国家出版基金项目
NATIONAL PUBLICATION FOUNDATION

A Dictionary of Seed Plant Names

Vol. 5 English, Japanese and Russian Indices

种子植物名称

卷5 英日俄名称索引

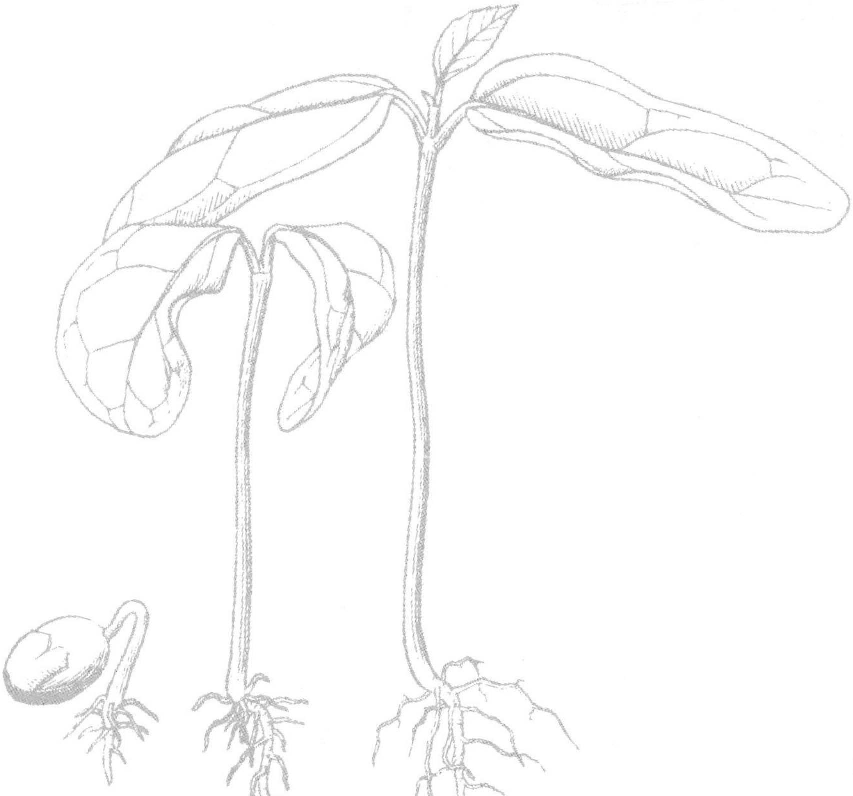

尚衍重　编著

中国林业出版社

图书在版编目(CIP)数据

种子植物名称. 卷5,英日俄名称索引 / 尚衍重编著. —北京:
中国林业出版社, 2012.6

ISBN 978-7-5038-6659-3

Ⅰ. ①种… Ⅱ. ①尚… Ⅲ. ①种子植物-专有名称-索引-英、日、俄
Ⅳ. ①Q949.4-61

中国版本图书馆 CIP 数据核字(2012)第139122号

中国林业出版社·自然保护图书出版中心

出 版 人:金 旻

策划编辑:温 晋

责任编辑:刘家玲 温 晋 周军见 李 敏

出版 中国林业出版社(100009 北京市西城区刘海胡同7号)
网址 http://lycb.forestry.gov.cn
E-mail wildlife_cfph@163.com 电话 010-83225836

发行 中国林业出版社
营销电话:(010)83284650 83227566

印刷 北京中科印刷有限公司

版次 2012年6月第1版

印次 2012年6月第1次

开本 889mm×1194mm 1/16

印张 92.75

字数 7613千字

印数 1~2000册

定价 566.00元

目　录

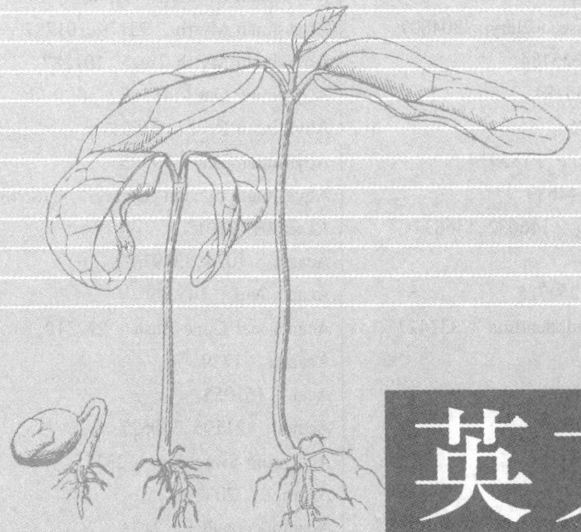

英文名称索引

English

A Dictionary of Seed Plant Names
种子植物名称

A

Aali 135192
Aandal Beadtree 7190
Aanteel-poprosie 108885
Aaron's Beard 5442,15876,116736,201787,273531,349936,372075
Aaron's Flannel 405788
Aaron's Rod 11549,37015,368480,389558,405788,405796
Aaron's-beard 107300,116736,201787,349936,372075
Aaronsbeard St. John's Wort 201787
Aba Angelica 24298
Aba Buttercup 325976
Aba Gentian 173183
Aba Sagebrush 35085
Abaca 260275
Abaca Banana 260275
Abactiana Louro-canfora 268699
Abacubeadbean 402149,402153
Abacus Indosasa 206881
Abage Palm 337883
Abajo Fleabane 150415,150530
Abarco 76832,76840
Abarco Wood 76832,76840
Abary Hemp 194779
Abassian Box 64345
Abassian Boxwood 64345
Abata Cola 99158
Abbespine 109857
Abbey 311208
Abbo Rubber 165263
Abby 311208
Abdominea 82
Abeel 311208
Abel Fig 164609
Abele 311208
Abelia 91
Abelia Multiflower 125
Abelmoschus 212
Abem 52843
Abeokuta Coffee 98941
Aberconway Rhododendron 330018
Aberrant Ardisia 31348
Aberrant Coral Holly 203653
Abert's Buckwheat 151794
Abert's Wild Buckwheat 151794
Abir 187468
Abiu 238111,313350
Ablamoth 133478
Able-tree 311208
Ablfgromwell 239167
Abnomal Milkvetch 42977
Abnormal Ardisia 31348
Abnormal Mallotus 243319
Abnormalleaf Clovershrub 70800
Abnormalumbrella Ribseedcelery 304807
Abolin Goosecomb 335184
Abolin Roegneria 335184
Abor Agapetes 10286
Aboriginal Prickly Apple 185914
Aboriginal Saltbush 44689
Aboriginal Willowherb 146632,146633
Aboudikro 145959
Aboutfish Cliffbean 254725
Above the Clouds Rhododendron 331427
Abraham 321643
Abraham's Balm 411189
Abraham's Blanket 405788
Abraham's Oak 324309
Abraham-isaac-and-jacob 393961
Abraham-isaac-jacob 57103,393959
Abrams Cypress 114650
Abricok 34475
Abrojo 273025
Abroma 672
Abronia 687
Abrus 738
Absinth 35090
Absinth Sagewort 35090
Absinth Sage-wort 35090
Absinth Wormwood 35090
Absinthe 35090
Absinthium 35090
Absolmsia 817
Absolute Entire Camellia 69151
Abundant Chinacane 162689
Abundant Magnoliavine 351077
Abundant-flower Bambusa 47347
Abundent Garcinia 171091
Abura Mitragyna 256120
Abutilon 837
Abutilonleaf Grewia 180666
Abutilon-leaved Grewia 180666
Abyssiana Dragon Tree 137463
Abyssinia Bellcalyxwort 231256
Abyssinia Coralbean 154620
Abyssinia Harrisonia 185932
Abyssinia Jujube 418145
Abyssinia Melhania 248823
Abyssinian Bamboo 278646
Abyssinian Banana 145846,260224
Abyssinian Basil 268526
Abyssinian Coffee 98857
Abyssinian Feather-top 289303
Abyssinian Gladiolus 4534
Abyssinian Millet 143551
Abyssinian Mustard 59348
Abyssinian Myrrh 93179,101287
Abyssinian Myrrh Tree 101287
Abyssinian Oat 45366
Abyssinian Rose 336917
Abyssinian Tea 79395
Abyssinian Yunnan-bamboo 278646
Acacallis 1018
Acacia 1024,334976
Acacia Nut 145899
Acacia-leaf Cone Bush 227318
Acaena 1739
Acai 161055
Acajou 21195,380527
Acalycine Sweetgum 232551
Acampe 2032
Acanthocephalus 2133
Acanthochlamys 2157
Acanthochlamys Family 2152
Acantholepis 2206
Acanthonema 2322
Acanthopanax 143575
Acanthophippium 2071
Acanthu-leaved Mahonia 242597
Acanthus 2657
Acanthus Bristle Thistle 73281
Acanthus Family 2060
Acanthus-leaved Carline Thistle 76989
Acanthuslike Bristlethistle 73281
Acapulco Grass 57925
Acaroid Resin 415172,415181
Acaulescent Parrya 224212
Acaulescent Pegaeophyton 287957
Acaulous Date 295453
Acceptable Rhododendron 330566
Aceituna 323553
Acer 2769
Acerola 243526
Acerose Melaleuca 248073
Acha 130544
Achasma 155395
Achenes Varnished 23802
Achimenes 4070
Achira 71169
Achnatherum 4113
Achorn 324335
Achyranthes 4249,4273
Achyrodes 220295
Achyrophorus 202387
Achyrospermum 4456

Acianthus 4511
Acicular Pricklythrift 2209
Acicular Rose 336848
Acicular Thorowax 63579
Acicularsplitleaf Spunsilksage 360878
Aciculate Chrysopogon 90108
Aciculate Eritrichium 153421
Acid Begonia 49593
Acid China-laurel 28310
Acid Fleshcoral 346995
Acid Jujube 418175
Acid Oxalis 277664
Acid Treebine 92587
Acid Woodsorrel 277664
Acidanthera 4523
Acid-leaf Begonia 49596
Acido 157164
Acid-water Arrowhead 342330
Acineta 4648
Ackee 55596,114581
Ackermann Nopalxochia 266648
Acland Cattleya 79527
Acmena 4877
Acoma Fleabane 150425
Acomastylis 4986
Acompes 279485
Aconite 5014,5442
Aconite Buttercup 325516
Aconite Monkshood 5442
Aconite Monks-hood 5442
Aconiteleaf Ampelopsis 20338
Aconiteleaf Snakegrape 20338
Aconiteleaf Syneilesis 381814
Aconite-leaved Begonia 49597
Aconite-leaved Buttercup 325516
Aconitum 5014
Aconitum-leaf Cowpea 408829
Acoop of Cock 182122,182126
Acorn 131501,263270,324335
Acorn Banksia 47645,47661
Acorn Corydalis 105639
Acorn Tree 324335
Acorn Yam 131501
Acorus 5793
Acrachne 5866
Acranthera 5884
Acrid Lettuce 219609
Acrid Lobelia 234860
Acriopsis 5947,5949
Acrocarpus 5980
Acrocephalus 5989
Acroceras 6099
Acrocomia 6128
Acroglochin 6160
Acronema 6196
Acronychia 6233,6251
Acroptilon 6272
Across Aster 40557
Acrossvein Arrowwood 408187
Actaea Rubra 61497
Actephila 6466
Actinialike Vineclethra 95473
Actinia-like Vineclethra 95473
Actinidia 6513
Actinidia Family 6728
Actinocarya 6739
Actinodaphne 6761
Actinoleaf Schefflera 350648
Actino-leaved Schefflera 350648
Actino-spiny Barberry 51271
Actinostemma 6888
Action Plant 255098
Acuate 219795
Aculeate Bittersweet 80130
Aculeate Euonymus 157282
Aculeate Meconopsis 247104
Aculeate Rockfoil 349627
Aculeate Staff Tree 80130
Acule-sepal China Laurel 28343
Acuminate Acmena 4879
Acuminate Aster 39920
Acuminate Atalantia 43719
Acuminate Baissea 94397
Acuminate Banana 260199
Acuminate Barberry 51275
Acuminate Barrenwort 146960
Acuminate Basketvine 9419
Acuminate Bauhinia 48993
Acuminate Cephaelis 81948
Acuminate Ceriman 258165
Acuminate Changed Sagebrush 35345
Acuminate Corydalis 105563
Acuminate Cowparsnip 192230
Acuminate Cymaria 116710
Acuminate Daphne 122526
Acuminate Delavay Meadowrue 388480
Acuminate Ehretia 141595
Acuminate Elatostema 142596
Acuminate Epigeneium 146529
Acuminate Epimedium 146960
Acuminate Eurya 160406
Acuminate Fairybells 134418
Acuminate Fimbristylis 166163
Acuminate Firmiana 166650
Acuminate Fluttergrass 166163
Acuminate Glochidion 177184
Acuminate Goosefoot 86892
Acuminate Gordontea 179731
Acuminate Hiptage 196822
Acuminate Ixora 211032
Acuminate Kitefruit 196822
Acuminate Lacquertree 393482
Acuminate Lasianthus 222085
Acuminate Leaved Chinkapin 79032
Acuminate Litse 6762
Acuminate Longbeak Eucalyptus 155518
Acuminate Maesa 241735
Acuminate Mosquitotrap 117339
Acuminate Ochrosia 268352
Acuminate Olax 269643
Acuminate Ottelia 277364
Acuminate Pomatocalpa 310798
Acuminate Raspberry 338082
Acuminate Rhubarb 329309
Acuminate Roughleaf 222085
Acuminate Sand Willow 343194
Acuminate Serrate Hydrangea 200089
Acuminate Sophora 368975
Acuminate Spapweed 204762
Acuminate Spiked Gingerlily 187469
Acuminate Stairweed 142596
Acuminate Swallowwort 117339
Acuminate Synotis 381919
Acuminate Thrixspermum 390487
Acuminate Touch-me-not 204762
Acuminate Twayblade 232074
Acuminate Twoformflower 128478
Acuminate Violet 409630
Acuminate Windhairdaisy 348104
Acuminate-bract Cryptandra 113339
Acuminateleaf Brassaiopsis 59188
Acuminate-leaf Cupular Willow 343259
Acuminate-leaf Eurya 160406
Acuminateleaf Firmiana 166650
Acuminateleaf Honeysuckle 235633
Acuminate-leaf Machilus 240551
Acuminate-leaf Newlitse 264018
Acuminateleaf Rhynchosia 333133
Acuminateleaf Solomonseal 308495
Acuminate-leaved Atalantia 43719
Acuminate-leaved Eurya 160406
Acuminate-leaved Firmiana 166650
Acuminate-leaved Honeysuckle 235633
Acuminate-leaved Neolitsea 264018
Acuminate-leaved Newlitse 264018
Acumite-leaf African Myrsine 261602
Acutangular Mananthes 244278
Acutangule Clematis 94700
Acutate Monkshood 5016
Acutatesepal Monkshood 5016
Acute Asiaglory 32602
Acute Bedstraw 170180
Acute Closedspurorchis 94464
Acute Cobwebcalyx 302796
Acute Dock 339905
Acute Fieldcitron 400518
Acute Gentianella 174103
Acute Gongshan Maple 3066
Acute Ho Schefflera 350710
Acute Lockhartia 235132
Acute Maple 2771

Acute Platycrater 302796
Acute Sedge 73589
Acute Sida 362478
Acute Turpinia 400518
Acuteangle Hedyotis 187501
Acuteangular Anisodus 25439
Acuteangular Cragcress 97969
Acuteangular Jute 104057
Acuteangular Scurvyweed 97969
Acuteangular Spikesedge 143042
Acute-contracted Fargesia 162636
Acute-dentate Ampelopsis 20294
Acutedentate Nordhagen Larkspur 124407
Acuteflower Heliotrope 190542
Acutefruit Rattanpalm 65755
Acutefruit Speedwell 407267
Acutefruit Woodbetony 287490
Acuteglume Garnotia 171487
Acuteglume Ricegrass 276017
Acuteknife Dendrobium 125188
Acuteleaf Alyxia 18509
Acuteleaf Beautyberry 66732
Acuteleaf Box 64371
Acuteleaf Caper 71679
Acuteleaf Cryptocarya 113424
Acuteleaf Japanese Spiraea 371965
Acuteleaf Purplepearl 66732
Acuteleaf Sageretia 342195
Acuteleaf Thickshellcassia 113424
Acute-leaved Aalyxia 18509
Acute-leaved Beautyberry 66732
Acute-leaved Box 64371
Acute-leaved Camellia 68888
Acute-leaved Caper 71679
Acute-leaved Common Barberry 52325
Acute-leaved Cryptocarya 113424
Acute-leaved Randia 12530
Acute-leaved Sageretia 342195
Acute-leaved Spindle-tree 177985
Acutelip Neottia 264651
Acutelobed Iris 208434
Acutemargin Shellfish Pricklyash 417217
Acutepetal Indigo 205619
Acutesepal Beautyberry 66838
Acutesepal Camellia 68882,69056
Acute-sepal Camellia 69080
Acutesepal Eurya 160418
Acutesepal St. John's Wort 201710
Acutesepal Syzygium 382535
Acute-sepaled Camellia 69056
Acute-sepaled Eurya 160418
Acute-sepaled Syzygium 382468
Acuteserrate Leatherleaf Buckthorn 328659
Acute-serrated Tutcheria 400685
Acutesprout Bamboo 297191
Acute-sproutted Bamboo 297191
Acutesrrate Stairweed 142598
Acutest Basketvine 9418
Acutest Camellia 69035
Acutestlike Camellia 69671
Acutest-like Camellia 69671
Acutetepal Elatostema 142599
Acutetepal Stairweed 142599
Acutidentate Ampelopsis 20294
Acutidentate Pellionia 288708
Acutidentate Snakegrape 20294
Acutifid Licuala Palm 228731
Acutifid Licualapalm 228731
Acutifoliate Jacaranda 211226
Acutifoliate Podocarpium 200742
Acutiligular Sinobambusa 364784
Acutilobate Angelica 24282
Acutiperulate Anlong Camellia 68907
Acutiperulate Camellia 68907
Acutipetal Indigo 205619
Acutiserrate Camellia 68888
Acutiserrate Tutcheria 400685
Ada 6988
Ada-jamir 93327
Adam and Eve 29365,37015
Adam And Eve Orchid 29365
Adam's Apple 93403,93534,260253,382763
Adam's Fig 260253
Adam's Flanne 405729
Adam's Flannel 405788
Adam's Hood 164608
Adam's Needle 416535,416538,416584,
416595,416649
Adam's Needle Yucca 416649
Adam's Needle-and-thread 416584,416649
Adam's Plaster 309570
Adam's Tree 167558
Adam's-needle 350425,416535,416584,
416607
Adam's-needle Yucca 416584
Adam-and-eve 5442,29359,29365,37015,
220346,273469,273531,321636,321643
Adams Sagebrush 35095
Adang Pinanga 299651
Adanson Polystachya 310311
Adansonia 7018
Adarf Buckeye 9693
Added Randia 162196
Adder's Cotton 115008
Adder's Eyes 21339
Adder's Flower 273531
Adder's Food 37015
Adder's Grass 273531,293709
Adder's Meat 28044,37015,250614,374916,
383675
Adder's Mouth 242993,273531
Adder's Poison 383675
Adder's Spit 232950,374916
Adder's Tongue 4005,37015,174877,232950,
273531,342400
Adder's Tongue Spearwort 326148
Adder's Violet 179571
Adder's-meat 374916
Adder's-mouth 306883
Adder's-mouth Orchid 242993,243084,254094
Adder's-tongue 154895,242672
Adder's-tongue Spearwort 326148
Adder-and-snake Plant 364193
Addermonth Orchid 242993,243149
Adders 272398
Adderslily 154895
Addersmeat 374916
Adderwort 37015,308893
Addington's St. John's Wort 201717
Addison Brown's Clematis 94703
Addison Brown's Leather-flower 94703
Addison's Leather-flower 94703
Addison's Virgin's-bower 94703
Adelostemma 7087
Aden Gum 1024
Aden Senna 78227
Adenandra 7113
Adenandra-leaf Sweetleaf 381154
Adenia 7220
Adenium 7333
Adenocaulon 7427
Adenopode Treebine 92596
Adenosma 7974
Adenostemma 8004
Adhatoda 214308
Adherent Bristle-grass 361686
Adhesive Litse 233928
Adhesive Rehmannia 327435
Adhesivecalyx Catchfly 364181
Adhesivehair Spiradiclis 371781
Adhib 160239
Adhibitgemma Azalea 332080
Adiantumleaf Corydalis 105566
Adina 8170
Adinandra 8205
Adjoin Clearweed 298864
Adjoin Clinopodium 96981
Adjoin Coldwaterflower 298864
Adjoin Wildbasil 96981
Adjoining Premna 313611
Adlai 99124
Adlay 99124
Adlumia 8298
Adnate Elder 345561
Adnate Perezia 290725
Adnate Solomonseal 308496
Adnate Treebine 92598
Adobe Hills Thistle 92305
Adobe-lily 168520
Adonis 8324,8387
Adonis Flower 8332

Adorned Peperomia 290398
Adorned Rhododendron 330329
Adoxalike Sage 344831
Adpresedleaf Elatostema 142852
Adpresedleaf Stairweed 142852
Adpressed Knotweed 308788
Adpressedhairy Wildberry 362341
Adrews Broadlily 97090
Adria Bellflower 70228
Adriatic Oak 323745
Adrossow Tamarisk 383431
Adrue 118520
Adsuki Bean 408839
Adun Corydalis 105630
Adun Ladybell 7601
Adunz Rockfoil 349085
Adunzi Gentian 173272
Adunzi Woodbetony 287023
Adzuki Bean 408839
Aechmanthera 8538,8540
Aechmea 8544,8555
Aegean Dock 340019
Aegean Wallflower 86427
Aegiceras 8645
Aegilops 8660
Aeginetia 8756
Aegypt Sage 344832
Aellenia 8848
Aeluropus 8855
Aeonium 9021,9023
Aerial Pseudosasa 318281
Aerial Yam 131501
Aerides 9271
Aerjin Dandelion 384439
Aerjin Windhairdaisy 348108
Aeroplane Propeller 109264
Aeroplanes 3462
Aerva 9361
Aeschynomene 9489
Aethiopia Sage 344836
Aethusa 9831
Aetna Barberry 51280
Afar Tree 386942
Afara 386652
Affadil 262441
Affined Arrow-wood 407672
Affined Conehead 385067
Affined Cystacanthus 120796
Affined Devil Rattan 121522
Affined Inga 206921,206942
Affined Microtoena 254242
Affined Salacca 342571
Affined Sterculia 376066
Affined Tinctorial Wendlandia 413831
Affinity Neillia 263143
Affodill 39328
Affrodile 262441
Afghan Azalea 330039
Afghan Barberry 51282
Afghan Bluegrass 305287
Afghan Erysimum 154521
Afghan Pine 299828
Afghan Poplar 311204
Afghan Rhododendron 330039
Afghan Rockfoil 349051
Afghan Yellow Rose 336872
Afghanistan Barberry 51282
Afo Nut 306777
Afraca Arrowroot 382923
Africa Aloe 16555
Africa Antiaris 28243
Africa Asparagus 38983
Africa Black Wood 121750
Africa Cedar 213863
Africa Daisy 175172
Africa Fingergrass 88352
Africa Greenheart 116569
Africa Jointfir 178525
Africa Juniper 213863
Africa Malcolmia 243164
Africa Mussaenda 260451
Africa Myrsine 261601
Africa Nettletree 80750
Africa Nittatree 284445
Africa Persimmon 132087
Africa Rosewood 121750
Africa Sphaeranthus 370958
African Afzelia 10106,10129
African Agapanthus 10245
African Albizzia 13476
African Amaranth 18781
African Antiaris 28243
African Apple 90008
African Apricot 268331
African Arctotis 31294
African Arrowroot 71163
African Asparagus 38983
African Asparagus Fern 38938
African Avodire 400622
African Barberry 51283
African Basil 268518
African Beech 162999
African Bent 12164
African Bermudagrass 117875
African Bird's-eye Bush 268226
African Black Wattle 288884
African Blackwood 121750,288884
African Blood Lily 184428,350312
African Blue Lily 10271
African Bonebract 354338
African Bowstring Hemp 346047,346100
African Box Thorn 239048
African Boxthorn 239048
African Boxwood 261601
African Breadfruit 394605
African Breadfruit-tree 394605
African Bristlegrass 361902
African Bur-grass 394380
African Canarium 71021
African Cedar 80080,213863
African Celtis 80570,80628,80750
African Cherry 316203,391608
African Coralbean 154622
African Coralwood 320330
African Coral-wood 320330
African Corn Lily 210689,210816
African Cornflag 86125
African Corn-lily 210689
African Cornlily Ixia 210689
African Couchgrass 130408
African Crabwood 72650
African Cubebs 300374
African Cucumber 256797
African Cypress 414058
African Cyrtococcum 120642
African Daisy 31178,31236,31294,31305,131147,131150,131191,175172,235503,235504,276556,276625,402754
African Dayflower 101103
African Dodder 114992
African Dogstooth Grass 117886
African Dropseed 372579
African Ebony 121750,132030,132117,132299,132353
African Eggplant 366920,367344
African Elemi 57521,71021
African Elephant's Foot 131693
African Elodea 219776
African Eremopogon 148395
African Erythrophleum 154964
African Evergreen 381861
African False Elm 80655
African Falseelm 80655
African Falsenettle 56206
African Feathergrass 289159
African Fern Pine 306435
African Finger Millet 143523
African Fountain Grass 289248
African Fountain-grass 289234,289248
African Golden Walnut 237927
African Hackberry 80750
African Hemp 346094,370130,370136
African Honey Suckle 184875
African Honeysuckle 184875
African Horned Cucumber 114221
African Iris 130292,130299,208524
African Jasmine 393657
African Jointfir 178525
African Juniper 213863
African Kino 320293
African Laburnum 68132,78219,78484

African Lily 10244,10245
African Linden 256113,256120
African Liverseed Grass 402533
African Locust 284456,284473
African Locust Bean 284456
African Lotus 418169,418177
African Love Grass 147612
African Lovegrass 147645
African Love-grass 147612
African Mahogany 216194,216196,216205,216214
African Malcolmia 243164
African Mammey Apple 243981
African Mammy Apple 268331
African Mangosteen 171129
African Marigold 131186,383090
African Mask 16488
African Milk Bush 381480,381481
African Milk Tree 160001
African Milkbush 381491
African Milkweed 179032
African Millet 143522,289116
African Mustard 59651,243164
African Myrrh 101290
African Myrrh Tree 101290
African Myrsine 261601
African Nettle Tree 80655
African Nettletree 80655
African Nitta Tree 284445,284475
African Nutmeg 257653
African Oak 88506,236334,270054
African Oil Palm 142257
African Oilpalm 142257
African Olive 270089,270100
African Oplismenus 272614
African Padauk 320330
African Padouk 320330
African Peach 346674
African Peanut 409069
African Pear 244583
African Peltophorum 288884
African Pencil Cedar 213863
African Pepper 415762
African Pepperwort 225283
African Persimmon 132087
African Piassava 326640
African Piptadenia 300609
African Plume 49094
African Rattlebox 112641
African Red Alder 114566
African Rhun Palm 57118
African Rosemallow 194689
African Rosewood 121750,181771,320293
African Rubber 169229,220814
African Rue 287978
African Sandalwood 372350
African Sandal-wood 161637
African Satinwood 417235
African Sausagetree 216324
African Scurf Pea 319224
African Scurf-pea 319224
African Senna 360431
African Sheepbush 290222,290248
African Signalgrass 402499
African Sparmannla 370136
African Sphaeranthus 370958
African Spinach 18734
African Star Chestnut 376067
African Star-apple 90008
African Sumac 332685
African Syngonium 381861
African Tamarisk 383414
African Teak 88506,254493,270054,320293
African Tragacanth 376199
African Tulip Tree 370351,370358
African Tuliptree 370358
African Tulip-tree 370358
African Valerian 163040
African Violet 342482,342493
African Walnut 108127,237927,352493
African Walnut Albizzia 13703
African Wheat 398853
African White Wood 25819
African Whitewood 25816,25819,145141
African Wild Date 295479
African Willow 343726
African Windflower 370136
African Wintersweet 4965
African Women's Hat 256116
African Wood Oil-nut 334411
African Yellow-wood 145141,306395
African-Daisy 31176
Africanlily 10244
Africanteak 254493
Afromelia 216194,216214
Afrormosia 10051,290876
After Szechuan Woodbetony 287425
Afternoon Iris 208524
Afzel Rose 336411
Afzel Strychnos 378639
Afzelia 10105,10106
Agalawood 29973
Agalloch 29973
Agallocha 161632,161639
Agalmyla 10188
Agalwood 29973
Aganonut Persian Peach 20948
Aganopersian Peach 20950
Aganosma 10214
Agapanthus 10244,10245,10271
Agapetes 10284
Agarito 52267
Agarwood 29977
Agate Pomegranate 321768
Agati 361384
Agati Sesbania 361384,361385
Agave 10778,10806,10852,10945
Agave Azul 10945
Agave Cactus 227755,227757
Agave Family 10775
Agavelike Living-rock Cactus 33207
Agba 179846
Agboin 300655
Agelaea 10975
Ageratum 11194,11199
Aggermony 11549
Aggie-land Gayfeather 228457
Agglomerate Flower Newlitse 264054
Agglomerate Newlitse 264054
Agglutinated Azalea 331479
Aggregate Barberry 51284
Aggregate Gynostemma 182999
Aggregateleaf Monkshood 5112
Agilawood 29973
Aging Leek 15734
Aging Onion 15734
Aglaia 11273
Aglaia Tree 11300
Aglaonema 11329,11339
Aglet 106703
Aglet-tree 106703
Agnus Castus 411189
Agnus-castus 411189
Agoseris 11408
Agosihis Cuming's Fig-tree 164864
Ag-paper 405788
Agreeable Draba 137053
Agreeable Elm 401447
Agreeable Oak 324200
Agreeable Tanoak 233285
Agreeable Whitlowgrass 137053
Agreen 359158
Agricola Barberry 51287
Agrimonia 11542
Agrimony 11542,11549,11558,11598,158318
Agriophyllum 11609
Agritos 52267
Agrostemma 11938
Agrostophyllum 12442
Aguacate 291494
Aguaje 246672
Aguassu 273241
Ague Root 14486
Ague Weed 158062,174148
Ague-root 14486
Ague-tree 347407
Agueweed 158250,174148
Ague-weed 174148,174149
Aguirre 150327
Agworm Flower 374916
Agworm-flower 374916

Agyoku Bittercress 72721
Ahart's Nailwort 284736
Ahart's Sulphur Flower 152566
Ahern Eugenia 156126
Ahernia 12483,12484
Ahikatea 306420
Ahun 18034
Aiara Terminaila 386652
Aibika 229
Aigreen 358062
Aik 324335
Aiktan 234269
Ailanthus 12558,12559
Ailanthus Family 364387
Ailanthus Prickly-ash 417139
Ailanthusleaf Evodia 161307
Ailanthusleaf Pricklyash 417139
Ailanthus-like Pricklyash 417139
Ailanthus-like Prickly-ash 417139
Ailanto 12559
Ailes 45566
Ails 45566
Ainee 36921
Ainsliaea 12599,12739
Ainsliifolious Cacalia 283787
Ainsl-leaved Lagarosolen 219788
Ainu Onion 15450
Aipi Cussawa 244512
Air Pine 8558
Air Plant 61572,391966,392015,392020
Air Potato 131501
Air Potato Yam 131501
Air Yam 131501
Airbell 70271
Airbell Hairbell 70271
Airbroom 54439
Aircraft Grass 89299
Airess 170193
Airid 170193
Airis 13595
Airplane Plant 88553
Airplant 61568,391631
Air-plant 61572
Air-potato 131501
Airpotato Yam 131501
Airup 170193
Aisatic Plantain 302068
Aischen 167955
Aishen-tree 167955
Ait Skeiters 24483
Aitchison Barberry 51288
Aitchison Habenaria 183402
Aitchison Iris 208435
Aitchorn 324335
Aiten 213702
Aitnagh 213702
Aiverin 338243
Aizoon 12926
Aizoon Rockfoil 349052
Aizoon Rockjasmine 23110
Aizoon Saxifrage 349054
Aizoon Stonecrop 294114
Ajamente 38129
Ajan Knotweed 308717
Ajan Ligusticum 391904
Ajania 12988
Ajaniopsis 13045
Ajipo 279737
Ajo Mountain Scrub Oak 323624
Ajo Rock Daisy 291298
Ajowan 393937,393940
Ajowan Caraway 393940
Ajuga 13174
Ajuga 'Bronze Beauty' 13174
Ajuga 'Burgundy Glow' 13174
Akakura 252610
Akar Dani 324677
Akeake 135215,270187
Ake-ake 135215,270214
Akebi Birthwort 34268
Akebia 13208,13225
Akebono Flowered Cherry 316933
Akee 55592,55596,114581
Akee Apple 55592,55596,114581
Akepuka 181267
Akesu Swallowwort 117542
Aketao Eritrichium 153476
Akiraho 270206
Akle 13475
Ako 28248
Ako Lindera 231301
Ako Spice Bush 231301
Ako Spicebush 231301
Ako Spice-bush 231301
Akoko 265879
Akom 386652
Akomu 321967
Aksai Sagebrush 35100
Aksaichin Sandwort 31738
Aksu Milkvetch 41935
Aktao Larkspur 124015
Akund Calotrope 68088
Akund Fibre 68095
Akyr 324335
Alabama 336284
Alabama Azalea 330053
Alabama Bellowsfruit 297852
Alabama Coneflower 339520
Alabama Croton 112832
Alabama Larkspur 124016
Alabama Leather-flower 95318
Alabama Ninebark 297852
Alabama Rhododendron 330053
Alabama Snow Wreath 265857
Alabama Snowwreath 265857
Alabama Supplejack 52466
Alabama Warbonnet 211554
Ala-compame 207135
Alaghez Nepeta 264862
Alai Kengyilia 215902
Alaishan Milkvetch 43004
Alajja 13334
Alal Licorice 177875
Alameda County Thistle 92337
Alamillo 311324
Alamo 302596
Alamo Vine 250777
Alan 267648,362188
Alang-alang 205497
Alangium 13345
Alangium Family 13343
Alashan Calligonum 66993
Alashan Cornulaca 104910
Alashan Figwort 355041
Alashan Gagea 169378
Alashan Goosecomb 144157
Alashan Herminium 192810
Alashan Hippolytia 196690
Alashan Kneejujube 66993
Alashan Lanceolate Sedge 75049
Alashan Lilac 382248
Alashan Madder 337931
Alashan Manaraceme Sweetvetch 188048
Alashan Meadowrue 355349
Alashan Medic 247244
Alashan Milkvetch 41941
Alashan Onion 15297
Alashan Panzerina 282478
Alashan Peppergrass 225289
Alashan Pepperweed 225289
Alashan Poplar 311207
Alashan Rockcress 30162
Alashan Rockjasmine 23117
Alashan Roegneria 144157
Alashan Saltbush 44303
Alashan Seepweed 379582
Alashan Windhairdaisy 348112
Alashan Woodbetony 286980
Alashan Youngia 416393
Alasja Birch 53549
Alaska Arnica 34780
Alaska Beauty 94294
Alaska Black Currant 334064
Alaska Blue Willow 343937
Alaska Bog Willow 343409
Alaska Bog-willow 343409
Alaska Cedar 85301
Alaska Currant 333928
Alaska Cypress 85301
Alaska Dandelion 384426
Alaska Fleabane 150944

Alaska Knotweed 309194
Alaska Larch 221904
Alaska Orach 44431
Alaska Pine 399905
Alaska Spruce 298435
Alaska Starwort 374728
Alaska Yellow Cedar 85301
Alaska-cedar 85301
Alaskan Bugseed 104818
Alaskan Pussytoes 26414
Alaskan Rush 212850
Alata Crocus 111483
Alata Mountain Crocus 111483
Alatai Chickweed 374729
Alatai Goosecomb 144159
Alatai Kengyilia 215903
Alatai Monkshood 5019
Alatao Crazyweed 279097
Alatao Milkvetch 41942
Alatao Sawwort 360985
Alatao Speedwell 317907
Alatao Willow 342969
Alatav Craneknee 221629
Alatav Stickseed 221629
Alataw Mountain Willow 342969
Alataw Speedwell 317907
Alate Nightshade 367726
Alate Sonerila 368870
Alatinerved Waterdropwort 269294
Alba Semi-plena Rose 336344
Alba-grey Simon Poplar 311495
Alban Onion 15030
Albania Lily 229794
Albanian Spurge 158640
Albany Bottlebrush 67298
Albany Bottle-brush 67298
Albany Gall-sick Bush 103508
Albany Pitcher Plant 82560,82561
Albany Woollybush 7199,7203
Albarco 76832
Albardine 239310
Albaspyne 109857
Albert Bellflower 69881
Albert Buttercup 325558
Albert Gagea 169379
Albert Honeysuckle 235649
Albert Pricklythrift 2213
Albert Rose 336345
Albert Skullcap 355351
Albert Willow 342995
Alberta Fleabane 151021
Albertisia 13447
Albertson's Rhododendron 330054
Albescent Camellia 68892
Albespeine 109857
Albiu 238111
Albizia 13474,13578
Albizzia 13474,13578
Albopilose Elatostema 142600
Albopubescent Coldbamboo 172737
Albopubescent Gelidocalamus 172737
Alchemilla filicaulis 14025
Alchemilla glabra 14037
Alchemilla glomerulans 14039
Alchemilla gracilis 14042
Alcimandra 14230,14231
Alcock Spruce 298225,298226
Alcock's Spruce 298225,298226
Alcove Rock Daisy 291333
Alcove Thistle 92355
Aldenham Barberry 51289
Aldenham Mahonia 242459
Alder 16309,16352
Alder Birch 53342
Alder Buckthorn 328592,328603,328701
Alderdraught 192358
Alder-leaf Begonia 49606
Alder-leaf Buckthorn 328603
Alder-leaf Clethra 96457
Alderleaf Elatostema 142602
Alder-leaf Mountain Mahogany 83820
Alderleaf Raspberry 338116
Alderleaf Serviceberry 19241
Alder-leaf Sida 362490
Alderleaf Stairweed 142602
Alderleaf Waxpetal 106629
Alderleaf Winterhazel 106629
Alder-leaved Buckthorn 328603
Alder-leaved Raspberry 338116
Alder-leaved Service-berry 19241
Alder-leaved Sida 362490
Alder-leaved Winterhazel 106629
Aldern 345631
Alderne 345631
Alderney Sea Lavender 230670
Alderney Sealavender 230670
Alderney Sea-lavender 230694
Aldertot 192358
Aldrovanda 14271,14273
Alecost 383712
Alecost Costmary 383712
Alectryon 14359
Alegampane 207135
Alehoof 176812,176824
Alemon 93558
Alemow 93558
Alepp Euphorbia 158414
Aleppo Avens 175378
Aleppo Galls 324314
Aleppo Grass 369652
Aleppo Oak 324050
Aleppo Pine 299964
Aleppo-grass 369652
Alerce 45239,166720,166726,299132,386942
Alerch 166720
Aletris 14471
Aleurites 14538
Aleutian Brome 60620
Aleutian Mouse-ear Chickweed 82642
Aleutian Phyllodoce 297016
Aleutian Selfheal 316136
Aleutian Wormwood 35111
Alexander Palm 321164
Alexander Rhubarb 329311
Alexander's Foot 21266
Alexander's Palm 321164
Alexanders 366739,366743
Alexandra 31013
Alexandra Catchfly 363165
Alexandra King Palm 31013
Alexandra Palm 31013
Alexandran King-palm 31013
Alexandria Cycad 145221
Alexandria Laurel 67860,122124
Alexandrian Laurel 67860,122121,122124
Alexandrian Parsley 366743
Alexandrian Senna 78212,78227
Alfafa-like Rattlebox 112396
Alfalfa 115008,247239,247456,247457
Alfalfa Dodder 114961
Alfilaria 153767
Alfilerea 153767
Alfred Combretum 100321
Alfred Loosestrife 239547
Alfred Stonecrop 356512
Alfred Willow 342997
Alfred Windmill 100321
Alfredia 14593
Alga Pondweed 312076
Algae-like Pondweed 312076
Algal-leaved Pondweed 312076
Algaroba 83527,315547,315551,315557,315568
Algarobilla 47089
Algarreba 315560
Algarroba 200844
Algarrobo 315560
Algarrobo Negro 315566
Algeria Cypress 114679
Algeria Eucalyptus 155474
Algeria Fir 435
Algerian 93429
Algerian Ash 167952
Algerian Butterfly-orchid 302246
Algerian Cedar 80080
Algerian Fibre 85671
Algerian Fir 435
Algerian Fir Iris 208863
Algerian Fir Oak 323732
Algerian Grass 376931
Algerian Iris 208863,208914

Algerian Ivy 187197,187200
Algerian Oak 323732
Algerian Oat 45424
Algerian Sealavender 230732
Algerian Winter Iris 208914
Algerita 51694,52267
Algid Primrose 314106
Algid Sawwort 360987
Algodones Dunes Sunflower 189011
Alhagi 14623
Alice 309570
Alice Areca Palm 31678
Alice Eastwood's Fleabane 150443
Alice Palm 31678
Alick 366743
Alicoche 140216,140288
Aligator Apple 25852
Aligator-apple 25852
Aligopane 207135
Alisanders 366743
Alishan Actinidia 6530
Alishan Acuminate Eurya 160535
Alishan Arisaema 33264
Alishan Astilbe 41827
Alishan Azalea 331565,331926
Alishan Barberry 51811,51926
Alishan Bentgrass 12000
Alishan Bulbophyllum 62985
Alishan Calanthe 65886
Alishan Camellia 69705
Alishan Chickweed 374757
Alishan Daisy 124772
Alishan Daphne 122377
Alishan Elm 401642
Alishan Entireliporchis 261470
Alishan Fig 165623
Alishan Forkfruit 128923
Alishan Gentian 173260
Alishan Greenbrier 366238
Alishan Holly 203575
Alishan Hornbeam 77314
Alishan Irisorchis 267927
Alishan Japenese Fig 165004
Alishan Kiwifruit 6530
Alishan Landpick 308616
Alishan Liparis 232322
Alishan Machilus 240707
Alishan Magnolia Vine 351012
Alishan Magnoliavine 351012
Alishan Mahonia 242608
Alishan Oberonia 267927
Alishan Oxtongue 298606
Alishan Rosewood 121621
Alishan Sabia 341575
Alishan Sage 345076
Alishan Sedge 73755,75418
Alishan Skimmia 365919
Alishan Slenderstalk Deutzia 126950
Alishan Small-flowered Sweet Spire 210407
Alishan Solomonseal 308616
Alishan Southstar 33264
Alishan Spotleaf-orchis 179577
Alishan Stonebean-orchis 62985
Alishan Supplejack 52403
Alishan Sweetleaf 381272
Alishan Sweetspire 210407
Alishan Swertia 380117
Alishan Tea 69705
Alishan Thistle 91753
Alishan Twayblade 232322
Alishan Zeuxine 417779
Alisma 14760
Alisma Commun 14788
Alisma Family 14790
Alisma Graminoide 14734
Alisma Subcorde 14785
Aliso 16425,302595
Alison 18313,45192
Alisson 235090
Alizanders 366743
Alizarin 338038
Alkakeng 297643
Alkali Aster 33039,398134
Alkali Barnyardgrass 140414
Alkali Belvedere 217372
Alkali Bulrush 56641,353689
Alkali Buttercup 184717,325749
Alkali Chalice 161027
Alkali Dropseed 372581
Alkali Goldfields 222618
Alkali Grass 134840,134842,321220,
321248,417880
Alkali Jimmyweed 209563
Alkali Mariposa-lily 67628
Alkali Muhly 259634
Alkali Rayless Aster 58265,58266
Alkali Sacaton 372581
Alkali Sand-spurrey 370624
Alkali Swainsonpea 371161
Alkali Yellowtops 166818
Alkaliaster 398134
Alkali-blite 87147
Alkaligrass 321220,321248
Alkali-grass 19766,321220,417890
Alkaline Water-nymph 262067
Alkali-sink Goldfields 222594
Alkanet 14836,14841,21914,21933,21949,
21959,289503
Alkekengi 297643,297645,297720
Alkekengy 297643
Alksuth Barberry 51290
All Gold Bamboo 297220
All Good 86970
Alla 277648
Allaeanthus 61096
Allamanda 14871,14873
Allar 16352
Allard Lavender 223249
All-bone 374916
Allegany Barberry 51422
Allegany Chinkapin 78807
Allegany Foamflower 391522
Alleghanies 377
Alleghany Barberry 51422
Alleghany Plum 316204
Alleghany Sloe 316204
Alleghany Spurge 279743
Allegheny Blackberry 338111
Allegheny Chinkapin 78807
Allegheny Monkey Flower 255237
Allegheny Monkey-flower 255237
Allegheny Moss 334969
Allegheny Serviceberry 19276
Allegheny Shadblow 19276
Allegheny Spurge 279754
Allegheny Stonecrop 200809
Allegheny Vine 8302
Allegheny-vine 8302
Alleizettella 14927
Alleluia 173082,277648,277688
Alleman Buttercup 325560
Allemande 14871
Allen-tree 16352
Aller 16352
Allerheiligenholz 184518
Aller-tree 16352
Allgood 86970
Allheal 373346,404316
Allhoove 176824
Allhose 176824
Alliaria 14953
Allicampane 207135
Allicoche Cactus 140280
Allienne 35090
Alliff 176824
Alliform Sedge 73636
Alligator Alternanthera 18128
Alligator Apple 25828,25876
Alligator Bonnet 267730
Alligator Juniper 213826,213847
Alligator Pear 291494
Alligator Pepper 9923
Alligator Redwood 181559
Alligator Tree 232565
Alligator Weed 18128
Alligator-apple 25828
Alligator-lily 200948
Alligator-pear 291494
Alligatorweed 18128
Allionia 14992
Allium 15018,15020,15057

Allmania 15910
Allocheilos 15956,15957
Allohairy Kudrjaschevia 218276
Allolida 277648
Allomorphia 16006
Allom-tree 401593
Allophylus 16043
Alloplectus 16174,16178
Allostigma 16212
Alloteropsis 16219
Allroundflower 280877,280880
Alls-bush 16352
Allscale 44609
Allseed 44576,307724,308816,325039,325041
Allseed Goosefoot 87130
Allsgheny Juneberry 19276
Allspice 68306,68310,68313,299325
Allspice Family 68300
Allspice Jasmine 172778
Allspice Pimenta 299325
Allthom Acacia 1297
Allthorn 217399
Alma Stonecrop 200778
Alman Radice 326616
Almon 362189
Almon Lauan 362203
Almond 20885,20890,386500
Almond Geranium 288459
Almond Gum 20890
Almond Pear 323086
Almond Willow 343013,344211
Almondette 61670
Almond-leafed Pear 323086
Almond-leaved Euonymus 157318
Almond-leaved Spindle-tree 157318
Almondleaved Willow 343013
Almond-leaved Willow 344211
Almond-scented Geranium 288459
Almond-scented Orchid 269055
Almond-shaped Pear Free 323086
Almond-shaped Pear-tree 323086
Almont Cherry 83208
Almum Sorghum 369593
Alocasia 16484,16512
Aloe 16540,16605,16687,16847,17017,
17165,17381
Aloe Family 17424
Aloe Wood 29973,29977
Aloe Woodbetony 286986
Aloe Yucca 416538
Aloeleaf Cymbidium 116781
Aloeleaf Dendrobium 124980
Aloes-wood 29973
Aloeswood Eaglewood 29973
Alone Themeda 389377
Along Newlitse 264020
Alonsoa 17475
Alopeculus-like Kudzuvine 321420
Alopecurus 17497
Alopecurus Brome 60621
Alopecurus Woodbetony 286988
Alp Anthyllis 28179
Alp Awngrass 255886
Alp Biebersteinia 54198
Alp Cephalanthera 82031
Alp Corydalis 106543
Alp Crazyweed 278700
Alp Dendrobium 125196
Alp Hairorchis 148754
Alp Knotweed 291663
Alp Leaflessorchis 29213
Alp Lily 234214
Alp Listera 264659
Alp Maximowicz Hillcelery 276754
Alp Milkvetch 41962
Alp Mouseear 83045
Alp Orchis 310943
Alp Peashrub 72208
Alp Pipewort 151219
Alp Platanthera 302502
Alp Scurrula 355300
Alp Spunsilksage 360866
Alp Veinyroot 237481
Alp Wendlandia 413825
Alp White Azalea 330385
Alp Wildsenna 389507
Alp Windhairdaisy 348121
Alp Yaccatree 306509
Alpam Root 390413
Alpbean 391535,391539
Alpenclock 367762
Alpencress 198480
Alpenrose 330695
Alphitonia 17613
Alphonse Karr Bamboo 47346
Alphonsea 17623
Alphonse-karr Hedge Bamboo 47346
Alpine 312863,357203
Alpine Acronema 6197
Alpine Aletris 14472
Alpine Alopecurus 17506
Alpine Alpinegold 199211
Alpine Anemone 321658
Alpine Anneslea 25774
Alpine Arctic-cudweed 270579
Alpine Arrowbamboo 162691
Alpine Ash 155558
Alpine Asiabell 98278
Alpine Aster 40016
Alpine Aussiepoplar 197623
Alpine Autumn-croctus 99296
Alpine Avens 175433,363062
Alpine Azalea 235283
Alpine Baeckea 46433
Alpine Barberry 51811
Alpine Barrenwort 146961
Alpine Bartsia 373608
Alpine Bastardtoadflax 389600
Alpine Bearberry 31142,31314
Alpine Bellflower 69889
Alpine Betony 373107
Alpine Bird's Foot Trefoil 237481
Alpine Birdtongue Chickweed 375124
Alpine Bistort 309954
Alpine Bittercress 72689,72749
Alpine Blue Sow-thistle 90836
Alpine Bluebasin 350090
Alpine Bluegrass 305303
Alpine Blue-grass 305303
Alpine Brassaiopsis 59192
Alpine Brook Saxifrage 349851
Alpine Brownish Sedge 74885
Alpine Brownleea 61170
Alpine Buckthorn 328604
Alpine Bugle 13149
Alpine Bulrush 353693
Alpine Burnet 345821
Alpine Buttercup 325562
Alpine Butterwort 299737
Alpine Campion 364088
Alpine Cat's Tail 294962
Alpine Cat's-tail 294962
Alpine Catchfly 363167,363168,364088,
410950
Alpine Cerastium 82646
Alpine Chaulmoogratree 199742
Alpine Chickweed 82646
Alpine China Ligusticum 229393
Alpine Chinese Buckthorn 328888
Alpine Chorispora 88997
Alpine Chrysanthemum 89419,227457,322646
Alpine Cinquefoil 312472,312852
Alpine Circaea 91508
Alpine Clematis 94714,95392
Alpine Clover 396818,396819
Alpine Club-rush 396005
Alpine Colt's-foot 197944
Alpine Coltsfoot 197944
Alpine Columbine 29997
Alpine Cotoneaster 107696
Alpine Cotton-grass 396005
Alpine Crazyweed 278700
Alpine Cretin Rhodiola 329862
Alpine Cryptandra 113337
Alpine Cudweed 224772
Alpine Currant 333904
Alpine Cystopetal 320240
Alpine Dandelion 384795
Alpine Delavay Holly 203756
Alpine Dewdrograss 91508
Alpine Dock 339914

Alpine Dogwood 380427
Alpine Dualbutterfly 398266
Alpine Dustymaidens 84490
Alpine Dwarf Sagebrush 35341
Alpine Edelweiss 224772
Alpine Elder 345660
Alpine Enchanter's-nightshade 91508
Alpine Epimedium 146961
Alpine Eryngo 154280
Alpine Euphorbia 158417
Alpine Eyebright 160231
Alpine Fargesia 162691
Alpine Fescue 164152
Alpine Fescuegrass 164152
Alpine Figwort 355142
Alpine Fir 402
Alpine Flax 231863
Alpine Fleabane 150446,150450,150454,150508
Alpine Fleeceflower 291663
Alpine Forget-me-not 260747
Alpine Forkstamenflower 416503
Alpine Fortune Plumyew 82511
Alpine Foxtail 17506
Alpine Fringed Pink 127870
Alpine Fritillary 168405
Alpine Gagea 169451
Alpine Galanthus 169706
Alpine Gentian 173198,173666,173872
Alpine Goatsmell Rockfoil 349458
Alpine Goatweed 8811
Alpine Golden Wild Buckwheat 152068
Alpine Goldenaster 194241
Alpine Goldenrod 368064
Alpine Goosecomb 335536
Alpine Grevillea 180568
Alpine Groundsel 279934
Alpine Hair Grass 126026
Alpine Hair-grass 126026
Alpine Hawkweed 195455
Alpine Heath 149147
Alpine Hedysarum 187769
Alpine Hemlock 399916
Alpine Herminium 192812
Alpine Heronbill 153903
Alpine Himalayas Ladybell 7662
Alpine Holly 203829
Alpine Homalanthus 197623
Alpine Hundredstamen 389600
Alpine Hyacinth 60339
Alpine Hyacinth Orchid 34896
Alpine Hypecoum 201600
Alpine Jerusalemsage 295054
Alpine Knotweed 291663,309954
Alpine Kobresia 217260
Alpine Laburnum 218757
Alpine Lady's Mantle 13959
Alpine Lady's-mantle 13959
Alpine Ladybell 7662
Alpine Ladymantle 13959
Alpine Lake False Dandelion 266825
Alpine Larch 221913
Alpine Larkspur 124024,124194
Alpine Laurel 215402
Alpine Leafybract Aster 380892
Alpine Lettuce 90836
Alpine Leucospermum 228031
Alpine Little Mouseear Cress 270461
Alpine Liver-balsam 151102
Alpine Lomatogoniopsis 235428
Alpine Loosestrife 239549
Alpine Lysionotus 239967
Alpine Marsh Violet 410347
Alpine Mazus 246965
Alpine Meadow Grass 305303
Alpine Meadow Rue 388404
Alpine Meadow Stonecrop 357044
Alpine Meadow-grass 305303
Alpine Meadowrue 388404
Alpine Meadow-rue 388404
Alpine Melic 249100
Alpine Merrilliopanax 250874
Alpine Milk Vetch 41962
Alpine Milkvetch 41962
Alpine Milk-vetch 41962
Alpine Mint Bush 315597
Alpine Minuartia 255506
Alpine Monkshood 5421
Alpine Mountain Sorrel 278577
Alpine mountain-sorrel 278577
Alpine Mouse Ear 82646
Alpine Mouseear 82646
Alpine Mouse-ear 82646
Alpine Mouse-ear Chickweed 82646
Alpine Muhly 259631
Alpine Nanbar Rattan Palm 65742
Alpine Neottia 264695
Alpine Nestorchid 264695
Alpine Oak 324373
Alpine Parnassia 284512
Alpine Pearlwort 342290
Alpine Peashrub 72208
Alpine Pea-shrub 72208
Alpine Penny Cress 390218
Alpine Pennycress 390242
Alpine Penny-cress 390208,390218,390242
Alpine Pepperbush 385091
Alpine Phyllodoce 297017
Alpine Pincushion 84490
Alpine Pine 299842,299886
Alpine Pine Dwarfmistletoe 30915
Alpine Pine Dwarf-mistletoe 30915
Alpine Pine Parasite 30915
Alpine Pink 127581,364088
Alpine Plantain 301853
Alpine Ploughpoint 401150
Alpine poker 217043
Alpine Pondweed 312037
Alpine Poppy 282506
Alpine Pricklyash 417282,417294
Alpine Prickly-ash 417282
Alpine Ptarmigan Berry 31314
Alpine Ptarmiganberry 31314
Alpine Ptarmigan-berry 31314
Alpine Pussytoes 26310
Alpine Puya 321937
Alpine Ragwort 359565
Alpine Red Catchfly 410950
Alpine Rice Flower 299305
Alpine Rock Cress 30164
Alpine Rock Jasmine 23121
Alpine Rockcress 30164
Alpine Rock-cress 30164
Alpine Rockjasmine 23182
Alpine Rock-jasmine 23121
Alpine Roegneria 335536
Alpine Roscoea 337066
Alpine Rose 330695,336348,336995
Alpine Rosularia 337297
Alpine Rush 212822
Alpine Sagewort 35993
Alpine Sandwort 32085,255573
Alpine Saussurea 348121
Alpine Savory 4662,65535
Alpine Sawwort 348121
Alpine Saw-wort 348089,348121
Alpine Saxifrage 349700
Alpine Scabious 350090
Alpine Scurvy-grass 98020
Alpine Sea-holly 154280
Alpine Sedge 73652
Alpine Serpentweed 392652,392653
Alpine Sheep Sorrel 340192
Alpine Shooting Star 135154
Alpine Silene 363167
Alpine Silverspikegrass 227952
Alpine Smartweed 308903
Alpine Smelowskia 366121
Alpine Snowbell 367763
Alpine Snow-gum 155667
Alpine Sow Thistle 90836
Alpine Sow-thistle 90836
Alpine Spapweed 204766,205184
Alpine Speedwell 406979
Alpine Spiraea 371798
Alpine Spirea 235651,371798,371945
Alpine Spruce 399916
Alpine Squill 352886
Alpine Stitchwort 255548
Alpine Stolon Willow 344176
Alpine Stonecrop 356988

Alpine Strawberry 167653,167657
Alpine Sun Rose 188586
Alpine Sun-rose 188586
Alpine Swampy Gentianopsis 174226
Alpine Sweet Grass 27963
Alpine Sweetgrass 27963
Alpine Sweet-grass 27963
Alpine Sweetvetch 187769
Alpine Tassel Flower 367766
Alpine Thermopsis 389507
Alpine Thickribcelery 279645
Alpine Thistle 76990,91931
Alpine Timothy 294962
Alpine Toadflax 230887
Alpine Totara 306509
Alpine Touch-me-not 204766,205184
Alpine Trigonotis 397464
Alpine Tripterospermum 398266
Alpine Turnflowerbean 98067
Alpine Typhonium 401150
Alpine Valerian 404292
Alpine Vanda 399700
Alpine Vernalgrass 27971
Alpine Violet 115931,409681,410161
Alpine Wallflower 154363,154497
Alpine Wattle 1037
Alpine Wendlandia 413825
Alpine Whitebark Pine 299786
Alpine Whitepearl 172059,172089
Alpine Willow Weed 146599
Alpine Willowherb 146609
Alpine Willowweed 146599
Alpine Wintergreen 172059,172089
Alpine Woundwort 373107
Alpine Yarrow 3921,36162
Alpine Yellow Gum 155754
Alpine Yucca 416560
Alpine-azalea 235279,235283
Alpine-bog Swertla 380302
Alpine-flames 323013
Alpinegold 199210
Alpine-loving Draba 137156
Alpine-loving Whitlowgrass 137156
Alpine-meadow Stonecrop 357044
Alpinia 17639
Alplily 234214
Alp-lily 234214
Alpparsley 98772,98786
Alps Alangium 13346
Alps Anthyllis 28179
Alps Honeysuckle 235651
Alps Onion 15493
Alps Plantain 301853
Alps Willow 343416
Alps Wormwood 36437
Alps Yarrow 3946
Alruna 244346
Alsace Broomrape 274941
Alsatian Clover 396930
Alsem 93868
Alseodaphne 17789
Alshinders 366743
Alsike Clover 396930
Alsine Starwort 374731
Alsine-like Evolvulus 161418
Alsophila Dogwood 380429
Alstonia 18027,18033,18034
Alstroemeria 18060,18062,18074
Altai Birch 53444
Altai Bluebasin 350111
Altai Bluegrass 305316
Altai Buttercup 325563
Altai Caladium 65269
Altai Catchfly 363169
Altai Centaurium 81523
Altai Chamaerhodos 85647
Altai Chickweed 374926
Altai Columbine 30038
Altai Cortusa 105489
Altai Craneknee 221765
Altai Crazyweed 278701
Altai Dandelion 384438
Altai Daphne 122365
Altai Doronicum 136339
Altai Draba 136918
Altai Euphorbia 158420
Altai Fakestellera 375209
Altai Falsehellebore 405604
Altai Falseoat 398442
Altai Fescuegrass 163802
Altai Flax 231864
Altai Furrowcress 383990
Altai Gagea 169383
Altai Galatella 169763
Altai Gentian 150457
Altai Globeflower 399491
Altai Goldenray 228977
Altai Gooseberry 333894
Altai Goosecomb 335200
Altai Hawthorn 109522
Altai Herelip 220053
Altai Heteropappus 193918
Altai Honeysuckle 235699
Altai Jurinea 193918
Altai Koeleria 217421
Altai Lagochilus 220053
Altai Leontice 182632
Altai Lettuce 219210
Altai Lymegrass 335200
Altai Milkaster 169763
Altai Milkvetch 41942
Altai Minorrose 85647
Altai Monkshood 5031,5581
Altai Mountain Centaurium 81523
Altai Mountain Greenbrier 366238
Altai Mountain Hawthorn 109522
Altai Mountain Thistle 92052
Altai Mountain Thorowax 63695
Altai Mountain Tulip 400116
Altai Mountains Desertcandle 148533
Altai Onion 15042
Altai Palm 246672
Altai Peony 280152
Altai Poppy 282666
Altai Pyrola 322810
Altai Rhubarb 329312
Altai Roegneria 335200
Altai Rush 212819
Altai Sagebrush 36174
Altai Salsify 394260
Altai Scabious 350111
Altai Scotch Rose 336971
Altai Sedge 73662
Altai Skullcap 355357
Altai Smilax 366238
Altai Starwort 374926
Altai Tansy 383714
Altai Taphropermum 383990
Altai Thistle 92052
Altai Thorowax 63695
Altai Thyme 203080
Altai Trisetum 398442
Altai Tulip 400116
Altai Vetch 408466
Altai Violet 409682
Altai Whitlowgrass 136918
Altai Wildryegrass 335200
Altai Windflower 23701
Altai Woodbetony 286991
Altai Yarrow 3966
Altaian Falsehellebore 405604
Altar Lily 334314,417092,417093
Altelnateleaf Goldsaxifrage 90325
Altelnateleaf Goldwaist 90325
Altenstein Encephalartos 145220,145221
Altered Andesite Wild Buckwheat 152433
Alternanthera 18089
Alternate Archiclematis 30955
Alternate Rhodiola 329835
Alternate Water-milfoil 261339
Alternate Wulingshan Ladybell 7895
Alternate-flowered Spartina 370156
Alternate-flowered Water-milfoil 261339
Alternateleaf Archiclematis 30955
Alternateleaf Butterfly Bush 61971
Alternateleaf Chessbean 116587
Alternateleaf Dogwood 104947
Alternate-leaf Golden Saxifrage 90325
Alternateleaf Mastixia 246210
Alternateleaf Peplis 290478
Alternateleaf Redsandplant 327195

Alternateleaf Solomonseal 308499
Alternateleaf Stringbush 414125
Alternate-leaved Butterfly-bush 61971
Alternate-leaved Dogwood 104947
Alternate-leaved Golden-saxifrage 90325
Alternate-leaved Loosestrife 239550
Alternate-leaved Redsandplant 327195
Alternate-leaved Stringbush 414125
Alternatetooth Hoarhound 245726
Althaea 18155,18173,195269
Althea 18155,195276,195278,195288
Alticupula Oriental Oak 323647
Altiligulate Bitter Bamboo 303997
Altiluda 277648
Altingia 18190
Altrot 192358
Alum 381035
Alum Bark Tree 378959
Alum Root 194411,194413,194428,194433,194437,194439
Alum-bark Tree 378959
Alum-bloom 174717
Aluminium Plant 298885
Aluminum Plant 298929
Alumroot 194411,194412,194413
Alum-root 174717,194411,194413
Alva Marina 418392
Alveolate Begonia 49607
Alyce Clover 18257,18273,18289
Alyceclover 18289
Alysicarpus 18257
Alysiclover 18257
Alyssum 18313,18374,18479,235090
Alyssus-like Asterothamnus 41748
Alyxia 18491,18516
Alyxialeaf Ardisia 31354
Alyxia-leaved Ardisia 31354
Am Aloe 10788
Amabile Lily 229723
Amaible Briggsia 60269
Amaible Michelia 252810
Amalilis Fir 280
Amalocalyx 18560
Amamastla Dock 339985
Amami Pine 299789
Amancay 18062
Amanis Gum 200844
Amanoa 18575
Amapola 290709
Amara 244397
Amaranth 18652,18687,18788,288863
Amaranth Cladostachys 123559
Amaranth Family 18646
Amaranthus 18652
Amareile Cherry 83361
Amaryllis 18862,196439,196443
Amaryllis Family 18861
Amaryllis-leaf Screwpine 280974
Amatnngulu 76864
Amatungula 76864,76926
Amaxon Grape 313262
Amazakoue 181773
Amazon Blue 276901
Amazon Chondrorhyncha 88855
Amazon Lily 155857,155858
Amazon Sprangletop 226022
Amazon Water Lily 408734
Amazon Water-platter 408734
Amazonian Zebra Plant 8554
Amazon-lily 155858
Amazonvine 376554
Amba 244397
Ambal 296554
Ambarella 372467
Ambari Hemp 194779
Ambary Hemp 194779
Ambatjang Mango 244393
Ambay Pumpwood 80004
Amber Bell 154899
Amber Groundsel 358240
Amber Lily 27394
Amber Pine 300221
Amber Tea Rose 336819
Amber Tree 27804
Amberbell 154899
Amberbloom Rhododendron 330704
Amber-bloom Rhododendron 331161
Amberboa 18946,18966
Ambiguous Barberry 51295
Ambiguous Cacalia 283789
Ambiguous Mono Maple 3174
Amblynotus 19042
Ambo Flindersia 166973
Amboi Phaius 293488
Amboin Coleus 99515
Amboin Trema 394628
Amboin Wildjute 394628
Amboina Dendrobium 124985
Amboina Pine 10496
Amboina Pitch Tree 10496
Amboina-berry 372467
Amboyna Wood 320269,320301
Ambrette Seed 235
Ambrevade 65146
Ambroma 672,675
Ambrose 388275
Ambrosia 19140,35132,139678,139682
Ambrozy Barberry 51559
Ambush 9531
Ambutch 9531
Amdo Sandwort 31740
Amee 19672
Amelachier 34854
Amelia's Sand Verbena 690
Amendoeira 386500
Amentotaxus 19356
Ameos 19672
Amer 380146
America Alder 16479
America Ash 167897,168032,168057
America Basil 268429
America Beech 162375
America Bellowsfruit 297842
America Bittersweet 80309
America Black Ash 168032
America Burnweed 148153
America Cow Slip 135153
America Elder 345580
America Elm 401431
America Ginsen 280799
America Jute 1000
America Larch 221904
America Mountainash 369313
America Overmallow 243880
America Persimmon 132466
America Planetree 302588
America Purplepearl 66733
America Red-onion 143571
America Sloughgrass 49540
America Smoketree 107315
America Spurge 159286
America Tallowwood 415558
America Trumpetcreeper 70512
America Waterwort 142558
America Wolftailgrass 289116
American Adder's Tongue 154899
American Aeschynomene 9495
American Agave 10787
American Alder 16368,16479
American Alkanet 233707
American Aloe 10787
American Alumroot 194413
American Angelica 24303
American Angelica Tree 30760
American Apple Mint 250362
American Arborvitae 390587
American Arbor-vitae 390587,390645
American Arrowhead 342363,342397
American Ash 167897,168032,168057
American Aspen 311547
American Avocado 291494
American Avocado-tree 291494
American Balsam Poplar 311237
American Barberry 51422
American Barnyard Grass 140451
American Basil 268429
American Basket Flower 303082
American Basketflower 303082
American Basswood 391649
American Beachgrass 19773
American Beach-grass 19773

American Beak Grain 128013
American Beakgrain 128013,128021
American Beauty Berry 66733
American Beauty Bush 66733
American Beautyberry 66733
American Beauty-berry 66733
American Beech 162375
American Bellbine 68746
American Bellflower 69892
American Birch 53338
American Bird Cherry 316918
American Bistort 54772
American Bittersweet 80285,80309
American Black Alder 16368
American Black Ash 168032
American Black Birch 53505
American Black Currant 333916
American Black Nightshade 366934
American Black Poplar 311292
American Black Spruce 298349
American Black Walnut 212631
American Black Willow 343302
American Blackberry 338281
American Blackcurrant 333916
American Bladder Nut 374110
American Bladdernut 374110
American Bluebell 69892
American Blue-eyed Grass 365782
American Bramble 338917
American Brooklime 406983
American Buckeye 9694
American Buckwheat-vine 61371
American Bugbane 91000
American Bugleweed 239186
American Bugseed 104753
American Bulrush 352249
American Burnet 345834
American Burnweed 148153
American Burn-weed 148153
American Burreed 370025
American Bur-reed 370025
American Butterflybush 61973
American Catalpa 79243
American Cherry 83311,316787
American Cherry Birch 53505
American Cherry Laurel 316285
American Chestnut 78790
American Christmas-mistletoe 295583
American Columbine 30007
American Columbo 167869,380146
American Coralbean 154646
American Cotton 179890
American Cowbane 90924
American Cow-parsnip 192278
American Cowslip 135153,135165
American Cow-wheat 248185,248186
American Crabapple 243583
American Cranberry 403894
American Cranberry Bush 408003,408188
American Cranberry Bush Viburnum 408191
American Cranberrybush 408002
American Cranberry-bush 408001,408188
American Crane's-bill 174717
American Cranesbill 174717
American Creeper 399606
American Crees 47963
American Cress 47931
American Crinum 111158
American Crowngrass 285419
American Cudweed 178367
American Cupscale 341993
American Cupscale-grass 341993
American Curatella 114809
American Currant 333916,334189
American Date Plum 132466
American Date-plum 132466
American Dendropanax 125592
American Dewberry 338225,338399
American Dittany 114527
American Dog Tooth Violet 154899
American Dog Violet 409851
American Dog-tooth Violet 154899
American Dog-violet 409851
American Dogwood 105023,105193
American Dragon's-head 137624
American Dragonhead 137624
American Dune Grass 228369
American Dwarf Birch 53594
American Ebony 61427
American Eelgrass 404536
American Elaeocarpus 142271
American Elder 345580
American Elderberry 345580,345646,345647
American Elemi 64068,64085
American Eleutherine 143571
American Elm 401431
American Ephedra 146134
American Euphorbia 159286
American Evergreen 381861
American Everlasting 21596
American False Hellebore 405643
American False Pennyroyal 187187
American Falsehellebore 405643
American False-hellebore 405643
American Falsemallow 243880
American Feverfew 285083,285088
American Figwort 355165
American Filbert 106698
American Fly Honeysuckle 235710
American Fly-honeysuckle 235709
American Forget-me-not 153420
American Fox-sedge 76735
American Fringetree 87736
American Frogbit 230240
American Frog-bit 230237,230240
American Fumitory 169076
American fustic Chittam 107315
American Gentian 173328
American Germander 388026,388032,388033
American Ginsen 280799
American Globe Flower 399512
American Globe-flower 399512
American Golden Dock 340060
American Golden Saxifrage 90335
American Gooseberry 333945
American Grass 391487
American Grass-of-parnassus 284545
American Great Burnet 345834
American Great Laurel 331207
American Great Rhododendron 331207
American Great-burnet 345834
American Greek Valerian 307241
American Green Alder 16479
American Gromwell 233753
American Groundnut 29298
American Hackberry 80698
American Hawthorn 109850
American Hazel 106698
American Hazelnut 106698
American Hellebore 405643
American Hemp 29473
American Highbush Cranberry 408188
American Hog-peanut 20560
American Holl 204114
American Honeylocust 176901,176903
American Hop 199387
American Hop Hornbeam 276818
American Hop-hornbeam 276818
American Hornbeam 77263,77264
American Horse Chestnut 9679
American Horse Mint 257184
American Hybrid Lilac 382171
American Hymenocallis 200941
American Ipecac 159150,175760,175763,175766
American Ipecacuanha 175763
American Ivy 20361,285136
American Jacob's Ladder 307245
American Jointvetch 9495
American Jointweed 308668
American Judas Tree 83757
American Knapweed 303082
American Knotweed 308824
American Lady's Tresses 372242
American Land Cress 47963
American Larch 221904
American Laurel 215386,215395
American Licorice 177922
American Lilac 81768
American Lily 263270
American Lily-of the-valley 102871

American Lime 391649
American Linden 391649
American Liquorice 177922
American Littorella 234147
American Liverleaf 192126
American Liverwort 192148
American Long-leaved Bamboo 181476
American Lop-seed 295984
American Lotus 263270
American Lotus-lily 263270
American Mahogany 380524,380527
American Mammea 243982
American Mandrake 306636
American Mangrove 329761
American Manna Grass 177610
American Marigold 383090
American Marram 19773
American Marsh Pennywort 200257
American Marsh Willow-herb 146766
American Mastic 350990
American Millet Grass 254515
American Milletgrass 254515
American Millet-grass 254515
American Mint 250331,257161
American Mistletoe 295575,295578,295581,295583
American Moor Rush 213478
American Mountain Ash 369313
American Mountain-ash 323083,369313
American Mulberry 259180,259186
American Muskwood 181578
American Nailwort 284737
American Narrow-leaved Bamboo 181474
American Narthecium 262574
American Needle-grass 376855
American Nelumbo 263270
American Nightshade 298094,366934
American Northern Dewberry 338399
American Oak 323625
American Oil Palm 142136,142264
American Oil-nut 323072
American Oilpalm 142264
American Olive 270061
American Onion 15857
American Orange-flower 88700
American Ox-eye 190518
American Palmetto 341422
American Pasqueflower 23979,321703,321708
American Pawpaw 38338
American Pennisetum 289116
American Pennyroyal 187187
American Persea 291494
American Persimmon 132466
American Pistachio 301004
American Pitch 299928
American Pitch Pine 300128
American Pitcherplant Family 347166
American Pitcher-plant Family 347166
American Plane 302588
American Plane Tree 302588
American Plane-tree 302588
American Plantain 302159
American Plum 316206
American Pogonia 306883
American Pokeberry 298094
American Pondweed 312042,312096
American Potato Bean 29298
American Potatobean 29298
American Pumpkin 114278
American Raspberry 338557
American Rattan Palm 126688
American Red Birch 53437
American Red Currant 334240
American Red Elder 345660
American Red Elderberry 345677
American Red Gum 232565
American Red Oak 320278,323894,324344
American Red Plum 316206
American Red Raspberry 338600,339310
American Red Spruce 298423
American Red-currant 334240
American River Sulphur Flower 152597
American Round-leaved Wintergreen 322782
American Rowan-tree 369313
American Rubber Plant 290395
American Sanicle 194413
American Sarsaparilla 30712,250221
American Sassafras 347407
American Saw-wort 348139
American Scurfpea 114425
American Sea Lavender 230566
American Sea Rocket 65185
American Sea-rocket 65175,65185
American Senna 360453
American Shoreweed 234151
American Silver Fir 282
American Skunk-cabbage 239517
American Sloe 408061
American Slough Grass 49540
American Sloughgrass 49540
American Slough-grass 49540
American Smoke Tree 107315
American Smoketree 107315
American Snowbell 379304
American Snowdrop Tree 184740
American Spatterdock 267276
American Speedwell 406983,407275
American Spider 242699
American Spikenard 30748,242699
American Spinach 298094
American Spurred-gentian 184694
American Squawroot 102003
American Squirrel Corn 128291
American Squirrel-corn 128291
American Starflower 396785
American Starwort 374741
American Stickseed 184300
American Storax 232565,379304
American Strawberry 167594
American Sugarberry 80673
American Sumach 64982
American Swamp Lily 348085
American Sweet Chestnut 78790
American Sweet Crabapple 243583
American Sweet Gum 232565
American Sycamore 302588
American Tallowwood 415558
American Tallow-wood 415558
American Tare 408257
American Tear-thumb 309796
American Threefold 399397
American Thrift 230566
American Trail-plant 7429
American Trout Lily 154943
American Trout-lily 154899
American Trumpet Creeper 70512
American Tulip Tree 232609
American Tupelo 267867
American Turk's Cap Lily 230044
American Turk's-cap Lily 230044
American Turkey Oak 324087
American Twinleaf 212299
American Umbrellaleaf 132678
American Upland Cotton 179906
American Valerian 120421,120432
American Valley Sulphur Flower 152581
American Vervain 405844
American Vetch 408257
American Wake Robin 33544
American Wake-robin 33544
American Walnut 212631
American Water Lily 267730
American Water Plantain 14760
American Water Weed 143920
American Water-horehound 239186
American Waterlily 267730
American Water-lily 267730
American Water-plantain 14760,14785
American Waterweed 143920
American Water-willow 214320
American Waterwort 142558
American Wayfaring 302068
American Wayfaring Tree 407674
American White Birch 53586
American White Elm 401431
American White Hellebore 405643
American White Oak 323625
American White Water-lily 267730,267777
American White Wood 391649
American Whitewood 232609

American Wild Celery 404536
American Wild Ginger 37574
American Wild Hazel 106698
American Wild Mint 250294,250331
American Wild Olive 276250
American Wild Plum 316206
American Wild Sensitive-plant 360437
American Wild Vanilla 228495
American Willowherb 146587,146664
American Willow-herb 146663,146664
American Winter Cress 47963
American Wintergreen 322782
American Wisteria 414565,414576
American Witch Hazel 185141
American Witch-hazel 185141
American Wood-lily 397540
American Woolly-fruit Sedge 75065,75066
American Wormseed 86934,139678
American Wormwood 19145
American Yard-grass 143553
American Yellow-rocket 47949
American Yellowwood 94022,94026
American Yellow-wood 94022,94026
American Yew 385342,385344
American-largeleaved Linden 391714
American-like Mint 257169
American-littleleaved Linden 391716
Ames Hemipilia 191583
Ames Pleurothallis 304905
Ames Vanda 197254
Ames' Rhododendron 330081
Amesiodendron 19417
Amethyst Aster 40047,380835
Amethyst Eryngo 154281
Amethyst Fescue 163811
Amethyst Flower 61118,61119
Amethyst Primrose 314117
Amethyst Shooting Star 135155
Amethystea 19458
Amiable Masdevallia 246102
Amilbed 93628
Amioki 296554
Amischophacelus 19484
Amischotolype 19490
Amitostigma 19498,19499
Amlexicaul Swertia 380212
Ammania 19550
Ammann Glorybind 102926
Ammi 19659
Ammobium 19690
Ammodendron 19728
Ammoniac Plant 136259
Ammopiptanthus 19776
Amnual Pink-sorrel 278062
Amoena Phlox 295246
Amoise 367357
Amole 244374
Amole Plant 244371
Amomum 19817,19930
Amoora 19952
Amorous Apple 239157
Amorpha 19989,20005
Amorpha Honey Locust 176862
Amorpha Honey-locust 176862
Amorphophallus 20037
Amourette 61059
Ampelocalamus 20192
Ampelopsis 20280
Ampfield Cotoncaster 107549
Amphibia Knotweed 308739
Amphibious Bistort 308739
Amphibious Knotweed 308739
Amphibious Marsh Cress 336172
Amphibious Marsh-cress 336172
Amphibious Persicaria 308739
Amphibious Water-cress 336172
Amphibola Willow 343008
Amphibolous Onion 15056
Amphicarpaea 20558
Ample Nymphaea 267654
Ample Onosma 271731
Ample Water Lily 267654
Amplectant Mahonia 242473
Amplectant Thrixsperm 390489
Amplexicaul Amitostigma 19500
Amplexicaul Burreed 370027
Amplexicaul Camellia 68903
Amplexicaul Cowwheat-like Loosestrife 239733
Amplexicaul Draba 136929
Amplexicaul Ixora 211038
Amplexicaul Jasmine 211812
Amplexicaul Jinggu Chickweed 375137
Amplexicaul Marshweed 230289
Amplexicaul Monocladus 56952
Amplexicaul Mosquitotrap 117364
Amplexicaul Rockcress 30182
Amplexicaul Rockfoil 349259
Amplexicaul Serrate Glorybower 96340
Amplexicaul Spapweed 204770
Amplexicaul Swallowwort 117364
Amplexicaul Tea 68903
Amplexicaul Thorowax 63711
Amplexicaul Thrixspermum 390489
Amplexicaul Touch-me-not 204770
Amplexicaul Whitlowgrass 136929
Amplexicaule Knotweed 308769
Amplexicaule Water Hymenacue 200827
Amplexicaul-leaf White-dragon Storax 379338
Amplexifoliate Camellia 68903
Amplexifoliate Eurya 160420
Amplexifolious Acanthocephalus 2134
Amplexifolious Glorybower 96340
Ampule Ascocentrum 38188
Amra 372478
Amsinckia 20840
Amsinkia 20843
Amsonia 20844
Amundson's Rhododendron 330083
Amur Adonis 8331
Amur Ampelopsis 20348
Amur Anemone 23705
Amur Angelica 24291
Amur Arisaema 33245
Amur Arrowwood 407677
Amur Barberry 51301
Amur Bellowsfruit 297837
Amur Bird Cherry 280028
Amur Birdcherry 280028
Amur Broomrape 274952
Amur Buttercup 325571
Amur Choke Cherry 280028
Amur Chokecherry 280028
Amur Choke-cherry Manchurian Bird-cherry 280028
Amur Cinquefoil 312659
Amur Columbine 29998
Amur Cork Tree 294231
Amur Corktree 294231
Amur Cork-tree 294231
Amur Corydalis 105594
Amur Cypressgrass 118491
Amur Deutzia 126841
Amur Erysimum 154380
Amur Eyebright 160134
Amur Falsespiraea 369264
Amur Flax 231865
Amur Galingale 118491
Amur Goosecomb 335204
Amur Grape 411540
Amur Grape Vine 411540
Amur Grapevine 411540
Amur Honeysuckle 235929
Amur Ladybell 7599
Amur Lilac 382277
Amur Lily 228565
Amur Lime 391661
Amur Linden 391661
Amur Maackia 240114
Amur Maple 2984
Amur Meadowsweet 166110
Amur Mountain Ash 369484
Amur Mountainash 369484
Amur Ninebark 297837
Amur Peppervine 20348
Amur Pepper-vine 20348
Amur Pink 127588
Amur Poplar 311229
Amur Poppy 282622
Amur Privet 229435,229558,229562

Amur Roegneria 335204
Amur Saddletree 240114
Amur Saussurea 348141
Amur Silver Grass 255873,394943
Amur Silvergrass 255873,394943
Amur Silver-grass 394943
Amur Smallflower Deutzia 126841
Amur Snakegrape 20348
Amur Southstar 33245
Amur Thyme 391078
Amur Valerian 404225
Amur Vetch 408278
Amur Violet 409687
Amur Waterchestnut 394417
Amur Whitevillous Monkshood 5665
Amur Willow Weed 146605
Amur Willowweed 146605
Amur Windflower 23705
Amurland Grape 411540
Amuyon Goniothalamus 179405
Amy 19664
Amydrium 20866
Amygdalata Leaved Tanoak 233104
Amygdaline Attalea 44802
Amygdaline Cryptocarya 113426
Amygdaline Horsfieldia 198508
Amygdaline Thickshellcassia 113426
Amyroot 29473
Ana Tree 1035,1037
Anaba 165423
Anabasis 21023
Anacacho Orchid Tree 49161
Anacahuita 104159
Anacampseros 21068
Anacamptis 21162
Anadendrum 21311
Anagyris-like Rattle-box 112405
Anami Gum 200844
Ananas 21472,21479
Ananbeam 157429
Anancia 175454
Anaphale Marguerite 21596
Anatolian Marsh Orchid 121369
Anatolian Orchid 273324
Anatto 54859,54862
Anatto Tree 54859,54862
Anattotree 54859
Anatto-tree 54859,54862
Anaxagorea 21847
Ancathia 21886,21888
Ancestral Tworow Barley 198365
Anchietea 21889
Anchor Bush 99814
Anchor Collabium 89939
Anchor Plant 99812,99814
Anchor Waterchestnut 394546
Anchored Water Hyacinth 141807
Anchorstyleorchis 130032
Anchovy Pear 181047
Anchusa 14836,21914
Ancient Hookime 113577
Ancient Pine 300040
Ancient Spikenard 262497
Ancient-coins Falsepimpernel 231553
Ancients Euphorbia 158456
Ancients Spurge 158456
Ancistrochilus 22040
Ancistrocladus 22048
Ancistrocladus Family 22047
Ancylostemon 22158
Anda Crazyweed 278707
Anda-Assy Oil 212380
Andaman Edony 132292
Andaman Gurjun 133553
Andaman Macaranga 240229
Andaman Marble 132244
Andaman Marble Wood 132244
Andaman Mombin 372478
Andaman Padauk 320288
Andaman Pyinma 219932
Andaman Redwood 320301
Andaman Zebra-wood 132244
Andean Barberry 51309
Andean Begonia 49614
Andean Sage 345005
Andean Silver Leaf 345005
Andean Wax Palm 84326
Andean Weeping Bamboo 90630
Anderson Alseodaphne 17790
Anderson Bulbophyllum 62557
Anderson Cherry-laurel 223094
Anderson Cirrhopetalum 62557
Anderson Clematis 94690
Anderson Curlylip-orchis 62557
Anderson Cyrtopodium 120706
Anderson Edelweiss 224800
Anderson Habranthus 184257
Anderson Lycium 239003
Anderson Rockfoil 349059
Anderson Wolfberry 239003
Anderson's Clematis 95299
Anderson's Hawksbeard 110991
Anderson's Larkspur 124033
Anderson's Mountaincrown 274125
Anderson's Thistle 91739
Andes Berry 338475
Andes Berry Bush 338475
Andes Podocarpus 316084
Andes Rose 49586,50646
Andes Yaccatree 316084
Andiroba 72640
Andley White Cedar 85376
Andorra Creeping Juniper 213799
Andorra Juniper 213786
Andoung 257730
Andrachne 22226
Andrachnelike Skullcap 355367
André Barberry 51310
Andrea's Broom 121005
Andréan Pitcairnia 301102
Andreas Broom 121001
Andreson Windhairdaisy 348144
Andréw's Gentian 173222
Andréws Eucalyptus 155480
Andréws' Bog Orchid 302251
Andrews' Lady's-slipper 120293
Andréws' Rose-purple Orchid 183412
Andrieux Mahonia 242475
Androcorys 22320
Andrographis 22400
Andromeda 22418,298703
Andropogonoid Yushania 416748
Androsaemme 202086
Androssov Skullcap 355368
Andrzejowsk Rose 336356
Andurion 158062
Andys Juniper 213763
Aneghany Black-berry Allegheny Blackberry 338105
Aneilima 260102
Anemarrhena 23669,23670
Anemoclema 23680
Anemone 23696,23724,23846,23917,77140,321653
Anemone Buttercup 325577
Anemone Clematis 95127
Anemoneflower Rose 336357
Anemone-flowered Rose 336357
Anemoneleaf Conebush 210235
Anemone-like Isopyrum 210256
Anemonelike Paraquilegia 283721
Anemonella 24133
Anemones 23767
Anemony 23696
Anemonyflower Rose 336357
Anet 24213
Anethum Conebush 210236
Aneurocarp Sedge 73688
Aneys 299351
Anfu Maple 3594
Angel 174877
Angel Face Rose 336285
Angel Flower 3978
Angel Gabriel 230058
Angel Trumpets 4779
Angel Wing Begonia 49587
Angel Wing Jasmine 211927
Angel Wings 65214,65218,65223
Angel's Eyes 407072
Angel's Fishing Rod 130187,130229
Angel's Fishing Rods 130229

Angel's Fishingrod 130187,130229
Angel's Fishing-rod 130187,130225
Angel's Hair 36225
Angel's Tears 54440,54454,123076,130229,262473,262474,274823,367772,407072
Angel's Trumpet 61230,61231,61235,61236,61239,61242,123061,123065,123076,123077
Angel's Trumpets 61242,123036,123054
Angel's Turnip 29468
Angel's Wing 65230
Angel's Wing Begonia 49729
Angel's-eyes 407072
Angel's-tears 61242
Angel's-trumpet 61235,61242,123061
Angel's-wing Begonia 49729
Angel's-wings 272980
Angelica 24281,24303,24475,24503,30587,30932,98772
Angelica Tree 30587,30604,30760,125592
Angelicaleaf Ligusticum 229292
Angelicalike Hogfennel 292770
Angelicalike Pleurospermum 304757
Angelicalike Ribseedcelery 304757
Angelica-tree 30587,135082
Angelico 229298
Angelim-rosa 302798
Angelin 22218
Angelin Tree 22211
Angelin-tree 22211,22218
Angelique 129488
Angelita Daisy 201296,387349,387355
Angelonia 24519
Angels 3462
Angels' Wings 65218,65228
Angels-and-devil 37015
Angels-and-devils 37015
Angels-tears 262473
Angels-trumpet 61231
Angel-tears Datura 61242
Angelwing Begonia 49729
Angel-wing Begonia 49729
Angelwing Jasmine 211927
Angel-wings 65214
Angico Gum 283669
Angle Bittersweet 80141
Angle Onion 15060
Angle-berry 222820
Angled Agapetes 10287
Angled Bittersweet 80141
Angled Pea 222676
Angled Solomon's Seal 308613
Angled Spike-rush 143316
Angled-seed Evodia 161380
Angledtwig Magnoliavine 351078
Angle-fruit Stewartia 376465
Anglehelmet Woodbetony 287003
Angle-pod 117552,179470
Angler's Flower 355061
Anglestem 80141
Angle-stem Beak Sedge 333498
Anglestem Primrose Willow 238180
Angle-stem Wild Buckwheat 151816
Angleton Bluestem 128571
Angle-twig Poplar 311231
Anglewing Begonia 49629
Anglojap 385384
Anglo-jap Yew 385384
Angola Afrormosia 10053
Angola Afzelia 10127
Angola Alangium 13384
Angola Pea 65146
Angola Protea 315718
Angophora 24606
Angostura 170165
Angostura Bark Tree 170167
Angostura-bark Tree 170167
Angraecum 24687
Anguineal Cactus 244064
Angular Agapetes 10287
Angular Barberry 51311
Angular Begonia 49617
Angular Fruit Fig 165658
Angular Holly 203543
Angular Leucas 227550
Angular Sea-fig 77446
Angular Skullcap 355727
Angular Solomon's Seal 308616
Angular Solomon's-seal 308613,308616
Angular Spapweed 204873
Angular Staff-tree 80141
Angular Touch-me-not 204873
Angular Willowweed 146612
Angular Woodbetony 287003
Angular-branch Holly 203543
Angularflower Elaeagnus 142011
Angular-flowered Elaeagnus 142011
Angularfruit Iris 208589
Angularfruit Swordflag 208589
Angular-stem Holboellia 197248
Angularstem Thin Raspberry 338777
Angulata Poplar 311292
Angulate Branchlet Poplar 311231
Angulate Clearweed 298854
Angulate Densehead Mountainash 369305
Angulate Poplar 311231
Angulate Torricellia 393108
Angulated Onion 15060
Anguloa 25134
Anguria 114122
Angustate Ainsliaea 12605
Angustate Rabbiten-wind 12605
Anguste Bitterbamboo 304006
Anguste Kobresia 217118
Angustem Bittersweet 80141
Angustibracteate Raspberry 338134
Angustibracted Linden 391669
Angustifoliate Bousigonia 57912
Angustifoliate Camellia 68904,69077
Angustifoliate Dogwood 124882
Angustifoliate Elaeagnus 141930
Angustifoliate Eritrichium 153424
Angustifoliate Fargesia 162642
Angustifoliate Gardenia 171440
Angustifoliate Holly 203801
Angustifoliate Ledum 223909
Angustifoliate Michelia 252813
Angustifoliate Privet 229436
Angustifoliate Scabrous Deutzia 127077
Angustifoliate Spice-bush 231303
Angustifoliate Stringbush 414129
Angustifoliate Sweetleaf 381145
Angustifoliate Water-plantain 14721
Angustifolius Pittosporum 301218
Angustilobed Cranesbill 174463
Angustin Syzygium 382677
Angustine Syzygium 382677
Angustine Wendlandia 413771
Angustior Sedge 73696
Angusti-sepal Anisetree 204475
Anhispid Ampelocalamus 20194
Anhui Asyneuma 43571
Anhui Barberry 51314,51454
Anhui Bittercress 72684
Anhui Bittergrass 404537
Anhui Boea 56053
Anhui Chickweed 374743
Anhui Chloranthus 88266
Anhui Clematis 94816
Anhui Crape-myrtle 219911
Anhui Fritillary 168344
Anhui Lily 229727
Anhui Lycoris 239256
Anhui Maple 2791
Anhui Primrose 314344
Anhui Rhododendron 331177
Anhui Sedge 74130
Anhui Skullcap 355375
Anhui Solomonseal 308503
Anhui Stonecrop 356542,356850
Anhui Stonegarlic 239256
Anhui Stringbush 414133
Anhui Touch-me-not 204775
Anhui Tulip 400117
Anhui Vetchling 222679
Anhui-Hubei Sage 345289
Anhwei Primrose 314344
Anhwei Skullcap 355375
Anil-de-pasto 206626
Animated Oat 45586
Anime 200848

Anio Mulberry 259097
Anisachne 25265
Anisadenia 25272,25273
Anise 24213,235090,242274,261585, 299351
Anise Burnet Saxifrage 299351
Anise Chervil 261585
Anise Gianthyssop 10407
Anise Hyssop 10402,10407
Anise Primrose 314122
Anise Root 276472
Anise Shrub 204474
Anise Tree 204474,204603
Aniseed 204474,299351
Aniseed Star 204603
Aniseed Tree 46335
Aniseed-tree 46335,204474,204603
Anise-hyssop 10407
Aniseia 25315
Aniseroot 276455
Anise-root 276472
Anise-scented Goldenrod 368290
Anise-scented sage 345072
Anisetree 204474
Anise-tree 204474
Anisetree-like Cinnamon 91338
Anisetreelike Pittosporum 301294
Anisetree-like Pittosporum 301294
Anisetree-like Seatung 301294
Anisette 167156,299351
Anisochilus 25379
Anisodus 25438
Anisolip Rabdosia 324728
Anisomeles 25466
Anisopapus 25488
Anisophyllous Beautyberry 66735
Anisophyllous Jasmine 212068
Anisophyllous Purplepearl 66735
Anisopterous Desertcandle 148534
Aniso-sepaled Honeysuckle 235667
Anisosepalous Persimmon 132045,132046
Ankle-aster 207393
Anlong Camellia 68906
Anlong Germander 388000
Anlong Mycetia 260611
Anlong Oilfruitcamphor 381790
Anlong Photinia 295621
Anlong Stonecrop 357253
Anlong Syndiclis 381790
Anlung Germander 388000
Anlung Photinia 295621
Ann Barberry 51315
Anna 25751,25754
Annabelle 199807
Annam Chaulmoogratree 199743
Annam Mahonia 242477
Annam Papeda 93430,93560
Annam Plum 166789
Annam Pouteria 313338
Annam Supplejack 52402
Annam Supple-jack 52402
Annam Thrixsperm 390490
Annam Thrixspermum 390490
Annam Vanilla 404978
Annamocarya 25759
Annaseed 261585
Annatto 54862
Anneslea 25771
Anneslia 25771
Anning Litse 233843
Anning Sedge 73707
Annona 25828
Annona Family 25909
Annual Agoseris 11460
Annual Aster 67314
Annual Baby's-breath 183191
Annual Bastard Cabbage 326833
Annual Bastard-cabbage 326833,326835
Annual Bedstraw 170193
Annual Bellflower 70001
Annual Blanket-flower 169603
Annual Bluegrass 305334
Annual Blue-grass 305334
Annual Bugloss 21920
Annual Bursage 19141
Annual Bur-sage 19145
Annual Bush Sunflower 364512
Annual Buttonweed 107755
Annual Canarygrass 293729
Annual Candyleaf 376409
Annual Candytuft 203184,203249
Annual Chrysanthemum 89466,89481
Annual Clary 345474
Annual Coreopsis 104598
Annual Delphinium 102833
Annual desert trumpet 151944
Annual Dry-flower 415317
Annual False Foxglove 45175,45181,45184
Annual Fescue 412407,412465
Annual Fleabane 150464
Annual Fountain Grass 289248
Annual Foxtail 17521
Annual Gentian 150464,173598
Annual Golden Eye 409148
Annual Greenhead Sedge 218618
Annual Ground Cherry 297720
Annual Gypsophila 183191
Annual Hair Grass 12789,170167
Annual Hairgrass 12816
Annual Hair-grass 12789
Annual Hedgenettle 373110
Annual Honesty 238371
Annual Hop 199382
Annual Knawel 353987
Annual Koeleria 217518
Annual Lavatera 223393
Annual Locoweed 42810
Annual Lonas 235504
Annual Majoram 274224
Annual Mallow 223393
Annual Meadow Grass 305334
Annual Meadow-grass 305334
Annual Mercury 250600
Annual Milkweed 37888
Annual Mugwort 35132
Annual Nettle 403039
Annual Oxeyedaisy 89466
Annual Pearlwort 342226
Annual Phlox 295267
Annual Poinsettia 159046
Annual Rabbit's-foot Grass 310125
Annual Rabbitsfoot Grass 310125
Annual Rock Rose 399957
Annual Rock-cress 30189
Annual Rockrose 399957
Annual Ryegrass 235334,235335
Annual Sage-wort 35132
Annual Salmarsh Aster 41317
Annual Saltmarsh Aster 256310
Annual Sandwort 255561
Annual Scorpion-vetch 105310
Annual Sea-blite 379553
Annual Sea-purslane 361660
Annual Silver Hairgrass 12816
Annual Skeleton-weed 362162
Annual Sow-thistle 368771
Annual Spike Rush 143261
Annual Spikerush 143158
Annual Stink Groundpine 70461
Annual Stock 246478,246479
Annual Stonecrop 356544
Annual Toadflax 230931,231084,267347
Annual Trampweed 161919
Annual Valerian 81751
Annual Vernal Grass 27934
Annual Vernalgrass 27934
Annual Vernal-grass 27934
Annual Vetch 222680
Annual Wallrocket 133286
Annual Wall-rocket 133286
Annual Wheatgrass 148420
Annual Wild Buckwheat 151827
Annual Wild Rice 418080
Annual Wildrice 418080
Annual Willow-herb 146638
Annual Woolly Bean 378502
Annual Wormwood 35132
Annual Yellow Clover 249218
Annual Yellow Sweetclover 249218
Annual Yellow Woundwort 373110
Annual Yellow-woundwort 373110

Annuciation Lily 229789
Annular Begonia 49622
Annular Cyclobalanopsis 116054
Annular Elegant Evergreenchinkapin 78964
Annular Holostemma 197476
Annular Oak 116054
Annular Qinggang 116054
Annular Scissorsvine 197476
Annular Tylophora 400879
Annular Willow 343028
Annularline Willow 343028
Annulate Bambusa 47187
Annulate Fig 164633
Annulate Gelidocalamus 172738
Annulate Tupistra 70594
Annunciation Lily 229789
Annuul Sage 345340
Anny 299351
Annyle 299351
Anoda 25922,25923
Anodendron 25927
Anoectochilus 25959
Anogeissus 26018
Anomal Mouseear 82664
Anomalous Crinum 111168
Anomalous Hawthorn 109526
Anomalous Mouse-ear Chickweed 82796
Anomalous Orinus 274250
Anomalous Sunflower 188921
Anona Blanca White Anona 25850
Anopyxis 26186
Anotom Figmarigold 251085
Anseikan 93470
Ansellia 26273
Ansellia Orchid 26274
Anserine Liana 61371
Anshan Willow 343677
Anshun Litse 233956
Anson's Sallow 343776
Ansu Apricot 34477
Ant Osbeckia 276130
Ant Palm 217918
Ant Tree 397933
Antao Nutmeg 261457
Antarctic Barley 198343
Antarctic Beech 266864,266875
Antarctic False Beech 266864
Antarctic Magellan Barberry 51402
Antecrenate Willow 343035
Antelope Brush 321886
Antelope Bush 321878
Antelope Grass 140490
Antelope Horn 38171
Antelope Orchid 125031
Antelope Sage 152160,152167
Antenoron 26622
Antflower 252771
Anthelmintic Ironweed 406102
Anthericum 26950
Antheroporum 27464
Anthocephalus 263983
Anthony Peak Larkspur 124044
Anthosachne Roegneria 144168
Anthosachnelike Goosecomb 144168
Anthrywood 400376,400377
Anthurium 28071
Anthyllis 28143
Antiaris 28242
Anticosti Aster 380838
Antidiarrhealtree 197175
Antidy Senteric Pulicaria 321554
Antidysenteric Fleaweed 321554
Antifebrile Dichroa 129033
Antigonon 28418
Antilles Fanpetals 362500
Antilles Fear 90019
Antilles Pear 90019
Antilles Pine 300125
Antilogous Mussaenda 260383
Antioquien Passionflower 285615
Antiotrema 28547
Antipdo Gumihan 36914
Antirhea 28563
Antlerorchis 310787,310797
Antlerpilose Grass 257537,257543
Antlerpilose Qinggang 116204
Antlevine 88888,88891
Antony Barberry 51316
Antshape Micrechites 25937
Antu Plantainlily 198601
Antu Woodrush 238676
Antuco Barberry 51317
Antwerp Hollyhock 13934
Anu 399611
Anuping 261457
Anyaran 134776
Anzacwood 310752
Anze Bouquet Larkspur 124243
Aoap Willow 344261
Apache Beads 24178
Apache Beggarticks 53905
Apache Pine 299931
Apache Plume 162554,162556
Apache-plume 162554,162556
Apalachicola Aster 160684
Apalachicola Doll's-daisy 56701
Apalanche 204376
Apama 28900
Aparejo Grass 259715
Apargidium 253916
Ape's Earring 301117,301128
Ape's-earring 301117
Apearleaf Rhododendron 332155
Apennine Anemone 23707
Apennine Globularia 177039
Apennine Rockrose 188590
Apependiculate Corydalis 105622
Apera 28969
Apetalous Manyleaf Paris 284362
Apetalous Monkshood 5042
Apetalous Paris 284362
Apetalous Pearlwort 342226
Apetalous Sandwort 256456
Apetalous Sterculia 376075
Apetalus Footcatkin Willow 343657
Aphanamixis 28993
Aphananthe 29004
Aphania 29031
Aphanopleura 29085
Aphelandra 29121
Aphragmus 29164
Aphrodita Moth Orchid 293582
Aphyllorchis 29212
Apiculate Barberry 51318
Apiculate Elaeocarpus 142273
Apiculate Onosma 271732
Apiculate Ormosia 274378
Apiculate Osmanther 276252
Apiculate Osmanthus 276252
Apiculate Plicate Woodbetony 287522
Apiculate Rockvine 387746
Apiculate Violet 410278
Apiculate-lobed Beautyberry 66838
Apiesdoring 1238
Apimpe 143530
Apio 34914
Apio Arracacia 34914
Apios 29295,29305
Apitong 133563
Aplder Orchid 272395
Apluda 29422,29430
Apocopis 29449
Apodytes 29581
Apoi Birch 53353
Apollo Fir 306,308
Aponogeton Family 29696
Apophyllum 29701
Apopo 237927
Aporosa 29759
Aporoseila 29727
Aposeris 29785
Apostasia 29787,29790
Apostasia Family 29796
Apostle Plant 264161,264163
Apostles 274823
Apothecaries' Rose 336581
Apothecary's Rose 336593
Apothecaries' Rose 336581
Appalachian Arrowhead 342317
Appalachian Bugbane 91037
Appalachian Bunch-flower 248662

Appalachian Goldenrod 368058
Appalachian Mock-orange 294479
Appalachian Nailwort 284905
Appalachian Rosinweed 364327
Appalachian Smooth Phlox 295275
Appalachian Stitchwort 255481
Appalachian Tea 203847,204391,407737
Appalachian Twayblade 232966
Appalachian White Snakeroot 11157
Apparent Rhododendron 331636
Appendageless Arisaema 33328
Appendicula 29807
Appendiculate Cremastra 110502
Appendiculate Galanga Galangal 17678
Appendiculate Sage 344854
Appendiculate Slugwood 50471
Appendiculate Spapweed 204778
Appendiculate Touch-me-not 204778
Appendiculate Waterhawthorn 29652
Apple 243543,243675,243711
Apple Bamboo 249620
Apple Banana 260253
Apple Berry 54432,54437
Apple Blossom 78336,78338
Apple Blossom Grass 172201
Apple Box 155507
Apple Dumpling 54437
Apple Family 242848
Apple Fig 165398
Apple Geranium 288396
Apple Mango 244406
Apple Mint 250294,250439,250457
Apple of Jerusalem 256797
Apple of Love 239157
Apple of Peru 123077,265940,265944
Apple of Sodom 44708,366906,366934,
367233,367481,367620
Apple Pelargonium 288396
Apple Pie 36474,72934,85875,146724
Apple Quince 116547
Apple Ring Acacia 1035,1037
Apple Ringie 35088
Apple Rose 336286,337006
Apple Serviceberry 19260
Apple Shrub 413591
Apple Tree 243543,243675
Appleberry 163502
Appleblossom Cassia 78316
Apple-blossom Cassia 78336
Applebox 155507
Applecactus 185912
Apple-flower Deutzia 127011
Apple-fruited Granadilla 285668
Apple-fruited Rose 337006
Appleleaf Arrowwood 407765
Apple-leaf Hawthorn 109670
Appleleaf Raspberry 338784
Appleleaf Viburnum 407765
Apple-leaved Raspberry 338784
Apple-leaved Willow 343664
Apple-mint 250457,250468
Apple-of-Peru 265940,265944
Apple-ring Acacia 1035
Apple-scented Mint 250439
Apple-topped Box 155481
Apple-tree 243543
Appressed Arrowbamboo 162637
Appressed Fargesia 162637
Appressed-hair Lasianthus 222090
Appressed-hair Roughleaf 222090
Appressipubescent Michelia 252818
Approximate Barberry 51319
Aprecock 34475
Apricock 34475
Apricot 34425,34475,34477,316166,316230
Apricot Geranium 288503
Apricot of San Domingo 243982
Apricot Plum 316813
Apricot Vine 285654
Apricotleaf Ladybell 7853
Apricotleaf Pear 323093
Apricotleaf Pimpinella 299376
Apricotleaf Tanoak 233104
Apricot-leaved Pear 323093
Apricot-leaved Tanoak 233104
Apricotsmell Rabbiten-wind 12635
Apricot-yellow Gongora 179289
Apricotyellow Paphiopedilum 282779
Aprono 244715
Apse 311537
Apsen-tree 311537
Aptandra 29842
Aptenia 29853
Apterosperma 29897,29898
Apuan Willow 343254
Apuleia 29956
Aquatic Barnyardgrass 140477
Aquatic Chikusichloa 87353
Aquatic Galangal 17646
Aquatic Gullygrass 87353
Aquatic Mannagrass 177629
Aquatic Milkweed 38058
Aquatic Morning Glory 207590
Aquatic Panicgrass 281342,282031
Aquatic Rattan Palm 65641
Aquatic Screwpine 280982
Aquatic Sedge 73732
Aquatic Soda Apple 367661
Aquatic Sunflower Jurinea 207136
Aquatic Tillaea 391923
Aquilegia 29994
Ara Wood 65002
Arab Bird's Foot Trefoil 237491
Arab Coffee 98857
Arab Gumtree 1572
Arab Jurinea 211990
Arab Speedwell 407287
Arabia Schismus 351163
Arabia Themeda 389361
Arabia Yellowback Grass 389361
Arabian Acacia 1427
Arabian Coffee 98857
Arabian Coffee Plant 98857
Arabian Euphorbia 158428
Arabian Horseradish 258936
Arabian Jasmine 211990
Arabian Juniper 213849
Arabian Pea 54841
Arabian Primrose 34600,34606,34635,
315087
Arabian Rue 185676
Arabian Schismus 351163
Arabian Senna 78227
Arabian Star Flower 274513
Arabian Star of Bethlehem 274513
Arabian Star-flower 274513
Arabian Tea 79395
Arabian Violet 161531
Arabian Wolfberry 239005
Arabian-primrose 34606,34609
Arabian-tea 79395
Arabic Acacia 1572
Arabic Gum 1572
Arabic Gum Tree 1572
Arabic Gumtree 1572
Arabic Star-of-bethlehem 274513
Arabic Tree 1572
Arabican Coffee 98857
Arach 44468
Arachnis 30526
Arachnislike Ornithoboea 274466
Arachnoid Everlasting 21534
Arachnoid Haworthia 186269
Arachnoid Leptospermum 226452
Arachnoid Pearleverlasting 21534
Araeostachya Willow 343040
Arage 44468,44576
Arakud Iguanura 203502
Aralia 30587,135082,350688
Aralia-leaf Raspberry 338967
Aramina Fibre 402245
Arar 386942
Arar Tree 386942
Arara Nut 212380
Arara Nut Tree 212380
Araroba 22212
Ara-root 245014
Arar-tree 386942
Araucaria 30830,263707
Araucaria Family 30857
Araujia 30860,30862

Arauria-da-caledomia 30839
Arb Rabbit 174877
Arbale 311208
Arbeal 311208
Arber Bean 294056
Arbeset 30888
Arbol Del Pito 155134
Arbor Sancta 248895
Arboreous Brugmansia 61231
Arboreous Fleshseed Tree 346980
Arborescent Aloe 16592
Arborescent Angel's-tears 61245
Arborescent Ardisia 31454
Arborescent Buchanania 61666
Arborescent Calligonum 67001
Arborescent Ceratoides 218121
Arborescent Evodia 161384
Arborescent Glochidion 177103
Arborescent Hakea 184591
Arborescent Heliotrope 190559
Arborescent Kneejujube 67001
Arborescent Pipturus 300835
Arborescent Pricklypear 272950
Arborescent Seepweed 379514
Arborous Beautyberry 66738
Arborous Ironweed 406109
Arborous Timonius 392088
Arborvitae 302718,390567,390661
Arbor-vitae 390567,390587,390645
Arbuteleaf Chosenia 89160
Arbute-leaved Chosenia 89160
Arbute-tree 30888
Arbuti Tree 334976
Arbutus 30877
Arby-root Arby 34536
Arcane Bamboo 297195
Arcang. sia 30897
Arcfruit Milkvetch 43100
Archangel 24281,30932,170025,220346,
220416,373455
Archangel Fir 300223
Archangelica 30929
Archarde 324335
Arched Leaf Mouseear Cress 317344
Archedleaf Mouseear Cress 317344
Archer Echidnopsis 140019
Archiatriplex 30946,30947
Archiboehmeria 30950
Archiclematis 30953
Archileptopus 30987,30988
Arching Barberry 51395
Arching Dewberry 339160
Archiphysalis 30998
Arctic Alpine Fleabane 150698
Arctic Aster 160681
Arctic Avens 138453
Arctic Bearberry 31314
Arctic Bellflower 70351
Arctic Birch 53530
Arctic Bladderwort 403353
Arctic Bluegrass 305362
Arctic Blue-grass 305362
Arctic Bramble 338142
Arctic Brome 60793
Arctic Butterbur 292354
Arctic Campion 363579,364088,410950
Arctic Cassiope 78630
Arctic Chrysanthemum 89424
Arctic Claytonia 94299
Arctic Creeping Willow 344000
Arctic Daisy 31027,31028
Arctic Dock 339937
Arctic Draba 136969,136995
Arctic Dwarf Birch 53530,53531
Arctic Fleabane 151036
Arctic Gentian 151036
Arctic Gray Willow 343425
Arctic Hairgrass 126029
Arctic Heather 298721
Arctic Iris 208819
Arctic Larkspur 124073
Arctic Marshmarigold 68160
Arctic Meadow-rue 388404
Arctic Minuartia 255442
Arctic Mouse-ear 82673
Arctic Peariwort 174347,342290
Arctic Poppy 282676
Arctic Pyrola 322830
Arctic Raspberry 338142
Arctic Rhododendron 331057
Arctic Rockcress 30434
Arctic Rush 212844,212852
Arctic Sagebrush 35153,35505
Arctic Sandwort 32103,255442
Arctic Saxifrage 349700
Arctic Scurvyweed 97977
Arctic Stitchwort 255442
Arctic Sweet Coltsfoot 292354
Arctic Sweet-colt's-foot 292354
Arctic Whitlowgrass 136995
Arctic Willow 343045,343592,343937
Arctic Willowweed 146623
Arctic Wintergreen 322830
Arctic Wood Rush 238571
Arctic Woodrush 238572
Arctic-cudweed 270571
Arcticflower 231676,231679
Arctogeron 31079,31080
Arctosepal Spapweed 204781
Arctotis 31176
Arctous 31311
Arcuate Calanthe 65881
Arcuate Larkspur 124046
Arcuate Monkshood 5043
Arcuate Peashrub 72184
Arcuate Pea-shrub 72184
Arcuate Seepweed 379495
Arcuate Spiraea 371814
Arcuate Trigonella 397192
Arcuate Waterchestnut 394421
Arcuate-fruit Milkvetch 43100
Arcuatenerve Drypetes 138581
Arcuatestigma Sedge 74051
Arcuscaul Burreed 370034
Ardisia 31347
Ardisia-like Blueberry 403729
Ardisia-like Holly 203571
Ard-losserey 176824
Arebian-tea 79387
Areca 31677,31680
Areca Nut 31680
Areca Palm 31677,31680,89360
Areca-nut Palm 31680
Arecapalm 31677
Arecastrum 31702
Areche 44468
Aren 32343
Arenarous Sainfoin 271172
Areng Palm 32343
Arenga 32330
Arere 398004
Aress 170193
Arethusa 32371,32373
Argan Tree 32403
Argans 274237
Argan-tree 32403
Argemone 312360
Argemony 32411,32423
Argent Asiabell 98279
Argent Bluebellflower 115340
Argent Cinquefoil 312389
Argentate Southstar 33276
Argentate Sweetleaf 381423
Argentatic Parthenium 285076
Argenteous Azalea 331824
Argenteous Krameria 218076
Argentill 13966
Argentina 312360
Argentina Rhododendron 330789
Argentina Senna 360429
Argentine 270705,312360
Argentine Black Walnut 212590
Argentine Cocklebur 415053
Argentine Dock 340008,340059
Argentine Evening-primrose 269468
Argentine Fingergrass 160992
Argentine Fleabane 103438
Argentine Habenaria 183391
Argentine Mesquite 315550
Argentine Pepperweed 225305
Argentine Pepperwort 225305

Argentine Screwbean 315573
Argentine Sprangletop 225990
Argentine Tecoma 385477
Argentine Trumpet Vine 97471
Argentine Vervain 405812
Argentine Walnut 212590
Argerian Cottonthistle 271672
Argid Jurinea 214033
Argostemma 32518
Argun Dragonhead 137551
Argun Greenorchid 137551
Argun Groundsel 358292
Argus Paphiopedilum 282778
Argute Barberry 51325
Argy Elaeagnus 141942
Argy Sagebrush 35167
Argy Sedge 73751
Argy's Wormwood 35167
Argyle Apple 155529
Argyll's Tea Tree 239011
Argyranthemum 32548
Argyrate Windmill-palm 181803
Argyreia 32600
Argyroderma 32689
Argyrolobium 32769
Argyrotrichia Oak 116055
Argyrotrichia Qinggang 116055
Arhar 65146
Aricanga Shadow Palm 174347
Ariculate Bamboo 297213
Arid Pear 323346
Arid Sedge 73754
Aridicolous Hibiscus 194717
Arid-stem Begonia 49633
Arietinous Lady's Slipper 120299
Arikury Palm 33197
Arillate Mayten 246810
Arisaema 33234,33540
Arisan Bentgrass 12000
Arisan Cherry 316874
Arishan Actinidia 6530
Arishan Elm 401642
Arishan Hornbeam 77314
Arishan Oak 116200
Arishan Sage 345076
Arishan Sedge 73755
Aristate Aloe 16605
Aristate Barberry 51327
Aristate Bellflower 69905
Aristate Biflorgrass 128571
Aristate Dali Jerusalemsage 295106
Aristate Dichanthium 128571
Aristate Fallacious Sugarcane 262550
Aristate Gentian 173261
Aristate Goosefoot 139681
Aristate Pfeiffer root 276455
Aristate Shrubbamboo 388748
Aristate Treebine 92614
Aristate Umbrella Bamboo 388748
Aristato-serrulate Barberry 51331
Aristea 33608,33643
Aristocrat Plant 186341
Aristolochia 34097
Aristotelia 34394
Aristulate Mahonia 242486
Aristulate Rockfoil 349069
Arizina Elder 345627
Arizona Agave 10802
Arizona Agoseris 11467
Arizona Alder 16425
Arizona Ash 168127,168136
Arizona Baccharis 46261
Arizona Barrel Cactus 163483
Arizona Black Walnut 212620
Arizona Blue Curls 396435
Arizona Blue-eyes 161428
Arizona Boxelder Maple 3225
Arizona Bramble 338150
Arizona Bugbane 91001
Arizona Claret-cup Cactus 140213
Arizona Cork Fir 403
Arizona Cottonrose 235255
Arizona Cottontop 130475
Arizona crested coral-root 194525
Arizona Cypress 114652,114692
Arizona Dock 340084
Arizona Fescue 163818
Arizona Fishhook Cactus 244087
Arizona Fleabane 150474
Arizona Fuschsia 146648
Arizona Goldenbush 150309
Arizona Grape Ivy 93007
Arizona Hedgehog Cactus 140213
Arizona Jewelflower 377634
Arizona Kidneywood 161862,161863
Arizona Longleaf Pine 299931
Arizona Lupine 238434
Arizona Madrone 30881
Arizona Mesquite 315575
Arizona Necklacepod 368972
Arizona Nut Pine 299926
Arizona Oak 323668
Arizona Pine 299797,299851,300153
Arizona Plane 302596
Arizona Planetree 302596
Arizona Ponderosa Pine 299797
Arizona Poppy 215381
Arizona Queen-of-the-night 288986
Arizona Rabbitbrush 90500
Arizona Rabbit-tobacco 317729
Arizona Rainbow Hedgehog Cactus 140313
Arizona Red Raspberry 338150
Arizona Rosewood 405223
Arizona Rrainbow Cactus 140313
Arizona Rubberweed 201332
Arizona Sandmat 158472
Arizona Siltbush 418479
Arizona Skeletonplant 239330
Arizona Slimpod 20851
Arizona Smooth Cypress 114652
Arizona Snakecotton 168683
Arizona Snake-cotton 168683
Arizona Snakeweed 182094
Arizona Sneezeweed 188400
Arizona Sophora 368971
Arizona Sunflower 188925
Arizona Sycamore 302596
Arizona Thistle 91756
Arizona Valerian 404228
Arizona Walnut 53530,212620
Arizona Water Willow 214403,214410
Arizona White Oak 323668
Arizona Wild Buckwheat 151841
Arizona Wild Fuchsia 417406
Arizona Wild Grape 411556
Arizona Willow 343541
Arizona-giant 77077
Arjinshan Alkaligrass 321231
Arjinshan Bluegrass 305369
Arjinshan Leymus 228340
Arjun 386462
Arjun Terminalia 386462
Arkansas Beard-tongue 289321
Arkansas Calamint 65538
Arkansas Doze-daisy 29106
Arkansas Gayfeather 228452
Arkansas Ironweed 406114
Arkansas Lazydaisy 29106
Arkansas Oak 323669
Arkansas Rose 336361,336363
Arkansas Yucca 416552
Arl 16352
Armand Clematis 94748
Armand Pine 299799
Armand's Pine 299799
Armed Barberry 51332
Armenia Epidendrum 146380
Armenia Grepe-hyacinth 260295
Armenia Pocktorchid 282779
Armenia Poppy 282517
Armenia Pricklythrift 2216
Armenia Skullcap 355381
Armenian Crane's-bill 174847
Armenian Grape Hyacinth 260295
Armenian Hawthorn Barberry 51511
Armenian Holly y 204114
Armenian Oak 324301
Armenian Plum 316356
Armeria 34488
Armgrass 58052,58088
Armillate Cactus 244004

Armodorum 34575
Armoise Absinthe 35090
Armoise Annuelle 35132
Armoise Bisannuelle 35598
Armoise Douce 35505
Armour Persimmon 132051
Armstrong 308816
Armstrong Gasteria 171622
Arnberry 338557
Arnebia 34600
Arnell Sedge 73767
Arnhem Land Bamboo 47189
Arnica 34655,34730,34752
Arnica Root 34752
Arnicks 325666
Arnold Hawthorn 109536,109850
Arnold Milkvetch 42017
Arnott Meliosma 249360
Arnott Pricklyash 417176
Arnott Spapweed 204785
Arnott Touch-me-not 204785
Arntree 345631
Arnut 63475,102727
Aroeira Blanca 233795
Aroid Palm 65239
Arolla Pine 299842
Aroma 129136
Aromatic Aster 40948,380937
Aromatic Bamboo 87633
Aromatic Blumea 55687
Aromatic Catchfly 363840
Aromatic Chondrorhyncha 88856
Aromatic Dendranthema 124813
Aromatic Epidendrum 146381
Aromatic Hamiltonia 370869
Aromatic Houlletia 198688
Aromatic Illigera 204612
Aromatic Knotweed 309951
Aromatic Litse 233882
Aromatic Lycaste 238733
Aromatic Pleurospermum 273937
Aromatic Ribseedcelery 273937
Aromatic Sage 344838
Aromatic Sumac 332477,332908
Aromatic Sumach 332477
Aromatic Syzygium 382477
Aromatic Turmeric 114859
Aromatic Wintergreen 172047,172089
Aromaticorchis 185753,185755
Aromaticroot 407579,407585
Aromcress 94232
Aron 37015
Arpent 357203
Arpent-weed 357203
Arracacha 34914
Arracacia 34909
Arrach 44576
Arradan 192358
Arran Service Tree 369493
Arran Service-tree 369493
Arrayan 291363
Arrow Arum 288800,288808
Arrow Aster 381007
Arrow Bamboo 304098,318277,318307
Arrow Broom 173051
Arrow Cowlily 267326
Arrow Fimbristyle 166411
Arrow Grass 397159
Arrow Head 342308
Arrow Pod Grass 397165
Arrow Pod-grass 397165
Arrow Primrose 314396
Arrow Vine 309796
Arrow Wood 407953,408081
Arrowanther Cheirostylis 86702
Arrowanther Lagotis 220202
Arrow-arum 288800
Arrowbamboo 162635,162695,162722,162752
Arrow-bearing Tree 16359,215442,345620
Arrowfeather 34000
Arrow-feather 34000
Arrow-feather Three-awn 34000
Arrowgrass 397142
Arrow-grass 350861,397142,397159,397165
Arrowgrass Family 212753
Arrow-grass Family 212753
Arrowhead 342308,342338,342363,342393,342397,342400,342424,381855
Arrowhead Family 14790
Arrowhead Groundsel 360240
Arrowhead Meadowrue 388667
Arrowhead Rattle-box 112636
Arrowhead Sweet Coltsfoot 292360
Arrowhead Sweet-colt's-foot 292360
Arrowhead Vine 381861
Arrowhead Violet 410527
Arrowheadlike Wildginger 37713
Arrow-leaf 342363
Arrowleaf Abelmoschus 242
Arrowleaf Alocasia 16508
Arrowleaf Aster 381007
Arrowleaf Balsam-root 47147
Arrowleaf Butterfly Pea 81877
Arrowleaf Clover 397130
Arrowleaf Cremanthodium 110456
Arrowleaf Gerbera 175195
Arrowleaf Goldenray 229168
Arrowleaf Greyawngrass 372425
Arrow-leaf Groundsel 360240
Arrowleaf Knotweed 309796
Arrowleaf Monochoria 257570
Arrowleaf Nutantdaisy 110456
Arrowleaf Orange 93492
Arrowleaf Philodendron 294835
Arrow-leaf Pond-lily 267326
Arrowleaf Sedge 73875
Arrowleaf Snakeroot 313878
Arrowleaf Sweet Coltsfoot 292360
Arrow-leaf Sweet-colt's-foot 292360
Arrow-leaf Tearthumb 291932
Arrow-leaf Wild Buckwheat 151934
Arrowleafbetony 252549,252550
Arrowlear False Rattlesnakeroot 283694
Arrow-leaved Aster 381007
Arrow-leaved Kumquat 93492
Arrow-leaved Tear-thumb 291932,309796
Arrow-leaved Tearweed 309796
Arrow-leaved Violet 410412,410527,410528
Arrowlike Parabaena 283060
Arrowpetal Stonecrop 357121
Arrowpoison Strophanthus 378445
Arrow-poison Strophanthus 378445
Arrowpole Coldbamboo 172743
Arrowroot 3978,37015,47147,192358,245013,245014,388387
Arrow-root 245013,245014
Arrowroot Family 245039
Arrowroot Tacca 382912,382923
Arrowscale 44606
Arrowshaft Poplar 311414
Arrowshaft Singlebamboo 56956
Arrow-shaped Fig 165609
Arrowshaped Habenaria 184053
Arrowshaped Tinospora 392269
Arrow-shaped Tinospora 392269
Arrow-vine 291932
Arrowweed 305134
Arrow-weed 305134,386764
Arrowwood 407645,407662,407688,407778,407779,407781,407785,407795,407943,408065
Arrow-wood 294462,328603,407645,407662,407778,408079,408080,408081
Arrowwood Viburnum 407778
Arroyo Lupine 238489
Arroyo Willow 343597
Arsenic Plant 99297
Arsenicke 309199
Arsesmart 309199,309570,322724
Artemisia 35084,35090,35837,36225
Arternateleaf Didymocarpus 129875
Artfulttea 79387,79395
Arthraxon 36612
Arthrophytum 36813
Arthrostemma 36880
Artichoke 117770,117787
Artichoke Betony 373439
Artichoke Betony Chorogi 373439
Artichoke Cactus 268086
Artichoke Thistle 117770

Articulate Pholidota 295511
Articulation-bearing 372406
Artillery Clearweed 298974
Artillery Plant 298848,298885,298974,299070
Artillery Weed 298974
Artocarpus 36902
Arts 403916
Arugula 154019
Arum 36967,37015,137744
Arum Family 30491
Arum Fern 65239
Arum Lily 36967,37015,417093
Arum-leaved Arrowhead 342315,342325
Arundina 37109
Arundinella 37350
Arvi Arva 374968
Arym 355202
Arytera 37521,37522
Asa's Foot 400675
Asafoetida 163574,163612
Asafoetida Giant Fennel 163580
Asahikan 93326
Asamela 290876
Asarabacca 37616,175454,207087
Asarabaeca 37556
Asarina 37548
Asarum-leaved Wintergreen 322787
Asbestus Yushanbamboo 416786
Ascendent Crabgrass 130489
Ascendent Cranesbill 174478
Ascendent Saxifrage 349076
Ascending Basil 268421
Ascending Wild Basil 96966
Ascending-branched Aster 40557
Ascension 360328
Ascent Leucadendron 227227
Ascesion Lily 229789
Ascherson's Orchardgrass 121264
Ascidiform Sweetleaf 381450
Ascidium Sweetleaf 381450
Ascinode Bluestem 22516
Ascitesgrass 407449,407498
Asclepias 37811,37979
Ascocentrum 38187
Ascotian Begonia 49634
Ascyrion 202179
Ash 167893,167940,167955,382329
Ash Barberry 242478
Ash Family 270182
Ash Gourd 50998
Ash of Jerusalem 209229
Ash Pumpkin 50998
Ash Tanoak 233147
Ashanti Blood 260414
Ashanti Gum 386448
Ashanti Pepper 300397,300401
Ashby's Banksia 47629
Ash-colored Smoke Tree 107309
Ash-colored Smoketree 107309
Ash-coloured Holly 203636
Ash-coloured Syzygium 382508
Ash-coloured Tanoak 233147
Ashe Birch 53630
Ashe Juniper 213620
Ashe Magnolia 241979
Ashe's Juniper 213620
Ashe's Magnolia 241979
Ashen Tree 167955
Ashen-tree 167955
Ash-grey Covered Rhododendron 331966
Ashgrey Lily 228584
Ashgrey Loosestrife 239617
Ash-grey Mussaenda 260433
Ashgrey Sedge 74135,75357
Ashhairy Pearleverlasting 21541
Ashing-tree 167955
Ashland Thistle 91880
Ashleaf Evodia 161333
Ashleaf Hiptage 196831
Ashleaf Kitefruit 196831
Ashleaf Lomatia 235407
Ashleaf Maple 3218
Ash-leaf Raspberry 338435
Ashleaf Stringbush 414208
Ash-leaved Maple 3218,3228
Ash-leaved Negundo 3218
Ash-leaved Raspberry 338433
Ash-leaved Stringbush 414208
Ashly Goldenrod 368250
Ashplant 227919
Ashrobe Azalea 331966
Ashrobe Rhododendron 331966
Ashthroat 405872
Ashwagandha 414601
Ashy Cinquefoil 312676
Ashy Crane's-bill 174542
Ashy Dichrostachys 129136
Ashy Hakea 184594
Ashy Limberbush 212121
Ashy Pipewort 151268
Ashy Poker 302103,302104
Ashy Sunflower 189001
Ashycoloured Ironweed 115595
Ashy-grey Sinobambusa 364799
Ashy-greyflower Lovegrass 148007
Asia Beech 162395
Asia Bell 98273
Asia Bell Tree 325002
Asia Belltree 325002
Asia Birdcherry 280004
Asia Bittergrass 404555
Asia Bluebells 250895
Asia Bushbeech 178026
Asia Cocklebur 11552
Asia Crinum 111167
Asia Currant 334232
Asia Dandelion 384628
Asia Dryas 138460
Asia Dunegrass 392095
Asia Forget-me-not 260762
Asia Ginsen 280741
Asia Glory 32600
Asia Grewia 180682
Asia Helictotrichon 190159
Asia Hepatica 192138
Asia Juneberry 19248
Asia Junegrass 217427
Asia Lily 228599
Asia Longleaved Pine 300186
Asia Medic 247257
Asia Minor Bluegrass 310109
Asia Mint 250323
Asia Neottia 264651
Asia Osmanther 276257
Asia Plantain 301871
Asia Poppy 335624
Asia Pouchvine 8299
Asia Rosewood 121631
Asia Sandwort 31758
Asia Selfheal 316097
Asia Shadbush 19248
Asia Starjasmine 393616
Asia Striga 377973
Asia Sweet Pink 127608
Asia Toddalia 392559
Asia Tree Cotton 179876
Asia Yarrow 3934
Asiabell 98273
Asiabigflower Sagebrush 35857
Asiaglory 32600
Asia-glory 32600
Asian Baneberry 6414
Asian Barberry 51333
Asian Bell 98273,325002
Asian Bell Tree 325002
Asian Bittersweet 80260
Asian Black Birch 53400
Asian Broomrape 274968
Asian Bulrush 353576
Asian Bushbeech 178026
Asian Butterflybush 61975
Asian Butterfly-bush 61975
Asian Cable Creper 25928
Asian Colubrina 100061
Asian Firethorn 322480
Asian Flatsedge 118491
Asian Fly Honeysuckle 235970
Asian Ghostweed 294344
Asian Goatsbeard 37056
Asian Grewia 180682

Asian Indian Mallow 853
Asian Koeleria 217427
Asian Lopseed 295986
Asian Marshweed 230323
Asian Mazus 247025
Asian Meadowsweet 372117
Asian Melastome 248732
Asian Mint 250323,309482
Asian Molucca Balm 256761
Asian Mountainfringe 8299
Asian Mustard 59651
Asian Nakedwood 100061
Asian Osmanthus 276257
Asian Pear 323268
Asian Pigeonwings 97203
Asian Pussy Willow 343185
Asian Selfheal 316097
Asian Service Berry 19248
Asian Serviceberry 19248
Asian Skunk-cabbage 239518
Asian Snakevine 100061
Asian Spiderwort 260102
Asian Spikesedge 218628
Asian Summerlilic 61975
Asian Taro 16518
Asian Tetracera 386897
Asian Thyme 391351
Asian Ticktrefoil 126389,126395
Asian Toddalia 19248,392559
Asian Virginsbower 94910
Asian Waterwort 142556
Asian White Birch 53483,53572
Asiatic Agrimonia 11552
Asiatic Apple 243702
Asiatic Ardisia 31578
Asiatic Barberry 51333
Asiatic Barringtonia 48510
Asiatic Beech 162395
Asiatic Bird Cherry 280003
Asiatic Bittersweet 80260
Asiatic Blister-cress 154533
Asiatic Bush-beech 178026
Asiatic Butter-fly-bush 61975
Asiatic Button-bush 82107
Asiatic Centella 81570
Asiatic Colubrina 100061
Asiatic Corn 417417
Asiatic Currant 334232
Asiatic Dayflower 100961
Asiatic Dewflower 260116
Asiatic Dragonflyoechis 302513
Asiatic Eight-petals Dryas 138460
Asiatic False Hawksbeard 416437
Asiatic Ginsen 280741
Asiatic Ginseng 280741
Asiatic Globe Flower 399494
Asiatic Greenhead Sedge 218547
Asiatic Hepatica 192138
Asiatic Jasmine 393616
Asiatic Longleaved Pine 300186
Asiatic Maple 3736
Asiatic Moonseed 250228
Asiatic Orach 44468
Asiatic Orache 44468
Asiatic Pennywort 81570,200259
Asiatic Plantain 301864,301871
Asiatic Rosewood 121631
Asiatic Serviceberry 19248
Asiatic Service-berry 19248
Asiatic Star-jasmine 393616
Asiatic Striga 377973
Asiatic Sweetleaf 381341,381355
Asiatic Tearthumb 309564
Asiatic Toddalia 392559
Asiatic Tree Cotton 25882,179876
Asiatic Tree-cotton 179876
Asiatic Witchweed 377973
Asiatic Yarrow 3934
Asidefield Cypressgrass 118982
Asidefield Galingale 118982
Asidiform Viburnum 408198
Askew Dishspurorchis 171896
Askew Gastrochilus 171896
Askew Vetch 408648
Askewfoot Dragonbamboo 125509
Asmanian Peppermint 155478
Asmanian Snow Gum 83060
Asmy Barberry 51335
Asna 386452
Asoca Tree 346504
Asoka 346504
Asp 311537
Asparagus 38904,39120
Asparagus Bean 319114,409086,409096
Asparagus Broccoli 59545
Asparagus Bush 137446
Asparagus Cowpea 409096
Asparagus Fern 38983,38985,39120,39171,39195
Asparagus Lettuce 219485,219488
Asparagus Pea 237771,319114,409096
Asparagus-fern 38904,39195
Asparaguslike Gentian 173265
Aspe 311537
Aspel-like Teasel 133454
Aspen 311131,311335,311398,311537,311547
Aspen Poplar 311537
Asper False Fairybells 134403
Asperge 38904
Asperulous Ehretia 141606
Asphodel 39300,39435,39438,39449
Asphodelus 39454
Aspidistra 39516,39532
Aspidocarya 39607
Aspidopterys 39666,39676
Ass Ear 381035
Ass Parsley 9832
Assacu 199465
Assagai Wood 114918
Assai 161055
Assai Palm 161055
Assam Buckeye 9678
Assam Caper 71694
Assam Cherry Laurel 223104
Assam Cherry-laurel 223106,316489
Assam Crotalaria 111915
Assam Dense-leaved Mahonia 242623
Assam Glochidion 177105
Assam Hairy Bamboo 297200
Assam Horse-chestnut 9678
Assam Hymenacue 200829
Assam King Begonia 50232
Assam Knotweed 308804
Assam lemon 93552
Assam Medinilla 247537
Assam Meranti 362190
Assam Onion 15375,15680
Assam Palm 130057
Assam Pennywort 200308
Assam Plectocomia 303088
Assam Plum Yew 82523
Assam Raspberry 338158
Assam Rattlebox 111915
Assam Rhododendron 330235
Assam Rosewood 121623
Assam Rubber 164925
Assam Rubber Tree 164925
Assam Rubber-tree 164925
Assam Sloanea 366037
Assam Tea 69644
Assam Treebine 92616
Assam Twayblade 232083
Assam Water Hymenacue 200829
Asse's Box Tree 238996
Ass-ear 381035
Assear Arisaema 33442
Assear Southstar 33442
Assegai Tree 114915,114918
Assegai Wood 386638
Assi Crazyweed 278724
Assion Lily 229789
Ass-parsley 9832,28044
Assurgent-flower Treemallow 223357
Assyrian Plum 104218
Asta 278159
Astelia 39874
Aster 16287,39910,39911,70948,155799,160645,160673,193125,268681,380833,380866
Aster Family 101642

Aster Pubescent 135274
Aster Rude 160676
Asteranthera 41551
Asterbush 41747
Asterlike Boltonia 56706
Asterothamnus 41747
Asthma Plant 159069
Asthma Weed 234547
Asthmaweed 103438,103617
Astilbe 41786,41789
Astilbileaf Pimpinella 299357
Astilboides 41870,41872
Astrachan Hawksbeard 110738
Astrantia 43306,43309
Astriate Holly 203788
Astridia 43320
Astrocarpus 361274
Astrocaryum 43372
Astroloba 43428
Astrologia 308893
Astronia 43457
Astrucarpus 43372
Asturian Daffodil 262354
Astyleorchis 19498,19500,19501,19512
Asymmetrical Leaf Actinidia 6617
Asymmetrical Leaf Kiwifruit 6617
Asyneuma 43569
Asystasia 43596
Asystasiella 43679
Atalantia 43718
Atamasco Lily 417608
Atamascolily 417608
Atamasco-lily 417608
Atchern 324335
Athel 383437
Athel Tamarisk 383437
Athel Tree 383437
Atherlus 176824
Atherton Palm 218792
Atis Root 5272
Atkinson Goldenray 228990
Atkinson Pyrethrum 322653
Atlantic Camas 68806
Atlantic Cedar 80080
Atlantic Cordgrass 370156
Atlantic Date Palm 295454
Atlantic Goldenrod 367965
Atlantic Ivy 187276
Atlantic Leatherwood 133656
Atlantic Ninebark 297849
Atlantic Pigeonwings 97195
Atlantic Pistache 300977
Atlantic Sedge 73784
Atlantic Star Sedge 73782,73784
Atlantic White Cedar 85375
Atlantic White-cedar 85375
Atlantic Yam 131896
Atlas Cedar 80080
Atlas Mastic 300977
Atlas Mastic Tree 300977
Atlas Mountain Cedar 80080
Atlas Pistache 300977
Atlas Poppy 282519
Atllspice 68313,383712
Atractylodeleaf Groundsel 381922
Atractylodes 44192
Atractylodes-leaf Synotis 381922
Atropa 44704
Atropanthe 44717,44719
Atropurple Lymegrass 144188
Atropurple Wildryegrass 144188
Atsa Rhodiola 329841
Attalea 44801
Attalea Cohune 273244
Attani 93753
Attenuate Elatostema 142610
Attenuate Eurya 160593
Attenuate Heteropanax 193901
Attenuate Ironweed 406123
Attenuate Knotweed 308808
Attenuate Osmanthus 276261
Attenuate Rockcress 30188
Attenuate Spikesedge 143055
Attenuate St. John's Wort 201761
Attenuate Stairweed 142610
Attenuated Haworthia 186303
Attenunate-leaf Spicebush 231424
Atterlothe 44708
Attopev Xanthophytum 415155
Attoto Yam 131516
Atuberculate Camellia 68925,69731
Atuntsi Woodbetony 287023
Atuntsuen Rhodiola 329842
Atuntze Barberry 51944
Atwood's Wild Buckwheat 152520
Atylosia 44824
Aubergine 367370
Aubert Blyxa 55913
Aubert Knotweed 162509
Aubespyne 109857
Aubit Dichorisandra 128989
Aubretia 44867
Aubrey Purple 201733
Aubrieta 44867
Aubrietia 44867,44868
Aucher Saltbush 44324
Aucher-eloy Iris 208704
Auckland Azalea 330794
Auckland Rhododendron 330794
Aucklandia 44878
Aucuba 44887,44915
Aucuricled Hedyotis 187510
Aucutesepal St. John's Wort 201711
Augers 344244
August Lily 18865,198639
August Plum 316206
August Squarebamboo 87616
Augustin St. John's Wort 201765
Augustine Henry's Rhododendron 330165
Augustine Rhododendron 330165
Augustine St. John's-wort 201765
August-lily 198639
Aul 16352
Aulacolepis 25331
Aulieat Pricklythrift 2217
Aulne 16352
Aulne Vert 16479
Aum 401593
Aum-tree 401593
Aunt Hannah 30164
Aunt Lucy 143886
Aunt Mary's Tree 203545
Aunt-eliza 111470
Auntie Polly 315101,315106
Aurahuaku-taquina Barberry 51343
Aurantiaceous Matucana 379638
Auratiaceous Lampranthus 220483
Aureate Rockfoil 349440
Aureatelike Rockfoil 349441
Aureo-coma Sasa 347197
Aureoglobose Chinese Juniper 213636
Aureus Epipremnum 147338
Aureus Kylinleaf 147338
Auricled Arisaema 33272
Auricled Greenbrier 366518
Auricled Millettia 254614
Auricled Twayblade 232893
Auricledleaf Mosquitotrap 117640
Auricledleaf Swallowwort 117640
Auricle-leaved Fig 164661
Auricula 314143
Auricula Tree 68095
Auriculate Acacia 1067
Auriculate Adinandra 8208
Auriculate Azalea 330186
Auriculate Bashania 48707
Auriculate Blastus 55140
Auriculate Cacalia 283792
Auriculate Combretum 100344
Auriculate Dichocarpum 128924
Auriculate Dichrocephala 129068
Auriculate Dysophylla 306960
Auriculate Eurya 160452
Auriculate False Foxglove 10140
Auriculate Fisheyeweed 129068
Auriculate Forkfruit 128924
Auriculate Hardhair Indocalamus 206769
Auriculate Hygrophila 200597
Auriculate Indocalamus 206769
Auriculate Leaf Actinidia 6617
Auriculate Leaf Kiwifruit 6617

Auriculate Lettuce 219221
Auriculate Liparis 232086
Auriculate Macaranga 240233
Auriculate Millettia 254614
Auriculate Mosquitotrap 117390
Auriculate Oreocharis 273840
Auriculate Paraboea 283109
Auriculate Perilepta 290919
Auriculate Pogostemon 306960
Auriculate Rhododendron 330186
Auriculate Rockcress 30189
Auriculate Rockfoil 349089
Auriculate Senna 78236
Auriculate Southstar 33272
Auriculate Spapweed 204790
Auriculate Stick Touch-me-not 204859
Auriculate Swallowwort 117390
Auriculate Touch-me-not 204790
Auriculate Twayblade 232086
Auriculate Watercandle 306960
Auriculate Willow 343063
Auriculate Windmill 100344
Auriculate-dentate Perilla 290947
Auriculateleaf Aster 40095
Auriculate-leaf Begonia 49642
Auriculate-leaf Blueberry 403971
Auriculate-leaf Ixora 211044
Auriculateleaf Paraprenanthes 283681
Auriculate-leaf Senna 78236
Auriculate-leaved Rhododendron 330186
Auriculatum Elatostema 142611
Auriform Eurya 160427
Auritus Fargesia 162643
Aurtcled Panicgrass 281362
Aruncus 37060
Aussiepoplar 197619
Austral Akebia 13240
Austral Arecastrum 31707
Austral Bladderwort 403110
Austral Indigo 205704
Austral Maackia 240118
Austral Rhubarb 329314
Austral Sarsaparilla 366246
Australasian Crane's-bill 174658
Australasian Geranium 174658
Australia Beech 155700
Australia Bird's Foot Trefoil 237506
Australia Blackwood 1384
Australia Burnweed 148163
Australia Dracaena 104366
Australia Gymnema 182384
Australia Nightshade 367289
Australia Veinyroot 237506
Australia Yaccatree 306424
Australian 10505
Australian Acacia 1384
Australian Acalypha 1790
Australian African Violet 56063
Australian Almond 386584
Australian Amaranth 18764
Australian Asthma-weed 159069
Australian Bamboo 47189
Australian Banyan 165274
Australian Baobab 7028
Australian Bean Flower 215978
Australian Beech 155700,266875,385530
Australian Beefwood 79157
Australian Black Wood 1384
Australian Blackwood 1384
Australian Blackwood Family 255141
Australian Bluebell 141258,368541,412691,412876
Australian Bluebell Creeper 368540,368541
Australian Bluestem 57548
Australian Bottlcbrush 67253
Australian Bottle Plant 212202
Australian Boxthorn 64064
Australian Brome 60632
Australian Brown Alder 4772
Australian Brush-cherry 382481
Australian Brush-cherry Eugenia 382481
Australian Bur-grass 394378
Australian Burnweed 148163
Australian Burr Grass 394378
Australian Butterflybush 61984
Australian Cabbage Palm 234166
Australian Cabbage-palm 234166
Australian Carrot 123180
Australian Cedar 392829
Australian Celtis 80708
Australian Cheesewood 301433
Australian Chestnut 79075
Australian Cinnamon 91388
Australian Couch 144453
Australian Currant 227962
Australian Cypress Pine 67405
Australian Daisy Bus 270207
Australian dendrobium 125206
Australian Desert Lime 148262
Australian Dogwood 211279
Australian Dropseedgrass 372659
Australian Earth Chestnut 143354
Australian Elder 345569
Australian Eriachne 148820
Australian Everlasting 329797
Australian Fan Palm 234166
Australian Fanpalm 234166
Australian Fan-palm 234166
Australian Feather Palm 321164
Australian Finger Lime 253285
Australian Fingergrass 88414
Australian Finger-lime 253285
Australian Firewheel Tree 375512
Australian Flannel Flower 6939
Australian Flatsedge 119713
Australian Frangipani 201240
Australian Fuchsia 105387,105394,105463,146067
Australian Gastrochilus 171951
Australian Ginger 17654
Australian Gooseberry 259585
Australian Grass 182942
Australian Grass-tree 415172
Australian Gum 155589
Australian Gymnema 182384
Australian Hareball 412750
Australian Hazel Nut 240210
Australian Heath 146067,146071
Australian Helmet Orchid 105528
Australian Hickory 1302
Australian Hollyhock 223384
Australian Honeysuckle 47626
Australian Indigo 205704
Australian Iris 285788
Australian Ivorywood 365271
Australian Ivy Plam 350648
Australian Kino 155468,155478,155517,155720
Australian Laurel 145320
Australian Lilly-pilly 156357
Australian Listera 232894
Australian Lovegrass 147773
Australian Macrozamia 241449
Australian Mahogany 139648
Australian Maple 166976
Australian Mint 315595
Australian Mountain Ash 155719
Australian Mountain Oak 155719
Australian Native Broom 409324
Australian Native Crepe-myrtle 219912
Australian Native Leek 62356
Australian Native Mangosteen 171140
Australian Native Quince 14374,71693
Australian Native Rosella 194919
Australian Native Tamarind 132992
Australian Native Violet 410058
Australian Native Water-lily 267687
Australian Nightshade 367289
Australian Nut 240209,240210
Australian Olive 270155
Australian Orache 44661
Australian Panicgrass 146026
Australian Pavetta 286104
Australian Pea 218721
Australian Pine 15953,30848,79157,79161,299807
Australian Pitcher Plant 82560,82561
Australian Posidonia 311971
Australian Quinine 18036
Australian Red Cedar 392829
Australian Redcedar 392826,392829

Australian Rhodes-grass 88339
Australian Rhododendron 331123
Australian River Oak 79157
Australian Rosemary 413914
Australian Rosewood 14369,139648
Australian Round Wild Lime 253288
Australian Saltbush 44641
Australian Sandalwood 155790,346212
Australian Sandarac 67417
Australian Sarcochilus 346691
Australian Sarsaparilla 185764
Australian Sassafras 43972
Australian Senna 360459
Australian Smut-grass 372666
Australian Snow Gum 155668
Australian Sprangletop 225993
Australian Stork's Bill 153796
Australian Sundew 138266
Australian Swamp Stonecrop 109059
Australian Sword Lily 25216
Australian Tallowwood 155654
Australian Tamarind 132989
Australian Tea Tree 226460
Australian Teak 166974
Australian Teatree 226460
Australian Tea-tree 226450,226460
Australian Tobacco 266058
Australian Toona 392826
Australian Tristellateia 398684
Australian Tulip 400229
Australian Umbrella Tree 350648
Australian Violet 410058
Australian Wallaby Grass 20549
Australian Walnut 145320
Australian Waterbuttons 107755
Australian White Ash 155585
Australian Wild Cotton 179882
Australian Wild Lime 253284
Australian Willow 1552,172478
Australian Willow Myrtle 11399
Australian Windmill Grass 88416
Australian Wisteria 254768
Australian-beech 155700
Australian-pine 79155
Australianyew 45317,45318
Australis Sweetleaf 381121
Australnut 240207
Austral-yunnan Box 64239
Austria Pine 300117
Austria Rose 336563
Austrian Briar 336563
Austrian Briar Rose 336563
Austrian Brier Rose 336563
Austrian Broom 120928
Austrian Bush Briar 336563
Austrian Chamomile 26751
Austrian Copper Briar 336563,336565
Austrian Copper Rose 336565
Austrian corsican pine 300117
Austrian Digitalis 130369
Austrian Dragon's-head 137555
Austrian Dragonhead 137555
Austrian Field Cress 336182
Austrian Fieldcress 336182
Austrian Field-cress 336182
Austrian Flax 231872
Austrian Hedge Maple 2843
Austrian Hogfennel 292779
Austrian Leopard's-Bane 136344
Austrian Milkvetch 42048
Austrian Mint 250326
Austrian Ononis 271337
Austrian Pea 301055
Austrian Pine 79155,300117
Austrian Pleurospermum 304765
Austrian Rocket 365406
Austrian Rose 336563
Austrian Sagebrush 35188
Austrian Snowbell 367764
Austrian Valerian 404365
Austrian Wormwood 35188
Austrian Yellow Cress 336182
Austrian Yellow Rose 336563
Austrian Yellowcress 336182
Austrian Yellow-cress 336182
Australian Silky-oak 73529,73530
Austrotaxus 45317
Ausubo 244530
Autograph Tree 97262
Autumn 99297
Autumn Adonis 8332
Autumn and Winter Squash 114288
Autumn Bellflower 173728
Autumn Bells 173728,174106
Autumn Bent Grass 12240
Autumn Blooming Cherry 83112
Autumn Cattleya 79546
Autumn Cherry 83112
Autumn Clovershrub 70910
Autumn Coralroot 104009
Autumn Coral-root 104009
Autumn Crocus 99293,99297,111566
Autumn Cudweed 178218
Autumn Daisy 227452,227455
Autumn Damask 336514,336515
Autumn Damask Rose 336515
Autumn Dandelion 384799
Autumn Elaeagnus 142214
Autumn Elm 401626
Autumn Flame Red Maple 3508
Autumn Flos Adonis 8332
Autumn Flowering Rhododendron 330845
Autumn Gastrodia 171909
Autumn Gentian 174106
Autumn Goldenrod 368432
Autumn Goosecomb 335497
Autumn Grape 411890
Autumn Hawkbit 224655
Autumn Heronbill 153909
Autumn Hubei Anemone 23854
Autumn Hupeh Anemone 23854
Autumn Kamaon Larkspur 124310
Autumn Lady's Tresses 372259
Autumn Lady's-tresses 372259
Autumn Ladytress 372259
Autumn Larkspur 124310
Autumn Lovegrass 147517
Autumn Lycoris 239280
Autumn Maple Tree 54620
Autumn Millet 281561
Autumn Moor Grass 361614
Autumn Olive 142214,142236
Autumn Onion 15765
Autumn Pleione 304278
Autumn Purple Ash 167900
Autumn Rain-lily 417612
Autumn Roegneria 335497
Autumn Sage 344822,345067
Autumn Sedge 73839,166190
Autumn Sneezeweed 188402
Autumn Snowdrop 169729
Autumn Snowflake 227858
Autumn Splendor Buckeye 9676
Autumn Squash 114288
Autumn Squill 352885
Autumn Tresess 372192
Autumn Violet 173728
Autumn Water Starwort 67361
Autumn Waterstarwort 67361
Autumn Water-starwort 67361
Autumn Willow 344093
Autumn Zephyr Lily 417608
Autumn Zephyrlily 417612
Autumn's Goodbye 284591
Autumn's Welcome 380930
Autumnal Fimbristylis 166190
Autumnal Meadow Saffron 99297
Autumnal Squill 352885
Autumnal Star Hyacinth 352885
Autumnal Starwort 67361
Autumnal Water-starwort 67361
Autumnblooming Laelia 219671
Autumn-Croctus 99293
Autumn-crocus 99297
Autumn-dandelion 224655
Autumn-flower Nutlikeseed Holly 204095
Autumn-flowering Elder 345580
Autumn-olive 142214
Autumnpurple Rhododendron 330691
Autumn-purpled Rhododendron 330691
Avalanche Lily 154933

Avalanche-lily 154933
Avance 175454
Avandrill 262441
Avaram Bark 78236
Avaram Senna 78236
Avaram Senna Bark 78236
Ave Grace 341064
Avecote 409085
Avels 198376
Avena Grass 45363
Avens 175375,175454
Average Bladder-senna 100182
Averill 262441
Averin Averen 338243
Averoyne 35088
Averrhoa 45721,45725
Avian Goosefoot 87045
Avicenna Pricklyash 417173
Avicenna Prickly-ash 417173
Avicennia 45740
Avignon Berry 328737,328849
Avocado 291491,291494
Avocado Pear 291494
Avocado-pear 291494
Avondale Redbud 83769
Avron 358062
Avrons 338243
Award Wound Weed 210543
Awarf Chinese Juniper 341719
Awarra 43379
Awarra Palm 43379
Awbel Aubel 311208
Aweet White Violet 409751
Awglen 109857
Awinhoe Raspberry 339338
Awkeotsang 165518
Awl 258882
Awl Aster 380953,380955
Awlbillorchis 333725,333731
Awlfruit 179153,179156
Awl-fruited Oval Sedge 76577
Awlleaf Arrowhead 342417
Awlleaf Chosenia 89160
Awlleaf Edelweiss 224952
Awlleaf Gentian 173498
Awl-leaf Pearlwort 32301
Awlleaf Pink 127676
Awlleaf Windhairdaisy 348935
Awl-leaf Windhairdaisy 348824
Awl-leaved Pearlwort 32301
Awlnettle 307040,307045
Awloleaved Pearlwort 32301
Awlquitch 131006,391440,391445
Awlshape Windhairdaisy 348551
Awlshaped Bulrush 353840
Awl-shaped Cryptostylis 113862
Awl-shaped Evergreen Chinkapin 79034
Awlshaped Rockjasmine 23215
Awlshaped Stenotaphrum 375773
Awl-shape-leaved Canary Tree 71024
Awl-shap-leaved Fig 165719
Awlstem Bulbophyllum 63006
Awlstem Stonebean-orchis 63006
Awltooth Bulbophyllum 62765
Awltooth Curlylip-orchis 62765
Awlwort 379649,379650
Awl-wort 379650
Awms 198376
Awned Bedstraw 170215
Awned Bent Grass 12088
Awned Canary-grass 293750
Awned Cypressgrass 119208
Awned Duckbeakgrass 209259
Awned Galingale 119208
Awned Graceful Sedge 74261
Awned Gyperus 245351
Awned Haworthia 186368
Awned Oval Sedge 76388
Awned Sawgrass 245351
Awned Serration Barberry 51331
Awned Umbrella-bamboo 388748
Awned Wheat Grass 144498
Awnedglume Goosecomb 144183
Awnedglume Leymus 228351
Awnedglume Roegneria 144183
Awned-leaved Barberry 51327
Awnedscale Sawgrass 245351
Awngrass 255827,255886
Awnlemma Bistamengrass 127468
Awnles Blueberry 403824
Awnless 215947
Awnless Brome 60760
Awnless Bromegrass 60760
Awnless Brome-grass 60760
Awnless Bush Sunflower 364509
Awnless Chikusichloa 87355
Awnless Cleistogenes 94631
Awnless Composite Oplismenus 272638
Awnless Duckbeakgrass 209345
Awnless Fescue 164095
Awnless Fescuegrass 164095
Awnless Garnotia 171516
Awnless Goatgrass 8691
Awnless Graceful Sedge 74595
Awnless Gullygrass 87355
Awnless Hairgrass 126054
Awnless Hideseedgrass 94631
Awnless Lazy Daisy 84937
Awnless Longbeaked Sedge 75190
Awnless Lymegrass 144487
Awnless Roegneria 335444
Awnless Sedge 74418
Awnless Wild-rye 144487
Awnless Wildryegrass 144487
Awnspine Azalea 331894
Awntlings 198376
Awntooth Barberry 52265
Awn-tooth Milkvetch 42301
Awntooth Primrose 314633
Awnut 63475
Awordtooth Ophiorrhiza 272201
Awun 18034
Awusa Nut 386838
Axcalyx Habenaria 183515
Axe Sandcress 321484
Axe-shaped ugionium 321484
Ax-flowered Sea-lavender 230643
Axial Paris 284293
Axile Arrow 381809
Axileflower 153396
Axilflower Pellacalyx 288683
Axilflower Wildmangrove 288683
Axilflower Woodbetony 287026
Axillary Acrocephalus 5995
Axillary Ascitesgrass 407453
Axillary Balm 249494
Axillary Biond's Magnolia 241998
Axillary Choerospondias 88691
Axillary Curvedstamencress 238006
Axillary Daphne 122380
Axillary Flowers Fetterbush 228155
Axillary Goldenrod 368003
Axillary Melodinus 249651
Axillary Petunia 292751
Axillary Poisonnut 378651
Axillary Polyspora 179729
Axillary Rattan Palm 65643
Axillary Strychnos 378651
Axillary Veronicastrum 407453
Axillaryflower Childgrass 340337
Axillaryflower Devilpepper 327062
Axillaryflower Eritrichium 153427
Axillaryflower Fieldorange 249651
Axillaryflower Glorybower 96023,96028
Axillaryflower Hooktea 52413
Axillaryflower Monkshood 5573
Axillaryflower Parryodes 285044
Axillaryflower Poisonnut 378651
Axillaryflower Rabbiten-wind 12695
Axillaryflower Rockjasmine 23126
Axillaryflower Rungia 340337
Axillaryflower Skullcap 355385
Axillaryflower Twistedstalk 377941
Axillary-flowered Supplejack 52413
Axillary-hair Meliosma 249448
Axillarytomentose Whiteroot Acanthopanax 2436
Axseed 105324
Axwort 105324
Axyris 45843
Ayan 134776

Aye 376174
Ayegreen 358062
Aye-no Bent 235334
Ayk 324335
Ayous 398004
Ayr Astrocaryum 43374
Ayrshire Rose 336364
Ayver 235334
Aza Catchfly 363225
Azalea 235279,330017,331345,331420,331546,331839
Azalealeaf Buckthorn 328841
Azalealeaf Willow 344265
Azania 'Daybreak' 172341
Azara 45952
Azarole 109545
Azarole Hawthorn 109545
Azed 417540
Azedarach 248895
Azima 45971
Azob 236336
Azon Lily 155858
Azorean Jasmine 211740
Azorella 56620
Azores Bellflower 70362
Azores Blueberry 403793
Azores Butterfly-orchid 302436
Azores Forget-me-not 260766
Azores Jasmine 211740
Aztec Clover 396820
Aztec Dahlia 121561
Aztec Lily 372912,372913
Aztec Marigold 383090
Aztec Pine 300277
Aztec Sweet Herb 232487,296121
Aztec Tabacco 266053
Azteclily 372912,372913
Aztekium 45993,45994
Azure Aster 380943
Azure Blue Sage 344885,344887
Azure Bluet 187526
Azure Bluets 187526,198708
Azure Ceanothus 79916
Azure Dualbutterfly 398263
Azure Monkshood 5193
Azure Red Fir 411
Azure Sanicle 345944
Azure Tripterospermum 398263
Azzy-tree 109857

B

B. H. Hedgson's Rhododendron 330877
B. Ririe's Rhododendron 331660
B. Scott's Rhododendron 331427
Baa Lamb's Tails 106703
Baa-lambs 106703,397043
Baa-Lambs' Tails 106703
Baalem's Smite 374916
Babaco 76809
Babacu 273241
Babai 120774
Babassu 273247
Babassu Palm 273241
Babbleleaf Azalea 330556
Babe-and-cradle 169126
Babe-in-the-cradle 37015,169126,355061
Baber Evodia 161311
Babes-in-the-wood 374897
Babiana 46022
Babies' Bells 262441
Babies'-breath 183225
Babington's Orache 44414
Baboen 261461
Baboon Flower 46022
Baboon Root 46060,46105
Baboso 279485
Babul Acacia 1427
Babul Bark 1427
Babul Tree 1427
Baby Blue Eyes 45984,263505
Baby Blue-eyes 263501
Baby Cabbage 59539
Baby Cakes 174711
Baby Corn 417417
Baby Echeveria 139986
Baby Faurax Rose 336307
Baby Goldenrod 368262
Baby Orchid 90711,269069
Baby Pepper 334897
Baby Pondweed 312236
Baby Primrose 314408,314613
Baby Rose 336783
Baby Rubber Plant 290395
Baby Rubber-plant 290317
Baby Ruber Plant 290317,290319
Baby Sage 345213
Baby Snapdragon 231084
Baby Sun-rose 29855
Baby Swiss Geranium 153903
Baby Toes 163328,168330
Baby's Bonnet 222789
Baby's Bonnets 222789
Baby's Breath 183157,183225,183238
Baby's Cradle 271280
Baby's Cradles 271280
Baby's Pet 50825
Baby's Pinafore 174877
Baby's Rattle 13174,248193,329521
Baby's Shoes 13174,30081
Baby's Tears 290418,367771,367772
Baby's-breath 170654,183157,183191,183225
Baby's-breath Gypsophila 183225
Baby's-tears 367771
Baby's-toes 163328,163329
Baby-blue Eyes 263505
Baby-blue-eyes 263505
Baby-blue-eyes Nemophila 263505
Babycom 417417
Babyin Cradle 25134
Babylon Weeping Willow 343070
Babylon Willow 342982,342993
Bacaba Palm 269373,269374,269376
Bacan 234033
Bacca Pearbamboo 249617
Bacca Sedge 73842
Baccamelia 91479,91480
Baccanisa 112952
Baccate Tallow-tree 346372
Baccaurea 46176
Baccharis 46210
Baccharis Bacchar 207087
Baccharis-leaf Penstemon 289323
Bacciferous Melocanna 249617
Baccobolts 401112
Baccy Lambs 106703
Baccybolts 401112
Baccy-plant 400675
Bach Rhododendron 330193
Bachang Mango 244393
Bachelor's Button 80892,81020,179236,363673
Bachelor's Buttons 4005,11947,30081,31051,50825,68189,81020,81237,81356,82098,174711,174877,179236,216085,216753,227533,248411,322724,325516,325518,325666,326287,350125,363461,363673,374916,379660,383874,401711,409339
Bachelor's Pear 336419,367357
Bachelor's-button 81020
Bachman Cape-cowslip 218840
Bachofen Dutch Beech 311218
Bachofen Poplar 311218
Bachofen's Speedwell 407022
Back's Sedge 73846

Back's Spindle-tree 157338
Backache Root 228529
Backeberg Cob Cactus 140850
Backer Elatostema 142613
Backer Stairweed 142613
Backglaucousleaf Dichocarpum 128935
Backlie Catnip 265071
Backlie Skullcap 355786
Back-rib Spindle-tree 157416
Back-to-back 410677
Backwardpilose Solms-laubachia 368578
Backwort 381035
Baclin 53816
Bacon 336419
Bacon And Eggs 123266
Bacon-and-eggs 231183,237539,262429,325580
Bacon-weed 86901
Bacopa 379882
Bacopa 'Snowflake' 379883
Bactard Mahogany 155769
Bacu 76832
Bacury 302621,302622
Bad 212636
Bad Man's Oatmeal 28044,72038,101852
Bad Man's Plaything 3978
Bad Man's Posies 220416
Bad Man's Posy 220416
Bad Pennisetum 289130
Badam 386613
Badashan Nepeta 264879
Badger's Bane 5389,5676
Badger's Flower 15861
Badger's-bane 5389
Badgersbane 5676
Badgeworth Buttercup 326148
Badi 262804
Badiane 204603
Badinjan 367370
Badminnie 252662
Badong Actinidia 6709
Badong Arrowwood 407882
Badong Ceropegia 84082
Badong Corydalis 105970
Badong Elaeagnus 141979
Badong Kiwifruit 6709
Badong Loosestrife 239776
Badong Manglietia 244464
Badong Monkshood 5287
Badong Pendentlamp 84082
Badong Summerlilic 61968
Badong Willow 343356
Badong Windhairdaisy 348369
Badong Wintergreen Barberry 51802
Badong Woodlotus 244464
Bad-smell Symplocarpus 381073
Badusan Sida 362672
Bady Blue-eyes 263505
Bady Sun Rose 29855
Baeckea 46423
Bael Fruit 8787
Baeltree 8787
Baerluke Fritillary 168354
Baeuerlen's Gum 155488
Bag Flower 96387
Bag Onion 15690
Bagac 133563
Bagasse 46514,46515
Bagcalyx Milkvetch 42255,42994
Bagcalyx Peashrub 72257
Bagfruit Milkvetch 42991
Bagger's Needle 153869
Bagpetaltree 342084
Bag-petal-tree 342084
Bagpod 361453
Bagras Kumurere Mindanao Gum 155557
Bagshaped Dendrobium 125087
Bag-shaped Dutchmanspipe 34318
Bagspur Corydalis 105647
Bagtikan 283900,283902
Bagtikan White Lauan 283900
Bagtikann 283900
Baguanhe Cycas 115800
Bahama Dildo 299264
Bahama Grass 117859
Bahama Hemp 10940
Bahama Pine 299837
Bahama Pitchpine 299836
Bahama Sachsia 342109
Bahama Sophora 369137
Bahama Tea 221238
Bahama Whitewood 71132
Bahaman Aster 380995
Bahamas Pitch Pine 299836
Bahan Spapweed 204795
Bahan Touch-me-not 204795
Bahera Nut 386473
Bahia 46527
Bahia Grass 285469
Bahia Lovegrass 147519
Bahia Piassava 44807
Bahia Rosewood 121770
Bahia Rose-wood 121770
Bahia Wood 65002
Bahiagrass 285469,285470
Bahia-grass 285469
Baia 270150
Baib Grass 156513
Baical Anemone 23709
Baical Feathergrass 376710
Baical Hogfennel 292780
Baical Monkshood 5048
Baical Muhly 259636
Baical Needlegrass 376710
Baical Windflower 23709
Baigoudashan Nightshade 367488
Baihe Dahur Cranesbill 174561
Baihe Willow 344296
Baihuashan Grape 411564
Baihuashan Roegneria 335550
Baikal Betony 373139
Baikal Cudweed 178090
Baikal Meadowrue 388432
Baikal Skullcap 355387
Baikal Vetch 408566
Bailan 252809
Bailang Monkshood 5049
Bailey Acacia 1069
Bailey Greasewood 346608
Bailey Grey Mockorange 294474
Bailey Primrose 314146
Bailey Redtwig Dogwood 104963
Bailey Red-twig Dogwood 104963
Bailey Rhododendron 330194
Bailey Sikkim Barberry 52153
Bailey Stonecrop 356562
Bailey Willow 343083
Bailey's Dewberry 338175
Bailey's Mimosa 1069
Bailey's Rabbitbrush 236496
Bailey's Stringybark 155489
Bailey's Wild Buckwheat 151848
Bailey's woolly Wild Buckwheat 151852
Bailing Magnolia 241991
Bailingshan Camellia 68935
Baillie Nicol Jarvie's Poker 217055
Baillom Miliusa 254547
Baillon Paramichelia 283501
Baillon Veratrilla 405567
Baima Barberry 51944
Baima Sage 344888
Baima Sedge 73847
Baimashan Milkvetch 42229
Baimashan Rockfoil 349093
Baimashan Stringbush 414137
Bain Hoodia 198040
Baines Lotononis 237244
Baiposhan Sedge 73848
Bairnwort 50825
Bairu Raspberry 338338
Baise Elatostema 142615
Baise Premna 313709
Baise Sloanea 366044
Baise Solomonseal 308591
Baise Spiradiclis 371744
Baise Stairweed 142615
Baise Willow 343117
Baisha Rosewood 121789
Baishan 326168
Baishan Bittercress 72694
Baishan Cranesbill 174487

Baishan Eightangle 204538
Baishan Illicium 204538
Baishanzu Fir 293
Baishanzu Yushanbamboo 416752
Baishanzu Yushania 416752
Baishou Chirita 87823
Baissea 94391
Baitalin Kengyilia 215905
Baitoa 297536
Baitoushan Sedge 75737
Baivcale Vetch 408302
Baiyan-flower 50825
Baiyao Pipewort 151268
Baiyu Greyawngrass 372402
Baiyu Spodiopogon 372402
Baiyushan Rose 336365
Baizhi Angelica 24325
Baja Bush Snapdragon 170850
Baja Bush Sunflower 145201
Baja Dalea 121880
Baja Elephant Tree 279505
Baja Evening Primrose 269514
Baja Fairy Duster 66665
Baja Fig 165423
Baja Ocotillo 167561
Baja Ruellia 339784
Baja-lily 398816
Bajotierra 103766
Bajoura Citron 93550
Bajri 289116
Bakain 248895
Bakain Cedar 248895
Bakain Mahogany 248895
Bakaurea Juniper 213644
Bakeapples 338243
Baked-apple Berry 338243
Baker Afzelia 10109
Baker Arborvitae 302724
Baker Begonia 49647
Baker Cotoneaster 107356
Baker Cypress 114662
Baker Iris 208456
Baker Lily 229742
Baker Platanthera 302259
Baker Sophora 368974
Baker Vetch 408303
Baker's Cord Grass 370158
Baker's Garlic 15185
Baker's Goldfields 222592
Baker's Hawksbeard 110745
Baker's Larkspur 124050
Baker's Wild Buckwheat 151835
Bakphul 361384
Baku 391608
Bakuanho Cycas 115800
Bakupari 329271
Bakupari Rheedia 329271
Bakuri Guiana Orange 302622
Bakury 302621
Balabar Ramie 56214
Balaca 46743,46744
Balaca Palm 46743
Balackstamen Actinidia 6666
Balackstamen Kiwifruit 6666
Balack-stamened Actinidia 6666
Balan 362205
Balang Azalea 330197
Balang Rhododendron 330197
Balang Willow 344133
Balanga Orange 93708
Balang-like Willow 344134
Balangshan Angelica 24305
Balangshan Buttercup 325635
Balangshan Draba 136946
Balangshan Lady's Slipper 120419
Balangshan Ladyslipper 120419
Balangshan Rockfoil 349095
Balanophora 46810
Balanophora Family 46888
Balans Miliusa 254548
Balansa Casearia 78099
Balansa Clover 396976
Balansa Cylindrokelupha 116589
Balansa Dacrydium 121086
Balansa Elaeocarpus 142287
Balansa Ormosia 274379
Balansa Pittosporum 301221
Balansa Rosewood 121629
Balansa Seatung 301221
Balansa's Litse 233848
Balansae Allomorphia 16007
Balanse Fissistigma 166651
Balanse Michelia 252817
Balanse Rattan Palm 65644
Balanse Tan Oak 233117
Balanse Tanoak 233117
Balanse Xanthophytopsis 415156
Balasier 66173
Balata 139904,244520,244530
Balata Gum 244530
Balau 362186
Balbis Banana 260206
Balck Acanthus 2699
Balck Oak 324071
Balcony Petunia 292747
Balcooa Bamboo 47203
Bald 252662
Bald Acacia 1422
Bald Brome 60745,60912
Bald Cypress 385253,385262,385281
Bald Cypress Family 385250
Bald Daisy 150528
Bald Man's Bread 102727
Bald Ruprecht Honeysuckle 236077
Bald Seatung 301366
Bald Spike-rush 143076,143143
Bald Willowherb 146663
Baldar Herb 18752
Baldare 18687
Baldcypress 385253,385262
Bald-cypress 385253,385262
Baldcypress Family 385250
Baldellia 46930
Baldemayne 174106
Baldemoyne 174106,252662
Balderry 273469,273531
Baldeyebrow 26768
Bald-fruit Fleabane 151056
Baldhead False Buttonweed 370758
Bald-leaved Felt Thorn 387079
Bald-leaved Felt-thorn 387079
Baldmoney 252660
Baldpistil Elaeocarpus 142327
Bald-rush 333640
Baldseed Hibiscus 194969
Baldshuan Lallemantia 220282
Baldshuan Skullcap 355389
Baldstem Eriocycla 151749
Baldstipe Metalleaf 296400
Baldwin's Nailwort 284778
Baldwin's Sedge 118673
Baldwin's Spike-rush 143062
Balearic Box 64240
Balearic Pearlwort 31767
Balearic Sandwort 31767
Bale-persicaria Curlytop Knotweed 309298
Balete Fig 165698
Balfour Azalea 330198
Balfour Corydalis 105640
Balfour Indigo 205709
Balfour Milkvetch 42063
Balfour Mosquitotrap 117395
Balfour Persimmon 132061
Balfour Pine 299795,299813
Balfour Polyscias 310237
Balfour Rockfoil 349094
Balfour Spruce 298248
Balfour Willow 343087
Balfour's Meadow Grass 305551
Balfour's Meadow-grass 305551
Balfour's Rhododendron 330198
Balfour's Spruce 298331
Balfour's Touch-me-not 204797
Balikun Buttercup 325637
Balincolong 93406
Baling Pink 127788
Baling Rattan Palm 65646
Baliosperm 46962
Baliospermum 46962
Balisier 189992,189995,190005
Balkan Acanthus 2664

Balkan Blue-grass 361621
Balkan Catchfly 363385
Balkan Clary 345246
Balkan Crabapple 243615
Balkan Maple 3013,3029
Balkan Pincushions 350216
Balkan Pine 299968,300144
Balkan Plum 316294
Balkan Toadflax 230978
Balkans Maple 3013
Ball Cactus 267027
Ball Clover 179226
Ball Denseflower Milkwort 308077
Ball Densi-flowered Milkwort 308077
Ball Eelgrass 418176
Ball Gentian 173479
Ball Honey Myrtle 248109
Ball Moss 392039
Ball Mustard 265553
Ballagan 221787
Ballam 316382
Ballard Spapweed 204798
Ballard Touch-me-not 204798
Ballart 161710
Ballast Eryngo 154313
Ballbag Milkvetch 43075
Ballbullate Milkvetch 43076
Ballcactus 267027
Ballcalyx Catchfly 363318
Balldaisy 56693
Ballerina Flower 413481
Ballflower 399504
Ballflower Bulbophyllum 63031
Ballflower Conehead 178861
Ballflower Cotoneaster 107467
Ballflower Crazyweed 278854
Ball-flower Dendrobium 125387
Ballflower Elatostema 142734
Ballflower Globethistle 140789
Ballflower Michelia 252959
Ballflower Stonebean-orchis 63031
Ballflower Windhairdaisy 348337
Ballflower Woodbetony 287238
Ballfruit Dyetree 302676
Ballfruit Rattlebox 112776
Ballfruit Slatepentree 322612
Ballfruit Tubergourd 390132
Ballhead Onion 15752
Ballhead Rush 213576
Ballhead Sandwort 31824
Ball-headed Mixed-flower 298065
Ball-headed Onion 15752
Ball-moss 392039
Ballmustard 265553
Ball-mustard 265550,265553
Ballnode Bitterbamboo 304029
Ballnut 212636
Ballocks Ballock-grass 273531
Ball-of-the-earth 81065
Ballon Pea 380032
Ballon-root 46022
Balloon Cottonbush 179096
Balloon Flower 289367,302747,302753
Balloon Vine 73207
Balloonbush 147445
Balloonflower 302747,302753
Balloon-flower 302747,302753
Balloonplant 38062
Balloon-vine 73207
Balloonvine Heartseed 73207
Ballota 46989
Balloxe 273531
Ballpit Arrowwood 408056
Ballraceme Mountainash 369400
Ballroom Bindersia 166974
Ballroom Flindersia 166974
Ballseed Cliffbean 66651
Ballshape Sagebrush 35546
Ballspike Cattail 401131
Ballspike Flatsedge 322233
Ballspike Knotweed 309616
Ballspike Pepper 300532
Ballspur Amitostigma 19525
Ballspur Astyleorchis 19525
Ballstem Begonia 50378
Ballweed 81237
Balm 249491,249504,311237
Balm Gentle 249504
Balm Mint 249491,250385,250420
Balm of Gilead 282,80064,101391,232562,
249542,311171,311237,311259,311554
Balm of Gilead Fir 282
Balm of the Warrior's Wound 202086
Balm Tree 261549,261559,407907
Balme 249504
Balmleaf Gomphostemma 179194
Balmleaf Melittis 249542
Balm-leaved Archangel 220412,249542
Balm-leaved Deadnettle 220412
Balm-leaved Figwort 355237
Balm-of-gilead 311186,311259,311360
Balm-of-gilead Poplar 311259
Balm-of-warrior 202086
Balmony 86783
Balmsamiferous Blumea Ai 55693
Balmtree 261549
Balm-tree 261549
Baloonbush 246659
Balotin Bergamot 93407
Balour Habenaria 183433
Balsa 268341,268344
Balsa Corkwood 268344
Balsa Wood 268344
Balsam 377,204756,204799,311237
Balsam Amyris 21005
Balsam Apple 97262,140527,239157,
256786,256797
Balsam Balmtree 261561
Balsam Bog 45982
Balsam Cuphea 114598
Balsam Family 47077
Balsam Fig 97262
Balsam Fir 282
Balsam Groundsel 279935
Balsam Herb 383712
Balsam Marshweed 230279
Balsam of Peru 261559
Balsam Pear 256797
Balsam Poplar 311171,311237,311259
Balsam Ragwort 279935,359697
Balsam Rhododendron 330200
Balsam Root 47147
Balsam Shrub 21003
Balsam Spurge 158509
Balsam Syzygium 382485
Balsam Tree 99980,103716
Balsam Willow 343964,343968
Balsam-apple 140527,256780
Balsam-bog 45982
Balsamina Family 47077
Balsamine Balsam 205175
Balsamint 383712
Balsam-leaf Calycera 68332
Balsam-of-Peru 261554
Balsampear 256797
Balsam-pear 256797
Balsam-poplar 311237
Balsamroot 47117
Balsam-root 47139
Balsam-tree 103691,261549
Balsamweed 204841
Balsour Axilflower Woodbetony 287027
Baltic Gentian 173283
Baltic Orchis 273333
Baltic Pine 300223
Baltic Redwood 300223
Baltic Red-wood 300223
Baltic Rush 212844,212852,212924
Baltic White Wood 298191
Baltic Whitewood 298191
Baltic Yellow Deal 300223
Balucanat 14544
Baluchistan Barberry 51344
Balustine Flowers 321764
Bambara Groundnut 409069
Bamboo 37127,47174,47190,87547,
162635,206765,297188,297194,297246,
297449,297499,303994,318277,318307,
347191,347268,347334,357932,362121,
362131,364631,416742
Bamboo Cycad 83681

Bamboo Eccoilopus 372403
Bamboo Elaeagnus 141946
Bamboo Grass 86140
Bamboo Leaf Grass 272625
Bamboo Leaf Oak 116156
Bamboo Muhly 259652
Bamboo Oilawn 372403
Bamboo Orchid 37115
Bamboo Palm 85420,85423,85428,85437,89360,326649,329176
Bamboo Shrub 82442
Bamboo Sugarcane 341909
Bamboo Vine 366428
Bamboo Violet 409711
Bamboobranch Dendrobium 125349
Bamboobroom 416823
Bambooforest Skullcap 355390
Bamboo-leaf Birthwort 34118
Bambooleaf Camellia 68937
Bambooleaf Clearweed 298873
Bambooleaf Coldwaterflower 298873
Bambooleaf Eria 148630
Bambooleaf Fig 165429
Bamboo-leaf Fig 165429
Bambooleaf Galangal 17647
Bambooleaf Gugertree 350906
Bambooleaf Hairorchis 148630
Bambooleaf Litse 234043
Bambooleaf Luckyweed 370407
Bambooleaf Nanmu 295360
Bambooleaf Oak 116156
Bamboo-leaf Osmanthus 276264
Bambooleaf Pepper 300347
Bambooleaf Pondweed 312171
Bambooleaf Pricklyash 417161
Bambooleaf Qinggang 116156
Bambooleaf Quitch 254036
Bambooleaf Raspberry 338177
Bambooleaf Seseli 361538
Bambooleaf Syzygium 382619
Bambooleaf Thorowax 63745
Bamboo-leaved Cyclobalanopsis 116156
Bamboo-leaved Gugertree 350906
Bamboo-leaved Holboellia 197241
Bamboo-leaved Litse 234043
Bamboo-leaved Oak 116153,116156
Bamboo-leaved Pepper 300347
Bamboo-leaved Prickly-ash 417161
Bamboo-leaved Raspberry 338177
Bambooshoot Dendrobium 125349
Bamboosifolious Clawtea 338177
Bambusa 47174
Bambusa-like Dendrocalamus 125466
Bamenda Cola 99161
Banadle 121001
Banak 410922
Banana 260198,260199,260253
Banana Family 260284
Banana Fig 165473
Banana Lily 267801
Banana Magnolia 252869
Banana Michelia 252869
Banana Passion Flower 285672
Banana Passion Fruit 285615,285672
Banana Passionflower 285672
Banana Passionfruit 285672
Banana Poka 285672,285710
Banana Shrub 232609,242096,252803,252869
Banana Water-lily 267714
Banana Yucca 416556
Bananaforest Loxostigma 238041
Banana-passion Fruit 285615,285672
Banane 260198
Banat campion 363222
Banat Silene 363222
Banathal 121001
Banbury Garcinia 171062
Bancaran Ramin 179533
Bancouloilplant 14544
Bancroft's Red Gum 155491
Bancroft's Wattle 1071
Band Plant 409335
Ban-dakai 219
Bandana-of-the-everglades 71171
Banded Arrowroot 245022
Banded Bullrush 352278
Baneberry 6405,6433,6441,6448
Banebind 50825,102933,325864,326034,326340
Banfenghe 357968,357969
Bangalay 155492,155505
Bangalow 31015
Bangalow Palm 31012,31015
Banga-wanga 19026,19029
Bange Buttercup 325638
Bangikat Poplar 311276
Bangkirai 362209
Bangladesh Bamboo 249617
Banglang 219966
Bangong Willow 343092
Banguo Buttercup 325638
Banian 165553
Banji Pat 104103
Banjo Fig 165426
Bank Catclaw 1522
Bank Cress 47964,365564
Bank Thistle 73430,92485
Bank Thyme 391365,391397
Bankan Parachampionella 283211
Bankok Giantarum 20049
Bankokai 219
Banks Centrolepis 81814
Banks Greenhoods 320872
Banks Grevilll 180570
Banks Pine 299818
Banks Rose 336366,336367
Banks Silk Oak 180570
Banks' Grevillea 180570
Banks' Pepper 300348
Banks' Rose 336366
Banksia 47626,47660
Banksia Rose 336366
Banksia-like Rose 336379
Banksian Pine 299818
Banksian Rose 336366,336367
Banlao Greenbrier 366250
Banma Azalea 330201
Banma Fritillary 168391
Banma Rhododendron 330201
Banma Sagebrush 35194
Banna Habenaria 183855
Banna Miliusa 254549
Banna Ophiorrhiza 272215
Banna Spapweed 204801
Banna Vatica 405184
Bannal 121001
Bannell 121001
Bannergowan 50825
Bannet-tree 212636
Bannit 212636
Bannock Wild Buckwheat 151872
Bannott 212636
Bannut-tree 212636
Banokan 93471
Banquet Herb 114884
Bantu Camphor 138548
Banwood 50825
Banwort 50825
Banyalla 301225
Banyan 164684,165553
Banyan Fig 164684
Banyan Tree 164684
Banzhuan Yam 131476
Banzuan Yam 131476
Baobab 7018,7022
Baobab Cream of Tartar Tree 7022
Baobabtree 7018,7022
Baobab-tree 7018,7022
Baohsing Gleadovia 176798
Baohua Machilus 240669
Baohua Magnolia 416727
Baohuashan Hornbeam 77353
Baohuashan Mucuna 259531
Baohuashan Sedge 73849
Baoji Pennisetum 289150
Baoji Wolftailgrass 289150
Baojing Barrenwort 146963
Baojing Epimedium 146963
Baokang Kinostemon 216462
Baokang Peony 280144

Baokang Primrose 314693
Baoshan Argyreia 32611
Baoshan Lysionotus 239984
Baoshan Monkshood 5424
Baoting Actinodaphne 6807
Baoting Adinandra 8239
Baoting Ardisia 31362
Baoting Dendropanax 125627
Baoting Eargrass 187513
Baoting Euonymus 157806
Baoting Fig 165800
Baoting Forkstyleflower 374490
Baoting Goldlineorchis 26004
Baoting Goniothalamus 179411
Baoting Hedyotis 187513
Baoting Holly 203980
Baoting Liparis 232096
Baoting Maesa 241755
Baoting Newlitse 264056
Baoting Persimmon 132359
Baoting Reevesia 327359
Baoting Slugwood 50475
Baoting Suoluo 327359
Baoting Treerenshen 125627
Baoting Twayblade 232096
Baotingflower 413758
Baotou Milkvetch 42066
Baoxing Acanthopanax 143577
Baoxing Aster 40880
Baoxing Azalea 331283
Baoxing Blueberry 403912
Baoxing Broomrape 275145
Baoxing Cranesbill 174746
Baoxing Currant 334104,334109
Baoxing Dogwood 380501
Baoxing Gentian 173284
Baoxing Gleadovia 176798
Baoxing Jerusalemsage 295169
Baoxing Larkspur 124588
Baoxing Little Willow 343690
Baoxing Loosestrife 239557,239800
Baoxing Parnassia 284551
Baoxing Pendentlamp 84196
Baoxing Primrose 314462,314676
Baoxing Raspberry 338938
Baoxing Sage 345288
Baoxing Sedge 75430
Baoxing Stonecrop 357003
Baoxing Whitlowgrass 137124
Baoxing Willow 343723
Baoxing Windflower 23714
Baoxing Woodbetony 287436
Baphicacanthus 47842
Bapo Cyclobalanopsis 116063
Bapo Greenbrier 366252
Bapo Oak 116063
Baptist Greenhoods 320873
Bar Clover 247246
Bar Crowfoot 326287
Bar Harbor Juniper 213787
Bara-bara 132466
Barand Barberry 51345
Barassu Palm 273243
Barb Grass 184581
Barba de Burro 418331
Barbaded Dendrocalamus 125467
Barbadine 285694
Barbados Almond 386500
Barbados Aloe 16792,17381
Barbados Cedar 213623
Barbados Cherry 156385,243525,243526
Barbados Cotton 179884
Barbados Flower-fence 65055
Barbados Gooseberry 290701
Barbados Heather 114605
Barbados Lily 196439,196443,196450,196458
Barbados Mastic 362940
Barbados Nut 212127
Barbados Pride 65055
Barbados Shrub 290701
Barbados Snowdrop 184257,417638
Barbados-lily 196439
Barbadosnut 212127
Barbadospride 65055
Barbados-pride 7190
Barbara's Buttons 245886,245887
Barbarealeaf Rorippa 336183
Barbarealeaf Yellowcress 336183
Barbareus 52322
Barbaric Indocalamus 206771
Barbary 52322
Barbary Fig 272891
Barbary Gum 1024,1266
Barbary Matrimony-vine 238996
Barbary Nut 208839
Barbary Squash 114292
Barbary Wolfberry 239011
Barbasco 235513
Barbat Coelogyne 98612
Barbat Fingergrass 88320
Barbate Alangium 13347
Barbate Aster 40120
Barbate Cyclea 116010
Barbate Deadnettle 220359
Barbate Delavay Taxillus 385195
Barbate Diplectria 132878
Barbate Griffith Lacquertree 393458
Barbate Indocalamus 206771
Barbate Lacquertree 332647
Barbate Mallotus 243327
Barbate Mongolian Pulsatilla 321660
Barbate Paraboea 283110
Barbate Petrocosmea 292548
Barbate Pink 127607
Barbate Ringvine 116010
Barbate Sandwort 31768
Barbate Singnalgrass 58202
Barbate Skullcap 355391
Barbate Spapweed 204802
Barbate Stonebutterfly 292548
Barbate Sumac 332647
Barbate Wildtung 243327
Barbate-veined Maple 2798
Barbatimao Alum Bark Tree 378960
Barbatimao Alum-bark Tree 378960
Barbato-axillate Cinnamon 91272
Barbeau 81020
Barbed Alangium 13347
Barbed Bristle Grass 361930
Barbed Bristlegrass 361930
Barbed Dendrocalamus 125467
Barbed Diplectria 132878
Barbed Gentianopsis 174205
Barbed Goatgrass 8732
Barbed Leaf Cinquefoil 312351
Barbed Lime 391738
Barbed Linden 391738
Barbed Panic-grass 281371
Barbed Rattlesnakeroot 313799
Barbed Skullcap 355391
Barbed Wire Cereus 2147
Barbedvein Maple 2798
Barbed-wire Plant 400795
Barbel Palm 2550,2553
Barbellate Milkwort 307946
Barber Pleurothallis 304906
Barber's Brushes 133478
Barber's Diascia 128032
Barber's Hawksbeard 110992
Barberry 51268,51302,51694,52013,52322
Barberry Family 51261
Barberry Hawthorn 109555
Barberry Mayten 182664
Barberryleaf Hawthorn 109555
Barberry-leaf Mahonia 242522
Barberryleaf Photinia 295640
Barberryleaf Rose 336849
Barberryleaf Willow 343101
Barberry-leaved Hawthorn 109691
Barberry-leaved Mahonia 242522
Barberry-leaved Photinia 295640
Barberry-leaved Rose 336849
Barberry-leaved Willow 343101
Barberry-like Mayten 182664
Barberton Daisy 175107,175172
Barbey Barberry 51346
Barbey Embelia 144721
Barbey Stonecrop 356566
Barbey's Larkspur 124054
Barbgrass 184581

Barbiferous Hooktea 52405
Barbiferous Supplejack 52405
Barbiferous Supple-jack 52405
Barbine 102933
Barboranne 52322
Barbwire Grass 117240
Barbwire Russian Thistle 344658
Barbwire Russian-thistle 344658
Barcelona Nut 106703
Barcoo Grass 209469
Bard Goatgrass 8732
Bardana 31051
Bardane 31051
Bardfield Oxlip 314333
Bare Elaeocarpus 142327
Bare Styles Rhododendron 331863
Bare Vrydagzynea 412388
Bare-anthered Ternstroemia 386696
Bare-buded Phoebe 295410
Bared Branches Rhododendron 330824
Bared Foot Rhododendron 330617
Bared Stamens Rhododendron 330463
Bare-eared Bambusa 47201
Bareet Grass 223984
Bare-flowered Beautyberry 66879
Bare-flowered Tetrameles 387297
Bare-flowered Trewia 394783
Barefoot Panicgrass 282286
Barefruit Groundsel 365050
Barefruit Hairyflower Actinidia 6591
Barefruit Hairyflower Kiwifruit 6591
Bare-fruited Pepper 300467
Bareleaf Rockfoil 349922
Bare-leaved Ormosia 274419
Bare-receptacled Wintersweet 87520
Bare-sheathed Fargesia 162749
Bare-stalked Metalleaf 296400
Bare-stamened Rhododendron 330463
Barestem Larkspur 124583
Barestem Teesdalia 385555
Bare-stemmed Tick-trefoil 126477
Bare-stigma Willow 343932
Bare-styled Rhododendron 331863
Baretwig Neststraw 379165
Bargeman's Cabbage 59603
Bargusinian Sagebrush 35197
Barilla 184947,184951,344425,379486
Barilla Plant 344585
Barilla-plant 344718
Bariloche Barberry 51347
Bark Cloth Tree 28243
Barkam Trigonotis 397382
Barkam's Rhododendron 330204
Bark-cloth Tree 28243,165344
Barker Notylia 267215
Barker's Hebe 186934
Barkgrass 317222
Barkley's Ragwort 359959
Barkly's Tavaresia 385149
Barksdale Trillium 397619
Barleria 48080,48272
Barley 198260,198376
Barleylike Perotis 291403
Barleyshaped Brome 60745
Barlike Fritillary 168591
Barm-leaf Barm 249504
Barn Vetch 408493
Barna 110235
Barna Tree 110227
Barnaby Star-thistle 81382
Barnaby's Thistle 81382
Barnados Gooseberry 297720
Barneby Rabbit-tobacco 127933
Barneby's Aster 193131
Barneby's Thistle 91793
Barnet 212636
Barnum Iris 208459
Barnut 212636
Barnwort 50825
Barnyard Aster 26768
Barnyard Grass 140367,140451
Barnyard Millet 140423
Barnyardgrass 140351,140367
Barnyard-grass 140367,140423
Baroba Nut 132963
Baron Ash 167926
Baron Oak 323695
Baron Pachypodium 279673
Baron Rosewood 121632
Baron's Lady's Flower 261585
Baron's Mercury 250600
Baroni Largeflower Deutzia 126846
Baros Camphor 138548
Barra Bean 145899
Barrel Bottle Tree 58348,376180
Barrel Bottle-tree 58348
Barrel Cactus 140111,140127,140872,163417,163483
Barrel-cactus 163417
Barrelhead Gayfeather 228454
Barrelier Bugloss 21928
Barrelier's Bugloss 21928
Barren Brome 60985
Barren Chickweed 83075
Barren Hemp 71218
Barren Myrtle 31142
Barren Strawberry 312549,313025,413030,413031
Barrens Claw Flower 68067
Barrens Silky Aster 380965
Barren-strawberry 413031
Barrenwort 146953,146961
Barren-wort 146961
Barres Pruinose Barberry 52070
Barrington Lycaste 238734
Barringtonia 48508,48518
Barringtonia Climber 178564
Barringtonia Family 48524
Barron's Wild Buckwheat 152477
Bar-room Plant 39532
Barrow Rose 336851
Barrow-rose 336851
Barstow Woolly Sunflower 152825
Barter Schefflera 350663
Barthea 48551
Bartlett Monkshood 5214
Bartlett Smalllobed Clematis 95204
Bartlett Yam 131477
Bartolona 273035
Bartonia 48585
Bartram Commersonia 101220
Bartram's Airplant 391978
Bartram's Ixia 68495
Bartram's Juneberry 19250
Bartram's Oak 324005
Bartsia 48601,268957
Barundi Dropseed 372803
Barus Camphor 138548,138550
Barweed 102933
Barwood 320278,320293,320330
Barwood Cam-wood 47788
Basal Pricklypear 272805
Basal-leaved Rosinweed Prairie-dock 364317
Basal-round Woodnettle 125541
Basalt Fleabane 150492
Basam 67456,121001
Base Broom 173082
Base Rocket 327879
Base Vervain 407072
Baseball Plant 159485
Baseflower Milkvetch 42067,42926
Baseflower Shortcuspidate Sedge 73931
Baseheartleaf Coldwaterflower 298874
Baseleaf Amitostigma 19501
Baseleaf Dichocarpum 128927
Basella Family 48698
Basellate Oak 323700
Baselvein Machilus 240576
Basel-veined Machilus 240576
Basford Willow 343094
Bash Daisy 32563
Bashan Beech 162396
Bashan Clematis 95214
Bashan Euonymus 157774
Bashan Grape 411572
Bashan Indocalamus 206772
Bashan Loosestrife 239680
Bashan Maple 3370
Bashan Paris 284296
Bashan Pine 299972
Bashan Sasa 206772

Bashan Snowbell 379310
Bashan Torreya 393058
Bashan Wildginger 37570
Bashanbamboo 48706,48711
Bashania 48706
Bashels 384714
Bashful Brier 255102
Bashful Bulrush 396019
Bashful Trillium 397546
Bashfulgrass 254979
Basicordateleaf Clearweed 298874
Basifoliate Forkfruit 128927
Basihirsute Dendrocalamopsis 125439
Basil 97060,268418,268438
Basil Balm 4666,96964,257160
Basil Thyme 4661,4666,65540,96964,
96969,97060
Basilicum 48724
Basillike Bruise-wort 346433
Basil-like Bruise-wort 346433
Basil-like Soapwort 346433
Basil-smelling Mint 250341,250420
Basil-thyme 4666
Basilweed 97060
Basin Fleabane 150903
Basin Goldenrod 368430
Basin Sagebrush 36400
Basin Saltbush 44411
Basin Sneezeweed 188398
Basindaisy 302914
Basinleaf Palmleaftree 59200
Basisagittate Bittercress 72696
Basker Ash 168032
Basket 302034
Basket Ash 168015
Basket Asparagus 39185
Basket Elm 401486
Basket Flower 7202,200945
Basket Flower American Star-thistle 303082
Basket Grass 272614
Basket Ivy 116732
Basket Oak 324166
Basket of Gold 45192
Basket Plant 9417
Basket Vine 9417
Basket Willow 145504,343937,344244
Basket With 393240
Basketflower 303080
Basket-flower 200913,200945,303082
Basketflower Centaurea 303082
Basket-of-gold 45192
Basketplant 67145
Basketshape Hymenocallis 200913
Baskettree 414809
Basketvine 9417
Basom 121001
Basralocus 129488
Bass Wood 391649
Bassam 67456
Bassam Broom 121001
Bassel 268438
Bassia 48753
Bassine Fibre 57122
Bassinet 174832,325518
Bassinet Geranium 174832
Bassom 67456
Bassweed 143920,143931
Basswood 391646,391649,391691
Basswood Leaved Mallotus 243452
Bast 391691
Bastaed Yellowwood 306426
Bastard 415558
Bastard Acacia 334976
Bastard Acorus 208771
Bastard Agrimony 11194,11199,31721,
54158,158062
Bastard Alkanet 233700
Bastard Almond 386500
Bastard Anemone 321721
Bastard Balm 249540,249542
Bastard Box 307983
Bastard Burr 415053
Bastard Cabbage 326833
Bastard Cabbage Tree 350795
Bastard Cedar 80015,90605,181618,
228639,369934
Bastard Cinnamon 91302
Bastard Clover 396930
Bastard Cobas 120108
Bastard Cress 390202,390213
Bastard Daffodil 154899
Bastard Daisy 50809,50841
Bastard Dittany 129618
Bastard Elm 80698,80728
Bastard Evening-primrose 269404
Bastard Flag 5793
Bastard Flower De Luce 208771
Bastard Flower-de-Luce 208771
Bastard Foxglove 246658
Bastard Hellebore 190959
Bastard Hemp 123028,158062,170078
Bastard Indigo 20005,386257
Bastard Ipecac 37888
Bastard Ipecacuanha 37888
Bastard Jasmine 84409
Bastard Jesuit's Bark 112884
Bastard Killer 213702
Bastard Lignum-vitae 181506
Bastard Lucerne 247507
Bastard Lupine 396956
Bastard Mahogany 72640,155492,155505,
155614,392829
Bastard Marjoram 274237
Bastard Mustard 182915
Bastard Myrobalan 386473
Bastard Narcissus 168476
Bastard Navelwort 116736
Bastard Nigella 11947
Bastard Oak 324426
Bastard Oleander 389978
Bastard Olive 62157
Bastard Parsley 79664
Bastard Pellitory 4005
Bastard Pellitory-of-Spain 205535
Bastard Pennyroyal 396439
Bastard Pine 300264
Bastard Rhubarb 340178,388506
Bastard River Yellowwood 306425
Bastard Rocket 364557
Bastard Rosewood 381965
Bastard Saffron 77748
Bastard Sage 152663
Bastard Sago 78056
Bastard Sago Palm 78045
Bastard Senna 100153,105317
Bastard Service Tree 369429
Bastard Service-tree 369297,369429
Bastard Speedwell 317964,317966
Bastard Teak 64148,320307
Bastard Toadflax 100227,100229,389592,
389727
Bastard Vervain 405853
Bastard White Oak 324426
Bastard Wild Poppy 282515
Bastard Wild Rubber 169227
Bastard Yellowwood 306425
Bastard's Fumitory 169013
Bastard-alkanet 233700
Bastard-sage 152663
Bastardtoadflax 389592
Bastard-toadflax 100229
Bastard-toadflax Family 346181
Bastardtoadflax-like Swallowwort 117721
Baster Kobas 120108
Basterkobas 120108
Baste-tree 391705
Basthag-bwee 89704
Bast-tree 391705
Bastwort 292957
Basu Crazyweed 278736
Basu Sandwort 31770
Basuto Kraal Aloe 17303
Bat Face 114604
Bat Flower 382912
Bat Leaf Passion Flower 285634
Bat's-wing Coral Tree 154740
Bat's-wing Coral-tree 154740
Bataan 362217
Bataan Mahogany 362217
Batai 162473
Batai Wood 162473

Batalin Larkspur 124056
Batalin Tulip 400125
Batalin Wildryegrass 215905
Batan Climbing Screw-pine 168251
Batan Woodnettle 125546,221534
Batan Wood-nettle 125546,221534
Batang Aster 40123
Batang Barberry 51348
Batang Catchfly 363237
Batang Cotoneaster 107511
Batang Cranesbill 174488
Batang Everlasting 21524
Batang Larkspur 124057
Batang Milkvetch 42068
Batang Mosquitotrap 117398
Batang Pearleverlasting 21524
Batang Primrose 314155
Batang Rhodiola 329968
Batang Rhododendron 331514
Batang Stonecrop 356782
Batang Threeawngrass 33771
Batang Ungeargrass 36616
Batang Woodbetony 287033
Batangas Mandarin 93728
Batavia Cinnamon 91282
Batavian Shorea 362238
Bate Haworthia 186542
Bateman Rodriguezia 335162
Bateman's Wild Buckwheat 151856
Bat-faced Cuphea 114604
Batflower 382912
Bat-flower 382912
Bath Asparagus 274746,274748
Bath Sponge 238261
Bathurst Bur 415053
Baticuling 234033
Batiki Blue Grass 209321
Bat-in-the-belfry 70331
Bat-in-water 250291
Batkudze 250218
Batoka Plum 166773
Batoko Plum 166773,166776,166786
Bat-poll 146108
Batrachium 48904
Batrachium-like Pennywort 200261
Bats in the Belfry 70331
Bats-in-the-belfry 70331
Batter Dock 292372,340151
Batter-dock 292372
Battinan Crapemyrtle 219963
Batty Gum 26027
Batweed 89201,89209
Bauer Acronychia 6237
Bauer Galeandra 169866
Bauer Oncidium 270796
Bauer Rhopalostylis 332436
Bauer's dracaena 104345
Bauer's Grevillea 180572
Bauhinia 48990
Bauhinia-like Rhododendron 330209
Baum Cactus 244008
Baumann Begonia 49652
Baume 250450
Baum-leaf 249542
Bauno 244390
Baur Hesperantha 193239
Bause Dieffenbachia 130096
Bause Tuftroot 130096
Bavarian Gentian 173287
Bavi Allomorphia 16008
Bavi Caryodaphnopsis 77957
Bavi Pinanga 299652
Bavi Syzygium 382487
Bawang Kumquat 167490
Bawang Newlitse 264028
Bawang Oak 323703
Bawangling Lipocarpha 232390
Bawangling Milkwort 307948
Bawangling Oak 323703
Bawchan Seed 114427
Bawd 212636
Bawd's Penny 252662
Bawdmoney 174106,252662
Bawdringie 252662
Bawei Cogflower 154142
Bawei Ervatamia 154142
Bawm 249504
Baw-tree 345631
Baxter Banksia 47632
Baxter's Banksia 47632
Bay 122359,223163,223203
Bay Berry 261132
Bay Biscayne 371233
Bay Cherry 223116
Bay Family 223006
Bay Forget-me-not 260813
Bay Lambs 300223
Bay Laurel 223163,223203
Bay Oak 324278
Bay Rum Tree 299323,299328
Bay Tree 223163,223203
Bay Willow 85875,343858
Bayard's Malaxis 243006
Bayberry 223203,261120,261137,261162,
261202,299323
Bayberry Box 64328
Bayberry Family 261243
Bayberry Waxmyrtle-fruit 386585
Bayberry Waxmyrtlefruit Terminalia 386585
Bayberry Willow 343738,343739
Baylaurel 223203
Bay-leaf Willow 343858
Bay-leaved Willow 343858
Baynt Tree 320828
Bayonet Grass 56641,353689
Bayonet Plant 4747,4748,346158
Bayonet Rush 213291
Bayoneta 416654
Bayonet-grass 56641
Bayrum 299328
Bay-rum 299328
Bay-rum Tree 299328
Bayrum-tree 299328
Baywood 380527,380528
Bazell 268438
Bazhi Fissistigma 166660
Bazier Basier 314143
Bazzies 31051
Bazzock 364557
Bblack Snakeroot 345986
Bbride's-bonnet 97098
Bdsde-grass 361877
Bea Nettle 170078,220346
Beach Aster 150665
Beach Chickweed 374941
Beach Clotbur 415057
Beach Clover 396900
Beach Cocklebur 415057
Beach Creeping Oxeye 285360
Beach Evening Primrose 269420
Beach Fleabane 150665
Beach Goldenaster 194246
Beach Grass 19765,19766
Beach Heath 198953
Beach Heather 198953,198955
Beach Knotweed 309525
Beach Leek 15552
Beach Morning Glory 68737
Beach Morningglory 208067
Beach Naupaka 350344
Beach Onion 15552
Beach Palm 46379
Beach Pancake 704
Beach Pea 222735,222743,222745,408435
Beach Pine 299869
Beach Plum 316543
Beach Rose 336901
Beach Sage 344839,344878
Beach Sage-wort 36314
Beach Salvia 344878
Beach Sand Verbena 710
Beach Screw Pine 281138
Beach Sheoak 79161
Beach She-oak 79161
Beach Speedwell 317927
Beach Spiderlily 200941
Beach Spiny-clubpalm 46379
Beach Star 327664
Beach Starwort 374941
Beach Strawberry 167604,167605
Beach Sunflower 188941

Beach Three-awn 34062
Beach Wood Sedge 75084
Beach Wormwood 35779,36314
Beach-head Iris 208628,208819
Beach-primrose 269420
Beachwort 48885
Beacon Bush 196599
Beaconweed 86901
Bead Lily 97086,97091
Bead Plant 265356,265358
Bead Sedge 370050
Bead Tree 7178,248893,248895
Bead Vine 765
Bead-bind 383675
Beadcress 264603
Beadgress 264639
Beadle Mock Orange 294460
Bead-like Sedge 75625
Beadplant 265353,265358
Bead-plant 265358
Beadruby 242668
Bead-ruby 242668
Beads Coldwaterflower 298981
Beads Sandwort 32083
Beadtree 7178,248893
Bead-tree 7178,45908,248893
Beadtree Ormosia 274416
Beagle 315082
Beak Grain 128011
Beak Grass 128013
Beak Rush 333477
Beak Splitmelon 351761
Beak Tickseed 104841
Beak Willow 343095
Beakcalyx Coldwaterflower 299062
Beakchervil 28013
Beaked Agrimony 11597
Beaked Chervil 28013
Beaked Dodder 115131
Beaked Hawk's-beard 111073
Beaked Hawksbeard 111073,111088
Beaked Hawkweed 195643
Beaked Hazel 106724,106771
Beaked Hazelnut 106720,106724
Beaked Panic Grass 281327
Beaked Parsley 28023,28024,28044,261585
Beaked Rush 333501,333585
Beaked Sanicle 346007
Beaked Sedge 76068
Beaked Skeleton-weed 362161,362162
Beaked Spike-rush 143331
Beaked Tasselweed 340476
Beaked Trout-lily 154945
Beaked Tubergourd 390141
Beaked Willow 343095
Beaked Yucca 416640,416654
Beaked-fruit Gynostemma 183032
Beakedfruit Spine olive 415236
Beakedleaf Hookvine 401777
Beakflower 332988
Beakfruit Euonymus 157843
Beakfruit Millettia 66653
Beakfruit Ormosia 274378
Beak-fruited Tailgrape 35052
Beak-fruited Xantolis 415236
Beakgrain 128011
Beakhair Woodbetony 287632
Beakleaf Ash 168086
Beak-leaved Ash 168086
Beakless Gastrodia 171908
Beakovary Habenaria 184039
Beakpetal Bruguiera 61268
Beakpod Cliffbean 66653
Beakpod Eucalyptus 155722
Beakrush 333477
Beak-rush 333477
Beaksac Crazyweed 279126
Beak-sedge 333477
Beak-seed Rhynchodia 333078
Beak-seeded Chonemorpha 333078
Beak-seeded Knotweed 308698
Beakseedvine 333076,333078
Beak-shaped Breynia 60080
Beak-shaped Tutcheria 400717
Beak-shaped Twayblade 232312
Beakstyle Marsdenia 245831
Beakstyle Milkgreens 245831
Beak-styled Condorvine 245831
Beaktooth Woodbetony 287629
Beale Mahonia 242487
Beale's Barberry 242487
Beam Tree 369322
Bean 293966,294056,408393
Bean Barberry 51351
Bean Broomrape 275029
Bean Caper 418585,418636
Bean Curd 177750
Bean Family 282926
Bean Herb 347506
Bean Meliosma 249357
Bean of India 263272
Bean Pear 323116
Bean Sprouts 409025
Bean St. John's Wort 201773
Bean St. John's-wort 201773
Bean Thermopsis 389540
Bean Tree 79075,79243,83527,83809,218754,352500
Bean Trefoil 21442,21445,218761,250502
Bean's Rhododendron 330210
Beanbroomrape 244635
Beancaper 418585
Bean-caper 418585
Beancaper Family 418580
Bean-caper Family 418580
Beanleaf Jasminorange 260164
Bean-leaf Wattle 1235
Beans 409025
Beantree 79075,79242,79260
Bean-tree 79243,218761
Beanweed 299759
Bear Bilberry 31142
Bear Brush 171552
Bear Ears 362726
Bear Grass 68794,122898,415441,416584,416603,416649
Bear Huckleberry 172262
Bear Oak 324039
Bear River Fleabane 151044
Bear Sedge 73742,73743
Bear Skeiters 24483,192358
Bear Valley Sandwort 32298
Bear Valley Wild Buckwheat 152613
Bear Whortleberry 31142
Bear's Beard 405788
Bear's Breech 2657,2664,2695,192358
Bear's Breeches 2657,2695,2711
Bear's Ear 314143,349936
Bear's Ears 314143,349936
Bear's Foot 5442,14065,190936,190959
Bear's Garlic 15861
Bear's Grape 31142,298094
Bear's Grass 416535
Bear's Paw 108048
Bear's-breech 2657,2695
Bear's-breech Family 2060
Bear's-ear Sanicle 105488
Bear's-foot 190936,190959
Bear's-garlic 15861
Bearbed Bulbostylis 63216
Bearbed Sallstyle 63216
Bearberry 31102,31142,328838
Bearberry Cotoneaster 107416
Bear-berry Cotoneaster 107416
Bearberry Honeysuckle 235874
Bearberry Vaccinium 403728
Bearberry Willow 344236
Bearbind 68713,95431,102933,162519,236022
Bearbine 102933
Beard Aster 40120
Beard Eria 148633
Beard Flower 306847
Beard Gentian 173365
Beard Grass 23077,182602,310102,310125
Beard Milkwort 307946
Beard Orchid 67543
Beard Tongue 289314,289352
Beard Valerian 404230
Beard Woodbetony 287773
Bearded Atylosia 65149

Bearded Bellflower 69916
Bearded Chirita 87825
Bearded Couch 144228
Bearded Creeper 113214
Bearded Darnel 235373
Bearded Dichelachne 128862
Bearded Diplectria 132878
Bearded Duckbeakgrass 209261
Bearded Fescue 412405,412409
Bearded Fig 164813
Bearded Flat Sedge 245351
Bearded Flowers Rhododendron 330087,330900
Bearded Goat Grass 8672
Bearded Grass-pink 67883
Bearded Heath 227964,227967
Bearded Iris 208583
Bearded Oat 45389
Bearded Oat-grass 45448
Bearded Oncidium 270795
Bearded Paphiopedilum 282787
Bearded Penstemon 289324
Bearded Pink 127607
Bearded Primrose 314150
Bearded Protea 315876
Bearded Rye-grass 235373
Bearded Schismus 351167
Bearded Setaria 361709
Bearded Shorthusk 58450,58452
Bearded Skeleton Grass 182602
Bearded Sprangletop 226005
Bearded Sprangle-top 226005,226008
Bearded Style Rhododendron 330941
Bearded Tinctorial Wendlandia 413832
Bearded Wheat Grass 144498,144501
Bearded Wheatgrass 144228
Beardedlip Twayblade 232094
Beardednode Arundinella 37356
Bearded-stalk Panicgrass 281370
Beard-flower 306847
Beardflower Azalea 330087
Beardflower Delavay Larkspur 124176
Beard-flowered Rhododendron 330087
Beardgrass 310102
Beard-grass 22476,310125
Beardless Actinidia 6668
Beardless Barley 198376,198396
Beardless Barnyardgrass 140360,140403
Beardless Chinchweed 286890
Beardless Dragonhead 137599
Beardless Duckbeakgrass 209264
Beardless Greenorchid 137599
Beardless Indian Lovegrass 147888
Beardless Kiwifruit 6668
Beardless Lovegrass 147888
Beardless Rabbit's-foot Grass 310134
Beardless Rabbitsfoot Grass 310134
Beardlip Liparis 232094
Beardsepal Corydalis 105646
Beardtongue 289314,289315,289324
Beard-tongue 289381
Beardtongue Penstemon 289328
Beard-tongue Penstemon 289328
Beard-tree 106703
Beargrass 10937,266482,266485,266493,266504,266506,415441,416535,416584,416595,416603
Bear-grass 122898,122910,415441
Bearing Gnaled Branches Rhododendron 330068
Bearing Hairs Rhododendron 330477,331997
Bearing Necklace Rhododendron 332155
Bearing Progeny Pleurothallis 304925
Bearing Runners Tournefortia 393267
Bearing Spines Rhododendron 331877
Bearing-vesicle Rhododendron 332059
Bearsear 314143
Bearsfoot 276724
Bearsfoot Hellebore 190936
Bearskin-grass 164312
Bearsolid Pseudosasa 318344
Bearwort 252662
Beatiful Erythropsis 155006
Beatiful Smallreed 127323
Beatifulwing Mosquitotrap 117414
Beatifulwing Swallowwort 117414
Beatley's Wild Buckwheat 152435
Beaton Currant 333922
Beaty Eyes 410677
Beaty Habenaria 183549
Beatyflower Pleurothallis 304918
Beaty-serrated Lime 391680
Beautiful Abrus 768
Beautiful Afzelia 10110
Beautiful Aralia 30626,30748
Beautiful Arisaema 33517
Beautiful Arrowbamboo 162657
Beautiful Aster 193128
Beautiful Bamboo 297338
Beautiful Banksia 47662
Beautiful Barberry 51297
Beautiful Bauhinia 49254
Beautiful Bigleaf Bredia 296397
Beautiful Bluebell 115353
Beautiful Bluebellflower 115353
Beautiful Bottlebrush 67298
Beautiful Bougainvillea 57868
Beautiful Briggsia 60287
Beautiful Bushclover 226977
Beautiful Buttercup 326261
Beautiful Carrionflower 373947
Beautiful Centaurium 81517
Beautiful Chinacane 364673
Beautiful Clarkia 94140
Beautiful Clematis 94969
Beautiful Cliffbean 66648
Beautiful Clover 397084
Beautiful Coelogyne 98743
Beautiful Colored Rhododendron 330650
Beautiful Condorvine 245835
Beautiful Coryanthes 105523
Beautiful Corydalis 106466
Beautiful Cotoneaster 107340
Beautiful Crapemyrtle 219980
Beautiful Crape-myrtle 219980
Beautiful Crazyweed 278738
Beautiful Cremanthodium 110449
Beautiful Currant 334166
Beautiful Cyclobalanopsis 116064
Beautiful Daffodil Orchid 208324
Beautiful Dendrobium 125152
Beautiful Deutzia 127049
Beautiful Dogwood 105059
Beautiful Euonymus 157789
Beautiful Falsetamarisk 261290
Beautiful Fan Palm 234193
Beautiful Fanpalm 234193
Beautiful Fargesia 162657
Beautiful Feathergrass 376890
Beautiful Felwort 235432
Beautiful Fir 280
Beautiful Fleabane 150648
Beautiful Flowering Dogwood 105027
Beautiful Floweringquince 84573
Beautiful Flowering-quince 84573
Beautiful Fritillary 168614
Beautiful Galangal 17774
Beautiful Gentian 173386,173455
Beautiful Gnetum 178533
Beautiful Goatwheat 44259
Beautiful Gooseberry 334166
Beautiful Grape 411573
Beautiful Hairs Rhododendron 331321
Beautiful Hairy-palmate Maple 3486
Beautiful Hawksbeard 110965
Beautiful Hawthorn 109969
Beautiful Head Synotis 381927
Beautiful Hogfennel 293053
Beautiful Hulsea 199226
Beautiful Iguanura 203512
Beautiful Indocalamus 206779
Beautiful Iris 208695
Beautiful Kohleria 217740
Beautiful Lady Rhododendron 330369
Beautiful Laelia 219692
Beautiful Large-spiny Barberry 51896
Beautiful Leafflower 296725
Beautiful Leaf-flower 296725
Beautiful Leea 223923
Beautiful Leopardflower 373947
Beautiful Leptodermis 226125

Beautiful Leptosiphonium 226436
Beautiful Leschenaultia 226672
Beautiful Lespedeza 226977
Beautiful Lijiang Woodbetony 287357
Beautiful Livistona 234193
Beautiful Lovegrass 147922
Beautiful Lyonia 239376
Beautiful Mazus 247021
Beautiful Meadowrue 388634
Beautiful Meconopsis 247193
Beautiful Milkgreens 245835
Beautiful Milkwort 308347
Beautiful Millettia 66648
Beautiful Mint 65581
Beautiful Mitragyna 256125
Beautiful Mullein 405779
Beautiful Muntries 218395
Beautiful Neoregelia 264426
Beautiful Nephelaphyllum 265122
Beautiful Nerve Eugenia 382533
Beautiful Newlitse 264088
Beautiful Nomocharis 266551
Beautiful Nutantdaisy 110449
Beautiful Oak 116064
Beautiful Odontoglossum 269083
Beautiful Oncidium 270837
Beautiful Orange Rhododendron 330426
Beautiful Osbeckia 276144
Beautiful Pansy Orchid 254937
Beautiful Pawpaw 123575
Beautiful Pellionia 288760
Beautiful Persimmon 132082
Beautiful Phalaenopsis 293634
Beautiful Phyllodium 297013
Beautiful Pieris 298722
Beautiful Pittosporum 301378
Beautiful Pleurospermum 304756
Beautiful Pommereschea 310829,310830
Beautiful Porana 396761
Beautiful Primrose 314159
Beautiful Purplecup Mockorange 294417
Beautiful Pyrostegia 322950
Beautiful Qinggang 116064
Beautiful Rattlebox 112682
Beautiful Renanthera 327694
Beautiful Rhododendron 330283,330723,331581
Beautiful Ribseedcelery 304756
Beautiful Rockfoil 349817
Beautiful Rodriguezia 335164,335170
Beautiful Rosarypea 768
Beautiful Sage 345207
Beautiful Sainfoin 271255
Beautiful Sandwort 31787,31903,255573
Beautiful Scabrous Wendlandia 413823
Beautiful Seatung 301378
Beautiful Sedge 74164
Beautiful Senna 360485
Beautiful Shrubcress 368571
Beautiful Sobralia 366781
Beautiful Solms-Laubachia 368571
Beautiful Sophora 368972
Beautiful Southstar 33517
Beautiful Spapweed 205263
Beautiful Spatholobus 370417
Beautiful Spicebush 231423
Beautiful Spice-bush 231423
Beautiful Spider Flower 95796
Beautiful Spindle-tree 157789
Beautiful Spiraea 371821
Beautiful St. John's Wort 201775
Beautiful St. John's-wort 201775
Beautiful Sulphur Flower 152603
Beautiful Sweetgum 232557
Beautiful Sweetleaf 381423
Beautiful Sweetspire 210360
Beautiful Swertia 380115
Beautiful Tanoak 233128
Beautiful Touch-me-not 205263
Beautiful Underleaf Pearl 296725
Beautiful Vanda 404648
Beautiful Wendlandia 413823
Beautiful Wightia 414114
Beautiful Windhairdaisy 348690
Beautiful Woodbetony 287034,287794
Beautiful Woolly Sunflower 152800
Beautiful Xikang Azalea 331829
Beautiful Xikang Rhododendron 331829
Beautiful Yellow Rhododendron 330312
Beautiful-colored Rhododendron 330650
Beautiful-covering Rhododendron 330292
Beautiful-flower Barberry 51416
Beautifulflower Cranesbill 174512
Beautiful-flower Greenbrier 366542
Beautifulflower Primrose 314201,314858
Beautifulflower Rhodiola 329849
Beautifulflower Saussurea 348690
Beautifulflower Tickclover 126276
Beautifulflower Uraria 402144
Beautiful-flowered Barberry 51416
Beautiful-flowered Tickclover 126276
Beautiful-flowered Uraria 402144
Beautifulform Goldenbutton 371645
Beautifulhead Edelweiss 224811
Beautifulhead Tailanther 381927
Beautifulleaf Belltree 324996
Beautifulleaf Clematis 95409
Beautiful-leaf Enkianthus 145675
Beautifulleaf Greenorchid 137567
Beautiful-leaf Kadsura 214951
Beautifulleaf Larkspur 124095
Beautiful-leaf Mucuna 259488
Beautiful-leaf Pendent-bell 145675
Beautiful-leaf Pouzolzia 313420
Beautifulleaf Sagebrush 35233
Beautiful-leaves Rhododendron 330281
Beautiful-nerve Maple 2829
Beautiful-nerved Maple 2829
Beautiful-nerved Syzygium 382533
Beautiful-perianth Azalea 330292
Beautiful-perianth Rhododendron 330292
Beautiful-perianthed Rhododendron 330292
Beautiful-raceme Barberry 51417
Beautiful-spiked Dendrocalamus 125471
Beautiful-spiked Rattan Palm 65663
Beautifulspikegrass 407458
Beautiful-stachys Dendrocalamus 125471
Beautiful-tooth Camellia 68964
Beautiful-toothed Camellia 68964
Beautify Gochnatia 178749
Beauty Berry 66727,66733,66743,66913
Beauty Bush 66727,217785,217787
Beauty Garishspiderorchis 155265
Beauty Goniolimon 179380
Beauty Leaf 67846,67860
Beauty Leaf Mastwood 67860
Beauty of Glen Retreat 93692
Beauty of the Night 255711
Beauty Small Reed 127323
Beautyberry 66727,66743,66910
Beauty-berry 66727
Beautyberryleaf Meliosma 249369
Beauty-berry-leaved Meliosma 249369
Beautyberrylike Gomphostemma 179177
Beautybract Tanoak 233127
Beautybush 217787,217789
Beauty-bush 217787
Beautyflower Bulbophyllum 63048
Beautyflower Curlylip-orchis 63048
Beautyflower Dendrobium 125233
Beautyflower Paphiopedilum 282845
Beautyflower Satyrium 347911
Beautyful Lilyturf 272112
Beautyhair Michelia 252835
Beautyhead 47155
Beautyhead Flickingeria 166958
Beautyleaf 67846,67860
Beauty-leaf 67846
Beautyleaf Hemipilia 191590
Beauty-leaf Mastwood 67860
Beautyleaf Reekkudzu 313420
Beauty-of-the-night 255711
Beautystyle 67449,67451
Beauverd Barberry 51352
Beauverd Photinia 295630
Beauverd Stonecrop 356569
Beauverd's Widow's-thrill 215097
Beaver Poison 90924,101852
Beaver Tree 242341
Beaver Wood 80698
Beaver-root 192278

Beavertail 356944
Beavertail Cactus 272805
Beavertail Pricklypear 272805
Beavertail-grass 67575
Beaverwood 80698,80728
Beaweed 359158
Bebb Willow 343095
Bebb's Oak 323582,323704
Bebb's Oval Sedge 73859
Bebb's Sedge 73859
Bebb's Willow 343095
Becaming Red Rhododendron 331700
Beccabunga Speedwell 407029
Beccar Dacrydium 121087
Beccar Halophila 184969
Beccar Pinanga 299654
Beccar Saltgrass 184969
Beccarinda 49394
Bechtel Crab 243637
Bechtel's Crab 243637
Beck Photinia 295636
Beck's Beggarticks 247882
Beck's Water-marigold 247881
Beckbean 250502
Becker Hawthorn 109552
Beckett Cryptocoryne 113528
Beckett's Water Trumpet 113528
Beckmann's-grass 49540
Becky-leaves 407029
Becoming Green Rhododendron 332070
Becoming Paler Rhododendron 331434
Becoming Red Ophiorrhiza 272298
Becoming Yellow Rhododendron 331159
Becoming-black Skullcap 355622
Becomingnaked Delavay Clematis 94875
Becomingred Marianthus 245272
Bed Furze 401390,401395
Bedbug Alloteropsis 16222
Bedda Nut 386473
Beddeome Spapweed 204806
Beddeome Touch-me-not 204806
Bedding Begonia 50298
Bedding Dahlia 121559
Bedding Geranium 288297
Bedding Pansy Violet 409858
Bedding Viola 409623
Bedding Violet 409624
Beddywind 102933
Bed-flower 170743
Bedford Willow 344250
Bed-furze 401390,401395
Bedigo Wax Flower 153336
Bedlam Cowslip 321643
Bedsfoot Flower 97060
Bedstraw 170175,170193,170246,170350,170490,170548
Bedstraw Asperula 39347
Bedstraw Bellflower 69899
Bedstraw Broomrape 274985
Bedstraw Family 338045
Bedstraw St. John's-wort 201888
Bedstruw Woodruff 39347
Bedvine 95431
Bedwen Bedewen 53563
Bedwind 68713,95431,102933
Bedwine 95431,102933,162519
Bee Balm 249504,257157,257161,257163,257169
Bee Bee Tree 161323
Bee Bread 57101,397019
Bee Catcher 130383
Bee Feed 152056
Bee Flower 86427,272398
Bee Larkspur 124194
Bee Nettle 170025,170076,170078,220346,220416
Bee Ophrys 272398
Bee Orchid 272398,272399
Bee Orchis 272398
Bee Sage 344853
Bee Wattle 1613
Bee Wort 5793
Bee's Nest 123141,192358
Bee's Rest 68189
Beead-and-cider 109857
Beebalm 257157,257158,257160,257161
Bee-balm 257161
Bee-bee Tree 161302
Beebrush 17600
Beech 162359,162386,162400,162407,162416
Beech Drops 146520
Beech Family 162033
Beech Tree 162359
Beech Vitex 411430
Beechdrops 146520
Beech-drops 146520
Beechey Bamboo 125441
Beechey Bambusa 125441
Beechey Dendrocalamopsis 125441
Beechey Fig 164671
Beechey Greenbamboo 125441
Beech-leaf Begonia 49699
Beechleaf Glochidion 177178
Beech-leaved Flemingia 297013
Beechnut 162375
Beechwheat 162312
Beedy's Eyes 410677
Beef Apple 244605
Beef Plant 208349
Beef Steak Plant 2011
Beef Suet Tree 362090
Beef Wood 79155,79161,79185,180651,375511
Beefsteak 47647,49818,79161,180651,290940,375511
Beefsteak Geranium 49587,50235
Beefsteak Plant 2011,208349,290932,290940
Beefsteakplant 290940
Beefsteak-plant 290940
Beef-suet Tree 362090
Beef-tree 181531
Beefwood 15953,79155,79157,79161,180651
Beef-wood 181531
Beefwood Family 79193
Beehive 130383
Beehive Cactus 107076,155228
Bee-in-a-bush 5442
Beelzebub 175444,321643
Bee-orchid 272398
Beering Mouseear 82732
Beerseem Clover 396817
Bees Barberry 51353
Bees Beakflower 332989
Bees Jasmine 211749
Bees Larkspur 124058
Bees Leek 15111
Bees Onion 15111
Bees Primrose 314157
Bees Red Nettle 220416
Bees Rhododendron 330211
Beesia 49575
Beesian Larkspur 124058
Beesom 121001
Beestebul 108818
Bee-swarm Orchid 120705,120715
Beet 53224,53249,53254,53267
Beet Broomwort 354660
Beet Chard 53257
Beetle-grass 225983
Beetleweed 169819
Beetlewood 169819
Beet-like Mandrake 119974
Beetraw 53254
Bee-tree 391649,391681,391727
Beetree Linden 391727
Beetrie 53254
Beetroot 53249,53267
Beet-root 53269
Beewood 40256
Beggar Brushes 95431
Beggar Lice 170193,221770
Beggar Tick 53755,54048
Beggar Ticks 53858,53879,53913,53976,54048,54158,54182
Beggar Weed 126649
Beggar's Basket 321643
Beggar's Blanket 405788
Beggar's Brush 95431

Beggar's Bur 31051
Beggar's Burr 31051
Beggar's Buttons 31051,123215,170193,332086
Beggar's Lice 123215,170193,184320,221727,221755
Beggar's Needle 350425
Beggar's Plant 95431
Beggar's Stalk 405788
Beggar's Ticks 247881
Beggar's-lice 184320,221711
Beggar's-needle 350425
Beggar-lice 170193,221770
Beggarlnan Cakes 18173
Beggarticks 53755,53797,53913
Beggar-ticks 53755,53797,53913,54048
Beggarweed 115008,115031,170193,192358,308816,370522
Beggary 169126
Begger Rose 336380
Begger's Tick 53976
Begh 53563
Beginners' Pondweed 312079
Begonia 49587
Begonia Family 50433
Begonia Treevine 92692,92777
Begonia Tree-vine 92692
Begonialeaf Cacalia 283795
Begonialeaf Petrocosmea 292549
Begonialeaf Sonerila 368896
Begonialeaf Stonebutterfly 292549
Begonialeaf Tremacron 394674
Begonialeaf Zippelia 418076
Begonifoliate Spapweed 204807
Begoon 367370
Beibei Ardisia 31364
Beibei Maple 3392
Beibeng Elatostema 142617
Beibeng Stairweed 142617
Beichatian Gastrodia 171948
Beichatian Platanthera 302481
Beichuan Arisaema 33447
Beichuan Southstar 33447
Beijing Barberry 51355
Beijing Batrachium 48929
Beijing Cabbage 59600
Beijing Catalpa 79247
Beijing Cortusa 105501
Beijing Corydalis 105920
Beijing Cotoneaster 107325
Beijing Hideseedgrass 94605
Beijing Hogfennel 292792
Beijing Honeysuckle 235766
Beijing Jijigrass 4150
Beijing Lilac 382278
Beijing Mockorange 294509
Beijing Mountainash 369379
Beijing Oak 323647
Beijing Peashrub 72310
Beijing Poplar 311143
Beijing Rockfoil 349895
Beijing Skullcap 355669
Beijing Speargrass 4150
Beijing Spurge 159540
Beijing Stonecrop 357011
Beijing Sweet Flag 5818
Beijing Thorowax 63597
Beijing Violet 410395
Beikiang Sedge 75618
Beimushan Bentgrass 12011
Beipei Fig 164682
Beipiao Dahurian Bushclover 226754
Beipiao Hawthorn 109553
Beipiao Truncate-leaved Maple 3717
Bejuco De Santiago 34361
Beketov Milkvetch 42072
Bekkabung 407029
Bel 8787
Bel Bael 8787
Bel Bor 300223
Bel Buttons 68189
Bela Movra 300144
Belah 79156,79177
Belangeran 362195
Bel-apple 285680
Bel-buttons 68189
Beldairy 273531,273545
Belder-root 269307
Beldrum 269307
Belene 201389
Belgaum Walnut 14544
Belgian Endive 90901
Belgian Orach 44668
Belgian Red Willow 342984
Belian Ironwood 160967,160968
Beliconia 189986
Belilaj 386473
Bell 8787
Bell Bottle 145487
Bell Catchfly 363285
Bell Flambeau Tree 370358
Bell Flambeautree 370358
Bell Flambeau-tree 370358
Bell Flower 69870
Bell Flower Hesperaloe 193225
Bell Gardenia 337500
Bell Heath 150131
Bell Heather 149205,150131
Bell Ling 149205
Bell Pepper 72068,72070,72075
Bell Rose 262441
Bell Thistle 92485
Bell Tree 121556,168767,184726,324993
Bell Tree Dahlia 121556
Bell Woodbind 68713
Bell Wort 403665
Bell's Honeysuckle 235674
Bella Sombra Tree 298103
Bella Umbra 298103
Belladonna 44708
Belladonna Atropa 44708
Belladonna Lily 18865,196443
Belladonnalily 18865
Bell-apple 285660
Bellard Kobresia 217121
Bellard's Smartweed 309531
Bellbind 68713,102933
Bellbinder 68713
Bellbine 68713
Bellbloom 262441
Bell-bottle 145487
Bellbract Sawwort 361029
Bellcalyx Camellia 68966
Bellcalyx Lasianthus 222293
Bellcalyx Pulsatilla 321665
Bellcalyx Roughleaf 222293
Bellcalyx Scape-like Sage 345368
Bell-calyxed Lasianthus 222293
Bellcalyxwort 231254
Bellcrown Mosquitotrap 117399
Belle Isle Cress 47963
Belle Rhododendron 330214
Belleyache Bush 212144
Bellflower 1000,68713,69870,69892,69942,70234,70255,86427,412568
Bell-flower 69870
Bellflower African Cornlily 210723
Bellflower Agapetes 10354
Bellflower Azalea 330303
Bell-flower Cherry 83158
Bellflower Clematis 95243
Bellflower Dahuria Gentian 173374
Bellflower Family 70372
Bellflower Gentian 173656
Bellflower Honeysuckle 235741
Bellflower Knotweed 308941
Bellflower Leek 15397
Bellflower Rhododendron 330303
Bellflower Sabia 341490
Bellflower Smartweed 308941
Bellflower Tulip 400229
Bellflowered Cherry 83158
Bell-flowered Cherry 83158
Bell-flowered Honeysuckle 235741
Bellflowered Onion 15397
Bell-flowered Rhododendron 330303
Bell-flowered Squill 199553
Bell-fruit Mallee 155705
Bell-fruited Mallee 155705
Bellfruit-tree 98256
Bell-heather 149205

Bellidiflorus 339419
Bellir Terminalia 386473
Bellisima Grande 28421
Belllike Oneflowerprimrose 270728
Bellota 376075
Bellows-flower 220729
Bellowsfruit 297833
Bellragges 336172
Bellrose 262441
Bell-rose 262441
Bells Ivy 68189
Bells of Ireland 256761
Bellsepal Tea 68966
Bellshaped Falselily 266898
Bell-shaped Rhododendron 330303
Bellshaped Spapweed 204837
Bellshaped Sterculia 376089
Bellshaped Touch-me-not 204837
Bellshaped Wintersweet 87515
Bell-shaped Wintersweet 87515
Bells-of-Ireland 256761
Belltree 324993
Bellum 316382
Bell-weed 81237,265944
Bellwind 68713,102933
Bellwinder 68713
Bellwort 98273,403665,403670,403672
Bellyache Bush 212144,212145
Belly-ache Bush 212144
Bellywind 95431
Belmore Palm 198806
Belmore Sentry Palm 198806
Belmores Howeia 198806
Belostemma 50931
Belosynapsis 50937
Beloved Rhododendron 330082,330626
Beltleaf Actinidia 6598
Beltleaf Bulbophyllum 63117
Beltleaf Chirita 87935
Beltleaf Curlylip-orchis 63117
Beltleaf Herminium 192838
Beltleaf Peristylus 291196
Beltleaf Perotis 291196
Beltleaf Platanthera 302388
Beltleaf Pocktorchid 282839
Beltleaf Primrose 315099
Beltleaf Tongoloa 392800
Beltleaf Windhairdaisy 348501
Beltlike Spicegrass 239896
Beltlip Perotis 291218
Beltpetal Platanthera 302517
Bel-tree 8787
Belumut Rattan Palm 65648
Belvedere 217361
Belvedere Cypres 217361
Belvedere Summer Cypres 217361
Belweed 81237
Ben 258945,363671,364193
Ben Lomand Spineflower 89106
Ben Lomand Wild Buckwheat 152321
Ben Nut 258945
Benchfruit 279743,279744
Bend 67456
Bendang 10496
Bendback Platanthera 302297
Bendbeak Milkvetch 42135
Bendehair Redcarweed 288768
Bendflower Milkvetch 42371
Bendhair Stairweed 142819
Bendhair Yushanbamboo 416776
Bending Bitter-cress 72749
Bending Hollow-bamboo 82454
Bendizao Mandarin 93732,93748
Bendleaf Figmarigold 177469
Bendo Eucalyptus 155578
Bendock 269307
Bendpole Arrowbamboo 162756
Bendstalk Amoora 19968
Bendstem Bellflower 69934
Bendtooth Milkvetch 42133
Bendtwig Peashrub 72184
Bendy Tree 389946
Bene 361317
Bene of Egypt 263272
Benecke Twinballgrass 209048
Benedict's Herb Blessed Herb 175454
Beneken Brome 60643
Benewith 236022
Bengal Arundinella 37357
Bengal Bamboo 47493
Bengal Bean 259559
Bengal Canary Tree 70992
Bengal Canarytree 70992
Bengal Cardamom 19824
Bengal Clock Vine 390787
Bengal Clockvine 390787
Bengal Clock-vine 390787
Bengal Coffee 98866
Bengal Dayflower 100940
Bengal Fig 164684
Bengal Gambir-plant 401757
Bengal Gamboge 171215
Bengal Gram 90801
Bengal India Rubber Tree 164925,165168
Bengal India-Rubber Tree 164925
Bengal Kino 64148
Bengal Loquat 151132
Bengal Meriandra 250656
Bengal Olive 70992
Bengal Quince 8787
Bengal Rinorea 334589
Bengal Rose 336485
Bengal Rush 212897
Bengal Trumpet 390787
Bengal Waterdropwort 269300
Bengal Wild Fig 164684
Bengalese Cyperus-like Sawgrass 245395
Benge 181765
Benghal Hiptage 196824
Benghal Kitefruit 196824
Benikoji Mandarine 93737
Benin Camwood 47806
Benin Mahogany 216194,216202,216203
Benin Pepper 300397,300401
Benin Walnut 237927
Benin Wood 216202,216203
Benjamin 231306
Benjamin Banyan 164688
Benjamin Bush 231306
Benjamin Fig 164688
Benjamin Tree 164688,379311,379312
Benne 361317
Bennel 359158
Bennergowan 50825
Bennert 50825
Bennet 299523,302034,302068
Bennettiodendron 51024
Benniseed 361317
Ben-nut Tree 258945
Ben-oil Tree 258945
Benoist Barberry 51357
Benom Rattan Palm 65649
Benquet Pine 299986
Benson Dendrobium 125014
Benson Vanda 404625
Bent 67456,302034
Bent Chandler Pieris 298735
Bent Eulalia 318150
Bent Fruits Rhododendron 330311
Bent Goatgrass 8674
Bent Grass 11968,19766
Bent Larkspur 124217
Bent Leymus 228361
Bent Milk Vetch 42299
Bent Ovary Rhododendron 330314
Bent Petal Paphiopedilum 282822
Bent Spike Roegneria 335284
Bent Spikerush 143158
Bent Trillium 397563
Bent Two-whorl Buckwheat 375583
Bentawn Muhly 259646
Bentawn Plumegrass 148865
Bentawned Roegneria 335323
Bentaxis Burreed 370061
Bent-axis Pricklyash 417329
Bentbeaked Sedge 75076
Bentbranch Tickseed 104778
Bentcalyx Honeysuckle 236194
Bent-culmed Fargesia 162756
Bentdentate Milkvetch 42133
Bentdentate-like Milkvetch 42132

Bent-ear Peashrub 72254
Bent-ear Pea-shrub 72254
Bentflower Bugle 13068
Bentflower Conehead 320117
Bent-foot Speedwell 407049
Bentfruit Chirita 87847
Bentfruit Tanoak 233168
Bent-fruited Tanoak 233168
Bentgrass 11968,11972
Bent-grass 11968
Benthair Feng Monkshood 5187
Bentham Anodendron 25933
Bentham Butterflygrass 392886
Bentham Cable Creper 25933
Bentham Eelvine 25933
Bentham Erysimum 154398
Bentham Fisheyeweed 129070
Bentham Oreocharis 273842
Bentham Photinia 295637
Bentham Portuguese Cypress 114664
Bentham Rosewood 121634
Bentham Sophora 368975
Bentham Staff-tree 257504
Bentham Torenia 392886
Bentham Yam 131485
Bentham's Cornel 124891
Bentham's Dichrocephala 129070
Bentham's Pipewort 151247
Bentinck Palm 51163
Bentinckia 51163
Bent-knees Microstegium 254010
Bentley's Coral-root 103979
Bentnerved Euonymus 157947
Bentricose Goatgrass 8746
Bent-seeded Hop Sedge 76613
Bentshoot Asparagus 38928
Bentshoot Greenbrier 366416
Bentshoot Sealavender 230626
Bentspike Dogtoothgrass 117881
Bentspike Fountaingrass 289176
Bentspike Sealavender 230610
Bentspikegrass 131214
Bentspur Larkspur 124098
Bentstalk Ardisia 31420
Bent-stalk Greenbrier 366231
Bentwood 187222
Benweed 359158
Benwood 187222
Benzhao Spikesedge 143305
Benzilan Monkshood 5062
Benzilan Rockfoil 349097
Benzoin 231306,379311
Benzoin Laurel 379311
Benzoin Sweetgum 232565
Bequaert Malouetia 243506
Bequilla 361380
Ber 418184
Berber 52322
Berberis 51268,52322
Berberry 242478
Berbine 405872
Berchemiella 52477
Berd's Grass 11972
Bere Bear 198376
Bereft Scales Rhododendron 331683
Berg Cypress 414060
Berg Lily 170828
Berg's Panicgrass 281381
Bergaalwyn 17017
Bergamot 93408,257157,257161,257169
Bergamot Mint 250291,250341,250362
Bergamot Orange 93408
Bergenia 52503
Berger Barberry 51359
Berger Photinia 295641
Berger Pricklypear 272812
Berger Stonecrop 356570
Berggarten Sage 345271
Bergia 52587
Bergmann Barberry 51360
Bergmann Elm 401454
Bering Mouse-ear Chickweed 82732
Bering Sea Sorrel 339952
Beringia Sorrel 339952
Beringian Spring Beauty 94369
Berk Apple 300223
Berk-apple 300223
Berkman Arborvitae 302725
Berlandier Reed 295896
Berlandier Wolfberry 239016
Berlandier's Acacia 1080
Berlandier's Goosefoot 86958
Berlandier's Hedgehog Cactus 140216
Berlin Poplar 311144
Berlinia 52843
Bermuda 245014
Bermuda Arrowroot 245014
Bermuda Buttercup 278008
Bermuda Cedar 213606,213623
Bermuda Falseolive 142458
Bermuda Grass 117859
Bermuda Juniper 213623
Bermuda Lily 229900,229902
Bermuda Oil-wood 142458
Bermuda Palm 341405
Bermuda Palmetto 341405
Bermuda Red Cedar 213623
Bermuda Red-cedar 213623
Bermuda Sisyrinchium 365747
Bermuda-buttercup 278008
Bermudagrass 117859
Bermuda-grass 117852,117859
Bernard's Goldenrod 367984
Bernard's Lily 88527
Bernard's Violet 409715
Bernardina Rabbitbrush 150337
Berneuxia 52928
Berneuxine 52928
Berolin Poplar 311144
Berry Bamboo 249617
Berry Bearing Oak 323769
Berry Catchfly 114057,114060
Berry Heath 149054
Berry Holly 203545
Berry Holm 203545
Berry Russianthistle 344538
Berry Saltbush 44641
Berrya 52954
Berry-bearing Aspidocarya 39608
Berry-bearing Campion 114060
Berry-bearing Chickweed 114060
Berry-bearing Cipadessa 91480
Berry-bearing Glossy Buckthorn 328701
Berry-bearing Poplar 311237,311398,311537
Berry-bush 166769,334250
Berry-girse 145073
Berrylike Oak 323777
Berry-like Sapium 346372
Berry-shaped Sauropus 348043,381916
Berry-tree 166769
Berseem 396817
Bertam 156116
Berteroella 53045
Berteron's Rattlebox 111949
Bertery 345631
Bertolon Timothy 294968
Bertoloni's Bee-orchid 272412
Bertram 21266,322724
Berula 53154
Besenkraut 217361
Besleria 53205
Besom 67456,121001
Besom Arrowbamboo 137851
Besom Heath 150027,150131
Besomweed 390202
Besomweed Pennycress 390202
Besser Centaurea 80965
Bessey Cherry 83157
Bessey's Cherry 83157
Bessie Banwood 50825
Bessy Bairnwort 50825
Be-still Tree 389978,389983
Beswind 68713
Beswlne 68713,102933
Betel 31680,300354
Betel Nut 31680
Betel Palm 31680
Betel Palms 31677
Betel Pepper 300354
Betelnut Palm 31680
Betel-nut Palm 31680

Betes 308893
Betfy-go-to-bed-at-noon 274823
Bethlehem Cowslip 321643
Bethlehem Lungwort 321647
Bethlehem Sage 321643,321647
Bethlehemsage 283603
Bethlem Star 274823
Bethmont Currant 333923
Bethroot 68713,95431,102933,162519,397556,397570
Bethwine 95431
Beticolor Russianthistle 344470
Betle-nut Palm 31680
Betlnutpalm Tanoak 233113
Betlnut-palm Tanoak 233113
Beton 53304
Betonica 53293
Betonicalike Jerusalemsage 295064
Betonicleaf Cinquefoil 312408
Betony 53304,373085,373166
Betonyleaf Mazus 247039
Betonyleaf Meconopsis 247109
Betony-leaf mistflower 101951
Betonylike Galangal 17759
Betonylike Morina 258859
Bettywind 102933
Beutiful Curvefruit Azalea 330312
Beverly Hills Arborvitae 302726
Bewitch Azalea 330046
Beyrich Galeandra 169867
Bezors 314143
Bhaji 18652
Bhang 71220
Bharbur Grass 156513
Bhendi Tree 389946
Bhesa 53678
Bhotan Dubyaea 138765
Bhotan Milkvetch 42074
Bhudan Small-flowered Barberry 51920
Bhudanese Rhododendron 331859
Bhutan Azalea 330794
Bhutan Bamboo 125508
Bhutan Barberry 51684
Bhutan Catchfly 363569
Bhutan Corydalis 106059
Bhutan Cremanthodium 110350
Bhutan Cypress 114668,114680,114775
Bhutan Liparis 232106
Bhutan Microula 254337
Bhutan Nutantdaisy 110350
Bhutan Pine 300297
Bhutan Sophora 368976
Bhutan Twayblade 232106
Bialynick's Mouse-ear Chickweed 82738
Biangular Spapweed 204816
Biangular-stalk Ternstroemia 386691
Biannual Wormwood 35598
Bianthela Silvergrass 127477
Biantheralga 184938,184943
Biaristate Huodendron 199449
Biasong 93634
Biba Pepper 300446
Bibiru-bark Tree 263063
Bible Flower 201732
Bible Frankincense 57516
Bible Hyssop 274226
Bible Leaf 158433,201732
Bible Wormwood 35600
Bibolo 237927
Bibract Diversifolious Osmanther 276313
Bibracteolate Elytranthe 241316
Bibracteolate Macrosolen 241316
Bicallus Gentian 156572
Bicallus Oncidium 270797
Bicanal Cowparsnip 192245
Bicapsular Senna 360426
Bicarinate Eulophia 156572
Bichy Tree 99158
Bicicatrical Dendrocalamopsis 47211
Bicknell's Crane's-bill 174495
Bicknell's Oval Sedge 73869
Bicknell's Rock-rose 188605
Bicknell's Sedge 73869
Bicolor Artocarpus 36946
Bicolor Barberry 51363
Bicolor Bulbophyllum 379773
Bicolor Bushclover 226698
Bicolor Bush-clover 226698
Bicolor Chickweed 374858
Bicolor Chirita 87826
Bicolor Cranesbill 174767
Bicolor Curlylip-orchis 379773
Bicolor Cymbidium 116796
Bicolor Dumasia 138929
Bicolor Fanmustard 265308
Bicolor Hairgrass 126071
Bicolor Kiwifruit 6587
Bicolor Lacaena 218768
Bicolor Lespedeza 226698
Bicolor Magnoliavine 351016
Bicolor Magnolia-vine 351016
Bicolor Mongolian Willow 343709
Bicolor Notylia 267216
Bicolor Onosma 242396
Bicolor Oriental Buckthorn 328667
Bicolor Patrinia 285849
Bicolor Peashrub 72190
Bicolor Pertybush 292052
Bicolor Rhododendron 330565
Bicolor Roughleaf 222101
Bicolor Sabia 341479
Bicolor Sage 345381
Bicolor Serrate Hydrangea 200093
Bicolor Spapweed 204904
Bicolor Sterculia 376082
Bicolor Sulphur Flower 152580
Bicolor Vanda 404626
Bicolor Velvetplant 183066
Bicolor Vetch 408377
Bicolor Willow 342995
Bicolor Woodbetony 287041
Bicolor Yam 131488
Bicolored Arisaema 33337
Bicolored Dendranthema 124781
Bicolored Everlasting 21525
Bicolored Onosma 242396
Bicolored Pearleverlasting 21525
Bicolored Southstar 33337
Bicoloured Ardisia 31403
Biconvex Halbertleaf Knotweed 308886
Bicorn Woodbetony 287042
Bicornute Hemiboea 191349
Bicoronate Genianthus 172850
Bicuhyba 53741
Bicuhyba Fat 53741
Bicusped Milkvetch 42078
Bicuspidate Falsenettle 56098
Biddy 237539
Biddy Bush 78608
Biddy's Eye 409703,410677
Biddy's Eyes 21339,72934,160236,175708,176824,237539,248312,260760,260842,289503,314366,342279,349044,374916,374968,407029,407072,407144,407256,409703,410677
Biddy-biddy 1752
Bident Germander 388012
Bidentate Holly 203589
Bidentate Onion 15113
Bidentate Woodbetony 287043
Bidgee-widgee 1739,1752
Bidibid 1752
Bidi-bidi 1739,1752
Bidwell Coralbean 154632
Bidwell's Knotweed 308890
Bidwill Incense-cedar 228636
Bidwill's Araucaria 30835
Bidwilli Sterculia 376084
Bieberstein Bentgrass 12015
Bieberstein Sainfoin 271181
Bieberstein Sandwort 31848
Bieberstein Toadflex 230910
Bieberstein Tulip 400131
Bieberstein's Carlina 76994
Bieberstein's Crocus 111603
Bieberstein's Mouse-ear Chickweed 31848
Biebersteinia 54194,54196
Biennial Hawksbeard 110752
Biennial Bee-blossom 172187
Biennial Campion 363385
Biennial Cork Oak 324247

Biennial Gaura 172187
Biennial Moonwort 238371
Biennial Motherwort 85058
Biennial Sage-wort 35204
Biennial Spinach Beet 53249
Biennial Wormwood 35204,35598
Biermann Lasianthus 222102
Biermann Roughleaf 222102
Biermannia 54210
Biet Woodbetony 287044
Bifid Bladderwort 403119
Bifid Deinanthe 123655
Bifid Draws Ear Grass 403119
Bifid Hempnettle 170060
Bifid Hemp-nettle 170060
Bifid Meliosma 249366
Bifid Passionflower 285613
Bifidcalyx Sage 344904
Biflor Crazyweed 278744
Biflorgrass 128565
Biflorous Abelia 416841
Biflorous Camellia 68939
Biflorous Litosanthus 233811
Biflorous Lycianthes 238942
Biflorous Microtropis 254284
Biflorous Morningglory 25316
Biflorous Raspberry 338183
Biflorous St. John's-wort 201892
Biflous Rose 336388
Biflower Begonia 49658
Biflower Daffodil 262417
Biflower Dendrobium 125158
Biflower Dogtoothgrass 117862
Biflower Listera 264660
Biflower Litse 233854
Biflower Minuartwort 255447
Biflower Nightshade 238942
Biflowered Rose 336388
Bifoil 232950
Bifoliate Ammodendron 19729
Bifora 54237
Bifork Ptilagrostis 321011
Bifork Tickclover 126307
Biform Ladybell 7605
Biformbean 20558
Biforme Microstegium 253987
Biformleaf Dendropanax 125639
Biformleaf Treerenshen 125639
Biformleaf Yam 131489
Biform-leaved Dendropanax 125639
Bifrenaria 54243
Bifruit Holly 203762
Bifurcate Cinquefoil 312412
Big Adinandra 8233
Big Ammi 19664
Big Angularfruit Swordflag 208508
Big Ant Palm 217922
Big Arrowleaf Knotweed 309036
Big Baby's-breath 183158
Big Basil 268438
Big Bend Aster 240469
Big Bend Beardtongue 289351
Big Bend Cactus 107018
Big Bend Cory Cactus 107058
Big Bend Hop-hornbeam 276805
Big Bend Woody-aster 415858
Big Bend Yucca 416640
Big Betony 373232
Big Bird-of-paradise 377563
Big Bird-of-paradise Flower 377563
Big Blue Lilyturf 232631
Big Blue Lobelia 234707
Big Bluegrass 305323,305567
Big Bluestem 22681
Big Blue-stem 22681
Big Bluethistle 141177
Big Bract Loranthus 392715
Big Bristlegrass 361743
Big Brome 60834
Big Broomrape 275128
Big Buckeye 9692
Big Buttercup 68189
Big Caltrap 215383
Big Chari Primrose 315061
Big Chickweed 82826
Big Cigar 114611
Big Clouds Azalea 330676
Big Conehead 378152
Big Copper Knotweed 309127
Big Cord-grass 370160
Big Cottonwood 311292
Big Cowparsnip 192303
Big Cypress 114691
Big Daisy 227533
Big Dammar Pine 10505
Big Devil's Beggar-ticks 54182
Big Dodder 115031
Big Dogwood 124914
Big Duckweed 372300
Big Eelgrass 418183
Big Elwes Woodbetony 287186
Big Eucalyptus 155598
Big Flower Tellima 385720
Big Galleta 304593
Big Garnotia 171504
Big Glomerate Arrowwood 407870
Big Glomerate Viburnum 407870
Big Glossodia 177314
Big Greenbamboo 47302
Big Gypsophila 183158
Big Hairfruit Microula 254375
Big Hairyleaf Alyxia 18532
Big Helicia 189933
Big Heliotrope 190559
Big Honeysuckle 25258
Big Hookleafvine 303088
Big Horseshoe Lake Dewberry 339306, 339307
Big Jokul Sandwort 32102
Big Jute-bamboo 125482
Big Larkspur 124371
Big Laurel 242135
Big Leaf Mahogany 380527
Big Leaf Periwinkle 409335
Big Marigold 383090
Big Maximowicz Hillcelery 276755
Big Monkshood 5413
Big Mountain Palm 188191
Big Mouseear 82918
Big Mouse-ear Chickweed 82826
Big Muhly 259676
Big Narrowflower Aniseia 25328
Big Nettle 402886
Big Nitraria 266365
Big Nod Bamboo 206868
Big Odontoglossum 269069
Big Peripterygium 73188
Big Periwinkle 409335
Big Phreatia 295960
Big Pink Deutzia 127068
Big Pipe Tanoak 233158
Big Plantain 302068
Big Pollia 307322
Big Pulsatilla 321681
Big Purple Achimenes 4078
Big Purple Roscoea 337101
Big Quakegrass 60379
Big Quaking Grass 60379
Big Quaking-grass 60379
Big Racemose Supplejack 52461
Big Rattanpalm 65682
Big Rattlesnake Plantain 179694
Big Red Sage 345299
Big Redbrownbristle Cottonsedge 152777
Big Reed Grass 65303
Big Sage 36400
Big Sagebrush 36400,360884
Big Salsify 394312
Big Salt Bush 44490
Big Scentless Mock Orange 294463
Big Scentless Mock-orange 294463
Big Scorpiongrass 175877
Big Shellbark 77905
Big Shellbark Hickory 77905
Big Silkpalm 413307
Big Skymallow 357920
Big Spapweed 204992
Big Stinkinggrass 249105
Big Stonecrop 356916
Big Striga 378024
Big Taiwan Meadowrue 388711

Big Tooth Maple 3000
Big Touch-me-not 204992
Big Tree 360575,360576
Big Tree of California 360561
Big Tree Plum 316553
Big Trefoil 237713
Big Trumpet Rhododendron 330666
Big Truncate Twinballgrass 209138
Big Virgate Lespedeza 226684
Big Wallich Rhodiola 329974
Big Wand-flower 369996
Big White Pelargonium 288264
Big White Trillium 397570
Big Whitethorn 266365
Big Wingseedtree 320835
Big Wood 360576
Big Yellowflower Sage 345030
Big Yellowflowered Violet 410279
Big Yunnan Dregea 137803
Biganther Narrowglume Leymus 228344
Biganther Pipewort 151503
Bigarade 93332,93408
Bigbagdzi 60230
Bigball Mucuna 259533
Bigbeak Monkshood 5403
Big-beak Rattan Palm 65730
Bigbeak Rattanpalm 65730
Bigbell Azalea 331660
Bigberry Creeping Mahonia 242627
Bigberry Manzanita 31116
Bigbill Woodbetony 287404
Bigbillorchis 346807,346808
Bigbloom Magnoliavine 351044
Big-botrys Franchet Barberry 51841
Bigbract Adinandra 8238
Bigbract Blueberry 403910
Bigbract Bulbophyllum 62672
Bigbract Camellia 69104
Bigbract Chamaesium 85719
Bigbract Chirita 87831
Bigbract Corydalis 106431
Bigbract Dayflower 101112
Bigbract Dendrobium 125412
Bigbract Garcinia 171069
Bigbract Gugertree 350916
Bigbract Heliotrope 190673
Bigbract Hemiboea 191372
Bigbract Macaranga 240239
Bigbract Monkshood 5525
Bigbract Ophiorrhiza 272209
Bigbract Pink 127733
Bigbract Pleurospermum 304822
Bigbract Ribseedcelery 304822
Bigbract Rockjasmine 23227
Bigbract Stonebean-orchis 62672
Bigbract Stonecrop 356530
Bigbract Thorowax 63622
Bigbract Willow 343927
Big-bracted Adinandra 8238
Big-bracted Blueberry 403910
Big-bracted Camellia 69108
Big-bracted Gugertree 350916
Big-bracted Keiskea 215816
Big-bracted Macaranga 240239
Bigbracted Saltbush 44349
Big-bracted Willow 343927
Bigbractginger 79726
Big-bud Hickory 77934
Big-budded Rhododendron 330756
Bigbulb Lycaste 238746
Bigcalyx Asiabell 98365
Bigcalyx Bluebell 115386
Bigcalyx Bluebellflower 115386
Bigcalyx Bubbleweed 297879
Bigcalyx Chelonopsis 86815
Big-calyx Currant 334080
Bigcalyx Deutzia 126856
Bigcalyx Glorybower 95979
Bigcalyx Gugertree 350927
Bigcalyx Kudzuvine 321426
Bigcalyx Lagotis 220164
Bigcalyx Litse 233849
Big-calyx Lyonia 239387
Bigcalyx Mazus 247028
Bigcalyx Microtoena 254256
Bigcalyx Paraboea 283128
Bigcalyx Phoebe 295382
Bigcalyx Rabdosia 209767
Bigcalyx Rhododendron 331212
Bigcalyx St. John's Wort 202004
Big-calyxed Camellia 68921,69331
Big-calyxed Lyonia 239387
Bigcarp Dutchmanspipe 34262
Bigcatkin Willow 343444
Big-catkined Willow 343444
Bigcavor Peliosanthes 288646
Bigcone Balsam Fir 290
Bigcone Douglas Fir 318579
Big-cone Douglas Fir 318579
Bigcone Douglas-fir 318579
Bigcone Pine 299876
Big-cone Pine 299876
Bigcone Spruce 298376,318579
Big-coned Spruce 298376
Bigcone-spruce 318579
Big-cupula Tanoak 233231
Bigdcale Dodder 115068
Bigdentate Ampelopsis 20360
Big-dentate Ampelopsis 20360
Bigdentate Snakegrape 20360
Big-discolour Deutzia 126914
Big-ear Supplejack 52426
Bigear Yushanbamboo 416750
Big-eared Bambusa 47331
Bigelow Beargrass 266485
Bigelow Nolina 266485
Bigelow Sagebrush 35206
Bigelow Sneezeweed 188410
Bigelow's Amaranth 18835
Bigelow's Bottlebrush Grass 144343
Bigelow's Clematis 94769
Bigelow's False Thoroughwort 89295
Bigelow's False Willow 46213
Bigelow's Fleabane 150499
Bigelow's Rabbitbrush 150338
Bigelow's Rubberweed 201307
Biger Leucanthemum 227512
Biger Whitedaisy 227512
Bigeye Bamboo 47266
Bigflower Allamanda 14874
Big-flower Anisetree 204550
Bigflower Anzacwood 310763
Bigflower Apios 29306
Bigflower Armand Clematis 94751
Bigflower Asiabell 98381
Bigflower Aster 40836
Bigflower Astyleorchis 19526
Bigflower Awlbillorchis 333729
Bigflower Barberry 51943
Big-flower Basketvine 9455
Bigflower Beancaper 418732
Bigflower Beautiful Lespedeza 226801
Bigflower Beijing Skullcap 355671
Bigflower Betonica 53301
Bigflower Bicolor Patrinia 285852
Bigflower Bigpanicle Snowgall 191946
Bigflower Birdlingbean 87248
Bigflower Birds-foot Violet 410386
Big-flower Bleeding-heart 203357
Bigflower Box 64259
Bigflower Branchy Milkvetch 42896
Bigflower Broadpetal Buttercup 326236
Bigflower Butterflybush 62000
Big-flower Caducouspetal Camellia 69236
Bigflower Cajan 65150
Bigflower Camellia 69333
Bigflower Carissa 76926
Bigflower Catchfly 363519
Bigflower Celandine 86761
Bigflower Centranthera 81727
Bigflower Cephalanthera 82038
Bigflower Cheirostylis 86683
Bigflower Chervil Larkspur 124042
Big-flower Childvine 400889
Bigflower Clematis 94865
Big-flower Clematis 95217
Bigflower Clinopodium 97017
Bigflower Clover 396975
Bigflower Clubfruitcress 376320
Bigflower Colewort 108613
Bigflower Colicweed 203357

Big-flower Common White Jasmine 211947
Bigflower Coreopsis 104509
Bigflower Cotyledon 107952
Bigflower Crazyweed 278862
Bigflower Cymbidium 116905
Bigflower Daphne 122510
Bigflower Darkred Gentian 173834
Bigflower Delavay Clethra 96492
Bigflower Delavay Microtoena 254248
Bigflower Dishspurorchis 171832
Bigflower Duabanga 138722
Bigflower Dwarf Mitrasacme 256163
Bigflower Dzungar Skullcap 355765
Bigflower Eightangle 204550
Bigflower Elaeagnus 142074
Big-flower Eurya 160541
Bigflower Eutrema 161127
Bigflower Eyebright 160190
Bigflower Falsehairfruit Microula 254362
Bigflower Falsehellebore 405598
Bigflower Farrer Asiabell 98313
Bigflower Felwort 235454
Bigflower Figwort 355097
Big-flower Fimbriate Orostachys 275365
Bigflower Fleshy Lobelia 234885
Bigflower Gaillardia 169586
Bigflower Galangal 17772
Bigflower Gansu Corydalis 105739
Bigflower Gardenia 171333
Bigflower Gentian 173951
Bigflower Giantarum 20125
Bigflower Gilia 99858
Bigflower Graceful Clematis 94960
Bigflower Greenbrier 366453
Bigflower Greenorchid 137591
Bigflower Greenvine 204625,204626
Bigflower Ground Ivy 176844
Bigflower Groundsel 359461
Bigflower Guangxi Aganosma 10227
Bigflower Gugertree 350914
Bigflower Habenaria 183731
Bigflower Hawkweed 195770
Bigflower Hemsleia 191941
Bigflower Herelip 220066
Bigflower Heteropappus 193968
Bigflower Horsfieldia 198511
Bigflower Houseleek 358044
Bigflower Huize Corydalis 106127
Bigflower Hyssop 203092
Bigflower Illigera 204625
Bigflower Indigo 206744
Bigflower Jasmine 211848
Bigflower Javatea 275689
Bigflower Junegrass 217494
Bigflower Jurinea 193968
Bigflower King Corydalis 106037
Bigflower Kiwifruit 6623
Bigflower Knotweed 291835
Bigflower Lady's Slipper 120396
Bigflower Ladyslipper 120396
Bigflower Leaflessorchis 29215
Bigflower Lily 229888
Bigflower Linociera 87728
Bigflower Liolive 87728
Bigflower Liparis 232139
Bigflower Listera 264752
Bigflower Longleaf Deutzia 127000
Bigflower Longstalk Corydalis 105716
Bigflower Loquat 151142
Bigflower Low Torularia 264615
Bigflower Luisia 238315
Bigflower Magnoliavine 351044
Bigflower Mamdaisy 246315
Bigflower Mattam Milkvetch 42686
Bigflower Mazus 247005
Bigflower Meconopsis 247126
Bigflower Mockorange 294463
Bigflower Morningglory 208196
Bigflower Morningstar Lily 229816
Bigflower Mountainash 369450
Big-flower Myrsine-leaf Syzygium 382620
Bigflower Neillia 263151
Bigflower Nestorchid 264700
Big-flower Oleander Agapetes 10346
Bigflower Ophiorrhiza 272249
Bigflower Oreocharis 273871
Bigflower Oreorchis 274052
Big-flower Oxalis 277702
Bigflower Papery Daphne 122573
Bigflower Paraquilegia 283721
Big-flower Pencilwood 139669
Bigflower Peristylus 291200
Bigflower Perotis 291200
Bigflower Pleione 304257
Bigflower Porana 396761
Bigflower Pricklepoppy 32423
Bigflower Primrose 314468
Bigflower Purslane 311852
Bigflower Qinjiu 173617
Bigflower Rhodiola 329860
Bigflower Rockfoil 349934
Bigflower Roscoea 337086
Bigflower Rose Stonecrop 357090
Bigflower Royle Woodbetony 287641
Big-flower Sage 345064
Bigflower Sagebrush 35859
Bigflower Sandwort 32229
Bigflower Sauraiua 347981
Bigflower Saxifrage 349934
Bigflower Scabious 350272
Bigflower Selfheal 316104
Bigflower Semeiandra 357888
Bigflower Siphocranion 365193
Bigflower Smilax 366453
Bigflower Snowbell 379357,379404
Bigflower Snowgall 191941
Bigflower Spicegrass 239663
Bigflower Splitmelon 351770
Bigflower Spotleaf-orchis 179580
Bigflower Spring Mouseear 82819
Bigflower Stauranthera 374447
Bigflower Stephania 375887
Bigflower Sterigmostemum 376320
Bigflower Stonecrop 356913
Bigflower Storax 379357
Bigflower Summerlilic 62000
Bigflower Swallowwort 117598
Bigflower Sweetvetch 187918
Bigflower Taiwan Toadlily 396594
Bigflower Taperleaf Camellia 69035
Big-flower Tickseed 104509,104510
Big-flower Tylophora 400889
Bigflower Vanda 404629
Bigflower Vetch 408327,408419
Bigflower Vinegentian 110295
Bigflower Whitedrumnail 307670
Bigflower Wildbasil 97017
Bigflower Wildorchis 274052
Bigflower Willow Weed 146763
Bigflower Wingstalk Hilliella 196271
Bigflower Woodbetony 287259
Bigflower Wright Cunealglume 29458
Bigflower Yunnan Corydalis 106614
Bigflower Zeuxine 417745
Big-flowered Actinidia 6623
Big-flowered Anise-tree 204550
Big-flowered Box 64259
Big-flowered Brome 60737,60894
Big-flowered Broom 85403
Big-flowered Buckthorn 328718
Big-flowered Cajan 65150
Bigflowered Cinnamon 91368
Big-flowered Clematis 94865
Big-flowered Deutzia 126956
Big-flowered Duabanga 138722
Big-flowered Elaeagnus 142074
Big-flowered Euonymus 157547
Big-flowered Forkstyleflower 374493
Big-flowered Gerardia 45169
Big-flowered Heterostemma 194160
Big-flowered Hippolytia 196703
Big-flowered Holboellia 197229
Big-flowered Illigera 204625
Big-flowered Java Tea 275781
Big-flowered Lacquer-tree 393455
Big-flowered Loquat 151142
Big-flowered Magnolia-vine 351044
Big-flowered Oleander-leaved Agapetes 10346
Big-flowered Pea-shrub 72241
Big-flowered Pottsia 313227

Big-flowered Pouteria 313369
Big-flowered Rhododendron 331183
Big-flowered Saurauia 347981
Big-flowered Selfheal 316104
Big-flowered Storax 379357
Big-flowered Styrax 379402
Bigfoot 175270
Bigfruir Hemlock 399916
Bigfruit Agapetes 10337
Bigfruit Ailanthus 12563
Bigfruit Altai Taphropermum 383990
Bigfruit Astilbe 41826
Bigfruit Atalantia 43726
Bigfruit Azalea 331855
Bigfruit Bastardtoadflax 389743
Big-fruit Begonia 50046
Bigfruit Blachia 54875
Bigfruit Burma Gynostemma 183001
Bigfruit Buttercup 326067
Bigfruit Cachrys 64788
Big-fruit Calcareus Pricklyash 417254
Bigfruit Camellia 69336
Bigfruit Carissa 76926
Bigfruit Catasetum 79342
Bigfruit Cheilotheca 86390
Bigfruit Cherryred Cotoneaster 107477
Bigfruit Chinese Hemlock 399889
Bigfruit Cinquefoil 313061
Bigfruit Clearweed 298965
Bigfruit Cleistocalyx 94574
Bigfruit Cogflower 154180
Bigfruit Coldwaterflower 298965
Bigfruit Cotoneaster 107558
Bigfruit Cypress 114723
Bigfruit David Rose 336520
Bigfruit Dendropanax 125598
Big-fruit Dodder 115077
Bigfruit Draba 137100
Bigfruit Dragonplum 137718
Bigfruit Dragon-plum 137718
Bigfruit Dwarfnettle 262225
Bigfruit Elaeocarpus 142316
Big-fruit Eleutharrhena 143558
Bigfruit Elm 401556
Bigfruit Ervatamia 154180
Bigfruit Euptelea 160345
Bigfruit Eurya 160445
Bigfruit Exbucklandia 161610
Bigfruit Faber Maple 2955
Bigfruit Falsepistache 384373
Bigfruit Fieldcitron 400539
Bigfruit Figwort 355179
Bigfruit Franchet Maple 2974
Bigfruit Furrowcress 383990
Bigfruit Garcinia 171164
Bigfruit Gaura 172187
Bigfruit Giantfennel 163658
Big-fruit Glabripetal Elaeocarpus 142323
Bigfruit Gouania 179948
Bigfruit Greenbrier 366452
Big-fruit Greenbrier 366452
Bigfruit Groundcherry 297699
Big-fruit Gynostemma 183001
Bigfruit Hao Poplar 311343
Big-fruit Hawthorn 109823
Bigfruit Hemsleia 191939
Bigfruit Holly 204001
Bigfruit Honeysuckle 235846
Bigfruit Hooktea 52425
Bigfruit Hundredstamen 389743
Bigfruit Hydrangea 199953
Bigfruit Hydrangeavine 351814
Bigfruit Leptospermum 226464
Bigfruit Litse 233970
Bigfruit Loropetal 237197
Bigfruit Loropetalum 237197
Bigfruit Manchurian Apricot 34445
Bigfruit Manglietia 244444
Bigfruit Manyflower Cotoneaster 107585
Bigfruit Megacarpaea 247786
Bigfruit Minuartwort 255514
Bigfruit Mongolian Oak 324195
Bigfruit Mountainash 369457
Bigfruit Mucuna 259535
Bigfruit Mycetia 260636
Big-fruit Myrica-like Distylium 134939
Big-fruit Myrica-like Mosquitoman 134939
Bigfruit Nakedflower Tubergourd 390167
Bigfruit Nanmu 295379
Bigfruit Nannoglottis 262225
Bigfruit Nenga 263550
Bigfruit Nucleonwood 291484
Bigfruit Oiltea Camellia 69015
Bigfruit Osmanther 276359,276371
Bigfruit Perrottetia 291484
Bigfruit Phoebe 295379
Bigfruit Pholidocarpus 295502
Bigfruit Potanin Larch 221933
Bigfruit Primrose 314630
Bigfruit Privet 229451
Bigfruit Pygeum 322401
Bigfruit Qinggang 116186
Bigfruit Redden Woodrush 238691
Bigfruit Rhodiola 329907
Bigfruit Rhodoleia 332209
Bigfruit Rockspray Cotoneaster 107558
Bigfruit Rockvine 387808
Big-fruit Sawtooth Oak 323618
Bigfruit Sedge 75284
Bigfruit Semiserrate Camellia 69618
Bigfruit Seseli 361460
Bigfruit Skeletonweed 88788
Bigfruit Slender Spicebush 231361
Bigfruit Slender Woodbetony 287247
Bigfruit Smalscale Sedge 74642
Bigfruit Smilax 366452
Bigfruit Snowbell 379403
Bigfruit Snowgall 191939
Bigfruit Sophora 369060
Bigfruit Storax 379403
Bigfruit Supplejack 52425
Bigfruit Tickseed 104803
Bigfruit Turpinia 400539
Bigfruit Waterelm 417563
Bigfruit Waterfig 94574
Bigfruit Woodlotus 244444
Bigfruit Xichou Rockvine 387854
Bigfruit Yua 416526
Big-fruited Agapetes 10337
Big-fruited Buckeye 9688
Big-fruited Camellia 69015,69336
Big-fruited Carissa 76926
Big-fruited China-laurel 28372
Big-fruited Cleistocalyx 94574
Big-fruited Common Juniper 213731
Big-fruited Cyclobalanopsis 116186
Big-fruited Dammar Pine 10505
Big-fruited Dendropanax 125598
Big-fruited Drimycarpus 138044
Big-fruited Elm 401556
Big-fruited Eurya 160445
Big-fruited Exbucklandia 161610
Big-fruited Falsepistache 384373
Big-fruited Hodgsonia 197072
Big-fruited Holly 204001
Big-fruited Litse 233970
Big-fruited Manglietia 244444
Big-fruited Mountain-ash 369457
Big-fruited Mucuna 259535
Big-fruited Oak 116186
Big-fruited Perrottetia 291484
Big-fruited Pygeum 322401
Big-fruited Rehder-tree 327418
Big-fruited Rhododendron 330763
Big-fruited Rhodoleia 332209
Big-fruited Rockvine 387808
Big-fruited Storax 379403
Big-fruited Supple-jack 52425
Big-fruited Tabernaemontana 154180
Big-fruited Turpinia 400539
Bigfuit Date 418183
Bigg 198376
Bigger Cyclorhiza 116372
Bigger R. Fourtunei Rhododendron 330676
Bigginseng 241160,241164
Big-ginseng 241160
Bigglume Kengyilia 215937
Bigglume Threeawngrass 376989
Bigglume Triawn 376989
Biggoty Lady 205175
Bighead Ainsliaea 12675

Bighead Camomile　26814
Bighead Carpesium　77178
Bighead Dandelion　384484
Bighead Dustymaidens　84511
Big-head Elachanthemum　141900
Bighead Galangal　17753
Bighead Gerbera　175193
Bighead Knapweed　81183
Big-head Knapweed　81183
Bighead Mongol Jerusalemsage　295152
Bighead Rabbiten-wind　12675
Big-head Rabbit-tobacco　127932
Bighead Sinacalia　364521
Bighead Straitjackets　253811
Bighead Threeleft Spunsilksage　360846
Bighead Windhairdaisy　348354
Bigheart Rockfoil　349620
Big-horn Euphorbia　158981
Bighorn Fleabane　150445
Big-horned Euphorbia　158981
Biginflorescens Summerlilic　62114
Bigkey Sycamore Maple　3471
Bigland Rock Willow　344021
Biglandular Bredia　59831
Biglandular Spapweed　204820
Biglandular Touch-me-not　204820
Biglangulose Bolbostemma　56657
Bigleaf Achyranthes　4308
Bigleaf Adinandra　8248
Bigleaf Agapetes　10332
Bigleaf Albizzia　13611
Bigleaf Alishan Chickweed　374758
Bigleaf Allophylus　16060
Bigleaf American Hackberry　80700
Bigleaf Angelica　24406
Bigleaf Antlevine　88893
Bigleaf Ardisia　31412
Bigleaf Argyreia　32668
Bigleaf Asiabell　98276
Bigleaf Aster　40817,160670
Big-leaf Beauty Berry　66864
Bigleaf Begonia　50071
Bigleaf Bennettiodendron　51049
Bigleaf Bergia　52597
Bigleaf Bitter Tanoak　233339
Bigleaf Bloodberry Cotoneaster　107608
Bigleaf Blueberry　403964
Bigleaf Bon Pepper　300362
Big-leaf Box　64284
Bigleaf Bracketplant　88598
Bigleaf Bredia　296396
Bigleaf Bristlegrass　361828
Bigleaf Bushclover　226764
Bigleaf Caesalpinia　65033
Bigleaf Calathea　66185
Bigleaf Canthium　71498
Bigleaf Carallia　72395
Big-leaf Caudate-leaf Blueberry　403817
Bigleaf Cayratia　79878
Bigleaf Cherry-laurel　223150
Bigleaf Chickweed　374832
Big-leaf Chinese Aspen　311200
Big-leaf Chinese Helwingia　191149
Bigleaf Chirita　87914
Bigleaf Chonemorpha　88893
Bigleaf Cinnamon　91341
Big-leaf Cleistanthus　94522
Bigleaf Clethra　96505
Bigleaf Coldwaterflower　298967
Bigleaf Corallodiscus　103955
Bigleaf Corydalis　106511
Big-leaf Creeper　285168
Bigleaf Creeping Mallotus　243440
Bigleaf Cylindrokelupha　116603
Bigleaf Damnacanthus　122064
Bigleaf Daphniphyllum　122705
Bigleaf David Maple　2923
Bigleaf Dayflower　101169
Bigleaf Desmos　126721
Bigleaf Devil Rattan　121513
Bigleaf Drymonia　138524
Bigleaf E. H. Wilson Pygeum　322414
Bigleaf Elaeocarpus　142287
Bigleaf Embelia　144812
Bigleaf Eritrichium　184295
Bigleaf Eucalyptus　155476,155722
Bigleaf Evergreenchinkapin　78987
Bigleaf Falsenettle　56206
Bigleaf Falsepimpernel　231543
Bigleaf Fang Eurya　160471
Bigleaf Fanpalm　234191
Bigleaf Fig　165841
Bigleaf Firmiana　166669
Bigleaf Fourtapeparsley　387899
Bigleaf Fragile Blueberry　403834
Bigleaf Fullmoon Maple　3050
Bigleaf Garcinia　171221
Bigleaf Goldenray　229005,229104
Bigleaf Goldsaxifrage　90405
Bigleaf Goldwaist　90405
Bigleaf Goniolimon　179372
Bigleaf Grewia　180918
Bigleaf Groundsel　359419
Bigleaf Hao Poplar　311344
Bigleaf Hartia　376490
Bigleaf Henry Raspberry　338510
Bigleaf Herminium　192858
Bigleaf Hibiscus　194986
Bigleaf Ho Schefflera　350711
Bigleaf Holly　203969
Big-leaf Hooker Deutzia　126972
Bigleaf Hooktea　52429
Bigleaf Hydrangea　199956
Big-leaf Hydrangea　199956
Bigleaf Japanese Beech　162370
Bigleaf Japanese Spindle-tree　157617
Big-leaf Jasmine　211739
Bigleaf Jasminorange　260170
Bigleaf Jinping Camellia　69725
Bigleaf Katsura Tree　83743
Bigleaf Katsuratree　83743
Bigleaf Ladyslipper　120349
Bigleaf Laurel Cherry　223150
Bigleaf Leea　223943
Big-leaf Leptodermis　226111
Bigleaf Licorice　177926
Bigleaf Linden　391817
Big-leaf Linden　391817
Bigleaf Linearleaf Thistle　92139
Bigleaf Lupine　238457
Big-leaf Lupine　238481
Bigleaf Machilus　240605
Bigleaf Magnolia　198700,242193
Big-leaf Magnolia　242193
Bigleaf Mahogany　380527
Bigleaf Maple　3127
Big-leaf Maple　3127
Bigleaf Markingnut　357881
Bigleaf Meadowrue　388495
Bigleaf Meliosma　249403
Bigleaf Microtropis　254310
Bigleaf Mitreola　256212
Bigleaf Motherweed　231543
Bigleaf Mountain Leech　126359
Bigleaf Mussaenda　260454
Bigleaf Myrioneuron　261325
Bigleaf Narrow-petal Hydrangea　199794
Bigleaf Nettletree　80664
Bigleaf Newlitse　264039
Bigleaf Nittatree　284464
Bigleaf Oak　323979
Bigleaf Obviousvein Tea　69060
Bigleaf Paraboea　283131
Bigleaf Paravallaris　216232
Big-leaf Paravallaris　216232
Bigleaf Parnassia　284571
Bigleaf Paspalum　285461
Bigleaf Pearweed　239864
Bigleaf Pellionia　288745
Bigleaf Periwinkle　409335
Big-leaf Periwinkle　409335
Bigleaf Petrocosmea　292558
Bigleaf Photinia　295731
Bigleaf Pilose Germander　388182
Bigleaf Pittosporum　301253,301255
Bigleaf Poplar　311371
Bigleaf Pricklyash　417278
Bigleaf Purplepearl　66864
Bigleaf Pygeum　322402
Bigleaf Qinggang　116120
Bigleaf Qiongzhu　87574

Bigleaf Ramie 56195
Bigleaf Redcarweed 288745
Bigleaf Resurrectionlily 215019
Bigleaf Rhodiola 329910
Bigleaf Rhododendron 330283
Bigleaf Robust Veronicastrum 407481
Bigleaf Rose 336725
Bigleaf Rush 212834
Bigleaf Rustyhair Raspberry 339165
Bigleaf Sage 345065
Bigleaf Sandwort 256450
Bigleaf Sanqi 280793
Bigleaf Schefflera 350737
Bigleaf Seatung 301253,301255
Bigleaf Senna 360434
Bigleaf Shamrue 185658
Bigleaf Siebold Elder 345682
Bigleaf Skullcap 355600
Bigleaf Skyblue Larkspur 124086
Bigleaf Slenderlobe Maple 3638
Bigleaf Slenderstyle Acanthopanax 143642
Bigleaf Smallligulate Aster 40009
Bigleaf Snowbell 379357,379358
Bigleaf Spapweed 205109,205319
Bigleaf Spiradiclis 371745
Bigleaf Square Bamboo 87560
Bigleaf St. John's Wort 202102
Bigleaf Stephania 375845
Bigleaf Sterculia 376127,376138
Bigleaf Stonebutterfly 292558
Bigleaf Stonecrop 356685
Big-leaf Storax 379414
Bigleaf Streblus 377542
Bigleaf Strigose Hydrangea 200114
Bigleaf Supplejack 52429
Bigleaf Sweetleaf 381217
Bigleaf Tanoak 233312
Bigleaf Thickhorn Elatostema 142774
Bigleaf Thistelike Windhairdaisy 348188
Bigleaf Thorowax 63728
Bigleaf Tickclover 126424
Bigleaf Tickseed 104542
Bigleaf Tigerthorn 122064
Bigleaf Tonkin Fishvine 126006
Bigleaf Tonkin Jewelvine 126006
Bigleaf Touch-me-not 205109,205319
Bigleaf Uvaria 403521
Bigleaf Violet 409895,410710
Bigleaf Waterelm 417552
Bigleaf White Mulberry 259083
Big-leaf Wildclove 226111
Bigleaf Willow 343458,343496
Big-leaf Willow 343458
Bigleaf Yam 131610
Bigleaf Yaoshan Blueberry 404064
Bigleaf Yushan Dewberry 339188
Big-leafed Snowbell 379358
Big-leafed Storax 379414
Bigleafstink Pricklyash 417278
Bigleafvine 392222,392223
Big-leaved Agapetes 10332
Big-leaved Ampelopsis 20420
Big-leaved Aphanamixis 28996
Big-leaved Aster 40817
Big-leaved Avens 175426,175429
Big-leaved Beautyberry 66864
Big-leaved Blueberry 403964
Big-leaved Box 64284
Big-leaved Bramble 338130
Big-leaved Bredia 296396
Big-leaved Buckeye 9715
Big-leaved Caesalpinia 65033
Big-leaved Carallia 72395
Big-leaved Cayratia 79878
Big-leaved Chaste-tree 209688
Big-leaved Cherry Laurel 223150
Big-leaved Chonemorpha 88893
Big-leaved Cinnamon 91341
Big-leaved Cleistanthus 94522
Big-leaved Clethra 96505
Big-leaved Cotoneaster 107541
Big-leaved Daphniphyllum 122705
Bigleaved Desmodium 126359
Big-leaved Desmos 126721
Big-leaved Dysoxylum 139655
Big-leaved Evergreen Chinkapin 78987
Big-leaved False-nettle 56206
Big-leaved Fig 165057
Big-leaved Flemingia 166880
Big-leaved Glochidion 177149
Big-leaved Grewia 180918
Big-leaved Hydrangea 199953
Big-leaved Kabatalia 216232
Big-leaved Lacquer-tree 393461
Big-leaved Lime 391788
Big-leaved Linden 391788
Big-leaved Livistona 234191
Big-leaved Magnolia 198700
Big-leaved Marking-nut 357881
Big-leaved Marshweed 230322
Big-leaved Microtropis 254310
Big-leaved Miliusa 254560
Big-leaved Newlitse 264069
Big-leaved Pepper 300435
Big-leaved Pittosporum 301253,301255
Big-leaved Pondweed 312044
Big-leaved Prickly-ash 417278
Big-leaved Psychotria 319606
Big-leaved Rose 336725
Big-leaved Schefflera 350737
Bigleaved Slenderlobe Maple 3638
Big-leaved Spine-mulberry 377542
Big-leaved Supple-jack 52429
Big-leaved Sweetleaf 381217
Big-leaved Tanoak 233233,233312
Big-leaved Wilkes Copperleaf 2017
Biglip Achasma 155402
Biglip Germander 388127
Biglip Large Rhinanthus-like Woodbetony 287619
Biglip Largeflower Listera 264668
Biglip Viscid Germander 388306
Biglip Woodbetony 287419
Big-lobed Milkvetch 42655
Biglomerate Elatostema 142619
Big-lowered Prickly-ash 417261
Bignay 28316
Bignay China Laurel 28316
Bignay Chinalaurel 28316
Bignay China-laurel 28316
Bignay Meytea 28316
Bigneck Gentian 173614
Big-nodded Shibataea 362147
Bignode Dendrobium 125301
Bignonia 54292
Bignonia Family 54357
Bignose Rhinacanthus 329471
Big-nosed Rhinacanthus 329471
Bigold 89704
Bigpanicle Snowgall 191965
Bigpanicled Rabdosia 209774
Bigpetal Clematis 95100
Bigpetalcelery 357905,357912
Big-petaled Clematis 95100
Bigpod Ceanothus 79954
Big-pod Vetch 408478
Big-pyrene Holly Taiwan Holly 203822
Bigred Raspberry 338374
Big-red Raspberry 338374
Bigrooot Statice 230674
Big-root 61497,207938,383675
Bigroot Chirita 87915
Bigroot Cranesbill 174715
Bigroot Geranium 174715
Big-root Geranium 174715
Bigroot Lady's Thumb 309428
Bigroot Morning Glory 208050
Bigroundleaf Primrose 314909
Big-sage 373298
Bigscale Azalea 330887
Bigscale Rhodiola 329908
Bigseed Bugle 13137
Bigseed Catchfly 364080
Big-seed Citrus 93563
Big-seed Crateva 110234
Bigseed Cycas 115852
Bigseed Fieldcitron 400528
Bigseed Jointfir 178555
Bigseed Merremia 250833
Big-seed Pepper-weed 225342
Bigseed Redflower Snakegourd 396265

Bigseed Sagebrush 36286
Bigseed Salacia 342686
Bigseed Swertia 380264
Big-seed Torreya 393075
Bigseed Turpinia 400528
Big-seeded Actinidia 6659
Big-seeded False Flax 68860
Big-seeded Gnetum 178534
Big-seeded Turpinia 400528
Bigsepal Camellia 69331
Bigsepal Gugertree 350927
Big-sepal Hydrangea 200019
Bigsepal Susanna Stonecrop 357195
Bigsepal Tea 69331
Bigsepal Tubergourd 390134
Big-sepaled Epimedium 147023
Big-sepalous Litse 233849
Big-sheath Pondweed 312285,378994
Bigsheath-pondweed 312285
Bigspike Alopecurus 17548
Bigspike Arundinella 37367
Bigspike Bellcalyxwort 231265
Bigspike Brome 60980
Bigspike Cliffbean 254762
Bigspike Drug Sweetflag 5800
Bigspike Garnotia 171511
Bigspike Japan Lawngrass 418427
Bigspike Microstegium 254002
Bigspike Ricegrass 300716
Bigspike Yam 131693
Big-spiked Butterfly-bush 62114
Big-spiked Dendrocalamus 125500
Big-spiked Hornbeam 77420
Big-spiked Tanoak 233313
Big-spine Honey Locust 176891
Bigspine Honeylocust 176901
Big-spine Honey-locust 176891
Bigspine Thistle 92035
Bigspine Waterchestnut 394440
Bigspiny Barberry 51533
Bigspur Violet 410192
Bigstachys Fruticose Birch 53448
Bigstachys Peliosanthes 288650
Bigstachys Wildryegrass 144195
Bigstalk Roegneria 335419
Bigstalk Waterchestnut 394485
Big-stigma Gaulteria 291366
Big-stigma Holly 204013
Bigstigma Rockfoil 349580
Bigsting Nettle 402886
Big-sting Nettle 402886
Bigstipular Coldwaterflower 298851
Bigstipule Milkvetch 43094
Big-stipulete Clearweed 298851
Bigstyle Shadegrass 352873
Big-styled Vine 247967
Bigstylevine 247967
Big-subkorean Raspberry 339317
Bigtail Heliotrope 190651
Big-thrse Styrax 379404
Bigthyrse Hemsleia 191965
Bigthyrse Rabdosia 209774
Bigthyrsiferous Porana 396761
Bigtooth Aspen 311335
Big-tooth Aspen 311335
Bigtooth Chirita 87913
Bigtooth Elatostema 142670
Bigtooth Goldenray 229103
Bigtooth Hillcelery 276748
Bigtooth Lycianthes 238968
Bigtooth Maple 3000
Big-tooth Maple 3000
Bigtooth Ophiorrhiza 272250
Bigtooth Oreocharis 273869
Bigtooth Rabdosia 209692
Bigtooth Red Silkyarn 238968
Bigtooth Rehder Mountainash 369503
Bigtooth Skullcap 355462,355591
Bigtooth Stairweed 142670
Bigtooth Suoluo 327376
Bigtooth Yunnan Croton 113062
Big-toothed Aspen 311335
Big-toothed Euphorbia 158982
Big-toothed Lycianthes 238968
Bigtoothed Primrose 314240
Bigtop Love Grass 147718
Bigtree 360576
Big-tree 360561
Big-tree Camellia 68911
Bigtree Hawthorn 109555
Bigtree Plum 316553
Big-tree Plum 316553
Bigtube Lagotis 220186
Bigtube Micromelum 253627
Bigtube Skullcap 355592
Bigtube Woodbetony 287405
Bigtuft Lymegrass 144383
Bigtuft Roegneria 335417
Bigumbel Rockvine 387806
Bigumbrella 59187
Bigwing Beancaper 418702
Big-wing Euonymus 157705
Bigwing Milkvetch 42657
Bigwing Mono Maple 3191
Bigwing Pterolobium 320603
Bigwing Tigerthorn 320603
Big-winged Pterolobium 320603
Bigword-shaped Rockfoil 349499
Bihi 325187
Bijiang Azalea 330227,331244
Bijiang Bittercress 72699
Bijiang Cephalanthera 82033
Bijiang Corydalis 105650
Bijiang Elatostema 142620
Bijiang Euonymus 157770
Bijiang Gingerlily 187422
Bijiang Pinkflush Rhododendron 331244
Bijiang Rhododendron 330227
Bijiang Rockfoil 349102
Bijiang Sinocarum 364894
Bijiang Stairweed 142620
Bijie Whytockia 413998
Bijugate Mahonia 242493
Bikou Willow 343107
Bilabiate Honeysuckle 235821
Bilberry 31142,403915,403916,404027
Bilberry 403710
Bildairy 273531,273545
Bilders 29341,192358,262722,269307
Bileaf Platanthera 302280
Bilimb Carambola 45724
Bilimbi 45724
Biling 45724
Bilinga 262804
Biliott Iris 208463
Bill Azalea 330644
Bill Buttons 175444,409339
Bill Sagebrush 36400
Bill Splitmelon 351761
Bill Williams Mountain Giant Hyssop 10412
Billadi Nitraria 266356
Billadi Whitethorn 266356
Billard Spirea 371834
Billard's Bridewort 371790
Billard's Spiraea 371834
Billardiera 54432
Billbergia 54439
Billbergialike Nidularium 266150
Billberry 403710
Billers 29341,192358,262722,269307
Billfruit Gynostemma 183032
Billhair Woodbetony 287632
Billiard Spirea 371790
Billington's Gentian 173296
Billion Dollar Grass 140367,140423
Billion-dollar Grass 140367,140423
Billtooth Woodbetony 287629
Billwalnut 25759
Billy Bachelor's Buttons 374916
Billy Beattie 284152
Billy Bright Eyes 407072
Billy Bright-eye 407072
Billy Buster 364193
Billy Buttons 11947,13934,31051,50825,68189,108610,108695,174877,175444,216753,227533,243823,243840,248312,248411,325518,349419,363461,363673,374916,409339
Billy Clippe 102933
Billy Clipper 102933
Billy White's Buttons 374916

Billy's Buttons 175444
Billy-goat Weed 11199
Billy-o'-buttons 68189
Bilobate Baliospermum 46965
Bilobate Darkred Gentian 173831
Bilobe Calanthe 65891
Bilobe Crazyweed 278745
Bilobed Grewia 180700
Bilobed Poorspikebamboo 270424
Bilobed Speedwell 407040
Bilobedpetal Camellia 69109
Bilobed-petaled Camellia 69109
Bilocular Dendropanax 125593
Bilocular Treerenshen 125593
Bilolo 93641
Bilow Breynia 60057
Bilow Rattanpalm 65673
Bilsted 232565
Biltmore Ash 167897,167927
Biltmore Hawthorn 109609
Bilumbing 45724
Biluo Nomocharis 266552
Biluoshan Rush 213470
Bilushan Buckthorn 328826
Bilushan False-nettle 56258
Bilushan Knotweed 309598
Bimaculate Corydalis 105651
Bimbila Barberry 51780
Bimble Box 155704
Bimucronate Mimosa 254997
Binarrowleaf Hoodorchis 264759
Binate Kengyilia 215908
Binate Paspalidium 285381
Binchuan Clematis 95236
Binchuan Deutzia 126856
Binchuan Linden 391849
Binchuan Mountainash 369474
Binchuan Spindle-tree 157794
Binchuan Swertia 380132
Binchuan Violet 409743
Bind 68713,236022
Bindang 10496
Bind-corn 102933,162312,162519
Binder 95431
Bindweed 68671,102903,102933,162505,162519,236022,359158,408423
Bind-weed 102903
Bindweed Family 102892
Bindweed Nightshade 91557
Bindweedlike Blinkworthia 55605
Bindweed-like Blinkworthia 55605
Bindwood 187222,236022
Bine 199384
Bine Lily 68713,102933
Binf-lily 68713,102933
Binian Ironwood 160967,160968
Binjai 244390
Binnwood 236022
Binode Bluegrass 305403
Binot Begonia 49659
Bintangor 67846
Binuang 268819
Binweed 359158
Biolett's Fleabane 150500
Bioletti's Rabbit-tobacco 317733
Biond Goldsaxifrage 90340
Biond Goldwaist 90340
Biond Hackberry 80580
Biond Nettletree 80580
Biond Willow 343109
Biond's Magnolia 416686
Biondia 54515
Biondia Mosquitotrap 117400
Biophytum 54532
Bioritsu Holly 203590
Biota 302721
Bipinnate Clematis 94857
Bipinnate Desmostachys 126741
Bipinnate Dragonhead 137557
Bipinnate Gold-rain Tree 217613
Bipinnate Greenorchid 137557
Bipinnate Vicatia 408241
Bipinnatifid Chiritopsis 87995
Bipinnatisect Aster 40138
Bipod Senna 360426
Bipore Porolabium 311713
Biradiateleaf Arisaema 33295
Biradiateleaf Southstar 33295
Birch 53309,53563
Birch Family 53651
Birch-bark Cherry 83312
Birchet 53563
Birch-leaf 323110
Birchleaf Anzacwood 310757
Birchleaf Arrowwood 407696
Birch-leaf Black Poplar 311405
Birchleaf Buckthorn 328619
Birch-leaf Buckthorn 328619
Birchleaf Grape 411576
Birchleaf Mountain-mahogany 83819
Birchleaf Pear 323110
Birch-leaf Pear 323110
Birchleaf Spiraea 371823
Birchleaf Viburnum 407696
Birch-leafed Mountain Mahogany 83817
Birch-leaved Fourstamen Maple 3678
Birch-leaved Grape 411576
Birch-leaved Pear 323110
Birch-leaved Spirea 371823
Birch-leaved Viburnum 407696
Bird 's Tongue 274857
Bird Beak Oncidium 270832
Bird Briar 336419
Bird Cactus 287853
Bird Cherry 83155,83274,279998,280003,280044,316166
Bird Crazyweed 278729
Bird Eagle 109857
Bird Eagles 109857
Bird Eye 66733
Bird Flower 111868
Bird Knotgrass 308816
Bird Nest Cypress 85292
Bird Nest Wild Buckwheat 152304
Bird of Paradise 5442,65055,377576
Bird of Paradise Bush 65014
Bird of Paradise Flower 377550
Bird Pears 109857,336419
Bird Pepper 72070,72077,72088,72095
Bird Rape 59603
Bird Spiderflower 95746
Bird Thistle 92485
Bird Tree 9701
Bird Vetch 408352,408356
Bird's Bannock 277648
Bird's Bill 135165
Bird's Brandy 221284
Bird's Bread 356468
Bird's Bread-and-cheese 277648
Bird's Candles 405788
Bird's Cherries 109857
Bird's Cherry 109857
Bird's Claws 237539
Bird's Eegle 109857
Bird's Egg 109857,364193
Bird's Eggle 109857
Bird's Eggs 109857,364193
Bird's Eye 21339,72934,160236,174877,175708,176824,237539,248411,260760,260842,314366,342279,349918,374916,374968,406966,407029,407144,407256,407287,410677
Bird's Eye Bush 268261
Bird's Eye Bush Ochna 268261
Bird's Eye Maple 3549
Bird's Eye Primrose 314366
Bird's Foot 237539,274879
Bird's Foot Clover 237539,397259
Bird's Foot Fenugreek 397259
Bird's Foot Millet 143522
Bird's Foot Trefoil 237476,237539
Bird's Foot Trigonella 397259
Bird's Foot Violet 410380
Bird's Knotgrass 308816
Bird's Knot-grass 308816
Bird's Meat 109857,302068
Bird's Nest 123141,232950
Bird's Nest Bromeliad 266155
Bird's Nest Orchid 264709
Bird's Nest Plant 266154
Bird's Rape 59575

Bird's Tongue 2837,21339,167955,285741,308816,359660,374916,377576
Bird's Tongue Flower 377550
Bird's-eye 407072,407256,407287
Bird's-eye Bush 268132,268202
Bird's-eye Cress 260577
Bird's-eye Gilia 175708
Bird's-eye Pearl-wort 342279
Bird's-eye Primrose 314366,314563,314652
Bird's-eye Primula 314366
Bird's-eye Speedwell 407072,407287
Bird's-foot 237539,274879
Bird's-foot Clover 237539,397259
Bird's-foot Deervetch 237539
Bird's-foot Deer-vetch 237539
Bird's-foot Fenugreek 397259
Bird's-foot Sedge 73782
Bird's-foot Trefoil 237476,237539,387239,397259
Bird's-foot Violet 410380,410385,410388
Bird's-foot-trefoil 237476,237539
Bird's-nest 123141,202767
Bird's-nest Family 258049
Bird's-nest Orchid 264650,264709
Bird's-nest Orehis 264650
Bird's-thot Clover 396994
Bird's-tongue 274857
Bird-beak Milkvetch 42834
Birdbill Dayflower 100989
Birdcage Evening-primrose 269429
Birdcage Plant 269429
Bird-card Catchbird Tree 81933
Bird-catcher 81933
Bird-catcher Tree 81933
Birdcatching Sedge 401804
Birdcherry 279998
Bird-drinking-at-a-fountain 266225
Birdeye Pearlwort 342279
Birdeye Speedwell 407287
Birdfoot Buttercup 326422
Birdfoot Gentian 173709
Bird-foot Grevill 180627
Birdfootorchis 347690
Bird-in-a-bush 106453
Birdleg Awlquitch 131018
Birdleg Dimeria 131018
Birdlike Crazyweed 278731
Bird-lime Flower 210689
Birdlime Mistletoe 410960
Birdlime Tree 81933,300955
Bird-lime Tree 104175,399433
Birdlingbran 87232
Bird-nest Neottia 264709
Bird-of-paradise 65014,377550,377576
Bird-of-paradise Flower 377550,377576
Bird-of-paradise Shrub 65014
Bird-on-a-thorn 106453
Birds 3462
Bird's Foot 274888
Birds On the Bough 169126
Birds On the Bush 169126
Birds' Wings 3462
Birdseed 302068,360328,364557
Birdseye Gilia 175708
Bird's-foot 274888
Birdsfoot Trefoil 237539
Birdsnest Spruce 298203
Birdtongue Chickweed 374731
Birdwood Grass 80850
Biriba 335829
Biriba Tree 335829
Birk 53563
Birthroot 397540,397556,397570
Birthwort 34097,34149
Birthwort Dutchman's-pipe 34149
Birthwort Family 34384
Biscay Heath 149398,149703
Bischofia 54620
Biscuit 174877,312520
Biscuit-flower 174877
Biscuit-leaves 162400
Biscuit-root 292851
Biscutella 54634,54690
Biseed Sedge 74342
Biseeded Pristimera 315356
Biseminal Poiret Barberry 52049
Biserrate Angelica 24306
Biserrate Begonia 49661,50130
Bisexual Cucumis 114128
Bisexual Muskmelon 114128
Bisexual-flowered Crowberry 145062
Bisharptooth Milkvetch 42078
Bishop Clove 396854
Bishop Lily 229870
Bishop Pine 300098
Bishop Weed 8826
Bishop's Beard 28044
Bishop's Cap 43509,43526,256010,256015
Bishop's Cap Cactus 43509,43526
Bishop's Cap Weed 8826
Bishop's Flower 53304
Bishop's Goutweed 8826
Bishop's Gout-weed 8826
Bishop's Hat 146953,146961,147054
Bishop's Hood 43509
Bishop's Leaf 355061
Bishop's Miter 43509
Bishop's Mitre 43173
Bishop's Mitre Cactus 43520
Bishop's Posy 227533
Bishop's Weed 8810,8826,19664,19672
Bishop's Wig 30164
Bishop's Wort 53304,266225
Bishop's-cap 256010,256015
Bishop's-weed 19672
Bishops Weed 8810
Bishopweed 54620,250291
Bishopwood 54618
Bishop-wood 54618
Bishopwort 250291
Bismarck Iris 208464
Bismarck Palm 54738,54739
Bismarckia 54738
Bismark Begonia 49662
Bisnaga 163439
Bispine Randia 325295
Bispinose Bigstalk Waterchestnut 394486
Bissell's Violet 409745
Bisselon 216214
Bisset's Bamboo 297244
Bissy Nut 99240
Bistamengrass 127467
Bistort 54757,308739,308893
Bistort Knotweed 308893
Bistyle Chickweed 374764
Bistyle Willow 343110
Biswarea 54834
Bitate Undulateleaf Oplismenus 272688
Bitchwood 157429
Biter Gentian 173598
Biternate Beggarticks 53801
Bitferwort 174106,174122,174221,384714
Bithwind 102933
Bithwine 102933
Bithynian Vetch 408322
Bithywind 95431,102933
Bitigma Bulrush 353374
Biting Persicaria 309199
Biting Stone Crop 356468
Biting Stonecrop 356468
Bitinga Rubber 326783
Bitny Bidny 53304
Bitooth Holly 203589
Bittedeaf 406987
Bitter Aks 384714
Bitter Almond 20892
Bitter Aloe 16817
Bitter Aloes 16540
Bitter Amur Maple 2986
Bitter Apple 93297,93304,256797,366906,367481
Bitter Ash 298510
Bitter Bamboo 303994,304000
Bitter Bark 18036
Bitter Buckwheat 162335
Bitter Bush 8411,158364
Bitter Buttons 383874
Bitter Candy Honeysuckle 235798
Bitter Candytuft 203184
Bitter Cassava 244507
Bitter Cherry 316411

Bitter Chinquapin 79017
Bitter Condalia 101734
Bitter Cress 72674,72927
Bitter Cryptandra 113338
Bitter Cucumber 93304,256797
Bitter Damson 364382
Bitter Dock 340151
Bitter Evergreen Chinkapin 79017
Bitter Evergreenchinkapin 79017
Bitter Fennel 167170
Bitter Flax 231884
Bitter Fleabane 150419
Bitter Gentian 150419,174221
Bitter Ginger 418031
Bitter Gourd 93304,256797
Bitter Greenbamboo 125439
Bitter Herb 81065,86783
Bitter Kola 171121
Bitter Kudzuvine 321459
Bitter Lady's Smock 72679
Bitter Land Cress 47964
Bitter Larkspur 124653
Bitter Leaf 406083
Bitter Leaf-flower 296471
Bitter Lettuce 219609
Bitter Medicine 345631
Bitter Melon 93307,256786,256797
Bitter Milkvetch 42550
Bitter Milkwort 308275,308276,308363
Bitter Momordica 256797
Bitter Mustard 59438
Bitter Nightshade 367120
Bitter Night-shade 367120
Bitter Nut 77889
Bitter Orange 93332,93408,93492,93534
Bitter Pea 123266
Bitter Pea Vine 222860
Bitter Peach 77881
Bitter Pecan 77876,77881
Bitter Poplar 311375
Bitter Quassia 364382
Bitter Raspberry 339390
Bitter Root 29468,29473,228298
Bitter Rroot 228256
Bitter Rubberweed 201323
Bitter Sagebrush 35090
Bitter Snakewood 101734
Bitter Sneezeweed 188397,188437
Bitter Sweet 367120
Bitter Tea Tree 110258
Bitter Thistle 73318
Bitter Tomato 367233
Bitter Vetch 222758,408392,408522
Bitter Vetchling 222758
Bitter Water Hickory 77910
Bitter Weed 19145,188397
Bitter Willow 343937
Bitter Winter Cress 47964
Bitter Yam 131567
Bitteraweetleaf Poisontree 366016
Bitteraweetleaf Sladenia 366016
Bitteraweet-leaved Sladenia 366016
Bitterbamboo 303994,304000
Bitter-bamboo 303994,304000
Bitterbean Pagodatree 368962
Bitter-berry 367544
Bitterbrush 321901
Bitterbush 298494
Bittercress 72674,262722
Bitter-cress 72674,72934
Bittercress Ragwort 279895
Bitter-dull Melandrium 248440
Bitterfigwort 298525
Bitter-flower 345631
Bittergiseng 369010
Bittergrass 404531,404555
Bitterhorsebean 371160,371161
Bitterleaf 406040,406083,406180,406246
Bitter-medicine 345631
Bitternut 77889
Bitternut Hickory 77889
Bitter-nut Hickory 77889
Bitterroot 228256,228298
Bitter-root 228298
Bitterseed Caper 71933
Bittersgall 243711
Bitterspine Nightshade 367087
Bittersweed-like Melodinus 249654
Bittersweet 80129,80141,80260,80309,166125,367120,367322
Bittersweet Nightshade 367120
Bittervetch 222758,408522
Bitter-vetch 408522
Bitterweed 19145,103446,188397,188402,188437,201293,201323,298642,311208,387348
Bitter-weed 188397
Bitterwood 364382,395466
Bittner Bulbophyllum 62588
Bittner Stonebean-orchis 62588
Bity Tongue 309199
Biuar-berry 338307
Biuisikul 93695
Bi-whiteflower Paraphlomis 283608
Bixa 54859
Bixa Family 54863
Bizarre Forsythia 167451
Biznaga De Dulce 163458
Biznaga Gigante 140150
Blab 334250
Blacck Persimmon 132431
Blacebergan 338447
Blach Barberry 51643
Blachfibre Shrimpalga 297183
Blachia 54874
Black 267867
Black Acacia 1384,334976
Black Afara 386563
Black Alder 16352,16368,204376,328603
Black Aller 16352,328603
Black Alpine-sedge 73788
Black Anthers Heath 149750
Black Apple 313344
Black Apricot 34431
Black aralia 310204
Black Archangel 47013
Black Arctous 31314,31316
Black Arisaema 33475
Black Ash 167923,167955,168032
Black Asper 175594
Black Austrian Pine 300117
Black Ballota 47013
Black Bamboo 175594,297188,297367,297476
Black Baneberry 6448
Black Barked 311398
Black Bead 301117,301162
Black Bean 79075,215981
Black Bean Tree 79075
Black Bean-tree 79075
Black Bearberry 31314
Black Beauty Elderberry 345631
Black Beech 266885
Black Begonia 50108
Black Bent 12118,17548
Black Bent Grass 12118
Black Bent Seed 203064
Black Bindweed 162519,383675
Black Birch 53437,53505,53536,53540
Black Bitter Vetch 222782
Black Bitter Vetchling 222782
Black Blueberry 403731
Black Bog Rush 352408
Black Bog-rush 352408
Black Bout 302034
Black Bowour 338447
Black Bowwower 338447
Black Box 155499,155620
Black Boy 415172,415173,415178
Black Boyd 338447
Black Boys 302034,401111
Black Bristle Palm 337059
Black Bristlegrass 361840
Black Broodhen Bamboo 297499
Black Broom 120983
Black Brush 99457
Black Bryony 383674,383675
Black Bryony Family 131918
Black Bulbophyllum 62936
Black Bulrush 353231
Black Bums 338447

Black Bush 242736
Black Butt 155696
Black Cabbage 59524
Black Calabash 145126
Black Caladium 99903
Black Calla 37024
Black Camellia 68923
Black Canary Tree 71015
Black Canarytree 71015
Black Canary-tree 71015
Black Cap 338909
Black Catechu 1120
Black Centaurea 81237
Black Chaetinii 8546
Black Cherry 44708,83311,316787
Black Chinese Olive 71015
Black Chokeberry 34864,34866,295732
Black Chokecherry 34869
Black Cinnamon 299323
Black Cohosh 6438,6441,6448,90994, 91034
Black Coral Pea 215981
Black Cosmos 107168
Black Cotoneaster 107543
Black Cottonwood 311233,311347,311554
Black Cowpea 409065
Black Cremanthodium 228991
Black Crowberry 145073
Black Cumin 266241
Black Currant 334117
Black Cutch 1120
Black Cyperus 118928
Black Cypress 67417
Black Cypress Pine 67417
Black Dalea 121888
Black Dammar 70988
Black Datura 123065
Black Dhup 71023
Black Doctor 355061
Black Dogwood 280003,328603
Black Dotted Leaves Blueberry 404017
Black Dye-tree 302684
Black Ebony 132117,132349,155967
Black Eentaurea 81237
Black Elder 158062,345631,345636,404316
Black Elderberry 345626,345632,345635, 345641
Black Elm 401489
Black Eurya 160539
Black False Hellebore 405618
Black Falsehellebore 405618
Black False-hellebore 405618
Black Fennel 167169
Black Flower Onion 15804
Black Fruit Madder 337925
Black Fruit Wolfberry 239106
Black Gamboge 171149
Black Garlic 15520
Black Giantgrass 175597
Black Gin 216415
Black Glands Rhododendron 331331
Black Gooseberry 334117
Black Gram 408982
Black Grama 57926
Black Grass 17553,213127,247366
Black Greasewood 346610
Black Guava 181731
Black Gum 155471,155680,267849,267867
Black Gypsy 302034
Black Haw 407917,408061,408062,408095
Black Haw Viburnum 408061
Black Hawberry 109684
Black Hawthorn 109684
Black Hay 247366
Black Headed Grass 238583
Black Heath 149205
Black Hellebore 43309,147212,190944
Black Hemlock 399916
Black Henbane 201389
Black Henna 206669
Black Hickory 77894,77932
Black Hickory Nut 212631
Black Highbush Blueberry 403731
Black Hills Spruce 298286
Black Holly 203581
Black Home Tanoak 233307
Black Honeysuckle 235874,235982
Black Horehound 46989,47013,47016,47017
Black Huckleberry 172249
Black Hurs 403916
Black Indian Hemp 29473
Black Iris 163516
Black Ironbox 155717
Black Ironwood 270070,270138
Black Italian Poplar 311163
Black Ivy 187222
Black Jack 53797,54048,302034
Black Jack Oak 324155
Black Jessie 301162
Black Jetbead 332282,332284
Black Jet-bead 332284
Black jewel Orchid 238135
Black Juniper 213807
Black Jurinea 201389
Black Kangaroo Paw 241206
Black Kangaroo-paw 241204,241206
Black Kauri 10501
Black Kauri Pine 10501
Black Kite 338447
Black Knapweed 81237
Black Knotweed 309525
Black Larch 221904
Black Laurel 179761,228161
Black Leaved Clover 397043
Black Leptodermis 226104
Black Lily 168361
Black Litse 233844
Black Locust 334946,334976
Black Locust Tree 334976
Black Lovage 366743
Black Magic 148077
Black Magnoliavine 351073
Black Man 302034
Black Man's Flower 316127
Black Man's Posy 220416
Black Mango 289476
Black Mangrove 45743,45746
Black Maple 3236,3549,3565
Black Marginate Groundsel 358748
Black Margine Groundsel 359586
Black Marlock 155718
Black Mary's Hand 273531
Black Masterwort 43309
Black Matfellon 81237
Black Meddick 247366
Black Medic 247366
Black Medicine 231298
Black Medick 247366
Black Merry 83155,280003
Black Mesquite 315566
Black Milkwort 42417
Black Mint 250423,250431
Black Mixed-flower 298064
Black Mondo Grass 272122,272123
Black Morrel 155648
Black Mountain Psychrogeton 319929
Black Mulberry 259180
Black Mullein 405737
Black Mustard 59515
Black Needlerush 213424
Black Nightshade 367416,367455,367535
Black Nigritella 266260
Black Nisewort 190936
Black Nonesuch 247366
Black Norway Spruce 298216
Black Nutantdaisy 228991
Black Oak 79177,323871,323873,324335, 324539
Black Oat 45602
Black Oat Grass 376707
Black Olive 61909,71015
Black Onion 15520
Black Ornamental Raspberry 339325
Black Palm 266678
Black Pansy 409858
Black Pea 222782
Black Pea Vine 222782
Black Pepper 300464
Black Pepper Mint 250420
Black Peppermint 155476,155478,155730, 250420,250429

Black Persimmon 132132,132322
Black Phyllary Edelweiss 224902
Black Pine 67411,299992,300117,300195,300281,306518,316087,379766
Black Plantain 302034
Black Pogostemon 306990
Black Poker 401112
Black Ponfianak 362186
Black Poplar 311398
Black Potherb 366743
Black Poui 211239
Black Protea 315852
Black Pudding 401112
Black Purple Flower Milkvetch 42919
Black Pussy Willow 343683
Black Quitch 17548
Black Raspberry 338741,338909
Black Rasp-berry 338909
Black Rattlepod 47861
Black Rockfoil 349082
Black Root 407434
Black Roseau 46379
Black Roseau Palm 46379
Black Rosewood 121684,121730,121770
Black Sage 19179,35998,345210
Black Sagebrush 35998,36400
Black Sally 155750,343151
Black Salsify 354870
Black Saltwort 176775
Black Sand Spikerush 143269
Black Sanguineous-scale Pycreus 322349
Black Sanicle 43309,345980
Black Sapote 132132
Black Sapote-tree 132132
Black Sarana 168361
Black Sassafras 43972
Black Savin 213807
Black Saxoul 185064
Black Sea Lily 229996
Black Sea Sideritis 362786
Black Sea Skullcap 355693
Black Sea Walnut 212636
Black Sedge 73788,73812,352408
Black Serpentaria 91034
Black Sesame 203064
Black Sesamum 203064
Black She-oak 15950
Black Siris 13635
Black Sloe 316890
Black Snakeroot 90994,91000,91034,345940,345945,345980,346002,417887
Black Snake-root 91034
Black Soap 81237,216753
Black Southstar 33475
Black Spice 338447
Black Spine Honey Locust 176881
Black Spinestemon 306990
Black Spots Rhododendron 331332
Black Spruce 298349
Black Stanhopea 373700
Black Sticks 401112
Black Stinkwood 268691
Black Stink-wood 268691
Black Sugar Maple 3236
Black Sugar-maple 3236
Black Sumach 332529
Black Swallowwort 117575,409394
Black Swallow-wort 117575
Black Swamp Ash 168032
Black Tanoak 233315
Black Tea-tree 248077,248101
Black Thistle 92125,92285,92485
Black Thorn 316819
Black Thorold Kengyilia 215943
Black Thorowax 63831
Black Thyme 391397
Black Titi 96881
Black Toothbrushes 180603
Black Top 81356,229662
Black Trefoil 247366
Black Tulip 185900
Black Tupelo 267867
Black Twinberry 235874
Black Vanilla Orchid 266260
Black Velnet Bean 259545
Black Vetch 408488
Black Vetchling 222782
Black Vine 383675
Black Walnut 53057,212582,212631
Black Wattle 1168,1373,1380,1384,1401,66975,66976
Black Wattle Acacia 1380
Black Whort 403916
Black Whortle 403916
Black Widow 174811
Black Wildclove 226104
Black Willow 343445,343767,343858,344077
Black Wood 1384,121730,288884
Black Woodbetony 287101,287457
Black Zedoary 114875
Black Zinger 417997
Blackacacia 334976
Blackalga 200184,200190
Blackamoor's Beauty 216753
Blackamore 401112
Blackanther Cranesbill 174734
Blackanther Goosecomb 215943
Blackanther Roegneria 215943
Blackanther Sandwort 32066
Black-anthered Heath 149750
Black-banded Rabbit Brush 90556
Blackbanded Rabbitbrush 90556
Blackbark 386563
Blackbark China Pine 300251
Blackbark Chinese Pine 300251
Blackbark Linociera 87726
Blackbark Liolive 87726
Black-bark Persimmon 132319
Blackbark Willow 343606
Black-barked Linociera 87726
Black-barked Persimmon 132319
Blackbead 301117,301134
Black-bead 301117
Blackbead Elder 345626
Black-beard Haworthia 186323
Blackbern 338447
Black-berried Eotoneaster 107543
Black-berried Honeysuckle 235982
Blackberry 145073,334117,338059,338105,338189,338265,338281,338307,338412,338447,403916
Blackberry Lily 50667,50669
Blackberry Mouchers 338447
Blackberry Rose 336896
Blackberry Token 338209
Blackberryleaf Primrose 314915
Blackberrylily 50667,50669
Blackberry-lily 50667,50669
Black-bindweed 162519
Blackbleg 338447
Blackblotch Tritonia 399067
Blackbowwowers 338447
Blackboy 415172,415174,415178
Blackbract Edelweiss 224902
Blackbract Milklettuce 259757
Blackbract Pyrethrum 322699
Blackbract Windhairdaisy 348522
Black-branch Milkvetch 43249
Blackbranch Tanoak 233315
Blackbranch Willow 343854
Black-branched Milkvetch 43249
Black-branched Tanoak 233315
Black-branched Willow 343854
Blackbrown Groundsel 358322
Blackbrush 1532,99457
Blackbrush Acacia 1532
Blackbud Bamboo 297202
Black-budded Bamboo 297202
Blackbulb Gagea 169472
Blackbulbil Yam 131706
Blackburn 341406
Blackburn Palmetto 341406
Blackbutt 155696
Black-butt 155696
Blackcalyx Crazyweed 279005
Black-calyx Mockorange 294457
Blackcalyx Primrose 314924
Blackcap 238583,338741,338909,401112
Black-cap 338909
Blackcap Raspberry 338808,338909

Blackcapsule 248691
Black-caudexed Daphne 122484
Blackcentral Rockfoil 349623
Blackcurrant 334117
Blackdamar 71023
Blackdisk Medick 247407
Blackdown Bottlebrush 67285
Blackdown Fan Palm 234179
Blackedge Gentianella 174120
Black-edge Sedge 75513
Blacken Archiboehmeria 30951
Blacken Catchfly 363810
Blacken Cytisus 120983
Blacken Eria 148658
Blacken Ophiorrhiza 272266
Blacken Saussurea 348557
Blacken Stonebean-orchis 62936
Blacken Syzygium 382626
Blackend Mosquitotrap 117385
Blackend Swallowwort 117385
Blackened Litse 233844
Blackened Rockfoil 349082
Blackening Teeth 263095,263096
Blackeye Root 383675
Black-eyed Bean 409086
Black-eyed Clockvine 390701
Black-eyed Dais 339512
Black-eyed Heath 149750
Black-eyed Heather 149138
Blackeyed Pea 409086
Black-eyed Pea 409086
Black-eyed Susan 765,339512,339539,
339556,339567,339593,390701
Blackeyed Susan Vine 390701
Black-eyed Susan Vine 390692,390701
Blackfellows 54048
Blackfibre Palm 32343
Blackflower Catchfly 363737
Blackflower Jerusalemsage 295146
Blackflower Notoseris 267174
Blackflower Purpledaisy 267174
Blackflowered Sedge 75326
Black-flowered Spindle-tree 157713
Blackfoot 407434
Blackfoot Daisy 248141
Blackfoot Greenbamboo 125447
Black-fringe Bindweed 162517
Black-fringed Giant Bamboo 175607
Blackfruit Aralia 30707
Blackfruit Arrowwood 407950
Blackfruit Bladderwort 403916
Blackfruit Buttercup 326070
Blackfruit Dandelion 384664
Black-fruit Elder 345626
Blackfruit Galangal 17728
Blackfruit Greenbrier 366343
Blackfruit Hawthorn 109597
Black-fruit Honeysuckle 235982
Blackfruit Litse 233844
Blackfruit Mountainash 369459
Blackfruit Nertera 265359
Blackfruit Qinggang 116073
Black-fruit Raspberry 338880
Blackfruit Skimmia 365970
Blackfruit Smilax 366343
Blackfruit Viburnum 407950
Blackfruit Wolfberry 239106
Blackfruit Yellowquitch 194040
Black-fruited Barberry 51336
Black-fruited Blueberry 403721
Blackfruited Cotoneaster 107543
Black-fruited Cotoneaster 107543
Black-fruited Greenbrier 366343
Black-fruited Hawthorn 109597
Black-fruited Holly 203581
Black-fruited Honeysuckle 235982
Black-fruited Lasianthus 222302
Black-fruited Litse 233844
Black-fruited Skimmia 365970
Black-fruited Spike-rush 143208
Black-fruited Wolfberry 239106
Black-galingale 169548
Black-girdled Wool-grass 353228
Blackgland Azalea 331331
Blackgland Elaeocarpus 142279
Blackgland Rockfoil 349696
Blackglandule Elaeocarpus 142279
Black-glandule Elaeocarpus 142279
Blackglandule Pearweed 239671
Black-grape Cotoneaster 107496
Blackgrass 17553,61727
Black-grass 17548,213127
Blackgreen Catnip 264876
Blackgreen Embelia 144768
Black-green Embelia 144768
Blackgreen Nepeta 264876
Blackgum 267867
Black-gum 267867
Blackhair Branchy Milkvetch 42898
Blackhair Crazyweed 279007
Blackhair Dendrobium 125415
Blackhair Dogwood 124940
Blackhair Four-involucre 124940
Black-hair Giant Bamboo 175607
Blackhair Giantgrass 175607
Blackhair Goldenray 229162
Blackhair Holly 204034
Blackhair Milkvetch 42949
Blackhair Narrowgalea Woodbetony 287699
Blackhair Windhairdaisy 348384
Black-haired Giant-grass 175607
Black-haired Holly 204034
Black-haired Persimmon 132189
Black-hairs Persimmon 132189
Blackhatflower 122924
Blackhaw 407917,408061
Black-haw 408061
Blackhaw Viburnum 408061
Black-haw Viburnum 408061
Blackhead 81237
Black-head Fleabane 150775
Blackhead Goldenray 229105
Blackhead Rush 212875
Blackhead Sedge 352348
Blackheadleaf Cowpea 408829
Blackheart Nanmu 295410
Blackheart Skullcap 355623
Blackhearts 403916
Blackie Toppers 401112
Blacking Plant 195149
Blackingirse 166125
Blackish Buckthorn 328795
Blackish Centaurea 81243
Blackish Paphiopedilum 282881
Blackish-fruited Barberry 51963
Blackite 338447
Blackjack 324155
Blackjack Oak 323873,324155,324225
Blackjack Pine 300153
Blackland Thistle 91949
Blackle Tops 302034
Blackleaf Alocasia 16494
Black-leaf Artocarpus 36936
Blackleaf Bamboo 47501
Blackleaf Caper 71846
Blackleaf Casearia 78138
Blackleaf Evergreenchinkapin 78994
Blackleaf Gendarussa 172831
Blackleaf Greenbrier 366485
Black-leaf Greenbrier 366485
Blackleaf Holly 204033
Black-leaf Indigo 206299
Blackleaf Machilus 240642
Black-leaf Memecylon 250028
Blackleaf Phoebe 295393
Blackleaf Podocarp 306506
Blackleaf Smilax 366485
Black-leaf Syzygium 382676
Black-leaved Caper 71846
Black-leaved Evergreen Chinkapin 78994
Black-leaved Indigo 206299
Black-leaved Memecylon 250028
Black-leaved Syzygium 382611,382676
Blackleg 338447
Blackline Dogwood 105121
Black-long-leaf Syzygium 382611
Blackmangrove 45746
Black-mangrove 45746
Blackmarginate Groundsel 358748
Blackmargine Groundsel 359586
Black-moss 392052

Blackmountain Aster 40909
Black-punctated Spicegrass 239753
Blackpurple Bluegrass 305773
Blackpurple Fringed Sagebrush 35509
Blackpurple Rhodiola 329935
Blackroot 381035
Blackscale 248497,248500
Blackscale Azalea 331332
Blackscale Gagea 169472
Black-scaled Longscourge Rattanpalm 65695
Blackseed 247366
Blackseed Feathergrass 376849
Blackseed Juniper 341760
Blackseed Kyllingia 218562
Blackseed Ricegrass 276058
Black-seed Sand-spurrey 370614
Blackseed Savin 341760
Blackseed Tubergourd 390191
Blackseed Water-centipede 218562
Blackseed Winged Euonymus 157298
Blackseed Winged Spindle-tree 157298
Blackseeded Juniper 341760
Black-seeded Mountain Rice 276058
Black-seeded Plantain 302159
Black-seeded Rice Grass 276058
Blacksetose Devil Rattan 121516
Black-she Oak 15950
Black-sheath Bamboo 297202
Black-spike Arrowbamboo 162714
Blackspike Cypressgrass 119256
Blackspike Milkvetch 42694
Blackspike Papposedge 375807
Blackspike Stephanachne 375807
Black-spike Umbrella Bamboo 162714
Black-spiked Fargesia 162714
Blackspiked Lovegrass 147844
Black-spiked Sedge 73788
Blackspine Honeylocust 176881
Blackspine Neoporteria 264346
Black-spine Prickly Pear 272970
Black-spined Pricklypear 272970
Blackspines Copiapoa 103745
Blackspinose Neoporteria 264362
Blackspot Ginger 418007
Black-spot Horn Poppy 176728
Blackspot Hornpoppy 176728
Black-spot Horn-poppy 176728
Black-spotted Ardisia 31639
Black-spotted Catasetum 79331
Black-stalk Masdevallia 246113
Black-stalked Japan Aucuba 44922
Blackstemmed Hydrangea 199977
Blackstrap 308816
Blackthorn 64064,83331,316819
Blackthorn May 316819
Blackthorn Plum 316819
Blackthroad Callalily 417108
Black-throated Arum 417097
Blackthyrse Goldenray 229106
Blacktiger Resurrectionlily 214959
Blacktine Corydalis 106189
Blacktongue Bulbophyllum 62900
Blacktongue Curlylip-orchis 62900
Blacktoothed Primrose 314134
Blacktop Corydalis 106189
Black-twig Willow 343743
Blackvein Cymbidium 116994
Blackvein Orchis 116994
Blackviolet Goldenray 228991
Blackweed 159634
Blackwhite Callalily 417108
Blackwindvine Fissistigma 166684
Blackwing Broomsedge 217347
Blackwing Summercypress 217347
Blackwood 1384,1474,121730,184518
Blackwood Acacia 1384
Blackwort 381035,403916
Blacky-Moor's Beauty 216753,350099
Blacky-more 401112
Bladder Bottle 364193
Bladder Campion 114057,363673,364193
Bladder Catchfly 364193
Bladder Cherry 297643,297645,297678
Bladder Flower 30860
Bladder Gentian 174034
Bladder Hawksbeard 111073
Bladder Hibiscus 195326
Bladder Ketmia 194680,195326
Bladder Nut 374087,374107
Bladder of Lard 364193
Bladder Pod 227017,234547,361453
Bladder Sage 355602
Bladder Sedge 74904,76677
Bladder Seed 228244,297967,297970
Bladder Senna 100149,100153
Bladder Silene 364158
Bladder Soapwort 346434
Bladder Vine 30862
Bladder Wort 403095
Bladderbean 100173
Bladdercampion 114057,114060
Bladder-campion 364193
Bladder-fruited Tanoak 233168
Bladderherb 297643,297645
Bladder-herb 297643
Bladderlikecalyx Skullcap 355678
Bladdernut 132472,374087,374089,374103,374107,374110
Bladder-nut 374087
Bladdernut Family 374114
Bladder-nut Family 374114
Bladderpod 18307,200414,227017,227018,361453
Bladderpod Locoweed 42231
Bladderpod Mustard 18385
Bladderpod Spiderflower 95709
Bladdersage 355602
Bladderseed 228250,297967,297970
Bladdersenna 100149,100153
Bladder-senna 100149,100153
Bladder-stem 152160
Bladderweed 364193
Bladderwort 297643,403095,403110,403382
Bladderwort Family 224499
Bladdery Oil-nut 323067
Bladdery Persimmon 132210
Blade Apple 290699
Blade Apple Cactus 290701
Bladseed Decaspermum 123450
Bladsheath Arrowbamboo 162749
Bladstem Rockfoil 349986
Blaeberry 403915,403916,404027
Blaebow 232001
Blaekbeetle Poison 220346
Blaekie Top 302034
Blagg 338447
Blaine Fishhook Cactus 354288
Blainvillea 55066
Blainy Eye 159221
Blak Rongcheng Bamboo 297245
Blake Bauhinia 49017
Blake Cyclobalanopsis 116065
Blake Oak 116065
Blake Qinggang 116065
Blake's Aster 268683
Blake's Knotweed 308698
Blakeley's Eucalyptus 155501
Blakeley's Red Gum 155501
Blak-leaf Schefflera 350660
Blakley's Spineflower 89057
Blanchard's Dewberry 339160
Blanchard's Hawthorn 109775
Blanco Beautyleaf 67853
Blanco Kibatalia 216230
Bland Ironweed 406150
Bland Pfeiffer Cicely 276466
Blandder Campion 363671
Blandder Catchfly 363671
Blanket 405788
Blanket Flower 169568,169571,169575,169586,169588,169603
Blanket Leaf 405788
Blanket Mullein 405788
Blanketflower 169575
Blanket-flower 169568,169575,169586,169608
Blanket-leaf 373166,405788
Blaspheme Vine 366428
Blastus 55135
Blatterdock 292372
Blaver 70271,81020,282546

Blaverole 81020
Blawort 70271,81020
Blaw-weary 316127
Blaxland's Strinybark 155502
Blayberry 403916
Blazing Star 14486,85505,228433,228439,228454,228461,228506,228509,228511,228519,228529,228534,228541,250481,250483,250485,250486,250489
Blazing Stars 228529
Blazing-star 85505,250481
Blazon Milkvetch 42384
Blazon Nervilia 265373
Blazon Orchis 310873
Blazon Taroorchis 265373
Blechum 55202
Bledewort 282685
Bleeberry 403916
Bleeding Grass 302068
Bleeding Heart 86427,128288,220729,270563,410677
Bleeding Heart Glorybower 96387
Bleeding Heart Tree 197625,270563
Bleeding Heart Vine 96387
Bleeding Nun 115949
Bleeding Tree 15953
Bleeding Warrior 86427
Bleeding Willow 273545
Bleeding Wire 86427
Bleedingheart 128288
Bleeding-heart 128288,128297,220729
Bleedingheart Glorybower 96387
Bleeding-heart Tree 197625
Bleeding-hearted Glorybower 96387
Bleedingheartleaf Asiabell 98310
Bleedy Warrior 86427
Bleedy War-rior 86427
Bleets 86970
Blessed Mary's Thistle 364360
Blessed Milk Thistle 364360
Blessed Milkthistle 364360
Blessed Thistle 73318,97581,97599,364360
Blethard 340249
Bletherweed 364193
Bletilla 55559,55575
Bleuet 81020
Blewart 407072
Blib Blob 68189
Blib-blob 68189
Blimbing 45724,45725
Blimbing Kamia 45724
Blin Conyza 103436
Blin Falsenettle 56358
Blin Photinia 295642
Blin Plagiopetalum 301664
Blin Sedge 73892
Blin Willowweed 146634
Blin's Conyza 103436
Blind Eyes 282546,282685
Blind Flower 407072
Blind Grass 379271
Blind Man 282685
Blind Man's Buff 95431
Blind Man's Hand 13174
Blind Nettle 170078,220346,373455
Blind Prickly Pear 273038
Blind Pricklypear 273038
Blind Tongue 170193
Blind Withy 309570
Blindeyes 282546
Blinding Tree 161632,161639
Blinding-tree 161639
Blindweed 72038
Blind-your-eye 161639
Blind-your-eyes 161639
Blind-your-eye-tree 161639
Blinking Chickweed 258248
Blinking-chickweed 258248
Blinks 258237,258248,258277
Blinks Family 311954
Blinkworthia 55604,55605
Blinn's Raspberry 338728
Blinn's Shaggy Raspberry 338728
Blister Begonia 50209
Blister Bush 92967
Blister Buttercup 326340,326344
Blister Creeper 73207
Blister Cup 325518
Blister Plant 325518
Blister Sedge 76677
Blistercress 154363,154466
Blister-cress 154363
Blistered Barberry 51398
Blistered Nautilocalyx 262897
Blisterwort 326281,326340
Blite 86891,86970
Blite Goosefoot 86984
Blithran 312360
Blob 130383,267294,334250
Blochman's Fleabane 150503
Blochman's Larkspur 124465
Block Broom 120942
Blockwheat 162312
Bloewood Condalia 101737
Blogga 68189
Blolly 181531
Blood Amaranth 18696
Blood Berry 334897
Blood Currant 334189,334260
Blood Flower 37888,68066,350312
Blood Iris 208806
Blood Leaf 208344
Blood Lily 184345,184372,184428,350312,350313
Blood Milkwort 308331
Blood of Adonis 8332
Blood of Mercury 405872
Blood of St. John 195858
Blood on the Snow 200085
Blood Orange 93765
Blood Pink 127698
Blood Plum 184503
Blood Polygala 308331
Blood Red Heath 149297
Blood Red Tassel Flower 66673
Blood Root 184527,312520,345805
Blood Staunch 103446
Blood Tongue 170193
Blood Vine 85875
Blood Wall 86427
Blood Wood 106788,155722,320269
Blood Wood Tree 184518
Blood Wort 184526
Bloodbark Maple 3003
Bloodbasin Sage 344947
Bloodbead Barberry 51982
Bloodberry 334897
Bloodberry Cotoneaster 107607
Blood-berry Cotoneaster 107607
Blood-colored Polygala 308331
Blooddisperser 297624,297632
Blood-drop Emlet 255218
Blood-drop-emlets 255218
Blooddrops 8332
Bloodflower 37888,350313
Blood-flower 37888,184345,350313,350321
Bloodflower Milkweed 37888
Bloodfruit Dandelion 384778
Bloodgood Japanese Maple 3303
Bloodleaf 208344,208349,208354,208357
Blood-leaf 208344
Bloodlike Azalea 330808
Blood-like Rhododendron 330808
Bloodlily 184345
Blood-lily 184345,184372,350321
Bloodred Azalea 331726
Bloodred Broughtonia 61091
Blood-red Cranesbill 174891
Bloodred Cryptochilus 113515
Bloodred Cyrtosia 120764
Bloodred Dendrobium 125351
Bloodred Disa 133932
Blood-red Eulophia 157110
Bloodred Euonymus 157855
Blood-red Euonymus 157855
Bloodred Flower 184345
Blood-red Geranium 174891
Bloodred Iris 208806
Bloodred Lycaste 238740
Bloodred Melastoma 248778
Blood-red Melastoma 248778

Bloodred Neoregelia 264414
Blood-red Rhododendron 331726,331743
Blood-red Spindle-tree 157855
Bloodred Swordflag 208806
Blood-red Tassel Flower 66673
Blood-red Tea-tree 226502
Bloodroot 219026,219030,345803,345805
Blood-root 345803,345805
Bloodroot Red Root 175393
Bloodstrange 260940
Bloodtwig Dogwood 380499
Blood-twig Dogwood 380499
Bloodtwig Shrub Dogwood 380499
Blood-veined Dock 340249
Bloodvine 347077,347078
Bloodvine Spatholobus 370421
Bloodweed 37888
Bloodwing Dogwood 380499
Blood-wing Dogwood 380499
Bloodwood 106805,155757,320278
Bloodwood Tree 184513,184518,320278
Bloodwort 3978,174877,184526,308816, 309199,340082,340249,345594,345805, 345860,345881
Bloodwort Dock 340249
Bloodwort Family 184524
Bloody Bells 130383
Bloody Bones 145487,273531
Bloody Butcher 81768,273317,273531, 397551,397599
Bloody Crane's Bill 174891
Bloody Crane's-bill 174891
Bloody Cranesbill 174891
Bloody Dock 340214,340249
Bloody Fingers 37015,130383,273531
Bloody Geranium 174891
Bloody Heath 149297
Bloody Man's Fingers 37015,130383, 273531,273545
Bloody Man's Hand 273531
Bloody Mary 174877
Bloody Noses 397599
Bloody Rod 380499
Bloody Sage 344980
Bloody Thumbs 60382
Bloody Triumph 396934
Bloody Twig 380499
Bloody Wall 86427
Bloody Wallier 86427
Bloody Walls 86427
Bloody Warrior 86427
Bloody William 363370
Bloody Wires 86427
Bloody-bark Maple 3003
Bloody-warrior 86427
Bloomer's Fleabane 150504
Bloomer's Goldenbush 150311
Bloom-fell 237539
Blooming Boxes 61575
Blooming Sally 85875,146724,240068
Blooming Spurge 158692
Blooming Willow 85875
Bloomingrush 64177,64179
Bloomy Down 127607
Bloomy Poplar 311433
Bloomy-down 127607
Bloomy-stem Rose 336762
Bloontleaf Screwtree 190107
Blossfeld Cactus 244011
Blossfeldia 55656
Blossom 109857
Blotchflor Azalea 331176
Blotchleaf Azalea 331585
Blotchleaf Violet 410721
Blotchstem Vinegentian 110321
Blotchy Mint Bush 315609
Blouaalwyn 17299
Bloubos 132275
Bloudow Iris 208466
Blowball 384714
Blow-ball 81020
Blower 384714
Blow-flower 81020
Blown-grass 12008
Blow-pine Bamboo 47254
Blowpipe Bamboo 47250,47254
Blow-pipe Tree 382329
Blowytree Wendlandia 413775
Bluc Moor-grass 361611
Blue Acacallis 1020
Blue African Lily 10245
Blue Ainsliaea 12619
Blue Alfalfa 247456
Blue Algerian Fir 436
Blue Aloe 16847
Blue Alpbean 391549
Blue Alpine Daisy 40016
Blue Altai Lily 210999
Blue Amaryllis 414779,414784
Blue and White Daisybush 276599
Blue Anemone 23707,23766,23975
Blue Appleberry 54436
Blue Ash 168065,168079,168106,382329
Blue Atlantic Cedar 80083
Blue Atlas Cedar 80083
Blue Ball 140776
Blue Ball Sage 344973
Blue Bamboo 137849
Blue Barrel Cactus 140133,163443
Blue Basin 174832
Blue Bead-lily 97091
Blue Bean 238473
Blue Bean Shrub 123367,123371
Blue Beard 345474
Blue Beargrass Tree 266505
Blue Beech 77263,77264
Blue Beli 368541
Blue Beli Creeper 368541
Blue Bell 141258,175990,199560,250883, 412568
Blue Bells 69870,70271
Blue Bells of Scotland 70271
Blue Berry 54436,127515
Blue Berry Elder 345578
Blue Betsy 409335,409339
Blue Bindweed 367120
Blue Birch 53369
Blue Bird 195276
Blue Bird Nemesia 263390
Blue Bird's Eye 407072
Blue Bird's Eyes 407072
Blue Bladderwort 403132
Blue Blaver 70271
Blue Blossom 79973
Blue Blossom Ceanothus 79973
Blue Blow 81020
Blue Bluesnow 83646
Blue Boneset 101955,383636
Blue Bonnets 81020,145487,211672,379660
Blue Boronia 57260
Blue Bottle 80892,81020,260333
Blue Bow 81020
Blue Box 155493
Blue Boys 322159
Blue Bramble 338209
Blue Brocade 39317
Blue Broomrape 275006
Blue Broomrupe 275173
Blue Bugle 13104
Blue Bugleweed 13104
Blue Bugloss 21914
Blue Bush 1151,155634,242737
Blue Butcher 272398,273531
Blue Butterfly 102833,337435
Blue Butterfly Bush 96417
Blue Butterflygrass 392905
Blue Buttons 81020,174832,211672, 216752,350125,379660,409335,409339
Blue Caladenia 65204
Blue Camassia 68803
Blue Camomile 398134
Blue Candle 261706
Blue Canyon Sage 345186
Blue Cap 81020
Blue Cardinal Flower 234707,234774, 234813
Blue Carpet Juniper 213944
Blue Cat's Tail 141347
Blue Cat's Tails 141347
Blue Catmint 264920
Blue Ceanothus 79902

Blue Cedar 80083
Blue Ceratostigma 83646
Blue Chalksticks 360027
Blue Chinese Poppy 247131
Blue Chip Juniper 213788
Blue Cohosh 79730,79732,79733
Blue Cohush 79733
Blue Colorado Bee Plant 95791
Blue Colorado Spruce 298401,298406
Blue Comfrey 381029,381038
Blue Coral Pea 185760
Blue Corn 417417
Blue Corn-lily 33608,33643
Blue Couch 130524
Blue Cowslip 321636
Blue Cow-wheat 248189
Blue Cowwheat 248189
Blue Cranesbill 174846
Blue Crazyweed 278757
Blue Crocus 385520
Blue Crown Passion Flower 285623
Blue Crown Passion-flower 285623
Blue Cupid's-dart 79276
Blue Cupidone 79271,79276
Blue Cupids-dart 79276
Blue Curls 208524,316127,396434,396437,396439,396440
Blue Daisy 86027,163131,177053,211672,398134
Blue Dalea 121880
Blue Dandelion 90901
Blue Dawn Flower 207891,207934
Blue Daze 161439,161447
Blue Devil 141087,141347,154334,380943
Blue Diamond Cholla 116668
Blue Dicks 128870
Blue Didymocarpus 129889
Blue Dogbane 20860
Blue Dogwood 104947
Blue Douglas Fir 318574,318580,318585
Blue Dracaena 104359
Blue Elder 345578,345579,345591,345604
Blue Elderberry 345578,345591,345627,345648
Blue Endive 90901
Blue Eryngo 154281,154317
Blue Evergreen Hydrangea 129033
Blue Eyes 407072,407287
Blue False Indigo 47859
Blue Fan Palm 59096
Blue Fenugreek 397208
Blue Fescue 163996,164059,164126,164164
Blue Fescuegrass 164164
Blue Field Gilia 175677
Blue Field Madder 362097
Blue Fieldmadder 362097
Blue Fingers 409335
Blue Fir 317
Blue Flag 208583,208924,208932,208933
Blue Flag Iris 208924
Blue Flax 231917,231933,231942
Blue Fleabane 150419,150454
Blue Flower Gueldenstaedtia 181635
Blue Flower Tibetia 391549
Blue Flowered Fenugreek 397208
Blue Fly Honeysuckle 235691
Blue Forest Juniper 213789
Blue Forget-me-not 260892
Blue Foxglove 70331
Blue Freesias 46022
Blue Fruits Rhododendron 330488
Blue Funnel-lily 23381
Blue Gem 110196
Blue Gentianella 174138
Blue Giant Hyssop 10407
Blue Giant Sequoia 360578
Blue Ginger 128997
Blue Ginseng 79733
Blue Girl Golly 203533
Blue Girsf 408352
Blue Globe-thistle 140660
Blue Glorybower 96417
Blue Goggles 145487
Blue Gowen Cypress 114695,114699
Blue Grama 57930,88882
Blue Granfer Greygle 145487
Blue Granfer-greygles 145487
Blue Grape 411551,411578
Blue Gress 163778,305274
Blue Gromwell 233772
Blue Gum 155500,155589,155756
Blue Hair-grass 217473
Blue Haw 407737
Blue Hawthorn 109559
Blue Haze 318424
Blue Haze Tree 211239
Blue Head 379660
Blue Hearts 61732
Blue Heath 297015,297019,297021
Blue Heaven Juniper 213923
Blue Hesper Palm 59096
Blue Hibiscus 18254,195269
Blue Holly 203533
Blue Honeysuckle 235691
Blue Hooded Violet 410306
Blue Horizon Juniper 213790
Blue Horse Mint 257180
Blue Huckleberry 172257,403932,403957
Blue Indianlotus 267723
Blue Indigo 47859
Blue Innocence 198708
Blue Iris 208819,208854
Blue Jacaranda 211239
Blue Jack 81020,81237,409335,409339
Blue Jack Oak 324045
Blue Jacket 394053
Blue Jackets 307197
Blue Jasmine 94870
Blue Joints 144661
Blue Kiss 379660
Blue Koeleria 217473
Blue Lace Flower 393883
Blue Lace-flower 393883
Blue Lagoon Juniper 213740
Blue Largeleaf Hydrangea 200007
Blue Larkspur 124104
Blue Latan 222634
Blue Latan Palm 222634
Blue Lawn Grass 171524
Blue Lawngrass 171524
Blue Lawson Cypress 85272,85293
Blue Leadwood 83646
Blue Leadwort 305172
Blue Leaf Sundrops 269441
Blue Leaf Wattle 1553
Blue Lechenauhia 223714
Blue Lettuce 219444,219458,219532,219609,259772
Blue Lillypilly 382636
Blue Lily 10245
Blue Lily-turf 232631
Blue Lips 99836
Blue Liriope 232631
Blue Lobelia 234774
Blue Lotus 267658
Blue Lotus of Egypt 267658
Blue Lotus of India 267723
Blue Lotus-of-the-nile 267658
Blue Lungwort 321643
Blue Lupin 238422
Blue Lupine 238422,238450,238467
Blue Lyme Grass 144144,228346
Blue Mahoe 194850
Blue Mallee 155701
Blue Marble Tree 142326
Blue Marguerite 163131
Blue Marlin Hibiscus 195274
Blue Marsh Bellflower 70347
Blue Marsh Violet 410306
Blue Marsh-bellflower 70347
Blue Mat Juniper 213791
Blue Maurandya 246649
Blue Medusa Rush 213165
Blue Melilot 397208
Blue Men 216753
Blue Mice 410493
Blue Milkvetch 42126
Blue Mist 77993
Blue Mist Shrub 77992,77993,77995,77996
Blue Mistflower 101955
Blue Money 321721

Blue Moonwort 367762,367763
Blue Moor Grass 361615
Blue Moorgrass 361615
Blue Moor-grass 361615
Blue Morning Glory 207839
Blue Moss Cypress 85347
Blue Mountain Ash 155679
Blue Mountain Heath 297019
Blue Mountain Juniper 213986
Blue Mountain Larkspur 124180
Blue Mountain Mahogany 155670
Blue Mountain Mallee 155753
Blue Mountain Tea 368106
Blue Mountainheath 297019
Blue Mountains Malice-ash 155753
Blue Muck 145487
Blue Mustard 89020
Blue Myrtle 409339
Blue Needlejade 297019
Blue Nest Spruce 298351
Blue New Nymph Flower 264159
Blue Nipliporchis 154872
Blue Noble Fir 443
Blue Nutateflower Crazyweed 279073
Blue Oak 323845,323849,323927,324141
Blue Oat Grass 190129,190196
Blue Oat-grass 190196
Blue Oats Grass 190196
Blue Onion 15220
Blue Orchid 193134,193154,404629
Blue Oxalis 284650
Blue Pacific Juniper 213741
Blue Palm 59096
Blue Palo Verde 83747,284481
Blue Paloverde 83747
Blue Panicum 281336
Blue Passion Flower 285623
Blue Passionflower 285623
Blue Passion-vine 285623
Blue Pea 97203,198762,319224
Blue Pea Bush 319224
Blue Penstemon 289344
Blue Peter 5442
Blue Peter Crocus 111519
Blue Petrocosmea 292552
Blue Phlox 295257,295258
Blue Pig-ear 231503
Blue Pigmy Water Lily 267671
Blue Pimpernel 21339,21340,21383,21396
Blue Pine 300297
Blue Pipe 382329
Blue Plantain Lily 198652
Blue Plantainlily 198652
Blue Plantain-lily 198652
Blue Poison 229662
Blue Poppy 81020,247099,247109
Blue Porterweed 373507
Blue Potato Bush 238975,367549
Blue Potato-bush 367549
Blue Prairie Violet 410297
Blue Quandong 142326
Blue Rabbiten-wind 12619
Blue Ray-floret Fleabane 151058
Blue Rhododendron 330442
Blue Ridge Blueberry 403952
Blue Ridge Goldenrod 368434
Blue Ridge Sedge 75226
Blue Ridge Wakerobin 397617
Blue Robin 57101
Blue Rock Palm 59095
Blue Rocket 5442,5676,145487
Blue Rose-of-sharon 195275,195286
Blue Rug Juniper 213792,213803
Blue Runner 176824
Blue Sage 148087,344885,344887,345023
Blue Sailor 90901
Blue Sailors 90901
Blue Salvia 344885,345292
Blue Scarlet Pimpernel 21340
Blue Scorpion-grass 260885
Blue Sedge 74542,74547,74656
Blue Shamrock 284650
Blue Shrub 79973
Blue Skullcap 355412,355440,355449,
355563,355851
Blue Smock 409339
Blue Snakeweed 373507
Blue Solanum Shrub 367212
Blue Sow Thistle 90851
Blue Sow-thistle 90835,90836,90851
Blue Spanish Fir 443
Blue Spiked Rampion 298056
Blue Spiraea 77993,78012
Blue Spoonleaf Anemone 23759
Blue Spruce 298401,298406
Blue Spur Flower 303318
Blue Spurge 159417
Blue Star 20857,20860,407072
Blue Star Creeper 210317,223040,223053,
313555,313568
Blue Star Juniper 213943,213945
Blue Stars 20844
Blue Stem 351309
Blue Stock Gilliflower 246478
Blue Stonebutterfly 292552
Blue Stonecrop 357114
Blue Stoneweed 233772
Blue Succory 90901
Blue Sugar Maple 3570
Blue Sugarbush 315900
Blue Syzygium 382200
Blue Tannia 415202
Blue Tar Fitch 408352
Blue Tar-fitch 408352
Blue Taro 415202
Blue Tea Bush 79905
Blue Thimble Flower 175677
Blue Thistle 92485,141347
Blue Throatwort 393608
Blue Tibetan Poppy 247099,247109
Blue Tinsel Lily 66374
Blue Toadflax 230931,231105,267347,
267349
Blue Top 81237,379660
Blue Torch Cactus 299253
Blue Torenia 392905
Blue Touch-me-not 204884
Blue Tree 155589
Blue Trigonella 397208
Blue Trumpet 145487
Blue Trumpet Vine 390787
Blue Trumpetcreeper 390822
Blue Trumpet-vine 390787
Blue Umber-lily 398789
Blue Umbrella-sedge 168839
Blue Utterfly-bush 96417
Blue Valley Orach 44435
Blue Vase Juniper 213639
Blue Velvet Honeysuckl 235904
Blue Venus' Pride 198708
Blue Vervain 405844
Blue Vetch 408352
Blue Victoria Salvia 345023
Blue Vine 97203,117552
Blue Vine Sage 344923
Blue Violet 174049,409648,410351,410493
Blue Virginia Rose 336308
Blue Warrior 174832
Blue Waterlily 267723
Blue Water-lily 267679
Blue Water-speedwell 406992
Blue Wattle 1165
Blue Waxweed 114614
Blue White Cedar 85383
Blue Whortleberry 403932
Blue Wild Indigo 47859
Blue Wild Rye 144318
Blue Wings 392905
Blue Wisteria 414576
Blue Wood Anemone 23707
Blue Wood Aster 40256,381007
Blue Wood Sedge 74555
Blue Woodruff 39314,39383
Blue Woodrush 39383
Blue Yucca 416556,416639
Blueanther Knotweed 309032
Blue-bark Schoepfia 352438
Bluebasin 350086
Bluebead 97091
Bluebead Greenbrier 366460
Blue-bead Greenbrier 366460

Bluebead Lily 97091
Bluebead Smilax 366460
Bluebead-lily 97091
Blue-bead-lily 97091
Bluebeard 77992,77993,78012,266225
Blue-beard 77992
Blue-beard Butterfly-bush 61992
Bluebeardleaf Butterflybush 61992
Bluebeardleaf Summerlilic 61992
Bluebeard-like Skullcap 355399
Blue-beech 77263,77264
Bluebelia 234447
Bluebell 30081,69870,70271,95243,115338,145487,199550,199560,250914,273531,307241,352880,409339,412632
Blue-bell 368539
Bluebell Barleria 48140
Bluebell Bellflower 70271
Bluebell Comastoma 100268
Bluebell Creeper 368539,368541
Bluebell Falsecuckoo 48140
Bluebell Throathair 100268
Bluebellflower 115338,115418
Bluebells 250883,250914
Bluebells-of-scotland 70271
Blue-berried Honeysuckle 235691
Blueberry 54436,104949,109559,172248,338209,403710,403780,403904,403916,403957
Blueberry Ash 142378
Blueberry Azalea 332046
Blue-berry Currant 334232
Blueberry Elder 345579,345591,345604,345648
Blueberry Family 403706
Blueberry Flax Lily 127517
Blueberry Root 79733
Blueberry Tree 142303
Blueberry Willow 343776
Blueberryash 142378
Blueberry-ash 142378
Blueberryleaf Box 64377
Blueberryleaf Knotweed 309933
Blueberryleaf Willow 344237
Blueberry-leaved Willow 344237
Blueberrylike Persimmon 132454
Blueberry-like Persimmon 132454
Blueberry-like Rhododendron 332046
Blueberry-like Willow 343745
Blue-bird-crocus 111518
Blueblack Arrowwood 407685
Blueblack Viburnum 407685
Blueblaw 81020
Blueblossom 79973
Blue-blossom 79973
Bluebonnets 81020
Bluebottle 5442,70271,80892,81020,81237,141347,145487,409339
Blue-bottle 81020
Bluebowls 175703
Bluebract Onion 15088
Blue-branched Ephedra 146188
Bluebunch Wheat Grass 144677
Bluebunch Wheat-grass 11883
Bluebush 132275,184762,242737
Blue-buttons 216753
Bluecalyx Japanese Rabdosia 209717
Blue-candle 261706
Bluecolored Chinese Fir 114540
Bluecrown Passionflower 285623
Blue-crown Passionflower 285623
Blue-curls 396437
Bluedevil 141348
Blue-devil 141347
Bluedick 128884
Bluedicks 128873
Blue-dicks 128870,128884
Blue-downy Lyme Grass 228346
Blueeargrass 115521
Blue-eyed African-daisy 31294
Blue-eyed Babies 198708
Blue-eyed Beauty 409335
Blue-eyed Grass 365678,365685,365695,365696,365700,365747
Blue-eyed Iris 208839
Blue-eyed Mary 99840,270677,270723
Blue-eyed-grass 365678
Blue-field Gilia 175678
Blueflag Iris 208924
Blueflag Swordflag 208924
Blue-flame 261706
Blueflower Catnip 264900
Blueflower Chinese Broomrape 275210
Blue-flower Common Lilac 382370
Blue-flower Cranesbill 174846
Blueflower Leadwort 305172
Blueflower Nepeta 264900
Blueflower Potanin Iris 208766
Blueflower Potanin Swordflag 208766
Blueflower Ricebag 181635
Blueflower Salt-loving Iris 208607
Blueflower Salt-loving Swordflag 208607
Blueflower Skullcap 355440
Blueflower Spapweed 204884
Blueflower Teasel 133464
Blueflower Vanda 404630
Blueflower Youngia 416405
Blue-flowered Germander 388086
Blue-flowered Milkweed 400750
Blue-flowered Torch 392026,392027
Blue-Flowered Vine 400844
Blue-flowered Woodrush 39314
Blueflowervine 292517
Bluefruit Azalea 330488
Bluefruit Bladderwort 403775
Blue-fruit Currant 334232
Bluefruit Elder 345604
Bluefruit Glycosmis 177828
Bluefruit Honeysuckle 235691,235746
Bluefruit Lasianthus 222124
Blue-fruit Memecylon 249937
Bluefruit Roughleaf 222124
Bluefruit Snakegrape 20297
Blue-fruited Dogwood 104951
Blue-fruited Honeysuckle 235746
Blue-fruited Lasianthus 222124
Blue-fruited Memecylon 249937
Bluegrass 305274,305703
Blue-grass 305274,305334,361619
Bluegrass-shaped Alkaligrass 321366
Blue-gray Barley 198328
Bluegray Maple 2826
Blue-gray Maple 2826
Blue-green Moor Grass 361621
Bluegreen Saltbush 44542
Bluegrey Azalea 330272
Blue-grey Fox Tail 17535
Bluegrey Gentian 173312
Blue-grey Milkvetch 42409
Blue-grey Underneath Rhododendron 330137
Bluegrey Willow 343133
Bluegum Eucalyptus 155589
Blueish-grey Rattanpalm 65656
Bluejack Oak 324045
Blue-jacket 394053
Blue-jasmine 94870
Bluejoint 22681,65303
Blue-joint Grass 65303
Bluejoint Reedgrass 65303
Blueleaf Acacia 1159
Blueleaf Honeysuckle 235903
Blue-leaf Litse 233872
Blueleaf Sotol 122902
Blueleaf St. John's Wort 201886
Blueleaf Wattle 1159
Blue-leaf Willow 343738,343739
Blue-leaved Wattle 1159
Blue-lily 398789
Blue-lips 99836
Bluemink 11206
Blue-mist Shrub 77993
Bluemony 321721
Blue-petal Milkvetch 42126
Bluepodded Rhododendron 330488
Blue-poppy 81020
Blue-ridge Buckbean 389558
Blue-runner 176824
Blue-sage 148087
Blue-sailors 90901
Blue-sky Vine 390787
Bluesnow 83641

Bluespath Onion 15088
Blue-star Amsonia 20860
Bluestem 351309
Blue-stem Prickly Poppy 32415
Bluestem Wheat Grass 144675
Bluestem Wheat-grass 144675
Bluestemlike Yushanbamboo 416748
Blue-stemmed Goldenrod 368003
Bluet 80892,81020,198708,403710
Bluet Rhododendron 330935
Bluethistle 141087,141347
Bluetongue Gentian 151058
Bluetop Eryngo 154280
Blue-top Eryngo 154280
Bluets 187526,198703,198708
Blue-tulip 258613
Blueweed 141087,141347,188933,383636
Blue-weed 141347
Blueweed Sunflower 188933
Bluewhite Larkspur 124018
Blue-white Trigoneila 397208
Bluewings 392905
Bluff Oak 323859
Bluff Rock Daisy 291297
Bluffs Gayfeather 228467
Blugga Bludda 68189
Blughtyns 68189
Bluish-gray Mahonia 242497
Bluish-grey Rattan Palm 65656
Bluish-purple Rhododendron 331713
Bluish-purple Tephrosia 386257
Bluleaf Noble Fir 454
Bluleek 15048
Blume Bambusa 47213
Blume Knotweed 308907
Blume Rattan Palm 65650
Blume Schizostachyum 351848
Blume Spine Bamboo 47213
Blume Spiraea 371836
Blumea 55669
Blumeopsis 55860
Blumer's Dock 340166
Blunderbuss 127635
Blunderbuss Lily 229900
Blundleaf Grewia 180938
Blunt Bog Orchid 302455
Blunt Broom-sedge 76577
Blunt Drypetes 138670
Blunt Leaves Azalea 330082
Blunt Pansy Orchid 254925
Blunt Plantain Lily 198599
Blunt Plantainlily 198599
Blunt Rhododendron 331370
Blunt Spike-rush 143239
Blunt Spikesedge 143239
Bluntbract Goldenrod 368297
Blunt-broom Sedge 76577
Bluntflower Rush 213487
Blunt-flowered Rush 213487
Blunt-fruited Water-starwort 67373
Bluntleaf Asiaglory 32638
Bluntleaf Bedstraw 170520
Blunt-leaf Bedstraw 170520
Blunt-leaf Bitter Pea 123269
Bluntleaf Cotton 179878
Bluntleaf Dock 340151
Bluntleaf Fishwood 110233
Bluntleaf grass 375764
Bluntleaf Grevill 180624
Blunt-leaf Heath 146074
Bluntleaf Hepatica 23703
Bluntleaf Litse 234095
Bluntleaf Liverleaf 23703
Bluntleaf Milkweed 37827
Bluntleaf Onewayflor 275482
Blunt-leaf Orchid 183911
Bluntleaf Protea 315907
Blunt-leaf Rhubarb 329365
Blunt-leaf Sandwort 256444
Bluntleaf Stairweed 142764
Bluntleaf Syzygium 382508
Blunt-leaved Dock 340151
Blunt-leaved Everlasting 178358
Blunt-leaved Milkweed 37827
Blunt-leaved Orchid 302455
Blunt-leaved Orchis 302455
Blunt-leaved Pondweed 312199
Blunt-leaved Privet 229558
Blunt-leaved Sandwort 256444
Blunt-leaved Silk Oak 180624
Blunt-leaved Spurge 159488
Blunt-leaved Willow 344004
Blunt-leaves Java Rhododendron 331634
Bluntooth Opithandra 272597
Bluntpetal Sinocarum 364882
Bluntscale Bulrush 352252
Blunt-scale Bulrush 352266,352269
Blunt-scaled Oak Sedge 73626
Blunt-scaled Wood Sedge 73633
Blunt-sepaled Starwort 375026
Blunt-sickle 81020
Blunttooth Chirita 87933
Blunttooth Clematis 94741
Blunttooth Stairweed 142763
Blurred Rhododendron 330935
Blush Daisy Bush 270203
Blush Honeysuckle 236144
Blush Red Arisaema 33325
Blush Tea Rose 336816
Blush Wood 200690
Blush-grey Leaves Rhododendron 332030
Blush-grey-leaved Rhododendron 332030
Blushing Azalea 331408
Blushing Bride 361189
Blushing Bromeliad 264409,266154
Blushing Philodendron 294805
Blushing Rhododendron 331408
Blushing Wild Buckwheat 152616
Blushing-bride 361189
Blush-red Bamboo 297448
Blushred Gentian 173828
Blush-red Maple 3209
Blushred Rabdosia 209811
Blush-red Rabdosia 209811
Blushwort 9417
Bluspine Craneknee 221645
Bly 338447
Blysmus 55889
Blyxa 55908
Buckley Centaury 80984
Bo 165553
Bo Thistle 92485
Bo Tree 165553
Boa Wood 132466
Boab 7028
Boadshape Spapweed 204887
Boadshape Touch-me-not 204887
Boar Thistle 91770,92485
Boar's Ear 314143
Boar's Ears 314143
Boar-thistle 91770,92485
Boat Arrowbamboo 162640
Boat Lily 394076
Boat Sterculia 376184
Boatcress 385129
Boat-lily 394076
Boats 2837
Boatshap Woodbetony 287138
Boatshaped Ant Palm 217925
Boat-shaped Bugseed 104815
Boat-shaped Haworthia 186370
Boat-shaped Skullcap 355617
Boatshapedbract Larkspur 124401
Boatshapedleaf Goldenray 229013
Boatshapedstalk Clematis 94879
Bob Ginger 309199
Bob Robert 174877
Bob Robin 248411
Bobberty 407072
Bobbies 174877,302034
Bobbin And Joan 37015
Bobbin Joan 37015,363461
Bobbin-and-joan 37015
Bobbins 37015,267294,302034
Bobby Hood 248411
Bobby Rose 397043
Bobby's Butions 68189,81020,81237,170193
Bobby's Buttons 31051,68189,81020,81237,170193
Bobby's Eye 407072

Bobby's Eyes 407072
Bobby-and-joan 37015
Bobbyns 53563
Bob-grass 60853
Bobrov Adonis 8340
Bobtail Barley 198278
Boccone's Sand-spurrey 370615
Boccone's Sandspurry 370615
Bock Adinandra 8209
Bock Elaeagnus 141947
Bock Greenbrier 366262
Bock Ironweed 406150
Bock Lilyturf 272057
Bock Rockfoil 349895
Bock Smilax 366262
Bock's Ironweed 406153
Bockwheat 162312
Bod 212636
Boddle 89704
Boddome Begonia 49654
Bodhi 165553
Bodinier Ampelopsis 20297
Bodinier Beautyberry 66743
Bodinier Benchfruit 279750
Bodinier Bittercress 72701
Bodinier Box 64243
Bodinier Buckthorn 328623
Bodinier Caper 71699
Bodinier Cinnamon 91277
Bodinier Clethra 96469
Bodinier Eargrass 187517
Bodinier Elsholtzia 143967
Bodinier Fairybells 134420
Bodinier Fleshspike 346936
Bodinier Hackberry 80739
Bodinier Handeliodendron 185263
Bodinier Hawksbeard 110756
Bodinier Hedyotis 187517
Bodinier Leaf-flower 296497
Bodinier Mahonia 242494
Bodinier Michelia 252825
Bodinier Monkeyflower 255193
Bodinier Pachysandra 279750
Bodinier Palmleaftree 185263
Bodinier Photinia 295643
Bodinier Puberulent Premna 313718
Bodinier Schefflera 350667
Bodinier Sedge 73893
Bodinier Spapweed 204823
Bodinier Starjasmine 393626
Bodinier Star-jasmine 393626
Bodinier's Saussurea 348617
Bodinier-like Beautyberry 66749
Bodinifer Pertybush 292045
Bodnant Viburnum 407648
Bodnier Lilyturf 272059
Boea 56044
Boed Living-rock Cactus 33220
Boedeker Cactus 244014
Boeica 56365,56367
Boer Bean 352485
Boer Beans 352485
Boerhavia 56397
Boerlagiodendron 56526
Boesenbergia 56532
Bog Adder's-mouth 185200
Bog Angelica 24429
Bog Arum 66593,66600
Bog Asphodel 262572,262574,262583
Bog Aster 40906,268684
Bog Bachelor's Buttons 219040
Bog Bean 250493,250502
Bog Bedstraw 170454
Bog Bell 22445
Bog Bilberry 404027
Bog Birch 53454,53594
Bog Black Spruce 298349
Bog Blueberry 404027
Bog Bluegrass 305803,306109
Bog Bulrush 352223
Bog Chickweed 374731,375036
Bog Cinquefoil 313087
Bog Clearweed 298917
Bog Cotton 152769,152791
Bog Crane's-bill 174799
Bog Daisy 68189
Bog Featherfoil 198680
Bog Goldenrod 368338,368454
Bog Hair-grass 126104
Bog Heather 150131
Bog Hop 250502
Bog Horn 292372,302091
Bog Hyacinth 273531
Bog Kalmia 215393,215405
Bog Laurel 215405
Bog Lobelia 234562
Bog Marsh Cress 336250
Bog Marshcress 336250
Bog Marsh-cress 336250
Bog Moss 246765
Bog Muhly 259713
Bog Myrtle 261162
Bog Nut 250502
Bog Orchid 185199,185200,302241
Bog Panicled Sedge 74304
Bog Pimpernel 21427
Bog Pondweed 312228
Bog Rattanpalm 65759
Bog Rhubarb 292372
Bog Rockfoil 349436
Bog Rorippa 336250
Bog Rosemary 22418,22436,22445
Bog Rush 212797,213476,213478,352348
Bog Sage 345449
Bog Sandwort 255582
Bog Smartweed 291958
Bog Speedwell 407349
Bog Spicebush 231443
Bog Spruce 298349
Bog St. John's-wort 394810
Bog Starwort 374731
Bog Stitchwort 374731
Bog Strawberry 100257
Bog Thistle 92285
Bog Trefoil 250502
Bog Twayblade 232229
Bog Valerian 404383
Bog Violet 299736,299759,410347
Bog White Violet 410166
Bog Whortleberry 404027
Bog Willow 343845,343964
Bog Willow Herb 146766
Bog Willow-herb 146766
Bog Wintergreen 322787,322922
Bog Yellow Cress 336250
Bog Yellow-cress 336250
Bog Yellow-eyed-grass 416045,416110
Boga Medaloa 385985
Bogang 95972
Bog-asphodel 262572
Bogbane 250502
Bogbean 250493,250502,314366
Bog-bean 250502
Bogbean Family 250492
Bogberry 100257,278366
Bogbutton 354438
Bog-cotton 152734
Bogdaisy 146042,146044
Bogdan Barley 198264
Bogeda Milkaster 169769
Bog-flower 72934
Boggart Posy 250614
Boggart-flower 250614
Boggy Spike Sedge 143271
Boggy Spike-sedge 143271
Bog-laurel 215405
Bog-mat 414684
Bog-moss Family 246767
Bog-myrtle 261120
Bog-myrtle Family 261243
Bogodosch Crazyweed 278747
Bogor Thin-walled Bamboo 47195
Bogorchid 158021
Bogorchid Jurinea 138892
Bogorchis 242993,243084
Bogota Kohleria 217741
Bogrose Orchid 32373
Bog-rosemary 22418,22436,22445
Bog-rosemary Andromeda 22445
Bogrush 352348
Bog-rush 352348,352408

Bogrushlike Crypsis 113316
Bogrushlike Pricklegrass 113316
Bog-sedge 75165
Bogspinks 72934
Bogwort 278366
Bohemia Henbane 201377
Bohemian Cranesbill 174503
Bohemian Knotweed 162508,162515,309940
Bohemian Serpent Root 354873
Bohemian Serpent-root 354873
Boholawn 359158
Bois Evergreen Chinkapin 78864
Bois Evergreenchinkapin 78864
Bois Pea-shrub 72191
Bois Syzygium 382490
Bois' Merremia 250760
Bois-chene 79256
Bois-jaune 232609
Boissier Barberry 51366
Boissier Iris 208467
Boissier Mandarin Orchid 295876
Boissier Rose 336402
Boissiera 56560
Bojers Spurge 159363
Bojiei Yushania 416754
Boke's Button Cactus 147423
Bokhara Clover 249203
Bokhara Vine 162509
Bola De Nieve 211032
Bolander Dandelion 293575
Bolander Sagebrush 35269
Bolander's Indian Pink 363553
Bolander's Knotweed 308913
Bolander's Ragwort 279890
Bolander's Sunflower 188929
Bolas 316470
Bolax 56633
Bolbonac 238371
Bolbostemma 56656
Boldingh's Dodder 114978
Boldo 293092
Bole Fritillary 168506
Boleweed 81237
Bolgan Leaves 221787
Boliaun 359158
Bolimonge 162312
Bolivia Barberry 51367
Bolivian Bark 91059
Bolivian Begonia 49664
Bolivian Black Walnut 212591
Bolivian Leaf 155067
Bolivian Panicgrass 281392
Bolivian Rubber 346360
Bollane-bane 36474
Bollas 316470
Bollea 56688
Bolleana Poplar 311208
Bolloms 316819
Bollwood 81237
Bolo Bolo 94107
Bolocephalus 56693
Boloflower 205546
Bolting Dewberry 338241
Boltonia 56700
Bolts 399504
Boluo Ormosia 274380
Bolus Loving-rock 304340
Bolwarra 160328,160330
Bomavist Bean 218721
Bombax 56770,56802
Bombax Family 56761
Bombay Aloe 10816
Bombay Arrowroot 114858
Bombay Black Wood 121730
Bombay Blackwood 121730,360481
Bombay Ebony 132137,132308
Bombay Hemp 112269
Bombay Mastic 300977
Bombay Mix 393940
Bombay Nepeta 264884
Bombay Rosewood 121730
Bombay Sumbul 136259
Bombay White Cedar 139663
Bomi Azalea 331519
Bomi Barberry 51691
Bomi Bastardtoadflax 389619
Bomi Birch 53366
Bomi Bluegrass 305405
Bomi Bulbophyllum 62590
Bomi Corydalis 106291
Bomi Curlylip-orchis 62590
Bomi Deutzia 126850
Bomi Gentian 173300
Bomi Hundredstamen 389619
Bomi Larkspur 124505
Bomi Mahonia 242597
Bomi Milkvetch 42090
Bomi Monkshood 5494
Bomi Primrose 314173
Bomi Rattlesnake Plantain 179584
Bomi Rhododendron 331519
Bomi Rockfoil 349443
Bomi Rockjasmine 23254
Bomi Sagebrush 36359
Bomi Sandwort 31774
Bomi Sedge 75854
Bomi Sickleleaf Primrose 314361
Bomi Spicebush 231412
Bomi Spice-bush 231412
Bomi Spotleaf-orchis 179584
Bomi Willow 343907
Bomi Windhairdaisy 348169
Bomi Woodbetony 287050
Bomi Woodrush 238580
Bon Banbury Garcinia 171068
Bon Clovershrub 70790
Bon Crucianvine 356268
Bon Pepper 300361
Bon Sauropus 348045
Bon Secamone 356268
Bon Xerospermum 415520
Bon's Machilus 240558
Bona Conyza 103438
Bonat Cliffbean 66617
Bonat Clovershrub 70788
Bonat Millettia 66617
Bonat Yunnan Felwort 235446
Bonavist 218721
Bonduc 64971
Bonduc Nut 64971
Bonduc Seed 64973
Bone-bract 354336
Boneedge Cystopetal 320211
Bone-flower 50825,177053
Boneseed 89394
Boneset 158021,158250,159911,381035
Boneweed 197486,197491
Bonewort 410187
Bonfire 248411
Bongard Peashrub 72193
Bongard Pea-shrub 72193
Bongassi 134776
Bongay 9701
Bongosei 236336
Boni Rattan Palm 65651
Boni Seed 203064
Bonin Azalea 330233
Bonin Island Juniper 213864
Bonin Isles Juniper 213864
Bonin Pandanus 280993
Bonin Rhododendron 330233
Bonin Spindle-tree 157343
Boniodendron 56970
Bonker Hedgehog 140220
Bonnace Tree 122621
Bonnafous Eight Trasure 200782
Bonnafous Stonecrop 200782
Bonnet Bellflower 98273,98290
Bonnet Strings 12323
Bonnet Tanoak 233119
Bonnets 30081
Bonnet-strings 12323
Bonny Bird Een 314366
Bonny Bird's Eye 72934
Bonny Bird's Eyes 72934
Bonny Rabbits 28617
Bonpland's Croton 112845
Bonsai Crassula 109353
Bonsamdua 134776
Bontaalwyn 16866
Bonvalot Azalea 330234

Bonvalot Larkspur 124507
Bonvalot Potanin Larkspur 124507
Bonvalot Rhododendron 330234
Bonwort 50825
Bonyberry 276533
Bony-berry 276533
Bonytip Fleabane 150724
Booag 336419
Boobialla 260717
Boobtalla 260706
Boobyalla 260706
Boodle 89704
Booin 359158
Boojum Tree 167557,203445
Boojum-tree 167557
Book Leaf 63805,201732
Bookara Gum 155574
Bookleaf Mallee 155616
Boomerang Wattle 1042
Boon Alstonia 18033
Boon Tree 345631
Boonaree 14369
Boonery 14369
Boonery Boonaree 14369
Boon-tree 345631
Boor's Mustard 390213
Boot's Rattlesnakeroot 313801
Booth Eastern Arborvitae 390590
Booth Epidendrum 146386
Booth's Rhododendron 330235
Boothill Eupatorium 101958
Boothriospermumlike Microula 254339
Bootlace Oak 184618
Bootlace Plant 299306
Bootlace Tree 184618
Bootry 345631
Boots 68189,347156
Boots-and-shoes 5442,28617,30081,120310,167955,237539
Boots-and-stockings 302103
Boott's Goldenrod 367968
Booyong 192477
Bopple Nut 240212
Bopplefnuts 240207
Bor Corydalis 105657
Borage 57095,57101
Borage Family 57084
Borassodendron 57114,57115
Borassus 57116
Borbor Oak 323635
Bordekin Plum 304197
Borden Littlespike Willow 343692
Border Forsythia 167435
Border Milkvetch 42009
Border Palo Verde 83747
Border Penstemon 289345
Border Privet 229558,229562
Border Sweetvetch 187984
Border White Pine 300210
Bordered Sinocrassula 218355
Borderhair Woodbetony 287118
Bordering 235090
Borderline Summerlilic 62108
Bordzilowski Sedge 73903
Bore's Ear 314143
Bore's Ears 314143
Boreal Bog Orchid 183571,230348
Boreal Bog Sedge 75281,75282
Boreal Broadlily 97091
Boreal Catchfly 364069
Boreal Chickweed 82739
Boreal Fleabane 150508
Boreal Manna-grass 177576
Boreal Panic-grass 281394
Boreal Sage 35211
Boreal Small Reed 127199
Boreal Starwort 374767
Boreal Stitchwort 255573
Boreau's Fumitory 169111,169112
Borecole 59524
Borecole Cole 59493
Boree 1465
Bores 31051
Bore-tree 345631,355202
Boriti Poles 329765
Borlotti Bean 294056
Borlotti Beans 294056
Bornean Rosewood 248529
Borneo Camphor 138547,138548
Borneo Cedar 362212
Borneo Climbing Bamboo 131287
Borneo Coelogyne 98615
Borneo Ironwood 160967,160968
Borneo Jasmine 211846
Borneo Mahogany 67860
Borneo Rubber 414418
Borneo Scrambling Bamboo 131287
Borneo Teak 207015
Borneo Wintergreen 172059
Bornet Cowlily 267280
Bornmueller Autumn-crocus 99310
Bornmueller Cinquefoil 312427
Bornmueller Fir 299
Bornmueller Maple 2806
Borod Columbine 30003
Borodin Euphorbia 158560
Borodin Milkvetch 42091
Borodin Pricklythrift 2220
Borodin Prickly-thrift 2220
Boronia 57257,57261,57262,57263,57273
Borotal Spunsilksage 360820
Borral 345631
Borrer's Saltmarsh Grass 321275
Borreria 57289
Borr-tree 345594
Borsczowia 57386
Borsig Phoenicophorium 295439
Borthwickia 57389
Bortree 345631
Bor-tree 345631
Bory's Orchid 273343
Borzi Cactus 32361
Bosc's Panic-grass 281398
Boschan Barberry 51941
Boschniakia 57429
Bose Maple 2807
Bose Premna 313709
Bose Willow 343117
Bosheng Elatostema 142618
Bosheng Stairweed 142618
Bosian Pine 300006
Bosistoa's Box 155504
Bosnia Iris 208468
Bosnian Maple 3425
Bosnian Pine 299968,300036
Bosqueia 57472
Bossell 89704
Bostock's Montia 258240
Boston Ivy 20455,285157,285159
Bostrychanthera 57503
Boswell 89704
Botany Bay 199956
Botcher's Blood 248411
Botel Tobago Cinnamon Tree 91357
Botham 89704
Bothen 89704,322724
Botherum 89704,407072,407144
Bothery-tree 345631
Bothriochloa 57545
Bothriospermum 57675
Bothrocaryum 57693
Botle Tobago Endiandra 145316
Botree 165553
Bo-tree 165553
Botree Fig 165553
Botryoidal Grepe-hyacinth 260301
Botterboom 400786
Bottery 345631
Bottle 89704,114245
Bottle Brush 67248,196818,248072,248104,370522
Bottle Crazyweed 278705
Bottle Gentian 173222,173998
Bottle Gourd 219821,219827,219843
Bottle Palm 49343,49345,201356,201359,201360
Bottle Plant 186156
Bottle Pleione 304273
Bottle Sedge 76068
Bottle Stopper 152160
Bottle Tree 7028,58334,376063,376102,

376205
Bottlebrush 67248,67253,67267,248114
Bottle-brush 67248,196818,248072,248104,252607,370522
Bottlebrush Buckeye 9719
Bottle-brush Buckeye 9719
Bottlebrush Bulrush 210012
Bottlebrush Grass 144340,203124
Bottle-brush Grass 203124
Bottlebrush Sedge 74862
Bottlebrush-grass 203124
Bottle-of-all-sorts 81020,321643
Bottlepitted Sweetleaf 381450
Bottles-of-wine 220729
Bottletree 166627
Bottle-tree 58334,376063
Bottom Shellbark 77905
Bottom White Pine 299954
Bottomland Aster 380939
Botton Oak 324354
Bottry 345631
Boudier Swallowwort 117408
Bougainvillea 57852,57857,57868
Bougainvillea Goldraintree 217613
Bougainvillea Goldraintree Tree 217613
Bough Elm 401512
Bouin 359158
Boulevard Cypress 85345,85347
Boulevard Falsecypress 85347
Bouncing Bess 81768,404238
Bouncing Bet 346433,346434,346436
Bouncing Betty 81768,346434
Bouncingbet 346434
Bouncing-bet 346434
Boundary Ephedra 146143
Boundary Goldenbush 150312
Boundary Mark 104367
Boundry Ephedra 146230
Bountree 345631
Boun-tree 345631
Bouquet 93376
Bouquet Aster 160672
Bouquet Larkspur 124237
Bour Tree 345631
Bourbon 417417
Bourbon Cotton 179924
Bourbon Dombeya 135957
Bourbon Lily 229789
Bourbon Ochrosia 268354
Bourbon Palm 201359,222626,234170
Bourbon Rose 336403
Bourbon Tea 24860
Bourbon Tea Orchid 24687
Bourbon Tea-orchid 24687
Bourgeau's Nutrush 354026
Bourgeau's Pepper-weed 225442
Bourne Confused Mahonia 242538
Bourne Honeysuckle 235681
Bourne Phoebe 295349
Bournea 57898
Bournemouth Pine 300146
Bourrier 31056
Bourtree 345631
Bousigonia 57911
Bout 68189
Boutrey 345631
Bouvardia 57948
Bovisand Sailor 81768
Bovisand Sailors 81768
Bow Bells 23925
Bow Kail 59532
Bow Thistle 92485
Bow Wood 240828
Bowedconical Rabdosia 209761
Bowed-conical Rabdosia 209761
Bowedinflorescense Conehead 320112
Bowedseed Mango 244392
Bowedsepal Spiraea 371910
Bowed-sepaled Spiraea 371910
Bowedstalk Greenbrier 366231,366544
Bowedstalk Smilax 366231,366544
Bowel-hive 13966
Bowel-hive Grass 13966
Bowen 359158
Bower Actinidia 6515
Bower Heteropappus 193927
Bower Jurinea 193927
Bower Kiwifruit 6515
Bower Plant 281225
Bower Vine 281225
Bower Wattle 1139
Bowerplant of Australia 281225
Bowfruit Rhododendron 330311
Bow-fruited Rhododendron 330311
Bowfruitvine 393525,393549
Bowhead Cock 71919
Bowie's Oxalis 277702
Bowie's Wood-sorrel 277702
Bow-kail 59532
Bowlbract Sawwort 361015
Bowles' Golden Sedge 74436
Bowles' Mint 250469
Bowley Sage 344910
Bowlocks 36474,359158
Bowman Root 407434
Bowman's Root 29473,175760,175766
Bowmann Tuftroot 130097
Bowmans Root 175766
Bown-tree 345631
Bowring Begonia 49986
Bowring Cattleya 79530
Bowringia 58005,58006
Bowstalk Corydalis 105925
Bowstring Hemp 68088,346047,346094,346100,346117,346173
Bowstringhemp 346047
Bowstring-hemp 346047
Bow-weed 81237
Bowwingpaesley 31325,31328
Bowwood 240806,240828
Bow-wood 81237,81356,240828
Bowyer's Mustard 225447,390213
Box 64235,155468
Box Blueberry 403944
Box Elder 3218,3228,3233
Box Elder Maple 3218
Box Family 64224
Box Hard-leaf 296172
Box Holly 340990
Box Honeysuckle 235984
Box Huckleberry 172253
Box Knotweed 308824
Box Pussytoes 26331
Box Thorn 64064,238996
Box Tree 64235
Boxberry 172135
Box-berry 172135
Boxelder 3218
Boxelder Maple 3218
Box-elder Maple 3218
Boxfruitvine 103362,250749
Boxifoliateoid Syzygium 382498
Boxing Glove Cholla 116657
Boxing Gloves 237539
Boxleaf Agapetes 10296
Boxleaf Atalantia 43721
Boxleaf Azara 45955
Box-leaf Azara 45954,45955
Boxleaf Basketvine 9431
Boxleaf Camellia 68960
Boxleaf Coronilla 105275
Boxleaf Dendrotrophe 125782
Boxleaf Distylium 134919
Boxleaf Eugenia 382499,382619
Boxleaf Hebe 186940
Boxleaf Leptodermis 226078
Boxleaf Mosquitoman 134919
Boxleaf Parasiticvine 125782
Box-leaf Pittosporum 301311
Boxleaf Scolopia 354597
Box-leaf Seatung 301311
Boxleaf Syzygium 382499
Boxleaf Tea 68960
Box-leaf Wattle 1104
Boxleaf Wildclove 226078
Box-leaved Agapetes 10296
Box-leaved Atalantia 43721
Box-leaved Barberry 51401
Box-leaved Basketvine 9431
Box-leaved Camellia 68960
Box-leaved Coronilla 105275

Boxleaved Cotoneaster 107370
Box-leaved Cotoneaster 107370
Box-leaved Dendrotrophe 125782
Box-leaved Distylium 134919
Box-leaved Eugenia 382499
Box-leaved Holly 203670
Box-leaved Honeysuckle 236035
Box-leaved Milkwort 307983
Box-leaved Pittosporum 301311
Box-leaved Scolopia 354597
Box-leaved Syzygium 382499
Box-like Holly 203606
Box-of-matches 2837
Boxoi Crazyweed 278736
Box-orange 362032
Boxthorn 238996,239003,239092
Box-thorn 238996,239011,239046
Box-thorn Barberry 51891
Box-thorn-like Barberry 51890
Boxton Gum 155547
Boxweed 6888
Boxwood 64235,64345,105023,155468
Box-wood 64235
Boxwood Family 64224
Box-wood Family 64224
Boxwood Knotweed 308824
Boy's Bacca 95431,192358
Boy's Love 35088
Boy's Mercury 250600
Boy's-love 35088
Boyce-Thompson Hedgehog 140221
Boyd 338447
Boyd's Pearlwort 342236
Boykinia 58015
Boynton Oak 323716
Boys-and-girls 250600,315082,315101
Boysenberry 338761
Boyton Thistle 92454
Bozen 89704
Bozzel 89704
Bozzom 89704,227533
Braam Butea 64145
Brabantica Poplar 311256
Brabtree 57122
Bracaatinga 255111
Brace's Aster 381001
Bracelet Honey Myrtle 248076
Brachanthemum 58038
Brachiate Olive 270066
Brachiton 58335
Brachy Porandra 311651
Brachyactis 58261
Brachybotrys 58299
Brachychiton 58334,376102
Brachycorythis 58373
Brachyfruit Beadcress 264604
Brachyfruit Torularia 264604
Brachystelma 58813
Brachystemma 58965
Brachytome 59002
Bracket Plant 88553
Bracketplant 88527,88546
Bracket-plant 88546
Brackfollicle Tylophora 400856
Brackish Water Crowfoot 325644
Brackish-water Crowfoot 325644
Bracksquamose Everlasting 21519
Bracksquamose Pearleverlasting 21519
Bract Amplexicaul Draba 136930
Bract Amplexicaul Whitlowgrass 136930
Bract Furnished Rhododendron 330251
Bract Hyparrhenia 201470
Bract Milkweed 37846
Bract Pholidota 295523
Bractacious Elatostema 142625
Bractacious Stairweed 142625
Bractbean 362292,362298
Bract-bearing Clearweed 298878
Bractbearing Erysimum 154563
Bractbearing Mulgedium 259752
Bractbearing Sugarmustard 154563
Bracteate Acanthochlamys 2158
Bracteate Alumroot 194415
Bracteate Barberry 51378
Bracteate Basketvine 9427
Bracteate Beadcress 264605
Bracteate Beautyleaf 67854
Bracteate Chelonopsis 86809
Bracteate Coeloglossum 98586
Bracteate Coleus 99530
Bracteate Corydalis 105663
Bracteate Eria 148639
Bracteate Garcinia 171069
Bracteate Goatwheat 44251
Bracteate Hyptianthera 203029
Bracteate Iris 208470
Bracteate Jerusalemsage 295074
Bracteate Leea 223930
Bracteate Macaranga 240239
Bracteate Machilus 240723
Bracteate Morina 258836
Bracteate Mycetia 260615
Bracteate Nepeta 264890
Bracteate Onosma 271741
Bracteate Orach 44644
Bracteate Premna 313605
Bracteate Pricklythrift 2221
Bracteate Primrose 314180
Bracteate Rattlebox 111968
Bracteate Rush 212923
Bracteate Spicate Woodbetony 287690
Bracteate Thistle 91814
Bracteate Torularia 264605
Bracteate Trigonotis 397384
Bracted Balsam Fir 287
Bracted Elaeocarpus 142292
Bracted Fir 453
Bracted Indigo 205739
Bracted Lousewort 287056
Bracted Plantain 301868
Bracted Poppy 282522
Bracted Races Blueberry 403738
Bracted Spapweed 204866
Bracted Spiderwort 394001
Bracted Strawflower 415386
Bracted Tick-trefoil 126301
Bracted Touch-me-not 204866
Bracted Utah Fleabane 150970
Bracted Viburnum 407709
Bracteolate Fissistigma 166652
Bracteolate Mahonia 242497
Bracteole Chosenia 89160
Bracteose Gugertree 350931
Bracteose Monkshood 5073
Bracteose Verbena 405817
Bracteous Ainsliaea 12685
Bracteous Glorybower 95968
Bracteous Rabbiten-wind 12685
Bractgoosefoot 47680,47681
Bractleaf Azalea 330251
Bractleaf Gentian 173614
Bractleaf Milklettuce 259752
Bractleaf Rabdosia 209779
Bractleaf Rhubarb 329311
Bractleaf Saussurea 348596
Bractleaf Stonecrop 356530
Bractleaf Thistle 92473
Bractleaf Windhairdaisy 348596
Bractleass Arrowbamboo 162677
Bractless Bastardtoadflax 389677
Bractless Cattail 401120
Bractless Chickweed 374872
Bractless Dilophia 130958
Bractless Elatostema 142653
Bractless Lady's Slipper 120302
Bractless Ladyslipper 120302
Bractless Parsnip-flower Wild Buckwheat 152140
Bractless Stairweed 142653
Bractless Stephania 375846
Bractlet grass 389318
Bract-persisted Sweetleaf 381422
Bractpersistent Sweetleaf 381422
Bractquitch 201454,201470
Bracts Tubergourd 390165
Bractscale 44644
Bradbract Emei Sage 345275
Bradbury Beebalm 257158
Brade Begonia 49675
Bradford Pear 323116
Brady's Hedgehog Cactus 287882

Bragge 235373
Brahea Palm 59095
Brahmi 46362
Brain Cactus 140566,244062,375494,375495
Brain Umbellate Barberry 52290
Brainerd Hawthorn 109561
Bramble 338059,338265,338447,338608,338945,339419
Bramble Acacia 1690
Bramble Breer 338447
Bramble Vetch 408624
Bramble Wattle 1690
Brambleberries 338447
Bramblekite 338447
Brambles 338059
Brammelkite 338447
Brammle 338447
Bran Zamia 417010
Branch Hookspur Corydalis 105957
Branch Reactiongrass 54539
Branch Thorn 151078
Branched British Inula 207065
Branched Broomrape 275184
Branched Bur-reed 370028,370050
Branched Cushion Wild Buckwheat 152363
Branched Fenugreek 397275
Branched Gayfeather 228457
Branched Goldenweed 271916
Branched Leymus 228380
Branched Panicled Aster 380986
Branched Plantain 301858,301864
Branched Porterweed 373494
Branched Rockfoil 349162
Branched Sawwort 361004
Branched Spodiopogon 372425
Branched St. Bernard's Lily 27320
Branched Wheat 398839,399003
Branched-setose Melastoma 248747
Branchhair Azalea 331604
Branchhair Melastoma 248747
Branching Asphodel 39454
Branching Blue-eyed Grass 365700
Branching Centaury 81517
Branching Draba 137202
Branching Fleabane 150925
Branching Larkspur 102833,124141
Branching Montia 258247
Branching Palm 202321
Branching Rush 213237
Branching Toadflex 231091
Branching Whitlowgrass 137202
Branchlet Craneknee 221726
Branchlet Flower Vetch 408563
Branchy Ainsliaea 12717
Branchy Ajania 13027
Branchy Apostasia 29791
Branchy Azalea 331512
Branchy Bastardtoadflax 389840
Branchy Bistamengrass 127479
Branchy Cointree 280555
Branchy Crazyweed 279117
Branchy Flatsedge 119410
Branchy Gentian 173459
Branchy Greenorchid 137635
Branchy Leek 15649
Branchy Marestail 196819
Branchy Mazus 247025
Branchy Milkvetch 42893
Branchy Onion 15649
Branchy Paliurus 280555
Branchy Pink 127808
Branchy Pycreus 119410
Branchy Rabbiten-wind 12717
Branchy Rugged Corydalis 106410
Branchy Sawwort 361004
Branchy Shadegrass 352871
Branchy Stairweed 142812
Branchy Tamarisk 383595
Branchy Yushanbamboo 416803
Branchy-eared Wheat 398839,399003
Brandegee Hesper Palm 59097
Brandegee's claytonia 94359
Brandegee's Rubberweed 201309
Brandegee's Siltbush 418478
Brandegee's Wild Buckwheat 151866
Brandis Agapetes 10291
Brandis Barberry 51381
Brandis Dendrocalamus 125469
Brandis Dragonbamboo 125469
Brandisia 59121,59127
Brandy Borrle 267274
Brandy Bottle 15861,267294,336419
Brandy Bottles 15861,267294,336343
Brandy Mazzard 83155
Brandy Mint 250420,250450
Brandy See 411979
Brandy Snap 231183,374916
Brandy Snaps 374916
Brandy-bottle 267294
Brank 162312
Brank Ursine 2695
Brank-Ursin 2695
Brannel Kite 338447
Brarted Green Orchid 98586
Brasenia 59144
Brashlach 364557
Brashlagh 364557
Brasil Parodia 284659
Brasiletto 64965,65060,184518,288882
Brasilian Araucaria 30831
Brass Buttons 107747,107766
Brassaiopsis 59187
Brassavol Epidendrum 146388
Brassavola 59254
Brassbuttons 107747
Brass-buttons 107766
Brass-coloured Dropseedgrass 372576
Brassica 59278
Brassica Broomrape 274972
Brassicalike Chirita 87832
Brassicalike Goldenray 228997
Brassick 364557
Brassock 364557
Bratiful Fir 280
Braue River Sage 227922
Braun Bedstraw 170282
Brauna 248570
Brauns Argyroderma 32696
Brave Bassinet 68189
Brave Bassinets 68189
Brave Celandine 68189
Braveness Azalea 330651
Bravistyle Elaeagnus 141980
Brawlins 31142,404051
Braya 59727
Brazil Arrowroot 207623,244507
Brazil Arundinella 37360
Brazil Barbados Cotton 179885
Brazil Bashfulgrass 255053
Brazil Begonia 49676
Brazil Bougainvillea 57868
Brazil Calaba 67855
Brazil Chastetree 411341
Brazil Cherry 156118,156148,156292,156385
Brazil Cholla 272818
Brazil Copai 200844
Brazil Cress 4866
Brazil Eugenia 156148
Brazil Groundsel 358432
Brazil Jacaranda 211226,211228
Brazil Jointfir 178552
Brazil Large-leaved Bamboo 181475
Brazil Leathern Barberry 51498
Brazil Melocactus 249597
Brazil Mimosa 255053
Brazil Nut 53057
Brazil Nutmeg 113473,261409
Brazil Pepper 350980,350995
Brazil Peppertree 350995
Brazil Pepper-tree 350995
Brazil Raintree 61304
Brazil Redwood 61062,65002
Brazil Rosewood 121607
Brazil Rubber Tree 194453
Brazil Rubbertree 194453
Brazil Satin-wood 161190,161191
Brazil Tea 204135
Brazil Trithrinax Palm 398826
Brazil Tulipwood 121663

Brazil Wood 65002,65060
Braziletto 65002
Braziletto Wood 288897
Brazilian Abutilon 957
Brazilian Angel's Trumpets 61242
Brazilian Araucaria 30831
Brazilian Arrowroot 71179,207623,244507
Brazilian Bachelor Button 81785
Brazilian Bamboo 250746
Brazilian Beefwood 79165
Brazilian Begonia 49924
Brazilian candles 286663
Brazilian Cashew 21195
Brazilian Catsear 202394
Brazilian Cedar 80030
Brazilian Cherry 156385
Brazilian Cocoa 285930
Brazilian Copal 200844
Brazilian Copperlily 184255
Brazilian Coral Tree 154656,154721
Brazilian Cress 4866
Brazilian Dwarf Morning-glory 161439
Brazilian Edelweiss 327324,364746
Brazilian Elemi 316023
Brazilian Firecracker 244364
Brazilian Firecracker Vine 244364
Brazilian Firetree 351732
Brazilian Flame Bush 66689
Brazilian Floss Silk Tree 88973
Brazilian Ginger 128997
Brazilian Glorybush 391577
Brazilian Glorytree 391569
Brazilian Gloxinia 364754
Brazilian Golden Vine 376554
Brazilian Guava 318734,318744
Brazilian Gunnera 181951
Brazilian Ironwood 65002,65008
Brazilian Jalap 250850
Brazilian Jasmine 211825
Brazilian Joyweed 18097
Brazilian Lucerne 379245
Brazilian Mahogany 380532
Brazilian Maximiliana 246733
Brazilian Milktree 255312
Brazilian Morning-glory 208181
Brazilian Myrocarpus 261527
Brazilian Nightshade 367596
Brazilian Nutmeg 113473
Brazilian Oak 311992
Brazilian Orchid tree 49093
Brazilian Orchid-tree 49093
Brazilian Parrot Feather 261342
Brazilian Parrot's Feather 261342
Brazilian Parrot-feather 261342
Brazilian Pepper 350995
Brazilian Pepper Tree 350995
Brazilian Peppertree 350995,350997
Brazilian Pepper-tree 350995
Brazilian Pine 30831
Brazilian Pitcairnia 301111
Brazilian Plume 214410
Brazilian Plume Flower 214410
Brazilian Potato Tree 367749
Brazilian Pricklypear 272818
Brazilian Red Cloak 247930,247931
Brazilian Redwood 61062,64971,64975,65002
Brazilian Rhatany 218076
Brazilian Rosewood 121770,211239
Brazilian Sassafras 25196,268700
Brazilian Satintail 205482
Brazilian Satinwood 161190,161191
Brazilian Sky Flower 139075
Brazilian Snapdragon 276901
Brazilian Tulipwood 121663,121683
Brazilian Umbrella Plant 181951
Brazilian Verbena 405812
Brazilian Vervain 405819
Brazilian Water Weed 141569
Brazilian Waterhyssop 46356
Brazilian Water-meal 414656
Brazilian Waterweed 141569
Brazilian Wax Palm 103737
Brazilian Wild Petunia 339709
Brazillian Galingale 118447
Brazil-nut 53055,53057
Brazil-nut Tree 53055,53057
Brazilredwood 65002
Brazilwood 61062,64971,64975,65002,65060
Brazil-wood 64975,65002
Brazzock 364557
Bread Fruit 36913
Bread Fruit Tree 36902,36913
Bread Palm 115794
Bread Wheat 398839
Bread-and-butter 72934,231183,312360
Bread-and-cheese 231183,237539,243840,301162,312360,336419,339887
Bread-and-cheese Tree 109857,336419
Bread-and-cheese-and-cider 23925,243840,277648
Bread-and-cheese-and-kisses 243840
Bread-and-cider 109857
Bread-and-marmalade 364557
Bread-and-milk 72934,277648
Breadfruit 36913,36920,258168,281089
Bread-fruit 36902,36913
Bread-fruit Tree 36913
Breadfruit-tree 61053
Bread-leaved Hedge Mustard 365503
Breadnut 36913,61053,287932
Breadnut Tree 61053
Breadroot 287932,319178
Breadroot Scurf-pea 287932
Break Jack 374897,374916
Break-basin 407072
Breakbones 374916
Breaking Spikes Rhododendron 331872
Breakingfruit Chinquapin 78936
Breakingfruit Evergreenchinkapin 78936
Breaking-fruited Evergreen Chinkapin 78936
Break-Jack 374897,374916
Breakstone 299523,342279
Breakstone Parsley 13966
Break-your-mother's-heart 28044,101852
Bream 121001
Breastwort 192941
Breath-of-heaven 131966
Breath of Heaven 131966
Breck Speedwell 407297
Breckland Catchfly 363855
Breckland Mugwort 35237
Breckland Thyme 391365
Breckland Wild Thyme 391365
Brecldand Speedwell 407297
Brecteate Acanthochlamys 2157
Bredia 59822
Breed Calf's Snout 28617
Breeders Gladioli 176222
Breeders Gladiolus 176222
Breekhout 13432
Breem 121001
Breer 336419
Brennende Liebe 363312
Brere 336419
Breslinge Strawberry 167674
Breton Fescue 163810
Bretschneider Barberry 51303
Bretschneider Dogwood 380434
Bretschneider Pear 323114
Bretschneidera 59965
Bretschneidera Family 59968
Brevicaulinary Ardisia 31371
Breviligulate Sicklebamboo 137844
Brevilimbate Monkshood 5081
Breviped Beadcress 264607
Breviped Pachypterygium 279704
Breviped Torularia 264607
Brevipetalous Monkshood 5082
Breviracemose Neillia 263146
Brevistylar Windflower 23735
Brewer Oak 323931
Brewer Pine 298265
Brewer Spruce 298265
Brewer's Aster 155803
Brewer's Fleabane 150516
Brewer's Hop 199384
Brewer's Mountain-heather 297018
Brewer's Ragwort 279892
Brewer's Saltbrush 44495

Brewer's Spineflower 89061
Brewer's Spruce 298265
Brewer's Thistle 91918
Brewer's Weeping Spruce 298265
Breynia 60049
Briancon Apricot 34430
Briar 149017,336419
Briar Balls 336419
Briar Root 149017
Briar Rose 336419,336892
Briar-bush 336419
Briar-rose 336419
Briar-tree 336419
Bribie Island Pine 67414
Brickberry Cotoneaster 107710
Brick-budded Rhododendron 330848
Brickellbush Aster 155821
Brick-red Azalea 331391
Brid Breer 336364,336851
Brid Rose 336851
Bridal Bouquet 311657
Bridal Broom 328237
Bridal Flower 245803
Bridal Heath 149069
Bridal Veil 175488
Bridal Wreath 70247,167733,167735,
245803,371785,371815,372050,372126
Bridal Wreath Spiraea 372050
Bridal-spray 371815
Bridal-veil-grass 376732
Bridalwreath 371785,372050
Bridal-wreath 371785,372126
Bridal-wreath Spiraea 372050
Bridalwreath Spirea 372050,372126
Bridal-wreath Spirea 372050
Bride's Feathers 37060,37061
Bride's-feathers 37060,37062
Bridelia 60162,60167
Bridelialeaf Litse 233930
Brideweed 231183
Bridewort 166125,231183,371785,371791,
372075
Bridewort Spiraea 372075
Bridges' Catchfly 363262
Bridges' Triteleia 398781
Bridget in Her Bravery 363312
Brier 338447
Brier Bush 336419
Brier Grape 411646
Brier Rose 339194,339195
Brier-berry 338307
Brigalow 1184,1274
Briggsiopsis 60289,60290
Brigham Pleurothallis 304907
Brigham Tea 146262,146265
Brigham Young Tea 146262,146265
Bright 325820
Bright Arrowbamboo 162718
Bright Balanophora 46876
Bright Birch 53510
Bright Dischidia 134145
Bright Eye 325820,407072
Bright Eyes 407072
Bright Green 182177
Bright Green Spike-rush 143155
Bright Meadow 68189
Bright Primrose 314210
Bright Puccinellia 321331
Bright Rattlebox 112405
Bright Snakemushroom 46876
Bright Sweetvetch 188129
Bright Swertia 380358
Bright Wormwood 35519
Bright Yellow Iris 208507
Bright Yellow-leaved Sycamore Maple 3470
Bright Yushanbamboo 416785
Brightbead Cotoneaster 107462
Bright-bead Cotoneaster 107462
Brightberry Cotoneaster 107714
Bright-berry Cotoneaster 107714
Bright-blue Speedwell 407362
Brightbract Ajania 13015
Brightbract Dandelion 384618
Bright-colored Pricklyash 417250
Bright-colored Prickly-ash 417250
Brightcolored Sedge 75036
Bright-coloured Blueberry 403880
Bright-coloured Sonerila 368887
Brighteye 325820,407072
Brightflower Goosecomb 335399
Brightfruit Balang Willow 344134
Bright-green Italian Cypress 114763
Bright-green Rhododendron 331037
Brighthair Azalea 331243
Brighthair Ormosia 274436
Brighthair Violet 410186
Brighthead Elatostema 142712
Brighthead Stairweed 142712
Brightleaf Anemone 23944
Brightleaf Arrowbamboo 162694
Brightleaf Aster 40488
Brightleaf Azalea 332055
Brightleaf Beech 162390
Brightleaf Bladderwort 403884
Brightleaf Forrest Larkspur 124342
Brightleaf Gentian 173634
Brightleaf Glycosmis 177839
Brightleaf Heteropanax 193912
Brightleaf Michelia 252879
Brightleaf Parakmeria 283421
Brightleaf Pondweed 312168
Brightleaf Primrose 314486
Brightleaf Qinggang 116177
Brightleaf Ramie 56192
Brightleaf Stairweed 142704
Brightleaf Willow 343833
Bright-leaved Pondweed 312034
Bright-leaved Yushania 416782
Bright-red Agapetes 10343
Brightred Azalea 330054
Brightred Bauhinia 49053
Bright-red Dandelion 384772
Brightred Fireflower 246793
Brightred Mayodendron 246793
Bright-red Mayodendron 246793
Brightridge Goosecomb 335402
Brightviolet Skullcap 355554
Brightwhite 313776
Bright-yellow Eastern Arborvitae 390610
Brightyellow Goldenray 228998
Brigssia 60251
Brihywine 95431
Brilliant Azalea 331636
Brilliant Campion 363472
Brilliant Lily 230036
Brilliant Lychnis 363472
Brilliant Sagebrush 36306
Brilliant Star 215102
Brilliant Sunset 188757
Brilliant-leaf Washingtonia 413307
Brimbel 338447
Brimble 338447
Brimmle 336419,338447
Brimstone 247366
Brimstone-wood 386563
Brimstonewort 292957,292966
Bringal 367370
Brinjal 367370
Brinjaul 367370
Brinton Root 407434
Briny 53304
Brisbane 301207
Brisbane Box 236462
Brisbane Lily 315499,315501
Brisbane Tristania 236462
Brisbanebox Tristania 236462
Brisewort 50825
Bristcone Pine 299795
Bristl Pennisetum 289205
Bristl Wolftailgrass 289205
Bristle Bent 12073
Bristle Cherry 83322
Bristle Cinquefoil 312399
Bristle Club-rush 210080
Bristle Cone Fir 507
Bristle Dahurian Rose 336526
Bristle Elatostema 142836
Bristle Goldenray 228974
Bristle Grass 361680
Bristle Hairgrass 126104
Bristle Maxillaria 246726

Bristle Meconopsis 247108
Bristle Nailwort 284898
Bristle Oat 45602
Bristle Philodendron 294844
Bristle Rhododendron 330203
Bristle Sedge 75361
Bristle Stairweed 142836
Bristle Sweetvetch 188100
Bristle Tamarisk 383517
Bristle Tubergourd 390180
Bristlebearing 143146
Bristle-bearing Rhododendron 331808
Bristlecone Fir 306,507
Bristlecone Pine 299795
Bristle-cone Pine 299795,300040
Bristled Umbrella Sedge 119576
Bristlefruit Hedgeparsley 392993
Bristlegrass 361680
Bristle-grass 361680
Bristlehead 77240
Bristleleaf Bulrush 210080
Bristleleaf Chaffhead 77233
Bristleleaf Lovegrass 147573
Bristle-leaf Sedge 74421
Bristleless Bogrush 352409
Bristleless Dark-green Bulrush 353434
Bristleless Spikesedge 143178
Bristle-pointed Iris 208819
Bristle-pointed Oat 45602
Bristle-seed Sand-spurrey 370627
Bristleseed Sandspurry 370627
Bristle-spine Hamatocactus 185158
Bristle-stalked Sedge 75120
Bristle-stem Hemp Nettle 170078
Bristlestem Hempnettle 170078
Bristle-stem Hemp-nettle 170078
Bristlethistle 73278
Bristle-thistle 73278
Bristletips 278602
Bristleweed 186881
Bristly Aralia 30681,30793
Bristly Aster 380970
Bristly Black Currant 334058
Bristly Blackberry 339262
Bristly Branchlets Rhododendron 331809
Bristly Buttercup 325942,325943,325948,326217
Bristly Canthium 71387
Bristly Crowfoot 326217
Bristly Desertstar 257834
Bristly Dewberry 338528
Bristly Dogstail Grass 118319
Bristly Fargesia 162755
Bristly Fiddle-neck 20843
Bristly Flatsedge 119015
Bristly Foxtail 358371,360318,361709,361930,361935
Bristly Foxtall Grass 361847
Bristly Globethistle 140786
Bristly Greenbrier 366592
Bristly Hawk's-beard 111017
Bristly Hawkbit 224668,224698
Bristly Hawksbeard 111017
Bristly Kohleria 217749
Bristly Lady's-thumb 309345
Bristly Locust 334965
Bristly Lovegrass 93978
Bristly Nama 262123
Bristly Oxtongue 298584
Bristly Ox-tongue 298584
Bristly Persimmon 132414
Bristly Poppy 282574
Bristly Raspberry 338859
Bristly Rhododendron 331808
Bristly Rockfoil 349725
Bristly Rose 336320,336322,336848
Bristly Rush 213446
Bristly Sarsaparilla 30681
Bristly Sedge 74157
Bristly Shibataea 362146
Bristly Speargrass 300769
Bristly Spikenard 30681
Bristly Spikesedge 143092
Bristly Sunflower 188980
Bristly Thistle 92033
Bristly Yushania 416758
Bristly-leaved Barberry 52142
Bristly-spiked Sedge 76234
Bristol Barberry 51393
Bristol Rock Cress 30429
Bristol Rock-cress 30429
Bristol Weed 250614
Briswort 381035
Britbean Loosestrife 239564
British Alkaligrass 321378
British Arrowroot 37015
British Columbia Wild Ginger 37582
British Columbian Pine 318580
British Dock 339957
British Honduras Yellow Wood 306437
British Inula 207046
British Myrrh 261585
British Timothy 294988
British Yellowhead 207046
Britle Kobresia 217177
Briton Indianpine 258008
Brittle Basketgrass 272614
Brittle Bent 12073
Brittle Bush 145188
Brittle Cactus 272898
Brittle Falsepimpernel 231503
Brittle Gum 155642
Brittle Motherweed 231503
Brittle Najad 262089
Brittle Prickly Pear 272898
Brittle Pricklypear 272898
Brittle Prickly-pear 272898
Brittle Sandwort 255545,255547
Brittle Spineflower 89058
Brittle Thatch Palm 390460
Brittle Waternymph 262089
Brittle Willow 343396
Brittlebush 145187,145188
Brittle-bush 145188
Brittlestem 317364
Brittle-stem Hemp-nettle 170060
Brittle-thatch 390462
Brittlewood 267370
Britton Ruellia 339684
Brittons 34536
Britton's Beargrass 266490
Britton's Spike-rush 143069
Britton's Wild Petunia 339684,339697
Britz Emarginate Barberry 51587
Brivet 229662
Briza Jocks 102727
Broad Bean 71042,408393
Broad Calf's Snout 28617
Broad Caucalis 400473
Broad Caucatis 400473
Broad Clover 397019
Broad Cuneateleaf Buttercup 325744
Broad Egyptian Privet 223454
Broad Fig 164763
Broad Fleabane 150748
Broad Helleborine 82038,147149
Broad Kelk 192358
Broad Kirilow Rhodiola 329898
Broad Leaf Eria 148707
Broad Leaf Pepperwort 225398
Broad Leaved Ragwort 358902
Broad Loose-flower Sedge 75084
Broad Path 18089
Broad Petioles Rhododendron 331497
Broad Sepal Onosma 242399
Broad Smooth-leaved Willowherb 146782
Broad Spike Pycreus 322210
Broad Spiny Rose 336857
Broad Spurge 159599
Broad Tube Rhododendron 330664
Broad Waxpetal 106670
Broadanther Illigera 204614
Broad-anthered Illigera 204614
Broadauriculate Raspberry 338731
Broad-auriculated Raspberry 338731
Broadbeak Woodbetony 287345
Broadbeaked Mustard 59512
Broadbean 408393
Broadbell Azalea 330211
Broadbill Woodbetony 287345
Broadbract Aster 40715

Broadbract Cousinia 108369
Broadbract Crazyweed 278954
Broad-bract Cryptandra 113342
Broadbract Dwarfnettle 262224
Broadbract Falsetamarisk 261250
Broadbract Groundsel-like Aster 41229
Broadbract Henry Rockjasmine 23194
Broadbract Jerusalemsage 295217
Broadbract Manyleaf Gentian 150797
Broadbract Microula 254369
Broadbract Monkshood 5072
Broadbract Nannoglottis 262224
Broadbract Potanin Larkspur 124509
Broadbract Shady Larkspur 124665
Broadbract Skullcap 355765
Broadbract Touch-me-not 205068
Broadbract Vetch 408453
Broadbract Vinegentian 108369
Broad-bracteate Mahonia 242522
Broad-bracted False-tamarisk 261250
Broadbracteole Potanin Larkspur 124509
Broadcalyx Herminium 192876
Broadcalyx Jaeschkea 211400
Broadcalyx Lily 228590
Broadcalyx Metalleaf 296392
Broadcalyx Onosma 242399
Broad-claw Vetch 408456
Broadcorymb Tarenna 384976
Broadcorymb Threevein Aster 39967
Broadcup Azalea 331850
Broadfilament Christolea 89270
Broadflower Anodendron 25929
Broad-flower Butterflybush 62104
Broadflower Corydalis 106058
Broadflower Cranesbill 174820
Broadflower Dendrocalamus 125482
Broad-flower Pincushion 84524
Broadflower Rabdosia 324802
Broadflower Spapweed 205070
Broad-flower Tarenna 384976
Broadflower Touch-me-not 205070
Broadflower Wallich Dragonhead 137692
Broadflower Wallich Greenorchid 137692
Broad-flowered Dragonbamboo 125482
Broadfruit Ash 168068
Broad-fruit Bur-reed 370058
Broadfruit Dactylicapnos 128316
Broadfruit Dichrostachys 129140
Broadfruit Dicranostigma 129572
Broadfruit Favusheadflower 129572
Broad-fruit Glochidion 177161
Broadfruit Hedinia 187363
Broadfruit Rhodiola 329870
Broadfruit Shrubcress 368552
Broadfruit Solms-Laubachia 368552,368570
Broad-fruit Yellowwood 94030
Broad-fruited Ash 168068
Broad-fruited Bur-reed 370058
Broad-fruited Cornsalad 404472
Broad-fruited Glochidion 177161
Broadglume Leymus 228391
Broadglume Roegneria 335393
Broad-grass 397019,397043
Broadhastate Goldenray 229086
Broadhorn Elatostema 142792
Broadhorn Stairweed 142792
Broadinflorescence Childvine 400917
Broad-inflorescence Millettia 66629
Broadinflorescence Tylophora 400917
Broad-inflorescence Tylophora 400917
Broadkidneyleaf Cranesbill 174825
Broad-lcaved Elm 401512
Broadleaf Acriopsis 5951
Broadleaf Actinidia 6648
Broadleaf Addermonth Orchid 130185
Broadleaf Adenostemma 8013
Broadleaf Ainsliaea 12663
Broadleaf Alangium 13366
Broadleaf Anadendrm 21312
Broadleaf Arborvitae 390673
Broadleaf Arborvitae Hiba 390680
Broadleaf Arnica 34730
Broad-leaf Arrowbamboo 162709
Broadleaf Arrowhead 342363
Broadleaf Arrowwood 407676
Broadleaf Ascitesgrass 407470
Broadleaf Aucuba 44906
Broadleaf Azalea 331496,331870
Broadleaf Bamboo 347268
Broadleaf Barbed Gentianopsis 174227
Broadleaf Blacken Catchfly 363812
Broadleaf Blolly 181535
Broadleaf Bluegrass 305436
Broadleaf Bogorchis 130185
Broad-leaf Bottle-tree 58337
Broadleaf Bracteate Goatwheat 44254
Broadleaf Buchanania 61671
Broadleaf Bunge Bedstraw 170297,170678
Broadleaf Bushclover 226892
Broadleaf Butomopsis 64179
Broadleaf Caldesia 66343
Broadleaf Camellia 69262
Broadleaf Caryodaphnopsis 77964
Broad-leaf Cat Tail 401112
Broadleaf Catchfly 363894
Broad-leaf Cathay Poplar 311267
Broadleaf Cattai 401112
Broadleaf Ceratozamia 83683
Broad-leaf Chasmanthium 86140
Broadleaf Childvine 400854
Broadleaf China Plumyew 82540
Broadleaf China Stachyurus 373535
Broadleaf Chinese Plumyew 82540
Broadleaf Chloranthus 88282
Broadleaf Cinnamon 91409
Broadleaf Cinquefoil 312541
Broadleaf Clematis 94705
Broadleaf Clovershrub 70829
Broadleaf Clubfilment 179189
Broadleaf Cogongrass 205524
Broadleaf Coldbamboo 172742
Broadleaf Collabium 99776
Broadleaf Combretum 100571
Broadleaf Commersonia 101220
Broadleaf Common Crapemyrtle 219951
Broadleaf Common Valerian 404325
Broadleaf Cortiella 105482
Broadleaf Corydalis 105817
Broadleaf Cottonrose 166016
Broadleaf Cottonsedge 152765
Broadleaf Crinum 111214
Broadleaf Cudweed 178061
Broad-leaf Cumbungi 401129
Broadleaf Curculigo 256582
Broad-leaf Currant 334061
Broadleaf Dampsedge 66343
Broad-leaf Deceptive Barberry 51605
Broadleaf Dendrochilum 125532
Broadleaf Devilpepper 327022
Broadleaf Dichroa 129042
Broadleaf Diels Euonymus 157678
Broadleaf Diffuse Cypressgrass 118754
Broadleaf Dock 340151
Broadleaf Dry Catchfly 363613
Broadleaf Dzungar Milkaster 169799
Broadleaf Earliporchis 277246
Broadleaf East Yunnan Sabia 341580
Broadleaf Eelgrass 418381
Broad-leaf Eightangle 204508
Broadleaf Elaeagnus 142054,142075
Broadleaf Elaeocarpus 142389
Broadleaf Elatostema 142794
Broadleaf Elder 345620
Broad-leaf Enchanter's-nightshade 91557, 91562
Broadleaf Epigeneium 146530
Broadleaf Epipactis 147149
Broadleaf Epiphyllum 147291
Broadleaf Eritrichium 153473,184295
Broadleaf Eucalyptus 155476
Broad-leaf Eucalyptus 155472
Broadleaf Euphorbia 159599
Broadleaf Eyebright 160203
Broadleaf False Buttonweed 370817
Broadleaf False Carrot 400473
Broadleaf Falsealyssum 170171
Broadleaf Falsepimpernel 231553
Broadleaf Falsetamarisk 261285
Broad-leaf Fewflower Lysionotus 239974
Broadleaf Filbert 106732
Broadleaf Fishvine 10210

Broadleaf Fissistigma 166655
Broad-leaf Flemingia 166872
Broadleaf Forget-me-not 260812
Broadleaf Galanthus 169725
Broadleaf Garden Burnet 345892
Broad-leaf Gee-bung 292029
Broadleaf Gelidocalamus 172742
Broadleaf Glandweed 284096
Broadleaf Globethistle 140732
Broad-leaf Glossy Privet 229529
Broadleaf Goatweed 8824
Broadleaf Goldenray 229030
Broadleaf Goldenrod 368100
Broadleaf Gomphostemma 179189
Broadleaf Goosecomb 144428
Broadleaf Graperfern-like Cystopetal 320207
Broadleaf Greenbrier 366559
Broadleaf Groundsel 358902,359769
Broadleaf Gymnema 182375
Broadleaf Hawthorn 109670
Broadleaf Hedge Glorybind 68721
Broadleaf Helleborine 147149
Broadleaf Herminium 192868,192883
Broadleaf Hogfennel 292908
Broadleaf Holboellia 197232
Broadleaf Holly 203961
Broadleaf Hypolytrum 202724
Broad-leaf Illicium 204508
Broadleaf Indocalamus 206800
Broad-leaf Japan Roughleaf 222178
Broad-leaf Japanese Lasianthus 222178
Broadleaf Jewelvine 10210
Broad-leaf Jewelvine 10210
Broadleaf Jointfir 178543
Broadleaf Kauri 10498
Broadleaf Kauri Pine 10498
Broadleaf Kiwifruit 6648
Broadleaf Knotweed 309601
Broad-leaf Knotweed 309394
Broadleaf Landpick 308558
Broadleaf Lavender 223298
Broadleaf Lilac 382220
Broadleaf Lilyturf 272126
Broadleaf Lime 391817
Broadleaf Liparis 232220
Broadleaf Liriope 232631
Broadleaf Litse 233866
Broadleaf Longstyle Pogostemon 306981
Broadleaf Mahonia 242487
Broadleaf Mananthes 244289
Broadleaf Manaraceme Sweetvetch 188049
Broadleaf Mannagrass 177618
Broadleaf Maple 2783,3127
Broad-leaf Maple 3127
Broadleaf Medick 247454
Broadleaf Microtropis 254306
Broadleaf Milkvetch 42888
Broadleaf Milkweed 37984
Broadleaf Monkeyflower 255254
Broadleaf Motherweed 231553
Broadleaf Mulberry 259152
Broad-leaf Mullein 405754
Broadleaf Murdannia 260100
Broad-leaf Myxopyrum 261898
Broadleaf Nettle 402946
Broadleaf Nodding Woodbetony 287073
Broadleaf Northern Bedstraw 170264
Broadleaf Northern Rockjasmine 23295
Broadleaf Onion 15389
Broadleaf Orchis 273501
Broadleaf Oriental Salsify 394322
Broad-leaf Paperbark 248114
Broadleaf Paper-bark 248114
Broad-leaf Paper-bark 248114
Broadleaf Paris 284369
Broadleaf Parnassia 284529
Broadleaf Patenthairy Anemone 23791
Broadleaf Pearlsedge 354255
Broad-leaf Pea-shrub 72235
Broadleaf Peppergrass 225398
Broadleaf Peppertree 350985
Broadleaf Pepperweed 225398
Broadleaf Pernettya 291367
Broadleaf Phacelurus 293215
Broadleaf Phlox 295248
Broadleaf Pine 306493
Broadleaf Pipewort 151452
Broad-leaf Plantain 302068
Broadleaf Platea 302601
Broadleaf Podocarpium 200740
Broadleaf Podocarpus 306493
Broadleaf Poisoncelery 90936
Broadleaf Poplar 311426
Broadleaf Premna 313667
Broadleaf Rabbitbrush 150370
Broadleaf Rabbiten-wind 12663
Broadleaf Rabdosia 209741
Broadleaf Rain Tree 61306
Broadleaf Raintree 61306
Broadleaf Raspberry 338731
Broadleaf Rattanpalm 65751,65774
Broadleaf Resin Hogfennel 293038
Broadleaf Rice 275932
Broadleaf Rockfoil 349979
Broadleaf Rockliving Beancaper 418754
Broadleaf Roegneria 144428
Broadleaf Rosularia 337307
Broadleaf Rush 213379
Broad-leaf Rush 213379
Broadleaf Sagebrush 35789
Broadleaf Salacia 342589
Broadleaf Satintail 205524
Broadleaf Saxifrage 349859
Broad-leaf Saxifrage 349859
Broadleaf Scaligeria 350374
Broadleaf Sealavender 230655
Broadleaf Sedge 76264
Broad-leaf Sedge 75833
Broadleaf Shorthair Aster 40163
Broadleaf Shorttail Hornbeam 77337
Broad-leaf Shrubby Peashrub 72235
Broadleaf Shrubcress 368552
Broadleaf Siberian Crabapple 243574
Broad-leaf Sichuan Acanthopanax 2488
Broadleaf Sichuan Woodbetony 287727
Broadleaf Slender Serpentroot 354927
Broadleaf Small Reed 127206
Broadleaf Small-leaved Willow 343519
Broadleaf Solms-Laubachia 368552
Broadleaf Solomon's Seal 308558
Broadleaf Solomonseal 308558
Broadleaf Speedwell 407018
Broad-leaf Speedwell 407018,407019
Broadleaf Spicebush 231380
Broad-leaf Spider-lily 200938
Broadleaf Spike Grass 86140
Broadleaf Sporoxeia 372895
Broadleaf Spring Knotweed 309578
Broadleaf Spurge 159223,159599
Broadleaf Stachyurus 373535
Broadleaf Stairweed 142794
Broadleaf Statice 230655
Broadleaf Stonecrop 294117,356760,357084
Broadleaf Striga 377973
Broadleaf Sweetflag 5815
Broadleaf Sweetgrass 177618
Broadleaf Sweetvetch 188049
Broadleaf Syzygium 382610
Broadleaf Tainia 383151
Broadleaf Taraike Willow 344186
Broadleaf Tenagocharis 64179
Broadleaf Thevetia 389970
Broadleaf Thyme 391247
Broadleaf Toadflex 231148
Broadleaf Toadlily 396601
Broadleaf Touch-me-not 205071
Broadleaf Tropidia 399634
Broadleaf Turgenia 400473
Broadleaf Twayblade 232220
Broadleaf Twinflower Violet 409729
Broadleaf Valerian 404316,404325
Broadleaf Variousforms Sisymbrium 365586
Broadleaf Vernalgrass 27957
Broadleaf Veronicastrum 407470
Broadleaf Vetch 408262
Broadleaf Vetchling 222750
Broadleaf Viburnum 407676
Broad-leaf Viburnum 407919
Broadleaf Vine Catchfly 363971
Broadleaf Waterbamboo 260100
Broadleaf Waterleaf 200484

Broadleaf Waterparsnip 365851
Broadleaf Wattle 1516
Broad-leaf Wild Buckwheat 152510
Broadleaf Wild Leek 15048,15091
Broadleaf Wild Smallreed 127206
Broad-leaf Wilhelms Willow 344274
Broadleaf Willowcelery 121046
Broadleaf Willowweed 85884
Broadleaf Windmill 100571
Broadleaf Woodnettle 273923
Broadleaf Woodsorrel 277925
Broadleaf Woolly Draba 137135
Broadleaf Yunnan Draba 137301
Broadleaf Yunnan Whitlowgrass 137301
Broadleaf Yushanbamboo 416797
Broadleaf Zamia 417015
Broadleaf Zeuxine 417700
Broad-leafed Hakea 184601
Broad-leafed Meryta 250948
Broad-leafed Native Cherry 161713
Broad-leafed Paperbark 248113,248125
Broad-leafed Privet 229529
Broad-leafed Scribbly Gum 155603
Broad-leafed Stringybark 155513
Broad-leaved Actinidia 6648
Broad-leaved Adinandra 8248
Broad-leaved Amur Barberry 51307
Broad-leaved Anemone 23846
Broad-leaved Arbor-vitae 390673
Broad-leaved Arrowhead 342363
Broad-leaved Bamboo 347268
Broad-leaved Barberry 52047
Broad-leaved Birch 53503
Broad-leaved Bluets 187652
Broad-leaved Borreria 57333
Broad-leaved Buchanania 61671
Broad-leaved Bunchflower 248655
Broad-leaved Cabbage-tree Mountain Cabbage Tree 104359
Broad-leaved Carpetgrass 45827
Broad-leaved Caryodaphnopsis 77964
Broad-leaved Cat-tail 401112
Broad-leaved Chinese Photinia 295777
Broad-leaved Cinnamon 91409
Broad-leaved Clovershrub 70829
Broad-leaved Cockspur-thorn 109928
Broad-leaved Combretum 100571
Broad-leaved Coral Tree 154681
Broad-leaved Cotton Grass 152765
Broad-leaved Cotton-grass 152765
Broad-leaved Croton 112891
Broad-leaved Cudweed 166025
Broadleaved Dandelion 384813
Broad-leaved Dock 340151
Broad-leaved Drumsticks 210235
Broad-leaved Elaeocarpus 142389
Broad-leaved Endive 90894
Broad-leaved Epipactis 147149
Broad-leaved Euonymus 157668
Broad-leaved Everlasting-pea 222750
Broad-leaved False-nettle 56206
Broad-leaved False-tamarisk 261285
Broad-leaved Fig 164626
Broad-leaved Fissistigma 166655
Broad-leaved Flemingia 166872
Broad-leaved Garlic 15861
Broad-leaved Gelidocalamus 172742
Broad-leaved Gilia 175688
Broad-leaved Glaucous Spurge 159417
Broad-leaved Golden Aster 90233
Broad-leaved Golden Rod 368100
Broadleaved Goldenrod 368100
Broad-leaved Goldenrod 368206
Broad-leaved Gomphostemma 179189
Broad-leaved Gromwell 233753
Broad-leaved Groundsel 358902
Broad-leaved Gymnema 182375
Broad-leaved Harebell 7739
Broad-leaved Hedgehog Parsley 400473
Broad-leaved Helleborine 147149
Broad-leaved Hemp-nettle 170066
Broad-leaved Holboellia 197232
Broad-leaved Holly 203961
Broad-leaved Hooker Barberry 51737
Broad-leaved Indocalamus 206800
Broad-leaved Ironbark 155741
Broad-leaved Ivy 215395
Broad-leaved Keck 192358
Broad-leaved Kindling-bark 155553
Broad-leaved Lady's Tresses 372227
Broad-leaved Laurel 215395
Broad-leaved Lavender 223298
Broad-leaved Leopard Tree 166977
Broad-leaved Lillypilly 4883
Broad-leaved Lime 391817
Broad-leaved Lucuma 313384
Broad-leaved Lupin 238459
Broad-leaved Mahogany 216202
Broad-leaved Mallotus 243418
Broad-leaved Maple 2783
Broad-leaved Marsh Dandelion 384813
Broad-leaved Marsh Orchid 273529
Broad-leaved Meadow-grass 305436
Broad-leaved Meadowsweet 371795
Broad-leaved Montia 258243
Broad-leaved Mouse Ear Chickweed 82849
Broad-leaved Mrs. E. H. Wilson's Barberry 52373
Broad-leaved Nipliporchis 154881
Broad-leaved Orchid 273529
Broad-leaved Orchis 273501
Broad-leaved Osier 342946
Broad-leaved Panic Grass 128515
Broad-leaved Panic-grass 281818
Broad-leaved Parakeelya 65832
Broad-leaved Pepper Grass 225398
Broad-leaved Peppermint 155564
Broad-leaved Pepperweed 225398
Broad-leaved Pepper-weed 225398
Broad-leaved Pepperwort 225398
Broad-leaved Plantain 302068,302159
Broad-leaved Platea 302601
Broad-leaved Podocarpus 306457
Broad-leaved Pondweed 312044,312190
Broad-leaved Premna 313667
Broad-leaved Purple Coneflower 140081
Broad-leaved Ragwort 358902
Broad-leaved Rattan Palm 65774
Broad-leaved Rhododendron 331496
Broad-leaved Rosebay 382763
Broad-leaved Rush 213379
Broad-leaved Sabia 341580
Broad-leaved Salacia 342589
Broad-leaved Sally 155524,155526
Broad-leaved Sea-lavender 230530
Broad-leaved Sedge 75833
Broad-leaved Signal-grass 402546
Broadleaved Silver Holly 203549
Broad-leaved Small Pondweed 312236
Broad-leaved Smalscale Sedge 74641
Broad-leaved Solomon's-seal 308558
Broad-leaved Sorrel 277925
Broad-leaved Spice-bush 231380
Broad-leaved Spindle 157668
Broadleaved Spindle-tree 157668
Broad-leaved Sporoxeia 372895
Broad-leaved Spring Beauty 94305
Broad-leaved Spurge 159599
Broadleaved Stairweed 142793
Broadleaved Texas Boxelder Maple 3231
Broad-leaved Thevetia 389970
Broad-leaved Thrift 34501
Broad-leaved Toothwort 72739
Broad-leaved Twayblade 232904
Broad-leaved Viburnum 407676
Broad-leaved Warted Spurge 159599
Broad-leaved Water Parsnip 365851
Broad-leaved Waterleaf 200484
Broad-leaved Whitebeam 369446
Broad-leaved Wild-rye 144541
Broad-leaved Wildryegrass 144428
Broad-leaved Willowherb 146782
Broad-leaved Wood Sedge 75833
Broad-leaved Wood Violet 410587
Broad-leaved Woolly Sedge 75740
Broad-leaved Yucca 416556
Broad-leaved Yushania 416797
Broad-leaves Rhododendron 331496
Broad-ligular Bamboo 297391
Broadlily 97086
Broadlinearleaf Erysimum 154451

Broadlip Galangal　17739
Broadlip Herminium　192850
Broadlip Hyssop　203088
Broadlip Liparis　232221
Broadlip Nephelaphyllum　265121
Broadlip Twayblade　232221,232289
Broad-lipped Twayblade　232904
Broadlobe Arecapalm　31687
Broadlobe Corydalis　106061
Broadlobe Cranesbill　174821
Broadlobe Jasmine　211999
Broadlobed Bauhinia　49004
Broad-lobed Bauhinia　49004
Broadlobed Kusnezoff Monkshood　5339
Broadlobed Whorledleaf Woodbetony　287806
Broadlobed Wormwood　35897
Broadpetal Barrenwort　147036
Broad-petal Biond's Magnolia　242001
Broadpetal Blume Spiraea　371841
Broadpetal Buttercup　326235
Broadpetal Catchfly　363924
Broadpetal Crazyweed　279079
Broadpetal Entireliporchis　261479
Broadpetal Epimedium　147036
Broad-petal Farges Paris　284323
Broadpetal Gloriosa　177244
Broad-petal Homalium　197689
Broadpetal Irisorchis　267965
Broadpetal Madder　337977
Broadpetal Michelia　252839
Broadpetal Oberonia　267965
Broadpetal Pearweed　239791
Broadpetal Rockfoil　349537
Broadpetal Spapweed　205072
Broadpetal Winterhazel　106670
Broad-petaled Homalium　197689
Broad-petaled Michelia　252839
Broad-petaled Winterhazel　106670
Broadpetiole Cacalia　283838
Broadpetiole Jointed Cinquefoil　312397
Broad-petioled Camellia　69264
Broadphyllaries Aster　40715
Broadpod Medic　247422
Broadpurpleleaf Violet　409797
Broadraceme Cliffbean　66629
Broadscale　44555
Broadscale Lepionurus　225648
Broadscale Rabbitbrush　150346
Broadscale Urobotrya　402477
Broad-scaled Urobotrya　402477
Broadseed Buttercup　326237
Broadseed Hakea　184626
Broad-sepal Brautiful St. John's Wort　201986
Broadsepal Briggsia　60271
Broadsepal Clematis　95362
Broadsepal Delavay Meadowrue　388481
Broadsepal Epimedium　147015
Broadsepal Hemiboea　191368
Broad-sepal Melastoma　248779
Broad-sepal Metalleaf　296392
Broad-sepal Mockorange　294557
Broadsepal Spapweed　205231
Broadsepal Stonecrop　357029
Broadsepal Threeflower Clematis　95362
Broadsepal Touch-me-not　205231
Broadsepal Tubergourd　390158
Broadsepal Violet　410017
Broadsepale Petrocosmea　292574
Broadsheath Greenbrier　366311
Broad-sheath Greenbrier　366311
Broadsheath Lily　228564
Broadsheath Sinocarum　364895
Broadsheath Smilax　366311
Broadspathe Leek　15602
Broadspathe Onion　15602
Broadspike Ainsliaea　12669
Broadspike Flatsedge　322210
Broadspike Melic　249060
Broadspike Rostellularia　337246
Broadspike Stinkinggrass　249060
Broad-spiked Millettia　66629
Broadspine Craneknee　221719
Broadspine Herelip　220085
Broadspine Lagochilus　220085
Broadspine Rattan Palm　65769
Broad-spine Rattan Palm　65770
Broadspine Rattanpalm　65770
Broadspine Rose　336857
Broad-spined Rose　336857
Broadsplit Primrose　314562
Broadspur Gastrochilus　171880
Broadspur Spapweed　205228
Broadspur Touch-me-not　205228
Broadspurred Trichophore Larkspur　124648
Broadstalk Azalea　331497,331672
Broadstalk Crazyweed　279078
Broadstalk Windhairdaisy　348653
Broad-stalked Orange　93529
Broad-stalked Rhododendron　331497
Broadstamen Burnet　345832
Broadstamen Rockjasmine　23166
Broadstipe Clematis　95191
Broadstipe Goldenray　229087
Broadstipule Cranesbill　175003
Broadstipule Eargrass　187647
Broadstipule Hedyotis　187647
Broadstyle Box　64270
Broadstyle Iris　208684
Broadstyle Onion　15161
Broadstyle Swordflag　208684
Broadstyle Willow Weed　146840
Broad-styled Box　64270
Broadtongue Gerbera　175183
Broadtongue Goldenray　229140
Broadtooth Cinquefoil　312904
Broadtooth David Woodbetony　287148
Broadtooth Lady's Mantle　14149
Broadtooth Lagochilus　220070
Broadtooth Strictflower Betony　373450
Broadtooth Triumfetta　399244
Broadtooth Unifolious Buttercup　326094
Broadtooth Whored Cinquefoil　313101
Broadtube Azalea　330664
Broadtube Gentian　173219
Broadtube Lagotis　220181
Broadtube Woodbetony　287346
Broad-tubed Rhododendron　330664
Broad-veined Chirita　87894
Broadweed　192358,216753
Broadwing Axe-shaped Pugionium　321488
Broadwing Craneknee　221722
Broadwing Crazyweed　278953
Broadwing Eritrichium　153548
Broadwing Everlasting　21589
Broadwing Goldenray　229152
Broadwing Oblongleaf Maple　3264
Broadwing Pearleverlasting　21589
Broadwing Rockcress　30406
Broadwing South China Barthea　48553
Broadwing Tailanther　381920
Broadwing Tickseed　104831
Broadwing Woad　209241
Broadwinged Iris　208437
Broatongue Nutantdaisy　110347
Brocade Bambusa　47462
Brocadebeldflower　413570,413591
Broccilo　59545
Broccoli　59529,59545
Brock　59532
Brodiaea　60431,60436,398806
Brodleaf Kinding Bark　155553
Brodleaf Osmanther　276295
Brodleaf Pfeiffer Osmanthus　276295
Brod-leaf Simon Poplar　311497
Brod-leaved Ribbon Gum　155553
Brodspur Cuphea　114606
Broken Blood Windhairdaisy　348099
Broken Bones Plant　275403
Broken Heart　96387
Broken Hearts　96387
Broken Pennywort　200261
Broken Rice Holly　203855
Brokenriceflower Azalea　331872
Brome　25280,60566,60612,83466
Brome Fescue　412407
Brome Grass　25280,60566,60612,60963,83466
Bromeform Melic　248966
Bromegrass　60612
Brome-grass　60612

Bromelia 60544
Bromelia Family 60557
Bromeliad 684,8558
Bromeliad Family 60557
Brome-like Duthiea 139123
Brome-like Sedge 73942
Brompton Stock 246478
Bromus 60612
Bromus-like Feathergrass 376721
Bromus-like Tripogon 398068
Bronckart Dendrobium 125020
Bronx Forsythia 167472
Bronze Beauty 257816
Bronze Fennel 167156
Bronze Hibiscus 194689
Bronze Lily 229758
Bronze Loquat 151144
Bronze Paper Tree 101408
Bronze Pirri-pirri-bur 1752
Bronze-bells 375456
Bronzecolour Braya 59759
Bronze-headed Oval Sedge 74579
Bronze-leaf Begonia 50243
Bronzeleaf Rodgersflower 335151
Bronzy Willowherb 146753
Bronzybasin Ardisia 31540
Bronzy-yellow Eastern Arborvitae 390595
Broodwing Broomsedge 217346
Broodwing Summercypress 217346
Broohweed 345733
Brook Alder 16404
Brook Betony 355061
Brook Cinquefoil 312936,312937
Brook Cress 72927
Brook Euonymus 157315
Brook Five-fingers 312936,312937
Brook Grass 79198,177629
Brook Leek 137747
Brook Lobelia 234562
Brook Saxifrage 349851
Brook Tongue 90932
Brook Wake-robin 397607
Brookbean 250502
Brooker's Gum 155508
Brook-feather 415166
Brookgrass 79194,79198
Brook-grass 79198
Brookleek 137747
Brooklem 407029
Brooklet Anemone 24024
Brooklet Astilbe 41841
Brooklet Azalea 331661
Brooklet Rhododendron 331661
Brooklet Sonerila 368898
Brooklet Windflower 24024
Brooklime 29341,262722,407029
Brookmint 250291
Brookside Avens 175444
Brookside Ophiorrhiza 272218
Brookside Rockfoil 349851
Brook-tongue 90932
Brookweed 345728,345729,345731,345733
Brookweedleaf Gentian 173835
Broom 67456,120903,121001,150131,
170743,172898,370191
Broom Baccharis 46255
Broom Bamboo 137851
Broom Beard Grass 351309
Broom Brome 60957
Broom Chinese White Poplar 311535
Broom Cluster Fig 165724
Broom Corn 369600,369700,369720
Broom Cupflower 266208
Broom Cypres 217361,217373
Broom Cypress 217361
Broom Dalea 319358
Broom Deer-weed 237749
Broom Groundsel 360081
Broom Heath 150131
Broom Honey Myrtle 248124
Broom Knotweed 308791
Broom Ladybell 7828
Broom Meadowrue 388716
Broom Milkvetch 43024
Broom Millet 281916
Broom Pine 300128
Broom Rape 274922,275010
Broom Reed 142942
Broom Sedge 74490,76207
Broom Simon Poplar 311494
Broom Snakeroot 20486
Broom Snakeweed 182103,182107
Broom Spurge 20480
Broom Teatree 226481
Broom Tea-tree 226481
Broom Wallflower 86461
Broom Wattle 1116
Broombeere 338281
Broombush False Willow 46219
Broomcom Millet 281916
Broomcorn 281916,369636,369700,369720
Broom-corn 369600
Broomcorn Millet 281917
Broom-corn Millet 281916,281917
Broomcorn Panicgrass 281916
Broomdashers 83361
Broomflower Corydalis 105798
Broomhead Ramie 56357
Broomjute Sida 362617
Broom-jute Sida 362617
Broomleaf Toadflax 230976
Broom-leaved Toadflax 230976
Broom-like Blueberry 403987
Broomlike Bluestem 351309
Broom-like Cassiope 78628
Broom-like Milkvetch 42933
Broom-like Priotropis 315263
Broomlike Serpentroot 354918
Broomlike Spunsilksage 360873
Broomrape 274922,275010,275121,275135,
275143
Broomrape Family 274921
Broomrap-like Striga 378027
Broomsedge 23077,217324,217361
Broom-sedge 23077,76207
Broomsedge Bluestem 23077
Broomshape China Pine 300255
Broomshape Chinese Pine 300255
Broomshaped Chinese Pine 300255
Broomshiped Cephalocereus 82210
Broomshoot Catnip 265103
Broomstraw 23077
Broomweed 20486,182103,182107,243893
Broomwort 354658,354662,390213
Broomy Rabdosia 209819
Brosdclaw Milkvetch 42593
Brosdleaf 181264
Brosdleaf Ajania 13006
Brosdleaf Goldband Lily 229734
Brosdleaf Villose Hawkweed 196082
Brosewort 201389
Brosh-tipped Sotol 122899
Brosplakkie 8471
Brotherwort 250432,391365
Broughtonia 61087
Brousea Tea 403728
Broussa Mullein 405674
Brow Willow 344265
Browallia 61118,61131,61132
Browbean 409092
Brow-coloured Rhododendron 331705
Browhair Michelia 252881
Brown African Padauk 320278
Brown Ash 168032
Brown Asian White Birch 53573
Brown Barrel 155579
Brown Beak Sedge 333576
Brown Beakrush 333495
Brown Beak-rush 333576
Brown Beaks 239421
Brown Beak-sedge 333576
Brown Bee Orchis 272435
Brown Beech 113447
Brown Beetle-grass 226005
Brown Bent 12025,12413
Brown Bent Grass 12025
Brown Birch 53590
Brown Bog-rush 352385
Brown Boronia 57267
Brown Bugle 13174
Brown Catchfly 363224

Brown Cress 262722
Brown Durra 369640
Brown Ebony 61427
Brown Eelgrass 418160
Brown Eulalia 156485
Brown Euonymus 157917
Brown Fasciated Haworthia 186432
Brown Flatsedge 118928
Brown Fox Sedge 76718,76735
Brown Galingale 118928
Brown Gentian 173701
Brown Hair Synotis 381940
Brown Haplosphaera 185705
Brown Head Synotis 381936
Brown Heterostemma 194157
Brown Indianhemp 194779
Brown Knapweed 81150
Brown Kneejujube 67011
Brown Lepisanthes 225670
Brown Lily 229754
Brown Mallet 155485
Brown Mint 250450
Brown Monkshood 5084
Brown Mustard 59438,59515
Brown Olive 270100
Brown Oval Sedge 73604
Brown Ovate Tylophora 400945
Brown Pine 306398,306424
Brown Pipewort 151441
Brown Pussytoes 26615
Brown River Finger 253292
Brown Salvia 344878
Brown Sedge 73945,73959,74354
Brown Shillers 106703
Brown Sily Oak 180569
Brown Sotvine 194157
Brown Spiderwort 362726
Brown Spiradiclis 371758
Brown Sterculia 376174
Brown Stringybark 155494,155525
Brown Sugar 339887
Brown Top 12034
Brown Tulip-oak 192500
Brown Tylophora 400945
Brown Wide-lip Orchid 232225
Brown Willow 343506
Brownback Azalea 330068
Brown-back Syzygium 382576
Brownback Willow 343275
Brown-backed Camellia 69150
Brown-backed Syzygium 382576
Brownbarrel Eucalyptus 155579
Brownbeard Rice 275957
Brown-bearded Sugarbush 315960
Brown-berried Cedar 213841
Brownberry Cotoneaster 107337
Brown-berry Cotoneaster 107337
Brownblotch Medic 247246
Brownbract Sagebrush 36076
Brownbract Threebibachene 397949
Brownbract Threeribfruit 397949
Brownbract Yam 131761
Brownbract Yarrow 3953
Brown-bracted Pussytoes 26615
Brown-branch Camellia 69482
Brown-branched Camellia 69482
Browndot Mountainash 369352
Browndot Mountain-ash 369352
Browne's Blechum 55207
Browne's Dock 339960
Brownet 355061,355202
Brown-eyed Susan 339556,339622
Brownflower E. Sichuan Rush 213053
Brownflower Lady's Slipper 120316
Brownflower Ladyslipper 120316
Brownflower Saussurea 348649
Brownflower Thomson Rush 213053
Brownflower Trillium 397556
Brownflower Vanilla 405015
Brownflower Windhairdaisy 348649
Brown-fruited Juniper 213841
Brown-fruited Rush 213365
Browngreen Sedge 76384
Browngreenutricle Sedge Sedge 76384
Brownhaie Sweetleaf 381272
Brownhair Aster 40495
Brownhair Azalea 332095
Brownhair Bird Cherry 316268
Brownhair Bristlegrass 361847
Brownhair Clematis 94933
Brownhair Cremanthodium 110353
Brownhair Currant 333983
Brownhair Dunbaria 138968
Brownhair Elaeocarpus 142311
Brownhair Four-involucre 105060
Brownhair Globethistle 140697
Brownhair Goldenray 229156
Brownhair Goldquitch 156485
Brownhair Inula 207142
Brown-hair Mallotus 243390
Brownhair Mountainash 369475
Brownhair Nutantdaisy 110353
Brownhair Oxtongue 298590
Brownhair Przewalsk Sage 345331
Brownhair Rattlebox 112446
Brownhair Seatung 301277
Brownhair Tephrosia 386316
Brownhair Thistle 91987
Brownhair Vetch 408527
Brownhair Wildtung 243390
Brownhair Willow 343400
Brownhair Woodbetony 287181
Brown-haired Dogwood 380449
Brown-haired Eustigma 161021
Brown-haired Korean Sorbus 369366
Brown-haired Pittosporum 301277
Brown-haired Rhododendron 330752
Brownhairy Bird Cherry 280015
Brown-hairy Bird Cherry 280015
Brownhairy Birdcherry 280015
Brownhairy Eustigma 161021
Brownhairy Goldenray 228974
Brownhairy Hairyflower Actinidia 6590
Brown-hairy Willow 343400
Brownhead Sagebrush 35178
Brownhead Tailanther 381936
Brown-headed Fox Sedge 73639,73641
Brown-headed Knapweed 81150
Brownie Lady's Slipper 120349
Brownie Lady's-slipper 120348
Brownie Thistle 92337
Brownies 354574
Brownish Anzacwood 310758
Brownish Beak Sedge 333504
Brownish Phalaenopsis 293605
Brownish Sedge 73959,73963,74884
Brownishpurple Aspidistra 39557
Brown-Jolly 367370
Brownleaf Achyranthes 4269
Brownleaf Azalea 331571
Brown-leaf Good Report Rhododendron 330661
Brownleaf Mischocarp 255951
Brownleaf Qinggang 116201
Brown-leaved Watercress 336232
Brown-leaves Rhododendron 330661
Brownleea 61169
Brownlip Coelogyne 98628,98660
Brownnedge Yarrow 3973
Brown-net 355061,355202
Brownplume Wirelettuce 375978
Brownpubescent Spapweed 205067
Brown-puctate Cherry-laurel 223117
Brownpunctate Cherry-laurel 223117
Brown-ray Knapweed 81150
Brown-rayed Knapweed 81150
Brown-red Juniper 213901
Brownred Sterculia 376179
Brown-red-spotted Cattleya 79541
Brown's Honeysuckle 235684
Brown's Larkspur 124233
Brown's Lily 229758
Brown's Myrobalan 386488
Brown's Yellowtops 166817
Brownscale Centaurea 81150
Brown-scale Centaurea 81150
Brownscale Small Rattanpalm 65645
Brownseed Paspalum 285481
Brownsheath Buttercup 326374
Brownsheath Corydalis 105668
Brownsheath Lilyturf 272073

Brown-shillers 106703
Brown-spine Hedgehog 140343
Brown-spined Pricklypear 273003
Brownspot Goldwaist 90365
Brownspotted Arisaema 33410
Brownspotted Southstar 33410
Brownstachys Leymus 228355
Brownstipule Violet 410017
Brown-top Millet 58155
Browntop Signalgrass 402567
Brown-top Stringybark 155672
Brownturbans 243517,243518
Brown-twig Poplar 311557
Browntwig Tea 69482
Brownvein Elatostema 142629
Brownvein Stairweed 142629
Brownvein Yellowhair Maple 2980
Brownwarty Petal Paphiopedilum 282797
Brownweed 182103
Brown-winged Orchid 273604
Brown-woolly Fig 164916
Brownwort 316127,355061,355202
Brown-yellow Iris 208697
Brownyellow Oncidium 270824
Brown-yellow Platanthera 302390
Brown-yellow Windhairdaisy 348603
Brucea 61199,61208
Bruennow Oreocereus 273815
Brugmansia 61230
Bruguiera 61250
Bruiseroot 176738
Bruise-root 176738
Bruisewort 50825,346434,381035
Brumble Kite 338447
Brumleyberry Bush 338447
Brummel 173082,338447,383433
Brummel Kite 338447
Brummelkite 338447
Brun Wormwood 35663
Brunei Teak 138547
Brunei Thin-walled Bamboo 216392
Brunel 316127
Brunell 316127
Brunet's Milk-vetch 42116
Brunnera 61364,61366
Brunnet 355061,355202
Brunon Deutzia 127108
Brunon Rose 336408
Brunon Stauntonvine 374390
Brunon Staunton-vine 374390
Brunon Veronicastrum 407458
Brunswickgrass 285468
Brush 81020,81237,121001,133478,
196818,273469
Brush Apple 313344
Brush Arisaema 33450
Brush Box 236462,398666
Brush Buxus 236462
Brush Cherry 382481,382604,382644
Brush Pepperbush 385090
Brush Wattle 283887
Brush-and-comb 133478
Brushbox 236462,398666
Brush-box 236462
Brush-box Tree 236462
Brush-chery 382481
Brushes 81020
Brushes-and-combs 133478
Brushlike Ajaniopsis 13046
Brushwattle 283887
Brussels Sprout 59539
Brussels Sprouts 59539
Brussel's Sprouts 59539
Bruyere 149017
Brych-chery Eugenia 382481
Brylkinia 61441
Brylocks 403916
Bryocarpum 61449
Bryoleaf Bulbophyllum 63035
Bryoleaf Curlylip-orchis 63035
Bryoleaf Sandwort 31778
Bryonopsis 61549
Bryony 61464,68713
Bryophytelike Gentian 173303
Bryouy 61497
Bryswort 50825
Bsitoa 297536
Btoad-leaf Palm Lily 104362
Buah Keras 14544
Buak Salak 342580
Buaze 356369
Bubble Gum Mint 10405
Bubblebamboo 318485,318487
Bubblefruit Whitethorn 266380
Bubbleleaf Azalea 330613
Bubbleweed 297875,297882
Bubby Water 90032
Bubescent Manchurian Alder 16403
Bubinga 103723,181771
Bubinga Didelotia 129732
Bubu Rattan Palm 65653
Bucalaun 359158
Bucayo 154660
Buccaneer Palm 347076
Buccinator Goblin Orchid 258991
Buch Clockvine 390759
Buch Vetch 408611
Buchanan Honeysuckle 235687
Buchanan Mayten 182667
Buchanan Barberry 51396
Buchanan Clematis 94789
Buchanan Cryptolepis 113577
Buchanania 61665,61666
Buchanania Meliosma 249472
Buchanania-leaf Meliosma 249472
Buchar Skullcap 355395
Bucharic Larkspur 124079
Buchholz Jointfir 178527
Buchnera 61727
Buchtorm Lily 228568
Buchtorm Milkvetch 42119
Buchu 10557,10575,10598,48472
Buck 162312,162400
Buck Bean 250493,250502
Buck Berry 403710
Buck Breer 336419
Buck Brush 22265,92485,296716,380748,
380749
Buck Bush 344585
Buck Eye 9675
Buck Spinifex 397769
Buck Thistle 73430,92485
Buckbean 250493,250502
Buck-bean 250493
Buckbean Family 250492
Buckberry 172262
Buckbrush 79921,380749
Buck-brush 380749
Buckbrush Ceanothus 79921
Buckbrush Maidenbush 22265
Buckbush 344585
Bucket Orchid 105518
Buckeye 9675,9692,9701,9719,77256
Buckeye Bottlebrush 9719
Buckeye Horse Chestnut 9701
Buckhorn 301868,302034
Buck-horn 301868
Buckhorn Cholla 116639,116640,272780
Buckhorn Plantain 301910,302034
Buck-horn Plantain 301910,302034
Buckie-faalie 315101
Buckie-lice 336419
Buckinghamia 61917
Buckitorn 302034
Bucklandia-like Chunia 90616
Buckle 336419
Buckler Mustard 54690
Buckler Thorn 328642
Bucklershaped Bittercress 72971
Buckler-shaped Sorrel 340254
Buckley Hickory 77932
Buckley Oak 323725
Buckley's Goldenrod 368002
Buckley's Oak 323725
Buckley's Yucca 416575
Buckmast 162400
Buckrams 15861,37015
Buck's Horn 301910
Buck's Horn Plantain 301910
Buck's Horn Tree 332916
Buck's Horn Weld 79276

Buck's-beard 37053,37060
Buck's-horn Plantain 301910,302091
Buck's-horn Weld 79276
Buckthorn 17614,316819,328592,328595,328625,328642,328755
Buckthorn Bumelia 63425
Buckthorn Family 328543
Buckthornleaf Mountainash 369504
Buckthorn-leaved Mountain-ash 369504
Buckwheat 151792,162301,162304,162312
Buckwheat Family 308481
Buckwheat Tree 96881
Buckwheat Vine 61378
Buckwheat-tree 96881
Bucky 336419
Bud Bamboo 297447
Bud Sagebrush 35779
Budbearing Rockfoil 349374
Budde Lice 336419
Budded Bittersweet 80193
Budded Daphne 122437
Buddha Bamboo 47498,47508
Buddha Belly Bamboo 47508
Buddha Belly Plant 212202
Buddha Common Bamboo 47520
Buddhabelly Bamboo 47508
Buddhagut Tree 212202
Buddhahalo Sage 345418
Buddhamallow 402238,402245,402267
Buddhanail 356884
Buddhanailm 356467
Buddhapalm Fig 165111
Buddha's Belly Bamboo 47508,47498
Buddha's Coconut 320987
Buddha's Common Bamboo 47516
Buddha's Hand 93604
Buddha's Head Citron 93603
Buddha's Lamp 260483
Buddh-belly Tree 212202
Buddhist Bauhinia 49247
Buddhist Pine 306457
Budding Primrose 314423
Budding Schoenorchis 352303
Buddleia 61953
Buddleia-leaved Begonia 49685
Buddlejaleaf Viburnum 407713
Buddy-bud 31051
Buddy-buss 31051
Budge Gum 64085
Budgerigar Flower 38135
Bud-growing Rockfoil 349376
Bud-producing Rockfoil 349376
Budsage 298676
Buenos Aires Conyza 103438
Buenos Ayres Verbena 405812
Buerger Bird Cherry 280016
Buerger Birdcherry 280016
Buerger Bush-clover 226721
Buerger Figwort 355067
Buerger Holly 203600
Buerger Lespedeza 226721
Buerger Maple 2811
Buerger Pipewort 151257
Buerger Rough Gentian 173852
Buerger's Adnate Elder 345709
Buerger's Elder 345709
Buerger's Raspberry 338205
Buff Fleabane 150825
Buffalo Bean 42156,42231,389547
Buffalo Berry 362089,362090
Buffalo Bill's Sulphur Flower 152600
Buffalo Bur 367567
Buffalo Clover 397041
Buffalo Currant 333919,334071,334127
Buffalo Gourd 114278
Buffalo Grass 58128,61723,61724,281479,375774
Buffalo Juniper 213902,213904
Buffalo Nut 323072
Buffalo Thorn 418190
Buffalo Wattle 1328
Buffalo Weed 19191
Buffalo Wood 63903,63904
Buffaloberry 362089,362090,362091
Buffalo-berry 362090
Buffalo-bur 367567
Buffalobur Nightshade 367567
Buffalo-bur Nightshade 367567
Buffalo-currant 334127
Buffalograss 61723,61724
Buffalo-grass 61724
Buffalo-weed 19191
Buffel Grass 80815,361725
Buffelgrass 80815
Bug Bean 250502,314366
Bug Hairgrass 16222
Bug Orchid 273380
Bugbane 6448,90994,91011,91034
Buggie Flower 364146
Bugheal 395685
Bugle 13051,13174,21920,141347
Bugle Arrowbamboo 162686
Bugle Azalea 330592
Bugle Dendrobium 125232
Bugle Lily 413315,413338
Bugle Weed 13104,239248
Bugle-bloom 236022
Bug-leg Goldenweed 323030
Buglelily 413315
Bugle-lily 413315,413328
Bugleweed 13051,13174,239184,239193,239248
Bugle-weed 13051
Bugloss 21914,21920,21933,141087,141258,260760,260892
Bugloss Cowslip 321643
Buglossoides 62294
Bugseed 104750,104753,104824
Bugwort 91034
Buk Choy 59595
Bukhara Fleeceflower 162509,162512
Bukhara Iris 208473
Bukkum Wood 65060
Bukkum-wood 65060
Bulangshan Dendrobium 125055
Bulb Buttercup 325666
Bulb Panicgrass 418487,418488
Bulb Red-onion 143572
Bulb-bearing Loose Strife 239879
Bulb-bearing Rockfoil 349137
Bulb-bearing Water Hemlock 90921
Bulbees 8418
Bulbiferous Corydalis 105670
Bulbiferous Giantarum 20055
Bulbiferous Lily 229766
Bulbiferous Stonecrop 356590
Bulbiferous Wand-flower 369980
Bulbiferous Wood Sorrel 277709
Bulbiferous Woodnettle 221538
Bulbiferous Wood-nettle 221538
Bulbil Bugle-lily 413369
Bulbil Cremanthodium 110354
Bulbil Flaxuose Corydalis 105882
Bulbil Hemsley Monkshood 5255
Bulbil Knotweed 309840,309954
Bulbil Lily 229766,229771,229772
Bulbil Loosestrife 239879
Bulbil Monkshood 5087
Bulbil Moxanettle 221538
Bulbil Nutantdaisy 110354
Bulbilferous Corydalis 106433
Bulbilifer Gagea 169392
Bulbiliferous Lovegrass 147609
Bulbiliferous Rockfoil 349166
Bulbilliferous Monkshood 5087
Bulbine 62334,62404
Bulblet Water-hemlock 90921
Bulbophyllum 62530,62984
Bulbose Chervil 84731
Bulbostyle 63204
Bulbostylis 63204
Bulbous Barley 198275
Bulbous Bitter-cress 72708
Bulbous Bluegrass 305419
Bulbous Buttercup 325666
Bulbous Canarygrass 293708
Bulbous Canary-grass 293708
Bulbous Corydalis 105721
Bulbous Crowfoot 325666
Bulbous Foxtail 17516
Bulbous Fumitory 106453

Bulbous Meadow-grass 305419
Bulbous Oat Grass 34920,34935
Bulbous Oatgrass 34935
Bulbous Primrose 314457
Bulbous Rush 212955
Bulbous Saxifrage 349419
Bulbous Violet 169719,227875,409788
Bulbous Wood Rush 238583
Bulbstem Yucca 416581
Bulbule Meadowrue 388447
Bulgefruit Crazyweed 279180
Bulgefruit Licorice 177915
Buljing Willow 343130
Bulking Willow 343130
Bull Apple 285660
Bull Banksia 47644
Bull Bay 242135,242136
Bull Beef 336419,338447
Bull Brier 366559
Bull Buttercup 68189
Bull Cottonthistle 271705
Bull Cup 68189
Bull Dairy 273469,273531
Bull Daisy 227533
Bull Face 126039
Bull Faces 126025
Bull Flower 18173,68189
Bull Grape 411893
Bull Haw 109857
Bull Haws 109857
Bull Head 81237
Bull Heather 150131
Bull Horn Thom 1150
Bull Jumpling 399504
Bull Mallee 155495
Bull Mallow 243805
Bull Nettle 97741,212228,367014,367139
Bull Oak 15951,79161,79181
Bull Pates 126025,126039
Bull Pine 299992,300153,300189,300264
Bull Rattle 248411,363673,364193
Bull Rush 68189
Bull Seg 273531,273545,401112
Bull Thistle 81237,92033,92112,92485,364360
Bullace 316470
Bullace Grape 411893
Bullace Plum 316382,316470
Bullate Ancylostemon 22161
Bullate Blueberry 403750
Bullate Corallodiscus 103938
Bullate Cotoneaster 107366
Bullate Dragonhead 137564
Bullate Euonymus 157344
Bullate Macaranga 240312
Bullate Microula 254363
Bullate Oilnut 323067
Bullate Primrose 314186
Bullate Spindle-tree 157344
Bullate Tangut Corydalis 106507
Bullate-leaf Blueberry 403750
Bullateleaf Windhairdaisy 348178
Bullate-leaved Blueberry 403750
Bullatiform Monkshood 5088
Bullberry 362090,403916
Bullbine 95431
Bullbleleaf Bladderwort 403750
Bullbrier 366264,366559
Bulldock 31051
Bulldogs 28617,68189
Bullem 316819
Bullen Paphiopedilum 282795
Bullers 192358,316470
Bullesse 316470
Bullet 411893
Bullet Wood 244530
Bulletwood 255260,255312
Bulley Asiabell 98284
Bulley Blueberry 403751
Bulley Corydalis 105678
Bulley Dali Corydalis 105678
Bulley Herminium 192830
Bulley Larkspur 124080
Bulley Primrose 314189
Bulley Rabdosia 209639
Bulley Rockfoil 349138
Bulley Rockjasmine 23133
Bulley Tali Corydalis 105678
Bulleyia 63388,63389
Bullfoot 400675
Bullgrass 259654
Bull-head Pond-lily 267333
Bull-horn Acacia 1150
Bull-horn Thom 1150
Bullhorn Wattle 1150
Bullies 316470,316819
Bullimong 162312
Bullins 316470,316819
Bullions 316470
Bullison 316470,316819
Bullister 316819
Bullnut 77934
Bullock Syzygium 382497
Bullock Threewingnut 398322
Bullocks 37015
Bullock's Eye 356468
Bullock's Eyes 85875,174877,198680,260892,356468,358062,407072,407287
Bullocks Heart 25882
Bullock's Heart 25882
Bullock's Lungwort 405788
Bullock's Saussurea 348179
Bullocksheart 25882
Bullock's-heart 25882
Bullock's-heart Custard Apple 25882
Bullocks-heart Custard-apple 25882
Bullock's-heart Tree 25882
Bulloe 316470
Bullpoll 126039
Bullpull 126039
Bullrushes 68189,133478
Bulls 37015,109857
Bull's Bags 273531,273545
Bull's Cocks 37015
Bull's Coral-drops 53221
Bull's Ear 405788
Bull's Ears 405788
Bull's Eye 160772
Bull's Eyes 9701,18173,68189,202086,227533,243840,248312,248411,267294,282685
Bull's Foot 400675
Bull's Horn Acacia 1611
Bull's Pease 329521
Bulls-and-cows 37015
Bulls-and-wheys 37015
Bull's-horn Acacia 1613
Bullslop 315106
Bulltree 345631
Bullums 316470,316819
Bullum-tree 316470
Bullweed 81237
Bullwort 19659,19664,19672,355061
Bully Head 81237
Bully-blooms 316470
Bully-mung 162312
Bulnesia 63405
Bulrose 262441
Bulrush 352206,352223,353179,353375,353699,353701,353713,401085,401112
Bulrush Millet 289116
Bulwand 36474,340151
Bum Pipe 384714
Bumalda Bladder-nut 374089
Bumbil 71808
Bumble Tree 71808
Bumble-bee 272398,316127
Bumble-bee Flower 220416
Bumble-bee Orchid 272416
Bumblebee-orchid 272416
Bumbleberry 336419,338447
Bumblekite 338447
Bumbo 122151
Bumbum Mango 244406
Bumelia-leaved Barberry 51399
Bumlykite 338447
Bummalkyte 399504
Bummel 338447
Bummelberries 338447
Bummelkite 338447,339240
Bummeltykite 316127

Bum-meltykite 338447
Bummull 338447
Bum-pipe 384714
Bumpy Ash 166984
Bumweed 148153
Bun 28044
Buna 162368
Bunch Berry 85581
Bunch Flower 248677
Bunch Grape 411531
Bunch Grass 351309
Bunch of Keys 167955,315082,401388
Bunchberry 85581,104917
Bunch-berry 339240
Bunchberry Dogwood 85581
Bunchberry Elder 345628
Bunched Plantain 302009
Bunchflower 248677
Bunch-flower 248638
Bunch-flowered Daffodil 262457
Bunch-flowered Narcissus 262382,262457
Bunchgrass 266509
Bunch-grass 351309
Bunching Pearl Onion 15048
Bunchlihe Suoluo 327382
Bunch-like Ardisia 31369
Bunch-like Reevesia 327382
Bunch-like Trigonostemon 397373
Bunch-o'-daisies 3978
Bunch-of-grapes 20435,130383
Bunch-of-keys 85581,167955,315082,339240,401388
Bund 81237,379660
Bundle Flower 126177
Bundleflower 126168
Bundlefruit Gooseberry 334230
Bundweed 81237,192358,359158,379660
Bundy 155596
Bunela Plum 166777
Bunewand 192358
Bunge Ash 167931
Bunge Batrachium 48907
Bunge Bedstraw 170289
Bunge Buckthorn 328626
Bunge Buttercup 48907
Bunge Cherry 83220
Bunge Cinquefoil 312431
Bunge Corydalis 105680
Bunge Dense-flower Barberry 51542
Bunge Desertcandle 148537
Bunge Feathergrass 376723
Bunge Gentian 173860
Bunge Hackberry 80596
Bunge Indigo 205752
Bunge Milkvetch 42120
Bunge Mosquitotrap 117412
Bunge Needlegrass 376723
Bunge Pea-shrub 72199
Bunge Pine 299830
Bunge Prickly Ash 417180
Bunge Pricklyash 417180
Bunge Prickly-ash 417180
Bunge Rockjasmine 23215
Bunge Rose 336409
Bunge Spindle-tree 157345
Bunge Swallowwort 117412
Bunge Toadflex 230925
Bunge Vetch 408327
Bunge Ziziphora 418123
Bunge's Smartweed 308922
Bungor Raya 219966
Bungu 83667
Bungua 31696
Bungua Areca Palm 31696
Bungua Arecapalm 31696
Bunias 63443
Bunks 68713,90901,101852
Bunnel 359158
Bunnen 192358
Bunnerts 192358
Bunnies' Ears 373166
Bunnle 192358
Bunny Ears 272980
Bunny Mouth 28617,28631
Bunny Rabbits 28617,116736,130383,231183,237539
Bunny Rabbit's Ear 237539
Bunny Rabbit's Ears 237539
Bunny Rabbit's Mouth 28617,116736,130383
Bunny-ears 272980
Bunny-ears Pricklypear 272980
Bunny's Ear 373166,405788
Bunny's Ears 373166,405788
Buntan 93579,93590
Buntan Shaddock 93468
Bunwand 192358
Bunwede 102933,162519,359158
Bunweed 192358
Bunwort 192358
Bunya Bunya 30835
Bunya Pine 30835
Bunya-bunya 30835
Bunyabunya Araucaria 30835
Bunya-bunya Araucaria 30835
Bunya-bunya Pine 30835
Bunya-pine 30835
Bupleuroid Pondweed 312066
Bupleurum-leaf Alysicarpus 18259
Bur 9701,92485,199384,300223
Bur Beaked Chervil 28023
Bur Beggarticks 54158
Bur Beggar-ticks 54158
Bur Bristlegrass 361930
Bur Buttercup 83443,326437
Bur Chervil 26905
Bur Clover 247246,247425,396951
Bur Crowfoot 326287
Bur Cucumber 114122,362460,362461
Bur Daisy 68079,68082
Bur Dock 31044
Bur Flag 370050
Bur Forget-me-not 221627,221711
Bur Gherkin 114122
Bur Grass 80803,394372
Bur Head 140536
Bur Ironweed 406266
Bur Khaki 18137
Bur Marigold 53755,53773,53797,53912,54158,247881
Bur Medick 247381,247425
Bur Nut 395064,395146
Bur Oak 324114,324141
Bur Parsley 28023,79664,392992
Bur Ragweed 19190
Bur Reed 370021,370102
Bur Rose 336885
Bur Sage 19162,169313
Bur Sedge 74711
Bur Thistle 31051,92485
Bur Thrissel 92485
Bur Thrissil 92485
Bur Tree 31051,345631
Bur Weed 370021,414990,415057
Bur-beaked Chervil 28023
Burbrush 399325
Burchard Caralluma 72425
Burclover 247425
Bur-cucumber 362461
Burdekin Plum 304197
Burdett's Eucalypt 155509
Burdett's Gum 155509
Burdock 31044,31051,31056,332086,415003
Burdock Clover 396951
Bureav's Rhododendron 330265
Bureja Gooseberry 333929
Burford Coral Holly 203651
Burgan 218392
Burging Croton 113039
Burgmott 257161
Burgrass 394372,394390
Bur-grass 80855,394372
Burgundian Rose 336477
Burgundy Cabbage Rose 336477
Burgundy Cabbage-rose 336477
Burgundy Hay 247456
Burgundy Pitch 298191
Burhead 140536,170193
Burhead Sedge 278309
Buri Palm 106986,307539

Buried Pepper 300420
Burileek 407029
Burk 53563
Burke's Goldfields 222590
Burke's Lupine 238437
Burkili Rattan Palm 65654
Burkwood Daphne 122393
Burkwood Osmanthus 276243
Burkwood Viburnum 407652
Burkwood's Viburnum 407652
Burler's Teasel 133478
Burley 262722
Bur-like Mallotus 243382
Burma Agapetes 10295
Burma Anise-tree 204490
Burma Bamboo 47225
Burma Bamboo Shrub 82441
Burma Bambusa 47225
Burma Barberry 51400
Burma Bauhinia 49181
Burma Bean 294010
Burma Blackwood 121645
Burma Butterflyfruit 94403
Burma Cedar 392829
Burma Cleisostoma 94429
Burma Clematis 94798
Burma Closedspurorchis 94429
Burma Coast Padauk 320301
Burma Coast Padouk 320301
Burma Conehead 290921
Burma Dendrocalamus 125470
Burma Dendropanax 125596
Burma Dragonbamboo 125470
Burma Fagraea 162348
Burma Gastrochilus 171832
Burma Gentian 173309
Burma Goldenray 229004
Burma Gugertree 350932
Burma Gynostemma 183000
Burma Holly 204490
Burma Ixora 211167
Burma Lancewood 197728
Burma Mahogany 289789
Burma Mangrove 61263
Burma Padauk 320306
Burma Pantling Woodbetony 287500
Burma Paradombeya 283305
Burma Pennywort 200267
Burma Pine 300305
Burma Pyingndo 415693
Burma Rattan Palm 65687
Burma Rattanpalm 65687
Burma Reed 265923
Burma Rhododendron 330267
Burma Rosewood 121645
Burma Sabia 341484
Burma Square Bamboo 87551
Burma Square-bamboo 87551
Burma Tailanther 381925
Burma Teak 385531
Burma Toon 392829
Burma Treerenshen 125596
Burma Tulipwood 121784
Burmacoast Padauk 320301
Burma-coast Padauk 320301
Burmaflame Rhododendron 331866
Burmamahogany 289408
Burman Arrowwood 407717
Burman Bamboo 125490
Burman Viburnum 407717
Burmann Bushmint 203043
Burmann Cinnamon 91282
Burmann Oplismenus 272619
Burmann Sundew 138273
Burmannia 63949
Burmannia Family 64000
Burmann's Cassia 91282
Burmareed 265915,265923
Bur-marigold 53755,53976
Burmese Bamboo 82452,87632,249620
Burmese Fishtail Palm 78047
Burmese Lacquer 177545
Burmese Lacquer-tree 177545
Burmese Plumbago 83643
Burmese Rosewood 320301
Burmese Rose-wood 320301
Burmese Synotis 381925
Burmese Timber Bamboo 47394
Burmese Weavers' Bamboo 47225
Burmuda Butter Cup 278008
Burn Plant 17381,17383
Burner Rose 336969
Burnet 345816,345860,345881,345917
Burnet Bloodwort 345881
Burnet Ragwort 279947
Burnet Rose 336851,336854
Burnet Saxifrage 299342,299466,299523
Burnet-saxifrage 299342,299523
Burnett Rose 336851
Burning Bush 48786,95431,100627,107313,129615,129618,157268,157285,157315,157327,157793,217361,217373
Burning Climber 95431
Burning Fire 384714
Burning Nettle 403039
Burning Sinobambusa 364826
Burningbush 129615,157285,157327,217361
Burning-bush 157268,157327
Burning-bush Euonymus 157327
Burn-nose Tree 122621
Burnt Orchid 273696
Burnt Sedge 73604
Burnt-stick Orchid 273696
Burnt-tip Orchid 273696
Burnut 395146
Bur-nut 395146
Burnweed 148148
Burr Bristlegrass 361686
Burr Chervil 28023
Burr Medic 247381
Burr Oak 324141
Burr Rose 336885
Burra Range Grevillea 180586
Burracoppine Mallee 155510
Burrawang 241449,241463,241465
Burrdaisytree 166700
Burreed 370021,370102
Bur-reed 370021,370025,370030,370050
Burreed Family 370018
Bur-reed Family 370018
Bur-reed Sedge 76307
Burren Myrtle 31142
Burretiodendron 64024
Burridge Burrage 57101
Burr-nut 395082
Burro Bush 201013,201015
Burro Fat 95709
Burro Tail 356944
Burro Weed 19163
Burrobush 19186
Burrograss 354494,354495
Burro's Tail 356944
Burro's Tails 356944
Burrow Rose 336851
Burroweed 14941,209584
Burro-weed Strangler 275123
Burrowing Clover 397106
Burrow-rose 336851
Burrow's Tail 356944
Burrow's Tails 356944
Burrow-weed 185489
Burr-reed 370021
Burrweed 368527
Burry Vervain 405861
Bursage 19163
Bursage Mugwort 35502
Bursaria 64063
Burscale 44299
Burseed 221711
Bursera 64068
Bursera Family 64088
Burser's Saxifrage 349139
Burst-bellies 127793
Burst-belly Pink 127793
Bursting Heart 157315
Burstwort 192926,192941,192965
Burtons 31051,133478,170193,243803,322724
Burvine 64454,64466
Burweed 170193,399295,415053,415057,415063

Burying Box 64345
Busch 64345
Busch Corydalis 105682
Busch Ladymantle 13987
Büser Gentian 173305
Bush Allamanda 14881
Bush Anemone 77141
Bush Arrowleaf 304628
Bush Banana 245784
Bush Barberry 51577
Bush Basil 97060,268566
Bush Bean 294056
Bush Beard Grass 22688
Bush Briar 336364
Bush Butter-tree 121114
Bush Calceolaria 66273
Bush Caper Berry 71693
Bush Chinkapin 79019
Bush Chinquapin 79019
Bush Cinquefoil 289713,312574,312577
Bush Clematis 95358
Bush Clover 226677,226698,226699,226720,226890,226903,226908,226970,227005,227012
Bush Coffee 99008
Bush Croton 112894
Bush Currant 96650,253003
Bush Daisy 160853
Bush Fig 165724
Bush Flax 39875,295600
Bush Germander 388085
Bush Golden Chinquapin 89973
Bush Grape 411784
Bush Greens 18687
Bush Groundsel 46227
Bush Honeysuckle 130242,130252,130254,235929,235970
Bush Kidney Bean 294057
Bush Ladybell 7789
Bush Lawyer 338169
Bush Lemon 93505
Bush Lily 39875
Bush Mango 104319
Bush Monkey Flower 255186,255192
Bush Morning Glory 102998,207792,207938
Bush Muhly 259687
Bush Nasturtium 399604
Bush Nut 240210
Bush Oak 88506
Bush Okra 104103
Bush Palmetto 341421
Bush Pea 389505,389540,389558
Bush Pea-shrub 72233
Bush Penstemon 289319
Bush Pepper 96457,300397
Bush Poppy 125585
Bush Protea 315907
Bush Pumpkin 114277,114301
Bush Red Pepper 72070,72100
Bush Redpepper 72070,72100
Bush Red-pepper 72100
Bush Sage 93202
Bush Seepweed 379573
Bush Spiderling 56441
Bush Strawflower 415386
Bush Sunflower 364507
Bush Tea 268614
Bush Tree 64345
Bush Trumpet 14881
Bush Vetch 408352,408611
Bush Violet 61131
Bush Windmill-palm 393806
Bushbean 294056
Bushbeech 178024
Bush-beech 178024
Bushclover 226677
Bush-clover 226677,226936
Bushe Turcoman Barberry 52285
Bushful Rhododendron 331578
Bush-honeysuckle 130242
Bushkiller 79850
Bushman Candle 64086,346648
Bushman's Candle 346644
Bushman's Candles 346644
Bushman's Clothes-pegs 180600
Bushman's Poison 4954,4966,4968,4981
Bushman's River Cycad 145243
Bushman's-poison 4954
Bushmint 203036,203066
Bush-pea 321726
Bushraan's Tea 79395
Bush's Goosefoot 86962
Bush's Oak 323583
Bush's Poppy Mallow 67126
Bush's Purple Coneflower 140080
Bush's Sedge 73976
Bush's Skullcap 355396
Bushsunflower 364507
Bush-wood Barberry 51577
Bushwood Parrya 284990
Bushy Aster 380866,380869
Bushy Beard 95431
Bushy Broom Grass 22688
Bushy Cinquefoil 313039
Bushy Eastern Arborvitae 390598
Bushy Glasswort 342858
Bushy Hawkweed 196057
Bushy Knotweed 309683
Bushy Machilus 240579
Bushy Mint 250362,250368
Bushy Penstemon 289336
Bushy Pigweed 18655
Bushy Pinweed 223705
Bushy Rhodamnia 329794
Bushy Rhododendron 330603
Bushy Schizostachyum 351853
Bushy Seedbox 238160
Bushy Vetchling 222854
Bushy Wallflower 154530
Bushy Wild Buckwheat 152502
Bushy Yate 155535,155621
Bushy-pondweed 262015
Busse Acacia 1102
Bussu Palm 244497
Bustard Teak 64148,320307
Buster Cup 325518
Busters 127793
Bustic Willow 132656
Busy Lizzie 204756,205444
Butcher 273531,363673
Butcher Boy 21176,273531
Butcher Flower 273531
Butcher's Blood 248411
Butchers Broom 340989
Butcher's Broom 340989,340990,340994
Butcher's Prick Tree 157327
Butcher's Prick-tree 157429,328603
Butcher's Prickwood 157327,328603
Butchersbroom 340989,340990
Butcher's-broom 340989,340990
Butchersbroom Family 340541
Butea 64144
Butia Palm 64159,64161
Butomus Family 64173
Buton Barringtonia 48511
Butt Bougainvillea 57854
Butte County Fritillary 168395
Buttefly Gardenia 382765
Butter 325820
Butter And Eggs 231183
Butter Basket 399504
Butter Bean 294010
Butter Blob 68189,399504
Butter Bump 399504
Butter Bur 292342
Butter Bush 82098
Butter Churn 267294,325518,325666,326287
Butter Cress 325518
Butter Cup 325498
Butter Daisy 227533,325518,325666,326287,405936
Butter Dock 31051,292372,339914,340109,340151
Butter Docken 339914
Butter Fingers 28200
Butter Flower 68189,237539,325518,325666,326287
Butter Flyiris 258529
Butter Fruit 132351
Butter Haw 109857

Butter Jags 237539
Butter Leaves 44468,339914
Butter Nut 77947,212599
Butter Orchid 302241
Butter Pat 409804
Butter Pats 409804
Butter Pea 81872,97203
Butter Plant 299759
Butter Print 1000
Butter Root 299759
Butter Rose 315082,315101,325518,325666,326287
Butter Seed 64203
Butter Tree 48990,109236,241514,289475,289476,400786
Butter Weed 38147
Butter Winter 87498
Butter Wood 132466
Butter Wort 299736,299759
Butter-and-bread 109857
Butter-and-cheese 243840,277648,325518,325666,325820,326287
Butter-and-eggs 208771,227533,227875,231183,237539,262405,262438,262441,277648
Butter-and-sugar 231183
Butterblob 68189
Butter-blob 68189,399504
Butterbumps 325518,325666,326287,399504
Butterbur 292342,292372,415003
Butter-bur 292342
Butterbur Coltsfoot 292372
Butterbur-like Cacalia 283857
Butterbush 301369
Butterchops 325820
Buttercreese 325666,326287
Buttercup 68189,312360,325498,325518,325612,326326
Buttercup Crowfoot 325518
Buttercup Family 325494
Buttercup Ficaria 325820
Buttercup Oxalis 278008
Buttercup Pennywort 200354
Buttercup Primrose 314405
Buttercup Tree 98113
Buttercup Waxpetal 106669
Buttercup Winter Hazel 106669
Buttercup Winterhazel 106669
Buttercup Winter-hazel 106669
Buttercupleaf Larkspur 124287
Buttercup-like Clematis 95272
Buttercup-like Globeflower 399539
Buttercuplike Metanemone 252498
Butterdock 292372
Buttered Eggs 90423,231183,237539,262405
Buttered Fingers 28200
Buttered Haycock 231183
Buttered Haycocks 231183
Butterfluflower Arrowwood 407878
Butterfly 3462,32563,222789
Butterfly Amitostigma 19523
Butterfly Arrowwood 408034
Butterfly Astyleorchis 19523
Butterfly Bush 49247,61953,61957,62019,62020,62021,62024,62028,62054,62099,62110,62118,337435,360426
Butterfly Dendrobium 125304
Butterfly Flower 37811,49173,351343,351345,410677
Butterfly Gaura 172201
Butterfly Ginger 187432
Butterfly Iris 130292,208740,208854,258380
Butterfly Ladies 282685
Butterfly Lady 282685
Butterfly Lily 187432
Butterfly Mariposa 67635
Butterfly Milkweed 38147
Butterfly Mist 11203
Butterfly Orchid 145298,270834,302241,302260,302280,319386
Butterfly Palm 31012,89345,89360
Butter-fly Pansy Orchid 254934
Butterfly Passionflower 285686
Butterfly Pea 97184,97195,97203
Butterfly Pincushion 350119
Butterfly Plant 200802
Butterfly Rose 336491
Butterfly Stonecrop 200802
Butterfly Swordflag 208640
Butterfly Tree 49211
Butterfly Tulip 67554,67634,67635
Butterfly Vanilla 405016
Butterfly Viola 409858
Butterfly Violet 410351,410587
Butterfly Weed 38147
Butterflybush 61953
Butterfly-bush 61953,62019,62035
Butter-fly-bush 61953
Butterfly-bush Family 62208
Butterfly-flower 351343
Butterflyflower Azalea 330018
Butterflyfruit 94401,94402
Butterflygrass 392880
Butterfly-lily 67554,187417,187432
Butterfly-orchid 270834,273569,302241
Butterflyorchis 293577,293582
Butterflypea 81872
Butterfly-pea 81872
Butterfly-tree 49211
Butterfly-weed 38147
Butterfruit 64203,64223,132351
Butter-fruit 291494
Butterknife Bush 114566
Butternhole Flower 159313
Butternut 77889,77946,77947,212599
Butter-nut 53055,77946
Butter-nut Tree 77946
Butternut Walnut 212599
Butterprint 1000
Butter-print 1000
Butterpump 267294
Butter-tree 64223
Butterweed 103446,279913,313881,358159,358757
Butterwort 299736,299759,345957
Butterworth's Wild Buckwheat 151887
Buttery 345631
Buttery Eggs 262429
Buttery-entry 410677
Button Burrweed 368528
Button Bush 82091,82098
Button Cactus 147422,147425
Button Clover 247407
Button Eryngo 154355
Button Everlasting 189761
Button Flower 194661
Button Grass 182616
Button Mangrove 101933
Button on a String 109266
Button Pink 127753
Button Snakeroot 154284,154355,228433,228511,228529
Button Weed 370736
Button Wood 101933,220233,302574
Button Wood Buttonwood 302588
Button Zehneria 417488
Buttonball 302588
Button-ball 302588
Buttonbush 82091,82098,82107
Buttonbush Dodder 114993
Buttonflower 404508,404523
Buttongrass 121306
Button-grass 121281,121306
Buttonhole Plant 94349
Button-rods 151208
Buttons 383874
Buttonweed 81237,107747,107766,131358,131361,192358
Button-weed 131358
Buttonwood 3462,101933,302574,313344
Button-wood 220233,302588
Butulang Canthium 71348
Butuo Cranesbill 174510
Butyrospermum 64203
Buwi Camellia 68970
Buxbaum's Sedge 73977,76490
Buxbaum's Speedwell 407287
Buxifoliate Cherry 156152
Buzzy Lizzy 205444

Byass Palm 271001
Byfield Fern 57974
Byrsonima 64425
Bysmale 18173
By-the-wind 95431
Byttneria 64454
By-Yu 241463
Byzant Oat 45424
Byzantine Gladiolus 176122

C

C. American Rubber 79110
C. B. Clarke Asian Barberry 51334
C. B. Clarke Begonia 49727
C. B. Clarke Blumea 55711
C. B. Clarke Boea 56050
C. B. Clarke Corydalis 105744
C. B. Clarke Gentian 173343
C. B. Clarke Kobresia 217142
C. B. Clarke Leucas 227575
C. B. Clarke Nepeta 264899
C. B. Clarke Primrose 314253
C. B. Clarke Trisetum 398453
C. B. Clarke Underleaf Pearl 296521
C. B. Clarke Uraria 402112
C. B. Clarke Willowweed 146668
C. B. Clarke Woodbetony 287097
C. B. Clarke's Evergreenchinkapin 78897
C. B. Clarke's Leafflower 296521
C. China Scurrula 236974
C. K. Schneid. Violet 410541
C. K. Schneid. Zelkova 417552
C. Shih Raspberry 339266
C. Shihmien Primrose 314945
C. Shihu Evodia 161376
C. Xizang Rockfoil 349909
Ca Jan Bean 65146
Caa-ehe 376415
Cabage Palm 337883
Cabage Tree 337883
Caballera Palo 273993
Cabaret 37616
Cabbage 59278,59520,59532,99086,326812
Cabbage Angelin Tree 22218
Cabbage Angelin-tree 22218
Cabbage Coneflower 339585
Cabbage Daisy 399504
Cabbage Family 113144
Cabbage Flower 364557
Cabbage Gum 155476,155688
Cabbage Lettuce 219489
Cabbage Palm 104334,104339,104359,161057,234162,234166,337883,341422
Cabbage Palmetto 341422
Cabbage Palmetto Palm 341422
Cabbage Rose 280231,336474
Cabbage Royal Palm 337883
Cabbage Seed 364557
Cabbage Thistle 92268
Cabbage Tree 22211,22218,27578,104334,104339,104350,104359,115190,115217,337883,406040,406252
Cabbage Tree Palm 234166
Cabbage-leaf Coneflower 339585
Cabbage-palm 341422
Cabbage-rose 336474
Cabbage-tree 22211,22218,104339
Cabbish 59532
Cabeca De Frade 249582
Cabelhida 261040
Cabeza De Viejo 82211,140313
Cabin Pogostemon 306964
Cabin Potchouli 306964
Cabinet Cherry 83311,316787
Cabomba Family 64503
Cabosu 93805
Cabrera Barberry 51410
Cabreuva 261527
Cabul Toadflex 230928
Cabuya 169242
Cabuya Fibre 169243
Cabuyao 93492,93559
Cacalia 283781
Cacalia Muhlenbergii 34809
Cacaliform Rhubarb 329345
Cacao 389399,389404
Cacao Tree 389404
Cacaotree 389399,389404
Cacao-tree 389404
Cachana 207457
Cache Valley Wild Buckwheat 152233
Cachibou 66173
Cachris 64776
Cachrys 64776
Cackle Buttons 31051
Cackle Dock 31051
Cacomite 391627
Cacti 272856
Cactuform Cissus 92640,411868
Cactus 243994
Cactus Apple 272873
Cactus Bossiaea 57501
Cactus Dahlia 121557
Cactus Family 64822
Cactus Tree 272924
Cadaga 106824,155763
Cadaghi 155763
Cadagi 106824
Caddell 192358
Caddie-me-to-you 410677
Cadia 64932
Cadillo 402245
Cadillo-leaved Grewia 181006
Cadlock 326609,364557
Caducouspetal Camellia 69235
Caducous-petaled Camellia 69235
Cadweed 192358
Caelospermum 98808
Caesalpinia 64965
Caesalpinia Family 65084
Caesar Weed 402245
Caesar's Weed 402245
Caesarweed 402245
Caesian Raspberry 338209
Caesious Acacia 1107
Caesious Caesalpinia 64979
Caespitose 346420
Caespitose Cousinia 108249
Caespitose Craneknee 221640
Caespitose Cystopetal 320209
Caespitose Garnotia 171490
Caespitose Leek 15145
Caespitose Onion 15145
Caespitose Vinegentian 108249
Caffir Lime 93492
Caffra Devilpepper 327001
Caffre Lime 93492
Caffron Gaoliang 369619
Caffron Sorghum 369619
Cafta 79387,79395
Cagayan Nutmeg 261412
Cage Arrowbamboo 162658
Cagge 208771
Cahaba Fleabane 150987
Cahaba Torch 228497
Cahaba-lily 200921
Cahum 169242
Cai Begonia 50374
Cai Helicia 189957
Cain-and-abel 30081,102727,273469
Cainito 90032
Cainito Star Apple 90032
Cainito Star-apple 90032
Cairo Morning Glory 207659
Cairo Morningglory 207659
Caise 101852
Caja 88691
Caja Fruit 372477

Cajan 65143,65146
Cajeput 248072,248104
Cajeput Tree 248104,248105,248114
Cajeputtree 248104
Cajeput-tree 248114
Cajun Bean 65146
Cajuput 248072,248104
Cajuput Oil 248081
Cajuput Tree 248104
Cake Tree 83736
Cakers 192358
Cakeseed 101852,192358
Cakezie 101852
Calaba 67856
Calaba Balsam 67871
Calabar Bean 297996,298004,298009
Calabar Ebony 132299
Calabarbean 298004
Calabash 111104,219821,219827,219843,383006,383009
Calabash Cucumber 219843
Calabash Gourd 219843
Calabash Jujube 418172
Calabash Nutmeg 257653
Calabash Sweetleaf 381423
Calabash Tree 111100,111104
Calabash-fruited Sweetleaf 381423
Calabashleaf Dutchmanspipe 34158
Calabashtree 111100,111104
Calabash-tree 111100,111104
Calabrian Black Pine 300117
Calabrian Pine 299827
Calabrian Soapwort 346421
Calabura 259910
Caladenia 65202
Caladium 65214,65218
Caladium Star Light 65229
Calaf of Persia Willow 342967
Calamagrostis Speargrass 4120
Calamander Wood 132365
Calambac 29977
Calamint 65528,65538,65540,96960
Calamint Balm 65540
Calamint Savory 65603
Calamintha 65528
Calamondin 93566,93636,93639
Calamondin Orange 93639
Calamus Root 5793
Calamus Sweetflag 5793
Calanthe 65862,65921
Calathea 66157,66159,66185
Calathian Violet 173728
Calathodes 66219
Calcalary 120310
Calcarate Hornflower 83393
Calcarate Moulmain's Rhododendron 331281
Calcarate Violet 409796
Calcareous Indigo 205765
Calcareous Pleurospermum 304770
Calcareus Draba 137126
Calcareus Pricklyash 417188
Calcareus Whitlowgrass 137126
Calcarious Michelia 252832
Calcarious Spiraea 371854
Calceolaria 66252,66271
Calceolate Thrixspermum 390496
Calcicole Lacquertree 393446
Calcicole Lacquer-tree 393443
Calcicole Ornithoboea 274467
Calcicole Pittosporum 301234
Calcicole Rabdosia 209641
Calcicole Sabia 341485
Calcicolous Amoora 19955
Calcicolous Bellflower 69930
Calcicolous Birch 53371
Calcicolous Corydalis 105687
Calcicolous Dimocarpus 131060
Calcicolous Evodia 161316
Calcicolous Machilus 240563
Calcicolous Manglietia 244424
Calcicolous Pittosporum 301234
Calcicolous Seatung 301234
Calcicolous Stachyurus 373519
Calcicolous Woodlotus 244424
Calciphile Casearia 78153
Calcisteppe Barberry 51413
Calciumsoil Goosecomb 335251
Calciumsoil Roegneria 335251
Calcutta Bamboo 125510,297367
Calder Primrose 314197
Caldesia 66337
Caleana 66368
Caleb Saltbush 44341
Calendula 66380,66445
Caley Banksia 47634
Caley Pea 222726
Caley's Grevillea 180579
Caley's Ironbark 155512
Calfkill 215387
Calf's Foot 15876,37015,364557,400675
Calf's Snout 28617
Calf's-foot 400675
Calgaroo 155684
Calgary Carpet Juniper 213908
Calicicolous Buckthorn 328628
Calicicolous Cryptocarya 113435
Calicicolous Thickshellcassia 113435
Calico Aster 40704,40706,40708,40709,380922
Calico Bush 215395
Calico Cactus 272891
Calico Flower 34246
Calico Hearts 8512
CalicoAster 380922
Calicoplant 18095
Califolian Cypress 114698
California 330277
California Agoseris 11465
California Alder 16435
California Amaranth 18682
California Aster 104652
California Barrel Cactus 163432,163434
California Bayberry 261132
California Big Tree 360576
California Black Currant 333928
California Black Oak 324071
California Black Walnut 212592
California Blackberry 339466
California Blue Bell 293179
California Blue Oak 323849
California Blue Sage 344973
California Bluebell 293158
California Blue-eyed Grass 365714
California bog-asphodel 262575
California Boxelder Maple 3227
California Brickellia 60108,61039
California Buckeye 9679
California Buckthorn 328630,328635,328638,328639
California Buckwheat 152056
California Bulrush 352169
California Bur Clover 247425
California Burclover 247425
California Bur-clover 247425
California Calla 417117
California Chicory 325074
California Cholla 116650
California Christmas-tree 80087
California Clubrush 352169
California Coneflower 339525
California Cottonrose 235261
California Currant 333932
California Dandelion 384486
California Desertdandelion 242926
California Dodder 114983
California Dutchman's Pipe 34133
California Elder 345604
California Ephedra 146144
California Falseindigo 19994
California Fan Palm 413298
California Fawnlily 154903
California Fawn-lily 154903
California Fern 101852
California Fescue 163853
California Fetid Adder's-tongue 354574
California Filbert 106725
California Fishhook Cactus 244054
California Flannel Bush 168216
California Fremontia 168216
California Fuchsia 417407
California Fuschia 146651

California Gilia 175669,175670
California Glorybind 103095
California Goldenbush 150322
California Goldenrod 368471
California Goldfields 222591
California Goosefoot 86982
California Gray Rush 213360
California Hazel 106725
California Hazelnut 106725
California Holly 193839,193841
California Holly Mahonia 242617
California Huckleberry 403944
California Hyacinth 60431
California Incense Cedar 228639
California Indian Pink 363282
California Joint Fir 146144
California Juniper 213625
California Lady's-slipper 120317
California Laurel 401672
California Lilac 79902,79945
California Live Oak 323621
California Mahonia 242504,242617
California Manyseed 307733
California Mogollon Ceanothus 79946
California Mountain-pincushion 275284
California Nutmeg 393047,393057
California Nutmeg Yew 393057
California Oak 324114
California Orach 44331
California Palm 413298
California Pepper 350990
California Pepper Tree 350990
California Peppertree 350990
California Pepper-tree 350990
California Pincushion 244243
California Pink 363651
California Pitcher 122758
California Pitcher Plant 122758,122759
California Pitcherplant 122758
California Plane 302595
California Plane Tree 302595
California Plumeseed 325074
California Poplar 311554
California Poppy 155170,155173,155180
California Privet 229569
California Rabbit-tobacco 317734
California Rape 364557
California Rayless Fleabane 150930
California Red Fir 409
California Redbud 83753,83798
California Redwood 360565
California Rhododendron 330277
California Rockcress 30200
California Rose-bay 330277
California Rosewood 405223
California Sagebrush 35231
California Sandwort 255451
California Saxifrage 349145
California Scrub Oak 323706,323849
California Sea-blite 379504
California Snakeroot 34133
California Snowbell 379319
California spicebush 68329
California Spineflower 259473
California Strawberry 167653
California Sunflower 188931
California Swamp Pine 300098
California Swamp Thistle 91917
California Sycamore 302595
California Thistle 92252
California Threefold 399397
California Thrift 34536
California Torreya 393057
California Tree Mallow 223357
California Tree Poppy 335870
California Tree-poppy 335870
California Valley Oak 324114
California Wakerobin 397549
California Walnut 212592
California Washington Palm 413298
California Washingtonia 413298
California Wax Myrtle 261132
California Wax-myrtle 261132
California White Fir 317
California White Oak 324114
California White Sage 344853
California Wild Grape 411594
California Wild Pine 300082
California Willow Dock 339977
California-holly 193841
Californian Alder 16435
Californian Allspice 68329
Californian Amaranth 18682
Californian Barberry 52046
Californian Bay 401670,401672
Californian Bayberry 261132
Californian Big Tree 360576
Californian Bird of Paradise 64980
Californian Black Oak 324071
Californian Bluebell 263501,293171,293179
Californian Brome 60659
Californian Buckeye 9679
Californian Buckthorn 328838
Californian Buttercup 325690
Californian Calocedar 228639
Californian Elder 345604
Californian Everlasting 178156
Californian False Hellebore 405581
Californian Fever-bush 171552
Californian Fir 407
Californian Fuchsia 146586,334212,417405,
417407
Californian Gum Plant 181161
Californian Harebell 70230
Californian Hazel 106772
Californian Honeysuckle 235857,235874
Californian Hyacinth 128870,128884,398806
Californian Laurel 401672
Californian Lilac 79902,79973
Californian Line Oak 323762
Californian Lobelia 136872,136873
Californian Mountain Pine 300082
Californian Mugwort 36489
Californian Nutmeg 393057
Californian Olive 401672
Californian Panther Lily 229962
Californian Pinyon Pine 300171
Californian Plum 316761
Californian Poison Oak 332570
Californian Poppy 155170,155173,302964,
335869,335870
Californian Privet 229569
Californian Red Fir 409
Californian Redwood 360565
Californian Rose Bay 330277
Californian Sassafras 401672
Californian Slippery Elm 168216
Californian Spearmint 139678
Californian Violet 410394
Californian Whispering Bells 145020
Californian White Cedar 228639
Californian Yew 385342
California-nutmeg 393057
California-poppy 155173
Californica Fluffweed 235261
Calisaya 91059
Calk Maple 3100
Calla 66593,66600,417092
Calla Family 30491
Calla Lilies 417092
Calla Lily 66593,68877,334310,417092,
417093,417118
Callalily 417092,417093
Callards 59532
Callery Pear 323116,323133
Calliandra 66660
Callianthemum 66697
Callicarpaeleaf Indianmulberry 258880
Callier Cinquefoil 312433
Callier Tulip 400138
Calligonum 66988
Calliopsis 104598
Callistemon 'Little John' 67306
Callitris 67405
Call-me-to-you 410677
Callock 364557
Callose Actinidia 6530
Callose Agrostophyllum 12445
Callose Fig 164743
Callose Kiwifruit 6530
Callous Grape 411636

Callous Photinia 295651
Callous Sabia 341487
Callous Stachyurus 373520
Callous-fruited Water Dropwort 269348
Calm Lilac 248895
Calochilus 67541
Calogyne 67686
Caloncoba 67704,270897
Calonesia Euphorbia 159159
Calophaca 67771
Calophanoides 67816
Calophyllous Dolomiaea 135724
Calophyllous Kadsura 214951
Calophyllous Mucuna 259488
Calopogon 67882
Calopogon Grass Pink 32373
Calospatha 68014
Calotis 68079
Calotrope 68086
Calpocalyx 68113
Calscalary 120310
Calthrop 395146
Caltilage Bladderwort 403837
Caltivated European Grape 411987
Caltra 80981
Caltrap 80981,395064,395146
Caltrop 68189,325518,325666,326287,
394500,395064,395146,418585
Caltrop Family 418580
Caltropleaf Milkvetch 43178
Caltroplike Evergreen Chinkapin 79054
Caltroplike Evergreenchinkapin 79054
Caltroplike Milkvetch 43180
Caltrops 80981,395146
Caltropus Milkvetch 43178
Calumba 212095
Calumba Root 212095
Calunga-bark 364370
Calvary Clover 247246,247282,247317
Calvepetal 177982
Calvert Glorybind 102971
Calvous Rhododendron 330300
Calycanthus 68306
Calycanthus Family 68300
Calyciform Clemati 78874
Calyciform Evergreen Chinkapin 78874
Calycinate Pimpinella 299373
Calycose Mille Graines 269878
Calycular Cassandra 85414
Calycular Chamaedaphne 85414
Calycular Willow 343136
Calyculate Aechmea 8550
Calypso 68531,68538
Calypso Orchid 68538,68541
Calyptrate Glyptic-petal Bush 177986
Calyptrocalyx 68614
Calysasa 91059
Calystegia 68671
Calyx Brachystemma 58966
Calyx Smelowskia 366122
Calyxform Figwort 355071
Calyx-leaved Sweetleaf 381291
Calyxless Sweetgum 232551
Calyxshape Kydia 218444
Calyx-shape Kydia 218444
Calyx-shaped Baliospermum 46966
Calyxshaped Carrierea 77644
Calyx-shaped Carrierea 77644
Calyx-shaped Daphniphyllum 122669
Calyxshaped Goathorntree 77644
Calyx-shaped Kydia 218444
Calyx-shaped Tigernanmu 122669
Camarid Maxillaria 246707
Camarotis 68783
Camas 68789
Camash 68789,417880
Camass 68789,68794,68803
Camassia 68789
Cambie-leaf 267294
Cambodia Apodytes 29587
Cambodia Decaspermum 123454
Cambodia Dracaena 137353
Cambodia Dragonbood 137353
Cambodia Eightangle 204491
Cambodia Mastixia 246218
Cambodia Paperelder 290376
Cambodia Persimmon 132193
Cambodia Polyosma 310061
Cambodia Pristimera 315359
Cambodia Rockvine 387752
Cambodia Thin Evodia 161351
Cambodia Whitlowwort 204491
Cambodia Whitlow-wort 204491
Cambodian Polyosma 310061
Cambodian Rockvine 387752
Cambogia Garcinia 171076
Cambridge Milk-parsley 357780,357789
Cambridge Oak 324553
Cambridge Parsley 357789
Cambuck 101852
Camchaya 68830
Camden Woollybutt 155633
Camel 3978
Camel Bush 395788
Camel Thorn 1206,1245,14631
Camelia Roja 290711
Camelina 68839
Camellia 68877,69156
Camellia Flower 69156
Camellia-flowered Balsam 204836
Camellia-like Rhododendron 330302
Camel's Foot 48990
Camel's Foot Tree 49211
Camel's-thorn 14638
Camelthorn 14623,14631
Camel-thorn 14638
Camel-tooth Actinidia 6534
Cameron Hibiscus 194774
Cameroon Cardamom 9904
Cameroon Jointfir 178527
Cameroon Persimmon 132231
Camerouns Cardamom 9904
Camfield's Hop Bush 135203
Camias 45724
Camlick 192358
Cammick 202086,271563,359158
Cammil 3978
Cammock 3978,237539,271347,271563,
292957,321554,350425,359158
Cammock Ononis 271563
Camomile 26738,246307,246396
Camomile Goldins 397970
Camomileleaf Soliva 368528
Camomileleaf Woodbetony 287009
Camomill 85526
Camomille Maroute 26768
Camomine 85526
Camoroche 312360
Camote 207623
Camovyne 26768,85526
Camowyne 85526
Campanelle 68713
Campanilla 14873
Campanula 69870
Campanula glomerata 70054
Campanulate Cremanthodium 110358
Campanulate Knotweed 308941
Campanulate Magnific Deutzia 127010
Campanulate Nutantdaisy 110358
Campanulate Pink Deutzia 127064
Campanulate Sabia 341490
Campanulate Sage 344926
Campanulate-sepaled Camellia 68966
Campanumoea 70386
Campbell Barberry 51418
Campbell Lycaste 238735
Campbell Magnolia 416688
Campbell Maple 2836
Campbell Sagebrush 35234
Campbell's Magnolia 416688
Campeachy Wood 184518
Campeche 184518
Campeche Wood 184518
Camperdown Elm 401512,401513
Campernelle 262429
Campernelle Jonquil 262429
Campestre Palm 398827
Camphire 111347,223454
Camphor 91287,383712
Camphor Arrowbamboo 137842
Camphor Basil 268550

Camphor Laurel 91287
Camphor of Borneo 138550
Camphor of Malaysia 138550
Camphor Plant 383712
Camphor Pluchea 305083
Camphor Tree 91252,91287
Camphor Weed 194195,194197,194254,194256,305083
Camphor Wood 91287,384877
Camphor Wormwood 35266
Camphora-tree 91287
Camphorfume 70460
Camphor-fume 70460
Camphorleaf Elaeagnus 141953
Camphorleaf Pepper 300484
Camphorleaf Stairweed 142791
Camphorsmell Adenosma 7980
Camphortree 91287
Camphor-tree 91287
Camphortreeleaf Pepper 300484
Camphorweed 194256
Camphorwood 384877
Campion 363141,363300,363411,363477,363671,364146
Campion of Constantinople 363312
Campos-porto Barberry 51419
Camproot 175452
Camptotheca 70574,70575
Campylotropous Barberry 51420
Camus Cyclobalanopsis 116069
Camus Lily 68789
Camwood 47685,47788,320278,320330
Can Dock 267294,267648
Cana 71181,341410
Cana Brava 37467
Canach 152769
Canada Anemone 23740
Canada Ash 167897,168032,168057
Canada Azalea 330321
Canada Balsam 282
Canada Barberry 51422
Canada Bead-ruby 242675
Canada Birch 53338
Canada Blackberry 338225
Canada Blueberry 403758,403915
Canada Bluegrass 305451
Canada Blue-grass 305451
Canada Brome 60654,60900
Canada Cinquefoil 312435
Canada Cocklebur 415003
Canada Crook-neck Squash 114292
Canada Dewberry 338225
Canada Ducklingcelery 113872
Canada Garlic 15149
Canada Gentian 103446
Canada Goldenrod 368013
Canada Gooseberry 334137
Canada Grasscoal 85581
Canada Hawkweed 195511,195677
Canada Hawthorn 109582
Canada Hemlock 399867
Canada Horse-balm 99844
Canada Lettuce 219249
Canada Lily 229779
Canada Lime-grass 144216
Canada Lymegrass 144216
Canada Manna-grass 177577
Canada Mayflower 242675,242676
Canada Moonseed 250221
Canada Mountain Rice Grass 276000
Canada Nettle 221547
Canada Onion 15149
Canada Parrot Feather 261379
Canada Parrot-feather 261379
Canada Parrot's-feather 261379
Canada Pea 408352
Canada Pedicularis 287063
Canada Plum 34431,316206,316591
Canada Poplar 311146
Canada Potato 189073
Canada Redbud 83757
Canada Reed-grass 65303
Canada Rush 212970
Canada Sand-spurry 370619
Canada Serviceberry 19251
Canada Shadbush 19251
Canada Snake-root 37574
Canada Soappod 182527
Canada Spruce 298286
Canada Thistle 91770
Canada Tick Clover 126278
Canada Tickclover 126278
Canada Toadflax 230931,267347
Canada Tree 172047
Canada Verbena 405820
Canada Vervain 405820
Canada Violet 409798
Canada Waterweed 143920
Canada Wild Rice 418080
Canada Wildginger 37574
Canada Wildrye 144216
Canada Wild-rye 144216
Canada Wood Nettle 221547
Canada Yew 385344
Canadapitch 282
Canadelabra Cactus 159204
Canadian Anemone 23740
Canadian Antennaria 26425
Canadian Arrowhead 342397
Canadian Aspen 311335
Canadian Beadruby 242675
Canadian Birch 53338
Canadian Bitter-root 228298
Canadian Black Currant 334037
Canadian Black Snakeroot 345945
Canadian Black Spruce 298349
Canadian Brome 60900
Canadian Burnet 345834
Canadian Clearweed 299024
Canadian Columbine 30007
Canadian Dogwood 105130
Canadian Elder 345580
Canadian Enchanter's-nightshade 91522
Canadian Eyebright 160143
Canadian False Lily-of-the-valley 242675
Canadian Fleabane 103446
Canadian Garlic 15149
Canadian Germander 388026
Canadian Golden Rod 368013
Canadian Goldenrod 368013,368029,368032,368035
Canadian Hemlock 399867
Canadian Hemp 29473
Canadian Honewort 113872
Canadian Horseweed 103446
Canadian Hybrid Lilac 382251
Canadian Juniper 213703,213706
Canadian Lettuce 219249
Canadian Lily 229779
Canadian Lousewort 287063
Canadian Maple 3505
Canadian Mayflower 242675
Canadian May-lily 242675
Canadian Milk Vetch 42137,42156
Canadian Milk-vetch 42137
Canadian Mint 250331
Canadian Moonseed 250221
Canadian Pine 300180
Canadian Plum 34431
Canadian Pondweed 143920
Canadian Poplar 311146,311237
Canadian Pussytoes 26425
Canadian Pussy-toes 26425
Canadian Red Oak 320278
Canadian Red Pine 300180
Canadian Reedgrass 65303
Canadian Rice Grass 276000
Canadian River Wild Buckwheat 151807
Canadian Rush 212970
Canadian Sagebrush 35273
Canadian Sand-spurrey 370619
Canadian Sanicle 345945
Canadian Serviceberry 19251
Canadian Setose Iris 208821
Canadian Small Reed 127219
Canadian Snakeroot 37574
Canadian Spruce 298286
Canadian St. John's-wort 201788
Canadian Tea 172135
Canadian Thistle 91770
Canadian Tick-trefoil 126278

Canadian Violet 409798,409799
Canadian Walnut 212631
Canadian Water-thyme 143920
Canadian Waterweed 143920
Canadian White Birch 53549
Canadian White Violet 409798
Canadian Wild Ginger 37574
Canadian Wild-ginger 37574
Canadian Wood-nettle 221547
Canadian Yellow Birch 53338
Canadian Yellow Pine 300211
Canadian Yew 385344
Canaigre 340084
Canaigre Dock 340084
Canal Grass 117245
Canaliculate Roegneria 335405
Canaliculate Sansevieria 346063
Canaliculate Sedge 74001
Cananga 70954
Canaomile Brassbuttons 107752
Canari Island Lavender 223271
Canari Lavender 223271
Canarian Date Palm 295474
Canaries Statice 230552
Canari-nut-tree 70996
Canarium 70988
Canary Ash 50474
Canary Balm 80063
Canary Banana 260250
Canary Bellflower 70974
Canary Bird Bush 111868
Canary Bird Flower 399606
Canary Broad-leaf 302068
Canary Broom 172922,173015
Canary Creeper 399606
Canary Date 295459
Canary Date-palm 295459
Canary Flower 302068
Canary Food 302068,360328
Canary Grass 293705,293729,293731
Canary Holly 204157
Canary Island Bellflower 70974
Canary Island Daisy 41617
Canary Island Date 295459
Canary Island Date Palm 295459
Canary Island Date-palm 295459
Canary Island foxglove 210231
Canary Island Geranium 174798
Canary Island Ivy 187200
Canary Island Juniper 213629
Canary Island Laurel 223178
Canary Island Pine 299835
Canary Island Sage 344936
Canary Island Spurge 158603
Canary Island St. Johnswort 201792
Canary Island Strawberry Tree 30882
Canary Island Tamarisk 383468
Canary Island Viburnum 408092
Canary Islands Banana 260250
Canary Ivy 187197,187200
Canary Jasmine 211740
Canary Laurel 223168,223178
Canary Mayten 246821
Canary Nasturtium 399606
Canary Oak 323732
Canary Palm 295459
Canary Phalaris 293729
Canary Pine 299835
Canary Rhododendron 331159
Canary Seed 360328
Canary Tree 70988,71023
Canary Whitewood 232609
Canary Wood 258882
Canary-bird Bush 111868
Canarybird Flower 399606
Canary-clover 136740
Canarygrass 293705,293729
Canary-grass 293705,293729
Canarytree 70988
Canary-tree 70988
Canavan 152769
Canby's Bulrush 352183
Cancellate Trigonella 397206
Cancer 248411
Cancer Bush 380032
Cancer Corn 102003
Cancer Jalap 298094
Cancer Root 216262
Cancer-root 102002,146520,275050,275234,275255,298094
Cancerweed 345187
Cancerwort 216262,216292,407256
Cancharana Cabralea 64507
Cancrinia 71085
Candalabra Aloe 16592
Candalabra Spruce 298368
Candelabar Tree 30831
Candelabra Cactus 159204,272915
Candelabra Dahlia 121556
Candelabra Flower 61390,61407
Candelabra Plant 16592,159204
Candelabra Tickseed 104755
Candelabra Tree 30831,159140
Candelabra-cactus 159204
Candelabro 279493
Candelabrum Larkspur 124099
Candele-berry 261137
Candelilla 158458
Candelilla Wax 158458
Candenat Rosewood 121647
Candicant Raspberry 338232
Candied Peel 93603
Candies 216753
Candle Anemone 23776
Candle Berry 14544,261137
Candle Delphinium 124194
Candle Larkspur 124194
Candle Nut 14544
Candle Nut Tree 14544,406019
Candle Orchid 25137
Candle Plant 129618,216654,303539,358306
Candle Tree 284497,289476
Candle Wood 21005
Candle Yucca 416607
Candlebark 155725
Candlebark Gum 155725
Candle-bark Gum 155725
Candleberry 14544,261137,261162,261176,261202,356206
Candleberry Myrtle 261132
Candle-berry Myrtle 261137
Candleberry Tree 14544,346408
Candlefruit 8645
Candlegostes 273531
Candlemas Bells 23925,169719
Candlemas Cap 23925
Candlemas Caps 23925
Candlenu Tree 14544
Candlenut 14544
Candle-nut Oil Tree 14544
Candlenut Tree 14544
Candlenut-oil Tree 14544
Candlenutree 14544
Candles 9701,167955
Candlestick 16241
Candle-stick Lily 229828
Candlestick Senna 360409
Candlestick Tickseed 104755
Candlesticks 174877,273531
Candletree 261202
Candle-tree 284497
Candleweek Flower 405788
Candlewick Flower 405788
Candlewick Statice 320002
Candlewood 21005,167562,261202
Candlybark 155772
Candolle Eriolaena 152684
Candolle Thorowax 63588
Candy Barrel Cactus 163483
Candy Barrel-cactus 163483
Candy Cactus 197762
Candy Cane Bamboo 196347
Candy Carrot 43795,43807
Candy Corn Vine 244367
Candy Flower 94362
Candy Mustard 9785,203184
Candy Palm 380555
Candy Tuft 203181,203239
Candyleaf 376405
Candystick 16241
Candy-stripe Bamboo 196347

Candytuft 203181,203184,203239,203249
Cane 37127
Cane Acanthopanax 143628
Cane Apple 30888
Cane Ash 167897
Cane Cholla 116672,272924,273067
Cane Grass 265923
Cane Palm 89360
Cane Palms 65637
Cane Tibouchina 391570
Canebrake 37127
Canella 71129,71132
Canella Family 71138
Canellini Bean 294056
Canereed Spiral Flag 107271
Canereed Spiralflag 107271
Canescent Altai Heteropappus 193919
Canescent Falsenettle 56207
Canescent Ptilotrichum 321089
Canescent Straight-raceme Barberry 51985
Canescent Stringbush 414143
Canescent Willowherb 146684
Canescent Willowweed 146684
Cangbaishan Barbedveiin Maple 2799
Cangjiang Apple 243661
Cangjiang Arisaema 33283
Cangjiang Jerusalemsage 295055
Cangjiang Newcinnamon 263764
Cangjiang Peashrub 72266
Cangjiang River Apple 243661
Canglangshan Barberry 52396
Cangshan Azalea 330590
Cangshan Bladderwort 403795
Cangshan Buttercup 325696
Cangshan Goldenray 229234
Cangshan Indigo 206044
Cangshan Jerusalemsage 295102
Cangshan Monkshood 5142
Cangshan Rabdosia 209639
Cangshan Rhododendron 330590
Cangshan Rockfoil 350004
Cangshan Sedge 74030
Cangshan Spapweed 205408
Cangshan Touch-me-not 205408
Cangshan Vinegentian 110336
Cangshan Whitepearl 172061
Cangshan Wintergreen 172061
Cangshan Woodbetony 287783
Cangwu Ophiorrhiza 272286
Cangyuan Agapetes 10318
Cangyuan Alphonsea 17632
Cang-zhu Atractylodes 44208
Canihua 87121
Canine Goosecomb 144228
Canine Roegneria 144228
Canistel 238117
Canistrum 71149,71150
Canker 282685,336419,384714
Canker Lettuce 322872
Canker Root 103850,216262,230566
Canker Rose 282685,336419
Canker Weed 359158
Canker-balls 336419
Cankerberry 103824
Canker-berry 336419
Canker-flower 336419
Canker-root 103850
Cankerweed 313778,313881
Cankerwort 359158
Canna 71156,71179
Canna Family 71235
Canna Lily 71173
Cannabis 71218
Canna-grass 152769
Cannaleaf Myrosma 261531
Canna-leaf Spathiphyllum 370336
Cannalike Dona 136081
Cannell 91446
Cannon-ball Plant 39532
Cannonball Tree 108193,108195
Cannon-ball Tree 108193,108195
Canoe Birch 53549,53563
Canoe Cedar 390645
Canoe-cedar 390645
Canoe-wood 232609
Canon Grape 411556
Canopytree 244485,244486
Canotia 71264
Canra Rubber 244509
Canscora 71270
Cansjera 71289
Cantala 10816,169245
Cantala Fibre 10816
Cantaloup Melon 114199
Cantaloupe 114189,114199
Cantaque 366580
Canterbury Bell 70164
Canterbury Bells 70054,70331,72934,177523
Canterbury Jack 61497
Canterburybells Gloxinia 177523
Canthium 71313
Cantley Dilochia 130946
Canton Abrus 747
Canton Ampelopsis 20327
Canton Buttercup 325697
Canton Caper 71707
Canton Cassia 91302
Canton Darkbrown Fluttergrass 166331
Canton Fairybells 134425
Canton Fibre 260235
Canton Lemon 93546
Canton Linen 56229
Canton Mombin 372471
Canton Salomonia 344347
Canton Snakegrape 20327
Canton Sonerila 368874,368895
Canton Water Pine 178019
Cantonese Buttercup 326365
Cantonese Fairy Bells 134425
Cantor Pseudosasa 318294
Cant-robin 336851
Cany Yellow Rhododendron 331862
Canyon Birch 53437
Canyon Grape 411556
Canyon Live Oak 323762
Canyon Maple 3000
Canyon Oak 323762
Canyon Penstemon 289370
Canyon Ragweed 19142
Canyon Sunflower 405382,405383
Cany-yellow Rhododendron 331862
Caoguoshan Raspberry 339504
Caorthann 369339
Caoutchouc 164925,194453
Caoutchouc Tree 164925,194453
Cap Docken 292372
Capa de Oro 366871
Capa Prieto 104153
Caparrosa 263097
Capcupule Oak 323986
Cap-docken 292372
Cape African-Queen 25420
Cape Alkanet 21933
Cape Aloe 16817,17381
Cape Ash 141863
Cape Asparagus 29660
Cape Aster 163121,163131,163132
Cape Bean 294010
Cape Beech 326537
Cape Bladder Pea 380032
Cape Blue Water Lily 267665
Cape Blue Waterlily 267665
Cape Blue Water-lily 267665
Cape Box 64277,179387
Cape Box Wood 179387
Cape Bramble 339194
Cape Broom 173015
Cape Bugle-lily 413328
Cape Bugloss 21933
Cape Cheesewood 301440
Cape Chestnut 67666,67667
Cape Clover 396811
Cape Coast Lily 111250
Cape Cow Slip 218824
Cape Cowslip 218824,218830
Cape Cudweed 166037,178481
Cape Daisy 276608
Cape Dandelion 31162
Cape Ebony 155967,194655
Cape Everlasting 293274,293275

Cape False Olive 142440
Cape Falseolive 142440
Cape Fig 164751,165724
Cape Figwort 218830,296108,296112,296113
Cape Flower 265259
Cape Forget-me-not 21933
Cape Fuchsia 296109,296110,296112
Cape Gardenia 171253
Cape gooseberry 297640
Cape Gooseberry 297658,297711,297720
Cape Grape 332324
Cape Gum 1297,1324
Cape Heath 149141
Cape Holly 204060
Cape Honey Flower 248944
Cape Honey Suckle 385508
Cape Honey-flower 248936,248944
Cape Honeysuckle 385505,385508
Cape Hyacinth 170828
Cape Ivy 359415
Cape Jasmine 171253,337500
Cape Krause Sorrel 340099
Cape Laburnum 111992
Cape Leadwort 305172
Cape Leeuwin Wattle 283887
Cape Lilac 45908,248895,410878
Cape Lily 111152,405356
Cape Mahogany 395488
Cape Mallow 25415,25420,25434
Cape Marigold 31162,131147,131150,131186,131191,131193
Cape Mistletoe 410991
Cape Myrtle 261601
Cape of Good Hope Yacca 306425
Cape of Good Podocarpus 306425
Cape Oil-wood 142440
Cape Oxalis 278137
Cape Pea 294056
Cape Pigweed 18685
Cape Pittosporum 301440
Cape Plumbago 305172,305175
Cape Pondweed 29643,29660
Cape Primrose 377652,377804,377818
Cape Province Pygmyweed 109174
Cape Ricegrass 4122
Cape Rush 88842
Cape Sable False Thoroughwort 89296
Cape Sable Whiteweed 11215
Cape Shamrock 277644
Cape Smilax 38938
Cape Sorrel 278008
Cape Spinach 144877
Cape Stock 190261
Cape Sundew 138275,138354
Cape Sweet Pea 218721
Cape Teak 378699
Cape Thorn 418190
Cape Treasure Flower 172326
Cape Tulip 184345,184372,197823,197830,197836
Cape Virgilia 410878
Cape Waterhawthorn 29660
Cape Water-lily 267665
Cape Weed 31162
Cape Willow 343726
Cape Yellowwood 306425
Cape York Palm 234188
Cape York Red Gum 155506
Cape-cowslip 218824,218830
Cape-gooseberry 297711
Cape-Ivy 123750
Cape-jasmine 171253
Cape-pondweed 29660
Cape-pondweed Family 29696
Caper 71676,71762,71871
Caper Bush 71676,71762,71871,159222
Caper Euphorbia 159222,159252
Caper Family 71671
Caper Plant 159222
Caper Spurge 159222,159252
Caper Tree 159222
Caperbush 71676
Caper-bush 71762
Caperoiles 222758
Caper-tree 71762
Capetown Grass 395024
Capetown Pea 276967
Capeweed 31160,31162
Capform Raspberry 339050
Capfruit Euonymus 157722
Capilatus Baron Oak 323695
Capillaceous Sedge 74023
Capillar Hairgrass 12789
Capillary Bentgrass 12323
Capillary Feathergrass 376730
Capillary Panic-grass 281438
Capillary Sagebrush 35282
Capillary Sedge 74032
Capillary Tailstachys Kobresia 217140
Capillary Wormwood 35282
Capillaryculm Sedge 74023
Capillaryleaf Kobresia 217131
Capillaryleaf Sedge 74044
Capillary-pedicel Barberry 51943
Capillary-pedicel Clovershrub 70793
Capillarystem Sedge 74032
Capillipedium 71602
Caping Flowers Rhododendron 330169
Capitate Acrocephalus 6039
Capitate Adenosma 7985
Capitate Aletris 14478
Capitate Amitostigma 19504
Capitate Argyreia 32618
Capitate Astyleorchis 19504
Capitate Azalea 330324
Capitate Box 64248
Capitate Bushclover 226729
Capitate Bushmint 203044
Capitate Calligonum 67008
Capitate Catchfly 363295
Capitate Cyathula 115704
Capitate Dodder 114989
Capitate Eargrass 187532
Capitate Eight Trasure 200805
Capitate Gentian 173321
Capitate Germainia 175242
Capitate Goosefoot 86984
Capitate Hedyotis 187532
Capitate Knotweed 308953
Capitate Late Juncellus 212782
Capitate Mint 250291
Capitate Mucuna 259489
Capitate Osbeckia 276089
Capitate Peppergrass 225322
Capitate Pepperweed 225322
Capitate Phrynium 296052
Capitate Plantain Lily 198593
Capitate Primrose 314212
Capitate Rattlebox 112383
Capitate Rhododendron 330324
Capitate Rush 212988
Capitate Salsify 394268
Capitate Schefflera 350671
Capitate Sedge 74046
Capitate Soft-hairy Hedyotis 187534
Capitate Spike-rush 143155,143158
Capitate Stonecrop 200805
Capitate Stringbush 414147
Capitate Tubergourd 390115
Capitate Willow 343150
Capitate-flower Asiaglory 32618
Capitateflower Speedwell 407059
Capitate-racemose Stringbush 414148
Capitatestigma Willow Weed 146606
Capitatestigma Willowweed 146606
Capitellate Micromeria 253646
Capleaf Cowparsnip 192309
Capon's Feather 30081
Capon's Feathers 30081,404316
Capon's Tail 30081,81768,404316
Capon's Tails 30081,81768,404316,404337
Cappadocian Maple 2849
Capper 71871
Capricorn Star Cactus 43495,43496,43497,43498,43499
Caprifoly 235713
Caprifoy 236022
Capron Strawberry 167627
Cap-shaped Rhododendron 330568
Capshaped Rose 336418

Capsiciform Amomum 19829
Capsicum 72068,367528
Captain Cook's Araucaria 30838
Captain Cook's Pine 30838
Captain-over-the-garden 5676
Capuchin's Beard 301910
Capucine Rose 336563
Capul 350573
Capuli 316918
Capulin 83311,316280,316760,316787
Capulin Black Cherry 316760,316792
Capulin Cherry 316760
Capurana-da-terra-firme 48842
Capus Willow 343167
Carabao Lime 93693
Caracas Rattlebox 112405
Caracus Wigandia 414099
Caragana 72180
Caraguata Fibre 154336
Caraguatatuba Begonia 49696
Carallia 72391
Caralluma 72402,72548
Caramaile 222758
Caraman Pine 300111
Caramba 45725
Carambola 45721,45725
Carambole 45725
Caramel Grevillea 180634
Caranda 398825
Caranda Palm 103731
Carandai Palm 398826
Caraway 77766,77784
Caraway Seed 77784
Caraway Thyme 391201
Card Teasel 133478
Card Thistle 133478
Cardamom 9873,19817,143499,143502
Cardamon 19836,143499,143502
Cardaria 73085
Cardboard Palm 417010
Cardiandra 73119
Cardinal Airplant 392001
Cardinal Climber 323457,323465,323470
Cardinal Creeper 323457
Cardinal Flower 234275,234277,234351,364741
Cardinal Gladiolus 176105
Cardinal Guard 269100
Cardinal Monkeyflower 255197
Cardinal Penstemon 289329
Cardinal Sage 345043
Cardinal Shrub 413570
Cardinal Spear 154668
Cardinal Star Glory 323470
Cardinal Star Ipomoea 323470
Cardinal Wood 61062
Cardinalflower 234351
Cardinal-flower 234351
Cardinal's Guard 279768
Cardinal's-guard 279771
Cardio Crinum 73157
Cardiocrinum 73154,73157
Cardiopteris Family 73181
Cardioteucris 73220
Cardoncillo 288989
Cardoon 117765,117770,117787
Cardwell Lily 315499
Careless Weed 18787,18822
Carelin Tansy 383780
Careluck 364557
Care-tree 369339
Carett 123141
Carex-like Kobresia 97769
Carey's Heart's-ease 308948
Carey's Sedge 74053
Carey's Smartweed 291716,308948
Carey's Wood Sedge 74053
Carib Grass 151695
Carib Heliconia 189992
Carib Potato 131501
Carib Royal Palm 337883
Carib Yam 325187
Caribbean Applecactus 185918
Caribbean copper Plant 158700
Caribbean Fingergrass 160984
Caribbean Pine 299835,299836,299928
Caribbean Pitch Pine 299836
Caribbean Royal Palm 337883
Caribbean Spikesedge 143087
Caribbean Trumpet Tree 382707
Caribbean Trumpet-tree 382707
Caribbee Royal Palm 337883
Carica Family 76820
Caricature Plant 180278
Caricature-plant 180278
Carier Currant 333935
Cariin Heath 149205
Carilla 259001
Carinate Batrachium 48910
Carinate Coelogyne 98623
Carinate Dendrobium 125037
Carinate Dichocarpum 128928
Carinate Metasasa 252536
Carissa 76854
Carl Hemp 71218
Carland Larkspur 124127
Carldod 302034
Carldoddie 92022
Carlemannia 76976,76978
Carles Arrowwood 407727
Carles Blueberry 403760
Carles Indigo 205782
Carles Viburnum 407727
Carlese Chinquapin 78877
Carlese Evergreen Chinkapin 78877
Carlese Evergreenchinkapin 78877
Carlesia 76986
Carlicups 68189
Carlin Heather 149205
Carlin Spurs 172905
Carlina 76988
Carline Thistle 76988,77022
Carline-thistle 76988
Carlin-heath 149205
Carlin-heather 149205
Carlin-spur 172905
Carlock 364557
Carlock Cups 68189
Carlowrightia 77035
Carludovica 77038
Carmale 222758
Carman Willow 343171
Carmel 222758
Carmel Ceanothus 79934
Carmel Creeper 79934
Carmel Daisy 235491
Carmel Valley Malacothrix 242956
Carmichael's Monkshood 5100
Carmile 222758
Carmine Barberry 51429
Carmine Begonia 49697
Carmine Pink Deutzia 127065
Carmine Scabrous Deutzia 127087
Carmine Thistle 92353
Carmineflower Specious Lily 230040
Carmona 77063
Carmylie 222758
Carnaba 103736
Carnadine 127635
Carnation 127573,127635,127779
Carnation Grass 74542,74547,74656
Carnation Leaf Gentian 173327
Carnation of India 154162
Carnation Poppy 282717
Carnation Sedge 75684
Carnation-gladiolus 176109
Carnauba Palm 103737
Carnauba Wax 103737
Carnauba Wax Palm 103737
Carnea Water Lily 267640
Carneous Kalanchoe 215109
Carniola Lily 229793
Carniolan Buckthorn 328697
Carnival Bush 268261
Carn-nock 271563
Carnose Deutzia 126862
Carnose Epithema 147429
Carnose Euonymus 157355
Carnose Goldsaxifrage 90343
Carnose Goldwaist 90343
Carnose Rockfoil 349154

Carnose Spindle-tree 157355
Carnose-flowered Euonymus 157355
Carnosflower Euonymus 157355
Caroa Fibre 263893
Carob 83523,83527
Carob Bean 83527
Carob Seed Gum 83527
Carob Tree 83527
Carobe-tree 83809
Carob-tree 83527,211243
Carol Barberry 51433
Carole Mackie Daphne 122394
Carolin Ash 167934
Carolin Elephantfoot 143456
Carolina 68313,172781
Carolina Allspice 68306,68310,68313,68314
Carolina Anemone 23745
Carolina Ash 167934
Carolina Basswood 391681
Carolina Bean 294010
Carolina Black Poplar 311231
Carolina Buckthorn 328641,328642,362957
Carolina Bush Pea 389558
Carolina Buttercup 325709
Carolina Cherry 316285
Carolina Cherry Laurel 316285
Carolina Cherrylaurel 316285
Carolina Clover 396859
Carolina Cranesbill 174524
Carolina Crane's-bill 174524,174526
Carolina Dayflower 100960
Carolina Doll's-daisy 56711
Carolina Fanwort 64496
Carolina Foxtail 17521
Carolina Geranium 174524,174526
Carolina Goldenrod 368336
Carolina Hemlock 399874
Carolina Hickory 69942,77885,77922
Carolina Holly 203542
Carolina Horse-nettle 367014
Carolina Ipecacuanha 159150
Carolina Iris 208614
Carolina Jasmine 172781
Carolina Jessamine 172781
Carolina Jointtail 98796
Carolina Larkspur 124104,124106
Carolina Laurel Cherry 316285
Carolina Laurelcherry 316285
Carolina Laurel-cherry 316285
Carolina Leek 15161
Carolina Lily 229938
Carolina Love Grass 147871
Carolina Lupine 389558
Carolina Magnoliavine 351029
Carolina Moonseed 97900
Carolina Neoregelia 264409
Carolina Nettle 367014
Carolina Nightshade 367014
Carolina Nut Grass 354189
Carolina Onion 15161
Carolina Palmetto 341422
Carolina Parnassia 284515
Carolina Phlox 295255
Carolina Pink 371616
Carolina Pluchea 305085
Carolina Poplar 311146,311149,311292
Carolina Puccoon 233708,233709
Carolina Rhododendron 330327
Carolina Rose 336462
Carolina Silverbell 184731,184746
Carolina Snailseed 97900
Carolina Spring Beauty 94305
Carolina Spring-beauty 94305
Carolina Statice 230566
Carolina Tanoak 233130
Carolina Tea 204391
Carolina Thermopsis 389558
Carolina Thistle 91843
Carolina Vetch 408336
Carolina Water Shield 64496
Carolina Whitlow-grass 137205
Carolina Wild Petunia 339687
Carolina Wild Pink 363300
Carolina Willow 343172
Carolina Yellow-eyed-grass 416032
Carolina-allspice 68313
Carolina-lily 417608
Carolina-poplar 311237
Caroline Ivory Nut Palm 252633
Carolinea-leaf Begonia 49698
Carolines Ivory Nut Palm 252633
Carolinian Laurel 215391
Carosella 167170
Carpathian Beech 162400
Carpathian Burl Elm 401593
Carpathian Current 334149
Carpathian Pussytoes 26362
Carpathian Walnut 212636
Carpatian Bellflower 69942
Carpatian Catisfoot 26362
Carpenter Grass 3978,316127
Carpenter's Apron 221787
Carpenter's Chips 262722
Carpenter's Herb 3978,13174,316127
Carpenter's Square 355184,355202
Carpenter's Weed 3978
Carpenter's-square 355184
Carper Spurge 159222
Carpesium 77145,77146
Carpet Begonia 49941
Carpet Bentgrass 12323
Carpet Bent-grass 12323
Carpet Bugle 13051,13174
Carpet Bugleweed 13174
Carpet Burrweed 368537
Carpet Cinquefoil 313039
Carpet Foxtail cactus 107067
Carpet Grass 45820
Carpet Juniper 213785
Carpet Plant 147384
Carpet Urbinea 139974
Carpet Weed 215382,256727
Carpet Weed Bedstraw 170490
Carpetgrass 45820,45827
Carpet-like Rhododendron 331933
Carpetweed 256689,256719,256727
Carpet-weed 256689,256727
Carpet-weed Bedstraw 170490
Carpetweed Family 12911
Carpet-weed Family 256684
Carquinez Goldenbush 209567
Carr Care 369339
Carrag 302068,340151
Carran 370522
Carretilla 131343
Carriere Hawthorn 109584
Carriere Spindle-tree 157499
Carrierea 77643
Carring Falls Grevillea 180641
Carrion Flower 366317,366364,366397,366427,366535,373719,373821,374013
Carrion Plant 373808,373821,373827,373836
Carrionflower 373719
Carrion-flower 273206,366364
Carrion-flower Form Ceropegia 84263
Carrion-flower Greenbrier 366364
Carrizo 37467
Carrizo Creek Ragwort 279949
Carrot 101852
Carrot Broomrape 275131
Carrot Burr Parsley 79664
Carrot Family 19681,29230
Carrot Plant 155173
Carrot Tops 302912
Carrot Wood 114585
Carrotleaf Woodbetony 287144
Carrot-shaped Semiaquilegia 357919
Carrotwood 114585
Carr's Rattlesnakeroot 313803
Carruth Wormwood 35307
Carruthers' Falseface 317253
Carry Me Seed 296471
Carrying Rockvine 387837
Carsinal's Hat 243951
Carskin 106703
Carslope 315082
Carsons 262722
Cartaphilago 178465
Carter's Orchid 48736
Carthage Crape Jasmine 382789

Carthagena Bark 91074
Carthaginian Apple 321764
Carthamine 77748
Carthusian Pink 127634
Cartilagegrass 263020
Cartilaginous Peppergrass 225326
Cartilaginous Pepperweed 225326
Cartwheel 101852
Cartwheel Flower 192302
Cartwheel-flower 192394
Carunclegrass 256441
Caruwaie 77784
Carver 77784
Carver's Tree 391706
Carveseed 178010
Carvie 77784
Carvy 77784
Caryer 77874
Caryodaphnopsis 77956,77969
Caryoleaf Martinezia 12772
Caryota Ruffle Palm 12772
Casa Banana 362419
Casca Bark 154973
Cascade Aster 155815
Cascade Azalea 330057
Cascade Canada Goldenrod 368091
Cascade Fleabane 150538
Cascade Knotweed 308951
Cascade Oregon Grape 51959
Cascade Penstemon 289376
Cascade Petunia 292747
Cascades Fir 280
Cascades Mahonia 242602
Cascalote 64982
Cascalote Caesalpinia 64978
Cascara 328838
Cascara Buckthorn 328838
Cascara Sagrada 328838
Cascara Tree 328838
Cascarilla 112884
Cascarilla Bark 112888
Cas'ear Boea 56063
Casearia 78095
Casearialeaf Microdesmis 253413
Casearia-leaved Microdesmis 253413
Case's Fireweed 105703
Case's Lady's-tresses 372191
Caseweed 72038
Casewort 72038
Cashemir Treemallow 223361
Cashes 28044,101852
Cashew 21191,21195
Cashew Apple 21195,244405
Cashew Family 21190
Cashew Gum 21195
Cashew Nut 21195
Cashew Tree 21195
Cashew-leaved Drimycarpus 138044
Cashew-nut 21195
Cashew-tree 21195
Cashionshape Cassiope 78642
Cashionshape Dropseedgrass 372810
Cashiony Dropseed 372810
Cashmer Jerusalemsage 295083
Cashmere Bouquet 95978
Casimiroa 78172
Caskets 59532
Casnation Rose 336559
Casp Bamboo 297191
Caspian Giantfennel 163586
Caspian Halostachys 185038
Caspian Honey Locust 176867
Caspian Honey-locust 176867
Caspian Iris 208555
Caspian Karelinia 215567
Caspian Locust 176867
Caspian Lotus 263268
Caspian Perovskia 291428
Caspian Saltclaw 215324
Caspian Saltspike 185038
Caspian Sea Inula 207075
Caspian Sea Kalidium 215324
Caspian Sea Willow 343174
Caspian Willow 342961
Caspiansea Willow 343174
Cassabully 47964
Cassandra 85412
Cassareep 244507
Cassava 244501,244507
Casse Fistuleuse 78300
Cassena 204391
Casse-weed 72038
Cassia 78204,91252,91302
Cassia Bark 91302
Cassia Bark Tree 91302
Cassia Bean 78300
Cassia Buds 91302
Cassia Cinnamon 91302
Cassia-bark Tree 91302
Cassiabarktree 91302
Cassia-bark-tree 91302
Cassia-flower-tree 91366
Cassiasistre 78300
Cassidony 223334
Cassie 1219
Cassie Flower 1219
Cassie-oil-plant 1219
Cassinia 78605
Cassiope 78622
Cassocks 144661
Cassowary Tree 79165
Cassumar 418017
Cassumar Ginger 418017
Cassumunar Ginger 418005
Cast Iron Plant 39532
Castanha De Curia 108124
Castilla 79108
Castilloa Rubber 79110
Castillon Bamboo 297215,297218
Castings 316819
Cast-iron Plant 39532
Castle 1219
Castle Gilliflower 246478
Castle Valley Saltbush 44408
Cast-me-down 223334
Castocks 59532
Castor 334435
Castor Aralia 215420,215442
Castor Bean 334432,334435
Castor Bean Plant 334435
Castor Oil Bean 212127
Castor Oil Plant 162879,334435
Castor Plant 334435
Castor Wood 242341
Castorbean 334432,334435
Castor-bean 334432,334435
Castor-bean Begonia 50243
Castorbean-oilplant 334432
Castor-oil Bean 334432
Castor-oil Plant 334432,334435
Castoroll Bean 212127
Castrate Carpgrass 36625
Cast-the-spear 368480
Castus Geranium 288209
Casual Arrowwood 407894
Casuarina 79155
Casuarina Cuscuta 115142
Casuarina Family 79193
Cat 284493
Cat Bed 81768
Cat Brier Catbrier 366559
Cat Choops 336419
Cat Claw Acacia 1264
Cat Germander 388147
Cat Grape 411834,411896
Cat Greenbrier 366341
Cat Gut 386365
Cat Haw 109857
Cat Haws 109857
Cat Heather 149205,150131
Cat Hep 336851
Cat Herb 46991
Cat Hip 336419
Cat Hips 336419
Cat Jug 336419
Cat Mint 264859
Cat Oak 2837
Cat O'nine Tails 106703,401112
Cat Pea 157429,237539,408352
Cat Posy 50825
Cat Puddish 237539

Cat Rose 336364,336419,336851
Cat Rush 157429
Cat Spruce 298286
Cat Tail 401112
Cat Tall Flag 401112
Cat Thyme 388147,388184
Cat Tiger Jaws 162914
Cat Trail 404316
Cat Whin 172905,271563,336419,336851,401390,401395
Catalina Ceanothus 79911
Catalina Ironwood 239411,239413
Catalina Mariposa-lily 67569
Catalina Mountain Lilav 79911
Catalonia Jasmine 211848
Catalonia Saffron 77748
Catalonian Jasmine 211848
Catalpa 79242,79243,79257,79260
Catalpa Leaf Paulownia 285956
Catalpa Maple 2870
Catalpaleaf Maple 2870
Catalpaleaf Paulownia 285956
Catalpa-leaved Maple 2870
Catalpa-leaved Paulownia 285956
Catamar Parodia 284662
Catapodium 79295
Catapuce 159222
Cataria 264897
Catasetum 79329
Catawba 79242,79243,79260
Catawba Rhododendron 330331
Catawba River Rhododendron 330331
Catawba Tree 79260
Catawbarebe 411764
Catberry 263498,263499,334250
Cat-berry 263498,263499
Catbird Grape 411834
Catbrier 366230,366264,366341,366559,366592
Catbrier Family 366141
Catbuw Blow 192358
Catch 1120
Catch Me if You Can 1779
Catch Skullcap 355804
Catchfly 28617,363141,363300,363368,363477
Catch-fly 363467
Catchfly Grass 223990
Catchgrass 170193
Catchrogue 170193
Catchthorn 418145
Catch-trap 138348
Catchweed 170193
Catchweed Bedstraw 170193
Catclaw 1264
Catclaw Acacia 1264
Catclaw Buttercup 326431
Catclaw Mimosa 254981,255089
Catclaw Sensitive Brier 255102
Catclawvine 240368
Catcluke 237539
Catdaisy 202387
Catear Metalleaf 296423
Catechu 1120,31680
Catechu Acacia 1120
Catechu Palm 31680
Catechu Tree 1490
Caten-aroes 227533
Caterpilar Plant 354755,354761
Caterpillar Amaranth 18696
Caterpillar Greasewood 346610
Caterpillar Plant 370316
Caterpillar Tree 79243
Caterpillar-plant 354755,354761
Catesby Lily 229797
Catesby Oak 324087
Catesby's Lily 229797
Catesby's Trillium 397546
Cateye Spurge 158857
Catfish-horn 313753
Catfoot 317761
Cat-foot 178318
Catfoot Poplar 311398
Catgut 386365
Cathartic Flax 231884
Cathaw-blow 192358
Cathay Hickory 77887
Cathay Knotweed 308953
Cathay Lily 73157
Cathay Poisonnut 378675
Cathay Poplar 311265
Cathay Semiliquidambar 357969
Cathay Silver Fir 79437,79438
Cathaya Hedyotis 187539
Cathaya Japan Rose 336792
Cathaya Japanese Rose 336792
Cathaya Manyflowered Rose 336792
Cathaya Starjasmine 393626
Cathayan Anemone 23746
Cathayan Begonia 49701
Cathayan Starjasmine 393626
Cathayan Windflower 23746
Cathayanthe 79441
Cathcart Begonia 49702
Cathcart Birthwort 34135
Cathcart Dutchman's-pipe 34135
Cathcart Landpick 308517
Cathcart Solomonseal 308517
Cathcart Spapweed 204842
Cathcart Touch-me-not 204842
Cathcart Twayblade 232116
Cathead Crazyweed 278680
Catheart Pothos 313195
Cathedral Bells 61572,97757
Cathedral Windows 49642
Catherine Wheel 184345,228043
Catherine's Pincushion 228043
Catiang Bean 409092
Catiang Cowpea 409092
Cat-in-clover 157429
Cativo 315255
Catjang 65146
Catjang Cow-pea 409092
Catkin Bush 171546
Catkin Yew 19357
Catleaf Stephanandra 375817
Cat-leaved Stephanandra 375817
Catluke 237539
Cat-mil 18687
Catmint 176824,264859,264897,265005,265064,388275
Catmon Dillenia 130923
Catnap 264897
Catnep 264859,264897
Catnip 264859,264897,264920,265005,265064
Cat-nip 264859,264897,264920,265005
Catnip Giant Hyssop 10410
Catnip Giant-hyssop 10410
Catnut 102727
Caton Rorippa 336187
Cat-o'-nine-tails 401085
Catrup 264897
Catrush 157429
Cat's Claw 301162
Cat's Claw Trumpet 240368
Cat's Claw Vine 240368
Cat's Claws 1264,28200,157429,237539,326287,334974,338447
Cat's Clover 237539
Cat's Ear 11947,26299,26385,67582,202387,202432,298589
Cat's Ears 26299
Cat's Eye 407072
Cat's Eyes 85875,131062,174877,198680,260892,407072,407287
Cat's Face 410677
Cat's Foot 26385,26490,176824,178358
Cat's Hair 159069,370522
Cat's Head 395146
Cat's Hips 336419
Cat's Hoof 176824
Cat's Jugs 336419
Cat's Keys 167955
Cat's Love 81768,404316
Cats' Love 404316
Cat's Lugs 314143
Cat's Milk 117112,158433,159027
Cat's Moustache 275639
Cat's Paw 25219,26385,237539,176824
Cat's Pea 121001

Cat's Peas 121001
Cat's Tail 5442,18687,62471,196818,294968,401085,401112
Cat's Tail Aloe 16696
Cat's Tail Grass 217501,294960
Cat's Tail Millet 361877
Cat's Tails 5442,18687,106703,141347,152769,196818,212636,294968,343396,401112
Cat's Tall 106703,141347,152769,212636,343396,401112
Cat's Tongue 400493
Cat's Valerian 404316
Cat's Whiskers 95759,182915,275639,275781,382912
Cat's Wort 264897
Cats-and-eyes 167955
Cats-and-keys 2837,3462
Cats-and-kittens 106703
Catsclaw 301162
Cat's-claw 240368
Cat's-claw Vine 240368
Cat's-claws 1264
Cat's-ear 202387,202432
Cat's-ears 26385,67554,202387
Cat's-eye 407072
Cat's-foot 26299,26385,26490,178318,317742,317756
Cat's-paw 25219
Cat's-tail 294960,401112
Cat's-tail Orchid 9271
Cat's-tail-grass 294992
Cat's-whiskers 182915
Cattail 401085
Cat-tail 18687,135308,401085,401112
Cattail Dolichandrone 135308
Cat-tail Dolichandrone 135308
Cattail Family 401140
Cat-tail Family 401140
Cat-tail Flag 208924,401085,401112
Cattail Gayfeather 228511
Cattail Grass 361847
Cat-tail Grass 217501
Cattail Millet 289116
Cat-tail Millet 289116
Cat-tail Sedge 76623
Cat-tail Tree 135299,135308
Cattail Uraria 402118
Cattailleaf Iris 208913
Cattailleaf Swordflag 208913
Cattaimillet 289116
Cattea-clover 157429
Catten Clover 237539
Catteridge Tree 380499
Cat-thorn 354345
Cattijugs 336419
Cattikeyns 167955
Cattle Beet 53254
Cattle Keys 167955
Cattle Spinach 44609
Cattle-beet 53249
Cattle-keys 167955
Cattle-spinach 44609
Cattley Guava 318737
Cattleya 79526,79543
Catton Ball 155296,155300,317685
Catton-ball 155296,155300,317685
Cat-trail 404316
Cat-tree 157429,380499
Catty-tree 157429
Catuaba Herbal 155051,203527,253776,387129
Catwood 157429,380499
Cauassu 66173
Caucas Larkspur 124110
Caucasia 325715
Caucasia Anemone 23750
Caucasia Atropa 44709
Caucasia Bluebasin 350119
Caucasia Broomrape 275036
Caucasia Caraway 77788
Caucasia Chorispora 89007
Caucasia Collectivegrass 381029
Caucasia Columbine 30060
Caucasia Corydalis 105719
Caucasia Cyclamen 115981
Caucasia Dittany 129628
Caucasia Fir 429
Caucasia Herniary 192940
Caucasia Iris 208481
Caucasia Jacaranda 211231
Caucasia Koeleria 217434
Caucasia Medic 247298
Caucasia Mint 250339
Caucasia Mockorange 294422
Caucasia Poppy 282526
Caucasia Rockcress 30220
Caucasia Sagebrush 35314
Caucasia Sandthorn 196760
Caucasia Sedge 74068
Caucasia Silverspikegrass 227950
Caucasia Sweetvetch 187823
Caucasia Yam 131515
Caucasian Alder 16463
Caucasian Ash 168043
Caucasian Beet 53248
Caucasian Bladdernut 374098
Caucasian Bluestem 57548
Caucasian Centaurea 81257
Caucasian Comfrey 381029
Caucasian Cow Parsnip 192343,192398
Caucasian Cow-parsnip 192343
Caucasian Crane's-bill 174660
Caucasian Daphne 122407
Caucasian Elm 417540
Caucasian Filbert 106724
Caucasian Fir 429
Caucasian Hackberry 80607
Caucasian Hornbeam 77271,77276
Caucasian Iris 208481
Caucasian Leopard's-bane 136346
Caucasian Lily 229944
Caucasian Lime 391682,391705
Caucasian Linden 391682
Caucasian Maple 2849
Caucasian Mock-orange 294422
Caucasian Nettle Tree 80607
Caucasian Oak 324140
Caucasian Penny-cress 279625
Caucasian Peony 280169
Caucasian Pyrethrum 322665
Caucasian Rhododendron 330334
Caucasian Snowdrop 169707
Caucasian Spikenard 207121
Caucasian Spruce 298391
Caucasian Spurge 22243
Caucasian Stonecrop 357169
Caucasian Whortleberry 403728
Caucasian Wing Nut 320356
Caucasian Wingnut 320356
Caucasian Zelkova 417540
Caucasus Catmint 264939
Caucasus Feathergrass 376737
Caucasus Mock Orange 294422
Caucasus Roegneria 335255
Caucasus Wingnut 320356
Cauchao 19948
Caudare-leaf Hedyotis 187540
Caudata Spicebush 231424
Caudata-leaflet Jewelvine 125947
Caudata-leafleted Jewelvine 125947
Caudate Adinandra 8212
Caudate Allophylus 16059
Caudate Ardisia 31378
Caudate Ash 167935
Caudate Brylkinia 61442
Caudate Camellia 68970
Caudate Cherry 83162
Caudate Chinese Neillia 263164
Caudate Copper-leaf 1811
Caudate Fig 164981
Caudate Greenvine 204646
Caudate Hairthroat Gentian 173438
Caudate Japanese Xylosma 415898
Caudate Mandarin Orchid 295878
Caudate Maple 2874
Caudate Masdevallia 246104
Caudate Mastixia 246213
Caudate Membranaceus Madder 337994
Caudate Microtropis 254287
Caudate Milkwort 307982

Caudate Mockorange 294423
Caudate Neocinnamomum 263754
Caudate Newcinnamon 263754
Caudate Ocotea 268696
Caudate Olive 270075
Caudate Osmanther 276274
Caudate Osmanthus 276272
Caudate Pectinated Maple 3380
Caudate Pentasacme 289925
Caudate Quisqualis 324671
Caudate Rhamnella 328546
Caudate Rose 336471
Caudate Rourea 337702
Caudate Semiliquidambar 357974
Caudate Styrophyton 379481
Caudate Sugarpalm 32333
Caudate Sweetleaf 381423
Caudate Taiwan Adinandra 8221
Caudate Themeda 389318
Caudate Tickclover 269613
Caudate Two-seeded Palm 32333
Caudate Vetchling 222692
Caudate Wildginger 37585
Caudate Wormwood 35315
Caudate-bracted Iris 208793
Caudate-leaf Barberry 51643
Caudate-leaf Blackbark Linociera 87710
Caudate-leaf Blueberry 403816
Caudateleaf Cinnamon 91327
Caudate-leaf Hydrangea 199852
Caudateleaf Jasmine 211781
Caudate-leaf Leptopus 226328
Caudateleaf Leptosiphonium 226513
Caudate-leaf Liolive 87710
Caudateleaf Mountainash 369527
Caudateleaf New Eargrass 262970
Caudate-leaf Pearlwood 296448
Caudate-leaf Phyllanthodendron 296448
Caudateleaf Raspberry 338238
Caudate-leaf Rhododendron 332041
Caudateleaf Rockvine 387754
Caudate-leaf Sixleaves Stauntonvine 374430
Caudateleaf Skullcap 355402
Caudateleaf Tanoak 233136
Caudate-leaf Waxplant 198874
Caudate-leaved Allomorphia 16032
Caudate-leaved Chinquapin 78877
Caudate-leaved Cinnamon 91327
Caudate-leaved Copper-leaf 1771
Caudate-leaved Crape-myrtle 219915
Caudate-leaved Hydrangea 199852
Caudate-leaved Maple 2872
Caudate-leaved Raspberry 338238
Caudate-leaved Rhododendron 332041
Caudate-leaved Rockvine 387754
Caudate-leaved Tanoak 233136
Caudatelobed Larkspur 124112
Caudate-spiked Sedge 74072
Caudatilobed Mussaenda 260392
Caudute Aechmea 8552
Caudute Maple 2870
Caughnawaga Thorn 110065
Caul 59532
Caulescent Mandrake 244343
Caulescent Miterwort 256014
Caulescent Wildginger 37590
Caulialate Dutchmanspipe 34137
Cauliflorous Brandisia 59122
Cauliflorous Bulbophyllum 62627
Cauliflorous Helicia 189913
Cauliflorous Stonebean-orchis 62627
Cauliflory Holly 203613
Cauliflory Rockvine 387755
Cauliflous-like Helicia 189914
Cauliflower 59529,345631
Cauli-flowered Rockvine 387755
Cautleya 79749
Cavaierie Premna 313606
Cavalearie Primrose 314222
Cavaler Fissistigma 166654
Cavaler Sweetleaf 381423
Cavaleri Nervedflower Phoebe 295390
Cavalerie Barberry 51437
Cavalerie Begonia 49703
Cavalerie Biond Hackberry 80582
Cavalerie Blastus 55143
Cavalerie Camellia 69494
Cavalerie Cleidiocarpon 94402
Cavalerie Clethra 96481
Cavalerie Fevervine 280070
Cavalerie Fishvine 125948
Cavalerie Goldsaxifrage 90344
Cavalerie Goldwaist 90344
Cavalerie Hemiboea 191350
Cavalerie Jewelvine 125948
Cavalerie Loxostigma 238034
Cavalerie Lysionotus 239936
Cavalerie Metalleaf 296371
Cavalerie Michelia 252838
Cavalerie Monkshood 5110
Cavalerie Mosla 259281
Cavalerie Paperelder 290316
Cavalerie Rhododendron 330335
Cavalerie Spicegrass 239572
Cavalerie Stauntonvine 374391
Cavalerie Staunton-vine 374391
Cavalerie Tailanther 381929
Cavalerie Trigonotis 397391
Cavalerie Velvety Sophora 369154
Cavalerie Wendlandia 413777
Cavalerie Willow 343182
Cavalerie's Machilus 240566
Cavalerie's Nothaphoebe 266789
Cavalerie's Synotis 381929
Cavate-nerved Blueberry 403764
Cave 59532
Cave Corydalis 106106
Cavea 79784
Cavea Primrose 314223
Cavea Rockfoil 349160
Caveler Corymbtree 160722
Caveler Euryocrymbus 160722
Caven 1128
Cavendish Banana 260250
Cavenia 1128
Cavernous-stone Holly 204092
Cavicolous Aspidistra 39525
Caw Hairs Spikesedge 143029
Cawdor Primrose 314224
Caxacubri 299270
Caxes 28044,101852,123141
Caxlies 192358
Caxsine 78540
Cay-cay Fat 208999
Cayenne Cherry 156385
Cayenne Incense 316022
Cayenne Jasmine 79418
Cayenne Pepper 72070,72077
Cayenne-cherry 156385
Cayratia 79831
Ceanothus 79902,79948,79962
Ceanothus Snowbush 79919
Ceara Rubber 244501,244509
Ceara Rubber Plant 244509
Ceara Rubber Tree 244509
Ceara Rubber-plant 244509
Ceara Rubbertree 244509
Cearарubber Tree 244509
Ceara-rubber Tree 244509
Cearyleaf European Grape 411985
Cebolleta 417617,417624,417626
Cedar 80030,80074,85271,213606,213984,390587
Cedar Acacia 283890
Cedar Breaks Goldenbush 150382
Cedar Breaks Wild Buckwheat 152370
Cedar Elm 401486
Cedar Gum 106805
Cedar of Goa 114714
Cedar of Lebanon 80100
Cedar Pine 299954
Cedar Sage 345350
Cedar Wattle 1195,1646
Cedarglade St. Johnswort 201886
Cedar-of-goa 114714
Cedar-of-Lebanon 80100
Cedars 390567
Cedars Fairy Lantern 67623
Cedar-wood 35088
Cedrat 93603
Cedrate Guarea 181552

Cedrela 80015
Cedro Blanco 114652
Cedron 323546,364368
Cedros Island Verbena 405864
Ceheng Begonia 49708
Ceheng Dogwood 124946
Ceheng Fig 164787
Ceheng Machilus 240702
Ceiba 80114
Celandine 86733,86755,204841,325820
Celandine Corydalis 105735
Celandine Crocus 111548
Celandine Meadowrue 388447
Celandine Poppy 86733,86755,379219,379220
Celandine Saxifrage 349229
Celandineleaf Meconopsis 247116
Celandine-poppy 379220
Cele Dandelion 384777
Cele Mouseear Cress 30138
Celebes Citrus 93422
Celebes Papeda 93422
Celebes Pepper 300473
Celeocollar Fargesia 196345
Celeriac 29329
Celery 29312,29327,29328
Celery Cabbage 59600
Celery Lettuce 219492
Celery Pine 296956
Celery Sagebrush 35308
Celery Wood 310189
Celery Wormwood 35308
Celerycress 366115,366122
Celeryleaf Aralia 30589
Celery-leaf Buttercup 326340
Celeryleaf Heronbill 153767
Celery-leaved Buttercup 326340
Celery-leaved Crowfoot 326340
Celery-pine 296956
Celery-seed 340151
Celery-top Pine 296957,299804
Celestial Lily 263310
Celestial-lily 68495,263307
Celia Stonecrop 356615
Cellender 104690
Cellophane Plant 140539
Celon Sphenoclea 371409
Celosia 80395
Cels Tulip 400143
Celsia 80495,405657
Celsia Mullein 405679
Celsus Buckinghamia 61918
Celtic Nard 404238
Celtic Spikenard 404238
Celtic Valerian 404238
Celtuce 219485
Cembran Pine 299842
Cembrian Pine 299842
Cencilla 361667
Cencleffe 262441
Cengialt Iris 208483
Cengshien Rhododendron 330381
Ceniza 227919
Cenizo 44334,227922,227926
Cenocentrum 80871
Cenolophium 80877
Centaurea 80892,81020
Centaurium 81485
Centaury 81020,81065,81198,81485,81490,81530,81544
Centaury Gentian 81065
Centaury Knapweed 80892
Centaury-like Slantdaisy 301620
Centiped Euphorbia 159069
Centipeda 81681
Centipedaplant 197769,197772
Centipede Bossiaea 57500
Centipede Closedspurorchis 288605
Centipede Grass 148249,148254
Centipede Plant 197772
Centipede Tongavine 147349
Centipedegrass 148249,148251
Centotheca 81692
Central American Cedar 80025
Central American Mahogany 380527
Central American Rubber 79110
Central American Rubber Tree 79110
Central Asia Asterbush 41749
Central Asia Asterothamnus 41749
Central Asia Gentian 173552
Central Asia Milkvetch 43247
Central Asia Peashrub 72364
Central Asia Saltbush 44347
Central Asian Peashrub 72364
Central Asian Pea-shrub 72364
Central Australian Cabbage Palm 234185
Central Australian Fan Palm 234185
Central Australian Ghost Gum 106789
Central China Bittercress 73029
Central China Clearweed 298856
Central China Clematis 95255
Central China Coldwaterflower 298856
Central China Cotoneaster 107686
Central China Hawthorn 110114
Central China Hogfennel 292924
Central China Holly 203614
Central China Scurrula 236974
Central Mexico Organ Pipe 279486
Central Sagebrush 35897
Central Shandong Willow 343645
Central Xizang Corydalis 105613
Central-Asia Birch 53629
Central-asian Arthraxon 36657
Central-south Raspberry 338487
Centranth 81743
Centranthera 81714
Centrantheropsis 81739,81740
Centre-of-the-sun 81065
Centro 81876
Centrol Asia Bedstraw 170588,170716
Centrol Asia Dandelion 384491
Centrol Asia Dock 340202
Centrol Asia Doronicum 136379
Centrol Asia Giantfennel 163633
Centrol Asia Henbane 201401
Centrol Asia Lymegrass 144517
Centrol Asia Salsify 394300
Centrol Asia Sophiopsis 368945
Centrol Asia Spurge 160022
Centrol Asia Swordflag 208466
Centrol Asia Tickseed 104783
Centrol Asia Waterparsnip 365854
Centrolepis 81812
Centrolepis Family 81811
Centrosema 81872
Centrostemma 81909,198821
Century 81495
Century Plant 10778,10787,10788,10793,10919,10933
Century Plant Agave 10787
Centuryplant 10778,10787
Centuryplant Agave 10787
Cephaelis 81945
Cephalanoplos 82017
Cephalanthera 82028,82113
Cephalaria 82113,82139
Cephalocereus 82204
Cephalomappa 82238
Cephalorrhynchus 82411
Cephalostigma 82464
Cephalotus 82560
Cerasoid Cherry 83165
Cerastium 82629
Cerated Bamboo 47230
Cerated Bambusa 47230
Ceratocarpus 83418
Ceratocephalus 83438,83443
Ceratoides 218116
Ceratolobus 83519
Ceratopetal Rockvine 387756
Ceratophore Spapweed 204845
Ceratostigma 83641,83646
Ceratostylis 83651
Ceratozamia 83680
Cerberus Tree 83690,83698
Cerberustree 83690
Cerberus-tree 83690
Cercidiphyllum 83734
Cercidiphyllum Family 83732
Cercis 83752
Cereal Rye 356237

Cerejeira 19228
Cereus 83855,83893
Cereusform Echidnopsis 140021
Ceriferous Maple 2883
Ceriferous Parmentiera 284497
Ceriman 258162,258168
Ceriops 83939,83946
Ceriscoides 83953,83954
Ceropegia 83975
Cerrado Pineapple 21474
Cerulean Flaxlily 127505
Cerulean Flax-lily 127507
Cervantes Odontoglossum 224350
Cervetano 290716
Cestrum 84400
Cevadilla 352124
Ceylan Pouzolzia 313483
Ceylon Ardisia 31437
Ceylon Azalea 332159
Ceylon Bamboo 318643
Ceylon Barberry 51443
Ceylon Beauty Leaf 67856
Ceylon Beauty-leaf 67856
Ceylon Boxwood 171350
Ceylon Caper 71935
Ceylon Cardamom 143503
Ceylon Cedar 248895
Ceylon Cinnamon 91446
Ceylon Cinnamon Tree 91446
Ceylon Date Palm 295490
Ceylon Ebony 132137
Ceylon Elaeocarpus 142382
Ceylon Ellipanthus 143871
Ceylon Erythrospermum 155037
Ceylon Fagraea 162346
Ceylon Fig 165758
Ceylon Gooseberry 136846
Ceylon Greenbrier 366624
Ceylon Gurjun 133590
Ceylon Homalium 197652
Ceylon Hound's Tongue 117965
Ceylon Houndstongue 117965
Ceylon Hunteria 199440
Ceylon Iron Wood 252369
Ceylon Ironwood 252370
Ceylon Leadwort 305202
Ceylon Leucas 227739
Ceylon Luisia 238329
Ceylon Mahogany 248895,248913
Ceylon Morningglory 250850
Ceylon Morning-glory 250850
Ceylon Mottled Reed Bamboo 268121
Ceylon Naravelia 262337
Ceylon Newlitse 264116
Ceylon Oak 76799,351968
Ceylon Ochna 268202
Ceylon Olive 142382
Ceylon Osbeckia 276159
Ceylon Persimmon 132137
Ceylon Piassava 78056
Ceylon Plantain 302217
Ceylon Pygeum 322415
Ceylon Redseedtree 155037
Ceylon Reed Bamboo 268120
Ceylon Rhododendron 332159
Ceylon Rose 265327
Ceylon Rosewood 13621,13635
Ceylon Sansevieria 346173
Ceylon Satinwood 88663
Ceylon Sophora 368984
Ceylon Spinach 48689,383345
Ceylon Streblus 377546
Ceylon Swamplily 111280
Ceylon Syzygium 382686
Ceylon Tarenna 385049
Ceylon Tea Tree 142449
Ceylon Tigertaillily 346173
Ceylon Willow 164688
Ceylon Wood 252370
Ceylonolive 142382
Chaalky Albizzia 13510
Chaconia 413269
Chacuob 357740
Chadamu Kneejujube 67087
Chaenomelea 84555
Chaetotropis 85016
Chaff Azalea 331868
Chaff Panicgrass 281390
Chaff Windhairdaisy 348618
Chaffanjon Ampelopsis 20331
Chaffanjon Snakegrape 20331
Chaff-flower 4249,18089
Chaffless Saw-wort 348590
Chaffseed 352751
Chaffweed 21335,21394
Chahomiiia 345040
Chahuiqui 10908
Chaidamu Leymus 228377
Chaidamu Milkvetch 42170
Chaidamu Russianthistle 344777
Chain Cactus 329705
Chain Flowered Redbud 83807
Chain Fruit 18522
Chain Plant 394047
Chain Rhipsalis 329705
Chain Tree 218754
Chainfruit Cholla 272906
Chain-fruit Cholla 116657
Chainfruit Jumping Cholla 272906
Chain-link Cactus 272915
Chain-link Cholla 116652
Chainpodpea 18257
Chainpodtree 274337,274343
Chair Dragonbamboo 125466
Chair Elm 401512
Chair Poplar 311568
Chairmaker's Rush 352249
Chair-Maker's Rush 352249
Chairs-and-tables 64345
Chaishou Thorowax 63593
Chaiturus 85054
Chaituruslike Motherwort 224977
Chaix Bluegrass 305436
Chaix Bulbil Lily 229767
Chaka's Wood 378699
Chakky Cheese 243840
Chalang Woodbetony 287784
Chalcedonian Lily 229806
Chalcedonian Lychnis 363312
Chalice Cup Narcissus 262405
Chalice Flower 262441
Chalice Lily 262441
Chalice Vine 366863,366871
Chalice-vine 366863
Chalk False-brome 58604
Chalk Maple 3100
Chalk Milkwort 307969
Chalk Plant 183157,183221,183225
Chalk Thistle 91713
Chalkleaf Barberry 51555
Chalk-leaved Barberry 51555
Chalkplant 183157
Chalky Anabasis 21035
Chalky Bluestem 23077
Chalky Pleurospermum 304779
Chalky Ribseedcelery 304779
Chalong Sandwort 31859
Chamabainia 85106
Chamac Michelia 252841
Chamae Iris 208698
Chamaedaphne 85412,85414
Chamaedorea 85420
Chamaedrys Germander 388042
Chamaemelum 85510
Chamaerhodos 85646
Chamaesciadium 85696
Chamaesium 85711
Chamaesphacos 85730
Chamber Bitter 296801
Chamberlain Paphiopedilum 282799
Chamberlayn Bignonia 54306
Chambeyronia Palm 85842
Chamburo 76815
Chameleon 198745
Chameleon Plant 198743,198745
Chamfuta 10127
Chamisa 44334
Chamise 8010,8031,8032
Chamise Greasewood 8010
Chamisso Arnica 34693
Chamisso Honeysuckle 235721

Chamisso's Cotton-grass 152749
Chamisso's Montia 258241
Chamisso's Starwort 374846
Chamizo 44334
Chamock 271563
Chamois Cress 198483
Chamois Ragwort 358753
Chamois-cress 198480
Chamomile 26738,85510,85520,85526,246307
Chamomile Sunray 190796
Chamomile Tea 85526
Chamomile-leaved Milfoil 3935
Chamomiletea 85526
Champa 252841
Champac Michelia 252841
Champaca 252841
Champacany-puti 252809
Champak 252841
Champereia 85932
Champin Grape 411611
Champion Azalea 330358
Champion Bauhinia 49046
Champion Birthwort 34139
Champion Burmannia 63960
Champion Cyclobalanopsis 116071
Champion Daphne 122410
Champion Diploprora 133145
Champion Dutchmanspipe 34139
Champion Dutchman's-pipe 34139
Champion Fig 165830
Champion Holly 203616
Champion Magnolia 232593
Champion Millettia 66620
Champion Mucuna 259493
Champion Oak 116071,324158,324344
Champion Qinggang 116071
Champion Rhodoleia 332205
Champion Slatepentree 400722
Champion Syzygium 382506
Championella 85937,85952
Champion's Rhododendron 330358
Champman Azalea 330361
Champman's Rhododendron 330361
Chan 362186
Chanal 174326
Chanar 174326
Chandelier Plant 215129,215278
Chanet Daisy 124778
Chanet Orostachys 275358
Chanet Rockcress 30222
Chanet's Dendranthema 124778
Chanet's Sawwort 361015
Chaney Ash 218761
Chang Azalea 330362
Chang Camellia 68933
Chang Dogtongueweed 365038
Chang Monkshood 5114
Chang Rhododendron 330362
Chang Sloanea 366041
Changart Milkvetch 42304
Changbai Crazyweed 278710
Changbai Honeysuckle 236076
Changbai Jerusalemsage 295128
Changbai Monkshood 5649
Changbai Pyrola 322921
Changbai Rockfoil 349529
Changbai Scotch Pine 300222
Changbai Iris 208706
Changbaishan Alpparsley 98784
Changbaishan Cacalia 283860
Changbaishan Corydalis 105726
Changbaishan Dogtongueweed 385913
Changbaishan Dry Catchfly 363615
Changbaishan Fescue 164334
Changbaishan Fescuegrass 164334
Changbaishan Goldenray 229065
Changbaishan Ladybell 7739
Changbaishan Mouseear 82728
Changbaishan Poppy 282683
Changbaishan Pyrola 322921
Changbaishan Rush 213272
Changbaishan Swordflag 208706
Changbaishan Tephroseris 385913
Changbaishan Willow 343780
Changbaishan Windhairdaisy 348854
Changdu Corydalis 105725
Changdu Crazyweed 279026
Changdu Fescue 163861
Changdu Fescuegrass 163861
Changdu Goosecomb 335193
Changdu Leek 15183
Changdu Milkvetch 42172
Changdu Onion 15183
Changdu Peashrub 72204
Changdu Pea-shrub 72204
Changdu Poplar 311452
Changdu Roegneria 335193
Changdu Sagebrush 36029
Changdu Sandwort 31813
Changeable Bamboo 47376,297288
Changeable Medic 247507
Changeable Rose 195040
Changeable Scorpion-grass 260791
Changeable Scorpionsenna 105324
Changeable Velvetberry 373510
Changeable Windhairdaisy 348547
Changeables 199956
Changecolor Pinangapalm 299659
Changed Acronema 6207
Changed Barberry 51482
Changed Magnolia 242217
Changed Sagebrush 35343
Changed Wormwood 35343
Changelobe Windhairdaisy 348914
Change-of-the-weather 21339
Changhua Bedstraw 170517
Changhua Hilliella 196275
Changhua Hornbeam 77401
Changhua Maple 2884
Changing Forget-me-not 260791
Changing Rose 195040
Changingcolour Caper 71919
Changing-coloured Rhododendron 331838
Changium 85971
Changjiang Cycas 115810
Changjiang Dendrobium 278636
Changjiang Deutzia 127088
Changjiang Ehretia 141614
Changjiang Paraboea 283112
Changmu Milkvetch 42173
Changmu Sedge 74090
Changning Camellia 68978
Changning Chien's Willow 343205
Changning Hemsleia 191900
Changning Sedge 75336
Changning Snowgall 191900
Changpai Mountains Ladybell 7739
Changpai Scotch Pine 300222
Changpeishan Tephroseris 385913
Changputong Cliffbean 254645
Changputong Millettia 254645
Changputong Rockfoil 349169
Changruicaoia 85976
Chang's Elm 401479
Chang's Persimmon 132097
Chang's Tephroseris 365038
Changsha Bamboo 297478
Changsha Firmbamboo 297478
Changshan Orange 93425
Changshanhuyou Tangelo 93688
Changshou Kumquat 167511
Changyang Mahonia 242638
Chang-zhu Atractylodes 44218
Channel Centaury 81418
Channel Island Oak 324489
Channelled Fargesia 162651
Channelled Heath 149138
Channelled Waterplantain 14725
Chansonette Camellia 69594
Chantin Aechmea 8554
Chantrier Tacca 382912
Chao'an Azalea 330363
Chao'an Rhododendron 330363
Chaofang's Sedge 74091
Chaotum Meadowrue 388520
Chaotung Actinidia 6694
Chaozhou Sweetleaf 381272
Chapa Ainsliaea 12621
Chapa Blueberry 10363
Chapa Greenbrier 366283

Chapa Liparis 232118
Chapa Michelia 252844
Chapa Rabbiten-wind 12621
Chapa Smilax 366283
Chaparal Broom 46241
Chaparral Beargrass 266492
Chaparral Ceanothus 79952
Chaparral Cluster-lily 60486
Chaparral Currant 334082
Chaparral Fleabane 150833
Chaparral Goldenweed 150312
Chaparral Honeysuckle 235873
Chaparral Lily 230025
Chaparral Nolina 266492
Chaparral Pea 298467
Chaparral Penstemon 289352
Chaparral Prickly Pear 272998
Chaparral Sage 345169
Chaparral Sumac 332751
Chaparral Wild Buckwheat 151978
Chaparro Prieto 1532
Chapistle 290720
Chaplash 36910
Chaplet Flower 245803
Chapman Bluegrass 305440
Chapman Oak 323750
Chapman White Oak 323750
Chapman's Gayfeather 228450
Chapman's Goldenrod 368291
Chapman's Rhododendron 330365
Chapote 132431
Charcoal Oak 324528
Charcoal Tree 394656
Chard 53224,53257
Chardon 73278
Chards 117787,394327
Chare 86427
Chari Primrose 315060
Chariot Tree 277462
Chariot-and-horses 5442,5676
Charity 307197
Charles' Sceptre 287657
Charleston Mountain Goldenbush 150314
Charleston Mountain Pussytoes 26595
Charleston Pussytoes 26595
Charleston Rabbitbrush 114517
Charleston Sandwort 31826
Charleston Tansy 371146
Charlesworth Paphiopedilum 282802
Charlick 364557
Charlock 326609,364557
Charlock Mustard 364557
Charming Barberry 52305
Charming Begonia 50359
Charming Paphiopedilum 282911
Charming Rhododendron 330046,331001
Charming Woody-aster 415856
Charnock 364557
Charruana Feathergrass 376744
Chartaceous Fig 164796
Chartaceous Gurjun 133557
Chartaceous-leaf Cinnamon 91309
Chartaceous-leaved Cinnamon 91309
Chartolepis 86046
Chasalia 86192
Chasam Island Akeake 270214
Chasbol 282685
Chase-the-devil 266225
Chasm Figwort 355085
Chasmanthe 86120,86122
Chaste Nut 78811
Chaste Tree 411187,411189,411362
Chaste Willow 411189
Chaste-lamb Tree 411189
Chastetree 411187,411189,411362,411373
Chaste-tree 411187,411189,411378
Chastey 78811
Chat 79387,79395
Chatham Island Forget-me-not 260734,
260735
Chatham Island Lily 260735
Chatham Isle Forget-me-not 260735
Chatham Isle Lily 260735
Chatham Sophora 368985
Chatianshan Liparis 232334
Chatianshan Twayblade 232334
Chats 3462,300223,367696
Chatterbox 28617,147142,174877
Chatterjee Pittosporum 301237
Chatts 300223
Chaudoc Glyptic-petal Bush 177987
Chaulmoogra Tree 199740
Chaulmoogratree 199740,199744
Chaulmoogra-tree 199740,199744
Chaulmugra 182962
Chausse-trappe 80981
Chautle Living Rock 33209
Chavalier Dendropanax 125604
Chawalong Azalea 330865
Chawalong Barberry 52281
Chawalong Larkspur 124134
Chawalong Meadowrue 388702
Chawalong Monkshood 5114
Chawalong Platanthera 302274
Chawalong Rhododendron 330865
Chawalong Tomentellate Honeysuckle 236179
Chawalong Woodbetony 287784
Chaw-leaf 369322
Chaws 109857
Chay Root 270028
Chaya 97739
Chaya Milkvetch 42169
Chaydaia 86313
Chayota 356352
Chayote 356350,356352
Chayu Arrowbamboo 162768
Chayu Aster 41523
Chayu Azalea 331487
Chayu Barberry 52397
Chayu Birdlingbran 87244
Chayu Catchfly 364226
Chayu Chesneya 87244
Chayu Clovershrub 70784
Chayu Corydalis 106557
Chayu Dubyaea 138781
Chayu Fescuegrass 163862
Chayu Fir 309
Chayu Machilus 240567
Chayu Milkvetch 43288
Chayu Monkshood 5116
Chayu Mountainash 369572
Chayu Mountain-ash 369572
Chayu Nettle 403048
Chayu Onosma 271849
Chayu Rex Woodbetony 287609
Chayu Rhododendron 331487
Chayu Robust Prive 229603
Chayu Rockfoil 350050
Chayu Rockjasmine 23347
Chayu Sagebrush 36567
Chayu Speedwell 407076
Chayu St. John's Wort 201969
Chayu Swertia 380419
Chayu Umbrella Bamboo 162768
Chayu Vetch 408303
Chayu Willow 344307
Chayu Woodbetony 287842
Chayu Yinshancress 416371
Chayu Yinshania 416371
Cheadle 250614
Cheadle Dock 359158
Cheal's Barberry 52121
Cheat 60612,60912,60963,235373
Cheat Grass 60989,60992
Cheatgrass 60989
Cheatgrass Brome 60989
Chebula Terminalia 386504
Chebule 386500
Chebulic Myrobalan 386504
Checker Fiddleneck 20843
Checker Mallow 362702,362705
Checker Tree 369541
Checkerberry 172089,172135,256002
Checkerberry Whitepearl 172135
Checkerberry Wintergreen 172135
Checkerbloom 362704
Checkered Fritillary 168476
Checkered Lily 168476
Checkered Rattlesnake-plantain 179707
Checkered-like Fritillary 168469
Checkered-lily 168476

Checker-lily 168335
Checker-Mallow 362702
Checker-tree Mountain Ash 369541
Checker-tree Mountain-ash 369541
Checkwood 154962
Cheescweed 201015
Cheddar Pink 127623,127728
Cheddies 367696
Chee Barringtonia 48509
Chee Reed Grass 65330
Chee Reedbentgrass 65330
Chee Reedgrass 65336
Chee Reed-grass 65330
Chee Woodreed 65330
Cheeful Qinggang 116080
Cheeful Tanoak 233103
Cheerful Blueberry 403880
Cheerful Chinese Sinobambusa 364822
Cheerful Tangbamboo 364822
Cheese 237539,243823,243840,277648,
280231,312360,336419,339887,383874
Cheese Alstonia 18033
Cheese Flower 243840
Cheese Log 243840
Cheese Rennet 170743
Cheese Tree 177130
Cheese Wood 301225
Cheese-and-bread 109857,277648
Cheese-and-butter 325820
Cheese-cake 28200,237539
Cheese-cake Flower 243840
Cheese-cake Grass 237539
Cheesecups 325820
Cheese-log 243840
Cheeseman Rhopalostylis 332437
Cheeseplant 258168
Cheese-rennet 170743
Cheese-running 170743
Cheeses 223350,243738,243823
Cheeseweed 201012,243810
Cheeseweed Mallow 243810
Cheesewood 18033,301207,301225,301433,
301440
Chef's Cap Correa 105392
Cheilotheca 86383
Cheiridopsis 86480,86496
Cheirostylis 86667
Cheiry 86427
Chekiang Chelonopsis 86810
Chekiang Clematis 94811
Chekiang Glorybower 96157
Chekiang Hackberry 80612
Chekiang Pycreus 322196
Chekiang Sinojohnstonia 364962
Chekiang Skullcap 355406
Chekiang Swallowwort 117424
Chekiang Willow 343200
Cheli Argyreia 32620
Cheli Goniothalamus 179406
Cheli Greenstar 307491
Cheli Sonerila 368878
Chelone 86781,86783
Chelonopsis 86805
Chelungkiang Fescuegrass 163863
Chemise 68713
Chempedak 36924
Chen Arundinella 37363
Chen Bulrush 353310
Chen Fakesnaketailgrass 193992
Chen Goosecomb 335256
Chen Indigo 205793
Chen Larkspur 124129
Chen Mnesithea 256338
Chen Monkshood 5117
Chen Mou Bulrush 352170
Chen Primrose 314234
Chen Roegneria 335256
Chen Sedge 74094
Chenault Coralberry 380741,380742
Cheng Azalea 330850
Cheng Blueberry 403770
Cheng Cypress 114669
Cheng Dysosma 139608
Cheng Euonymus 157362
Cheng Hemsley's Rhododendron 330850
Cheng Spicebush 231319
Chengbu Holly 203620
Chengbu Longstalk Maple 3117
Chengde Forget me not 260769
Chengingleaf Tree 98177,98190
Chengiopanax 86853
Chengjiang Cogflower 154147,382743
Chengjiang Ervatamia 154147,382743
Chengkou Actinidia 6552
Chengkou Barberry 51518
Chengkou Bulbophyllum 62632
Chengkou Calanthe 66057
Chengkou Cinnamon 91409
Chengkou Cowparsnip 192262
Chengkou Curlylip-orchis 62632
Chengkou Holly 203621
Chengkou Kiwifruit 6552
Chengkou Mahonia 242637
Chengkou Parnassia 284516
Chengkou Pimpinella 299411
Chengkou Primrose 314359
Chengkou Rose 336484
Chengkou Sedge 75227
Chengkou Smoketree 107308
Chengkou Stonecrop 356577
Chengkou Tongoloa 392797
Chengkou Trigonotis 397393
Chengkou Wildginger 37596
Chengkou Windhairdaisy 348320
Chengmai Begonia 49717
Chengmai Fimbristylis 166215
Chengmai Fluttergrass 166215
Cheng's Mountainash 369359
Chengshuzhi Litse 233861
Chengtu Clematis 94859
Chengxian Woodbetony 287088
Chenile 201389
Chenille Copperleaf 1894
Chenille Copper-leaf 1894
Chenille Honey Myrtle 248098
Chenille Plant 1894,139993
Chenille Prickly Pear 272785
Chen-mo Bulrush 352170
Chenmou Euonymus 157364
Chennault Barberry 51445
Chenook Liquorice 238461
Chen's Oak 323751
Chequen 238339
Chequer Tree 369541
Chequer Vanda 404679
Chequerboard Juniper Alligator Bark
Juniper 213847
Chequered Daffodil 168476
Chequered Lily 168476
Chequered Tulip 168476
Chequer-fruit Razorsedge 354260
Chequers 369541
Chequer-shape Indocalamus 206833
Chequer-shaped Evergreen Chinkapin 79005
Chequer-shaped Gelidocalamus 172750
Chequer-shaped Indocalamus 206833,347314
Chequer-tree 369541
Chequer-wood 369541
Cherimoya 25828,25835,25836
Cherimoya-of-the-lowlands 25850
Cherimoya-tree 25836
Cherimoyer 25836
Cherisaunce 86427
Cherokec Tickclover 126649
Cherokee Bean 154668
Cherokee Rose 336675
Cherokee Sedge 74095
Cherokee-bean 154668
Cheronjee 61670
Cherry 83110,83155,83234,83284,83361,
155067,156385,279998,316166
Cherry Ballart 161712
Cherry Bay 223116,316518
Cherry Birch 53437,53505
Cherry Elaeagnus 142088
Cherry Emu Plum 277640
Cherry Guava 318737
Cherry Gum 83361
Cherry Honeysuckle 235797
Cherry Laurel 223088,223116,316285,
316518

Cherry Mahogany 391608
Cherry of the Rio Grande 156126
Cherry Orange 93269
Cherry Palm 347074
Cherry Pepper 72072,72075
Cherry Pie 85875,146724,190540,190559,190622,190649,190702,404251,404316
Cherry Plum 316166,316294
Cherry Prinsepia 315171,315176,365002
Cherry Red Sage 344980
Cherry Sage 345067
Cherry Silverberry 142088
Cherry Tomato 239158
Cherry Tree 83361
Cherry Woodbine 235651
Cherrybark Oak 323896
Cherry-berried Cotoneaster 107729
Cherryberry Cotoneaster 107729
Cherry-berry Cotoneaster 107729
Cherry-coloured Rhododendron 330343
Cherry-currant 334179
Cherryflower Azalea 330343
Cherryflower Camellia 69348
Cherryfruit Hackberry 80610
Cherry-fruited Hackberry 80610
Cherrylaurel 223088
Cherry-laurel 223088,223116,316518
Cherryleaf Arrowwood 408061
Cherry-leaf Elaeocarpus 142374
Cherry-leaved Elaeocarpus 142374
Cherryleaved Eurya 160439
Cherry-leaved Eurya 160439,160585
Cherrylike Greenstar 307489
Cherry-like Greenstar 307489
Cherry-of-the-rio Grande 156125
Cherry-pie 85875,146724,190622,404251,404316
Cherry-plum 316294
Cherry-red Barberry 51441
Cherryred Cotoneaster 107474
Cherry-red Cotoneaster 107474,107729
Cherrystone Juniper 213827
Cherry-wood 407989
Cherry-woodbine 235651
Cherson Bedstraw 113135
Chervell 28024,236022
Cherviileaf Corydalis 105620
Chervil 28013,28023,28024,28044,84723,84768
Chervil Larkspur 124039
Chervil-like Violet 409816
Chesapeake Panicgrass 282031
Chesbol 282685
Chesboul 282717
Chesbow 282685,282717
Chesbowl 282717
Chesneya 87232,87239,87271
Chesniella 87268
Chess 60612,60963,235373
Chess Apple 369322
Chess Brome 60963
Chess-apple 369322
Chessbean 116585
Chessboard Flower 417880,417925
Chesses 280231
Chestnus Taxillus 385187
Chestnut 9701,78766,78790,78802,78811
Chestnut Dioon 131429
Chestnut Fruit 84854
Chestnut Mucuna 259491
Chestnut Oak 324166,324201,324204
Chestnut Rattan Palm 65658
Chestnut Rose 336885
Chestnut Rush 212990
Chestnut Sage 344941
Chestnut Sedge 74066,166447,166449,166490
Chestnut Setafruit 84854
Chestnut Taxillus 385187
Chestnut Vine 387853
Chestnut Woodland Sedge 74066
Chestnut-brown Neoporteria 264337
Chestnut-colored Sedge 73952
Chestnutflower Rush 212990
Chestnutfruit 84852,84854
Chestnut-fruit 84852,84854
Chestnutleaf False Croton 71581
Chestnutleaf Trumpetbush 385472
Chestnut-leafed Oak Chestnut-leaf Oak 323737
Chevalier Adenia 7234
Chevalier Chessbean 116591
Chevalier Cyclobalanopsis 116073
Chevalier Oak 116073
Chevalier Premna 313607
Chevalier's Cylindrokelupha 116591
Chevisaunce 86427
Chevorell 28024
Chewbark 401512
Chewinggum-tree 244605
Chewing's Fescue 164243,164253,164272
Chew-root 384606
Chhnese Horned Holly 203660
Chia 344983,345083
Chia Sage 344983
Chia Seeds 344821,344983,345083
Chiacha Monkshood 5119
Chian Leptopus 226320
Chian Turpentine 301002
Chian Turpentine Tree 301002
Chianshan Vetch 408338
Chiapas Mahonia 242506
Chiapas Sage 344952
Chiastophyllum 87315,87318
Chibai Spindle-tree 157365
Chibbal 15170,15289,15621
Chibble 15170,15289,15621
Chibenh Rattan Palm 65659
Chibou 64085
Chiboul 15170,15289,15621
Chica 34902
Chicago Lustre 407781
Chiche Clearweed 299000
Chichelings 222844,408571
Chichen's Toes 103976,104019
Chichester Elm 401529
Chichibe 307170
Chichipe 307170
Chichituna 307170
Chichling Vetch 408452
Chick Pea 90797,90801,408257
Chick Wittles 374968
Chickasaw Plum 316219,316220
Chicken Corn 369600,369720
Chicken Grape 411631,411887,411979
Chicken Spike 371409
Chickenclaw Grass 274250
Chicken-claws 342859
Chickenleg Violet 409630
Chickens 349044
Chicken's Evergreen 374968
Chicken's Meat 374968
Chickensage 371141
Chickenspike 371409
ChickenVine 138934
Chickenweed 83064,311921,360328,374968
Chickenwood 364767,364768
Chickenwort 374968
Chickintestine Chickweed 375007
Chickling Pea 222832
Chickling Vetch 222832
Chicknyweed 374968
Chick-pea 90801
Chickpea Milk-Vetch 42193
Chick-pea Vetch 42193
Chickrassy 90605
Chicks-and-hens 358062
Chickweed 82629,82646,82742,82991,83020,220355,374724,374968
Chickweed Baby's-breath 183183
Chickweed Starry Puncturebract 362996
Chickweed Willowherb 146601
Chickweed Wintergreen 396785,396786
Chickweedlike Lepyrodiclis 226626
Chickweedlike Thincapsulewort 226626
Chickweed-wintergreen 396780
Chicle 244605
Chicory 90889,90894,90901
Chicoryleaf Wirelettuce 375958
Chicot 182527
Chicuta Palm 208373

Chiddies 367696
Chie Acronema 6202
Chieh-gua 51000
Chien Beech 162366
Chien Bigginseng 241161
Chien Big-ginseng 241161
Chien Briggsia 60259
Chien Primrose 314236
Chien Sinosenecio 365039
Chien Spicebush 231321
Chien Spice-bush 231321
Chienning Monkshood 5121
Chien's Chinese Groundsel 365039
Chien's Maple 2886
Chien's Willow 343203
Chie-Pei David Keteleeria 216123
Chifu Tanoak 233140
Chigger Flower 38147
Chigger-flower 38147
Chigger-plant 38147
Chigger-weed 26768
Chihli 59600
Chihshui Skullcap 355408
Chihsin Azalea 330384
Chihsin Rhododendron 330384
Chihuahua Ash 168051
Chihuahua Flax 232010
Chihuahua Oak 323754
Chihuahua Pine 299861,300033,300035
Chihuahuan Evening Primrose 269514
Chihuahuan Fishook Cactus 176701
Chihuahuan Mouse-ear Chickweed 83056
Chihuahuan Sage 227922
Chiiding Sweet William 292665
Chiken Eye Parnassia 284628
Chiken Heart grass 284521
Chikpea 90797,90801
Chiku 244605
Chikusichloa 87352
Chi-lan Shan Lasianthus 222289
Chilauni 350949
Chilblain-berry 383675
Childgrass 340336,340367
Childing Cudweed 166038
Childing Pink 292660,292664,292665
Children of Israel 243209,321643
Children-of-israel 321643,349044
Children's Clock 384714
Children's Daisy 50825
Children's Tomato 366947
Children's Tomatoes 366947
Childs Gladiolus 176116
Childseng 318489,318501
Childvine 400844,400942
Chile 72070
Chile Barberry 51446
Chile Bells 221340,221341
Chile Chaetotropis 85017
Chile De Perro 273025
Chile Dodder 115141
Chile Florifenoua Barberry 51624
Chile Hazel 175469
Chile Lantern Tree 111146
Chile Mayten 246817
Chile Mayten Tree 246796
Chile Nettle 234251
Chile Nut 30832,175469
Chile Nut Tree 30832
Chile Pepper 72070
Chile Pine 30832
Chile Pine Family 30857
Chile Pyrrhocactus 264340
Chile Satinflower 365803
Chile Strawberry 167604
Chile Tarweed 241543
Chile Wine Palm 212557
Chile Yaccatree 316084
Chilean Bamboo 90630,90632
Chilean Beech 266879
Chilean Beet 53227,53257
Chilean Bell Flower 221341
Chilean Bellflower 221340,221341,266470
Chilean Blue Crocus 385520
Chilean Bluegrass 305442
Chilean Cedar 45238,45239,299132
Chilean Chess 60645
Chilean Cranberry 401346
Chilean Crocus 385520,385521
Chilean Currant 333985
Chilean Evening-primrose 269509
Chilean Fig 251025
Chilean Firebush 144847,144848
Chilean Fire-bush 144848
Chilean Giant-rhubarb 181958
Chilean Glory Flower 139934,139936
Chilean Glory Vine 139934
Chilean Glory Vive 139936
Chilean Guava 261726,261765,401346
Chilean Gum Box 155141
Chilean Gum-box 155141
Chilean Gunnera 181958
Chilean Hazel 175469
Chilean Incense Cedar 45238,45239
Chilean Incense-cedar 45239
Chilean Iris 228625,228628
Chilean Jasmine 242324,244327
Chilean Jessamine 84419
Chilean Laurel 223011,223014,223015
Chilean Mesquite 83527,315551
Chilean Monkey-flower 255198
Chilean Myrtle 156133
Chilean Needlegrass 262615
Chilean Nightshade 367170
Chilean Nut 175469
Chilean Palo Verde 174326
Chilean Pfeiffer Cicely 276463
Chilean Pine 30832
Chilean Plantain 301974
Chilean Potato Tree 367062
Chilean Rabbitsfoot Grass 310103
Chilean Ricegrass 4124
Chilean Rium Dacrydium 121099
Chilean Sophora 369065
Chilean Strawberry 167604
Chilean Tarweed 241543
Chilean Totara 306510
Chilean Waterwort 142574
Chilean Weeping Bamboo 90634
Chilean Willow 343207
Chilean Wine Palm 212557
Chilean Yellow-sorrel 278136
Chilean Yew-pine 306517
Chile-bells 221340,221341
Chilen Fire Bush 144848
Chilen Fire Tree 144848
Chile-tree 244605
Chilgoza Pine 299953
Chili 72070
Chili Clover 126596
Chili Pepper 72071
Chili Pine 30832
Chili Strawbwrry 167604
Chilian Gloryflower 139936
Chilian Lily 18065
Chilicote 114278
Chilien Moutain Monkshood 5123
Chilienshan Milkvetch 42176
Chilita 243994
Chillan Barberry 51448
Chilli 72070
Chilli Pepper 72077
Chilly Tinospore 392252
Chilosehista 87456
Chilte 97738,97740
Chilte Rubber 97740
Chiltern Barberry 51450
Chiltern Gentian 174133
Chimbo Barberry 51451
Chimen Skullcap 355409
Chimney Bellflower 70247
Chimney Plant 70247
Chimney Smock 23925
Chimney Sweep 81237,238583,302034,302103
Chimney Sweeper 238583,302034,401112
Chimney Sweepers 238583
Chimney Sweeper's Brush 401112
Chimneyflower 254892
Chin Cactus 182419,182444
China Abelia 113
China Abutilon 989

China Acanthopanax 2395
China Acronema 6205
China Adinandra 8214
China Aeginetia 8767
China Aeluropus 8884
China Alangium 13353
China Allspice 68308
China Aloe 17383
China Altingia 18192
China Amesiodendron 19418
China Amomum 19833
China Amsonia 20850
China Amydrium 20868
China Andrographis 182815
China Anise 204603
China Anisopapus 25504
China Aporosa 29761
China Apple 243667
China Aralia 30604
China Arborvitae 302721
China Ardisia 31380
China Armand Pine 299799
China Arrowwood 407939
China Artichoke 373100,373439
China Ash 167940
China Aspen 311193
China Aster 67312,67314
China Asyneuma 43577
China Asystasiella 43682
China Atractylodes 44208
China Atropanthe 44719
China Aucuba 44893
China Awlquitch 131030
China Axileflower 153398
China Bagpetaltree 342086
China Banana 260201
China Batkudze 250225
China Bayberry 261212
China Beakrush 333518
China Begonia 49890
China Bellflower 302753
China Berry 248895
China Berry Tree 248895
China Betony 373173
China Bhesa 53680
China Billwalnut 25762
China Biondia 54516
China Birch 53379
China Bitterbamboo 304044
China Bladderwort 403899
China Bletilla 55574
China Blooddisperser 297635
China Bluestem 22569
China Bogorchid 158070
China Bonyberry 276544
China Bournea 57902
China Box 260173
China Box Jasmine Orange 260165
China Box Jasminorange 260173
China Bretschneidera 59966
China Briar 366264
China Brier 366284
China Brocadebeldflower 413613
China Bromegrass 60972
China Broomrape 275209
China Buckeye 9683
China Buckthorn 328883
China Buddhamallow 402249
China Bushbeech 178028
China Bushclover 226739
China Buttercup 325718
China Cacalia 283869
China Caesalpinia 252754
China Calanthe 66073
China Calophanoides 67821
China Cane 162635,329173,364631
China Carlesia 76987
China Cedar 113721
China Cephalomappa 82240
China Cheirostylis 86670
China Cherry 83238
China Chestnut 376144
China Chest-nut 376144
China Chickweed 374802
China Childgrass 340344
China Chirita 87966
China Chlamydoboea 88166
China Cinquefoil 312450
China Cistanche 93079
China Cliffbean 254812
China Clubfilment 179178
China Cluster Mallow 243866
China Coca 155088
China Cockclawthorn 278323
China Cockehead-yam 152888
China Cockle 346434
China Cogflower 154150,382766
China Coinmaple 133599
China Cointree 280548
China Coir Palm 393809
China Collabium 99773
China Congea 101776
China Corktree 294238
China Corydalis 105730
China Cotton 179878
China Cowlily 267329
China Crab Apple 243630
China Crabdaisy 413514
China Cranesbill 174918
China Crinum 111171
China Crosnes 373100
China Crucianvine 356324
China Cryptocoryne 113541
China Cryptolepis 113612
China Cryptomeria 113721
China Curculigo 114849
China Curlylip-orchis 62629
China Cypress 178014,178019
China Dandelion 384473
China Date-plum 418169
China Decumaria 123545
China Dendrobium 125359
China Desmos 126717
China Didissandra 129833
China Dischidia 134032
China Dishspurorchis 171894
China Dodder 114994
China Dogliverweed 129243
China Dogwood 104994
China Douglas Fir 318591
China Dove Tree 123245
China Dregea 137795
China Drumprickle 244938
China Dullcolored Carpesium 77193
China Dunnia 138986
China Dwarf Cherry 83220
China Dyschoriste 139494
China Eaglewood 29983
China Eargrass 187539
China Edelweiss 224938
China Eelgrass 418169
China Elaeocarpus 142298
China Elder 345586
China Entireliporchis 261469
China Ephedra 146253
China Eria 148775
China Euonymus 157745
China Eurya 160442
China Evergreen 11329,11348
China Faberdaisy 161889
China Fagraea 162346
China Fescue 164321
China Fevervine 280097
China Figwort 355181
China Filbert 106716
China Fir 114532,114539
China Flatspike 55897
China Fleece Vine 162509
China Flowering Crabapple 243702
China Fluttergrass 6865
China Forkstyleflower 374495
China Fortunearia 167488
China Foxglove 327434
China Fritillary 168564
China Galangal 17656
China Gastrochilus 171894
China Gentian 173336
China Ghostfluting 228333
China Giantarum 20136
China Gianthyssop 236267
China Globeflower 399500

China Glorybind 102933
China Glycosmis 177849
China Golden Larch 317822
China Golden Rose 336636
China Goldthread 103828
China Goldwaist 90346
China Goniothalamus 179407
China Gooseberry 6553
China Goosecomb 335505
China Grand Crinum 111171
China Grass 56094,56229,335649
China Grassleaf Chickweed 374899
China Greenbrier 366284
China Greenstar 307493
China Greentwig 352432
China Grumewood 211022
China Gugertree 350943
China Gymnotheca 182848
China Hairorchis 148775
China Hairtea 28566
China Halfstyleflower 360700
China Hanceola 185252
China Hartia 376467
China Hat Plant 197368
China Hawthorn 109933
China Heartleaf Hornbeam 77278
China Helwingia 191141
China Hemlock 399879
China Henbane 201392
China Herbclamworm 398071
China Heteropanax 193899
China Heterosmilax 194110
China Hibiscus 195233
China Hideseedgrass 94591
China High Smallreed 127297
China Hippeophyllum 267994
China Holcoglossum 197268
China Holly 204272
China Honeylocust 176901
China Honeysuckle 236180
China Hoofrenshen 133077
China Hooktea 52468
China Hookvine 401781
China Hopea 198140
China Hornfennel 201612
China Hornsage 205580
China Horsetailtree 332329
China Hosiea 198576
China Houndstongue 117908
China Houses 99838
China Hundredstamen 389638
China Hydrangea 199853
China Hydrangeavine 351802
China Hypecoum 201612
China Incense Cedar 67529
China Indocalamus 206829
China Indosasa 206900
China Irisorchis 267933
China Ivy 187307
China Ixeris 210525
China Ixora 211067
China Jellygrass 252242
China Jobstears 99134
China Juneberry 19292
China Jurinea 212004
China Jute 1000
China Katsuratree 83736
China Keiskea 215817
China Kellogia 215847
China Knotweed 308965
China Ladybell 7811
China Ladytress 372247
China Lakemelongrass 232391
China Lantern 297643
China Larch 221930,221934
China Largelily 73157
China Laurel 28309
China Lawngrass 418445
China Lerchea 226644
China Leucas 227572
China Leymus 228356
China Ligusticum 229390
China Lilac 382132,382256
China Lilyturf 272134
China Linden 391685
China Lizardtail 348087
China Lobelia 234363
China Loropetal 237191
China Lotusdaisy 111826
China Machilus 240570
China Macrocarpium 104994
China Madder 337916
China Magnolia 232594,279250
China Magnoliavine 351021
China Mahonia 242534
China Mallow 243833
China Maple 3615
China Marshweed 230286
China Meliosma 249463
China Mezoneuron 252754
China Miliusa 254558
China Milkgreens 245845
China Milkvetch 42177,43053
China Monimopetalum 257433
China Monkey Earrings Pea 301147
China Morina 258840
China Mosla 259282
China Mosquitoman 134922
China Mosquitotrap 117425
China Mountain Cypress 114690
China Mountainsorrel 278583
China Mucilagefruit 101236
China Mulberry 259122
China Munronia 259895
China Mustard 59444
China Mycetia 260650
China Naiad 262024
China Narcissus 262468
China Nardostachys 262497
China Neillia 263161
China Nertera 265361
China Nettletree 80739
China Nightshade 367322
China Oat 45431
China Oilnut 323073
China Oldham Meliosma 249426
China Olive 70988
China Onion 15185
China Onosma 271828
China Ophiorrhiza 272190
China Orange 93765
China Orange Gingerlily 187467
China Orchidantha 273269
China Orchis 117042
China Orthocarpus 397905
China Osbeckia 276090
China Osmanther 276256
China Pagodatree 368986
China Paper Birch 53335
China Paradombeya 283307
China Paris 284367
China Parnassia 284517
China Pear 323268
China Pearleaf Crabapple 243551
China Peashrub 72342
China Pendent-bell 145693
China Pennywort 200274
China Pentaphragma 289696
China Penthorum 290134
China Pepper 300370,300373
China Persimmon 132089
China Pertybush 292078
China Pholidota 295518
China Photinia 295630
China Pinangapalm 299657
China Pine 300243
China Pink 127654
China Pistachio 300980
China Platanthera 302404
China Plum 316761
China Plumyew 82545
China Polemonium 307215
China Poplar 311371
China Porana 311659
China Pothos 313197
China Prettybean 67774
China Pride 248895
China Primrose 314975
China Privet 219933,229523
China Pseudopogonatherum 318153
China Pulsatilla 321672

China Purple Willow 344121
China Purplepearl 66762
China Purplestem 376478
China Quassia 298509
China Randia 278323
China Rattlebox 112013
China Red Bud 83769
China Red Silkyarn 238967
China Redbud 83769
China Rhabdothamnopsis 328443
China Rhynchosia 333182
China Ricegrass 276001
China Root 183124,366284,366592
China Roscoea 337106
China Rose 195149,336485
China Roughleaf 222115
China Saddleorchis 2079
China Saddletree 240128
China Sage 344957
China Saltclaw 215331
China Sandspike 148539
China Sandthorn 196766
China Sanicle 345948
China Saraca 346502
China Sasa 347391
China Savin 213634
China Sawwort 361018
China Scaleseedsedge 225572
China Schefflera 350675
China Schizostachyum 351850
China Scolopia 354598
China Scrofella 355034
China Scurrula 355317,385192
China Seaberry 184997
China Sealavender 230759
China Sedge 74098
China Sericocalyx 360700
China Serpentroot 354952
China Shortia 362257
China Sida 362514
China Silkvine 291082
China Sindechites 364698
China Sinofranchetia 364947
China Siphonostegia 365296
China Siris 13515
China Slantwing 301691
China Sloanea 366077
China Snakegrape 20447
China Snowbell 379325,379474
China Snowberry 380758
China Snowgall 191903
China Soapberry 346338
China Sotvine 194166
China Spicebush 231324
China Spinestemon 306966
China Spiraea 371884
China Spiralflag 107246
China Spotseed 57678
China Sprangletop 225989
China Spruce 298232
China Spurge Pachysandra 279744
China Squash 114292
China St. John's Wort 202021
China Stachyurus 373524
China Starjasmine 393657
China Stephanandra 375812
China Stephania 375899
China Stonebutterfly 292581
China Sumac 332513,332662
China Sweetgrass 177582
China Sweetleaf 381341
China Sweetspire 210364
China Sweetvetch 187831
China Swordflag 208480,208667
China Syzygium 382504
China Tailanther 381953
China Tailleaftree 402616
China Tallowtree 346408
China Tarenna 385026
China Teasel 133463
China Themeda 389358
China Thorny Bamboo 47445
China Thorowax 63594
China Tickseed 104849
China Tineanther 5885
China Tinospore 392274
China Toadflex 231187
China Toona 392841
China Torreya 393061
China Touch-me-not 204848
China Trapella 394560
China Tree 217626,248895
China Tree Family 248929
China Triawn 33797
China Tricarpweed 395372
China Trumpetcreeper 70507
China Tuliptree 232603
China Tupelo 267864
China Tupistra 70600
China Twigrush 93913
China Umbrellaleaf 132685
China Uraria 402147
China Vetch 408339
China Vetchling 222710
China Violet 410360
China Walnut 212595
China Waterdropwort 269360
China Waxgourd 50998
China Waxmyrtle 261212
China Waxpetal 106675
China Weasel-snout 170021
China White Pine 299799
China White Poplar 311530
China Whitepearl 172150
China Whitesilkgrass 87776
China Wildclove 226106
China Wildginger 37597
China Wildmedia 28999
China Wildsenna 389518
China Willow 343179
China Willowweed 146883
China Windhairdaisy 348199
China Wingnut 320377
China Winterberry Currant 333972
China Witchgrass 226230
China Witchhazel 185130
China Wolfberry 239021
China Wolftailgrass 289015
China Woodbetony 287090,287478
China Wood-oil Tree 406018
China Woodrush 238615
China Woodsorrel 278288
China Yaccatree 306469
China Yelloweyegrass 416008
China Yellowflower Willow 344117
China Yellowtung 145417
China Yellowwood 94020
China Yew 385405
China-aster 67312
Chinabell 16291
Chinabells 16291
Chinaberry 248893,248895,346336
China-berry 248893
Chinaberry Evodia 161356
Chinaberry Family 248929
Chinaberry Tree 248895
China-berry Tree 248895
Chinaberrytree 248895
Chinaberry-tree 248895
Chinacane 162635,364631
Chinafir 114532
China-fir 114532,114539,114548
China-grass 56229
Chinagreen 11329,11348
Chinalaurel 28309
China-laurel 28309
Chinaman's Breeches 220729
China-paper Birch 53335
Chinaroot Green Brier 366284
Chinaroot Greenbrier 366284
Chinaroot Smilax 366284
Chinarose Campion 363265
China-shrub 31607
China-tree 217626,248895
Chinaure 56380
China-wood 406018
Chincherinchee 274810
Chinchone 360328
Chinchow Spiraea 372038
Chinchweed 286897
Chinense Aphanamixis 28999

Chines Glaucous Bluegrass 305380
Chinese Abelia 113
Chinese Abutilon 989
Chinese Acanthopanax 2395
Chinese Acanthophippium 2079
Chinese Aconite 5335
Chinese Acranthera 5885
Chinese Actinidia 6553
Chinese Adinandra 8214
Chinese Aeginetia 8767
Chinese Aeluropus 8884
Chinese Alangium 13353
Chinese Albizia 13515
Chinese Albizzia 13515
Chinese Allium 15649
Chinese Allspice 68308,231298
Chinese Aloe 17383
Chinese Alpine Juniper 213857
Chinese Altelnateleaf Goldsaxifrage 90346
Chinese Altingia 18192
Chinese Alyxia 18529
Chinese Amaranth 18836
Chinese Amesiodendron 19418
Chinese Ampelopsis 20447
Chinese Amplexicaule Knotweed 308772
Chinese Amydrium 20868
Chinese and Japanese Chrysanthemum 124790
Chinese Andrachne 226320
Chinese Angelica 24475
Chinese Angelica Tree 30604
Chinese Anisatum 204575
Chinese Anise 204575,204603
Chinese Anise Shrub 204526
Chinese Anisopapus 25504
Chinese Annamocarya 25762
Chinese Anthocephalus 263983
Chinese Antirhea 28566
Chinese Apes-earring 301147
Chinese Aporosa 29761
Chinese Apple 243667
Chinese Apple Pear 323268
Chinese Aralia 30604
Chinese Arborvitae 302718,302721,390587
Chinese Arbor-vitae 302718,302721
Chinese Archiphysalis 31002
Chinese Ardisia 31380
Chinese Arrowhead 342424
Chinese Arrowroot 263272
Chinese Artichoke 373100,373439
Chinese Arundina 37115
Chinese Ash 167940,320377
Chinese Asparagus 38960
Chinese Aspen 311193
Chinese Aster 41278,67312,67314
Chinese Astilbe 41795,41847
Chinese Asyneuma 43577
Chinese Asystasiella 43682
Chinese Atropanthe 44719
Chinese Attenuate Bluegrass 305985
Chinese Aucuba 44893
Chinese Azalea 331257
Chinese Bagpetaltree 342086
Chinese Bag-petal-tree 342086
Chinese Bamboo 347391,362123
Chinese Banana 260201,260250
Chinese Bandoline Wood 240669
Chinese Banyan 165307,165630
Chinese Baotingflower 413759
Chinese Barbed Gentianopsis 174208
Chinese Barberry 51360,51452,51802
Chinese Bastardtoadflax 389638
Chinese Bayberry 261212
Chinese Bead Plant 265361
Chinese Beakrush 333518
Chinese Beauty Bush 66913
Chinese Beautyberry 65670,66762
Chinese Beauty-berry 66913
Chinese Beech 162372
Chinese Bellflower 913,69959,302753
Chinese Bellfower 302747
Chinese Berry 248895
Chinese Betony 373173
Chinese Biondia 54516
Chinese Birch 53335,53379
Chinese Bishopwood 54620
Chinese Bitter Sweet 80203
Chinese Bitterbamboo 304044
Chinese Bittersweet 80203,80260
Chinese Black Olive 71015
Chinese Bladdernut 374103
Chinese Bladder-nut 374103
Chinese Bluestem 22569
Chinese Blysmus 55897
Chinese Bog Trollius 399500
Chinese Bonyberry 276544
Chinese Bony-berry 276544
Chinese Bothriospermum 57678
Chinese Bottle Tree 166627
Chinese Bournea 57902
Chinese Box 64285,64369,260165,260173
Chinese Box Thorn 239011
Chinese Boxrange 43721
Chinese Box-thorn 239011
Chinese Bracketplant 88550
Chinese Bramble 339392
Chinese Bredia 59881
Chinese Bretschneidera 59966
Chinese Briar 366264
Chinese British Inula 207059
Chinese Brome 60972
Chinese Broomrape 275209
Chinese Buckeye 9683
Chinese Buckthorn 328883
Chinese Buckwheat 162309
Chinese Bugle Weed 13079
Chinese Bulbophyllum 62629
Chinese Bush Cherry 83208,83238
Chinese Bush Clover 226742
Chinese Bush Fruit 83347
Chinese Bushbeech 178028
Chinese Bush-beech 178028
Chinese Bushcherry 83238
Chinese Bushclover 226739
Chinese Bush-clover 226739
Chinese Buttercup 325718
Chinese Butternut 212595
Chinese Buttonbush 8199
Chinese Cabbage 59438,59595,59600
Chinese Cadillo 402249
Chinese Caesalpinia 252754
Chinese Calligonum 67010
Chinese Calogyne 67690
Chinese Calophaca 67774,67785
Chinese Calophanoides 67821
Chinese Calycanthus 68308
Chinese Camp Brodiaea 60512
Chinese Cane 364820
Chinese Cardiocrinum 73157
Chinese Casearialeaf Microdesmis 253414
Chinese Catalpa 79257
Chinese Cedar 80015,113633,113721,114539,302721,392841
Chinese Cedrela 392841
Chinese Celery 269326
Chinese Cephalomappa 82240
Chinese Chaste Tree 411362
Chinese Chengiopanax 86854
Chinese Cherry 83284
Chinese Chestnut 78802
Chinese Chest-nut 376144
Chinese Chi 393491
Chinese Chickweed 374899
Chinese Chionographis 87776
Chinese Chirita 87966
Chinese Chives 15534,15843
Chinese Chlamydoboea 88166
Chinese Cinnamon 91302
Chinese Cinquefoil 312450
Chinese Cistanche 93079
Chinese Cladium 93913
Chinese Cleistogenes 94591
Chinese Clematis 94814
Chinese Clinopodium 96970
Chinese Clover 43053
Chinese Clover Shrub 70842
Chinese Clovershrub 70833
Chinese Cluster Mallow 243866
Chinese Cocaine Tree 155088
Chinese Coffeetree 182526
Chinese Coir Palm 393809

Chinese Collabium 99773
Chinese Common Toadflex 231187
Chinese Condorvine 245845
Chinese Congea 101776
Chinese Cordia 104247
Chinese Coriaria 104694,104701
Chinese Cork Oak 324532
Chinese Cork Tree 294238
Chinese Corktree 294238
Chinese Cork-tree 294238
Chinese Cornel 104994
Chinese Corydalis 105730
Chinese Cotton 179876
Chinese Cow Lily 267329
Chinese Cow's Tail Pine 82507
Chinese Cow's Tall Pine 82496
Chinese Cow's-tail Pine 82507
Chinese Cowtail Pine 82507
Chinese Crab Apple 243630
Chinese Crab-apple 243702
Chinese Crabgrass 130481
Chinese Cracker Flower 322950
Chinese Cranesbill 174918
Chinese Creeper 285102
Chinese Crinum 111171
Chinese Crossostephium 111826
Chinese Cryptocarya 113436
Chinese Cryptocoryne 113541
Chinese Cryptolepis 113612
Chinese Cryptomeria 113721
Chinese Cunninghamia 114539
Chinese Cup Grass 151702
Chinese Curculigo 114849
Chinese Cymbidium 117042
Chinese Cypress 114680,178019
Chinese Cystopetal 320239
Chinese Dandelion 384473
Chinese Date 418169,418173
Chinese Date-plum 418169
Chinese Datetree 418169
Chinese Day-lily 191266
Chinese Deadnettle 170021
Chinese Decaspermum 123451
Chinese Deciduous Cypress 178014,178019
Chinese Decumaria 123545
Chinese Dendrobium 125359
Chinese Dendrocalamus 125509
Chinese Desert Candle 148539
Chinese Desertcandle 148539
Chinese Desert-thorn 239021
Chinese Desmos 126717
Chinese Dicliptera 129243
Chinese Didissandra 129833
Chinese Dimeria 131030
Chinese Dipentodon Dipentodon 132611
Chinese Diplopanax 133077
Chinese Dipteronia 133599
Chinese Dischidia 134032
Chinese Distylium 134922
Chinese Dodder 114994
Chinese Dogwood 104994,124923,124925
Chinese Douglas Fir 318591
Chinese Dove-tree 123245
Chinese Dregea 137795
Chinese Dunnia 138986
Chinese Dwarf Bamboo 47348
Chinese Dwarf Cherry 83220
Chinese Dwarf Chinquapin 78823
Chinese Dwarfmistletoe 30908
Chinese Dwarf-mistletoe 30908
Chinese Dyschoriste 139494
Chinese Eagle Wood 29983
Chinese Eaglewood 29973,29983
Chinese Edelweiss 224938
Chinese Edible Bamboo 297194,297263
Chinese Eelgrass 418445
Chinese Egg-gooseberry 6693
Chinese Elaeocarpus 142298
Chinese Elatostema 142840
Chinese Elder 345586,345708
Chinese Elderberry 345679
Chinese Elkwood 413843
Chinese Elm 401581,401583
Chinese Endospermum 145417
Chinese Enkianthus 145693
Chinese Ephedra 146253
Chinese Epidendrum 146395
Chinese Eriosema 152888
Chinese Erismanthus 153398
Chinese Ervatamia 154150,382766
Chinese Euonymus 157745,157954
Chinese Eupatorium 158070
Chinese Eurya 160442,160519
Chinese Evergreen 11329,11332,11340,11348
Chinese Evergreen Azalea 331839
Chinese Evergreen Chinkapin 78889
Chinese Evergreen Magnolia 232595
Chinese Evergreenchinkapin 78889
Chinese Everlasting 21682
Chinese Faberdaisy 161889
Chinese Fagraea 162346
Chinese Falconer Rhododendron 331850
Chinese False-pistache 384371
Chinese False-quince 317602
Chinese Fan Palm 234170,393809
Chinese Fanpalm 234170
Chinese Fan-palm 234170
Chinese Fescuegrass 164321
Chinese Fevervine 280097
Chinese Few-flowered Spindle-tree 157954
Chinese Fig 164763
Chinese Fig Hazel 380602
Chinese Fighazel 380602
Chinese Figwort 355181
Chinese Filbert 106716
Chinese Fimbristylis 6865
Chinese Fir 114532,114539
Chinese Firethorn 322451,322465
Chinese Fish-rod Bamboo 297194
Chinese Fishtail Palm 78052
Chinese Fivefruit Larkspur 124582
Chinese Fixed-petal Vine 257433
Chinese Flame Tree 217613
Chinese Flowering Apple 243702
Chinese Flowering Ash 168100
Chinese Flowering Chestnut 415096
Chinese Flowering Crab Apple 243702
Chinese Flowering Crabapple 243702
Chinese Flowering Crab-apple 243702
Chinese Flowering Quince 84573,317602
Chinese Floweringquince 317602
Chinese Flowering-quince 317602
Chinese Foldwing 129243
Chinese Forgetmenot 117908
Chinese Forget-me-not 117900,117908
Chinese Forkstyleflower 374495
Chinese Fortunearia 167488
Chinese Fountain Bamboo 162752
Chinese Fountain Palm 234170
Chinese Fountain-bamboo 162718
Chinese Fountaingrass 282366,289015
Chinese Fountain-palm 234170
Chinese Fox Glove 327434
Chinese Foxglove 327429,327434
Chinese Foxtail 361743
Chinese Fragrant Bamboo 87633
Chinese Fragrant Fern 35132
Chinese Fragrant Sarcococca 346745
Chinese Freyn Cinquefoil 312571
Chinese Fringe Tree 87729
Chinese Fringelily 391501
Chinese Fringetree 87729
Chinese Fringe-tree Tasseltree 87729
Chinese Funeral-cypress 114690
Chinese Galan 17733
Chinese Galangal 17656,17733
Chinese Galeobdolon 170021
Chinese Gambirplant 401781
Chinese Gambir-plant 401781
Chinese Garlic 15843
Chinese Giant Hyssop 10414
Chinese Gingerlily 187467
Chinese Glacial Stonecrop 357154
Chinese Globe Flower 399500
Chinese Globeflower 399500
Chinese Glory Bower 96009
Chinese Glorybower 96009
Chinese Glycosmis 177849,177851
Chinese Goat Willow 344117
Chinese Golden Larch 317822

Chinese Goldenrain Tree 217613
Chinese Gold-rain Tree 217610
Chinese Goldsaxifrage 90452
Chinese Goldthread 103828
Chinese Goldwaist 90452
Chinese Gomphostemma 179178
Chinese Goniothalamus 179407
Chinese Gooseberry 6513,6553,6558,45725
Chinese Grass 56229
Chinese Green Indigo 328592
Chinese Greenstar 307493
Chinese Grewia 180700
Chinese Grey Bamboo 87643
Chinese Greytwig 352432
Chinese Grey-twig 352432
Chinese Ground Orchid 55575
Chinese Groundsel 365035,365068,365073
Chinese Gugertree 350943,350945
Chinese Guger-tree 350945
Chinese Gutta Percha 156041
Chinese Gymnotheca 182848
Chinese Hackberry 80739
Chinese Hanceola 185252
Chinese Hartia 376467
Chinese Hat Plant 197368
Chinese Hawthorn 109781,109924,109933,109936,295773
Chinese Hazel 106716,106736
Chinese Hazelnut 106716
Chinese Heartleaf Hornbeam 77278
Chinese Hedyotis 187539
Chinese Helwingia 191141
Chinese Hemigraphis 191486
Chinese Hemlock 399879
Chinese Hemlock-parsley 101823
Chinese Hemp 837,1000
Chinese Hemsleia 191903
Chinese Heptacodium 192168
Chinese Heteropanax 193899
Chinese Heterosmilax 194110
Chinese Heterostemma 194166
Chinese Hibiscus 195149,195233
Chinese Hickory 25762,77887
Chinese High Small Reed 127297
Chinese Himalayahoneysuckle 228333
Chinese Himalaya-honeysuckle 228333
Chinese Hispid Knotweed 308969
Chinese Holly 203625,203660,204217,204272
Chinese Honey 93717
Chinese Honey Locust 176901
Chinese Honey-locust 176901
Chinese Honeysuckle 235885,236180
Chinese Hooktree 337291
Chinese Hopea 198140
Chinese Hornbeam 77291,77303,77411,77420
Chinese Horse Chestnut 9683
Chinese Horsechestnut 9683
Chinese Horse-chestnut 9683
Chinese Hosiea 198576
Chinese Hound's-tongue 117908
Chinese Houses 99838
Chinese Hydrangea 199853
Chinese Hydrangeavine 351797,351802
Chinese Hydrangea-vine 351802
Chinese Incarvillea 205580
Chinese Incense Cedar 67529
Chinese Indigo 205876,209229,309893
Chinese Indigo Bush 206140
Chinese Indocalamus 206829
Chinese Indosasa 206900
Chinese Iris 208480,208667
Chinese Ivy 187307,393613
Chinese Ixeris 210525
Chinese Ixonanthes 211022
Chinese Ixora 211067
Chinese Jade Plant 311956
Chinese Japanese Cranberry 403872
Chinese Japanese Sabia 341514
Chinese Jasmine 211864,212004,393613,393657
Chinese Joint Fir 146253
Chinese Judas Tree 83769
Chinese Jujube 418169,418173
Chinese Juniper 213634
Chinese Juniper Dwarfmistletoe 30912
Chinese Juniper Dwarf-mistletoe 30912
Chinese Jute 837
Chinese Kadsura 351021
Chinese Kagwort 364522
Chinese Kale 59520,59526
Chinese Kalidium 215331
Chinese Katsura Tree 83740
Chinese Katsuratree 83736,83740
Chinese Keiskea 215817
Chinese Kellogia 215847
Chinese Kidney Bean 414576
Chinese Knotweed 308965,309966
Chinese Kousa Dogwood 124925
Chinese Kurrimia 218422
Chinese Lacquer 332938
Chinese Lacquer Tree 393491
Chinese Lacquer-tree 393495
Chinese Ladiesstresses 372247
Chinese Ladybell 7811
Chinese Lantern 837,913,957,975,267620,267621,297640,297643,345782
Chinese Lantern Flower 145693
Chinese Lantern Plant 297643,297645
Chinese Lantern-plant 297643
Chinese Lanterns 129136,174877,297640,297643
Chinese Larch 221930,221934,317822
Chinese Lasianthus 222115
Chinese Laurel 28309,28316
Chinese Leptocanna 351850
Chinese Leptodermis 226106,226120
Chinese Leptopus 226320
Chinese Lerchea 226644
Chinese Lespedeza 226739,226742
Chinese Leucas 227572
Chinese Leymus 228356
Chinese Ligusticum 229390
Chinese Lilac 382132,382134
Chinese Lilyturf 272134
Chinese Lime 391685,391795,397878
Chinese Linden 391685
Chinese Lipocarpha 232391
Chinese Liquidambar 232557
Chinese Litse 233880
Chinese Livistona 234170
Chinese Lizardtail 348087
Chinese Lobelia 234363
Chinese Locust 176901
Chinese Long Bean 409086
Chinese Longstigma Meadowrue 388552
Chinese Lophanthus 236267
Chinese Lophather 236295
Chinese Loropetalum 237191
Chinese Lotus 263272
Chinese Lovegrass 148028
Chinese Lycianthes 238967
Chinese Lycoris 239259
Chinese Maackia 240128
Chinese Macaranga 240322
Chinese Machilus 240570
Chinese Macrocarpium 104994
Chinese Madder 337916
Chinese Magnolia 198699,232594,279250,416682
Chinese Magnolia Vine 351021
Chinese Magnoliavine 351021
Chinese Magnolia-vine 351021
Chinese Mahogany 392841
Chinese Mahonia 242534
Chinese Mallow 243833,243862
Chinese Mammy Apple 268335
Chinese Manglietiastrum 244486
Chinese Mannagrass 177582
Chinese Mapania 244938
Chinese Maple 3615
Chinese Marsdenia 245845
Chinese Marshweed 230286
Chinese Mastixia 246219
Chinese Matrimony Vine 239011,239021
Chinese Matrimonyvine 239021
Chinese Matrimony-vine 239021
Chinese May Apple 365009
Chinese May-apple 365005,365009
Chinese Meadow Rue 388477

Chinese Meadow-rue 388477
Chinese Medick 247453
Chinese Medlar 151162
Chinese Meliosma 249463
Chinese Melodinus 249657
Chinese Meranti 362197
Chinese Merrilliopanax 250876
Chinese Mesona 252242
Chinese Mezoneuron 252754
Chinese Miliusa 254558
Chinese Milk Vetch 43053
Chinese Milkvetch 42177,43053
Chinese Milkwort 308081
Chinese Millettia 254812
Chinese Mock-orange 294471
Chinese Money Maple 133599
Chinese Monimopetalum 257433
Chinese Monkshood 5185
Chinese Moonseed 250225
Chinese Morina 258840
Chinese Mosla 259282
Chinese Mountain-loving Draba 137157
Chinese Mountain-loving Whitlowgrass 137157
Chinese Moxanettle 221538
Chinese Mugwort 35167,36454
Chinese Mulberry 259122
Chinese Munronia 259895
Chinese Musk-mallow 229
Chinese Muskroot 364523,364524
Chinese Mustard 59348,59438,59444,59595
Chinese Mycetia 260650
Chinese Myrmechis 261469
Chinese Narcissus 262382,262468
Chinese Nardostachys 262497
Chinese Necklace Poplar 311371
Chinese Neillia 263161
Chinese Nertera 265361
Chinese Nettle-tree 80739
Chinese Newlitse 264069
Chinese Nightshade 367322
Chinese Nutmeg 393061
Chinese Nuttall Rhododendron 331855
Chinese Oat 45431
Chinese Oberonia 267933
Chinese Oblique-wing 301691
Chinese Oeder Woodbetony 287478
Chinese Oilfruitcamphor 381791
Chinese Oilnut 323073
Chinese Oil-nut 323073
Chinese Olive 70988,70989,71015
Chinese Onion 15185,15289
Chinese Onosma 271828
Chinese Ophiorrhiza 272190
Chinese Orange 167503
Chinese Orchidantha 273269
Chinese Oriental Blueberry 403739
Chinese Oriental Photinia 295813
Chinese Orophea 275322
Chinese Orthocarpus 397905
Chinese Osbeckia 276090
Chinese Osmanthus 276256
Chinese Pachygone 279560
Chinese Pachysandra 279744
Chinese Paliurus 280548
Chinese Paperbark Maple 3003
Chinese Paradombeya 283307
Chinese Parashorea 362197
Chinese Parasol 166627
Chinese Parasol Tree 166612,166627
Chinese Parasoltree 166627
Chinese Parasol-tree 166627
Chinese Parnassia 284517
Chinese Parsley 104690
Chinese Patchouly 254262
Chinese Pea Tree 72342
Chinese Pear 323268,323330
Chinese Pear Apple 323268
Chinese Pear-apple 323268
Chinese Pearl bloom Tree 307291
Chinese Pearlbloomtree 307291
Chinese Pearlbloom-tree 307291
Chinese Pearleaf Crab-apple 243551
Chinese Pear-leaved Crab-apple 243551
Chinese Pearleverlasting 21682
Chinese Peashrub 72342
Chinese Pea-shrub 72342
Chinese Peliosanthes 288651
Chinese Pennisetum 289015
Chinese Pennywort 200274
Chinese Pentaphragma 289696
Chinese Penthorum 290134
Chinese Peony 280147,280213
Chinese Pepper 300370,300373
Chinese Perfume Plant 11300
Chinese Persimmon 132089,132219
Chinese Perty-bush 292078
Chinese Petrocosmea 292581
Chinese Phoebe 295353
Chinese Pholidota 295518
Chinese Photinia 295773,377444
Chinese Physaliastrum 297635
Chinese Pieris 22445
Chinese Pinangapalm 299657
Chinese Pinanga-palm 299657
Chinese Pine 300243
Chinese Pink 127654
Chinese Pistache 300980
Chinese Pistachio 300980
Chinese Platanthera 302404
Chinese Plum 20970,316761
Chinese Plum Yew 82507
Chinese Plumbago 83650
Chinese Plumyew 82507,82545
Chinese Plum-yew 82507,82545
Chinese Podocarpus 306469
Chinese Pogostemon 306966
Chinese Poisonnut 378675
Chinese Polemonium 307215
Chinese Poplar 311371
Chinese Poppy 247116
Chinese Porana 311659
Chinese Poranopsis 311607
Chinese Portia Tree 389941
Chinese Portiatree 389941
Chinese Potato 131772
Chinese Pothos 313197
Chinese Preserving Melon 50998
Chinese Primrose 314975
Chinese Primula 314975
Chinese Privet 229529,229617
Chinese Psilopeganum 318947
Chinese Pulsatilla 321672
Chinese Pummelo 93803
Chinese Pumpkin 50998
Chinese Purple Roscoea 337106
Chinese Purple Willow 344121
Chinese Pusley 190597
Chinese Pussy Willow 343444
Chinese Pyrola 322801
Chinese Quassia Wood 298509
Chinese Quassiawood 298509
Chinese Quince 84573,317602
Chinese R. Falconer Rhododendron 331850
Chinese R. Grande Rhododendron 331851
Chinese R. Nuttall Rhododendron 331855
Chinese Radish 326622
Chinese Rag'wort 364517
Chinese Raisin 198769
Chinese Raisin Tree 198769
Chinese Raisin-tree 198769
Chinese Randia 278323
Chinese Raspwort 179435,184997
Chinese Rattan Vine 52468
Chinese Rattlebox 112013
Chinese Razorsedge 354050
Chinese Red Birch 53335
Chinese Red Bud 83769
Chinese Red Pine 300054,300243
Chinese Red-barked Birch 53335
Chinese Redbud 83769
Chinese Renanthera 327683
Chinese Rhabdothamnopsis 328443
Chinese Rhododendron 330727,331257,331855
Chinese Rhoiptelea 332329
Chinese Rhubarb 329366,329372
Chinese Rhynchosia 333182
Chinese Ricegrass 276001
Chinese Roegneria 335505
Chinese Root 366284

Chinese Rose 195149,336485
Chinese Rose Mallow 402249
Chinese Rostrinucula 337291
Chinese Rubber Tree 156041
Chinese Runcinate Knotweed 309716
Chinese Rungia 340344
Chinese Sacred Bamboo 262189
Chinese Sacred Lily 262382,262457
Chinese Sage 344957,345393
Chinese Sand Pear 323268
Chinese Sanicle 345948
Chinese Sapium 346408
Chinese Saraca 346502
Chinese Sasa 347391
Chinese Sasamorpha 347391
Chinese Sassafras 347413
Chinese Savory 96970
Chinese Sawwort 361018
Chinese Scaleseed Sedge 225572
Chinese Scapose Larkspur 124583
Chinese Scarlet Egg-plant 366920
Chinese Schefflera 350675
Chinese Schizostachyum 351850
Chinese Scholar Tree 369037
Chinese Scholatree 369037
Chinese Scolopia 354598
Chinese Screwpine 281089,281143
Chinese Scrofella 355034
Chinese Scurrula 385192
Chinese Sea Buckthorn 196766
Chinese Seabuckthorn 196766
Chinese Sealavender 230759
Chinese Secamone 356324
Chinese Sedge 74098
Chinese Sedum 357225
Chinese Semiliquidambar 357969
Chinese Sericocalyx 360700
Chinese Serpentroot 354952
Chinese Serviceberry 19292
Chinese Service-berry 19292
Chinese Shibataea 362123
Chinese Shortia 362257
Chinese Shrub 78608
Chinese Shrubalthea 195233
Chinese Sida 362514
Chinese Silk Plant 56229,56236
Chinese Silk Vine 291082
Chinese Silkvine 291082
Chinese Silver Grass 255886
Chinese Silvergrass 255886
Chinese Sindechites 364698
Chinese Sinobambusa 364820
Chinese Sinofranchetia 364947
Chinese Siphonostegia 365296
Chinese Slender Woodbetony 287248
Chinese Sloanea 366077
Chinese Snowball 407939
Chinese Snowball Bush 407939
Chinese Snowball Tree 407939
Chinese Snowball Viburnum 407939
Chinese Snowbell 379474
Chinese Snowberry 380758
Chinese Snowdrop Tree 87525
Chinese Snowdrop-tree 87525
Chinese Soapberry 346338
Chinese Soap-pod Locust 176901
Chinese Sophora 368986
Chinese Sorbaria 369272
Chinese Sour Cherry 316279
Chinese Spapweed 204848
Chinese Spatholobus 370421
Chinese Spicebush 231324
Chinese Spice-bush 231324
Chinese Spinach 18836
Chinese Spiny Bamboo 47445
Chinese Spiraea 371884
Chinese Spiranthes 372247
Chinese Spirea 371884
Chinese Sprangletop 225989
Chinese Spruce 298232
Chinese St. John's Wort 202021
Chinese St. John's-wort 202021
Chinese Stachyurus 373524
Chinese Star Jasmine 393657
Chinese Star Wort 374802
Chinese Star-anise 204603
Chinese Starjasmine 393657
Chinese Star-jasmine 393657
Chinese Statice 230759
Chinese Stauntonia 374392
Chinese Stauntonvine 374392
Chinese Staunton-vine 374392
Chinese Stellera 375187
Chinese Stephanandra 375812
Chinese Stephania 375899
Chinese Stewartia 376442,376473,376478
Chinese Stonegarlic 239259
Chinese Storax 379325,379474
Chinese Stranvaesia 377444
Chinese Strawberry 261212
Chinese Strychnos 378675
Chinese Sugar Maple 369700
Chinese Sugarcane 341909
Chinese Sumac 12559,332509,332662
Chinese Sumach 12559,332509,332662
Chinese Supplejack 52468
Chinese Supple-jack 52468
Chinese Swallow Wort 117425
Chinese Swallowwort 117425
Chinese Swamp Cypress 178019
Chinese Sweet Gum 232562
Chinese Sweet Shrub 68308
Chinese Sweetgum 232551
Chinese Sweetleaf 381341
Chinese Sweetshrub 68308
Chinese Sweetspire 210364
Chinese Sweetvetch 187831
Chinese Syndiclis 381791
Chinese Synotis 381953
Chinese Syzygium 382504
Chinese Tallow 346408
Chinese Tallow Tree 346408
Chinese Tallowtree 346408
Chinese Tallow-tree 346408
Chinese Tamarisk 383469
Chinese Tamarix 383469
Chinese Tarenna 385026
Chinese Taxillus 385192
Chinese Tea 69634
Chinese Tea-plant 239021
Chinese Teasel 133463
Chinese Textile Bamboo 47475
Chinese Textile Bambusa 47475
Chinese Themeda 389358
Chinese Thermopsis 389518
Chinese Thickshellcassia 113436
Chinese Thistle 91864
Chinese Thorny Bamboo 47445
Chinese Thorny Bambusa 47445
Chinese Thorowax 63594
Chinese Thuja 302721
Chinese Thyme 391284
Chinese Timber Chinquapin 78795
Chinese Tinospora 392274
Chinese Tirpitzia 392371
Chinese Toon 392822,392841
Chinese Toona 392841
Chinese Torreya 393061
Chinese Trapella 394560
Chinese Tree 248895
Chinese Tree Lilac 382278
Chinese Tree Peony 50341,280182,280286
Chinese tree-of-Heaven 12559
Chinese Tricarpelema 395372
Chinese Trichilia 395538
Chinese Trigonostemon 397359
Chinese Tripogon 398071
Chinese Tripterospermum 398262
Chinese Trumpet Creeper 70507
Chinese Trumpet Flower 70507
Chinese Trumpet Lily 229758
Chinese Trumpet Vine 70507
Chinese Trumpetcreeper 70507
Chinese Trumpet-creeper 70507
Chinese Trumpet-vine 70507
Chinese Tsoong's Tree 399856
Chinese Tuan Linden 391850
Chinese Tulip Tree 232603
Chinese Tuliptree 232603
Chinese Tulip-tree 232603
Chinese Tupelo 267864

Chinese Tupistra 70600
Chinese Umbrellaleaf 132685
Chinese Uraria 402147
Chinese Urophyllum 402616
Chinese Varied Litse 234092
Chinese Varnish Tree 332775
Chinese Vegetable Tallow 346408
Chinese Verrucose Spindle-tree 157954
Chinese Vetch 408339
Chinese Viburnum 407939,408118
Chinese Violet 43624,385752
Chinese Virginia Creeper 285110
Chinese Wallich Palm 413090
Chinese Wallichia 413090
Chinese Walnut 212595,212621,212636
Chinese Wampee 94207
Chinese Water Chestnut 143122,143123
Chinese Water Pine 178019
Chinese Water-chestnut 143122
Chinese Wattle 1134
Chinese Wax Gourd 50998
Chinese Wax Myrtle 261212
Chinese Waxgourd 50998
Chinese Waxmyrtle 261212
Chinese Wayfaring Tree 408210
Chinese Wedelia 413514
Chinese Weeping Cypress 114690
Chinese Weigela 413613
Chinese Wenchengia 413759
Chinese Wendlandia 413843
Chinese White Pine 299799
Chinese White Poplar 311530
Chinese Whitebark Pine 299830
Chinese Wild Ginger 37732
Chinese Wildbasil 96970
Chinese Willow 343179,343366
Chinese Willow Pattern Tree 217626
Chinese Windhairdaisy 348199
Chinese Windmill Palm 393809
Chinese Wing Nut 320377
Chinese Wingleaf Pricklyash 417161
Chinese Wingnut 167940,320377
Chinese Wing-nut 320346
Chinese Winter Hazel 106675
Chinese Winterberry Currant 333972
Chinese Wintergreen 172150
Chinese Winterhazel 106675
Chinese Winter-hazel 106675
Chinese Wisteria 414576
Chinese Witch Hazel 106675,185130
Chinese Witchgrass 226230
Chinese Witchhazel 185130
Chinese Wolfberry 238996,239021
Chinese Wolftailgrass 289015
Chinese Woodbetony 287090
Chinese Woodnettle 221538
Chinese Wood-oil Tree 406018
Chinese Woollychin Woodbetony 287341
Chinese Xanthoceras 415096
Chinese Yam 131577,131645,131747,131772
Chinese Yellowcress 336187
Chinese Yellowwood 94020
Chinese Yellow-wood 94018,94020
Chinese Yew 385405,385407
Chinese Yua 416527
Chinese Zanthoxylum 417180
Chinese Zelkova 417552,417563
Chinese-decorated Gentian 173900
Chinese-hat Plant 197368
Chinese-hat-plant 197361,197368
Chinese-houses 99838
Chinese-lantern 297640,297643,297645
Chinese-lantern Lily 345783
Chinese-quince 317602
Chineses Threeawngrass 33797
Chinese-weeping Cypress 114690
Chinfu Mountain Lathraea 222642
Chinfu Mountain Square Bamboo 87625
Chinfushan Square-bamboo 87625
Ching Arrowwood 407743
Ching Banfenghe 357977
Ching Barberry 51454
Ching Blueberry 403852
Ching Caper 71715
Ching Cryptocarya 113437
Ching Greenbrier 366295
Ching Henry Blueberry 403852
Ching Herelip 220061
Ching Jadeleaf and Goldenflower 260394
Ching Jijigrass 4127
Ching Lagochilus 220061
Ching Lanceolate Draba 137074
Ching Lanceolate Whitlowgrass 137074
Ching Lilyturf 272064
Ching Lysionotus 239939
Ching Magnolia 242030
Ching Manglietia 244427
Ching Maple 2887
Ching Ophiorrhiza 272192
Ching Photinia 295657
Ching Redbud 83778
Ching Rehmannia 327433
Ching Sagebrush 35330
Ching Semiliquidambar 357977
Ching Sloanea 366044
Ching Smilax 366295
Ching Speargrass 4127
Ching Stemonurus 178923
Ching Tailanther 381931
Ching Thickshellcassia 113437
Ching Viburnum 407743
Ching Willow 343209
Ching Windhairdaisy 348200
Ching Woodbetony 287094
Ching Yam 131521
Ching Yushanbamboo 416760
Ching Yushania 416760
Chingan Tickseed 104758
Chingdao Bulrush 353901
Chinghai Larkspur 124632
Chinghai Mandrake 244344
Chinghai Pleurospermum 304851
Chinghai Rhubarb 329372
Chinghai Russianthistle 344493
Chinghai Stonecrop 357255
Chingho Buttercup 325719
Chingho Jerusalemsage 295087
Chingiacanthus 209876
Chingling Crazyweed 278783
Chingling Edelweiss 224846
Chingma 837
Chingma Abutilon 1000
Ching's Clematis 94823
Ching's Ironweed 406222
Ching's Milkvetch 42178
Ching's Mussaenda 260394
Ching's Sage 344959
Ching's Scorpiongrass 175868
Ching's Scurrula 355297
Ching's Sweetspire 210376
Ching's Synotis 381931
Chingsi Galangal 17699
Chingtung Clematis 95465
Chingtung Primrose 314499
Chingtung Stonecrop 356619
Chinkapin 78766,78807,78848,324204
Chinkapin Oak 324204
Chinks 274810
Chinling Fir 311
Chinling Ladybell 7761
Chinling Maple 3724
Chinling Mountain Columbine 30042
Chinling Mountains Burger Figwort 355068
Chinling Mountains Fir 311
Chinling Mountains Speedwell 407422
Chinling Rhododendron 331588
Chinling Swallowwort 117400
Chinnampo Saussurea 348201
Chinnampo Windhairdaisy 348201
Chino Bitterbamboo 304014
Chinois 93765
Chinotto 93643
Chinotto Orange 93389
Chinping Alphonsea 17625
Chinquapin 78766,78807,78848,89968,89969,324204
Chinquapin Oak 324204
Chinquapin Water Chestnut 263270
Chinshan Viburnum 407751
Chinsin Tallow-tree 346376

Chiogenes 87668
Chionocharis 87746
Chionodoxa 87757
Chionographis 87775
Chiotilla 155249
Chipple 15289
Chir Pine 300128,300186
Chiretta 380157
Chiricahua Dock 340166
Chiricahua Fleabane 150739
Chiricahua Mountain Tansy-aster 33039
Chirinole 375522
Chirita 87815
Chiritopsis 87994,88002
Chirms 68189
Chishui Altingia 18202
Chishui Begonia 49715
Chishui Bredia 59839
Chishui Camellia 68901
Chishui Elatostema 142853
Chishui Honeysuckle 236182
Chishui Indocalamus 206774
Chishui Skullcap 355408
Chishui Spapweed 204849
Chishui Touch-me-not 204849
Chisocheton 88111
Chisos Agave 10862,10866
Chisos Bluebonnet 238449
Chisos Hop-hornbeam 276805
Chisos Mountains Wild Buckwheat 152136
Chisos Oak 323972
Chisos Red Oak 323978
Chisos Rosewood 405224
Chitah 228298
Chitalpa 88158
Chit-chat 369339
Chitra 51327
Chittagong 90605
Chittagong Chickrassy 90604,90605
Chittagong Ehickrassy 90605
Chittagong Wood 90605
Chittam 107315
Chittam Wood 107315
Chittamwood 107315
Chittick 220310
Chittim Wood 362953
Chiubei Clematis 94824
Chiukiang Milkvetch 42183
Chiupei Camellia 68983
Chive 15018,15709
Chive Garlic 15709
Chive Onion 15709
Chivegrass 416127
Chive-like Onion 15708
Chiveorchis 254226,254238
Chives 15709
Chivi Tree 181768
Chivken Mulberry 259097
Chiwu Leek 15187
Chiwu Onion 15187
Chiwu Sedge 74103
Chixiaodou Cowpea 409085
Chlamydites 88158
Chlamydoboea 88164
Chlamydoleaf Pearleverlasting 21539
Chloraea 88257
Chloranth 88264
Chloranth Family 88263
Chloranthus 88264
Chloranthus Family 88263
Chloranthus-like Euonymus 157373
Chloris 88316
Chlorophytum 88527
Cho-Cho 356352
Chochocala Mahonia 242507
Chock Cheese 243840
Chock-cheese 243840
Choco 356352
Chocolate 389404
Chocolate Bean 389404
Chocolate Berry 411398
Chocolate Cosmos 107160
Chocolate Daisy 52829
Chocolate Family 376218
Chocolate Flower 52829,174717
Chocolate Lily 36833
Chocolate Nut 389404
Chocolate Plant 107160,317247
Chocolate Pudding Tree 132132
Chocolate Root 175393,175444
Chocolate Tree 389399,389404
Chocolate Vine 13225
Chocolate Weed 249633
Chocolate-lily 128983,168355
Chocolatetree 389399
Chocolate-tree 389399,389404
Chocolate-vine 13225
Chocolateweed 249633
Chocon Palm 194610
Choctaw Root 29473
Choerospondias 88690,88691
Choi Sum 59595
Choisy 192148
Choisy St. John's Wort 201811
Choisy's St. John's-wort 201811
Choju Kinkan 167511
Choke Cherry 316918
Choke Pear 323133,369541
Chokeberry 34845,295609,295732
Chokecherry 316918
Chokeweed 275135
Chokky-cheese 243840
Choko 356352
Cholla 116636,272775,272856
Cholla Cactus 272775
Chomonque 178750
Chondrilla 88776
Chondrorhyncha 88854
Chonemorpha 88888
Chong'an Sage 344967
Chongqing Camellia 68993
Chongras 298094
Chongyang Pondweed 312071
Chonnocks 59575
Chonta Palm 12772
Chooky-pigs 28617,273531
Choop Rose 336419
Choop-rose 336419
Choops 336419
Choop-tree 336419
Chop Nut 298009
Chopped Eggs 231183
Chorisis 88975
Chorispora 88996
Chorogi 373439
Choroki 373439
Chorro Creek Bog Thistle 91981
Chosenia 89159
Chou Bourache 31056
Choumutov Milkvetch 42186
Choups 336419
Chou's Burreed 370104
Chow Chow 356352
Chow Windhairdaisy 348209
Chowlee 409096
Chowps 336419
Choy Sum 59595
Choyita 140320,244087
Choyote 356352
Christ Paphiopedilum 282837
Christ Plant 159363
Christ-and-the-apostles 285623
Christensen Ceropegia 84040
Christensen Pendentlamp 84040
Christia 89201
Christian 316470
Christina Loosestrife 239582
Christisonia 89217
Christlings 316470
Christmas Begonia 49711,49712
Christmas Bell 55114,55115,55116
Christmas Bells 55114,345782,345783
Christmas Berry 193839,193841,203545,
203611,295766,350995
Christmas Bush 14177,14182,83533,83535,
89299,159675,315601
Christmas Cactus 116665,351993,351994,
351995,351997,418532
Christmas Camellia 69594
Christmas Candles 9701
Christmas Candlestick 224600

Christmas Cassia 360471
Christmas Cherry 367522,367528
Christmas Cholla 272948
Christmas Daisy 380900
Christmas Flower 148104,159675
Christmas Gambol 334869
Christmas Grass 389312
Christmas Heath 149138
Christmas Heather 149138
Christmas Heliconia 189989
Christmas Lily 229900
Christmas Oriental Bitter Sweet 80285
Christmas Palm 405258
Christmas Pepper 72070
Christmas Peppertree 350995
Christmas Pride 339761,339779
Christmas Protea 315732
Christmas Rose 148104,190925,190944,190949
Christmas Senna 360471
Christmas Spruce 298191
Christmas Tree 272,9701,203545,252613,267409,298191
Christmasberry 193841
Christmas-berry-tree 350995
Christmasbush 14177
Christmas-bush 14177
Christmas-flowering Iris 208437
Christmas-rose 190944
Christmas-tree of New Zealand 252627
Christmastree Plant 215174
Christmas-vine 311657
Christmasvine Porana 311657
Christmas-vine Poranopsis 311657
Christophine 356352
Christplant 159363
Christ's Eye 345458
Christ's Eyes 68189,345458
Christ's Heel 302034,302068
Christ's Herb 190944,308454
Christ's Ladder 81065,202086
Christ's Thorn 76873,159363,203545,280559,418216
Christ's Wort 190944
Christ's-thorn 76873,280559
Christ's-thorn Paliurns 280559
Christthorn Cointree 280559
Christthorn Paliurus 280559
Christ-thorns 280559
Chritmas Berry 88029
Chritton Paliurus 280559
Chroesthes 89282
Chromeflower Barberry 52052
Chromeflowered Barberry 52052
Chrysalidocarpus 89345
Chrysanihemum Greens 89481
Chrysanthemum 89402,89531,124790,124826
Chrysanthemum Wood 49046
Chrysanthemumleaf Rhodiola 329850
Chrysanthemum-like Groundsel 359253
Chrysocephalous Leek 15190
Chrysocephalous Onion 15190
Chrysoglossum 89938
Chrysohairy Newlitse 264038
Chrysopogon 90107
Chu Bulrush 353308
Chu Edelweiss 224820
Chu Monkshood 5132
Chu Newlitse 264039
Chu Sedge 74129
Chu Stringbush 414154
Chuan Corktree 294238
Chuanching Buttercup 325722
Chuan-Dian Buckthorn 328707
Chuan-Dian Dutchmanspipe 34144
Chuan-Dian Hemsleia 191907
Chuan-Dian Sanicle 345941
Chuan-Dian Snowgall 191907
Chuandian Soapberry 346333
Chuan-Dian Stonecrop 356753
Chuan-Dian Valerian 404339
Chuandian Windhairdaisy 348331
Chuandian Woodlotus 244433
Chuangan Dandelion 384646
Chuangan Onion 15223
Chuanigan Windhairdaisy 348103
Chuanlong Yam 131734
Chuanminshen 90600,90601
Chuannan Rhododendron 331864
Chuan-Qian Glorybower 96014
Chuanqing Buttercup 325722
Chuan-Qing Milkvetch 42877
Chuanshan Pricklyash 417298
Chuanshan Windhairdaisy 348491
Chuanxi Hairystyle Mockorange 294551
Chuanxiong Ligusticum 229309
Chuan-Zang Rabdosia 209794
Chuanzang Small Reed 127285
Chuanzang Smallreed 127285
Chuchupate 229374
Chuckenwort 243840,374968
Chucky-cheese 109857
Chueky Cheese 109857,243840
Chufa 118822,118837
Chufa Flat Sedge 118822
Chufa Nut 118822
Chulan Aglaia 11300
Chulan Tree 11300
Chu-lan Tree 88301
Chulta 130919
Chumb Woodbetony 287095
Chumbi Felwort 235439
Chumlangma Larkspur 124135
Chun Azalea 330398
Chun Bambusa 47233
Chun Blachia 54876
Chun Blueberry 403775
Chun Buckeye 9688
Chun Bulrush 353310
Chun Camellia 68995
Chun Croton 112862
Chun Deutzia 126868
Chun Evergreen Chinkapin 78896
Chun Evergreenchinkapin 78896
Chun Grewia 180732
Chun Helicia 189937
Chun Holly 203632
Chun Hornbeam 77274
Chun Horse-chestnut 9688
Chun Indigo 205806
Chun Jasminanthes 211694
Chun Lasianthus 222120
Chun Litse 233865
Chun Maple 2888
Chun Miliusa 254548
Chun Parabarium 283067
Chun Persimmon 132104
Chun Pleione 304227
Chun Redbud 83779
Chun Rhododendron 330398
Chun Roughleaf 222120
Chun Spicebush 231322
Chun Spice-bush 231322
Chun Spine Bamboo 47233
Chun Stephanotis 211694
Chun Sweetleaf 381366
Chun Syzygium 382507
Chun Viburnum 407752
Chunan Barberry 51467
Chunechites 90612
Chung Bamboo 47232
Chung Bambusa 47232
Chung China Cane 364641
Chung Chinacane 364641
Chung Cyclobalanopsis 116077
Chung Distylium 134927
Chung Grape 411614
Chung Hydrangea 199860,199931
Chung Microcos 253367
Chung Mosquitoman 134927
Chung Oak 116077
Chung Qinggang 116077
Chung Sedge 74130
Chungba Larkspur 124136
Chungdian Bluebellflower 115343
Chungien Primrose 314246
Chungning Milkvetch 42813
Chungtien Cotoneaster 107525
Chungtien Groundsel 358550
Chungtien Loosestrife 239588

Chungtien Skullcap 355411
Chunia 90615
Chuniophoenix 90619
Chunnian Azalea 330399
Chunnien Rhododendron 330399
Chunpi Rockfoil 349179
Chun's Ironweed 406226
Chuosijia Monkshood 5134
Chupa Sangre 242719
Chupachupa 139936
Chupadilla 120617
Chuparosa 25262,214398
Church Bells 116736,381035
Church Broom 133478
Church Lily 229900
Church Steeple 11549,158062
Church Steeples 11549,158062
Churchman's Greeting 410960
Churchmouse Three-awn 33830
Church-mouse Three-awn 33830,33831
Churchsteeples 11549
Church-steeples 11549
Churchwort 250432
Churchyard Elder 72038
Churchyard Yew 385314
Churee 159739
Churl 86427
Churl Head 81356
Churl Hemp 71218
Churlick 364557
Churl's Cress 225315
Churl's Head 81237
Churl's Treacle 15698
Churn 262441,267294,325666,326287
Churnstaff 159027,231183
Churnwood 93267
Churras 71220
Churrs Head 81237
Chu's Euonymus 157374
Chusan Palm 266677,266678,393803,
393809
Chusquea 90629,90633
Chuxiong Moth Orchid 293590
Chuxiong Stonecrop 356622
Chuxiong Trigonotis 397394
Chysis 90711
Ciant Fishtail Palm 78050
Cibbols 15170
Cibol 15289
Ciboul Onion 15289
Ciboule 15170,15289
Cibourroche 31056
Cicadawingvine 356361,356367
Cicatrical Machilus 240573
Cicatricosous Rapanea 261617
Ciccifer-like Oak 323777
Cicely 28044,261585
Cicer 42193
Cicer Milkvetch 42193
Cich 90801
Cichlings 408571
Cichorium Family 90886
Cichou Paperbush 141471
Cickenwort 374968
Cider Eucalyptus 155602
Cider Gum 106805,155468,155602
Cider Tree 155602
Ciderage 309199
Cidra 114277
Cien Elaeagnus 142013
Cigar Flower 114606
Cigar Orchid 120715
Cigar Plant 114606
Cigar Tree 79243,79260
Cigarbox Cedar 80030
Cigar-box Cedar 80025,80030
Cigarbox Cedrela 80030
Cigar-box Cedrella 80030
Cigarette Plant 114606
Cigar-flower 114606
Cigartree 79243,79260
Cigar-tree 79260
Cigu Biondia 54529
Cikai Monkshood 5589
Cikai Pleurospermum 304835
Cikou Woodbetony 287785
Cilantro 104690
Cilff Net Bush 68065
Ciliabractbean 172497,172509
Ciliare Epidendrum 146396
Ciliate Agapetes 10300
Ciliate Aloe 16709
Ciliate Arundinella 37389
Ciliate Azalea 330402
Ciliate Barberry 51468
Ciliate Bedstraw 170251
Ciliate Belosynapsis 50939
Ciliate Bergenia 52510
Ciliate Birdfootorchis 347844
Ciliate Blue Star 20848
Ciliate Blueberry 403776
Ciliate Brachyactis 58265
Ciliate Brassaiopsis 59197
Ciliate Brome-grass 60667
Ciliate Bugle 13079
Ciliate Calyx Mock Azalea 250507
Ciliate Cassiope 78625
Ciliate Cat's Ear 202400
Ciliate Cayratia 79839
Ciliate Centipede-grass 148251
Ciliate Cliff Taihangia 383118
Ciliate Crazyweed 278789
Ciliate Cryptocoryne 113530
Ciliate David Gooseberry 333951
Ciliate Diplocyatha 132949
Ciliate Duckbeakgrass 209321
Ciliate Elatostema 142748
Ciliate Euaraliopsis 59197
Ciliate Eurya 160446
Ciliate Falsepimpernel 231496
Ciliate Fig 164810
Ciliate Fuirena 168831
Ciliate Galinsoga 170146
Ciliate Garnotia 171492
Ciliate Gastrochilus 171836
Ciliate Girald Spindle-tree 157543
Ciliate Girardinia 175892
Ciliate Goldsaxifrage 90394
Ciliate Goldwaist 90394
Ciliate Goosecomb 335258
Ciliate Heath 149195
Ciliate Heteropappus 193930
Ciliate Honeysuckle 235739
Ciliate Leucas 227574
Ciliate Longleaf Chickweed 374944
Ciliate Lovegrass 147598
Ciliate Lycaste 238737
Ciliate Microstegium 253992
Ciliate Motherweed 231496
Ciliate Nepeta 264898
Ciliate Palmleaftree 59197
Ciliate Pearlsedge 354052
Ciliate Pepper 300505
Ciliate Platanthera 183501
Ciliate Pleurothallis 304908
Ciliate Poplar 311276
Ciliate Rattan Palm 65660
Ciliate Rattlesnake Plantain 347844
Ciliate Razorsedge 354052
Ciliate Rhododendron 330402
Ciliate Roegneria 335258
Ciliate Salomonia 344351
Ciliate Scale Wormwood 35207
Ciliate Sedge 74193
Ciliate Shrubcress 368571
Ciliate Small Reed 127202
Ciliate Smithia 366654
Ciliate Solms-Laubachia 368571
Ciliate Spoonleaf Gentian 173914
Ciliate Steironema 239589
Ciliate Stelis 374704
Ciliate Taihang flower 383118
Ciliate Thick-leaved Slugwood 50591
Ciliate Toona 392829
Ciliate Torenia 392893
Ciliate Trowelpetal Stonecrop 357248
Ciliate Wild Smallreed 127202
Ciliate Yu Woodbetony 287840
Ciliatecalyx Azalea 330403
Ciliatecalyx Rhododendron 330403
Ciliate-calyxed Rhododendron 330403

Ciliateflower Twinballgrass 209047
Ciliateflowered Twinballgrass 209047
Ciliatescale Wormwood 35207
Ciliato-toothed Winged Spindle-tree 157293
Cilicia Fir 314
Cilicia Iris 208644
Cilician Fir 314
Cilicica Abies 314
Cilicica Fir 314
Cilicle Syrenia 382107
Ciliolate Habenaria 183503
Ciliolate Paphiopedilum 282804
Ciliolate Sandwort 31819
Ciliospiny Holly 203633
Cilmbing Raspberry 338407
Cimarron Wild Buckwheat 151809
Cinchon 91055
Cinchona 91055
Cinchona Tree 91055,91075
Cinderella-slippers 364752
Cinderlike Sweetleaf 381148
Cinder-like Sweetleaf 381148
Cineraria 91107,290814,290821,358395
Cinereous Spindle-tree 157375
Cingulate Onosma 271747
Cinkefield 312925
Cinnabar Agapetes 10341
Cinnabar Begonia 49721
Cinnabar Cuphea 114613
Cinnabarhair Saurauia 347967
Cinnabar-haired Saurauia 347967
Cinnabar-red Rhododendron 330409
Cinnabarroot 31396
Cinnamomum Hackberry 80767
Cinnamomum Leaves Rhododendron 330127
Cinnamon 91252,91446
Cinnamon Bark Clethra 96456
Cinnamon Clethra 96456
Cinnamon Iris 5793
Cinnamon Rhododendron 330749
Cinnamon Root 207087
Cinnamon Rose 336498,336733
Cinnamon Sedge 5793
Cinnamon Vine 131747,131772
Cinnamon Wattle 1348
Cinnamon Willow Herb 146670
Cinnamon Willow-herb 146670
Cinnamon Wood 347407
Cinnamon-bark 71132
Cinnamonbark Clethra 96456
Cinnamonleaf Arrowwood 407756
Cinnamonleaf Bladderwort 403814
Cinnamonleaf Dogwood 380480
Cinnamon-leaf Elaeagnus 141953
Cinnamonleaf Hackberry 80618,80767
Cinnamonleaf Hibiscus 194907
Cinnamonleaf Indianmulberry 258881
Cinnamon-leaf Jasmine 211777
Cinnamonleaf Maple 2891
Cinnamonleaf Nettletree 80767
Cinnamonleaf Passionflower 285660
Cinnamonleaf Sloanea 366041
Cinnamon-leaf Sterculia 376094
Cinnamonleaf Viburnum 407756
Cinnamon-leaved Elaeagnus 141953
Cinnamon-leaved Hackberry 80767
Cinnamon-leaved Jasmine 211777
Cinnamon-leaved Schefflera 350761
Cinnamon-leaved Sterculia 376094
Cinnamon-leaved Viburnum 407756
Cinnamonroot Inula 207087
Cinnamon-tree 91302,91446
Cinquefoil 100254,312322,312670,312812,312814,312925
Cinquefoil False Sagebrush 371148
Cinquefoil Geranium 174830
Cinquefoil-like Raspberry 339092
Cinquuefoli-like Daisy 124845
Ciote Bur 31051
Cipadessa 91479
Ciper Nut 63475
Circaea 91499
Circaea-like Bittercress 72721
Circaeashape Pearweed 239591
Circaeaster Family 91598
Circaester 91596
Circassian Bean 7190
Circinal Dunbaria 138967
Circinate Arrowbamboo 162653
Circinate Glazegrass 367811
Circinate Hemsley Monkshood 5251
Circinate Solenanthus 367811
Circled Gentianopsis 174212
Circlefruit Hedinia 187365
Circular Actinidia Ballfruit Kiwifruit 6620
Circular Fruit Actinidia 6620
Circumlobed Begonia 49722
Circumpilose Bamboo 297256
Circus Triggerplant 379077
Cirio 167557,203445
Cirnela 88691
Cirrhaea 91611
Cirtonella 117136
Cirtusleaf Glycosmis 177845
Ciruelas 317171
Cismontane Minuartia 255456
Cissampelopsis 92513
Cissampelos 92523,92555
Cissus 92586,92777,93007
Cissusform Hemsleia 191909
Cissusform Snowgall 191909
Cissusleaf Acanthopanax 143582
Cistanche 93047,93075
Cistancheherb 57429
Cistsage 93202
Cistus 93120
Citerne Barberry 51475
Cithara Wild Buckwheat 151927
Cithernwood 35088
Citherwood 35088
Citrate Actinodaphne 6771
Citrine Fawnlily 154908
Citrine Melaleuca 248085
Citrine Oncidium 270800
Citroflower Morning-star Lily 229815
Citron 93539,93603
Citron Daylily 191266
Citron Rrain-lily 417616
Citron Zephyrlily 417616
Citron-coloured Gesneria 175298
Citronella 22836,99844
Citronella Grass 117218
Citronella Oil 117218
Citronella-grass 117218
Citrullus 93286
Citrus 93314
Citrusform Groundsel 358561
City Avens 175454
City Goosefoot 87186
Cives 15709
Civet 15709
Civet Bean 294010
Civet Durian 139092
Civet-cat Fruit 139092
Ckecker Bloom 362704
Clabgourd 263567
Clackamas Iris 208885
Cladanthus 93882
Claden 170193
Clader 170193
Cladium 93901
Cladocalyx Gum 155531
Cladogynos 93951
Cladopus 93968
Cladostachys 123558
Claggers 170193
Claiton 170193
Clammy Azalea 332086
Clammy Campion 364179,410953,410957
Clammy Cherry 104220
Clammy Chickweed 82849
Clammy Cudweed 317753
Clammy Cuphea 114614
Clammy Everlasting 178155,317753
Clammy False Foxglove 45175,45181,45184
Clammy False Oxtongue 317578
Clammy Goosefoot 87142
Clammy Ground Cherry 297678
Clammy Ground-cherry 297678
Clammy Hedge Hyssop 180319
Clammy Hedge-hyssop 180297,180319

Clammy Hop Seed Bush 135215
Clammy Hopseed Bush 135215
Clammy Hop-seed bush 135215
Clammy Hopseedbush 135215
Clammy Locust 335013
Clammy Plantain 301864
Clammy Weed 95791
Clammyweed 307132,307141
Clamoun 215395
Clamshell Orchid 145290
Clam-shell Orchid 315616
Clamshell Pricklyash 417216
Clanwilliam Cedar 414058,414059
Claoxylon 94051,94055
Clappedepouch 72038,170193
Clapweed 364193
Clap-weed 146136,146234
Claret Ash 167913
Claret Cup 140231,140291,140332
Claret Vine 411981
Claret-cup Cactus 140231,140332
Clarimond Tulip 400206
Clarke Arachnis 155267
Clarke Blumea 55711
Clarke Broomrape 275002
Clarke Clematis 94840
Clarke Garishspiderorchis 155267
Clarke Leaf-flower 296521
Clarke Lilyturf 272067
Clarke's Leptopus 226324
Clarkia 94132,94138,94144
Clary 344821,345379,345447,345458
Clary Sage 345379
Clasping Arnica 34729
Clasping Coneflower 339518
Clasping Cress 225428
Clasping Dogbane 29514
Clasping False Solomonseal 242684
Clasping Goldenrod 367980
Clasping Heliotrope 190550
Clasping Iris 208865
Clasping Leaf Pondweed 312220
Clasping Milkweed 37827
Clasping Mullein 405747
Clasping Pepperweed 225428
Clasping Pepper-weed 225428
Clasping Swertia 380212
Clasping Twistedstalk 377930
Clasping Venus'-looking-glass 397759
Clasping Yellowtops 166819
Clasping-leaf Coneflower 339518
Clasping-leaf Doll's-daisy 56712
Claspingleaf Goldenrod 367980
Clasping-leaf Pondweed 312220
Clasping-leaved Dogbane 29473
Clasping-leaved St. John's-wort 201906
Clasping-leaved Streptopus 377930
Claspingstem Solena 367775
Claspleaf Pennycress 390249
Claspleaf Twistedstalk 377912,377930
Clasp-leaf Twisted-stalk 377905,377930
Classical Fenugreek 397229
Clathrate Willow 343228
Clatter Malloch 397019
Clatterclogs 400675
Claus Cowparsnip 192252
Clausia 94232
Claussen Barberry 51476
Clavate Arisaema 33292
Clavate Beech 162387
Clavate Calanthe 65906
Clavate Lobelia 234372
Clavate Parnassia 284579
Clavate Southstar 33292
Clavate Sporoxeia 372890
Clavate-flower Syzygium 382512
Clavate-glandular Rockfoil 349199
Clavatehair Fairy Lantern 67570
Clavatehair Osbeckia 276148
Clavateleaf Lilyturf 272068
Clavateleaf Oberonia 267936
Clavatespur Sporoxeia 372890
Clavatestamen Rockfoil 349186
Clavatetail Spapweed 204856
Clavellina 181459
Claver 170193,397019,397043
Claver Devil 115008
Claver Sorrel 277648
Claver-grass 170193
Claver-sorrel 277648
Claviculate Cherry 316334
Claviflower Bauhinia 49179
Claviform Aspidistra 39526
Claviform-styled Rhododendron 330468
Clavisepal Paraboea 283113
Clavver-grass 170193
Claw Cactus 418532
Claw Rhubarb 329401
Clawless Rockfoil 349279
Clawpetal Caper 71762
Clawpetal Rockfoil 350020
Clawpetal Stonecrop 356986
Clawy Acanthophippium 2084
Clawy Rockfoil 350012
Clay Hill Wild Buckwheat 152656
Clay Sand-verbena 693
Clay Wattle 1255
Claybush Wattle 1255
Clay-bush Wattle 1255
Clay-loving Wild Buckwheat 152383
Clayton Pfeiffer root 276466
Claytonia 94292
Clayton's Bedstraw 170320
Clayton's Pfeiffer-root 276466
Clayweed 400675
Claywort 400675
Clean Azalea 330559
Clean Rhododendron 330559
Cleansing Grass 159222
Clear Creek Fleabane 150932
Clear-eye 345379,345458
Clear-eyes 345379,345458
Clearing Nut 378860
Clearvein Ardisia 31353
Clearweed 298848,298917,298974,299024
Clearweed Elegant 298996
Clearweed Wild 298945
Clearweedlike Lecanthus 223685
Cleat 292372,400675
Cleat-leaves 292372
Cleaved Grass 170193
Cleavers 31051,133478,170193
Cleden 170193
Cleeve Pink 127728
Cleft Phlox 295252
Cleft Rosewood 121805
Cleft Violet 410340
Cleftedcalyx Saunders Cinquefoil 312958
Clefted-leaf Potanin Cinquefoil 312898
Cleftleaf Groundsel 279952
Cleftpetal Largeflower Larkspur 124252
Cleft-wing Milkvetch 42566
Cleidiocarpon 94401
Cleidion 94404
Cleisostoma 94425,94465
Cleistanthus 94498,94539
Cleistes 94545
Cleistocalyx 94571
Cleistocatus 94551
Cleistogenes 94585
Cleland's Evening-primrose 269422
Clematis 94676
Clematis 'Prince Charles' 95450
Clematis Agdestis 10973
Clematis Asiabell 98290
Clematis Corydalis 105749
Clematis Iris 208543
Clemens Planchonella 301770
Clemens Wildolive 301770
Clementine 93717
Clementine Mandarin 93429
Clementine's Rhododendron 330427
Clench 325612
Cleome 95606
Cleopatra Mandarin 93716
Clereye 345379
Clerodendranthus 95860
Clerodendrum 95978
Cletheren 170193
Clethra 96454
Clethra Family 96561

Clethra Loosestrife 239594
Cleveland Sage 344973
Cleveland's Desertdandelion 242928
Cleveland's Ragwort 279897
Cleveland's Spineflower 89067
Clever 170193
Cleyera 96576,386696
Clianthus 96633
Clide 170193
Clidemialike Ramie 56110
Cliden 170193
Clider 170193
Cliff Aletris 14527
Cliff Alstonia 18054
Cliff Ancylostemon 22174
Cliff Anemone 24046
Cliff Argostemma 32524
Cliff Arisaema 33485
Cliff Azalea 331755
Cliff Barrel 163436
Cliff Begonia 50018
Cliff Blueberry 403988
Cliff Bottlebrush 67263,67264
Cliff Cherry 83308
Cliff Chickweed 375136
Cliff Cinquefoil 312946
Cliff Cudweed 317770
Cliff Currant 334203
Cliff Date Palm 295485
Cliff Desertdandelion 242954
Cliff Draba 137147
Cliff Elm 401618
Cliff Elsholtzia 144088
Cliff Ephedra 146247
Cliff Fendlerbush 163315
Cliff Goldenaster 194276
Cliff Goldenbush 150317
Cliff Goldenrod 368383
Cliff Honeysuckle 236070
Cliff Jamesia 211550,211551
Cliff Mallee Ash 155549
Cliff Maple 243437
Cliff Monkshood 5520
Cliff Oak 323596
Cliff Oresitrophe 274156
Cliff Penstemon 289373
Cliff Pink 127728
Cliff Poorspikebamboo 270448
Cliff Rose 34536
Cliff Snowflake 32524
Cliff Southstar 33485
Cliff Spurge 159378
Cliff Spurrey 370705
Cliff Stenosolenium 375726
Cliff Stickseed 184298,184300
Cliff Taihangia 383117
Cliff Thistle 92456
Cliff Whitlowgrass 137147
Cliff Wild Buckwheat 152457
Cliff Windflower 24046
Cliffbean 66626,66637,254596,254779
Cliffbush 211550,211551
Cliffliving Eurya 160597
Cliffliving Pleurospermum 304844
Cliffortia-like Barberry 51477
Cliff-plum Rhododendron 331710
Cliffrose 108497
Cliffrose Azalea 330485
Cliffrose Rhododendron 330485
Cliffvine 13447
Cliiff Patrinia 285854
Climber Floscopa 167040
Climbers 95431
Climbing Acacia 1308
Climbing Aloe 16709
Climbing Amaryllis 56755
Climbing Apple Bamboo 249605,249608
Climbing Asparagus 38950
Climbing Asparagus Fern 39195
Climbing Asparagus-fern 39195
Climbing Aster 20188
Climbing Bamboo 351852
Climbing Bauhinia 49098,49226,49227
Climbing Bellflower 98290,234139
Climbing Bindweed 162552
Climbing Birthwort 34149
Climbing Bittersweet 80129,80309
Climbing Boneset 254443
Climbing Bredia 59855
Climbing Buckwheat 162519
Climbing Butcher's Broom 357898,357899
Climbing Butcher's-broom 357899
Climbing Cactus 147283,147288
Climbing Cassia 360467,360471
Climbing Cobaea 97757
Climbing Corydalis 83412,105745
Climbing Dahlia 195396
Climbing Dogbane 393638
Climbing Drepanostachyum 20209
Climbing Embelia 144807
Climbing Entada 145899
Climbing Erythropalum 154958
Climbing Euonymus 157473,157872
Climbing False Buckwheat 162528,162552, 309757
Climbing False-buckwheat 162552
Climbing Fig 165515,165624
Climbing Fishvine 125998
Climbing Fumitory 8302,105745,169125
Climbing Gambirplant 401778
Climbing Gambir-plant 401778
Climbing Gazania 260530
Climbing Glorypea 96640
Climbing Grass Nut 128886
Climbing Groundsel 359980
Climbing Harrisia 185920
Climbing Hedyotis 187666
Climbing Hempweed 254414,254443
Climbing Hirsute Bredia 59855
Climbing Hydrangea 123544,199800, 200059,200077
Climbing Ilang-Ilang 35016
Climbing Jewelvine 125998
Climbing Knotweed 162552
Climbing Lignum 259585
Climbing Lily 177230,177251
Climbing Lysionotus 239939
Climbing Mikania 254443
Climbing Milkweed 117552,246268,246269, 246271,246273,347008
Climbing Nasturtium 399601
Climbing Nightshade 367120,367130
Climbing Onion 57984
Climbing Pagodatree 369157
Climbing Pandanus 168247
Climbing Parrotbeak 96640
Climbing Parrotbill 96640
Climbing Philodendron 294838
Climbing Porandra 311652
Climbing Pothos 313207
Climbing Prairie Rose 336942,336943
Climbing Premna 313751
Climbing Pricklyash 417329
Climbing Prickly-ash 417329
Climbing Prinsepia 315174
Climbing Rhaphidophora 328997
Climbing Rose 336942
Climbing Sailor 116736
Climbing Sarcodum 96640
Climbing Sarmienta 347105
Climbing Screw Pine 168239
Climbing Screw-pine 168238,168245
Climbing Seedbox 238211
Climbing Sicklebamboo 20209
Climbing Smartweed 162542
Climbing Sophora 369156,369157
Climbing Tibetan Hydrangea 199800
Climbing Tribisee 222047
Climbing Velvetplant 183122
Climbing Vineclethra 95543
Climbing Violet 409898
Climbing wortclub 101266
Climbing Yellowwood 94031
Climbing Ylang-Ylang 35016
Climbing-bamboo 249605,249612
Climbinglike Asparagus 39223
Climborchis 193070
Clime 170193
Clinacanthus 96901
Cling Fingers 273531
Clinger 114245

Cling-fingers 273531
Clinging Sweethearts 170193
Cling-rascal 170193
Clingstone Nectarine 20955
Clinker Beech 266886
Clinopodium 96960,96983
Clinopodium Calamintha 97060
Clintonia 97086,97091
Clintonis 97086
Clinton's Bulrush 396012
Clinton's Club-rush 353318
Clinton's Lily 97086,97091,97096
Clip-me-dick 31051,158730,162519
Clitch Buttons 31051,170193
Clite 31051,170193
Clithe 170193
Clitheren 170193
Clithers 170193
Clits 170193
Cliven 170193
Clivers 170193
Clivia 97218
Clobweed 81237
Clock 302034,329521,384714
Clock Needle 350425
Clock Posy 384714
Clock Vine 390692,390701,390842
Clock-flower 384714
Clock-needle 350425
Clock-posy 384714
Clocks-and-watches 384714
Clockvine 390692
Clock-vine 390692,390787
Clockvineleaf Natsiatopsis 262784
Clockvine-leaved Natsiatopsis 262784
Clod Bur 31051
Clod-burr 31051
Clodweed 166038,216753
Cloffing 326340
Clog Plant 16179,202452
Cloggirs 170193
Clogweed 31051,192358,216753
Clokey Thistle 91930
Clokey's Fleabane 150548
Clokey's Wild Buckwheat 152128
Close Sciences 31051,267294,292372,405788
Close to Chive 15708
Close-bracted Jointfir 178529
Closed Gentian 173188,173222,173836,173998
Closedbract Jointfir 178529
Closedfruit Craneknee 221713
Closedfruit Stickseed 221713
Closedfruit Tanoak 233148
Closed-fruited Tanoak 233148
Closed-sheath Cotton-grass 152744
Closedspurorchis 94425
Closeflower Ainsliaea 12622
Closeflower Rabbiten-wind 12622
Close-flowered Orchid 264599
Close-headed Alpine-sedge 75529
Closeleaf Eargrass 187552
Close-relative Tanoak 233349
Closes Sciences 267294
Closesheath Tanoak 233162
Clot Bur 11549
Clotbur 31044,31051,415053,415063
Clot-burr 11549,31051,415003,415037,415053,415057
Clote 170193,292372,405788
Clote-burr 31051
Cloth Alarm 411420
Cloth of Gold 46580,111612
Cloth of Gold Crocus 111612
Clothed Colquhounia 100048
Clothed Dicliptera 129268
Clothed Embelia 144821
Clothed Glandularstipe Sweetleaf 381095
Clothed Kadsura 214968
Clothed Leptodermis 226139
Clothed Rhododendron 331808
Clothed Taxillus 385246
Clothed Wampee 94219
Clothed Wildclove 226139
Clothed Willow 344243
Clothes Pegs 130383,273531
Clothes Brush 133478
Clothing Meadowsweet 166131
Cloth-leaved Sallow 343829
Cloth-of-gold-crocus 111612
Cloth-tree 28243
Cloud Chesneya 87251
Cloud Garcinia 171086
Cloud Glochidion 177160
Cloud Nootka Cypress 85312
Cloud Plant 183157
Cloud Sedge 75544
Cloudberry 338243
Clouded Honeysuckle 235991
Clouded Iris 208946
Cloudgrass 12217
Cloud-grass 12217,12496
Cloudland Holly 204091
Cloudland Rhododendron 330913
Cloudleaforchis 265115,265123
Cloud-loving Hardy Ice Plant 123932
Cloudmist Gentian 173667
Cloudmist Glochidion 177160
Clouds Azalea 330727
Cloudy Bluegrass 305783
Cloudy Buttercup 326120
Cloudy Honeysuckle 235991
Cloudy Windhairdaisy 348158
Cloudy Woodbetony 287779
Clout 31051
Clove 382477
Clove Azalea 330687
Clove Bark Tree 129718
Clove Bush 233700
Clove Currant 334071,334127
Clove Gilliflower 127635,127793
Clove July Flower 127635
Clove July-flower 127635
Clove Pink 127635,127793
Clove Tongue 190944
Clove Tree 382477
Clove-bark Tree 129718
Clove-lip Toadflax 230911
Clovenfoot Plumegrass 128862
Clovenlip Toadflax 230911
Cloven-lip Toadflex 230911
Clover 157429,396806,397097
Clover Broomrape 275135
Clover Dodder 115008,115146
Clover Eagle-claw Cactus 354292
Clover Rape 275135
Clover Rose 397019
Clover Shrub 70781
Clover Trefoil 396806
Clover-knob 81237
Clove-root 175454
Clove-Root Plant 100132
Clover-rape 275135
Clover's Fishhook Ccactus 354292
Cloverscented Broomrape 274985
Clovershrub 70781,70833
Cloves 382477
Clove-scented Broomrape 274985
Clove-tongue 190944
Clove-tree 382477
Clovewort 175454,325518
Clowes Pansy Orchid 254928
Clowes Tulip Orchid 25137
Clowing-flower Hippeastrum 196456
Clown Fig 165438
Clowns 299759
Clown's Allheal 373346,373455
Clown's Lungwort 222651,405788
Clown's Mustard 34590,203184
Clown's Treacle 15698
Clown's Woundwort 373346
Clown's-mustard 203184
Club Awn Grass 106924
Club Awn-grass 106924
Club Bunches 407989
Club Foot 279685
Club Gourd 396151
Club Palm 104334,104339
Club Sedge 76490

Club Wheat 398877
Clubawngrass 106924
Clubbedspike Tanoak 233357
Clubbed-spike Tanoak 233357
Clubby Fig 165711
Club-cholla 181450,181453
Clubed Begonia 49755
Cluber 397043
Clubfilment 179174
Club-flowered Bauhinia 49179
Club-flowered Syzygium 382512
Clubfoot Amaranth 18692
Clubfruit Snowgall 191940
Clubfruitcress 376314
Club-haired Osbeckia 276148
Club-hairy Osbeckia 276148
Clubhead Cutgrass 223984
Clubleaf Fox-tail Orchid 9315
Club-moss Cypress 85321
Clubrush 352206
Club-rush 143089,209945,352158,352206,
353179,396003
Clubshaped Cherry 83197
Club-shaped Epidendrum 146398
Club-shaped Flower Eugenia 382672
Club-shaped Gourd 219843
Clubshaped Hemsleia 191940
Club-shaped Holly 203851
Club-shaped Sweetleaf 381221
Club-spur Orchid 302287
Clubspur Sporoxeia 372890
Clubweed 81237
Cluckenweed 374968
Cluckenwort 374968
Cluckweed 374968
Clumn Echidnopsis 140021
Clump Bulbophyllum 63116
Clump Knotweed 309624
Clump Speedwell 317927
Clump Stonebean-orchis 63116
Clump Verbena 405820
Clump-forming Bamboo 'Damarapa' 196347
Clumping Bamboo 162635,162679
Clumping Bamboo 'Damarapa' 196347
Clumping Giant Timber Bamboo 125453
Clumpy Catchfly 363278
Clumpybranch Hornsage 205585
Clumpybranch Incarvillea 205585
Clury Sage 345379
Cluse Crocus 111595
Clusia 97260
Clusia Family 97266
Clusia-leaved Peperomia 290317
Clusius Gentian 173344
Cluster Aster 380883,380887
Cluster Bean Guvar 115323
Cluster Cherry 280003
Cluster Clover 396917
Cluster Eelgrass 418382
Cluster Fescue 164197
Cluster Fig 165541
Cluster Holly Grape 52046
Cluster Liparis 232117
Cluster Mahonia 242617
Cluster Mallow 243862,243877
Cluster Palm 321170
Cluster Pea 131306,131319
Cluster Peppers 72074
Cluster Pine 300146
Cluster Redpepper 72074
Cluster Rose 336856
Cluster Yam 131501,131567
Cluster-Amaryllis 239253
Clusterberry 404051
Cluster-cherries 280003
Clustered Agapetes 10371
Clustered Aspidistra 39523
Clustered Barberry 51284
Clustered Beak-rush 333504,333589
Clustered Bellflower 70054,70055
Clustered Black Snakeroot 345969
Clustered Bracted Sedge 74077
Clustered Broom-rape 275050
Clustered Bur-reed 370066
Clustered Camellia 69068
Clustered Casearia 78120
Clustered Clover 396917
Clustered Coriandrifolious Cinquefoil 312471
Clustered Corydalis 105685
Clustered Dock 339989
Clustered Draba 137004
Clustered Field Sedge 75860
Clustered Fish-hook Pincushion 244245
Clustered Fishtail Palm 78047
Clustered Fishtail-palm 78047
Clustered Flatsedge 118646
Clustered Flower Actinidia 6594
Clustered Goldenrod 368123
Clustered Goldenweed 323039
Clustered Green Dock 339989
Clustered Halogeton 184951
Clustered Helleborine 147212
Clustered Hippolytia 196696
Clustered Ivy 187233
Clustered Jacot Edelweiss 224809
Clustered Knotweed 308939
Clustered Lady's-slipper 120348
Clustered Leptodermis 226088
Clustered Lilyturf 272153
Clustered Mallow 243862
Clustered Marshmarigold 68167
Clustered Mouse Ear Chickweed 82849
Clustered Peliosanthes 288656
Clustered Poppy Mallow 67131
Clustered Redcurrant 333970
Clustered Sacseed 390968
Clustered Saltlivedgrass 184951
Clustered Sedge 73986,74225
Clustered Snakeroot 345971
Clustered Thistle 91822
Clustered Thorowax 63611
Clustered Twayblade 232117
Clustered Wax Flower 245803
Clustered Whitlowgrass 137004
Clustered Yellowtops 166827
Clusteredflower Argyreia 32649
Clusteredflower Sweetleaf 381366
Clustered-flowered Actinidia 6594
Clusterflower Peliosanthes 288656
Clusterflower Syzygium 382538
Cluster-flowered Argyreia 32649
Cluster-flowered Sweetleaf 381366
Clusterflowered Swertia 380208
Clusterhead 127634
Clusterhead Pink 127634
Cluster-head Pink 127634
Clusterleaf Gayfeather 228483
Cluster-leaf Tick-trefoil 126366
Cluster-lily 60431
Clusterpalm 6849
Clutch 308816
Clutch Buttons 170193
Clutch-button 170193
Cly 170193
Clyder 170193
Clyte 170193
Clythers 170193
Cmianas 831
Cmnese Narcissus 262382
Cmpet Pink Pincushion 350086
Cnesmone 97499
Cnestis 97510
Cnicus 97581,97599
Cnidium 97697
Cnidiumleaf Pimpinella 299384
Cnoutberry 338243
Coach Horse 410677
Coach-and-horses 5442
Coach-horses 410677
Coachman's Buttons 216753
Coach-whip 167562
Coachwood 83534,156068
Coahuila Oak 324299
Coahuila Orach 44301
Coahuila Pine-leaved Mahonia 242616
Coahuila Sage 344975
Coal Oil Brush 387079
Coalseleaf Azalea 331756
Coanther Petrocosmea 292553
Coanther Stonebutterfly 292553
Coarse Verbena 405885

Coarse-edge Spiradiclis 371779
Coarsefruit Corydalis 106541
Coarsefruit Tanoak 233396
Coarseleaf Eargrass 187697
Coarseleaf Wendlandia 413817
Coarse-leafed Mallee 155599
Coarse-serrate Mongolian Oak 324190
Coarseserrate Oak 324190
Coast Amaranth 18808
Coast Azalea 330160
Coast Banksia 47647
Coast Barnyard Grass 140509
Coast Beard Heath 227968
Coast Beefwood 15953
Coast Blite 87147
Coast Casuarina 15953,79182
Coast Cinquefoil 312860
Coast Cockspur 140509
Coast Coral Tree 154636
Coast Cottonwood 195311
Coast Douglas Fir 318580
Coast Douglas-fir 318580
Coast Goldenrod 368414
Coast Gray Box 155658
Coast Grewia 180925
Coast Grey Box 155504
Coast Juniper 213776
Coast Larkspur 124091,124168
Coast leucothoe 228155
Coast Licuala Palm 228755
Coast Licualapalm 228755
Coast Lily 229921
Coast Live Oak 323621
Coast Live-forever 138832
Coast Madrone 334976
Coast Myall 1082
Coast Nettle 402869
Coast Oak 79161
Coast Persimmon 132291
Coast Pignut Hickory 77895
Coast Pine 299869
Coast Range Agoseris 11469
Coast Range Claytonia 94319
Coast Range Triteleia 398809
Coast Range Wild Buckwheat 151839
Coast Red Elder 345579
Coast Red Elderberry 345579
Coast Redwood 360565
Coast Rosemary 413914
Coast Sedge 74478
Coast She Oak 15953,79178
Coast She-oak 15953
Coast Silk Tassel 171546
Coast Silktassel 171546
Coast Silver Oak 58489
Coast Spruce 298435
Coast Tarweed 241543
Coast Violet 409785
Coast Wattle 1610
Coast West Cedar 85375
Coast Whistling Thorn 1712
Coast Whitethorn 79944
Coast Whitethorn Ceanothus 79944
Coast Willow 343496
Coast Willow-weed 146663
Coast Wollflower 154419
Coastal Avicennia 45746
Coastal Azalea 330160
Coastal Banksia 47647
Coastal Barrel Cactus 163479
Coastal Blackroot 320393
Coastal Blue Sage 345364
Coastal Blumea 55814
Coastal Bulrush 353569
Coastal Burnweed 148161
Coastal California Buckwheat 152056,152060
Coastal Carolina Spider-lily 200922
Coastal Century Plant 10938
Coastal Cholla 116670
Coastal False Asphodel 392632
Coastal Galenia 169997
Coastal Gem 180610
Coastal Glehnia 176923
Coastal Goldfields 222615
Coastal Gumplant 181180
Coastal Heritiera 192477,192494
Coastal Joint-weed 308669
Coastal Jugflower 7198
Coastal Leucothoe 228155
Coastal Mallow 217958
Coastal Myall 1083
Coastal Plain Flat-topped Goldenrod 161100
Coastal Plain Goldentop 161074
Coastal Plain Joepyeweed 161170
Coastal Plain Willow 343172
Coastal Plain Yellowtops 166816
Coastal Prickly Pear 272962,273061
Coastal Redcedar 213998
Coastal Redwood 360565,360566
Coastal Rosemary 413914
Coastal Sage Scrub Oak 323850
Coastal Sagewort 36127
Coastal Salt Grass 134842
Coastal Sedge 74478
Coastal Silverpuffs 253915
Coastal Silvertree 192494
Coastal Sneezeweed 188411
Coastal Squill 352960
Coastal Star Sedge 74478
Coastal Tea Tree 226460
Coastal Tea-tree 226460
Coastal Waterhissop 46362
Coastal Wattle 1161,1610
Coastal Wild Buckwheat 151926
Coastal Willow 343496
Coastal Woollybush 7203
Coastalplain Goldenaster 90261
Coastal-plain Nailwort 284854
Coastal-plain Thistle 92347
Coastal-plain Willow 343628
Coast-blite 87147
Coat Flower 292656,292668
Coatbuttons 396697
Coatedleaf Birthwort 34141
Coat's Thorn 41906
Cob cactus 107072
Cob Cactus 140248,155201,155208,155225,234900
Cob Nut 106703
Cobaea 97752
Cobaea Beardtongue 289332
Cobaea Beard-tongue 289332
Cobaea Penstemon 289332
Coban Tree 407907
Cobb Dendrochilum 125528
Cobbedy-cut 106703
Cobbler's Pegs 53797,54048,103446
Cobbly-cut 106703
Cobin Tree 407907
Cobnut 106696,106703
Cobnut Hazel 106703
Cobnuts 106696
Cobra Lily 33234,33517,122759
Cobra Plant 122759
Cobresia 217117
Cobweb Houseleek 358021
Cobwebby Amur Linden 391663
Cobwebby Halogeton 184950
Cobwebby Saltlivedgrass 184950
Cobwebby Thistle 92249
Cobwebcalyx 302795
Coca 155051,155067
Coca Family 155048
Coca Shrub 155067
Cocaine 155067
Cocaine Family 155048
Cocaine Plant 155067
Cocaine Tree 155051
Cocaine-plant 155067
Cocainetree 155051,155067
Cocayams 415201
Coccineus Colquhounia 100031
Cocculua Indicus 21459
Coccus Wood 61427
Cocdare-ovate-leaf Hanceola 185246
Cochichina Bog Rattanpalm 65760
Cochichina Centranthera 81718
Cochichina Cheirostylis 86675
Cochichina Momordica 256804
Cochichinense Asparagus 38960
Cochil Sapote 78175

Cochin Goraka 171221
Cochinchina Afzelia 10129
Cochinchina Ardisia 31642
Cochin-China Begonia 50314
Cochinchina Blastus 55147
Cochin-China Blastus 55147
Cochinchina Cordia 104164
Cochin-China Cordia 104164
Cochin-China Cratoxylum 110251
Cochinchina Cudrania 240813
Cochin-china Cudrania 240813
Cochin-China Cyclobalanopsis 116059
Cochinchina Elytranthe 241317
Cochinchina Glaucous Oil-wood 142450
Cochinchina Glycosmis 177822
Cochin-China Glycosmis 177822
Cochinchina Gymnopetalum 182575
Cochinchina Helicia 189918
Cochin-china Helicia 189918
Cochinchina Holly 203642
Cochin-China Holly 203642
Cochinchina Homalium 197659
Cochin-China Homalium 197659
Cochinchina Indianmulberry 258885
Cochin-China Indian-mulberry 258885
Cochinchina Ixonanthes 211023
Cochin-China Ixonanthes 211023
Cochinchina Leafflower 296522
Cochin-China Leaf-flower 296522
Cochin-China Litse 234089
Cochinchina Meranti 362200
Cochinchina Nosema 266759
Cochinchina Ormocarpum 274343
Cochin-China Ormocarpum 274343
Cochinchina Passionflower 285677
Cochin-China Pleurostylia 304898
Cochinchina Randia 12519
Cochin-China Randia 12519
Cochinchina Raspberry 338265
Cochin-China Raspberry 338265
Cochinchina Sweetleaf 381143
Cochin-China Sweetleaf 381143
Cochinchina Thaumastochloa 388909
Cochinchina Turpinia 400521
Cochinchina Walsura 413125
Cochin-China Walsura 413125
Cochin-ChinaTurpinia 400521
Cochinchinense Osage Orange 240813
Cochinchinese Camellia 68999
Cochinchinese Elytranthe 241317
Cochinchinese Excoecaria 161647
Cochinchinese Macrosolen 241317
Cochin-chinese Macrosolen 241317
Cochineal Cactus 272775
Cochineal Fig 272830
Cochineal Nopal Cactus 272830
Cochineal Nopal-cactus 272830
Cochineal Plant 272830
Cochise Foxtail cactus 107061
Cochise Pincushion 155218
Cochise Pincushion Cactus 107061
Cochise Rock Daisy 291305
Cochlear-like Bunias 63446
Cochleate Falsepanax 266912
Cochlianthus 98064
Cochlioda 98085
Cock 248411
Cock Battler 302034
Cock Bramble 336419,338447
Cock Brumble 338447
Cock Fighters 302034
Cock Foot 86755
Cock Robin 248411,363461
Cock Roise 282685
Cock Sorrel 339887
Cock Thistle 271666
Cockatoo Bush 260717
Cock-battler 302034,336419,338447
Cockbillthorn 252754
Cockbone Elsholtzia 144023
Cockbone's Aroma 112871
Cockbum Primrose 314256
Cockburn Raspberry 338267
Cockclawflower 66219,66220
Cockclawthorn 278314
Cock-drunks 369339
Cockehead-yam 152853
Cockens 282685
Cockerel 11947
Cockeyeweed 218344,218347
Cock-fighter 302034,336419,338447
Cock-flower 273531
Cockfoot 86755
Cock-grass 235373,302034,329521
Cockies' Tongues 385812
Cockiloorie 50825
Cockle 11938,11947,31051,203789, 235373,237539,346434,363673,364193, 409335
Cockle Bells 31051
Cockle Bur 31051
Cockle Buttons 31051
Cockle Dock 133478
Cockle Orchid 145290
Cockle Shells 409339
Cockle-bell 31051
Cocklebur 11542,11549,11572,31051, 414990,415057
Cockle-bur 11549,31051,415057
Cockleburleaf Pogostemon 307013
Cockleburleaf Spinestemon 307013
Cocklebur-like Villous Amomum 19933
Cockle-buttons 31051
Cockleford 11947
Cockle-shell 409339
Cockly Bur 11549,31051
Cockly-bur 31051
Cocknest Spine Bamboo 47281
Cockroach Plant 185692
Cockrose 282546
Cock-rose 282685
Cocks 302034,302068
Cock's Comb 18752,18788,80395,133478, 154647,190651,268991,271280,273469, 273531,282685,287494,287721,329521, 363461
Cock's comb 80378
Cock's Foot 30081
Cock's Foot Waldheimia 413021
Cock's Head 28617,271280,282515, 282546,282685,302034,397019
Cocks-and-hens 175444,237539,302034
Cockscomb 80378,80381,80395,154647
Cockscomb Coralbean 154647
Cockscomb Coral-tree 154647
Cockscomb Hedgehog Cactus 140256
Cockscomb Japanese Cedar 113692
Cockscomb Pondweed 312083
Cockscomb Rattleweed 329521
Cock's-comb Rhinanthus 329543
Cockscomb Sainfoin 271191
Cock's-eggs 344369,344372
Cocksfoot 121226,121233
Cock's-foot 121233
Cocksfoot-grass 121226,121233
Cock's-grass 121226
Cock's-head 271280
Cocksomb Yunnanclave 238108
Cock-sorrel 339887
Cockspur 1150,81198,140351,140367, 300944
Cockspur Coral Bean 154647
Cockspur Coral Tree 154647
Cockspur Coralbean 154647
Cockspur Coral-bean 154647
Cockspur Coral-tree 154647
Cockspur Flower 303324
Cockspur Grass 140367,140451
Cockspur Hawthorn 109636
Cockspur Thorn 1191,109636,240813
Cockspur-grass 140367
Cockspur-thorn 109636
Cock-upon-perch 231183,237539
Cockweed 11947,225315
Cocky Apple 301791
Cocky Baby 37015
Cocky-baby 37015
Cocldebur 415042
Coco 99910
Coco Grass 118642,119503
Coco Magnolia 232594

Coco Nut Palm 98136
Coco Plum 89812
Coco Yam 99910
Cocoa 389404
Cocoa Tree 389404
Cocoa-buttons 31051
Cocoa-root 65218
Cocobolo 121607,121803
Coco-de-mer 235159
Coco-grass 119503
Coconut 98134,98136
Coco-nut 98134
Coconut Geranium 288273
Coconut Palm 98134,98136
Coconut Tree 98136
Cocoon Plant 359026
Cocoonhead 222527
Cocoplum 89816
Coco-plum 89816
Cocoron Elaeodendron 142476
Cocos Palm 31705,89816
Cocos Wood 61427
Cocoswood 61426
Cocowort 72038
Cocoxochitl 121561
Cocoyam 99910
Coco-yam 99910
Cocozelle 114300,219843
Cocum Coakum 298094
Cocumbe Palm 283408
Cocus Wood 61427
Cocuswood 61426,61427
Codariocalyx 98152
Codded Trefoil 387251
Codded Willowherb 146724
Coddled Apples 146724
Codlins 146724,243711
Codlins and Cream 146724
Codlins-and-cream 146724
Codonacanthus 98220
Codonanthe 98228,98232
Codweed 81237
Coelachne 98446
Coelodiscus 98555
Coeloglossum 98560,98586
Coelogyne 98601
Coelogyne Epigeneium 146534
Coelomate Veins Rhododendron 330441
Coelonema 98762
Coelospermum 98808
Coerulescent Pternandra 320199
Coeruleus Larkspur 124083
Coffeate Jasmine 211783
Coffee 98844,98857,339887
Coffee Bean Tree 411398
Coffee Berry 328630
Coffee Fence 95936
Coffee Flower 381035
Coffee Jurinea 211783
Coffee Plant 98857
Coffee Root 90901
Coffee Senna 360463
Coffee Shrub 98872
Coffee Tree 57443,98857,182524,213883,295386,310204
Coffee Weed 360461,360463
Coffee Wood 213883,295386
Coffeebarry 328630
Coffeebean 361378,361453
Coffeesenna 360463
Coffeetree 182524
Coffee-tree 98857,182524
Coffeeweed 360463,361390
Coffee-weed 360461
Coffin Juniper 213883,213889
Coffin Nail 21195
Coffin-nail 21195
Cogflower 154139
Cognate Campion 363337
Cogniaux Blastus 55141
Cogon Grass 205497
Cogon Satin Tail 205497
Cogon Satintail 205497
Cogon Satin-tail 205497
Cogongrass 205470,205497
Cogon-grass 205497
Cogweed 247246,325612
Cogwood 79902,418155
Cogwood Jujube 418155
Cohesive-styled St. John's-wort 201818
Cohosh 79730,90994
Cohune Nut 273244
Cohune Palme 273244
Cohune Tree 273244
Cohunenut 273244
Coif Azalea 331254
Coigue 266868
Coil Fieldhairicot 138967
Coilbract Windhairdaisy 348772
Coiled Fargesia 162653
Coiled Pondweed 340472
Coiled Rhododendron 331984
Coiledanther Southstar 33328
Coiledflower Nightshade 367632
Coiledleaf Everlasting 21549
Coiledleaf Pearleverlasting 21549
Coil-flowered Nightshade 367632
Coilhair Azalea 330420
Coilleaf Azalea 331675
Coilmargin Willow 344124
Coilstem Heshouwu 162519
Coin Larch 317822
Coin Maple 133596
Coinbag Milkwort 307923
Coinleaf Whitepearl 172127
Coinmaple 133596
Coin-shaped Blueberry 403920
Coinshaped Saltbush 44542
Cointree 280539
Coir Palm 393809
Cokar-nut 98136
Coke 155067
Coke Pintel 37015
Coker Nut 98136
Coker's Gayfeather 228451
Cokeweed 11947
Col 104690
Cola 99157,99158
Cola Lagarto 290718
Cola Nut 99157,99158,99240
Colander 104690
Colchester Lily 229758
Colchicum 99293,99297
Colchis Bladdernut 374098
Colchis Ivy 187205
Cold Buttercup 325886
Cold Crazyweed 278835
Cold Dogwood 380429
Cold Euonymus 157522
Cold Milkvetch 42384
Cold Mountain Crazyweed 278764,278766
Cold Orchis 116926
Cold Rhodiola 329877
Cold Spindle-tree 157522
Cold Umbrella Bamboo 162691
Coldbamboo 87581,172736,172748
Coldenia 99358
Coldish Dandelion 384823
Coldland Stonecrop 356995
Coldwaterflower 298848,298989
Coldwater-flower-like Falsestairweed 223685
Cole 59278,59532,108627
Colebrookea 99418
Colebrookia 99418
Coleostephus 60095
Colesat 59493
Coleseed 59493
Coleslaw 59520
Coleus 99506,99509,99711,367870
Coleusleaf Skullcap 355843
Colewort 59493,59532,108591,108627,175454
Colgajito 140320
Coliander 104690
Colic Root 14486,131896
Clickwort 13966
Colicroot 228534
Colic-root 14471,14486,131896
Colicweed 128288,220729
Coliledfruit Acacia 1615
Coling 243711

Colins' Arabis 30230
Colioba 21308
Coliseum Maple 2849
Coliseum-ivy 116736
Collabium 99770
Collarbind 30081
Collard 59493,59524
Collards 59493,59524
Collectivegrass 381024
Colletia 99804
Collets 59532
Collett Barberry 51479
Collett Iris 208496
Collett Rosewood 121860
Collett Tanoak 233151
Collett Yam 131529
Collett-like Barberry 51478
Colliguaji Bark 99822
Collimamol 238337
Collinhood 282685
Collins Hackberry 80712
Collins Nettletree 80712
Collinsia 99833
Collomia 99850,99859
Collumella Raspberry 338269
Collumellaelike Raspberry 338269
Colluts 59532
Colmenier 127607
Colocasia 99899
Colocasiifoliate Slender Wildginger 37630
Colo-colo 93707
Colocynth 93297,93304
Colombia Dense Barberry 51539
Colombia Hemlock 399912
Colombia Net-veined Barberry 52113
Colombian Ball Cactus 414107
Colombian Barberry 51480
Colombian Bark 91074
Colombian Bluestem 351269
Colombian Box 78145
Colombian Boxwood 78145
Colombian Mahogany 76840,380527
Colombian Wax Palm 84328
Colombian Waxweed 114599
Colona 99964
Colona-like Euonymus 157380
Colona-like Spindle-tree 157380
Colonial Bent 12034
Colonial Bent Grass 12034
Colonial Bentgrass 12034
Colonial Bent-grass 12034
Colonial Oak Sedge 74155
Colony Broodhen Bamboo 297284
Colony Hydrangea 199862
Coloquintida 93304
Color Brome 60679
Color Marking Dracaena 137401
Color Marking Dragonbood 137401
Coloradillo 185169
Colorado 213922
Colorado Barberry 51612
Colorado Blue Spruce 298401
Colorado Bluestem 144675
Colorado Bristlecone Pine 299795
Colorado Columbine 30051
Colorado Corn Cockle 403687
Colorado Currant 333940
Colorado Desert Wild Buckwheat 152001
Colorado Douglas Fir 318585
Colorado Fir 317
Colorado Four-o'clock 255732
Colorado Grass 282304
Colorado Green Thread 389154
Colorado Juniper 213922
Colorado Millet 282304
Colorado Pine 299926
Colorado Pinyon 299926
Colorado Pinyon Pine 299926
Colorado Red Cedar 213922
Colorado River Hemp 361380
Colorado Root 229324
Colorado Rubberweed 201325,201326
Colorado Sage 35505
Colorado Spruce 298401,298406
Colorado Stonecrop 357164
Colorado Thistle 92363
Colorado White Fir 317
Colorado Wild Buckwheat 151932
Colorata Water Lily 267671
Colorate Chinese Loropetalum 237194
Colorate Loropetal 237194
Colorcalyx 89282
Colored Fortune Spindle-tree 157509
Colored Gentian 173725
Colored Maxillaria 246711
Colored Mistletoe 410992
Colored Peristrophe 291138
Colored Pole Bambusa 47409
Coloredfruit Spapweed 204905
Colorful Azalea 331580
Colorful Forkliporchis 332340
Colorines 154658
Colorless Trillium 397555
Coloroured Rattlesnake Plantain 179594
Colorus Actinidia 6584
Colorus Kiwifruit 6584
Colory Bamboo 47178
Colory Bambusa 47178
Colory Woodbetony 287793
Colosseum Ivy 116736
Colourbine 30081
Coloured Ardisia 31383
Coloured Bellflower 69972
Coloured Currant 333940
Coloured Erythropsis 155000
Coloured Geniosporum 172858
Coloured Nymphaea 267671
Coloured Panicgrass 281479
Coloured Sedge 76551
Coloured Sterculia 155000
Coloured Waterlily 267671
Coloured-flower Vetch 408377
Colpodes Skullcap 355414
Colquhounia 100030
Colt-herb 400675
Colts Foot 400671,400675
Colt's Foot 400675
Colt's Tail 103446,196818
Colt's Tails 103446,196818
Coltsfoot 37574,292342,400671,400675
Colts-foot 400675
Colubrina 100058
Columar Italian Cypress 114768
Columba's Herb 202086
Columbia Goldenweed 150377
Columbia Gorge Larkspur 124055
Columbia Hawthorn 109621
Columbia Lily 229810
Columbia R. Giant Dogwood 105131
Columbia River goldenrod 100086
Columbia Windflower 23784
Columbian Black Walnut 212601
Columbian British Pine 318580
Columbian Goldenbush 150377
Columbian Monk's-hood 5136
Columbian Pine 318580
Columbian Rodriguezia 335165
Columbian Spruce 298273
Columbian Stitchwort 255445
Columbian White-topped Aster 360713
Columbine 29994,30007,30034,30041,30081,405872
Columbine Meadow Rue 388423
Columbine Meadow-rue 388423
Columbine Olympicus 30060
Columbineleaf Spiraea 371807
Columbine-leaved Meadow Rue 388423
Columbine-leaved Spiraea 371807
Columbo 380146
Columbus Grass 369593
Columcarp Begonia 49760
Column Ascocentrum 38191
Column Duckbeakgrass 209314
Column Fruit Actinidia 6577
Column of Pearls 186341
Column Spapweed 204863
Column Touch-me-not 204863
Columnar 43510
Columnar Alaska Cedar 85305
Columnar Araucaria 30838,30839
Columnar Coneflower 326960

Columnar English Oak 324338
Columnar English Yew 385311
Columnar European Hornbeam 77257
Columnar Norway Maple 3446
Columnar Ornament Pear 323118
Columnar Phaius 293491
Columnar Plum-yew 82525
Columnar Podocarpus 306468
Columnar Poplar 311226
Columnar Red Maple 3520
Columnar Silver Fir 273
Columnar Sugar Maple 3550
Columnar Yew 385390
Columnea 100110,100115
Columnumbel Hippolytia 196694
Coluria 100132
Colvill Gladiolus 176121
Colza 59493,59603
Comanche Western-daisy 43296
Comanthosphace 100232
Comastoma 100263
Comb 350425
Comb Corydalis 105769
Comb Five-fingers 312648
Comb Grevillea 180604
Comb Hedgehog 140282
Comb Pondweed 378991
Comb Toothed Rhododendron 331455
Comb Windmill Grass 88383
Comb-and-brush 133478
Comb-and-hairpins 384714
Combbush 292621
Comber Lasiococca 222385
Comber's Barberry 51481
Combform Wheatgrass 11699
Combined Spicebush 231298
Combined Spice-bush 231298
Combose Tabernaemontana 382766
Combretum 100309
Combretum Family 100296
Comb's Crowngrass 285393
Combs-and-hairpins 384714
Combshapeleaf Cassiope 78641
Come-and-kiss-me 410677
Come-quickly Death 174877
Comfort Root 194919
Comfrey 50825,381024,381035
Comfrey Consound 381035
Commandra 100225
Commelina 100960
Commercial Cranberry 403894
Commercial Vanilla 405017
Commerson Dallisgrass 285508
Commerson Paspalum 285508
Commersonia 101219
Commerson's Latan Palm 222626
Commom Sinochasea 364899
Common Acacia 334976
Common Achyranthes 4259
Common Achyrophorus 202400
Common Acroceras 6109
Common Acroglochin 6164
Common Acronema 6224
Common Adder's-tongue 154899
Common Adelostemma 7088
Common Adenostemma 8012
Common Aechmanthera 8540
Common Aerva 9396
Common Aeschynomene 9560
Common African Violet 342482,342493
Common Afzelia 10106
Common Agalinis 10157
Common Aganosma 10229
Common Agapetes 10344
Common Agrimony 11549,11558
Common Ahernia 12484
Common Alcimandra 14231
Common Alder 16352,16445
Common Alfredia 14596
Common Alkanet 21959
Common Allamanda 14873
Common Allmania 15912
Common Almond 20890
Common Alpine-lily 234234
Common Alplily 234234
Common Alstonia 18055
Common Alyxia 18516
Common Amaranth 18810
Common Amberboa 18966
Common Amblynotus 19045
Common Ambroma 675
Common Amentotaxus 19357
Common American Columbine 30007
Common Amischophacelus 115528
Common Ammania 19566
Common Ancathia 21888
Common Androcorys 22323
Common Andrographis 22407
Common Aniseia 25316
Common Anisochilus 25381
Common Anisodus 25447
Common Anneslea 25773
Common Annual Candytuft 203184
Common Anodendron 25928
Common Anotis 262966
Common Anthurium 28119
Common Antiaris 28248
Common Aotus 28874
Common Apea-earring 301128
Common Apluda 29430
Common Appendicula 29809
Common Apple 243550,243675
Common Apple Tree 243675
Common Apricot 34425,34475
Common Apterosperma 29898
Common Arbor-vitae 390587
Common Arctogeron 31080
Common Argyreia 32602
Common Arnebia 34624
Common Arrowbamboo 162656
Common Arrowhead 342363,342400
Common Artichoke 117787
Common Arundina 37115
Common Arundinella 37352
Common Arytera 37522
Common Ascocentrum 38188
Common Ash 167955
Common Asparagus 39120
Common Asparagus-fern 39195
Common Aspidistra 39532
Common Aspidopterys 39676
Common Asterbush 41748
Common Astilboides 41872
Common Asystasia 43610
Common Atractylodes 44208
Common Atropa 44708
Common Aubrieta 44868
Common Aucklandia 44881
Common Aulacolepis 25343
Common Australian Buttercup 326239
Common Autumn Crocus 99297
Common Avens 175454
Common Averrhoa 45725
Common Axyris 45844
Common Baby's-breath 183225
Common Baccaurea 46198
Common Balata 244530
Common Baldcypress 385262
Common Balldaisy 56695
Common Balloon Vine 73207
Common Balm 249504
Common Bamboo 47516
Common Banana 260253
Common Baphicacanthus 47845
Common Barberry 52322
Common Barley 198376
Common Basil 268438
Common Bean 294056
Common Beancaper 347066
Common Bedstraw 170193
Common Beech 162400
Common Beet 53249
Common Beggarticks 54182
Common Beggar-ticks 53913
Common Begonia 50168
Common Bennettiodendron 51039
Common Bent 12034
Common Bent Grass 12034,12323
Common Bentgrass 12025
Common Bentspikegrass 131224
Common Bergia 52592

Common Betony 53304
Common Bigginseng 241164
Common Bindweed 102933
Common Birch 53563,53590
Common Bird's-foot 274896,274895
Common Bird's-foot Trefoil 237539
Common Bird's-foot-trefoil 237539
Common Bistort 308893
Common Bitterfigwort 298527
Common Black Poplar 311398
Common Blackberry 338105
Common Blackcap 338808
Common Bladder 100153
Common Bladder Senna 100153
Common Bladdersenna 100153
Common Bladder-senna 100153
Common Bladderwort 403382,403385
Common Blainvillea 55067
Common Blanket-flower 169575
Common Bleeding Heart 220729
Common Bleeding-heart 220729
Common Bletilla 55575
Common Bloodlily 350312
Common Blossfeldia 55658
Common Blue Heart-leaved Aster 40256
Common Blue Sow-thistle 90851
Common Blue Squill 199560
Common Blue Violet 410351,410587
Common Blue Wood Aster 40256,380856
Common Bluebeard 78012
Common Blue-beard 78012
Common Blueeargrass 115589
Common Blue-eyed Grass 365747,365782
Common Blue-eyed-grass 365680,365695
Common Bluets 198708
Common Bog Arrow-grass 397159
Common Bog Bean 250502
Common Bogbean 250502
Common Bogrush 352382
Common Bolocephalus 56695
Common Bombax 56802
Common Boneset 158250
Common Borage 57101
Common Borneo Camphor 138548
Common Borszczowia 57387
Common Bowringia 58006
Common Box 64345
Common Boxweed 6907
Common Boxwood 64345
Common Brachybotrys 58300
Common Bramble 338265
Common Brassavola 59259,333102
Common Brassbuttons 107766
Common Bredia 59855
Common Briar 336419
Common Brighteyes 327473
Common Bristlethistle 73430
Common Broadlily 97093
Common Broom 121001
Common Broomrape 275135
Common Bruguiera 61263
Common Bryocarpum 61450
Common Buckbean 250502
Common Bucket Orchid 105522
Common Buckthom 328642
Common Buckwheat 162312
Common Buffalograss 61724
Common Bugle 13174
Common Bugloss 21959,57101
Common Bugseed 104861
Common Bulrush 352284,353231,353434,
353686
Common Burdock 31056,45982
Common Burmannia 63971
Common Burreed 370102
Common Bur-reed 370058
Common Burstwort 192965
Common Bushweed 167096
Common Butter Wort 299759
Common Buttercup 325518
Common Butterfly Bush 62019
Common Butterwort 299759
Common Button Cactus 147425
Common Buttonbush 82098,82107
Common Cabbage Tree 115239
Common Calabash Tree 111104
Common Caladium 65218
Common Calamint 96966,97027
Common Calanthe 65921
Common Caleana 66369
Common California Poppy 155173
Common Calla 417093
Common Calla Lily 417093
Common Callalily 417093
Common Callianthemum 66713
Common Calochilus 67542
Common Calotis 68081
Common Calypso 68538
Common Camarotis 68787
Common Camas 68794,68803
Common Camash 68803
Common Camass 68794,68803
Common Camassia 68803
Common Camchaya 68833
Common Camellia 69156
Common Camptotheca 70575
Common Canary Grass 293729
Common Canarytree 70996
Common Cancrinia 71088
Common Canscora 71279
Common Caper 71762,71871
Common Capers 71871
Common Caraway 77784
Common Cardaria 73090
Common Carline 77022
Common Carpesium 77146
Common Carpetgrass 45821,45833
Common Carrion Flower 366427
Common Carrion-flower 366364
Common Cashew 21195
Common Cassava 244507
Common Cassiope 78645
Common Cat Tail 401112
Common Catalpa 79243
Common Catbrier 366559
Common Catchfly 363477
Common Catdaisy 202400
Common Caterpilar 354778
Common Cat's Ear 202432
Common Cat's-ear 202432
Common Cattail 401112
Common Cat-tail 401112
Common Cattleya 79546
Common Celandine 86755
Common Centaurium 81495,81546
Common Centaury 81495
Common Centipedegrass 148254
Common Centotheca 81694
Common Cephalanoplos 82022
Common Ceratocephalus 83443
Common Ceratoides 218122
Common Cerberustree 83698
Common Cerberus-tree 83698
Common Ceriops 83946
Common Ceropegia 84287
Common Chamomile 85526
Common Championella 85952
Common Chaulmoogra Tree 199744
Common Chaulmoogra-tree 199744
Common Cherry 83361
Common Cherry-laurel 223116
Common Chestnut 78811
Common Chickweed 82818,82826,342874,
374968,374975,375033
Common Chicory 90901
Common China Aster 67314
Common China Fir 114539
Common China-aster 67314
Common Chinchweed 286897
Common Chinese Bamboo 304000
Common Chinese Calophaca 67785
Common Chinquapin 78807
Common Chirita 130071
Common Chloraea 88259
Common Choke Cherry 316918
Common Chonemorpha 88891
Common Christia 89209
Common Christisonia 89219
Common Christolea 89229
Common Chunechites 90613
Common Cineraria 290821

Common Cinquefoil 312993
Common Cissampelos 92555
Common Cladopus 93975
Common Claoxylon 94068
Common Clary 345379
Common Cleavers 170193
Common Cleistanthus 94539
Common Closedspurorchis 94465
Common Clotbur 415063
Common Clubfruitcress 273971
Common Club-rush 352206
Common Cnestis 97552
Common Cockle 11947
Common Cocklebur 11549,415057,415063
Common Cockscomb 80395
Common Codariocalyx 98159
Common Coelogyne 98711
Common Coffee 98857
Common Coleostephus 99491
Common Coleus 99711
Common Colewort 108627
Common Coltsfoot 400675
Common Comfrey 381035
Common Comh'ey 381035
Common Commersonia 101220
Common Conandron 101681
Common Conehead 47845
Common Coral Bean 154643
Common Coral Tree 154647,154687
Common Coralbean 154643
Common Coral-bean 154643
Common Corallodiscus 103941
Common Coral-tree 154643,154647
Common Cordgrass 370157
Common Cord-grass 370157
Common Corn Cockle 11947
Common Corncockle 11947
Common Cornsalad 404439
Common Corybas 105531
Common Corydalis 105846
Common Corymborkis 106882,106883
Common Cosmos 107161
Common Cotoneaster 107504
Common Cotton-grass 152769
Common Cottonrose 166038
Common Cottonthistle 271666
Common Cotton-weed 168687
Common Cottonwood 311292
Common Couch 144661
Common Cowpea 409086
Common Cow-pea 409086
Common Cow-wheat 248193
Common Crabgrass 130745
Common Crape Myrtle 219933
Common Crapemyrtle 219933
Common Crape-myrtle 219933
Common Cress 225450
Common Crocosmia 111464
Common Crocus 111625
Common Crowberry 145073
Common Crupina 113214
Common Crypsis 113306
Common Cryptochilus 113513
Common Cryptodiscus 113554
Common Cryptomeria 113685
Common Cuckoo-orchis 110502
Common Cudweed 166038
Common Curculigo 114838
Common Currant 334179,334201
Common Custard Apple 25882
Common Cutgrass 223987
Common Cyanotis 115589
Common Cyclamen 115965
Common Cymbidiella 116757
Common Cypress 114753
Common Cypress-pine 67433
Common Cyrtosperma 120773
Common Cystopetal 320248
Common Daffodil 262441
Common Dahlia 121561
Common Daisy 50825,227533
Common Danceweed 98159
Common Dandelion 384714,384717
Common Daphniphyllum 122679
Common Date Palm 295461
Common Day Flower 100961
Common Dayflower 100961
Common Devilpepper 327069
Common Devil-pepper 327069
Common Devil's Claws 315434
Common Dewberry 338399
Common Dholl 65146
Common Dicerma 128372
Common Dichapetalum 128675
Common Dichotomanthus 129026
Common Dickinsia 129196
Common Didymostigma 130071
Common Dill 24213
Common Dimocarpus 131059
Common Dimorphostemon 131141
Common Diplachne 226005
Common Diplacrum 132789
Common Diplazoptilon 132862
Common Diplomeris 133042
Common Disanthus 134010
Common Diuranthera 135059
Common Dobinea 135099
Common Dodder 115008,115040
Common Dog Fennel 26768
Common Dogbonbavin 133185
Common Dogmustard 154091
Common Dog-mustard 154091
Common Dogoak Maple 2837
Common Dog-violet 410493
Common Dogweed 391042
Common Dogwood 380461,380499
Common Dolicholoma 135372
Common Douglas Fir 318580
Common Dracaena 104367
Common Dracunculus 137747
Common Dragonhead 137651
Common Dragon's Plant 137747
Common Drywheatgrass 148420
Common Ducksmeat 372300
Common Duckweed 224375
Common Dumasia 138942
Common Dutchman's-pipe 34337
Common Dwarf Lupine 238469
Common Dwarfnettle 262218
Common Dyer's Greenwood 173082
Common Dysophylla 139602
Common Dysosma 139629
Common Eastern Fleabane 150980
Common Eastern Poison-Ivy 393470,393471
Common Eastern Wild-rye 144528
Common Eberhardtia 139782
Common Edelweiss 224772,224893
Common Eelgrass 418392
Common Eelvine 25928
Common Eggplant 367370
Common Egretgrass 135059
Common Eight Trasure 200784
Common Elachanthemum 141898
Common Elaeocarpus 142303,142396
Common Elder 345580,345631
Common Elderberry 345580
Common Eleutherine 143573
Common Elm 401593
Common Elsholtzia 143974
Common Elytranthe 144595
Common Embelia 144752
Common Emu Bush 148363,148364
Common Enchanter's-nightshade 91557
Common Endive 90894
Common Engelhardia 145517
Common English Ivy 187222
Common Enkianthus 145714
Common Epidendrum 146378
Common Epilasia 146573
Common Epipactis 147149
Common Epipogium 147323
Common Epipremnum 147349
Common Eremopyrum 148420
Common Eremostachys 148497
Common Eriachne 148818
Common Ethulia 155365
Common Eulalia 156499
Common Eulaliopsis 156513
Common Eurya 160503
Common Euscaphis 160954
Common Eustachys 160993

Common Evening Primrose 269404,269522
Common Evening-primrose 269404,269510
Common Everlasting 178358
Common Eversmannia 161295
Common Evolvulus 161418
Common Exbucklandia 161607
Common Fagraea 162346
Common Fairybells 134467
Common Fairyvine 164458
Common Falcaria 162465
Common False Fairybells 134404
Common False Flax 68860
Common False Foxglove 10157
Common Fargesia 162656
Common Feather Cockscomb 80395
Common Featherlip Orchis 274488
Common Fedtschenkiella 163089
Common Felwort 235435
Common Fennel 167156
Common Fibraurea 164458
Common Fiddleneck 20840
Common Field Melilot 249232
Common Field Speedwell 407287
Common Field-speedwell 407287
Common Fig 164763
Common Figwort 355202
Common Filbert 106703
Common Finger-lime 253285
Common Fish Geranium 288297
Common Fishhook Cactus 244243
Common Fishtail Palm 78052
Common Fishtailpalm 78052
Common Flat Pra 302850
Common Flat-topped Goldenrod 161081
Common Flax 232001
Common Fleabane 150862,321554
Common Flowering Quince 84573,84583
Common Floweringquince 84573
Common Fluffweed 166038
Common Fog Fruit 296121
Common Footstyle-orchis 125535
Common Fordia 167284
Common Forget-me-not 260868
Common Forkstamenflower 416508
Common Four-o'clock 255711
Common Fox Sedge 76375
Common Foxglove 130383
Common Foxtail 17521
Common Freesia 168162,168183
Common Fringe-myrtle 68761
Common Fritillary 168476
Common Frog Fruit 296121
Common Frogbit 200230
Common Fuchsia 168750
Common Fumitory 169126
Common Furrowlemma 25343
Common Gaillardia 169575
Common Gapfruityam 351383
Common Garcinia 171197
Common Garden Canna 71173
Common Garden Cosmos 107161
Common Garden Fuchsia 168750
Common Garden Honeysuckle 235878
Common Garden Parsley 292694,292698
Common Garden Peony 280213
Common Garden Petunia 292745
Common Garden Sunflower 188908
Common Garden Thyme 391397
Common Garden Tulip 400162
Common Garden Verbena 405852
Common Garden Vervain 405852
Common Gardenia 171253
Common Garishtaro 65218
Common Gaultheria 172100
Common Gendarussa 172832
Common Geranium 288297
Common German Flag 208583
Common Giant Fennel 163590
Common Ginger 418010
Common Ginger Lily 187432
Common Gladiolus 176122
Common Glasswort 342859
Common Globe Amaranth 179236
Common Globe Flower 399504
Common Globe Thistle 140789
Common Globe-amaranth 179236
Common Globedaisy 177053
Common Globeflower 399504
Common Globethistle 140789
Common Globe-thistle 140789
Common Glochidion 177174
Common Gloxinia 364754
Common Goat's-roe 169935
Common Goatsrue 169935
Common Golden Alexanders 418109
Common Golden Star 55648
Common Goldenrod 368013,368035,
368073,368268,368480
Common Goldenthistle 354651
Common Goldentop 161081
Common Goldfields 222607
Common Gold-star 202870
Common Goosecomb 144353
Common Gorse 203789,401388
Common Granadilla 285694
Common Grape 411979
Common Grape Hyacinth 260301
Common Grape-hyacinth 260301
Common Grassleaf orchis 12445
Common Great Angelica 24303
Common Greenbrier 366554,366559
Common Greenhoods 320877
Common Greenorchid 137651
Common Greentwig 352438
Common Grey Twig 352438
Common Greytwig 352438
Common Grey-twig 352438
Common Gromwell 233760
Common Ground Cherry 297694
Common Ground Ivy 176824
Common Ground-ivy 176824
Common Groundnut 29298
Common Groundsel 360328
Common Guarri 155977
Common Guava 318742
Common Guava goyave 318742
Common Guihaiothamnus 181807
Common Gum Cistus 93161
Common Gurjun 133588
Common Gurjun Oil Tree 133588
Common Gurjunoiltree 133588
Common Gurjun-oiltree 133588
Common Gypsophila 183191
Common Gypsyweed 407256,407257
Common Gypsy-weed 407256
Common Haberlea 184225
Common Hackberry 80698
Common Halogeton 184951
Common Hardgrass 354403
Common Hawkbit 224731
Common Hawksbeard 110785
Common Hawkweed 195708,196093
Common Hawthorn 109857
Common Hazel 106703
Common Heath 146071
Common Heather 67456
Common Heathgrass 122210
Common Hedgenettle 53304
Common Hedge-nettle 373459
Common Hedgeparsley 393004
Common Hedyotis 187545
Common Heliotrope 190559,190702
Common Hemlock 399867
Common Hemp 71218
Common Hemp Nettle 170078
Common Hemp-nettle 170060,170078
Common Henbane 201389
Common Hepatica 192134
Common Hepauca 192135
Common Herminium 192837,192862
Common Heron's Bill 153914
Common Hetaeria 332337
Common Hoarhound 245770
Common Hog's-fennel 292957
Common Holarrhena 197199
Common Holly 203545
Common Honey Locust 176903
Common Honeylocust 176901,176903
Common Honey-locust 176903
Common Honeysuckle 236022
Common Hoodshaped Orchid 264779

Common Hoofcelery 129196
Common Hop 199384,199387,199389,199390
Common Hop Sedge 75233
Common Hop Tree 320071
Common Hops 199384
Common Hoptree 320071
Common Hop-tree 320071
Common Horaninowia 198238
Common Horehound 245770
Common Hornbeam 77256
Common Hornwort 83545
Common Horse Chestnut 9701
Common Horse Gentian 397844
Common Horsechestnut 9701
Common Horse-chestnut 9701
Common Houlletia 198685
Common Hound's Tongue 118008
Common Hound's-tongue 118008
Common Houseleek 358062
Common Huntleya 199444
Common Hyacinth 199583
Common Hyacinthus 199583
Common Hydrangea 199956
Common Hyssop 203095
Common Iceplant 251265
Common Ice-plant 251265
Common Ichnocarpus 203336
Common Immortal 415317
Common Immortelle 415317
Common Indian Mulberry 258923
Common Indian-mulberry 258923
Common Iris 208583
Common Ironweed 406325,406631
Common Ivory Palm 298051
Common Ivy 187222
Common Jack Bean 71042
Common Jack-bean 71042
Common Jacob's-rod 39438
Common Jasmine 211940
Common Jasminefruit 283993
Common Jasminorange 260173
Common Jasmin-orange 260173
Common Javatea 275810
Common Jerusalem Sage 295111
Common Jimmyweed 209569
Common Jointfir 178549
Common Jujuba 418169
Common Jujube 418173
Common Juniper 213702,213706
Common Jurinea 211848,214121
Common Kadsura 214972
Common Kidneyvetch 28200
Common Kidney-vetch 28200
Common Kingidium 293585
Common Knapweed 81237
Common Knight's Star 196456
Common Knot-grass 308816
Common Knotweed 308816,308827,309192,309602
Common Knotweedtree 397932
Common Knotweed-tree 397932
Common Kochia 217361
Common Kopsia 217867
Common Kylinleaf 147349
Common Laburnum 218761
Common Ladies' Tresses 372192
Common Ladymantle 14065
Common Lady's-mantle 14166
Common Lady's-tresses 372259
Common Lagedium 219806
Common Lagenophora 219900
Common Lake Sedge 75034
Common Lamb's-quarters 86901
Common Lamiophlomis 220336
Common Lantana 221238
Common Larch 221855
Common Latouchea 222872
Common Laurel 223116
Common Laurel Cherry 223116
Common Laurel-cherry 223116
Common Lavender 223251
Common Lavender-cotton 346250
Common Leaf-flower 296801
Common Leaflessorchis 29216
Common Lecanthus 223679
Common Leibnitzia 224110
Common Lentil 224473
Common Leopardbane 34657
Common Lepidagathis 225179
Common Lepironia 225661
Common Leptopyrum 226341
Common Lespedeza 218347
Common Leucanthemella 227453
Common Leucomeris 178749
Common Leucosceptrum 228001
Common Leymus 228384
Common Licorice 177893,177897
Common Lignum Vitae 181505
Common Lilac 382329
Common Limaciopsis 230090
Common Lime 93329,391706
Common Linaria 231183
Common Linden 391706
Common Liolive 231753
Common Listera 264727
Common Locust 176903,334976
Common Loeseneriella 235224
Common Loosestrife 240068
Common Lophather 236284
Common Lousewort 287063
Common Lungwort 321643
Common Lyonia 239391
Common Macaranga 240325
Common Machilus 240707
Common Macrocarpium 105146
Common Macropodium 241249
Common Madder 338038
Common Madia 241528
Common Malleola 243289
Common Mallow 243823,243840
Common Mango 244397
Common Manzanita 31123
Common Maoutia 244893
Common Maple 2837
Common Mappianthus 244985
Common Mare's-tail 196818
Common Marigold 66445,383090
Common Marjoram 274237
Common Marsh Bedstraw 170529
Common Marsh Mallow 18173
Common Marsh Marigold 68189
Common Marsh Water-cress 336250
Common Marsh-bedstraw 170529
Common Marshmallow 18173
Common Marshmarigold 68189
Common Marsh-pink 341466
Common Matrimony Vine 239055
Common Matrimonyvine 239046
Common Mattiastrum 246578
Common May Apple 306636
Common Meadow Buttercup 325518
Common Meadow Grass 305855
Common Meadow-rue 388506
Common Mecopus 247205
Common Medinilla 247579
Common Mediterranean Grass 351167
Common Megacodon 247844
Common Melanolepis 248500
Common Melanosciadium 248543
Common Melastoma 248732
Common Melhania 248857
Common Melicope 249161
Common Mermaid-weed 315540
Common Mesquite 315553,315557,315560
Common Mesua 252370
Common Michaelmas-daisy 41190
Common Micrechites 203336
Common Microtoena 254252
Common Mignonette 327896,327904
Common Milfoil 3978
Common Milkpea 169660
Common Milkweed 38135,159971
Common Milkwort 308357,308454
Common Millet 281916,361794
Common Miniwhitedaisy 227453
Common Mistletoe 410960
Common Miterwort 256015
Common Mitragyna 256122
Common Mock Orange 294426
Common Monkey Earrings Pea 301128,

301164
Common Monkey Flower 255210
Common Monkeyflower 255210,255218
Common Monkshood 5100
Common Moonflower 67736
Common Moonseed 250221
Common Moonwort 238371
Common Morina 258859
Common Morning Glory 208120
Common Morning-glory 208120
Common Motherwort 224976
Common Mountain Mint 321964
Common Mouse Ear 82826
Common Mouse Ear Chickweed 82758
Common Mouse-ear 82818,82826
Common Mouse-ear Chickweed 82818
Common Mouseear Cress 30146
Common Mouse-eared Chickweed 82818
Common Mugwort 36474
Common Mulberry 259180
Common Mulgedium 259772
Common Mullein 405788
Common Muntingia 259910
Common Muskmelon 114201
Common Mustard 364557
Common Myoporum 260711
Common Myripnois 261399
Common Myrrh 101474
Common Myrrh Tree 101474
Common Myrtle 261739
Common Nabalus 261926
Common Nandina 262189
Common Nannoglottis 262218
Common Nasturtium 399601
Common Natsiatum 262787
Common Needlegrass 329049
Common Needle-grass 329049
Common Neomicrocalamus 324933
Common Nepenthes 264846
Common Net Brush 68064
Common Nettle 402886
Common Newlitse 264021
Common Night-blooming Cereus 200713
Common Nightblooming-cereus 200713
Common Nightshade 367416
Common Ninebark 297842
Common Nipplewort 221784,221787
Common Nobillorchis 197447
Common Nomocharis 266570
Common Nonea 266597
Common Nospurflower 374076
Common Nucleonwood 291486
Common Nutmeg 261424
Common Nut-meg 261424
Common Oak 324335
Common Oak Sedge 75744
Common Oat 45566
Common Oleander 265327
Common Oligochaeta 270301
Common Olive 270099
Common Onion 15165
Common Ophiorrhiza 272260
Common Opilia 272555
Common Opithandra 272601
Common Orache 44576
Common Orange Daylily 191284,191304
Common Orange-lily 229766
Common Orchardgrass 121233
Common Oreorchis 274058
Common Origanum 274237
Common Ormocarpum 274343
Common Oscularis 251194
Common Osier 344244
Common Otophora 277309
Common Ottochloa 277430
Common Owl's-clover 275499
Common Oxmuscle 129026
Common Oxwood 110251
Common Oxygraphis 278473
Common Oxyspora 278602
Common Pachygone 230090
Common Padritree 376266
Common Paedicalyx 280138
Common Panicgrass 281438
Common Panisea 282417
Common Papaw 38338
Common Paper Mulberry 61107
Common Paper-mulberry 61107
Common Papposedge 375809
Common Paraprenanthes 283695
Common Parastyrax 283993
Common Parthenium 285086
Common Passion Flower 285623
Common Passionflower 285623
Common Pawpaw 38338
Common Peach 20935
Common Peach-seven 365009
Common Pear 323133
Common Pearl Bush 161749
Common Pearlbush 161749,161755
Common Pearl-bush 161749
Common Pearleverlasting 21596
Common Pearly Everlasting 21596
Common Pecteilis 286836
Common Pedilanthus 287853
Common Peganum 287978
Common Pemphis 288927
Common Pencil-wood 139644
Common Pendentlamp 84287
Common Pennywort 200388
Common Penstemon 289327
Common Pentapanax 289632
Common Pentaphylax 289702
Common Penthea 290118
Common Peony 280213,280253
Common Peppergrass 225475
Common Perennial Gaillardia 169575,169586
Common Perilla 290940
Common Periwinkle 409335,409339
Common Perotis 291221
Common Persicaria 309570
Common Persimmon 132466
Common Petrocodon 292537
Common Petrocosmea 292554
Common Petunia 292742,292745
Common Phacelia 293162
Common Phaius 293537
Common Philydrum 294895
Common Phreatia 295958
Common Phylacium 296130
Common Phyllanthodendron 296445
Common Physochlaina 297882
Common Picria 298527
Common Pileostegia 299115
Common Pimpernel 21339
Common Pipewort 151232
Common Pipsissewa 87498
Common Piptadenia 300621
Common Piptanthus 300667
Common Pistache 301005
Common Pitcher Plant 264846,347156
Common Pitcherplant 347156
Common Pitcher-plant 347156
Common Plantain 302068
Common Pleione 304218
Common Plum 316382
Common Plumegrass 148906
Common Poacynum 29506
Common Pockberry 298094
Common Podocarpium 200739
Common Poinsettia 159675
Common Poison Bush 4968
Common Poison Ivy 393470
Common Poison-ivy 393470
Common Pokeberry 298094
Common Poker-plant 217055
Common Polemonium 307197
Common Pomatosace 310807
Common Pomegranate 321764
Common Pondweed 312188,312190
Common Poolmat 417053
Common Poppy 282685,282713,282717
Common Portulaca 311852
Common Pratia 313564
Common Pricklegrass 113306
Common Prickly Ash 417157,417231
Common Prickly Pear 272987,273103
Common Pricklyash 417157
Common Prickly-ash 417157
Common Pricklypear 272987
Common Pricklythrift 2211

Common Prickly-thrift 2211
Common Primrose 315101
Common Princess Palm 129684
Common Privet 229569,229662
Common Pterocypsela 320515
Common Pulai 18032
Common Purple Beech 162415
Common Purslane 311817,311890
Common Pussytoes 26385
Common Pygmyweed 391923
Common Pyrethrum 322665,322669
Common Pyrola 322815
Common Qiongzhu 87555
Common Quaking Grass 60382
Common Quaking-grass 60382
Common Quick-weed 170146
Common Quietvein 244985
Common Quince 116532,116546
Common Rabdosia 209625,303129
Common Ragweed 19145
Common Ragwort 290821,359158
Common Ramping-fumitory 169111
Common Raphistemma 326801
Common Raspberry 338557
Common Rathairdaisy 146573
Common Rattan Palm 65769,65782
Common Rattle 329521
Common Red Clover 397019
Common Red Currant 334201
Common Reed 295881,295888
Common Reed Grass 295888
Common Reevesia 327371
Common Renanthera 327688
Common Restharrow 271563
Common Rhododendron 331520
Common Ricegrass 300733
Common Rock Pueslane 65843
Common Rockrose 188757
Common Rock-rose 188610,188757
Common Rockvine 387822
Common Rockycabbage 394838
Common Roegneria 144353
Common Rose Mallow 195032
Common Rosemary 337180
Common Rue 341064
Common Rush 213021,213036,213066
Common Russian Thistle 344585
Common Russianthistle 344496
Common Rye-grass 235334
Common Ryssopterys 341226,341227
Common Saddleorchis 2081
Common Sage 345271,345310
Common Sagebrush 35090,36400
Common Sago Palm 252636
Common Sainfoin 271280
Common Salttree 184838
Common Saltwort 344585
Common Sand Aster 104652
Common Sandbur 80821
Common Sandspike 148497
Common Sandwhip 317004
Common Sanvitalia 346304
Common Saraca 346504
Common Sassafras 347407,347413
Common Satin Grass 259660
Common Saurauia 347993
Common Sauropus 348041
Common Saussurea 348146
Common Sawgrass 245556
Common Sawwort 361011
Common Scaletail 225648
Common Schefflera 350706
Common Schizocapsa 351383
Common Schizomussaenda 351741
Common Schizonepeta 265001
Common Schizostachyum 351862
Common Schoenorchis 352303
Common Schumannia 352703
Common Sciaphila 352875
Common Scopolia 25439,25442
Common Scrambling Bamboo 131288
Common Screw Pine 281166
Common Screwpine 281166
Common Screw-pine 281166
Common Screwtree 190108
Common Scurvy Weed 98020
Common Scurvygrass 98020
Common Scurvy-grass 98020
Common Scurvyweed 98020
Common Scyphiphora 355909
Common Sea Buckthoru 196757
Common Seagrape 97865
Common Sea-lavender 230815
Common Seamulberry 368915
Common Secamone 356279
Common Securidaca 356367
Common Sedge 74155,75508
Common Seepweed 379531
Common Sehima 357365
Common Selfheal 316127
Common Self-heal 316127
Common Sensitivebrier 352637
Common Seradella 274985
Common Serapias 360591
Common Serpentroot 354813
Common Serradella 274895
Common Serviceberry 19243
Common Sesbania 361370
Common Sevenseedflower 192168
Common Shamrockpea 284650
Common Sheep Sorrel 339897
Common Shuteria 362298
Common Shuttleleaffruit 139782
Common Silvianthus 364339
Common Sinia 364733
Common Sinodielsia 247716
Common Sinojohnstonia 364966
Common Sinolimptichtia 364970
Common Skapanthus 365905
Common Skeletonweed 88796
Common Skullcap 355449
Common Small Reed 127197
Common Smartweed 309199,309542
Common Smithorchis 366725
Common Smoke Bush 107303
Common Smoke Tree 107300,107303
Common Smoketree 107300
Common Smoke-tree 107300
Common Snail-seed Pondweed 312091
Common Snapdragon 28617
Common Sneezeweed 188402
Common Snowball 407989
Common Snowberry 380736,380749
Common Snowdrop 169719
Common Soapwort 346434,346436
Common Solisia 368526
Common Solomon's-seal 308529
Common Sonneratia 368915
Common Sorrel 339887
Common Souliea 369895
Common Sow Thistle 368771
Common Sowthistle 368771
Common Sow-thistle 368771
Common Spatterdock 267276
Common Speedwell 407014,407256
Common Speirantha 370490
Common Spenceria 370503
Common Spiderflower 182915
Common Spiderlily 200941
Common Spiderwort 394053,394088
Common Spike Sedge 143271
Common Spike-rush 143271
Common Spikesedge 143123,143391
Common Spike-sedge 143271
Common Spikeweed 81832
Common Spinach 371713
Common Spindle 157429
Common Spindle-tree 157429
Common Spirorrhynchus 372335
Common Splitmelon 351762
Common Spoongrass 222872
Common Spotted Orchid 121382
Common Spruce 298191
Common Squill 353057
Common St. John's Wort 202086
Common St. Johnswort 202086
Common St. John's-wort 202086
Common St. Paulswort 363090
Common Stapfiophyton 374076
Common Star-grass 202870
Common Star-of-bethlehem 274823

Common Starwort 67393,374897
Common Stelmatocrypton 375237
Common Stenocoelium 375552
Common Stephanachne 375809
Common Sterculia 376144
Common Sterigmostemum 273971
Common Stictocardia 376536
Common Stiff Sedge 76532
Common Stimpsonia 376683
Common Stinging Nettle 402886
Common Stink Dragon 137747
Common Stitchwort 374897
Common Stock 246478
Common Stonebutterfly 292554
Common Stonecrop 200784,356468
Common Stork's Bill 153764
Common Stork's-bill 153767
Common Strawberry 167597,167663
Common Strawberry Tree 30888
Common Streptanthera 377626
Common Strophioblachia 378474
Common Sulphur Flower 152550
Common Sumac 332608
Common Sumbaviopsis 379748
Common Summer Cypres 217373
Common Sunflower 188908
Common Sunrose 188757
Common Sun-rose 188757
Common Suoluo 327371
Common Sweet Shrub 68313
Common Sweetleaf 381423
Common Sweetshrub 68313,68314
Common Sweet-shrub 68313
Common Swisscentaury 375274
Common Swollennoded Cane 87555
Common Symphyllocarpus 380792
Common Syncalathium 381648
Common Tabacco 266060
Common Tabog 84983
Common Tall Meliot 249205
Common Tamarisk 383491
Common Tansy 383874
Common Tare 408571
Common Tea 69634
Common Teak 385531
Common Teasel 133478,133482,133517
Common Teatree 226471,226472
Common Teramnus 386410
Common Tetraplasia 387487
Common Thellungiella 389200
Common Theropogon 389568
Common Thistle 92485
Common Thoracostachyum 390378
Common Thorn Apple 123077
Common Threebibachene 397964
Common Threelobed Hepatica 192134
Common Three-square 352249
Common Three-square Bulrush 352249
Common Threewingnut 398322
Common Thrift 34536
Common Thrixsperm 390498
Common Thrixspermum 390498
Common Thunia 390917
Common Thyme 391397
Common Thyrsia 391445
Common Tiger Flower 391627
Common Tigerflower 391627
Common Tigerheadthistle Schmalhausenia 352006
Common Tigernanmu 122679
Common Timothy 294992
Common Toadflex 231183
Common Tomato 239157
Common Torulinium 393194
Common Tree Daisy 270186
Common Tricalysia 133185
Common Trichilia 395480
Common Trichurus 396492
Common Tridenworth 394838
Common Triplostegia 398024
Common Tripterospermum 398264
Common Triribmelon 141462
Common Triteleia 398806
Common Triumfetta 399312
Common Trumpet Creeper 70512
Common Trumpetcreeper 70512
Common Trumpet-creeper 70512
Common Ttule 352165
Common Tubeorchis 106883
Common Tulip 400162
Common Tupidanthus 400377
Common Turczaninovia 400467
Common Turmeric 114871
Common Tussock Sedge 74435,76399
Common Tutcheria 400722
Common Twayblade 232950
Common Umbrella Bamboo 162752
Common Uncifera 401796
Common Unicorn Flower 315434
Common Valerian 404316
Common Vanilla 404990,405017
Common Varrea 413246
Common Velvet Grass 197301
Common Velvetgrass 197301
Common Velvet-grass 197301
Common Verbena 405872
Common Vervain 405872
Common Vetch 408352,408571,408578
Common Vetchling 222820,222828
Common Vine 411979
Common Vineclethra 95543
Common Violet 410320,410351,410587,410730
Common Violet Cuckoo's Shoes 410493
Common Viper's Bugloss 141347
Common Viper's-bugloss 141347
Common Virginia Creeper 285127
Common Vladimiria 135739
Common Wall Cress 30146
Common Wallflower 86427
Common Walnut 212636
Common Waltheria 413149
Common Washington Geranium 288206
Common Water Hemlock 90924
Common Water Hyacinth 141808
Common Water Milfoil 261360
Common Water-crowfoot 325580
Common Water-dropwort 278559
Common Waterhawthorn 29673
Common Water-hemlock 90924
Common Water-horehound 239186
Common Waterhyacinth 141808
Common Water-hyacinth 141808
Common Water-lily 267294,267730
Common Water-meal 414659
Common Water-milfoil 261360
Common Water-plantain 14760,14785
Common Watershield 59148
Common Waterstarwort 67376
Common Water-starwort 67376,67389,67393
Common Waterweed 141569,143920
Common Wax Plant 198827
Common Waxplant 198827
Common Wax-plant 198827
Common Webseedvine 235224
Common Wendlandia 413842
Common Wheat 398839
Common White Jasmine 211940
Common White Quebracho 39700
Common White Snakeroot 11154
Common Whitebeam 369322
Common Whitlow-grass 153964
Common Whitlow-grass 137205
Common Wild Lupine 238473
Common Wild Sorghum 369597
Common Wildmedia 28997
Common Wild-oat-grass 122292
Common Wildorchis 274058
Common Willow 344244
Common Windmilltung 147356
Common Window Orchid 113742
Common Winter Aconite 148104
Common Winter Cress 47964
Common Winterberry 204376
Common Wintercreeper 157515
Common Winter-creeper Euonymus 157515
Common Wintergreen 172135,322853
Common Witch Grass 281438
Common Witch Hazel 185141
Common Witchgrass 281438
Common Witch-grass 281438

Common Witchhazel 185141
Common Woadwaxen 173082
Common Woad-waxen 173082
Common Wombgrass 365905
Common Wood Rush 238647
Common Wood Sedge 73887
Common Wood Sorrel 277973
Common Wood-reed 91236
Common Woodruff 170477
Common Wood-rush 238583
Common Wood-sorrel 277994
Common Woolly Sunflower 152809
Common Woollybush 7200
Common Wormwood 35090,36474
Common Wrightia 414813
Common Xanthopappus 415132
Common Xylocarpus 415716
Common Xylopia 415826
Common Yam 131501,131772
Common Yarrow 3978,3981
Common Yellow Azalea 331161
Common Yellow Flax 231923,231924
Common Yellow Lake Sedge 76653
Common Yellow Loosestrife 239898
Common Yellow Oxalis 278099
Common Yellow Stonecrop 356468
Common Yellow Trefoil 396882
Common Yellow-cress 336250
Common Yellowflax 327553
Common Yellow-flax 327553
Common Yellowwort 88248
Common Yellow-wort 54896
Common Yew 385302
Common Zanonia 417081
Common Zenia 417574
Common Zinnia 418043
Commonage Eargrass 187545
Common-cudweed 155874
Comose Bauhinia 49056
Comose Desertcandle 148540
Comose Grepe-hyacinth 260308
Comose Skullcap 355416
Compact Asparagus-fren 39149
Compact Begonia 49730
Compact Brome 60820
Compact Ceratoides 218124
Compact Chickensage 371146
Compact cobwebby Thistle 92254
Compact Cranberry Bush 407991
Compact Cryptomeria 113690
Compact Deutzia 126871
Compact Dodder 114996
Compact Eastern Arborvitae 390593
Compact Edelweiss 224848
Compact Eutrema 161132
Compact Evodia 161320
Compact Fargesia 162750
Compact Gelidocalamus 172747
Compact Golden-striped Bamboo 304011
Compact Hornsage 205554
Compact Incarvillea 205554
Compact Innocence 263414
Compact jatropha 212160
Compact Lemoine Deutzia 126990
Compact Onion 15876
Compact Oregon Grape 242480
Compact Pee Gee Hydrangea 200048
Compact Rhododendron 330447
Compact Rockspray 107490
Compact Rush 213021
Compact Sawgrass 245376
Compact Sedge 76464
Compact White Fir 319
Compact White-striped Bamboo 304109
Compact Winged Euonymus 157286
Compacte Globose Zo-asa 209062
Compacte Strawberry Bush 30889
Compact-flower Anzacwood 310774
Compactleaf Dandelion 384501
Compass Barrel 163483
Compass Barrel Cactus 163432
Compass Lettuce 219507
Compass Plant 219507,337180,364255,
364290
Compass-plant 219497,364290,364301
Compass-weed 337180
Comper Centaurea 81010
Complanate Phyllophyton 245691
Completcalyx Gentian 173586
Complete Zornia 418339
Complex Rhododendron 330449
Composite Dragonhead 137693
Composite Family 101642
Composite Greenorchid 137693
Composite Henry Monkshood 5269
Composite Oplismenus 272625
Composite Sedge 74162
Composite Spike Sawgrass 245585
Composites 101642
Compound-serrated Grewia 180748
Compressed Pipewort 151276
Compress Cypressgrass 118642
Compress Galingale 118642
Compress Parasite 293190,293191
Compressed Bambusa 47234
Compressed Common Juniper 213703
Compressed Fargesia 162660,162661
Compressed Fig 165126
Compressed Flat-sedge 118642
Compressed Milkvetch 42211
Compressed Pincution 244037
Compressed Rush 213008
Compressed Scales Rhododendron 330150
Compressed Sedge 73743
Compressed Spike-rush 143096
Compte Lyonia 239376
Compte Spapweed 204865
Comtie 417006
Comwood 320330
Cona Stonecrop 357256
Conandron 101680
Concave Aspidopterys 39670
Concave Chinese Privet 229622
Concave Eria 299617
Concave Hairorchis 299617
Concave Leaf Aster 41148
Concaveleaf Aster 41148
Concaveleaf Fissistigma 166685
Concaveleaf Magnolia 242234
Concaveleaf Nicolsonia 265997
Concave-leaved Fissistigma 166685
Concave-leaved Stachyurus 373568
Concavelip Addermonth Orchid 110641
Concavelip Bogorchis 110641
Concavelip Liparis 232200
Concavelip Pholidota 295520
Concavelip Twayblade 232200
Concavelip Woodbetony 287128
Concavenerve Raspberry 338615
Concavepetal Ancylostemon 22159
Concavepetal Parnassia 284575
Concavepetal Rockfoil 349669
Concavevein Bladderwort 403861
Concavevein Maesa 241752
Concavevein Sterculia 376125
Concavevein Tea 69147
Concave-veined Aucuba 44893
Concave-veined Blueberry 403861
Concave-veined Maesa 241752
Concealed Maclurolyra 240857
Concentrate Neoregelia 264412
Concentric Spapweed 205015
Concentric Touch-me-not 205015
Conch 144624
Conchilinque 244185
Conchita 81872
Concinnity Fargesia 162657
Concolor Actinidia 6672
Concolor Fir 317
Concolor Flickingeria 166960
Concolor Greenbrier 366340
Concolor Grewia 180738
Concolor Kiwifruit 6672
Concolor Oncidium 270801
Concolor Paphiopedilum 282806
Concolor Vancouveria 404614
Concolor Wintersweet 87526
Concolorleaved Falsenettle 56240
Concolorous Barberry 51486
Concolorous Napal Lily 229960
Concolorous Torenia 392895

Concolorous Vanda 404632
Concolour Grewia 180738
Condalia 101730
Condederate Rose 195040
Condensed Schizachyrium 351269
Condensed Thorowax 63611
Condlewood 167556
Condor Vine 245798
Condorvine 245777
Conduplicate Garnotia 171493
Conduplicate Gentian 173346
Condurango 245798
Cone Bush 227369,227401
Cone Flower 140066,339512
Cone Pepper 72075
Cone Plant 102213
Cone Redcedar 390645
Cone Redpepper 72073
Cone Sticks 292621,292625
Cone Wheat 398986
Conebearing Gomphostemma 179205
Conebearing Woodbetony 287106
Conebush 210234
Cone-dealapple 298191
Coned-seeded Jointfir 178528
Coneflower 140066,140081,326960,339512,339518,339574,339585
Cone-hair Wallrocket 133304
Conehead 47845,115794,323064,378079,378172
Conehead Thyme 390987
Conejo Buckwheat 151968
Conelike Redpepper 72073
Cone-like Silene 363368
Coneshape Tanoak 233368
Cone-shape Tanoak 233368
Coneshaped Fruit Hemsleia 191971
Coneshaped Galangal 17761
Cone-shaped Halocnemum 184933
Conespike Flemingia 166897
Cone-spiked Flemingia 166897
Conevine 39607,39608
Coney Parsley 28044
Coney's Parsley 28044
Confederate Daisy 189033
Confederate Jasmine 393657
Confederate Rose 195040
Confederate Vine 28422
Confederate Violet 410351,410439,410587
Confederate-jasmine 393657
Confederate-vine 28418
Confert Tickseed 104762
Conferted Flower Helixanthera 190852
Conferted Flower Mistletoe 411052
Conferted Jerusalemsage 295091
Conferted Marshweed 230288
Conferted Milkvetch 42216
Conferted Sawgrass 245376
Confertedflower Sarcococca 346725
Confertedfruit Dendropanax 125602
Confetti 86901
Confetti Bush 99467
Confetti Tree 246927
Confined Brome 60687
Confined Dimocarpus 131058
Confluent Woodbetony 287105
Confrte Girardinia 175889
Confuse Gastrodia 171913
Confuse Goosecomb 335278
Confuse Melicope 249166
Confuse Privet 229450
Confuse Roegneria 335278
Confuse Sweetleaf 381355
Confused Bridewort 371790
Confused Cactus 244042
Confused Calospatha 68015
Confused Canary-grass 293722
Confused Dendropanax 125602
Confused Eria 299603
Confused Fescue 164048
Confused Iris 208498
Confused Licuala Palm 228733
Confused Licualapalm 228733
Confused Mahonia 242522
Confused Michaelmas-daisy 40923
Confused Ochrosia 268356
Confused Oiltea Camellia 69414
Confused Privet 229450
Confused Storax 379327
Confused Swordflag 208498
Confused Yushania 416765
Confusing Trillium 397615
Cong 67867
Congaberry 76854
Congdon's Wild Buckwheat 151943
Congea 101774
Conger 114245
Congested Cotoneaster 107392
Congested Eria 299600
Congested Sagewort 35550
Congested Sedge 73614
Congested Snakeroot 11174
Congested Xylosma 415869
Congested-flowered Syzygium 382516
Congest-flowered Barberry 51492
Conghua Eurya 160546
Conghua Galangal 17658
Congling Dandelion 384769
Congling Sedge 73620
Conglobate Edelweiss 224822
Congo 114245
Congo Acacia 289439
Congo Bean 65146
Congo Coffee 98872,98896
Congo Copai 181771
Congo Goober 409069
Congo Jute 402245
Congo Pea 65146
Congo Rubber 165263
Congo Senna 78227
Congo Snake 346158
Congo Stick 78811
Congo Wood 237927
Congopea 65146
Congowood 121750
Congregate Hairorchis 299600
Congregate Primrose 314439
Congyuan Pepper 300537
Conic Gymnadenia 182230
Conic Sugar Maple 3556
Conical Arborvitae 302727
Conical Catchfly 363368
Conical Eastern Arborvitae 390594
Conical Greenbrier 366271
Conical Silene 363363
Conical Smilax 366271
Conical White Fir 324
Conical-beaked Cornera 104890
Conicalfruit Slatepentree 400728
Conicalfruit Ternstroemia 386692
Conicalfruit Tutcheria 400728
Conical-fruited Ternstroemia 386692
Conical-fruited Tutcheria 400728
Conifer Cotton-grass 152788
Conifer Saxifrage 349198
Conifer Willow 342927,343236
Conium 101845
Conker Tree 9701
Conkerberry 76854,336419
Conkers 9701,114245,302034,336419
Conks 9701
Conlike Woodbetony 287106
Connarus 101860
Connarus Family 101858
Connarus-like Trichilia 395480
Connate Beggarticks 53845
Connate Clematis 94853
Connate Gerbera 175139
Connate Hedyotis 187552
Connate Nepeta 264902
Connate Woodbetony 287107
Connatestyle Camellia 69007
Connaught Marsh Orchid 121389
Connectedfruit Stonecrop 356629
Connectedsepal Kalanchoe 215269
Connemara Heath 121066
Connivent Wing Mono Maple 3157
Conobea 228127
Conophor Nut 386838
Conquer-more 384714
Conqueror-flower 302034

Conquerors 9701
Conradina Cherry 83177
Conringia 102808
Consanfuinity Eargrass 187548
Consanfuinity Hedyotis 187548
Consanguineous Greenstar 307517
Consanguineous Maesa 241756
Consimilar Epipactis 147121
Consolation Flower 95431
Consolida 102827
Consound 13174,50825,381035
Consperseflower Ardisia 31386
Consperse-flowered Ardisia 31386
Conspicuous Rhododendron 330312
Conspicuous-veined Spatholobus 370420
Consprout 9489,9560
Constantinople Hazel 106724
Consumption Weed 46227
Consumption-weed 46227
Contaminated Cape-cowslip 218853
Continent Aralia 30618
Continent Calvepetal 177988
Continent Ervatamia 154152,382766
Continent Long-pedicellate Spindle-tree 177988
Continental Aralia 30618
Continental Cogflower 154152,382766
Continental Glyptopetalum 177988
Continental Longpedicel Holly 204147
Continuous Sedge 74183
Contorted Filbert 106703
Contorted Meadowrue 388460
Contorted Mulberry 259098
Contorted Tanglehead 194020
Contorted Tetracme 386949
Contorted Willow 343670
Contorted Yellowquitch 194020
Contra Costa Goldfields 222596
Contra Costa Jimmyweed 209567
Contra Hierba 136470
Contra Yerba 136470
Contract Cinnamon 91316
Contract Flower Cassia 91316
Contracte Bamboo 47234
Contracted Bentgrass 12237
Contracted Euonymus 157386
Contracted Ladybell 7628
Contracted Reed-grass 65438
Contracted Sawwort 361131
Contracted Spindle-tree 157386
Contract-flowered Cinnamon 91316
Contrayerva 34286,136470
Contrayerva Root 136470
Controcted Bentgrass 12310
Conval Lily 102867
Convallarialike Odontoglossum 269062
Convally 102867
Convex Gingerlily 187431
Convexa Japanese Holly 203672
Convexed Veins Rhododendron 330874
Convexhole Habenaria 183393
Convex-leaved Barberry 52131
Convex-nerved Pittosporum 301259
Convexpetal Ancylostemon 22163
Convextine Azalea 331851
Convexutricle Sedge 75769
Convexvein Actinidia 6521
Convexvein Holly 203779
Convexvein Kiwifruit 6521
Convexvein Trigonotis 397411
Convex-veined Holly 203779
Convex-veined Rhododendron 330874
Convey Redbud 83759
Convict Grass 81768
Convolule Square-bamboo 87557
Convolute Mahonia 242460
Convolute-leaved Rhododendron 331675
Convolvulate Asiabell 98292
Convolvulate Knotweed 162519
Convolvulus 102903,103340
Convulsion Root 258044
Conway Straight-raceme Barberry 51986
Conyza 103402
Conyzalike Windhairdaisy 348227
Conzatt Rattail Cactus 29714
Cooba 1528,1552,1553
Cooburn Eucalyptus 155499
Cook Pine 30838
Cooking Banana 260253
Cook's Scurvy Grass 225426
Cook's Scurvy-grass 225426
Cook's Triteleia 398798
Cooktown Orchid 125015
Cool Tankard 57101
Coolabah 155655
Coolabah Apple 24611
Coolabah Coolibah 155537
Coolabar Grass 147474
Coolamon 382616
Cooley's Meadow-rue 388462
Coolgardie Gum 155765
Coolibah 155655
Coolibah Coolabar 155655
Coolwort 391522
Coolwort Foamflower 391525
Coomail 83578
Coombs 350425
Coon Root 192148,345805
Coondi 72640
Coon's-tail 83545
Coontail 83540,83545,83558,261348
Coontie 417002,417006,417013,417017
Coonties 417013
Cooper Asparagus 38972
Cooper Barberry 51496
Cooper Begonia 49736
Cooper Diplazoptilon 132861
Cooper Haworthia 186354
Cooper Osmanther 276276
Cooper Osmanthus 276276
Cooper Rush 213026
Cooper's Bitterweed 201310
Cooper's Goldenbush 150315
Cooper's Hardy Ice Plant 123854
Cooper's Milk-vetch 42783
Cooper's Paperflower 318978
Cooper's Pondweed 312078
Cooper's Rrush 213026
Cooper's Rubberweed 201310
Coosil 400675
Cooslip 315082
Cootamundra Wattle 1069,1070,282546,282685
Cop Rose 282685
Copaiba 103719
Copaiba Balsam 103718,103719
Copaiba Copal Tree 103719
Copaiba Copal-tree 103719
Copaiba Oil 103718
Copaifera 103691,103696
Copal 64077,64080
Copal Tree 12559,103691
Copalehi Bark 112972
Copalquia 196419
Copalquin 279505
Copal-tree 12559,103691
Copao 157164
Copao De Philippi 157166
Copeland Addermonth Orchid 243028
Copeland Bogorchis 243028
Copenhagen Hawthorn 109609
Copernicia 103731
Copernicus Palm 103731
Copetal Corydalis 105987
Copey 97262
Copiapoa 103739
Copiapoa De Bridges 103740
Copiapoa De Carrizal 103763
Copiapoa De Philippi 103745
Copihue 221340,221341
Copper Beard 67542
Copper Beech 162404,162406,162415
Copper Canyon Daisy 383096
Copper False Chestnut 78903
Copper Iris 208578
Copper Knotweed 309474
Copper Leaf 1769,1931,2011
Copper Rose 282685
Copper Zyphyr-lily 417624
Copper-buff 354435
Copper-cash Wildginger 37608

Coppercolor Azalea 332078
Copperhammer and Jadebelt 313564
Copperhammwer grass 277776
Copperleaf 1769,1790,2011,18089
Copper-leaf 1769,1869,2011
Copper-lily 184257
Copperoze 282685
Copperpod 288897
Coppertip 111455,111458,111466
Coppertone Loquat 151162
Coppertone Sedum 356968
Copperweed 210463,278657
Copper-weed 278657
Coppery Mesemb 243253
Coppery Monkeytlower 255198
Coppery St. John's-wort 201833
Coppery-red Iris 208578
Coppice Rhododendron 331124
Coprosma 103782
Copse Bindweed 162528
Copse Laurel 122492
Copse-bindweed 162519,162528
Coptide 103824
Coptosapelta 103861
Coquiila Palm 44807
Coquimbano 103749
Coquimbo Barberry 51497
Coquina Nut 44807
Coquito 212557
Coquito Palm 212557
Coracan 143522
Corakan 143522
Coral 104009
Coral Aloe 17299
Coral Aphelandra 29126
Coral Ardisia 31396,31502
Coral Bark Japanese Maple 3328
Coral Beads 97894,97900
Coral Bean 154617,154632,154668,154734,369116
Coral Beauty 380749
Coral Beauty Cotoneaster 107417
Coral Begonia 49739
Coral Bells 70331,97757,130383,194411,194437,321721
Coral Berry 8562,31408,380749
Coral Bower 154617
Coral Broom 104029
Coral Bulbophyllum 62661
Coral Bush 212186,341020,385812
Coral Cactus 329692
Coral Creeper 28422,215978
Coral Drops 53213,53218
Coral Flower 194437
Coral Fountain 341020
Coral Gem 237509
Coral Gem Trefoil 237509
Coral Ginger 417974
Coral Gum 155765
Coral Heath 146073
Coral Hibiscus 195216
Coral Holly 203648
Coral Honeysuckle 236089
Coral Ixora 211113
Coral Lily 230005,230009,230019
Coral Moss 265356
Coral Necklace 204469
Coral Nettletree 80661
Coral Pea 765,7190,185764,215978
Coral Peadtree 7190
Coral Peony 280231
Coral Plant 51267,212186,334189,341020,385812
Coral Root 104009,104023,125836
Coral Rose 336918
Coral Sealavender 230586
Coral Spurge 158690
Coral Stonebean-orchis 62661
Coral Sumach 252595
Coral Tree 154617,154618,154620,154636,154643,154647,154734,212186
Coral tree 154632
Coral Vine 28418,28422,97900
Coral Winterberry 204379
Coral Zinger 417974
Coral-bark Maple 3328
Coral-bark Willow 342975,342982
Coral-bead Plant 765
Coral-beads 97894,97900
Coralbean 154617,154734,368958
Coral-bean 154617,154647
Coral-bean Tree 154643
Coralbells 194411
Coral-bells 194437
Coralbells Alumroot 194437
Coralberry 6405,8562,31396,31408,380734,380749
Coral-berry 31396,380749
Coralblow 341016
Coral-blow 341016
Coralbush 212186
Coral-drops 53213,53218
Coralflower 120525
Coralgreens 176921,176923
Coralhead Plant 765
Coralliformis Hinoke Cypress 85315
Coralline Honeysuckle 235730
Coralline Persimmon 132112
Corallita 28422
Corallodiscus 103937
Corallorhiza 103976
Coral-necklace 204443,204469
Coralroot 103976,125836
Coral-root 103976
Coralroot Bittercress 125836
Coralroot Orchid 103976,104019
Coral-root Orchid 104019
Coral-tree 154617,154643,154647
Coralvine 28418,28422
Coral-vine 28418
Coralwood 7190
Coralwort 125836
Coral-wort 125836
Coraseena 403916
Corazon-de-jesua 49845
Corbier Pyrola 322813
Corcassian Walnut 212636
Corchoropsis 104039
Cord Grass 370155
Cordare Halfummer 299719
Cordare Pinellia 299719
Cordate Arisaema 33299
Cordate Camellia 69010
Cordate Carpesium 77155
Cordate Circaea 91537
Cordate Dewdrograss 91537
Cordate Drymaria 138482
Cordate Esquirol Bredia 59842
Cordate Flemingia 166854
Cordate Japanese Staunton-vine 374428
Cordate Lepisanthes 225669
Cordate Odontoglossum 269063
Cordate Ornamental Rhododendron 330525
Cordate Panicgrass 281984
Cordate Pearleaf Raspberry 339068
Cordate Peppergrass 225333
Cordate Pepperweed 225333
Cordate Primrose 315036
Cordate Rhododendron 330150
Cordate Sepal Gaultheria 172061
Cordate Shortia 362255
Cordate Southstar 33299
Cordate Spapweed 204872
Cordate Stachyurus 373537
Cordate Sweetshell Azalea 330525
Cordate Telosma 385752
Cordate Touch-me-not 204872
Cordate Truncate-leaved Maple 3715
Cordate Urochloa 402506
Cordatelaef Primrose 314521
Cordate-leaed Maple 2901
Cordateleaf Ainsliaea 12616
Cordateleaf Bluebellflower 115345
Cordate-leaf Bredia 59842
Cordateleaf Butterflygrass 392899
Cordateleaf Cardioteucris 73221
Cordateleaf Caudate Wildginger 37581
Cordateleaf Cayratia 79840
Cordateleaf Ceratoides 218129
Cordateleaf Clearweed 298897
Cordate-leaf Cryptocoryne 113531

Cordateleaf Dumasia 138933
Cordateleaf Eight Trasure 200796
Cordate-leaf Fig 165602
Cordateleaf Greenvine 204615
Cordateleaf Hedge Sageretia 342203
Cordateleaf Illigera 204615
Cordateleaf Jasmine 211997
Cordate-leaf Jasmine 211960
Cordateleaf Ladybell 7629
Cordateleaf Leek 15559
Cordateleaf Lepisanthes 225669
Cordateleaf Lycianthes 238960
Cordateleaf Maple 2901
Cordateleaf Pleurospermum 304839
Cordateleaf Rockfoil 349152
Cordateleaf Shortia 362255
Cordateleaf Sida 362523
Cordateleaf Slatepentree 322588
Cordateleaf Sloanea 366046
Cordateleaf Snailseed 97902
Cordateleaf Snakegourd 396162
Cordateleaf Stonecrop 200796
Cordateleaf Strophioblachia 378477
Cordateleaf Torenia 392899
Cordateleaf Tubergourd 390119
Cordateleaf Tutcheria 322588
Cordateleaf Twayblade 232125
Cordateleaf Waxpetal 106660
Cordateleaf Winterhazel 106660
Cordateleaflet Asiabell 98308
Cordate-leaved Ceratoides 218129
Cordate-leaved Illigera 204615
Cordate-leaved Lepisanthes 225669
Cordate-leaved Mint 250468
Cordate-leaved Sida 362523
Cordate-leaved Sloanea 366046
Cordate-leaved Spring Beauty 94311
Cordgrass 370155
Cord-grass 370155,370166,370182
Cordia 104147
Cordifoliate Zygia 30963
Cordiform-based Rhododendron 331401
Cordipetal Catchfly 363297
Cord-root Sedge 74118
Cord-rooted Sedge 74118
Corduroy 113440
Cordyline 104334,104339
Cordyline Terminalis 278124
Corean Pine 300006
Corenate Sawwort 361023
Coreopsis 104429,104444,104509,104531,104598,104616
Coriaceous Bambusa 47236
Coriaceous Briggsia 60278
Coriaceous Cotoneaster 107411
Coriaceous Dutchman's-pipe 34323
Coriaceous Endiandra 145316
Coriaceous Euonymus 157678
Coriaceous Fleabane 150955
Coriaceous Galangal 17660
Coriaceous Glochidion 177117
Coriaceous Indian Madder 337932
Coriaceous Knotweed 309014
Coriaceous Leaf Cissampelopsis 92518
Coriaceous Phyllaries Groundsel 358606
Coriaceous Sinacalia 364518
Coriaceousleaf Actinidia 6693
Coriaceousleaf Briggsia 60278
Coriaceousleaf Dutchmanspipe 34323
Coriaceousleaf Glomerulate Brassaiopsis 59212
Coriaceousleaf Kiwifruit 6693
Coriaceousleaf Lagotis 220156
Coriaceousleaf Maple 2905
Coriaceousleaf Pimpinella 299386
Coriaceousleaf Sabia 341495
Coriaceousleaf Stringbush 414254
Coriaceousleaf Sweetspire 210378
Coriaceous-leaved Maple 2905
Coriaceous-leaved Sabia 341495
Coriaceous-leaved Stringbush 414254
Coriaceous-leaved Sweetspire 210378
Coriander 104687,104690
Coriandrifolious Cinquefoil 312470
Coriaria 104691
Coriaria Family 104710
Cori-leaf Rose 336504
Corispermumlike Trigonotis 397402
Cork Bark Fir 402
Cork Bark Rhododendron 332155
Cork Camellia 69486
Cork Elm 401430,401475,401549,401618,401633,401638
Cork Fir 402,403
Cork Lady's Tresses 372242
Cork Oak 324453,324532
Cork Tree 184640,268344,284248,294230,324453
Cork Wood 145991,184586,224276,261100
Corkbark Fir 402,403
Cork-bark Honey Myrtle 248120
Corkbark Tree 184614
Cork-barked Fir 403
Corkbarked Litse 234070
Cork-barked Litse 234070
Cork-barked Smooth-leaved Elm 401475
Corkbush 157285
Cork-bark Smooth-leaved Elm 401633
Corkleaf Snowbell 379457
Corkleaf Storax 379457
Corkleaf Wingseedtree 320849
Cork-leaved Actinidia 6706
Cork-petaled Brassaiopsis 59245
Cork's Foot 30081
Corkscrew Flower 378399
Corkscrew Grass 376687
Corkscrew Hazel 106703,106705,106759
Corkscrew Rush 213067
Corkscrew Willow 343668,343670
Corktree 157793,294230,389946
Cork-tree 294230,324453
Cork-wing Spindle-tree 157793
Corkwood 138744,145321,145990,145991,184614,224275,224276,260287,268344
Cork-wood 25852,145990,145991
Corkwood Duboisia 138744
Corkwood Family 224279
Corky Creeper 285151
Corky Macrosolen 241319
Corky Spindletree 157285
Corkyfruit Waterdropwort 269348
Corky-fruited Water-dropwort 269348
Corkyleaf Actinidia 6706
Corky-leaved Storax 379457
Corkystem Passionflower 285703
Corkywing Euonymus 157793
Corky-winged Euonymus 157793
Corme 369541
Cormeille 222758
Cormele 222758
Corn 417412,417417
Corn Bedstraw 170694
Corn Bellflower 224067
Corn Bind 102933
Corn Bindweed 162519
Corn Blinks 81020
Corn Bluebottle 81020
Corn Bottle 81020
Corn Brome 60977
Corn Bugloss 21920
Corn Bur Parsley 392968
Corn Buttercup 325612
Corn Campion 11947
Corn Carraway 292711
Corn Centaury 81020
Corn Chamomile 26746
Corn Champion 11947
Corn Chervil 350425
Corn Chrysanthemum 89704
Corn Cleavers 170694
Corn Cob Euphorbia 159305
Corn Cockle 11938,11947
Corn Corsican Hellebore 190929
Corn Crowfoot 325612
Corn Feverfew 397970
Corn Flag 175990,208771
Corn Flag Gladiolus 176281
Corn Flakes 417417
Corn Floor 417417
Corn Flower 81020,210816
Corn Gilliflower 224067

Corn Gold 89704
Corn Grass 28972
Corn Gromwell 233700
Corn Honewort 292711
Corn Kale 364557
Corn Lettuce 404439
Corn Lily 68713,97091,102933,210689,405581
Corn Marigold 89704
Corn Marigold of Candy 89481
Corn Mayweed 397970
Corn Mignonette 327904
Corn Mint 4666,250294,250362
Corn Mustard 364557
Corn of Candy Marigold 89481
Corn On The Cob 417417
Corn Panicgrass 140360
Corn Pansy 409703,410677
Corn Parsley 292711,397731
Corn Pheasant's Eye 8332
Corn Pimpernel 21339
Corn Pink 11947,224067
Corn Pinks 81020
Corn Plant 137367,137397
Corn Pop 364193
Corn Poppy 282546,282685
Corn Rocket 63443,63450
Corn Rose 11947,282685,336364
Corn Salad 404394,404439,404468,404484
Corn Silk 417417
Corn Snapdragon 28631
Corn Sow Thistle 368635
Corn Sow-thistle 368635
Corn Speedwell 407014,407144
Corn Spurry 370522
Corn Thistle 91770
Corn Violet 224067
Corn Woundwort 373125
Cornalee 380499
Cornberry 278366
Corn-bin 102933
Cornbind 102933
Corn-bind 102933,162519
Corn-bine 102933
Corn-bottle 81020
Corncalyx Elatostema 142860
Corncarp Begonia 49709
Corncockle 11938,11947
Corndaisy 89704
Cornel 104917,105111,380425,380499
Cornel Cherry 105111
Cornelia Tree 105111
Cornelian 380499
Cornelian Cherry 105111
Cornelian Cherry Dogwood 105111
Cornelian Dogwood 105111
Cornelian-cherry Dogwood 105111
Cornel-leaf Whitetop Aster 135257
Cornel-leaved Aster 40625
Cornel-leaved Whitetop Aster 135257
Cornemellagh 222758
Corneously-leaved Barberry 51442
Corner Eria 148656
Corner Licuala Palm 228734
Corner Licualapalm 228734
Corner Plectocomiopsis 303097
Corner Rattan Palm 65665
Cornera 104889
Corneredcalyx Larkspur 124116
Corneredcalyxlike Larkspur 124115
Cornets 271563
Cornfield Crowfoot 325612
Cornfield Knot-grass 308836
Cornflag 175990,176122
Corn-flag 175990
Cornflower 11947,80892,81020,81356,216753,282685
Cornflower Aster 377287
Cornflower Bachelor's Button 81020
Cornflower Centaurea 81020
Cornflower-aster 377287
Cornfruit Wild Saltbush 44401
Corn-gold 89704
Corn-grass 28972,128494
Corniculate Aegiceras 8646
Corniculate Albizia 13525
Corniculate Albizzia 13525
Corniculate Bamboo 47235
Corniculate Bambusa 47235
Corniculate Candlefruit 8646
Corniculate Cayratia 79841
Corniculate Clematis 94863
Corniculate Falsestairweed 223675
Corniculate Lecanthus 223675
Corniculate Siris 13525
Corniculate Spurgentian 184691
Cornifle 83540
Cornifle Nageante 83545
Cornifoliate Helicia 189922
Cornish Bellflower 69885
Cornish Elm 401485,401573,401631
Cornish Fumitory 169125
Cornish Fuzz 401390
Cornish Golden Elm 401469
Cornish Heath 150201
Cornish Heather 150201
Cornish Mallow 223364
Cornish Money Wort 362406
Cornish Moneywort 362402,362406
Cornish Smooth-leaved Elm 401485
Cornish Whitebeam 369429
Corn-leaf 401711
Corn-lily 97091,210689,405570
Corn-marigold 89704
Corn-on-the-cob 417417
Cornsalad 404394,404439,404454
Corn-salad 404394,404454
Cornt Violet 409834
Cornucopia Floripondia 123065
Cornudentate Ardisia 31387
Cornulaca 104909
Cornute Milkvetch 42221
Cornuted Pugionium 321481
Cornuted Sandcress 321481
Cornutia 105229
Cornviolet 224067
Cornwood 380499
Coroba 350986
Corokia 105240
Corokio 105239
Corolla Hairorchis 148658
Coromadel Coast Falsemallow 243893
Coromandel Ebony 132298
Coromandel Lannea 221144
Corona Regis 249232
Coronarious Gingerlily 187432
Coronate Mountainash 369371
Coronation 127635
Coronilla 105269
Corozo 6136
Corozo Nut Palm 298051
Corozo Palm 142264
Corpse Flower 222651
Corpse Plant 258044
Corpus Christi Fleabane 150890
Corr 222758
Corrameille 222758
Correa 105387
Correhuela 97904
Correll's Wild Buckwheat 151946
Corrietree Azalea 331640
Corrietree Rhododendron 331640
Corrie-tree Rhododendron 331640
Corrigiole 105410
Corringin Grevillea 180645
Corrot 123135,123141,123164
Corrugate Chinese Dregea 137796
Corrugate Dregea 137796
Corsica mint 250437
Corsican Heath 150127
Corsican Hellebore 190929,190942,190949
Corsican Lily 280896
Corsican Mint 250437
Corsican Pearlwort 32301,342298
Corsican Pine 300017,300114,300117,300120
Corsican Sandwort 31767
Corsican Speedwell 407309
Corsican Thyme 391201
Corsican Toadflax 116735
Corsiopsis 105448,105449

Corticeira-da-serra 154656
Cortiella 105481
Corts 123141
Cortusa 105488
Cortuseleaf Rockfoil 349204
Corulwort 125836
Corumb Rose 336419
Cory Barberry 51503
Cory Ephedra 146151
Cory Iris 208499
Corybas 105524
Corydalis 105557,105795,106387
Corydalisleaf Meadowrue 388539
Corydalis-like Woodbetony 287109
Corylopsis 106628
Corylus-leaf Hydrangeavine 351783
Corymb Ardisia 31389
Corymb Bistamengrass 127471
Corymb Bladderwort 403780
Corymb Corydalis 105758
Corymb Deutzia 126879
Corymb Everlasting 21555
Corymb Milklettuce 259776
Corymb Nepal Pearleverlasting 21644
Corymb Pearleverlasting 21555
Corymb Szechwan Deutzia 127094
Corymb Tansy 383857
Corymb Wood Sorrel 277776
Corymb Woodbetony 287110
Corymbate Osmanther 276277
Corymbate Osmanthus 276277
Corymbe Polycarpaea 307656
Corymbflower Tansy 383737
Corymblike Coelogyne 98633
Corymborkis 106871
Corymbose Abelia 119
Corymbose Barberry 51504
Corymbose Bauhinia 49058
Corymbose Beakrush 333530
Corymbose Coelogyne 98633
Corymbose Corydalis 105758
Corymbose Deutzia 126879
Corymbose Eargrass 187555
Corymbose Epaulettetree 320880
Corymbose Epaulette-tree 320880
Corymbose Hedyotis 187555
Corymbose Largeleaf Hydrangea 200014
Corymbose Larkspur 124147
Corymbose Mosquitotrap 117433
Corymbose Nivenia 266403
Corymbose Premna 313740
Corymbose Pyrethrum 322723
Corymbose Sichuan Deutzia 127094
Corymbose Squill 353001
Corymbose Swallowwort 117433
Corymbose Viburnum 407764
Corymbous Ardisia 31389
Corymbous Arrowwood 407764
Corymbtree 160720
Corypha 106984,106999
Coryphantha 107004
Cory's Chinese Privet 229623
Cory's Dutchman's-pipe 34156
Cory's Ephedra 146151
Cos Lettuce 219492
Coses Sciences 31051,292372,405788
Coslseleaf Milkvetch 43012
Cosmetic Bark 260173
Cosmopolite Ruellia 339731
Cosmos 107158,107161,107170,107172
Cosmostigma 107174
Cossack Asparagus 401112
Cossatot Mountain Leafcup 310028
Cost Rica Columnea 100115
Costa Pica 387932
Costa Rican Butterfly Vine 121921
Costa Rican Guava 318740,318747
Costa Rican Nightshade 367745
Costa Rican Weeping Bamboo 90631
Costamen Greenbrier 366307
Costantin Stonecrop 356638
Costarica Varrea 413244
Costate Camellia 69012
Costate Tea 69012
Coste Tea 69014
Costmary 383712
Costmary Chrysanthemum 383712
Costulate Barberry 51506
Costyle St. John's Wort 201818
Cosyr Statice 230588
Cotine 179032
Coto Bark 25198
Coto Nectandra 263052
Cotoneaster 107319,107321,107325,107766
Cotoneaster Anzacwood 310759
Cotswoid Penny Cress 390249
Cottage Pink 127635,127793
Cottage Rose 336343
Cottagers 130383
Cottam's Wild Buckwheat 151875
Cotton 179865,179900,179906
Cotton Abroma 675
Cotton Ambroma 675
Cotton Arrowbamboo 162692
Cotton Ball 155300
Cotton Batting Plant 178119
Cotton Bean 294010
Cotton Burdock 31068
Cotton Chickweed 374936
Cotton Daisy 80374
Cotton Deer-grass 396005
Cotton Flower 302103
Cotton Grass 152734,152769,152796,
205497
Cotton Gum 267850,267867
Cotton Hippolytia 196697
Cotton Lavender 346243,346250
Cotton Palm 413298,413307
Cotton Root 179900
Cotton Rose 195040
Cotton Saussurea 348344
Cotton Sedge 152734,152769,152791
Cotton Thistle 91954,271663,271666
Cotton Thorn 387085
Cotton Tree 56802,80120
Cotton Weed 276917
Cotton Willow 343602
Cotton Wood 311131,311292,311554
Cotton Wool Grass 205497
Cottonbatting Cudweed 178119
Cotton-batting-plant 317772
Cottondaisy 293408
Cottonflower 179850
Cottonfruit Milkvetch 43046
Cotton-grass 152734,152765,152769
Cotton-grass Sedge 152788
Cottonhead Argyreia 32621
Cottonhead Windhairdaisy 348294,348465
Cottonheads 263283
Cottonleaf Physic 212144
Cotton-leaved Nettle Spurge 212144
Cottonrose 195040,235253
Cottonrose Hibiscus 195040
Cotton-rose Hibiscus 195040
Cottonsedge 152734
Cotton-sedge 152734,152791
Cottonseed 253841
Cotton-seed Tree 46227
Cottonspine 312317
Cottonthistle 271663
Cottontop 253845
Cottontop Cactus 140152
Cotton-tree 311292
Cottontube Camellia 69257
Cottonweed 165906,165915,168678,
168691,235253,276916,276917
Cotton-weed 168691
Cottonwood 311131,311237,311292,311324
Cotton-wood 311237
Cottonwood Tree 195311
Cottony Anisochilus 25385
Cottony Antlerpilose Grass 257542
Cottony Aster 40587
Cottony Azalea 330710,331048
Cottony Burdock 31068
Cottony Dandelion 384619
Cottony Goldenaster 90206
Cottony Goldenray 229242
Cottony Hoarhound 245747
Cottony Jujube 418184
Cottony Leucas 227625

Cottony Licuala Palm 228746
Cottony Licualapalm 228746
Cottony Milkvetch 42580
Cottony Sage 345151
Cottony Wild Buckwheat 152099
Cottonycalyx Basketvine 9444
Cottonyflower Basketvine 9443
Cottonyfruit Corydalis 106055
Cottonymouth Leucas 227595
Cottonystalk Lycaste 238744
Cotula 107810
Cotylanthera 107835,107836
Cotyledon 87315,107840,107952,107976
Coubaril 200844
Couch 117859,144144,144661
Couch Grass 117859,144661
Couchgrass 144661
Couch-grass 11628,144661
Couchy Bent 12323
Cough Bush 305085
Cough Root 235417
Coulter Mock Orange 294445
Coulter Pine 299876
Coulter Willow 344122
Coulter's Brickellia 60113
Coulter's Fleabane 150571
Coulter's Globemallow 370943
Coulter's Goldfields 222604
Coulter's Lupine 238487
Coulter's Matilija Poppy 335870
Coulter's Mock-orange 294445
Coulter's Orach 44365
Coulter's Pine 299876
Coulter's Thistle 92255
Couma Rubber 108147
Couming Erythrophleum 154968
Council Tree 164626
Counsellors 92485
Counter Wood 254493
Counter-wood 88506
Country Almond 386500
Country Gooseberry 45725
Country Lawyers 338447
Country Pellitory 280347
Country Pepper 356468
Countryman's Treacle 15698,341064,404316
Countryman's Weatherglass 21339
Coupleflower 134010
Courgette 114300
Coursetia 108199
Courtois Elaeagnus 141964
Courtoisia 108213
Courtoisina 108213
Courtship and Matrimony 166125
Courtship-and-matrimony 166125,239092
Cous root 235415
Cousinia 108217
Couve Tronchuda 59522
Cove Keys 315082
Cove-keys 315082
Covenna 128870
Coven-tree 407907
Coventry Bells 70331,321721
Cover Savin 213902
Covered Begonia 50393
Covered Bractbean 362306
Covered Congea 101780
Covering Ancistrocladus 22067
Covexvein Bladderwort 404020
Covexvein Blueberry 404020
Covexvein Holly 203947
Covexvein Oak 116088
Covex-veined Blueberry 404020
Coville Wild Buckwheat 151965
Coville's Barrel Cactus 163439
Coville's Rush 213028
Coville's Sulphur Flower 152576
Cow Basil 346434,403687
Cow Bells 364193
Cow Belly 192358
Cow Berry 404051
Cow Blinder 273038
Cow Breadnut Tree 61057
Cow Breadnut-tree 61057
Cow Bumble 192358
Cow Cakes 192358,285741
Cow Chervil 28044,261585
Cow Clogweed 192358
Cow Clover 396939,396967,397019
Cow Cockle 346434,403687
Cow Crane 68189
Cow Cranes 68189
Cow Creek Spider-lily 200928
Cow Cress 29341,225315,407029
Cow Eyes 227533
Cow Flop 130383,262441,285741
Cow Foot 346434
Cow Garlic 15876
Cow Grass 302068,308816,325864,396967,397019
Cow Herb 403687
Cow Horn Agave 10813
Cow Itch 259558,309199,336419,393470
Cow Itch Tree 220228
Cow Keeks 192358
Cow Keeps 192358
Cow Lily 68189,267274,267276,267645
Cow Mack 363673,364193
Cow Mumble 28044,84768,192358
Cow Oak 324166
Cow Okra 284493
Cow Paigle 315082
Cow Paps 364193
Cow Parsley 9832,28013,28044,84768
Cow Parsnip 192227,192278,192303,192312,192358
Cow Pea 409096
Cow Peggle 315082
Cow Pops 297650,366941
Cow Quakers 60382
Cow Quakes 60382
Cow Rattle 363673,364193
Cow Serl 339897
Cow Sinkin 315106
Cow Slip 250914,315082
Cow Soapwort 346434,403687
Cow Stripling 315082
Cow Stropple 315082
Cow Strupple 315082
Cow Tamarind 345525
Cow Tongue Prickly Pear 272876
Cow Tree 61057,108153
Cow Tree Tabernaemontana 382852
Cow Vetch 408352,408356
Cow Wheat 248155,248185,248193
Cowage 259558
Cowage Velvet Bean 259558
Cowage Velvet-bean 259558
Cowan's Rhododendron 330467
Cow-bana 90932
Cowbane 90918,90932,269307,278559
Cowbean 90924
Cowbell 363671
Cowbells 403670
Cowberry 100257,403916,404051
Cowbind 61497
Cowboy's Red Whiskers 272785
Cowcockle 403687
Cowcumber 114245
Cow-cummer 114245
Cower-slop 315082
Cow-fat 403687
Cowflock 68189
Cowflop 45566,130383,192358,285741,315082
Cowfoot 359158
Cow-grass 397019
Cowhage 243529,259558
Cowheave 400675
Cow-heave 400675
Cowherb 346434,403686,403687
Cow-herb 403687
Cowherb Soapwort 403686,403687
Cowhorn Orchid 120715
Cowitch 259558
Cow-Itch 70512
Cowlily 267274,267294
Cow-lily 267274,267333
Cowll 106703
Cowmack 363673,364193
Cow-parsley 28044

Cowparsnip 192227
Cow-parsnip 192227,192278,192303,192358
Cowparsnipleaf Bugbane 91023
Cowparsnipleaf Corydalis 105974
Cowparsnipleaf Pleurospermum 304806
Cowparsnipleaf Ribseedcelery 304806
Cowpea 408825,409086
Cow-pea 408825,409086
Cowpea Witchweed 378003
Cowpeatree 325001
Cow-poison 124654
Cowpops 297650
Cowquake 370522
Cowrie Pine 10494
Cowryleafpalm 106984,106999
Cows 401711
Cows And Calves 37015
Cow's Ear 405788
Cow's Eyes 227533
Cow's Hair-felt Spikesedge 143029
Cow's Horn Euphorbia 158981
Cow's Lick 61497
Cow's Lungwort 190944,405788
Cow's Mouth 315082
Cow's Parsley 321643
Cow's Parsnip 37015
Cow's Tail 103446
Cow's Tail Pine 82521
Cow's Tails 103446
Cow's Tall Pine 82524
Cow's Thistle 91770
Cow's Tongue Pricklypear 272877
Cow's Weatherwind 373455
Cow's Weather-wind 373455
Cow's Withwind 373455
Cow's Withywind 373455
Cow's Wort 287494
Cows-and-bulls 37015
Cows-and-calves 37015
Cows-and-kies 37015
Cowskin Azalea 330182
Cowslip 23925,68189,130383,168476,250914,262441,273531,314083,314143,315082,325518,325666,326287
Cowslip Bush 106682
Cowslip Lungwort 321636
Cowslip Primrose 315082
Cowslip-scented Orchid 404620
Cowslop 68189,130383,315082
Cowslop Cowslap 315082
Cow's-tail Pine 82523,82524
Cowtail Pine Family 82493
Cowthwort 224976
Cow-tree Tabernaemontana 382852
Cow-weed 28044,81356,261585,325580
Cowwheat 248155
Cow-wheat 248155,248161,248185,248187,248193,329521
Cowwheatflower Woodbetony 287421
Cowwheat-like Loosestrife 239732
Cow-wort 175454
Cox Barberry 51507
Cox Juniper 213889
Cox Savin 213889
Coxcomb Coral-tree 154647
Cox's Mariposa-lily 67579
Coyaye 182107
Coynes 116546
Coyo Avocado 291609
Coyoli 6134
Coyoli Palm 6134
Coyote Brush 46234
Coyote Bush 46210,46241
Coyote Melon 114299
Coyote Mint 257197
Coyote Thistle 154325
Coyote Tobacco 266038
Coyote Willow 343360
Coyure Palms 12771
Craa's Foot 237539
Craa's Foot Craa-taes 237539
Crab 243711,295227
Crab Apple 243675,243711
Crab Beancaper 418721
Crab Cactus 351993,418529,418532
Crab Cherry 83155
Crab Grass 130404,130489,130745,143512,281274,308816,342859
Crab Oil 72640
Crab Stock 243711
Crab Wood 72634,72640
Crab Zygophyllum 418721
Crabapple 243543,243667,243675,243711
Crab-apple 243543,243675,243711
Crabapple-like Actinidia 6661
Crabapple-like Kiwifruit 6661
Crabapples 243635
Crabcactus 418529
Crab-claw Plum 412343
Crabdaisy 413490
Crabgrass 130404,130537,143512
Crab-grass 130404,130745
Crab's Claws 309570,377485
Crab's Eye 765
Crab's Eyes 765
Crab's Stone 765
Crab's-eye Vine 765
Crabweed 162868,308816
Craches 374968
Crack Nut 106703
Crack Wheat 398900
Crack Willow 343396
Crackerberry 85581
Crackflower 322948,322950
Crack-nut 106703
Cradle Dock 359158
Cradle Orchid 25137
Cradle-dock 359158
Crae-nels 28200
Crafty Stonecrop 356677
Cragcress 97967,98020
Crags Rhododendron 331772
Craib Glycosmis 177825
Craib Milkvetch 42228
Craib Rhynchosia 333263
Craib Tanoak 233161
Craibiodendron 108574
Craid 396854
Craigia 108588
Crain 325820
Craisey 313025
Crake Needle 350425
Crakeberry 145069,145073
Crakefeet 145487,273531,273545
Cramantree 181558
Crambe 108591
Crambling Rocket 327866
Crammick 271563
Cramp Bark 407989
Cramp Thistle 91770
Crampbark 407989
Cramp-bark 408002
Cramp-weed 312360
Cranberry 31142,278351,278366,403710,403894,403944,404051,408188
Cranberry Bush 407989,408188
Cranberry Chickweed 375030
Cranberry Cotoneaster 107345
Cran-berry Cotoneaster 107345
Cranberry Tree 407989,408002
Cranberry Viburnum 407989,408001
Cranberry Wire 278366,404051
Cranberrybush 407989
Cranberrybush Viburnum 408188
Cranberry-bush Viburnum 407989
Cranberry-tree 407675,407989
Cranberry-wire 278366,404051
Cran-commer 343994
Crane 145073,278366
Crane Catchfly 363467
Crane Flower 377576
Crane Ginseng 302386
Crane Johnny 68189
Crane Lily 377576
Crane Woodbetony 287261
Crane-beaked Plantain Lily 198619
Craneberry 278366
Craneberry Gourd 639,641
Craneberry Wire 31142
Cranefly Orchid 392348
Crane-fly Orchis 392346

Cranehead Woodbetony 287261
Craneknee 221627
Crane's Bill 174440
Cranesbill 153903,174551,174569,174719,174793
Cranes-bill 288045
Crane's-bill 174717,174945
Crane's-bill Cranesbill 174440
Crane's-bill Crow Foot 174832
Crane's-bill Family 174431
Cranesbill Geranium 174891
Craniospermum 108661
Craniotome 108667
Crannaberry 278366
Crannock 401388
Cranops 162312,235334,235373,325666,326287,326609,364557
Cranston Spruce 298195
Crape Flower 219933
Crape Jasmine 382763
Crape Jasmine Tabernaemontana 154162
Crape Myrtle 219908,219933
Crape Myrtle of the North 192168
Crape-leaf Creeper 285164
Crapeleaf Kiwifruit 6724
Crapemyrtle 219933
Crape-myrtle 219908,219933
Crash 262722
Craspedolobium 108698
Crassicauliferous Monkshood 5148
Crassilimbatus Barberry 51509
Crassilobed Spapweed 204878
Crassipediceled Dyeing-tree 346480
Crassitube Chirita 87845
Crassleaf Hakea 184622
Crassula 108775,109229,109452
Crassula Family 109505
Crassulaleaf Gentian 173356
Crataegifoliate Groundsel 358637
Crater Lake Sandwort 32156
Crateva 110205,110235
Cratlock 165541
Crato Passion Fruit 285630
Cratoxylum 110238
Craw 325666
Craw Ash 166974
Craw Berry 145073
Craw Crowfoot 325666
Craw Foot 121281,325498
Craw Foot Grass 143530
Craw Pea 222820
Crawcraw 360409
Craw-craw 360409
Crawcrooks 145073
Crawcroups 145073
Crawe's Sedge 74195
Crawfoot 145487,273531,326287
Crawford's Oval Sedge 74196
Crawford's Sedge 74196
Crawfurdia 110292
Crawlers 98020
Crawley 104009
Crawley Barberry 52122
Crawling Lignariella 228961
Crawnberry 278366
Craw-shaw 250502
Craw-taes 145073,145487,273531,278366,325518,326287
Crayfery 321643
Crayfish 136365
Crayfish Leopard's Bane 136365
Crazy 68189,267294,325518,325666,325820,326287
Crazy Bess 68189
Crazy Bet 68189,227533,267294,325518,325666,325820,326287
Crazy Bets 68189,227533,267294,325518,325666,325820,326287
Crazy Betsy 68189
Crazy Betty 68189
Crazy Cup 325820
Crazy Lily 68189
Crazy Weed 278940,325518,325666,326287
Crazy-leafpinetorum 50160
Crazymar 326287
Crazy-moir 326287
Crazy-more 326287
Crazy-pates 68189
Crazyweed 278679,278940,279174
Creagh Antflower 252773
Creagh Mezzettiopsis 252773
Cream Albizia 13480
Cream Avens 175459
Cream Bush 197404
Cream Cactus 244107,244137
Cream Clematis 94910
Cream Cup 302964
Cream Cups 302961,302964
Cream Gentian 173193
Cream Narcissus 262457
Cream Nut 7028,53057
Cream Pea-vine 222787
Cream Pincushions 350207
Cream Sisyrinchium 365803
Cream Violet 410602
Cream White Indigo 47861
Cream Wild Indigo 47862
Cream-and-butter 325820
Creambush 197404
Cream-colored Avens 175459
Cream-colored Vetchling 222787
Creamcups 302961,302964
Cream-flower Rock-cress 30306
Creamish Stonecrop 356979
Cream-nut 53055
Creamy Butterbur 292374
Creamy Scabious 350207
Creamy Vetch 222787
Creamy Vetchling 222787
Creamy Violet 410602
Creamy Wh. Gypsy Rose 216086
Creaseleaf Camellia 69027
Creaseleaf Tea 69027
Creashak 31142
Creed 224375
Creek Dogwood 104977
Creek Sandpaper Fig 164853
Creek Senecio 358757,358882
Creekside Wirelettuce 375971
Creep Ivy 68713
Creep Stonebean-orchis 63032
Creeper 68713,285097
Creeping Acanthopanax 143654
Creeping Acroptilon 6278
Creeping Amaranth 18668
Creeping Aster 160685
Creeping Azalea 330626,330724
Creeping Baby's Breath 183238
Creeping Banksia 47664
Creeping Bastardtoadflax 389848
Creeping Bellflower 70255
Creeping Bent 12323
Creeping Bent Grass 12323,12344
Creeping Bentgrass 12323
Creeping Bent-grass 12323
Creeping Berries 216637
Creeping Bladderwort 403191
Creeping Blue Blossom 79974
Creeping Blueberry 403790
Creeping Bluesnow 83644
Creeping Bluets 187627
Creeping Boeica 56371
Creeping Boobialla 148358,260720
Creeping Bramble 339188
Creeping Broad Leafed Sedge 76264
Creeping Broad-leaf Sedge 76264
Creeping Buds 326287
Creeping Bugle 13174
Creeping Bugleweed 13174
Creeping Bulbophyllum 63032
Creeping Burhead 140543
Creeping Bush Clover 226943
Creeping Bush-clover 226943
Creeping Buttercup 326287,326288,326290
Creeping Cactus 273028
Creeping Catchfly 363958
Creeping Cedar 213785
Creeping Ceratostigma 83644
Creeping Charley 239755,298991
Creeping Charlie 176824,298991,356468
Creeping Chirita 87890

Creeping Chorisis 88977
Creeping Cinquefoil 312925
Creeping Comfrey 381031
Creeping Corydalis 106380
Creeping Cotoneaster 107336
Creeping Crazy 326287
Creeping Crowfoot 326287,326303
Creeping Cucumber 249843
Creeping Currant 334245
Creeping Curvedstamencress 72953,238021
Creeping Dentella 125872
Creeping Devil 240512,375522
Creeping Devil Cactus 240512,375522
Creeping Dichondra 128964
Creeping Dipteracanthus 133541
Creeping Dogwood 85581
Creeping Draba 137183,137237
Creeping Edelweiss 224947
Creeping Eria 148754
Creeping Falsepimpernel 231483
Creeping Falsetamarisk 261288
Creeping False-tamarisk 261288
Creeping Farges Meehania 247727
Creeping Ficus 165515
Creeping Fig 165515
Creeping Fordiophyton 167316
Creeping Forget-me-not 260862,260874,270723
Creeping Foxtail 17512
Creeping Gardenia 171258
Creeping Germander 388042
Creeping Ginsen 280793
Creeping Glasswort 346771
Creeping Gromwell 233772
Creeping Groundsel 358875
Creeping Gypsophila 183238
Creeping Heliotrope 190550
Creeping Holly Grape 52104
Creeping Honeysuckle 235745
Creeping Indigo 205950
Creeping Ixeris 88977
Creeping Jack 356468
Creeping Jacob's Ladder 307241
Creeping Jane 239755
Creeping Jenny 20435,68713,90423,102933,116736,170743,239755,239898,312925,356468,357114
Creeping Juniper 213702,213785,213864
Creeping Knotweed 309084
Creeping Lady's Sorrel 277747
Creeping Lady's-tresses 179685
Creeping Lantana 221284
Creeping Launaea 222988
Creeping Leadwort 83644
Creeping Lepturus 226599
Creeping Lespedeza 226943
Creeping Lettuce 210673
Creeping Licorice 177937
Creeping Lignariella 228961
Creeping Ligusticum 229381
Creeping Lilyturf 232640,272131
Creeping Lily-turf 232640
Creeping Lip Plant 296121
Creeping Lippia 232476
Creeping Liriope 232640
Creeping Lobelia 234363
Creeping Loosestrife 239755,239813
Creeping Love Grass 147733,147941
Creeping Love-grass 147733
Creeping Mahonia 242625
Creeping Mallotus 243437
Creeping Mannagrass 177560
Creeping Marshweed 230319
Creeping Marshwort 29345
Creeping Mazus 247008,247030
Creeping Meadow Foxtail 17512
Creeping Mint 250437
Creeping Mirrorplant 103798
Creeping Motherweed 231483
Creeping Muhly 259700
Creeping Myoporum 260720
Creeping Nailwort 284897
Creeping Navel Seed 270723
Creeping Navel-seed 270723
Creeping Ninenode 319833
Creeping Oregon Grape 52104
Creeping Oxalis 277747
Creeping Oxeye 413559
Creeping Peliosanthes 288651
Creeping Pellionia 288767
Creeping Pennisetum 289067
Creeping Pentapterygium 289750
Creeping Phlox 295321
Creeping Pine 253178,253179,299786
Creeping Pink 127809
Creeping Plantain 179685
Creeping Polemonium 307241
Creeping Potentilla 312813
Creeping Pothos 313206
Creeping Pratia 313555
Creeping Primrose 314113
Creeping Primrose Willow 238203
Creeping Psychotria 319833
Creeping Quackgrass 144661
Creeping Raspberry 339012
Creeping Raspwort 179438,184995
Creeping Rattlesnake Plantain 179685
Creeping Rattlesnake-plantain 179685
Creeping Red Fescue 163816,164243
Creeping Red Thyme 391365
Creeping Redflush 220660
Creeping Redhairgrass 333024
Creeping Restharrow 271563
Creeping Rhododendron 330724
Creeping Rhynchelytrum 333024
Creeping Rockcress 30263
Creeping Rockfoil 349936
Creeping Rosemary 337180,337201
Creeping Rostellularia 337253
Creeping Rush 213417,213490
Creeping Sage 345418
Creeping Sailor 116736,349936,356468
Creeping Sainfoin 271191,271272
Creeping Saltbush 44627,44641
Creeping Sandwort 31957
Creeping Sanicle 345993
Creeping Sanvitalia 346304
Creeping Savin 213864
Creeping Saxifrage 349936
Creeping Sebastiania 356201
Creeping Sedge 74118
Creeping Signal Grass 402544
Creeping Signalgrass 402553
Creeping Skyflower 139062
Creeping Sky-flower 139062
Creeping Snapdragon 28617,37554
Creeping Snowberry 172085,380746
Creeping Softgrass 197306
Creeping Soft-grass 197301,197306
Creeping Sow Thistle 368635
Creeping Spearwort 325864,325870,326303
Creeping Speedwell 407124,407241,407309
Creeping Spiderling 56497
Creeping Spikesedge 143271
Creeping Spotflower 4864
Creeping spring beauty 94358
Creeping Spurge 159417
Creeping St. John's Wort 202133
Creeping St. John's-wort 201842,201930,202133
Creeping St. Johnswort 201787
Creeping Stem 256001,256002
Creeping Stonecrop 357169
Creeping Strawberry Bush 157651
Creeping Swamp Red Currant 334245
Creeping Tailgrass 402553
Creeping Tansy 312360
Creeping Telosma 385759
Creeping Thistle 91770
Creeping Thyme 391080,391365,391397
Creeping Tickle Grass 12323,12344
Creeping Tinctorial Fig 165766
Creeping Tom 356468
Creeping Tormentil 312357
Creeping Treebine 92920
Creeping Trefoil 397043
Creeping Twinballgrass 209121
Creeping Urochloa 402553
Creeping Velvet Grass 197306
Creeping Velvetgrass 197306
Creeping Vervain 405815

Creeping Violet 410422
Creeping Wart Cress 105350
Creeping Wartcress 105350
Creeping Water Forget-me-not 260874
Creeping Water Parsnip 365861
Creeping Water Rocket 336277
Creeping Water Scorpion-grass 260874
Creeping Waxweed 114618
Creeping Wheat-grass 144661
Creeping Whitepearl 172085
Creeping Whitlowgrass 137183
Creeping Wild Rye 144661
Creeping Willow 343046,343994
Creeping Wintergreen 172089,172135
Creeping Winter-green 172047
Creeping Wire Vine 259589
Creeping Wolftailgrass 289067
Creeping Woodbetony 287583,287792
Creeping Woodsorrel 277747
Creeping Wood-sorrel 277747
Creeping Yellow Cress 336277
Creeping Yellow Wood-sorrel 277747
Creeping Yellowcress 336277
Creeping Yellow-cress 336277
Creeping Zinnia 346300,346304,418038
Creepingbract Primrose 314195
Creepingbranch St. John's wort 202133
Creeping-charlie 176824,176835,239755
Creeping-cudweed 155873
Creeping-jenny 239755
Creeping-oxeye 371233
Creeping-root Violet 409799
Creeping-snowberry 172085
Creep-ivy 68713
Creepy Mallotus 243437
Creese 262722
Creevereegh 111347
Creggans 401388
Creivel 315082
Cremanthodium 110343,110368
Cremastra 110500
Creme-de-menthe Mint 250437
Crenate Bittercress 72879
Crenate Chinese Helwingia 191146
Crenate Deutzia 126891
Crenate Dew Rabdosia 324786
Crenate Elatostema 142635
Crenate Euonymus 157395
Crenate Firethorn 322458
Crenate Fleshspike 346939
Crenate Gugertree 350912
Crenate Hedge Buckthorn 328692
Crenate Kalanchoe 215122
Crenate Largeleaf Bittercress 72879
Crenate Pride-of-Rochester 126891
Crenate Procris 315465
Crenate Reddish Beautyberry 66915
Crenate Reddish Purplepearl 66915
Crenate Sichuan Bittercress 72829
Crenate Spapweed 204879
Crenate Spindle-tree 157395
Crenate Stairweed 142635
Crenate Szechuan Bittercress 72829
Crenate Touch-me-not 204879
Crenate Willow 343035,343255
Crenate Woodbetony 287121
Crenate Yadong Poplar 311580
Crenateflower Calceolaria 66263
Crenateleaf Heteropappus 193932
Crenateleaf Jurinea 193932
Crenate-leaved Ardisia 31396
Crenulate Barberry 51514
Crenulate Bellflower 69979
Crenulate Cinquefoil 312475
Crenulate Jasminorange 260163
Crenulate Jasmin-orange 260163
Crenulate Leaves Rhododendron 330475
Crenulate Orchis 310885
Crenulate Spapweed 204880
Crenulate Violet 409881
Crenulate Woodbetony 287125
Creole Cotton 179884
Creole Tea 348975
Creosote Bush 221975,221979,221983
Creosote Rush 221980
Creosotebush 221983
Creosote-bush 221975,221983
Creosote-bush Family 418580
Crepe 154162
Crepe flower 219933
Crepe Jasmine 154162
Crepe Myrtle 219933
Crepe-myrtle 219908,219933
Crepidate Dendrobium 125082
Crepidiastrum 110602
Crepp Sicklebamboo 20199
Crescent Ophrys 272454
Crescent Pax Maple 3378
Crescent-shaped Euphorbia 158857
Cress 72674,225279,225450,406992
Cress Rocket 336277
Cressel 355061
Cresset 355061
Cressil 355061
Cress-rocket 336277
Crest Iris 208502
Crest Sandcress 321482
Crestate Elsholtzia 143993
Crested Arrowhead 342324
Crested Buckwheat 162552
Crested Bunias 63450
Crested Cabbage-rose 336475
Crested Coral Root 194524
Crested Coral-root 194524
Crested Corydalis 105770
Crested Cow-wheat 248168
Crested Dock 340006
Crested Dog's-tail 118315
Crested Dog's-tail Grass 118315
Crested Dogstail Grass 118315
Crested Dogtail 118315
Crested Dogtailgrass 118315
Crested Dwarf Iris 208502
Crested Field-speedwell 407092
Crested Floating Heart 267818
Crested Floatingheart 267818
Crested Gentian 173886
Crested Goosefoot 86997
Crested Greenhead Sedge 218628
Crested Hair Grass 217494
Crested Hair-grass 217494
Crested Hakea 184598
Crested Iris 208502
Crested Irises 208875
Crested Latesummer Mint 143974
Crested Late-summer Mint 143974
Crested Moss-rose 336475
Crested Neriumleaf Euphorbia 159457
Crested Nerium-leaved Euphorbia 159457
Crested Oleander Cactus 159457
Crested Onion 15210
Crested Oval Sedge 74209
Crested Ovary Rhododendron 331139
Crested Panicgrass 281512
Crested Philippine Violet 48140
Crested Pipewort 151281
Crested Prickly Poppy 32449
Crested Pugionium 321482
Crested Sedge 74209
Crested Ventilago 405446
Crested Wartycabbage 63450
Crested Wheat Grass 11696
Crested Wheatgrass 11696
Crested Wheat-grass 11696
Crested Wingdrupe 405446
Crested Yellow Orchis 302296
Crestedgalea Woodbetony 287112
Cresthair Woodbetony 287378
Crestless Corydalis 105843
Crestless Tall Corydalis 105852
Crestmarine 111347
Crestpetal-tree 236414
Crestrib Morning Glory 207720
Cret Betony 373185
Cretaceous Dendrobium 125083
Cretaceousleaf Actinidia 6667
Cretaceousleaf Kiwifruit 6667
Cretan Barberry 51515
Cretan Bears-tail 405659
Cretan Bee-orchid 272429
Cretan Bryony 61485,61497

Cretan Crownvetch 356383
Cretan Date Palm 295489
Cretan Dittany 274210
Cretan Maple 3588
Cretan Meadow Foxtail 17522
Cretan Mullein 405691
Cretan Palm 295489
Cretan Rockrose 93137
Cretan Rose 93161
Cretan Spikenard 404331
Cretan Tulip 400220
Cretan Viper's Bugloss 141145
Cretanweed 187718
Crete Tulip 400220
Cretin Rhodiola 329861
Crewel 315082
Crex 316470
Criapateleaf Newlitse 264027
Cricket Bat Willow 342980
Cricket Rhaphis 90125
Cricket-bat Willow 342973,342977,342980,343167
Cricks 316470
Cricksey 316470
Cricktb Hakea 184626
Crilly-greens 59524
Crimean Bumet-saxifrage 299551
Crimean Burnet Saxifrage 299551
Crimean Candytuft 203248
Crimean Comfrey 381039
Crimean Cotton-thistle 271705
Crimean Golden Drop 271835
Crimean Golden-drop 271835
Crimean Ivy 187337
Crimean Juniper 213767
Crimean Lime 391705
Crimean Linden 391705
Crimean Melick 248975
Crimean Pine 300111
Crimean Snowdrop 169726
Crimped Bellflower 70370
Crimson Agapetes 10343
Crimson Bells 194437
Crimson Bottlebrush 67253
Crimson China Rose 336495
Crimson Clover 396934
Crimson Creeper 193747
Crimson Cup Ixia 210723
Crimson Flag 351883,351884
Crimson Foliage Maple 3529
Crimson Fountain Grass 289248
Crimson Fountaingrass 289248
Crimson Glory Vine 411623
Crimson Glory-vine 411623
Crimson Grevill 180638
Crimson Ixora 211068
Crimson Lady 127635
Crimson Mallee 155619
Crimson Monkey-flower 255197
Crimson Ochrosia 268355
Crimson passion Flower 285717
Crimson Pitcher Plant 347149
Crimson Pitcherplant 347152
Crimson Pygmy Japanese Barberry 52227
Crimson Shoes 222784
Crimson Spotted Rhododendron 331707
Crimson Thyme 391373
Crimson Trefoil 396934
Crimson Tulip 400150
Crimson Turkey Bush 148365
Crimson Vetchling 222784
Crimson Woodsorrel 277905
Crimsonband Lily 229733
Crimsonberry 298094
Crimson-cup Neoregelia 264416
Crimson-eyed Rose Mallow 195032
Crimson-eyed Rosemallow 195032
Crimson-flower Eucalyptus 106800
Crimsonleaved Maple 2984
Crimsons 246479
Crimson-spot Rockrose 93161
Crimson-spotted Rhododendron 331707
Crincle Passionflower 285645
Crincled Passionflower 285645
Crincled Passion-flower 285645
Crinite Abelmoschus 217
Crinite Anemone 23773
Crinite Matucana 246593
Crinite Pogonatherum 306832
Crinite Windflower 23773
Crinkle Bush 235412
Crinkle Leaf Plant 8429
Crink-leaved Glyptopetalum 178001
Crinklebush Lomatia 235412
Crinkled Hair Grass 126074
Crinklefruit Camellia 69569
Crinklehair Cowpea 409028
Crinkleleaf Spiderling 56420
Crinkleroot 72739,125841
Crinkle-root 72739
Crinkly-leaf Geranium 288179
Crinum 111152
Crinum Lily 111152,111250
Crinum-Lilies 111152
Crippsi Golden Cypress 85314
Crisitate Gentian 173363
Crislings 316470
Crisp Fox-tail Orchid 9281
Crisp Starwort 374812
Crispate Abutilon 192470
Crispate Camellia 69027
Crispate Dock 339994
Crispate Ottelia 277365
Crispate Plantainlily 198598
Crispate Rhubarb 329321
Crispateleaf Ardisia 31408
Crispateleaf Leea 223931
Crispateleaf Snakegourd 396171
Crispate-leaved Ardisia 31408
Crispate-leaved Leea 223931
Crisped Ardisia 31408
Crisped Barberry 51516
Crisped Bretschneider Dogwood 104972
Crisped Bunch-flower 248657
Crisped Corydalis 105763
Crisped Dock 339994
Crisped Elaeagnus 141965
Crisped Goldenbush 150316
Crisped Iris 208831
Crisped Larkspur 124153
Crisped Mint 250347
Crisped Mock Vervain 176691
Crisped Odontoglossum 269065
Crisped Oncidium 270802
Crispedflower Wildginger 37605
Crisped-leaf Mustard 59444
Crisped-stalked Yushania 416767
Crispleaf Amaranth 18694
Crispleaf Mint 250347
Crisp-leaf Wild Buckwheat 151947
Crisp-leaved Amaranth 18694
Crispy-haired Sweetleaf 381446
Crisscross Aspidistra 39527
Crisscross Chirita 87846
Crisscross Milkvetch 42236
Cristaldre 81065
Cristate Begonia 49750
Cristate Blueeargrass 115541
Cristate Caesalpinia 64983
Cristate Coelogyne 98634
Cristate Cryptomeria 113692
Cristate Cyanotis 115541
Cristate Floatingheart 267809
Cristate Geissaspis 172509
Cristate Lepidagathis 225151
Cristate Nephelaphyllum 265123
Cristate Odontoglossum 269066
Cristate Oncidium 270803
Cristate Pleurospermum 304783
Cristate Ribseedcelery 304783
Cristate Satyrium 347755
Cristate Skullcap 355420
Cristate Vanda 404633
Cristate Woodbetony 287127
Cristate Youngia 416404
Cristateless Dali Corydalis 105836
Cristateless Kashmir Corydalis 105843
Cristateless Tali Corydalis 105836
Cristens 316470
Cristensen Cranesbill 174536
Cristmas Box 346723

Critical Meadow-rue 388459
Crnekelty Bur 31051
Crocea Iceplant 243253
Crockelty-bur 31051
Crocks-and-kettles 64345
Crocodile 95431,203545
Crocosmia 111455,111464,111468
Crocus 99297,99345,111480,111589
Crocus Autumn 99297
Crocus Japonica 216085
Crocus Tritonia 399062
Crocus Tulip 400213
Crocusleaf Primrose 314279
Crocus-like Iris 208659
Crofton Weed 11153
Croizat Macaranga 240229
Croizat's Barbate Mallotus 243331
Crokers 367696
Cromwell 233760
Croneberry 278366,403916
Croneshanks 309570
Cronquist Paris 284307
Cronquist's Fleabane 150575
Cronquist's Wild Buckwheat 151970
Cronquist's Woody-aster 415848
Cron-reish 367120
Crooked Aster 380966
Crooked Holly 203744
Crooked Yellow Stonecrop 357114
Crookedhead Sawwort 361110
Crookedstem Aster 380966
Crooked-stem Aster 380966
Crooked-stemmed Aster 380966
Crook-neck Squash 114292
Crook-nerved Blood-red Spindle-tree 157860
Croomia 111674
Croos Tree 111101
Crop 162312,235334,235373
Cropleaf Ceanothus 79924
Cropweed 81237
Crosby's Wild Buckwheat 151971
Crosnes 373100
Cross Cleavers 170316
Cross Flower 86427,145487,273531,308454
Cross Gentian 173364
Cross Milkwort 308014
Cross of Jerusalem 363312
Cross Vine 54292,54303
Cross Wort 113135
Crossandra 111693,111724
Crossberry 180897
Crossbow Arrowbamboo 162738
Crossbow Fargesia 162738
Crossed Heath 150131
Crossflower 89020
Cross-flower 86427
Crossleaf Goldenray 229231
Crossleaf Heath 150131
Cross-leaf Honey-myrtle 248090
Cross-leaf Milkwort 308014
Crossleaf Pondweed 312220
Cross-leaved Bedstraw 170246
Cross-leaved Gentian 173364
Cross-leaved Heath 150131
Cross-leaved Milkwort 308014
Crossorchis Habenaria 184053
Crossostephium 111823
Cross-shaped Chloris 88331
Cross-shaped Milkvetch 42236
Crossvine 54303
Cross-vine 54303
Crosswort 113110,113130,113135,158250,170335
Cross-wort Gentian 173364
Crotalaria 111851
Croton 98177,98190,112829,113039
Croton Bicolor 161643
Croton-oil Plant 113039
Crotonvine 108698
Crottle 234269
Crouched Gasteria 171650
Croupans 145073
Crow Bells 145487,262441,325666
Crow Broom 383670
Crow Claws 325612,326287
Crow Corn 14486
Crow Crane 68189
Crow Flower 68189,145487,174951,273531,325518,363461
Crow Foot 145487,237539
Crow Garlic 15876
Crow Killer 21459
Crow Leek 145487,340990
Crow Ling 67456,145073,149205,150131
Crow Needle 350425
Crow Onion 15876
Crow Packle 325666
Crow Parsnip 384714
Crow Pea 145073
Crow Peas 145073,408611
Crow Peck 72038,273531,325612,350425
Crow Persimmon 132371
Crow Pightle 325666,325820
Crow Poison 128870,266959,417887
Crow Potato 239193
Crow Rocket 158062
Crow Soap 346434
Crow Toe 145487,237539,273531,326287
Crow Toes 145487,237539,273531,326287
Crow Vetch 408352
Crow's Ash 166974
Crow's Claw Cactus 163456
Crow's Flower 273531
Crow's Foot 325666,374916,400675
Crow's Foot Plantain 301910
Crow's Legs 145487
Crow's Nest 123141
Crow's-foot Plantain 301910
Crowa 21479
Crowbars 31142
Crowberry 145054,145073,298094,403916,404051
Crow-berry 145073
Crowberry Family 145052
Crowberry-leaved Barberry 51590
Crowcup 168476
Crowd-ball Smooth-leaved Barberry 51437
Crowded Astilbe 41832
Crowded Barberry 51487
Crowded Chive 15198
Crowded Jerusalemsage 295091
Crowded Milkvetch 42216
Crowded Onion 15198
Crowded Oval Sedge 74225
Crowded Sideritis 362768
Crowded-flower Salacia 342611
Crowded-flowered Salacia 342611
Crowded-leaf Wattle 1144
Crowdipper 299724
Crowdy-kit-o'-the-wall 356468
Crowed-leaved Mahonia 242508
Crow-flower Mayten 246826
Crow-flowered Mayten 246826
Crowflower-leaf Primrose 314247
Crowfoot 125846,145487,174440,174717,237539,238583,255223,273524,273531,273545,301910,325498,325518,325666,325864,359158
Crowfoot Cranesbill 174832
Crowfoot Family 325494
Crow-foot Fox Sedge 74219
Crowfoot Geranium 288465
Crowfoot Grass 121285,143530
Crowfoot Plantain 301910
Crow-foot Rock Daisy 291307
Crowfoot Violet 410380
Crowfoot Waybread 301910
Crowfootgrass 121281
Crowfoot-like Echinodorus 140556
Crowfruit 403832
Crowgarlic 15876
Crown Anemone 23767
Crown Bark 91082
Crown Beard 6833,405911
Crown Begonia 49774
Crown Cactus 327250
Crown Campion 363235
Crown Daisy 89402,89481
Crown Daisy Chrysanthemum 89481
Crown Fritillary 168430
Crown Gum 244534

Crown Imperial 168430
Crown Imperials 168430
Crown Mayweed 246332
Crown of Gold Tree 78276
Crown of Thorns 159363,159364,159366,159371,266225,285623,418216
Crown of Thorns Euphorbia 159363
Crown of the Field 11947
Crown Pea 301070
Crown Pink 363370
Crown Vetch 105269,105324,356377
Crown Waterchestnut 394480
Crownbeard 405911
Crowncactus 327250
Crowndaisy 89481
Crowndaisy Chrysanthemum 89481
Crown-daisy Chrysanthemum 89481
Crowndaisy Oxeyedaisy 89481
Crowned Goldfields 222597
Crowned-fruit Honeysuckle 236118
Crown-fruited Honeysuckle 236118
Crown-jewels 236245,236246
Crownless Flex Corydalis 106478
Crownless Indian Horseorchis 215345
Crownless Waterchestnut 394478
Crownless Wingpetal Corydalis 106333
Crown-of-thorns 159363
Crownplant 68088
Crowns' Treacle 15698
Crownscale 44361
Crownseed Pectis 286877
Crownvetch 105269,105324
Crown-vetch 105324
Crown-vetch Coronilla 105324
Crow-Poison 19468
Crows 315082
Crow-spur Sedge 74219
Crowther Soap 346434
Crowtoes 325518,325666,326287
Crozier Cycas 115812
Crucia Lovegrass 147995
Crucianvine 356259,356279
Cruciate Blackgrass 61766
Cruciate Buchnera 61766
Cruciate Dysophylla 139556
Cruciate Rockvine 387762
Cruciate Sedge 74211
Cruciate Watercandle 139556
Crucifix Orchid 146429
Crucifix-flower 86427
Crucifixion Thorn 71264
Cruciform Crabgrass 130509
Cruciform Hemipilia 191598
Cruciform Rockvine 387762
Cruel 315082
Cruel Hawthorn 109635
Cruel Plant 30862
Cruel Vine 30862
Cruian Gall 241817
Cruise's Goldenaster 90207
Crumble Lily 230058
Crummock 365871
Crumple Lily 229922
Crunchweed 364557
Crupina 113205,113214
Crusader's Spear 352960
Crusaders' Spears 352960
Crushed-rice Flat-sedge 119041
Cruss-leaved Bedstraw 170246
Crust 295227
Cry-baby 21339,154647,174877
Cry-baby Crab 21339,174877
Crybabytree 154647
Crybe 113230,113231
Crypsis 113305
Cryptandra 113335
Cryptandrous Dropseed 372642
Cryptantha 113350
Cryptanthus 113364
Crypteronia 113397
Crypteronia Family 113402
Cryptic Fruit Tanoak 233162
Cryptic-fruited Tanoak 233168
Cryptocarya 113422
Cryptochilus 113511
Cryptocoryne 113527
Cryptodiscus 113551
Cryptolepis 113566
Cryptomeria 113633,113685
Cryptomeria-like Taiwania 383189
Crypto-nerved Camellia 69029
Cryptonerved Nankun Rhododendron 331304
Cryptonervius Camellia 69029
Cryptopetalous Burmannia 63966
Cryptospora 113815
Cryptostachys Sedge 74224
Cryptostegia 113820
Cryptostylis 113846,113848
Cryspateleaf Pittosporum 301249
Cryspateleaf Seatung 301249
Cryspate-leaved Pittosporum 301249
Crystal 316470
Crystal Anthurium 28093
Crystal Dendrobium 125089
Crystal Springs Lessingia 227098
Crystal Tea 223901
Crystal Tea Ledum 223901
Crystal Wendlandia 413795
Crystaleaf Figmarigold 251265
Crystalline 251265
Crystalline iceplant 251265
Crystalloid Dendrobium 125089
Crystlline Mitreola 256201
Ctenolepis 113980
Cuachilote 284493,284498
Cuan-Dian Birdlingbran 87257
Cuan-gan Fritillary 168372
Cuaromo 80009
Cuba Bean 409086
Cuba Devilpepper 327011
Cuba Devil-pepper 327011
Cuba Eightangle 204496
Cuba Hemp 169249
Cuba Mahogany 380528
Cuba Pine 300289
Cuba Pink Tree 382720
Cuba Pink Trumpet Tree 382720
Cuba Royalpalm 337885
Cuban Bast 194850
Cuban Holly 49753
Cuban Lily 353001
Cuban Mahogany 380528
Cuban Petticoat Palm 103737
Cuban Pine 299836
Cuban Raintree 61309
Cuban Royal Palm 337880,337885
Cuban Walnut 212617
Cuban Zephyrlily 417630
Cubanlily 353001
Cuban-lily 353001
Cubeb 300377
Cubeb Litse 233882
Cubeb Pepper 300377
Cubebs 300377
Cubitt Millettia 254663
Cubitt Rhododendron 330481
Cuckenwort 374968
Cuckle 31051
Cuckle Buttons 31051
Cuckle Dock 31051
Cucklemoors 31051
Cuckold 31051,53755
Cuckold Buttons 31051
Cuckold Dock 31051
Cuckold's Cap 5442
Cuckoldy Buttons 31051
Cuckoo 13174,23925,70271,72934,145487,273531,273545,300223,315082,363461
Cuckoo Babies 37015
Cuckoo Baby 37015
Cuckoo Bitter-cress 72934
Cuckoo Boots 145487
Cuckoo Bread-and-cheese 277648
Cuckoo Buttons 31051,31056,92485
Cuckoo Cheese 277648
Cuckoo Cheese-and-bread 277648
Cuckoo Cock 37015,273531
Cuckoo Flower 72674,72934,363461
Cuckoo Gilliflower 363461
Cuckoo Hitter Cress 72934
Cuckoo Hood 81020

Cuckoo Lily 37015
Cuckoo Orchid 273531
Cuckoo Orchis 273531
Cuckoo Pint 36967,37015,72934,273531
Cuckoo Pintel 37015,72934
Cuckoo Point 37015
Cuckoo Potato 102727
Cuckoo Potatoes 102727
Cuckoo Rose 262441
Cuckoo Sorrel 277648,339887,339897
Cuckoo Sorrow 339887
Cuckoo Spice 72934,277648
Cuckoo Spit 23925,37015,72934,248411
Cuckoo's 273531
Cuckoo's Beads 109857
Cuckoo's Boots 145487
Cuckoo's Bread 277648
Cuckoo's Bread-and-cheese 277648
Cuckoo's Bread-and-cheese Tree 109857
Cuckoo's Cap 5442,5676
Cuckoo's Clover 277648
Cuckoo's Eye 174877
Cuckoo's Eyes 174877
Cuckoo's Meat 174877,277648,339887,339897,374916
Cuckoo's Shoes-and-stockings 72934
Cuckoo's Sorrel 277648
Cuckoo's Sourock 277648
Cuckoo's Sourocks 277648
Cuckoo's Stockings 70271,145487,237539,410493
Cuckoo's Victuals 174877,277648,374916
Cuckoo's-meat 277688
Cuckoo-bread 72934,277648,302068
Cuckoo-buds 72934,273531,325518,325666,326287
Cuckoo-buttons 31051,31056,92485
Cuckoo-cheese 277648
Cuckoo-cock 37015,273531
Cuckooflower 72934,363461
Cuckoo-flower 23925,72934,72939,363461
Cuckoo-grass 238583
Cuckoo-hood 81020
Cuckoo-meat 277648,339887
Cuckoo-orchis 110500
Cuckoopint 37015
Cuckoo-pint 37015,68189,72934,248411,273531
Cuckoo-pintle 37015,72934
Cuckoo-point 37015
Cuckoo-sorrow 339887
Cuckoosour 277648
Cuckoo-spice 72934,277648
Cuckoo-spit 23925,37015,248411
Cuckow Pint 37015
Cuctirating Teasel 133505
Cucubalus 114057
Cuculate Caesalpinia 252745
Cuculate Mezoneuron 252745
Cucullate Begonia 49755
Cucullate Greenvine 204657
Cucullate Roxie's Rhododendron 331676
Cucullate Tanoak 233164
Cucumber 114108,114245,208771
Cucumber Magnolia 241962
Cucum-ber Root 247228
Cucumber Tree 45724,125750,216324,241951,241962,242333
Cucumberleaf Sunflower 188936
Cucumber-leaf Sunflower 188941
Cucumber-leaved Sunflower 188936,188941
Cucumber-tree 45724,241951,241962
Cucumis 114108
Cucummer 114245
Cucuzzi 219843
Cuddapah Almond 61670
Cuddie's Lungs 405788
Cuddle-me 410677
Cuddle-me-to-you 410677
Cuddy's Lugs 405788
Cuddy's Lungs 405788
Cudgerie 166984
Cudjoe Wood 211374
Cudjoe-wood 211369
Cudo Clinopodium 97044
Cudrania 114319,240842
Cudrania Tree 240842
Cudweed 26385,165906,166038,178056,178062,178323,235267,317742,317756,415386
Cudweed Everlasting 189661
Cudweedlike Fleaweed 321561
Cudweedlike Pulicaria 321561
Cudweed-like Saussurea 348342
Cudweed-like Windhairdaisy 348342
Cud-weed-old Balsa 178358
Cudwort 166038
Cuekold's Cap 5442
Cuesta Ridge Thistle 92256
Cuichunchulli 199702
Cuija 272818
Cuipo 79780
Cuirn 369339
Cuirn Sorbus 44960
Cukoo-pint 37015
Culcate Sasa 347308
Culciform Milkvetch 42017
Culen 319186
Culerage 309199
Culilawan 91317
Culinary Asparagus 39120
Cullavine 30081
Cullenbeam 30081
Cull-me-to-you 410677
Culm Spikesedge 143149
Culnbar Bean 298009
Culrache 309199
Culrage 309199
Cultiform Oat 45437
Cultiva Vetch 408571
Cultivate Mountainash 369381
Cultivate Teasel 133505
Cultivate Vetchling 222832
Cultivated Apple 243675
Cultivated Beet 53243
Cultivated Camelina 68860
Cultivated Currant 334179
Cultivated Endive 90894
Cultivated Falseflax 68860
Cultivated Fenugreek 397211
Cultivated Flax 232001
Cultivated Garlic 15698
Cultivated Hop 199384
Cultivated Knotweed 309616
Cultivated Licorice 177893
Cultivated Mint 250376
Cultivated Oats 45566
Cultivated Olive 270111
Cultivated Onion 15165
Cultivated Plum 316382
Cultivated Radish 326616
Cultivated Raspberry 338375
Cultivated Tobacco 266060
Culver's Root 407434,407449,407507
Culver's-physic 407449,407507
Culver's-root 407507
Culverfoot 174551
Culverkeys 30081,145487,167955,273531,315082,408620
Culvers 145487
Culverwort 30081
Cumberfield 308816
Cumberland 308816
Cumberland Azalea 330196,330484
Cumberland Hawthorn 369322
Cumberland Rosemary 102805
Cumberland Rosinweed 364267
Cumberland Stitchwort 255457
Cumbungi 401085
Cumbungi Reed 401085
Cumfirie 50825
Cumfurt 381035
Cumin 114502,114503
Cumin Royal 19672
Cuming Cordia 104203
Cuming Croton 112853
Cuming Drypetes 138603
Cuming Eulalia 156458
Cuming Fig 164864
Cuming Ironweed 406272
Cuming Linociera 231739

Cuming Pencilwood 139641
Cuming Pencil-wood 139641
Cuming Peristrophe 291148
Cuming Threeawngrass 33817
Cuming Triawn 33817
Cuming's Lovegrass 147609
Cuming's Medinilla 247549
Cuming's Screw Pine 168244
Cuming's Whitepearl 172100
Cuming's Wintergreen 172100
Cuminum 114502
Cummin 114502
Cummins Devilpepper 327012
Cumquat 167503
Cunealglume 29449,29454
Cunealleaf Birdlingbran 87235
Cunealleaf Chesneya 87235
Cunealleaf Dregea 137787
Cunealleaf Jerusalemsage 295094
Cunealleaf Spiraea 371860
Cuneare-base-leaves Leucothoe 228175
Cuneate Base Leaves Leucothoe 228175
Cuneate Bigfruit Mountainash 369458
Cuneate Cinquefoil 312483
Cuneate Daisy 124832
Cuneate Elatostema 142637
Cuneate Eurya 160449
Cuneate Girald Currant 333987
Cuneate Goldenray 229010
Cuneate Kobresia 217146
Cuneate Larkspur 124156
Cuneate Lespedeza 226742
Cuneate Redstem Skullcap 355859
Cuneate Sibbaldia 362344
Cuneate Staff-tree 80168
Cuneate Sterculia 376122
Cuneate Trillium 397551
Cuneate Wildberry 362344
Cuneate-leaf Actinidia 6595
Cuneateleaf Azalea 330485
Cuneateleaf Buttercup 325743
Cuneateleaf Dendranthema 124832
Cuneateleaf Forkstyleflower 374487
Cuneate-leaf Gesneria 175299
Cuneate-leaf Goldsaxifrage 90385
Cuneateleaf Jerusalemsage 295094
Cuneate-leaf Kiwifruit 6595
Cuneateleaf Knotweed 309909
Cuneateleaf Meadowrue 388594
Cuneateleaf Meliosma 249373
Cuneateleaf Miliusa 254551
Cuneateleaf Pepperweed 225338
Cuneateleaf Physospermopsis 297946
Cuneateleaf Swertia 380172
Cuneateleaf Windhairdaisy 348523
Cuneate-leaved Arrowleaf 342325
Cuneate-leaved Meliosma 249373
Cuneate-leaved Miliusa 254551
Cuneate-wing Peashrub 72213
Cuneate-winged Caunealwing Peashrub 72213
Cuniform Stairweed 142637
Cunjevoi 16493,16512
Cunningham Araucaria 30841
Cunningham Beefwood 79157
Cunningham Casuarina 79157
Cunningham Centipeda 81683
Cunningham Didiciea 392347
Cunningham Primrose 314288
Cunningham Seaforthia 31015
Cunningham She-oak 79157
Cunningham's Araucaria 30841
Cunningham's Beefwood 79157
Cuntblows 85526
Cunure 66689
Cuona Arrowbamboo 162696
Cuona Chickweed 374827
Cuona Cremanthodium 110366
Cuona Doronicum 136350
Cuona Fargesia 162696
Cuona Larkspur 124139
Cuona Rockfoil 349090
Cuona Sagebrush 35355
Cuona Spapweed 204883
Cuona Stonecrop 357256
Cuona Swertia 380164
Cuona Tomentellate Honeysuckle 236178
Cuona Touch-me-not 204883
Cuona Windhairdaisy 348347
Cup and Saucer 70166
Cup and Saucer Plant 197368
Cup and Saucer Vine 97757
Cup Flower 266200
Cup Grass 151655
Cup Gum 155543
Cup Kalanchoe 215269,215270
Cup of Gold 366871
Cup of Gold Vine 366871
Cup Omphalotrigonotis 270761
Cup Oxalis 278137
Cup Plant 364301
Cup Rosinweed 364301
Cup Thicked Woodbetony 287759
Cup-and-ladle 324335
Cup-and-saucer 68189,70166,401711,407072
Cup-and-saucer Canterbury Bell 69931
Cup-and-saucer Canterbury Bells 69931
Cup-and-saucer Flower 97757
Cup-and-saucer Plant 197368
Cup-and-saucer Vine 97757
Cupania 114574
Cupanis-like Tulip Wood 185895
Cupanis-like Tulipwood 185895
Cupcalyx Azalea 331503
Cup-calyx Honeysuckle 235868
Cupcalyx Milkvetch 42251
Cup-calyx Pencilwood 139642
Cupcalyx Seamulberry 368913
Cupcalyx Sonneratia 368913
Cupcalyx Tea 69025
Cup-calyx Two-colored Rhododendron 330568
Cup-calyxed Sonneratia 368913
Cupdaisy 115639
Cupepreous Holly 203743
Cupflower 266194,266199
Cup-flower 266194
Cupflower Aspidistra 39528
Cupflower Leek 15222
Cupflower Onion 15222
Cupgrass 151655
Cup-grass 151655
Cuphea 114595,114606
Cupi-coloured Holly 203743
Cupi-coloured Michelia 252807
Cupi-coloured Rhododendron 330304
Cupid Peperomia 290426
Cupid's Bower 4070
Cupid's Dart 79271,79276
Cupid's Darts 79276
Cupid's Flower 410677
Cupid's-bower 4070
Cupid's-dart 79271,79276
Cupid's-darts 79271
Cupidone 79271,79276
Cupids-dart 79271
Cupilar-calyx Milkvetch 42251
Cupiliform Bluspine Craneknee 221647
Cup-leaf Desert Ceanothus 79933
Cupleaf Passionflower 285635
Cupleaf Woodbetony 287135
Cupleaflike Woodbetony 287136
Cup-leaved Passionflower 285635
Cup-of-gold 366871
Cup-of-wine 385302
Cup-plant 364301
Cuprea Bark 327661,327662
Cupri-coloured Clethra 96528
Cup-rose 282685
Cupscale 341937
Cupseed 68370,68371
Cup-shaped Evergreen Chinkapin 78874
Cupshapedstalk Clematis 94859
Cupshaped-stalk Clematis 94859
Cupstalk Agapetes 10358
Cupstalk Clematis 94859
Cupstem Balanophora 46812
Cupstem Snakemushroom 46812
Cupuacu 389403,389405
Cupula Oak 323986
Cupular Biondia 54523

Cupular Pencil-wood 139642
Cupular Willow 343258
Cupularcalyx Camellia 69025
Cupular-calyxed Camellia 69025
Cupular-calyxed Rhododendron 331503
Cupulate Dodder 114961
Curacao Aloe 17381,17383
Curacao Apple 382660
Curage 309199
Curagua 21485
Curare 378914
Curatella 114808
Curb Bushclover 227001
Curcas Bean 212127
Curculigo 114814,114838
Curcuma 114852
Curdly-greens 59387
Curds-and-cream 349044
Cure-all 175444,249504
Cure-for-all 305085
Curiosity Plant 83891
Curious Dutchmanspipe 34225
Curious Iris 208601
Curious Kadsura 214967
Curious Pseudanthistria 317194
Curlans 102727
Curldoddy 50825,216753,273531,302034,
302068,379660,397043
Curled Basil 290952
Curled Dock 339994
Curled Kale 59387
Curled Laelia 219675
Curled Lettuce 219490
Curled Mallow 243738,243771,243803,
243862
Curled Maple 3505
Curled Mint 250347,250420
Curled Mustard 59438
Curled Odontochilus 269018
Curled Parsley 292694
Curled Petal Little Man Orchid 178894
Curled Pondweed 312028,312079
Curled Tansy 383878
Curled Thistle 73337
Curled Tinospore 392252
Curled Wickerware Cactus 329695
Curled-leaved Kale 59493,59524
Curled-leaved Weeping Willow 343027
Curledsepal Snakegourd 396164
Curlew Berry 145073
Curlewberry 145073
Curley-doddies 273531,359158
Curlick 364557
Curl-leaf Cercocarpus 83819
Curl-leaf Mountain Mahogany 83819
Curl-leaf Mountain-mahogany 83818,83819
Curlock 326609
Curltop Lady's-thumb 309298
Curly Begonia 49723
Curly Bristle Thistle 73337
Curly Bristlethistle 73337
Curly Butterflybush 62009
Curly Butterfly-bush 62009
Curly Clematis 94870
Curly Dock 339994,339995
Curly Hairs Rhododendron 331992
Curly Mallow 243771,243862
Curly Mesquite 196197
Curly Mint 250347
Curly Motherweed 231504
Curly Muckweed 312079
Curly Nuts 102727
Curly Oat Grass 122292
Curly Paela 65053
Curly Palm 198806
Curly Plumeless Thistle 73337
Curly Pondweed 312079
Curly Poplar 311261
Curly Sedge 79746
Curly Sentry Palm 198806
Curly Summerlilic 62009
Curly Water-thyme 219776
Curly Waterweed 219768,219776
Curly Wig 79746
Curlybract Cottonthistle 271678
Curly-cup Gum-weed 181174,181178
Curly-head Goldenweed 323019
Curly-heads 95173
Curlyleaf Barberry 52105
Curly-leaf Mint 250450
Curly-leaf Pondweed 312079
Curly-leaf Reckrose 93141
Curly-leaved Barberry 52105
Curly-leaved Pondweed 312079
Curly-leaved Rock Rose 93141
Curlylip-orchis 62530
Curly-locks 392045
Curlypalm 198806
Curly-styled Wood Sedge 76056
Curlytop Gumweed 181174
Curly-top Knotweed 309298
Curna Palm 273248
Curnberries 334117
Curn-flower 11947
Curranbine 30081
Curran-petris 123141
Currant 333893,333928,334003,334062,
411979
Currant Bush 226234
Currant Dumpling 146724
Currant Grape 411924
Currant Neillia 263158
Currant Tomato 239165,367503
Currant-berry 334179
Currant-dumpling 146724
Currant-of-texas 52267
Currawong Acacia 1184
Curry 114871
Curry Bush 202134
Curry Leaf 260168
Curry Plant 189131,189458
Curry Sugarpalm 32349
Curry Tree 260168
Curryleaf Jasminorange 260168
Curryleaftree 260168
Curry-leaved Jasmin-orange 260168
Curry-plant 189131
Curse of Barbados 221238
Cursed Crowfoot 326340,326344
Cursed Thistle 73318,91770
Curt Rose 282685
Curtain Plant 61572
Curtis Aster 380977
Curtis Meranti 362202
Curtis Paphiopedilum 282809
Curtis' Gentian 173366
Curtis' Goldenrod 368057
Curtis' Lasianthus 222123
Curtis' Mouse-ear 82991
Curtiss Grape 411933
Curtiss' Milkwort 308014
Curtiss' Three-awn 33831
Curuba 362419
Curv Arisaema 33540
Curvate Wildryegrass 144306
Curvateflower Panicgrass 281532
Curvateleaf Kobresia 217150
Curve Leaf-tip Rhododendron 330028
Curve Lovegrass 147612
Curvebeak Monkshood 5096
Curvebract Gerbera 175145
Curvebract Goldenray 229011
Curvecalyx Bulbophyllum 63014
Curvecalyx Stonebean-orchis 63014
Curved Ardisia 31420,31422
Curved Drypetes 138581
Curved Hard-grass 283656
Curved Indian Nightshade 367249
Curved Sandwort 255564
Curved Sedge 75304
Curved Sichuan Square Bamboo 87618
Curved Sicklegrass 283656
Curved Wood Rush 238572
Curved Wood-rush 238572
Curvedawn Fimbristylis 166499
Curvedawn Fluttergrass 166499
Curvedflower Chasalia 86207
Curvedflower Corydalis 105778
Curvedflower Phlogacanthus 295022
Curvedflower Rockjasmine 23141
Curved-flowered Chasalia 86207

Curvedfruit Speedwell 407049
Curvedhair Shortstalk Monkshood 5070
Curvedleaf Ascocentrum 38189
Curved-leaf Holly 203744
Curvedleaf Iris 208510
Curved-leaf Spanish Dagger 416610
Curvedleaf Swordflag 208510
Curvedleaf Wampee 94191
Curved-prickle Rose 336380
Curved-prickled Rose 336380
Curvedsepal St. John's Wort 201826
Curve-dsepaled St. John's-wort 201826
Curved-stalk Pratt Barberry 52062
Curvedstamencress 72944,238005,238017
Curvedstipe Primrose 314327
Curvedstyle Meadowrue 388708
Curvedstyle Solomonseal 308528
Curvedtooth Shieldfruit 391420
Curvedtooth Thyrocarpos 391420
Curvedtube Woodbetony 287132
Curvedutricle Sedge 74339
Curveflower Corydalis 105778
Curvefruit Azalea 330311
Curvefruit Fenugreek 397192
Curvefruit Herb Naiad 262042
Curvefruit Naiad 262018
Curvefruit Sweetvetch 187813
Curvehair Falsepimpernel 231504
Curvehair Stairweed 142636
Curve-knife Gasteria 171609
Curveleaf Gentian 173369
Curveleaf Yucca 416610
Curve-leaf Yucca 416610
Curverwell Currant 333941
Curveseed Butterwort 83443
Curve-seed Butterwort 83443,326437
Curveshoot Rosewood 121647
Curvestalk Corydalis 106295
Curvestyle Azalea 330314
Curvethrum Gentian 173367
Curvibranch Monkshood 5195
Curvicalvarate Larkspur 124517
Curvihaired Rhododendron 330420
Cuscus 407585
Cuscuta Family 115155
Cushag 359158
Cushaw 114273,114292
Cushaw Squash 114292
Cush-cush 131880
Cush-cush Yam 131880
Cush-cush Yampee 131880
Cushia 192358
Cushings Curshins 34536
Cushion 34536,216753
Cushion Baby's-breath 183221
Cushion Bush 67533,227940
Cushion Cactus 107076
Cushion Calamint 97060
Cushion Cassiope 78642
Cushion Euphorbia 159615
Cushion Foxtail Cactus 107007
Cushion Gypsophila 183221
Cushion Honeysuckle 236002
Cushion Miner's-lettuce 94357
Cushion Pink 34536,363144
Cushion Rockfoil 349819
Cushion Rockjasmine 23313
Cushion Spurge 159615
Cushion Wild Buckwheat 152346
Cushion Wildberry 362367
Cushion Willow 343119
Cushionbury Wild Buckwheat 152365
Cushionshape Hippolytia 196701
Cushionshaped Bush Cinquefoil 312613
Cushion-shaped Cassiope 78642
Cushionshaped Ceratoides 218124
Cushionshaped Echeveria 140004
Cushion-shaped Jasmine 211932
Cushion-shaped Parrya 285014
Cushion-shaped Sibbaldia 362367
Cusick Camas 68791
Cusick Camass 68791
Cusick's Aster 380859
Cusick's Hawksbeard 110746
Cusick's Pincushion 84487
Cusick's Sunflower 188939
Cusick's Wild Buckwheat 151974
Cusickii Camass 68794
Cusp Dodder 115003
Cusp Tangbamboo 364826
Cusparia Bark 24642
Cuspid Chamabainia 85107
Cuspidate Banfenghe 357976
Cuspidate Chinese Stachyurus 373532
Cuspidate Cypressgrass 118679
Cuspidate Distylium 134928
Cuspidate Elatostema 142640
Cuspidate Galingale 118679
Cuspidate Gnetum 178531
Cuspidate Hyssop 203084
Cuspidate Inula 138890
Cuspidate Mosquitoman 134928
Cuspidate Mountainash 369375
Cuspidate Olive 270089
Cuspidate Semiliquidambar 357976
Cuspidate Smallleaf Elatostema 142785
Cuspidate Spindle-tree 157400
Cuspidate Stairweed 142640
Cuspidate Tea 69033
Cuspidated Gladiolus 176144
Cuspidateleaf Acanthopanax 2370
Cuspidateleaf Bladderwort 403792
Cuspidateleaf Kalidium 215325
Cuspidateleaf Saltclaw 215325
Cuspidate-leaved Acanthopanax 2370
Cuspidate-leaved Jacaranda 211235
Cuspidate-leaved Kalidium 215325
Cuspid-leaf Blueberry 403792
Cussonia 114040
Custard Apple 25828,25836,25838,25882,25898
Custard Cheese 243840
Custard Cup 146724
Custard Squash 114301
Custardapple 25828,25882
Custard-apple 25828,25836,25882,25898
Custardapple Family 25909
Custard-apple Family 25909
Custard-cheese 243840
Custard-cups 146724
Custin 316470
Cut Cowparsnip 192393
Cut-and-come-again 131516
Cutandia 115252
Cutberdill 2695
Cutberdole 2695
Cutch-tree 1120
Cuted Marsdenia 245819
Cuted Milkgreens 245819
Cut-eye Bean 71042
Cutfinger 409335
Cut-finger 355202,401711,404316,409335,409339
Cut-finger Leaf 404316
Cutfruit Tanoak 233401
Cutgrass 223973
Cut-grass 223973,223992,302068
Cut-heal 404316
Cution Bush 227938
Cution Pink 363144,363399
Cut-leaf banksia 47660
Cutleaf Barberry 51470
Cut-leaf Beech 162431
Cut-leaf Birch 53564
Cutleaf Blackberry 338695
Cut-leaf Blackberry 338695
Cutleaf Brickellia 60128
Cutleaf Burnweed 148152
Cutleaf Bushclover 226742
Cutleaf Coneflower 339574,339582
Cut-leaf Coneflower 339574
Cutleaf Crabapple 243722
Cutleaf Daisy 145542,145543
Cut-leaf Elderberry 345636
Cut-leaf Evening-primrose 269446,269456
Cutleaf False Oxtongue 55768
Cut-leaf Full-moon Maple 3035
Cutleaf Geranium 174569
Cutleaf Germander 388015
Cut-leaf Germander 388015,388128
Cutleaf Gilia 175687

Cut-leaf Goldenrod 367965
Cutleaf Goosefoot 87095
Cut-leaf Goosefoot 139689
Cutleaf Ground Cherry 297650
Cutleaf Groundcherry 297650
Cutleaf Guinea Flower 194664
Cut-leaf Ironplant 414980
Cutleaf Lilac 382195
Cut-leaf Lilac 382195
Cutleaf Medick 247337
Cut-leaf Mignonette 327866
Cutleaf Mono Maple 3179
Cutleaf Nightshade 367689
Cut-leaf Nightshade 367689
Cutleaf Philodendron 294796
Cut-leaf Philodendron 294839
Cutleaf Self Heal 316116
Cutleaf Selfheal 316116
Cutleaf Silverpuffs 253927
Cutleaf Staghorn Sumac 332917
Cut-leaf Staghorn Sumac 332917
Cutleaf Stephanandra 375818
Cut-leaf Stephanandra 375817
Cut-leaf Stephanandre 375817
Cutleaf Stork's Bill 153825,153902
Cutleaf Teasel 133494
Cutleaf Vervain 176686
Cutleaf Vipergrass 354881
Cut-leafed Mint Bush 315600
Cut-leaved Anemone 23891
Cut-leaved Barberry 51470
Cut-leaved Beech 162431
Cut-leaved Blackberry 338695
Cut-leaved Bugbane 91026
Cut-leaved Bugleweed 239186
Cut-leaved Coneflower 339574
Cut-leaved Crabapple 243722
Cut-leaved Crane's-bill 174569
Cut-leaved Cranesbill 174569
Cut-leaved Dead Nettle 220392
Cut-leaved Deadnettle 220392
Cut-leaved Dead-nettle 220392
Cut-leaved Evening Primrose 269456
Cut-leaved Evening-primrose 269456
Cut-leaved Geranium 174569
Cut-leaved Germander 388015
Cutleaved Japanese Maple 3309
Cut-leaved Lavender 223308
Cut-leaved Lettuce 219494
Cut-leaved Maple 2770
Cut-leaved Nightshade 367689
Cut-leaved Selfheal 316116
Cut-leaved Self-heal 316116
Cut-leaved Silverpuffs 253927
Cut-leaved Teasel 133494
Cut-leaved Toothwort 72723,125846
Cut-leaved Verbena 405809
Cut-leaved Violet 410380
Cut-leaved Walnut 212639
Cut-leaved Water-parsnip 53157
Cut-leaved Whitebeam 369432
Cutler's Alpine Goldenrod 368209
Cut-lobed Spurge 159788
Cut-petaled Hibiscus 195216
Cut-scale Canary Grass 293764
Cut-scale Canary-grass 293764
Cutstyle Flower 351883
Cutting Lettuce 219490
Cutting Rockjasmine 23153
Cuttonguetree 413124
Cutwafed Stephanandra 375817
Cuviera 115278
Cuyamaca Cypress 114652
Cuyamaca Larkspur 124282
Cyaennepepper 72068
Cyamopsis 115315
Cyananthus 115338
Cyanofruit Memecylon 249937
Cyanolimne Sand Willow 343262
Cyanotis 115521
Cyathocline 115639
Cyathostemma 115690
Cyathula 115697
Cycad 115794,115915,131436,131441,145238,145240
Cycad Family 115793
Cycas 115794,115884
Cycas Family 115793
Cyclam Leaf Chinese Groundsel 365041
Cyclamen 115931,115949,115953,115965
Cyclamen Cherry 83184
Cyclamen Poppy 146054
Cyclamen-flowered Daffodil 262395
Cyclamenleaf Sinosenecio 365041
Cyclamenlike Cremanthodium 110372
Cyclamenlike Nutantdaisy 110372
Cyclamen-primula 135165
Cyclamineus Daffodil 262395
Cyclamineus Narcissus 262395
Cyclanthera 115990
Cyclanthus Family 115989
Cycle Childvine 400879
Cycle Primrose 314116
Cyclea 116008
Cyclebill Woodbetony 287137
Cyclecalyx Spapweed 204886
Cyclelike Tylophora 400879
Cycleshape Conehead 378116
Cyclicscale Horny Tanoak 233160
Cyclicscale Pipe Tanoak 233160
Cyclobalanopsis 116044,116177
Cyclocarya 116239,116240
Cyclo-fruited Mucuna 259501
Cyclops Acacia 1161
Cyclorhiza 116370
Cyderach 309199
Cydonia 84549
Cylinder Colletia 99815
Cylinder Immortelle 415326
Cylinderflower Lovegrass 147615
Cylinder-fruit Bruguiera 61258
Cylindric Bruguiera 61258
Cylindric Disa 133757
Cylindric Fruit Slugwood 50496
Cylindric Kiwifruit 6577
Cylindric Lymegrass 144266
Cylindric Spike Kyllingia 218518
Cylindric Spiradiclis 371747
Cylindric Wildryegrass 144266
Cylindric Willowweed 146677
Cylindric Wolfberry 239029
Cylindric Yam 131523
Cylindrical Actinidia 6577
Cylindrical Blazing-star 228454
Cylindrical Camellia 69038
Cylindrical False Solomonseal 366200
Cylindrical Flower Leucothoe 228178
Cylindrical Lovegrass 147614
Cylindrical Poppy 282539
Cylindrical Pricklypear 272849
Cylindrical Pterygiella 320916
Cylindrical Sansevieria 346070
Cylindrical Slugwood 50496
Cylindrical Wolfberry 239029
Cylindrical-flower Leucothoe 228178
Cylindric-branch Pyrenocarpa 322634
Cylindric-branched Pyrenocarpa 322634
Cylindric-inflorescence Addermonth Orchid 130181
Cylindricleaf Leek 15819
Cylindricleaf Onion 15819
Cylindri-flower Eucalyptus 155551
Cylindrokelupha 116585
Cylindrycculm Spikesedge 143257
Cymaria 116709
Cymbaeform Monkshood 5153
Cymbalaria Halerpestes 184703
Cymbalaria Soda Buttercup 184703
Cymbaria 116745
Cymbidiella 116755
Cymbidium 116769,116915
Cymbidium Orchid 116769
Cymbidium-like Epigeneium 146535
Cyme Goldenray 229014
Cyme Metalleaf 296377
Cyme Rose 336485
Cymligerous Deutzia 126902
Cymophora 117316,117318
Cymos Fimbristylis 166236
Cymos Fluttergrass 166236
Cymose Billardiera 54435

Cymose Larkspur 124336
Cymose Metalleaf 296377
Cymose Olinia 270468
Cymous Cuminum 114503
Cynocramba Family 389181
Cynomorium 118127
Cynomorium Family 118124
Cynop Plantain 301945
Cynthia 218202
Cyp 104153
Cypella 118393
Cyperus Sedge 75895
Cyperus-like Sawgrass 245392
Cyperuslike Sedge 73894
Cyphel 255576,255578,358062
Cyphokentia 119963
Cypholophus 119968
Cyphotheca 120275
Cypre 104153
Cypress 85263,85329,114649,178019,253164,385253,385262
Cypress Euphorbia 158730
Cypress Family 114645
Cypress Grass 118428
Cypress Knees 385262
Cypress Lavender Cotton 346250
Cypress Lavender-cotton 346250
Cypress Oak 324338
Cypress of Babylon 114871
Cypress of India 114871
Cypress of Malabar 114871
Cypress Pine 67405,67433,386942,414060
Cypress Poplar 311400
Cypress Shrub 223454
Cypress Spurge 158730
Cypress Vine 207546,208120,255398,323450,323465
Cypress Wormwood 36088
Cypressgrass 118428
Cypress-grass 118428
Cypress-knees 385262
Cypress-like Sedge 75895
Cypress-pine 67405,67432
Cypress-root 119125
Cypressvine 323465
Cyprian Cedar 80085
Cyprian Golden Oak 323649
Cypripedium 120287
Cyprus Bee-orchid 272449
Cyprus Cedar 80085
Cyprus Helleborine 147245
Cyprus Iris 208511
Cyprus Oak 324134
Cyprus Pine 299964
Cyprus Plane Tree 302580
Cyprus Turpentine 301002
Cyrillo 374968
Cyrtandra 120501,120508
Cyrtanthera 120525
Cyrtanthus 120532
Cyrtococcum 120624
Cyrtosia 120758
Cyrtosperma 120768
Cystacanthus 120794
Cystoid Kobresia 217177
Cystoid Mosquitotrap 117675
Cystoid Zornia 418335
Cystoidflower Swordflag 208921
Cystoidglandula Felwort 235464
Cystopetal 320203
Cytinus 120875
Czetyrkinin Peashrub 72253
Czetyrkinin Pea-shrub 72253
Czimganic Roegneria 335287

D

Daa Nettle 220416
Daa-nettle 220416
Dabanshan Azalea 331561
Dabanshan Milkvetch 42257
Dabanshan Rhododendron 331561
Dabbery 334250
Dabieshan Arundinella 37407
Dabieshan Dutchmanspipe 34160
Dabieshan Fritillary 168344
Dabieshan Hickory 77892
Dabieshan Holly 203745
Dabieshan Photinia 295664
Dabieshan Pine 299939
Dabieshan Sage 344995
Dabieshan Sedge 74252
Dabieshan Wildginger 37607
Dabieshan Willow 343264
Daboma 300655
Dachang Camellia 69686
Dachang Tea 69686
Dacheng Azalea 330493
Dacheng Rhododendron 330493
Dacrydium 121084
Dacts Tree 295461
Dactylicapnos 128288,128318
Dactylis 121226
Dactyloideus Violet 409878
Dadap 154726,154734
Daddy Man's Beard 95431
Daddy's Beard 95431
Daddy's Whiskers 95431
Daddy's White Shirt 68713
Dadu Monkshood 5208
Daekyellow Lycaste 238742
Dafang Snakegourd 396178
Daff Lily 262441
Daffadilly 262441
Daffadoondilly 262441
Daffadowndilly 262441
Daffany 122515
Daffidowndilly 262441
Daffodil 122515,168476,262348,262404,262417,262441
Daffodil Azalea 331749
Daffodil Garlic 15509
Daffodil Orchid 208323,208324
Daffodilly 262441
Daffodowndilly 262441
Daffy Downdilly 262441
Daffydowndilly 262441
Daft Berries 44708
Daft Lily 262441
Dagga 71220
Dagger 99297
Dagger Cholla 181453
Dagger Cocklebur 415053
Dagger Flower 208771,208924,244728
Dagger Hakea 184646
Dagger Plant 234437,416538
Dagger Weed 415053,416603
Dagger-flower 244728
Daggerfruit Hakea 184630
Daggerleaf Cottonrose 235262
Dagger-leaf Rush 213093
Dagger-leaves Heath 150049
Daggerweed 415053
Daghestan Skullcap 355422
Daghestan Sweetclover 249232
Daguan Spapweed 204888
Daguan Willow 343267
Dahai Rockfoil 349233
Dahai Southstar 33307
Dahai Woodbetony 287738
Daheba Kengyilia 215944
Daheba Roegneria 335432
Daheishan Camellia 69688
Daheshan Holly 204306
Dahlberg Daisy 391050
Dahlia 121539,121561
Dahlia Fair 121561
Dahlia Hedgehog Cactus 140293

Dahlia-rooted Cereus 288989
Da-Ho 28044
Dahoba Roegneria 335432
Dahoma 300655
Dahomey Rubber 165263
Dahoon 203611,204391
Dahoon Holly 203611
Dahur Asparagus 38977
Dahur Birch 53400
Dahur Bugbane 91008
Dahur Cranesbill 174559
Dahur Cymbaria 116747
Dahur Deerdrug 242679
Dahur Falsehellebore 405585
Dahur Fescuegrass 163885
Dahur Goldenrod 368502
Dahur Hawthorn 109662
Dahur Helictotrichon 190145
Dahur Lily 229828
Dahur Lymegrass 144271
Dahur Milkvetch 42258
Dahur Mint 250353
Dahur Mouseear 82786
Dahur Pulsatilla 321676
Dahur Pyrola 322814
Dahur Sakebed 97703
Dahur Savin 213752
Dahur Speedwell 317909
Dahur Sweetvetch 187849
Dahur Willowweed 146680
Dahuria Calystegia 68676
Dahuria Cranesbill 174559
Dahuria Cymbaria 116747
Dahuria Falsehellebore 405585
Dahuria Gentian 173373
Dahuria Glorybind 68676
Dahuria Hawthorn 109662
Dahurian Asparagus 38977
Dahurian Birch 53400
Dahurian Bluegrass 305380
Dahurian Buckthorn 328680,328684
Dahurian Bugbane 91008
Dahurian Bushclover 226751
Dahurian Bush-clover 226751
Dahurian Currant 334147
Dahurian False Solomonseal 242679
Dahurian False Tamarix 261254
Dahurian Falsetamarisk 261254
Dahurian Glorybind 103013
Dahurian Goatwheat 44260
Dahurian Goldenrod 368502
Dahurian Hawthorn 109662
Dahurian Helictotrichon 190145
Dahurian Juniper 213752
Dahurian Ladybell 7850
Dahurian Larch 221866
Dahurian Lespedeza 226751
Dahurian Lily 229828
Dahurian Loosestrife 239902
Dahurian Milkvetch 42258
Dahurian Mint 250353
Dahurian Moonseed 250228
Dahurian Oxtongue 298580
Dahurian Patrinia 285859
Dahurian Pulsatilla 321676
Dahurian Pyrola 322814
Dahurian Rhododendron 330495
Dahurian Rose 336522
Dahurian Sedge 74253
Dahurian Speedwell 317909
Dahurian Spiraea 371895
Dahurian Sweetvetch 187849
Dahurian Thyme 391167
Dahurian Wild Rye 144271
Dahurian Wildrye 144271
Dahuricum Bedstraw 170340
Dahursk Buckthorn 328680
Dai Dendrocalamopsis 47302
Daibu Anisetree 204501
Daibu Eightangle 204501
Daibuzan Spotleaf-orchis 179600
Daidai 93368
Daidaihua 93368
Daikon 326616,326622
Dailing Angelica 24334
Dailing Monkshood 5066
Dailing Sedge 74256
Dailing Willow 343269
Daily Bread 336419
Daily-dew 138261
Daimio Oak 323814
Daimyo Oak 323814
Dainty Barberry 51483
Daintypetal Rose 336615
Dairity Moraea 258682
Daisha 361380
Daishan Rose 336513
Daisy 50808,50825,124767,258191,359158
Daisy Bush 270184,270194
Daisy Desertstar 257833
Daisy Dichrocephala 129075
Daisy Family 101642
Daisy Fisheyeweed 129075
Daisy Fleabane 150464,150724,150862,
150925,150980,150989
Daisy Gentian 173842
Daisy Goldins 227533
Daisy Tree 270207,306223
Daisybush 270184
Daisy-leaf Goodenia 179557
Daisyleaf Goosefoot 139693
Daisyleaf Parrya 224210
Daisyleaf Primrose 314165
Daisyleaf Statice 230543
Daisyleaf Storkbill 288468
Daisy-leaved Bittercress 72698
Daisy-leaved Chickweed 256711
Daisy-leaved Toadflax 21765
Daisylike Groundsel 359253
Daiyunshan Arrowwood 408163
Daiyunshan Hard-leaf Holly 203811
Daiyunshan Rhododendron 331203
Daiyunshan Viburnum 408163
Dajian Windhairdaisy 348850
Dajianlu Gentian 173959
Dajianlu Rockfoil 349981
Dajianlu Sedge 76496
Dajin Buttercup 326257
Dajin Corydalis 105802
Dajin Dogwood 380446
Dajin Fritillary 168389
Dakota Mock Vervain 176686
Dakota Potato 29298
Dakota Vervain 176686,405809
Dakua 10500
Dakua Tree 10509
Dalai Sagebrush 35364
Dalat Cylindrokelupha 116593
Dale Cup 68189
Dalea 121871
Dalechampia 121912
Dale-cup 68189,325518,325666,326287
Dalhousiae's Rhododendron 330501
Dali Asparagus 39225
Dali Azalea 331921
Dali Barberry 52214
Dali Begonia 50354
Dali Bentgrass 12365
Dali Betony 373456
Dali Buckthorn 328679
Dali Butterflybush 62178
Dali Butterfly-bush 62178
Dali Cacalia 283872
Dali Camellia 69691
Dali Cinquefoil 313024
Dali Corallodiscus 103972
Dali Corybas 105533
Dali Corydalis 106500
Dali Daphne 122432
Dali Dragonhead 137671
Dali Elaeagnus 142203
Dali Euonymus 157318,157907
Dali Falsehellebore 405638
Dali Fishvine 125970
Dali Gentian 173956
Dali Greenorchid 137671
Dali Helmet Orchid 105533
Dali Iris 208495
Dali Jerusalemsage 295105
Dali Jewelvine 125970
Dali Knotweed 309839

Dali Larkspur 124618
Dali Lily 230048
Dali Lobelia 234816
Dali Madder 337989
Dali Oak 323778
Dali Paris 284310
Dali Pearweed 239874
Dali Persimmon 132061
Dali Petrocosmea 292556
Dali Pricklyash 417294
Dali Primrose 315030
Dali Rabdosia 324890
Dali Rhododendron 331921
Dali Roscoea 337084
Dali Sedge 76083
Dali Slipper 120336
Dali Sow-thistle 90856
Dali Spapweed 205355
Dali Speedwell 407127
Dali Stenoseris 375714
Dali Stonebutterfly 292556
Dali Summerlilic 62178
Dali Swordflag 208495
Dali Tea 69691
Dali Teasel 133466
Dali Willow 343273
Dali Woodbetony 287741
Daliangshan Aster 41338
Dalibarda 339171
Dallid Fissistigma 166680
Dallis Grass 285423
Dallisgrass 285390,285423
Dalmais Milkwort 308016
Dalmatian Broom 172928
Dalmatian Clover 396876
Dalmatian Insect Flower 322660
Dalmatian Laburnum 292736,292737
Dalmatian Pine 300113
Dalmatian Pyrethrum 322660
Dalmatian Toadflax 230978
Dalmatin Iris 208744
Dalton Willow 343275
Dalton Woodbetony 287140
Dalton's Glochidion 177117
Dalung Willow 343277
Daluokou Spiraea 372094
Dalxell Spapweed 204889
Dalxell Touch-me-not 204889
Dalziel Opithandra 272593
Dalziel Pteroptychia 320793
Damaling Ginkgo 175816
Damar Minyak 10496
Damar Pine 10492
Damara Tree 10492
Damarapa Bamboo 196347
Damar-pine 10492
Damascisa 176941,176949
Damascus Rose 336514
Damascus Saltwort 344758
Damask Fennel 266225
Damask Rose 336514,336581
Damask Violet 193417
Damasonium 121993
Damayeshu 112874
Dambala 319114
Dame's Gilliflower 193417
Dame's Rocket 193387,193417
Dame's Violet 193417
Dame's-violet 193387,193417
Dameihe Ginkgo 175817
Dames Rocket 193417
Dames Violet 193417
Damiana 400489
Damianita 89337
Damianita Daisy 89337
Damiao Mountain Pepper 300381
Damiaoshan Bauhinia 49063
Damiaoshan Camellia 69690
Damiaoshan Pepper 300381
Damiaoshan Snakegourd 396179
Damiaoshan Synotis 381933
Damiaoshan Tailanther 381933
Damiaoshan Tanoak 233169
Daming Bitterbamboo 304032
Daming Dichroa 129032
Daming Mananthes 244286
Daming Pine 300270
Damingshan Chinkapinga 78918
Damingshan Cyclobalanopsis 116078
Damingshan Dichroa 129032
Damingshan Evergreenchinkapin 78918
Damingshan Fig 164891
Damingshan Oak 116078
Damingshan Pine 300270
Damingshan Platanthera 302299
Damingshan Qinggang 116078
Damingshan Square Bamboo 87558
Damingshan Square-bamboo 87558
Dammar Pine 10492,10494,10496
Dammar-pine 10492
Damnacanthus 122027
Damoceen 90019
Dampsedge 66337,66343
Dams Chin Cactus 182439
Damsel 316470
Damsel of the Wood 53563
Damsil 316470,316819
Damson 316470
Damson Plum 316382,316470
Damson-leaved Sallow 343278
Damson-plum 90082
Dan's Cabbage 359279
Danba Azalea 330505
Danba Corydalis 105950
Danba Rhododendron 330505
Danba Yellowhair Maple 2978
Danceweed 98152
Dancing Grass 98152
Dancing Lady 168735
Dancy Tangerine 93837
Dandelion 68189,110710,384420,384519,384714
Dandelion Hawksbeard 110990
Dandelionleaf Windhairdaisy 348849
Dandong Dandelion 384514
Dandong Figwort 355148
Dandy Goshen 273545
Dandy Goslings 273531,273545
Dandy Gussets 273529
Dandy-gusset 273529
Dane's Blood 70054,70055,321721,345594
Dane's Elder 345594
Dane's Flower 321721
Dane's Weed 321721
Dane's Wood 345594
Dane's-blood 70054
Dane's-blood Bellflower 70054
Daneball 345594
Danesblood 70055
Danesblood Bellflower 70054
Daneweed 154301,345594
Danewort 345594
Danford Iris 208514
Dangle Bells 102867
Dangleberry 172257
Danglepod 361380
Dangling Bells 102867
Danica Arborvitae 390594
Daniell 381982
Daniell Evodia 161323
Danish Beech 162400
Danish Scurvy-grass 97981
Dan-shen 345214
Dansia Phoenix-tree 166617
Danta 265586
Danthonia 122180
Danube Gentian 173701
Danube Grass 295888
Danxiashan Phoenix Tree 166617
Danya 354345
Danzhai Camellia 69039
Danzig Fir 300223
Danzig Pine 300223
Danzig Vetch 408337
Dao 137715
Daocheng Apple 243588
Daocheng Arisaema 33308
Daocheng Barberry 51519
Daocheng Crabapple 243588
Daocheng Cremanthodium 110374
Daocheng Draba 136974

Daocheng Gentian 173377
Daocheng Groundsel 358677
Daocheng Incarvillea 205549
Daocheng Indigo 205868
Daocheng Nutantdaisy 110374
Daocheng Rockfoil 349235
Daocheng Sharptooth Hornsage 205549
Daocheng Spiraea 371896
Daocheng Woodbetony 287141
Daofu Azalea 330517,331754
Daofu Barberry 51527
Daofu Catchfly 363400
Daofu Rabdosia 209653
Daofu Rhododendron 331754
Daofu Rockfoil 349575
Daofu Stonecrop 356768
Daoxian Citrus 93437
Daoxian Linden 391734
Daoxian Wild Mandarin 93437
Daozhen Machilus 240575
Dapao Woodbetony 287744
Dapaoshan Rockfoil 349726
Dapaoshan Stonecrop 356707
Daphne 122359,122515
Daphne Cneorum 122515
Daphne Flowered Rhododendron 331700
Daphne Heath 58528
Daphne Pittosporum 301253
Daphne Rhododendron 331700
Daphne Willow 343280
Daphneflower Valerian 404248
Daphne-like Willow 343280
Daphniphyllum 122661
Daphniphyllum Family 122656
Dapieh Mountain Pine 299939
Dapingzi Premna 313757
Dapingzi Sedge 76487
Dapingzi Skullcap 355791
Dapplebloom Rhododendron 331480
Dapu Aster 40637
Dapu Azalea 331916
Dapu Rhododendron 331916
Dapu Roughhaired Holly 203580
Daqingshan Bellflower 70060
Daqingshan Crazyweed 278805
Daqingshan Kobresia 217152
Daqingshan Milkvetch 42263
Darbottle 81237
Darele 340082
Darf Altai Draba 136925
Darf Altai Whitlowgrass 136925
Darin's Wild Buckwheat 151941
Darjeeling Edgaria 141462
Darjeeling Guger Tree 350949
Darjeeling Gugertree 350949
Darjeeling Guger-tree 350949
Dark Eugeissoma 156116
Dark Golden Rhododendron 331477
Dark Green Rhododendron 330162
Dark Grey Mouse Ear 82793
Dark Hair Crowfoot 48933
Dark Hawksbeard 110739
Dark Linear Japanese Maple 3301
Dark Mullein 405737
Dark Purple Thinleaf Stringbush 414204
Dark Rampion 298066
Dark Red Helleborine 147112
Dark Red Lauan 362217
Dark Red Meranti 362202,362216
Dark Red Seraya 362216
Dark Tea 68923
Dark-bloodcoloured Cinquefoil 312393
Dark-blue Viburnum 407685
Dark-brown Blueberry 403836
Darkbrown Fimbristylis 166329,166420
Darkbrown Fluttergrass 166329
Dark-brown Leaves Azalea 331705
Dark-brown Leaves Rhododendron 331705
Darkbrown-hairy Rhododendron 330752
Darkcolor Catchfly 363267
Darkcoloured Carpesium 77190
Dark-fruit Barberry 51336
Dark-golden Rhododendron 331477
Darkgray-leaved Cyclobalanopsis 116118
Darkgreen Azalea 330162
Dark-green Barberry 51340
Dark-green Bulrush 353231
Darkgreen Carludovica 77039
Darkgreen Corydalis 106146
Darkgreen Dandelion 384516
Dark-green Eastern Arborvitae 390613
Darkgreen Elatostema 142608
Dark-green Encephalartos 145237
Darkgreen Helicia 189941
Darkgreen Maizailan 11305
Darkgreen Milkvetch 42350
Dark-green Mouse-ear Chickweed 82793
Darkgreen Norway Maple 3442
Darkgreen Plantain Lily 198637
Dark-green Rhododendron 330162
Dark-green Sedge 76669
Darkgreen Stairweed 142608
Darkgreen Woodbetony 287022
Darkleaf Azalea 330083
Darkleaf Litse 233844
Dark-leaf Machilus 240642
Darkleaf Nanmu 295393
Dark-leaf Phoebe 295393
Dark-leaved Litse 233844
Dark-leaved Machilus 240642
Dark-leaved Phoebe 295393
Dark-leaved Rhododendron 331705
Dark-leaved Willow 343776
Dark-prasine Barberry 51340
Darkpurple Aralia 30591
Darkpurple Azalea 330161
Dark-purple Chinese Waxmyrtle 261214
Darkpurple Columbine 30001
Dark-purple Common Barberry 52331
Dark-purple Cymbidium 117042
Darkpurple Dubyaea 138764
Dark-purple Epidendrum 146383
Dark-purple European Spindle-tree 157441
Darkpurple Hemsley Monkshood 5248
Dark-purple Horned Violet 409860
Dark-purple Japanese Barberry 52227
Darkpurple Jerusalemsage 295060
Dark-purple Liparis 232085
Dark-purple Lysionotus 239930
Darkpurple Pleurospermum 304760
Dark-purple Rhododendron 330161
Darkpurple Risleya 334782
Darkpurple Sage 344866
Darkpurple Swertia 380121
Darkpurple Teasel 133460
Darkpurple Trachydium 304807
Darkpurple Twayblade 232085
Darkpurple Whiter Daphne 122547
Darkpurpleflower Gongora 179290
Darkpurpleflower Window Orchid 113740
Dark-purple-flowered Indigo 205698
Dark-red Barberry 51287
Darkred Caraway 77774
Darkred Gentian 173828
Darkred Ginger 417966
Darkred Premna 313720
Darkred Sage 344867
Dark-red Straight-raceme Barberry 51987
Darkred Torchflower 100031
Darkred Zinger 417966
Dark-red-stem Barberry 52124
Darkrose Fawnlily 154944
Dark-scale Cotton-grass 152798
Dark-spined Prickly Pear 272797
Darkspinose Neoporteria 264334
Darkviolet Pimpinella 299359
Dark-winged Orchid 273696
Dark-winged Orchis 273696
Darkyellow Selenipedium 357752
Darlahad Barberry 51327
Darley Dale Heath 148942
Darling April 315101
Darling Lily 111195
Darling Ofapril 315101
Darling Pea 380067,380068
Darlington Oak 324002,324095
Darmell 273531
Darmell Goddard 37015,68189,72934,145487,174569,174739,248411,273469,273531,273545,277648,349419,363461,363673,374916

Darnel 235373
Darnel Lovegrass 147482
Darnel Rye Grass 235373
Darnel Ryegrass 235373
Darnel Rye-grass 235373
Darning Old Wife's Needle 350425
Darris Smilax 366308
Darris Tarphochlamys 385068
Darrow's Wild Buckwheat 151977
Dartford Cotuneaster 107607
Dartnell 167955
Dart's Gold Ninebark 297843
Darwas Iris 208516
Darwin Barberry 51520
Darwin Stringy-bark 155761
Darwin's Berberis 51520
Dash Bagger 86901
Dashan Swertia 380391
Dashan Woodbetony 287737
Dasheen 99910
Dashel 91770,92485,384714
Dashel Flower 384714
Dashen 99899,99910,99919,99936
Daso 99699
Dassel 91770
Dasymaschalon 122924
Datch 408571
Date 295451,295461
Date Palm 267838,295451,295461
Date Plum 132030,132264,132466
Date Sugar Palm 295487
Datepalm 295451,295461
Date-palm 295451,295461
Datepalm Machilus 240671
Date-plum 132030,132219,132264,132466
Dateplum Persimmon 132264
Date-plum Persimmon 132264
Datil 416556
Datil Yucca 416556
Datilillo 416664
Datisca 123027,123028
Datisca Family 123033
Datong Larkspur 124545
Datong Milkvetch 42268
Datoucai Leaf-mustard 59461
Dattock 126805
Datun Wildginger 37674
Datung Milkvetch 42268
Datunshan Rhododendron 331305
Datura 123036,123065,123077
Dau 133559
Daughter-before-mother 99297
Dauke 123141
Daur Batkudze 250228
Daur Moonseed 250228
Daur Shamrue 185635
Daurian Haplophyllum 185635
Daurian Peashrub 72288
Daurian Pea-shrub 72288
Davall Sedge 74258
Davall's Sedge 74258
Davazamc Pea-shrub 72216
David Adonis 8353
David Anemone 23779
David Arrowwood 407774
David Astilbe 41806
David Azalea 330515
David Barberry 51526
David Barrenwort 146977
David Bushclover 226764
David Bush-clover 226764
David Calanthe 65915
David Cattail 401120
David Christmas bush 14184
David Christmasbush 14184
David Christmas-bush 14184
David Corydalis 105808
David Currant 333950,334073
David Cystopetal 320215
David David 204891
David Elaeagnus 141977
David Elm 401489
David Epimedium 146977
David False Panax 266913
David False-panax 266913
David Fritillary 168390
David Gentian 173378
David Goldsaxifrage 90349
David Goldwaist 90349
David Gooseberry 333950
David Greenbrier 366309
David Hemiptelea 191629
David Hydrangea 199865
David Keteleeria 216120,216125
David Larkspur 124164
David Lily 229829
David Lobelia 234398
David Maple 2920
David Milkvetch 42269
David Ostryopsis 276846
David Parnassia 284522
David Peach 20908
David Pine 299799
David Pleurospermum 304768
David Poplar 311281
David Primrose 314295
David Pseudostellaria 318493
David Ribseedcelery 304768
David Rose 336519
David Sedge 74259
David Sinacalia 364519
David Smilax 366309
David Sophora 368994
David Stranvaesia 377444
David Tubergourd 390122
David Vetchling 222707
David Viburnum 407774
David Windflower 23779
David Woodbetony 287145
David Xmas Bush 14184
David's Harp 308564,308607
David's Keteleeria 216120
David's Lily 229829
David's Maple 2920
David's Mountain Laurel 368994
David's Peach 20908
David's Pine 299799
David's Rhododendron 330515
David's Root 80309
David's Seal 308607
David's Spurge 158738
David's Viburnum 407774,407775
Davidia Family 123253
Davids Harp 308529
Davidson Azalea 330516
Davidson Photinia 295675
Davidson Rhododendron 330516
Davidson's Indian-lettuce 94346
Davidson's Orach 44645
Davidson's Plum 123261
Davidson's Wild Buckwheat 151979
Davil Devilrattan 121514
Davis Alleghany Plum 316205
Davis Begonia 49762
Davis Cholla 116653
Davis Mountains Mock Vervain 176696
Davis' Desertdandelion 242951
Davis' Fleabane 150576
Davis' Hedgehog Cactus 140238
Davis' Sedge 74261
Davis' Thistle 92050
Davur Buckthorn 328680
Davur Windhairdaisy 348243
Davurian Buckthorn 328680
Dawdle-grass 60382
Daweishan Begonia 49763
Daweishan Elatostema 142644
Daweishan Hopea 198143
Daweishan Lady's Slipper 120339
Dawn Cypress 252540
Dawn Redwood 252538,252540,252541
Dawo Rabdosia 209653
Dawo Rhododendron 330517
Dawo Rose 336527
Dawson Magnolia 416692
Dawu Anise-tree 204583
Dawu Aspidistra 39529
Dawu Hemlock 399882
Dawu Rhododendron 331754
Dawu Sandwort 31847
Dawushan Litse 233975

Dawushan Newlitse 264043
Dawyck Beech 162409,162410
Daxikeng Burmannia 63967
Daxin Begonia 49765
Daxing'anling Monkshood 5155
Day Cestrum 84408
Day Daisy 227533
Day Flower 93161,100892
Day Jessamine 84408
Day Lily 191261,191284,191302,191304,384714
Day Nettle 170078,220346
Day Paphiopedilum 282810
Day Spiderwort 394088
Day's Eye 50825
Day's Eyes 50825
Dayam Cereus 83862
Dayao Arrowbamboo 162710
Dayao Camellia 69692
Dayao Skullcap 355807
Dayaoshania 123296,123297
Dayberry 334250
Day-blooming Jessamine 84408
Dayflower 100892,100954,100961,100990,101008,101193
Day-flower 100892
Dayflower Family 101200
Dayflower-leaf Addermouth Orchid 243025
Dayflowerlike Spapweed 204864
Dayflowerlike Touch-me-not 204864
Dayi Azalea 330518
Dayi Rhododendron 330518
Daylily 191261,191278,191284,191302
Day-lily 191261,191272,191284
Dayllower 100960
Dayong Bamboo 4596
Dayong Hornbeam 77285
Dayong Indocalamus 206777
Dayong Sedge 74263
Dayuan Fir 331
Dazeg 50825
Dazzle 91770,384714
Dazzling Rose 336868
Dcrub Palmetto 341421
Dea Nettle 170078,220346,373346
Dead Creepers 61497
Dead Finish 1647
Dead Man 130383,275135,287494
Dead Man's Bellows 13174,130383,287494
Dead Man's Bells 130383,168476,364146
Dead Man's Bones 231183,374916
Dead Man's Creesh 269307
Dead Man's Finger 273531
Dead Man's Fingers 37015,130383,174106,237539,269307,273469,273531,273545
Dead Man's Flesh 28044
Dead Man's Flourish 101852
Dead Man's Grief 364146
Dead Man's Hand 273469,273531
Dead Man's Thimbles 130383
Dead Man's Thumb 174106,273531
Dead Man's Tongue 269307
Dead Nettle 220345,220401,220416
Dead Sea Apple 324134,367620
Dead Tongue 269307
Deadly Calabar Bean 298009
Deadly Calabarbean 298009
Deadly Carrot 388868
Deadly Dwale 44708
Deadly Nightshade 44704,44708,367120,367416
Dead-netde 220345
Deadnettle 220345
Dead-nettle 220355
Dead-nettle Family 218697
Deadnettleleaf Leucas 227621
Deadnettleleaf Nepeta 264974
Deadnettle-like Nepeta 264975
Deadnettlelike Woodbetony 287333
Deadweed 103446
Deadwort 345594
Deaf Nettle 170078,220346,220416
Deaf-and-dumb 170025
Deal Apple 298191,300223
Deal Wood 300223
Deal-apple 298191,300223
Dealbate Cactus 244049
Dealies 300223
Dealseys 300223
Deal-tree 300223
Deam's Coneflower 339541
Deam's Oak 323585
Deam's Verbena 405835
Deam's Vervain 405835
Deane's Gum 155555
Deane's Wirelettuce 375967
Death Alder 157429
Death Bell 168476
Death Camas 417880,417891,417908,417920,417928
Death Camash 417920
Death of Man 90924
Death Valley Ephedra 146176
Death Valley Goldeneye 46552
Death Valley Joint-fir 146176
Death Valley Sage 345044
Death Warrant 61497
Death's Flower 169719
Death's Herb 44708
Death-camus 417880
Death-flower 3978
Deathin 90932,269299
Deathweed 44708
Deathwort 345594
Debao Cycas 115819
Debao Holly 203750
Deberry 334250
Debregeasia 123316
Decaisne Angelica Tree 30628
Decaisne Angelica-tree 30628
Decaisnea 123367
Decalvate Greyblue Deutzia 126936
Decapetalous Caesalpinia 64990
Decary Pricklyash 417208
Decaschistia 123439
Decaspermum 123443,123451
Decastamen Bigginseng 241163
Decayed Flower Amomum 19908
Decayedflower Amomum 19908
Deccan Hemp 194779
Deccusate Rockfoil 349240
Deceitful Pussytoes 26531
Deceiving Goatwheat 44259
Deceiving Mosquitotrap 117443
Deceiving Swallowwort 117443
Deceiving Trillium 397552
Deceiving Yam 131546
Deceptious Mussaenda 260399
Deceptive Barberry 51526,51603
Deceptive Fan Palm 234175
Deceptive Fanpalm 234175
Deceptive Mahonia 242512
Deceptive Microtropis 254292
Deceptive Rhododendron 330522
Dechang Banana 260219
Dechang Delavay Meadowrue 388479
Dechang Yushanbamboo 416763
Dechang Yushania 416763
Dechen Monkshood 5462
Deciduosbark Elm 401550
Deciduos-bark Elm 401550
Deciduous Atylosia 65156
Deciduous Azalea 331257
Deciduous Cypress 385254,385262
Deciduous Fargesia 162650
Deciduous Fig 165721
Deciduous Holly 203752
Deciduous Leaf Eria 148675
Deciduous Manglietia 244432
Deciduous Mazus 246969
Deciduous Swamp Cypress 385253
Deciduous Xylobium 415707
Deciduous-swamp-cypress 385253
Deciduous-yew-cypress 385253
Deckenia Palm 123497
Declinate Bluegrass 306052
Declinate Tickseed 104765
Decline Cherry 83258
Declined Crinum 111169
Declined Trillium 397563
Declivious Sagebrush 35397

Declivious Wormwood 35397
Declivous Fargesia 162669
Decodon 123525
Decolor Phyllophyton 245693
Decorated Euonymus 157405
Decorated Gonatanthus 179260
Decorated Kinostemon 216458
Decorative Rattan Palm 65753
Decorative Swertia 380176
Decoratted Bauhinia 49181
Decored Stauntonvine 374396
Decumaria 123543
Decumbens Hymenolobus 198485
Decumbent Bluebellflower 115365
Decumbent Bugle 13091
Decumbent Corydalis 105810
Decumbent Evolvulus 161422
Decumbent Gentian 173387
Decumbent Goldenrod 368398
Decumbent Hogfennel 292825
Decumbent Kingiella 216437
Decumbent Lagotis 220166
Decumbent Longstyle Azalea 331137
Decumbent Longstyle Rhododendron 331137
Decumbent Mountainash 369499
Decumbent Mountain-ash 369499
Decumbent Trillium 397554
Decurrent Archangelica 30931
Decurrent False Aster 56712
Decurrent Incense-cedar 228639
Decurrent Pearweed 239608
Decurrent Rhaphidophora 328997
Decurved Elaeocarpus 142412
Decurved Fargesia 162670
Decussate Olive 270066
Dedekens Edelweiss 224823
Dee Nettle 220416
Dee Pegwood 157429
Dee-nettle 220416
Deep Rosy Flowering Dogwood 105025
Deep Yellow-wood 332256
Deepblue Honeysuckle 235691
Deep-blue Honeysuckle 235691
Deepbrown Sedge 73814
Deepbrown-like Sedge 73819
Deepfid Roundlobe Buttercup 325790
Deepgreen Wildginger 37704
Deeplobe Agapetes 10330
Deeplobe Vilmorin Monkshood 5668
Deep-lobed Clematis 95388
Deeplobed Grape 411543
Deep-lobed Largeleaved Linden 391823
Deeppurple Candolle Thorowax 63589
Deeppurple Lysionotus 239930
Deeppurple Mrs. E. H. Wilson's Azalea 331065
Deeppurple Square Bamboo 87603
Deeppurple Teasel 133460
Deep-purpleflower Indigo 205698
Deepred Agapetes 10341
Deepred Javatea 275759
Deepred Jurinea 190832
Deepred Primrose 314914
Deepred Vermilion Azalea 330416
Deepsplit Dandelion 384818
Deepsplit Sawwort 361039
Deepsplit Windhairdaisy 348632
Deeptooth Primrose 314631
Deeptust Ballspike Flatsedge 322235
Deepviolet Gentian 173271
Deepviolet Rush 213319
Deepviolet Stairweed 142607
Deer Brush 79945
Deer Cabbage 265197
Deer Goldenbush 150313
Deer Grass 259695
Deer Lodge Wild Buckwheat 152388
Deer Nut 364450
Deer Oak 324356
Deer Pea Vetch 408468
Deer Vine 231682
Deer's Ear 380336
Deer's Ears 380336
Deer's Hair 353272
Deer's Milk 158433
Deer's Tongue 228495
Deer's-foot Bindweed 102933
Deerberry 172135,403753,404012
Deerbrush 79945
Deerdrug 366142
Deerfoot 4105
Deer-foot 4103,4105
Deer-grass 145073,329427,396003
Deer-hair 353272
Deer-hair Brush 353272
Deerhair Bulrush 353272
Deerhorn Azalea 331065
Deerhorn Cedar 390680
Deerhorn Chinese Juniper 213655
Deerhorn Rhododendron 331065
Deer-horn Rhododendron 331065
Deering Velvet Bean 259559
Deering Velvet-bean 259559
Deeronion Stonegarlic 239280
Deertongue 128494
Deer-tongue Grass 128494
Deervetch 237476,387239
Deer-vetch 237733,237781
Deerweed 237749
Defe Nettle 220346
Definite Stonecrop 200807
Deflective-fruited Nightshade 367087
Deflex Waterstarwort 67349
Deflexed Asparagus 39186
Deflexed Bostrychanthera 57504
Deflexed Bottlebrush Sedge 76014
Deflexed Brachiaria 58079
Deflexed Crazyweed 278809
Deflexed Empetrum 297023
Deflexed Enkianthus 145694
Deflexed Erysimum 154442
Deflexed Glorybower 96034
Deflexed Needlejade 297023
Deflexed Pendent-bell 145694
Deflexed Primrose 314298
Deflexed Sugarmustard 154442
Deflexed Vetch 408374
Deflexedpod Tickclover 126373
Defoliate Cymbidium 116820
Defoliate Orchis 116820
Deformedfruit Craneknee 221632
Deg Corydalis 105815
Deg Milkvetch 42275
Degame 68416
Degame Lancewood 68416
Degami 68416
Dege Parnassia 284524
Degenerate Nospurflower 167298
Degenerate Stapfiophyton 167298
Degron's Rhododendron 330532
Dehaasia 123598,123606
Dehiscent Common Flax 232003
Dehiscentfruit Privet 229613
Dehong Camellia 69646
Dehong Holly 203753
Dehong Tea 69646
Deinanthe 123654,123659
Deinbollia 123661
Deinocheilos 123722,123724
Deinostemma 123727
Deith Bell 168476
Deith-bell 168476
Deity Cup 326287
Deji Cheirostylis 86677
Dekata Rhododendron 330545
Del Norte County Iris 208639
Delaet Aloe 16756
Delanay Lily 229745
DeLaney's Goldenaster 90188
Delavai Myriactis 261065
Delavay 262220,287679
Delavay Acacia 1176
Delavay Ampelopsis 20335
Delavay Anemone 23782
Delavay Apios 29303
Delavay Arisaema 33310
Delavay Azalea 330546
Delavay Barberry 52029
Delavay Bauhinia 49064
Delavay Bellflower 69987
Delavay Birch 53405

Delavay Birthwort 34163
Delavay Bittercress 72732
Delavay Bladdersenna 100173
Delavay Bluebell 115347
Delavay Bluebellflower 115347
Delavay Blueberry 403795
Delavay Bushbeech 178029
Delavay Bush-beech 178029
Delavay Butterflybush 62045
Delavay Butterfly-bush 62045
Delavay Calanthe 65917
Delavay Catchfly 363403
Delavay Chamaesium 85712
Delavay Chinquapin 78920
Delavay Cinquefoil 312494
Delavay Clematis 94874
Delavay Clethra 96487
Delavay Clovershrub 70799
Delavay Corydalis 105816
Delavay Cotoneaster 107422
Delavay Cranesbill 174567
Delavay Cremanthodium 110380
Delavay Curvedstamencress 238008
Delavay Cyclobalanopsis 116079
Delavay Cystopetal 320217
Delavay Dallisgrass 285421
Delavay Dobinea 135098
Delavay Dryquitch 148393
Delavay Dysosma 139610
Delavay Edelweiss 224826
Delavay Elaeagnus 141978
Delavay Elegant Pouzolzia 313425
Delavay Elegant Reekkudzu 313425
Delavay Eremopogon 148393
Delavay Evergreen Chinkapin 78920
Delavay Evergreenchinkapin 78920
Delavay Everlasting 21561
Delavay Evodia 161328
Delavay False Manyroot 283763
Delavay Figwort 355097
Delavay Fir 333
Delavay Flatsedge 322202
Delavay Fritillary 168391
Delavay Gentian 173388
Delavay Gerbera 175147
Delavay Goldsaxifrage 90351
Delavay Goldwaist 90351
Delavay Greenleaf Elaeagnus 142242
Delavay Gueldenstaedtia 181638
Delavay Habenaria 183553
Delavay Helictotrichon 190146
Delavay Hippolytia 196693
Delavay Holly 203754
Delavay Honey Locust 176883
Delavay Honeylocust 176883
Delavay Honeysuckle 236104
Delavay Indigo 205888
Delavay Iris 208519
Delavay Lacquertree 393449
Delavay Lacquer-tree 393449
Delavay Larkspur 124171
Delavay Ligusticum 229314
Delavay Magnolia 232595
Delavay Mazus 247025
Delavay Meadowrue 388477
Delavay Meconopsis 247119
Delavay Megacarpaea 247770
Delavay Microtoena 254246
Delavay Mockorange 294451
Delavay Monkshood 5156
Delavay Munronia 259875
Delavay Nannoglottis 262220
Delavay Neocinnamomum 263759
Delavay Nothopanax 266918
Delavay Nutantdaisy 110380
Delavay Oak 116079
Delavay Oneflowerprimrose 270729
Delavay Oreocharis 273851
Delavay Osmanther 276283
Delavay Osmanthus 276283
Delavay Oxygraphis 278469
Delavay Pararuellia 283763
Delavay Paris 284312,284313
Delavay Parnassia 284525
Delavay Paspalum 285421
Delavay Pearleverlasting 21561
Delavay Pearweed 239614
Delavay Peony 280182
Delavay Physospermopsis 297947
Delavay Privet 229454
Delavay Pycreus 322202
Delavay Raspberry 338317
Delavay Rhododendron 330546
Delavay Rhubarb 329324
Delavay Ricebag 181638
Delavay Rockjasmine 23156
Delavay Rockvine 387763
Delavay Sandwort 31849
Delavay Sauropus 348049
Delavay Schefflera 350680
Delavay Scurrula 236646
Delavay Sedge 74283
Delavay Silver Fir 333
Delavay Skullcap 355426
Delavay Snakegrape 20335
Delavay Snowgall 191913
Delavay Soapberry 346333
Delavay Solena Telfairia 367775
Delavay Southstar 33310
Delavay Spapweed 204894
Delavay St. John's Wort 201829
Delavay Stephania 375836
Delavay Stringbush 414160
Delavay Summerlilic 62045
Delavay Swordflag 208519
Delavay Taxillus 385194
Delavay Teaolive 276283
Delavay Touch-me-not 204894
Delavay Tupistra 70602
Delavay Vinegentian 110306
Delavay Violet 409888
Delavay Wildginger 37610
Delavay Willow 343286
Delavay Windflower 23782
Delavay Windhairdaisy 348249
Delavay Wingnut 320350
Delavay Yellowwood 94020
Delavay's Fir 333
Delavaya 123765
Delavigne Koeleria 217462
Delei Ludlow Barberry 51943
Delevay Swertia 380179
Delicacy Vetch 408539
Delicate Asparagus-fren 39148
Delicate Bamboo 87633
Delicate Bess 81744,404238
Delicate Bodinier Fleshspike 346938
Delicate Corydalis 105822
Delicate Cyclobalanopsis 116080
Delicate Everlasting 170919
Delicate Falsepimpernel 231507
Delicate Fleabane 151053
Delicate Fragrant-bamboo 87633
Delicate Gentian 173390
Delicate Ionopsis 207448
Delicate Lily 88216
Delicate Rabbit-tobacco 317756
Delicate Roegneria 335403
Delicate Trigonotis 397406
Delicate Twayblade 232133
Delicate Twinballgrass 209098
Delicatefoot Spapweed 205083
Delicatefoot Touch-me-not 205083
Delicatefragrance Gingersage 253655
Delicatefragrance Micromeria 253655
Delicatehair Agapetes 10300
Delicatespur Columbine 30051
Delicious Actinidia 6558
Delicious Chinese Torreya 393062
Delicious Kiwifruit 6558
Delicious Statice 230598
Delight Oak 116080
Delightful Dendrobium 124988
Delightful Osmanthus 276399
Delima 321764
Delisle Ceanothus 79923
Dell 300223
Dellcup 325518
Delmatian Bellflower 70228
Delosperma 123819
Delosperma Nubigenum 89959

Delphinium 124008,124194,124301
Delseed 300223
Delta Arrowhead 342393
Delta Bulrush 352178
Delta Post Oak 324422,324443
Delta Woolly Marbles 318803
Deltaleaf Goldthread 103832
Deltleaf Nettle 403032
Deltoialeaf Eutrema 161126
Deltoid Asiabell 98309
Deltoid Balsam-root 47124
Deltoid Goldenray 229063
Deltoid Goldthread 103832
Deltoid Loosestrife 239615
Deltoid Oscularis 220523
Deltoid Synurus 382069
Deltoid Yam 131550
Deltoidleaf Cacalia 283807
Deltoid-leaf Groundsel 360244
Deltoidleaf Saussurea 348256
Deltoid-leafed Dewplant 220523
Deltoid-leaved 220464
Deltoid-leaved Dew-plant 220523
Delt-orach 44576
Delt-orache 44576
Delty-cup 326287
Delziel Arrowwood 407771
Delziel Viburnum 407771
Demater Iris 208521
Demerara Pinkroot 371614
Denali Fleabane 150587
Dendranthema 124767,124783
Dendrobenthamia 124880
Dendrobium 124957
Dendrobium Orchid 125365
Dendrocalamopsis 125435
Dendrocalamus 125461
Dendrochilum 125526
Dendrochilum Orchid 125526
Dendrolobium 125574
Dendropanax 125589
Dendrophthoe 125661
Dendrotrophe 125781
Dene Peashrub 72216
Denegri Obregonia 268086
Denflower Euonymus 157386
Dengchuan Larkspur 124634
Denhamia 125802
Denison Hibiscus 194830
Denison Spapweed 204896
Denison Touch-me-not 204896
Denison Vanda 404635
Denmoza 125817
Dennes Dendrobium 125005
Dense Agapetes 10371
Dense Arrowwood 407758
Dense Blazing Star 228529
Dense Calligonum 67018
Dense China Pine 300244
Dense Chinese Everlasting 21690
Dense Chinese Pearleverlasting 21690
Dense Cotton-grass 152784
Dense Crazyweed 278810
Dense Fargesia 162658
Dense Flower Synotis 381928
Dense Gay-feather 228529
Dense Gerard Ephedra 146184
Dense Grayhair Everlasting 21543
Dense Grayhair Pearleverlasting 21543
Dense Hybrid Yew 385385
Dense Jacot Edelweiss 224865
Dense Jadeleaf and Goldenflower 260401
Dense Japanese Yew 385359
Dense Kadsura 214968
Dense Leaflets Crazyweed 278813
Dense Linearleaf Kobresia 217132
Dense Logwood 415869
Dense Nerve Eugenia 382530
Dense Pennywort 200275
Dense Pink Deutzia 127067
Dense Pondweed 312086
Dense Primrose 314303
Dense Raceme Heath 149776
Dense Salacca 342572
Dense Sandwort 31850
Dense Scales Rhododendron 331474
Dense Sedge 74198,74225,74289
Dense Senna 360422
Dense Shrubby St. Johnswort 201831
Dense Silkybent 28971
Dense Silky-bent 28971
Dense Sisymbrium 365433
Dense Small Reed 127224
Dense Smallreed 127224
Dense Spike Primrose 56555
Dense Tumorfruitparsley 393862
Dense Viburnum 407758
Denseball Ramie 56126
Densebract Eria 299611
Densebract Galangal 17664
Densebract Hairorchis 299611
Densebract Irisorchis 268000
Densebract Oberonia 268000
Densebranch Azalea 330691
Densebranch Craneknee 221634
Densebranch Kaschdaisy 215617
Densebranch Kaschgaria 215617
Dense-branched Maple 3493
Densebranched Russianthistle 344575
Densebranchlet Juniper 341724
Densebranchlet Savin 341724
Denseciliate Buttercup 325768
Denseclump Crazyweed 278810
Dense-column Cedar 390655
Densefimbriate Sedge 74286
Denseflor Spiradiclis 371752
Denseflower Acacia 1516
Denseflower Actinidia 6699
Denseflower Actinodaphne 6774
Denseflower Aucuba 44905
Dense-flower Barbate Mallotus 243328
Dense-flower Barberry 51540
Denseflower Bedstraw 170572
Denseflower Billardiera 54438
Denseflower Bladderwort 403299
Dense-flower Bluebeardleaf Butterflybush 61994
Denseflower Blumea 55722
Denseflower Bulbophyllum 62955
Denseflower Calanthe 65920
Dense-flower Carnose Deutzia 126863
Denseflower Catchfly 363404
Denseflower Catnip 264909
Denseflower Chee Woodreed 65333
Denseflower Chickweed 374805
Denseflower Childgrass 340346
Denseflower Chiritopsis 87996
Denseflower Cliffbean 66623
Denseflower Conehead 378122
Denseflower Cordgrass 370161
Denseflower Corydalis 105753
Denseflower Crazyweed 279009
Denseflower Cryptocarya 113441
Denseflower Dendrobium 125109
Denseflower Denseflower Sealavender 230585
Dense-flower Deutzia 127020
Dense-flower Drypetes 138601
Denseflower Ehretia 141626
Denseflower Elaeagnus 141956
Denseflower Elsholtzia 144002
Denseflower Eria 148667
Denseflower Eritrichium 153438
Denseflower Euonymus 157732
Denseflower Falsenettle 56125
Denseflower Fargesia 162673
Denseflower Firethorn 322462
Denseflower Gentianella 174132
Denseflower Gingerlily 187438
Denseflower Glaucous Bauhinia 49099
Denseflower Goldenray 229007
Denseflower Groundsel 381928
Denseflower Habenaria 183643
Denseflower Hairorchis 299639
Denseflower Hippeophyllum 196463
Denseflower Holly 203644
Denseflower Hoodorchis 264768
Denseflower Hoodshaped Orchid 264768
Denseflower Indigo 206661
Dense-flower Jasmine 211779
Dense-flower Johnston Pittosporum 301303
Dense-flower Johnston Seatung 301303

Denseflower Jurinea 190852
Denseflower Kalanchoe 215130
Denseflower Kudzuvine 321420
Denseflower Larkspur 124177
Denseflower Lemongrass 117158
Denseflower Lily 228571
Denseflower Longstamen Onion 15451
Denseflower Loosestrife 239597
Denseflower Mayten 246826
Denseflower Microtropis 254295
Denseflower Milkvetch 42280
Denseflower Milkwort 308416
Denseflower Millettia 66623
Denseflower Monkshood 5496
Denseflower Mullein 405694
Dense-flower Mullein 405694
Denseflower Mussaenda 260401
Denseflower Mycetia 260619
Denseflower Neillia 263147
Denseflower Nepeta 264909
Denseflower Onosma 271748
Dense-flower Pencilwood 139643
Denseflower Peppergrass 225340
Denseflower Pepperweed 225340
Denseflower Phreatia 295954
Denseflower Plantain 301878
Denseflower Platanthera 302355
Denseflower Pleurothallis 304909
Denseflower Poisonnut 378845
Denseflower Pollia 307342
Denseflower Quisqualis 324672
Denseflower Rattan Palm 65668
Denseflower Reedbentgrass 65333
Denseflower Reevesia 327375
Denseflower Resembling Sedge 76284
Denseflower Rotala 337330
Denseflower Rungia 340346
Denseflower Rush 213202
Denseflower Sedge 74171
Dense-flower Shortfruit Bauhinia 49025
Denseflower Speedwell 407106
Denseflower Spikesedge 143100
Denseflower Striga 377994
Denseflower Strychnos 378845
Denseflower Suoluo 327375
Denseflower Sweetleaf 381154
Denseflower Tamarisk 383444
Denseflower Tanoak 233176
Denseflower Thickshellcassia 113441
Denseflower Thorowax 63615
Denseflower Wallichia 413091
Denseflower Wendlandia 413803
Denseflower Woodbetony 287103
Denseflower Xylosma 415881
Denseflower Yunnan Felwort 235447
Denseflower Yunnan Rattanpalm 65817
Dense-flowered 334144
Dense-flowered Bamboo 58698
Dense-flowered China Laurel 28345
Dense-flowered Climbing-bamboo 249608
Dense-flowered Dock 340016
Dense-flowered Drypetes 138601
Dense-flowered Elaeagnus 141956
Denseflowered False-nettle 56125
Dense-flowered False-nettle 56125
Dense-flowered Fargesia 162736
Dense-flowered Fishvine 126003
Dense-flowered Fumitory 169052
Dense-flowered Maple 3492
Dense-flowered Millettia 66623
Dense-flowered Mullein 405694
Dense-flowered Mycetia 260619
Dense-flowered Neillia 263147
Dense-flowered Orchid 264595,264599
Dense-flowered Rush 213309
Dense-flowered Sweetleaf 381154
Dense-flowered Taiwan Epithema 147432
Dense-flowered Tamarisk 383444
Dense-flowered Viburnum 407758
Dense-flowered Wallichia 413091
Dense-flowered Wendlandia 413803
Dense-flpwer Knotweed 309043
Densefruit Dittany 129629
Dense-fruit Indigo 205892
Densefruit Pricklyash 417237
Densefruit Treerenshen 125602
Dense-fruited Cornera 104892
Dense-fruited Indigo 205892
Dense-fruited Prickly-ash 417237
Dense-glandule Hornbeam 77300
Densehair Aster 41482
Densehair Bigpetalcelery 357910
Densehair Blastus 55171
Densehair Bredia 296391
Densehair Clematis 95269
Densehair Dockleaved Knotweed 309312
Densehair Duplicatefruit Craneknee 221662
Densehair Galangal 17659
Densehair Ginger 417975
Densehair Helicia 189958
Densehair Italian Jasmine 211867
Densehair Microula 254348
Densehair Paphiopedilum 282920
Densehair Sedge 74290
Densehair Shortplume Clematis 94783
Densehair Windhairdaisy 348353
Dense-haired Blastus 55171
Dense-haired Lasianthus 222090
Dense-haired Metalleaf 296391
Densehaired Rockjasmine 23268
Dense-haired Veinfruit Seabuckthorn 196756
Dense-hairs Metalleaf 296391
Densehairs Snailseed 97934
Densehairy Anemone 23794
Densehairy Aucuba 44914
Densehairy Chiritopsis 88001
Densehairy Lycianthes 238950
Dense-hairy Podocarpium 200725
Densehairy Red Silkyarn 238950
Densehairy Rose 336802
Dense-head Mountain Ash 32988
Densehead Mountain-ash 32988
Densehead Tailanther 381928
Densehead Tansy 383740
Densehead Woodland Sagebrush 36355
Densehead Woodland Wormwood 36355
Dense-headed Mountain-ash 32988
Densehispid Metalleaf 296391
Denselanose Aster 40700
Denseleaf Aster 41131
Denseleaf Azalea 330553
Denseleaf Cardiandra 73132
Denseleaf Crazyweed 278813,279029
Denseleaf Dwarf Mitrasacme 256162
Denseleaf Gentian 173347
Denseleaf Haworthia 186341
Denseleaf Holly 203748
Denseleaf Larkspur 124319
Denseleaf Meadowrue 388598
Denseleaf Nepeta 265081
Denseleaf Newlitse 264041
Denseleaf Paraboea 283139
Denseleaf Paramignya 283509
Denseleaf Peashrub 72221
Denseleaf Poplar 311524
Denseleaf Pussytoes 26379
Denseleaf Rhododendron 330553
Denseleaf Rockfoil 349243
Denseleaf Sageretia 342189
Denseleaf Sedge 75311
Denseleaf Senna 360457
Denseleaf Sinocrassula 364917
Denseleaf Stonelotus 364917
Denseleaf Vanda 404671
Dense-leaved Baeckea 46429
Dense-leaved Barberry 51546
Dense-leaved Elodea 141569
Dense-leaved Holly 203748
Dense-leaved Jasmine 211779
Denseleaved Maple 2899
Dense-leaved Maple 2899
Dense-leaved Pea-shrub 72221
Dense-leaved Poplar 311524
Denseleaved Porana 396765
Dense-leaved Rhododendron 330553
Dense-leaved Senna 360457
Dense-leaved Vine-lime 283509
Denseleaved Violet 409842
Dense-leaved Yew 385376
Denselobate Primrose 314872
Denselraf Porana 396765

Denselutinous Tickclover 126250
Densely-flowered Elaeocarpus 142362
Denselyspine Raspberry 339328
Densenerve Hartia 376463
Densenerve Seatung 301257
Dense-net-leaved Euonymus 157402
Dense-net-leaved Spindle-tree 157402
Densenode Smithia 366655
Densepod Indigo 205892
Densepubescent Cherry 83281
Dense-pubescent Cherry 83281
Densepubescent Glorybind 68701
Densepulvinate Draba 136977
Densepulvinate Whitlowgrass 136977
Dense-pyramid Eastern Arborvitae 390596
Denseraceme Monkshood 5517
Dense-root Milkvetch 42958
Densescale Ardisia 31423
Denseseta Greenbrier 366571
Densesetaceous Greenbrier 366571
Dense-setaceous Greenbrier 366571
Dense-shrub Common Yew 385303
Densespicate Galangal 17662
Dense-spikated Willow 343961
Densespike Corydalis 105824
Densespike Galangal 17754
Densespike Imbricate Galingale 119018
Densespike Knotweed 308713
Densespike Rhubarb 329318
Densespike Sedge 75941
Densespike Woodbetony 287156
Dense-spiked Knotweed 308713
Densespine Coryphantha 107054
Densespine Raspberry 339328
Dense-spines Coryphantha 107054
Dense-spiny Raspberry 339328
Densespurred Larkspur 124543
Dense-squamaceous Ardisia 31423
Densestriped Maesa 241761
Densethorn Evergreenchinkapin 78921
Densethorn Kneejujube 67018
Dense-thorned Greenbrier 366310
Densethorny Greenbrier 366310
Densetomentose Ajania 13034
Densetooth Elatostema 142805
Densetooth Grewia 180668
Densetooth Half-capitate Hemiboea 191388
Densetooth Stairweed 142805
Dense-toothed Folgner Mountainash 369390
Dense-toothed Grewia 180668
Densetoothed Willow 343192
Dense-toothed Willow 343192
Denseumbel Groundsel Groundsel 358848
Denseumbel Pennywort 200349
Densevein 261322
Dense-vein Bigleaf Clethra 96508
Densevein Clovershrub 70790
Densevein Indocalamus 206822
Densevein Ninenode 319500
Densevein Ophiorrhiza 272196
Densevein Pittosporum 301257
Densevein Psychotria 319500
Densevein Schefflera 350799
Dense-veined Pittosporum 301257
Dense-veined Schefflera 350799
Dense-veins Bridelia 60223
Densewart Greenbrier 366283
Densewart Smilax 366283
Densicorrugate Prostrate Bulrush 353848
Densiflowered Aucuba 44905
Densi-flowered Cryptocarya 113441
Densiflowered Firethorn 322462
Densiflowered Holly 203644
Densiflowered Milkwort 308416
Densiflowered Reevesia 327375
Densiflowered Sarcococca 346725
Densifolious Eria 396472
Densihaired Helicia 189958
Densileaf Mahonia 242508
Densi-leaved Newlitse 264041
Densileaved Sageretia 342189
Densispike Willow 343961
Densispikeed Willow 343961
Densispined Evergreen Chinkapin 78921
Densithorned Evergreen Chinkapin 78921
Densiwart Greenbrier 366283
Dent Corn 417416,417417
Dentate Beautyberry 66767
Dentate Calyx Rhododendron 330554
Dentate Chinese Goat Willow 344118
Dentate Dock 340019
Dentate Dontostemon 136163
Dentate Engler Stonecrop 356704
Dentate Figwort 355098
Dentate Ixeris 210527
Dentate Jerusalemsage 295096
Dentate Membraneroot Leek 15362
Dentate Nepeta 264910
Dentate Nettle 402949
Dentate Paniculate Ladybell 7731
Dentate Rhubarb 329325
Dentate Rockvine 387766
Dentate Scopolia 25443
Dentate Sophora 368997
Dentate Yunnan Bastard Cress 390272
Dentate-calyxed Rhododendron 330554
Dentated Melilot 249212
Dentateleaf Banksia 47639
Dentateleaf Tubergourd 390123
Dentate-leaved American Linden 391656
Dentedleaf Milkvetch 42976
Dentelion 384714
Dentella 125868
Denthtooth Asparagus 39087
Denticulate Chamabainia 85108
Denticulate Chinese Groundsel 365042
Denticulate Gentian 173636
Denticulate Holly 203759
Denticulate Ixeris 283388
Denticulate Japanese Kerria 216096
Denticulate Macaranga 240250
Denticulate Oreocharis 273841
Denticulate Primrose 314310
Denticulate Willow 343180
Dentiferous Dendropanax 125604
Dentilinn 384714
Dentipetal Catchfly 363405
Dentylion 384714
Denudate Cactus 244053
Denudate Rhododendron 330556
Denya 116569
Deodar 80087
Deodar Cedar 80087
Depauperate Cryptocarya 113442
Depauperate Fake Needlegrass 318201
Depauperate Pseudoraphis 318201
Depauperate Senecio 359697
Depauperate Spapweed 204898
Depauperate Touch-me-not 204898
Dependent Cirrhaea 91612
Dependent Skullcap 355428
Dependent Staff-tree 80273
Depgul 220796
Deplanche's She-oak 182836
Deppe's Lycaste 238741
Depressa Threeawngrass 33825
Depressa Triawn 33825
Depressed Ardisia 31424
Depressed Bottle Gourd 219848
Depressed Chickweed 374833
Depressed Clearweed 298905
Depressed Gaultheria 172067
Depressed Lily 228577
Depressed Oak 323830
Depressed Orach 44566
Depressed Orange 93448
Depressed Plantain 301952
Depressed Sedge 74277
Deptford Pink 127600
Deqin Arrowbamboo 162757
Deqin Aster 41351
Deqin Azalea 331307
Deqin Barberry 51495
Deqin Bastardtoadflax 389672
Deqin Bulbophyllum 62963
Deqin Dwarf Woodbetony 287571
Deqin Eritrichium 153446
Deqin Everlasting 21588
Deqin Hackberry 80597
Deqin Hornbeam 77397
Deqin Hundredstamen 389672

Deqin Largeleaf Rabdosia 209690
Deqin Lovegrass 147631
Deqin Monkshood 5462
Deqin Parnassia 284528
Deqin Pearleverlasting 21588
Deqin Pimpinella 299446
Deqin Poplar 311340
Deqin Rhodiola 329842
Deqin Rockfoil 349245
Deqin Rose 336528
Deqin Rush 213184
Deqin Sandwort 31759
Deqin Sedge 74295
Deqin Sinocrassula 364934
Deqin Spapweed 204899
Deqin Stonebean-orchis 62963
Deqin Stonecrop 357294
Deqin Stonelotus 364934
Deqin Stringbush 414266
Deqin Woodbetony 287162
Derived Silverpuffs 374543
Derong Barberry 51547
Derong Clovershrub 70792
Derong Rose 336529
Derris 125958
Dese Serrate Groundsel 358708
Deseret Sulphur Flower 152578
Desert Agave 10843
Desert Alyssum 225367
Desert Anemone 24101
Desert Apricot 20919,316412
Desert Ash 168136
Desert Aster 39916
Desert Baccharis 46257
Desert Bahia 46527
Desert Baileya 46580
Desert Barberry 51640,51694
Desert Bearclaw-poppy 31081
Desert Beard Tongue 289370
Desert Bedstraw 170645
Desert Bell 293158
Desert Bird of Paradise 65014
Desert Blue Bells 293158
Desert Blueeargrass 115571
Desert Box 155655
Desert Broom 46255
Desert Buckwheat 152001,152056
Desert Calico 221035
Desert Candle 79698,122905,148531
Desert Carpet 1522
Desert Cassia 78230,360459
Desert Catalpa 87449
Desert Ceanothus 79932
Desert Chalkplant 183187
Desert Chicory 325075
Desert Christmas Cactus 116665,272948
Desert Christmas Cholla 116665
Desert Cinquefoil 312497
Desert Cudweed 170938
Desert Cyanotis 115571
Desert Dale 46767
Desert Dandelion 242923,242926,242942,
384683
Desert Date 46768
Desert Death Camas 417884
Desert Desertcandle 148541
Desert Elderberry 345627
Desert Eucalyptus 155727
Desert Everning Primrose 269429
Desert False Indigo 20005
Desert Fan Palm 413298
Desert Fanbract 379164
Desert Fern 239541
Desert Fig 165423,165471
Desert Five-spot 243906
Desert Four-o'clock 255732
Desert Geranium 288592
Desert Giantfennel 163727
Desert Globemallow 370935
Desert Gold 230863
Desert Gold Poppy 155179
Desert Goosefoot 87133
Desert Grape 411710
Desert Grevillea 180595
Desert Gum 155727
Desert Hackberry 80707,80708
Desert Hakea 184621
Desert Heron's Bill 153921
Desert Holly 5862,44471
Desert Hollyhock 370935
Desert Honeysuckle 25262
Desert Horse-purslane 394903
Desert Indian Wheat 302119
Desert Indigo-bush 20005
Desert Ironwood 270527
Desert Juniper 213625
Desert Kumquat 148262
Desert Kurrajong 58340
Desert Lamb's Quarters 87020
Desert Larkspur 124104
Desert Lavender 203050
Desert Lily 193463
Desert Lime 93465
Desert Madwort 18374,18375
Desert Mahonia 242536
Desert Mallow 370935
Desert Marigold 46578,46580
Desert Mariposa 67589
Desert Mariposa Lily 67589
Desert Milkweed 37911,38129
Desert Mistletoe 295577
Desert Mountains Thistle 91761
Desert Night-blooming Cereus 288986
Desert Nut Pine 300079
Desert Oak 79158,79165,324556
Desert Olive 167323,167324,167339
Desert Onion 15807
Desert Paintbrush 79116,79119
Desert Palm 413307
Desert Peach 223094
Desert Peashrub 72333
Desert Pepperweed 225367
Desert Petunia 339684,339784
Desert PhloxSanta Catalina Mountain
Phlox 295327
Desert Pincushion 84495,84524
Desert Plume 373709
Desert Poinsettia 158840
Desert Poppy 215381
Desert Pot-herb 65830
Desert Prenanthella 313776
Desert Primrose 269416
Desert Privet 290382
Desert Pussytoes 26580
Desert Rabbitbrush 150339
Desert Raisin 367143
Desert Range Almond 316413
Desert Rock Daisy 291317
Desert Rock Pea 237741
Desert Rock-nettle 156002
Desert Rod 148466,148494
Desert Rose 7333,7343,7353,336974,
405223
Desert Rose Mallow 194818
Desert Rosemallow 194818
Desert Rose-mallow 194818
Desert Rosemary 307282
Desert Ruellia 339784
Desert Sage 344999,345011
Desert Sagebrush 35373
Desert Saltbush 44609
Desert Sand Verbena 714
Desert Sand-verbena 714
Desert Scrub Oak 324067
Desert Senna 360430
Desert Spider Flower 180637
Desert Spoon 122912
Desert Spunsilksage 360870
Desert Sunflower 174427,188956
Desert Surprise 289380
Desert Tansy-aster 33034,33035
Desert Tea 146262
Desert Thistle 92222
Desert Thorn 238996,239046,239092
Desert Tobacco 266048,266063
Desert Trumpet 152160
Desert Trumpet Flower 123071
Desert Twinbugs 129476,129478
Desert Valley Fishhook Cactus 354319
Desert Velvet 317093
Desert Wallflower 154390

Desert Wheat Grass 11710
Desert Wheatgrass 11710
Desert Willow 87446,87449
Desert Wind Flower 24101
Desert Windflower 24101
Desert Wirelettuce 375981
Desert Wishbone Bush 255679
Desert Wormwood 35373
Desert Yaupon 350573
Desert Yellow Fleabane 150758
Desert Yellow-head 416304
Desert Zinnia 418036
Desertbroom 46255
Desertcandle 148531
Desert-candle 148531
Desertdandelion 242923
Desert-gold 174427
Desert-lily 193462
Desertliving Asparagus 39032
Desertliving Cistanche 93054
Desertpeony 5859
Desertstar 257832
Desert-sunflower 174426
Desett Pea 380066
DesmondMallee 155559
Desmos 126716
Desmostachys 126738
Despain Footcactus 287891
Despain's Pincushion Cactus 287891
Dessert Banana 260253
Detenna 268729
Detling's Silverpuffs 253928
Deutery Dewtry 123077
Deutsa 220729
Deutzia 126832,126834,126879
Deutzianthus 127129
Devil Cholla 181454,181456
Devil Club-cholla 181454
Devil Daisy 26768,227533,322724
Devil Maple 2928
Devil Pepper 326995,327058
Devil Rattan 121486,121514
Devil Tanoak 233132
Devil Thorn 395146
Devil Tree 18055,154643
Devil Walking Stick 30760
Devil Wood 276250
Devil's Apple 123077,244346,366906
Devil's Apple Tree 159027
Devil's Apron 402886
Devil's Backbone 61568,287853
Devil's Bane 201924,321658,321721,
358062,405872
Devil's Beard 321658,321721,358062
Devil's Beggar-ticks 53913
Devil's Berries 44708,201732,383675
Devil's Bit 81237,85505,195858,216753,
379660
Devil's Bit Cotton 675
Devil's Bit Fairy-wand 85505
Devil's Bit Guts 114947
Devil's Bit Paintbrush 195478
Devil's Bit Scabious 379660
Devil's Bite 228529,379660
Devil's Blanket 405788
Devil's Boot 347144
Devil's Boots 347156
Devil's Bread 102727
Devil's Burtons 175444,216753
Devil's Buttercup 326340
Devil's Button 175444,216753,379660,
417093
Devil's Candles 244346,417093
Devil's Candlesticks 271280
Devil's Cherries 44708,61497,367120
Devil's Cherry 44708,61497,367120
Devil's Churn 159557
Devil's Citurnstaff 159027
Devil's Claw 185846,315434,315438
Devil's Claw Pisonia 300944
Devil's Claws 1264,237539,315430,325612
Devil's Clover 247246
Devil's Club 272707
Devil's Coachwheel 325612
Devil's Coach-wheel 325612
Devil's Corn 374916
Devil's Cotton 675
Devil's Cup-and-saucer 158433
Devil's Cups And Saucers 158433
Devil's Curry-comb 325612
Devil's Cut 95029,95431
Devil's Daisy 227533,322724
Devil's Darning Needle 95426,350425
Devil's Darning-needle 95426,350425
Devil's Dew 266225
Devil's Dung 163612
Devil's Dye 206534
Devil's Ear 33544,374916
Devil's Elshins 350425
Devil's Entrails 102933
Devil's Eye 201389,374916,407072
Devil's Eyes 201389,374916,407072
Devil's Fiery Poker 217055
Devil's Fig 32429
Devil's Fingers 237539,311398,325612
Devil's Flight 202086
Devil's Flower 101852,208771,248312,
370522,374916,407072
Devil's Garter 68713,102933
Devil's Garters 68713,102933
Devil's Gillofer 86427
Devil's Guts 68713,95431,102933,114947,
115008,115031,326287,379660
Devil's Hand 273531
Devil's Hate 369339,405872
Devil's Hatties 364146
Devil's Head 140133,231183,302034
Devil's Herb 44708,123077,305194
Devil's Horn 315431
Devil's Horsewhip 4259,4268
Devil's Ivy 147338,147349
Devil's Ivyarum 147338
Devil's Kirnstaff 159027
Devil's Ladies And Gentlemen 37015
Devil's Ladies-and-gentlemen 37015
Devil's Leaf 402886
Devil's Lingels 308816
Devil's Meal 28044
Devil's Meat 28044,28045
Devil's Men And Women 37015
Devil's Men-andwomen 37015
Devil's Milk 86755,159027,159557
Devil's Milk Plant 384714
Devil's Milkpail 384714
Devil's Milkplant 384714
Devil's Net 115008
Devil's Nettle 3978
Devil's Nightcap 5442,68713,102933,
374916,392992
Devil's Oatmeal 102727,192358,292694
Devil's Oats 101852
Devil's Painkiller 195478
Devil's Pain-killer 195478
Devil's Paintbrush 195478
Devil's Parsley 28044
Devil's Paw 267047
Devil's Pinch 309570
Devil's Plague 123141
Devil's Plaything 3978,53304,402886
Devil's Poker 217055,401112
Devil's Posy 15861,15876,208771
Devil's Potato 367014
Devil's Rattle 3978
Devil's Rhubarb 44708,292372
Devil's Ribbon 28617,231183
Devil's River 417826
Devil's Rush 213447
Devil's Shirt Buttons 374916
Devil's Shoelaces 70512
Devil's Shoestrings 105324,386365,407674
Devil's Spineflower 89111
Devil's Spit 81237
Devil's Spoons 14760,312190
Devil's Tether 162519
Devil's Thorn 395146
Devil's Thread 95431,115008
Devil's Threads 95431,115008
Devil's Tobacco 192358
Devil's Tomato 367014
Devil's Tongue 163456,282685
Devil's Tongue Barrel 163456

Devil's Torch 217055
Devil's Tree 56784,213902
Devil's Trumpet 123065,123077
Devil's Turnip 61497
Devil's Twine 78738,95431,102933
Devil's Vine 68713
Devil's Walking Stick 30760
Devil's Walkingstick 30760
Devil's Walking-stick 30760
Devil's Wand 9832
Devil's Wood 345631
Devil's Wort 148104
Devil's Yam 95431
Devil's-apple 306636
Devil's-backbone 61568,287853
Devil's-bit 85505,379660
Devil's-bit Scabious 379655,379660
Devil's-claw 1264,300944,315431
Devil's-claw Cactus 354307
Devil's-claws 315430
Devil's-club 272704
Devil's-darning-needle 95426
Devil's-guts 115008
Devil's-head 197762
Devil's-head Cactus 140133
Devil's-lettuce 20843
Devil's-paintbrush 195478
Devil's-tail Tearthrumb 309564
Devil's-thorn 144880
Devil's-tongue 20037,20132,272915
Devil's-tree 18055
Devil's-walking-stick 30760
Devil-among-the-tailors 266225
Devil-and-angels 37015
Devildums 359158
Deviled Goldenray 229038
Devil-flower 382912
Devil-in-a-bush 81020,266215,266216,266225,284387
Devil-in-a-fog 266225
Devil-in-a-frizzle 266225
Devils-in-a-hedge 266225
Devil-in-a-mist 266225
Devil-in-church 57101
Devil-in-thebush 81020
Devil-in-the-pulpit 394088
Devil-may-care 15861
Devil-on-all-sides 325612
Devil-on-both-sides 325612
Devilpepper 326995,327069
Devil-pepper 326995
Devilrattan 121486,121514
Devil's Candlesticks 176824
Devils Coachwheel 325612
Devils Currycomb 325612
Devil's Ivy 147347
Devil's Root 275135
Devils-and-angels 37015,302034
Devilsbit 379660
Devilsclaw 1264
Devilsclaw Pisonia 300944
Devils-tongue 20132
Deviltree 18046
Devil-tree 18046,18055
Deviltree Alstonia 18046
Devil-tree Alstonia 18046
Devil-weed 102933
Devilwood 276242,276250
Devilwood Osmanther 276250
Devilwood Osmanthus 276250
Devirs-club 272704
Devon Dendrobium 125112
Devon Eaver 235373
Devon Galeandra 169870
Devon Pride 81768
Devon Stanhopea 373687
Devon Violet 410320
Devonshire Myrtle 261162
Dew Azalea 330939
Dew Bean 408829
Dew Corydalis 106392
Dew Cup 14166,325518,325666,326287
Dew Flower Ice Plant 138233
Dew Mantle 13959
Dew Plant 138261,138348
Dew Rabdosia 209711
Dewberry 334250,338059,338209,338243,338399,338557,339407,339466
Dewberryleaf Cinquefoil 312550
Dewcovered Rhododendron 330939
Dew-covered Rhododendron 330939
Dewcup 14166,325518,325666,326287
Dewdrograss 91499
Dewdrop 169719,339171
Dewdrop Bittercress 72721
Dewdrop Rabdosia 209711
Dewey's Sedge 74297
Dew-plant 29855,134394,138261,220464
Dew-plant Family 12911
Dew-thread 138289
Dewy Pine 138376
Deye Nettle 170078,373455
Dhaincha 361367
Dhak 64148
Dhak Tree 64148
Dhak-tree 64148
Dhal 65146
Dhaman 180995
Dhani 104690
Dhawa 26027
Dhob 117859
Dhub 117859
Dhup 70998,284421
Dhupa Fat 405154
Diabol Strychnos 378704
Diabolo Rattlebox 111879
Dialium 127375
Dialypetalous Carolina Silverbell 184733
Dialysepalous Lady's Slipper 120430
Diamond 343095
Diamond Cholla 116671
Diamond Flower 212474,212476
Diamond Leaf Oak 324095
Diamond Mountain Buckthorn 328689
Diamond Plant 251025
Diamond Willow 343095,343343,343649
Diamond-flower 212476
Diamond-leaf Laurel 301390
Diamondleaf Oak 324095,324240
Diamond-leaf Oak 324240
Diamondleaf Persimmon 132371
Diamond-leaf Willow 343894
Dian Egerton Swan Orchid 116520
Dian Spine olive 415241
Diana Hibiscus 195279
Diana Rose-of-sharon 195279
Dianbai Rattan Palm 65669
Dianbai Rattanpalm 65669
Dian-chuan Windhairdaisy 348394
Diandra Sandspurry 370624
Diandrous Sedge 74304
Dianella 127504
Diangui Holly 203761
Dian-Qian Ramie 56297
Dian-Shu Sandwort 31856
Dianthus 127573,127654
Dianthus Pinks 127793
Dianxi Eargrass 187563
Dianxi Hedyotis 187563
Dianxibei Rockfoil 349247
Dianxiong 297942,297947
Dian-Zang Sandwort 32096
Dian-Zang Spotleaf-orchis 179689
Diaoluoshan Bamboo 47240
Diaoluoshan Bambusa 47240
Diaoluoshan Cyclobalanopsis 116211
Diaoluoshan Devilpepper 327068
Diaoluoshan Mayten 182791
Diaoluoshan Mud Bamboo 47240
Diaoluoshan Oak 116211
Diaoluoshan Qinggang 116211
Diapensia 127910,127916
Diapensia Family 127926
Diapensia Pricklythrift 2227
Diapensia Rockfoil 349249
Diapensia-like Prickly-thrift 2227
Diaphonous Barberry 51548
Diarthron 128023
Diastema 128144
Dicentra 128288
Dicercoclados 128369,128370

Dicerma 128371,128372
Dichaea 128397
Dichanthium 128565
Dichapetalium 128589
Dichapetalum 128589
Dichapetalum Family 128588
Dichelachne 128861
Dichocarpum 128919
Dichondra 128955,128964,128972
Dichorisandra 128987
Dichotoma Silene 363409
Dichotomal Chickweed 374836
Dichotomanthus 129025
Dichotomous Anemone 23796
Dichotomous Batrachium 48913
Dichotomous Cordia 104175
Dichotomous Cymaria 116711
Dichotomous Fimbristylis 166248
Dichotomous Fluttergrass 166248
Dichotomous Gingerbread Palm 202275
Dichotomous Knotweed 309051
Dichotomous Ptilagrostis 321011
Dichotomous Sobralia 366782
Dichotomous Star Wort 374836
Dichotomous Tabernaemontana 327577
Dichotomous Tickclover 126307
Dichotomous Windflower 23796
Dichotomous Woodbetony 287163
Dichotomous-like Fimbristylis 166275
Dichotomous-like Fluttergrass 166275
Dichroa 129030
Dichrocephala 129062
Dichromatic Willow 343297
Dichromic Monomeria 257710
Dichrostachys 129127,129136
Dichrotrichum 129181
Dickasonia 129191
Dickins Barbed Cinquefoil 312498
Dickins Sedge 74309
Dickinsia 129195
Dickson Ehretia 141629
Dickson's Golden Elm 401469
Dicky Birds 3462,169126
Dicky Daisy 50825
Dicky Dilver 409335,409339
Dicky Primrose 314312
Dicliptera 129220,129231,129308,129325
Dicolor Broadleaf Woodnettle 273924
Dicolorflower Crazyweed 278814
Dicranostigma 129563
Dicyrta 129721
Didder Grass 60382
Diddery Dock 60382
Diderrich Fatheadtree 262804
Diderrich Nauclea 262804
Didgeridoo Bamboo 47189
Didiciea 129768
Didier's Tulip 400162
Didierea 129776
Didissandra 129790
Didymocarpus 129868
Didymophylla Devil Rattan 121492
Didymoplexis 130037
Didymostigma 130069
Didymous Milkwort 308024
Didymousleaf Bauhinia 49067
Dieck Maple 2933
Dieffenbach's Hebe 186948
Dieffenbachia 130083
Diels Abelia 122
Diels Angelica 24342
Diels Ardisia 31415
Diels Barberry 51563
Diels Calligonum 67019
Diels Cherry 83191
Diels Childvine 400887
Diels Coral Ardisia 31415
Diels Cotoneaster 107424
Diels Euonymus 157407
Diels Hogfennel 292833
Diels Indigo 205709
Diels Kneejujube 67019
Diels Leptodermis 226081
Diels Ligusticum 229315
Diels Millettia 66624
Diels Moxanette 221539
Diels Pennywort 200282
Diels Rhodiola 329863
Diels Rhynchosia 333215
Diels Rockfoil 349252
Diels Sabia 341497
Diels Sealavender 230608
Diels Stephania 375840
Diels Stonecrop 356674
Diels Thorowax 63616
Diels Trema 394637
Diels Vetchling 222710
Diels Waterdropwort 269309
Diels Wildjute 394637
Diels Woodbetony 287166
Diels Wood-nettle 221539
Diels' Cotoneaster 107424
Diepenhorst Rattan Palm 65671
Dietinguished Rhododendron 330562
Dietrich Begonia 49780
Difengpi Anise-tree 204504
Difengpi Eightangle 204504
Diferttooth Milkvetch 42487
Differbract Dandelion 384572
Differcentre Corydalis 105982
Differcolor Thrixsperm 390500
Differcolor Thrixspermum 390500
Differcolor Wildjute 394656
Differcolor Windhairdaisy 348176
Different Colours Star Wort 374858
Different Craneknee 221687,221707
Different Leaf Bittercress 72799
Different Leaves Pseudostellaria 318501
Different Leaves Small Cowpea 408973
Different Violetleaf Bittercress 73038
Different Willow 343310
Different-appearance Barberry 51720
Differentbract Aster 40565
Differentcalyx Gentianella 174114
Differentcalyx Honeysuckle 235667
Differentcolor Rhynchotechum 333735
Differentcolor Sage 345079
Differentcolor Taro 99921
Differentcolour Milkvetch 42298
Differentflower Indigo 206061
Different-flowered Indigo 206061
Differentfruit Dendrobium 125188
Differentfruit Millettia 66626
Differentfruit Tickclover 126389
Differenthair Rockfoil 349442
Differentleaf Araucaria Pine 30848
Differentleaf Clovershrub 70800
Differentleaf Corydalis 106368
Different-leaf Crazyweed 278816
Differentleaf Elsholtzia 144037
Differentleaf False Rattlesnakeroot 283693
Differentleaf Goldenray 229046
Differentleaf Hogfennel 292858
Differentleaf Honeysuckle 235843
Differentleaf Jasmine 212068
Differentleaf Schefflera 350674
Different-leaf Tickclover 126396
Differentleaved Arisaema 33349
Differentleaves Eutrema 161132
Differentlip Rabdosia 324728
Differentshoot Bladderwort 403220
Differentspike Fingergrass 88394
Different-spike Sedge 74801
Differentspine Craneknee 221685
Differentstamen Gentian 173507
Differentthrum Gentian 173231
Different-tooth Milkvetch 42487
Differenttooth Rhodiola 329881
Differflower Ainsliaea 12653
Differflower Rabbiten-wind 12653
Differfruit Corydalis 105976
Differhair Munronia 259884
Differing Breviligulate Sicklebamboo 137846
Differleaf Coldwaterflower 298858
Differleaf Pondweed 312138
Differleaf Stairweed 142747
Differleaf Triratna Tree 397363
Differscale Azalea 331241
Differsepal Flax 231909
Differsplit Windhairdaisy 348399
Differtooth Corydalis 105985

Diffferleaf Ramie 56095
Difformed Epidendrum 146407
Difformed Galingale 118744
Diffrantflower Parasiticvine 125788
Diffranther Awlquitch 131015
Diffranther Dimeria 131015
Diffrantleaf Yam 131489
Diffrent Eritrichium 184302
Diffrentfruit Eritrichium 153459,184302
Diffuse Barberry 51564
Diffuse Boerhavia 56425
Diffuse Corydalis 105716
Diffuse Cypressgrass 118749
Diffuse Dayflower 100990
Diffuse Emei Woodbetony 287486
Diffuse Erysimum 154418
Diffuse Galingale 118749
Diffuse Hedyotis 187565
Diffuse Iguanura 203506
Diffuse Javatea 275667
Diffuse Knapweed 81034
Diffuse Leucas 227587
Diffuse Marshweed 230293
Diffuse Microula 254342
Diffuse Millet-grass 254515
Diffuse Orychophragmus 275866
Diffuse Rostellularia 337235
Diffuse Schizostachyum 351852
Diffuse Small Reed 127228
Diffuse Smallreed 127228
Diffuse Spiderling 56425
Diffuse Spineflower 89072
Diffuse Star Wort 374944
Diffuse Sugarmustard 154418
Diffuse Wallflower 154418
Diffuse White Falsenettle 56112
Diffuseflower Meadowrue 388485
Diffusicallose Rockfoil 349253
Diffusive Woodbetony 287168
Diflugossa 130315
Digger Pine 300189
Digger Speedwell 283372
Digger's Speedwell 283372
Digitalis 130344
Digitate Begonia 49781
Digitate Bothriochloa 57567
Digitate Habenaria 183568
Digitate Oncidium 270799
Digitate Phacelurus 293211
Digitate Polytoca 310703
Digitatelaef Eria 148735
Digitgrass 130537
Digitstyleorchis 376576,376578
Diglyphosa 130851,130852
Digua Fig 165759
Digynian Caesalpinia 64997
Dijon Mustard 59438
Dika 208990
Dika Butter 208990
Dika Fat 208990
Dika Nuts 208990
Dike Rose 336419
Dike-rose 336419
Dikuscha Currant 333957
Dilatate Arisaema 33311
Dilatate Clematis 94879
Dilatate Linearleaf Speedwell 317925
Dilatate Southstar 33311
Dilated Beadruby 242680
Dilem 306994
Dilienia Family 130928
Diliporchis 130037
Dill 24178,24208,24213,392992,408423,408611
Dill Daisy 32563
Dill-cup 68189,325518,325612,325666,325820,326287
Dillenia 130913
Dillenialeaf Litse 233891
Dillenia-leaved Litse 233891
Dillenia-leaved Meliosma 249378
Dillenius Sauerldee 278099
Dillenius' Oxalis 277803
Dillenius' Speedwell 407114
Dillenius' Tick-trefoil 126363,126508
Dillens Tick-trefoil 126508
Dillflowers 267294
Dillies 60382
Dillleaf Corydalis 105615
Dillleaf Sagebrush 35119
Dillleaf Wormwood 35119
Dill-like Sagebrush 35128
Dill-like Wormwood 35128
Dillon Vanilla 404989
Dilly 262441
Dilly Boolen 277638
Dilly Daff 262441
Dilly Dally 262441
Dilochia 130945
Dilophia 130956
Diluted Swertia 380184
Dime-a-bottle Plant 175766
Dimeria 131006
Dimgreen Sagebrush 35185
Diminashed Gnetum 178532
Dimmit Sunflower 189035
Dimocarpus 131057
Dimorphantis 30604
Dimorphic Crabgrass 130592
Dimorphicleaf Net-veined Maple 3496
Dimorphicscale Sedge 74327
Dimorphleaf Spapweed 204908
Dimorphleaf Stonecrop 356675
Dimorphocalyx 131103
Dimorphostemon 131132
Dimorphous Allomorphia 16076
Dimorphous Pondweed 312267
Dimpled Trout-lily 154949
Dimplewort 401711
Dimyellow Clovershrub 70809
Dina Red Ivorywood 297044
Dindle 368771,384714
Dindygul Peperomia 290305
Dinetus 131242
Ding'an Fig 164910
Ding'an Machilus 240577
Ding-dong 70271
Dinggongvine 154233
Dinghu Azalea 331980
Dinghu Beautyberry 66939
Dinghu Chirita 87865
Dinghu Clematis 95380
Dinghu Eargrass 187568
Dinghu Fishvine 10207
Dinghu Jewelvine 10207
Dinghu Macaranga 240320
Dinghu Oak 116081
Dinghu Oligostachyum 270444
Dinghu Poorspikebamboo 270444
Dinghu Purplepearl 66939
Dinghu Qinggang 116081
Dinghu Rhododendron 331980
Dinghu Wildginger 37679
Dinghushan Cyclobalanopsis 116081
Dinghushan Opithandra 272594
Dinghushan Rhododendron 331980
Dingjie Bomi Deutzia 126851
Dingjie Buttercup 325783
Dingjie Milkvetch 42296
Dingjie Osmanther 276285
Dingjie Osmanthus 276285
Dingjie Rush 212995
Dingle Bells 169719
Dingleberry 403822
Dingle-dangles 169719
Dingri Fritillary 168563
Dingri Milkvetch 43159
Dingy Chamaesaracha 85692
Dinkel Wheat 398890
Dinklage Stephania 375841
Dinner Plate Fig 164892
Dinnerplate Thistle 92361
Dinochloa 131277
Dinosaur Food 181958
Dioecious Balanophora 46826
Dioecious Ephedra 146216
Dioecious Olive 270094
Dioecious Sedge 74330,76368
Dioecious Snakemushroom 46826
Dioecism Nettle 402886
Diogenes' Lantern 67560

Dion 131427
Dioon 131427
Dioscorea-leaved Bauhinia 49069
Diosma 99458,131938
Diosma-leaved Honey Myrtle 248091
Diospyros Ferrea 132153
Dipelta 132598
Dipentodon 132608
Diphylax 132668,132672
Diplachne 132720,226005
Diplacrum 132787,132789
Dipladenia 244329
Diplarche 132828,132829
Diplazoptilon 132860
Diplectria 132874
Diploclisia 132928,132929
Diplocyatha 132947
Diploknema 133006
Diplomeris 133038
Diplomorpha Stringbush 414257
Diplopanax 133076
Diploprora 133143
Dipodium 133395
Dipoma 133410
Dippel Deutzia 127100
Dipper Gourd 219843
Dipsacuslike Fimbristylis 166293
Dipsacuslike Fluttergrass 166293
Dipteracanthus 133532
Dipterocarpus Family 133551
Dipteronia 133596
Diptychocarpus 133640
Diqing Monkshood 5162
Dirt-a-bed 384714
Dirtflower Windhairdaisy 348800
Dirtweed 86901
Dirty Dick 86901
Dirty Dora 118744
Dirty Groundsel 360096
Dirty Jack 86901
Dirty John 86901,87200
Dirty Melandrium 248423
Dirty-brown Currant 334078
Dirtyflower Onosma 242398
Dirtyhair Everlasting 21660
Dis Grass 20268
Disa 133684
Disanthus 134009
Disc Camomile 246375
Disc Goldenray 229019
Disc Heteropappus 193979
Disc Mayweed 246375
Disc Water-hyssop 46371
Dischidanthus 134025
Dischidia 134027,134146
Disc-noded Bamboo 20206
Discocleidion 134143
Discoid Beggarticks 53879
Discoid Cancrinia 71090
Discoid Cherry 83196
Discoid Cremanthodium 110382
Discoid Heteropappus 193937
Discoid Nutantdaisy 110382
Discoid-stigma Blinkworthia 55605
Discolor Addermonth Orchid 243036
Discolor Agapetes 10303
Discolor Argostemma 32520
Discolor Brandisia 59123
Discolor Catasetum 79338
Discolor Cinquefoil 312502
Discolor Conehead 323062
Discolor Dendrobium 125113
Discolor Epipactis 147149
Discolor Gansu Fritillary 168523
Discolor Ludisia 238135
Discolor Microtropis 254291
Discolor Nervilia 265389
Discolor Rhodiola 329865
Discolor Rhynchotechum 333735
Discolor Smallligulate Aster 40003
Discolor Spapweed 204909
Discolor Spatholobus 370413
Discolor Steudnera 376383
Discolor Ternstroemia 386693
Discolor Tipularia 392348
Discolor Touch-me-not 204909
Discolor Willow 343302
Discolored Actinidia 6533
Discolored Kiwifruit 6533
Discolored Skullcap 355429
Discolorleaved Mahonia 242608
Discolour Barberry 51565
Discolour Clinopodium 96987
Discolour Deutzia 126904
Discolour Dragonhead 137575
Discolour Haemaria 238135
Discolour Homogyne 197945
Discolour Wildbasil 96987
Discoloured Agapetes 10303
Discolour-leaved Mahonia 242608
Discontinuous Moxanettle 221568
Discostigma 134181
Discretesepal Ladyslipper 120430
Discretestamen Camellia 69270
Discretus L. H. Bailey 338491
Disc-shaped Cherry 83196
Discus-fruit Hydrangea 199867
Disected Ladypalms 329186
Disepal Comastoma 100270
Disepal Throathair 100270
Disguised St. John's-wort 201834
Dish Mustard 390213
Dish Oak 116172
Dish Stairweed 142650
Dishalaga 400675
Dishcalyx Azalea 331446
Dishcloth Gourd 238257,238261
Dish-cloth Gourd 238257,238261
Dishclothgourd 238257
Dishcupula Oak 116082
Dishcupula Qinggang 116082
Dish-cupule Cyclobalanopsis 116082
Dish-cupule Oak 116082
Dishflower Dzungar Milkaster 169798
Dishflower Lily 266573
Dishflower Milkaster 169767
Dishflower St. John's Wort 201717
Dishflower Tupistra 70624
Dishfruit Seepweed 379533
Dishfruit Tickseed 104829
Dishgland Chessbean 116599
Dishilago 400675
Dishlago 400675
Dish-like Stairweed 142649
Dishrag Gourd 238261
Dish-rag Gourd 238261
Dish-shaped Oak 116172
Dishshaped Syncalathium 381647
Dishspurorchis 171825,171835,171881,342015
Dishstyle Aspidistra 39517
Dishylagie 400675
Diskfruit Wingnut 116240
Disk-leaved Hebe 186967
Disperis 134258
Dispermous Big-ginseng 241164
Dispermous Jasmine 211795
Dispersalstyle Schefflera 350718
Dispersed Humid-shaped Barberry 51745
Diss 20268
Dissect Cousinia 108268
Dissect Vinegentian 108268
Dissecte Eryuan Cystopetal 320234
Dissected Cremanthodium 110383
Dissected Crenate Woodbetony 287122
Dissected Dandelion 384524
Dissected Elatostema 142650
Dissected Garishcandle 28137
Dissected Gentian 173376
Dissected Giantfennel 163602
Dissected Leaf Woodbetony 287171
Dissected Lettuce 219303
Dissected Merremia 250777
Dissected Nutantdaisy 110383
Dissected Rock Daisy 291308
Dissected Rockjasmine 23159
Dissected Sinolimptichtia 364971
Dissected Violet 409914
Dissected Woodbetony 287170
Dissectedleaf Graceful Clematis 94958
Dissectedleaf Sawwort 361039

Dissectleaf Woodbetony 287171
Dissimilar Willow 343310
Distaff Thistle 44114,77684,77716,77748
Distant Nepeta 264912
Distant Phacelia 293162
Distant Sedge 74348
Distantflower Small Reed 127207
Distanttooth Tea 69547
Distichous Angraecum 24815
Distichous Calanthe 65952
Distichous Cotoneaster 107595
Distichous Dishspurorchis 171842
Distichous Eragrostiella 147464
Distichous Euonymus 157414
Distichous Gastrochilus 171835,171842
Distichousleaf Eurya 160453
Distichous-leaved Eurya 160453
Distichous-leaved Wallichpalm 413093
Distinchous Rattan Palm 65673
Distinchous Rattanpalm 65673
Distinchous Vineclethra 95487
Distinct Camellia 69059
Distinct Gardneria 171442
Distinct Gladiolus 176273
Distinct Hawthorn 109678
Distinct Pondweed 312090
Distinct Rattan Palm 65712
Distinct Yunnan-bamboo 278649
Distinctia Schefflera 350725
Distinctive Schefflera 350725
Distinctstyle Monkshood 5605
Distinctstylelike Monkshood 5604
Distinctvein Ardisia 31353
Distinctvein Barberry 52029
Distinctvein Bauhinia 49103
Distinctvein Bedstraw 170445
Distinctvein Cylindrokelupha 116593
Distinctvein Cymbidium 117017
Distinctvein Homalium 197710
Distinctvein Illigera 204638
Distinctvein Inflated Dendropanax 125619
Distinctvein Monkshood 5658
Distinctvein Stauntonvine 374393
Distinctvein Taxillus 385211
Distinctvein Thyme 391302
Distinctvein Viburnum 407964
Distinctvein Wild Quince 374393
Distinct-veined Barberry 51982
Distinct-veined Homalium 197710
Distinct-veined Thyme 391302
Distingchus Vineclethra 95487
Distingct Tomentose Pentapanax 289666
Distinguished Biondia 54522
Distinguished Spapweed 205031
Distinguished Touch-me-not 205031
Distylium 134918
Dita 18055
Dita Bark 18055
Ditch Bur 31051,415003,415057
Ditch Dallisgrass 285471
Ditch Fimbry 166471
Ditch Grass 340476
Ditch Moss 143920
Ditch Paspalum 285471
Ditch Polypogon 310109,310113
Ditch Stonecrop 290138
Ditchgrass 340476
Ditch-grass 340470
Ditch-grass Family 340502
Ditch-moss 143914
Ditfuse Coptosapelta 103862
Dithering Grass 60382
Dithery Dother 60382
Ditinguished Rhododendron 331538
Dittander 225398
Dittany 114527,129615,129618,225398,
274197,274210
Dittany Candy 274210
Dittany of Candy 274210
Dittany of Crete 274210
Dittany-of-crete 274210
Ditten 225398
Diuranthera 135055,135059
Diuris 135063
Divaricate Acanthopanax 143588
Divaricate Barberry 51567
Divaricate Bassia 48762
Divaricate Batrachium 48914
Divaricate Bentgrass 12080
Divaricate Bluebeard 78002
Divaricate Carpesium 77156
Divaricate Cleistes 94547
Divaricate Cogflower 154162
Divaricate Cranesbill 174570
Divaricate Daphniphyllum 122676
Divaricate Epaltes 146099
Divaricate Ervatamia 154162
Divaricate Houndstongue 117956
Divaricate Ironweed 406284
Divaricate Jadeleaf and Goldenflower 260403
Divaricate Knotweed 309056
Divaricate Listera 264673
Divaricate Mussaenda 260403
Divaricate Oncidium 270806
Divaricate Oxtongue 298583
Divaricate Pentanema 289566
Divaricate Phlox 295257
Divaricate Rockfoil 349255
Divaricate Saposhnikovia 346448
Divaricate Serpentroot 354846
Divaricate Spapweed 204911
Divaricate Starjasmine 393640
Divaricate Strophanthus 378386
Divaricate Sunflower 188958
Divaricate Swertia 380188
Divaricate Thespis 389962
Divaricate Tigernanmu 122676
Divaricate Typhonium 401156
Divaricate Velvetplant 183082
Divaricate Willow 343314
Divaricatebranch Kangding Monkshood 5627
Divaricated-veined Sweetleaf 381148
Divaricate-vein Sweetleaf 381148
Divergent Ironweed 406284
Divergent Wild Buckwheat 152004
Divergentstyle Willow 343318
Divergent-styled Willow 343318
Divers Jadeleaf and Goldenflower 260382
Divers Mussaenda 260382
Diverscolour Sage 345079
Diverse Male Willow 343482
Diverse Mesquitilla 66663
Diverse Mussaenda 260382
Diverse Periwinkle 409327
Diverse Sagebrush 35136
Diverse Willow 343480
Diverse Wormwood 35136
Diversecolor Catnip 264911
Diversecolor Ligusticum 229318
Diversecolor Maple 2936
Diversecolor Meliosma 249416
Diversecolor Microtropis 254291
Diversecolor Nepeta 264911
Diversecolor Pinangapalm 299659
Diversecolor Pinanga-palm 299659
Diversecolor Sabia 341498
Diverse-colored Maple 2936
Diverse-colored Sabia 341498
Diversecolour Skullcap 355474
Diverse-cymose Mayten 246830
Diverse-foliage Gueldenstaedtia 181642
Diversefolious Rockfoil 349256
Diversefolius Cystopetal 320221
Diverse-leaf Achimenes 4079
Diverseleaf Adenia 7263
Diverseleaf Ainsliaea 12634
Diverseleaf Alajja 13335
Diverseleaf Alangium 13364
Diverseleaf Asiabell 98320
Diverseleaf Bean 293999
Diverseleaf Biond Hackberry 80584
Diverseleaf Deutzia 126970
Diverseleaf Dragonhead 137577
Diverseleaf Fedde Elsholtzia 144015
Diverseleaf Fescue 164006
Diverseleaf Fescuegrass 164006
Diverseleaf Fig 165099
Diverseleaf Greenorchid 137577
Diverseleaf Hedgeparsley 392985
Diverseleaf Honeysuckle 235843
Diverseleaf Knotweed 308816

Diverseleaf Lambsquarters 86903
Diverseleaf Mayten 182683,246853
Diverseleaf Nettle Spurge 212148
Diverseleaf Pertybush 292044
Diverseleaf Pseudopyxis 318187
Diverseleaf Rabbiten-wind 12634
Diverseleaf Ricebag 181642
Diverseleaf Rotala 337325
Diverseleaf Sawwort 361059
Diverseleaf Schefflera 350674
Diverseleaf Southstar 33349
Diverseleaf Spapweed 204912
Diverseleaf Thistle 92022
Diverseleaf Touch-me-not 204912
Diverseleaf Trigonostemon 397363
Diverseleaf Willowweed 146842
Diverseleaf Wingseedtree 320831
Diverseleaf Youngia 416416
Diverse-leaved Honeysuckle 235843
Diversestamen Oreocharis 273865
Diverseumbel Pleurospermum 304807
Diversianther 43765,43768
Diversianther Lilyturf 272083
Diversibract Alpine Aster 40019
Diversi-calyx Persimmon 132045,132046
Diversicolored Rhododendron 330609
Diversicymose Mayten 246830
Diversiflorous Dendrotrophe 125788
Diversiflorous Indigo 206061
Diversiflorous Star Wort 374859
Diversifoliaceous Barberry 51748
Diversifoliolate Schefflera 350674
Diversifolious Blooddisperser 297627
Diversifolious Buckthorn 328726
Diversifolious China Starjasmine 393657
Diversifolious Chinese Starjasmine 393657
Diversifolious Creeper 285117
Diversifolious Elatostema 142747
Diversifolious Goldsaxifrage 90434
Diversifolious Harrysmithia 185939
Diversifolious Hemiphragma 191574
Diversifolious Lettuce 219304,283695
Diversifolious Lysionotus 239947
Diversifolious Osmanther 276313
Diversifolious Osmanthus 276313
Diversifolious Patrinia 285834
Diversifolious Physaliastrum 297627
Diversifolious Poplar 311308
Diversifolious Pricklyash 417290
Diversifolious Pterospermum 320836
Diversifolious Schefflera 350674
Diversifolious Tulip 400174
Diversifolious Whytockia 413999
Diversifolious Youngia 416433
Diversifolius Ampelopsis 20354
Diversi-hair Munronia 259884
Diversileaf Artocarpus 36920
Diversi-leaf Clematis 94880
Diversileaf Creeper 285105
Diversileaf Elaeagnus 142023
Diversileaf Firethorn 322469
Diversileaf Marshweed 230304
Diversileaf Pimpinella 299395
Diversisepal Shagvine 97505
Diversisifolious Alangium 13364
Diversityleaf Evergreenchinkapin 78922
Divided Sedge 74360
Divided-by-the-brook 247025
Dividedcalyx Strawberry 167613
Dividivi 64982
Divi-divi 64982
Divine Cactus 236425
Dixie Daisy 291336
Dixie Goldenrod 368001
Dixie Iris 208614
Dixie Rosemallow 195040
Dixie Signalgrass 58155
Dixie Spider-lily 200923
Dixie Stitchwort 255530
Dixie white-topped aster 360716
Diyu Camellia 69046
Djave 46607
Djeroek Balik 93674
Djeroek Hunje 93513
Djeroek Limoe 93322
Djeroek Papaya 93685
Dlender-leaf Begonia 50363
Dobbin-in-the-ark 5442
Dobinea 135097
Dobo Lily 120542
Dock 34590,339880,339887,339989,
339994,340019,340084,340178,340269
Dock Cress 221780,221787
Dock Flower 86901,308739
Dock Sorrel 339887
Dock Vine 390692
Dock's Meat 224375
Dock's Mouth 130383
Docken 339887
Dockerene 221787
Dockleaf Goldenray 229085,229166
Dockleaf Knotweed 309298
Dockleaved Knotweed 309298
Dock-leaved Smartweed 309298
Dockmackie 407662
Docko 36474
Dockseed 339887
Doctor Doodles 65055
Doctor's Gum 252595
Doctor's Love 170193
Doctor's Medicine 338447,340151
Doctorbush 305202
Docynia 135113
Dod 401112
Dodartia 135132
Dodder 102933,114947,114996,115008,
162519,370522,401112
Dodder Family 115155
Dodder Grass 60382
Dodder-cake Plant 68860
Dodderformed Rockjasmine 23152
Doddering Dickies 60382
Doddering Dillies 60382
Doddering-dillies 60382
Dodder-laurel 78726
Dodder-of-thyme 115008
Doddle Grass 60382
Dodge's Hawthorn 109680
Dodger 91770
Dodgeweed 182103
Dodgill-reepan 273469
Dodo Cloth 382821
Dodol 171136
Dodonaea 135192
Dodonaea-leaf Tanoak 233181
Doellingeria 135253
Does-my-mother-want-me 235334
Dog Almond 22218
Dog Apple 38334
Dog Bane 29465
Dog Banner 26768
Dog Berries 44708,336419,369339
Dog Binder 26768
Dog Bobbins 37015
Dog Breer 336419
Dog Briar 336419
Dog Caper 71706
Dog Cherry 380499
Dog Cholla 181459
Dog Choops 336419
Dog Clover 247366
Dog Cocks 37015
Dog Couch-grass 144498
Dog Daisy 3978,26768,50825,227533,
397970
Dog Drake 229662
Dog Eller 407989
Dog Fennel 26738,26768,158031,158066,
292957,292966
Dog Figwort 355075
Dog Finkle 26768
Dog Flower 227533,250614
Dog Gowan 397970
Dog Grass 144661
Dog Heather 67456
Dog Hip 336419
Dog Hippan 336419
Dog Hobble 228162
Dog Job 336419
Dog Jump 336419
Dog Laurel 44915

Dog Leek 274823
Dog Lime 365424
Dog Liver 215122
Dog Mint 388275
Dog Mouth 28617
Dog Mustard 154091
Dog Nettle 170078,220416,403039
Dog Oak 2837
Dog Parsley 28044
Dog Pinanga 299656
Dog Pine 121088,184913
Dog Plum 141863
Dog Poison 9832
Dog Posy 384714
Dog Rise 157429
Dog Rose 48977,336419,407989
Dog Rowan 407989
Dog Senna 78436,360444
Dog Snout 28617
Dog Spear 37015
Dog Stalk 359158
Dog Standard 359158
Dog Standers 359158
Dog Stinker 26768
Dog Stinkers 26768
Dog Stinks 26768
Dog Tansy 312360
Dog Thistle 91770
Dog Timber 157429,407907
Dog Tooth Berry 157429
Dog Tooth Violet 125836,154909
Dog Tree 16352,105023,157429,345631,380499,407989
Dog Violet 409804,409851,410161,410493
Dog's Airach 87200
Dog's Arrach 87200
Dog's Briar 336419
Dog's Cabbage 389184
Dog's Camomile 26768,397970
Dog's Camovyne 397970
Dog's Carvi 28044
Dog's Cherries 61497
Dog's Cherry 61497
Dog's Cole 250614
Dog's Dibble 37015
Dog's Dogger 273531
Dog's Ear 36474
Dog's Ears 36474
Dog's Fennel 26738,26768
Dog's Finger 130383
Dog's Fingers 130383
Dog's Hair 143365,143369
Dog's Leek 169460
Dog's Lugs 130383
Dog's Medicine 250614
Dog's Mercury 250614
Dog's Mouth 28617,231183
Dog's Nose 28617
Dog's Onion 274823
Dog's Orach 87200
Dog's Orache 87200
Dog's Paise 28200
Dog's Parsley 9832,28044
Dog's Pennies 329521
Dog's Rib 302034
Dog's Siller 329521
Dog's Tansy 312360
Dog's Tassel 37015
Dog's Tausle 37015
Dog's Thistle 37015,368771
Dog's Thorn 336419
Dog's Timber 157429,380499
Dog's Toe 174877
Dog's Tongue 118008
Dog's Tooth 154909
Dog's Tooth Violet 154895,154909
Dog's-tail 118307,118315
Dog's-Tail Grass 118307
Dog's-tail-grass 118307
Dog's-tongue 317233
Dog's-tooth Violet 154895,154909
Dog's-tooth-violet 154895
Dog-and-bobbin 37015
Dog-and-bobbins 37015
Dogbane 29465,29473,29514
Dogbane Family 29462
Dogberry 31142,333945,369290,380499,407907,407989
Dogbonbavin 133181,395164
Dogbone Ardisia 31348
Dogfennel 158066
Dog-fennel 26768
Doggies 231183
Doghobble 228155,228162
Dogleg Ash 167981
Doglike Sage 344993
Dogliverweed 129220
Dogmint 264897
Dogmouth 28617
Dog-mustard 154091
Dogrise 157429
Dogs-and-cats 396828
Dogsbane 29468
Dogstail Grass 118307,118315,118319
Dogstones 273531
Dogstooth Grass 117852
Dogtail 61975
Dogtailgrass 118307
Dogteeth Glorybower 96059
Dogtongueweed 385884,385903
Dog-tooth 154899
Dogtooth Fawn Lily 154909
Dogtooth Lily 154936
Dog-tooth Lily 154936
Dogtooth Pea 222832
Dogtooth Violet 154909
Dog-tooth Violet 154909
Dog-tooth Violet Fawnlily 154909
Dog-toothed Violet 125836
Dogtoothgrass 117852,117859
Dog-violet 409804,410483
Dogweed 139710,391042
Dogwood 104917,105146,124880,157429,323083,328603,367120,380425,380444,380499,407907,407989
Dogwood Family 104880
Dogwood Navelseed 270684
Dogwood Navel-seed 270684
Doi's Sweetleaf 381104
Doken 339880
Doleful Bells 44708
Doleful Bells of Sorrow 168476
Dolichandrone 135299,135308
Dolichang Triangle Oak 167344
Dolicho Vigna 409008
Dolicholoma 135371
Dolichopetalum 135389
Dolichos 135399
Dolichos Bean 218721
Dolichothele 135686
Dolium Rehder Willow 343986
Doll Cheese 243823
Doll Cheeses 243823
Doll's Eyes 6433,6448
Doll's-daisy 56700
Doll's-eyes 6405,6433
Doll's-head 219795
Dollar Orchid 315614
Dollar Plant 238371
Dollarleaf 126586
Dollar-leaf 322872
Dollarleaf Rhynchosia 333377
Dollarplant 238371
Dolls-eyes 6441
Dolly Soldiers 174739
Dolly Varden 386365
Dolly Winter 248312
Dolly's Apron 174877
Dolly's Bonnets 30081
Dolly's Nightcap 174877
Dolly's Pinafore 174877
Dolly's Shoes 30081,174877
Dolomiaea 135721,135739
Dolores River Skeletonplant 239332
Dolphin-flower 102833
Dombeya 135753,136031
Domestic Langsat 221227
Dominica Begonia 49789
Dominican Palm 341410
Dominican Palmetto 341410
Don Cranesbill 174572

Don Lobelia 234436
Don Quixote's Lace 416659
Don River Sainfoin 271273
Don's Twitch 144228
Dona 136072
Donald Honeysuckle 235808
Donald Magnoliavine 351041
Donegal 361317
Dong 346209
Dong Bells 262441
Dong Fritillary 168587
Dong Nut 346209
Dong'an Camellia 69735
Dongbazi Milkvetch 43195
Dongbei Serpentroot 354897
Dongbei Spunsilksage 360832
Dongchuan Barberry 51571,51903
Dongchuan Cacalia 283876
Dongchuan Manyvein Buckeye 9725
Dongchuan Milkvetch 43196
Dongchuan Rockfoil 349274
Dongchuan Spapweed 204822
Dongchuan Yellowish Woodbetony 287390
Dongfang Cleistanthus 94507
Dongfang Firmiana 166692
Dongfang Sedge 76619
Dongfang Slugwood 50621
Donggou Willow 343327
Dongjiu Goldenray 229229
Dongle Remarkable Barberry 51771
Dong-quai 24475
Dongtinghuang Ginkgo 175818
Dongtou Falsenettle 56208
Dongwang Rockfoil 349275
Dongxing Bamboo 47235
Dongxing Briggsia 60262
Dongxing Camellia 69149
Dongxing Helicia 189925
Dongxing Machilus 240713
Dongya Goosefoot 87187
Dongyang Bamboo 297484
Dongyang Greenbark Bamboo 297484
Dongzhi Stonecrop 356679
Donkey 31051,170193
Donkey Ears 215150
Donkey Eye Bean 259573
Donkey Orchid 135063
Donkey Tail 159417
Donkey's Ear 215277,302034,373166,405788
Donkey's Ears 215277,302034,373166,405788
Donkey's Oats 339887,340151
Donkey's Rhubarb 328345
Donkey's Tai 356944
Donkey's Tails 356944
Donkey's Thistle 133478
Donkey's-taill 356944
Donkey-eye Bean 259573
Donkeytail 159417
Donkey-tail 356944
Donner Pass Sulphur Flower 152608
Donnhove 400675
Donninethell 170078
Dontostemon 136156
Doob 117859
Doom Bark 154973
Doon 362186
Doonhead Clock 384714
Doon-head Clock 384714
Doorweed 308816
Dooryard Dock 340109
Door-yard Dock 340109
Dooryard Knotweed 308816
Door-yard Knotweed 308788
Dooryard Plantain 302068
Door-yard Violet 410587
Dooryard Weed 308816
Dopatricum 136209
Doplopod 148249
Dorea Palm 85420
Dorest Heath 149195
Doritis 136312
Dorj Corydalis 105831
Dornbloom 1297
Dornel 235373
Doronicum 136333
Doronoki 311389
Dorrigo Waratah 16253
Dorsay's Buttonhole Rose 337007
Dorsiflorous Phaulopsis 294084
Dorstenie 136417
Dorward Cliffbean 66628
Dorward Millettia 66628
Dosheng La Rhododendron 330597
Dossinia 136784
Dotalu Palm 237972
Dother 115008,162519,370522,408423
Dother Grass 60382
Dothering 60382
Dothering Dick 60382
Dothering Dickies 60382
Dothering Dillies 60382
Dothering Dock 60382
Dothering Grass 60382
Dothering Jockies 102727
Dothering Nancy 60382
Dotherum 407072,407144
Dothery 60382
Dotless Adinandra 8219
Dotroa 123077
Dots-and-dashes 349044
Dotted Ash 168077
Dotted Beebalm 257184
Dotted Blazing Star 228506
Dotted Bridalveil 175489
Dotted Button Snakeroot 228506
Dotted Camellia 69531
Dotted Combretum 100728
Dotted Gayfeather 228506
Dotted Gay-feather 228506
Dotted Hawthorn 109970
Dotted Horsemint 257188
Dotted Lancepod 235567
Dotted Loosestrife 239804
Dotted Melaleuca 248099
Dotted Persimmon 132364
Dotted Rockfoil 349691
Dotted Rockjasmine 23109
Dotted Sclerachne 353974
Dotted Sedge 75931
Dotted Smartweed 309654,309658,309666
Dotted Thorn 109970
Dotted Wild Coffee 319784
Dotty Azalea 330427
Dotty Rhododendron 331480
Double Almond 316405
Double Apple Rose 336861
Double Bladder Campion 364146
Double Bubble Mint 10405
Double Buttercup 325518
Double Buttercup Daisy 326290
Double California Rose 336414
Double Camomile 85526
Double Cinnamon Rose 336733
Double Coconut 235151,235159
Double Common Lilac 382371
Double Common Oleander 265335
Double Common Snowdrop 169720
Double Cream Blackberry 339195
Double Daisy 50825
Double Dumb Nettle 47013
Double Eaglеclawbean 370203
Double Fingers-and-thumbs 28200
Double Gean 316247
Double Hydrangea 199912
Double Impatiens 205444
Double Lady's Fingers-and-thumbs 28200,237539
Double Madonna Lily 229792
Double Magnolia-vine 351077
Double Meadow Buttercup 325519
Double Orange Daylily 191291
Double Pincushion 28200
Double Queen of Meadowsweet 166127
Double Reeves Spirea 371868
Double Roxburgh Rose 336891
Double Rugose Rose 336730
Double Saxifrage Tunicflower 400357
Double Scabrous Deutzia 127074
Double Sea Campion 364148

Double Soapwort 346435
Double Tongue 340994
Double Virginia Rose 336308
Double Weaversbroom 370203
Double Weeping Cherry 316636
Double White Banksia Rose 336367
Double White Crowfoot 325516
Double White Spirea 371868
Double Yellow Rocket 47967
Doubleanther Dropseed 372650
Doublecorolla Aster 40315
Doubleear Southstar 33278
Double-ear Spatholobus 370412
Doublefile Viburnum 408027,408034,
408035,408036
Doubleflower Autumn-crocus 99305
Doubleflower Cherokee Rose 336674
Doubleflower Chinese Hibiscus 195180
Doubleflower Chinese Lilac 382136
Doubleflower Clematis 94914
Doubleflower Common Lilac 382371
Doubleflower Cottonrose Hibiscus 195042
Doubleflower Daylily 191291
Doubleflower Flowering Plum 316884
Doubleflower Fragrant Plantainlily 198642
Doubleflower Goldband Lily 229731
Doubleflower Golden Rhododendron 330185
Doubleflower Guangdong Rose 336670
Doubleflower Hibiscus 195042
Doubleflower Japan Buttercup 325983
Doubleflower Juteleaf Raspberry 338284
Doubleflower Kerria 216088
Doubleflower Magnoliavine 351077
Doubleflower Nippon Hawthorn 109652
Double-flower Pomegranate 321776
Doubleflower Reeves Spiraea 371874
Doubleflower Shortfruited Rhododendron
330242
Doubleflower Siberia Fritillary 168506
Doubleflower Smallfruit Rose 336510
Doubleflower Tigwer Lily 230061
Doubleflower Tuoli Fritillary 168591
Doubleflower Xinjiang Fritillary 168624
Double-flowered Chinese Hibiscus 195149
Double-flowered Fragrant
Mockorange 294492
Double-flowered Gorse 401389
Double-flowered Gypsy Rose 216088
Double-flowered Horsechestnut 9701
Double-flowered Pure White Mockorange
294588
Double-flowered Salmonberry 339286
Double-fowered Cherry 316247
Doublefurrow Rockjasmine 23129
Doublegee 144877
Doublehead Elatostema 142648
Doublehead Stairweed 142648
Doublehead Woodbetony 287165
Doublehorn Reinorchis 182218
Doublehorned Crabgrass 130454
Double-leaf 232950
Double-leaf Platanthera 302260
Doubleligulate Groundsel 358400
Doublelip Dendrobium 125183
Double-lobe Meliosma 249366
Doublelobed Smallleaf Bittercress 72898
Double-perianth Azalea 330228
Doublepetal Cream Clematis 94914
Doublepetal Roseleaf Raspberry 339195
Doublepetalous White Pomegranate 321769
Double-petalous White Rugose Rose 336905
Doubleriver Pipewort 151210
Doubleseed Sugarpalm 32333
Double-serrate Meliosma 249378
Double-serrate Windhairdaisy 348425
Doublespine Sesbania 361367
Double-spines Pincution 244081
Doublestamen Bergia 52617
Doublestegine Bamboo 47246
Double-stigma Bulrush 353374
Doubletail Orchis 412385,412388
Doubleteeth Figwort 355100
Double-teeth Maple 2942
Doubleteethed Bittercress 72882
Doubletine Ramie 56098
Doubletooth Grewia 180748
Doubletoothed Girald Rose 336608
Double-toothed Maple 2942
Doubletoothed Rose 336535
Double-toothed Rose 336535
Doublewave Pimpinella 299364
Doublewhite Banks Rose 336367
Double-white Thimbleberry 338971
Doublewing Globba 176996
Doubt Bamboo 297261
Doubt Rattlebox 112094
Doubtful Azalea 330079
Doubtful Barberry 51295,51572
Doubtful Cacalia 283789
Doubtful Camellia 69052
Doubtful Chickweed 82796
Doubtful China Tree 248913
Doubtful Chinaberry 248913
Doubtful Cotoneaster 107339
Doubtful Currant 333914
Doubtful Hairystyle Mockorange 294550
Doubtful Knight's Spur 102833
Doubtful Knight's-spur 102833
Doubtful Langsat 221227
Doubtful Rhododendron 330079
Doubtful Rush 212832
Doubtful Sinocrassula 364916
Doubtful Skullcap 355432
Doubtful Sophora 368967
Doudlar 250502
Dougal Grass 209968
Dough Fig 164763
Douglas Blue-eyed-grass 365726
Douglas Coreopsis 104480
Douglas Fir 318562,318580
Douglas Groundsel 358757
Douglas Hackberry 80626
Douglas Hawthorn 109684
Douglas Iris 208528
Douglas Juniper 213793
Douglas Pine 299920
Douglas Rabbit Brush 90492
Douglas Rabbitbrush 90492
Douglas Ragwort 358757
Douglas Rocky Mountain Maple 2993
Douglas Sagewort 35404
Douglas Spirea 371902
Douglas Spruce 318580
Douglas Spurred Lupin 238445
Douglas Squaw-weed 358757
Douglas Tree 318580
Douglas Violet 409937
Douglas' Aster 380993
Douglas' Catchfly 363419
Douglas' Dustymaiden 84488
Douglas' Falsewillow 46221
Douglas' Knotweed 309063
Douglas' Meadow Foam 230210
Douglas' Silverpuffs 253921
Douglas' Spineflower 89074
Douglas' Stitchwort 255463
Douglas' Thistle 91917
Douglas' Wild Buckwheat 152006
Douglas-fir 318562,318580
Douglas-spruce 318580
Doulwnny Sophora 369134
Doum Dom Palm 202321
Doum Palm 202256,202321
Doumer Crabapple 243610
Dovaston Yew 385305
Dove Dock 400675
Dove Flower 5442,291127,291129
Dove Foot 174739
Dove Orchid 291127,291129
Dove Pincushions 350125
Dove Scabious 350125
Dove Tree 123244,123245
Dove's Dung 274708,274823
Dove's Foot 30081,174717
Dove's Foot Cranesbill 174739
Dove's Plant 30081
Dove's-foot Crane's-bill 174739
Dove's-foot Cranesbill 174739
Dove's-foot Geranium 174739
Dovedale Moss 349496
Dove-dock 400675

Dove-foot 174739
Dovefoot Geranium 174739
Dove-plant 30081
Dove-plum 97862
Doves-at-the-fountain 30081
Dovesfoot Cranesbill 174739
Doves-in-the-ark 5442,30081
Dovesround-aodish 30081
Dovetree 123244,123245
Dove-tree 123244,123245
Dove-weed 113021,113037,148238,159634
Dovewood 14177
Dovyalis 136832
Dowerin Rose 155712
Down Dilly 262441
Down Hawthorn 109850
Down Thistle 91954,271666
Down Tree 268344
Downhair Azalea 330560
Downivine 95431
Downscwob 68189
Downweed 166038
Downy Agrimony 11595
Downy Alpine Oatgrass 190187
Downy American Snowbell 379306
Downy Apple 123071
Downy Arrow Wood 408079
Downy Arrowwood 408079
Downy Arrow-wood 408079,408080
Downy Arrowwood Viburnum 408079
Downy Aster 380872
Downy Birch 53563,53590
Downy Brome 60977,60989,60992
Downy Cherry 83347
Downy Chess 60989,60992
Downy Cinquefoil 312683
Downy Currant 334214
Downy Daisy Bush 270213
Downy False Foxglove 45175,175093
Downy Gentian 173763
Downy Golden Rod 368329
Downy Goldenrod 368329
Downy Grape 411615
Downy Green Sedge 76463
Downy Ground Cherry 297720
Downy Ground-cherry 297676,297720
Downy Hawthorn 109850
Downy Heath 149894
Downy Hedge Nettle 170064
Downy Hemp Nettle 170064,170073
Downy Hemp-nettle 170073,170075
Downy Holly 204184
Downy Japanese Maple 3034
Downy Jasmine 211867,211912
Downy Juneberry 19243
Downy Lady's Slipper 120421,120432
Downy Lens Grass 285525
Downy Leucothoe 228155
Downy Lightyellow Sophora 369015
Downy Lobelia 234707
Downy Magnolia-vine 351089
Downy Manzanita 31140
Downy Milkpea 169654
Downy Mussaenda 260483
Downy Oak 324314
Downy Oat-grass 45558,190187
Downy Pagoda-plant 55523
Downy Pagodatree 369149
Downy Paintbrush 79135
Downy Painted Cup 79135
Downy Painted-cup 79135
Downy Phlox 295312
Downy Poplar 311208,311347
Downy Prairie Dalea 121898
Downy Prairie-clover 121908
Downy Qinggang 116118
Downy Ragged Goldenrod 368312
Downy Rattlesnake Plantain 147209,179680
Downy Rattlesnake-plantain 179680
Downy Rhododendron 331575
Downy Rose 336993
Downy Rose Myrtle 332221
Downy Rosemyrtle 332221
Downy Rose-myrtle 332221
Downy Safflower 77716
Downy Sedge 75065
Downy Serviceberry 19243,19251
Downy Skullcap 355490
Downy Snowbell 379306
Downy Solomon's-seal 308633
Downy Speedwell 407196
Downy Sunflower 189001
Downy Swamp Blueberry 403731
Downy Sweetleaf 381143
Downy Thorn Apple 123061
Downy Thorn-apple 123065,123071
Downy Thyme 391394
Downy Trailing Lespedeza 226936
Downy Tree-of-heaven 12559
Downy Viburnum 408065,408079
Downy Wattle 1508
Downy Wild Rye 144524
Downy Wild-rye 144524
Downy Willow 343595
Downy Willowherb 146779,146895
Downy Willow-herb 146895
Downy Wood Mint 55523
Downy Woundwort 373224
Downy Yellow Painted-cup 79135
Downy Yellow Violet 410466
Downy Zieria 417868
Downybranched Eurya 160490
Downy-branched Eurya 160490
Downy-capsule Willow 343893
Downy-flowered Caper 71830
Downy-fruit Milkvetch 42477
Downy-fruited Box 64258
Downy-fruited Field Maple 2844
Downy-fruited Sedge 75065
Downy-leaved Camellia 69520
Down-you-go 407674
Dowo Barberry 51527
Doyonon Raspberry 338338
Dr. Cowan's Rhododendron 330467
Dr. Hu's Rhododendron 330886
Dr. Kendrick's Rhododendron 331002
Dr. M. Albrecht's Azalea 330058
Dr. Martius' Wood-sorrel 277776
Dr. Weych Azalea 332104
Draba 136899,137180
Drabalike Coelonema 98764
Drablike Rockfoil 349277
Dracaena 104334,135389,137330,137382
Dracaena Fig 165502
Dracaena Palm 104334
Dracaena-palm 104339
Dracena 104359,137330
Draftgoldquitch 156511,156513
Dragance 137747
Dragge 235373
Dragon 35411,137382
Dragon Arum 33234,33314,137744,137747
Dragon Bamboo 125477
Dragon Blood Tree 137382
Dragon Bones 159204
Dragon Chinese Juniper 213647
Dragon Claw 104009
Dragon Dracaena 137382
Dragon Gum 41883,41885
Dragon Head 137545
Dragon Heath 137732
Dragon Juniper 213647
Dragon Mallow 366902,367416
Dragon Plum 137714
Dragon Root 33314
Dragon Sage-wort 35411
Dragon Savin 213647
Dragon Spruce 298232
Dragon Tree 137330,137382
Dragon Wormwood 35411
Dragon's Arum 137747
Dragon's Blood 174877
Dragon's Blood Palm 121507
Dragon's Blood Sedum 357169
Dragon's Blood Tree 137382
Dragon's Eye 131061
Dragon's Female 137747
Dragon's Head 28617,137545
Dragon's Mouth 28617,32373,130383,198424,198426
Dragon's Mugwort 35411

Dragon's Teeth 387251
Dragon's-blood 137382,320291
Dragon's-blood Padauk 320291
Dragon's-claw Willow 343670
Dragon's-eye Pine 299894
Dragon's-head 137545,137651
Dragon's-mouth 28617,32371,32373,198426
Dragon's-mouth Orchid 146371
Dragon's-teeth 237476,369707
Dragonbamboo 125461
Dragonblood 320291
Dragonblood Padauk 320291
Dragonbood 137330,137382,137463,137487
Dragonclaw Copiapoa 103749
Dragonclaw Date 418170
Dragonclaw Elm 401609
Dragoneye Machilus 240660
Dragon-eye Machilus 240660
Dragon-eye Pine 299894
Dragon-eyes Tanoak 233297
Dragonflies 3462
Dragon-flower 208771,208924,220412
Dragonflyoechis 302513,400280
Dragonhead 137545,137624,297984
Dragon-head 137545
Dragonhead Arrowbamboo 162679
Dragonhead Bamboo 47516
Dragon-head Bamboo 47516,162679
Dragon-head Fargesia 162679
Dragonhead Gladiolus 176171
Dragonheadflower 372912,372913
Dragonheadsage 247720
Dragonmouth 198424,198426
Dragonpearl 399995,399998
Dragonplum 137714
Dragon-plum 137714
Dragon-root 33314,33544,91557
Dragons 35411,37015,137747,308893
Dragons Month 32373
Dragontooth Coralbean 154643
Dragontree 137382
Dragon-tree 137382
Dragonwort 35411,137747,308893
Drake 235373
Drake Corydalis 105832
Drake's Foot 273531
Draketree 264039
Drakeye Sunflower 188926
Dralyer 102933
Drangon-claw Hankow 343670
Drank 235373
Drank Drunk 235373
Drant 154019
Draper's Teasel 133478
Dravick 235373
Drawk 11947,235373
Drawn By Doves 5442
Draws Ear Grass 403095
Dream Primrose 314758
Dregea 137774
Dregea Rhynchelytrum 333006
Dregea Statice 230609
Dregonheadsage Skullcap 355598
Drejerella 137813
Drepaniflower Laelia 219682
Drepanophore Spapweed 204914
Drepanostachyum 137841
Dress' Goldenaster 90231
Dried Calophanoides 67831
Driftless Area Goldenrod 368383
Drilsepal Chirita 87977
Drilthrum Chiritopsis 88004
Drimycarpus 138043,138044
Drimys 138051
Drink Corn 198376
Drisag 338447
Droguetia 138068,138077
Droke 235373
Dromedary 81237,81356
Droopflower Billardiera 54436
Droopflower Primrose 314835
Droopfruit Monkshood 5480
Droophead Holly 203840
Drooping Asparagus 38974
Drooping Avens 175444
Drooping Bambusa 47394
Drooping Barberry 51531
Drooping Bells 169719
Drooping Bells of Sodom 168476
Drooping Birch 53563
Drooping Branches Agapetes 289750
Drooping Brome 60989
Drooping Broomrape 274987
Drooping Carpesium 77149
Drooping Catchfly 363882
Drooping Clinacanthus 96904
Drooping Coneflower 326964
Drooping Cypress-pine 67432
Drooping Fargesia 162671
Drooping Fimbristylis 166421
Drooping Head 169719
Drooping Heads 169719
Drooping Juniper 213773,213883
Drooping Leucothoe 228159,228162
Drooping Lily 169719
Drooping Lymegrass 144414
Drooping Melaleuca 248076
Drooping Pentanema 289564
Drooping Pimpinella 299509
Drooping Prickly Pear 273103
Drooping Primrose 314713
Drooping Razorsedge 354175
Drooping Rockfoil 349695
Drooping Savin 213883
Drooping Saxifrage 349162
Drooping Sedge 75556,75865
Drooping She-oak 15953
Drooping Silene 363882
Drooping Star of Bethlehem 274708
Drooping Star-of-bethlehem 274708,274823
Drooping Stringbush 414224
Drooping Tickclover 126339
Drooping Timber Bamboo 297281
Drooping Trillium 397563
Drooping Tulip 168476
Drooping Wendlandia 413809
Drooping Wildrye 144459
Drooping Wildryegrass 144414
Drooping Willow 218761,342993
Drooping Woodland Sedge 73743
Drooping Woodreed 91239
Drooping Wood-reed 91239
Droopingflower Cinquefoil 312876
Droopingflower Felwort 235467
Droopingflower Greenhoods 320876
Droopingflower Pleurothallis 304917
Droopingflower Raspberry 339025
Droopingflowered Onion 15173
Droopingfruit Sisymbrium 365482
Droopingleaf Landpick 308528
Droopingspike Pearlsedge 354175
Droopraceme Pokeweed 298094
Drooptwig Bitterbamboo 304003
Droopy Rhubarb 329399
Drophip Rose 336348
Dropin Cathay Poplar 311269
Dropin Scabrous Wendlandia 413818
Dropin Snailseed 97937
Dropin-flower Myrioneuron 261328
Dropin-fruit Barberry 51975
Droping Agapetes 10347
Dropingcalyx St. John's Wort 201826
Dropining Myrioneuron 261328
Drop-leaf Wild Buckwheat 152054
Dropptogue 351146
Drops 177893
Drops of Blood 21339
Drops of Snow 23925
Dropseed 372574,372666
Dropseedgrass 372574,372666
Drops-of-snow 23925
Dropwort 166071,166132
Drosophyllum 138375
Drosophyllum Family 138374
Drought Panicgrass 281532
Droughtdysentery Antidiarrhealtree 197199
Droughtdysentery Holarrhena 197199
Druee's Pearlwort 342275
Drug Centaurium 81544
Drug Eyebright 160233,160268
Drug Eye-bright 160268

Drug Fumitory 169126,169127
Drug Fustic Tree 88514
Drug Fustic-tree 88514
Drug Hedge Hyssop 180321
Drug Lion-ear 224585
Drug Rhubarb 329366
Drug Snowbell 379419
Drug Solomon's-seal 308613
Drug Speedwell 407256
Drug Sweet Flag 5793
Drug Sweetflag 5793
Druggists'Bark 91055
Drum Sticks 210236
Drum-heads 308014
Drummer Boy 81237
Drummer Boys 81237
Drummer Daisy 227533
Drummer Head 81237
Drummond Aster 380863
Drummond Pennyroyal 187179
Drummond Phlox 295267
Drummond Red Maple 3521
Drummond Wattle 1187
Drummond Waxmallow 243932
Drummond's Arabis 30252
Drummond's Aster 40330,380863
Drummond's Catchfly 363427
Drummond's Clematis 94887
Drummond's Cypress Pine 67416
Drummond's Evening Primrose 269432
Drummond's Goldenrod 368080
Drummond's Half-chaff Sedge 232396
Drummond's Ironweed 406593
Drummond's Iron-weed 406593
Drummond's Jimmyweed 209572
Drummond's Mountain-avens 138450
Drummond's Nailwort 284836
Drummond's Phlox 295267
Drummond's Rock-cress 30252
Drummond's Rush 213056
Drummond's Side-saddle Flower 347149
Drummond's Snake-cotton 168686
Drummond's Soapberry 346336
Drummond's Stitchwort 255465
Drummond's Thistle 91920
Drummond's Anemone 23800
Drumprickle 244897
Drumstick Primrose 314310
Drumstick Primula 314310
Drumstick Tree 78300,258945
Drumsticks 15752,81237,81356,108693,
210234,210235,258945,345881
Drumstick-tree 78300
Drunk 235373
Drunkard's Dream 186156
Drunkard's Nose 81768
Drunkards 18173,68189,81768,172135,
174877,248411,363461
Drunken Elm 401512
Drunken Plant 235373
Drunken Sailor 81768,324677
Drunken Slots 404316
Drunken Willy 81768
Drunkits 81768
Drupe-fruited Fig 164916
Drupetea 322564
Druprez Cypress 114679
Dry Catchfly 363606
Dry Celery 29327
Dry Cuckoo 349419
Dry Falsepimpernel 231568
Dry Kesh 192358
Dry Motherweed 231568
Dry Shrubcress 368579
Dry Solms-Laubachia 368579
Dry Strawberry 413031,413034
Dry Woods Sedge 73626
Dryad 138445
Dryadleaf Primrose 314320
Dryander 138423
Dryander's Grevillea 180591
Dryandra-leaved banksia 47640
Dryas 138445
Dryas Chinese Groundsel 365044
Dryas Palm 130057
Dryasalike Sinosenecio 365044
Drybirdbean 87268
Dryground Spiraea 372086
Dry-ground Spiraea 372086
Dry-hot Barberry 51326
Dryish Spunsilksage 360874
Dryland Barnyardgrass 140438
Dry-land Bitter-cress 72920,72922
Dryland Blueberry 403952,404040
Dry-land Blueberry 403952
Dryland Chelonopsis 86836
Dryland Cuckoo 349419
Dryland Elm 401519
Dryland Rosemallow 194717
Dryland Sandwort 32315
Dryland Scout 192358
Dryland Sedge 73754
Dryland Willow 343668
Dryliving Chelonopsis 86836
Dryliving Olive 270172
Dryloving Rabdosia 209870
Drymaria 138470,138482
Drymary 138470
Drymoda 138515
Drymonia 138523
Drymophloeus 138532
Dryophyla Sweetleaf 381188
Drypetes 138575
Dryquitch 148392
Dryslope Dewberry 338807
Dry-spiked Sedge 76261
Drywheatgrass 148398
Dry-zone Mahogany 216214
Dschangar Milkvetch 42304
Dschungar Centaurea 81049
Dsharkent Milkvetch 42306
Dshimen Milkvetch 42494
Duabanga 138721
Dualbutterfly 398251,398262
Dualflower Primrose 314311
Dualflower St. John's Wort 201892
Dualgland Bredia 59831
Dualplicated Coelogyne 98744
Dualsheathorchis 85448
Dualspike Dallisgrass 285426
Dualspike Paspalum 285426
Dual-split Meliosma 249366
Dualtooth Meliosma 249378
Dubious Barberry 51572
Dubious Crazyweed 278702
Dubious Plectocomiopsis 303098
Dubious Thundermelia 327558
Dubjansky Thyme 391176
Duboisia 138739,138744
Dubyaea 138762
Duc Van Thol Tulip 400227
Duchang Indocalamus 206776
Duchartre Lily 229835
Duchartre's Lily 229835
Duchess Protea 315795
Duck 263270
Duck Acorn 263270
Duck Lettuce 277369
Duck Oak 324225
Duck Orchid 66370
Duck Plant 380032
Duck Potato 342335,342359,342363
Duck Willow 342973
Duck Woodbetony 286999
Duck's Bill 208771,220729,382329
Duck's Eyes 107766
Duck's Foot 14065,306636
Duck's Mouth 130383
Duck's-meat 372295
Duckbeakgrass 209249,209288
Duckbill Woodbetony 287014
Duckess Grass 285507
Duckgrass 121226,121233,151481
Duckhead Woodbetony 286999
Ducklettuce 277369
Duck-lettuce 277369
Ducklinggrass 263020
Duckmeat 372300
Ducknut 212636
Duckpalm Bredia 59881
Duck-potato 342363

Ducks-and-drakes 273531
Ducksmet 372295
Ducktongue Monochoria 257583
Duckweed 224359,224360,224375,224380,224385,224399,224400,414649
Duckweed Family 224403
Ducky Daisy 50825
Ducloux Albizia 13533
Ducloux Albizzia 13533
Ducloux Angelica 24344
Ducloux Begonia 49793
Ducloux Bigleaf Albizzia 13612
Ducloux Blueberry 403810
Ducloux Chastetree 411261
Ducloux Chaste-tree 411261
Ducloux Cherry 83198,316400
Ducloux Chinese Neillia 263165
Ducloux Corydalis 105836
Ducloux Cypress 114680
Ducloux Cypressgrass 118781
Ducloux Dinetus 131247
Ducloux Elaeocarpus 142311
Ducloux Fig 165622
Ducloux Galingale 118781
Ducloux Jasmine 211806
Ducloux Mahonia 242517
Ducloux Manglietia 244433
Ducloux Monkshood 5169
Ducloux Paperelder 290328
Ducloux Pimpinella 299401
Ducloux Podocarpium 200734
Ducloux Porana 131247
Ducloux Pterygiella 320917
Ducloux Round-leaf Poplar 311462
Ducloux Sichuan Taxillus 385235
Ducloux Spapweed 204915
Ducloux Spodiopogon 372411
Ducloux Stauntonvine 374398
Ducloux Staunton-vine 374398
Ducloux Tailanther 381934
Ducloux Wild Quince 374398
Ducloux Woodbetony 287179
Ducloux's Synotis 381934
Ducumbent Currant 334157
Ducumbent Garnotia 171533
Dudaim Melon 114150
Dudgeon 64345
Dudley Willow 343767
Dudley's Brome-grass 60704
Dudley's Rush 213060,213066,213507
Dudley's Triteleia 398787
Duffin Bean 294010
Duffle 405788
Duguey Stonecrop 356692
Duhat 382522
Duiker Tree 317810
Duila Monkshood 5620
Duilongdeqing Fritillary 168563
Duke Cherry 83121
Duke of Argyll's Tea Plant 239055
Duke of Argyll's Teaplant 239011
Duke of Argyll's Tea-tree 239011
Duke Tea Tree 239011
Dukou Terminalia 386523
Dulan Milkvetch 42198
Duliche 138918
Dulichium 138919
Dull Lanceolateleaf Smilax 366419
Dull Ophrys 272435
Dullbrown Deerdrug 242686
Dullbrown False Solomonseal 242686
Dull-brown Indianmulberry 258874
Dullcolor Greenbrier 366419
Dullcolored Carpesium 77190
Dullcolour Woodbetony 287779
Dull-coloured Lime 391840
Dullish Blue Leaves Rhododendron 330272
Dullish Blue-leaved Rhododendron 330272
Dullooa Bamboo 263946
Dull-seed Cornbind 162519
Dulong Agapetes 10304
Dulong Anoak 233183
Dulong Arisaema 33317
Dulong Azalea 331001
Dulong Barberry 52216
Dulong Beadruby 242681
Dulong Calanthe 65940
Dulong Cyclobalanopsis 116084
Dulong Elatostema 142652
Dulong Eurya 160616
Dulong Greenbrier 366477
Dulong Gugertree 350942
Dulong Hydrangea 200122
Dulong Jerusalemsage 295103
Dulong Liparis 232348
Dulong Litse 234077
Dulong Mahonia 242650
Dulong Maple 3655
Dulong Milkvetch 42312
Dulong Monkshood 5625
Dulong Oak 116084
Dulong Ophiorrhiza 272200
Dulong Paris 284316
Dulong Pearlsedge 354076
Dulong Pentapanax 289658
Dulong Raspberry 339355
Dulong Razorsedge 354076
Dulong Rhaphidophora 328998
Dulong Schefflera 350804
Dulong Spapweed 205359
Dulong Stairweed 142652
Dulong Touch-me-not 205359
Dulong Twayblade 232348
Dulong Woodbetony 287180
Dulong Yushania 416774
Dulongjiang Corydalis 105837
Dulongjiang Hemsleia 191916
Dulongjiang Snowgall 191916
Dulongjiang Yushanbamboo 416773
Dumasia 138928
Dumb Cammock 271563
Dumb Cane 130083,130135
Dumb Nettle 170025,220346,220416
Dumb Plant 130093
Dumbcane 130096
Dumble Dor 272398
Dumbweed 86901
Dummies 292372
Dummy Nettle 220346
Dumpling Cactus 236425
Dumplings 54437
Dumpy Centaury 80989
Dun Daisy 227533
Dun Nettle 220346
Dun Pea 301055,301070
Dunal's Blueberry 403814
Dunbar's Hawthorn 109551
Dunbaria 138966
Duncan's Pincushion Cactus 107020
Duncecap Larkspur 124233
Dunce-nettle 92213
Dunch 220346
Dunch Nettle 220346,220416
Dunchi Fiber 361367
Dundathu Pine 10505
Dunder Daisy 227533
Dundle Daisy 227533
Dune Ephedra 146142
Dune Fan Flower 350325
Dune Fescue 412422
Dune Gentian 174164
Dune Goldenrod 368402
Dune Grass 19773,228346
Dune Helleborine 147142,147168
Dune Knotweed 309525
Dune Larkspur 124465
Dune Malacothrix 242944
Dune Manzanita 31133
Dune Milkvetch 42203
Dune Myrtle 156155
Dune Poison Bush 4965
Dune Sandbur 80855
Dune Squincywort 39380
Dune Thistle 92307
Dune Three-awn Grass 34062
Dune Willow 343343
Dunedelion 242944
Dunegrass 19773,392091
Dune-groundnut 269631
Dungbeetle Cajan 65156
Dunhua Monkshood 5171

Dunkeld Larch 221830
Dunkfruit Kneejujube 67035
Dunks 418169
Dunn Anisetree 204507
Dunn Anise-tree 204507
Dunn Antiotrema 28548
Dunn Azalea 330854
Dunn Blastus 55149
Dunn Bushclover 226777
Dunn Bush-clover 226777
Dunn Distylium 134929
Dunn Ehretia 141640
Dunn Giantarum 20071
Dunn Goathorntree 77645
Dunn Holly 203777
Dunn Hornbeam 77645
Dunn Liparis 232145
Dunn Litse 233894
Dunn Mallotus 243353
Dunn Milkwort 308028
Dunn Mosquitoman 134929
Dunn Mountainash 369382
Dunn Mountain-ash 369382
Dunn Oak 323857
Dunn Paris 284317
Dunn Raspberry 338340
Dunn Snakegourd 396183
Dunn Sophora 369002
Dunn Starjasmine 393642
Dunn Star-jasmine 393642
Dunn Tainia 383132
Dunn Tongoloa 392786
Dunn Twayblade 232145
Dunn Wampee 94185
Dunn Willow 343332
Dunn's Henry Azalea 330854
Dunn's Henry Rhododendron 330854
Dunn's Hooker Lasianthus 222164
Dunn's Hooker Roughleaf 222164
Dunn's Mariposa-lily 67581
Dunn's Pearlwood 296451
Dunn's Phyllanthodendron 296451
Dunn's Redflower Greenvine 204651
Dunn's Redflower Illigera 204651
Dunn's Rhododendron 330854
Dunnia 138985
Dunny Nettle 47013,170025,220346,400675
Dunny-leaf 400675
Dunny-nettle 47013,170025,220346
Dunny-weed 400675
Dunse Nettle 220346,220416
Duoxiongshan Gentian 173411
Duperre Crestpetal-tree 236415
Duperrea 139029
Duplicate Ovate-fruit Rose 336621
Duplicate Sasa 347211
Duplicatefruit Craneknee 221660
Duplicatehair Azalea 331538
Duplicatetooth Figwort 355100
Duplicate-tooth Rose 336535
Duppy Basil 268565
Duppy Needles 54048
Durand Oak 323859,324426
Durango Pine 299924
Durango Root 123029
Duraznillo 272950
Duraznillo Blanco 272950
Durban Crowfootgrass 121285
Durban Grass 121288
Durgan Mai Yang 133588
Durham Mustard 364557
Durian 139089,139092
Durian-tree 139092
Durietz Taiwan Eyebright 160148
Durkens 300223
Durmast Oak 323805,324278
Durobby 382616
Durobriv Barberry 51578
Durra 369600,369640,369720
Durum Wheat 398900
Dushan Photinia 295800
Dushan Sinobambusa 364787
Dushan Spicegrass 239625
Dushan Spiraea 371876
Dushan Tangbamboo 364787
Duskblotch Rhododendron 331176
Dusky Coral Pea 215983
Dusky Crane's-bill 174811
Dusky Cranesbill 174877
Dusky Dogfennel 85526
Dusky Pear 323259
Duskybloom Rhododendron 330314
Dusky-bloomed Rhododendron 330314
Duskyleaf Raspberry 338451
Dusky-leaved Raspberry 338451
Duskyred Raspberry 338452
Dusky-red Raspberry 338452
Dusted Vriesea 412355
Dustpan Willow 344161
Dusty Bob 358395
Dusty Daisy Bush 270207
Dusty Foxglove 130376
Dusty Husband 30164,83060
Dusty Miller 35779,81004,81103,83060,314143,358159,358395,363370
Dusty Zenobia 417592
Dusty-miller 363370
Dusty-miller Grape 411980
Dusty-miller Sage-wort 36314
Dutch 397043
Dutch Agrimony 53816,158062
Dutch Arbel 311208
Dutch Barley 198376
Dutch Beech 311208
Dutch Box 64356
Dutch Cheese 243823
Dutch Clover 397019,397043
Dutch Crocus 111625
Dutch Daffodil 39328
Dutch Dill 252662
Dutch Elm 401522
Dutch Flax 68850
Dutch Grass 58128,143530
Dutch Honeysuckle 236022
Dutch Hyacinth 199583
Dutch Hyacinths 199583
Dutch Iris 208626,208946
Dutch Lavender 223251
Dutch Medlar 252305
Dutch Mezerion 122515
Dutch Mice 222851
Dutch Morgan 227533
Dutch Myrtle 261162
Dutch Osier 342982
Dutch Pink Plant 327879
Dutch Tonka Bean 133629
Dutch Tonka-bean 133629
Dutch White Clover 397043
Dutch Yellow Crocus 111558
Dutchman's Breeches 128288,128294,220729
Dutchman's Pipe 34097,34133,34193,34246,34262,34264,34337,34356,210329
Dutchman's Pipe Cactus 147291
Dutchman's Trousers 220729
Dutchman's-breeches 128294
Dutchman's-pipe 34097,34149,34264,34337,147291,202767
Dutchman's-pipe Family 34384
Dutchmanspipe 34097
Dutchmanspipe Family 34384
Duthie Barberry 51579
Duthie Spapweed 204917
Duthie Speargrass 4131
Duthie Tibet Draba 137270
Duthie Tibet Whitlowgrass 137270
Duthie Touch-me-not 204917
Duthie Xizang Whitlowgrass 137270
Duthiea 139120
Duvalia 139133
Duwny Bridelia 60211
Dwaef Poinciana 65055
Dwale 44708,190936,367416
Dwarf Acomastylis 4990
Dwarf Ainsliaea 12686
Dwarf Alberta Spruce 298286
Dwarf Alberta White Spruce 298288
Dwarf Alder 16348,167537
Dwarf Aletris 14514
Dwarf Aloe 16894

Dwarf Alpine Hawksbeard 110916
Dwarf Alpine Oak 324199
Dwarf Alpine Onion 15271
Dwarf Alpinegold 199221
Dwarf Amazon Sword Plant 140547
Dwarf American Iris 208922
Dwarf Amorpha 20023
Dwarf Ancylostemon 22166
Dwarf Anisacanthus 25259
Dwarf Apple 24605,243675
Dwarf Apples 60072
Dwarf Arctic Birch 53530
Dwarf Arctic Ragwort 279899
Dwarf Ardisia 31471
Dwarf Arrowbamboo 162671
Dwarf Arrowhead 342396
Dwarf Ascocentrum 38194
Dwarf Ash 167922
Dwarf Asiatic Elm 401602
Dwarf Asparagus-fren 39150
Dwarf Astilbe 41851
Dwarf Autumnal Furze 401395
Dwarf Axyris 45848
Dwarf Azalea 330160,331248,331556
Dwarf Baccharis 46241
Dwarf Balsam Fir 282,291
Dwarf Bamboo 297352,304086
Dwarf Banana 260250
Dwarf Basketvine 9441
Dwarf Bayberry 261195
Dwarf Bean 294056
Dwarf Beanbroomrape 244636
Dwarf Bearclaw-poppy 31083
Dwarf Bearded Iris 208781,208874
Dwarf Bedstraw 170767
Dwarf Bentgrass 12176
Dwarf Bernard's Lily 88537
Dwarf Bifurcate Cinquefoil 312417
Dwarf Bigbract Chamaesium 85719
Dwarf Bigleaf Rush 213313
Dwarf Bilberry 403754,403765
Dwarf Birch 53454,53470,53528,53530,53594
Dwarf Bird's Foot Trefoil 237485
Dwarf Black Elder 345640
Dwarf Black Spruce 298351
Dwarf Blue Plumbago 83646
Dwarf Blue Scotch Pine 300230
Dwarf Blueberry 403754,403768
Dwarf Bog Birch 53454
Dwarf Bothriochloa 57573
Dwarf Bottlebrush 67306
Dwarf Box 64356
Dwarf Bramble 338718
Dwarf Broadleaf Arborvitae 390681
Dwarf Brodiaea 60505,60525
Dwarf Broom 173030
Dwarf Buckeye 9719
Dwarf Buckthorn 328788,328835
Dwarf Bunge Indigo 205754
Dwarf Bur Biebersteinia 54168
Dwarf Burhead 140563
Dwarf Burnet 345845
Dwarf Burning Bush 157740
Dwarf Burreed 370088
Dwarf Bushclover 226812
Dwarf Bush-honeysuckle 130252
Dwarf Buttercup 326273
Dwarf Butterfly Orchid 315618
Dwarf Calligonum 67070
Dwarf Canadian Primrose 314652
Dwarf Cape Gooseberry 297720
Dwarf Cape-gooseberry 297716
Dwarf Carline Thistle 91713
Dwarf Carpesium 77166
Dwarf Cassiope 78639
Dwarf Catalpa 79246
Dwarf Catchfly 363796
Dwarf Centaurea 81378
Dwarf Centranthera 81736
Dwarf Chaparral False Willow 46241
Dwarf Cheilotheca 258069
Dwarf Cherry 83202,83258,83361,316735,316739
Dwarf Chestnut 78807
Dwarf Chestnut Oak 324304
Dwarf Chickling Pea 222694
Dwarf Chickweed 256470
Dwarf Childseng 318507
Dwarf Childvine 400936
Dwarf Chilean Beech 266884
Dwarf Chin Cactus 182425
Dwarf Chinese Fig 165515
Dwarf Chinese Mountain Ash 369499
Dwarf Chinese Waxmyrtle 261215
Dwarf Chinkapin 324304
Dwarf Chinquapin 324304
Dwarf Cholla 181458
Dwarf Christolea 89259,126162
Dwarf Chuniophoenix 90622
Dwarf Cinquefoil 312435
Dwarf Clearweed Dwarf 299006
Dwarf Coldwaterflower 299006
Dwarf Coleus 99711
Dwarf Cornel 85581,105204
Dwarf Corydalis 105994,106345
Dwarf Cotoneaster 107587
Dwarf Cotton 178465
Dwarf Cottonrose 235259
Dwarf Cowlily 267320
Dwarf Coyote Brush 46241
Dwarf Crazyweed 278897
Dwarf Crested Iris 208502
Dwarf Crimean Iris 208698
Dwarf Cryptomeria 113697,113701
Dwarf Cudweed 127927,178444,178463
Dwarf Cultivar Crapemyrtle 219933
Dwarf Cup Flower 266195
Dwarf Cupflower 266199,266201
Dwarf Currant 334038
Dwarf Curry 189460
Dwarf Cushion Wild Buckwheat 152355
Dwarf Cutleaf Stephanandra 375818
Dwarf Cymbidium 116863
Dwarf Cypress 114698
Dwarf Cyrtosia 120763
Dwarf Dandelion 218199,218206,218208,218210,218212,218213
Dwarf Date 295475,295479
Dwarf Date Palm 295453,295479,295484
Dwarf Daylily 191314
Dwarf Delavay Megacarpaea 247776
Dwarf Dendrobium 125013
Dwarf Denseflower Elsholtzia 144003
Dwarf Desert Knotweed 309180
Dwarf Deutzia 126946
Dwarf Devilpepper 326999
Dwarf Dichotomous Fimbristylis 166251
Dwarf Didymocarpus 129937
Dwarf Differmed Galingale 118745
Dwarf Dishspurorchis 171873
Dwarf Dock 339897
Dwarf Dogwood 85581
Dwarf Draba 137013
Dwarf Dracaena 137512
Dwarf Dragonbood 137512
Dwarf Dwostle 250432
Dwarf Eargrass 187637
Dwarf Eastern White Pine 300218
Dwarf Edelweiss 224913
Dwarf Eelgrass 262281,418389,418396
Dwarf Eel-grass 418396
Dwarf Elder 345594
Dwarf Elderberry 345594
Dwarf Elm 401602
Dwarf Elsholtzia 144085
Dwarf Enchanter's-nightshade 91508
Dwarf Encyclia 315618
Dwarf Entireliporchis 261476
Dwarf Epidendrum 315618
Dwarf Eritrichium 153461
Dwarf Erythrina 154670
Dwarf Euonymus 157740
Dwarf Euphorbia 159095
Dwarf European Cranberry Bush 407992
Dwarf European Fly Honeysuckle 236227
Dwarf European Spindle-tree 157448
Dwarf Evax 193375
Dwarf Fan Palm 85665,85671,234188
Dwarf Fan-palm 85671
Dwarf Fascicular Sagebrush 35230

Dwarf Fennel 167142
Dwarf Fimbristylis 166418
Dwarf Flannel Flower 6940
Dwarf Flatsedge 322329
Dwarf Flatsheath Fimbristylis 166226
Dwarf Flax 231884
Dwarf Fleabane 103580,150591,150807
Dwarf Flemingia 166889
Dwarf Flowering Almond 20970,83208
Dwarf Flowering Cherry 83208,83238
Dwarf Flowering-quince 84556
Dwarf Fothergilla 167540
Dwarf French Rose 336595
Dwarf Furze 401390,401395
Dwarf Galangal 17744
Dwarf Garland Spirea 371787
Dwarf Gastrochilus 171873
Dwarf Genista 173000
Dwarf Gentianella 174147
Dwarf Germander 388161
Dwarf Giantarum 20120
Dwarf Ginsen 280817
Dwarf Ginseng 280817
Dwarf Globe Flower 399524
Dwarf Globeflower 399507,399524
Dwarf Globethistle 140722,140753
Dwarf Globular Spike Pycreus 322253
Dwarf Glorybind 103340
Dwarf Goats-beard 37063
Dwarf Golden Arborvitae 302721,302722
Dwarf Golden Star 55653
Dwarf Goldenbush 150331
Dwarf Goldenrod 368262,368404
Dwarf Goldsaxifrage 90439
Dwarf Gonatanthus 179262
Dwarf Goosecomb 144337
Dwarf Gorse 401390,401395,401396
Dwarf Gragontree 137512
Dwarf Gray Willow 343506
Dwarf Grease-bush 177364
Dwarf Greenbrier 366480,366536
Dwarf Green-leaved Pleioblastus 304109
Dwarf Greenwood 326999
Dwarf Grevillea 180589
Dwarf Grewia 180825
Dwarf Ground Rattan 329176
Dwarf Ground-rattan 329176
Dwarf Hackberry 80760
Dwarf Halopeplis 184965
Dwarf Hawksbeard 110916
Dwarf Hawthorn 110098
Dwarf Heavenly Bamboo 262202
Dwarf Hedyotis 187637
Dwarf Heliotrope 190753
Dwarf Herbclamworm 398082
Dwarf Herminium 192835
Dwarf Himalayan Edelweiss 224856
Dwarf Hinoki Cypress 85323
Dwarf Hippeophyllum 196460
Dwarf Hoarhound 245754
Dwarf Hoarypea 386252
Dwarf Hogfennel 292950
Dwarf Holeglumegrass 57573
Dwarf Holly 203615
Dwarf Honeysuckle 105204,235859,236226
Dwarf Hooker Sarcococca 346733
Dwarf Hooktea 52413
Dwarf Horned Holly 203661
Dwarf Horsechestnut 9719
Dwarf Horse-chestnut 9719
Dwarf Houseleek 357114
Dwarf Huckleberry 172256,403765
Dwarf Humboldt Lily 229866
Dwarf Hydrangea 199980
Dwarf Indian Cress 399604
Dwarf Indica Azalea 330675
Dwarf Indigo 206590
Dwarf Indocalamus 206819
Dwarf Iris 208514,208623,208632,208698,
208781,208788,208874,208922
Dwarf Japanese Cedar 113706
Dwarf Japanese Floweringquince 84556
Dwarf Japanese Flowering-quince 84556
Dwarf Japanese Garden Juniper 213866
Dwarf Japanese Spirea 371945
Dwarf Japanese Yew 385361
Dwarf Jasmine 211956
Dwarf Jerusalem Sage 295129
Dwarf Jerusalemsage 295154,295180
Dwarf Juneberry 19241
Dwarf Juniper 213702,213712
Dwarf Jurinea 214033
Dwarf Kaffirboom 154670
Dwarf Kengyilia 215906
Dwarf Knotweed 309198
Dwarf Kobresia 217195
Dwarf Korean Lilac 382200
Dwarf Kowhai 369111
Dwarf Kurrajong 58338
Dwarf Ladypalm 329184
Dwarf Ladypalms 329184
Dwarf Laelia 219689
Dwarf Lagotis 220176
Dwarf Lake Iris 208671
Dwarf Largecalyx Cinquefoil 312468
Dwarf Largeleaf 147160
Dwarf Largeleaf Hydrangea 199991
Dwarf Larkspur 124180,124536,124650
Dwarf Laurel 122515,215387
Dwarf Leek 15064
Dwarf Leptospermum 226506
Dwarf Lessingia 227125
Dwarf Lilac 382200
Dwarf Lily 229952
Dwarf Lilyturf 272090
Dwarf Lily-turf 272090
Dwarf Liriope 232630
Dwarf Loblolly Pine 300265,300266
Dwarf Longstalk Gentian 174064
Dwarf Longstem Cystopetal 320229
Dwarf Loosestrife 239803
Dwarf Lucky Bean Tree 154670
Dwarf Lymegrass 144337
Dwarf Madonna Lily 370345
Dwarf Magellan Barberry 51406
Dwarf Mahonia 242622
Dwarf Maiden Grass 255892
Dwarf Mallow 243823
Dwarf Mannagettaea 244636
Dwarf Manycleft Cinquefoil 312802
Dwarf Manyflowered May-apple 139612
Dwarf Maple 2991
Dwarf Marigold 351909,351917
Dwarf Marsh Orchid 273605
Dwarf Marshmarigold 68206
Dwarf Marshweed 230317
Dwarf Masterwort 43311
Dwarf Mazus 246985
Dwarf Meadow Rue 388573
Dwarf Meadow-rue 388404
Dwarf Meadowsweet 166111
Dwarf Medinilla 247597
Dwarf Megadenia 247852
Dwarf Microcaryum 253204
Dwarf Milfoil 3997
Dwarf Milkvetch 43082
Dwarf Milkweed 38045
Dwarf Milkwort 307905
Dwarf Minnesota Pure White Mockorange
294573
Dwarf Miscanthus 255892
Dwarf Mistletoe 30906
Dwarf Mitrasacme 256161,256162
Dwarf Mondo Grass 272091
Dwarf Montane Rockfoil 349660
Dwarf Montia 258246
Dwarf Morning Glory 68742,103340
Dwarf Morning-glory 102884,103340
Dwarf Mountain Cinquefoil 312940
Dwarf mountain Fleabane 150555
Dwarf Mountain Laurel 215397
Dwarf Mountain Palm 85426
Dwarf Mountain Pine 300086
Dwarf Mountainash 369488
Dwarf Mountain-ash 369488
Dwarf Mountain-loving Draba 137160
Dwarf Mountain-loving Whitlowgrass 137160
Dwarf Mountain-pine 300086
Dwarf Mouse-ear 82991,82994
Dwarf Mousetail Cupscale 341979
Dwarf Murdannia 260116

Dwarf Nasturtium 399604
Dwarf Nealie 1105
Dwarf Nealing 1704
Dwarf Nenga 263551
Dwarf Nettle 403039
Dwarf Nightshade 367432
Dwarf Ninebark 297846
Dwarf Nipplewort 221792
Dwarf Nippon Hawthorn 109656
Dwarf Nitid Sagebrush 35399
Dwarf North China Monkshood 5302
Dwarf Norway Spruce 298393
Dwarf Oak 324199
Dwarf Obedient Plant 297987
Dwarf Oncidium 270838
Dwarf Ophiorrhiza 272282
Dwarf Orach 44625
Dwarf Orchid 273696
Dwarf Oregon Grape 242625
Dwarf Oreorchis 274057
Dwarf Palmetto 341411,341421
Dwarf Panicle Hydrangea 200045
Dwarf Pansy 410140
Dwarf Papyrus 119422
Dwarf Paris 284371
Dwarf Parnassia 284548,284584
Dwarf Parrotfeather 261350
Dwarf Pawpaw 38330,38333
Dwarf Pea Shrub 72334
Dwarf Peashrub 72187,72322
Dwarf Pea-shrub 72322
Dwarf Pedilanthus 287854
Dwarf Periwinkle 409335,409339
Dwarf Pernettya 291372
Dwarf Pieris 31038
Dwarf Pine 254092,300086
Dwarf Pink Astilbe 41796
Dwarf Pink Geranium 174891
Dwarf Pink Hibiscus 195097
Dwarf Pinnaflower 4990
Dwarf Plain-green Leaf Pleioblastus 304109
Dwarf Plane 407989
Dwarf Plantain 301888,302103,302209
Dwarf Plantain Lily 198655
Dwarf Plateaucress 89259,126162
Dwarf Pleione 304267
Dwarf Pleurospermum 304826
Dwarf Plumbago 83646
Dwarf Pomegranate 321771
Dwarf Pondweed 312188
Dwarf Post Oak 324153
Dwarf Prairie Rose 336361
Dwarf Przewalsk Woodbetony 287548
Dwarf Pseudostellaria 318507
Dwarf Puccinellia 321324
Dwarf Punctulate Rockfoil 349828
Dwarf Pure White Mockorange 294576
Dwarf Purple Orchid 273605
Dwarf Purple Rhododendron 330913
Dwarf Purple Willow 343941
Dwarf Purplebract Iris 208801
Dwarf Purplebract Swordflag 208801
Dwarf Pycreus 322329
Dwarf Qiongpalm 90622
Dwarf Rabbitbrush 150349
Dwarf Rabbiten-wind 12686
Dwarf Ragweed 19184
Dwarf Raspberry 338142,338264,338962,339124
Dwarf Rattlebox 112449
Dwarf Rattlesnake-plantain 179685
Dwarf Red Buckeye 9722
Dwarf Red Fountain Grass 289248
Dwarf Red Raspberry 339124
Dwarf Red Rattle 287721
Dwarf Red-leaved Barberry 52227,52231
Dwarf Redtip Dogwood 105166
Dwarf Red-tipped Dogwood 105166
Dwarf Reedmace 401128
Dwarf Rex Woodbetony 287607
Dwarf Rhodiola 329886
Dwarf Rhododendron 330160,331248,331556,331584
Dwarf Rhubarb 329380,329383
Dwarf Ribseedcelery 304826
Dwarf Ricegrass 300717
Dwarf Rockfoil 349189
Dwarf Rose 336879
Dwarf Rostellularia 337244
Dwarf Rue 341075
Dwarf Ruellia 339736
Dwarf Rush 212988,213292
Dwarf Russian Almond 20931,316859
Dwarf S. China Milkwort 308082
Dwarf Sandholly 19778
Dwarf Sanguineous-scale Pycreus 322346
Dwarf Sarcococca 346733
Dwarf Saw-wort 348590
Dwarf Scorzonera 354873
Dwarf Scotch Pine 300237
Dwarf Screwpine 281114
Dwarf Screwtree 190108
Dwarf Screw-tree 190108
Dwarf Sea-buckthorn 196758
Dwarf Sedge 74842,74849,75930
Dwarf Serbian Spruce 298387
Dwarf Servian Spruce 298389
Dwarf Serviceberry 19293
Dwarf Sheareria 362083
Dwarf Shibataea 362146
Dwarf Shortstyle Lily 229752
Dwarf Siberian Pine 300163
Dwarf Silky Willow 343994
Dwarf Silver Fir 274
Dwarf Silver Santolina 346251
Dwarf Smoketree 107314
Dwarf Smoke-tree 107314
Dwarf Snapdragon 84596,84604
Dwarf Snowbell 367767
Dwarf Solomon's Seal 308562
Dwarf Solomon's-seal 242672
Dwarf Sour Bush 88921
Dwarf South China Milkwort 308082
Dwarf Southern Magnolia 242141
Dwarf Spanish Heath 150187
Dwarf Spapweed 205122
Dwarf Speedwell 407306
Dwarf Spider-lily 200951
Dwarf Spiderwort 394040,394083
Dwarf Spike Sedge 143289
Dwarf Spike-rush 143094,143289
Dwarf Spikesedge 143289
Dwarf Spike-sedge 143289
Dwarf Spotflower 4874
Dwarf Spruce 298404
Dwarf Spurge 158869
Dwarf Spurless Touch-me-not 205122
Dwarf St. John's-wort 202029
Dwarf Staff-tree 80168,80265
Dwarf Stalegrass 182621
Dwarf Stitchwort 255561
Dwarf Stone Pine 300163
Dwarf Stringbush 375187
Dwarf Sumac 332529
Dwarf Sumach 332529
Dwarf Summer-sweet 96458
Dwarf Sundew 138271
Dwarf Sunflower 189038
Dwarf Superior Rockfoil 349289
Dwarf Supplejack 52413
Dwarf Sweetflag 5803
Dwarf Swertia 380222
Dwarf Swiss Mountain Pine 300095
Dwarf Swordflag 208874
Dwarf Tanbark 233177
Dwarf Tangut Dragonhead 137674
Dwarf Tangut Greenorchid 137674
Dwarf Tatarian Statice 230790
Dwarf Tephrosia 386252
Dwarf Thelasis 389128
Dwarf Thermopsis 389555
Dwarf Thickspurred Larkspur 124299
Dwarf Thistle 76990,91713
Dwarf Thistle Short-stemmed Thistle 91920
Dwarf Thomson Rhododendron 330351
Dwarf Thorowax 63852
Dwarf Thread Sawara Falsecypress 85356
Dwarf Toitoi 87752,87753,87754,87756
Dwarf Trillium 397584,397594
Dwarf Trilobe Buttercup 325938
Dwarf Tripogon 398108

Dwarf Tuftroot 130102
Dwarf Turk's Cap Cactus 249593
Dwarf Twinflower 139438
Dwarf Two-seeded Palm 130065
Dwarf Tylophora 400936
Dwarf Umbrella Grass 118749
Dwarf Umbrella Tree 350654
Dwarf Umbrella-sedge 168876
Dwarf Uvaria 403441
Dwarf Valerian 404365
Dwarf Vanda 404661
Dwarf Variegata Thujopsis 390682
Dwarf Variegated Bamboo 304007
Dwarf Vegetablepalm 341421
Dwarf Vetchling 222727
Dwarf Villous Amomum 19931
Dwarf Violet Iris 208922
Dwarf Wakerobin 397594
Dwarf Water Gum 398669
Dwarf Waterbamboo 260116
Dwarf Water-milfoil 261374
Dwarf Wax Myrtle 261195
Dwarf Waxmyrtle 261195
Dwarf White Flowering Trillium 397594
Dwarf White Trillium 397584
Dwarf White Wake-robin 397584
Dwarf White Wood Lily 397584
Dwarf White Wood-lily 397584
Dwarf Whitebeam 369358
Dwarf Whitebracteole Bugle 13134
Dwarf Whitepearl 172124
Dwarf Whitestripe Bamboo 304110,347214
Dwarf White-stripe Bamboo 304109
Dwarf White-stripe-leaved Bamboo 304110
Dwarf Whitlowgrass 137013
Dwarf Wild Buckwheat 151882
Dwarf Wild Iris 208502
Dwarf Wild Pine 300086
Dwarf Wildorchis 274057
Dwarf Wildsenna 389555
Dwarf Willow 343476,343506,343994,
343995
Dwarf Willowherb 146805
Dwarf Willowweed 146805
Dwarf Windhairdaisy 348698
Dwarf Windmillpalm 393814
Dwarf Winteraconite 148111
Dwarf Wintercreeper 157486
Dwarf Wintergreen 172124
Dwarf Wintergreen Barberry 51802
Dwarf Witch Alder 167540
Dwarf Wobamboo 362146
Dwarf Woodbetony 287284,287570
Dwarf Woodsorrel 278044
Dwarf Woollyheads 318802
Dwarf Yarrow 3997
Dwarf Yaupon 204391
Dwarf Yaupon Holly 204392
Dwarf Yellow Daylily 191312
Dwarf Yellow Fleabane 150543
Dwarf Yellow Lily 229956
Dwarf-cape-jasmine 171374
Dwarfdandelion 218199,218206
Dwarf-dubmoss Cassiope 78645
Dwarf-flower 139690
Dwarfflower Loosestrife 239594
DwarfFragrant Sumac 332478
Dwarf-globe Eastern Arborvitae 390606
Dwarfich Sasa 347279
Dwarfindigo Amorpha 20023
Dwarfish Rhododendron 331584
Dwarfish Vetchling 222825
Dwarfjasmine Rockjasmine 23144
Dwarfmistletoe 30906
Dwarf-mistletoe 30906
Dwarfnettle 262217,262243,262247
Dwarf-oak 324166
Dwarf-pear Cymbidium 116993
Dwarf-pear Orchis 116993
Dwarf-pyramid Eastern Arborvitae 390597
Dwart Morning Glory 103340
Dwayoberry 44708
Dweller in Thicketts Rhododendron 330603
Dweller on Trees Rhododendron 330550
Dwf Varieg Sweet Flag 5805
Dwinkle 409339
Dwraf China Acronema 6206
Dwyer's Red Gum 155572
Dybowsk Cephalocereus 82206
Dyckia 139254
Dye Euonymus 157926
Dye Tree 302684
Dye Wendlandia 413829
Dyebark Evodia 161356
Dyed Morinda 258920
Dyeing Barberry 52257
Dyeing Carthamus 77748
Dyeing Indian-mulberry 258920
Dyeing Peristrophe 291138
Dyeing Silvergrass 255918
Dyeing Tree 346471
Dyeingtree 346471
Dyeing-tree 346471
Dyer Bedstraw 39413
Dyer Gurjun 133561
Dyer Macrozamia 241452
Dyer Rosewood 121670
Dyer's Alkanet 14841
Dyer's Broom 173082
Dyer's Bugloss 14841,289503
Dyer's Camomile 26900
Dyer's Chamomile 26900
Dyer's Croton 89330
Dyer's Grape 298094
Dyer's Grapes 298094
Dyer's Green Weed 173082
Dyer's Greening Weed 173082
Dyer's Greening-weed 173082
Dyer's Greenweed 173082,209229,327879
Dyer's Herb 209229
Dyer's Knotgrass 309893
Dyer's Madder 338038
Dyer's Mulberry 88514
Dyer's Oak 324050,324539
Dyer's Plumeless Saw-wort 361139
Dyer's Rocket 327879
Dyer's Root 338038
Dyer's Saffron 77748
Dyer's Savory 361139
Dyer's Sawwort 361139
Dyer's Saw-wort 361139
Dyer's Weed 39413,173082,209229,327879
Dyer's Weld 327879
Dyer's Woad 209229
Dyer's Woodruff 39413
Dyer's Woodrush 39413
Dyer's Yellow 327879
Dyer's Yellow-weed 173082
Dyer's-oak 324539
Dyer's-rocket 327866
Dyer's-weed Goldenrod 368268
Dyer's Bugloss 21984
Dyers Garcinia 171221
Dyers Serratula 361139
Dyersweed 368268
Dyetree 302674,302684
Dye-tree 302674,302684
Dyeweed 173082
Dyewood 172898,173082
Dypsis 139300
Dyschoriste 139433
Dysentery-herb 258096
Dysophylla 139544,306973
Dysophyllalike Anisochilus 25383
Dysosma 139606,139629
Dysoxylum 139637
Dzevanovsky Thyme 391177
Dzungar Bedstraw 170635
Dzungar Buckthorn 328861
Dzungar Buttercup 326377
Dzungar Catchfly 364072
Dzungar Centaurea 81049
Dzungar Chickweed 375100
Dzungar Clematis 95319
Dzungar Cotoneaster 107688
Dzungar Crazyweed 279169
Dzungar Cynomorium 118133
Dzungar Elaeagnus 142195
Dzungar Euphorbia 159851
Dzungar Garliccress 365589
Dzungar Giantfennel 163712

Dzungar Globeflower 399503
Dzungar Goldenray 229197
Dzungar Hawthorn 110048
Dzungar Hogfennel 292937
Dzungar Jurinea 214074
Dzungar Knotweed 309808
Dzungar Leaflessbean 148447
Dzungar Milkaster 169796
Dzungar Milkvetch 42395
Dzungar Monkshood 5585
Dzungar Mullein 405777
Dzungar Peashrub 72345
Dzungar Pink 127839
Dzungar Primrose 314700
Dzungar Redsandplant 327226
Dzungar Rhodiola 329890
Dzungar Ricegrass 300747
Dzungar Russianthistle 344524
Dzungar Salsify 394346
Dzungar sand Sagebrush 36304
Dzungar Sedge 76299
Dzungar Serpentroot 354954
Dzungar Skullcap 355765
Dzungar Spunsilksage 360849
Dzungar Stink Groundpine 70466
Dzungar Sweetvetch 188116
Dzungar Swordflag 208848
Dzungar Thistle 91723
Dzungar Tulip 400222
Dzungar Willow 344127
Dzungar Woodbetony 287684
Dzungaria Cotoneaster 107688
Dzungaria Monkshood 5585
Dzungarian Mullein 405777

E

E. Africa Cordia 104148
E. Africa Gum 1186
E. African Elemi 57521
E. African Rubber 220814
E. Australian Blackwood 1384
E. Bureav's Rhododendron 330265
E. China Clearweed 298998
E. China Coldwaterflower 298998
E. China Globethistle 140716
E. China Grape 411863
E. China Hillcelery 276749
E. China Indigo 205995
E. China Ladybell 7762
E. China Marshmarigold 68204
E. China Walnut 212608
E. H. Wilson Acanthopanax 143701
E. H. Wilson Anemone Clematis 95145
E. H. Wilson Angelica 24476
E. H. Wilson Aralia 30794
E. H. Wilson Arisaema 33565
E. H. Wilson Arrowwood 408221
E. H. Wilson Banana 260281
E. H. Wilson Begonia 50409
E. H. Wilson Berchemiella 52481
E. H. Wilson Bird Cherry 280061
E. H. Wilson Buckeye 9738
E. H. Wilson Buckthorn 328900
E. H. Wilson Butterflyorchis 293648
E. H. Wilson Chokecherry 280061
E. H. Wilson Cinnamon 91449
E. H. Wilson Clovershrub 70911
E. H. Wilson Corktree 294256
E. H. Wilson Corydalis 106604
E. H. Wilson Cymbidium 117102
E. H. Wilson Daphne 122632
E. H. Wilson Dendrobium 125417
E. H. Wilson Dogwood 380520
E. H. Wilson Douglas Fir 318599
E. H. Wilson Edelweiss 224961
E. H. Wilson Elaeagnus 142207
E. H. Wilson Elm 401647
E. H. Wilson Euonymus 157966
E. H. Wilson Gentian 174072
E. H. Wilson Glochidion 177190
E. H. Wilson Goldenray 229250
E. H. Wilson Grape 412004
E. H. Wilson Hawthorn 110114
E. H. Wilson Holly 204401
E. H. Wilson Hookettea 52481
E. H. Wilson Horsechestnut 9738
E. H. Wilson Hydrangea 200148
E. H. Wilson Indigo 206744
E. H. Wilson Indocalamus 206839
E. H. Wilson Iris 208941
E. H. Wilson Juniper 213857
E. H. Wilson Ladybell 7893
E. H. Wilson Leptodermis 226141
E. H. Wilson Lily 230076
E. H. Wilson Litse 234108
E. H. Wilson Loosestrife 239904
E. H. Wilson Lysionotus 239991
E. H. Wilson Maddenia 241507
E. H. Wilson Magnoliavine 351118
E. H. Wilson Maple 3749
E. H. Wilson Melaleuca 248126
E. H. Wilson Michelia 252974
E. H. Wilson Mountainash 369565
E. H. Wilson Nepeta 265106
E. H. Wilson Orchis 117102
E. H. Wilson Pagodatree 369162
E. H. Wilson Pearlbush 161745
E. H. Wilson Pentapanax 289672
E. H. Wilson Plumyew 82550
E. H. Wilson Poplar 311568
E. H. Wilson Primrose 315123
E. H. Wilson Pygeum 322413
E. H. Wilson Raspberry 339475
E. H. Wilson Rockjasmine 23343
E. H. Wilson Sophora 369162
E. H. Wilson Southstar 33565
E. H. Wilson Spindle-tree 157966
E. H. Wilson Spiraea 372135
E. H. Wilson St. John's Wort 202220
E. H. Wilson Stinkcherry 241507
E. H. Wilson Stonecrop 357299
E. H. Wilson Sumac 332969
E. H. Wilson Supplejack 52481
E. H. Wilson Swordflag 208941
E. H. Wilson Tulip 400253
E. H. Wilson Viburnum 408221
E. H. Wilson Wildclove 226141
E. H. Wilson Willow 344277
E. H. Wilson Yarrow 4062
E. H. Wilson Youngia 416495
E. H. Wilson's Notoseris 267180
E. India Copal 10496
E. India Vatica 405164
E. Indian Gum 1024
E. Indian Walnut 13595
E. London Box 64277
E. Sichuan Fritillary 168579
E. Sichuan Gentian 173254
E. Sichuan Giantarum 20109
E. Sichuan Petrocosmea 292566
E. Sichuan Rush 213052
E. Sichuan Sage 345192
E. Sichuan Stonebutterfly 292566
E. Sichuan Swertia 380175
E. Sichuan Windhairdaisy 348267,348309
E. Tianshan Milkvetch 42091
E. Xinjiang Rhodiola 329965
E. Xizang Azalea 331410
E. Xizang Melic 249089
E. Xizang Rhubarb 329367
E. Xizang Rockfoil 349500
E. Xizang Sedge 74052
E. Xizang Silene 364224
E. Yunnan Sagebrush 36030
E. Yunnan Woodbetony 287319
E. Zhejiang Fritillary 168587

Eacor 324335
Eagle Vine 245798
Eagle Wood 29972
Eagle's-claw Maple 3435
Eaglebeak Clearweed 299079
Eaglebeak Coldwaterflower 299079
Eaglebeak Woodbetony 287014
Eagleclaw 34995
Eagle-claw Cactus 140111,354287
Eagle-claw Norway Maple 3435
Eagleclawbean 370191,370202
Eagleclaws 140133
Eaglewood 29972,29973
Eak 324335
Ear Leaf Groundsel 358554
Ear Willow 343063
Earache 282685
Earanther 276911
Ear-anther 276911
Earbase Blastus 55140
Earbase Clearweed 298870
Earbase Coldwaterflower 298870
Earbract Dayflower 100930
Eardrop Vine 61371
Ear-drops 220729
Eared Coneflower 339520
Eared Coreopsis 104444
Eared False Foxglove 10140
Eared Gentian Gentian 173692
Eared Goldenrod 367980
Eared Indian Plantain 283792
Eared Osier 342948
Eared Sallow 343063
Eared Strangler Fig 164661
Eared Willow 343063
Earedwing Cacalia 283853
Earfruit Macaranga 240233
Eargrass 187497,187510
Earl of Bute 376428
Ear-laef Synotis 381923
Earlaef Tailanther 381923
Earleaf Acacia 1067
Earleaf Ammania 19554
Earleaf Azalea 330186
Earleaf Bladderwort 403971
Earleaf Conehead 290919
Earleaf Groundsel 358554
Earleaf Hakea 184592
Earleaf Knotweed 309374
Earleaf Nightshade 367364
Earleaf Pearweed 239556
Earleaf Roscoea 337067
Earleaf Saussurea 348552
Earleaf Sealavender 230700
Earleaf Tanoak 233233
Earleaf Tasselflower 144903
Earleaf Thorowax 63832
Earleaf Windhairdaisy 348552
Ear-leafed Magnolia 242110
Ear-leaved Acacia 1067
Ear-leaved Blueberry 403971
Ear-leaved Brome 60624,60793
Ear-leaved Gerardia 10140
Ear-leaved Stenoseris 375711
Ear-leaved Willow 343063
Earless Dinochloa 131283
Earless Drepanostachyum 137848
Earless Poorspikebamboo 270425
Earless Sinobambusa 364788
Earless Vinebamboo 131283
Earlike Gentian 173694
Earlimart Orach 44357
Earline Thistle 77022
Earliporchis 277239,277248
Early Anemone 23891
Early Azalea 331546
Early Bamboo 297483
Early Black Wattle 1168
Early Blue Violet 410340,410612
Early Blueberry 403932,403957
Early blue-top Fleabane 151057
Early Bugloss 21928
Early Buttercup 325812
Early Columbine 30061
Early Coralroot 104019
Early Coral-root 104019
Early Crocus 111542,111617
Early Cudweed 170926
Early Dark-green Bulrush 353451
Early Daylily 191272
Early Deutzia 126956
Early Dock 340203
Early Dog-violet 410483
Early Dutch Honeysuckle 236023
Early Everlasting 26299
Early Fen Sedge 74195
Early Field Scorpion-grass 260800
Early Figwort 355165
Early Flowering Iris 208874
Early Flowering Lilac 382171
Early Fly Honeysuckle 235710
Early Forget-me-not 260800,260855
Early Forsythia 167452
Early Garden Bamboo 297401
Early Gentian 173224,174112
Early Golden Rod 368106
Early Goldenrod 368106,368188
Early Hair Grass 12839
Early Hairgrass 12839
Early Hoarhound 245760
Early Horse-gentian 397830
Early Jessamine 84412
Early Lilac 382220
Early Lilac Hybrid 382116
Early Lorget-me-not 260779
Early Low Blueberry 403716
Early Low-blueberry 403716
Early Low-bush Blueberry 403952
Early Madonna Lily 229788,
Early Marsh Orchid 273469
Early Meadow Rue 388486
Early Meadow-grass 305599
Early Meadow-rue 388486,388712
Early Medick 247437
Early Millet 254542
Early Mushroom 292372
Early Nancy 414878
Early Oak Sedge 76630
Early Orache 44619
Early Perennial Phlox 295326
Early Pink Flowering Peach 20935
Early Pleione 304291
Early Purple Orchid 273531
Early Red Clover 397066
Early Rose 315101
Early Russianthistle 344674
Early Sandgrass 252785
Early Sand-grass 252783,252785
Early Saxifrage 350032
Early Scorpion-grass 260905
Early Sedge 75744
Early Spider-orchid 272490
Early Spring Rhododendron 331535
Early Spring Shoot Bamboo 297483
Early Star-of-bethlehem 169390
Early Tamarisk 383581
Early Tulip 400206
Early Water Avens 175457
Early Water Grass 140466
Early White Lily 229788
Early White-top Fleabane 151055
Early Wild Rose 336389
Early Winter Cress 47947,47963
Early Wood Buttercup 325506
Early Yellow Violet 410509
Early Yellowrocket 47963
Earlybloom Honeysuckle 236040
Earlybloom Medic 247437
Earlybloom Primrose 314830
Early-blossoming Honeysuckle 236040
Early-coning Spruce 298192
Earlyflower Gerbera 175131
Early-flower Honeysuckle 236040
Earlyflower Raspberry 339096
Early-flowered Hydrangea 200056
Early-flowering Yellow Rocket 47963
Early-flushed Rhododendron 330477
Earlymadder 226237,226243
Early-purple Orchid 273531
Early-ripe Medic 247437
Early-ripe Tickseed 104832

Earlyspring Azalea 331535
Earlyspring Primrose 314837
Earning Grass 299759
Earpetal Crazyweed 278727
Earpetal Melandrium 248287
Earpetiolate Chinese Groundsel 365047
Earpetiole Aster 40000
Earpod 145992,145994
Earpod Tree 145992
Ear-pod Tree 145992
Ear-pod Wattle 1067
Earpodtree 145992,145993
Ear-rings 218761
Earshape Catchfly 363855
Earshape Indocalamus 206769
Earshape Stairweed 142611
Ear-shaped Eurya 160427
Ear-shaped Wild Buckwheat 152320
Ear-shapeleaf Sealavender 230700
Earshapeleaf Thorowax 63832
Ear-shape-leaved Rhododendron 330186
Earstalk Corydalis 105638
Earstalk Loosestrife 239765
Ear-stalk Synotis 381946
Earstipe Sinosenecio 365047
Earth Almond 118822,118837,189073
Earth Apple 115949,189073
Earth Bark 312520
Earth Chestnut 63475,102727,222851
Earth Gall 169126
Earth Ivy 176824
Earth Nut 30498,102727
Earth Pea 30498
Earth Quakes 60382
Earth Star 113364,113374
Earth Stars 113364
Earthbeet 53249
Earthgall 54896,81065
Earthgallgrass 143453,143464
Earthloving Wildginger 37626
Earthnut 30498,102727,269348
Earth-nut 30498,63475,102720,102727
Earthnut Pea 222851
Earth-nut Pea 222851
Earth-silkworm Betony 373222
Earthsmoke 169126,169183
Earthwormfruit Beadcress 264609
Eartooth Catchfly 363859
Earwig 102933
Easrern Hornbeam 77359
East Africa Antiaris 28248
East African Bombax 56779
East African Camphorwood 268734
East African Cedar 213863
East African Doum Palm 202268
East African Fig 165065
East African Juniper 213863
East African Olive 270129
East African Sandalwood 276878,276893
East African Yellow Wood 9995
East African Yellowwood 306490
East African Yellow-wood 306435
East Anglian Elm 401500
East Asia Goosefoot 87187
East Asian Anise-tree 204598
East China Globethistle 140716
East China Hydrangea-vine 351797
East China Ladybell 7762
East China Walnut 212608
East Guangxi Rhododendron 331747
East Himalayan Fir 480
East Himalayan Spruce 298437
East India Rosewood 121730
East Indian Almond 386585
East Indian Arrowroot 114858,382923
East Indian Basil 268518
East Indian Bristlegrass 361709
East Indian Coreltree 154734
East Indian Crabgrass 130768
East Indian Dill 24213
East Indian Ebony 132308
East Indian Hygrophila 191326
East Indian Island Pea 89209
East Indian Lily 263272
East Indian Lotus 263272
East Indian Mahogany 320288
East Indian Plum Tree 166777
East Indian Rhubarb 329372
East Indian Rosebay 382763
East Indian Rosewood 121730
East Indian Satinwood 88663
East Indian Walnut 13595
East Indian Wine Palm 78056
East Indies Arrowroot 114858
East Indies Bluestem 57567
East Indies Hemp 112269
East London Box 64277
East London Boxwood 64277
East Siberian Fir 427
East View Rubberweed 201308
East Xizang Rhododendron 331410
East Yunnan Calophanoides 67834
East Yunnan Corydalis 106126
East Yunnan Sabia 341579
East Yunnan Woodbetony 287319
Eastafrican Fig 165065
East-Asia Low Meadowrue 388583
East-Asia Privet 229561
East-chinese Manyflowered May-apple 139623
Easter Bell 374916
Easter Bells 374916
Easter Broom 121013
Easter Cactus 186153,329681,351999
Easter Daisy 39910,393396
Easter Flower 23925,159675,262429,321721,356366,374916
Easter Giants 308893
Easter Hedges 308893
Easter Heraldtrumpet 49359
Easter Herald-trumpet 49359
Easter Ledger 308893
Easter Ledgers 308893
Easter Ledges 308893
Easter Lily 154936,229900,229902,262441
Easter Lily Cactus 140876
Easter Lily Vine 49359
Easter Mangiants 308893
Easter Mantgions 308893
Easter Orchid 139743
Easter Qingmingflower 49359
Easter Rose 262429,262441,315101
Easter Sedge 308893
Easter-bell 374916
Easter-bell Starwort 374916
Easterlily 140872
Easter-lily 417608
Eastern Annual Saltmarsh Aster 41317
Eastern Arborvitae 390587,390617
Eastern Arbor-vitae 302721,390587
Eastern aromatic Aster 40951
Eastern Baccharis 46227
Eastern Balsam Poplar 311237
Eastern Black Currant 333916
Eastern Black Walnut 212631
Eastern Bladdernut 374110
Eastern Blazing Stars 228519
Eastern Bleeding-heart 128296
Eastern Bluebells 250914
Eastern Blue-eyed-grass 365695
Eastern Blue-eyed-mary 99840
Eastern Bluestar 20860
Eastern Bog Laurel 215405
Eastern Borage 393961
Eastern Bottlebrush Grass 144340,144343
Eastern Brisdecone Pine 299795
Eastern Burning Bush 157327
Eastern Burningbush 157327
Eastern Camas 68806
Eastern Camass 68806
Eastern Cedar 213984
Eastern Chokecherry 316918
Eastern Columbine 30007
Eastern Coneflower 339539,339614
Eastern Coral Bean 154668
Eastern Cottonwood 311237,311292
Eastern Cyclamen 115942
Eastern Daisy Fleabane 150464
Eastern Dewberry 338929
Eastern Dog-violet 410246
Eastern Dwarf Mistletoe 30917

Eastern Elaeagnus 141936
Eastern Feathergrass 376859
Eastern Feverfew 285088
Eastern Figwort 355184
Eastern Fir 282
Eastern Fire-berry Hawthorn 110003
Eastern Flowering Dogwood 105023
Eastern Gama Grass 398139
Eastern Gamagrass 398139
Eastern Giants 308893
Eastern Gladiolus 176122
Eastern Green-violet 199681
Eastern Groundsel 360306
Eastern Heloniopsis 191065
Eastern Hemlock 399867
Eastern Hemlock-spruce 399867
Eastern Hooded Helleborine 82043
Eastern Hop Hornbeam 276818
Eastern Hophornbeam 276818
Eastern Hop-hornbeam 276818
Eastern Joe-pye weed 158106
Eastern Juniper 213984
Eastern Larch 221904
Eastern Larkspur 102838
Eastern Leatherwood 133656,156068
Eastern Leopard's-bane 136349
Eastern Lined Aster 380912
Eastern Manna Grass 177655
Eastern Mistletoe 295581
Eastern Narrow-leaved Sedge 73675,74715
Eastern Ninebark 297842
Eastern Nit-grass 171814
Eastern Parthenium 285088
Eastern Pasque Flower 321703
Eastern Pasque-flower 321703
Eastern Penstemon 289353
Eastern Planetree 302592
Eastern Poison Oak 393468
Eastern Poplar 311237,311292
Eastern Prairie Fringed Orchid 302390
Eastern Prickly Gooseberry 333945
Eastern Prickly Pear 272915
Eastern Pricklypear 272915
Eastern Prickly-pear Cactus 272915
Eastern Purple Bladderwort 403296
Eastern Purple Coneflower 140081
Eastern Rabbit-tobacco 317761
Eastern Red Cedar 213979,213984
Eastern Red Elderberry 345660
Eastern Red Oak 320278
Eastern Redbud 61166,83757
Eastern Redcedar 213984
Eastern Red-cedar 213984
Eastern Redcedar-ouge 213984
Eastern Rocket 365568
Eastern Rough Sedge 76160
Eastern Savin 213767
Eastern Shooting-star 135165
Eastern Showy Aster 160683
Eastern Silvery Aster 380854
Eastern Spring Beauty 94373
Eastern Spruce 298391,298423
Eastern St. Paul's Wort 363090
Eastern Star 285623
Eastern Star Sedge 75959
Eastern Stork's-bill 153794
Eastern Straw Sedge 76388
Eastern Strawberry-tree 45724
Eastern Swamp Saxifrage 349771
Eastern Tansy 383768
Eastern Teaberry 172135
Eastern Thuja 390587
Eastern Violet 410326
Eastern Violet Helleborine 147119
Eastern Wahoo 157327
Eastern Western-daisy 43297
Eastern White Beard-tongue 289366
Eastern White Cedar 390587
Eastern White Oak 323625
Eastern White Pine 300211
Eastern White-cedar 390587
Eastern Whorled Milkweed 38160
Eastern Wild Ginger 37574
Eastern Willow-herb 146670
Eastern Witch Hazel 185141
Eastern Woodland Sedge 73887
Eastern Woody-pear 415730
Eastindian Cycad 115812
Eastindian Persimmon 132137
East-Indian Lotus 263272
East-Liaoning Oak 324100,324559
Eastward Marjoram 274237
Eastwood Lily 229962
Eastwood's Goldenbush 150324
Eastwood's Larkspur 124466
Eastwood's Sandwort 31861
Eastwood's Wild Buckwheat 152014
Eatable Corydalis 105846
Eaten Away Rhododendron 330638
Eatin Berries 213702
Eating Pine 299926
Eaton's Aster 380870
Eaton's Bur-marigold 53887
Eaton's Fleabane 150603
Eaton's Thistle 91929
Eau De Cologne 93332
Eau-de-cologne Mint 250341
Eaver 235334,235373
Eba Ekki Lophira 236336
Ebbing's Silverberry 141981
Ebble 311208,311537
Ebei Fritillary 168344
Eberhard Scutia 355879
Eberhard Twinspinevine 355879
Eberhardt Blumea 55729
Eberhardt Chinabells 16293
Eberhardt Cylindrokelupha 116595
Eberhardt Hyparrhenia 201489
Eberhardt's Passionflower 285636
Eberhardtia 139781
Ebian Aspidistra 39531
Ebian Azalea 330608
Ebian Privet 229629
Ebian Rhododendron 330608
Ebian Shing-leaf Beech 162391
Ebi-lake Calligonum 67022
Ebilake Kneejujube 67022
Ebinger's Wild-rye 144295
Eble 345594
Eboe 133631
Ebony 132030,132137
Ebony Blavkbead 301136
Ebony Coccuswood 61427
Ebony Family 139763
Ebony Heart 142288
Ebony Monkey Earrings Pea 301136
Ebony Persimon 132137
Ebony Tree 132137,132298
Ebonyshoot Mistletoe 411016
Ebony-shooted Mistletoe 411016
Ebor 133631
Eccle Grass 299759
Eccle-grass 299759
Eccoilopus 139905
Ecdysanthera 139945
Echeveria 139970,139986,139987
Echeveria 'Set-Oliver' 139972
Echidnopsis 140017
Echinacea 140066
Echinaria 140099
Echinate Arisaema 33318
Echinate Cactus 244063
Echinate Euonymus 157418
Echinate Indigo 206311
Echinate Keeled Barberry 51428
Echinate Southstar 33318
Echinate-fruit Euonymus 157418
Echinocactus 140111
Echinocereus 140209
Echinochloaeform Sedge 74429
Echinocodon 140517,140520
Echinodorus 140536
Echinofossloсactus 140576
Echium 141087
Echiumlike Andrographis 22401
Echiumlike Craniospermum 108665
Echiumlike Onosma 271756
Echte Schliisselblume 315082
Echuan Ginger 417979
Eciliate Superior Rockfoil 349287
Ecjinate Licorice 177885

Eckberry 280003
Eclipta 141365
Ecuadoor Large-flowered Barberry 51679
Ecuador 140877
Ecuador Balmtree 261559
Ecuador Begonia 49800
Ecuador Laurel 104153
Ecuador Rocky Barberry 52133
Ecuador Sage 345092
Ecuador Slightly Barberry 51749
Ecuador Walnut 212613
Edder 345631
Edderwort 137747
Eddick 31051
Eddo 99910
Eddoes 99910
Eded Cinquefoil 312510
Edeltanne 272
Edelweiss 224767,224772
Edelweisslike Windhairdaisy 348481
Edentate Abutilon 990
Edentate Barberry 51580
Edgaria 141461
Edge Falls Anemone 23802
Edgehair Goldenray 229092
Edgehair Woodbetony 287118
Edgeweed 269299
Edgeworth Amphicarpaea 20566
Edgeworth Barberry 51581
Edgeworth Biformbean 20566
Edgeworth Figwort 355109
Edgeworth Madder 337953
Edgeworth Sagebrush 35456
Edgeworth Sandwort 31865
Edgeworth Spapweed 204918
Edgeworth Supple-jack 52413
Edgeworth Touch-me-not 204918
Edgeworth's Rhododendron 330613
Edging 34536,235090,349044
Edging Box 64356
Edging Candytuft 203239
Edging Lobelia 234363,234447
Edible Abelmoschus 219
Edible Bamboo 125447,297266
Edible Banana 260199,260253
Edible Begonia 49802
Edible Brachystelma 58858
Edible Burdock 31051
Edible Cana 71181
Edible Canna 71169
Edible Casearia 78117
Edible Casimiroa 78175
Edible Chesmut 78811
Edible Chinacane 4599
Edible Cyperus 118822
Edible Date 295461
Edible Debregeasia 123322
Edible Deepblue Honeysuckle 235702
Edible Dendrocalamopsis 125447
Edible Dragon Plum 137717
Edible Dragonplum 137717
Edible Fargesia 162682
Edible Fig 164763
Edible Figmarigold 251357
Edible Fruit Blueberry 403832
Edible Fruit Gaultheria 172057
Edible Fruit Whitepearl 172057
Edible Garcinia 171100,171101
Edible Granadilla 285637
Edible Ironweed 406311
Edible Kudzuvine 321431
Edible Lannea 221154
Edible Mountain Ash 369466
Edible Murdannia 260095
Edible Nightshade 367696
Edible Oilnut 323069
Edible Oil-nut 323069
Edible Oxystelma 278618
Edible Pangium 281246
Edible Pokeweed 298093
Edible Prickly Pear 272891
Edible Salacca 342580
Edible Sandcress 321431
Edible Snake Gourd 396151
Edible Snakegourd 396151
Edible Sour Bamboo 4599
Edible Stairweed 142654
Edible Thistle 91940
Edible Tulip 18569
Edible Valerian 404254,404255
Edible Vladimiria 135730
Edible Waterbamboo 260095
Edible Yam 131577
Edinam 145950
Edle Clivie 97223
Edle Sanchezia 345770
Edmundo Begonia 49801
Eduard Leek 15259
Eduard Onion 15259
Edum 254493
Edwards Chickweed 374951
Edwards Condalia 101738
Edwards Eutrema 161129
Edwards Hole-in-the-sand Plant 265988
Edwards Lazy Daisy 84928
Edwards' Plateau Larkspur 124368
Eegleclaw 401786
Eel-beds 48918
Eelgrass 404531,404563,418377,418392
Eel-grass 404536,418377,418392
Eelgrass Family 418405
Eel-grass Family 418405
Eelvine 25927
Eelweed Eelware 325580
Eerduosi Youngia 416460
Eerie 3978
Eerjisi Poplar 311188
Eever 235334
Eevy 187222
Effuse Baliospermum 46968
Effuse Mannagrass 177673
Effuse Rabdosia 324757
Effuse-flower Ixora 211085
Effusive Baliospermum 46968
Effusive Hedyotis 187568
Efoveotate Microula 254343
Efwatakala 249303
Efwatakala Grass 249303
Eg Climbing Hydrangea 299115
Egerton Odontoglossum 269068
Egerton Swan Orchid 116518
Egg Fruit 367370
Egg Orchid 82038
Egg Plant 367370,380756
Egg Squash 114310
Egg Thickshellcassia 113462
Egg Tree 171221
Egg-and-cheese 277648
Eggbag Sedge 75178
Eggberry 280003
Eggers 109857
Eggert's Sunflower 188962
Eggfruit 238117
Egg-fruit 238110,238117
Eggfruit Camellia 69426
Eggfruit Rhubarb 329358
Eggfruit Spruce 298307
Eggfruit Tree 238117
Eggfruit-lucuma 238117
Egg-in-the-pan 45192
Eggle 109857
Eggleaf Birthwort 34288
Eggleaf Dutchmanspipe 34288
Eggleaf Rhododendron 331420
Eggleaf Silktassel 171555
Eggleaf Slatepentree 322599
Eggleaf Spurge 159486
Egg-leaved Skullcap 355643
Egg-peg Bush 316819
Eggplant 366902,367370
Egg-plant 366902
Eggplant Ironweed 406817
Eggplant Wormwood 367605
Eggplant-flower Spicegrass 239858
Eggplantleaf Mazus 247037
Eggplantleaf PinnateGroundsel 263516
Eggplantleaf Rehmannia 327456
Eggremunny 11549
Eggs and Bacon 237539
Eggs and Bacon Pea 130937
Eggs-and-bacon 231183,237539,325612,

325666,325929,338209,367696
Eggs-and-butter 231183,262438,325518,325666
Eggs-and-collops 231183,237539
Eggs-eggs 109857
Egg-shaped Leaves Rhododendron 331420
Egg-shaped Twayblade 232950
Eggspike Sedge 74406
Egg-yolk-colour Epidendrum 146503
Eglantine 236022,336538,336741,336892
Eglantine Rose 336538,336892
Eglon Eglet 109857
Egremoine 11549
Egremounde 11549
Egretgrass 135055
Egri Mony 11549
Egusi 114106,256797
Egyp Broomrape 274927
Egyp Carissa 76894
Egypt Conyza 103407
Egypt Trefoil 396817
Egyptian 175444
Egyptian Bean 218721,263272
Egyptian Broomrape 274927
Egyptian Carissa 76894
Egyptian Clover 396817
Egyptian Cotton 179884
Egyptian Cottonthistle 271670
Egyptian Doum Palm 202321
Egyptian Fig 165726
Egyptian Geum 175444
Egyptian Granny's Cap 175444
Egyptian Grass 121285
Egyptian Henbane 201385
Egyptian Kidney Bean 218721
Egyptian Lettuce 219507
Egyptian Lotus 267658,267698
Egyptian Lovegrass 147475
Egyptian Lupin 238421
Egyptian Lupine 238444
Egyptian Mallow 243810
Egyptian Marjoram 274227
Egyptian Millet 143515,369720
Egyptian Mimosa 1427
Egyptian Mint 250402,250439
Egyptian Multiplier Onion 15169
Egyptian Myrobalan 46768
Egyptian Onion 15166,15167
Egyptian Paper Plant 119347
Egyptian Paper Reed 119347
Egyptian Paper Rush 119347
Egyptian Paper-reed 119347
Egyptian Papyrus 119347
Egyptian Paradise Seed 19858
Egyptian Paspalidium 285381
Egyptian Privet 223454
Egyptian Reed 119347
Egyptian Riverhemp 361430
Egyptian Rose 216753,350099
Egyptian Sesban 361430
Egyptian Sour Bread 7022
Egyptian Star 289831
Egyptian Starcluster 289831
Egyptian Star-cluster 289831
Egyptian Sycamore 165726
Egyptian Thorn 1427,1680,322450
Egyptian Trefoil 396817
Egyptian Walking Onion 15165
Egyptian Water Grass 285381
Egyptian Water Lily 267698
Egyptian Willow 342967
Egyptian Yarrow 3916
Egyptian's Herb 239202
Ehe Goatwheat 44264
Ehe Poplar 311188
Ehrenberg Bulrush 353381
Ehrenberg Mahonia 242519
Ehretia 141592
Ehretia-like Persimmon 132140
Eichlam Cactus 244060
Eichler Tulip 400154,400247
Eichwald Ammodendron 19732
Eichwald Heliotrope 190610
Eidin Docken 292372
Eight Ribbed Memecylon 250036
Eight Treasure 200775
Eight Treasures Tea 157809
Eightangle 204474,204571,204603
Eightangle Bethlehemsage 283629
Eightangle Paraphlomis 283629
Eightcornut Corydalis 106546
Eight-day Healing Bush 235007
Eight-flowered Fescue-grass 412474
Eightleaves Argyroderma 32733
Eight-males Rhododendron 331386
Eight-nerved Lasianthus 222245
Eightnerved Premna 313705
Eight-nerved Premna 313705
Eightpetal Begonia 50120
Eightpetal Camellia 69015
Eightpetal Mountain-avens 138461
Eight-petaled Camellia 69015
Eight-petals Dryas 138461
Eight-ribbed Memecylon 250036
Eightstamen Azalea 331386
Eightstamen Mastixia 246215
Eightstamen Waterwort 142570
Eight-stamened Rhododendron 331386
Eight-stamened Waterwort 142570
Eightvein Premna 313705
Eight-veins Lasianthus 222245
Eight-veins Roughleaf 222245
Eike-tree 324335
Eincorn 398924
Eisch Keys 167955
Eisch-keys 167955
Eithin 401388
Ejow 32343
Ejow Palm 32343
Ekar Ekra 341842
Eker 262722
Eke-tree 324335
Ekhimi 300655
Ekkel-girse 299759
Ekkern 324335
Ekki 236334
Ekpogoi 52843
Eksis-girse 384714
Elachanthemum 141897
Elaeagnus 141929,142214
Elaeagnus Pear 323165
Elaeagnus Willow 343336
Elaeagnus-leaf Tanoak 233194
Elaeagnus-leaved Tanoak 233194
Elaeagnuslike Distylium 134930
Elaeagnus-like Distylium 134930
Elaeagnuslike Mosquitoman 134930
Elaeocarpus 142269,142303
Elaeocarpus Family 142267
Elaeopian Olive 141932
Elagant Falsetamarisk 261259
Elagant False-tamarisk 261259
Elagant Greenbrier 366319
Elagant Ironweed 405445
Elagant Ventilago 405445
Elangate Magnoliavine 351034
Elangate Mussaenda 260410
Elangate Polypogon 310108
Elastic Grass 148003
Elaterium Fruit 139852
Elatior Begonia 49921
Elatior Hybrid Primroses 314085
Elatostema 142594,142842
Elavated-midrib Cyclobalanopsis 116088
Elburs Iris 208520
Elden 292372
Elden Eldin 292372
Elder 16352,345558,345580,345631
Elder Rose 407989
Elderberry 345558,345580,345631,345660
Elderbush 300382
Elder-flower Orchid 273623
Elderflower Photinia 295767
Elder-flowered Orchid 273623
Elderleaf Ghost lampstand 335153
Elderleaf Rodgersflower 335153
Eldern 345631
Elder-trot 192358
Eldin 292372
Eldin-docken 292372,339925
Eldon 345631

Eldorad Cattleya 79549
Eldrot 28044,101852
Elecampane 207025,207135,207213
Elecampane Inula 207135
Elecampane Jurinea 207135
Election Pink 331546
Election Pink Azalea 331546
Electleaf Tanoak 233264
Elederberry Panax 310236
Elegance Gladiolus 176531
Elegans Bladdernut 374088
Elegans Boxwood 64350
Elegans Cryptomeria 113695
Elegant Ainsliaea 12630
Elegant Aralia 30626
Elegant Ardisia 31435
Elegant Arrowbamboo 162683
Elegant aster 155806
Elegant Barberry 51483
Elegant Begonia 49733,49769
Elegant Brodiaea 60460
Elegant Cactus 244061
Elegant Camas 417890
Elegant Clarkia 94144
Elegant Cluster-lily 60460
Elegant Codonanthe 98233
Elegant Coldwaterflower 298996
Elegant Colquhounia 100037
Elegant Confused Storax 379329
Elegant Corydalis 105750
Elegant Crassula 108985
Elegant Cudweed 170923
Elegant Death-camas 417890
Elegant Dendrobenthamia 105059
Elegant Deutzia 126919
Elegant Dicliptera 129255
Elegant Dipelta 132600
Elegant Dogliverweed 129255
Elegant Dontostemon 136169
Elegant Dutchman's Pipe 34246
Elegant Duvalia 139149
Elegant Epidendrum 146409
Elegant Eulophia 156732
Elegant Evergreen Chinkapin 78963
Elegant Evergreenchinkapin 78963
Elegant Everlasting 21564
Elegant Evodia 161329
Elegant Falseginsen 280791
Elegant Fawn-lily 154913
Elegant Four-involucre 105059
Elegant Gayfeather 228461
Elegant Goldenrod 368211
Elegant Groundsel 279918,358788
Elegant Habenaria 183619
Elegant Hawksbeard 110796
Elegant Hedge Bamboo 47347
Elegant Herminium 192847
Elegant Hogfennel 292843
Elegant Hymenocrater 201040
Elegant Ixeris 210538
Elegant Lady's Slipper 120344
Elegant Ladyslipper 120344
Elegant Larkspur 124691
Elegant Licuala Palm 228736,228759
Elegant Lockhartia 235134
Elegant Lophanthus 236269
Elegant Maple 2944
Elegant Meadowrue 388492
Elegant Meconopsis 247118
Elegant Michelia 252863
Elegant Milkwort 308032
Elegant Mountain Leech 126329
Elegant Orchis 116827
Elegant Panicgrass 281591
Elegant Pearleverlasting 21564
Elegant Philodendron 294804
Elegant Phreatia 295958
Elegant Phyllodium 297008
Elegant Platanthera 302318
Elegant Pleurothallis 304912
Elegant Pouzolzia 313424
Elegant Ptilagrostis 321008
Elegant Pussytoes 26564
Elegant Rabbiten-wind 12630
Elegant Raspberry 338121
Elegant Rattan Palm 65664,65676
Elegant Reekkudzu 313424
Elegant Rhodiola 329842
Elegant Rhododendron 330615,330675
Elegant Rush 213014
Elegant Sagebrush 35635
Elegant Schismatoglottis 351149
Elegant Sedge 74164
Elegant Shrubalthea 195295
Elegant Silverpuffs 253924
Elegant Sourbamboo 4618
Elegant Spapweed 204923
Elegant Spindle-tree 157405
Elegant Spiraea 371903
Elegant Stephania 375847
Elegant Stitchwort 255466
Elegant Tanoak 233196
Elegant Tickclover 126329
Elegant Torchflower 100037
Elegant Touch-me-not 204923
Elegant Twayblade 232150
Elegant Vanda 404652
Elegant Vetch 408539
Elegant Waterstarwort 67382
Elegant Wild Buckwheat 152038
Elegant Windhairdaisy 348284
Elegant Zigadenus 417890
Elegant Zinnia 418061
Elegantest Euonymus 157420
Elegantest Spindle-tree 157420
Elegant-leaved Remarkable Barberry 51768
Elegent Azalea 330369
Elegent Rhododendron 330369
Elegentflower Azalea 330366
Elegentflower Rhododendron 330366
Elegent-flowered Rhododendron 330366
Elem 401593
Elemi Frankincense 57521
Elephant Apple 130919,163502
Elephant Arisaema 33319
Elephant Bush 311956
Elephant Cactus 279490
Elephant Cattail 401107
Elephant Climber 32642
Elephant Creeper 32642
Elephant Ear 49587,415201
Elephant Ears 99910
Elephant Foot 20132,131573
Elephant Garlic 15048
Elephant Grass 289218
Elephant Head Masdevallia 246107
Elephant Hedge 352500
Elephant Mucuna 259512
Elephant Orange 378888
Elephant Pumpkin 114288
Elephant Southstar 33319
Elephant Tree 57532,64073,64080,64081,279505
Elephant Wood 56735
Elephant Yam 20057,20125
Elephant's Ear 16484,16512,49587,52514,99899,145993,99910,145993,145994,294802
Elephant's Ear Begonia 49602
Elephant's Ear Tree 145994
Elephant's Earpodtree 145994
Elephant's Ears 52503
Elephant's Food 311956
Elephant's Foot 7291,49343,131573,131863,143456
Elephant's Trunk 279685
Elephant's-ear 16484,99899,99910,145994
Elephant's-ear Begonia 50284
Elephant's-ears 52503,52514,187206
Elephant's-foot 131451,131573,143453
Elephant's-trunk 245977,245979
Elephant-ear 52503
Elephantfoot 143453
Elephant-foot Tree 49343
Elephantgrass 289218
Elephant-grass 289218
Elephant-leaved Saxifrage 52532
Elephantorange 163499,163502
Elephants Tooth 107023
Elephanttooth Coryphantha 107023
Elephant-tree 64080

Elephat's-ear 65214
Eleusine 143512
Eleutharrhena 143557
Eleutherine 143570
Eleuthro-petal Neoregelia 264415
Elevated Yushania 416770
Elevatednerve Pittosporum 301259
Elevatednerve Seatung 301259
Elevatehead Saussurea 348798
Elevatinode Riparial Silverreed 394928
Eleven o'clock Flower 274823,311852
Eleven o'clock Lady 274823
Elevenleaf Hemsleia 191922
Elevenleaf Snowgall 191922
Elf Dock 207135
Elf-dock 207135
Elfin Wands 130187
Elf-shot 14065
Elf-wort 207135
Elgins 339925
Elgon Olive 270074,270176
Elicompane 207135
Eliff Cinquefoil 312946
Elim Heath 149979
Elizabeth Primrose 314335
Elk Nut 323072
Elk Sedge 73829,74626,93913,93919
Elk Thistle 91977,92360
Elk's Lip 68181
El-kantara 93403
Elk-bark 242341
Elk-grass 415441
Elkhorn 159204
Elkhorns 332359
Elkweed 380302
Elkwood 242333
Ell Docken 292372
Ellangowan Poison Bush 148359
Ellarne 345631
Ell-docken 292372
Ellem 401593
Ellen 16352,345631
Ellen-tree 16352,345631
Eller 16352,345631
Ellern 345631
Ellern Aul 345631
Ellet 345631
Elliott Barberry 52258
Elliott Love Grass 147655
Elliott Monkshood 5172
Elliott Woodbetony 287184
Elliott Yellow-eyed-grass 416054
Elliott's Aster 380871
Elliott's Blueberry 403820
Elliott's Goldenrod 368207
Elliott's Huckleberry 403820
Elliott's Love Grass 147655
Elliott's Broomsedge 22704
Ellipanthus 143867
Ellipseleaf Clearweed 298913
Ellipseleaf Coldwaterflower 298913
Ellipsoid Lanceleaf Litse 233968
Ellipsoidal-fruit Litse 233968
Elliptic Biondia 54525
Elliptic Casearia 78155
Elliptic Elaeocarpus 142396
Elliptic Epidendrum 146410
Elliptic Fishvine 125958
Elliptic Gambirplant 401752
Elliptic Garcinia 171201
Elliptic Ginger 417981
Elliptic Heliotrope 190614
Elliptic Indian Skullcap 355504
Elliptic Jadeleaf and Goldenflower 260409
Elliptic Jewelvine 125958
Elliptic Leaves Rhododendron 330617
Elliptic Mussaenda 260409
Elliptic Myrsine 261619
Elliptic Nepeta 264914
Elliptic Oldham Persimmon 132335
Elliptic Oreocharis 273851
Elliptic Pondweed 312095
Elliptic Primrose 314336
Elliptic Raspberry 338347
Elliptic Rhynchotechum 333739
Elliptic Rockfoil 349292
Elliptic Rose 336650
Elliptic Roughleaf 222130
Elliptic Shin-leaf 322823
Elliptic Smallligulate Aster 40007
Elliptic Spike-rush 143127
Elliptic Stauntonvine 374399
Elliptic Tall Dock 340040
Elliptic Tarenna 385040
Elliptic Yellowwood 268357
Elliptical Euonymus 157423
Elliptical Fruit Newlitse 264046
Elliptical Mussaenda 260409
Elliptical Myrsine 261619
Elliptical Newlitse 264046
Elliptical Ochrosia 268357
Elliptical Ormosia 274386
Elliptical Prain Kobresia 217251
Elliptical Raspberry 338347
Elliptical Staunton-vine 374399
Elliptical Wildbasil 96989
Elliptical-fruited Newlitse 264046
Elliptic-buds Magnolia 416701
Ellipticfruit Corydalis 105855
Ellipticfruit Draba 136989
Ellipticfruit Hemsleia 191917
Ellipticfruit Meconopsis 247116
Ellipticfruit Snowgall 191917
Ellipticfruit Spikesedge 143193
Ellipticfruit Whitlowgrass 136989
Elliptic-fruited Maackia 240122
Ellipticleaf Aglaia 11291
Ellipticleaf Bennettiodendron 51032,51040
Elliptic-leaf Blueberry 403970
Ellipticleaf Camellia 69082
Ellipticleaf Debregeasia 123323
Ellipticleaf Dinggongvine 154237
Ellipticleaf Gentian 173208
Ellipticleaf Hapaline 185305
Elliptic-leaf Hydrangeavine 351786
Ellipticleaf Indigo 205783
Ellipticleaf Knotweed 309078
Elliptic-leaf Magnolia 242082
Ellipticleaf Maizailan 11291
Ellipticleaf Meranti 362214
Elliptic-leaf Siberian Crabapple 243578
Ellipticleaf Spurgentian 184696
Elliptic-leaf St. John's Wort 201844
Elliptic-leaf Stachyurus 373520
Ellipticleaf Waternettle 123323
Elliptic-leaved Aglaia 11291
Elliptic-leaved Blueberry 403970
Elliptic-leaved Camellia 69082
Elliptic-leaved Indigo 205783
Elliptic-leaved Magnolia 242082
Elliptic-leaved Spindle-tree 157804
Elliptic-leaved St. John's-wort 201844
Elliptic-ovate Bract Larkspur 124196
Elliptoid Milkvetch 42322
Ellisia 143886
Ellisiophyllum 143894
Ellliptic Beccarinda 49395
Ellliptic-leaf Beccarinda 49395
Ellow Bartsia 284098
Ell-shinders 16352,345631,359158
Ellum 345631,401593
Ellwanger Hawthorn 109688
Ellwood Cypress 85277
Elm 401425,401593,401602
Elm Family 401415
Elm Leaf Actinidia 6714
Elm Zelkova 417540
Elmer Macrozamia 241766
Elmer's Fleabane 150615
Elmerrill Tanoak 233201
Elmin Elmen 401593
Elm-leaf Begonia 50379
Elmleaf Blackberry 339419,339427
Elm-leaf Blackberry 339419
Elmleaf Bramble 339419
Elmleaf Kiwifruit 6714
Elm-leaved Actinidia 6714
Elm-leaved Blackberry 339419
Elm-leaved Goldenrod 368461
Elm-leaved Spiraea 371877
Elm-leaved Sumach 332533

Elmlike Meadowsweet 166125
Elm-wych 401512
Elodea 143914,143920,143931
Elodea-like St. John's Wort 201845
Elongate Barrenwort 146987
Elongate Dropseed 372659
Elongate Elytrigia 144643
Elongate Epimedium 146987
Elongate Geniosporum 172861
Elongate Hetaeria 193602
Elongate Imbricate Galingale 119020
Elongate Litse 233899
Elongate Paulownia 285958
Elongate Peduncle Cinquefoil 312867
Elongate Primrose 314337
Elongate Serpentroot 354853
Elongate Snow White Cinquefoil 312821
Elongate Swertia 380195
Elongate Tickseed 104771
Elongated Cymaria 116712
Elongated Lovegrass 147656
Elongated Mustard 59380
Elongated Screwtree 190100
Elongated Screw-tree 190100
Elongated Sedge 74442
Elongated Spapweed 204926
Elongated Touch-me-not 204926
Elongateleaf Cherry-laurel 223135
Elongateleaf Tanoak 233264
Elongate-raceme Dense-flower Barberry 51544
Elongatespur Hemsley Monkshood 5253
Elonghate Wombgrass 365907
Elphamy 61497
Elren 345631
Elsedock 207135
Elsholtzia 143958,143974,143993
Elsholtzia-like Keiskea 215807
Elsinore Wild Buckwheat 152345
Eltrot 8826,28044,192358,192367,269307
Eluanbi Senna 78303
Eluanbi Small-fruit Fig 165317
Elven 401593
Elwes Forkliporchis 269022
Elwes Gentian 173415
Elwes Monkshood 5176
Elwes Odontochilus 269022
Elwes Primrose 270731
Elwes Woodbetony 287185
Elytranthe 144593
Elytrigia 144624
Elytrophorus 144696
Ema Azalea 331384
Emaginate Bulbophyllum 62712
Emaginate Cirrhopetalum 62712
Emaginate Curlylip-orchis 62712
Emarginate Amaranth 18670
Emarginate Amaranthus 18670
Emarginate Azalea 330622
Emarginate Barberry 51586
Emarginate Blueberry 403821
Emarginate Bractleaf Stonecrop 356533
Emarginate Erycibe 154237
Emarginate Eurya 160458
Emarginate Holly 203616
Emarginate Mayten 182687
Emarginate Merremia 250779
Emarginate Ormosia 274388
Emarginate Rhododendron 330622
Emarginate Roscoea 337111
Emarginate Roureopsis 337772
Emarginate Sabia 341501
Emarginate Sageretia 342177
Emarginate Stonecrop 356702
Emarginate Wampee 94188
Emarginate-leaf Indigo 205945
Emarginateleaf Ormosia 274388
Emarginate-leaf Silkwormthorn 114325
Emarginate-leaved Indigo 205945
Emasculate Scabrous Hydrangea 199818
Embelia 144716
Embero 237927
Emblic 296554
Emblic Leaf Flower 296554
Emblic Leafflower 296554
Emblic Leaf-flower 296554
Emblic Underleaf Pearl 296554
Embothrium 144847
Embrert 96140
Embroidered Sheath Bamboo 297215
Embroiderhair Cliffbean 66647
Embul 260253
Emei Actinodaphne 6806
Emei Amitostigma 19507
Emei Ancylostemon 22169
Emei Angelica 24424
Emei Arisaema 33441
Emei Arrowwood 407988
Emei Aspidistra 39567
Emei Aster 41470
Emei Astyleorchis 19507
Emei Aucuba 44898
Emei Azalea 330141
Emei Bamboo 47264
Emei Bambusa 47264
Emei Barberry 51279
Emei Begonia 49808
Emei Birch 53628
Emei Blueberry 403929
Emei Broad-leaved Sabia 341580
Emei Calanthe 65943
Emei Camellia 69423
Emei Chestnut 78796
Emei Chickweed 375028
Emei Cliffbean 66640
Emei Clinopodium 97041
Emei Closedfruit Tanoak 233150
Emei Cyclea 116033
Emei Cymbidium 116853
Emei Daphne 122424
Emei Deutzia 127044
Emei Dogwood 124893
Emei Dualbutterfly 398264
Emei Dubyaea 138776
Emei Edelweiss 224924
Emei Elatostema 142767
Emei Euonymus 157755
Emei Eupatorium 158243
Emei False Ironculm 317193
Emei Falsehellebore 405626
Emei Four-involucre 124893
Emei Fritillary 168385
Emei Gentian 173680
Emei Germander 388170
Emei Ginger 418013
Emei Gingerlily 187442
Emei Globba 176990
Emei Goldlineorchis 25978
Emei Goldsaxifrage 90381
Emei Goldthread 103840
Emei Goldwaist 90381
Emei Greenbrier 366326
Emei Gymnadenia 182244
Emei Helwingia 191194
Emei Hemiboea 191377
Emei Hemsleia 191921
Emei Henry Spiraea 371930
Emei Holly 204112
Emei Honeysuckle 236106
Emei Hooktea 52446
Emei Hornbeam 77358
Emei Indocalamus 206782
Emei Kadsura 214975
Emei Kudzuvine 321457
Emei Ladybell 70202
Emei Landpick 308624
Emei Larkspur 124428
Emei Leek 15543
Emei Lily 229837
Emei Linden 391797
Emei Litse 234004
Emei Loosestrife 239762
Emei Lysionotus 239960
Emei Machilus 240646
Emei Madder 337987
Emei Maple 2946
Emei Mazus 247018
Emei Meadowrue 388607
Emei Microtoena 254261
Emei Millettia 66640
Emei Mischobulbum 383137

Emei Morningglory 207775
Emei Mountain Winterhazel 106668
Emei Muping Corydalis 105892
Emei Nanmu 295394,295408
Emei Nepal Saurauia 347971
Emei Oblongleaf Maple 3266
Emei Onion 15543
Emei Ophiorrhiza 272191
Emei Orchis 116853,310923
Emei Osbeckia 276381
Emei Osmanther 276381
Emei Parakmeria 283422
Emei Parnassia 284587
Emei Pennsylvania Bittercress 72812
Emei Pepper 300385
Emei Pericampylus 290845
Emei Phoebe 295394,295408
Emei Pipewort 151301
Emei Pittosporum 301352
Emei Ploughpoint 401170
Emei Potanin Birch 53628
Emei Primrose 314552
Emei Pseudanthistria 317193
Emei Rhododendron 330141
Emei Rhynchoglossum 333092
Emei Rhynchotechum 333747
Emei Rockjasmine 23252
Emei Rose 336825
Emei Sage 345274
Emei Sagebrush 35463
Emei Sageretia 342182
Emei Sandwort 32114
Emei Seatung 301352
Emei Sedge 75602
Emei Skullcap 355634
Emei Sloanea 366049
Emei Smilax 366326
Emei Snowgall 191921
Emei Solomonseal 308624
Emei Sowthistleleaf Primrose 314995
Emei Spapweed 205197
Emei Spicebush 231419
Emei Spiradiclis 371755
Emei Splitmelon 351771
Emei Stairweed 142767
Emei Supplejack 52446
Emei Sweetspire 210403
Emei Swertia 380197
Emei Taillessfruit 100144
Emei Tainia 383137
Emei Tanakaea 383890
Emei Tanoak 233329
Emei Teasel 133457
Emei Thistle 91962
Emei Tilia 391797
Emei Tinospora 392270
Emei Tinospore 392270
Emei Touch-me-not 205197
Emei Trigonotis 397447
Emei Tumorous Primrose 315017
Emei Tupistra 70604
Emei Typhonium 401170
Emei Viburnum 407988
Emei Vineclethra 95529
Emei Violet 409954
Emei Violetbloom Rhododendron 331345
Emei Waxpetal 106668
Emei Whytockia 414009
Emei Wildbasil 97041
Emei Wildberry 362355
Emei Willow 343804
Emei Winterhazel 106668
Emei Woodbetony 287485
Emei Zinger 418013
Emeishan Bogorchid 158243
Emeishan Euonymus 157755
Emeishan Spindle-tree 157755
Emeishan Tropidia 399643
Emerald Creeper 378353
Emerald Gaiety Euonymus 157479
Emerald Green Arborvitae 390587
Emerald Lily 229988
Emerald Ripple 290315
Emerald-fern 38985
Emeri 386563
Emermann Cactus 244072
Emero 237927
Emersed Pondweed 312096
Emetic Crucianvine 356303
Emetic Devilpepper 327080
Emetic Devil-pepper 327080
Emetic Mosquitotrap 117743
Emetic Root 158692
Emetic Weed 234547
Emien 18033
Emil Barberry 51588
Emin Begonia 49809
Emmel Emmal 401593
Emmenopterys 145023
Emmer Wheat 398890
Emmet's Stalk 240068
Emmons' Sedge 73627
Emod Onosma 242398
Emodi Pine 300186
Emory Barrel 163439
Emory Oak 323873
Emory Sedge 74453
Emory's Barrel Cactus 163439
Emory's Rock Daisy 291309
Emory's Sedge 74452,74453
Emperator Rhododendron 330914
Emperor's Sulphur Flower 152579
Emping 178535
Empress Tree 285951,285958,285981
Empress-of-Germany 266651
Empress-tree 285981
Empty Arrowbamboo 162682
Empty Pseudosasa 318281
Emptygrass 98792
Emptyhorn Peristylus 291199
Emptyhorn Perotis 291199
Emptystalk Yushanbamboo 416759
Emptystem Sweetvetch 187873
Emu Apple 277638
Emu Berry 306423
Emu Bush 148354,148360,148366
Emu Plum 277636
Emu-bush 148354
Emulative Leymus 228339
Emur Goatwheat 44264
Emwi Barberry 51653
Enancia 175454
Enares Pansy Orchid 254930
Encephalartos 145217
Enchanter's Nightshade 91499,91557,244340
Enchanter's-nightshade 91499,91524,91557
Enchanting Bauhinia 49254
Encina 323621
Encinilla 112894
Encinitas Baccharis 46268
Encinitas False Willow 46268
Endiandra 145315
Endive 90894,90901
Endive Daisy 328520
Endlicher Callitris 67417
Endlicher Grevill 180593
Endod 298104
Endophyllous Fig 164885
Endospermum 145413
Endres Stelis 374705
Endress' Cranesbill 174581
Enem 23925
Enemion 145492
Enemy Anenemy 23925
Enerv Sedge 74453
Enervious Barberry 51967
Eng 133587
Eng Gurjun Oil Tree 133588
Engelhardia 145504
Engelmann Aster 40368
Engelmann Daisy 145542,145543
Engelmann Fishhook Cactus 354319
Engelmann Oak 323874
Engelmann Pine 299931
Engelmann Spruce 298273
Engelmann's Aster 155807
Engelmann's Daisy 145542,145543
Engelmann's Fleabane 150618
Engelmann's Goldenweed 271912
Engelmann's Hedgehog Cactus 140242
Engelmann's Knotweed 309086

Engelmann's Pine 299931
Engelmann's Prickly Pear 272891
Engelmann's Pricklypear 272873
Engelmann's Spike-rush 143132
Engelmann's Spruce 298273
Engelmann's Thistle 91949
Engelmann's Vervain 405840
England Sedge 76780
Engler Abelia 123
Engler Barberry 51592
Engler Beech 162372
Engler Begonia 49810
Engler Bittercress 72742
Engler Milkvetch 42327
Engler Oak 323875
Engler Rockjasmine 23164
Engler Spicegrass 239626
Engler Stonecrop 356703
Engler Sugarpalm 32336
English Alder 16352
English Arrowroot 367696
English Bean 408393
English Beech 162372,162400
English Birch 53563
English Bluebell 199560
English Box 64345
English Boxwood 64345,64356
English Bryony 61497
English Camomile 85526
English Catch 363477
English Catchfly 363477
English Cherry 223116
English Cinquefoil 312357
English Cost 383874
English Costmary 383874
English Daffodil 262441
English Daisy 50808,50825
English Earigold 66445
English Elm 401468,401508,401593
English Flytrap 138348
English Forcing Cucumis 114246
English Forget-me-not 260868
English Gooseberry 334250
English Greenweed 172905
English Hare's Bell 145487
English Hare's Bells 145487,145488
English Harebell 145487
English Hawthorn 109790,109857
English Holly 203545,203562
English Hyacinth 145487
English Iris 208444,208679,208944
English Ivy 187195,187222
English Laurel 223116
English Lavender 223251,223291
English Mandrake 61497
English Maple 2837
English Marjoram 274237
English Marquery 86970
English Masterwort 8826
English Meadowsweet 166125
English Mercury 86970
English Mulberry 259180
English Oak 324278,324335
English Orris 404316
English Palm 343151
English Passion Flower 37015
English Peony 280231
English Planetree 302582
English Plantain 302034,302068
English Primrose 314820,315101
English Rhubarb 329388,388506
English Rose 336419
English Rye Grass 235334
English Ryegrass 235334
English Sage 344907
English Sandwort 32104
English Sarsaparilla 312520
English Scurvy Grass 97976
English Scurvy-grass 97976
English Sea Grape 342859
English Serpentary 308893
English Sorrel 339887
English Stonecrop 356537
English Sundew 138267
English Tamarisk 383433
English Thyme 391365,391397
English Treacle 388042,388272,388275
English Violet 410320
English Wallflower 86427
English Walnut 212636
English Water Grass 177629
English Wild Daffodil 262441
English Yew 385302
Englishman's Foot 302068
Engraved Cotoncaster 107499
Enhalus 145642
Enisomeles 147082
Enkianthus 145673,145719
Enkokorutanga 16882
Enneaphyllous Caesalpinia 252748
Enneapogon 145756
Enormous Euphorbia 159140
Ensete 145823,145846,260208,260224
Enshi Barrenwort 146988
Enshi Epimedium 146988
Enshi St. John's Wort 201852
Enshi Teasel 133471
Ensiform Knifebean 71042
Entada 145857
Entadalike Cliffbean 254688
Entada-like Millettia 254688
Entangled Holly 203923
Enteropogon 145999
Entire Barberry 51730
Entire Beautyberry 66804
Entire Butterflybush 62073
Entire Butterfly-bush 62073
Entire Byttneria 64469
Entire Callery Pear 323122
Entire Camellia 69151,69675
Entire Catchfly 363551
Entire Central Clematis 95256
Entire Cherry-laurel 223113
Entire Chloranthus 88284
Entire Clematis 95029
Entire Cudrania 114332
Entire Dianxiong 297943
Entire Elatostema 142692
Entire Fiddleleaf Fig 165430
Entire Fig 165430
Entire Firethorn 322451
Entire Garliccress 365532
Entire Indian Hibiscus 194937
Entire Kalanchoe 215174
Entire Lagotis 220179
Entire Leaf Vineclethra 95475
Entire Linden 391734
Entire Meconopsis 247139
Entire Micromelum 253628
Entire Oberonia 267954
Entire Ochna 268200
Entire Pellionia 288732
Entire Peppergrass 225364
Entire Pepperweed 225364
Entire Photinia 295706
Entire Platanthera 302373
Entire Purplepearl 66804
Entire Shortbranch Willow 343121
Entire Showy Woodbetony 287035
Entire Snakegourd 396238
Entire Steudnera 376384
Entire Summerlilic 62073
Entire Velvetplant 183096
Entire Viburnum 407895
Entire Willow 343529
Entire Woodnettle 125548
Entirelea Decorated Kinostemonf 216460
Entireleaf Arrowwood 407895
Entireleaf Banksia 47647
Entireleaf Bird Cherry 280025
Entireleaf Birdcherry 280025
Entireleaf Burvine 64469
Entireleaf Clabgourd 417081
Entireleaf Clematis 95029
Entireleaf Corydalis 106380
Entireleaf Dontostemon 136173
Entireleaf Dragonhead 137601
Entireleaf Eutrema 161134
Entireleaf Germander 388110
Entireleaf Goldenmelon 182580
Entireleaf Goldenray 229109

Entireleaf Goldraintree 217615
Entireleaf Greenorchid 137601
Entireleaf Hawksbeard 110860
Entireleaf Hemiboea 191367
Entire-leaf Hydrangea 199910
Entireleaf Indigo 205945
Entireleaf Lagarosolen 219790
Entireleaf Lagotis 220180
Entireleaf Ligusticum 229340
Entireleaf Mustard 59438
Entireleaf N. China Eight Trasure 200808
Entireleaf Neoalsomitra 417081
Entireleaf Petunia 292749
Entireleaf Physospermopsis 297943
Entireleaf Pricklyash 417247
Entireleaf Purpledaisy 267172
Entireleaf Raphiolepis 329086
Entireleaf Redcarweed 288732
Entireleaf Runcinate Saussurea 348744
Entireleaf Sawwort 360987
Entireleaf Snakegourd 396238
Entireleaf Spiritweed 8775
Entireleaf Stairweed 142692
Entireleaf Tacca 382920
Entireleaf Tatarinow Stonecrop 200808
Entire-leaf Tianshan Mountainash 369537
Entireleaf Tridenwort 394837
Entireleaf Valerian 404280
Entireleaf Windhairdaisy 348389
Entireleaf Woodbetony 287296
Entireleaf Woodnettle 273908
Entireleaflet Amesiodendron 19419
Entire-leafleted Amesiodendron 19419
Entire-leaved Adinandra 8242
Entire-leaved Adonis 8363
Entire-leaved Aporosa 29776
Entire-leaved Bird Cherry 280025
Entire-leaved Burvine 64469
Entire-leaved Cherry Laurel 223113
Entire-leaved Cotoneaster 107504
Entire-leaved Daisy 199237,199238
Entire-leaved Drypetes 138636
Entire-leaved False Foxglove 10146
Entire-leaved Fourleaf Ladybell 7851
Entire-leaved Groundsel 359138
Entire-leaved Hackberry 80586
Entire-leaved Linden 391734
Entire-leaved Medick 247462
Entire-leaved Micromelum 253628
Entire-leaved Prickly-ash 417247
Entire-leaved Raphiolepis 329086
Entire-leaved Salmon Barberry 51284
Entire-leaved Sloanea 366058
Entire-leaved Woodnettle 273908
Entire-leaved Yellow Mahonia 242517
Entirelip Cheirostylis 86725
Entirelip Ginger 417988
Entirelip Irisorchis 267954
Entirelip Lecanorchis 223658
Entirelip Zeuxine 417750
Entirelip Zinger 417988
Entireliporchis 261467
Entirelobe Indian Hibiscus 194937
Entire-marginal Barberry 51730
EntireRockycabbage 394837
Entire-sepal Barleria 48204
Entire-sepal Falsecuckoo 48204
Eowler's Service 369339
Eoxflop 130383
Epacris 146067
Epaltes 146096
Epaulette Tree 320878,320880,320886
Epaulettetree 320878,320886
Epaulette-tree 320878
Epazote 86934,139678
Epephant Bush 311956
Ephedra 146122
Ephedra Family 146270
Ephedra Wild Buckwheat 152042
Epidendrum 146371
Epidendrum Orchid 146371,146451
Epigeal Stephania 375848
Epigeneium 146528,146532
Epigynum 146558
Epilasia 146565
Epilose Jingdong Primrose 314500
Epimedium 146953,146958,147010
Epimeredi 147082
Epipactis 147106
Epiphany 115008
Epiphyllum 147283
Epiphyte Blueberry 403799
Epiphyte Currant 333914
Epiphytic Azalea 330550
Epiphytic Medinilla 247535
Epiphytic Mountainash 369383
Epiphytic Mountain-ash 369383
Epipogium 147303,147323
Epipremnum 147336
Epiprinus 147353
Episcia 147384
Episia 147384
Epithema 147427
Eprinose Mahonia 242543
Epruinose Primrose 314331
Eps 311537
Epunctate Swertia 380206
Equal Alopecurus 17498
Equalbract Aster 40589
Equalglume Ricegrass 300702
Equal-leaved Knotgrass 308788
Equallip Figwort 355040
Equallyflower Primrose 314472
Equallypedicel Primrose 314533
Equalpetal Corydalis 105987
Equalphyllaries Aster 40589
Equal-sepal Barberry 52009
Equalsepal Bulbophyllum 63175
Equalsepal Curlylip-orchis 63175
Equalsepal Stonecrop 357238
Equaltoothed Cinquefoil 312997
Equal-winged Begonia 49950
Equatorial Papillose Barberry 52004
Equeal-sided Casearia 78097
Equestrian Star-flower 196443
Eragrostiella 147463,147465
Eragrostis Fimbristylis 166304
Eragrostis Fluttergrass 166304
Eranthemum 148053
Erduosi Sunrose 188771
Erduosi Sun-rose 188771
Ere Sedge 74459
Erect Achimenes 4076
Erect Angraecum 24839
Erect Ascitesgrass 407468
Erect Asiabell 98301
Erect Averts 175446
Erect Bamboo 47467,297271
Erect Bedstraw 170372
Erect Beefwood 79185
Erect Berula 53157
Erect Betony 373122
Erect Brome 60707
Erect Bugle 13104
Erect Bulbophyllum 63160
Erect Bur-head 140539
Erect Cephalanthera 82044
Erect Chamaerhodos 85651
Erect Cinquefoil 312520,312912
Erect Clematis 95277
Erect Climbing-bamboo 249606
Erect Common Juniper 213725
Erect Common Yew 385310
Erect Cone Pine 299827
Erect Corydalis 106488
Erect Craneknee 221758
Erect Cranesbill 174865
Erect Cryptostylis 113853
Erect Cudweed 178101
Erect Curlylip-orchis 63160
Erect Dayflower 101008
Erect Dilepyrum 58452
Erect Dysophylla 139558
Erect Ephedra 146136
Erect Erysimum 154467
Erect Evax 193379
Erect Fescue 164333
Erect Fig 164671,164947
Erect Filmcalyx 292658
Erect Flower Cassiope 78640
Erect Glorybind 103206

Erect Goldenrod 368095
Erect Green Cedar 85278
Erect Hedgeparsley 392992
Erect Hedge-parsley 392992
Erect Heterosmilax 366588
Erect Hiddenflower Woodbetony 287130
Erect Hybrid Yew 385390
Erect Hypecoum 201605
Erect Japanese Barberry 52234
Erect Knotgrass 308820
Erect Knotweed 309094
Erect Lobelia 234442
Erect Magnific Deutzia 127007
Erect Marshweed 230294
Erect Meadowrue 388683
Erect Metalleaf 296381
Erect Milkvetch 42594
Erect Minorrose 85651
Erect Mountain Clematis 94709
Erect Nepeta 264915
Erect Norway Maple 3432
Erect Pacific Dogwood 105131
Erect Periwinkle 409329
Erect Pimpinella 299539
Erect Primrose Willow 238166
Erect Pseudosasa 318335
Erect Raspberry 339295
Erect Rattanpalm 65677
Erect Rest Harrow 271347
Erect Rhododendron 330691
Erect Rockjasmine 23165
Erect Seaberry 185002
Erect Sedge 75984
Erect Silky Leather-flower 95173
Erect Skullcap 355346
Erect Spiderling 56437
Erect St. John's Wort 201853
Erect Stickseed 221758
Erect Tunicflower 292658
Erect Twocolored Monkshood 5026
Erect Vinegentian 110328
Erect Wallgrass 284145
Erect Wildcoral 170044
Erectcalyx Gentian 173420
Erectcalyx Skullcap 355641
Erect-cone Aleppe Pine 299827
Erect-cone Aleppo Pine 299827
Erectflower Cassiope 78640
Erectfruit China Spiraea 371885
Erecthair Crabgrass 130783
Erecthair Swertia 380199
Erect-leaved Plantain Lily 198643
Erectsepal Rockfoil 349296
Erectspine Sanicle 345988
Erectstem Rhodiola 329932
Erectthrum Chirita 87937
Erecttop Wintercress 47949
Eremophilous Krylovia 218245
Eremopogon 148392
Eremopyrum 148398
Eremosparton 148441,148444
Eremostachys 148466
Eremotropa 258055
Erewort 358062
Erguna Sedge 73750
Erhai Spikesedge 143142
Erhizomatose Spikesedge 143058
Eria 148619
Eriachne 148808
Erianthous Cactus 244071
Erianthous Kirilowia 216487
Erianthous Persimmon 132145
Erica 150229
Ericaleaf Lambertia 220307
Erie Viburnum 407787
Erigeron 150414
Erima 268819
Eringo 154276,154327
Erinus-like Koellikeria 217598
Erinuslike Rockfoil 349297
Eriocalyx Fleabane 150621
Eriocalyx Gentian 150621
Eriocarpous Glochidion 177123
Eriocarpous Jewelvine 125959
Eriocarpous Morningglory 207780
Eriocarpous Willow 343342
Eriocaulon 151208
Eriocephalous Argyreia 32621
Eriocladous Willow 343344
Eriocycla 151746
Eriodes 151767,151768
Erioglossum 151782
Eriogonum 151792
Eriolaena 152683
Eriophyllum 152799
Eriophyton 152838
Eriopodium Chinese Groundsel 365046
Eriosema 152853
Eriosolena 153101
Eriostachys Elsholtzia 144011
Eriostachys Grevill 180595
Eriostamen Cranesbill 174589
Erismanthus 153396
Eritrichium 153420
Erlangshan Larkspur 124206
Erlangshan Sagebrush 35472
Erlian Peashrub 72222
Erlian Pea-shrub 72222
Erman's Birch 53411
Erna Blastus 55153
Ernest Fir 346
Ernest Milkvetch 42348
Ernest Sweetleaf 381291
Ernest Willow 343351
Ernst Spapweed 204935
Ernut 63475
Eroded Rockfoil 349298
Eroded Saxifrage 349298
Erodium 153711
Erose Azalea 330638
Erose Jadeleaf and Goldenflower 260413
Erose Martinezia 12773
Erose Mussaenda 260413
Erose Primrose 314347
Erose Rhododendron 330638
Erose Viburnum 407802
Erose Wuyishan Clethra 96556
Erosepetalous Stonecrop 356658
Erostrate Wintergreen 172136
Eroton Oil Plant 334435
Errick 31051
Errie 3978
Errif 170193
Errucose Largeleaf Trigonotis 397432
Ersmart 309199
Ertix River Poplar 311188
Eru 178525
Erubescent Miner's-lettuce 94356
Ervatamia 154139,382727
Ervatamialike Glorybower 96059
Ervatamia-like Glorybower 96059
Erycibe 154233
Eryngo 154276,154280
Erysimum 154363
Erythea 59095,154571
Erythropalum 154956
Erythropalum Family 154952
Erythrophleum 154962
Erythropsis 154998
Erythrorchis 155011,155012
Erythrospermum 155032
Eryuan Barberry 51856
Eryuan Corydalis 106473
Eryuan Cystopetal 320231
Eryuan Goldenray 229082
Eryuan Larkspur 124543
Eryuan Rockfoil 349785
Eryuan Sage 345155
Esa African Celtis 80794
Escabon 85395
Escallonia 155130,155141
Escallonia Family 155155
Escarole 90894
Eschallot 15170
Eschauzier's Pincushion 244013
Eschscholtz Saxifrage 349300
Escoba Palm 113292
Escobilla 182107
Escobita 275499
Esetose Rose 336887
Eskimo 408203

Esmeralda Rubber 346393
Esmeralda Wild Buckwheat 152050
Esp 311537
Espadin 10942
Esparcet 271166,271280
Esparsette 271280
Esparto Grass 239310,376931
Esparto Needle Grass 376931
Esparto Needle-grass 376931
Esparto-grass 239310,376931
Esperance Waxflower 85852
Espercet 187842
Espibawn 227533
Espin 311537
Espina De Santo Antonio 290714
Espina Del Guanaco 242719
Espino 1128
Espino-caven 1128
Esquiral Tolypanthus 392713
Esquirol Aspidopterys 39674
Esquirol Bauhinia 49079
Esquirol Biophytum 54539
Esquirol Bredia 59839
Esquirol Buckthorn 328695
Esquirol Burretiodendron 64026
Esquirol Clethra 96481
Esquirol Coleus 99573
Esquirol Corydalis 105863
Esquirol Daphne 122426
Esquirol Decaspermum 123451
Esquirol Deutzia 126924
Esquirol Dischidia 134035
Esquirol Fig 164964
Esquirol Flamenettle 99573
Esquirol Heterostemma 194158
Esquirol Indigo 205962
Esquirol Jadeleaf and Goldenflower 260415
Esquirol Leptopus 226328
Esquirol Macaranga 240256
Esquirol Mayte 182692
Esquirol Mussaenda 260415
Esquirol Nymphaea 267682
Esquirol Ottelia 277390
Esquirol Parnassia 284531
Esquirol Persimmon 132147
Esquirol Plagiopetalum 301667
Esquirol Pogostemon 306971
Esquirol Pricklyash 417226
Esquirol Prickly-ash 417226
Esquirol Primrose 314351
Esquirol Slantpetal 301667
Esquirol Sotvine 194158
Esquirol Spinestemon 306971
Esquirol Tetragoga 387151
Esquirol Tickclover 126329
Esquirol Water Lily 267682
Esquirol Yam 131580
Esquirol's Mallotus 243355
Essia 292466
Esterledges 308893
Esteve Pincushion 84524
Esteve's Pincushion 84524
Esthwaite Waterweed 200184,200190
Estoril Thrift 34560
Estragon 35411
Estrella Begonia 49820
Estrigose Featherleaf Ghost Lampstand 335149
Estrigose Featherleaf Rodgersflower 335149
Estuarine Sedge 75984
Estuary Sea-blite 379521
Etenanthe 113944
Eteng 321967
Eternal Flower 189814
Ethiopia Chasmanthe 86121
Ethiopia Eggplant 366920
Ethiopian Bamboo 278646
Ethiopian Banana 145846
Ethiopian Climbing Bamboo 278646
Ethiopian Cumin 19672
Ethiopian Dogstooth Grass 117853
Ethiopian Egg-plant 366920
Ethiopian Niger Seed 181873
Ethiopian Niger-seed 181873
Ethiopian Pepper 415762
Ethiopian Rape 59348,59438
Ethiopian Rattlebox 111970,111972
Ethiopian Sour Bread 7022
Ethiopian Wheat 398836
Ethulia 155354
Etnagh-berries 213702
Etruscan Honeysuckle 235771
Ettley 402886
Etuoke Dragonhead 137609
Etuoke Greenorchid 137609
Etuoke Milkvetch 42831
Euaraliopsis 59187
Eucador Laurel 104185
Eucalypt 24603,106788,155468
Eucalyptus 155468,155589,155682,155749
Eucalyptus Indian Kino 320307
Eucalyptus Raspberry 338366
Eucalyptusleaved Maple 2950
Eucalyptuslike Maple 2950
Eucalyptus-like Maple 2950
Eucharis 155857
Eucharis Lily 155857,155858
Euchresta 155888
Euchresteleaf Jasminorange 260164
Euchrest-leaved Jasmin-orange 260164
Euclidium 155984
Eucommia 156040,156041
Eucommia Family 156042
Eucryphia 156058
Eucrypta 156073
Eufah 148356
Eugeissoma 156113
Eugene Poplar 311146,311149
Eugenia 156118,382465
Eugenia Ormosia 274389
Eugenialeaf Ormosia 274389
Eugenia-leaved Leptospermum 226468
Eugenia-leaved Ormosia 274389
Eugh 385302
Eula's Aster 380881
Eulalia 156439,156471,255827,255886,255891
Eulalia Grass 255886,341887
Eulalia-like Silvergrass 255843
Eulaliopsis 156511
Eulophia 156521
Eumong 1623
Euodia 161302
Euonymus 157268,157473
Euonymus Leaf Alyxia 18516
Euonymusleaf Azalea 330622
Euonymusleaf Camellia 69058
Euonymusleaf Rhododendron 330622
Euonymusleaf Syzygium 382532
Euonymus-leaved Rhododendron 330622
Euonymus-leaved Syzygium 382532
Eupator's Agrimony 11549
Eupatorium 101955,158021
Eupatorium-like Inula 138892
Euphorbia 158385,158640,159319,159997
Euphorbium 159710
Euphrasia Poplar 311308
Euphrasy 160128,160236
Euphrates Aspen 311308
Euphrates Balsam Tree 311308
Euphrates Balsam-tree 311308
Euphrates Poplar 311308
Euptelea 160338
Euptelea Family 160349
Eurabbie 155500
Eurasia Centaurea 81345
Eurasia Greenbrier 366240
Eurasia Scurrula 236704
Eurasia Stink Groundpine 70463
Eurasian Catchfly 363827
Eurasian Chestnut 78811
Eurasian Greenbrier 366240
Eurasian Mountain-ash 369339
Eurasian Nailwort 284840
Eurasian Smilax 366240
Eurasian Smoke Bush 107300
Eurasian Solomon's-seal 308607
Eurasian Water Milfoil 261364
Eurasian Watermilfoil 261364
Eurasian Water-milfoil 261364
Eurasian Water-plantain 14760

Eurasian White Adder's-mouth 243084
Eurasian Wild Plum 316470
Eurasiatic Spiraea 372006
Eureka 123555
Eureka Lily 229967
Euro-asiatic Centaurea 81345
Europe 305183
Europe Arrowwood 407989
Europe Barberry 52322
Europe Biglef Poplar 311259
Europe Birdcherry 280003
Europe Bladdernut 374107
Europe Box 64345
Europe Bugbane 91010
Europe Cherry 83155
Europe Cowlily 267294
Europe Cowparsnip 192344
Europe Currant 334250
Europe Dodder 115031
Europe Dune Lymegrass 228346
Europe Eight Trasure 200812
Europe Euonymus 157429
Europe Forsythia 167430
Europe Globeflower 399504
Europe Glorybind 102933
Europe Goldenrod 368480
Europe Grape 411979
Europe Groundsel 360328
Europe Hop 199384
Europe Ironwood 276803
Europe Lily 229922
Europe Linden 391706
Europe Lymegrass 228346
Europe Madwort 18327
Europe Mountainash 369339
Europe Nettle 403039
Europe Nightshade 367120
Europe Pennywort 200388
Europe Pine 300036
Europe Plantain 302092
Europe Plum 316382
Europe Poplar 311398,311537
Europe Privet 229662
Europe Sage 345379
Europe Sainfoin 271280
Europe Smallreed 127257
Europe Smoothleaf Elm 401468
Europe Spruce 298191
Europe Star Flower 396786
Europe Starflower 396786
Europe Strawberry 167653
Europe Sweetgrass 177648
Europe Toadflex 231183
Europe Vervain 405872
Europe Vetch 408666
Europe Vetchling 222800
Europe Waterhemlock 90932
Europe White Birch 53563
Europe White Elm 401549
Europe White Waterlily 267648
Europe Willow 343937
Europe Winteraconite 148104
Europe Yew 385302
European Agrimonia 11549
European Agrimony 11549
European Alder 16352,16368
European Alder Buckthorn 328603,328701
European Alkali Grass 321248,321249
European Aper Birch 53563
European Asarum 37616
European Ash 167955
European Aspen 311537
European Barberry 52322
European Beach Grass 19766
European Beachgrass 19766
European Beach-grass 19766
European Bedstraw 170597
European Beech 162400
European Bellflower 69921,70255
European Birch 53563,53590
European Bird Cherry 280003,280044
European Bird-cherry 280003
European Bishop 54240
European bistort 54798
European Bistort 308893
European Bitter Night Shade 367120
European Bittersweet 367120
European Black Alder 16352
European Black Currant 334117
European Black Elderberry 345631
European Black Hawthorn 109868
European Black Pine 300117
European Black Poplar 311398
European Blackberry 338447,339446
European Bladdernut 374107
European Blue Lupin 238450
European Blue Lupine 238450
European Blueberry Willow 343982
European Box 64345
European Boxthorn 239046
European Boxwood 64345
European Box-wood 64345
European Brome 60680,60707
European Brooklime 407029
European Broom 121001
European Buckeye 9701,64345
European Buckthorn 328642
European Bugbane 91010
European Bugle 13174
European Bugleweed 239202
European Bur-grass 394390
European Burning Bush 157429
European Bur-reed 370044
European Centaury 81495
European Cherry 83155
European Chestnut 78811
European Chickweed 82994,375036
European Chrysopogon 90125
European Columbine 30081
European Corn Salad 404439
European Cornsalad 404402,404472
European Cotoneaster 107504
European Cow Lily 267294,267320
European Cowlily 267320
European Cow-lily 267294
European Crabapple 243711
European Cranberry 278351,278366,407989
European Cranberry Bush 407989
European Cranberry Viburnum 407989
European Cranberrybush 407989
European Cranberry-bush 407989
European Cranberrybush Viburnum 407989
European Craneknee 221711
European Currant 334250
European Cyclamen 115949
European Date Plum 132264
European Date-plum 132264
European Dewberry 338209
European Dodder 114972
European Dogwood 380499
European Dune Leymus 228346
European Dune Wild Rye 228346
European Dune Wildrye 228346
European Elder 345631
European Elm 401593
European Euonymus 157429
European Fan Palm 85665,85671
European Fan-palm 85671
European Feather Grass 376877
European Featherfoil 198680
European Feathergrass 376877
European Feather-grass 376877
European Field Pansy 409703
European Filbert 106703
European Fir 272
European Firethorn 322454
European Fireweed 146724
European Fly Honeysuckle 236226
European Fly-honeysuckle 236226
European Forsythia 167430
European Frog-bit 200230
European Frogs-bit 200230
European Frostweed 399957
European Ginger 37616
European Globe Flower 399504
European Globeflower 399504
European Glorybind 102933
European Golden Ball 167430
European Goldenrod 368480
European Goldilocks 231844
European Gooseberry 334250,334257

European Gorse 401388
European Grape 411979
European Green Alder 16472
European Gromwell 233760
European Ground Cherry 83202
European Gymnadenia 182216
European Hackberry 80577
European Hawkweed 195708,195710
European Hawthorn 109790
European Hazel 106703
European Hazelnut 106703
European Heliotrope 190622
European Highbush Cranberry 407989
European Highbush-cranberry 407989
European Holly 203545
European Honeysuckle 236022
European Hop 199384
European Hop Hornbeam 276803
European Hop-hornbeam 276803
European Hornbeam 77256
European Horse Chestnut 9701
European Horse-chestnut 9701
European Knotweed 308782
European Lady's Slipper 120310
European Lady's-slipper 120310
European Ladyslipper 120310
European Lady-slipper 120310
European Larch 221855
European Leadwort 305183
European Lily of the Valley 102867
European Lily-of-the-valley 102867
European Lime 391706
European Linden 391691,391706
European Little-leaved Linden 391691
European Madwort 18327
European Mannagrass 177648
European Marshwort 29341
European Meadow Rush 213165
European Meadowsweet 166125
European Milkvetch 42461
European Mistletoe 410960
European Mock-orange 294426
European Monkshood 5442
European Motherwort 224976
European Mountain Ash 369339
European Mountainash 369339
European Mountain-ash 369339
European Needle Grass 376877
European Needle-grass 376877
European Nettle Tree 80577
European Nettle-tree 80577
European Oak 324335
European Olive 270099
European Osier 343937
European Pasque Flower 321721
European Pasqueflower 321721
European Pasque-flower 321653
European Pear 323133
European Pedicularis 287494
European Pepperwort 225398
European Planetree 302582
European Plum 316382
European Plum Gristlings 316470
European Potato 367696
European Privet 229569,229662
European Pussywillow 343151
European Pyrola 322872
European Raspberry 338209,338557
European Red Elder 345660
European Red Elderberry 345660
European Red Raspberry 338557
European Sanicle 345957
European Scopolia 354686
European Scurrula 236704
European Seaheath 167813
European Searocket 65185
European Seaside Plantain 302091,302092
European Shoreweed 234150
European Shrubby Horsetail 146154
European Silver Fir 272
European Slough-grass 49540
European Small Reed 127257
European Smoketree 107300
European Snakeroot 34149
European Snowball 407989
European Speedwell 407029
European Spindle 157429
European Spindle Tree 157429,157435,
157436
European Spindletree 157429
European Spindle-tree 157429
European Spruce 298191
European Starflower 396786
European Stickseed 221711,221755
European Stonecrop 356979
European Stoneseed 233760
European Strawberry 167653
European Sumach 332533
European Swallow-wort 117670
European Swamp Thistle 92285
European Tea 407256
European Thimbleweed 23925
European Touch-me-not 205175
European Turk's-cap Lily 229922
European Turkey Oak 323745
European Umbrella Milkwort 392672
European Venus' Looking Glass 224073
European Verbena 405872
European Vervain 405872
European Vetch 408378
European Violet-willow 343280
European Wallflower 154467
European Walnut 212636
European Wand Loosestrife 240100
European Water Hemlock 90932
European Water Plantain 14760
European Water-milfoil 261364
European Waterplantain 14760
European Waythorn 328642
European Weeping Birch 53563
European White Birch 53563,53590
European White Elm 401549
European White Hellebore 405571
European White Lime 391836
European White Poplar 311208
European White Water Lily 267648
European White Willow 342973
European Wild Ginger 37616
European Wildginger 37616
European Willow 342973
European Wolfberry 239046
European Wood Anemone 23705,23925
European Wood Rush 238640
European Wood Sorrel 277648
European Wood-anemone 23925
European Woodland Sedge 76465
European Yellow Iris 208771
European Yellow Lupin 238463
European Yellow Lupine 238463
European Yellow Violet 410187
European Yellowrattle 329543
European Yew 385302
Eurya 96576,160405
Euryale 160635
Euryaleaf Holly 203791
Euryaleaf Sweetleaf 381199
Euryaleaf Tea 69064
Eurya-leaved Sweetleaf 381199
Euryalike Camellia 69064
Eurya-like Camellia 69064
Euryal-leaved Holly 203791
Eurycoma 160716
Euryocrymbus 160720,160722
Euryodendron 160741
Euryops 160853
Eurysolen 160916
Euscaphis 160946,160954
Euscaphis Euonymus 157453
Euscaphis Spindle-tree 157453
Eustachys 160979
Eustigma 161020
Euterpe Palm 161053
Eutrema 161120
Evacuate Arrowbamboo 162661
Evans Begonia 49886
Evax 161220
Eve's Apple 154161,327577
Eve's Apple Tree 154161
Eve's Cup 347156
Eve's Cups 347156
Eve's Cushion 349496

Eve's Needle 273067
Eve's Tears 169719
Evelynia Keteleeria 216134
Even Leea 223922
Evening Campion 363671,363673,363814
Evening Christia 89209
Evening Close 363673
Evening Lychnis 363673
Evening Pride 236022
Evening Primrose 159069,269394,269404,269422,269443,269445,269448
Evening Primrose Family 270773
Evening Snow 175683
Evening Star 103674,269404,269441
Evening Trumpet Flower 172781
Evening Twilight 23925,277648
Eveningprimrose 269394
Evening-primrose 68502,269394,269404
Eveningprimrose Family 270773
Evening-primrose Family 270773
Ever 101852
Everard Catnip 264918
Everard Nepeta 264918
Everbearing Grape 411811
Everbearing Strawberry 167653
Everblooming Acacia 1528
Everblooming China Rose 336495
Everblooming Chinese Rose 336495
Everblooming French Heath 149363
Everblooming Gardenia 171259
Everblooming Honeysuckle 235828
Ever-blooming Horned Violet 409990
Ever-blooming Iris 208797
Everest Barberry 51597
Ever-flowering Gladiolus 176601
Everflowering Rose 336485
Everglade Milkvetch 43204
Everglade Windhairdaisy 348892
Everglades Palm 4949,4951
Evergold Striped Weeping Sedge 75636
Evergreen Ailanthus 12570
Evergreen Alder 16315,16396
Evergreen Alkanet 289503
Evergreen Arrowwood 408123
Evergreen Ash 168124
Evergreen Azalea 331088
Evergreen Bayberry 261169
Evergreen Blackberry 338695
Evergreen Blueberry 403794,403914
Evergreen Buckthorn 328595
Evergreen Bugloss 289503
Evergreen Burnet 345881
Evergreen Candytuft 203239
Evergreen Cherry 316452
Evergreen Chinese Chickweed 374903
Evergreen Chinkapin 78848
Evergreen Chinkapiu 78848
Evergreen Clematis 94748,95431
Evergreen Clivers 338007
Evergreen Coral Bean 369116
Evergreen Deutzia 127049
Evergreen Dogwood 124891
Evergreen Elm 401552
Evergreen Enkianthus 145714
Evergreen Euonymus 157601
Evergreen Four-involucre 124891
Evergreen Grape 332324,411600
Evergreen Helictotrichon 190196
Evergreen Honeysuckle 236089
Evergreen Huckleberry 403944
Evergreen Laburnum 300691
Evergreen Magnolia 242135
Evergreen Maple 3255
Evergreen Mock Orange 294485
Evergreen Mockorange 294500
Evergreen Mortonia 259055
Evergreen Mucuna 259566
Evergreen Oak 324027,324515
Evergreen Pear 323207
Evergreen Pendent-bell 145714
Evergreen Pistache 300994
Evergreen Privet 229613
Evergreen Pussytoes 26605
Evergreen Raspberry 338656
Evergreen Rose 336925
Evergreen Saxifrage 349054
Evergreen Spindle 157601
Evergreen Spindle-tree 157601
Evergreen Stonecrop 294122
Evergreen Sumac 332850,332956,332957
Evergreen Sweetspire 210384
Evergreen Tamarisk 383437
Evergreen Tree Viburnum 407977
Evergreen Viburnum 407756,408056,408089,408123
Evergreen Violet 410558
Evergreen White Oak 323874
Evergreen Winterberry 203847
Evergreen Wisteria 66643
Evergreenchinkain Mallotus 243425
Evergreenchinkapin 78848
Everlasting 21506,178056,189093,190791,317756,415386
Everlasting Fire Thorn 322454
Everlasting Flower 189093,415386
Everlasting Friendship 170193
Everlasting Grape 411811
Everlasting Grass 271280
Everlasting Neststraw 379160
Everlasting Pea 222671,222724,222750,222844
Everlasting Peavine 222844
Everlasting Persimmon 132089
Everlasting Pussytoes 26605
Everlasting Sand Flower 19691
Everlasting Thorn 322454
Evermann's Fleabane 150627
Evermann's Pincushion 84494
Everocks 338243
Everred Japanese Maple 3309
Eversmann Cinquefoil 312530
Eversmann's Tamarisk 383484
Eversmannia 161294
Everspringtree 250873,250876
Every-grass 235334
Everywhere St. John's Wort 202217
Eves-necklace 368958
Eveweed 193417
Evil Windhairdaisy 348509
Evodia 161302
Evodialeaf Acanthopanax 2387
Evodia-leaved Acanthopanax 2387
Evolvulus 161416
Evonymus 157268
Evonymusleaf Syzygium 382532
Evrard Phreatia 295958
Evron 338243
Ewan's Larkspur 124275
Ewbank Iris 208553
Ewe 385302
Ewe Bramble 338447
Ewe Daisy 312520
Ewe Gollan 50825
Ewe Gowan 50825
Ewe-bramble 338447
Ewe-daisy 312520
Ewe-gollan 50825
Ewe-gowan 50825
Ewers Stonecrop 200788
Ewfrace 160236
Ewfras 160236
Ewgh 385302
Exacum 161528,161531
Exalate Hopea 198189
Exaltate Ophiuros 272350
Exalted Snaketailgrass 272350
Exaristate Blueberry 403824
Exbucklandia 161605
Excavated Gasteria 171666
Excellence Azalea 331532
Excellent Azalea 330452,331245
Excellent Barberry 52059
Excellent Clearweed 298942
Excellent Coldwaterflower 298942
Excellent Goniolimon 179374
Excellent Rhododendron 330666,331532,331824
Excelsa Rose 336289
Excentricflower Indigo 206057
Excentricflower Primrose 314936
Excentricflowers Rabdosia 209824

Excentric-vein Eargrass 187634
Excisalike Rabdosia 209663
Excoecaria 161628
Excurrent Bristlegrass 361737
Exellent Begonia 49804
Exellent Paphiopedilum 282887
Exellent Woodbetony 287197
Exeter Elm 401514
Exile Tree 389978
Expanded Lobster Claw 190024
Expanded Lobsterclaw 190024
Expanded Panicgrass 281918
Expandtwig Knotweed 309531
Expanse Gentian 173428
Expansile Privet 229458
Explanate Ironweed 406320
Explorer's Gentian 173316
Exposedpetal Lily 266549
Exposite Fargesia 162685
Exquisite Rhododendron 331411
Exsert Acianthus 4512
Exserted Hanceola 185247
Exserted Indian Paintbrush 79124
Exserted Onosma 271758
Extend Gentian 173740
Extendedtwig Indosasa 206895
Extensed Wingfruitvine 261388
Extensed Wing-fruit-vine 261388
Extension-flowered Elegant Barberry 51485
Extented Clematis 94879
Extented Sunken Sedge 74883
Extracuspidate Tea 69477
Extralongstalk Tea 69309
Extremeeast Feathergrass 4134
Extremeeast Peashrub 72237
Extremely Azalea 330936
Extremely Rhododendron 330936
Extremi-oriental Fescue 163950
Extremioriental Jijigrass 4134
Extremioriental Speargrass 4134
Exvellent Rhododendron 330668
Eyan 237927
Eye of Christ 407072
Eye of Day 50825
Eye Poppy 282515
Eyebane 159286
Eye-bane 85779
Eyeberry 339124
Eyebright 21339,21959,85875,146834,160128,160133,160233,160236,198708,234547,282685,374916,374944,407072,410677
Eyebright Cow-wheat 268991
Eyebrightlike Striga 377999
Eyebrow Bambusa 47494
Eyebrow Leaves Cassiope 78625
Eyed Bee-orchid 272404
Eye-form Callalily 197794
Eyeglasses 300223
Eyelash Begonia 49669
Eyelash Bulbophyllum 63185
Eyelash Crabgrass 130489
Eyelash Curlylip-orchis 63185
Eyelash Grass 57930
Eyelash Rockjasmine 23146
Eyelashleaf Stonecrop 356572
Eye-of-the-sun Tulip 400200
Eyeseed 345458
Eyeshaped Dendrobium 125135
Eye-spot Kohleria 217747
Eye-spotted Achimenes 4085
Eyong 376150
Eyre Chinquapin 78927
Eyre Evergreen Chinkapin 78927
Eyre Evergreenchinkapin 78927
Eyries Sea-urchin Cactus 140860
Ezo-yama-hagi 226698

F

F. C. How Ixora 211090
F. C. How Trigonostemon 397364
F. G. Webster Rhododendron 332098
F. P. Metcalf Cryptocarya 113470
F. P. Metcalf Maple 3143
F. P. Metcalf Persimmon 132300
F. P. Metcalf Thickshellcassia 113470
Faasen's Catmint 264920
Fabe 334250
Faber Alangium 13362
Faber Ardisia 31443
Faber Aster 40410
Faber Bamboo 297273
Faber Barberry 51601
Faber Bauhinia 49022
Faber Bluegrass 305513
Faber Bristlegrass 361743
Faber Chinquapin 78932
Faber Clethra 96496
Faber Cymbidium 116851
Faber Densevein 261326
Faber Deutzia 126926
Faber Dragonheadsage 247722
Faber Eulophia 156651
Faber Evergreen Chinkapin 78932
Faber Evergreenchinkapin 78932
Faber Fir 349
Faber Fordiophyton 167299
Faber Juteleaf Raspberry 338285
Faber Litse 233901
Faber Maple 2951
Faber Meadowrue 388495
Faber Meehania 247722
Faber Oak 323881
Faber Orchis 116851
Faber Parnassia 284532
Faber Phoebe 295360
Faber Primrose 314358
Faber Rapanea 261621
Faber Raspberry 338384
Faber Rattan Palm 65682
Faber Rattanpalm 65682
Faber Rattlesnakeroot 313820
Faber Rhododendron 330672
Faber Snowbell 379336
Faber Spapweed 204942
Faber St. John's Wort 201874
Faber Storax 379336
Faber Vineclethra 95489
Faber Wildcoral 170035
Faber's Carpesium 77160
Faber's Groundsel 358848
Faberdaisy 161881
Faberia 161881
Faberiopsis 161893
Faberry 334250
Fabiana 161896,161897
Fabirama 99699
Face-and-hood 410677
Face-in-hood 5442
Fachno 290704
Faddy-tree 3462
Fadou Burmannia 63972
Fadou Oak 116069
Fadou Qinggang 116069
Fadyen's Silktassel 171549
Faeberry 334250
Fagaropsis 162178
Fagerlindia 162193,162196
Fagonia 162216
Fagopyrum 162312
Fagraea 162343
Faham 24860
Faham Tea 24860
Fair Days 312360
Fair Gentian 173386
Fair Grass 312360,325666
Fair Indigo 205876

Fair Jerusalemsage 295165
Fair Lady 44708
Fair Maid of France 4005,227533,325516,349419,363461
Fair Maid of Kent 325516
Fair Maids 169719,349419
Fair Maids of February 169719
Fair Maids of France 4005,325516,349419,363461
Fair Maids of Kent 325516
Fair Maids-of-france 325516
Fair Masdevallia 246103
Fair Meconopsis 247190
Fair Weather Thistle 76990
Fairenther Rinorea 334618
Fairgreen Sea-urchin Cactus 140852
Fairies 37015
Fairleaf Addermonth Orchid 110640
Fairway 11696
Fairway Wheatgrass 11696
Fairy Basin 315082,325518,325666,326287
Fairy Bells 70271,102867,130383,134417,134442,262441,277648,315082
Fairy Boats 267294
Fairy Cap 130383
Fairy Caps 130383
Fairy Cheese 243823,243840
Fairy Cheesecake 247366
Fairy Cheese-cake 247366
Fairy Clock 384714
Fairy Cups 159027,159557,315082
Fairy Dell 159027,159557
Fairy Dresses 130383
Fairy Duster 66669
Fairy Fan Flower 350332
Fairy Fanflower 350324
Fairy Fans 94134
Fairy Fingers 130383
Fairy Flax 160236,231884,248411
Fairy Foxglove 151098,151102
Fairy Grass 60382
Fairy Hair 115008
Fairy Hat 130383
Fairy Hats 130383
Fairy Horse 359158
Fairy Lamps 37015
Fairy Lantern 67554,67555,134417,134439,134494
Fairy Lanterns 67555,134494
Fairy Lily 230038,417606,417608,417612
Fairy Lint 231884
Fairy Loeseneriella 235213
Fairy Petticoats 130383
Fairy Pops 397019
Fairy Potato 102727
Fairy Potatoes 102727
Fairy Primrose 314613
Fairy Queen 410677
Fairy Ringers 70271
Fairy Rose 336485,336490,336542
Fairy Rose Bush 336490
Fairy Soap 308454
Fairy Table 200388
Fairy Tables 200388
Fairy Thimbles 69970,70271,130383
Fairy Trumpet 68713
Fairy Trumpets 68713,236022
Fairy Wallflower 154363
Fairy Wand 85505
Fairy Wax Flower 294869
Fairy Webseedvine 235213
Fairy Weed 130383
Fairy Wings 146954
Fairy Wives' Distaff 401112
Fairy Woman's Flax 231884
Fairy's Basin 315082,325518,325666,326287
Fairy's Bath 175444
Fairy's Bed 355202
Fairy's Beds 355202
Fairy's Bells 145487
Fairy's Broom 133478
Fairy's Cap 70271
Fairy's Clock 8399
Fairy's Corn 222758
Fairy's Cup 70271,315082
Fairy's Dresses 130383
Fairy's Fire 84573,133478
Fairy's Gloves 130383
Fairy's Lamp 37015
Fairy's Lanterns 231183
Fairy's Paintbrush 409339
Fairy's Paintbrushes 409339
Fairy's Petticoats 130383
Fairy's Thimbles 69970,70271
Fairy's Umbrella 102933
Fairy's Umbrellas 102933
Fairy's Wand 11549,405788
Fairy's Wind Flower 23925
Fairy's Windflower 23925
Fairy's Winecup 102933
Fairy's Winecups 102933
Fairybells 134417,134425
Fairy-bells 315524
Fairy-carpet Begonia 50392
Fairy-duster 66669
Fairy-lantern 315530
Fairy-slipper 68531
Fairy-slipper Orchid 68538,68541
Fairyvine 164456
Fairy-wand 85505
Faith Rhododendron 330676
Fake Akebia 211694
Fake Basket Willow 344071
Fake Boxleaf Syzygium 382498
Fake Bubbleleaf Bladderwort 403969
Fake Chestnut 366055
Fake Earthgall 317231,317233
Fake Flabellateleaf Willow 343830
Fake Glaucous Poplar 311438
Fake Hankow Willow 343678
Fake Lobelia 234291
Fake Needlegrass 318193
Fake Orientvine 245860
Fake Pruinashoot Willow 344028
Fake Pubescent Willow 343926
Fake Sawwort 361023
Fake Simon Poplar 311436
Fake Sterculia 376130
Fake Tang Willow 343928
Fake Wallich Willow 343929
Fake Woloho Willow 343930
Fake Woodbean 125574,125580
Fake Woollypistil Willow 343918
Fake-macro-petiole 203775
Fakesnaketailgrass 193988
Fakestellera 375207
Falcaria 162457
Falcate Acacia 1217
Falcate Cephalanthera 82051
Falcate Ceratocephalus 83439
Falcate Clematis 95332
Falcate Crateva 110215
Falcate Crazyweed 278825
Falcate Cryptospora 113817
Falcate Decorated Kinostemon 216459
Falcate Dimeria 131008
Falcate Fox-tail Orchid 9286
Falcate Gonggashan Monkshood 5368
Falcate Grewia 180773
Falcate Hare's-ear 63625
Falcate Helicia 189930
Falcate Leaf Aster 40416
Falcate Ligusticum 229321
Falcate Micromelum 253627
Falcate Milkvetch 42355
Falcate Monkshood 5182
Falcate Paraserianthes 162473
Falcate Pigeonwings 97188
Falcate Pogostemon 306973
Falcate Sedge 74483
Falcate Solomonseal 308538
Falcate Spapweed 204943
Falcate Spurge 158876
Falcate Tickseed 104776
Falcate Watercandle 306973
Falcate Yellowwood 306439
Falcate-auriculate Yushania 416773
Falcated-leaved Asparagus 39009
Falcateleaf Asparagus 38997
Falcateleaf Clearweed 299043

Falcateleaf Gagea 169417
Falcateleaf Helicia 189930
Falcateleaf Iris 208537
Falcateleaf Kobresia 217164
Falcateleaf Madder 337955
Falcateleaf Sarcochilus 346697
Falcate-leaved Albizzia 162473
Falcate-leaved Blueberry 404019
Falcate-leaved Rhododendron 330678
Falcatepetal Oncidium 270808
Falcate-pod Milkvetch 42021
Falcon's Claw Acacia 1490
Falconer Dendrobium 125128
Falconer Hepatica 192131
Falconer Milkvetch 42357
Falconer Sweetvetch 187868
Falconer Wildcoral 170036
Falconer's Mock Orange 294459
Falconer's Mock-orange 294459
Falconer's Rhododendron 330680
Falfalaries 168476
Fall Adonis 8332
Fall Coralroot 104009
Fall Coral-root 104009
Fall Crocus 99297
Fall Daffodil 376356
Fall Dandelion 224655
Fall Grape 411575
Fall Hawkbit 224655
Fall Knotweed 309813
Fall Meadow Rue 388621
Fall Meadow-rue 388621
Fall Panic Grass 281561
Fall Panicgrass 281561
Fall Panicum 281561
Fall Phlox 295288
Fall Poison 158305
Fall Witch Grass 130501
Fallacious Sugarcane 262548
Fallacious Tickclover 200740
Fall-daffodil 376356
Fall-dandelion 224655
Falleneaves 351146,351150
Fall-flowering ixia 263310
Falling Star 83545,111455,111458
Falling Stars 83545,111458,111466
Fallingcoin 183402
Fallingleaf Bulbophyllum 62779
Fallingleaf Stonebean-orchis 62779
Fallows Buddleja 62054
Falmeflower 295018,295022
Falscher Jasmin 294426
False Acacia 334946,334976
False African Violet 377818
False Agave 187137
False Agoseris 266823
False Alkanet 21928
False Aloe 17293,244377
False Alp Corydalis 106293
False Alpine Aster 40021
False Alpine Oak 324322
False Alpine Windhairdaisy 348676
False Alyssum 53034
False Amur Adonis 8378
False Anemone 24166,24167
False Apple-mint 250439
False Aquatic Gentian 173751
False Aralia 135079,135082,350688
False Arborvitae 390673,390680
False Arbor-vitae 390673
False Asphodel 392598,392608,392630
False Aster 56704,56705,56706,56713,
215358
False Baby's-breath 170490
False Baby-stars 175672
False Banana 145830
False Banks Rose 336379,336875
False Banyan 164626
False Bashfulgrass 265228,265234
False Bayleaf Willow 343834
False Beardgrass 90107
False Beardsepal Corydalis 106296
False Beautyberry 399851,399852
False Beech 266861
False Beech Drops 202767
False Beechdrops 202767
False Beech-drops 202767
False Bentfruit Tanoak 233235
False Bentspurred Larkspur 124517
False Biba Pepper 300495
False Bigstigma Rockfoil 349599
False Bigthyrse Rabdosia 324822
False Bird of Paradise 190052
False Bird-of-paradise 189986
False Bird-of-paradise Family 190073
False Bittersweet 80309
False Black Barberry 51747
False Blue Japanese Oak 116181
False Bluebell 250914
False boatshaped Hyparrhenia 201556
False Bog Rush 333504
False Boneset 60116,60119
False Boxleaf Syzygium 382498
False Boxwood 182199
False Bracteate Chelonopsis 86832
False Bristly Sedge 75895
False Broadbean Sweetvetch 188177
False Broadspike Flatsedge 322327
False Broadspike Pycreus 322327
False Brome 58572
False Brome Grass 58572
False Buck's-beard 41786,41817
False Buckthorn 362953
False Buckwheat 162519,162552,309796
False Bugbane 91034,394583,394584
False Bullate Blueberry 403969
False Bulrush 401112
False Burmaflame Rhododendron 331868
False Cactus 159204
False Calabash 383007
False Calumba Root 107120
False Camas 417899
False Camomile 397970
False Caper 159222
False Carraway 77825
False Carrot 79614
False Castor-off Plant 162879
False Chamae Iris 208778
False Chamomile 56700,246396,397964
False Changdu Corydalis 105813
False Chelone 266840
False Chestnutflower Rush 213403
False Chia Bugle 13165
False Chickweed 318489
False China Indocalamus 206821
False China Whie Poplar 311443
False China-root Greenbrier 366271
False Chinese Swertia 380319
False Chinese Taxillus 385225
False Chinese Whie Poplar 311443
False Christmas Cactus 418532
False Clearweed 299021
False Cleavers 170193,170205
False Coffee 133185
False Coldwaterflower 299021
False Columbine Corydalis 105623
False Corianda 368528
False Corn-flag 176122
False Cottony Olgaea 270244
False Cowwheat Woodbetony 287560
False Crane Catchfly 363932
False Crested Corydalis 106298
False Crusiate Rockvine 387838
False Curvetube Woodbetony 287557
False Cuspidate Elatostema 142799
False Cuspidate Stairweed 142799
False Cypress 85263,85266,85375
False Daisy 50809,141374
False Dalbergia 290877
False Dandelion 11408,11445,202432,
218202,253918,266823,322997
False Date-palm 295454
False Dayflower 392133
False Diamondflower 212476
False Dittany 46990,47018,129618,245770
False Dragonhead 297973,297974,297978,
297981,297984
False Dragon-head 297984
False Drake Corydalis 106301
False Duhat 382677
False Ebony 218761

False Elaeocarpus 142310
False Elephant's Foot 317231
False Elm 80698,80728
False Fairybells 134400
False Fan-shaped Willow 343830
False Farges Corydalis 106302
False Fennel 24213,158197,334477,334478
False Fern-like 287206
False Fern-like Woodbetony 287206
False Figleaf Holly 204285
False Flax 68839,68860
False Fleabane 321509
False Flexuose Bittercress 72764
False Foxglove 45168,45174,289337,375411
False Fox-sedge 75641
False Freesia 168176
False Freesia Lapeyrouse 221353
False Gapleaf Corydalis 106311
False Garlic 15876,266957,266959
False Giant Woodbetony 287558
False Glaucous Poplar 311438
False Goat's Beard 41786,41793
False Goatsbeard Astilbe 41793
False Golden Sedge 74626
False Goldeneye 190248,190253
False Goldenray 229261,229262
False Goldenrod 368432
False Grape 20334,20361
False Grass-poly 240053
False Grey-blue Cyclobalanopsis 116181
False Griffith Agapetes 10358
False Gromwell 271855,271861
False Harpetiole Willow 343918
False Headflower Woodbetony 287556
False Heath 161897
False Heather 114605,198955
False Hellebore 8332,8387,405570,405581,405643,405649
False Hemp 112269
False Henry Loosestrife 239797
False Henry St. John's Wort 202106
False Himalayan Berry 338456
False Holly 276313
False Holly-leaved Barberry 52081
False Hookspur Corydalis 106306
False Hop Sedge 75232
False Horn-fruited Spindle-tree 157523
False Houseleek 357144
False Impatiens Corydalis 106310
False Incised Monkshood 5577
False Indian Almond 264222
False Indigo 19989,19994,20005,47854,47856,47859,47931,205611,206061,206140,206445
False Indigo Bush 20005
False Ipecac 175760
False Ipecacuanha 334338
False Iris 208854
False Iroko 28243,28248
False Ironculm 317192
False Jacob's Ladder 307241
False Jagged-ckickweed 226621
False Jalap 255711
False Japan Listera 264726
False Japanese Elm 401600
False Jasmine 172781
False Jerusalem Cherry 367110,367528
False Juncus Corydalis 106312
False Kola 171121
False Lacteum Rhododendron 331640
False Largeflower Motherwort 225005
False Leafy Corydalis 106313
False Lemon 93707
False Lettuce 219321
False Lily of the Valley 242675,242699
False Lily-of-the-valley 242668,242675,242680
False Littleleaf Photinia 295750
False Little-leaved Photinia 295750
False Lobate Buttercup 326257
False Locust 187088
False London Rocket 365529
False Londonpride 349797
False London-rocket 365529
False Longstalk Corydalis 106314
False Loosestrife 238150,238171,239780
False Lubbers Begonia 50199
False Lupin 389505,389547
False Lupine 389505,389558
False Lychee 374392
False Lyrateleaf Loosestrife 239726
False Machilus-leaved Holly 204183
False Mallow 243877,362702
False Manyroot 283760
False Many-tooth Camellia 68908
False Maple 133596
False Maximowicz Poplar 311441
False Mayflower 242706
False Mayweed 397964,397970
False Medlar 369358
False Melic Grass 351251,351254
False Mercury 86970
False Mermaid 166994
False Mermaid-weed 166994
False Mesquite 66669
False Milk Parsley 357789
False Milkvetch 43123
False Milkvetch Sweetvetch 188062
False Miterwort 391522
False Mitrewort 391520,391525
False Morning Glory 190592
False Mountain Hairy Willow 343926
False Muscicolous Woodbetony 287561
False Muskroot Corydalis 106291
False Naked Wild Buckwheat 152314
False Nakedstem Corydalis 106312
False Nettle 56023,56094,56119,221547
False Nut Sedge 119624
False Nutgrass 119624
False Nutmeg 257653,261423,321973
False Nutsedge 119624
False Oat 34930,398437,398531
False Oat Grass 34930
False Oat-grass 34920,34930
False Obtuse-leaved Fig 165402
False Olive 142422
False Onion 266957
False Oroko 28248
False Pakchoi 59556
False Papillate Bambusa 47406
False Pappograss 397509
False Pareira Root 92523,92555
False Parsley 9832,90924
False Pellitory 4005,280347
False Pellitory-of-Spain 205535
False Pennyroyal 187176,396436
False Peristylus 291207
False Perotis 291207
False Peyote 33209
False Pimpernel 231473,231476,231514
False Pineaple 280993
False Pistache 384369
False Pleasing Barberry 52080
False Poinciana 361422
False Pothos 313169,313170
False Psilate Primrose 314851
False Queen Anne's Lace 19664
False Ragweed 210485
False Rampion 70255
False Rattlesnakeroot 283680,283695
False Reddish Beautyberry 66902
False Robust Vetch 408551
False Rockcress 44868
False rosemary 102804
False Rosewood 389946
False Roxie's Rhododendron 331681
False Rubber Tree 169227
False Rue Anemone 145494,145498,210260
False Rue-anemone 145492,210260
False Saffron 77684,77748
False Sagebrush 371141,371147
False Sago 115812
False Sainfoin 408518
False Sand Evening-primrose 269400
False Sandalwood 260722
False Sarsaparilla 30712,185764
False Saxifrage 349856
False Scaleleaf Corydalis 106396
False Seaonion 274556
False Seatung 301205,301206
False Shagbark 77912

False Shagbark Hickory 77894
False Shamrock 278124
False Shea 236336
False Shorttooth Elatostema 142798
False Shorttooth Stairweed 142798
False Sickleleaf Tickseed 104833
False Silkleaf Corydalis 106303
False Simon Poplar 311436
False Sisal 10841
False Slatepentree 400711
False Small-flowered Illigera 204646
False Smallpanicl Rabdosia 324745
False Soap Willow 343929
False Solomon's Seal 242699,366142,366198
False Solomon's-seal 366142
False Sonneratia 368911
False Sowthistle 327478
False Spectacular Penstemon 289370
False Spider 242699
False Spider-orchid 272402
False spikeflower 197347
False Spikenard 242699,366198
False Spinulose Barberry 52082
False Spiraea 41789,41795,41816,41852,369259,369279
False Spirea 41786,369272
False Spotted Staunton-vine 374435
False Spring Elm 401600
False Spruce 318580
False Stapf Monkshood 5504
False Starwort 56706,56713
False Steininger Woodbetony 287562
False Strawberry 138796
False Sunflower 188402,190505,190525,295341
False Tacheng Larkspur 124513
False Tadehagi 383007
False Tamarisk 261246,261262
False Tang Willow 343928
False Tarragon 35407,35411
False Tatarian Ixiolirion 211000
False Thorow-wax 63842
False Threadleaf 351909
False Thyme 391334
False Thyres Corydalis 105763
False Tick Trefoil 98156
False Timothy 294989
False Tinctorial Marsdenia 245834
False Toadflax 100229,174290
False Tomentose Poplar 311443
False Tongol Corydalis 106326
False Tutcheria 400711
False Umbellate Barberry 52084
False Valerian 358336
False Vanda 404745
False Versicolor Woodbetony 287564
False Vervain 405844
False Vetch 408551
False Virginia Creeper 285127
False Wallflower 154390
False Wallich Willow 343929
False Water Pepper 291792
False Water-pepper 291792
False Waterwillow 22401
False Westsichuan Larkspur 124530
False Weymouth Pine 300162
False Whiskbranch Gentian 173939
False White Cedar 390587
False Willow 46272
False Winter's Bark 91250
False Wintergreen 322872
False Woloho Willow 343930
False Woolly-leaved Rhododendron 331044
False Xizang Barberry 52083
False Xizang Tanoak 233352
False Yarrow 84488,84524
False Yellowflower Monkshood 5041
False Yellowhead 135051
False Yew 318539
False Zeshan Gentian 150897
False-agave 187137
Falsealyssum 53034
Falseareca 31012,31013
Falseasterostigma Corydalis 106294
Falsebanksia Rose 336875
Falsebeautiful Meconopsis 247169
False-beautiful Rhododendron 331580
Falsebeautyberry 399851,399852
Falsebetony 373068,373071
Falsebigtail Heliotrope 190709
Falsebox 286749
Falsebracteate Chelonopsis 86832
Falsebranch Brome 60896
Falsebroadsepal Larkspur 124528
Falsebrome 58572
False-brome 58572
False-brome Grass 58572
False-buckwheat 162505
False-bullated Blueberry 403969
Falsecactus Euphorbia 159641
Falsecapform Raspberry 339116
False-capform Raspberry 339116
Falsecatch Skullcap 355700
Falsecherryplum Eurya 160585
Falsechervil 84721,84722
Falsechirita 317537
Falsecoeruleus Larkspur 124516
Falsecristate Junegrass 217494
Falsecuckoo 48080
Falsecypress 85263
False-cypress 85263
False-dandelion 218202,218204
Falsedeadnettle 283435,283436
Falsedenticulate Primrose 314849
Falsedivaricate Monkshood 5499
Falsefalcate Tickseed 104833
Falseflax 68839
False-fragrant Scurrula 236974
Falseginsen 280778
Falseglacial Larkspur 124521
False-gold Groundsel 279939,279941
Falsegrip Skullcap 355700
Falsegrniculate Monkshood 5500
Falsehairfruit Microula 254361
Falsehairstalk Loosestrife 239798
Falsehairyfruit Microula 254361
Falsehellebore 405570,405618
False-hellebore 405570
Falsehelleboreleaf Calanthe 66101
Falsehelleboreleaf Neuwiedla 265840
Falsehelleborelike Swertia 380400
Falsehorsebean Sweetvetch 188177
False-Indian Tea Rose 336814
Falseindigo 19989,19997,20005
False-indigo 19989,20005
False-larch 317821
Falselily 266893,266896
Falselittleleaf Corydalis 106316
False-loosestrife 238160,238210
False-lupine 389547
Falsemallow 243877
Falsemanynerved Eurya 160586
Falsenetted Grape 411863
False-netted Grape 411863
Falsenettle 56094
False-nettle 56094
Falsenettleleaf Actinidia 6718
Falsenettleleaf Pepper 300357
Falsenettle-like Elatostema 142623
Falseoat 398437
Falseolive 142422
Falseoneflower Azalea 331457
Falseoneflower Rhododendron 331457
Falsepanax 266907
False-panax 266907
Falsepectinate Hatchet Cactus 288617
False-pennyroyal 396436
Falsepimpernel 231473,231504,231507,231508
Falsepimpernelleaf Gentian 174035
Falsepinnate Hypersilvery Cinquefoil 312672
Falsepistache 384369,384371
False-pistache 384369
False-raceme Bennettiodendron 51025,51056
False-raceme Mayten 182710
False-racemose Glycosmis 177849
Falserambutan 283529
Falsereed Reedbentgrass 65464
Falsereed Woodreed 65464
False-scabrous-leaf Milkvetch 42930

Falsescale Gentian 173758
Falseseapurslane 394879,394903
False-sepal Devil Rattan 121523
Falsesessileflower Monkshood 5536
False-sharpleaf Gambirplant 401777
False-short-keel Milkvetch 42923
False-sootep Spindle-tree 157811
Falsesour Cherry 83284
Falsespiny Meconopsis 247165
Falsespiraea 369259
False-spiraea 369259
Falsespiralspur Larkspur 124080
False-spotted Stauntonvine 374435
Falsestairweed 223674,223679
Falsestarwort 318489
Falsestonecrop 318388
Falsetamarisk 261246
False-tamarisk 261246
Falsetaro 99917
Falsetea Ehretia 77067,141611
False-timothy 113316
False-tinctorial Condorvine 245834
Falsetinctorial Milkgreens 245834
Falsetonkin Lilyturf 272127
Falsetufuted Spiradiclis 371776
False-underground Milkvetch 42926
False-uniflorous Rhododendron 331457
False-vagant Spindle-tree 157813
Falsevalerian 373490
False-varied-colour Milkvetch 42938
False-verticillate Daphniphyllum 122726
False-yellowflower Milkwort 308050
False-yellow-flowered Milkwort 308050
Falseyunnan Camellia 69783
Faluohai Cowparsnip 192236
Fame Flower 383293,383304
Fameflower 294318,383293,383345
Faminterry 169126
Famous Heliconia 190021
Famous Hepatica 192134
Famous Leafflower 245221
Fan Aloe 17175
Fan Columbine 30031
Fan Flowe 350339
Fan Flower 350323,350324,350332
Fan Forrest Woodbetony 287219
Fan Leaf Actinidia 6716
Fan Licuala Palm 228741
Fan Licualapalm 228741
Fan Maple 2964
Fan Palm 85665,106999,228762,234162,
234194,315376,329176,341406,393803,
393809,413289,413298,413307
Fan Wattle 1039
Fanbract Corydalis 106388
Fanbract Tacca 382931
Fanchen Leafflower 296560
Fanchen Leaf-flower 296560
Fanchen Underleaf Pearl 296560
Fanching Mountain Greenbrier 366617
Fancy 410677
Fancy Geranium 288206
Fancy Leaved Caladium 65228
Fancy Washington Geranium 288206
Fancy-leaved Caladium 65218,65223
Fanflower 350323
Fang Aucuba 44898
Fang Barrenwort 146990
Fang Bashanbamboo 37171
Fang Bashania 37171
Fang Cane 37174
Fang Canebrake 37174
Fang Chirita 87856
Fang Epimedium 146990
Fang Eurya 160469
Fang Goldenray 229032
Fang Henry Pentapanax 289637
Fang Hornbeam 77290
Fang Intermediate Holly 203921
Fang Monkshood 5183
Fang Paralagarosolen 283434
Fang Pratia 234467
Fang Primrose 314362
Fang Sinosenecio 365048
Fang Tanoak 233207
Fang's Chinese Groundsel 365048
Fang's Cotoneaster 107442
Fang's Poplar 311320
Fang's Thistle 91962
Fangcheng Anodendron 25935
Fangcheng Azalea 330682
Fangcheng Camellia 69067
Fangcheng Eelvine 25935
Fangcheng Rhododendron 330682
Fangcheng Tea 69067
Fangchi 34176
Fangding Archileptopus 30988
Fangding Ophiorrhiza 272205
Fangfeng 346447,346448
Fanghsien Small Reed 127246
Fangshan Bouquet Larkspur 124244
Fangshan Corydalis 105868
Fangshan Oak 323899
Fangxian Milkvetch 42360
Fangxian Smallreed 127246
Fangxian Willow 344008
Fanjing Edelweiss 224832
Fanjing Mountain Greenbrier 366617
Fanjing Primrose 314363
Fanjingshan 204944
Fanjingshan Actinidia 6592
Fanjingshan Aster 40419
Fanjingshan Azalea 331310
Fanjingshan Burmareed 265918
Fanjingshan Chinese Groundsel 365049
Fanjingshan Dendrobium 125129
Fanjingshan Edelweiss 224832
Fanjingshan Fir 354
Fanjingshan Greenbrier 366617
Fanjingshan Kiwifruit 6592
Fanjingshan Microtoena 254274
Fanjingshan Monkshood 5184
Fanjingshan Persimmon 132149
Fanjingshan Rhododendron 331310
Fanjingshan Rockjasmine 23228
Fanjingshan Sinosenecio 365049
Fanjingshan Smilax 366617
Fanjingshan Thistle 91963
Fanjingshan Touch-me-not 204944
Fanjingshan Yushanbamboo 416764
Fanjingshan Yushania 416764
Fanleaf Ancylostemon 22164
Fanleaf Buttercup 325816
Fanleaf Corydalis 105877
Fan-leaf Fleabane 150633
Fanleaf Gentian 173418
Fanleaf Geranium 174570
Fanleaf Hawthorn 109707
Fan-leaf Hawthorn 109707
Fanleaf Maple 3035
Fanleaf Primrose 314744
Fanleaf Redhair Rockfoil 349862
Fanleaf Trachydium 304809
Fanleaf Willow 343381
Fanleaflike Willow 343830
Fan-leaved Buttercup 325838
Fan-leaved Water-crowfoot 48919
Fanlike Primrose 314398
Fanlip Hemipilia 191602
Fanlip Liparis 232339
Fanlip Nailorchis 9289
Fanpalm 234162,234170,413298
Fan-palm 234162
Fansel 31680
Fanshape Rattan Palm 65691
Fanshaped Cogflower 154162,154166
Fanshaped Columbine 30031
Fan-shaped Corallodiscus 103944
Fanshaped Ervatamia 154162,154166
Fan-shaped Hawthorn 109707
Fanshaped Iris 208939
Fanshaped Salacca 342574
Fanshaped Skullcap 355439
Fanshaped Swordflag 208939
Fanshaped Umbrellasedge 118477
Fanshaped-bract Corydalis 106388
Fantail Willow 344232
Fanvein Rabdosia 209669
Fanweed 390213
Fanwort 64491,64495,64496
Fape 334250

Far-eastern Heteropappus 193973
Farenut 102727,166125,359158
Farer Beakrush 333560
Farewell Summer 40989,346434
Farewell to Spring 94133
Farewell-summer 40989,284591,346434,368480
Farewell-to-spring 94132,94133,94139,269413,269527
Farfugium 162612
Farges Angelica 24347
Farges Aralia 30659
Farges Arisaema 33330
Farges Balanophora 46823
Farges Barrenwort 146991
Farges Bashania 48711
Farges Birch 53436
Farges Bittercress 72746
Farges Calanthe 65946
Farges Cane 37176,48711
Farges Canebrake 37176,48711
Farges Catalpa 79250
Farges Chengiopanax 86854
Farges China Wildginger 37598
Farges Chinquapin 78933
Farges Clethra 96501
Farges Corydalis 105869
Farges Decaisnea 123371
Farges Dichocarpum 128931
Farges Dishspurorchis 171844
Farges Dragonheadsage 247723
Farges Epigeneium 146536
Farges Epimedium 146991
Farges Evergreen Chinkapin 78933
Farges Evergreenchinkapin 78933
Farges Evodia 161330
Farges Figwort 355116
Farges Filbert 106731
Farges Fir 356
Farges Forkfruit 128931
Farges Gastrochilus 171844
Farges Goldenray 229033
Farges Habenaria 183619
Farges Harlequin Glorybower 96398
Farges Hazel 106731
Farges Holboellia 197213
Farges Holly 203798
Farges Honeysuckle 235776
Farges Hornbeam 77291
Farges Indocalamus 37176
Farges Isometrum 210179
Farges Lily 229840
Farges Meadowrue 388497
Farges Meehania 247723
Farges Mistletoe 411023
Farges Neocinnamomum 263761
Farges Newcinnamon 263761
Farges Oreorchis 274037
Farges Parasite 293192
Farges Paris 284319
Farges Paulownia 285959
Farges Phacellaria 293192
Farges Pine Taxillus 385190
Farges Rhododendron 331408
Farges Sedge 74487
Farges Snakemushroom 46823
Farges Southstar 33330
Farges Spapweed 204945
Farges Speedwell 407120
Farges Stringbush 414172
Farges Thistle 91964
Farges Torreya 393058
Farges Twayblade 232156
Farges Wildginger 37617
Farges Windhairdaisy 348309
Farges Woodbetony 287198
Farges' Currant 333969
Farges' Eightangle 204513
Farges' Illicium 204513
Farges' Nothaphoebe 266791
Farges' Willow 343366
Fargesia 162635
Farinose Azalea 330686
Farinose Barberry 51607
Farinose Honeysuckle 235860
Farinose Kiwifruit 6593
Farinose Primrose 314366
Farinose Single Bamboo 47232
Farinoseback Schefflera 350725
Farinosefruit Azalea 330897
Farinosefruit Bladderwort 403953
Farinoseleaf Habenaria 183662
Farinoseleaf Kiwifruit 6615
Farinosestalk Primrose 314634
Farkleberry 403724
Farkleberry Tree 403724
Farmer's Clock 384714
Farmer's Plague 8826
Farmer's Ruin 370522
Farmer's Weatherglass 21339
Farnambuck 64971
Farnetto 323917
Faro 122160
Farrain 192358
Farrer Asiabell 98314
Farrer Aster 40420
Farrer Barberry 51608
Farrer Callianthemum 66706
Farrer Cranesbill 174605
Farrer Globeflower 399507
Farrer Isometrum 210180
Farrer Onion 15223
Farrer Onosma 271759
Farrer Parnassia 284536
Farrer Primrose 314385
Farrer Rhododendron 330687
Farrer Rose 336551
Farrer Viburnum 407837
Farsetia 162813
Farthing Rot 200388
Farthing Stephania 375836
Farwell's Milfoil 261347
Farwell's Water-milfoil 261347
Fasciate Aechmea 8558
Fasciated Haworthia 186422
Fasciated Slugwood 50511
Fasciatedpeduncle Slugwood 50511
Fascicled Anemone 23815
Fascicled Beaksedge 333563
Fascicled Branches Rhododendron 330273
Fascicled Clematis 94894
Fascicled Japanese Korthalsella 217907
Fascicled Microstegium 254004
Fascicled Redbead Barberry 51313
Fascicled Redpepper 72074
Fascicled-branched Rhododendron 330273
Fascicledflower Kiwifruit 6594
Fascicled-flower Sabia 341504
Fascicled-flowered Sabia 341504
Fascicle-flower Bushclover 226789
Fascicleflower Chirita 87857
Fascicleflower Daphne 122429
Fascicleflower Hippolytia 196694
Fascicleflower Ophiorrhiza 272206
Fascicleflower Primrose 314386
Fascicleflower Rush 212832
Fascicleflower Wintersweet 87513
Fascicleleaf Rockfoil 349194
Fascicular Agapetes 10371
Fascicular Bladderwort 403754
Fascicular Cactus 244073
Fascicular Gentian 173974
Fascicular Ophiorrhiza 272206
Fascicularhair Antlevine 88892
Fascicularia 162868
Fasciculate Alyxia 18498
Fasciculate Bluebell 115349
Fasciculate Bluebellflower 115349
Fasciculate Brome 60718
Fasciculate ephedra 146157
Fasciculate Hemarthria 191232
Fasciculate Machilus 240583
Fasciculate Osier-like Rattan Palm 65810
Fasciculate Pycnarrhena 321981
Fasciculate Sandthorn 196747
Fasciculate Tea 69068
Fascicullate Yunnan Sandwort 32322
Fascileaf Aspidistra 39538
Fascinate Bluegrass 305518
Fascinate Fescuegrass 163955
Fascinate Honda Roegneria 335350

Fascite Saurauia 347957
Faselles 294056
Fassett's Locoweed 278766
Fast Parodia 284668
Fastenblume 315082
Fastigiate Fortune Spindle-tree 157511
Fastigiate Gaillardia 169582
Fastigiate Japan Plumyew 82525
Fastigiate Japanese Plumyew 82525
Fastigiate Japanese Spindle-tree 157630
Fastigiate Milkaster 169777
Fastigiate Mouse-ear Chickweed 82810
Fastigiate Poplar 311400
Fastigiate Rabdosia 209819
Fastigiate Sedge 74490
Fastigiate Tulip Tree 232611
Fastigiate Windhairdaisy 348923
Fastigiate Woodbetony 287199
Fastigiate-branch Willow Weed 146696
Fat Albert Spruce 298405
Fat Bellies 364193
Fat Bitterbamboo 304005
Fat Duckweed 224366
Fat Goose 86970
Fat Grass 406992
Fat Head Tree 262795
Fat Hen 36474,44576,72038,86901,86970, 87200,89704,162312,170193,176824, 309570,357203
Fat Hen Saltbush 44576
Fat Pine 300128
Fat Pork Tree 97262
Fat Primrose 314740
Fat Solomon 242700
Fat Touch-me-not 205188
Fata 367696
Fatch 408571
Fatheadtree 262795
Fat-hen 44576,44621,86901
Father Bigface 73430
Father David's Maple 2920
Father Hugo Rose 336636
Father Hugo's Rose 336636
Father of Heath 150131
Father Time 95431
Father-of-heath 150131
Fathomlong Euonymus 157339
Fatleaf Seepweed 379543
Fatoua 162868,162872
Fatox Chirita 87881
Fatsia 162876,162879
Fatsialike Brassaiopsis 59200
Fatsia-like Brassaiopsis 59200
Fatsialike Euaraliopsis 59200
Fat-spiked Yard-grass 143539
Fattahs 109857
Fatuous Nutmug 261423
Fatweed 167289
Faucimaculate Dutchmanspipe 34177
Faulhaber Dendrobium 124963
Faurie Avens 175410
Faurie Beakgrain 128015
Faurie Championella 85942
Faurie Crapemyrtle 219922
Faurie Lovegrass 147670
Faurie Microstegium 254005
Faurie Monkshood 5185
Faurie Poppy 282563
Faurie Sagebrush 35481
Faurie Valerian 404316
Fauroie Spiraea 371906
Faury Bamboo 259652
Fausse Camomille 26746
Faustrimedin 253285
Fava Bean 408393
Faveolate Birthwort 34187
Faverel 153949
Faverole 137747
Favose Mrs. E. H. Wilson's Barberry 52372
Favusheadflower 129563,129571
Fawn Lily 154895,154899,154903,154909, 154936
Fawn Lily Trout 154899
Fawnlily 154895
Fawn-lily 154895
Faxon Barberry 51610
Faxon Fir 362
Faxon Yucca 416583
Faxonian-like Willow 343371
Faxons Fir 362
Faxy-red Cherry 83295
Faya Tree 261160
Fayberry 334250,403916
Fea 334250
Feabe 334250
Feaberry 278366,334250
Feap 334250
Feapberry 334250
Feasil 294056
Feather Amaranth 80395
Feather Amsonia 20846
Feather Bells 375453
Feather Bladderwort 403885
Feather Bush 239541
Feather Cactus 244192
Feather Cinquefoil 312890
Feather Cockscomb 80381,80400
Feather Dalea 121887
Feather Fern 371944
Feather Finger Grass 88421
Feather Finger-grass 88344
Feather Flower 414823
Feather Geranium 139682
Feather Grape-hyacinth 260310
Feather Grass 65254,225983,376687,376877
Feather Hyacinth 260310
Feather Lovegrass 147995
Feather Peabush 121887
Feather Plume 121887
Feather Reed Grass 65254,65264
Featherbells 375453
Feather-bells 375453
Featherbow 322724
Featherbract Ligusticum 229312
Feathered Cinquefoil 312890
Feathered Columbine 388423
Feathered Geranium 139682
Feathered Hoarhound 245759
Feathered Hyacinth 260301
Feathered Pink 127793
Featherfall 322724
Featherfew 322724
Feather-fingergrass 88344
Featherfleece 375453
Feather-fleece 375453
Feather-flower 407529
Featherfoe 322724
Featherfoil 198676,198679,198680,261337, 322724
Featherfold 322724
Featherfowl 322724
Featherfoy 322724
Featherfull 322724
Feather-geranium 139682
Feathergrass 376687
Feather-grass 376687
Feather-head Knapweed 81436
Feather-leaf Fleabane 150870
Featherleaf Ghost lampstand 335147
Featherleaf Pepperweed 225433
Featherleaf Rodgersflower 335147
Featherleaf Sanqi 280722
Featherlip Orchis 274485
Featherpappo Cottonthistle 271693
Feathers 18687,182942
Feathertop 65330,289303
Feather-top 289303
Featherweed 170941
Featherwheelie 322724
Feathery Bamboo 47516
Feathery Cassia 360419
Feathery Lanaria 220790
Feathery Pentapanax 289662
Feathery Plum 182942
Feathery Plume 182942
Feathery Rhodes-grass 387512
Feathery Woodrush 238685
Featherycleft Horsegentian 397847
Feathery-sepaled Raspberry 339055
Feathyfew 322724
Feberry 334250

February Daphne 122515
February Fair Maid 169719
February Fair Maids 169719
February Fair-maid 169719
February Rhododendron 331535
February Spicebush 231415
February Spice-bush 231415
Fechter 302034
Fedde Barberry 51611
Fedde Calvepetal 177989
Fedde Corydalis 105870
Fedde Elsholtzia 144014
Fedde Glyptic-petal Bush 177989
Fedde Glyptopetalum 177989
Fedde Mahonia 242528
Fedde Pimpinella 299414
Fedde Raspberry 338393
Fedde Sagebrush 35483
Fedde Stonecrop 356732
Fedde Wormwood 35483
Fedderfew 322724
Fedschenko Stonecrop 356733
Fedtschenk Rose 336554
Fedtschenkiella 163086,163089
Fedtschenko Bluegrass 305521
Fedtschenko Corydalis 105872
Fedtschenko Ephedra 146159
Fedtschenko Madwort 18381
Fedtschenko Rose 336554
Fedtschenko Valerian 404267
Fedtschenko Willow 343373
Feeble Devil Rattan 121501
Feeneh Saugh 85875
Feijoa 163116,163117
Feited Cranesbill 174715
Felicia 163121
Fellbloom 237539,247366
Fell-bloom 237539
Fell-fields Claytonia 94334
Fellon-berry 61497
Fellon-grass 174877,190936,190944,205535
Fellon-herb 36474,82758,190959,195858
Fellon-weed 359158
Fellon-wood 205535,367120
Fellon-wort 36474,86755,174877,205535,
367120,367416
Fellwort 81065,174106,174221
Felon Herb 36474
Felonewood 367120
Felon-herb 35084
Felt Bush 215097
Felt Oak 323754
Felt Plant 215097,215102
Felted Hydrangea 199861
Felthair Aster 41473
Feltleaf Ceanothus 79911
Feltleaf Willow 342970
Feltwort 405788
Felwort 174122,235431,380302
Female Bamboo 47203
Female Cornel 380499
Female Dogwood 104949
Female Dragons 66611
Female Fluella 216292
Female Fluellin 216292
Female Handed Orchid 273531
Female Horsetail 196818
Female Nettle 402886
Female Pimpernel 21339
Female Regulator 358336
Female Satyrion 273531
Female Speedwell 216292
Femble 71218
Fen Bedstraw 170717
Fen Betony 287335
Fen Grass-of-parnassus 284545
Fen Groundsel 359660
Fen Lobelia 234562
Fen Orchid 232072,232229,232364
Fen Panicled Sedge 75864
Fen Pondweed 312073
Fen Ragwort 359660
Fen Rue 388506
Fen Sedge 93913
Fen Sow-thistle 368781
Fen Star Sedge 76368
Fen Violet 410401
Fen Willow-herb 146766
Fen Woodrush 238675
Fen Wood-rush 238677
Fenberry 278366
Fence Morning-glory 208028
Fenceline Dewberry 338381
Fendler Barberry 51612
Fendler Bush 163312
Fendler Ceanothus 79926
Fendler Meadowrue 388499
Fendler Palicourea 280483
Fendler Penstemon 289340
Fendler's Aster 380889
Fendler's Bladderpod 227018
Fendler's Brickellbush 60159
Fendler's Desertdandelion 242934
Fendler's Globemallow 370947
Fendler's Hedgehog Cactus 140258
Fendler's Meadow-rue 388499
Fendler's Ragwort 279908
Fendler's Sundrops 68507
Fendler's Waterleaf 200485
Fendler's Sandwort 31875
Fenestraria 163326
Fenestrate Tanoak 233209
Feng Barberry 51613
Feng Begonia 49826
Feng Broadleaf Sporoxeia 372896
Feng Caper 71744
Feng Glyptopetalum 177990
Feng Maple 3067
Feng Medinilla 247562
Feng Monkshood 5539
Feng Persimmon 132152
Feng Rock Monkshood 5539
Feng Schefflera 350693
Feng Spindle-tree 177990
Feng Willow 343375
Feng's Clematis 94896
Feng's Cyclobalanopsis 116090
Feng's Elaeocarpus 142295
Feng's Oak 116090
Feng's Qinggang 116090
Feng's Trigonostemon 397360
Feng's Triratna Tree 397360
Feng's Woodbetony 287201
Fengcheng Cliffbean 66639
Fengcheng Euonymus 157711
Fengcheng Millettia 66639
Fengcheng Oak 323904
Fenghuai Spapweed 204946
Fenghuai Touch-me-not 204946
Fenghuang Aspidistra 39539
Fenghuangshan Aucuba 44901
Fenghuangshan Bulbophyllum 62725
Fenghuangshan Sedge 74496
Fenghuangshan Stonebean-orchis 62725
Fengjie Beggarticks 53903
Fengjie Elaeocarpus 142315
Fengjie Wildginger 37695
Fengkai Atalantia 43725
Fengkai Azalea 331031
Fengkai Bellcalyxwort 231258
Fengkai Fan Palm 234177
Fengkai Fanpalm 234177
Fengkai Honeysuckle 235780
Fengkai Rhododendron 331031
Fengqin Didymocarpus 129949
Fengqing Grape 411668
Fengqing Hackberry 80636
Fengqing Holly 203807
Fengqing Kadsura 214969
Fengqing Premna 313622
Fengshan Begonia 49714
Fengshan Mitreola 256216
Fengxian Willow 342998
Fengxiang Primrose 314985
Fenhe Cypressgrass 119243
Fenhe Galingale 119243
Fenkel Fenckell 167156
Fennel 167141,167146,167156
Fennel Finkle 167156
Fennel Flower 266225,266241
Fennel Giant-hyssop 10407

Fennel Pondweed 378991
Fennel Primrose 314122
Fennel-finkle 167156
Fennelflower 266215,266231
Fennel-flower 266215
Fennelleaf Pondweed 378991
Fennel-leaved Peony 280318
Fennel-leaved Pondweed 378991
Fenqihu Clearweed 298919
Fenqihu Coldwaterflower 298919
Fenqing Camellia 69069
Fenugreek 86755,397186,397229,397259
Fenugreek Trigonella 397229
Fenzel Engelhardia 145512
Fenzel Maple 2962
Fenzel Milkvetch 42363
Fenzel Pine 299938
Fenzel Tanoak 233213
Ferdinand Honeysuckle 235781
Ferdinand-coburg Saxifrage 349306
Ferdinandi-coburg Barberry 51614
Ferdinandi-coburg Haberlea 184224
Fergan Beancaper 418648
Fergan Craneknee 221675
Fergan Elytrigia 144648
Fergan Fritillary 168401
Fergan Russianthistle 344531
Fergan Spunsilksage 360831
Fergan Sweetvetch 187870
Ferges Mahonia 242638
Ferges Violet 409980
Fern Acacia 1051
Fern Asparagus 39171,39195
Fern banksia 47633
Fern Buttercup 312360
Fern Cycad 373673
Fern Cycas 115812
Fern Grass 126201,354487
Fern Mint 257169
Fern Nut 102727
Fern Palm 115812
Fern Pine 9994,306435
Fern Pondweed 312254
Fern Tree 211237,211239
Fernald Primrose 314389
Fernald's False Manna Grass 321278
Fernald's Iris 208557
Fernald's Manna-grass 177597
Fernald's Oval Sedge 75340
Fernald's Sedge 75340
Fernald's Yellow-cress 336251
Fernandez Lactoris 219205
Fernandez Sophora 369006
Fernando Rib Begonia 49828
Fernandoa 163396
Fernbush 85142,85146
Ferngrass 354487
Fern-grass 79295,354487
Fernhemp Cinquefoil 312360
Fernleaf Acacia 1394,1473
Fernleaf American Elder 345645
Fern-leaf Aralia 310195
Fern-leaf Bamboo 37127
Fern-leaf Banksia 47655
Fernleaf Bayberry 261126
Fern-leaf Beech 162422,162431
Fern-leaf Begonia 49844
Fern-leaf Buckthorn 328702
Fernleaf Clematis 95320
Fern-leaf Coreopsis 104616
Fernleaf Dichocarpum 128929
Fern-leaf False Foxglove 10143
Fern-leaf Fleabane 150553
Fernleaf Forkfruit 128929
Fern-leaf Grevillea 180566,180579,180636
Fernleaf Groundsel 359827
Fernleaf Hedge Bamboo 47347
Fern-leaf Hopbush 135199
Fernleaf Lavender 223308
Fernleaf Ligusticum 229377
Fernleaf Maple 3035
Fernleaf Mountainash 369495
Fernleaf Mulgedium 283692
Fernleaf Nanchuan Sage 345243
Fern-leaf Nitta Tree 284456
Fernleaf Polyscias 310195
Fernleaf Red Elderberry 345665
Fernleaf Sage 345112
Fernleaf Sweet Gale 261126
Fern-leaf Tree 166056,166059
Fernleaf Waxmyrtle 261126
Fernleaf Woodbetony 287203
Fernleaf Yarrow 3948
Fern-leaf Yellow False Foxglove 45175,45181,45184
Fern-leaved Beech 162403,162431
Fern-leaved Black Elder 345636
Fern-leaved Clematis 94835
Fern-leaved Corydalis 105730
Fern-leaved Daisy 397970
Fern-leaved Elder 345636,345637
Fern-leaved False Foxglove 45175
Fern-leaved Lily 230005,230009
Fern-leaved Mountain-ash 369495
Fern-leaved Parsley 292697
Fern-leaved Phacelia 293154
Fern-leaved Polyscias 310195
Fernlike Asparagus 39014
Fernlike Cystopetal 320219
Fern-like Kobresia 217166
Fernlike Sedge 74513
Fern-like Sedge 76421
Fernlike Sinocarum 364887
Fern-like Woodbetony 287204
Fernspray Cypress 85316
Fernstem Dichocarpum 128939
Fern-tree 211225
Ferocious Blue Cycad 145229
Feroniella 163505,163506
Ferre Fir 363
Ferrer Lady's Slipper 120347
Ferrer Ladyslipper 120347
Ferris' Goldfields 222600
Ferris' Sandwort 32044
Ferris' Sulphur Flower 152606
Ferrocalamus 163567
Ferrousscale Sedge 76226
Ferrugineous Evodialeaf Acanthopanax 2388
Ferrugineous Peridiscus 290924
Ferruginous Clerodendrum 96400
Ferruginous Elm 401476
Fertile Euonymus 157459
Fertile Knotweed 309105
Ferty Lilac 382302
Ferula 163574
Ferworth 380105
Fescue 163778,163806,412394
Fescue Like Scolochloa 354583
Fescue Oval Sedge 74503
Fescue Sandwort 31791
Fescue Sedge 73933,74503
Fescuegrass 163778,163950,164295
Fescue-grass 163778
Fescueleaf Hairgrass 126039
Festike Nut 301005
Festuelike Sandwort 31895
Fetch 408571
Feterita 369600
Fetferfoe 322724
Fetid Adder's-tongue 354573
Fetid Asiabell 98316
Fetid Buckeye 9694
Fetid Clary Sage 345379
Fetid Currant 334159
Fetid Dragonhead 137581
Fetid Glorybower 96078
Fetid Groundsel 360321
Fetid Gymnema 182366
Fetid Larkspur 124218
Fetid Marigold 139710
Fetid Passionflower 285639
Fetid Peppergrass 225447
Fetid Securinega 167096
Fetid Trillium 397564
Fetid-marigold 139710
Fetish Tree 216324
Fetisow Garlic 15278
Fetisow Onion 15278
Fetissow Woodbetony 287202
Fetter Bush 228162,228180,239385,298721
Fetterbush 228154,228155,228162,228180,

228181,239385,298721
Fetter-bush 228155,298721
Fetterbush Lyonia 239385
Fever Bark 18036,112829
Fever Bush 171546,171552,231297,231306
Fever Leaf 268671
Fever Nut 64971
Fever Plant 3978,268671
Fever Root 90924
Fever Tree 1708,155589,232494,299701,
299702,299703,417190
Fever Vine 280063
Fever Weed 405889
Fever Wort 397844
Fever-bark 18036
Feverberry 334250
Feverfew 81065,285072,322724
Feverfew Chrysanthemum 322724
Feverfoullie 322724
Feverfuge 81065
Fevertory 169126
Fever-twig 80309
Fevervine 280063
Feverwort 81065,158250,397828,397844
Few Flower Blastus 55174
Few Flower Crazyweed 279071
Few Flower Thorowax 63816
Fewanther Eightangle 204567
Few-bracted Beggar-ticks 53879
Fewdentate Pellionia 288752
Fewflower Affinity Neillia 263144
Fewflower Alangium 13354
Fewflower Aletris 14519
Fewflower Argyreia 32625
Fewflower Arrowwood 407987
Fewflower Aucuba 44908
Fewflower Barrenwort 147032
Fewflower Beautyberry 66889
Fewflower Bedstraw 170545
Fewflower Bladderwort 403191
Fewflower Blastus 55174
Fewflower Bluegrass 305437
Fewflower Bulrush 353696
Fewflower Caltropus Milkvetch 43179
Few-flower Caper 71749
Fewflower Childvine 400944
Fewflower Christolea 89265,126164
Fewflower Cinnamon 91397
Fewflower Clearweed 298999
Fewflower Codonacanthus 98223
Fewflower Coldwaterflower 298999
Fewflower Conehead 85949
Fewflower Corydalis 106206,106255
Fewflower Cotoneaster 107611
Fewflower Crimson Flag 351884
Fewflower Cryptocarya 113442
Fewflower Daphne 122421
Fewflower Diflugossa 130322
Fewflower Diplarche 132830
Fewflower Droguetia 138077
Fewflower Elaeocarpus 142286
Fewflower Elsholtzia 143978
Few-flower Embelia 144779
Fewflower Enkianthus 145709
Fewflower Epimedium 147032
Fewflower Figwort 355213
Few-flower Fumitory 169153
Fewflower Funnel Flower 130223
Fewflower Gagea 169480
Fewflower Gardenia 171456
Few-flower Gayfeather 228500
Fewflower Gesneria 175300
Fewflower Gingerlily 187463
Fewflower Glycosmis 177843
Fewflower Goosecomb 335457
Fewflower Greenbrier 366253
Few-flower Griffith Elaeagnus 142017
Few-flower Gueldenstaedtia 181693
Fewflower Hairycalyx Lobelia 234697
Fewflower Hawkweed 196159
Few-flower Ichnocarpus 203330
Fewflower Indianmulberry 258903
Fewflower Ironweed 406226
Fewflower Jerusalemsage 295145
Fewflower Lobelia 234398,234697
Fewflower Loosestrife 239874
Fewflower Lopseed 296043
Fewflower Lysionotus 239967
Fewflower Madder 338004
Few-flower Memecylon 250041
Fewflower Microtoena 254263
Fewflower Milkvetch 43179
Fewflower Milkweed 37979
Fewflower Mosla 259319
Fewflower Mouseear 82955
Fewflower Neillia 263144
Fewflower Nepal Cranesbill 174759
Fewflower Ovate-leaf Metalleaf 296406
Fewflower Oxtongue 298633
Fewflower Pencilwood 139668
Fewflower Pendent-bell 145709
Fewflower Phaius 293525
Fewflower Pilose Galingale 119381
Fewflower Pimpinella 299522
Fewflower Pittosporum 301360
Fewflower Plateaucress 89265,126164
Fewflower Premna 313707
Fewflower Pseudosasa 318337
Fewflower Purplepearl 66889
Few-flower Raspberry 339284
Fewflower Red Thorowax 63816
Fewflower Redbud 83802
Fewflower Roegneria 335457
Fewflower Sage 345295
Fewflower Sarcococca 346739
Fewflower Saussurea 348606
Fewflower Seatung 301360
Fewflower Skeletonweed 88793
Fewflower Slugwood 50586
Fewflower Smilax 366253
Fewflower Snailseed 97933
Few-flower Spike Galingale 118758
Fewflower Spikesedge 143319
Fewflower St. John's Wort 202217
Fewflower Thick-leaf Deutzia 126889
Fewflower Trisetum 398514
Fewflower Twoleaved Swertia 380127
Fewflower Viburnum 407987
Fewflower Waistbone Vine 203330
Few-flower Wild Buckwheat 152379
Fewflower Wildrice 418095
Fewflower Windhairdaisy 348606
Fewflower Woodbetony 287482
Fewflower Woodrush 238671
Fewflower Yelloweyegrass 416127
Fewflower Yellowleaftree 415145
Fewflower Youngia 416486
Few-flowered Bog Sedge 75711
Few-flowered Cinnamon 91397
Few-flowered Cotoneaster 107593,107611
Few-flowered Cranberry-tree 407905
Few-flowered Elaeocarpus 142286
Few-flowered Fargesia 162733
Few-flowered Fumitory 169183
Few-flowered Leek 15585
Few-flowered Maple 3371
Few-flowered Mayten 246894
Few-flowered Meadow-rue 388673
Few-flowered Microtropis 254315
Few-flowered Nut Grass 354173
Few-flowered Panic Grass 128535,128537
Few-flowered Pencil-wood 139668
Few-flowered Plantain 302117
Few-flowered Sageretia 342186
Few-flowered Sarcococca 346739
Few-flowered Sedge 75711,75980,76540
Few-flowered Senecio 359682
Few-flowered Shooting Star 135178
Few-flowered Slugwood 50586
Few-flowered Spike Sedge 143397
Few-flowered Spike-rush 143319
Few-flowered Spike-sedge 143397
Few-flowered Spindle-tree 157777
Fewflowered Swertia 380417
Few-flowered Tick Clover 126514
Few-flowered Verrucose Spindle-tree 157777
Few-flowered Wildryegrass 144200
Few-flowered Wirelettuce 375978
Few-flowered Yellow-leaved Tree 415145
Fewfruit Cotoneaster 107612
Fewfruit Elaeagnus 142207

Fewfruit Gentian 173534
Few-fruit Illicium 204574
Fewfruit Lycianthes 238974
Fewfruit Maple 3274
Fewfruit Red Silkyarn 238974
Few-fruit Rhododendron 331395
Fewfruit Stonecrop 356984
Fewfruit Taillessfruit 100143
Few-fruited Gray Sedge 75589
Fewfruited Maple 3274
Few-fruited Pittosporum 301294
Fewhair Beccarinda 49398
Fewhair Betony 373092
Fewhair Maple 3423
Fewhair Scorpiothyrsus 354747
Fewhair Shade Clearweed 299076
Fewhair Zuogong Rockfoil 350061
Fewhairs Hairycalyx Blueberry 403974
Few-hairs Radde's Willow 343978
Fewhairy Cherry 83171
Few-headed Blazing-star 228454
Few-headed Grass-leaved Goldenaster 301511
Fewhir Chirita 87870
Fewleaf Absolmsia 400942
Fewleaf Aphania 225676
Fewleaf Bluegrass 305817
Fewleaf Deerdrug 242702
Fewleaf Devil Rattan 121520
Fewleaf False Solomonseal 242702
Fewleaf Gentian 173678
Fewleaf Ginger 418011
Fewleaf Schnabelia 352060
Fewleaf Thistle 92342
Fewleaved Bluegrass 306034
Few-leaved Bluegrass 305360
Few-leaved Hawkweed 195803
Few-leaved Sunflower 189017,189028,
189030
Few-leaves Mountainash 369427
Few-leaves Stauntonvine 374433
Few-male Holly 204567
Fewnerne Bauhinia 49193
Fewnerve Bowfruitvine 393542
Fewnerve Camellia 69421
Fewnerve Garcinia 171163
Fewnerve Linden 391807
Fewnerve Spapweed 205192
Fewnerve Touch-me-not 205192
Fewnerve Toxocarpus 393542
Few-nerved Cotton-grass 152788
Few-nerved Lime 391807
Few-nerved Linden 391807
Few-nerved Saurauia 347975
Few-nerved Sweetleaf 381353
Few-nerved Toxocarpus 393542
Few-nerved Wood Sedge 75126
Fewpair Emei Rose 336827
Fewpetal Camellia 69463
Fewroot Ducksmeat 221007
Fews 358062
Fewscale Clearweed 299052
Fewscale Cotylanthera 107836
Fewseed Corydalis 106207
Fewseed Milkwort 308238
Fewseed Speedwell 407259
Few-seeded Hop Sedge 75591
Few-seeded Sedge 75591
Fewsetose Bethlehemsage 283641
Fewsetose Paraphlomis 283641
Fewspike Cypressgrass 118758
Fewspike Greyawngrass 372423
Fewspike Hypolytrum 202731
Fewspike Ricegrass 300717
Fewspike Sedge 74431
Fewspike Spodiopogon 372423
Few-spiked et Slender-stem Bulrush 353413
Fewspikelet Fimbristylis 166471
Fewspikelet Fluttergrass 166471
Few-spiny Fragrant-bamboo 87598
Fewspoke Sinocarum 364889
Fewstamen Grewia 180900
Fewstamen Meadowrue 388605
Few-stamened Grewia 180900
Few-strigose Milkvetch 42358
Fewstrobile Hypolytrum 202731
Fewteeth Mountainash 369477
Few-teethed Mountain-ash 369477
Fewtooth Raspberry 339005
Fewtooth Skullcap 355599,355632
Fewtooth SW. China Clematis 95258
Fewtooth Willowleaf Spiraea 372076
Few-toothed Barberry 52021
Fewtube Shorthair Cowparsnip 192318
Fewtwig Asparagus 39126
Fewvein Camellia 69421
Fewvein Litse 234023
Fewvein Sageretia 342185
Fewvein Saurauia 347975
Fewvein Seatung 301350
Fewvein Skullcap 355633
Fewveined Acronychia 6249
Few-veined Acronychia 6249
Few-veined Bauhinia 49193
Few-veined Microtropis 254318
Few-veined Shaanxi Hornbeam 77389
Few-veined Soulie Barberry 52170
Fewveined Sweetleaf 381353
Few-veins Bauhinia 49193
Few-veins Coronate Mountainash 369372
Fiber Flax 232001
Fiber Lily 295594
Fiber Optics Plant 353290
Fibraurea 164456
Fibre Mauritia 246672
Fibre Roegneria 335307
Fibre-tree 356368
Fibrous Crabgrass 130550
Fibrous Tussock-sedge 73729
Fibrous-root Sedge 74155
Ficarialeaf Buttercup 325832
Fickeisen Pincushion Cactus 287905
Fickeisen's Navajo Cactus 287905
Ficus 164608
Ficus Benjamina 164688
Fiddle 68189,123141,355061,355202
Fiddle Dock 340214
Fiddle Leaf Ficus 165426
Fiddle-back Fig 165426
Fiddle-cases 329521
Fiddle-grass 146724
Fiddleleaf Aster 40988
Fiddleleaf Dubyaea 138777
Fiddleleaf Fig 165426
Fiddle-leaf Fig 165267,165426
Fiddleleaf Hawksbeard 110990
Fiddle-leaf Horsfieldia 198518
Fiddle-leaved Fig 165429
Fiddle-leaved Flock 340214
Fiddleneck 20827,20837,20840,20842,
20843
Fiddle-neck 20827,20840
Fiddler's Trumpet 347149
Fiddle-shaped Leaf Nan 295397
Fiddle-shaped Leaf Raspberry 338952
Fiddle-shaped-leaf Waxplant 198882
Fiddle-shape-leaved Raspberry 338952
Fiddlesticks 355061,355202
Fiddlestrings 302068,355061
Fiddlewood 93237,93239,93247,292482,
355061
Fiebrig Barberry 51617
Field Ash 369339
Field Aster 215343
Field Balm 96969,176824,264946
Field Basil 97060
Field Bastardtoadflax 389611
Field Bean 294056,408393
Field Beet 53254
Field Betony 373125
Field Bindweed 102903,102933
Field Blackcapsule 14297
Field Brome 60634
Field Brome Gras 60634
Field Brome-grass 60634
Field Bugloss 21920
Field Bur Parsley 392968
Field Burrweed 368536
Field Cabbage 59603
Field Camomile 26746
Field Caraway 77777
Field Cerastlum 82758

Field Chickweed 82680,82688
Field Circaester 91597
Field Clover 396854
Field Copperleaf 1788
Field Cottonrose 165915
Field Cottonweed 168687
Field Cowpea 409113
Field Cow-wheat 248161
Field Cranesbill 174832
Field Cress 225315,390213
Field Cypress 13073
Field Daisy 227533,322724
Field Dock 340209
Field Dodder 114986,115099
Field Dragonbamboo 125507
Field Duckbeakgrass 209356
Field Edelweiss 224818
Field Elm 401468,401508,401593
Field Eryngo 154301
Field Eulophia 156608
Field Euphorbia 158407
Field Fennelflower 266216
Field Fescue 163828
Field Fleawort 385899,385903
Field Forget-me-not 260760
Field Foxtail 17548
Field Garlic 15540,15876
Field Gentian 156608,173318,173473,174122
Field Glycosmis 177822
Field Goldenrod 368268,368270
Field Gromwell 233700
Field Groundsel 359138,359632,385903
Field Hawkweed 195506
Field Hedgeparsley 392968
Field Hedge-parsley 392968
Field Hop 3978
Field Hundredstamen 389611
Field Indian Paintbrush 79118
Field Jerusalemsage 295051
Field Kale 364557
Field Lacquertree 393479
Field Lacquer-tree 393479
Field Lady's Mantle 13966
Field Lady's-mantle 13966
Field Larkspur 102833,102841,124141
Field Madder 362095,362097
Field Maple 2837
Field Marigold 66395,89704
Field Melasma 14297
Field Milk Thistle 368635
Field Milk-vetch 41925
Field Milkwort 308331,308449
Field Milky Thistle 368635
Field Mint 250294,250331,250362,250370,264897
Field Morning Glory 102933
Field Mouse Ear 82680
Field Mouse-ear 82680
Field Mouse-ear Chickweed 82680
Field Mugwort 35237
Field Muskmelon 114194
Field Mustard 59575,59603,364557
Field Nepeta 264895
Field Nettle 373125
Field Nut Sedge 118822,118829
Field Oval Sedge 75399
Field Pansy 409703,409728,410140,410480,410677
Field Parsley Piert 13966
Field Paspalum 285456
Field Pea 301070
Field Pen 301055
Field Penny Cress 390213
Field Pennycress 390213
Field Penny-cress 390213
Field Pepper Grass 225315
Field Pepper-grass 225315
Field Pepperweed 225315
Field Pepper-weed 225315
Field Pepperwort 225315,225316
Field Poppy 282546,282685
Field Pumpkin 114300
Field Pussytoes 26490
Field Pussy-toes 26427,26490
Field Restharrow 271333
Field Rose 336364
Field Sagebrush 35237
Field Sagewort 35237
Field Sage-wort 35237,35244
Field Sandbur 80830
Field Sandspurry 370696
Field Scabious 81356,216753
Field Scorpion-grass 260760
Field Snake-cotton 168687
Field Sorrel 339897
Field Southernwood 35237
Field Sow Thistle 368635
Field Sowthistle 368635
Field Sow-thistle 368635
Field Speedwell 406973,407014,407287
Field Storax 379301
Field Thistle 91770,91914
Field Thlaspi 390213
Field Violet 409703
Field Wallflower 326609
Field Wood Rush 238583
Field Woodrush 238583
Field Wood-rush 238583
Field Wormwood 35237,35244
Field Woundwort 373125
Field Yellow-cress 336182
Fieldcitron 400515,400530
Fieldcress 225315
Fieldhaircot 138966
Fieldhove 400675
Fieldia 165892
Fielding Fox-tail Orchid 9288
Fieldjasmine 199448
Fieldnettle Betony 373125
Field-nettle Betony 373125
Fieldorange 249648,249657
Fieldthistle 91914
Fieldwort 81065,174106,199624
Fiery Love-flower 148071
Fiery Thorn 322454
Fig 164608,164763,165307
Fig Buttercup 325820
Fig Marigold 136397,177449,251025,251357
Fig Mulberry 165726
Fig Nut 212127
Fig Tree 164763
Figerleaf Morningglory 207988
Fighazel 380586,380602
Fighead Stairweed 142658
Fightee Cocks 302034
Fighting Cocks 302034
Figleaf Brassaiopsis 59202
Figleaf Dendropanax 125597
Figleaf Euaraliopsis 59202
Figleaf Goosefoot 87007
Figleaf Gourd 114277
Fig-leaf Gourd 114277
Figleaf Grape 411669,411930
Figleaf Holly 203814
Figleaf Palmleaftree 59202
Figleaf Spindle-tree 157463
Fig-leaved Brassaiopsis 59202
Fig-leaved Dendropanax 125597
Fig-leaved Goosefoot 87007
Fig-leaved Grape 411669
Fig-leaved Holly 203814
Fig-leaved-like Grape 411682
Figlike Biflorgrass 128574
Fig-like Holly 203814
Figmarigold 220674,251025
Fig-marigold 77429,220464
Fig-marigold Family 12911
Fig-mulberry 165726
Figo Michelia 252869
Figroot Buttercup 325820
Fig-tree 164608,164763
Figwort 325820,355039,355061,355165,355184,355201,355202
Figwort Family 355276
Figwort Giant Hyssop 10420
Figwortflower Picrorhiza 264318
Fiji Arrowroot 382923
Fiji Fan-palm 315386
Fiji Kauri 10509

Fiji Longan 310823
Fijian Banfenghe 357971
Fijian Kauri 10500
Fijian Kauri Pine 10509
Fijian Semiliquidambar 357971
Filaera 158062,404316
Filamental Flowering Crab 243623
Filamentary Clematis 95096
Filamentary Meadowrue 388503
Filamentless Gingerlily 187439
Filamentose Sedge 74043
Filamentous-leaf Bulbostylis 63253
Filaree 153767,153869
Filaree Storks-bill 153767
Filarious Pyrethrum 322641
Filbeard 106703
Filberd 106703
Filbert 106696,106703,106757
Filbert Family 106623
Filbert Nut-tree 106703
Filberts 106696
Filbest-leaved Sida 362668
Filbord 106703
Filchner Parnassia 284537
Fildchive 294892,294895
Fildchive Family 294886
Filewort 235267
Filicaudate Aucuba 44907
Filicauline Eyebright 160286
Filicauline Rockfoil 349308
Filiculm Feathergrass 376772
Filifolious Kobresia 217169
Filifolious Pilostemon 299278
Filifolium 166065
Filiform Ardisia 31446
Filiform Begonia 49834
Filiform Cleisostoma 94438
Filiform Closedspurorchis 94438
Filiform Cotton-grass 152788
Filiform Dendrochilum 125529
Filiform Gagea 169420
Filiform Ladypalm 329178
Filiform Ladypalms 329178
Filiform Leptaleum 225803
Filiform Luisia 238311
Filiform Pondweed 378986
Filiform Rootless Vine 78731
Filiform Rush 213114
Filiform Sedge 74694
Filiformed Pedicel Speedwell 407125
Filiformis Rockjasmine 23170
Filiformis-leaved Russianthistle 344694
Filiformleaf Buttercup 326117
Filiformleaf Carum 77779
Filiform-leaf Corydalis 105873
Filiformleaf Pipewort 151524
Filiformleaf Sallstyle 63253
Filiform-stem Milkvetch 42366
Filiformstyle Willow 343490
Filimoto 166788
Filipedicelled Lasianthus 222132
Filipedunculate Sedge 74523
Filisect Halerpestes 184707
Filisect Soda Buttercup 184707
Filistalk Aster 40433
Filistalk Madder 337956
Filistyle Willow 343490
Filler Azalea 330508
Fillet Rattlesnake Plantain 179714
Fillyfindillan 166132
Film Azalea 331965
Filmbract Chalkplant 183182
Filmbract Windhairdaisy 348174
Filmcalyx 292656,292668
Filmcalyx Sandwort 32069
Film-ear Rush 213278
Filmleaf Ainsliaea 12682
Filmleaf Caper 71899
Filmleaf Childvine 400929
Filmleaf Habenaria 183431
Filmleaf Nanmu 295384
Filmleaf Nettle 402968
Filmleaf Rabbiten-wind 12682
Filmleaf Woodbetony 287422
Filmleaf Woodnettle 273902
Filmsepal Tea 69610
Filmy Angelica 24493
Filmy Seatung 301410
Fimble 71218
Fimbriate Aspidistra 39540
Fimbriate Begonia 49835
Fimbriate Catasetum 79339
Fimbriate Chondrorhyncha 88859
Fimbriate Climbing Apple Bamboo 249609
Fimbriate Climbing-bamboo 249609
Fimbriate Coelogyne 98651
Fimbriate Dallisgrass 285435
Fimbriate Fenugreek 397226
Fimbriate Fragrant Bamboo 87636
Fimbriate Fragrant-bamboo 87636
Fimbriate Gentian 173702
Fimbriate Jerusalemsage 295100
Fimbriate Orostachys 275363
Fimbriate Paspalum 285435
Fimbriate Platanthera 302324
Fimbriate Sandwort 31900
Fimbriate Spapweed 204949
Fimbriate Spindle-tree 157466
Fimbriate Touch-me-not 204949
Fimbriate Trigonella 397226
Fimbriate Tupistra 70607
Fimbriate Wild Sage 345454
Fimbriatebract Touch-me-not 204826
Fimbriate-ligular Bamboo 297280
Fimbriate-petal Corydalis 105875
Fimbriatesepal Chirita 87859
Fimbriate-stipulate Begonia 49837
Fimbribracteate Blueberry 403826
Fimbribriate-bracted Blueberry 403826
Fimbribriate-calyxed Blueberry 403827
Fimbricalyx Blueberry 403827
Fimbricalyx Leafflower 296562
Fimbriligulate Bamboo 297280
Fimbriligulate Membranaceous Dendrocalamus 125491
Fimbriligulate Yellow Dragonbamboo 125491
Fimbrisepal Loxostigma 238035
Fimbristylis 166156
Finckle 167156
Findley Dendrobium 125140
Fine Bent 12034
Fine Bent-grass 12375
Fine Cut-leaf Geranium 174661
Fine Euonymus 157546
Fine Leaves Rhododendron 331830
Fine Mountainash 369388
Fine Mountain-ash 369388
Fine Pinanga-palm 299662
Fine Ringvine 116017
Fine Stem Metalleaf 301682
Fine Stephania 375856
Fine Stonecrop 357189
Fine Windhairdaisy 348348
Fine Woodbetony 287246
Finebristle Blastus 55175
Fine-bristle Bluegrass 305967
Fine-caudated Aucuba 44907
Fineflower Buckthorn 328758
Fineflower Groundsel 358985
Fineflower Nephelaphyllum 265123
Fineflower Ringvine 116013
Fineflower Syzygium 382597
Fineflowered Alkaligrass 321397
Fine-flowered Buckthorn 328758
Fine-flowered Syzygium 382597
Finehair Duckbeakgrass 209321
Finehairgoosefoot 235614,235618
Finehead Rattlesnakeroot 313844
Fineleaf 410320
Fineleaf Bittercress 73018
Fineleaf Fumitory 169136
Fine-leaf Goosefoot 87070
Fineleaf Maackia 240133
Fineleaf Maxillaria 246727
Fineleaf Miyabe Spiraea 372015
Fineleaf Saddletree 240133
Fineleaf Schizonepeta 265078
Fineleaf Serpentroot 354926
Fineleaf Sheep Fescue 163958
Fineleaf Shrubcress 368566
Fineleaf Solms-Laubachia 368566

Fineleaf Waterdropwort 269299
Fineleaf West China Spiraea 371992
Fine-leaved Clover 396823
Fine-leaved Fumitory 169136
Fine-leaved Goldilocks 89870
Fine-leaved Heath 149205
Fine-leaved Heather 149205
Fine-leaved Hedge Mustard 126127
Fine-leaved Maackia 240133
Fine-leaved Mugwort 35088
Fine-leaved Peony 280318
Fine-leaved Pondweed 378986,378988,378989
Fine-leaved Sandwort 255494,255586
Fine-leaved Sheep's-fescue 163958
Fine-leaved Sneezeweed 188437
Fine-leaved Toothwort 125855
Fine-leaved Vetch 408624
Fine-leaved Water Dropwort 269299,269347
Fine-leaved Water-dropwort 269299
Finelydivided 312765
Finelydivided Phtheirospermum 296072
Fine-ribbed China-laurel 28323
Finerostrate Pterocypsela 320522
Finery Windhairdaisy 348695
Finescale Azalea 330214
Fineshoot Spiraea 372023
Fineshoot Syzygium 382476
Fine-shooted Spiraea 372023
Fine-shooted Syzygium 382476
Finespike Dragonbamboo 125471
Finespike Roegneria 335552
Finesplit Woodbetony 287754
Finespur Corydalis 106516
Finestalk Bulrush 210080
Finestalk Notoseris 267171
Finestalk Pearlsedge 354197
Finestalk Razorsedge 354197
Finestem Feathergrass 376934
Finestem Uncifera 401798
Finestem Woodbetony 287753
Finestipe Poorspikebamboo 270431
Finet Addermonth Orchid 110642
Finet Bogorchis 110642
Finet Clematis 94899
Finet Habenaria 183624
Finet Meadowrue 388505
Finet Monkshood 5192
Finet Platanthera 302325
Finetail Lasianthus 222285
Fine-tail Lasianthus 222285
Finetail Roughleaf 222285
Finetooth Holly 204239
Finetooth Tea 68964
Finetop Salt Grass 372581
Finetube Woodbetony 287354
Finevein Ligusticum 229350
Finger Bowl Geranium 288179
Finger Cactus 107070
Finger Cap 130383
Finger Cherry 332220
Finger Citron 93604
Finger Coreopsis 104566
Finger Euphorbia 159975
Finger Gloves 130383
Finger Grass 88316,130404
Finger Hakea 184601
Finger Hut 130383
Finger Millet 143522
Finger Nail Plant 264426
Finger of God 8582
Finger Orchid 273531
Finger Poppy Mallow 67127
Finger Speedwell 407421
Finger Sprangletop 225995
Finger Tickseed 104566
Finger Tree 159975
Fingerberry 338281
Finger-cap 130383
Finger-cone Pine 300082
Fingered Anthurium 28095
Fingered Citron 93604
Fingered Lemon 93604
Fingered Saxifrage 349999
Fingered Sedge 75954
Fingered Speedwell 407421,407422
Finger-flower 130383
Finger-gloves 130383
Fingergrass 88316,88421,130404
Finger-grass 88316,130404,285390
Finger-hut 130383
Finger-leaf 208051
Fingerleaf Ghost Lampstand 335142
Fingerleaf Hairorchis 148735
Fingerleaf Rodgersflower 335142
Finger-leaved Gourd 114276
Finger-leaved Ivy 187286
Finger-lime 253284,253285
Fingernail Plant 264426
Finger-root 130383
Fingers 130383
Fingers-and-thumbs 28200,106117,120310,130383,196630,222820,231183,237539,247239,247366,248411,401388,408352,408611
Fingers-and-toes 28200,237539
Fingershapesplit Sagebrush 36398
Finger-tips 130383
Finite Rockfoil 349314
Finkle 167156
Finland Pink 127596,127598
Finland Spruce 298282
Finlayson Aralia 30664
Finngerleaf Merremia 250821
Finnigig 397259
Finnish Dock 340209
Finnish Whitebeam 369429
Finocchio 167146
Finweed 271563
Finzach 308816
Fique 169242
Fir 269,298189
Fir Apple 298191,300223
Fir Bob 300223
Fir Deal Tree 300223
Fir Deal-tree 300223
Fir Helictotrichon 190130
Fir-apple 298191,300223
Fir-ball 300223
Fir-bob 300223
Fircwhcel Tree 375510
Fir-dale Tree 300223
Firda-tree 386504
Fire Birch 53586
Fire Bush 185169,217361,322454
Fire Catchfly 364175
Fire Chalice 417405
Fire Cherry 83274
Fire Cracker 114618
Fire Cracker Flower 111724
Fire Cracker Plant 341020
Fire Crowncactus 327297
Fire Cypress 85310
Fire Dahlia 121561
Fire Evax 193376
Fire Flag 388387
Fire Flags 388387
Fire Lights 410320
Fire Lily 120532,196450,229766,230058
Fire Moss 78648
Fire Pink 364175
Fire Poppy 282524
Fire Power Nandina 262192
Fire Red Tiger Lily 230064
Fire Screen 399608
Fire She Oak 79157
Fire She-oak 79157
Fire Star Orchid 146470
Fire Thorn 322448,322454,322465
Fire Tree 261160,267409,375512
Fire Weed 146614,359423
Fire Wheel Tree 375512
Fire Willow 344077,344085
Fire-ball Lily 350308
Fireberry Hawthorn 109598,109599,109603,110003
Firebush 144847,185169,217373
Fire-cherry 83274
Firecracker Bamboo 341016,341020
Firecracker Cactus 94556
Firecracker Flower 59982,59985,111724

Fire-cracker Flower 59982,59985
Firecracker Penstemon 289339
Firecracker Plant 9720,60481,114606,341020
Fire-cracker Plant 114606
Firecracker Vine 244362,255398
Firecracker-flower Azalea 331877
Firecracker-plant 9720
Fire-flag 388387
Fireflout 282685
Fireflower 246792
Firefly 114620
Fire-grass 13966,302034
Fire-king Begonia 49879
Fire-leaf 302034,302103
Fireman's Cap 154632
Fire-o'-gold 68189
Fire-on-the-mountain 158727,159046
Fire-pink 364175
Fire-plant 305185
Fire-red Azalea 331247
Firerope 152683
Fire-screen 399608
Firespike 269094,269096,269100,269102
Firesticks 179032
Firetail 1955
Firethorn 322448,322450,322454,322465
Firethornleaf Eurya 160589
Firetree 261160
Fire-tree 267409
Fireweed 85875,85880,103446,123077,146586,146587,146614,146619,148148,148153,217361,302034,302103
Fire-weed 85875
Firewheel 169575,169603,180663
Firewheel Pincushion 228111
Firewheel Tree 375512
Fire-wheel Tree 375510,375512
Fire-wheels 169603
Firewood Cinnamon 91432
Firewood Teasel 295202
Firforest Monkshood 5015
Firm Peach-leaved Eurya 160580
Firm Qiongzhu 87609
Firm Threevein Aster 39956
Firmb Bamboo 297486
Firmbamboo 297188
Firmbark Starjasmine 393642
Firmculm Capillipedium 71603
Firmleaf Hornbeam 77296
Firm-peachleaf Eurya 160580
Firmstip Primrose 314394
Fir-rape 202767
Firrope Bauhinia 49007
First of May 349419
First of the First Rhododendron 331557
First Rose 315101
First Sagebrush 36097
First Wormwood 36097
Fir-top 300223
Fischer Cenolophium 80879
Fischer Germander 388078
Fischer Goldenray 229035
Fischer Monkshood 5193
Fischer Spurge 158895
Fischer Stonecrop 356743
Fischer's Lily 229934
Fischer's Mouse-ear Chickweed 82813
Fischer's Orchid 121359
Fiscus Rubber 164925
Fish Belly 91954,92022
Fish Berries 21459
Fish Bones 9701,235387,312360
Fish Creek Fleabane 150874
Fish Geranium 288297
Fish Grass 64491,64496
Fish Killer 21459
Fish Lake Thistle 91881
Fish Mint 250291,250450
Fish Poison Bean 386368
Fish Storkbill 288297
Fish Tail Palm 78052,78056
Fishbelly Bamboo 47284
Fishberry 21458,21459,97900
Fishbone Cactus 147286
Fishbone Cotoneaster 107486
Fishbone Thistle 97618
Fish-cluster Millettia 254725
Fisher Bambusa 47412
Fisherman Bambusa 47412
Fisheye Coldwaterflower 298964
Fisheyeweed 129062
Fish-grass 64496
Fishhook Barrel Cactus 163483
Fishhook Barrel-cactus 163483
Fishhook Cactus 244087,244152,354287
Fishhook Pincushion Cactus 244087
Fish-killer Tree 48510
Fishook Cactus 163417
Fishpoison 386257
Fishpole Bamboo 297203
Fish-poly Bamboo 297203
Fishscale Bamboo 297296
Fishtail Palm 78045,78047,78053,78056
Fish-tail Palm 78045,78056
Fish-tail Palms 78045
Fishtail Wallichpalm 413086
Fishtailpalm 78045,78052
Fishvine 125937
Fishwood 110205
Fissistigma 166649,166682
Fist Knotwood 44258
Fistular Fig 164985
Fistular Onion 15289
Fistular Onosma 271767
Fistular Sweetvetch 187873
Fistular-onion-like Rush 213014
Fistulose Blumea 55736
Fistulose Onosma 271767
Fistulous Goat's-beard 394281
Fistulous Marshmarigold 68196
Fit Plant 258044
Fitch's spikeweed 81826
Fitchacks 408352,408571
Fitches 266241
Fitsroot 42417,258044
Fitted Onion 15410
Fitweed 105703,154316
Fitzgerald Sarcochilus 346698
Fitzroya 166720
Five Finger 312435,312993
Five Fingers 237539,280799,312322,312520,312925,315106,318062,381856
Five Fingers of Mary 312925
Five Fingers Syngonium 381856
Five Sisters 159027
Five Spot 263504
Five Spot Nemophila 263504
Five Stamems Mock Azalea 250528
Five Stamens Daphniphyllum 122713
Five Stamens Pittosporum 301363
Five Stamina Gaultheria 172148
Fiveangle Fimbry 166454
Five-angle Raspberry 339029
Fiveangle Snakegourd 396254
Fiveangle Woodbetony 287509
Five-angled Dodder 115099
Five-angled Raspberry 339029
Five-angular Ardisia 31578
Fiveangular Bulrush 352162
Fiveangular Desmodium 126307
Fiveangular Fimbristylis 166402,166454
Fiveangular Fluttergrass 166402,166454
Fiveangular Gymnema 182377
Five-angular Gymnema 182377
Five-angular Holly 204155
Fiveangular Prestonia 313976
Fivebract Sida 362611
Fivecalyx Coldwaterflower 298877
Fivecalyx Persimmon 132112
Five-celled Eriolaena 152692
Five-corners 379273
Fived Camellia 69466
Five-faced Bishop 8399
Five-finger 312322
Five-finger Blossom 312925
Fivefinger Chickensage 371148
Five-finger Grass 312520,312925
Fivefinger Swan Orchid 116522
Five-fingered Root 269307
Fivefingers 381855,381856

Five-fingers 237539,280799,312520,315106,318062
Fiveflower Ardisia 31380
Five-flower Rock Daisy 291328
Five-flowered Ardisia 31380
Five-flowers Enkianthus 145714
Fivefoliolate Clematis 95271
Fivefoliolate Corydalis 106349
Fivefruit Larkspur 124582
Fivehel Hedgehog Cactus 140288
Fivehook Bassia 48769
Five-hook Bassia 48769
Five-hooked Bassia 48769
Fivehorn Smotherweed 48769
Five-horned Spindle-tree 157393
Fivelaeves Twoflower Cinquefoil 312411
Five-leaf 13225
Fiveleaf Akebia 13225
Five-leaf Akebia 13225
Fiveleaf Allomorphia 16076
Fiveleaf Aralia 143677
Fiveleaf Azalea 331462
Fiveleaf Bedstraw 170584
Fiveleaf Carpetweed 256719
Fiveleaf Chaste Tree 411420
Fiveleaf Chastetree 411420
Fiveleaf Goldthread 103843
Fiveleaf Gynostemma 183023
Fiveleaf Landpick 308495
Fiveleaf Maple 3396
Fiveleaf Raspberry 339149
Fiveleaf Snakegourd 396255
Fiveleaf Sneezeweed 188397
Fiveleaf Sorrel 278019
Fiveleaf Sumac 332766
Five-leaf Sumac 332766
Fiveleaf Tubergourd 390153
Fiveleaf Whitehair Cinquefoil 312724
Fiveleaf Wildberry 362358
Fiveleaf Yam 131759
Fiveleaflet Strawberry 167645
Fiveleaflets Oldham Sping Raspberry 339149
Five-leaved Chaste-tree 411420
Five-leaved Glycosmis 177819
Five-leaved Grass 153767,168476,312520,312925
Five-leaved Ivy 20361,285136
Fiveleaved Japanese Raspberry 338997
Five-leaved Lacquer-tree 393469
Five-leaved Lady's Mantle 13959
Five-leaved Mahonia 242624
Fiveleaved Raspberry 339075
Five-leaved Raspberry 339075,339149
Five-leaved Rhododendron 331462
Five-leaved Sumach 332766
Fiveleaved Whiteleaf Raspberry 338633
Five-leaved Wood-sorrel 278019
Fiveleaved Woollystyle Raspberry 338722
Five-leaves Azalea 331593
Fiveleaves Gentian 174056
Fiveleaves Japanese Pagodatree 369045
Five-leaves Rhododendron 331593
Fiveleaves Sibbaldia 362358
Fiveleaves Snakegourd 396255
Fiveleaves Tubergourd 390153
Fivelobe Cacalia 283862
Fivelobe Goldthread 103845
Fivelobed Acute Maple 2772
Fivelobed Ainsliaea 12679
Fivelobed Bur Biebersteinia 54166
Fivelobed Griffith Bittercress 72794
Fivelobed Makino's Rhododendron 331189
Fivelobed Raspberry 338758
Five-lobed Raspberry 338758
Fivelobed Yam 131804
Fivelocular Brassaiopsis 59232
Fivelocular Camellia 69686
Five-locular Camellia 68927
Fivelocular Eurya 160590
Five-locular Eurya 160590
Five-locules Camellia 68927
Fiveloculous Bigumbrella 59232
Five-needled Fetid Marigold 391042
Fivenerve Mistletoe 411061
Five-nerve Muli Maple 3213
Fivenerved Gongora 179296
Five-nerved Mistletoe 411061
Fivenerved Obliquecalyxweed 237960
Fivenodes Awngrass 255849
Five-parted Toothwort 72723
Five-partite Rhododendron 330532
Five-petal Camellia 69468
Fivepetal Eremotropa 148530
Fivepetal Helixanthera 190848
Fivepetal Mischocarpus 255951
Fivepetal Pearweed 239781
Fivepetal Scurrula 190848
Fivepistil Sandwort 32148
Five-point Bishop's-cap 256037
Fiverib Elatostema 142806
Fiverib Eria 299633
Fiverib Hairorchis 299633
Fiverib Stairweed 142806
Fiveribbed Thyme 391347
Five-ribbed Thyme 391347
Fivering Qinggang 116176
Fiveroom Firerope 152692
Fiveroom Tea 69669
Fivesepal Clearweed 298877
Fivespined Hornwort 83577
Fivesplit Goldthread 103845
Five-spot Baby 263504
Five-spot Nemophila 263504
Fivestamen Chickweed 83020
Five-stamen Chickweed 83020
Fivestamen Dendrophthoe 125667
Five-stamen Dendrophthoe 125667
Fivestamen Glutinousmass 179495
Fivestamen Mastixia 246217
Fivestamen Miterwort 256037
Five-stamen Mouse-ear Chickweed 83020
Fivestamen Pygmyweed 391953
Fivestamen Sphenodesma 371421
Fivestamen Tamarisk 383469,383591
Five-stamen Tamarisk 383469,383595
Fivestamen Tillaea 391953
Fivestamen Whitepearl 172148
Fivestamen Wintergreen 172148
Five-stamened Belltree 325001
Five-stamened Wintergreen 172148
Five-stamens China Laurel 28381
Five-stamens Laurel 28378
Fivestyle Camellia 69474
Fivestyle Eurya 160578
Fivestyle Schefflera 350759
Fivestyle Tea 69474
Five-styled Camellia 69474
Fivetooth David Woodbetony 287147
Fivetooth Mount Willow 343813
Fivevein Meconopsis 247175
Fivevein Melaleuca 248114
Fivevein Vetchling 222828
Fiveveine Rattanpalm 65775
Five-veined Hairyleaf Maple 3634
Five-veined Lovegrass 147931
Five-veined Rattan Palm 65775
Fivewing Agapetes 289750
Five-winged Macleania 240758
Fixed-petal Vine 257432
Fixweed 126112
Fizgerald Little Cone 254091
Fizz-gigg 359158
Flabby Dock 130383
Flabby Small Reed 127234
Flabby Smallreed 127234
Flabellate Hemipilia 191602
Flabellate Nengella 263555
Flabellate Rattan Palm 65692
Flabellate Willow 343381
Flabellateleaf Alumroot 194420
Flabellateleaf Maple 2964
Flabellateleaf Myrothamnus 261542
Flabellateleaf Oxalis 277841
Flabellateleaf Rockfoil 349862
Flabellateleaf Willow 343381
Flabellateleaf Woodsorrel 277841
Flabellate-leaved Maple 2964
Flaccid Anemone 23817
Flaccid Aster 40440
Flaccid Bentgrass 12104
Flaccid Bluegrass 305530

Flaccid Canna 71171
Flaccid Coelogyne 98653
Flaccid Conehead 47845
Flaccid Corydalis 105878
Flaccid Everlasting 21567
Flaccid Hymenodictyon 201066
Flaccid Jurinea 214086
Flaccid Knotweed 309116
Flaccid Leaf Yucca 416595
Flaccid Leucas 227597
Flaccid Pearleverlasting 21567
Flaccid Pennisetum 289096
Flaccid Rockfoil 349317
Flaccid Spapweed 204954
Flaccid Touch-me-not 204954
Flaccid Valerian 404269
Flaccid Woodbetony 287208
Flaccidest Mayten 246841
Flaccidleaf Chinese Fir 114541
Flaccidleaf Larkspur 124372
Flaccid-leaf Yucca 416595
Flacourtia 166761
Flacourtia Family 166793
Flag 208433,208583,208744
Flag Grass 372629,372630
Flag Iris 208583
Flag Lily 208771,208924
Flag Root 5793
Flag Sedge 208771
Flagelibranch Valerian 404270
Flageliforme Typhonium 401160
Flagellar Cinquefoil 312540
Flagellaria 166795
Flagellaria Family 166801
Flagellate Ant Palm 217921
Flagellate Bittersweet 80189
Flagellate Dewberry 338399
Flagellate Goldwaist 90362
Flagellate Peristylus 291211
Flagellate Perotis 291211
Flagellate Rockfoil 349319
Flagellate-flowered Raspberry 338407
Flagelliflory 146108
Flagellum Rattan Palm 65694
Flageolet 294056
Flageolet Bean 294056
Flagliporchis 218294,218296
Flagons Flaggers 208771
Flag-pawpaw 38327,38329,38334
Flagroot 5793
Flags 68713,208583,401112
Flagstaff Sulphur Flower 152575
Flake Flower 81020
Flaky Fir 485
Flaky Juniper 213943
Flam Buttercup 326303
Flambeau Tree 370351,370358
Flambeautree 370351,370374
Flambeau-tree 370351,370358
Flamboyant 123801,123811
Flamboyant Tree 123801,123811
Flamboyanttree 123801,123811
Flame Adonis 8362
Flame Anemone 23823
Flame Anisacanthus 25261
Flame Azalea 330276,331161
Flame Bottle Tree 58335
Flame Bottle-tree 58335
Flame Buckeye 9729
Flame Climber 399608
Flame Coral Tree 154643,154646
Flame Coral-tree 154646
Flame Creeper 100627,399608
Flame Flower 211068,217055,322950,383304,410677
Flame Gold-tips 228061
Flame Grass 255889,357174
Flame Grevillea 180588
Flame Heath 43452,149446,379276
Flame Ixora 211037
Flame Kurrajong 58335
Flame Lily 97213,177251,229984
Flame Nasturtium 399608
Flame Nettle 99506,99711,367870
Flame of the Forest 64148,123811,260414,370358
Flame of the Wood 211068
Flame of the Woods 211068
Flame Pea 89134,89135,89136
Flame Plant 28119
Flame Ragwort 279914
Flame Tasselflower 144903
Flame Tree 58335,123811,125937,159675,267409,330111,370358
Flame Vine 322950
Flame Violet 147384,147386,147391,147393
Flame Vriesea 412350
Flamecolored Groundsel 385892
Flamecoloured Adonis 8362
Flame-coloured Inflorescence Helicia 189947
Flamecoloured Pyrostegia 322950
Flameflower 294318
Flame-flower 216923,217055
Flamegold Tree 217620
Flamegoldrain Tree 217620
Flamegold-tree 217620
Flamegrass 79113,79130
Flameleaf Sumac 332529
Flame-leaf Sumac 332529
Flame-leaved Crowfoot 325864
Flamenettle 99506
Flame-of-the-forest 64148,123811,370358
Flame-of-the-woods 211029,211068
Flameorchis 327681,327685
Flameray Gerbera 175172
Flame-ray Gerbera 175172
Flamered Goblin Orchid 258997
Flame-tree 58335,267409
Flamevine 322950
Flame-violet 147386
Flaming Katy 215102
Flaming Sword 217055,412371
Flamingo Boxelder 3221
Flamingo Celery 269326
Flamingo Flower 28071,28084,28119
Flamingo Lily 28084
Flamingo Plant 28071,28119
Flamingo Primrose 314664
Flamtree 370351,370374
Flamy 410677
Flander's Poppy 282685
Flanders Poppy 282685
Flanged Brome 60807
Flannel 405788
Flannel Bush 168215,168216
Flannel Flower 6938,6939,168215,296283,405788
Flannel Jacket 405788
Flannel Leaf 43550,405788
Flannel Mullein 405788
Flannel Petticoat 405788
Flannel Petticoats 405788
Flannel Plant 405788
Flannel-bush 168216
Flannel-leaved Mullein 405788
Flap Dock 130383
Flap-a-dock 130383
Flapdick 130383
Flapdock 130383
Flapper Bags 31044,31051
Flapper Dock 130383,292372
Flappy Dock 130383
Flare's Eyes 248312
Flash Camellia 69319
Flask-shaped Orange 93323
Flat Buerger Maple 2816
Flat Camellia 69400
Flat Crown 13476
Flat Crown Tree 13569
Flat Gentian 173395
Flat Milkvetch 42208
Flat Pea 222844
Flat Pea Vine 222844
Flat Peach 20952
Flat Peavine 222844
Flat Pea-vine 222844
Flat Pra 302848
Flat Sedge 118428,118822,118829,393205
Flat Siebold Walnut 212602
Flat Spine Prickly-ash 417340

Flat Tree Dahlia 121553
Flat Trident Maple 2816
Flat Vervain 407072
Flatawndaisy 413007,413021
Flatball Liparis 232151
Flatball Twayblade 232151
Flatbract Dolomiaea 135734
Flatbranch Biond's Magnolia 242008
Flat-crown Tree 13569
Flat-culm Oat 45466
Flatculm Sedge 75827
Flatenfruit Machilus 240675
Flaten-fruited Machilus 240675
Flatfruit Buttercup 326286
Flatfruit Camellia 69002
Flatfruit Gynostemma 183004
Flatfruit Meadowrue 388511
Flatfruit Oak 116072
Flat-fruit Oppositeleaf Fig 165126
Flatfruit Qinggang 116072
Flatfruit Sawtooth Oak 323617
Flat-fruited Cyclobalanopsis 116072
Flat-fruited Meadow Rue 388673
Flat-fruited Oak 116072
Flat-fruited Supple-jack 52410
Flatfruited Yellowwood 94030
Flat-fruited Yellowwood 94030
Flathair Aster 40178
Flathead Cousinia 108323
Flathead Larkspur 124066
Flathead Vinegentian 108323
Flat-horn Orach 44636
Flatleaf Eryngo 154337
Flat-leaf Palm 198808
Flatleaf Stonecrop 357026
Flatleaf Vanilla 405017
Flat-leaved Bladderwort 403220
Flat-leaved Indocalamus 206820
Flat-leaved Palm 198808
Flat-leaved Willow 343894
Flatlobed Pittosporum 301370
Flat-lobed Pittosporum 301370
Flatlobed Seatung 301370
Flatnode Riparial Silverreed 394939
Flatpetal Gloriosa 177247
Flatpetiolate Basketvine 9473
Flat-petiole Spindle-tree 157797
Flatpod 203398
Flat-pod Pea Vine 222694
Flatpodded Conringia 102815
Flatraceme Milkvetch 42211
Flatrund Petrocosmea 292573
Flatrund Stonebutterfly 292573
Flatscape Irisorchis 267978
Flatsedge 118428,322179,393205
Flat-sedge 55889,55892,118428
Flatsedge Courtoisia 108215
Flatsedgeleaf Cymbidium 116813
Flatsedge-like Courtoisia 108215
Flatsheath Fimbristylis 166223
Flatsheath Fluttergrass 166223
Flatshoot Mistletoe 410979
Flat-shooted Mistletoe 410979
Flatspike 55889,55892
Flatspike Oxwhipgrass 191228
Flatspine Evergreenchinkapin 79004
Flatspine Peashrub 72191
Flatspine Prickly Ash 417340
Flat-spine Prickly Ash 417180
Flatspine Pricklyash 417340
Flatspine Prickly-ash 417340
Flatspine Rose 336981
Flat-spined Evergreen Chinkapin 79004
Flatstalk Bulrush 353707
Flatstalk Greenbrier 366523
Flat-stalk Greenbrier 366523
Flatstalk Lilyturf 272069
Flatstalk Smilax 366523
Flat-stalked Pondweed 312123
Flatstalkgrass 220281
Flatstem Dutchmanspipe 34151
Flatstem Liparis 232083
Flatstem Milkvetch 42208
Flatstem Pondweed 312299
Flat-stem Pondweed 312075,312299
Flatstem Rockvine 387837
Flatstem Rush 213008
Flat-stemmed Meadow-grass 305451
Flat-stemmed Pondweed 312075,312299
Flat-stemmed Rockvine 387837
Flat-stemmed Spike Rush 143096
Flat-stemmed Spike-rush 143096
Flatstiped Corydalis 106170
Flatstylebase Spikesedge 143145
Flattend Oat Grass 122207
Flattened Blysmus 55892
Flattened Cactus 244003
Flattened Culm Bluegrass 305875
Flattened Epidendrum 146377
Flattened Fig 165126
Flattened Hemarthria 191228
Flattened Laelia 219670
Flattened Meadow-grass 305451
Flattened Pansy Orchid 254926
Flattened Phacellaria 293191
Flattened Wild-oat-grass 122207
Flattened Yelloweyegrass 416037
Flattened-axis Pricklyash 417329
Flattened-fruit Supplejack 52410
Flattenedleaf Little Man Orchid 178896
Flatterbaw 292372
Flatterdock 267294,267648
Flattering Anemone 23724
Flat-top Acacia 1026,1027
Flattop Ageratum 11203
Flat-top Aster 41448
Flat-top Pussytoes 26375
Flat-top Whiteweed 11203
Flat-topped Aster 41448
Flat-topped Skeleton Weed 151981
Flat-topped Viburnum 408210
Flat-topped White Aster 41448
Flattopped Yate 155674
Flat-tops 406631
Flatutricle Sedge 74187
Flatweb Sandwort 31822
Flatweed 202432
Flaunt Azalea 331215
Flaveria 166814
Flavescent Polystachya 310365
Flavescet Hooktea 52416
Flavescet Supplejack 52416
Flavescet Supple-jack 52416
Flaviflower Michelia 252871
Flaw-flower 23925,321721
Flax 231856,231884,232001
Flax Dodder 115007,115008
Flax Family 230854
Flax Lily 127504,127505,127511,127517,295594,295600
Flax Spurge 159526
Flax Wattle 1355
Flaxaster 231818
Flaxfield Catchfly 363690
Flaxfield Rye-grass 235353
Flaxflower Gilia 175691
Flax-flowered Ipomopsis 208320
Flaxleaf Alyssum 18407
Flax-leaf Ankle-aster 207397
Flaxleaf Aster 207397
Flaxleaf Buchea 57827
Flaxleaf Diarthron 128026
Flaxleaf Fanpetals 362575
Flaxleaf Iris 208689
Flaxleaf Knema 216851
Flaxleaf Mask Flower 17481
Flaxleaf Navelseed 270705
Flaxleaf Paperbark 248105
Flaxleaf Pimpernel 21383
Flaxleaf Seepweed 379547
Flaxleaf Stringbush 414211
Flaxleaf Whitetop 207397
Flax-leafed Paperbark 248105
Flax-leaved Aster 40767
Flax-leaved St. John's Wort 201992
Flax-leaved Stringbush 414211
Flaxlike Gentian 173599
Flax-lily 127504
Flaxseed 325039,325041
Flaxseed Plantain 301864
Flax-seed Plantain 301864

Flaxtail 401112
Flaxuose Corydalis 105881
Flaxweed 68850,220790,231183
Flea Dock 292372
Flea Mint 250432
Flea Nit 301864,359158
Flea Nut 301864,359158
Flea Sedge 75929
Fleabane 103402,103446,135049,150414,150419,150444,150457,150594,150655,150658,151005,207025,305072,321509,321554,363214
Fleabane Daisy 150414,150594
Fleabane-mullet 321554
Fleabite 248312
Fleas-and-lice 116736
Fleaseed 301864
Fleaweed 170743,321509,321595
Fleawood 261162
Fleawori 385899
Fleawort 103446,207087,301864,359158,385884
Flecampane Inula 207135
Fleece Flower 162505,308690,308769
Fleece Vine 162509
Fleece-bearing Rhododendron 331503
Fleeceleaf Azalea 330809
Fleeceleaf Rhododendron 330809
Fleecy Rhododendron 331195,332053
Flee-dod 359158
Fleischer Willowweed 146700
Fleme Red Rhododendron 331247
Fleme-red Rhododendron 331247
Flemingia 166850
Flesg-coloured Lycoris 239262
Flesh Ching Arrowwood 407744
Flesh-and-blood 312520
Fleshcolor Apios 29301
Fleshcolor Azalea 330326,330911
Fleshcolor Pincushion 84535
Fleshcolored Rhododendron 330326
Flesh-colored Rhododendron 330326
Fleshcolour Apios 29301
Fleshcolour Habenaria 183486
Flesh-colour Pholidota 295517
Flesh-coloured Bamboo 297327
Fleshcoloured Bitter Bamboo 304043
Flesh-coloured Bitter-bamboo 304043
Flesh-coloured Buckeye 9680
Flesh-coloured Caladenia 65205
Fleshcoloured Condorvine 245791
Flesh-coloured Manyflowered Rose 336789
Fleshcoloured Milkgreens 245791
Fleshcoral 346993
Fleshfingered Citron 93604
Fleshflower Dendrobium 125039
Fleshflower Snowgall 191899
Fleshleaf Eargrass 187549
Fleshleaf Hedyotis 187549
Fleshred Coelogyne 98624
Fleshseed Tree 346979
Flesh-seed Tree 346979
Fleshspike 346935,346936
Fleshspike Tree 346979
Fleshstem Ophiorrhiza 272188
Fleshtepal 346720,346721
Fleshy Coralflower 120526
Fleshy Cyrtanthera 120526
Fleshy Hawthorn 110065
Fleshy Honeysuckle 235714
Fleshy Lobelia 234807,234884
Fleshy Lysionotus 239934
Fleshy Pincushion 84535
Fleshy Rhododendron 331180
Fleshy Scales Rhododendron 331089
Fleshy Stitchwort 374807
Fleshy Willow Dock 339993
Fleshy Yellow-sorrel 277959
Fleshycolored Orchis 273469
Fleshy-flower Hemsleia 191899
Fleshyfruit Embelia 144728
Fleshy-fruited Embelia 144728
Fleshy-fruited Yucca 416556
Fleshyleaf Actinidia 6545
Fleshyleaf Bassia 48790
Fleshyleaf Coleus 99539
Fleshyleaf Flamenettle 99539
Fleshyleaf Kiwifruit 6545
Fleshy-leaved Actinidia 6545
Fleshy-leaved Honeysuckle 235714
Fleshyorchis 346899
Fleshyroot Buttercup 326241
Fleshy-stalked Slugwood 50555
Fleshy-trunk Treebine 120108
Fletcher Cypress 85280
Fletcher Woodbetony 287213
Fletcher's Rhododendron 330708
Flett's Fleabane 150637
Flett's Ragwort 279910
Flettened Litse 233888
Fleur-de-lis 208433,208583
Fleur-de-lys 208583
Fleur-de-paradis 123811
Fleury Camellia 69075
Fleury Cyclobalanopsis 116092
Fleury Elaeocarpus 142316
Fleury Evergreen Chinkapin 78939
Fleury Evergreenchinkapin 78939
Fleury Oak 116092
Fleury Podocarpus 306429,306435
Fleury Yaccatree 306429,306435
Flex Corydalis 106477
Flex Lovegrass 147938
Flex Rochelia 335107
Flex Rush 213165
Flexbract Ironweed 406827
Flexbract Liparis 232360
Flexbract Twayblade 232360
Flexed Hanceola 185248
Flexed-flower Milkvetch 42371
Flexedstem Rabdosia 209673
Flexhair Gyrocheilos 183320
Flexible Pine 299942
Flexiblebranch Primrose 314394
Flexible-stemmed Rabdosia 209673
Flexlip Liparis 232101
Flexlip Platanthera 302300
Flexlip Twayblade 232101
Flexpetal Primrose 314093
Flexspike Lemongrass 117171
Flexstalk Ardisia 31589
Flexstem Dendrobium 125146
Flexstem Gentian 173453
Flextine Azalea 330028
Flexuose Bamboo 47275,297281
Flexuose Barberry 51621
Flexuose Bittercress 72749
Flexuose Bluegrass 305569
Flexuose Casearia 78118
Flexuose Hanceola 185248
Flexuose Rockfoil 349337
Flexuose Vallesia 404528
Flexuose-like Bittercress 73023
Flexuosus Leptopus 296565
Flexuous Meliosma 249388
Flexuous Tickclover 126348
Flexuouse Hawksbeard 110806
Flibberty-gibbet 243840
Flicate Sweet-grass 177648
Flickingeria 166954,166959
Flig 374968
Fliggers 208771
Flimbract Everlasting 21584
Flinders Grass 209469
Flinders Range Wattle 1314
Flindersia 166971
Flint Corn 417431
Flint Maize 417431
Flint Wheat 398900
Flintwood 155696
Flirtwort 322724
Flixweed 126127
Flix-weed 126127
Flixweed Tansy Mustard 126127
Flixweed Tansy-mustard 126127
Flixwort Flixweed 126127
Floadng Hearts 267674
Floating Bogbean 267825
Floating Bur-reed 370030,370062
Floating Buttercup 326114
Floating Club-rush 210002,353158,353179

Floating Dock 267294,267648
Floating Foxtail 17498,17527
Floating Heart 267800,267825
Floating Hearts 267800,267801,267806
Floating Knotweed 308739
Floating Leaf Pondweed 312190
Floating Manna Grass 177655
Floating Marsh Marigold 68186
Floating Marshmarigold 68186
Floating Marsh-marigold 68186
Floating Orchid 184021
Floating Ottelia 277389
Floating Pennywort 200354
Floating Pondweed 312188,312190
Floating Primrose Willow 238203
Floating Primrose-willow 238204
Floating Sweet-grass 177599
Floating Water Plantain 238510,238511
Floating Water-plantain 238510,238511
Floatingheart 267800
Floatingleaf Pondweed 312190
Floating-leaf Pondweed 312188,312190
Floating-leaved Bur-reed 370062
Floatleaf Arrowhead 342385
Floatleaf Pondweed 312190
Flobby Dock 130383
Floccose Beautyberry 66843
Floccose Catnip 264924
Floccose Chonemorpha 88892
Floccose Nepeta 264924
Floccose Purplepearl 66843
Floccose Willow 343385
Floccosous Speedwell 407019
Flock Nematanthus 263325
Flockyhead Bethlehemsage 283653
Floderus Willow 343388
Flodman's Thistle 91976
Flooded Box 155655
Flooded Gum 155598,155727,155756
Flop 130383
Flop Dock 130383
Flop Docken 130383
Flop Poppy 130383
Flop-a-dock 130383
Flop-dock 130383
Flopdocken 130383
Flopper 61568
Floppers 61572
Flop-poppy 130383
Floppy Dock 130383
Flop-top 130383
Flor Impia 289684
Flora's Clock 394333
Flora's Paintbrush 144976
Floramor 18687,18836
Floramour 18836
Floras-paintbrush 144903
Floren Tine Iris 208565
Florence Court Yew 385302,385314
Florence Fennel 167146,167164
Florence Flower 266241
Florence Whisk 369600
Florentine Iris 208565
Florentine Tulip 400229
Flores De Palo 295575
Floribunda Rose 336641
Florida Adder's-mouth 243130
Florida Amaranth 18717
Florida Anise Tree 204517
Florida Anisetree 204517
Florida Anise-tree 204517
Florida Arrowroot 417013,417017
Florida Azalea 330190
Florida Bean 259573
Florida Beargrass 266484
Florida Bellwort 403669
Florida Betony 373209
Florida Bitterbush 298500
Florida Box 350574
Florida Boxwood 350574
Florida Butterfly Orchid 145298
Florida Cedar 213623
Florida Celestial 263310
Florida Cherry 156385
Florida Chinchweed 286893
Florida Clamshell Orchid 315616
Florida Cockspur 140472
Florida Coneflower 339580
Florida Crinum 111158
Florida Dock 340056
Florida Dollar-orchid 315614
Florida Dutchman's Pipe 34271
Florida Dutchman's-pipe 34271
Florida Eightangle 204517
Florida False Sunflower 295342
Florida Flame Azalea 330190
Florida Gambirplant 401753
Florida Gayfeather 228496
Florida Goldenaster 90198
Florida Grayback Grape 411618
Florida Hopbush 135215
Florida Hopseedbush 135215
Florida Keys Umbrella Thoroughwort 217113
Florida Lettuce 219321
Florida Leucothoe 228155
Florida Linden 391715
Florida Lobelia 234487
Florida Longleaf Pine 300128
Florida Maple 2797,2966
Florida Milkweed 37997
Florida Moss 392052
Florida Mudmigdet 414684
Florida Paintbrush 77228
Florida Palmetto 341422
Florida Panhandle Spider-lily 200919
Florida Paspalum 285437
Florida Pigeon Grape 411618
Florida Pine 300128
Florida Pinxter 330322
Florida Pondweed 312115
Florida Rhododendron 330190
Florida Rosemary 83398
Florida Royal Palm 337879,337885
Florida Semaphore Cactus 102821,102826
Florida Silver Palm 97885
Florida Silverbell 184749
Florida snake-cotton 168687
Florida Spider-lily 200957
Florida Strangler Fig 164659
Florida Sugar Maple 2797,2966,3562
Florida Sunflower 188967
Florida Swamp-lily 111158
Florida Tasselflower 144916
Florida Tassel-flower 144916
Florida Thatch Palm 390463,390467
Florida Thistle 92036
Florida Thoroughwort 158042
Florida Torreya 393089
Florida Trema 394652
Florida Velvet Bean 259558,259559
Florida Velvet Mucuna 259559
Florida Waltheria 413149
Florida water Aster 380897
Florida Water Cress 336197
Florida Water-hemp 18717
Florida Willow 343391
Florida Yellowtops 166821
Florida Yellowtrumpet 385493
Florida Yew 385375
Florida Zamia 417006
Florida-anise 204517
Florida-dandelion 52836
Florimer 18687,18836
Florin 11972,12323
Florind Primrose 314407
Floripondio Datura 61231
Florist Chrysanthemum 124826
Florist Gayfeather 228529
Florist's Calla 417093
Florist's Candytuft 203249
Florist's Chrysanthemum 124767
Florist's Cyclamen 115965
Florist's Daisy 89619
Florist's Gloxinia 364754
Florist's Hydrangea 199956
Florist's Pyrethrum 322665
Florist's Spiraea 41817
Florists Chrysanthemum 124785
Florists Cineraria 290821,358642
Florists Daisy 124785
Florists Dendranthema 124785,124826

Florists Dendranthemum 124826
Florists Pyrethrum 322665
Florists Violet 410320
Florists' Chrysanthemum 124826
Florists' Cineraria 290821
Florists' Larkspur 124301
Florists' Pyrethrum 322665
Florrush 64180,64184
Flos Adonis 8325,8332
Floscopa 167010,167040
Flosculous Greenstar 307497
Flosdocken 130383
Floss 213021
Floss Azalea 331430
Floss Bethlehemsage 283609
Floss Bushclover 226989
Floss Butea 64145
Floss Cliffbean 254876
Floss Docken 130383
Floss Flemingia 166867
Floss Flower 11194,11206
Floss Greyawngrass 372410
Floss Milkvetch 42483
Floss Mountain Leech 126663
Floss Ostodes 276788
Floss Paraphlomis 283609
Floss Photinia 295771
Floss Rhodoleia 332206
Floss Rhynchosia 333389
Floss Seaves 152769
Floss Silk Tree 88973
Floss Spodiopogon 372410
Flossback Azalea 330902
Flossback Windhairdaisy 348919
Floss-docken 130383
Flossflower 11194
Flossfruit Crazyweed 278824
Flossfruit Partridgeflower 395538
Flosshead Elatostema 142656
Flosshead Stairweed 142656
Flossleaf Cousinia 108321
Flossleaf Dragonhead 137689
Flossleaf Greenorchid 137689
Flossleaf Vinegentian 108321
Flosspetiole Clovershrub 70902
Floss-seaves 152769
Floss-silk Tree 80118,88972,88973
Flossstalk Chickweed 374937
Flote Grass 177599
Flour Corn 417414
Flour Maize 417414
Flous Docken 130383
Flous Taiwania 383189
Flous-docken 130383
Flower Amor 18836
Flower Arachnis 30532
Flower Armour 18687
Flower Axistree 177845
Flower Bamboo 297359
Flower De Luce 208771,229789
Flower Dust Plant 215238
Flower Fence 65055
Flower Flames 399608
Flower Gentle 18836
Flower Hedge 26950
Flower of a Day 394088
Flower of an Hour 195326
Flower of Bristol 363312
Flower of Bristow 363312
Flower of Constantinople 363312
Flower of Death 409339
Flower of Dunluce 174832
Flower of Heaven 83545
Flower of Jove 363462
Flower of Spring 50825
Flower of St. Sebastian 79574
Flower Spurge 158692
Flower Velure 18687
Flower-deluce 208771
Flower-de-Luce 229789
Floweret Azalea 330717
Floweret Rhododendron 330717
Flowerfence Poinciana 65055
Flower-fence Poinciana 65055
Flowering Almond 20970,83208,316210
Flowering Apricot 34448
Flowering Ash 167949,167953,168041
Flowering Banana 260217,260252
Flowering Box 404051
Flowering Bran 293985
Flowering Cabbage 59524
Flowering Cassia 360429
Flowering Cherry 83314,83365
Flowering Crab 243657
Flowering Crabapple 243691
Flowering Crab-apple 243617,243623
Flowering Currant 333919,334189
Flowering Dogwood 105023,105130
Flowering Flax 231907
Flowering Gum 106800
Flowering Kerria 216085
Flowering Lignum 148370
Flowering Malpe 837,913
Flowering Maple 957
Flowering Moss 323108,323366
Flowering Nutmeg 228321
Flowering Peach 20942
Flowering Pear 323116
Flowering Plum 20970,316294,316302
Flowering Polygala 308253
Flowering Quince 84549,84573
Flowering Raspberry 338917
Flowering Reed 71156
Flowering Rush 64180,64184
Flowering Rush Family 64173
Flowering Sally 240068
Flowering Spurge 158692
Flowering Stones 233459
Flowering Tobacco 266031,266035,266054,266059
Flowering Willow 87449
Flowering Wintergreen 308253
Flowering Withy 85875
Flowering-crab 243543
Flowering-malpe 837,913
Flowering-quillwort 229687
Floweringquince 84549
Flowering-quince 84549,84573
Flowering-quince Willow 343185
Flowering-rush 64180,64184
Floweringrush Family 64173
Flowering-Rush Family 64173
Flowering-wintergreen 308253
Flower-of-a-day 394088
Flower-of-an-hour 195326,243840
Flower-of-an-hour Hibiscus 195326
Flower-of-Jove 363462
Flower-of-Jupiter 363462
Flower-of-the-hour 195326
Flower-of-the-Incas 71547
Flower-of-the-sun 261739
Flower-of-the-western-wind 417612
Flower-river Tylophora 400849
Flowers Acemes Distylium 134946
Flowery Acacia 1528
Flowery Aspidopterys 39675
Flowery Azalea 330335
Flowery Bogorchid 158038
Flowery Bushclover 226793
Flowery Caper 71749,71809
Flowery Chickweed 375018
Flowery Chirita 87863
Flowery Clovershrub 70866
Flowery Coelogyne 98753
Flowery Corallodiscus 103964
Flowery Corydalis 106218
Flowery Cotoneaster 107583
Flowery Cranesbill 174829
Flowery Crazyweed 278831
Flowery Dendrobium 125109
Flowery Docken 86970
Flowery Fragileorchis 2046
Flowery Hairgrass 126095
Flowery Jasminanthes 245803
Flowery Kitefruit 196843
Flowery Lecanorchis 223656
Flowery Leptoboea 225896
Flowery Magnolia 416710
Flowery Marsh Felwort 235463
Flowery Michelia 252872
Flowery Milkvetch 42280,42374

Flowery Nailorchis 9310
Flowery Orchis 116863
Flowery Purplevine 414554
Flowery Rabdosia 324845
Flowery Rush 213297
Flowery Saddletree 240124
Flowery Senna 78301,360429
Flowery Silkvine 291060
Flowery Sphenodesma 371418
Flowery Spicebush 231339
Flowery Spice-bush 231339
Flowery Stonebean-orchis 62955
Flowery Summerlilic 62127
Flowery Sweetleaf 381381
Flowery Tupelo 267867
Flowery Violet 410464
Flowery Wallichpalm 413091
Flowery Whiteheadtree 171562
Flowery Willow 344095
Flowery Woodrush 238647
Flowery Woolly Buttercup 326073
Flowery Zinnia 418058
Flowster Docken 130383
Flowster-docken 130383
Flox 130383
Flue Arrowwood 408202
Flue Brush 401112
Flue Brushes 401112
Flueggea 167067
Fluellen 216239,216262,216292,407256
Fluellin 216239
Fluff Grass 152719,397772
Fluff-grass 152719
Fluffweed 165906,165915,235253,235261,405788
Fluffy Buttons 343151
Fluffy-puffy 384714
Fluted Pumpkin 385654
Fluttergrass 166156
Fluttering Elm 401549
Fluvial Hemigraphis 360701
Fluvial Ninenode 319548
Fluvial Pipewort 151306
Fluvial Psychotria 319548
Fluvial Syzygium 382534
Fluxweed 126127,396436
Fly 363477
Fly Flower 316127
Fly Honey Suckle 236226
Fly Honeysuckle 235710,235874,236224,236226
Fly Marshelder 210474
Fly Ophrys 272466
Fly Orchid 272466
Fly Orchis 272444
Fly Poison 19471
Fly-angels 3462
Flyaways 3462
Flycatcher 37015,138348,299759,347156
Fly-catcher Rhododendron 330280
Fly-dod 359158
Fly-honeysuckle 236226
Flying Dragon Trifoliate-orange 310852
Flying Dutchman 3462
Flying Dutchmen 3462
Flying Moth Tree 2775
Flying-angels 3462
Flylike Primrose 314684,314686
Fly-orchid 272444,272466
Fly-Poison 19468
Flytrap 29468,138348,347156
Fly-trap 138348
Fly-wing Tickclover 126651
Foal's Foot 400675
Foal's-foot 400675
Foalfoot 37616,325820,400675
Foal-foot 400675
Foal's Foot 176824
Foalsfoot 400675
Foalswort 400675
Foam Dock 346434
Foam Flower 391520,391522
Foam of May 371815
Foamflower 391520,391522
Foam-flower 391522
Foam-of-may 371815
Foamy Gentian 173237
Focke Raspberry 338418
Fodder Beet 53249,53262
Fodder Burnet 345866
Fodder Pea 222680
Fodder Vetch 408571,408693
Foeecy Hair Rhododendron 330809
Foefid Camomile 26768
Foeniy-tree 166627
Foerstermann Calanthe 66001
Foetid Blueberry 403828
Foetid Cranesbill 174877
Foetid Elder 345708
Foetid Eryngo 154316
Foetid Fevervine 280078
Foetid Goosefoot 87045,139693
Foetid Horehound 47013
Foetid Momordica 256822
Foetid Nothapodytes 266805
Foetid Viburnum 407844
Foetid Wild Gourd 114278
Foetid Yew 393047,393089
Foetld Cassia 360493
Fog Crocus 99297
Fog's Hemistepta 191666
Fogfruit 232485,232496
Foggy Gentian 173667
Fogs Cat 336419
Fogweed 44391
Fogwort 325820
Fohai Bulrush 352185
Foile Foot 400675
Foile-foot 400675
Fokien Cypress 167200,167201
Fokien Microtropis 254293
Foldcalyx Azalea 330188
Foldleaf Daylily 191316
Foldwing 129308
Folesfoth 176824
Folgner Mountainash 369389
Folgner Mountain-ash 369389
Folhado 96464
Foliage Beet 53257
Foliage Flower 60057,296475
Foliate Bethlehemsage 283613
Foliate Paraphlomis 283613
Foliated Kalidium 215327
Foliole Bugbane 91013
Foliose Cleistogenes 94609
Foliose Goosefoot 87016
Foliose Kitagawa Hideseedgrass 94609
Foliose Thistle 91977
Foliosous Parnassia 284539
Follan-fing 170611
Followers 170193,338447
Follow-my-lad 170193
Folly's Flower 30081
Fomes Heritiera 192491
Fonio 130544
Fontanesia 167230
Fontinal Pondweed 312120
Foo 358062
Foochow Yam 131597
Food Inga 206928
Food of the Gods 163683
Food Pangium 281246
Fool Gooseberry 321643
Fool's Ballocks 273545
Fool's Cap 30081,367120
Fool's Cherry 280003
Fool's Cicely 9832
Fool's Cress 365861
Fool's Huckleb Erry 250513
Fool's Huckleberry 250513
Fool's Onion 128870,398794
Fool's Parsley 9831,9832
Fool's Stones 273531,273545
Fool's Watercress 29341
Fool's Water-cress 29341
Fool's-parsley 9832
Fool's-parsley Aethusa 9832
Foolproof Vase Plant 54456
Foolproofplant 54456
Foolwort 400675
Foonchew Machilus 240584

Fooning Greenbrier 366334
Fooning Lilyturf 272077
Foorflower Ricebag 181693
Foorflower Spathoglottis 370398
Foorflower Sunipia 379779
Foose 358062
Foot Cactus 287881
Footcatkin Willow 343656
Foot-catkin Willow 343656
Foothill Arnica 34714
Foothill Ash 167953
Foothill Bladder-pod 227022
Foothill Death Camas 417921
Foothill Fleabane 150773
Foothill Palo Verde 284482
Foothill Penstemon 289352
Foothill Sunburst 317364
Foothill Yellow Palo Verde 83748
Foothills Nolina 266493
Foothills Paloverde 83748
Footleaf Anthurium 28113
Footstool Palm 234162,234190
Footstool-palm 234190
Footstyle-orchis 125526
Forage Kochia 217353
Foramenbearing Sedge 74590
Foraminatiform Sedge 74591
Forbes Alyxia 18519
Forbes Bristlegrass 361750
Forbes Cattleya 79538
Forbes Euonymus 157675
Forbes Garcinia 171104
Forbes Notopterygium 267152
Forbes Primrose 314408
Forbes Spindle-tree 157675
Forbes Wildginger 37622
Forbes' Saxifrage 349771
Forbestown Rush 213034
For-bete 379660
Forbidden Fruit 327577
Forbitten More 379660
Forchhammeria 167275
Forcipate Herminium 291215
Ford Ailanthus 12570
Ford Ardisia 31450
Ford Barrel Cactus 163441
Ford Begonia 49847
Ford Birthwort 34185
Ford Breynia 60060
Ford Bridelia 60182
Ford Bushclover 226796
Ford Bush-clover 226796
Ford Checkwood 154971
Ford Cherry 223101
Ford Cherry Laurel 223101
Ford Cherry-laurel 223101
Ford Chinalaurel 28329
Ford China-laurel 28329
Ford Chinquapin 78941
Ford Chirita 87864
Ford Crapemyrtle 219927
Ford Crape-myrtle 219927
Ford Elytranthe 241317
Ford Erythrophleum 154971
Ford Evergreen Chinkapin 78941
Ford Evergreenchinkapin 78941
Ford Fatweed 167300
Ford Fimbristylis 166326
Ford Fishvine 125963
Ford Fordiophyton 167300
Ford Jewelvine 125963
Ford Lasianthus 222133
Ford Licuala Palm 228738
Ford Licualapalm 228738
Ford Loosestrife 239642
Ford Mahonia 242532
Ford Manglietia 244434
Ford Meliosma 249390
Ford Metalleaf 59844
Ford Millettia 66631
Ford Mosquitotrap 117468
Ford Nepeta 264925
Ford Ormosia 274391
Ford Osmanther 276287
Ford Osmanthus 276287
Ford Premna 313631
Ford Roughleaf 222133
Ford Slugwood 50513
Ford Swallowwort 117468
Ford Sweetleaf 381205
Ford Tanoak 233218
Ford Torenia 392904
Ford Woodlotus 244434
Ford Yam 131592
Fordia 167282
Fordiophyton 167289
Ford's Elm 401514
Forebit 379660
Forefather's Cup 347156
Foreign Feathergrass 376689
Foreign Needlegrass 376689
Foreign Rattan Palm 65767
Foreign Rhododendron 331469
Foreign Sealavender Statice 230711
Foreign Statice 230711
Foreign Toona 392839
Foreigntoona 80015
Forest Angelica 24483
Forest Bedstraw 170316,170317
Forest Bell Bush 240736
Forest Belonging Rhododendron 330897
Forest Betony 373455
Forest Blueberry 403986
Forest Bluegrass 305743,305946
Forest Bride's-bush 286312
Forest Cabbage Tree 350795
Forest Calanthe 66086
Forest Chroesthes 89283
Forest Colorcalyx 89283
Forest Conehead 320118
Forest Corydalis 106185
Forest Cotoneaster 107686
Forest Didymocarpus 129961
Forest Dombeya 136009
Forest Elder 267378
Forest European Grape 411983
Forest False Rattlesnakeroot 283697
Forest Globethistle 140814
Forest Goldsaxifrage 90400
Forest Goldwaist 90400
Forest Grape 93030,411887
Forest Grass Tree 415173
Forest Grey Gum 155756
Forest Helicia 189952
Forest Indigo 206044
Forest Ironweed 406854
Forest Jerusalemsage 295221
Forest Lily 405358
Forest Lobelia 234841
Forest Lousewort 287063
Forest Madder 338032
Forest Mangrove 1375
Forest Meadow-grass 305946
Forest Muhly 259707
Forest Nettle 402946
Forest Nuxia 267378
Forest Oak 15950,15952,79161,79170,79185
Forest Orania Palm 273125
Forest Pansy Redbud 83760
Forest Pea 222854,222860
Forest Phlox 295257,295258
Forest Rabdosia 209834
Forest Red Gum 155756
Forest Redgum 155756
Forest Rhododendron 331320,331360
Forest River Gum 155756
Forest Sedge 75226
Forest She Oak 79189
Forest She-oak 15952,79189
Forest Skullcap 355643
Forest Slaty Gum 155756
Forest Sunflower 188953
Forest Teasel 133508
Forest Wild Elder 267378
Forest Willow 343328
Forest Wood Rush 238640
Forest Woodbetony 287721
Forest Wood-rush 238565
Forest Yellowcress 336277
Forestloving Sage 345093

Forest-she Oak 15952
Forget me not 260737
Forget Nautilocalyx 262898
Forgetmenot 260737
Forget-me-not 13073,260735,260737,260760,260868,260892,407072,410677
Forget-me-not Family 57084
Forgetmenotlike Microula 254355
Forgetmenot-like Trigonotis 397441
Fork Cordia 104183
Fork Euonymus 157414
Fork Gentian 173407
Fork Mouseear 82834
Fork Rush 213237
Forkbract Monkshood 5152
Forkbranch Sagebrush 35400
Forked Aster 40489,160660
Forked Beard-grass 18668
Forked Catchfly 363409,363427
Forked Cerastium 82834
Forked Chickweed 82789,284782,284845,284905
Forked Duckweed 224396
Forked Fanwort 64499
Forked Mayten 246941
Forked Mouse-ear Chickweed 82789
Forked Nailwort 284869
Forked Nightshade 367168
Forked Panic Grass 128505
Forked Scale-seed 370892
Forked Silene 363409
Forked Swertia 380181
Forked-chickweed 284782,284845
Forkedflower Dendrobium 125232
Forked-leaved Sundew 138270
Forkedlike Corymborkis 106877
Forkedlip Calanthe 66101
Forkedpilose Cragcress 97989
Forkedpilose Scurvyweed 97989
Forkers' Coralbean 154659
Forkfruit 128919
Forkful Stairweed 142796
Forkglome Stairweed 142619
Forking Larkspur 102833,102841,124141
Forkleaf Cycas 115854
Forkleaf Maple 3497
Fork-leaved Maple 3497
Forklip Dendrobium 125373
Forklip Herminium 192851
Forklip Luisia 238322
Forklip Nobillorchis 197448
Forkliporchis 25959
Forkpaniclegrass 209876
Forkshoot Meadowrue 388656
Forkshoot Seseli 361595
Forkspike Rhubarb 329379
Forkspike Sweetleaf 381291
Fork-spiked Sweetleaf 381291
Forkspine Bamboo 47425
Forkstalk Gagea 169409
Forkstamenflower 416501,416508
Forkstem Sagebrush 35622
Forkstem Wormwood 35622
Forkstyle Tofieldia 392604
Forkstyle Willow 343318
Forkstyleflower 374468,374476
Forktip Three-awn 33768
Fork-tip Three-awn Grass 33768
Forktogue Cremanthodium 110465
Forktogue Nutantdaisy 110465
Fork-toothed Ookow 128873
Forktwig Milkvetch 42405
Forky Knotweed 309906
Forman's Mallee 155583
Formania 167339
Formosa Aglaia 11293
Formosa Caper 71751
Formosa Ehretia 141691
Formosa Falsenettle 56140
Formosa Firethorn 322471
Formosa Gum Tree 232557
Formosa Juniper 213775
Formosa Lily 229845
Formosa Nato Tree 280440
Formosa Natotree 280440
Formosa Pieris 298734
Formosa Pyracantha 322471
Formosa Supple Jack 52423
Formosa Viburnum 407853
Formosa Wendlandia 413784
Formosan Actinidia 6530
Formosan Alder 16345
Formosan Alder Mistletoe 410992
Formosan Amentotaxus 19363
Formosan Apple 243610
Formosan Argyreia 32623
Formosan Ash 167982
Formosan Bamboo 297282
Formosan Barthea 48552
Formosan Beauty-berry 66779
Formosan Blue Berry 403796
Formosan Blueberry 404061
Formosan Bramble 338425
Formosan Breynia 60061
Formosan Buckthorn 328700
Formosan Butter-fly-bush 62015
Formosan Cans 65697
Formosan Condorvine 245805
Formosan Cypress 85268
Formosan Dandelion 384681
Formosan Date 295465
Formosan Date Palm 295465
Formosan Douglas Fir 318599
Formosan Ehretia 141691
Formosan Elaeagnus 141991
Formosan Elegant Pouzolzia 313424
Formosan Excoecaria 161657
Formosan False Cypress 85268
Formosan False-nettle 56140
Formosan Fig Hazel 380590
Formosan Fir 396
Formosan Gold-rain Tree 217620
Formosan Gooseberry 333978
Formosan Gum 232557
Formosan Holly 203821
Formosan Honeylocust 176870
Formosan Koa 1145
Formosan Lasianthus 222138
Formosan Lettuce 320513
Formosan Medinilla 247564
Formosan Michelia 252849
Formosan Ormosia 274392
Formosan Pachycentria 279479
Formosan Raspberry 338425
Formosan Rattan Palm 65697
Formosan Rattanpalm 65697
Formosan Red Bark Slugwood 50514
Formosan Red Cypress 85268
Formosan Reevesia 327362
Formosan Rose 336793
Formosan Seashore Rose 336705
Formosan Snowbell 379347
Formosan Spruce 298375
Formosan Staunton-vine 374425
Formosan Sugarpalm 32336
Formosan Supple-jack 52423
Formosan Swallowwort 117469
Formosan Sweet Gum 232557
Formosan Sweet Spire 210402
Formosan Sweetgum 232557
Formosan Sweetleaf 381148,381206
Formosan Thistle 92066
Formosan Trachelospermum 393626
Formosan Turpinia 400522
Formosana Juniper 213775
Formosana Sinopanax 364997
Forrest Abelia 127
Forrest Abutilon 907
Forrest Acronema 6209
Forrest Actinodaphne 6779
Forrest Agapetes 10309
Forrest Angelica 24352
Forrest Asiabell 98293
Forrest Barberry 51629
Forrest Bedstraw 170383
Forrest Begonia 49851
Forrest Bladdernut 374102
Forrest Bladder-nut 374102
Forrest Bluebeard 78008
Forrest Blue-beard 78008
Forrest Braya 59738

Forrest Briggsia 60264
Forrest Bugle 13102
Forrest Bulbophyllum 62740
Forrest Bushclover 226812
Forrest Bush-clover 226812
Forrest Buttercup 325817
Forrest Butterflybush 62059
Forrest Butterfly-bush 62059
Forrest Cacalia 283814
Forrest Camellia 69079
Forrest Chelonopsis 86815
Forrest Chirita 87866
Forrest Common Pentapanax 289633
Forrest Conehead 320122
Forrest Cranesbill 174615
Forrest Cremanthodium 110390
Forrest Curlylip-orchis 62740
Forrest Daylily 191283
Forrest Deerdrug 242684
Forrest Dianxiong 297963
Forrest Dolomiaea 135731
Forrest Douglas Fir 318571
Forrest Dragonhead 137584
Forrest Dumasia 138934
Forrest Dutchmanspipe 34186
Forrest E. H. Wilson Arisaema 33566
Forrest E. H. Wilson Southstar 33566
Forrest Edelweiss 224840
Forrest Fescuegrass 163970
Forrest Fir 369
Forrest Garuga 171564
Forrest Gingerlily 187449
Forrest Goldleaf 108576
Forrest Goldwaist 90364
Forrest Greenorchid 137584
Forrest Gugertree 350914
Forrest Hemlock 399903
Forrest Holly 203824
Forrest Incarvillea 205561
Forrest Inula 138895
Forrest Iris 208575
Forrest Jerusalemsage 295102
Forrest Lady's Slipper 120354
Forrest Ladyslipper 120354
Forrest Larkspur 124219
Forrest Leaf-flower 296521
Forrest Leek 15302
Forrest Leptodermis 226084
Forrest Litse 233912
Forrest Lysionotus 239941
Forrest Manglietia 244439
Forrest Maple 2970
Forrest Meconopsis 247123
Forrest Microula 254345
Forrest Milkvetch 42382
Forrest Monkshood 5205
Forrest Mosquitotrap 117473
Forrest Nomocharis 266556
Forrest Nutantdaisy 110390
Forrest Oneflower Amitostigma 19521
Forrest Oneflower Astyleorchis 19521
Forrest Onion 15302
Forrest Oreocharis 273861
Forrest Paris 284329
Forrest Pennywort 200308
Forrest Peristylus 291218
Forrest Persimmon 132163
Forrest Phoebe 295362
Forrest Physospermopsis 297963
Forrest Piptanthus 300674
Forrest Pleione 304248
Forrest Podocarpus 306433
Forrest Primrose 314415
Forrest Pyrola 322826
Forrest Rhamnella 328549
Forrest Rhodiola 329875
Forrest Rhododendron 330724
Forrest Rhodoleia 332206
Forrest Rhubarb 329330
Forrest Rockfoil 349343
Forrest Rockjasmine 23174
Forrest Rose 336573
Forrest Sagebrush 35498
Forrest Sandwort 31908
Forrest Sedge 74599
Forrest Silkvine 291061
Forrest Sinocrassula 364920
Forrest Skullcap 355443
Forrest St. John's Wort 201883
Forrest St. John's-wort 201883
Forrest Stonecrop 356753
Forrest Summerlilic 62059
Forrest Swallowwort 117473
Forrest Syzygium 382536
Forrest Tea 69079
Forrest Threewingnut 398316
Forrest Trachydium 297963
Forrest Tremacron 394675
Forrest Underleaf Pearl 296571
Forrest Violet 410001
Forrest Vladimiria 135731
Forrest Wallflower 154455
Forrest Whiteheadtree 171564
Forrest Whitepearl 172074
Forrest Wildclove 226084
Forrest Wintergreen 172074
Forrest Woodbetony 287218
Forrest Woodlotus 244439
Forrest Yaccatree 306433
Forrest's Birch 53441
Forrest's Day Lily 191284
Forrest's Day-lily 191284
Forrest's Ironweed 406338
Forrest's Sawwort 361052
Forrest's Scabrous Wendlandia 413819
Forskohl Coleus 99583
Forskohl Flamenettle 99583
Forsskal Barberry 51630
Forsteck Cypress 85281
Forster Stephania 375852
Forster Woodrush 238619
Forster's Wood Rush 238619
Forster's Wood-rush 238619
Forsteronia 167405
Forsythia 209,167427,167435,167441
Forsythia Sage 345191
Forsythia-of-the-wilds 231306
Fort Bragg Manzanita 31126
Fort Mohave Wild Buckwheat 152342
Fort Sheridan Hawthorn 109528
Fortnight Lily 130292,130298,130299
Fortunat Actinidia 6598
Fortunat Kiwifruit 6598
Fortune 393809
Fortune Apios 29304
Fortune Bamboo 347214
Fortune Bitterbamboo 347214
Fortune Bogorchid 158118
Fortune Cape Jasmine 171330
Fortune Chinabell 16295
Fortune Chloranthus 88276
Fortune Cyrtandra 120503
Fortune Euonymus 157473
Fortune Firethorn 322465
Fortune Fontanesia 167233
Fortune Forsythia 167432
Fortune Glochidion 177170
Fortune Indigo 205995
Fortune Japanese Spiraea 371971
Fortune Keteleeria 216142
Fortune Loosestrife 239645
Fortune Mahonia 242534
Fortune Meadowrue 388514
Fortune Meadowsweet 371971
Fortune Osmanther 276290
Fortune Osmanthus 276290
Fortune Palm 393809
Fortune Plantainlily 198608
Fortune Plumyew 82507
Fortune Plum-yew 82507
Fortune Rhododendron 330727
Fortune Rockfoil 349344
Fortune Rose 336577
Fortune Sasa 347214
Fortune Spathoglottis 370402
Fortune Spindle-tree 157473
Fortune's Chinabells 16295
Fortune's Creeping Spindle 157473,157515
Fortune's Double Yellow 336814
Fortune's Double Yellow Tea Rose 336814
Fortune's Eupatorium 158118

Fortune's Fontanesia 167233
Fortune's Keteleeria 216142
Fortune's Paulownia 285960
Fortune's Plum Yew 82507
Fortune's Windmill Palm 393809
Fortunearia 167487
Fortunei Bamboo 304110
Fortune's Osmanthus 276290
Fortune-teller 384714
Fosberg Tasselflower 144916
Fosberg's Pualele 144916
Fossil Tree 252540
Fossulate Eritrichium 153553
Foster Aechmea 8561
Foster Canistrum 71151
Foster Iris 208576
Foster Palm 198808
Foster Sentry Palm 198808
Foster Tulip 400159
Foster Vriesea 412351
Foster's Caryodaphnopsis 77960
Foster's Rock Daisy 291310
Fothergilla 167537
Foues Gentian 173596
Fouets 358062
Foul-grass 144661
Foully Operculina 103362
Foulrush 157429
Fountain Bamboo 162635,162718,364631
Fountain Buddleja 61971
Fountain Butterflybush 61971
Fountain Butterfly-bush 61971
Fountain Dracaena 104359
Fountain Flower 84235
Fountain Grass 289011,289013,289015,289162,289248,289250
Fountain Humea 67698
Fountain Palm 234162,234170
Fountain Plant 18836,341020
Fountain Summerlilic 61971
Fountain Thistle 91979
Fountain Tree 15953,370358
Fountainbush 341020
Fountain-grass 289234,289248
Fountain-palm 234170
Four Corners 180897
Four Corners Thistle 91757
Four Flower Synotis 381955
Four Muskroot 387068,387069
Four Seasons Rose 336515
Four Seoals Gaultheria 172162
Four Sisters 308454
Four Stamens Metalleaf 296421
Four-anglar Bredia 59873
Fourangle Sweetleaf 381291
Fourangle Threeribfruit 397986
Fourangle Treebine 92983
Fourangled Bredia 59873
Four-angled Circaea 91524
Four-angled Microtropis 254329
Four-angled Serrate Plagiopetalum 301668
Four-angled Stem St. John's Wort 202168
Fourangledculm Spikesedge 143374
Fourangleseed Threebibachene 397986
Fourangular Begonia 50365
Fourangular Cassiope 78648
Fourangular Dendrobium 125386
Fourangular Greenbrier 366524
Four-angular Greenbrier 366524
Fourangular Hollowed Wampee 94193
Fourangular Microtropis 254329
Four-angular Microtropis 254329
Fourangular Mockorange 294558
Fourangular Olive 270171
Fourangular Padritree 376266
Four-angular Padritree 376266
Fourangular Rattlebox 112745
Fourangular Sauropus 348059
Fourangular St. John's-wort 202179
Fourangular Syzygium 382675
Four-angular Syzygium 382675
Fourangular Willowweed 146905
Fourarista Stonecrop 357225
Four-armed Grass 58183
Fourbearing Singnalgrass 58183
Fourcalyx Actinidia 6708
Fourcalyx Kiwifruit 6708
Four-cat Purslane 311921
Fourcolor Cape-cowslip 218832
Fourcolor Cattleya 79553
Four-corners Orach 44608
Fourcorners Stanhopea 373701
Fourfile Germander 388247
Fourfinger Rattan Palm 65800
Fourfingers Rattan Palm 65800
Fourfingers Rattanpalm 65800
Four-fingers-like Rattanpalm 65799
Fourflower Litse 233907
Fourflower Sedge 75954
Fourflower Vetch 408636
Four-flowered Litse 233907
Four-flowers Jasminorange 260178
Fourfruit Wildtung 243452
Four-fruited Larkspur 124638
Four-fruits Mallotus 243452
Fourgland Hetaeria 193589
Fourglandlike Willow 343838
Fourhead Tailanther 381955
Four-involucre 124880,124925
Fourleaf Devilpepper 327067
Fourleaf Dysophylla 139580
Fourleaf Gentian 173970
Fourleaf Kidney Vetch 28196
Fourleaf Ladybell 7850
Fourleaf Lepisanthes 225682
Fourleaf Manyseed 307755,307756
Fourleaf Milkweed 38080
Fourleaf Paperelder 290439
Fourleaf Paris 284387
Fourleaf Peperomia 290439
Fourleaf Stonecrop 357066
Fourleaf Tulip 400241
Four-leaved Allseed 307724,307755
Fourleaved Australnut 240212
Four-leaved Clover 278124
Four-leaved Devil-pepper 327067
Four-leaved Grass 284387
Four-leaved Loosestrife 239810
Four-leaved Macadamia 240212
Four-leaved Milkweed 38080
Four-leaved Senna 78210
Four-leaved Truelove 284387
Fourleaves Falserobust Vetch 408558
Fourleaves Madder 338027
Fourleaves Stonecrop 357066
Fourleves Bedstraw 170289
Fourlobe Catchfly 363950
Four-lobe Sandwort 32173
Fourlobed Skullcap 355706
Four-needle Pyron 300171
Four-nerve Daisy 387380
Four-nerved Daisy 201314
Fournerved Eulalia 156490
Fournerved Leucosyca 228147
Four-o'clock 255672,255694,255711,269404,274823,384714
Four-o'clock Family 267421
Four-o'clock Flower 255672,255711
Four-o'clock Milkweed 38039
Four-o'clock Plant 255672,255711,255727
Fourpetal Aglaia 19970
Fourpetal Amoora 19970
Fourpetal Purslane 311921
Fourpetal Rockfoil 349657
Fourpetal Rose 336930
Fourpetal St. John's-wort 202179
Fourpetale Mangrove 329765
Fourpetaled Mangrove 329765
Four-petaled Pawpaw 38337
Four-petaled Pussypaws 93096
Four-petaled Rose 336930
Fourpistil Nettletree 80764
Four-point Evening-primrose 269497
Fourrange Gentian 173971
Fourridge Greenbrier 366524
Fourridgy Rattlebox 112745
Fourseasons Poorspikebamboo 270438
Fourseed Conehead 85952
Fourseed Vetch 408638
Four-seed Vetch 408638
Fourseed Willow 344199

Four-seed Willow 344199
Four-seeded Vetch 408638
Foursepal Anemone 24082
Foursepal Windflower 24082
Four-sepaled Wintergreen 172162
Foursplit Rhodiola 329930
Four-spotted Orchid 273607
Fourstamen Alangium 13388
Fourstamen Buttercup 326437
Fourstamen Chickweed 82793
Fourstamen Gomphandra 178923
Four-stamen Gomphandra 178923
Fourstamen Hackberry 80764
Fourstamen Maple 3677
Fourstamen Metalleaf 296421
Fourstamen Parrotfeather 261375
Fourstamen Prismatomeris 315332
Fourstamen Stephania 375904
Four-stamen Tamarisk 383581
Four-stamened Prismatomeris 315332
Fourstamens Sibbaldia 138411
Fourtain Bamboo 297203
Fourtapeparsley 387893
Fourteeth Schnabelia 352063
Fourtepal Elatostema 142875
Fourtepal Stairweed 142875
Fourtimes Swertia 380385
Fourtooth Sandwort 32173
Fourtooth Tetracme 386953
Fourvein Goldquitch 156490
Fourvein Leucosyca 228147
Fourverticalrank Buchnera 61871
Fourwing Eveningprimrose 269519
Fourwing Greenbrier 366336
Four-wing Greenbrier 366336
Four-wing Mallee 155759
Fourwing Saltbush 44334
Four-wing Saltbush 44334
Fourwing Smilax 366336
Four-wing Sophora 369130
Four-winged Mallee 155760
Fourwings Cliffbean 254857
Four-wings Millettia 254857
Fourwings Pagodatree 369130
Fourwings Sophora 369130
Fouse 358062
Fouze 358062
Fovealip Neottia 264718
Fovealip Nestorchid 264718
Foveate Jasmine 211832
Foveolate Fig 165002
Foveolate Litse 233914
Foveolate Michelia 252874
Foveolate Spicebush 231341
Foveolate Spice-bush 231341
Fowl Bluegrass 305804
Fowl Blue-grass 305804
Fowl Manna Grass 177662
Fowl Meadow Grass 177662,305804
Fowl Meadow-grass 305804
Fowler's Knotweed 309129
Fowler's Service 369339
Fowlfoot 143530
Fox And Cubs 195478
Fox Berry 66733
Fox Brush Orchid 9271
Fox Docken 130383
Fox Finger 130383
Fox Fingers 130383
Fox Geranium 174877
Fox Glove 130344,130383
Fox Grape 411764
Fox Grass 174877
Fox Nut 160637
Fox Plum 31142
Fox Poison 122492
Fox Rose 336851
Fox Sedge 76716,76718,76735
Fox Tail 17497
Fox Tail Agave 10806
Fox Tail Lily 148531
Fox Tail Millet 361794
Fox Tail Orchid 333725
Fox Ter-leaves 130383
Fox Valley Dwarf River Birch 53536
Fox's Brush 81768
Fox's Clote 31051
Fox's Glove 130383
Fox's Meat 277648
Fox's Mouth 5442
Fox-and-cubs 195478
Fox-and-hounds 231183
Fox-and-leaves 130383
Foxbane 5676
Foxberry 404051
Fox-brush Orchid 9288
Foxcolor Ancylostemon 22177
Foxglove 70119,130344,130383,405788
Foxglove Beardtongue 289337
Foxglove Beard-tongue 289337
Foxglove Pecteilis 285960
Foxglove Tree 285981
Foxglove-like Sage 345002
Foxglove-tree 285981
Fox-leaves 130383
Fox-sedge 76716
Foxtaail-like Kudzuvine 321420
Foxtail 17497,106703,374968
Fox-tail 196818
Foxtail Amaranth 18687
Foxtail Barley 198311
Fox-tail Blackroot 320393
Foxtail Brisde-grass 361794
Foxtail Bristle Grass 361794
Foxtail Bristlegrass 361794
Foxtail Brome 60943
Fox-tail Cactus 244196
Foxtail Chess 60943
Foxtail Clover 397059
Foxtail Dalea 121893
Foxtail Fern 315645
Foxtail Fescue 412465
Foxtail Flatsedge 118473
Foxtail Grass 17497,17527
Foxtail Knotweed 308726
Foxtail Lily 148531
Foxtail Milkvetch 41961,43236
Foxtail Millet 361794
Foxtail Mint 250468
Fox-tail Orchid 9271,9302,30532,333725
Fox-tail Orchids 9271
Foxtail Palm 414642,414643
Foxtail Pine 299795,299813
Foxtail Pricklegrass 113310
Foxtail Restharrow 271300
Foxtail Sedge 73639,73641
Fox-tail Wild Buckwheat 152401
Foxtailgrass 361680
Foxtail-like Falsetamarisk 261250
Foxtail-like Milkvetch 43236
Foxtail-like Sophora 368962
Fox-taoil Orchis 9271
Foxter 130383
Foxter-leaves 130383
Fox-tongued Melastoma 248778
Foxtree 130383
Foxwort 325820
Foxy 130383
Foxy-leaves 130383
Foxy-red Fargesia 162743
Foxy-red Mayten 182772
Fozhi Ginkgo 175820
Frachet Groundcherry 297645
Frachet's Currant 333982
Fra-fra Potato 99699
Fraghan 403916
Fragile Almond 20899
Fragile Amomum 19851
Fragile Blueberry 403832
Fragile Dragonhead 137586
Fragile Ginger 417984
Fragile Glasswort 342890
Fragile Honeysuckle 235795
Fragile Oat 172034
Fragile Pentachaeta 289425
Fragile Prickly Pear 272898
Fragile Prickly-pear 272898
Fragile Willow 343396
Fragileorchis 2032
Fragile-shell Almond 20899
Fragile-stem Aster 40479

Fragile-stemmed Aster 40480
Fragrabt Pavetta 286388
Fragrans Brassavola 59257
Fragrant Abelia 152
Fragrant Acanthophippium 2081
Fragrant Agrimonia 11594
Fragrant Agrimony 11594
Fragrant Ainsliaea 12635
Fragrant Albizia 13635
Fragrant Albizzia 13635
Fragrant Amomum 19923
Fragrant Angelica 24327
Fragrant Apostasia 29790
Fragrant Arrowwood 407837
Fragrant Ash 167949
Fragrant Asparagus 38974
Fragrant Azalea 330405
Fragrant Balm 257161
Fragrant Bamboo 87630,87633
Fragrant Barberry 51633
Fragrant Bedstraw 170708
Fragrant Begonia 50121
Fragrant Bffyerweed 201323
Fragrant Bluebeard 78000
Fragrant Blue-beard 78000
Fragrant Bogorchid 89299
Fragrant Brachyactis 58264
Fragrant Bugle-lily 413366
Fragrant Bulbophyllum 62552
Fragrant Bursera 64073
Fragrant Calanthe 66022
Fragrant Cananga 70960
Fragrant Candytuft 203229
Fragrant Carpet 313555
Fragrant Champaca 252841
Fragrant Chervil 84726
Fragrant Cinnamon 91287,91392,91429
Fragrant Citrus 93515
Fragrant Clematis 94901
Fragrant Coelogyne 98710
Fragrant Cudweed 178318,317761
Fragrant Currant 333972
Fragrant Cymbidium 117067
Fragrant Cyperus 393205
Fragrant Dracaena 137397
Fragrant Dragonbood 137397
Fragrant Dragonhead 137613
Fragrant Dutchman's Pipe 34286
Fragrant Eagleclaw 35013
Fragrant Elsholtzia 143991
Fragrant Epaulette Tree 320886
Fragrant Epidendrum 146421,146453
Fragrant Erythrophleum 154983
Fragrant Eupatorium 89299
Fragrant Evening Primrose 269417,269509
Fragrant Evening-primrose 269509
Fragrant Everning Primrose 269475
Fragrant False Garlic 266960
Fragrant False Indigo 20023
Fragrant Fieldorange 249679
Fragrant Flowers Rhododendron 330405
Fragrant Fritillary 168454
Fragrant Gaillardia 169610
Fragrant Galangal 17661
Fragrant Garland-flower 187427
Fragrant Gaultheria 172078
Fragrant Giant Hyssop 10407
Fragrant Giantarum 20123
Fragrant Gingersage 253655
Fragrant Gladiolus 4534
Fragrant Glorybower 96009
Fragrant Golden Rod 368134
Fragrant Goldenrod 368290
Fragrant Greenorchid 137613
Fragrant Greentwig 352435
Fragrant Greytwig 352435
Fragrant Grey-twig 352435
Fragrant Gymnadenia 182261
Fragrant Gynocardia 182962
Fragrant Hairorchis 148705
Fragrant Haraella 185755
Fragrant Heliotrope 190746
Fragrant Heteropanax 193900
Fragrant Hibiscus 194881
Fragrant Holly 204280
Fragrant Hopea 198176
Fragrant Horseegg 182962
Fragrant Hosta 198639
Fragrant Landpick 308613
Fragrant Lathyrus 222789
Fragrant Lavender 223288
Fragrant Leaf-stem Iris 208600
Fragrant Liparis 232261
Fragrant Litse 233882,234026
Fragrant Loosestrife 239640
Fragrant Loquat 151155
Fragrant Lovegrass 147986
Fragrant Luculia 238102
Fragrant Machilus 240700,240726
Fragrant Magnolia 232599
Fragrant Manglietia 244423
Fragrant Manjack 104175
Fragrant Maple 232557
Fragrant Marshweed 230275
Fragrant Melodinus 249679
Fragrant Michelia 252885
Fragrant Micromeria 253655
Fragrant Mimosa 254999
Fragrant Mockorange 294426,294491
Fragrant Moraea 258468
Fragrant Muping Corydalis 105883
Fragrant Myall 1292
Fragrant Nailorchis 9302
Fragrant Nanmu 12515,12517
Fragrant New Nymph Flower 264163
Fragrant Odontoglossum 269080
Fragrant Olive 276291
Fragrant Orange 93515
Fragrant Orchid 182213,182230
Fragrant Orchis 117067
Fragrant Osbeckia 276399
Fragrant Padritree 376292
Fragrant Pancratium 280893
Fragrant Paracryphia 283251
Fragrant Pink 127713
Fragrant Plagiopteron 301691
Fragrant Plantain Lily 198639
Fragrant Plantainlily 198639
Fragrant Plantain-lily 198639
Fragrant Pleione 304231
Fragrant Plum Rose 336287
Fragrant Poikilospermum 307045
Fragrant Popcorn-flower 301625
Fragrant Poplar 311509
Fragrant Premna 313706
Fragrant Prickly-apple 185918
Fragrant Pterocaulon 320395
Fragrant Rabbit-tobacco 317732
Fragrant Raspberry 338252
Fragrant Rhododendron 331387
Fragrant Rodriguezia 335169
Fragrant Rondeletia 336104
Fragrant Rosewood 121782
Fragrant Sand Verbena 700
Fragrant Sandalwood 346219
Fragrant Sarcococca 346743
Fragrant Savin 213857
Fragrant Schoenorchis 352302
Fragrant Schoepfia 352435
Fragrant Siris 13635
Fragrant Snakeroot 11164
Fragrant Snowbell 379414
Fragrant Sobralia 366783
Fragrant Solomonseal 308613
Fragrant Spicebush 231342
Fragrant Spice-bush 231342
Fragrant Star-lily 227819
Fragrant Sterculia 376110
Fragrant Stixis 377125
Fragrant Stock 246521
Fragrant Stonebean-orchis 62552
Fragrant Stonegarlic 239262
Fragrant Storax 379414
Fragrant Sumac 332477,332478,332483,
332502
Fragrant Sumach 332477
Fragrant Syzygium 382633
Fragrant Tailgrape 35063
Fragrant Tea 69654
Fragrant Tea Olive 276291
Fragrant Telosma 385752

Fragrant Thimbleberry 338917
Fragrant Thistle 92330
Fragrant Twayblade 232261
Fragrant Unguiculate Iris 208914
Fragrant Viburnum 407646,407647
Fragrant Violet 410320
Fragrant Wampee 94214
Fragrant Wandflower 369987,369990
Fragrant Water Lily 267730
Fragrant Waterlily 267730
Fragrant Water-lily 267730
Fragrant Waxpetal 106643
Fragrant Waxplant 198869
Fragrant White Sand-verbena 700
Fragrant Whitepearl 172078
Fragrant Winter Honeys 235798
Fragrant Wintergreen 172078
Fragrant Winterhazel 106637,106643
Fragrant Wintersweet 87525
Fragrant Witer Hazel 106637
Fragrant Witer-hazel 106637
Fragrant Woodlotus 244423
Fragrant Woolly Cactus 185918
Fragrant Yunnanclave 238102
Fragrant Zeuxine 417772
Fragrantbamboo 87630,87633
Fragrant-bamboo 87630
Fragrantflower Arachnis 30533
Fragrantflower Dragonhead 137677
Fragrantflower Greenorchid 137677
Fragrantflower Primrose 314128
Fragrantflower Wampee 94214
Fragrant-flowered Rhododendron 330405
Fragrantfruit Newlitse 264046
Fragrant-like Marshweed 230277
Fragrant-root Iris 208565
Frail Corydalis 105878
Frail Woodbetony 287291
Frailea 167689
Frailecillo 212144
Framboise 338557
Framboise Bush 338557
Framboys 338557
Framlington Clover 316127
Franceschi Palm 59100
Franchet Arisaema 33335
Franchet Barberry 51634
Franchet Barrenwort 146993
Franchet Buttercup 325882
Franchet Corydalis 105914
Franchet Cotoneaster 107451
Franchet Cowparsnip 192266
Franchet Dichocarpum 128933
Franchet Dicranostigma 129564
Franchet Edelweiss 224842
Franchet Epimedium 146993
Franchet Favusheadflower 129564
Franchet Figwort 355122
Franchet Forkfruit 128933
Franchet Goldenray 229040
Franchet Holly 203833
Franchet Lady's Slipper 120355
Franchet Landpick 308544
Franchet Leaf-flower 296573
Franchet Leptopus 296573
Franchet Linden 391717
Franchet Loosestrife 239666
Franchet Maple 2972
Franchet Monkshood 5208
Franchet Oak 323920
Franchet Peashrub 72227
Franchet Pea-shrub 72227
Franchet Photinia 295688
Franchet Pleurospermum 304795
Franchet Ribseedcelery 304795
Franchet Sealavender 230627
Franchet Skullcap 355448
Franchet Solomonseal 308544
Franchet Sophora 369018
Franchet Southstar 33335
Franchet Spurge 158915
Franchet Staff-tree 80191
Franchet Stonecrop 356758
Franchet Terminalia 386537
Franchet Underleaf Pearl 296573
Franchet Vineclethra 95495
Franchet Wild Stonecrop 356758
Franchet Woodbetony 287222
Franchet's Cotoneaster 107451
Francis' Water-gum 382537
Franciscan Raintree 61295
Franciscan Thistle 91740
Francisco Willowweed 146705
Franck Spurry 370522
Francke Spurry 370522
Francking Spurwort 370522
Francoa 167733,167735
Frangipani 305206,305213,305221,305225
Frangipani Tree 305226
Frangrant Birch 53479
Frangrant Daphne 122532
Frangula-like Rhamnella 328550
Franke 370522
Frankenia 167765
Frankenia Family 167827
Frankincense 57508,57537
Frankincense Pine 300264
Frankincense Tree 57537
Franklin Tree 167835,179761
Franklin's Lady's-slipper 120426
Franklin's Sandwort 31915
Franklin's Tree 179734
Franklinia 167834,167835
Franklinia Tree 167835
Franzenbach Rhubarb 329331
Frap 130383
Fraser Balsam Fir 285,377
Fraser Beefwood 79164
Fraser Cedar 85282
Fraser Fir 377
Fraser Magnolia 242110
Fraser Photinia 295612
Fraser Sophora 369019
Fraser's Balsam Fir 377
Fraser's Fir 377
Fraser's Magnolia 242110
Fraser's Marsh St. John's-wort 394810
Fraser's Photinia 295612
Fraser's Sedge 117324
Fraser's She Oak 79164
Fraser's She-oak 79164
Frata 367696
Fraughan 403916
Fraughans 403916
Fraw-cup 168476
Frawn 403916
Frawns 403916
Fraxinell 308607
Fraxinella 129615,129618
Fray 401388
Freckle Face 202595,202608
Freckled Face 314083,315082
Freckled Milk-vetch 42599
Freckle-face 202595
Frecnch Geranium 288206
Fredonia Wild Buckwheat 152292
Free Flowering Rhododendron 330714
Free Holly 203545
Free-flowered Waterweed 143931,200190
Free-flowering Andromeda 298721
Free-flowering Barberry 51622
Freeman Magnolia 241952
Freeman's Maple 2976
Freesia 168153,168174
Free-stamen Camellia 69270
Freestone Nectarine 20955
Free-style Camellia 69272
Free-style-like Camellia 69274
Freeway Sedge 75860
Frei Jo 104190
Freijo 104190
Freiser 167653
Fremont Cottonwood 311324
Fremont Holly Grape 51640
Fremont Holly-grape 51640
Fremont Lycium 239052
Fremont Mahonia 242536
Fremont Pincushion 84495
Fremont Poplar 311324
Fremont Silk-tassel 171552
Fremont Thornbush 239052

Fremont's Chaffbush 20689
Fremont's Cottonwood 311324
Fremont's Death Camas 417899
Fremont's Goldfields 222601
Fremont's Leather Flower 94918
Fremont's Pepper-grass 225367
Fremont's Poplar 311324
Fremont's Wild Buckwheat 152328
Fremont's-gold 382060
Fremontia 168216,168218
French 271280
French Alpine Juniper 213969
French Artichoke 117787
French Ash 218761
French Asparagus 274748
French Bartsia 268968
French Bay 85875
French Bean 293985,294056
French Beech 162400
French Berries 328737
French Bladdernut 374099
French Box 64356
French Broom 173015
French Brum 218761
French Catchfly 363477
French Chestnut 78811
French Cotton 68095
French Cowslip 314143
French Cranesbill 174581
French Cress 47964
French Endive 90901
French Fennel 167146
French Figwort 355075
French Flax 231996
French Fuzz 401388
French Garlic 15726
French Grass 271280,273236,274748
French Guava 360409
French Hales 323292,369446,369447
French Hardhead 81150
French Hawksbeard 110929
French Hay 271280
French Heath 149544
French Honey Suckle 187842
French Honeysuckle 187753,187842,188028
French Hybrid Ceanothus 79923
French Hybrid Lilac 382171
French Hydrangea 199956,199957,199958,199959,199961,199966,199976,199977,199983
French Jasmine 211940
French Jujube 418169
French Lace Lavender 223308
French Lavender 223284,223334,346250
French Leek 15621
French Lilac 169935,382329
French Lungwort 195803
French Mallow 223378,243805,243840
French Marigold 383090,383103
French Meadow-rue 388423
French Mercury 250600
French Mignonette 127654,350011
French Millet 281916
French Mulberry 66727,66733
French Mustard 154424
French Nettle 220416
French Nut 78811,212636
French Oak 324278,324335
French Oat-grass 172032,172034
French Onion 352960
French Paigle 321643
French Physic Nut 212127
French Physicnut 212127
French Pink 34536,127654
French Plantain 260253
French Poppy 405788
French Pussy Willow 343151
French Rocket 365459
French Rose 336581
French Rye Grass 34930
French Rye-grass 34930
French Sally 343858
French Saugh 85875
French Silene 363477
French Snowdrop 274823
French Sorrel 277648,340249,340254
French Sparrow-grass 274748
French Spikenard 404238
French Spinach 44468,340254
French Spotted Marigold 383107
French Sumac 104700
French Tamarisk 383491
French Tarragon 35411
French Thyme 391402
French Toadflax 230895
French Tree 383491
French Turpentine 300146
French Wheat 162312
French Whisk 90125
French Wild Guava 360463
French Willow 85875,344211,389978
French Wormwood 35868,36088
French's Shooting Star 135160
French's Shooting-star 135160
French-and-English Grass 302034
French-and-English Soldiers 302034
French-and-English Weed 302034
French-berries 328642
Frenchweed 390213
Freneh Marjoram 274231
Freshwater Mangrove 48509
Freshwater Soldiers 377485
Freshyellow Azalea 332125
Freycinetia 168238
Freyn Ash 167955
Freyn Cinquefoil 312565
Freyn Sagebrush 35503
Freyns Pink 127714
Frez 401388
Friar's Balsam 379300
Friar's Cap 5442
Friar's Cowl 33593,34600,37015
Friar's Crown 91954
Fribourg Begonia 49855
Fried Candlesticks 273531
Fried Egg 227533
Fried Eggs 227533
Friendship Plant 54454,298885,298943
Friendship Tanoak 233294
Friendship Tree 109229
Fries Pondweed 312123
Fries' Pondweed 312123
Fries' Pussytoes 26413
Fries' Sandwort 31931
Frijol 294056
Frijoles 294056
Frijolillo 369116
Frijolko 369116
Frikart Barberry 51641
Frilled Fan 159204
Fringe Bell 351439
Fringe Cups 385719,385720
Fringe Flower 87694,237191,237195,351343
Fringe Myrtle 68755,68761
Fringe Tree 87694,87736
Fringecups 385719,385720
Fringed Amaranth 18714
Fringed Amazonvine 376557
Fringed Bindweed 162517
Fringed Black Bindweed 162517
Fringed Bleeding Heart 128296
Fringed Blue Aster 380852
Fringed Blue Star Flower 20848
Fringed Brome 60599,60667
Fringed Calyx Rhododendron 330403
Fringed Campion 363904
Fringed Centipede Grass 148251
Fringed Chirita 87859
Fringed Dodder 115141
Fringed Dropseed 372679
Fringed Everlasting 189186
Fringed False Hellebore 405592
Fringed False Pimpernel 231496
Fringed Fescue 412409
Fringed Gentian 173181,173342,174215
Fringed Grass-of-parnassus 284538
Fringed Heath 149195
Fringed Heath Myrtle 253740
Fringed Heath-myrtle 253740
Fringed Hibiscus 195216

Fringed Houstonia 198707
Fringed Iris 208640
Fringed Jointweed 308672
Fringed Knotweed 309291
Fringed Lily 391500
Fringed Loosestrife 239589
Fringed Milkwort 308253
Fringed Onion 15289
Fringed Orchid 302241
Fringed Pearlwort 342239
Fringed Peony 280318
Fringed Phacelia 293165
Fringed Pink 127772,127852
Fringed Polygala 308253
Fringed Puccoon 233749
Fringed Quickweed 170146
Fringed Rhododendron 330402
Fringed Rock-cress 30292
Fringed Rosemallow 195216
Fringed Rue 341047
Fringed Rupturewort 192975,192979
Fringed Sage 35505
Fringed Sagebrush 35505
Fringed Sage-wort 35505
Fringed Sandwort 31816
Fringed Sedge 74200
Fringed Sneezeweed 188416
Fringed Spiderflower 95781
Fringed Spineflower 89077
Fringed Water Lily 267825
Fringed Water-lily 267800,267825
Fringed Wattle 1224
Fringed Wormwood 35505
Fringed-calyx Azalea 330935
Fringed-calyx Rhododendron 330935
Fringed-lily 391500
Fringe-flower 87694,351343
Fringeleaf Ruellia 339736
Fringe-leaf Ruellia 339736
Fringelily 391500,391501
Fringe-myrtle 68755
Fringe-myrtles 68755
Fringe-scaled Thistle 92345
Fringetree 87694,87736
Fringe-tree 87694,87736
Frisco Wild Buckwheat 152469
Frith 109857
Fritillaria 168332
Fritillary 168332,168476
Fritsch Briggsia 60278
Fritschiana Spirea 371915
Fritsch Nematanthus 263324
Frivald's Frog Orchid 318212
Friz 401388
Frizzle New Eargrass 262957
Frizzle Tanoak 233216
Frocious Barberry 51616
Frocken 403916
Frockup 168476
Froebel Begonia 49856
Froebel's Four-o'clock 255732
Frog Bit 200225,200228
Frog Bites 200230
Frog Cheese 243840
Frog Grass 342859
Frog Orchid 98560,98586,98596,125694
Frog's Foot 325666,325820
Frog's Meat 37015
Frog's Mouth 28617,255218,273531
Frog's-bit 230240
Frogbit 200225,200228,200230
Frog-bit 200225,200230
Frogbit Family 200235
Frogflower 325498
Frogfoot 405872
Frog-orchis 98560
Frogs-bit Family 200235
Frogwort 273531,273545
Frolov Windhairdaisy 348328
Frond Woodbetony 287565
Frontier Milkvetch 42009
Frontier Peashrub 72193
Forechap 168476
Frost Aster 380953,380955
Frost Boil Pussytoes 26415
Frost Flower 40479
Frost Flowers 39910
Frost Grape 411631,411887,411979,411991
Frost Grass 372428
Frost Weed 188610,405973
Frost Weed Aster 380953
Frost-blite 86902
Frostbloom Bethlehemsage 283637
Frosted Curls 73632
Frosted Hawthorn 109949,109957,109960
Frosted Orache 44668
Frosted Wild Buckwheat 152155
Frost-mat 4414
Frostweed 188610,399957
Frostwort 111397
Frosty Cinquefoil 312630
Frosty Wattle 1502
Frothy Poppy 364193
Froz 401388
Frughan 403916
Fruit Salad Plant 163117
Fruit Sheath Plant 370316
Fruitful Arrowbamboo 162689
Fruitful Fargesia 162689
Fruitful Giantgrass 175598
Fruitful Maple 3461
Fruitful Pricklyash 417237
Fruitful Rattanpalm 65814
Fruitful Rhynchosia 333134
Fruitfulgrass 307724,307748
Fruiting Duckweed 265358
Fruiting Elderberry 345581,345583
Fruiting Myrtle 156118
Fruity Teucrium 388100
Fruog 403916
Frutescent Cladostachys 123559
Frutescent Montanoa 258199
Fruticose Birch 53444
Fruticose Breynia 60064
Fruticose Cordyline 104350
Fruticose Coriaria 104350
Fruticose Cudrania 240820
Fruticose Dracaena 104350
Fruticose Eritrichium 153453
Fruticose Falsepanax 266927
Fruticose Glorybind 103053
Fruticose Gomphocarpus 179032
Fruticose Naiheadfruit 179032
Fruticose Peristrophe 291148
Fruticose Polyscias 310199
Frying-pans 155182
Ftanley's Wash Tub 20057,20125
Fu Stonecrop 356760
Fu's Stonecrop 356760
Fubao Sweetleaf 381423
Fuchs Groundsel 359565
Fuchsia 168735,168740,168749,168767,168788,296108,417405
Fuchsia Begonia 49844,49845
Fuchsia Flowered Rhododendron 330740
Fuchsia Gooseberry 334212
Fuchsia Grevillea 180574
Fuchsia Gum 155584
Fuchsia Heath 146072
Fuchsia Mallee 155584
Fuchsia Tree 352493
Fuchsiaflowered Currant 334212
Fuchsia-flowered Currant 334212
Fuchsia-flowered Gooseberry 334212
Fuchsia-flowered Rhododendron 330740
Fuchsialeaf Jurinea 211837
Fuchsia-leaved Jasmine 211837
Fuchsrebe 411764
Fuchstraube 411764
Fuchuan Rhododendron 331571
Fufangteng Euonymus 157473
Fugong Arrowbamboo 162663
Fugong Barberry 51767
Fugong Dendrocalamus 125476
Fugong Dragonbamboo 125476
Fugong Magnolia 242279
Fugong Neillia 263148
Fugong Rockvine 387764
Fuhai Draba 136997
Fuhai Sterigmostemum 376318
Fuighans 403916

Fuirena 168819
Fuji Cherry 83224
Fujian Azalea 331832
Fujian Barberry 51642
Fujian Bedstraw 170511
Fujian Birthwort 34188
Fujian Camellia 69083
Fujian Cassiope 78629
Fujian Clearweed 299087
Fujian Coldwaterflower 299087
Fujian Crapemyrtle 219956
Fujian Elaeagnus 142132
Fujian Euscaphis 160949
Fujian Groundsel 358924
Fujian Holly 203835
Fujian Loosestrife 239651
Fujian Lovegrass 147673
Fujian Machilus 240587
Fujian Maddencherry 241499
Fujian Maddenia 241499
Fujian Manyfruit Idesia 203424
Fujian Mazus 246979
Fujian Michelia 252878
Fujian Microtropis 254293
Fujian Moxanettle 221561
Fujian Mucuna 259501
Fujian Nanmu 295349
Fujian Oplismenus 272645
Fujian Pagodatree 369018
Fujian Photinia 295686
Fujian Poorspikebamboo 270426
Fujian Primrose 314722
Fujian Raspberry 338449
Fujian Rhododendron 331832
Fujian Ringvine 116025
Fujian Schefflera 350695,350726
Fujian Shibataea 362141
Fujian Sisymbrium 365469
Fujian Sour Bamboo 4611
Fujian Sour-bamboo 4611
Fujian Stinkcherry 241499
Fujian Sweetleaf 381210
Fujian Tonkin Pseudosasa 318284
Fujian Uraria 402141
Fujian Vinegentian 110325
Fujian Wildginger 37624
Fujian Witchgrass 130564
Fujian Wobamboo 362141
Fujiancypress 167200,167201
Fujian-Guangxi Machilus 240647
Fujisan Clematis 94818
Fujisan St. John's-wort 201865
Fujiyama Rhododendron 330240
Fukang Giantfennel 163618
Fukien Crape-myrtle 219956
Fukien Cypress 167200,167201
Fukien Elaeagnus 142132
Fukien Euscaphis 160949
Fukien Groundsel 358924
Fukien Loosestrife 239651
Fukien Machilus 240587
Fukien Mazus 246979
Fukien Mucuna 259501
Fukien Photinia 295686
Fukien Rhododendron 331832
Fukien Tea 77067,141611
Fukien Wildginger 37624
Fukiencypress 167200,167201
Fukulai-mikan 93460
Fukure-mikan 93859
Fukushu Kumquat 167511
Fukutome Monkshood 5214
Fukuyama Bedstraw 170387
Fukuyama Bentgrass 12115
Fulgent Ixora 211091
Fulham Yew 385310
Fuliginos Oak 116095
Fuling Teasel 133475
Full Moon Maple 3034
Fuller's Grass 346434
Fuller's Hawthorn 109723
Fuller's Herb 346434
Fuller's Teasel 133478,133505
Fuller's Thistle 133478
Fullmoon Maple 3003,3034
Full-moon Maple 3034
Fullstem 85166,85167
Fulvescent Dogwood 380449
Fulvileaf Pittosporum 301277
Fulvotomentose Honeysuckle 235801
Fulvous Daylily 191284
Fulvous Day-lily 191284
Fulvous Dogwood 380449
Fulvous Fig 165023
Fumariaeleaf Corydalis 105917
Fumaria-like Cragcress 97986
Fumaria-like Scurvyweed 97986
Fumarileaf Hilliella 196282
Fumeterre 168964
Fumewort 169126
Fumin Ladybell 7663
Fumin Trifoliate-orange 310849
Fumiterre 169126
Fumitory 106117,168964,169126
Fumitory Corydalis 106350
Fumitory Family 169194
Fumitoryleaf Corydalis 105917
Funadoko 93461
Fundi 130544
Funeral Cypress 114690,114753
Funeral Palm 115794
Funereal Cypress 114690
Fung Jujube 418160
Fung Waxplant 198843
Funghom Bamboo 47281
Funghom Bambusa 47281
Funghom Schizostachyum 351855
Fungiform Tupistra 400394
Fungiliving Woodbetony 287446
Fungistyle Aspidistra 39542
Fungivorous Woodbetony 287446
Fungohom Cane 37189
Fungohom Canebrake 37189
Fungose-pithed Fargesia 162692
Funing Alyxia 18513
Funing Bulbophyllum 62745
Funing Curlylip-orchis 62745
Funing Earlymadder 226239
Funing Elaeocarpus 142312
Funing Funingvine 284100
Funing Greenbrier 366334
Funing Hackberry 80638
Funing Lilyturf 272077
Funing Maple 3297
Funing Mosquitotrap 117738
Funing Oak 324409
Funing Oilfruitcamphor 381792
Funing Parepigynum 284100
Funing Pellionia 288724
Funing Redcarweed 288724
Funing Rockvine 387775
Funing Sedge 74614
Funing Spicegrass 239641
Funing Swallowwort 117738
Funing Syndiclis 381792
Funing Trigonotis 397412
Funingvine 284099
Funiushan Arrowbamboo 162693
Funiushan Dendrobium 125157
Funiushan Magnolia 242117
Funiushan Redbud 83782
Funiushan Spicebush 231348
Funkia 198589
Funnel Bubbleweed 297878
Funnel Chirita 87886
Funnel Didissandra 25753
Funnel Flower 130187
Funnel Primrose 314312
Funneled Physochlaina 297878
Funnel-fruited Gum 155533
Funnel-lily 23378
Funnel-shaped Crossandra 111724
Funnelshaped Dendrobium 125196
Funny Face 399601
Funny-face 399601,410677
Fuping Adonis 8379
Fuping Corydalis 106602
Fuping Yellowhair Maple 2979
Fura-fura Potato 99699
Furbish's Pedicularis 287223
Furcate Cordia 104183

Furcate Craniotome 108668
Furcate Milkwort 308065
Furcate Roundleaf Sundew 138350
Furcate Screwpine 281031
Furcatebract Monkshood 5152
Furcatefid Buttercup 325884
Furcillate Spapweed 204969
Furcillate Touch-me-not 204969
Furfuraceous Fruit Slugwood 50519
Furfuraceous Slugwood 50519
Furfuraceous Woodbetony 287224
Furfuraceousfruit Camellia 69084
Furfuraceous-fruited Camellia 69084
Furfuraceous-leaf-like Camellia 69430
Furfurasceous Knema 216811
Furin Holly 203840
Furnace Creek Wash Wild Buckwheat 152147
Furra 401388
Furrow Arrowbamboo 162651
Furrowcress 383989
Furrowed Caterpilar 354774
Furrowed Melilot 249251
Furrowed Saxifrage 349303
Furrowedstone Cherry 83275
Furrowed-stone Cherry 83275
Furrowglume 357361,357365
Furrowlemma 25331
Furrowstem Milkvetch 43112
Furrow-weed 169126
Furry Feather Calathea 66203
Furry Kittens 115587
Furry Willow 342965
Furturaceous Syndiclis 381793
Furze 271563,401373,401388
Furze Hakea 184649
Furzen 401388
Fuscescent Beadruby 242685
Fuscifolius Raspberry 338451
Fuscous Begonia 49862
Fuscous Cypressgrass 118928
Fuscous Galingale 118928
Fuscous Sprangletop 226005
Fuscous-spotted Begonia 49863
Fusiform Fieldorange 249664
Fusiform Melodinus 249664
Fusong Monkshood 5217
Fussy Cats 106703,343151
Fustic 88514,107300,240834
Fustic Tree 88500
Fustic-tree 88500
Fusui Fig 165027
Futi 211233
Futterer Gentian 173466
Futui 211233
Fuxiong Ligusticum 229307
Fuyang Bamboo 297366
Fuyang Broodhen Bamboo 297366
Fuyuan Azalea 330753
Fuyuan Rhododendron 330753
Fuyuan Small-fruit Fig 165312
Fuyun Currant 333984
Fuzhou Falsenettle 56142
Fuzhou Maple 3103
Fuzhou Persimmon 132090
Fuzhou Yam 131597
Fuzz 401388
Fuzzen 401388
Fuzzy Deutzia 127072
Fuzzy Flatsedge 119378
Fuzzy Magnoliavine 351089
Fuzzy Mock Orange 294560
Fuzzy Mountain Silverbell 184748
Fuzzy Pride-of-Rochester 127072
Fuzzy Sedge 74812
Fuzzy Wuzzy Sedge 74805
Fuzzy-weed 35407
Fyfield Pea 222851
Fynbos Aloe 17320
Fynel 167156
Fyrrys 401388

G

G. Forrest's Rhododendron 330724
G. Houlston's Rhododendron 330592
G. Houlston's Rhododendron Azalea 330592
G. Medd's Rhododendron 331209
G. Nakahara's Rhododendron 331305
G. Sears' Rhododendron 331776
G. W. Trall's Rhododendron 331987
Gabon Bombax 56786
Gabon Ebony 132353
Gabon Symphonia 380697
Gabon Yellow-wood 289439
Gaboon Chocolate 208990
Gaboon Mahogany 44883,44884
Gaboon Nut 108127
Gabriel's Trumpet 366870
Gad Rouge 157429
Gadgevraw 227533
Gadjerwraw 227533
Gadrise 157429,380499,407989
Gaelder Rose 407989
Gaerthera 169313
Gagea 169375
Gagnepain Barberry 51643
Gagnepain Begonia 49867
Gagnepain Rockjasmine 23175
Gagnepain Woodbetony 287227
Gagnepain's Barberry 51643
Gagroot 234547
Gagued Protea 315803
Gahnia 169548
Gaillardia 169568,169571,169586
Gaitberry 338447
Gaiter 380499
Gaiterberry 338447
Gaiter-tree 380499
Gaize Crazyweed 278840
Gaize Sandwort 31922
Gajanimma 93693
Gal 261162
Galanga 17639,214995,215011
Galanga Galangal 17677
Galanga Resurrectionlily 215011
Galangal 17639,17677,17733
Galangale 119125
Galanthus 169705
Galatella 169756
Galax 169817,169819
Galba 67855
Galbanum 163619,163705
Galcier Pink 127790
Gale 261162
Gale Gowan 89704
Galé's Epidendrum 146372
Galeandra 169865
Galeate Lomatogoniopsis 235429
Galeate Monkshood 5708
Galeate Spokeflower 235429
Galeate Woodbetony 287228
Galenia 169953
Galeobdolon 170015
Galeott Cactus 244174
Galericulate Skullcap 355449
Gales 261162
Galgal Large 93557
Galgal Lemon 93706
Galiccressleaf Nightshade 367617
Galiccress-like Sophiopsis 368951
Galilean Orchid 273442
Galingale 17677,118428,119125
Galingale Flat Sedge 119125
Galinglelike Sawgrass 245392
Galinsoga 170137
Galium Circaezans 271563
Gall 261162
Gall Oak 324050

Galla 381035
Gallant Soldier 170142
Gallant-soldier 170137,170142,170146
Gallberry 203847
Galligaskins 315082
Gallinita 246022
Gallion 170193,170743
Gallipoli Rose 93202
Gallito 287850
Gall-of-the Earth 81065
Gall-of-the-earth 81065,174148,313778, 313881,313901
Gallon 292372
Galloway Whin 172905
Gallow-grass 71218
Gallpea 170164
Galluc 381035
Gallwood 36474
Galndularhairy Aletris 14495
Galpin Acacia 1238
Galtonia 170827
Gama Grass 398138,398139
Gamagrass 398138
Gama-grass 57922,398138,398139
Gamalu 320307
Gamar 178025
Gamba Grass 22675
Gambel Oak 323927
Gambel Primrose 314422
Gambel's Oak 323927
Gambia Pods 1427
Gambian Bush Tea 232503
Gambian Dayflower 101027
Gambian Kino 320293
Gambian Mahogany 216214
Gambier 401757
Gambier Cutch 401757
Gambier Plant 401757
Gambier-plant 401757
Gambinn Bush Tea 232503
Gambir 401757
Gambir Plant 401737,401773
Gambirplant 171046,401737
Gambir-plant 401737
Gamble Barberry 51417
Gamble Cyclobalanopsis 116097
Gamble Embelia 144739
Gamble Machilus 240588
Gamble Oak 116097
Gamble Olive 270123,270165
Gamble Qinggang 116097
Gamble-weed 345982
Gambo Hemp 194779
Gamboge 171046,171221
Gamboge Tree 171144
Gambosa Tree 171144
Gamma-grass 57922
Gamosepal Hemiboea 191357
Gamosepal Lysionotus 239942
Gamotepalous Goosefoot 86990
Gamp 363235
Gampi 414257
Ganagra 340084
Gan-Chuan Rush 213220
Gandaria 57847
Gander Cholla 116659
Gander's Cholla 116659
Gander's Ragwort 279912
Gander-gauze 273531
Ganderglass 273531
Gandergoose 273545,359158
Gander-gosse 273531
Gander-grass 312360
Gandigoslings 273531
Ganeschin Bluegrass 305543
Gangaw 252370
Gangba Felwort 235451
Gangchai Riparial Silverreed 394929
Gangdise Catchfly 363771
Gangera Amaranth 18836
Ganges Asystasia 43624
Ganges Primrose 43624
Gang-flower 308454
Gangkou Gugertree 350946
Gangrenbuchi Milkvetch 43137
Gangrenbuqi Milkvetch 43137
Ganja 71220
Ganjanimma 93693
Ganlanfoshow Ginkgo 175821
Ganluo Yinshancress 416336
Ganluo Yinshania 416336
Gannan Azalea 330755
Gannan Barberry 51775
Gannan Corydalis 106449
Gannan Deutzia 126912
Gannan Fritillary 168523
Gannan Gentian 173467
Gannan Primrose 314348
Gannan Rhodiola 329876
Gannan Rhododendron 330755
Gannan Willow 344035
Ganning Crazyweed 278838
Ganpin Clematis 95265
Ganpin Shorthair Clematis 95265
Ganping Mahonia 242538
Ganping Woodbetony 287230
Ganqing Needlegrass 376887
Gansu Apple 243642
Gansu Aralia 30691
Gansu Arrowwood 407900
Gansu Asparagus 39048
Gansu Azalea 331528
Gansu Baical Windflower 23711
Gansu Barberry 51805
Gansu Beancaper 418681
Gansu Birdlingbran 87239,87271
Gansu Butterflybush 62149
Gansu Cacalia 283815
Gansu Chesniella 87271
Gansu Clearweed 298922
Gansu Coldwaterflower 298922
Gansu Columbine 30065
Gansu Corydalis 105738
Gansu Cowparsnip 192275
Gansu Crabapple 243642
Gansu Craneknee 221678
Gansu Crazyweed 278919
Gansu Crenate Firethorn 322459
Gansu Deadnettle 220407
Gansu Deutzia 126839
Gansu Doronicum 136355
Gansu Drybirdbean 87271
Gansu Dwarf Childvine 400937
Gansu Dwarf Tylophora 400937
Gansu Fescuegrass 164029
Gansu Figwort 355151
Gansu Forkstamenflower 416506
Gansu Fritillary 168523
Gansu Gooseberry 334220
Gansu Hawthorn 109781
Gansu Honeysuckle 235895
Gansu Houndstongue 117967
Gansu Indigo 206421
Gansu Iris 208747
Gansu Jerusalemsage 295124
Gansu Jurinea 211865
Gansu Kansu 87239,87271,181649
Gansu Kneejujube 67010
Gansu Kobresia 217198
Gansu Larkspur 124312
Gansu Liriope 232627
Gansu Maple 3130
Gansu Milkvetch 42617
Gansu Mockorange 294484
Gansu Nettle 402889
Gansu Panzerina 282481
Gansu Parnassia 284544
Gansu Peach 20921
Gansu Peashrub 72256
Gansu Pea-shrub 72256
Gansu Poplar 311182
Gansu Raspberry 338453,339231
Gansu Rhodiola 329894
Gansu Rhododendron 331528
Gansu Ricebag 181649
Gansu Sagebrush 35468,35523
Gansu Sandwort 31971
Gansu Sedge 74973
Gansu Skullcap 355713
Gansu Spurge 159171
Gansu Stonecrop 357015

Gansu Summerlilic 62149
Gansu Swordflag 208747
Gansu Tamarisk 383514
Gansu Viburnum 407900
Gansu Willow 343368
Gansu Windhairdaisy 348418
Gansu Withania 414601
Gansu Woodbetony 287302
Gansui Spurge 159172
Gansu-Mongol Peashrub 72303
Gansu-Mongol Pea-shrub 72303
Gan-Xin Beadcress 264618
Ganyazi Goldenray 229071
Ganymede's Cup 262473
Gan-Zang Buttercup 325895
Ganzi Ladybell 7673
Ganzi Larkspur 124314
Ganzi Unibract Fritillary 168606
Gaoliang 369590,369720
Gaoligong Asiabell 98325
Gaoligong Cymbidium 116873
Gaoligong Platanthera 302347
Gaoligong Spapweed 204847
Gaoligong Tanoak 233223
Gaoligong Touch-me-not 204847
Gaoligongshan Sedge 74625,74678
Gaoligongshania 171016,171017
Gaopo Hanceola 185249
Gaoxiong Aspidistra 39519
Gaoxiong Fig 165203
Gaoxiong Galangal 17707
Gaoxiong Gentian 173548
Gaoxiong Leafflower 296779
Gaoxiong Naiad 262021
Gaoxiong Taxillus 385225
Gaozhou Oil Camellia 69092
Gape False-olive 142440
Gapfruityam 351380
Gaping Dutchman's Pipe 198802
Gaping Jack 231183
Gaping Marianthus 245273
Gaping Maxillaria 246721
Gapleaf Corydalis 106004
Gap-mouth 28617,130383,231183,255210
Garad Fennelflower 266229
Garambullo 261708
Garanbi Clematis 95361
Garanbi Spurge 158935
Garanli Fig 165444
Garapatilla 14976
Garbanzo 90801
Garbanzo Bean 90801
Garbanzo Beans 90801
Garbeen 106822
Garcia Barberry 51659
Garcinia 171046
Garcinia Family 97266
Garck Woodbetony 287231
Garclive 11549
Garden Abutilon 913
Garden Aconite 5442
Garden Alternanthera 18095
Garden Anchusa 21949
Garden Anemone 23767,23846
Garden Angelica 30897,30899,30932
Garden Arabis 30220
Garden Asparagus 38985,39120
Garden Aster 67314
Garden Baby's-breath 183243
Garden Balsam 204799,249208
Garden Basil 268438
Garden Bean 294056
Garden Beet 53259,53269
Garden Black Currant 334117
Garden Burner 345881
Garden Burnet 345881
Garden Candytuft 203249
Garden Carrot 123164
Garden Catmint 264920
Garden Cat-mint 264920,265005
Garden Celery 29328
Garden Chervil 28024
Garden Chimneyflower 254900
Garden Chrysanthemum 89481,89619,124785
Garden Clover 249208
Garden Columbine 30081
Garden Coneflower 339581
Garden Coreopsis 104598
Garden Cornflower 81020
Garden Cosmos 107161
Garden Cress 225450
Garden Croton 98190
Garden Cucumber 114245
Garden Currant 334179,334201
Garden Cypress 35868,346250
Garden Dahlia 121561
Garden Daisy 50825
Garden Dock 340178
Garden Egg 367370
Garden Eggplant 367370
Garden Euphorbia 159069
Garden Evening-primrose 269443
Garden Everlasting Pea 222750
Garden Fennel Flower 266225
Garden Fennelflower 266241
Garden Flag 208583
Garden Forget-me-not 260794,260892
Garden Gates 174877
Garden Geranium 288045,288297
Garden Ginger 418010
Garden Gold 66445
Garden Grepe-hyacinth 260295
Garden Heliotrope 190559,404316
Garden Hoary 246478
Garden Honeysuckle 235876
Garden Huckleberry 367256
Garden Hyacinth 199583
Garden Hydrangea 199956
Garden Laovage 228250
Garden Larkspur 102834
Garden Lavender 223291
Garden Leek 15621
Garden Lemon 114200
Garden Lettuce 219485
Garden Lobelia 234447
Garden Loosestrife 239804
Garden Lovage 228250
Garden Lupin 238481
Garden Marjoram 274224
Garden May 408167
Garden Mercury 250600
Garden Mignonette 327896
Garden Millingtonia 254900
Garden Mint 250287,250450
Garden Monkshood 5442
Garden Mum 89619
Garden Myrrh 261585
Garden Nasturtium 399601
Garden Nightshade 367416
Garden Onion 15166
Garden Orach 44468
Garden Orache 44468
Garden Orpine 200784
Garden Pansy 409624,410677
Garden Parsley 77828,292694
Garden Parsnip 285741
Garden Patience 339914,340178
Garden Pea 301053,301070
Garden Pear 323133
Garden Penstemon 289345
Garden Petunia 292742
Garden Phlox 295288
Garden Pink 127793
Garden Pink-sorrel 277925
Garden Plum 316382
Garden Poppy 282717
Garden Primrose 269443,314856
Garden Privet 229569
Garden Purslane 311899
Garden Radish 326616
Garden Ranunculus 325614
Garden Red Currant 334179
Garden Rhubarb 329308,329386,329388
Garden Rocket 102841,153998,154019,154041,193417
Garden Rue 341064
Garden Sage 345271
Garden Sagebrush 35088
Garden Serviceberry 19288
Garden Snapdragon 28617
Garden Solomon's-seal 308564

Garden Sorrel 339887,340279
Garden Spapweed 204799
Garden Speedwell 317927,406973
Garden Spotted-leaf 98190
Garden Stonecrop 200784
Garden Storkbill 288206
Garden Strawberry 167597
Garden Succory 90894
Garden Sunflower 188908
Garden Sword-like Iris 208547
Garden Thyme 391397
Garden Tickseed 104598
Garden Tomato 239157,239158,367322
Garden Touch-me-not 204799
Garden Trifoly 249208,277747
Garden Valerian 404316
Garden Verbena 405852
Garden Veronica 317927
Garden Vervain 405812,405852
Garden Vetch 408571,408578
Garden Violet 410320
Garden Wormwood 36088
Garden Yellow Loosestrife 239898
Garden Yellowrocket 47964
Garden Yellow-rocket 47964
Garden Zinnia 418061
Garden's Plague 8826
Garden-cress 225450
Garden-cress Peppergrass 225450
Gardencress Pepperweed 225450
Garden-crowfoot 325614
Gardener's Delight 363370
Gardener's Eye 363370
Gardener's Garter 293710
Gardener's Garters 293709,293720
Gardengate 309494
Garden-heliotrope 404316
Garden-huckleberry 367585
Gardenia 154162,171238,171246,171253
Gardenia-flowered Jasmine 211887
Garden-sorrel 339887
Garder Onion 15165
Garder's Delight 363312
Gardner Angelonia 24523
Gardner Goniothalamus 179409
Gardner Neogyna 263911
Gardner Spapweed 204974
Gardner Sweetleaf 381213
Gardner Touch-me-not 204974
Gardner's Saltbush 44406
Gardneria 171439
Garget 298094
Garget-plant 298094
Gargut-root 190936
Garhadiolus 171475
Garhwal Petioled Barberry 52025
Gari 244507
Gariand Spirea 371815
Gariand-lily 68031
Garish Coleus 99711
Garish Flamenettle 99711
Garishcandle 28071
Garishleaf Pyrola 322779
Garishspiderorchis 155264
Garishtaro 65214
Garland Chrysanthemum 89481
Garland Cock's Head 187842
Garland Crab Apple 243583
Garland Crabapple 243583
Garland Daphne 122416
Garland Flower 122359,122416,187417,
187432
Garland Flower Daphne 122416
Garland Gold Yucca 416585
Garland Rhododendron 330875
Garland Thorn 280559
Garland-tree 243583
Garlete 15698
Garlic 15018,15698
Garlic Chives 15843
Garlic Germander 388272,388275
Garlic Lily 230030
Garlic Mustard 14953,14964
Garlic Onion 15480
Garlic Pear 110205
Garlic Penny-cress 390205
Garlic Sage 388275
Garlic Tree 354735
Garlic Vine 54303,116530,317477
Garlic Wood 78657
Garliccress 365386,365469,365482
Garlicfruit 242983,242984
Garlicleaf Germander 388272
Garlicleaf Salsify 394327
Garlicmustard 14953
Garlic-scented Vine 317477
Garlicsmell Germander 388272
Garlicvine 244710
Garliek Germander 388272,388275
Garlock 364557
Garnesie Violet 246478
Garnet Berry 334179
Garnetberry 334179
Garnet-berry 334117
Garnier Crestedgalea Woodbetony 287114
Garnotia 171486,171509
Garou Bush 122448
Garrett Litse 233990
Garrett Loosestrife 239652
Garrett Mallotus 243362
Garrett Obtuseleaf Acacia 1383
Garrett Sauropus 348050
Garrett Yam 131598
Garrett's Fleabane 150651
Garrett's Litse 233920
Garrett's Manglietia 244440
Garrett's Saltbush 44413
Garrett's Tanoak 233224
Garrettia 171540
Garrocha 385477
Garroway's Australian Wild Lime 253289
Garry Oak 323931
Garter-berry 338447
Garth Cress 225450
Garuga 171561
Gas Plant 129615,129618
Gaskins 83155,315082,334250
Gasparillo 155256
Gasparrin Fig 165037
Gasplant 129618
Gas-plant 129618
Gas-plant Dittany 129618
Gasso Nut 244658
Gasteria 171607
Gastrochilus 171825,342015
Gastrodia 171905,171918
Gat's-toes 26385
Gate Canyon Wild Buckwheat 152154
Gaterfruit Gynostemma 182999
Gates Cactus 244080
Gatfridge Tree 157429
Gather Primrose 314992
Gathercupule Tanoak 233378
Gatten-tree 157429,380499,407989
Gatter-berry 338447
Gatter-bush 157429,380499
Gatteridge-tree 157429,380499,407989
Gatter-tree 338447,380499
Gattinger's Goldenrod 368104
Gattinger's Panic-grass 281638
Gattinger's Prairie Clover 121889
Gattridge-berry 157429
Gaub 132286
Gaub Tree 132286
Gauceous Lampranthus 220564
Gaudalup Cypress 114702
Gaudichaud Camellia 69094
Gaudichaud Heterosmilax 194114
Gaudichaud Polycarpaea 307670
Gaudineess Windhairdaisy 348222
Gaudrangular Bishop's Cap 43520
Gaugau 252370
Gaul 261162
Gaul Nut 212636
Gaule 261162
Gaul-nut 212636
Gaultheria 172047
Gaumachil Apea-earring 301134
Gaumachil Monkey Earrings Pea 301134
Gaund 292372
Gaura 172185,172201

Gaura bisannuellc 172187
Gaura Smallflower 172203
Gaussen Douglas Fir 318572
Gaussen Elm 401511
Gaussen Juniper 213780
Gauze Tree 219986
Gavelstyle Orchis 243287
Gay Feather 228433,228439,228454,228509,228511,228519,228534,228541
Gay Feathers 228519,228529
Gay Flower 169603
Gayanum Painted Daisy 227491
Gayfeather 228433,228506,228529
Gay-feather 228433,228529
Gaywings 308253
Gay-wings 308253
Gazania 172276,172348
Gazania 'Fiesta Red' 172342
Gazania 'Sun Gold' 172343
Gazania 'Yellow Trailing' 172344
Gazel 109857,334117,334179
Gazle 109857,334117,334179
Gdgson 248411,363461
Geagles 192358
Geal Gowan 89704
Geal-gowan 89704
Geal-seed 89704
Gean 83155
Geanucanach's Pipes 324335
Gear 282717
Gebang 106986,107000
Gebang Palm 106986,107000
Gebauer Abutilon 896
Geckdor 170193
Gecko Beancaper 418721
Geckowood 348038,348041
Gedang Rockfoil 349372
Gedu Nohor 145950
Geebung 292025,292028,292031
Geen 83155
Geese-and-gullies 343151
Geiger Tree 104245
Geiser's Fleabane 150652
Geissaspis 172497
Geji Corydalis 106165
Gelbe Zeitlose 315082
Gelders Rose 407989
Geldilocks 40781
Gelidocalamus 172736
Gellalfred 86427
Gelonium-leaved Glyptic-petal Bush 177991
Gelonium-leaved Glyptopetalum 177991
Gelrica Poplar 311150
Gelsemine 211940
Gelsemium 172778
Gelu Grape 411686
Geminate Bauhinia 49067
Geminate Kengyilia 215908
Geminate Roegneria 335318
Geminate Speedwell 407109
Geminate Woodbetony 287046
Geminatepetal Osmanthus 276284
Geminate-petaled Osmanther 276284
Gemmate Daphne 122437
Gemmate Staff-tree 80193
Gemmiparous Pondweed 312126
Gemsbuck Bean 49077
Gendarussa 172812,172832
General Leucanthemum 227533
General Whitedaisy 227533
Genet 121001
Geneva Bugle 13104
Geneva Bugleweed 13104
Geneva Plant 213702
Gengma Euonymus 157649
Gengma Forkliporchis 417735
Gengma Rattlebox 111880
Gengma Spatholobus 370414
Genianthus 172848
Genichi Euonymus 157355
Geniculate Bedstraw 170391
Geniculate Buttercup 326121
Geniculate Cayratia 79847
Geniculate Elaeagnus 141996
Geniculate Fig 165040
Geniculate Franchet Monkshood 5209
Geniculate Holly 203840
Geniculate Microstegium 254010
Geniculate Monkshood 5219
Geniculate Spikesedge 143158
Geniculateclaw Distinctstyle Monkshood 5608
Geniculatelike Monkshood 5500
Geniculatepetal Monkshood 5219
Geniosporum 172852
Geniostoma 172881
Genip 172888,172890,201647,249126
Genip Tree 172888,172890,201647
Genipa 172888,172890,201647
Genipa Fruit 172890
Genipa Tree 172890
Genipap 172888,172890,201647
Genipapo 172890
Genip-tree 172888,201647,249125
Genista 173000,173082
Genista Anglica 401395
Genista Broom 173082
Genonggang 110246
Genshokan 93463
Genshokan Citrus 93463
Gentian 81065,150414,154895,156521,173181,173264,173318
Gentian Blue Gromwell 233434
Gentian Family 174100
Gentian Rockbell 412750
Gentian Sage 345292
Gentian Speedwell 407133
Gentianella 173184,174101
Gentianopsis 174204
Gentian-root 173610
Gentian-wood 334373,334374
Gentiawood 334373,334374
Gentil Begonia 49868
Gentil Cliffbean 66633
Gentil Millettia 66633
Gentile Birthwort 34191
Gentili's Delavay Ampelopsis 20340
Gentili's Delavay Snakegrape 20340
Gentle Dock 308893
Gentle Teasel 133488
Gentle Thistle 91916
Gentleman John 410677
Gentleman Tailor 410677
Gentleman's Buttons 379660
Gentleman's Finger 37015
Gentleman's Fingers 37015
Gentleman's Pincushion 216753
Gentleman's Purse 72038
Gentleman's Tormenters 170193
Gentleman's Tormentors 170193
Gentlemen's and Ladies' Fingers 37015
Gentlemen's Buttons 379660
Gentlemen's Cap-and-frills 325820
Gentlemen's-and-ladies' Fingers 37015
Gentlemen-and-ladies 37015
Gentlemen-and-ladies' Fingers 37015
Genturyplant Family 10775
Genuine Mahogany 380527
Genuine Mapleleaf Excoecaria 161632
Genuine Storax Tree 232562
Geocarpon 174284,174285
Geonomaform Iguanura 203508
Geophila 174355
Georg Dolomiaea 135732
George 404575
George Lily 120546,404575,404581
George Monkshood 5225
George Rockfoil 349588
George's Fir 379
George's Holly 203844
Georgia Aster 380899
Georgia Bark 299703
Georgia Bark Tree 299702
Georgia Beargrass 266496
Georgia Buckeye 9693
Georgia Bulrush 353434
Georgia Ceanothus 79906
Georgia Centaurea 81112
Georgia Gayfeather 228499
Georgia Gooseberry 333943
Georgia Hackberry 80760
Georgia Oak 323935

Georgia Pine 300128
Georgia Rush 213126
Geotube Woodbetony 287235
Geraflour 86427
Gerald Rockcress 30277
Geraldton Carnation Weed 159947
Geraldton Wax 122786
Geraldton Wax Flower 85852,122786
Geraldton Waxflower 122786
Geranium 174440,174715,174739,288045
Geranium Aralia 310204
Geranium Family 174431
Geranium of Florists 288045
Geranium of House-plants 288045
Geranium Oil 117204,288045
Geranium Rockjasmine 23176
Geranium Saxifrage 349380
Geranium-leaf Aralia 310204
Geranium-leaf Begonia 49869
Geranium-like Begonia 49869
Gerard Ephedra 146183
Gerard Pea-shrub 72239
Gerard Pine 299953
Gerard Rush 213127
Gerard Vetch 408352
Gerard's Pine 299953
Gerard's Rock Cress 30277
Gerardia 10142
Gerbera 175107
Gerbera Daisy 175172
Gereken 114122
Geriba Palm 31705
Gerland Rose 336940
Germainia 175241
German Asphodel 392601
German Betony 373224
German Broom 120997
German Campion 410957
German Catchfly 364179,410957
German Celery 29329
German Chamomile 246396
German Filago 166038
German Fluffweed 166038
German Garlic 15734
German Gentian 173473
German Greenweed 172976
German Hedgenettle 373224
German Honeysuckle 236022
German Houndstongue 117968
German Iris 208583
German Ivy 123750
German Jurinea 207117
German Knotgrass 353987
German Lilac 81768
German Madwort 39298,39299
German Millet 361794
German Pellitory 21230
German Pink 127634
German Sausage Tree 216322,216324
German Spurge Olive 122515
German Statice 179384
German Sundrops 269404
German Swordflag 208583
German Tamarisk 261262
German Thyme 391366
German Velvet Grass 197306
German Velvetgrass 197306
German Velvet-grass 197306
German Vetch 408381
German Woad-waxen 172976
German-and-English 326340
Germander 387989,388026,388042,388279
Germander Chickweed 406973,407072
Germander Meadowsweet 371877,371883
Germander Sage 344950
Germander Speedwell 407072
Germander Spiraea 371877
Germanderleaf Speedwell 407072
German-ivy 123746,123750
German-madwort 39299
Germany False Tamarisk 261262
Germany Falsetamarisk 261262
Gerneter 321764
Gernut 102727
Gero 289116
Geroge Oreocharis 273862
Geroggong 110240
Gerrard's Aristea 33651
Gerrard's Ephedra 146177
Gerrard's Rock-cress 30277
Gerze Crazyweed 278840
Gesmine 211940
Gesnera 327322
Gesneria 175295,327322
Gesneria Family 175306
Gesnerialike Rabdosia 209678
Gesse 211940
Gessemine 211940
Gethsemane 37015,211940,273531
Geuky Flower 248312,273531
Geum 175375
Geutian 174101
Gevuina 175468
Gew-gog 334250
Geyer Onion 15312
Geyer Waterplantain 14731
Geyer's Aster 380910
Geyer's Larkspur 124226
Geyer's Sand-mat 85749
Geyer's Spurge 85749
Gezan Monkshood 5228
Gezlins 343151
Gharab 311308
Ghasal Combretum 100481
Ghasal Windmill 100481
Ghent Gladiolus 176222
Gherkin 114122,114245
Gholson's Gayfeather 228467
Ghost candle 294988
Ghost Flower 256526,258044
Ghost Grass 182942
Ghost Gum 106788,106814,155688
Ghost Gum-bark White 106814
Ghost Iris 208441
Ghost Kex 24483
Ghost lampstand 335141,335151
Ghost Orchid 125694,147303,147307
Ghost Plant 180269
Ghost Poppy 282647
Ghost Thistle 91935
Ghost Tree 123245
Ghost-arrow Peashrub 72249
Ghostflower 258044
Ghostfluting 228319,228321
Ghost-kex 24483
Ghost-tree 123244,123245
Ghost-weed 159313
Ghstplant Sagebrush 35770
Ghstplant Wormwood 35770
Giam 198126
Gianifennel-like Ligusticum 229323
Giant Adder's Tongue 154936
Giant Aechmea 8564
Giant Airplant 392001
Giant Allium 15445
Giant Alocasia 16512
Giant Anisetree 204553
Giant Arborvitae 390645
Giant Arbor-vitae 390645
Giant Arrowhead 342381,342383
Giant Arum 20037,20145
Giant Asphodel 148531
Giant Aspidistra 39553
Giant Azalea 330789
Giant Bamboo 47203,125461,125465,125477,175588
Giant Bell 276795
Giant Bellflower 70119
Giant Bell-flower 276795,276796
Giant Bentgrass 12118
Giant Bird of Paradise 377563
Giant Bird's-nest 320853
Giant Blue Iris 208588
Giant Blue Wild Rye 228379
Giant Bothrocaryum 57695
Giant Brunsvigia 61398
Giant Bulrush 352169,352187
Giant Burmese Honeysuckle 235846
Giant Burrawang 241458
Giant Burreed 370058
Giant Bur-reed 370058

Giant Bushman's River Hybrid Cycad 145230
Giant Butterbur 292374
Giant Buttercup 325518,326049
Giant Buttonflower 404518
Giant Cabomba 64492
Giant Cactus 77077
Giant Cactus of Arizona 77077
Giant Cactus of Califoruia 77077
Giant Caladium 16496
Giant Cane 37191,37467
Giant Cardiocrinum 73159
Giant Cardon 279490
Giant Catmint 264927
Giant Cedar 390645
Giant Chickweed 260920,260922,374713,374749,375067
Giant Chin Cactus 182476
Giant Chinkapin 89969
Giant Chinquapin 89969
Giant Claret Cup 140292
Giant Climbing Tearthumb 309564
Giant Coneflower 339585
Giant Coral-root 194521
Giant Corydalis 105929
Giant Coryphantha 107008
Giant Cowslip 314407
Giant Cutgrass 418103
Giant Daisy Giantdaisy 227455
Giant Death Camas 417895
Giant Dendrocalamus 125477
Giant Desertcandle 148554
Giant Devil Rattan 121497
Giant Dodder 115123
Giant Dogwood 57695
Giant Dracaena 104339
Giant Dragonbamboo 125477
Giant Duckweed 221007,372300
Giant Dumbcane 130093
Giant Eightangle 204553
Giant Elephant's Ear 16512
Giant Elodea 141569
Giant False Sensitive Plant 255020
Giant False Spiraea 369272
Giant Feather Grass 376779
Giant Feathergrass 376779
Giant Fennel 163574,163580,163590,167145
Giant Fescue 163993,350625
Giant Fescuegrass 163993
Giant Filbert 106757
Giant Fir 384
Giant Forget-me-not 118031,260734
Giant Four-involucre 124914
Giant Four-o'clock 255741
Giant Fox Tail Grass 361826
Giant Foxtail 361743,361826
Giant Garlic 15048,15726
Giant Goblin Orchid 258993
Giant Golden Bamboo 47518
Giant Golden Chinkapin 89969
Giant Goldenrod 368106
Giant Granadilla 285694
Giant Green Foxtail 361954
Giant Green-striped Bamboo 47518
Giant Groundsel 46227,125716
Giant Gum 155478,155719
Giant Hairysheath Edible Bamboo 297306
Giant Heath 149017
Giant Helleborine 147142
Giant Hemsleia 191925
Giant Herb-Robert 174719
Giant Hesperaloe 193228
Giant Hogweed 192302,192306
Giant Homalomena 197791
Giant Honey-myrtle 248076
Giant Hyssop 10402,10407,10410,10422,236262
Giant Jewel Plant 17447
Giant Kalanchoe 215097
Giant Knotweed 162549,309723,328345
Giant Larkspur 124233
Giant Lily 73154,73159,169245
Giant Lily Turf 272086
Giant Lombardy Poplar 311399
Giant Mallow 194680
Giant Mannagrass 177629
Giant Meadowsweet 166089
Giant Mexican Cereus 279490
Giant Mexican Lily 169242
Giant Milkweed 68088
Giant Miscanthus 255849
Giant Montbretia 111468
Giant Mountain Fishtailpalm 78051
Giant Mountain Rattan 303087,303094
Giant Mucuna 259512
Giant Mugo Pine 300223
Giant Mullein 405788
Giant Needleleaf 307840
Giant Nettle 125540
Giant Nolina 266506
Giant Onion 15445
Giant Orchid 48411,147142,180116,180126,196379
Giant Papyrus 119347
Giant Pine 300011
Giant Pineapple Flower 156028
Giant Pineapple Lily 156028
Giant Pinedrops 320853
Giant Plantain 302101
Giant Potato Creeper 367745
Giant Potatocreeper 367745
Giant Prairie Lily 417617
Giant Protea 315776
Giant Pumpkin 114288
Giant Pussy Willow 343185
Giant Ragweed 19191
Giant Raisin 180811
Giant Rattanpalm 65698
Giant Rattlesnake Plantain 179664
Giant Rattlesnake-plantain 179664
Giant Red Paintbrush 79129
Giant Redbud 83785
Giant Redwood 360576
Giant Reed 37450,37467,295888
Giant Reed-bamboo 47192
Giant Rescue 163993
Giant Rhododendron 330789
Giant Rotang Palm 121497
Giant Rush 213131
Giant Rye 398939
Giant Sacaton 372888
Giant Saltbush 44542
Giant Scabious 82113,82139
Giant Sensitive Plant 255089
Giant Sensitive-plant 255020
Giant Sequoia 360565,360575,360576
Giant Snowdrop 169710
Giant Solomon's Seal 308508,308527
Giant Solomon's-seal 308508
Giant Spaniard 4747
Giant Spanish Dagger 416572
Giant Spider Flower 95796
Giant Spider Plant 95796
Giant St. John's Wort 201743,201982
Giant St. John's-wort 201743
Giant Summer-hyacinth 170828
Giant Sunflower 188968
Giant Sweet Chervil 261585
Giant Sweet Sultan 81116
Giant Tabacco 266062
Giant Taro 16496,16512,99919
Giant Tea Rose 336817
Giant Themeda 389333
Giant Thevetia 389983
Giant Thistle 73398,92127
Giant Thorny Bamboo 47190,47448
Giant Throatwort 70119
Giant Thuya 390645
Giant Tick Clover 126278
Giant Timber Bamboo 297215
Giant Tree 360575,360576
Giant Tree of California 360576
Giant Trillium 397549
Giant Typhonium 401161
Giant Valerian 404337
Giant Vallaris 404518
Giant Variegated Mediterranean 37469
Giant Vetch 408407
Giant Water Gum 382537
Giant Water Lily 408734
Giant Waterweed 141569

Giant White Bush 335873
Giant Wild Rye 144255
Giant Wild-pine 392053
Giant Wildrye 228358
Giant Woodbetony 287293
Giant Yellow Mulberry 261105
Giant Yellow-cress 336172
Giant Yucca 416581
Giant's Rattle 145899
Giant's Viper's-bugloss 141255
Giantarum 20037
Giantbract Monkshood 5409
Giantdishtree 166971,166973
Giantfennel 163574,163580
Giantflower Mischobulbum 255937
Giantflower Snowgall 191925
Giant-flowered Broom 77060
Giantgrass 175588
Giant-grass 175588
Gianthead Elatostema 142739
Gianthead Stairweed 142739
Gianthyssop 10402,236262
Giantic Epipactis 147142
Giantic Pine 300011
Giantleaf Ardisia 31455
Giant-leaf Holly 204161
Giantleaf Plantain 302068
Giant-leaved Ardisia 31455
Giant-leaved Marking-nut 357881
Giant-panda Fodder-bamboo 162717
Giantpetal Pocktorchid 282791
Giantreed 37450,37467
Giant-rhubarb 181958
Giant-rhubarb Family 181960
Giant-timbered Bamboo 297215
Gibbles 15289
Gibboseflower Rabdosia 209680
Gibbosity Bambusa 47283
Gibbosity-like Bambusa 47284
Gibbous Bamboo 47283
Gibbous Bredia 59846
Gibbous Chin Cactus 182444
Gibbous Duckweed 224366
Gibbous Euonymus 157539
Gibbous Fig 165762
Gibbous Sophora 369024
Gibbous Spindle-tree 157539
Gibbs Barberry 51663
Gibbs Firethorn 322451
Giblets 15289
Gibraltar Candytuft 203209
Gibraltar Mint 250432
Gibs 343151
Gibson Bushmint 203052
Gibson Dendrobium 125165
Gichtblume 315082
Gicks 192358
Giddy Gander 273531,273545
Giddy-gander 273531,273545
Gidgee 1292
Gidgee Myall 1292
Gigant Bittergrass 404555
Gigant Cactus 244083
Gigant Eucalyptus 155558
Gigant Goosefoot 87025
Gigant Rattan Palm 65698
Gigante Indocalamus 206880
Gigantic Bamboo 125477
Gigantic Bulbil Lily 229770
Gigantic Carrionflower 373821
Gigantic Dendrobenthamia 124914
Gigantic Eulophia 156723
Gigantic False Vanda 404747
Gigantic Leopardflower 373821
Gigantic Lycaste 238743
Gigantic Rhododendron 331558
Gigantic Typhonium 401161
Gigantic Wightia 414114
Giggary 262441
Gil Cup 325518
Gila Rock Daisy 291311
Gila Sophora 368972
Gilashan Willow 343413
Gilawfer 246478
Gilbralter Azalea 331010
Gilcup 68189,325518,325666,325820,326287
Gildcup 68189,325518,325666,326287
Giles Knotweed 162314
Gilg Barberry 51665
Gilg Buckthorn 328707
Gilgit Rhamnella 328552
Gilia 175668,175696
Gill 176824
Gill Gowan 89704
Gill Hen 176824
Gill Oak 323936
Gill Soroseris 369855
Gill-ale 176812,176824
Gill-creep-by-the-ground 176824
Gillgo-by-ground 176824
Gill-go-by-the-hedge 176824
Gill-go-on-the-ground 176824
Gillies Caesalpinia 65014
Gilliflower 34536,86427,127573,154363,246438,307197
Gilliver 86427
Gillman's Goldenrod 368402
Gilloflower 72934
Gill-over-the-ground 176824,176835,264946
Gill-over-the-hill 176824
Gill-run-along-the-ground 176824
Gill-run-bythe-ground 176824
Gill-run-by-the-street 346434
Gilly 86427
Gillyfer 86427
Gillyflower 127635,246478
Gilly-flower 86427
Gilman's Wild Buckwheat 152091
Gilrumbithground 176824
Gilt Cup 325820
Gilted Cup 325518,325666
Gilty Cup 68189,325518,325666,325820
Gilty-cup 68189,325518,325666,325820,326287
Gilver 86427
Gimlet 155734
Gimlet Gum 155734
Ginfoshan Mountain False Solomonseal 194046
Gingelly 361317
Ginger 356468,357114,383874,417964,417968,418010
Ginger Bread Palm 202321
Ginger Bread Plum 263715
Ginger Bush 387032
Ginger Family 418034
Ginger Ginsen 280821
Ginger Grass 117204
Ginger Lily 187417,187432
Ginger Mint 250362,250368
Gingerbread Palm 202256,202321
Gingerbread Plum 263715
Ginger-bread Plum 284244
Gingerbread Plum-tree 284244
Gingerbush 286703
Gingerleaf 238941
Ginger-leaf Morning-glory 207603
Gingerlike Lilyturf 272161
Gingerlike Sanqi 280821
Gingerlily 187417,187432
Ginger-lily 17639,187417
Ginger-pine 85271
Ginger-plant 383874
Gingersage 253631
Ginger-wort 187417
Gingham Golf Ball 159485
Gingham Golf-ball 159485
Gingham Golf-ball Euphorbia 159485
Gingseng Childvine 400913
Ginkgo 175812,175813
Ginkgo Family 175839
Ginny-hen Flower 168476
Ginsen 280712,280741
Ginseng 143677,280712,280799
Ginseng Family 30800
Gippsland Fountain Palm 234166
Gippsland Fountain-palm 234166
Gippsland Palm 234166
Gippsland Waratah 385737
Gipsoy Onion 15861

Gipsy Rose 336292
Gipsygirl Crocus 111521
Gipsywort 239202,407256
Giradol 89330
Giraffe Thorn 1206
Girald Acanthopanax 143596
Girald Actinidia 6519
Girald Ailanthus 12571
Girald Aster 40507
Girald Barberry 52128
Girald Beautyberry 66789
Girald Bluegray Maple 2827
Girald Bodinier Beautyberry 66789
Girald Chelonopsis 86816
Girald Corydalis 105934
Girald Crazyweed 278841
Girald Daphne 122444
Girald Edelweiss 224846
Girald Euonymus 157540
Girald Forsythia 167433
Girald Jasmine 211822
Girald Kiwifruit 6519
Girald Knotweed 320912
Girald Larkspur 124229
Girald Lilac 382164
Girald Mosquitotrap 117482
Girald Pleurospermum 304796
Girald Primrose 314434
Girald Pteroxygonum 320912
Girald Purplepearl 66789
Girald Ribseedcelery 304796
Girald Rose 336607
Girald Sagebrush 35528
Girald Sandwort 31923
Girald Sanicle 345969
Girald Spindle-tree 157540
Girald St. John's Wort 201996
Girald Swallowwort 117482
Girald Trigonotis 397415
Girald Woodbetony 287236
Girald's Currant 333986
Girald's Supplejack 52418
Girardinia 175865
Girasole 189073
Gireoud Spider Orchid 59270
Girgensohnia 175904
Girigiri 371540
Girl's Curly Love 35088
Girl's Delight 35088
Girl's Love 337180
Girl's Mercury 250600
Gironniera 175912
Girps Mercury 250600
Girt Ox-eye 227533
Gisekia 175930,175943
Gith 11947,235373,266225,266241
Githago Agrostemma 11947
Githcorn 122492
Gixy Gix 101852
Gjant Vriesea 412358
Glabose Date 418176
Glabous Alumroot 194422
Glabrate Bamboo 297284
Glabrate Chinese Linden 391687
Glabrate Henry Linden 391726
Glabrate Indosasa 206881
Glabrate Jadeleaf and Goldenflower 260420
Glabrate Mussaenda 260420
Glabrate Rockfoil 349686
Glabrate Skullcap 355455
Glabrate Wrinkledleaf Buckthorn 328845
Glabratefruit Anemone 23710
Glabrescent Aerva 9370
Glabrescent Alumroot 194421
Glabrescent Ampelopsis 20454
Glabrescent Apple 243725
Glabrescent Benchfruit 279747
Glabrescent Coronate Mountainash 369373
Glabrescent Ehretia 141667
Glabrescent Eriolaena 152687
Glabrescent Ernest Willow 343352
Glabrescent Firecracker-flower Azalea 331879
Glabrescent Firerope 152687
Glabrescent Hairypistil Photinia 295722
Glabrescent Hooktea 52431
Glabrescent Kydia 218447
Glabrescent Larkspur 124311
Glabrescent Leptodermis 226115
Glabrescent Oresitrophe 274157
Glabrescent Pachysandra 279747
Glabrescent Pittosporum 301417
Glabrescent Seatung 301417
Glabrescent Storax 379321
Glabrescent Supplejack 52431
Glabrescent Velvety Spiraea 372130
Glabrescent Waxpetal 106678
Glabrescent Western Arrowwood 407960
Glabrescent Wildclove 226115
Glabrescent Winterhazel 106678
Glabrescent Yam 131547
Glabrescent Yunnan Brachytome 59005
Glabriflorous Willowshoot Rhododendron 332073
Glabripetal Elaeocarpus 142320
Glabripetaled Elaeocarpus 142320
Glabros Betony 373440
Glabrous Acoop of Cock 182128
Glabrous Ainsliaea 12640
Glabrous Aletris 14494
Glabrous Amomum 19855
Glabrous Arabis 400644
Glabrous Aridicolous Hibiscus 194718
Glabrous Artocarpus 36937
Glabrous Ascitesgrass 407504
Glabrous Atriplex 44414
Glabrous Bamboo 297284
Glabrous Barbate Deadnettle 220359
Glabrous Barrenwort Epimedium 147050
Glabrous Beardtongue 289358
Glabrous Begonia 49874
Glabrous Bigfruit Elm 401557
Glabrous Bigleaf Cayratia 79880
Glabrous Birchleaf Spiraea 371830
Glabrous Black Holly 203582
Glabrous Bougainvillea 57857
Glabrous Box 64325
Glabrous Brandisia 59125
Glabrous Braya 59763
Glabrous Butterflygrass 392884
Glabrous Camellia 69097,69111
Glabrous Canon Grape 411557
Glabrous Chestnut Sage 344942
Glabrous Childvine 400891
Glabrous Chinese Madder 337921
Glabrous Chinese Privet 229627
Glabrous Chinese Sasa 347393
Glabrous Chinese Sumac 332512
Glabrous Chingiacanthus 209897
Glabrous Cinquefoil 312634
Glabrous Columnea 100114
Glabrous Common Trichilia 395481
Glabrous Coneshaped Galangal 17762
Glabrous Craib Glycosmis 177826
Glabrous Crapemyrtle 219928
Glabrous Crape-myrtle 219928
Glabrous Crazyweed 278842
Glabrous Creeping Tailgrass 402554
Glabrous Creeping Urochloa 402554
Glabrous Croton 112931
Glabrous Curculigo 114827
Glabrous Curvicalvarate Larkspur 124518
Glabrous Custardapple 25852
Glabrous Custard-apple 25852
Glabrous Dahurian Rose 336524
Glabrous Daisy 124784
Glabrous Date-plum 132267
Glabrous Decaspermum 123450
Glabrous Delavay Ampelopsis 20338
Glabrous Dendranthema 124784
Glabrous Denseciliate Buttercup 325769
Glabrous Deutzia 126933,127028
Glabrous Dichotomanthus 129027
Glabrous Discocleidion 134145
Glabrous Draba 136998,137232
Glabrous Dunn Raspberry 338341
Glabrous Elaeagnus 141999
Glabrous Elsholtzia 144034
Glabrous Eurya 160474
Glabrous Fake Maximowicz Poplar 311442
Glabrous Falsecapform Raspberry 339117
Glabrous Field Pepperweed 225316

Glabrous Fieldcitron 400531
Glabrous Fig 165053
Glabrous Fleabane 150750
Glabrous Fleshseed Tree 346984
Glabrous Fleshspike Tree 346984
Glabrous Ford Premna 313632
Glabrous Forrest Holly 203825
Glabrous Foxglove-like Sage 345003
Glabrous Franchet Terminalia 386538
Glabrous Garuga 171567
Glabrous Gentian 150750
Glabrous Goosecomb 144314
Glabrous Gouania 179941
Glabrous Greenbrier 366338
Glabrous Greyleaf Holly 204335
Glabrous Gueldenstaedtia 181670
Glabrous Hainan Pencilwood 139666
Glabrous Hairyleaf Spiraea 372017
Glabrous Hairystem Clearweed 299090
Glabrous Hastate Cacalia 283825
Glabrous Henry Honeysuckle 235634
Glabrous Henry Spiraea 371929
Glabrous Henry Wilsontree 365094
Glabrous Herbcoral 346527
Glabrous Homalium 197693
Glabrous Honghe Holly 204020
Glabrous Hooktea 52428
Glabrous Horsfieldia 198508
Glabrous Illigera 204623
Glabrous Indigo 206295,206612
Glabrous Indocalamus 206881
Glabrous Itoa 210443
Glabrous Japanese Spiraea 371972
Glabrous Keel Roegneria 335402
Glabrous Kurz Machilus 240608
Glabrous Ladymantle 14037
Glabrous Lagochilus 220064
Glabrous Lambert Raspberry 338703
Glabrous Lanceleaf Arthraxon 36707
Glabrous Largeflower Deutzia 126958
Glabrous Largeleaf Ehretia 141667
Glabrous Largeleaf Tickclover 126463
Glabrous Larkspur 124229
Glabrous Lasianthus 222146
Glabrous Leaf Actinidia 6614
Glabrous Leaf Pittosporum 301279
Glabrous Leaf Sophora 369135
Glabrous Licuala Palm 228739
Glabrous Licualapalm 228739
Glabrous Longstamen Willow 343632
Glabrous Longtail Holly 203989
Glabrous Low Torularia 264613
Glabrous Macaranga 240260
Glabrous Malpighia 243526
Glabrous Manchurian Apricot 34446
Glabrous Margin Rockfoil 349686
Glabrous Mattfeld Sagebrush 35893
Glabrous Mattfeld's Wormwood 35893
Glabrous Meconopsis 247176
Glabrous Meyer Clematis 95117
Glabrous Michelia 252886
Glabrous Mistletoe 411071
Glabrous Miyabe Spiraea 372013
Glabrous Monkshood 5233
Glabrous Mountainash 369397
Glabrous Multiradiate Fleabane 150802
Glabrous Net-veined Persimmon 132370
Glabrous Newlitse 264035
Glabrous Northeastern Corydalis 105608
Glabrous North-Xinjiang Fleabane 150898
Glabrous Orach 44414
Glabrous Ormosia 274393
Glabrous Oxmuscle 129027
Glabrous Palmate Meadowsweet 166114
Glabrous Paper-leaf Holly 203619
Glabrous Paulownia 285985
Glabrous Pendentfruit Rockcress 30389
Glabrous Pentapanax 289635
Glabrous Pertybush 292055
Glabrous Pittosporum 301366
Glabrous Pogostemon 306978
Glabrous Polemonium 307248
Glabrous Primrose 314435
Glabrous Protected Rockvine 387825
Glabrous Przewalsk Sage 345329
Glabrous Pubescent Box 64325
Glabrous Pubescent Holly 204185
Glabrous Purplebranch Willow 343479
Glabrous Qiujiang Mountainash 369442
Glabrous Rabbiten-wind 12640
Glabrous Rabdosia 209627
Glabrous Rattlebox 329527
Glabrous Red Raspberry 338593
Glabrous Redbud 83787
Glabrous Reticulate Leafflower 296735
Glabrous Reticulate Underleaf Pearl 296735
Glabrous Rice 275926
Glabrous Rockcress 400644
Glabrous Roegneria 144314
Glabrous Rough Pricklyash 417282
Glabrous Roughleaf 222146
Glabrous Rush Sandwort 31969
Glabrous Saffron-coloured Flower Raspberry 338922
Glabrous Sarcandra 346527
Glabrous Schneider Spiraea 372080
Glabrous Serpentroot 354913
Glabrous Seseli 361495
Glabrous Shadbush 19276
Glabrous Shady Raspberry 338812
Glabrous Sharpfruit Aphragmus 29171
Glabrous Shuteria 362299
Glabrous Sibbaldia 362346
Glabrous Sichuan Fritillary 168575
Glabrous Sichuan Willow 343513
Glabrous Silky Raspberry 338753
Glabrous Silky Rose 336933
Glabrous Silverleaf Azalea 330143
Glabrous Silverleaf Rhododendron 330143
Glabrous Singnalgrass 58204
Glabrous Slenderprickle Holly 203902
Glabrous Smilax 366338
Glabrous Sophora 368989
Glabrous South China Xylosma 415878
Glabrous Spicegrass 239627
Glabrous Spinestemon 306978
Glabrous Stenotaphrum 375767
Glabrous Stephania 375854
Glabrous Stewartia 376444
Glabrous Stringbush 414181
Glabrous Subshrub Nightshade 367382
Glabrous Sudden Corydalis 106015
Glabrous Supplejack 52428
Glabrous Sweetgrass 27942
Glabrous Taishan Willow 344171
Glabrous Tanoak 233228
Glabrous Tan-oak 233228
Glabrous Tephrosia 386264
Glabrous Ternate Waldsteinia 413035
Glabrous Ternstroemia 386696
Glabrous Thistle 91991
Glabrous Thomson Aralia 30780
Glabrous Tomentose Sophora 369135
Glabrous Tongol Milkvetch 43167
Glabrous Torenia 392884
Glabrous Towercress 400644
Glabrous Tree Falsespiraea 369267
Glabrous Tubergourd 390170
Glabrous Turpinia 400531
Glabrous Tylophora 400891
Glabrous Undulateleaf Oplismenus 272689
Glabrous Unexpected Corydalis 106015
Glabrous Unlovely Willow 343525
Glabrous Veronicastrum 407504
Glabrous Vineclethra 95489
Glabrous Wedge-leaf Eurya 160450
Glabrous Wedgeleaf Fig 165792
Glabrous Whitlowgrass 137232
Glabrous Whitlow-grass 153951
Glabrous Wild Oat 45454
Glabrous Wildberry 362346
Glabrous Willowleaf Agapetes 10323
Glabrous Willowleaf Honeysuckle 235912
Glabrous Wilsontree 365094
Glabrous Windhairdaisy 348333
Glabrous Wingseedtree 320834
Glabrous Woodbetony 287237
Glabrous Yalu Monkshood 5294
Glabrous Yam 131547,131601
Glabrous Yellow Bluebell 115352
Glabrous Yunnan Holly 204417

Glabrous Yunnan Rockvine 387873
Glabrous Yushania 416809
Glabrousbract Ajania 13015
Glabrousbract Notoseris 267177
Glabrousbract Willow 344191
Glabrous-bracted Willow 344191
Glabrousbranched Eurya 160443
Glabrous-branches Holly 204098
Glabrous-callyx Common Pricklythrift 2251
Glabrous-calyx 229563
Glabrous-calyx Chickweed 374933
Glabrous-calyx Currant 333989
Glabrouscalyx Littlerhomboidleaf Bluebellflower 115398
Glabrous-calyx Pavetta 286096
Glabrous-calyx Photinia 295811
Glabrous-calyx Rockfoil 349292
Glabrous-calyx Taizhong Arrowwood 407857
Glabrous-calyx Taizhong Viburnum 407857
Glabrouscalyx Viscid Germander 388304
Glabrous-calyx Wild Honeysuckle 235743
Glabrouscalyx Woodbetony 287540
Glabrousflower Helictotrichon 190166
Glabrousflower Roegneria 335399
Glabrousfruit Corchoropsis 104051
Glabrousfruit Greyleaf Willow 344143
Glabrous-fruit Milkvetch 42127
Glabrousfruit Peony 280325
Glabrous-fruit Pilose Poplar 311424
Glabrousfruit Raspberry 338467
Glabrous-fruit Shiquan Willow 344105
Glabrous-fruit Viburnum 407915
Glabrousfruit Willow 343562,344134
Glabrous-fruited Medick 247429
Glabrous-fruited Raspberry 338467
Glabrous-fruited Sedge 76561
Glabrousinflorescence Indian Quassiawood 298519
Glabrousinflorescence Larkspur 124309
Glabrousinflorescence Monkshood 5352
Glabrousleaf Asiabell 98286
Glabrousleaf Beautyberry 66837
Glabrousleaf Begonia 50202,50416
Glabrousleaf Bluebellflower 115352
Glabrousleaf Cherry 83204
Glabrousleaf China Corktree 294240
Glabrousleaf Chinese Corktree 294240
Glabrous-leaf Clematis 94945
Glabrous-leaf Currant 333990
Glabrousleaf Dinggongvine 154260
Glabrousleaf Ellipanthus 143868
Glabrousleaf Epaulettetree 320886
Glabrousleaf Erycibe 154260
Glabrousleaf Esquirol Buckthorn 328696
Glabrous-leaf Evodia 161335
Glabrousleaf Eyebright 160286
Glabrousleaf Flatawndaisy 14917
Glabrousleaf Greenbrier 366304
Glabrousleaf Hirsute Roegneria 335344
Glabrousleaf Leea 223924,223936
Glabrous-leaf Manchurian Apricot 34446
Glabrous-leaf Marsdenia 245807
Glabrous-leaf Meliosma 249374
Glabrousleaf Meranti 362209
Glabrous-leaf Milkgreens 245807
Glabrousleaf Moraea 258607
Glabrous-leaf Myrsine 261657
Glabrousleaf Persimmon 132134
Glabrousleaf Pittosporum 301279
Glabrousleaf Pondweed 312168
Glabrousleaf Primrose 314790
Glabrousleaf Purplepearl 66837
Glabrous-leaf Scorpiothyrsus 354745
Glabrousleaf Seatung 301279
Glabrousleaf Siris 13609
Glabrous-leaf Stem-flower Lepisanthes 225672
Glabrousleaf Sunkenvein Schefflera 350724
Glabrousleaf Trema 394635
Glabrous-leaf Tucher Azalea 332033
Glabrous-leaf Tucher Rhododendron 332033
Glabrousleaf Uvaria 403430
Glabrousleaf Violet 410615
Glabrousleaf Waldheimia 14917
Glabrousleaf Willow 343833
Glabrous-leaved Beautyberry 66837
Glabrous-leaved Cherry 83204
Glabrous-leaved Cylindrokelupha 116596
Glabrous-leaved Ellipanthus 143868
Glabrous-leaved Epaulette-tree 320886
Glabrous-leaved Erycibe 154260
Glabrous-leaved Fargesia 162694
Glabrous-leaved Mulberry 259156
Glabrous-leaved Myrsine 261657
Glabrous-leaved Pentapanax 289635
Glabrous-leaved Persimmon 132134
Glabrous-leaved Pittosporum 301279
Glabrous-leaved Trema 394635
Glabrous-leaved Uvaria 403430
Glabrous-leaved Willow 343833
Glabrouslocular Camellia 69111
Glabrous-locular Camellia 69111
Glabrous-midrib Camellia 69099
Glabrouspetal Paphiopedilum 282904
Glabrouspod Bashfulgrass 254997
Glabrousraceme Spinehairy Blueberry 404024
Glabrousseed Hibiscus 194969
Glabrous-seeded Hibiscus 194969
Glabroussepal Apple 243648
Glabroussepal Glabrate Rockfoil 349687
Glabroussepal Rattlebox 112757
Glabrous-sheath Bamboo 47285
Glabrous-sheath Bambusa 47285
Glabrous-spiked Lymegrass 144314
Glabrous-stalk Franchet Barberry 51634
Glabrousstalk Maple 3096
Glabrousstalk Trilobe Buttercup 325935
Glabrous-stem Actinidia 6629
Glabrousstem Bluebell 115363
Glabrousstem Bluebellflower 115363
Glabrousstem Larkspur 124221,124229
Glabrous-stem Pepper 300358
Glabrousstemmed Buttercup 325895
Glabroustwig Nanmu 295392
Glabroustwig Phoebe 295392
Glabrous-twigged Phoebe 295392
Glacial Crazyweed 278848
Glacial Currant 333991
Glacial Knotweed 309143
Glacial Larkspur 124230
Glacial Pastinacopsis 285749
Glacial Rockfoil 349386
Glacial Stonecrop 357154
Glacial Windhairdaisy 348334
Glacialform Buttercup 325897
Glacier Alpenclock 367763
Glacier Buttercup 325898
Glacier Crowfoot 325898
Glacier Draba 136999
Glacier Knotweed 309143
Glacier Lily 154915
Glacier Rhododendron 331346
Glacier Whitlowgrass 136999
Glacier-lily 154915
Gladden 5793
Gladdon 5793,208566
Glade Lily 229984
Glade Mallow 262286
Gladen 208771
Gladene 352960
Glading Root 208566
Gladiola 175990
Gladioli 175990
Gladiolus 175990,176260
Glads 175990
Gladulose Cinquefoil 312651
Gladwin 208566,208771
Gladwyn 208566
Gladwyn Gladwin 208566
Gladwyn Iris 208566
Gladyne 208566
Glamerated Synotis 381937
Glamerated Tailanther 381937
Glamerule Whitetree 379800
Gland 283682
Gland Bellflower 7596
Gland Bell-flower 7596
Gland Olive 270155
Gland Styles Yellow Rhododendron 330439
Gland Yellowhair Azalea 331708
Glandanther Azalea 330439

Glandbeak Habenaria 184038
Gland-bearing Azalea 330026
Glandbearing Cryptocarya 106640
Gland-bearing Rhododendron 330026
Glandbearing Waxpetal 106640
Glandbract Chinese Groundsel 365051
Glandbract Dogtongueweed 385887
Glandcalyx Astilbe 41847
Glandcalyx Azalea 330193
Glandcalyx Bladderwort 403972
Glandcalyx Catchfly 363151
Glandcalyx Didymocarpus 129871
Glandcalyx Hemiboea 191358
Glandcalyx Maximowicz Rose 336739
Glandcalyx Raspberry 338471
Glandfloss Azalea 331094
Glandflower Ajania 12990
Gland-flower Azalea 330024
Glandflower Butterflybush 61963
Gland-flower Rhododendron 330024
Glandfruit Jurinea 214030
Glandhai Meadowrue 355345
Glandhair Campion 364219
Glandhair Chickweed 375008
Glandhair Cremanthodium 110391
Glandhair Didymocarpus 129904
Glandhair Epimedium 146994
Glandhair Fixweed 126131
Glandhair Flemingia 166866
Glandhair Hairystem Larkspur 124291
Glandhair Hemiboea 191383
Glandhair Keiskea 215810
Glandhair Metalleaf 296422
Glandhair Murdannia 260115
Glandhair Nutantdaisy 110391
Glandhair Purpledaisy 267169
Glandhair Tailanther 381950
Glandhair Waterbamboo 260115
Glandhair Windhairdaisy 348335
Glandleaf Clethra 96502
Glandleaf Dolomiaea 135732
Glandleaf Lily 223117
Glandleaf Maesa 241804
Glandleaf Premna 313638
Glandleaf Rose 336659
Glandleaf Tea 69465
Glandless Denseleaf Larkspur 124321
Glandless Eucalyptus Raspberry 338368
Glandless Evodia 161333
Glandless Five-angle Raspberry 339030
Glandless Greywhitehair Raspberry 339361
Glandless Licorice 177886
Glandless Rockfoil 349285
Glandless St. Paulswort 363094
Glandless Star-jasmine 393642
Glandless Tsang Raspberry 339410
Glandpetal Rockfoil 350035
Glandpistil Azalea 330026
Gland-scaleleafruit Sophora 369034
Glandspot Sweetleaf 381422
Glandstalk Rose 336558
Glandstalk St. Paulswort 363093
Gland-stem Blackberry 338470
Glandstem Willowweed 146709
Glandstyle Azalea 330799
Gland-style Rhododendron 331407
Glanduiferous Purplebract Saussurea 348397
Glandular Adinandra 8260
Glandular African Myrsine 261606
Glandular Anodendron 25945
Glandular Atyle Azalea 330030
Glandular Atyle Rhododendron 330030
Glandular Bbaby's-breath 183243
Glandular Birch 53454
Glandular Blackberry Rose 336897
Glandular Blumea 55678
Glandular Branches Rhododendron 330044
Glandular Burnet 345890
Glandular Calyx Raspberry 338471
Glandular Calyx Rhododendron 330193
Glandular Cape Marigold 131191
Glandular Cinquefoil 312406
Glandular Clethra 96502
Glandular Clockvine 390696
Glandular Denselyspine Raspberry 339329
Glandular Dentate Dontostemon 136163
Glandular Dimorphostemon 131137
Glandular Dotted Pearweed 239656
Glandular Eelvine 25945
Glandular Emei Rose 336826
Glandular Fennelflower 266230
Glandular Flower Japanese Raspberry 338993
Glandular Flowers Phyllodoce 297026
Glandular Fourpetal Rose 336934
Glandular Fruit Piso Tree 300944
Glandular Globe-thistle 140789
Glandular Glorybower 96011
Glandular Grape 411526
Glandular Greenstar 307527
Glandular Hair Gaultheria 172052
Glandular Hair Monkshood 5120
Glandular Hair Whitepearl 172052
Glandular Isometrum 210181
Glandular Keiskea 215810
Glandular Korean Rose 336661
Glandular Lambert Raspberry 338704
Glandular Largeflower Larkspur 124245
Glandular Largeleaf Rose 336728
Glandular Leaf Cherry 223117
Glandular Machilus 240598
Glandular Meliosma 249393,249425
Glandular Notoseris 267169
Glandular Oak 323942
Glandular Olive 270155
Glandular Onosma 271728
Glandular Pedicel Rhododendron 330028
Glandular Phtheirospermum 317400
Glandular Phyllaries Tephroseris 385887
Glandular Pinnatesepal Raspberry 339057
Glandular Plantain 301834
Glandular Premna 313638
Glandular Rhododendron 330763
Glandular Rose 336659,336995
Glandular Sandwort 31925
Glandular Saturate Rose 336919
Glandular Senna 360455
Glandular Setose Dunn Blastus 55151
Glandular Shady Raspberry 338814
Glandular Skullcap 355456
Glandular Slimtop Meadowrue 388670
Glandular Spapweed 204982
Glandular Speedwell 407096
Glandular Spinycalyx Raspberry 338102
Glandular Strawberryleaf Raspberry 338430
Glandular Styles Rhododendron 330799
Glandular Sweetspire 210382
Glandular Sweginzow Rose 336982
Glandular Synotis 381950
Glandular Touch-me-not 204982
Glandular Twoflower Raspberry 338185
Glandular Wand Goldenrod 368479
Glandular Wild Buckwheat 152094
Glandular Wilford Cranesbill 175010
Glandular Willmott Rose 337023
Glandular Willowherb 146709
Glandular Willow-weed 146664
Glandular Windhairdaisy 348336
Glandular Winterhazel 106640
Glandular Yellownerve Raspberry 339487
Glandular-apiculate Spapweed 205100
Glandularcalyx Redfruit Raspberry 338362
Glandular-calyxed Blueberry 403972
Glandular-colored Butterfly-bush 61963
Glandular-dotted Sweetleaf 381422
Glandularflower Butterflybush 61963
Glandularflower Onosma 271728
Glandularflower Rabdosia 209617
Glandularflower Summerlilic 61963
Glandular-flowered Rhododendron 330024
Glandularfruit Aromcress 317565
Glandularfruit Clausia 317565
Glandularfruit Hodgsonia 197073
Glandularfruit Jurinea 214030
Glandularfruit Oilresiduefruit 197073
Glandularfruit Prickly Rose 336328
Glandularfruit Raspberry 338473
Glandular-fruited Pisotree 300944
Glandularhair Bluebeard 78035
Glandularhair Columbine 30059
Glandularhair Indigo 206510
Glandularhair Lungwort 321639

Glandularhair Porana 131247
Glandularhair Siphonostegia 365298
Glandular-haired Raspberry 338088
Glandular-hairy Heath 149493
Glandularhairy Meadowrue 388510
Glandularhairy Raspberry 338088
Glandularleaf Cherry 223117
Glandularleaf Prickly Rose 336326
Glandularleaf Rabdosia 303129
Glandularleaf Rose 336659
Glandularleaf Sweetleaf 381094
Glandular-leaved Adinandra 8260
Glandular-leaved Olive 270155
Glandular-leaved Rabdosia 303129
Glandular-leaved Sweetleaf 381094
Glandular-lepidote Ledum 223894
Glandularnerve Pepper 300349
Glandular-nerved Pepper 300349
Glandular-ovared Rhododendron 330026
Glandular-ovary Rhododendron 330026
Glandular-pediceled Rhododendron 330028
Glandular-punctated Sweetleaf 381422
Glandular-stalked Rose 336558
Glandularstipe Sweetleaf 381095
Glandular-stiped Sweetleaf 381095
Glandular-styled Rhododendron 330799
Glandular-styled Yellow Rhododendron 330439
Glandular-toothed Ardisia 31387
Glandulary-dotted Raspberry 339194
Glandulate Bigred Raspberry 338376
Glandulate Ovate-fruit Rose 336622
Glandule Chiritopsis 87998
Glandule Flowering-quince Willow 343188
Glandule Smallligulate Aster 40004
Glandule Willow 343417
Glandule-hair Blueberry 403839
Glanduleless Adinandra 8219
Glanduleless Dontostemon 136173
Glanduleless Glabrousfruit Raspberry 338468
Glanduleless Gongshan Raspberry 338478
Glanduleless Hedge Gooseberry 333902
Glanduleless Larkspur 124193
Glanduleless Lindley's Eupatorium 158202
Glanduleless Maire Lovegrass 147795
Glanduleless Sachalin Raspberry 339230
Glanduleless Smallfruit Spicegrass 239737
Glanduleless Starjasmine 393642
Glanduleless Sterigmostemum 273968
Glandulifer Nettle Spurge 212139
Glandulifering Winterhazel 106640
Glanduliferous Balsam 204982
Glanduliferous Fennelflower 266230
Glandulipilose Babylon Weeping Willow 343078
Glandulose Belltree 324997
Glandulose Cranesbill 174620
Glandulose Creeping Catchfly 363970
Glandulose Dimorphostemon 131137
Glandulose Filbert 106754
Glandulose Forrest Rose 336574
Glandulose Henry Rose 336626
Glandulose Licorice 177901
Glandulose Pinnatesepal Rose 336855
Glandulose Willowleaf Blueberry 403866
Glandulose-leaved Elaeocarpus 142397
Glandulous Cissampelopsis 92520
Glandwin 208566
Glareosous Buttercup 325900
Glass Mountain Coral-root 194523
Glass Mountain Rock Daisy 291339
Glass Saltwort 342859
Glasswort 342847,342859,344425,344585,346761
Glastonbury Thorn 109858,109863
Glatfelter Willow 343424
Glatfelter's Willow 343424
Glaubrous Pinnatifid Parrya 285007
Glaucescent Barberry 51669
Glaucescent Chinese Hydrangeavine 351806
Glaucescent Corydalis 105936
Glaucescent Diploclisia 132931
Glaucescent Elm 401519
Glaucescent Erycibe 154243
Glaucescent Fissistigma 166659
Glaucescent Ligusticum 229330
Glaucescent Mosquitotrap 117486
Glaucescent Motherwort 224981
Glaucescent Petrosimonia 292721
Glaucescent Swallowwort 117486
Glaucescent Thomson Yua 416531
Glaucescent Winterhazel 106643
Glauco-fruited Barberry 51670
Glauco-reded Winter-greenleaf Blueberry 403838
Glaucos Fruits Blueberry 403952
Glaucosleaf Cotoneaster 107462
Glaucous Acacia 1249
Glaucous Allspice 68321,231355
Glaucous Antheroporum 27465
Glaucous Bamboo 297285
Glaucous Barberry 51668
Glaucous Bauhinia 49098
Glaucous Bittersweet 80195
Glaucous Broadleaf Poplar 311430
Glaucous Bulrush 352278
Glaucous Campion 363385
Glaucous Canna 71179
Glaucous Cape-cowslip 218868
Glaucous Cassia 360490
Glaucous China-fir 114540
Glaucous Deutzia 126937
Glaucous Dichaea 128398
Glaucous Dinggongvine 154243
Glaucous Dog-rose 336411
Glaucous Echeveria 139990
Glaucous Encephalartos 145229
Glaucous Forkfruit 128935
Glaucous Gastrodia 171923
Glaucous Gendarussa 172817
Glaucous Glasswort 342879
Glaucous Goosefoot 87029
Glaucous Guangxi Falsechirita 317539
Glaucous Hair Grass 217416
Glaucous Honeysuckle 235760
Glaucous King-devil 195857
Glaucous Kobresia 217179
Glaucous Koeleria 217473
Glaucous Lagotis 220167
Glaucous Larkspur 124232
Glaucous Leaf Machilus 240591
Glaucous leaf Nanmu 295365
Glaucous Leaf Slugwood 50525
Glaucous Leaves Heath 149499
Glaucous Leaves Jewelvine 283284
Glaucous Leptodermis 226122
Glaucous Meadow-grass 305551
Glaucous Michaelmas-daisy 380907
Glaucous Moraea 258508
Glaucous Motherwort 224981
Glaucous Newlitse 264025
Glaucous Oil-wood 142449
Glaucous Pendentfruit Rockcress 30393
Glaucous Persimmon 132121
Glaucous Pleurospermum 304798
Glaucous Poplar 311333
Glaucous Rattlesnakeroot 313871
Glaucous Rockcress 30437
Glaucous Rush 213165
Glaucous Saxifrage 349144
Glaucous Sedge 74542,74547,74656,74664
Glaucous Senna 78308
Glaucous Separatestyle Acanthopanax 2383
Glaucous Sour Bamboo 4594
Glaucous Spapweed 204985
Glaucous Spiraea 371887
Glaucous Stauntonvine 374402
Glaucous Sweetleaf 381217
Glaucous Swellpod 27465
Glaucous Touch-me-not 204985
Glaucous Trifolialate Mahonia 242655
Glaucous Turkey Grape 411783
Glaucous Walking Iris 264160
Glaucous White Lettuce 313871
Glaucous White-lettuce 313871
Glaucous Wildclove 226122
Glaucous Willow 343425
Glaucousback Actinodaphne 6780
Glaucousback Ampelopsis 20298
Glaucousback Honeysuckle 235860
Glaucousback Slugwood 50525

Glaucousback Smoketree 107310
Glaucousback Snakegrape 20298
Glaucousback Tanoak 233254
Glaucousback Threewingnut 398317
Glaucousback Waxpetal 106641
Glaucousback Winterhazel 106641
Glaucous-backed Tanoak 233254
Glaucous-dorsal Honeysuckle 235860
Glaucousleaf Cyclea 116019
Glaucousleaf Dandelion 384563
Glaucous-leaf Dunbaria 138972
Glaucousleaf Habenaria 183662
Glaucousleaf Hoary Spiraea 371862
Glaucousleaf Hydrangea 199881
Glaucous-leaf Manglietia 244442
Glaucousleaf Paphiopedilum 282824
Glaucous-leaf Phoebe 295365
Glaucousleaf Roegneria 335326
Glaucousleaf Salacia 342646
Glaucousleaf Smoketree 107310
Glaucousleaf Wattle 1499
Glaucous-leaved Cyclea 116019
Glaucous-leaved Jewelvine 283284
Glaucous-leaved Machilus 240591
Glaucous-leaved Phoebe 295365
Glaucous-leaved Salacia 342646
Glaucous-leaved Senna 78308
Glaucous-leaved Staff-tree 80195
Glaucous-like Slugwood 50527
Glaudular Eurya 160475
Glaudular Mycetia 260623
Glaudulose Mycetia 260623
Glazegrass 367804
Glaziov Barberry 51671
Glaziov Cassava 244509
Glaziov Cedrela 80021
Glaziov Foreigntoona 80021
Glaziov Ocotea 268706
Gleadovia 176792
Gleaming Star 349044
Gledhill Pavonia 286585
Glehn's Spruce 298298
Glehnia 176921
Glennies 325518,325666,326287
Glens 262441
Glenwoodgrass 341960
Gleome 95606
Glinus 176941,176949
Glistening Leaf Aster 40488
Glittering Prickly-ash 417282
Glittering Scales Rhododendron 330841
Glitter-scaled Rhododendron 330841
Gllted Cup 325518,325666
Globate Rockjasmine 23179
Globate Rorippa 336200
Globate Yellowcress 336200
Globate-flower Deutzia 126938
Globba 176985
Globe Amaranth 18752,179221,179236,179240
Globe Artichoke 117787
Globe Butterfly-bush 62064
Globe Cactus 243994
Globe Candytuft 203249
Globe Cardoon 117787
Globe Centaurea 81183
Globe Chamomile 270993
Globe Crow's Foot 399504
Globe Crowfoot 399504
Globe Cucumber 114231
Globe Daisies Family 177054
Globe Daisy 177022,177023
Globe Eastern Arborvitae 390602
Globe Euphorbia 158958
Globe Everlasting 179236
Globe Flat Sedge 118784
Globe Flower 82098,177053,399484,399501,399504
Globe Flower Adina 8192
Globe Gilia 175677
Globe Lily 67555
Globe Mallow 370934,370935,370942,370947
Globe Orchid 394580
Globe Ranunculus 399504
Globe Rhododendron 331400
Globe Thistle 140652,140776
Globe Tulip 67554
Globe White Fir 325
Globe Willow 343676
Globeamaranth 179221,179236
Globe-amaranth 179221
Globedaisy 177022,177052
Globe-daisy 177022
Globe-daisy Family 177054
Globeflower 399484,399485,399504
Globe-flowered Magnolia 279248
Globe-flowers 82098
Globemallow 370934
Globe-mallow 370934
Globepea 371160
Globe-podded Hoary Cress 73098
Globe-podded Hoary-cress 73098
Globes 399504
Globe-thisde 140776
Globethistle 140652
Globe-thistle 140652,140704,140789
Globe-tulip 67554
Globigemmate Evergreenchinkapin 78944
Globosa Japanese Pine 299893
Globose Adjoin Wildbasil 96983
Globose Chinese Juniper 213642
Globose Euphorbia 158958
Globose Galangal 17681
Globose Glans Cyclobalanopsis 116110
Globose Glans Oak 116110
Globose Head Hat-sedge 118744
Globose Marsdenia 245810
Globose Milkgreens 245810
Globose Norway Maple 3449
Globose Phaenosperma 293296
Globose Raspberry 339290
Globose Red Maple 3523
Globose Sinosenecio 365050
Globose Spurge 158958
Globose Zo-sasa 209059
Globose-flower Michelia 252959
Globose-flower Syzygium 382516
Globose-fruit Manglietia 244443
Globose-fruit Microtropis 254327
Globose-fruit Nitraria 266380
Globosefruit Shiquan Willow 344106
Globose-fruit Woodlotus 244443
Globose-fruited Barberry 51672
Globose-node Bitter Bamboo 304029
Globose-noded Bitter-bamboo 304029
Globose-stamen Magnoliavine 351098
Globose-urceolate Heath 149503
Globular Coneflower 326964
Globular Firmiana 166660
Globular Head Largeflower Onion 15752
Globular Pepper 300460
Globular Spike Pycreus 322233
Globular Woodbetony 287712
Globularflower Catasetum 79340
Globularflower Rockfoil 349390
Globularflower Saussurea 348337
Globular-flowered Cotoneaster 107467
Globular-fruit Banksia 47670
Globularfruit Tubergourd 390132
Globularspike Fimbristylis 166550
Globularspike Fluttergrass 166550
Globularspike Sedge 74672
Globularstyle Sedge 74669
Globule-bearing Naravelia 262335
Globulebearing Pothos 313204
Glochid Craneknee 221682
Glochidiate Houndstongue 117970
Glochidion 177096
Gloden-eyed Grass 365714
Glomerata Juniper 213794
Glomerate Arrowwood 407869
Glomerate Barberry 51673
Glomerate Burreed 370066
Glomerate Condorvine 245812
Glomerate Cypressgrass 118957
Glomerate Deutzia 126940
Glomerate Falsenettle 56155,56214
Glomerate Galingale 118957
Glomerate Hechtia 187132
Glomerate Holly 203851

Glomerate Leptodermis 226087
Glomerate Milkgreens 245812
Glomerate Mouseear 82849
Glomerate Onosma 271771
Glomerate Photinia 295697
Glomerate Sedge 73617
Glomerate Soroseris 369860
Glomerate Viburnum 407869
Glomerate Wildclove 226087
Glomerate-flower Onosma 271790
Glomerate-flower Sweetleaf 381221
Glomeratelike Bellflower 70065
Glomerulate Bigumbrella 59207
Glomerulate Brassaiopsis 59207
Glomerulate Mountainash 369400
Glomerulate Mountain-ash 369400
Glomerulate Suregada 379800
Glomerule Cotoneaster 107467
Glomerule Schefflera 350699
Glomerule Sweetleaf 381221
Glomerulose Schefflera 350699
Glonerate Leek 15324
Glonerate Onion 15324
Gloomy Gahnia 169561
Gloomy Laelia 219695
Gloria De La Manana 207792
Glorida Chinkapin 78807
Gloriless 8399
Gloriosa 177230
Gloriosa Daisy 339512,339556
Gloriosa Lily 177230
Gloriosa Water Lily 267628
Glorious Bollea 56689
Glorious Clethra 96520
Glorous Laelia 219679
Glory Acacia 1142
Glory Bower 95934,96009,96398
Glory Bush 391566,391569,391573,391571,391577
Glory Flower 95978,139935,139936,391577
Glory Jadeleaf and Goldenflower 260383
Glory Lily 177230,177244,177251
Glory Michelia 252900
Glory of the Mountains Rhododendron 331406
Glory of the Snow 87759
Glory of the Sun 227813,227814
Glory Pea 96633,96637,96639,96890,380065,380066
Glory Tree 95934,96009
Glory Vine 96633
Glory Vive 139936
Glory Wattle 1612
Glorybind 68671,102903
Glorybind Begonia 49735,50278
Glorybind Family 102892
Glory-bind-like Blinkworthia 55605
Glorybower 95934,96362,96387
Glory-bower 95934
Glorybush 391577
Glorylily 177230
Glory-lily 177230,177244
Glory-of-texas 389205
Glory-of-the-blue 87762
Glory-of-the-pink 87764
Glory-of-the-snow 87757,87759
Glory-of-the-sun 227814
Glory-of-the-white-snow 87763
Glorypea 96633,96637
Glory-tree 95934
Gloryvine 411623
Glossodia 177313
Glossostigma 177407
Glossy Abelia 135
Glossy Adina 8199
Glossy Begonia 50111
Glossy Buckthorn 328603,328701
Glossy Cotoneaster 107448
Glossy Encephalartos 145244
Glossy Laurel 113458
Glossy Magnolia 242220,283421
Glossy Privet 229529
Glossyleaf China Cane 162718
Glossyleaf Chinacane 162718
Glossy-leaved Aster 380890,380970
Glossyshower Senna 360490
Glossy-shower Senna 360490
Glove-bark 129718
Glovewort 102867
Glow Vine 347092
Glowing Bambusa 47433
Glowvine 347092
Glowwort 347183
Gloxinia 177515,364737,364754
Gluey Bark Litse 233928
Gluey Litse 233928
Glumeless Wild-rye 144340
Glumelike Dendrochilum 125531
Glume-like Pricklythrift 2237
Glustered Rose 336612
Glusterflower Deutzia 126938,126940
Gluster-flowered Deutzia 126940
Gluten Tofieldia 392608
Glutinous Bartsia 268966
Glutinous Bluebeard 78010
Glutinous Blue-beard 78010
Glutinous Coral-blow 341021
Glutinous False Asphodel 392608
Glutinous Flemingia 166866
Glutinous Nepeta 264935
Glutinous Sage 345057
Glutinous Spapweed 205437,205438
Glutinous Touch-me-not 205437,205438
Glutinousmass 179485,179488
Glyceria 177556
Glycosmis 177817
Glycyphyllous Milkvetch 42417
Glycyphyllous-like Milkvetch 42416
Glyptic-petal Bush 177982
Glyptopetalum 177982
Gmelin Alpparsley 98777
Gmelin Birch 53461
Gmelin Buttercup 325901
Gmelin Euphorbia 158965
Gmelin Globe Thistle 140712
Gmelin Globethistle 140712
Gmelin Ladybell 7649
Gmelin Lymegrass 335547
Gmelin Onosma 271772
Gmelin Rockjasmine 23182
Gmelin Sagebrush 35560
Gmelin Sealavender 230631
Gmelin Sedge 74676
Gmelin Sweetvetch 187912
Gmelin Teasel 133486
Gmelin Vetchling 222720
Gmelin's Buttercup 325901
Gmelin's Orach 44430
Gmelin's Wormwood 35560
Gmelina 178024
Gnapperts 222758
Gnat Flower 81237
Gnat Orchid 4511,4513,182230
Gnawedpetal Corydalis 106190
Goa Bean 319114
Goa Butter 171119
Goa Ipecacuanha 262543
Goabean 319101,319114
Goasey-goosey-gander 273531
Goat Apple 123077
Goat Grass 8660
Goat Leaf Honeysuckle 235713
Goat Nut 364450
Goat Pea 356387
Goat Pepper 72070
Goat Stones 273531
Goat Tree 236022
Goat Willow 343151
Goat's Airach 87200
Goat's Arrach 87200
Goat's Bean 250502
Goat's Beard 37053,37060,37085,90894,166125,202086,394281,394327,394333
Goat's Cullions 196376
Goat's Foot 8826,188757,394333
Goat's Foot Ipomaea 208067
Goat's Herb 8826
Goat's Horns Cactus 43495
Goat's Leaf 236022
Goat's Marjoram 274237
Goat's Nut 364450

Goat's Rue 169919,169935,386365
Goat's Weed 11199
Goat's Willow 343151
Goat's-beard 37053,37060,37062,37085, 41786,166125,394257,394333
Goat's-foot Ipomaea 208067
Goat's-horn cactus 43495
Goat's-rue 169919,169935,386365
Goat's-wheat 44247
Goatbean 169919,169935
Goatgrass 8660,8727
Goat-grass 8660
Goathorn Disa 133728
Goathorntree 77643
Goat-leaf Honeysuckle 235713
Goatnut 364450
Goatroot 271464
Goatroot Ononis 271333,271464
Goatsbeard 37053,37060,37085,394257
Goat-scented Tutsan 201924
Goatsmell Rockfoil 349452
Goatsrue 169919
Goatweed 8810,8826,11199,162519,202086
Goatwheat 44247
Gobblety-guts 339887
Gobbo 219
Gobernadora 221983
Gobi Beancaper 418667
Gobi Brachanthemum 58042
Gobi Calligonum 67028
Gobi Kengyilia 215909
Gobi Kneejujube 67028
Gobi NW. China Spunsilksage 360858
Gobi Peach 316695
Gobi Tianshan Feathergrass 376937
Goblet Aster 40704,40706,40708,40709
Goblin Orchid 258987
Goblin's Gloves 130383
Goblin's Thimbles 130383
Gobo 219,31051
Gochnatia 178748
Go-cup 68189,325518,325820
God Aimighty's Fingers-and-thumbs 28200
God Almighty's Bread-and-cheese 277648
God Almighty's Flower 237539
God Almighty's Thumb-and-finger 237539
God Almighty's Thumbs-and-fingers 237539
God Senna 78338
God's Eye 345379,407072
God's Eyes 345379,407072
God's Fingers-and-thumbs 169126,237539
God's Flower 189814
God's Meat 109857
God's Stinking Tree 345631
God's Tree 35088,203545
Godalming Cotoneaster 107462
Goddard 349044
Godetia 94132,94133,94139
Godfathers And Godmothers 410677
Godfrey's Gayfeather 228505
Godfrey's Goldenaster 90205
Godfrey's Spider-lily 200931
Godfrey's Stitchwort 255483
Godfrey's Thoroughwort 158138
Godon Hoodia 198053
Goering Cymbidium 116880
Goering Lemongrass 117181
Goggles 334250
Goiners 189073
Goktschaic Hoarhound 245740
Gold 66445,89704,261162
Gold And Silver Chrysanthemum 89659
Gold Apple 239157
Gold Arborvitae 390618
Gold Basket 45192
Gold Bell 262441
Gold Bells 262441
Gold Birch 53411
Gold Bloom 66445
Gold Buckwheat 162309
Gold Cephalanthera 82051
Gold Chain 356468
Gold Ciliate Poplar 311277
Gold Coast Jasmine 211795
Gold Coin 269133,280584
Gold Cone Juniper 213708
Gold Cup 325518,325666,326287,366871
Gold Drop Shrubby Cinquefoil 312580
Gold Dust 45192,356468
Gold Dust Plant 44915
Gold Dust Shrub 44887,44919
Gold Dust Tree 44915
Gold Flame Honeysuckle 235828
Gold Flower 178330,189814,201709
Gold Fruit 93314
Gold Hedgehog Holly 203552
Gold Knob 325518,326287
Gold Knop 81020,81237,81356,325518, 325666,326287
Gold Knots 325518
Gold Lace Cactus 244062
Gold Lily 229730
Gold Locket-and-chain 218761
Gold Lockets-and-chains 218761
Gold Marble bidens 53905
Gold Medal Rose 336288
Gold Medallion Tree 78368
Gold Nugget Ice Plant 123853
Gold of Pleasure 68839,68860
Gold Poppy 155170
Gold Ring Barberry 52235
Gold Rose 336814
Gold Shower 390542
Gold Silver Nightshade 367735
Gold Star 175454
Gold Star Juniper 213644
Gold Stargrass 202812
Gold Tooth Aloe 16541,17038,17089,17289
Gold Tree 79260,115784
Gold Watch 201787
Gold Watch-and-chain 218761
Gold Watchchain 218761
Gold Watches 176738,201787
Gold Winterberry 204379
Gold Wire 201819
Gold Withy 261162
Goldaster 90151
Goldbach Cinquefoil 312645
Goldbachia 178814
Goldball Corydalis 105658
Goldband Lily 229730
Gold-banded Lily 229730
Goldbeard Iris 208562
Goldblotch Gladiolus 176431
Gold-crap 325518,325666
Goldcrop 325518
Goldcup 325666
Goldcup Oak 323762
Gold-dust 1029,45192
Gold-dust Dracaena 137409
Gold-dust Dragonbood 137409
Gold-dust Wattle 1029
Goldear Dendrobium 125191
Goldedge Kerria 216091
Goldedge Osmanther 276314
Goldedge Osmanthus 276314
Golden Adonis 8343
Golden Alexander 418113
Golden Alexanders 418107,418109
Golden Alison 45192
Golden Alpine-lily 234237
Golden Alyssum 18465
Golden American Elder 345633
Golden Angel's Trumpet 61233
Golden Apple 8787,88691,372467
Golden Apple Mint 250362
Golden Arborvitae 302721,302723
Golden Arum 417104
Golden Arum Lily 417104
Golden Ash 167958,167963
Golden Aspen 311547
Golden Aster 59056,90151,90220,90233, 90294,194195,194197,194254,194256, 194261,194264
Golden Awngrass 255886
Golden Azalea 331711
Golden Baby Goldenrod 367941
Golden Baker Lily 229744
Golden Ball Butterfly Bush 62064
Golden Ball Cactus 140127,267041,284673
Golden Ball-cactus 267041

Golden Bamboo 47516,47518,297203,297215,297464
Golden Banner 389547
Golden Barberry 52169
Golden Barrel 140127
Golden Barrel Cactus 140127
Golden Barrel-cactus 140127
Golden Bead Tree 139062
Golden Bean Kumquat 167516
Golden Bearclaw-poppy 31082
Golden Begonia 50413
Golden Bell 167427
Golden Bellapple 285660
Golden Bells 167471,315082
Golden Bristlegrass 361877
Golden Brodiaea 398796
Golden Bunch 111484
Golden Butter 325864
Golden Buttercup 68189
Golden Butterfly Palm 89360
Golden Buttons 383874
Golden Cabomba 64492
Golden Calla 417093,417104
Golden Callalily 417104
Golden Camellia 69480
Golden Camomile 26900
Golden Candles 279769
Golden Cane Palm 89345,89360
Golden Cane-palm 89360
Golden Canna 71171
Golden Cap-and-frills 325820
Golden Carpet 152246,175786,356468
Golden Cassia 78284,85977
Golden Cassia Bark 78284
Golden Cassidony 189814
Golden Cereus 52572
Golden Chain 218754,218761
Golden Chain Tree 218755,218761
Golden Chamomile 26900
Golden Champaca 252841
Golden Chervil 84728
Golden Chested Beehive Cactus 107059
Golden Chestnut 89969,90180
Golden Chinese Arborvitae 302733
Golden Chinese Juniper 213635
Golden Chinkapin 89969
Golden Cholla 116654
Golden Chysis 90712
Golden Cinquefoil 312406
Golden Clematis 95345
Golden Clover 397096
Golden Club 275295,275298
Golden Clubcactus 52572
Golden Columbine 30017
Golden Column 140888,413475
Golden Common Bamboo 47518
Golden Common Yew 385304
Golden Coneflower 339539
Golden Coppertip 111458
Golden Coreopsis 104598
Golden Cornflower 89704
Golden Corydalis 105631
Golden Cotton Wood 78610
Golden Cottonwood 78610
Golden Creeping Juniper 213797
Golden Crest 236322,236324
Golden Crocus 111515,111535
Golden Crosswort 170335
Golden Crownbeard 405936
Golden Crown-beard 405936
Golden Cup 68189,199407,199408,325518,325820
Golden Currant 333919,334127
Golden Cuvely Dock 340116
Golden Daffodil 262356
Golden Daimyo Oak 323815
Golden Daisy 89704,325820
Golden Dawyck 162407
Golden Dead Nettle 170025
Golden Deodar 80088
Golden Desert Snapdragon 256525
Golden Dew Drop Skyflower 139062
Golden Dewdrop 139062,139065
Golden Dock 340060,340116
Golden Dog's-tail 220295,220298
Golden Dollar 280584
Golden Drinking Cup 325820
Golden Drop 271769
Golden Drops 218761,315082
Golden Dryandra 138435
Golden Duckbeakgrass 209276
Golden Dust 45192,356468,368480
Golden Dwarf Hinoki Cypress 85322
Golden Dyssodia 391042
Golden Eardrops 128292,141590
Golden Eargrass 187501
Golden Easter Lily Cactus 140847
Golden Eastern Arborvitae 390589
Golden Egerton Swan Orchid 116519
Golden Elaeagnus 142156
Golden Elder 345631,345633
Golden Elderberry 345633
Golden Eulalia 156442
Golden European Ash 167958
Golden Evening Primrose 269416
Golden Evergreen Raspberry 338347
Golden Evergreenchinkapin 89969
Golden Everlasting 415386
Golden Eye 409147,409153
Golden Fairy Lantern 67560
Golden Fairy Lanterns 67560
Golden False Beardgrass 90108
Golden Fawn-lily 154938
Golden Feather 383822
Golden Fenestraria 163329
Golden Fernleaf 85329
Golden Fig 164659
Golden Fimbristylis 166359
Golden Fir 409
Golden Fishpole Bamboo 297203
Golden Flat Juniper 213704
Golden Fleabane 150485
Golden Fleece 391042,391050
Golden Fleece Daisy 391050
Golden Flower Velure 18775
Golden Flowering Currant 333919
Golden Fluttergrass 166359
Golden Foxtail-grass 17554
Golden Fringe Myrtle 68757
Golden Fullmoon Maple 3036,3600
Golden Garlic 15480
Golden Gilia 175674
Golden Globe Tulip 67560
Golden Globe-amaranth 179239
Golden Globes 239597,239794
Golden Globe-tulip 67560
Golden Glory Pea 179163,179164
Golden Glow 339574,368480
Golden Gorse 401388
Golden Grain 405788
Golden Gram 409025
Golden Grass 184677
Golden Grevillea 180580
Golden Groundsel 279888,358336
Golden Guinea 325820
Golden Guinea Flower 194661
Golden Gymnopetalum 182575
Golden Hairy Tan Oak 233143
Golden Hakonechloa 184658
Golden Hardhack 289713,312542
Golden Hawksbeard 110741
Golden Heather 198953
Golden Hedgehog 140275
Golden Hedgehog Cactus 140275
Golden Hedge-hyssop 180297
Golden Henbane 201401
Golden Hinoke Cypress 85314
Golden Hurricane Lily 239254,239257
Golden Ice Plant 220485
Golden Indian Bean-tree 79244
Golden Inside-out Flower 404612
Golden Iris 208719
Golden Irish Yew 385312,385322
Golden Ironlock 317035
Golden Japanese Red Pine 299892
Golden Kangaroo Paw 25221
Golden Kangaroo-paw 25221,25223
Golden Kingcup 68189
Golden Knee 89959
Golden Knob 68189,325689
Golden Korean Fir 399

Golden Lampranthus 220485
Golden Larch 317821,317822
Golden Lawson Cypress 85289
Golden Leaf 90007
Golden Lily of Japan 229730
Golden Linden 391708
Golden Lobster Claw 190024
Golden Locks 218761
Golden Loosestrife 239898
Golden Lungwort 195803
Golden Lycoris 239257
Golden Margin Agave 10788
Golden Margin Boxelder Maple 3226
Golden Margin Sansevieria 346165
Golden Marguerite 26900,107295
Golden Marjoram 274203
Golden Meadow Foxtail 17497,17554,17556
Golden Meadow-parsnip 418109
Golden Medal Rose 336288
Golden Melilot 249205
Golden Metalleaf 296369
Golden Midnight 237539
Golden Millet Grass 254516
Golden Mimosa 1069
Golden Monkey Flower 255218
Golden Monkeyflower 255218
Golden Monkey-flower 255218
Golden Mos Cypress 85352
Golden Mothwort 178330
Golden Mouse Ear 195478
Golden Mouse-ear 195437,195478
Golden Mugget 170335
Golden Muguet 170335
Golden Narcissus 262356
Golden Nigger 188908
Golden Nordmann Fir 431
Golden Oak 324335,324337
Golden Oak of Cyprus 323649
Golden Oak-of-cyprus 323649
Golden Oat Grass 398518
Golden Oat-gras 398518
Golden Oatgrass 398445
Golden Oats 376779
Golden Osier 261162,342982
Golden Paper Daisy 415386
Golden Parrot Tree 180636
Golden Pea 389547
Golden Penda 415206
Golden Pfeiffer Osmanther 276292
Golden Pfeiffer Osmanthus 276292
Golden Pfitzer 213611
Golden Phoenix Tree 123811
Golden Pine 180642,317822
Golden Ploughpoint 401172
Golden Plumose Sawara False Cypress 85349
Golden Pogonatherum 306834
Golden Poplar 311140
Golden Pothos 147338,147349
Golden Powder Puff 284663
Golden Prince's Plume 373709
Golden Privet 229569,229571,229663
Golden Pulsatilla 321662
Golden Purslane 311899
Golden Pussies 343151
Golden Ragwort 279888,279939,358280,
358336,358750
Golden Rain 217626,218754,218761
Golden Rain Tree 217610,217626
Golden Rain Wattle 1496
Golden Rain-tree 217626
Golden Rambutan 265126
Golden Rat Tail 94569
Golden Ray 228969
Golden Rhododendron 330182,331161,
331711
Golden Rod 11549,121001,202006,202179,
367939,368480,405788
Golden Rose 315101,336636
Golden Rose of China 336636
Golden Samphire 230109
Golden Sam-phire 230109
Golden Sawara False Cypress 85346
Golden Saxifrage 90317
Golden Scales Rhododendron 330397
Golden Sea Lavender 230524
Golden Seal 200173,200175
Golden Sealavender 230524
Golden Sedge 73829
Golden Sedum 356475
Golden Senecio 358336
Golden Shower 78300,218761,322950
Golden Shower Tree 78300
Golden Shrimp Plant 279769
Golden Slipper 120310,237539
Golden Snakecactus 52572
Golden Sovereigns 312360
Golden Spaniard 4743
Golden Spapweed 204885
Golden Sphaerical Corydalis 105658
Golden Spider Flower 95757
Golden Spider Lily 239254,239257
Golden Spleenwort 90423
Golden Spoon 64430
Golden Spreading English Yew 385318
Golden Star 55644,89952,89959,315101,
325820
Golden Star Boxelder Maple 3223
Golden Stars 55644,244062
Golden Stonecrop 356468
Golden Stonegarlic 239257
Golden Sulphur Flower 152569
Golden Summer Daylily 191263
Golden Sun 384714
Golden Sunrose 188760
Golden Swan 111468
Golden Sword 240481
Golden Swordflag 208562
Golden Tassel 156000
Golden Tauhium 78610
Golden Thistle 354648,354651,354655
Golden Thread 103850
Golden Threadleaf Sawara Cypress 85355
Golden Thryallis 390542
Golden Thyme 391372
Golden Tickseed 104598
Golden Timothy Grass 361902
Golden Tip 179567
Golden Top 220298
Golden Torch Cereus 140888
Golden Touch-me-not 204885
Golden Tree 180636
Golden Tree Mallow 223379
Golden Trefoil 192130
Golden Trumpet 14873,262441
Golden Trumpet Bush 14880
Golden Trumpet Flower 14873
Golden Trumpet Tree 382711,382712
Golden Tuft 45192
Golden Turk's Cup Lily 229857
Golden Turk's Lily 229857
Golden Variegated Boxelder Maple 3219
Golden Vine 376557
Golden Wattle 1159,1516
Golden Wave 104448,104482
Golden Wavy Hair-grass 126075
Golden Weeping Willow 342926,342978,
342982
Golden Western Arborvitae 390647
Golden White Fir 322
Golden Wild Buckwheat 151950
Golden Willow 1159,342975,342982
Golden Wings 368480
Golden Wire 356468
Golden Withy 261162
Golden Wonder 360431
Golden Wonder Senna 360486
Golden Woodbetony 287024
Golden Yarrow 152802,152803,152809
Golden Yew 385304,385321
Golden Zephyrlily 417610
Golden Zizia 418109
Golden-apple 8787
Golden-apple Tree 372467
Goldenaster 59054,194189,207393
Golden-back Argyreia 32630
Golden-back Bauhinia 49135
Golden-back Przewalsk Azalea 331562
Golden-back Przewalsk
Rhododendron 331562
Golden-ball 140127
Golden-ball Lead Tree 227434

Goldenball Leadtree 227434
Golden-banded Lily 229730
Golden-bark Clementine's Rhododendron 330428
Golden-bark Dotty Azalea 330428
Golden-bean 389547
Goldenbean Kumquat 167516
Golden-bean Kumquat 167499,167516
Golden-bell 167456
Goldenbell Forsythia 167471
Goldenbells 167456
Golden-bells 167427
Golden-blossom Tree 48076,48078
Goldenbowl Mariposa-lily 67578
Golden-bowl-crocus Crocus 111520
Goldenbush 150306
Goldenbutton 4870,371627
Golden-buttons 383874
Goldencalyx Azalea 330394
Goldencalyx Rhododendron 330394
Golden-calyxed Rhododendron 330394
Golden-carpet 356468
Goldencattail Narenga 262548
Golden-chain 218754,218761
Goldenchain Laburnum 218761
Golden-chain Laburnum 218761
Goldenchain Tree 218754,218761
Golden-chalice Vine 366871
Golden-club 275295,275298
Golden-coloured Newlitse 264021
Golden-crest 236322,236324
Goldencup Metalleaf 296369
Goldencup Oak 323762
Golden-cup St. John's-wort 202070
Golden-dewdrop 139062
Goldendotted Eurya 160536
Golden-drop 271727,271835
Goldenearring 37653
Goldeneye 409147
Golden-eye 190248
Golden-feather Palm 89360
Goldenfleece 150308
Goldenfleece Epimedium 147035
Goldenflower Actinidia 6568
Goldenflower Baker Lily 229744
Goldenflower Dendrobium 125048
Goldenflower Kiwifruit 6568
Goldenflower Michelia 252892
Goldenflower Milkwort 308164
Golden-flower Rhododendron 330182
Golden-flowered Actinidia 6568
Golden-flowered Agave 10824
Golden-flowered Camellia 69480
Golden-flowered Daphne 122378
Golden-flowered Norway Willow 343217
Golden-flowered-like Camellia 69704
Golden-fruit Fissistigma 166657
Golden-fruit Hawthorn 109598
Golden-fruited Fissistigma 166657
Golden-fruited Sedge 73829
Goldenglandular Arrowwood 407752
Golden-glandular Bluebeard 77999
Golden-glandular Blue-beard 77999
Golden-glow 339574
Golden-groove Bamboo 297204
Golden-guinea Everlasting 189158
Golden-hair Anzacwood 310772
Goldenhair Bamboo 297373
Goldenhair Cephalocereus 82208
Goldenhair Elaeocarpus 142281
Golden-hair Fig 165023
Golden-hair Grass 306832
Goldenhair Hollowbamboo 82457
Golden-hair Newlitse 264038
Goldenhair Tanoak 233143
Golden-haired Bamboo 82457
Golden-haired Crowfoot 325628
Golden-haired Elaeocarpus 142281
Golden-haired Newlitse 264038
Goldenhairgrass 306828,306834
Goldenhairy Tanoak 233143
Golden-hairy Tanoak 233143
Goldenhead 2049
Golden-head Common Juniper 213724
Goldenhook 318661
Golden-larch 317821,317822
Goldenleaf 90007
Goldenleaf Chinkapin 89969
Goldenleaf Eurya 160565
Goldenleaf Hedge Maple 2839
Goldenleaf Indianmulberry 258883
Goldenleaf Michelia 252874
Goldenleaf Osmanther 276315
Goldenleaf Osmanthus 276315
Goldenleaf Philodendron 294792
Golden-leaf Sweet Mockorange 294427
Golden-leafed European Cranberry Bush 407990
Golden-leaved Catalpa 79244
Golden-leaved Elderberry 345582
Golden-leaved European Viburnum 407990
Golden-leaved Eurya 160565
Golden-leaved Japanese Barberry 52228
Golden-leaved Laburnum 218762
Golden-leaved Mexican Orange 88701
Golden-leaved Mockorange 294427
Golden-leaved Queen-of-the-meadow 166126
Golden-leaved Weeping Forsythia 167463
Golden-light Addermonth Orchid 243077
Golden-lily 239257
Goldenlip Thrixsperm 390502
Goldenlip Thrixspermum 390502
Golden-looking Azalea 331865
Golden-marginata Japanese Spindle-tree 157615
Golden-marginated Common Barberry 52332
Golden-marked Japanese Spindle-tree 157616
Goldenmelon 182574,182575
Golden-nerve Yerbadetajo 141375
Goldenpearl Maesa 241808
Golden-pert 180297
Goldenpoppy 379219,379228
Golden-rain 218761
Goldenrain Tree 217613,217626
Golden-rain Tree 217610,217626
Golden-rain Wattle 1496
Goldenraintree 217610
Goldenray 228969
Golden-rayed Lily 229730
Golden-rayed Lily-of-Japan 229730
Golden-rayed Pentachaeta 289421
Goldenraylike Sinosenecio 365064
Goldenrod 367939,367965,368002,368013,368080,368104,368106,368162,368179,368240,368312,368353,368432,368480
Goldenrod Goldenray 229246
Goldenrod Tree 57455
Goldenrod-like Synotis 381954
Goldenrod-tree 57455
Goldenroot 103850
Golden-samphire 230109
Golden-saxifrage 90317,90335
Goldenseal 200173,200175
Golden-seal 200173,200175
Golden-shower 78300
Goldenshower Senna 78300
Golden-shower Senna 78300
Golden-shower Tree 78300
Goldenshrub 181932,264233
Goldensilk Grass 306832
Golden-sphaeroidal Barberry 51466
Goldenspine Parodia 284657
Golden-spined Prickly Pear 272801
Goldenstar 89959
Golden-star Cactus 244062
Goldenstar Rockfoil 349926
Goldentainui Anzacwood 310761
Goldenthistle 354648
Goldenthread Meadowrue 388519
Golden-tip Broadleaf Arborvitae 390682
Goldentop 161069,220295,220298
Golden-top 220295,220298
Goldentop Grass 220298
Golden-top Wattle 1650
Goldentuft 45191,45192
Golden-tuft 45192
Goldentuft Alyssum 45192
Golden-tuft Alyssum 45192
Goldentwig Dogwood 105198
Golden-variegata Black Elder 345638
Goldenvein Iris 208487

Goldenvein Swordflag 208487
Goldenwave Coreopsis 104448
Goldenweed 150313,150378,185566,209562,265638,323011,414976
Goldenweed Goldenweed 375786
Goldenwing Milkvetch 42189
Golden-winter-crocus Crocus 111515
Golden-wreath Wattle 1553
Goldenyellow Ancylostemon 22159
Golden-yellow Bauhinia 49007
Golden-yellow Besleria 53208
Golden-yellow Broomrape 275126
Goldenyellow Crocus 111558
Golden-yellow Dolichandrone 135321
Goldenyellow Epidendrum 146505
Goldenyellow Hawksbeard 110775
Golden-yellow Horned Violet 409861
Goldenyellow Iris 208507
Goldenyellow Laelia 219698
Goldenyellow Oncidium 270793
Goldenyellow Rambutan 265126
Goldenyellow Thorowax 63571
Goldenyellow Woodbetony 287024
Goldfieds Bottlebush 248086
Goldfields 222588
Goldfields Grevillea 180592
Goldfinger Shrubby Cinquefoil 312581
Goldfish Plant 16179,100115,263325,263330
Goldfishflower 255394
Goldflame Honeysuckle 235828
Goldflower Tea 69480
Goldfussia 378086
Goldhair Eargrass 187544
Goldhair Hedyotis 187544
Goldhead Cudweed 178125
Goldic Dracaena 137409
Goldicup 68189
Goldie Dracaena 137416
Goldie Dracena 137416
Goldie Dragonbood 137416
Goldie's Starwort 374951
Goldilocks 18775,68189,189814,231844,325628,399504,405788
Goldilocks Aster 40781
Goldilocks Buttercup 325628
Golding Cup 325820
Gold-in-green 89959
Goldings 66445,89704,227533
Goldlarch 317821
Goldleaf 108574
Gold-leaf Sedum 356918
Goldleaftree 90007,90062
Gold-leaved Barberry 52228
Gold-leaved Bellflower 70216
Gold-leaved English Elm 401597
Gold-leaved Forsythia 167438
Gold-leaved Lamium 220403
Gold-leaved Ninebark 297847
Gold-leaved Red Osier Dogwood 105196
Goldlip-orchis 89938,89948
Goldlock 364557
Goldmat Azalea 330182
Goldmoss 356468
Gold-moss 356468
Goldmoss Sedum 356468
Goldmoss Stonecrop 356468
Gold-moss Stonecrop 356468
Goldnet Honeysuckle 235879
Goldnyellow Sagebrush 35187
Goldnyellow Wormwood 35187
Gold-of-pleasure 68839,68841,68850,68860
Goldpoppy 155170
Goldquitch 156439,156499
Goldquitchlike Bistamengrass 127473
Gold-rain Tree 217610
Goldraintree 217610
Gold-rain-tree 217610
Goldsaxifrage 90317
Goldsaxifrageleaf Bellflower 69962
Gold-saxifrage-leaf Rockfoil 349178
Gold-shining Rambutan 265126
Goldshower Thryallis 390542
Goldspot Pacific Dogwood 105132
Goldstripe Heliconia 189992
Goldthread 103824,103850
Goldthreadweed 26622,26624
Goldwaist 90317
Goldweed 325518,325612,325666,326287
Goldwool Clematis 94830
Goldwort 66445
Goldy 175454,325518,325666,326287
Goldy Knob 325518,325820
Goldyellow Ophiorrhiza 272180
Goldyellow Tea 68927
Gole 89704,261162
Golf Ball Cactus 244126
Golf Ball Pincushion Cactus 244126
Golfob 334250
Golland 68189,89704,227533,325518,325666,326287,399504
Gollin 68189
Gollywog 282685
Golmu Milkvetch 42419
Gololen Barberry 52371
Gombalan Cinquefoil 312646
Gombo 219
Gomphandra 178916
Gomphia 70712,70750
Gomphocarpus 178974
Gomphococcous Attalea 44808
Gomphogyne 179153
Gomphostemma 179174
Gomuti 32343
Gomuti Palm 32343
Gomuti Sugar Palm 32343
Gomuti Sugarpalm 32343
Gomuti Sugar-palm 32343
Gonatanthus 179258
Goncalo Alves 43475
Gongbala Franchet Barberry 51637
Gongbogyanda Barberry 51823
Gongbu Azalea 331016
Gongbu Barberry 51823
Gongbu Groundsel 359237
Gongbu Larch 221901
Gongbu Monkshood 5327
Gongbu Primrose 314547
Gongbu Rhodiola 329927
Gongbu Rhododendron 331016
Gongbu Woodbetony 287316
Gongbu-like Monkshood 5503
Gongdong Photinia 295713
Gongga Amitostigma 19511
Gongga Astyleorchis 19511
Gongga Goldenray 229079
Gongga Larkspur 124234,124298
Gongga Milkvetch 42432
Gongga Poplar 311334
Gongga Sedge 74679
Gonggashan Azalea 330783
Gonggashan Monkshood 5367
Gonggashan Rhododendron 330783
Gonggashan Rockfoil 349410
Gongjia Acronychia 6249
Gongnaisi Catchfly 363646
Gongora 179288
Gongronema 179301
Gongshan Actinidia 6680
Gongshan Ampelopsis 20359
Gongshan Artocarpus 36919
Gongshan Asiabell 98325
Gongshan Azalea 330785
Gongshan Birch 53466
Gongshan Bulbophyllum 62756
Gongshan Coelogyne 98664
Gongshan Cowparsnip 192276
Gongshan Curlylip-orchis 62756
Gongshan Currant 334008
Gongshan Deerdrug 242687
Gongshan Dishspurorchis 171856
Gongshan Elatostema 142673
Gongshan Eurya 160482
Gongshan False Solomonseal 242687
Gongshan Falsehellebore 405597
Gongshan Falsepanax 266916
Gongshan Fargesia 162695
Gongshan Fleabane 150738
Gongshan Gastrochilus 171856
Gongshan Gentian 150738
Gongshan Goatsbeard 37079

Gongshan Griffith Currant 334008
Gongshan Hackberry 80645
Gongshan Himalayahonerysuckle 228334
Gongshan Holly 203891
Gongshan Hornbeam 77421
Gongshan Jadeleaf and Goldenflower 260503
Gongshan Kiwifruit 6680
Gongshan Litse 233932
Gongshan Longstalk Willow 343138
Gongshan Machilus 240592
Gongshan Maple 3065
Gongshan Meliosma 249479
Gongshan Monkshood 5333
Gongshan Mulberry 259123
Gongshan Mussaenda 260503
Gongshan Newlitse 264104
Gongshan Oak 324079
Gongshan Platanthera 302345
Gongshan Plumyew 82539
Gongshan Primrose 314972
Gongshan Raspberry 338477
Gongshan Rehdertree 327413
Gongshan Rehder-tree 327413
Gongshan Rhododendron 330785
Gongshan Rockfoil 349411,349503
Gongshan Rockvine 387874
Gongshan Sagebrush 35573
Gongshan Scurrula 355306
Gongshan Sedge 74680
Gongshan Sinocarum 364874
Gongshan Spapweed 204987,205076
Gongshan Spicebush 231331
Gongshan Spice-bush 231331
Gongshan Stairweed 142673
Gongshan Syzygium 382545
Gongshan Thistle 91953
Gongshan Touch-me-not 204987,205076
Gongshan Umbrella Bamboo 162695
Gongshan Whytockia 414002
Gongshan Willow 343375
Gongshan Woodbetony 287243
Gongshan-like Stairweed 142782
Goniolimon 179368
Goniostemma 179400
Goniothalamus 179403
Gonocaryum 179441
Gonostegia 179485
Gonshan Begonia 49895
Goober 30494,30498
Goober Pea 30498
Goocoo-flower 72934
Good Banksia 47643
Good Friday 8399
Good Friday Flower 8399,285623
Good Friday Grass 238583
Good Friday Plant 321643
Good Henry 86970
Good King Harry 86970
Good King Henry 86970,340151
Good Luck 277648
Good Luck Flower 351778
Good Luck Leaf 61572
Good Luck Money Tree 279381
Good Luck Plant 278124
Good Luck Tree 389978
Good Neighbourhood 81768,86970
Good Neighbours 81768
Good Report Rhododendron 330651
Goodding Willow 343437,343767
Goode 227533
Goodenia 179556
Goodenia Family 179563
Gooding Ash 167980
Good-king-henry 86970
Good-king-henry Goosefoot 86970
Good-luck Leaf 278124
Good-luck Palm 85425
Goodluck Plant 104350
Good-luck Plant 104367,278124
Goodman's Sulphur Flower 152586
Goodman's Unarmed Wild Buckwheat 152159
Good-night-at-noon 195326,243840
Good-tongue Eulophia 156691
Goody's Eye 345379
Goody's Eyes 345379
Goodyer's Elm 401573
Gookoo Buttons 72934
Gools 66445,68189,89704,364557
Goose Bill 170193,174877
Goose Chicks 343151
Goose Cleavers 170193
Goose Flop 130383,262441
Goose Foot 86901
Goose Grass 143530,170193,170743,308816,308893,312360,383874
Goose Gray 312360
Goose Grease 170193
Goose Heiriff 170193
Goose Lake Wild Buckwheat 152494
Goose Leek 262441
Goose Nest 202767
Goose Plum 316206
Goose Starwort 260922
Goose Tansy 312360
Goose Tongue 3978,4005,170193,325864,383712
Goose Withy 343151
Goose-and-gander 248411,408352
Goose-and-goslings 273531,273539,273545,343151
Goose-and-gubblies 343151
Gooseberry 333893,334250
Gooseberry Family 181339
Gooseberry Fig 77430
Gooseberry Geranium 288273
Gooseberry Gourd 114225
Gooseberry Mallee 155516
Gooseberry Pie 146724,271563,363673,381035,404251
Gooseberry Pudding 146724
Gooseberry Tomato 297658
Goosechite 11549
Goosecomb 335183
Goosefeather Habenaria 183559
Goosefeather Wobamboo 362123
Goosefoot 3393,86891,86901,139682,312360
Goosefoot Corn-salad 404404
Goosefoot Family 86873
Goosefoot Maple Stripe Bark 3393
Goosefoot Mercury 86970
Goosefoot Moonpo 19720
Goosefoot Plant 381861
Goosefootleaf Skullcap 355407
Goosefootlike Pearweed 239579
Goosegob 334250
Goosegog 334250
Goosegrass 5871,143512,143530,170193
Goose-grass 143530,170193
Goose-grass Bedstraw 170193
Goosehare 170193
Goosehareth 170193
Goosehead Woodbetony 287089
Gooseheriff 170193
Gooseneck Loosestrife 239594
Gooseneck Yellow Loosestrife 239594
Goosepalm Windflower 23817
Goose-share 170193
Goose-tongue 3978,4005,170193,325864,383712
Gooseweed 170193,312360,371409
Goosewhite Eria 299641
Goosewhite Hairorchis 299641
Goosewort 312360
Goosey Gander 145487,273531,273545
Goosey-gander 145487,273531,273545
Goosey-goosey-gander 145487,273531
Gopher Plant 159222,159726
Gopher Wood 299780
Gopher-berry 38333
Gopherwood 114753,393089
Goraka 171076
Gorbunov Crazyweed 278857
Gordeiev Medic 396920
Gordejev Clover 396920
Gordejev Willow 343438
Gordjev Siberian Crabapple 243559
Gordolaba 3978
Gordolobos 358881
Gordon Currant 334003

Gordon Euryale 160637
Gordon Mock Orange 294462
Gordon Mockorange 294462
Gordon's Bladderpod 18385
Gordon's Mock-orange 294462
Gordon's Wild Buckwheat 152098
Gordonia 179729
Gordontea 179729
Gore Thetch 408571
Gore-thetch 408571
Gorge Fleabane 150832
Gorge Goldenrod 368096
Gorgeous Gentian 173900
Gorgeous Rhododendron 330651
Gorgon Plant 160637
Gorgon Waterlily 160637
Gorgon's Head 158970
Gorilla Grass 159046
Goris Stonecrop 356772
Gorli Oil 270897
Gorman's Aster 155814
Gorodkov Mouseear 82859
Gorongowe Cycad 145239
Gorse 401373,401388
Gorse Bitter Pea 123271
Gorse Hakea 184649
Gorst 213702,401388
Gorstberry 334250
Gortschakov Anemone 23832
Gortschakov Corydalis 105939
Gortschakov Glorybind 103065
Gortschakov Windflower 23832
Gosford Wattle 1496
Goshiki False Holly 276313
Gosler 334250
Gosling Grass 170193
Gosling Scratch 170193
Gosling Tree 343151
Goslings 273531,343151,343858
Gosling-weed 170193
Goss 271563,401112,401388
Gost 401388
Go-to-bed-at-noon 394333
Go-to-sleep-at-noon 99297
Gottridge 407989
Gouania 179940,179947
Goud 209229
Goudot Barberry 51675
Gouget Statice 230633
Gough Spapweed 204990
Gough Touch-me-not 204990
Goulan 66445,68189,89704
Gould Remarkable Barberry 51769
Goule 261162
Gouls 89704
Goumi 142088,142102
Gourd 114271,219821,219843
Gourd Family 114313
Gourd grass 87690
Gourd-tree 7028
Gourian Clematis 94953
Gourka 171092
Gourlins 102727
Gourou Nut 99240
Goury Tree 7028
Gout Ivy 13073
Gout Plant 212202
Gout Weed 8810,8826
Goutou Orange 93334
Goutweed 8826
Gout-weed 8826
Goutwort 8826
Gouty Houseleek 12511
Gouty Vine 120019,411979
Goutystalk Nettlespurge 212202
Gouyahua 154163
Govan Corydalis 105943
Govan Nepeta 264937
Goven Cypress 114698
Governor Plum 166773
Governor's Plum 166773,166786,166788
Governorsplum 166773
Gow 261162
Gowan 50825,66445,68189,89704,325518,384420,384714
Gowen 50825
Gowen Cypress 114698
Gowk Meat 273531,273545,277648,339887
Gowk's Clover 277648
Gowk's Hose 70119,70271,145487
Gowk's Meat 273531,273545,277648,339887
Gowk's Shillings 329521
Gowk's Siller 329521
Gowk's Sixpences 329521
Gowk's Thumb 70271
Gowlan 50825,66445,89704,227533
Gowland 89704
Gowles 66445,68189,89704
Gowlins 102727
Goyle 261162
Gozill 334179,334250
Gozzle-grass 170193
Grab 243711
Grab-apple 243711
Grabappleleaf Sonerila 368896
Grabrous-fruit Taohe Willow 344182
Grace of God 174832,202086,203102,301910
Grace Valerian 404384
Graceful Aster 40907
Graceful Azalea 331411
Graceful Bamboo 297270
Graceful Bedstraw 170354
Graceful Bush-pea 321729
Graceful Camellia 69057
Graceful Clematis 94957
Graceful False Flax 68859
Graceful Flowers Rhododendron 330366
Graceful Grama Grass 57930
Graceful Gypsophila 183191
Graceful Honey Myrtle 248115
Graceful Jessamine 172779
Graceful Love Grass 147650
Graceful Lovegrass 147650
Graceful Nightshade 367183
Graceful Pondweed 312205
Graceful Pouzolzsbush 313483
Graceful Primrose 314262
Graceful Sedge 74694
Graceful Trillium 397569
Graceful Veined Jasmine 211919
Graceful Wattle 1167
Graceful-clump Iris 208879
Gracile 29305
Gracile Clearweed 298926
Gracile Cochlianthus 98065
Gracile Haworthia 186448
Gracile Knotweed 162316
Gracile Maple 2996
Gracile Mint 250368
Gracile Mycetia 260624
Gracile Potatobean 29305
Gracile Sainfoin 271210
Gracile Willow 343440
Gracilebranch Amaranthus 18724
Gracilepeduncle Evodialeaf Acanthopanax 2391
Gracilestem Woodbetony 287245
Gracile-style Willow 343444
Gracillis Hinoke Cypress 85318
Gracy Daisy 50825,262441
Gracy Day 262441
Graebner Honeysuckle 235817
Graebner Piratebush 61932
Graessner Ballcactus 267035
Graessner Parodia 59163
Graffith Larch 221881
Gragon Palm 393814
Gragontongue Ottelia 277369
Graham Sedge 74696
Graham's Club-cholla 181455
Graham's Fishhook Cactus 244087
Graham's Manihot 244510
Graham's Nipple Cactus 244087
Graham's Prickly Pear 272915
Grain Sorghum 369600
Grain Tree 323769,324027
Grainfield Spurge 159802
Grains 224375
Grains of Paradise 9923,19858,143502

Grains-of-paradise 9923
Grain-spotted Cattleya 79540
Grainy Cinquefoil 312651
Gram Chick Pea 90801
Gram Chick-pea 90801
Gram Chikpea 90801
Grama 57922,57930
Grama-grass Cactus 354306
Gramineous Barberry 51677
Gramineous Plants 180078
Graminileaf Ixeris 210541
Graminoid Bitter-bamboo 304032
Gramma 57922
Gramma Grass 57922
Gramma Grass Cactus 354306
Grammatophyllum 180116
Grammer Greygle 145487,236022
Gramophone 399601
Gramophone Horn 236022
Gramophone Horns 236022,285623,285637
Grampian Eucalyptus 155475
Grampian Stringybark 155475
Grampians Bauera 48978
Grampians Fringe Myrtle 68756
Grampians Fringe-myrtle 68760
Grampians Grevillea 180563,180582
Grampians Gum 155475
Grampians Thryptomene 390549
Granada 321764,321776
Granadilla 61427,285607,285623,285637,
285694
Granadilla Vine 285694
Granadillo 121698,121803
Granberry 404051
Granberry-bushes 407917
Grand Adinandra 8233
Grand Bamboo 47516
Grand Canyon Beavertail Pricklypear 272808
Grand Canyon Evening-daisy 90559
Grand Canyon Glowweed 90559
Grand Canyon Rock Daisy 291306
Grand Crinum 111167,111171
Grand Devil's-claws 300955
Grand Eucalyptus 155598
Grand Fir 384
Grand Helicia 189933
Grand Lake Blackberry 338871
Grand Lake Gentian 173487
Grand Licuala Palm 228741
Grand Myrtle 219933
Grand Paspalum 285492
Grand Red-stem 19621
Grand Torreya 393061
Grand Valley Desert Trumpet 152084
Grand Valley Wild Buckwheat 151945
Grand White Fir 384
Grande Fire Pink 363891
Grande Passerage 225403
Grandfather Griggle 273531
Grandfather's Beard 95431,175452
Grandfather's Buttons 68189
Grandfather's Weatherglass 21339
Grandfather's Whiskers 95431
Grandfy's Beard 95431
Grandibracted Cliff Monkshood 5522
Grandiflower Wintersweet 87528
Grandiflower-concolor Wintersweet 87527
Grandifoliate Lanceolate Elaeagnus 142045
Grandligulate Sedge 74705
Grandmother 227533
Grandmother's Bonnet 5442
Grandmother's Bonnets 5442,24525,30081
Grandmother's Darning Needle 350425
Grandmother's Darning Needles 350425
Grandmother's Hair 164006
Grandmother's Needle 404316
Grandmother's Nightcap 5442,68713,364146
Grandmother's Pincushion 216753
Grandmother's Slipper 237539
Grandmother's Spectacles 238371
Grandmother's Toenails 237539
Grandsir-greybeard 87736
Granebill Family 174431
Granfer Goslings 273531
Granfer Greygle 273531
Granfer Grigg 145487,273531
Granfer Griggle 145487,248312,273531
Granfer Griggle-sticks 145487,273531,
384714
Granfer Jan 248312
Granfer-goslings 273531
Granfer-gregor 145487
Granfer-greygles 248312,273531
Granfer-griddle-goosey-gander 273531
Granfer-griggles 145487,248312,273531
Granfer-grizzle 28200
Grang Bassam 216205
Grange Cestrum 84404
Grangea 180163
Granite Bottlebrnsh 248092
Granite Gooseberry 333943
Granite Honey-myrtle 248092
Granite Mountain Wild Buckwheat 152511
Granite Pink 127727
Granite-living Draba 137011
Granite-living Whitlowgrass 137011
Granjeno 80708
Grannies Ringlets 113703
Granny Bonnets 30081
Granny Griggle 145487
Granny Hood 30081
Granny Threads 326287
Granny's Bonnet 5442,28617,30081,68713,
102833,130383,174832,175444,297643
Granny's Bonnets 5442,28617,30081,
68713,102833,130383,174832,175444,
297643
Granny's Cap 30081,175444
Granny's Face 410677
Granny's Gloves 130383
Granny's Needle 174877
Granny's Night Bonnet 68713
Granny's Night-bonnets 68713
Granny's Nightcap 5442,23925,28617,
30081,57101,68713,102833,102933,
174877,175444,191304,363673,367120,
374897,374916
Granny's Shoes 5442
Granny's Slipper 5442,237539
Granny's Slippers 5442,237539
Granny's Slipper-sloppers 222820
Granny's Tears 70271
Granny's Thimble 30081
Granny's Thimbles 30081
Granny-bonnets 29994,30081
Granny-griggles 145487
Granny-jump-out-of-bed 5442,30081
Granny-threads 326287
Granny-thread-the-needle 23925,174877
Grannyvine 208250
Grant Barberry 51680
Grant Lemon 93631
Grant's Rattlebox 112801
Granter Periwinkle 409335
Granular Crabgrass 130407
Granular Gagea 169443
Granular Mountainash 369374
Granular Mountain-ash 369374
Granular Skullcap 355460
Granulate Clematis 95116
Granulate Craneknee 221684
Granulate Parasiticvine 125786
Granyagh 370522
Grape 411521,411979
Grape Cactus 244202
Grape Clematis 95448
Grape Corydalis 106584
Grape Eucalyptus 155505
Grape Family 411165
Grape Flower 260301
Grape Fruit 93690
Grape Grevillea 180574
Grape Honeysuckle 236066
Grape Hyacinth 260295,260299,260301,
260302,260308,260310,260333
Grape Ivy 92586,92859,92977,411885
Grape Leather-flower 95448
Grape Pomelo 93690
Grape Vine 411979
Grape Vine Ivy 411885
Grape Woodbine 285164

Grapefruit 93690
Grapefruit Citrus 93690
Grape-fruit Citrus 93690
Grape-hyacinth 260301,260333
Grapeleaf Actinidia 6724
Grapeleaf Anemone 24116
Grape-leaf Anemone 24116
Grapeleaf Begonia 49802,50229
Grape-leaf Begonia 49790,50275,50395
Grapeleaf Geranium 288588
Grapeleaf Larkspur 124584
Grapeleaf Moxanettle 221604
Grapeleaf Windflower 24116
Grape-leaf Woodnettle 221604
Grape-leaved Actinidia 6724
Grape-leaved Geranium 288588
Grape-like Eucalyptus 155505
Grape-like Falsepimpernel 231595
Grape-of-mount-Ida 404051
Graperfern-like Cystopetal 320206
Grapes 356468
Grape-scented Sage 345209
Grapevine 411979
Grape-vine 411521
Grapevine Begonia 50404
Grape-vine Family 411165
Grape-vine Ivy 411885
Grapewort 6448,61497
Graphistemma 180211
Grapple Plant 185846
Graptopetalum 180265
Graptophyllum 180270
Grass 5793,71218,376687
Grass Corn 293729
Grass Family 180078
Grass Gum 415172,415178
Grass Iris 208600
Grass Myrtle 5793
Grass Nettle 373455
Grass Nut 398806
Grass Nuts 128870
Grass of Parnassus 284500,284591
Grass Pea 222784,222832,222837
Grass Pea Vine 222832
Grass Peavine 222832
Grass Pink 67882,67894,127793
Grass Pink Orchid 67894
Grass Poly 240047
Grass Sedge 74935
Grass Tree 216415,415172,415173,415174,415178,415180
Grass Triggerplant 379079
Grass Vetch 222784
Grass Vetchling 222784
Grass Water Plantain 14734
Grass Waterplantain 14734
Grass Water-plantain 14734
Grass Widow 270533,365726
Grass Wrack 418392
Grasscoal 85579
Grasses 180078
Grassland Agoseris 11454
Grassland Azalea 331944
Grass-land Blueberry 403968
Grassland Buttercup 326019
Grassland Chalkplant 183185
Grassland Cherry 83202
Grassland Cranesbill 174836
Grassland Dogtongueweed 385915
Grassland Gagea 169504
Grassland Gaultheria 172134
Grassland Gentian 173734
Grassland Hawksbeard 110957
Grassland Hogfennel 293022
Grassland Japanese Buttercup 325987
Grassland Lily 228601
Grassland Medic 247291
Grassland Milkvetch 42260
Grassland Monkshood 5466
Grassland Rose 336650
Grassland Sedge 74366
Grassland Silverpuffs 374543
Grassland Spunsilksage 360872
Grassland Whitepearl 172134
Grassland Whitlowgrass 98764
Grassland Windhairdaisy 348669
Grasslaving Glorybind 103310
Grassleaf Acronema 6211
Grassleaf Arundina 37115
Grass-leaf Barbara's Button 245889
Grassleaf Bladderwort 403199
Grassleaf Blue Orchid 193154
Grassleaf Catchfly 363516
Grassleaf Chickweed 374897
Grassleaf Coelogyne 98754
Grassleaf Coneflower 339551
Grassleaf Draws Ear Grass 403199
Grassleaf Eria 299619
Grass-leaf Gayfeather 228502
Grassleaf Goldaster 301505
Grassleaf Gymnadenia 182250
Grassleaf Hairorchis 299619
Grassleaf Hakea 184608
Grassleaf Kobresia 217183
Grassleaf Landpick 308553
Grassleaf Liriope 232623
Grassleaf Orchis 12442
Grassleaf Pepperweed 225369
Grassleaf Primrose 314443
Grass-leaf Sandwort 31933
Grassleaf Saussurea 348350
Grassleaf Screwpine 281037
Grassleaf Solomonseal 308553
Grassleaf Sorrel 340068
Grassleaf Spurge 158978
Grassleaf St. John's Wort 201903
Grassleaf Stonecrop 356775
Grassleaf Sweetflag 5803
Grass-leaf Sweetflag 5803
Grassleaf Windhairdaisy 348350
Grass-leaved Arrowhead 342338
Grass-leaved Buttercup 325912
Grass-leaved Day Lily 191312
Grass-leaved Daylily 191312
Grass-leaved Day-lily 191312
Grass-leaved Golden Aster 194223
Grass-leaved Goldenaster 301505
Grass-leaved Goldenasters 301497
Grass-leaved Goldenrod 161081,161086,368134
Grass-leaved Ladies'-tresses 372277
Grass-leaved Nerine 265269
Grass-leaved Orache 44510
Grass-leaved Pea 222784
Grass-leaved Pondweed 312126
Grass-leaved Rock Goldenrod 292505
Grass-leaved Rush 213254
Grass-leaved Sagittaria 342338
Grass-leaved Sandwort 31933
Grass-leaved Scabious 350156
Grass-leaved Sorrel 340068
Grass-leaved Starwort 374897
Grass-leaved Stitchwort 374897
Grassleaved Sweetflag 5803
Grass-leaved Water Plantain 14734
Grass-leaved Water-plantain 14734
Grass-like Arrowleaf 342338
Grass-like Bamboo 304032
Grasslike Flatsedge 322376
Grass-like Flat-sedge 322376
Grasslike Sedge 75684
Grasslike Starwort 374897
Grass-like Starwort 374897
Grassnut 30499
Grass-nut 398806
Grass-of-parnassus 284498,284509,284545,284546,284591
Grass-pink Orchid 67882,67894
Grass-wrack Pondweed 312075
Grassy Arrowhead 342338
Grassy Coneflower 339588
Grassy Death Camas 417930
Grassy Pondweed 312199
Grassy Rush 64184
Grassy Screwpine 281037
Grassy St. Johnswort 201903
Grassy-leaved Sweet Flag 5803
Gratiole 180292,180321
Gravel Azalea 330445
Gravel Bind 367762
Gravel Crazyweed 278852

Gravel Root 158264
Gravel Sedge 74654
Gravel-bind 367762
Gravel-ghost 44296
Gravelly Meadowrue 388679
Graves Oak 323978
Graves' Blackberry 338111
Graveyard Spurge 158730
Graviola 25866
Grawmpy Griggle 145487
Gray Alder 16368,16440
Gray Barleria 48094
Gray Beardtongue 289328
Gray Birch 53338,53586
Gray Bird Cherry 280023
Gray Bog Sedge 74004,74007
Gray Box 155658
Gray Camomile 26764
Gray Chickensage 371144
Gray Chickweed 82742
Gray Chokecherry 280023
Gray Choke-cherry 280023
Gray Clubawn Grass 106932
Gray Dogwood 105039,105169
Gray Echeveria 139990
Gray Elm 401431,401620
Gray Everlasting 170923,279349
Gray Fescuegrass 163867
Gray Fig 165841
Gray Ghost Organ Pipe 375530
Gray Globe Mallow 370948
Gray Goddess 59096
Gray Goldenrod 368268
Gray Green Washingtonia 413298
Gray Gum 155711,155756
Gray Hair Grass 106932
Gray Hakea 184594
Gray Hawksbeard 110934
Gray Honey Myrtle 248100
Gray Ironbark 155682
Gray Japanese Helwingia 191181
Gray Loco 42130
Gray Mangrove 45748
Gray Mouse-ear Chickweed 82742
Gray Mulga 1090
Gray Myrtle 46338
Gray Nightshade 367233
Gray Oak 320278,323982,324344
Gray Pine 299818,300189
Gray Poplar 311261
Gray Raspberry 338487
Gray Rock Daisy 291304
Gray Rockcress 30220
Gray Rock-cress 30220
Gray Rupturewort 192981
Gray Sage 345169
Gray Sagebrush 44334
Gray Scurf-pea 319270
Gray Sedge 74715
Gray Sheepbur 1744
Gray Sheoak 79165
Gray She-oak 79165
Gray Spider Flower 180578
Gray Thistle 92461
Gray Thorn 418196
Gray Umbrellaleaf 132682
Gray Willow 343060,343221
Gray's Aster 193127
Gray's Bur Sedge 74711
Gray's Catchfly 363521
Gray's Lily 229855,229934
Gray's Sedge 74711,118970
Gray's Wild Buckwheat 152624
Grayback Grape 411615
Gray-bark Grape 411615
Graybark Sargent Spruce 298260
Graybark Spruce 298260
Gray-budded Snakebark Maple 3529
Grayflower Korean Pulsatilla 321669
Gray-glaucous Acaena 1744
Graygreen Thistle 91901
Gray-green Wood Sorrel 278099
Gray-hair Actinidia 6604
Grayhair Aster 41057
Grayhair Elaeocarpus 142358
Grayhair Everlasting 21541
Gray-hair Kiwifruit 6604
Grayhair Maple 3026
Gray-haired Actinidia 6604
Gray-haired Elaeocarpus 142358
Gray-haired Maple 3026
Grayhairy Bellflower 69934
Grayhead Prairie Coneflower 326964
Gray-headed Coneflower 326964
Grayish Onion 15338
Gray-leaf Cedar 85283
Gray-leaf Pine 300189
Gray-leaved Willow 343425
Graylocks Rubberweed 201314
Graysheath Aletris 14480
Grayspatted Giantarum 20116
Gray-stemmed Goldenrod 368268
Gream Fruit 378401
Gream Rhododendron 331034
Grease-bush 177361
Greasewood 8032,46255,346607,346610
Great American Laurel 331207
Great Angelica 24303
Great Aspen 311208
Great Aster 380900
Great Astilbe 41812
Great Barberry 51670
Great Basin Blue Sage 345011
Great Basin Bristlecone Pine 300040
Great Basin Brittle Spineflower 89060
Great Basin Claytonia 94371
Great Basin Desert Wild Buckwheat 152002
Great Basin Fishhook Cactus 354314
Great Basin Naturalist 255725,393406
Great Basin Rabbitbrush 150351
Great Basin Sagebrush 36400
Great Basin Wild Buckwheat 152280
Great Bear Azalea 331532
Great Bearclaw-poppy 31084
Great Bellflower 70119
Great Bind 68713
Great Bindweed 68713,68746
Great Bistort 308893
Great Bladderwort 403382
Great Blue Lobelia 234774,234779,234813
Great Bougainvillea 57868
Great Britain Blackberry 339375
Great Broad-leaved Mullein 405788
Great Brome 60696
Great Broomrape 275192
Great Bulbous Iris 208944
Great Bulrush 352164,352192,352206,352278
Great Bur 31051
Great Burdock 31051,332086
Great Burmese Honeysuckle 235846
Great Burnet 345881
Great Buttercup 325916
Great Butterwort 299744
Great Candlesticks 363312
Great Cat-tail 401112
Great Chervil 261585
Great Ciote Bur 31051
Great Coneflower 339585
Great Consound 381035
Great Cowslip 315106
Great Crow Leek 15861
Great Daisy 227533
Great Darnel 235334
Great Desert Poppy 31084
Great Dragon 37015
Great Duckweed 372300
Great Earth Nut 63475
Great Earthnut 63475
Great False Leopardbane 136365
Great Fen Ragwort 359660
Great Fen-sedge 93901,93913
Great Fir 384
Great First of the First Rhododendron 331558
Great Fleabane 207087
Great Forget-me-not 61364,61366
Great Furze 401388
Great Galangal 17677
Great Globe Thistle 140789
Great Globethistle 140789
Great Globe-thistle 140789

Great Grass 401112
Great Ground Ivy-leaved Chickweed 220355
Great Gunnera 181951
Great Hawkweed 298589
Great Hedge 170490
Great Hedge-nettle 373179
Great Horse Knob 81356
Great Indian Cress 399597
Great Indian Plantain 34809
Great Indian-plantain 34809
Great Juniper 386942
Great Knapweed 81183,81356
Great Lakes Corn-salad 404404
Great Lakes Dewberry 339040
Great Lakes Gentian 173836
Great Lakes Sand Cherry 316735,316739
Great Laurel 331207
Great Laurel Magnolia 242135
Great Laurel Rhododendron 331207
Great Lead Tree 227420
Great Leopard's Bane 136365
Great Leopard's-bane 136365
Great Lettuce 219609
Great Lobelia 234774,234813
Great Lunary 238371
Great Maple 3462
Great Marsh Bristlethistle 73523
Great Marsh Thistle 73454
Great Masterwort 43309
Great Millet 369600
Great Morel 44708
Great Mouse-ear Chickweed 82918
Great Mullein 405788
Great Murray Pine 67433
Great Nettle 402886
Great Northern Aster 70949
Great Orchid 273541
Great Oriental Bellflower 276796
Great Park Lily 102867
Great Periwinkle 409335
Great Pignut 63467,63475
Great Pilewort 355202
Great Plains Canadian Goldenrod 368026
Great Plains Cottonwood 311465
Great Plains False Willow 46253
Great Plains Flat Sedge 119139,119140
Great Plains Flat-topped Goldenrod 161086
Great Plains Fringed Gentian 174228
Great Plains Gayfeather 228484
Great Plains Goldentop 161086
Great Plains Ladies' Tresses 372229
Great Plains lady's-tresses 372229
Great Plains late Goldenrod 367953
Great Plains Ragwort 279954
Great Plains Sand Sedge 119549
Great Plains Yucca 416603
Great Plantain 302068
Great Poppy 282522
Great Prickly Lettuce 219609
Great Quaking Grass 60379
Great Ragweed 19191
Great Raifort 34590
Great Reed 37450
Great Reedmace 401112
Great Rose Mellow 194901
Great Round-headed Garlic 15048
Great Round-leaved Willow 343151
Great Sallow 343151
Great Sallow Willow 343151
Great Saniicle 14065
Great Scarlet Poppy 282522
Great Silver Bell 184731
Great Soft-rush 213351
Great Solomon's Seal 102867,308508,
308527
Great Solomon's-seal 308527
Great Spanish Bastard Rhubarb 388423
Great Spicebush 231385
Great Spotted Iris 208870
Great Spurge 334435
Great Squash 114288
Great St. John's Grass 202179
Great St. John's Wort 201743,202179
Great St. John's-wort 201743
Great Stonecrop 356928,401711
Great Sundew 138267
Great Thunderbolts 14760
Great Trefoil 247456
Great Turkey Garlic 15726
Great Turkey-garlic 15726
Great Water Cress 336172
Great Water Dock 339957,340082,340161
Great Water-dock 339957,340082,340161
Great Water-leaf 200483
Great Waybread 302068
Great White Lettuce 313809
Great White Trillium 397570
Great Wild Tare 222844
Great Willow 342973,343151
Great Willow Herb 85875
Great Willowherb 85875,146724
Great Willow-herb 85875,85880,146614,
146619
Great Willowherb Willowweed 85875
Great Wood Rush 238709
Great Woodrush 238709
Great Wood-rush 238709
Great Wood-sorrel 277973
Great Woolly Nutmeg 216831
Great Yellow Gentian 173610
Great Yellow Pansy 410187
Great Yellowcress 336172
Great Yellow-cress 336172
Greata's Aster 380902
Greater Alpenclock 367766
Greater Ammi 19664
Greater Arctic Campion 363579
Greater Bird's-foot Trefoil 237713
Greater Bird's-foot-trefoil 237713
Greater Bistort 308893
Greater Bittercress 72749
Greater Bladder Sedge 74904
Greater Bladderwort 403110,403382
Greater Breomrape 275128
Greater Broomrape 275192
Greater Brown Sedge 73952
Greater Bur Parsley 400473
Greater Burdock 31051
Greater Burnet Saxifrage 299467
Greater Burnet-saxifrage 299467
Greater Bur-parsley 400473
Greater Butterfly Orchid 302280
Greater Butterfly-orchid 302280
Greater Calamint 97000
Greater Cat's Tail 401112
Greater Cat's Tails 401112
Greater Celandine 86733,86755
Greater Centaury 54896,81356
Greater Chickweed 342877,375007
Greater Churmel 54896
Greater Coreopsis 104542
Greater Cuckoo Flower 72948
Greater Daisy 227533
Greater Dodder 115031
Greater Duckweed 372295,372300
Greater Fringed Gentian 174215
Greater Galangal 17677
Greater Hawkbit 224698
Greater Henbit 220355
Greater Indian Cress 399601
Greater Knapweed 81356
Greater Malayan Chestnut 89972
Greater Meadow Rue 388423,388564
Greater Musk-mallow 243746
Greater Nan 240615
Greater Periwinkle 409335,409339
Greater Pimpinella 299466,299467
Greater Plantain 302068
Greater Pond Sedge 76044
Greater Pond-sedge 76044
Greater Quaking Grass 60379
Greater Quaking-grass 60379
Greater Red-hot Poker 217017
Greater Rockjasmine 23227
Greater Salad Burnet 345881
Greater Sand Goat's-beard 394281
Greater Sand-spurrey 370675
Greater Sea Rush 212808
Greater Sea Spurrey 370675
Greater Sea-kale 108599
Greater Sea-spurrey 370675

Greater Snowdrop 169710
Greater Spearwort 326034
Greater St. John's-wort 202010
Greater Stitchwort 374916
Greater Straw Sedge 75527
Greater Swinecress 105352
Greater Thrumwort 14760
Greater Tickseed 104542
Greater Trefoil 237713
Greater Tussock-sedge 75686
Greater Water Dock 339957
Greater Water Parsnip 365851
Greater Water-parsnip 365835,365851
Greater Waterwort 142574
Greater Wintergreen 322852
Greater Woodrush 238709
Greater Wood-rush 238709
Greater Yellow Cress 336172
Greater Yellow Rattle 329510
Greater Yellow-rattle 329510
Greater-pleasure Gladiolus 176097
Great-flowered Cornel 105023
Great-head Garlic 15048
Greatleaf Goldenray 229104
Greatleaf Rhododendron 331851
Great-leaved Rhododendron 331851
Great-leaved Sundew 138267
Great-spurred Violet 410547
Grecian Ceder 213767
Grecian Fir 306
Grecian Foxglove 130369
Grecian Hay 86755
Grecian Juniper 213767
Grecian Laurel 223203
Grecian Scabious 350225
Grecian Silk Vine 291063
Grecian Silk-vine 291063
Grecian Strawberry Tree 30880
Grecille Wolfberry 239053
Gree Jewel Touch-me-not 205346
Greece Fir 306
Greece Mountainash 369403
Greece Silkvine 291063
Greek Alfalfa 247456
Greek Anemone 23724
Greek Bellflower 70005
Greek Dock 340006,340019
Greek Fir 306
Greek Fritillary 168415
Greek Hayseed 397259
Greek Juniper 213767
Greek Mallow 362702,362704
Greek Maple 3013
Greek Micromeria 253666
Greek Myrtle 261739
Greek Nettle 402992
Greek Oregano 274237
Greek Sand Spurrey 370615
Greek Sea-spurrey 370615
Greek Spiny Spurge 158391
Greek Strawberry Tree 30880,403728
Greek Thimbleweed 23724
Greek Valerian 307190,307197,307241
Greek Yarrow 3918
Greek-valerian 307241
Greek-valerian Polemonium 307197
Green Adder's-mouth 243142
Green Aglaia 11305
Green Ailanthus 12570
Green Alder 16472,16479
Green Alkanet 289500,289503
Green Allomorphia 16164
Green Aloe 169245
Green Amaranth 18734,18773,18804,18848
Green and Gold 89952,89959
Green Apple 123077
Green Ardisia 31639
Green Arrow 3978
Green Arrow-arum 288808
Green Artichoke 117787
Green Ash 168057,168065,168134
Green Asiabell 98289
Green Azalea 332077
Green Back Osmanthus 161657
Green Bamboo 125453,297486
Green Basil 268671
Green Basom 121001
Green Bayberry 261122
Green Bean 409025
Green Besom 121001
Green Bird-lime Flower 210955
Green Bitterbamboo 304043
Green Blotchleaf Violet 410724
Green Blumea 55840
Green Bog Sedge 73959,73963
Green Bottlebrush 67308
Green Bottle-brush 67294
Green Briar 366559
Green Bristle Grass 361935,361954
Green Bristlegrass 361935,361954
Green Bristle-grass 361935
Green Broadwing Everlasting 21590
Green Broadwing Pearleverlasting 21590
Green Broom 121001
Green Buckwheat 162335
Green Bull Rush 213066
Green Bur-reed 370037
Green Button Protea 315954
Green Calla 36979
Green Cardamom 143502
Green Carpet Plant 192965
Green Carpetweed 256727
Green Carpet-weed 256727
Green Celery 29319
Green Cestrum 84419
Green Cloud Texas Ranger 227920
Green Condalia 101742
Green Cotton-grass 152798
Green Death Camas 417932
Green Desert Spoon 122899
Green Dock 339989
Green Douglas Fir 318580
Green Dracaena 104339
Green Dragon 33314
Green Earth Star 113365
Green Ebony 132077,211225,211239,211241
Green Ebony Persimmon 132102
Green Endive 219507,219609
Green Ephedra 146265
Green Epidendrum 146501
Green Everlasting 21726
Green Excoecaria 161657
Green Eye Rhododendron 330390
Green Eyes 52823,52836,52837,353987
Green False Hellebore 405643
Green Falsehellebore 405643
Green False-hellebore 405643
Green Feathergrass 376960
Green Fescue 164126
Green Field Speedwell 406973
Green Field-speedwell 406973
Green Fig Marigold 33183
Green Figwort 355261
Green Five Corners 379278
Green Five-corners 379278
Green Flat Sedge 119750
Green Flower Nipple Cactus 244255
Green Fountain Grass 289248
Green Foxtail 361935,361954
Green Fringed Orchid 302382
Green Gastrodia 171928
Green Gentian 380146,380302
Green Giant Fox Tail Grass 361935
Green Ginger 35090,36474
Green Goosefoot 87179
Green Gram 409025
Green Grass 88316
Green Greenstar 307535
Green Grevillea 180606
Green Grower 158433
Green Habenaria 184186
Green Hawthorn 110107
Green Heather 150027
Green Hedge-nettle 373179
Green Hellebore 190959,405643
Green Helleborine 147204
Green Hillcelery 276780
Green Holly 204388
Green Honey Myrtle 248091
Green Hoods 320862,320871

Green Hooker Barberry 51738
Green Hound's Tongue 117968
Green Hound's-tongue 117968
Green Hygro 191326
Green Imperial Begonia 49942
Green Iris 208931
Green Jessamine 84419
Green Ladyslipper 120363
Green Lavender 223347
Green Lavender Cotton 346271
Green Lemongrass 117150
Green Lily 190959
Green Litse 234106
Green Machilus 240718
Green Magnolia-vine 351014
Green Malice 155773
Green Mallee 155773
Green Man Orchid 3777
Green Manzanita 31132
Green Maple 3591
Green Marke 269307
Green Mexican-hat 326966
Green Milkweed 38168,38171
Green Millet 361935
Green Mint 250450
Green Monocelastrus 80348
Green Mormon Tea 146265
Green Mound Juniper 213864
Green Mountain Sallow 343017
Green Needle Grass 376960
Green Nightshade 367499,367583
Green Oblongleaf Maple 3258
Green One-seed Staff-tree 80348
Green Onion Bamboo 318309
Green Osier 104947,105023
Green Panic Grass 402531
Green Panicgrass 281459
Green Parrot's Feather 261357
Green Pearleverlasting 21726
Green Pellionia 288781
Green Pepper 72068,72070,72075
Green Pigweed 18734
Green Pinanga 299680
Green Pinangapalm 299680
Green Pinanga-palm 299680
Green Pitaya 140342
Green Pop 130383
Green Poplar 311265
Green Poppy 130383
Green Pops 130383
Green Pricklyash 417330
Green Protea 315954
Green Purslane 311890
Green Pyrola 322804
Green Rabbit Bush 150594
Green Rabbitbrush 150379
Green Radish 34590
Green Ramie 56241
Green Redcarweed 288781
Green Ricegrass 276068
Green Rock-cress 30358
Green Rose 336486,337010
Green Rose Buds 9029
Green Sage 35244
Green Sagebrush 36469
Green Sandy Peashrub 72218
Green Sandy Pea-shrub 72218
Green Santolina 346267,346271
Green Sapote 313408
Green Sauce 277648,339887,339897
Green Shrimp Plant 55207
Green Singleseed Bittersweet 80348
Green Smokebush 107300
Green Snob 277648,339887
Green Sorrel 277648,339887
Green Sotol 122910
Green Souce 339887
Green Spider-lily 200932
Green Spike Trisetum 398566
Green Spikerush 143155
Green Spike-rush 143155
Green Spindle-tree 157962
Green Spotleaf-orchis 179709
Green Spotted Orchid 273580
Green Sprangletop 225997
Green Staghorn Sumac 332922
Green Stem Forsythia 167471
Green Sterculia 376114
Green Strawberry 167674
Green Stripe Common Bamboo 47518
Green Striped Bamboo 47518
Green Sulphur Bamboo 297469
Green Tanoak 233416
Green Tansy Mustard 126120,126121
Green Taro 99910,99922
Green Thistle 91864
Green Thorn-apple 123077
Green Thread 389153
Green Threerib Flickingeria 166969
Green Trillium 397631,397633
Green Twayblade 232229
Green Violet 199678,199681
Green Wandering Jew 394021
Green Wattle 1168,1384
Green Wattle Acacia 1168
Green Waves 250614
Green Waxmyrtle 261122
Green Weeping Beech 162413
Green Weeping Willow 343070
Green Wildginger 37704
Green Windhairdaisy 348167
Green Woodland Orchid 302287
Green Wood-orchis 302287
Green Wormwood 36469
Green Yellow Sedge 76701
Greenaepal Glabrousstemmed Buttercup 325896
Green-and-Gold 89959
Green-back Adinandra 8221
Greenback Fortune Plumyew 82512
Green-back Hypoglaucous Deutzia 126979
Green-back Redback Christmasbush 14222
Greenback Schefflera 350717
Greenback Tanoak 233256
Greenback Whitepearl 172090
Green-backed Wintergreen 172090
Greenbamboo 125435,125453
Green-banded Mariposa-lily 67601
Greenbark Ceanothus 79972
Greenbean 409025
Greenbird Flower 112047
Greenbract Galangal 17651
Greenbract Protea 315954
Greenbract Sagebrush 36468
Greenbract Serrate Arisaema 33502
Greenbract Spiralflag 107287
Greenbract Thistle 91872
Greenbranch Sweetleaf 381450
Greenbranch Tamarisk 383516
Greenbranched Sweetleaf 381450
Green-branched Sweetleaf 381450
Greenbriar 366240,366341
Greenbrier 366230,366264,366408,366559
Greenbrier Family 366141
Greenbrierleaf Snakegourd 396273
Greenbrown Tainia 383179
Greencalyx Camellia 69749
Greencalyx Mumeplant 34455
Greencalyx Tea 69749
Green-calyxed Camellia 69749
Green-carpet 192965
Greencenetr Haworthia 186494
Greencolored Chinese Maple 3619
Greencone Norway Spruce 298214
Greencovering Sideritis 362765
Greendot Azalea 331776
Green-dragon 33314
Greene Mountainash 369519
Greene's elegant Gayfeather 228466
Greene's Goldenbush 150326
Greene's Goldfields 222599
Greene's Larkspur 124236
Greene's Mariposa-lily 67588
Greene's Narrow-leaf Fleabane 150681
Greene's Rabbitbrush 90496
Greene's Rush 213140
Greene's Thistle 92049
Greene's Wild Buckwheat 152498
Greeneye Primrose 314354
Green-felt Twigs Rhododendron 332113
Greenfleshy Hawthorn 109597

Greenflower Akebia 13235
Greenflower Alumroot 194416
Greenflower Asiabell 98428
Green-flower Asiatic Currant 334264
Greenflower Azalea 331218
Greenflower Bauhinia 49255
Greenflower Catchfly 363315,364178
Greenflower Chamaesium 85724
Greenflower China Rose 336486
Greenflower Chinese Rose 336486
Greenflower Claw Rhubarb 329403
Greenflower Cleisostoma 94478
Greenflower Cliffbean 66620
Greenflower Closedspurorchis 94478
Greenflower Coelogyne 98701,98722
Greenflower Columbine 30077
Green-flower Currant 334264
Greenflower Cyrtopodium 120719
Greenflower Herminium 192835
Greenflower Houndstongue 118033
Greenflower Japan Aucuba 44927
Greenflower Lily 229840
Green-flower Melaleuca 248125
Greenflower Morina 258842
Greenflower Parnassia 284627
Greenflower Peristylus 291221
Greenflower Pyrola 322804
Greenflower Sarcochilus 346711
Greenflower Sinocrassula 364924
Greenflower Spapweed 205436
Greenflower Sterculia 376114
Greenflower Stonelotus 364924
Greenflower Swertia 380402
Greenflower Tainia 25188
Greenflower Touch-me-not 205436
Greenflower Tulip 400251
Greenflower Twayblade 232339
Greenflower Wartybark-like Euonymus 157950
Green-flower Willowleaf Bottlebrush 67294
Green-flowered Cold Orchis 116930
Green-flowered Helleborine 147204
Green-flowered Milkweed 38171
Green-flowered Pepper Grass 225340
Green-flowered Rabbitear Orchis 116941
Green-flowered Sterculia 376114
Green-flowered Verruca-like Spindle-tree 157950
Green-flowered Wintergreen 322804
Green-fly Orchid 146439
Green-fruit Burreed 370037
Greenfruit Fig 165821
Green-fruit Olga Bay Larch 221927
Green-fruit Variable Fig 165821
Greengage 316333,316391
Green-ginger 36088,36474
Greengroove Bamboo 297207
Green-grooved Bamboo 297207
Greenhair Arisaema 33291
Greenhair Poplar 311479
Greenhair Southstar 33291
Greenhead Sedge 74105,218457
Green-headed Coneflower 339574,339582
Greenheart 88430,263063
Greenhood 320871
Greenhoods 320871
Greenhorn Fritillary 168359
Greening-weed 173082
Greenish Barrel Cactus 163479
Greenish Bluegrass 306146
Greenish Euonymus 157769
Greenish Helictotrichon 190211
Greenish Lily 229765
Greenish Pseudosasa 318353
Greenish Purple Heath 150246
Greenish Ramie 56241
Greenish Sedge 76701
Greenish Wintergreen 322804
Greenish-flowered Pyrola 322804
Greenish-white Sedge 75197
Greenisland Ministalkgrass 71607
Greenland Bluegrass 305551
Greenland Stitchwort 255486
Greenland Wood Rush 238624
Green-leaf Ardisia 31639
Greenleaf Bushclover 226721
Greenleaf Elaeagnus 142241
Greenleaf Elaeocarpus 142407
Greenleaf Goatwheat 44266
Greenleaf Japanese Maple 3300
Greenleaf Knotwood 44266
Green-leaf Machilus 240718
Greenleaf Magnoliavine 351014
Greenleaf Manzanita 31132
Green-leaf Manzanita 31132
Green-leaf Peashrub 72294
Green-leaf Pea-shrub 72294
Greenleaf Piptanthus 300667
Greenleaf Tanoak 233416
Greenleaf Willow 343687
Greenleaf Zeuxine 417702
Green-leaved Ardisia 31639
Green-leaved Elaeagnus 142241
Green-leaved Goatwheat 44266
Green-leaved Helleborine 147168
Green-leaved Hound's-tongue 117968
Green-leaved Machilus 240718
Green-leaved Piptanthus 300667
Green-leaved Pruinose Barberry 52355
Green-leaved Rattlesnake-plantain 179664
Green-leaved Willow 343360,343687
Green-lily 352122
Greenman Litse 233934
Greenman's Cock's-comb 194521
Greenman's Hexalectris 194521
Green-marked Japanese Spindle-tree 157638
Greenmound Juniper 213865
Greenorchid 137545,137688
Greenpeel Caper 71851
Greenpetal Stonecrop 357042
Greenpole Bamboo 47494
Green-pyramid Eastern Arborvitae 390619
Green-ribbed Sedge 73878
Greenridgy Rockjasmine 23223
Greens 59278,109863,224375
Greens Royalpalm 337883
Greensauce Dock 339887
Green-sepal Pink Deutzia 127070
Greensepal Spapweed 204851
Greensepal Touch-me-not 204851
Greenskin Bamboo 47475
Greensnake Vine 291046
Greenspike Falseoat 398566
Greenspike Goosecomb 144527
Greenspike Roegneria 144527
Greenspike Sedge 74032,74111
Green-spiked Wildryegrass 144527
Greenspine Rattan Palm 65812
Greenspire Euonymus 157601
Green-stamen Erythrophleum 154967
Greenstar 307484
Greenstem Coldwaterflower 298935
Greenstem Elatostema 142890
Green-stem Elatostema 142890
Greenstem Forsythia 167471
Green-stem Forsythia 167471
Greenstem Fritillary 168614
Greenstem Mistletoe 411071
Greenstem Stairweed 142890
Greenstem Yellowflower Fritillary 168621
Greenstem Yunnan Sida 362696
Green-stemmed Forsythia 167471
Green-stemmed Joe-pye Weed 158264
Green-stemmed Joe-pye-weed 158264
Greenstripe 18653
Greenstripe Amaranth 18653
Greenstripe Common Bamboo 47518
Green-striped Bamboo 47518
Green-striped Red-culmed Bamboo 196347
Green-styled Shortflower Rhododendron 330238
Green-sulphur Bamboo 297486
Green-thrum Epimedium 146968
Greentip Kaffir Lily 97223
Greentip Kaffir-lily 97223
Greentwig 352430
Green-veined Orchid 273545
Greenvine 204609
Green-violet 199681
Green-wattle Acacia 1168
Greenwax Golden Bamboo 297485

Greenweed 172898,173082,327879
Green-winged Meadow Orchid 273545
Green-winged Orchid 273545
Greenwood 121001,173082
Greenwood's Goldenbush 150328
Greenyellow Crazyweed 279236
Green-yellow Fritillary 168563
Greenyellow Stephania 375907
Green-yellow Stephania 375907
Greenyellow Stonecrop 356905
Green-yellowflower Checkered-like Fritillary 168469
Greet-wort 99297
Grefsheim Spirea 371786,371788
Gregariousflower Mistletoe 411052
Gregariousleaf Hornsage 205575
Gregariousleaf Incarvillea 205575
Gregg Ash 167981
Gregg Catclaw 1264
Gregg Catdaw 1264
Gregg Hawthorn 109742
Gregg Mortonia 259053
Gregg Pine 299957
Gregg's Amaranth 18733
Gregg's Campion 363652
Gregg's Mexican Pink 363652
Gregg's Wild Buckwheat 152114
Greggle 145487
Gregory 262441
Greig Tulip 400171
Greigia 180535
Greins 9923
Greins of Paris 9923
Grenadille 285623,285637
Grenadillo 320291
Grenadine 321764
Grepe Hyacinth 260293
Grepe-hyacinth 260293,260295
Gressitt Screwpine 281039
Grevill 180561
Greville Barberry 51682
Grevillea 180561,180575,180652,180656
Grevillea de Banks 180570
Grewia 180665
Grewialeaf Hibiscus 194907
Grewia-leaved Hibiscus 194907
Grey Alder 16368,16421
Grey Alpine Groundsel 359105
Grey Bamboo 297379
Grey Beard 95431
Grey Beneath Rhododendron 331939
Grey Birch 53586
Grey Bird Vetch 408360
Grey Box 155652
Grey Buloke 79165
Grey Bulwand 36474
Grey Cinquefoil 312459,312676
Grey Club-rash 352278
Grey Copiapoa 103745
Grey Crazyweed 278767
Grey Desert Senna 360422
Grey Felt Thorn 387075
Grey Felt-thorn 387075
Grey Field Speedwell 407292
Grey Field-speedwell 407292
Grey Fragrant Bamboo 87643
Grey Germander 388035,388288
Grey Giantfennel 163585
Grey Goose 312360
Grey Goosefoot 87112
Grey Gum 155711
Grey Hair-grass 106924,106932
Grey Heath 149205
Grey Ironbark 155682
Grey Leaf Pearleverlasting 21700
Grey Lemongrass 117150
Grey Mahogany 155529
Grey Mallee 155659
Grey Manglietia 244441
Grey Mangrove 45746
Grey Marshweed 230283
Grey Mat 44611
Grey Millet 233760
Grey Mint 250463
Grey Mockorange 294471
Grey Mock-orange 294471
Grey Mouse-ear 82742
Grey Myle 233760
Grey Nicker 64971
Grey Owl Juniper 213989
Grey Pea 301055
Grey Pine 300189
Grey Plum 284215
Grey Poplar 311208,311261,311262
Grey Rhododendron 330485
Grey Rockfoil 349144,349184
Grey Sallow 343060,343221
Grey Saussurea 348183
Grey Scabious 350117
Grey Sedge 74004,74366
Grey Skullcap 355486
Grey Statice 230558
Grey Tanoak 233246
Grey Teak 178024
Grey Thistle 92001
Grey Tubergourd 390116
Grey Twig 352430
Grey Violet 409800
Grey Willow 343221
Grey Woodbetony 287096
Grey Woodlotus 244441
Greyawngrass 372398,372428
Greyback Azalea 330871
Greyback Cranesbill 175018
Greyback Forkstyleflower 374482
Greyback Oak 324382
Greyback Tanoak 233254
Grey-beard 95431
Grey-blue Acacia 1249
Greyblue Anplectrum 55147
Grey-blue Bamboo 297485
Grey-blue Cyclobalanopsis 116100
Greyblue Deutzia 126935
Grey-blue Deutzia 126935
Greyblue Ensete 145833
Greyblue Fishvine 283284
Greyblue Jewelvine 283284
Greyblue Leafflower 296589
Grey-blue Leaf-flower 296589
Greyblue Pericampylus 290843
Grey-blue Pericampylus 290843
Grey-blue Poplar 311333
Greyblue Sagebrush 35538
Grey-blue Slugwood 50525
Greyblue Sophora 369149
Greyblue Spicebush 231355
Grey-blue Spice-bush 231355
Grey-blue Staunton-vine 374402
Greyblue Tabacco 266043
Greyblue Underleaf Pearl 296589
Greyblue Wormwood 35538
Grey-blue Yushania 416778
Greyblueleaf Lilyturf 272066
Grey-blue-leaved Caesalpinia 64979
Grey-blue-leaved Spiraea 371887
Greybract Goldenray 229001
Greybranch Aster 41056
Grey-branch Holly 203636
Greybranch Loeseneriella 235219
Grey-branch Loeseneriella 235219
Greybranch Webseedvine 235219
Greybranched Serpentroot 354950
Greycupule Tanoak 233388
Grey-cupuled Tanoak 233388
Grey-dorsal Oak 324382
Greyfruit Dandelion 384664
Greyfruit Tanoak 233388
Greygle 145487,273531
Greygole 145487
Greygreen Corydalis 105572
Greygreen Cremanthodium 110392
Greygreen Gentian 174084
Greygreen Knotweed 308697
Greygreen Lecanorchis 223668
Greygreen Nutantdaisy 110392
Grey-green Yushanbamboo 416757
Greyhair Arnebia 34618
Greyhair Asiabell 98285
Greyhair Baccamelia 91482
Greyhair Bellflower 69934
Greyhair Bluebeard 78008

Greyhair Bluebell 115365
Greyhair Briggsia 60253
Greyhair Chastetree 411222
Greyhair Cipadessa 91482
Greyhair Cliffbean 66621
Greyhair Clubfruitcress 376322
Greyhair Clusteredflower Argyreia 32650
Greyhair Curvedstamencress 238013,238015
Greyhair Dolomiaea 135740
Greyhair Eritrichium 153433
Grey-hair Fernleaf Mountainash 369496
Greyhair Flaccid Aster 40442
Greyhair Glorybower 95984
Grey-hair Honeysuckle 235740
Greyhair Indigo 205782,205798
Greyhair Jadeleaf and Goldenflower 260433
Greyhair Kangding Milkvetch 43135
Greyhair Millettia 66621
Grey-hair Mussaenda 260433
Greyhair Pagodatree 369149
Greyhair Poppy 282525
Greyhair Raspberry 338641
Greyhair Sterigmostemum 376322
Greyhair Trigonotis 397474
Greyhair Youngia 416402
Greyhair Yunnan Skullcap 355365
Grey-haired Chaste-tree 411222
Grey-haired Cipadessa 91482
Grey-haired Eurya 160479
Grey-haired Glorybower 95984
Grey-haired Honeysuckle 235740
Grey-haired Indigo 205798
Grey-haired Japanese Cayratia 79851
Grey-haired Millettia 66621
Grey-haired Raspberry 338641
Greyhairy Arnebia 34618
Greyhairy Eurya 160479
Greyhairy Lily 228603
Greyhairy Michelia 252876
Grey-hairy Newlitse 264079
Greyhairy Speedwell 407050
Greyia 181043
Greyish Anzacwood 310760
Greyish Cowparsnip 192251
Greyish Crazyweed 278790
Greyish Dichrostachys 129136
Greyish Largeleaf Falsenettle 56207
Greyish Primrose 314248
Greyishbark Poplar 311414
Greyishgreen Corydalis 105572
Greyish-green Yushania 416757
Greyish-white Broomsedge 217355
Greyishwhite Nelsonia 263242
Greyish-white Summercypress 217355
Greyleaf Actinidia 6616
Greyleaf Bittersweet 80195
Greyleaf Blueberry 403842
Greyleaf Corydalis 105936
Greyleaf Crazyweed 278790
Grey-leaf Deutzia 126870
Greyleaf Dogwood 380488
Greyleaf Everlasting 21700
Grey-leaf Fewflower Lysionotus 239970
Greyleaf Greenbrier 366244
Grey-leaf Greenbrier 366244
Grey-leaf Henry Mockorange 294466
Greyleaf Hippolytia 196710
Greyleaf Maesa 241753
Greyleaf Maple 3457
Greyleaf Milkvetch 42298
Greyleaf Mountainash 369478
Greyleaf Onion 12992
Greyleaf Ophiorrhiza 272186
Greyleaf Opithandra 272592
Greyleaf Pearweed 239660
Grey-leaf Peashrub 72286
Grey-leaf Pea-shrub 72286
Greyleaf Pimpinella 299422
Greyleaf Pine 300189
Greyleaf Pyrethrum 322732
Greyleaf Rockfoil 349389
Greyleaf Smilax 366244
Greyleaf Trigonotis 397396
Greyleaf Willow 344140
Greyleafback Oak 324382
Grey-leaved Actinidia 6616
Grey-leaved Blueberry 403842
Grey-leaved Cistus 93124
Grey-leaved Dogwood 380488
Grey-leaved Maesa 241753
Grey-leaved Manglietia 244442
Grey-leaved Maple 3457
Grey-leaved Mountain-ash 369478
Grey-leaved Willow 344140
Greymill 233760
Greynerve Sedge 76547
Greypole Bamboo 47413
Greyshoot Serpentroot 354950
Greytwig 352430
Grey-twig 352430
Grey-white Blueberry 403841
Greywhite Lallemantia 220283
Grey-white Raspberry 339360
Greywhite Rhododendron 330757,332157
Grey-white Rhododendron 330757,332157
Greywhite Saltbush 44333
Greywhitehair Melhania 248861
Greywhitehair Plectranthus 324781
Greywhitehair Raspberry 339360
Greywhitehair Toadflex 231008
Gribble 243711
Grierson's Rhododendron 330793
Griffith Agapetes 10312
Griffith Anemone 23834
Griffith Antlevine 88895
Griffith Arisaema 33338
Griffith Ash 167982
Griffith Barberry 51683
Griffith Bellcalyxwort 231261
Griffith Bittercress 72790
Griffith Bluesnow 83643
Griffith Calanthe 65972
Griffith Ceratostigma 83643
Griffith Chonemorpha 88895
Griffith Cinquefoil 312652
Griffith Cliffbean 254713
Griffith Cockclawthorn 278319
Griffith Combretum 100500
Griffith Condorvine 245814
Griffith Currant 334007
Griffith Dogwood 380478
Griffith Dutchman's-pipe 34200
Griffith Dysophylla 139562
Griffith Elaeagnus 142015
Griffith Elsholtzia 144035
Griffith Euphorbia 159002
Griffith Fighazel 380593
Griffith Fleshseed Tree 346982
Griffith Fleshspike Tree 346982
Griffith Glabrous Artocarpus 36938
Griffith Glorybower 96111
Griffith Goldsaxifrage 90371
Griffith Goldwaist 90371
Griffith Goniothalamus 179412
Griffith Hedgehyssop 180306
Griffith Holly 203864
Griffith Hydrobryum 200203
Griffith Iris 208602
Griffith Lacquertree 393457
Griffith Lacquer-tree 393457
Griffith Leadwort 83643
Griffith Leucothoe 228175
Griffith Loxostigma 238038
Griffith Mahonia 242597
Griffith Marsdenia 245814
Griffith Marshweed 230302
Griffith Milkgreens 245814
Griffith Millettia 254713
Griffith Neonauclea 264219
Griffith Nightshade 367190
Griffith Oak 323979
Griffith Pine 300297
Griffith Plumyew 82541
Griffith Primrose 314446
Griffith Randia 278319
Griffith Rhododendron 330794
Griffith Sandwort 31936
Griffith Saraca 346503
Griffith Saurauia 347961,347962
Griffith Sedge 74713
Griffith Sophora 369075

Griffith Southstar 33338
Griffith Spapweed 204995
Griffith Spinestemon 306980
Griffith Streptocaulon 377856
Griffith Touch-me-not 204995
Griffith Vanilla 405028
Griffith Watermoss 200203
Griffith Waxpetal 106648
Griffith Whitepearl 172080
Griffith Windflower 23834
Griffith Windmill 100500
Griffith Wintergreen 172080
Griffith Wood Sorrel 277878
Griffith Woodsorrel 277878
Griffith's Butterflybush 62115
Griffith's Pondweed 312134
Griffith's Waxplant 198847
Griffiths' Saltbush 44676
Grig 67456,150131
Griggle 145487,273531
Griglans 67456
Griglings 67456
Griglum 67456
Grijs Elaeagnus 142018
Grijs Machilus 240595
Grim the Collier 195478
Grimson Star Glory 255398
Grim-the-collier 195478
Grindstone Apple 243711
Gringel 141347
Grinning Swallow 360328
Grinsel 360328
Grip-grass 170193
Gris Palm 46374
Griscom's Arnica 34720
Grisebach False Boxwood 182201
Grisebach Saxifrage 349425
Grisons Sallow 343460
Gristleleaf Rockfoil 349158
Gristleleaf Saxifrage 349158
Gristly Orostachys 275356
Grizzle 334250
Grizzly Bear Cactus 273016
Grizzly Bear Pricklypear 273016
Groat Bulrush 352206
Groats 45566
Groby Pleurothallis 304914
Grodtmann Barberry Barberry 51686
Groff Eurya 160480
Groff Sweetleaf 381232
Gromaly 233760
Gromvel 233760
Gromwell 233692,233731,233760,250894
Gromwell Corn Cockle 233700
Gromwellleaf Mouseear 82911
Grondbox Milkwort 307983
Gronovius Dodder 115040
Groot Kersbos 346648
Groove Bamboo 297204
Grooved Agrimony 11598
Grooved Elaeocarpus 142350
Grooved Flax 231984
Grooved Oligostachyum 270451
Grooved Orange 93820
Grooved Waxplant 198852
Grooved Yellow Flax 231984
Grosart 334250
Grosel 334250
Groser 334250
Grosert 334250
Grosier 334250
Gross Brome 60739
Gross Bulrush 352187
Gross Skullcap 355461
Grossberry 334250
Grossedentate Ampelopsis 20360
Grossedentate Clematis 94964
Grossek Bellflower 70073
Grosser Maple 3004
Grosseserrate Carpesium 77189
Grosset 334250
Grossheim Iris 208604
Grossheim's Iris 208604
Grosstooth Rabdosia 209692
Grosstooth Skullcap 355462
Grosstooth Willowleaf Spiraea 372077
Grosvenor Momordica 365332
Groud Seeking Rhododendron 330356
Groudnut 29295
Ground Ash 8826,24483
Ground Bamboo 304084
Ground Bean 20560,241427
Ground Birch 53605
Ground Box 307983
Ground Cedar 213702
Ground Cherry 83202,297640,297641,297658,297664,297665,297669,297678,297703
Ground Chilli Thyme 391347
Ground Clematis 95277
Ground Cypress 346250
Ground Daisy 393417
Ground Elder 8810,8826,24483,345594
Ground Eller 8826
Ground Enell 350425
Ground Flax 231884
Ground Furze 271563
Ground Glutton 360328
Ground Hemlock 385344
Ground Holly 87480,172135
Ground Honeysuckle 237539
Ground Iris 208701
Ground Ivvin 176824
Ground Ivy 13073,68713,102933,176812,176824,264946,409339
Ground Lemon 306636
Ground Lilac 81768
Ground Lily 102933,397540,397556
Ground Mat Cholla 181457
Ground Morning Glory 103234
Ground Mulberry 338243
Ground Needle 153869
Ground Nut 29298,30498,63475,102727,177689
Ground Oak 388042
Ground Orchid 55559,55575
Ground Pea 30498
Ground Phlox 295324
Ground Pine 13073
Ground Pink 295324
Ground Plum 42156,42231
Ground Protea 315930
Ground Rattan-cane 329176
Ground Soap 346434
Ground Tea 172135
Ground Thistle 91713
Ground Virginsbower 95277
Ground Willow 308739
Groundcherry 297640,297643
Ground-cherry 83202,297640,297650,297686,297711
Groundcherrylike Bladderwort 278366
Ground-elder 8810,8826
Groundfig Spurge 158632,159634
Ground-hale 233760,407256
Ground-heele 407256
Ground-hemlock 385344
Groundie Swallow 360328
Groundie-swallow 360328
Ground-ivy 176812,176824,176835
Groundnut 29298,30498,280817
Groundnut Pea Vine 222851
Groundnut Peavine 222851
Ground-pine 13073
Ground-plum 42231
Ground-plum Milk-Vetch 42231
Groundsel 46255,279956,358159,359980,360328
Groundsel Bush 46227
Groundsel Heath 360172
Groundsel Hippolytia 196708
Groundsel Tree 46227
Groundsel-like Aster 41228
Groundsel-tree 46210,46227
Ground-sill 176824
Groundstar 22027
Groundswell 360328
Groundwill 360328
Group Sandwort 32145
Groupflower 263982,263983
Groupflower Azalea 330094

Groupflower Newlitse 264054
Groupflower Sweetleaf 381221
Groupflower Syzygium 382516
Grouse Whortleberry 403987
Grouseberry 403987
Grove Hawthorn 109809
Grove Marjoram 274237
Grove Meadow-grass 305314
Grove Sandwort 32004,256444
Grove Wood Rush 238663
Grove Wood-rush 238663
Grove-nut 102727
Groves 224375
Grovy Barberry 51957
Growing on Tree Rhododendron 330552
Grozen 224375
Grozer 334250
Grozet 334250
Grozzle 334250
Grubov Crazyweed 278864
Grubov Kudrjaschevia 218277
Grubov's Lily 228582
Gru-gru 6136,6137
Grugru Acrocomia 6129
Gru-gru Nut 6136
Grugru Palm 6128,6137
Gru-gru Palm 6128,6136
Gruie 277638
Gruisplakkie 108960
Grum Gentian 173492
Grumbell 233760
Grumewood 211021
Grumichama 156148
Grummel 233760
Grumsel 384714
Grundavy 176824
Grundswaith 359158
Grundswathe 359158
Grundy Swallow 360328
Grundy-swallow 360328
Grunnishule 360328
Grunnistule 360328
Grunnut 102727
Grunsel 359158
Grunsil 359158,360328
Grunty Ash 8826
Gruzel 334250
Gruzzle 334250
Guaangxi Spiraea 371988
Guaba 206952
Guacalla 105437
Guacimilla 394652
Guaco 254434
Guadalupe Cypress 114702
Guadalupe Erythea 59099
Guadalupe Lazy Daisy 84936
Guadalupe Palm 59099
Guadalupe Rabbitbrush 236502
Guadalupe Water-nymph 262044,262048
Guadeloupe Blackbead 99155
Guadeloupe Erythca 154574
Guadeloupe Naias 262044
Guado 96643
Guaiac 63409
Guaiac Wood 181505
Guaiacum 181505
Guaiacum Wood 181505
Guajacum Tree 181505
Guajacumwood 181496
Guaje 227422
Guajillo 1080
Guajilote 284498
Gualapo Cacalia 283836
Guama Venezolano 206940,206941
Guanabana 25866,25868
Guanabana Custard-apple 25868
Guanabana Soursop 25868
Guanacaste 145994
Guanacaste Earpodtree 145994
Guandishan Chinese Goat Willow 344119
Guang Pseudosasa 318317
Guangdong Abrus 747
Guangdong Actinodaphne 6784
Guangdong Ampelopsis 20327
Guangdong Ardisia 31491
Guangdong Atalantia 43730
Guangdong Azalea 331031
Guangdong Beautyberry 66833
Guangdong Bentgrass 12058
Guangdong Birthwort 34176
Guangdong Blueberry 403844
Guangdong Browndot Mountainash 369353
Guangdong Bushclover 226796
Guangdong Caesalpinia 64983
Guangdong Camellia 69246,69352
Guangdong Chilosehista 87457
Guangdong Cinnamon 91358
Guangdong Cleisostoma 94468
Guangdong Cliffbean 66631
Guangdong Closedspurorchis 94468
Guangdong Craibiodendron 108583
Guangdong Crapemyrtle 219927
Guangdong Cryptocarya 113457
Guangdong Dehaasia 123608
Guangdong Dendrobium 125417
Guangdong Dishspurorchis 171857
Guangdong Euonymus 157656
Guangdong Galangal 17715
Guangdong Galeobdolon 170024
Guangdong Gastrochilus 171857
Guangdong Gleadovia 176794
Guangdong Glorybower 96167
Guangdong Goldleaf 108583
Guangdong Goldsaxifrage 90382
Guangdong Goldwaist 90382
Guangdong Grewia 180835
Guangdong Gugertree 350923
Guangdong Gymnema 182374
Guangdong Habenaria 183718
Guangdong Hairyvein Maple 3484
Guangdong Hartia 376490
Guangdong Helicia 189937
Guangdong Hemiboea 191385
Guangdong Hilliella 196277
Guangdong Holly 203953
Guangdong Hydrangea 199936
Guangdong Indocalamus 206786
Guangdong Jadeleaf and Goldenflower 260439
Guangdong Juncellus 212783
Guangdong Lasianthus 222123
Guangdong Late Juncellus 212783
Guangdong Leaf-flower 296596
Guangdong Lilyturf 272129
Guangdong Liparis 232217
Guangdong Litse 233966
Guangdong Loosestrife 239702
Guangdong Machilus 240610
Guangdong Mananthes 244291
Guangdong Manglietia 244451
Guangdong Meiogyne 248031
Guangdong Metalleaf 296380
Guangdong Moneygrass 126623
Guangdong Mouretia 259411
Guangdong Mussaenda 260439
Guangdong New Eargrass 262968
Guangdong Newlitse 264062
Guangdong Paris 284368
Guangdong Passionflower 285659
Guangdong Pearlsedge 354126
Guangdong Pigeonwings 97190
Guangdong Pine 300009
Guangdong Pricklyash 417170
Guangdong Primrose 314549
Guangdong Purplepearl 66833
Guangdong Rattlesnake Plantain 179637
Guangdong Rehdertree 327416
Guangdong Rehder-tree 327416
Guangdong Rhododendron 331031
Guangdong Rockyellowtree 415157
Guangdong Rose 336667
Guangdong Roughleaf 222123
Guangdong Sasa 347218
Guangdong Scolopia 354627
Guangdong Sedge 73603
Guangdong Sonerila 368874
Guangdong Spicebush 231377
Guangdong Spice-bush 231377
Guangdong Spiradiclis 371759
Guangdong Spotleaf-orchis 179637
Guangdong Stonebean-orchis 62832

Guangdong Syzygium 382591
Guangdong Tallculm Cypressgrass 118847
Guangdong Tallculm Galingale 118847
Guangdong Tea 69246,69352
Guangdong Ternstroemia 386712
Guangdong Thickshellcassia 113457
Guangdong Twayblade 232217
Guangdong Vanda 404642
Guangdong Waterstarwort 67384
Guangdong Weasel-snout 170024
Guangdong Wendlandia 413788
Guangdong Willow 343685
Guangdong Woodlotus 244451
Guangdong Wrightia 414807
Guangdong Xanthophytopsis 415157
Guangdong-Guangxi Actinidia 6654
Guangdong-Guangxi Adinandra 8229
Guangdong-Guangxi Guihaia 181805
Guangdong-Guangxi Holly 204104
Guangdong-Guangxi Kiwifruit 6654
Guangfeng Chirita 87919
Guangfu Azalea 330592
Guangfu Rhododendron 330592
Guangnan Azalea 330797
Guangnan Camellia 69243
Guangnan Holly 203866
Guangnan Homalium 197709
Guangnan Maple 3068
Guangnan Mayten 246849
Guangnan Rhododendron 330797
Guangnan Tanoak 233260
Guangnan Tea 69243
Guangtoushan Bittercress 72742
Guangtoushan Sedge 75027
Guanguangtree 399855,399856
Guangxi Acrocarpus 5984
Guangxi Actinidia 6669
Guangxi Aganosma 10239
Guangxi Agapetes 10313
Guangxi Ailanthus 12579
Guangxi Alangium 13374
Guangxi Allocheilos 15958
Guangxi Allomorphia 16129
Guangxi Allostigma 16213
Guangxi Alstonia 18048
Guangxi Amomum 19871
Guangxi Archiphysalis 297625
Guangxi Arrowbamboo 162668
Guangxi Aspidistra 39573
Guangxi Aspidopterys 39669
Guangxi Awlquitch 131014
Guangxi Azalea 331026
Guangxi Bamboo 47303,297333
Guangxi Bambusa 47303
Guangxi Basketvine 9426
Guangxi Bedstraw 170356
Guangxi Begonia 49714,49912
Guangxi Bell Tree 324997
Guangxi Belltree 324997
Guangxi Bigleafvine 392225
Guangxi Bigumbrella 59221
Guangxi Birthwort 34238
Guangxi Bitter Bamboo 304051
Guangxi Bitterbamboo 304051
Guangxi Blueberry 404001
Guangxi Brandisia 59128
Guangxi Brassaiopsis 59221
Guangxi Breynia 60066
Guangxi Briggsia 60288
Guangxi Buckthorn 328753
Guangxi Calophanoides 67825
Guangxi Camellia 69244
Guangxi Chastetree 411310
Guangxi Chaste-tree 411310
Guangxi Chinese Groundsel 365052
Guangxi Chirita 317538
Guangxi Clearweed 298973
Guangxi Coffeetree 182529
Guangxi Cogflower 154173,382766
Guangxi Coldwaterflower 298973
Guangxi Common Reevesia 327372
Guangxi Conehead 320126
Guangxi Craigia 108589
Guangxi Cryptocoryne 113539
Guangxi Currant 334039
Guangxi Cyclobalanopsis 116126
Guangxi Damnacanthus 122036
Guangxi Dendrobium 125170
Guangxi Dendropanax 125623
Guangxi Denseflower Holly 203645
Guangxi Dimeria 131014
Guangxi Dolichopetalum 135390
Guangxi Dracaena 135390
Guangxi Dutchman's-pipe 34238
Guangxi Dutchmanspipe 34238
Guangxi Dysosma 139621
Guangxi Elaeocarpus 142347
Guangxi Eriolaena 152690
Guangxi Ervatamia 154173,382766
Guangxi Erythropsis 155003
Guangxi Eurya 160574
Guangxi Falleneaves 351148
Guangxi Falsechirita 317538
Guangxi Fernandoa 163402
Guangxi Fevervine 280091
Guangxi Fewtooth Raspberry 339006
Guangxi Fieldjasmine 199457
Guangxi Fig 165079
Guangxi Firerope 152690
Guangxi Fissistigma 166667
Guangxi Galangal 17713
Guangxi Gapfruityam 351381
Guangxi Garcinia 171123
Guangxi Gentian 173569
Guangxi Ginger 417985
Guangxi Gingerlily 187455
Guangxi Glorybower 96029
Guangxi Gordonia 179758
Guangxi Gordontea 179758
Guangxi Greenbrier 366407
Guangxi Gueldenstaedtia 181661
Guangxi Gynostemma 183008
Guangxi Helixanthera 190836
Guangxi Hemipilia 191610
Guangxi Heterostemma 194167
Guangxi Hogfennel 292873
Guangxi Holly 204355
Guangxi Homalium 197690
Guangxi Hopea 198155
Guangxi Huodendron 199457
Guangxi Hydrangea 199931
Guangxi Irisorchis 267964
Guangxi Ironweed 406222
Guangxi Island Ringvine 116021
Guangxi Jadeleaf and Goldenflower 260438
Guangxi Jasmine 211852
Guangxi Jasminorange 260169
Guangxi Jasmin-orange 260169
Guangxi Jurinea 190836
Guangxi Kadsura 214964
Guangxi Kiwifruit 6669
Guangxi Linociera 87708,231747
Guangxi Liolive 87708,231747
Guangxi Litse 233965
Guangxi Lobelia 234405
Guangxi Local Milkvetch 266464
Guangxi Loosestrife 239547
Guangxi Lovegrass 147703
Guangxi Lycoris 239260
Guangxi Lysionotus 239953
Guangxi Mallotus 243346
Guangxi Mayten 246850
Guangxi Mosquitotrap 117551
Guangxi Mussaenda 260438
Guangxi Nanmu 295372
Guangxi Newlitse 264061
Guangxi Oak 116126
Guangxi Oilfruitcamphor 381795
Guangxi Olive 270126
Guangxi Ophiorrhiza 272233
Guangxi Orophea 275320
Guangxi Padritree 376268
Guangxi Parashorea 362197
Guangxi Parsonsia 285060
Guangxi Pavieasia 286576
Guangxi Peristrophe 291155
Guangxi Phoebe 295372
Guangxi Photinia 295715
Guangxi Pittosporum 301309
Guangxi Platanthera 302380
Guangxi Pricklyash 417249

Guangxi Pseudosasa 318297
Guangxi Pubescent Holly 204187
Guangxi Purplehammer 313575
Guangxi Qinggang 116126
Guangxi Rapanea 261626
Guangxi Raphistemma 326800
Guangxi Raspberry 338687
Guangxi Rattan Palm 65702
Guangxi Rattanpalm 65702
Guangxi Redbud 83779
Guangxi Rhododendron 331026
Guangxi Ricebag 181661
Guangxi Rockfoil 349527
Guangxi Rockvine 387794
Guangxi Sasa 347219
Guangxi Schefflera 350728
Guangxi Schizocapsa 351381
Guangxi Seatung 301309
Guangxi Sedge 75026
Guangxi Single Bamboo 47303
Guangxi Sinosenecio 365052
Guangxi Slatepentree 322592
Guangxi Smilax 366407
Guangxi Soappod 182529
Guangxi Sotvine 194167
Guangxi Spicebush 231366
Guangxi Spice-bush 231366
Guangxi Stephania 375880
Guangxi Sterculia 376116
Guangxi Stonegarlic 239260
Guangxi Storax 379411
Guangxi Strophioblachia 378475
Guangxi Suoluo 327372
Guangxi Swallowwort 117551
Guangxi Sweetleaf 381272
Guangxi Sweetspire 210391
Guangxi Syndiclis 381795
Guangxi Syzygium 382549
Guangxi Tailleaftree 402624
Guangxi Tarenna 384972
Guangxi Tea 69244
Guangxi Thickfilmtree 163402
Guangxi Tigerthorn 122036
Guangxi Tinomiscium 392225
Guangxi Tinospora 392256
Guangxi Tinospore 392256
Guangxi Tonkin Maple 3696
Guangxi Treerenshen 125623
Guangxi Triratna Tree 397366
Guangxi Turmeric 114868
Guangxi Twinballgrass 209065
Guangxi Urophyllum 402624
Guangxi Vatica 405168
Guangxi Vineclethra 95500
Guangxi Wampee 260169
Guangxi Waxplant 198847,198895
Guangxi Wendlandia 413770
Guangxi Wildtung 243322,243346
Guangxi Woodnettle 273911
Guangxi Zinger 417985
Guangxi-Guizhou Lysionotus 239927
Guangzhou Abrus 747
Guangzhou Caper 71707
Guangzhou Darkbrown Fimbristylis 166331
Guangzhou Darkbrown Fluttergrass 166331
Guangzhou Dropseed 372696
Guangzhou Hedyotis 187529
Guangzhou Ophiorrhiza 272187
Guangzhou Salomonia 344347
Guangzhou Speranskia 370512
Guangzhou Yellowcress 336187
Guano 268344
Guano Palm 97885,97886
Guano-weed 370522
Guanshan Bedstraw 170453
Guanshan Listera 232925
Guanshan Orchis 273492
Guanxi Maple 3696
Guanxian Leek 15339
Guanxian Maple 3008
Guanxian Mountainash 369406
Guanxian Mountain-ash 369406
Guanxian Muping Corydalis 105886
Guanxian Onion 15339
Guapiles Balsa 268342
Guapilla 187132
Guar 115315,115323
Guara 114574
Guaramaco 61159
Guaran Colorado 385477
Guarana 285930
Guarana Paullinia 285930
Guarani 285930
Guard-robe 337180
Guarea 181544
Guarri 155919
Guashacks 31142
Guashicks 31142
Guatambu Moroti 46955
Guatemala Fir 386,458
Guatemala Grass 398140
Guatemala-grass 398140
Guatemalan Bird of Paradise 190067
Guatemalan Fir 386
Guatemalan Yellow Wood 306437
Guatemale Rhubarb 212202
Guava 163117,318733,318742
Guava Berry 382601
Guava Tree 318742
Guaxima 402245
Guayacan 181496
Guayal 405224
Guayamochil 301134
Guaymochil 301128
Guayule 285076
Guayule Parthenium 285076
Guayule Rubber 285076
Guazuma 181616
Gubgub 409086,409096
Gucheng Sedge 75018
Guckoo-flower 145487
Guckoos 145487
Guebrianth Willow 343461
Gueldenstaedtia 181626,181649
Gueldenstaedtia-like Crazyweed 278866
Guelder Rose 407989
Guelder-rose 407989
Guere Palm 43372
Guernsey Barberry 52093
Guernsey Centaury 161525,161526
Guernsey Chickweed 307755
Guernsey Coltsfoot 292353
Guernsey Elm 401632
Guernsey Fleabane 103617
Guernsey Flower 265294
Guernsey Lily 265254,265294
Guernsey Star of Bethlehem 15509
Guernsey Star-of-Bethlehem 15509
Guest Tree 216642
Guger Tree 350904,350945
Gugertree 350904,350945
Guger-tree 350904
Guggul 101591
Guhtzun Mrs. E. H. Wilson's Barberry 52373
Guiana Arrowhead 342347
Guiana Crabwood 72640
Guiana-chestnut 279381
Guibei Swertia 380220
Guichow Manyflowered May-apple 139621
Guide Kengyilia 215917
Guiding Azalea 330740
Guiding Hornbeam 77316
Guidong Rhododendron 331747
Guihai Azalea 330798
Guihaia 181802
Guihaiothamnus 181806
Guihaitree 181806,181807
Guilache Barberry 51688
Guild 52322,66445,68189,89704
Guild Tree 52322
Guild Weed 89704
Guildford Grass 336031
Guile 89704
Guilfoyle Polyscias 310204
Guilielm Skullcap 355464
Guilin Chirita 87876
Guilin Chiritopsis 88003
Guilin Elatostema 142672
Guilin Kiwifruit 6624
Guilin Maple 3070
Guilin Parnassia 284547

Guilin Sapium 346376
Guilin Stairweed 142672
Guiling Crapemyrtle 219931
Guiling Crape-myrtle 219931
Guiling Sedge 74128
Guillaumin Atalantia 43726
Guills 89704
Guilty Cup 325518
Guilty-cup 325518
Guimauve 18173
Guinan Crazyweed 278867
Guinan Fissistigma 166662
Guinan Galangal 17684
Guinan Milkvetch 42452
Guinan Stephania 375879
Guinan Syzygium 382574
Guinan Willow 343555
Guind 83155
Guinea Abutilon 905
Guinea Chestnut 279381
Guinea Corn 369590,369600,369720
Guinea Cubeb 300397
Guinea Cubebs 300397,300401
Guinea Erythrophleum 154973
Guinea Eulophia 156744
Guinea Flower 168476,194661,194668,
216085
Guinea Gold Vine 194661,194668
Guinea Grains 9923
Guinea Grass 281887,369652,402531
Guinea Hen 168476
Guinea Hen Weed 292490
Guinea Nutmeg 261456
Guinea Pea 765
Guinea Peach 346674
Guinea Pepper 9923,72070,300397,415762
Guinea Pepper-tree 415762
Guinea Plant 216085
Guinea Plum 284215
Guinea Sorrel 195196
Guinea Syzygium 382567
Guinea Yam 131516,131814
Guineaflower 192768,192786
Guinea-flower 168476,194661
Guineafowl Asiabell 98372
Guineagrass 281887,402531
Guinea-grass 281887,369652
Guinea-hen-flower 168476
Guinea-hen-weed 292490
Guinean Syzygium 382567
Guinean Trema 394642
Guinea-pods 72070
Guinea-wing Begonia 49603
Guiping Giantarum 20063
Guiqian Lysionotus 239927
Guirado's Goldenrod 368144
Guisaro 318744
Guishan Begonia 49892
Guitar Plant 235413
Guixi Arisaema 33340
Guiyang Betony 373274
Guiyang Championella 85946
Guiyang Greenbrier 366356
Guiyang Hornbeam 77318
Guiyang Parnassia 284606
Guiyang Persimmon 132147
Guiyang Pyrola 322813
Guizhou Aalyxia 18509
Guizhou Actinodaphne 6785
Guizhou Aganosma 10234
Guizhou Alyxia 18506
Guizhou Ampelocalamus 20196
Guizhou Ancylostemon 22170
Guizhou Azalea 330800
Guizhou Bamboo 297289
Guizhou Barberry 51690
Guizhou Bauhinia 49155
Guizhou Beautyberry 66799
Guizhou Begonia 49966
Guizhou Bigflower Azalea 331183
Guizhou Bridelia 60208
Guizhou Bristlegrass 361777
Guizhou Buckthorn 328695
Guizhou Calanthe 66107
Guizhou Calophanoides 67824
Guizhou Camellia 69014,69248
Guizhou Carpesium 77160
Guizhou Childvine 400961
Guizhou Chinese Groundsel 365053
Guizhou Clematis 95063
Guizhou Cliffbean 254741
Guizhou Cogflower 154176,382766
Guizhou Corydalis 105627,106252
Guizhou Cremastra 110510
Guizhou Currant 333973
Guizhou Cycas 115830
Guizhou Densevein 261326,261327
Guizhou Dogtongueweed 385916
Guizhou Dragonbamboo 125515
Guizhou Drepanostachyum 137847
Guizhou Dysosma 139621
Guizhou Elaeagnus 142021
Guizhou Ervatamia 154176,382766
Guizhou Eurya 160525
Guizhou Evergreenchinkapin 78969
Guizhou Evodia 161375
Guizhou Fieldorange 249676
Guizhou Fig 165082,165796
Guizhou Fissistigma 166694
Guizhou Flemingia 166871
Guizhou Fritillary 168491
Guizhou Fuirena 168841
Guizhou Gentian 173421
Guizhou Goldenray 229091
Guizhou Hanceola 185245
Guizhou Handbamboo 20196
Guizhou Hartia 376435
Guizhou Hazelnut 106748
Guizhou Hickory 77904
Guizhou Holly 203868
Guizhou Honeysuckle 236014
Guizhou Hornbeam 77315
Guizhou Leafflower 296497
Guizhou Linden 391751
Guizhou Liparis 232155
Guizhou Lobelia 234377
Guizhou Loosestrife 239669
Guizhou Maple 3009
Guizhou Mayten 182692
Guizhou Mazus 247000
Guizhou Meliosma 249397
Guizhou Melodinus 249676
Guizhou Milkwort 308028
Guizhou Millettia 254741
Guizhou Mosquitotrap 117727
Guizhou Mucuna 259571
Guizhou Myrioneuron 261326
Guizhou Neomartinella 161149
Guizhou Notoseris 267172
Guizhou Oak 116055
Guizhou Oneleaf Munronia 259900
Guizhou Oreocharis 273845
Guizhou Ornithoboea 274469
Guizhou Ottelia 277412
Guizhou Pecan 77904
Guizhou Petrocosmea 292550
Guizhou Phaius 293505
Guizhou Pholidota 295535
Guizhou Photinia 295643
Guizhou Physaliastrum 297632
Guizhou Pittosporum 301310
Guizhou Pricklyash 417226
Guizhou Prickly-ash 417226
Guizhou Primrose 314550
Guizhou Purplepearl 66799
Guizhou Pyrola 322851
Guizhou Qinggang 116055
Guizhou Rehdertree 327417
Guizhou Rehder-tree 327417
Guizhou Rhododendron 330800,331183
Guizhou Rockjasmine 23208
Guizhou Rose 336671
Guizhou Rosewood 121829
Guizhou Sage 344945
Guizhou Scurrula 236755
Guizhou Seatung 301310
Guizhou Sinosenecio 365053
Guizhou Slatepentree 400701
Guizhou Slugwood 50542
Guizhou Snakegourd 396188
Guizhou Spapweed 204997

Guizhou Spindle-tree 157658
Guizhou Spiraea 371989
Guizhou St. John's Wort 201974
Guizhou St. John's-wort 201974
Guizhou Stonebutterfly 292550
Guizhou Swallowwort 117727
Guizhou Sweetspire 210413
Guizhou Swertia 380245
Guizhou Synotis 381938
Guizhou Tailanther 381938
Guizhou Tengia 385838
Guizhou Thorowax 63697
Guizhou Touch-me-not 204997
Guizhou Tutcheria 400701
Guizhou Twayblade 232155
Guizhou Underleaf Pearl 296497
Guizhou Ungeargrass 36644
Guizhou Uvaria 403503
Guizhou Vineclethra 95502
Guizhou Violetcress 161149
Guizhou Wendlandia 413777
Guizhou Willow 343581
Guizhou Windflower 23977
Guizhou-Guangxi Machilus 240569
Gulashan Willow 343583
Gulch Grape 411967
Gules 66445,68189,89704
Gules-gowan 89704
Gulf Barnyardgrass 140415
Gulf Coast Camphor-daisy 327155
Gulf Coast Gayfeather 228435
Gulf Cockspur 140415
Gulf Cockspur Grass 140415
Gulf Cord Grass 370179
Gulf Cypress 385262
Gulf Leaf-flower 296574
Gulf Muhly 259642
Gulf of St. Lawrence Aster 380924
Gulf of St. Lawrence Dandelion 384625
Gulf Rabbit-tobacco 127935
Gulf Stream Nandina 262194
Gulf triandra 228692
Gulfcost Sageretia 342181
Gulin Ginger 417986
Gulin Hemsleia 191958
Gulin Sandwort 32028
Gulin Snowgall 191958
Guling Ampelopsis 20356
Guling Hooktea 52435
Guling Mockorange 294542
Guling Raspberry 338683
Guling Snakegrape 20356
Guling Stringbush 414245
Guling Supplejack 52435
Guling Vetch 408449
Gulinqin Begonia 49894
Gull 89704
Gull-grass 170193
Gulls 343151
Gully Azalea 331906
Gully Gum 155745
Gully Peppermint 155745
Gullygrass 87352
Gully-root 292490
Gulped Stonecrop 356971
Gulseck-girse 388275
Gulty Cup 325518
Gulty-cup 325518
Gum 24603,106788,155468
Gum Acacia 1572
Gum Arabic 1572
Gum Arabic Tree 1427
Gum Benjamin 379311
Gum Bumelia 63422
Gum Cistus 93161
Gum Copal 103696
Gum Dragon 41883
Gum Elemi 64085
Gum Euphorbia 159710
Gum Plant 181092,181161,381035
Gum Rockrose 93161
Gum Sarcoculla 41885
Gum Senegal 1572
Gum Succory 88759,88776
Gum Thistle 159710,271666
Gum Tragacanth 41883
Gum Tree 79108,155468
Gum Vine 220814
Gum Weed 181149,181174
Gumarabic Acacia 1572
Gum-arabic Acacia 1572
Gum-arabic Tree 1427,1572,1588
Gumari 178025
Gumball Sweetgum 232570
Gum-barked Coolabah 155611
Gum-bearing Cactus 244091
Gumbo 219
Gumbo Limbo Tree 64085
Gumbolimbo 64085
Gumbo-limbo 64085
Gum-elastic 362953
Gumhar 178025
Gumhead 182628,182629
Gumi 142088
Gum-lac Tree 351968
Gummy Love Grass 147611
Gumplant 181092
Gum-plant 181092,181161,181174,181178
Gum-top Ironbark 155556
Gum-topped Box 155658
Gum-tree 1024,79108,155468,386942
Gumvine 220814
Gumweed 181092,181174,241534
Gum-weed 181174,181178
Gumweed Aster 414976
Gundabluey 1690
Gungurru 155511
Gunn's Beech 266872
Gunn's Eucalyptus 155602
Gunnera 181946,181951,181958
Gunny 104056
Gunpowder Plant 298974
Gunpowder Tree 394656
Guo Spikesedge 143186
Guodong Raspberry 338476
Guoluo Azalea 330782
Guoluo Rhododendron 330782
Guoxun Brome 60981
Guoxun Small Reed 127225
Guoxun Smallreed 127225
Gurdner's Eyes 363370
Gurjun 133552,133563,133588
Gurjun Balsam 133552,133553
Gurjun Family 133551
Gurjun Oil 133552
Gurjun Oil Tree 133552
Gurjun Oil Tree Family 133551
Gurjun-oil Tree 133587,133588
Gurjunoiltree 133552
Gurjun-oiltree 133552
Gurjunoiltree Family 133551
Gurjun-oiltree Family 133551
Gurney's Claret-cup 140234,140334
Gushan Waxplant 198829
Gushanlong 30897
Gushanlung 30899
Gussets 273531
Gussips 273531
Gutta Percha 280437
Gutta Percha Tree 156041
Gutta-percha Mayten 246908
Gutta-percha Tree 156041,280441
Guttate Drejerella 66725
Gutzlaffia 182122
Guyana Arrowroot 131458
Guyana Arrow-root Greater Yam 131458
Guyanese Arrowhead 342346
Guyuan Cyclobalanopsis 116104
Guzmania 182149,182181
Guzmannia 182149
Guzzleberry 334250
Gwar 115323
Gyaca Monkshood 5119
Gyaca Rockfoil 349890
Gyala Barberry 51691
Gye 11947,170193,282685,325612,326287
Gyirong Thermopsis 389525
Gymea Lily 136729,136730
Gymea-lily 136730
Gymnadenia 182213
Gymnanthera 182324

Gymnaster 256301
Gymnaster Lixian 182347
Gymnema 182361
Gymnocalycium 182419
Gymnocarpos 182491
Gymnopetalum 182574
Gymnosperms 182628
Gymnotheca 182847
Gympie 125540
Gympie Messmate 155532
Gynocardia 182961
Gynostemma 182998
Gynura 183097
Gypsies' Daisy 227533
Gypsophila 183157,183221
Gypsophila Cerastioidesnone 174715
Gypsum Cactus 287908,354318
Gypsum Phacelia 293169
Gypsum Wild Buckwheat 152115
Gypsum-loving Larkspur 124268
Gypsy 174832,174877,238583,302034, 359158
Gypsy Comb 31051
Gypsy Curtains 28044,101852
Gypsy Daisy 216753,227533,397970
Gypsy Flower 9832,28044,91770,101852, 118008,174832,174877,216753,248312, 363461,397970
Gypsy Lace 170529
Gypsy Laces 28044
Gypsy Maids 81768
Gypsy Nut 109857,336419
Gypsy Nuts 109857,336419
Gypsy Onion 15861
Gypsy Pea 408571
Gypsy Peas 408571
Gypsy Pink 127635
Gypsy Rose 216084,216753,336290, 336364,350099,379660
Gypsy Violet 410493
Gypsy's Bacca 95431,339887,388275
Gypsy's Comb 31051,133478
Gypsy's Curtains 101852
Gypsy's Daisy 227533
Gypsy's Gibbles 15861
Gypsy's Hat 102933
Gypsy's Lace 192358
Gypsy's Money 68189
Gypsy's Parsley 28044,174877
Gypsy's Rhibarb 292372
Gypsy's Rhubarb 31051,292372
Gypsy's Sage 388275
Gypsy's Tobacco 95431,339887
Gypsy's Treacle 345631
Gypsy's Umbrella 28044
Gypsyflower 118008
Gypsy-flower 118008
Gypsyweed 239202,239248,407256
Gypsywort 47013,239184,239202
Gypsy-wort 239202
Gyrocheilos 183315,183316
Gyrogyne 183324,183325

H

H. Collett's Rhododendron 330443
H. F. Hance's Rhododendron 330816
H. Falconer's Rhododendron 330680
H. I. Harding's Rhododendron 330084
H. R. Fletcher's Rhododendron 330708
H. Shepherd's Rhododendron 331002
H. T. Tsai Begonia 50374
H. T. Tsai Blastus 55186
H. T. Tsai Damnacanthus 122084
H. T. Tsai Eightangle 204601
H. T. Tsai Helicia 189957
H. T. Tsai Hornbeam 77400
H. T. Tsai Illicium 204601
H. T. Tsai Marsdenia 245821
H. T. Tsai Neonauclea 264226
H. T. Tsai Rockvine 387863
H. T. Tsai Tigerthorn 122084
H. T. Tsai Urophyllum 402646
H. T. Tsai's Rhododendron 332009
H. W. Kung Honeysuckle 235905
Ha'uowi 405865
Haag Saxifrage 349428
Haage Campion 363265
Haasballetjies 21095
Haassuring 21095
Haba Catchfly 363531
Haba False Bigstigma Rockfoil 349600
Haba Lily 229856
Haba Monkshood 5240
Haba Mountain Milkvetch 42457
Haba Rockfoil 349429
Habahe Kengyilia 215918
Habashan Milkvetch 42457
Habashan Woodbetony 287272
Habenaria 183389
Haberlea 184222
Hab-nabs 109857
Habranthus 184249
Hachuela 288623
Hack 280003
Hackberries 80569
Hackberry 80569,80580,80661,80673, 80698,80723,80771,280003
Hackberry-leaf Grewia 180724
Hackberry-leaved Grewia 180724
Hackel Cleistogenes 94601
Hackel Hideseedgrass 94601
Hackel Roegneria 335371
Hackel Spain Barberry 51727
Hacker 280003
Hackmatack 221904,311237
Hackwood 280003
Hackymore 81237
Hacquetia 184326
Hadder 67456
Haddyr 67456
Haedhead Bamboo 297477
Haemaria 184471
Haenke Barberry 51695
Hag 109857
Hag Bush 109857
Hag Haw 109857
Hag Rope 95431
Hag Taper 405788
Hag Tree 109857
Hag's Taper 405788
Hagag 109857
Hagberry 83155,109857,280003,369541
Hager Tulip 400172
Haggas 109857
Haggil 109857
Hag-haw 109857
Hagisses 336419
Hag-leaf 405788
Hag-rope 95431
Hag-taper 405788
Hagthorn 109857
Hague 109857
Hagworm Flower 374916
Ha-Ho 28044,192358
Hahs-bush 109857
Haiari 235513
Haiball Crazyweed 279218
Haichow Elsholtzia 144093
Haiey-branch Holly 203747
Haiey-twig Holly 203747
Haig 109857
Haigh 109857
Haihan Macaranga 240265
Haikang Pseuderanthemum 317260
Haila Maple 3019
Hailar Crazyweed 278869
Hailar Spiraea 371926

Hailweed 115031
Hainan Acacia 1268
Hainan Addermonth Orchid 110644
Hainan Adina 292038
Hainan Adinandra 8235
Hainan Aerva 9371
Hainan Aganosma 10234
Hainan Albizia 13493
Hainan Albizzia 13493
Hainan Alocasia 16498
Hainan Alphonsea 17626
Hainan Alseodaphne 17800
Hainan Altingia 18203
Hainan Alyxia 18516
Hainan Amomum 19878
Hainan Amydrium 20867
Hainan Anneslea 25775
Hainan Antlevine 88901
Hainan Apama 28902
Hainan Ardisia 31579
Hainan Argostemma 32521
Hainan Arisaema 33341
Hainan Arrowbamboo 162698
Hainan Arrowwood 407877
Hainan Aspidistra 39544
Hainan Aspidopterys 39688
Hainan Awlbillorchis 333729
Hainan Azalea 330815
Hainan Balanophora 46851
Hainan Bamboo 37229,47222
Hainan Bambusa 47305,47458
Hainan Basketvine 9463
Hainan Batweed 89205
Hainan Bauhinia 49116
Hainan Begonia 49897
Hainan Bell Tree 324998
Hainan Belltree 324998
Hainan Birthwort 34201
Hainan Blachia 54876
Hainan Bladderwort 403113
Hainan Blueberry 403845
Hainan Bogorchis 110644
Hainan Bowfruitvine 393531
Hainan Box 64253
Hainan Brachytome 59003
Hainan Buckthorn 328720
Hainan Bulbophyllum 62767
Hainan Bulrush 353448
Hainan Bushbeech 178031
Hainan Bush-beech 178031
Hainan Butterflyorchis 293607
Hainan Calophanoides 67823
Hainan Calvepetal 177990
Hainan Cane 37229
Hainan Canebrake 37229
Hainan Caper 71757
Hainan Casearia 78097
Hainan Cayratia 387836
Hainan Ceratostylis 83653
Hainan Championella 85948
Hainan Chaulmoogra Tree 199748
Hainan Chaulmoogratree 199748
Hainan Chaulmoogra-tree 199748
Hainan Cherry 83215
Hainan Childvine 400896
Hainan Chinalaurel 28334
Hainan China-laurel 28334
Hainan Chinese Groundsel 365054
Hainan Chonemorpha 88901
Hainan Christia 89205
Hainan Chuniophoenix 90620
Hainan Claoxylon 94063
Hainan Clematis 94983
Hainan Clockvine 390793
Hainan Clubfilment 179185
Hainan Cnesmone 97501
Hainan Cogflower 154169
Hainan Coldwaterflower 299074
Hainan Condorvine 245815
Hainan Copperleaf 1888
Hainan Copper-leaf 1888
Hainan Cosmostigma 107175
Hainan Crabgrass 130768
Hainan Croton 112933
Hainan Cryptocarya 113451
Hainan Cycas 115833
Hainan Damnacanthus 122037
Hainan Daphniphyllum 122717
Hainan Dehaasia 123606
Hainan Dendrobium 125173
Hainan Dendropanax 125611
Hainan Devilpepper 327070
Hainan Dichapetalum 128726
Hainan Dichroa 129038
Hainan Dinggongvine 154244
Hainan Dishspurorchis 171858
Hainan Dolichos 135641
Hainan Draws ear grass 403113
Hainan Drypetes 138624
Hainan Dutchman's-pipe 34201
Hainan Dutchmanspipe 34201
Hainan Dyeingtree 346481
Hainan Dyeing-tree 346481
Hainan Dysophylla 139588
Hainan Eagleclaw 35015
Hainan Eggplant 367518
Hainan Ehretia 141648
Hainan Elaeocarpus 142328
Hainan Elatostema 142674
Hainan Endiandra 145318
Hainan Engelhardia 145515
Hainan Eria 125573
Hainan Ervatamia 154169
Hainan Erycibe 154244
Hainan Euonymus 157556
Hainan Euphorbia 159018
Hainan Eurya 160484
Hainan Evergreen Chinkapin 78947
Hainan Evergreenchinkapin 78947
Hainan False Manyroot 283767
Hainan Falsenettle 56257
Hainan False-nettle 56257
Hainan Falserambutan 283531
Hainan Fargesia 162698
Hainan Fewflower Pittosporum 301364
Hainan Fig 165084
Hainan Fimbristylis 166346
Hainan Fishvine 283285
Hainan Flemingia 166875
Hainan Fluttergrass 166346
Hainan Forkstyleflower 374480
Hainan Galangal 17685
Hainan Gardenia 171301
Hainan Garnotia 171513
Hainan Gastrochilus 171858
Hainan Glorybower 96114
Hainan Glycosmis 177841
Hainan Glyptic-petal Bush 177990
Hainan Glyptopetalum 177990
Hainan Gomphandra 178923
Hainan Gomphostemma 179185
Hainan Goniothalamus 179413
Hainan Gordonia 179753
Hainan Gordontea 179753
Hainan Graptophyllum 180282
Hainan Greenstar 307506
Hainan Gymnema 182370
Hainan Heath 125573
Hainan Helicia 189934
Hainan Herbcoral 346532
Hainan Hetaeria 193589
Hainan Holly 203869
Hainan Homalium 197685
Hainan Homalomena 197792
Hainan Honeysuckle 235709
Hainan Hopea 198157
Hainan Hornbeam 77336
Hainan Hornstedtia 198471
Hainan Hornstyle 83653
Hainan Horny Tanoak 233157
Hainan Horsfieldia 198512
Hainan Humpvine 250868
Hainan Hypolytrum 202717
Hainan Indianmulberry 258891
Hainan Indigo 206740
Hainan Indocalamus 206821
Hainan Ironweed 405446
Hainan Ivyarum 353127
Hainan Ixora 211103
Hainan Jadeleaf and Goldenflower 260423
Hainan Jasmine 211893

Hainan Javatea 275761
Hainan Jewelvine 283285
Hainan Joined Prismatomeris 315322
Hainan Jointfir 178538
Hainan Keteleeria 216147
Hainan Koilodepas 217760
Hainan Kopsia 217870
Hainan Kumquat 167498
Hainan Lasianthus 346491
Hainan Leafflower 296599
Hainan Leaf-flower 296599
Hainan Lepidagathis 225168
Hainan Lepisanthes 225674
Hainan Leptopus 226330
Hainan Linociera 87710,231749
Hainan Liolive 87710,231749
Hainan Liparis 232183
Hainan Litse 233981
Hainan Lobelia 234520
Hainan Lovegrass 147708
Hainan Lysionotus 239945
Hainan Madhuca 241513
Hainan Mahonia 242550
Hainan Mallotus 243364
Hainan Manglietia 244435
Hainan Maple 3010
Hainan Marsdenia 245815
Hainan Masson Pine 300056
Hainan Mayten 246851
Hainan Medinilla 247569
Hainan Meiogyne 87344
Hainan Memecylon 249976
Hainan Merremia 250788
Hainan Merrillanthus 250868
Hainan Metalleaf 296388
Hainan Meyna 252705
Hainan Milkgreens 245815
Hainan Milkwort 308095
Hainan Mischocarp 255949
Hainan Mischocarpus 255949
Hainan Mosquitotrap 117526
Hainan Mucuna 259519
Hainan Munronia 259877
Hainan Mussaenda 260423
Hainan Mycetia 260625
Hainan Myxopyrum 261895
Hainan Nanmu 295367
Hainan Naville Spicegrass 239748
Hainan Newlitse 264052
Hainan Nightshade 367518
Hainan Ninenode 319569
Hainan Oak 116204
Hainan Olive 270128
Hainan Oncodostigma 87344
Hainan Ophiorrhiza 272210
Hainan Orchidantha 273271
Hainan Ormosia 274428
Hainan Orophea 275325
Hainan Osbeckia 276111
Hainan Osmanther 276311
Hainan Osmanthus 276311
Hainan Parabarium 283069
Hainan Paraboea 283118
Hainan Paradrupetea 283713
Hainan Paramignya 283510
Hainan Paranephelium 283531
Hainan Pararuellia 283767
Hainan Parsonsia 285056
Hainan Pellionia 288754
Hainan Pencilwood 139665
Hainan Pencil-wood 139665
Hainan Pepper 300407
Hainan Peristrophe 291153
Hainan Persimmon 132186
Hainan Pertusadina 292038
Hainan Phaius 293507
Hainan Phoebe 295367
Hainan Phoenix Tree 166618
Hainan Phoenix-tree 166618
Hainan Phrynium 296033
Hainan Pittosporum 301364
Hainan Plumyew 82521
Hainan Plum-yew 82521
Hainan Podocarpus 306402
Hainan Pointedflowervine 334363
Hainan Poisonnut 378761
Hainan Poisonrat 128726
Hainan Premna 313642
Hainan Psychotria 319569
Hainan Purpleleaf 180282
Hainan Purslane 311857
Hainan Pyrenocarpa 322633
Hainan Rambutan 265139
Hainan Randia 278319
Hainan Rattanpalm 65703
Hainan Rattlebox 112207
Hainan Rhodamnia 329795
Hainan Rhododendron 330815
Hainan Rhodoleia 332211
Hainan Richella 334363
Hainan Rockvine 387836
Hainan Rockyellowtree 415155
Hainan Rosewood 121702
Hainan Sagebrush 35701
Hainan Salacia 342647
Hainan Sanqi 215035
Hainan Sarcandra 346532
Hainan Sarcococca 346749
Hainan Sasa 347220
Hainan Sawgrass 245437
Hainan Scaevola 350331
Hainan Schefflera 350703
Hainan Schizostachyum 351857
Hainan Schoenorchis 352304
Hainan Seamulberry 368917
Hainan Seatung 301364
Hainan Sedge 74731
Hainan Shagvine 97501
Hainan Sicklehairicot 135641
Hainan Single Bamboo 47305
Hainan Sinosenecio 365054
Hainan Skullcap 355467
Hainan Sloanea 366053
Hainan Slugwood 50627
Hainan Snakegourd 396175
Hainan Snakemushroom 46851
Hainan Snowbell 379361
Hainan Snowflake 32521
Hainan Sonerila 368885
Hainan Sonneratia 368917
Hainan Sophora 369134
Hainan Southstar 33341
Hainan Spapweed 204998
Hainan Spicebush 231430
Hainan Spice-bush 231430
Hainan Spiradiclis 371760
Hainan Spurge 159018
Hainan Stairweed 142674
Hainan Stauntonvine 374403
Hainan Staunton-vine 374403
Hainan Stephania 375859
Hainan Sterculia 376119
Hainan Stonebean-orchis 62767
Hainan Storax 379361
Hainan Stringbush 414186
Hainan Strychnos 378761
Hainan Sunipia 379778
Hainan Swallowwort 117526
Hainan Sweetleaf 381233
Hainan Syzygium 382569
Hainan Tailgrape 35015
Hainan Tallculm Cypressgrass 118844
Hainan Tallculm Galingale 118844
Hainan Tarenna 385039
Hainan Terminalia 386589
Hainan Ternstroemia 386702
Hainan Teyleria 388336
Hainan Thickshellcassia 113451
Hainan Thottea 390414
Hainan Thrixspermum 390504
Hainan Tigerthorn 122037
Hainan Tinospora 392257
Hainan Tinospore 392257
Hainan Touch-me-not 204998
Hainan Toxocarpus 393531
Hainan Treebine 92919
Hainan Treerenshen 125611
Hainan Twayblade 232183
Hainan Twinballgrass 209067
Hainan Tylophora 400896
Hainan Umbrella Bamboo 162698

Hainan Underleaf Pearl 296599
Hainan Ungeargrass 36625
Hainan Velvet Bean 259519
Hainan Ventilago 405446
Hainan Viburnum 407877
Hainan Wampee 94196
Hainan Waxplant 198881
Hainan White-flower Trigonostemon 397369
Hainan Whitetung 94063
Hainan Wild Kat 93480
Hainan Wild Orange 93480
Hainan Wild Quince 374403
Hainan Wildtung 243364
Hainan Wingdrupe 405446
Hainan Woodlotus 244435
Hainan Woodnettle 125550,273928
Hainan Wrinkle-seed Spiderflower 95782
Hainan Xanthophyllum 415144
Hainan Yaccatree 306402
Hainan Yellow Leaf Tree 415144
Hainan Yellowleaftree 415144
Hainan Yellow-leaved Tree 415144
Hainania 184575
Hainanlinden 184575
Hainberry 338557
Haines Barberry 51696
Hainryfruit Croton 112927
Hainry-fruited Croton 112927
Hainryleaf Ardisia 31632
Haipin Tanoak 233236
Hair Ailanthus 12571
Hair Beak-rush 333501
Hair Beneath Rhododendron 330902
Hair Capitate Eargrass 187534
Hair Carp 158318
Hair Clovershrub 70822
Hair Corydalis 106537
Hair Cowpea 409008
Hair Crabgrass 130496
Hair Dogbonbavin 133186
Hair Dogwood 380514
Hair Dragonbamboo 125514
Hair Drymaria 138498
Hair Duckbeakgrass 209255
Hair Dwarfnettle 262249
Hair Fescuegrass 163958
Hair Fimbriatesepal Chirita 87860
Hair Glycosmis 177825
Hair Goosecomb 335471
Hair Grass 12139,12776,12816,126025,
217416,259630,259642,398437
Hair Groundcherry 297720
Hair Hartia 376487
Hair Henry Clematis 94992
Hair Hogfennel 292988
Hair Hymenodictyon 201080
Hair Jujube 418152
Hair Kadsura 214968
Hair Knotweed 308877
Hair Ladyslipper 120355
Hair Lilac 382302
Hair Lupine 238482
Hair Metalleaf 296398
Hair Mnesithea 256342
Hair Mycetia 260626
Hair Ningpo Comanthosphace 100239
Hair Osmanther 276403
Hair Palm 85665,85671
Hair Panicgrass 282316
Hair Parabarium 402190
Hair Pipewort 151243
Hair Purslane 311915
Hair Safflower 77716
Hair Sedge 63229,74032
Hair Skunk Bugbane 91021
Hair Spike Newstraw 317084
Hair Spine Bamboo 47244
Hair Spiradiclis 371783
Hair Themeda 389376
Hair Triumfetta 399212
Hair Twig Cassia 91265
Hair Viscid Germander 388308
Hair Wild Pepper 300454
Hair Yushanbamboo 416753
Hairander Tea 69036
Hairanther Bulbophyllum 62958
Hairanther Camellia 69548
Hairanther Curlylip-orchis 62958
Hairanther Oreocharis 273844
Hairaxi Sweetleaf 381423
Hairaxle Arundinella 37418
Hairaxle Orchis 310916
Hairback Nabilous Azalea 330352
Hairback Windhairdaisy 348687
Hairbase Elatostema 142683
Hairbase Stairweed 142683
Hair-beard 238583
Hairboad Woodbetony 287771
Hairbract Elliott Monkshood 5175
Hairbract Gentian 150740
Hairbract Thin Hemiboea 191360
Hairbranch Callose Actinidia 6540
Hairbranch Hemsley Monkshood 5263
Hairbranch Lysionotus 239979
Hairbranch Whitepearl 172164
Hairbrush 133478
Hairbubble Crazyweed 279217
Haircalyx Azalea 330195
Haircalyx Catchfly 363942
Haircalyx Dualbutterfly 398271
Haircalyx Elsholtzia 144009
Haircalyx Gyrocheilos 183318
Haircalyx Hemsley Monkshood 5257
Haircalyx Monoflorcress 287958
Haircalyx Primrose 314152
Haircalyx Sandwort 32021
Haircalyx Spapweed 205400
Haircalyx Sporoxeia 372892
Haircalyx Touch-me-not 205400
Haircalyx Tripterospermum 398271
Haircalyx Violet 410650
Haircalyx Wildcoral 170044
Haircalyx Yunnan Stephania 375909
Haircockclaw Maple 3485
Haircolumn Azalea 332054
Haircorol Azalea 331069
Hair-covered Cactus 244202
Hairculm Goosecomb 335474
Hairdragonhead Arrowbamboo 162670
Haired British Inula 207066
Haired Scapose Sedge 76172
Haired Sloanea 366080
Haired Stygma 396454
Haireve 170193
Hairflower 409642
Hairflower Azalea 330900
Hairflower Bunge Bedstraw 170294
Hairflower Cheilotheca 86392
Hairflower Cremanthodium 110399
Hairflower Dallisgrass 285423
Hairflower Daphne 122386
Hairflower Gentian 173766
Hairflower Glandhair Didymocarpus 129905
Hairflower Hairfruit Microula 254374
Hairflower Hawkweed 195856
Hairflower Honeysuckle 236189
Hairflower Indianpine 202767
Hairflower Manyleaf Fleabane 150798
Hairflower Nutantdaisy 110399
Hairflower Oreocharis 273849
Hairflower Ovateleaf Microula 254359
Hairflower Trigonotis 397419
Hairflower Violet 410456
Hairflower Wood Didymocarpus 129905
Hairfoot Dragonbamboo 125468
Hairfruit Azalea 331795
Hairfruit Baoxing Whitlowgrass 137127
Hairfruit Bigfruit Tickseed 104811
Hairfruit Bigpetalcelery 357906
Hairfruit Billfruit Gynostemma 183031
Hairfruit Blueberry 403886
Hairfruit Bogorchid 158318
Hairfruit Caper 71908
Hairfruit Christmasbush 14210
Hairfruit Dilophia 130960
Hairfruit Ebony 132436
Hairfruit Eritrichium 153472
Hairfruit Flatawndaisy 14913
Hairfruit Graceful Clematis 94959
Hairfruit Grewia 180768
Hairfruit Gynostemma 183024

Hairfruit Heliotrope 190625
Hairfruit Hemiboea 191354
Hairfruit Henry Monkshood 5270
Hairfruit Involucrate Draba 137050
Hairfruit Involucrate Whitlowgrass 137050
Hairfruit Ladygin Draba 137066
Hairfruit Ladygin Whitlowgrass 137066
Hairfruit Largeleaf Trigonotis 397431
Hairfruit Liaoxi Tickseed 104768
Hairfruit Madder 338039
Hairfruit Microula 254373
Hairfruit Mouping Draba 137127
Hairfruit Netseed 332374
Hairfruit Nettle 403035
Hairfruit Ophiorrhiza 272287
Hairfruit Orychophragmus 275882
Hairfruit Pamir Tickseed 104827
Hairfruit Pearlsedge 354141
Hairfruit Peony 280214
Hairfruit Poppy 282618
Hairfruit Rattlebox 111968
Hairfruit Red-ligulate Tephroseris 385918
Hairfruit Roughleaf Bedstraw 170220
Hairfruit Serpentroot 354855,354876
Hairfruit Snakegourd 396279
Hairfruit Sopubia 369217
Hairfruit Spring Bittercress 72835
Hairfruit Stenocoelium 375555
Hairfruit Tanoak 233351
Hairfruit Underleaf Pearl 296593
Hairfruit Waldheimia 14913
Hairfruit Willow 343893
Hairfruit Windhairdaisy 348686
Hairfruit Woad 209205
Hairfruit Woodlotus 244475
Hairfruit Xing'an Tickseed 104761
Hairfruit Xizang Tickseed 104857
Hairfruit Xmas Bush 14210
Hairglume Arundinella 37437
Hairgrass 12776,12816,126025
Hair-grass 12776,126025,217416
Hairhead Elatostema 142706
Hairhead Stairweed 142706
Hairhound 47013
Hairhull Qinggang 116167
Hairiff 170193
Hairitch 170193
Hairleaf Aeluropus 8877
Hairleaf Allomorphia 16160
Hairleaf Amur Saddletree 240114
Hairleaf Aspidopterys 39685
Hairleaf Asterbush 41757
Hairleaf Awlnettle 307042
Hairleaf Azalea 331601
Hairleaf Big-leaved Fig 165058
Hairleaf Blackhair Milkvetch 42950
Hairleaf Convolvulate Asiabell 98299
Hairleaf Copperleaf 1919
Hairleaf Cordia 104218
Hairleaf Cremanthodium 110448
Hairleaf Cyclobalanopsis 116122
Hairleaf Cystopetal 320251
Hairleaf Dandelion 384676
Hairleaf Datura 123061
Hairleaf Elatostema 142746
Hairleaf Goosecomb 335267
Hairleaf Hairy-callus Roegneria 335227
Hairleaf Japanese Spiraea 371979
Hairleaf Jurinea 211822
Hairleaf Lyonia 239406
Hairleaf Magnolia 279248
Hairleaf Magnoliavine 351089
Hairleaf Mastixia 246221
Hairleaf Milkvetch 42586
Hairleaf Mulberry 259157
Hairleaf Nutantdaisy 110448
Hairleaf Oak 324150
Hairleaf Oat 45592
Hairleaf Olive 71024
Hairleaf Pennywort 200329
Hairleaf Pepper 300489
Hairleaf Phacelurus 293225
Hairleaf Plumegrass 148924
Hairleaf Qinggang 116122
Hairleaf Resupinate Woodbetony 287588
Hairleaf Rosewood 121822
Hairleaf Sandwort 31787,32283
Hairleaf Scurrula 355337
Hairleaf Siris 13621
Hairleaf Skullcap 355613
Hairleaf Sonerila 368905
Hairleaf Stairweed 142746
Hairleaf Stink Asiabell 98321
Hairleaf Sunflower 189001
Hairleaf Taroorchis 265418
Hairleaf Tea 69520
Hairleaf Vinegentian 110327
Hairleaf Wendlandia 413850
Hairleaf Whitepearl 172165
Hairleaf Xianbeiflower 362399
Hairleass Bluegrass 305501
Hair-leaved Cordia 104218
Hairless Aristulate Rockfoil 349071
Hairless Bigseed Bugle 13138
Hairless Blue Sow-thistle 90857
Hairless Catmint 265016
Hairless Chinese Groundsel 365070
Hairless Chinese Hawthorn 109937
Hairless Cinquefoil 312531
Hairless Cremanthodium 110446
Hairless Eucalyptus Raspberry 338367
Hairless Fairy Lantern 67608
Hairless Fiddle-shaped Leaf Raspberry 338953
Hairless Fleabane 150991
Hairless Garliccress 365416
Hairless Gingerlily 187453
Hairless Goldwaist 90367
Hairless Greenvine 204623
Hairless Hipericumlike Rockfoil 349490
Hairless Ladymantle 14037
Hairless Lamiophlomis 220337
Hairless Laxfruit Tickclover 126374
Hairless Leaf Milkvetch 43065
Hairless Leptinella 107770
Hairless Licorice 177893
Hairless Listera 264743
Hairless Littleflower Raspberry 338416
Hairless Milkvetch 43040
Hairless Mountain Jerusalemsage 295164
Hairless Nutantdaisy 110446
Hairless Oxtail Sagebrush 36340
Hairless Panicled Raspberry 338957
Hairless Pashi Pear 323253
Hairless Pearlwort 342290
Hairless Peduncle Cinquefoil 312869
Hairless Primrose 314342
Hairless Rhododendron 330644
Hairless Rockspray Cotoneaster 107560
Hairless Sagebrush 35530
Hairless Silverleaf Cinquefoil 312715
Hairless Smallleaf Cinquefoil 312770
Hairless Spicebush 231375
Hairless Spikesedge 143159
Hairless Stalk Maple 3096
Hairless Whorlleaf Blueberry 404037
Hairless Willow 343416
Hairless Yellow Spiderflower 34407
Hairlessleaf Loxostigma 238037
Hairlesss Rockvine 387825
Hairless-stalked Maple 3096
Hairlike Pondweed 312282
Hair-like Pondweed 312282
Hairlike Randia 12533
Hairlike Sedge 74032
Hair-like Sedge 74032
Hair-like Stenophyllus 63229
Hairlip Eulophia 156751
Hairlip Gentian 156751
Hairlip Habenaria 183703
Hairlip Liparis 232206
Hairlip Nervilia 265395
Hairlip Pleione 304261
Hairlip Taroorchis 265395
Hairlip Twayblade 232206
Hairliporchis 395855,395871
Hairlrss Jadegreen Sasa 304009
Hairnode Herelip 220076
Hairnode Squarebamboo 87563
Hairorchis 148619
Hairough 170193

Hairpalmleaf Peashrub 72272
Hairpetal Azalea 330510
Hairpetal Crazyweed 279157
Hairpetal Grewia 180686
Hairpetal Lady's Slipper 120346
Hairpetal Ladyslipper 120346
Hairpetal Milkvetch 42585
Hairpetal Monkshood 5488
Hairpetal Tea 69528
Hairpetal Violet 410673
Hairpetal Whitethorn 266360
Hairpetiole Willow 343598
Hairpin Banksia 47671
Hairpistil Caper 71830
Hairpod Medic 247285
Hairpod Senna 360439
Hairpole Yushanbamboo 416780
Hairraceme Crazyweed 279216
Hairreceptacle Buttercup 326442
Hairrib Azalea 330165
Hairring Chinacane 364799
Hairring Fishscale Bamboo 297213
Hairrode Arundinella 37356
Hairrostra Azalea 331999
Hairscale Rattanpalm 65802
Hairscape Habenaria 183503
Hairscape Longlegorchis 320449
Hairscape Pteroceras 320449
Hairseed Camellia 69715
Hairseed Tea 69490
Hairsepal Camellia 69530
Hairsepal Hemsley Monkshood 5261
Hair-sheath Bambusa 47419
Hairsheath Broodhen Bamboo 297256
Hairshoot False Rattlesnakeroot 283689
Hairshoot Purpledaisy 267180
Hairshooy Milklettuce 283692
Hairslume Fescuegrass 164229
Hairspike Falsehellebore 405606
Hairspike Leymus 228373
Hairspike Maesa 241779
Hairspike Woodbetony 287143,287343
Hairspine Peashrub 72362
Hairstalk Azalea 332047
Hairstalk Dandelion 384536
Hairstalk Dendrobium 125036
Hairstalk Milkvetch 42145
Hairstalk Serpentroot 354930
Hairstalk St. Paulswort 363080
Hairstalk Stairweed 142803
Hairstalk Stapf Monkshood 5599
Hairstalk Tea 69082
Hairstalk Violet 410073
Hairstamen Gentian 173876
Hairstem Auriculatum Elatostema 142612
Hairstem Clearweed 299089
Hairstem Coldwaterflower 299089
Hairstem Corydalis 106337
Hairstem Earshape Stairweed 142612
Hairstem Gentian 173764
Hairstem Skullcap 355596
Hair-stemmed Dwarf Rush 212987
Hairstipe Sinosenecio 365046
Hairstyle Azalea 331062
Hairstyle Clematis 95113
Hairstyle Milkvetch 42488
Hairstyle Poisonnut 378833
Hairtea 28563
Hairthroat Azalea 330338
Hairthroat Gentian 173437
Hairthroat Milkvetch 41925
Hairthrum Broomrape 275154
Hairtooth Crazyweed 279215
Hairtwig Azalea 331859
Hairtwig Elm 401449
Hairtwig Tanoak 233357
Hairtwig Tea 69713
Hairtwig Threevein Aster 39966
Hairumbel Seseli 361487
Hairup 170193
Hair-vein Agrimony 11572
Hairvein Alpine Oak 324322
Hairvein Aster 41398
Hairvein Azalea 331576
Hairvein Chloranthus 88286
Hairvein Gentian 173910
Hairvein Greenstar 307535
Hairvein Listera 264696
Hairvein Trigonotis 397434
Hairvine Yam 131648
Hairvinebean 67897,67898
Hairweed 115008,115031,170193
Hairwood 3462
Hairy Abrus 763
Hairy Abutilon 900
Hairy Acuminate Raspberry 338083
Hairy Adinandra 8237
Hairy Agapetes 10300
Hairy Ailanthus 12571
Hairy Alder 16368,16400
Hairy Alopecurus Woodbetony 286989
Hairy Alpen Rose 330875
Hairy Alumroot 194439
Hairy Angelica 24503
Hairy Apple 123065,243649
Hairy Arabis 30292
Hairy Arnica 34748
Hairy Arrowbamboo 162736
Hairy Arundinella 37421
Hairy Aster 40194,380953,380955
Hairy Astridia 43342
Hairy Auriculate Dichocarpum 128926
Hairy Awnedglume Roegneria 335214
Hairy Azalea 331518
Hairy Bamboo 297188,347334,347361
Hairy Bambooleaf Pricklyash 417162
Hairy Bamboo-shoot Giant-grass 175600
Hairy Banana 260279
Hairy Barberry 52266
Hairy Basketvine 9468
Hairy Bassia 48767
Hairy Bayberry 261155
Hairy Bead Grass 285522,285524,285525
Hairy Beardtongue 289353,289356
Hairy Beard-tongue 289353
Hairy Bedstraw 170557
Hairy Beggarticks 54048
Hairy Begonia 49924
Hairy Beijing Cotoneaster 107720
Hairy Bellsepal Sage 344931
Hairy Bigfruit Tickseed 104811
Hairy Bigthyrse Rabdosia 209776
Hairy Bind 115008
Hairy Bindweed 68676
Hairy Birch 53590
Hairy Bird's Eye 14375
Hairy Bird's-foot Trefoil 237762
Hairy Bird's-foot-trefoil 237762
Hairy Birdsfoot Trefoil 237762
Hairy Bitter Cress 72802
Hairy Bittercress 72802
Hairy Bitter-cress 72802
Hairy Boronia 57274
Hairy Bougainvillea 57855
Hairy Bracted Tick-trefoil 126302
Hairy Bridalwreath Spiraea 372053
Hairy Broadleaf Holly 203962
Hairy Broadstamen Burnet 345833
Hairy Brome 60680,60918
Hairy Brome Grass 60918
Hairy Brome-grass 60918
Hairy Broom 121026
Hairy Broomrape 275113
Hairy Brucea 61211
Hairy Bugseed 104861
Hairy Bunge Bedstraw 170293
Hairy Burstwort 192975
Hairy Bush Cinquefoil 312599
Hairy Bush Clover 226833
Hairy Buttercup 326326
Hairy Caltrop 215382
Hairy Camellia 69527
Hairy Campion 363370
Hairy Canton Sonerila 368875
Hairy Caper 71727
Hairy Carrion-flower 366427
Hairy Cat's Ear 202432
Hairy Cat's-ear 202432
Hairy Catchfly 363409
Hairy Catsear 202432
Hairy Ceanothus 79956

Hairy Chaffhead 77232,397529
Hairy Chelonopsis 86820
Hairy Cherry-laurel 223104,223137
Hairy Chervil 84743
Hairy Chess 60680,60912
Hairy Chestnut 78802
Hairy China Buckthorn 328885
Hairy Chinese Buckthorn 328885
Hairy Chinese Viburnum 407943
Hairy Chittagong Chickrassy 90606
Hairy Cinnamon 91265
Hairy Cinquefoil 313029
Hairy Clematis 95016
Hairy Clovershrub 70822
Hairy Cochinchina Sweetleaf 381151
Hairy Cockspur-thorn 110060
Hairy Combretum 100637
Hairy Common White Jasmine 211951
Hairy Conebush 210244
Hairy Corktree 294249
Hairy Correa 105389
Hairy Corydalis 106106
Hairy Cowpea 409008
Hairy Crab Grass 130745
Hairy Crabgrass 130745
Hairy Crab-grass 130745
Hairy Crabweed 162872
Hairy Crazyweed 279216
Hairy Cress 73090
Hairy Crinklemat 392360
Hairy Crocus 111577
Hairy Cucumber 114136
Hairy Culm Roegneria 335474
Hairy Cup Grass 151702
Hairy Cupgrass 151702
Hairy Daisy 124857
Hairy Date-plum 132268
Hairy Datura 123061
Hairy Delavay Ampelopsis 20339
Hairy Dendranthema 124857
Hairy Desert Wheatgrass 11712
Hairy Desert-sunflower 174427
Hairy Dewflower 138183
Hairy Divaricate Mussaenda 260404
Hairy Dog-rose 336410
Hairy Draba 137019
Hairy Dropseed 372803
Hairy Dropseedgrass 372803
Hairy Drymonia 138525
Hairy Duckbeakgrass 209255
Hairy Dutchmanspipe 34275
Hairy E. H. Wilson Buckthorn 328901
Hairy Elaeagnus 142018
Hairy Elephantfoot 143460
Hairy Elm 401466,401511
Hairy Entire Micromelum 253629
Hairy Euphorbia 160047
Hairy Evening-primrose 269522,269525
Hairy False Golden-aster 194261,194264
Hairy Falsebetony 373070
Hairy Falsenettle 56326
Hairy Farges' Willow 343367
Hairy Feathergrass 376866
Hairy Fig 165111
Hairy Fighead Stairweed 142662
Hairy Fimbristylis 166447,166449
Hairy Fimbry 166447,166449
Hairy Finger Grass 130554
Hairy Finger-grass 130745
Hairy Flat-top Aster 41453
Hairy Flax 231911
Hairy Fleabane 150901
Hairy Fleshseed Tree 346983
Hairy Fleshspike Tree 346983
Hairy Florida Maple 2969
Hairy Flower Agapetes 10359
Hairy Flower Rhododendron 331991
Hairy Flowers Heath 149551
Hairy Flueggea 167068
Hairy Fly-honeysuckle 236206
Hairy Forked Nailwort 284845
Hairy Forked Nail-wort 284845,284847
Hairy Forrest Braya 59740
Hairy Four-o'clock 255708
Hairy Fragrant Arrowwood 407839
Hairy Fruit Gynostemma 183024
Hairy Garlic 15781
Hairy Gastrodia 171925
Hairy Gayfeather 228480
Hairy Gentian 173342
Hairy Gerbera 175203
Hairy Germander 388166
Hairy Ginger 417990
Hairy Globba 176987
Hairy Goatgrass 8670
Hairy Goldaster 90294
Hairy Golden Aster 90175,194264,194270
Hairy Goldenaster 194261
Hairy Goldeneye 190249
Hairy Goldenrod 368162,368163
Hairy Goldsaxifrage 90433
Hairy Goldwaist 90429
Hairy Gooseberry 334030
Hairy Graceful Bedstraw 170361
Hairy Grama Grass 57931
Hairy Grape 411735
Hairy Greenweed 173030
Hairy Ground Cherry 297720
Hairy Guangdong Indocalamus 206787
Hairy Gynostemma 183026
Hairy Hawkbit 224731
Hairy Hawkweed 195643,195760
Hairy Head 81237
Hairy Heart-leaved Aster 40330
Hairy Hedge Nettle 373244
Hairy Helictotrichon 190187
Hairy Hemsleia 191906
Hairy Herelip 220068
Hairy Himalayan Currant 334028
Hairy Homalium 197703
Hairy Honeysuckle 235847
Hairy Hooker Sweetleaf 381241
Hairy Hopbush 135199
Hairy Hop-bush 135199
Hairy Hopea 198174
Hairy Huckleberry 403853,403854
Hairy Huodendron 199456
Hairy Hygrophila 200637
Hairy Indian Madder 337940
Hairy Indian Mallow 900
Hairy Indocalamus 206789
Hairy Inland Ceanothus 79941
Hairy Ironwort 362825
Hairy Jadeleaf and Goldenflower 260404, 260461
Hairy Japanese Maple 2942
Hairy Japanese Yam 131648
Hairy Jimsonweed 123061
Hairy Kengyilia 215922
Hairy Kerr Petrocosmea 292563
Hairy Keteleeria 216150
Hairy Knight's-spur 102839
Hairy Knotweed 308877
Hairy Korean Litse 233878
Hairy Lady's Smock 71106,71107,72802
Hairy Lady's-mantle 14090
Hairy Laevigate Elatostema 142705
Hairy Lagochilus 220068
Hairy Largefruit Holly 204005
Hairy Laurel 215394
Hairy Lavenderleaf Dendranthema 124812
Hairy Lazydaisy 29102
Hairy Leaf Mountain Mint 321960
Hairy Leaves Gaultheria 172165
Hairy Lens Grass 285524
Hairy Leptocodon 226064
Hairy Leptodermis 226133
Hairy Lespedeza 226833
Hairy Lettuce 84960
Hairy Leucas 227731
Hairy Lilac 382257,382312
Hairy Linden 391738
Hairy Littleleaf Supplejack 52452
Hairy Little-leaved Willow 343891
Hairy Loosestrife 239589
Hairy Lotus 237776
Hairy Lovegrass 147891
Hairy Lupine 238496
Hairy Lymegrass 144521
Hairy Lyonia 239406
Hairy Maddertree 12533

Hairy Maesa 241819
Hairy Maire Porana 131243
Hairy Mandarin 315525
Hairy Manschurian Currant 334085
Hairy Manzanita 31110
Hairy Marsdenia 245853
Hairy Marshmallow 18166
Hairy Meadow-parsnip 388898
Hairy Medicinal Indianmulberry 258907
Hairy Medick 247393,247425
Hairy Melic Grass 248975
Hairy Merremia 250794
Hairy Michaelmas-daisy 380930
Hairy Micromelum 253629
Hairy Milkgreens 245853
Hairy Milkvetch 42209
Hairy Milkweed 37968
Hairy Mint 250291,250468
Hairy Mnesithea 256342
Hairy Mongolian Wheatgrass 11797
Hairy Mountain Laurel 215394
Hairy Mountain Mint 321960
Hairy Mouthed Rhododendron 331999
Hairy Mussaenda 260461
Hairy Mycetia 260626
Hairy Narrowleaf Dogwood 124884
Hairy Narrowleaf Rockvine 387787
Hairy Nautilocalyx 262902
Hairy Needle on Both Sides 417282
Hairy Nepal Carpesium 77183
Hairy Nepeta 264997
Hairy Nightshade 367295,367583,367726
Hairy Nut Grass 354052
Hairy Oat 45558
Hairy Oliver Linden 391796
Hairy Oneflowerprimrose 270741
Hairy Orbicular Snailseed 97934
Hairy Orientvine 364989
Hairy Osbeckia 276403
Hairy Oxtail Greenbrier 366554
Hairy Pachilan Bamboo 47403
Hairy Pagoda-plant 55524
Hairy Pale-spike Lobelia 234787
Hairy Panic Grass 128478
Hairy Parabarium 402190
Hairy Parsley 235418
Hairy Paulownia 285981
Hairy Pea 763,222726
Hairy Pearleaf Raspberry 339069
Hairy Peking Cotoneaster 107720
Hairy Pendentfruit Rockcress 30391
Hairy Pennywort 200336
Hairy Pentanema 289575
Hairy Pepper 300415,300491
Hairy Persimmon 132414
Hairy Petals Rhododendron 330510
Hairy Pfeiffer Cicely 276466
Hairy Phacelia 293166
Hairy Pinweed 223703,223708
Hairy Plantain 302019
Hairy Poorspikebamboo 270443
Hairy Pottsia 313234
Hairy Pricklyash 417137
Hairy Prickly-ash 417340
Hairy Primrose Willow 238180
Hairy Pristimera 315371
Hairy Privet 229461
Hairy Pseudosasa 318305
Hairy Puccoon 233708,233709
Hairy Purpleanther Sandwort 31964
Hairy Purplepearl 66873
Hairy Purslane Speedwell 407283
Hairy Ramie 56339
Hairy Randia 12533
Hairy Raspberry 338130
Hairy Rattlesnake-plantain 179680
Hairy Red Maple 3526
Hairy Reed 295907
Hairy Rhododendron 330875
Hairy Rhombicleaf Rockvine 387862
Hairy Rhynchotechum 333749
Hairy Rigid Kengyilia 215956
Hairy Rock Cress 30292
Hairy Rock Rose 93137
Hairy Rock-cress 30292,30298,30299,30306
Hairy Rocket 59407,154073,154091
Hairy Rockrose 93152,93154,93155
Hairy Roegneria 335471
Hairy Rosarypea 763
Hairy Rose Mallow 194963
Hairy Rough Melic 249085
Hairy Ruellia 339687,339736
Hairy Rupturewort 192975
Hairy Rupture-wort 192975,192979
Hairy Safflower 77716
Hairy Sand-spurrey 370724
Hairy Sandspurry 370724
Hairy Sandwort 31816,32281
Hairy Scabrous Wendlandia 413820
Hairy Seablite 48767
Hairy Sedge 74806,74810,75785
Hairy Sepal Honeysuckle 236195
Hairy Sepal Rocket 94241
Hairy Serpentroot 354922
Hairy Serrate Hydrangea 200092
Hairy Serrulate Rockvine 387817
Hairy Seseli 361480
Hairy Sharpleaf Four-involucre 124884
Hairy Shasta Wild Buckwheat 152419
Hairy Sheep's Scabious 211672
Hairy Sheep's-bit Scabious 211672
Hairy Shinyleaf Cliffbean 254778
Hairy Shinyleaf Millettia 254778
Hairy Showy Daisy 150992
Hairy Sierra Sedge 74812
Hairy Signalgrass 58199
Hairy Simple-leaf Evodia 161379
Hairy Skullcap 355433
Hairy Skyblue Adonis 8349
Hairy Slendertail Elatostema 142870
Hairy Sloanea 366080
Hairy Smallleaf Willow 343891
Hairy Small-leaved Tick-trefoil 126291
Hairy Smartweed 291777
Hairy Smotherweed 48767
Hairy Snowberry 380746
Hairy Solomon's Seal 308508,308633
Hairy Solomon's-seal 308633
Hairy Sphenodesma 371420
Hairy Spicebush 231398
Hairy Spice-bush 231398
Hairy Spike Leymus 228373
Hairy Spiraea 372095
Hairy Spotflower 4873
Hairy Spurge 158659,159069,160047
Hairy St. John's Wort 201924
Hairy St. John's-wort 201924
Hairy Stalk Violet 410073
Hairy Star of Bethlehem 169524
Hairy Starwort 40194
Hairy Stonecrop 357282
Hairy Stranvaesia 377436
Hairy Sunflower 188980
Hairy Swamp Milkweed 37968
Hairy Sweetgrass 196153
Hairy Sweetleaf 381232
Hairy Swollennoded Cane 87600
Hairy Sycamoreleaf Snowbell 379432
Hairy Taiwan Roegneria 335316
Hairy Taiwan Storax 379350
Hairy Tare 408423
Hairy Tea 69652
Hairy Teasel 133502
Hairy Think Skullcap 355430
Hairy Thistle 92111
Hairy Thorn-apple 123065
Hairy Threelobed Spiraea 372120
Hairy Toadlily 396598
Hairy Tongue Lycaste 238745
Hairy Toothed She Oak 79191
Hairy Touch-me-not 205067
Hairy Tree Falsespiraea 369268
Hairy Tricalysia 133186
Hairy Trichilia 395498
Hairy Triumfetta 399212
Hairy Tube Metalleaf 296413
Hairy Twiggy Buckthorn 328894
Hairy Twigs Gaultheria 172171
Hairy Twigs Whitepearl 172171
Hairy Uraria 402118
Hairy Urceola 402190

Hairy Vetch 408423,408693
Hairy Vetchling 222726
Hairy Viburnum 408075
Hairy Violet 409843,410066
Hairy Wallaby Grass 341250
Hairy Wattle 1508,1689
Hairy Waxmyrtle 261155
Hairy White Oldfield Aster 380953
Hairy Whitetop 73098
Hairy White-top 73098
Hairy Whitlowgrass 137019
Hairy Whitlow-grass Nailwort 153964
Hairy Wild Buckthorn 328901
Hairy Wild Petunia 339736
Hairy Wild Sunflower 189001
Hairy Wildclove 226133
Hairy Wild-rye 144524
Hairy Wildryegrass 144521
Hairy Willow Weed 146724
Hairy Willowherb 146724
Hairy Willow-herb 146663,146664,146724
Hairy Willowweed 146724
Hairy Wireweed 308670
Hairy Wood Mint 55524
Hairy Wood Rush 238565,238681
Hairy Wood Sedge 74810,74816
Hairy Wood Violet 410587
Hairy Woodbetony 287434,287773
Hairy Woodbrome 60807
Hairy Woodland Brome 60900
Hairy Woodrush 238681
Hairy Woodsorrel 277889
Hairy Wormwood 36460
Hairy Xikang Spruce 298336
Hairy Yam 131546
Hairy Yellow Vetch 408429
Hairy Yellow-vetch 408429
Hairy Yunnan Blueberry 403813
Hairy Yushania 416753
Hairy Zinger 417990
Hairyanther Honeysuckle 236095
Hairy-anthered Drypetes 138658
Hairy-anthered Honeysuckle 236095
Hairy-back Cherry 34440
Hairy-back High Holly 203793
Hairyback Korean Litse 233878
Hairyback Mountainash 369335
Hairyback Rhamnella 328557
Hairy-back Rhamnella 328557
Hairyback Sharptooth Buckthorn 328615
Hairy-backed Cherry Laurel 223104
Hairy-backed Mountain-ash 369335
Hairy-bamboo 297188
Hairybract Siberian Goldenray 229182
Hairy-bracted Oligostachyum 270448
Hairy-bracted She-oak 79191
Hairy-branch Blueblack Viburnum 407688
Hairybranch Camellia 69713
Hairybranch Cathercupule Tanoak 233381
Hairybranch Cherry-laurel 223124
Hairy-branch Chun Arrowwood 407754
Hairy-branch Chun Viburnum 407754
Hairy-branch Evergreen Arrowwood 408124
Hairy-branch Evergreen Viburnum 408124
Hairy-branch Fujian Holly 203836
Hairy-branch Japanese Xylosma 415899
Hairy-branch Lasianthus 222137
Hairy-branch Mockorange 294458
Hairybranch Mulgedium 283692
Hairybranch Oak 116113
Hairybranch Qinggang 116113
Hairybranch Rockvine 387819
Hairy-branch Roughleaf 222137
Hairybranch Sageretia 342172
Hairy-branch Taizhong Arrowwood 407859
Hairy-branch Taizhong Viburnum 407859
Hairybranch Wintergreen 172164
Hairy-branched Camellia 69713
Hairy-branched Cyclobalanopsis 116113
Hairy-branched Milkvetch 42883
Hairy-branched Oak 116113
Hairybranched Paraprenanthes 283689
Hairy-branched Rockvine 387819
Hairy-branched Sallow 343494
Hairy-branched Ventilago 405442
Hairy-branched Willow 343493
Hairy-branched Wingdrupe 405442
Hairy-branched Wintergreen 172164
Hairybroom 121026
Hairybud Eurya 160552
Hairy-callus Roegneria 335224
Hairy-calyx Basketvine 9444
Hairycalyx Blueberry 403973
Hairy-calyx Crape Myrtle 219913
Hairycalyx Crapemyrtle 219913
Hairycalyx Crazyweed 279215
Hairycalyx Elsholtzia 144009
Hairycalyx Glaucosleaf Cotoneaster 107466
Hairycalyx Himalayan Cherry 83297
Hairy-calyx Himalayan Currant 334027
Hairycalyx Indigo 205775
Hairycalyx Jasmine 211785
Hairycalyx Lobelia 234696
Hairycalyx Manyflower Monkshood 5491
Hairy-calyx Mockorange 294448
Hairycalyx Parrya 224211
Hairycalyx Photinia 295752
Hairy-calyx Rose 336682
Hairycalyx Storax 379385
Hairycalyx Stranvaesia 377436
Hairycalyx Swertia 380226
Hairycalyx Tarenna 384974
Hairy-calyxed Basketvine 9444
Hairy-calyxed Blueberry 403973
Hairy-calyxed Crape-myrtle 219913
Hairy-calyxed Indigo 205775
Hairy-calyxed Jasmine 211785
Hairy-calyxed Stranvaesia 377436
Hairycapsule Oak 116075
Hairycup Statice 230804
Hairy-cupuled Cyclobalanopsis 116075
Hairycupuled Qinggang 116075
Hairycyme Woodbetony 287771
Hairydish Goosecomb 335224
Hairydisk Meconopsis 247120
Hairydisk Mockorange 294539
Hairyfilament Camellia 69036
Hairy-filamented Camellia 69036
Hairyflower Actinidia 6588
Hairyflower Agapetes 10359
Hairyflower Ancylostemon 22176
Hairy-flower Arrowwood 407772
Hairy-flower Axispalm 228735
Hairy-flower Basketvine 9443
Hairyflower Broomsedge 217344
Hairy-flower Caper 71830
Hairy-flower Cherry 83297
Hairyflower Chroesthes 89283
Hairyflower Clematis 94871
Hairyflower Colorcalyx 89283
Hairyflower Earlymadder 226238
Hairyflower Eriosolena 153102
Hairy-flower Fewflower Lysionotus 239948
Hairyflower Kiwifruit 6588
Hairy-flower Leptodermis 226137
Hairy-flower Licualapalm 228735
Hairyflower Lovegrass 148021
Hairyflower Maple 2947
Hairyflower Mukia 259739
Hairyflower Narrowleaf Pittosporum 301375
Hairyflower Ovateleaf Microula 254359
Hairy-flower Papery Daphne 122571
Hairyflower Rhododendron 330087,330900
Hairy-flower Sour Bamboo 4607
Hairyflower Spiraea 371897
Hairyflower Stringbush 414244
Hairy-flower Viburnum 407772
Hairy-flower Wild Buckwheat 152144
Hairyflower Wild Petunia 339693
Hairy-flower Wildclove 226137
Hairy-flowered Actinidia 6588
Hairy-flowered Agapetes 10359
Hairy-flowered Basketvine 9443
Hairy-flowered Chroesthes 89283
Hairy-flowered Eriosolena 153102
Hairy-flowered Licualapalm 228735
Hairy-flowered Lovableflower 148086
Hairy-flowered Maple 2947
Hairy-flowered Milkvetch 42264
Hairy-flowered Rhododendron 330900
Hairy-flowered Sour-bamboo 4607

Hairy-flowered Spiraea 371897
Hairy-flowered Stringbush 414244
Hairy-flowered Viburnum 407772
Hairyfruit Actinodaphne 6820
Hairyfruit African Malcolmia 243164
Hairyfruit Anemone Clematis 95131
Hairy-fruit Beautiful Spiraea 371822
Hairyfruit Bedstraw 170764
Hairyfruit Blume Spiraea 371846
Hairyfruit Boatcress 385132
Hairyfruit Boatshapedbract Larkspur 124402
Hairyfruit Box 64258
Hairyfruit Buttercup 326425
Hairyfruit Camellia 69711
Hairyfruit Cancrinia 71095
Hairyfruit Chienning Monkshood 5122
Hairyfruit Chingan Tickseed 104761
Hairyfruit Clethra 96522
Hairyfruit Clustered Whitlowgrass 137005
Hairyfruit Craneknee 221695
Hairyfruit Draba 137005
Hairyfruit Dyers Woad 209236
Hairyfruit Elm 401520
Hairy-fruit Emarginate Azalea 330623
Hairy-fruit Emarginate Rhododendron 330623
Hairyfruit Embelia 144745
Hairyfruit Eurya 160618
Hairy-fruit Eurya 160479
Hairyfruit Fig 164918
Hairy-fruit Fig 165778
Hairy-fruit Fineshoot Spiraea 372024
Hairyfruit Forrest Monkshood 5206
Hairy-fruit Gaultheria 172094
Hairyfruit Glaucescent Elm 401520
Hairyfruit Grewia 180768
Hairyfruit Hackberry 80771
Hairyfruit Hairyfruit 125959
Hairyfruit Holly 204339
Hairyfruit Honeysuckle 236183
Hairy-fruit Hophornbeam 276817
Hairyfruit Impatient Bittercress 72835
Hairyfruit Ironweed 405441
Hairyfruit Ironwood 276817
Hairyfruit Jilin Monkshood 5319
Hairyfruit Jujube 418152
Hairyfruit Kirin Monkshood 5319
Hairyfruit Lacquer-tree 393488
Hairy-fruit Lake Sedge 76582
Hairyfruit Longstamen Azalea 331883
Hairyfruit Longstamen Rhododendron 331883
Hairyfruit Lyonia 239397
Hairy-fruit Manglietia 244475
Hairyfruit Mazus 247038
Hairy-fruit Meliosma 249470
Hairyfruit Micrechites 203336
Hairyfruit Microula 254373
Hairyfruit Milkvetch 42588
Hairyfruit Mongolian Draba 137118
Hairyfruit Mountain-loving Draba 137158
Hairyfruit Mountain-loving Whitlowgrass 137158
Hairyfruit Musella 260359
Hairyfruit Neoalsomitra 263570
Hairyfruit Netseedgrass 332374
Hairyfruit Obtusesepal Clematis 95230
Hairyfruit Pale Tickseed 104768
Hairyfruit Pamir Tickseed 104827
Hairyfruit Pendentfruit Rockcress 30391
Hairy-fruit Porana 131256
Hairyfruit Qianning Monkshood 5122
Hairyfruit Raisin Tree 198786
Hairyfruit Raspberry 339122
Hairyfruit Razorsedge 354141
Hairyfruit Rehdertree 327417
Hairy-fruit Sedge 76582
Hairyfruit Serpentroot 354876
Hairyfruit Shortbranch Willow 343123
Hairy-fruit Short-hair Clematis 95267
Hairy-fruit Silkhairy Willow 343640
Hairyfruit Smallfruit Fig 165319
Hairyfruit Snakegourd 396279
Hairyfruit Speedwell 407117
Hairyfruit Sweetdaisy 71095
Hairyfruit Tall Monkshood 5576
Hairyfruit Tangut Monkshood 5623
Hairyfruit Tanoak 233351
Hairyfruit Tauscheria 385132
Hairy-fruit Tetracera 386900
Hairyfruit Tickseed 104858
Hairy-fruit Tree Currant 334110
Hairyfruit Turnjujube 198786
Hairyfruit Valerian 404281
Hairyfruit Veitch's Litse 234096
Hairyfruit Ventilago 405441
Hairyfruit Violet 409834
Hairy-fruit Whitepearl 172094
Hairyfruit Willow 344221
Hairyfruit Wingdrupe 405441
Hairyfruit Yunnan Blueberry 403811
Hairy-fruited Actinodaphne 6820
Hairy-fruited Broom 121015
Hairy-fruited Camellia 69711
Hairy-fruited Cape 71908
Hairy-fruited Christmas-bush 14210
Hairy-fruited Cornsalad 404422
Hairy-fruited Cushion Willow 343893
Hairy-fruited Dinetus 131256
Hairy-fruited Embelia 144745
Hairy-fruited Eurya 160618
Hairy-fruited Guangdong Beautyberry 66834
Hairy-fruited Honeysuckle 236183
Hairy-fruited Hop-hornbeam 276817
Hairy-fruited Jujube 418152
Hairy-fruited Lacquer-tree 393488
Hairy-fruited Leaf-flower 296593
Hairy-fruited Milkvetch 42588
Hairyfruited Nightshade 367295
Hairy-fruited Photinia 295752
Hairy-fruited Raisin-tree 198786
Hairy-fruited Raspberry 339122
Hairyfruited Taiwan Toadlily 396599
Hairy-fruited Tanoak 233351
Hairy-fruited Tetracera 386900
Hairyfruited Twonectary Lepistemon 225712
Hairy-fruited Ventilago 405441
Hairy-fruited Willow 344221
Hairyfruitgrass 222352,222353
Hairygalea Woodbetony 287772
Hairyglume Bluegrass 305909
Hairyglume Jijigrass 4155
Hairyglume Sloughgrass 49544
Hairyglume Speargrass 4155
Hairy-head Argyreia 32621
Hairyhead Sagebrush 35779
Hairyhead Wormwood 35779
Hairy-inflorescence Anzacwood 310765
Hairy-inflorescens Peashrub 72291
Hairy-inflorescens Pea-shrub 72291
Hairy-jointed Meadow-parsnip 388898
Hairyleaf Aeonium 9066
Hairyleaf Alyxia 18532
Hairy-leaf Ampelopsis 20428
Hairyleaf Antenoron 26625
Hairyleaf Antlevine 88902
Hairyleaf Aphanopleura 29086
Hairyleaf Apricot 34436
Hairyleaf Aralia 30623
Hairyleaf Arrowwood 408096
Hairyleaf Asparagus 39239
Hairyleaf Assam Treebine 92617
Hairyleaf Asterothamnus 41757
Hairyleaf Batrachium 48933
Hairyleaf Bedstraw 170577
Hairyleaf Bird Cherry 280007
Hairyleaf Bluebell 115367
Hairyleaf Bluebellflower 115367
Hairyleaf Bunge Pricklyash 417181
Hairy-leaf Camellia 69520
Hairy-leaf Canton Mombin 372472
Hairyleaf Caper 71831
Hairyleaf Cherry 83189
Hairy-leaf Chinese Clematis 94821
Hairyleaf Chonemorpha 88902
Hairyleaf Cinnamon 91381
Hairyleaf Cogflower 154192,154194,382822
Hairyleaf Cotoneaster 107697
Hairy-leaf Delavay Clethra 96489
Hairyleaf Draba 137083
Hairyleaf Elm 401455,401553
Hairyleaf Engelhardia 145510

Hairyleaf Ervatamia 154192,154194,382822
Hairyleaf Eucalyptus 155763
Hairyleaf Evergreenchinkapin 78885
Hairyleaf Flowering Quince 84553
Hairyleaf Floweringquince 84553
Hairyleaf Fullmoon Maple 3043
Hairyleaf Ganpin Clematis 95266
Hairyleaf Gansu-Mongol Peashrub 72306
Hairyleaf Gansu-Mongol Pea-shrub 72306
Hairyleaf Girald Rose 336609
Hairyleaf Goldenray 229153
Hairyleaf Grape 411735
Hairyleaf Greenbrier 366231
Hairy-leaf Greywhite Rhododendron 332158
Hairyleaf Guangdong Rose 336669
Hairy-leaf Henry Rockvine 387781
Hairyleaf Hornbeam 77280
Hairyleaf Hymenolyma 201157
Hairyleaf Ichnocarpus 203326
Hairyleaf Ironweed 405448
Hairyleaf Japanese Cayratia 79862
Hairyleaf Japanese Cherry 83319
Hairy-leaf Knotweed 309508
Hairyleaf Korean Raspberry 338295
Hairyleaf Kydia 218448
Hairy-leaf Lambert Raspberry 338710
Hairyleaf Larkspur 124292
Hairyleaf Litse 233996
Hairyleaf Looseflower Rose 336687
Hairyleaf Lyonia 239406
Hairy-leaf Lysionotus 239949
Hairy-leaf Maack Euonymus 157703
Hairyleaf Magnolia 279248
Hairy-leaf Manyfruit Idesia 203428
Hairyleaf Maple 3632
Hairyleaf Marshweed 230307
Hairyleaf Martin Spiraea 372004
Hairy-leaf Mastixia 246221
Hairy-leaf Milkvetch 42586
Hairyleaf Mitreola 256212
Hairyleaf Monkshood 5106
Hairyleaf Moyes Rose 336780
Hairy-leaf Mycetia 260652
Hairyleaf Nervilia 265418
Hairyleaf Newlitse 264111
Hairyleaf Obtusesepal Clematis 95228
Hairy-leaf Ostryopsis 276845
Hairyleaf Padritree 376283
Hairyleaf Paperelder 290442
Hairyleaf Peduncled Sinosideroxylon 365085
Hairyleaf Peony 280247
Hairy-leaf Porana 131254
Hairyleaf Raspberry 339084
Hairyleaf Reticulate Meadowrue 388637
Hairyleaf Rockfoil 349519
Hairyleaf Rose 336731
Hairyleaf Sageretia 342206
Hairyleaf Scabious 350132
Hairyleaf Shorthair Clematis 95266
Hairyleaf Sibiraea 362399
Hairy-leaf Sichuan Holly 204320
Hairyleaf Skullcap 355613
Hairy-leaf Slatepentree 322585
Hairy-leaf Smallfruit Rose 336511
Hairyleaf South Ailanthus 12587
Hairyleaf South China Pricklyash 417171
Hairyleaf Spiraea 372016
Hairyleaf Taxillus 385221
Hairyleaf Tianshanyarrow 185260
Hairy-leaf Tutcheria 322585
Hairyleaf Ventilago 405448
Hairyleaf Viburnum 408096
Hairyleaf Wampee 94219
Hairyleaf West China Spiraea 371991
Hairy-leaf Whiteface Azalea 332158
Hairyleaf Whitlowgrass 137083
Hairyleaf Wingdrupe 405448
Hairyleaf Wingless Begonia 49595
Hairyleaf Wintergreen 172165
Hairy-leaf Wuhan Grape 412006
Hairy-leaved Albizia 13621
Hairy-leaved Allomorphia 16160
Hairy-leaved Ampelopsis 20428
Hairy-leaved Aralia 30623
Hairy-leaved Calliandra 66669
Hairy-leaved Caper 71831
Hairy-leaved Cherry 83189
Hairy-leaved Chonemorpha 88902
Hairy-leaved Cinnamon 91381
Hairy-leaved Cotoneaster 107697
Hairy-leaved Custardapple 25835
Hairy-leaved Dewberry 339082
Hairy-leaved Engelhardia 145510
Hairy-leaved Eucalyptus 155763
Hairy-leaved Evergreen Chinkapin 78885
Hairy-leaved Evodia 161373
Hairy-leaved Flowering-quince 84553
Hairy-leaved Lake Sedge 73782
Hairy-leaved Lasianthus 222158
Hairy-leaved Litse 233996
Hairy-leaved Lyonia 239406
Hairy-leaved Magnolia 279248
Hairy-leaved Maple 3632
Hairy-leaved Mastixia 246221
Hairy-leaved Mockorange 294560
Hairy-leaved Newlitse 264111
Hairy-leaved Ostryopsis 276845
Hairy-leaved Padritree 376283
Hairy-leaved Pepper 300489
Hairy-leaved Raspberry 339084
Hairy-leaved Rush 213496
Hairy-leaved Sedge 74463
Hairy-leaved South Ailanthus 12587
Hairy-leaved Spiraea 372016
Hairy-leaved Sweetscented Basil 268523
Hairy-leaved Taxillus 385221
Hairy-leaved Wintergreen 172165
Hairy-leaves Enkianthus 145694
Hairylip Sage 345317
Hairylobed Butterbur 292404
Hairylraf Rockvine 387826
Hairy-microphyllous Willow 343891
Hairy-nerve China Sweetspire 210387
Hairynerve Choerospondias 88693
Hairy-nerve Clustered Casearia 78121
Hairynerve Falmeflower 295024
Hairy-nerve Fuzzy Magnoliavine 351091
Hairy-nerve Indochinese Sweetspire 210387
Hairynerve Magnoliavine 351091
Hairy-nerve Pendent-bell 145721
Hairynerve Phlogacanthus 295024
Hairynerve Rockvine 387839
Hairy-nerve Serrulate Enkianthus 145721
Hairy-nerve Storax 379391
Hairynerve Tarenna 385016
Hairy-nerved Adina 8196
Hairy-nerved Fig 165513
Hairy-nerved Rockvine 387839
Hairy-nerved Sweetspire 210387
Hairy-nerved Syzygium 382681
Hairy-nerved Tarenna 385016
Hairynerved Wallich Combretum 100853
Hairy-nerves Pearfruit Fig 165532
Hairynode Hairy-callus Roegneria 335229
Hairynode Lagochilus 220076
Hairynode Roegneria 335463
Hairy-noded Oat 45526
Hairy-palmate Maple 3485
Hairypedicel Heartleaved Violet 409840
Hairypedicel Larkspur 124295
Hairy-pedicel Osbeckia 276389
Hairypedicel Smallfruit Holly 204053
Hairypedicel Stairweed 142681
Hairypedicel Tali Larkspur 124623
Hairy-pediceled Osbeckia 276389
Hairy-pedunecle Cherry-laurel 223125
Hairypetal 300698
Hairypetal Ascitesgrass 407456
Hairy-petal Camellia 69528
Hairypetal Clethra 96534
Hairypetal Cliffbean 254797
Hairypetal Cogflower 154155,382766
Hairypetal Ervatamia 154155,382766
Hairypetal Fairy Lantern 67568
Hairypetal Habenaria 183425
Hairypetal Lovely Monkshood 5511
Hairypetal Meconopsis 247192
Hairy-petal Milkvetch 42585
Hairy-petal Nitraria 266360
Hairypetal Photinia 295778
Hairypetal Plantain 302026

Hairypetal Rockfoil 349181
Hairypetal Soapberry 346344
Hairypetal Veronicastrum 407456
Hairy-petaled Indigo 206046
Hairy-petaled Rhododendron 330510
Hairy-petaled Soapberry 346344
Hairypetallike Rockfoil 349668
Hairy-petiolate Maple 3487
Hairy-petiole Maple 3487
Hairypetiole Pennywort 200281
Hairypetiole Wuyuan Maple 3757
Hairypink 292660
Hairypistil Acanthopanax 143625
Hairypistil Actinidia 6712
Hairypistil Kiwifruit 6712
Hairypistil Photinia 295721
Hairypistil Splitmelon 351766
Hairy-pistiled Acanthopanax 143625
Hairy-pistiled Actinidia 6712
Hairy-pistiled Photinia 295721
Hairy-pitted Stork's-bill 153743
Hairypod Fishvine 125959
Hairy-pod Medic 247285
Hairypunjab Sumac 332790
Hairy-pyrene Holly 203979
Hairy-rachilla Littleleaf Willow 343515
Hairyrachis Himalayan Stachyurus 373540
Hairyrachis Sweetleaf 381423
Hairy-rachis Xing'an Poplar 311351
Hairy-rachis Yadong Poplar 311582
Hairy-rachised Sweetleaf 381423
Hairy-rhachis Engelhardia 145519
Hairy-ribbed Tutcheria 322609
Hairy-root Luisia 238327
Hairy-scale Bulbostylis 63327
Hairy-scaled Rattan Palm 65802,65804
Hairyseed Bahia 46527
Hairy-seed Camellia 69490,69715
Hairyseed Stonecrop 357242
Hairy-seeded Camellia 69490,69715
Hairyseeded Hainanlinden 184578
Hairy-seeded Haonania 184578
Hairy-sepal Blackbark Linociera 87726
Hairy-sepal Clematis 95382
Hairysepal Franchet Monkshood 5211
Hairy-sepal Jinping Camellia 69726
Hairy-sepal Jinping Tea 69726
Hairy-sepal Liolive 87726
Hairy-sepal Metalleaf 296413
Hairysepal Rabdosia 209661
Hairysepal Raspberry 338254
Hairysepal Rocket 94241
Hairysepal Rose 336682
Hairy-sepaled Honeysuckle 236195
Hairy-sepaled Rabdosia 209661
Hairy-sepaled Raspberry 338254
Hairysheath Duckbeakgrass 209265
Hairy-sheath Indocalamus 206791
Hairy-sheathed Hedge Bamboo 47367
Hairyshoot Elm 401449
Hairy-shooted Elm 401449
Hairyspathe Devil Rattan 121509
Hairysphere Bluebeard 78042
Hairyspherical Blue-beard 78042
Hairyspike Japanese Germander 388115
Hairyspike Nepeta 264916
Hairy-spike Pepper 300535
Hairyspike Roegneria 335532
Hairyspike Woodbetony 287143
Hairy-spiked Maesa 241779
Hairystalk Bluegrass 305641
Hairy-stalk Clematis 94868
Hairystalk Faber Maple 2958
Hairy-stalk Metalleaf 296366
Hairy-stalk Newlitse 264078
Hairystalk Pearweed 239885
Hairystalk Rhododendron 331430
Hairystalk Serpentroot 354930
Hairystalk Spapweed 204841
Hairystalk Spicebush 231471
Hairystalk Spicegrass 239571
Hairystalk Subfalcate Clematis 95333
Hairystalk Tinospora 392248
Hairystalk Tinospore 392248
Hairystalk Touch-me-not 204841
Hairy-stalk Yunnan Rockvine 387875
Hairy-stalked Clovershrub 70793
Hairy-stalked Spice-bush 231471
Hairy-stalked Tinospore 392248
Hairystamen Primrose 314169
Hairy-stamen Rockvine 387847
Hairystamen Tulip 400148
Hairystamen Wolfberry 239030
Hairy-stamened Wolfberry 239030
Hairystem Elatostema 142803
Hairystem Gooseberry 334030
Hairy-stem Gooseberry 334030
Hairystem Larkspur 124288
Hairystem Rockfoil 349803
Hairy-stemed Veiny Stairweed 142802
Hairystipt Osmanther 276389
Hairystomate Monkshood 5347
Hairy-stomatic Rhododendron 331999
Hairy-stone Holly 203979
Hairystrum Mockorange 294524
Hairystyle Catchfly 363516
Hairystyle Cherry 83277,316705
Hairy-style Delavay Rhododendron 330548
Hairy-style Milkvetch 42488
Hairystyle Mockorange 294549
Hairy-style Rusty Rhododendron 331685
Hairy-styled Camellia 69489
Hairy-styled Cherry 83277,316705
Hairy-styled Mockorange 294549
Hairy-styled Water Nightshade 367334
Hairy-style-like Milkvetch 42719
Hairythroat Marsdenia 245822
Hairythroat Milkgreens 245822
Hairy-throated Condorvine 245822
Hairytongue Woodbetony 287329
Hairytongued Woodbetony 287329
Hairy-toothed She-oak 79191
Hairyturg Largeleaf Spicebush 231388
Hairy-twig Box 64336
Hairy-twig Rhododendron 331992
Hairytwig Willow 343283
Hairy-twigged Cinnamon 91265
Hairy-twigged Rhododendron 330508,331992
Hairyvein 85880
Hairyvein Agrimonia 11572
Hairyvein Ardisia 31567
Hairy-vein Ash 168076
Hairyvein Childseng 318506
Hairyvein China Shortia 362258
Hairy-vein Creeper 285103
Hairyvein Dock 340064
Hairyvein Heshouwu 162516
Hairyvein Jujube 418201
Hairyvein Lettuce 320520
Hairyvein Maesa 241802
Hairyvein Maple 3482
Hairyvein Oak 116212
Hairyvein Ophiorrhiza 272178
Hairyvein Pseudostellaria 318506
Hairyvein Pterocypsela 320520
Hairyvein Qinggang 116212
Hairyvein Samaradaisy 320520
Hairyvein Spindle-tree 157561
Hairyvein Starjasmine 393664
Hairyvein Sweetspire 210387
Hairy-vein Syzygium 382681
Hairy-veined Ardisia 31567
Hairy-veined Cyclobalanopsis 116212
Hairy-veined Jujube 418201
Hairy-veined Maesa 241802
Hairy-veined Maple 3482
Hairywinged Fruit Sophora 369076
Haiscale Sallstyle 63327
Haish 167955
Haitang Blueberry 403846
Haitian Catalpa 79256
Haitian Pine-pink 55548
Haizhou Elsholtzia 144093
Haizhou Sagebrush 35481
Haizishan Sandwort 31941
Hajastan Nepeta 264941
Hakea 184586
Hakerne 324335
Hakone Dichocarpum 128934
Hakone Grass 184656
Hakone Muhly 259669

Hakone Smallreed 127243
Hala Screw Pine 281089
Halabula Siberia Fritillary 168506
Halberd Willow 343467
Halberdleaf Saltbush 44621
Halberd-leaf Tearthumb 291684
Halberd-leaved Orache 44621
Halberd-leaved Rose Mallow 194955
Halberd-leaved Rose-mallow 194955
Halberd-leaved Tear Thumb 309877
Halberd-leaved Tearthumb 309877
Halberd-leaved Tear-thumb 291684,309877
Halberd-leaved Violet 410056
Halberd-leaved Willow 343467
Halbert Dualsheathorchis 85456
Halbertleaf Dock 340075
Halcup 68189
Haldina 184677,184678
Haldu 184678
Halehouse 176824
Hale-men 279685
Halenut 106703
Halerpestes 184702
Hales 106703,109857
Hale-weed 115008,115031
Half Lotus 355391
Halfa 376931
Half-and-half 109857
Halfawn Quakegrass 60390
Halfbearded Larkspur 124572
Half-capitate Hemiboea 191387
Half-clavate Larkspur 124573
Half-cordate Fig 165655
Half-crescent-shaped Rhododendron 331218
Half-exserted Hamilton Spindle-tree 157569
Half-flower 350344
Half-globe Juniper 341765
Half-globe Savin 341765
Half-green Bunge Spindle-tree 157353
Half-hairy Leaved Rhododendron 330848
Halfmoon Loco 42002
Half-moon Loco 42002
Halfmoon Stonecrop 357143
Halfnaked Sinobambusa 364675
Halfpanicle Microtropis 254324
Halfpennies-and-pennies 401711
Halfpouch Corydalis 105969
Halfribpurse 191525
Halfsagittate Wingseedtree 320848
Halfserrate Eurya 160600
Half-serrated Eurya 160600
Halfshrub Securinega 167092
Half-skilt Daffodil 262405
Halfspreading Ironweed 406667
Halfstyle Hairorchis 148656
Halfstyleflower 191486,360701
Halftooth Camellia 69613
Halftwisting Woodbetony 287669
Halfummer 299717,299724
Halfwinged Alloteropsis 16231
Halfwood 95431,239021,367120
Halgania 184760
Halimium Wild Buckwheat 152486
Halimocnemis 184810
Haliwort 106117
Hall Barberry 51699
Hall Crab Apple 243623
Hall Violet 410051
Hall's Aster 380903
Hall's Bulrush 352191
Hall's Catchfly 364030
Hall's Crabapple 243623
Hall's Fleabane 150436
Hall's Hawksbeard 110994
Hall's Honeysuckle 235878
Hall's Rush 213147
Hall's Thistle 91940
Hall's Totara 306438
Hall's Woolly Sunflower 152816
Hallelujah 277648
Haller Pulsatilla 321682
Haller Sedge 74746
Halocnemum 184930
Halogeton 184947,184951
Halopeplis 184962
Halophila 184967
Halophilous Birch 53468
Halophilous Wormwood 35595
Halostachys 185034
Halse 106703,401512
Halse-bush 16352
Halstand Rheedia 329274
Halvah 361317
Halve 109857
Halved Apodytes 29587
Hamabo Hibiscus 194911
Hamakua Pamakani 11171
Hamate Calanthe 65966
Hamate Larkspur 124271
Hamate Mimosa 255040
Hamatipetalous Monkshood 5242
Hamatocactus 185155
Hamburg Parsley 292695,292705
Hame Flower 322950
Hame Vine 322950
Hamelia 185164,185169
Hami Buttercup 325921
Hami Goldenray 229044
Hami Madwort 18415
Hami Milkvetch 42460
Hamilla 52957
Hamilton Barberry 51701
Hamilton Blumea 55749
Hamilton Dendrocalamus 125479
Hamilton Euonymus 157559
Hamilton Ramie 56165
Hamilton Spindle-tree 157559
Hamilton Sweet Dragonbamboo 125479
Hamilton Violet 410052
Hamilton's Spindletree 157559,157699
Hamiltonia 370863
Hammer Sedge 74806
Hammerfruit Zehneria 417517
Hammerwort 284152
Hammock Maple 2797
Hammock Sweet Azalea 332087
Hamose Milkvetch 42461
Hamosespine Cactus 244084
Hampshire Purslane 238192
Hampshire-purslane 238150,238192
Hanaupah Rock Daisy 291338
Hanayu 93869
Hanbury Garcinia 171113
Hance Ampelopsis 20352
Hance Ardisia 31462
Hance Brandisia 59127
Hance Croton 112905
Hance Date 295465,295475
Hance Datepalm 295465
Hance Didymocarpus 129910
Hance Dropseedgrass 372696
Hance Fig 165086
Hance Fishvine 283286
Hance Holly 203871
Hance Jewelvine 283286
Hance Lobelia 234291
Hance Pepper 300408
Hance Raspberry 338498
Hance Rosewood 121703
Hance Sagebrush 35596
Hance Snakegrape 20352
Hance Syzygium 382570
Hance Tanoak 233238
Hance Viburnum 407878
Hance's Rhododendron 330816
Hanceola 185244
Hancock Azalea 330817
Hancock Calanthe 65974
Hancock Clematis 94985
Hancock Dendrobium 125176
Hancock Everlasting 21580
Hancock Grape 411719
Hancock Gromwell 233743
Hancock Indigo 206044
Hancock Luisia 238312
Hancock Mahonia 242553
Hancock Milkvetch 42470
Hancock Mosquitotrap 117606
Hancock Rhododendron 330817
Hancock Sedge 74752
Hancock Spapweed 205000

Hancock Swallowwort 117606
Hancock Violet 410053
Hancock's Curse 328345
Hancock's Pearleverlasting 21580
Hancock's St. John's Wort 201920
Hancockia 185254,185256
Hand of God 221787
Hand Orchid 273531
Hand Orchis 273531
Hand Plant Tree 87809
Hand Satyrion 273531
Handbamboo 20192
Handel Alangium 13370
Handel Arisaema 33342
Handel Asiabell 98397
Handel Begonia 49898
Handel Camellia 69121
Handel Epipactis 147121
Handel Goatweed 8821
Handel Hairycalyx Lobelia 234699
Handel Larkspur 124272
Handel Monkshood 5243
Handel Pennywort 200301
Handel Pleurospermum 304804
Handel Rhodiola 329878
Handel Southstar 33342
Handel Syzygium 382571
Handel Tea 69121
Handel Thistle 92009
Handel Thorowax 63803
Handel Wendlandia 413835
Handel's Tanoak 233240
Handeliodendron 185262
Handflower 86427
Handful Stringbush 414162
Handkerchief Plant 260418
Handkerchief Tree 123245
Handro Begonia 49901
Hand-shaped Polystachya 310372
Hands-in-pockets 20435
Handsome Barberry 51356
Handsome Cactus 244009
Handsome Flat Pra 302849
Handsome Mackaya 240736
Handsome Paphiopedilum 282898
Handsome Parnassia 284626
Handsome Pussytoes 26321
Handsome Sedge 74595
Handsome-harry 329427
Hang Milkvetch 42283
Hang'ai Sagebrush 35414
Hangchow Mosla 259295
Hangdowns 243711
Hanging Bellflower 145714
Hanging Down Bean 293989
Hanging Geranium 288430,349936
Hanging Heliconia 190052
Hanging Rhododendron 331458
Hanging Sedge 75742
Hanging-down Milkvetch 42283
Hanging-flower Euonymus 157761
Hanging-flowered Euonymus 157761
Hanging-garden Rock Daisy 291333
Hangskirt Rattlebox 112615
Hangzhou Azalea 330820
Hangzhou Bitter Bamboo 304002
Hangzhou Bitterbamboo 304002
Hangzhou Mosla 259295
Hangzhou Rhododendron 330820
Hangzhou Stonecrop 356781
Hanilton Yam 131614
Hankow Aster 40160
Hankow Willow 343668,343669
Hanmi Raspberry 338551
Hanous Chuniophoenix 90622
Hanry Atalantia 43727
Hanry Blackhatflower 122933
Hanry Dasymaschalon 122933
Hanry Sinojackia 364953
Hanry Weigttree 364953
Hans and Roosters 410380
Hans Nodding Broomrape 274995
Hansen's Bush Fruit 83347
Hansen's Larkspur 124273
Hanson's Lily 229857
Hao Poplar 311340
Hao Stringbush 414188
Hao Willow 343465
Hao Windhairdaisy 348365
Hao's Currant 334014
Haoge Zinnia 418038
Haohan Prickly-thrift 2249
Haoping Fourstamen Maple 3685
Haoryanther Baoshan Monkshood 5438
Haoryanther Paoshan Monkshood 5438
Hap Pennies-and-pennies 401711
Hapaline 185304
Haplanthoides 185337
Haplophyllum 185615
Haplophyllum-like Edelweiss 224850
Haplosphaera 185703
Hapotato 99699
Happenies-and-pennies 401711
Happy Plant 137397
Haps 336419
Haps Tree 311537
Haraella 185753
Hara-giri 215442
Harber 77256
Harberdleaf Violet 409716
Harberdlip Stonebean-orchis 62685
Harbertleaf Knotweed 309877
Harbin Elm 401521
Harbin Green Poplar 311274
Harbin Poplar 311274
Harbin Sharphead Sagebrush 36033
Harbinger Cactus 244099
Harbinger of Spring 150412
Harbinger-of-spring 150412
Harbison Hawthorn 109748
Harbor Cactus 244101
Harbour Dwarf 262195
Harbs 109857
Harbur 77256
Harcher 86427
Hard Arrowbamboo 162681
Hard Bambusa 47321
Hard Beech 266886
Hard Birch 53338
Hard Bluegrass 306003
Hard Catchfly 363451
Hard Elm 401618
Hard Eremopyrum 148400
Hard Fargesia 162681
Hard Fescue 163850,164159,164364
Hard Fescuegrass 163918
Hard Gentian 173814
Hard Giantfennel 163584
Hard Giant-timbered Bamboo 297215
Hard Goosecomb 144455
Hard Grass 354400,354403
Hard Hack 372106
Hard Hay 202179
Hard Heads 81237
Hard Ironbark 155571
Hard Maple 3236,3549,3576
Hard Meadow Grass 354477
Hard Melandrium 363451
Hard Milkvetch 42982
Hard Parasite 293195
Hard Pine 300181
Hard Quandong 142365
Hard Rockfoil 349147
Hard Rockjasmine 23266
Hard Roegneria 144455
Hard Rush 213165
Hard Sheep Fescue 164364
Hard Sikkim Sweetvetch 188108
Hard Sorghum 369640
Hard Thistle 91770
Hard Wallflower 154467
Hard Wheat 398900
Hard Woodbetony 287633
Hard Yelloweyegrass 416037
Hardback 99844
Hardbean 241408,241440
Hardbract Cousinia 108400
Hardbract Vinegentian 108400
Hardbract Windhairdaisy 348231
Hardcalyx Arnebia 34609
Hardculm Gaoliang 369640

Hardculm Goosecomb 335483
Hardee Peppertree 350994
Hardenbergia 185759
Hardened Ormosia 274403
Hardflower Craibiodendron 108582
Hardflower Goldleaf 108582
Hard-flowered Craibiodendron 108582
Hardfruit Calvepetal 178004
Hardfruit Glyptopetalum 178004
Hardfruit Halimocnemis 184824
Hardfruit Sedge 76205
Hardgrass 354400,354403,385816,385818
Hard-grass 354400,354579
Hardhack 81237,371902,372106
Hard-hack 372106,372107
Hardhack Spirea 372106
Hardhair Bittersweet 80205
Hardhair Bractbean 362297
Hardhair Cajan 65149
Hardhair Clubfilment 179204
Hardhair Crazyweed 278828
Hardhair Dumasia 138936
Hardhair Evodia 161342
Hardhair Gentian 173521
Hardhair Giantarum 20090
Hardhair Grewia 180948
Hardhair Indocalamus 206792
Hardhair Knotweed 309188
Hardhair Lacquertree 393460
Hardhair Locust 334965
Hardhair Pellionia 288780
Hardhair Redcarweed 288780
Hard-hair Sophora 369007
Hardhair Stairweed 142682
Hardhair Violet 410066
Hardhair Whitefruit Whitepearl 172102
Hardhairy Bethlehemsage 283620
Hardhairy Lacquer-tree 393460
Hardhay 202086,202179
Hardhead 4005,6278,11947,81237,81356,
192358,235373,302034,302068,345881,
379660
Hardhead Bitterbamboo 304056
Hardhead Knapweed 81237
Hardheads 6278,81150,81237,81356
Hard-heads 81356
Hardhow 66445
Hardine 81237,325612
Harding Grass 293708,293768
Harding's Grass 293764
Harding's-grass 293764
Hardleaf Arundinella 37373
Hardleaf Azalea 331939
Hardleaf Bicolor Orchis 116797
Hardleaf Camellia 69094
Hardleaf Chirita 87958
Hardleaf Cinnamon 91309
Hardleaf Cymbidium 116797
Hardleaf Elytrigia 144675
Hardleaf Golden Rod 368354
Hard-leaf Holly 203810
Hardleaf Mycetia 260620
Hardleaf Orchis 116915
Hardleaf Oreorchis 274051
Hardleaf Pipewort 151474
Hardleaf Pocktorchid 282867
Hardleaf Sedge 75724
Hardleaf Tanoak 233162
Hardleaf Wildorchis 274051
Hardleaf Willow 344072
Hardleaf Windhairdaisy 348211
Hard-leaf Yunnan Holly 204419
Hard-leafed Monkey Plum 132386
Hardleaf-like Willow 344075
Hard-leaved Mycetia 260620
Hard-leaved Willow 344072
Hardlemma Barnyardgrass 140430
Hardock 31051
Hardoke 31051
Hardpith Maesa 241737
Hard-pithed Maesa 241737
Hardshape Woodbetony 287634
Hardsheath Yushanbamboo 416758
Hardspike Fescuegrass 163914
Hardspine Azalea 330203
Hard-spinous Rhododendron 330203
Hardspiny Cocklebur 415057
Hardstem Bulrush 352164
Hard-stem Bulrush 352164
Hardstem Spikesedge 143073
Hard-tack 83820
Hardway 202179
Hardwick Cucumis 114249
Hardwick Valerian 404272
Hardwood Sweetleaf 381188
Hardy Ageratum 101955
Hardy Banana 260198,260208
Hardy Begonia 49886
Hardy Blue Bamboo 162718
Hardy Canna 388385
Hardy Catalpa 79260
Hardy Chinese Ash 167931
Hardy Cluster Amaryllis 239280
Hardy Crinum 111180,111217
Hardy Dragon Bamboo 162679
Hardy Eucryphia 156064
Hardy Fuchsia 168767
Hardy Garden Pink 127747
Hardy Geranium 174715
Hardy Giant Timber Bamboo 297215
Hardy Gloxinia 205546,205548,205557
Hardy Hibiscus 194680
Hardy Hummingbird Trumpet 146648
Hardy Ice 89959
Hardy Ice Plant 123819,123854,123932
Hardy Kiwi 6515
Hardy Orange 310850
Hardy Orchid 55575
Hardy Pampas Grass 148906
Hardy Pink Ice Plant 123854
Hardy Plumbago 83641
Hardy Purple Sage 345472
Hardy Rubber Tree 156040,156041
Hardy Rye Grass 235353
Hardy Ryegrass 235353
Hardy Rye-grass 235353
Hardy Timber Bamboo 297215
Hardy Yellow Ice Plant 123932
Hare Figwort 355165
Hare Nut 102727
Hare's Ballocks 196376
Hare's Beard 405788
Hare's Colewort 368771
Hare's ear 63553,63805,102811
Hare's Ear Cabbage 102811
Hare's Ear Mustard 102811
Hare's Eye 248312
Hare's Foot Clover 396828
Hare's Foot Dalea 121893
Hare's Foot Plantain 302028
Hare's Foot Trefoil 396828
Hare's Lettuce 260596,368771
Hare's Meat 277648
Hare's Palace 368771
Hare's Parsley 28044,101852
Hare's Tail 152769,152791
Hare's Tail Grass 220254,220252
Hare's Tails 152769
Hare's Thistle 368771
Hare's-ear 63553,102808
Hare's-ear Cabbage 102811
Hare's-ear Mustard 102808,102811
Hare's-foot 396828
Hare's-foot Clover 396828
Hare's-foot Fern 122898
Hare's-foot Plantain 302028
Hare's-foot Sedge 75030,75116
Hare's-foot Trefoil 396828
Hare's-tail 152791,220252,220254
Hare's-tail Cotton-grass 152791
Hare's-tail Grass 220252,220254
Harebell 69870,70271,130383,145487,
199560,412568
Harebell Flax 231882
Harebell Hungarian Speedwell 407299
Harebell Poppy 247175
Harebottle 81237
Hareburr 31051
Hare-ear Mustard 102811
Harefoot 175454,396828
Harenut 102727

Harestail 152791
Harestail Grass 220254
Harestones 273531
Harestrong 292957
Haretail Uraria 402132
Harewell Speedwell 407299
Harewood 3462
Harewort 225398
Harford's Ragwort 279891
Harford's Wild Buckwheat 152325
Harger's Goldenrod 368029
Haricot 218715
Haricot Bean 294056
Harif 170193
Haritch 170193
Harland Balanophora 46829
Harland Box 64254
Harland Boxwood 64254
Harland Fig 164985,165231
Harland Pearlsedge 354110
Harland Razorsedge 354110
Harland Sedge 74758
Harland Snakemushroom 46829
Harland Tanoak 233241
Harlequin Blue Flag 208924
Harlequin Butterfly Bush 62025
Harlequin Flower 369975,370009
Harlequin Fuchsia Bush 148360
Harlequin Glorybower 96398
Harlequin-flower 106430
Harley Meconopsis 247127
Harlock 31051,364557
Harmal 287978
Harmal Peganum 287978
Harmand Aganosma 10220
Harmand Antheroporum 27466
Harmand Elaeocarpus 142330
Harmand Fig 165092,165231
Harmand Glyptic-petal Bush 177994
Harmand Rockvine 387838
Harmand Spatholobus 370415
Harmand Swellpod 27466
Harmand's Blueberry 403850
Harms Clovershrub 70819
Harms Honeysuckle 236180
Harofe 170193
Haropito 318624,318625
Harp Onefruit 185355
Harpachne 185828,185830
Harpaper Stairweed 142790
Harpaper Treerenshen 125630
Harper's Beauty 185875
Harper's Trout-lily 154901
Harper's Wild Buckwheat 152243
Harping Johnny 357203
Harpur-Crewe's Leopard's-bane 136335
Harriff 170193
Harrington Plum Yew 82523
Harrington Plumyew 82523
Harris' Goldenrod 367970
Harrisia 185912
Harrison Bifrenaria 54246
Harrison Yellow Rose 336291
Harrison's Barberry 51704
Harrisonia 185931
Harrow Gentian 173497
Harrow Jewelvine 125970
Harrow Mountainash 369407
Harrow Mountain-ash 369407
Harrow's Butterflybush 62070
Harrow-rest 271563
Harrul 16352
Harry Dobs 127635
Harry Lauder's walking stick 106703
Harry Nettle 53304
Harrya Odontoglossum 269071
Harrysmith Cotoneaster 107473
Harrysmith Sedge 74762
Harrysmithia 185935
Harsculm Ministalkgrass 71603
Harsh Downy-rose 336993
Harsh Hakea 184629
Harstrong 292957
Harsy 109857
Hart Claver 249232
Hart Lasianthus 222153,222172
Hart Mint 250450
Hart Roughleaf 222153,222172
Hart's Clover 249205,249232
Hart's Ragwort 279915
Hart's Thorn 328642
Hart's Tree 249232
Hart's-thorn 328642
Hartberry 403916
Hartia 185943
Hartmann Sarcochilus 346701
Hartmann Sedge 74763
Harts 403916
Hartsease 410677
Hartshorn 105352,301910,328642
Hartshorn Plantain 301910
Hartsthorn 328642
Hartweg Iris 208608
Hartweg Lupine 238448
Hartweg Mahonia 242554
Hartweg Pine 299967
Hartweg's Golden Sunburst 317363
Hartweg's Sisymbrium 365480
Hartwegs Penstemon 289350
Hartwig Crabapple 243628
Hartwort 361526,392863,392875
Harves 109857
Harvest Bells 70271,173728
Harvest Brodiaea 60447,60460
Harvest Daisy 227533,397970
Harvest Flower 89704,321554
Harvest Lice 11549,170193
Harvest Lily 68713
Harvest-bells 173840
Harvest-lice 11571
Harvey's Beaked Rush 333604
Harvey's Buttercup 325922
Harvies 109857
Haryhound 245770
Haselberg Ballcactus 59164
Hashish 71218,71220
Hash-khuli 93451
Hasill Tree 106703
Haskett 2837,106703
Haskwort 70119,70331
Haspen 311537
Hassaku-mikan 93482
Hassock 213021
Hassock Grass 126039
Hastate Atriplex 44621
Hastate Butterflybush 62071
Hastate Butterfly-bush 62071
Hastate Cacalia 283816,283826
Hastate Calanthe 66020
Hastate Clematis 94986
Hastate Indian-plantain 186137
Hastate Knotgrass 309877
Hastate Merremia 415299
Hastate Mosquitotrap 117420
Hastate Orache 44621
Hastate Rhubarb 329338
Hastate Sugarpalm 32337
Hastate Swallowwort 117420
Hastate Treebine 92741
Hastate Verbena 405844
Hastate Violet 410056
Hastate Willow 343467
Hastateleaf Ixeris 210547
Hastate-leaf Japanese Mulberry 259103
Hastateleaf Mulberry 259103
Hastateleaf Paraprenanthes 283684
Hastateleaf Raspberry 338499
Hastateleaf Sage 344918
Hastateleaf Sedge 74766
Hastateleaf Skullcap 355469
Hastateleaf Treebine 92741
Hastate-leaved Dock 340075
Hastate-leaved Raspberry 338499
Hastatelip Bulbophyllum 62685
Hastatelip Odontoglossum 269072
Hastileleaf Summerlilic 62071
Hasty Peach 34475
Hasty Roger 221787,355202
Hasty Sergeant 221787
Hatchet Cactus 288615,288616
Hatchet Vetch 356377

Hatchets-and-billhooks 2837
Hatch-horn 324335
Hate's Ballocks 196376
Hate's Beard 405788
Hate's Colewort 368771
Hate's Thistle 368771
Hather 67456
Hat-pins 151208,219040
Hatpins Pipewort 151276
Hatsushima Ladybell 7660
Hattie's Pincushion 43306
Hattor Sedge 74768
Haughty Acineta 4653
Haul 106703
Haumachil 301128
Haun Barberry 51706
Hausa Potato 99670,99699,303289,303624
Haussknecht's Sulphur Flower 152587
Hautbois 167631
Hautbois Strawberry 167614,167627,167631
Hautboy 167627,167631
Hautboy Strawberry 167627
Hauth 401388
Hav 45566
Havana Snakeroot 11163
Havard Agave 10866
Havard Oak 323998
Havard's False Willow 46229
Havard's Wild Buckwheat 152120
Havels 198376
Haverdril 262441
Havers 45448
Haves 109857,336419
Haw 45566,109857,336419,408061
Haw Bush 109857
Haw Gaw 109857
Haw Tree 109857
Hawai Woodnettle 221568
Hawaian Blue Eyes 161447
Hawaii Ticktrefoil 126596
Hawaii Yelloweyed Grass 416037
Hawaiian Arrowroot 382918,382923
Hawaiian Cherry 382722
Hawaiian False Nettle 56164
Hawaiian Giant Taro 16512
Hawaiian Hibiscus 195149
Hawaiian Ti Plant 104350
Hawaiian White Hibiscus 194722
Hawallan Elf Schefflera 350654
Hawberry 109857,109996
Hawk Nut 63467,102727
Hawk's Beard 110710,111017,111050
Hawk's Bill Bramble 338447
Hawk's Bit 224651
Hawk's Eye 160837
Hawk's Foot 30081
Hawk's Weed 195437
Hawk's-beard 110710,110767,111050
Hawk's-bit 224655
Hawkberry 83155,280003
Hawkbit 224651
Hawkes Pleurothallis 304915
Hawkins' Oak 323586
Hawknut 63467
Hawk-nut 102727
Hawks Beard 110965
Hawksbeard 110710
Hawks-beard 110710
Hawksbeard Velvetplant 108736
Hawkweed 195437,195643,195708,195780,195948
Hawkweed Gowan 384714
Hawkweed Leaf Synotis 381939
Hawkweed Oxtongue 298589
Hawkweed Ox-tongue 298589
Hawkweed Saxifrage 349444
Hawkweed Wild Buckwheat 152137
Hawkweedleaf Blumea 55754
Hawkweedleaf Erysimum 154467
Hawkweedleaf Sugarmustard 154467
Hawkweedleaf Tailanther 381939
Hawkweed-like Saussurea 348370
Hawk-your Mother's-eyes-out 407072
Hawk-your-Mother's-eyes-out 407072
Haworth Groundsel 359026
Haworth's Aeonium 9047
Haworthia 186251,186792
Haworth's Aeonium 9048
Hawp 336419
Hawth 401388
Hawthorn 109509,109513,109541,109559,109674,109700,109729,109743,109746,109749,109782,109788,109790,109818,109821,109822,109832,109857,109870,109876,109909,109942,109995,110009,110086,329056
Hawthorn Barberry 51510
Hawthorn Cherry 83182
Hawthorn Currant Tree 334137
Hawthorn Maple 2910
Hawthornleaf Cherry 83182
Hawthornleaf Maple 2910
Hawthornleaf Raspberry 338300
Hawthorn-leaved Cherry 83182
Hay Beech 77256,162400
Hay Gob 162519
Hay Hoa 176824
Hay Plant 313528,364193
Hay Rattle 329521,364193
Hay Rope 95431
Hay Sedge 74579,76261
Hay Shakers 60382
Hayata Barberry 51708,51926
Hayata Begonia 49626
Hayata Cleyera 96593
Hayata Corkleaf Storax 379348
Hayata Cranesbill 174645
Hayata Elaeocarpus 142400
Hayata Eurya 160553
Hayata Fig 165762
Hayata Glochidion 177184
Hayata Greenbrier 366361
Hayata Holly 203883
Hayata Leaf-flower 296695
Hayata Litsea 233832
Hayata Medinilla 247571
Hayata Natotree 280440
Hayata Spiraea 371927
Hayata Storax 379348
Hayata Strawberry 167620
Hayata Tainan Greyawngrass 372435
Hayata Uvaria 403519
Hayata Wildginger 37646
Hayata Yellow Cranesbill 174949
Hayata's Snowbell 379348
Hayatagrass 186856
Hayden's Mountaincrown 274126
Hayden's Sedge 74774
Hayfield Tarweed 191710
Hayhouse Hayhove 176824
Haymaidens 176824
Haymaids 176824
Haynald Paphiopedilum 282834
Hayne Matucana 246595
Hayrattle 329543
Hayriff 162519,166125
Hay-rough 170193
Haythorn 109857
Hazal Family 106623
Hazel 106696,106703,109857,234269
Hazel Alder 16440,16445,16447
Hazel Bottle Tree 376110
Hazel Catkins 106703
Hazel Crottle 234269
Hazel Dodder 114999
Hazel Pine 232565
Hazel Sterculia 376110
Hazelnut 106696,106698,106703
Hazelnut Family 106623
Hazelnuts 106696
Hazel-raw 234269
Hazelwort 37616
Hazle 109857
Hazzle 106703
He Broom 218761
He Bulkishawn 359158
He Dishspurorchis 171859
He Gastrochilus 171859
He Heather 67456
He Holly 203545
He Peony 280231

Head Betony 287063
Head Cabbage 59520
Head Echinaria 140105
Head Elsholtzia 143973
Head Flowers Rhododendron 330324
Head Garden Lettuse 219489
Head Lettuce 219489
Head Pellionia 288717
Head Rattlebox 112383
Head Redcarweed 288717
Head Rhubarb 329337
Headache 72934,174877,282546,282685,
374916
Headache Tree 313740,401670,401672
Headache Tree Premna 313702
Headacher 282685
Head-dog 235090
Headed Cabbage 59532
Headed Flowers Rhododendron 330338
Headed Ixora 211065
Headflower Bellflower 70065
Headflower Elsholtzia 143970
Headflower Gentian 173333
Headflower Knotweed 308946
Headflower Roscoea 337074
Headflower Woodbetony 287069
Head-flowered Elsholtzia 143970
Head-form Kunming Holly 203951
Headfort Rhododendron 330830
Headfruit 108661,108665
Headleaf Loosestrife 239786
Headlike Bulrush 396025
Headlike Chalkplant 183180
Head-like Leucas 227571
Headman 302034
Headshape Kneejujube 67008
Headstyle Rush 212994
Headwarke 282685
Headwork 282685
HeaHeart-leaved Leptopus 226325
Heal All 316116
Heal-all 37015,99844,316127,357203,
404316
Heal-bite 235090
Heal-dog 235090
Healing Blade 302068,358062
Healing Leaf 302068,357203,358062
Heardtwig Buckwheat 162336
Heare-nut 102727
Hearer Hayver 235334
Hearfruit Cystopetal 320210
Heart Balata 244605
Heart Calligonum 67013
Heart Clover 247246,384714
Heart Fever Grass 384714
Heart Leaf 294838
Heart Leaf Adina 184678
Heart Leaved Aster 381007
Heart Liverleaf 192123
Heart Liver-leaf 192123
Heart Medick 247246
Heart Nut 212624
Heart of Flame 60546
Heart Pansy 410677
Heart Pea 73207
Heart Pertybush 292047
Heart Rockfoil 349202
Heart Seed 73207
Heart Snakeroot 37754
Heart Tree 83736
Heart Trefoil 247246
Heart Vandellia 231479
Heart Vine 84305
Heart's Dea 73201
Heart's Delight 690
Heart's Ease 309570,410677
Heart's Seed 73201
Heart's-ease 309298,309570,410677
Heart-based Globe Rhododendron 331401
Heart-denseleaf Poplar 311525
Heart-flowered Orchis 360591
Heart-flowered Serapias 360591
Heartfruit Milkwort 308122
Heartfruit Speedwell 407063
Hearth French Heath 149565
Heartleaf 194569
Heart-leaf 29853,171969,192123,192148
Heartleaf Actinidia 6518
Heartleaf Adenia 7233
Heartleaf Ampelopsis 20334
Heartleaf Aptenia 29855
Heartleaf Ardisia 31509
Heartleaf Arnica 34699
Heartleaf Arrowwood 407759
Heartleaf Asiabell 98308
Heartleaf Aster 380856
Heart-leaf Aster 40256
Heart-leaf Bbirch 53387
Heartleaf Begonia 49971
Heart-leaf Belostemma 50932
Heartleaf Bergenia 52514
Heartleaf Berrya 52957
Heartleaf Birch 53387
Heart-leaf Birch 53388
Heartleaf Bittercress 275869
Heartleaf Bluebell 115345
Heartleaf Boesenbergia 56535
Heartleaf Buckwheat 162314
Heartleaf Carpesium 77155
Heartleaf Catnip 264925
Heart-leaf Cayratia 79839
Heartleaf Chirita 87842
Heartleaf Chiritopsis 87997
Heartleaf Collectivegrass 381030
Heartleaf Convolvulate Asiabell 98295
Heartleaf Coreopsis 104468
Heartleaf Crabapple-like Actinidia 6663
Heartleaf Dumasia 138933
Heartleaf Euphorbia 159854
Heartleaf Evening-primrose 269423
Heartleaf Falsebetony 373069
Heartleaf Falsetamarisk 261290
Heartleaf Felwort 235441
Heartleaf Fieldcitron 400545
Heartleaf Figmarigold 29855
Heartleaf Foamflower 391522
Heart-leaf Foam-flower 391522
Heartleaf Fordiophyton 167297
Heartleaf Four o'clock 255735
Heartleaf Goldeneye 409152
Heartleaf Goldenray 229107
Heartleaf Greygreen Gentian 174085
Heartleaf Hartia 376435
Heartleaf Hemipilia 191591
Heartleaf Heterostemma 194165
Heartleaf Hibiscus 195000
Heartleaf Hornbeam 77276
Heartleaf Horsenettle 367012
Heartleaf Houndstongue 118023
Heartleaf Houttuynia 198745
Heartleaf Iceplant 29855
Heart-leaf Ice-plant 29855
Heartleaf Jewelflower 377638
Heartleaf Jurinea 211960
Heartleaf Kiwifruit 6518
Heartleaf Ladybell 7629
Heartleaf Lily 73158
Heartleaf Liparis 232125
Heartleaf Madeiravine 26265
Heartleaf Mananthes 244284
Heartleaf Merremia 250772
Heartleaf Mischobulbum 383127
Heartleaf Mock Orange 294426
Heartleaf Morning-glory 207714
Heartleaf Myriactis 261085
Heartleaf Ophiorrhiza 272193
Heart-leaf Orach 44356
Heartleaf Oreocharis 273848
Heartleaf Ottelia 277382
Heartleaf Oxeye 385616
Heartleaf Parnassia 284520
Heartleaf Penstemon 289333
Heartleaf Pentapanax 289664
Heartleaf Peppervine 20334
Heart-leaf Philodendron 294798
Heartleaf Plantain 301907
Heartleaf Primrose 314774
Heartleaf Rabbiten-wind 12616
Heartleaf Red Silkyarn 238960
Heartleaf Rhubarb 329309
Heartleaf Ribseedcelery 304839

Heartleaf Saussurea 348228
Heartleaf Sloanea 366046
Heartleaf Snakegourd 396162
Heartleaf Snakegrape 20334
Heartleaf Snakeroot 37754
Heartleaf Sotvine 194165
Heartleaf Speedwell 317909
Heart-leaf Speedwell 407022
Heartleaf Spicegrass 239601
Heartleaf Spiradiclis 371751
Heartleaf Sunflower 189003
Heartleaf Swertia 380166
Heartleaf Synotis 381932
Heartleaf Tailanther 381932
Heartleaf Tailgrass 402506
Heartleaf Tea 69010
Heart-leaf Tickclover 126349
Heartleaf Truncate Twinballgrass 209134
Heartleaf Tubergourd 390119
Heart-leaf Twayblade 232905
Heartleaf Urariopsis 402153
Heartleaf Vetch 408348
Heartleaf Viburnum 407964
Heart-leaf Vineclethra 95483
Heartleaf Waxplant 198836
Heart-leaf Waxplant 198835
Heart-leaf Willow 343343
Heartleaf Windhairdaisy 348228
Heartleaf Woodnettle 221541
Heartleaf Zygia 30963
Heart-leafed Aster 40256
Heart-leafed Beard Heath 227963
Heart-leafed Flame Pea 89135
Heart-leafed Flame-pea 89135
Heart-leaved Adenia 7233
Heart-leaved Adina 184678
Heart-leaved Alexanders 418107
Heart-leaved Begonia 49740
Heart-leaved Bennettiodendron 51031
Heart-leaved Bergenia 52514
Heart-leaved Berrya 52957
Heart-leaved Bitter-cress 72724
Heart-leaved False-tamarisk 261290
Heart-leaved Globe Daisy 177035
Heart-leaved Globe-daisy 177035
Heart-leaved Golden Alexanders 418107
Heart-leaved Groundsel 279888,279939,279941
Heart-leaved Hartia 376435
Heart-leaved Heterostemma 194165
Heart-leaved Honeysuckle 236120
Heart-leaved Hornbeam 77276
Heart-leaved Jasmine 211960
Heart-leaved Meadow Parsnip 418107
Heart-leaved Meadow-parsnip 418107
Heart-leaved Mock Orange 294425
Heart-leaved Mock-orange 294426
Heart-leaved Oxeye 385616
Heart-leaved Pentapanax 289664
Heart-leaved Philodendron 294838
Heart-leaved Plantain 301907,301909
Heart-leaved Silver Gum 155538
Heart-leaved Skullcap 355643
Heart-leaved Snakeroot 91007
Heart-leaved Stachyurus 373537
Heartleaved Stonebreak 52514
Heart-leaved Twayblade 232905
Heart-leaved Umbrellawort 255735
Heart-leaved Umbrella-wort 255735
Heart-leaved Urariopsis 402153
Heartleaved Violet 409837,409840
Heart-leaved Willow 343343
Heart-leaved Zygia 30963
Heartlip Addermonth Orchid 110653
Heartlip Bogorchis 110653
Heart-liver 247246
Heartmint 250450,403916
Heartnut 212602
Heart-o'-the-earth 316127
Heart-of-flame 60546
Heart-of-Jesus 65218
Heart-ovateleaf Hanceola 185246
Heartovateleaf Oreocharis 273839
Heart-pea 73207
Heartpetal Corydalis 106190
Heart-podded Hoary Cress 73090
Heart-podded Hoary-cress 73090
Hearts 403916
Hearts And Honey Vine 323457
Hearts at Ease 410677
Hearts Ease 410677
Hearts-and-honey 323457
Hearts-and-honey Vine 323457
Heartsease 86427,309570,316127,381035,409624,409703,410513,410677
Heartsease Wild Pansy 410677
Heartseed 73194,73207,309542,410677
Heart-seed 73194,73207
Heartseed Gynostemma 183002
Heartseed Horseegg 183002
Hearts-entangled 84305
Heartshape False Pickerelweed 257583
Heartshape Kneejujube 67013
Heartshape Mikania 254424
Heart-shaped Aster 40256
Heartshaped Calanthe 65901
Heart-shaped Neohouzeaua 351851
Heart-shaped Orchid 121370
Hearts-on-strings 220729
Heartstipular Clearweed 298899
Heartstipular Coldwaterflower 298899
Heartweed 309570
Heartwing Sorrel 340075
Heartwort 34149,249232,309570
Heartychoke 117787
Heath 67456,145073,148619,148941,383491
Heath Aster 380872
Heath Banksia 47641
Heath Bedstraw 170611
Heath Bellflower 70271
Heath Bramble 338209
Heath Broom 150131
Heath Cudweed 178465
Heath Dog Violet 409804
Heath Dog-violet 409804
Heath False Brome 58604
Heath Family 150286
Heath Grass 122210
Heath Groundsel 360172
Heath Lobelia 234860,234867
Heath Melaleuca 248093
Heath Milkwort 100871,308357
Heath Myrtle 46425
Heath Orchid 121382
Heath Pea 222758
Heath Pearlwort 32301,342298
Heath Rush 213471
Heath Sedge 74542,74547,74656
Heath Speedwell 407256
Heath Spotted Orchid 273531
Heath Stitchwort 374897
Heath Urts 145073
Heath Vetch 222758
Heath Violet 409804
Heath White Cedar 85377
Heath Woodrush 238610
Heath Wood-rush 238647
Heathbell 70271
Heathberry 145073,403916
Heather 67455,67456,114604,149205
Heather Bell 70271,150131
Heather Family 150286
Heather Goldenweed 150322
Heather Whin 172905,173082
Heatherberry 403916
Heatherbun Chamaecyparis 85378
Heather-fue 152769
Heather-leaf Wild Buckwheat 152047
Heath-goldenrod 150328
Heath-grass 122180,122210
Heathleaf Banksia 47641
Heath-leaf Banksia 47641
Heath-leaved Banksia 47641
Heathliving Rabdosia 209870
Heaven's Eye 260842
Heaven's Eyes 260842
Heavenly Bamboo 262188,262189
Heavenly Rhododendron 330440
Heavenly Wild Buckwheat 152352
Heavenly-bamboo 262189

Heaver Hayver 235334
Heavespike Loosestrife 239558
Heavy Sedge 74709
Heavysnow Cymbidium 116984
Heavysnow Orchis 116984
Heavy-wooded Pine 300153
Hebakou Ceropegia 84259
He-balsam 298423
He-barfoot 190936
Hebe 186925
Hebei Caraway 77775
Hebei Chalkplant 183263
Hebei Cortusa 105501
Hebei Goldenray 229058
Hebei Mockorange 294538
Hebei Oak 324012
Hebei Orchis 169906
Hebei Pear 323194
Hebei Poplar 311349
Hebei Reed 295914
Hebei Sedge 76485
Hebei Speedwell 407078
Hebei Walnut 212614
Hebei Whitethroat Monkshood 5360
Hebei Willow 344172
Hebetate Goldwaist 90375
Hebon 385302
He-brimmel 338447
He-broom 218761
Hebukesair Dragonhead 137597
Hebukesair Greenorchid 137597
He-bulkishawn 359158
Hech-How 101852
Hechi Chirita 87885
Hechi Pepper 300414
Hechtia 187129
Heckberry 280003
Heckel Stonecrop 356782
Heckrialeaf Rockfoil 349433
Hector Willowweed 146722
Hedder 67456
Hedehog 325612
Heder's Hippolytia 196698
Hedera Spindle-tree 157576
Hedge Acacia 227430
Hedge Apple 240828
Hedge Bamboo 47345
Hedge Bambusa 47345
Hedge Barberry 52169
Hedge Basil 97060
Hedge Bedstraw 170490
Hedge Bells 68713,102933
Hedge Bindweed 68699,68705,68713,
68732,162528,207590
Hedge Buckthorn 328691
Hedge Bur 170193
Hedge Cactus 83896
Hedge Calamint 97060
Hedge Calystegia 68713
Hedge Caper 71851
Hedge Clivers 170193
Hedge Dead Nettle 373455
Hedge Euphorbia 159457,159458
Hedge False Bindweed 68713
Hedge Feathers 95431
Hedge Garlic 14953,14964
Hedge Garliccress 365564
Hedge Glorybind 68713
Hedge Gooseberry 333900
Hedge Grape 61497
Hedge Hyssop 180292,180304,180319,
180321,180338,203102,308454,355609
Hedge Ivy 383675
Hedge James Barberry 51782
Hedge Knotweed 162528
Hedge Lily 68713
Hedge Lime 365424
Hedge Lover 174877
Hedge Maple 2837
Hedge Mustard 126127,365398,365424,
365500,365564,365568
Hedge Nettle 170078,373085,373126,
373163,373253,373346,373455,373459
Hedge Nut 106703
Hedge Parsley 392959,392992
Hedge Picks 316819
Hedge Pigs 316819
Hedge Pink 346434
Hedge Poppy 130383,405788
Hedge Prinsepia 315177,365003
Hedge Privet 229617
Hedge Ragwort 58477
Hedge Sageretia 342198
Hedge Sisymbrium 365564
Hedge Speak 316819,336419
Hedge Stitchwort 374916
Hedge Taper 405788
Hedge Thorn 76864,109857
Hedge Veronica 186925,186926
Hedge Vetch 408611
Hedge Vine 61497,95431
Hedge Violet 410493
Hedge Woundwort 373455
Hedge-apple 240828
Hedge-ball 240828
Hedgebavin 392368,392371
Hedgeberry 280003
Hedgeheriff 170193
Hedgehog Agave 10942
Hedgehog Barberry 51715
Hedgehog Broom 151077,151078
Hedgehog Cactus 140209,140242,140842,
287881
Hedgehog Clover 396885
Hedgehog Club Rush 118784
Hedgehog Coneflower 140081
Hedgehog Dog's-tail 118319
Hedgehog Dogtailgrass 118319
Hedgehog Felicia 163188
Hedgehog Fir 442
Hedgehog Gourd 114147
Hedgehog Grass 80803
Hedgehog Hakea 184606
Hedgehog Holly 203553
Hedgehog Juniper 213707
Hedgehog Norway Spruce 298197
Hedgehog Parsley 79664
Hedgehog Prickly Poppy 32424
Hedgehog Pricklypoppy 32453
Hedgehog Raspberry 338342
Hedgehog Rose 336901
Hedgehogcactus 140842
Hedgehog-like Raspberry 338342
Hedgehogs 31051,170193,325612,350425
Hedgehogweed 203124,203127
Hedge-htssopklike Marshweed 230307
Hedgehyssop 180292
Hedge-hyssop 180292,180321
Hedgekog-spiny Crazyweed 278899
Hedgemaids 176824
Hedgemustard 365564
Hedge-mustard 365564
Hedgenettle 373085
Hedge-nettle 373346,373348,373365
Hedge-nettle Betuny 373110
Hedgeparsley 392959
Hedge-parsley 392959
Hedgeparsleyleaf Hogfennel 293046
Hedgepegs 316819
Hedgepicks 316819
Hedgerow Crane's-bill 174856
Hedgerow Cranesbill 174856
Hedgerow Geranium 174856
Hedgerow Rose 336901
Hedge-row Rose 336901
Hedgespeaks 316819,336419
Hedgespecks 109857,316819
Hedge-taper 405788
Hedge-thorn 76861
Hedgeweed 365564
Hedgy-pedgy 336419
Hedin Milkvetch 42478
Hedin Pleurospermum 304805
Hedin Pricklythrift 2239
Hedin Prickly-thrift 2239
Hedin Reedbentgrass 65362
Hedin Ribseedcelery 304805
Hedin Sagebrush 35598
Hedin Woodreed 65362
Hedin's Wormwood 35598
Hedinia 187361

Hedionda-macho 78301
Hediondo 57455
Hedley's Jamie Clover 316127
Hedychium 187417
Hedyosmum 187491
Hedyotidous Hedyotis 187581
Hedyotis 187497
Hedyotis Eargrass 187666
Hedysarum 187753
Heel Trot 285741
Heermann's Coast Range Wild Buckwheat 152131
Heermann's Great Basin Wild Buckwheat 152130
Heermann's grooved Wild Buckwheat 152134
Heermann's Rough Wild Buckwheat 152127
Heermann's Wild Buckwheat 152121
Heermann's Woolly Wild Buckwheat 152129
Heethen-berry 109857
Hefei Vetch 408425
Hefeng Anemone 23820
Hefeng Chirita 87835
Hefeng Fritillary 168614
Hefeng Windflower 23820
Heg-beg 402886
Hegberry 83155,280003
Heg-peg 109857,316819
Heg-peg Bush 109857,316819
He-heather 67456
He-holly 203545
He-huckleberry 239384
Heidenreich Cinquefoil 312656
Heihow 176824
Heil's Wild Buckwheat 151956
Heilongjiang Catnip 264990
Heilongjiang Dock 339918
Heilongjiang Fleabane 150733
Heilongjiang Germander 388298
Heilongjiang Poplar 311229
Heilongjiang Sedge 76008
Heilongjiang Thyme 391078
Heilongjiang Vetch 408278
Heilongjiang Windhairdaisy 348141
Heimia 188263
Heiriff 170193
Heiriffe 170193
Heishan Sweetleaf 381235
Heishui Azalea 330837
Heishui Bittercress 72796,365090
Heishui Euphorbia 159026
Heishui Ligusticum 229357
Heishui Rhododendron 330837
Heishui Willow 343473
Heizhugou Azalea 330839
Heizhugou Rhododendron 330839
Hejiang Azalea 330840
Hejiang Pittosporum 301374
Hejiang Rhododendron 330840,330933
Hejiang Seatung 301374
Hejiang Square Bamboo 87562
Hejiang Square-bamboo 87562
Hejing 187369
Hejing Buttercup 325925
Hejing Kengyilia 215919
Hejing Milkvetch 42481
Hekou Alseodaphne 17801
Hekou Anemone 23845
Hekou Aspidistra 39545
Hekou Begonia 49910
Hekou Camellia 69123
Hekou Dragonbood 137424
Hekou Elatostema 142675
Hekou Grape 411721
Hekou Ormosia 274395
Hekou Paraboea 283119
Hekou Poisonnut 378908
Hekou Raspberry 339025
Hekou Salacia 342697
Hekou Saurauia 347994
Hekou Spiradiclis 371756
Hekou Stairweed 142675
Hekou Sterculia 376183
Hekou Strychnos 378908
Hekou Tea 69123
Hekou Whytockia 414003
Hekou Windflower 23845
Helanshan Buttercup 325556
Helanshan Catchfly 363160
Helanshan Chickweed 374727
Helanshan Childseng 318495
Helanshan Corydalis 105582
Helanshan Crazyweed 278888
Helanshan Kobresia 217188
Helanshan Nettle 402930
Helanshan Pseudostellaria 318495
Helanshan Rockcress 30162
Heldreich Hawthorn 109750
Heldreich Jancaea 211560
Heldreich Pine 299968,300036
Heldreich's Alfalfa 247300
Heldreich's Maple 3013
Helen Barberry 51709
Helen Rose 336620
Helen Skullcap 355473
Helen's Elecampane 207135
Helen's Flower 188393,188402
Helena Bulbophyllum 62771
Helena Curlylip-orchis 62771
Helen-flower 188393
Heleninm 188393
Helep Gaoliang 369652
Helfer Bluntleaf grass 375770
Helfer Dysophylla 139563
Helfer Leucas 227607
Helfer Marshweed 230303
Helfer Persimmon 132193
Helfer Stenotaphrum 375770
Helianthemum 188583
Helianthus 188902
Helichrysum 189093
Helicia 189910
Helicia Nut 189924
Heliciopsis 189966
Helicoid-stamenal Parsonsia 285056
Heliconia 189986
Heliconia Family 190073
Heliconias 189986
Helictotrichon 190129
Heliophilous Blastus 55136
Heliopsis 190502
Heliotrope 190540,190559,190597,190622,190649
Heliotrope Cherry 190559
Heliotrope Ehretia 141595
Helix Androcorys 22329
Helixanthera 190830
Hell Vine 70512
Hell's Curse 18788
Hellbind 115008
Helleboraster 190936
Hellebore 190925,405643
Helleborine 82028,147106,147149
Helleborine Orchid 147149
Helleborus 190925,190949
Heller's Cudweed 317742,317756
Heller's Gayfeather 228479
Heller's Japanese Holly 203675
Heller's Plantain 302000
Heller's Rabbit-tobacco 317742
Heller's Rosette Grass 128535
Hellfetter 366592
Hellotropeleaf Trigonotis 397419
Hellroot 275135,366743
Helltrot 192278
Hellweed 68713,102933,115008,325612
Hellycompane 207135
Helm 401512
Helmet Flower 5442,105518,355340,364741
Helmet Orchid 105524,108650
Helmetbeard Woodbetony 287378
Helmet-flower 5442,220346,355449
Helmetpetal Stonecrop 357184
Heloniopsis 191059
Helophilic Gentian 173502
Helrut 366743
Heltrot 192358
Helver 203545
Helwingia 191139
Helxine 367772
Hemarthria 191225
Hemiadelphis 191325

Hemiboea 191347
Hemiboeopsis 191396,191397,239956
Hemigraphis 360700
Hemilophia 191525
Hemiphragma 191573
Hemipilia 191582,191598
Hemiptelea 191626
Hemisphaeric Horny Tanoak 233158
Hemisphaerical Brassbuttons 107776
Hemisphere Eritrichium 153458
Hemispherical Eritrichium 153458
Hemispherical Rockfoil 349437
Hemistepta 191665
Hemitory 169126
Hemlock 28044,101845,101852,192358,
399860,399867
Hemlock Chervil 392968,392992
Hemlock Chervill 392992
Hemlock Fir 399905
Hemlock Panic Grass 128480
Hemlock Panic-grass 128480
Hemlock Parsley 101820
Hemlock Rosette Grass 128553
Hemlock Spruce 399860,399867,399905
Hemlock Stork's-bill 153767
Hemlock Storksbill 153767
Hemlock Water Dropwort 269307
Hemlock Water-dropwort 269307
Hemlock Waterparsnip 365872
Hemlock Water-parsnip 365872
Hemlock-fir 399905
Hemlock-leaved Cranesbill 153767
Hemlockparsley 101820,101823
Hemlock-parsley 101823
Hemlock-spruce 399860
Hemming-and-sewing 3978
Hemp 13073,71209,71218,123141,123215,
393883
Hemp Agrimony 53816,158021,158062
Hemp Bogorchid 158062
Hemp Broomrape 275184
Hemp Dogbane 29473
Hemp Eupatorium 158062
Hemp Family 71202
Hemp Hibiscus 194779
Hemp Melhania 248835
Hemp Nettle 170056,170078
Hemp Palm 393803,393809
Hemp Tree 411362
Hemp-agrimony 158021,158062
Hemp-dogbane 29473
Hemper 329521
Hempleaf Copperleaf 1912
Hempleaf Croton 112874
Hempleaf Groundsel 358500
Hempleaf Monkshood 5099
Hempleaf Mountainash 369384
Hempleaf Negundo Chastetree 411373
Hempleaf Nettle 402869
Hempleaf Ragwort 358500
Hempleaf Spiraea 371868
Hemp-leaved Croton 112874
Hemp-leaved Dead Nettle 170078
Hemp-leaved Deadnettle 170078
Hemp-leaved Hibiscus 194779
Hemp-leaved Hollyhock 18160
Hemp-leaved Mountain-ash 369384
Hempnettle 170056
Hemp-nettle 170056,170058,170060,170078
Hemp-plant 10940
Hempseed 158062
Hemptree 411189
Hemp-tree 411189
Hemsle Loosestrife 239667
Hemsle Storax 379364
Hemsle Woodbetony 287275
Hemsleia 191894
Hemsley Actinidia 6625
Hemsley Azalea 330849
Hemsley Barberry 51710
Hemsley Begonia 49911
Hemsley Biondia 54519
Hemsley Buckthorn 328722
Hemsley Corydalis 105970
Hemsley Cowparsnip 192270
Hemsley Dogwood 380451
Hemsley Euonymus 157578
Hemsley Greenbrier 366363
Hemsley Honeysuckle 235830
Hemsley Jasmine 211918
Hemsley Kiwifruit 6625
Hemsley Macaranga 240265
Hemsley Melodinus 249665
Hemsley Metaplexis 252513
Hemsley Monkeyjoy 366055
Hemsley Monkshood 5247
Hemsley Mountainash 369414
Hemsley Mountain-ash 369414
Hemsley Pricklyash 417139
Hemsley Rockvine 387779
Hemsley Setose Vineclethra 95505
Hemsley Sloanea 366055
Hemsley Smilax 366363
Hemsley Spindle-tree 157578
Hemsley Vineclethra 95505
Hemsley Yam 131616
Hemsley's Rhododendron 330849
Hemsley's Spicebush 231427
Hen and Chicken Houseleek 212541,358060
Hen and Chicks 139971,139987,140008,
212541
Hen Apple 369322
Hen Bell 201389
Hen Bells 201389
Hen Cress 72038
Hen Drunks 369339
Hen Flower 168476
Hen Gorse 268991,271563
Hen Pea 417417
Hen Peas 417417
Hen Plant 302034
Hen's Bill 271280
Hen's Comb 273531,329521
Hen's Evergreen 374968
Hen's Eyes 31396
Hen's Foot 79664,105745
Hen's Kames 273531
Henahena Palm 68614
Henan Actinidia 6627
Henan Azalea 330851
Henan Barberry 51732
Henan Birch 53469
Henan Bulbophyllum 62773
Henan Catnip 264948
Henan Crabapple 243629
Henan Curlylip-orchis 62773
Henan Dendrobium 125179
Henan Dicranostigma 129566
Henan Favusheadflower 129566
Henan Forkleaf Maple 3498
Henan Kiwifruit 6627
Henan Knotweed 309187
Henan Larkspur 124294
Henan Magnolia 242152
Henan Meadowrue 388533
Henan Nepeta 264948
Henan Paulownia 285952
Henan Rhododendron 330851
Henan Sage 345086
Henan Skullcap 355480
Henan Spapweed 205003
Henan Spicebush 231370
Henan Spiny Ailanthus 12589
Henan Tickseed 104832
Henan Touch-me-not 205003
Henan Woodbetony 287282
Hen-and-chickens 28200,30081,50825,
50826,64184,72038,116736,174877,
176824,237539,262441,322724,349044,
349936,356468,358060,358062
Henbane 201370,201389
Hen-bell 201389
Henbit 47013,220345,220355
Henbit Dead Nettle 220355
Henbit Deadnettle 220355
Henbit Dead-nettle 220355
Henbit Nettle 220355
Henchum Pencil-wood 139655
Hencress 72038
Hendel Wendlandia 413789
Henderson Allamanda 14874

Henderson Ascocentrum 38190
Henderson Corydalis 105971
Henderson Fawnlily 154923
Henderson Milkvetch 42483
Henderson Rattan Palm 65705
Henderson Shooting Star 135162
Henderson's Aster 380904
Henderson's claytonia 94301
Henderson's Fawn-lily 154923
Henderson's Shooting Star 135162
Henderson's Triteleia 398791
Hendl Acronema 6212
Hendriksen Oreocereus 273820
Henep 13073
Henequen 10854
Henequen Agave 10854
Hengchun Bayberry 261122
Hengchun China Laurel 28341
Hengchun Copperleaf 1923
Hengchun Eugenia 382589
Hengchun Euonymus 157769
Hengchun Forkliporchis 25986
Hengchun Gambirplant 401768
Hengchun Holly 204025
Hengchun Honeylocust 176900
Hengchun Hookvine 401768
Hengchun Lettuce 219409
Hengchun Liparis 232180
Hengchun Loquat 151144
Hengchun Mayer Woodnettle 125546
Hengchun Mucuna 259512
Hengchun Nagi 306431
Hengchun Ormosia 274396
Hengchun Parakmeria 283418
Hengchun Pencilwood 139655
Hengchun Pepper 300432
Hengchun Premna 313643
Hengchun Sweetleaf 381217
Hengchun Syzygium 382589
Hengchun Twayblade 232180
Hengduanshan Clovershrub 70840
Hengduanshan Crabgrass 130589
Hengduanshan Fascicle-flower Bushclover 226790
Hengduanshan Rockfoil 349438
Hengduanshan Spapweed 205004
Hengduanshan Stonecrop 356783
Hengduanshan Touch-me-not 205004
Hengduanshan Valerian 404278
Henghe Terminalia 386505
Hengshan Arrowwood 407881
Hengshan Bluegrass 305575
Hengshan Indocalamus 206802
Hengshan St. John's Wort 201916
Hengshan Viburnum 407881
Hengxian Azalea 331111
Hengxian Rhododendron 331111
Hengxian Slugwood 50532
Henkam 201389
Henkel Yellowwood 306439
Henna 223451,223454
Henning Pondweed 312136
Henol Lemongrass 117165
Henon Bamboo 297373
Henpen 201389
Henpenny 201389,248193
Henric Lily 229860
Henric Meconopsis 247128
Henrieville Woody-aster 415847
Henry Acanthopanax 2412,143610
Henry Acoop of cock 283361
Henry Actinidia 6537,6628
Henry Actinodaphne 6781
Henry Alder 16358
Henry Alstonia 18043
Henry Angelica 24369
Henry Anisetree 204526
Henry Anise-tree 204526
Henry Aralia 30680
Henry Argyreia 32629
Henry Asiabell 98331
Henry Aspidopterys 39679
Henry Asystasia 43659
Henry Azalea 330853
Henry Balanophora 46835
Henry Bamboo 297294
Henry Barberry 51711
Henry Begonia 49914
Henry Biondia 54520
Henry Blueberry 403851
Henry Blumea 55751
Henry Brachycorythis 58393
Henry Buckthorn 328725
Henry Camellia 69127
Henry Caryodaphnopsis 77961
Henry Chestnut 78795
Henry Chinkapin 78795
Henry Chloranthus 88282
Henry Clematis 94989
Henry Clovershrub 70820
Henry Crabgrass 130590
Henry Cragcress 416338
Henry Craibiodendron 108578
Henry Cranesbill 174885
Henry Creeper 285110
Henry Currant 334016
Henry Damnacanthus 122038
Henry Deerdrug 242688
Henry Dendrobium 125182
Henry Dragonheadsage 247728
Henry Dunbaria 138970
Henry Dyeingtree 346482
Henry Dyeing-tree 346482
Henry Eightangle 204526
Henry Elaeagnus 142022
Henry Emmenopterys 145024
Henry Euphorbia 159029
Henry Eurya 160492
Henry Evodia 161340
Henry False Solomonseal 242688
Henry Falseoat 398482
Henry Fieldhairicot 138970
Henry Fieldorange 249658
Henry Fig 165097
Henry Figwort 355135
Henry Fluttergrass 166348
Henry Galangal 17686
Henry Gerbera 175148
Henry Glorybower 96116
Henry Goatweed 8822
Henry Goldleaf 108578
Henry Goldsaxifrage 90376
Henry Goldwaist 90376
Henry Grewia 180806
Henry Gueldenstaedtia 181667
Henry Gutzlaffia 283361
Henry Heliciopsis 189968
Henry Hemiboea 191387
Henry Hemipilia 191609
Henry Hepatica 192132
Henry Hogfennel 292880
Henry Holarrhena Conessi 197199
Henry Honeysuckle 235633
Henry Hornbeam 77303
Henry Illicium 204526
Henry Illigera 204630
Henry Indigo 206057
Henry Iris 208613
Henry Ixora 211105
Henry Jadeleaf and Goldenflower 260425
Henry Kiwifruit 6537,6628
Henry Lady's Slipper 120363
Henry Larkspur 124278
Henry Lasianthus 222155
Henry Leek 15344
Henry Lettuce 267173
Henry Lily 229862
Henry Linden 391725
Henry Loosestrife 239668
Henry Loquat 151160
Henry Macaranga 240268
Henry Magnolia 232598
Henry Magnoliavine 351054
Henry Magnolia-vine 351054
Henry Mazus 246982
Henry Meehania 247728
Henry Meliosma 249397
Henry Melodinus 249658
Henry Microtropis 254297
Henry Milkvetch 42484
Henry Mockorange 294465

Henry Monkshood 5268
Henry Munronia 259881
Henry Mussaenda 260425
Henry Ninenode 319576
Henry Onion 15344
Henry Oreocharis 273863
Henry Ormosia 274399
Henry Osmanther 276312
Henry Osmanthus 276312
Henry Paphiopedilum 282836
Henry Paragutzlaffia 283361
Henry Passionflower 285649
Henry Pecteilis 286832
Henry Pentapanax 289636
Henry Pimpinella 299427
Henry Pine 299972
Henry Pipewort 151322
Henry Piratebush 61935
Henry Pittosporum 301290
Henry Pleione 304218
Henry Plum 83216
Henry Pocktorchid 282836
Henry Porana 311607
Henry Premna 313644
Henry Privet 229466
Henry Psychotria 319576
Henry Purpledaisy 267173
Henry Pygeum 322397
Henry Rabdosia 209694
Henry Randia 325341
Henry Raspberry 338508
Henry Rattan Palm 65706
Henry Rattlesnake Plantain 179631
Henry Rehmannia 327448
Henry Retuse Ash 167996
Henry Rhodiola 329879
Henry Rhododendron 330853
Henry Rhodoleia 332207
Henry Ricebag 181667
Henry Ricegrass 276014
Henry Rockjasmine 23191
Henry Rockvine 387780
Henry Rose 336625
Henry Rosewood 121705
Henry Roughleaf 222155
Henry Rungia 340352
Henry Sageretia 342176
Henry Sanicle 345988
Henry Sarmentose Fig 165623
Henry Saruma 347186
Henry Scolopia 354613
Henry Scurvyweed 416338
Henry Seatung 301290
Henry Sedge 74790
Henry Sindechites 364701
Henry Sinobambusa 364797
Henry Smallspine Caper 71760
Henry Snakemushroom 46835
Henry Snakewood 378675
Henry Spapweed 205005
Henry Speedwell 407150
Henry Spiraea 371928
Henry St. John's Wort 201919
Henry St. John's-wort 201919
Henry Staff-tree 80204
Henry Sterculia 376121
Henry Steudnera 376382
Henry Sweetleaf 381291
Henry Swordflag 208613
Henry Taillessfruit 100136
Henry Tailsacgrass 402652
Henry Tangbamboo 364797
Henry Tanoak 233246
Henry Thistle 92019
Henry Tigerthorn 122038
Henry Toona 392832
Henry Touch-me-not 205005
Henry Tubergourd 390138
Henry Twayblade 232188
Henry Urophysa 402652
Henry Viburnum 407882
Henry Wampee 94197
Henry Waxpetal 106649
Henry Willowleaf Cotoneaster 107674
Henry Wilsontree 365093
Henry Wilson-tree 365093
Henry Winterhazel 106649
Henry Woodbetony 287276
Henry Yam 131548
Henry Yinshancress 416338
Henry Yinshania 416338
Henry Youngia 416432
Henry's Ainsliaea 12648
Henry's Caper 71760
Henry's Crabgrass 130489
Henry's Erycibe 154245
Henry's Honeysuckle 235831
Henry's Ironweed 406402
Henry's Lily 229862
Henry's Maple 3015
Henry's Notoseris 267173
Henry's Spider-lily 200932
Henryana Poplar 311151
Henry-flowered Storax 379331
Henryleaf Callose Actinidia 6543
Henryleaf Callose Kiwifruit 6543
Hens 273531
Hens and Chicks 358062
Hep Briar 336419
Hep Tree 336419
Hepatica 23706,192120,192126,192134,192135
Hepaticaleaf Corydalis 105973
Heping Raspberry 339194
Heptacodium 192166
Hep-thorn 109857
Heqing Ceropegia 84254
Heqing Clematis 94752
Heqing Cotoneaster 107484
Heqing Cowparsnip 192345
Heqing Kobresia 217126
Heqing Magnoliavine 351118
Heqing Meadowrue 388553
Heqing Rattlebox 112824
Heqing Rocky Aster 40974
Heqing Windhairdaisy 348461
Herald's Trumpet 49359
Herald-trumpet 49355
Herb Bennett 101852,175454,404316
Herb Bifoil 232950
Herb Bonnet 175454
Herb Canary-clover 136746
Herb Carpenter 13174
Herb Christopher 6448,321554
Herb Corydalis 106482
Herb Dendrobium 125075
Herb Dragon's blood 309507
Herb Eaglewood 161630
Herb Eargrass 187582
Herb Eve 13073,105352,301910
Herb Gerard 8826
Herb Git 266225
Herb Grace 341064
Herb Ive 13073,105352,301910
Herb Ivy 13073,105352,301910
Herb Knotweed 309018
Herb Lily 18060,18062
Herb Margaret 50825
Herb Mary 383712
Herb Mercury 250600,250614
Herb Naiad 262040
Herb of Gilead 80063
Herb of Grace 341064,405872
Herb of Mary 321643
Herb of Memory 337180
Herb of Repentance 341064
Herb of the Cross 405872
Herb of the Trinity 410677
Herb of Vine 39328
Herb Paralysy 315082
Herb Paris 284289,284387
Herb Patience 340178
Herb Peter 315082
Herb Rober Cranesbill 174877
Herb Robert 174877,344980
Herb Robert Geranium 174877
Herb Robin 174877
Herb Sagapene 163590
Herb Screwpine 280986
Herb Seatung 350323,350344
Herb Sophia 126127

Herb Squinantyke 39328
Herb Stephania 375860
Herb Tree Mallow 223393
Herb Treemallow 223393
Herb Tree-mallow 223393
Herb Trinity 192130,410513,410677
Herb Truelove 284387
Herb True-love 284387
Herb Twopence 239755,240068
Herb Weiling Ascitesgrass 407485
Herb William 19672
Herb Willow 239898
Herba Impia 165906
Herbaceous Begonia 49917
Herbaceous Blackberry 338660
Herbaceous Elder 345594
Herbaceous Geophila 174366
Herbaceous Glorytree 391570
Herbaceous Haworthia 186472
Herbaceous Hedyotis 187582
Herbaceous Indian Madder 337934
Herbaceous Peony 280213
Herbaceous Periwinkle 409335
Herbaceous Pygmaeopremna 322423
Herbaceous Seepweed 379553
Herbaceous Stephania 375860
Herb-a-grass 341064
Herbal Bonnet 175454
Herbal Hibiscus 194973
Herbclamworm 398064,398068
Herbcoral 346525
Herb-flower 13174
Herbgrass 341064
Herb-mercury 250600
Herb-of-grace 341064
Herb-of-the-seven-cures 3978
Herb-Paris 284387
Herb-Robert 174877
Herbs 109857
Herb-scented Geranium 288550
Herb-sophia 126127
Herbst Bloodleaf 208349
Herbst's Bloodleaf 208349
Herbygrass 341064
Herby-grass 341064
Hercules Club 30760,158603
Hercules' Club 30760,417199
Hercules' Woundwort 316127
Hercules'-club 30760,417199
Hercules-club 30634
Herd's Grass 294992
Herd's-grass 294960,294992
Herder Leek 15345
Herder Onion 15345
Herdreich Lily 229858
Herdreich Onion 15346
Herds-grass 11972
Herelip 220050
Hereroa 192427
Herif 170193
Herissantia 192470
Heritiera 192477
Herklots Indocalamus 206788
Hermann Cob Cactus 234928
Hermann's Dwarf Rush 213148
Herminium 192809
Hermit Cactus 354311
Hermitgold Clematis 95295
Hernandez Stanhopea 373693
Hernandez's Spineflower 89056
Hernandia 192909
Hernandia Family 192921
Hernandia-leaed Stephania 375861
Hernandialeaf Stephania 375861
Herniaria-leaf Spurge 159042
Herniary 192926,192941,192965
Heron's Bill 153711,153767,153869,153903
Heron's-Bill 153711
Heronbill 153711,153914
Herpetospermum 193057
Herpysma 193070
Herrer Cactus 244104
Herringbone Cotoneaster 107486
Herringbone Plant 245023
Hers Barberry 51713
Hers Maple 3006
Herteclowre 388042
Hertfordshire Weed 345631
Herts 403916
Hertwort 167955
Hervey Mahonia 242462
Hervey's Aster 160665
Heshouwu 162505,162542
Heshuo Sedge 74794
Hesper Palm 59095,59096,154571
Hesperantha 193232
Hess' Fleabane 150683
Hessei 107321
Hessel 106703
Hester's Foxtail Cactus 107028
Hester's Pincushion Cactus 107028
Hetaeria 193581
Hetao Rhubarb 329342
Heteraeia 193657
Heterocarpous Mountain Leech 126389
Heterocarpous Tickclover 126389
Heterocaryum 193736
Heterocentre Corydalis 105982
Heterochetous Bulrush 353457
Heterochromia Willow 343477
Heteroclite Rhododendron 330861
Heterocycle Bamboo 297301
Heteroflower Childseng 318496
Heteroflower Pseudostellaria 318496
Heterohair Honeysuckle 235938
Heterohead Cremanthodium 110398
Heterohead Nutantdaisy 110398
Heterolamium 193815
Heteroleaf Chinese Privet 229624
Heteroleaf Jilin Monkshood 5320
Hetero-leaf Sichuan Holly 204317
Heteroleaf Spicebush 231405
Heteroleaf Threevein Aster 39964
Heteroleaved Chun Maple 2889
Heterolepidote Sedge 74797
Heterolobe Pellionia 288729
Heterolobe Redcarweed 288729
Heteromerous Willow 343480
Heteromorphic Fig 165099
Heteropanax 193897
Heteropappus 193914
Heteropetalous Tulip 400173
Heterophullous Earlymadder 226241
Heterophyllaries Aster 40565
Heterophyllous Arrowleaf 342349
Heterophyllous Artocarpus 36920
Heterophyllous Bittercress 72799
Heterophyllous Bogorchid 158149
Heterophyllous Camellia 69552
Heterophyllous Chastetree 411374
Heterophyllous Elsholtzia 144037
Heterophyllous Eupatorium 158149
Heterophyllous Fig 165100
Heterophyllous Mountain Leech 126396
Heterophyllous Negundo Chastetree 411374
Heterophyllous Tickclover 126396
Heterophyllous Trigonostemon 397363
Heterophyllous Wingseedtree 320836
Heterophyllous Wing-seed-tree 320836
Heteroplexis 194009,194014
Heteroscale Sedge 74797
Heterosepal Cogflower 154204,382766
Heterosepal Ervatamia 154204,382766
Heterosepal Spapweed 205006
Heterosmilax 194106
Heterostachys Sedge 74801
Heterostalk Leek 15347
Heterostalk Onion 15347
Heterostamen Willow 343482
Hetero-stamenned Willow 343480
Heterostemma 194155
Heterostemonous Biebersteinia 54196
Heterotropalike Wildginger 37635
Heterotuberous Yam 131611
Hetian Buttercup 325930
Hetian Dandelion 384814
Hetian Milkvetch 42500
Hetian Serpentroot 354872
Hettle 402886
Hetz's Midget Eastern Arborvitae 390604
Heucher Figwort 355137

Heuchera 194411
Heucherialeaf Didymocarpus 129913
Heudes Sedge 74802
Heuffel Saffron 111540
Heuffel's Saffron 111540
Hevea 194451
Hever 101852
Hewittia 194459
Hexagon Coldwaterflower 298934
Hexapetalous Tailgrape 35016
Hexaphyllous Staunton-vine 374409
Hexastichous Pinanga-palm 299663
Hexham Scout 249218
Hexi Giantfennel 163628
Hexian Tupistra 70600
Hexinia 194600,194601
Heyde Primrose 314650
Heyde Skullcap 355476
Heyder Cactus 244105
Heyder's Nipple Cactus 244105
He-yew 213902
Heyhove 176824
Heyhown 176824
Heyne Anisomeles 25473
Heyne Lobelia 234527
Heyne Patchouli 306994
Heyne Wingseedtree 320837
Heyne's Grape 411735
Heyriffe 170193
Hezhang Rose 336629
Hezzle 106703
Hhammock Spider-lily 200946
Hiang 133573
Hiba 390680
Hiba Arbor-vitae 390673,390680
Hiba Cedar 390680
Hiba Cedarvitae 390680
Hiba False Arborvitae 390680
Hiba False Arbor-vitae 390680
Hiba False-arborvitae 390680
Hibbered Barberry 52069
Hibbert Bean 294010
Hibbin 187222
Hibiscus 194680,194804,194919,195149,
195269
Hicberry 280003
Hiccup Nut 100359
Hichel Evergreenchinkapin 78932
Hicken 369339
Hickman's Knotweed 309181
Hickory 77874,77920
Hickory Elm 401618,401638
Hickory Pine 299795,300168
Hicks Mahonia 242555
Hicks Monkshood 5273
Hickymore 81237
Hicorier 77874
Hidalg Cactus 244111
Hidalgo 373103
Hidcote Lavender 223254
Hidden Drop-seed 372625
Hidden Euphorbia 158649
Hidden Homalomena 197794
Hidden Lily 114852
Hidden Neoporteria 263748
Hidden Panicgrass 128494
Hidden Sedge 76630
Hidden Stonecrop 356612
Hiddenaxle Balanophora 46815
Hiddenaxle Snakemushroom 46815
Hiddenflower Eargrass 187560
Hiddenflower Hedyotis 187560
Hiddenflower Woodbetony 287129
Hidden-fruited Bladderwort 403190
Hidden-lily 114852
Hiddenpetal Catchfly 248255
Hiddenspur Bladderwort 403824
Hiddenstem Rockfoil 349507
Hiddenvein Microtropis 254314
Hidefruit Goldwaist 90319
Hideleaf Galangal 17660
Hideseedgrass 94585
Hidgy-pidgy 402886
Hidian Fountain-bamboo 37136
Hieransan Drypetes 138627
Hierba De La Hormiga 14992
Hierba Del Soldado 249641
Hierba Limpia 235259
Hierba Loca 112894
Hierbamora 57455
Hieronymus Barberry 51722
Hig Taper 405788
Higan Cherry 83334
Higgins Mahonia 242556
Higgis Taper 405788
High Asterbush 41751
High Bluegrass 305315
High Bush Blueberry 403780
High Bush-cranberry 408188
High Coffee 98914
High Cranberry 408002,408188
High Crapemyrtle 219920
High Crape-myrtle 219920
High Currant 333913
High Deerdrug 242671
High False Solomonseal 242671
High Figwort 355110
High Holly 203792
High Larkspur 124194
High Leymus 228342
High Lymegrass 144303
High Mallow 243840
High Monkeyflower 255255
High Mountain Juniper 213943
High Mountain Larkspur 124504
High Mountain Trigonotis 397464
High Sanicle 345952
High Schefflera 350687
High Seabuckthorn 196766
High Stem Aster 41080
High Taper 405788
High Wildryegrass 144303
High-arctic Dandelion 384583
Highbush Blackberry 338105,338444,
338843,338928,338933
Highbush Blueberry 403780
High-bush Blueberry 403780
Highbush Cranberry 408002,408188
High-bush Cranberry 407989,408001
Highbush Huckleberry 403780,404012
High-bush Huckleberry 403780
Highbush-cranberry 407917
Highclere Holly 203566
High-climbing Jointfir 146128
Highcristate Henderson Corydalis 105972
Higher-hairy-scaled Rattan Palm 65711
Highest Sisymbrium 365398
Highfront Woodbetony 286992
High-land Barberry 51599
Highland Barley 198293
Highland Bent 12038
Highland Breadfruit 164892
Highland Conehead 320119
Highland Cudweed 178311,178465
Highland Fleabane 150508
Highland Hogfennel 229296
Highland Hookleafvine 303091
Highland Kobresia 217258
Highland Live Oak 324556
Highland Madder 337923
Highland Meadowrue 388465
Highland Ploughpoint 401158
Highland Rockjasmine 23346
Highland Rosewood 121860
Highland Saxifrage 349851
Highland Typhonium 401158
Highlands Goldenaster 90216
Highligule Bitterbamboo 303997
Highly Prized Rhododendron 330094
High-moutain Oak 324322
High-plains Goldenrod 367949
Highsheath Arisaema 33276
Highslope Hanceola 185249
Highstalk Pearlsedge 354254
Highstalk Razorsedge 354254
Highstem Pearlsedge 354254
Highstem Razorsedge 354254
Highstem Sedge 73658
Hight Habenaria 183408
High-taper 405788
Highveld Cabbage Tree 115232

Highwater Shrub 210478
Hig-taper 405788
Hi-How 28044
Hiiranshan Hawthorn 329073
Hila Ducloux Mahonia 242517
Hila Maple 3019
Hildaberry 338059
Hildebrand Eulophia 156756
Hildebrand Honeysuckle 235846
Hildebrandt Encephalartos 145228
Hill Beautyberry 66765
Hill Blackgum 267867
Hill Cherry 83234
Hill Combretum 100397
Hill Cranesbill 174548
Hill Euonymus 157376
Hill Gosseberry 332221
Hill Lemon 93706
Hill Lily 229730
Hill Lotus 237652
Hill Monkshood 5422
Hill Mustard 63452
Hill Palm 78056
Hill Poppy 130383,405788
Hill Spindle-tree 157376
Hill Strawberry 167610
Hill's Oak 323871
Hill's Pondweed 312140
Hill's Thistle 92027,92328
Hill's Weeping Fig 165314
Hillbubble Crazyweed 278958
Hillburn's Silver Globe Juniper 213924
Hillcelery 276743,276768
Hill-growing Beautyberry 66765
Hill-growing Combretum 100397
Hill-growing Purplepearl 66765
Hill-growing Windmill 100397
Hilliella 196267
Hillier Rhododendron 330865
Hillman's Orach 44319
Hillock Bush 248099
Hillock Ironweed 406243
Hills Oak 323871
Hills of Snow 199808
Hillside Blueberry 403952
Hillside Creeper Pine 300232
Hillside Sedge 76261
Hillside Strawberry 167658
Hills-of-snow Hydrangea 199807
Hilly Eulalia 156450
Hilly Ironweed 406243
Hilly Sandwort 31772
Hilly Small Reed 127204
Hilo Holly 31396
Hilograss 285417
Himalaya Berry 338326
Himalaya Blueberry 403814
Himalaya Cassiope 78628
Himalaya Cinquefoil 312393
Himalaya Cypress 114775
Himalaya Honeysuckle 228319,228321
Himalaya Knotweed 309084
Himalaya Pheasantberry 228321
Himalaya Plectocomia 303091
Himalaya Roscoea 337066
Himalaya Sapria 346470
Himalaya Sarcococca 346731
Himalaya Spruce 298436
Himalaya Tanoak 233247
Himalaya-honeysuckle 228319
Himalayan Actinidia 6697
Himalayan Adenocaulon 7433
Himalayan Alder 16421
Himalayan Alumroot 194423
Himalayan Andromeda 298722
Himalayan Anise-tree 204524
Himalayan Ash 167975
Himalayan Aster 40568
Himalayan Aucuba 44911
Himalayan Balsam 204982
Himalayan Bamboo 137849,416747
Himalayan Bamboo 'Damarapa' 196347
Himalayan Barberry 51311,51724
Himalayan Bastardtoadflax 389718
Himalayan Berry 338189
Himalayan Birch 53635
Himalayan Black Juniper 213807
Himalayan Blackberry 338151,338326,339101
Himalayan Bladdernut 374100
Himalayan Blue Bamboo 196350
Himalayan Blue Pine 300297
Himalayan Blue Poppy 247109
Himalayan Bluegrass 305579
Himalayan Boschniakia 57434
Himalayan Box 64380
Himalayan Bramble 339392
Himalayan Brome 60741
Himalayan Camellia 69049,69411
Himalayan Caper 71762
Himalayan Carpet Bramble 338877
Himalayan Cassiope 78628
Himalayan Cedar 80087
Himalayan Ceratostylis 83654
Himalayan Cherry 83295
Himalayan Chir Pine 300186
Himalayan Christolea 89233,126154
Himalayan Clematis 95127
Himalayan Comfrey 271819
Himalayan Coral Bean 154626
Himalayan Coralbean 154626
Himalayan Coral-bean 154626
Himalayan Cotoneaster 107454,107687
Himalayan Cowslip 314407,314961
Himalayan Crane's-bill 174653
Himalayan Cranesbill 174653
Himalayan Creeper 285144
Himalayan Cremanthodium 110376
Himalayan Currant 333962,334025
Himalayan Cypress 114775
Himalayan Daphniphyllum 122691
Himalayan Desertcandle 148543
Himalayan Diapensia 127913
Himalayan Dogwood 124891
Himalayan Drooping Juniper 213883
Himalayan Edelweiss 224855
Himalayan Eragrostiella 147466
Himalayan Eriosema 152943
Himalayan Euonymus 157547
Himalayan Fan Palm 393813
Himalayan Figwort 355138
Himalayan Fir 480
Himalayan Firethorn 322458
Himalayan Fleabane 150690
Himalayan Fleeceflower 308713
Himalayan Foamflower 391524
Himalayan Fritillary 168405
Himalayan Gentian 150690,173988
Himalayan Ginger 337065
Himalayan Gueldenstaedtia 391539
Himalayan Gymnadenia 182252
Himalayan Hazel 284966,284968
Himalayan Helwingia 191157
Himalayan Hemlock 399896
Himalayan Holly 203765
Himalayan Honeysuckle 228321
Himalayan Hornbeam 77420,77421
Himalayan Horsegentian 397836
Himalayan Hound's Tongue 118006
Himalayan Hound's-tongue 118006
Himalayan Hydrangea 199891
Himalayan Illicium 204524
Himalayan Indigo 206005,206061
Himalayan Inula 207213
Himalayan Ivy 187292
Himalayan Jasmine 211864
Himalayan Juniper 213883,213943
Himalayan Knotweed 292007,308941
Himalayan Kobresia 217265
Himalayan Laburnum 369070
Himalayan Lady's Slipper 120364
Himalayan Ladybell 7661
Himalayan Larch 221881,221891
Himalayan Lilac 382153
Himalayan Lily 229958
Himalayan Longleaf Pine 300186
Himalayan Long-leaf Pine 300186
Himalayan Maddencherry 241500
Himalayan Maddenia 241500
Himalayan Manna Ash 167975
Himalayan Maple 3255

Himalayan May Apple 365009
Himalayan May-apple 365009
Himalayan Medinilla 247572
Himalayan Mouseear Cress 113148
Himalayan Musk Rose 336408
Himalayan Musk-rose 336408
Himalayan Pear 323251
Himalayan Peony 280197
Himalayan Pheasantberry 228319
Himalayan Pieris 298722,298729
Himalayan Pine 300186,300297
Himalayan pittosporum 301266
Himalayan Plantain 302002
Himalayan Plectocomia 303091
Himalayan Poppy 247109
Himalayan Primrose 314310
Himalayan Prinsepia 315179
Himalayan Rhubarb 329314,329327
Himalayan Rhynchosia 333262
Himalayan Rockfoil 349134
Himalayan Rockjasmine 23281
Himalayan Rockvine 387842
Himalayan Rush 213153
Himalayan Savin 213883
Himalayan Siberian Crabapple 243571
Himalayan Silver Fir 439
Himalayan Small Reed 127247
Himalayan Snakegourd 396193
Himalayan Sorbaria 369288
Himalayan Speedwell 407151
Himalayan Spiraea 371860
Himalayan Spruce 298436
Himalayan Spurge 160074
Himalayan Stachyurus 373540
Himalayan Stickseed 221688
Himalayan Tanoak 233247
Himalayan Teasel 133454
Himalayan Tibetia 391539
Himalayan Touch-me-not 204982
Himalayan Toxocarpus 393532
Himalayan Tree Cotoneaster 107454
Himalayan Trevesia 394757
Himalayan Trigonella 397216
Himalayan Trisetum 398545
Himalayan Viburnum 407958
Himalayan Virginia Creeper 285144
Himalayan Waxpetal 106650
Himalayan Weeping Juniper 213883
Himalayan Whitebeam 369375
Himalayan Whorlflower 258831
Himalayan Willow 343487
Himalayan Willow Weed 146870
Himalayan Yew 385404
Himalayan-berry 338189
Himalayas Adenocaulon 7433
Himalayas Alopecurus 17536
Himalayas Aucuba 44911
Himalayas Birch 53635
Himalayas Birdcherry 280018
Himalayas Bowfruitvine 393532
Himalayas Bromegrass 60741
Himalayas Catchfly 363544
Himalayas Cherry 83295
Himalayas Childseng 318503
Himalayas Cistancheherb 57434
Himalayas Cockehead-yam 152943
Himalayas Coralbean 154626
Himalayas Craneknee 221688
Himalayas Creeper 285144
Himalayas Currant 333962
Himalayas Cypress 114775
Himalayas Diapensia 127913
Himalayas Edelweiss 224855
Himalayas Falseoat 398545
Himalayas Fenugreek 397216
Himalayas Filbert 106733
Himalayas Foamflower 391524
Himalayas Fox Tail 17536
Himalayas Gentian 174047
Himalayas Helwingia 191157
Himalayas Hornstyle 83654
Himalayas Horsegentian 397836
Himalayas Hundredstamen 389718
Himalayas Hydrangea 199891
Himalayas Jurinea 207213
Himalayas Kobresia 217265
Himalayas Ladybell 7661
Himalayas Ladyslipper 120364
Himalayas Larch 221891
Himalayas Lemongrass 117234
Himalayas Medinilla 247572
Himalayas Mouseear Cress 113148
Himalayas Muhly 259670
Himalayas Nettle 402854
Himalayas Nutantdaisy 110376
Himalayas Peony 280197
Himalayas Pieris 298722
Himalayas Pimpinella 299561
Himalayas Pine 300297
Himalayas Plateaucress 89233,126154
Himalayas Prinsepia 315179
Himalayas Pseudostellaria 318503
Himalayas Rhodiola 329882
Himalayas Rhubarb 329414
Himalayas Rhynchosia 333262
Himalayas Rockfoil 349134
Himalayas Rockjasmine 23281
Himalayas Rush 213153
Himalayas Sapria 346470
Himalayas Smallreed 127247
Himalayas Speedwell 407151
Himalayas St. John's Wort 201922
Himalayas Stachyurus 373540
Himalayas Stinkcherry 241500
Himalayas Thorowax 63686
Himalayas Wildginger 37642
Himalayas Willow 343487
Himalayas Willowweed 85886,146870
Himalayas Xylanche 57434
Himalrandia 196351
Himalyan Knotweed 309616
Hime Kan Suga 74174
Hime Lemon 93546
Himekitsu 93483
Hinahina 249175
Hinau 142308
Hinau Elaeocarpus 142308
Hinckley Oak 324006
Hinckley's Columbine 30040
Hinckley's Golden Columbine 30019
Hind's Black Walnut 212612
Hind's Walnut 212612
Hindberry 338557
Hindheel 383874
Hinds Bittersweet 80203
Hinds Cane 318303
Hinds Kumquat 167499
Hinds Pseudosasa 318294
Hinds Staff-tree 80203
Hinds Willow 343491
Hinds' Pseudosasa 318303
Hindsacred Lotus 263272
Hindu Cowpea 409092
Hindu Datura 123065
Hindu Laurel 97919
Hindu Lotus 263272
Hineberry 338557
Hinge Bell 351439
Hinnysickle 236022
Hinojo 63809
Hinoke Cypress 85328
Hinoki Cedar 85310
Hinoki Cypress 85310,85311
Hinoki False Cypress 85310
Hinoki Falsecypress 85310
Hip 336419
Hip Bramble 336419
Hip Briar 336419
Hip Haw 109857
Hip Rose 336419
Hip Tree 336419
Hipericumlike Rockfoil 349484
Hippan 336419
Hippeastrum 196439
Hippeophyllum 196459
Hipperty Haw 109857
Hippo Grass 140500
Hippo Root 158692
Hippobroma 196487,196490
Hippocratea 196501
Hippocratea Family 196616

Hippolytia 196688
Hippoman Cupflower 266199
Hipson 336419
Hiptage 196821,196824
Hiptypips 336419
Hipwort 401711
Hirane Indian Hawthorn 329073
Hiroshima-natsuzabon 93485
Hirse 281916
Hirsute Abutilon 910
Hirsute Adinandra 8237
Hirsute Allomorphia 16095
Hirsute Altai Heteropappus 193920
Hirsute Altai Jurinea 193920
Hirsute Anotis 262960
Hirsute Barbate Deadnettle 220359
Hirsute Bauhinia 49122
Hirsute Bayberry Waxmyrtlefruit Terminalia 386586
Hirsute Begonia 49922
Hirsute Belostemma 50934
Hirsute Cajan 65149
Hirsute Camellia 69627
Hirsute Chian Leptopus 226321
Hirsute Chinacane 416753
Hirsute Cliffbean 66639
Hirsute Cointree 280549
Hirsute Corneredcalyx Larkspur 124119
Hirsute Dali Larkspur 124621
Hirsute Delavay Windhairdaisy 348250
Hirsute Densefimbriate Sedge 74287
Hirsute Dichroa 129036
Hirsute Dumasia 138936
Hirsute Eremopyrum 148413
Hirsute Feathergrass 376876
Hirsute Frankenia 167792
Hirsute Gambirplant 401761
Hirsute Glochidion 177141
Hirsute Gomphostemma 179204
Hirsute Goosecomb 335341
Hirsute Grama 57931
Hirsute Grewia 180813
Hirsute Ground Ivy 176837
Hirsute Hedge Calystegia 68722
Hirsute Hogfennel 292983
Hirsute Holly 203889
Hirsute Hooker Onosma 271779
Hirsute Indigo 206081
Hirsute Jadeleaf and Goldenflower 260426
Hirsute Koeleria 217479
Hirsute Lancea 220795
Hirsute Landpick 308556
Hirsute Lasianthus 222158
Hirsute Leucas 227609
Hirsute Lomatia 235408
Hirsute Marshweed 230286
Hirsute Millettia 66639
Hirsute Mockorange 294468
Hirsute Mussaenda 260426
Hirsute Nanchuan Trigonotis 397424
Hirsute Orophea 275326
Hirsute Paliurus 280549
Hirsute Paphiopedilum 282839
Hirsute Paraphlomis 283620
Hirsute Passed Over Rhododendron 331534
Hirsute Petiole Rhododendron 330874
Hirsute Photinia 295701
Hirsute Pinnatifid Parrya 285008
Hirsute Pleurothallis 304916
Hirsute Przewalsk Woodbetony 287546
Hirsute Raspberry 338516
Hirsute Rockcress 30292
Hirsute Roegneria 335341
Hirsute Roughleaf 222158
Hirsute Saurauia 347990
Hirsute Screwtree 190103
Hirsute Screw-tree 190103
Hirsute Senna 360439
Hirsute Shiny Bugleweed 239222
Hirsute Silvervein Creeper 285114
Hirsute Small Reed 127205
Hirsute Snailseed 97914
Hirsute Solomonseal 308556
Hirsute Soroseris 369861
Hirsute Spiraea 371933
Hirsute Sporoxeia 372892
Hirsute St. John's Wort 201924
Hirsute Staff-tree 80205
Hirsute Tali Larkspur 124621
Hirsute Tartar Heteropappus 193984
Hirsute Tartar Jurinea 193984
Hirsute Torenia 392911
Hirsute Twinballgrass 209071
Hirsute Twinflower Violet 409729
Hirsute Vetch 408423
Hirsute Vineclethra 95550
Hirsute Wild Smallreed 127205
Hirsute Woodbetony 287280
Hirsutecalyx Asiabell 98354
Hirsuteflower Angelica 24372
Hirsute-flower Leptodermis 226091,226092
Hirsute-flower Wildclove 226091
Hirsuteleaf Asiabell 98299
Hirsutepetiole Azalea 330874
Hirsute-poled Yushania 416780
Hirsute-stalk Chillan Barberry 51449
Hirtellous Anemone 23819
Hirtellous Windflower 23819
Hirtellous-pedicel Barberry 51725
Hirtelous Hairyleaf Batrachium 48939
Hirtose Crazyweed 278881
Hirtose Oxalis 277889
Hirtose-sheathed Indocalamus 206791
Hiry Brdock 31068
Hiryleaf Martagon Lily 229928
Hiryu Azalea 330082,331370
Hisauch Bluegrass 305584
Hi-scented Sumach 332908
Hislop Polystachya 310432
Hispan Fennelflower 266231
Hispania Oak 324010
Hispanic Hyacinthoides 199553
Hispaniolan Palmetto 341406
Hispaniolan Palmetto Palm 341410
Hispid Acanthopanax 2495
Hispid Actinidia 6558,6559
Hispid Alfredia 14595
Hispid Althaea 18166
Hispid Amischotolype 19493
Hispid Arthraxon 36647
Hispid Aster 90220
Hispid Azalea 330876
Hispid Baikal Betony 373147
Hispid Bamboo 297325,362129
Hispid Bambusa 47244
Hispid Barnyardgrass 140438
Hispid Begonia 49925
Hispid Bethlehemsage 283621
Hispid Blackberry 338528
Hispid Bonvalot Larkspur 124205
Hispid Borreria 57321
Hispid Bothriospermum 57680
Hispid Bottle Gourd 219851
Hispid Brassaiopsis 59219
Hispid Brier Grape 411650
Hispid Buchnera 61795
Hispid Buttercup 325942,325943,325948
Hispid Calabash 219851
Hispid Cathayan Anemone 23749
Hispid Chiogenes 87673
Hispid Clovershrub 70795
Hispid Copperleaf 1894
Hispid Crabgrass 130768
Hispid Cranesbill 174656
Hispid Dubyaea 138771
Hispid Dysosma 139616
Hispid Elatostema 142682
Hispid Euaraliopsis 59219
Hispid Fieldcitron 400517
Hispid Fig 165671
Hispid Flaccid Knotweed 308969
Hispid Girald Acanthopanax 143598
Hispid Greenbrier 366592
Hispid Hainan Milkwort 308096
Hispid Hedyotis 187697
Hispid Heteropappus 193939
Hispid Holly 203889
Hispid Honeysuckle 235852
Hispid Hooktea 52427
Hispid Indocalamus 206792
Hispid Indosasa 206883

Hispid Jigongshan Pseudosasa 318322
Hispid Jurinea 193939
Hispid Kiwifruit 6559
Hispid Knotweed 309831
Hispid Kohleria 217749
Hispid Kunming Buttercup 326011
Hispid Ligusticum 229335
Hispid Locust 334965
Hispid Malcolmia 243193
Hispid Mallow 18166
Hispid Marsh Mallow 18166
Hispid Marsh-cress 336254
Hispid Marsh-mallow 18166
Hispid Metalleaf 296383
Hispid Microula 254348
Hispid Mock Vervain 176690
Hispid Morningglory 207862
Hispid Navelgrass 171904
Hispid Oat 45602
Hispid Oxyspora 278611
Hispid Palmleaftree 59219
Hispid Paraphlomis 283621
Hispid Plectranthus 209700
Hispid Rabdosia 209700
Hispid Rattlesnake Plantain 179634
Hispid Rose 336632
Hispid Roughleaf Bedstraw 170221
Hispid Sage 345324
Hispid Selfheal 316109
Hispid Shady Larkspur 124667
Hispid Shellfish Pricklyash 417218
Hispid Shibataea 362129
Hispid Slenderstyle Acanthopanax 143642
Hispid Small Bugbane 91024
Hispid Starbur 2612
Hispid Supplejack 52427
Hispid Supple-jack 52427
Hispid Trachydium 393843
Hispid Trifoliate Acanthopanax 143675
Hispid Turpinia 400517
Hispid Vineclethra 95543
Hispid Willd Barley 198272
Hispid Windhairdaisy 348768
Hispid Yam 131630
Hispid Yellow-cress 336254
Hispid Yunnan Gaultheria 172102
Hispid Yunnan Whitepearl 172102
Hispidcalyx Pogostemon 306985
Hispidfruit Poison-ivy 393471
Hispidless Spikesedge 143179
Hispid-node Square Bamboo 87563
Hispid-noded Square-bamboo 87563
Hispidocostate Lasianthus 222310
Hispidocostate Roughleaf 222310
Hispidpeduncle Girald Acanthopanax 143702
Hispidstem Morningglory 207981
Hispidulous Crazyweed 278882
Hispidulous Stephania 375870
Hispidulous Twoflower Micromeria 253641
Hispidutricle Sedge 74870
Hispid-villous Begonia 49927
Hissar Skullcap 355478
Hissing Tree 284248
Hisute Acanthopanax 30755
Hisute Arundinella 37379
Hisute Crazyweed 278883
Hisute Scape-like Sage 345369
Hisute Shuteria 362297
Hisute Tutcheria 322585
Hitchcock's Sedge 74816
Hitchcock's Wood Rush 238626
Hittefwood 395466
Hitter Thistle 73318
Hjelrnquvist's Cotoneaster 107488
Hmnble Bee 272398
Ho Groundsel 92522
Ho Primrose 314471
Ho Twinballgrass 209075
Ho Wood 91287
Ho's Longbeaked Sedge 75191
Ho's Rhododendron 330878
Ho's Schefflera 350708
Hoar Withy 369322
Hoarhound 245723,245770
Hoarhound-like Chaiturus 85058
Hoarhound-like Leucas 227631
Hoarless Chalk-leaved Barberry 51559
Hoarstrange 292957
Hoarstrong 292957
Hoary Alder 16368
Hoary Alison 53034,53038
Hoary Alyssum 18327,53038
Hoary Azalea 330322,331667
Hoary Basil 268429,268438,268460
Hoary Beaked Orchid 375700
Hoary Bitter Cress 72802
Hoary Bluebellflower 115367
Hoary Caper 71769
Hoary Chamaerhodos 85648
Hoary Cherry 83161
Hoary Cherryred Cotoneaster 107476
Hoary Cinquefoil 312389
Hoary Cistus 93154
Hoary Clematis 94925
Hoary Clubawngrass 106932
Hoary Cress 73090,73098
Hoary Erysimum 154418
Hoary False Alyssum 53038
Hoary False Madwort 53038
Hoary Falsealyssum 53038
Hoary Fleabane 150534
Hoary Frostweed 188605
Hoary Goatwheat 44256
Hoary Goldenaster 194197
Hoary Groundsel 279958,358828
Hoary Manzanita 31109
Hoary Mock Orange 294488,294517
Hoary Mock-orange 294488
Hoary Mountain-mint 321957
Hoary Mugwort 35779
Hoary Mullein 405754
Hoary Mustard 196950,196952
Hoary Nightshade 367233
Hoary Pea 385926,386365
Hoary Pepperwort 73090
Hoary Pincushion 84488
Hoary Plantain 302103,302209
Hoary Puccoon 233707
Hoary Ragwort 358828
Hoary Rattlebox 112238
Hoary Rock Rose 188613
Hoary Rockrose 188613
Hoary Saltbush 44334
Hoary Sinobambusa 364799
Hoary Skullcap 355490
Hoary Spiraea 371860
Hoary Stock 246478
Hoary Stringbush 414143
Hoary Sun Rose 188613
Hoary Sunray 190795
Hoary Sunrose 188613
Hoary Tick Clover 126280
Hoary Tick-trefoil 126280
Hoary Verbena 405889
Hoary Vervain 405889
Hoary Whitlow-grass 137039,137069
Hoary Wild Buckwheat 152450
Hoary Willow 342973,343142,343335,343336
Hoary Willowherb 146834
Hoary-alyssum 53038
Hoary-aster 130272
Hoaryhair Minorrose 85648
Hoaryleaf Ceanothus 79920
Hoarypea 385926
Hobamboolike Indosasa 206898
Hobble Bush 407674
Hobblebush 407674
Hobble-bush 407674
Hobblebush Arrowwood 407674
Hobblebush Viburnum 407674
Hobble-gobbles 37015,68189
Hobbly Honker 9701
Hobbly-flower 9701
Hobbly-honkers 9701
Hobei Reed 295914
Hobgoblin 259000
Hoblionker 9701
Hoblionkers 9701
Hobson Barberry 51728
Hobson Lignariella 228957

Hobson Rhodiola 329884
Hoburn Saugh 218761
Hochreutiner Barberry 51729
Hock 13934,243840
Hockerie-topner 358062
Hockherb 243840
Hock-holler 13934
Hockle Elderberry 21459
Hock-leaf 243840
Hockweed 285741
Hodge Cotoneaster 107531
Hodgin's Arizona Cypress 114693
Hodgson Azalea 330877
Hodgson Goldenray 229049
Hodgson Rhododendron 330877
Hodgson Talauma 383246
Hodgsonia 197069
Hodrod 315082
Hod-the-rake 326287
Hoe Nightshade 367499
Hoefft Heron's Bill 153883
Hofeng Anemone 23820
Hoferia 386689
Hoffmann's Wild Buckwheat 152145
Hoffmannsegg Cactus 244112
Hoffmeister Woodruff-like Bedstraw 170225
Hog Apple 306636
Hog Bean 201370
Hog Birch 53454
Hog Brake 19145
Hog Cherry 280003
Hog Cranberry 31142
Hog Doctor Tree 64085
Hog Fennel 278559
Hog Gum 252595,259023
Hog Haw 109857
Hog Millet 281916
Hog Nut 77894,102727
Hog Peanut 20560
Hog Plum 316206,316751,316890,372459,
372467,372477,372479,415558
Hog Weed 103446,192358
Hog Weed Giant 192302
Hog's Bean 398134
Hog's Beans 201389,398134
Hog's Fennel 292758,292957,292966
Hog's Garlic 15861
Hog's-Fennel 292758
Hog-a-back 379660
Hogail 109857
Hogan Western Arborvitae 390651
Hog-apple 109636
Hogarve 109857
Hogasses 109857
Hogazel 109857
Hogbean 201389
Hog-bean 201370
Hog-berry 109857,280003
Hogbrake 19145
Hogfennel 292758,292966,292982
Hoggan 109857
Hoggin 109857
Hog-gosse 109857
Hog-grass 105352
Hog-haghes 109857
Hog-haw 109857
Hog-Hazel 109857
Hognut 77894,77934,102720,102727
Hognut Broom Hickory 77894
Hogpeanut 20558
Hog-peanut 20560
Hog-plum 88691,372478
Hogwallow Starfish 193378
Hogweed 103446,192227,192358,282685,
308816,368781,392992,400675
Hogweed Cow Parsnip 192358
Hogweed Cow-parsley 192358
Hogwort 112852,113037
Hog-wort 112852
Hohenacheir Pricklythrift 2241
Hohenacker's Tamarisk 383519
Hohokam Agave 10910
Hokkaiolo Hemlock 399895
Hokou Anemone 23845
Hoky-poky 402886
Hola 133590
Holarrhena 197175
Holboell's Arabis 30308
Holboellia 197206
Holcoglossum 197253,197254,197265
Holcus 197278
Holdfast 271563
Hold-me-tight 4259,4263
Holdstem Thorowax 63711
Holdtight Medinilla 247636
Hole 106703
Holeanther Boeica 56370
Holeglumegrass 57545,57579
Hole-in-the-sand Plant 265987
Holenacker Glochidion 177142
Holene 203545
Hole-white Cinquefoil 312667
Holewort 8399,106560
Holey Naiad 262035
Holford Pine 299978
Holger Juniper 213947
Holiday Cactus 351994,351995
Holiday Tree 203545
Holigold 66445
Holland Iris 208626
Holland Smock 68713
Hollen 203545
Holleyleaf Chamaesphacos 85731
Hollick 358062
Hollin 203545
Hollin-traie 154327
Holliock 13934
Holloak 13934
Hollock 386585
Hollond 203545
Hollow Bamboo 82440,82442,82445
Hollow Cress 174122
Hollow Joe-pye weed 158113
Hollow Joepyeweed 161171
Hollow Yushania 416759
Hollowbamboo 82440,82445
Hollow-bamboo 82440,82445
Hollow-cress 174122
Hollow-culm Spikesedge 143149
Hollowed Wampee 94191
Hollowleaf Annual Lupine 238489
Hollow-root 8399,105721,106560
Hollowroot Birthwort 105721
Hollow-rooted Thistle 92027
Hollowstem Burnet Saxifrage 299467
Hollowstem Iris 208474
Hollowstem Marsh Marigold 68196
Hollowstem Marshmarigold 68196
Hollowstem Swordflag 208474
Hollowstem Thorowax 63713
Hollow-wort 8399,106560
Holly 203527,203545,203566,203625,
203660,203855,204394
Holly Barberry 51755
Holly Family 29968
Holly Flame Pea 89136
Holly Flame-pea 89136
Holly Grape 242458,242534,242625
Holly Grevillea 180564,180605,180661
Holly Mahonia 242478
Holly Oak 13934,324027,324316
Holly Olive 276242,276313
Holly Osmanthus 276313
Holly Tea Olive 276313
Holly Tree 203545
Holly Wood 301390
Hollyanders 13934
Hollyberry Cotoneaster 107362,107363
Holly-berry Cotoneaster 107362
Holly-dandelion 178012
Hollygreen Barberry 52069
Holly-green Barberry 52069
Hollyhock 13911,13934,18155,18163,
130383,223355
Hollyhock Begonia 49885
Hollyhock Hibiscus 195032
Hollyhock Mallow 243746
Hollyhock-like Yam 131462
Hollyhocks 13934
Hollyleaf Acanthus 2684
Holly-leaf Banksia 47646

Hollyleaf Barberry Hollyleaf Mahonia 242478
Hollyleaf Begonia 49753
Holly-leaf Begonia 49598
Hollyleaf Buckthorn 328668
Holly-leaf Bur Sage 19169
Holly-leaf Bur-sage 19169
Hollyleaf Calvepetal 177995
Hollyleaf Camellia 69144
Holly-leaf Ceanothus 79968
Hollyleaf Cherry 316452
Holly-leaf Cherry 316452
Hollyleaf Cherry-laurel 223096
Hollyleaf Euonymus 177985
Hollyleaf Glyptopetalum 177995
Hollyleaf Herelip 220069
Hollyleaf Lagochilus 220069
Hollyleaf Lomatia 235409
Hollyleaf Mayten 246859
Holly-leaf Mayten 246859
Hollyleaf Mutisia 260541
Hollyleaf Oak 323657
Hollyleaf Osmanther 276313
Holly-leaf Pea 278495
Hollyleaf Red Berry 328734
Hollyleaf Redberry Buckthorn 328734
Hollyleaf Streblus 377540
Hollyleaf Sweetspire 210384
Holly-leaf Sweetspire 210384
Holly-leafed Cherry 316452
Holly-leafed Fuchsia Bush 180277
Hollyleafed Lomatia 235409
Holly-leafed Mirbelia 255776
Hollyleaf-like Oak 323657
Holly-leaved Acanthus 2684
Hollyleaved Barberry 242478
Holly-leaved Barberry 242458,242478
Holly-leaved Begonia 49753
Holly-leaved Camellia 69144
Holly-leaved Cherry Laurel 223096
Holly-leaved Euonymus 177995
Holly-leaved Glyptopetalum 177995
Holly-leaved Grevill 180564
Holly-leaved Mangrove 2672
Holly-leaved Naiad 262067
Holly-leaved Osmanthus 276327
Holly-leaved Sweetspire 210384
Holly-lenf Spurge 386965
Hollysedge 154327
Hollywood Juniper 213634,213943
Holm 401593
Holm Oak 324027
Holme 401512
Holmen's Dandelion 384577
Holmes' Hawthorn 109754
Holmstrup Arborvitae 390605
Holocalyx Woodbetony 287281
Holocheila 197386
Hologlandular Machilus 240598
Holo-haired Actinidia 6631
Holohairy Actinidia 6631
Holohairy Kiwifruit 6631
Hololeaf Fig 165430
Hololip Oblique Rhynchoglossum 333089
Holostemma 197475
Holostyl 49889
Holrod 315082
Holtrot 192358
Holttum Rattan Palm 65709
Holway's Sage 345085
Holy Basil 268614,268644
Holy Clover 271166,271280
Holy Flax 346243,346271
Holy Ghost 24483,30932
Holy Ghost Orchid 291127,291129
Holy Gosta Orchid 291127
Holy Grass 27967,196155
Holy Hay 247456
Holy Herb 203095,405872
Holy Innocents 109857
Holy Lagoseris 220145
Holy Mallow 13934
Holy Orchid 273625
Holy Plant 24483
Holy Poker 401112
Holy Rope 158062,170078
Holy Rose 336917
Holy Sagebrush 36177
Holy Thistle 73318,364360
Holy Thorn 52322,109863
Holy Thorn of Christmas 343302
Holy Vervain 405872
Holy Wood 181505
Holyclover 271280
Holy-grass 27967,196116
Holyn 203545
Holyoke 13934
Holywood 181506
Holywood Lignumvitae 181506
Holywood Lignum-vitae 181506
Homalanthus 197619,197624
Homalium 197633
Homalomena 197785,197794
Home 203545
Homer's Garlic 15520
Homewort 358062
Hominy 417417
Homocodon 197894,197895
Homoflower Mamdaisy 246375
Homogyne 197943
Homomallous Gambir-plant 401762
Homonoia 197960
Honan Actinidia 6627
Honan Crabapple 243629
Honan Forkleaf Maple 3498
Honan Larkspur 124294
Honan Meadowrue 388533
Honan Sage 345086
Honan Skullcap 355480
Honan Woodbetony 287282
Honda Common Juniper 213728
Honda Duckbeakgrass 209259
Honda Hiba False Arborvitae 390684
Honda Roegneria 335350
Hondapara 130919
Hondapara Tree 130919
Hondel Milkvetch 42471
Hondo Spruce 298303,298307,298323
Honduras Mahogany 380527,380528
Honduras Pine 299838,300124
Honduras Rosewood 121830
Honesty 95431,238369,238371,238373
Honesty Plant 238369
Honewort 13966,113872,365357,365362,397729,397731
Honey Bee 220346
Honey Blob 334250
Honey Box 155649
Honey Bush 248944
Honey Cap 287494
Honey Eucalyptus 155649
Honey Flower 220308,248944,315937
Honey Garlic 263085,263087
Honey Locust 176860,176903
Honey Mesquite 315557,315560
Honey Myrtle 248072
Honey Palm 212557
Honey Plant 198821,198827
Honey Plantain 302103
Honey Protea 315937
Honey Sage 345210
Honey Spurge 159340
Honey Suckle 235630
Honey Tree 351968
Honey Wort 83923
Honeyball 62064,82098
Honey-balls 82098
Honey-balm 249542
Honey-bear Begonia 50020
Honey-bee 220346
Honey-bee Flower 272398
Honeybell Bush 168262
Honey-bell Rhododendron 330311
Honey-bells 266957,266965,266966
Honeyberry 249124,249126
Honeybind 236022
Honey-blob 334250
Honeybottle 67456,150131,401388
Honeybush 248944
Honey-cap 287494
Honeycomb Litse 233914
Honeycup 417592

Honeyed Acacia 1386
Honeyflower 220308,248936,315937
Honey-flower 166125,220346,248936,248944,272398,315937,397019
Honeyflower Azalea 331218
Honey-flower Rhododendron 331218
Honeyfruit Evodia 161320
Honeygrass 249270,249303
Honey-grass 249303
Honeylocust 176860,176903,176904
Honey-locust 176860,176903
Honey-plant 249504
Honey-plantain 302103
Honeypot Protea 315776
Honeysookies 287721
Honeystalk 397019,397043
Honeystick 95431
Honeysuchle-like Vineclethra 95521
Honeysuck 236022,397019
Honeysuckle 30007,68713,105204,214398,220311,220346,228320,235630,235730,235812,236022,236089,237539,287494,287721,329521,331352,385505,397019,397043
Honeysuckle Azalea 331138
Honeysuckle Banksia 47650
Honeysuckle Family 71955
Honeysuckle Fuchsia 168789
Honeysuckle Trefoil 397019
Honeysuckle-leaf Holly 203992
Honeysuckle-leaf Scurrula 355319
Honey-suckle-leaved Holly 203992
Honeysuckley Spider Flower 180608
Honeysuklelike Jurinea 211926
Honeysweet 166125,391603
Honeyweed 225009
Honeywort 83918,83923,170335
Hong Kong Lily 229755
Hong Kong Machilus 240570
Hong Kong Mucuna 259547
Hong Kong Orchid 49017,49247
Hong Kong Orchid Tree 49247
Hong Kong Orchidtree 49017
Hong Kong Orchid-tree 49017
Hong Kong Rose 332205
Hong Wintercress 47946
Hongay Hopea 198159
Hongcaoguo Amomum 19863
Hongcone-fir 318562
Honghe Butterflyorchis 293609
Honghe Cliffbean 254663
Honghe Giantarum 20087
Honghe Holly 204019
Honghe Litse 233940
Honghe Manglietia 244447
Honghe Orange 93486
Honghe Schefflera 350708
Hongjinma Hibiscus 194683
Hongjinma Rosemallow 194683
Hongkong Azaleastrum 330879
Hongkong Birthwort 34372
Hongkong Bulbophyllum 63155
Hongkong Burmannia 63962
Hongkong Camellia 68920,69135
Hongkong Cliffbean 254792
Hongkong Croton 112905
Hongkong Cudrania 114325
Hongkong Curlylip-orchis 63155
Hongkong Dendrobenthamia 105056
Hongkong Disperis 134324
Hongkong Dogwood 105056
Hongkong Dualbutterfly 398290
Hongkong Dutchman's-pipe 34372
Hongkong Dutchmanspipe 34372
Hongkong Eagleclaw 35018
Hongkong Eargrass 187696
Hongkong Elaeagnus 142213
Hongkong Eria 148687
Hongkong Fissistigma 166693
Hongkong Four-involucre 105056
Hongkong Gaerthera 399833
Hongkong Giantarum 20124
Hongkong Glochidion 177192
Hongkong Goldlineorchis 26014
Hongkong Gordonia 179745
Hongkong Gordontea 179745
Hongkong Habenaria 183524
Hongkong Hairorchis 148687
Hongkong Hawthorn 329068
Hongkong Hedyotis 187696
Hongkong Kumquat 167499,167516
Hongkong Lily 229754
Hongkong Loosestrife 239549
Hongkong Maackia 240122
Hongkong Magnolia 232593
Hongkong Milkwort 308108
Hongkong Millettia 254792
Hongkong Oak 116071
Hongkong Paphiopedilum 282889
Hongkong Pavetta 286264
Hongkong Pearlsedge 354219
Hongkong Pencilwood 139653
Hongkong Pencil-wood 139653
Hongkong Raphiolepis 329068
Hongkong Rattlesnake Plantain 179719
Hongkong Razorsedge 354219
Hongkong Rhaphidophora 329000
Hongkong Rosewood 121756
Hongkong Saddletree 240122
Hongkong Sasa 347307
Hongkong Spapweed 205019
Hongkong Spotleaf-orchis 179719
Hongkong Sumac 332646
Hongkong Tailgrape 35018
Hongkong Tainia 383145
Hongkong Tea 68920
Hongkong Tetrathyrium 387937
Hongkong Touch-me-not 205019
Hongkong Tripterospermum 398290
Hongkong Tsiangia 399833
Hongkong Wild Kumquat 167499
Hongkong Wildginger 37645
Hongkong Willd Kumquat 167499
Hongmao Mountain Phoebe 295371
Hongmaoshan Nanmu 295371
Hongmaoshan Phoebe 295371
Hongping Apricot 34439
Hongqiao Sage 345320
Hongshuihe Jasmine 211862
Hongtou Beautyberry 66831
Hongya Arisaema 33354
Hongya Ploughpoint 401165
Hongya Southstar 33354
Hongya Typhonium 401165
Hongyao Chirita 87905
Hongyuan Honda Roegneria 335354
Hongyuan Sedge 74823
Honiton Lace 28044,101852,392992
Honjiso Mandarin 93748
Honohonokukui 272614
Honolulu Creeper 28422
Honolulu Rose 96009
Honolulu-queen 200713
Honshu Draba 137143
Honshu Spruce 298307
Honshu Whitlowgrass 137143
Hontauyu Copperleaf 1997
Honytoothed Ardisia 31387
Honyybag Leek 15791
Honyybag Onion 15791
Hood Canarygrass 293750
Hood Orchids 320862,320871
Hooded Arrowhead 342383
Hooded Bindweed 68713
Hooded Coralroot 104016,104019
Hooded Hakea 184599
Hooded Helleborine 82037
Hooded Ladies'-tresses 372242
Hooded Lady's-tresses 372242
Hooded Milfoil 403382
Hooded Orchid 376391
Hooded Pitcher Plant 347153
Hooded Pitcherplant 347153
Hooded Pitcher-plant 347153
Hooded Skullcap 355449
Hooded Violet 410306
Hooded Willow 355449
Hooded Willowherb 355449
Hoodia 198035,198079
Hood-leafed Hakea 184599
Hoodorchis 264757,264779

Hoodshape Orchis 264757
Hoodshaped Alocasia 16495
Hood-shaped Caesalpinia 252745
Hoodshaped Oncidium 270804
Hoodshaped Orchid 264757
Hood-shaped Tanoak 233164
Hoodwort 355449,355563
Hooey Bush 248936
Hoof Ploughpoint 401176
Hoof Typhonium 401176
Hoofcelery 129195
Hooffruit 316018,316031
Hoofleaf Goldenray 229123
Hoofrenshen 133076
Hoofs 400675
Hoog Iris 208627
Hoog Tulip 400176
Hoogendorp Livistona 234181
Hoohked Eritrichium 153553
Hook Thorn 1117
Hook Thorn Tree 1117
Hookcyme Chirita 87880
Hooked Agrimony 11558
Hooked Bristle Grass 361930
Hooked Bristlegrass 361930
Hooked Bristle-grass 361930
Hooked Buttercup 326281
Hooked Clearweed 299026
Hooked Clematis 95396
Hooked Crowfoot 326281
Hooked Dendrobium 124963
Hooked Deutzia 126963
Hooked Dock 339960
Hooked Fortune Spindle-tree 157493
Hooked Groundstar 22028
Hooked Kobresia 217304
Hooked Needlewood 184645
Hooked Rattlebox 112776
Hooked Robiquetia 335036
Hooked Rostrinucula 337290
Hooked Sageretia 342170
Hooked Spine Hamatocactus 185157
Hooked Tongueleaf 177469
Hooked Violet 409648
Hookedhair Fig 164652
Hookedhair Madder 338002
Hookedhairy Beautyberry 66895
Hookedhairy Purplepearl 66895
Hookedhairypod Tickclover 126359
Hookedsepal 267077,267078
Hookedspine Bittersweet 80189
Hookedspur Larkspur 124271
Hooked-spur Violet 409648
Hookedstyle Meadowrue 388706
Hooker Acronema 6213
Hooker Amischotolype 19495
Hooker Arundinella 37400
Hooker Azalea 330880
Hooker Barberry 51733
Hooker Basketvine 9440
Hooker Begonia 50298
Hooker Bellcalyxwort 231262
Hooker Bentgrass 12133
Hooker Bittersweet 80207
Hooker Bluebell 115355
Hooker Bluebellflower 115355
Hooker Cephalostigma 412702
Hooker Chionocharis 87747
Hooker Christisonia 89219
Hooker Clearweed 298937
Hooker Cortiella 105487
Hooker Corydalis 105991
Hooker Croton 112910
Hooker Cuphea 114604
Hooker Denseflower Blumea 55724
Hooker Deutzia 126971
Hooker Dragonhead 137598
Hooker Elaeocarpus 142336
Hooker Elatostema 142685
Hooker Eriolaena 152689
Hooker Fedtschenkiella 163088
Hooker Fig 165135
Hooker Goblin Orchid 258996
Hooker Goldenray 229055
Hooker Greenorchid 137598
Hooker Hawksbeard 369855
Hooker Holly 203891
Hooker Iris 208629
Hooker Knotweed 309188
Hooker Lacquertree 393461
Hooker Lacquer-tree 393461
Hooker Landpick 308560
Hooker Lasianthus 222163
Hooker Leaf-flower 296632
Hooker Leek 15351
Hooker Mallotus 243368
Hooker Manglietia 244448
Hooker Maple 3020
Hooker Mayten 246856
Hooker Murdannia 260099
Hooker Onion 15351
Hooker Onosma 271777
Hooker Paphiopedilum 282843
Hooker Pearlsedge 354119
Hooker Pennywort 200308
Hooker Pepper 300416
Hooker Pleurospermum 304808
Hooker Plumegrass 148883
Hooker Primrose 314474
Hooker Raffia Palm 326640
Hooker Razorsedge 354119
Hooker Rhaphidophora 329001
Hooker Rhododendron 330880
Hooker Ribseedcelery 304808
Hooker Rockfoil 349477
Hooker Rockjasmine 23199
Hooker Rocktaro 327674
Hooker Roughleaf 222163
Hooker Sarcococca 346731
Hooker Sharptooth Photinia 295626
Hooker Shortawn Ungeargrass 36624
Hooker Solomonseal 308560
Hooker Soroseris 369862
Hooker Spapweed 205020
Hooker Speargrass 397506
Hooker Staff-tree 80207
Hooker Stairweed 142685
Hooker Sugarpalm 32338
Hooker Sweet Vernalgrass 27945
Hooker Sweetleaf 381240
Hooker Themeda 389352
Hooker Touch-me-not 205020
Hooker Trikeraia 397506
Hooker Tubergourd 390144
Hooker Vanda 404646
Hooker Vernalgrass 27945
Hooker Waterdropwort 269323
Hooker Whinghead 320435
Hooker Whinghead Flower 320435
Hooker Whitepearl 172086
Hooker Wildtung 243368
Hooker Willow 343496
Hooker Wintergreen 172086
Hooker's Barberry 51733
Hooker's Ceropegia 84124
Hooker's Evening Primrose 269434
Hooker's Evening-primrose 269450
Hooker's Fleabane 150924
Hooker's Gaultheria 172086
Hooker's Hebe 186936
Hooker's Indian Pink 363552
Hooker's Larkspur 124233
Hooker's Manzanita 31118
Hooker's Onion 15020
Hooker's Orchid 183700,302358
Hooker's Orchis 183700
Hooker's Rhododendron 330880
Hooker's Sandwort 31948
Hooker's Scratchdaisy 111687
Hooker's St. John's Wort 201925
Hooker's St. John's-wort 201925
Hooker's St. Johnswort 201925
Hooker's Thistle 92030
Hooker's Whing Head 320435
Hooker's Wild Buckwheat 152149
Hooker's Willow 343496
Hookettea 52477
Hook-haired Beautyberry 66895
Hook-hairy-podded Tickclover 126359
Hook-heal 316127
Hookleafvine 303087,303092

Hookpetal Monkshood 5242
Hooks-and-hatchets 2837
Hook-shap Flowers Rhododendron 332037
Hookship-flowered Rhododendron 332037
Hookspur Calanthe 65966
Hookspur Corydalis 105957
Hook-spur Violet 409648,409650
Hookstalk Uraria 402146
Hooktea 52400
Hooktooth Buckthorn 328754
Hooktree 337289,337290
Hookvine 401737,401761
Hookweed 316127
Hoop Ash 80698,168032
Hoop Petticoat 262441
Hoop Petticoat Daffodil 262361
Hoop Petticoats 262441
Hoop Pine 30841
Hoopash 168032
Hoop-petticoat Daffodil 262361
Hoopvine 396455
Hoover's Desert Trumpet 151929
Hoover's Wirelettuce 375965
Hooves 400675
Hop 61497,199376,199379,199384,
199387,276818
Hop Bitter Pea 123268
Hop Bush 135214,135215
Hop Clover 247366,396854,397096
Hop Family 71202
Hop Goodenia 179559
Hop Hornbeam 276801,276803
Hop Marjoram 250432,274210
Hop Medic 247366
Hop Medick 247366
Hop Medick Shamrock 247366
Hop Plant 274210
Hop Sedge 75233
Hop Seed Bush 135192
Hop Tree 320065,320071
Hop Trefoil 396854
Hopbush 135192,135215
Hop-bush 135192,135215
Hop-clover 396806,397096
Hope And Charity Faith 321643
Hopea 198126
Hopeh Speedwell 407078
Hopei Oak 324012
Hopei Pear 323194
Hopei Poplar 311349
Hopei Walnut 212614
Hopes 246478
Hophead Philippine Violet 48241
Hophornbeam 276818
Hop-hornbeam 276801
Hop-horn-beam 276801
Hopi Cotton 179908
Hopleaf Ampelopsis 20400
Hopleaf Snakegrape 20400
Hop-leaved Ampelopsis 20400
Hopley's Purple Oregano 274223
Hoplike Sedge 75232
Hop-like Sedge 75232
Hop-o'-my-thumb 174877,237539
Hopp's Figwort 355077
Hops 199379,199382,199384
Hopsage 180470
Hopsedbush-leaved Tanoak 233181
Hopseed Bush 135192,135215
Hop-seed Bush 135192
Hopseedbush 135192
Hop-shaped Dodder 115066
Hoptensia 199956
Hoptree 320065,320071
Hop-tree 320065
Hopweed 345269
Hora 133590
Horaninowia 198237
Hore Calamint 65540
Horehound 47013,245723,245770
Horehound Motherwort 85058
Horestrange 292957
Horestrong 292957
Horewort 166038
Horizontal Italian Cypress 114760
Horizontal Juniper 213785
Horizontalwing Maple 3764
Hormigo 397931
Horn Bamboo 297280,297296
Horn Beech 77256
Horn Birch 401512
Horn Corydalis 105754
Horn Flower 417093
Horn Milkvetch 42166
Horn Nut 394500
Horn of Plenty 123065
Horn Poppy 176724,176738
Horn Wort 83540,83545
Hornbeam 2868,77252,77256,77263,
77264,276818,401512
Hornbeam Maple 2868
Hornbeam Mountain-ash 369354
Hornbeamleaf Elm 401468
Hornbeamleaf Mountainash 369570
Hornbeam-leaved Maple 2868
Hornbeam-leaved Sallow 343173
Hornbean Maple 2943
Hornbract Dandelion 384816
Horncap Eucalyptus 155756
Horncone 83680,83682
Horned Beech 77256
Horned Bladderwort 403147
Horned Clover 247366,387251
Horned Cucumber 114221
Horned Dandelion 384493
Horned Dock 339961
Horned Holly 203660
Horned Maple 2928
Horned Melon 114221
Horned Milkwort 308006
Horned Pansy 409858
Horned Pleurothallis 304910
Horned Pondweed 417045,417053
Horned Poppy 176724,176735,176738
Horned Rampion 298054,298068
Horned Rush 333527,333625
Horned Sea Poppy 176738
Horned Sea-blite 379503
Horned Spindle-tree 157391
Horned Tulip 400112
Horned Violet 409858
Hornedcalyx Chirita 87843
Hornedhead Elatostema 142669
Hornedhead Stairweed 142669
Horned-pondweed 417045,417053
Horned-pondweed Family 417073
Horned-poppy 176724,176728,176738
Horned-wing Birch 53376
Horneman Willow Weed 146734
Hornemann's Willowweed 146734
Hornet Orange 93561
Hornet Rose 336969
Hornfennel 201597,201605
Hornflower 83392
Horn-fruited Spindle-tree 157361
Hornicks Hornecks 102727
Hornkraut 82629
Hornleaf Maple 3650
Horn-leaved Maple 3650
Hornless Dandelion 384529
Hornlike Euonymus 157391
Hornlike Milkvetch 42166
Hornlikepetal Habenaria 183488
Horn-nut 394500
Horn-of-plenty 123065,163040
Hornpoppy 176724,176735
Horn-poppy 176724
Hornpoppyleaf Anemoclema 23682
Hornrim Saxifrage 349608
Horns 198376
Hornsage 205546
Hornseed 83443
Hornstedtia 198470
Hornstyle 83651
Hornthorn-leaf Raspberry 338300
Hornthorn-leaved Raspberry 338300
Hornwort 83540,83545,83558,83578
Hornwort Family 83539
Hornwort Pondweed 83545
Horny Bamboo 47236
Horny Bird Cherry 280018

Horny Conebush 210242
Horny Euonymus 157361,157391
Horny Holly 203660
Horny Tanoak 233153
Hornyfruit Seepweed 379509
Horny-like Euonymus 157523
Horoeka 318063
Horone 245770
Horrid Greenbrier 366392
Horrid Thistle 92033
Horse Apple 240828
Horse Balm 99843,99844
Horse Bean 71042,408393,408394
Horse Beech 77256
Horse Blob 68189
Horse Blobs 68189,227533
Horse Bramble 336419
Horse Breath 271563
Horse Brier 366559
Horse Brush 387071
Horse Buckle 315082
Horse Buttercup 68189
Horse Buttons 243840
Horse Cassia 78316
Horse Cassia Bark 78316
Horse Chestnut 9675,9679,9701,77256
Horse Chestnut Family 196494
Horse Cress 407029
Horse Crippler 197762
Horse Daisy 26768,227533,397970
Horse Dashel 92485
Horse Elder 207135
Horse Fennel 196723
Horse Flower 248211
Horse Gentian 397828,397833
Horse Gog 316819
Horse Gold 325612,325666
Horse Gowan 227533,384714
Horse Gram 241440
Horse Hardhead 81237
Horse Hoof 68189,400675
Horse Knob 81237
Horse Knop 81020,81237,81356
Horse Knot 81237,81356
Horse May 401593
Horse Mint 10422,250291,250385,250439,257169,257184
Horse Nettle 10422,367014,367663
Horse Nicker 64971
Horse Nop 81237
Horse Parsley 192358,366743
Horse Pease 408522
Horse Pen 329521
Horse Pennies 227533,329521
Horse Pepper 24483
Horse Peppermint 13174
Horse Plum 34431,316206
Horse Purslane 394903,416898
Horse Radish 34590
Horse Radish Tree 258945
Horse Saving 213702
Horse Shoe Vetch 196630
Horse Shoes 3462
Horse Snap 81237
Horse Sorrel 340082
Horse Sugar 381441
Horse Thistle 92485,219609
Horse Thyme 97060,391397
Horse Violet 410066,410493,410677
Horse Weed 19191,103446
Horse Well Cress 407029
Horse Well-cress 262722,407029
Horse Wellgrass 407029
Horse Wicky 298770
Horse's Hooves 68189
Horse's Mouth 28617
Horse's Tooth 249232
Horse-and-hounds 13174
Horsebalm 99844
Horsebane 269299
Horsebean 408393
Horsebean Caltrop 418641
Horsebladder 316953
Horse-breath 271563
Horsebrier 366559
Horse-brier Riparian Greenbrier 366554
Horsebush 288889
Horse-cane 19191
Horsechestnut 9675,9679,9701
Horse-chestnut 9675,9701
Horsechestnut Family 196494
Horse-Chestnut Family 196494
Horsechire 388042
Horse-crippler 197762
Horseegg 182961
Horse-eye Bean 71042,259573
Horseflesh Mahogany 239540
Horse-fly Weed 47880
Horsefly's Eye 136214
Horsefly-weed 47880
Horsefoot 400675
Horsegentian 397828
Horse-gentian 397828
Horse-gog 316819
Horsegram 241440
Horseheel 207135
Horsehelne 207135
Horsehoof Bamboo 47318
Horsehoof Waterwort 142570
Horsekidneyfruit 11313
Horselene 207135
Horsemint 250385,257157,257169,257184,257188
Horse-Mint 257157
Horse-nettle 367014
Horse-nut Tree 9701,366743
Horse-purslane 394879
Horseradish 34582,34590,258935,258945
Horse-radish 34590
Horseradish Family 258961
Horseradish Tree 258945
Horse-radish Tree 258945
Horseradishtree 258945
Horseradish-tree 98258,258945
Horses-and-chariot 60382
Horses-and-chariots 60382
Horse-seed Bush 135215
Horseshoe Geranium 288297,288594
Horseshoe Pelargonium 288594
Horse-shoe Pelargonium 288594
Horseshoe Vetch 196621,196630
Horseshoe-orchid 272432
Horseshoes 3462
Horseshoe-vetch 196630
Horse-strong 292957
Horsesugar 381441
Horsetail Beefwood 79161
Horsetail Milkweed 38160
Horsetail Paspalum 285496
Horsetail Pine 300054
Horsetail She-oak 79162
Horsetail Spike Rush 143141
Horsetail Spike-rush 143141
Horsetail Thorowax 63756
Horsetail Tree 79161
Horsetails-like Spikesedge 143217
Horsetailtree 332328
Horsetailtree Family 332330
Horsetops 92213
Horseweed 99844,103402,103446
Horseweed Fleabane 103446
Horseweed Fleahane 103446
Horsewhip Buckthorn 328617
Horse-whip Buckthorn 328617
Horsfield Euchresta 155892
Horsfieldia 198507
Horshelne 207135
Horshone 176824
Horst Beech 77256
Hortensia 199956,203545
Hortleberry 403916
Horts 403916
Hortulan Plum 316444
Hortulana Plum 316444
Hose-doup 252305
Hosie Indigo 206099
Hosie Ormosia 274401
Hosiea 198574
Hosoba Kan Suge 76503
Hosokawa Thistle 92037
Hospital Canyon Larkspur 124092

Host's Sedge 74830
Hosta 198589
Hosts Saxifrage 349481
Hot Arsesmart 309199
Hot Cross Bun 157429
Hot Pepper 72070
Hot Pink Silene 363300
Hot Springs Sagebrush 35150
Hot Water Plant 4070,4078
Hotan Milkvetch 42500
Hot-dogcactus 216654
Hot-dry Iris 208439
Hoth 401388
Hotmint Lemongrass 117190
Hot-scented Milkvetch 42444
Hotspring Larkspur 124697
Hottentot Bean 352493
Hottentot Bedding 189276
Hottentot Bread 131573
Hottentot Cherry 246681
Hottentot Fig 77431,77439,77441,251265,251357
Hottentot Tea 189772
Hottentot's Bread 131573
Hottentot-bread 131573
Hottentot-fig 77429,77439,77441
Hot-water Plant 4070
Houdapu Tanoak 285233
Houghton Currant 334036
Houghton's Cyperus 119004
Houghton's Flat Sedge 119004
Houghton's Flatsedge 119004
Houghton's Goldenrod 368170
Houghton's Sedge 74832
Houghton's Woolly Sedge 74832
Houhe Sarsaparilla 30682
Houhere 197154
Houlletia 198684
Hound's Fennel 26768
Hound's Head 28631
Hound's Mie 118008
Hound's Mile 118008
Hound's Piss 118008
Hound's Tongue 117900,118008,221770,373346
Hound's Tooth 144661
Hound's Tree 380499
Hound's-tongue 117900,117971,118008
Houndberry Tree 380499
Houndsbane 245770
Houndsberry 44708,367416
Houndstongue 117900
Houpara 318069
Houpu 198699
House Bean 408393
House Geranium 288297
House Leek 358015,358062
House Leeks 358062
House Lime 370130,370136
House Pine 30848
House's Rose 336635
House's Stitchwort 255513
Housegreen 358062
Household Bean 408393
Houseleek 358015,358021,358062
Houserock Cactus 287902
Houston camphor-daisy 327154
Houstonia 198703
Houstonia Bluets 198703
Houttuynia 198743,198745
Hove 176824
How Adinandra 8239
How Anodendron 25940
How Begonia 49934
How Birthwort 34210
How Croton 112911
How Dutchman's-pipe 34210
How Dutchmanspipe 34210
How Eelvine 25940
How Elaeocarpus 142337
How Indianmulberry 258892
How Ixora 211090
How Ophiorrhiza 272216
How Ormosia 274402
How Parsonsia 285056
How Persimmon 132202
How Portia Tree 389938
How Portiatree 389938
How Raspberry 338537
How Syzygium 382572
How Tanoak 233251
How Trigonostemon 397364
How Triratna Tree 397364
How's Dichapetalum 128675
How's Kurz Lasianthus 222199
How's Kurz Roughleaf 222199
How's Mycetia 260633
How's Newlitse 264056
Howard's Rabbitbrush 150368
How-doup 252305
Howell Allomorphia 16023
Howell Anemone 23848
Howell Clovershrub 70823
Howell Cmas 68799
Howell Coldwaterflower 298939
Howell Fawnlily 154924
Howell Indigo 206101
Howell Windflower 23848
Howell's Fleabane 150696
Howell's Montia 258256
Howell's Pussytoes 26424
Howell's Pussy-toes 26425,26427,26428
Howell's Rush 213159
Howell's Silverpuffs 253926
Howell's Spineflower 89079
Howell's Stitchwort 255493
Howell's Wild Buckwheat 152150
Howell's-buckwheat 212397
Howes 109857
Howitt's Wattle 1299
Howka 252662
Howler 16352
How-like Begonia 50335
Hoya 198821
Hoya Shooting Star 81911
Hraerleaf Baliosperm 46971
Hris Plant 16525
Hsiang Cheng 93515
Hsiangyang Cherry 83184
Hsien Mu 161622
Hsienchu Bitter Bamboo 304037
Hsienmu 161621,161622
Hsienmu Burretiodendron 161622
Hsinba Milkvetch 42501
Hsinchu Leafflower 296695
Hsing'an Larkspur 124296
Hsingan Poplar 311350
Hsingan Willow 343497
Hsintucho Monkshood 5633
Hsinyi Sterculia 376193
Hsitao Thick-leaf Hydrangeavine 351785
Hsueh Fargesia 162699
Hsuyun Barberry 51740
Hu Childvine 400905
Hu Grape 411742
Hu Larkspur 124298
Hu Loosestrife 239673
Hu Maple 3024
Hu Rustyhair Raspberry 339163
Hu Storax 379370
Hu Tylophora 400905
Hu's Cyclobalanopsis 116114
Hu's Euonymus 157582
Hu's Holly 203896
Hu's Nanmu 295369
Hu's Oak 116114,324014
Hu's Phoebe 295369
Hu's Raspberry 339163
Hu's Rhododendron 330886
Hu's Sichuan Holly 204319
Huachuca or Rutter's Goldenaster 194243
Huaco 244371,244374
Huading Rhododendron 330885
Huajillo 186238
Hualian Ainsliaea 12694
Hualian Bogorchid 158153
Hualian Copperleaf 1997
Hualian False-nettle 56175
Hualian Photinia 295773
Hualian Privet 229587
Hualian Rabbiten-wind 12694

Hualian Tainia 383149
Hualian Willow 344167
Hualien Privet 229587
Huan Pine 121100
Huanghe Dahurian Bushclover 226753
Huangho Tickseed 104785
Huanghuahaosu 35132
Huanghuali 121701,121702
Huangjingtiao Peashrub 72233
Huanglianshan Alseodaphne 17802
Huanglianshan Begonia 49738
Huanglin Masson Pine 300058
Huanglong Willow 343611
Huangmushu Trigonostemon 397365
Huangmushu Triratna Tree 397365
Huangping Maple 3023
Huangping Raspberry 338538
Huangshan Ash 167989
Huangshan Azalea 330097
Huangshan Bittercress 72718
Huangshan Cacalia 283828
Huangshan Eurya 160524
Huangshan Fritillary 168491
Huangshan Magnolia 416690
Huangshan Monkshood 5105
Huangshan Mountain Magnolia 416690
Huangshan Mountain Monkshood 5105
Huangshan Mountain Pine 300269
Huangshan Mountainash 369309
Huangshan Mountain-ash 369309
Huangshan Oak 324452
Huangshan Pentapanax 289639
Huangshan Pine 300269
Huangshan Redbud 83778
Huangshan Rhododendron 330097
Huangshan Rose 336638
Huangshan Sage 344953
Huangshan Skullcap 355481
Huangshan Snakegourd 396258
Huangshan Windhairdaisy 348379
Huangyan Orange 93735
Huangyan Spapweed 205022
Huangzhong Larkspur 124297
Huaning Gymnema 182366
Huanita 57905
Huanjiang Aspidistra 39546
Huanjiang Ophiorrhiza 272217
Huanoco Cocaine-tree 155067
Huanuco Barberry 51705
Huanuco Cocaine Tree 155067
Huanuco Leaf 155067
Huapei Scabious 350265
Huashan Ash 168090
Huashan Chalkplant 183208
Huashan Leptodermis 226093
Huashan Maackia 240129
Huashan Milkvetch 42474
Huashan Mountain Maackia 240129
Huashan Newstraw 317079
Huashan Pine 299799
Huashan Primrose 314481
Huashan Psathyrostachys 317079
Huashan Saddletree 240129
Huashan Sedge 74836
Huashan Wildclove 226093
Huashan Windhairdaisy 348376
Huata 213625
Huaxi Childvine 400849
Huaxi Dogwood 124928
Huaxi Four-involucre 124928
Huaxi Tylophora 400849
Hubbard Squash 114288
Hubei Actinidia 6632
Hubei Anemone 23850
Hubei Angelica 24318
Hubei Aralia 30685
Hubei Arrowwood 407886
Hubei Arundinella 37402
Hubei Ash 167991
Hubei Beautyberry 66795
Hubei Bedstraw 170421
Hubei Buckthorn 328731
Hubei Bushclover 226840
Hubei Camellia 69142
Hubei Catchfly 363559
Hubei Chickweed 374914
Hubei Childvine 400960
Hubei Chloranthus 88283
Hubei Clematis 95023
Hubei Clovershrub 70847
Hubei Corydalis 105564
Hubei Crabapple 243630
Hubei Cranesbill 174885
Hubei Dualbutterfly 398267
Hubei Early Lilac 382232
Hubei Euonymus 157587
Hubei Eurya 160499
Hubei Evergreenchinkapin 78952
Hubei Evodia 161340
Hubei Fritillary 168491
Hubei Glaucous Bauhinia 49101
Hubei Glochidion 177190
Hubei Grape 411923
Hubei Greenbrier 366511
Hubei Hawthorn 109760
Hubei Hornbeam 77306
Hubei Indianmulberry 258893
Hubei Jurinea 207145
Hubei Kiwifruit 6632
Hubei Ladybell 7665
Hubei Landpick 308665
Hubei Ligusticum 229394
Hubei Lily 229862
Hubei Lime 391730
Hubei Linden 391730
Hubei Liriope 232648
Hubei Litse 233942
Hubei Lycianthes 238954
Hubei Maesa 241774
Hubei Mahonia 242522
Hubei Mallotus 243332
Hubei Mountain-ash 369420
Hubei Oligostachyum 270435
Hubei Orychophragmus 275879
Hubei Osmanther 276323
Hubei Oxalis 277897
Hubei Pondweed 312141
Hubei Poorspikebamboo 270435
Hubei Primrose 314717,314748
Hubei Purplepearl 66795
Hubei Red Silkyarn 238954
Hubei Redbud 83787
Hubei Rehmannia 327448
Hubei Rosewood 121714
Hubei Sage 345091
Hubei Sasa 347226
Hubei Slenderstalk Starjasmine 393616
Hubei Small Reed 127251
Hubei Smallreed 127251
Hubei Solomonseal 308665
Hubei Spapweed 205249
Hubei Spindle-tree 157587
Hubei Splitmelon 351765
Hubei Spurge 159101
Hubei St. John's Wort 201928
Hubei Thistle 92040
Hubei Toona 392845
Hubei Touch-me-not 205249
Hubei Trigonotis 397436
Hubei Tripterospermum 398267
Hubei Tubergourd 390170
Hubei Viburnum 407886
Hubei Wildtung 243332
Hubei Willow 343509
Hubei Windflower 23850
Hubei Windhairdaisy 348368
Hubei Wingnut 320358
Hubei Woollystyle Raspberry 338724
Huckleberry 172248,172249,367368,
403710,403716,403916
Huckleberry Lily 229984
Huckleberry Oak 324530
Hudang Rattan Palm 217920
Huds 340082
Hudson Bay Rose 336389
Hudson Currant 334037
Hudson Fir 283
Hudsonian Club-rush 396005
Hudsonian knotweed 309130
Hueco Rock Daisy 291316
Huegel Barberry 51741

Huegel's She-oak 79166
Huernia 198969
Huff Cap 144661
Huff-cap 144661
Huge Indocalamus 206884
Hugel Muhly 259671
Huggaback 408352
Huggan 336419
Huggy-me-close 170193
Hugher's Trillium 397551
Hughes Juniper 213795
Hugo Rose 336636
Hugon Bentgrass 12136
Hugonia 199137
Hugonis Rose 336636
Huguang Euonymus 157584
Hu-Guang Euonymus 157584
Huidong Azalea 330888
Huidong Lily 229864
Huidong Petrocosmea 292567
Huidong Rhododendron 330888
Huidong Stringbush 414191
Huigel Waldheimia 413011
Huili Mahonia 242638
Huili Monkshood 5283
Huili Primrose 314785
Huili Small Reed 127298
Huili Smallreed 127298
Huining Milkvetch 42503
Huisache 1219
Huisache Dulce 1219
Huiyang Azalea 331980
Huiyang Rhododendron 331980
Huizache 1473
Huizache Chino 1560
Huize Arisaema 33307
Huize Corydalis 106126
Huize Hogfennel 292764
Huize Monkshood 5284
Hulidao Dendranthema 124795
Huller Nut 212599
Hullin Hull 203545
Hulst 203545
Hulver 203545
Hulver Archaic 203545
Hulver Oak 324027
Hulwort 388184
Huma Willow 343504
Humack 336419
Humbert Barberry 51743
Humble 396854
Humble Gilia 175681
Humble Plant 255098
Humble-bee 272398
Humble-plant 255098
Humblot Cymbidiella 116759
Humblot Vanilla 405002
Humboldt Cactus 244113
Humboldt Laelia 219683
Humboldt Lily 229865
Humboldt Willow 343207
Humboldt's Willow 343505
Humea 67698,199295
Humid Rattlebox 112773
Humid-shaped Barberry 51744
Humifuse Paraixeris 283397
Humifuse Pricklypear 272915
Humifuse Speedwell 407362
Humifuse Spurge 159092
Humildito 103759
Humillis Juniper 213796
Humlick 101852
Humlock 28044,101852
Humly 101852
Humming Bird's Trumpet 417407
Hummingbird Bush 25261,129325,180655,185169
Hummingbird Grevillea 180655
Hummingbird Tree 86783
Hummingbird Trumpet 146651
Hummingbird's Trumpet 417407
Hummock Sedge 74435,76399
Humpbacks 410320
Humped Bladderwort 403191
Humpvine 250867
Humpy-scrumple 192358
Humus Corydalis 105995
Humutiatenga 93497
Hunan Arisaema 33353
Hunan Aspidistra 39579
Hunan Aster 40598
Hunan Azalea 330891
Hunan Babylon Weeping Willow 343079
Hunan Barnyardgrass 140423
Hunan Barrenwort 147008
Hunan Chinese Groundsel 365058
Hunan Christmasbush 14197
Hunan Christmas-bush 14197
Hunan Clethra 96551
Hunan Currant 334039
Hunan Cyclobalanopsis 116115
Hunan Elatostema 142686
Hunan Elsholtzia 144039
Hunan Epimedium 147008
Hunan Groundsel 358177
Hunan Hackberry 80739
Hunan Hickory 77900
Hunan Hilliella 196291
Hunan Holly 203897
Hunan Indocalamus 206796
Hunan Ladybell 7763
Hunan Lime 391704
Hunan Linden 391704
Hunan Litse 233941
Hunan Maple 3217
Hunan Mountainash 369418
Hunan Mountain-ash 369418
Hunan Munronia 259886
Hunan Nanmu 295370
Hunan Nettletree 80739
Hunan Oak 116115
Hunan Ophiorrhiza 272219
Hunan Phoebe 295370
Hunan Ploughpoint 401167
Hunan Qinggang 116115
Hunan Qiongzhu 87595
Hunan Raspberry 338544
Hunan Rhododendron 330891
Hunan Sida 362525
Hunan Sinosenecio 365058
Hunan Skullcap 355483
Hunan Southstar 33353
Hunan Spapweed 205024
Hunan Stairweed 142686
Hunan Swollennoded Cane 87595
Hunan Tanoak 233253
Hunan Thrixsperm 390490
Hunan Thrixspermum 390490
Hunan Touch-me-not 205024
Hunan Trabeculate Alder 16471
Hunan Typhonium 401167
Hunan Violet 410091
Hunan Xmas Bush 14197
Hunan Yushanbamboo 416775
Hunan Yushania 416775
Hunangemoho Grass 87751
Hunangemoho-grass 87750
Hunan-Guangdong Rhododendron 330887
Hunan-Guangxi Newlitse 264057
Hunan-Guangxi Persimmon 132476
Hunan-Guangxi Rhododendron 330887
Hunaniopanax 199396,199397
Hunan-Kwangsi Mulberry 259202
Hunchy Sedge 74644
Hunderedlobe Windhairdaisy 348194
Hundred Holes 202086
Hundred Leaved Thistle 154301
Hundred Thistle 154301
Hundred-dente Spindle-tree 157360
Hundred-flower Barberry 51440
Hundred-leaved Grass 3978
Hundreds-and-thousands 116736,190622,243209,321643,339887,349044,356468,358062
Hundredstamen 389592
Hung Sha 221930
Hungarian 107300
Hungarian Acanthus 2683
Hungarian Brome 60760
Hungarian Chrysanthemum 89695
Hungarian Climber 95029

Hungarian Clover 396999
Hungarian Fustic 107300
Hungarian Gentian 173701
Hungarian Iris 208919
Hungarian Lilac 382180
Hungarian Linden 391747
Hungarian Milkvetch 42218
Hungarian Millet 361794
Hungarian Mullein 405779
Hungarian Oak 323917
Hungarian Sainfoin 271172
Hungarian Speedwell 407198
Hungarian Spurge 158857
Hungarian Thorn 109868
Hungarian Vetch 408527
Hungary Deadnettle 220412
Hungdredtooth Euonymus 157360
Hungerweed 325612
Hunger-weed 17548,325518,325612
Hungry Grass 17548
Hungry Rice 130544
Hungry Tickclover 126471
Hungta Beautyberry 66800
Hunteria 199412
Hunters 339887
Huntingdon Elm 401522,401529
Huntingdon Willow 342973
Huntleya 199443
Huntsman's Cap 355061,355262
Huntsman's Cup 347156
Huntsman's Horn 347150,347156
Huntsman's Trumpet 347150
Huocheng Catchfly 363558
Huocheng Crazyweed 278786
Huodendron 199448
Huolushan Sedge 74854
Huon Dacrydium 121100
Huon Pine 121084,121100,219792
Huonpine 121084
Huonpine Yaccatree 306420
Huoshan Dendrobium 125192
Huoshan Germander 388107
Huoshan Holly 203899
Huozhou Bird Rape 389518
Hupeh Anemone 23850
Hupeh Aralia 30685
Hupeh Bauhinia 49101
Hupeh Beautyberry 66795
Hupeh Chloranthus 88283
Hupeh Clematis 95023
Hupeh Clovershrub 70847
Hupeh Crab 243630
Hupeh Crab Apple 243630
Hupeh Crabapple 243630
Hupeh Evergreen Chinkapin 78952
Hupeh Fritillary 168491
Hupeh Hawthorn 109760
Hupeh Hornbeam 77306
Hupeh Inula 207145
Hupeh Lycianthes 238954
Hupeh Maesa 241774
Hupeh Monkshood 5552
Hupeh Mountain-ash 369420
Hupeh Primrose 314748
Hupeh Rosewood 121714
Hupeh Rowan 369420
Hupeh Sage 345091
Hupeh Solomonseal 308665
Hupeh Splitmelon 351765
Hupeh Vine 6553
Hupeh Wingnut 320358
Hupoci Raspberry 338305
Hurdreve 81065
Hurds 403916
Huron Green Orchid 302360
Hurr Bur 31051
Hurrah-bush 298770
Hurrahgrass 285524
Hurr-burr 31051
Hurricane Palm 321170
Hurricanegrass 166493
Hurricane-grass 166493
Hurrivane Palm 129684
Hurt Sickle 81020,81237
Hurtful Crowfoot 326340
Hurtleberry 403916
Hurts 403916
Hurtsickle 81020
Hurt-sickle 81020,81237
Huru 199465
Husbandman's Dial 66445
Husbandman's Tree 167955
Husbandman's Weather Warner 21339
Husbandman's Weatherglass 21339,373346,373455
Husbandman's Weather-warner 373346,373455
Husbandman's Woundwort 373346,373455
Husbeech 77256
Husk Corn 417434
Husk Ground-cherry 297676,297720
Husk Tomato 297640,297642,297716,297720
Husked Nut 78811
Husk-tomato 297640,297712,297720
Huslock 358062
Hutchinsia 198479,198483,199480
Hutchinson's Delphinium 124300
Hutchinson's Larkspur 124300
Hutera wrightii 99094
Hutton Cockscomb 80396,80422
Huxian Windhairdaisy 348860
Huzhu Azalea 331563
Huzhu Eritrichium 153462
Huzhu Rhododendron 331563
Hwalien False-nettle 56175
Hwalien Hornbeam 77302
Hwolute Sedge 76056
Hyacinth 176281,199565,199583,208771,260301
Hyacinth Bean 218721
Hyacinth Bletilla 55575
Hyacinth Brodiaea 398794
Hyacinth Dolichos 218721
Hyacinth Falselily 266896
Hyacinth Like Eria 148699
Hyacinth Lilac 382171
Hyacinth of Peru 353001
Hyacinth Orchid 34895,133399
Hyacinth Squill 352915
Hyacinthbean 218721
Hyacinth-flowered Larkspur 102833
Hyacinth-of-Peru 353001
Hyacinthus 199565
Hyalea 199620
Hyaline Feathergrass 376801
Hybrid Abutilon 913
Hybrid Actaea 6427
Hybrid Alfalfa 247507
Hybrid Aloe 16756
Hybrid Amaranth 18734
Hybrid Amaranthus 18734
Hybrid Anemone 23856
Hybrid Axyris 45853
Hybrid Black Poplar 311146
Hybrid Black Willow 343424
Hybrid Cardinal Flower 234277,234784
Hybrid Cattail 401108
Hybrid Cat-tail 401086
Hybrid Cattleya 79543
Hybrid Chokeberry 34869,295732
Hybrid Cinquefoil 312781
Hybrid Columbine 30041
Hybrid Coneflower 339572
Hybrid Coral Tree 154632
Hybrid Crabapples 243635
Hybrid Crack Willow 344046
Hybrid Crack-willow 344046
Hybrid Dead Nettle 220392
Hybrid Dogbane 29483
Hybrid Dombeya 135804
Hybrid Fleabane 150699
Hybrid Flowering Tobacco 266054
Hybrid Fuchsia 168750
Hybrid Gaillardia 169588
Hybrid Goosecomb 335543
Hybrid Jonquil 262429
Hybrid Knapweed 80893
Hybrid Knotweed 162508,309940
Hybrid Kurrajong 58347
Hybrid Larch 221830

Hybrid Lavender 223291
Hybrid Leathery-leaf Maple 2908
Hybrid Lobelia 234541
Hybrid Lupine 238456
Hybrid Magnolias 241954
Hybrid Maple 3025
Hybrid Medic 247507
Hybrid Milkwort 308116
Hybrid Minuartia 255494
Hybrid Montebretia 111464
Hybrid Oat 45481
Hybrid Pagodatree 369044
Hybrid Pertybush 292058
Hybrid Plume-poppy 240764
Hybrid Poplar 311146
Hybrid Poppy 282574
Hybrid Roegneria 335543
Hybrid Sage 344824
Hybrid Scabious 350168
Hybrid Serapias 360625
Hybrid Soundbean 100182
Hybrid Spathiphyllum 370339
Hybrid Stonecrop 294122
Hybrid Strawberry Tree 30878
Hybrid Sumac 332787
Hybrid Thorn 110047
Hybrid Trefoil 396930
Hybrid Tuberous Begonia 50378
Hybrid Verbena 405852
Hybrid Water-cress 336169
Hybrid Witch Hazel 185093
Hybrid Witchhazel 185093
Hybrid Woundwort 373086
Hybrid Yellow-cress 336173
Hybrid Yew 385384
Hybrids Dahlia 121541
Hydnophytum 199759
Hydrangea 199787,199807,199956
Hydrangea Family 200160
Hydrangea Vine 351780
Hydrangeavine 351780
Hydrangea-vine 351780
Hydrilla 200184,200190
Hydrobryum 200201
Hydrocera 200213
Hydrolea 200411,200424,200429
Hydrolea Family 200476
Hyères Tree-mallow 223378
Hygrochilus 200581,200584
Hygrometric Boea 56063
Hygrophila 200589
Hygrophilous Bittercress 72828
Hylodesmum 200724
Hylomecon 200753
Hylotelephium 200775
Hymenacue 200822
Hymenocallis 200906
Hymenocarpous Caesalpinia 65020
Hymenocrater 201038
Hymenodictyon 200848,201057,201066
Hymenolobus 201135
Hymenolyma 201155
Hymenopogon 263958
Hyokan 93323
Hyospathe 201433
Hyparrhenia 201454
Hypecoum 201597
Hypericifoliate Spiraea 371938
Hypericum 201705
Hypersilvery Cinquefoil 312671
Hypocrites 410493
Hypocyrta 202450
Hypodulloured Litse 233945
Hypoestes 202495
Hypoglaucous Bauhinia 49136
Hypoglaucous Deutzia 126976
Hypoglaucous Grape 411745
Hypoglaucous Greenbrier 366395
Hypoglaucous Hydrangea 199881,199904
Hypoglaucous Pearlwood 296453
Hypoglaucous Phyllanthodendron 296453
Hypoglaucous Smilax 366395
Hypoglaucous Stairweed 142688
Hypoglaucous Storax 379471
Hypoglaucous Yam 131531
Hypogomphia 202649
Hypogynousflower Wildginger 37646
Hypoleuca Ajania 13002
Hypolytrum 202702
Hypopitys 202761
Hypotomentose Rhododendron 330902
Hypotrichous Cherrylaurel 223104
Hypserpa 203018,203023
Hyptianthera 203028
Hyrcania Germander 388104
Hyrcanian Maple 3029
Hyssop 10402,10407,53304,203079,203095
Hyssop Hedge Nettle 373126,373253
Hyssop Hedge-nettle 373253
Hyssop Loosestrife 240047
Hyssop Lythrum 240047
Hyssop of the Bible 10402,274226
Hyssop Skullcap 355519
Hyssop-leaf Fleabane 150702
Hyssopleaf Leucas 227611
Hyssopleaf Pepperweed 225283
Hyssopleaf Thoroughwort 158154
Hyssop-leaved Boneset 158154
Hyssop-leaved Bugseed 104786
Hyssop-leaved Eupatorium 158154
Hyssop-leaved Fleabane 150702
Hyssop-leaved Thoroughwort 158154
Hyssop-leaved Thorough-wort 158154
Hyssop-leaved Tickseed 104786
Hystrix Devil Rattan 121500
Hyuganatsu 93834
Hyven 187222
Hyvin 187222

I

Iady's Lint 374916
Iasmine-flowered Rhododendron 330955
Iberi Iris 208635
Iberian Geranium 174660
Iberian Iris 208635
Iberian Knapweed 81112
Iberian Lallemantia 220284
Iberian Spiderflower 95704
Iberian Spirea 371938
Iberian Star Knapweed 81112
Iberian Star Thistle 81112
Iberis 203181
Ibex Wild Buckwheat 151812
Iboga 382865
Ibota Liguster 229558
Ibota Privet 229558
Ibul Palm 273123
Icacina Family 203303
Icaco 89816
Icaco Coco Plum 89816
Icaco Coco-plum 89816
Icacos 89816
Ice Pink 127725
Ice Plant 29855,77429,138183,177446,177465,220485,220660,220674,243253,251265,276202
Ice Rockfoil 349386
Iceboggrass 350859,350861
Iceboggrass Family 350864
Icecold Anemone 24038
Ice-cream Bean 206928
Icegrass 294907
Ice-grass 294907
Iceland Phyllophyton 245695
Iceland Poppy 282536,282617
Iceland Purslane 217648

Iceland Watercress 336250
Icelandknotweed 217640,217648
Iceland-purslane 217640,217648
Iceplant 251025
Ice-plant 29853,136394,200802,251265,
258044,349936,401711
Ichang Bitter Orange 93495
Ichang bitterorange 93495
Ichang Bramble 338554
Ichang Clovershrub 70833
Ichang Lemon 93495,93869
Ichang Lily 229890
Ichang Monkshood 5287
Ichange Ihih 93495
Ichant Papeda 93495
Ichnocarpus 203321
Idaeuslike Raspberry 338556
Idaho Fleabane 150481
Idaho Goldenweed 397748
Idaho Hawksbeard 110747
Idaho White Pine 300082
Idesia 203417
Idewa 185697
Idigbo 386563
Idria Wild Buckwheat 152623
Ifafa Lily 120532
Igaganga 121120
Ignat Poisonnut 378761
Ignatius Bean 378761
Igneous Coloured Rhododendron 330911
Igneous-coloured Rhododendron 330911
Ignorant Barberry 51754
Igorot lemon 93857
Iguanatail 346100
Iguanura 203501
Igusa 213066
Iigiri Tree 203422
Iironwood 276811
Ikoma Woodbetony 287288
Ikonnikov Serpentroot 354876
Ikonnikov Skullcap 355711
Ikonnikovia 203518
Ilama 25850
Ilang-Ilang 70960
Ilan-ilan 70960
Ilb 418216
Iles 138348,198376
Ilex 203527
Ilex Oak 324027
Ilexleaf Osmanther 276327
Ilex-leaved Osmanthus 276327
Ilex-leaved Spindle-tree 177995
Ili Barberry 51757
Ili Beancaper 418677
Ili Bubbleweed 297876
Ili Gagea 169450
Ili Larkspur 124302
Ili Poplar 311355
Ili Sweetvetch 187940
Ili Tulip 400178
Ili Willow 343523
Iliau 414323
Iliecebrum 204443
Iljin Goosefoot 87053
Iljinia 204434
Illawara Flame-tree 58335
Illawara Pine 67432
Illawarra Flame Tree 58335,376067
Illawarra Palm 31015
Ill-coloured Azalea 331710
Ill-coloured Rhododendron 331710
Illicit Verbena 405853
Illicium Family 204471
Illigera 204609
Illinois Bundleflower 126176
Illinois Bundle-flower 126176
Illinois Carrion-flower 366397
Illinois Greenbrier 366397
Illinois Hawthorn 109948
Illinois Horse-gentian 397833
Illinois Pondweed 312143
Illinois Rose 336942,336943
Illinois Tick Clover 126407
Illinois Tick-trefoil 126407
Illipe 241510,362186
Illipe Butter 362186
Illipe Butter Tree 241514
Illipe Nut 241510,362186
Ill-scented Sumach 332908
Ill-scented Trillium 397556
Illyarrie 155574
Illyarrie Red-cup Gum 155574
Illyrian Buttercup 325971
Illyrian Cottonthistle 271692
Illyrian Pancratium 280896
Illyrian Thistle 271692
Ilomba 321967
Ilsemann Philodendron 294813
Imbe 171129
Imbreke 358062
Imbricate Anemone 23858
Imbricate Chickweed 374923
Imbricate Cinquefoil 312674
Imbricate Corydalis 105998
Imbricate Cypressgrass 119016
Imbricate Dragonhead 137600
Imbricate Galingale 119016
Imbricate Greenorchid 137600
Imbricate Mahonia 242561
Imbricate Melastoma 248753
Imbricate Podocarpus 121079
Imbricate Pricklypear 272924
Imbricate Sandwort 31896
Imbricate Windflower 23858
Imbricate Yaccatree 121079
Imbricateleaf Stairweed 142823
Imbu 372459,372482
Imitating Cymbidium 116781
Imitative Rhododendron 331245,331856
Immaculate Skullcap 355489
Immersed Crazyweed 278901
Immortelle 189131,415315,415317,415386
Immortelle Blanche 21596
Imo 99910
Imodi Pine 300186
Impala Lily 7340
Impalalily 7343
Impatiens 204756,205444
Impatient Bittercress 72829
Impatient Corydalis 105999
Impatient Lady's Smock 72829
Imperate Crocus 111542
Imperfect Fleabane 150771
Imperfect Wild Buckwheat 152251
Imperforate St. John's-wort 202006
Imperforete St. John's Wort 202006
Imperial Begonia 49941
Imperial Crown-fritillary 168430
Imperial Frltillary 168430
Imperial Japanese Morning Glory 208016
Imperial Lily 168430
Imperial Morning Glory 208016
Imperial Tuftroot 130103
Imphee 369600
Impresednerve Photinia 295704
Impresed-nerved Photinia 295704
Impressed Cleyera 96586
Impressed Eurya 160501
Impressed Newlitse 264058
Impressed Sarmentose Fig 165624
Impressed-leaf Newlitse 264058
Impressednerve Raspberry 338615
Impressedvein Newlitse 264058
Impressvein Dutchmanspipe 34212
Impressvein Heteroplexis 194010
Impudent Lawyer 231183
Impudentlawyer 231183
Inaba Cheirostylis 86707
Inca Lily 18062
Inca queen 214800
Inca Wheat 18687
Incanous Begonia 49943
Incanous Stock 246479
Incaparina 179865
Incarnate Bamboo 297327
Incarnate Begonia 49944
Incarnate Swallowwort 37967
Incarnation 127635
Incarnation Rose 336343
Incarvillea 205546
Incense 57508

Incense Bush 67698
Incense Cedar 67526,228635,228639
Incense Juniper 213969
Incense Machilus 240726
Incense Nanmu 240726
Incense Plant 67698
Incense Rose 336872
Incense Tree 57508,64085,71021,316023
Incense-bearlng Jumper 213969
Incense-cedar 228635,228639
Incense-tree 57508,64085
Incensier 337180
Incertitude Bluegrass 305595
Incertitude Spindle-tree 157593
Inch Plant 67145,394021
Inch Plum 316553
Inch Worm 359712
Inch-high Rush 213559
Inchplant 394093
Inchworm 359712
Incienso 145188
Incisa Waterchestnut 394459
Incise Primrose 314505
Incise Rabdosia 209663
Incised Corydalis 106004
Incised Elsholtzia 144095
Incised Figwort 355145
Incised Fumitoryleaf Corydalis 105919
Incised Groundsel 359985
Incised Horseorchis 215339
Incised Japanese Mulberry 259104
Incised Kalimeris 215339
Incised Larkspur 124303
Incised Leucospermum 228080
Incised Monkshood 5288
Incised Notopterygium 267154
Incised Primrose 314107
Incised Windflower 23871
Incisedleaf Nemosenecio 263513
Incisedleaf Skullcap 355492
Incisedleaf Spiraea 371974
Incise-serrate Begonia 49945
Incisoserrate Stinkcherry 241504
Inclined Cinquefoil 312676
Incongspiuous Roughleaf 222169
Inconspicuous Honeysuckle 235868
Inconspicuous Larkspur 124304
Inconspicuous Lasianthus 222169
Inconspicuous Pea 222731
Inconspicuous Sandwort 31959
Inconspicuous Spapweed 205029
Inconspicuous Touch-me-not 205029
Inconstant Torenia 392945
Incrassate Dehaasia 123607
Incurvate Membranaceus Madder 337995
Incurvate Oncidium 270816
Incurvate Psilurus 319064
Incurved Javatea 275701
Incurved Parapholia 283656
Incurved Skullcap 355493
Incurved Star Wort 374924
Incurvedfpur Spapweed 205317
Indehiscentfruit Spicegrass 239633
Inder Euphorbia 159129
India Abutilon 934
India Adenosma 7985
India Aeginetia 8760
India Aromatic Bamboo 87560
India Atalantia 43731
India Azalea 330917
India Ber 418184
India Boltonia 215343
India Canna 71181
India Carallia 72393
India Catchfly 363568
India Cedar 80087
India Charcoal Trema 394664
India Charcoal-trema 394656
India Childvine 400855
India Chrysanthemum 124790
India Cliffbean 254810
India Copperleaf 1900
India Cowparsnip 192243
India Cupscale 341960
India Daisy 124790
India Dandelion 384587
India Date 295487
India Date-palm 295487
India Dillenia 130919
India Docynia 135120
India Dogliverweed 129235
India Duck-beak 209321
India Epimeredi 147084
India Fieldcress 336211
India Fig 164684
India Flagellaria 166797
India Floatingheart 267819
India Fruitfulgrass 307748
India Galangal 17655
India Glorybower 96139
India Green Star 307509
India Green-star 307509
India Hibiscus 194936
India Horseorchis 215343
India Iphigenia 207497
India Ivygourd 97801
India Jointfir 178541
India Jujube 418164
India Lasiococca 222384
India Leadwort 305185
India Leea 223943
India Lettuce 320515
India Locust 200843,200844
India Love Grass 147883
India Lovegrass 147883
India Love-grass 147883
India Macaranga 240272
India Madder 337925
India Maesa 241775
India Marshweed 230307
India Melothria 417470
India Minigrass 253276
India Mockstrawberry 138796
India Morning glory 207891
India Mustard 59438
India Myrrhtree 101591
India Nightshade 367241
India Orangevine 238536
India Oreorchis 274047
India Pentanema 289568
India Perotis 291405
India Persimmon 132434
India Pluchea 305100
India Plum 166786
India Plumyew 82541
India Pockberry 298093
India Pokeweed 298093
India Polyscias 310199
India Poon 67860
India Pristimera 315361
India Quassia Wood 298516
India Quassiawood 298516
India Ramontchi 166773
India Reed Bamboo 268116
India Rosewood 121730
India Rubber Plant 165168
India Rubber Tree 164925
India Rubber Vine 113820,113823
India Rush 213000
India Sagebrush 35634
India Schefflera 350673
India Scutia 355872
India Sedge 74875
India Sesbania 361430
India Seseli 361483
India Skullcap 355494
India Snakemushroom 46839
India Sphaeranthus 370989
India Spruce 298436
India Spurge 159102
India Squill 352920
India Stonecayenne 327553
India Striped Bamboo 196350
India Sundew 138295
India Sweetclover 249218
India Teasel 133454
India Tigerthorn 122040
India Trumpet Flower 275401
India Trumpetflower 275401,275403
India Trumpet-flower 275401
India Tulip 400146

India Waltheria 413149
India Wheat 162335
India Whitetung 94068
India Wildorchis 274047
India Yellowcress 336211
India Zehneria 417470
India-almond 386500
India-Burma Stranvaesia 377462
India-currant 380749
India-currant Snowberry 380749
India-form Sedge 74877
Indian Abelmoschus 228
Indian Abutilon 934
Indian Aconite 5188
Indian Acrocephalus 6039
Indian Adenosma 7985
Indian Aeginetia 8760
Indian Agueweed 158250
Indian Ague-weed 158250
Indian Albizia 13595
Indian Almond 33544,386500
Indian Apple 123071
Indian Arrow Wood 157327
Indian Arrowroot 114858
Indian Arrowwood 294493
Indian Arrow-wood 157327
Indian Aster 215343
Indian Atalantia 43731
Indian Azalea 331839
Indian Balanophora 46839
Indian Balm 397556
Indian Balmony 380157
Indian Balsam 204982
Indian Bamboo 47190
Indian Banyan 164684
Indian Barberry 51327,51333
Indian Barringtonia 48510
Indian Bead-tree 248895
Indian Bean 79242,79243,79247,79260,154658,218721
Indian Bean Tree 79243
Indian Bean-tree 79243
Indian Beautyberry 66835
Indian Beech 254596,310962
Indian Begonia 49613
Indian Berry 21459
Indian Bindweed 208120
Indian Birch 53635
Indian Black Wood 121730
Indian Blackwood 121750
Indian Blanket 79120,169568,169603
Indian Bluegrass 305294
Indian Borage 99515
Indian Boxwood 171350
Indian Breadroot 287932,319178
Indian Buckeye 9706
Indian Butter Bean 218721
Indian Butter-tree 133007
Indian Cabbage 416603
Indian Cajan 65146
Indian Camphor Tree 91330
Indian Camphortree 91330
Indian Camphorweed 305100
Indian Cane 71181
Indian Canna 71181
Indian Canyon Fleabane 151042
Indian Caoutchouc Tree 164925
Indian Carallia 72393
Indian Cassia 91432
Indian Catalpa 79242
Indian Cedar 80087
Indian Cherry 328641,328674
Indian Chickweed 256719,256727
Indian Chinquapin 78959
Indian Chir Pine 300186
Indian Chocolate 175444
Indian Cinnamon 91436
Indian Claoxylon 94068
Indian Clematis 94856
Indian Cliffbean 254810
Indian Colza 59344
Indian Coral Bean 154734
Indian Coral Tree 154647,154734
Indian Coralbean 154734
Indian Coral-tree 154647
Indian Corn 417412,417417
Indian Cotton 179878
Indian Cottowood 56802
Indian Cow Parsnip 192278
Indian Crabgrass 130635
Indian Cress 399601
Indian Cucumber Root 247228
Indian Cucumber-root 247225,247228
Indian Cup 347144,347156,364301
Indian Cupscale 341960
Indian Currant 66733,380749
Indian Currant Coralberry 380749
Indian Cycas 115888
Indian Damnacanthus 122040
Indian Date 383407
Indian Date Palm 295485
Indian Dendranthema 124790
Indian Derris 125974
Indian Dicliptera 129235
Indian Dill 24213
Indian Dillenia 130919
Indian Dock 340019
Indian Docynia 135120
Indian Dodder 115049
Indian Dropseed 372650
Indian Drypetes 138634
Indian Duckbeakgrass 209321
Indian Ebony 132298,132434
Indian Egony 132318
Indian Elm 401509
Indian Epimeredi 147084
Indian Euphorbia 159102
Indian Evergreen Chinkapin 78959
Indian Evergreenchinkapin 78959
Indian Eye 127793
Indian Eyes 127793
Indian Fig 164684,165553,272775,272891
Indian Fig Pear 272891
Indian Filbert 31680
Indian Fire 344980
Indian Flagellaria 166797
Indian Floatingheart 267819
Indian Flowering Reed 71181
Indian Fog 356770,357114
Indian Fountain Bamboo 416747
Indian Fountain-bamboo 36940
Indian Frankincense 57539
Indian Fumitory 169076
Indian Galangal 17655
Indian Gardenia 171350
Indian Gentian 380157
Indian Giantfennel 163683
Indian Ginger 17655,37574
Indian Glorybower 96139
Indian Goose Grass 143530
Indian Gooseberry 296466
Indian Goosegrass 143530
Indian Grass 104072,182942,369575,369582
Indian Gugertree 350920
Indian Gum 1024,26018,100309,230461
Indian Hawhtorn 329056
Indian Hawthorn 329060,329068
Indian Hay 71218
Indian Head 107048
Indian Hedgemustard 365568
Indian Hedge-mustard 365568
Indian Heliotrope 190651
Indian Hellebore 405643
Indian Hemp 29465,29473,29514,71209,71218,71220,112269
Indian Henbane 266060
Indian Hibiscus 194936
Indian Hog Potato 197113
Indian Horse Chestnut 9706
Indian Horse-chestnut 9706
Indian Hyacinth 128870
Indian Ink 86984
Indian Ipecacuanha 113542
Indian Iphigenia 207497
Indian Ironweed 405450
Indian Ivygourd 97801
Indian Jack-in-the-pulpit 33544
Indian Jalap 207839
Indian Jointfir 178541
Indian Jointvetch 9560
Indian Jujube 418164,418184

Indian Kale 65230
Indian Kalimeris 215343
Indian Katon 345786
Indian Kidney Tea 275781
Indian Knot-grass 309001
Indian Krathon 345786
Indian Laburnum 78300
Indian Laurel 67860,165307,386452
Indian Laurel Fig 165307,165316
Indian Lavender 64082
Indian Leadwort 305185
Indian Leafflower 245220
Indian Leea 223943
Indian Lettuce 258237
Indian Licorice 765
Indian Lilac 219933,248895
Indian Liquorice 765
Indian Locust 200843
Indian Long Pepper 300446
Indian Lotus 263272
Indian Love Grass 147883
Indian Lovegrass 147883
Indian Luvunga 238536
Indian Macaranga 240272
Indian Madder 270028,337925
Indian Madhuca 241514
Indian Maesa 241775
Indian Mahogany 369934
Indian Mallow 837,932,967,1000,402238
Indian Mangrove 45746
Indian Marshweed 230307
Indian May Apple 365009
Indian May-apple 365009
Indian Microchloa 253276
Indian Mikania 254426
Indian Millet 140367,289116,369590
Indian Mitrasacme 256154
Indian Mock Strawberry 138796
Indian Mockstrawberry 138796
Indian Mock-strawberry 138796
Indian Moss 349496,357114
Indian Mulberry 258871,258882,258938,259137
Indian Murainagrass 209321
Indian Mustard 59348,59438
Indian Naiad 262054
Indian Nard 262497
Indian Night Jasmine 267430
Indian Nightshade 367241,367733
Indian Oak 48509
Indian Oleander 265339
Indian Olibanum 57539
Indian Olive 270089
Indian Orangevine 238536
Indian Orris 44881
Indian Paint 86984,233698,233707,233717,345805
Indian Paint-Brosh 79113
Indian Paintbrush 38147,79113,79116,79120,79127
Indian Paper Birch 53362
Indian Paspalum 285507
Indian Pavetta 286275
Indian Pea 222832
Indian Pear 323251
Indian Peashrub 72239
Indian Pedalium 286930
Indian Pennywort 200259
Indian Pepper 300330,417301
Indian Perotis 291405
Indian Persimmon 132434
Indian Pharbitis 207891
Indian Physic 29473,175763,175766,212127
Indian Physic-plant 175760
Indian Pink 127654,127663,234351,248411,363282,363461,364175,371616
Indian Pipe 258006,258044
Indian Pipeweed 152160
Indian Plant 306982
Indian Plantain 34799,34802,64704,64731,283781,301864
Indian Pluchea 305100
Indian Plum 166788,269291
Indian Plumyew 82541
Indian Pockberry 298093
Indian Poke 298093,405643
Indian Pokeweed 298093
Indian Polycarpon 307748
Indian Polyscias 310199
Indian Poppy 176738,247113
Indian Posy 178358
Indian Potato 29298,188968,208050,367696,396790
Indian Primrose 314310
Indian Pristimera 315361
Indian Prune 166788
Indian Pseudostreblus 377541
Indian Quamash 68791,68794
Indian Quassia Wood 298516
Indian Quassiawood 298516
Indian Ramontchi 166773
Indian Rattlebox 111915
Indian Red Water Lily 267756
Indian Redwood 90605,369934
Indian Rhododendron 248732,248762,330917,331839
Indian Rhubarb 122766,329314,349770
Indian Rice 418080,418095
Indian Rice Grass 4139
Indian Ricegrass 4139,276022
Indian Rice-grass 276022
Indian Root 34370,37888
Indian Rorippa 336211
Indian Rose Chestnut 252369,252370
Indian Rosewood 121730,121750,121810,121826
Indian Rotala 337351
Indian Rubber 164925
Indian Rubber Fig 164925
Indian Rubber Plant 164925
Indian Rubber Tree 164925
Indian Rubberplant 164925
Indian Rubbertree 164925
Indian Saffron 114871
Indian Sage 158250
Indian Sago Palm 78056
Indian Sandalwood 346209
Indian Sandburr 80810
Indian Saraca 346504
Indian Scabious 350099
Indian Schefflera 350673
Indian Scutia 355872
Indian Senna 78300
Indian Sesbania 361430
Indian Shamrock 397556
Indian Shell-flower 17639
Indian Shot 71156,71181
Indian Shot Plant 71156
Indian Silk Cotton Tree 56784
Indian Silver Fir 480
Indian Sinocrassula 364919
Indian Skullcap 355494
Indian Snakeroot 327058
Indian Sphaeranthus 370989
Indian Spinach 48689
Indian Spruce 298436
Indian Spurgetree 159457
Indian Strawberry 138796
Indian Streblus 377541
Indian Stringbush 414193
Indian Sumac 107300
Indian Sumach 107300
Indian Sundew 138295
Indian Sunflower 188908
Indian Swampweed 191326
Indian Swede 386500
Indian Sweet-clover 249218
Indian Teak-tree 385531
Indian Teasel 133505
Indian Thistle 91822,91940
Indian Tobacco 26530,26551,234547,266040
Indian Tobacco Lobelia 234547
Indian Tobacco-lobelia 234547
Indian Toothcup 337351
Indian Traveller's Joy 94953
Indian Trumpet Flower 275401,275403
Indian Trumpetflower 275401,275403
Indian Turnip 31142,33234,33544,319251,386500
Indian Turnsole 190651

Indian Underleaf Pearl 245220
Indian Valerian 262497
Indian valley brodiaea 60448
Indian Ventilago 405450
Indian Walnut 14544
Indian Walnut Tree 14544
Indian Wampee 94203
Indian Water Navelwort 200259
Indian Wattle 1308
Indian Wheat 162312,417417
Indian White Mahogany 70998
Indian Wikstroemia 414193
Indian Wild Citrus 93497
Indian Wild Date 295487
Indian Wild Fig 164684
Indian Wild Flacourtia 166777
Indian wild Orange 93497
Indian Wild Pear 323251
Indian Willow 307509
Indian Wood Oats 86140
Indian Woodoats 86140
Indian Wormseed 139678
Indian Yam 131880
Indian Yelloweyegrass 416082
Indian Yellowflax 327553
Indian Zehneria 417470
Indian's Pipe 258044
Indian's Virgin's Bower 95127
Indiana Bergen 26768
Indiana Dalbergia 121826
Indian-apple 306636
Indian-basket-grass 415441
Indian-bean 79243,79260
Indian-charcoal Trema 394656,394664
Indian-cherry 328641
Indian-currant 380749
Indian-currant Coralberry 380749
Indian-currant Snowberry 380749
Indian-fig 272891
Indian-fig Pricklypear 272891
Indiangrass 369575
Indian-lettuce 94343
Indianmulberry 258871
Indian-mulberry 258871,258882
Indian-mulberryoid Psychotria 319696
Indian-paint 86984
Indianpine 258006,258044
Indianpink 364175
Indian-Pipe 258006
Indian-pipe 258044
Indianpipe Hypopitys 202767
Indian-plum Family 166793
Indian-potato 29298
Indian-rhubarb 122764,122766
Indianroot 34370
Indian-root 30748
Indian-rubber Tree 164925
Indian-shot 71181
Indian-shot Plant 71181
Indian-tabacco 234547
Indiantree Spurge 159975
Indian-wood 184518
Indiapoon Beautyleaf 67860
India-rubber Fig 164925
India-rubber Plant 164925
India-rubber Tree 164925
India-wheat 162335
IndiaWingdrupe 405450
Indica Azalea 330917
Indies Goa Bean 319114
Indies Goabean 319114
Indies Goa-bean 319114
Indigenous Cinnamon Tree 91389
Indigo 205611,206626
Indigo Broom 47880
Indigo Bush 20005,121901,121906,205876,206061,206445
Indigo Flower 83646
Indigo Plant 206669
Indigo Vine 235527
Indigo Woad 209232
Indigoblue Woad 209232
Indigo-broom 47880
Indigobush 20005
Indigo-bush 20005
Indigobush Amorpha 20005
Indigo-bush Amorpha 20005
Indigofera 205639
Indish Pepper 72070
Indistinct Corydalis 106013
Indo Mitrasacme 256154
Indo Peashrub 72239
Indocalamus 206765
Indochina Actinidia 6635
Indo-China Actinidia 6635
Indochina Asparagus 38960
Indo-China Camellia 69148
Indochina Dragon Plum 137716
Indochina Dragonplum 137716
Indochina Dragon-plum 137716
Indo-China Ebony 132457
Indo-China Euonymus 157539
Indochina Fieldcitron 400526
Indochina Goldenhook 318662
Indochina Gurjun 133559,133561
Indochina Kiwifruit 6635
Indochina Persimmon 132312
Indochina Pseuduvaria 318662
Indo-China Rehder-tree 327415
Indo-China Senna 78338
Indo-China Sweetspire 210386
Indochinese Actinidia 6635
Indochinese Camellia 69148
Indochinese Chinquapin 78855
Indochinese Evergreenchinkapin 78855
Indo-chinese Excoecaria 161647
Indochinese Feroniella 163508
Indochinese Oak 116059
Indochinese Sweetspire 210386
Indochinese Turpinia 400526
Indofevillea 206847
Indomalayan Caladium 16501
Indonesian Asper 125465
Indonesian Gum 155557
Indonesian Gurjun 133562
Indopacific Strand 104251
Indosasa 206868,206877
Indy 248312,363461
Indy Corn 417417
Indy Pink 127635
Ine 15165
Inebriate Jijigrass 4142
Inebriate Speargrass 4142
Inerm Lambertia 220310
Ineyun 15165
Infant Onion 15709
Infant Streptocaulon 377856
Infattyation Heath 149337
Infirmary Clock 8399
Inflared Peduncle Groundsel 359811
Inflatable-fruited Senna 360483
Inflate Kobresia 217197
Inflate Licorice 177915
Inflate Mayten 246862
Inflate Ormosia 274404
Inflatecalyx Bluebell 115376
Inflated Bladderwort 403214
Inflated Bluebellflower 115376
Inflated Calyx Crazyweed 278680
Inflated Campanumoea 70393
Inflated Cylindrokelupha 116610
Inflated Dendrobium 125400
Inflated Dendropanax 125613
Inflated Habenaria 184211
Inflated Hornwort 83562
Inflated Jadeleaf and Goldenflower 260434
Inflated Landpick 308565
Inflated Leopard 70393
Inflated Lobelia 234547
Inflated Mussaenda 260434
Inflated Sedge 76677
Inflated Solomonseal 308565
Inflated Stonecrop 357075
Inflated Treerenshen 125613
Inflated Wildginger 37651
Inflatedbract Iris 208921
Inflatedcalyx Woodbetony 287517
Inflatedfruit Senna 360483
Inflatedstyle Sedge 73673
Inflated-tubed Mussaenda 260434
Inflatefruit Milkvetch 43197

Inflatfruit Thermopsis 389527
Inflatted Hymenocrater 201039
Inflexed Lygisma 239319
Inflexed Rabdosia 209702
Infracted Chickweed 374924
Infrapurple Wildginger 37674
Infundibular Angraecum 24896
Inga 206918
Ingan 15165
Inin 15165
Ining 15165
Inion 15165
Ink Root 230566,230815
Ink Vine 285703
Inkberry 84408,203847,298094,298119
Inkberry Hawthorn 109864
Inkbush 379616
Inkfruit Barberry 51698
Inkweed 298116,379610
Inkwood 138493
Inland Bluegrass 305602,305604
Inland Boxelder Maple 3228
Inland Ceanothus 79940,79957
Inland Golden Star 55651
Inland Jersey Tea 79957
Inland Juneberry 19272
Inland Marsh Fleabane 305083
Inland Muhly 259666
Inland New Jersey Tea 79940
Inland Panicled Aster 380919
Inland Rush 213173
Inland Salt Grass 134842
Inland Sea Oats 86140
Inland Sedge 74894
Inland Serviceberry 19272
Inland Star Sedge 74894
Inner Coast Range Chaenactis 84502
Inner Mongol Catchfly 363851
Inner Mongol Crazyweed 279032
Inner Mongol Sagebrush 36557
Inner Mongol Seseli 361508
Inner Mongolia Pondweed 312155
Inner Mongolia Spikesedge 143169
Inner Mongolian Milkvetch 42704
Inner Mongolian Poplar 311356
Innermongol Goosecomb 335360
Inner-mongolia Barley 198305
Inner-mongolia Naiad 262084
Inner-mongolian Ajania 12991
Innermongolian Ceratoides 83501
Innion 15165
Innocence 187526,187648,198708
Innocent 50825,102867
Innocent Nidularium 266155
Innocent-weed 80830
Inodorate Mockorange 294478
Inodorous Bifrenaria 54247
Inoi Nut 306777
Inon 15165
Inornate Humid-shaped Barberry 51841
Inornate Sandwort 31960
Inoy Nut 306777
Inrygertjie 109349
Insane-root 201389
Insect Powder Plant 322660
Insertedstyle Woodbetony 287648
Inside-out Flower 404610
Inside-out-flower 404615
Insignificant Primrose 314838
Insignis Nouelia 267222
Insignis Pine 300173
Insignis Pulicaria 321570
Insignis Willow 343528
Insignis Wingnut 320359
Insignis Woodbetony 287294
Insipid Bulbophyllum 62806
Insipid Curlylip-orchis 62806
Insipid Stonecrop 357147
Insleay Odontoglossum 269073
Insula Childvine 400945
Insular Alyxia 18504
Insular Ash 167994
Insular Bambusa 47312
Insular Bridelia 60189
Insular Cyclea 116020
Insular Gurjun 133565
Insular Holly 203860
Insular Osmanther 276336
Insular Osmanthus 276336
Insular Purslane 311890
Integerleaf Artocarpus 36924
Integral Holly 203908
Integriflious Drypetes 138636
Integrifolious Barberry 51782
Integrifolious Chloranthus 88284
Integrifolious Globethistle 140724
Integrifolious Horseorchis 215350
Integrifolious Hydrangea 199910
Integrifolious Kalimeris 215350
Integrifolious Lovely Barberry 51292
Integrifolious Viburnum 407895
Integrifolius Ainsliaea 12676
Integrifolius Goldraintree 217615
Integripetal Rhodiola 329945
Inter Grape 411575
Intergrading Ragweed 19171
Interior Live Oak 324556
Interior Rush 213173
Interior Wild Rice 418096
Intermediary Spike-rush 143165
Intermediate Agapetes 10319
Intermediate Angledtwig Magnoliavine 351080
Intermediate Angulate Torricellia 393109
Intermediate Asian Starjasmine 393621
Intermediate Bambusa 47313
Intermediate Bethlehemsage 283622
Intermediate Bitter Bamboo 304044
Intermediate Bitter-bamboo 304044
Intermediate Bladderwort 403220
Intermediate Bothriochloa 57548
Intermediate Box 64373
Intermediate Brome 60773
Intermediate Bulrush 353182
Intermediate Chartolepis 86049
Intermediate Cinquefoil 312683
Intermediate Citrus 93501
Intermediate Clinopodium 97019
Intermediate Deadnettle 220373
Intermediate Devil Rattan 121502
Intermediate Dogbane 29483
Intermediate Dysophylla 139590
Intermediate Elytrigia 144652
Intermediate Ephedra 146192
Intermediate European Spindle-tree 157444
Intermediate Evening-primrose 269437
Intermediate Forrest Skullcap 355444
Intermediate Goldwaist 90372
Intermediate Holly 203919
Intermediate Inula 207149
Intermediate Laggera 220025
Intermediate Lavender 223291
Intermediate Lepistemon 225708
Intermediate Lilyturf 272085
Intermediate Loosestrife 240050
Intermediate Luculia 238106
Intermediate Mahonia 242463
Intermediate Meadowsweet 166088
Intermediate Melastoma 248754
Intermediate Miterwort 256022,256023
Intermediate Nanchuan Sage 345241
Intermediate Nipplewort 221794
Intermediate Northern Bedstraw 170256
Intermediate Oplismenus 272612
Intermediate Orange 93501
Intermediate Paraphlomis 283622
Intermediate Paspalum 285451
Intermediate Patrinia 285839
Intermediate Peashrub 72258
Intermediate Pea-shrub 72258
Intermediate Pennisetum 289155
Intermediate Peristrophe 291152
Intermediate Periwinkle 409327
Intermediate Pinweed 223699
Intermediate Pond-lily 267324
Intermediate Primrose 314501
Intermediate Purplesepal Rabdosia 324769
Intermediate Rattlesnake-plantain 179664
Intermediate Rigid Kengyilia 215955
Intermediate Sedge 74899,75320,75527
Intermediate Selfheal 316112

Intermediate Sinobambusa 364800
Intermediate Sixleaves Stauntonvine 374429
Intermediate Slugwood 50536
Intermediate Spike Microtoena 254268
Intermediate Spikesedge 143168
Intermediate Stephania 375868
Intermediate Stickseed 221689
Intermediate Sundew 138299
Intermediate Swollennoded Cane 87564
Intermediate Syzygium 382577
Intermediate Water-starwort 67359
Intermediate Whentgrass 144652
Intermediate Wintercress 47947
Intermediate Wintergreen 322852
Intermediate Wintersweet 87532
Intermediate Yellow Rocket 47947
Intermixed Bristlegrass 361789
Intermountain Aster 380839
Intermountain Bitterweed 201315
Intermountain Bristlecone Pine 300040
Intermountain Miner's-lettuce 94350
Intermountain Rubberweed 201315
Interpositional Barberry 51776
Interrupeded Woodnettle 221568
Interrupt Mucuna 259525
Interrupted Apera 28971
Interrupted Brome 60774
Interrupted Cinquefoil 312893
Interrupted Pepper 300422
Interrupted Premna 313664
Interrupted Rabdosia 209709
Interrupted Rhododendron 331307
Interrupted Sophora 369036
Interspike Cupscale 341967
Interwoven Rhododendron 330449
Intricate Clematis 95031
Intricate Gurjun 133566
Intricate Holly 203923
Intricate Oak 324056
Intricate Terminalia 386540
Introduced Sage 345321
Intromongolian Roegneria 335360
Inul 207135
Inula 207025
Inun 15165
Inundated Beakrush 333608
Inundated Sweetvetch 187946
Inusitate Japanese Mulberry 259106
Inverse Hookvine 401766
Inversley Heart-shaped Aucuba 44944
Inverted Castillon Bamboo 297239
Invert-hearted Aspidopterys 39687
Invincible Rhododendron 330936
Invisible-veine Anzacwood 310766
Invisible-veined Rhododendron 332025
Involucral Primrose 314503
Involucrate Balanophora 46841
Involucrate Begonia 49948
Involucrate Buttercup 326372
Involucrate Cudweed 178232
Involucrate Draba 137049
Involucrate Elatostema 142694
Involucrate Evax 193378
Involucrate Flemingia 166870
Involucrate Fritillary 168433
Involucrate Hydrangea 199911
Involucrate Local Tianqi 373635
Involucrate Lysionotus 239952
Involucrate Microula 254349
Involucrate Oreomyrrhis 273979
Involucrate Sensitivebrier 352633
Involucrate Snakemushroom 46841
Involucrate Sphenodesma 371419
Involucrate Stahlianthus 373635
Involucrate Whitlowgrass 137049
Involucred Cudweed 178232
Involucrete Symphorema 380713
Involute Thuarea 390553
Inyo Lily 229870
Inyo Rabbitbrush 90566
Inyo Rock Daisy 291318
Inyo Wild Buckwheat 152205
Inyun 15165
Iochroma 207342
Iodes 207354
Iodine Bush 14940,14941,243306,243308
Iodine Weed 379610
Ione Wild Buckwheat 151832
Iow Bamboo 304040
Iowa Crab 243636
Iowa Crabapple 243636
Iowii Orchid 282768
Ipecac 81970
Ipecacuanha 81970
Ipecacuanha Spurge 159150
Ipheion 207487
Ipheion Uniflorum 207489
Iphigenia 207492
Ipo 77809
Ipoh 28248
Ipomea Quamoclit 323465
Ipomoea 207546
Iran Asparagus 39141
Iran Broomsedge 217379
Iran Crazyweed 279146
Iran Speedwell 407287
Iran Stock 246448
Iran Summercypress 217379
Iranian Box 64345
Iranian Knapweed 81033
Iranian Stork's Bill 153806
Irby-dale Grass 159027
Iriartea 208366
Iridescent Rosy Coneflower 140084
Irion County Wild Buckwheat 152300
Iris 208433
Iris Family 208377
Iris Moraea 258529
Irish Bladderwort 403220
Irish Broom 121001
Irish Butterwort 299744
Irish Common Juniper 213709
Irish Daisy 384714
Irish Fleabane 207219
Irish Heath 121066,149398
Irish Heather 121066
Irish Hill Wild Buckwheat 151833
Irish Ivy 187276
Irish Juniper 213709
Irish Lady's Tresses 372242
Irish Lady's-tresses 372242
Irish Mahogany 16352
Irish Marsh Orchid 273529,273681
Irish Moss 32301
Irish Orchid 264599
Irish Potato 367696
Irish Queen Anne's Jonquil 262354
Irish Sandwort 31816
Irish Saxifrage 349856
Irish Shamrock 277648
Irish Spurge 159058
Irish St. John's Wort 201788
Irish St. John's-wort 201788
Irish Strawberry Tree 30888
Irish Tea 316819
Irish Tutsan 202106
Irish Vine 236022
Irish Worts 121066
Irish Yew 385314,385324
Irisleaf Oberonia 267973
Irisleaf Satinflower 365767
Irisleaf Saussurea 348731
Irisleaf Windhairdaisy 348731
Iris-like Cymbidium 116919
Iris-like Orchis 116919
Irisorchis 267922,267973
Irkut Brome 60776
Irkutsk Anemone 23701
Iroko 88506,254493
Iroko Fustic Wood 88506
Iroko Fustic-tree 88506
Iron Bamboo 163567,163570
Iron Crabapple 159363
Iron Cross Begonia 50067
Iron Grass 235385
Iron Head 81237
Iron Holly 204217
Iron Hopea 198189
Iron Knob 81237
Iron Oak 323849,324244,324443
Iron Pear 369322

Iron Persimmon 132153
Iron Ringvine Iron Vine 116031
Iron Straw-sandal 198885
Iron Sweetleaf 381372
Iron Tree 252607,284961
Iron Weed 406040,406311
Iron Wood 79161,252369
Iron Woundwort 362728
Ironaxle Germander 388247
Ironbark 155468,155743
Iron-chopstick 190925,190956
Iron-cross Begonia 50067
Ironculm 317194
Iron-flower 15861,211672
Iron-grass 308816
Iron-hard 81237,240068,405872
Iron-head 81237
Ironheart Tree 380083
Ironhook pole 407504
Iron-knob 81237
Ironolive 365082
Iron-pear 369322
Ironpole Rose 336870
Iron-requiring Persimmon 132398
Iron-root 44576
Iron-sheet Dendrobium 125288
Iron-tree 252607
Ironweed 405438,406040,406114,406130,
406365,406566,406593,406631,407254
Iron-weed 81237,141347,233700,406631
Ironweed Thistle 92472
Ironwhip Buckthorn 328617
Ironwhip Bushclover 226924
Ironwood 1213,46338,77252,77256,77263,
77264,79155,88930,103691,118050,
160967,160968,252370,254711,270058,
270155,270527,276801,276811,276818,
284961,315548,362957,415693
Ironwort 362728
Iroquois Breadfruit 33544
Iroquois Breadroot 33544
Irrawadi Camellia 69153
Irregular Garcinia 171054
Irregular Milkvetch 42977
Irregularteeth Skullcap 355521
Irregulartooth Skullcap 355521
Irrorate Rabdosia 209711
Irwin Barberry 51777
Irwin's Tanoak 233260
Isabellarebe 411764
Isano Oil 271133
Isanu 399611
Ischnogyne 209404
Island Addermonth Orchid 110645
Island Ash 167994
Island Bladderwort 404060
Island Bloodwood 106811
Island Bogorchis 110645
Island Bristleweed 186884
Island Bush Poppy 125585
Island Bush Snapdragon 170893
Island Canary Juniper 213629
Island Desertdandelion 242963
Island Dyeing-tree 346491
Island Glorybush 394703,394704
Island Holly 203860
Island Koenigia 217648
Island Linden 391733
Island Live Oak 324489
Island Manzanita 31120
Island Milkwort 308119
Island Oak 324489
Island Rhode Bent 12025
Island Ringvine 116020
Island Sagebrush 35980
Island Scrub Oak 324138
Island Snapdragon 170852
Island Speedwell 407292
Island Torrey Pine 299986
Island-poppy 155191
IslandTree-poppy 125584
Islandvine 182324
Islay 316452
Isle Belle Cress 47963
Isle of Wight Vine 61497,383675
Isle-of-man Cabbage 99089
Isle-of-wight Hellehorine 147204
Ismene 200906
Isocarp Milkwort 308122
Isoloma 217738
Isometrum 210176
Isopyrum 210251,210260
Ispahan Nepeta 264957
Isphaghul Plantain 302119
Isphaghul Seeds 302119
Israelites 243209
Issykkol Spunsilksage 360843
Istria Iris 208621
Isu Tree 134946
Ita Palm 246672
Itala Palm 202268
Itali Stairweed 141202
Itali Toadflex 231014
Italian Alder 16323
Italian Alkanet 21949
Italian Arum 37005
Italian Aster 40038
Italian Bellflower 70093
Italian Bluebell 199555
Italian Broccoli 59545
Italian Buckthorn 328595
Italian Buck-thorn 328595
Italian Bugloss 21925,21949
Italian Catchfly 363583
Italian Chestnut 78811
Italian Clematis 95449
Italian Clover 396934
Italian Corn Salad 404422
Italian Cornsalad 404422
Italian Corn-salad 404422
Italian Crabapple 243615
Italian Cranesbill 174715
Italian Cypress 114753
Italian Fennel 167170
Italian Garlic 15595
Italian Gladiolus 176281
Italian Hawksbeard 110764
Italian Hedgenettle 373329
Italian Honeysuckle 235713
Italian Jasmine 211848,211863,211864
Italian Lavender 223334
Italian Leather Flower 95449
Italian Lords and Ladies 37005
Italian Lords-and-ladies 37005
Italian Maple 3107,3286
Italian Millet 361794
Italian Narcissus 262381
Italian Oak 323917
Italian Parsley 292700
Italian Plumeless Thistle 73467
Italian Poplar 311400
Italian Ray Grass 235315
Italian Rock-cress 30229,30363
Italian Rocket 327866
Italian Rye 235315
Italian Rye Grass 235334,235335
Italian Ryegrass 235315,235334,235335
Italian Rye-grass 235315
Italian Sainfoin 187842
Italian Senna 78436,360444,360461
Italian Silene 363583
Italian Squill 199555
Italian Starwort 40038
Italian Stone Pine 300151
Italian Thistle 73467
Italian Timothy 295001
Italian Toadflax 116738,230892,231014
Italian Turnip Broccoli 59619
Italian Turnipbroccoli 59619
Italian Upright Celery 29327
Italian Vetch 169935
Italian Viper's Bugloss 141202
Italian Windgrass 28971
Italian Woodbine 235713
Italian Yellow Jurinea 211863
Italy Catchfly 363583
Italy Cypress 114753
Italy Gladiolus 176281
Italy I-poplar 311152
Italy Jurinea 211863
Italy Poplar 311152

Italy Ryegrass 235315
Italy Squill 199555
Italy Stone Pine 300151
Itch 175813
Itchgrass 337544,337555
Itching Berry 336419
Itchweed 405643
Itchy Crabgrass 130768
Itchy-backs 336419
Itea 210421
Iteoleaf Michelia 252894
Ithaca Blackberry 338647
Ithuriel's Spear 398806
Ito Burmannia 63977
Ito Keisuke's Rhododendron 330998
Ito Wintergreen 172059
Itoa 210441
Ityo 175813
Itzershan Gentian 173537
Ivanov Hairgrass 126042
Ivary Nut Palm 298050,298051
Ivary Palm 298050
Ivary Primrose 314329
Iver 235334
Ivery 187222
Ivin 187222
Ivory 187222
Ivory Bamboo 47518
Ivory Cane Palm 299664
Ivory Coast Almond 386563
Ivory Coast Checkwood 154976
Ivory Coast Erythrophleum 154976
Ivory Coast Khaya 216205
Ivory Coast Mahogany 216205
Ivory Curl Tree 61918
Ivory Lady's-slipper 120375
Ivory Nut 298051
Ivory Nut Palm 252631,252633
Ivory Palm 298051
Ivory Thistle 154317
Ivory Wood 365271
Ivory's Hakea 184614
Ivory-fruited Hartwort 392869
Ivory-nut Palm 98549,252631,298050,298051
Ivory-white Chirita 87855
Ivorywhite Cymbidium 116823
Ivorywhite Stanhopea 373688
Ivorywood 46955,365270
Ivray 235373
Ivvy 187222
Ivy 187195,187222,215395
Ivy Arum 353121
Ivy Begonia 49877,50012
Ivy Bells 68189
Ivy Bindweed 162519
Ivy Broomrape 275086
Ivy Bush 215395
Ivy Campanula 412697,412698
Ivy Chickweed 407144
Ivy Crowfoot 325923
Ivy Duckweed 224396
Ivy Flower 157429,192130
Ivy Geranium 288430
Ivy Glorybind 68686
Ivy Gourd 97801
Ivy Halo Dogwood 104920
Ivy League Wild Buckwheat 151837
Ivy Leaved Toadflax 116736
Ivy of Uruguay 92977
Ivy Peperomia 290354
Ivy Tree 350706
Ivy Treebine 92586
Ivyarum 353121
Ivy-Arum 353121
Ivybush 215395
Ivygourd 97781,97801
Ivyleaf Cyclamen 115965
Ivy-leaf False Thoroughwort 89297
Ivyleaf Geranium 288430
Ivy-leaf Morning Glory 207839,208016
Ivyleaf Morning-glory 207839
Ivy-leaf Morning-glory 207839
Ivy-leaf Peperomia 290354
Ivyleaf Saxifrage 349229
Ivyleaf Speedwell 407144
Ivy-leaf Violet 410058
Ivy-leaved Antirrhinum 116736
Ivy-leaved Bellflower 412697
Ivy-leaved Beltfower 412568
Ivyleaved Crowfoot 325923
Ivy-leaved Crowfoot 325923
Ivyleaved Cyclamen 115953
Ivy-leaved Cyclamen 115965
Ivyleaved Duckweed 224396
Ivy-leaved Duckweed 224396
Ivy-leaved Geranium 288430
Ivy-leaved Ground Cherry 297677
Ivy-leaved Harebell 412697
Ivyleaved Maple 2896
Ivy-leaved Morning Glory 207839
Ivy-leaved Morning-glory 207839,208016
Ivy-leaved Pelargonium 288430
Ivy-leaved Scurvy-grass 97981
Ivy-leaved Snapdragon 116736
Ivy-leaved Speedwell 407144
Ivy-leaved Toadflax 116732,116736
Ivy-leaved Violet 410058
Ivy-leaved Water Buttercup 325923
Ivy-like Calystegia 68686
Ivylike Euonymus 157576
Ivy-like Euonymus 157576
Ivy-like Fig 165096
Ivylike Merremia 250792
Ivy-of-uruguay 92977
Ivywort 116736
Iwaikan 93503
Iwaikan Citrus 93503
Iwarancusa Grass 117190
Ixbut 159213
Ixeris 210553
Ixia 210689
Ixiolirion 210988,210999
Ixonanthes 211021
Ixonanthes Family 211020
Ixora 211029,211068
Ixtle De Jaumave 10857
Ixtle Fibre 10778
Ixtli Fibre 10778
Iyo-mikan 93504

J

J. C. Williams' Rhododendron 332110
J. E. Smith's Rhododendron 331859
J. H. Edgar Azalea 330612
J. H. Edgar's Rhododendron 330612
J. Lindley's Rhododendron 331109
JabilyElephant Tree 271956
Jabon 93579
Jaboncillo 346336
Jaborand Pepper 300425
Jaborandi 299167
Jaborandi Pepper 300446
Jaborosa 211219,211220
Jaboticaba 156169,261253,261294
Jaboticabe 261047,261049
Jaca 36920
Jacamatraca 288989
Jacaranda 121770,211225,211239
Jacaranda Rosa 121683
Jacareuba 67855
Jacinth Irisorchis 267976
Jacinth Oberonia 267976
Jack 36920,127635
Jack Bean 71036,71042,71050
Jack Beanstalk 97757

Jack Cheese 243840
Jack Common Juniper 213729
Jack Frost Siberian Bugloss 61366
Jack Fruit 36920
Jack in Prison 266225
Jack in the Pulpit 33295,33544
Jack Nut 36920
Jack Oak 323871,324155
Jack Pine 299799,299818
Jack Snaps 374916
Jack Sprat 374916
Jack Straw 302034
Jack Torreya 393074
Jack Tree 36920
Jack Wingseedtree 320838
Jack's Cheeses 243840
Jack's Ladder 409085
Jack's Poplar 311186,311360
Jack's Wild Buckwheat 152249
Jack-abed-at-noon 394333
Jack-an-apes-on-horseback 50825,66445
Jack-and-the-beanstalk 308454
Jackass Clover 414518
Jack-at-the-hedge 170193
Jackbean 71036
Jack-bean 71036
Jack-behind-the-garden-gate 410677
Jack-by-the-ground 176824
Jack-by-the-hedge 14964,47964,84596,174877,176824,248312,394333
Jack-durnals 102727
Jack-flower 174877
Jackfruit 36920
Jack-fruit 36920
Jack-go-to-bed-at-noon 21339,274823,394307,394333
Jack-horner 174877
Jackie's Saddle 290290
Jack-in-a-lantern 297658
Jack-in-Prison 266225
Jack-in-the-box 37015,192913
Jack-in-the-bush 221787,401711
Jack-in-the-buttery 356468
Jack-in-the-green 8332,37015,315106
Jack-in-the-hedge 61497,176824,248312,374916
Jack-in-the-lantern 248312,374916
Jack-in-the-pulpit 33234,33270,33544,37015
Jack-jennets 102727
Jack-jump-about 8826,24483,30932,237539
Jack-jump-up-and-kiss-me 410677
Jackman Clematis 95036
Jack-o'-both-sides 325612
Jack-o'-lantern 23925,158062
Jack-o'-lanterns 23925
Jack-o'-twosides 325612
Jack-of-the-buttery 356468
Jack-of-the-hedge 14964
Jackpine 299818
Jack-run-along-by-the-hedge 14964
Jack-run-in-the-country 68713,102933
Jack-run-in-the-hedge 68713
Jack-run-up-dyke 170193
Jack-snaps 374916
Jacksonbrier 366580
Jacksonia 211273
Jack-straws 302034
Jacktree 364951,364957,364959
Jackweed 325612
Jackwood 113447,280003
Jacky Cheese 243840
Jacky's Cheese 243840
Jacky-jurnals 102727
Jacob's Chariot 5442
Jacob's Coat 2011,18095,18101,18150
Jacob's Groundsel 359158
Jacob's Ladder 5442,13934,28617,44708,86755,102833,102867,205175,307190,307195,307197,307241,307245,308607,357203,366364
Jacob's Rod 39435,39438
Jacob's Staff 39438,167562,405788
Jacob's Stee 44708
Jacob's Tears 102867
Jacob's Walking Stick 307197
Jacob's Walking-stick 307197
Jacob's Word 208771
Jacob's-ladder 307190,307197,366364
Jacob's-ladder Family 307186
Jacob's-wort 208771
Jacobaea 211295,358788
Jacobean Amaryllis 372913
Jacobean Lily 372912,372913
Jacobeanlily 372912,372913
Jacobinia 211338,214403
Jacobite Rose 336343
Jacobo-ladder 307197
Jacobsen Groundsel 359162
Jacobs-rod 39435
Jacoby Fleawort 359158
Jacot Edelweiss 224858
Jacquemont Hazelnut 106784
Jacquemont Roegneria 335364
Jacquemont Speargrass 4143
Jacu Rose 336653
Jacub Kudrjaschevia 218278
Jade Cicada Swordflag 208543
Jade Cymbidium 116967
Jade Groundsel 359516
Jade Mountain Aster 40878
Jade Mountain Fleabane 150788
Jade Orchid 416694
Jade Orchis 116967
Jade Plant 108818,109229,311956
Jade Tree 108818,109229
Jade Vine 378353
Jadebean 388335,388336
Jadeedge Kerria 216092
Jadegreen Sasa 347279
Jadeleaf and Goldenflower 260375,260483,260484
Jadeolive 179441,179444
Jaeschke Barberry 51779
Jaeschkea 211399
Jaffrey Primrose 314507
Jagged Bugle 13104
Jagged Chickweed 197486,197491
Jagged Cranesbill 174569
Jagged Mallow 13934
Jagged Sea Orache 44668
Jagged-leaved Fig Marigold 9780
Jaggery Palm 78056
Jaggy Nettle 402886
Jagua 172890
Jail Germander 388112
Jak 36920
Jakfruit 36920
Jakfrujt 36924
Jak-in-a-box 192909,192913
Jak-in-a-box Family 192921
Jakkalbessie 132299
Jakut Fescuegrass 164023
Jalap 207546,207901,208117,255711
Jalap Bindweed 208117
Jalap Purge 208117
Jalapa Plane Tree 302586
Jalisco Pine 299974,300125
Jalisco Rabbit-tobacco 317746
Jalu Goldenray 229063
Jam Tarts 169126,174739,174877
Jam Tree 259910
Jamacar Cereus 83885
Jamaica 418155
Jamaica Allspice 299325
Jamaica Bark 161821
Jamaica Bloodwood 221527
Jamaica Caper Tree 71726
Jamaica Carludovica 77041
Jamaica Cherry 259910
Jamaica Cobnut 270622
Jamaica Dogwood 300937
Jamaica Ebony 61427
Jamaica Falsevalerian 373507
Jamaica Flower 195196
Jamaica Honeysuckle 285660
Jamaica Honey-suckle 285660
Jamaica Horse Bean 71042
Jamaica Ironwood 155055
Jamaica Linden Hibiscus 194850
Jamaica Linden-hibiscus 194850
Jamaica Liquorice 38135

Jamaica Mignonette 158730,223454
Jamaica Mountain Sage 221238
Jamaica Pepper 299325
Jamaica Pimento 299325
Jamaica Plum 88691,372477
Jamaica Quassia Wood 298510
Jamaica Quassiawood 298510
Jamaica Sarsaparilla 366542
Jamaica Senna 78436
Jamaica Sorrel 195196
Jamaica Thatch Palm 390463
Jamaica Vervain 373507,405857
Jamaican Ebony 61427
Jamaican Feverplant 395082
Jamaican Forget-me-not 61119
Jamaican Horse Bean 71042
Jamaican King 97865
Jamaican Kino 97865
Jamaican Licorice 765
Jamaican Navel-spurge 270619
Jamaican Oak 79256
Jamaican Ormosia 274405
Jamaican Red Cedar 80030
Jamaican Rose 55080
Jamaican Satin Wood 417230
Jamaican Sorrel 195196
Jamaican Thatch Palm 390463
Jamaica-wood 184518
Jaman 382522
Jamba 154019
Jamberberry 297686,297712
Jambhiri Orange 93505
Jambolan 382522
Jambolanplum 382522
Jambolan-plum 382522
Jambos 382582,382606
Jambosa 382660
Jambosleaf Raspberry 338650
Jamboslike Raspberry 338650
Jambu 156118,382465,382582
Jambu Batu 318742
Jambu Biji 318742
Jambul 382522
James Barberry 51782
James Gentian 173540
James Goldenray 229065
James Nailwort 284855
James' Antelope Sage 152167
James' Clammy-weed 307149
James' Cristatella 307149
James' Cryptantha 113356
James' Monkey-flower 255206
James' Nailwort 284855
James' Prairie Clover 121891
James' Rubberweed 201318
James' Sedge 74935
James' Waterlily 267694
James' Weed 359158
Jamesia 211550
Jameson Barberry 51785
Jamestown Lily 123077
Jamestown Weed 123077
Jamiaca Nightshade 367262
Jamie Hedley's Clover 316127
Jammy-mouth 341165
Jan the Crowder 363461
Jan's Catchfly 363794
Jananese Fir 365
Jancaea 211559
Janczewsky Currant 334048
Janet-flower 68189
Janka Lily 229795
January Jasmine 235796
Janusia 211594
Japan Acanthopanax 143685
Japan Alder 16386
Japan Alopecurus 17537
Japan Anemone 23854
Japan Angelica 24380
Japan Arborvitae 390660
Japan Ardisia 31477
Japan Arrowroot 321441
Japan Arrowwood 407897,407980
Japan Artichoke 373100,373439
Japan Ash 167970,167999
Japan Atractylodes 44205
Japan Aucuba 44915
Japan Avens 175417
Japan Banana 260208
Japan Barberry 52225
Japan Barnyardgrass 140384
Japan Barrenwort 147021
Japan Bauhinia 49140
Japan Beakrush 333628
Japan Bean 177750
Japan Beech 162382
Japan Benchfruit 279757
Japan Betony 373262
Japan Birch 53520
Japan Black Pine 300281
Japan Bladdercampion 114066
Japan Bladderwort 403868
Japan Blooddisperser 297628
Japan Bluebasin 350175
Japan Bogorchid 158161
Japan Brocadebeldflower 413611
Japan Bromegrass 60777
Japan Buckeye 9732
Japan Bugbane 91024
Japan Bugle 13146
Japan Bulbophyllum 62813
Japan Burreed 370072
Japan Bushclover 226849
Japan Butterbur 292374
Japan Buttercup 325981
Japan Camellia 69156
Japan Catnip 264958
Japan Cayratia 79850
Japan Cedar 85310,113633,113685
Japan Centaurium 81502
Japan Championella 85944
Japan Chestnut 78777
Japan Chiogenes 172092
Japan Chloranthus 88289
Japan Cinnamon 91351
Japan Citrus 93823
Japan Clethra 96466
Japan Cleyera 96587
Japan Clover 218347
Japan Columbine 30043
Japan Comanthosphace 228002
Japan Conehead 85944,387136
Japan Conyza 103509
Japan Coriaria 104696
Japan Cottonsedge 353606
Japan Cowlily 267286
Japan Crazyweed 278910
Japan Creeper 285157
Japan Crinum 111170
Japan Croomia 111677
Japan Cryptomeria 113685
Japan Cudweed 178237
Japan Curlylip-orchis 62813
Japan Currant 333970
Japan Cutgrass 224002
Japan Dallisgrass 285530
Japan Dandelion 384591
Japan Deerdrug 242691
Japan Dishspurorchis 171864
Japan Dock 340089
Japan Dodder 115050
Japan Dogliverweed 291161
Japan Douglas Fir 318575
Japan Dualbutterfly 398273
Japan Dualsheathorchis 85455
Japan Ducklingcelery 113879
Japan Edelweiss 224872
Japan Elaeocarpus 142340
Japan Entireliporchis 261474
Japan Euchresta 155897
Japan Euonymus 157601
Japan Eurya 160503
Japan Eutrema 161149
Japan Falsehellebore 405601
Japan Farfugium 162616
Japan Fatsia 162879
Japan Fescue 164025
Japan Fir 365
Japan Fleeceflower 328345
Japan Floweringquince 84556
Japan Four-involucre 124923

Japan Furrowlemma 25346
Japan Galangal 17695
Japan Galls 332662
Japan Garliccress 365515
Japan Gastrochilus 171864
Japan Gentian 154926
Japan Ginger 418002
Japan Globeflower 399510
Japan Glorybind 68692,103095
Japan Glorybower 96147
Japan Goldenray 229066
Japan Goldthread 103835
Japan Goosecomb 335370
Japan Greenbrier 366486
Japan Greens 59444
Japan Greygreen Gentian 174086
Japan Ground Ivy 176821
Japan Gymnaster 41201
Japan Hedge Glorybind 68692,68724
Japan Hedgehyssop 180308
Japan Helwingia 191173
Japan Hemlock 399923
Japan Heterosmilax 194120
Japan Hirsute Rockcress 30301
Japan Hogfennel 292892
Japan Holly 203670
Japan Honeylocust 176881
Japan Honeysuckle 235878
Japan Hookettea 52478
Japan Hop 199392
Japan Hornwort 83575
Japan Horsechestnut 9732
Japan Hosiea 198575
Japan Hylomecon 200754
Japan Iris 208543
Japan Irisorchis 267958
Japan Ironwood 276808
Japan Ivy 187318
Japan Junesnow 360933
Japan Kadsura 214971
Japan Knotweed 309262
Japan Korthalsella 217906
Japan Kumquat 167503
Japan Lacquer 332938
Japan Ladyslipper 120371
Japan Landpick 308613
Japan Larch 221894
Japan Lawngrass 418426
Japan Lecanorchis 223647
Japan Leek 15381,15822
Japan Leopard 70398
Japan Lilac 382276
Japan Lily 229730,229758,229872
Japan Lime 391739
Japan Linden 391739
Japan Liparis 232197
Japan Listera 232919
Japan Litse 233949
Japan Loosestrife 239687
Japan Loquat 151162
Japan Lovegrass 147746
Japan Machilus 240604
Japan Maesa 241781
Japan Magnificent Fig 165717
Japan Maple 3300
Japan Mazus 247025
Japan Medlar 151162
Japan Meytea 28341
Japan Microstegium 254020
Japan Microtropis 254303
Japan Milkwort 308123
Japan Mint 250381
Japan Monkshood 5298
Japan Morning Glory 208016
Japan Mosla 259300
Japan Mosquitotrap 117537
Japan Mountainash 33008
Japan Muhly 259672
Japan Mulberry 259097
Japan Nailorchis 356457
Japan Nakeaster 41201
Japan Nightshade 367262
Japan Nucleonwood 291483
Japan Onion 15381,15822
Japan Opithandra 272598
Japan Orixa 274277
Japan Orostachys 275375
Japan Ottelia 277397
Japan Oxtongue 298618
Japan Pagodatree 369037
Japan Paperelder 290365
Japan Paris 284340
Japan Parsley 113872
Japan Pear 323268
Japan Pendent-bell 145710
Japan Peony 280207
Japan Pepper Tree 417301
Japan Peristrophe 291161
Japan Persimmon 132216
Japan Photinia 295691
Japan Pieris 298734
Japan Pine 299890
Japan Pink 127739
Japan Pipsissewa 87485
Japan Plantain 302014
Japan Plantainlily 198618
Japan Platanthera 302377
Japan Plum 316761
Japan Plumyew 82523
Japan Pogonia 306870
Japan Pokeweed 298114
Japan Pollia 307324
Japan Poplar 311389
Japan Premna 313692
Japan Primrose 314508
Japan Privet 229485
Japan Pseudosasa 318307
Japan Purplepearl 66808
Japan Pyrola 322842
Japan Quince 84556,84854
Japan Rabdosia 209713
Japan Ramie 56181
Japan Raspberry 338985
Japan Rose 336783
Japan Roughleaf 222172
Japan Sabia 341513
Japan Sagebrush 35674
Japan Sakebed 97711
Japan Sedge 74941
Japan Sedirea 356457
Japan Seepweed 379542
Japan Smithia 366670
Japan Snailseed 97947
Japan Snakegourd 396203
Japan Snakemushroom 46848
Japan Snowbell 379374
Japan Southstar 33369
Japan Spicebush 231403
Japan Spruce 298442
Japan St. John's Wort 201942
Japan Stemona 375343
Japan Stephania 375870
Japan Stephanotis 211695
Japan Stone Pine 300163
Japan Stonecrop 356841
Japan Summerlilic 62099
Japan Sweetgrass 177560
Japan Sweetspire 210389
Japan Swertia 380234
Japan Tallowtree 346390
Japan Teasel 133490
Japan Theligon 389186
Japan Themeda 389355
Japan Thistle 92066
Japan Threeflower Gentian 174007
Japan Thrixsperm 390506
Japan Toad-lily 396598
Japan Toothwort 222645
Japan Torreya 393077
Japan Triadenum 394811
Japan Triumfetta 399261
Japan Tropidia 399650
Japan Tupistra 400397
Japan Turnjujube 198769
Japan Twigrush 93927
Japan Twinballgrass 209105
Japan Umbrella Pine 352854,352856
Japan Umbrellaleaf 132682
Japan Valerian 404284
Japan Varnish Tree 12559
Japan Velvetplant 183097

Japan Vernalgrass 27971
Japan Vetch 408435
Japan Vetchling 222735
Japan Waterstarwort 67367
Japan Wax Tree 332865
Japan Weed 210623
Japan White Birch 53483
Japan White Pine 300130
Japan Whitebractcelery 266975
Japan Wildeggplant 58462
Japan Wildginger 37659
Japan Wildtung 243371
Japan Windflower 23854
Japan Wood Oil Tree 406016
Japan Wood-oil Tree 406016
Japan Woodorchis 200771
Japan Yam 131645
Japan Yew 385355
Japan Yoania 416382
Japanes Knotweed 328328
Japanese Abelia 177
Japanese Acanthopanax 143685
Japanese Achyranthes 4275
Japanese Actinidia 6523
Japanese Adinandra 8244
Japanese Alder 16341,16386
Japanese Allspice 87525
Japanese Alopecurus 17537
Japanese Alpine Cherry 83267
Japanese Ampelopsis 20408
Japanese Andromeda 298734
Japanese Anemone 23850,23854,23856
Japanese Angelica 30634,30636
Japanese Angelica Tree 30634
Japanese Angelica-tree 30634
Japanese Anise Tree 204575
Japanese Anisetree 204533,204575
Japanese Anise-tree 204575
Japanese Apricot 34448
Japanese Aralia 30634,162879
Japanese Arborvitae 390660
Japanese Arbor-vitae 390660
Japanese Ardisia 31477
Japanese Arrow Bamboo 318307
Japanese Arrowhead 342400,342401,342421
Japanese Arrowroot 321441
Japanese Arrowwood Viburnum 407802
Japanese Artichoke 373100,373439
Japanese Ash 167970,167999,168018,168023
Japanese Aspen 311480
Japanese Aster 215334,215358
Japanese Astilbe 41817
Japanese Asyneuma 43580
Japanese Atractylodes 44205
Japanese Aucuba 44915
Japanese Aulacolepis 25346
Japanese Avens 175417
Japanese Azalea 331258,331370
Japanese Balanophora 46848
Japanese Bamboo 328345,362131
Japanese Banana 260208
Japanese Barberry 52225,52235,52240
Japanese Barnyard Millet 140423
Japanese Basil 290952
Japanese Bastard Cress 390237
Japanese Bauhinia 49140
Japanese Beakgrain 128017
Japanese Beakrush 333628
Japanese Bean 294031
Japanese Beauty Berry 66808
Japanese Beauty Bush 66808
Japanese Beautyberry 66808
Japanese Beauty-berry 66808
Japanese Bedstraw 170445
Japanese Beech 162368,162382
Japanese Bellflower 302753
Japanese Bentawn Muhly 259646
Japanese Berchemiella 52478
Japanese Betony 373262
Japanese Big-leaf Magnolia 198698
Japanese Big-leaved Magnolia 198698
Japanese Bindweed 68686
Japanese Birch 53388,53520,53572
Japanese Bird Cherry 280023
Japanese Bird's Foot Trefoil 237554
Japanese Bitter Orange 8787,310850
Japanese Black Crowberry 145073
Japanese Black Pine 300281
Japanese Bladdernut 374089
Japanese Bladderwort 403222
Japanese Blood Grass 205470,205497,205509,205517
Japanese Blue Beech 162382
Japanese Blue Oak 116100
Japanese Blueberry 403868
Japanese Bluegrass 305776
Japanese Blyxa 55922
Japanese Bog Rosemary 22448
Japanese Box 64285,64303
Japanese Boxwood 64369
Japanese Bristle Grass 361743
Japanese Bristlegrass 361743
Japanese British Inula 207151
Japanese Brome 60777
Japanese Buckeye 9732
Japanese Buckthorn 328740
Japanese Bugbane 91024
Japanese Bugle 13146
Japanese Bule Vetch 408361
Japanese Bunching Onion 15289
Japanese Butterbur 292374
Japanese Buttercup 325981
Japanese Butterfly Bush 62099
Japanese Butterflybush 62099
Japanese Butter-flybush 62099
Japanese Butterfly-bush 62093,62099
Japanese Caesalpinia 64993
Japanese Callicarpa 66808
Japanese Camellia 69156
Japanese Camphor 91287
Japanese Cassia 91351
Japanese Catalpa 79257
Japanese Cayratia 79850
Japanese Cedar 85310,113633,113685
Japanese Cedar Cypress 85310
Japanese Centaurium 81502
Japanese Chaff Flower 4274,4275
Japanese Cheesewood 301413
Japanese Chengiopanax 86855
Japanese Cherry 83314
Japanese Cherry Birch 53465
Japanese Chess 60777
Japanese Chestnut 78777
Japanese Chestnut Oak 323611
Japanese China-berry-tree 248903
Japanese Chinalaurel 28341
Japanese China-laurel 28341
Japanese Chinquapin 78903
Japanese Chloranthus 88289
Japanese Cinnamon 91351
Japanese Cladium 93927
Japanese Clematis 95358
Japanese Clethra 96466
Japanese Cleyera 96587
Japanese Clover 218347
Japanese Cocklebur 11566
Japanese Coelachne 98451
Japanese Coltsfoot 292374,292403
Japanese Columbine 30043
Japanese Comanthosphace 228002
Japanese Common Barberry 52339
Japanese Conehead 85944,387136
Japanese Conyza 103509
Japanese Coriaria 104696
Japanese Corktree 294233
Japanese Cornal 105146
Japanese Cornal Dogwood 105146
Japanese Cornelian Cherry 105146
Japanese Cornelian-cherry 105146
Japanese Cow-lily 267286
Japanese Cowslip 314508
Japanese Crab 243617,243717
Japanese Crabapple 243617
Japanese Cranberry 403868
Japanese Crapemyrtle 219922
Japanese Crazyweed 278910
Japanese Creeper 285157,285159
Japanese Crimson Glory-vine 411623
Japanese Crinum 111170
Japanese Croomia 111677

Japanese Crowberry 145073
Japanese Cryptomeria 113685
Japanese Cucumber Tree 198698,283421
Japanese Cudweed 178237
Japanese Cultivar Maple 3300
Japanese Curely Dock 340089
Japanese Currant 333970
Japanese Cutgrass 224002
Japanese Cypress 85310
Japanese Dahurian Buckthorn 328684
Japanese Dandelion 384591
Japanese Dappled Willow 343530
Japanese Date Palm 132219
Japanese Date-plum 132275
Japanese Dendrobenthamia 124923
Japanese Dichocarpum 128938
Japanese Dicliptera 291161
Japanese Dock 340089
Japanese Dodder 115050
Japanese Dogwood 105146,124923,124925
Japanese Doublefile Viburnum 408034
Japanese Douglas Fir 318575
Japanese Draba 137052
Japanese Dragon-head 137551
Japanese Dwarf Iris 208657
Japanese Dwarf Swordflag 208657
Japanese Edelweiss 224872
Japanese Eightangle 204533
Japanese Elaeocarpus 142340
Japanese Elatostema 142759
Japanese Elm 401490,417558
Japanese Enkianthus 145710
Japanese Epimedium 147021
Japanese Erysimum 154485
Japanese Euonymus 157473,157601
Japanese Eupatorium 158161
Japanese Eurya 160503
Japanese Euscaphis 160954
Japanese Evergreen Oak 323599
Japanese Evodia 161336,161356
Japanese False Bindweed 68686
Japanese False Brome 58604
Japanese False Cypress 85310
Japanese False Oak 233193
Japanese False Solomonseal 242691
Japanese Falsebrome 58604
Japanese False-brome 58604
Japanese Falsecypress 85345
Japanese Falsehellebore 405601
Japanese Fatsia 162879
Japanese Fawnlily 154928
Japanese Fescuegrass 164025
Japanese Filbert 106777
Japanese Flag 208543
Japanese Fleece Flower 328345
Japanese Fleeceflower 309018,328345
Japanese Fleece-flower 328345
Japanese Flowering Cherry 83314,83365
Japanese Flowering Crab Apple 243617
Japanese Flowering Crabapple 243617
Japanese Flowering Crab-apple 243617
Japanese Flowering Dogwood 124923
Japanese Flowering Quince 84556,84573
Japanese Floweringquince 84556
Japanese Forest Grass 184656
Japanese Forsythia 167444
Japanese Fox-taoil Orchis 356457
Japanese Fritillary 168434
Japanese Gagea 169452
Japanese Galangal 17695
Japanese Garden Anemone 23857
Japanese Garden Juniper 213864
Japanese Gentian 173662,173847
Japanese Germander 388114
Japanese Ginsen 280793
Japanese Glabrous Lithocarpus 233193
Japanese Globeflower 399510
Japanese Glorybower 96147
Japanese Goldenray 229066
Japanese Goldsaxifrage 90383
Japanese Goldwaist 90383
Japanese Greenbrier 366486
Japanese Greens 59444
Japanese Hawk's Beard 416437
Japanese Hazel 106777
Japanese Hazelnut 106777
Japanese Hedgehyssop 180308
Japanese Hedgeparsley 392992
Japanese Hedge-parsley 392992
Japanese Heloniopsis 191065
Japanese Helwingia 191173
Japanese Hemarthria 191234,191251
Japanese Hemlock 399895,399923
Japanese Hemlock Fir 399923
Japanese Heterosmilax 194120
Japanese Hibiscus 195216
Japanese Hill Cherry 83234
Japanese Hinoki Chamaecyparis 85310
Japanese Hogfennel 292892
Japanese Holly 203670
Japanese Honey Locust 176881
Japanese Honeylocust 176881
Japanese Honey-locust 176881
Japanese Honeysuckle 235878,235984,
413570
Japanese Hop 199382,199392
Japanese Hop Hornbeam 276808
Japanese Hop-hornbeam 276808
Japanese Hops 199382
Japanese Hornbeam 77312
Japanese Horse Chestnut 9732
Japanese Horsechestnut 9732
Japanese Horse-chestnut 9732
Japanese Hyacinth 272049
Japanese Hydrangea 200087,200088
Japanese Hydrangea Vine 351793
Japanese Hylomecon 200754
Japanese Illicium 204533
Japanese Indosasa 206898
Japanese Inula 207151
Japanese Iris 208543
Japanese Ivy 20455,187318,285157
Japanese Jasmine 211905
Japanese Juniper 213864
Japanese Kadsura 214971
Japanese Katsura Tree 83736
Japanese Katsura-tree 83736
Japanese Keiskea 215811
Japanese Kerria 216085
Japanese Knotweed 162532,308675,309262,
328345
Japanese Korthalsella 217906
Japanese Kummerowia 218345
Japanese Kumquat 167503
Japanese Lacquer 332938,393491
Japanese Lacquer Tree 393491
Japanese Lady's Slipper 120371
Japanese Ladymantle 14065
Japanese Lantern 195216
Japanese Lanterns 297643
Japanese Larch 221894
Japanese Larch Tree 221894
Japanese Large-leaved Podocarp 306457
Japanese Lasianthus 222172
Japanese Late Cherry 83316
Japanese Lathraea 222645
Japanese Laurel 44915
Japanese Lawn Grass 418426
Japanese Lawngrass 418426
Japanese Lawn-grass 418426
Japanese Lecanorchis 223647
Japanese Leek 15289
Japanese Lespedeza 218347,226849
Japanese Lilac 382277
Japanese Lily 229730,229872,230040
Japanese Lime 391739
Japanese Linden 391739,391773
Japanese Listera 232919
Japanese Litse 233949
Japanese Little-leaf Box 64303
Japanese Locust 176881
Japanese Loosestrife 239687
Japanese Loquat 151162
Japanese Lovegrass 147995
Japanese Machilus 240604
Japanese Maesa 241781
Japanese Magnolia 198698,242169,242333
Japanese Mahonia 242563
Japanese Maize 417420
Japanese Mallotus 243371
Japanese Manchurian Ash 168025

Japanese Mannagrass 177560
Japanese Maple 2846,2868,3034,3127,3300,3302
Japanese Mat Rush 213066
Japanese Mazus 247025
Japanese Meadowsweet 166117,371944
Japanese Medlar 151162
Japanese Metaplexis 252514
Japanese Microstegium 254020
Japanese Microtropis 254303
Japanese Milkwort 308123
Japanese Millet 140367,140421,140423,140503
Japanese Mint 250331,250376
Japanese Miterwort 256024
Japanese Morning Glory 208016
Japanese Mosla 259300
Japanese Mountain Ash 33008,369454
Japanese Muhly 259672
Japanese Mulberry 259097
Japanese Myrmechis 261474
Japanese Nanocnide 262247
Japanese Nepeta 264958
Japanese Nervilia 265412
Japanese Nightshade 367262
Japanese Nipplewort 221784
Japanese Nucleonwood 291483
Japanese Nutmeg 393077
Japanese Nutmeg Yew 393077
Japanese Oak 116153,323611,324173
Japanese Oberonia 267958
Japanese Olive 276313
Japanese Onion 15734
Japanese Ophiorrhiza 272221
Japanese Oplopanax 272712
Japanese Orange 93717
Japanese Orixa 274277
Japanese Oxtongue 298618
Japanese Pachysandra 279757
Japanese Pagoda 369037
Japanese Pagoda Tree 369037
Japanese Pagodatree 369037
Japanese Panicgrass 281390
Japanese Paris 284401
Japanese Paspalum 285530
Japanese Peace Palm 329176
Japanese Pear 323268
Japanese Pearlwort 342251
Japanese Pellionia 288729
Japanese Pennisetum 289135
Japanese Peony 280207
Japanese Peperomia 290365
Japanese Pepper 300427,417301
Japanese Peppercorns 417301
Japanese Peristrophe 291161
Japanese Persimmon 132219
Japanese Photinia 295691
Japanese Phtheirospermum 296068
Japanese Phyllodoce 297030
Japanese Physaliastrum 297628
Japanese Pieris 298734
Japanese Pine 299890,300163
Japanese Pink 127654,127663,127739
Japanese Pink Lily 229872
Japanese Pipsissewa 87485
Japanese Pittosporum 301413
Japanese Plantain Lily 198618,198624
Japanese Plantainlily 198624
Japanese Platanthera 302377
Japanese Plum 151162,316761
Japanese Plum Yew 82523,82524
Japanese Plume Grass 394943
Japanese Plumyew 82523
Japanese Plum-yew 82499,82523,393077
Japanese Pockberry 298114
Japanese Podocarp 306469
Japanese Podocarpium 200742
Japanese Pogonia 306870
Japanese Poinsettia 287853
Japanese Poisoncelery 90937
Japanese Pollia 307324
Japanese Poplar 311389
Japanese Premna 313692
Japanese Prickly Ash 417301
Japanese Prickly-ash 417301
Japanese Primrose 314508
Japanese Privet 229485,229558,229569
Japanese Ptarmiganberry 31316
Japanese Pussywillow 343185
Japanese Pyrola 322842
Japanese Quince 84549,84556,84573
Japanese Rabdosia 209713
Japanese Radish 326622
Japanese Raisin 198769
Japanese Raisin Tree 198769
Japanese Raisintree 198769
Japanese Raisin-tree 198769
Japanese Randia 325393
Japanese Raspberry 338985
Japanese Ray Lily 229730
Japanese Red Cedar 113685
Japanese Red Oak 323599
Japanese Red Pine 299890
Japanese Red-cedar 113633,113685
Japanese Rhododendron 331341
Japanese Roegneria 335370
Japanese Roof Iris 208875
Japanese Rose 216084,216085,336783,336901
Japanese Rose Cowwheat 248201
Japanese Rowan 369362,369363,369364,369365
Japanese Rubber Plant 109229
Japanese Rush 5803,212890
Japanese Sabia 341513
Japanese Sage 345108,345252
Japanese Sago Cycad 115884
Japanese Sago Plum 115884
Japanese Sapium 346390
Japanese Sarmentose Fig 165628
Japanese Sasa 318307
Japanese Saussurea 348403
Japanese Saxifrage 349520
Japanese Scabious 350175
Japanese Scopolia 354691
Japanese Sedge 74941,75003,75421,76264
Japanese Sedirea 356457
Japanese Seepweed 379542
Japanese Serissa 360926
Japanese Serisse 360933
Japanese Sharp-leaved Spindle-tree 157764
Japanese Shibataea 362131
Japanese Shorthusk 58462
Japanese Shrub Lespedeza 226708
Japanese Silver Fir 365
Japanese Silver Tree 264090
Japanese Silvergrass 255849
Japanese Silver-grass 255886
Japanese Singleseed Chickweed 375002
Japanese Sisymbrium 365515
Japanese Skimmia 365929,365974
Japanese Skullcap 355760
Japanese Small-leaved Box 64303
Japanese Smithia 366670
Japanese Snailseed 97947
Japanese Snakegourd 396170
Japanese Snakegrape 20408
Japanese Snowball 408027,408034
Japanese Snow-ball 408027
Japanese Snowball Tree 408027,408034
Japanese Snowbell 379374
Japanese Snowdrop Tree 379374
Japanese Snowflower 126946
Japanese Soaptree 346390
Japanese Solomonseal 308613
Japanese Speedwell 407252
Japanese Spice Bush 231403
Japanese Spicebush 231403
Japanese Spice-bush 231403
Japanese Spicebush Lindera 231403
Japanese Spikesedge 143300
Japanese Spindle 157601
Japanese Spindle Tree 157601
Japanese Spindletree 157601
Japanese Spiraea 371852,371944
Japanese Spirea 371951,371953,371956,372101
Japanese Spruce 298442
Japanese Spurge 279757
Japanese Spurge Shrub 243315
Japanese St. John's Wort 201942

Japanese Star Anise 204575
Japanese Star Jasmine 393616
Japanese Statice 230797
Japanese Staunton Vine 374409
Japanese Staunton-vine 374409
Japanese Stemona 375343
Japanese Stephania 110227,375870
Japanese Stewarta 376466
Japanese Stewartia 376457,376466
Japanese Stilt Grass 254046
Japanese Stone Pine 300163
Japanese Stonecrop 200798,356841
Japanese Strawberry 124923,167635
Japanese Strawberry Tree 124923,124925
Japanese Strippedbark Maple 2846
Japanese Styrax 379374
Japanese Summersweet 96466
Japanese Surfgrass 297183
Japanese Sweet Coltsfoot 292374
Japanese Sweet Flag 5803
Japanese Sweetflag 5803
Japanese Sweet-flag 5803
Japanese Swertia 380234
Japanese Tachibana 93823
Japanese Tallow 393479
Japanese Tallow Tree 332865,346390
Japanese Tallow-tree 346390
Japanese Tanoak 233193
Japanese Tea Party 23850
Japanese Teasel 133490
Japanese Temple Bamboo 357939
Japanese Ternstroemia 386696
Japanese Theligonum 389186
Japanese Themeda 389355
Japanese Thimbleweed 23850
Japanese Thistle 92066
Japanese Threestamen Willow 344213
Japanese Thrixspermum 390506
Japanese Thuja 390680
Japanese Thyme 391225
Japanese Timber Bamboo 297215
Japanese Tipularia 392349
Japanese Toadflex 231015
Japanese Tofieldia 392622
Japanese Torreya 393077
Japanese Tree 393491
Japanese Tree Lilac 382178,382276,382277
Japanese Triadenum 394811
Japanese Tripterospermum 398273
Japanese Triumfetta 399261
Japanese Tubeflower 96147
Japanese Tubocapsicum 399998
Japanese Tupistra 400397
Japanese Turk's Cup Lily 229857
Japanese Twayblade 232197
Japanese Twinballgrass 209105
Japanese Twotooth Achyranthes 4275
Japanese Udo Salad 30619
Japanese Umbrella Pine 352854,352856
Japanese Undulateleaf Oplismenus 272693
Japanese Valerian 404284
Japanese Vanilla-leaf 4104
Japanese Velvet Bean 259520
Japanese Viburnum 407897,407980
Japanese Vinegentin 398273
Japanese Violet 409942
Japanese Virnish Tree 166627
Japanese Virnish-tree 166627
Japanese Walnut 212621,212626
Japanese Water Iris 208543
Japanese Waterchestnut 394463
Japanese Waterparsnip 365858
Japanese Wax 332865,346390
Japanese Wax Tree 332865,393479
Japanese Weigela 413611
Japanese White Birch 53483
Japanese White Pine 300130,300137
Japanese White Spirea 371797
Japanese Whitebractcelery 266975
Japanese Wikstroemia 414271
Japanese Wild Sumac 393479
Japanese Willow 343546
Japanese Windhairdaisy 348403
Japanese Wineberry 339047,339048
Japanese Wing Nut 320372
Japanese Wingnut 320372
Japanese Winter Berry 204239
Japanese Winter-berry 204239
Japanese Wistaria 414554
Japanese Wisteria 414554,414556
Japanese Witch Hazel 185104
Japanese Witchhazel 185104
Japanese Wood Oil Tree 406016
Japanese Wood-oil Tree 406016
Japanese Woodrush 238629
Japanese Wormwood 35674
Japanese Wych Elm 401542
Japanese Xylosma 415869
Japanese Yam 131645
Japanese Yellow Loosestrife 239687
Japanese Yellowwood 94030
Japanese Yellow-wood 94030
Japanese Yew 306457,385355
Japanese Youngia 416437
Japanese Zelkova 417558
Japanese-aralia 162879
Japanese-lantern 195216,297640,297643,
297645
Japanese-rose 216085
Japaneses Arisaema 33369
Japaneses Vernalgrass 27971
Japenese Fig 165002
Japenese Plumyew 82499
Japenese Podocarpus 306493
Japenese Sweetleaf 381143,381252
Japgarden Juniper 213864
Japgarden Savin 213864
Japonese Boykinia 288848
Japonese Bush Spruce 298354
Japonese Didiciea 392350
Japonese Erman's Birch 53421
Japonese Ligusticum 229343
Japonese Monochasma 257538
Japonica 84549,84556,84573
Jaquemont's Birch 53481
Jar Osbeckia 276135
Jaragua Grass 201563
Jaraguagrass 201563
Jarcalyx Bubbleweed 297886
Jarcalyx Physochlaina 297886
Jarcalyx Woodbetony 287791
Jarilla 48074
Jarrah Eucalyptus 155644
Jarritos 279486
Jarshape Osmanther 276402
Jasione 211639
Jasioneleaf Ladybell 7673
Jasmin De Rosa 325333
Jasmin Orange 260158
Jasmin rose 211749
Jasmine 172781,211715,211740,211834,
211940,211963
Jasmine Alyxia 18516
Jasmine Confederate 393657
Jasmine Nightshade 367265
Jasmine Orange 260158,260173
Jasmine Pandorea 281225
Jasmine Tobacco 266035
Jasmine Tree 197184,305226
Jasminefruit 283992
Jasmine-like Heptacodium 192168
Jasminodora Greytwig 352438
Jasminodora Grey-twig 352438
Jasminorange 260158,260173
Jasmin-orange 260158
Jasninelike Azalea 330956
Jasnine-like Rhododendron 330956
Jasperbells Barberry 51863
Jasquemont Arisaema 33368
Jasquemont Southstar 33368
Jata De Guanbacoa 103737
Jatamans Valerian 404285
Jau 79161
Jaubert Sophora 369053
Jaumarve Istle 10857
Jaumave Fibre 10857
Jaunders-berry 52322
Jaunders-tree 52322
Jaundice-berry 52322
Jaundice-tree 52322
Jausar Mahonia 242571

Java Acanthophippium 2074
Java Acriopsis 5950
Java Almond 71004
Java Almond Canary-tree 70996
Java Almond Tree 70996
Java Amomum 19836
Java Apple 382660
Java Bean 294010
Java Bethlehemsage 283623
Java Bishopwood 54620
Java Bishop-wood 54620
Java Black-galingale 169555
Java Brucea 61208
Java Campanumoea 70396
Java Cardamom 19888
Java Cassia 91353
Java Cinnamon 91282,91353
Java Cleidion 94408
Java Clematis 95042
Java Cocainetree 155067
Java Cymbidium 116921,116938
Java Cyrtosia 120761
Java Devilpepper 327058
Java Devil-pepper 327058
Java Engelhardia 145505
Java Eria 148705
Java Eugenia 382660
Java Fan Palm 234190
Java Fan-palm 234190
Java Fig 164682,164688
Java Gahnia 169555
Java Galangal 17677
Java Gastrodia 171939
Java Glorybower 96356
Java Gouania 179944
Java Graceful Bedstraw 170357
Java Grass 294981,310722
Java Indigo 205678,206669,245854
Java Koilodepas 217759
Java Leopard 70396
Java Leptospermum 226458
Java Meadowrue 388541
Java Mukia 259738
Java Olive 376110
Java Paphiopedilum 282855
Java Paraphlomis 283623
Java Parkia 284473
Java Pennywort 200312
Java Pepper 300377
Java Plum 382522
Java Podocarpus 121079
Java Rattan Palm 65713
Java Sanicle 345977
Java Sawgrass 245453
Java Shagvine 97503
Java Sida 362569
Java Skullcap 355531
Java Soja 264241
Java Speedwell 407164
Java Stereosandra 376244
Java Stonewood 192492
Java Tasselflower 144976
Java Tea 275781
Java Treevine 92777
Java Waterdropwort 269326
Java Willow 165841
Java-almond 70996
Java-almond Canary Tree 70996
Java-almond Tree 70996
Javan Adenosma 7986
Javan Brucea 61208
Javan Grape 387740
Javan Loosestrife 239696
Javan Meadowrue 388541
Javan Podocarpus 121079
Javan Quassia Wood 298513
Javan Quassiawood 298513
Javan Spindle-tree 157639
Javan Tupelo 267856
Javanese Azalea 330957
Javanese Bishopwood 54620
Javanese Flatsedge 245453
Javanese Glorybower 96356
Javanese Treebine 92859
Java-plum 382522
Javatea 275624
Javawood 54620
Javelina Bush 101733
Jayweed 26768
Jazmin De Trapo 211825
Jealousy 329935
Jeckfrult 36924
Jeelico 24483
Jeetrym-jees 196818
Jeffer Hemlock 399912
Jeffersonie 212296
Jeffrey Aster 40639
Jeffrey Pine 299992
Jeffrey's Pine 299992
Jegidi 225678
Jeheb Nut 104144
Jehol Jerusalemsage 295123
Jehol Ligusticum 229344
Jelly Palm 31705,64159,64161,64163,
98136
Jelly Stock 86427
Jelly-beans 357111
Jelly-fig 165518
Jelly-flower 86427,246478
Jellygrass 252241
Jelly-stock 86427
Jelski Barberry 51794
Jelutong 139271
Jelutong Bukit 139271
Jelutong Paya 139271
Jembolan Plum 382522
Jeneper 213702
Jeneru Tenga 93649
Jenkins Cherry Laurel 223106
Jenkins Coffee 98927
Jenkins Dendrobium 125202
Jenkins Irisorchis 267960
Jenkins Livistona 234183
Jenkins Oberonia 267960
Jenny Creeper 239755
Jenny Flower 174877
Jenny Greenteeth 224375
Jenny Hood 174877
Jenny Nettle 402886
Jenny Stone Crop 357114
Jenny Stonecrop 357114
Jenny Wren 174877
Jenny's Stonecrop 357114
Jenny-run-by-the-ground 176824
Jensen Cinnamon 91354
Jepson Ceanothus 79905,79950
Jepson's Woolly Sunflower 152808
Jequerity Seeds 765
Jequirity 765
Jequirity Bean 765
Jequirity Rosary Pea 765
Jequirity Rosarypea 765
Jequirity Rosary-pea 765
Jequitiba 76832
Jerdon Spapweed 205038
Jerdon Touch-me-not 205038
Jerdonia 212335
Jericho Balsam 46768
Jericho Potato 367233
Jericho Resurrection Mustard 21805
Jericho Resurrection-mustard 21805
Jerlin 235373
Jeroffleris 86427
Jerry-Jerry 19602
Jerrymander 407072
Jersalem Cowslip 321643
Jersev Thrift 34536
Jersey Butrercup 325838
Jersey Buttercup 326174
Jersey Cudweed 178265,317748
Jersey Elm 401575,401623,401632
Jersey Forget-me-not 260878
Jersey Knapweed 81263
Jersey Lily 18862,18865,404581
Jersey Live-long 178265
Jersey Love-grass 147883
Jersey Orchid 273513
Jersey Orchis 273565
Jersey Pigmyweed 109287
Jersey Pine 300296
Jersey Pink 127723

Jersey Rabbit-tobacco 317748
Jersey Tea 79940,172135
Jersey Thistle 80981
Jersey Thrift 34501,34560
Jersey Toadflax 231074
Jeruju 2672
Jeruk Hunje 93513
Jerusale Thorn 280539
Jerusalem Artichoke 189073
Jerusalem Ash 209229,327879
Jerusalem Cherry 367139,367522
Jerusalem Cowslip 321643
Jerusalem Cross 363312
Jerusalem Cucumber 114122
Jerusalem Date 49173
Jerusalem Nettle 220401
Jerusalem Nightshade 367522
Jerusalem Oak 139682
Jerusalem Pine 299964
Jerusalem Primrose 321643
Jerusalem Rye 398939
Jerusalem Sage 295049,295111,321643
Jerusalem Seeds 321643
Jerusalem Star 83060,201787,358395
Jerusalem Tea 139678
Jerusalem Thorn 280559,284477,284479
Jerusalem Wormseed 139682
Jerusalem-artichoke 189073
Jerusalemcherry 367522
Jerusalem-cherry 367522
Jerusalem-oak 139682
Jerusalem-oak Goosefoot 139682
Jerusalemsage 295049,295215
Jerusalem-sage 295111
Jerusalemthorn 284477,284479
Jeshal 158062
Jessamine 84400,84408,172778,211715,
211940,273531
Jessamy 211940,273531
Jesse 211940
Jesse's Flower 211940
Jessica's Aster 380906
Jesuit Tea 139678,204135
Jesuit's Bark Tree 91055
Jesuit's Brazil Tea 204135
Jesuit's Cress 399601
Jesuit's Nut 394500
Jesuit's Tea 319186
Jesuit's-nut 394436
Jesuits' Bark 91082
Jesuits' Cress 399601
Jesuits' Tea 276969
Jesup's Hawthorn 109779
Jetbead 332282,332284
Jetbead Barberry 51336
Jetbead Cotoneaster 107648
Jetberry Bush 332284
Jetty Palm 64161
Jew Plum 372467
Jew's Apple 367370
Jew's Beard 358062
Jew's Ear 220729,239157
Jew's Mallow 104103,216085
Jew's Myrtle 340990
Jew's Plum 372467
Jewel Orchid 25959,25976,108650,108654
Jewel Orchids 179571
Jewel Vine 125937
Jewel Weed 204756,204841
Jeweled Aloe 17041
Jeweled Distaff Thistle 77734
Jeweled Rocket 365406
Jeweled Shooting-star 135155
Jewel-leaf Plant 180266
Jewelled Shooting-star 135178
Jewels of Opar 383324
Jewelvine 125937
Jewelweed 204756,204838,205204
Jewel-weed 204841,205204
Jews Mallow 104103
Jews-mallow 104103
Jezo Spruce 298307
Jiabaotai Cyclobalanopsis 116111
Jiacha Monkshood 5119
Jiacha Rockfoil 349890
Jiacha Swertia 380221
Jiacha Windhairdaisy 348362
Jiadifengpi Anisetree 204534
Jiadifengpi Anise-tree 204534
Jiadifengpi Eightangle 204534
Jiagedaqi Wormwood 35467
Jiajiang Ardisia 31481
Jiala Rockfoil 349427
Jiala Woodbetony 287313
Jialasa Clematis 95043
Jian'ou Sour-bamboo 4594
Jianchuan Leek 15184
Jianchuan Monkshood 5243
Jianchuan Onion 15184
Jiande Gentian 173626
Jiande Mosla 259296
Jiande Swertia 380239
Jianfeng Evergreenchinkapin 78962
Jianfeng Lasianthus 222221
Jianfeng Passionflower 285656
Jianfeng Rattlebox 112263
Jianfeng Roughleaf 222221
Jianfeng Syzygium 382586
Jianfengling Cyclobalanopsis 116133
Jianfengling Evergreen Chinkapin 78962
Jianfengling Machilus 240649
Jiang Woodbetony 287786
Jiangcheng Lilyturf 272099
Jiangcheng Pearlsedge 354123
Jiangda Catnip 264961
Jiangda Nepeta 264961
Jiangda Willow 343462
Jiange Cypress 114709
Jiangjie Willow 343560
Jiangkou Actinidia 6638
Jiangkou Kiwifruit 6638
Jiangle Pseudosasa 318315
Jiangnan Peonygrass 182633
Jiangnan Stephania 375850
Jiangnan Stonecrop 356860
Jiangshan Shibataea 362122
Jiangshan Wobamboo 362122
Jiangsu Arisaema 33316
Jiangsu Lycoris 239261
Jiangsu Sedge 74993
Jiangsu Stonegarlic 239261
Jiangxi Amomum 19865
Jiangxi Ampelopsis 20421
Jiangxi Azalea 331006
Jiangxi Barberry 51454,51795
Jiangxi Cliffbean 66634
Jiangxi Cystopetal 320222
Jiangxi Deinocheilos 123723
Jiangxi Dogwood 124908
Jiangxi Elaeagnus 142036
Jiangxi Falsepimpernel 231537
Jiangxi Field Lacquertree 393483
Jiangxi Four-involucre 124908
Jiangxi Glorybower 96161
Jiangxi Hemiboea 191386
Jiangxi Holly 203943
Jiangxi Longleaf Pyrola 322821
Jiangxi Maple 3062
Jiangxi Michelia 252896
Jiangxi Millettia 66634
Jiangxi Mosla 259283
Jiangxi Motherweed 231537
Jiangxi Paper-mulberry 61100
Jiangxi Pearweed 239698
Jiangxi Raspberry 338490,338657
Jiangxi Rhododendron 331006
Jiangxi Sage 345142
Jiangxi Smallleaved Box 64371
Jiangxi Spiraea 371872
Jiangxi Twinballgrass 209106
Jiangxi Violet 410136
Jiangxi Woodbetony 287314
Jiangyong Pseudosasa 318323
Jiangzi Chickweed 374906
Jiangzi Everlasting 21563
Jiangzi Monkshood 5382
Jiangzi Pearleverlasting 21563
Jiangzi Sagebrush 35584
Jiangzi Sandthorn 196750
Jiangzi Seabuckthorn 196750
Jiankun Greenbrier 366405

Jianlei Tabernaemontana 382743
Jianmen Catchfly 364138
Jianning Euscaphis 160957
Jianning Goldsaxifrage 90386
Jianning Goldwaist 90386
Jianshan Elatostema 142699
Jianshi Spapweed 205322
Jianshu Hopea 198174
Jianshui Dendrocalamus 125481
Jianshui Dragonbamboo 125481
Jianshui Maple 2785
Jianwei Buckthorn 328839
Jiaobai Wildrice 418095
Jiaodong Birch 53492
Jiaodong Euonymus 157651
Jiaodong Linden 391746
Jiaodong Sedge 74954
Jiaodong Spindle-tree 157651
Jiaodong Stonecrop 356849
Jiaoling Holly 203928
Jiaonan Sinobambusa 364675
Jiaonan Tangbamboo 364675
Jiaozhou Corydalis 106034
Jiaozhou Euonymus 157651
Jibbles 15289
Jicama 279732
Jicana 279734
Jiculi 236425
Jienfun Syzygium 382586
Jieshan Maple 2819
Jieshan Trident Maple 2819
Jifengshan Milkvetch 42551
Jiggle-joggles 60382
Jigong Indigo 206126
Jigong Mountain Willow 343206
Jigong Willow 343206
Jigongshan Pseudosasa 318321
Jigongshan Willow 343206
Jihua Primrose 314478
Jijigrass 4113,4163
Jilaffer 86427
Jilin Bulrush 352205
Jilin Monkshood 5318
Jilin Poplar 311330
Jilin Sedge 75001
Jilin Stonecrop 294124
Jilin Wakerobin 397574
Jilin Windhairdaisy 348823
Jill 176824
Jilliver 86427
Jilloffer 86427,246478
Jilloffer Stock 246478
Jilly Offers 86427
Jilong 306005
Jilong Arisaema 33295
Jilong Arrowbamboo 196346
Jilong Barberry 51666
Jilong Beautyberry 66808
Jilong Bogorchid 158182
Jilong Catchfly 363529
Jilong Chickweed 374907
Jilong Ciliate Poplar 311278
Jilong Clematis 94763
Jilong Deadnettle 220387
Jilong Epipremnum 147344
Jilong Eupatorium 158182
Jilong Fargesia 196346
Jilong Fritillary 168380
Jilong Gentian 173493
Jilong Homalomena 197793
Jilong Honeysuckle 235890
Jilong Kobresia 217148
Jilong Larkspur 124616
Jilong Leptodermis 226094
Jilong Ligusticum 229333
Jilong Meconopsis 247159
Jilong Microula 254350
Jilong Monkshood 5304
Jilong Peashrub 72228
Jilong Pea-shrub 72228
Jilong Pennywort 200317
Jilong Privet 229463
Jilong Rockcress 30458
Jilong Rush 213228
Jilong Sagebrush 35719
Jilong Scurrula 236814
Jilong Small Reed 127240
Jilong Smallreed 127240
Jilong Southstar 33295
Jilong Thermopsis 389525
Jilong Umbrella Bamboo 196346
Jilong Wildclove 226094
Jilong Wildsenna 389525
Jilong Willow 343463
Jilong Windhairdaisy 348145
Jilong Woodbetony 287270
Jilong Woodrush 238631
Jilung Microula 254350
Jim Hill Mustard 365398
Jim Sage 344973
Jimmy Weed 185489
Jimmyweed 209562
Jimpson Weed 123077
Jimson Weed 123077
Jimsonweed 123036,123077,277747
Jimson-weed 123036,123077
Jimson-weed Datura 123077
Jimsy-weed 123077
Jimunai Calligonum 67034
Jimunai Fritillary 168614
Jimunai Kneejujube 67034
Jimunai Yellowflower Fritillary 168616
Jinan Lily 228586
Jinchang Azalea 330959
Jinchang Rhododendron 330959
Jinchuan Cranesbill 174673
Jinchuan Peristylus 291233
Jinchuan Perotis 291233
Jinchuan Sweetvetch 187949
Jinchuan Willow 343552
Jinding Azalea 330672
Jinfoshan Azalea 330285
Jinfoshan Barberry 51797
Jinfoshan Beautiful Rhododendron 330285
Jinfoshan Clematis 94977
Jinfoshan Cranesbill 174501
Jinfoshan Deerdrug 194046
Jinfoshan Dutchmanspipe 34216
Jinfoshan Euonymus 157641
Jinfoshan False Fairybells 134406
Jinfoshan False Solomonseal 194046
Jinfoshan Hemsleia 191959
Jinfoshan Lathraea 222642
Jinfoshan Lily 229877
Jinfoshan Notoseris 267175
Jinfoshan Purpledaisy 267175
Jinfoshan Raspberry 338659
Jinfoshan Sasa 206839
Jinfoshan Sedge 74956
Jinfoshan Snowgall 191959
Jinfoshan Square Bamboo 87625
Jinfoshan Tubergourd 390176
Jinfoshan Viburnum 407751
Jingbo Helictotrichon 190162
Jingbohu Batrachium 48940
Jingbohu Pink 127665
Jingdong Aralia 30674
Jingdong Arisaema 33375
Jingdong Casearia 78103
Jingdong Childvine 400868
Jingdong Clematis 95045,95465
Jingdong Cudrania 240808
Jingdong Elaeagnus 142037
Jingdong Epigeneium 146539
Jingdong Eurya 160523
Jingdong Gentian 173547
Jingdong Holly 203846
Jingdong Indigo 206127
Jingdong Mahonia 242612
Jingdong Maple 3055
Jingdong Millettia 254815
Jingdong Monkshood 5127
Jingdong Mosquitotrap 117546
Jingdong Nanmu 295416
Jingdong Persimmon 132238
Jingdong Phoebe 295416
Jingdong Pimpinella 299459
Jingdong Primrose 314499
Jingdong Rockfoil 349511
Jingdong Rockvine 387792
Jingdong Small-leaf Cliffbean 254815

Jingdong Small-leaf Millettia 254815
Jingdong Southstar 33375
Jingdong Spicegrass 239699
Jingdong Stenoseris 375713
Jingdong Stephania 375834
Jingdong Sterculia 376127
Jingdong Stonecrop 356619
Jingdong Swallowwort 117546
Jingdong Tanoak 233267
Jingdong Tylophora 400868
Jingdong Wendlandia 413791
Jingdong Willow 343553
Jingdong Wingseedtree 320839
Jingdong Wing-seed-tree 320839
Jinggang Grape 411756
Jinggang Spapweed 205039
Jinggang Touch-me-not 205039
Jinggangshan Actinidia 6563
Jinggangshan Azalea 330960
Jinggangshan Chinacane 364635
Jinggangshan Cordateleaf Maple 2902
Jinggangshan Euonymus 157643
Jinggangshan Holly 203929
Jinggangshan Kiwifruit 6563
Jinggangshan Neillia 263153
Jinggangshan Rhododendron 330960
Jinggangshan Snakegourd 396207
Jinggu Chickweed 375136
Jinghong Argyreia 32620
Jinghong Begonia 49786
Jinghong Dendrobium 125126
Jinghong Drupetea 322575
Jinghong Goniothalamus 179406
Jinghong Greenstar 307491
Jinghong Helixanthera 190832
Jinghong Pyrenaria 322575
Jinghong Sonerila 368878
Jinghong Waxplant 198834
Jingmen Bulrush 353487
Jingning Magnolia 242291
Jingping Begonia 49653
Jingxi Cyclobalanopsis 116074
Jingxi Galangal 17699
Jingxi Oak 116074
Jingxi Ottelia 277366
Jingxi Qinggang 116074
Jingxi Rockvine 387759
Jingxi Soft Sloanea 366066
Jingyuan Corydalis 106025
Jinjiang Azalea 330615
Jinjiang Colquhounia 100034
Jinjiang Dutchmanspipe 34215
Jinmen Galingale 119513
Jinny Joes 384714
Jinny Lily 230005,230009
Jinping Alphonsea 17625
Jinping Azalea 330961
Jinping Birch 53494
Jinping Camellia 69724
Jinping Cyclobalanopsis 116121
Jinping Dendrocalamus 125503
Jinping Dragonbamboo 125503
Jinping Elatostema 142700
Jinping Esquirol Pogostemon 306972
Jinping Esquirol Spinestemon 306972
Jinping Eurya 160621
Jinping Forest Azalea 331320
Jinping Forkstamenflower 416505
Jinping Hairstyle Azalea 331489
Jinping Litse 233864
Jinping Oak 116121
Jinping Pencilwood 139654
Jinping Qinggang 116121
Jinping Rattlebox 112563
Jinping Rhododendron 330961,331489
Jinping Schefflera 350762
Jinping Stairweed 142700
Jinping Tanoak 233189
Jinping Tea 69724
Jinping Yushanbamboo 416754
Jinsha Bittercress 72847
Jinsha Larkspur 124371
Jinsha Lemongrass 117188
Jinsha Tailgrass 402521
Jinsha Urochloa 402521
Jinshajiang Arisaema 33376
Jinshajiang Bluebeard 78019
Jinshajiang Camellia 69226,69552
Jinshajiang Rungia 340353
Jinshajiang Southstar 33376
Jinshajiang Stemona 375344
Jinshajiang Taxillus 385238
Jinshan Arrowwood 407751
Jinshan Azalea 331135
Jinshan Long-stalked Rhododendron 331135
Jinshan Rhododendron 331135
Jinshan Tupistra 70609
Jinshan Viburnum 407751
Jinshanpalm 31705,380553
Jinshoei Tanoak 233360
Jinshuiying Tanoak 233360
Jinta Tamarisk 383521
Jinxian Awngrass 255855
Jinxian Buckthorn 328898
Jinxian Crabapple 243640
Jinxian Pulsatilla 321675
Jinxiu Azalea 330962
Jinxiu Begonia 49878
Jinxiu Rhododendron 330962
Jinxiu Rockvine 387793
Jinyang Dubyaea 138773
Jinyang Euonymus 157645
Jinyang Holly 203931
Jinyang Mayten 182723
Jinyang Monkshood 5306
Jinyun Ardisia 31483
Jinyun Beautyberry 66790
Jinyun Camellia 69224
Jinyun Daphne 122472
Jinyun Dogwood 124911
Jinyun Euonymus 157373
Jinyun Four-involucre 124911
Jinyun Holly 203932
Jinyun Maple 3747
Jinyun Purplepearl 66790
Jinyun Skullcap 355820
Jinyunshan Euonymus 157373
Jinyunshan Tea 69686
Jinzhai Clematis 95046
Jinzhai Daphne 122474
Jinzhai Grape 411757
Jinzhou Peashrub 72276
Jinzhou Spiraea 372038
Jinzhula Rockfoil 349509
Jio 100940
Jiotilla 155249
Jiqi 290570
Jiquitiba 76832
Jisha Arisaema 33320
Jisha Southstar 33320
Jishan Peony 280212
Jishan Sedge 74491
Jishi Willow 343554
Jitang Sagebrush 35585
Jitterpappel 311537
Jiuding Mountain Willow 343008
Jiufengshan Potanin Larkspur 124508
Jiufengshan Roegneria 335374
Jiuhua Chinese Groundsel 365059
Jiuhuashan Falsepimpernel 231536
Jiuhuashan Merremia 250832
Jiuhuashan Motherweed 231536
Jiuhuashan Orostachys 275377
Jiuhuashan Sedge 75290
Jiuhuashan Sinosenecio 365059
Jiujiang Maple 2818
Jiulong Arrowbamboo 162701
Jiulong Barberry 51799
Jiulong Birch 53495
Jiulong Cacalia 283831
Jiulong Daphne 122625
Jiulong Fargesia 162701
Jiulong Monkshood 5307
Jiulong Spapweed 204850
Jiulong Touch-me-not 204850
Jiulong Umbrella Bamboo 162701
Jiulongshan Grand Torreya 393069
Jiulongshan Spapweed 205040
Jiulongshan Stonecrop 356852
Jiulongshan Touch-me-not 205040
Jiuquan Milkvetch 42539

Jiuwanshan Chirita 87887
Jiuwanshan Holly 203933
Jiuxianshan Raspberry 339497
Jiuxianshan Sedge 74961
Jiuyishan Camellia 69230
Jiuyishan Tea 69230
Jixin Rhododendron 330384
Jizhuang Sedge 74962
Jizushan Violet 410114
Jnjerto 313408
Joan's Ribbon 293709
Joan's Silver Pin 282685
Joanna's Thistle 92086
Job's Tears 99124,308607,321554
Job's-tears 99124
Jobarbe 358062
Jobo 88691,372477
Jobstears 99105,99124
Jockies 60382
Jocky Jurnals 102727
Joe Stanley 174877
Joepye Thoroughwort 161170
Joe-pye Weed 158021,158113,158219,
158264
Joepyeweed 161168
Joe-pye-weed 158219
Joey Hooker 170142
Joey Palm 212402
Johan 202086
Johann Barberry 51800
John Day Ppincushion 84515
John George 68189
John Georges 68189
John Hood 174877
John's Cabbage 200488
John's Feast-day Wort 36474
John's Flower 174711
John's Plant 364193
John's Wort 202086
John's-cabbage 200488
John-go-to-bed-at-noon 21335,274823,
394333
Johnny Cocks 273531
Johnny Cranes 68189
Johnny Jumpup 410677
Johnny Mac Gorey 109857
Johnny O'neele 86901,86970
Johnny Prick-finger 133478
Johnny Woods 248312
Johnny-go-to-bed 394333
Johnny-jump-up 409728,410340,410394,
410513,410677
Johnny-jump-well 410480,410677
Johnny-run-the-street 410677
Johnsmas Flowers 302034,302068
Johnsmas Pairs 302034,302068
Johnson Grass 369652
Johnson Pleurothallis 304919
Johnson's Grevillea 180607
Johnsongrass 369652
Johnson-grass 369652
Johnston Begonia 49952
Johnston Cactus 244118
Johnston Mahonia 242572
Johnston Pittosporum 301302
Johnston River hardwood 46336
Johnston Seatung 301302
Johnston's Gum 155613
Johnston's Wild Buckwheat 152278
John's-wort 201705
John-that-goes-to-bed-at-noon 21339,394333
Johol Siberian Thorowax 63828
Joined Prismatomeris 315321
Joint Fir 146122,178524
Joint Grass 98796
Joint Rush 213325
Joint Vetch 9489,9629
Joint Wood 78425
Jointed Bog Rush 213291
Jointed Charlock 326609
Jointed Cinquefoil 312396
Jointed Flatsedge 118520
Jointed Glasswort 342859
Jointed Goatgrass 8672
Jointed Rush 212866,213325
Jointed Spikerush 143171
Jointed Spike-sedge 143141
Jointed-culm Blysmus 55898
Jointfir 146122,146154,178524,178549
Joint-fir 146122,146136,146154,178524
Jointfir Ephedra 146154
Joint-fir Ephedra 146154
Jointfir Family 178522
Joint-fir Family 146270,178522
Joint-flowered Knotweed 309602
Jointgrass 285390
Joint-grass 170743,196818
Joint-pine 146154,146163,146168,146262
Joint-tail Grass 337544
Joint-tall-grass 337544
Joint-vetch 9489
Jointweed 308667,308668,308690
Jointwood 78425
Jointwood Senna 78425
Joinwood 78425
Jo-Jo Weed 368536
Jojoba 364447,364450
Jojoba Family 364451
Jokul Arrowbamboo 162702
Jokul Azalea 330040
Jokul Sandwort 32209
Jolkin Euphorbia 159159
Jolkin Spurge 159159
Jolly Soldiers 273531
Jolmi Larkspur 124135
Jolmo Lungma Asiabell 98310
Jolmolugma Didymocarpus 129994
Jolmolungma Eritrichium 153513
Jolmolungma Gentian 173924
Jolmolungma Groundsel 359925
Jolmolungma Primrose 314191
Jones' Aster 160668
Jones' Fleabane 150717
Jones' Goldenaster 194221
Jones' Nailwort 284860
Jones' Penstemon 289338
Jones' Saltbush 44409
Jones' Wild Buckwheat 152179
Jonette 68189
Jongkong Merubong Medang 121280
Joni Azalea 330964
Joni Rhododendron 330964
Jonocostle 375533
Jonquil 262410,262429
Jonquil-scented Jasmine 211938
Joo Violet 410116
Jopi Weed 158264
Jordan Almond 20897
Joseph Bank Pine 299818
Joseph Begonia 49954
Joseph Milkvetch 42540
Joseph Rockfoil 349350
Joseph Sage 345474
Joseph Tipularia 392351
Joseph's Coat 18089,18095,18823,18836,
18839,321643
Joseph''s Coat 18103
Joseph's Flower 394333
Joseph's Walking Stick 307197
Joseph's Walkingstick 307197
Joseph-and-Mary 321636,321643
Josephine's Lily 61398
Joshua Tree 416535,416563
Joshua-tree 416563
Josiah Amaranth 18733
Jove's Beard 358062
Jove's Flower 363462
Jove's Nut 324335
Jovellana 212529,212532
Jowar 369600
Joy 325612
Joy Weed 18095
Joyful Disa 133821
Joyful Embelia 144752
Joyful Knotweed 309269
Joyful Osmanther 276399
Joyful Primrose 314525
Joynson's Remedy Cheese 383874
Joy-of-the-ground 409339
Joy-of-the-mountain 274237
Joyous Iris 208443

Joyweed 18089,18124,18150
Juanulloa 212553
Juba's Bush 208347
Jubaea 212556
Jubarbe 358062
Jubard 358062
Jubilee Blueberry 403780
Jubilee Hunter 338209
Juburishan Willow 343555
Jucaro Oxhorn Bucida 61909
Juchaihu Aster 40646
Juda's Bush 208357
Judaes Tree 83809
Judas Pence 238371
Judas Tree 83757,83809,345631
Judas' Bag 7022
Judas-tree 83752,83757,83798,83809
Judd Viburnum 407657
Judean Sage 345137
Jug Orchid 303020
Jugly Onosma 242397
Jugly Oreocharis 273874
Jugor Poppy 282581
Jujube 418144,418169
Jujube Lotus 418177
Jujube-leaved Birch 53463
Jujube-tree 418144
Jujubevine 143557,143558
Jujubileaf Kydia 218449
Jujubi-leaved Kydia 218449
Juldus Onion 15385
Julia's Goldenrod 368187
Julian 262441
Julian's Hackberry 80661
Juliana Lilac 382265
Julie Camellia 68878
July Gold 123555
July-flower 86427
Jum 235373
Jumbie Balsam 268565
Jumbie Tree 56784
Jumble Balsam 268565
Jumble Beads 765
Jumble Beans 7190,154617,227420,274375
Jumby-bubby 367357
Jump-about 8826
Jumping Betty 205175
Jumping Cholla 116670,272814
Jumping Chotla 272906
Jumping Jack 204756
Jumping Wattle 1202
Jumpsecd 393363
Jumpseed 309940
Jump-up-and-kiss-me 311915,410677
Junak's Desertdandelion 242948
Junatov Ptilagrostis 321013
Juncagina Family 212753
Juncellus 212763
Junceus Calligonum 67035
Junc-like Sandwort 31967
Junco Espinoso 288988
Junction Vine 34286
Juncus Corydalis 106027
Jundzill's Rose 336735
June Berry 19243
June Grape 411887,411996
June Grass 60989,60992,217494,305855
June Lily 229789
Juneberry 19240,19241,19243,19251,19273,19276,19279,19288,19290
June-berry 19240
Junebush 19241
June-flower 28044
Junegrass 217416,217441,217494,217550
June-grass 217473
Junesnow 360923,360933,360935
Junfandaisy 364967,364968
Jungfrau Saxifrage 349216
Junggar Pea-shrub 72345
Junghuhn Yellow-wood Barberry 52386
Jungle Corydalis 106185
Jungle Flame 211068
Jungle Holly 377540
Jungle Plum 362902
Jungle Rice 140360,140367,140423
Jungle Whitepearl 172069
Jungleplumleaf Persimmon 132398
Junglerice 140360
Juniper 213606,213702,341696
Juniper Bush 213984
Juniper Family 114645
Juniper Globemallow 370944
Juniper Grevill 180609
Juniper -leaf Grevillea 180609
Juniper Mistletoe 295580
Juniper Mountain Wild Buckwheat 152272
Juniper Myrtle 11399,11400
Juniper Parasite 30912
Juniper Savin 213902
Juniper Saxifrage 349514
Juniper Sedge 74968
Juniper Tamarisk 383469
Juniper Thrift 34522
Juniper Thyme 391253
Juniper Wild Buckwheat 151931
Juniperforest Buttercup 325995
Juniper-leaf Minuartia 255501
Juniperwoods Buttercup 325995
Junlian Hemsleia 191961
Junlian Snowgall 191961
Juno Iris 208816
Juno's Rose 229789
Juno's Tears 405872
Junshan Riparial Silverreed 394938
Junshan Sweetleaf 381255
Jupar Rhodiola 329892
Jupati Palm 326649
Jupiter's Bean 201389
Jupiter's Beard 28200,81743,81768,321721,358062
Jupiter's Distaff 345057
Jupiter's Eye 358062
Jupiter's Eyes 358062
Jupiter's Staff 405788
Jupiter's-beard 28151,81768
Jupiter's-beard Centranthus 81768
Jura Turpentine 298191
Jurinea 190830,190848,193914,207151,214028
Jur-nut 102727
Justiceweed 158198
Justicia 214304
Jute 104056,104072,104103
Jute Greenbamboo 47451
Juteleaf Melochia 249633
Juteleaf Raspberry 338281
Juteleaf Spapweed 204871
Juteleaf Touch-me-not 204871
Jute-leaved Raspberry 338281
Jutt Cissus 120108
Juvia-nut 53057
Juxian Bitter-bamboo 304045
Juzepczuk Skullcap 355536

K

K. Heyne Cogflower 154172
K. Heyne Ervatamia 154172
Kabong 32343
Kabuchii 93520
Kabul Psychrogeton 319928
Kachin Blueberry 403874
Kachin Elsholtzia 144046
Kachina Fleabane 150719
KachinFleshseed Tree 346983
Kachirachia Magnolia 283418
Kadam 263983
Kadam Tree 263983

Kadle Dock 28044,292372,359158
Kadsura 214947,214972
Kadsura Pepper 300427
Kaempfer Grape 411759
Kaempfer Paper-mulberry 61101
Kaempfer Taxillus 385202
Kaempfer's Glorybower 96147
Kaempfer's Goldenray 162616
Kaempfer's Rhododendron 330966
Kaempferi Azalea 330966
Kaffir Bean Tree 352485
Kaffir Boom 154636
Kaffir Bread 145217
Kaffir Corn 369600,369720
Kaffir Date 185867
Kaffir Fig 77441,251357
Kaffir Lily 97213,351883,351884
Kaffir Nut 185867
Kaffir Orange 378888
Kaffir Plum 166773,185867
Kaffir Potato 99571,99583
Kaffir Thorn 238999
Kaffir Tree 154636
Kaffirboom 154636
Kaffirboom Coral Tree 154636
Kaffirlily 97213
Kaffir-lily 97213
Kafir Corn 369600
Kafir-lily 351883,351884
Kager 123141,285741
Kagné Butter 14895
Kahikatea 121076,306420
Kahili Flower 180570
Kahiliflower 180570
Kahua Kahua Bark 386462
Kai Choy 59575
Kaibab Pincushion Cactus 287902
Kaibab Pussytoes 26586
Kaibab Rock Daisy 291306
Kaido Crab-apple 243657
Kaili Azalea 332102
Kaili Corydalis 106028
Kaili Rhododendron 332102
Kainjul 54620
Kaivum Fibre 190105
Kaiyer 123141,285741
Kaizuca Juniper 213647
Kaka 101852
Kakabeak 96633
Kaka-beak 96639
Kakabo Monkshood 5311
Kakaha 39875
Kakezie 101852
Kaki 132219
Kaki Persimmon 132219
Kaku Oil 236334
Kakuda Figwort 355148
Kala Monkshood 5316
Kaladana 207839
Kalakunlun Christolea 89236
Kalakunlun Plateaucress 89236
Kalamazoo Dewberry 339437
Kalanchoe 215086,215102
Kalanchoe Stonecrop 61570
Kalanhoe 61561
Kalatao Spunsilksage 360847
Kalaw 199746
Kale 59493,59524,59551,108591
Kales 59524
Kalgan Boronia 57265
Kali 344585
Kali Catchfly 363629
Kaliang 369671
Kalidium 215322
Kalimandan Jointfir 178532
Kalimantan Meranti 362195
Kalimeris 215334,215358
Kaliphora 215373,215374
Kalkliving Bittercress 72712
Kalkora Mimosa 13586
Kalm Cowlily 267291
Kalm's Brome 60793
Kalm's Hawkweed 195677
Kalm's Lobelia 234562
Kalm's St. John's-wort 201953
Kalmia 215386,215395
Kalmia Deutzia 126983
Kalmiopsis 215412
Kalmuk Corydalis 106453
Kalo 99910
Kalofilum 67860
Kalofilum Kathing 67860
Kalonji 266241
Kalopanax 215420
Kalpi 93439
Kama Cranesbill 174514
Kama Willow 343557
Kamahi 413702
Kamala 243427
Kamala Tree 243427
Kamalatree 243427
Kamala-tree 243427
Kamani 67860
Kamaon Larkspur 124309
Kamarere 155557
Kamassi 179387
Kambroo 167123
Kamchatka Bugbane 91038
Kamchatka Meadowsweet 166089
Kamchatka Pleurospermum 304772
Kamchatka Ribseedcelery 304772
Kamchatka Thistle 92089
Kamchatka Trillium 397574
Kammick Kamics 271563
Kamole 238188,309043
Kamoon Yam 131655
Kamschatka Sedum 294123
Kamtschatic Rockcress 30345
Kamtschatka Alkaligrass 321309
Kamtschatka Azalea 389572
Kamtschatka Bedstraw 170434
Kamtschatka Blueberry 403967
Kamtschatka Bluegrass 305835
Kamtschatka Daphne 122477
Kamtschatka Fir 383
Kamtschatka Fleabane 150422
Kamtschatka Fritillary 168361
Kamtschatka Gentian 150422
Kamtschatka Monkshood 5313
Kamtschatka Neottia 264665
Kamtschatka Nestorchid 264665
Kamtschatka Northern Bedstraw 170259
Kamtschatka Rhododendron 389572
Kamtschatka Rosewood 121720
Kamtschatka Spikesedge 143178
Kamtschatka Spruce 298319
Kamtschatka Stonecrop 294123
Kamtschatka Woodrush 238573
Kan Chiku 87581
Kanara Ebony 132318
Kanari-nut-tree 70996
Kanashira Alkaligrass 321313
Kanashiro Sagebrush 35723
Kanayan-tinik 47213
Kandelia 215508,215510
Kanehira Azalea 330990
Kanehira Caelospermum 98810
Kanehira Coelospermum 98810
Kanehira Holly 204344
Kanehira Mitrastemon 256182
Kang Primrose 314533
Kanga Butter 289476
Kangaron Thorn 1272
Kangaroo 92607
Kangaroo Apple 367289
Kangaroo Grass 389307,389315,389358
Kangaroo Paw 25216,25218
Kangaroo Thorn 1060
Kangaroo Treebine 92607
Kangaroo Vine 92607
Kangaroo-apple 367289
Kangaroo-grass 389307,389315
Kangaroo-paw 25216
Kangba Cotoneaster 107684
Kangba Willow 343573
Kangding Acanthopanax 143625
Kangding Angelica 24383
Kangding Bedstraw 170568
Kangding Bristly Rhododendron 331601
Kangding Bugle 13069
Kangding Buttercup 325774

Kangding Cherry 83343
Kangding Chickweed 375101
Kangding Chirita 87981
Kangding Cinquefoil 313062
Kangding Cowparsnip 192357
Kangding Currant 334263
Kangding Dogwood 380503
Kangding Eritrichium 153471
Kangding Falsecapform Raspberry 339118
Kangding Falseoat 398454
Kangding Fritillary 168563
Kangding Goldenray 229074
Kangding Holly 203833
Kangding Honeysuckle 235841
Kangding Indigo 206564
Kangding Jerusalemsage 295203
Kangding Landpick 308632
Kangding Larkspur 124630
Kangding Litse 233950
Kangding Milkvetch 43132
Kangding Monkshood 5626
Kangding Mosquitotrap 117566
Kangding Parnassia 284549
Kangding Poplar 311362
Kangding Rabdosia 209724
Kangding Reddrop Barberry 51279
Kangding Regel Eyebright 160255
Kangding Rhubarb 329345
Kangding Rockcress 365534
Kangding Rockjasmine 23216
Kangding Rush 213183
Kangding Sage 345322
Kangding Spapweed 205327
Kangding Spruce 298368
Kangding Spurge 159167
Kangding St. John's Wort 202003
Kangding Stonecrop 356904
Kangding Swertia 380357
Kangding Teasel 133492
Kangding Touch-me-not 205327
Kangding Trisetum 398454
Kangding Willow 343559,343834
Kangding Windhairdaisy 348195
Kangding Woodbetony 287301,287746
Kangding Youngia 416447
Kangkong 207590
Kangling Bamboo 297329
Kangmar Sagebrush 35724
Kangnan Azalea 332116
Kangpu Rush 213185
Kangra Buckwheat 162335
Kangting Fritillary 168563
Kangting Larkspur 124630
Kangting Woodbetony 287301
Kangxian Cacalia 283832
Kangzang Mountainash 369533
Kanluang 262823
Kannghmu 25612
Kanniedood Aloe 17136
Kan-non-chiku 329176
Kanooka 398675
Kanran 116926
Kansas antelope Sage 152175
Kansas Arrowhead 342313
Kansas Feather 228511
Kansas Feathers 228511
Kansas Gay Feather 228449,228511
Kansas Gayfeather 228449,228511
Kansas Gay-feather 228511,228529
Kansas Hawthorn 109612
Kansas Sunflower 189032
Kansas Thistle 367567
Kansi 93415
Kanska 934
Kansu Anemone 23711
Kansu Aralia 30691
Kansu Asparagus 39048
Kansu Cacalia 283815
Kansu Columbine 30065
Kansu Corydalis 105738
Kansu Cowparsnip 192275
Kansu Crab Apple 243642
Kansu Crabapple 243642
Kansu Crazyweed 278919
Kansu Crenate Firethorn 322459
Kansu Doronicum 136355
Kansu Figwort 355151
Kansu Hawthorn 109781
Kansu Houndstongue 117967
Kansu Iris 208747
Kansu Jerusalemsage 295124
Kansu Larkspur 124312
Kansu Liriope 232627
Kansu Maple 3130
Kansu Milkvetch 42617
Kansu Panzerina 282481
Kansu Peach 20921
Kansu Rhodiola 329894
Kansu Sagebrush 35523
Kansu Sandwort 31971
Kansu Saussurea 348418
Kansu Sedge 74973
Kansu Viburnum 407900
Kansu Willow 343368
Kansu Woodbetony 287302
Kansu Wormwood 35468
Kansu-mongolian Peashrub 72303
Kantokei Zeuxine 417752
Kanuka 218392
Kanuka Box 398675
Kanuro-Zasa 304011
Kao Osmanther 276343
Kao Osmanthus 276343
Kao Peliosanthes 288644
Kao Scurrula 236788
Kao Yam 131668
Kao-hsiung Leafflower 296779
Kapa 61107
Kapok 80114,80115,80120
Kapok Bush 151562
Kapok Ceiba 80120
Kapok Tree 80120
Kapokkie Heath 149894
Kapok-tree 56770,80120
Kapong 387297
Kapuka 181264
Kapur 138547,138548,138550
Kaqin Blueberry 403874
Karaczukur Catchfly 363620
Karaka 106945
Karaka Nut 106944,106945
Karakor Sandwort 31975
Karamu 103791,103806
Karamyle 222758
Karanda 76873
Karanda Carissa 76873
Karanda Nuts 142288
Karanja 310962
Karaping Drypetes 138641
Karasberg Aloe 17300
Karatav Skullcap 355538
Karateg Larkspur 124316
Karaya 98113,376205
Karaya Gum 98113
Karee 332685
Karee Moer 395831
Karekir Catchfly 363621
Karela 256797
Karelin Craneknee 221691
Karelin Fritillary 168438
Karelin Halimocnemis 184815
Karelin Leymus 228363
Karelin Pricklythrift 2244
Karelin Saltbane 184815
Karelin Willow 343565
Karelin's Tamarisk 383526
Karelinia 215566
Kari Spicebush 231374
Kari Spice-bush 231374
Kari Woodbetony 287308
Kariyat 22407
Karka Reed 295916
Karkaral Skullcap 355540
Karker Milkvetch 42544
Karl Foerster Feather Reed Grass 65261
Karlong Sedge 74980
Karna Khatta 93517
Karnip 37015
Karo 301248
Karo Nut 99240
Karo's Marsh Woodbetony 287495

Karoo Aloe 16991
Karoo Conebush 227331
Karoo Nut 99240
Karoo Rose 221518
Karoub 83527
Karr Bamboo 47345
Karri 155562,155563
Karri Eucalyptus 155562
Karri Gum 155562
Karro Otaheite Potato 131501
Karroo Thorn 1324
Karroo Violet 29919
Karsten Crowded Barberry 51488
Karta Barberry 51808
Karum Tree 310962
Karwinsk Begonia 49956
Karwinsky's Fleabane 150724
Kasachstanian Cudweed 178240
Kaschdaisy 215616
Kaschgar Beancaper 418683
Kaschgar Ephedra 146239
Kaschgar Herelip 220073
Kaschgar Hippolytia 196699
Kaschgar Jurinea 214106
Kaschgar Kengyilia 215929
Kaschgar Lagochilus 220073
Kaschgar Lymegrass 144357
Kaschgar Overlord 418683
Kaschgar Reaumuria 327212
Kaschgar Redsandplant 327212
Kaschgar Rhodiola 329896
Kaschgar Sealavender 230648
Kaschgar Spunsilksage 360848
Kaschgar Tamarisk 383517
Kaschgar Windhairdaisy 348424
Kaschgaria 215616
Kaschghar Pyrethrum 322696
Kashgar Christolea 89238
Kashgar Plateaucress 89238
Kashi Barberry 51809
Kashi Corydalis 106032
Kashi Holly 203625
Kashi Juniper 213877
Kashi Savin 213877
Kashimir Catchfly 363305
Kashmir Barberry 51810
Kashmir Corydalis 105710
Kashmir Cotoneaster 107385
Kashmir Cypress 114668
Kashmir Fenugreek 397204
Kashmir Fescuegrass 164031
Kashmir Iris 208652
Kashmir Minuartwort 255502
Kashmir Plume 309616
Kashmir Rowan 369356
Kashmir Sage 295083
Kashot Syzygium 382649
Kasi Sawgrass 245559
Kaso Rhododendron 330992
Kassod Tree 360481
Kat 79395
Kataka 378860
Katamba 386500
Katang 54620
Kataya Gum 376205
Katharine Blood Lily 350313
Katherine Bloodlily 350313
Katherine Blood-lily 350313
Katherine's Flower 266225
Kathing 67860
Katkin Grevillea 180653
Kato 19954
Kato Taro 99928
Katon 345788
Katong Milkvetch 43016
Katsumada Galangal 17702
Katsura 83736
Katsura Tree 83734,83736,83740
Katsura Tree Family 83732
Katsuratree 83734,83736
Katsura-tree 83734,83736
Katsuratree Family 83732
Katsura-tree Family 83732
Katty Keys 167955
Katu-kitul Palm 270998
Katune 344496
Katus 78955
Kaufmann Desertcandle 148546
Kaufmann Gentian 173552
Kaufmann Ikonnikovia 203519
Kaulfussia 86027
Kauri 10492,10494
Kauri Dammar Pine 10494
Kauri Dammar-pine 10494
Kauri Pine 10492,10494
Kauripine 10494
Kaushue Holly 203938
Kava 241210,300458
Kava Pepper 300458
Kav-apple 136839
Kawa Pepper 300458
Kawabata 93404
Kawachi Spindle-tree 157647
Kawaka 228649
Kawakami Asiabell 98341
Kawakami Azalea 330993
Kawakami Barberry 51811
Kawakami Chinquapin 78966
Kawakami Evergreen Chinkapin 78966
Kawakami Evergreenchinkapin 78966
Kawakami Excoecaria 161666
Kawakami Falsebrome 58592
Kawakami Fir 389,396
Kawakami Honeysuckle 235897
Kawakami Hornbeam 77314
Kawakami Hydrangea 199817
Kawakami Maple 3031,3058
Kawakami Paulownia 285966
Kawakami Pepper 300432
Kawakami Rhododendron 330993
Kawakami Sagebrush 35732
Kawakami Spicebush 231355
Kawakami Spicy Tree 231355
Kawakami Sweetleaf 381256
Kawakami Tanoak 233269
Kawakami Thistle 92096
Kawakami's Gambirplant 401761
Kawakami's Raspberry 338664
Kawakawa 241210
Kawa-Kawa 241210
Kawal 360461
Kaweah Brodiaea 60481
Kaweah Lakes Fawn-lily 154941
Kawla 93435
Kaworthialike Aloe 16878
Kaya 393077
Kaya Nut 393077
Kaya Torreya 393077
Kaye's Wild Buckwheat 151842
Kayea 215663,252369
Kays 3462,167955
Kazinoki Paper-mulberry 61103
Keaki 417558
Kearney Wild Buckwheat 152332
Kearney's Three-awn 33915
Keatlegs 273531,273545
Keblock 59575,364557
Keck 28044,192358
Keckers 192358
Kecklock 364557
Kecksy 101852
Ked 17614
Keddle Dock 359158
Keddle-dock 359158
Kedlack 364557
Kedlet 364557
Kedlock 24483,28044,192358,326609,
359158,364557
Kedondong 70988
Keedle Dock 359158
Keedle-dock 359158
Keek Legs 273531
Keek-legs 273531
Keeks 101852
Keel Brome 60659
Keele 59532
Keeled Barberry 51427
Keeled Corn Salad 404402
Keeled Cornsalad 404402
Keeled Garlic 15160
Keeled Goosefoot 139684

Keeled Odontoglossum 269059
Keeled Onion 15160
Keeled Thelasis 389123
Keeled-fruited Cornsalad 404402
Keeled-leaf Thrixspermum 390497
Keeledsiliclecress 302792
Keenan 215749
Keer 369339
Keeshion 28044
Keeslip 170743
Kegang Himalayan Kobresia 217267
Keggas 192358
Keggers 123141
Keglus 24483,192358
Kei Apple 136839
Kei Cycad 145241
Keicer Keice 101852
Keik-kecksy 101852
Keirn 369339
Keiry 86427
Keisak Murdannia 260102
Keisak Waterbamboo 260102
Keish 101852
Keiske Sagebrush 35733
Keiske Wormwood 35733
Keiskea 215805,215818
Keissler Mountainash 369437
Keissler Mountain-ash 369437
Keissleri Barberry 51815
Keitao Stauntonvine 374417
Keitao Staunton-vine 374417
Kekui Oil Plant 14544
Keladi 99910
Kelch-grass 351162,351167
Keliage 309199
Kelimu Milkvetch 43139
Kelimu Rockfoil 349505
Kelk 9832,28044,101852,364557
Kelk-kecksy 24483,101852
Kellas 102727
Keller Broomrape 275102
Kelley Lily 229879
Kellock 28044,326609,364557
Kellogg Lily 229880
Kellogg Oak 324071
Kellogg's Knotweed 309610
Kellogg's Spurred Lupin 238438
Kellogglily 229880
Kellogia 215845
Kelly 102727
Kelly Grass 337555
Kelong Bluegrass 306005
Kelp 344585
Kelring Lily 229881
Kelso Water-crowfoot 325500
Kelubi Palm 342572
Kelung Christmas-bush 14199
Kelung Rockcress 30458
Kelway Chamomile 26804
Kemaman Licuala Palm 228742
Kemaman Licualapalm 228742
Kemaman Rattan Palm 65716
Kemang 244390
Kemaon Iris 208653
Kembang 95972
Kemmick 271563
Kemng 310962
Kempas 217855
Kemps 302034,302068
Kempseed 302034
Kenaf 194779
Kenaf Hemp 194779
Kenaf Hibiscus 194779
Kenai Birch 53498,53549
Kenanga 70960
Kenaph 194779
Kendal Green Weed 173082
Kendrick Azalea 331002
Kendrick's Rhododendron 331002
Keng Alkaligrass 354404
Keng Blueberry 403875
Keng Holly 203940
Keng Lymegrass 144360
Keng Reedbentgrass 65384
Keng Stiffgrass 354404
Keng Woodreed 65384
Kenguel Seed 364360
Kengyilia 215901
Kenia Alpine Groundsel 125741
Kenilworth Ivy 116732,116736
Kenilworth-Ivy 116736
Kennedia 215978
Kennedy Hippolytia 196701
Kennedy's Bentgrass 12368
Kennedy's Blackberry 338667
Kennedy's Wild Buckwheat 152184
Kennin Herb 325612
Kenning Heal 325612
Kenning-wort 86755
Kenry 364557
Kensui False Harpetiole Willow 343919
Kentia 198804
Kentia Palm 198808
Kentia Palm Howeia 198804
Kentish Balsam 250614
Kentish Clover 397043
Kentish Cob 106757
Kentish Milkwort 307905
Kentucky 94026
Kentucky Blue Grass 305855
Kentucky Bluegrass 305855
Kentucky Blue-grass 305855
Kentucky Coffee 182527
Kentucky Coffee Tree 182527
Kentucky Coffeetree 182527
Kentucky Coffee-tree 182527
Kentucky Starwort 374889
Kentucky Viburnum 407953
Kentucky Wisteria 414571
Kentucky Yellow Wood 94026
Kentucky Yellowwood 94022
Kentucky Yellow-wood 94026
Kenworthy Begonia 49959
Kenya Burr Grass 394383
Kenya Hyacinth 346124
Kenyan Shower 78219
Keonla 93435
Keppel 374688
Keraji 93519
Kerala Reed Bamboo 268118,268119,268122
Keranda Nut 142288
Keratto 10778
Kerchov Arrowroot 245026
Kerguelen Cabbage 315154
Kerk 24483
Kerlack 364557
Kerleck 364557
Kerlick 364557
Kerlock 364557
Kerm Oak 324027
Kermadec Pohutukawa 252618
Kermes Bush 298094
Kermes Oak 323769,324027
Kermes Oak-leaved Barbados Cherry 243523
Kermode Willowweed 146744
Kern 369339
Kern Beavertail Pricklypear 272810
Kern County Larkspur 124276
Kern River Fleabane 150795
Kernelwort 355202
Kerner Bristle Thistle 73377
Kerner Iris 208654
Kerner's Dock 340096
Kerner's Mullein 405721
Kernera 216074
Kerosene Bush 279347,279348,334357
Kerosene Weed 279348
Kerqin Poplar 311367
Kerr Barberry 51816
Kerr Bauhinia 49187
Kerr Brachystelma 58886
Kerr Clematis 95340
Kerr Copperleaf 1907
Kerr Cylindrokelupha 116599
Kerr Fevervine 280081
Kerr Large-wing Citrus 93561
Kerr Lasianthus 222265
Kerr Nightshade 367596
Kerr Oak 116122
Kerr Obovate Waxplant 198850,198877

Kerr Persimmon 132235
Kerr Petrocosmea 292562
Kerr Pittosporopsis 301206
Kerr Pittosporum 301305
Kerr Plectocomia 303092
Kerr Pothos 313200
Kerr Roughleaf 222265
Kerr Seatung 301305
Kerr Snakegourd 396208
Kerr Stonebutterfly 292562
Kerr Tephrosia 386129
Kerr Treebine 92791
Kerr Tylophora 400913
Kerr's Thailand Papeda 93561
Kerria 216084,216085
Kerroon 83155
Kerroon Coroon 83155
Kerry Lily 364407,364409
Kerse 225450,262722
Kerslins 316470
Kersouns 262722
Kersting's Groundnut 241427
Keruing 133552
Kervanh 19870
Kesh 8826,101852,192358
Kesk 24483,28044,192358
Keslings 316470
Kesselring Lily 229881
Ketaki 281138
Keteleeria 216114
Keteleeria Parasite 30908
Ketlack 364557
Ketlock 326609,364557
Ketmen Crazyweed 278925
Ketsa-shupfu 93495
Kett 144661
Kettle Cap 273531
Kettle Case 273531
Kettle Dock 28044,292372,340151,359158
Kettle Smock 102933,236022,248411
Kettle-case 273531,292372
Kettlecupule Tanoak 233188
Kettle-cupuled Tanoak 233188
Kettle-dock 28044,340151,359158,364557
Kettle-pad 273531
Kettles-and Crocks 64345
Kettle-smocks 102933,236022,248312
Ketty Keys 2837,167955
Ketty-keys 2837,167955
Kew Barberry 51817
Kew Begonia 49961
Kew Cactus 244121
Kew Primrose 314084
Kew Weed 170142
Kewsies 24483,28044,101852,192358
Kex 28044,101852,123141,192358,316819
Kexies 101852
Key Lime 93329
Key Palm 390460
Key Tree Cactus 299266
Keyaki 417558
Keyamasakura 316913
Key-balls 300223
Keyesia 178012
Key-flower 315082
Keyn 167955
Keys 2837,3462
Keys of Heaven 315082
Keys' Rhododendron 331003
Keys-of-heaven 315082
Keywort 315082
Khair 1120
Khair Gum 1120
Khair Tree 1120
Khair-tree 1120
Khaki Bur 18137
Khaki Weed 18137
Khans Addermonth Orchid 110646
Khans Bogorchis 110646
Khart Ajania 13004
Khas Aletris 14520
Khas Catchfly 363631
Khas Clearweed 298953
Khas Coldwaterflower 298953
Khas Gugertree 350920
Khas Lemongrass 117195
Khas Milkwort 308136
Khas Podochilus 306538
Khas Pricklyash 417248
Khas Schefflera 350726
Khas Streptolirion 377885
Khas Thelasis 389127
Khas Whitetung 94069
Khasi Nightshade 366910
Khasi Panicgrass 281791
Khasi Papeda 93529
Khasi Pine 299986,299999
Khasi Prickly-ash 417248
Khasi Rostellularia 337245
Khasi Skullcap 355542
Khasi Trichodesma 395764
Khasia Barberry 51818
Khasia Bauhinia 49146
Khasia Berry 107687
Khasia Claoxylon 94069
Khasia Fieldorange 249667
Khasia Fighazel 380593
Khasia Glochidion 177146
Khasia Gugertree 350920
Khasia Indofevillea 206848
Khasia Ladybell 7674
Khasia Melodinus 249667
Khasia Mnesithea 256336
Khasia Parasite Pentapanax 289660
Khasia Schefflera 350726
Khasian Bluegrass 305617
Khasian Illigera 204631
Khasy Bulbophyllum 62825
Khasy Stonebean-orchis 62825
Khasya Bauhinia 49146
Khasya Pine 299986,299999
Khasyan Litse 233953
Khat 79387,79395
Khate Tree 79387
Khatta 93517
Khaya 216194,216205
Khayer 1120
Khella 19672
Khesara 222832
Khing'an Monkshood 5032
Khingan Fir 427
Khus Khus 407585
Khus-Khus 407585
Kiaat 320278
Kiah Devil Rattan 121505
Kiah Licuala Palm 228743
Kiah Licualapalm 228743
Kiangsi Falsepimpernel 231537
Kiangsi Glorybower 96161
Kiangsi Millettia 66634
Kiangsi Violet 410136
Kiangsi Woodbetony 287314
Kiangxi Sage 345142
Kiaometi Sage 345143
Kiawe 315568
Kiawe Bean 315560
Kibatalia 216228
Kibo Rose 336292
Kiboh Rose 336292
Kicking Colt 204841
Kicks 9832
Kicksy 101852
Kickxia Rubber 167954
Kidney Bean 294056
Kidney Begonia 49776,49818
Kidney Cotton 179893
Kidney Liverleaf 192126
Kidney Palegreenvine 230090
Kidney Saxifrage 349473
Kidney Shaped Crazyweed 279120
Kidney Sorrel 278577
Kidney Vetch 28143,28200,196630
Kidney Vetch Anthyllis 28200
Kidney Weed 128964
Kidney Wood 161863
Kidneyauricle Sinobambusa 364806
Kidneyauricle Tangbamboo 364806
Kidney-form Begonia 50229
Kidneyleaf Bellflower 70194
Kidneyleaf Bittercress 72721,72827
Kidneyleaf Chinese Groundsel 365057

Kidneyleaf Didymocarpus 129953
Kidneyleaf Gentian 173357
Kidney-leaf Grass-of-parnassus 284509
Kidneyleaf Habenaria 184019
Kidneyleaf Mountain Leech 126574
Kidneyleaf Mountainsorrel 278577
Kidneyleaf Pennywort 200390
Kidneyleaf Pimpinella 299510
Kidneyleaf Primrose 314521
Kidneyleaf Pulsatilla 321703
Kidneyleaf Pyrola 322870
Kidney-leaf Rosinweed 364268
Kidney-leaf Sinosenecio 365057
Kidneyleaf Synotis 381949
Kidneyleaf Tickclover 126574
Kidney-leaf Twayblade 232966
Kidney-leaf Wild Buckwheat 152428
Kidneyleaf Wildginger 37709
Kidneyleaf Windhairdaisy 348098
Kidney-leaved Buckwheat 152428
Kidney-leaved Violet 410484
Kidneypetal Crazyweed 279120
Kidney-shaped Caldesia 66343
Kidneyshaped Leaf Calystegia 68737
Kidneyshaped Seed Milkvetch 43063
Kidney-shaped Whitfordiodendron 9852
Kidney-shaped Wild Buckwheat 152501
Kidneyweed 128955,128964,401711
Kidneywort 192148,349932,401711
Kidnyfruit Actinidia 6565
Kidnyfruit Milkwort 308024
Kiele 171404
Kieran 369339
Kif 71220
Kifonsan Milkvetch 42551
Kikuchi Caper 71775
Kikudaidai 93417
Kikuyu Grass 289067
Kikuyugrass 289067
Kikuyu-grass 289067
Kilk 364557
Kill Wart 86755
Kill Wood 285985
Killarney Strawberry-tree 30888
Kill-bastard 213902
Kill-bastard Bastard Killer 213702,213902
Killimore 102727
Killridge 309199
Kill-wart 86755
Kill-your-mother-quick 28044
Kilmarnock Willow 343151,343152,343153
Kilmarnock's Willow 343152
Kiloneedleleaf Azalea 331515
Kilung Clematis 94763
Kimball Holcoglossum 197260
Kimball Spathoglottis 370399
Kimball Vanda 197260
Kimberley Gray Box 155484
Kimberley Heath 68759
Kindlingweed 182107
King Alps 153487
King Anthurium 28138
King Azalea 331635
King Begonia 50232
King Billy 44050
King Billy Pine 44050
King Boris' Fir 298
King Ceratolobus 83520
King Cob 68189,325518,325612,325666,
326287
King Corallodiscus 103957
King Corydalis 106036
King Cup 68189
King Devil 195478,195506,195857,195873
King Fan Palm 234184
King Fanpalm 234184
King Fig 164630
King Fingers 273531,273545,277648
King Fisher Daisy 163151
King Horsfieldia 198514
King Island Melilot 249218
King Kong 68189
King Larkspur 124319
King Licuala Palm 228744
King Licualapalm 228744
King Litse 233954
King Mandarin 93649
King of Snowdrops 169711
King of the Alps 153487
King of the Bromeliads 412357
King of the Meadow 388621
King of the Woods 30748
King Ophiorrhiza 272229
King Orange 93649,93717
King Palm 31012,31013,31015
King Primrose 314536
King Protea 315776
King Rhododendron 331635
King Rosewood 121721
King Snakegourd 396170
King Solomon's-seal 308508
King Solomonseal 308572
King Tangor 93649
King William Billy Pine 44050
King William Pine 44050
King Wood 43475,121650
King Woodbetony 287605
King's Alpine Holly 203830
King's Aster 193130
King's Bush 401388
King's Catchfly 363636
King's Clover 249205,325518,325666,
326287
King's Cob 68189,325518,325612,325666,
326287
King's Consound 102833
King's Cross 86427
King's Crown 129325,214410,249232,
397019,407989
King's Elwand 130383
King's Evil 325820
King's Feather 350011
King's Feathers 350011
King's Fingers 237539,273469,273531
King's Flower 156018
King's Hood 174951
King's Knob 325518,325666
King's Knobs 325518,325666
King's Mantle 390759
King's Mountain Gayfeather 228548
King's Nobs 325666
King's Rod 262583
King's Rush 213188
King's Sandwort 31979
King's Saussurea 348428
King's Spear 39195,39328,39438,39454,
262441
King's Taper 405788
King's-spear 148531
King-cob 68189,325612
Kingcup 68157,68189,325518,325666,
325820,326287,399504
Kingcup Cactus 140332
Kingcups 68189
Kingdevil 195857,195873
King-devil 195506,195603
King-devil Hawkweed 195857
Kingdon Corydalis 106035
Kingdon Willowweed 146749
Kingdonia 216398
Kingdon-ward Blueberry 403876
Kingdon-ward Mahonia 242503
Kingdon-ward Rhododendron 332093
Kingdon-ward Trachydium 393841
King-finger 273531,273545,277648
Kingfischer Daisy 163151
King-fischer Daisy 163151
Kingfisher 273545
Kingfisher Daisy 163131
Kingfisher Ophrys 272481
Kingfisher Woodbetony 287667
Kingfishershape Woodbetony 287667
Kingfowshan Clematis 94977
Kingia 216414
Kingidium 216424
Kingiella 216436
Kingnut 77905
King-nut 77905
King-of-bromeliads 412357
King-of-the-bromeliads 412357
King-of-the-meadow 388628

King-of-the-night 357748
Kingpalm 31012,31013
King-palm 31012
King-root 301907,301909
Kings-and-queens 37015
Kingsbloom 280253
Kingscote Mallee 155728
Kingwood 121650
Kinkle 364557
Kinkoji 93672
Kinnickinnick Dewberry 338856
Kinnikinnick 31142
Kinnikinnik 31142,104949,105193
Kino 320307
Kino Eucalyptus 155720
Kino Gum 155720
Kinokuni 93724
Kinokuni Citrus 93724
Kinokuni Mandarin 93724
Kinostemon 216454
Kinukawa 93464
Kinzu 167499
Kioshan Vetch 408446
Kippernut 63475,222758
Kirai Orchis 273481
Kirais Windhairdaisy 348430
Kirengeshoma 24167,216473
Kirghis Sweetvetch 187953
Kirialov Giantfennel 163645
Kirilow Clematis 95051
Kirilow Falsespiraea 369272
Kirilow False-spiraea 369272
Kirilow Figwort 355152
Kirilow Indigo 206140
Kirilow Rhodiola 329897
Kirilow's Groundsel 385903
Kirilow's Indigo 206140
Kirilow's Tephroseris 385903
Kirilowia 216486,216487
Kirin Monkshood 5318
Kirishima Azalea 331370
Kirk Crinum 111212
Kirk Meliosma 249401
Kirk Ochna 268206
Kirk Pine 121102
Kirki Pine 184917
Kirnstaff 159027
Kisidwe 14886
Kiskies 192358
Kiss Camellia 69235
Kiss Me Over the Garden Gate 309494
Kiss-and-go 410960
Kiss-and-look-up 410677
Kiss-at-the-garden-gate 410677
Kiss-behind-the-garden-gate 349044
Kiss-behind-the-pantry-door 81768
Kisses 31051,410677
Kiss-I'-my-corner 35088
Kissing Kind 81768
Kiss-me 81768,174877,349044,410677
Kiss-me-and-go 35088
Kiss-me-at-the-garden-gate 410677
Kiss-me-behind-the-garden-gate 410677
Kiss-me-dick 158730
Kiss-me-ere-I-rise 410677
Kiss-me-John-at-the-garden-gate 410677
Kiss-me-love 81768,174877,349044,410677
Kiss-me-love-at-the-garden-gate 174877, 349044,410677
Kiss-me-love-behind-the-gardengate 349044
Kiss-me-not 349044
Kiss-me-over-the-garden-gate 309494,410677
Kiss-me-quick 18687,31056,35088,61306, 81768,166125,170193,174877,349044, 364193,410677
Kiss-me-quick-and-go 35088
Kiss-me-twice-before-I-rise 266225
Kiss-tile-garden-door 81768
Kit Keys 2837,167955
Kit Willow 344211
Kitagawa Cleistogenes 94608
Kitagawa Hideseedgrass 94608
Kitamura Veronicastrum 407468
Kitchen Sage 345271
Kitchli 93565
Kite Keys 167955
Kite Tree 267378
Kite's Legs 99297,273531
Kite's Pan 273531
Kitefruit 196821
Kitembilla 136846
Kit-run-the-fields 410677
Kitten's Tails 106703
Kitten's-tails 53221
Kittens'tails 106703
Kittitas Sulphur Flower 152589
Kittool 78056
Kittool Palm 78056
Kittool-palm 78056
Kitty Keys 2837,167955,369339
Kitty-come-down-the-lane-jumpand-kiss-me 37015
Kitty-run-the-street 410677
Kitty-two-shoes 237539
Kitul 78056
Kitul Palm 78056
Kiujiang Fleabane 150729
Kiukiang Hornbeam 77421
Kiukiang Maple 3063
Kiukiang Mountainash 369439
Kiukiang Mountain-ash 369439
Kiwano 114221
Kiwi 6513,6553,6558,6639,6666
Kiwi Fruit 6513,6553,6558
Kiwi Rose 385738
Kiwi Vine 6639
Kiwifruit 6513
Kiwifruit Family 6728
Kiwifruitlike Vineclethra 95473
Kix 101852,316470
Kjelman Draba 137058
Klaes Galeandra 169872
Klain Anopyxis 26189
Klamath Arnica 34771
Klamath Coneflower 339573
Klamath Fawn-lily 154931
Klamath Fleabane 150730
Klamath Mountain Wild Buckwheat 152143
Klamath Plum 316833
Klamath Thistle 92346
Klamath Weed 202086
Klamath-weed 202086
Klarke's Sikkim Clematis 95300
Klatt Primrose 314540
Klaver Primrose 314541
Klebelsberg Cowwheat 248171
Kleberg's Bluestem 128568
Kleim's Hardy Gardenia 171309
Klein Cholla 116664
Klein Cinquefoil 312690
Klein Spapweed 205057
Klein Touch-me-not 205057
Kleinaalwyn 16656
Kleine Bisamhyazinthe 260301
Kleine Traubenhyazinthe 260301
Kleinhovia 216641
Klementz Calligonum 67038
Klemenz Feathergrass 376940
Klickitat Aster 155811
Klimu Cotoneaster 107701
Klimu Desertcandle 148559
Klimu Hogfennel 293030
Klimu Iris 208874
Klimu Mouseear 83049
Klimu Onosma 271835
Klinegrass 281479
Klinki Pine 30850,30852
Klipnoors 159485
Klissing Cactus 244122
Klondike Cosmos 107172
Klonger 336419
Kloss Mahonia 242573
Klugia 216733
Kluk Rose 336658
Klunger 336419
Knap 397019
Knap Bottle 364193
Knap-bottle 282685,364193
Knappers 324335
Knapperts 222758
Knapperty 222758

Knapweed 80892,81035,81150,81183,81237,81243
Knapweed Broomrape 275044
Knapweed Harshweed 81150
Knapwort Harshweed 81150
Knauperts 145073
Knautia 216752,216769
Knawel 353986,353987
Knee Grass 281561
Knee Holly 340990
Knee Holme 340990
Knee Hulver 340990
Knee Hulyer 340990
Kneed Devil Rattan 121496
Kneed Holly 340990
Kneeholy 340990
Kneejujube 66988
Kneelike Microstegium 254010
Knee-pine 300086
Knema 216783
Knickier Tree 182527
Knife Acacia 1156,1157
Knife Leaf Wattle 1156,1157
Knife-and-fork 174877
Knifebean 71036,71050
Knife-leaf Acacia 1156
Knife-leaf Condalia 101741
Knife-leaf Waggle 1156
Knife-leaf Wattle 1156
Knigbt's Pondweed 377485
Knight Barberry 51820
Knight Cross 363312
Knight Portuguese Cypress 114720
Knight Star Lily 196439
Knight's Milfoil 3978
Knight's Pondweed 377485
Knight's Spur 102833
Knight's Star 196439
Knight's Star Lily 18865,196439
Knight's-spur 102839,102841
Knight's-star 18862,196439
Knights-and-ladles 37015
Kniphofia 216923
Knipper Nut 222758
Knitback 381035
Knitbone 381035
Knives-and-forks 3462,174877
Knob 68189
Knob Sedge 74908
Knob Thorn 1424
Knobbed Crowfoot 325666
Knobbed Hop Sedge 75232
Knobby Club-rush 210042
Knobcone Pine 299806
Knob-cone Pine 300290
Knob-root 99844
Knob-styled Dogwood 104949
Knobthorn 417132,417190
Knob-thorn 1424
Knobweed 81020,81237,81356,99844
Knobwood 417132,417190
Knoll 59575
Knolles 59493,59575
Knollige Tillandsie 391984
Knolls 59493,59575
Knoop 338243
Knopweed 81020,81237,81356
Knorring Nepeta 264962
Knot Oat-grass 34930
Knotberry 338243
Knotbone 381035
Knotgrass 81237,105410,162505,285426,291651,308690,308816
Knot-grass 105416,308690,308816
Knotroot Bristlegrass 361753
Knotroot Bristle-grass 361856
Knot-root Bristle-grass 361753
Knotroot Foxtail 361856
Knot-sheath Sedge 76014
Knot-stem Bastard-sage 152671
Knotted Bur Parsley 392996
Knotted Clover 397095,397097
Knotted Crane's-bill 174763
Knotted Cranesbill 174763
Knotted Figwort 355202
Knotted Geranium 288248
Knotted Hedge Parsley 392996
Knotted Hedge-parsley 392996
Knotted Majoram 274224
Knotted Marjoram 274224
Knotted Pearlwort 342271
Knotted Persicaria 309468
Knotted Rush 213325
Knotted Spike-rush 143141
Knotted Spurry 342271
Knotted Storksbill 288248
Knottedflower Phyla 296121
Knotty Birch 53563
Knotty Meal 102727
Knotty Senna 78425
Knotty-leaved Rush 212800
Knotweed 81020,81237,81356,162505,291651,308690,308698,308788,308816,308824,308832,308946,309685
Knotweed Blue 309893
Knotweed Dodder 115099,115116
Knotweed Family 308481
Knotweed Spineflower 89097
Knotweed Threewingedcalyx 224045
Knotweed Violet 410758
Knotweedleaf Begonia 50176
Knotweedleaf Milkwort 308259
Knotweedleaf Ottelia 277407
Knotweedleaf Pondweed 312228
Knotweedleeaf Violet 410758
Knotweedlike Microula 254360
Knotweed-like Torenia 224045
Knotweedtree 397931,397932
Knotweed-tree 397931
Knotwood 44247,44260
Knotwort 308816
Knoutberry 338243
Knowlton Hophornbeam 276811
Knowlton Ironwood 276811
Knowlton's Cactus 287893
Knowlton's Hop-hornbeam 276811
Knowlton's Minute Cactus 287893
Knowsley Begonia 49963
Knowtweedleaf Windhairdaisy 348658
Knuckle-bleeders 9701
Knuppelplakkie 8427
Knutsford Devil 68713
Knysna Lily 404581
Ko Childvine 400914
Ko Condorvine 245821
Ko Lasianthus 222190
Ko Marsdenia 245821
Ko Milkwort 308138
Ko Tylophora 400914
Koa 1334
Koa Acacia 1334
Koa Haole 227430
Koali-pehu 207570
Kobayashi Alkaligrass 321296
Kobresia 217117
Kobus Magnolia 242169
Koch Beautyberry 66829
Koch Iris 208658
Koch Willow 343569
Kochang Lily 229834
Kochia 48753,217324,217361
Kock Clematis 95058
Kodo Millet 285507
Kodo Wood 141595
Kodomillet 285507
Kodo-millet-like Panicgrass 282041
Koehne Barberry 51821
Koehne Callery Pear 323123
Koehne Dogwood 380459
Koehne Honeysuckle 235732
Koehne Mountainash 369443
Koehne Mountain-ash 369443
Koehne Spindle-tree 157560
Koeler Grass 217441,217494,217550
Koeleria 217416
Koellikeria 217597
Koelpinia 217600
Koelz Figwort 355153
Koenig Amomum 19868
Koenig Jasminorange 260168
Koenigia 217640

Koesnaatjie 108905
Kofa Mountain Barberry 51704
Kohekobe 143076
Kohekohe 139673
Kohleria 217738
Kohlrabi 59541
Kohl-rabi 59541
Kohuhu 301398
Koidzum Firethorn 322471
Koie-Yan 162569
Koilodepas 217758
Koji 93530
Kojima Monkshood 5323
Kok Corydalis 106043
Kokam 171119
Kokand Rose 336659
Kokanic Pricklythrift 2249
Kokerbom 16768
Kokerite 246733
Kokko 13595,13649
Kokni 93556
Kokonor Orinus 274253
Kokoon 217775
Kokrodua 290876
Koksaghyz 384472
Kok-saghyz Rubber 384472
Kokum 171119
Kokum Butter 171119
Kola 99157,99158,99240
Kola Nut 99157
Kolhol Gum 1572
Koling 243711
Kolitachibana 93648
Kolkwitzia 217785
Kolomikta Actinidia 6639
Kolomikta Kiwifruit 6639
Kolomikta Vine 6639
Kolomikta-vine 6639
Kolomikta-vine Actinidia 6639
Kolomikta-vined Actinidia 6639
Kolorkow Arum 37013
Komalov Bulrush 352205
Komalov Goosecomb 335389
Komalov Mosquitotrap 117548
Komalov Pimpinella 299447
Komalov Roegneria 335389
Komalov Russianthistle 344606
Komalov Swallowwort 117548
Komarnitzk Windhairdaisy 348437
Komarov Apple 243645
Komarov Bellflower 70104
Komarov Bittercress 72851
Komarov Bluegrass 305623
Komarov Cacalia 283835
Komarov Catchfly 363640
Komarov Currant 334055
Komarov Dragonhead 137606
Komarov Hedgehogweed 203136
Komarov Juniper 213815
Komarov Kaschdaisy 215618
Komarov Koeleria 217484
Komarov Lilac 382190
Komarov Maple 3064
Komarov Nitraria 266358
Komarov Plantain 302024
Komarov Raspberry 338679
Komarov Savin 213815
Komarov Spruce 298316
Komarov Thorowax 63693
Komarov Vetchling 222747
Komarov Waterchestnut 394526
Komarov Whitethorn 266358
Komarov's Fleabane 150732
Komarov's Kaschgaria 215618
Komarov's Photinia 295712
Kombe 378417
Kombo False Nutmeg 321967
Komper's Orchid 101634
Kong Croton 112920
Kongba Willow 343573
Kongbo Groundsel 359237
Kongbo Larch 221901
Kongbo Rhododendron 331016
Kongbo Rockfoil 349522
Kongbo Woodbetony 287316
Kongkang 329745
Kongmu Willow 343585
Kongpo Monkshood 5327
Kongu 198178
Konishi China Fir 114548
Konishi China-fir 114548
Konishi Cryptocarya 113438
Konishi Fig 165823
Konishi Machilus 266792
Konishi Neolitsea 264059
Konishi Newlitse 264059
Konishi Sweetleaf 381148
Konishi Tanoak 233278
Konishi Taro 99918
Konishi's Nothaphoebe 266792
Konishi-fir 114548
Konjaku 20132
Konka Monkshood 5367
Kontum Cyclobalanopsis 116189
Koochla Tree 378836
Kooker Dragonbood 137425
Kooyah 404255
Kopetdagh Leymus 228364
Kopetdagh Nepeta 264966
Kopsia 217864,217867
Korakan 143522
Korarima Cardamom 9890
Korea Alkaligrass 321236
Korea Angelica 24358
Korea Arborvitae 390579
Korea Arisaema 33260
Korea Atractylodes 44202
Korea Barberry 51824
Korea Barrenwort 147013
Korea Buckthorn 328752
Korea Bugleweed 239194
Korea Catchfly 363641
Korea Clematis 95061
Korea Cocklebur 11546
Korea Cranesbill 174682
Korea Dandelion 384505
Korea Daylily 191268
Korea Gentian 174029
Korea Hornbeam 77283
Korea Lawngrass 418426
Korea Litse 233876
Korea Lovegrass 147671
Korea Minuartwort 255516
Korea Monkshood 5144
Korea Pimpinella 299448
Korea Pine 300006
Korea Poplar 311368
Korea Pulsatilla 321667
Korea Purplepearl 66824
Korea Raspberry 338292
Korea Rhubarb 329320
Korea Rose 336660
Korea Sinosenecio 365060
Korea Southstar 33260
Korea Speedwell 317947
Korea Spiraea 372114
Korea Trigonotis 397462
Korea Twistedstalk 377919
Korea Willow 343574
Korean Abelia 117,210
Korean Agrimonia 11546
Korean Angelica 24358,24386
Korean Arborvitae 390579
Korean Arbor-vitae 390579
Korean Arrowwood 407905
Korean Astilbe 41822
Korean Azalea 331289
Korean Barberry 51824
Korean Beautyberry 65670,66824
Korean Bittersweet 80189
Korean Blueberry 403878
Korean Bluegrass 306059
Korean Box 64285
Korean Buckthorn 328752
Korean Bush-clover 218345
Korean Chinese Groundsel 365060
Korean Chrysanthemum 89723
Korean Clematis 95061
Korean Clover 218345
Korean Daphne 122486
Korean Deutzia 126875

Korean Dewberry 338292
Korean Dogwood 124923,124925,380444
Korean Early Lilac 382224
Korean Eelgrass 418426
Korean Elder 345592
Korean Epimedium 147013
Korean Euodia 161323
Korean Evodia 161323
Korean Fir 396,398
Korean Forsythia 167452,167473
Korean Hackberry 80664
Korean Jerusalemsage 295128
Korean Larch 221919
Korean Lawn Grass 418426
Korean Lawngrass 418453
Korean Lespedeza 218345,226956
Korean Leymus 228349
Korean Lilac 382268
Korean Lily 229723,230063
Korean Litse 233876
Korean Lovegrass 147671
Korean Monkshood 5144
Korean Mountain Ash 32988
Korean Mulberry 259097
Korean Oxtongue 298625
Korean Pearl Bush 161755
Korean Pine 300006
Korean Poplar 311368
Korean Pulsatilla 321667
Korean Purplebloom Maple 3480
Korean Raspberry 338292,338300
Korean Rhododendron 331289
Korean Rockfoil 349351
Korean Rose 336660
Korean Sandwort 31991
Korean Sophora 369054
Korean Sorbus 369362
Korean Speedwell 317947
Korean Spice Viburnum 407727,407728,407729
Korean Spice-viburnum 407727
Korean Spiraea 371915,372114
Korean Spruce 298323
Korean Stewartia 376446
Korean Sweetheart Tree 160954
Korean Thuja 390579
Korean Trigonotis 397462
Korean Twistedstalk 377919
Korean Velvet Grass 418453
Korean Viburnum 407729,407905
Korean Violet 409665
Korean Weeping Willow 343918
Korean Weigela 413576
Korean Wheel Lily 230065
Korean Willow 343574
Koreanspice Viburnum 407727,407728
Koriba Rattan Palm 65717
Korintji Cinnamon 91333
Koriyanagi Willow 343579
Kork-fruited Camellia 69484
Korkhull Camellia 69484
Korky Eucalyptus 155497
Korolkov Torularia 264618
Korolkov Tulip 400184
Korolkow Iris 208662
Korolkow Leek 15402
Korolkow Onion 15402
Korolkow Tamarisk 383530
Koromiko 186925,186972
Korovin Arthrophytum 36823
Korovin Nodosetree 36823
Korsakow Monochoria 257572
Korshinsk Peashrub 72258
Korshinsk Pea-shrub 72258
Korshinsky Broomrape 275018
Korshinsky Kudrjaschevia 218279
Korshinsky Larkspur 124325
Korshinsky Licorice 177916
Korshinsky Russianthistle 344607
Korshinsky Sedge 75012
Korthalsella 217901
Korunca 189559
Koshewnikov Sedge 75013
Koshun Galangal 17707
Kosinsk's Rose 336663
Koso 184565
Kossinsky Hornwort 83573
Kossinsky Lovegrass 147754
Kossinsky Seepweed 379543
Koster's Curse 96651
Koster's Hinoke Falsecypress 85320
Kosum Oil 351968
Kosumba 351968
Kotchy's Crocus 111549
Koto Alderleaf Raspberry 338435
Koto Newlitse 264112
Koto Nightshade 238942
Koto Purplepearl 66831
Kotokan 93526
Kotokan Citrus 93526
Kotschy Broomrape 275105
Kotschy Colewort 108621
Kotschy Crocus 111549
Kotschy Goatgrass 8684
Kotschy's Azalea 331018
Kotschy's Rhododendron 331018
Kotukutuku 168735,168746
Kou 104251
Kouga Aloe 17167
Kous 101852
Kousa 124923,124925
Kousa Dogwood 124923,124925
Koushe 101852
Koushle 101852
Kou-shuipaper Mulberry 61107
Kousso 59794,184565
Kowhai 369067,369130
Koyama Spruce 298323
Koyama's Spruce 298323
Kozlov Dandelion 384609
Kozlov Peashrub 72266
Kozlov Pea-shrub 72266
Kozlov Spurge 159193
Kozo 61103
Kraalaalwyn 16715
Krabak 25605,25608
Krabao 199744
Kral's Goldenrod 368194
Kralik Fumitory 169078
Kramer's Lily 229872
Krameria 218074,218076,218085
Kramics 271563
Krantz Aloe 16592
Krapovickasia 218097
Kraschenninkov Groundsel 359242
Krasev Skullcap Helmit Skullcap 355449
Krasnovia 218146
Kraus Haworthia 186494
Krause Sorrel 340099
Krempf Pine 300007
Kricksies 101852
Krim-saghyz Rubber 384666
Krishna Bor 165210
Krislings 316470
Krobonko 385654
Kroneburg Milkvetch 42556
Kronenburg Newstraw 317083
Kronenburg Psathyrostachys 317083
Krotkov's Goldenrod 368195
Kruse's Mallee 155616
Krylov Broomsedge 217343
Krylov Buttercup 326093
Krylov Crazyweed 278934
Krylov Dragonhead 137607
Krylov Feathergrass 376910
Krylov Foot Trefoil 237664
Krylov Gentian 150737
Krylov Gianthyssop 236271
Krylov Lophanthus 236271
Krylov Summercypress 217343
Krylov Sweetvetch 187959
Krylov Vetchling 222748
Krylov's Fleabane 150737
Krylovia 218242
Krylow Broomrape 275107
Krysna Crassula 109097
Kuang Ostodes 276788
Kuanzhao Croton 112911
Kuanzhao Spiradiclis 371761
Kubilan 234033
Kubili Nut 114048

Kucha Camellia 68917
Kuche Knotweed 309620
Kuchugouzhui Evergreen Chinkapin 78850
Kuding Cratoxylum 110258
Kudingcha 110258,203938
Kudingcha Holly 203938
Kudo Melastoma 248754
Kudo Photinia 295713
Kudoacanthus 218272
Kudrjaschevia 218274
Kudzu 321441,321453
Kudzu Bean 321419
Kudzu Vine 321419,321441
Kudzubean 321419
Kudzuvine 321419,321441
Kudzu-vine 321441
Kuerle Calligonum 67039
Kuerle Kneejujube 67039
Kuhl Palm 299664
Kuhl Wildcoral 170040
Kukaipua'a-Uka 130825
Kukui Nut 14544
Kuling Diversifolius Ampelopsis 20356
Kuling Raspberry 338683
Kuling Supple-jack 52435
Kullam Nut 46784
Kulu Azalea 330031
Kulu Rhododendron 330031
Kuma Bamboo 362131
Kuma Sasa 347324
Kuma Shibataea 362131
Kuma Zasa 347324
Kumang Sandwort 31992
Kumaon Barberry 51281,51825
Kumar 207623
Kumara 207623
Kumbanzo 197199
Kumbaru 174326
Kumbuk 386462
Kumen Swordflag 208653
Kummerowia 218344
Kumon Sweetvetch 187962
Kumquat 167489,167503,167506
Kumson Forsythia 167474
Kumurere 155557
Kumurere Gum 155557
Kunai 205470,205497
Kunamur Lagotis 220183
Kunawar Violet 410149
Kunawur Barberry 51826
Kunda-oll Tree 72650
Kunenbo 93660
Kungia 218354
Kungshan Monkshood 5333
Kungshan Oak 324079
Kungshan Plum-yew 82539
Kungshan Thistle 91953
Kunish Gelidocalamus 172741
Kunishi Cane 172741
Kunlun Buttercup 326007
Kunlun Catnip 264970
Kunlun Crabgrass 130780
Kunlun Juniper 213630
Kunlun Mountain Juniper 213630
Kunlun Needlegrass 376903
Kunlun Nepeta 264970
Kunlun Savin 213630,213810,213932
Kunlun Savin Juniper 213810,213932
Kunlun Spunsilksage 360850
Kunlun Windhairdaisy 348213,348264
Kunlunshan Bentgrass 12390
Kunlunshan Goldenray 229080
Kunlunshan Groundsel 359248
Kunlunshan Larkspur 124327
Kunlunshan Nettle 402944
Kunlunshan Pepperweed 225395
Kunlunshan Sagebrush 35975
Kunlunshan Sedge 75024
Kunlunshan Wildryegrass 144210
Kunming Acacia 1177
Kunming Asparagus 39082
Kunming Barberry 51827
Kunming Bentgrass 12163
Kunming Buttercup 326008
Kunming Cacalia 283878
Kunming Dutchmanspipe 34237
Kunming Elm 401480
Kunming Gentian 173412
Kunming Hackberry 80668
Kunming Holboellia 197245
Kunming Holly 203950
Kunming Indigo 206342
Kunming Jurinea 212019
Kunming Ladybell 7833
Kunming Mockorange 294486
Kunming Mosquitotrap 117747
Kunming Pipewort 151343
Kunming Pittosporum 301308
Kunming Ploughpoint 401168
Kunming Rabdosia 209732
Kunming Rhodiola 329901
Kunming Rhododendron 330602
Kunming Rhynchosia 333286
Kunming Rocky Aster 40973
Kunming Roscoea 337093
Kunming Rose 336665
Kunming Savin 213780
Kunming Seatung 301308
Kunming Thorowax 63696
Kunming Typhonium 401168
Kunstler Devil Rattan 121506
Kunstler Licuala Palm 228745
Kunstler Licualapalm 228745
Kunth Begonia 49970
Kunth Microchloa 253277
Kunth Minigrass 253277
Kunth Nutmeg 261453
Kunth Onion 15424
Kunth Twinballgrass 209081
Kunth's Beaksedge 333615
Kunth's Evening Primrose 269455
Kuntz Whiteleaf Raspberry 338631
Kuntze Cactus 244123
Kuntze Larkspur 124314
Kunze's Club-cholla 181456
Kuo's Longfruit Oak 116141
Kuo's Longfruit Qinggang 116141
Kuocangshan Broadsepal Spapweed 205232
Kuomei Maple 3067
Kupong 284473
Kura Clover 396821
Kurakkan 143522
Kuranjan 5983
Kurara 22218
Kurchi Bark 197199
Kurdee 77748
Kurdish Bee-orchid 272450
Kurile Larch 221866
Kurkuti Licuala Palm 228756
Kurly Hairs Rhododendron 330420
Kurogane Holly 204217
Kuroiwa Grass 390551,390553
Kurrajong 376102
Kurrajong Bottle Tree 376102
Kurrat 15048
Kurrimia 218418
Kurro Picrorhiza 298670
Kurssanov Onion 15407
Kurssanov Skullcap 355553
Kurume Azalea 331370
Kurwini Mango 244406
Kurz Alangium 13369
Kurz Allaeanthus 61105
Kurz Casearia 78133
Kurz Chaulmoogratree 199751
Kurz Chaulmoogra-tree 199748
Kurz Fig 165211
Kurz Lasianthus 222193
Kurz Macaranga 240278
Kurz Paper-mulberry 61105
Kurz Phyllodium 297010
Kurz Roughleaf 222193
Kurz Uvaria 403502
Kusaihu Dali Larkspur 124626
Kusam 351968
Kusamaki 306457,306469
Kusano Bitter Sweet 80221
Kusano Chinkapin 78967
Kusano Holly 203952
Kusano Namu 240605
Kusano Persimmon 132245

Kusano Staff-tree 80221
Kusano Willow 343587
Kuschakewicz Pink 127749
Kuschakewicz Sagebrush 35754
Kusnezoff Monkshood 5335
Kusnezow Bothriospermum 57681
Kusnezow Gentian 173192
Kusnezow Spotseed 57681
Kusso Tree 184565
Kusso-tree 184565
Kutira Gum 98113,376205
Kutki 298670
Kuye Island Goldenray 229167
Kuye Raspberry 339227
Kuye Rhodiola 329944
Kuye Wildginger 37635
Kuyedao Mint 250443
Kwangfu Rhododendron 330592
Kwangnan Maple 3068
Kwangs Parashorea i 362197
Kwangsi Actinidia 6669
Kwangsi Alangium 13374
Kwangsi Archiphysalis 297625
Kwangsi Azalea 331026
Kwangsi Bamboo 297333
Kwangsi Belltree 324997
Kwangsi Bitterbamboo 304051
Kwangsi Blueberry 404001
Kwangsi Brandisia 59128
Kwangsi Brassaiopsis 59221
Kwangsi Camellia 69244
Kwangsi Chastetree 411310
Kwangsi Conehead 320126
Kwangsi Cyclobalanopsis 116126
Kwangsi Dolichopetalum 135390
Kwangsi Dutchman's-pipe 34238
Kwangsi Dutchmanspipe 34238
Kwangsi Eriolaena 152690
Kwangsi Erythropsis 155003
Kwangsi Falleneaves 351148
Kwangsi Falsechirita 317538
Kwangsi Fevervine 280091
Kwangsi Fissistigma 166667
Kwangsi Galangal 17713
Kwangsi Ginger 417985
Kwangsi Gingerlily 187455
Kwangsi Glorybower 96029
Kwangsi Gordonia 179758
Kwangsi Greenbrier 366407
Kwangsi Helixanthera 190836
Kwangsi Ironweed 406222
Kwangsi Jasminorange 260169
Kwangsi Lobelia 234405
Kwangsi Local Milkvetch 266464
Kwangsi Lysionotus 239953
Kwangsi Maple 3696
Kwangsi Mussaenda 260438
Kwangsi Oak 116126
Kwangsi Oilfruitcamphor 381795
Kwangsi Ophiorrhiza 272233
Kwangsi Orophea 275320
Kwangsi Padritree 376268
Kwangsi Pavieasia 286576
Kwangsi Phoebe 295372
Kwangsi Photinia 295715
Kwangsi Pricklyash 417249
Kwangsi Prickly-ash 417249
Kwangsi Pseudosasa 318297
Kwangsi Rapanea 261626
Kwangsi Raspberry 338687,339006
Kwangsi Rockfoil 349527
Kwangsi Seatung 301309
Kwangsi Slatepentree 322592
Kwangsi Spicebush 231366
Kwangsi Spiraea 371988
Kwangsi Stephania 375880
Kwangsi Strophioblachia 378475
Kwangsi Suoluo 327372
Kwangsi Swallowwort 117551
Kwangsi Sweetleaf 381272
Kwangsi Syzygium 382549
Kwangsi Tarenna 384972
Kwangsi Tinospore 392256
Kwangsi Turmeric 114868
Kwangsi Umbrella Bamboo 162668
Kwangsi Vatica 405168
Kwangsi Waxplant 198847
Kwangsi Wendlandia 413770
Kwangsi Wildtung 243322
Kwangsi Woodnettle 273911
Kwangtung Atalantia 43730
Kwangtung Bauhinia 49140
Kwangtung Beautyberry 66833
Kwangtung Bentgrass 12058
Kwangtung Blueberry 403844
Kwangtung Browndot Mountainash 369353
Kwangtung Bulbophyllum 62832
Kwangtung Bushclover 226796
Kwangtung Camellia 69246
Kwangtung Cliffbean 66631
Kwangtung Craibiodendron 108583
Kwangtung Dehaasia 123608
Kwangtung Galeobdolon 170024
Kwangtung Gastrochilus 171857
Kwangtung Gleadovia 176794
Kwangtung Glorybower 96167
Kwangtung Grewia 180835
Kwangtung Gugertree 350923
Kwangtung Gymnema 182374
Kwangtung Helicia 189937
Kwangtung Hemiboea 191385
Kwangtung Holly 203953
Kwangtung Hydrangea 199936
Kwangtung Indocalamus 206786
Kwangtung Lasianthus 222123
Kwangtung Lilyturf 272129
Kwangtung Litse 233966
Kwangtung Loosestrife 239702
Kwangtung Machilus 240610
Kwangtung Mananthes 244291
Kwangtung Manglietia 244451
Kwangtung Maple 3484
Kwangtung Meiogyne 248031
Kwangtung Moneygrass 126623
Kwangtung Mouretia 259411
Kwangtung Mussaenda 260439
Kwangtung New Eargrass 262968
Kwangtung Pigeonwings 97190
Kwangtung Pine 300009
Kwangtung Pricklyash 417170
Kwangtung Primrose 314549
Kwangtung Razorsedge 354126
Kwangtung Rehdertree 327416
Kwangtung Rhododendron 331031
Kwangtung Rose 336667
Kwangtung Scolopia 354627
Kwangtung Sedge 73603
Kwangtung Spicebush 231377
Kwangtung Spiradiclis 371759
Kwangtung Syzygium 382591
Kwangtung Ternstroemia 386712
Kwangtung Thickshellcassia 113457
Kwangtung Twayblade 232217
Kwangtung Waterstarwort 67384
Kwangtung Wendlandia 413788
Kwangtung Willow 343685
Kwangtung Wrightia 414807
Kwangtung Xanthophytopsis 415157
Kwangtung-kwangsi Actinidia 6654
Kwangtung-kwangsi Adinandra 8229
Kwangtung-kwangsi Kiwifruit 6654
Kwangxi Garcinia 171123
Kwasho Pepper 300434
Kweek 117859
Kweek Grass 117859
Kwei 276291
Kweichou Fissistigma 166694
Kweichou Lime 391751
Kweichou Linden 391751
Kweichou Physaliastrum 297632
Kweichou Primrose 314550
Kweichou Slugwood 50542
Kweichow Clematis 95063
Kweichow Densevein 261327
Kweichow Evergreen Chinkapin 78969
Kweichow Hanceola 185245
Kweichow Hornbeam 77315
Kweichow Mazus 247000
Kweichow Mucuna 259571
Kweichow Sage 344945
Kweichow Thorowax 63697

Kweilin Maple 3070
Kweiyang Betony 373274
Kweizhou Scurrula 236755
Kweme Nut 385655
Kwigga 144661
Kwila 207015
Kwort 321643
Kyasuma Grass 289195,289196
Kyaw's Rhododendron 331032
Kydia 218443,218444
Kyerlic 364557
Kyetpaung 402189
Kylinleaf 147336
Kyllingia 218457
Kyor 342400
Kyushu Azalea 331010
Kyushu Linden 391749

L

L. B. Stewart's Rhododendron 331890
L. C. Chia Bugle 13078
L. Ramsden's Rhododendron 331606
La Campana 405872
La Graciosa Thistle 92362
La Marjolaine 274237
Labiate Beanbroomrape 244639
Labiate Mannagettaea 244639
Lablab 218721
Lablab Bean 218721
Labord Hibiscus 194954
Labord Woodbetony 287324
Laborde Begonia 49971
Laborde Spapweed 205063
Labrador Deerdrug 242706
Labrador False Solomonseal 242706
Labrador Marsh Bedstraw 170454
Labrador Milk-vetch 42565
Labrador Tea 223888,223901,331980
Labrador Violet 410161
Labrador Woodbetony 287326
Labrador-tea 223888,223894,223895,223901
Labrose Spiderorchis 30534
Labruscan Vineyard Grape 411764
Labuleng Azalea 331033
Labuleng Rhododendron 331033
Laburnum 218754,218755,218761
Labyrinth Pondweed 378991
Lac Tree 351968
Lacaena 218767
Laccospadix 218792
Lace Agapetes 10322
Lace Aloe 16605
Lace Bark Elm 401581
Lace cactus 140224,244062
Lace Curtains 9832
Lace Fern 39195
Lace Flower 17953,19664,147390
Lace Flower Vine 147390
Lace Grass 147566
Lace Hedgehog Cactus 140307
Lace Michelia 252900
Lace Pine 299830
Lace Shrub 375811,375817,375818
Lace Tree 219986
Lacebark 172245,197154,219986
Lace-bark 197154,219986
Lacebark Kurrajong 58339,376102
Lacebark Pine 299830
Lace-bark Pine 299830
Lace-bark Tree 197154,219986
Lacecap Hydrangea 200086
Lace-fern 39195
Lace-flower 101852,147390,392992
Lace-leaf 29326,29663,29676
Lace-leaved Rock Daisy 291299
Lacepod 391472
Lacerate Anemone 23871
Lacerate Begonia 49973
Lacerate Blazing-star 228439
Lacerate Radde Anemone 24017
Lacerate Yushania 416781
Laceratedlip Habenaria 291234
Lacerate-leaved Blumea 55766
Laceratelobed Larkspur 124640
Laceshrub 375817
Lacespine Nipple Cactus 244126
Lacewood 302582,302588
Lacewort 29643
Lachnanthes 219026
Lachnoloma 219056,219057
Lachow Lily 229830
Laciate Kalanchoe 215183
Lacinate Spineflower 89078
Laciniate Begonia 49974
Laciniate Blumea 55768
Laciniate Cremanthodium 110404
Laciniate Currant 334057
Laciniate Devil Rattan 121508
Laciniate Elm 401542
Laciniate Falsecuckoo 48219
Laciniate Kalanchoe 215110
Laciniate Nutantdaisy 110404
Laciniate Rhubarb 329347
Laciniate Rockfoil 349529
Laciniate Selfheal 316116
Laciniateleaf Primrose 314555
Laciniate-leaved Lilac 382256
Lacinose Primrose 314555
Lacinoseleaf Madwort 18407
Lackner Pommereschea 310829
Lackschewitz Fleabane 150741
Lacomte Neocinnamomum 263763
Lacomte Newcinnamon 263763
Lacoocha 36928
Lacost Larkspur 124330
Lacoste Windhairdaisy 348458
Lacquer 177541
Lacquer Tree 393441,393491
Lacquered Wine Cup 8561
Lacquerleaf Meliosma 249445
Lacquertree 332646,393441,393491
Lacquer-tree 332646,393441
Lacteous Carnose Deutzia 126864
Lactistriate Square Bamboo 87565
Lactoris 219204
Lactucarium 219609
Lacustrine Iris 208671
Lacy Germander 388128
Lacy Nut 77971
Lacy Phacelia 293176
Lacy Tree Philodendron 294839
Lacy Wirevine 259591
Lad Savour 35088
Lad's Love 35088
Ladanol Nepeta 264972
Ladanum 93154,93161
Ladd's Favorite 249053
Ladder Love 81020
Ladder of Christ 202086
Ladder To Heaven 102867,307197,308607
Ladder Wild Buckwheat 152053
Ladder-to-heaven 102867,307197
Laddervein Bladderwort 404018
Ladder-veins Ardisia 31594
Ladder-veins Blueberry 403984
Ladie's Ear 168767
Ladie's Eardrops 168735,168767
Ladies Mantle 14065
Ladies Slipper 199699
Ladies Tobacco 26299
Ladies' Bedstraw 170175
Ladies' Delight 410677
Ladies' Eardrops 61378
Ladies' False Fleabane 321515
Ladies' Flower 72934,410677
Ladies' Lords 37015
Ladies' Smock 72934

Ladies' Tobacco 26299,26530,26551
Ladies'-eardrops 61371
Ladies'-tresses 372176
Ladies-and-gentlemen 37015,410677
Ladies-in-a-ship 5442
Ladies-in-the-shade 266225
Ladies-in-white 349044
Ladino Clover 397043
Ladles' Thistle 73318
Lad-love-lass 35088
Ladon-shrub 93161
Ladoo 93692
Ladsavvur 35088
Ladu 93692
Lady Banks Rose 336366
Lady Banks' Rose 336366,336367
Lady Bell 7596
Lady Bells 7682,7833
Lady Betty 81768
Lady Birch 53563
Lady Cakes 277648
Lady Cues 315082
Lady Cushion 34536
Lady Dove And Her Coach And Pair 5442
Lady Eleven o'clock 274823
Lady Finger Cactus 140288
Lady Fingers 244062
Lady Garten Berries 338447
Lady Jane 174877
Lady Laurel 122515
Lady Lavinia's Dove Carriage 5442
Lady Lily 229789
Lady Lupine 238496
Lady Mary's Tears 321643
Lady Nut 78811,145899
Lady of Spring 384714
Lady of the Lake 267648
Lady of the Meadow 166125
Lady of the Night 61295,84417,255694
Lady of the Night Cactus 83882
Lady of the Woods 53563
Lady Ofthe Woods 162400
Lady Orchid 273604
Lady Orchis 273604
Lady Palm 85671,329173,329176,329184
Lady Poplar 311208,311398,311400
Lady Slipper 120287,282839,295875
Lady Smock 72934
Lady Traces 372259
Lady Tulip 400145
Lady Washington Geranium 288206
Lady Washington Pelargonium 288206
Lady Whin 271563
Lady's Ball 81237
Lady's Bed 170743
Lady's Bedstraw 170335,170743
Lady's Bell 7596
Lady's Bonnet 30081
Lady's Bonnets 30081
Lady's Boots 120310,237539
Lady's Bouquet 170708
Lady's Bower 95431
Lady's Bowler 95431,266225
Lady's Brush 133478
Lady's Brush-and-comb 133478
Lady's Brushes 133478
Lady's Brushes-and-combs 133478
Lady's Bunch of Keys 315082
Lady's Bunch-of-keys 315082
Lady's Buttons 374916
Lady's Candle 405788
Lady's Candles 405788
Lady's Candlesticks 315106
Lady's Chain 218761
Lady's Cheese 243840
Lady's Chemise 23925,68713,374916
Lady's Cloak 72934
Lady's Clover 277648
Lady's Comb 350425
Lady's Cowslip 169460
Lady's Cushion 28200,30220,34536,81237,90423,216753,237539,349496
Lady's Delight 410513
Lady's Double 28200
Lady's Double Fingers-and-thumbs 237539
Lady's Eardrops 168735,168767
Lady's Ear-drops 220729
Lady's Embroidery 374916
Lady's Featherbeds 349419
Lady's Fingers 219,28200,37015,130383,196630,218761,222820,236022,237539,273531,312925,315082,325612,360328
Lady's Fingers-and-thumbs 237539
Lady's Flannel 405788
Lady's Flock 72934
Lady's Foxglove 405788
Lady's Garten-berry 338447
Lady's Garter-berry 338447
Lady's Garters 293709,338447
Lady's Glove 72934,130383,169126,207087,237539,237713
Lady's Gloves 72934,130383,169126,207087,237539,237713
Lady's Golden Bedstraw 170743
Lady's Grass 293709
Lady's Hair 60382
Lady's Hand 60382
Lady's Hands 60382
Lady's Hatpins 216753
Lady's Hearts 220729
Lady's Keys 3462,37015,315082
Lady's Lace 3978,9832,28044,101852,115008,392992
Lady's Laces 115008,293709
Lady's Leek 15173
Lady's Lint 374916
Lady's Locket 2837,169126,220729,238371,308607
Lady's Looking-glass 224067
Lady's Maid 35323
Lady's Mantle 8399,13951,14065,14089,72934,302034
Lady's Meat 109857
Lady's Mii 321643
Lady's Milk 364360
Lady's Milkcans 23925
Lady's Milking Stools 374897
Lady's Milking-stools 374897
Lady's Milksile 72934,321643
Lady's Milkwort 321643
Lady's Mint 250450
Lady's Navel 401711
Lady's Needle 350425
Lady's Needlework 14964,28044,81768,101852,235090,350099,374916,392992
Lady's Nightcap 23925,68713,102933,208120
Lady's Parasol 123141
Lady's Petticoats 23925,30081
Lady's Pincushion 28200,34536,106117,216753,237539,321643,350099
Lady's Posy 397019
Lady's Pride 72934
Lady's Purse 23925,30081,72038,220729
Lady's Ribands 293709
Lady's Ruffles 166132,262441
Lady's Seal 61497,308529,308607,383675
Lady's Shakes 60382
Lady's Shimmey 23925,68713
Lady's Shimmy 23925,68713
Lady's Shoes 30081,72934,169126,237539
Lady's Shoes-and-stockings 237539
Lady's Signet 308607,383675
Lady's Slipper 5442,28200,28617,30081,37015,120287,121001,130383,196630,222820,231183,237539,282768,325518
Lady's Slipper Orchid 120287,120310,282768
Lady's Smock 23925,35770,37015,68713,72934,102933,374916
Lady's Sorrel 277644,277648,277747,278099
Lady's Sunshade 102933
Lady's Taper 405788
Lady's Tears 102867,367772
Lady's Thimble 70271,130383,145487,374916,407072
Lady's Thimbles 70271,130383,145487,374916,407072
Lady's Thistle 364360
Lady's Thumb 309494,309570

Lady's Thumbs-and-fingers 237539
Lady's Tobacco 26385,26551
Lady's Tresses 37015,60382,170743,
232950,372176,372259
Lady's Tresses Orchid 372176
Lady's Two-shoes 237539
Lady's Umbrella 68713,102933,367120
Lady's White Petticoats 374916
Lady's-Bell 7596
Lady's-comb 350425
Lady's-erdrops 168735
Lady's-finger 219
Lady's-glove 130383
Lady's-mande 14166
Lady's-mantle 13951,14089
Lady's-seal 383675
Lady's-slipper 120287,120288
Lady's-slipper Orchid 120287
Lady's-smock 72934
Lady's-sorrel 278099
Lady's-thumb 309494
Lady's-tresses 372176,372246,372259
Ladybell 7596,7830
Ladybells 7596,7789
Ladybird 21339
Ladybird Beads 765
Ladycoot Beadtree 7187
Ladydaisy 400466,400467
Ladyfinger Cactus 140288
Ladygin Crazyweed 278939
Ladygin Draba 137065
Ladygin Whitlowgrass 137065
Ladygin Windhairdaisy 348459
Lady-in-the-bath 220729
Lady-in-the-boat 220729
Lady-in-the-bower 95431,266225
Ladykiller Crocus 111522
Ladymantle 13951
Ladymantle-like Rockjasmine 23120
Lady-my-lord 37015
Lady-of-the-night 59259,61295,84417,
255694
Ladypalm 329173
Ladypalms 329173,329176
Lady-pea 222789
Ladyslipper 120287
Lady's-slipper 120350
Ladytress 372176
Laege-leaved Nearsmooth Barberry 52201
Laelia 219668
Laelia Orchid 219668
Laevicalyx Hawthorn 109663
Laevigate Bowfruitvine 393534
Laevigate Elatostema 142704
Laevigate Fig 165225
Laevigate Lacquertree 393534
Laevigate Leptospermum 226460
Laevigate Parameria 283492
Laevigate Poppy 282585
Laevigate Premna 313666
Laevigate Seepweed 379544
Laevigate Toxocarpus 393534
Lafaldense Chin Cactus 182450
Lafoensia 219729
Lagarosolen 219787
Lagedium 219805
Lageniform Jujube 418172
Lageniforme Actinidia 6564
Lageniforme Kiwifruit 6564
Lagenophora 219892,219900
Lagetto 219986
Laggera 219990
Lagochilus 220050
Lagopsis 220117,220126
Lagos Ebony 132299
Lagos Rubber 169229
Lagos Rubber Tree 169229
Lagos Rubbertree 165263
Lagos Silkrubber 167954
Lagos Silk-Rubber 169229
Lagos Silk-Rubber Tree 169229
Lagoseris 220135
Lagotis 220155
Lagurus 220252
Lagwort 292372
Lahontan Basin Wild Buckwheat 152442
Lahsa Rhododendron 331001
Lahule Bluegrass 305295
Laifeng Chirita 87891
Laiguan Maple 3080
Laikwan Maple 3080
Laisect Beesian Larkspur 124060
Laister 208771
Laisu Maple 3081
Laiyuan Goosecomb 335198
Laiyuan Roegneria 335198
Laizhou Rhaphidophora 329002
Lake Athabasca starwort 374952
Lake Champlain Hawthorn 109591
Lake Cress 34583,34587
Lake Huron Tansy 383768
Lake Hygrophila 200607
Lake Loosestrife 239879
Lake Louise Arnica 34740
Lake Mistassini Primrose 314652
Lake Sallow 343590
Lake Sedge 75034
Lake St. John Aster 40642
Lake Tahoe Serpentweed 392774
Lake Washington Claytonia 94376
Lake Willow Dock 340100
Lake-marshes Kobresia 217203
Lakemelongrass 232380
Lakepypie 413369
Lakeshore Bentgrass 12265
Lakeshore Bulrush 352206
Lakeshore Nutrush 354129
Lakeshore Sedge 75106
Lakeside Sunflower 188932
Lakeweed 308739,309199,309570
Lakoocha 36928
Lalang 205497
Lalang Grass 205497,205506
Lalashan Holly 204204
Lalashan Platanthera 302383
Lalinde Bollea 56690,56691
Lallemantia 220281
Lama Catchfly 363664
Lamance Iris 208471
Lamarck Serviceberry 19279
Lamarck's Bedstraw 170350
Lamarck's Evening-primrose 269443
Lamark Ladybell 7678
Lamashan Willow 343591
Lamb Lettuce 404439
Lamb Mint 250420,250450
Lamb Tongue 87186
Lamb's Cress 72802
Lamb's Ear 220416,302103,312360,
373166,379660
Lamb's Ears 373166
Lamb's Foot 14065,28200,237539,302068
Lamb's Lakens 37015,72934
Lamb's Lettuce 221787,345881,404439,
404454
Lamb's Pummy 14964
Lamb's Quarters 44576,86901,87096,
397556
Lamb's Quarters Goosefoot 86901
Lamb's Succory 34815,34816
Lamb's Sucklings 237539,397019,397043
Lamb's Tails 16352,28200,106703,107975,
302034,302068,343151,343396
Lamb's Toe 28200,237539,247366
Lamb's Tongue 86901,250294,302034,
302103,309570,329521,373166
Lamb's Tongue Plantain 302103
Lamb's-ear 373166
Lamb's-lettuce 404414,404439
Lamb's-quarters 86901
Lamb's-tail 87318
Lamb's-tail Cactus 140321
Lamb's-tongue 302103,373166
Lamb's-tongue Plantain 302103
Lambert Barberry 51829
Lambert Cranesbill 174690
Lambert Hawthorn 109796
Lambert Nut 106757
Lambert Pine 300011
Lambert Scurrula 236814
Lambert's Filbert 106757

Lambert's Raspberry 338698
Lambertia 220306
Lamb-in-a-pulpit 37015
Lambkill 215387,215391
Lamb-kill 215387
Lambkill Kalmia 215387
Lambkins 106703
Lamb-lakins 37015,250450
Lambs 9701
Lambsquarters 86901,86921
Lambs-quarters 86901
Lamb-tongue 87186
Lamellate Mucuna 259531
Lamellate Velvet Bean 259531
Lamellosehair Cystopetal 320224
Lamer Pachypodium 279681
Laminate Cyclobalanopsis 116130
Laminate Oak 116130
Laminated Sanicle 345978
Laminate-like Cyclobalanopsis 116128
Lamiophlomis 220335
Lamium 220345
Lammie Sou Rock 339887
Lammint 250450
Lamont Chinquapin 78970
Lamont Evergreen Chinkapin 78970
Lamont Evergreenchinkapin 78970
Lampflower 363312
Lamp-flower 363141
Lampoil Bittersweet 80273
Lampranthus 220464
Lamps of Scent 236022
Lampshade Poppy 247139
Lampstand Bladderwort 403751
Lampstand Rabbiten-wind 12676
Lampstand Rockfoil 349150
Lampstandtree 57693,57695
Lampwick Grass 213021
Lamula Milkvetch 42574
Lamuta 118060
Lamy Butter 289476
Lamy Willow Weed 146905
Lamy Willowweed 146905
Lamy's Willowweed 146905
Lanaria 220788
Lanate Alder 16400
Lanate Betony 373166
Lanate Cactus 244124
Lanate Hippolytia 196697
Lanate Oreocereus 273875
Lanate Russianthistle 344609
Lanate Saussurea 348197
Lanate Serpentroot 354831
Lanatehead Saussurea 348465
Lanatous Calophanoides 67818
Lancang Albizzia 13589
Lancang Aralia 30694
Lancang Begonia 49994
Lancang Bellflower 70167
Lancang Buckeye 9712
Lancang Cliffbean 254743
Lancang Climbing Apple Bamboo 249606
Lancang Conehead 320131
Lancang Elaeocarpus 142353
Lancang Firethorn 322470
Lancang Larkspur 124639
Lancang Lilyturf 272101
Lancang Loxostigma 238040
Lancang Meadowrue 388547
Lancang Millettia 254743
Lancang Mucilagefruit 101253
Lancang Oak 324073
Lancang Onosma 271808
Lancang Pearbamboovine 249606
Lancang Podocarpium 200726
Lancang Premna 313687
Lancang Sandwort 31996
Lancang Saurauia 347965
Lancang Sedge 75046
Lancang Senna 78362
Lancang Stringbush 414160
Lancang Tanoak 233314
Lancang Waxplant 252998
Lancangjian Toxocarpus 393548
Lancangjiang Cayratia 79895
Lancangjiang Loxostigma 238040
Lancangjiang Notopterygium 267151
Lancangjiang Spapweed 205248
Lancashire Asphodel 262583
Lancaster St. John's Wort 201981
Lance Asiabell 98343
Lance Herminium 192851
Lance Medinilla 247579
Lance Self-heal 316146
Lance Tree 235521
Lancea 220794
Lancealeaf Leucas 227620
Lance-bract Kopsia 217867
Lance-cleft Cinquefoil 312701
Lancedentate Paraphlomis 283633
Lance-fruited Oval Sedge 76207
Lancelate Ragweed 19150
Lanceleaf Anneslea 25779
Lanceleaf Arnica 34728
Lanceleaf Arthraxon 36673
Lanceleaf Azara 45954
Lanceleaf Bennettiodendron 51033
Lance-leaf Buckthorn 328755
Lanceleaf Candytuft 203216
Lanceleaf Chastetree 411219
Lanceleaf Coreopsis 104531
Lanceleaf Cottonwood 311142
Lance-leaf Crotalaria 112307
Lanceleaf Eightangle 204544
Lance-leaf Figwort 355165
Lance-leaf Fog-fruit 296118
Lance-leaf Gee-bung 292028
Lance-leaf Goldenrod 368201
Lance-leaf Goldenweed 323032
Lanceleaf Greenbrier 366358
Lanceleaf Greenstar 307505
Lanceleaf Hackberry 80601
Lanceleaf Halerpestes 184709
Lanceleaf Hygrophila 200652
Lanceleaf Illicium 204544
Lanceleaf Illigera 204631
Lanceleaf Kalimeris 215351
Lanceleaf Kitefruit 196834
Lance-leaf Lasianthus 222208
Lanceleaf Lemongrass 117199
Lanceleaf Lily 228587,230058
Lanceleaf Litse 233967
Lanceleaf Marguerite 32562
Lanceleaf Mask Flower 17476
Lanceleaf Meconopsis 247143
Lanceleaf Mockorange 294510
Lanceleaf Pentapanax 289646
Lance-leaf Persimmon 132248
Lanceleaf Piptanthus 300677,300691
Lanceleaf Poplar 311142
Lanceleaf Ragweed 19150
Lanceleaf Rattlebox 112307
Lance-leaf Roughleaf 222208
Lance-leaf Sage 345153,345340
Lance-leaf SandalwoodPlum Wood 346212
Lanceleaf Soda Buttercup 184709
Lanceleaf Stachyurus 373573
Lanceleaf Sterculia 376129
Lanceleaf Stringbush 414201
Lanceleaf Sumac 332686
Lanceleaf Sweetleaf 381279
Lanceleaf Tarenna 384974
Lanceleaf Tatarian Statice 230789
Lanceleaf Thermopsis 389533
Lanceleaf Thoroughwort 158191
Lanceleaf Thorow Wax 63699
Lance-leaf Tickseed 104531,104532
Lance-leaf Tiger Lily 230058
Lanceleaf Viburnum 407906
Lanceleaf Vineclethra 95511
Lanceleaf Water Plantain 14745
Lanceleaf Waxplant 198847
Lance-leaf Wild Buckwheat 152203
Lanceleaf Wildsenna 389533
Lanceleaf Wingseedtree 320840
Lance-leafed Waxberry 261233
Lance-leaved Anemone 23873
Lance-leaved Anise-tree 204544
Lanceleaved Arrowhead 342359
Lance-leaved Aster 380912
Lance-leaved Basketvine 9442

Lance-leaved Buckthorn 328755,328756
Lance-leaved Chaste-tree 411219
Lance-leaved Eurya 160526
Lance-leaved Goldenrod 368134
Lance-leaved Greenstar 307505
Lance-leaved Ground-cherry 297740
Lance-leaved Hamilton Spindle-tree 157561
Lance-leaved Lily 230036
Lance-leaved Litse 233967
Lance-leaved Loosestrife 239679,239703
Lance-leaved Parrya 284994
Lance-leaved Pentapanax 289646
Lance-leaved Piptanthus 300677
Lance-leaved Plantain 302034
Lance-leaved Reevesia 327364
Lance-leaved Sage 345340
Lance-leaved Shibataea 362136
Lance-leaved Steironema 374667
Lance-leaved Sterculia 376129
Lance-leaved Thistle 92112
Lance-leaved Tickseed 104531
Lance-leaved Trillium 397576
Lance-leaved Violet 410166
Lance-leaved Wild Licorice 170457
Lance-leaved Wing-seed-tree 320840
Lance-leaved-tickseed 104531
Lance-leved Gambirplant 401766
Lancelike Spike Rush 143187
Lancelimbate Basketvine 9442
Lanceolare-leaved Mahonia 242575
Lanceolate Actinidia 6647
Lanceolate Anogeissus 26022
Lanceolate Anzacwood 310767
Lanceolate Barberry 52302
Lanceolate Beautyberry 66844
Lanceolate Birdcherry 280030
Lanceolate Blumea 55769
Lanceolate Brome 60797
Lanceolate Bromegrass 60797
Lanceolate Callery Pear 323124
Lanceolate Caper 71779
Lanceolate Cavea Rockfoil 349161
Lanceolate Chinese Stachyurus 373534
Lanceolate Clematis 95065
Lanceolate Clovershrub 70836
Lanceolate Cuckoo-orchis 110615
Lanceolate Cymbidium 116938
Lanceolate Dendrolobium 125577
Lanceolate Dichotomous Star Wort 374841
Lanceolate Draba 137069
Lanceolate Dualbutterfly 398285
Lanceolate Dubyaea 138774
Lanceolate Elaeagnus 142044
Lanceolate Elaeocarpus 142351
Lanceolate Eutrema 161133
Lanceolate Fake Woodbean 125577
Lanceolate Figwort 355165
Lanceolate Forkliporchis 269033
Lanceolate Fritillary 168444
Lanceolate Glochidion 177150
Lanceolate Greenvine 204631
Lanceolate Herminium 192875
Lanceolate Hiptage 196834
Lanceolate Jasmine 211884
Lanceolate Kalanchoe 215106
Lanceolate Kauri 10497
Lanceolate Kauri Pine 10497
Lanceolate Kiwifruit 6647
Lanceolate Lasianthus 222209
Lanceolate Leptodermis 226096
Lanceolate Lily 230001
Lanceolate Longpetiole Hydrangea 199948
Lanceolate Madder 337910
Lanceolate Maple 3082
Lanceolate Memecylon 249989
Lanceolate Microstegium 254025
Lanceolate Monbeig Deutzia 127017
Lanceolate Nanmu 295373
Lanceolate Oak 324093
Lanceolate Oligostachyum 270436
Lanceolate Orange Rockfoil 349087
Lanceolate Osmanther 276344
Lanceolate Osmanthus 276344
Lanceolate Palaktree 280445
Lanceolate Paris 284342
Lanceolate Parnassia 284552
Lanceolate Pentas 289831
Lanceolate Phoebe 295373
Lanceolate Piratebush 61935
Lanceolate Platanthera 302385
Lanceolate Poikilospermum 307042
Lanceolate Purplepearl 66844
Lanceolate Raphiolepis 329095
Lanceolate Rattlebox 112307
Lanceolate Red Ash 168065
Lanceolate Rhaphidophora 329003
Lanceolate Screwtree 190106
Lanceolate Screw-tree 190106
Lanceolate Sedge 75048
Lanceolate Spindle-tree 157663
Lanceolate Swallowwort 117698
Lanceolate Tarenna 384974
Lanceolate Tripterospermum 398285
Lanceolate Violet 410166
Lanceolate Waterplantain 14745
Lanceolate Whitlowgrass 137069
Lanceolate Wildclove 226096
Lanceolate Wrinkleleaf Cinquefoil 312352
Lanceolatecalyx Camellia 69253
Lanceolatecalyx Tea 69253
Lanceolated Milkwort 37979
Lanceolate-foliolate Grape 411772
Lanceolateleaf Arrowwood 407906
Lanceolateleaf Basketvine 9442
Lanceolateleaf Camellia 69253
Lanceolateleaf Dock 340209
Lanceolateleaf Gardneria 171451
Lanceolate-leaf Glochidion 177150
Lanceolateleaf Hornbeam 77336
Lanceolateleaf Mazus 247001
Lanceolateleaf Pentapanax 289646
Lanceolate-leaf Phoebe 295373
Lanceolateleaf Smilax 366414
Lanceolate-leaf Spathiphyllum 370344
Lanceolateleaf Tibet Lyonia 239398
Lanceolateleaf Windhairdaisy 348807
Lanceolate-leaved Camellia 69255
Lanceolate-leaved Gardneria 171451
Lanceolate-leaved Greenbrier 366414
Lanceolate-leaved Poplar 311363
Lanceolate-leaved Sweetleaf 381279
Lanceolate-leaved Viburnum 407906
Lanceolatelobe Thistle 92057
Lanceolate-petal Rockfoil 349907
Lanceolatetooth Bethlehemsage 283633
Lanceoleaf Bugbane 91002
Lanceoleaf Camellia 69252
Lanceo-leaf Ceropegia 84079
Lanceoleaf Cinchona 91074
Lanceoleaf Cuphea 114609
Lanceoleaf Cupri-coloured Clethra 96530
Lanceoleaf Euonymus 157862
Lanceoleaf Frangipani 305215
Lanceoleaf Kangding Willow 343835
Lanceoleaf Northern Bedstraw 170269
Lanceoleaf Ormosia 274407
Lanceoleaf Reevesia 327364
Lanceoleaf Sandwort 31997
Lanceoleaf Trigonostemon 397374
Lanceolobed Rustyhair Raspberry 339164
Lancepod 235513
Lance-sepal Camellia 69297
Lance-sepaled Camellia 69297
Lancewood 138879,263050,278159,318063
Lancileaf Fewflower Lysionotus 239971
Lanci-leaf Poplar 311363
Lancinate Yushanbamboo 416781
Lanciniate Cinquefoil 312701
Lancipetal Adinandra 8246
Lanci-petaled Adinandra 8246
Lanci-scaled Sedge 75058
Lancisepal Spapweed 205065
Land Cress 47963,47964,72679,72802
Land Stonecrop 356768
Land Whin 271563
Landpick 308493
Landsberg Houlletia 198687
Laney's Hickory 77875,77906
Lang De Beef 298584
Lang Jurinea 211887
Langao Aster 40699

Lange Loose-flowered Barberry 51836
Lange's Spiraea 371850
Langgong Azalea 332004
Langgong Rhododendron 332004
Langkazi Buttercup 326132
Langkok Fig 165231
Langley Beef 298584
Langong Snowbell 379400
Langsat 221224,221226,221789
Langsdorff Small Reed 65473
Langsdorff Smallreed 65473
Langshan Chirita 87893
Langshan Crazyweed 278946
Langshan Feathergrass 376786
Langshan Seseli 361514
Langtaodian Flatutricle Sedge 74188
Langwort 405571
Langxian Corydalis 106349
Langxian Milkvetch 42770
Langxian Rockfoil 349690
Langya Elm 401481
Langya Mountain Elm 401481
Langya Solomonseal 308578
Langyan Stonecrop 356900
Langyashan Rose 336681
Langyu Elm 401581
Langzi Larkspur 124398
Lanhan Hornbeam 77381
Laniculate Gentian 173572
Laniferous Leptospermum 226462
Lanigera Grevill 180610
Lanise Cactus 244124
Lankao Pecteilis 285958
Lankong Catchfly 363665
Lankong Lily 229835,229885
Lankong Sage 345155
Lannea 221130
Lanoschin Woodbetony 287340
Lanose Fleabane 150744
Lanose Gentian 150744
Lanose Photinia 295720
Lanose Saurauia 347961
Lanose Thistle 92111
Lanose Tritonia 399091
Lanose Willow 343594
Lanoseovary Azalea 330675
Lanping Arrowbamboo 162721
Lanping Elaeagnus 142053
Lanping Honeysuckle 235949
Lanping Maple 3083
Lanping Mountainash 369445
Lanping Woodbetony 287337
Lanshan Yushanbamboo 416783
Lantana 221229,221238,221241,221244,
221272
Lantanaphyllum Viburnum 407660
Lantern Agapetes 10322
Lantern Dogwood 57695
Lantern Fruit 297711
Lantern Seedbox 238188
Lantern Tree 111145,111146
Lantern Wildginger 37651
Lantern-leaves 326287
Lanterns 77256
Lantern-tree 111146
Lanternweed 297678
Lanthorn Lily 262441
Lantsang Millettia 254743
Lantsang Waxplant 252998
Lanuginose Broomrape 275113
Lanuginose Clematis 95067
Lanuginose Crazyweed 278947
Lanuginose Fevervine 280083
Lanuginose Inula 138888
Lanuginose Rockjasmine 23211
Lanuginous Buttercup 326018
Lanveleaf Clinopodium 97021
Lanyu Addermonth Orchid 110637
Lanyu Albizia 13653
Lanyu Alderleaf Raspberry 338435
Lanyu Alyxia 18504
Lanyu Ardisia 31519
Lanyu Beautyleaf 67853
Lanyu Begonia 49827
Lanyu Boerlagiodendron 56530
Lanyu Bogorchis 110637
Lanyu Canthium 71381
Lanyu Caper 71779
Lanyu China-laurel 28402
Lanyu Chisocheton 88122
Lanyu Cinnamon 91357
Lanyu Cogflower 154192,382822
Lanyu Elaeocarpus 142402
Lanyu Eugenia 382664,382672
Lanyu Euonymus 157539
Lanyu Excoecaria 161666
Lanyu Fig 165442
Lanyu Fig-tree 165819
Lanyu Garcinia 171128
Lanyu Gymnosporia 182687
Lanyu Hackberry 80709
Lanyu Holly 203952
Lanyu Indiapoon Beautyleaf 67859
Lanyu Jasminorange 260163,260173
Lanyu Jasmin-orange 260163
Lanyu Jewelvine 283282
Lanyu Litse 233920
Lanyu Litsea 233920
Lanyu Long-leaved Beauty-berry 66840
Lanyu Lysionotus 239968
Lanyu Maesa 241795
Lanyu Maoutia 244895
Lanyu Memecylon 249892
Lanyu Mucuna 259541
Lanyu Nagi 306419
Lanyu Neolitsea 264112
Lanyu Nervilia 265406
Lanyu Newlitse 264112
Lanyu Ninenode 319477
Lanyu Pencilwood 139641
Lanyu Pencil-wood 139639
Lanyu Pepper 300335
Lanyu Persimmon 132241
Lanyu Pinanga 299679
Lanyu Pinangapalm 299679
Lanyu Pinanga-palm 299679
Lanyu Pittosporum 301340
Lanyu Planchonella 301771
Lanyu Podocarpus 306419
Lanyu Pouteria 301771
Lanyu Psychotria 319477
Lanyu Pterospermum 320843
Lanyu Raspberry 338435
Lanyu Rattan Palm 65793
Lanyu Rattlesnake Plantain 179721
Lanyu Rockvine 387796
Lanyu Snowbell 379357
Lanyu Spotleaf-orchis 179721
Lanyu Starjasmine 393616
Lanyu Sterculia 376092
Lanyu Swallowwort 117555
Lanyu Sweetleaf 381149
Lanyu Tabernaemontana 327577
Lanyu Taroorchis 265406
Lanyu Timonius 392088
Lanyu Tylophora 400945
Lanyu Velvetplant 183086
Lanyu Yaccatree 306419
Lanyu Yam 131586
Lanzhou Cistanche 93061
Lanzhou Lily 228588,229830
Lao Ginger 417994
Lao Zinger 417994
Laojunshan Azalea 331055
Laojunshan Barberry 51831
Laojunshan Michelia 252828
Laojunshan Rhododendron 331055
Laojunshan Spapweed 205066
Laos Camellia 69612
Laos Caryodaphnopsis 77963
Laos Cylindrokelupha 116601
Laos Elaeocarpus 142356
Laos Euchresta 155894
Laos Homalium 197653
Laos Mitrephora 256227
Laos Mytilaria 261885
Laos Staff-tree 80225
Laos Syzygium 382594
Laos Tanoak 233286
Laoshan Buckthorn 328757
Laoshan Deutzia 126931

Laoshan Ophiorrhiza 272238
Laoshan Pear 323328
Laovage 228244,228250
Laoyu Maple 3089
Laoyu Sedge 74648
Lapacho 382704,382706,382718
Lapachol 382704,382718
Lapanshan Sandwort 32003
Lapegrass 404531
Lapidicolous Narrow-leaved Rhododendron 330108
Laping Pennisetum 289055
Lapland Azalea 331057
Lapland Buttercup 326020
Lapland Cornel 105204
Lapland Dandelion 384620
Lapland Diapensia 127916
Lapland Marsh Orchid 121394
Lapland Mountain Sorrel 340103
Lapland Pedicularis 287335,287338
Lapland Pine 300223
Lapland Rhododendron 331057
Lapland Rose Bay 331057
Lapland Rosebay 331057
Lapland Rose-bay 331057
Lapland Sedge 75062
Lapland Sorrel 340103
Lapland Willow 343595
Lapleaf Creeping Mahonia 242629
Laplove 102933,162519
Lapp Heather 297019
Lapper Gowan 399504
Lapper-gowan 399504
Lappon Crazyweed 278948
Lapsana 221787
Lapulan Milkvetch 42582
Lapweed 162519
Laramie Chickensage 371151
Laramie Sourock 339887,339897
Laran 263983
Larch 221829,221855,221904
Larch-leaf Minuartia 255505
Larchleaf Russianthistle 344611
Larchleaf Sandwort 32004
Larch-leaved Russianthistle 344611
Lardfruit 289475,289476
Lardizabala 221815
Lardizabala Family 221818
Lareabell 188908
Largancillo 1148
Large Aglaia 11291
Large Alfalfa Dodder 115047
Large Anemone Clematis 95132
Large Angularfruit Iris 208508
Large Anise-tree 204553
Large Apple Mint 250459
Large Autumn-crocus 99304
Large Banksia 47644
Large Barnyard Grass 140367
Large Beard-tongue 289349
Large Beggarticks 53913
Large Bellwort 403670
Large Berried Juniper 213843
Large Bilberry 404027
Large Bindweed 68732
Large Bird's-foot Trefoil 237713
Large Bird's-foot-trefoil 237713
Large Bittercress 72679
Large Bitter-cress 72679
Large Bladder Ground Cherry 297745
Large Blazing Stars 228519
Large Blue Alkanet 21949
Large Blue Hair Grass 217473
Large Blue Lobelia 234813
Large Buds Rhododendron 330756
Large Bullwort 19664
Large Button-snakeroot 228519
Large Buttonweed 131361
Large Calyx Rhododendron 330590,331212
Large Camas 68803
Large Chessbean 116595
Large Chinese Hawthorn 109936
Large Clammy-weed 307143
Large Compact Sawgrass 245378
Large Coneflower 339585
Large Coralroot 103990
Large Crab-grass 130745
Large Cranberry 403894
Large Cuckoo-pint 37009
Large Dahurian Bushclover 226751
Large Dicky Daisy 227533
Large Drimycarpus 138044
Large Duckweed 221007
Large Elsholtzia 144076
Large Evening Primrose 269441
Large False Solomon's-seal 242699
Large Feathergrass 376795
Large Field Gentian 174133
Large field Mouse-ear Chickweed 83075
Large Field Speedwell 407287
Large Flamingo Flower 28094
Large Flower Actinidia 6623
Large Flower Atylosia 65150
Large Flower Deutzia 126956
Large Flower Koeleria 217494
Large Flower-of-an-hour 383304
Large Forget-me-not 260868
Large Fothergilla 167541
Large Fruit Fig 165297
Large Fruit Hodgsonia 197072
Large Gardenia 171393
Large Grass-leaved Rush 212901
Large Gray Willow 343060,343221
Large Head Aster 40836
Large Head Groundsel 359461
Large Headed Tickseed 104509
Large Hedge Nettle 170076
Large Hemp Nettle 170076
Large Hop Clover 396854
Large Hop Trefoil 397096
Large Hop-clover 396854
Large Hoptrefoil 397096
Large Houstonia 198717
Large Indian Cress 399601
Large Iris 208587
Large Jadeleaf and Goldenflower 260454
Large Juniper 213841
Large Kangaroo Apple 367289
Large Kingidium 293598
Large Lady Palm 329176
Large Leaf Aster 40817
Large Lilyturf 272080
Large Lineolate Elatostema 142719
Large Magnolia 242135
Large Marsh-pink 341469
Large Merry-bell 403670
Large Moonflower 67736
Large Mouse-ear 232483
Large Naiad 262067
Large Needlegrass 376795
Large Ox-eye 227512
Large Periwinkle 409335
Large Pheasant's Eye 8362
Large Pheasant's Eyes 8362
Large Phoebe 295379
Large Pink Nerine 265259
Large Plantain Lily 198639
Large Poroporo 367289
Large Purple Fringed Orchid 302324
Large Quaking Grass 60379
Large Quaking-grass 60379
Large Rampion 269404
Large Reedbentgrass 65419
Large Rhinanthus-like Woodbetony 287617
Large Rhododendron 330111,330789,331212
Large Roegneria 335336
Large Round Kumqat 167493
Large Round-leaf Orchid 302472
Large Round-leaved Orchid 302401,302472
Large Sahmarsh Aster 380999
Large Sand Catchfly 363368
Large Scrub Oak 324556
Large Seed Wirelettuce 375968
Large Selfheal 316104
Large Self-heal 316104
Large Shortfilament Flower 221412
Large Silverleaf Cotoneaster 107616
Large Skullcap 355519
Large Speedwell 407018
Large Stemona 375355
Large Stonecrop 357203

Large Strict Roegneria 335517
Large Sweet Basil 268438
Large Terebinth 300977
Large Thickroot Iris 208898
Large Thyme 391346
Large Toadshade 397551
Large Tongoloa 392787
Large Tongue Orchid 113862
Large Tooth Cress 125852
Large Toothwort 72895,125852
Large Trefoil 397096
Large Twayblade 232225
Large Vanilla 405027
Large Visnaga 140126
Large Wand-flower 369996
Large Water Starwort 67366
Large Water-starwort 67366
Large White Groundcherry 227932
Large White Helleborine 82038
Large White Petunia 292751
Large White Plantain Lily 198639
Large White Trillium 397570
Large White-flowered Anemone 24021
Large Whorled 210336
Large Whorled Pogonia 210336
Large Whorled Pogonia Orchid 210336
Large Wild Thyme 391346
Large Woodreed 65419
Large Yellow Foxglove 130348,130365
Large Yellow Lady's Slipper 120310
Large Yellow Lady's-slipper 120432
Large Yellow Loosestrife 239804
Large Yellow Moraea 258650
Large Yellow Pond Lily 267645
Large Yellow Rest Harrow 271464
Large Yellow Restharrow 271464
Large Yellow Sedge 74557
Large Yellow Stonecrop 357114
Large Yellow Vetch 408419
Large Yellow Wood Sorrel 277877
Large Yellowflag Iris 208773
Largeangular Trichocereus 395645
Large-anther Bentgrass 12178
Largeanther Senna 360490
Large-anthered Senna 360490
Largeawn Roegneria 335413
Largebract Didissandra 129795
Largebract Eargrass 187524
Largebract Flemingia 166897
Largebract Hedyotis 187524
Large-bract Helitrope 190673
Large-bract Indigo 205739
Largebract Iris 208476
Largebract Liparis 232171
Largebract Marsdenia 245828
Largebract Murdannia 260090
Largebract Rhodiola 329837
Largebract Saltbush 44349
Largebract Spapweed 204796
Largebract Swordflag 208476
Largebract Thorowax 63622
Largebract Touch-me-not 204796
Largebract Tubergourd 390119
Largebract Twayblade 232171
Largebract Waterbamboo 260090
Largebract Willow 343927
Large-bracteate Barberry 51678
Large-bracted Plantain 301868
Large-bracted Tick-trefoil 126301,126302
Large-bracted Vervain 405815
Largebud Azalea 330756
Largebugel Azalea 330666
Largecalyx Azalea 331212
Largecalyx Bluegrass 305686
Largecalyx Breynia 60053
Largecalyx Bubbleweed 297879
Largecalyx Cinquefoil 312467
Largecalyx Nanmu 295382
Large-calyx Phoebe 295382
Largecalyx Physochlaina 297879
Large-calyx Sweetleaf 381423
Largecalyx Tea 68921
Largecalyx Trichodesma 395736
Large-calyxed Brunfelsia 61297,61310
Large-calyxed Phoebe 295382
Large-calyxed Rhododendron 330590
Large-cluster Plantainlily 198612
Large-cone Douglas Fir 318579
Large-coned Douglas Fir 318579
Largedentate Ainsliaea 12647
Largedentate Meadowrue 388525
Largedentate Rabbiten-wind 12647
Large-eaered Quaking Grass 60379
Large-eaf Flemingia 166880
Largeear Bamboo 47331
Largeear Windhairdaisy 348506
Large-eye Bambusa 47266
Large-eyed Bambusa 47266
Largefibre Torpalm 181805
Largeflower Aletris 14513
Largeflower Amitostigma 19526
Largeflower Arrowwood 407876
Largeflower Arundinella 37378
Largeflower Aspidistra 39578
Large-flower Aucuba 44909
Largeflower Basketvine 9462
Large-flower Basketvine 9462
Largeflower Beautiful Lespedeza 226801
Largeflower Bellcalyxwort 231259
Largeflower Bellwort 403670
Largeflower Bleedingheart 203357
Largeflower Broomrape 275133
Large-flower Bucket Orchid 105521
Largeflower Buckthorn 328718
Largeflower Butterflybush 62000
Largeflower Caltrop 395082
Largeflower Canna 71173
Largeflower Carissa 76926
Largeflower Carrionflower 373827
Largeflower Caudate Rose 336472
Largeflower Cephalanthera 82038
Largeflower Chinese Iris 208670
Largeflower Chinese Spiraea 371886
Largeflower Coneflower 339552
Largeflower Corydalis 106121
Largeflower Cryptostegia 113823
Largeflower Curvedstamencress 238015
Largeflower Delavay Megacarpaea 247774
Largeflower Deltoialeaf Eutrema 161127
Largeflower Dendrobium 125366
Largeflower Deutzia 126956
Largeflower Dragonhead 137591
Largeflower Dwarf Mitrasacme 256163
Largeflower Eargrass 187506
Large-flower Echinodorus 140544
Largeflower Epimedium 146995
Largeflower Euonymus 157547
Largeflower Fairybells 134452
Largeflower Falsehellebore 405598
Largeflower Fameflower 383304
Largeflower Farrer Asiabell 98313
Largeflower Fawnlily 154915
Largeflower Flax 231907
Largeflower Gardenia 171333
Largeflower Gentian 173951
Largeflower Gentianopsis 174218
Largeflower Gladiolus 176314
Large-flower Goldenweed 323014
Largeflower Goniothalamus 179412
Largeflower Greenstar 307919
Large-flower Gurjun 133563
Largeflower Habenaria 183828
Largeflower Hawksbeard 110934
Largeflower Heartleaf 194585
Largeflower Heterostemma 194160
Largeflower Holboellia 197229
Largeflower Honeysuckle 235935
Largeflower Hoodia 198044
Largeflower Hoodorchis 264762
Largeflower Hoodshaped Orchid 264762
Largeflower Hygrophila 200628
Largeflower Incarvillea 205571
Largeflower Jerusalemsage 295144
Largeflower Kaempfer Taxillus 385203
Largeflower Kobresia 217216
Largeflower Lacquertree 393455
Largeflower Laelia 219681
Largeflower Lagochilus 220066
Largeflower Larkspur 124237
Largeflower Leek 15446
Largeflower Leptospermum 226457

Largeflower Lilyturf 272114
Largeflower Listera 264752
Largeflower Luculia 238101
Largeflower Magnific Deutzia 127009
Largeflower Mamdaisy 246315
Largeflower Manuka Tea-tree 226499
Largeflower Maxillaria 246713,246723
Largeflower Meadowrue 388526
Largeflower Meconopsis 247126
Largeflower Melandrium 363519
Largeflower Melic 249015
Largeflower Mexican Clover 334323
Largeflower Microula 254362
Largeflower Mischobulbum 383157
Largeflower Montanoa 258201
Largeflower Morina 258863
Largeflower Morningstar Lily 229816
Largeflower Motherwort 224996
Largeflower Mouseear Chickweed 82819
Largeflower Narrowwing Milkvetch 42276
Largeflower Oncidium 270827
Largeflower Onion 15446
Largeflower Panicled Hydrangea 200055
Largeflower Pansy Orchid 254929
Largeflower Pashi Pear 323252
Largeflower Pencilwood 139669
Largeflower Perotis 291416
Largeflower Petrocosmea 292557
Largeflower Phaius 293545
Largeflower Pleione 304257
Largeflower Pottsia 313227
Largeflower Pretty Crocus 111606
Largeflower Pricklyash 417261
Largeflower Pricklythrift 2300
Largeflower Primrose Willow 238172
Large-flower Primrose-willow 238172
Largeflower Purslane 311852
Largeflower Queen-of-night 83874
Largeflower Rockcress 30198
Large-flower Sandwort 255513
Largeflower Sansevieria 346089
Large-flower Saurauia 347981
Largeflower Scabious 350272
Largeflower Shrubalthea 195296
Largeflower Skeletonplant 239329
Largeflower Small Reed 127271
Largeflower Smallreed 127271
Largeflower Sobralia 366786
Largeflower Solomonseal 308592
Largeflower Sotvine 194160
Largeflower Spapweed 204773
Largeflower Spurgentian 184697
Largeflower Stanhopea 373691
Large-flower Star Glory 323462
Largeflower Stinkinggrass 249015
Largeflower Stonebutterfly 292557
Largeflower Stork's Bill 153921
Largeflower Tainia 383157
Largeflower Tavaresia 385151
Largeflower Theligon 389187
Largeflower Theligonum 389187
Largeflower Touch-me-not 204776
Largeflower Trefoil 396893
Largeflower Triplostegia 398025
Largeflower Tulip Orchid 25136
Largeflower Twayblade 232139
Largeflower Twinginseng 398025
Largeflower Uvaria 403481
Largeflower Vanda 404629
Largeflower Viburnum 407876
Largeflower Waxmallow 243944
Largeflower Wildginger 37674
Largeflower Woodbetony 287259,287417
Largeflower Woodsorrel 277702
Large-flower Yili Milkvetch 42517
Largeflower Yunnan Corydalis 106614
Large-flowered Abelia 135
Large-flowered Andromeda 22457
Large-flowered Aster 380900
Large-flowered Beardtongue 289349
Large-flowered Beard-tongue 289349
Large-flowered Bellflower 70218
Large-flowered Bellwort 403670
Large-flowered Butterfly-bush 62000
Large-flowered Butterwort 299744
Large-flowered Calamint 65581
Large-flowered Catchfly 363438
Large-flowered Combretum 100495
Large-flowered Cut-leaf Evening-primrose 269446
Large-flowered Dendrocalamus 125482
Large-flowered Epimedium 146995
Large-flowered Evening-primrose 269443
Large-flowered Fairy-bells 315530
Large-flowered Fuchsia 168737
Large-flowered Gaura 172202
Large-flowered Ground-cherry 227932
Large-flowered Hedge Nettle 170076
Largeflowered Hedyotis 187506
Large-flowered Hemp Nettle 170076
Large-flowered Hemp-nettle 170076
Large-flowered Honeysuckle 235935
Large-flowered Indian-lettuce 94344
Large-flowered Leafcup 310035
Largeflowered Lychnis 363235
Large-flowered Meadow-grass 305506
Large-flowered Mullein 405788,405796
Large-flowered Neillia 263151
Large-flowered Ochna 268149
Large-flowered Orlaya 274288
Large-flowered Pink-sorrel 277776,277787
Largeflowered Purslane 311852
Largeflowered Sandwort 31934
Large-flowered St. John's Wort 201787
Large-flowered Star-tulip 67634
Large-flowered Storksbill 153921
Largeflowered Styrax 379402
Largeflowered Tavaresia 385151
Large-flowered Tick Clover 126366
Large-flowered Tickseed 104509,104510
Large-flowered Tick-trefoil 126371
Large-flowered Trillium 397570
Largeflowered Trisetum 398547
Large-flowered Triteleia 398789
Large-flowered Uvaria 403481
Large-flowered Vetch 222724
Large-flowered Viburnum 407876
Large-flowered Waterweed 141568,141569
Large-flowered Willowherb 146724
Large-flowered Yellow False Foxglove 45169,45171
Large-flowered Yellow Oleander 389983
Largeflower-like Honeysuckle 235941
Largefoottube Eria 299625
Largefoottube Hairorchis 299625
Largefruit Amaranth 18701
Largefruit Azalea 330763
Largefruit Buttercup 326067
Largefruit Camellia 69336
Largefruit Coral Holly 203654
Large-fruit Cranberry 403894
Largefruit Cranberryf 278366
Largefruit Dendropanax 125598
Largefruit Dragonplum 137718
Largefruit Eleutharrhena 143558
Largefruit Evening Primrose 269464
Largefruit Fairy Lantern 67601
Largefruit Groundcherry 297699
Largefruit Holly 204001
Largefruit Iodes 207356
Large-fruit Ivary Palm 298050
Large-fruit Mucuna 259535
Largefruit Murdannia 260107
Largefruit Mycetia 260636
Largefruit Oak 116186
Largefruit Oilresiduefruit 197072
Large-fruit Pachira 279387
Largefruit Pedilanthus 287850
Largefruit Pygeum 322401
Largefruit Rehder Tree 327418
Largefruit Rehdertree 327418
Largefruit Rockjasmine 23227
Largefruit Sedge 75284
Largefruit Smalscale Sedge 74642
Largefruit Syrenia 382104
Largefruit Ternstroemia 386703
Large-fruit Thorn 109970
Largefruit Treerenshen 125598
Largefruit Waterbamboo 260107
Large-fruited Black Snakeroot 346007
Large-fruited Buffalo Currant 334128

Large-fruited Bushwillow 100867
Large-fruited Iodes 207356
Large-fruited Malabar-chestnut 279387
Large-fruited Mallee 155712
Large-fruited Red Mahoany 155689
Large-fruited Rose-apple 382606
Large-fruited Sand-verbena 703
Large-fruited Sandwort 255514
Large-fruited Star Sedge 74423
Large-fruited Tea-tree 226464
Large-fruited Ternstroemia 386703
Large-fruited Tsarong Barberry 52281
Large-fruited Whitebeam 369457
Largeglume Bluegrass 305693
Largeglume Goosecomb 335333
Largeglume Roegneria 335333
Largehairyfruit Eurya 160545
Large-hairy-fruited Eurya 160545
Largehead Atractylodes 44218
Largehead Blumea 55783
Largehead Celery Wormwood 35311
Largehead Sinacalia 364521
Largehead Sweet Sagebrush 35134
Largehead Sweet Wormwood 35134
Largehead Windhairdaisy 348163
Largehead Wormwood 35859
Large-headed Rush 213275
Large-headed Sedge 74076
Largeinflorescence Butterflybush 62114
Largekeel Vetch 408484
Largeleaf Adenosma 7987
Largeleaf Adinandra 8255
Largeleaf Aeonium 9076
Largeleaf Alfafa-like Rattlebox 112397
Largeleaf Alliaria 275869
Largeleaf Allophylus 16060
Largeleaf Anemone 23945
Largeleaf Aphanamixis 28996
Large-leaf Appressed-hair Lasianthus 222090
Largeleaf Ardisia 31412
Largeleaf Ash 168086
Large-leaf Australian wild lime 253290
Largeleaf Avens 175426
Largeleaf Beancaper 418701
Largeleaf Beautyberry 66864
Largeleaf Birthwort 34219
Largeleaf Bittercress 72876
Largeleaf Blackberry 338779
Largeleaf Box 64284
Largeleaf Brunnera 61366
Largeleaf Bubbleweed 297880
Largeleaf Buckeye 9715
Largeleaf Bulbophyllum 62555
Largeleaf Buttercup 325915
Largeleaf Cacalia 283812
Largeleaf Caesalpinia 65033
Largeleaf Canthium 71498
Large-leaf Chinese Ash 168086
Largeleaf Cleisostoma 94461
Largeleaf Closedspurorchis 94461
Largeleaf Clovershrub 70818,70867
Largeleaf Corydalis 106511
Largeleaf Cowlily 267296
Large-leaf Cryptocoryne 113535
Largeleaf Curculigo 114819
Largeleaf Curlylip-orchis 62555
Largeleaf Dendrobium 125245
Largeleaf Densevein 261325
Largeleaf Distichous Cotoneaster 107599
Largeleaf Distylium 134937
Largeleaf Dogwood 380461
Largeleaf Ehretia 141629
Largeleaf Epipactis 147176
Largeleaf Evergreenchinkapin 78987
Largeleaf Falselily 266901
Largeleaf Falsenettle 56206
Largeleaf Fasciated Haworthia 186431
Largeleaf Fig 165274,165698
Largeleaf Fourpetal Aglaia 19971
Largeleaf Fritsch Spiraea 371916
Largeleaf Gambirplant 401770
Largeleaf Garlicmustard 275869
Largeleaf Geigertree 104245
Largeleaf Gentian 173615
Largeleaf Globeflower 399508
Largeleaf Glochidion 177149
Largeleaf Griffith Bittercress 72792
Largeleaf Gymnema 182374
Largeleaf Henry Raspberry 338510
Largeleaf Hibiscus 194986
Largeleaf Hollyberry Cotoneaster 107366
Largeleaf Hookvine 401770
Large-leaf Hop-bush 135214
Largeleaf Horsechestnut 9715
Largeleaf Hydrangea 199956
Largeleaf Ironweed 406931
Largeleaf Japanese Beech 162370
Largeleaf Japanese Spindle-tree 157617
Largeleaf Jasminefruit 283994
Largeleaf Jerusalemsage 295142
Largeleaf Jurinea 207132
Largeleaf Kalanchoe 61568
Largeleaf Kauri 10500
Largeleaf Kauri Pine 10500
Largeleaf Lacquertree 393461
Largeleaf Landpick 308600
Largeleaf Leea 223946
Largeleaf Lepisanthes 225670
Largeleaf Licorice 177926
Largeleaf Lilyturf 272103
Largeleaf Linden 391788,391817
Largeleaf Litse 233891
Largeleaf Loropetal 237198
Largeleaf Loropetalum 237198
Largeleaf Lupine 238457
Largeleaf Lycaste 238747
Largeleaf Maesa 241800
Largeleaf Manglietia 244431,244455
Large-leaf Manuka Tea-tree 226498
Largeleaf Markingnut 357881
Largeleaf Marsh Pennywort 200265
Largeleaf Marshpennywort 200265
Largeleaf Microtropis 254310
Largeleaf Milkwort 308149
Largeleaf Mosquitoman 134937
Largeleaf Mustard 59438
Largeleaf Newlitse 264069
Largeleaf Oak 116120
Largeleaf Ophiorrhiza 272288
Largeleaf Orthoraphium 275613
Largeleaf Parastyrax 283994
Largeleaf Paxistima 286747
Largeleaf Peliosanthes 288645
Largeleaf Pendent Inflorescence Indigo 206379
Largeleaf Peony 280227
Largeleaf Pepper 300435
Largeleaf Photinia 295731
Largeleaf Physochlaina 297880
Largeleaf Pine 299931
Largeleaf Poacynum 29506
Large-leaf Podocarpus 306457
Largeleaf Pondweed 312044
Largeleaf Pouteria 313369
Largeleaf Primrose 314605
Largeleaf Rabdosia 209688,209769
Largeleaf Raphiolepis 329097
Largeleaf Rattlesnakeroot 313849
Largeleaf Rhaphidophora 329007,329009
Largeleaf Robiquetia 335039
Largeleaf Rockjasmine 23231
Largeleaf Rose 336725
Largeleaf Rosemallow 194986
Largeleaf Roupala 337678
Largeleaf Sansevieria 346170
Largeleaf Sibbaldia 362361
Large-leaf Small-flowered Sweet Spire 210407
Largeleaf Snakegrape 20420
Largeleaf Sobralia 366789
Largeleaf Solomonseal 308600
Largeleaf Spathoglottis 370397
Largeleaf Spicebush 231385
Largeleaf Statice 230672
Largeleaf Stink Pricklyash 417278
Largeleaf Supplejack 52429
Largeleaf Sweetspire 210397
Largeleaf Swollennoded Cane 87574
Largeleaf Tanoak 233233
Largeleaf Tea 69651
Largeleaf Tickclover 126462

Largeleaf Tickseed 104542
Largeleaf Trigonotis 397430
Largeleaf Vetch 408266,408551
Largeleaf Vineclethra 95515
Largeleaf Wendlandia 413813
Largeleaf Whiten Eriocycla 151748
Largeleaf Wildginger 37680
Largeleaf Wildmedia 28996
Largeleaf Windhairdaisy 348355
Large-leaf Wireweed 308679
Largeleaf Woodlotus 244431,244455
Large-leaf Yellow Teatree 226466
Largeleaf Yushan Dewberry 339188
Large-leafed Guarri 155957
Large-leafed Lime 391817
Large-leafed Rock Fig 164611
Large-leaved American Linden 391652
Large-leaved Antennaria 26402
Large-leaved Aster 40817
Large-leaved Avens 175426,175429
Large-leaved Bamboo Shrub 82446
Large-leaved Barberry 51902
Large-leaved Beggarticks 53913
Large-leaved Brunfelsia 61306
Large-leaved Canthium 71498
Large-leaved Crinum 111214
Large-leaved Cucumber Tree 242193
Largeleaved Cucumber-tree 242193
Large-leaved Cyclobalanopsis 116120
Large-leaved Distylium 134937
Largeleaved Dogwood 380461
Large-leaved Dogwood 380461
Large-leaved Elaeagnus 142075
Large-leaved Epipactis 147149
Large-leaved Germander 388275
Large-leaved Giant Solomon's-seal 308592
Large-leaved Goldenrod 368226,368233
Large-leaved Hollow Bamboo 82446
Large-leaved Holly 203961,204064
Large-leaved Ironweed 406931
Large-leaved Japanese Spindle-tree 157617
Large-leaved Kalanchoe 215150
Largeleaved Lime 391817
Large-leaved Lime 391817
Large-leaved Linden 391817
Large-leaved Lindera 231385
Large-leaved Lupine 238481
Large-leaved Machilus 240605
Large-leaved Maple 3127,3462
Large-leaved Meadowsweet 371795
Large-leaved Mucuna 259512
Large-leaved Mussaenda 260454
Large-leaved Nanmu 240605
Large-leaved Oak 116120
Large-leaved Orchis 183830
Large-leaved Osmanther 276370
Large-leaved Photinia 295731
Large-leaved Pondweed 312044
Large-leaved Raphiolepis 329097
Large-leaved Sandwort 32058
Large-leaved Shin-leaf 322823
Large-leaved Spice-bush 231385
Large-leaved Spindle 157668
Large-leaved Square-bamboo 87560
Large-leaved Sweetspire 210397
Large-leaved Swollennoded Cane 87574
Large-leaved Tickclover 126462
Largeleaved Tuftroot 130111
Large-leaved Vagrant Spindle-tree 157944
Large-leaved Variegated Bamboo 362124, 362132
Large-leaved Wild Indigo 47857
Large-leaves Tickclover 126359
Largelily 73154,73159
Largelip Germander 388127
Largelip Woodbetony 287419
Large-lobed Dandelion 384623
Largenode Bamboo 297336
Largepanicle Bentgrass 12192
Largepanicle Bluegrass 305723
Largepanicle Raspberry 338486
Large-panicled Raspberry 338486
Largepaniculate Raspberry 338486
Largepediole Holly 204008
Largepetal Onion 15447
Largepetal Purpleflower Sibbaldia 362370
Largepetiole Elaeocarpus 142330
Large-pod Pinweed 223699
Larger Bindweed 68713
Larger Blue Flag 208924
Larger Blue Iris 208924
Larger Bur-marigold 53976
Larger Canada St. John's-wort 202010
Larger Canadian St. John's-wort 202010
Larger Duckweed 372300
Larger Scented Orchid 182243
Larger Sunshade 68713
Larger Wintergreen 322872
Larger Yellow Lady's Slipper 120432
Largeraceme Ophiorrhiziphyllon 272311
Largeroot Orchis 116975
Largescale Chosenia 89160
Large-scaled Chosenia 89160
Largeseed Actinidia 6659
Largeseed Corydalis 106140
Large-seed False Flax 68860
Largeseed Goosefoot 87080
Large-seed Hemsleia 191943
Largeseed Jointfir 178566
Largeseed Kiwifruit 6659
Largeseed Snowgall 191943
Large-seeded Hawthorn 109823
Large-seeded Mercury 1829
Large-seeded Nodding Campion 364154
Largesepal Acomastylis 4994
Largesepal Adinandra 8232
Large-sepal Barberry 51897
Largesepal Cinquefoil 312740
Largesepal Pinnaflower 4994
Largesepal Rabdosia 209767
Largesepal St. John's Wort 202004
Largesepal Tubergourd 390134
Largesepal Whiteleaf Raspberry 338632
Largeserrate Mosla 259292
Large-serrete Windhairdaisy 348360
Largespike Bluegrass 306010
Large-spike Bristlegrass 361825
Largespike Garnotia 171511
Largespike Hornbeam 77420
Largespike Japan Lawngrass 418427
Large-spike Jointfir 178547
Largespike Lawngrass 418431
Largespike Millettia 254762
Largespike Ricegrass 300716
Largespike Sedge 75274,76029
Largespike Woodnettle 221552
Large-spiked Millettia 254762
Large-spiked Moxanettle 221552
Large-spiked Mycetia 260637
Large-spiked Woodnettle 221552
Large-spiny Barberry 51895
Large-spiny Dense-flower Barberry 51543
Largestalk Slugwood 50555
Large-stipule Milkvetch 43094
Largestipule Rattlebox 112682
Largestipule Vetchling 222819
Largetooth Aspen 311335
Largetooth Chirita 87888
Largetooth Didymocarpus 129907
Largetooth Maple 3142
Largetooth Skullcap 355591
Largetoothed Aspen 311335
Large-toothed Aspen 311335
Large-toothed Maple 3142
Large-tree Rhododendron 331558
Largetrifoliolious Bugbane 91023
Largetube Didissandra 129817
Largewing Bean 241258
Large-wing Citrus 93559
Largewing Euonymus 157705
Large-wing Iris 208837
Largewing Spindle-tree 157705
Largewing Wingnut 320364
Large-wing Wingnut 320364
Largewingbean 241258,241263
Large-winged Begonia 50072
Large-winged Orange 93559
Largeyellow Gardenia 171393
Largeyellow Iris 208736
Largh-leafed Guarri 155977
Largipedioled Holly 204008

Largrflower Nardostachys 262497
Largrflower Oncidium 270790
Largrfruit Euonymus 157732
Lark's Claws 102833
Lark's Eye 407072,410677
Lark's Eyes 407072,410677
Lark's Heel 102833
Lark's Toe 102833
Larkdaisy 81781,81787
Larkseed 302068
Larkspur 102827,102833,124008,124168,124301,124583,231183
Larkspur Corydalis 105823
Larkspur Lettuce 219458
Larkspur Violet 410392
Larkspur-leaved Cacalia 283806
Larkspurlike Corydalis 105823
Larp Mallee 155609
Larpente Plumbago 83646
Laryngitis Summerlilic 62138
Lasa Catchfly 363683
Lasa Corydalis 106070
Lasa Creeping Leadwort 83645
Lasa Cudweed 178182
Lasa Dandelion 384800
Lasa Groundsel 359345
Lasa Heteropappus 193938
Lasa Jurinea 193938
Lasa Larkspur 124267
Lasa Longfruit Speedwell 407080
Lasa Milkvetch 42583
Lasa Rhubarb 329349
Lasa Thickribcelery 279649
Lasa Tickseed 104801
Lasa Wallflower 86477
Lasa Windhairdaisy 348428
Laserwort 222020
Lasia 222038
Lasia-like Cyrtosperma 120773
Lasiandra 391566
Lasianthus 222081
Lasianthus-leaf Syzygium 382595
Lasianthus-leaved Syzygium 382595
Lasiocarpous Craneknee 221695
Lasiocarpous Eritrichium 153472
Lasiococca 222383
Lasiopetalous Millettia 254745
Lasiopetiole Willow 343598
Lasiostyle Adinandra 8247
Laspur Milkvetch 42590
Lasser Galangal 17733
Lasua 28997
Latan Palm 222625
Latania 222625
Latanier 341410
Latarier Balai 97886
Late Aster 41293,380947
Late Black Wattle 1380
Late Boneset 158314
Late Coral Root 104009
Late Coralroot 104009
Late Cotoneaster 107520
Late Dutch Honeysuckle 236026
Late Eupatorium 158314
Late Figwort 355184
Late Golden Rod 367986
Late Goldenrod 367950,368106
Late Gypsophila 183245
Late Horse-gentian 397844
Late Juncellus 212778
Late Leucanthemum 227526
Late Lilac 382268,382312
Late Low Blueberry 403716,404040
Late Low-blueberry 404040
Late Meadow-rue 388628
Late Michaelmas-daisy 41480
Late Ovalleaf Primrose 315042
Late Purple Aster 380947
Late Snakeweed 182111
Late Spider Orchid 272434
Late Spider-orchid 272434
Late Tamarisk 383595
Late Tulip 400162,400183
Late Yellow Daylily 191318
Late-bloom Plantainlily 198627
Latebloom Tulip 400240
Late-blooming Eupatorium 158314
Lateblooming Ophrys 272434
Lateflower Gerbera 175217
Lateflower Pendent-bell 145719
Lateflower Tulip 400183
Late-flowered Cockspur-thorn 109612
Lateflowered Dwarf Willow 343810
Late-flowered Enkianthus 145719
Lateflowered Willow 343807
Late-flowered Willow 343807
Late-flowering Boneset 158314
Late-flowering Dewberry 339354
Late-flowering Goosefoot 87171
Lateflowering Thoroughwort 158314
Late-flowering Thoroughwort 158314
Late-flowering Yellowrattle 329510
Latentpistil Azalea 330935
Latentvein Azalea 332025
Latentvein Camellia 69029
Latentvein Elatostema 142762
Latentvein Stairweed 142762
Later Sichuan Woodbetony 287425
Lateral Flowered Rhododendron 331063
Lateral Flowers Heath 149639
Lateral Podocarpium 200727
Lateral Prostrate Sophora 369145
Lateral Ricegrass 300724
Lateral Viburnum 407914
Lateralflower Agriophyllum 11613
Lateralflower Caruncelgrass 256444
Lateralflower Hoodorchis 264783
Lateral-flowered Indigo 206057
Lateral-flowered Rhododendron 331063
Lateralleaf Arrowwood 407914
Lateralleaf Moehringia 256444
Lateralspiked Spapweed 205069
Lateralspiked Touch-me-not 205069
Lateralstem Goldenray 229143
Laterflower Azalea 331063
Lateripening Bartsia 268999
Latexplant 259044
Lather Leaf 100061
Latherleaf 100061
Latherwort 346434
Lathraea 222641
Latifoliate Fissistigma 166669
Latin American Fleabane 150724
Latin American Mock Vervain 176695
Latin Globemallow 370940
Latipetiolate Camellia 69264
Latisepal Kirilow Clematis 95214
Latisepalous Larkspur 124528
Latouche's Chinese Groundsel 365062
Latouche's Groundsel 365062
Latouchea 222871
Latrobe Acacia 1029
Latter Pine 300030
Lattice-leaf 29326,29663,179571
Latua 222887
Lau Bauhinia 49257
Lau Cephaelis 81975
Lau Jujube 418176
Lau Meliosma 249402
Lau Ninenode 81975
Lau Tarenna 384977
Lau Temple-bells 366716
Lauan 283891,362186
Laucel Magnolia 242135
Laucel-leaved Magnolia 242135
Laudan 93161
Laudanum 93161
Laughing Parsley 321721
Laughter-bringer 21339
Launaea 222907
Launceley 302034
Laura Ninenode 319632
Laurel 122492,215386,223010,223116,223163,223203,316518,401672
Laurel Bay 223203
Laurel Cherry 223116,316166,316285
Laurel Clockvine 390822
Laurel Crown of Caesar 340994
Laurel Family 223006
Laurel Fig 165307
Laurel Greenbrier 366428

Laurel Magnolia 242135,242341
Laurel Negro 104153
Laurel Oak 324002,324041,324095,324240
Laurel Poplar 311375
Laurel Rockrose 93170
Laurel Sabino 242313
Laurel Skimmia 365969
Laurel Sumac 332687
Laurel Sweetleaf 381148
Laurel Willow 343858
Laurel-bay 223203
Laurel-cherry 223116
Laurell Fleshseed Tree 346985
Laurelleaf Albertisia 13456
Laurelleaf Cliffvine 13456
Laurelleaf Currant 334062
Laurelleaf Fighazel 134909
Laurelleaf Genianthus 172850
Laurel-leaf Grevillea 180611
Laurelleaf Poplar 311375
Laurelleaf Sloanea 366041
Laurelleaf Snailseed 97919
Laurel-leaf Snailseed 97919
Laurelleaf Sweetleaf 381148
Laurel-leaf Willow 343858
Laurel-leaved Albertisia 13456
Laurel-leaved Cistus 93170
Laurel-leaved Currant 334062
Laurel-leaved Dogwood 380480
Laurel-leaved Fighazel 134909
Laurel-leaved Poplar 311375
Laurel-leaved Rock Rose 93170
Laurel-leaved Snailseed 97919
Laurel-leaved Thunbergia 390822
Laurel-leaved Willow 342937,343858
Laurel-like Barberry 51834
Laurellike Fleshseed Tree 346985
Laurellike Fleshspike Tree 346985
Laurelwood 67860
Laurel-wood 122492
Laurent Millettia 254747
Laurentia 223053
Laureole 122492
Laurere 223203
Lauriel 122492
Laurier Rose 265327
Laurier Tulipier 242135
Lauro Cordia 104261
Lauro Pardo 104261
Laurocerasus 223088
Laurusleaf Jasmine 211891
Laurus-leaved Jasmine 211891
Laurustine 408167
Laurustinus 408167
Laurustinus Viburnum 408167
Laury 122492
Laus Tibi 262438
Lauter 223203
Lauyu Maple 3089
Lava Ankle-aster 207394
Lava Fig 165455
Lava rayless Fleabane 150708
Lavable Pseudosasa 318283
Lavalle Corktree 294234
Lavandin 223290,223291
Lavatera 223350
Lavelle Hawthorn 109800
Lavender 223248,223251,223290,223318, 309570
Lavender Blue Willow 343133
Lavender Corn 346250
Lavender Cotton 346243,346250
Lavender Gentle 223334
Lavender Giant Hyssop 10407
Lavender Grevill 180612
Lavender Grevillea 180612
Lavender Musk 255237
Lavender Oldfield Aster 380969
Lavender Pebbles 180266
Lavender Popcorn 221307
Lavender Scallops 61570
Lavender Snips 116736
Lavender Spice 307285
Lavender Star Flower 180897
Lavender Starflower 180897
Lavender Thrift 230566,230815
Lavender Water Musk 255237
Lavender-cotton 346243,346250
Lavenderleaf Aster 40728
Lavenderleaf Daisy 124806
Lavenderleaf Dendranthema 124806
Lavenderleaf Sagebrush 35794
Lavenderleaf Wormwood 35794
Lavender-leaved Aster 40728
Lavender-scallops 61570
Laver 208771,311736
Laverock 277648
Laveroek's Lint 231884
Lavrencko Windhairdaisy 348476
Law Camellia 69266
Law Spapweed 205073
Law Touch-me-not 205073
Lawn Bastardtoadflax 389805
Lawn Bburweed 368536
Lawn Bladderwort 403968
Lawn Craneknee 221725
Lawn Daisy 50825
Lawn Grass 418424
Lawn Hundredstamen 389805
Lawn Marshpennywort 200366
Lawn Orchid 417795
Lawn Pennywort 200366
Lawn Prunella 316127
Lawn Silaus 363128
Lawn Trefoil 396854
Lawn Vetchling 222820
Lawndaisy 50825
Lawngrass 418424
Lawn-grass 418424
Lawnleaf 128955,128964,128972
Lawrell 122492
Lawrence Cattleya 79558
Lawrence Paphiopedilum 282856
Lawrence Polystachya 310471
Lawson Cypress 85263,85271
Lawson Euonymus 157669
Lawson False Cypress 85271
Lawson Pine 300031
Lawson's Cypress 85271
Lawson's Spindle-tree 157669
Lawson's Cypress 85295
Lawsons-cypress 85271
Lawyers 336419,338447
Lawyer-weed 325518,325666,326287
Lax Chinese Everlasting 21694
Lax Chinese Pearleverlasting 21694
Lax Cowwheat 248175
Lax Eritrichium 153474
Lax Falsetamarisk 261278
Lax False-tamarisk 261278
Lax Germander 388130
Lax Goldwool Clematis 95074
Lax Horseegg 183012
Lax Liriope 232636
Lax Marshweed 230310
Lax Meadowrue 388548
Lax Puccinellia 321248
Lax Rubgrass 398140
Lax Sea Lavender 230643
Lax Viper's-bugtoss 141287
Laxative Brome-grass 60900
Laxative Ram 328642
Laxbract Bastardtoadflax 389846
Laxcyme Euonymus 157674
Laxenode Loosestrife 239821
Laxflower Aletris 14506
Laxflower Andrographis 22403
Laxflower Bamboo 47427
Laxflower Beautyberry 66830
Laxflower Brome 60928
Laxflower Bromegrass 60928
Laxflower Buttercup 326063
Laxflower Clinopodium 97011
Laxflower Corydalis 105835
Laxflower Curculigo 114828
Laxflower Dendrobium 125182
Laxflower Eargrass 187624
Laxflower Embelia 144780
Laxflower Euonymus 157675
Laxflower Fevervine 280084
Laxflower Fishvine 125978

Laxflower Galangal 17722
Laxflower Goniolimon 179370
Laxflower Hairysepal Rabdosia 209661
Laxflower Hedyotis 187624
Laxflower Jadeleaf and Goldenflower 260445
Laxflower Jewelvine 125978
Laxflower Knotweed 309054
Laxflower Lambsquarters 86919
Laxflower Larkspur 124594
Laxflower Leaflesslotus 292673
Laxflower Licorice 177919
Laxflower Lilyturf 272135
Laxflower Loosestrife 239821
Laxflower Maple 3090
Laxflower Meadowrue 388485
Laxflower Pertybush 292049
Laxflower Petrosavia 292673
Laxflower Podocarpium 200728
Laxflower Pottsia 313229
Laxflower Spapweed 205074
Laxflower Speedwell 407200
Laxflower SW. China Ungeargrass 36732
Laxflower Tickclover 126424
Laxflower Tupistra 70600
Laxflower Wall Heshouwu 162530
Laxflower Wild Smallreed 127207
Laxflower Woodbetony 287347
Lax-flowered Embelia 144780
Lax-flowered False-tamarisk 261280
Lax-flowered Jewelvine 125978
Lax-flowered Mayten 182710
Lax-flowered Mockorange 294489
Lax-flowered Mussaenda 260445
Lax-flowered Orchid 273513
Lax-flowered Tickclover 126424
Laxfruit Tickclover 126373
Laxhair Capitate Eargrass 187533
Laxhair Capitate Hedyotis 187533
Laxhair Crazyweed 279076
Laxhair Hillcelery 276766
Laxhair Narrowflower Didymocarpus 129972
Laxhair Ramie 56257
Laxhair Rockfoil 349963
Laxhair Stairweed 142600
Laxhairy Falsenettle 56257
Lax-hairy False-nettle 56257
Laxicymose Spindle-tree 157674
Laxiflorous Kalanchoe 215193
Laxiflorous Maesa 241796
Laxiflorous Olive 270140
Laxiflower Kengyilia 215935
Laxiflowered Euonymus 157675
Laxiflowered Maple 3090
Laxiflowered Pottsia 313229
Laxiflowered Pygeum 322400
Laxiflowered Rose 336685
Laxifoliolate Licorice 177903
Laxinflorescence Meadowrue 388548
Laxity-hair Corydalis 105835
Laxleaf Alangium 13371
Laxleaf Angelica 24389
Laxleaf Birdlingbran 87254
Laxleaf Chesneya 87254
Laxleaf Yellow Silver Pine 225602
Laxlip Dendrobium 125372
Laxmann Bugle 13130
Laxnode Leymus 228387
Laxpaniculated Chastetree 411368
Laxpaniculated Negundo Chastetree 411368
Laxpilose Handel Monkshood 5245
Laxspiculate Cypressgrass 118765
Laxspiculate Galingale 118765
Laxspike Bigglume Kengyilia 215937
Laxspike Kengyilia 215936
Laxspike Kobresia 217206
Laxspike Thinstalk Knotweed 162320
Laxspike Woodbetony 287349
Laxspine Bittersweet 367413
Laxspine Nightshade 367413
Laxtooth Gugertree 350940
Laxtooth Pendent-bell 145698
Laxumbrella Elatostema 142710
Laxumbrella Stairweed 142710
Laxymose Larkspur 124336
Lay-a-bed 384714
Laylock 72934,382329
Layne's Ragwort 279920
Lazane 386169
Lazarus Bell 168476
Lazarus-bell 168476
Lazy Daisy 84921
Lazy-bones 313025
Lazydaisy 29099
Le Conte's Thistle 92125
Le Conte's Violet 409654
Lea Cranesbill 174832
Leachi Barberry 51838
Leachian Kalmiopsis 215411
Lead Plant 19997
Lead Tree 227420,227430
Lead Wort 305167
Leading Size Rhododendron 331532
Leadplant 19997
Lead-plant 19997
Leadplant Amorpha 19997
Leadtree 227420,227430
Lead-tree 227420
Leadwood 218236
Leadwort 83641,83646,255598,305167,
305172,305194
Leadwort Family 305165
Leadwortleaf Tabacco 266051
Leaf Beet 53257
Leaf Between Flower 57685
Leaf Blade Longitudinally Green-and-white
Striped 182178
Leaf Blade of One Color 182177
Leaf Cactus 147283,290699
Leaf Cassia 360457
Leaf Chervil 28024
Leaf Cup 310021,310030
Leaf Flower 296465
Leaf Mustard 59438
Leaf-beet 53257
Leafbract Lobelia 234489
Leafbract Loosestrife 239666
Leafbud Rockcress 30272
Leafcalyx Gentian 173722
Leafcalyx Sweetleaf 381291
Leafcroton 98177
Leaf-croton 98177
Leafcup 310021
Leaf-flower 296465,296515
Leafflower Stonecrop 357022
Leaf-flowering Cactus 147283
Leafless Anabasis 21028
Leafless Beaked Orchid 375690
Leafless Bentspur Orchid 70672
Leafless Bossiaea 57500
Leafless Calligonum 66998
Leafless Caper 71689
Leafless Epipogium 147307
Leafless Fimbristylis 166182
Leafless Fluttergrass 166182
Leafless Gentian 157110
Leafless Hawk's-beard 110956
Leafless Iris 208447
Leafless Kneejujube 66998
Leafless leadwort 305172
Leafless Mosla 259288
Leafless Orchid 147307
Leafless Spurge 158460
Leafless Stalegrass 182621
Leafless Taeniophyllum 383064
Leafless Tamarisk 383437
Leafless Vanilla 404979,404989
Leaflessbean 148441
Leaflesslotus 292671,292674
Leaflessorchis 29212
Leaf-like Bracteal Aster 40618
Leaflikebract Aster 40618
Leaflikebract Azalea 389577
Leaflikebract Sagebrush 36077
Leaflikecalyx Swertia 380140
Leaflikefruit Honeysuckle 235797
Leaflikesheath Goldenray 229136
Leaflikestiped Cremanthodium 110429
Leaflikestiped Nutantdaisy 110429
Leaf-mustard 59438
Leaf-scaled Cacalia 283858
Leafspike Rabdosia 209798

Leafspine Brain Cactus 140579
Leafstipe Rabdosia 209796
Leaftraceflower Spapweed 204874
Leaftraceflower Touch-me-not 204874
Leafy Adder's-mouth Orchid 185200
Leafy Arnica 34693
Leafy Aster 380892
Leafy Azalea 330208
Leafy Belltree 324996
Leafy Bell-tree 324996
Leafy Bossiaea 57493
Leafy Broom 173073
Leafy Butea 64148
Leafy Cactus 290701
Leafy Catchfly 363463
Leafy Coldbamboo 172745
Leafy Corydalis 106280
Leafy Crazyweed 279029
Leafy Desertdandelion 242936
Leafy Dwarf Knotweed 309394
Leafy Elm 401508
Leafy Ephedra 146148
Leafy Euphorbia 160062
Leafy Fleabane 150641
Leafy Goldenweed 271913
Leafy Goosefoot 87016
Leafy Hairy-callus Roegneria 335226
Leafy Hawkweed 196057
Leafy Jadeleaf and Goldenflower 260418
Leafy Javatea 275812
Leafy Knotweed 309121
Leafy Lagotis 220157
Leafy Lespedeza 226746
Leafy Lousewort 287217
Leafy Lupine 238481
Leafy Meadowsweet 166097
Leafy Muehlenbergia 259656
Leafy Paris 284358
Leafy Peashrub 72311
Leafy Pinellia 299723
Leafy Pondweed 312117
Leafy Pricklyash 417277
Leafy Primrose 314419
Leafy Rabdosia 324856
Leafy Rock Daisy 291330
Leafy Rockfoil 349736
Leafy Rush 213116
Leafy Satin Grass 259679
Leafy Spotleaf-orchis 179610
Leafy Spurge 158857
Leafy Thistle 91977
Leafy Vanilla 405015
Leafy Vetch 408503
Leafy White Orchis 183571,230348
Leafybract Altai Draba 136923
Leafybract Altai Whitlowgrass 136923
Leafy-bract Aster 380892
Leafybract Corydalis 105910
Leafybract Dwarf Rush 212988
Leafy-bracted Aster 380892
Leafy-bracted Beggarticks 53842
Leafy-bracted Goldenweed 271913
Leafycalyx Cinquefoil 289720
Leafycalyx Gentian 173722
Leafyflower 57852,57868
Leafy-fruited Nightshade 367583
Leafyinvolucre Saussurea 348643
Leafy-stemmed Plantain 301864
Lean Woodbetony 287400
Leanshan Milkvetch 42596
Leaping Cucumer 139852
Leap-up-and-kiss-me 410677
Leari's Morningglory 207934
Least Bell 367765
Least Bur Reed 370088
Least Burreed 370088
Least Bur-reed 370088
Least Cactus 244066
Least Cinquefoil 362361
Least Cudweed 235267
Least Duckweed 224378,224385,414653
Least dwarf rush 213149
Least Evening Primrose 269481
Least Hop Clover 396882
Least Houseleek 356468
Least Letruce 219480
Least Mallow 243810
Least Marshwort 29327
Least Mouse Ear Chickweed 83020
Least Neopicrorhiza 264317
Least Pepperwort 225475
Least Primrose 314646,314650
Least Snowbell 367765
Least Spike Rush 143022
Least Stouecrop 356907
Least Trillium 397594
Least Water Pepper 309395
Least Water-lily 267320
Least Waterwort 142572
Least Willow 343476,344045
Least Yellow Trefoil 396977
Least Yellow Water Lily 267320
Least-daisy 207396
Leastflower Abdominea 83
Leath Saxifrage 52514
Leather 208771
Leather Bergenia 52514
Leather Flower 95243,95423
Leather Jacket 48076,48078
Leather Leaved Ash 168127
Leather Oak 323865
Leather Spineflower 222586
Leatherbract Groundsel 358606
Leatherbract Windhairdaisy 348230
Leather-coatplum 372477
Leather-colored Ardisia 31353
Leathercress 378287,378290
Leatherflower 95423
Leatherflower Clematis 95423
Leatherjacket Eucalyptus 155711
Leatherleaf 85412,85414
Leather-leaf 85414,85416
Leatherleaf Acacia 1153
Leatherleaf Arrowwood 408089
Leatherleaf Azalea 330462
Leather-leaf Bodinier Clethra 96470
Leatherleaf Buckthorn 328658
Leatherleaf Dutchmanspipe 34323
Leatherleaf Elatostema 142633
Leatherleaf Euonymus 157678
Leatherleaf Gentian 150955,173882
Leather-leaf Grape 411636
Leatherleaf Luchun Raspberry 338770
Leatherleaf Mahonia 242487
Leather-leaf Mahonia 242487
Leatherleaf Millettia 66643
Leatherleaf Mycetia 260620
Leatherleaf Rhododendron 330462
Leatherleaf Sedge 73967,76600
Leatherleaf Sinacalia 364518
Leatherleaf Stairweed 142633
Leatherleaf Stringbush 414254
Leatherleaf Subsoja 272371
Leatherleaf Viburnum 408089
Leatherleaf Windhairdaisy 348663
Leather-leaved Ash 168131
Leather-leaved Buckthorn 328658
Leather-leaved Mahonia 242487
Leather-leaved Millettia 66643
Leather-leaved Rhododendron 330462
Leather-leaved Rose 336504
Leatherleaved Triflower Maple 3713
Leather-leaved Viburnum 408089
Leather-lef Holly 203642
Leatherlraf Knotweed 309014
Leathern Barberry 51500
Leatherstem 212130
Leatherwood 120485,120488,120501,
133654,133656,156062,156066
Leatherwood Titi 120488
Leathery Begonia 49741
Leathery Holboellia 197223
Leathery Knotweed 308698
Leathery Litse 234067
Leatherybractdaisy 400035,400036
Leatheryleaf Cremanthodium 110347
Leathery-leaf Litse 234067
Leatheryleaf Nutantdaisy 110368
Leathryleaf Primrose 314233
Leavenworth Eryngo 154325
Leavenworth's Bracted Sedge 75095

Leavenworth's Eryngo 154325
Leavenworth's Goldenrod 368208
Leavenworth's Sedge 75095
Leaves Entire Prickly-ash 417247
Leazel 133505
Lebanon Barberry 51858
Lebanon Buckthorn 328765
Lebanon Cedar 80100
Lebanon Oak 324104
Lebanon Onion 15431
Lebanon Squill 321892,321894,321896
Lebanonzeder 80100
Lebbek Albizia 13586
Lebbek Albizzia 13586,13595
Lebbek Siris 13595
Lebbek Tree 13595
Lebrun Greenbrier 366429
Lebrun Smilax 366429
Lebu Arrowbamboo 162687
Lecanorchis 223634
Lecanthus 223674
Lechang Calanthe 65993
Lechang Michelia 252844
Lechleri Barberry 51840
Lechuguilla 10881
Lechuguilla Verde 10813
Lecler Euonymus 157459
Lecler Spindle-tree 157678
Leclere Windhairdaisy 348477
Lecocia 223728
Lecockia 223728
Lecomte Barberry 51841
Lecomte Bushbeech 178035
Lecomte Bush-beech 178035
Lecomte Mazus 247002
Lecomte Woodbetony 287350
Lecteous Cotoneaster 107520
Lecythis Family 223762
Ledebour Chive 15420
Ledebour Globe Flower 399509
Ledebour Globeflower 399515
Ledebour Lily 229886
Ledebour Onion 15420
Ledebour Seseli 361519
Ledebour Yarrow 3966
Ledger Bark 91059,91075
Ledger Cinchona 91075
Ledgerbark Cinchona 91075
Leding's Hedgehog Ccactus 140268
Ledong Alyxia 18516
Ledong Bulbophyllum 62841
Ledong Evergreen Chinkapin 78974
Ledong Jadeleaf and Goldenflower 260449
Ledong Magnolia 283419
Ledong Mussaenda 260449
Ledong Oak 324131
Ledong Oilfruitcamphor 381796
Ledong Parakmeria 283419
Ledong Passionflower 285661
Ledong Skullcap 355586
Ledong Stonebean-orchis 62841
Ledong Syndiclis 381796
Ledu Milkvetch 42601
Leduce's Thistle 92126
Ledum 223888,223901
Ledum Boronia 57266
Leea 223919,223943
Leech Lime 93492
Leechee 233077,233078
Leechee Litchi Nut 233078
Leechwort 345957
Leek 15018,15621,15843
Leek Cress 14964
Leek Orchid 313550
Leek-cress 14964
Leek-green Sedge 75865
Leek-leaved Salsify 394327
Leemers 106703
Legendre Monkshood 5351
Legendre Nanmu 295377
Legendre Phoebe 295377
Legendre Woodbetony 287351
Legnephora 224051
Legume Family 224085
Legumes 224085
Legwort 325820
Lehman Sedge 75099
Lehmann Banksia 47648
Lehmann Barberry 51842
Lehmann Crazyweed 278957
Lehmann Lachnoloma 219057
Lehmann Lovegrass 147768
Lehmann Milkvetch 42597
Lehmann Rockjasmine 23213
Lehmann Seseli 361521
Lehmann Sophora 369056
Lehmann Spunsilksage 360851
Lehmann Tulip 400190
Lehmann's Gum 155621
Lehmann's Love Grass 147768
Lehui Machilus 240624
Lei Cryptocarya 113462
Leiberg's Panic Grass 128518
Leibo Azalea 331079
Leibo Barberry 51839
Leibo Chinese Groundsel 365063
Leibo Deutzia 126987
Leibo Didymocarpus 129921
Leibo Duckweed 224370
Leibo Gleadovia 176796
Leibo Holly 203971
Leibo Kunming Buttercup 326009
Leibo Landpick 308588
Leibo Maple 3097
Leibo Monkshood 5502
Leibo Pricklyash 417254
Leibo Prickly-ash 417254
Leibo Qiongzhu 87575
Leibo Rhododendron 331079
Leibo Sinosenecio 365063
Leibo Solomonseal 308588
Leibo Vineclethra 95519
Leichhardt 262823
Leichhardt Bean 78250
Leichhardt Tree 262823
Leichtlin Camass 68800
Leichtlin Lily 229887
Leichtlin's Lily 229887
Leigongshan Maple 3095
Leipzig Poplar 311153
Leishan Azalea 331085
Leishan Daphne 122496
Leishan Euonymus 157681
Leishan Fragrant-bamboo 87566
Leishan Paraprenanthes 283685
Leishan Rhododendron 331085
Leishan Snakegourd 396215
Leishan Square-bamboo 87566
Leiwuqi Monkshood 5353
Leko 336232
Lely 229717
Lemee Spapweed 205078
Lemmarosa 93408
Lemmon Penstemon 289362
Lemmon's Bitterweed 201320
Lemmon's Candyleaf 376408
Lemmon's Catchfly 363681
Lemmon's Fleabane 150751
Lemmon's Rock Daisy 291319
Lemmon's Rubberweed 201320
Lemmon's Sedge 75102
Lemmon's Snakeroot 11166
Lemmon's Syntrichopappus 382062
Lemmon's Triteleia 398807
Lemmon's Wild Buckwheat 152224
Lemmons Ceanothus 79951
Lemoine Barberry 51844
Lemoine Deutzia 126989
Lemoine Gladiolus 176308
Lemoine Mockorange 294491
Lemoine's Deutzia 126989
Lemon 93539,93546
Lemon Bacopa 46352
Lemon Balm 249504
Lemon Basil 268438,268470
Lemon Beebalm 257159
Lemon Beebrush 232483
Lemon Bottlebrush 67253,67282
Lemon Bottle-brush 67253,67275
Lemon Dalea 121884
Lemon Day Lily 191284

Lemon Daylily 191304
Lemon Day-lily 191304
Lemon Eucalyptus 106793
Lemon Flower 11549
Lemon Flowered Mallee 155776
Lemon Geranium 288179
Lemon Grass 117136,117153
Lemon Guava 318749
Lemon Gum 155776
Lemon Lily 191304,229980
Lemon Mint 249504,250291,250341,257159,257184
Lemon Monarda 257184
Lemon Peony 280236
Lemon Pistol Bush 139210
Lemon Plant 232483
Lemon Sabia 341523
Lemon Sagewort 35909
Lemon Scented Gum 106793
Lemon Sumac 332477
Lemon Sumach 332477
Lemon Tea 69148
Lemon Tea Tree 226471
Lemon Thyme 391152,391346,391366
Lemon Tree 232483
Lemon Verbena 17605,187179,232483
Lemon Verbena Lippia 232483
Lemon Vine 290701
Lemonade 11549
Lemonade Berry 332658,332908
Lemonade Sumac 332658
Lemonade Sumach 332908
Lemonade Tree 332658,332916
Lemonbalm 249504
Lemon-flowered Gum 155776
Lemonfragrant Hillcelery 276745
Lemongrass 117136,117153
Lemon-grass 117153
Lemonleaf Whitepearl 172149
Lemon-like Camellia 69148
Lemonlike Citrus 93546
Lemon-like Citrus 93546
Lemonlike Orange 93546
Lemonnier Iris 208719
Lemon-scent Bottlebrush 67253
Lemon-scented Cucumber 114150
Lemon-scented Darwinia 122780
Lemon-scented Geranium 288179,288396
Lemon-scented Gum 106793
Lemon-scented Ironbark 155748
Lemon-scented Myrtle 46337
Lemon-scented Spotted Gum 106793
Lemon-scented Tea-tree 226471
Lemon-scented Thyme 391152
Lemon-scented Verbena 232483
Lemon-verbena Lippia 232483
Lemonweed 286877
Lemonwood 301261
Lemon-wood 415955
Lemon-yellow Cattleya 79532
Lemon-yellow False Golden Aster 194195
Lemonyellow Flowered Rhododendron 330422
Lemonyellow Iris 208561
Lemonyellow Rosemallow 194772
Lemperg's Barberry 51845
Lenga 266884
Lengthened Jasmine 211812
Lengthened Mussaenda 260410
Lengua De Vaca 272798,272830,272853
Lenormand's Water Crowfoot 326147
Lens Coldwaterflower 299005
Lenscale 44490
Lenspod Whitetop 73090
Lens-podded Hoary Cress 73090
Lent Cocks 262441
Lent Cup 262441
Lent Lily 262441
Lent Pitcher 262441
Lent Rose 262417,262441,315101
Lent-cocks 262441
Lenten 391706
Lenten Rose 190925,190949
Lenten-rose 190949
Lenticel Evodia 161349
Lenticel Loeseneriella 235223
Lenticel Pencilwood 139658
Lenticel Webseedvine 235223
Lenticel-bearing Embelia 144773
Lenticellate Embelia 144822
Lenticellate Evodia 161349
Lenticellate Indigo 206165
Lenticellate Machilus 240614
Lenticellate Pencil-wood 139658
Lenticellate Rockvine 387798
Lenticular Sedge 75106
Lenticules 224359
Lentiginous Coelogyne 98685
Lentigo 224375,224473,262441
Lentil 224471,224473
Lentil Vetch 408638
Lentisc 300994
Lentisc Pistache 300994
Lentisco 300994,301004,332956
Lentiscus Pistachio 300994
Lentisk 300994
Lentisk Lentisco 300994
Lentisk Pistache 300994
Lenty Cup 262441
Lenty Cups 262441
Lenty Lily 262441
Leny Cocks 262441
Lenycocks 262441
Leonard's Skullcap 355565,355667
Leonotis 224558
Leontice 182634,224625
Leontopodium 224767
Leopard 70386
Leopard Begonia 50060
Leopard Bulbophyllum 62648
Leopard Camphor 234048
Leopard Flower 50667,50669
Leopard Lily 50669,130083,130135,168476,229797,229978,346158
Leopard Longlegorchis 320451
Leopard Orchid 26273,26274
Leopard Plant 162616,162617,228969,229016
Leopard Pteroceras 320451
Leopard Rockfoil 349749
Leopard Staurochilus 374460
Leopard Stonebean-orchis 62648
Leopard Tree 65008
Leopard Wood 61059
Leopard's Bane 34657,34752,136333,136366,284387
Leopard's Head 168476
Leopard's-bane 34752,136333,136365
Leopardess Water Lily 267629
Leopardflower 373719
Leopard-flower 50669
Leopard-palm 20132
Leopardspot Odontoglossum 269081
Leopard-spot Thrixspermum 390521
Leopardtrree 65008
Leopardwood 166980
Leopardwood Tree 166980
Leopold Sycamore Maple 3465
Leopold's Tuftroot 130107
Leopoldia 225027
Lepechinella 225088
Lepidagathis 225134
Lepidobotrys 225480
Lepidoid Brassaiopsis 59223
Lepidotestoloniform Rockfoil 349540
Lepidozamia 225622
Lepin Snakegourd 396217
Lepionurus 225644
Lepironia 225659
Lepisanthes 225668
Lepistemon 225699
Leprosy Gourd 256797
Leprous Tree 212098,212127
Leptaleum 225802
Leptinella 107821
Leptoboea 225895
Leptocanna 225910
Leptocarp Yu Maple 3762
Leptocodon 226058
Leptodermis 226075
Leptophyllous Stringbush 414203

Leptopus 226315
Leptopyrum 226339
Leptosiphonium 226435
Leptospermum 226450
Leptunis 226569,226571
Lepturus 226585,226599
Lepturus Pricklythrift 2254
Lepyrodiclis 226615
Lerchea 226633,226639
Lerp 155568
Leschenault Clematis 95077
Leschenault Goldquitch 156471
Leschenault Loosestrife 239709
Leschenault Mahonia 242597
Leschenault Rush 213213
Leschenault Spapweed 205084
Leschenault Touch-me-not 205084
Leschenaultia 226669
Leser White Water Lily 267649
Leshan Aspidistra 39549
Lespedeza 226677
Lesser Alpine Pearlwort 342268
Lesser Bindweed 102933
Lesser Bladderwort 403252
Lesser Bottlebrush 67286
Lesser Bougainvillea 57857
Lesser Broomrape 275135
Lesser Brown Sedge 73604
Lesser Bugloss 21920
Lesser Bulrush 401094
Lesser Bur 31056
Lesser Burdock 31056
Lesser Burnet 299523,345860
Lesser Bur-reed 370025
Lesser Butter And Eggs 231140
Lesser Butterfly Orchid 302260
Lesser Butterfly-orchid 302260
Lesser Calamint 96969
Lesser Canadian St. John's-wort 201788
Lesser Canary Grass 293743
Lesser Canary-grass 293743
Lesser Candelabra Tree 158688
Lesser Cat's Foot 26427
Lesser Caucasian Stonecrop 357181
Lesser Celandine 325820
Lesser Centaury 81065,81517,199624
Lesser Chickpea 222694
Lesser Chickweed 375033
Lesser Chilean Iris 228627
Lesser Churmel 81065
Lesser Clearweed 298917
Lesser Cord-grass 370166
Lesser Curmel 81065
Lesser Daisy Fleabane 150464
Lesser Dandelion 384614
Lesser Dodder 115008
Lesser Duckweed 224360,224375
Lesser Evening Primrose 269404
Lesser Flowering Quince 84556
Lesser Fringed Gentian 174228
Lesser Furze 401395
Lesser Glory-of-the-snow 87770
Lesser Goat's-beard 394333
Lesser Gold-of-pleasure 68850
Lesser Gorse 401395
Lesser Hairy-brome 60643
Lesser Hawkbit 224731,224732
Lesser Hemlock 9832
Lesser Henbit 407144
Lesser Honcywort 83929
Lesser Honeywort 83923
Lesser Hop Trefoil 396882
Lesser Knapweed 81237
Lesser Knotweed 308941
Lesser Londonpride 349227
Lesser Marigold 383105
Lesser Marsh Dandelion 384824
Lesser Marshwort 29332
Lesser Meadow Rue 388572
Lesser Meadow-rue 388572
Lesser Mexican Stonecrop 356633
Lesser Milk-vetch 42815
Lesser Mohavea 256525
Lesser New Zealand Flax 295598
Lesser Old-man 140239
Lesser Orach 44567
Lesser Panicled Sedge 74304
Lesser Periwinkle 409339
Lesser Pond Sedge 73597
Lesser Pond-sedge 73597
Lesser Pondweed 312236
Lesser Purple Fringed Orchid 302491
Lesser Quaking-grass 60383
Lesser Quick-weed 170142
Lesser Rattlesnake Plantain 147192
Lesser Rattlesnake-plantain 179685
Lesser Red Knot-grass 308782
Lesser Reedmace 401094
Lesser Sarsaparilla 159221
Lesser Sea Spurrey 370661
Lesser Sea-fig 148562,148579
Lesser Sea-spurrey 370661,370706
Lesser Shepherd's Cress 385552
Lesser Shepherdscress 385552
Lesser Skullcap 355609
Lesser Snapdragon 28631
Lesser Solomon's-seal 308613
Lesser Spearwort 325864
Lesser St. John's Wort 201942
Lesser Star-thisde 81035
Lesser Stitchwort 374897
Lesser Sunflower 189032
Lesser Swine's Cress 105340,105352
Lesser Swinecress 105340
Lesser Swine-cress 105340
Lesser Texas Silverleaf 227923
Lesser Toadflax 84596
Lesser Tree Mallow 223364
Lesser Trefoil 396882
Lesser Tttfted Vetch 408693
Lesser Tussock-sedge 74304
Lesser Twayblade 232905
Lesser Valerian 404251
Lesser Water Parsnip 53157
Lesser Water Plantain 46931
Lesser Water-parsnip 53154,53157
Lesser Waterplantain 46931
Lesser Water-plantain 46930,46931
Lesser Wintergreen 322853
Lesser Yam 131577
Lesser Yellow Trefoil 396882
Lesser Yellow-sorrel 277832
Lessert Mulgedium 259757
Lessflower Beautyberry 66881
Lessing Mediterranean Camphorfume 70464
Lessing Plantain 302050
Lessing Seseli 361524
Lesther-leaf Deutzia 126876
Lest-we-forget 327866
Lesueur's Rush 213218
Lether Flower 94676
Letherleaf Monkshood 5145
Letter Flower 108775
Letter Plant 180126
Letterwood 61059
Lettnee-leaf Basil 290952
Lettuce 219206,219485
Lettuce Tree 300955
Lettuce-leaf Basil 290952
Lettuceleaf Favusheadflower 129570
Lettuce-like Dicranostigma 129570
Lettucelike Primrose 314557
Letung Bulbophyllum 62841
Leucaena 227420,227430
Leucanthemella 227452
Leucanthemum 227467
Leucas 227537
Leucomeris 178748
Leucorchis 318209
Leucosceptrum 227999
Leucosyca 228144,228147
Leucosyke 228144
Leucothoe 228154,228181
Levant Berry 21459
Levant Cotton 179900
Levant Galls 324314
Levant Garlic 15048
Levant Madder 338007
Levant Nut 21459
Levant Rose 336563
Levant Scammony 103255

Levant Sea Holly 154301
Levant Storax 232562
Levant Wormseed 35333,35868,360825
Levant-wormseed 360825
Leveille Mahonia 242494
Leveille Peashrub 72272
Leveille Pea-shrub 72272
Leveille Willow 343603
Levelled Leea 223922
Lever 208771
Lever Wood 276818
Leverwood 276818
Lever-wood 276818
Levigated Yushania 416785
Levine Aalyxia 18509
Levine Bulbophyllum 62852
Levine Rhododendron 331103
Levine Stonebean-orchis 62852
Levine Syzygium 382599
Levine's Newlitse 264069
Levver 208771,401112
Lewis Cactus 244130
Lewis Devil Rattan 121511
Lewis Flax 231917
Lewis Mock Orange 294493
Lewis Mockorange 294493
Lewis Syringa 294493
Lewis' Groundsel 279907
Lewis' Hebe 186959
Lewis' Monkeyflower 255216
Lewis' Syringa 294493
Lewisia 228256,228270
Lewiston Cornsalad 404439
Ley 127635
Leybold Jointfir 178545
Leye Aspidistra 39550
Leye Camellia 69268
Leye Trigonotis 397427
Leymus 228337
Lezabyz 219914
Leza-wood 219976
Lezzory 369541
Lhasa Groundsel 359345
Lhorong Corydalis 106071
Li Trigonostemon 397370
Li's Litse 233975
Li's Mallotus 243384
Li's Privet 229523
Li's Waxplant 198864
Lian Bamboo 47312
Lian Mallotus 243384
Liancheng Sedge 75148
Liang Cinnamon 91359
Liang Holly 203980
Liang Stringbush 414206
Liang Waxplant 198863
Liangguang Dogwood 105095
Liangguang Globularspike Fimbristylis 166340
Liangguang Ophiorrhiza 272239
Liangguang Rostellularia 337250
Liangshan Anemone 24095
Liangshan Azalea 330886
Liangshan Dutchmanspipe 34244
Liangshan Fescuegrass 164051
Liangshan Gentian 173587
Liangshan Groundsel 359346
Liangshan Larkspur 124347
Liangshan Leek 15429
Liangshan Lemongrass 117201
Liangshan Monkshood 5362
Liangshan Osmanther 276347
Liangshan Osmanthus 276347
Liangshan Pagodatree 368996
Liangshan Primrose 315009
Liangshan Rabdosia 209744
Liangshan Sophora 368996
Liangshan Tupistra 70612
Liangshan Windflower 24095
Liangshui Willow 343975
Lianhua Eurya 160591
Lianhuachi Eurya 160591
Lianhuashan Milkvetch 42734
Lianhuashan Violet 410178
Lianshan Camellia 69277
Lianshan Eargrass 187603
Lianshan Hedyotis 187603
Lianshan Milkvetch 42596
Lianxian Chirita 87899
Liao Linden 391760
Liaodong Birch 53507
Liaodong Lilac 382383
Liaodong Monkshood 5363
Liaodong Pink 127668
Liaodong Rorippa 336226
Liaodong Sagebrush 36453
Liaodong Sedge 74650
Liaodong Violet 410531
Liaodong Yellowcress 336226
Liaodongapanese Aralia 30649
Liaoning Apricot 34473
Liaoning Seepweed 379545
Liaoning Simon Poplar 311498
Liaotung Rorippa 336226
Liaoxi Azalea 331105
Liaoxi Melicope 249193
Liaoxi Rhododendron 331105
Liaoxi Tickseed 104767
Liaoxi Violet 410181
Liatris 228433
Libanon Periwinkle 409334
Libanotis 228556
Libbard's Bane 5442
Libbardine 5442
Libearleaf Sagebrush 36348
Liberated Stauntonvine 374421
Liberated Wild Quince 374421
Liberbaileya 228620
Liberia Coffee 98941
Liberia Ebony 132382
Liberia Persimmon 132382
Liberia Sansevieria 346109
Liberian Cherry 342116
Liberian Coffee 98941
Libertia 228625
Liblong 357203
Libo Camellia 69286,69577
Libo Chirita 87898
Libo Clematis 95084
Libo Dendrocalamus 125489
Libo Distylium 134935
Libo Dragonbamboo 125489
Libo Elaeagnus 142055
Libo Elatostema 142715
Libo Euonymus 157687
Libo Indosasa 206888
Libo Machilus 240621
Libo Mosquitoman 134935
Libo Mulberry 259154
Libo Pricklyash 417258
Libo Raspberry 338742
Libo Stairweed 142715
Libo Tea 69286
Libo Waxplant 198867
Libosheng's Litse 233974
Licania 228663
Licebane 124604
Licent Amur Barberry 51713
Licent Pea-shrub 72273
Licent's Saussurea 348491
Lichee 233077,233078
Lichiang Dactylicapnos 128302
Lichuan Arrowhead 342372
Lichuan Dysosma 139618
Lichuan Falsepistache 384370
Lichuan False-pistache 384370
Lichuan Hilliella 196278
Lichuan Hydrangea 199939
Lichuan Machilus 240620
Lichuan Maple 3101
Lichuan Nanmu 295378
Lichuan Phoebe 295378
Lichuan Raspberry 338744
Lichuan Tupistra 70614
Lichuan Yellowwood 94024
Lichwort 284152
Lick 15621
Lickorice 177873
Licorice 42417,177873,177893
Licorice Bedstraw 170316,170317
Licorice Milkvetch 42417

Licorice Mint 10407
Licorice Plant 189659,189661
Licorice Vine 765
Licorice-plant 189659
Licuala Palm 228729,228735,228766
Licualapalm 228729
Licualapalm Pumila 228741
Lidded Cleistocalyx 94576
Lide Lily 262441
Lidongshan Raspberry 339185
Lie-abed 360328
Liechtenstein Barberry 52058
Lien Hua Chih Helicia 189948
Lietze Calathea 66169
Lietzia 228802
Lieven Falsestonecrop 318397
Life Everlasting 21596,26385,26551
Life Plant 54555,215150
Life-everlasting 21596
Life-long 357203
Life-of-man 30748,191304,394088
Life-plant 61572
Life-root 358336
Liffleraceme Cherry-laurel 223133,223136,316922
Lift-up-your-head-and-I'll-Kissyou 220729
Light Leaf Aster 40911
Light Poppy Mallow 67125
Light Red Meranti 362187
Lightblack Porpax 311735
Lightblue Primrose 314257
Lightcolorbell Azalea 331044
Lightcolour Woodrush 238675
Lightgreen Leaf Actinodaphne 6780
Lightgreen Sedge 74540
Lightgreen Yushanbamboo 416782
Light-green-leaved Actinodaphne 6780
Lightgrey Linden 391840
Lightgrey Willow 343133
Light-grey Willow 343133
Lightning 282685
Lightning-flower 282685
Lightred Azalea 331647
Lightred Honeysuckle 235633
Lightred Meranti 362210
Light-red Rhododendron 331705
Lightred Rhynchosia 333392
Lightting Rockfoil 349577
Lightviolet Catnip 265107
Lightwave Podocarpium 200747
Lightwood 1302
Lightwood Hickory 1302
Lightwood Scented Satinwood 83534
Lightwort 250894
Lightyellow Azalea 330704
Lightyellow Broomrape 275214
Lightyellow Crazyweed 279055
Lightyellow Hoodorchis 264772
Lightyellow Lily 230077
Lightyellow Rockjasmine 23172
Lightyellow Sagebrush 35527
Lightyellow Skullcap 355583
Lightyellow Sophora 369010
Light-yellow Sophora 369010
Lightyellow Spapweed 205175
Lightyellow Touch-me-not 205175
Lightyellow Woodbetony 287386
Lightyellowflower Ecdysanthera 139959
Light-yellow-flowered Larkspur 124417
Lign-aloes 29977
Lignariella 228955
Lignum Vitae 181496,181505,181506
Lignum-nephriticum 161863
Lignumvitae 181496,181505
Lignum-vitae 181496,181505
Liguelate Spapweed 205089
Liguelate Touch-me-not 205089
Ligularia 228969
Ligularia-like Chinese Groundsel 365064
Ligulate Cissampelopsis 92517
Ligulate Largelip Woodbetony 287420
Ligulate Sedge 75161
Ligulate Small Reed 127208
Liguleleaf Goldwaist 90368
Ligulelobe Sage 345172
Ligurian Yarrow 3970
Ligusticum 229283
Ligustrum 229529,229617
Liguulate Ricegrass 300705
Liheng Bittercress 72864
Liheng Stairweed 142716
Lihsien Monkshood 5365
Lihsien Rabdosia 209746
Lijiang Acronema 6222
Lijiang Angelica 24391
Lijiang Anisodus 25450
Lijiang Arisaema 33388
Lijiang Asparagus 39156
Lijiang Aster 40754
Lijiang Barberry 51860
Lijiang Bauhinia 49019
Lijiang Bluebell 115382
Lijiang Bluebellflower 115382
Lijiang Cacalia 283840
Lijiang Catchfly 363684
Lijiang Ceropegia 83995
Lijiang Chelonopsis 86817
Lijiang Chun Litse 233867
Lijiang Clematis 94745
Lijiang Copperleaf 1975
Lijiang Cotoneaster 107526
Lijiang Cowparsnip 192299
Lijiang Crucianvine 356299
Lijiang Dactylicapnos 128302
Lijiang Dandelion 384515
Lijiang Deerdrug 242693
Lijiang Deutzia 126943
Lijiang Dolichos 135412
Lijiang Draba 137088
Lijiang Ephedra 146203
Lijiang Euonymus 157683
Lijiang Eurya 160486
Lijiang Everlasting 21592
Lijiang Evodia 161328
Lijiang False Solomonseal 242693
Lijiang Felwort 235450
Lijiang Flatsedge 322282
Lijiang Forsythia 167449
Lijiang Gerbera 175187
Lijiang Goldenray 229094
Lijiang Groundsel 359357
Lijiang Gutzlaffia 283361
Lijiang Hemlock 399903
Lijiang Hemsleia 191934
Lijiang Himalrandia 196352
Lijiang Holly 203755
Lijiang Indigo 205709
Lijiang Jerusalemsage 295132
Lijiang Kiwifruit 6655
Lijiang Knotweed 309335
Lijiang Lady's Slipper 120383
Lijiang Ladyslipper 120383
Lijiang Larkspur 124348
Lijiang Leptodermis 226081
Lijiang Ligusticum 229350
Lijiang Lily 229898,229976
Lijiang Lime 391759
Lijiang Linden 391759
Lijiang Litse 233867
Lijiang Loosestrife 239712
Lijiang Meadowrue 388719
Lijiang Meconopsis 247123
Lijiang Milkvetch 42134
Lijiang Milkwort 308163
Lijiang Mockorange 294419
Lijiang Mosquitotrap 117576
Lijiang Oak 323840
Lijiang Onion 12990
Lijiang Onosma 271789
Lijiang Parnassia 284559
Lijiang Pearleverlasting 21592
Lijiang Platanthera 302391
Lijiang Pleurospermum 229350,304793
Lijiang Pycreus 322282
Lijiang Randia 196352
Lijiang Rhodiola 329902
Lijiang Rhubarb 329350
Lijiang Ribseedcelery 304793
Lijiang Rockfoil 349552
Lijiang Rose 336690
Lijiang Roughfruitparsley 393952

Lijiang Secamone 356299
Lijiang Sedge 74310
Lijiang Sicklehairicot 135412
Lijiang Skullcap 355570
Lijiang Snowgall 191934
Lijiang Southstar 33388
Lijiang Spiraea 371997
Lijiang Spruce 298331
Lijiang Stringbush 414207
Lijiang Swallowwort 117576
Lijiang Swertia 380179
Lijiang Synotis 381942
Lijiang Tailanther 381942
Lijiang Teasel 133495
Lijiang Thick-tooth Clematis 94965
Lijiang Thistle 92131
Lijiang Tillaea 391937
Lijiang Tubergourd 390156
Lijiang Whitlowgrass 137088
Lijiang Wildclove 226081
Lijiang Windhairdaisy 348492
Lijiang Woodbetony 287356
Like Bone Larkspur 124439
Like E. Bureau's Rhododendron 330265
Like Empetrum Phyllodoce 297024
Like Euonymus Leaves Rhododendron 330622
Like Furfuraceousfruit Camellia 69430
Like Gibbous Bamboo 47284
Like Hongkong Camellia 68921
Like Hypnum Cassiope 78630
Like Iteo Leaf Blueberry 403863
Like Littlespike Willow 343198
Like Lovely Rhododendron 330790
Like Lycopodum Cassiope 78631
Like Phillyrea Leavea Pieris 298788
Like R. Chrysanthum Rhododendron 331565
Like R. concinnum Rhododendron 330451
Like R. Faber's Rhododendron 330672
Like R. Roxieanum Rhododendron 331681
Like R. Sperabile Rhododendron 331868
Like Santalum Gaultheria 172074
Like Saura Larkspur 124563
Like Selago Cassiope 78645
Like Slendertail Elatostema 142868
Like Smallsepal Camellia 69459
Like Stellera Cassiope 78646
Like Strachey Corydalis 106479
Like Tendrilleaf Solomonseal 308522
Like Vaccinium Rhododendron 332046
Like Vagrant Spindle-tree 157941
Like Winteraconite Anemone 23807
Likest Spiraea 371793
Likiang Aster 40754
Likiang Bluebellflower 115382
Likiang Dolichos 135412
Likiang Draba 137088
Likiang Ephedra 146203
Likiang False Solomonseal 242693
Likiang Jerusalemsage 295132
Likiang Knotweed 309335
Likiang Larkspur 124348
Likiang Lily 229976
Likiang Loosestrife 239712
Likiang Onion 12990
Likiang Pleurospermum 304793
Likiang Rhodiola 329902
Likiang Rhubarb 329350
Likiang Saussurea 348492
Likiang Secamone 356299
Likiang Skullcap 355570
Likiang Spruce 298331
Likiang Swallowwort 117576
Likiang Woodbetony 287356
Likiu Cheirostylis 86698
Lilac 72934,382114,382117,382329
Lilac Chaste Tree 411189
Lilac Chastetree 411189
Lilac Chaste-tree 411189
Lilac Clematis 95216
Lilac Croton 112999
Lilac Dandelion 384633
Lilac Daphne 122438
Lilac Globeflower 399518
Lilac Hibiscus 18254
Lilac Honeysuckle 236073
Lilac Melaleuca 248090
Lilac Orchid Vine 246020
Lilac Penstemon 289346
Lilac Pink 127852
Lilac Pualele 144975
Lilac Sage 345471
Lilac Snapdragon 28584
Lilac Spapweed 205090
Lilac Sunbonnet 175701
Lilac Tasselflower 144975,144976
Lilac Vine 185764
Lilac Windhairdaisy 348668
Lilacbell 208264
Lilacbush 44868
Lilac-flower 250291
Lilac-flowered Wild Hyacinth 398808
Lilac-leaved Honeysuckle 235995
Lilies-of-the-field 23767
Liliput Gasteria 171688
Liljestrand Monkshood 5368
Lillo Barberry 51861
Lilly Pilly 4885
Lilly-convally 102867
Lillypilly 4885,156357,382481
Lilly-pilly 4885
Lilly-pilly Tree 4885
Lily 37015,68713,102933,162519,208566,228556,229717,229862,230019,262441,267730,404575
Lily Azalea 331107
Lily Bind 102933
Lily Can 267294
Lily Family 229700
Lily Leek 15480
Lily Magnolia 416707
Lily Oak 382329
Lily of Nile 417093
Lily of Paradise 283300
Lily of the Incas 18060,18074
Lily of The Nile 10245
Lily of the Nile 10271
Lily of the Valley 102856,102867
Lily of the Valley Shrub 298754
Lily of the Valley Tree 96464
Lily of the Valley Vine 344372
Lily Potato 39438
Lily Thorn 79376
Lily Tree 416694
Lily Turf 232617,232631,272049
Lily Twayblade 232225
Lily-among-thorns 236022
Lily-bind 68713,102933
Lily-can 267294
Lily-confancy 102867
Lily-constancy 102867
Lily-flower 68713
Lily-flowered Magnolia 416707
Lily-grass 37015
Lilyleaf Ladybell 7682
Lilyleaf Twayblade 232225
Lilyleaf-like Ladybell 7685
Lily-leaved Lady Bell 7682
Lily-leaved Ladybell 7682
Lily-leaved Twayblade 232225
Lily-leek 15480
Lily-like Rhododendron 331107
Lily-oak 382329
Lily-of-China 335760
Lily-of-the-incas 18060,18074
Lily-of-the-mountain 308607
Lily-of-the-nile 10244,10245,417093
Lily-of-the-valley 102856,102867
Lily-of-the-valley Bush 298734
Lily-of-the-valley Tree 96464,278389,298734
Lily-of-the-valley-like Eria 148650
Lilypad Begonia 50104
Lily-pad Begonia 49943,50104
Lily-royal 230044,250432
Lily-tree 416694
Lily-tuff 272049,272090
Lilyturf 232617,272049,272086
Lily-turf 232617
Lima Bean 294006,294010
Lima Wood 64975

Limaciopsis 230089
Limao 93423
Limba 386652
Limbate Bluegrass 305661
Limbate Otostegia 277339
Limber Honeysuckle 235760
Limber Pine 299942
Limberbush 212116
Limbricht Begonia 50013
Limbricht Cinquefoil 312728
Limbricht Cranesbill 174702
Lime 93329,93534,391646,391705,391706
Lime Berry 397878
Lime Family 391860
Lime Mountainash 369389
Lime Prickly Ash 417228
Lime Pricklyash 417228
Lime Tree 391646,391706
Limeberry 397878
Lime-fruit Tree 93329
Limeixing Apricot 34442
Lime-leaved Mallotus 243453
Lime-leaved Maple 2939
Lime-loving Ampelocalamus 20196
Lime-loving Stachyurus 373519
Limestone 131060
Limestone Bedstraw 170648
Limestone Bitter-cress 72740
Limestone Bugheal 395688
Limestone Bugle 13167
Limestone Calamint 65538
Limestone Cassia 360421
Limestone Cleidion 94405
Limestone Corydalis 105687
Limestone Cranesbill 174619
Limestone Fleabane 150984
Limestone Hawksbeard 110861
Limestone Indigo 205765
Limestone Longan 131060
Limestone Macaranga 240256
Limestone Meadow Sedge 74706
Limestone Michelia 252832
Limestone Ruellia 339819
Limestone Sedge 73990
Limestone Shagvine 97505
Limestone Skullcap 355443
Limestone Spiraea 371854
Limestone Swamp Bedstraw 170284
Limestone Waxpetal 106631
Limestone Wild Buckwheat 152044
Limestone Woundwort 373107
Lime-tree 93329
Limetta 93534
Limette 93534
Limewort 363214,407029
Limit Butterflybush 62108
Limitaneous Subalpine Arrowwood 408147
Limitaneous Subalpine Viburnum 408147
Limitas 332908
Limitedleaf Haworthia 186501
Limitless Rhododendron 330099
Limnophila 230323
Limodore 230378
Limon Balsa 268346
Limon Real 93456
Limonia 93546,230089
Limpernscrimp 192358
Limper-scrimp 192358
Limpet-scrimp 224375
Limpia Evening Primrose 269498
Limpo Grass 191227
Limpograss 191227
Limprecht Rockjasmine 23216
Limprichet Swallowwort 117566
Limpricht Bentgrass 12171
Limpricht Bittercress 275869
Limpricht Buttercup 326030
Limpricht Calanthe 65995
Limpricht Clematis 94876
Limpricht Daphne 122505
Limpricht Goldenray 229095
Limpricht Habenaria 183794
Limpricht Hemipilia 191612
Limpricht Milkvetch 42619
Limpricht Orchis 310914
Limpricht Rockfoil 350018
Limpricht Spicebush 231382
Limpricht Spice-bush 231382
Limpricht Storax 379400
Limpricht Taxillus 385208
Limpricht Windhairdaisy 348495
Limpricht Woodbetony 287359
Limpricht's Willow 343606
Limpwort 407029
Limy Albizia 13510
Limy Cinnamon 91285
Limy Rhododendron 330137
Lin 232001,391706
Lin Garcinia 171128
Lin Madder 337981
Lin Wei-Fang Garcinia 171128
Lin'an Maple 3102
Lin's Litse 233840
Lin's Physaliastrum 297625
Linaigrette 152734
Linaloa Oil 64068
Linaloe 64068
Linan Chen's Oak 323751
Linan Sikoku Pipewort 151495
Linarfruit Lignariella 228960
Linary 231183
Linaw Dendrobium 125221
Lincang Blueberry 403888
Lincang Casearia 78124
Lincang Fargesia 162702
Lincang Lovegrass 147775
Lincang Marsdenia 245869
Lincang Milkgreens 245869
Lincang Rockvine 387800
Lincang Stephania 375883
Lincang Umbrella Bamboo 162702
Linch Bulbophyllum 62900
Linch Curlylip-orchis 62900
Lincolnshire Asparagus 86970
Lincolnshire Spinach 86970
Lind 391706
Lindelofia 231243
Lindemann Broom 120970
Linden 391646,391649,391681,391706,
391727
Linden Arrowwood 407785
Linden Blood-leaf 208354
Linden Canistrum 71152
Linden Family 391860
Linden Hibiscus 195311,195321
Linden Viburnum 407785
Linden's Angurek 125694
Linden's Bloodleaf 208354
Linden-leaf Asiaglory 376536
Linden-leaf Christmasbush 14216
Lindenleaf Cowparsnip 192388
Linden-leaf Grewia 180995
Linden-leaf Mallotus 243453
Lindenleaf Sage 345432
Lindenleaf Torricellia 393110
Lindenleaf Wildtung 243453
Lindenleaf Xmas Bush 14216
Linden-leaved Grewia 180995
Linden-leaved Torricellia 393110
Linder-leaved Christmas-bush 14216
Lindernia 231527
Lindheimer Hackberry 80687
Lindheimer Senna 360451
Lindheimer's Beargrass 266498
Lindheimer's Croton 112941
Lindheimer's Muhly 259676
Lindheimer's Panic Grass 128483
Lindheimer's Rock Daisy 291320
Lindley Aster 40777
Lindley Azalea 331109
Lindley Barberry 51862
Lindley Begonia 50015
Lindley Bogorchid 158200
Lindley Butterfly Bush 62110
Lindley Butterflybush 62110
Lindley Butterfly-bush 62110
Lindley Dendrobium 125224
Lindley False Spirea 369288
Lindley Glorybower 96180
Lindley Odontoglossum 269075
Lindley Pear 323217

Lindley Pleurospermum 304817
Lindley Summerlilic 62110
Lindley Willow 343607
Lindley's Aster 40777,380852
Lindley's Butterfly Bush 62110
Lindley's Butterflybush 62110
Lindley's Clerodendrum 96180,96181
Lindley's Cotoneaster 107440
Lindley's Eupatorium 158200
Lindley's Rhododendron 331109
Lindley's Saltbush 44505
Lindley's Silver Puff 402587
Line 232001,391706
Line Fragrant Spicebush 231343
Linear Alpine Kobresia 217261
Linear Arisaema 33395
Linear Basketvine 9450
Linear Biformbean 20572
Linear Bitter-bamboo 304054
Linear Calyx Azalea 331110
Linear Calyx Rhododendron 331110
Linear Crazyweed 278829
Linear Cupulate Rhododendron 331111
Linear Desert Sagebrush 35380
Linear Kobresia 217156
Linear Koelpinia 217604
Linear Leaf Milkvetch 43247
Linear Leaves Rhododendron 331113
Linear Lobed Calyx Azalea 331115
Linear Lobed Calyx Rhododendron 331115
Linear Nepeta 264983
Linear Primrose 315024
Linear Rapanea Forgeirontree 261629
Linear Roxie's Rhododendron 331679
Linear Skullcap 355574
Linear Southstar 33395
Linear Spapweed 205092
Linear Stole Bulrush 353557
Linear Stonecrop 356884
Linear Thickleaf Chickweed 374809
Linear Touch-me-not 205092
Linear Waxplant 198865
Linearbract Crazyweed 278969
Linearbract Milkvetch 42856
Linearbract Thinleaf Carpesium 77171
Linearbracted Crazyweed 278969
Linearcalyx Lasianthus 222215
Linearcalyx Roughleaf 222215
Linearcalyx Silvianthus 364344
Linear-cordate Yam 131683
Linear-fruit Parabarium 402192
Linear-fruited Urceola 402192
Linearifolious Groundsel 358773
Linearifolious Hedyotis 187612
Linearifolious Loosestrife 239715
Linearifolious Melaleuca 248105
Linearifolious Pinellia 299725
Linearifolious Rockfoil 349980
Linearifolious Shamgoldquitch 318154
Linearifolious Solms-laubachia 59769
Linearifolious Tuberous Merremia 250798
Linearifolious Vine Catchfly 363973
Linearilobe Corydalis 105599
Linearleaf Alyssum 18403
Linearleaf Argyreia 32633
Linearleaf Aspidistra 39551
Linear-leaf Bamboo 304054
Linearleaf Bedstraw 170461
Linearleaf Bowwingpaesley 31327
Linearleaf Box 64271
Linearleaf Buckwheat 162321
Linearleaf Bugle 13131
Linearleaf Carpetweed 256695
Linearleaf Chalkplant 183213
Linearleaf Chirita 87901
Linearleaf Clearweed 298959
Linearleaf Coldwaterflower 298959
Linearleaf Corydalis 106085
Linearleaf Cremanthodium 110407
Linearleaf Cryptandra 113341
Linearleaf Cymbidium 116887
Linearleaf Delavay Holly 203757
Linearleaf Dendrobium 125003
Linearleaf Dichotomous Fimbristylis 166250
Linearleaf Dichotomous Star Wort 374842
Linearleaf Dontostemon 136173
Linearleaf Draba 137091
Linearleaf Dragonhead 137608
Linearleaf Dutchmanspipe 34282
Linearleaf Dysophylla 139569
Linearleaf Elatostema 142857
Linearleaf Euonymus 157684
Linearleaf Euphorbia 159249
Linearleaf Flemingia 166876
Linearleaf Gentian 173432,173625
Linearleaf Gladiolus 176023
Linearleaf Glorybind 103110
Linearleaf Greenorchid 137608
Linear-leaf Grevillea 180615
Linearleaf Habenaria 183797
Linear-leaf Holboellia 197241
Linearleaf Houndstongue 270705
Linearleaf Indigo 206181
Linearleaf Inula 207165
Linearleaf Japanese Windhairdaisy 348410
Linearleaf Jurinea 207165
Linearleaf Knotweed 162321
Linearleaf Kobresia 217131
Linearleaf Lasianthus 222216
Linearleaf Ligusticum 229368
Linearleaf Madwort 18403
Linearleaf Milkvetch 42784
Linearleaf Mosquitotrap 117527
Linearleaf Nutantdaisy 110407
Linearleaf Pseudopogonatherum 318151
Linear-leaf Raspberry 338748
Linearleaf Rattlebox 112340
Linearleaf Rhodiola 329897
Linearleaf Rockfoil 349555
Linearleaf Roughleaf 222216
Linearleaf Sandwort 32154
Linearleaf Sharpperianth Lily 229912
Linearleaf Shrubcress 368563
Linearleaf Skeletonweed 88771
Linearleaf Small Cowpea 408975
Linearleaf Snapdragon 28631
Linearleaf Soja 177780
Linearleaf Solms-Laubachia 368563
Linearleaf Sophora 369057
Linearleaf Soybean 177780
Linearleaf Speedwell 317924
Linearleaf Spoonleaf Anemone 23763
Linearleaf Spurge 159249
Linearleaf Squarebamboo 87567
Linearleaf Stairweed 142857
Linearleaf Swallowwort 117527
Linearleaf Synstemon 136173
Linearleaf Syzygium 382583
Linearleaf Thistle 92132
Linearleaf Thorowax 63562
Linearleaf Tickseed 104777
Linearleaf Tulip 400191
Linearleaf Vetchling 222807
Linearleaf Watercandle 139569
Linearleaf Waterdropwort 269335
Linearleaf Waterstarwort 67361
Linearleaf Whitlowgrass 137091
Linearleaf Willow 343111,343880,344273
Linearleaf Xizang Braya 59782
Linearleaf Yunnan Sandwort 32323
Linear-leaved Agapetes 10328
Linear-leaved Argyreia 32633
Linear-leaved Bamboo 304054
Linear-leaved Barberry 51863
Linear-leaved Basketvine 9446
Linear-leaved Box 64271
Linear-leaved Discoid Gumweed 150330
Linear-leaved Euonymus 157684
Linear-leaved False Willow 46236
Linear-leaved Flemingia Linearleaf Wurrus 166876
Linear-leaved Gardneria 171440
Linear-leaved Holboellia 197241
Linear-leaved Panic Grass 128525
Linear-leaved Spindle-tree 157684
Linear-leaved Sundew 138304
Linear-leaved Willow 343111
Linear-lobe Cinquefoil 312453
Linearlobe Corydalis 106370
Linearlobe Japanese Maple 3356
Linearlobe Yunnan Glorybower 96450

Linearlobed Pleurospermum 304818
Linear-lobed Rhododendron 331604
Linearlobed Ribseedcelery 304818
Linearpetal Bulbophyllum 62766
Linearpetal Habenaria 183637
Linearpetal Stonebean-orchis 62766
Linearpetap Catchfly 363688
Linearsegmented Corydalis 106083
Linearsepal Gambirplant 401768
Linearsepal Hookvine 401768
Linearsepal Mosquitotrap 117570
Linearsepal Rhodiola 329915
Linearsepal Spiny Raspberry 339134
Linearsepal Swallowwort 117570
Linear-sepaled Gambir-plant 401768
Linear-sepaled Rhododendron 331115
Linearspike Cypressgrass 119111
Linearspike Galingale 119111
Linearspike Sedge 75488
Linearstalk Bedstraw 170323
Linearstalk Pepper 300483
Linearstripe Rabdosia 209753
Linear-stypulate Willow 343610
Linearstypule Willow 343610
Lineartongue Aster 40135
Lineartooth Rabdosia 209632
Linear-toothed Rabdosia 209632
Lineate Basketvine 9450
Lineate Eargrass 187612
Lineate Hooktea 52436
Lineate Jackbean 71055
Lineate Raspberry 338751
Lineate Supplejack 52436
Lineate Supple-jack 52436
Lineate Syzygium 382601
Lineate Woodbetony 287361
Linedaisy 166065
Line-leaf Agapetes 10328
Line-leaf Ardisia 31504
Line-leaf Basketvine 9446
Line-leaf Eurya 160529
Line-leaf Greencalyx Camellia 69750
Line-leaf Leptopus 226333
Lineleaf Leucas 227626
Line-leaf Pond Cypress 385255
Line-leaf Privet Honeysuckle 236037
Lineleaf Rockvine 387801
Linen Buttons 102867
Linen-buttons 102867
Lineolate Elatostema 142717
Lineolate Yushania 416786
Line-parted Japanese Mulberry 259108
Linepetal Orchis 116781
Linetooth Rabdosia 209632
Ling 67455,67456,383491,401388
Ling Goldenray 229096
Ling Gowland 89704
Ling Groundsel 359362
Ling Heather 67456
Ling Maple 3103
Ling Nut 394426
Ling Sedge 75173
Ling's Azalea 331651
Ling's Hydrangea 199938
Ling's Rhododendron 331651
Ling's Sowthistle 368745
Lingbao Azalea 330852
Lingbao Henan Rhododendron 330852
Lingbao Larkspur 124351
Lingbao Rhododendron 330852
Lingberry 145073,403916,404051
Lingchuan Indosasa 206887
Lingelsheim Woodbetony 287364
Linglan Euonymus 157395
Lingling Chirita 87902
Lingnan Ailanthus 12587
Lingnan Artocarpus 36939
Lingnan Azalea 331202
Lingnan Loosestrife 239639
Lingnan Masson Pine 300059
Lingnan Mombin 372471
Lingnan Schizostachyum 351858
Lingonberry 404051,404052,404060
Lingshan Heterostemma 194167
Lingshui Cycas 115844
Lingshui Greenstar 307515
Lingshui Herbcoral 346534
Lingshui Pepper 300442
Lingua Gasteria 171690
Linguaramie 30950,30951
Lingue 291563
Lingui Hydrangea 199939
Lingui Hydrangeavine 351782
Lingui Kiwifruit 6656
Lingui Photinia 295656
Lingui Rhododendron 331117
Lingui Spicegrass 239717
Lingulate Epipactis 147149
Lingulate Holcoglossum 197262
Lingulate Windhairdaisy 348497
Lingwort 30932,405571
Lingyun Arisaema 33396
Lingyun Bauhinia 49157
Lingyun Clematis 95089
Lingyun Southstar 33396
Lingyun Toxocarpus 393542
Linhua Sage 345093
Linifolious Oligomeris 270343
Linjiang Corydalis 106086
Link Moss 357114
Link Plant 329705
Link's Blackberry 338754
Linking Together Rhododendron 332123
Linn 232001,391706
Linociera 231729
Linociera-like Slugwood 50550
Linosyris 231818
Linseed 232002
Lin-spurs Cap 172905
Linstow Corydalis 106087
Lint 232001,408571
Lint Bells 232001
Lint Bennels 232001
Lint Lacebark 219986
Lint Lacebark Tree 219986
Lint Lacebark-tree 219986
Lint Spurge 158857
Lint-bow 232001
Lintels 408423
Lintin 408571
Lint-spurge 158857
Liny Nan 295351
Liny Nanmu 295351
Linyun Ginger 417997
Linzhi Azalea 331363,331929
Linzhi Catchfly 364209
Linzhi Denseciliate Buttercup 325771
Linzhi Gentian 173673
Linzhi Goldenray 229122
Linzhi Meconopsis 247153
Linzhi Oxygraphis 278470
Linzhi Pimpinella 299486
Linzhi Primrose 314695
Linzhi Rhododendron 331363,331929
Linzhi Rockfoil 349506
Linzhi Sedge 74027
Linzhi Small Reed 127281
Linzhi Smallreed 127281
Linzhi Spapweed 205094
Linzhi Spruce 298345
Linzhi Touch-me-not 205094
Linzhi Vinegentian 110323
Linzhi Woodbetony 287464
Linzhou Windhairdaisy 348502
Liodendron 138611
Liolive 231729
Lion Citrus 93705,93711
Lion's Claw 111868
Lion's Claws 111868
Lion's Cudweed 224772
Lion's Ear 224558,224585,224600,224976
Lion's Foot 14065,14166,190944,224767,224772,313787
Lion's Heart 220285,297984
Lion's Herb 30081
Lion's Mouth 28617,28631,130383,176824,231183
Lion's Paw 14065
Lion's Snap 28617,170025,220355
Lion's Tail 10409,224585,224976
Lion's Tails 10409,224585,224976

Lion's Teeth 260596,384714
Lion's Tongue 231183
Lion's-beard 95016
Lion's-ear 224558,224585
Lion's-foot 313881
Lion's-Heart 297973
Lion's-Leaf 224625
Lion's-tail 85058,224976
Liongrass 191325
Lionhead Thelocactus 389219
Lion's Mane Maple 3334
Liontail Rhaphidophora 329000
Liontongue Masdevallia 246111
Liou Cuevedprickle Rose 336381
Liou Monkshood 5370
Liou Sedge 75175
Liou's Currant 334144
Liou's Milkvetch 42862
Liou's Willow 343611
Liparis 232072
Lipfern Stevenia 376389
Lipfernleaf Woodbetony 287084
Lipflower Honeysuckle 236131
Lipflower Larkspur 124127
Liping Begonia 50016
Liping Camellia 69284
Liping Cleyera 96605
Liping Photinia 295658
Lipo Indosasa 206888
Lipocarpha 232380,232403
Lipped Calanthe 65990
Lipped-flower Honeysuckle 236131
Lippia 232457,296121
Lipsky Aster 40782
Lipsky Craneknee 221698
Lipsky Larkspur 124352
Lipsky Lophanthus 236272
Lipsky Nepeta 264984
Lipsky Rex Woodbetony 287606
Lipsky's Carpesium 77172
Lipsky's Jurinea 214121
Lipstick Plant 9062,9451,9475,9476,54862
Lipsticktree 54862
Liqing Brome 60927
Liqing Sedge 75177
Liquid Amber 232565
Liquid Storax 232565
Liquidambar 232565
Liquorice 177873,177893
Liquorice Plant 271563
Liquorice Vetch 42417
Liquoriceleaf Corydalis 105938
Liquory Knot 222758
Liquory-knot 222758
Liquory-stick 271563
Liricon-fancy 102867
Liricum-fancy 102867
Liriope 232617,232631
Liry-confancy 102867
Lisamoo 192358
Lisbon False Sun-rose 184792
Lishan Machilus 240707
Lishan Monkshood 5214
Lishan Obtusesepal Clematis 95227
Lishizhen Barrenwort 147017
Lishizhen Epimedium 147017
Lishui Bitter Bamboo 304068
Lishui Bitterbamboo 304068
Lishui Bitter-bamboo 304068
Lishui Raspberry 338755
Lisianthus 161027
Lister Agapetes 10329
Lister Everspringtree 250876
Lister Merrilliopanax 250876
Lister Primrose 314587
Lister's Tanoak 233294
Listera 232890
Lit Tlegood 159027
Litang Barberry 52276
Litang Honeysuckle 235924
Litang Ligusticum 229351
Litang Rockfoil 349557
Litang Sandthorn 196752
Litceeleaf Azalea 331120
Litchi 233077,233078
Litchi Leaves Rhododendron 331120
Litchi Sage 345310
Litchi-leaf Ormosia 274434
Litchi-leaved Rhododendron 331120
Lithe Primrose 315045
Lithewort 260760,345594,407907
Lithodora 233434
Lithological Corydalis 106405
Lithophilous Bamboo 297334
Lithophilous Violet 410513
Lithuania Forgetmenot 260823
Lithuania Mannagrass 177626
Lithwort 345594
Lithy-tree 407907
Li-ti-ping Rhododendron 332093
Litle Kurz Lasianthus 222202
Litle Kurz Roughleaf 222202
Litosanthus 233809
Litse 233827
Litsea 233827
Litsealeaf Holly 203984
Litsea-leaved Cyclobalanopsis 116134
Litsea-leaved Greenstar 307507
Litsea-leaved Holly 203984
Litsea-leaved Oak 116134
Litseileaf Greenstar 307507
Litseleaf Elatostema 142721
Litseleaf Indianmulberry 258896
Litseleaf Machilus 240623
Litseleaf Maple 3105
Litseleaf Oak 116134
Litseleaf Qinggang 116134
Litseleaf Stairweed 142721
Litseleaf Tanoak 233295
Litseleaf Wendlandia 413794
Litse-leaved Machilus 240623
Litse-leaved Maple 3105
Litse-leaved Tanoak 233295
Littedale Ligusticum 229352
Little 360328
Little Addermonth Orchid 268004
Little Anther Small Reed 127280
Little Bachelor's Buttons 174877
Little Baliosperm 46969
Little bargepole grass 260113
Little Barley 198344
Little Beavertail Pricklypear 272807
Little Blue Cypress 85290
Little Blue Stem 341421
Little Bluebell 409804
Little Bluestem 351309
Little Blue-stem 351309,351312
Little Bodnier Lilyturf 272060
Little Boxleaved Cotoneaster 107380
Little Bristlegrass 361877
Little Brush 133478
Little Burdock 118008
Little Burreed 370096
Little Cabbage Tree 115232
Little Candles 244198
Little Cattail 401128
Little Centaury 81065
Little Chickweed 342279
Little Cich 222832
Little Club-spear Orchid 302287
Little Cokernut 212557
Little Common Alplily 234238
Little Cone 254090
Little Cottonrose 235267
Little Cranesbill 174877
Little Croton 113040
Little Darling 327866
Little Desert Trumpet 152536
Little Dlender Iris 208715
Little Duckbeakgrass 209341
Little Duckweed 224380
Little Eelgrass 418197
Little Epaulettetree 320880
Little Fair One 121001
Little False Solomon's-seal 242701
Little Fantasy 290315
Little Flame of God 384714
Little Floating Bladderwort 403302
Little Flower Alangium 13362
Little Flower Dipodium 133397
Little Flower Prickly-ash 417271

Little Flower Rhodoleia 332210
Little Forget-me-not 260760
Little Fragrant Bamboo 87634
Little Fruit Japanese Yew 385355
Little Fruit Privet 229549
Little Fruit Sophora 369066
Little Gagea 169479
Little Gem Arborvitae 390608
Little Gem Magnolia 242135
Little Glossodia 177315
Little Gold Poppy 155186
Little Golden Zinnia 418046
Little Goosefoot 87087
Little Gossips 273531
Little Green Sedge 76701
Little Groundcherry 297703
Little Hao Poplar 311345
Little Hogweed 311890
Little Honeysuckle 397019
Little Hop Clover 396882
Little Houseleek 356468
Little Ironweed 115595
Little Jack 144880,174877
Little Jan 174877
Little Jane 21339
Little Jen 174877
Little John 174877,374916
Little John Robin Hood 174877
Little Kurrajong 58338
Little Ladies' Tresses 372275
Little Ladies'-tresses 372274
Little Larkspur 124066
Little Leaf Astilbe 41828
Little Leaf Boxwood 64369
Little Leaf Cordia 104228
Little Leaf Rourea 337729
Little Lie-abed 360328
Little Love Grass 147815
Little Lovegrass 147815
Little Love-grass 147815
Little Lupine 238467
Little Mallow 243810
Little Man Orchid 178893
Little Marigold 383099
Little Medic 247381
Little Millet 282286
Little Mountain Anemone 23882
Little Mouse Ear 26385,83020
Little Mouse-ear 83020
Little Mouseear Cress 270462
Little Mugo Pine 300095
Little Mullein 148238
Little Muscadine Grape 411811
Little Naiad 262089
Little Nanophyton 262260
Little Nipple Cactus 244105
Little Oat 45402
Little Open Star 50825
Little Oreorchis 274040
Little Peep 21339
Little Peepers 21339
Little peganum 287983
Little Petelot's Camellia 69481
Little Phaseolus 294019,408975
Little Pink 349044
Little Pipsissewa 87489
Little Plantain 302107
Little Platanthera 302439
Little Pricklypear 272898
Little Prickly-pear 272898
Little Pyrethrum 322730
Little Quaking Grass 60383
Little Quakinggrass 60383
Little Quaking-grass 60383
Little Rattanpalm 65700
Little Rattlepod 42156
Little Red Robin 174877
Little Redflower Scurrula 355318
Little Rhynchosia 333323
Little River Black-eyed Susan 339554
Little Robin 174852,174877
Little Rockfoil 349756
Little Schizonepeta 264869
Little Seed Spruce 298307
Little Serradella 274888
Little Shellbark Hickory 77920
Little Shin-leaf 322853
Little Silverbell 184749
Little Smooth Speedwell 407358
Little Sophora 369003
Little Sour Bamboo 4604
Little Speargrass 4123
Little Spike Willow 343691
Little Spurflower 303561
Little Spurge 158857
Little Square Bamboo 87557
Little Star 50825,409339
Little Starwort 374897
Little Statice 230681
Little Stonecrop 356468,356733,356736
Little Sunflower 188564,189038
Little Sur Manzanita 31114
Little Sweet Betsy 397551
Little Trumpet 152536
Little Valerian 404439
Little Walnut 212627,212646
Little Washington Lily 230071
Little White Bells 102867
Little White Bird's-foot 274888
Little Wild Sage 35505
Little Wildorchis 274040
Little Window-palm 327540
Little Yellow-rattle 329543
Little-and-pretty 11947,243209,349044,407072
Little-anther Alkaligrass 321337
Littleanther Russianthistle 344627
Littlebeaked Peonygrass 182634
Little-caalyx Hartia 376447
Littlecalyx Belltree 325000
Littlecalyx Woodbetony 287428
Little-candles 244198
Littlecone Veitch Fir 499
Littleconical Rabdosia 209646
Littledalea 234122
Littleflower Agrimonia 11568
Littleflower Alangium 13362
Littleflower Alumroot 194426
Littleflower Appendicula 29818
Littleflower Basil 268565
Littleflower Betony 295045,295048
Little-flower Bigginseng 241179
Little-flower Bodinier Clethra 96473
Littleflower Bugleweed 239235
Little-flower Camellia 69437
Littleflower Catnip 264995
Littleflower Chelonopsis 86816
Littleflower Chickweed 374986
Littleflower Cinchona 91079
Littleflower Citrus 93634
Littleflower Clovershrub 70797
Little-flower Combretum 100322
Littleflower Currant 334076
Littleflower Cyrtanthus 120542
Littleflower Dontostemon 136179
Littleflower Dry Catchfly 363609
Littleflower Drymonia 138526
Littleflower Dutchmanspipe 34377
Littleflower Edelweiss 224825
Littleflower Elaeagnus 142081
Littleflower Epaulettetree 320885
Littleflower Figwort 355245
Littleflower Fritillary 168538
Littleflower Fuchsia 168780
Littleflower Gentian 130922
Littleflower Germander 388178
Littleflower Globeflower 399520
Littleflower Glycosmis 177845
Littleflower Greenstar 307497
Littleflower Gugertree 350937
Littleflower Hartia 376455
Littleflower Hiptage 196842
Littleflower Hoarhound 245756
Little-flower Holboellia 197247
Little-flower Holly 204558
Littleflower Jellygrass 252244
Littleflower Lady's Slipper 120414
Littleflower Ladybell 7697
Little-flower Linden 391749
Littleflower Lockhartia 235135
Littleflower Machilus 240668

Little-flower Magnoliavine 351069
Littleflower Marshweed 230311
Littleflower Melaleuca 248110
Littleflower Meliosma 249429
Littleflower Mesona 252244
Littleflower Milkwort Corydalis 106279
Littleflower Morina 258865
Littleflower Mosquitotrap 117454
Littleflower Nepeta 264995
Littleflower Nettle 402854
Littleflower Ophiorrhiza 272241
Littleflower Orange 93634
Littleflower Ottochloa 277431
Littleflower Panzerina 282487
Littleflower Paraphlomis 283638
Littleflower Parviflorous Whitlowgrass 137180
Littleflower Pearlwort 342252
Littleflower Phoebe 295384
Little-flower Photinia 295772
Littleflower Plumbagella 305163
Littleflower Pricklyash 417271
Little-flower Pterolobium 320604
Little-flower Quickweed 170142
Little-flower Quiekweed 170142
Littleflower Raspberry 338413,338886
Littleflower Rhodoleia 332210
Littleflower Rockfoil 349977
Littleflower Rose 336741
Littleflower Rosularia 337318
Littleflower Silene 363259
Littleflower Small Reed 127276
Littleflower Smallreed 127276
Littleflower Sophora 369109
Littleflower Spapweed 205141
Little-flower Square Bamboo 87587
Littleflower Squaregrass 226369
Littleflower Stelis 374710
Littleflower Stemona 375349
Littleflower Sterculia 376142
Littleflower Stringbush 414216,414235
Littleflower Sundrops 269481
Little-flower Sweetspire 210407
Little-flower Tamarisk 383581
Little-flower Tigerthorn 320604
Littleflower Touch-me-not 205141
Littleflower Twistedstalk 377933
Littleflower Urophyllum 402635
Littleflower Uvaria 403569
Littleflower Valerian 404309
Littleflower Wendlandia 413808
Littleflower Whitlowgrass 137180
Little-flower Wild Buckwheat 152327
Littleflower Woodbetony 287427
Little-flowered Gugertree 350937
Little-flowered Melaleuca 248110
Little-flowered Raspberry 338413
Little-flowered Stringbush 414216
Little-flowered Uvaria 403569
Littlefruit Aspidopterys 39684
Littlefruit Bittersweet 80206
Littlefruit Canton Salomonia 344349
Littlefruit Chinense Maple 3622
Littlefruit Clethra 96546
Littlefruit Coriaceousleaf Maple 2907
Littlefruit Cranberry 278360
Littlefruit Elegant Raspberry 338123
Littlefruit Forkstamenflower 416504
Littlefruit Gardenia 171374
Littlefruit Grape 411568
Littlefruit Griffith Lacquertree 393459
Little-fruit Hairyvein Maple 3483
Littlefruit Hooker Lacquertree 393462
Littlefruit Illigera 204639
Littlefruit Jujube 418197
Littlefruit Lyonia 239392
Littlefruit Machilus 240645
Littlefruit Mahonia 242494
Littlefruit Mallotus 243391
Littlefruit Markingnut 357882
Littlefruit Pittosporum 301356
Littlefruit Purplebloom Maple 3480
Littlefruit Raspberry 338123
Littlefruit Rockfoil 349642
Littlefruit Seatung 301356
Littlefruit Tanoak 233317
Littlefruit Trichilia 395482
Littlefruit Trigonotis 397434
Littlefruit Tutcher Maple 3731
Littlefruit Tutcheria 322596
Littlefruit Uvaria 403521
Littlefruit Wallich Lacquertree 393496
Littlefruit Winterhazel 106658
Little-fruited Machilus 240645
Little-fruited Pittosporum 301356
Little-fruited Sophora 369066
Little-fruited Tanoak 233317
Little-fruited Tutcheria 322596
Littleglume Helictotrichon 190193
Little-goodie 159027
Little-guid 159027
Littlehair Michelia 252932
Littlehead Japan Edelweiss 224880
Littlehip Hawthorn 110050,110054
Littleleaf Abelia 200
Little-leaf Ailanthus 12561
Littleleaf Alangium 13365
Littleleaf Alpine Meadowrue 388410
Little-leaf Angelica 357789
Littleleaf Aralia 30666
Littleleaf Ash 167981
Littleleaf Bamboo 297385
Littleleaf Banfenghe 357973
Littleleaf Beancaper 418703
Littleleaf Bedstraw 170222
Littleleaf Bigbloom Magnoliavine 351049
Littleleaf Black Thorowax 63833
Littleleaf Blueberry 403956
Little-leaf Blume Spiraea 371843
Littleleaf Box 64375
Littleleaf Bredia 59862
Littleleaf Buckthorn 328816
Littleleaf Bulbophyllum 63139
Littleleaf Bushbeech 178029
Littleleaf Buttercup 326084
Little-leaf Buttercup 325506
Littleleaf Canarytree 71014
Littleleaf Chikpea 90821
Littleleaf Childvine 400887
Little-leaf Chu-lan Aglaia 11302
Littleleaf Clearweed 298974
Littleleaf Climbing Fig 165520
Littleleaf Coldwaterflower 298974
Littleleaf Coliseum Maple 2866
Littleleaf Confused Storax 379328
Littleleaf Cordia 104227
Littleleaf Cotoneaster 107549
Littleleaf Crazyweed 279014
Little-leaf Creeper 285167
Littleleaf Dandelion 384564
Littleleaf Deervetch 237768
Littleleaf Denseflower Milkwort 308419
Littleleaf Diels Cotoneaster 107425
Littleleaf Distichous Cotoneaster 107600
Littleleaf Dogwood 380486
Littleleaf Elaeocarpus 142373
Littleleaf Elm 401602
Littleleaf Emarginate Eurya 160460
Littleleaf Epimedium 147031
Littleleaf Euphorbia 159358
Littleleaf Field Lacquertree 393484
Littleleaf Fieldhairicot 138973
Little-leaf Franchet Holly 203834
Little-leaf FritschSpiraea 371918
Littleleaf Greenbrier 366466
Little-leaf Greenbrier 366466
Littleleaf Hainan Munronia 259880
Littleleaf Hairorchis 396476
Littleleaf Hartia 185970
Littleleaf Helwingia 191150
Littleleaf Herb seatung 350331
Littleleaf Hillcelery 276775
Littleleaf Holly 204418
Little-leaf Holly 204571
Littleleaf Honeysuckle 235952
Littleleaf Hooktea 52450
Littleleaf Horsebrush 387079
Little-leaf Horsebrush 387079
Littleleaf Indianmulberry 258910
Little-leaf Indocalamus 206894
Little-leaf Indosasa 206894

Littleleaf Italian Jasmine 211872
Littleleaf Japan Eurya 160508
Littleleaf Japanese Germander 388118
Littleleaf Jasminorange 260171
Littleleaf Lead Tree 227434
Littleleaf Leadtree 227434
Littleleaf Leptodermis 226099
Little-leaf Leptopus 226335
Littleleaf Leucaena 227434
Littleleaf Lilac 382266
Littleleaf Linden 391691
Little-leaf Linden 391691,391697
Littleleaf Lysionotus 239959
Littleleaf Madder 338024
Littleleaf Maesa 241815
Little-leaf Mahonia 242590
Littleleaf Maple 3099
Littleleaf Memecylon 250065
Littleleaf Meranti 362215
Little-leaf Meyer Lilac 382205
Littleleaf Mitragyna 256121
Littleleaf Mitreola 256215
Littleleaf Mountainash 369464
Littleleaf Muddy Snowbell 379328
Littleleaf Mulberry 259164
Little-leaf Mulberry 259164
Littleleaf Myoporum 260720
Littleleaf Negundo Chastetree 411379
Littleleaf Olive 71014
Littleleaf Oriental Grape 411686
Littleleaf Ormosia 274412
Little-leaf Osmanthus 276373
Littleleaf Palo Verde 284482
Littleleaf Paphiopedilum 282788
Littleleaf Paraquilegia 283726
Littleleaf Pea Shrub 72285
Littleleaf Pearweed 239775
Littleleaf Peashrub 72285
Littleleaf Pepper 300338,300533
Littleleaf Pertybush 292063
Littleleaf Phoebe 295383
Littleleaf Photinia 295747
Littleleaf Pittosporum 301359
Littleleaf Pocktorchid 282788
Littleleaf Pondweed 312083
Littleleaf Porandra 311650
Littleleaf Premna 313710
Little-leaf Pricklythrift 2227
Littleleaf Privet 229525
Littleleaf Procumbent Mallow 402269
Littleleaf Pussytoes 26470
Littleleaf Pyrola 322852
Littleleaf Qinggang 116153
Littleleaf Qinghai Willow 343973
Littleleaf Rabdosia 209792
Littleleaf Raspberry 338886
Littleleaf Ratany 218077
Littleleaf Redflower Cotoneaster 107662
Littleleaf Robiquetia 335040
Littleleaf Rock Cotoneaster 107490
Littleleaf Rourea 337729
Littleleaf Scaevola 350331
Littleleaf Schefflera 350745,350757
Littleleaf Sebaea 356122
Littleleaf Semiliquidambar 357973
Littleleaf Sicklebamboo 137861
Littleleaf Sida 362494
Littleleaf Siebold Maple 3608
Littleleaf Smilax 366466
Little-leaf Sophora 369067
Littleleaf Spiradiclis 371772
Littleleaf Stairweed 142784
Littleleaf Stephania 375902
Littleleaf Stonebean-orchis 63139
Littleleaf Sugarpalm 32340
Little-leaf Sumac 332716
Littleleaf Supplejack 52450
Littleleaf Tea 69443
Littleleaf Ternstroemia 386720
Little-leaf Toothless Vetch 408386
Littleleaf Tree Millettia 254814
Littleleaf Ungeargrass 36682
Littleleaf Viburnum 407802,408019
Littleleaf Wendlandia 413793,413801
Littleleaf Wildclove 226099
Littleleaf Willow 343516
Littleleaf Woodbetony 287104
Little-leaf Wrinkledfruit Camellia 69570
Littleleaf Yunnan Holly 204418
Littleleaf Yunnan Ostryopsis 276850
Littleleaf Yunnan Poplar 311585
Little-leafed Sage 345213
Little-leaved Abelia 200
Little-leaved Amur Linden 391666
Little-leaved Aralia 30666
Little-leaved Dogwood 380486
Little-leaved Horse Bean 83748
Little-leaved Indian-mulberry 258910
Little-leaved Lime 391691
Little-leaved Maesa 241815
Little-leaved Memecylon 250065
Little-leaved Mint 250368
Little-leaved Mock Orange 294501
Little-leaved Mock-orange 294501
Little-leaved Mountain Mahogany 83818
Little-leaved Pea Shrub 72285
Little-leaved Perty-bush 292063
Little-leaved Pittosporum 301359
Little-leaved Premna 313710
Little-leaved Rabdosia 209792
Little-leaved Rourea 337729
Little-leaved Scaevola 350331
Little-leaved Stephania 375902
Little-leaved Ternstroemia 386720
Little-leaved Tunic-flower 292665
Littlelip Woodbetony 287429
Littlemouth Onosma 242400
Littlepod False Flax 68850
Little-pod False Flax 68850
Littlerhomboidleaf Bluebellflower 115396
Littleroot Woodbetony 287382
Littleseed Canarygrass 293743
Littleseed Densebranchlet Savin 341724
Littleseed Jaeschkea 211404
Littleseed Jointfir 178548
Little-seed Juniper 213748
Littleseed Spruce 298307
Little-seeded Canary Grass 293743
Little-seeds Beadtree 7187
Littlesepal Stonecrop 356937
Little-serrated Willow 343840
Littlespike Plectocomia 303094
Little-spike Statice 230777
Littlespike Willow 343691
Little-spiked Plectocomia 303094
Littlespur Corydalis 105622
Littlespur Spapweed 205143
Littlespur Touch-me-not 205143
Littlespur Violet 410647
Littletail Clearweed 298968
Littletail Coldwaterflower 298968
Little-tooth Basketvine 9433
Littletooth Hydrangeavine 351802
Littletree Willow 343042
Little-tree Willow 343042
Littlewale 233760
Littonia 234130
Littoral Oak 116133
Littoral Spinifex 371724
Littoral Spurge 158490
Litvinov Fescuegrass 164056
Litvinowii Junegrass 217487
Litwinov Colewort 108626
Litwinov Petrosimonia 292723
Litwinov Poppy 282597
Litwinow Broom 120972
Litwinow Chalkplant 183215
Litwinow Peashrub 72276
Litwinow Pea-shrub 72276
Litwinow Rhodiola 329905
Litwinow Skullcap 355576
Litwinowia 234154
Liu Onosma 271792
Liu Raspberry 338756
Liu Spikesedge 143193
Liu's Euphorbia 159251
Liugu 74760
Liujiang Aspidistra 39569
Liujiang Chirita 87903
Liukiu Buckthorn 328767
Liukiu Greenbrier 366484

Liukiu Japan Pieris 298758
Liukiu Privet 229524
Liukiu Slimstem Lily 229776
Liupanshan Claw Rhubarb 329402
Liupanshan Crazyweed 279035
Liupanshan Sweetvetch 188042
Liuqiu Arrowwood 407867
Liuqiu Azalea 331759
Liuqiu Bitterbamboo 304054
Liuqiu Box 64373
Liuqiu Buckthorn 328767
Liuqiu Cowpea 409036
Liuqiu Euonymus 157698
Liuqiu Glochidion 177153
Liuqiu Greenstar 307508
Liuqiu Juniper 213962,213963
Liuqiu Ninenode 319671
Liuqiu Persimmon 132291
Liuqiu Pine 300044
Liuqiu Raphiolepis 329096
Liuqiu Skimmia 365965
Liuqiu Slimstem Lily 229776
Liuqiu Starjasmine 393656
Liuqiu Wedelia 413507
Liuzhou Elaeagnus 142057
Live Forever 138829,138928,200802
Live Long 356467
Live Oak 323762,323873,324544
Live-forever 26385,138836,178481,200784,
200812,357203
Live-for-ever 356467
Live-forever Sedum 357203
Liveforver 358015
Live-in-Idleness 410677
Live-long 21596,221150,357203,404924
Livelong Saxifrage 349054
Live-long Saxifrage 349054
Live-longand-love-long 357203
Livelong-love-long 357203
Liver 68189,208771
Liver Leaf 192120
Liver Lily 208924
Liver-Balsam 151098
Liverleaf 192120,192126,192134
Liver-leaf 192123,192126,192148
Liver-leaf Hepatica 192134,192148
Liver-leaf Kidney 192126
Liver-leaf Wintergreen 322787
Livermore Nailwort 284878
Livermore Sandwort 32029
Liversidge Leptospermum 226463
Liverwort 11549,158062,192120,192126,
192134,321643
Livid Amaranth 18670
Livid Sedge 75184
Living Idols 410677
Living Rock 25179,33206,33209,33217,
304340
Living Stones 233459,233500
Living tree Bladderwort 403799
Livingbaseball 159485
Living-baseball 159485
Living-rock Cactus 25179,33206
Livingston Garcinia 171129
Livingstone Daisy 136394,136395,251255
Living-stones 233459,233624
Livistona 234162
Lixian Azalea 331994
Lixian Cranesbill 174844
Lixian Dogwood 380506
Lixian Fritillary 168605
Lixian Goldsaxifrage 90401
Lixian Goldwaist 90401
Lixian Monkshood 5365
Lixian Nakeaster 182347
Lixian Rabdosia 209746
Lixian Rhododendron 331994
Lixian Rockfoil 349559
Liyuying's Litse 233982
Lizard Arum 348022
Lizard Flower 196376
Lizard Orchid 64018,196376
Lizard Orchis 196372
Lizard Plant 215129
Lizard Tail 24178,152832
Lizard's Tail 109180,348082,348085
Lizard's-tail 348082,348085
Lizard's-tail Family 348077
Lizards Tail 348082,348085
Lizardtail 348082,348085
Lizard-tail 172203
Lizardtail Family 348077
Lizardtaro 348022
Lizzie-by-the-hedge 176824
Lizzie-in-the-hedge 170193
Lizzie-run-the-hedge 176824
Lizzie-run-up-the-hedge 176824
Lizzory 369541
Lleyn Cotoneaster 107720
Lloyd Cactus 244131
Lloyd's Hedgehog Cactus 140316
Lloyd's Mariposa Cactus 140140,354302
Lloyd's Raspberry 338757
Lloydii Poplar 311154
Llume Palm 172225
Lluslow's Peony 280222
Lluvia De Orquideas 101778
Loaves-of-bread 201389,243840
Lobate Buttercup 326036
Lobate Campion 363337
Lobate Goldfishflower 255398
Lobate Larkspur 124675
Lobate Nanocnide 262249
Lobate Tibet Neillia 263175
Lobate Wildcotfon 402245
Lobate Yam 131686
Lobateleaf Ladybell 7686
Lobate-leaf Storax 379464
Lobb Barberry 51870
Lobb Californiapoppy 155182
Lobb Ceanothus 79953
Lobb Cornera 104891
Lobb Dwarf Cryptomeria 113698
Lobb Gonocaryum 179444
Lobb Oak 116136
Lobb's Climbing Nasturtium 399605
Lobb's Gooseberry 334070
Lobb's Wild Buckwheat 152231
Lobbiana False Pothos 313170
Lobbiana Pothoidium 313170
Lobe Agapetes 10330
Lobe Nettle 402952
Lobeapiculate Beautyberry 66838
Lobe-bracted Blumea 55780
Lobed Actinostemma 6907
Lobed Addermonth Orchid 110635
Lobed Amur Linden 391667
Lobed Arisaema 33397
Lobed Asiatic Currant 334233
Lobed Bogorchis 110635
Lobed Bract Blumea 55780
Lobed Brassaiopsis 59218
Lobed Bugle 13132
Lobed Cudweed 35837
Lobed Currant 334057
Lobed David Gooseberry 333952
Lobed Densehead Mountainash 369306
Lobed Drummond Red Maple 3522
Lobed Euaraliopsis 59218
Lobed Fleabane 150761
Lobed Garcinia 171125
Lobed Heliciopsis 189972
Lobed Hibiscus 194973
Lobed Kudzuvine 321441
Lobed Large Moonflower 67738
Lobed Maple 3500
Lobed Monkshood 5371
Lobed Mountain Ash 369306
Lobed Palmleaftree 59218
Lobed Southstar 33397
Lobed Spatulateleaf Willow 344130
Lobed Spiraea 371999
Lobed Thin Fruit-cup Litse 234090
Lobed Violet 410612
Lobedbeak Woodbetony 287665
Lobedbract Blumea 55780
Lobedfruit Tacca 351383
Lobedleaf Arrowwood 407919
Lobedleaf Bluebell 115383
Lobedleaf Bluebellflower 115383
Lobedleaf Elm 401542

Lobedleaf Epimedium 147018
Lobed-leaf Hopbush 135206
Lobedleaf Lagotis 220191
Lobedleaf Lepistemon 225710
Lobedleaf Nettle 402916
Lobedleaf Photinia 295702
Lobedleaf Primrose 314462
Lobedleaf Raspberry 338258,338759
Lobedleaf Siraitia 365330
Lobedleaf Speedwell 407432
Lobedleaf Starfruit 41733
Lobedleaf Starfruitestraw 41733
Lobedleaf Viburnum 407919
Lobedleaf Whinghead 320426
Lobedleaf Whinghead flower 320426
Lobed-leaf White Mulberry 259077
Lobed-leaved Fig 165035
Lobedsepal Bellsepal Sage 344929
Lobedsepal Raspberry 339081
Lobel Maple 3107
Lobel's Catchfly 363214
Lobel's Maple 3107
Lobeleaf Groundsel 279925
Lobeleaf Incarvillea 205559
Lobeleaf Pennywort 200282
Lobeleaf Yichang Arrowwood 407819
Lobe-leaved American Linden 391650
Lobelia 234275,234447,234867
Lobelia Family 234890
Lobelip Laelia 219685
Loberg Norway Maple 3436
Lobgawl Dandelion 384643
Lob-grass 60853
Lobinlly Tree 114574
Lobivia 234936
Loblolly 18687,242135,404439
Loblolly Bay 179729,179761
Loblolly Magnolia 242135
Loblolly Pine 300264
Lobster Claw 189986,190005,190052
Lobster Claws 96639,154647,189986,190031,412343
Lobster Flower 159675
Lobster Plant 214379
Lobster's Claw 96639
Lobster-claw 96639,154647,190018,190031
Lobster-claw Family 190073
Lobster-claws 189986
Lobulate Chiritopsis 88000
Lobulate Fourstamen Maple 3686
Lobulate Spiraea 371999
Lobulate Winteraconite 148107
Lobulate-leaved Raspberry 338759
Local Bambusa 47311
Local Milkvetch 266463
Local Tianqi 373633
Localginseng 383293
Loch Leven Spearwort 325501
Locheng Photinia 295727
Lock Elm 401500
Lockan Gowan 399504
Locken Gowan 399504
Locken Gowlan 399504
Locker Goulon 399504
Lockergoulon 399504
Locket-and-chain 220729
Lockets-and-chains 220729
Lockhartia 235131
Lockin-ma-gowan 399504
Lock-leaved Salsify 394327
Lockrea Gowlan 399504
Lockren Gowlan 399504
Locks-and-keys 2837,3462,145487,167955,220729,273531,336419,391706
Lockyer-golden 399504
Loco 41906
Locoweed 278679,278757,279156,279174
Loco-weed 278940
Locust 43480,64425,83527,176860,176903,200843,334946,334976
Locust Bean 43480,64425,79243,79247,83527,176860,200843,284444,334946
Locust Been Gum 83527
Locust Berry 64429
Locust Gum 83527
Locust Tree 83527,334976
Locust Wood 43475
Locustleg 261398,261399
Locusts 329521
Locust-tree 83523,83527
Locust-weed 78284
Loddges' Latania 222634
Loddiges Cattleya 79560
Loddiges Dendrobium 125233
Loddon Pondweed 312194
Lodewort 325580
Lodgepole Pine 299869,299874
Lodh Bark 381379
Loebner Magnolia 241956
Loeoweed 41906
Loesel Garliccress 365529
Loesel Liparis 232229
Loesel Twayblade 232229
Loesel Waterplantain 14749
Loesel's Liparis 232229
Loesel's Twayblade 232229
Loesener Bittersweet 80296
Loesener Staff-tree 80296
Loeseneriella 235202
Lofty Fescuegrass 163922
Lofty Fig 164626
Lofty Hopea Big-flower Hopea 198189
Lofty Pine 300297
Lofu Reevesia 327365
Lofu Sterculia 376192
Lofu Suoluo 327365
Lofushan Bamboo 297336
Loganberry 338761
Logania Family 235251
Loggerheads 81214,81237,81356
Loggerum 81237
Logwood 184513,184518
Lohan Pine 306457
Lohfau Holly 203987
Loho Dendrobium 125235
Lohui Machilus 240624
Loja Barberry 51880
Lokao Buckthorn 328713
Lokis Wool 152769
Lokki's-oo 152769
Lolagbola 278631
Loliondo 270074,270176
Lollilop Plant 279769
Lolly Bush 96076
Lolog Barberry 51871
Lomaria-leaved Mahonia 242608
Lomatia 235402
Lomatogoniopsis 235427
Lombardy Poplar 311398,311400
Lomboy Blanco 212121
Lomonosow's Olgaea 270239
Lompoc Manzanita 31135
Lonachies 144661
Lonas 235503
Londen Plane 302582
Londen Planetree 302582
London Basket 175444,175454
London Bells 68713
London Bob 127607,175444
London Bobs 127607,175444
London Bottle 316127
London Bottles 316127
London Bur-marigold 53845
London Daisy 227533
London Flag 208583
London Green Sauce 339887
London Pink 174877
London Plane Tree 302582
London Pretty 349044
London Pride 35088,127607,138348,349044,350011,356468,363312
London Pride Saxifrage 350011
London Rocket 365503
London Tuft 127607,349044
Londonpride 350011
Londonpride Saxifrage 350011
London-rocket 365503
Lone Fleabane 150540,151033
Lone Manzanita 31124
Lone Mountain Serpentweed 392775
Lonely Lily 148264

Lonesome Lady 72934
Long Acuminate Gaultheria 172080
Long and Narrow Ironweed 406123
Long Arrowleaf Knotweed 291775
Long Auricle Indocalamus 206801
Long Awned Taiwan Roegneria 335314
Long Beaks 350425
Long Birthwort 34251
Long Bract Persimmon 132260
Long Broadbract Crazyweed 278955
Long Calyx Bredia 59857
Long Calyx Rhododendron 331128
Long Cherry 105111
Long Cyperus 119125
Long Drupe Actinodaphne 233830
Long Drupe Litsea 233830
Long Drupetea 322605
Long Falcate-fruit Rhododendron 331129
Long Feathergrass 376814
Long Flower Heliophilous Blastus 55138
Long Flowered Potatobean 29306
Long Glans Oak 116139
Long Harbertleaf Knotweed 309356
Long Heartwort 34251
Long Heath 67456
Long Inflorescence Cryptocarya 113470
Long Jack 166985,397931
Long John 397934
Long Knight's-spur 102843
Long Leaf Pondweed 312194
Long Leaf Wax Flower 294868
Long Legs 315082
Long Linear-leaved Barberry 51865
Long Lovegrass 147656
Long Mamma 107031
Long Mammilated Cactus 135689
Long Marked Azalea 331839
Long Marked Rhododendron 331839
Long Monkshood 5167
Long Moss 392052
Long Peasen 294056
Long Pedicel Chesneya 87234
Long Peduncle Kadsura 214972
Long Peduncled Aster 40326
Long Pepper 300446,300495
Long Petiolate Beech 162386
Long Petiole Beech 162386
Long Petioled Kalimeris 215353
Long Plantain 302034
Long Pondweed 312232
Long Pricklyhead Poppy 282515
Long Prickly-headed Poppy 282515
Long Purples 37015,130383,240068,273531
Long Quackgrass 144666
Long Racemose Gooseberry 334072
Long Rape 59493
Long Rayed Triteleia 398806
Long Rough-headed Poppy 282515
Long Scales Rhododendron 331125
Long Sedge 74587
Long Smooth-headed Poppy 282546
Long Spike Tamarisk 383479
Long Spine olive 415239
Long Stalk Fordiophyton 167308
Long Stalk Metalleaf 296375
Long Stephania 375884
Long Style Rhododendron 331990
Long Trumpet 152318
Long Yellow Daylily 191266
Long'an Begonia 50023
Long'an Lilyturf 272117
Long's Sedge 75197
Longacuminate Basketvine 9418
Long-acuminateleaf Copperleaf 1916
Long-ament Chinese Poplar 311372
Long-ament Dali Willow 343274
Longan 131057,131061
Longan Aspidistra 39552
Long-and-love-long 357203
Longan-like Tanoak 233297
Longanoid Tanoak 233297
Longanshape Tanoak 233297
Longanther Azalea 330501
Longanther Eight Trasure 200802
Longanther Lilyturf 272120
Longanther Stonecrop 200802
Longaristate Kobresia 217214
Long-aristate Metalleaf 296393
Longarm Bulbophyllum 62861
Longarm Curlylip-orchis 62861
Longauricle Indocalamus 206801
Long-auricled Indocalamus 206801
Long-auriculate Yushania 416787
Longawn Alopecurus 17541
Longawn Barnyardgrass 140356
Long-awn Duckbeakgrass 209271
Longawn Fox Tail 17541
Longawn Goosecomb 335291
Longawn Herbclamworm 398091
Longawn Metalleaf 296393
Longawn Roegneria 335291
Longawn Tripogon 398091
Long-awned Bracted Sedge 74709
Long-awned Spineflower 89099
Long-awned Wood Grass 58450,58452
Longaxis Azalea 331606
Longbag Hairorchis 148763
Longbarb Arrowhead 342374
Long-bard Begonia 50025
Longbeak Arrowhead 342375
Longbeak Buttercup 326417
Longbeak Chervil 28029
Longbeak Eucalyptus 155517
Longbeak Evodia 161389
Longbeak Gerbera 175178
Longbeak Leek 15563
Longbeak Magnolia Houpu 198700
Longbeak Meadowrue 388559
Longbeak Monkshood 5225
Longbeak Onion 15563
Longbeak Ranalisma 325270
Longbeak Rattlebox 112353
Longbeak Stork's Bill 153738
Longbeak Woodbetony 287404
Long-beaked Bald Sedge 333687
Long-beaked Bald-rush 333687
Long-beaked Evodia 161389
Long-beaked Oak Sedge 75226
Long-beaked Pavlov Milkvetch 43266
Longbeaked Sedge 75189
Long-beaked Sedge 76333
Long-beaked Water-crowfoot 326045
Longbeakorchis 399819,399820
Long-beard Hawkweed 195760
Longbeard Metalleaf 296393
Longbeard Peristylus 291197
Longbeard Perotis 291197
Longbian Pine 300041
Longbill Crazyweed 279020
Longbill Heronbill 153764
Longbract Aletris 14507
Longbract Alpine Oak 323907
Longbract Amitostigma 19508
Longbract Astyleorchis 19508
Longbract Beautyberry 66839
Longbract Broomrape 275213
Longbract Bugbane 91016
Longbract Catnip 264985
Longbract Cattail 401105
Longbract Cephalanthera 82058
Longbract Coeloglossum 98593
Longbract Corydalis 106091
Longbract Crazyweed 278974
Longbract Daylily 191311
Longbract Dendranthema 124823
Longbract Diphylax 132669
Longbract Elatostema 142723
Longbract Eria 299624
Longbract Eulophia 156589
Longbract Fir 379
Longbract Gentian 156589
Longbract Greenbrier 366433
Longbract Hairorchis 299624
Longbract Hawksbeard 110961
Longbract Irisorchis 267966
Longbract Kunming Roscoea 337094
Longbract Landpick 308532
Longbract Largeleaf Primrose 314608
Long-bract Lilyturf 272105
Longbract Linden 391849
Longbract Liparis 232196

Longbract Man Orchid 3782
Longbract Mosla 259313
Longbract Nepeta 264985
Longbract Oak 323907
Longbract Oberonia 267966
Longbract Osmanther 276351
Longbract Peliosanthes 288649
Longbract Philippine Pipewort 151371
Longbract Pink 127670
Longbract Pipewort 151288
Longbract Purplepearl 66839
Longbract Purplestem 376449
Longbract Rush 213223
Longbract Salsify 394294
Longbract Sedge 74481
Longbract Skunk Bugbane 91016
Longbract Solomonseal 308532
Longbract Spiradiclis 371765
Longbract Spotleaf-orchis 179682
Longbract Spring Orchis 117086
Longbract Stairweed 142723
Longbract Stewartia 376449
Longbract Tongol Milkvetch 43171
Longbract Twayblade 232196
Longbract Upright-spiked Cypressgrass 119320
Longbract Upright-spiked Galingale 119320
Longbract Uraria 402137
Longbract Viscid Germander 388305
Longbract Yellowflower Crazyweed 279054
Long-bracteate Mahonia 242583
Long-bracted 98593
Long-bracted Beautyberry 66839
Long-bracted Cymbidium 117086
Long-bracted Green Orchid 98586,98596
Long-bracted Hemlock 399914
Long-bracted Linden 391849
Long-bracted Oak 323907
Long-bracted Orchid 98560,98586,302264
Long-bracted Persimmon 132260
Long-bracted Sedge 74481
Long-bracted Spiderwort 394001
Long-bracted Stewartia 376449
Long-bracted Tussock Sedge 73732,73739
Long-bracted Uraria 402137
Long-bracted Wild Indigo 47861,47862
Longbranch Arthrophytum 36822
Longbranch Bambusa 47250
Long-branch Frostweed 188610
Longbranch Monkshood 5378
Longbranch Nodosetree 36822
Long-branch Pinanga 299666
Longbranch Trigonotis 397429
Long-branched Bamboo 47250
Long-branched Bambusa 47250
Long-branched Pinanga-palm 299666
Longbreach Monkshood 5375
Longbreast Dolichothele 135689
Longbud Spiraea 372000
Longcalyx Antlevine 88897
Longcalyx Ardisia 31605
Longcalyx Azalea 331128
Long-calyx Basketvine 9453,9478
Long-calyx Blueberry 403787
Long-calyx Bredia 59857
Longcalyx Camellia 69297
Longcalyx Chonemorpha 88897
Longcalyx Cockeyeweed 218345
Longcalyx Crazyweed 279069
Longcalyx Dendropanax 125638
Long-calyx Duhat 382524
Longcalyx Flax 231895
Longcalyx Gentian 173409
Longcalyx Hairyfruit Raspberry 339123
Longcalyx Jerusalemsage 295135
Longcalyx Lasianthus 222110
Longcalyx Licuala Palm 228747
Longcalyx Licualapalm 228747
Longcalyx Lobelia 234381
Longcalyx Lysionotus 191397,239956
Long-calyx Michelia 252912
Longcalyx Microtoena 254254
Longcalyx Milkvetch 42629
Longcalyx Ophiorrhiza 272252
Longcalyx Pieris 298797
Longcalyx Pink 127756
Longcalyx Pyrola 322849
Longcalyx Rattlebox 111987
Longcalyx Roughleaf 222110
Longcalyx Slatepentree 322628
Longcalyx Snakegourd 396213
Longcalyx Spicegrass 239681
Longcalyx Stonecrop 356761,357302
Longcalyx Stonelotus 364916
Longcalyx Tea 69297
Longcalyx Violet 410270
Longcalyx Woodbetony 287366
Long-calyxed Basketvine 9478
Longcalyxed Camellia 69297
Long-calyxed Chonemorpha 88897
Long-calyxed Dendropanax 125638
Long-calyxed Lasianthus 222113
Long-calyxed Pieris 298797
Long-calyxed Rhododendron 331128
Longcap Cleisostoma 94451
Longcap Closedspurorchis 94451
Longcapsule Azalea 331884
Long-capsule Burretiodendron 161624
Longcapsule Primrose 314422
Long-capsule Xianmu 161624
Longcapsuled Falsepimpernel 231479
Longcapsuled Motherweed 231479
Longcarp Begonia 50026
Longcarp Little Flower Alangium 13363
Longcatkin Poplar 311438
Longcaudate Blueberry 403891
Longcaudate Camellia 69301
Longcaudate Delavay Falsepanax 266917
Long-caudate Holly 203988
Longcaudate Maple 2788
Long-caudate Mulberry 259172
Longcaudate Nothopanax 266917
Long-caudated Blueberry 403891
Long-caudated Camellia 69301
Long-chawed Peashrub 72336
Long-chawed Pea-shrub 72336
Longchi Primrose 314601
Longclaw Dendrobium 125044
Longclaw Epigeneium 146547
Longclaw Orchid 144123
Long-cluster Japanese Wistaria 414556, 414572
Long-cluster Japanese Wisteria 414565
Longcluster Sanicle 345955
Longcomb Corydalis 106052
Longcorn Crestedgalea Woodbetony 287116
Longcorn Elatostema 142641
Longcorn Stairweed 142641
Longcorolla Bladderwort 403850
Longcorollatube Woodbetony 287152
Longcoronate Sage 345315
Longcusp Camellia 69303
Long-cusp Camellia 69303
Longcusp Tea 69303
Long-cusp Vetch 408467
Longcyme Alyxia 18527
Long-cymed Alyxia 18527
Longdisk Sneezeweed 188435
Longear Hymenacue 200834
Long-ear Rhododendron 330188
Longear Rush 212878
Longear Water Hymenacue 200834
Longear Yushanbamboo 416787
Long-ear Yushania 416787
Long-eared Bamboo 297213
Longeared Epigynum 146559
Long-eared Epigynum 146559
Longeared Ladybell 7832
Longeared Pentastelma 290056
Long-eared Pentastelma 290056
Long-eared Rhododendron 330188
Longeared Simaovine 146559
Longehairy Cactus 244132
Longelanceolate Maesa 241797
Longen 131061
Longepetal Stock 246494
Longerpeduncle Greenbrier 366301
Longer-peduncle Greenbrier 366301
Longerpeduncle Smilax 366301
Longest Camellia 69309
Longestapex Beautyberry 66853

Longestapex Purplepearl 66853
Longest-apexed Beautyberry 66853
Longest-pediceled Camellia 69309
Longevity Kalanchoe 215102
Longeyelash Honeysuckle 235739
Long-famele Camellia 69305
Longfeather Needlegrass 376814
Long-fimbriate Bitter-bamboo 304056
Longfloss Fieldhairicot 138978
Longflower Alyxia 18519
Longflower Amomum 19846
Longflower Asiabell 98420
Longflower Asparagus 39066
Longflower Astridia 43332
Longflower Awnless Bromegrass 60767
Longflower Basketvine 9434
Longflower Bastardtoadflax 389768
Long-flower Bee-blossom 172202,172203
Longflower Bentgrass 12134
Longflower Bluebell 115385
Longflower Bluebellflower 115385
Longflower Brome 60767
Longflower Bushman's Poison 4961
Longflower Bushman's-poison 4961
Long-flower Catchfly 363706
Longflower Cleistogenes 94614
Longflower Cochichina Centranthera 81722
Longflower Conehead 85947
Longflower Creeping Dentella 125873
Longflower Dutchmanspipe 34260
Longflower Ehretia 141662
Longflower Engler Milkvetch 42331
Longflower Evening-primrose 269397
Longflower Fescuegrass 163909
Longflower Four-o'clock 255727
Longflower Greyawngrass 372415
Longflower Hemarthria 191247
Longflower Hideseedgrass 94614
Long-flower Honeysuckle 235926
Longflower Hundredstamen 389768
Longflower Indianmulberry 258898
Long-flower Indigo 206101
Long-flower Ixora 211111
Longflower Japanese Youngia 416443
Longflower Lawngrass 418450
Longflower Lily 229900
Longflower Limpricht Taxillus 385210
Long-flower Linociera 87716,231763
Longflower Liolive 87716,231763
Longflower Mayten 246938
Longflower Milkvetch 42633
Longflower Mycetia 260632
Longflower Neomartinella 264165
Longflower Onosma 271777,271782
Longflower Oxwhipgrass 191247
Longflower Pilose Deutzia 127042
Longflower Pollia 307337
Longflower Protea 315860
Longflower Sage 345008
Longflower Sandwort 32031
Longflower Small Reed 127260
Longflower Smalllobed Clematis 95207
Longflower Smallreed 127260
Longflower Spodiopogon 372415
Long-flower Starjasmine 393626
Longflower Stonelilac 263960
Longflower Stringbush 414162
Longflower Stringbush Bashful 414162
Long-flower Syzygium 382601
Longflower Tabacco 266048
Long-flower Thin-branch Leptodermis 226089
Long-flower Wildclove 226089
Longflower Wildginger 37672
Longflower Willow 343619
Longflower Woodbetony 287172
Longflower Youngia 416450
Long-flowered Actephila 6469
Long-flowered Bamboo 364820
Long-flowered Basketvine 9434
Long-flowered Ehretia 141662
Long-flowered Heterosmilax 194122
Long-flowered Honeysuckle 235926
Long-flowered Hymenopogon 263960
Long-flowered Linociera 87716,231763
Long-flowered Mycetia 260632
Long-flowered rabbitbrush 90493
Long-flowered Snowberry 380743
Long-flowered Stringbush 414162
Long-flowered Tobacco 266048
Longflowered Veldtgrass 141761
Long-flowered Willow 343619
Longfoot Pholidota 295528
Longfruit Actinodaphne 233830
Longfruit Amomum 19843
Longfruit Amplexicaul Draba 136933
Longfruit Amplexicaul Whitlowgrass 136933
Longfruit Astilbe 41825
Longfruit Asyneuma 43579
Longfruit Beancaper 418680
Longfruit Bombax 56795
Longfruit Camellia 69298
Longfruit Cleyera 96607
Long-fruit Cleyera 96605
Longfruit David Rose 336521
Longfruit Endiandra 145317
Longfruit Erysimum 154498
Longfruit Fewflower Pittosporum 301362
Longfruit Fewflower Seatung 301362
Long-fruit Gordonia 179763
Longfruit Gordontea 179763
Longfruit Hemsleia 191937
Longfruit India Iphigenia 207498
Longfruit Indigo 206002
Longfruit Korolkov Torularia 264620
Longfruit Litse 234011
Long-fruit Manyfruit Idesia 203427
Longfruit Meconopsis 247119
Longfruit Mountainash 369571
Longfruit Oak 116139
Longfruit Osmanther 276353
Long-fruit Pittosporum 301324
Longfruit Pleurospermum 304819
Longfruit Pyrenaria 322605
Longfruit Qinggang 116139
Longfruit Raspberry 338334
Longfruit Rattlebox 112307
Long-fruit Seatung 301324
Longfruit Shrubcress 368552
Longfruit Sinojackia 85969
Longfruit Snowgall 191937
Longfruit Solms-Laubachia 368552
Longfruit Speedwell 407079
Longfruit Sugarmustard 154498
Longfruit Sugarpalm 32339
Longfruit Tea 69298
Longfruit Thermopsis 389520
Longfruit Weigttree 85969
Longfruited Actinodaphne 233830
Long-fruited Anemone 23776
Long-fruited Bombax 56795
Long-fruited Camellia 69298
Long-fruited Cleyera 96607
Long-fruited Croton 112918
Long-fruited Cyclobalanopsis 116139
Long-fruited Endiandra 145317
Long-fruited Gordonia 179763
Long-fruited Honeylocust 176889
Longfruited Horsebean Caltrop 418643
Long-fruited Japanese Ash 168003
Long-fruited Jute 104103
Longfruited Kiwifruit 6657
Long-fruited Knotweed 309683
Long-fruited Raspberry 338334
Long-fruited Sanide 346007
Long-fruited Sinojackia 85969
Long-fruited Snakeroot 346007
Longfruited Youngia 416475
Longfunicle Stonecrop 356898
Longgang Birthwort 34258
Longgang Chirita 87904
Longgang Hemiboea 191369
Longgang Leaf-flower 296456
Longgang Pearlwood 296456
Longgang Perilepta 290925
Longgearistate Roegneria 335404
Longgland Cherry 83197
Longgland Greywhitehair Raspberry 339365
Longgland Platanthera 302399
Longglandular Eyebright 160169
Long-glosse Feathergrass 376845
Longglume Bluegrass 305846

Longglume Fescue 164061
Longglume Fescuegrass 164061
Longglume Goosecomb 335409
Longglume Kengyilia 215904
Longglume Oat 45495
Longglume Roegneria 335409
Longgynophore Willow 343332
Longhair Aristulate Rockfoil 349073
Longhair Azalea 331991
Longhair Begonia 49747,50394
Longhair Chickweed 375060
Longhair Christolea 89231,126151
Long-hair Clematis 94761
Longhair Clubfilment 179180
Longhair Didymocarpus 129989
Longhair Draba 136981
Longhair Eritrichium 153555
Longhair Germander 388181
Longhair Goldenray 229245
Longhair Gomphostemma 179180
Longhair Hemsleia 191904
Long-hair Honeysuckle 235739
Longhair Kangding Buttercup 325776
Longhair Milkvetch 42747
Longhair Milkwort 308457
Longhair Nanmu 295362
Longhair Plateaucress 89231,126151
Longhair Ricegrass 276022
Longhair Saurauia 347966
Longhair Schultzia 352690
Longhair Small Reed 127322
Longhair Smallligulate Aster 40011
Longhair Snowgall 191904
Long-hair Spiraea 372001
Longhair Sweetleaf 381187
Longhair Whitlowgrass 136981
Longhair Wildginger 37706
Longhair Windhairdaisy 348370
Longhair Yushanbamboo 416788
Long-haired Adinandra 8231
Long-haired Ardisia 31629
Long-haired Eurya 160577
Long-haired Gomphostemma 179203
Long-haired Hawkweed 195760
Long-haired Mekong River Rhododendron 331217
Long-haired Saurauia 347966
Long-haired Star-tulip 67593
Long-haired Sweetleaf 381187
Long-haired Willowleaf Hygrophila 200653
Long-haired Yushania 416788
Longhairlike Clubfilment 179203
Longhairlike Gomphostemma 179203
Long-hairs Cassiope 78649
Longhairy Antenoron 26624
Longhairy Ardisia 31629
Longhairy Eurya 160577
Longhairy Lycaste 238739
Longhairy Rockfoil 349864
Longhairy Speedwell 317921
Long-hairy-pod Milkvetch 42747
Longhastate Goldenray 229098
Longhead Japanese Windhairdaisy 348411
Longhead Poppy 282546
Longhead Prairie Coneflower 326960
Longhead Ramie 56131
Longhead Raspberry 338334
Long-headed Anemone 23776
Long-headed Coneflower 326960
Long-headed Poppy 282546
Longhook Triumfetta 399296
Longhorn Catchfly 363699
Longhorn Dandelion 384773
Longhorn Disa 133829
Longhorn Ophiorrhiza 272243
Longhorn Spapweed 205097
Longhorn Touch-me-not 205097
Longhorned Elatostema 142841
Long-horned Habenaria 184008
Longhushan Begonia 50380
Longibracteate Ferre Fir 364
Longibracteate Mahonia 242583
Longibracted Shrubalthea 195302
Longibuded Spiraea 372000
Longicarcarate Larkspur 124186
Longicaudate Clematis 94695
Longiciate Begonia 50027
Longifalcate-fruited Rhododendron 331129
Longifimbriate Bitter Bamboo 304056
Longifruited Indigo 206002
Longiinternode Gelidocalamus 172743
Longileaved Sarcococca 346736
Longiligulate Ginger 417999
Longilobate Monkshood 5375
Longinflorescence Amomum 19852
Longinflorescence Falsepimpernel 231541
Longinflorescence Motherweed 231541
Longinflorescence Raspberry 338246
Longinflorescence Small Reed 127195
Long-inflorescence Vanda 404681
Long-internoded Fargesia 162706
Long-internoded Gelidocalamus 172743
Longipedicel Monkshood 5376
Longipediceled Holly 204138
Longipediceled Supple-jack 52437
Longipediceleed Mango 244404
Longipedicellate Larkspur 124354
Longipedicellate Maple 3115
Longipedicellate Monkshood 5376
Longipeduncle Rehder Acanthopanax 143652
Longipedunculate Hairycalyx Storax 379390
Longipedunculate Holly 203692
Longipedunculate Machilus 240632,240712
Longipetiolate Monkshood 5377
Longipetiole Beech 162386
Longipetiole Sauropus 348054
Longipetioled Beech 162386
Longipetioled Elaeocarpus 142371
Longipetioled Helicia 189940
Longipetioled Sarcococca 346737
Longiracemed Daphniphyllum 122698
Longispiculate Bamboo 47330
Longispiked Barberry 51524
Longispiked Croton 112952
Longispined Evergreen Chinkapin 78924
Longispiny Barberry 51877
Longistalked Sichuan Sabia 341562
Longistiped Burretiodendron 64031
Longistyled Daphniphyllum 122715
Longistyled Mahonia 242517
Longitudinally Striped Hakea 184652
Longitudinally-striped Barbadoslily 196458
Longitudinalwing Seepweed 379590
Longiveined Sabia 341530
Longjin Loosestrife 239834
Longkeel Milkvetch 42664
Longleaf 162457,162465
Longleaf Acanthus 2691
Longleaf Actinodaphne 6762
Longleaf Alfredia 14597
Longleaf Appendicula 29821
Longleaf Ardisia 31640
Longleaf Arnica 34735
Longleaf Arrowhead 342310
Longleaf Arthrophytum 36824
Longleaf Asiabell 98362
Longleaf Aster 40325
Longleaf Bastardtoadflax 389769
Longleaf Batrachium 48922
Longleaf Beautyberry 66784,66840
Longleaf Beefwood 79165
Longleaf Bethlehemsage 283630
Longleaf Bicolored Everlasting 21527
Longleaf Bicolored Pearleverlasting 21527
Longleaf Bifurcate Cinquefoil 312418
Longleaf Bluegrass 305675
Longleaf Bodinier Photinia 295644
Longleaf Briggsia 60273
Longleaf Burnet 345894
Longleaf Burreed 370075
Longleaf Bushclover 226736
Longleaf Butchersbroom 340994
Longleaf Buttercup 326034
Longleaf Caelospermum 98811
Longleaf Cajan 65153
Long-leaf Camellia 68963
Longleaf Campanumoea 70403
Long-leaf camphorweed 305109
Longleaf Carpesium 77177
Longleaf Catasetum 79341
Longleaf Cephalanthera 82060

Longleaf Ceratozamia 83684
Longleaf Ceropegia 84154
Longleaf Chasalia 86208
Longleaf Chastetree 411219
Long-leaf Cherry-laurel 223100
Longleaf Chickweed 374944
Long-leaf Childvine 400921
Longleaf Cinquefoil 312734
Longleaf Claoxylon 94073
Longleaf Cleisostoma 94444
Longleaf Closedspurorchis 94438,94444
Longleaf Cocklebur 11586
Longleaf Codariocalyx 98159
Longleaf Coelospermum 98811
Longleaf Coin-shaped Blueberry 403921
Long-leaf Corkwood 184640
Longleaf Cowpea 408950
Longleaf Cylindrokelupha 116587
Longleaf Cymbidium 116827,116847
Longleaf Dallisgrass 285460
Longleaf Date-plum 132265
Longleaf Debregeasia 123330
Longleaf Dendrochilum 125533
Longleaf Deutzia 126992
Longleaf Dock 340109
Longleaf Dryander 138432
Longleaf Dutchmanspipe 34139
Long-leaf Dwarf Peashrub 72324
Longleaf Edelweiss 224899
Longleaf Elaeagnus 142075
Longleaf Embelia 144759
Long-leaf Ephedra 146262
Longleaf Epidendrum 146437
Long-leaf Evergreenchinkapin 78877
Longleaf Faber Maple 2953
Longleaf False Fairybells 134410
Longleaf False Goldeneye 190251
Longleaf Falsenettle 56195,56254
Longleaf Fawnlily 154948
Long-leaf Fig-tree 164877
Long-leaf Fleabane 150570
Long-leaf Flemingia 166895
Longleaf Glorybower 96183
Longleaf Glorytree 391574
Longleaf Glycosmis 177838
Longleaf Goldenray 229097
Longleaf Goniothalamus 179409
Longleaf Greenbrier 366418
Longleaf Greenhoods 320875
Longleaf Greenstar 307509
Longleaf Habenaria 183812
Longleaf Hairflower Honeysuckle 236194
Longleaf Halerpestes 184711
Longleaf Halimocnemis 184818
Longleaf Hawksbeard 110717
Longleaf Herb Screwpine 280987
Longleaf Himalayan Aucuba 44912
Longleaf Holly Grape 51959
Longleaf Hundredstamen 389769
Longleaf Hydrangea 199945
Longleaf India Pine 300186
Longleaf Indian Madder 337910
Longleaf Indian Pine 300186
Longleaf Ixora 211136
Longleaf Japan Aucuba 44924
Longleaf Koilodepas 217761
Longleaf Kumquat 167513
Longleaf Ladybell 7789
Longleaf Lambertia 220309
Longleaf Laxflower Maple 3091
Longleaf Ledum 223908
Longleaf Leea 223945
Longleaf Leopard 70403
Longleaf Lettuce 219306
Longleaf Leucas 227629
Longleaf Lilyturf 272107
Longleaf Litse 233970
Longleaf Luckyweed 370407
Longleaf Luisia 238330
Longleaf Lupine 238462
Longleaf Machilus 240625
Longleaf Madder 337952
Longleaf Magnolia 242244
Longleaf Mahonia 242602
Long-leaf Mammea 243987
Longleaf Mandarin Orchid 295879
Longleaf Manyspike 309618
Long-leaf Maple 2997
Long-leaf Marginate Osmanthus 276365
Longleaf Meconopsis 247143
Longleaf Michelia 252907
Longleaf Milkwort 308170
Longleaf Mitreola 256205
Longleaf Morina 258853
Longleaf Mycetia 260634
Longleaf Nodosetree 36824
Longleaf Notoseris 267167
Longleaf Oak 116137,324200
Longleaf Oncidium 270822
Longleaf Orchis 116847
Long-leaf Osmanther 276365
Longleaf Paraphlomis 283630
Longleaf Parasiticvine 125793
Longleaf Paspalum 285460
Longleaf Pendentlamp 84079
Longleaf Peppertree 350989
Longleaf Periwinkle 79413
Long-leaf Philodendron 294820
Longleaf Photinia 295644
Longleaf Phyllodium 297012
Longleaf Pine 300128
Long-leaf Pine 300128
Longleaf Plantain 302034
Longleaf Pluchea 305092
Longleaf Podocarpus 306457
Longleaf Pondweed 312166,312194
Long-leaf Pondweed 312194
Longleaf Poplar 311575
Longleaf Primrose-willow 238183
Longleaf Privet 229444
Longleaf Purpledaisy 267167
Longleaf Purplepearl 66840
Longleaf Purplestem 376461
Longleaf Pycnarrhena 321980
Longleaf Pyrola 322820
Longleaf Qinggang 116137
Longleaf Rabdosia 209861
Longleaf Ramie 56254
Longleaf Raspberry 338336
Longleaf Rhaphidophora 329007
Longleaf Rockjasmine 23219
Long-leaf Rush 213251
Longleaf Rustyhair Raspberry 339166
Longleaf Saltbane 184818
Longleaf Sarcococca 346736
Longleaf Saussurea 348500
Longleaf Saxifrage 349562
Longleaf Shellfish Pricklyash 417219
Longleaf Silver Maple 3535
Longleaf Skullcap 355572
Long-leaf Slender Honeysuckle 236194
Longleaf Sloanea 366037
Longleaf Soda Buttercup 184711
Longleaf South China Screwpine 280987
Longleaf Speedwell 317927
Longleaf Spicegrass 239705
Longleaf Spiradiclis 371774
Longleaf Spotleaf-orchis 179582
Longleaf Stewartia 376461
Longleaf Sunflower 188995
Longleaf Tailgrass 402525
Longleaf Taillessfruit 100139
Longleaf Tarenna 385047
Longleaf Tick Clover 126301
Longleaf Tiger Jaws 162932
Longleaf Torreya 393074
Longleaf Tubergourd 390159
Long-leaf Tylophora 400921
Longleaf Urochloa 402525
Long-leaf Villose Hydrangeavine 351817
Longleaf Walking Iris 264162
Longleaf Waternettle 123330
Longleaf Waveleaf Rhubarb 329411
Long-leaf Waxflower 153333
Long-leaf Waxplant 198868
Longleaf Wheelstamentree 399435
Longleaf Whiteback Qinggang 116181
Longleaf Whitetung 94073
Long-leaf Wild Buckwheat 152240
Longleaf Wildorchis 274037
Longleaf Wildtung 243355

Longleaf Willow 343881
Longleaf Windhairdaisy 348500
Longleaf Wurrus 166880
Long-leaf Wurrus 166895
Longleaf Xylosma 415891
Longleaf Yellow Bedstraw 170751
Longleaf Yellow Pine 300128
Long-leafed Wax Flower 153333
Long-leafed Yellowwood 306439
Long-leaved Argyle Apple 155528
Long-leaved Aster 40923,380839,380926
Long-leaved Beautyberry 66840
Long-leaved Beauty-berry 66853
Longleaved Beefwood 79165
Long-leaved Blue Aster 380926
Long-leaved Bluets 187617,198709
Long-leaved Box 64276
Long-leaved Bush-clover 226736
Long-leaved Cherry Laurel 223100
Long-leaved Claoxylon 94073
Long-leaved Cyclobalanopsis 116137
Long-leaved Cylindrokelupha 116587
Long-leaved Debregeasia 123330
Long-leaved Deutzia 126992
Long-leaved Dock 340109
Long-leaved Drypetes 138651
Long-leaved Embelia 144759
Longleaved Felwort 235452
Long-leaved Glorybower 96183
Long-leaved Grass Nut 60445
Long-leaved Ground-cherry 297694,297697
Long-leaved Guadua 181476
Long-leaved Helleborine 82060
Long-leaved Houstounia 198709
Longleaved Indian Pine 300186
Long-leaved Ixora 211136
Long-leaved Japanese Spindle-tree 157631
Long-leaved Joint Fir 146262
Long-leaved Joint-fir 146262
Long-leaved Lasianthus 222208
Longleaved Laxflower Maple 3091
Long-leaved Leea 223945
Long-leaved Magnolia 242244
Long-leaved Magnolia-vine 351066
Long-leaved Mallotus 243405
Long-leaved Milkvetch 42302
Long-leaved Oak 324200
Long-leaved Paintbrush 79128
Long-leaved Paspalum 285460
Long-leaved Phlox 295281
Long-leaved Phyllodium 297012
Longleaved Pine 300128
Long-leaved Podocarpus 306457
Long-leaved Pondweed 312194,312294
Long-leaved Poplar 311575
Long-leaved Privet 229444
Long-leaved Pruinose Barberry 52074
Long-leaved Raspberry 338336
Long-leaved Reed 37467
Long-leaved Reed Grass 65633
Long-leaved Sage 35835
Long-leaved Sallow 344126
Long-leaved Speedwell 317927
Long-leaved Starwort 374944,374951
Long-leaved Stitchwort 374944,374946
Long-leaved Sundew 138307
Long-leaved Tarenna 384983
Long-leaved Traill Rhododendron 331988
Long-leaved Water Crowfoot 48918
Long-leaved Willow 343881
Long-leaved Xylosma 415891
Longlegorchis 320446
Long-legs 315082
Longlemma Bluegrass 305492
Longlemma Fake Needlegrass 318194
Longlemma Helictotrichon 190183
Longlemma Pseudoraphis 318194
Longlife Daffodil 262410
Longlife Narcissus 262410
Long-ligulate Groundsel 358284
Long-ligulated Giant-grass 175605
Long-ligule Pseudosasa 318317
Longlin Chirita 87909
Longlin Eurya 160538
Longlin Mahonia 242597
Longlin Mayten 246872
Longlin Reevesia 327367
Longlin Suoluo 327367
Longling Camellia 69313
Longling Elaeocarpus 142359
Longling Eria 299620
Longling Hairorchis 299620
Longling Holly 203622
Longling Newlitse 264072
Longling Rockvine 387804
Longling Smilax 366438
Longling Taxillus 385233
Longling Woodbetony 287383
Longlip Liparis 232278
Longlip Listera 232930
Longlip Siphocranion 365196
Longlip Twayblade 232278
Long-lipped Serapias 360642
Longlive Flower Kalanchoe 215102
Longliving Rockjasmine 23292
Longlobe Bucklershaped Bittercress 72980
Longlobe Garcinia 171125
Longlobe Milkvetch 42634
Longlobe Oldworld Arrowhead 342421
Long-lobe Violet 410184
Long-lobed Arrowhead 342383
Longlobed Chinese Maple 3621
Longlobed Daphne 122367
Longlobed Elaeagnus 142060
Long-lobed Elaeagnus 142060
Longlobed Krasnovia 218148
Longlobed Milkvetch 42634
Longlobed Paraprenanthes 283686
Longlobed Sowthistle 368675
Longlobed Taihangdaisy 272586
Longlower Woodbetony 287368
Longmai Cyclobalanopsis 116146
Longmai Oak 116146
Longmai Qinggang 116146
Longmamma Nipple Cactus 244231
Long-marked Rhododendron 331839
Longmucuo Christolea 89244
Longmucuo Plateaucress 89244
Longnan Clematis 95192
Longnan Opithandra 272591
Longneck Crinum 111230
Longneck Iris 208838
Longnode Chinacane 162706
Longnode Eargrass 187691
Long-node Fragrant Bamboo 87640
Long-noded Fragrant-bamboo 87640
Long-nut Oak 116139
Longovata Larkspur 124196
Longpanicle Stringbush 414212
Longpaniculate Cinnamon 91363
Longpaniculate Yunnan Lacquertree 393498
Long-paniculated Cinnamon 91363
Longpanla Sedge 75220
Longpedicel Aletris 14523
Longpedicel Bittercress 72867
Longpedicel Bittergrass 404554
Longpedicel Brassaiopsis 59214
Longpedicel Buttercup 326207
Longpedicel Calvepetal 177998
Long-pedicel Cathay Poplar 311268
Longpedicel Childvine 400922
Longpedicel Chinese Bushcherry 83241
Long-pedicel Chinese Peashrub 72343
Long-pedicel Chinese Pea-shrub 72343
Long-pedicel Cleistanthus 94529
Longpedicel Corydalis 106099
Longpedicel Dipentodon 132609
Longpedicel Falsepanax 266922
Longpedicel Hoary Clematis 95430
Long-pedicel Holly 204138
Longpedicel Homocodon 197896
Longpedicel Hooktea 52438
Longpedicel Incarvillea 205551
Longpedicel Korean Willow 343577
Longpedicel Ladybell 7687
Longpedicel Landpick 308590
Long-pedicel Largeflower Euonymus 157549
Longpedicel Larkspur 124354
Longpedicel Leek 15513
Longpedicel Liriope 232629
Long-pedicel Lushan Hibiscus 195090

Long-pedicel Lysionotus 239955
Longpedicel Machilus 240628
Longpedicel Mango 244404
Long-pedicel Michelia Michelia 252905
Longpedicel Newlitse 264071
Longpedicel Onion 15513
Longpedicel Ophiorrhiza 272245
Longpedicel Pennywort 200350
Longpedicel Pentapanax 289657
Long-pedicel Privet 229526
Long-pedicel Purple Willow 343953
Longpedicel Raspberry 339220
Longpedicel Sedge 74675
Longpedicel Sharptooth Hornsage 205551
Longpedicel Solomonseal 308590
Longpedicel Spiradiclis 371766
Longpedicel Stachyurus 373581
Longpedicel Supplejack 52438
Longpedicel Tanoak 233300
Longpedicel Ternstroemia 386713
Longpedicel Tupelo 267878
Longpedicel Tupistra 70618
Longpedicel Tylophora 400922
Long-pedicel Viburnum 407927
Longpedicel Wendlandia 413797
Long-pedicel Willow 343630
Long-pedicel Wonderflower Willow 343059
Long-pedicel Woodnettle 273916
Long-pedicel Yunnan Poplar 311587
Long-pedicele Camellia 69307
Long-pediceled Camellia 69307
Long-pediceled Machilus 240628
Long-pediceled Millettia 66636
Long-pediceled Miquel Linden 391776
Long-pediceled Newlitse 264071
Long-pediceled Tanoak 233300
Long-pediceled Ternstroemia 386713
Long-pediceled Tylophora 400891
Longpedicellate Acanthopanax 2496
Longpedicellate Angelica 24394
Long-pedicellate Glyptopetalum 177998
Long-pedicellate Spindle-tree 177998
Longpedicelled Dogwood 124938
Longpedicelled Goldsaxifrage 90336
Longpedicelled Goldwaist 90336
Longpedicelled Kidneyleaf Mountain Leech 126575
Longpedicelled Kidneyleaf Tickclover 126575
Long-pedicelled Pepper 300459
Longpediculate Aspidistra 39554
Longpediole Amomum 19879
Longpediole Cinnamon 91365
Longpediole Lacquertree 393456
Long-pedioled Cinnamon 91365
Longpeduncle Calvepetal 177997
Longpeduncle Clearweed 298964
Long-peduncle Clovershrub 70838
Longpeduncle Gentian 174062
Longpeduncle Holly 204004
Longpeduncle Holocheila 197387
Longpeduncle Kadsura 214972
Longpeduncle Lysionotus 239955
Longpeduncle Merremia 250803
Long-peduncle Millettia 66636
Longpeduncle Pellionia 288744
Longpeduncle Redcarweed 288744
Longpeduncle Spicebush 231383
Longpeduncle Wholelipflower 197387
Longpeduncle Wrinkly-leaf Glyptopetalum 178002
Longpeduncled Alder 16326
Long-peduncled Alder 16326
Long-peduncled Crazyweed 278975
Longpeduncled Indigo 206194
Long-peduncled Kadsura 214972
Longpeduncled Larkspur 124355
Long-peduncled Pentapanax 289657
Long-peduncled Pondweed 312232
Long-peduncled Portia Tree 389949
Long-peduncled Spice-bush 231383
Long-peduncled Synotis 381941
Longpedunculate Cliffbean 66636
Longpedunculate Indigo 206194
Longpeicel Rockvine 387803
Long-peniculate Stringbush 414212
Longperula Azalea 331839
Long-perula Rhododendron 331132,331839
Longpes Turpinia 400544
Longpetal Aspidistra 39555
Longpetal Chickweed 374780
Longpetal Daphne 122507
Longpetal Dichapetalum 128726
Longpetal Eargrass 187620
Longpetal Exbucklandia 161606
Longpetal Glabrous Cinquefoil 312635
Longpetal Globeflower 399519
Longpetal Grewia 180855
Long-petal Hedyotis 187620
Longpetal Herminium 192868
Longpetal Hipericumlike Rockfoil 349486
Longpetal Iris 208692
Longpetal Paphiopedilum 282814
Longpetal Parnassia 284561
Longpetal Pocktorchid 282814
Longpetal Reed 295888
Longpetal Rockfoil 349566
Longpetal Serapias 360619
Longpetal Yucca 416556
Longpetal Yunnan Globeflower 399554
Long-petaled Daphne 122507
Long-petaled Exbucklandia 161606
Long-petaled Grewia 180855
Long-petaled Iris 208692
Longpetal-like Parnassia 284565
Longpetiolate Horseorchis 215353
Longpetiolate Speedwell 407215
Longpetiole Elatostema 142725
Longpetiole Glorybower 96263
Longpetiole Helicia 189940
Longpetiole Hydrangea 199946
Longpetiole Reevesia 327366
Longpetiole Sandwort 32034
Longpetiole Slugwood 50552
Longpetiole Speedwell 407215
Longpetiole Stairweed 142725
Longpetiole Tea 69307
Longpetiole Woodbetony 287375
Longpetiole Yellowish Woodbetony 287388
Long-petioled Hydrangea 199946
Long-petioled Privet 229526
Long-petioled Reevesia 327366
Long-petioled Slugwood 50552
Long-petioled Trillium 397592
Long-petioled Viburnum 407927
Long-petioled Wendlandia 413797
Long-petioled Wintergreen 172068
Longpilose Beautyberry 66897
Longpilose Cushionshaped Ceratoides 218127
Longpilose Elsholtzia 144001
Longpilose Purplepearl 66897
Long-pilosed Beautyberry 66897
Longpinnate Radish 326622
Longpistil Camellia 69305
Longpod Milkvetch 42654
Long-pod Mucuna 259533
Long-pod Poppy 282546
Long-pod Stitchwort 255514
Longpod Tickclover 126679
Long-podded Mucuna 259533
Long-podded Tickclover 126679
Longqi Sedge 74100
Longqishan Bitterbamboo 304063
Longquan Ascitesgrass 407475
Longquan Grape 411786
Longquan Stonecrop 356903
Longquan Veronicastrum 407475
Long-raceme Barberry 51524
Longraceme Currant 334072
Longraceme Daphniphyllum 122698
Longraceme Elm 401504
Long-raceme Elm 401504
Longraceme Goldenray 229021
Longraceme Indigo 206101
Longraceme Milkwort 308080
Longraceme Tigernanmu 122698
Long-racemed Akebia 13219
Long-racemed Currant 334072
Long-racemed Raspberry 338246
Long-racemose Euchresta 155905
Longracemose Lovely Monkshood 5512
Longracemose Vetch 408341

Long-racemose Whitepearl 172109
Long-racemose Wintergreen 172109
Longradiate Hedyotis 187570
Long-radiate Hooker Mayten 246857
Longradiate Red Thorowax 63815
Long-radiate Viburnum 407928
Long-radiated Viburnum 407928
Long-rayed Brodiaea 398813
Long-root 255451
Longroot Chive 15868
Longroot Onion 15868
Long-root Smartweed 308750
Longroot Woodbetony 287175
Long-rooted Cat's Ear 202432
Long-rooted Hawkweed 202432
Long-rooted Onion 15868
Long-rooted Turmeric 114871
Longrui Sabia 341526
Longrunner Panicgrass 282200
Longscale Azalea 331125
Long-scaled Green Sedge 73976
Long-scaled Rhododendron 331125
Long-scaled Tussock Sedge 74774
Longscape Jurinea 214167
Long-scape Milkvetch 42637
Longscapus Milkvetch 42637
Longscourge Rattanpalm 65694
Longsdorff Bellflower 70116
Longseed Osmanther 276356
Longseed Willow Weed 146849
Longseed Willowweed 146849
Longsepaied Treerenshen 125638
Longsepal Ardisia 31582
Longsepal Clematis 95350
Longsepal Five-angle Raspberry 339031
Longsepal Hancock Azalea 330818
Long-sepal Hancock Rhododendron 330818
Longsepal Hemiboea 191370
Longsepal Honeysuckle 235927
Long-sepal Lasianthus 222113,222219
Longsepal Lysionotus 191397,239956
Longsepal Maxillaria 246714
Longsepal Microtoena 254254
Longsepal Raspberry 338785
Long-sepal Roughleaf 222219
Longsepal Snakegourd 396213
Longsepal Stonecrop 356761
Longsepal Tubergourd 390160
Longsepal Violet 410100,410270
Long-sepaled Blueberry 403787
Long-sepaled Clematis 95350
Longsepaled Rattlebox 111987
Long-sepaled Threeleaf Akebia 13241
Long-sepaled Tutcheria 400703
Longseta Knotweed 309345
Longseta Milkvetch 42301
Longseta Needle Spikesedge 143026
Longsetose Rattan Palm 65725
Longsetose Sandwort 32035
Longshag Azalea 330477
Longsheath Angelica 24313
Longsheath Cremanthodium 110356
Longsheath Pondweed 312250
Long-sheath Pseudosasa 318319
Longsheath Yushanbamboo 416791
Long-sheathed Anemone 321721
Long-sheathing Yushania 416791
Longsheng Camellia 69321
Longsheng Hogfennel 292915
Longsheng Isometrum 210187
Longsheng Maple 3125
Longsheng Parnassia 284567
Longsheng Persimmon 132263
Longsheng Rabdosia 209763
Longsheng Rockfoil 349570
Longsheng Sedge 75221
Longsheng Spicebush 231384
Longsheng Spice-bush 231384
Longshoot Balanophora 46820
Long-shoot Bamboo 47250
Longshoot Monkshood 5378
Longshoot Snakemushroom 46820
Longshoot Willow 343650
Long-shooted Willow 343650
Longshoulderhair Yushanbamboo 416821
Longshoulderhair Yushania 416821
Longsickle Azalea 331129
Longsilk Stonecrop 356570
Long-sleeved Begonia 50059
Longspathe Rattan Palm 65726
Long-spicate Indocalamus 206890
Longspicate Semnostachya 358001
Long-spiculate Bambusa 47330
Longspike Ardisia 31560
Longspike Ascitesgrass 407472
Longspike Bamboo 47330
Longspike Birch 53394
Longspike Blueberry 403819
Longspike Bog Rattanpalm 65761
Longspike Bogrush 352357
Longspike Cacalia 283841
Longspike Chastetree 411402
Longspike Clearweed 667
Longspike Coldwaterflower 667,298971
Longspike Conehead 358001
Longspike Crazyweed 278983
Long-spike Evening-primrose 269497
Longspike Falseoat 398453
Longspike Fimbristylis 166386
Longspike Fluttergrass 166386
Longspike Hedgenettle 373294
Longspike Hoarypea 386199
Longspike Indianmulberry 258894
Longspike Indosasa 206890
Longspike Loosestrife 239580
Longspike Lovegrass 147783
Longspike Luisia 238313
Longspike Mosla 259314
Longspike Oak 324129
Longspike Pearweed 239580
Longspike Pennisetum 289150
Longspike Peristylus 291240
Longspike Perotis 291240
Longspike Rabbiten-wind 12648
Longspike Sedge 74374
Longspike Senna 360431
Longspike Shrubby Elsholtzia 144027
Longspike Smallreed 127195
Longspike Tallculm Cypressgrass 118846
Longspike Tallculm Galingale 118846
Longspike Tamarisk 383479
Longspike Tickseed 104771
Longspike Tridens 396710
Longspike Tupistra 400404
Longspike Valerian 404272
Longspike Veronicastrum 407472
Longspike Willow 343979
Longspike Wolftailgrass 289150
Longspike Woodbetony 287177
Long-spiked Ardisia 31560
Long-spiked Birch 53394
Long-spiked Blueberry 403819
Long-spiked Chaste-tree 411402
Long-spiked Croton 112952
Longspiked Enteropogon 146002
Long-spiked Glasswort 342858
Long-spiked Habenaria 183813
Long-spiked Indosasa 206890
Long-spiked Oak 324129
Long-spiked Sedge 75213
Long-spiked Tamarisk 383479
Long-spiked Thorn 109818
Long-spiked Willow 343979
Longspikelet Cypressgrass 118756
Longspikelet Galingale 118756
Longspine Caper 71760
Long-spine Hawthorn 110065
Longspine Russian Thistle 344658
Longspine Xantolis 415239
Long-spined Xantolis 415239
Long-spiny Barberry 51877
Long-spiny Evergreenchinkapin 78924
Longspiny Hookedsepal 267079
Longspiny Sainfoin 271228
Longsplit Irisorchis 267925
Longsplit Oberonia 267925
Longspur Amitostigma 19506
Longspur Angraecum 25063
Longspur Astyleorchis 19506
Longspur Barrenwort 146995
Longspur Calanthe 66086

Longspur Columbine 30054
Long-spur Columbine 30047,30054
Longspur Corydalis 106092,106414
Longspur Crestless Corydalis 105844
Longspur Dali Larkspur 124619
Longspur Dendrobium 125237
Longspur Droopfruit Monkshood 5481
Longspur Epimedium 146995
Longspur Gentian 156651
Longspur Golden Columbine 30021
Longspur Green Habenaria 184187
Longspur Habenaria 183540
Longspur Hemipilia 191607
Longspur Platanthera 302395
Longspur Tianmushan Spapweed 205383
Longspur Tianmushan Touch-me-not 205383
Longspur Toadflex 231031
Longspurlike Larkspur 124186
Longspurred Dendrobium 125237
Longspurred Fairybells 134423
Longspurred Larkspur 124634
Long-spurred Orchid 273520
Long-spurred Pansy 409796
Longspurred Tali Larkspur 124619
Long-spurred Violet 410502
Long-stachys Falsenettle 56131
Longstachys Monkshood 5167
Long-stachys Willow 343979
Longstaik Allomorphia 16111
Long-staiked Allomorphia 16111
Longstalk Actinidia 6570
Long-stalk Adinandra 8218
Longstalk Ainsliaea 12718
Longstalk Almond 20933
Long-stalk Anemone Clematis 95136
Longstalk Arrowwood 407927
Longstalk Aster 40326
Longstalk Azalea 331134
Long-stalk Badong Willow 343357
Longstalk Banfenghe 357979
Longstalk Barbate Mallotus 243334
Longstalk Bauhinia 49160
Longstalk Beancaper 418722
Longstalk Bearberry Cotoneaster 107420
Long-stalk Bennettiodendron 51046
Long-stalk Biondia 54524
Longstalk Birdlingbran 87234
Longstalk Bladderwort 403232
Longstalk Brigssia 60275
Longstalk Buckthorn 328768
Longstalk Calamintha 65590
Longstalk Calvous Rhododendron 330301
Longstalk Camelina 68850
Longstalk Catchfly 363941
Longstalk Chinchweed 286896
Longstalk Chinese Bastardtoadflax 389639
Longstalk Clearweed 298857
Longstalk Clinopodium 97014
Longstalk Coelogyne 98690
Longstalk Comastoma 100276
Longstalk Cordateleaf Maple 2904
Longstalk Corydalis 106099
Longstalk Cranesbill 174551
Longstalk Crazyweed 278973,278975,
279080
Long-stalk Creeper 285108
Longstalk Cymbidium 116966
Long-stalk David Poplar 311288
Long-stalk Densevein 261332
Longstalk Devil Rattan 121512
Longstalk Didissandra 129814
Longstalk Dipentodon 132609
Longstalk Draws ear grass 403232
Longstalk E. H. Wilson Poplar 311572
Longstalk Elaeagnus 141978
Longstalk Elatostema 142724
Longstalk Eritrichium 153477
Long-stalk Euonymus 157415
Longstalk Falsehellebore 405624
Longstalk Fieldhairicot 138974
Long-stalk Fig 164982
Longstalk Figwort 355116
Longstalk Fimbristylis 166389
Long-stalk Gaultheria 172068
Longstalk Gentian 174062
Longstalk Glinus 176955
Longstalk Greenbrier 366435
Longstalk Groundsel 358848
Longstalk Gueldenstaedtia 181665
Longstalk Gynostemma 183013
Long-stalk Hainan Drypetes 138625
Longstalk Helicia 189959
Longstalk Heritiera 192484
Longstalk Holly 204138
Long-stalk Holly 203770,204138
Longstalk Kiwifruit 6570
Longstalk Landpick 308540
Longstalk Lasianthus 222132
Longstalk Licuala Palm 228748
Longstalk Licualapalm 228748
Longstalk Lilyturf 272108
Longstalk Litse 233912
Longstalk Loosestrife 239631,239722
Longstalk Low Meadowrue 388584
Long-stalk Machilus 240634
Long-stalk Mahonia 242584
Longstalk Maple 3116
Long-stalk Maple 3115
Longstalk Marsdenia 245826
Longstalk Martin Spiraea 372003
Longstalk Mazus 246982
Longstalk Meadowrue 388554,388627
Longstalk Meliosma 249409
Longstalk Michelia 252905
Longstalk Microula 254352
Longstalk Milkgreens 245826
Longstalk Monkshood 5377
Longstalk Nematanthus 263326
Long-stalk Oak 324127
Long-stalk Oblong-leaf Agapetes 10349
Longstalk Omphalothrix 270758
Longstalk Oncidium 270823
Longstalk Orchis 116966
Longstalk Ormosia 274409
Longstalk Paraprenanthes 283683
Longstalk Peach 20933
Longstalk Pepper 300524
Longstalk Petrocosmea 292565
Longstalk Phaius 82088
Longstalk Photinia 295708
Longstalk Phyllodium 297012
Longstalk Pollia 307339
Longstalk Portiatree 389938
Longstalk Primrose 314599
Longstalk Rabbiten-wind 12718
Longstalk Redcarweed 288740
Longstalk Redfruit Spicebush 231335
Long-stalk Rhododendron 331134
Long-stalk Rhynchotechum 333744
Longstalk Ricebag 181665
Longstalk Rockyellowtree 415156
Longstalk Roscoea 337081
Longstalk Roughleaf 222132
Longstalk Roundpetal
Coldwaterflower 298857
Longstalk Sandwort 32033
Longstalk Saussurea 348272
Long-stalk Sedge 75733
Longstalk Shedhair Azalea 330301
Longstalk Sida 362521
Longstalk Silvertree 192484
Longstalk Sinocarum 364886
Longstalk Slatepentree 322620
Longstalk Smallleaf Falseflax 68850
Longstalk Solomonseal 308540
Longstalk Sophora 369058
Longstalk Spapweed 205098
Longstalk Speedwell 407103
Longstalk Spiny-fruited Spindle-tree 157272
Long-stalk Spiny-fruited Spindle-tree 157270
Longstalk Splitmelon 351769
Longstalk Spreading Hedyotis 187523
Longstalk Stairweed 142724
Longstalk Stephania 375886
Longstalk Stonebutterfly 292565
Long-stalk Sugerok Holly 204301
Longstalk Suoluo 327366
Longstalk Sweetvetch 187992
Longstalk Swertia 380256
Longstalk Tailanther 381941
Longstalk Taiwan Violet 409995

Longstalk Tanoak 233300
Longstalk Ternstroemia 386715
Longstalk Throathair 100276
Longstalk Tickclover 200732
Long-stalk Tickseed 104531,104532
Long-stalk Tonkin Myrioneuron 261332
Longstalk Touch-me-not 205098
Long-stalk Trigonostemon 397373
Longstalk Trigonotis 397428,397433
Longstalk Tubergourd 390179
Longstalk Tutcheria 322620
Longstalk Valvate Kiwifruit 6719
Longstalk Variegate Azalea 330610
Longstalk Violet 409976
Longstalk White-snaketotue-grass 187523
Longstalk Wild Buckwheat 162333
Longstalk Wildbasil 97014
Longstalk Willow 343136
Longstalk Windhairdaisy 348272
Longstalk Woodbetony 287374
Longstalk Woodnettle 273916
Longstalk Youngia 416451
Longstalk Yunnan Parnassia 284643
Long-stalked Adinandra 8218
Long-stalked Almond 20933
Long-stalked Aster 380866,380869
Long-stalked Bambooleaf Fig 165691
Long-stalked Bauhinia 49160
Long-stalked Buckthorn 328768
Long-stalked Condorvine 245826
Long-stalked Crane's-bill 174551
Long-stalked Cranesbill 174551
Long-stalked Deutzia 127121
Long-stalked Diversicolored Rhododendron 330610
Long-stalked Geranium 174551
Long-stalked Heritiera 192484
Longstalked Herpysma 193072
Long-stalked Holly 203770,204138
Long-stalked Indigo 206368
Long-stalked Machilus 240634
Long-stalked Maple 3116
Long-stalked Meliosma 249409
Long-stalked Metalleaf 296375
Long-stalked Oleaster 142088
Long-stalked Orache 44513
Long-stalked Ormosia 274409
Long-stalked Panic Grass 128547
Long-stalked Peach 20933
Longstalked Pondweed 312232
Long-stalked Rape 59380
Long-stalked Rhododendron 331134
long-stalked Sedge 75733
Long-stalked Sida 362521
Long-stalked Sitchwort 374951
Long-stalked Starwort 374951
Long-stalked Stephania 375886
Long-stalked Supple-jack 52438
Long-stalked Sweetleaf 381310
Long-stalked Tutcheria 322620
Long-stalked Velvety Sophora 369155
Long-stalked Yellow Sedge 76702,76706
Longstalked Yew 306395
Longstalked Yew Family 306345
Longstalk-pod 247204,247205
Longstamen Azalea 331882
Long-stamen Barberry 51437
Longstamen Burnet 345894
Longstamen Champereia 85934
Longstamen Epimedium 146983
Longstamen Garlic 15450
Longstamen Litse 233984
Longstamen Michelia 252912
Longstamen Onion 15450
Longstamen Pearweed 239720
Longstamen Rhodiola 329958
Longstamen Rhododendron 331882
Longstamen Rush 213233
Longstamen Willow 343631
Long-stamened Litse 233984
Long-stamened Michelia 252912
Long-stamened Rhododendron 331882
Long-stamened Willow 343631
Long-stem Basketvine 9454
Longstem Briggsia 60272
Longstem Buttercup 326263
Long-stem Clearweed 298961
Longstem Climborchis 193072
Longstem Coldwaterflower 298961
Longstem Cystopetal 320227
Longstem Fleabane 150616
Longstem Gentian 150616
Longstem Jacot Edelweiss 224863
Longstem Liparis 232364
Longstem Mapania 244909
Longstem Phaius 293534
Longstem Renanthera 327686
Longstem Rooted Pellionia 288763
Longstem Salsify 394284
Longstem Sandwort 32030,32032
Longstem Thorowax 63707
Longstem Twayblade 232364
Longstem Violet 410017
Long-stem Waterwort 142574
Long-stem Wild Buckwheat 152040
Longstem Wildginger 37671
Longstem Woodbetony 287367
Longstem Xylobium 415706
Long-stemmed Basketvine 9454
Long-stemmed Cushion Wild Buckwheat 152361
Longstigma Baikal Meadowrue 388435
Longstigma Monkshood 5165
Longstipal Corallodiscus 103966
Longstipe Beautyberry 66851
Longstipe Burretiodendron 64031
Longstipe Carallia 72400
Longstipe Corydalis 105756
Longstipe Disanthus 134012
Longstipe Dubyaea 138779
Longstipe Dunbaria 138974
Longstipe Fissistigma 166676
Longstipe Fordiophyton 167308
Longstipe Primrose 314596
Longstipe Privet 229526
Longstipe Purplepearl 66851
Longstipe Simpleleaf Turpinia 400544
Longstipe Waterwort 142556
Longstipe Whitepearl 172068
Longstipe Winteraconite 148109
Longstipe Woodbetony 287376
Long-stiped Beautyberry 66851
Long-stiped Carallia 72400
Longstiped Cremanthodium 110427
Long-stiped Disanthus 134012
Longstiped Nutantdaisy 110427
Longstipular Stairweed 142727
Longstock Drumprickle 244909
Longstock Rockjasmine 23161
Long-stolon Sedge 74886,74887
Longstraw Pine 300128
Longstyle Aspidistra 39530
Longstyle Azalea 331136
Longstyle Bittercress 72870
Longstyle Catchfly 363718
Longstyle Chirita 87907
Longstyle Ciliatecalyx Azalea 330404
Longstyle Clematis 95095
Longstyle Daphniphyllum 122715
Longstyle Deerdrug 242695
Longstyle Duperrea 139030
Longstyle Eargrass 187570
Long-style Echinodorus 140548
Long-style Eurya 160533
Longstyle False Solomonseal 242695
Longstyle Figwort 355250
Longstyle Fritillary 168563
Longstyle Gentianopsis 174219
Longstyle Glochidion 177146
Long-style Grevillea 180616
Long-style Hemsle Dogwood 380456
Longstyle Hilliella 196279
Longstyle Hipericumlike Rockfoil 349493
Longstyle Ladybell 7816
Longstyle Leek 15440
Longstyle Lindelofia 231252
Long-style Mahonia 242517
Longstyle Marshmarigold 68200
Longstyle Meadowrue 388555
Long-style Michelia 252913
Longstyle Monkshood 5165

Longstyle Mosquitotrap 117573
Long-style Oldham Daphniphyllum 122715
Longstyle Oneflowerprimrose 270738
Longstyle Onion 15440
Longstyle Phyllodium 297010
Longstyle Pogostemon 306980
Longstyle Purplestem 376469
Longstyle Rhododendron 331136
Longstyle Rush 213234,213400
Longstyle Russianthistle 344731
Longstyle Sandwort 32036
Longstyle Sinocrassula 364925
Longstyle Speedwell 407387
Longstyle St. John's Wort 201995
Longstyle Stonelotus 364925
Long-style Straight Twigs Rhododendron 331415
Longstyle Straighttwig Azalea 331415
Longstyle Sundew 138321
Longstyle Swallowwort 117573
Long-style Thistle 92162
Long-style Tigernanmu 122715
Longstyle Tupistra 400395
Longstyle Wallich Willow 344262
Longstyle Willow 343342
Longstyle Woodbetony 287173,287714
Longstyle Yunnan Sida 362694
Longstyled Clematis 95095
Long-styled Duperrea 139030
Long-styled Dwarf Rush 213542
Long-styled Pfeiffer Cicely 276455,276472
Long-styled Pfeiffer-root 276472
Long-styled Rhododendron 331136,331928
Long-styled St. John's-wort 201995
Long-styled Stewartia 376469
Longstyled Sweetleaf 381423
Long-styled Sweetleaf 381423
Longsurculose Draba 137258
Longsurculose Whitlowgrass 137258
Longtail Bladderwort 403891
Longtail Camellia 69301
Longtail E. H. Wilson Maple 3752
Longtail Elegant Maple 2945
Longtail Holly 203988
Longtail Lasianthus 222174
Longtail Maple 2874
Longtail Mastixia 246213
Longtail Milkvetch 41957
Longtail Narrowleaf Eurya 160605
Longtail Qinggang 116202
Longtail Roughleaf 222174
Longtail Rourea 337702
Longtail Spicebush 231449
Longtail Steward Cyclobalanopsis 116202
Longtail Steward Oak 116202
Longtail Swordflag 208793
Longtail Tea 68970
Longtail Wantien Camellia 69515
Long-tailed Fourstamen Maple 3680
Long-tailed Lasianthus 222174
Long-tailed Wild Ginger 37582
Longtailleaf Angelica 24393
Longtassel Dendrobium 125021
Longtassel Gentian 173489
Longthorn Dock 340293
Longtine Tea 69035
Longtogue Pseudosasa 318328
Longtongue Cremanthodium 110444
Longtongue Fragrant Bamboo 87639
Long-tongue Fragrant-bamboo 87639
Longtongue Melic 249036
Longtongue Needlegrass 376845
Longtongue Nutantdaisy 110444
Longtongue Sasa 347241
Longtongue Sour Bamboo 4613
Longtongue Wild Smallreed 127208
Longtongue Woodbetony 287174
Longtongued Bamboo 347241
Longtongued Sasa 347241
Long-tongued Sasa 347241
Longtooth Manyleaf Rabdosia 209803
Longtooth Monkshood 5373
Longtooth Narrowpod Milkvetch 43086
Longtooth Plumegrass 341876
Longtooth Rose 336692
Longtooth Thyme 391172
Longtooth Tiger Jaws 162930
Longtooth Tongol Milkvetch 43169
Longtooth Vetch 408467,408516
Longtooth Wendlandia 413795
Long-toothed Basketvine 9418
Long-toothed Indigo 205918
Long-toothed Lake Sedge 75037
Long-toothed Rose 336692
Long-toothed Wendlandia 413795
Longtooth-leaf Sichuan Deutzia 127095
Longtouhu Epimedium 147048
Longtube Azalea 332031
Longtube Batweed 89204
Longtube Bluebells 250888
Longtube Camellia 69057
Long-tube Camellia 69311
Long-tube Christia 89204
Longtube Daphne 122509
Longtube Daylily 191286
Longtube Fringed Pink 127756
Long-tube Gold-drop Onosma 271835
Long-tube Golddrop Onosmo 271835
Longtube Ground Ivy 176839
Long-tube Henry Privet 229467
Long-tube Honeysuckle 235927
Long-tube Iris 208886
Long-tube Jasmine 211896
Longtube Lycoris 239264
Longtube Microula 254353
Longtube Milkvetch 42619
Longtube Nepeta 264986
Long-tube Privet 229467
Longtube Rabdosia 209748
Long-tube Rhododendron 332031
Longtube Stonegarlic 239264
Longtube Tea 69057
Longtube Woodbetony 287176
Long-tubed Gilia 175675
Long-tubed Honeysuckle 235927
Long-tubed Huernia 199031
Long-tubed Lily 229900
Long-tubed Privet 229467
Long-tuft Leaves Dracaena 137478
Longtwig Bamboo 47250
Longtwig Pinangapalm 299666
Longumbell Angelica 24395
Longumbrella Red Thorowax 63815
Longutricle Sedge 74758
Longvein Sabia 341530
Longvelvety Chirita 87989
Longvine Mazus 247004
Longwalk Alone 71679
Longwhip Rhodiola 329873
Longwing Begonia 50024
Longwing Crazyweed 278973
Long-wing Hydrangea 199943
Longwing Maple 3114
Longwing Spapweed 205096
Longwing Touch-me-not 205096
Long-winged Maple 3114
Longwort 21266,405788
Longxi Azalea 331158
Longxi Corydalis 106109
Longxi Hanceola 185250
Longxi Primrose 314601
Longxi Rhododendron 331158
Longxishan Sedge 75222
Longyen 131061
Longzhou Amesiodendron 19419
Longzhou Begonia 50092
Longzhou Blachia 54877
Longzhou Camellia 69325
Longzhou Chastetree 411421
Longzhou Chirita 87911
Longzhou Dye Tree 302678
Longzhou Elatostema 142729
Longzhou Evergreen Chinkapin 78982
Longzhou Evergreenchinkapin 78982
Longzhou Fieldorange 249673
Longzhou Fig 164762,165262
Longzhou Grape 411682
Longzhou Hemiboea 191371
Longzhou Holly 203991
Longzhou Melodinus 249673
Longzhou Ophiorrhiza 272246

Longzhou Pearlwood 296447
Longzhou Perilepta 290926
Longzhou Phyllanthodendron 296447
Longzhou Pothos 313194
Longzhou Schefflera 350732
Longzhou Spiradiclis 371767
Longzhou Stairweed 142729
Longzhou Tea 69325
Longzhou Wendlandia 413804
Longzi Azalea 330545
Longzi Milkwort 308162
Longzi Stringbush 414213
Longzxle Arrowwood 407928
Lonicera 235630
Lonicera Flowers Rhododendron 331138
Lonicera-flowered Rhododendron 331138
Lonsplit Chickweed 374965
Lontar 57122
Loof 37024
Loofah 238257,238261
Look up Windhairdaisy 348798
Looking Glass Tree 192494
Looking-glass 374916
Looking-glass Plant 103798
Looking-glass Tree 192494
Look-vp-and-kiss-me 349044,410677
Loongwoolly Actinidia 6651
Loongwoolly Kiwifruit 6651
Loose Covering Rhododendron 330644
Loose Goldwool Clematis 95074
Loose Magnific Deutzia 127006
Loose Raspberry 338732
Loose Scaled Rhododendron 330264
Loose Scales Rhododendron 331163
Loose Schneider Deutzia 127089
Loose Serrulate Enkianthus 145698
Loose Silky Bent 28972
Loose Silkybent 28972
Loose Silky-bent 28972
Loose Skinned Orange 93446
Loose Speargrass 4128
Loose Spicegrass 239707
Loose Spike Roegneria 335525
Loose Spike Shortfilament Flower 221418
Loose Spreading Cyrtococcum 120632
Loose Strife 240025
Loose Sweetgrass 196142
Loosebract Beautifulhead Edelweiss 224812
Looseflower Alpine Sweetvetch 187770
Loose-flower Ash 167951
Loose-flower Beautiful Azalea 330287
Loose-flower Beautiful Rhododendron 330287
Looseflower Bentgrass 12133
Looseflower Burnet 345842
Loose-flower Cherry 83246,280027
Loose-flower Cinnamon 91412
Looseflower Dyckia 139261
Loose-flower Erycibe 154253
Loose-flower Faber Clethra 96498
Loose-flower Falsetamarisk 261280
Looseflower Goosecomb 335395
Looseflower Gynostemma 183011
Looseflower Helictotrichon 190202
Looseflower Kangding Rockjasmine 23217
Looseflower Little Man Orchid 178895
Looseflower Lobelia 234592
Looseflower Longleaf Rockjasmine 23220
Looseflower Maesa 241796
Looseflower Mazus 247017
Looseflower Meadowrue 388673
Looseflower Milkvetch 42496
Loose-flower Mockorange 294489
Loose-flower Moellendorff Cardiandra 73137
Loose-flower Mussaenda 260445
Looseflower Needlegrass 376875
Loose-flower Neillia 263171
Looseflower Orchis 273513
Loose-flower Peachleaf Rhododendron 330085
Loose-flower Podocarpium 200728
Loose-flower Pygeum 322400
Looseflower Rockjasmine 23212
Looseflower Roegneria 335395
Looseflower Rose 336685
Looseflower Shortstalk Monkshood 5071
Looseflower Small Reed 127229
Looseflower Smallreed 127229
Looseflower Tianmu Sanicle 346006
Looseflower Toona 392834
Looseflower Water Willow 209922
Loose-flower Wendlandia 413792
Loose-flowered Barberry 51835
Looseflowered Bentgrass 12133
Loose-flowered Blood-red Spindle-tree 157864
Loose-flowered Feathergrass 376875
Looseflowered Jewelvine 125978
Loose-flowered Olive 270140
Loose-flowered Orchid 273513
Loose-flowered Phacelia 293154
Loose-flowered Rattan Palm 65721
Loose-flowered Rhododendron 330085
Loose-flowered Spindle-tree 157675
Loose-flowers Coalseleaf Azalea 331757
Loose-flowers Rough-leaves Rhododendron 331757
Loosefruit Pepper 300422
Loosehairy Leucas 227657
Loosehead Sandwort 32260
Loose-headed Bracted Sedge 76307
Loose-headed Oval Sedge 75879
Loose-leaf Cliffbean 254813
Loose-leaf Millettia 254813
Looseleaf Pricklyash 417329
Loose-leaf Wild Buckwheat 151879
Looseleaved Lettuce 219490
Looselobe Woodbetony 287749
Loosely Woodrush 238614
Looselyflower Anodendron 25929
Looselyflower Eelvine 25929
Loose-node Sophora 369101
Loose-noded Sophora 369101
Loosescale Gayfeather 228534
Loose-scaled Rhododendron 330264
Looseskinned Orange 93717
Loosespike Elsholtzia 143984
Loosespike Oplismenus 272676
Loosespike Small Wild Buckwheat 162320
Loose-spiny Shinyleaf Pricklyash 417282
Loosestrife 239543,239679,239703,239755,239795,239804,239809,240025,240100
Loose-strife 239898,240025
Loosestrife Family 240020
Loose-strife Family 240020
Loosestrifeleaf Mosquitotrap 117576
Loosestrifeleaf Swallowwort 117576
Loose-tooth Lemperg's Barberry 51846
Loose-toothed Camellia 69547
Loosetwig Rhubarb 329345
Loosevein Bigginseng 241175
Loose-vein Bigleaf Clethra 96506
Looswspike Lovegrass 147876
Lopez Root 392559
Lopgrass 60745
Lop-grass 60853
Lophanthus 236262
Lophather 236281
Lophatherum 236281,236284
Lophira 236334
Lophotocarpus 236478
Loppedleaf Raspberry 339378
Lopper Gowan 399504
Lopper-gowan 399504
Loppy-major 31051
Lops-and-lice 336419
Lopsang Dalai Rhododendron 331142
Lopseed 295978,295984,295986
Lopseed Family 296006
Lopside Oat 45602
Lopsided Oat 45602
Loquat 151130,151162
Loquatleaf Arrowwood 408089
Loquatleaf Aucuba 44906
Loquatleaf Beautyberry 66829
Loquat-leaf Machilus 240558
Loquatleaf Purplepearl 66829
Loquatleaf Sweetleaf 381412
Loquat-leaf Sweetleaf 381412
Loquatleaf Tanoak 233204
Loquat-leaved leaf Aucuba 44906
Loquat-leaved Tanoak 233204

Loranth 236507
Lord Adson's Blue Pea 222781
Lord Howe Mountain Rose 252621
Lords and Ladies 36967,37015
Lords' And Ladies'fingers 37015
Lords-and-ladies 36967,37015,273531, 302034,302103
Lords-and-ladies Family 30491
Lords-and-ladies' Fingers 37015
Lordwood 232562
Lorel 122492
Lorelon Pride 127607,349044
Lorentz Loxopterygium 237997
Lorentz Red Quebracho 323563
Lorer 223203
Lore-tree 80577
Loropetal 237190
Loropetalum 237190
Lorraine Begonia 49712
Lorry 223203
Lorts-and-Ladies 37015
Lorty Seepweed 379492
Los Angeles Sunflower 189014
Losefruit 371237
Lost Creek Wild Buckwheat 152286
Lost Thistle 92312
Lotebush Condalia 101739
Lotien Skullcap 355579
Lotonbean 237219,237244
Lotononis 237219
Lot-tree 369322
Lotung Mussaenda 260449
Lotung Parakmeria 283419
Lotungvine 90612,90613
Lotus 237476,263267,263270,263272, 418177
Lotus Bulbophyllum 62780
Lotus Bush 418177
Lotus Curlylip-orchis 62780
Lotus Flower 64184
Lotus Fruit 418177
Lotus Magnolia 242135
Lotus Mosquitotrap 117666
Lotus Sweetjuice 176949
Lotus Tree 266355,418177,418216
Lotusberry 80577
Lotusdaisy 111823
Lotusleaftung 192909,192913
Lotusleaftung Family 192921
Lotus-lily Family 263281
Loudon Barberry 51879
Lougb Lily 267648
Louis Edmunds 31108
Louise's Swallow-wort 117575
Louisiana Broom-rape 275121
Louisiana Cypress 385262
Louisiana Flatsedge 322342
Louisiana Goldenrod 368224
Louisiana Marsh Spider-lily 200940
Louisiana Sage-wort 35837
Louisiana Sedge 75223
Louisiana Trillium 397577
Louisiana Wormwood 35837
Loulan Peppergrass 225409
Loureir Angularflower Elaeagnus 142012
Loureir Beauty-berry 66829,66856
Loureir Crinum 111221
Loureir Elaeagnus 142071
Loureir Gentian 173606
Loureir Spathulate Sundew 138357
Loureiro Clematis 95096
Loureiro Murdannia 260105
Loureiro Pothos 313206
Lourier Cassava 244511
Louro 268688
Louro Inamui 268700
Louro Preto 268688
Loury 122492
Louse Bur 31056
Lousebane 124604
Louseberry 21459
Louseberry Tree 157429
Louse-herb 124604
Lousewort 124604,190959,286972,287063, 287494,287721
Lousy Arnot 63475,102727
Lousy Arnut 63475,102727
Lousy Bed 248312
Lousy Beds 248312
Lousy Grass 370522
Lousy Soldier's Buttons 248411
Lovable Oreocharis 273832
Lovable Panicgrass 281323
Lovable Paphiopedilum 282773
Lovable Spunsilksage 360812
Lovableflower 148053,148087
Lovache 229384
Lovage 228250,229298,229384,366743
Love Apple 146724,239157,244346
Love Bean 765
Love Creeper 100872,177707
Love Dodder 115040
Love Entanggle 356468
Love Grass 90108,147468,147699
Love Idol 410677
Love in a Puff 73207
Love Links 357114
Love Nut 765
Love Pea 765
Love Tree 83809
Love Troth 284387
Love Vine 28422,78731,114996,115008, 115040,115047
Love's Test 26551
Loveable Otophora 277308
Loveache 326340
Love-and-Idleness 410677
Love-and-Idols 410677
Love-and-tangle 396854,397019
Love-apple 239157,284289
Lovebamboo 47264,264510
Lovebeecelery 105467,105469
Love-bind 95431
Love-creeper 100872
Love-entangle 266225
Love-entangled 95431,237539,266225, 356468
Lovegrass 147468
Love-grass 147468,147650
Love-Idol 410677
Love-in-a-chain 357114
Love-in-a-hedge 167156,266225,285639
Love-in-a-mist 266215,266225
Love-in-a-puff 73207
Love-in-a-puzzle 266225
Love-in-a-tangle 356468
Love-in-idle 410677
Love-in-idleness 410513,410677
Love-in-idleness Ife 385302
Love-in-the-mist 266225
Love-in-vain 410677
Love-in-winter 87498
Lovejoy 147384
Love-leaves 31051
Love-lies-bleeding 8332,18687,18810, 220729,238371,410677
Love-links 357114
Love-long 357203
Lovely Achnatherum 4163
Lovely Azalea 331581
Lovely Barberry 51292
Lovely Brigssia 60269
Lovely Chirita 87971
Lovely Coelachne 98455
Lovely Crape-myrtle 219909
Lovely Dogtailgrass 118325
Lovely Evergreen Chinkapin 78852
Lovely Fir 280
Lovely Glorylily 177251
Lovely Golden Larch 317822
Lovely Golden-larch 317822
Lovely Hemsleia 191896
Lovely Jijigrass 4163
Lovely Knotweed 309269
Lovely Kohleria 217739
Lovely Lily 229872
Lovely Monkshood 5510
Lovely Pricklythrift 2300
Lovely Rhododendron 330626,331581, 331641
Lovely Snowgall 191896

Lovely Stringbush 414220
Lovely Tanoak 233103
Lovely-shaped Rhododendron 330279
Loveman 95431,170193
Love-man 95431,170193
Love-me 95431,260760
Love-me-not 407072
Love-me-not Love-me 235334
Lover's Kisses 170193
Lover's Knots 170193
Lover's Links 401711
Lover's Pride 309570
Lover's Steps 235373
Lover's Thoughts 410677
Lover's Wanton 273469,273531
Love-restoring Stonecrop 356535
Love-root 208565
Lovers' Kisses 170193
Lovers' Knots 170193
Lovers' Links 401711
Lovers' Pride 309570
Lovers' Thoughts 410677
Lovesalt Milkvetch 43000
Lovesand Milkvetch 41976
Love-seed 99042
Lovesnow Milkvetch 42796
Love-stone 187222
Lovestone Milkvetch 42878
Love-tree 83809
Love-troth 284387
Love-true 410677
Love-vine 78731,114955,114996,115040,
115047,115097
Lovi-Lovi 166776
Loving Andréws 174832
Loving Lidols 410677
Loving Lydles 410677
Loving-rock 304336
Low Ardisia 31471
Low Baby's-breath 183221
Low Bamboo 304084
Low Belia 234547
Low Bigfruit Elm 401556
Low Bindweed 68742
Low Birch 53594
Low Blueberry 403957
Low Box 64376
Low Broom 173082
Low Bulrush 396020
Low Calamint 65538
Low Chamaesium 85718
Low Chinese Box 64376
Low Cornel 85581
Low Cornflower 81033
Low Cremanthodium 110403
Low Cudweed 178481
Low Cyperus 118734
Low Cypressgrass 119461
Low Dutch Heath 150131
Low Erysimum 154534
Low Euonymus 157737
Low Eyebright 160286
Low False Bindweed 68742
Low Flatsedge 322329
Low Flax 232001
Low Forked Chickweed 284845,284847
Low Galingale 119461
Low Geniculatepetal Monkshood 5220
Low Geranium 174854
Low Germander 388069
Low Ground Rattan 329184
Low Ground-rattan 329184
Low Ground-rattan Palm 329184
Low Hackberry 80723
Low Hairstalk Serpentroot 354935
Low Hemarthria 191250
Low Honeysuckle 235859
Low Hop Clover 396854
Low Hop-clover 396854
Low Huckleberry 403987
Low Indocalamus 206819
Low Japanese Barberry 52247
Low Japanese Fleeceflower 309262
Low Jasmine 211863
Low Juniper 213706
Low Kalanchoe 215160
Low Larkspur 124066
Low Leptodermis 226126
Low Lily 230009
Low Liriope 232644
Low Livistona 234182
Low Love Grass 147871
Low Lupin 238483
Low Mallow 243823
Low Meadow Rue 388572
Low Meadowrue 388572
Low Northern Sedge 74164
Low Nut Grass 354274
Low Nutantdaisy 110403
Low Nut-rush 354274
Low Opithandra 272600
Low Or Two-form Or Cushion
Pussytoes 26380
Low Pectis 286880
Low Peliosanthes 288658
Low Peppergrass 225440
Low Poppy Mallow 67128
Low Rabbitbrush 150369
Low Rhododendron 331584
Low Rockfoil 349685
Low Roegneria 144337
Low Rough Aster 160676
Low Running Blackberry 339103
Low Sagebrush 35147
Low Sedge 74849
Low Serpentroot 354956
Low Service Berry 19264
Low Shadblow 19290
Low Shadbush 19264
Low Shoebutton 31471
Low Shortleaf Kyllingia 218490
Low Slender Iris 208885
Low Slender Tulip 400201
Low Spear-grass 305334
Low Spearwort 326271
Low Spike-rush 143289
Low Spindle-tree 157737
Low Stanhopea 373695
Low Starburr 2613
Low Starwort 374919
Low Sugarmustard 154534
Low Sweet Blueberry 403716
Low Tick Trefoil 126586
Low Torularia 264609
Low Water-milfoil 261350
Low Water-parsnip 53157
Low Whorled Milkweed 38076
Low Wildclove 226126
Low Youngia 416415
Low Yushan Gentian 173872
Low's Caladium 16509
Low's White Fir 407
Lowanther-orchis 85136,85139
Lowbush Blueberry 403716,403718,403952
Low-bush Blueberry 403716,403718
Lowbush Penstemon 289343
Lowdaphne Stringbush 414150
Low-daphne Stringbush 414150
Lowell Ash 168021
Lower Jasmine 212019
Lowflower Dasymaschalon 122933
Low-growing Knotweed Knotweed 309198
Low-growing Leymus 228393
Low-growing Melocanna 249620
Low-growing Ophiorrhiza 272218
Lowhill Banksia 47637
Lowi Paphiopedilum 282857
Lowi Phalaenopsis 293616
Lowia Family 237937
Lowland Fir 384
Lowland Rotala 337384
Lowland Spruce 298435
Lowland White Fir 384
Lowland Yellow Loosestrife 239679
Low-level Oak 323611
Low-leveled Oak 323611
Lowly Ricegrass 300717
Lowly Small Reed 127294
Lowndes Rhododendron 331145
Lowrie's Aster 40792
Lowry 122492

Lowsweet Blueberry 403716
Loxa Bark 91082
Loxabark 91082
Loxbrct Fedde Elsholtzia 144017
Loxococcus 237971
Loxopterygium 237996
Loxostigma 238030
Loyar Loyak 228729
Ltarshweed 81356
Lttleflower Crapemyrtle 219959
Luahan Viburnum 407795
Luanda Greenbrier 366537
Luanda Purplepearl 66904
Luanda Sageretia 342191
Luandashan Blueberry 403976
Luandashan Cinnamon Tree 91429
Luanguofoshow Ginkgo 175822
Luanta Sageretia 342191
Luanta-fir 114548
Luantashan Blueberry 403976
Lubber-lub 250502
Lubbers Begonia 50034
Lubricant Barberry 51881
Lubricous Oligostachyum 270438
Lucens Peridiscus 290897
Lucerne 247456,271280
Luchu Azalea 331759
Luchu Greenbrier 366484
Luchu Pine 300044
Luchuan Stonecrop 356901
Luchuen Suzukia 380045
Luchun Elatostema 142728
Luchun Glaucousleaf Wattle 1500
Luchun Irisorchis 267924
Luchun Medinilla 247585
Luchun Memecylon 250003
Luchun Oberonia 267924
Luchun Ophiorrhiza 272247
Luchun Paraprenanthes 283687
Luchun Raspberry 338769
Luchun Rhaphidophora 329005
Luchun Stairweed 142728
Luchun Suzukia 380045
Luchun Tanoak 233303
Luchun Yushanbamboo 416756
Luchun Yushania 416756
Luciana Pink Beauty Water Lily 267630
Lucid Albizia 13609
Lucid Camellia 69319
Lucid Clubfilment 179191
Lucid Gomphostemma 179191
Lucid Peridiscus 290897
Lucid Scolopia 354616
Lucid Spindle-tree 157695
Lucida Water Lily 267631
Lucidleaf Childvine 400945
Lucidleaf Fieldcitron 400543
Lucidleaf Galangal 17693
Lucidleaf Maple 3124
Lucidleaf Sageretia 342179
Lucidleaf Turpinia 400543
Lucid-leaved Daphne 122386
Lucid-leaved Maple 3124
Lucid-leaved Rose 336714
Lucid-leaved Sageretia 342179
Lucid-leaved Turpinia 400543
Lucidstem Actinidia 6696
Lucidstem Kiwifruit 6696
Lucifer 247456
Lucifer Matches 365564
Luck 28200
Luckan Gowan 399504
Lucken Golland 68189
Lucken Gowan 399504
Luck-herb 202117
Luckie Gowan 399504
Luckie's Mutch 5442
Lucky Bean 738,765,10127,154617,389967
Lucky Bean Coral Tree 154687
Lucky Bean Tree 10127,154620,154687
Lucky Beans 765
Lucky Clover 278124
Lucky Gowan 399504
Lucky Lilyturf 272086
Lucky Moon 401711
Lucky Nut 389978
Luckynut Thevetia 389978
Lucky-nut Thevetia 389978
Luckyweed 370405
Lucretia Dewberry 339190
Luculia 238100
Lucuma 238110
Lucy Arnut 63475,102727
Lucy Locket 72934
Lucyarnut 102727
Ludao Ixora 211096
Ludao Pachystoma 279858
Ludian Anemone 24097
Ludian Holly 203997
Ludian Windflower 24097
Luding Draba 137096
Luding Gentian 173607
Luding Greyawngrass 372421
Luding Rose 336716
Luding Spodiopogon 372421
Luding Willow 343642
Ludisia 238133
Ludlow Azalea 331150
Ludlow Barberry 51943
Ludlow Corydalis 106114
Ludlow Cotoneaster 107533
Ludlow Dandelion 384645
Ludlow Lady's Slipper 120387
Ludlow Ladyslipper 120387
Ludlow Monkshood 5382
Ludlow Peony 280189
Ludlow's Rhododendron 331150
Ludovician Oat 45495
Ludweig Woodbetony 287382
Ludwig Begonia 50036
Lueddemann Phalaenopsis 293617
Luehea 238246
Luffa 238261
Lug 208771
Lugu Begonia 50038
Luhuo Azalea 331151
Luhuo Barberry 51888
Luhuo Rhododendron 331151
Luisia 238301
Lujula 277648
Luk Pencil-wood 139661
Lukes 59575
Lukiang Rhododendron 331152
Lukin Gowan 399504
Lulang Azalea 331157
Lulang Rhododendron 331157
Lulo 367543
Luma 238336
Lumbago 83646
Lumbang Oil 14544
Lumbangoil Plant 14544
Lumboltz Pine 300048
Lumie 93554
Lumnitzera 238345,238354
Lumper-scrump 192358
Lumpho 10109
Lunan Nut 225673
Lunan Ottelia 277367
Lunan Touch-me-not 205100
Lunar Crater Howell's-buckwheat 212398
Lunary 238371
Lunate Indosasa 206891
Lunate Irisorchis 267968
Lunate Oberonia 267968
Lunate Sundew 138331
Lundy Cabbage 99094
Lung Moss 234269
Lungan 131061
Lungchi Primrose 314601
Lung-flower 173728
Lungling Greenbrier 366438
Lungling Newlitse 264072
Lungling Woodbetony 287383
Lungsheng Maple 3125
Lungsheng Rabdosia 209763
Lungtsuan Stonecrop 356903
Lungwort 222651,234269,321633,321636,321643,321647
Lungwort Comastoma 100278
Lungwort Crottle 234269
Lungwort Sticta 234269

Lungwort Throathair 100278
Luning Monkshood 5383
Luntai Serpentroot 354892
Lunumidella 248895
Lunumidena Cedar 248895
Lunumidena Mahogany 248895
Luocheng Grape 411787
Luocheng Hornbeam 77340
Luocheng Photinia 295727
Luodian Ampelocalamus 20199
Luodian Aspidistra 39556
Luodian Begonia 50181
Luodian Bethlehemsage 283618
Luodian Drepanostachyum 20199
Luodian Fig 164964
Luodian Glyptic-petal Bush 177989
Luodian Handbamboo 20199
Luodian Hiptage 196837
Luodian Kitefruit 196837
Luodian Paraphlomis 283618
Luodian Pararuellia 283762
Luodian Privet 229625
Luodian Skullcap 355579
Luoding Mallotus 243385
Luoding Wildtung 243385
Luoergai Buttercup 326048
Luofu Fluttergrass 166326
Luofu Jointfir 178546
Luofu Maple 2951,3627
Luofu Mountain Jointfir 178546
Luofu Purplepearl 66881
Luofu Reevesia 327365
Luofu Sterculia 376192
Luofushan Jointfir 178546
Luofushan Raspberry 338762
Luohanfruit 365325,365332
Luohanguo Momordica 365332
Luohanguo Siraitia 365332
Luohe Dendrobium 125235
Luolong Corydalis 106071
Luolong Milkvetch 42616
Luoping Corydalis 106111
Luoping Thistle 92066
Luopingshan Spapweed 205244
Luotian Atractylodes 44209
Luotian Magnolia 416712
Luoxiang Elaeagnus 142072
Lupin 238419
Lupinaster 396956
Lupine 238419,238467
Lupine Clover 396956
Luping-like Thermopsis 389540
Lupinlike Corydalis 106115
Luqiu Greenbrier 366484
Luquan Corydalis 106116
Luquan Onosma 271796
Luquan Paris 284346
Luquan Stonecrop 356901
Lurgadish 250432
Lurgeydish 250432
Lurid Ophiorrhiza 272248
Luristan Ophrys 272480
Lurkeydish 250432
Lurk-in-ditch 250432
Lus Na Laoch 329935
Lushan Arundinella 37398
Lushan Ash 168100
Lushan Banana 260244
Lushan Barberry 52320
Lushan Blushred Rabdosia 324862
Lushan Bulrush 353576
Lushan Elatostema 142844
Lushan Epimedium 147030
Lushan Germander 388179
Lushan Goldsaxifrage 90403
Lushan Goldwaist 90403
Lushan Grape 411742
Lushan Hibiscus 195089
Lushan Honeysuckle 235965
Lushan Loosestrife 239822
Lushan Pseudosasa 318305
Lushan Sedge 75240
Lushan Solomonseal 308582
Lushan Spiny-fruited Spindle-tree 157696
Lushan Stairweed 142844
Lushan Teasel 133497
Lushan Violet 410598
Lushan Windhairdaisy 348179
Lushan Yushanbamboo 416820
Lushan Yushania 416820
Lushi Blushred Rabdosia 324863
Lushi Gymnaster 256307
Lushi Nakeaster 256307
Lushui Arrowbamboo 162707
Lushui Bentgrass 12219
Lushui Fargesia 162707
Lushui Greenbrier 366439
Lushui Lilyturf 272110
Lushui Mockorange 294495
Lushui Plantain Calanthe 66032
Lushui Rockfoil 349576
Lushui Smilax 366439
Lushui Umbrella Bamboo 162707
Lushun Alder 16458
Lusitanian Cypress 114714
Lusitanian Heath 149703
Lusitanian Oak 324134
Lusmore 130383
Luson Pine 299986
Luss-ny-ollee 299759
Lusterleaf Holly 203961
Luster-leaf Holly 203961
Lustwort 138348
Lus-y-volley 170743
Lutao Copperleaf 1811
Lutao Eugenia 382649
Luteous Begonia 50040
Luteous Silver Maple 3536
Lutescent Pycnospora 322078
Lutescent Seedybean 322078
Lutien Anemone 24097
Lutqua 46198
Lutz Stonecrop 356904
Luvunga 238530
LUX Poplar 311293
Luxi Elaeagnus 142073
Luxi Elatostema 142730
Luxi Nearsmooth Barberry 52201
Luxi Stairweed 142730
Luxi Thorowax 63742
Luzhai Begonia 50042
Luzhi Swollennoded Cane 87570
Luzon Anaxagorea 21857
Luzon Arrowwood 407930
Luzon Canarytree 71004
Luzon Dualbutterfly 398287
Luzon Eugenia 156126
Luzon Fruit 378761
Luzon Gonocaryum 179442
Luzon Hackberry 80688
Luzon Illigera 204634
Luzon Nightshade 367317
Luzon Pennywort 200263
Luzon Schefflera 350745
Luzon Skullcap 355584
Luzon Strychnos 378761
Luzon Tephrosia 386167
Luzon Vanda 404655
Luzon Viburnum 407930
Luzon Wendlandia 413798
Ly Purplepearl 66787
Ly's Beautyberry 66787
Ly's Paragutzlaffia 283362
Ly's Rhododendron 330404
Lyaall Hollyleaf Mahonia 242484
Lyall's Larch 221913
Lyall's Serpentweed 392778
Lycaste 238732
Lycaste Orchid 238732
Lychee 233077,233078
Lycheeshape Longan 131058
Lychnis 363141
Lycian Hawthorn Barberry 51513
Lycianthes 238939
Lycoctonifoliate Monkshood 5387
Lycoperdon Tanoak 233305
Lycopodium-like Dacrydium 121104
Lycopsis 239167
Lycoris 239253
Lydia Broom 173000
Lygisma 239318

Lying Dround Rhododendron 331722
Lying-flat Crazyweed 279087
Lyme Grass 144144,228337,228346
Lymegrass 144144
Lyme-grass 144144,228337,228346
Lynch Nautilocalyx 262899
Lynch's Alloplectus 16177
Lynd 391706
Lyne 232001
Lynn Haven Goldenaster 90227
Lynndyl Fourwing 44340
Lynx Flower 373693
Lyon Epigeneium 146541
Lyon's Pentachaeta 289426
Lyon's Turtlehead 86793
Lyonia 239369,239384
Lyrate Bittercress 72873
Lyrate Hemistepta 191666
Lyrate Rockcress 30342
Lyrate Rock-cress 30342
Lyrate Woodbetony 287397
Lyrateleaf Loosestrife 239763
Lyrateleaf Meconopsis 247150
Lyrateleaf Windhairdaisy 348504
Lyre Flower 220729
Lyre-flower 220729
Lyreleaf Jewelflower 377637
Lyreleaf Parthenium 285078
Lyre-leaf Sage 345187
Lyre-leaved Rock Cress 30342
Lyre-leaved Rock-cress 30342
Lyre-leaved Sage 345187
Lyre-shaped Rockcress 30342
Lyrestyle Sage 345252
Lyre-tree 232609
Lysidice 239516,239525
Lysionotus 239926,239967
Lytchee 233078
Lythewale 233760
Lythewale Lichwale 233760
Lythrum 240025
Lyver 401112

M

M. Degron's Rhododendron 330532
M. O. Albertson's Rhododendron 330054
M. P. Edgeworth's Rhododendron 330613
M. S. Ames' Rhododendron 330081
M. Smirnow's Rhododendron 331858
Ma Bamboo 125482
Ma's Windhairdaisy 348507
Maack Aster 40805
Maack Euonymus 157699
Maack Falsehellebore 405606
Maack Honeysuckle 235929
Maack Larkspur 124360
Maack Laurel Cherry 280028
Maack Onion 15444
Maack Pondweed 312170
Maack Sedge 75252
Maack Thistle 92167
Maack's Honeysuckle 235929
Maackia 240112
Maanshan Ceropegia 84276
Maass Parodia 284674
Maaximowicz's Lily 229888
Mabian Elatostema 142731
Mabian Figwort 355183
Mabian Maple 3131
Mabian Stairweed 142731
Mabian Ternstroemia 386719
Mabian Yushanbamboo 416794
Mabian Yushania 416794
Macadamia 240207,240212
Macadamia Nut 240207,240209,240210,240212
Macadamla 240210
Macao Nightshade 367334
Macaranga 240226,240325
Macariney Rose 336405
Macaroni 398900
Macaroni Wheat 398900
Macarthur Clusterpalm 6854
MacArthur Feather Palm 321170
Macarthur Palm 321170
Macary Bitter 298494,298495
Macasaar Oil 351968
Macassa 8998
Macassar 261402
Macassar Ebony 132093,132379
Macassar Oil Tree 70960
Macaw Bush 367357
Macaw Fat 142257
Macaw Palm 6136,12773
Macaw-bush 367357
Macaw-fat 142257
Macawflower 189992
Macawood 302860
Macclure Michelia 252915
Macclure Uvaria 403519
Macdougal Cactus 244137
Macdougall Begonia 50045
Mace 261424,324335
Mace Flinwort 227533
Mace Sedge 74711
Macedonia Pine 300144
Macedonian Camomile 26813
Macedonian Oak 324498
Macedonian Parsley 366743
Macedonian Pine 300144
Macedonian Toadflax 231033
Macey 324335
Macey-tree 324335
Macgregor Silverbell 184743
MacGregor's Wild-rye 144380
Mach Red Scales Rhododendron 331427
Machete 154631
Machfragrant Magnolia 232599
Machhair Flatawndaisy 413011
Machiloid Holly 204183
Machilus 240550
Machilus Leaf Litse 233987
Machilusleaf Litse 233987
Machilus-leaved Holly 203999
Machilus-leaved Litse 233987
Machilus-leaved Maple 3126
Mach-interest Azalea 331890
Machure Sweetleaf 381094
Machuria Semiaquilegia 210281
Mack Mullein 405737
Mackay Bean 145875,145899
Mackay's Heath 149709
Mackaya 240735
Mackenzie Willow 343649
Mackerel Mint 250450
Mackerel Plant 17136
Mackinnon Addermonth Orchid 110647
Mackinnon Bogorchis 110647
Maclaren St. John's-wort 202003
Macleania 240750
Macleay Laurel 26185
Maclelland Fig 165269
Maclure Ardisia 31509
Maclure Championella 85948
Maclure China-laurel 28358
Maclure Cryptocarya 113467
Maclure Fissistigma 166670
Maclure Galangal 17714
Maclure Gonocaryum 179444
Maclure Ivyarum 353127
Maclure Persimmon 132279
Maclure Premna 313679
Maclure Thickshellcassia 113467
Maclure Tolypanthus 392715
Maclurolyra 240856
MacNab Cypress 114721
MacNab's Cypress 114721
Macoun's Blue-joint 65307

Macoun's Rabbit-tobacco 317753
Macoun's Reed Grass 65307
Macoun's Spike-rush 143194
Macoun's Thistle 91941
Macrae Bulbophyllum 62873
Macrae Curlylip-orchis 62873
Macrae's Hakea 184619
Macranthum Azalea 330917
Macrodont Lagochilus 220079
Macrofloral Honeysuckle 235941
Macrophyllous Dogwood 380461
Macrophyllous Ehretia 141629
Macrophyllous Gambir-plant 401770
Macrophyllous Hibiscus 194986
Macrophyllous Hydrangea 199956
Macrophyllous Knotweed 309360
Macrophyllous Mussaenda 260454
Macrophyllous Parastyrax 283994
Macropiper 241208
Macropodium 241248,241249
Macropodous Daphniphyllum 122700
Macropodous Landpick 308593
Macropodous Pepper 300449
Macropodous Solomonseal 308593
Macropodous Tigernanmu 122700
Macrosepal Gugertree 350927
Macrosepal Largeleaf Hydrangea 200009
Macrosolen 241313
Macrostachys Eulophia 156947
Macrotyloma 241408
Macrozamia 241448
Maculate Achyrophorus 202424
Maculate Basketvine 9456
Maculate Bitter Bamboo 37271
Maculate Bubei Spindle-tree 157589
Maculate Dayflower 101085
Maculate Indosasa 206892
Maculate Japan Aucuba 44920
Maculate Rhododendron 331176
Maculate Sedge 75279
Maculate Stauntonvine 374423
Maculate-fruited Cryptocarya 113468
Mad Apple 123077,367370
Mad Dog Skullcap 355563
Mad Dog Weed 14760
Mad Dog's Berries 367120
Mad Nip 61497
Mad River Fleabane 150772
Mad Woman's Milk 159027
Madagascan Ocotillo 16262
Madagascar Bean 294010
Madagascar Beautyleaf 67871
Madagascar Burmareed 265916
Madagascar Calanthe 66003
Madagascar Cardamom 9878
Madagascar Chaplet Flower 245803
Madagascar Clove 10550
Madagascar Cloves 10550
Madagascar Copal 200848
Madagascar Crabgrass 130661
Madagascar Cryptostegia 113825
Madagascar Dragonbood 137447
Madagascar Dragon-tree 137447
Madagascar Ebony 132188,132435
Madagascar Ebony Ebony Veine 132192
Madagascar Flatsedge 119370
Madagascar Gragon Tree 137447
Madagascar Jasmine 211693,245803,376022
Madagascar Mahogany 216209
Madagascar Nutmeg 10550
Madagascar Olive 266704
Madagascar Ordeal Tree 83706
Madagascar Palm 89345,89360,279681
Madagascar Pepper 300464
Madagascar Periwinkle 79418
Madagascar Plum 83224,166761,166773,166786
Madagascar Raffia Palm 326639
Madagascar Raffia-palm 326639
Madagascar Ragwort 359423
Madagascar Reed Bamboo 268114,268117
Madagascar Rubber 113820,220814,245777,246030
Madagascar Rubbervine 113825
Madagascar Traveler's-tree 327095
Madagascar Travelerstree 327095
Madagascar Widow's-thrill 215102
Madagascar Yellowwood 306485
Madagascarpalm 89345,89360
Madake Bamboo 297215
Madam Gorgon 4524
Madam Trip 357114
Madam Ugly 168476
Mad-apple 123077,367370
Madar 68086,68088
Madarin 397970
Madberry 405571
Madden Cherry 241498,241502
Madden Rhododendron 331179
Maddencherry 241498,241502
Madden-cherry 241498,241502
Maddenia 241498
Madder 26768,337906,338038
Madder Family 338045
Maddern 26768
Madders 26768
Maddertree 12515,12519
Madderwort 39299,235090
Mad-dog Skullcap 355563
Madegascar Butterflybush 62118
Madegascar Butterfly-bush 62118
Madegascar Summerlilic 62118
Madeila Barberry 51901
Madeira Bay 291543
Madeira Bay Persea 291543
Madeira Bay-persea 291543
Madeira Bluethistle 141129
Madeira Broom 173078,173129
Madeira Dyer's Greenweed 173002
Madeira Holly 204157
Madeira Ivy 187200
Madeira Marrow 356352
Madeira Palm 4951
Madeira Vine 26264,26265
Madeira Viper's Bugloss 141129
Madeira Whortleberry 403951
Madeiran Lady's-tresses 179647
Madeiran Orchid 121381
Madeiravine 26264,26265
Madeira-vine 26264,26265,48688
Madeira-vine Family 48698
Maden 413706
Maderaspat Leaf-flower 296647
Madern 227533
Maderwort 36474
Madflower 27664
Madhkankur 93614
Madhuca 241510
Madia 241523
Madnef 192358
Madonna Lily 229789,370332
Madras Bean 241440
Madras Grangea 180177
Madras Hemp 112269
Madras Thorn 301128,301134
Madras-thorn 301134
Madre 176970
Madre de Cacao 176970
Madrean miner's-lettuce 94351
Madrean Rosemary 307285
Madrit Brome 60820
Madron 322724
Madrona 30877,30885
Madrona Laurel 30885
Madrone 30877,30885,30893
Madrono 30885
Madrono Laurel 30885
Madura Rattlebox 112381
Madweed 47013,245770
Madwoman's Milk 159027
Mad-woman's-milk 159027
Madwort 18313,39298,39299,45192,47013,235090
Maedakan 93457
Maerkaang Willow 343655
Maerkang Azalea 330204
Maerkang Onosma 271799
Maerkang Thorowax 63744
Maershan Gentian 173618
Maershan Monkshood 5156
Maerua 241572

Maesa 241734
Mafu 162179
Mafura Tallow 395488
Mafurra 395488
Magan Mananthes 244292
Mage Arrowbamboo 162643
Magellan Barberry 51401
Magellan Crowberry-leaved Barberry 51591
Magellan Darwin Barberry 51522
Magellan Fuchsia 168767
Magellan Ragwort 360057
Magennis' Dogstooth Grass 117874
Magenta Brush Cherry 382644
Magenta Cherry 382481,382644
Mageroum 274237
Maghet 322724
Magic Flower 4082,71547
Magic Flower of the Incas 71544
Magic Herb 123077
Magic Lily 239266,239280
Magician Deutzia 126833
Magician's Cypress 213902
Magnein Tanoak 233307
Magnific Deutzia 127005
Magnific Rhododendron 331182
Magnific Tigridiopalma 391632
Magnific Torchflower 100034
Magnifica Rose 336906
Magnifica Septemlobate Kalopanax 215443
Magnificent Basketvine 9482
Magnificent Clethra 96520
Magnificent Fieldorange 249670
Magnificent Fig 165721
Magnificent Mahonia 242586
Magnificent Melodinus 249670
Magnificent Mint Bush 315603
Magnificent Willow 343656
Magnificent Woodbetony 287719
Magnified Clethra 96520
Magnol Melic 249037
Magnolia 241951
Magnolia Azalea 331361
Magnolia Family 242370
Magnolia Laurel 242341
Magnolia Leaf Litse 234060
Magnolia Taxillus 385208
Magnolia Vine 351011,351021
Magnolia Water Lily 267777
Magnolia Water-lily 267777
Magnolialeaf Litse 234060
Magnolia-leaved Litse 234060
Magnoliavine 351011
Magnolia-vine 351011
Magpie Gladiolus 4534
Maguan Actinodaphne 6821
Maguan Azalea 331184
Maguan Begonia 50145
Maguan Camellia 69346
Maguan Cymbidium 116978
Maguan Fragrant Bamboo 87641
Maguan Fragrant-bamboo 87641
Maguan Manglietia 244454
Maguan Michelia 252940
Maguan Orchis 116978
Maguan Primrose 314232
Maguan Rhododendron 331184
Maguan Tea 69346
Maguey 10778,10787,10816,10933,10948,169245
Maguey Agave 10933
Maguey Lechugilla 10797
Maguey Pajarito 10914
Maguey Tecolote 10899
Maguey Verde 10891
Maguire's Fleabane 150768
Magweed 227533
Ma-hai 143122
Mahaleb 83253
Mahaleb Cherry 83253
Mahaw 109857
Mahoe 194850,195311,249175,389946
Mahoe Wao 249174
Mahogany 216194,380522,380528,395488
Mahogany Bean 10106,10127
Mahogany Birch 53505
Mahogany Family 248929
Mahon Stock 243209
Mahong Spapweed 204793
Mahonia 242458
Mahoschan Milkvetch 42665
Mahot Coton 56764
Mahua 241514,241516
Mahwa 241516
Mai Yang 133588
Maid of the Mead 166125
Maid of the Meadow 166125
Maid's Hair 170743
Maid's Love 35088
Maiden Blue-eyed Mary 99839
Maiden Eucalyptus 155640
Maiden Grass 255827,255887,255888
Maiden Hair 170193
Maiden Hair Tree 175813
Maiden Mercury 250600
Maiden Pink 127600,127698,127700
Maiden Sorrel 340136
Maiden's Australian Wild Lime 253291
Maiden's Blush 256797,366039
Maiden's Breath 183225
Maiden's Delight 35088
Maiden's Gum 155640
Maiden's Hair 60382,95431,115008,170335,170743,262583
Maiden's Head 345881
Maiden's Heads 345881
Maiden's Honesty 95431,238371
Maiden's Quiver Tree 17215
Maiden's Ruin 35088
Maiden's Tears 364193
Maiden's Wattle 1373
Maiden's Wreath 167735
Maiden's-tears 364193
Maiden-grass 255886
Maidenhair 170193,176824,388573
Maidenhair Barry 87678
Maidenhair Meadowrue 388506
Maidenhair Tree 175812,175813
Maidenhair Tree Family 175839
Maidenhair Vine 259591
Maidenhairleaf Sage 344848
Maidenhairtree 175812,175813
Maidenhair-tree 175812,175813
Maidenhairtree Family 175839
Maidenhair-Tree Family 175839
Maidens Tears 364193
Maidenstears 364193
Maidentile-meadow 325666
Maidenwort 36474
Maid-in-the-meadow 325666
Maid-in-the-mist 401711
Maid-of-the-mead 166125
Maid-of-the-meadow 166125
Maid-of-the-mist 388499
Maids 322724
Maids'love 35088
Maidsweet 166125
Maidu 320306,320307
Maidweed 26768
Maidwort 35088
Maihop 47413
Maikin Maiken 208771
Maikoa 61231
Maile-hohono 11199
Mailing Toadflax 84604
Mainbridge's Rhododendron 330195
Maine Sedge 75286
Maingay Mitrephora 256231
Maingay Silverhook 256231
Maingay Sterculia 376139
Maingay's Oak 324147
Mainland Torrey Pine 300288
Mainlin Poplar 311383
Maior Madden's Rhododendron 331179
Maiqishan Sedge 74623
Maiquel Mazus 247008
Maire Ainsliaea 12681
Maire Alstonia 18047
Maire Ancylostemon 22168
Maire Anisodus 25451
Maire Asparagus 39082
Maire Ballpit Arrowwood 408058

Maire Barberry 51903
Maire Bluebell Falsecuckoo 48142
Maire Briggsia 60277
Maire Brome 60835
Maire Bromegrass 60835
Maire Camellia 69338
Maire Ceropegia 84163
Maire Cnesmone 97504
Maire Conehead 182126
Maire Copperleaf 1919
Maire Cystopetal 320230
Maire Eugenia 156284
Maire Fargesia 162710
Maire Fescue 164076
Maire Fescuegrass 164076
Maire Fetid Gymnema 182366
Maire Greenbrier 366447
Maire Habenaria 183839
Maire Hornsage 205568
Maire Incarvillea 205568
Maire Jujube 418183
Maire Lilac 382197
Maire Lilyturf 272111
Maire Loose-node Sophora 369107
Maire Lovegrass 147793
Maire Mahonia 242517
Maire Microtoena 254255
Maire Nettle 402960
Maire Nomocharis 266570
Maire Onion 15453
Maire Osbeckia 276123
Maire Pagodatree 369107
Maire Paris 284347
Maire Pendentlamp 84163
Maire Peony 280229
Maire Porana 131243
Maire Rabbiten-wind 12681
Maire Rose 336731
Maire Sage 345192
Maire Sagebrush 35863
Maire Shagvine 97504
Maire Skullcap 355596
Maire Skunk Bugbane 91011
Maire Smilax 366447
Maire Sophora 369104,369107
Maire Spapweed 205115
Maire Spurge 159295
Maire Stemona 375346
Maire Strawberry 167634
Maire Tanoak 233308
Maire Tremacron 394676
Maire Trigonotis 397433
Maire Woodbetony 287410
Maire Wothflower Raspberry 339097
Maire Xingshan Viburnum 408058
Maire Yew 385407
Maire Youngia 416454
Mairehau 294106
Mairei Ephedra 146204
Maise 227533
Mait-banes 408393
Maiten 246817
Maiten Mayten 246817
Maithen 26768,227533
Maither 227533
Maithes 322724
Maitheweed 26768
Maizailan 11273,11300
Maize 417412,417417
Majestic Eremurus 148557
Majesty Palm 327111
Majevski Milkvetch 42672
Major Celandine 86755
Major Chinese Hawthorn 109936
Major Cholla 116641
Major Excellent Barberry 51437
Major Hemsleia 191946
Major Walnut 212620
Majorana 274224
Majorleaf Slenderstyle Acanthopanax 143642
Makaka 301612
Makalao 119073
Makeawning Porana 311655
Makebate 307197
Make-beggar 342279
Makebeggars 342279
Makestem Goldenray 229212
Makha Huakham 10129
Maki Podocarpus 306469
Makinboy 159058
Makino Bamboo 297337
Makino Fimbristylis 166393
Makino Fluttergrass 166393
Makino Liparis 232322
Makino Listera 232934
Makino Sedge 75287
Makino Stonecrop 356917
Makino's Lily 229919
Makino's Rhododendron 331187
Makole 265358
Makomako 34400
Makopa 382660
Makoy Calathea 66175
Makrut Lime 93492
Maku Sedge 75288
Makua 132305
Makua-plua 132305
Malabar Adenosma 7988
Malabar Almond 386500
Malabar Begonia 50256
Malabar Black Wood 121730
Malabar Blackwood 121730
Malabar Bombax 56802
Malabar Cardamom 143504
Malabar Glorylily 177251
Malabar Glory-lily 177251
Malabar Glory Lily 177251
Malabar Gourd 114277
Malabar Kino 320307
Malabar Melastome 248762
Malabar Nightshade 48688,48689
Malabar Nut 214308
Malabar Nutmeg 261444
Malabar Oil 117153,117171
Malabar Ottochloa 277428
Malabar Pencilwood 139663
Malabar Plum 382582
Malabar Randia 79595
Malabar Rosewood 121730
Malabar Simal Tree 56784,56802
Malabar Spinach 18836,48689
Malabar Squash 114277
Malabar Tallow 405154
Malabar Tree 159975
Malabarnum 214308
Malabartree Euphorbia 159975
Malabar-tree Euphorbia 159975
Malabartree Spurge 159975
Malacca Albizzia 162473
Malacca Cane 65637,65787
Malacca Cypressgrass 119160
Malacca Fishvine 125983
Malacca Galangal 17719
Malacca Galingale 119160
Malacca Iguanura 203509
Malacca Megistostigma 247968
Malacca Pseuderanthemum 317259
Malacca Sphaerocaryum 371090
Malacca Teak 10128,207017
Malachite Stonecrop 356924
Malagetta Pepper 19858
Malaguetta Pepper 9923
Malaisia 242974,242975
Malaja Licuala Palm 228749
Malaja Licualapalm 228749
Malambo 112953
Malanga 415190,415201
Malango 415190
Malania 242983
Malanvine 134025,134026
Malatti 211715
Malay Apple 382582,382606
Malay Arrow-poison 28248
Malay Blumea 55766
Malay Bushbeech 178025
Malay Bush-beech 178025
Malay Catchbird Tree 81933
Malay Ginger 107271
Malay Glycosmis 177822,177846
Malay Guttaperch Natotree 280441
Malay Guttapercha Nato Tree 280441

Malay Guttapercha Nato-tree 280441
Malay Kumquat 167513
Malay Nettle Tree 125550
Malay Nettle-tree 125550
Malay Palaktree 280441
Malay Pinanga 299668
Malay Rambutan 265136
Malay Strophanthus 378375
Malay Sweet Dragonbamboo 125465
Malaya Prismatomeris 315328
Malaya Rhododendron 331193
Malaya Windmilltung 147355
Malayan Blumea 55766
Malayan Date 295477
Malayan Fan Palm 234181
Malayan Ground Orchid 370401
Malayan Jasmine 393657
Malayan Kumquat 167513
Malayan Rice Paper 350344
Malayapple 382606
Malay-apple 382606
Malaysia Dacrydium 121093
Malaysian Apple 382606
Malaysian Aralia 310186
Malaysian Dacrydium 121093
Malaysian Eugenia 382606
Malaysian Mischocarpus 255954
Malaysian Mountain Bamboo 197602
Malaysian Palm Grass 361850
Malaysian Persimmon 132291
Malaysian Scurfpea 114427
Malaysian Thin-walled Bamboo 216393
Malaytea Scurfpea 114427
Malcolm Stock 243156
Malcolmia 243156
Malcon Thorowax 63744
Malden 26768
Malden Oak 324278
Malden's Delight 35088
Malden's Tears 364193
Maldive Nut 235159
Male 384714
Male Bamboo 47190,125510
Male Berry 239384
Male Cornel 105111
Male Dendrocalamus 125510
Male Dogwood 105111
Male Dragonbamboo 125510
Male Ginkgo 175815
Male Kola 171121
Male Lily 102867
Male Nettle 227533,243840,402992
Male Orchid 273604
Male Orchis 273531
Male Peony 280231
Male Pimpernel 21339
Male Speedwell 407256
Maleberry 239384
Malegueta Pepper 9923
Malesia Bitterorange 93559
Maleysia Jointfir 178536
Malheur Wirelettuce 375973
Mali Arrowbamboo 162711
Mali Fargesia 162711
Malibu Baccharis 46234
Maling Bamboo 47333
Maling Bambusa 47333
Malinta Habenaria 183844
Malipo Agapetes 10333
Malipo Alseodaphne 17804
Malipo Ardisia 31516
Malipo Argyreia 32635
Malipo Azalea 331194
Malipo Barberry 51904
Malipo Calanthe 65908
Malipo Coelogyne 98694
Malipo Common Greentwig 352440
Malipo Common Greytwig 352440
Malipo Dichocarpum 128936
Malipo Eurya 160542
Malipo Greenbrier 366448
Malipo Grey-twig 352440
Malipo Holly 204022
Malipo Ichnocarpus 203331
Malipo Lady's Slipper 120409
Malipo Loquat 151165
Malipo Lycianthes 238972
Malipo Michelia 252845
Malipo Micrechites 203331
Malipo Oak 324156
Malipo Oilfruitcamphor 381797
Malipo Paphiopedilum 282858
Malipo Pocktorchid 282858
Malipo Raspberry 338787
Malipo Red Silkyarn 238972
Malipo Rhododendron 331194
Malipo Schefflera 350739
Malipo Smilax 366448
Malipo Spapweed 205116
Malipo Spiradiclis 371769
Malipo Syndiclis 381797
Malkamong 10129
Malkang Roughfruitparsley 393954
Mallee 155468,155568
Mallee Cypress Pine 67434
Mallee Honey-myrtle 248074
Malleola 243287
Mallet Bark 155485
Mallet Flower 400377
Mallotus 243315
Mallotusleaf Raspberry 338790
Mallotus-leaf Storax 379406
Mallotus-leaved Raspberry 338790
Mallow 18155,194680,223350,243494,243738,243746,243840
Mallow Bellowsfruit 297840
Mallow Bindweed 102914
Mallow Family 243873
Mallow Hock 243840
Mallow-hock 243840
Mallowleaf Grewia 181006
Mallowleaf Primrose 314616
Mallow-leaved Bindweed 102914
Mallow-rock 358062
Mallow-rose 195032
Malls 86901
Mallus 243840
Mally 243840
Mally-gowl 66445
Maloga Bean 408935
Malong Privet 229539
Malope 243472
Malou Ardisia 31517
Malpighia 243519,243523
Malpighia Family 243530
Malta Cross 395146
Malta Lily 372913
Malta Orange 93765
Malta Thistle 81198
Maltese Clover 187842
Maltese Cockspur 81198
Maltese Cross 363312,363313,395146
Maltese Cross Campion 363312
Maltese Cross Rose 336930
Maltese Star 81198
Maltese Star Thistle 81198
Maltese Star-thistle 81198
Maltese Thistle 81198
Maltesecross 363312
Maltese-cross 363312
Maltifruited Pepper 300483
Malu Fargesia 162713
Malugay 310823
Malzev Awnless Bromegrass 60768
Mam Bamboo 47218
Mamalis 301363
Mamane Sophora 368988
Mamdaisy 246307
Mamee-sapote 67524
Mamey 243979,243982
Mamey Colorado 238115
Mamey Sapote 238115,313399
Mamillate Ardisia 31518
Mamillate Spike-rush 143198
Mamillate Spikesedge 143198
Mamillatestylebase Spikesedge 143198
Mamma's Milk 159027
Mammea 243979
Mammea Apple 243982
Mammee 243979
Mammee Apple 243982

Mammee Sapote 67524,313399
Mammee Zapote 313399
Mammee-apple 243982
Mammey 67524,243979,243982
Mammey Apple 285654
Mammey Sapote 67524
Mammillaria 243994,244265
Mammillary Holly 204017
Mammillopsis 243974
Mammoth Clover 396967
Mammoth Tree 360561,360565,360576
Mammoth Tree of California 360576
Mammoth Wildrye 228379
Mamoncillo 249125,249126
Mamoncillo Spunish-lime Honeyberry 249125
Mamora Pear 323223
Man Orchid 3774,3777,232950,273469,
273531,273545
Man's Motherwort 334435
Man's Naughty Cherry 44708
Man's Purse 72038
Mana Grass 117218
Manaca Raintree 61304
Manala Mats 79966
Manan Rattan Palm 65731
Mananthes 244276
Manaplant Alhagi 14638
Manaraceme Sweetvetch 188047
Manasi Motherwort 224998
Manasi Rush 213225
Manateagrass 117296,117308
Manatee-grass 382418
Manatee-grass Family 117311
Manatree Grape 411746
Manchester Poplar 311398,311405
Manchineel 196716
Manchineel Tree 196715,196716
Manchineel-tree Mancenillier 196716
Manchrian Elder 345624
Manchrian Elm 401496
Manchrian Glabrous Cinquefoil 312636
Manchrian Nepeta 264990
Manchu Cherry 83347
Manchu Filbert 106781
Manchu Rose 337029
Manchu Tuber Gourd 390128
Manchu Tuber-gourd 390128
Manchur Ash 168023
Manchur Groundsel 358241
Manchur Rockfoil 349607
Manchur Tubergourd 390128
Manchur Wildginger 37638
Manchuri Alder 16401
Manchuri Beakgrain 128019
Manchurian Alder 16359,16362,16401
Manchurian Apricot 34443
Manchurian Arisaema 33404
Manchurian Arrowwood 407715
Manchurian Ash 168023
Manchurian Beakgrain 128019
Manchurian Catalpa 79247
Manchurian Catawba 79247
Manchurian Cherry 280028
Manchurian Clematis 95366
Manchurian Cudweed 178278
Manchurian Dandelion 384719
Manchurian Dutchman's-pipe 34268
Manchurian Dutchmanspipe 34268
Manchurian Elm 401542
Manchurian Eritrichium 153479
Manchurian Fir 388,427
Manchurian Forsythia 167450
Manchurian Goatwheat 44269
Manchurian Gooseberry 6639
Manchurian Groundsel 358241
Manchurian Hazel 106751
Manchurian Honeysuckle 236076
Manchurian Hornwort 83568
Manchurian Jute 837,1000
Manchurian Largeleaf Hydrangea 200011
Manchurian Lilac 382259,382268,382277
Manchurian Lime 391760
Manchurian Linden 391760
Manchurian Monk's-hood 5659
Manchurian Monkshood 5659
Manchurian Ninebark 297837
Manchurian Peashrub 72279
Manchurian Pea-shrub 72279
Manchurian Rhododendron 331239
Manchurian Rockfoil 349607
Manchurian Rose 337029
Manchurian Saussurea 348510
Manchurian Schneider Buckthorn 328853
Manchurian Thyme 391264
Manchurian Tubergourd 390128
Manchurian Viburnum 407715
Manchurian Violet 410201
Manchurian Walnut 212621
Manchurian Water Rice 418080
Manchurian Whortleberry-like Willow 343747
Manchurian Wildginger 37638
Manchurian Wildrice 418095
Manchurian Woodbetony 287411
Manchurian Yellow Loosestrife 239558
Manchustriple Maple 3665
Manda's Woolly-bear Begonia 50007
Mandacaru 83885
Mandarin 93446,93717,377902
Mandarin Azalea 331742
Mandarin Blueberry 403899
Mandarin Ischnogyne 209405
Mandarin Jacker Bambusa 47493
Mandarin Orange 93649
Mandarin Orchid 295875
Mandarin Rhododendron 330592,331742
Mandarin's Hat Plant 197368
Mandarine Orange 93446,93717
Mandevilla 244320,244321,244329
Mandioca 244507
Mandragon 244346
Mandrake 37015,61497,91557,244340,
244346,306609,306636,383675
Maned Adinandra 8231
Maned Crabgrass 130616
Maned Cristesion 198311
Manelet 89704
Manettia 244358
Mang We 386638
Mangabeira Rubber 185258
Mangachapo Vatica 405178
Mangasinoro 362186
Mangdangshan Machilus 240640
Mangel 53245
Mangelong Croton 112954
Mangelwurzel 53245,53249
Mangel-wurzel 53249
Mange-tout 301080
Mangetti Tree 334417
Mangis 171136
Mangium 1375
Mangkang Ploughpoint 401169
Mangkang Typhonium 401169
Mangle Bobo 57012
Mangle Dulce 246908
Mangles Everlasting 329801
Manglietia 244420
Manglietiastrum 244485
Mango 244386,244397
Mango Ginger 114856
Mango Melon 114200
Mangoba 185258
Mangold 53245,53249,53257
Mangostan 171136
Mangosteen 171136
Mangosteen Garcinia 171136
Mangrove 45740,329744,329761,329765
Mangrove Date Palm 295477
Mangrove Family 329776
Mangrove Fan Palm 228766
Mangrove Hibiscus 195311
Mangrove Palm 267838
Mangrove Rubber Vine 328375
Mangrove Spider-lily 200938
Mangshan Aster 40830
Mangshan Camellia 69371
Mangshan Chirita 87917
Mangshan Citrus 93572
Mangshan Hydrangea 200023
Mangshan Pipewort 151362
Mangshan Wild Mandarin 93572
Manhao Begonia 50058

Manhao Elatostema 142735
Manhao Stairweed 142735
Manicon 44708
Manienie-Alii 143530
Manifestnerved Bigumbrella 59250
Manifest-nerved Bigumbrella 59250
Manifestnerved Brassaiopsis 59250
Manifestnerved Sibbaldia 362360
Manihot Caoutchouc 244509
Manila Bean 319114
Manila Champereia 85935
Manila Copal 10496
Manila Elemi 71004
Manila Grass 418432
Manila Hemp 260275
Manila Leea 223937
Manila Maguey 10816
Manila Padauk 320301
Manila Palm 405258
Manila Surinamcherry 156126
Manila Tamarind 301134
Manilagrass 418432
Manila-grass 418432
Manilla Tamarind 301128
Man-in-the Pulpit 37015
Man-in-the-ground 208050
Man-in-ti-the-pulpit 37015
Manio 306510,306517
Manio Podocarpus 306510
Manioc 244501,244507
Manioc Bean 279737
Maniot 244501
Manipue Mahonia 242597
Manipur Azalea 331197
Manipur Barberry 51905
Manipur Lily 229913
Manipur Lycianthes 238969
Manipur Maesa 241801
Manipur Red Silkyarn 238969
Manipur Rhododendron 331197
Manisoba 244509
Manitoba Maple 3218
Manju Mandarine 93844
Man-ketti Nut 334417
Mankich 93844
Mankieh 93844
Mankil Apple 382660
Mankil Rose-apple 382660
Mann Butterflyorchis 293620
Mann Calanthe 66005
Mann Meliosma 249412
Mann Phalaenopsis 293620
Mann Plumyew 82541
Mann Plum-yew 82541
Mann Tanoak 233309
Mann's Bamboo 297338
Mann's Hollowbamboo 82448
Manna 168041
Manna Ash 168041
Manna Eucalyptus 155772
Manna Grass 177556,177559,177599,177617
Manna Gum 155642,155772
Manna Oak 323745,324478
Manna Plant 383491
Manna-ash 168041
Mannagettaea 244635
Mannagrass 177556
Manna-grass 177556,177599
Mannatamarisk 383553
Manoao 121090,184915
Man-of-the-earth 208050
Man-of-war 302034
Manschurian Currant 334084
Manshuria Poplar 311386
Manshurian Birch 53572
Manshurian Filbert 106751
Manshurian Maple 3129
Manshurian Poplar 311386
Man-tie 308816
Mantspike Diffuse Cypressgrass 118755
Manuka 226481
Manuka Tea-tree 226481
Manx Gorse 401395
Many Diversifolious Pricklyash 417291
Many Flower Crapemyrtle 219924
Many Flower Fatweed 167312
Many Flower Garcinia 171145
Many Flower Glorybower 96028
Many Flower Tamarisk 383519
Many Flowered Clovershrub 70866
Many Flowers Fordiophyton 167312
Many Glands Rhododendron 330765
Many Knees 308607
Many Leaflets Indigo 206352
Many Leaves Vetch 408545
Many Prickle Acanthopanax 143657
Manyandra Heterosmilax 194124
Manyangular Euphorbia 159617
Many-anguled Begonia 50095
Manyawn Microstegium 254043
Manyball Gagea 169476
Manybract Bothriospermum 57683
Manybract Camellia 69375
Manybract Coldwaterflower 298878
Manybract Firmiana 166652
Many-bract Gongronema 179307
Manybract Groundsel 359529
Manybract Gugertree 350931
Manybract Ligusticum 229342
Manybract Primrose 314181
Manybract Rose 336781
Manybract Spotseed 57683
Manybracted Rose 336781
Manybracteole Bugle 13063
Manybranch Candolle Thorowax 63590
Manybranch Cleistogenes 94588
Manybranch Dragonhead 137635
Manybranch Elatostema 142796
Manybranch Gentian 173652
Manybranch Gutzlaffia 182134
Manybranch Hideseedgrass 94588
Manybranch Indian Skullcap 355500
Manybranch Indigo 206466
Manybranch Leymus 228370
Manybranch Meadowrue 388632
Manybranch Milkvetch 42893
Manybranch Monkshood 5526
Manybranch Muhly 259691
Manybranch Onosma 271807
Manybranch Rattan Palm 65739
Manybranch Toadflex 230927
Manybranch Wideword Parnassia 284596
Manybranch Willow 343906
Manybranch Willowweed 146696
Manybranch Yellowish Woodbetony 287389
Manybranched Ailanthus 12562
Manybranched Armgrass 58155
Manybranched Crazyweed 279025
Manybranched Elsholtzia 143982
Manybranched Goldsaxifrage 90442
Manybranched Goldwaist 90442
Manybranched Larkspur 124376
Manybranched Rattan Palm 65777
Many-branched Rhododendron 331512
Manybranched Ricegrass 276039
Manybranched Signalgrass 58155
Manybranches Angraecum 25020
Manybranches Brome 60918
Manybranches Odontoglossum 269084
Many-ciliate Microstegium 254033
Manycleft Cinquefoil 312797
Many-cleft Rosewood 121805
Many-coloured Rattlepodd 41956
Many-crenate Chinese Holly 203698
Manyculm Goosecomb 335436
Many-dissected Elaeocarpus 142350
Many-dissected Elatostema 142751
Many-feet 326287
Manyflorous Shrubcress 368556
Manyflower 363781
Manyflower Abolin Goosecomb 335190
Manyflower Acroglochin 6166
Manyflower Actinidia 6678
Manyflower Ajania 13011
Manyflower Alkaligrass 321285
Manyflower Alternateleaf Stringbush 414126
Manyflower Ammania 19602
Manyflower Apple 243617
Manyflower Ash 167975
Many-flower Aspidopterys 39675

Manyflower Astilbe 41844
Manyflower Azalea 331596
Manyflower Barrenwort 147027
Manyflower Bauhinia 49045
Manyflower Begonia 50313
Manyflower Bellcalyxwort 231277
Manyflower Bentgrass 12212
Manyflower Boxleaved Cotoneaster 107377
Manyflower Bugle 13143
Manyflower Bushclover 226793
Manyflower Buttercup 326242
Manyflower Butterflybush 62127
Manyflower Caper 71809
Manyflower Cautleya 79752
Manyflower Centrostemma 81911
Manyflower Childvine 400888
Manyflower Chin Cactus 182466
Manyflower Cladium 93925
Manyflower Clovershrub 70866
Manyflower Colona 99965
Manyflower Cotoneaster 107583
Manyflower Cuspidated Tulip 400211
Manyflower Cymbidium 116863
Manyflower Daylily 191313
Manyflower Deutzia 127013
Manyflower Dinggongvine 154249
Manyflower Diplarche 132829
Manyflower Dogwood 380494
Manyflower Dryander 138430
Manyflower Elaeocarpus 142317
Manyflower Embelia 144736
Manyflower Epidendrum 146419,146460
Manyflower Epimedium 147027
Manyflower Erycibe 154249
Manyflower Eupatorium 158038
Manyflower Figwort 355218
Manyflower Fissistigma 166684
Manyflower Four-o'clock 255732
Manyflower Fourstamen Prismatomeris 315334
Manyflower Fox-tail Orchid 9301
Manyflower Fritillary 168520
Manyflower Galangal 17740
Manyflower Garcinia 171145
Manyflower Gardneria 171455
Manyflower Gentian 173221,173936
Manyflower Giant Orchid 180125
Manyflower Girald Currant 333988
Manyflower Gladiolus 176206
Manyflower Glaucosleaf Cotoneaster 107463
Manyflower Glorybower 96028
Many-flower Griffith Elaeagnus 142016
Many-flower Gueldenstaedtia 181695
Manyflower Hairgrass 126095
Manyflower Hiptage 196843
Manyflower Hollyberry Cotoneaster 107364
Manyflower Hooktea 52418
Manyflower Inflated Dendropanax 125615
Manyflower Inflated Treerenshen 125615
Manyflower Jasmine 211963
Manyflower Jurinea 214138
Manyflower Kiwifruit 6678
Manyflower Ladybell 7768
Manyflower Lambertia 220311
Manyflower Landpick 308529
Manyflower Leek 15487
Manyflower Leptoboea 225896
Manyflower Maackia 240124
Manyflower Magnolia 416710
Manyflower Mallotus 243358
Manyflower Marshweed 230312
Manyflower Maximowicz Currant 334089
Manyflower Maximowicz Sage 345206
Manyflower Meliosma 249414
Manyflower Memecylon 250044
Many-flower Memecylon 249962
Manyflower Michelia 252872
Manyflower Microula 254344
Manyflower Milkvetch 42374
Manyflower Monkshood 5490
Manyflower Ochna 268261
Manyflower Onion 15487
Manyflower Pavetta 286418
Manyflower Pittosporum 301266
Manyflower Rabdosia 324845
Manyflower Red Fescuegrass 164262
Manyflower Rhododendron 331596
Manyflower Ricebag 181695
Manyflower Rockfoil 349597
Manyflower Rockvine 387753
Manyflower Roegneria 335190
Manyflower Rush 213297,213382
Manyflower Ryegrass 235315
Manyflower Sagebrush 35968
Manyflower Sealavender 230688
Manyflower Seatung 301266
Many-flower Sichuan Sabia 341560
Manyflower Sikkim Clematis 95298
Manyflower Silkstalk Elatostema 142664
Manyflower Silkstalk Stairweed 142664
Manyflower Silkvine 291060
Manyflower Silvergrass 255849
Manyflower Solomonseal 308529
Manyflower Sphenodesma 371418
Manyflower Starapple 90061
Manyflower Stilpnolepis 376651
Manyflower Stonecrop 294120
Manyflower Supine Chickweed 374830
Manyflower Supplejack 52418
Many-flower Syzygium 382651
Manyflower Tamarisk 383519
Many-flower Tangut Barberry 51525
Manyflower Thenardia 389398
Manyflower Tickclover 126471
Manyflower Tobacco 266032
Manyflower Trigonotis 397409
Manyflower Tulip 400133
Manyflower Twigrush 93925
Manyflower Tylophora 400888
Manyflower Vineclethra 95492
Manyflower Waterhissop 46359
Manyflower Waxpetal 106659
Manyflower Wendlandia 413834
Manyflower Wildryegrass 335190
Manyflower Wildtung 243358
Manyflower Willow 344095
Manyflower Winterhazel 106659
Manyflower Woodbetony 287216
Manyflower Woodrush 238647
Manyflower Yam 131726
Manyflower Yunnan Cherry 83370,316943
Many-flowerbuds Biond's Magnolia 242004
Many-flowered Agave 10909
Many-flowered Agrimony 11571
Many-flowered Aster 380872
Many-flowered Bamboo 47347
Many-flowered Begonia 50096
Many-flowered Bitter Sweet 80274
Manyflowered Bittersweet 80274
Manyflowered Cerberustree 83696
Manyflowered Cotoneaster 107583
Many-flowered Cotoneaster 107583
Manyflowered Daffodil 262440
Manyflowered Garcinia 171145
Many-flowered grass-pink 67885
Many-flowered Japanese Barberry 52249
Manyflowered May-apple 139606,139629
Manyflowered Narcissus 262440
Manyflowered Paniculate Staff-tree 80274
Many-flowered Phinaea 294906
Manyflowered Rose Many-flowered Rose 336783
Many-flowered Sikkim Clematis 95298
Many-flowered Stickseed 184305
Manyflowered Wheatgrass 11700
Many-flowered Wild Buckwheat 152296
Manyflowers Bittercress 72903
Manyflowers Canebrake 37252
Many-flowers Christmasbush 14187
Manyflowers Eria 148677
Manyflowers Masdevallia 246108
Manyflowers Melastoma 248727
Manyflowers Smallflower Deutzia 127036
Manyflowers Solms-Laubachia 368556
Manyflowersd Soapwort 346421
Manyfork Catchfly 363782
Manyfork Ladybell 7884
Manyfork Ramie 56294
Manyforked Primrose 314318
Manyform Dutchmanspipe 34299
Manyform Medic 247425

Manyfruit Affinity Neillia 263145
Manyfruit Anemone 24039
Many-fruit Bishopwood 54623
Manyfruit Calathodes 66224
Manyfruit Cockclawflower 66224
Manyfruit Fatsia 162895
Manyfruit Gongbu Monkshood 5328
Manyfruit Hemsleia 191962
Manyfruit Horny Tanoak 233156
Manyfruit Idesia 203422
Manyfruit Maple 3461
Manyfruit Mimosa 255092
Manyfruit Monkshood 5492
Many-fruit Newlitse 264087
Many-fruit Pepper 300483
Manyfruit Sapium 346401
Manyfruit Scaligeria 350375
Manyfruit Seatung 301376
Many-fruited Juniper 213859
Manyfurrow Elaeocarpus 142350
Manygap Cowparsnip 192254
Manygland Willow 343780
Manyglandular Mallotus 243397
Manyhair Begonia 50179
Manyhair Caper 71727
Manyhair Chinacane 364672
Manyhair Cranehead Woodbetony 287262
Manyhair Creeper 285166
Manyhair Crisped Corydalis 105763
Manyhair Flowery Milkvetch 42376
Manyhair Littleconical Rabdosia 209648
Manyhair Maxianshan Milkvetch 42667
Many-hair Mussaenda 260461
Manyhair Pearlbush 161756
Manyhair Sichuan Speedwell 407401
Manyhair Snowgall 191906
Many-hair Vetch 408527
Manyhairs Croton 112994
Manyhairs Indocalamus 206789
Many-hairs Rhododendron 331518
Manyhairy Ladybell 7807
Manyhairy Lovegrass 147891
Many-hairy-scale Rattanpalm 65804
Manyhead Cinquefoil 312796
Manyhead Clinopodium 97043
Manyhead Cousinia 108372
Many-head Desertdandelion 242939
Manyhead Groundsel 359545
Manyhead Headleaf Loosestrife 239787
Manyhead Ixeris 210654
Manyhead Pincution 244160
Manyhead Saussurea 348655
Manyhead Sawwort 361104
Manyhead Stairweed 142733
Manyhead Vinegentian 108372
Manyhead Visnaga 140152
Manyhead Windhairdaisy 348655
Many-headed Barrel Cactus 140152
Many-headed Dryandra 138439
Many-headed Groundsel 360081
Many-headed Oval Sedge 76464
Many-headed Sedge 76464
Many-heads Cinquefoil 312796
Manyhorn Spapweed 205239
Manyinflorescence Elatostema 142733
Manyinflorescenced Sweetvetch 188047
Manyleaf Aloe 17182
Manyleaf Altai Heteropappus 193921
Manyleaf Altai Jurinea 193921
Manyleaf Brachycorythis 58414
Manyleaf Cinquefoil 312892
Manyleaf Cleistogenes 94619
Manyleaf Clematis 95152
Manyleaf Cranehead Woodbetony 287263
Manyleaf Dwarf Globeflower 399532
Manyleaf Fleabane 150796
Manyleaf Gentian 150796
Manyleaf Goosecomb 335226
Manyleaf Hideseedgrass 94619
Manyleaf Iris 208569
Manyleaf Leek 15607
Manyleaf Lily 229994
Manyleaf Meadowrue 388513
Manyleaf Monkey Earrings Pea 301150
Manyleaf Onion 15607
Manyleaf Paphiopedilum 282879
Manyleaf Paris 284358
Many-leaf Pendent Inflorescence Indigo 206381
Manyleaf Primrose 314829
Manyleaf Pulsatilla 321691
Manyleaf Rabdosia 324856
Manyleaf Rockfoil 349736
Manyleaf Schefflera 350744
Manyleaf Sedge 75849
Manyleaf Sichuan-Xizang Rabdosia 324856
Manyleaf Skullcap 355691
Manyleaf Woodbetony 287447
Many-leaflets Sophora 369156
Many-leaved Gelidocalamus 172745
Many-leaved Lily 229947
Many-leaved Rhododendron 331513
Manyleaves Bittercress 72888
Manyleaves Cinquefoil 312892
Manyleaves Draba 137190
Manyleaves Lupine 238481
Many-leaves Pagodatree 369143
Manyleaves Peashrub 72311
Manyleaves Sophora 369100
Many-leaves Sophora 369143
Manyleaves Twayblade 232163
Manyleg Knotweed 309607
Manyleves Sedge 74583
Manyleves Whitlowgrass 137190
Manylobe Groundsel 359536
Manylobed Arisaema 33423
Many-lobed Ladypalms 329186
Manylobed North China Grape 411590
Manylobed Samaradaisy 320518
Manylobed Snakegourd 396226
Manylobed Southstar 33423
Manylobed Tentacle Catchfly 363535
Manylobed Water Nightshade 367035
Manymale Sedge 75848
Many-nerne Holly 204177
Many-nernes Camellia 68918
Manynerv Purple Willow 343955
Manynerve Bauhinia 49103
Manynerve Embelia 144773
Manynerve Hophornbeam 276815
Manynerve Hornbeam 77369
Manynerve Ironwood 276815
Manynerve Osmanther 276386
Manynerve Primrose 314821
Manynerve Spapweed 205240
Manynerve Touch-me-not 205240
Manynerved Altingia 18202
Manynerved Clethra 96508
Manynerved Greenstar 307521
Manynerved Loosefruit Pepper 300423
Many-nerved Mistletoe 411065
Many-nerves Eurya 160583
Manynode Anisachne 25268
Manynode Arundinella 37416
Manynode Capillipedium 71614
Manynode Dragonhead 137616
Manynode Germander 388155
Manynodes Brome 60891
Many-ovele Barberry 51949
Manyovule Gagea 169476
Manypair Smallleaf Cinquefoil 312395
Manypaired Obtuseleaf Rose 336941
Many-pedunded Bur-reed 370084
Many-petal Barberry 52056
Manypetal Begonia 50178
Manypetal Camellia 69516
Manypetal Cyclea 116031
Manypetal Marshmarigold 68213
Manypetal Oxygraphis 278472
Many-petaled Cape-jasmine 171308
Manypetaled Mangrove 61263
Manypetal-like Syzygium 382652
Manypin Neoporteria 264365
Many-pinnate Cycas 115865
Manypinnate Woodbetony 287471
Manypit Schefflera 350670
Manyprickle Acanthopanax 143657
Many-pyrene Holly 204179
Manyraceme Grevill 180634
Manyraceme Sagebrush 36087
Manyray Aster 380837

Manyrib Burreed 370086
Manyribbed Brain Cactus 140577
Manyroot 339658
Manyroot Buttercup 326245
Manyroot Leek 15615
Manyroot Monkshood 5314
Manyroot Onion 15615
ManyscaleAzalea 331514
Many-scales Rhododendron 331514
Manyscape Chirita 87942
Manyscape Dandelion 384685
Manyscape Fawnlily 154935
Manysect Leaf Mustard 59438
Manyseed 307724
Manyseed Alphonsea 342086
Manyseed Cladostachys 123569
Manyseed Goosefoot 87130,87131,87132
Many-seed Goosefoot 87130,87131
Manyseed Hemiadelphis 191326
Many-seed Idesia 203422
Manyseed Kadsura 214979
Manyseed Liongrass 191326
Manyseed Plantain 302132
Manyseed Tarenna 385015
Manyseeded Euptelea 160344
Many-seeded Goosefoot 87130
Many-seeded Hygrophila 191326
Manyseeds Salacia 342705
Manysepal Camellia 69384
Manysepal Chinese Pulsatilla 321674
Many-sepal Parapyrenaria 283713
Manysepal Tea 69384
Manyserrate Barberry 51950
Many-sheath Bluegrass 305851
Manysheath Windhairdaisy 348656
Manyshoot Asparagus 38986
Manyshoot Flower Hedge 27320
Manyshoot Gladiolus 176491
Manyshoot Moraea 258625
Manysmallleaf Skunk Bugbane 91013
Manyspike Anisochilus 25389
Manyspike Bulrush 353878
Manyspike Chloranthus 88292
Manyspike Galingale 119410
Manyspike Ginger 418016
Manyspike Grevill 180635
Manyspike Knotweed 309616
Manyspike Marshweed 230313
Manyspike Moraea 258613
Manyspike Podocarpus 306513
Manyspike Rabdosia 324847
Manyspike Rattan Palm 65772
Manyspike Rattanpalm 65651
Manyspike Tanoak 233348
Manyspike Tropidia 399653
Manyspike Yaccatree 306513
Manyspike Zinger 418016
Manyspiked Cupgrass 151690
Many-spiked Umbrella Sedge 119410
Manyspikeorchis 310309,310365
Manyspine Asparagus 39104
Manyspine Chalkplant 183250
Manyspine Eria 148676
Manyspine Fagerlindia 162194
Manyspine Rattan Palm 65799
Manyspine Rattanpalm 65799
Many-spined Opuntia 273014
Manyspiny Erect Raspberry 339296
Manyspliked Barkgrass 317226
Manyspliked Pseudochinolaena 317226
Manysplit Corydalis 106174
Manysplit Isometrum 210178
Manysplitted Monkshood 5493
Manyspot Iris 208556
Manyspot Masdevallia 246117
Manyspot Swordflag 208556
Many-stachys Tamarisk 383538
Manystalk Cinquefoil 312795
Manystalk Roegneria 335436
Many-stalked Spike-rush 143229
Manystamen Amoora 19959
Manystamen Balanophora 46868
Many-stamen Euptelea 160347
Manystamen Forchhammeria 167277
Many-stamen Heterosmilax 194124
Manystamen Megacarpaea 247790
Manystamen Pokeweed 298118
Manystamen Snakemushroom 46868
Manystamen St. John's Wort 201811
Manystamen Stockorange 310849
Many-stamen Trifoliate-orange 310849
Many-stamens Rhododendron 331510,331882
Many-stanen Henry Eightangle 204529
Many-stanen Henry Illicium 204529
Manystem Anemone 24033
Manystem Angelica 24416
Many-stem Ardisia 31532
Manystem Cremanthodium 229143
Manystem Cudweed 178355
Manystem Dragonhead 137614
Manystem Everlasting 21581
Manystem Hawksbeard 110906
Manystem Honeysuckle 236208
Manystem Lovegrass 147823
Manystem Nutantdaisy 229143
Manystem Pachypterygium 279714
Manystem Pearleverlasting 21581
Many-stem Rabbit-tobacco 127934
Manystem Stonecrop 356953
Manystem Vetch 408502
Manystem Wildrye 228370
Manystem Windflower 24033
Manystem Woolly Sunflower 152826
Many-stemmed Barberry 51946
Manytendril Snakegourd 396259
Manythorny Asparagus 39104
Manythrum Paris 284357
Manytooth Broomrape 275242
Manytooth Camellia 69513
Manytooth Euonymus 157360
Manytooth Lysionotus 239940
Manytooth Ministylenettle 85108
Manytooth Purplepearl 66767
Manytooth Raspberry 339088
Manytooth Woodbetony 287528
Manytoothed Actinidia 6630
Many-toothed Mahonia 242620
Manytoothed Woodbetony 287528
Manytube Ligusticum 229367
Manyumbell Giantfennel 163610
Manyumbell Thorowax 63595
Manyvein Arisaema 33301
Manyvein Buckeye 9724
Manyvein Conehead 378210
Manyvein Delavay Birch 53410
Manyvein Dendrotrophe 125789
Manyvein Dogwood 124942
Many-vein Elm 401476
Manyvein Four-involucre 124942
Manyvein Gugertree 350938
Manyvein Leucadendron 227408
Manyvein Machilus 240650
Manyvein Madder 338018
Manyvein Michelia 252948
Manyvein Parasiticvine 125789
Manyvein Saurauia 347977
Manyvein Schefflera 350749
Manyvein Skimmia 365971
Manyvein Southstar 33301
Manyvein Tengyue Arrowwood 408161
Manyvein Tengyue Viburnum 408161
Many-veined Begonia 50097
Manyveins Ardisia 31535
Many-veins Ardisia 31391
Many-veins Baccaurea 46193
Many-veins Chinese Buckthorn 328886
Many-veins Fig 165479
Many-veins Mussaenda 260465
Manyveins Oak 116152
Many-veins Pencilwood 139661
Many-veins Waxplant 198884
Manywing Groundsel 358985
Manzanilla 109968,415863
Manzanillo 415863
Manzanita 30877,31102,31123,31134,31142
Manzanote 270515
Maoershan Azalea 331200
Maoershan Birch 53423
Maoershan Rhododendron 331200
Maofeng Lime 391778
Maofeng Linden 391778

Maoling Roundleaf Azalea 331402
Maomian Azalea 331280
Maoming Bigumbrella 59226
Maoming Brassaiopsis 59226
Maoming Euaraliopsis 59226
Maoming Palmleaftree 59226
Maoping Ungeargrass 36687
Maori Bulrush 401094
Maori Celery 29319
Maori Fire 288995
Maori Flax 295600
Maori Holly 270195,270198
Maori Mint 250351
Maorshan Great Buttercup 325917
Maorshan Sedge 75300
Maoutia 244892,244893
Maowen Angelica 24399
Maowen Azalea 331201
Maowen Cacalia 283842
Maowen Leek 15454
Maowen Milkvetch 42677,42855
Maowen Monkshood 5412
Maowen Onion 15454
Maowen Rhododendron 331201
Maowen Yam 131508
Maoxian Larkspur 124373
Map Tree 158982
Mapania 244897
Mapian Figwort 355183
Mapien Maple 3131
Maple 2769,2837,2976,3127,3393,3462
Maple Family 3765
Maple Lemongrass 117273
Maple Pea 301055
Maple Service 369541
Maple Tree 369541
Mapleleaf Ainsliaea 12600
Mapleleaf Ampelopsis 20282
Mapleleaf Arrowwood 407662
Maple-leaf Bayur 320828
Mapleleaf Begonia 49782,49790,50404
Maple-leaf Begonia 49790,50122,50341,50404
Mapleleaf Buttercup 325515
Maple-leaf Clematis 94677
Mapleleaf Excoecaria 161630
Mapleleaf Goosefoot 87048
Mapleleaf Meadowsweet 166117
Maple-leaf Oak 323595
Mapleleaf Rabbiten-wind 12600
Maple-leaf Sterculia 58335
Mapleleaf Viburnum 407662
Mapleleaf Wingseedtree 320828
Mapleleafgrass 259730,259731
Maple-leaved Ampelopsis 20282
Maple-leaved Arrow-wood 407662
Maple-leaved Clematis 94677
Maple-leaved Excoecaria 161630
Maple-leaved Geranium 288050
Maple-leaved Goosefoot 87048,87159
Maple-leaved Pennywort 200263
Maple-leaved Pterospermum 320828
Maple-leaved Viburnum 407662
Maple-leaved Wing-seed-tree 320828
Maplin 2837
Ma-plua 132305
Mapney 131880
Mappianthus 244984
Maqu Crazyweed 278826
Maqu Kobresia 217223
Maqu Sedge 75301
Mar Querry 86970
Marabou-coronal Bulbophyllum 63084
Marabou-coronal Curlylip-orchis 63084
Marabutan 165307
Maracaibo Box 78145
Maracaibo Boxwood 78145
Maracaibo Lignum 63406
Maracock 285623
Maracuja 285694
Marang 36940
Maranon 21195
Maranta Family 245039
Marara 318620
Marasca Sour Cherry 316318
Maraschino Cherry 316318
Maraschino Marasco 316318
Maravilla 255672
Marble Arrowroot 261531
Marble Bamboo 87581
Marble Canyon Cactus 287882
Marble Gasteria 171703
Marble Leaf 77482,291158
Marble Mountain Bar Campion 363728
Marble Mrtagon Lily 229835
Marble Neoregelia 264421
Marble Pea 409092
Marble Treebine 92729
Marble Vine 61552
Marbled Bamboo 87581
Marbled Dossinia 136785
Marbled Kalanchoe 215206
Marbled Paris 284351
Marble-leaved Ivy 187238
Marbles Vine 131325
Marbleseed 271861
Marble-seed 271854,271855,271857,271861
Marblet Alumroot 194432
Marcaram 86970
March 29327
March And May 30164
March Daisy 50825
March Lily 18865
March Orchis 273501
March Qiongzhu 87595
March Violet 410320
Marchand Daphniphyllum 122706
Marchand Doellingeria 135258
March-and-may 30164
March-white Snow Marguerite 135359
Mardin Iris 208581
Mare Blob 68189
Mare Fart 359158
Mare's Fat 321554
Mare's Milk 159027
Mare's Tail 103446,158433,196818
Mare's Tails 103446,158433,196818
Mare's-tail 158433,196810,196818
Mare's-tail Family 196809
Mareer 104251
Marer 104251
Marestail 196810,196818
Marestail Family 196809
Marfim 46955
Marg 26768
Margan 26768
Margaret 227533
Margaret Rotang Palm 121514
Margarett's Hawthorn 109832
Margaric Devilrattan 121514
Margarite Haworthia 186534
Margery 86970,274237
Margil 97900
Margin Brome 60840
Marginate Agapetes 10335
Marginate Alyxia 18513
Marginate Eria 116621
Marginate Hairorchis 116621
Marginate Hakea 184620
Marginate India Rush 213001
Marginate Orostachys 218355
Marginate Osmanther 276362
Marginate Osmanthus 276362
Marginate Resurrectionlily 215021
Marginate Rockbell 412750
Marginate Sand Spurrey 370675
Marginate Sand-spurrcy 370675
Margincolored Cymbidium 117058
Margined Buttercup 326061,326062
Margined Craneknee 221702
Margined Fishvine 125986
Margined Henry Magnoliavine 351054
Margined Jewelvine 125986
Margined Pachycereus 279486
Margined Rose 336735
Margined Stickseed 221702
Marginvein Greenbrier 366483
Marginvein Smilax 366483
Margin-veined Greenbrier 366483
Margosa 45908
Margosa Tree 45908,248895

Margo-spreading Bennettiodendron 51027
Margret 227533
Marguerite 26738,32548,32563,227467,
227533
Marguerite Daisy 32563
Mari Eucalyptus 155514
Mari Gum 155514
Maria Bluegrass 305711
Maria Larkspur 124374
Maria Phalaenopsis 293621
Maria Woodbetony 287413
Marian 92485
Marian Thistle 364360
Marian Watson Flower 413369
Marianthus 245269
Marica 264158
Maricao 64432
Marie's Fir 414
Maries Ash 168100
Maries Bluegrass 305712
Maries Fir 414
Maries' Azalea 331203
Maries' Rhododendron 331203
Mariesi Viburnum 408035
Mariet 70164
Marigold 66380,66395,66445,68189,
89704,383087,383090,383103
Marigold Goldins 89704
Marigold of Peru 188908
Marigold-goldins 89704
Marihuana 71209,71218
Marijuana 71209,71218,71220
Marilandica Poplar 311155
Marin Knotweed 309375
Marine Batrachium 48926
Marine Ivy 92768
Marine Sedge 75303
Marine Vine 92768
Mariola 285087
Marionberry 338059
Marios Largeleaf Hydrangea 199974
Mariposa 67554
Mariposa Cactus 354302
Mariposa Lily 67554,67634,67635
Mariposa Pussypaws 93094
Mariposa Tulip 67554
Mariposa-lily 67554,67648
Marish Beetle 401112
Marisque 93901
Maritime Allomorphia 16155
Maritime Bulrush 353587
Maritime Dock 340116
Maritime Eryngo 154327
Maritime Eyebright 160284
Maritime Goldfields 222613
Maritime Larkspur 124467
Maritime Pea Vine 222735
Maritime Pine 300146
Maritime Rhubarb 329355
Maritime Thrift 34536
Maritime Wormwood 35868
Maritine Widgeonweed 340476
Marjolett Tulip 400192
Marjoram 274224,274237
Marjorana 242792,274197
Mark's Wild Buckwheat 152299
Markaang Willow 343655
Markang Catchfly 363726
Markang Mulberry 259167
Markang Pyrola 322850
Marked Pseudosasa 318332
Marked Willow 343528
Marked Woodbetony 287294
Marked-auriculate Yushania 416750
Markery 86970
Marking Nut 21195,357877,357878
Markingnut 357877
Marking-nut 21195,357877
Markweed 393470
Marlberry 31441,31477
Marlborough Rock Daisy 270196
Marl-grass 396967,397019
Marliac's Bamboo 297222
Marliac's Castillon Bamboo 297218
Marlock 155468
Marmalade Box 172890
Marmalade Bush 377951
Marmalade Plum 67524,313399
Mar-malade Plum 313399
Marmalade-plum 238115,244605
Marmaritan 280231
Marmorate Gasteria 171703
Maroola Plum 354345,354346
Maroon Gaillardia 169574
Maroon Tree Peony 280182
Maroon-leaved Japanese Maple 3335
Maroon-stemmed Banana 260207
Marram 19765,19766
Marram Grass 19766,19773
Marram Sea Grass 228346
Marram-grass 19766,19773
Marrom 19766
Marrow 114271,114300,114310
Marrow Cabbage 59524
Marrow Pea 301081
Marrow Squash 114304
Marrowfat 222832,301081
Marrow-type Pumpkin 114304
Marrube 245770
Marschall Campbell's Rhododendron 330127
Marschall Dock 340123
Marschall Onion 15465
Marschall Sagebrush 35888
Marschall Thyme 391274
Marsdenia 245777,245795
Marsh Alkali Aster 16287,16288
Marsh Andromeda 22445
Marsh Arrow Grass 397165
Marsh Arrowbamboo 162700
Marsh Arrow-grass 397165
Marsh Asphodel 262583
Marsh Aster 380912
Marsh banksia 47657
Marsh Barren Horsetail 196818
Marsh Bedstraw 170529
Marsh Beetle 401112
Marsh Beggartick 54012
Marsh Bellflower 69899,69900
Marsh Betony 373346,373365
Marsh Blazing-star 228529
Marsh Blue Violet 410306
Marsh Bluegrass 305803
Marsh Bristlegrass 361856
Marsh Bulrush 353276
Marsh Cinquefoil 100254,100257
Marsh Cistus 22445,223901
Marsh Claver 250502
Marsh Claytonia 94342
Marsh Cleavers 250502
Marsh Clematis 94870
Marsh Clover 250502
Marsh Collard 267274
Marsh Coneflower 339543
Marsh Corkwood 25876
Marsh Cress 336268
Marsh Crowfoot 326340
Marsh Cudweed 178481
Marsh Daisy 34536
Marsh Dandelion 384824
Marsh Dewflower 260102
Marsh Dock 340170
Marsh Elder 210462,210468,210478,
210485,407989
Marsh Everlasting 178481
Marsh Featherfoil 198680
Marsh Felwort 235461,380302
Marsh Figwort 355061
Marsh Five-finger 100257
Marsh Flawort 358287
Marsh Fleabane 150862,305083,305093,
358590
Marsh Fleawort 385908
Marsh Fox Tail 17527
Marsh Foxtail 17527
Marsh Gayfeather 228529
Marsh Gay-feather 228529
Marsh Gentian 173675,173728
Marsh Germander 388272
Marsh Gilliflower 363461
Marsh Goldenray 229081
Marsh Goldenrod 368454

Marsh Goldenweed 323036
Marsh Grass 370155
Marsh Grass-of-parnassus 284591,284601
Marsh Groundsel 358280,359660
Marsh Hawk's-beard 110946
Marsh Hawksbeard 110946
Marsh Hedge-nettle 373346,373348,373365
Marsh Helleborine 147195
Marsh Hibiscus 194804
Marsh Hog's Fennel 292957,292966
Marsh Holy Rose 22445
Marsh Horehound 239202
Marsh Horsegowl 68189
Marsh Knotweed 309524
Marsh Lavender 230815
Marsh Ledum 223901
Marsh Lily 68189
Marsh Loosewort 287494
Marsh Mallow 18155,18173,195032
Marsh Marigold 68157,68181,68189
Marsh Mermaid-weed 315539,315540
Marsh Milkwort 308014
Marsh Mint 250291
Marsh Mouse Ear Chickweed 260922
Marsh Muhly 259667,259689
Marsh Nanmu 295346
Marsh Oak 324262
Marsh Orchid 121358,273501
Marsh Parsley 29327,116384,292966
Marsh Pea 222800
Marsh Pea Vine 222800
Marsh Peavine 222800
Marsh Penny 200388
Marsh Pennywort 200257,200383,200388
Marsh Pepper 309199
Marsh Pestle 401112
Marsh Phlox 295275
Marsh Pilewort 325820
Marsh Pine 299928,300195
Marsh Pinegrass 65303
Marsh Pink 341439
Marsh Pitcher 188555
Marsh Plume Thistle 91916
Marsh Purslane 238192
Marsh Ragwort 358280,358590
Marsh Red Rattle 287494
Marsh Reedgrass 65303
Marsh Rhododendron 331944
Marsh Rose 275398
Marsh Roseau 230815
Marsh Rosemary 223888,223901,230508,
230559,230566,230815
Marsh Samphire 342847,342859
Marsh Sandwort 32118
Marsh Saxifrage 349452
Marsh Scheuchzeria 350861
Marsh Sedge 73597
Marsh Seed-box 238192
Marsh Sesbania 361394
Marsh Silverpuffs 253937
Marsh Skullcap 355435,355449
Marsh Sow Thistle 368781
Marsh Sow-thistle 368781
Marsh Speckled Climber 178564
Marsh Speedwell 407349
Marsh Spike-rush 143271
Marsh St. John's Wort 201847
Marsh St. John's-wort 201847,202207,
394813,394817
Marsh Stitchwort 375036
Marsh Straw Sedge 76508,76509
Marsh Sweetvetch 187946
Marsh Tea 223901
Marsh Thistle 92285
Marsh Trefoil 250502
Marsh Valerian 404251,404383
Marsh Vetchling 222800
Marsh Violet 410306,410347
Marsh Whortleberry 278366
Marsh Wild Pea 222800
Marsh Wild-timothy 259667
Marsh Willow Herb 146812
Marsh Willowherb 146812
Marsh Willow-herb 146812
Marsh Windhairdaisy 348890
Marsh Woodbetony 287494
Marsh Woundwort 373346
Marsh Yellow Cress 336250
Marsh Yellow-cress 336250
Marsh's Dutchman's-pipe 34294
Marshallia 245889
Marshcress 336166
Marsh-cress 336251
Marshcressleaf Woodbetony 287454
Marsh-elder 210462,210485
Marshes Rhododendron 331944
Marshfire Glasswort 342859
Marshhay Cordgrass 370171
Marshland Goosefoot 87151
Marshmallow 18173
Marsh-mallow 18155
Marshmarigold 68157
Marsh-marigold 68157,68189
Marshmarigoldleaf Primrose 314207
Marshmarigold-leaved Beesia 49576
Marshmarigold-leaved Goldenray 228999
Marshpepper Knotweed 309199
Marsh-pepper Knotweed 309199
Marshpepper Smartweed 309199
Marsh-pepper Smartweed 309199
Marsh-pink 341435,341463
Marsh-purslane 240065
Marshweed 230274
Marshwort 29312,29327,278366
Marshy Ant Palm 217923
Marshy Bedstraw 170717
Marshy Betony 373346
Marshy Buttercup 326278
Marshy Chickweed 375036
Marshy Germander 388270
Marshy Habenaria 184162
Marshy Honeysuckle 235649
Marshy Jewelvine 126007
Marshy Juncellus 212771
Marshy Ladybell 7728
Marshy Litse 234084
Marshy Mimulicalyx 255149
Marshy Orchis 273565
Marshy Panicgrass 282031
Marshy Pimpinella 299426
Marshy Pinanga 299665
Marshy Rattan Palm 65759
Marshy Rattanpalm 65759
Marshy Rhododendron 331346
Marshy Sedge 75165
Marshy Sowthistle 368781
Marshy Spapweed 205208
Marshy Spikesedge 143271
Marshy St. John's Wort 202199
Marshy Stylegrass 379082
Marshy Stylidium 379082
Marshy Tephroseris 385908
Marshy Threeribfruit 397969
Marshy Tofieldia 392629
Marshy Willow Weed 146812
Marshy Willowweed 146812
Marshy Woodbetony 287788
Mart Hemp Palm 393813
Martagon 230058,232950
Martagon Imperial 229922
Martagon Lily 229922
Martar Bamboo 297391
Martha 26746
Marthus 26746
Martian Philodendron 294822
Martijanov Crazyweed 278998
Martin Bedstraw 170476
Martin Clearweed 298967
Martin Lemongrass 117204
Martin Michelia 252924
Martin Pepper 300454
Martin Petrocosmea 292568
Martin Rhamnella 328562
Martin Spiraea 372002
Martin Stonebutterfly 292568
Martin Yam 131700
Martin's Fumitory 169153
Martin's Paraboea 283120
Martin's Prickly Pear 272975
Martin's Ramping-fumitory 169153
Martin's Rhododendron 331205

Martin's Thistle 92262
Martinez Cactus 244145
Martinique Plum 166776
Martinique Trimezia 397685
Martinoe 215334
Martius' Chusan Palm 393813
Martius' Windmillpalm 393813
Martynia 245977,245979
Martynia Family 245991
Maru-bushukan 93550
Marula 354345,354346
Marumi Kumquat 167503
Marum-leaf Wild Buckwheat 152252
Marupa 364382
Marvel 47013,245770
Marvel of Peru 255672,255694,255711
Marvelin Hair 232557
Marvel-of-Peru 255672,255711,255727
Mary Alone 221787
Mary Gold 66445
Mary Gooles 66445
Mary Gowan 50825
Mary Gowlan 66445,89704
Mary Jane 174877,248411
Mary Palmer 57868
Mary Spink 315101
Mary's Bean 250776
Mary's Clover 239750
Mary's Flower 21805
Mary's Gold 68189
Mary's Kidney 259573
Mary's Nut 145899
Mary's Rest 407072
Mary's Seed 368771
Mary's Taper 169719
Mary's Tears 321643
Mary's Thistle 364360
Mary's Titistle 364360
Mary-at-the-cottage-gate 374916
Marybout 68189
Marybud 66445,68189,325518,325666,326287
Maryjane 71218
Maryland Aster 90233
Maryland Dwarf American Holly 204121
Maryland Dwarf Morning-glory 161436
Maryland Figwort 355184
Maryland Golden Aster 90233
Maryland Goldenaster 90233
Maryland Meadow Beauty 329422
Maryland Pinkroot 371616
Maryland Sanicle 345980
Maryland Senna 360453
Maryland Tick Clover 126458
Marysvale Fleabane 150811
Masafuer Barberry 51908
Masaika Caper 71796
Masang Deutzia 126843
Mascadine 411893
Mascadine Grape 411893
Mascaren Grass 418453
Mascarene Grass 418453
Mascarene Island Leaf-flower 296785
Masdevallia 246101
Masdevallia Orchid 246101,246105
Maser 2837
Maser Tree 2837
Masertree 2837
Mashan Clematis 95106
Mashan Hogfennel 292922
Mashan Stephania 375888
Mash-corns 312360
Mashua 399611
Mashy Primrose 314453
Masindi 117886
Masindi Grass 117869
Masiro Gumi 142214
Mask 324335
Mask Flower 17475,17481,17490
Masked Cranesbill 153869
Maskert 373346
Mask-flower 255218
Maso Bamboo 297266
Mason Neststraw 379163
Mason Valley Cholla 116656
Mass 324335
Mass Polytoca 87690
Massang Arrowroot 245027
Massl 120432
Masslin 410960
Masso Cinnamon 91374
Masson Pine 300054
Massoy Bark 113429
Massy-cup Oak 323745
Mast Beech 162400
Master Wort 43306,43309
Master's Larch 221914
Masters Gomphostemma 179193
Masters Pine 299802
Mastersia 246175,246176
Masterwort 24303,30932,43306,43309,43310,192303,192358,205535
Masterwort Hog's-fennel 205535
Mastic 300994
Mastic Tree 300994
Mastick Thyme 391275
Mastic-tree 300994
Mastixia 246209
Masur 224473
Masur Wood 53563,53590
Mat Amaranth 18668
Mat Bean 408829
Mat Daisy 326513
Mat Grass 262537
Mat Muhly 259694
Mat Panic Grass 128482
Mat Reed 401112
Mat Rush 235386
Mat Sandbur 80830,80837
Mat Spurge 85795,158963
Mata Ajam 95972
Mata Kuching 131062
Mata Kucin 131062
Matac 37888
Matai 306518,316081,316089
Matai Podocarpus 306518
Matang Draba 137104
Matang Lily 229930
Matang Whitlowgrass 137104
Matapen Nolina 266502
Matarab Tea 78236
Matarique 316964
Mat-atriplex 44364
Match Wood 85271
Match-me-if-you-can 2011
Matchweed 182103
Maternity Plant 61568
Matfellon 81237,81356
Mat-forming Pernettya 291370
Matgrass 19766,262532,262537,308816
Mat-grass 262532,262537
Mat-grass Fescue 412493
Mather 227533
Mather Bambusa 47218
Mathern 26768,227533
Mathers 403932
Mati Milkvetch 42684
Matico 300333
Matico Pepper 300333
Matilija Poppy 335869,335870
Matilija-poppy 335869
Matricaria 89402
Matricary 246307
Matrimony Vine 238996,239011,239055,239092
Matrimonyvine 238996,239011
Matrimony-vine 238996,239011
Matrona Sedum 357203
Matrush 213036
Mat-saltbush 44364
Matscale 44364
Matsuda Addermonth Orchid 110648
Matsuda Arisaema 33406
Matsuda Bogorchis 110648
Matsuda Copperleaf 1923
Matsuda Dishspurorchis 171869
Matsuda Euonymus 157709
Matsuda Gastrochilus 171869
Matsuda Holly 203995
Matsuda Hornbeam 77382
Matsuda Osmanthus 276362

Matsuda Scurrula 385189
Matsuda Small Reed 127217
Matsuda Smallreed 127217
Matsuda Southstar 33406
Matsuda Spindle-tree 157709
Matsuda Tail Themeda 389319
Matsuda Taxillus 385189
Matsuda Winterhazel 106656
Matsumura Alder 16412
Matsumura Bentgrass 12186
Matsumura Drypetes 138658
Matsumura Leafflower 296803
Matsumura Spotleaf-orchis 179650
Matsumura Underleaf Pearl 296803
Matsumura White-dragon Storax 379408
Matsumura's Snowbell 379408
Matsyda Yam 131702
Mattae Wildginger 37626
Mattam Milkvetch 42685
Matte 265602
Matted Globularia 177035
Matted Pea Bush 321730
Matted Pearlwort 342279
Matted Sagewort 36069
Matted Sea Lavender 230543
Matted Sea-lavender 230543
Matted Spike-rush 143165
Matted Spikesedge 143165
Mattfeld Sagebrush 35892
Mattfeld's Wormwood 35892
Matthiol Cortusa 105496
Mattiastrum 246576
Mattipaul 12587
Mattop 285654
Matucana 246590
Matura Tea 78236
Mature Hawthorn 109837
Matweed 18668,19771,262537
Matweed Amaranth 18668
Mau 195311
Mau'u-Kukaepua'a 130768
Maudeline 383712
Maudia Michelia 252926
Maudlin 227533
Maudlin Daisy 227533
Maudlinwort 227533
Maul Oak 323762
Maule Maul 243840
Maule's Quince 84556,116546
Maunaloa 71040
Mauple 2837
Maurandia 246651
Maurit Pearweed 239727
Maurit Violet 410233
Mauritanian Grass 20268
Mauritia 246671
Mauritian Grass 29430
Mauritinus Hibiscus 194971
Mauritius 93492
Mauritius Aloe 169245
Mauritius Hemp 169242,169245
Mauritius Papeda 93492
Mauritius Prune 166788
Mauritius Raspberry 339194
Mauritius Thorn 64990
Maurole 245770
Maussarif Nepeta 264992
Mauther 26768
Mauthern 26768,227533
Mauve 243840
Mauve Catmint 264920
Mauve Dancing Ladies 177000
Mauve Sleekwort 232225
Mauvequeen Rhododendron 331683
Mavin 26768
Maw Seed 282717
Mawah 288045
Mawai Rattan Palm 65735
Mawroll 245770
Maws 243840
Mawseed 282717
Mawth 26768
Mawthem 26768
Max Chrysanthemum 227512
Maxburretia 246697
Maxianshan Milkvetch 42665
Maxillaria 246704
Maximilian Sunflower 188997
Maximilian's Sunflower 188997
Maximiliana 246731
Maximovicz Buckthorn 328781
Maximovicz Honeysuckle 235945
Maximovicz Rush 213272
Maximovicz Septemlobate Kalopanax 215443
Maximowicz Alder 16328
Maximowicz Ampelopsis 20321
Maximowicz Angelica 276752
Maximowicz Beadcress 264622
Maximowicz Berteroella 53047
Maximowicz Birch 53400
Maximowicz Cancrinia 71096
Maximowicz Cherry 83255
Maximowicz Cranesbill 174729
Maximowicz Currant 334089
Maximowicz Daisy 124824
Maximowicz Dendranthema 124824
Maximowicz Deutzia 127012
Maximowicz Euonymus 157711
Maximowicz Figwort 355148
Maximowicz Fritillary 168467
Maximowicz Hawthorn 109839
Maximowicz Hillcelery 276752
Maximowicz Japanese Barberry 52246
Maximowicz Ladybell 7695
Maximowicz Larkspur 124376
Maximowicz Linden 391766
Maximowicz Maple 3137
Maximowicz Nightshade 367365
Maximowicz Onion 15467
Maximowicz Peashrub 72283
Maximowicz Pea-shrub 72283
Maximowicz Poplar 311389
Maximowicz Primrose 314624
Maximowicz Pseudostellaria 318496
Maximowicz Rapanea 261632
Maximowicz Rhubarb 329356
Maximowicz Rose 336738
Maximowicz Sage 345205
Maximowicz Sagebrush 35895
Maximowicz Saltbush 44520
Maximowicz Sedge 75314
Maximowicz Snakegrape 20321
Maximowicz Spikesedge 143207
Maximowicz Spotleaf-orchis 179653
Maximowicz Spruce 298354
Maximowicz Sweetdaisy 71096
Maximowicz Waterchestnut 394496
Maximowicz Weigela 413616
Maximowicz Willow 343680
Maximowicz Windhairdaisy 348515
Maximowicz Woodland Bulrush 353676
Maximowicz's Birch 53520
Maximowicz's Cancrinia 71096
Maximowicz's Saltbush 44520
Maximowiz Bushclover 226892
Maximowiz Bush-clover 226892
Maximum Cattleya 79564
Maxionggou Rockfoil 349615
Maxon Melocactus 249594
Maxon Woodbetony 287415
Maxwell Spruce 298202
Maxwell's Norway Spruce 298202
May 3462,26768,109857,401593,408167
May Apple 285654,306609,306636
May Blob 68189,72934,326340,399504
May Blub 68189
May Bread-and-cheese Bush 109857
May Bread-and-cheese Tree 109857
May Bubble 68189
May Bush 109790
May Flower 109509
May Flowering Tasmanian Daisy Bush 270207
May Gollen 68189
May Goslings 343151
May Gowan 50825
May Grass 293731
May Hawthorn 110006
May Lily 102867,242668,242672,382329
May Maid 176824
May Night Sage 345246

May of the Meadow 166125
May Pink 127635
May Rose 336733
May Tassel 407989
May Thorn 109509,109857
May Tossel 407989
May Tossels 407989
May Tosty 407989
May Tree 109857
May Weed 26768
May Woodbetony 287416
Maya Corydalis 106133
Mayaca Family 246767
Mayapple 306636
May-apple 285654,306609,306636
May-ball 407989
Mayberry 109857
May-blob 68189,72934,326340,399504
May-blossom 102867
May-blub 68189
May-bubbles 68189
Maybuds 325518,325666,326287
May-bush 109790,109857
Mayday Tree 280003
Mayer Cinquefoil 312757
Mayer Rice 275937
Mayer Woodnettle 125545
Mayer Wormwood 35908
Mayflower 72934,146526,315082
May-flower 68189,72934,86755,109857,
315082,315101,374916,382329
Mayflower Lily 102867
May-flower Lily 102867
May-fruit 109857
May-goslings 343858
May-grass 374916
Mayhaw 109879
May-lily 102867,242680,382329
Mayll 1292
Mayodendron 246792
May-of-the-meadow 166125
May-pink 127635
Maypole 407989
Maypole-ing Tree 2837
Maypop 285654
Maypops 285654
Mayr Mono Maple 3198
May-rose 336498,407989
Mayrs Maple 3140
Maysor Melothria 417488
Maysor Zehneria 417488
May-spink 315101
May-tassels 407989
Mayten 246796,246817,246856
Maythem 26768
Maythig 26768
Maythorn 109857
May-thorn 109857
May-tree 3462,109857
Mayweed 26768,28044,227533,246307,
246396,322724,397948
Mayweed Camomile 26768
Maywort 36474,170335
Maza Begonia 50070
Mazalium 122515
Mazapan 243951
Mazari Palm 262234,262235
Mazar-tree 280003
Mazatlan Cactus 244146
Mazeerie 122515
Mazell 122515
Mazer 2837
Mazes 26768,227533
Mazus 246964
Mazzard 83155
Mazzard Cherry 83155
Mazzud 83155
Mber 201732,202086
Mbo-carya 6137
Mbocaya 6136
Mboga 361667
Mbura 284211
McCabe's Rhododendron 331164
McCart's Nailwort 284874
McDonald Oak 324138
Mcdougall's Yellowtops 166823
Mea Parsnip dow 285740
Meaberry 278366
Mead's Milkweed 38009
Mead's Sedge 75319
Mead's Stiff Sedge 75319
Meaden 26768
Meadow 72934
Meadow Alopecurus 17553
Meadow Anemone 23740,321721
Meadow Anemony 321721
Meadow Arrowleaf Knotweed 309801
Meadow Barley 198337,198354
Meadow Beauty 329419,329422,329427
Meadow Bistort 308893
Meadow Bittercress 72934
Meadow Bitter-cress 72934
Meadow Blue Violet 410351
Meadow Blueberry 403968
Meadow Bluegrass 306152
Meadow Bout 68189
Meadow Brome 60680,60707
Meadow Buttercup 325518
Meadow Cabbage 92268,381073
Meadow Campion 363461
Meadow Chickweed 82680
Meadow Clary 345321
Meadow Clover 396967,397019
Meadow Crane's Bill 174832
Meadow Crane's-bill 174832
Meadow Cranesbill 174715,174832
Meadow Cress 72934
Meadow Crocus 99297
Meadow Crowfoot 325518
Meadow Death Camas 417928
Meadow Drop-seed 372629,372630
Meadow Evening-primrose 269489
Meadow Eyebright 160233
Meadow False Fleabane 321554
Meadow Falsebrome 58608
Meadow Fatch 271280
Meadow Fescue 163819,164219,350631
Meadow Fescuegrass 164219
Meadow Foam 230205,230210
Meadow Fox Tail 17553
Meadow Foxtail 17553
Meadow Garlic 15149
Meadow Gentian 173318
Meadow Geranium 174832
Meadow Giantfennel 163644
Meadow Goldenrod 368013
Meadow Gowan 68189
Meadow Gras 305274
Meadow Hawksbeard 110994
Meadow Hawkweed 195506,195858
Meadow Honeysuckle 397019
Meadow Jerusalemsage 295172
Meadow Knapweed 80893,81026
Meadow Larkspur 124182
Meadow Leek 15149
Meadow Ligusticum 229346
Meadow Lily 229779,229789
Meadow Lychnis 363461
Meadow Maid 166125
Meadow Monkshood 5466
Meadow Oat Grass 45548
Meadow Oat-grass 45548
Meadow Orchid 273531,273545
Meadow Parsnip 192358,388895,388899,
418109,418113
Meadow Pasqueflower 23998
Meadow Pea 222820
Meadow Pea Vine 222820
Meadow Peavine 222820
Meadow Pea-vine 222820
Meadow Phlox 295282
Meadow Pinegrass 65303
Meadow Pink 72934,127698,127700,363461
Meadow Plagiobothrys 301627,301628
Meadow Pulsatilla 23998
Meadow Pussytoes 26331,26375
Meadow Queen 166125
Meadow Rattle 329521
Meadow Rice Grass 141784
Meadow Rocket 273469
Meadow Rockfoil 349419,349801

Meadow Rue 388397,388423,388506,388572,388638
Meadow Runagates 239755
Meadow Rye Grass 164219
Meadow Saffron 62518,99293,99297,99352
Meadow Sage 345246,345321
Meadow Salsify 394333
Meadow Saussurea 348129
Meadow Saxifrage 349419,361456,363123
Meadow Scabious 216753,379660
Meadow Serpentroot 354917
Meadow Silaus 363128
Meadow Soft-grass 197301
Meadow Sorrel 339887
Meadow Spink 363461
Meadow Starwort 375036
Meadow Succisa 379660
Meadow Sundrops 269489
Meadow Sweet 166125,371785
Meadow Thistle 91916,92360
Meadow Trefoil 397019
Meadow Valley Sandwort 32237
Meadow Vetchling 222820
Meadow Violet 410351,410474,410587
Meadow Willow 343868,343909
Meadow Windhairdaisy 348129,348858
Meadow Wintergreen 172134
Meadow Woollyheads 318806
Meadow-bout 68189
Meadowbright 68189
Meadow-cresses 72934
Meadow-cuckoo 72934
Meadow-fern 261162
Meadowflower 72934
Meadow-flower 72934
Meadow-foam 230205,230210
Meadow-foam Family 230183
Meadow-grass 305274,305334
Meadowgrass-like Koeleria 217494
Meadow-lychnis 363461
Meadowrue 388397
Meadow-rue 388397,388423,388506
Meadowrueleaf Asiabell 98419
Meadowruelike Bowwingpaesley 31329
Meadowruelike Isopyrum 210291
Meadows Hillcelery 276764
Meadows Ostericum 276764
Meadow-saffron-of-the-spring 99352
Meadowsoot 166125
Meadowsweet 166071,166110,166125,166132,371785,371795,372075
Meadow-weed 388628
Meadsweet 166125
Meaduart 166125
Meadwort 166125
Meager Pentachaeta 289423
Meagre Maesa 241798
Meagre-like Maesa 241799
Meal Plum 31142
Meal Tree 407907
Mealberry 31142
Mealie 417417
Mealie Heath 149850
Mealies 417417
Mealy Corydalis 105775
Mealy Cup Sage 345023
Mealy Dendrocalamus 125474
Mealy Guelder Rose 407907
Mealy Plum 31142
Mealy Primrose 314366,314563
Mealy Rhododendron 330686
Mealy Sage 345023
Mealy Sinobambusa 364790
Mealy Stringbark 155529
Mealy Tonkin Pseudosasa 318285
Mealycup Sage 345023
Mealy-cup Sage 345023
Mealy-plum 31142
Mealy-tree 407907
Mearded Protea 315876
Mearns Acacia 1380
Measle-flower 66445
Measure of Love 50825
Measures Giant Orchid 180119
Meat Nut 78811,102727
Mecaw Palm 6129
Mecca Aster 40235
Mecca Balsam 101391,101488
Mecca Galls 324314
Mecca Myrrh 101391,101488
Mecca Myrrh Tree 101488
Mecca Orach 44386
Mecca Woody-aster 415846
Meccaaster 40235
Meconopsis 247099
Mecopus 247204
Meda Iris 208709
Medake Bamboo 304098
Medeola 247225
Medewspike Bog Rattanpalm 65762
Media Mahonia 242463
Media Sandspurry 370675
Median Apple 93603
Median Sandspurry 370675
Medianfruit Coffee 98872
Median-fruited Coffee 98872
Mediate Privet 229544
Medic 247239
Medical Dogwood 105146
Medicinal Agrimony 11549
Medicinal Alkanet 21959
Medicinal Aloe 16792,17381
Medicinal Archangelica 30932
Medicinal Basketvine 9474
Medicinal Betonica 53304
Medicinal Breynia 60073
Medicinal Buckthorn 328642
Medicinal Changium 85972
Medicinal Chirita 87922
Medicinal Cinchona 91075,91082
Medicinal Citron 93603
Medicinal Citrus 93603
Medicinal Cogflower 382740
Medicinal Collectivegrass 381035
Medicinal Comfrey 381035
Medicinal Condorvine 245830
Medicinal Cornel 105146
Medicinal Cowherb 403693
Medicinal Cyathula 115731
Medicinal Damnacanthus 122077
Medicinal Devilpepper 327074
Medicinal Ervatamia 382740
Medicinal Evodia 161373
Medicinal Fat Head Tree 262822
Medicinal Fatheadtree 262822
Medicinal Fumitory 169126
Medicinal Garlicmustard 14964
Medicinal Holarrhena 197199
Medicinal Hyssop 203095
Medicinal Indianmulberry 258905
Medicinal Indian-mulberry 258905
Medicinal Jerusalemsage 295143
Medicinal Kopsia 217867
Medicinal Lavender 223251
Medicinal Lily 230038
Medicinal Magnolia 198699
Medicinal Mandrake 244346
Medicinal Marsdenia 245830
Medicinal Mescal Bean 369116
Medicinal Milkgreens 245830
Medicinal Nutmeg 261452
Medicinal Onion 15480
Medicinal Plantain 302103
Medicinal Rhubarb 329366
Medicinal Sage 345271
Medicinal Sakebed 229371
Medicinal Solomonseal 308613
Medicinal Speedwell 407256
Medicinal Swallowwort 117637
Medicinal Tigerthorn 122077
Medicinal Toona 392836
Medicinal Wallgrass 284167
Medicine Dandelion 384714
Medicine Stephania 375893
Medicine Terminalia 386504
Medick 247239,247391
Medick Fetch 271280
Medick Fitch 271280
Medick Fitchling 271280
Medick Fodder 247456
Medick-fetch 271280

Medick-fitch 271280
Medick-fitchling 271280
Medick-fodder 247456
Medick-vetchling 271280
Mediek Vetchling 271280
Medilong Rabdosia 209771
Medilung Rabdosia 209771
Medinilla 247526,247589
Mediostachys Broad-spine Rattan Palm 65771
Mediterranea Stink Groundpine 70464
Mediterranean Aloe 17381
Mediterranean Barley 198297
Mediterranean Beach Daisy 280584
Mediterranean Brome 60797
Mediterranean Broom 172995
Mediterranean Buckthorn 328595
Mediterranean Cabbage 59389
Mediterranean Camphorfume 70463
Mediterranean Chrysanthemum 89421
Mediterranean Clover 397086
Mediterranean Coriaria 104700
Mediterranean Crownvetch 105317
Mediterranean Cypress 114753
Mediterranean Fan Palm 85671
Mediterranean Figwort 355216
Mediterranean Germanier 388184
Mediterranean Glorybind 102914
Mediterranean Hairgrass 217518
Mediterranean Hair-grass 217518,337269
Mediterranea Hartwort 392869
Mediterranean Heath 149398
Mediterranean Herb Elder 345594
Mediterranean Herb-elder 345594
Mediterranean Lily 280900
Mediterranean Lineseed 50710,50717
Mediterranean Love Grass 147520
Mediterranean Lovegrass 147520
Mediterranean Medlar 109545
Mediterranean Mezereon 122448
Mediterranean Mullein 405774
Mediterranean Needle-grass 376728
Mediterranean Onion 15567
Mediterranean Palm 85665,85671
Mediterranean Pepperweed 225371
Mediterranean Polypogon 310121
Mediterranean Rabbitsfoot Grass 310121
Mediterranean Radish 326602
Mediterranean Redbud 83809
Mediterranean Restharrow 271451
Mediterranean Rocket 365459
Mediterranean Rose 336947
Mediterranean Rye-grass 235354
Mediterranean Saltbush 44446
Mediterranean Saltwort 344758
Mediterranean Sea Lavender 230815
Mediterranean Serpent Root 354881
Mediterranean Serpent-root 354881
Mediterranean Spurge 158640
Mediterranean Statice 230815
Mediterranean Stinkbush 21445
Mediterranean Stork's Bill 153839
Mediterranean Stork's-bill 153738
Mediterranean Sweet Lemon 93534
Mediterranean Sweetclover 249251
Mediterranean Tapeweed 311972
Mediterranean Tree Mallow 243797
Mediterranean Tree-mallow 243797
Mediterranean Turnip 59651
Meditteranean Grass 351167
Meditteranean Sage 344836
Medium Barberry 51910
Medium Roegneria 335425
Medium-pictured Japanese Spindle-tree 157608
Medlar 252291,252305
Medlar-bush 19288
Medlar-Hawthorn 110205
Medlar-tree 252305
Mednip 61497
Medog Agapetes 10335
Medog Basketvine 9460
Medog Cleisostoma 94452
Medog Clematis 95112
Medog Closedspurorchis 94452
Medog Glochidion 177157
Medog Glycosmis 177842
Medog Holly 204030
Medog Mahonia 242620
Medog Pencil-wood 139664
Medog Phoebe 295385
Medog Raspberry 338818
Medog Rhododendron 331211
Medog Spicebush 231396
Medog Square-bamboo 87586
Medong Barberry 51911
Medusa Saussurea 348521
Medusa Windhairdaisy 348521
Medusa's Head 158614
Medusahead 383034,383040
Medwart Meduart 166125
Meebold Rockfoil 349618
Meedles 44576
Meehan Barberry 51912
Meehan Carolina Silverbell 184734
Meehania 247720
Meehanialike Skullcap 355598
Meeks 308893
Meelalla 57122
Meet-her-in-the-entry-kiss-her-in-the-buttery 410677
Meeting House Seeds 24213
Meeting Houses 30007
Meeting Seeds 167156
Meeting Spunsilksage 360827
Meet-me-love 349044,410677
Meet-me-love-behind-the-garden-door 410677
Meg Many-feet 326287
Megacarpaea 247767
Megacodon 247843,247844
Megadenia 247850
Megaheaded Melaleuca 248107
Megaphyllous Adinandra 8255
Megaphyllous Manglietia 244431,244455
Megaphyllous Tickclover 126462
Megistostigma 247967
Meg-many-feet 326287
Megweed 366743
Mei 34448
Mei Flower 34448
Mei Hua 34448
Meifeng Listera 232936
Meigong Sage 345208
Meigu Cranesbill 174731
Meigu Rush 213276
Meihuashan Cyclobalanopsis 116147
Meihuashan Oak 116147
Meihuashan Qinggang 116147
Meiliqike Bluegrass 305684
Mei-nung Broadflower Dendrocalamus 125483
Mei-nung Dragonbamboo 125483
Meiogyne 248028
Meiostamen Grease-bush 177362
Meissner Banksia 47652
Meiwa Kumquat 167493
Meixian Apricot 34479
Meixian Crested Corydalis 105773
Meixian Sedge 75324
Mekilwort 44708
Mekkin 208771
Mekong Barberry 51913
Mekong Bousigonia 57913
Mekong Cayratia 79895
Mekong Chinquapin 78989
Mekong Evergreen Chinkapin 78989
Mekong Evergreenchinkapin 78989
Mekong Formania 167340
Mekong Giantarum 20114
Mekong Indigo 206234
Mekong Jinjiang Colquhounia 100035
Mekong Myriactis 261080
Mekong Neocinnamomum 263764
Mekong Premna 313687
Mekong River Rhododendron 331216
Mekong Sage 345208
Mekong Tanoak 233314
Mekong Wallich Scleropyrum 354503
Mel Grass 19766
Melabranch Burreed 370028
Melaleuca 248072,248114
Melampode 190944

Melampodium 248133,248140
Melancholy 3978
Melancholy Gentlemen 43309,193452
Melancholy Thistle 92012,92022
Melandrium 248235
Melandrium Sandwort 32067
Melandriumleaf Gentian 173630
Melanesian Papeda 93559
Melanocarpous Viburnum 407950
Melanohairy Dendrobenthamia 124940
Melanolepis 248497
Melanosciadium 248542
Melasma 248691
Melastoma 248726
Melastoma Family 248806
Melastomalike Canscora 71273
Mclastoma-like Metalleaf 296398
Melde 44576
Meldweed 86901
Melegueta Pepper 9923,19858
Mel-grass 19766
Melgs 86901
Melhania 248822
Melia 248893,248895
Melialeaf Evodia 161335
Melic 248949
Melic Grass 248949
Melica Manna-grass 177635
Melicgrass 248949
Melic-grass 248949
Melick 248949,249054
Meliclike Sicklebamboo 20200
Melic-oats 398499
Melicope 249127
Melilot 249232
Melilot Trefoil 249232
Melilot-trefoil 249232
Melindjo 178535
Melinis 249270
Meliosma 249355,249373
Meliosmaleaf Mountainash 369461
Meliosma-leaved Mountain-ash 369461
Meliot 249197
Melissaleaf Betony 373307
Melittis 249540
Mell Actinidia 6673
Mell Apple 243651
Mell Eargrass 187625
Mell Giantarum 20115
Mell Hedyotis 187625
Mell Kiwifruit 6673
Mell Lobelia 234628
Mell Sageretia 342180
Mellifluous Iris 208711
Melliodendron 249557
Melmont Berries 213702
Melmot Berries 213702
Melocactus 249578
Melocanna 249614
Melochia 249627
Melodinus 249648
Melon 114108,114189
Melon Cactus 249578,249590
Melon Euphorbia 159342
Melon Loco 29537
Melon Pear 367400
Melon Pumpkin 114288
Melon Spurge 159342
Melon Tree 76813
Meloncillo 285703
Melonleaf Begonia 49756
Melonleaf Birthwort 34158
Melonleaf Dutchmanspipe 34157
Melonleaf Mesquite 315553
Melonleaf Pertybush 292057
Melonleaf Snakegourd 396166
Melon-leaved Nightshade 367205
Melon-Loco 29537
Melonseed Hairorchis 396472
Melothria 249760
Mel-sylvestre 236022
Melton bluebird 412913
Membraanousleaf Primrose 314635
Membraceousleaf Woodbetony 287422
Membrana Arrowbamboo 137857
Membranacarp Begonia 49939
Membranaceous Beautyleaf 67864
Membranaceous Beauty-leaf 67864
Membranaceous Bethlehemsage 283636
Membranaceous Blumea 55784
Membranaceous Camellia 69353
Membranaceous Caper 71798
Membranaceous Casearia 78137
Membranaceous Dendrocalamus 125490
Membranaceous Marshmarigold 68201
Membranaceous Paraphlomis 283636
Membranaceousleaf Caper 71802
Membranaceous-leaved Caper 71802
Membranaceous-phyllaris Pearleverlasting 21584
Membranaceous-podded Caesalpinia 65020
Membranaceus Madder 337993
Membrana-leaf Camellia 69353
Membrana-leaf Pittosporum 301336
Membranaleaf Purplepearl 66869
Membrana-leaf Rehdertree 327420
Membrana-leaf Seatung 301336
Membrana-sepal Camellia 69610
Membrane Jadeleaf and Goldenflower 260458
Membranebract Cremanthodium 110461
Membranebract Nutantdaisy 110461
Membranecalyx Sterculia 376124
Membraneedge Swertia 380271
Membrane-flower Stelis 374709
Membranefruit Gentian 173528
Membraneleaf Catnip 264993
Membraneleaf Swertia 380278
Membraneleaf Tea 69353
Membraneous Begonia 50073
Membraneousbract Spikesedge 143298
Membraneousleaf Speedwell 407311
Membraneroot Leek 15361
Membraneroot Onion 15361
Membraneusleaf Nepeta 264993
Membranous Bract Iris 208813
Membranous Dendrocalamus 125490
Membranous Epimedium 146977
Membranous Garlicvine 244711
Membranous Merrilliopanax 250877
Membranous Milkvetch 42699
Membranous Rockvine 387810
Membranous-calyxed Sterculia 376124
Membranousleaf Camellia 69267
Membranousleaf Childvine 400929
Membranousleaf Clematis 94796
Membranousleaf Linden 391770
Membranousleaf Maesa 241804
Membranousleaf Mucuna 259541
Membranous-leaf Mucuna 259545
Membranousleaf Mussaenda 260458
Membranousleaf Rockvine 387810
Membranousleaf Woodnettle 273902
Membranous-leaved leaf Maesa 241804
Membranous-leaved Lime 391770
Membranous-leaved Linden 391770
Membranous-leaved Mucuna 259541
Memecylon 249886
Memecylonleaved Holly 204035
Memecylon-leaved Holly 204035
Memorial Rose 337015
Memphisgrass 115259
Memzies Tolmiea 392668
Men's Face 410677
Men's Faces 410677
Men-and-women 37015
Mendel Cactus 244149
Mendocino Cypress 114698,114749
Menel Aster 40839
Mengban Flemingia 166885
Mengda Milkvetch 42666
Menghai Bulbophyllum 62908
Menghai Caper 71750
Menghai Cherry Laurel 223115
Menghai Cherry-laurel 223115
Menghai Clearweed 298972
Menghai Cleisostoma 94453
Menghai Closedspurorchis 94453
Menghai Coldwaterflower 298972
Menghai Dendrobium 125360
Menghai Denseflower Elaeagnus 141959
Menghai Elaeagnus 141959

Menghai Galangal 17721
Menghai Gastrodia 171943
Menghai Giantarum 20051,20095
Menghai Ginger 418001
Menghai Grape 411796
Menghai Heterostemma 194161
Menghai Irisorchis 267971
Menghai Linociera 231771
Menghai Liolive 231771
Menghai Maple 3024
Menghai Oberonia 267971
Menghai Pepper 300372
Menghai Premna 313630
Menghai Sotvine 194161
Menghai Spicebush 231395
Menghai Spice-bush 231395
Menghai Stonebean-orchis 62908
Menghai Tanoak 233217
Menghai Yushanbamboo 416799
Menghai Yushania 416799
Menghua Germander 388144
Mengkulang 192477
Mengla Amomum 19893
Mengla Argyreia 32639
Mengla Cayratia 79874
Mengla Clematis 95111
Mengla Dracaena 137451
Mengla Dragonbood 137451
Mengla Drupetea 322595
Mengla Drypetes 138630
Mengla Ginger 418004
Mengla Irisorchis 267972
Mengla Macrosolen 241319
Mengla Newlitse 264073
Mengla Oberonia 267972
Mengla Podocarpium 200735
Mengla Pyrenaria 322595
Mengla Raspberry 338806
Mengla Rattan Palm 65715
Mengla Rattanpalm 65715
Mengla Schefflera 350742
Mengla Semnostachya 358002
Mengla Sheathflower 241319
Mengla Yam 131707
Menglian Begonia 50074
Menglian Brachycorythis 58404
Menglian Kamalatree 243428
Menglian Lilyturf 272115
Menglian Nettle-leaf Croton 113051
Menglian Pellionia 288747
Menglian Petrocosmea 292570
Menglian Redcarweed 288747
Menglian Stonebutterfly 292570
Menglian Sweetleaf 381304
Menglian Wildtung 243428
Menglong Alyxia 18514
Menglong Nanbar Rattan Palm 65745
Menglun Bulbophyllum 62909
Menglun Elatostema 142740
Menglun Slugwood 50478
Menglun Spicebush 231399
Menglun Stairweed 142740
Menglun Stonebean-orchis 62909
Menglun Triratna Tree 397370
Menglun Wingseedtree 320842
Menglun Wing-seed-tree 320842
Mengma Greenbrier 366458
Mengma Smilax 366458
Mengpeng Flemingia 166885
Mengshan Cranesbill 174984
Mengshan Hornbeam 77342
Mengshan Willow 343778
Mengtze Clovershrub 70872
Mengtze Waxplant 198875
Mengtzu Falsehellebore 405614
Mengxing Basketvine 9461
Mengzi Albertisia 13505
Mengzi Albizia 13505
Mengzi Alder Birch 53347
Mengzi Amomum 19895
Mengzi Aspidopterys 39679
Mengzi Azalea 331221
Mengzi Clematis 94930
Mengzi Clovershrub 70872
Mengzi Cotoneaster 107472
Mengzi Didymocarpus 129928
Mengzi Dinetus 131245
Mengzi Euonymus 157714
Mengzi Falsehellebore 405614
Mengzi Grape 411797
Mengzi Greenvine 204630
Mengzi Gutzlaffia 182127
Mengzi Illigera 204630
Mengzi Indigo 206240
Mengzi Lysionotus 239934
Mengzi Mayten 246878
Mengzi Paperelder 290357
Mengzi Petrocosmea 292560
Mengzi Porana 131245
Mengzi Premna 313644
Mengzi Rhododendron 331221
Mengzi Spapweed 205132
Mengzi St. John's Wort 201920
Mengzi Sterculia 376121
Mengzi Stonebutterfly 292560
Mengzi Sweetleaf 381291
Mengzi Swertia 380252
Mengzi Touch-me-not 205132
Mengzi Yushanbamboo 416793
Mengzi Yushania 416793
Meni Oil 236334
Meni Oil Tree 236336
Menslow Spapweed 205133
Menslow Touch-me-not 205133
Menten Twinballgrass 209054
Mentor Barberry 51914
Mentzelia 250481
Menyuan Buttercup 326075
Menyuan Kobresia 217225
Menzies Banksia 47653
Menzies Gooseberry 334093
Menzies Larkspur 124379
Menzies Nemophila 263505
Menzies Penstemon 289363
Menzies Peppergrass 225413
Menzies Pipsissewa 87489
Menzies Spruce 298435
Menzies' Catchfly 363741
Menzies' Giant Rattlesnake-plantain 179664
Menzies' Goldenbush 209576
Menzies' Larkspur 124379
Menzies' Rattlesnake Plantain 147126
Menziesia 250505,297019
Meon 252662
Meranti 362186
Merawan 198126,198167,198176
Merbau 207015,207017
Merch 29327
Merckia 414311
Mercury 44576,86970,250597
Mercury Docken 86970
Mercury Goosefoot 86970
Mercury's Moist Blood 405872
Mercury's Violet 70164,70331
Mercury-docken 86970
Merecrop 21339
Meriandra 250655
Merill Ormosia 274411
Merill's Dunbaria 138971
Merisier 53338
Merk Crazyweed 279009
Merkus Pine 300068
Mermaid Weed 315539
Mermaids 267648
Mermaidweed 315539
Merremia 250749
Merril Primrose 314637
Merrill Actephila 6476
Merrill Ardisia 31522
Merrill Chaulmoogratree 199754
Merrill Dyeingtree 346491
Merrill Hemiboea 191374
Merrill Herpysma 193073
Merrill Ixora 211065
Merrill Palm 405258
Merrill Pittosporum 301338
Merrill Randia 325377
Merrill Veitchia 405258
Merrill Wendlandia 413800
Merrill Woodbetony 287424
Merrill's Croton 112962

Merrill's Holly 204556
Merrillanthus 250867
Merrillia 250869,250870
Merrilliopanax 250873
Merry Bells 403665,403670
Merrybells 403665
Merry-bells 403670
Merry-go-rounds 66445
Merry-tree 83155
Mersawa 25605
Merten's Coral-root 104002
Merten's Saxifrag 349633
Mertensia 250914
Meru Oak 411307
Mesa Brodiaea 60486
Mesa Oak 323874
Mesa Sacahuista 266493
Mesa Tansyaster 240457
Mesa Verde Fishhook Cactus 354304
Mescal 10787,10843,10948
Mescal Agave 10921
Mescal Bean 369116
Mescal Button 236423,236425
Mescal Button Peyote 236425
Mescal Buttons 236425
Mescal Ceniza 10830
Mescalbean 369116
Mescal-button 236423,236425
Mescat Acacia 1148
Mesdow Rose 336389
Mesembryanthemum 251025
Meserve Hybrid Holly 203533
Meseta Cottonwood 311324
Mesh Mellice 18173
Mesny Jasmine 211905
Mesny Willow 343685
Mesona 252241
Mesopotamian Poplar 311308
Mespil 252291,252305
Mesquita 315547
Mesquite 315547,315553,315557,315560,
315565,315568
Mesquite Neststraw 379166
Mesquitilla 66669
Mess 324335
Mess-a-bed 384714
Messanius Fritillary 168483
Messerschmidia 252333
Messmate 155672
Messmate Stringbark 155672
Mestoklema 252361,252366
Mesua 252369
Metabriggsia 252386,252387
Metadina 252395
Metake 318307
Metalleaf 296365,296371
Metal-leaf Begonia 50076
Metallic Taro 16522
Metallic-leaf Begonia 50076
Metanemone 252497,252498
Metapetrocosmea 252506,252507
Metaplexis 252508
Metapolyanthous Barberry 51915
Metasasa 252533,252536
Metcalf Cryptocarya 113470
Metcalf Mallotus 243390
Metcalf Maple 3143
Metcalf Persimmon 132300
Metcalf Sabia 341491
Metcalf Spicebush 231392
Metcalf Spice-bush 231392
Metel 123065
Meteor Crater Wild Buckwheat 152413
Methonica Lily 177251
Methyst Flower 61131
Metterich Rhododendron 331227
Meu 252662
Mew Meu 252662
Mex Begonia 50077
Mexiaan Red Juniper 213826
Mexica Spruce 298357
Mexican Aalpinegold 199220
Mexican Aechmea 8576
Mexican Ageratum 11206
Mexican Alder 16315,16396,16425
Mexican Apple 78175
Mexican Aster 107158,107161
Mexican Astrocaryum 43376
Mexican Avocado 291495
Mexican Bald Cypress 385281
Mexican Baldcypress 385281
Mexican Bald-cypress 385281
Mexican Balsamea 417407
Mexican Bamboo 162532,328345
Mexican Barberry 51694,242654
Mexican Basketflower 303085
Mexican Bird Cherry 316760
Mexican Bird of Paradise 65037
Mexican Bloewood Condalia 101736
Mexican Blood Flower 134877
Mexican Blue Oak 324235
Mexican Blue Palm 59096
Mexican Blue Sage 344950
Mexican Bone-bract 354336
Mexican Brassbuttons 107789
Mexican Bread Fruit 258168
Mexican Breadfruit 258168
Mexican Buck Eye 401829
Mexican Buckeye 401829
Mexican Bush Sage 345167
Mexican Calceolaria 66281
Mexican Campion 363649
Mexican Catchfly 363649
Mexican Cedar 80025
Mexican Cedrela 80025
Mexican Ceratozamia 83685
Mexican Chickweed 374819
Mexican Chinchweed 286880
Mexican Cigar Plant 114606
Mexican Claret Cup 140231
Mexican Cliffrose 321885
Mexican Clover 334310,334338
Mexican Columbine 30075
Mexican Common Pine 300277
Mexican Coral Drops 53213
Mexican Creeper 28418,28422
Mexican Creeping Zinnia 346304
Mexican Crowngrass 285433
Mexican Crucillo 101744
Mexican Cypress 114714,385281
Mexican Daisy 150724,150792
Mexican Dayflower 100960
Mexican Day-flower 100960
Mexican Devil 11153,158024
Mexican Devilweed 41297,88253,88254
Mexican Dock 340128
Mexican Drooping Juniper 213773
Mexican Elder 345627
Mexican Eupatorium 158199
Mexican Evening Primrose 269504
Mexican Everning Primrose 269403
Mexican Fairy Lily 417613
Mexican Fan Palm 413307
Mexican Fanpalm 413307
Mexican Feather Grass 262621,376931,
376934
Mexican Fence Post Cactus 279683
Mexican Fern Palm 131429
Mexican Fir 458
Mexican Fire Barrel Cactus 163464,163465
Mexican Fire Bush 217361
Mexican Fire Plant 159046,159675
Mexican Fire-bush 217361
Mexican Firecracker 140009
Mexican Fire-plant 159046,159675
Mexican Fireweed 217361
Mexican Flag 128988
Mexican Flame Leap 159675
Mexican Flame Tree 159675
Mexican Flame Vine 358588
Mexican Flamebush 66689
Mexican Flamevine 317787
Mexican Flame-vine 358588
Mexican Fleabane 150724,150792
Mexican Foxglove 387346
Mexican Frangipani 305225,305226
Mexican Giant Cactus 279490
Mexican Giant Hyssop 10402,10409
Mexican Giant-hyssop 10402,10409
Mexican Gold Poppy 155183
Mexican Grain Amaranth 18788

Mexican Grape Herb 139678
Mexican Grass 211027
Mexican Grass Plant 122908
Mexican Grass Tree 122908,122909,266499
Mexican Ground Cherry 297686
Mexican Groundcherry 297712,297713
Mexican Ground-cherry 297712
Mexican Hat 326957,326960,339512
Mexican Hat Plant 61568
Mexican Hawthorn 109967
Mexican Heather 114605
Mexican Hens 140010
Mexican Hickory 77927
Mexican Holdback 65037
Mexican Honeysuckle 214817
Mexican Horncone 83685
Mexican Ivy 97757
Mexican Juniper 213773,213826
Mexican Leatherwood 133655
Mexican Lily 196439,196452
Mexican Lime 93329
Mexican Lobelia Bush 234592
Mexican Love Grass 147808
Mexican Mahogany 380526
Mexican Malpighia 243527
Mexican Marigold 383097,383107
Mexican Mint 303139
Mexican Mint Marigold 383097
Mexican Mock Orange 294499,294500
Mexican Mock-orange 294500
Mexican Mountain Pine 299875
Mexican Muehlenbergia 259679
Mexican Mugwort 35904
Mexican Muhly 259679
Mexican Mulberry 259164,259165
Mexican Nut Pine 299851
Mexican Oleander 389984
Mexican Orange 88696,88698,88700
Mexican Orange Blossom 88696,88700
Mexican Orange Flower 88700
Mexican Oregano 232492,307284
Mexican Oxalis 277925
Mexican Palmetto 341418
Mexican Palo Verde 284479
Mexican Paloverde 284479
Mexican Papaya 76813
Mexican Papyrus 118935
Mexican Pepper 300488
Mexican Persimmon 132431
Mexican Petunia 339684
Mexican Pine 299809,299851,300048,300103,300140
Mexican Pinyon 299851
Mexican Pinyon Pine 299851
Mexican Plane 302587
Mexican Plane Tree 302587
Mexican Plum 316553
Mexican Plume 214490
Mexican Pokeweed 298110
Mexican Poppy 32411,32423,32429,215381
Mexican Prickly Poppy 32423,32429
Mexican Primrosewillow 238188
Mexican Red Sage 345043
Mexican Redbud 83794
Mexican Ricegrass 4129
Mexican Rubber Tree 79110
Mexican Rush 212857,213289
Mexican Sage 345177,345212
Mexican Sagebrush 35837
Mexican Shell Flower 391624
Mexican Shell Flowers 391622
Mexican Shrubby Spurge 158700
Mexican Spangletop 226005
Mexican Star 254564
Mexican Star of Bethlehem 254564
Mexican Stone Pine 299851
Mexican Stonecrop 356934
Mexican Sun Flower 392432
Mexican Sunflower 392428,392432,392434,392436
Mexican sunflowerweed 392432
Mexican Swamp Cypress 385281
Mexican Tarragon 383097
Mexican Tea 86934,86946,139678,146122,146262
Mexican Teosinte 155878
Mexican Thistle 32429
Mexican Tiger-lily 391627
Mexican Tree Daisy 258195
Mexican Tree Ocotillo 167560
Mexican Tree-daisy 258195
Mexican Trixis 399398
Mexican Tulip Poppy 199407,199408,199409
Mexican Tulip-poppy 199407,199408
Mexican Turk's Cap 243932,243949,286663
Mexican Twinbloom 59715
Mexican Twist 236455
Mexican Vanilla 405010,405017
Mexican Violet 387346
Mexican Viper 246653
Mexican Washington Palm 413307
Mexican Washingtonia 413307
Mexican Weeping Bamboo 276923
Mexican Weeping Pine 300140
Mexican White Pine 299809,300210,300219
Mexican Whorled Milkweed 38012
Mexican Willow 340128
Mexican Wormwood 35842
Mexican Yellow Pine 300140
Mexican Zinnia 418047
Mexican-apple 78175
Mexican-blue 105232
Mexican-hat 326957
Mexican-tea 139678,146262
Mexico Ageratum 11206
Mexico Cypress 114714
Mexico Hickory 77927
Mexico Holly Mahonia 242560
Mexico Nut Pine 299851
Mexico Pine 299809,299851
Mexico Tea 139678
Mexico Weeping Pine 300140
Meyen Bluebells 250898
Meyen Bluegrass 305728
Meyen Duckbeakgrass 209272
Meyen Spapweed 205138
Meyen's Flatsedge 119192
Meyendorff Jurinea 193973
Meyer Bamboo 297344
Meyer Citrus 93631
Meyer Currant 334094
Meyer Juniper 213948
Meyer Lemon 93631
Meyer Lilac 382200,382201
Meyer Peppergrass 225414
Meyer Sedge 75348
Meyer Single-seed Juniper 213948
Meyer Spruce 298358
Meyer Tamarisk 383557
Meyer's Bamboo 297344
Meyer's Clematis 95113
Meyer's Currant 334094
Meyna 252704
Meytea 28309
Mezard 83155
Mezercon 122359
Mezereon 122515,391015
Mezereon Spurge 122515
Mezereon Tree 122515
Mezereum 122515
Mezereum Family 391030
Mezeroon Spurge 122515
Mezoneuron 252741
Mezzettiopsis 252771
Mfukufuku 60035
Mfungu 80381
Mi 241516
Miami Mist 293173
Miami Palmetto 341419
Miami-mist 293173
Mianhuaguo Ginkgo 175823
Mianning Azalea 331238
Mianning Barberry 51916
Mianning Dendrocalamus 20202
Mianning Dragonbamboo 20202
Mianning Hornbeam 77343
Mianning Mengzi Porana 131245
Mianning Monkshood 5351
Mianning Ostryopsis 276848
Mianning Rhododendron 331238

Mianning Tanoak 233316
Mianxian Sedge 76241
Mianyang Snakegourd 396222
Miaofengshan Negundo 411378
Miaoli Dumasia 138937
Miaoshan Cassia 91376
Miaoshan Holly 203627
Miaoshan Machilus Machilus 240643
Miaoshan Maple 3144
Miaoshan Persimmon 132301
Miaotai Maple 3145
Miaray 93633
Mibora 252783
Micangshan Primrose 314933
Mice's Mouth 231183
Michaelmas Crocus 99297
Michaelmas Daisy 39910,40989,322724,380930,398134
Michaelmas-daisy 39910
Michaemas Daisy 40923
Michanovich Chin Cactus 182459
Michaux Oak 324166
Michaux's Birch 53524
Michaux's Blue-eyed Grass 365785
Michaux's Lily 229938
Michaux's Orchid 184008
Michaux's Saxifrage 349634,350032
Michaux's Sedge 75349
Michaux's Stitchwort 255524
Michaux's Sumac 332714
Michay Barberry 51919
Michel Aster 40125
Michel Cypressgrass 119196
Michel Galingale 119196
Michel Sedge 75355
Michelia 252803,252860
Michella-like Hayatagrass 186858
Michelson Willow 343689
Michen 252662
Michigan Lily 229939
Michoaca Pine 300077
Michoacan Pine 300077
Micholitz Alocasia 16517
Micholitz Cycas 115854
Micholitzia 252996
Micken 252662
Mickey Mouse Plant 268261,268273
Mickey-mouse Plant 268261
Micocoulier occidental 80698
Miconia 253003,253007
Miconia-like Heptacodium 192168
Micrandra 253037
Micrechites 203321,253135
Micro Crazyweed 279107
Microanther Aeluropus 8874
Microanther Bluegrass 305729
Microcachrys 253178
Microcarp 253197
Microcarpaea 253197
Microcaryum 253203
Microcephaloid Adina 8186
Microchloa 253266
Microcitrus 253285
Microcordateleaf Maple 2903
Microcos 253361
Microdesmis 253409
Microflorous Crape-myrtle 219959
Microflower Stephania 375890
Microglossa 253450
Microgynoecium 253487,253488
Microhead Elatostema 142741
Microhead Stairweed 142741
Microleaf Fordia 167286
Microlobed Amalocalyx 18562
Micromelum 253626,253628
Micromered Rhododendron 331241
Micromeria 253631
Micrope 253841
Micropetalous Larkspur 124383
Microphyll Cinquefoil 312764
Microphylloua Rockjasmine 23234
Microphyllous Leptodermis 226099
Microphyllous Pea-shrub 72285
Microphyllous Schefflera 350745
Microsisymbrium 253949
Micro-spicated Willow 343691
Microstalk Corydalis 105846
Microstegium 253980,254034
Microstigma 254082
Microtis 254226,254238
Microtoena 254241
Microtooth Cherry-laurel 223123
Micro-tooth Cherry-laurel 223123
Microtropis 254283
Microula 254335
Microula-like Neckfruitgrass 252398
Mid-Asia Sedge 74408
Midday Flower 220523
Midday Marvel 275403
Midday Star 202825
Middayflower 289682
Midden Mylies 59493,86901,86970,285741
Middendorff Birch 53527
Middendorff Day Lily 191309
Middendorff Daylily 191309
Middendorff Stonecrop 294124
Middendorff's Day-lily 191309
Middetendril Pincution 244026
Middle Clearweed 298969
Middle Coldwaterflower 298969
Middle Comfrey 13174
Middle Consound 13174
Middle Ephedra 146192
Middle Fleabane 321554
Middle Hairgrass 126093
Middle Peashrub 72258
Middle Rockfoil 349616
Middle Trefoil 396967
Middlecover Skullcap 355601
Middle-east Tea 68880
Middling Agapetes 10319
Mide China fir 114557
Midge Orchid 173169
Midge Plant 36474
Midget Crab Apple 243657
Midget Crabapple 243657
Midgewort 36474
Midland Elm 401502
Midland Hawthorn 109790
Midland Sand Sedge 118435
Midnight Horror 275403
Midrib Gurjun 133559
Midshipman's Butter 291494
Midsouth China Raspberry 338487
Midsummer Daisy 227533,322724
Midsummer Fairmaid 34536
Midsummer Lily 229789
Midsummer Men 81768,275135,329935,357203
Midsummer Plantainlily 198604
Midsummer Silver 312360
Midsummer Thistle 81382
Midwestern Arrowhead 342319
Midwestern Blue Heart-leaved Aster 380983
Midwestern Tickseed-sunflower 53773
Midwestern White Heart-leaved Aster 40489
Midwestern White-lettuce 313809
Miethke Mahobarnerry 242453
Migao Cinnamon 91380
Mignonette 327793,327866,327896
Mignonette Family 327934
Mignonette Tree 223454
Mignonette Vine 26264,26265
Mignonettevine 26264,26265
Mignonette-vine 26265
Migo Indocalamus 206800
Migo Sweetleaf 381309
Migrescent Memecylon 250028
Migu Holly 204059
Migum 155623
Migwort 36474
Mikania 254414
Mikinor Epimedium 147026
Mikun Barberry 51925
Milashan Rush 213290
Mild Water pepper 291792
Mild Water-pepper 291792
Milder 86970
Mild-leaved Barberry 52128
Mildweed 37811
Mile 29327

Mile a Minute Vine 207659
Mile Groundsel 359076
Mile Swertia 380252
Mile Tree 79161
Mile-a-minute 162509,207659
Mile-a-minute Plant 162509
Mile-a-minute Vine 162512,218721,309564
Mile-a-minute Weed 309564
Mile-a-minute-weed 309564
Miles Iris 208712
Milfoil 3978,3981,261337
Milfoil Yarrow 3978
Milin Azalea 331185
Milin Buttercup 326057
Milin Chickweed 374963
Milin Corydalis 106115
Milin Gentian 173619
Milin Jerusalemsage 295149
Milin Larkspur 124578
Milin Milkvetch 42718
Milin Monkshood 5417
Milin Poplar 311383
Milin Rhododendron 331185
Milin Rockfoil 349991
Milin Spapweed 205185
Milin Touch-me-not 205185
Milin Violet 410196
Miling Jerusalemsage 295149
Miling Rockfoil 349991
Milion 114189
Military Orchid 272460,273541
Military Orchis 273541
Miliumlike Fimbristylis 166402
Miliumlike Fluttergrass 166402
Miliusa 254546
Milk Barrel 158625
Milk Bush 159975
Milk Chive 15310
Milk Fig-tree 164671
Milk Flower 363673
Milk Gowan 384714
Milk Maidens 315082,374916
Milk Onion 15310
Milk Parsley 292966,357780
Milk Pea 169654
Milk Purslane 85770,159102,159286,159911
Milk Rhododendron 330754
Milk Spurge 159286,159634
Milk the Cows 401711
Milk Thistle 364353,364360,368625,368771
Milk Tree 18055,108153,165089
Milk Vetch 41906,42417
Milkaster 169756,169790
Milkbean 169645,169660
Milkberry 87661
Milkbush 159975,381475
Milkcans 374916
Milkflower Cotoneaster 107520
Milk-flower Photinia 295716
Milk-fruit Actinidia 6546
Milk-fruit Kiwifruit 6546
Milkgrass 404439
Milkgreens 245777
Milkies Milkgirl 72934
Milking Maids 72934
Milkleaf Azalea 330754
Milklettuce 259746,259772
Milkline Squarebamboo 87565
Milkmaid's Eye 407072
Milkmaid's Eyes 407072
Milkmaids 14964,23925,68713,72934,
237539,308454,349419,363673
Milkpans 374916
Milkpea 169645,169666
Milk-purslane 85770,159286
Milksile 72934
Milkstools 64345
Milktree 255260
Milkvetch 41906,42699
Milk-vetch 41906,42137,42417
Milkvetch Huangchi 42699
Milkvetch-like Crazyweed 279212
Milkweed 29468,29473,37811,38135,
38147,158385,158692,159027,159557,
292966,368771
Milkweed Family 37806
Milkweed Gentian 173264
Milkweed Rubber 37811
Milkwhite Euphorbia 159204
Milk-white Milkvetch 42391
Milkwood 255260,255312
Milkwort 70271,159027,159557,307891,
307969,307983,308006,308173,308454,
368771
Milkwort Boronia 57276
Milkwort Corydalis 106278
Milkwort Family 308474
Milkwortlike Woodbetony 287527
Milky Bellflower 70107
Milky Dashel 368771
Milky Dassel 368771
Milky Dicel 368771
Milky Dickle 368771
Milky Disle 368771
Milky Draba 137063
Milky Maidens 72934
Milky Malden 72934
Milky Mangrove 161639
Milky Milkvetch 42391
Milky Rhododendron 330754,331034
Milky Stonegarlic 239255
Milky Tassel 368771
Milky Thrissel 364360
Milky White Indigo 47856,47857
Milky Whitlowgrass 137063
Milkyellow Azalea 331034
Milky-flowered Violet 410165
Milkywhite Everlasting 21586
Milkywhite Pearleverlasting 21586
Milkywhiteflower Columbine 30049
Milky-yellow Primrose 314101
Millboro Leather-flower 95448
Mille Flower 3978
Millen Swertia 380252
Miller 's Star 374916
Miller Gladiolus 176375
Miller's Delight 81020
Miller's Jerkin 405788
Millet 254496,281274,281916,361680,
369590,369600,369720
Millet Grass 254496
Millet Woodrush 238678
Milletformis Singnalgrass 58185
Milletgrass 254496
Millet-grass 254496,254515
Milletgrasslike Twinballgrass 209094
Millet-like Scargrass 203319
Millett Adinandra 8256
Millettia 254596
Millettia Vine 66643
Milliet Wildtung 243392
Millingtonia 254892
Million 114288
Million Bells 66531,66535
Millstream Palm 234163
Milner Flower 146724
Miltoniopsis 254941
Milula 254944,254945
Mimeng Summerlilic 62134
Mimicry Plant 304340,304344
Mimosa 1165,13578,254979
Mimosa Bush 1219,1395
Mimosa Family 255141
Mimosa Grevill 180619
Mimosa Thorn 1324
Mimosa Tree 13578
Mimosae-leaved Jacaranda 211239
Mimosalike Caesalpinia 65039
Mimosa-tree 13578
Mimose-leaf Jacaranda 211239
Mimoselike Caesalpinia 65039
Mimoselike Rosewood 121760
Mimose-like Rosewood 121760
Mimulicalyx 255148
Mimulus 255184
Mimusops 255260
Min Fir 455
Mina Climber 255398
Mina Lobata 255398
Minaret-flower 224585
Minaster 175454

Mindanao Gum 155557
Mind-your-own-business 367771,367772
Miner's Lettuce 94349
Miner's Tea 146224
Minfeng Reaumuria 327219
Minfeng Redsandplant 327219
Ming Aralia 310199,310237
Ming Asparagus 39171
Mingal Bamboo 262761
Mingwort 35090
Minhe Milkvetch 42722
Mini Azalea 331584
Mini Beccarinda 49397
Mini Cattail 401128
Mini Childgrass 340356
Mini Cowpea 408969
Mini Crazyweed 279107
Mini Eria 148750
Mini Gasteria 171705
Mini Gentian 173637
Mini Hairorchis 148750
Mini Marguerite 246665
Mini Nasturtium 399604
Mini Oreocharis 273873
Mini Petrocosmea 292571
Mini Primrose 314838
Mini Protea 315954
Mini Rattlesnake Plantain 179718
Mini Rhynchosia 333323
Mini Rockfoil 349757
Mini Rungia 340356
Mini Sandwort 32075
Mini Spotleaf-orchis 179718
Mini Stonebutterfly 292571
Mini Themeda 389357
Mini Woodbetony 287432
Minianther Eulalia 156473
Minianther Goldquitch 156473
Minianther Smallreed 127280
Miniature Beefsteak Plant 259284
Miniature Beefsteakplant 259284
Miniature Caladium 65229
Miniature Date 295484
Miniature Date Palm 295484
Miniature Date-palm 295484
Miniature Flatsedge 119422
Miniature Grape Ivy 92977
Miniature Holly 243523
Miniature Huntsman Horn 347163
Miniature Lupine 238436
Miniature Marigold 383100
Miniature Pine Tree 109452
Miniature Pond-lily Begonia 49938
Miniature Rhododendron 332278
Miniature Spearwort 325864
Miniature Wax Plant 198826
Miniball Milkwort 308077
Miniebush 250531
Miniflower Azalea 331249
Miniflower Corydalis 106159
Miniflower Kitefruit 196842
Miniflower Milkwort 307927
Miniflower Nanmu 295384
Miniflower Tea 69368
Minifruit Aspidopterys 39684
Minigrass 253266
Minihair Gyrocheilos 183319
Minihook Chirita 87925
Minileaf Camellia 69363
Minileaf Glutinousmass 179494
Minileaf Honeysuckle 235955
Minilip Woodbetony 287433
Minilobe Machilus 240648
Minima Arborvitae 390612
Minimal Oak 324167
Miniorchis 254192,254196
Minisepal Spapweed 205149
Minispotted Chirita 87926
Ministalkgrass 71602,71608
Ministylenettle 85106,85107
Minitooth Milkvetch 42437,42725
Minitooth Pondweed 312170
Miniwhitedaisy 227452
Minjiang Azalea 330893
Minjiang Barberry 52058
Minjiang St. John's Wort 201921
Minjiang Stonecrop 269606
Minjiang Sweetvetch 187871
Minjiang Willow 343696
Minjiri 360481
Minnesota Dwarf Trout-lily 154939
Minnesota Pure White Mockorange 294580
Minnesota Wild-rye 144292
Minnie Cedar 414066
Minnie Minnies 765
Minnieritchie 1665
Minnieroot Ruellia 339828
Minor Begonia 50082
Minor Caleana 66370
Minor Date Palm 295468
Minor Eastern Arborvitae 390615
Minor Fieldcitron 400540
Minor Habenaria 183871
Minor Hardgrass 385817
Minor Melianthus 248941
Minor Nasturtium 399604
Minor Platanthera 302439
Minor Turpinia 400540
Minorca Box 64240
Minor-leaf Largeflower Deutzia 126960
Minorrose 85646
Minqin Sedge 75377
Minqin Spunsilksage 360854
Minshan Barberry 51759
Minshan Dragonhead 137640
Minshan Kobresia 217232
Minshan Milkvetch 42724
Minshan Mono Maple 3199
Minshull Crab 252305
Mint 250287,250291,250362,250370
Mint Bush 144096,315595,315598,315602,315605,315608
Mint Family 218697
Mint Geranium 383712
Mint of Life 250291
Mint Shrub 143958,144096
Mintbush 315595
Mint-bush 144096,315595
Mintdrop 248411
Mint-geranium 383712
Mintleaf 303478
Mintleaf Bee Balm 257169
Mint-like Pogostemon 306988
Mint-like Spinestemon 306988
Mint-musk 144096
Mintodorous Brachyactis 58264
Mintweed 345340
Minuartia 255435
Minuartie de Dawson 255458
Minuartwort 255435,255565
Minute Duckweed 224385
Minute Haworthia 186301
Minute Honeysuckle 235953
Minute Primrose 314650
Minute Sedge 75378
Minute Toadflex 231046
Minute Woodbetony 287432
Minuteflower Microsisymbrium 203150,253956
Minuteflower Rhododendron 331249
Minuteflower Smallgarliccress 203150,253956
Minute-flowered Rhododendron 331249
Minute-flowered Secamone 356303
Minuteleaf Leptospermum 226465
Minute-leaved Honeysuckle 235955
Minutelip Woodbetony 287433
Minute-lobed Machilus 240648
Minutely Fringed Rhododendron 330871
Minutely-serrate Hornbeam 77346
Minutepuberula Hedyotis 187628
Minute-serrated Hornbeam 77314
Minute-stalk Gyala Barberry 51691
Minute-veined Holly 204371
Minuteyellowflower Iris 208715
Minutserrate Hornbeam 77346
Minwan Anise-tree 204562
Minwan Eightangle 204562
Minwan Illicium 204562
Minxian Sedge 75383
Minxian Spurge 159375

Min-Yue Groundsel 360104
Mioga Ginger 418002
Mioga Zinger 418002
Mip 59575
Miquel Elder 345709
Miquel Lime 391773
Miquel Linden 391773
Miquel Macrozamia 241457
Miquel's Mazus 247008
Miquel's Racemose Fig 165542
Miquelberry 172111
Mir Violet 410246
Miraa 79395
Mirabelle 316294,316394,316470
Mirabelle Plum 316394
Miracle Tree 413748
Miraculous Berry 381982
Miraculous Fruit 388921
Mirbeck Oak 323732
Mirbeck's Oak 323732
Mirbel Star Cactus 43527
Mirbelia 255772
Mire Blob 68189
Miriti Palm 246672
Miro 316086
Mirror Bush 103798
Mirror of Venus 272502
Mirror Orchid 272502
Mirror Plant 103798
Mirror Sedge 75769
Mirror-of-venus 272502
Mirror-orchid 272502
Mirrot 123141
Mis of Peru 272897
Misceldin 410960
Miscelto 410960
Mischievous Jack 374968
Mischobulbum 255933,383127
Mischocarp 255943,255954
Mischocarpus 255943
Mishmi Litse 233993
Misle 410960
Mislin-bush 410960
Miss Harp Rose 336311
Miss Jessop's Variety Rosemary 337183
Miss Kim Lilac 382259
Miss Modesty 50825,410320
Miss Scenty 410320
Miss Willmott's Lily 229831
Miss Willmott's Rose 337022
Miss Wilmott's Ghost 154317
Missanda 154973
Missel 410960
Misselto 410960
Misseltoe 410960
Misseltow 410960
Mission Bells 168355,168444
Mission Blue Spruce 298410
Mission Cactus 272891
Mission Grass 289205,289208
Mission Pricklypear 272891
Mississippi Arrowhead 342383
Mississippi Hackberry 80673
Mississippi Hawberry 109612
Mississippi Penstemon 289337
Mississippi Spangletop 226021
Mississippi Valley Goldentop 161088
Mississippi-valley Loosestrife 239679
Missletoe 410960
Missouri Bladderpod 297772
Missouri Cob Cactus 155215
Missouri Coneflower 339587
Missouri Currant 333919,334127
Missouri Dewberry 338839
Missouri Flag 208718
Missouri Foxtail Cactus 107036
Missouri Goldenrod 368240,368242
Missouri Gooseberry 333916,334099
Missouri Gourd 114278
Missouri Grape 411834
Missouri Iris 208718
Missouri Lily 207489
Missouri Milkvetch 42732
Missouri Orange Coneflower 339587
Missouri Pigweed 87090
Missouri Primrose 269464
Missouri River Willow 343343,344010
Missouri Rock Cress 30358
Missouri Rock-cress 30358
Missouri Spurge 85775
Missouri Viburnum 408079
Missouri Violet 410260
Missouri Water Lily 267632
Missy-Moosey 323083
Mist Birdlingbran 87251
Mist Flower 11171,101955
Mistassini Primrose 314652
Misteli Mistil 97060
Mistflower 11154,101955
Mist-flower 101955
Mistletoe 295580,295581,295583,295586,
410959,410960
Mistletoe Cactus 329683,329684,329689
Mistletoe Family 236503,410934
Mistletoe Fig 164900
Mistmaiden 335855
Mit 59575
Mita-nimbu 93536
Mitchell Grass 43316,43317
Mitchell's Caper 71808
Miterwort 256010,256015,391522
Mithridate Mustard 225315,390213
Mithridate Pepperwort 225315
Mithridatum 390213
Mitnan 391006
Mitracarpus 256082
Mitragyna 256107
Mitral Spindle-tree 157722
Mitraria 256142
Mitrasacme 256147
Mitrastemon 256175
Mitre Cress 260577
Mitredfruit Hemsleia 191948
Mitredfruit Snowgall 191948
Mitre-flower Aloe 17038
Mitreola 256198,256214
Mitrephora 256221
Mitrewort 256010,256015
Mitriform Rhododendron 331254
Mitsai Mockorange 294475
Mitsuba 113872
Mitsu-ba 113879
Mitsuharu 93640
Mitsumata 141470
Mitute Monoflorcress 287953
Mitute Pegaeophyton 287953
Mitzuba 113872
Miuntely Toothed Photinia 295758
Mix Crazyweed 278799
Mix Milkvetch 42207
Mixed Gaoliang 369593
Mixed Giant Timber Bamboo 297229
Mixed Helicia 189944
Mixed Herniary 193009
Mixed Sorghum 369593
Mixed-flower 298054
Mixi Beautyberry 66872
Mixi Purplepearl 66872
Miximoewicz Pearlwort 342259
Mix-together Milkvetch 42207
Miyabe Cinquefoil 312782
Miyabe Daphne 122519
Miyabe Maple 3149
Miyabe Sedge 75396
Miyabe Spiraea 372012
Miyabe Willow 343702
Miyagi Camellia 69370
Miyaluo Milkvetch 42732
Miyama Cherry 83255
Miyi Azalea 331256
Miyi Bamboo 47342
Miyi Loosestrife 239740
Miyi Microtoena 254257
Miyi Rhododendron 331256
Miyi Rose 336760
Miyi Rush 213295
Miyi Stephania 375892
Miyi Whitepearl 172112
Miyi Wintergreen 172112
Mizzeltoe 410960
Mkani Fat 14899

Mlanje Cedar 61339
Mlanji Widdringtonia 414066
Mlokose Peony 280236
Mmurrain Berries 61497
Mmyrtle 261168
Mnesithea 256331
Mnleery Tea 4005
Moa 241516
Moab Woody-aster 415852
Moabi 46607
Mobola 132351,284211
Mobola Persimmon 132351
Mobola Plum 284211
Mocassin Flower 120290
Moccasin Flower 120287
Moccasin-flower 120287,120290
Moch Orange 301433
Mochi-yu 93499
Mochorange 301433
Mock Azalea 250505,250513
Mock Bishop's Weed 321031,321032,
321033
Mock Bishop's-weed 321031
Mock Chervil 28044,261585,350425
Mock Cucumber 140526,140527
Mock Cypress 217324,217361
Mock Gilliflower 346434
Mock Goldenweed 265638,375786
Mock Lemon 11300
Mock Leopardbane 34704
Mock Liquorice 169935
Mock Locust 19994
Mock Mesquite 66669
Mock Olive 266783
Mock Orange 240828,294407,294426,
294468,294479,294485,294493,294517,
294556,301413,379304
Mock Orenge 260158
Mock Oyster 394327
Mock Pennyroyal 187181,187184
Mock Plane 3462
Mock Poppy 379220
Mock Privet 294761,294772
Mock Saffron 77748
Mock Strawberry 138790
Mock-cucumber 140526,140527
Mock-eel Root 90924
Mocker Nut 77934
Mockernut 77934
Mockernut Hickory 77934
Mocketprivet-like Oak 324283
Mocknut Hickory 77934
Mockorange 294407,294408,294468,
294471,294479,294560
Mock-orange 294407,294426,294493,
301413
Mock-orange Family 200160,294404
Mockstrawberry 138790
Mock-strawberry 138790,138796
Mockstrawberry Cinquefoil 312446
Mocucca Albizia 162473
Mocucca Siris 162473
Moderate Asiabell 98401
Moderate Bellflower 70173
Moderate Daphne 122520
Moderate Fescue 164083
Moderate Fescuegrass 164083
Moderate Honeysuckle 235964
Moderate Ligusticum 229367
Modern Rose 336640
Modest Licuala Palm 228751
Modest Licualapalm 228751
Modest Maiden 274823,410320
Modest Prickle Grass 113319
Modest Smallleaf Bedstraw 170702
Modest Spapweed 205151
Modest Touch-me-not 205151
Modingshan Barberry 52117
Modjadjis Palm 145242
Modoc Cypress 114662
Modoc Hawksbeard 110900
Modog Monkshood 5172
Modog Ophiorrhiza 272252
Moehringia 256441,256456
Moehringie 256441
Moellendorff Cardiandra 73136
Moellendorff Crazyweed 279018
Moellendorff Figwort 355191
Moellendorff Milkvetch 42733
Mogador Acacia 1266
Mogdad Coffee 360463
Moghat Root 177402
Mogollon Ceanothus 79945
Mogollon Death Camas 417917
Mogollon Larkspur 124225
Mogue Tobin 89704
Mogvurd 36474
Mohave Aster 39916
Mohave Buck Brush 79980
Mohave Buckbrush 79980
Mohave Buckwheat 152287
Mohave Coreopsis 104464
Mohave Firethorn 322449
Mohave Groundsel 359499
Mohave Loco 42737
Mohave Orach 44321
Mohave Pennyroyal 257195
Mohave Rose 336761
Mohave Rubberbrush 90531
Mohave Sage 345227
Mohave Saltbush 44654
Mohave Sandwort 32042
Mohave Sulphur Flower 152593
Mohave Woolly Sunflower 152825
Mohave Yucca 416646
Mohawk Viburnum 407655
Mohilevi Mallow 243801
Mohintli 214817
Mohlenbrock's Sedge 118971
Mohr Oak 324172
Mohr's Coneflower 339588
Mohr's Thoroughwort 158237
Mohri Eupatorium 158237
Mohur 123801
Mohwa 241516
Moidenke 216085
Moiken 252662
Moist Aster 40764
Moist Bank Pimpernel 231514
Moist Rockfoil 349604
Moistloving Low Torularia 264617
Moistureloving Bittercress 72828
Moisture-loving Fargesia 162700
Moitch Eucalyptus 155727
Moithern 26768
Mojave 150362
Mojave Aster 240463
Mojave Deer Vetch 237741
Mojave Desert California Buckwheat 152062
Mojave Desert Star 257832,257834
Mojave Desertstar 257834
Mojave Eagle-claw Cactus 354311
Mojave Indian-lettuce 94345
Mojave Lupine 238487
Mojave Mound Cactus 140332
Mojave Neststraw 379162
Mojave Pincushion 84511
Mojave Prickly Pear 273003
Mojave Pricklypear 273016
Mojave Rabbitbrush 150348
Mojave Sagebrush 36404
Mojave Sand Verbena 708
Mojave Spineflower 89115
Mojave Thistle 92204
Mojave Woody-aster 415854
Mojave Yucca 416646
Mojiang Flemingia 166854
Mojiang Snailseed 97934
Mokelumne Hill Wild Buckwheat 152248
Molasses Grass 249303
Molassesgrass 249303
Molasses-grass 249303
Molayne 405788
Moldavian Balm 137613
Moldavian Dragon's-head 137613
Moldavian Dragonhead 137613
Molded Wax Agave 139974
Mole Plant 159222,159252
Mole Weed 159222,159252
Moleery Tea 3978
Moleery-tea 3978,4005

Moleplant 159222
Mole-plant 159222
Moleroot Lily 229887
Mole-tree 159222
Moleweed 159222
Molikurikur 93865
Molinia 256597
Molishan Willow 343721
Moll Blob 68189,72934
Moll-blob 68189,72934
Molle 350985,350986
Moll-o'-the-woods 23925,37015
Molloy Red Box 155622
Mollugo Family 256684
Molly Blob 68189
Molly's Rhododendron 331274
Mollyblobs 68189
Molt Elm 401550
Moltebeere 338243
Moltke Linden 391780
Moltkia 256746
Molucca Albizzia 162473
Molucca Balm 256758,256761
Molucca Cypholophus 119969
Molucca Duabanga 138724
Molucca Licuala Palm 228764
Molucca Mallotus 243395
Molucca Palm 332377
Moluccan Bramble 339055
Moluccan Cabbage 300956
Moly 15480,15861
Mombin 88691,372459,372467
Momi Fir 365
Momiyama Duckbeakgrass 209274
Momordica 256780
Monadelphus Milkvetch 42740
Monal Bamboo 297328
Monarch Birch 53520
Monarch of The East 348036
Monarch of the Veldt 346304
Monarch Redstem 19566
Monarch Rosemallow 195132
Monarch-of-the-east 348036
Monarch-of-the-veld 31221
Monarch-of-the-veldt 346304
Monarda 257157
Monastery Bells 97757
Monbeig Catchfly 363769
Monbeig Deutzia 127016
Monbeig Dogwood 380465
Monbeig Euonymus 157725
Monbeig Hornbeam 77350
Monbeig Indigo 206270
Monbeig Milkvetch 42744
Monbeig Mountainash 369465
Monbeig Mountain-ash 369465
Monbeig Spindle-tree 157725
Monbeig Violet 409714
Monbeig Woodbetony 287435
Monbin 372477
Mondell Pine 299828
Mondo Grass 272049,272090
Mondurup Bell 122783
Moneses 257371
Money 220790,329521,356468
Money Bags 72038,205175
Money Bush 360426
Money Flower 238371
Money Grass 329521
Money Maple 133596
Money Penny 401711
Money Plant 109229,147350,238369,
238371,390213
Money Pockets 238371
Money Tree 109229,155710,390213
Money Wild Buckwheat 152332
Money-in-both-pockets 2837,11549,238371
Money-in-every-pocket 238371
Money-in-your-pocket 238371
Money-leaved Barberry 51968
Moneywort 21427,238371,239755
Mongil Falseoat 398559
Mongilian Trisetum 398559
Mongol Adonis 8371
Mongol Almond 20928
Mongol Bluebeard 78023
Mongol Bluegrass 305733
Mongol Broomrape 275142
Mongol Cistanche 93068
Mongol Cocklebur 415031
Mongol Cotoneaster 107573
Mongol Cottonspine 312318
Mongol Cymbaria 116750
Mongol Daisy 124825
Mongol Dandelion 384681
Mongol Drybirdbean 87273
Mongol Eight Trasure 200791
Mongol Ephedra 146155
Mongol Euphorbia 159387
Mongol Fescue 163886
Mongol Helictotrichon 190175
Mongol Hogfennel 292986
Mongol Horseorchis 215356
Mongol Jerusalemsage 295150
Mongol Jurinea 214133
Mongol Leek 15489
Mongol Linden 391781
Mongol Milkvetch 42704
Mongol Mulberry 259165
Mongol Needlegrass 376852
Mongol Oak 324173
Mongol Pulsatilla 321659
Mongol Pygmyweed 391939
Mongol Russianthistle 344573
Mongol Scotch Pine 300240
Mongol Serpentroot 354904
Mongol Shorttonguedaisy 58044
Mongol Spiraea 372018
Mongol Spunsilksage 360855
Mongol Sugarmustard 154451
Mongol Tamarisk 383559
Mongol Tickseed 104814
Mongol Violet 410269
Mongol Wheatgrass 11796
Mongol Whitlowgrass 137115
Mongol Wildsenna 389543
Mongol Windhairdaisy 348538
MongolArrowwood 407954
Mongol-Daimyo Oak 324198
Mongolia Chesneya 87273
Mongolia Motherwort 225003
Mongolia Sedge 75298
Mongolian Ammopiptanthus 19777
Mongolian Batrachium 48927
Mongolian Bigfruit Elm 401556
Mongolian Bluebeard 78023
Mongolian Blue-beard 78023
Mongolian Bluegrass 305733
Mongolian Brachanthemum 58044
Mongolian Broomrape 275142
Mongolian Calligonum 67050
Mongolian Cherry 83202
Mongolian Chesniella 87273
Mongolian Cocklebur 415031
Mongolian Cotoneaster 107573
Mongolian Cymbaria 116750
Mongolian Dandelion 384681
Mongolian Dendranthema 124825
Mongolian Deutzia 127032
Mongolian Draba 137115
Mongolian Ephedra 146155
Mongolian Feathergrass 376852
Mongolian Fescuegrass 163886
Mongolian Goldenray 229109
Mongolian Helictotrichon 190175
Mongolian Jerusalemsage 295150
Mongolian Jurinea 214133
Mongolian Kalimeris 215356
Mongolian Kneejujube 67050
Mongolian Lime 391781
Mongolian Linden 391781
Mongolian Milkvetch 42704
Mongolian Mosquitotrap 117606
Mongolian Mulberry 259165
Mongolian Oak 324173
Mongolian Onion 15489
Mongolian Pear 323330
Mongolian Piptanthus 19777
Mongolian Poplar 311509
Mongolian Potaninia 312318
Mongolian Pride-of-Rochester 127032

Mongolian Psammochloa 317004
Mongolian Ptilagrostis 321016
Mongolian Pulsatilla 321659
Mongolian Russianthistle 344573
Mongolian Sagebrush 35916
Mongolian Sandthorn 196763
Mongolian Saussurea 348538
Mongolian Scotch Pine 300240
Mongolian Seabuckthorn 196763
Mongolian Serpentroot 354904
Mongolian Skullcap 355614
Mongolian Snakegourd 396210
Mongolian Spiraea 372018
Mongolian Spruce 298360
Mongolian Stonecrop 200791
Mongolian Swallowwort 117606
Mongolian Sweetvetch 188007
Mongolian Tamarisk 383559
Mongolian Tetraena 387116
Mongolian Thermopsis 389543
Mongolian Thyme 391284
Mongolian Tickseed 104814
Mongolian Tillaea 391939
Mongolian Tugarinovia 400036
Mongolian Viburnum 407954
Mongolian Violet 410269
Mongolian Wheatgrass 11796
Mongolian Willow 343706
Mongolian Wormwood 35916
Mongolian-Daimyo Oak 324198
Mongolsandy Sagebrush 35739
Mongonogo 351005
Mongrel Larkspur 124301
Mongut Milkvetch 42746
Moniliform Dendrobium 125257
Moniliform Gentian 173643
Moniliform Ligusticum 229363
Moniliform Skullcap 355615
Monilla 401829
Monimopetalum 257432
Monjet Mango 244403
Monk Orchid 269222
Monk's Basil 268614
Monk's Beard 90901
Monk's Belly Bamboo 297203
Monk's Cowl 5442
Monk's Herb 250439
Monk's Hood 5442,43526
Monk's Pepper 411189
Monk's Pepper Tree 411189
Monk's Rhubarb 339914,340151,340178
Monk's-hood 5014,5442
Monk's-rhubarb 339914
Monkey 199465
Monkey Apple 25572,25828,25852,25876,90085,378633,378888
Monkey Balls 378888
Monkey Bells 68189
Monkey Blossom 255210
Monkey Bread 7022
Monkey Bread Tree 7022,7029
Monkey Chops 28617,176824
Monkey Cinnamon 91277
Monkey Cocoa 389401
Monkey Comb 301169
Monkey Cup 255210
Monkey Dinner-bell 199465
Monkey Earrings Pea 301117
Monkey Face 28617,231183,410677
Monkey Face Tree 243427
Monkey Faces 28617,231183,410677
Monkey Fiddle 159975
Monkey Flower 28617,176824,231183,255184,255186,255210,255237
Monkey Guava 132299
Monkey Hat 399601
Monkey Jack 36928,36945,255210
Monkey Jaw 116736
Monkey Jaws 116736
Monkey Kola 99181
Monkey Mint 248193
Monkey Mouth 28617,116736
Monkey Musk 28617,255184,255210,255218
Monkey Nose 28617
Monkey Noses 28617
Monkey Nut 30498,305334
Monkey Orchid 272487,273633
Monkey Plant 231183,255210,339762
Monkey Plum 132275,415558,415560
Monkey Pot 76832,223764
Monkey Puzzle 30830,30832,30835
Monkey Puzzle Tree 30830,30832
Monkey Star-apple 90085
Monkey Vine 208016
Monkey's Ear-rings 301117,301128
Monkey's-comb 54303
Monkey's-hand 87809
Monkeyblotch Azalea 330692
Monkeybread 7022
Monkey-bread 7022
Monkey-bread Tree 7022
Monkeyface Tanoak 233117
Monkey-face Tree 243427
Monkeyfaced Pansy 4070
Monkey-faced Pansy 4082
Monkeyflower 255184,255197,255237,255246
Monkey-flower 255184,255237
Monkeyhead Azalea 331832
Monkeyhead Rhododendron 331832
Monkey-head Rhododendron 331832
Monkey-nut 30498
Monkeypod 301134,345525
Monkey-pot 223778
Monkeypot Tree 223764
Monkey-pot Tree 223764,223771
Monkey-puzzle 30830,30832
Monkey-puzzle Araucaria 30832
Monkey-puzzle Family 30857
Monkey-puzzle Tree 30832
Monkeys Hood 5442,5676
Monkeys Rhododendron 331832
Monkeysoap 145994
Monkeytail Plant 12599
Monkeytlower 255210
Monkflower 79329
Monks' Pepper Tree 411189
Monks' Rhubarb 339914
Monkshood 5014,5100,5442,5654,43509
Monkshoodleaf Boykinia 58019
Monkshoodvine 20284
Monkshood-vine 20284
Monkswort 266593
Monnaghs Heather 145073
Monnier's Snowparsley 97719
Monnocs Heather 145073
Mono Maple 3154
Mono Wild Buckwheat 151813
Monocelastrus 257503
Monocephal Sandwort 32097
Monochasma 257537
Monochoria 257564,257572
Monocladus 56951,257595
Monodora 257630
Monoear Spatholobus 370423
Monoflorcress 287948,287957
Monoflower 135136,135137
Monoflower Bedstraw 170376
Monoflower Fenugreek 397254
Monoflower Honeysuckle 236130
Monoflower Leafflower 296675
Monoflower Panisea 282418
Monoflower Pertybush 292083
Monoflower Underleaf Pearl 296675
Monofruit Fissistigma 166678
Monofruit Tanoak 233350
Monogynous Alphonsea 17628
Monohead Elegant Rush 213017
Monoleaf Munronia 259898
Monoleaf Pholidota 295527
Monoleaf Pyrola 322857
Monomeria 257708
Monoorchid Herminium 192862
Monophyllous Rockvine 387812
Monoplant Alhagi 14638
Monopolize Orchis 116823
Monopterous Russianthistle 344635
Monopyle 257837
Monos Plum 261046
Monoseed Sandwort 32084

Monoseed Sawgrass 245484
Monosomic Camellia 69373
Monospermous Argyreia 32640
Monospermous Ephedra 146219
Monospike Biflorgrass 128574
Monospike Clethra 96526
Monospike Dichanthium 128574
Monospke Bulrush 353626
Monostachys Clethra 96526
Monostamen Patrinia 285841
Monostyle St. John's Wort 202021
Monotropa Family 258049
Monovein Rhubarb 329413
Monowing Russianthistle 344635
Monox 145073
Monox Heather 145073
Monro's Ragwort 58475
Monrovia Coffee 98941
Monsonia 258096
Monster Flower 325061
Monstera 258162,258168
Monsterflower 325061
Monsterflower Family 325065
Monstrous Goldenray 229128
Monstruocalamus 258183
Montain Peristrophe 291168
Montain Rose 315897
Montain Tea 172135
Montaine Giantfennel 163577
Montane Anadendrm 21313
Montane Aphyllorchis 29216
Montane Arecapalm 31690
Montane Azalea 331273,331411
Montane Baliosperm 46970
Montane Baliospermum 46970
Montane Bedstraw 170570
Montane Ceropegia 84176
Montane Cochlianthus 98067
Montane Coneflower 339590
Montane Conehead 320133
Montane Cyphotheca 120278
Montane Dandelion 384763
Montane Draba 376389
Montane Falsepimpernel 231546
Montane Fragrantbamboo 87642
Montane Germander 388154,388177
Montane Helicia 189917
Montane Kudzuvine 321453
Montane Lady's Slipper 120415
Montane Larch 221917
Montane Larkspur 124432
Montane Macaranga 240297
Montane Machilus 240663
Montane Meadowrue 388703
Montane Milkvetch 42752
Montane Monkshood 5422
Montane Motherweed 231546
Montane Pendentlamp 84176
Montane Pipsissewa 87490
Montane Platanthera 302443
Montane Poppy 282683
Montane Pratia 313562
Montane Rabdosia 209788
Montane Ramontchi 166782
Montane Rattan Palm 65750
Montane Rockfoil 349728,349913
Montane Sainfoin 271234
Montane Sandwort 32081
Montane Saurauia 347969
Montane Sideritis 362838
Montane Skullcap 355637
Montane Spicebush 231429
Montane Spice-bush 231429
Montane Stonecrop 356784
Montane Sweetgum 232558
Montane Sweetvetch 188102
Montane Tournefortia 393261
Montane Turpinia 400530
Montane Waxplant 198896
Montane Whitlowgrass 376389
Montane Willow 343814
Montane Windhairdaisy 348541
Montaneleader Azalea 331406
Montaneous Glycosmis 177841
Montanoa 258191
Montanous Baliospermum 46970
Montanous Bluebell 115351
Montanous Bluebellflower 115351
Montanous Muntries 218397
Montbretia 111455,111464,399043
Monte Baldo Anemone 23715
Monterey Cedar 114723
Monterey Cypress 114723
Monterey Manzanita 31118
Monterey Pine 300264
Monterey Spineflower 89103
Monterrey Cypress 114723
Monterrey Pine 300173
Montery Ceanothus 79970
Montevideo Escallonia 155143
Montezuma Bald Cypress 385281
Montezuma Baldcypress 385281
Montezuma Bald-cypress 385281
Montezuma Cypress 385262,385281
Montezuma Pine 300081
Monthly Rose 336514
Monthly Rosebush 336485
Montia 258237
Monticule Rhododendron 331273
Montpelier Broom 173015
Montpelier Cistus 93180
Montpelier Rock Rose 93180
Montpellier Maple 3206
Montpellier Pink 127772
Montro Rhododendron 331274
Monument Plant 167878,380146
Monuseed Sarsaparilla 250221
Monyul Mahonia 242592
Mooley Plum 277638
Mooli 326622
Moon Beam 154162
Moon Cactus 357735
Moon Carrot 361526
Moon Cereus 357735
Moon Chilosehista 87460
Moon Daisy 227455,227533
Moon Flower 23925,67734,109857,207570,227533,238371,374916
Moon Penny 227533
Moon Plant 346995
Moon Seed 250221
Moon Trefoil 247249
Moon's Eye 227533
Moon's Eyes 227533
Moonah 248101
Mooncactus 185922
Mooncarrot 361526
Moon-carrot 361456,361526
Moon-daisy 227533
Moonflower 67734,67736,207570
Moon-flower Vine 207570
Moonlight 28044
Moonlight Broom 121006
Moonlight Cactus 357735
Moonlight Holly 203554
Moonogs 145073,404051
Moons 238371
Moonseed 250218,250221
Moonseed Family 250217
Moonseed Sarsaparilla 250221
Moonseed Vine 250218
Moonshine 21596
Moonshine Yarrow 3913
Moonstones 279635,279639
Moon-toby 227533
Moonvine 67736,207570
Moonwort 22445,227533,238369,238371,250221,374916
Moon-wort 22445,367762
Moor 59575,67456
Moor Everlasting 26385
Moor Grass 256597,256601,361611
Moor Heath 150201
Moor King 287657
Moor Myrtle 261162
Moor Rush 213476,213478
Moor Whin 172905
Moorberry 278366,404027
Moor-berry 278366
Moorcroft Sage 345230
Moorcroft Sagebrush 35957

Moorcroft Sedge 75416
Moorcroft Sophora 369078
Moorcroft's Wormwood 35957
Moore Agapetes 10344
Moore Legnephora 224052
Moore Macrozamia 241458
Moore Mosquitotrap 117610
Moore Robiquetia 335037
Moore Swallowwort 117610
Moore Wallichpalm 413094
Moor-gloom 138348
Moor-grass 138348,256601,312360,361611
Moorhens 312360
Moorhouse Rattan Palm 65738
Moorish Mallow 18173
Moorland Spotted Orchid 273531
Moor-whin 172905
Moorwort 22436
Moose Elm 401509,401620
Moose Maple 3393,3629
Mooseberry 407674,407799
Moose-berry Viburnum 407799
Mooseflower 397570
Mooseform Sandwort 31778
Moosewood 3393,3629,133656,407674
Moosewood Viburnum 407799
Mop 81356
Mopane 99980,103716
Mophead Acacia 334995
Mop-headed Acacia 334982
Mopopaja Tree 376067
Mops Mugo Pine 300090
Moquinia 258360
Mor Angelica 24411
Moraea 258380
Moraine Buttercup 325562
Moral 259186
Morama Bean 49077
Moran Mahonia 242593
Morane 192358
Morbeam 259180
Morcrop 21339
More 123141
More Centaury 54896
More Flowers Rhododendron 332155
More Needle Like Leaves Rhododendron 331515
Morefield Neststraw 379162
Morefield's Clematis 95147
Morefield's leather-flower 95147
Morel 367416
Morelittle Primrose 314648
Morelle De Balbis 367617
Morello 83361,316318,316320
Morello Cherry 83361
Moreno Palm 85420,85432
Moreton Bay Ash 106822
Moreton Bay Bean 79075
Moreton Bay Chestnut 79073,79075
Moreton Bay Fig 165274
Moreton Bay Pine 30841
Morette 44708
Morgan 26768
Morgeline 407144
Mori Bedstraw 170503
Mori Clematis 95148
Mori Cyclobalanopsis 116150
Mori Hawkweed 195797
Mori Oak 116150
Mori Rhododendron 331565
Mori Sedge 75417
Mori Thistle 92210
Mori Willow 343321
Mori's Cleyera 96598
Mori's Raspberry 338850
Moriche Palm 246671
Morina 258831
Morinda Spruce 298349,298436
Moringa Family 258961
Moringo 258945
Morison's Spurry 370565
Mormon Tea 146122,146154,146224,146260,146262
Mormon-tea 146122,146260
Mormon-tea Family 146270
Morning Brides 84487
Morning Campion 363411
Morning Glory 68713,102903,102933,207546,207579,207659,207676,208016,208120,208250,293861,334869
Morning Glory Family 102892
Morning Light 413914
Morning Light Miscanthus 255888
Morning Star 374916
Morning Star Lily 229811
Morning Yellow-eyed-grass 415990,415993
Morningbride 350099
Morningflower Renanthera 327692
Morningglory 207546
Morning-glory 207546,208051
Morning-glory Begonia 49735
Morningglory Family 102892
Morning-glory Family 102888,102892
Morningstar Lily 229811
Moroccan Broom 120939
Moroccan Toadflax 231084
Morocco Glorybind 103234
Morocco Gum 1024,1266
Morocco Iris 208899
Morocco knapweed 412272
Morocco Toadflex 231084
Morot Stonecrop 356945
Morris Dichaea 128399
Morris Persimmon 132309
Morrison 407529
Morrison Actinodaphne 234000
Morrison Arrowwood 407696
Morrison Aster 40878
Morrison Bentgrass 12145
Morrison Cacalia 283847
Morrison Cotoneaster 107575
Morrison Elaeagnus 142207
Morrison Euonymus 157728
Morrison Feather Flower 407532
Morrison Ladybell 7702
Morrison Listera 264704
Morrison Mountain Spiraea 372021
Morrison Mouseear 82924
Morrison Pine 300083
Morrison Privet 229553
Morrison Rockcress 30345
Morrison Sagebrush 35964
Morrison Spiraea 372021
Morrison Spruce 298375
Morrison Stonecrop 356947
Morrison Stranvaesia 377457
Morrison Sweetleaf 381104
Morrison Viburnum 407696
Morrison's Barberry 51940
Morrison's Cypress Pine 67423
Morrito 111101
Morrow Honeysuckle 235970
Morrow Sedge 75421
Morrow's Honeysuckle 235970
Morrow's Sedge 75421
Morse Luisia 238318
Morse Melodinus 249673
Morse Spapweed 205154
Morse Touch-me-not 205154
Mortal 367120
Morthen 216753
Mortification Root 18173
Mortinia 403911
Mortinia Blueberry 403911
Mortonia 259051
Morub 309570
Mos Cypress 85353
Mos Phlox 295324
Mosaic Dichorisandra 128990
Mosaic Fig 165438
Mosaic Plant 166709
Moschate Nutmeg 261450
Moschatel 8396,8399
Moschatel Family 8406
Moschosma 387032
Moscow Salsify 394296
Mose Bamboo 297306
Moser Mahonia 242468
Moses' Blanket 405788
Moses-and-the-bulrushes 394088
Moses-in-the-boat 394076

Moses-in-the-bulrushes 342400,394088
Moses-in-the-cradle 394076
Moshan Naiad 262091
Mosla 259278
Moso Bamboo 297306
Mosquito Bulrush 353451
Mosquito Bush 268418,268565
Mosquito Flatsedge 119423
Mosquito Orchid 4511
Mosquito Plant 10405,268565
Mosquito Rush 213471
Mosquito Wood 259326,259327
Mosquito Wormwood 35282
Mosquitograss 122963
Mosquito-grass 57930
Mosquitoman 134918
Mosquitotrap 117334,117351,117415,117476,117488,117597,117598,117690
Moss Cabbage Rose 336476
Moss Cabbage-rose 336476
Moss Campion 363144,363399,364023
Moss Locust 334965
Moss Nymph 32373
Moss Phlox 295324
Moss Pink 295324,307190
Moss Pygmyweed 109463
Moss Rose 311852,336476
Moss Sandwort 255598,255600
Moss Saxifrage 349496
Moss Silene 363144
Moss Thistle 92285
Moss Verbena 176694,405841,405897
Moss Wytham 261162
Moss-berry 278366,403916
Moss-corns 312360
Moss-crop 152769,287494,312360
Moss-flower 287494
Mossgrowing Eria 148722
Mossleaf Corydalis 105973
Mossleaf Groundsel 358717
Mosslike Dwarf Rush 212928
Mosslike Edelweiss 224911
Mosslike Primrose 314686
Mosslike Rockjasmine 23237
Mosslike Stonecrop 357036
Moss-living Acronema 6218
Mossliving Brigssia 60279
Moss-loving Bluegrass 305416
Moss-millions 278366
Moss-mingin 278366
Mosspink 295324
Moss-pink 295324
Mossplant 78630
Moss-rose 311852
Mossstem Stonecrop 356692
Moss-whin 172905
Mossy Cup Oak 324141
Mossy Cyphel 255576
Mossy Gentian 173651
Mossy Hairorchis 148722
Mossy Pearl Wort 342279
Mossy Sandwort 31767,256452
Mossy Saxifrage 349496,349856
Mossy Stonecrop 109463,356468
Mossy-cup Oak 324141
Most Chin Cactus 182465
Most Divaricate Bentgrass 12080
Most Loftily Fanpalm 234164
Most Loose Rattan Palm 65722
Most-acuminate Rhynchosia 333134
Mostbeautiful Briggsia 60263
Mostbeautiful Rattan Palm 65795
Most-dwarf Milkvetch 42768
Most-slender Ant Palm 217927
Most-spiny Barberry 52175
Most-spiny Pincution 244234
Moswuito Flower 236247
Moth Bean 408829
Moth Fanpetals 362653
Moth Mullein 405667,405788
Moth Orchid 293577,293582,293633
Moth Plant 405788
Mothan 342279
Moth-bean 408829
Mother 227533
Mother Chrysanthemum 124790
Mother Daisy 227533
Mother Mary's Milk 308454
Mother of Hundreds 244037
Mother of Pearl Plant 180269
Mother of Thousands 116736,349936,367771,367772
Mother of Thyme 4666,391365
Mother of Wheat 407144
Mother Shimble's Snick-needles 374916
Mother Thyme 4666
Mother's Heart 72038
Mother's Nightcap 5442,68713
Mother's Tears 4070
Mother's Thimble 374916
Mother's Thimbles 374916
Mother-breaks-her-heart 407072
Mother-dee 248411,392992
Mother-die 3978,101852
Mother-in-law Plant 65214,130093,130135
Mother-in-law's Armchair 140127
Mother-in-law's Seat 140127
Mother-in-law's Tongue 346047,346158
Mother-of-millions 116736,174739
Mother-of-pearl Plant 180269
Mother-of-the-evening 193417
Mother-of-thousands 3978,106117,116736,243209,349044,349936,367772,392668
Mother-of-thyme 4666,391080,391329,391365,391397
Mother-of-wheat 407144
Mother-thread-my-needle 174877
Motherthyme 4666,391365
Motherweed 231473
Mother-will-die 109857
Motherwood 35088
Motherwort 36474,44576,193417,224969,224976,225009,239755,391397
Motherwortlike Hoarhound 245748
Moth-Orchid 293577
Mothweed 178330
Mothwort 178330,189814
Motlery Nageia 261983
Motlrer Thyme 391397
Moto Manglietia 244458
Moto Woodlotus 244458
Motse 1427
Mottlecah 155634
Mottled Bamboo 87581
Mottled Dutchman's Pipe 34239
Mottled Paris 284351
Mottled Red-leaved Barberry 52239
Mottled Spurge 159204
Mottled Toothedthread 269097
Mottleleaf Squill 223819
Motuo Agapetes 10336
Motuo Alseodaphne 17805
Motuo Arrowwood 407720
Motuo Aster 40879
Motuo Azalea 331211,331274
Motuo Basketvine 9460
Motuo Begonia 49903
Motuo Bulbophyllum 62716
Motuo Calanthe 66015
Motuo Catchfly 363795
Motuo Cayratia 79872
Motuo Chirita 87853
Motuo Clearweed 298970
Motuo Clematis 95112
Motuo Coldwaterflower 298970
Motuo Cyclobalanopsis 116151
Motuo Didymocarpus 129927
Motuo Elatostema 142737
Motuo Elytranthe 144607
Motuo Eria 148718
Motuo Fir 338
Motuo Gentian 173655
Motuo Glochidion 177157
Motuo Glycosmis 177842
Motuo Greenbrier 366353
Motuo Hairorchis 148718
Motuo Holboellia 197243
Motuo Holly 204030
Motuo Honeylocust 176893
Motuo Hooktea 52442
Motuo Jadeleaf and Goldenflower 260399

Motuo Lily 229936
Motuo Lilyturf 272116
Motuo Lysionotus 239958
Motuo Marsdenia 245829
Motuo Medinilla 247616
Motuo Milkgreens 245829
Motuo Monkshood 5172
Motuo Moxanettle 221576
Motuo Nanmu 295385
Motuo Oak 116151
Motuo Oxyspora 278595
Motuo Pearweed 239731
Motuo Pencilwood 139664
Motuo Phoebe 295385
Motuo Pricklyash 417275
Motuo Prickly-ash 417275
Motuo Qinggang 116151
Motuo Raspberry 338818
Motuo Rattanpalm 65688
Motuo Rhododendron 331211,331213,331274
Motuo Rockfoil 349617
Motuo Sedge 75429
Motuo Spapweed 205129
Motuo Spice-bush 231396
Motuo Square Bamboo 87586
Motuo Stairweed 142737
Motuo Stonebean-orchis 62716
Motuo Supplejack 52442
Motuo Tanoak 233331
Motuo Touch-me-not 205129
Motuo Tubergourd 390162
Motuo Viburnum 407720
Motuo Willow 343681
Mouchers 338447
Mouillac Barberry 51941
Moujean Tea 262601
Moulds 176824
Moulmain's Rhododendron 331280
Moulmein Cedar 392829
Moulmein Lancewood 197728
Moulmein Rhododendron 331445
Moulmein Teak 385531
Mound Lily 416607
Mound Lily Yucca 416607
Mound-lily 416607
Moundlily Yucca 416607
Mound-lily Yucca 416607
Mounrain Rye 356245
Mount Atlas Cedar 80080
Mount Atlas Daisy 21230,21249
Mount Atlas Mastic 300977
Mount Atlas Mastic Tree 300977
Mount Atlas Pistache 300977
Mount Atlas Pistachio 300977
Mount Broom 172901
Mount Caper 273469
Mount Cook Lily 326049
Mount Coper 273469
Mount Dellanbaugh Sandwort 31728
Mount Diablo Cottonseed 253843
Mount Etna Barberry 51280
Mount Etna Broom 172901
Mount Hood Bugbane 91026
Mount Katahdin Sedge 74984
Mount Lemmon Marigold 383096
Mount Lewis Palm 31016
Mount Lysionotus 239967
Mount Morgan Wattle 1488
Mount Morrison Barberry 51940
Mount Morrison Spruce 298375
Mount Omei Rose 336930
Mount Pinos Larkspur 124468
Mount St. Helena Ceanothus 79925
Mount St. Helena Fawn-lily 154922
Mount Tamalpais Thistle 92043
Mount Washington Bluegrass 305646
Mount Washington Blue-grass 305646
Mount Washington Dryad 138461
Mount Wellington Peppermint 155533
Mount Willow 343812
Mountai Gooseberry 334137
Mountain Agave 10891
Mountain Agoseris 11420
Mountain Alangium 13346
Mountain Alder 16368,16375,16459,16464,16472,16479
Mountain Alyssum 18433
Mountain Andromeda 298721
Mountain Anemone 23707,23873
Mountain Apple 382606
Mountain Arnica 34752
Mountain Ash 155719,311537,323083,369290,369313,369339,407737
Mountain Aster 39920,41333,268682
Mountain Atlas Cedar 80080
Mountain Avens 138445,138461,175433,175436
Mountain Azalea 235283,330322,331546,331667
Mountain Balm 65540,96966
Mountain Bamboo 37131
Mountain Bastard Cress 390265
Mountain Bearberry 31314
Mountain Beech 266866,266885
Mountain Bellwort 403673,403675
Mountain Bitter Bamboo 304022
Mountain Bladder-sedge 74696
Mountain Blue Gum 155555
Mountain Bluebell 250887
Mountain Blue-eyed Grass 365782,365784
Mountain Blue-eyed-grass 365782,365803
Mountain Bluegrass 305794
Mountain Bluet 81214
Mountain Bog-sedge 75980
Mountain Bottlebrush 67288
Mountain Box 31142
Mountain Bramble 338243
Mountain Bred Rhododendron 330150
Mountain Brome 60842
Mountain Brush 81214
Mountain Bugle 13163
Mountain Burmareed 265922
Mountain Bush 218387
Mountain Buttercup 326096
Mountain Butterwort 299737
Mountain Cactus 287909
Mountain Calamint 65540
Mountain Camellia 376465
Mountain Campion 363144,363548
Mountain Carpet Weed 159392
Mountain Catchfly 363770
Mountain Cedar 61339,228636
Mountain Cherry 316219
Mountain Chestnut Oak 324201
Mountain Chick-pea 42193
Mountain Cinnamon 91250
Mountain Cinnamon Tree 91351
Mountain Clematis 95127
Mountain Clover 396981
Mountain Common Juniper 213703,213934
Mountain Cornflower 81214
Mountain Coronilla 105299
Mountain Correa 105393
Mountain Cottonwood 311233
Mountain Cowparsnip 192337
Mountain Cowslip 314143
Mountain Cranberry 404051,404052,404060
Mountain Crane's-bill 174856
Mountain Cranesbill 174856,174951
Mountain Crowberry 145062
Mountain Crowfoot 325559
Mountain Currant 333904
Mountain Custardapple 25866
Mountain Cypress 61339
Mountain Damson 364382
Mountain Dandelion 11408,11445
Mountain Death Camas 417890,417891
Mountain Devil 220308
Mountain Dogwood 105130
Mountain Dwarf Rhododendron 330065
Mountain Dwarfdandelion 218210
Mountain Ebony 48990,49123,49247
Mountain Elder 345660
Mountain Elm 401512
Mountain Eryngo 154280
Mountain Everlasting 26299,26385
Mountain Fetterbush 298721
Mountain Fetter-bush 298721
Mountain Fig Dapasan 165370
Mountain Flax 81065,231884,295598,

295599,308347,370522
Mountain Fleabane 150831
Mountain Fleece 308769
Mountain Flower 174951
Mountain Fly Honeysuckle 235691,236206,236207
Mountain Fly-honeysuckle 236206
Mountain Forget-me-not 260892
Mountain Four o'clock 255736
Mountain Four-o'clock 255731
Mountain Foxglove 277596
Mountain Fragrant Bamboo 87642
Mountain Fragrant-bamboo 87642
Mountain Fringe 8302,35505
Mountain Fritillary 168585
Mountain Gentian 150831
Mountain Germander 388154
Mountain Globemallow 204423
Mountain Glory 337885
Mountain Golden Chain 218757
Mountain Golden Star 55651
Mountain Goldenrod 368058,368398
Mountain Grape 242478,411805,411900
Mountain Gray Gum 155552
Mountain Graygum 155596
Mountain Grevillea 180563
Mountain Grey Gum 155596
Mountain Groundsel 360172
Mountain Gum 155553,155604
Mountain Hairy Willow 343865
Mountain Hard Pear 270470
Mountain Hawksbeard 110905
Mountain Hawthorne 110090
Mountain Heath 146072,227970,297015,297019
Mountain Helicia 189917
Mountain Hemlock 229384,399905,399916
Mountain Hickory 1474
Mountain Holly 204064,263498,263499,270184,270195
Mountain Honeysuckle 235760
Mountain Horseorchis 215352
Mountain Houseleek 358052
Mountain Houstonia 187652
Mountain Immortelle 154704
Mountain Indian Physic 175766
Mountain Iris 208528,208720
Mountain Ironwort 362838
Mountain Jasmine 211918
Mountain Jerusalemsage 295163
Mountain Jewelflower 377639
Mountain Joint Fir 146265
Mountain Jujube 418189
Mountain Juniper 213934
Mountain Kalimeris 215352
Mountain Knapweed 81214
Mountain Krigia 218210
Mountain Lady's-mantle 13959
Mountain Larkspur 124180,124233,124391
Mountain Laurel 215395,332751,369116
Mountain Leatherwood 156067
Mountain Leech 126242
Mountain Lilac 79902
Mountain Lily 229730,229922
Mountain Lime 93868
Mountain Linden 391782
Mountain Lungwort 321640
Mountain Machilus 240663
Mountain Maesa 241808
Mountain Magnolia 242110,242267
Mountain Mahoe 194850
Mountain Mahogany 53505,83817,83819,83820
Mountain Mahonia 242478
Mountain Malaxis 243129
Mountain Maple 2991,3629
Mountain Marigold 359398,383096
Mountain Marygold 136365
Mountain Meadow Seseli 361526
Mountain Meadow Wild Buckwheat 152480
Mountain Meadow-rue 388450,388673
Mountain Meadow-saxifrage 361526
Mountain Meadow-seseli 361526
Mountain Melic 249054
Mountain Melick 249054
Mountain Mint 65540,96966,257161,321952,321962,321964
Mountain Misery 85141
Mountain Mist 67456
Mountain Morgan Wattle 1488
Mountain Muchuan Yushania 416772
Mountain Muhly 259684
Mountain Mulberry 259164
Mountain Myrtle 224196
Mountain Needlewood 184617
Mountain Neinei 137735
Mountain Neststraw 379160
Mountain Ninebark 297841
Mountain Northern Ladybell 7612
Mountain Nutmeg 261423
Mountain Organ Pipe 375529
Mountain Oxeye 190524
Mountain Panicgrass 281984,282200
Mountain Pansy 410187
Mountain Papaw 76811
Mountain Papaya 76809
Mountain Paper Birch 53549
Mountain Parsley 292961
Mountain Pasqueflower 23969
Mountain Pawpaw 76811,76815
Mountain Pea Vine 222758
Mountain Peavine 222758
Mountain Pepper 122515,138051,138052,138056
Mountain Pepperbush 96456
Mountain Persimmon 132308
Mountain Pfeiffer Osmanthus 276344
Mountain Phlox 295287
Mountain Pieris 298721
Mountain Pincushion 84479
Mountain Pine 121088,184913,300086,300168,300223
Mountain Pink 80966,80984,331546
Mountain Plantain 302112
Mountain Plum Pine 306453
Mountain Podocarpus 306493
Mountain Poisonnut 378834
Mountain Polyraphioid Rhododendron 331517
Mountain Pride 289363,289364,370308
Mountain Primrose 314668
Mountain Purplestem 376465
Mountain Rabbitbrush 150372
Mountain Radish 34590
Mountain Ramontchi 166782
Mountain Red Cedar 213922
Mountain Red Elderberry 345628
Mountain Red Oak 320278
Mountain Ribbonwood 172245,197152,197153
Mountain Rice 275991
Mountain Rice Grass 276055
Mountain Rimu 225602
Mountain Rock Cress 30164
Mountain Rockcress 30164
Mountain Rock-cress 30160,30299
Mountain Rocky Oak 324515
Mountain Rose 28422,336914,337026
Mountain Rose Bary 330331
Mountain Rosebay 330331
Mountain Rosebay Rhododendron 330331
Mountain Sage 321643,345341
Mountain Sagebrush 35505,36408
Mountain Sandwort 32081,32085,255447,255486,255573
Mountain Sanicle 43309,105496
Mountain Sannicle 43309
Mountain Satinflower 365782
Mountain Savory 347597
Mountain Saxifrage 349216
Mountain Scurvy-grass 98014,98020
Mountain Sedge 75409
Mountain Serviceberry 19250
Mountain She Oak 79172
Mountain She-oak 79172
Mountain Silverbell 184746
Mountain Silver-bell 184746
Mountain Sneezeweed 188393
Mountain Snow 30220,159313
Mountain Snowbell 79905,367766
Mountain Snowberry 380751
Mountain Snowdrop 227856

Mountain Snowdrop Tree 184746
Mountain Snowdrop-tree 184746
Mountain Snuff 34752
Mountain Sorrel 278576,278577
Mountain Soursop 25866
Mountain Soursop-tree 25866
Mountain Sow Thistle 90836
Mountain Sow-thistle 90836
Mountain Spapweed 205153
Mountain Speedwell 407238,407358
Mountain Spicy Litse 233882
Mountain Spicy Tree 233882
Mountain Spinach 44468
Mountain Spray 197404
Mountain Spruce 298273
Mountain Spurge 159392,279754
Mountain St. John 's-wort 202021
Mountain St. John's Wort 202026
Mountain Stewartia 376465
Mountain Stitchwort 255447
Mountain Stonecrop 357224
Mountain Strychnos 378834
Mountain Sumach 332529
Mountain Swamp Gum 155524
Mountain Tallow Tree 346379
Mountain Tallow-tree 346379
Mountain Tarweed 241533
Mountain Tassel 367766
Mountain Tassel Flower 367766
Mountain Tea 172135,250331
Mountain Thatch Palm 390463
Mountain Thistle 91931
Mountain Thyme 391365
Mountain Tobacco 34752,266038,367146
Mountain Triteleia 398811
Mountain Tubergourd 390163
Mountain Tulip 400197
Mountain Valerian 404310,404383
Mountain Viburnum 408056
Mountain Violet 409716
Mountain Watercress 72961
Mountain White Birch 53387,53388
Mountain White Pine 300082
Mountain Whitethorn 79919
Mountain Widow-wail 122416
Mountain Willow 343041,343400,343744,
343814,344085
Mountain Willow Weed 146782
Mountain Willowweed 146782
Mountain Wineberry 34395
Mountain Winterberry 204064
Mountain Wood Aster 160652
Mountain Wood-sorrel 277973
Mountain Wormwood 35841
Mountain Xyris 416110
Mountain Yam 131877
Mountain Yew 213702
Mountain-angelica 30587
Mountain-apple 382606
Mountainash 369290
Mountain-ash 369290,369379,369484
Mountainash Actinidia 6702
Mountain-ash False Spiraea 369279
Mountainash Falsespiraea 369279
Mountainash False-spiraea 369279
Mountainash Kiwifruit 6702
Mountainash Woodbetony 287685
Mountainashleaf Woodbetony 287685
Mountain-ash-leaved Actinidia 6702
Mountain-asphodel 415440
Mountain-aweet 79905
Mountain-bells 375456
Mountain-bluet 81214
Mountain-camellia 376465
Mountain-cedar 213620
Mountaincrown 274122
Mountaineer Willow 343814
Mountainfringe 8298
Mountainglary Primrose 314762
Mountainglory Primrose 314762
Mountainglory Rhododendron 331406
Mountain-glory Rhododendron 331406
Mountainheath 297015
Mountainliving Devil Rattan 121518
Mountain-loving Draba 137155
Mountain-loving Parnassia 284588
Mountain-loving Whitlowgrass 137155,
137170
Mountain-mahogany 83816,83819
Mountain-mint 321952,321957,321958
Mountainous Euonymus 157737
Mountainous Jerusalemsage 295163
Mountainous Leucas 227661
Mountainous Maesa 241808
Mountainous Milkvetch 42752
Mountainous Mulberry 259168
Mountainous Tubergourd 390163
Mountainrose Coralvine 28422
Mountain-rose Coralvine 28422
Mountain-rose Coral-vine 28422
Mountainsorrel 278576
Mountain-sorrel 278576
Mountaintop Thistle 91929
Mountainview Gentian 173681
MountainWillow-herb 146782
Mountainy Anadendrm 21313
Mountanain Turpinia 400530
Mountane Garcinia 171100
Mountane Sandwort 32116
Mountanous China-laurel 28368
Mountanous Nothapodytes 266801
Mounte Maesa 241808
Mount-trifoly 192130
Moupin Cotoneaster 107576
Moupin Litse 234003
Moupin Milkvetch 42757
Moupin Violet 410275
Mouping Draba 137124
Mouping Dutchman's-pipe 34276
Mouping Microtoena 254259
Mouping Palmatisect Cacalia 283855
Mouping Rhododendron 331283
Mouping Sedge 75430
Mouping Small Reed 127275
Mouping Smallreed 127275
Mouret Statice 230683
Mouretia 259410
Mournful Bells of Sodom 168476
Mournful Widow 174811,216753,350099
Mourning Bride 216753,350086,350099
Mourning Cypress 114690
Mourning Iris 192904,208870
Mourning Widow 174811,216753,350099
Mourning Widow Iris 192904
Mouse Barley 198327
Mouse Bloodwort 195858
Mouse Ear 82629,82758,195858,197486,
260737,373166
Mouse Ear Coreopsis 104444
Mouse Ear Cress 30146
Mouse Ear Everlasting 178358
Mouse Ear Hawkweed 195858
Mouse Ear Scorpion Grass 260760
Mouse Fox Tail 17548
Mouse Foxtail 17548
Mouse Milk 159027
Mouse Pea 222758,222820,408352,408423
Mouse Plant 33588
Mouse Tail 260940
Mouse Wild Buckwheat 152324
Mouse's Mouth 231183
Mouse's Pease 408352
Mouse-car Betony 373166
Mouseear 82680,82732
Mouse-ear 195858,197486,260737
Mouse-ear Betony 373224
Mouseear Chalkplant 183183
Mouseear Chickweed 82629
Mouse-ear Chickweed 82629,82818,82826,
82849
Mouseear Cress 30105,30146
Mouse-ear Cress 30146
Mouse-ear Everlasting 178358
Mouseear Gypsophila 183183
Mouse-ear Gypsophila 183183
Mouseear Hawkweed 195858,195859
Mouse-ear Hawkweed 195835,195858,
299215
Mouse-ear-cress 30146
Mouse-eared Chickweed 82793
Mousehole Tree 260718

Mouseplant 33588
Mousetail 260837,260930,260940,356468
Mouse-tail Cassiope 78638
Mousetail Cupscale 341977
Mousetail Indigo 206283
Mousetaillike Cassiope 78638
Mousetail-plant 33584,33588
Mousewood 133656
Moustache Plant 78012
Moutain Cypress Pine 67419
Moutainous Garcinia 171115
Moutan 280286
Moutan Peony 280286
Mouthroot 103850
Movingue 134776
Mowa 241516
Mowing Daisy 227533
Moxa 111826
Moxanettle 221529,221552
Moxie Plum 172085
Moya Grass 289124
Moyes Rose 336777
Moysey Licuala Palm 228752
Moysey Licualapalm 228752
Mozambique Ebony 121750,132188
Mozambique Ochna 268226
Mozhugongka Willow 343662
Mozugongka Willow 343662
Mr. Bidwell's Coral Tree 154632
Mr. Bidwill's Araucaria 30835
Mr. Elliott's Rhododendron 330616
Mr. Keys' Rhododendron 331003
Mr. Mainbridge's Rhododendron 330195
Mrs. Austin's knotweed 308814
Mrs. Britton's shadow witch 311021
Mrs. E. H. Wilson's Azalea 331065
Mrs. E. H. Wilson's Barberry 52371
Mrs. E. H. Wilson's Rhododendron 331065
Mrs. Farrer's Rhododendron 330687
Mrs. Johnston's Rhododendron 330963
Mrs. Lott's Vanilla 404989
Mrs. Parry's Rhododendron 331447
Mrs. Roxie's Rhododendron 331675
Mrs. Sargent Lily 230029
Mrs. Wilson's Barberry 52371
Mt. Albert Goldenrod 368398,368401
Mt. Amagi Azalea 330071
Mt. Atlas Daisy 21230
Mt. Atlas Mastic Tree 300977
Mt. Diablo Wild Buckwheat 152546
Mt. Etna Broom 172901
Mt. Hamilton Thistle 91980
Mt. Kyushu Rhododendron 331010
Mt. Lassen Fleabane 150745
Mt. Morgan Wattle 1488
Mt. Morrison Actinodaphne 234000
Mt. Nankotaisan Rhododendron 331311
Mt. Pontus Rhododendron 331520
Mt. Poukhan Azalea 331530
Mt. Rose Wild Buckwheat 152434
Mt. Shasta Arnica 34784
Mt. Shasta Snakeroot 11173
Mt. Taurus Pink 127620
Mtambara 82437
Mtn Sneezeweed 188393
Mtn. Avens 138445
Mu Oil Tree 406019
Mu Oiltree 406019
Mu Tree 406019
Mubian Cyclobalanopsis 116169
Mubian Oak 116169
Mubian Qinggang 116169
Mucaia Acrocomia 6136
Mucaja Acrocomia 6136
Muchbranching Statice 230732
Muchcolor Rhododendron 330609
Much-flowered Sweetleaf 381381
Muchform False Rattlesnakeroot 283688
Much-good 292813
Muchhair Jurinea 206789
Muchhigh Aster 41080
Muchlovable Dendrobium 125169
Muchnarrow Bamboo 47185
Muchnarrowleaf Sagebrush 35130
Muchslender Everlasting 21710
Muchslender Pearleverlasting 21710
Muchsmall Pipewort 151379
Muchsticky Sagebrush 36472
Muchuan Yushania 416771
Muchvein Ardisia 31535
Mucilagefruit 101230
Muck Sedge 75165
Muck Sunflower 189058
Muckhill Weed 86901
Muckies 336419
Muckle Kokkeluri 227533
Muckweed 86901
Mucronate Bulrush 352223
Mucronate Corydalis 106169
Mucronate Cotoneaster 107579
Mucronate Delavay Meadowrue 388482
Mucronate Hideseedgrass 94615
Mucronate Honeysuckle 235973
Mucronate Jasminanthes 211695
Mucronate Lepironia 225664
Mucronate Obovateleaf Woodnettle 273913
Mucronate Prettyleaf Silkvine 291050
Mucronate Rabdosia 209780
Mucronate Sedge 74383
Mucronate Stephanotis 211695
Mucronate Thickribcelery 229365
Mucronate Torenia 392921
Mucronatefruit Ervatamia 154185,154192,382822
Mucronipetale Muping Corydalis 105890
Mucronulate Windhairdaisy 348543
Mucuna 259479
Mud Bamboo 47283
Mud Crowfoot 326475
Mud Cudweed 178481
Mud Grass 99411,99413
Mud Knotweed 309337
Mud Plantain 193676,193683,193686,193689
Mud Rush 213127
Mud Sedge 75165
Mud Spine Bamboo 47242
Mud Tanoak 233209
Mudar 68086,68088
Mudar Fibre 68086
Mudbank Paspalum 285424
Muddy Bamboo 47242
Muddy Bambusa 47242
Muddy Snowbell 379327
Mud-Fish's Egg Climber 178564
Mudgee Wattle 1612
Mudgrass 99411,99413
Mud-grass 99413
Mudmats 177408
Mud-nut 177413
Mud-purslane 142572
Muduga Oil 64148
Mudweed 29327
Mudwort 230828,230831
Muehldorf Violet 410279
muehlenberg's Sedge 75435
Muehlenpfordt Cactus 244159
Mueller Callitris 67424
Mueller Mahonia 242595
Mueller Olive 270149
Mueller's Cypress Pine 67424
Mueller's Helleborine 147168
Mueller's Saltbush 44532
Mueller's Wattle 1410
Mueller's Hawkbit 224709
Muenscher's Water-nymph 262047
Mugga 155741,155743,155744
Mugga Ironbark 155743
Mugger 36474
Muggert 36474,359158
Muggert Kale 36474
Mugget 102867,407989
Mugget Rose 407989
Muggins 36474
Muggles 71218
Muggons 36474
Muggurth 36474
Muggwith 36474
Mugho Pine 300086
Mugo Pine 300086
Mugonga 8186

Mugongo 351005
Mugonione 29587
Mugs. Without-handles 70164
Muguet 102867
Mugweed 36474
Mugwood 36474
Mugwort 35084,35090,35407,35534,36474,113135,170335
Mugwort Wormwood 36474
Muhimbi 118050
Muhlenberg's Bracted Sedge 75435,75436
Muhlenberg's Cacalia 34809
Muhlenberg's Hairy Bead Grass 285524
Muhlenberg's Nut Grass 354225
Muhlenberg's Sedge 75436
Muhly 259630,259645
Muhly Grass 259630,259642
Muir's Fleabane 150794
Mukia 259733,259739
Muku Tree 29005
Mukumari 104148,104150
Mukunawanna 18147
Mula 326622
Mulberry 115008,259065,259180,338209,338447,369322,403916
Mulberry Family 258378
Mulberry Fig 165726
Mulberry-leaf Begonia 50091
Mulberryleaf Grape 411736
Mulberryleaf Raspberry 338664
Mulberryleaf Windhairdaisy 348542
Mulberry-leaved Raspberry 338664
Mulberry-weed 162872
Mule Fat 46251,46269
Mule Grass 259642
Mule Mountain False Brickellbush 37539
Mule's Fat 46251
Mule's-ears 414925
Mule-ear Oncidium 395616
Mule-ears 414924
Mule-fat 46269
Mule-grippler Cactus 140133
Mulei Milkvetch 42791
Mules-ears 414924
Mulga 1049
Mulga Acacia 1049
Mulgamulga 1049
Mulgedium 259746,283692
Muli Abelmoschus 238
Muli Acroglochin 6165
Muli Apple 243658
Muli Azalea 330031,331712
Muli Barberry 51943
Muli Begonia 50094
Muli Bentgrass 12208
Muli Bittercress 72901
Muli Catchfly 363778
Muli Ceropegia 84180
Muli Comastoma 100275
Muli Conyza 103539
Muli Corydalis 105679,106172
Muli Cotoneaster 107581
Muli Crabapple 243658
Muli Cremanthodium 110464
Muli Delavay Holly 203758
Muli Deutzia 127018
Muli Dianxiong 297957
Muli Diflugossa 130319
Muli Dubyaea 138775
Muli Elaeagnus 141948
Muli Everlasting 21637
Muli Fern-like Woodbetony 287205
Muli Goldenray 229113
Muli Groundsel 359527
Muli Indigo 206278
Muli Jerusalemsage 295153
Muli Kobresia 217307
Muli Larkspur 124394
Muli Leek 15352
Muli Ligusticum 229398
Muli Maple 3211
Muli Microtoena 254260
Muli Microula 254354
Muli Milkvetch 42758
Muli Monkshood 5484
Muli Mosquitotrap 117614
Muli Nutantdaisy 110464
Muli Oak 324383
Muli Onion 15352
Muli Pearleverlasting 21637
Muli Pendentlamp 84180
Muli Phtheirospermum 296070
Muli Physospermopsis 297957
Muli Primrose 314178
Muli Rabdosia 209782
Muli Rhododendron 330031,331712
Muli Sage 345074
Muli Sandwort 32089
Muli Sedge 75438
Muli Skullcap 355446
Muli Spapweed 205155
Muli Spiraea 372022
Muli Swallowwort 117614
Muli Synotis 381943
Muli Tailanther 381943
Muli Thickribcelery 279651
Muli Thistle 92211
Muli Throathair 100275
Muli Tree Currant 334109
Muli Tremacron 394679
Muli Trigonotis 397439
Muli Valerian 404311
Muli Willow 343734
Muli Windhairdaisy 348544
Mull 326622,397043
Mullein 405657,405694,405788
Mullein Dock 405788
Mullein Foxglove 10121
Mullein Nightshade 367115,367146
Mullein Pink 11947,363370
Mullein-foxglove 10121
Mullein-leaved Sage 345457
Mulleinlike Primrose 314225
Mullen 405788
Muller Oak 323791
Mullet Creek Grevillea 180647
Multa-mulla 321102
Multibract Raspberry 339080
Multibracteate Camellia 69375
Multibracted Gugertree 350931
Multibracted Raspberry 339080
Multi-bracted Rose 336781
Multibracteolate Groundsel 359529
Multibranch Thorowax 63785
Multibranched Leymus 228370
Multi-branched Rhododendron 331512
Multibranched Willow 343906
Multicapital Monkshood 5490
Multicarpous Monkshood 5492
Multicauliforous Barberry 51946
Multicaulinary Ardisia 31532
Multicaulis Cinquefoil 312795
Multicaulis Mulberry 259085
Multicellutate Clearweed 298983
Multicellutate Coldwaterflower 298983
Multicolor Azalea 331710
Multicoloured Dragonhead 137615
Multicostate Brain Cactus 140577
Multidente Euonymus 157360
Multifid Anemone 23906
Multifid Biebersteinia 54197
Multifid Cuneateleaf Buttercup 184718
Multifid Dali Teasel 133468
Multifid Gerbera 175123
Multifid Groundsel 359536
Multifid Hylomecon 200758
Multifid Ladypalm 329186
Multifid Lavender 223312
Multifid Monkshood 5493
Multifid peganum 287982
Multifid Pulsatilla 321708
Multifid Rosewood 121805
Multifid Turtschaninov Corydalis 106569
Multifid Windflower 23906
Multifid Youngia 416472
Multifid-leaved Nettle Spurge 212186
Multiflora Bean 409085
Multiflora Rose 336783
Multiflorous Anzacwood 310768
Multiflorous Ash 167975
Multiflorous Barberry 51440

Multiflorous Bauhinia 49045
Multiflorous Bush-clover 226793
Multiflorous Butterfly-bush 62127
Multiflorous Calycopteris 175369
Multiflorous Caper 71809
Multiflorous Clovershrub 70866
Multiflorous Colona 99965
Multiflorous Cotoneaster 107583
Multiflorous Cushion Willow 344095
Multiflorous Deutzia 127013
Multiflorous Diplarche 132829
Multiflorous Elaeocarpus 142317
Multiflorous Embelia 144736
Multiflorous Erycibe 154249
Multiflorous Garcinia 171145
Multiflorous Gardneria 171455
Multiflorous Glorybower 96028
Multiflorous Jasmine 211912
Multiflorous Maackia 240124
Multiflorous Melastoma 248727
Multiflorous Memecylon 249962
Multiflorous Michelia 252872
Multiflorous Pavetta 286418
Multiflorous Pink Deutzia 127069
Multiflorous Silkvine 291060
Multiflorous Sweetleaf 381381
Multiflorous Syzygium 382651
Multiflorous Tailgrape 35034
Multiflorous Tamarisk 383519
Multiflorous Tickclover 126471
Multiflorous Vineclethra 95492
Multiflorous Violet 409851
Multiflorous Waxplant 81911
Multiflorous Winterhazel 106659
Multiflorous Wisteria 414554
Multiflower Bauhinia 49045
Multiflower Bittercress 72903
Multiflower Knotweed 162542
Multiflower Longleaf Briggsia 60274
Multiflower Ratpalm Cranesbill 174910
Multiflower Smallbract Milkvetch 42855
Multiflower Waxplant 81911
Multiflowered Aspidopterys 39675
Multiflowered China-laurel 28358
Multiflowered Clover 396939
Multiflowered Euonymus 157732
Multiflowered Garuga 171563
Multiflowered Meliosma 249414
Multi-flowered Oxalis 277776
Multiflowered Supple-jack 52418
Multiflowers Jasmine 211912
Multiflrous Dogwood 380494
Multifolia Lacquertree 393451
Multifoliolate Incarvillea 205572
Multifoliolate Sage 345130
Multifolious Schefflera 350744
Multiform Bambusa 47413
Multiforme Ceheng Fig 164789
Multifruit Hemsleia 191957
Multifruit Snowgall 191957
Multifruited Bishop-wood 54623
Multifruited Maple 3461
Multifruited Tallow-tree 346401
Multijugate Sweetvetch 188016
Multijugous Mountainash 369467
Multijugous Mountain-ash 369467
Multijugous Prickly-ash 417277
Multileaf Onosma 271817
Multileaf Schefflera 350744
Multi-leaved Bluegrass 306004
Multileaved Gelidocalamus 172745
Multileaved Roegneria 335226
Multinerved Altingia 18202
Multinerved Catchfly 363366
Multinerved Dendrobenthamia 124942
Multi-nerved Holly 204177
Multinerved Hop-hornbeam 276815
Multinerved Hornbeam 77369
Multinerved Mistletoe 411065
Multi-nerved Mussaenda 260465
Multinerved Pepper 300518
Multinerved Schefflera 350749
Multinerved Thorowax 63761
Multinervous Buckeye 9724
Multinode Calophanoides 67827
Multi-noded Bluegrass 305846
Multiperulated Camellia 69613
Multipetalous Bruguiera 61263
Multipetalous Oxygraphis 278472
Multipinnate Cycas 115865
Multiple Violet 410432
Multiple-fractate Fargesia 137851
Multiplier Onion 15167
Multiprickeled Acanthopanax 143657
Multiradiate Deutzia 127019
Multiradiate Fleabane 150800
Multiradiate Gentian 150800
Multiramose Onosma 271807
Multisepal Paradrupetea 283713
Multiserrate Barberry 51950
Multiserrate Caudate Maple 2878
Multispiked Rattan Palm 65740
Multispinate Barberry 51533
Multistalk Rye 356241
Multivein Alphonsea 17632
Multivein Guizhou Primrose 314551
Multivein Machilus 240650
Multivein Schefflera 350749
Multiveined Cyclobalanopsis 116152
Multiveined Dendrotrophe 125789
Multi-veined Elm 401476
Mulun Magnolia 242215
Mum 89402,124826
Mume 34448
Mume Plant 34448
Mumeplant 34448
Mume-plant 34448
Mumleaved Violet 410633
Mumpstree 189976
Mums 124785
Munan Manglietia 244426
Munan Woodlotus 244426
Mundi Rose 336585
Mundt Cactus 244162
Mundu 171092
Mundulea 259816
Mung 409025
Mung Bean 408982,409025
Mung-bean 408825,408982
Mung-hean 409025
Mungo Bean 408982
Munguba 56805
Munich Grane's-bill 174441
Muninga 320278
Munj Sweetcane 341842
Munjeet 337925
Munro Black Bamboo 297367
Munro Hairyfruitgrass 222354
Munron Spapweed 205158
Munron Touch-me-not 205158
Munronia 259874
Munshook 404051
Munson Plum 316583
Muntingia 259907
Muntries 218387,218394
Munz Cholla 116669
Munz's Iris 208722
Munz's Mariposa-lily 67618
Munz's Shrub 259918
Munz's Sulphur Flower 152594
Mu-oil 406019
Mu-oil Tree 406019
Muoiltree 406019
Mu-oiltree 406019
Mupin Milkvetch 42757
Mupin Willow 343723
Muping Ainsliaea 12661
Muping Corydalis 105881
Muping Cotoneaster 107576
Muping Dutchman's-pipe 34276
Muping Dutchmanspipe 34276
Muping Rabbiten-wind 12661
Muping Sinojohnstonia 364964
Muping Spindle-tree 157899
Muping Violet 410275
Muping Willow 343723
Murchion River Oak 79166
Murdannia 260086
Murfeys 367696
Murg 26768
Muricate Amaranth 18781

Muricate Aspidistra 39562
Muricate Ballcactus 267044
Muricate Barnyard-grass 140451
Muricate Begonia 50098,50260
Muricate Camellia 69385
Muricate Eurya 160551
Muricate Slugwood 50567
Muricate-fruited Slugwood 50567
Muricateleaf Whiteroot Acanthopanax 2439, 143632
Muricatum Parrotfeather 261370
Muriculate Syzygium 382618
Muriel Bamboo 162717,162752,388761
Muriel Rose 336800
Muringa 104150
Murity Palm 246674
Murphy's Agave 10910
Murrain Berries 383675
Murrain-grass 355202
Murray Phasemy Bean 241263
Murray Red Gum 155517
Murray Redgum 155517
Murray River Red Gum 155520
Murray's Birch 53529
Murraya 260158
Murren 61497,374968
Murren-berries 61497,383675
Murtle-leaf Orange 93389
Murton Heraldtrumpet 49363
Murton Herald-trumpet 49363
Murton Qingmingflower 49363
Murumuru 43377
Murumuru Astrocaryum 43377
Murun 374968
Muscadine Grape 411893
Muscar Liriope 232631
Muscari 260308
Muscatel 411979
Muscatel Sage 345379
Muscicolous Elatostema 142752
Muscicolous Woodbetony 287438
Musciform Woodbetony 287439
Muscle Tree 77263
Musclewood 77256,77263,77264
Muscle-wood 77263
Muscovy 153869
Musella 260356
Musengera 9994
Musengerra Podocarpus 306435
Musengerra Yaccatree 306435
Musenna Albizzia 13488
Muses-in-the-cradle 394076
Mush 338447
Musha Actinodaphne 6800
Musha Machilus 240728
Mushan Actinodaphne 6800
Mushfragrant Milkvetch 42444
Mushquash Root 90924,101852
Mushroom Tanoak 233305
Musine 112957,112959
Musizi 241896
Musk 153869,255184,255223
Musk 'Maple' 313593
Musk Bristle Thistle 73430
Musk Bristlethistle 73430
Musk Bristle-thistle 73430
Musk Button 398134
Musk Buttons 398134
Musk Clover 153869
Musk Crow Flower 8399
Musk Crowflower 8399
Musk Dendrobium 125264
Musk Giantfennel 163681
Musk Grape Flower 260301
Musk Hyacinth 260301,260322,260333
Musk Larkspur 124076
Musk Lime 93636,93639
Musk Mallow 235,194684,243738,243803
Musk Maple 313692
Musk Melon 114189
Musk Milfoil 3978,3995
Musk Million 114189
Musk Monkeytlower 255223
Musk Okra 235
Musk Onion 15499
Musk Orchid 192862
Musk Orchis 192809,192862
Musk Plant 243803,255223
Musk Primrose 314675
Musk Rose 243803,336364,336408,336766
Musk Saxifrage 349671
Musk Stork's-bill 153869
Musk Storksbill 153869
Musk Strawberry 167627
Musk Thistle 73430
Musk Willow 342967
Musk Wood Crawfoot 8399
Musk-clover 153869
Musked Cranesbill 153869
Muskingum Sedge 75466
Muskmallow 235
Musk-mallow 194684,243803
Muskmallowleaf Primrose 314929
Muskmelon 114108,114189
Musk-plant 255223
Muskrat Weed 90924
Muskrat Wort 388397
Muskrat-weed 388628
Muskroot 8396,8399
Musk-root 8399,163681
Muskroot Family 8406
Muskrootleaf Corydalis 105567
Muskroot-like Sage 344831
Muskroot-like Semiaquilegia 357919
Muskroot-like Semiaquilegia Big 357920
Musk-rose 336408
Muskseed 194684
Muskweed 8399
Muskwood 13391,181544
Musky Dendrobium 125264
Musky Gourd 114292
Musky Hibiscus 195032
Musky Marshpennywort 200336
Musky Mint 203037
Musky Monkey-flower 255223
Musky Saxifrage 349680
Musky Stork's Bill 153868,153869
Musky Yarrow 3995
Musodo Manketti Nut 334411
Musodo Manketti-nut 334411
Musquash Root 90924
Mussaenda 260375,260472
Mussel-shell Pea 97203
Mussoorie-berry 104701
Mussot Swertia 380285
Mussot Woodbetony 287442
Mustang Clover 230865
Mustang Grape 411596
Mustard 59278,59524,196952,364544, 364545
Mustard Evening Primrose 69821
Mustard Family 113144
Mustard Greens 59438
Mustard Leaf Groundsel 358717
Mustard Treacle 154363
Mustard Tree 266043,344804
Mustard Valerian 404354
Mustard Weed 364557
Mustard Weld 364557
Mustardform Sisymbrium 365416
Mustardleaf Dandelion 384479
Mustard-tips 247366,396854
Muster John Henry 383100
Musulman Iris 208723
Mutabilis Rose 336491
Muticous Henbane 201388
Mutis' Burrweed 368531
Mutisia 260530
Mutter Pea 222832,301055
Mutton Chops 86901,170193
Mutton Dock 86970
Mutton Rose 397043
Mutton Tops 86901
Mutualroll Landpick 308499
Muuse-ear Scorpion-grass 260760
Muya Stonecrop 356956
Muyou Oiltung 406019
My Lady's Bell 166125
My Lady's Eardrops 168767
My Lady's Lace 28044
My Lady's Washing Bowl 346434

Myagrum 260569
Myall 1465
Myallwood 1292
Myall-wood 1465
Mycetia 260607,260623
Mycetialeaf Ophiorrhiza 272261
Myers Asparagus 38983,38985
Mylady's Smock 72934
Myles 86901
Mynaking 414193
Myoporum 260706,260711
Myoporum Family 260700
Myose-tail Elsholtzia 144062
Myosot 260920,260922
Myosotis Stickseed 221711
Myosotis-like Cassiope 78638
Myosurus Elsholtzia 144062
Myosuruslike Dysophylla 139577
Myriactis 261060
Myriad-flower Spindle-tree 157732
Myrialepis Palm 261090,261091
Myrica Asplenifolia 261162
Myrica Box 64328
Myrica-like Distylium 134938
Myrica-like Mosquitoman 134938
Myricaria 261262
Myrioneuron 261322
Myripnois 261398
Myrmechis 261467
Myrobalan 296554,316294,386448,386473,386500,386504
Myrobalan Family 100296,386668
Myrobalan Plum 316294
Myrobalans 316294,386448,386504
Myrobalanwood 386504
Myrocarpus 261525
Myrrh 93137,101285,101474,261576,261585
Myrrh Tree 101285,101474
Myrrhtree 101285,101474
Myrsina-leaved Cyclobalanopsis 116153
Myrsine 261598
Myrsine Family 261596
Myrsine-leaf Holly 204048
Myrsineleaf Oak 116153
Myrsine-leaf Oak 116153
Myrsine-leaf Syzygium 382619
Myrsine-leaved Syzygium 382619
Myrsine-like Azalea 331301
Myrsine-like Rhododendron 331301
Myrsine-like Willow 343743
Myrsin-like Paxistima 286748
Myrsinoid Leptospermum 226467
Myrt Tree 261739
Myrtle 19947,261162,261169,261726,261739,401672,409335,409339
Myrtle Beech 266867
Myrtle Blueberry 403916
Myrtle Cactus 261705
Myrtle Coriaria 104700
Myrtle Dahoon 204078
Myrtle Eoriaria 104700
Myrtle Euphorbia 159417
Myrtle Family 261686
Myrtle Flag 5793,5794,208771
Myrtle Grass 5793,208771
Myrtle Greenbrier 366476
Myrtle Hakea 184623
Myrtle Holly 204078
Myrtle Leaved Holly 203611
Myrtle Lime 397876,397878
Myrtle Luma 156133
Myrtle Milkwort 308209
Myrtle Oak 324210
Myrtle Sedge 5793
Myrtle Spurge 159222,159417
Myrtle Tree 266867
Myrtle Wattle 1413
Myrtle Whortleberry 403916
Myrtle Willow 343743
Myrtleleaf Coriaria 104700
Myrtleleaf Heimia 188264
Myrtleleaf Holly 204078
Myrtle-leaf Holly 204078
Myrtle-leaf Milkwort 308209
Myrtle-leaf Orange 93643
Myrtle-leaved Grass Pea 222778
Myrtle-leaved Heimia 188264
Myrtle-leaved Leafflower 296672
Myrtle-leaved Leaf-flower 296672
Myrtle-leaved Marsh Pea 222778
Myrtle-leaved Mimosa 1413
Myrtle-wood 401672
Mysor Sida 362590
Mysor Spapweed 205162
Mysor Touch-me-not 205162
Mysore Fig 165336,165338
Mysore Raspberry 338886
Mysore Sida 362590
Mysore Thorn 64990,64993
Mysorethorn 64990
Mysteria 99297
Mysterious Plant 122515
Mytilaria 261884
Myxopyrum 261893

N

N. America Piratebush 61931
N. Burma Asiabell 98288
N. China Bastardtoadflax 389636
N. China Bentgrass 12053
N. China Bluebasin 350265
N. China Columbine 30090
N. China Eight Trasure 200807
N. China Figwort 355191
N. China Grape 411589
N. China Hogfennel 292877
N. China Honeysuckle 236161
N. China Hundredstamen 389636
N. China Marshweed 230282
N. China Monkshood 5303
N. China Mouseear 82910
N. China Nightshade 367282
N. China Rhubarb 329331
N. China Sandwort 31939
N. China Sweetvetch 187912
N. China Swertia 380415
N. China Woodrush 238671
N. Desert Milkvetch 42051
N. Guangdong Tanoak 233140
N. Guangxi Spiradiclis 371768
N. Guizhou Barrenwort 146964
N. Guizhou Epimedium 146964
N. Hebei Larkspur 124585
N. Hubei Stringbush 414232
N. Japanese Hill Cherry 83301
N. S. W. Coral Heath 146077
N. Sichuan Primrose 314470
N. Sichuan Wildclove 226076
N. Vietnam Hookvine 401762
N. Wallich's Rhododendron 332091
N. Xing'an Sedge 73904
N. Xingjiang Bellflower 70058
N. Xingjiang Skeletonweed 88784
N. Xinjiang Eutrema 161146
N. Xinjiang Sedge 75618
N. Xizang Bluegrass 305406
N. Xizang Draba 137306
N. Xizang Kobresia 217211
N. Xizang Plateaucress 89227,126149
N. Xizang Whitlowgrass 137306
N. Yunan Monkshood 5290
N. Yunnan Bulbophyllum 63126
N. Yunnan Camellia 68941
N. Yunnan Clematis 95044
N. Yunnan Dandelion 384822
N. Yunnan Elaeocarpus 142290
N. Yunnan Raspberry 338194

N. Yunnan Stonebean-orchis 63126
Nabalus 261913
Nabilous Azalea 330351
Nabilous Yaccatree 306510
Naboom 158982,159140
Nachlinger's Catchfly 363794
Nada Uvaria 403519
Nadelkerbel 350425
Nadina Kudrjaschevia 218280
Naegelia 366717
Nagai Podocarpus 306493
Nagaio 260718
Nagami Kumquat 167506
Nagarze Sweetvetch 188021
Nagasa Bamboo 347324
Naga-sar 252369
Nagasawa Everlasting 21643
Nagasawa Raspberry 338864
Nagasawa St. John's Wort 202035
Nagasawa Violet 410289
Nagasawa's Pearleverlasting 21643
Nageia 261970
Nageia Family 261992
Nagens Groundsel 381944
Nagesi Groundsel 381944
Nagi 306493
Nagi Podocarpus 306493
Nagi Yaccatree 306493
Nagoya Elm 401545
Nagoya Lobedleaf Elm 401545
Nahanni Aster 380929
Na-How 101852
Naiad 262015
Naiad Family 262013
Naias Family 262013
Naias Tanoak 233321
Naibu Bamboo 20205
Naibun Drepanostachyum 20205
Naibun-like Drepanostachyum 137863
Naid 262015
Naidong Rockfoil 349244
Naiheadfruit 178974
Nailorchis 9271
Nails 50825
Nailwort 284735,284855,349999
Naio 260722
Naivansha Thorn 1708
Najue Barberry 51341
Nakahara Azalea 331305
Nakahara Buckthorn 328793
Nakahara Paperelder 290389
Nakai Cleistogenes 94604
Nakai Corydalis 106519
Nakai Goosecomb 335446
Nakai Hideseedgrass 94604
Nakai Jijigrass 4148
Nakai M. Degron's Rhododendron 330536
Nakai Podocarpus 306502
Nakai Poplar 311397
Nakai Roegneria 335446
Nakai Sagebrush 35972
Nakai Speargrass 4148
Nakai's Ochrosia 263199
Nakamura St. John's Wort 202037
Nakao Machilus 240652
Nakao Monkshood 5218
Nakao Shaped Rockfoil 349683
Nake Barley 198293
Nake Bothriochloa 57553
Nake Holeglumegrass 57553
Nake Violet 410303
Nakeaster 256301,256308
Naked Albizia 13511
Naked Bishop's Cap 43511
Naked Bladderwort 403147
Naked Boys 48918,99297,111566
Naked Broom-rape 275234
Naked Buried Pepper 300421
Naked Catchfly 363192
Naked Coral Tree 154643,154646
Naked coral-tree 154646
Naked Crabgrass 130517,130683
Naked Farinose Primrose 314382
Naked Flower Tetrameles 387297
Naked Flower Trewia 394783
Naked Flowers Rhododendron 331352
Naked Fruits Rhododendron 331240
Naked Gansu Apple 243644
Naked Hardleaf Azalea 331940
Naked Hawksbeard 110953
Naked Indian Tree 64085
Naked Jack 99297
Naked Ladies 72934,99297,111566
Naked Lady 18865,72934,196439
Naked Lady Lily 18865
Naked Leafyflower 57857
Naked Lily 239257,239266,239280
Naked Maidens 99297,169719
Naked Men 99297
Naked Mexican-hat 326962
Naked Microstegium 254036
Naked Mistletoe 411071
Naked Miterwort 256032
Naked Nannies 99297,273531
Naked Nepeta 265016
Naked Oat 198293
Naked Ormosia 274419
Naked Prairie Coneflower 326962
Naked Qiaojia Yushanbamboo 416813
Naked Qiaojia Yushania 416813
Naked Silverweed Cinquefoil 312369
Naked Stalk Metalleaf 296400
Naked Styles Azalea 331928
Naked Styles Rhododendron 331928
Naked Tatsienlu Rhododendron 331940
Naked Tick-trefoil 126477
Naked Virgin 99297
Naked Virgins 99297
Naked Whiterod 407970
Naked Wild Buckwheat 152318
Nakedanther Jijigrass 4153
Nakedanther Speargrass 4153
Nakedanther Ternstroemia 386696
Nakedaxil Osmanther 276376
Nakedbud Phoebe 295410
Nakedcaule Goldsaxifrage 90421
Nakedcaule Goldwaist 90421
Nakedflower Althaea 13932
Nakedflower Beautyberry 66879
Nakedflower Murdannia 260110
Nakedflower Purplepearl 66879
Nakedflower Tetrameles 387297
Nakedflower Trewia 394783
Nakedflower Tubergourd 390164
Naked-flowered Sneezeweed 188419
Naked-flowered Tick-trefoil 126477
Nakedfruit 182491,182498
Nakedfruit Dandelion 384560
Nakedfruit Gongshan Willow 343376
Nakedfruit Kobresia 217217
Nakedfruit Pellionia 288741
Nakedfruit Redcarweed 288741
Naked-fruit Shortspine Evergreenchinkapin 78925
Nakedfruit Tanoak 233235
Nakedfruit Tickseed 104852
Naked-fruited Tanoak 233235
Nakedhead Spicegrass 239664
Naked-lady 417608
Nakedleaf Ormosia 274419
Nakedleaf Photinia 295755
Nakedleaf Solitary Rose 336384
Nakedleaf Tanoak 233308
Nakedleaf Wildjute 394635
Naked-mouth Gloxinia 177520
Naked-pedicel Barberry 52085
Nakedscale Ricegrass 300754
Nakedspike Goosecomb 335448
Nakedspike Roegneria 335448
Naked-stalk Bistamengrass 127477
Naked-stalk Silvergrass 127477
Nakedstalk Siphocranion 365197
Naked-stalked Candytuft 385555
Naked-stalked Rock-cress 385555
Nakedstamen Azalea 330463
Nakedstamen Monkshood 5237
Nakedstem Bittercress 72964
Nakedstem Corydalis 106027
Nakedstem Dewflower 260110
Naked-stem Hawksbeard 110990
Nakedstem Parrya 285004

Nakedstem Rhubarb 329335
Naked-stemmed Bulrushes 352158
Naked-stemmed Daisy 145209
Naked-stemmed Sunflower 189017
Naked-stemmed Tick Clover 126477
Nakedstigma Willow 343932
Nakedstyle Azalea 331928
Nakedtwig Azalea 330824
Nakeear Bamboo 47201
Nakeflower Waterbamboo 260110
Nakefruit Ginger 418009
Nakefruit Pepper 300467
Nakefruit Sinosenecio 365070
Nakeovary Purplestem 376444
Nakepetal Pipewort 151314
Nakepetal Violet 410770
Nakephyllary Purpledaisy 267177
Nakepistil Tea 69111
Nakerue 318945,318947
Nakesepal Pipewort 151351
Nakespike Crabgrass 130786
Nakespread Violet 409911
Nakestalk Rockfoil 349114
Nakestalk Tea 69052
Nakestem Corydalis 105953
Nakestem Cystopetal 320235
Nakestem Poppy 282617
Nakestigma Willow 343932
Nal 295888
Nalin Onion 15816
Nalta Jute 104103
Namaqua Daisy 131150,346304
Namaqualand Daisy 131150,131191,346304
Namara Potato 189073
Nambu Everest Barberry 52009
Nameless Iris 208639
Namibian Partridge Breast Aloe 16774
Nam-Nam 118060
Namula Monkshood 5440
Nan Wade 405872
Nana 21479
Nana Gracilis Falsecypress 85324
Nana Lutea Chamaecyparis 85322
Nanachangney 93645
Nanan Wood 219953
Nan-au Azalea 330255
Nanba Rattanpalm 65741
Nanban Tanoak 233389
Nanbar Rattan Palm 65741
Nanbian Azalea 331222
Nanbula Rockfoil 349684
Nance 64430
Nanchang Euonymus 157423
Nanchang Spindle-tree 157423
Nanchang Waterchestnut 394532
Nanchuan Artocarpus 36935
Nanchuan Ash 168030
Nanchuan Barberry 51602
Nanchuan Beadruby 242694
Nanchuan Begonia 49778
Nanchuan Bogorchid 158238
Nanchuan Camellia 69389
Nanchuan Clearweed 298986
Nanchuan Clethra 96536
Nanchuan Coldwaterflower 298986
Nanchuan Comanthosphace 100236
Nanchuan Cudweed 178303
Nanchuan Deerdrug 242694
Nanchuan Deutzia 127022
Nanchuan Dishspurorchis 171872
Nanchuan Diuranthera 135058
Nanchuan Drepanostachyum 20200
Nanchuan Egretgrass 135058
Nanchuan Elaeagnus 142118
Nanchuan Elatostema 142755
Nanchuan Euonymus 157338
Nanchuan Eupatorium 158238
Nanchuan Faberdaisy 161887
Nanchuan Faberiopsis 161894
Nanchuan False Solomonseal 242694
Nanchuan Falsehellebore 405617
Nanchuan Gastrochilus 171872
Nanchuan Goldenray 229116
Nanchuan Gooseberry 334176
Nanchuan Helwingia 191164
Nanchuan Holly 204079
Nanchuan Ironweed 406153
Nanchuan Isometrum 210188
Nanchuan Lime 391784
Nanchuan Linden 391784
Nanchuan Listera 264706
Nanchuan Longstalk Maple 3119
Nanchuan Machilus 240653
Nanchuan Notoseris 267176
Nanchuan Oak 324214
Nanchuan Osmanther 276374
Nanchuan Paper-mulberry 61110
Nanchuan Parnassia 284505
Nanchuan Purpledaisy 267176
Nanchuan Sage 345240
Nanchuan Sedge 75474
Nanchuan Spiraea 372072
Nanchuan Stairweed 142755
Nanchuan Stonecrop 356958,357101
Nanchuan Tangtsinia 383911
Nanchuan Tea 69389
Nanchuan Trigonotis 397423
Nanchuan Veronicastrum 407499
Nanchuan Vineclethra 95524
Nanchuan Wildginger 37691
Nanchuan Willow 344039
Nanchuan Woodbetony 287451
Nanchun Bugbane 91032
Nancy 23925,262441,374916
Nancy Gladiolus 176393
Nancy Pretty 11947,243209,349044,
374916,407072
Nancy-none-so-pretty 349044
Nancy-pretty 11947,243209,349044,
374916,407072
Nandan Ophiorrhiza 272264
Nandan Sinobambusa 364804
Nandan Tangbamboo 364804
Nandan Trigonotis 397445
Nandi Coffee 98913
Nandi Flame 370358
Nandina 262188,262189,262195,262196
Nandina Heavenly Bamboo 262189
Nanfeng Milkvetch 42769
Nangchien Larkspur 124397
Nangka Langka 36920
Nangong Cleisostoma 94455
Nangong Closedspurorchis 94455
Nangqian Catchfly 363797
Nangqian Larkspur 124397
Nangqian Onosma 271810
Nangqian Rockfoil 349689
Nangxian Rockfoil 349690
Nanhai Elatostema 142654
Nanhai Stairweed 142654
Nanhu Azalea 331308
Nanhu Barberry 51708
Nanhu Buttercup 325879
Nanhu Orchis 273481
Nanhu Rattlesnake Plantain 179660
Nanhu Red Fescuegrass 164289
Nanhu Rhododendron 331308
Nanhu Spotleaf-orchis 179660
Nanhu Waterchestnut 394414
Nanhudashan Bedstraw 170513
Nanhudashan Willowweed 146787
Nanhushan Willow 344173
Nanjian Azalea 331309
Nanjian Rhododendron 331309
Nanjiang Milkvetch 42771
Nanjiang Wingnut 320368
Nanjing Azalea 330140
Nanjing Cherry 83347
Nanjing Figwort 355197
Nanjing Indigo 205793
Nanjing Linden 391773
Nanjing Pearlsedge 354167
Nanjing Petitmenginia 292487
Nanjing Razorsedge 354167
Nanjing Rhododendron 330140
Nanjing Willow 343757
Nankeen Lily 230055
Nanking Cherry 83347
Nanking Figwort 355197
Nanking Petitmenginia 292487
Nanko Bluegrass 305739

Nanko Podocarpus 306503
Nanko Yaccatree 306503
Nankou Peashrub 72380
Nankun Azalea 331303
Nankun Calanthe 66016
Nankun Hartia 376487
Nankun Rhododendron 331303
Nankung Nagi 306503
Nanling Azalea 331103
Nanling Goldlineorchis 269039
Nanling Indianmulberry 258902
Nanling Mananthes 244290
Nanling Purplestem 376459
Nanling Stewartia 376459
Nanmu 240550,295344,295386,295417
Nanmu Phoebe 295349,295386
Nanmu Scurrula 355322
Nanmu Wood 295386
Nanning Evergreenchinkapin 78852
Nanning Fimbristylis 166417
Nanning Fluttergrass 166417
Nanning Holly 204080
Nanning Ormosia 274417
Nanning Rose 336795
Nannoglottis 262217
Nanny Goat's Mouth 116736
Nannyberry 273531,407917
Nannyberry Sweet Viburnum 407917
Nannyberry Viburnum 407917
Nanocnide 262243
Nanophyton 262259
Nanous Ammopiptanthus 19778
Nanpie 280231
Nanping Acanthopanax 2445
Nanping Azalea 331314
Nanping Cane 37162
Nanping Canebrake 37162
Nanping Loosestrife 239746
Nanping Rhododendron 331314
Nanping Shibataea 362140
Nanping Syzygium 382622
Nanping Wildtung 243353
Nanping Wobamboo 362140
Nanrensahan Arisaema 33426
Nanrenshan Newlitse 264053
Nansho-daidai 93341
Nansi Youngia 416457
Nanto Actinodaphne 6762
Nanto Tanoak 233323
Nantou Actinodaphne 6762
Nantou Begonia 50100
Nantou Butterflygrass 392923
Nantou Camellia 69398,69706
Nantou Fairybells 134453
Nantou Fig 164635
Nantou Gordonia 179745
Nantou Gordontea 179745
Nantou Greenbrier 366481
Nantou Hedyotis 187630
Nantou Holly 203883
Nantou Iris 208725
Nantou Pipewort 151400
Nantou Tanoak 233323
Nantou Tea 69706
Nanzhang Arisaema 33427
Nanzhang Ironweed 406618
Nanzhang Southstar 33427
Nanzhao Chinese Aspen 311199
Nanzheng Taillessfruit 100145
Naoshichi 93833
Naoyu Millettia 254725
Napa Thistle 81198
Napal Catnip 264935
Napal Lily 229958
Napal Violet 409720
Nap-at-noon 274823,394327,394333
Nape 59493
Nape Urceola 402197
Napier 289218
Napier Grass 289218
Napier's Fodder 289218
Napiergrass 289218
Napier-grass 289218
Napiform Root Neoporteria 264359
Napkinring Wild Buckwheat 152166
Naples Garlic 15509
Naples Onion 15509
Napo Chirita 87930
Napo Clematis 95155
Napo Elatostema 142756
Napo Fig 165343
Napo Forbes Alyxia 18501
Napo Galangal 17727
Napo Litse 234015
Napo Mananthes 244288
Napo Manglietia 244468
Napo Mycetia 260612
Napo Ophiorrhiza 272265
Napo Ormosia 274418
Napo Schefflera 350752
Napo Spapweed 205165
Napo Spiradiclis 371773
Napo Stairweed 142756
Napo Touch-me-not 205165
Napoleon 396934
Napoleon's Plume 49173
Napple-root Napple Nappery 222758
Nappy Catnip 264964
Nappy Didymocarpus 87849
Nappy Jurinea 214115
Nappypod Ormosia 274422
Naranjilla 367543,367611
Naravelia 262333
Narbone Flax 231933
Narbonne Vetch 408505
Narcisse 262348
Narcissus 262348,262438
Narcissus Anemone 23901
Narcissus Iris 208727
Narcissus Onion 15508
Narcissus-flowered Anemone 23901
Narcissuslike Swordflag 208727
Nard 262497
Nard Grass 117218,262532,262537
Nardgrass 262532,262537
Nard-grass 117218
Nardostachys 262494
Nardus Lemongrass 117218
Narea St. John's Wort 202035
Narenga 262547,262552
Narenga Sugarcane 262552
Narihira Bamboo 357939
Narihira Cane 357939
Narinjin 93579
Narnara Potato 189073
Naroowbract Windhairdaisy 348400
Naroow-leaf Pendent Inflorescence Indigo 206384
Naroowleaf Perovskia 291429
Naroow-leaved Sweetspire 210370
Narra 2602
Narra Padauk 320307
Narras 2602
Narrawa Burr 367040
Narraw-leaved Cress 225447
Narrnw-lenved Woodrush 238619
Narroecalyx Mycetia 260651
Narroeleaf Crabapple 243585
Narroeleaf Knema 216785
Narroe-leafed Mallee 155483
Narroe-leaved Apple 24604
Narroeleved Marsh Dandelion 384824
Narroelip Bulbophyllum 62738
Narroelip Curlylip-orchis 62738
Narroelip Herminium 192828
Narroelobed Eupatorium 158041
Narroeseptate Pegaeophyton 287949
Narroespike Peristylus 291202
Narroespike Perotis 291202
Narroespike Platanthera 291202
Narrosepal Pearweed 239862
Narrow Aster 58265
Narrow Bambooleaf Thorowax 63748
Narrow Bent 228346
Narrow Blackflower Jerusalemsage 295147
Narrow Blue Flag 208767
Narrow Canarygrass 293706
Narrow Clover 396823
Narrow Corncockle 11941
Narrow Cudweed 235262
Narrow Dock 339994,340135

Narrow Dwarf Daylily 191272
Narrow False Oats 398531
Narrow Fluffweed 235262
Narrow Gambirplant 401747
Narrow Garden Lettuce 219487
Narrow Hare's Ear 63775
Narrow Helleborine 82060
Narrow Hollyleaf Camellia 69146
Narrow Indian Horseorchis 215348
Narrow Knifebean 71055
Narrow Leaf Blue Star 20848
Narrow Leaf Gimlet 155746
Narrow Leaf Mountain Mint 321961
Narrow Leaf Red Tephrosia 386001
Narrow Leafed Plantain Lily 198624
Narrow Leaved Cattail 401094
Narrow Lobed Meadowsweet 166073
Narrow Macrophyllous Knotweed 309362
Narrow Metalleaf 296415
Narrow Petal Paphiopedilum 282899
Narrow Plumegrass 148921
Narrow Pod Milkvetch 43085
Narrow Rhodiola 329836
Narrow Rockfoil 349063
Narrow Scabrousleaf Elatostema 142824
Narrow Schomburgk Globba 176997
Narrow Sedge 73696
Narrow Sheath Bamboo 297194
Narrow Siberian Messerschmidia 393273
Narrow Small-reed 65490
Narrow Spike Cupscale 341962
Narrow Spot-leaf Aucuba 44889
Narrow Water Starwort 67361
Narrow Waterlily 267770
Narrow Willowleaf Cotoneaster 107672
Narrowawn Roegneria 335440
Narrowbract Bulbophyllum 63058
Narrowbract Dandelion 384467
Narrowbract Diversefolious Rockfoil 349265
Narrowbract Goldenray 229060
Narrowbract Lagotis 220160
Narrow-bract Rabbitbrush 150367
Narrowbract Raspberry 338134
Narrowbract Shorthair Aster 40161
Narrowbract Stonebean-orchis 63058
Narrowbract Tickseed 104850
Narrowbract Whiteedge Dandelion 384751
Narrowbract Windhairdaisy 348814
Narrowbract Yunnan Aster 41521
Narrowcalyx Grewia 180678
Narrow-calyx Hartia 376441
Narrowcalyx Jerusalemsage 295219
Narrowcalyx Mazus 246966
Narrowcalyx Microtoena 254266
Narrowcalyx Primrose 315007
Narrowcase Woodbetony 287701
Narrowcelony Woodbetony 287701
Narrowcleft Petalformed Meadowrue 388617
Narrowcrown Chinese Arborvitae 302736
Narrowear Bamboo 47184
Narrow-eared Bambusa 47184
Narrowedge Dandelion 384764
Narrower Sedge 76408
Narrowest-eared Bambusa 47185
Narrowflower Aniseia 25325
Narrowflower Basketvine 9487
Narrowflower Corydalis 105617
Narrowflower Didymocarpus 129971
Narrowflower Gentianella 174113
Narrowflower Hirsute Twinballgrass 209073
Narrowflower Marsdenia 245849
Narrowflower Milkgreens 245849
Narrowflower Poison-nut 378648
Narrowflower Spapweed 204775,205333
Narrowflower Strychnos 378648
Narrowflower Touch-me-not 204773,205333
Narrow-flowered Condorvine 245849
Narrow-flowered Poisonnut 378648
Narrowfruit 221742
Narrowfruit Catchfly 363555
Narrowfruit Cornsalad 404413
Narrowfruit Currant 334220
Narrowfruit Draba 137242
Narrowfruit Fragileorchis 2039
Narrowfruit Jobstears 99137
Narrowfruit Sedge 73695
Narrowfruit Snakegourd 396170
Narrowfruit Spicate Woodbetony 287691
Narrowfruit Stickseed 221742
Narrowfruit Weigttree 364957
Narrowfruit Whitlowgrass 137242
Narrow-fruited Cornsalad 404413
Narrow-fruited Currant 334220
Narrow-fruited Sedge 73700
Narrow-fruited Water-cress 336232
Narrowgalea Woodbetony 287698
Narrowglume Bluegrass 305331
Narrowglume Goosecomb 335440,335513
Narrowglume Kengyilia 215965
Narrowglume Leymus 228343
Narrowglume Roegneria 335440,335513
Narrow-glumme Bluegrass 306019
Narrow-grooved Rhododendron 331884
Narrowhairleaf Eurya 160603
Narrowhead Everlasting 21702
Narrowhead Goldenray 229202
Narrowhead Pearleverlasting 21702
Narrowhead Windhairdaisy 348266
Narrowhelmet Monkshood 5036
Narrowhelmet Tall Monkshood 5575
Narrow-inflorescence Ceanothus 79947
Narrowlabiate Woodbetony 287007
Narrowlea Bag-shaped Dutchmanspipe 34319
Narrowlea Vetchling 222748
Narrowleaf Acanthopanax 2433,2505
Narrowleaf Acuminate Goosefoot 86893
Narrowleaf Aechmea 8547
Narrowleaf Agave 10797
Narrow-leaf Agave 10797
Narrowleaf Ainsliaea 12606
Narrowleaf Altingia 18204
Narrowleaf Alyxia 18523
Narrowleaf Angelica 24293
Narrowleaf Angelon 24522
Narrowleaf Angelonia 24522
Narrowleaf Anisetree-like Pittosporum 301301
Narrowleaf Anisetree-like Seatung 301301
Narrowleaf Ant Palm 217919
Narrowleaf Ardisia 31446,31635
Narrowleaf Arisaema 33258
Narrowleaf Arnica 34677
Narrowleaf Aucuba 44911
Narrowleaf Azalea 330107
Narrowleaf Azima 45972
Narrowleaf baccharis 46212
Narrowleaf Baikal Betony 373142
Narrowleaf Balansa Pittosporum 301222
Narrowleaf Balansa Seatung 301222
Narrowleaf Baliosperm 46963
Narrowleaf Balsam Poplar 311233
Narrowleaf Basket Willow 344245
Narrowleaf Basketvine 9423
Narrowleaf Batavia Cinnamon 91333
Narrowleaf Bayberry Box 64329
Narrowleaf Beautiful Shrubcress 368571
Narrowleaf Beautiful Solms-Laubachia 368571
Narrowleaf Beautyberry 66817
Narrowleaf Beauty-berry 66897
Narrowleaf Bedstraw 170355
Narrowleaf Bengal Loquat 151133
Narrowleaf Bethlehemsage 283624
Narrowleaf Betony 373109
Narrowleaf Bigumbrella 59194
Narrow-leaf Bird's-foot Trefoil 237617
Narrow-leaf Bitter Pea 123269
Narrowleaf Bittercress 72829
Narrowleaf Bittersweet 80252
Narrowleaf Bladdercampion 114062
Narrowleaf Bladderwort 403716
Narrowleaf Bluebasin 350129
Narrowleaf Bluebeard 78015
Narrowleaf Bluebell Creeper 368540
Narrowleaf Bluegrass 305759
Narrowleaf Blumea 55833
Narrowleaf Bock Lilyturf 272055
Narrow-leaf Bottlebrush 67278
Narrowleaf Bousigonia 57912
Narrow-leaf Bower Wattle 1139
Narrowleaf Box 64329,64378
Narrowleaf Bracteate Goatwheat 44252

Narrowleaf Brassaiopsis 59194
Narrowleaf Burmann Cinnamon 91333
Narrowleaf Burreed 370030,370100
Narrowleaf Bur-reed 370030
Narrowleaf Calanthe 65879
Narrowleaf Callianthemum 66701
Narrowleaf Camellia 68904,69077
Narrowleaf Campion 363691
Narrowleaf Cape-cowslip 218835
Narrowleaf Catnip 265052
Narrowleaf Cattail 401094
Narrowleaf Cayratia 79868
Narrowleaf Ceropegia 84269
Narrowleaf Chalkplant 183158
Narrowleaf Changbaishan Ladybell 7748
Narrow-leaf Cherry-laurel 223155
Narrowleaf Chinese Knotweed 308972
Narrowleaf Chinese Photinia 295775
Narrowleaf Chinese Vetch 408340
Narrowleaf Chloranthus 88265
Narrowleaf Cinquefoil 312358,313016
Narrowleaf Clark Clematis 94841
Narrowleaf Cliffbean 66646
Narrowleaf Clovershrub 70869
Narrow-leaf Clusterpalm 6853
Narrowleaf Coffee 99008
Narrowleaf Colona 99968
Narrowleaf Common Bluebeard 78015
Narrowleaf Composite Oplismenus 272678
Narrowleaf Concavepetal Ancylostemon 22160
Narrowleaf Corydalis 105819,106141
Narrowleaf Cottonrose 235262
Narrowleaf Cottonsedge 152769
Narrowleaf Cottonwood 311233
Narrowleaf Cowparsnip 192263
Narrowleaf Cowpea 408831
Narrowleaf Cremanthodium 110345
Narrowleaf Crimson Clover 396823
Narrowleaf Crucianella 113112
Narrow-leaf Cumbungi 401105
Narrow-leaf Daphne 122520
Narrowleaf Daphniphyllum 122664
Narrow-leaf Deepblue Honeysuckle 235701
Narrowleaf Delavay Lacquertree 393450
Narrowleaf Delavay Peony 280186
Narrowleaf Devil Rattan 121487
Narrowleaf Differentfruit Tickclover 126390
Narrowleaf Dishspurorchis 171865
Narrowleaf Dissimilar Willow 343311
Narrowleaf Dittany 129627
Narrowleaf Dock 340269
Narrowleaf Dogwood 124882
Narrowleaf Dracaena 137337
Narrowleaf Dragonbood 137337
Narrowleaf Dragonheadsage 247733
Narrow-leaf Drumsticks 210236
Narrowleaf Drymonia 138528
Narrow-leaf Dwarf Peashrub 72328
Narrow-leaf Dwarf Pea-shrub 72328
Narrowleaf Dzungar Milkaster 169797
Narrowleaf Earliporchis 277244
Narrowleaf Early Blueberry 403933
Narrowleaf Echinodorus 140538
Narrowleaf Elaeagnus 141930
Narrowleaf Elatostema 142717
Narrowleaf Elsholtzia 143999
Narrowleaf Eritrichium 153424
Narrowleaf Eurya 160602
Narrowleaf Evergreenchinkapin 78893
Narrowleaf Falsehellebore 405636
Narrowleaf Falsepimpernel 231544
Narrowleaf Falsevalcrian 373492
Narrow-leaf Farges Hornbeam 77292
Narrowleaf Fascicled Clematis 94895
Narrow-leaf Fewflower Lysionotus 239975
Narrowleaf Fiddleleaf Fig 165429
Narrowleaf Fieldcitron 400532
Narrowleaf Fire Thorn 322450
Narrowleaf Firethorn 322450
Narrow-leaf Fireweed 85875
Narrowleaf Flax 231880
Narrowleaf Fleabane 150763
Narrowleaf Flickingeria 166956
Narrowleaf Forkstyleflower 374496
Narrowleaf Fourleves Bedstraw 170290
Narrowleaf Foxglove 130369
Narrowleaf Galangal 17682
Narrowleaf Gardenia 171401,171440
Narrow-leaf Gardenia 171246
Narrowleaf Gastrochilus 171865
Narrow-leaf Geebung 292030
Narrowleaf Germander 388184
Narrowleaf Ginsen 280788
Narrowleaf Globemallow 370938
Narrowleaf Glomerulate Brassaiopsis 59208
Narrowleaf Glutinousmass 179500
Narrowleaf Goldenaster 90230
Narrowleaf Goldenbush 150329
Narrowleaf Goosefoot 86893,87133
Narrow-leaf Goosefoot 87133
Narrowleaf Graceful Bedstraw 170355
Narrowleaf Grape 411973
Narrowleaf Grassland Chalkplant 183186
Narrowleaf Green Bitterbamboo 304019
Narrowleaf Greenman Litse 233936
Narrowleaf Greyleaf Willow 344142
Narrow-leaf Gromwell 233698
Narrow-leaf Gueldenstaedtia 181687
Narrowleaf Gymnaster 256303
Narrowleaf Hawksbeard 111050
Narrowleaf Hawkweed 195736
Narrowleaf Hedyotis 187512
Narrowleaf Helwingia 191152
Narrowleaf Hempnettle 170058
Narrow-leaf Heritiera 192485
Narrowleaf Heterocarpus Mountain Leech 126390
Narrowleaf Hillcelery 276777
Narrowleaf Hoary Spiraea 371866
Narrowleaf Holboellia 197213
Narrow-leaf Holly 203801
Narrowleaf Homalium 197723
Narrowleaf Honeysuckle 235665
Narrowleaf Hongkong Milkwort 308109
Narrow-leaf Hongkong Milkwort 308109
Narrowleaf Hooker Whitepearl 172087
Narrowleaf Hooker Wintergreen 172087
Narrowleaf Horny Tanoak 233155
Narrow-leaf Hydrangea 200106
Narrowleaf Impatient Bittercress 72831
Narrowleaf Indian Horseorchis 215349
Narrowleaf Indian Madder 337945
Narrow-leaf Indianmulberry 258873
Narrowleaf Irisorchis 267935
Narrowleaf Ironbark Eucalyptus 155545
Narrow-leaf Italian Buckthorn 328600
Narrowleaf Ixeris 210540
Narrowleaf Japan Eurya 160504
Narrowleaf Japanese Buckthorn 328741
Narrow-leaf Kadsura 214950
Narrowleaf Kirilow Clematis 95052
Narrowleaf Knotweed 308777,308881
Narrow-leaf Knotweed 308832,308836
Narrowleaf Kozlov Spurge 159194
Narrowleaf Ladybell 7828
Narrowleaf Landpick 308647
Narrowleaf Large-keel Vetch 408484
Narrowleaf Laurel Sweetleaf 381148
Narrowleaf Lavender 223251,223345
Narrowleaf Leathery Litse 234069
Narrowleaf Leathery-leaf Litse 234069
Narrowleaf Ledum 223909
Narrowleaf Leopard Lily 229979
Narrowleaf Lepidagathis 225224
Narrowleaf Lilyturf 272147
Narrowleaf Litse 234069
Narrowleaf Lobelia 234376
Narrowleaf Longanther Eight Trasure 200804
Narrowleaf Loosespike Oplismenus 272678
Narrowleaf Loosestrife 239773
Narrowleaf Lovegrass 147979
Narrowleaf Low Torularia 264611
Narrowleaf Lungwort 321636
Narrowleaf Lupine 238422
Narrowleaf Lyonia 239398
Narrowleaf Magnoliavine 351066
Narrowleaf Mapleleaf Excoecaria 161631
Narrowleaf Maya Corydalis 106134
Narrowleaf Meehania 247733
Narrowleaf Melaleuca 248084

Narrowleaf Meliosma 249359
Narrow-leaf Metalleaf 296415
Narrowleaf Michelia 252813,252850
Narrowleaf Microula 254342,254368
Narrowleaf Milkaster 169764
Narrowleaf Milkvetch 41987
Narrowleaf Millettia 66646
Narrow-leaf Mint Bush 315602
Narrowleaf Mock Goldenweed 265640
Narrowleaf Morina 258854
Narrowleaf Mosquitotrap 117694
Narrowleaf Motherweed 231544
Narrowleaf Mountain Leech 126616
Narrowleaf Mouseear 82692
Narrowleaf Murdannia 260101
Narrow-leaf Myrtle 45301
Narrowleaf N. China Sweetvetch 187914
Narrowleaf Nakeaster 256303
Narrowleaf Nepal Lily 230000
Narrowleaf Nepeta 264947,265052
Narrowleaf Nettle 402847
Narrowleaf Nigritella 266259
Narrowleaf Northern Bedstraw 170250
Narrowleaf Nutantdaisy 110345
Narrowleaf Oak 116056,324449
Narrowleaf Oberonia 267935
Narrowleaf Oil Camellia 69252
Narrowleaf Olive 270152
Narrowleaf Oreorchis 274050
Narrowleaf Osbeckia 276091
Narrowleaf Osmanther 276261
Narrow-leaf Osmanther 276362
Narrow-leaf Osmanthus 276362
Narrowleaf Pair Vetch 408654
Narrow-leaf Paper-bark 248075
Narrowleaf Paraphlomis 283624
Narrowleaf Paris 284343,284378
Narrowleaf Patrinia 285819
Narrowleaf Peashrub 72351
Narrowleaf Pecteilis 286835
Narrowleaf Penduliracemed Indigo 206376
Narrowleaf Penstemon 289320
Narrowleaf Peony 280186
Narrow-leaf Pepperwort 225447
Narrowleaf Pertybush 292043
Narrowleaf Phacelurus 293216
Narrowleaf Phoebe 295346
Narrowleaf Photinia 295794
Narrowleaf Pink 127820
Narrowleaf Pipe Tanoak 233155
Narrowleaf Pittosporum 301373
Narrow-leaf Pittosporum 301218
Narrow-leaf Plantain 302034
Narrowleaf Plantain Lily 198624
Narrowleaf Plantain-lily 198624
Narrowleaf Platanthera 302522
Narrowleaf Plumegrass 148920
Narrowleaf Podocarpus 306467
Narrowleaf Pogostemon 306969
Narrowleaf Poisoncelery 90935
Narrowleaf Poplar 311233
Narrowleaf Potanin Leptodermis 226121
Narrowleaf Potanin Wildclove 226121
Narrowleaf Pothos 313189
Narrowleaf Potts Waxplant 198885
Narrowleaf Precocious Stachyurus 373563
Narrowleaf Pricklyash 417345
Narrowleaf Primrose 314983
Narrowleaf Privet 229436
Narrowleaf Pseuderanthemum 317256
Narrowleaf Pseudosasa 318330
Narrowleaf Purple Willow 343947
Narrow-leaf Purpleflower Euonymus 157802
Narrowleaf Purpleflower Isometrum 210185
Narrowleaf Purpleflower Oreocharis 273835
Narrowleaf Purplepearl 66817
Narrowleaf Pussytoes 26603
Narrowleaf Qinggang 116056
Narrowleaf Rabbiten-wind 12606
Narrowleaf Rabdosia 209628
Narrowleaf Rapanea 261626
Narrowleaf Rattlebox 112342,112475
Narrowleaf Red Hoarypea 386001
Narrowleaf Red Ironbark 155545
Narrowleaf Reddish Beautyberry 66914
Narrowleaf Reddish Purplepearl 66914
Narrowleaf Redflower Greenvine 204649
Narrowleaf Rhodiola 329897
Narrowleaf Rhubarb 329400
Narrowleaf Ribbed Vetch 408351
Narrowleaf Ricebag 181687
Narrowleaf Rockjasmine 23303
Narrowleaf Rockvine 387785
Narrowleaf Roegneria 335506
Narrowleaf Rose Cowwheat 248208
Narrowleaf Rustyhair Clematis 95077
Narrowleaf Sageretia 342160
Narrow-leaf Sand Verbena 691
Narrowleaf Sandwort 32318
Narrowleaf Satinflower 365747
Narrowleaf Scabious 350129
Narrowleaf Schefflera 350747
Narrowleaf Screwtree 190094
Narrowleaf Seatung 301373
Narrow-leaf Seatung 301218
Narrowleaf Selfheal 316146
Narrowleaf Seoul Catchfly 364049
Narrowleaf Sericeous Newlitse 264091
Narrowleaf seudostellaria 318515
Narrowleaf Shibataea 362136
Narrowleaf Short-stalked Indian-mulberry 258879
Narrowleaf Sibiraea 362394
Narrowleaf Silky Raspberry 338752
Narrowleaf Sisyrinchium 365747
Narrowleaf Skullcap 355711
Narrowleaf Smallligulate Aster 40005
Narrowleaf Smallreed 127196
Narrow-leaf Snake Herb 139469
Narrowleaf Solomonseal 308647
Narrowleaf Sophora 369125
Narrowleaf Southstar 33258
Narrowleaf Spicebush 231303
Narrowleaf Spiderling 56454
Narrow-leaf Spineflower 89053
Narrowleaf Spinestemon 306969
Narrow-leaf Spiny Goatwheat 44277
Narrowleaf Spiraea 371895
Narrowleaf Spring Bittercress 72831
Narrowleaf Squarebamboo 87550
Narrowleaf St. John's Wort 201710
Narrowleaf Stairweed 142717
Narrowleaf Stellera 375187
Narrowleaf Stonecrop 357123
Narrowleaf Stoneseed 233749
Narrow-leaf Strigose Hydrangea 200110
Narrowleaf Stringbush 414129,414261
Narrowleaf Stylegrass 379081
Narrowleaf Stylidium 379081
Narrowleaf Subundulate Pellionia 288773
Narrowleaf Sunflower 188906
Narrowleaf Sutured Pittosporum 301368
Narrowleaf Sutured Seatung 301368
Narrowleaf Swallowwort 117694
Narrowleaf Sweetleaf 381145
Narrowleaf Syrenia 382103
Narrowleaf Tabacco 266061
Narrowleaf Tainia 383120
Narrowleaf Taiwan Arisaema 33332
Narrowleaf Taiwan Fig 164996
Narrowleaf Tang Willow 344180
Narrowleaf Tanoak 233152
Narrowleaf Tea 68904
Narrowleaf Thorow Wax 63772
Narrowleaf Thorowax 63748
Narrowleaf Threevein Aster 39958
Narrowleaf Tickclover 126616
Narrowleaf Tigernanmu 122664
Narrowleaf Trefoil 396823
Narrowleaf Trema 394631
Narrowleaf Tremacron 394677
Narrowleaf Trigonotis 397392,397398
Narrowleaf Tulip 400225
Narrowleaf Turpinia 400532
Narrowleaf Undulateleaf Oplismenus 272691
Narrowleaf Uniflower Iris 208918
Narrowleaf Uniflower Swordflag 208918
Narrowleaf Uvaria 403543
Narrowleaf Veing Pepper 300519
Narrowleaf Vetch 408284
Narrowleaf Villose Paphiopedilum 282918

Narrowleaf Waterbamboo 260101
Narrowleaf Waterparsnip 53157
Narrowleaf Waterplantain 14725
Narrowleaf Waveedge Redcarweed 288773
Narrowleaf Waxplant 198847
Narrowleaf Whiteflower Michelia 252928
Narrowleaf Whiteroot Acanthopanax 143633
Narrow-leaf White-topped Aster 360709
Narrowleaf Wild Arrowhead 342421
Narrowleaf Wildjute 394631
Narrowleaf Wildorchis 274050
Narrowleaf Willow 343360
Narrow-leaf Willowherb 85875
Narrowleaf Wobamboo 362136
Narrowleaf Xianbeiflower 362394
Narrowleaf Yellowtops 166822
Narrowleaf Yunnan Devilpepper 327084
Narrowleaf Zamia 417003
Narrowleaf Zinnia 418038
Narrow-leafed Ash 167910
Narrow-leafed Black Peppermint 155666
Narrowleafed Firethorn 322450
Narrow-leafed Grass Tree 415175
Narrow-leafed Montia 258258
Narrow-leafed Mountain-mint 321961
Narrow-leafed Peppermint 155716
Narrowleaflet Milkvetch 41987
Narrow-leaflet Rosewood 121829
Narrow-leaved Alternanthera 18147
Narrow-leaved Altingia 18204
Narrow-leaved Alyxia 18523
Narrow-leaved Anzacwood 310754
Narrow-leaved Arrowhead 342417
Narrow-leaved Ash 167910
Narrow-leaved Atriplex 44510
Narrow-leaved Baliospermum 46963
Narrow-leaved Barberry 52169
Narrow-leaved Basketvine 9423
Narrow-leaved Beauty-berry 66817
Narrow-leaved Bellflower 70214
Narrow-leaved Bird's-foot-trefoil 237617
Narrow-leaved Bittercress 72829
Narrow-leaved Bitter-cress 72829
Narrow-leaved Blue-eyed Grass 365747
Narrow-leaved Blue-eyed-grass 365685
Narrow-leaved Bluegrass 305328
Narrow-leaved Blue-grass 305328
Narrow-leaved Bluets 187632
Narrow-leaved Bottlebrush 67278
Narrow-leaved Box 64378,155652
Narrow-leaved Brassaiopsis 59194
Narrow-leaved British Inula 207056
Narrow-leaved Brook Euonymus 157316
Narrow-leaved Bur-reed 370030,370044
Narrow-leaved Caper 71679
Narrow-leaved Cattail 401094
Narrow-leaved Cat-tail 401094
Narrow-leaved Centaury 81179
Narrow-leaved China-laurel 28390
Narrow-leaved Coffee 99008
Narrow-leaved Cold Spindle-tree 157523
Narrow-leaved Cotton Bush 179032
Narrow-leaved Cotton-grass 152769
Narrow-leaved Cow-wheat 248185,248186
Narrow-leaved Cudweed 235262
Narrow-leaved Cyclobalanopsis 116056,116200
Narrow-leaved Damnacanthus 122028
Narrow-leaved Daphniphyllum 122664
Narrow-leaved Day Lily 191312
Narrow-leaved Dayflower 101008
Narrow-leaved Dock 339994,340269
Narrow-leaved Dracaena 137337
Narrow-leaved Eelgrass 418380
Narrow-leaved Eel-grass 418387
Narrow-leaved Euonymus 157321
Narrow-leaved European Spindle-tree 157440
Narrow-leaved Eurya 160602
Narrow-leaved Evening-primrose 269401
Narrow-leaved Evergreen Chinkapin 78893
Narrow-leaved Everlasting Pea 222844
Narrow-leaved Everlasting Peavine 222844
Narrow-leaved Everlasting-pea 222844
Narrow-leaved Fevervine 280110
Narrowleaved Fieldorange 249649
Narrow-leaved Fire Thorn 322450
Narrow-leaved Firethorn 322450
Narrow-leaved Flax 231880
Narrow-leaved Forrest Barberry 51336
Narrow-leaved Fortune Spindle-tree 157496
Narrow-leaved Four-o'clock 255723
Narrow-leaved Gentian 173591
Narrow-leaved Goldenrod 368134
Narrow-leaved Grape 411973
Narrow-leaved Greenbrier 366324
Narrow-leaved Gromwell 233749
Narrow-leaved Guadua 181474
Narrow-leaved Guilfoyle Polyscias 310209
Narrow-leaved Gum-weed 181132
Narrow-leaved Hare's-ear 63826
Narrow-leaved Hawk's-beard 111050
Narrow-leaved Hawkweed 196057
Narrow-leaved Hedge Nettle 170058
Narrow-leaved Hedge Veronica 186972
Narrow-leaved Hedge-nettle 373459
Narrow-leaved Hedyotis 187681
Narrow-leaved Helleborine 82060
Narrow-leaved Hemp Nettle 170058
Narrow-leaved Hemp-nettle 170058
Narrow-leaved Homalium 197723
Narrow-leaved Honeysuckle 235665
Narrow-leaved Hydrangea 200106
Narrow-leaved Iceplant 101810
Narrow-leaved Indian-mulberry 258873
Narrowleaved Ironbark 155545
Narrow-leaved Jerusalem Cowslip 321636
Narrow-leaved Knotweed 308881
Narrow-leaved Lamb's Quarters 87070
Narrow-leaved Laurel 215387
Narrow-leaved Lavender 223251
Narrow-leaved Lily 230005,230009
Narrow-leaved Loosestrife 239809,240055
Narrow-leaved Lungwort 321636
Narrow-leaved Lupin 238422
Narrow-leaved Machilus 240726
Narrow-leaved Mahonia 242476
Narrow-leaved Marsh Dandelion 384824
Narrow-leaved Marsh Orchid 273681
Narrow-leaved Marsh Willowherb 146812
Narrow-leaved Meadow Rue 388421
Narrow-leaved Meadow-grass 305328
Narrow-leaved Meadow-rue 388421
Narrow-leaved Meliosma 249359
Narrowleaved Melodinus 249649
Narrow-leaved Melodinus 249649
Narrow-leaved Michaelmas-daisy Panicled Aster 380912
Narrow-leaved Milkweed 38125
Narrow-leaved Miner's Lettuce 258275
Narrow-leaved Monardella 257196
Narrow-leaved Mountain Mint 321961
Narrow-leaved Mountain Trumpet 99859
Narrow-leaved Mouse Ear 83064
Narrow-leaved Murdannia 260113
Narrow-leaved Namu 240726
Narrow-leaved Nanmu 240604
Narrow-leaved Oak 116200
Narrow-leaved Oleaster 141932
Narrow-leaved Olive 270152
Narrow-leaved Osmanther 276261
Narrow-leaved Oval Sedge 76508
Narrow-leaved Ovate-fruited Milkvetch 42447
Narrow-leaved Palm Lily 104366
Narrow-leaved Panic-grass 128525
Narrow-leaved Pea-shrub 72351
Narrowleaved Pendentlamp 84269
Narrow-leaved Pepperwort 225447
Narrow-leaved Phillyrea 294762
Narrow-leaved Phoebe 295346
Narrow-leaved Photinia 295794
Narrow-leaved Pinweed 223706
Narrow-leaved Plantain 302034
Narrow-leaved Pondweed 312046,312270
Narrow-leaved Prickly-ash 417345
Narrow-leaved Puccoon 233749
Narrow-leaved Purple Coneflower 140067
Narrow-leaved Pygmy Pussytoes 26475
Narrow-leaved Rag'wort 359103
Narrow-leaved Red Ironbark 155545
Narrow-leaved Reedmace 401094
Narrow-leaved Rhododendron 330107

Narrow-leaved Rock Daisy 291300
Narrow-leaved Rockvine 387785
Narrow-leaved Rosewood 121829
Narrow-leaved Sageretia 342160
Narrow-leaved Saw-wort 348146
Narrow-leaved Screwtree 190094
Narrow-leaved Screw-tree 190094
Narrow-leaved Senecio 358828
Narrow-leaved Sibiraea 362394
Narrow-leaved Sida 362478
Narrow-leaved Skeletonplant 375984
Narrow-leaved Small Pondweed 312241
Narrow-leaved Sneezeweed 188397
Narrow-leaved Soaproot 416603
Narrow-leaved Soapweed 416603
Narrow-leaved Sorrel 340279
Narrow-leaved Southern Rhododendron 331223
Narrow-leaved Speedwell 407349
Narrow-leaved Spindle-tree 157321
Narrow-leaved Spring Beauty 94373
Narrow-leaved Square-bamboo 87550
Narrow-leaved St. John's-wort 201710
Narrow-leaved Strawberry-bush 157316
Narrow-leaved Stylose Staff-tree 80330
Narrow-leaved Sundew 138299
Narrow-leaved Sunflower 188906
Narrow-leaved Sweet Briar 336650
Narrowleaved Swertia 380111
Narrow-leaved Tanoak 233152
Narrow-leaved Tetraplasia 122028
Narrow-leaved Tickclover 126616
Narrow-leaved Tick-clover 126620
Narrow-leaved Trema 394631
Narrow-leaved Umbrellawort 255723
Narrow-leaved Umbrella-wort 255723
Narrow-leaved Vervain 405888
Narrowleaved Vetch 408284
Narrow-leaved Vetch 408284,408578
Narrow-leaved Vetchling 222844
Narrow-leaved Viburnum 407895
Narrow-leaved Water Parsnip 53157
Narrow-leaved Water Primrose 238181
Narrow-leaved Water-dropwort 269358
Narrow-leaved Water-parsnip 53157
Narrow-leaved Water-plantain 14734,14745
Narrow-leaved White-topped Aster 360709
Narrow-leaved Wild Leek 15832
Narrow-leaved Willow Herb 146766
Narrow-leaved Willowherb 146766
Narrow-leaved Wood Sedge 74312
Narrow-leaved Woolly Sedge 75065,75066
Narrow-leaved Xingshan Barberry 51336
Narrow-leaved Yucca 416603
Narrow-leaves Camellia 69146
Narrowlef Meryta 250946
Narrowligule Dwarfnettle 262221
Narrowligule Nannoglottis 262221
Narrowlip Arachnis 30534
Narrowlip Epipogium 147324
Narrowlip Woodbetony 287006
Narrow-lipped Helleborine 147168
Narrowlobate Primrose 315046
Narrowlobe Cinquefoil 312359
Narrowlobe Curvedstamencress 238028
Narrowlobe Duvalia 139134
Narrowlobe Monkshood 5531
Narrowlobe Woodbetony 287007
Narrowlobed Aletris 14533
Narrow-lobed Callalily 417100
Narrowlobed False Rattlesnakeroot 283686
Narrowlobed Rattlesnakeroot 313792
Narrowlobed Shanxi Monkshood 5583
Narrowly Obovateleaf Stauntonvine 374426
Narrow-oblong Basketvine 9422
Narrowpanicle Murdannia 260117
Narrow-panicle Rush 212924
Narrowpanicle Waterbamboo 260117
Narrowpetal African Malcolmia 243164
Narrowpetal Argyroderma 32691
Narrowpetal Buttercup 326082
Narrow-petal Carnose Deutzia 126865
Narrowpetal Chayu Rockfoil 350051
Narrowpetal Crapemyrtle 219968
Narrowpetal Deerdrug 242704
Narrowpetal Deqin Sandwort 31762
Narrow-petal Epidendrum 146482
Narrowpetal Gentian 173226
Narrowpetal Habenaria 184085
Narrow-petal Hydrangea 199792
Narrowpetal Madonna Lily 229791
Narrow-petal Magnolia 242058
Narrowpetal Paphiopedilum 282859
Narrowpetal Parnassia 284506
Narrowpetal Platanthera 302519
Narrowpetal Pocktorchid 282859
Narrowpetal Rhodoleia 332211
Narrowpetal Rockfoil 349809
Narrow-petaled Hydrangea 199792
Narrow-petaled Trillium 397543
Narrowpod Milkvetch 43085
Narrow-raceme Fevervine 280109
Narrowraceme Meadowrue 388431
Narrow-scaled Felt-thorn 387087
Narrow-sealed Felt Thorn 387087
Narrow-sepal Basketvine 9481
Narrowsepal Ghostfluting 228326
Narrowsepal Grewia 180678
Narrowsepal Hartia 376441
Narrowsepal Himalayan Honeysuckle 228326
Narrow-sepal Lysionotus 239954
Narrowsepal Spapweed 205334
Narrowsepal Thickspur Larkspur 124446
Narrowsepal Thickspurred Larkspur 124446
Narrowsepal Touch-me-not 205334
Narrowsepal Tube-flower Basketvine 9486
Narrowsepal Xantolis 415241
Narrowsepaled Ardisia 31613
Narrow-sepaled Basketvine 9481
Narrow-sepaled Grewia 180678
Narrow-sepaled Hartia 376441
Narrow-sepaled Himalaya-honeysuckle 228326
Narrowsheath Bamboo 297194
Narrow-sheath Indosasa 206872
Narrow-sheathed Bamboo 297194
Narrow-sheathed Indosasa 206872
Narrowspathe Devil Rattan 121488
Narrowspiculate Galingale 119687
Narrow-spike Bambusa 47213
Narrowspike Barley 198369
Narrowspike Bedstraw 170610
Narrowspike Brome 60984
Narrowspike Cypressgrass 119687
Narrowspike Eight Trasure 200780
Narrowspike Goosecomb 335207
Narrowspike Larkspur 124036
Narrowspike Meliosma 249431
Narrowspike Needlegrass 376899
Narrowspike Pleurothallis 304929
Narrowspike Roegneria 335207
Narrowspike Sinocrassula 218362
Narrowspike Stonecrop 200780
Narrow-spiked Barberry 52187
Narrow-spiked Meliosma 249431
Narrowsplit Monkshood 5103
Narrowspur Corydalis 106043
Narrowstipe Primrose 315047
Narrowstipule Violet 409690
Narrowteeth Betony 373385
Narrowtepal Elatostema 142604
Narrowtepal Stairweed 142604
Narrowtongue Cremanthodium 110463
Narrowtongue Nutantdaisy 110463
Narrow-tongue Pansy Orchid 254938
Narrowtongue Pelatantheria 288602
Narrowtongued Doronicum 136375
Narrowtooth Rabdosia 324882
Narrowtube Honeysuckle 236117
Narrowtube Skullcap 355775
Narrowtube Woodbetony 287756
Narrow-utricle Sedge 74333
Narrowwing Cowparsnip 192381
Narrowwing Hookedstyle Meadowrue 388707
Narrowwing Milkvetch 42275
Narrowwing Saussurea 348329
Narrowwing Tailanther 381951
Narrowwing Twayblade 232089
Narrowwing Valerian 404359
Narrowwing Windhairdaisy 348329
Narrow-winged Ainsliaea 12614

Narrow-winged Girald Spindle-tree 157541
Narrow-winged Rabbiten-wind 12614
Narrow-winged Synotis 381951
Narthecium 262572
Naseberry 244605
Nashag 31142
Nashi Pear 323268
Nastal-cherry 136009
Nastrutium-leaf Begonia 49853
Nasturtium 399597,399601
Nasturtium Family 399596
Nasturtium Lobbianum 399605
Nasturtiumleaf Spapweed 205406
Nasturtiumleaf Touch-me-not 205406
Natal Bauhinia 49178
Natal Begonia 50101
Natal Bottlebrush 181040,181042,181043
Natal Chili 72070
Natal Chilli 72070
Natal Coral Tree 154670
Natal Fig 165344
Natal Flame Bush 13432
Natal Gardenia 171276
Natal Glorybower 96099
Natal Grass 249323,333024
Natal Guarri 155957
Natal Indigo 205678
Natal Ivy 359415
Natal Mahogany 216354,216355
Natal Orange 378888
Natal Plum 76926
Natal Primrose 390718
Natal Shellflower Bush 57996
Natal Vine 411885
Natal Wedding Flower 136009
Natal Wild Banana 377563
Natal-grass 333024
Natal-plum 76926
Nathalia Ramonda 325226
Nathaliella 262776
Native Apricot 301369
Native Australian Frangipani 201240
Native Bamboo 47311
Native Banana 260207
Native Bloodleaf 4263
Native Boronia 57277
Native Broom 409324
Native Carrot 174806
Native Cassia Bark Tree 91389
Native Cherry 161712
Native Daphne 153333
Native Frangipani 201240
Native Fuchsia 105395,146072
Native Hibiscus 194837
Native Hopbush 135214
Native Hops 135215
Native Jacob's Ladder 307241
Native Mulberry 300836
Native Orange 71808
Native Parsley 235412
Native Peren Sunflower 188997
Native Poplar 197625,270563
Native Potato 99571
Native Quince 14374
Native Rose 57277
Native Rosemary 413914
Native Tamarind 132992
Native Willow 343726
Native Wistaria 185760
Native Wormwood 35837
Nato Tree 280437
Natotree 280437
Natsiatopsis 262783,262784
Natsiatum 262786
Natsu Gumi 142152
Natsu-daidai 93340
Natural Grass 247366
Nauclea 262795
Naughty Man 36474
Naughty Man's Cherries 44708
Naughty Man's Oatmeal 28044
Naughty Man's Parsley 28044
Naughty Man's Plaything 72038,402886
Nautilocalyx 262896
Navajo Bridge Pricklypear 273019
Navajo ephedra 146153
Navajo Fleabane 150561
Navajo Pincushion Cactus 287904
Navajo Tea 194415
Navajo Thistle 91758
Nave Elm 401593
Navel of the Earth 401711
Navel Orange 93797
Navelgrass 171903
Navelseed 270677
Navel-seed 270677
Navelwort 118008,200253,270677,270705,
401678,401711
Naven 59603
Navet 59493
Navet-gentle 59493
Navew 59493,59575
Naville Spicegrass 239747
Navy Bean 294056
Nayong Maple 3216
Nayong Synotis 381945
Nayong Tailanther 381945
Nazingu 256107
NE. China Alopecurus 17543
NE. China Beautiful Currant 334167
NE. China Bulrush 353750
NE. China Burreed 370078
NE. China Cranesbill 174583
NE. China Cudweed 178278
NE. China Dandelion 384719
NE. China Eritrichium 153479
NE. China Forsythia 167450
NE. China Hornwort 83568
NE. China Isopyrum 210281
NE. China Knotwood 44269
NE. China Lady's Slipper 120289
NE. China Ladyslipper 120289
NE. China Lily 229834
NE. China Maple 3129
NE. China Neillia 263183
NE. China Peashrub 72279
NE. China Platanthera 302290
NE. China Sagebrush 35865
NE. China Schneider Buckthorn 328853
NE. China Speedwell 317949
NE. China Swertia 380302
NE. China Vetchling 222852
NE. China Violet 410201
NE. China Wildtung 243384
NE. China Willowweed 146663
NE. China Windhairdaisy 348510
Nea Zealand Lilac 190247
Nealley's Globe-amaranth 179241
Neapolitan Cyclamen 115953
Neapolitan Garlic 15509
Neapolitan Melon 114212
Neapolitan Poplar 311411
Neapolitan Star of Bethlehem 274708
Neaps Neeps 59575
Near Equal Bamboo 47462
Near Hoarhound 245761
Near Larkspur 124511
Near-entire Dark-fruit Barberry 51336
Nearly Entire Groundsel 360137
Nearly Entire Tephroseris 385920
Nearly Entire-leaved Mahonia 242645
Nearly Gaoliang 369689
Nearlyentireleaf Dogtongueweed 385920
Nearlyentireleaf Groundsel 360137
Nearly-fascicled Box-thorn Barberry 51893
Nearlyleathern Sinosenecio 365075
Near-muchsharp Tea 69671
Nearness Arrowbamboo 162761
Nearness Umbrella Bamboo 162761
Near-smallsepal Tea 69459
Nearsmooth Barberry 52201
Neat Mountain Leech 126296
Neat Parodia 284665
Neat Philippine Violet 410412
Neat Rhododendron 330452
Neat Spike-rush 143237
Neatlegs 273531,273545
Nebraska Fern 101852
Nebraska Sedge 75481,75488
Nebrod Ephedra 146221
Nebrown Caralluma 72548

Neckfruitgrass 252397
Necklace Clearweed 298981
Necklace Gladecress 223541
Necklace Leaf-flower 296679
Necklace Poplar 311237,311259,311292
Necklace Sedge 75879
Necklace Tree 274375,274385
Necklace Weed 6433,407275
Necklace-bearing Poplar 311237
Necklace-spike Sedge 75625
Necklace-spike Wood Sedge 75625
Neckweed 71218,407275
Neck-weed 407275
Neckwort 70331,262441
Nectandra 263044
Nectarberry 338059
Nectarean Whiteleaf Raspberry 338630
Nectarine 20955,20957,316667
Nectary Azalea 331523
Nectary Maesa 241753
Ned Ironbark 155545
Nedcushion 23925
Neddy Grinnel 336419
Ned-veins Rhododendron 330594
Née Tree 324353
Needdle Grass 376687
Needle and Thread Grass 376749
Needle Beak Sedge 333501
Needle Bush 184586,184616
Needle Chervil 350425
Needle Fir 388
Needle Furze 172905
Needle Gorse 172905
Needle Grass 376917,376931
Needle Green Weed 172905
Needle Greenweed 172905
Needle Indiangrass 369585
Needle Juniper 213896
Needle Needlebush 184586
Needle on both Sides 417282
Needle Palm 326669,329169,329170,416584,416649
Needle Pink 127578
Needle Points 350425
Needle Rush 213424,213569
Needle Sandwort 31730
Needle Spike Sedge 143022
Needle Spikerush 143022
Needle Spike-rush 143022
Needle Spikesedge 143022,143029
Needle Spike-sedge 143022
Needle Spined Pineapple Cactus 140614
Needle Stonecrop 356884
Needle Wattle 1275,1531
Needle Whin 172905
Needle Woad-waxen 172905
Needlealga 382418,382420
Needle-and-thread 376749
Needlebract Vetch 408454
Needlebractdaisy 395964,395965
Needlebush 177170,184609,184617,184646
Needle-bush Wattle 1531
Needle-cases 381035
Needlegrass 329042,376687,376732
Needle-grass 329042,376687,376917
Needlegrass Rush 213424
Needlejade 297015
Needleleaf 307830
Needleleaf Bluet 187499
Needleleaf Dendrobium 125323
Needle-leaf Fleabane 150812
Needleleaf Gentian 173501
Needleleaf Hakea 184589
Needleleaf Leek 15019
Needleleaf Sedge 74406,75610
Needle-leaved Bottlebrush 67301
Needlelike Asparagus 38906
Needlelike Rattlebox 111856
Needlepod Rush 213438
Needles 350425
Needles-and-pins 401388,410677
Needleshaped Currant 333894
Needle-shaped Currant 333894
Needleshaped Rattlebox 111856
Needle-tip Blue-eyed-grass 365785
Needletooth Myrsine 261650
Needletooth Woodbetony 287717
Needletwig Knotweed 309607
Needletwig Shamrue 185675
Needleweed 350425
Needlewood 184616
Needle-wood 184586
Neelgerr Sandwort 32098
Neem 45908
Neem Tree 45908,248895
Nees Feathergrass 376855
Nees Lovegrass 147839
Neesewort 4005,405571
Neesing Root 405571
Neglect Clovershrub 70872
Neglect Fevervine 280094
Neglect Magnoliavine 351071
Neglect Sweetvetch 188023
Neglected Antennaria 26490
Neglected Bluegrass 305740
Neglected Dropseed 372782
Neglected Goatgrass 8694
Neglected Iris 208729
Neglected Magnoliavine 351071
Neglected Spapweed 205168
Neglected Sunflower 189009
Neglected Sweetvetch 188023
Neglected Touch-me-not 205168
Neglectus Magnolia-vine 351071
Negrita 370934
Negro Coffee 360463
Negro Pepper 415762
Negro Yam 131516
Negro's Head Palm 298050,298051
Negro's Slippers 159417
Negundo 3218
Negundo Chaste Tree 411362
Negundo Chastetree 411362,411373,411374
Negundo Chaste-tree 411362
Neibun Sicklebamboo 20205
Neidpath Yew 385310
Neighbor Fleabane 151059
Neighbouring Tanoak 233349
Neighbouring Viburnum 408056
Neighbours 81768
Neiha Kinkan 167493
Neilgherry Lily 229957
Neillia 263142
Neined Argyreia 32642
Nejap Cactus 244165
Nekbudu 165263
Nele 11947
Nellie Cory Cactus 155213
Nellie's Pincushion Cactus 107035
Nelson Pine 300103
Nelson Pinyon 300103
Nelson's Bugle 13174
Nelson's Saltbush 44407
Nelson's Sulphur Flower 152595
Nelsonia 263238
Nelumbium 263267
Nelumbo 263267,263270
Nematanthus 263320
Nemesia 263374,263381
Neminy 23925
Nemophila 263501,263505
Nemoros Conehead 320132
Nemosenecio 263510
Nemoto Holly 204081
Nenga 263549
Nengella 263553
Nenggaoshan Camellia 69398
Nenggaoshan Holly 204356
Nenggaoshan Sweetleaf 381330
Nengka Camellia 69398
Nengkao Sweetleaf 381330
Nengkaoshan Rose 336871
Nengpula Qinggang 116157
Nenjiang Spruce 298325
Nenjiang Xing'an Smallreed 127322
Nenkao Sweetleaf 381330
Nenufar 59575,176824,223251,264897,267294,267648
Neoalsomitra 263567
Neocinnamomum 263753
Neodiocious Antennaria 26427

Neofinetia 263865
Neogyna 263910
Neohouzeaua 263944
Neohusnotia 263952
Neolitsea 264011,264021
Neomartinella 264164
Neomicocalamus 264175
Neonauclea 264218,264219
Neonotonia 264239,264241
Neopallasia 264248
Neopalmer Cactus 244169
Neopicrorhiza 264316
Neoporteria 264332
Neoregelia 264406
Neosinglelepistil Milkvetch 42788
Neosinocalamus 264510
Neo-sulphur-colored Epimedium 147073
Neottia 264650,264700
Neowilson Willow 343763
Nepal Aconite 5188
Nepal Adonis 8373
Nepal Agrimonia 11587
Nepal Alder 16421
Nepal Alpine Monkshood 5030
Nepal Arundinella 37414
Nepal Awngrass 255862
Nepal Barberry 51327,242563
Nepal Barley 198376
Nepal Begonia 50161
Nepal Bird Cherry 280036
Nepal Birdfootorchis 347842
Nepal Bistamengrass 127475
Nepal Bladdersenna 100185
Nepal Bluebeard 78010
Nepal Bracketplant 88602
Nepal Brome 60865
Nepal Buckthorn 328794
Nepal Burmannia 63987
Nepal Camphor Tree 91330
Nepal Camphortree 91330
Nepal Camphor-tree 91330
Nepal Cardamom 19923
Nepal Carpesium 77182
Nepal Catchfly 363800
Nepal Centranthera 81726
Nepal Chickweed 375017
Nepal Chokecherry 280036
Nepal Cinnamon 91330
Nepal Clematis 95154
Nepal Coriaria 104701
Nepal Corydalis 105971
Nepal Cowparsnip 192324
Nepal Cranesbill 174755
Nepal Cremanthodium 110417
Nepal Currant 333991
Nepal Cycas 115874
Nepal Dock 340141
Nepal Dualbutterfly 398308
Nepal Everlasting 21643
Nepal Fieldcitron 400533
Nepal Firethorn 322458
Nepal Fleshspike 346950
Nepal Fourtapeparsley 387896
Nepal Goldsaxifrage 90416
Nepal Goldwaist 90416
Nepal Gongronema 179308
Nepal Groundsel 359246
Nepal Holly 203908
Nepal Iris 208517
Nepal Ivy 187306,187307
Nepal Knotweed 309459
Nepal Kobresia 217236
Nepal Mahonia 242597
Nepal Mallotus 243399
Nepal Mango 244410
Nepal Meconopsis 247157
Nepal Monkeyflower 255253
Nepal Mountainash 369391
Nepal Mycetia 260640
Nepal Myriactis 261081
Nepal Nutantdaisy 110417
Nepal Osbeckia 276130
Nepal Paper 122386
Nepal Pearleverlasting 21643
Nepal Pegaeophyton 287954
Nepal Pennywort 200312
Nepal Pepper 300462
Nepal Persicaria 309459
Nepal Pipewort 151409
Nepal Piptanthus 300691
Nepal Pittosporum 301344
Nepal Rattlesnake Plantain 347842
Nepal Red Pouzolzia 313475
Nepal Red Reekkudzu 313475
Nepal Rhubarb 329362
Nepal Russianthistle 344642
Nepal Sassafras 91330
Nepal Saurauia 347968
Nepal Seatung 301344
Nepal Silvergrass 127475,255862
Nepal Small Reed 127279
Nepal Smallreed 127279
Nepal St. John's-wort 201845
Nepal Swordflag 208517
Nepal Tripterospermum 398308
Nepal Turpinia 400533
Nepal Twinballgrass 209030
Nepal Velvetplant 183108
Nepal Windhairdaisy 348555
Nepalese Blue Bamboo 137849
Nepalese Browntop 254046
Nepalese Firethorn 322458
Nepalese Ivy 187306
Nepalese Smartweed 309459
Nepalese White-thorn 322458
Nepentes 264829
Nepentes Family 264827
Nepenthes Southstar 33430
Nepeta 264859,264897
Nepetaleaf Leonotis 224600
Nepetaleaf Leucas 227665
Nephelaphyllum 265115
Nephroid-auriculate Sinobambusa 364806
Nephroidleaf Violet 410543
Nephrosperma 265202
Nephthytis 381861
Nept Nepe 264897
Nerine 265254,265283,265294
Nerium-leaf Rapanea 261648
Nero's Crown 382763
Neroli 93332
Nertera 265353
Nervate Ardisia 31535
Nervate Embelia 144773
Nervate Swertia 380287
Nervate-flowered Phoebe 295388
Nerve Plant 166709
Nerved Asiabell 98380
Nerved Catnip 265011
Nerved Cinquefoil 312811
Nerved Disa 133856
Nerved Flax 231936
Nerved Manna-grass 177662
Nerved Nematanthus 263327
Nerved Nepeta 265011
Nerved Petrocosmea 292572
Nerved Sorghum 369671
Nerved Stonebutterfly 292572
Nerved Trillium 397582
Nerved Twayblade 232252
Nerved Waxplant 198877
Nerved Willow Weed 146795
Nervedcalyx Bluebellflower 115386
Nervedflower Phoebe 295388
Nervedleaf Pterygopleurum 320976
Nervedrachis Acronema 6219
Nerveless Osmanther 276286
Nerveless Osmanthus 276286
Nerveless Woodland Sedge 75126
Nerve-root 120290
Nervilia 265366
Nervin's Mahonia 242605
Nervose Arrowwood 407964
Nervose Chessbean 116593
Nervose Clearweed 298960
Nervose Cremanthodium 110418
Nervose Elatostema 142727
Nervose Fig 165370
Nervose Gaoliang 369671
Nervose Greenvine 204638
Nervose Liparis 232252

Nervose Magnolia 232596
Nervose Megacodon 247846
Nervose Nutantdaisy 110418
Nervose Pencilwood 139661
Nervose Skimmia 365971
Nery Melocactus 249597
Neslia 265550
Nespite 65540
Nespoli 151162
Nest Arrowbamboo 162648
Nest Neoporteria 264360
Nest Windhairdaisy 348556
Nestorchid 264650
Neststraw 379154
Net Bush 68063
Net Cliffbean 66643
Netbract Dandelion 384556
Netcalyx Falsepimpernel 231508
Netde-tree 80698
Netfruit Bugle 13097
Netfruit Dock 339983
Netfruit Pearlsedge 354260
Netleaf Hackberry 80728
Net-leaf Hackberry 80728
Netleaf Oak 324353
Net-leaf Oak 324326
Netleaf Primrose 314894,314895
Net-leaf White Oak 324299
Netleaf Willow 344003
Net-leaf Willow 344003
Net-leafed Willow 344003
Net-like Rhododendron 331623
Net-nerved Drypetes 138679
Netseed 129690,129691
Netseed Mitreola 256218
Netseedgrass 129690
Netted Apple 25882
Netted Barberry 51553
Netted Boea 283115
Netted Camellia 69552
Netted Cotoneaster 107648
Netted Custard Apple 25882
Netted Custard-apple 25882
Netted Holboellia 197249
Netted Iris 208788
Netted Melon 114189,114214,114215
Netted Nut-rush 354225
Netted Sauropus 348063
Netted Willow 344003
Netted-vein Spicebush 231393
Netted-vein Spice-bush 231393
Netted-veined Leafflower 296734
Netted-veined Leaf-flower 296734
Nettle 402840,402875,402886,402916,
403039
Nettle Family 403050
Nettle Fibre 402886
Nettle Hemp 170078
Nettle Spurge 212098,212127,212202
Nettle Tree 80569,80577,80698,80728
Nettleeoot 373455
Nettleleaf Bellcalyxwort 231286
Nettleleaf Chirita 87985
Nettleleaf Clinopodium 97056
Nettleleaf Crabdaisy 413563
Nettle-leaf Croton 113050
Nettleleaf Dragonheadsage 247739
Nettleleaf Falsepimpernel 231589
Nettleleaf Figwort 355263
Nettleleaf Goosefoot 87096
Nettle-leaf Goosefoot 87096
Nettle-leaf Hyssop 10422
Nettle-leaf Leucas 227724
Nettleleaf Meehania 247739
Nettleleaf Microtoena 254271
Nettleleaf Motherweed 231589
Nettleleaf Motherwort 225020
Nettle-leaf Mullein 405680
Nettleleaf Obliquecalyxweed 237961
Nettle-leaf Porterweed 373511
Nettle-leaf Premna 313759
Nettleleaf Primrose 315074
Nettleleaf Skullcap 355855
Nettle-leaf Soldier 170150
Nettleleaf Spapweed 205423
Nettleleaf Touch-me-not 205423
Nettleleaf Velvetberry 373511
Nettleleaf Wedelia 413563
Nettleleaf Wildbasil 97056
Nettle-leaved Bellflower 70331
Nettle-leaved Croton 113050
Nettle-leaved Figwort 355216
Nettle-leaved Goosefoot 87096
Nettle-leaved Michaelmas-daisy 41217
Nettle-leaved Mullein 405680
Nettle-leaved Premna 313759
Nettle-leaved Vervain 373511,405903,
405904
Nettle-leaved Vervein 405903
Nettles Bull 212228,367014,367139
Nettles Burning 402992,403039
Nettletree 80569,80698,394626
Nettle-tree 80569,80577
Netty Raspberry 338260
Nettysepal Motherweed 231508
Netvein Bladderwort 403791
Netvein Boea 283115
Netvein Camellia 69552
Netvein Chickweed 375074
Netvein Cinnamon 91420
Netvein Crapemyrtle 219975
Netvein Fenugreek 397206
Netvein Goldenray 229018
Netvein Grape 412004
Netvein Microtropis 254321
Netvein Paraboea 283115
Netvein Snakegourd 396256
Netvein Tanoak 233179
Netvein Wild Mung Bean 126580
Netvein Wildtung 243442
Net-veined Actinidia 6579
Net-veined Bamboo 297435
Net-veined Barberry 51553,52108
Net-veined Blueberry 403791
Netveined Cotoneaster 107648
Net-veined Crapemyrtle 219975
Net-veined Crape-myrtle 219975
Net-veined Fortune Spindle-tree 157504
Net-veined Holly 204207
Net-veined Kiwifruit 6579
Net-veined Mahonia 242513,242631
Net-veined Maple 3495
Net-veined Microtropis 254321
Netveined Osmanther 276393
Net-veined Osmanthus 276393
Netveined Panicgrass 281863
Net-veined Persimmon 132369
Net-veined Rockvine 387840
Net-veined Sauropus 348063
Net-veined Spicebush 231393
Net-veined Spice-bush 231393
Net-veined Tanoak 233179
Net-veined Willow 344003
Neubert Mahobarnerry 242454
Neurada 265713
Neurolepis 265756
Neuropeltis 265783
Neurose Gugertree 350938
Neurose-fruited Sea-buckthorn 196755
Neutral Henna 418169
Neuwiedla 265836
Nevada Agave 10949
Nevada Arnica 34754
Nevada Bentgrass 12220
Nevada City Wild Buckwheat 152405
Nevada Dustymaidens 84513
Nevada Ephedra 146224
Nevada Fleabane 150605
Nevada Goldeneye 190256
Nevada Goldenrod 368430
Nevada Grease-bush 177363
Nevada Joint Fir 146224
Nevada Jointfir 146224
Nevada Joint-fir 146224
Nevada Nut Pine 300079
Nevada Rabbitbrush 150373
Nevada Rock Daisy 291322
Nevada Skeleton Weed 151993
Nevada Sulphur Flower 152596
Nevada Wild Almond 223094
Nevada-pondweed 378993

Neverdie 215117,215174
Never-die 61572,215122,221150,258945,404924
Nevin Bamboo 297354
Nevin Lovegrass 147842
Nevski Wildryegrass 144411
New Belgium Aster 40923
New Britain Island Screwpine 280989
New Caledonia Cypress Pine 67425
New Caledonia Pine 30838
New Chapel Flower 275135
New China Bulrush 353655
New Cocoyam 415201
New Dawn Rose 336293
New Day Rose 336294
New Deal Weed 46236
New Eargrass 262956,262969
New England Aster 40923,380930
New England Blackbutt 155480
New England Blue Violet 410301
New England Boxwood 105023
New England Hawkweed 195931
New England Hemlock 399867
New England Mahogany 83311
New England Northern Reed Grass 65438,65491
New England Oak Sedge 75539
New England Pine 300211
New England Sedge 75539
New England Serviceberry 19290
New England Tea-tree 226470
New Forest Crowfoot 325502,326130
New Guinea Asparagus 361850
New Guinea Clumping Bamboo 262761
New Guinea Creeper 259484
New Guinea Dacrydium 121107
New Guinea Falleneaves 351153
New Guinea Impatiens 205001
New Guinea Kauri 10502
New Guinea Kauri Pine 10502
New Guinea Rosewood 320301
New Guinea Schismatoglottis 351153
New Guinea Teak 411237
New Guinea Thin-walled Bamboo 47195
New Guinea Walnut 137714
New Guinea Wild Lime 253293
New Jersey Tea 79902,79905
New Mexican Alder 16425
New Mexican Feather Grass 193525
New Mexican Locust 334974
New Mexican Muhly 259686
New Mexico Agave 10920
New Mexico Evergreen Sumac 332517
New Mexico Fleabane 150813
New Mexico Groundsel 279927
New Mexico Larkspur 124409
New Mexico Locust 334974
New Mexico Love Grass 147808
New Mexico Olive 167324,167326
New Mexico Plumseed 325075
New Mexico Privet 167324
New Mexico Rock Daisy 291335
New Mexico Saltbush 44555
New Mexico Thistle 92222
New Mexico Tick-trefoil 126474
New Mexico Yucca 416612
New Mown Hay 166125,329521
New Nymph Flower 264158,264161
New Purple Willow 343763
New South Wales Christmas Bush 83535
New South Wales Dianella 127505
New South Wales Sassafras 136768
New South Wales Walnut 145322
New South Wales Waratah 385738
New Splitmelon 351760
New Veitch Spruce 298376
New Wilson Willow 343763
New World Trumpet Tree 382708
New Year's Gift 148104
New Year's Rose 190944
New York Aster 40923
New York Hawthorn 109668
New York Ironweed 406631,407254
New Zealand Beech 266870,266874,266885
New Zealand Black Pine 316089
New Zealand Bluebell 412581
New Zealand Brass Buttons 107821,225890
New Zealand Broadleaf 181260,181264
New Zealand Broom 77050
New Zealand Bur 1744,1752
New Zealand Cabbage Palm 104339
New Zealand Cabbage Tree 104339
New Zealand Cabbage-tree 104339
New Zealand Cedar 228636,228649
New Zealand Christmas Tree 252613
New Zealand Christmas-tree 252613
New Zealand Daphne 299313
New Zealand Dracaena 104339
New Zealand Edelweiss 227835
New Zealand Everlasting Flower 189188
New Zealand Fiber Lily 295600
New Zealand Flame Bush 186940
New Zealand Flax 231925,295594,295598,295600
New Zealand Geranium 174874
New Zealand Glory-pea Kaka Beak 96639
New Zealand Gloxinia 328446
New Zealand Hebe 186973
New Zealand Hemp 295600
New Zealand Holly 270198
New Zealand Honeysuckle 216922
New Zealand Iceplant 387221
New Zealand Iris 228630
New Zealand Jovellana 212531
New Zealand Kauri 10492,10494
New Zealand Kowhai 369130
New Zealand Laburnum 369130
New Zealand Lacebark 197153,197154
New Zealand Laurel 106945
New Zealand Lemonwood 301261
New Zealand Lilac 186957
New Zealand Mountain Pine 121088,184913
New Zealand Native Blueberry 126981
New Zealand Nightshade 367289
New Zealand Pennywort 200341
New Zealand Pincushion 326513
New Zealand Pygmyweed 109059
New Zealand Red Beech 266870
New Zealand Red Pine 121094
New Zealand Rium 121094
New Zealand Satin Flower 228629
New Zealand Scented Broom 77058
New Zealand Schefflera 350682
New Zealand Sedge 73543
New Zealand Silver Beech 266874
New Zealand Silver Pine 244664,244665
New Zealand Spinach 146644,387156,387221
New Zealand Tea 226481
New Zealand Teak 411322
New Zealand Tree Fuchsia 168746
New Zealand Violet 409875
New Zealand Walnut 50636
New Zealand White Podocarpus 306420
New Zealand Willowherb 146644
New Zeyland Bur 1739
New Zeyland Quintinia 324661
Newbroadtube Woodbetony 287456
Newcastle-thorn 109636
Newcinnamon 263753,263759
Newell Cestrum 84412
Newell Fascicled Cestrum 84413
New-England-staudenaster 380930
Newfoundland Dwarf Birch 53524
Newfoundland Mouse-ear Chickweed 83053
New-jersey-tea 79905
Newlitse 264011,264021
Newmarket Oregon-grape 242461
New-monadelphus Milkvetch 42788
Newmoon Azalea 331216
Newmoon Rhododendron 331218
New-mown Hay 166125
Newrussia Skullcap 355625
Newstraw 317073,317081
Newton Water Lily 267633
New-York Aster 40923
Ngai 55693
Ngai Camphor 55693
Ngaio Tree 260718
Ngali Nut 71000
Niagara Hawthorn 110021

Niangon 192501
Niaoguan Corydalis 106168
Niaouli Oil 248114
Nibung 68617
Nibung Oncosperma 271003
Nicaragua Pine 300126
Nicaragua Rosewood 121803
Nicaragua Wood 64965,121607,184513
Nicaraguan Cocao-shade 176970
Nicaraguan Fountaingrass 289069
Nicaraguan Pine 300126
Nice Hemp Nettle 170076
Nice Hemp-nettle 170076
Nichobar Bentinckia 51164
Nichol's Hedgehog 140275
Nichol's Hedgehog Cactus 140275
Nicholls Manuka Tea-tree 226500
Nickel's Coryphantha 107045
Nickels' Pincushion Cactus 107045
Nicker Bean 64971,145875,145899
Nicker Nut 64971,64973
Nicker Nut Caesalpinia 64983
Nickernut Caesalpinia 64971,64983
Nicker-nut Caesalpinia 64973,64983,64990,65078
Nicobar Bentinckia 51164
Nicodemia 62118
Nicola Milkvetch 42791
Nicotiana 266031,266035
Nicuri 380556
Nicuri Palm Nut 380556
Niddle 402886
Nidulant Thelocactus 389221
Nidularium 266148
Nielamu Buttercup 326131
Nielamu Chickweed 375024
Nielamu Corydalis 106248
Nielamu Cowparsnip 192330
Nielamu Gentian 173671
Nielamu Jerusalemsage 295156
Nielamu Monkshood 5454
Nielamu Raspberry 338899
Nielamu Rockfoil 349670
Nielamu Thickribcelery 279652
Nielamu Willow 343765
Nielamu Windhairdaisy 348594
Nienku Nightshade 367413
Niepa-bark Tree 345515
Nierembergia 266199
Nieuwland's Gayfeather 228521
Nig 23767
Nigella 266215,266225
Niger Plum 166769
Niger Seed 181872,181873
Niger Thistle 181872,181873
Nigerian Golden Walnut 237927
Nigerian Satin Wood 134776
Nigerian Satinwood 134776
Nigerian Stylo 379241
Niger-seed 181872,181873
Nigger 23767,302034
Nigger Toe 53057
Nigger's Head 302034,302103
Nigger's Heal 302034,302103
Niggerhead Beech 266875
Niggertoe 53057
Nigger-toe 53057
Nigh'r-bonnets 61497
Night and Afternoon 317289
Night Blooming Cereus 83890,288986,288988
Night Blooming Nightshade 366934
Night Bonnets 61497
Night Cestrum 84417
Night Jasmin 267429,267430
Night Jassamine 84417
Night Phlox 416917,416927
Night Scented Stock 246443,246494
Night Violet 302280
Nightblooming Catchfly 363815
Night-blooming Cereus 83874,200707,200713,357735
Nightblooming Cestrum 84417
Night-blooming Cestrum 84417
Night-blooming Epidendrum 146450
Nightblooming Jassamine 84417
Night-blooming Jassamine 84417
Nightblooming-cereus 200707
Nightcap 30081,68713,273531
Nightcaps 24004
Night-closing Flower 232594
Nightflower Melandrium 248382
Night-flowering Cactus 147288
Night-flowering Campion 363815
Night-flowering Catchfly 248382,363815
Night-flowering Cereus 83874,200713
Night-flowering Jasmine 267430
Night-flowering Petunia 292751
Nightflowering Silene 363815
Night-flowering Silene 363815
Nightingale 37015,174877,374916
Nightingale Flower 72934
Nightjasmine 267429,267430
Night-jasmine 267429,267430
Night-phlox 416917
Night-scented Jassamine 84417
Night-scented Stock 246443,246494
Night-scented Tobacco 266035
Nightshade 123077,366902,367120,367322,367535
Nightshade Bindweed 91557
Nightshade Family 366853
Nightshadeflower Snowflake 32525
Nightshadeleaf Ironweed 406817
Nightshadeleaf Mazus 247037
Nightshadeleaf Rehmannia 327456
Nightshade-leaved Ironweed 406817
Nightshade-like Vallaris 404523
Nightshirt 68713
Night-smelling Epidendrum 146450
Night-smelling Rocket 193417
Nigrescent Buckthorn 328795
Nigrescent Leptodermis 226104
Nigrescent Rosewood 121772
Nigrescent Sandwort 32100
Nigrescent Skullcap 355622
Nigrescentspike Galingale 119256
Nigritella 266257
Niijima Zeuxine 417770
Nikau Palm 332438
Nikau Rhopalostylis 332438
Nikko Euonymus 157567
Nikko Fir 388,389
Nikko Maple 3137,3243,3286
Nikko Silver Fir 389
Nikko Spindle-tree 157567
Nikko St. John's-wort 202047
Niko Bedstraw 170761
Niko Groundsel 263515
Nilagirian Sagebrush 35983
Nilagiris Helicia 189941
Nile Cabbage 301025
Nile Crossandra 111741
Nile Grass 6108
Nile Tamarisk 383563
Niles' Wild Buckwheat 151960
Nilghiri Barberry 51964
Nilgiri Nettle 175877
Nilgiris Lily 229957
Nim 45908
Nimble Will 259660,259696
Nimble-will 259696
Nimble-will Muhly 259696
Nimmon 266807
Nimu Bluegrass 305583
Ninde 8957
Nine Joints 308816
Nine-anther Prairie-clover 121886
Nine-anthered Prairie Clover 121886
Ninebark 297833,297842,297849,297850
Ninefoliolate Indigo 206182
Nineleaf Caesalpinia 252748
Nineleaflets Indigo 206182
Nine-leaved Bittercress 72743
Nine-leaved Caesalpinia 252748
Ninenode 81945,319392
Nineseed Green Violet 199686
Ninety Knot 308816
Ninety-knot 308816
Nine-wheel Primrose 315082
Ninewing Amomum 19888

Ning'an Hawthorn 109869
Ningbo Buerger Maple 2820
Ningbo Deutzia 127026
Ningbo Figwort 355201
Ningbo Indigo 205880
Ningbo Maple 2820
Ningbo Trident Maple 2820
Ningde Holly 204086
Ninggang Cyclobalanopsis 116162
Ninggang Oak 116162
Ninggang Qinggang 116162
Ningguo Fritillary 168491
Ninghua Chirita 87877
Ningjingshan Clematis 95158
Ninglang Gentian 173660
Ninglang Mosquitotrap 117545
Ninglang Swallowwort 117545
Ningming Slugwood 50573
Ningnan Hemsleia 191905
Ningnan Lily 229964
Ningnan Snowgall 191905
Ningnan Square-bamboo 87592
Ningpo Clematis 95067
Ningpo Comanthosphace 100238
Ningpo Deutzia 127026
Ningpo Figwort 355201
Ningqiang Oak 324228
Ningshan Linden 391810
Ningshan Poplar 311415
Ningsia Ladybell 7716
Ningwu Monkshood 5455
Ningxia Catchfly 363813
Ningxia Cistanche 93069
Ningxia Crazyweed 279035
Ningxia Fourleaf Paris 284390
Ningxia Fritillary 168655
Ningxia Ladybell 7716
Ningxia Poplar 311415
Ningxia Reedbentgrass 65447
Ningxia Spiraea 372027
Ningyuan Kobresia 217200
Ningyun Azalea 331337
Ningyun Rhododendron 331337
Ninnyver 267648
Niopo Snuff 21308
Niove 374378
Nipa Palm 267837,267838
Nipbone 381035
Niphaea 266307
Nipliporchis 154871
Nippernail 336419
Nippernut 222758
Nipple Azalea 331441
Nipple Beehive Cactus 107031
Nipple Cactus 107070,244015,244147
Nipple Gentian 173704
Nipple Heritiera 192496
Nipplebearing Embelia 144753
Nipplewort 221780,221787,401711
Nippon Chrysanthemum 89628,89646
Nippon Cinquefoil 312478
Nippon Common Juniper 213735
Nippon Daisy 89628,89646,266319
Nippon Greenbrier 366486
Nippon Hawthorn 109650
Nippon Juniper 213735
Nippon Lily 335758
Nippon Maple 3248
Nippon Spirea 372029,372030
Nippon-bells 362270
Nipponese Listera 232946
Nipponlily 335758,335760
Nirrhe 156064
Niruri 296530
Nisan Anisodus 25450
Nishiki Willow 343530
Nispero 151162,244605
Nissol's Vetchling 222784
Nit Clickers 68713
Nit Grass 171809,171814
Nit Sprangletop 226021
Niterhush 266355
Nit-grass 171809,171818
Nitid Azalea 331343
Nitid Euonymus 157745
Nitid Heteropanax 193912
Nitid Leptospermum 226469
Nitid Spindle-tree 157745
Nitida Rose 336807
Nitidcalyx Spotleaf-orchis 179631
Nitidfruit Tanoak 233328
Nitidleaf Aster 40911
Nitidleaf Croton 112931
Nitidspike Itchgrass 337583
Nitraria 266355
Nitre 266864
Nitre Goosefoot 87105
Nitrebush 266355
Nits 374916
Nits and Lice 201835
Nitta Tree 284444
Nittatree 284444
Nitta-tree 284444
Nivenia 266397
Nlunt-leaved Ardisia 31540
Narrow-leaved Crab Apple 243550
No Papillary Twigs Rhododendron 331441
Noah's Ark 5442,5676,30081,102833,
120421,120432
Nobillorchis 197446
Noble Agrimony 192130
Noble Aloe 16810,17089
Noble Bluethistle 141164
Noble Bottle-tree 376144
Noble Cactus 244082
Noble Chamaemelum 85526
Noble Corydalis 106453
Noble Cremanthodium 110419
Noble Dendrobium 125279
Noble Fir 453
Noble Gentian 173471
Noble Goldenrod 368422
Noble Hepatica 192134
Noble Laurel 223203
Noble Leafflower 245221
Noble Lily 229966
Noble Liverwort 192130
Noble Milkvetch 42801
Noble Nutantdaisy 110419
Noble Odontoglossum 269078
Noble Rhubarb 329363
Noble Sanchezia 345770
Noble Yarrow 3998
Nobly Mulberry 259182
Nobody's-flower 66445
Nocaudate Gentian 173414
Nodal Flower Groundsel 359594
Nodalflower Allmania 15912
Nodalflower Synedrella 381809
Nodalhead Groundsel 359594
Nodding Agrimony 53816
Nodding Arnica 34732,34757
Nodding Avens 175444
Nodding Barberry 51974
Nodding Beggarticks 53816
Nodding Beggar-ticks 53816
Nodding Billbergia 54454
Nodding Bristle-grass 361743
Nodding Brome 60977
Nodding Broomrape 274987
Nodding Buckwheat 151968
Nodding Bur Marigold 53816
Nodding Bur-marigold 53816
Nodding Campion 364149
Nodding Catasetum 79335
Nodding Catchfly 363827,363882
Nodding Chickweed 82935
Nodding Clerodendron 96439
Nodding Clover 396861
Nodding Cyrtandra 120506
Nodding Dichrostachys 129164
Nodding Dragonhead 137617
Nodding Enkianthus 145689
Nodding Fescue 164341
Nodding Fescue-grass 164109
Nodding Flowers Gaultheria 172111
Nodding Flowers Whitepearl 172111
Nodding Foxtail 361743
Nodding Galingale 119282
Nodding Gomphostemma 179197
Nodding Greenhood 320876

Nodding Hackelia 184298
Nodding Hypecoum 201618
Nodding Indian-grass 369582
Nodding Isabel 60382
Nodding Ladies' Tresess 372192
Nodding Lady's Tresses 372192
Nodding Lady's-tresses 372192
Nodding Leucas 227671
Nodding Lilac 382190,382251
Nodding Lily 229779,229803,230044
Nodding Mandarin 134450,315526
Nodding Meadowrue 388680
Nodding Melic 249054
Nodding Melick 249054
Nodding Microseris 253936
Nodding Mouse-ear Chickweed 82935
Nodding Muhly 259641
Nodding Onion 15173
Nodding Pendent-bell 145689
Nodding Pincushion 228048,228057
Nodding Plumeless Thistle 73430
Nodding Pogonia 397877,397892
Nodding Primrose 314228
Nodding Rattlesnakeroot 313809
Nodding Rattlesnake-root 313809
Nodding Rockfoil 349162
Nodding Roegneria 335451
Nodding Sage 345266
Nodding Sedge 74720
Nodding Silene 363827
Nodding Silverpuffs 253936
Nodding Sophronitis 369179
Nodding Sorghum 369601
Nodding Spurge 85779,159286
Nodding Star-of-bethlehem 274708
Nodding Stickseed 184298,184300
Nodding Thistle 73430
Nodding Tresess 372192
Nodding Trillium 397547
Nodding Wake-robin 397563
Nodding Water Hemp Agrimony 53816
Nodding Water-nymph 262028
Nodding Wild Buckwheat 151904
Nodding Wild Onion 15173
Nodding Wintergreen 275480,275486
Nodding Woodbetony 287072
Noddingflower 314715
Noddingflower Elsholtzia 144076
Nodding-nixie 29888,29889
Noddingspike Goosecomb 335451
Node China Flatspike 55898
Nodeflower Flax 231937
Nodeflower Xianmu 161627
Nodehair Windhairdaisy 348810
Nodeleaf Rush 213143
Nodfruit Rose 336944
Nod-fruited Rose 336944
Noding Cape-cowslip 218920
Nodose Elaeocarpus 142362
Nodose Greenorchid 137616
Nodose Hedgeparsley 392996
Nodose Landpick 308611
Nodose Ottochloa 277430
Nodose Pholidota 295511
Nodose Solomonseal 308611
Nodosetree 36813
Nodosewhip Galangal 17657
Nodostipe Elaeocarpus 142362
Nodular Pondweed 312194
No-eye Pea 65146
Nogal 212620,212627
Nogs 71218
Noisette Rose 336809
Noko St. John's Wort 202048
Noko Stonecrop 356964
Noko Sweetleaf 381330
Noko Tea 69398
Noko-mt St. John's Wort 202048
Nokye 181773
Nolana 266470
Noli-me-tangere 205175
Nolina 266482
Nomame Senna 78429
Nomocharis 266548
None Found 146451
Nonea 266593
None-so-pretty 178358,243209,315106,
349044,350011,363214
Nonesuch 247366
Nonesuch Daffodil 262405
Nonggang Camellia 69105
Nonggang Cyclea 116022
Nonggang Hemsleia 191935
Nonggang Pellionia 288742
Nonggang Ringvine 116022
Nonggang Yunnan Wampee 94227
Nonghua Elaeagnus 142126
Nongo 13567,13703
Nonnea 266593
Nonscales Rhododendron 330576
Nonsuch 363312
Noogoora Burr 415033
Noon Peepers 274823
Noon-day Flower 394333
Noon-flower 50825,394333
Noontide 394333
Noops 338243
Noosoora Bur 414990
Nootka Cedar 85301
Nootka Cypress 85301
Nootka False Cypress 85301
Nootka Lupin 238472
Nootka Rose 336810
Nootka Sound 390645
Nootka Sound Cypress 85301
Nopal 266637,272775,272830,272951
Nopal Blanco 272950
Nopal Cactus 272830
Nopal Camueso 272940
Nopal Crinado 273012
Nopal De Culebra 272853
Nopal De Tortuga 273024
Nopalxochia 266647
Nordhagen Larkspur 124406
Nordic Bladderwort 403353
Nordmann Christmas Tree 429
Nordmann Fir 429
Norfolk Betel Palm 332436
Norfolk Everlasting-pea 222725
Norfolk Island Hibiscus 220228
Norfolk Island Palm 332436
Norfolk Island Pine 30846,30848
Norfolk Palm 332436
Norfolk Pine 30848
Norfolk Reed 295888
Norfolk Skullcap 355469
Norfolk-Island Pine 30848
Noriake Azalea 331275
Noriake Rhododendron 331275
Normal Manchurian Rose 337034
Normal Roxburgh Rose 336888
Normal Sedge 75527
Normandy Cress 47963
Norrlin Hawkweed 195814
Norsveld Plakkie 108954
Nortern Willow 343116
North Africa Grass 405429
North African Catchfly 363930
North African Knapweed 81035
North America Peppergrass 225475
North America Platanthera 302263
North America Violet 409798
North American Ebony 132466
North American Fringe Tree 87736
North American Hackberry 80698
North American Ironwood 276818
North American Ox-eye 190502
North American Persimmon 132466
North American Pitcherplant 347144
North American Pitcher-plant 347144
North American White Adder's-
mouth 243085
North Butterwort 299757
North China Bentgrass 12053
North China Grape 411589
North China Red Raspberry 338588
North China Scabious 350265
North Chinapaper Birch 53336
North Cinnamon 91425
North Corydalis 106453
North Figwort 355067

North Galingale 118931
North Hopeh Larkspur 124585
North Hubei Arrowwood 407888
North Hubei Viburnum 407888
North Island Edelweiss 227837
North Japanese Magnolia 242171
North Lagotis 220161
North Magnoliavine 351063
North Onion 15432
North Pole Fruit 31311
North Roegneria 335230
North Sasa 347200
North Sawtooth Oak 323611
North Sichuan Leptodermis 226076
North Sichuan Wildginger 37597
North Snowy Rhododendron 331348
North Szechwan Wildginger 37597
North Tibet Christolea 89227,126149
North Viet Nam Gambirplant 401762
North Wild Smallreed 127199
North Yunan Monkshood 5290
North Yunnan Camellia 68941
North Yunnan Clematis 95044
North Yunnan Elaeocarpus 142290
North Yunnan Long Fruit Rhododendron 330645
North Yunnan Raspberry 338194
North's False Flag 264163
North-China Bittersweet 367282
North-China Nightshade 367282
Northeast Alkaligrass 321335
Northeast Bulrush 353750
Northeast Caudate Maple 3732
Northeast Crazyweed 278990
Northeast Crowberry 145073
Northeast Falsebrome 58596
Northeast Helictotrichon 190206
Northeast Melic 249012
Northeast Speedwell 407013
Northeaste Mannagrass 177672
Northeasten Neillia 263183
Northeastern Beard-tongue 289353
Northeastern Bedstraw 170574
Northeastern Bladderwort 403311
North-eastern China Isopyrum 210281
Northeastern Corydalis 105594
Northeastern Cutleaf Coneflower 339578
Northeastern Falsenettle 56318
Northeastern Fox Tail 17543
Northeastern Juniper 341781
Northeastern Podocarpium 126501
Northeastern Rosemaryleaf Willow 344036
Northeastern Sedge 74223
North-eastern Serpentroot 354897
Northeastern Willow-herb 146895
Northern Agrostis 12196,18652
Northern Anemone 23980
Northern Arnica 34735
Northern Arrowhead 342325
Northern Asparagus 38948
Northern Bayberry 261202
Northern Bedstraw 170246
Northern Beggarticks 53947
Northern Bilberry 404027
Northern Bitter Cress 72740
Northern Black Currant 334037
Northern Black Wattle 1067
Northern Bladderwort 403220
Northern Blazing Star 228446
Northern Blue Flag 208924
Northern Blue Violet 410560
Northern Bluebells 250902
Northern Bog Aster 40147,380842
Northern Bog Bedstraw 170454
Northern Bog Goldenrod 368454
Northern Bog Sedge 74722
Northern Bog Violet 410297
Northern Bog-violet 410297
Northern Boobialla 260708
Northern Bottle Tree 58337
Northern Bugleweed 239246,239248
Northern Bungalow Palm 31013
Northern Burgundy 407779
Northern Burma Pine 299999
Northern Bur-reed 370066,370071
Northern Bush-honeysuckle 130252
Northern Butterfly-orchid 302361
Northern California Walnut 212612
Northern Calypso 68538
Northern Catalpa 79260
Northern Catchfly 364202
Northern Chinese Pine 300243
Northern Cinquefoil 100257
Northern Clustered Sedge 73742,73743
Northern Comandra 174290
Northern Commandra 100226
Northern Coralroot 104019
Northern Coral-root 104019
Northern Cottonwood 311237,311292
Northern Crabgrass 130745
Northern Crane's-bill 174495
Northern Crowfoot 326351
Northern Cypress Pine 67421
Northern Dandelion 384738
Northern Daylily 191278
Northern Dead-nettle 220373
Northern Decurrentleaf Loosestrife 239851
Northern Dewberry 338399
Northern Dianthera 214320
Northern Dock 340109
Northern Dropseed 372698
Northern Drop-seed 372698
Northern Dutchmanspipe 34154
Northern Dwarf Cornel 105204
Northern Elegant Sedge 74164
Northern Enchanter's-nightshade 91508
Northern Enneapogon 145766
Northern Eritrichium 153430
Northern Evening Primrose 269481
Northern Evening-primrose 269481
Northern Eyebright 160219
Northern Fairy Candelabra 23294
Northern False Asphodel 392603
Northern Fog Fruit 296118
Northern Fringe Myrtle 68759
Northern Frog Fruit 296118
Northern Gayfeather 228519
Northern Gentian 173888,174103,174117
Northern Goldenrod 368258
Northern Gooseberry 334030,334137
Northern Green Orchid 302256,302361
Northern Green Orchis 302361
Northern Green Rush 212822
Northern Groundsel 279917
Northern Hackberry 80698
Northern Hawk's-beard 110854
Northern Hawksbeard 110904
Northern Hawkweed 195491
Northern Hawthorn 109531,109676
Northern Heart-leaved Aster 380852
Northern Hemlock 399867
Northern Horse-balm 99844
Northern Hounds-tongue 117926
Northern Inside-out Flower 404615
Northern Jack-in-the-pulpit 33522,33544
Northern Japan Aucuba 44932
Northern Japanese Hemlock 399895
Northern Jointweed 308669
Northern Juniper 213702
Northern Knot-grass 309858
Northern Knotweed 308823
Northern Ladybell 7609
Northern Listera 232905
Northern Long Sedge 74587
Northern Mahonia 242495
Northern Manna Grass 177576
Northern Manna-grass 177655
Northern Marsit Orchid 273605
Northern Meadow Groundsel 279935
Northern Meadow-rue 388459
Northern Meadow-rue 388712
Northern Medinilla 247623
Northern Microseris 253916
Northern Mountain Thistle 91934
Northern Mountain-ash 369377
Northern Najad 262028
Northern Oak Sedge 74277
Northern Panic Grass 128490
Northern Pin Oak 323871
Northern Pine 300181
Northern Pink 127809

Northern Pitch Pine 300181
Northern Pitcher Plant 347156
Northern Pitcherplant 347156
Northern Plains Blazing-star 228487
Northern Plains Gayfeather 228487
Northern Pogostemon 307003
Northern Prairie Rose 336389
Northern Prickly Ash 417157
Northern Pricklyash 417157
Northern Prickly-ash 417157
Northern Ragwort 279935
Northern Rata 252625
Northern Rattle 329512
Northern Red Currant 334179,334214,334223
Northern Red Lily 229984
Northern Red Oak 320278,324158,324344
Northern Red-currant 334179
Northern Reed Grass 65438,65490,65491
Northern Riverbank Wild-rye 144541
Northern Rock Sandwort 32245
Northern Rock-cress 30399
Northern Rockjasmine 23294
Northern Roughleaf Dogwood 105017
Northern Sandalwood 346209,346212
Northern Scurvy-grass 97996
Northern Sea Oats 86140
Northern Sedge 74277
Northern Shorewort 250894
Northern Shorthusk 58450
Northern Slender Lady's-tresses 372214,372247
Northern Snailseed Pondweed 312267
Northern Snail-seed Pondweed 312267
Northern Spider-lily 200946
Northern Spike-rush 143060
Northern Spinestemon 307003
Northern St. John's-wort 201781
Northern Starwort 374767,374783
Northern Stitchwort 374767
Northern Sugar Maple 3577
Northern Sumac 332787
Northern Swamp Dogwood 105169
Northern Swamp Groundsel 358590
Northern Sweet-colt's-foot 292354
Northern Tansy Mustard 365414
Northern Three-lobed Bedstraw 170696
Northern Tickseed-sunflower 53858
Northern Twayblade 232897
Northern Twinflower 231679
Northern Vancouveria 404615
Northern Vetch 408563
Northern Violet 410547
Northern Water Forget-me-not 260770,260883
Northern Water Gum 398674
Northern Water Plantain 14788
Northern Waterchestnut 394489
Northern Water-hemlock 90932
Northern Water-horehound 239246
Northern Water-meal 414654
Northern Water-nymph 262028
Northern Water-plantain 14788
Northern Water-starwort 67361
Northern Weak Sedge 74271
Northern White Cedar 390587
Northern White Pine 300211
Northern White Violet 410189
Northern White-cedar 390587
Northern Wild Comfrey 117926,118031
Northern Wild Raisin 407737
Northern Wild Rice 418096,418098
Northern Wild Senna 360437
Northern Wild-licorice 170434
Northern Wild-raisin 407971
Northern Willowherb 146587,146709
Northern Willowweed 146636
Northern Winter-cress 47949
Northern Wolfberry 239023
Northern Wood Rush 238610
Northern Wormwood 35211
Northern Yellow Lake Sedge 76068
Northern Yellow Locoweed 278766
Northern Yellow-cress 336250
Northern Yellow-eyed-grass 416110
Northern Yunnan Falsehellebore 405637
Northern-candleberry 261137
North-tailand China-laurel 28401
Northwest Cinquefoil 312647,312648
Northwest Melic 249066
Northwest Mugwort 35404
Northwest Sedge 74165
Northwest Territory Sedge 76653
Northwest Twayblade 232903
Northwest Willow 344099
Northwest Yunnan Raspberry 339389
North-western Lepyrodiclis 226621
Northwestern Paper Birch 53549
Northwestern Rabbit-tobacco 317773
North-Xinjiang Fleabane 150897
Norton Larkspur 124408
Norton Rockjasmine 23241
Norton Sagebrush 35992
Norvegian Cudweed 178311
Norway Draba 137147
Norway Fir 348,384,300223,318580
Norway Maple 3425
Norway Pine 300180,300223
Norway Rockfoil 349719
Norway Sedge 75529
Norway Spruce 298191
Norway Spruce Fir 298191
Norway Spruce-fir 298191
Norway Whitlowgrass 137147
Norwegian Arctic-cudweed 270574
Norwegian Cinquefoil 312828
Norwegian Draba 137147
Norwegian Grey Alder 16368
Norwegian Mugwort 35993
Norwegian Sandwort 32103
Norwegian Wintergreen 322861
Norwey Cudweed 178311
Nose Ticklers 399601
Nosebleed 3978,101852,322724,397556
Noseburn 394114,394129,394142,394200
Nosegay Frangipani 305225
Nosema 266758
Nose-smart 399601
Nose-tickler 399601
Nosetwitcher 399601
Nosewort 4005
Nospurflower 374070
Not Bearded Rhododendron 330912
Notable Azalea 330932
Notable Barberry 51966
Notable Linociera 231753
Notable Rhododendron 331636
Notable Stanhopea 373694
Notable Statice 230645
Notable Wattle 1444
Notable Wildginger 37653
Notchbract Waterleaf 200483
Notched Leaf Milkvetch 42976
Notched Wampee 94188
Notchleaf Sea Lavender 230761
Notch-leaf Sea-lavender 230761
Notchleaf Statice 230761
Notchleaf Willow 344004
Notchweed 87200
Note Dallisgrass 285469
Note Paspalum 285469
Noteworthy Mulberry 259183
Notha Iris 208858
Nothaphoebe 266677,266678,266787
Nothapodytes 266799
Nothern Draba 136951
Nothern Epipactis 147255
Nothern Groundsel 358769
Nothern Incense Cedar 67526
Nothern Whitlowgrass 136951
Nothing in Summer 105810
Nothoduguey Stonecrop 356965
Nothoscordum 266957
Noticeable Rhodiola 329912
Notocactus 267030
Notopterygium 267147
Notoseris 267166
Notothixos-like Scurrula 355315
Nottingham Catchfly 363827
Notylia 267213
Nouelia 267221

November Goldenrod 368106
November Lily 229900,229902
Nowallcress 373642
Ntcidleaf Gerbera 175212
Ntonga Nut 113460
Nub-berry Nub 338243
Nubbletwig Azalea 330153
Nubian Senna 78227
Nubicolous Holly 204091
Nubilous Honeysuckle 235991
Nuchen Germander 388165
Nucleonwood 291481
Nude Orach 44541
Nude Phyllostachys 297379
Nude-stem Magellan Barberry 51406
Nudicaulous Poppy 282617
Nuffin Idols 410677
Nujiang Arrowbamboo 162720
Nujiang Aster 41195
Nujiang Auriculateleaf Aster 40096
Nujiang Azalea 331720
Nujiang Bedstraw 170608
Nujiang Bladderwort 403316
Nujiang Camellia 69589
Nujiang Clearweed 299039
Nujiang Clematis 95160
Nujiang Coldwaterflower 299039
Nujiang Craibiodendron 108576
Nujiang Dolomiaea 135735
Nujiang Draws Ear Grass 403316
Nujiang Dutchmanspipe 34320
Nujiang Eurya 160620
Nujiang Fargesia 162720
Nujiang Fir 433
Nujiang Garcinia 171152
Nujiang Larch 221947
Nujiang Loquat 151171
Nujiang Maple 2886,2887
Nujiang Mountainash 369511
Nujiang Mountain-ash 369511
Nujiang Pennywort 200360
Nujiang Primrose 314611
Nujiang Raspberry 339233
Nujiang Sagebrush 36000
Nujiang Sandwort 32203
Nujiang Smoothleaf Maple 3078
Nujiang Spapweed 205302
Nujiang Taxillus 385246
Nujiang Umbrella Bamboo 162720
Nujiang Waxplant 198893
Nujiang Willow 343233,343779
Nujiang Winterhazel 106643
Nujiang Youngia 416459
Nukiang Clematis 95160
Nukiang Fir 433
Numerous-branched Flat-sedge 119410
Numerousflower Elaeocarpus 142285
Numman-idles 410677
Nummlar Leaves Gaultheria 172127
Nummular-leaved Wintergreen 172127
Nummularylike Wintergreen 172127
Num-num 76864
Numpinole 21339
Nun 273545
Nun's Scourge 18687
Nun's-hood Orchid 293537
Nunas Bean 294056
Nunate Watertowerflower 54454
Nuoqiang Rueleaf Sagebrush 36175
Nur Devil Rattan 121519
Nut Bush 106703
Nut Grass 118565,119015,119503
Nut Hall 106703
Nut Halse 106703
Nut Palm 106703,115852
Nut Pine 299842,299851,299854,299926,
300171,300189
Nut Rags 106703
Nut Rush 354052,354173,354189,354225,
354274
Nut Sedge 118897,119139,119549,393205
Nut Stowel 106703
Nut Stowell 106703
Nut Tree 106703
Nut Wood 386465
Nutall's Waterwced 200190
Nutant Dandelion 384700
Nutant Gaoliang 369601
Nutant Paraboea 283122
Nutant Raceme Indigo 206373
Nutant Racemelike Indigo 206386
Nutant Thrixsperm 390522
Nutant Thrixspermum 390522
Nutant Vanda 399700
Nutant Windhairdaisy 348936
Nutantbranch Tamarisk 383530
Nutantdaisy 110343
Nutantflower Anemone 23937
Nutantflower Densevein 261328
Nutantflower Forrest Sandwort 31909
Nutantflower Gentian 173670
Nutantflower Licorice 177929
Nutantflower Monkshood 5459
Nutantflower Mycetia 260640
Nutantflower Ophiorrhiza 272267
Nutantflower Primrose 314715
Nutantflower Spicegrass 239758
Nutantflower Windflower 23937
Nutantfruit Eritrichium 153505
Nutantfruit Flax 231940
Nutantfruit Violet 410396
Nutant-fruited Barberry 51975
Nutanthead Goldenray 229009
Nutanthook Azalea 332037
Nutantiflorous Monkshood 5459
Nutantleaf Sagebrush 35491
Nutantsa Japgarden Vin 341751
Nutantspike Gingerlily 187458
Nutantspike Lovegrass 147685
Nutantspike Primrose 314228
Nutantstem Phreatia 295953
Nutanttwig Savin 213854
Nutanttwig Wendlandia 413809
Nutate Bamboo 47394
Nutate Catchfly 363859
Nutateflower Crazyweed 279041
Nutateflower Cymbidium 116809
Nutateflower Orchis 116809
Nut-bearing Torreya 393077
Nutbrwon Sage 344941
Nutgall Oak 324050
Nutgall Tree 332509
Nutgrass 118820,119503
Nut-grass 119503
Nutgrass Cypressgrass 119503
Nut-grass Flat Sedge 119503
Nutgrass Flatsedge 119503
Nut-grass Flat-sedge 119503
Nutgrass Galingale 119503
Nutka Rose 336810
Nutlikeseed Holly 204092
Nutmeg 261402,261423,261424
Nutmeg Bush 387032
Nutmeg Family 261464
Nutmeg Flower 266241
Nutmeg Geranium 288239,288396
Nutmeg Hickory 77910
Nutmeg Melon 114214
Nutmeg Pelargonium 288396
Nutmeg-scented Geranium 288239
Nutmeg-tree 393047
Nutmug 261424
Nuttal Oak 324232
Nuttall 106703
Nuttall Oak 324479
Nuttall Povertyweed 257687
Nuttall Rhododendron 331361
Nuttall Violet 410305
Nuttall Willow 344077
Nuttall's Alkali Grass 321344
Nuttall's Death Camas 417920
Nuttall's Evening-primrose 269471
Nuttall's False Sagebrush 371143
Nuttall's Knotweed 309475
Nuttall's Larkspur 124411,124416
Nuttall's Lobelia 234660
Nuttall's Milkwort 308231
Nuttall's Oak 324479
Nuttall's Prairie-parsley 310674
Nuttall's Pussytoes 26536
Nuttall's rayless-goldenrod 54278

Nuttall's Sandwort 255542
Nuttall's Sedum 356969
Nuttall's Sensitive-briar 255082
Nuttall's Snapdragon 28630
Nuttall's Sunflower 189012
Nuttall's Thistle 92245
Nuttall's Wirelettuce 375962
Nuttall's Dogwood 105130
Nuuanu White Hibiscus 194722
Nux Vomica 378836
Nux-vomica 378836
Nux-vomica Poison Nut 378834
Nuxvomica Poisonnut 378836
Nux-vomica Poison-nut 378836
Nux-vomica Tree 378836
NW. China Cotoneaster 107729
NW. China Eutrema 161129
NW. China Minuartwort 255512
NW. China Spunsilksage 360857
NW. Sichuan Sedge 74454
NW. Tiger Lily 229810
Nyala Tree 415100
Nyanang Rockfoil 349670
Nyasaland Mahogany 216196,216211
Nyctelea 143886
Nyctocalos 267598,267599
Nye Fishhook Cactus 354305
Nylon Hedgehog Cactus 140342,140343
Nyman Cladopus 93975
Nymphaea 267627
Nymphae-leaf Echinodorus 140551
Nymphae-leaf Rhododendron 331364
Nypa 267837,267838
Nypa Palm 267837,267838
Nyssa Family 267881

O

Oahu Stringbush 414174
Oak 2837,116044,233246,323580,323582,323583,324278,324335,324384
Oak Atchern 324335
Oak Creek Ragwort 279944
Oak Creek Triteleia 398807
Oak Fern 182491
Oak Leaf Ivy 411885
Oak Lung 234269
Oak Mistletoe 295583
Oak Nut 324335
Oak of Cappadocia 19145
Oak of Jerusalem 388015
Oak of Mamre 302034,302103,316127,324309
Oak of Paradise 139682
Oak Rags 234269
Oak Sedge 73626,73627
Oak Walnut 113440
Oak-belt Gooseberry 334169
Oakberry 324335
Oakes Evening Primrose 269473
Oakes Pondweed 312195
Oakes' Evening-primrose 269473
Oakes' Pondweed 312195
Oak-forest Wood Rush 238640
Oakforest Woodrush 238640
Oak-forest Wood-rush 238640
Oak-inhabiting Mistletoe 410979
Oakland Star-tulip 67632
Oakleaf Ajania 13026
Oakleaf Azalea 331477
Oakleaf Banksia 47663
Oakleaf Bigumbrella 59239
Oakleaf Brassaiopsis 59239
Oak-leaf Dryandra 138442
Oak-leaf Fleabane 150922
Oakleaf Garden Geranium 288430
Oakleaf Geranium 288459
Oak-leaf Geranium 139682,288459
Oakleaf Goosefoot 87029
Oakleaf Hydrangea 200063,200064,200065
Oak-leaf Hydrangea 200063,200064,200067
Oak-leaf Ivy 411885
Oakleaf Loquat 151168
Oakleaf Mountain Ash 369429
Oak-leaf Mountain Ash 369429
Oakleaf Saussurea 348703
Oakleaf Tanoak 233354
Oakleaf Windhairdaisy 348703
Oak-leafed Hydrangea 200063
Oak-leafed Mountain Ash 369297
Oak-leaved Brassaiopsis 59239
Oak-leaved Geranium 288459
Oak-leaved Goosefoot 87029
Oak-leaved Hydrangea 200063
Oak-leaved Loquat 151168
Oak-leaved Tanoak 233354
Oakleeeh 45165
Oak-lefaf Grevillea 180640
Oak-like Sweetleaf 381188
Oak-macey 324335
Oak-of-cappadocia 19174
Oak-rag 234269
Oaksey Lily 168476
Oat 45363,45566
Oat Chestnut 78848
Oat Grass 34920,45363,190129
Oat Thistle 271666
Oatchestnut 78848,78889
Oatgrass 34920
Oat-grass 34930,190129
Oat-like Pricklythrift 2218
Oatmeal Grass 223990
Oats 45363,45566
Oaxaca Palmetto 341418
Oaxaca Pine 300124
Obada 165263
Obconic Slugwood 50575
Obconical Hsienmu 161624
Obconical Slugwood 50575
Obconic-fruit Hemsleia 191951
Obconicfruit Snowgall 191951
Obcordate Christia 89207
Obcordate Golden Evergreen Raspberry 338354
Obcordate Stauntonvine 374424
Obcordate Staunton-vine 374424
Obcordateleaf Aspidopterys 39687
Obcordateleaf Aucuba 44944
Obcordate-leaved Aucuba 44944
Obcordatelimb Larkspur 124418
Obcordate-shaped Aspidopterys 39687
Obcordifolious Larkspur 124418
Obeche 398004
Obedience 297984
Obedience Plant 245014
Obedient Plant 297740,297973,297984,297987
Oberonia 267922
Oberonia Orchid 267991
Obispo Pine 300098
Oblance-leaf Ardisia 31444
Oblanceleaf Aspidistra 39564
Oblanceleaf Aucuba 44913
Oblanceleaf Daphniphyllum 122716
Oblanceleaf Devilpepper 327072
Oblanceleaf Sweetleaf 381218
Oblanceleaf Syzygium 382629
Oblanceleaf Tigernanmu 122716
Oblance-leaved Brazil Leathern Barberry 51499
Oblanceolata Loose-flowered Barberry 51837
Oblanceolata Tickseed 104797
Oblanceolate Arisaema 33332
Oblanceolate Catchfly 363829
Oblanceolate Linear-leaved Basketvine 9448
Oblanceolate Pruinashoot Willow 344025
Oblanceolate Southstar 33332
Oblanceolate Stranvaesia 377466

Oblanceolate Sweetleaf 381218
Oblanceolate Syzygium 382629
Oblanceolate Tanoak 233329
Oblanceolate Willow 344188
Oblanceolate-leaf Devilpepper 327072
Oblanceolateleaf Syzygium 382629
Oblanceolate-leaved Barberry 52062
Oblanceolate-leaved Rhododendron 331365
Oblanceolate-leaved Staff-tree 80252
Oblanceolate-leaved Sweetleaf 381218
Oblanceoleaf Azalea 331365
Oblanceoleaf Rhododendron 331365
Oblanceotaeleaf Windhairdaisy 348562
Oblate Camellia 69400
Oblate Honeysuckle 235995
Oblate Listera 264714
Oblate Syzygium 382631
Oblionker 9701
Oblique Arrowbamboo 162669
Oblique Caudateleaf Skullcap 355403
Oblique Ceriman 258174
Oblique Erismanthus 153397
Oblique Eurya 160502
Oblique Fargesia 162723
Oblique Lagochilus 220063
Oblique Leaf Cordia 104220
Oblique Metalleaf 296408
Oblique Oreocharis 273877
Oblique Petal Metalleaf 296408
Oblique Photinia 295743
Oblique Pilose Cypressgrass 119379
Oblique Pilose Galingale 119379
Oblique Primrose 314737
Oblique Rhynchoglossum 333086
Oblique Wallich Lasianthus 222305
Oblique Wallich Roughleaf 222305
Oblique-beard Schizachyrium 351277
Obliquecalyxweed 237957
Obliqueflower Dutchmanspipe 34285
Obliquefruit Bladderwort 403257
Obliquefruit Draws Ear Grass 403257
Obliqueleaf Begonia 49641
Oblique-leaf Ceriman 258175
Obliqueleaf Eurya 160561
Obliqueleaf Eutrema 161140
Obliqueleaf Fig 165390
Obliqueleaf Thyme 391220
Oblique-leaflet Rosewood 121793
Oblique-leaved Eurya 160561
Oblique-leaved Pepper 300507
Obliquenerv Hedyotis 187634
Obliquenerved Greenstar 307523
Oblique-nerved Greenstar 307523
Obliquepedicellate Apple 243671
Obliqueswollen Malabarnum 172831
Obliquetooth Woodbetony 287683
Obliquevein Eucalyptus 155614
Obliquevein Lasianthus 222237
Obliquevein Microtropis 254313
Oblique-vein Microtropis 254313
Obliquevein Roughleaf 222237
Oblique-veined Lasianthus 222237
Oblique-wing 301688
Obliqui-veined Microtropis 254313
Oblong Arrowwood 407972
Oblong Bugle 13150
Oblong Burmannia 63988
Oblong Conescaled Keteleeria 216144
Oblong Daphniphyllum 122710
Oblong Evergreen Chinkapin 78996
Oblong Evergreenchinkapin 78996
Oblong Falsebetony 373071
Oblong Falsepimpernel 231555
Oblong Fishvine 283282
Oblong Gomphostemma 179198
Oblong Helwingia 191197
Oblong Hooktea 52420
Oblong Jadeleaf and Goldenflower 260410
Oblong Juniper 213833
Oblong Kumquat 167506
Oblong Leaves Azalea 331366
Oblong Maire Nettle 402961
Oblong Motherweed 231555
Oblong Motuo Elatostema 142738
Oblong Motuo Stairweed 142738
Oblong Niphaea 266308
Oblong Orbicular-leaved Rhododendron 331403
Oblong Pygeum 322404
Oblong Qruis Varied Litse 234093
Oblong Raspberry 338901
Oblong Skullcap 355627
Oblong Sophora 369089
Oblong Spiralflag 107261
Oblong Sunflower 188980
Oblong Supplejack 52420
Oblong Tupelo 267865
Oblong Villose Hawkweed 196083
Oblong Yushanbamboo 416805
Oblong Yushania 416805
Oblongate Blueberrylike Persimmon 132455
Oblongate Draba 137148
Oblongate Fishvine 283282
Oblongate Microula 254356
Oblongate Yunnan Rhodiola 329976
Oblongateleaf Cremanthodium 110420
Oblongateleaf Elaeocarpus 142364
Oblongateleaf Nutantdaisy 110420
Oblonge-leaf Jointfir 178553
Oblong-foliate Sasa 347261
Oblongfruit Hemsleia 191944
Oblong-fruit Holly 204103
Oblongfruit Largeseed Snowgall 191944
Oblongfruit Sedge 74191
Oblong-fruit Serviceberry 19250
Oblongfruit Woad 209217
Oblong-fruited Holly 204103
Oblongifoliate Lime 391789
Oblongifoliate Linden 391789
Oblongifruited Pygeum 322404
Oblongifruited Pyrenaria 322605
Oblongileaf Glaudular Eurya 160478
Oblongileaved Elaeocarpus 142364
Oblongileaved Euonymus 157748
Oblong-leaf Agapetes 10348
Oblongleaf Aspidistra 39566
Oblongleaf Betony 373321
Oblongleaf Blumea 55796
Oblongleaf Brownhair Aster 40496
Oblongleaf China Sweetspire 210370
Oblongleaf Chirita 87931
Oblongleaf Common Vineclethra 95517
Oblong-leaf Dahurian Birch 53400
Oblong-leaf Daphniphyllum 122710
Oblongleaf Dendrobium 125120
Oblong-leaf Dogwood 380474
Oblongleaf Doronicum 136362
Oblongleaf Dragonhead 137619
Oblongleaf Dumasia 138939
Oblongleaf Elaeocarpus 142364
Oblongleaf Elatostema 142761
Oblongleaf Embelia 144773
Oblongleaf Euonymus 157748
Oblongleaf Eurya 160562
Oblongleaf Eustigma 161023
Oblongleaf Falsenettle 56244
Oblongleaf Garcinia 171155
Oblongleaf Hackberry 80697
Oblongleaf Hetaeria 193617
Oblongleaf Heterostemma 194162
Oblong-leaf Holly 204000
Oblongleaf Hoodorchis 264777
Oblongleaf Hornbeam 77353
Oblongleaf Ironweed 405452
Oblong-leaf Japanese Mulberry 259111
Oblongleaf Kadsura 214974
Oblongleaf Linden 391789
Oblongleaf Litse 234048
Oblongleaf Lopseed 295986
Oblongleaf Luzon Viburnum 407936
Oblong-leaf Lysionotus 239964
Oblong-leaf Mallotus 243405
Oblongleaf Maple 3255
Oblongleaf Maximovicz Buckthorn 328782
Oblongleaf Michelia 252937
Oblongleaf Mountain Leech 126485
Oblongleaf Newlitse 264074
Oblong-leaf Onosma 271812
Oblong-leaf Orach 44554
Oblongleaf Orache 44554
Oblongleaf Ramie 56244

Oblongleaf Salomonia 344351
Oblongleaf Saltbush 44554
Oblong-leaf Sasa 347261
Oblongleaf Schefflera 350752
Oblongleaf Schnabelia 352060
Oblongleaf Sotvine 194162
Oblong-leaf Stachyurus 373550
Oblongleaf Stairweed 142761
Oblongleaf Swallowwort 117395
Oblongleaf Sweetspire 210370
Oblong-leaf Sweetspire 210370
Oblong-leaf Tigernanmu 122710
Oblongleaf Tupelo 267865
Oblongleaf Ventilago 405452
Oblongleaf Vetch 408269
Oblongleaf Windhairdaisy 348595
Oblongleaf Wingdrupe 405452
Oblong-leaved Agapetes 10348
Oblong-leaved Artocarpus 36917
Oblong-leaved Aster 40948,380937
Oblong-leaved Barberry 51978
Oblong-leaved Dogwood 380474
Oblong-leaved Embelia 144773
Oblong-leaved Epimedium 147051
Oblong-leaved Eurya 160562
Oblongleaved Eustigma 161023
Oblong-leaved Eustigma 161023
Oblong-leaved Garcinia 171155
Oblong-leaved Heterostemma 194162
Oblong-leaved Kadsura 214974
Oblong-leaved Maple 3255
Oblong-leaved Newlitse 264074
Oblong-leaved Stachyurus 373550
Oblong-leaved Sundew 138299
Oblong-leaved Vanilla 405015
Oblong-leaved Ventilago 405452
Oblong-leaved Wintergreen Barberry 51802
Oblonglipped Dendrobium 125221
Oblongpetal Parnassia 284553
Oblongsheath Arrowbamboo 162724
Oblongspikelet Juncellus 212774
Oblong-squamate Hemlock 399920
Obovata Acuminate Ehretia 141597
Obovate Agapetes 10351
Obovate Altingia 18203
Obovate Beak Grain 128021
Obovate Cleyera 96615
Obovate Cremanthodium 110422
Obovate Hairystyle Cherry 83279
Obovate Honeysuckle 235830
Obovate Hydrangea 199907
Obovate Indianmulberry 258926
Obovate Japanese Staunton-vine 374413
Obovate Kumquat 167511
Obovate Leaf Tephrosia 386210
Obovate Loosestrife 239760
Obovate Loquat 151166
Obovate Nutantdaisy 110422
Obovate Parnassia 284586
Obovate Peony 280241
Obovate Planchonella 301777
Obovate Rhynchotechum 333745
Obovate Shortstipe Beautyberry 66759
Obovate Staunton-vine 374425
Obovate Tanoak 233335
Obovate Waxpetal 106667
Obovate Waxplant 198850
Obovate Wildolive 301777
Obovate Winterhazel 106667
Obovate-leaed Tanoak 233330
Obovateleaf Acanthopanax 2451
Obovateleaf Actinidia 6675
Obovateleaf Actinodaphne 6804
Obovateleaf Agapetes 10351
Obovateleaf Arrowwood 408134
Obovateleaf Bennettiodendron 51048
Obovateleaf Blueberry 403745
Obovate-leaf Buckthorn 328620
Obovateleaf Camellia 69402
Obovateleaf Canthium 71349
Obovateleaf Daphne 122454
Obovate-leaf Dock 340150
Obovate-leaf Elaeagnus 142214
Obovateleaf Glochidion 177162
Obovateleaf Helicia 189943
Obovateleaf Hoarypea 386210
Obovateleaf Holly 204025
Obovateleaf Honeysuckle 235830
Obovateleaf Hornbeam 77356
Obovate-leaf Hydrangea 200031
Obovateleaf Kiwifruit 6675
Obovateleaf Machilus 240658
Obovate-leaf Manglietia 244461
Obovateleaf Photinia 295638,295721
Obovateleaf Primrose 314738
Obovateleaf Privet 229557
Obovateleaf Sabia 341531
Obovateleaf Shortstipe Beautyberry 66759
Obovateleaf Sida 362495
Obovateleaf Stachyurus 373554
Obovateleaf Stauntonvine 374425
Obovate-leaf Stauntonvine 374428
Obovateleaf Stringbush 414250
Obovateleaf Tanoak 233330
Obovateleaf Windhairdaisy 348756
Obovateleaf Woodnettle 273912
Obovate-leaved Acanthopanax 2451
Obovate-leaved Actinidia 6675
Obovate-leaved Actinodaphne 6804
Obovate-leaved Barberry 51979
Obovate-leaved Camellia 69402
Obovate-leaved Dock 340150
Obovate-leaved Helicia 189943
Obovate-leaved Machilus 240658
Obovate-leaved Peony 280241
Obovate-leaved Privet 229557
Obovate-leaved Sabia 341531
Obovate-leaved Senna 78436
Obovate-leaved Stachyurus 373554
Obovatelimb Salacia 342697
Obovate-limb Salacia 342697
Obovateutricle Sedge 76015
Obovoidberry Cherry Elaeagnus 142105
Obovoid-fruited Rattan Palm 65748
Obregonia 268085,268086
Obscure Begonia 50118
Obscure Morning-glory 208023
Obscure Nothapodytes 266808
Obscure Willow 343789
Obscure Woodbetony 287465
Obscurenerve Slugwood 50580
Obscurenerved Cleyera 96616
Obscure-nerved Cleyera 96616
Obscure-nerved Slugwood 50580
Obscure-nerved Staunton-vine 374432
Obscurevein Actinodaphne 6805
Obscurevein Camellia 69404
Obscurevein Leafflower 296596
Obscurevein Machilus 240659
Obscurevein Underleaf Pearl 296596
Obscure-veined Actinodaphne 6805
Obscure-veined Machilus 240659
Obscure-veined Microtropis 254314
Obscuricallose Rockfoil 349753
Obsolescent Begonia 50119
Obsoletedentate Mountainash 369474
Obsoletedentate Mountain-ash 369474
Obtriangular Oxalis 277985
Obtriangular Woodsorrel 277985
Obtuse Ardisia 31540
Obtuse Armgrass 58128
Obtuse Barleria 48272
Obtuse Bird Cherry 280040
Obtuse Ciliate Barberry 51469
Obtuse Clematis 94763
Obtuse E. H. Wilson Maple 3753
Obtuse Elaeagnus 142127
Obtuse Falsecuckoo 48272
Obtuse Figwort 355203
Obtuse Greyleaf Willow 344073
Obtuse Ground Cypress 85310
Obtuse Hairy-fruit Fig 165779
Obtuse Hardleaf Willow 344073
Obtuse Honeysuckle 236064
Obtuse Italian Maple 3289
Obtuse Leaf Rosewood 121779
Obtuse Longbeak Eucalyptus 155520
Obtuse Meranti 362213
Obtuse Pashi Pear 323255
Obtuse Peppergrass 225425
Obtuse Pepperweed 225425

Obtuse Pondweed 312199
Obtuse Ricegrass 300722
Obtuse Rush 213332
Obtuse Sphenopholis 371493
Obtuse Swertia 380291
Obtuse Taeniophyllum 383066
Obtuse Taiwan Adinandra 8228
Obtuse Twistedstalk 377930
Obtuse Water-cress 336279
Obtuse Yunnan Bittercress 73049
Obtusebract Gerbera 175222
Obtusebract Windhairdaisy 348557
Obtusecalyx Chickweed 374740
Obtusedentate Raoping Raspberry 339155
Obtuse-fruit Gomphocarpus 179096
Obtuseleaf Acacia 1381
Obtuseleaf Achyranthes 4263
Obtuse-leaf Actinidia 6578
Obtuse-leaf Adinandra 8228
Obtuseleaf Argyreia 32638
Obtuseleaf Azalea 331370
Obtuseleaf Camellia 69406
Obtuseleaf Cassia 91276
Obtuseleaf Cinnamon 91276
Obtuseleaf Corydalis 105966
Obtuseleaf Cowparsnip 192331
Obtuseleaf Cow-wheat 248203
Obtuseleaf Cryptocarya 113455
Obtuseleaf Curly Pondweed 378983
Obtuseleaf Cystopetal 320250
Obtuseleaf Dinggongvine 154252
Obtuseleaf Dolichandrone 135322
Obtuseleaf Drypetes 138670
Obtuse-leaf Elaeagnus 142127
Obtuseleaf Elatostema 142764
Obtuseleaf Eritrichium 153463
Obtuseleaf Erycibe 154252
Obtuseleaf Eurya 160564
Obtuseleaf Fighazel 134913
Obtuseleaf Fiveleaf Akebia 13234
Obtuseleaf Flowering-quince Willow 343187
Obtuseleaf Gasteria 171721
Obtuseleaf Groundsel 359608
Obtuseleaf Gurjun 133573
Obtuseleaf Impatient Bittercress 72840
Obtuseleaf Inula 207189
Obtuseleaf Jurinea 207189
Obtuse-leaf Kiwifruit 6578
Obtuseleaf Lilyturf 272054
Obtuseleaf Litse 233885
Obtuse-leaf Millettia 254779
Obtuseleaf Newlitse 264075
Obtuseleaf Noble Milkvetch 42802
Obtuseleaf Nothapodytes 266809
Obtuseleaf Onion 15602
Obtuseleaf Orangedred Raspberry 338167
Obtuseleaf Orostachys 275379
Obtuseleaf Peperomia 290395
Obtuseleaf Premna 313702
Obtuseleaf Rapanea 261629
Obtuse-leaf Rhododendron 331370
Obtuseleaf Rose 336940
Obtuseleaf Rosewood 121779
Obtuseleaf Russianthistle 344564
Obtuseleaf Screwtree 190107
Obtuseleaf Shortstyle Camellia 69406
Obtuseleaf Sinocrassula 364922
Obtuseleaf Skullcap 355628
Obtuseleaf Spring Bittercress 72840
Obtuseleaf Thickshellcassia 113455
Obtuseleaf Throughstem Clearweed 299027
Obtuseleaf Trefoil 396882
Obtuseleaf Wingseedtree 320844
Obtuseleaf Yunnan Bittercress 73049
Obtuse-leaved Acacia 1381
Obtuse-leaved Argyreia 32638
Obtuse-leaved Camellia 69406
Obtuse-leaved Cinnamon 91276
Obtuse-leaved Cryptocarya 113455
Obtuse-leaved Erycibe 154252
Obtuse-leaved Eurya 160564
Obtuse-leaved Everlasting 178318
Obtuse-leaved Fig 164874
Obtuse-leaved Grewia 180938
Obtuse-leaved Gugertree 350935
Obtuse-leaved Gurjun 133577
Obtuse-leaved Hiptage 196844
Obtuse-leaved Newlitse 264075
Obtuse-leaved Nothapodytes 266809
Obtuse-leaved Pileostegia 123545
Obtuse-leaved Pondweed 312199
Obtuse-leaved Premna 313702
Obtuse-leaved Primrose 314741
Obtuse-leaved Rapanea 261629
Obtuse-leaved Rose 336940
Obtuse-leaved Rosewood 121779
Obtuse-leaved Screw-tree 190107
Obtuse-leaved Tutcheria 400709
Obtuselob Coeruleus Larkspur 124087
Obtuselobate Sagebrush 36012
Obtuselobe Snakegourd 396210,396232
Obtuselobed Anemone 23939
Obtuselobed Coeruleus Larkspur 124087
Obtuselobed Nujiang Pennywort 200361
Obtuselobed Snakegourd 396210,396232
Obtuselobed Windflower 23939
Obtusepetal Gagea 169431
Obtusepetal Stonecrop 356974
Obtuseridge Pondweed 312203
Obtusescale Sedge 73826
Obtusesepal Chickweed 374913
Obtusesepal Chirita 87910
Obtusesepal Clematis 95226
Obtusesepal Stonecrop 356871
Obtusesepal Trigonotis 397452
Obtusetooth Clearweed 299003
Obtusetooth Coldwaterflower 299003
Obtuse-tooth Elatostema 142763
Obtusetooth Eyebright 160286
Obtusetooth Farges Meehania 247724
Obtusetooth Mountainash 369410
Obtuse-tooth Urophyllous Clematis 95401
Obtuse-toothed Eurya 160448
Obtuse-toothed Mountain-ash 369410
Obtusewing Ash 168068
Obtusilobed Deutzia 127030
Obuse Breynia 60078
Obvate Hartia 376463
Obvatefruit Rattanpalm 65748
Obvate-leaf Hartia 376463
Obvateleaf Oak 116164
Obvateleaf Qinggang 116164
Obvate-leaved Cyclobalanopsis 116164
Obvate-leaved Oak 116164
Obvious Barberry 52029
Obvious Cyclobalanopsis 116177
Obvious Newlitse 264082
Obviousnerve Newlitse 264082
Obvious-nerved Newlitse 264082
Obviousvein Tea 69059
Oca 277779,278130
Oca Quina 401412
Oca Quirts 401412
Occasional Larkspur 124305
Occidental Hackberry 80698
Occidental Milkvetch 42921
Occidental Pear 323133
Occidental Pot Barberry 51462
Occidental Sophora 369090
Occidental Swamp-lily 417608
Occult-nerved Holly 204104
Ocean Mountainash 369525
Ocean Spray 197404
Oceanblue Morning Glory 207891
Oceanspray 197401,197404
Ocean-spray 197401,197404
Ocellate Violet 410319
Ocellated Humboldt Lily 229867
Ochna 268132
Ochna Family 268277
Ochotsk Clematis 95304
Ochotsk Corydalis 106197
Ochotsk Knotweed 309478
Ochre Lily 229999
Ochrecoloured Coelogyne 98701
Ochreyellow Epidendrum 146452
Ochre-yellow Lily 230001
Ochre-yellow Mountain-ash 369475
Ochreyellow Rhododendron 331384
Ochre-yellow Rhododendron 331384
Ochrocarpus 268330

Ochrosia 268351
Oconee Azalea 330702,331865
Oconee Bells 362246
Ocote Chino 300033
Ocotea 268688
Ocotillo 167556,167562
Ocreate Greenbrier 366491
Ocreate Knotweed 309479
Ocreate Smilax 366491
October Clematis 94740
October Daphne 200798
October Glory Maple 3513
October Lady's-tresses 372235,372236
October Plant 200798
October-flower 308683
Octocrown Corydalis 106546
Octopus Agave 10956
Octopus Cactus 375519
Octopus Tree 129776,350648,350706
Octoraro Creek Chickweed 83076
Oculate Stanhopea 373698
Oddity Azalea 331444
Oddrod 315082
Ode 209229
Odessa Tamarisk 383595
Odhran 192358
Odier Neoporteria 263749
Odin's Grace 174832
Odiyal Flour 57122
Odoko 355004
Odollam Cerberustree 83700
Odollam Cerberus-tree 83698
Odontites 268986,268991,268999
Odontochilous Chelonopsis 86825
Odontoglossum 269055
Odontoglossum Orchid 269055
Odora Porophyllum 311716
Odorata Water Lily 267730
Odorate Schefflera 350756
Odorless Bayberry 261176
Odorless Dendrobium 124995
Odorless Wax-myrtle 261176
Odour Cassia-bark 91389
Odour-bark Cinnamon 91389
Odum 88506
Oeder Woodbetony 287468
Oenanthe 269292
Oenanthem Chin Cactus 182469
Oenocarpus 269373
Oenothera 269394
Oersted Columnea 100124
Oersted Trichilia 395518
Ofbicular Snailseed 97934
Ofbit 379660
Ofbiten 379660
Ofer Hollow Reedgrass 65462
Official Monkshead 5442
Officinal Acotite 5442
Officinal Breynia 60073
Officinal Bugloss 21959,321637
Officinal Cherry Laurel 223116
Officinal Magnolia 198699
Officinal Sage 345271
Officinal Squill 352960
Officinale Ivyarum 353130
Officinale Mallow 243806
Ofram 386652
Ofthe Field 70271
Ogata Achyranthes 4319
Ogawa Spiraea 372036
Ogbane 29473,265327
Ogden's Pondweed 312204
Ogea 122160
Ogeechee Lime 267859,267867
Ogeechee Tupelo 267859
Ogiera 143565
Ogilvie Claytonia 94341
Oglethorpe Oak 324248
Ogon-daidai 93328
Ogotoruk Creek Ragwort 279931
Ogre Lily 230058
Ogura Cowlily 267313
Ogura Naiad 262096
Ohelo Berry 403978
Ohi Spiderwort 394053
Ohi's Lehua 252623
Ohia 80750,382606
Ohio Buckeye 9694
Ohio Goldenrod 368294
Ohio Horse Mint 55523
Ohio Horsemint 55523
Ohio Suckeye 9694
Ohla 382606
Ohlendorf Eastern Arborvitae 390614
Ohw Vetch 408516
Ohwi Alishan Oxtongue 298607
Oi 373497,373507,373511,405865
Oil Arrowbamboo 162642
Oil Bean 289439
Oil Bitter Bamboo 37271
Oil Bitterbamboo 37271
Oil Camellia 69411
Oil Cinnamon 91363
Oil Flax 232002
Oil Grass 117136,407579
Oil Malania 242984
Oil Neststraw 379158
Oil Nut 212599
Oil Palm 142020,142255,142257
Oil Pine 300243
Oil Poppy 282717
Oil Sindora 364707
Oil Spine Bamboo 47318
Oil Tea 69411
Oil Vine 385650
Oilawn 139905,372406
Oilfruitcamphor 381788
Oilicica 228663
Oilleaf Tanoak 233278
Oilnut 212599,323066
Oil-nut 323066
Oilpalm 142255,142257
Oilplant 138348
Oilresiduefruit 197069
Oils 45566,198376
Oils Hoyles 45566
Oilseed 68860,361317
Oil-seed Rape 59493,59575
Oiltea Camellia 69411
Oil-tea Camellia 69411
Oiltung 406014,406018
Oil-vine 385650
Oil-wood 142422
Oily Bambusa 47467
Oily Bitter-bamboo 37271
Oily Malania 242984
Oily Persimmon 132339
Oily-grain Plant 361317
Oionut Tree 67860
Oisterloit 308893
Ojai Fritillary 168499
Oka Oxalis 277779,278130
Okame Zasa 362131
Okamoto Maple 3273
Okamoto Willow 343797
Okan 116569
Okennon's Anemone 23972
Ok-gue 165515
Okhotian Bugseed 104818
Okie Bean 218721
Okinawa Spinach 183066
Okkerdi 220416
Oklahoma Grass Pink 67886
Oklahoma Plum 316434
Oklahoma Redbud 83763
Okoume 44883
Okra 219
Okshit 8787
Okuro 13703
Olancha Peak Bastard-sage 152672
Olax 269642
Olax Family 269640
Old Baldy Sulphur Flower 152592
Old Coco Yam 99910
Old Coco-yam 99910
Old English Lavender 223251
Old English Poplar 311398
Old Fashioned Weigela 413596,413601
Old Field Birch 53586
Old Field Clover 396828
Old Field Pine 300264

Old Field Three-awn 33957
Old Folks' Herb 138348
Old Fustic 88514
Old Garden Dahlia 121561
Old Gold Juniper 213634
Old Lad's Corn 374916
Old Lady 35211
Old Lady Cactus 244097
Old Lady's Bonnets 30081
Old Lady's Lace 28044
Old Lady's Smock 68713
Old Maid 79418,86427
Old Maid's Basket 30081
Old Maid's Flower 410677
Old Maid's Last Friend 410677
Old Maid's Nightcap 174717
Old Maid's Pink 11947
Old Maid's Scent 322724
Old Man 21339,35088,95431,337180,358757
Old Man Banksia 47667
Old Man Beard 94887
Old Man Cactus 82211
Old Man of Mexico 82211
Old Man of the Andes 273813
Old Man of the Mountain 273826
Old Man Palm 97887
Old Man Saltbush 44542
Old Man Wormwood 35088
Old Man's Beard 35088,70271,79746,87736,94676,94747,94773,95431,145487,166125,196818,201787,266225,336419,349936,392052
Old Man's Beard 95085
Old Man's Bell 145487
Old Man's Bells 70271,145487
Old Man's Bread-and-cheese 243840
Old Man's Buttons 31051,68189,325518,325666,326287
Old Man's Clock 384714
Old Man's Face 28617,410677
Old Man's Flannel 405788
Old Man's Friend 21339
Old Man's Glass Eye 21339
Old Man's Looking Glass 21339
Old Man's Looking-glass 21339
Old Man's Love 35088
Old Man's Mustard 3978
Old Man's Nightcap 68713,102933,208120
Old Man's Pepper 3978,345881
Old Man's Pepper 4005
Old Man's Pepper-box 4005
Old Man's Plaything 299523
Old Man's Pulpit 37015
Old Man's Shirt 68713,374916
Old Man's Weatherglass 21339
Old Man's Whiskers 175452
Old Man's Woozard 95431
Old Pipewort 151479
Old Plainsman 201195
Old Sow 21596,249208
Old Uncle Harry 36474
Old Warrior 36088
Old Wife Hood 5442
Old Wife's Darning-needle 350425
Old Wife's Threads 326287
Old Witch Grass 281438
Old Wives' Mutch 5442
Old Wives' Mutches 5442
Old Wives' Tongue 311537
Old Wives' Tongues 311537
Old Woman 35090,35779,35868,36088,223334
Old Woman Threading the Needle 174877
Old Woman's Bonnets 30081,70164,72038,175444
Old Woman's Eye 409339
Old Woman's Eyes 409339
Old Woman's Needle 350425
Old Woman's Needlework 81768
Old Woman's Nightcap 5442,68713,70164
Old Woman's Orchid 273604
Old Woman's Penny 238371
Old Woman's Petticoats 282685
Old Woman's Pincushion 273531
Old Woman's Purse 205175
Old Woman's Toenails 237539
Old World Arrowhead 342400,342421
Old World Flatsedge 245392
Old World Locust 83527
Oldenlandia 187521,187695
Old-fashioned Bleeding Heart 220729
Oldfashioned Weigela 413591
Old-fashioned Weigela 413591
Oldfield Aster 380953
Old-field Balsam 317761
Oldfield Birch 53586
Old-field Blackberry 338119
Old-field Cinquefoil 312993
Old-field Cudweed 178318
Old-field Five-fingers 312993
Old-field Goldenrod 368268,368270
Oldfield Sneezeweed 188413
Old-field Three-awn 33957
Old-field Toadflax 267347
Old-field-balsam 178318
Oldham Bamboo 125453
Oldham Bambusa 125453
Oldham Bredia 59864
Oldham Chalkplant 183222
Oldham Chloranthus 88295
Oldham Daphniphyllum 122713
Oldham Dendrocalamopsis 125453
Oldham Elaeagnus 142132
Oldham Fissistigma 166675
Oldham Groundsel 365066
Oldham Gypsophila 183222
Oldham Habenaria 183921
Oldham Meliosma 249424
Oldham Persimmon 132335
Oldham Podocarpium 200736
Oldham Rhododendron 331391
Oldham Scolopia 354621
Oldham Sinosenecio 365066
Oldham Sping Raspberry 339135
Oldham Tigernanmu 122713
Oldham Titanotrichum 392424
Oldham Yam 131647
Oldham's Baby's-breath 183222
Oldham's Bambusa 125453
Oldham's Blueberry 403925
Oldham's Chinese Groundsel 365066
Oldham's Rhododendron 331391
Old-hook 300944
Oldlady Cactus 244097
Old-lady Cactus 244097,244124
Oldmaid 175943
Oldman 35090
Oldman and Woman 358062
Old-man Cactus 82211
Old-man Saltbush 44542
Old-man Wormwood 35088
Oldman's Beard 87736
Oldman's Cap 307656
Old-man's-whiskers 175452
Old-man-in-the-spring 360328
Old-man-of-the-andes 273813,273826
Old-man-of-the-andes Cactus 273826
Oldrot 28044,192358
Old-witch Grass 281438
Old-woman Cactus 244097
Oldworld Arrowhead 342400
Old-world Arrowhead 342400,342421,342424
Oleander 265310,265327
Oleander Agapetes 10345
Oleander Flowered Rhododendron 331321
Oleander Leaf Rapanea 261648
Oleander Podocarpus 306506
Oleander Rhododendron 331321
Oleander Spurge 159458
Oleander Wattle 1422
Oleanderleaf Allamanda 14881
Oleanderleaf Ardisia 31620
Oleanderleaf Camellia 69146
Oleanderleaf Nothoscordum 266970
Oleanderleaf Pittosporum 301286
Oleanderleaf Rapanea 261648
Oleanderleaf Seatung 301286
Oleanderleaf Yaccatree 306506
Oleander-leafed Protea 315900

Oleander-leaved Agapetes 10345
Oleander-leaved Ardisia 31620
Oleander-leaved Euphorbia 159458
Oleander-leaved Fig 165361
Oleander-leaved Protea 315900
Oleander-leaved Rapanea 261648
Oleaster 141929,141930,141932,142214,270099
Oleaster Family 141928
Oleaster Pear 323165
Oleraceous Woodnettle 221581
Olern Oler 16352
Olga Bay Larch 221919
Olga Desertcandle 148552
Olga Gagea 169475
Olgaea 270227,270239
Olibanum 57508
Oligandrous Anise-tree 204567
Oliganthous Beautyberry 66881
Oliganthous Bedstraw 170526
Oliganthous Microtropis 254315
Oligocarpous Hymenopogon 263959
Oligocarpous Maple 3274
Oligochaeta 270297
Oligodontous Dendropanax 125627
Oligodontous Holly 204111
Oligofloculous Ichnocarpus 203330
Oligoflorous Premna 313707
Oligoflorous Viburnum 407987
Oligolobed Flexhair Gyrocheilos 183321
Oligomeris 270326
Oligophyllous Aphania 225676
Oligophyllous Staunton-vine 374433
Oligospermous Corydalis 106207
Oligostachyum 270423
Oligo-vened Litse 234023
Oligrant Rhododendron 331395
Olimpic Draba 137154
Olimpic Whitlowgrass 137154
Olinia 270463
Olivacea Veitch Fir 497
Olivaceous Alyxia 18527
Olivaceous Ardisia 31545
Olivaceous Croton 112979
Olivaceous Maple 3275
Olivaceous Sweetleaf 381119
Olive 70988,141929,270058,270099
Olive Croton 112979
Olive Family 270182
Olive Grevillea 180588,180626
Olive Indosasa 206880
Olive Palm 138536
Olive Plum 142449
Olive Roughleaf 222209
Olive Sweetleaf 381119
Olive Tanoak 233332
Olive Tree 270099
Olive Willow 343336
Oliveblotch Calathea 66165
Olivecolor Ardisia 31545
Olive-coloured Ormosia 274421
Olivecoloured Twayblade 232261
Oliveform Combretum 100674,100805
Oliveform Rockvine 387834
Oliveform Windmill 100805
Olive-green Ardisia 31545
Olivegreen Ormosia 274421
Oliveleaf Grevill 180626
Oliveleaf Tanoak 233332
Oliveleaf Vineclethra 95527
Olive-leaved Sallow 343336
Olive-leaved Tanoak 233332
Olive-leaved Vineclethra 95527
Olive-like Combretum 100805
Olive-like Foliage Rhododendron 332071
Oliver Bark 91388
Oliver Cromwell's Creeping Companion 367772
Oliver Ligusticum 229372
Oliver Linden 391795
Oliver Maple 3277
Oliver Meconopsis 247155
Oliver Plumyew 82542
Oliver Plum-yew 82542
Oliver Scabious 350209
Oliver Taibamboo 391458
Oliver Thyrsostachys 391458
Oliver Tubergourd 390170
Oliver Woodbetony 287483
Oliver's Lime 391795
Oliver's Sea Holly 154281
Oliver's Touch Me Not 205326
Oliver's Touch-me-not 205326
Oliver-bark Tree 386500
Olive-shaped Kumquat 167506
Olivespot Rhododendron 330169
Olivier Odorant 276291
Olivillo 9844
Oliwa-ku-kahakai 61572
Ollar 16352
Ollick 15621,358062
Olluco 401412
Olney Three-square 352166
Olney's Hairy Sedge 74723
Olney's Three-square Bulrush 352166
Olona 393216
Olopua 276395
Olson Begonia 50123
Olulu 60292
Olympic Bellflower 70069,70121
Olympic Mountain Aster 155818
Olympic Mountain Larkspur 124231
Olympic Violet 409989
Oma Rhododendron 331384
Omato 239157
Ombrocharis 270587
Ombrocharis Broomrape 275154
Ombu 298103
Ombu Tree 298103
Ombutree Pockberry 298103
Omei Actinodaphne 6806
Omei Amplexicaul 239762
Omei Aucuba 44898
Omei Bittercress 72812
Omei Chestnut 78796
Omei Chickweed 375028
Omei Clinopodium 97041
Omei Closedfruit Tanoak 233150
Omei Cymbidium 116853
Omei Edelweiss 224924
Omei Eupatorium 158243
Omei Fritillary 168385
Omei Germander 388170
Omei Goldthread 103840
Omei Hemsleia 191921
Omei Henry Spiraea 371930
Omei Holly 204112
Omei Kudzuvine 321457
Omei Landpick 308624
Omei Lime 391797
Omei Lysionotus 239960
Omei Machilus 240646
Omei Mazus 247018
Omei Meadowrue 388607
Omei Microtoena 254261
Omei Millettia 66640
Omei Mountain Actinodaphne 6806
Omei Mountain Bamboo 47264
Omei Mountain Barberry 51279
Omei Mountain Blueberry 403929
Omei Mountain Hornbeam 77358
Omei Mountain Parakmeria 283422
Omei Mountain Pittosporum 301352
Omei Mountain Pollia 307333
Omei Mountain Rose 336825
Omei Mountain Tanakaea 383890
Omei Mountain Trigonotis 397447
Omei Nepal Saurauia 347971
Omei Orchis 310923
Omei Osmanther 276381
Omei Parakmeria 283422
Omei Pepper 300385
Omei Phoebe 295408
Omei Primrose 314552
Omei Rhynchotechum 333747
Omei Rose 336825
Omei Sage 345274
Omei Sageretia 342182
Omei Sibbaldia 362355
Omei Skullcap 355634
Omei Spapweed 205197

Omei Splitmelon 351771
Omei Supple-jack 52446
Omei Tanoak 233329
Omei Teasel 133457
Omei Viburnum 407988
Omei Violetbloom Rhododendron 331345
Omei Willow 343804
Omei Woodbetony 287485
Omeo Grevillea 180662
Omeo Round-leaf Gum 155664
Omeo Round-leaved Gum 155664
Ome-tree 401512,401593
Omikanto 93678
Omission Knotweed 309629
Omit Knotweed 309629
Omixochitl 307269
Omoto 335760
Omoto Nippon Lily 335760
Omoto Nipponlily 335760
Omphalodes 270684
Omphalogramma 270726
Omphalothrix 270757,270758
Omphalotrigonotis 270760
Oncidium 270787
Oncidium Orchid 270787
Oncoba 270883,270924
Oncodostigma 270973
Oncosperma 270997
One Blotch Rhododendron 331895
One Flower Pratt Milkvetch 42912
One Flower With Sevenleaves 284358
One Umbrella Southstar 33325
One-auricled Spatholobus 370423
Oneawn Goatgrass 8738
One-awn Spineflower 89125
One-berry 80728,256002,284387
One-blade 242672
One-coloured Lily 229811
One-day Flower 394088
Oneeared Eurya 160625
One-eared Eurya 160625
One-eye Bean 71042
One-eyed Sulphur Flower 152567
One-eyed Wild Buckwheat 151924
One-fathom Spindle-tree 157339
Oneflower 135136
Oneflower Altingia 18199
Oneflower Astyleorchis 19520
Oneflower Azalea 331270
Oneflower Brodiaea 60526
Oneflower Brome 61008
Oneflower Catchfly 364146
Oneflower Eargrass 187677
One-flower Fleabane 151036
Oneflower Hancockia 185256
Oneflower Hawthorn 110098
One-flower Hawthorn 110098
Oneflower Hedyotis 187677
Oneflower Kingdonia 216399
Oneflower Larkspur 124100
Oneflower Litse 135137
Oneflower Lycianthes 238957
Oneflower Macrotyloma 241440
One-flower Meyer's Clematis 95119
Oneflower Microstegium 254027
Oneflower Monkshood 5203
Oneflower Monochasma 257541
Oneflower Papilioorchis 282935
Oneflower Peony 280328
One-flower Pittosporum 301342
Oneflower Pycnoplinthus 322061
One-flower Raintree 61316
Oneflower Red Silkyarn 238957
Oneflower Rhododendron 331270
Oneflower Rose 337004
One-flower Seatung 301342
One-flower Sikkim Clematis 95301
Oneflower Snakegourd 396257
One-flower Stitchwort 255597
Oneflower Sweetleaf 381395
Oneflower Taillessfruit 100141
Oneflower Trigonella 397254
Oneflower Vetch 408493
One-flower Vetch 408493
One-flower Wichura Rose 337016
Oneflower Windhairdaisy 348895
One-flowered Bellflower 70351
Oneflowered Bluebeard 78025
One-flowered Bluebell 70351
One-flowered Broomrape 275234
One-flowered Broom-rape 275234
One-flowered Cancer-root 275234
One-flowered Cushion Saxifrage 349139
One-flowered Dewberry 338357
One-flowered Dropseed 372866
One-flowered Glasswort 342890
One-flowered Harebell 70351
One-flowered Horehound 239246
One-flowered Lambertia 220312
One-flowered Pyrola 257374
One-flowered Satin Grass 259713
One-flowered Sclerolepis 354442
One-flowered Sea-daffodil 280915
One-flowered Shin-leaf 257374
One-flowered Wintergreen 257371,257374
One-flowered Wood-nymph 257374
Oneflowerprimrose 270726
One-fruit Privetleaf Memecylon 249997
Onefruit Tanoak 233350
One-fruited Maple 3205
One-glumed Hard-grass 184575,184581
One-grain Wheat 398924
One-grained Wheat 398924
Onehead Emei Aster 41471
One-head Pertybush 292061
One-headed Rabbitbrush 150371
Onehorn Ploughpoint 401161
Oneida Viburnum 407789
One-I-eat 30888
One-leaf Dyetree 302679
Oneleaf Gueldenstaedtia 181673
Oneleaf Munronia 259898
One-leaf Orchid 273614
Oneleaf Pine 300079
One-leaf Pine 300079
Oneleaf Rockvine 387812
One-leaved Ash 167960
One-leaved Bamboo 297461
One-leaved Butterfly-orchid 302455
One-o'clock 274823,374916,384714,394333,400675
Onepetal Milkwort 308196
Onepistil Alphonsea 17628
Onerow Yellowcress 336232
One-row Yellowcress 336232
One-row Yellow-cress 262707,336232
One-rowed Watercress 336232
Oneseed Butea 64148
One-seed Croton 112965
Oneseed Ephedra 146219
Oneseed Hawthorn 109857
One-seed Juniper 213827
Oneseed Monocelastrus 257504
One-seed Oiltea Camellia 69416
One-seed Persimmon 132446
One-seed Rockvine 387772
One-seed Staff-tree 257504
One-seeded Bur-cucumber 362461
One-seeded Hawthorn 109857
One-seeded Juniper 213827
One-seeded Mercury 1930
One-seeded Plume Poppy 240770
One-side Racemes Leucothoe 228166
Oneside Vinegentian 110307
One-sided Aster 380922
One-sided Bottle-brush 68064
Onesided Galenia 170005
One-sided Pyrola 275486
One-sided Shinleaf 275486
One-sided Shin-leaf 275486
One-sided Wintergreen 275486
Onespike Guzmania 182177
Onespike Microstegium 254032
Onespiked Indosasa 206899
Onespikegrass 310713,310720
Onestachys Japan Pieris 298750
One-stamen Cocainetree 155093
Onestamen Stephanachne 375806
Onestamened Patrinia 285841
Onetoothed Eyebright 160286
One-two-three 384714

One-way Brome 60977
One-way Chess 60977
Onewayflor 275480,275486
Onion 15018,15165,15429,15689
Onion Cordia 104153
Onion Couch 34934,34935
Onion Grass 248949,335879
Onion Iris 192904
Onion Orchid 254226
Onion Stinker 15861
Onion Twitch 34934
Onion Weed 15839,39463,266960
Onion-flower 15861
Oniongrass 248949
Onion-grass 248949,336030
Onion-leaved Asphodel 39463
Onion-rooted Crowfoot 325666
Onions 15165
Onionweed 39463
Onion-weed 39463
Onondaga Viburnum 408100
Ononis 271290
Onosma 271727
Ontario Aster 380939
Ontario Goatsbeard 394258
Ontario Goldenrod 368405
Ontario Hawthorn 109870,109871
Ontario Lobelia 234562
Ontario Poplar 311171,311259
Ontario White Pine 300215
Onyx Flower 4414
Ookow 60516,128873
Ooleaf Amitostigma 19514
Ooleaf Astyleorchis 19514
Ooleaf Chickweed 375029
Ooleaf Hoodorchis 264778
Ooleaf Tainia 383156
Oolip Orchis 169887
Oolivaceous Bitterweed 201326
Oosepal Addermonth Orchid 110651
Oosepal Bogorchis 110651
Ooster-munath-jonnums 308893
Ooze Apple 123071
Opaque Parsnip 285738
Opaque-leaved Papillose Barberry 52005
Open Azalea 330169
Open Eyes 50825
Open Gowan 68189
Open Jaws 28617
Open Mouth 28617
Open Star 227533
Open-and-shut 274823
Open-arce 252305
Open-arse 252305
Open-ass 252305
Opencalyx Sage 344904
Opener 252305
Open-field Sedge 74180
Open-fruited Cotoneaster 107354
Open-gowan 68189
Open-jaws 28617
Open-mouth 28617
Opepe 262804
Operculate Cleistocalyx 94576
Operculate Waterfig 94576
Operculina 250749
Ophelia Rose 336295
Ophiglossumlike Androcorys 22323
Ophiorrhiza 272173
Ophiorrhiziphyllon 272308
Ophiuros 272345
Ophrestia 272360
Ophrys 272395
Ophthalmic Barberry 51327
Opien Bentgrass 12391
Opien Swollennoded Cane 87595
Opier 407989
Opilia 272553,272555
Opilia Family 272580
Opithandra 272589
Opium 282717
Opium Lettuce 219609
Opium Poppy 282713,282717
Ople-tree 407989
Oplismenus 272611
Oplopanax 272704
Opopanax 1219,101428,272730,272732
Oposite Pleurostylia 304898
Oposite Shieldstyle 304898
Opossum Wood 184731
Oppisiteleaf Ophiorrhiza 272269
Opple-tree 407989
Opposite Rockfoil 349200
Oppositebane 175904,175907
Oppositeflower Gladiolus 176419
Oppositeleaf Bushman's Poison 4968
Oppositeleaf Colebrookea 99420
Oppositeleaf Engelhardia 145526
Oppositeleaf Eritrichium 153509
Oppositeleaf Fig 165125
Oppositeleaf Girgensohnia 175907
Oppositeleaf Lady's Slipper 120341
Oppositeleaf Ladyslipper 120341
Oppositeleaf Landpick 308625
Oppositeleaf Rhodiola 329962
Oppositeleaf Rockfoil 349379
Oppositeleaf Russian Thistle 344718
Oppositeleaf Solomonseal 308625
Oppositeleaf Spapweed 205199
Oppositeleaf Stairweed 142840
Opposite-leaved Colebrookia 99420
Opposite-leaved Dwarfdandelion 218206
Opposite-leaved Fig 165125
Opposite-leaved Golden Saxifrage 90423
Opposite-leaved Pondweed 181285,181286
Oppositeleaved Tarweed 191693
Opposite-leaves Gaultheria 172131
Opposite-leaves Whitepearl 172131
Oppossum Wood 184731
Optical Fibre Plant 209968
Opuntia 272775,273029
Opyman 252305
Orach 44298,44468
Orache 44298,44317,44468,44576
Orage 44468
Orange 93314,93332,93717,93765
Orange Agoseris 11420
Orange Aphelandra 29123
Orange Azalea 330422
Orange Ball Tree 62064
Orange Ball-tree 62064
Orange Balsam 204838,204841
Orange Batalin Tulip 400129
Orange Bell Lily 229855
Orange Bells 385462,385464
Orange Bird's Foot 274890
Orange Bird's-foot 274890
Orange Bladder-senna 100182
Orange Blossom 294426
Orange Bulbil Lily 229766,229769
Orange Bulbine 62390
Orange Butterfly Bush 62064
Orange Canada Lily 229784
Orange Cestrum 84404
Orange Clock Vine 390791
Orange Clockvine 390791
Orange Coneflower 339539,339544,339614
Orange Cosmos 107172
Orange Cotoneaster 107451
Orange Cypress Vine 323459
Orange Daisy 150484
Orange Daphne 122378
Orange Daylily 191284
Orange Day-lily 191284
Orange Dendrochilum 125527
Orange Desertcandle 148535
Orange Dogtongueweed 385892
Orange Dwarfdandelion 218202
Orange Dwarf-dandelion 218202,218204
Orange Edelweiss 224803
Orange Erysimum 154414
Orange Eye 62019
Orange Eye Butterflybush 62019
Orange Fig 164655
Orange Fire Thorn 322450
Orange Firethorn 322450
Orange Fire-thorn 322450
Orange Fleabane 150484
Orange Floatingheart 267802
Orange Flower Tree 294426
Orange Foxtail 17498

Orange Gentian 150484
Orange Gladiolus 176039
Orange Globe Flower 399494
Orange Gooseberry 334156
Orange Groundsel 358510,385917
Orange Gum 155491
Orange Habenaria 184025
Orange Hawkweed 195478
Orange Honey Rose 336296
Orange Honeysuckle 235738
Orange Jasmine 260173
Orange Jessamine 84404,260158,260173
Orange Jewelweed 204838,204841
Orange King 51863
Orange Ligulate Tephroseris 385892
Orange Lily 21339,229718,229766,229769
Orange Luckynut Thevetia 389979
Orange Lycoris 239271
Orange Magnoliavine 351103
Orange Magnolia-vine 351103
Orange Milkweed 38147
Orange Milkwort 308173
Orange Mint 250341
Orange Mullein 405747
Orange Oreocharis 273836
Orange Plume Flower 214817
Orange Poppy 282536
Orange Pricklypear 272799
Orange Primrose 314139
Orange Puccoon 233707
Orange Raspberry 338166
Orange Rhododendron 330422
Orange River Lily 111180
Orange Rockfoi 349086
Orange Root 200173,200175
Orange Sage 344834,345396
Orange Salacia 342591
Orange Sneezeweed 188424
Orange Speciosum 229862
Orange Star 194671
Orange Stonecrop 294123
Orange Sugarmustard 154414
Orange Sunflower 190502
Orange Sunrose 188761
Orange Tea Rose 336814
Orange Thorn 93262
Orange Thyme 391191
Orange Touch-me-not 204838,204841
Orange Tree 93765
Orange Tremacron 394673
Orange Triumph Rose 336297
Orange Trumpet Vine 322950
Orange Ursinia 402803
Orange Wattle 1159,1553
Orange Willow 232483
Orange Wisteria Shrub 361422
Orange Zexmenia 413559,417826
Orange Zinnia 418038
Orange-ball-tree 62064
Orangebead Cotoneaster 107649
Orange-beaded Cotoneaster 107649
Orangeberry Nightshade 367292
Orangeberry Pittosporum 301433
Orange-berry Pittosporum 301433
Orange-bush Monkeytlower 255192
Orangecolour Sage 344834
Orangecoloured Hackberry 80664
Orange-coloured Peashrub 72187
Orange-coloured Pea-shrub 72187
Orange-cup Lily 229984
Orangedred Raspberry 338166
Orange-eye Butterfly Bush 62019
Orangeeye Butterflybush 62019
Orange-eye Butterfly-bush 62019
Orangeeye Summerlilic 62019
Orangeflower Cordia 104251
Orangeflower Cremanthodium 110364
Orangeflower Jerusalemsage 295111
Orangeflower Murdannia 260091
Orangeflower Nutantdaisy 110364
Orangeflower Tupistra 70596
Orangeflower Waterbamboo 260091
Orange-fruit Horse-gentian 397830,397833
Orangefruit Salacia 342591
Orange-fruited Horse-gentian 397830
Orange-fruited Salacia 342591
Orange-grass 201895
Orange-jessamine 260173
Orange-peel Clematis 95179,95345
Orangered Hawthorn 109544
Orange-red Hawthorn 109544
Orangeroot 200173,200175
Orange-sneezeweed 201316
Orangevine 238530
Orangeyellow Amomum 19900
Orangeyellow Irisorchis 267950
Orangeyellow Oberonia 267950
Orania 51164
Orania Palm 273120
Orbicular Addermonth Orchid 110650
Orbicular Bogorchis 110650
Orbicular Illigera 204640
Orbicular Medic 247407
Orbicular Primrose 314759
Orbicular Sedge 75618
Orbicular Snailseed 97933
Orbicular-leaf Blueberry 403920
Orbicularleaf Sida 362498
Orbicular-leaf Wild Buckwheat 151961
Orbicular-leaved Rhododendron 331400
Orbiculate Dunn's Redflower Greenvine 204652
Orbiculate Ginger 418014
Orbiculate Greenvine 204640
Orbiculate Mayten 182751
Orbiculate Primrose 314759
Orbiculate Shrubcress 368570
Orbiculate Solms-Laubachia 368570
Orbignya 273241
Orcanette 21959
Orchanet 21920,21959
Orchard Grass 121226,121233
Orchard Snakeweed 182106
Orchard Weed 28044
Orchardgrass 121226,121233
Orchard-grass 121226,121233
Orchid 273317
Orchid Bauhinia 49200
Orchid Bush 48993
Orchid Cactus 147283,266648
Orchid Canna 71193
Orchid Family 273267
Orchid Iris 208738
Orchid Pansy 4070
Orchid Rockrose 93199
Orchid Tree 11300,19463,48990,49173,49211,49247
Orchid Vine 376554,376557,376559
Orchidaceous Cape-cowslip 218909
Orchidantha 273268
Orchidflowered Canna 71193
Orchid-tree 49247
Orchis 98593,116769,121358,273317,310873
Orchis Morio 359158
Orchislike Spapweed 205200
Orcutt Cactus 244177
Orcutt Pachycereus 279488
Orcutt Spineflower 89090
Orcutt's Bristleweed 186886
Orcutt's Brodiaea 60511
Orcutt's Pincushion 84505
Orcutt's Woody-aster 415853
Ordeal Bean 298009
Ordeal Tree 154973,383907
Ordos Leptodermis 226109
Ordos Sagebrush 36023
Ordos Wildclove 226109
Ordos Wormwood 36023
Oread Rhododendron 331411
Orech 44576
Oregan Rockfoil 349724
Oreganillo 17607
Oregano 99515,232457,232492,274197,274229,274237
Orege 44576
Oregold Rose 336311
Oregon 385342
Oregon Agoseris 11415
Oregon Alder 16429,16437
Oregon Anemone 23976

Oregon Ash 168015,168039
Oregon Balsam Poplar 311554
Oregon Boxleaf 286749
Oregon Boxwood 286749
Oregon Catchfly 363847
Oregon Cedar 85271
Oregon Crab Apple 243622
Oregon Crabapple 243622
Oregon Evergreen Blackberry 338695
Oregon Fawnlily 154936
Oregon Fawn-lily 154936
Oregon Fetid Adder's-tongue 354575
Oregon Fir 384,318580
Oregon Flag 208879
Oregon Fleabane 150972
Oregon Goldenaster 194234
Oregon Grape 242478,242602
Oregon Grape Holly 242478
Oregon Holly Grape 242478
Oregon Hop 228321
Oregon Lily 229810
Oregon Maple 3127
Oregon Myrtle 401670,401672
Oregon Oak 323931
Oregon Pine 318580
Oregon Plum 269274,269291
Oregon Sunshine 152799,152809,152838
Oregon Trillium 397607
Oregon White Oak 323927,323931
Oregon White-topped Aster 360710
Oregon Woolly Marbles 318808
Oregon Woollyheads 318808
Oregon Yew 385342
Oregon-grape 51268,242458,242478
Oregon-myrtle 401672
Oregon-pine 318580
Orenge Grape 242458
Oreocereus 273809
Oreocharis 273830
Oreocnide 273899
Oreomyrrhis 273976
Oreorchis 274033
Oreosolen 274110
Oresitrophe 274155
Orgal 250432
Orgament 274237
Organ 250432,274237
Organ Mountain Giant Hyssop 10413
Organ Mountain Rock Daisy 291302
Organ Mts. Evening Primrose 269478
Organ Pipe Cactus 375534
Organ-herb 250432
Organillo 83874,181452
Organo 279486
Organ-pipe Cactus 245252
Organy 250432,274237
Orglon 109857
Orhamwood 401431
Orient Platea 302604
Oriental Ablfgromwell 21965
Oriental Acantholepis 2207
Oriental Alder 16431
Oriental Ampelopsis 20430
Oriental Apple 243662
Oriental Arborvitae 302721
Oriental Arbor-vitae 302721
Oriental Ardisia 31437
Oriental Arum 37022
Oriental Asperula 39383
Oriental Barberry 51981
Oriental Bastard-cabbage 326833,326835
Oriental Beech 162395
Oriental Berry 21459
Oriental Bittersweet 80260
Oriental Bladder Senna 100187
Oriental Bladdersenna 100187
Oriental Bladder-senna 100187
Oriental Blueberry 403738
Oriental Bubbleweed 297881
Oriental Buckthorn 328665
Oriental Bulbophyllum 62961
Oriental Bulrush 353510
Oriental Bunias 63452
Oriental Caper-bush 141470
Oriental Cashew Nut 357878
Oriental Catchfly 363852
Oriental Cattail 401129
Oriental Cherry 83312,83314
Oriental Chrysopogon 90133
Oriental Cinquefoil 312841
Oriental Cladogynos 93952
Oriental Clematis 95179
Oriental Cocaine Tree 155088
Oriental Cocainetree 155088
Oriental Cogflower 154190
Oriental Cudrania 240813
Oriental Currant 334132
Oriental Dock 340164
Oriental Dodartia 135134
Oriental Drywheatgrass 148415
Oriental Eremopyrum 148415
Oriental Ervatamia 154190
Oriental False Hawksbeard 416437
Oriental False Hawks-beard 416437
Oriental False Wheatgrass 148415
Oriental Falseolive 142461
Oriental Fatheadtree 262823
Oriental Fennelflower 266236
Oriental Figwort 355208
Oriental Fissistigma 166692
Oriental Fountain Grass 289185
Oriental Foxglove 130381
Oriental Garlic 15843
Oriental Garliccress 365568
Oriental Gentian 173864
Oriental Germander 388172
Oriental Goatbean 169940
Oriental Goosecomb 335423
Oriental Grape 411686
Oriental Hackberry 80577
Oriental Hawksbeard 416437
Oriental Hawthorn 109788
Oriental Henbane 201394
Oriental Hornbeam 77359
Oriental Horsebean Caltrop 418644
Oriental Iris 208740
Oriental Knight's-spur 102833
Oriental Lady's Thumb 309624
Oriental Lady's-thumb 308960,308961
Oriental Ladysthumb 309624
Oriental Larkspur 102838
Oriental Leafflower 296546
Oriental Leopard's-bane 136364
Oriental Lespedeza 226977
Oriental Lilies 229973
Oriental lily 230036
Oriental Lizardtail 348087
Oriental Lotus 263272
Oriental Lycopsis 21965
Oriental Mangrove 61263,61267
Oriental Maple 3294
Oriental Meadow Saffron 99313
Oriental Meadow-saffron 99313
Oriental Merremia 250854
Oriental Mullein 405681
Oriental Naiad 262098
Oriental Needlegrass 376859
Oriental Oak 116083,323630,324532
Oriental Oat 45533
Oriental Oil-wood 142461
Oriental Paper Bush 141470
Oriental Paperbush 141470
Oriental Pear 323268
Oriental Pennisetum 289185
Oriental Persimmon 132219
Oriental Photinia 295773,295808
Oriental Physochlaina 297881
Oriental Pickling Cucumis 114201
Oriental Pickling Melon 114201
Oriental Pink 127787
Oriental Plane 302592
Oriental Plane Tree 302592
Oriental Planetree 302592
Oriental Poison Sumac 393485
Oriental Poison-ivy 393466
Oriental Poison-oak 393466
Oriental Poppy 282496,282647,282668
Oriental Qinggang 116083
Oriental Rattan Palm 65751
Oriental Rhazya 329263
Oriental Roegneria 335423

Oriental Russianthistle 344652
Oriental Sagebrush 36028
Oriental Salsify 394321
Oriental Sarcococca 346738
Oriental Scariola 350507
Oriental Senna 360493
Oriental Sesame 361317
Oriental Shepherd's Purse 72054
Oriental Sida 362594
Oriental Silene 363361
Oriental Sisymbrium 365568
Oriental Skullcap 355638
Oriental Slugwood 50621
Oriental Spicebush 231303
Oriental Spiraea 372006
Oriental Spirea 372006
Oriental Spooncress 63452
Oriental Spruce 298391
Oriental Stephania 375833
Oriental Stonebean-orchis 62961
Oriental Strawberry 167641
Oriental Sweet Gum 232562
Oriental Sweetgum 232562
Oriental Thorn 109788
Oriental Threestamen Waterwort 142578
Oriental Tickseed 104819
Oriental Trema 394656
Oriental Trilobe Buttercup 325939
Oriental Underleaf Pearl 296546
Oriental Veriegated Coral-bean 154734
Oriental Vetch 408435
Oriental Violet 410326
Oriental Virginsbower 95179
Oriental Waterplantain 14754
Oriental Wendlandia 413837
Oriental White Oak 323630
Oriental Wolftailgrass 289185
Oriental Wood 145320
Oriental Woodruff 39383
Oriental Wormwood 36232
Oriental-cedar 302721
Orientale Bugle 13153
Orientale Giantfennel 163688
Orientvine 364985,364986
Origanum 274197
Origanumlike Dragonhead 137620
Origanumlike Greenorchid 137620
Original Gastrodia 171907
Orinoco Sassafras 268700
Orinoco Simaruba 364382
Orintal Debregeasia 123332
Orintal Hedyosmum 187495
Orinus 274248
Orixa 274276
Orl 16352
Ormocarpum 274337
Ormosia 274375,274399
Ornamemtal Cabbage 59520
Ornamental Aglaonema 11355
Ornamental Arum 37027
Ornamental Banana 260217,260256
Ornamental Brassicas 59520
Ornamental Cabbage 59372
Ornamental Catmint 264920
Ornamental Corn 417417,417420
Ornamental Giant Bamboo 47518
Ornamental Jewelweed 204982
Ornamental Kale 59520
Ornamental Leaf Beet 53257
Ornamental Maize 417417
Ornamental Millet 254515,289116
Ornamental Monkshood 43526
Ornamental Onion 15023,15188,15389,15445,15509,15550,15752
Ornamental Origano 274223
Ornamental Peach 20942
Ornamental Pepo 114310
Ornamental Pepper 72070
Ornamental Raspberry 338917,339321
Ornamental Rhododendron 330524
Ornamental Rhubarb 329372
Ornamental Thorn 109509
Ornamental Yam 131556
Ornate Dinetus 131243
Ornate Peperomia 290398
Ornate Porana 131243
Ornate Staunton-vine 374396
Ornduff's Goldfields 222617
Ornithoboea 274465,274471
Ornithogalum 274495
Orobanch 275135
Oroental Basketvine 9428
Orogbo Kola 171121
Orono Sedge 75626
Oront Snapdragon 28631
Orophea 275319
Orostachys 275347
Orphan John 357203
Orphine 200802
Orphyne 357203
Orpies 357203
Orpine 356467,357203,385629
Orpine Family 109505
Orpy-leaf 357203
Orrice 208565
Orrice-root 208565
Orris 208565
Orris Root 208565
Orthoraphium 275607
Orthotropic-spike Sedge 73782
Orychophragmus 275865
Oryelle 16352
Oryza-like Acroceras 6110
Osage 240828
Osage Orange 240806,240828
Osageorange 240828
Osage-orange 240806,240828
Osbeckia 276072,276135
Osceola's Plume 417887
Oscularis 251025
Osha 229324
Oshima Cherry 83331
Oshima Childvine 400944
Oshima Kan Suge 75636
Oshima Meadowrue 388609
Osier 342923,344244
Osier Willow 344244
Osier-like Rattan Palm 65809
Osier-like Rattanpalm 65809
Osmanther 276242,276264,276268,276270,276395
Osmanther Microtropis 254316
Osmanthus 276242,276323
Osmanthus Epidendrum 146455
Osmanthus-like Microtropis 254316
Osmaston Barberry 51990
Osmoxylon 276484
Oso Berry 269290,269291
Oso-berry 269290,269291
Ospin Pleurothallis 304923
Ossumwood 132466
Osterhout's Thistle 91884
Osterick 308893
Ostericum 276743
Osti Peony 280258
Ostodes 276786,276789
Ostrovsk Tulip 400202
Ostryopsis 276844
Oswego Tea 257161
Oswegotea 257161
Oswego-tea 257161
Osyris 276866
Otachibana 93679
Otaheite Apple 372459,372467,382606
Otaheite Chestnut 207000
Otaheite Gooseberry 60057,296466,296543
Otaheite Goose-berry 296466
Otaheite Gooseberryleaf Flower 296466
Otaheite Myrtle 356395
Otaheite Salap-plant 382923
Otaheite Walnut 14544
Otaheite Yam 131501
Otaheite-apple 372467
Otaheite-gooseberry Leal Flower 296466
Otanthera 276911
Otatea 276920
Oto 93680
Otophora 277306,277309
Otostegia 277326
Ottawa Barberry 51991
Ottelia 277361

Otter Bush 179214
Otto Luken Cherry Laurel 316506
Ottochloa 277426
Otton Polystachya 310534
Ottonis Ballcactus 267045
Otu 94654
Ouachita Mountains Goldenrod 368296
Ouininc Bush 108497
Ouizhou Aster 40839
Oulia Champ 252841
Ountain Banksia 47635
Our Lady's Basin 133478
Our Lady's Bedstraw 170743
Our Lady's Candle 405788
Our Lady's Candles 405788
Our Lady's Cushion 34536
Our Lady's Flannel 405788
Our Lady's Heart 220729
Our Lady's Milk Thistle 364360
Our Lady's Mint 383712
Our Lady's Nightcap 68713
Our Lady's Pincushion 72038
Our Lady's Seal 61497
Our Lady's Smock 68713
Our Lady's Tears 233760
Our Lady's Thimble 70271
Our Lady's Thimbles 70271
Our Lady's Thistle 73318
Our Lady's Vine 383675
Our Lady-in-aboat 220729
Our Lord's Candle 193534,193538
Our Lord's Candles 193538
Our Lord's Flannel 141347,405788
Our Saviour's Flannel 141347,405788
Our's Flannel 405788
Ouratea 277463
Ouricuri 380556
Ouricuri Palm 380556
Our-lady-in-a-boat 220729
Our-Lord's Candle 193538
Ourplish Wine Washington Lily 230072
Outcrop Pussytoes 26415
Outeniqua Yellowwood 9993,306426
Outeniqua Yellow-wood 306426
Outspread Rhododendron 331215
Ouw 200388
Ovaid Manglietia 244462
Oval Coelogyne 98711
Oval Halophila 184979
Oval Kumquat 167506
Oval Ladies' Tresses 372235
Oval Lady's-tresses 372235,372236
Oval Milkweed 38045
Oval Miterwort 256035
Oval Sedge 75644
Oval Spike-rush 143261
Oval Spiraea 372043
Ovalcone Pine 300126
Oval-coned Pine 300126
Ovale-leaved Dutchman's-pipe 34288
Ovalfruit Hakea 184604
Ovalglume Kengyilia 215907
Oval-headed Sedge 74078
Ovalleaf Anemone 23948
Ovalleaf Aspidistra 39580
Ovalleaf Birch 53544
Ovalleaf Bowfruitvine 393538
Ovalleaf Bugle 13154
Ovalleaf Camellia 69443
Oval-leaf Cassia 360421
Ovalleaf Cow-wheat 248204
Ovalleaf Dutchmanspipe 34348
Oval-leaf False Buttonweed 370802
Oval-leaf Grevillea 180620
Ovalleaf Japanese Spiraea 371977
Oval-leaf Knotweed 308827,308971
Ovalleaf Meranti 362214
Oval-leaf Mint-bush 315605
Ovalleaf Mistletoe 411085
Ovalleaf Primrose 314766
Ovalleaf Threevein Aster 39974
Ovalleaf Toxocarpus 393538
Ovalleaf Twistedstalk 377931
Oval-leafed Privet 229569
Oval-leaved Barberry 51993
Oval-leaved Birch 53544
Oval-leaved Huckleberry 403932
Oval-leaved Knotweed 308788
Oval-leaved Milkweed 38045
Oval-leaved Mistletoe 411085
Oval-leaved Red-root 79957
Oval-leaved Willow 343820
Ova-ova 258044
Ovate Aster 40976
Ovate Asteranthera 41554
Ovate Bay-leaf Willow 343862
Ovate Biond's Magnolia 242005
Ovate Catalpa 79257
Ovate Childvine 400945
Ovate Dillenia 130920
Ovate Dragonbamboo 125497
Ovate Eargrass 187636
Ovate Galangal 17734
Ovate Goatgrass 8674,8695
Ovate Gomphostemma 179199
Ovate Hedgebavin 392370
Ovate Hirsute Rockcress 30301
Ovate Lagurus 220254
Ovate Leaves Lyonia 239391
Ovate Leucas 227682
Ovate Leymus 228372
Ovate Littleleaf Willow 343821
Ovate Manchurian Linden 391763
Ovate Metalleaf 296405
Ovate Monsonia 258141
Ovate Periwinkle 79414
Ovate Poisonnut 378845
Ovate Sandwort 198010
Ovate Spike-rush 143261
Ovate Spikesedge 143261
Ovate Strychnos 378845
Ovate Talauma 383251
Ovate Thyme 391346
Ovate Tirpitzia 392370
Ovate Tylophora 400945
Ovate Windhairdaisy 348614
Ovatebract Eria 299625
Ovatebract Macaranga 240329
Ovate-bract Rattlebox 112094
Ovatebract Sweetleaf 381423
Ovate-bracted Sweetleaf 381423
Ovatecalyx Sanicle 345970
Ovatecordateleaf Rockfoil 349627
Ovateflower Mannagrass 177668
Ovateflower Sweetgrass 177668
Ovate-fruit Camellia 69426
Ovatefruit Craneknee 221715
Ovatefruit Galangal 17735
Ovatefruit Integral Holly 203916
Ovatefruit Reineckea 327528
Ovate-fruit Rose 336620
Ovatefruit Sevanlobed Nightshade 367606
Ovatefruit Slugwood 50583
Ovatefruit Stickseed 221715
Ovate-fruited Camellia 69426
Ovate-fruited Milkvetch 42446
Ovate-fruited Slugwood 50583
Ovateleaf Adenosma 7990
Ovateleaf Ainsliaea 12731
Ovateleaf Andrographis 22406
Ovateleaf Anemone 23718
Ovateleaf Arrowwood 408016
Ovateleaf Aster 40976
Ovateleaf Azalea 330279
Ovateleaf Bauhinia 49190
Ovateleaf Begonia 50126
Ovate-leaf Birch 53400
Ovateleaf Cinnamon 91421
Ovateleaf Clematis 95194
Ovateleaf Coelogyne 98704
Ovateleaf Copperleaf 1907
Ovateleaf Cottonwood 311284
Ovateleaf Cragcress 98023
Ovate-leaf Dutchmanspipe 34288
Ovateleaf Ehretia 141683
Ovateleaf Elaeagnus 142139
Ovateleaf Euonymus 157932
Ovateleaf Eurya 160573
Ovateleaf False Loosestrife 238191
Ovateleaf Fieldcitron 400536
Ovateleaf Fig 165412

Ovateleaf Flaccid Aster 40443
Ovateleaf Flexuose Bittercress 72770
Ovateleaf Forsythia 167452
Ovateleaf Gardenia 171339
Ovateleaf Gardneria 171459
Ovate-leaf Gaultheria 172133
Ovateleaf Gentian Croton 112856
Ovate-leaf Guangxi Azalea 331027
Ovate-leaf Guangxi Rhododendron 331027
Ovateleaf Hawkweed 195915
Ovateleaf Hedyotis 187636
Ovateleaf Hilliella 196297
Ovateleaf Holly 204217
Ovateleaf Homalium 197711
Ovateleaf Honeysuckle 235871
Ovateleaf Hooker Deutzia 126974
Ovateleaf India Tigerthorn 122056
Ovate-leaf Inula 207154
Ovateleaf Jerusalemsage 295218
Ovate-leaf Kurz Lasianthus 222203
Ovate-leaf Kurz Roughleaf 222203
Ovateleaf Leatherybractdaisy 400037
Ovateleaf Leek 15558
Ovateleaf Lepidagathis 225178
Ovateleaf Leptodermis 226110
Ovateleaf Leucas 227632
Ovateleaf Listera 232950
Ovateleaf Litse 234049
Ovateleaf Loosestrife 239768
Ovateleaf Madder 338003
Ovate-leaf Mayten 246899
Ovate-leaf Metalleaf 296405
Ovateleaf Microula 254358
Ovateleaf Milkwort 307925
Ovateleaf Mistletoe 410969
Ovateleaf Mosquitotrap 117470,117695
Ovateleaf Multiradiate Fleabane 150803
Ovateleaf Newlitse 264077
Ovateleaf Notopterygium 267155
Ovateleaf Onion 15558
Ovateleaf Oxystelma 385752
Ovate-leaf Pachygone 279558
Ovateleaf Peltate Clearweed 299002
Ovateleaf Peltate Coldwaterflower 299002
Ovate-leaf Peony 280269
Ovateleaf Pepper 300341
Ovateleaf Plantain 302119
Ovateleaf Pricklyash 417290
Ovateleaf Privet 229569
Ovateleaf Rhamnella 328565
Ovateleaf Rhododendron 330279
Ovate-leaf Rhododendron 330359
Ovateleaf Rhubarb 329370
Ovateleaf Rockjasmine 23249
Ovateleaf Sanicle 345994
Ovateleaf Scurvyweed 98023
Ovateleaf Seedbox 238191
Ovate-leaf Seedbox 238191
Ovateleaf Simon Poplar 311500
Ovate-leaf Smallflower Deutzia 127038
Ovateleaf Snakegourd 396236
Ovateleaf Swallowwort 117470,117695
Ovateleaf Swampy Gentianopsis 174227
Ovateleaf Sweetleaf 381272
Ovateleaf Swertia 380411
Ovateleaf Taiwan Toadlily 396596
Ovateleaf Tanoak 233330,233410
Ovateleaf Tugarinovia 400037
Ovateleaf Turpinia 400536
Ovateleaf Tutcheria 322599
Ovateleaf Twistedstalk 377931
Ovate-leaf Tylophora 400945
Ovateleaf Viburnum 408016
Ovateleaf Waxdaisy 189188
Ovateleaf Wildclove 226110
Ovateleaf Windflower 23718
Ovateleaf Windhairdaisy 348615
Ovateleaf Woodlotus 244461
Ovateleaf Woodruff 170558
Ovateleaf Yam 131603
Ovateleaf-like Jerusalemsage 295220
Ovate-leaved Bauhinia 49190
Ovate-leaved Camellia 69425
Ovate-leaved Campion 363860
Ovate-leaved Catchfly 363860
Ovate-leaved Cinnamon 91421
Ovate-leaved Debregeasia 123328
Ovate-leaved Eurya 160573
Ovate-leaved Fig 165412
Ovate-leaved Forsythia 167452
Ovate-leaved Gardneria 171459
Ovate-leaved Holboellia 197245
Ovate-leaved Holly 204217
Ovate-leaved Japan Aucuba 44937
Ovate-leaved Leptodermis 226110
Ovateleaved Lomatogoniopsis 235430
Ovate-leaved Lyonia 239391
Ovate-leaved Metalleaf 296405
Ovate-leaved Newlitse 264077
Ovateleaved Pepper 300341
Ovate-leaved Privet 229569
Ovate-leaved Rhamnella 328565
Ovate-leaved Rhododendron 331420
Ovateleaved Spokeflower 235430
Ovate-leaved Sweetleaf 381272
Ovate-leaved Turpinia 400536
Ovate-leaved Tutcheria 322599
Ovate-leaved Viburnum 408016
Ovate-leaved Violet 409983
Ovatelip Flickingeria 166966
Ovate-lobed Machilus 240664
Ovatepetal Chervil Larkspur 124043
Ovatepetal Rockfoil 349729
Ovatesepal Rhodiola 329914
Ovatesepal Strophanthus 378375
Ovate-sepaled Strophanthus 378375
Ovate-small-leaved Willow 343821
Ovate-spiked Sedge 75648
Ovatileaf Prostrate Axyris 45857
Ovatipetal Sandwort 31973
Oven Bush 103508
Ovens Acacia 1494
Ovens Wattle 1494,1528
Overcup Oak 324137
Overenyie 35088
Overhilldregon 351148
Overlong Arrowbamboo 162734
Overlook Bean 71042
Overlord 347066,418585
Overmallow 243877,243893
Ovest 324335
Ovoid Spike Sege 143261
Ovoid Spikesedge 143261
Ovoid Spike-sedge 143261
Ovoidfruit Pittosporum 301360
Ovoidfruit Seatung 301360
Ovoid-fruited Pittosporum 301360
Owala Oil Tree 289439
Owala-oil Tree 289439
Owan's Flatsedge 119332
Owatar Oplismenus 272625
Owd Lad's Peascods 218761
Owi 373497,373507,373511,405865
Owl Clover 275499
Owl's Claws 188424
Owl's Crown 166038,170931
Owl's Eye 21339
Owl's Eyes 244182
Owl's Eyes Cactus 244182
Owl's Socks 347156
Owl's-claws 201316
Owlder 16352
Owler 16352,311537
Owl-eyes 199091
Owl-fruit Sedge 76375
Owlorn 16352
Owls Eyes 21339
Owm 401593
Ownbey's Goatsbeard 394296
Ownbey's Thistle 92282
Owpa Socks 347156
Owusa Nut 305150
Owyhee Sage 36055
Ox Balm 99844
Ox Eye 190525
Ox Eyes 8387,21339,89704,227533
Ox Heel 190936
Ox Tongue 21920,21959,57101
Oxalis 277644,277877
Oxalis Family 277643
Ox-balm 99844

Oxberry 37015,383675
Oxblood Lilies 332184
Oxborn Tree 16293
Oxear Windhairdaisy 348938
Oxet Bamboo 47416
Oxeye 63524,63531,190502,190505,385614,385616
Ox-eye 8387,21339,63524,89704,190505,190525,227533
Ox-eye Camomile 26900
Oxeye Caper 71935
Ox-eye Chamomile 26900
Oxeye Daisy 227533
Ox-eye Daisy 227467,227533
Ox-eye Sunflower 190505
Oxeyedaisy 63524,89402
Ox-eyes 190505
Oxford and Cambridge Bush 360096
Oxford Barberry 51994
Oxford Cedar 85271
Oxford Groundsel 360096
Oxford Ivy 116736
Oxford Ragwort 360096
Oxford Rampion 298068
Oxford Weed 116736
Oxford-and-cambridge 260298
Oxford-and-cambridge Flower 260298
Ox-heal 190936
Ox-heel 190936
Oxhneedaisy 170137
Oxhorn Parabarium 402192
Oxley Rattan Palm 65754
Oxlip 314333
Oxlip Primrose 314333
Ox-louse Lovegrass 148028
Oxmouth Thistle 92386
Oxmuscle 129025
Oxmusclefruit 185931
Oxtail Greenbrier 366548
Oxtail Rabdosia 209844
Oxtail Rhubarb 329330
Oxtail Sagebrush 35430
Oxtongue 191013,298550,298584
Ox-tongue 21920,21959,57101,171767,298550,298584
Ox-tongue Broomrape 274958
Oxtongue Gasteria 171767
Ox-tongue Gasteria 171767
Ox-tongue Lily 184372
Oxus Locoweed 279127
Oxwhipgrass 191225
Oxwood 110238
Oxyacanthous Rose 336835
Oxycoccus Blueberry 278366
Oxygraphis 278466
Oxygraphisleaf Primrose 314768
Oxygraphislike Goldsaxifrage 90425
Oxygraphislike Goldwaist 90425
Oxygyne 278482
Oxylobus 278497
Oxypetalum 278546
Oxyspora 278590
Oxystelma 278614
Oxytoothed Brome 60871
Oyama Magnolia 279249
Oyster 300223,382329,394327
Oyster Bay Pine 67432
Oyster Flay Pine 67432
Oyster Leaf 250894
Oyster Nut 385655
Oyster Plant 2711,250883,250894,332293,354648,394076,394327,394333
Oyster-leaf 250894
Oysternut 385655
Oyster-nut Vine 385654
Oyster-plant 394076,394327
Ozark Blue Star Flower 20854
Ozark Bunch-flower 248679
Ozark Chestnut 78804,78807
Ozark Chinkapin 78804
Ozark Chinquapin 78807
Ozark Corn Salad 404459
Ozark Dropseed 372791
Ozark Green Trillium 397633
Ozark Ironweed 406114
Ozark Larkspur 124404
Ozark Milk Vetch 42299
Ozark Pignut Hickory 77932
Ozark Poverty Grass 372791
Ozark Spiderwort 394056
Ozark Sundrop 269464
Ozark Trillium 397633
Ozark Wake Robin 397594
Ozark Witch Hazel 185138
Ozark Witchhazel 185138
Ozark Witch-hazel 185138
Ozothamnus 279341

P

P. C. Tsoong Pellionia 288740
Pacara Earpod Tree 145993
Pacara Earpodtree 145993
Pacaya 85420
Pacaya Palm 85438
Pachakshir Mahonia 242620
Pachan Trichocereus 395649
Pachilan Bamboo 47402
Pachira 279380
Pachira-nut 279387
Pachycentria 279476
Pachycereus 279482
Pachycladous Fargesia 162726
Pachyclldous Dendrocalamus 125499
Pachygone 279556
Pachyleaf Holly 204133
Pachypodium 279666
Pachypterygium 279703
Pachysandra 279743
Pachystachys 279765
Pachystoma 279852,279859
Pacific Alpinegold 199211
Pacific Anemone 23891
Pacific Bayberry 261132
Pacific Bentgrass 12008
Pacific Bergenia 52530
Pacific Bleeding-heart 128297
Pacific Bugseed 104823
Pacific Chalkplant 183223
Pacific Circaea 91581
Pacific Coast Maple 3127
Pacific Coast Red Elder 345579
Pacific Crabapple 243622
Pacific Cudweed 170941
Pacific Dogwood 105130
Pacific False Buttonweed 370773
Pacific Fir 280
Pacific Flowering Dogwood 105130
Pacific Fringed Thistle 92345
Pacific Goldenrod 368297
Pacific Grape 411594
Pacific Gypsophila 183223
Pacific Hemlock 399905
Pacific Island Kauri 10500
Pacific Island Silvergrass 255849
Pacific Island Wild Buckwheat 152107
Pacific Jimmyweed 209576
Pacific Juneberry 19259
Pacific Knotweed 309505
Pacific Labrador Tea 223892
Pacific Madrone 30885,334976
Pacific Mahogany 380526
Pacific Maple 3127
Pacific Mountainash 369525
Pacific Orach 44562
Pacific Plum 316833
Pacific Raspberry 338942

Pacific Red Cedar 390645
Pacific Red Elder 345579
Pacific Redcedar 390645
Pacific Rhododendron 331171
Pacific Rush 213217
Pacific Serviceberry 19259,19272
Pacific Silver Fir 280
Pacific Skullcap 355674
Pacific Tea 223892
Pacific Walnut 137714
Pacific Waxmyrtle 261132,261162
Pacific White Fir 407
Pacific Willow 343596
Pacific Willow Dock 340289
Pacific Yew 385342
Paco-paco 414533
Paczosk Gagea 169477
Padang Cassia 91282
Padang Cinnamon 91282
Padang Rattan Palm 65757
Padauk 320269,320288,320301,320306,320330
Padda Foot 387993
Paddle Plant 215275
Paddlewood 39696
Paddock Cheese 39120
Paddock Flower 68189
Paddock's Pipe 196818
Paddock's Pipes 196818
Paddock's Spindle 273531,273545
Paddy 275958
Paddy Barnyardgrass 140466
Paddy Field Fimbristyle 166499
Paddy's Lucerne 362617
Paddy's Pipe 196818
Paddyfield Flat-sedge 118982
Paddyfield Prostrate Bulrush 352276
Paddy-field Spikesedge 143300
Padelion 14065
Pad-leaf 302358
Padouk 320269,320288,320306,320330
Padri 376266
Padri Tree 376257
Padritree 376257
Padshape Catchfly 363399
Padshaped Honeysuckle 236002
Padshaped Sagebrush 35912
Padue-leaf Blueberry 403951
Paederia 280063
Paederota 280122
Paedicalyx 280135,280138
Paeony 280143
Paeony-flowered Poppy 282655
Page Oak 393470
Paggle 315082
Pagle 315082
Pagnucc Grape 411853
Pagoda Dogwood 104947
Pagoda Flower 95972,96257,96356
Pagoda Maple 3147
Pagoda Plant 109254
Pagoda Savory 96981
Pagoda Tree 305221,305225,305226,368955,369037
Pagoda Wild Buckwheat 152432
Pagodalike Goldenray 229227
Pagodatree 368955,369037
Pahang Licuala Palm 228754
Pahang Licualapalm 228754
Pahautea 228636
Pahrump Orach 44320
Pahrump Valley Wild Buckwheat 151862
Pahudia 10105
Pahutan Mango 244388
Pahute Mesa Wild Buckwheat 152279
Pai Woodbetony 287491
Paich-Ha 157699
Paigle 72934,314333,315082,315101,325498,325518,325666,326287,374916
Paigya Bean 294010
Paigyabean 294010
Paiheng Maple 3297
Paiheng Tanoak 233339
Pained Nettle 99583
Paintbrush 81237,184347,394333
Painted Acanthophippium 2077
Painted Bamboo 47518
Painted Buckeye 9693,9716
Painted Caladium 65237
Painted Canarygrass 293710
Painted Copperleaf 2011
Painted Copper-leaf 2011
Painted Cup 79113,79120,79128,79130,287721
Painted Daisy 322640,322665
Painted Damask 336514,336516
Painted Damask Rose 336514
Painted Drop Tongue 11344
Painted Droptongue 11344
Painted Drop-tongue 11344
Painted Drymoda 138517
Painted Epidendrum 146422
Painted Euphorbia 159046
Painted Feather-flower 407533
Painted Gentian 173725
Painted Graphistemma 180215
Painted Grass 293709
Painted Kerria 216092
Painted Lady 127607,139986,174997,222789,322665,349044
Painted Leaf 158727
Painted Leaf Begonia 50232
Painted Leaves 99506
Painted Lily 397627
Painted Maple 3154
Painted Maxillaria 246718
Painted Melic 249072
Painted Mono Maple 3160,3398
Painted Net Leaf 166709
Painted Net-leaf 166709
Painted Nettle 99509,99711
Painted Orchis 273592
Painted Pink 127715
Painted Poinsettia 158727
Painted Skullcap 355679
Painted Spurge 159046
Painted Tongue 344383
Painted Trillium 397627
Painted Wood Lily 397627
Painted Wood-lily 397627
Paintedcup 79113,79130
Painted-leaf Begonia 50232
Painted-leaf Schismatoglottis 351154
Painted-tongue 344383
Painter's Palette 28084,28119
Paiqu Rockfoil 349735
Pair Vetch 408648
Paired Flower St. John's Wort 201892
Pairing Cryptodiscus 113555
Pairleaf Clematis 95468
Pairleaf Euphorbia 159853
Pairleaf Madder 338030
Pairleaf Milkvetch 43207
Pairleaf Touch-me-not 205199
Pairspot Eria 299608
Pairspot Hairorchis 299608
Paise Premna 313709
Paise Rhubarb 388506
Paishan Japanese Buttercup 326168
Paiute Cabbage 373709
Pajaro Manzanita 31131
Pajuil 21195
Pakchoi 59595
Pak-choi 59575
Pak-choi Cabbege 59595
Pakha Tanoak 233340
Pakim La Rhododendron 331002
Palace Purple 194426
Palafoxia 280413
Palaktree 280437
Palamut 324146
Palas 64148,228729
Palas Tree 64148
Palasa 64148
Palasan 265137
Palas-tree 64148
Palatebearing Pseuderanthemum 317267
Palay Rubber 113823
Palay Rubbervine 113823
Palczewsk Currant 334144
Palczewsk Gooseberry 334144

Paldao 137715
Pale Agoseris 11445
Pale Alyssum 18327,18344
Pale Arrowbamboo 162728
Pale Avens 175459
Pale Azalea 330757,332030
Pale Bamboo 47404
Pale Beard-tongue 289366
Pale Biddy-biddy 1756
Pale Bittersweet 80211
Pale Bladderwort 403268,403841
Pale Blue-eyed-grass 365680
Pale Bridewort 371795
Pale Broom 120909
Pale Bugseed 104825
Pale Bulrush 353686
Pale Butterflybush 62134
Pale Butterfly-bush 62134
Pale Butterwort 299751
Pale Cabbage 59651
Pale Catechu 401757
Pale Chickweed 375033
Pale Claytonia 94317
Pale Coneflower 140081
Pale Corydalis 106201,106430,169065
Pale Dew-plant 138135,138171
Pale Didymoplexis 130043
Pale Diliporchis 130043
Pale Diploclisia 132931
Pale Dock 339916,340135
Pale Dog-violet 410164
Pale Dogwood 104949,105133
Pale Dualbutterfly 398291
Pale Eulalia 156482
Pale Evening Primrose 269479
Pale Face 194831
Pale False Foxglove 10156
Pale False Manna Grass 321345
Pale Fargesia 162728
Pale Fawn-lily 154908
Pale Flax 231880
Pale Fleabane 150840
Pale Forget-me-not 260883
Pale Fragrant-bamboo 87643
Pale Galingale 322342
Pale Gentian 173193,173457
Pale Giant Hyssop 10411
Pale Globose-stamen Magnoliavine 351100
Pale Goldenrod 367986
Pale Goldquitch 156482
Pale grass-pink 67887
Pale Green Orchid 302326,302328
Pale Green Orchis 302326
Pale Green Sedge 75672
Pale Hawkweed 195603
Pale Heath Violet 410164
Pale Hickory 77926
Pale Indian Plantain 34802
Pale Indian-plantain 34802
Pale Iris 208744
Pale Laelia 219691
Pale Laurel 215405
Pale Madwort 18327
Pale Mahonia 242559
Pale Mallotus 243419
Pale Manna-grass 177643
Pale Meadow Violet 409661
Pale Meadow-beauty 329422
Pale Meadow-violet 409661
Pale Microtropis 254317
Pale Mint 250432
Pale Oxalis 277905
Pale Painted-cup 79130,79134
Pale Panic Grass 128562
Pale Pasqueflower 24109
Pale Pasque-flower 24109
Pale Peashrub 72317
Pale Perfoliate Honey Suckle 235713
Pale Persicaria 309298
Pale Pink Boronia 57264
Pale Pink-sorrel 277905
Pale Plantain 302159
Pale Poppy 282515
Pale Poppy Mallow 67125
Pale Primrose 314591
Pale Purple Coneflower 140074,140088
Pale Ragwort 359660
Pale Sage 345284
Pale Sedge 75672
Pale Senna 78308
Pale Serrate Azalea 330406
Pale Serrate Leaves Rhododendron 330406
Pale Smartweed 309298
Pale Speedwell 407096
Pale Sphenopholis 371497
Pale Spike Lobelia 234785
Pale Spikerush 143154
Pale Spike-rush 143154
Pale St. John's Wort 201842,202026
Pale St. John's-wort 201842,202021
Pale Staff-tree 80211
Pale Stiff Sedge 75184
Pale Stonecrop 356961
Pale Sunflower 188953
Pale Sweetleaf 381217
Pale sweetshrnb Pale Sweet Shrub 68310
Pale Sweetshrub 68313
Pale Sweet-shrub 68321
Pale Swordflag 208744
Pale Three-winged Nut 398317
Pale Threewingnut 398317
Pale Tickseed 104767
Pale Toadflax 231105
Pale Touch-me-not 205204
Pale Tripterospermum 398291
Pale Trumpets 208320
Pale Umbrella Bamboo 162728
Pale Umbrellawort 255674
Pale Umbrella-wort 255674
Pale Vanilla-lily 36837
Pale Vetch 408336
Pale Vetchling 222787
Pale Violet 410602
Pale Weasel's-snout 255984
Pale Wild Bergamot 257180
Pale Wildtung 243419
Pale Willow Dock 340167
Pale Willowherb 146859
Pale Willowweed 146859
Pale Willow-weed 309298
Pale Windhairdaisy 348183
Pale Wolfberry 239092
Pale Wood Violet 410483
Pale Woodrush 238675
Pale Yellow Rattan Palm 65727
Pale Yellow Trillium 397555
Pale Yellow-eyed Grass 365803
Pale Yellowflag Iris 208774
Pale Yucca 416630
Paleaceous Knotweed 309507
Paleback Azalea 330848
Paleback Bittersweet 80211
Paleback Fieldhairicot 138972
Paleback Ringvine 116019
Paleback Yam 131531
Pale-face 194830
Paleface Rosemallow 194831
Paleflower Henry Lily 229863
Pale-flowered Brown's Larkspur 124233
Pale-flowered Leaf-cup 310023
Pale-flowered Megacarpaea 247779
Pale-flowered Orchid 273564
Pale-flowered Parish's Larkspur 124459
Pale-flowered Red Maple 3524
Palegreen Bamboo 297285
Palegreen Euonymus 157375
Palegreen Magnoliavine 351041
Palegreen Spunsilksage 360830
Palegreen Tangbamboo 364790
Palegreenvine 279556,279560
Palehair Tangbamboo 364799
Pale-laurel 215405
Paleleaf Barberry 51424
Pale-leaf Goldenweed 209563
Pale-leaf Maple 3026
Paleleaf Reevesia 327363
Paleleaf Salacia 342646
Paleleaf Suoluo 327363
Paleleaf Wildtung 243362
Paleleaf Woodlotus 244442
Pale-leaved Common Linden 391707

Pale-leaved Crowfoot 326326
Pale-leaved Deer-weed 237675
Pale-leaved Reevesia 327363
Pale-leaved Spindle-tree 157769
Pale-leaved Sunflower 189063
Pale-leaved Woodland Sunflower 189063
Palepurple Maple 2775
Pale-purple Maple 2775
Palered Chinese Woodbetony 287093
Palermo Pondweed 312236
Paleroseflower Snakegourd 396276
Pale-seed Plantain 302209
Pale-seeded Plantain 302209
Palespike Lobelia 234785
Pale-spike Lobelia 234785,234787
Pale-stemmed Four-o'clock 255676
Palestine Anemone 23767
Palestine Bramble 339234
Palestine Buckthorn 328810
Palestine Iris 208452,208742
Palestine Nightshade 367233
Palestine Terebinth 301002,301003
Palestine Willow 343726
Palestinian Tumbleweed 21805
Palette Flower 28084
Paleyellow Armenia Pocktorchid 282786
Paleyellow Arrowwood 407929
Paleyellow Buckthorn 328699
Pale-yellow Buckthorn 328699
Paleyellow Elsholtzia 144053
Paleyellow Halfribpurse 191528
Paleyellow Iris 208771
Pale-yellow Iris 208771
Paleyellow Milkvetch 41949,42294
Paleyellow Monkshood 5199
Paleyellow Silvergrass 255847
Pale-yellow Tulip 400226
Paleyellow Viburnum 407929
Pale-yellow Viburnum 407929
Paliavana 280478
Palicourea 280481
Pali-mara 18055
Palimara Alstonia 18055
Palina-rosa 117204
Palipo Begonia 50056
Palisander 121607,121770,211225,240484
Palisandro 121803
Paliurus 280539
Pall Corydalis 106227
Pall Orchis 273564
Pallas Bugseeed 104824
Pallas Buttercup 326172
Pallas Crazyweed 279067
Pallas Garlic 15564
Pallas Lagotis 220190
Pallas Larkspur 124454
Pallas Milkvetch 42845
Pallas Onion 15564
Pallas Saffron 111573
Pallas Sainfoin 271241
Pallas Seseli 361547
Pallas Thyme 391311
Pallas' Buckthorn 328693,328812
Pallass Bluebells 250900
Pallescent Stonecrop 200793
Pallet Apocopis 29454
Pallid Anisochilus 25386
Pallid Barberry 51997
Pallid Fescuegrass 164189
Pallid Javatea 275739
Pallid Madder 338005
Pallid Mahonia 242610
Pallid Newlitse 264079
Pallid Rattan Palm 65758
Pallid Rattlebox 112491
Pallid Rattle-box 112491
Pallid Rockfoil 349736
Pallid Scargrass 203318
Pallid Skullcap 355655
Pallid Vernalgrass 28001
Pallid Wolfberry 239092
Pallida Rose 336298
Pallidflower Columbine 30062
Pallid-flower Huashan Milkvetch 42475
Pallid-green Galingale 118929
Pallidileaved Euonymus 157769
Pallid-leaf Ormosia 274435
Pallid-leaved Bambusa 47404
Pallidus Ceanothus 79962
Palm 64159,64345,106703,385302,393803,
393809
Palm Beachbells 215150
Palm Cabbage 142257
Palm Dulce 59098
Palm Family 280608
Palm Grass 361850
Palm Hearts 6136,161053,337873
Palm Lily 104339,104359,416607
Palm Newlitse 264059
Palm of Christian 334435
Palm Perotis 291215
Palm Rockjasmine 23176
Palm Sedge 75466,75775
Palm Springs Daisy 93884
Palm Syneilesis 381820
Palm Willow 343151
Palma Christi 273531,334432,334435
Palma Fibre 345752
Palma Istle 416572
Palma Pita 416659
Palma Real 337875
Palma Rosa 117204
Palma-christi 334435
Palmata-leaf Pinellia 299721
Palmate Bamboo 347268
Palmate Begonia 50131
Palmate Butterbur 292403
Palmate Carludovica 77043
Palmate Cassava 244512
Palmate Dragonhead 137621
Palmate Germander 338053
Palmate Hop Clover 397096
Palmate Jewelvine 125992
Palmate Kirengeshoma 216473,216474
Palmate Leaf Knotweed 309510
Palmate Manycleft Cinquefoil 312803
Palmate Meadowsweet 166110
Palmate Norway Maple 3437
Palmate Poisongourd 132957
Palmate Raspberry 338945
Palmate Sweet-coltsfoot 292403
Palmate Violet 410340
Palmated Larkspur 124604
Palmateileaved Jewelvine 125992
Palmateleaf Bigumbrella 59235
Palmateleaf Boerlagiodendron 56529
Palmateleaf Corydalis 105733
Palmateleaf Garishcandle 28113
Palmateleaf Monkshood 5067
Palmateleaf Primrose 314771
Palmateleaf Reddish Raspberry 339221
Palmateleaf Synotis 381947
Palmate-leaved Morningglory 207659
Palmatelobe Buttercup 326303
Palmate-lobed Violet 410340
Palmately Leaf Chinese Groundsel 365067
Palmatipartite Septemlobate
Kalopanax 215443
Palmatisect Cacalia 283854
Palmatisect Sinosenecio 365068
Palmbract Corydalis 106348
Palmcalyx 317399
Palmella 416578
Palmer 343151
Palmer Black Maple 3240
Palmer Ceanothus 79964
Palmer Cottonwood 311419
Palmer Oak 323857,324261
Palmer Penstemon 289367
Palmer Sagewort 36047
Palmer's Amaranth 18787
Palmer's Bloodleaf 208355
Palmer's Catchfly 363244
Palmer's Cockscomb 80448
Palmer's elm-leaf Goldenrod 368463
Palmer's Goldenbush 150359
Palmer's Goldenweed 150359
Palmer's Indian Mallow 967
Palmer's Mariposa-lily 67616
Palmer's Spineflower 89091
Palmer's Umbrella Thoroughwort 217108

Palmer's Wild Buckwheat 152368
Palmer's Zinnia 418053,418054
Palmerosa Oil Grass 117204
Palmer's Agave 10917
Palmers Slimpod 20856
Palmetto 341401,341421,341422
Palmetto Royal 341422
Palmgermander 338052,338053
Palmgrass 361680,361850
Palm-grass 114819,361850
Palmhair Azalea 330752
Palmier 315214
Palmier Royal 337885
Palmilla 416639,416646
Palmiste 337883
Palmlaef Cypress Vine 323470
Palmleaf Begonia 50148
Palm-leaf Begonia 50041
Palmleaf Bristlegrass 361850
Palmleaf Butterbur 292403
Palmleaf Buttercup 325717
Palmleaf Cinquefoil 312460
Palmleaf Corydalis 105733
Palmleaf Cremanthodium 110423
Palmleaf Dewberry 339270
Palmleaf Falsepanax 266918
Palmleaf Fishvine 125992
Palmleaf Goldenray 229151
Palmleaf Greenorchid 137621
Palmleaf Groundsel 359654
Palmleaf Knotweed 309510,309641
Palmleaf Lizardtaro 348036
Palm-leaf Marshmallow 18160
Palmleaf Merremia 250859
Palm-leaf Mistflower 101958
Palmleaf Motherwort 225008
Palmleaf Nutantdaisy 110423
Palmleaf Raspberry 338250,339029
Palmleaf Rhubarb 329372
Palmleaf Selenipedium 357753
Palmleaf Violet 409878
Palmleaf Xylobium 415708
Palmleaftree 59187,155445,185262
Palm-leaved Raspberry 338250
Palmleaved Rhubarb 329372
Palm-like Fig 165502
Palm-polly 125694
Palmsplit Sinosenecio 365067
Palmstipe Brassaiopsis 59229
Palmstipe Euaraliopsis 59229
Palmstipe Palmleaftree 59229
Palm-stiped Brassaiopsis 59229
Palmtelobe Buttercup 326310
Palmvein Catchfly 363218
Palmvein Windhairdaisy 348164
Palmyra 57122
Palmyra Palm 57122
Palo Adan 167558
Palo Alto Thistle 92312
Palo Amarillo 68418
Palo Blanco 1704,1705,80687
Palo Brea 83749,284483
Palo Calorado 238337
Palo Colorado 120488
Palo Del Muerto 207593
Palo Dulce 161862
Palo Madrono 19948
Palo Mortero 392345
Palo Prieto 405224
Palo Santo 63409
Palo Verde 83747,83748,284477,284479,284481
Palomar Mountain Wild Buckwheat 152102
Palosapis 25605,25613
Palouse Goldenweed 323035
Palouse Thistle 91820
Palp Peristylus 291274
Palp Perotis 291274
Palrymple Bean 408950
Palsy Plant 315101
Palsy-curer 73207
Palsywort 315082
Palsy-wort 315082
Palta 291494
Palu 244551
Paludal Burreed 370074
Paludal Oak 324262
Pameruon Bark 395516
Pamir Brome 60873
Pamir Chickweed 375144
Pamir Clematis 95199
Pamir Corydalis 106246
Pamir Crazyweed 279084
Pamir Eritrichium 153497
Pamir Goldenray 228976
Pamir Hairgrass 126047
Pamir Jurinea 214144
Pamir Kengyilia 215951
Pamir Milkvetch 42848
Pamir Nathaliella 262777
Pamir Nepeta 265020
Pamir Nitraria 266359
Pamir Parrya 224213
Pamir Perovskia 291433
Pamir Pondweed 312208
Pamir Poplar 311420
Pamir Rhodiola 329917
Pamir Saltbush 44564
Pamir Sedge 75678
Pamir Serpentroot 354912
Pamir Skullcap 355656
Pamir Solms-Laubachia 368569
Pamir Tarragon 35421
Pamir Tickseed 104826
Pamir Toadflex 231019
Pamir Trigonella 397263
Pamir Whitethorn 266359
Pampainin Honeysuckle 236014
Pampanin Bractbean 362300
Pampanin Indigo 206342
Pampanin Shuteria 362300
Pampanin Stonecrop 357002
Pampanin Stringbush 414232
Pampas Grass 105450,105456,148906,182941,182942
Pampas-grass 105456,182941
Pam's Pink Rose 336299
Pan American love Plant 21144
Pan Arisaema 33448
Panalero 167323,350573
Panama Crowngrass 285435
Panama Dichaea 128401
Panama Gum Tree 79110
Panama Gum-tree 79110
Panama Hat Palm 77043
Panama Hat Plant 77043
Panama Oncidium 270833
Panama Orange 93639
Panama Rubber 79110
Panama Rubber Tree 79110
Panama Rubbertree 79110
Panama Screw Pine 77043
Panama Tree 376075
Panama-hat Palm 77040,77043
Pan-american Friendship Plant 298943
Panamiga 298943
Panamint Daisy 145208
Panamint Mariposa-lily 67619
Panamint Mountain Wild Buckwheat 152369
Panamint Rock Goldenrod 114517
Panamint Sulphur Flower 152610
Panamint Wild Buckwheat 152282
Panax-scandent Hemsleia 191954
Panaxscandent Snowgall 191954
Pancake 401711
Pancake Plant 243840
Pancake Prickly Pear 272826
Pancake Pricklypear 272826
Pance 410677
Pan-chinese Bambusa 47402
Pancratium 280877
Panda Family 280960
Panda Plant 215277
Pandacaqu Tabernaemontana 154192,382822
Pandan Wangi 280974
Pandang 281089,281138
Pandani 334355
Pandanni 334355
Pandanny 334355
Pandanus 280968
Panderia 281173

Pandora Vine　281227
Pandorea　281223
Pandurate Conehead　320134
Pandurate Raspberry　338952
Pandurifoliate Nettle Spurge　212194
Panduriform Mananthes　244294
Panes　285741
Panga-Panga　254851
Pangium　281245
Panglandular Rabdosia　209790
Pangola Grass　130514,130537
Pangola-grass　130514
Pangpanga　254851
Panhandle Grape　411658
Paniala　166777
Panic Grass　128478,128494,128502,128505,128516,128518,128525,128535,281274,281394,281398,281495,281565,281704,281818,281873,282104,282159,282214,282250
Panic Liverseed Grass　402538
Panic Millet　281916
Panic Veldtgrass　141755
Panicgrass　281274
Panic-grass　281274
Panicgrasslike Tailgrass　402538
Panicgrasslike Urochloa　402538
Panicilate Microcos　253385
Panicle Anisochilus　25387
Panicle Chalkplant　183225
Panicle Closedspurorchis　94458
Panicle Flemingia　166887
Panicle Gypsophila　183225
Panicle Hydrangea　200038,200043,200044,200046,200047,200049,200050
Panicle Meconopsis　247157
Panicle Milkwort　308247
Panicle Neslia　265553
Panicle Ophiorrhiza　272270
Panicle Paraboea　283134
Panicle Poorspikebamboo　270442
Panicle Rattlesnakeroot　313895
Panicle Sabia　341537
Panicle Tanoak　233341
Panicle Willowweed　146831
Panicle-corymbose Barberry　51505
Panicled Aster　40989,256310,380912,380919,380986
Panicled Bellflower　70002
Panicled Bittersweet　80273
Panicled Bluebell　250902
Panicled Bulrush　353618
Panicled Dogwood　105169
Panicled False Solomonseal　366198
Panicled Fameflower　383324
Panicled Golden Rain Tree　217626
Panicled Golden-rain Tree　217626
Panicled Gold-rain Tree　217626
Panicled Goldraintree　217626
Panicled Hawkweed　195836
Panicled Hydrangea　200038
Panicled Knapweed　81263
Panicled Littleflower Stringbush　414217
Panicled Localginseng　383324
Panicled Mallotus　243420
Panicled Monk's-hood　5468
Panicled Monkshood　5468,5469
Panicled Pittosporum　301355
Panicled Raspberry　338954
Panicled Sedge　75686
Panicled Tick-trefoil　126363,126507,126508
Panicled Wildtung　243420
Panicleleaf Pond Cypress　385257
Paniculate　124456
Paniculate Abutilon　968
Paniculate Acronema　6220
Paniculate Amaranth　18788
Paniculate Amaranthus　18788
Paniculate Anodendron　25944
Paniculate Arrowwood　408077
Paniculate Barberry　51999
Paniculate Bedstraw　170533
Paniculate Bluebeard　78032
Paniculate Blue-beard　78032
Paniculate Bolbostemma　56659
Paniculate Butterflybush　62138
Paniculate Butterfly-bush　62138
Paniculate Caspian Sea Inula　207070
Paniculate Cephalostigma　82475
Paniculate Chisocheton　88121
Paniculate Cleisostoma　94458
Paniculate Connarus　101889
Paniculate Corydalis　106244
Paniculate Crypteronia　113401
Paniculate Cystacanthus　120797
Paniculate Dallisgrass　285476
Paniculate Decaspermum　123456
Paniculate Dinggongvine　154263
Paniculate Echinodorus　140552
Paniculate Epidendrum　146456
Paniculate Eria　148733
Paniculate Erycibe　154263
Paniculate False Solomonseal　242704
Paniculate Falsetamarisk　261284
Paniculate False-tamarisk　261284
Paniculate Flemingia　166887
Paniculate Glorybower　96257
Paniculate Grevill　180628
Paniculate Hairorchis　148733
Paniculate Inflated Treerenshen　125617
Paniculate Jasminorange　260173
Paniculate Jointfir　178554
Paniculate Ladybell　7618
Paniculate Lilyturf　272119
Paniculate Mahonia　242611
Paniculate Milkwort　308247
Paniculate Mosquitotrap　117643
Paniculate Oligostachyum　270442
Paniculate Onosma　271814
Paniculate Ostodes　276789
Paniculate Paspalum　285476
Paniculate Pimpinella　299559
Paniculate Raspberry　338954
Paniculate Rattlebox　112496
Paniculate Rockcress　30383
Paniculate Sabia　341537
Paniculate Singleseed Chickweed　375003
Paniculate Staff-tree　80273
Paniculate Stringbush　414217
Paniculate Swallowwort　117643
Paniculate Threeleaf Lacquertree　386678
Paniculate Tickclover　126485
Paniculate Trileaved Lacquer-tree　386678
Paniculate Viburnum　408077
Paniculateflower Raspberry　339067
Panicullate Spotflower　4870
Panicum　281274
Panicutate Pittosporum　301355
Panicutate Seatung　301355
Panir Littledalea　234124
Panirband　414590
Panisea　282409
Panjiang Camellia　69429
Panlanshan Barberry　52130
Panlongqi Hemsleia　191956
Panlongqi Snowgall　191956
Panma Sagebrush　36568
Pannicle Jointvetch　9609
Pannose Meliosma　249450
Pannose Pearleverlasting　21660
Pans-and-cakes　243840
Pansura Grewia　180888
Pansy　316127,409608,409624,409703,410677,410749
Pansy Butterflygrass　392897
Pansy Orchid　254924
Pansy Orchids　254924
Pansy Torenia　392897
Pansy Violet　410380,410677
Pansy-orchid　254924
Panther Lily　229978
Pantling Woodbetony　287497
Pantomorphic Rhododendron　331439
Panxian Herbclamworm　398110
Panzerina　282477
Panzhihua Cycas　115870
Panzhu Gingerlily　187459
Panzhuteng Childvine　400948
Paohsing Blueberry　403912
Paohsing Broomrape　275145
Paohsing Currant　334104

Paohsing Jerusalemsage 295169
Paohsing Larkspur 124588
Paohsing Primrose 314676
Paohsing Rhododendron 331283
Paohsing Sage 345288
Paohsing Stonecrop 357003
Paohsing Strawberry 167629
Paohsing Woodbetony 287436
Paoshan Monkshood 5424
Paosing Raspberry 338938
Paoxing Ceropegia 84196
Papain 76813
Papapsco 2769
Papauma 181264
Papaw 38322,38338,76813
Papaw Family 76820
Papaya 38338,76808,76813
Papaya Family 76820
Papaya Like Bigumbrella 59231
Papaya Like Brassaiopsis 59231
Papeda 93523
Paper Bark 248114
Paper Bark Maple 3003
Paper Beech 53563
Paper Birch 53533,53549,53563
Paper Bush 141466,141470
Paper Crop 356468
Paper Daisy 329801,415385,415386
Paper Flower 57857,57868,238371,318978
Paper Mulberry 61096,61107
Paper Mulberry Fargesia 162730
Paper Nailwort 284809
Paper Reed 119347
Paper Sedge 119347
Paper Spine Cactus 385873,385874,385875
Paper-bag Bush 355602
Paperbagbush 355602
Paperbark 248072,248114,377538
Paper-bark 248104,248114
Paperbark Albizia 13680
Paper-bark Birch 53549,53563
Paperbark False Thorn 13680
Paperbark Maple 3003
Paper-bark Maple 3003
Paperbark Tea-tree 226504
Paperbark Thorn 1601
Paper-bark Thorn 1601
Paperbark Tree 248114
Paper-bark Tree 248114
Paperbush 141466,141470
Paper-bush 141466,141470
Paperelder 290286
Paperflower 57857,318976,318978,390311
Paper-flower 57864
Paperleaf Azalea 332155
Paperleaf Bladderwort 403876
Paper-leaf Daphniphyllum 122673
Paper-leaf Holly 203618
Paperleaf Litse 233860
Paper-leaf Tigernanmu 122673
Paper-leaved Holly 203618
Paper-leaved Rhododendron 332155
Paperlike Clearweed 298894
Paperlike Coldwaterflower 298894
Paper-mulberry 61096,61107
Papershell Pinyon 300177
Paper-spined Cactus 354306
Paper-white Daffodil 262433
Paperwhite Narcissus 262433
Paper-white Narcissus 262433,262457
Papery Allophylus 16060
Papery Daphne 122567
Papery Daphniphyllum 122673
Papery Fig 164797
Papery Whitlow-wort 284809
Paperyflower Onion 15509
Paperysepal Globeflower 399499
Paphiopedilum 282768
Paphlagoni Barberrya 51453
Papilionaceous Plants 282926
Papilionaceus Violet 410351
Papilioorchis 282929,282934
Papillalaef Corydalis 106372
Papillary Holly 204017
Papillary Rhododendron 331441
Papillastalk Corydalis 106248
Papillate Aspidistra 39568
Papillate Bambusa 47405
Papillate Blueberry 403953
Papillate Forrest Rhododendron 330725
Papillate Helwingia 191188
Papillate Lily 229976
Papillate Nightshade 367357
Papillated Haworthia 186589
Papilliferous Barberry 52003
Papillose Barberry 52003
Papillose Dogwood 380482
Papillose Epipactis 147196
Papillose Fragileorchis 2042
Papillose Himalayan Helwingia 191166
Papillose Spiraea 372045
Papillose Stairweed 142780
Papillseleaf Knotwood 44261
Papilose Magellan Barberry 51407
Papoose Root 79730,79733
Papoose-root 79733
Papose Crabgrass 130666
Papov Pricklythrift 2273
Papov Prickly-thrift 2273
Papple 11947
Pappose Garhadiolus 171478
Papposedge 375805
Pappus Grass 282979
Pappus-grass 282979
Pappusless Groundsel 358784
Pappusless Nemosenecio 263511
Pappusless Pinnate Groundsel 263511
Paprika 72070,72077
Papua New Guinea Rosewood 320301
Papua Nutmeg 261402
Papua Sandalwood 346218
Papuan Walnut 137714
Papusform Speargrass 397509
Papustular Blueberry 403954
Papwort 250614
Papyrus 119347
Papyrus Plant 119347
Para Angelwood Para Angel-wood 129489
Para Caoutchouc Tree 194453
Para Cress 4866
Para Devilpepper 327049
Para Grass 58045,58128
Para Guava 70490
Para Nut 53057
Para Para 81933
Para Piassava 225039
Para Rhatany 218076
Para Rubber 194453
Para Rubber Tree 194453
Para Rubbertree 194453
Para Rubber-tree 194453
Parabaena 283059
Parabarium 283065,402194
Paraboea 283107,283127
Parachampionella 283207
Parachinense Ophiorrhiza 272271
Parachute Plant 44296,84235
Parachute Sedge 76562
Parachutes 70164
Paracrenate Gugertree 350935
Paracress Spotflower 4866
Para-cress Spotflower 4866
Paracryphia 283249
Paracuminate Stairweed 142781
Paracuspidate Elatostema 142855
Paracuspidate Stairweed 142855
Paradise Apple 239157,243549,243675,243682
Paradise Flower 65055,367745
Paradise Lily 282685
Paradise Nut 223764,223776,223778
Paradise Nut Tree 223768
Paradise Palm 198808
Paradise Plant 122515
Paradise Tree 323548,364384
Paradisiacal Apple 243682
Paradombeya 283304
Paradox Acacia 1272
Paradox Pinanga 299669
Paradox Sunflower 189023
Paradoxical Bedstraw 170534

Paradrupetea 283710,283713
Para-grass 58128
Paraguay Dock 340176
Paraguay Holly 204135
Paraguay Nightshade 238975,367549
Paraguay Palm 6136,6137
Paraguay Starbur 2609
Paraguay Tea 204135
Paraguayan Dock 340176
Paraguayan Purslane 311818
Paraguayan Siver Trumpet Tree 382706
Paraguayan Windmill Grass 88325,88326
Paraguaylignum 63409
Paragutzlaffia 283360
Paraguyaran Graptopetalum 180269
Paraisometrum 283380
Paraixeris 283385
Parakedya 65832
Parakeelya 65827
Parakeetflower 190038
Parakmeria 283417
Paralagarosolen 283433
Paralbizzia 283437
Parallelvein Micrechites 25937
Paralle-veined Anodendron 25937
Paramartin Paraboea 283123
Parameria 283489
Paramichelia 283500,283501
Paramicrorhynchus 283504
Paramignya 283508
Parana Araucaria 30831
Parana Pine 30831
Paranephelium 283529
Parang Bean 71050
Para-nut 53055,53057
Para-para 81933
Paraphlomis 283603
Parapholia 283654
Parapinnatevein Elatostema 142858
Parapinnatevein Stairweed 142858
Paraprenanthes 283680,283682
Parapruinose Barberry 52006
Parapteropyrum 283707
Parapyrenaria 283710
Paraquadrangular Eurya 160576
Paraquilegia 283720
Pararotary Jerusalemsage 295170
Pararotate Jerusalemsage 295170
Pararuellia 283760
Paraserianthes 283885
Parashorea 283891
Parashortiflowered Machilus 240667
Parasite 30906,293187
Parasite Helixanthera 190848
Parasite Pentapanax 289659
Parasite Scurrula 144607,190848
Parasitic Medinilla 247535
Parasitic Pentapanax 289659
Parasitic Salomonia 147368
Parasitic Scurrula 355317
Parasiticvine 125781
Parasol 102933,345881
Parasol Aster 41448,41453
Parasol Flower 197368
Parasol Leaf Tree 240325
Parasol Pine 352856
Parasol Plant 350654
Parasol Sedge 76630
Parasol Tree 350654
Parasol Whitetop 135268
Parasol Wild Buckwheat 151874
Paraspectacular Barberry 52007
Parastyrax 283992
Paratetragland Willow 343838
Paratinytail Stairweed 142868
Paravallaris 216228
Parched Pendentlamp 83995
Parchment-like Elatostema 142790
Parcifoliate Monkshood 5476
Parcitment Bark 301248
Pareira 88749
Pareirs Brava 88815
Pareirs Root 88815
Parepigynum 284099
Parfa Nut 53057
Paria Sunflower 190258
Parietary 284167
Paril Arrowroot 244507
Parinarium 284191
Paris 284289
Paris Bedstraw 170536
Paris Circaea 91557
Paris Daisy 32563,160772
Parish Chaenactis 84517
Parish Cholla 181457
Parish Cleisostoma 94459
Parish Closedspurorchis 94459
Parish Club-cholla 181457
Parish Cymbidium 116826
Parish Dendrobium 125296
Parish Dendrocalamus 125501
Parish Dragonbamboo 125501
Parish Paphiopedilum 282883
Parish Phtheirospermum 296071
Parish Rhododendron 331445
Parish Spapweed 205211
Parish Touch-me-not 205211
Parish Vandopsis 200584
Parish Western Bruning Bush 157752
Parish's Catchfly 363867
Parish's Fleabane 150843
Parish's Goldenbush 150363
Parish's Ironweed 406662
Parish's Larkspur 124458
Parish's Rabbitbrush 150363
Parish's Sunflower 189014
Parish's Wild Buckwheat 152371
Parishes' Orach 44570
Parisleaf Eargrass 187641
Parisleaf Hedyotis 187641
Parisshape Loosestrife 239770
Parit Grass 58045
Park Hedgehog Cactus 287902
Park Leaves 201732
Parka Rockfoil 349751
Parker Barberry 52010
Parker Christolea 89251
Parker Pipewort 151424
Parker Plateaucress 89251
Parker Raspberry 338964
Parker's Jasmine 211956
Parker's Pipewort 151424
Parkinson Pincution 244182
Parkman Crabapple 243624
Parkman's Hall Crabapple 243624
Parks' Jointweed 308681
Parlin's Pussytoes 26530
Parlin's Pussy-toes 26530,26531
Parlor Palm 85420,85426
Parlour Palm 85425,85426
Parma Violet 409661,410320
Parmentiera 284490
Parnassia 284500
Parnassia Rockfoil 349269
Parnassialeaf Rockfoil 349269
Parnassis-leaf Caldesia 66343
Parnell 407989
Parney Cotoneaster 107520
Parod Barberry 52011
Parodia 284655
Parple Viper's-bugloss 141258
Parque Cestrum 84419
Parquet Burr 399206
Parralena 391042
Parramatt Eucalyptus 155684
Parramatta Grass 372579
Parriette Fishhook Cactus 354291
Parriette Hookless Cactus 354325
Parritory 284152
Parrot Alstroemeria 18076
Parrot Beak 96639,190005
Parrot Bush 138443
Parrot Chastetree 411409
Parrot Coralbean 154626
Parrot Feather 261337,261342,261364
Parrot Feather Watermilfoil 261342
Parrot Gladiolus 176151
Parrot Head 347155
Parrot Leaf 18101
Parrot Pea 130934,130937
Parrot Pitcher Plant 347155

Parrot Pitcherplant 347155
Parrot Pitcher-plant 347155
Parrot's Beak 237509,237681
Parrot's Bill 96639
Parrot's Feather 261337,261342
Parrot's Flower 190038
Parrot's Plantain 190038
Parrot's-bill 96639
Parrot's-feather 261342
Parrot-beak 96633
Parrotbill 96633
Parrot-bill 96633
Parrotfeather 261337,261364
Parrot-feather 261337
Parrotheak 96633
Parrothill 96633
Parrotia 284957
Parrotleaf 18095
Parrot-lily 18076
Parrot-pea 130934
Parrots Heliconia 190038
Parrot-seed 77748
Parrotweed 56027
Parrot-weed 56027,154316
Parry Agave 10919
Parry Dalea 245292
Parry Nolina 266506
Parry Northern Mahonia 242517
Parry Petrocosmea 292552
Parry Pine 300171
Parry Pinyon 299857,300171
Parry Sagewort 35762
Parry Thistle 92288
Parry's Agave 10919
Parry's Alpinegold 199228
Parry's Arnica 34757
Parry's Beargrass 266506
Parry's Catchfly 363871
Parry's Centuryplant 10919
Parry's Fleabane 150844
Parry's Goldenweed 273897
Parry's Knotweed 309527
Parry's Larkspur 124464
Parry's Lazy Daisy 84940
Parry's Penstemon 289368
Parry's Rock Daisy 291327
Parry's Rush 213357
Parry's Saltbush 44570
Parry's Spineflower 89094
Parry's Wild Buckwheat 151865
Parry's Wirelettuce 375977
Parrya 224209,284969
Parrya-like Phaeonychium 293370
Parryodes 285043
Parsley 292685,292694
Parsley Breakstone 13966,14065,14166
Parsley Fern 383874
Parsley Hawthorn 109527,109836
Parsley Hawtitorn 109527
Parsley Peart 13966
Parsley Peat 13966
Parsley Perk 13966
Parsley Piercestone 13966
Parsley Pierce-stone 13966
Parsley Piert 13966,29018,29025,353987
Parsley Vlix 13966
Parsley Water Dropwort 269334,269348
Parsley Water-dropwort 269334
Parsley-leafed Thorn 109527
Parsley-piert 13951,29018
Parsley-piert Parsley Piert 13966
Parsnip 285731,285741,285742
Parsnip-flower Wild Buckwheat 152137
Parson's Billycock 37015
Parson's Nose 273539,273545
Parson-and-clerk 37015
Parson-in-hissmock 37015
Parson-in-his-smock 37015
Parson-in-the-pulpit 5442,5676,33544,37015
Parsons Barberry 52012
Parson's Pink Rose 336300
Parsonsia 285055
Parted Gentian 173376
Parthenium 285072,285086
Partially Bearded Rhododendron 331791
Particoloured Pinanga 299659
Parti-coloured Scorpion-grass 260791
Partitipilose Garnotia 171518
Partridge Berry 172047,172135,256002
Partridge Breast Aloe 17136
Partridge Feather 383746
Partridge Pea 78284,78421,301055,301070,
360453
Partridge Wood 22218
Partridgeberry 172135,256001,256002,
404051,404052,404060
Partridge-berry 172135,256002
Partridge-breast 17136
Partridge-breasted Aloe 17136,17375
Partridgeflower 194638,395466,395480
Partridgegrass 148808,148818
Partridge-pea 78284,301055,360453
Parture Tephroseris 385915
Parvicaudate Camellia 69433
Parviflorous Bittercress 72920
Parviflorous Dendropanax 125628
Parviflorous Giant-grass 175608
Parviflorous Magnolia-vine 351069
Parvifolia Rose 336477
Parvifoliate Abelia 200
Paschal Flower 321721
Pascuita 159240
Pashan Clematis 95214
Pashan Maple 3370
Pashan Paris 284296
Pashi Pear 323251
Pashia Pear 323251
Pash-leaf 302034
Pasment 285741
Pasmet 285741
Paspalidium 285376
Paspalum 285390,285491,285522
Paspalum Tailgrass 402543
Paspalum Urochloa 402543
Paspalum-flower Rattan Palm 65765
Pasque Flower 321653,321721
Pasqueflower 23696,321653,321708,321721
Pasque-flower 321721
Pasquier Madhuca 241519
Passed Over Rhododendron 331533
Passe-flower 321721
Passerage 225450
Passevelours 18687
Passing Lovely Rhododendron 331215
Passion 308893,340178
Passion Dock 308893,340178
Passion Flower 37015,285607,285612,
285623,285639,285654,285655,285666
Passion Fruit 285607,285637
Passion Vine 285612
Passionflower 285607,285623,285637,
285654
Passion-flower 285607
Passionflower Family 285725
Passionfruit 285637
Passment 285741
Passper 111347
Pastel 209229
Pastinacopsis 285748
Pastis 299351
Pasto Lacquer 141926
Pastural Willow 343909
Pasture Goatsbeard 394296
Pasture Gooseberry 333945
Pasture Hawksbeard 110947
Pasture Hawthorn 110054
Pasture Knapweed 80979
Pasture Milkvetch 42855
Pasture Rose 336462,336637
Pasture Sagebrush 35505
Pasture Spike Sedge 218488
Pasture Spikesedge 218547
Pasture Spruce 298286
Pasture Thistle 91914,92330,92461
Pastureweed 115738
Patagonian Bean 71042
Patagonian Cypress 166720,166726
Patagonian Fitzroya 166726
Patagonian Pilgerodendron 299132
Patagonian Plantain 302122
Patagoniau Brome 83468

Patana Oak 76799
Patang Primrose 314155
Patashte 389403
Pataua 269374
Pataxte 389403
Patchoul Microtoena 254262
Patchouli 306994
Patchouli Plant 306994
Patchouly 306964
Patchy Indumentum Rhododendron 331504
Pate 350682
Pated Indosasa 206895
Patelliform Cyclobalanopsis 116172
Patelliform Dragonbamboo 125502
Patelliform Tickseed 104829
Patenehairy Vilmorin Monkshood 5670
Patent Hamelia 185169
Patent Saltbush 44575
Patent Toxocarpus 393541
Patent Tulip 400203
Patentbeak Monkshood 5458
Patentbract Fleabane 150846
Patentbract Gentian 150846
Patentbranch Anabasis 21064
Patentflower Monkshood 5115
Patent-haired Melastoma 248765
Patenthairy Anemone 23785
Patent-hairy Melastoma 248765
Patenthairy Monkshood 5108
Patenthairy Windflower 23785
Patentpillose Manyroot Monkshood 5315
Patentpilose Curvebeak Monkshood 5097
Patentsetulose Bethlehemsage 283640
Patentsetulose Paraphlomis 283640
Paternoster Bean 765
Paternoster Beans 765
Paterson's Curse 141258
Pates Crazy 68189
Path Rush 213507
Pathfinder Juniper 213926
Patience 205444,308893,340178
Patience Dock 308893,340178
Patience Plant 205444
Patience-dock 308893
Patient Dock 340178
Patient Lucy 205444
Patridge Cane 329173
Patrin Figwort 355211
Patrinia 285814
Patrot's Bill 96637
Pattens-and-clogs 231183,237539
Pattern Wood 18034
Patterson Sagewort 36062
Patterson's Bluegrass 305803
Pattikeys 167955
Patty-carey 192130
Pattypan Squash 114301
Patty-pan Squash 114301
Patty-pans 267294
Patula Pine 300140
Patulous Anemone 23990
Patulous Chingiacanthus 209886
Patulous False Simon Poplar 311437
Patulous Forkpaniclegrass 209886
Patulous Garnotia 171509
Patulous Leathern Barberry 51502
Patung Actinidia 6709
Patung Elaeagnus 141979
Patung Loosestrife 239776
Patung Manglietia 244464
Pau Marfim 46955
Pau Rosa 121683
Pau Santo 216301
Pauciannulated Camellia 69460
Pauciflorous Beautyberry 66889
Pauciflorous Bedstraw 170545
Pauciflorous Blastus 55174
Pauciflorous Diplarche 132830
Pauciflorous Enkianthus 145709
Pauciflorous Erycibe 154253
Pauciflorous Greenbrier 366253
Pauciflorous Lysionotus 239967
Pauciflorous Memecylon 250041
Pauciflorous Pittosporum 301360
Pauciflorous Sagebrush 360835
Pauciflorous Sedge 75711
Pauciflorous Ungeargrass 36710
Paucilobed Begonia 50144
Paucinerved Big-ginseng 241175
Paucinerved Garcinia 171163
Paucinerved Sweetleaf 381353
Paucipilose Golmu Milkvetch 42420
Paucipunctated Camellia 69465
Paucisepaled Camellia 69462
Paucisete Vineclethra 95532
Paucity Flower Yellowleaftree 415145
Paucity Milkwort 308238
Paucity Seed Milkwort 308238
Pauderpuff 66660
Pauderpuff Tree 66660
Pauho Machilus 240669
Pauilinia 285928
Paul Begonia 50146
Paul Primrose 314784
Paul's Betony 407256,407358
Paul's Scarlet Hawthorn 109792
Pauldopia 285916,285917
Paulownia 285951,285981
Paulse Russian Thistle 344658
Paulsen Brome 60882
Paulsen Dragonhead 137625
Paulsen Greenorchid 137625
Paunce 410677
Paunsy 410677
Pauper's-tea 342198
Paurotis Palm 4951
Pau-santo 63409
Pavetta 286073
Pavieasia 286573
Pavlov Milkvetch 42860
Pavon Barberry 52018
Pavon Moonflower 67753
Pavonia 286580
Paw Paw 38322
Pawdered Scurrula 355324
Pawpaw 38322,38338,76808,76813
Paw-paw 76813
Pawpaw Family 76820
Pawpaw-tree 76813
Pawple 11947
Pax Ash 168054
Pax Daphniphyllum 122717
Pax Mahonia 242613
Pax Maple 3374
Pax Sedge 75720
Pax Stringbush 414242
Pax Tigernanmu 122717
Pax Woodbetony 287504
Pax's Mallotus 243424
Paxistima 286746
Paxiuba Palm 208368,366807
Paxman Parnassia 284604
Payena 286754
Payshun Dock 340178
Payshun-dock 340178
Payson's Groundsel 279905
Pea 222671,301053,301070
Pea Family 224085
Pea Gull 315082
Pea Mint 250362,250450
Pea Seven 329879
Pea Shrub 72168
Pea Thatch 237539
Pea Tree 72168,72180,218761
Pea Vetch 408543
Pea Vine 222671
Pea-bean 294056
Peabush 361357
Peace Lily 370332,370338,370344,370345
Peace Plant 33544
Peace Rose 336301
Peace-and-plenty 349044
Peace-lily 370332
Peach 20885,20935,77902,316166
Peach Bells 70214,256432
Peach Lilac 382241
Peach Nut 46376
Peach Oak 324282
Peach Palm 46376
Peach Protea 315810
Peach Springs Cholla 116638

Peach Thorn 239026
Peach Tiger Flower 391627
Peach Willow 343013
Peach Wood 65002,184516
Peach-flowered Rhododendron 331559
Peach-flowered Tea-tree 226503
Peachform Mango 244407
Peach-formed Mango 244407
Peachgrass 172185
Peachleaf Azalea 330084
Peachleaf Bellflower 70214
Peachleaf Buckthorn 328739
Peachleaf Eurya 160580
Peachleaf Oak 324282
Peachleaf Photinia 295757
Peachleaf Rhododendron 330084
Peachleaf Willow 343013
Peach-leaved Bellflower 70214
Peach-leaved Buckthorn 328739
Peachleaved Campanula 70214
Peach-leaved Dock 339916
Peach-leaved Eurya 160584
Peach-leaved Photinia 295757
Peach-leaved Rhododendron 330084
Peach-leaved Willow 343013
Peach-seven 365005
Peach-thorn 239026
Peachwood 65002,65060
Peachwort 309298,309570
Peacock Anemone 23992
Peacock Echeveria 140000
Peacock Fir 113721
Peacock Flower 4534,65055,123811,
258380,258606,391624,391627
Peacock Flower-fence 7190
Peacock Iris 258588,258606
Peacock Pine 113685
Peacock Plant 66175,147386
Peacock Poppy 282659
Peacock Sprangletop 225994
Peacock Tiger-flower 391627
Peacockflower 391627
Peacock-flower 123811
Peacock-plume Grass 88320
Peacocksplume 162473
Peacock-tigerflower 391627
Peafruit Popowia 311108
Pea-fruited Cypress 85345
Peagle 315082,315101,325612
Peakock Barnyardgrass 140415
Pealikefruit Popowia 311108
Pealike-fruited Popowia 311108
Peanut 30494,30498
Peanut Cactus 85163,85165,140855
Peanut Clover 397015
Peanut Madhuca 241519
Peanut Stitchwort 255569
Peanut Tree 376171
Peanutbrittle Begonia 49788
Pear 323076,323133
Pear Bamboo 249617
Pear Cashew 21195
Pear Gourd 114300
Pear Hawthorn 109573
Pear Jobstears 99146
Pear Machilus 240679
Pear Maesa 241817
Pear Thorn 110088
Pear Umbrella 31511
Pearbamboo 249614
Pearbamboovine 249605
Pearce Barberry 52019
Pearce Begonia 50147
Pearform Citrus 93712
Pearfruit Fig 165527
Pear-fruit Fig 165527
Pear-fruited Gockspur-thorn 109609
Pear-fruited Mallee 155712
Pearl Acacia 1488
Pearl Anthurium 28118
Pearl Arrowwood 407846
Pearl Barley 198376
Pearl Berry 245259
Pearl Bloom Tree 307290
Pearl Bush 161741,161748,161749
Pearl Cudweed 21596
Pearl Everlasting 21506,21596
Pearl Flower 193752
Pearl Fruit 245257,245259
Pearl Grass 34930
Pearl Haworthia 186534
Pearl Laceleaf 28118
Pearl Lupin 238468
Pearl Millet 289116
Pearl of Spain 260302
Pearl Onion 15048
Pearl Plant 186534,233760
Pearl Pyrola 322914
Pearl Russianthistle 344656
Pearl Viburnum 407846
Pearl Wattle 1488
Pearlberry 404529
Pearlbloomtree 307290
Pearlbloom-tree 307290
Pearlbush 161741,161744,161749
Pearl-bush 161741
Pearleaf Crabapple 243667
Pearleaf Goldenray 229158
Pear-leaf Holly 204198
Pearleaf Knotwood 44272
Pearleaf Microglossa 253476
Pearleaf Raspberry 339067
Pearleaf Sweetleaf 381378
Pear-leaf Vineclethra 95539
Pear-leaf Wintergreen 322872
Pear-leaved Crab Apple 243667
Pear-leaved Crabapple 243667
Pear-leaved Goatwheat 44272
Pear-leaved Holly 204198
Pear-leaved Microglossa 253476
Pear-leaved Raspberry 339067
Pear-leaved Sweetleaf 381378
Pearleverlasting 21506
Pearlflower 193752
Pearl-flowered Everlasting 21596
Pearl-flowered Life Everlasting 21596
Pearl-grass 60379
Pearlhead 209551
Pearlikeinflorescence Elatostema 142658
Pearlmillet 289116
Pearlsedge 353998
Pearlweed 342221,342279
Pearlweed Rockfoil 349870
Pearlweedlike Sandwort 32201
Pearlwood 296442,296445
Pearlwort 342221,342251,342279
Pearlwort Spurrey 370565
Pearly Everlasting 21596,21716
Pearly Globe-amaranth 179242
Pearly Immortelle 21596
Pearly Pussytoes 26321
Pearly-shing Maesa 241817
Pearoot Corydalis 106558
Pear-shaped Fruit Melocanna 249617
Pearweed 239543
Peasant's Clock 384714
Pease 301070
Pease Earthnut 222758
Pease Earth-nut 222758
Pease Everlasting 222750
Peashrub 72168
Pea-shrub 72168,72213,72257,72324
Peatpink 363300
Peatree 72168
Pea-tree 72168,72180,361350
Peaty Dewberry 338091
Peavine 222671
Pea-vine Clover 397019
Pebble Plant 102037,102213
Pebble Vetch 408571
Pebble-vetch 408571
Pecan 77902
Pecan Hickory 77902,204114
Pecan Nut 77902
Pecan-tree 77874,77902
Peck Neststraw 379165
Peck's Oak Sedge 75721
Peck's Sedge 75721
Peckled Nomocharis 266567
Pecos River Skeletonplant 239337
Pecos Sunflower 189023

Pecos Twistflower 377637
Pecteilis 286829
Pectinate Barberry 52020
Pectinate Boerlagiodendron 56530
Pectinate Brome 60883
Pectinate Bulbophyllum 62985
Pectinate Cassiope 78641
Pectinate Corydalis 105601
Pectinate Eyebright 160239
Pectinate Habenaria 183945
Pectinate Hedgehog Cactus 140282
Pectinate Love-grass 147871
Pectinate Neopallasia 264249
Pectinate Osmoxylon 276486
Pectinate Pinanga 299671
Pectinate Pondweed 378991
Pectinate Primrose 314787
Pectinate Raspberry 339013,339014
Pectinate Rockfoil 350010
Pectinate Rungia 340367
Pectinate Spartina 370173
Pectinate Stonebean-orchis 62985
Pectinate Woodbetony 287506
Pectinatebract Windhairdaisy 348635
Pectinatebracted Saussurea 348635
Pectinateciliate Eritrichium 153501
Pectinated Maple 3379
Pectinatehair Eritrichium 153501
Pectinatetooth Raspberry 339013
Pectinatilobe Buttercup 326200
Pectinato-edge Barberry 52021
Peculiar Arisaema 33309
Peculiar Kleinhovia 216642
Peculiar Southstar 33309
Pedalium 286925
Pedalium Family 286918
Pedate Cyclanthera 115995
Pedateleaf Snakegourd 396247
Pedatifid Begonia 50148
Pedatifid Buttercup 326201
Pedelion 190944
Pedgery 138742
Pediceld Flower Pennywort 200346
Pedicellaate Taohe Willow 344183
Pedicellate Actinodaphne 233945,234075
Pedicellate Cleistanthus 94529
Pedicellate Eulalia 156453
Pedicellate Gentian 173710
Pedicellate Hemiphragma 191576
Pedicellate Litse 233969
Pedicellate Sageretia 342187
Pedicelled Pepper 300475
Pedicelless Reticulinerve Barberry 52111
Pedicillate Wool-grass 353699
Pedicled Pepper 300475
Pedilanthus 287846
Pediocactus 393235
Pedisellate Gesneria 175301
Pediselless Rattlebox 112667
Pedlar's Basket 116736,349936
Peduncled Cinquefoil 312862
Peduncled Gomphostemma 179202
Peduncled Jerusalemsage 295171
Peduncled Sedge 75733
Peduncled Sinosideroxylon 365084
Peduncular Milkvetch 42864
Pedunculate Achimenes 4086
Pedunculate Acronychia 6251
Pedunculate Bluebell 115408
Pedunculate Bluebellflower 115408
Pedunculate Butterfly-bush 62108
Pedunculate Clubfilment 179202
Pedunculate Contracted Spindle-tree 157388
Pedunculate Fairybells 134457
Pedunculate Farges Meehania 247725
Pedunculate Garcinia 171164
Pedunculate Gesneria 175302
Pedunculate Herpetospermum 193062
Pedunculate Jerusalemsage 295171
Pedunculate Oak 324335
Pedunculate Sea-purslane 44601
Pedunculate Sedge 75733
Pedunculate Sprague Barberry 52313
Pedunculate St. John's Wort 202082
Pedunculate Trigonotis 397451
Pedunculate Water-starwort 67345
Pedunculate Waveribgourd 193062
Pedunculate Windhairdaisy 348639
Peduneled Loosestrife 239780
Pee Gee Hydrangea 200043,200049,200055
Pee-a-bed 384714
Pee-bed 384714
Peeble's Navajo Cactus 287904
Peegee Hydrangea 200038,200055
Pee-in-bed 384714
Peek-a-boo Plant 4866
Peelbark St. John's-wort 201875
Peelberry Poison Sumac 393488
Peeling Plne 268244
Peelu Extract 344804
Peep-o'-day 274823
Peeps 315082
Peepul 165553
Peepul Tree 165553
Peeseweep Grass 238583
Peeweets 160236
Pegaeophyton 287948
Peganum 287975
Peggall-bush 109857
Peggle 109857,315082
Peggy-ailes 109857
Peggyiles 109857
Pegia 287986
Peg-nut 102727
Pegu Windhairdaisy 348643
Pegunny 49123
Pegwood 157429,380499
Pegyll 315101
Pehpei Maple 3392
Pei Beautyberry 66805
Pei Begonia 50151
Pei Glorybower 96263
Pei Purplepearl 66805
Peiktushan Sedge 75737
Peilitory 284167
Peirson Pincushion 84486
Peirson's Lessingia 227114
Peirson's Mountaincrown 274129
Peirson's Serpentweed 236499
Pejibaye 46376
Pejivalle 46376
Pejivalle Pejibaye 46376
Pekea Nut 77946
Pekin Willow 343669
Peking Cabbage 59600
Peking Cleistogenes 94605
Peking Cortusa 105501
Peking Cotoneaster 107325
Peking Euphorbia 159540
Peking Flowering Crab 243667
Peking Honeysuckle 235766
Peking Lilac 382278
Peking Mock Orange 294509
Peking Mockorange 294509
Peking Mock-orange 294509
Peking Oak 323647
Peking Pea-shrub 72310
Peking Skullcap 355669
Peking Thorowax 63597
Peking Tree Lilac 382278
Peking Violet 410395
Peking Willow 343668
Pelargonium 288045
Pelatantheria 288598,288604
Peldate-leaved Macaranga 240227
Peletir 284152
Pelexia 288620
Pelican Dutchman's-pipe 34198
Pelican Flower 34193,34198
Pelicanflower 34198
Pelican-flower Fleurs 34198
Peliosanthes 288638
Pelitgrain Oil 93332,93765
Pellacalyx 288682
Pell-a-mountain 391397
Pellas 243823
Pellionia 288705,288768
Pellionifoliate Elatostema 142788
Pelliot Anabasis 21054
Pelliot Eriocycla 151751
Pelliot Ptilagrostis 321018

Pellitory 21266,284131,284174,322724
Pellitory of The Wall 284152
Pellitory-of-Spain 21266
Pellitory-of-the-wall 284131,284152
Pellucid Ophiorrhiza 272276
Pellucid Rockfoil 349768
Pellucid-leaved Spindle-tree 157784
Pellucidmargin Agapetes 10316
Peltate Briggsia 60275
Peltate Clearweed 299001
Peltate Clematis 95316
Peltate Coldwaterflower 299001
Peltate Dutchman's Pipe 34293
Peltate Globeflower 399555
Peltate Lallemantia 220285
Peltate Smilax-leaf Clematis 95316
Peltate Sundew 138329
Peltate Yam 131917
Peltateleaf Begonia 49703,50153
Peltateleaf Clematis 95316
Peltateleaf Halfummer 299722
Peltateleaf Meadowrue 388536
Peltateleaf Paraboea 283124
Peltate-leaf Petrocosmea 252507
Peltateleaf Raspberry 339024
Peltateleaf Storkbill 288430
Peltate-leaved Raspberry 339024
Peltatelike Pothomorphe 313178
Peltatelike Vinepepper 313178
Peltatescaled Windmill 100728
Peltatesepal Corydalis 106267
Peltate-stipule Cowpea 409065
Peltophorum 288882,288897
Peltostyle 288882
Peluskina 301055
Pelviform Tickseed 104830
Pemako Rhododendron 331457
Pembina 408002
Pemphis 288926,288927
Pen Gentian 174095
Pen Shamgoldquitch 318150
Penaga Lili 252370
Penang Lawyers 228729
Penang Rattan Palm 303089
Penard's Larkspur 124106
Pence 329521
Pencil Bush 159975
Pencil Cactus 140293
Pencil Cedar 213984,310227
Pencil Cholla 116647,116665,116671,
272793
Pencil Flower 379239
Pencil Juniper 213984
Pencil Pine 44046
Pencil Savin 213984
Pencil Southstar 33450
Pencil Tree 159975
Pencil Willow 343505
Pencilflower 379239
Pencil-flower 379239
Pencilled Crane's-bill 174997
Pencilled Cranesbill 174997
Pencil-pine 44047
Penciltree 159975
Pencilwood 139637,139644
Pencil-wood 139637
Pencuir Kale 308893
Pendant Barberry 51639
Pendate Clematis 94900
Penddulous European Aspen 311539
Pendent Agapetes 10353
Pendent Azalea 331458
Pendent Branches Agapetes 10353
Pendent Euonymus 157695
Pendent Goosecomb 335461
Pendent Inflorescence Indigo 206373
Pendent Japanese Pagodatree 369038
Pendent Roegneria 335461
Pendent Silver Linden 391816
Pendent Sliver-lime 391647
Pendent Wendlandia 413809
Pendent-bell 145673
Pendentbranch Bitter Bamboo 304003
Pendentbranch Chinese Juniper 213673
Pendentbranch David Poplar 311286
Pendentbranch Longbeak Eucalyptus 155521
Pendentbranch Siberian Elm 401608
Pendentflower Greenorchid 137617
Pendentfruit Rockcress 30387
Pendentlamp 83975
Pendentlamp Hibiscus 195216
Pendentseed Jointfir 178557
Pendent-seeded Jointfir 178557
Penduculate Daphne 122580
Penduculatum Meadowrue 388613
Pendular Cape-cowslip 218920
Pendular Silver Maple 3534
Pendulate Baron Oak 323695
Pendulate Thistle 92296
Pendulate Torenia 392948
Pendule Kengyilia 215952
Pendulicarpous Monkshood 5480
Penduliflory Raspberry 339025
Penduline Rose 336348
Penduliracemed Indigo 206373
Penduliracemeoid Indigo 206386
Pendulous Alder 16433
Pendulous Bammula Heath 149727
Pendulous Begonia 50154
Pendulous Bicolor Bushclover 226702
Pendulous Bottlebrush 67302
Pendulous Bunge Spindle-tree 157352
Pendulous Ceratostylis 83655
Pendulous China Nettletree 80740
Pendulous Cliffbean 254801
Pendulous Cymbidium 116815
Pendulous Dove Orchid 291130
Pendulous Euonymus 157761
Pendulous Flower Galangal 17751
Pendulous Fourstamen Hackberry 80765
Pendulous Ginkgo 175824
Pendulous Hall's Crabapple 243625
Pendulous Hankow Willow 343673
Pendulous Heads Groundsel 358765
Pendulous Higan Cherry 83337
Pendulous Hypecoum 201618
Pendulous Japanese Chestnut 78781
Pendulous Katsuratree 83744
Pendulous Keteleeria 216137
Pendulous Leymus 228375
Pendulous Millettia 254801
Pendulous Monkshood 5483
Pendulous Mulberry 259068
Pendulous Odontoglossum 269082
Pendulous Orchis 116815
Pendulous Oriental Arborvitae 302738
Pendulous Pagodatree 369038
Pendulous Peach 316636
Pendulous Racemose Distylium 134948
Pendulous Red Cedar 341778
Pendulous Sedge 75742
Pendulous Simon Poplar 311502
Pendulous Smoketree 107311
Pendulous Spapweed 205219
Pendulous Spindle-tree 157695
Pendulous Sweetleaf 381354
Pendulous Touch-me-not 205219
Pendulous Yucca 416610
Pendulousflower Agapetes 10347
Pendulousflower Galangal 17751
Pendulousflower Glorybower 96264
Pendulous-flowered Helleborine 147204
Pendulous-fruit Tickclover 126621
Pendulous-fruited Catchfly 363882
Pendulousshead Groundsel 358765
Pendulous-panicle Pratt Barberry 52062
Pendulus-bamboo 20192
Penelope Rose 336302
Penetrateleaf Canscora 71281
Penflower 379237
Penflower Stylosanthes 379245
Peng Willow Weed 146838
Pengcuo Sagebrush 36071
Penghu Soja 177707
Penghu Soybean 177707
Penglai Bambusa 47259
Penglai Yellow Bamboo 47259
Pengrass 318147,318150
Pengshui Sanicle 345995
Pengxian Hemsleia 191957
Pengxian Snowgall 191957

Pengzhou Monkshood 5523
Penicaud Gymnopetalum 182580
Penicillate Daphne 122582
Penicillate Rattan Palm 65766
Peninsula Hebei Pear 323195
Peniterry 284152
Penn Sedge 75744
Pennants 86120
Pennate Threeawngrass 377011
Pennate Triawn 377011
Pennies 238371
Pennies-and-happenies 239755,329521
Pennilabium 289007
Pennisetum 289011
Penniwinkle 409335
Pennsylvania Bittercress 72802,72927
Pennsylvania Bitter-cress 72927
Pennsylvania Blackberry 338843
Pennsylvania Buttercup 326217
Pennsylvania Hawthorn 109923
Pennsylvania Hitter Cress 72802
Pennsylvania Knotweed 309542
Pennsylvania Persicaria 309542
Pennsylvania Rush 213144
Pennsylvania Saxifrage 349771
Pennsylvania Sedge 75744
Pennsylvania Smartweed 309542
Pennsylvania Sumach 332608
Pennsylvania Tea 257161
Pennsylvania Wild Pink 363301
Pennsylvanian Pellitory 284174
Penny Box 329521
Penny Cake 401711
Penny Cakes 401711
Penny Cap 401711
Penny Cod 401711
Penny Cods 401711
Penny Cress 390213,390249
Penny Daisy 227533
Penny Flower 238371,401711
Penny Girse 329521
Penny Grass 329521
Penny Hat 401711
Penny Hats 401711
Penny Hedge 14964
Penny in the Hedge 14964
Penny Leaves 401711
Penny Mouchers 338447
Penny Mountain 391397
Penny Pie 401711
Penny Plate 401711
Penny Plates 401711
Penny Ratile 329521
Penny Rattle 329521
Penny Rot 200388
Pennycress 390202
Penny-cress 390202
Pennyeress 390202
Pennygrass 200388,401711
Pennyrinkle 409339
Pennyroyal 187187,250385,250432
Pennyroyal Mint 250432
Pennyroyal Thyme 391346
Pennywall 401711
Pennyweed 329521
Pennywinkle 409339
Pennywort 81564,116736,200253,200388,238371,239755,268079,362406,401711
Pennywort Begonia 49938
Pennywortleaf Goldsaxifrage 90380
Pennywortleaf Goldwaist 90380
Pensil Agapetes 10353
Penstemon 289314
Pensylvania Cudweed 178342
Pensylvanian Cudweed 178342
Pentace 289407
Pentachondra 289431
Pentaclethra 289434
Pentacme 289440
Pentacyclic Cyclobalanopsis 116176
Pentacyclic Oak 116176
Pentagland Sabia 341543
Pentagon Dodder 115099
Pentagonous Ardisia 31578
Pentagonous Dysophylla 139579
Pentagonous Watercandle 139579
Pentagynous Dillenia 130922
Pentagynous Eurya 160578
Pentaleaved Maple 3396
Pentalobe Yellowhair Maple 2981
Pentalocule Brassaiopsis 59232
Pentamerous Camellia 69466
Pentamerous Tea 69466
Pentandrous Bell Tree 325001
Pentandrous Dendrophthoe 125667
Pentandrous Fivestamen 246217
Pentandrous Sphenodesma 371422
Pentanema 289563
Pentapanax 289627
Pentapera 148941
Pentapetaled Mischocarp 255951
Pentapetes 289682
Pentaphragma 289693
Pentaphylax 289700
Pentaphylax Family 289699
Pentaphylax-like Camellia 68998
Pentapterygium 289744
Pentarhaphia 289759
Pentas 289784,289831
Pentasacme 289924
Pentastelma 290055
Penthea 290113
Penthorum 290133,290138
Pentitsao 93732
Pentopetia 290158,290197
Pentz Blachia 54878
Penwiper Plant 215206,267188
Penzig Ahernia 199033
Peonia 5861
Peony 280143,280213,280231,280253
Peony Rose 280197
Peony Shrubalthea 195297
Peonygrass 224625
Peonyleaf Angelica 24428
Pepal Barberry 51327
Pepelot's Artocarpus 36942
Pepermioides Clearweed 299005
Peperomia 290286
Pepilary 311398
Pepilles 290489
Pepino 367400
Pepino Dulce Melon 367400
Pepita 114300
Peplar 311398
Peplidium 290471
Peplis 290476
Peplis Purslane 240065
Peplis Spurge 159557
Peppe Tree 417330
Pepper 72070,300323,300427,300464
Pepper and Salt 150412
Pepper Box 282685
Pepper Bush 385090
Pepper Chili 72070
Pepper Cress 73090,225279,225398,225447,225450
Pepper Crop 356468
Pepper Face 290395
Pepper Family 300557
Pepper Grass 225340,225450,225475
Pepper Mint 250420
Pepper Mint Tree 11399
Pepper Plant 300464
Pepper Saxifrage 363123
Pepper Tree 216502,241210,318624,318625,350980,350990
Pepper Turnip 33544,267867
Pepper Vine 20296,300464
Pepper-and-salt 72038,150412
Pepper-box 282685,329521
Pepperbush 96457
Pepperbush Clethra 96457
Pepperbush Family 96561
Peppercom 350991
Pepper-cress 225279
Pepper-girse 4005
Peppergrass 225279,225295,225450,225475
Pepper-grass 72038,225279,225340,225450,365424
Pepperidge 267867
Peppermint 155476,250362,250420,250432,

250450,287721
Peppermint Geranium 288550
Peppermint Gum 155478,155533
Peppermint Scented Geranium 288550
Peppermint Tree 11399,155478,155675
Peppermint-scented Geranium 288550
Peppermint-tree Wattle 1646
Pepper-plant 72038,309199
Pepper-root 125841,125846
Pepper-saxifrage 363123,363128
Peppertree 350980,350991
Pepper-tree 241210,318624,350980
Pepper-tree Prickly-ash 417330
Peppervine 20296
Pepper-vine 20296
Pepperweed 225279,225295
Pepperweed Whitetop 73090
Pepper-weed White-top 73090
Pepperwood 401672
Pepperwort 225279,225315,225398,363123
Peppetree 417330
Peppilary 311398
Pequi 77943
Peracarpa 290572,290573
Perak Devilpepper 327050
Perak Pinanga 299672
Perak Rattan Palm 65764
Percely 292694
Percepier 13966
Perchment-like Leaf Slugwood 50592
Percuspidate Camellia 69477
Pere A. Gregestier's Rhododendron 330757
Pere David's Maple 2920
Pere Farges Rhododendron 331408
Pere L. F. Faurie's Rhododendron 330240
Pere Valentin's Rhododendron 332047
Peregrina 212160,212161
Peregrine Hoarhound 245757
Peregrine Saltbush 44661
Peregrine Thistle 91900
Perenial Sunflower 189044
Perennial 183225
Perennial Agrostis 12240
Perennial Bluegrass 305824
Perennial Bursage 19190
Perennial Candytuft 203236,203239
Perennial Centaury 81530
Perennial Conyza 103561
Perennial Cornflower 81214
Perennial Crabgrass 130800
Perennial Cupgrass 151694
Perennial Cup-grass 151694
Perennial Dontostemon 136184
Perennial Dove's Foot Cranesbill 174856
Perennial Dove's-foot Cranesbill 174856
Perennial Duckweed 224399
Perennial Felwort 235457
Perennial Flax 231942
Perennial Gaillardia 169575
Perennial Glasswort 342900,346761,346771
Perennial Goldfields 222593
Perennial Goosefoot 86970
Perennial Greenhead Sedge 218480
Perennial Honesty 238379
Perennial Knawel 353994
Perennial Lettuce 219444
Perennial Lovegrass 147875
Perennial Lupine 238473
Perennial Marjoram 274231
Perennial Pea 222671,222750
Perennial Pea Vine 222750
Perennial Peanut 30496
Perennial Pea-vine 222750
Perennial Pepper Grass 225398
Perennial Pepper-grass 73090
Perennial Pepperweed 225398
Perennial Phlox 295288
Perennial Pigweed 18701
Perennial Quakegrass 60382
Perennial Quaking Grass 60382
Perennial Quaking-grass 60382
Perennial Ragweed 19179
Perennial Rocket 365621
Perennial Rye Grass 235334
Perennial Ryegrass 235334
Perennial Rye-grass 235334
Perennial Saltmarsh Aster 380999
Perennial Snakeweed 182103
Perennial Sow Thistle 368635
Perennial Sow-thistle 368635
Perennial Soybean 264241
Perennial Stock 246478
Perennial Sundrops 269486
Perennial Sunflower 188987
Perennial Sweet Pea 222750
Perennial Swertia 380302
Perennial Teosinte 155880
Perennial Veldtgrass 141743
Perennial Wallrocket 133313
Perennial Wall-rocket 133313
Perennial Welsh Onion 15441
Perennial Yellow-woundwort 373389
Perenniel Pea 222844
Pereskia 290699
Perez Statice 230712
Perez's Sealavender 230712
Perezia 290724
Perfect Clearweed 298923
Perfect Coldwaterflower 298923
Perfoliate Alexanders 366744
Perfoliate Bellwort 403672
Perfoliate Blackfoot 248145
Perfoliate Greenbrier 366518
Perfoliate Honeysuckle 235713
Perfoliate Knotweed 309564
Perfoliate Loosestrife 239782
Perfoliate Mouseear 82968
Perfoliate Penny Cress 390249
Perfoliate Penny-cress 390249
Perfoliate Pepper Grass 225428
Perfoliate Pepperwort 225428
Perfoliate Pondweed 312220
Perfoliate Spineflower 259477
Perforate Fleeceflower 309564
Perforate St. John's Wort 202086
Perforate St. John's-wort 202086
Perforated Bothriochloa 57579
Perforated Haplophyllum 185658
Perforated Harrisonia 185934
Perforated Matricary 397964
Perforated Mayweed 397964
Perforated Oxmusclefruit 185934
Perforted Ceriman 258177
Perfumed Cherry 83253
Perfumed Fairy Lily 88216
Perfumed Passionflower 285717
Pergamentaceous Slugwood 50592
Pergularia 290739
Perianth Bristles Absent 352266
Perianth Rudimentary 352266
Pericampylus 290840,290843
Peridiscus 290896
Periflower Nematanthus 263328
Perilepta 290917
Perilla 290932,290940,290952
Perilla Mint 290940
Perilla-mint 290940
Peripterygium 73184,73190
Peristrophe 291134
Peristylus 291188
Periwinkle 79410,145487,340990,409325,409335,409339
Periwinkle Family 29462
Perky Sue 387355
Permanent Wave Tree 343670
Pernambuco 65002
Pernambuco Redwood 64971
Pernambuco Rubber 185258
Pernambuco Wood 64971,65002
Pernel 21339
Pernettya 291363
Perny False Fairybells 134412
Perny Germander 388179
Perny Holly 204162
Perny's Holly 204162
Peroba 284006
Peroba Rosa 39693
Peroffsky Lepidozamia 225624
Perofski's Hedge-mustard 154521
Perona Maple 3397
Perotis 291188,291396

Perovskia 291427
Perpetual Begonia 50298
Perpetual Spinach 53249
Perplexed Tick-trefoil 126363,126508
Perrier Cymbidiella 116763
Perrin Laelia 219688
Perrot Stonecrop 357015
Perrottetia 291481
Perry's Aloe 17163
Perry's Silver Weeping Holly 203548
Perry's Vervain 405880
Perrymedoll 70247
Persea 291491,291540
Persele 292694
Persetose Rose 336848
Persia Austrian Rose 336568
Persia Barrenwort 147034
Persia Epimedium 147034
Persia Fritillary 168514
Persia Sugarmustard 154522
Persia Walnut 212636
Persian Acacia 13578
Persian Apple 93603
Persian Berries 328737
Persian Berry 328849
Persian Berry Buckthorn 328737
Persian Box 64345
Persian Briar 336563
Persian Buttercup 325614,325954
Persian Calligonum 67059
Persian Carpet Flower 141485
Persian Cat-mint 265005
Persian Centaurea 81024
Persian Clover 397043,397050
Persian Cornflower 81024
Persian Cyclamen 115965
Persian Darnel 235351
Persian Everlasting Pea 222831
Persian Goatgrass 8671
Persian Helleborine 147203
Persian Hoarhound 245758
Persian Hyssop 390987
Persian Iris 208755
Persian Ironwood 284961
Persian Ivy 187205
Persian Kingcup 325614
Persian King-cup 325614
Persian Knotweed 308791
Persian Kobresia 217247
Persian Lilac 248895,382241
Persian Lily 168332,168430,168514
Persian Lime 93528
Persian Melic 249066
Persian Mulberry 259180
Persian Oak 324140
Persian Parrotia 284961
Persian Pearl Tulip 400215
Persian Pellitory 322665
Persian Ranunculus 325614
Persian Rose 336849
Persian Ryegrass 235351
Persian Sagebrush 36072
Persian Saxifrage 349229
Persian Saxoul 185072
Persian Shield 290921
Persian Silk Tree 13578
Persian Speedwell 407287
Persian Stone Cress 9798
Persian Stone-cress 9798
Persian Velvet Maple 3737
Persian Violet 115931,115965,161531
Persian Walnut 212636
Persian Wheat 398869
Persian Willow 85875
Persian Witch Hazel 284961
Persian Witch-hazel 284961
Persian Yellow Rose 336567,336568
Persicaria 291651,309570
Persicary 309570
Persimmon 132030,132219,132466
Persimmon-leaved Litse 233999
Persimmon-loving Mistletoe 411016
Persistent Barnyardgrass 140474
Persistent Buttercup 325897
Persistent Styles Rhododendron 330398
Persistent Sweetleaf 381422
Persistentleaf Woodbetony 287750
Persistentscale Azalea 330099
Persll 292694
Persoonia 292025
Persson Plantain 302129
Pertaining Violet 410483
Pertaining-cloudy Parnassia 284581
Pertusadina 292037
Pertybush 292040
Perty-bush 292040
Pertybush-like Ainsliaea 12695
Peru Balsam 261559
Peru Balsam-balm Tree 261554
Peru Balsam-tree 261559
Peru Barberry 52022
Peru Bark 91055
Peru Broad-leaved Barberry 51833
Peru Cotton 179884
Peru Daffodil 200906
Peru False Heath 161897
Peru Greentwig 352442
Peru Groundcherry 297711
Peru Haired Stygma 396456
Peru Nutmeg 223014
Peru Peppertree 350990
Peru Petioled Barberry 52048
Peru Rock Purslane 65855
Peru Rocky Barberry 52132
Perule Honeysuckle 236033
Perules Bird Cherry 280042
Peruvian Almond 77946,386594
Peruvian Apple Cactus 83890,83896
Peruvian Balsam 261554
Peruvian Bark 91055,91082
Peruvian Bark Tree 91086
Peruvian Black Salvia 345005
Peruvian Cereus 83890
Peruvian Cherry 297658
Peruvian Cotton 179884
Peruvian Daffodil 200906,200945
Peruvian Elder 345655
Peruvian Ground Cherry 297711
Peruvian Groundcherry 297711
Peruvian Ground-cherry 297711
Peruvian Heliotrope 190702
Peruvian Lily 18060,18062,18072,18073,18074,353001
Peruvian Marvel 255711
Peruvian Mastic 350990
Peruvian Mastic Tree 350990
Peruvian Mastic-tree 350990
Peruvian Mock Vervain 176693
Peruvian Nightshade 367494
Peruvian Nutmeg 223014
Peruvian Old Man 155303
Peruvian Old Man Cactus 155300
Peruvian Old-man Cactus 155300
Peruvian Parsnip 34914
Peruvian Paspalum 285495
Peruvian Pepper Tree 350990,350994
Peruvian Peppertree 350990
Peruvian Pepper-tree 350990
Peruvian Primrose Willow 238208
Peruvian Ragweed 19178
Peruvian Rhatany 218085
Peruvian Rock Pueslane 65855
Peruvian Scilla 271014
Peruvian Snowball 155300
Peruvian Squill 353001
Peruvian Swamp Lily 417612
Peruvian Swamp-lily 417612
Peruvian Torch 83891
Peruvian Turnsole 190622
Peruvian Verbena 176693
Peruvian Watergrass 238560
Peruvian Wild Petunia 339689
Peruvian Zinnia 418058
Peruviana Bark Tree 91090
Peruvian-daisy 170146
Peruvian-lily 18060
Peruvlan Leaf 155114
Pervenkle 409339
Pervinca 409339
Pervinkle 409339
Perwinkle 409339

Perwinkle Phlox 295244
Pest Pricklypear 273042,273061
Pestilence Wort 292372
Pestilence-wort 292372
Pestilentwort 292372
Petai 284472
Petalcalyx Azalea 330329
Petalformed Meadowrue 388614
Petalless Beautyleaf 67852
Petalless Chickweed 375033
Petalless Jujube 418149
Petalless Monkshood 5042
Petalless Rorippa 336193
Petalless Sibbaldia 362363
Petalless Viburnum Leaf Raspberry 339449
Petalless Yellowcress 336193
Petaloid Pussytoes 26428
Petasisecte Whored Cinquefoil 313102
Petasites 292342
Petch Henric Lily 229861
Petelot Actinidia 6679
Petelot Allomorphia 16129
Petelot Anise-tree 204574
Petelot Cleistanthus 94530
Petelot Clethra 96540
Petelot Eightangle 204574
Petelot Elatostema 142791
Petelot Fissistigma 166681
Petelot Greenstar 307520
Petelot Habenaria 183962
Petelot Kiwifruit 6679
Petelot Lecanthus 223682
Petelot Lysionotus 239977
Petelot Medinilla 247607
Petelot Microtropis 254319
Petelot Sikok Loosestrife 239783
Petelot Tanoak 233344
Petelot Wildginger 37702
Petelot Woodbetony 287510
Petelot's Camellia 69480
Peter Keys 167955
Peter Milkvetch 42877
Peter's Cress 111347
Peter's Grass 329521
Peter's Pence 238371
Peter's Staff 405788
Peterkin 315082
Peterson's Campion 363886
Peterson's Catchfly 363886
Petgery 138742
Petha 50998
Pethwine 95431
Petigree 340990
Petigrue 340990
Petiolar Hakea 184625
Petiolar Microstegium 254041
Petiolate Acianthus 232284
Petiolate Eritrichium 153506
Petiolate Liparis 232284
Petiolate Mitreola 256214
Petiolate Paniculate Ladybell 7733
Petiolate Paris 284325
Petiolate Primrose 314792
Petiolate Rocky Spindle-tree 157871
Petiolate St. John's Wort 202093
Petiolate Swertia 380307
Petiolate Trigonotis 397458
Petiolea St. John's Wort 202093
Petiolebore Protea 315918
Petioled Barberry 52023
Petioled Cuphea 114614
Petioled Fleabane 150859
Petioled Fountaingrass 289201
Petioled Garlicmustard 14964
Petioled Gentian 150859
Petioled Sunflower 189032
Petioled Twayblade 232284
Petiole-shorted Linden 391799
Petiolulate Thorowax 63778
Petit Barberry 52026
Petite Bardane 31056
Petite Flamboyant Bauhinia 49176
Petite Pink Scotch Rose 336303
Petitmengin Woodbetony 287511
Petitmenginia 292485
Petrea 292517
Petrocodon 292536
Petrocosmea 292547
Petrophila Spiradiclis 371775
Petropolit Vriesea 412362
Petrosavia 292671
Petrosimonia 292714
Petrov Sweetvetch 188042
Petrov Synstemon 382011
Petrov Windhairdaisy 348646
Pe-tsai 59595,59600
Petti Grue 340990
Petticoat Daffodil 262361
Petticoat Palm 413298
Pettigrew 203095
Petty Begonia 50140
Petty Cotton 178465
Petty Euphorbia 159557
Petty Morel 367416
Petty Morrel 30748
Petty Mugget 170743
Petty Mullein 315082
Petty Panick 293729
Petty Spurge 159557
Petty Whin 172905,401395
Petty Whir 271563,401395
Petunia 292741,292745
Pevit 229662
Peweep-tree 3462
Pewzow Sagebrush 36075
Peyote 236423,236424,236425
Peyote Verde 140267
Peyoti 236425
Peyotillo 288616
Peyotl 236425
Pfeiffer Anise 276472
Pfeiffer Cicely 276450,276466
Pfeiffer Leaf Nautilocalyx 262900
Pfeiffer Olive 276283,276291
Pfeiffer Osmanther 276291
Pfeiffer Osmanthus 276257,276291
Pfeiffer Root 276450
Pfeiffer Tea 276291
Pfeifferheart Ivy 187285
Phacelia 293151,293154,293158,293168,
293174,293176
Phacelia Family 200476
Phacelialeaf Woodbetony 287513
Phacellanthus 293181,293186
Phacellaria 293187,293190
Phacelurus 293207
Phaeanthus 293238
Phaenosperma 293293
Phaeonychium 293365
Phagnalon 293408
Phai-tong 125465
Phaius 293484,293537
Phalaenopsis 293577,293582
Phalsa 180682
Phanera 49058
Phantom Orchid 82028,82032
Phantom Tree 258946
Phar Sandwort 32138
Pharaoh's Date Palm 326649
Pharaoh's Fig 165726
Pharaoh's Pea 222844
Pharaoh's Peas 222844
Pharbitis 293861
Pharbitis Sees 207839
Phari Leek 15599
Phari Onion 15599
Phari Rhodiola 329929
Phasemy Bean 241263
Phaulopsis 294062
Phayung 121653
Pheasant Grass 376705
Pheasant Lily 168476
Pheasant's Eye 8324,8325,8332,21339,
127635,262438,289503
Pheasant's Eye Narcissus 262438
Pheasant's Eye Pink 127793
Pheasant's Eyes 8332,21339,127635,
262438,289503
Pheasant's Feathers 349044
Pheasant's Foot Geranium 288256
Pheasant's Head 168476

Pheasant's Tail 28091
Pheasant's Tail Grass 376705
Pheasant's Wings 17136
Pheasant's-eye 8324,8325,8332,262439
Pheasant's-eye Adonis 8325,8332
Pheasant's-eye Daffodil 262438
Pheasant's-eye Narcissus 262438
Pheasant-grass 376705,376728
Pheasants-eye 262438
Phebalium 294105
Phellem-bark Camellia 69486
Phellemy Camellia 69484
Phenomenal Berry 338059
Pheulpin Woodbetony 287514
Philadelphia Daisy 150862
Philadelphia Fleabane 150862
Philadelphia Lily 229984
Philadelphia Panic Grass 282067
Philadelphia Panic-grass 282067
Philadelphia Snowbell 379327
Philadelphia Witchgrass 282067
Philadelphus 294407
Philadelphus Family 294404
Philanthropos 11549
Philbrick's Desertdandelion 242938
Philip Barberry 52031
Philippin Epipremnum 147346
Philippin Homalomena 197795
Philippin Jade Vine 378353
Philippin Kadsura 214978
Philippin Litsea 234033
Philippin Olax 269663
Philippin Teak 385532
Philippine Acanthophippium 2075
Philippine Achyrospermum 4464
Philippine Agave 10816
Philippine Alcimandra 14549
Philippine Alphitonia 17621
Philippine Bellcalyxwort 231271
Philippine Blachia 54879
Philippine Boea 56069
Philippine Bushbeech 178039
Philippine Camarotis 68785
Philippine Cinnamon 91405
Philippine Cochinchina Sweetleaf 381149
Philippine Dehaasia 123600
Philippine Dichroa 129041
Philippine Dillenia 130923
Philippine Drypetes 138649
Philippine Eargrass 187645
Philippine Ebony 132351
Philippine Ehretia 141687
Philippine Eria 148738
Philippine Evergreen 11332,11339
Philippine Fig 164630,165502
Philippine Fighazel 380599
Philippine Flemingia 166888
Philippine Garcinia 171201
Philippine Glochidion 177165
Philippine Glorybower 96009
Philippine Ground Orchid 370401
Philippine Hackberry 80709
Philippine Hedyotis 187645
Philippine Holly 204575
Philippine Ixora 211153
Philippine Jadeleaf and Goldenflower 260478
Philippine Kauri 10503
Philippine Kauri Pine 10503
Philippine Kingiella 216439
Philippine Leea 223952
Philippine Lily 229988
Philippine Litse 233920
Philippine Lychee 233083
Philippine Mahogany 362212,362217
Philippine Mahonia 242614
Philippine Meadowrue 388618
Philippine Nato Tree 280448
Philippine Nettletree 80709
Philippine Nightshade 367706
Philippine Padauk 320340
Philippine Paphiopedilum 282886
Philippine Pennywort 200263
Philippine Pepper 300479
Philippine Persimmon 132351
Philippine Pine 299986
Philippine Pipewort 151369
Philippine Pittosporum 301363
Philippine Platycarpa 369098
Philippine Podocarpus 306512
Philippine Psychotria 319477
Philippine Scurrula 355289
Philippine Spikesedge 143309
Philippine Teak 219963
Philippine Tinomiscium 392224
Philippine Toona 392827
Philippine Violet 48140
Philippine Wild Coffee 319477
Philippine Wildtung 243427
Philippine Yaccatree 306512
Philippine Yellow Spathoglottis 370395
Philippines Scrambling Bamboo 82449
Philippines Sweet-shoot Bamboo 125508
Phillip Island Hibiscus 194938
Phillyrea 294761
Philodendron 294785,294798,294810
Philodendron-leaf Begonia 50385
Philodendronlike Begonia 50159
Philoxerus 294877
Philtrewort 91557
Philydrum 294892
Philydrum Family 294886
Phinaea 294905
Phlogacanthus 295018
Phlomis 295049
Phlox 295239,295288,295324
Phlox Family 307186
Phoebe 295344
Phoebe Scurrula 355322
Phoenicia Juniper 213849
Phoenician Juniper 213849
Phoenician Machilus 240671
Phoenician Rose 336850
Phoenicophorium 295437
Phoenix Haworthia 186430
Phoenix Machilus 240671
Phoenix Maple 2963
Phoenix Tree 166612,166627
Phoenix Violet 409880
Phoenix-tree 166612,166627
Phoenixtree Maple 2963
Phoenixtree Poplar 311441
Pholidocarpus 295501
Pholidota 295510
Photinia 295609,295773
Photinia-leaved Barberry 52032
Phragmipedium Orchid 295878
Phragmites 295888
Phreatia 295952
Phrynium 296018
Phtheirospermum 296061
Phu 404316
Phyla 296114
Phylacium 296129,296130
Phyllanthodendron 296442
Phyllanthus-like Mayten 246908
Phyllocalyx Sweetleaf 381291
Phyllodium 297007
Phylloid-spiked Rabdosia 209798
Phyllophyton 245687
Phyllostachys 297188
Physaliastrum 297624
Physalis 297640,297643,297645,297711
Physalislike Spicegrass 239788
Physic Nut 212127,212186,212202,214308
Physicnut 212127
Physic-nut 113039
Physochlaina 297875
Physospermopsis 297942
Phytocrenum 298082
Phytolacca 298103
Pia 382923
Pia Family 382934
Piairie Trillium 397599
Pianet 280231
Pianma Arisaema 33456
Pianma Arrowbamboo 162638
Pianma Asiabell 98393
Pianma Clematis 95234
Pianma Didymocarpus 129945
Pianma Fargesia 162638
Pianma Spapweed 205222

Pianma Swertia 380313
Piano Rose 280231
Piano-roses 280231
Pianowood 121770
Piasecki Pricklyash 417298
Piasecki Prickly-ash 417298
Piasezk Rehmannia 327454
Piasezky Grape 411849
Piassaba Attalea 44807
Piassaba Palm 225039
Pican 77874
Picart's Poplar 311262
Piccabeen 31013
Piccabeen Bangalow Palm 31015
Piccabeen Palm 31015
Piccol's Gymnaster 256308
Pichi 161897
Pichincha Barberry 52034
Pichon 135215
Pichurim Bean 263059
Pick Cheese 243840
Pick Needle 153869
Pick Purse 72038
Pickaback Plant 392668
Pick-a-back Plant 392668
Pick-a-cheese 243840
Pickerel Rush 310993
Pickerel Weed 310990,310993,310996,312190
Pickerelweed 94177,310990,310993
Pickerel-weed 310990,310993,325580
Pickerelweed Family 311019
Pickerel-weed Family 311019
Pick-folly 72934
Pickle Fruit 45724
Pickle Grass 113316
Pickle Plant 358689
Pick-needle 153869
Pickpocket 14964,72038,72934,116736,235334,316127,326287,356468,370522,374916,407287
Pick-purse 72038,370522
Picks 316819
Pick-your-mother's-eyes-out 407072
Pick-your-mother's-heart-out 72038
Picmoc 126689
Picnic Thistle 91713
Picotee 127635
Picotee Lilac 382361
Picria 298525
Picris Broomrape 274958
Picrorhiza 298668
Picrotoxin 21459
Pictou Disease 359279
Picturesque Nautilocalyx 262901
Piddly-bed 384714
Pie Apple 300223
Pie Cherry 83361
Pie Cress 29341
Pie Marker 1000
Pie Plant 329386,329388
Pie Print 1000
Pie Vanille Heliotrop 190559
Piecestone 14166
Pie-cherry 83361
Pie-cress 29341
Piedmont Azalea 330322,331248,331546
Piedmont Bedstraw 113137
Piedmont Butterfly-pea 81878
Piedmont Gayfeather 228548
Piedmont Primrose Willow 238163
Piedmont Primula 314789
Piedmont Ragwort 279923
Piedmont Rhododendron 331248
Piedmont Saxifrage 349763
Piemarker 1000
Pie-Marker 1000
Pie-nanny 280231
Pie-plant 329386,329388
Pie-print 1000
Pierce Rhododendron 331487
Pierce-snow 169719
Pierce-stone 111347
Pieris 22418,298703
Pierot Dogtongueweed 385914
Pierot Fimbristylis 166435
Pierot Fluttergrass 166435
Pierot Willow 343890
Pierot's Tephroseris 385914
Pierott Clematis 95201
Pierre Argyreia 32651
Pierre Bridelia 60208
Pierre Chastetree 411409
Pierre Chaste-tree 411409
Pierre Dacrydium 121105
Pierre Garuga 171567
Pierre Helixanthera 190852
Pierre Hopea 198182
Pierre Myxopyrum 261897
Pierre Sauropus 348058
Pierre Scyphellandra 334705
Pierre Whiteheadtree 171567
Pig Balsam 387130
Pig Daisy 26768,321554
Pig Dock 9832
Pig Face 77446
Pig Grass 308816,309570
Pig Haw 109857
Pig Laurel 215387
Pig Leaves 91916,271666
Pig Lily 37015
Pig Nut 77894,102727,197114,212127,364450
Pig Root 365678
Pig Rose 282685,336419
Pig Rush 308816
Pig Squeak 52514
Pig Thigh 198745
Pig Violet 410493
Pig Weed 18655,18810
Pig's Ailes 109857
Pig's Bubbles 192358
Pig's Chops 28617,231183
Pig's Cole 192358
Pig's Cress 26768,221787
Pig's Daisy 26768,321554
Pig's Dichrocephala 129068
Pig's Ear 107976,107985,356468
Pig's Eye 72934
Pig's Eyes 72934
Pig's Flop 192358
Pig's Flower 26768
Pig's Food 192358
Pig's Foot 237539
Pig's Grease 407030
Pig's Hales 109857
Pig's Haws 109857
Pig's Heels 109857
Pig's Hells 109857
Pig's Isles 109857
Pig's Mouth 28617,231183
Pig's Parsley 123141,392992
Pig's Parsnip 192358
Pig's Pear 109857
Pig's Pears 109857
Pig's Pettitoes 237539
Pig's Rhubarb 31051,332086
Pig's Rose 336419
Pig's Snout 28617
Pig-a-back Plant 392667,392668
Pigafettia 298830
Pigales 109857
Pig-ales 109857
Pigall 109857
Pigaul 109857
Pig-berry 109857
Pigbutt Three-awn 33830
Pigeon Bean 408394
Pigeon Berry 104947,139062,298094,334897
Pigeon Grape 411531,411615
Pigeon Grass 361680,361877
Pigeon Orchid 125087
Pigeon Pea 65146,218761
Pigeon Vetch 408423
Pigeon's Eye 72934
Pigeon's Eyes 72934
Pigeon's Foot 174739,342859
Pigeon's Grass 405872
Pigeon's Knee 73207
Pigeon's Meat 405872

Pigeonberry 139062,298094,298119,334897
Pigeon-berry 139062,139065,235874,298094,328630,334897
Pigeon-foot 342859
Pigeon-grass 361680
Pigeonpea 65143,65146
Pigeon-pea 65143
Pigeon-plum 97862
Pigeons 3462
Pigeonwings 97184
Pigeon-wings 97184,97195
Pigeonwood 187412
Pigeon-wood 181531
Pigface 77441,134394,134395
Piggle 315082
Piggyback Begonia 49926
Piggy-back Begonia 49926
Piggyback Plant 392667,392668
Piggy-back Plant 392668
Piggy-wiggy 28617
Pighau 109857
Pig-in-the-hedge 316819
Pigmentary Garcinia 171167
Pigmy Date Palm 295485
Pigmy Rush 213409
Pignon Pignoli 300151
Pignum Hickory 77912
Pignut 63475,77889,77894,102720,102727
Pig-nut 77889
Pignut Hickory 77889,77894,77926,77932
Pig-nut Hickory 77894
Pignut Palm 201356
Pig-o'-the-wall 28617
Pigshell 109857
Pigsy Pear 109857
Pigsy-pears 109857
Pigtail 170193
Pigtail Anthurium 28119
Pigtail Plant 28119
Pigtoes 237539
Pigweed 8826,18652,18655,18810,18848,86891,86901,86962,87000,87007,87029,87054,87107,87112,87120,87133,87142,87145,87147,87164,87171,87203,192358,308816,311890,311899,381035
Pigweed Family 18646
Pikeleaf Gentian 150763
Pikeleaf Ungeargrass 36673
Pikpoktsai 60225,60230
Pikpoktsai Bridelia 60230
Pileate Morningglory 208092
Pileate Raspberry 339050
Pileostegia 299109
Piles 198376
Pilewort 148153,164469,325820
Pilgrim Rose 336304
Pilgrim's Gourd 219843
Pilgrim's Tree 327095
Pili Nut 70988,71013
Piliferous Agapetes 10354
Piliferous Elytrigia 144680
Piliferous Michelia 252945
Piliferous Psychotria 319898
Piliol 391397
Pill Flower 279345
Pill Nettle 402992
Pill Nut 70998
Pill Sedge 75788
Pill's Heel 109857
Pillans Huernia 199049
Pillar Apple 243729
Pillarwood 78686
Pillerds 198376
Pill-headed Sedge 75788
Pillose Muli Monkshood 5485
Pillowlike Rockfoil 349226
Pillpod Sand-mat 159069
Pillus 45448
Pilocarpus 299167
Pilose Actinidia 6680
Pilose Actinodaphne 6811
Pilose Alternateleaf Solomonseal 308500
Pilose Armour Persimmon 132053
Pilose Asiabell 98395
Pilose Basil 268445
Pilose Beggarticks 54059
Pilose Bentgrass 12252
Pilose Bigleaf Bennettiodendron 51036
Pilose Biondia 54527
Pilose Bluebellflower 115386
Pilose Broadleaf Sedge 76268
Pilose Bushclover 226924
Pilose Byttneria 64471
Pilose Calogyne 67689
Pilose Cheng Blueberry 403771
Pilose Chinese Chickweed 374901
Pilose Chinese Sweetleaf 381137
Pilose Chinese Wendlandia 413845
Pilose Cinquefoil 312886
Pilose Combretum 100702
Pilose Currant 334155
Pilose Cypressgrass 119378
Pilose Deqin Monkshood 5464
Pilose Deutzia 127041
Pilose Elsholtzia 144077
Pilose Euphorbia 158659
Pilose Fatoua 162871
Pilose Ford's Fordiophyton 167302
Pilose Fourlobed Skullcap 355707
Pilose Galingale 119378
Pilose Germander 388181
Pilose Goldsaxifrage 90429
Pilose Halfribpurse 191530
Pilose Indianmulberry 258904
Pilose Japanese Abelia 182
Pilose Jasminanthes 211697
Pilose Laxcymose Larkspur 124337
Pilose Leaf Actinodaphne 6811
Pilose Leptodermis 226113
Pilose Leucas 227687
Pilose Longraceme Currant 334075
Pilose Maple 3423
Pilose Marsh Peavine 222803
Pilose Martagon Lily 229929
Pilose Meliosma 249417
Pilose Membranaceous Dendrocalamus 125492
Pilose Miyabe Spiraea 372014
Pilose Petrov Synstemon 382012
Pilose Pittosporumleaf Bittersweet 367503
Pilose Pittosporumleaf Nightshade 367503
Pilose Pleurospermum 304830
Pilose Poplar 311423
Pilose Pricklyash 417300
Pilose Prickly-ash 417300
Pilose Protected Rockvine 387826
Pilose Red Ninenode 319811
Pilose Red Psychotria 319811
Pilose Reeves Spiraea 371875
Pilose Seatung Nightshade 367503
Pilose Seguin Colquhounia 100045
Pilose Septemlobate Kalopanax 215451
Pilose Showy Hemilophia 191530
Pilose Sicklelobe 409008
Pilose Silverleaf Gentian 112854
Pilose Snakehead Dipoma 133415
Pilose St. Johnsedleaf Skullcap 355485
Pilose Stephanotis 211697
Pilose Sundrops 269489
Pilose Sweetleaf 381363
Pilose Syncalathium 381650
Pilose Synstemon 382012
Pilose Teasel 133502
Pilose Threevein Aster 39984
Pilose Triumfetta 399296
Pilose Vetch 408541
Pilose Watersnake 162871
Pilose White Chinaure 56384
Pilose Windmill 100702
Pilose Woodrush 238681
Pilose Xikang Rose 336953
Pilose Yellow Dragonbamboo 125492
Pilose-antlered Cyclobalanopsis 116204
Pilosecalyx Acaulescent Pegaeophyton 287958
Pilosefruit Anemone 24034
Piloseleaf Desmos 126720
Piloseleaf Eagleclaw 35045
Piloseleaf Girald Acanthopanax 143702
Piloseleaf Tailgrape 35045
Pilose-leaved Actinodaphne 6811

Pilose-leaved Barberry 52035
Pilose-leaved Desmos 126720
Pilose-little-leaf Willow 343891
Pilosemarginate Shortwoolled Goldwaist 90398
Pilose-nernes Grape 411855
Pilosestachys Woodbetony 287518
Pilose-styled Elaeagnus 142149
Pilose-sutural Delavay Willow 343290
Pilosocereus 299255
Pilostemon 299277
Pilostemonlike Jurinea 214148
Pilostyle Elaeagnus 142149
Pilostyle Laurel 122473
Pilot-weed 364290
Pilous Byttneria 64471
Pilous Zhongdian Monkshood 5487
Pilular Adina 8192
Pilushan False-nettle 56258
Pimenta 299325
Pimentary 249504
Pimento 72068,72070,72075,299321,299325
Pimento Royal 261162
Pimmerose 315101
Pimo Rock Palm 59103
Pimpernel 21335,21339,345881
Pimpernel Rose 336851
Pimpernel-like Speedwell 407003
Pimpinell 345866
Pimpinella 299342,299351
Pimpon 376144
Pimrose 315101
Pin Apple 300223
Pin Bur 170193
Pin Cherry 83274
Pin Clover 153767
Pin Grass 153767
Pin Oak 323859,324232,324262,324282
Pinaleno Mountains Rubberweed 201301
Pinan Brome 60889
Pinang 31680
Pinang Palm 299649
Pinang Tanoak 233113
Pinang Utan 68617
Pinanga 299649
Pinangapalm 299649
Pinanga-palm 299649
Pinaster 300146
Pin-burr 170193
Pinbush Wattle 1100
Pince Pine 300148
Pince Pinyon Pine 300148
Pinch-me-tight 273531
Pinchot Juniper 213853
Pinchuan Clematis 95236
Pinchweed 309570
Pinckneya 299701,299703
Pin-clover 153767
Pincushion 28200,34536,81020,84478,106117,133478,157429,216753,228029,228057,237539,265358,309570,350099,379660,407989
Pincushion Cactus 107004,107076,244001,244037,244081,244096
Pincushion Coneflower 210239
Pincushion Euphorbia 159681
Pincushion Flower 84495,216783,228029,350086,350099,350119,350125
Pincushion Hakea 184615
Pincushion Plant 107757
Pincushion Shrub 157429
Pincushion Tree 62064,184586,184615,407989
Pincushiontree 407989
Pincution 243994
Pincution Cactus 243994
Pindar 30498
Pinder 30498
Pindilic Barberry 52036
Pindo Palm 31705,31707,64161,98136
Pindrow Fir 439,480
Pine 299780
Pine Apple 298191,300223
Pine Barren Thoroughwort 158284
Pine Colonnaire 30838
Pine Family 299594
Pine Geranium 288192
Pine Goldenweed 150376
Pine Hill Flannel Bush 168217
Pine Lily 229797
Pine Mountain Grevillea 180606
Pine Spurge 158857
Pine Strawberry 167597,167610
Pine Sugar Protea 315732
Pine Taxillus 385189
Pineapple 21472,21479
Pine-apple 298191,300223
Pineapple Broom 120939
Pineapple Cactus 107063,107070,354311
Pineapple Family 60557
Pineapple Flower 156012,156015,156022
Pineapple Germander 388100
Pineapple Ginger 114855
Pineapple Guava 2751,163117,318737
Pineapple Lily 156018
Pineapple Mayweed 246375
Pineapple Mint 250439,250440,250458
Pineapple Sage 345016,345360
Pineapple Scented Sage 345016
Pineapple Shrub 68313
Pineapple Weed 246375
Pineapple-flower 156032
Pineapple-scented Sage 345016
Pineappleweed 246375
Pineapple-weed 246375
Pine-barren Death Camas 417887
Pine-barren Gentian 173276
Pine-barren Goldenrod 368097
Pine-barren Sandwort 31801,255451
Pine-barren Stitchwort 255451
Pinebarren Whitetop Aster 268685
Pinebush 150376
Pinecone Cactus 288618
Pine-cone Epidendrum 146484
Pine-drops 320853
Pineforest Azalea 331491
Pineforest Begonia 50163
Pineforest Kobresia 217248
Pineforest Saussurea 348651
Pineforest Sinocarum 364890
Pineforest Windhairdaisy 348651
Pineforest Woodbetony 287519
Pinehill Blue-stem 351312
Pine-hyacinth 94757
Pineland Chaffhead 77227,233823
Pineland Conehead 130321
Pineland False Sunflower 295343
Pineland Goldenaster 90229
Pineland Nailwort 284883
Pineland Rayless-goldenrod 54275
Pineland Scaly-pink 377074
Pineland Snakeherb 139438
Pineland Spapweed 205226
Pineland Sunflower 189041
Pineland Wild Buckwheat 152289
Pineleaf Asiabell 98327
Pineleaf Buttercup 326303
Pineleaf Eargrass 187646
Pineleaf Geebung 292033
Pine-leaf Geebung 292033
Pineleaf Hedyotis 187646
Pineleaf Knotweed 308695
Pineleaf Ladybell 7765
Pineleaf Milkweed 37991
Pineleaf Seseli 361603
Pine-leaved Bottlebrush 67287
Pine-leaved Goldenaster 301512
Pine-leaved Lily 229991
Pine-leaved Mahonia 242615
Pinellia 299717,299722
Pinemat 31125
Pinemat Manzanita 31125
Pine-mat Manzanita 31125
Pine-needle Gayfeather 228542
Pineneedle Milkweed 37991
Pineneedle Toadflax 231077
Pine-pink 55549
Pinesap 202767,258006
Pine-sap 202767,258044
Pine-strawberry 167610

Pineweed 201895
Pine-weed 201895
Pine-wodd 411782
Pinewood Cranesbill 174815
Pinewood Knotweed 309599
Pinewood Syzygium 382246
Pine-woods Clematis 94757
Pinewoods Coneflower 339522
Pinewoods Milkweed 37966
Pinewoods Oxeye 190504
Pinewoods Pussytoes 26419
Pinewoods Rhododendron 331491
Pinewoods Sedge 119014
Pinewoods-lily 17580
Piney Lilac 382246
Piney Tree 405154
Piney Varnish 405154
Pinfa Clethra 96505
Pinfa Raspberry 339468
Ping Juniper 213854
Ping Machilus 240673
Pingba Maple 3598
Pingba Mazus 246970
Pingba Spapweed 204973
Pingbian Adinandra 8267
Pingbian Ainsliaea 12703
Pingbian Anodendron 25932
Pingbian Arisaema 33458
Pingbian Azalea 331492
Pingbian Barberry 52037
Pingbian Begonia 50164
Pingbian Burmannia 63991
Pingbian Caper 71774
Pingbian Cheirostylis 86715
Pingbian Childgrass 340369
Pingbian China-laurel 28313
Pingbian Cinnamon 91407
Pingbian Cyclobalanopsis 116178
Pingbian Cypressgrass 118759
Pingbian Distylium 134942
Pingbian Dualbutterfly 398292
Pingbian Eelvine 25932
Pingbian Ehretia 141690
Pingbian Elaeocarpus 142393
Pingbian Evergreen Chinkapin 79000
Pingbian Evergreenchinkapin 79000
Pingbian Galingale 118759
Pingbian Giantarum 20127
Pingbian Ginsen 280814
Pingbian Goldlineorchis 25995
Pingbian Greenstar 307521
Pingbian Hairy-petiole Maple 3488
Pingbian Hemiboea 191380
Pingbian Hemsleia 191955
Pingbian Herbcoral 346535
Pingbian Hopea 198183
Pingbian Jadeleaf and Goldenflower 260480
Pingbian Lilyturf 272121
Pingbian Mosquitoman 134942
Pingbian Mussaenda 260480
Pingbian Newlitse 264084
Pingbian Oak 116178
Pingbian Oilfruitcamphor 381798
Pingbian Ophiorrhiza 272279
Pingbian Ormosia 274427
Pingbian Pepper 300481
Pingbian Qinggang 116178
Pingbian Rabbiten-wind 12703
Pingbian Rattanpalm 65818
Pingbian Rehdertree 327411
Pingbian Rhododendron 331492
Pingbian Rungia 340369
Pingbian Schefflera 350763
Pingbian Skullcap 355680
Pingbian Slatepentree 322609
Pingbian Snowgall 191955
Pingbian Sterculia 376158
Pingbian Syndiclis 381798
Pingbian Tanoak 233285
Pingbian Tripterospermum 398292
Pingbian Tupistra 400406
Pingbian Tutcheria 322609
Pingbian Wendlandia 413811
Pingbian Whitepearl 172104
Pingbian Whytockia 414004
Pingbian Wintergreen 172104
Pingchang Evergreen Chinkapin 79003
Pingdong Clematis 94708
Pingdong Copperleaf 1775
Pingdong Copper-leaf 1775
Pingdong Duckbeakgrass 209251
Pingdong Indigo 205764
Pingdong Litse 233832
Pingdong Litsea 233832
Pingdong Pricklyash 417372
Pingdong Rattlebox 112671
Pingdong Starjasmine 393669
Pingfa Briggsia 60282
Pingfa Greenbrier 366521
Pinggu Pinnate Clematis 95240
Pingguo Camellia 69491
Pingguo Tea 69491
Pinghe Holly 204173
Pingjiang Barberry 52320
Pingli Willow 343892
Pingnan Chinacane 364653
Pingnan Elaeagnus 142150
Pingnan Holly 204174
Pingnan Poorspikebamboo 270429
Pingpien Galingale 118759
Pingpien Hairy-petiole Maple 3488
Ping-pong 376144
Ping-pong Ball Cactus 147422
Pin-grass 153767
Pingshan Barberry 52039
Pingshan Beautyberry 66899
Pingshan Mosquitotrap 117654
Pingshan Purplepearl 66899
Pingshan Swallowwort 117654
Pingtung Copperleaf 1775
Pingtung Litsea 233832
Pingue 201325
Pingue Rubberweed 201325
Pinguin 60553
Pinguin Fibre 60553
Pingwing 201325
Pingwu Barberry 52040
Pingwu Beech 162366
Pingwu Corydalis 106273
Pingwu Deutzia 126998
Pingwu Fritillary 168563
Pingwu Vineclethra 95534
Pinha 25898
Pinheads 81768
Pini Jambu 382660
Pinjane 418392
Pink 11947,72934,79963,127573,127779,127793,292656
Pink ‘Forsythia’ 211
Pink Acacia 334965
Pink Allemande 113823
Pink and White Everlasting 190819
Pink and Yellow Corydalis 106430
Pink Arum 417118
Pink Azalea 330495,331352,331408
Pink Baby's Breath 383324
Pink Baby-breath 383324
Pink Ball Dombeya 135804
Pink Beauty 127607
Pink Beauty Bush 217788
Pink Bird's Eye 174877
Pink Bird's Eyes 174877
Pink Bird-lime Flower 210917
Pink Bloodwood 106807
Pink Bogbutton 354442
Pink Breath of Heaven 99467
Pink Bridewort Spirea 372075
Pink Broom 267183
Pink Butterfly Orchid 273569
Pink Butterfly-orchid 273569
Pink Buttons 218390
Pink Calla 417118
Pink Camellia 69532
Pink Campion 363958
Pink Candle 80381
Pink Cassia 78425
Pink Cedar 5983
Pink Cheirostylis 86692
Pink Cherokee Rose 336676
Pink Chirita 87955
Pink Cinquefoil 312816

Pink Corydalis 106430
Pink Cranesbill 174542
Pink Dandelion 110854,384488
Pink Deutzia 127061,127065
Pink Diosma 99467
Pink Dombeya 135795
Pink Easter Lily Cactus 140860
Pink Escallonia 155138
Pink Everlasting 26579
Pink Fairy Duster 66669
Pink Family 77980
Pink Fawn-lily 154943
Pink Fetterbush 239385
Pink Firecracker-flower Azalea 330602
Pink Flame Tree 58339
Pink Flax 231948
Pink Floss-silk Tree 80121
Pink Fritillary 168520
Pink Frog-orchid 264765
Pink Funnel-lily 23380
Pink Gaura 172201
Pink Hair Grass 259642
Pink Hawk's-beard 110985
Pink Heath 146071
Pink Hyacinth 199583
Pink Hyssop 203102
Pink Ice Plant 123854
Pink Japanese Snowbell 379377
Pink Jasmine 211963
Pink Kinostemon 216455
Pink Knotweed 309542
Pink Kopsia 217869
Pink Kunzea 218390
Pink Lady's Slipper 120290
Pink Lady's-slipper 120290
Pink Lady's-tresses 372247
Pink Largeleaf Hydrangea 199994
Pink Lime-berry 94191
Pink Masterwort 43309
Pink Meadowrue 388650
Pink Melaleuca 248108
Pink Milkwort 308118
Pink Mimosa 13578,254999
Pink Mink 315900
Pink Minor 126834
Pink Mocassin 120290
Pink Moccasin Flower 120290
Pink Moccasin-flower 120290
Pink Monkeytlower 255201
Pink Mountain Berry 227968
Pink Mountain Silverbell 184747
Pink Mountain-heather 297024
Pink Muhly 259642
Pink Needle 153767,153869
Pink of My John 410677
Pink Onion 15666
Pink Opal Water Lily 267634
Pink Oxalis 277849,278064
Pink Panda 167593
Pink Paper Daisy 190819
Pink Pasqueflower 24109
Pink Pearweed 239824
Pink Phlox 295254
Pink Phyllanthodendron 296460
Pink Pinafore 174877
Pink Pine 121090,184915
Pink Plumepoppy 240765
Pink Pokers Grevillea 180631
Pink Polka-dot Plant 202595
Pink Polygala 308275
Pink Poppy Mallow 67125
Pink Poui 382723
Pink Powder Puff 66673
Pink Pricklythrift 2224
Pink Primrose 315101
Pink Princess Escallonia 155131
Pink Prosperity Rose 336305
Pink Protea 315970
Pink Purslane 94297,94362
Pink Pussy Willow 343151
Pink Pussytoes 26579
Pink Pyrola 322787
Pink Queen 95796
Pink Rabbit-tobacco 317767
Pink Rabdosia 324722
Pink Reineckea 327525
Pink Reineckia 327525
Pink Rhipsalidopsis 329681
Pink Rice Flower 299307
Pink Rock Orchid 125206
Pink Rockrose 93154
Pink Rockrose 93212
Pink Rugose Rose 336908
Pink Sand Verbena 710
Pink Sand-verbena 710
Pink Saxifrage Tunicflower 400358
Pink Shepherd's-purse 72059
Pink Shin-leaf 322787
Pink Shower 78316,78336
Pink Siris 13578
Pink Siris Persian Acacia 13578
Pink Slender-thoroughwort 166837
Pink Smartweed 291696
Pink Snakeweed 373510
Pink Snowball 135804
Pink Snowball Tree 135804
Pink Snowberry 380741
Pink Spider Flower 180646
Pink Spike Hakea 184596
Pink Spineflower 89086
Pink Spiraea 371944
Pink Spot Begonia 50300
Pink Spotted Rhododendron 330893
Pink Stonecrop 357282
Pink Streptopus 377934
Pink Sundew 138276
Pink Sunrose 188759
Pink Tea-tree 226503
Pink Tecoma Tree 382722
Pink Three-flower 14992
Pink Tickseed 104577
Pink Tips 67293
Pink Torch Rhododendron 331878
Pink Tree Broom 77056
Pink Trumpet Tree 382715,382718,382723
Pink Trumpet Vine 306706
Pink Trumpet-tree 382723
Pink Trumpetvine 306706
Pink Tulip Tree 416688
Pink Tuliptree 416688
Pink Tulip-tree 416688
Pink Turtlehead 86793,86795
Pink Turtle-head 86793
Pink Velvet Bush 222450
Pink Vine 28422
Pink Walnut 145321
Pink Water Speedwell 407064
Pink Water-speedwell 407064
Pink Wax Flower 153328
Pink Wild Bean 378506
Pink Wild Onion 15765
Pink Wild Pear 135795
Pink Windmills 14992
Pink Wood Sorrel 277776,277787
Pink Woodruff 39411
Pink Woodrush 39411
Pink Woodsorrel 277776,277787
Pink Yun-Zang Azalea 331961
Pink-and-white Powderpuff 66686
Pinkblush Cotoneaster 107593
Pink-bud Rhododendron 331870
Pinkcolor Sandwort 32068
Pinkdactil Everlasting 21674
Pinkdactil Pearleverlasting 21674
Pink-dot 202595
Pink-dot Plant 202595
Pinkeney John 410677
Pink-eye 387914
Pink-eyed John 410677
Pinkflower Autumn-crocus 99347
Pinkflower Bindweed 103037
Pinkflower Burnet 345888
Pink-flower Camellia 69532
Pinkflower Corydalis 106340
Pinkflower Ecdysanthera 402201
Pinkflower Fawnlily 154943
Pinkflower Indigo 205639
Pink-flower Rhododendron 331647
Pinkflower Sandwort 32226
Pink-flowered Dogbane 29468
Pink-flowered Doughwood 249138

Pink-flowered Indigo 205639
Pink-flowered Korkwood 249138
Pink-flowered Urceola 402201
Pink-flowering Ironbark 155744
Pink-flowering Yellow Gum 155625
Pinkflush Rhododendron 331243
Pink-flushed Rhododendron 331243
Pinkfringe 36880
Pink-frtd Pernettya 291368
Pink-fruit Barberry 52068
Pink-head Knotweed 291714,308946
Pinkhead Smartweed 308946
Pinkies 397019
Pinklady 134707
Pinkleaf Chickweed 374834
Pinkleaf False Loosestrife 238207
Pinkleaf Gentian 173327
Pinkleaf Seedbox 238207
Pinklike Broomrape 274985
Pink-needle 153767,153869,350425
Pink-o'-my-John 410677
Pink-of-my-John 410677
Pink-queen 95796
Pinkray Fremont's-gold 382062
Pinkroot 371616
Pink-root 371612,371614
Pinks 127573,127654
Pinkscale Gayfeather 228461
Pinkshell Azalea 332050
Pink-shell Azalea 332050
Pink-shower 78316
Pinkshower Senna 78316
Pink-sorrel 277688
Pink-spotted Rhododendron 330893
Pinkster-flower 331352
Pinkvine 28422
Pinkweed 308816,309542
Pinkwood 121683,129718,156062,156066,
156068
Pinky Anisacanthus 25260
Pin-leaf Seepweed 379547
Pinnacle Primrose 314503
Pinnacle Rockfoil 349143
Pinnacles Wild Buckwheat 152313
Pinnaflower 4986
Pinnaleaf Knotweed 309711
Pinnan Galangal 17738
Pinnate Alishan Sage 345077
Pinnate Beggarticks 54001
Pinnate Biantheralga 184940
Pinnate Boronia 57275
Pinnate Braya 59784
Pinnate Candytuft 203234
Pinnate Chirita 87940
Pinnate Clematis 95238
Pinnate Coral-root 72799
Pinnate Dahlia 121561
Pinnate Dragonhead 137630
Pinnate Ellisiophyllum 143896
Pinnate False Threadleaf 351917
Pinnate Globethistle 140770
Pinnate Groundsel 263510,358746
Pinnate Hop Clover 396854
Pinnate Hopbush 135209
Pinnate Kalanchoe 61572
Pinnate Lavender 223318
Pinnate Lilac 382247
Pinnate Meliosma 249435
Pinnate Nyctocalos 267600
Pinnate Paraglycine 272371
Pinnate Peppergrass 225433
Pinnate Podocarpium 200736
Pinnate Polyscias 310201
Pinnate Pometia 310823
Pinnate Prairie Coneflower 326964
Pinnate Pretty Mountain 273805
Pinnate Pterocypsela 320518
Pinnate Rosewood 121793
Pinnate Serpentroot 354954
Pinnate Signdaisy 143896
Pinnate Sinacalia 364522
Pinnate Speedwell 317940
Pinnate Tansy Mustard 126120,126121
Pinnate Twiggy Cinquefoil 313105
Pinnate Windhairdaisy 348487,348652
Pinnatebract Angelica 24431
Pinnatebract Muping Corydalis 105894
Pinnatebract Notopterygium 267157
Pinnateleaf Aralia 30729
Pinnateleaf Corydalis 106274
Pinnateleaf Garuga 171568
Pinnateleaf Groundsel 358828
Pinnate-leaf Japanese Spiraea 371978
Pinnateleaf Lilac 382247
Pinnateleaf Meconopsis 247159
Pinnate-leaf Ophrestia 272371
Pinnateleaf Pellionia 288733
Pinnate-leaf Philodendron 294796
Pinnate-leaf Pond Cypress 385256
Pinnateleaf Primrose 314390
Pinnateleaf Redcarweed 288733
Pinnateleaf Saunders Cinquefoil 312960
Pinnateleaf Siebold Elder 345701
Pinnateleaf Whiteheadtree 171568
Pinnate-leaved Garuga 171568
Pinnate-leavedf Lilac 382247
Pinnately Azalea 330046
Pinnately Divided Sinacalia 364522
Pinnately Veined Rhododendron 330046
Pinnatenerve Christmasbush 14213
Pinnate-nerve Newlitse 264085
Pinnatenerve Trema 394650
Pinnatenerve Wildjute 394650
Pinnatenerve Xmas Bush 14213
Pinnate-nerved Newlitse 264085
Pinnate-nerved Trema 394650
Pinnate-sepal Raspberry 339055
Pinnatesepal Rose 336854
Pinnate-sepaled Raspberry 339055
Pinnatifid Chirita 87941
Pinnatifid Christolea 89254
Pinnatifid Colewort 108629
Pinnatifid Conehead 320135
Pinnatifid Conyza 103613
Pinnatifid Cremanthodium 110431
Pinnatifid Delavay Megacarpaea 247780
Pinnatifid Ducklingcelery 113885
Pinnatifid Elatostema 142749
Pinnatifid Globethistle 140770
Pinnatifid Mullein 405749
Pinnatifid Nutantdaisy 110431
Pinnatifid Parrya 285006
Pinnatifid Philodendron 294832
Pinnatifid Plateaucress 89254
Pinnatifid Primrose 314805
Pinnatifid Rhodiola 329923
Pinnatifid Sage 344911
Pinnatifid Synotis 381957
Pinnatifid Youngia 416461
Pinnatileaf Meliosma 249435
Pinnatilobate Isometrum 210189
Pinnatipartite Primrose 314491
Pinnatipartite Windflower 24070
Pinnatisect Cremanthodium 110432
Pinnatisect Nutantdaisy 110432
Pino Barda Pine 300048
Pino Chino 300033
Pino Prieto 300035
Pino Real 300035
Pinon 300079
Pinon Strangleroot 275050
Pinpatch 409339
Pin-pointed Clover 396926
Pins-and-needles 216753,349044,350425,
374916,401388
Pintelwort 37015,159222
Pinto Bean 294056
Pinto Beans 294056
Pintorum Knotweed 309599
Pintwater Rabbitbrush 90494
Pinweed 188868,223696,223699,223703,
223704,223706
Pinwheel Aeonium 9047
Pinwheel Flower 154162
Pinwheelflower 154162
Pinxter Flower 330322,331352
Pinxterbloom 331352
Pinxterbloom Azalea 331352
Pinxterflower 331352
Piny 280231
Pinyon 299851,299926

Pinyon Pine 299926
Pinyon Spineflower 89132
Pinyuan Bamboo 297388
Piony 280231
Pipal 165553
Pipal Tree 165553
Pipal Tree of the Hindoos 165553
Pipal-tree 165553
Pipe Privit 382329
Pipe Tanoak 233153
Pipe Tree 382329
Pipe Vine 34337,34606
Pipe Wort 151232
Pipebract Rattanpalm 65792
Pipe-privit 382329
Piper Mahonia 242619
Piper's Anemone 23995
Piper's Fleabane 150873
Piper's Wild Buckwheat 152076
Piper's Wood Rush 238684
Piperomia 290286
Pipestem 95071
Pipe-stem 95071
Pipe-tree 78300,382329
Pipevine 34356
Pipe-vine 34337,34356
Pipewort 142554,151208,151232,151257
Pipewort Family 151206
Pipili 138482
Pipinola 356352
Pippalatree 165553
Pipperidge Bush 52322
Pipple 311208,311537
Pipple Hogfennel 292973
Pipple Rush 213355
Piprige Piprage 52322
Pips 315082
Pipsissewa 87480,87498
Piptadenia 300607
Piptanthus 300660
Pipturus 300833,300835
Pipul 165553
Piqui 77943,77948
Pirageia 21889
Piratebush 61929
Pirri-pirri Bur 1752
Pirripirri-bur 1739
Pirri-pirri-bur 1752
Piscushion-flower 350086
Pisgie-flower 374916
Pishamoolag 384714
Pishan Leymus 228376
Piskies 192358,374916
Pismire 384714
Pisonia 300941
Pissabed 325689
Pissibed 384714
Pissimire 384714
Pisspot 68713
Pistache 300974
Pistache Tree 300974
Pistache-leaf Ash 168067
Pistache-leaf Pricklyash 417290
Pistachia Gall 300994
Pistachio 300974,301005
Pistachio Galls 301002
Pistachio Nut 301005
Pistavhe-leaved Ash 168084
Pistol 199465
Pistol Bush 139211
Pistol Plant 298974
Pita 169242
Pita Fibre 8571
Pitahaya 77077
Pitahaya Anaranjada 2147
Pitahaya De Queretaro 375531
Pitahaya Dulce 375534
Pitahaya Real 357748
Pitahayita 244087
Pitajaya 200711
Pitamoreal 146143
Pitanga 156118,156385
Pitanja Cherry 156385
Pitard Camellia 69494
Pitard Heraldtrumpet 49364
Pitard Herald-trumpet 49364
Pitard Qingmingflower 49364
Pitay Cinchona 91085
Pitaya 140209,140248,200707,272775,375516,375533
Pitaya Agria 375524
Pitaya Colorada 375529
Pitaya De Cerro 134232
Pitaya Hedgehog Cereus 140209
Pitayita 244229
Pitayo 375530
Pitayo De Mayo 375523
Pitcairnia 301099
Pitch Apple 97262
Pitch Pine 299876,300128,300153,300181
Pitch Trefofi 319139
Pitcher New-jersey-tea 79908
Pitcher Plant 264829,347144,347156
Pitcher Plants 122758,188554,264829
Pitcher Sage 225082,344887
Pitcher's Clematis 95243
Pitcher's Plants 347144
Pitcher's Stitchwort 255552
Pitcherflower Agapetes 10342
Pitcheri 138742
Pitcher-like Arisaema 33430
Pitcherplant 264829,347156,347165
Pitcher-plant 264829
Pitcherplant Family 347166
Pitchershaped-fruit Photinia 295705
Pitchery 138742
Pitchfork 53755,53797,54048,129525
Pitchfork Paspalum 285400
Piteira Hurcrea 169245
Pith Paper 387432
Pith Paper Plant 387432
Pith Plant 9489
Pithywind 95431
Pitomba 156279
Pitpit 361850
Pitscale Grass 184323
Pitseed Goosefoot 86958
Pit-seed Goosefoot 86958,86962,86966
Pittea Leaf Cryptocarya 113466
Pitted Beardgrass 57579
Pitted Onion 15410
Pitted-leaf Litse 233914
Pittedleaf Spicebush 231341
Pittle-bed 384714
Pittosporopsis 301205
Pittosporum 301207,301248,301398
Pittosporum Family 301203
Pittosporumleaf Bittersweet 367503
Pittosporumleaf Litse 234037
Pittosporumleaf Sweetleaf 381235
Pittosporum-leaved Bittersweet 367503
Pittosporum-leaved Eurya 160582
Pittosporum-leaved Litse 234037
Pittosporumlesf Eurya 160582
Pittosporum-like Cinnamon 91408
Pittosporum-like Nothapodytes 266810
Pituri 138742
Piute Wild Buckwheat 151867
Pivot 229662
Pixie Caps 4511
Pixie Mops 292623
Pixie Orchid 4513
Pixies 374916
Pixy Lily 374916
Pixy Pear 109857,336419
Pixy-grass 152769
Pixy-pear 109857,336419
Plackett 45566
Plage Qinggang 116133
Plaggis 315082
Plagiopetalum 301663
Plagiopteron 301688
Plagiopteron Family 301687
Plagiostachys 301721
Plague Flower 292372
Plain Harlequin Flower 369990
Plain Treasureflower 31160,31162
Plain-green Leaf Pleioblastus 304109
Plainleaf Pussytoes 26530,26551
Plains Acacia 1080
Plains Beebalm 257183

Plains Cactus 182459,287881
Plains Coreopsis 104598
Plains Cottonwood 311294
Plains Crazyweed 278764
Plains Eryngo 154337
Plains Evening Primrose 68508
Plains Fleabane 150782
Plains Gayfeather 228506
Plains Larkspur 124106
Plains Lazydaisy 29103
Plains Love Grass 147737
Plains Muhly 259651
Plains Oval Sedge 73933
Plains Poplar 311465
Plains Prickly Pear 272950,272972,273014
Plains Pricklypear 273079
Plains Prickly-pear 273014
Plains Puccoon 233708,233709
Plains Ragwort 279918
Plains Sagittaria 342313
Plains Sea-blite 379503
Plains Snake-cotton 168687
Plains Sunflower 188908,189032
Plains Three-awn Grass 33957
Plains Tickseed 104598
Plains Violet 410745
Plains Wallflower 154390
Plains Whorled Milkweed 38076
Plains Wild Indigo 47861,47862
Plains Yellow Primrose 68508
Plains Yucca 416540,416569
Plains Zinnia 418046
Plainsdaisy 128475
Plains-sandhill Ground-cherry 297683
Plaister Clover 249232
Plait-leaf Dewberry 339078
Plakkiebos 108940
Planain Cremanthodium 110384
Planain Nutantdaisy 110384
Planchonella 301760
Plane 3462,302574,302588
Plane Leatherleaf Mahonia 242491
Plane Maple 3425
Plane Tree 302574
Plane Tree Family 302229
Planeleaf Alangium 13376
Planeleaf Ramie 56343
Plane-leaved Alangium 13376
Planer Tree 301801,301804
Planertree 301804
Planetree 302574
Plane-tree 302574
Planetree Family 302229
Plane-tree Family 302229
Planetree Maple 3462
Plane-tree Maple 3462
Planetreeflower Clearweed 299017
Planetreeleaf Falsenettle 56260
Plant of Gluttony 105204
Plantago Family 301821
Plantago Singnalgrass 58139
Plantagoleaf Everlasting 21521
Plantagoleaf Pearleverlasting 21521
Plantain 260198,260253,301832,302034,302068
Plantain Banana 260198,260253
Plantain Calanthe 66031
Plantain False Leopard Bane 136366
Plantain Family 301821
Plantain Goldenweed 323055
Plantain Leaf Sedge 75831
Plantain Leaved Sedge 75831
Plantain Lily 198589,198616,198646
Plantain Pussy-toes 26530,26531,26551
Plantain Shore Weed 234150
Plantain Shoreweed 234150,234151
Plantain Signalgrass 402544
Plantain Thrift 34536,34550
Plantainleaf Ainsliaea 12705
Plantainleaf Primrose 314983
Plantainleaf Pussytoes 26551
Plantainleaf Rabbiten-wind 12705
Plantain-leaved Chickweed 256456
Plantain-leaved Everlasting 26551
Plantain-leaved Leopard's-bane 136366
Plantain-leaved Pondweed 312073
Plantain-leaved Pussy Toes 26551
Plantain-leaved Pussytoes 26551
Plantain-leaved Pussy-toes 26551
Plantain-leaved Sedge 75831
Plantain-leaved Thrift 34501
Plantain-leaved Wood Sedge 75831
Plantainlike Anisochilus 25388
Plantainlily 198589
Plantain-lily 198589
Plantainly-leaved Leek 15306
Plantainly-leaved Onion 15306
Plantainshaped Cremanthodium 110384
Plantaln-leaved Pussy Toe 26551
Plantine 302068
Plantree-leaf Begonia 50166
Platan 302574
Platanthera 302241
Platea 302599
Plateau Arisaema 33359
Plateau Arrowhead 342311
Plateau Bluegrass 305300
Plateau Cinquefoil 312852
Plateau Elsholtzia 144014
Plateau Fortune Spindle-tree 157507
Plateau Meadowrue 388465
Plateau Nettle 402938
Plateau Oak 323926
Plateau Ophiorrhiza 272299
Plateau Platanthera 302321
Plateau Rocktrumpet 241311
Plateau Southstar 33359
Plateau Spunsilksage 360836
Plateau Stephania 375901
Plateau Stonecrop 357049
Plateau Swordflag 208496
Plateau Threeawngrass 33753
Plateau Triawn 33753
Plateaucress 89229
Plateaucress Christolea 89223
Plateau-living Bluegrass 305300
Plateum Carpesium 77172
Platifilament Plateaucress 89270
Plat-petiole Basketvine 9473
Plat-stalked Basketvine 9473
Platte Groundsel 279936
Platte Thistle 91840
Platter Dock 267648
Platter Leaf 97865
Platycarya 302674
Platycraspedum 302792
Platycrater 302795
Platyglossey Bamboo 297391
Platy-stamened Holboellia 197239
Platystemma 302958
Platystemon 302961
Playboy Rose 336306
Playfair's Spicebush 231300
Playful Begonia 50035
Pleasant Azalea 331782
Pleasant Barley 198316
Pleasant Dendrobium 125217
Pleasant Lockhartia 235133
Pleasant Melocactus 249580
Pleasant Nepeta 264867
Pleasant Orange 93806
Pleasant Rhododendron 331782
Pleasant Strophanthus 378401
Pleasant Valley Mariposa-lily 67571
Pleasant-in-sight 72934,363461
Pleasing Barberry 51297
Pleasing King Azalea 331641
Pleasing King Rhododendron 331641
Pleaster Pear 323165
Pleasure Gladiolus 176051
Pleated Cereus 83885
Pleated Gentian 173188
Pleated Leaves 111455
Pleated Snowdrop 169726
Pleat-leaf Iris 192405
Pleat-leaf Knotweed 309866
Plebian Knotweed 309602
Plectocomia 303087
Plectocomiopsis 303096
Pleione 304204,304218
Pleiophyllous Pea-shrub 72311

Pleiophyllous Rabdosia 324856
Pleiosorbus 304329,304330
Pleiospermous Euptelea 160344
Plenioculm Fargesia 162735
Plenty 356468
Pleomele 137478
Plereau Cocklebur 415011
Pleteau Buttercup 326422
Pleuriflower Cotoneaster 107622
Pleurisy Root 38147
Pleurisy-root 38147
Pleurospermum 304752
Pleurostylia 304892
Pleurothallis 304904,304927
Pliant Bamboo 47344
Pliant Cactus 244128
Plicate Spathoglottis 370401
Plicate Woodbetony 287521
Plicatebract Everlasting 21662
Plicate-phyllaries Pearleverlasting 21662
Plotchleaf Arrowwood 408074
Plot's Elm 401592
Plough Share Wattle 1156
Ploughbracted Saltbush 44374
Ploughman's Mignonette 158730
Ploughman's Spider 207087
Ploughman's Spikenard 207087,292353
Ploughman's Spike-nard 207087
Ploughman's Weatherglass 21339
Ploughman's-spikenard 207087
Ploughpoint 401148,401156
Ploughshare Orchid 360642
Ploughshareleaf Violet 410195
Plover Eggs 8427
Plowman's-wort 305083
Pluchea 305072
Plukenet's Sedge 119391
Plukenet's Umbrella Sedge 119391
Plum 316166,316382,316761
Plum Aralia 310237
Plum Juniper 213763,213843
Plum Leaf Azalea 331560
Plum Leaf Viburnum 408061
Plum Leaved Apple 243667
Plum Mango 57847
Plum Nut Palm 201359
Plum Pine 306424
Plum Pudding 146724,248411,363673
Plum Purple 262196
Plum Yew 82496,82507,82523,316084
Plumajillo 3978
Plumas Mountaincrown 274127
Plumbagella 305162
Plumbago 83641,83643,83646,305167,
305172,309570
Plumboy 339124
Plum-bush 71411
Plume Acacia 1466
Plume Albizia 13578,283887
Plume Albizzia 283887
Plume Bleeding Heart 128296
Plume Cedar 113696
Plume Clematis 94901
Plume Cockscomb 80453
Plume Cryptomeria 113696
Plume Feathers 182942
Plume Fern 39195
Plume Grass 148855,255886,341829,394943
Plume Incense Cedar 228649
Plume Knapweed 81232,81443
Plume Leucadendron 227345
Plume Poppy 56023,240763,240765
Plume Tiquilia 392359
Plume-albizia 283887
Plumed Fruticose Polyscias 310201
Plumed Goldenrod 368319
Plumed Indian Polyscias 310201
Plumed Pentapanax 289662
Plumed Polyscias 310201
Plumed Thistle 91697
Plumegrass 128861,128863,148855,148915
Plume-grass 148855,148906
Plumeless Thistle 73278,73281
Plumepoppy 240763,240765
Plume-poppy 240763,240765
Plumeri Carludovica 77044
Plumeria 305206,305225
Plum-fir 316081
Plum-fruited Yew 316084
Plumier Skyflower 139065
Plumleaf Actinidia 6660
Plumleaf Azalea 331560
Plum-leaf Crab 243667
Plumleaf Crabapple 243667
Plumleaf Cratoxylum 110258
Plum-leaf Deutzia 126963
Plumleaf Elatostema 142797
Plumleaf Elm 401599
Plumleaf Eurya 160584
Plum-leaf Helwingia 191169
Plumleaf Kiwifruit 6660
Plumleaf Oxwood 110258
Plumleaf Spiraea 372050
Plumleaf Stairweed 142797
Plum-leaved Apple 243667
Plum-leaved Elm 401599
Plum-leaved Willow 343041
Plum-like Leaves Rhododendron 331560
Plummer's Baccharis 46245
Plummer's Candyleaf 376412
Plummer's Mariposa-lily 67621
Plummer's Siltbush 418480
Plumosa Compacta Juniper 213786
Plumosa Juniper 213657
Plumose Aralia 30729
Plumose False Cypress 85350
Plumose Feathergrass 376927
Plumose Goldenrod 368319
Plumose Lanaria 220790
Plum-pine 306424
Plumrocks 315101
Plumwood 156068,346212
Plumyew 82496
Plum-yew 82496,316084,393077
Plumyew Family 82493
Plum-yew Family 82493
Plunkett Mallee 155550
Pluristem Swertia 380282
Plush Anemone 23767
Plush Plant 140004
Plymouth Pear 323152
Plymouth Thistle 73467
Plyvens 397019
Po Mu 167201
Poa 305274
Poached Egg Flower 230210
Poached Egg Plant 230210
Poached-egg Daisy 261303
Poached-egg Flower 230210
Poached-egg Plant 82038,230210
Poaleaf Pondweed 312126
Poaleaf Primrose 314443
Poaleaf Rockjasmine 23185
Pocan 298094
Pochetfruit Naiheadfruit 179096
Pochote 80116
Pock Berry 298088
Pockberry 298094
Pocket Handker Chief Tree 123245
Pocket Tanoak 233164
Pocketbook Plant 66263
Pocketflower Honeysuckle 236078
Pocket-handkerchief Tree 123244,123245
Pocktorchid 282768
Pockwood Tree 181505
Pocosin Pine 300195
Pod Corn 417434
Pod Grass 397142
Pod Thistle 91713
Podded Mouse Ear 30146
Podder 115008,115031
Podgrass 397142,397165
Pod-grass 350861,397142
Podo 306435
Podoa Family 306337
Podocarileaf Pittosporum 301312
Podocarileaf Seatung 301312
Podocarp Family 306345
Podocarpifolious Photinia 295753
Podocarpifolius Blueberry 403966
Podocarpium 200724,200739

Podocarpus 306395,306469
Podocarpus Family 306345
Podocarpus-leaf Blueberry 403966
Podocarpus-like Bladderwort 403966
Podochilus 306534
Podograss 397142
Podranea 306704
Podunk Ragwort 279922
Poet Daffodil 262438
Poet's Cassia 276870
Poet's Daffodil 262438
Poet's Jasmine 211848,211940
Poet's Jessamine 211940
Poet's Laurel 122124,223203
Poet's Narcissus 262438
Poets Jasmine 211848
Pogge Begonia 50171
Pogonatherum 306828
Pogonflower Sandwort 32142
Pogonia 210336,306847
Pogostemon 306956
Poha 297658
Pohuashan Mountain Mountain-ash 369484
Pohuashan Mountainash 369484
Pohuashan Mountain-ash 369484
Pohutukawa 252613
Poi 48689
Poikilospermum 307040
Poilane Basketvine 9474
Poilane Bridelia 60210
Poilane Cyclobalanopsis 116180
Poilane Evergreen Chinkapin 79005
Poilane Evergreenchinkapin 79005
Poilane Firmiana 166682
Poilane Hypoestes 202597
Poilane Lily 229993
Poilane Oak 116180
Poilane Pycnarrhena 321983
Poilane Yam 131768
Poinciana 64965,123801,123811,307055
Poinsettia 159675,307076
Poinsettia Tree 299702
Point Leaves Rhododendron 331289
Point Reyes Creeper 79929
Point Sublime Wild Buckwheat 152678
Pointed Azalea 331280
Pointed Blue-eyed Grass 365747
Pointed Blue-eyed-grass 365685
Pointed Bulrush 352164
Pointed Cat's Tail 294989
Pointed Leaves Rhododendron 331280
Pointed Tick-trefoil 126366
Pointed Zelkova 417558
Pointedflowervine 334358
Pointed-leaved Tick-trefoil 126366
Pointleaf Indigo 206761
Pointleaf Manzanita 31134
Pointletted Rockfoil 349672
Pointreyes Ceanothus 79929
Pointvetch 278679
Point-vetch 278679
Poiret Barberry 52049
Poiret Pine 300116
Poison Apple 367481
Poison Ash 87736,393491
Poison Bay 204517
Poison Berry 21459,37015,208566,367120,369339,383675
Poison Bulb 111167
Poison Bush 4954,4966,385302
Poison Cup 326340
Poison Daisy 26768
Poison Dart 11332
Poison Devil's-pepper 327080
Poison Dogwood 393491
Poison Elder 332941,393491
Poison Fingers 37015
Poison Flag 208924
Poison Flagroot 208767
Poison Flower 367120
Poison Fool's Parsley 101852
Poison Hemlock 90924,101852
Poison Hog Meat 34198
Poison Ivy 393470
Poison Ivy Oak 332803
Poison Larkspur 124654
Poison Milkweed 38130
Poison Nut 378633,378836
Poison Oak 332570,332803,393453,393468,393470
Poison Parsley 101852
Poison Poppy 282685
Poison Primrose 314717
Poison Rhubarb 292372
Poison Root 37015
Poison Sanicle 345942
Poison Sego 417920
Poison Suckleya 379676
Poison Sumac 332959,393479,393491
Poison Sumach 393491
Poison Tree 4954
Poison Vine 393470
Poison Walnut 113446
Poison Wood 252595
Poisonbulb 111167
Poisonbush 171969
Poisoncelery 90918,90932
Poisondart 11329
Poisongourd 132954
Poisonhemlock 90918,101845,101852
Poison-hemlock 101852
Poisonhemlockleaf Vicatia 408244
Poisoning Berries 61497,367120
Poisonivy 332803
Poison-ivy 393470,393471
Poisonnut 378633,378836
Poison-nut 378633
Poisonnutleaf Greenbrier 366438
Poisonous Buttercup 326340
Poisonous Excoecaria 161680
Poisonous Flueggea 167096
Poisonous Giantarum 20125
Poisonous Lacquer-tree 393470
Poisonous Lettuce 219609
Poisonous Tea Plant 367120
Poisonous Woodnettle 125545
Poisonous Wood-nettle 125545
Poisonous-root Ironweed 406272
Poisonrat 128589,128675
Poisonrat Family 128588
Poisontree 366015
Poisonvetch 41906
Poisonweed 124226
Poisson Primrose 314815
Pojarkova Kudrjaschevia 218281
Pokaka 142336
Poke 298094,298119,405643
Poke Milkweed 37914
Poke Needle 350425
Poke Weed 298094
Pokeberry 298088,298094,298119
Poke-needle 350425
Poker 37015,217055,401112
Poker Alumroot 194417
Poker Plant 217055
Pokeroot 298094
Poke-root 298094
Pokerplant 216923
Poker-plant 217055,401112
Poketlike Cremanthodium 110370
Poketlike Nutantdaisy 110370
Poketlip Dendrobium 124996
Pokeweed 298088,298094,298119,355202,374968
Pokeweed Family 298124
Pokosola 268357
Poland15A Poplar 311156
Polanfd Larch 221929
Polanisia 307141
Polant 315106
Polar Bluegrass 305459
Polar Plant 337180
Polar Willow 343900
Pole Bambusa 47408
Pole Bean 294010
Pole Reed 295888
Pole Windhairdaisy 348221
Polecat Bush 332477,332908
Polecat Geranium 221284
Polecat Tree 204517
Polecat-bush 38327,38328

Polecat-tree 204517
Polecat-weed 381073
Polemonium 307190,307197
Polemonium Family 307186
Polenta 417417
Pole-reed 295888
Poley 250432
Poleye 388184
Poliane Dimorphocalyx 131104
Policeman's Buttons 68189
Policeman's Helmet 5442,204982
Poliman's Helmet 5442,204982
Polion 388184
Poliothyrsis 307291
Polish Ash 167955
Polish Larch 221929
Polish Oak 324278,324335
Polish Wheat 398939
Polished Cinnamon 91363
Polished Kobresia 217237
Polished Leaves Rhododendron 330070
Polished Pieris 298734
Polished Willow 343590
Polka Dot Cactus 272980
Polka Dot Plant 202595
Polka-dot Plant 202495,202595
Polkadot-plant 202595
Pollard-flowers 391706
Pollia 307312
Polluting Geranium 288310
Polly Andréws 315106
Polly Ann 315106
Polly Baker 363461
Polly Nut 78811
Polly Pods 238371
Pollyandice 315106
Polour Peashrub 72312
Polour Pea-shrub 72312
Polstead Cherry 280003
Poltate 367696
Polters 130383
Poltoratzk Larkspur 124502
Poluglandular Willow 343780
Polumorph Dutchmanspipe 34299
Polunin Barberry 52051
Polunin's Cotoneaster 107625
Poly Germander 388181
Polyadelphous Rosewood 121795
Polyandrous Trifoliate Orange 310849
Polyantha Rose 336783
Polyanthous Fissistigma 166684
Polyanthous Jasmine 211963
Polyanthus 314820,315106
Polyanthus Daffodil 262457
Polyanthus Narcissus 262382,262457
Polyanthus Primrose 314820,315079
Polyatha Grandiflora Rose 336859
Polybrached Willow 343906
Polybract Firmiana 166652
Polycarpaea 307642
Polycarpon 307724
Polycarpos Idesia 203422
Polycarpous Fatsia 162895
Polycarpous Newlitse 264087
Polycephalous Adina 252396
Polycephalous Ixeris 210654
Polyclade Rhododendron 331512
Polycnemum 307830
Polydentate Mahonia 242620
Polydentate Raspberry 339088
Polygala 307891,308014
Polygala Knotweed 309608
Polygala-like Indigo 206582
Polygamous Orchardgrass 121264
Polyglandular Eyebright 160286
Polygonum 308690
Polygonum-like Twinballgrass 209114
Poly-hairscale Rattanpalm 65804
Polyleaf Akebia 13225
Polymorph Indian Horseorchis 215347
Polymorphic Angelica 24433
Polymorphic Garliccress 365585
Polymorphic Morningglory 208104
Polymorphous Barberry 52054
Polymorphous Iguanura 203511
Polymorphous Pinanga 299673
Poly-mountain 97060,373608
Polynerve Fig 165479
Poly-nerved Eurya 160583
Poly-nerved Saurauia 347977
Polynesia Bonyberry 276534
Polynesia Ironwood 79161
Polynesian Arrowroot 382923
Polynesian Bonyberry 276534
Polynesian Chestnut 56033,207000
Polynesian Iron Wood 79161
Polynesian Ivory Nut Palm 252633
Polynesian Ohe 351856
Polyosma 310060
Polypetal Cyclea 116031
Polypetaloid Syzygium 382652
Polyphyllous Corydalis 106280
Polyphyllous Onosma 271817
Polyphyllous Whitlowgrass 137190
Polypodium 70119
Polypodium-like Embelia 144786
Polypogon 310102
Polyraphioid Rhododendron 331515
Polyscias 310169
Polyspermius Kadsura 214979
Polyspermous Salacia 342705
Polyspike Falsenettle 56294
Polystachous Knotweed 309616
Polystachous Rabdosia 324847
Polystachya 310309
Polystachyous Aphanamixis 28997
Polystachyus Basilicum 48727
Polystone Holly 204179
Polystone Schefflera 350670
Polystylous Camellia 69518
Polytoca 310699
Polytrichous Cherry 83281
Polytrichous Rhododendron 331518
Polytrichumlike Fimbristylis 166444
Polytrichumlike Fluttergrass 166444
Polyvein Firmiana 166651
Pomarack 382606
Pomarrosa 382582
Pomatocalpa 310787
Pomatosace 310806
Pome Larkspur 124505
Pome Monkshood 5494
Pome Rhododendron 331519
Pome Tanoak 233251
Pomegarnet 321764
Pomegranate 321763,321764
Pomegranate Family 321777
Pomegranate Mistletoe 411082
Pomelo 93314,93579,93690
Pomerac 382606
Pometia 310818
Pomi Spicebush 231412
Pomion 114300
Pomme Blanche 319178
Pomme-de-prairie 287932
Pommereschea 310828
Pompelmous 93579
Pompion 114300
Pompom Rhododendron 330741
Pompon 121539
Pompon Bush 121575
Pompon Cabbage Rose 336481
Pompon Cabbage-rose 336481
Pompon Lily 229995
Pompon Tree 121575
Poncirus 310848
Pond Apple 25852
Pond Apple-tree 25852
Pond Baldcypress 385253,385264
Pond Bald-cypress 385254
Pond Cypress 385254,385264
Pond Dock 339925
Pond Dogwood 82098
Pond Grass 100990
Pond Lily 49818,208771,267730
Pond Litse 233928
Pond Lovegrass 147746
Pond Nut 263270
Pond Nuts 263270
Pond Pine 300195
Pond Plantain 312190

Pond Sedge 138919
Pond Spice 233928
Pond Thatch 341422
Pond Top 341422
Pond Water Starwort 67393
Pond Water-crowfoot 326210
Pond Waterstarwort 67393
Pond Water-starwort 67393
Pond Weed 312028,312090
Pond-apple 25852
Pondcypress 385264
Ponderosa Lemon 93712
Ponderosa Pine 300153
Pond-lily 267627,267730
Pond-lily Begonia 50104
Pond-lily Cactus 147283,266651
Pondoland Palm 212561
Pond-spice 233831
Pondweed 312028,312090,312104
Pondweed Family 312300,417073
Pondweeds 312028
Poneshe Pepper 300485
Pongame Oiltree 254803
Pongamia 310950
Poninac 1219
Ponkan Orange 93717
Ponki Citrus 93698
Ponki Mandarin 93698
Pontefract Root 177893
Pontic Azalea 331161
Pontic Blue Sow-thistle 90839
Pontic Fritillary 168521
Pontic Helleborine 147206
Pontic Oak 324301
Pontic Rhododendron 331520
Pontic Wormwood 36088
Pontine Oak 324301
Pontus Sagebrush 36088
Pony Bee Balm 257184
Pony Foot 128964
Pony Grass 147733,147941
Pony Tail Plant 49343
Pony's Foot 128972
Pony's Tail 302068
Pony's Tails 302068
Ponyfoot 128972
Pony-foot 128959
Pony-tail 49343
Ponytail Palm 49343,49345
Pook-needle 11947,153767,350425
Pool Moss 246761
Poolmat 417045,417053
Poolmat Family 417073
Poolroot 345957
Poon 67860,67873
Poonen Citrus 93728
Poongaoil Pongamia 310962
Poonga-oil Tree 310962
Poor Jan's Leaf 358062
Poor Jane 174877,248312,363673
Poor Knight's Lily 415423
Poor Man's Baccy 400675
Poor Man's Beer 199384
Poor Man's Blanket 405788
Poor Man's Blood 273531
Poor Man's Brush 133478
Poor Man's Cabbage 308893,340178
Poor Man's Flannel 405788
Poor Man's Friend 95431
Poor Man's Geranium 349936
Poor Man's Herb 339887
Poor Man's Lignumvitae 276818
Poor Man's Mustard 14964
Poor Man's Orchid 208944,351343,351344,
351345
Poor Man's Parmacetty 72038
Poor Man's Pepper 225315,225398,225475,
345881,356468,404251
Poor Man's Pepper Grass 225475
Poor Man's Purse 72038
Poor Man's Remedy 404316
Poor Man's Rhubarb 388506
Poor Man's Salve 355061,355202
Poor Man's Tea 407072
Poor Man's Tobacco 400675
Poor Man's Treacle 14964,15698,15876,
388275
Poor Man's Weather Glass 21339
Poor Man's Weatherglass 21339
Poor Oat 45448
Poor Oats 45448
Poor Robert 174877
Poor Robin 174877,248411,268991,363461
Poor Robin's Plantain 150901,196077
Poor Robins Plantain 151054
Poor Sagebrush 35369
Poor White Mahogany 155527
Poor Widow 350099
Pooranther Microtropis 254315
Poorflower Amitostigma 19524
Poorflower Arrowbamboo 162733
Poorflower Astyleorchis 19524
Poorflower Azalea 331205
Poorflower Bulbophyllum 63079
Poorflower Calanthe 65976
Poorflower Clovershrub 70860
Poorflower Crazyweed 279071
Poorflower Epipactis 147121
Poorflower Euonymus 157675
Poorflower Falseoat 398514
Poorflower Glycosmis 177843
Poorflower Ophiorrhiza 272275
Poorflower Pencilwood 139668
Poorflower Pseudosasa 318337
Poorflower Rush 213362
Poorflower Stonebean-orchis 63079
Poorflower Thickshellcassia 113442
Poorflower Yellowleaftree 415145
Poorfruit Azalea 331395
Poorfruit Sagebrush 36020
Poorhair Oblongate Microula 254357
Poorhead Windhairdaisy 348609
Poor-joe 131358
Poor-land Daisy 227533
Poorleaf Azalea 330816
Poorleaf Licorice 177933
Poorleaf Murdannia 260109
Poorleaf Waterbamboo 260109
Poorlower Dendrobium 125295
Poor-man's Mustard 365564
Poorman's Orchid 351343,351345
Poorman's Pepper 225475
Poor-man's Pepperwort 225475
Poor-man's Treacle 15698
Poorman's Weatherglass 21339
Poor-man's Weather-glass 21339
Poor-man's-pepper 225475
Poorscale Azalea 332153
Poorspike Bashanbamboo 37314
Poorspikebamboo 270423,270451
Poorspine Japanese Aralia 30655
Poorstem Coelogyne 98745
Poortassel Gentian 173620
Poortooth Croton 112940
Poortwig Yushanbamboo 416808
Poorvein Microtropis 254318
Pop Ash 167934
Pop Ash Tree 167934
Pop Bells 130383
Pop Bladder 130383
Pop Corn 417417
Pop Dock 130383
Pop Glove 130383
Pop Jack 374916
Pop Saltbush 44467
Pop Vine 297720
Pop-a-dock 130383
Popash 167934
Pop-bells 130383
Pop-bladder 130383
Popcorn 417432
Popcorn Flower 301623
Popcorn Senna 360431
Popcorn Tree 346408
Popcorns 157429
Pop-dock 130383
Pope 282685
Pope's Ode 5676
Pope's Phacelia 293172
Popenoe Begonia 50180
Pop-gun 130383,364193,374916

Popilary 311398
Popillary 311398
Popille 11947
Popinac 1219
Pop-Jack 374916
Poplain 311208
Poplar 311131,311208,311547
Poplar Birch 53586
Poplar Box 155704
Poplar Eucalyptus 155472
Poplar Gum 155472
Poplar Leaves Leucothoe 228179
Poplar Pine 311400
Poplar-leaf Argyreia 32652
Poplarleaf Linden 391826
Poplarleaf Litse 234042
Poplarleaf Monbeig Dogwood 380469
Poplarleaf Saussurea 348666
Poplarleaf Sterculia 376165
Poplarleaf Vineclethra 95536
Poplarleaf Windhairdaisy 348666
Poplar-leaved Lime 391826
Poplar-leaved Linden 391826
Poplar-leaved Litse 234042
Poplar-leaved Rock Rose 93198
Poplar-leaved Rockrose 93194
Poplar-leaves Leucothoe 228179
Poplin 311208
Popolo 367620
Popolo-kikania 366934
Popossum Wood 324662
Popotillo 146224,146230
Popov Brome 60892
Popov Windhairdaisy 348665
Popowia 311043
Popper 130383,364193,374916
Poppet 282685
Poppilery 311398
Popple 11947,282685,311398,311547,342973
Poppy 11947,130383,282496,282519,282685,364193,374916
Poppy Anemone 23767
Poppy Dock 130383
Poppy Family 282751
Poppy Mallow 67124,67128,67129
Poppy of California 335873
Poppy Tree Peony 280307
Poppy-flowered Anemone 23767
Poppy-of-the-dawn 146053
Poppywort 247109
Pops 130383,297720,374916
Populage 68157
Popular Rhododendron 331523
Populose Blueberry 403954
Pop-up 99297
Pop-vine 297720
Porana 131242,131254
Porandra 311649,311651
Porcelain Berry 20348
Porcelain Flower 198827
Porcelain Heath 150214
Porcelain Vine 20280,20348
Porcelainberry 20348
Porcelain-berry 20348
Porcelane 311899
Porch Penstemon 289378
Porcupine Crazyweed 278899
Porcupine Cucumis 114171
Porcupine Euonymus 157332
Porcupine Flower 48295
Porcupine Grass 255890,376917
Porcupine Jointvetch 9556
Porcupine Plant 367289
Porcupine Pod Tree 81817
Porcupine Pricklypear 273017
Porcupine Sedge 74862
Porcupine Spindle-tree 157332
Porcupine Wood 98136
Porcupine-grass 203139
Porcupine-pod Tree 81817
Poreleaf 311715,311724
Porho 165541
Pork and Beans 357111
Porkbloodtree 160741
Porkkidneybean 9852,413975
Pork-wood 181531
Poro Poro 367289
Porolabium 311712
Poro-poro 367289
Porose Boeica 56370
Porpax 311733
Porrect Disa 133895
Porrect Spapweed 205242
Porrect Touch-me-not 205242
Porsild Spruce 298286
Porsild's Arctic Fleabane 150886
Porsild's Starwort 375061
Port Jackson 1553
Port Jackson Cypress 67432
Port Jackson Cypress Pine 67432
Port Jackson Fig 165586
Port Jackson Pine 67432
Port Jackson Willow 1159,1553
Port Macquarie Pine 67422
Port Orford Cedar 85271
Port Orford White Cedar 85271
Port Orford Wild Buckwheat 152326
Port Royal Senna 78436,360444
Port Wine Magnolia 232609,252869
Porter Begonia 50181
Porter Ligusticum 229374
Porter Mugwort 36090
Porter's Aster 380957
Porter's Groundsel 279937
Porter's Joy 207934
Porter's Sulphur Flower 152599
Porterandia 311790
Porterweed 373507
Portia Oil Nut 389946
Portia Oilnut 389946
Portia Tree 389928,389946
Portiatree 389928,389946
Port-Jackson Fig 165586
Portland Arrowroot 37015
Portland Sago 37015
Portland Spurge 159623
Portland Starch 37015
Port-orford-cedar 85271
Portugal Heath 149703
Portugal Laurel 316518
Portuguese Broom 120909,120979
Portuguese Cabbage 59522,59541
Portuguese Cherrylaurel 316518
Portuguese Cypress 114714
Portuguese Daisy 227506
Portuguese Fritillary 168461
Portuguese Heath 149703
Portuguese Laurel 316518
Portuguese Laurel Cherry 316518
Portuguese Laurel-cherry 316518
Portuguese Oak 324134
Portuguese Orange 93765
Portuguese Quince 116546
Portuguese Squill 353001
Portuguese Sundew 138376
Portulaca 311817,311852
Portulacaria 311955
Port-wine Magnolia 252869
Poshte 25884
Posh-te 25884
Posidonia 311969,311971
Possum Banksia 47631
Possum Grape 411567
Possum Haw 203752,407970
Possum Oak 324225
Possum Wood 132466
Possumhaw 203752,407970
Possum-haw 407662,407970,407971
Possumhaw Viburnum 407970
Possum-haw Viburnum 407970
Possumwood 132466,324662
Post Cedar 213620,228639
Post Oak 323931,324244,324443
Postoak Grape 411782
Post-oak Grape 411782
Posumbu Knotweed 309345
Posy 280231
Pot 71218
Pot Ash 8826
Pot Barberry 51461

Pot Herb Mustard 59438
Pot Marigold 66380,66445
Pot Marjoram 274224,274231,274237
Pot Purslane 311890
Potamot 378982
Potanin Ajania 13022
Potanin Barberry 52058
Potanin Birch 53587
Potanin Bushclover 226931
Potanin Buttercup 326250
Potanin Cinquefoil 312897
Potanin Clematis 95249
Potanin Corydalis 106283
Potanin Dandelion 384754
Potanin David Peach 20911
Potanin Falsealyssum 170171
Potanin Helictotrichon 190184
Potanin Indigo 206421
Potanin Iris 208765
Potanin Juniper 213972
Potanin Jurinea 214155
Potanin Ladybell 7789
Potanin Larch 221930
Potanin Larkspur 124506
Potanin Leptodermis 226120
Potanin Lily 230011
Potanin Monkshood 5496
Potanin Nutantdaisy 110443
Potanin Peashrub 72313
Potanin Pea-shrub 72313
Potanin Rhododendron 331528
Potanin Rush 213388
Potanin Sage 345320
Potanin Sealavender 230716
Potanin Spapweed 205243
Potanin Sumac 332775
Potanin Swordflag 208765
Potanin Wildclove 226120
Potanin Woodbetony 287535
Potanin's Cremanthodium 110443
Potaninia 312317
Pot-ash 8826
Potato 367696
Potato Bean 29295,29298
Potato Bush 367549
Potato Creeper 367596
Potato Crisps 367696
Potato Dandelion 218208
Potato Dwarfdandelion 218208
Potato Onion 15165,15167,15169
Potato Orchid 171951
Potato Tree 367146,367749
Potato Vine 208050,367265
Potato Yam 131501,131577
Potatobean 29295,29298
Potato-bean 29298
Potatoes-in-the-dish 158433,159027
Potato-yam 131577
Pot-bellied Palm 100029
Potchouli 306964
Potentil 312322
Potentilla 312322
Potentilla-like Dendranthema 124845
Pother 115008,115031
Potherb Jute 104103
Potherb Onion 15540
Pothoidium 313169
Pothomorphe 313172
Pothos 147338,147349,313186,353132
Pot-marigold 66445
Potmarigold Calendula 66445
Pot-of-gold Lily 229871
Potosi Pinyon 299879
Pots-and-kettles 64345
Potsheath Tanoak 233188
Potsheath Yushanbamboo 416795
Pott's Mammillaria 244196
Pott's Montbretia 111474
Pottage Iterb 59493
Pottinger Greenbrier 366528
Pottinger Heterosmilax 366528
Potts Flower 111474
Potts Waxplant 198885
Pottsia 313224
Pouch Bean 293985
Pouch Flower 66273
Pouch Nemseia 263459
Pouch Stephania 375839
Pouched Nemesia 263459
Pouchlikeflower Rabdosia 209680
Pouchvine 8298
Poulard Wheat 398986
Pounce 386942
Pound Needle 350425
Pound-garnet 321764
Pourret Bentgrass 12262
Pouteria 301777,313325
Pouzolzia 313411
Poverty 271563,307197,342279,370522
Poverty Brome 60985
Poverty Danthonia 122292
Poverty Drop-seed 372870
Poverty Grass 33830,122292,372782,372870
Poverty Oat Grass 122292
Poverty Panic Grass 128502
Poverty Pink 316127
Poverty Purse 72038
Poverty Rush 213507
Poverty Weed 257687
Poverty-grass 198955,302068,372870
Poverty-pink 316127
Poverty-purse 72038
Poverty-weed 227533,248161,329521
Powder Dragonbamboo 125504
Powder Linden 391795
Powder Primrose 314863
Powder Puff 66668
Powder Puff Cactus 244012
Powder Sourbamboo 4594
Powder Stephania 375904
Powder Sterculia 376108
Powder Yushanbamboo 416774
Powdered Bean 314366
Powdered Dendrocalamus 125504
Powdered Frankenia 167813
Powdered Hedge Maple 2840
Powdered Thalia 388385
Powderflower Birthwort 34357
Powderflower Dutchmanspipe 34357
Powderflower Spirea 371944
Powder-horn 82680
Powderleaf Goldenflower 260432
Powderleaf Mussaenda 260432
Powderleaf Plantainlily 198646
Powderleaf Spiraea 371887
Powderleaf Tanoak 233306
Powder-puff Cactus 244012
Powderpuff Lillypilly 382683
Powderpuff Lily 184428,350312
Powderpuff Thistle 303082
Powderpuff Tree 48518,66673
Powderpuff-tree 13578
Powdery Dudleya 138835
Powdery Hair Actinidia 6593
Powdery Live-forever 138835
Powdery Plantainlily 198591
Powdery Thalia 388385
Powdery-haired Actinidia 6593
Powdery-leaf Actinidia 6615
Powdery-leaved Actinidia 6615
Powdery-leaved Tanoak 233306
Powell Crinum 111250
Powell Lycaste 238748
Powell's Amaranth 18804
Powell's Orach 44617
Powell's Smooth Amaranth 18804
Power-wort 325820
Powk Needle 153767
Powk-needle 153767
Powton 285960
Poyang Naiad 262101
Poyi 121750
Palm 273241
Pracaxi Fat 289438,289439
Praemorse Violet 410436
Pragense Viburnum 407659
Prague Viburnum 407659
Prain Aster 88160
Prain Clovershrub 70794
Prain Kobresia 217250
Prain Microtoena 254264

Prain Neomicocalamus 264180
Prain Rhodiola 329925
Prain Sedge 75862
Prain Woodbetony 287539
Prain's Chlamydites 88160
Prain's Jasmine 211965
Prain's Ninenode 319771
Prain's Psychotria 319771
Prairie Acacia 1051
Prairie Alumroot 194433
Prairie Anemone 23745
Prairie Apple 319178
Prairie Aster 381005
Prairie Baccharis 46260
Prairie Beard Grass 351309
Prairie Bergamot 257159
Prairie Blazing Star 228511
Prairie Blazing Stars 228511
Prairie Blazing-star 228511
Prairie Blue-eyed Grass 365715
Prairie Blue-eyed-grass 365715
Prairie Brome 60793
Prairie Broom-rape 275121
Prairie Broomweed 20486
Prairie Bulrush 56641,353689
Prairie Bundle-flower 126176
Prairie Bush Clover 226815
Prairie Bush-clover 226876
Prairie Buttercup 325812,326234,326309
Prairie Camas 68790
Prairie Cherry 316434
Prairie Cinquefoil 312391,312795
Prairie Closed Gentian 173222
Prairie Clover 237483,292279
Prairie Coneflower 140074,326957,326960,326964,326966,339543
Prairie Cord Grass 370173,370174
Prairie Cord-grass 370173
Prairie Coreopsis 104566
Prairie Crab 243636
Prairie Crab Apple 243636
Prairie Crabapple 243636
Prairie Crab-apple 243636
Prairie Crocus 321721
Prairie Crowfoot 326309
Prairie Cup Grass 151668
Prairie Dandelion 253918
Prairie Desmanthus 126176
Prairie Dock 364317
Prairie Dogbane 29473
Prairie Dogtooth Violet 154932
Prairie Droopseed 372698
Prairie Drop-seed 372698
Prairie Evening Primrose 269398
Prairie False Dandelion 253918,266826
Prairie False Indigo 47871
Prairie False Willow 46260
Prairie Fame Flower 383327
Prairie Fame-flower 294330,294331
Prairie Flameleaf Sumac 332686
Prairie Flame-leaf Sumac 332530
Prairie Flax 231917
Prairie Fleabane 150980,150989
Prairie Fringed Orchis 302390
Prairie Gagea 169504
Prairie Gayfeather 228511
Prairie Gentian 161026,161027,173188,173763
Prairie Golden Aster 90175,194195
Prairie Goldenrod 368324,368422
Prairie Gray Sedge 74180
Prairie Ground Cherry 297691,297727
Prairie Ground-cherry 297683
Prairie Groundsel 279936
Prairie Hawkweed Hawkweed 195760
Prairie Heart-leaved Aster 380943
Prairie Hummock Sedge 76033
Prairie Indian-plantain 34808
Prairie Iris 208812,263311
Prairie Ironweed 406325
Prairie June Grass 217441,217494
Prairie Junegrass 217441,217494,217550
Prairie June-grass 217494
Prairie Koeleria 217441
Prairie Ladybell 7792
Prairie Larkspur 124104,124106,124677
Prairie Leek 15393
Prairie Lespedeza 226815,226876
Prairie Lettuce 219403
Prairie Mallow 362702,362704
Prairie Milkweed 37964,38132
Prairie Mimosa 126168,126176
Prairie Mouse-ear Chickweed 82680
Prairie Muhly 259651
Prairie Net-leaf 166709
Prairie Onion 15393,15765
Prairie Panic Grass 128518
Prairie Parsley 310674
Prairie Pepperweed 225340
Prairie Pepper-weed 225340
Prairie Phlox 295312
Prairie Pinweed 223705
Prairie Plum 316206
Prairie Potato 319178
Prairie Ragwort 279936
Prairie Red-root 79940
Prairie Rocket 154390
Prairie Rooster Bill 135165
Prairie Rose 336361,336363,336637,336942
Prairie Rose Gentian 341440
Prairie Rosinweed 364285,364317
Prairie Sage 35570
Prairie Sagebrush 35505
Prairie Sagewort 35505
Prairie Sage-wort 35505
Prairie Sand Reed 65633
Prairie Sandreed 65633
Prairie Sand-reed 65633
Prairie Satin Grass 259651
Prairie Sedge 75864
Prairie Senna 78284
Prairie Shoestring 19997
Prairie Skeletonplant 375978
Prairie Smoke 175452
Prairie Spiderwort 394050
Prairie Spurge 85775,159856
Prairie Star 233456
Prairie Straw Sedge 76420
Prairie Sumac 332686
Prairie Sundrops 269489,269491
Prairie Sunflower 189028,189032,189044
Prairie Sweet Pea 222860
Prairie Tea 112965
Prairie Thermopsis 389547
Prairie Thistle 91840,91914,91976,92027
Prairie Three-awn 33957
Prairie Tickseed 104566
Prairie Tick-trefoil 126407
Prairie Trefoil 237733,237781
Prairie Trout-lily 154932
Prairie Turnip 287932,319178
Prairie Verbena 176686
Prairie Violet 410392
Prairie Voneflower 326957
Prairie Wedge Grass 371493
Prairie Wedge Scale 371493
Prairie Wedgescale 371493
Prairie White Fringed Orchid 302390
Prairie White-fringed Orchis 302390
Prairie White-fringed-orchid 302390
Prairie Wild Lettuce 219403
Prairie Wild Onion 15765
Prairie Willow 343506,343508
Prairie Zinnia 418046
Prairie-crocus 321708
Prairiedawn 201333
Prairie-fire Pink 364084
Prairiemallow 362702
Prairie-mimosa 126176
Prairie-parsley 310674
Prairie-plum 42231
Prairie-smoke 23979,175452,321703,321708
Prairie-tea 112965
Prairie-turnip 287932
Praiseworthy Rhododendron 331069,331070
Prantt Leek 15626
Prantt Onion 15626
Prarie Willow 343909
Prase 367696
Pratia 223053,313552
Pratie 367696

Pratt Anemone 23999
Pratt Barberry 52062
Pratt Catnip 265028
Pratt Caudate Maple 2879
Pratt Corydalis 106286
Pratt Crabapple 243665
Pratt Gentian 173735
Pratt Kobresia 217254
Pratt Milkvetch 42909
Pratt Mountainash 369490
Pratt Mountain-ash 369490
Pratt Nepeta 265028
Pratt Pleurospermum 304831
Pratt Primrose 314834
Pratt Reddish Arrowwood 407833
Pratt Ribseedcelery 304831
Pratt Rockfoil 349802
Pratt Rose 336870
Pratt Solomonseal 308632
Pratt St. John's-wort 202102
Pratt Windflower 23999
Pratt Youngia 416467
Pratt's Rhododendron 330673
Pratt's Spicebush 231419
Pratt's Spice-bush 231419
Prattling Parnell 349044
Prayer Beads 765
Prayer Beans 765
Prayer Plant 66200,245022
Prayer-plant 245013
Prayer-plant Family 245039
Prazer Sophora 369104
Prazer Yam 131777
Preacher-in-the-pulpit 37015
Preaux Statice 230717
Precatory 765
Precatory Bean 765
Precatory-pea 765
Precipice Corydalis 106284
Precipice Crazyweed 279137
Precipice Woodbetony 287538
Precocious Apple 243664
Precocious Brocadebeldflower 413621
Precocious Matricary 246392
Precocious Mayweed 246392
Precocious Stachyurus 373556
Precocious Tree 34475
Pregnant Onion 274679
Preiss Callitris 67427
Preiss Cypress Pine 67427
Premna 313593
Premontane Germander 388242
Premorse Daphne 122425
Premorse Scabious 379660
Prenanth Hawkweed 195880
Prescot Chervil 84758
Pressedhairy Sibbaldia 362341
Prestoea 313952
Preston Lilac 382251,382254
Prestonia 313956
Prettiest Caesalpinia 65055
Pretty Anthurium 28105
Pretty Ardisia 31571,31578
Pretty Autumn-crocus 99345
Pretty Baby 81768
Pretty Ball Cactus 284665
Pretty Ballcactus 284665
Pretty Barberry 51483
Pretty Bedstraw 170325
Pretty Begonia 49605
Pretty Betsy 81768,349044,349918
Pretty Betty 349044
Pretty Broomrape 274951
Pretty Chrysoglossum 89948
Pretty Clematis 95409
Pretty Clinopodium 97025
Pretty Condorvine 245835
Pretty Corydalis 105568
Pretty Cratoxylum 110256
Pretty Crocus 111603
Pretty Dendrobium 125365
Pretty Didymocarpus 129950
Pretty Doritis 136320
Pretty Dryander 138431
Pretty Eargrass 187650
Pretty Face 60431,398777,398797,398806
Pretty Fairy Lantern 67635
Pretty Falseoat 398526
Pretty Fig 164833
Pretty Gentian 173290
Pretty Grepe-hyacinth 260335
Pretty Hedyotis 187650
Pretty Honeysucle 235674
Pretty Hyalea 199624
Pretty Lady 349044
Pretty Leaf Tan Oak 233128
Pretty Leptodermis 226125
Pretty Lindelofia 231251
Pretty Lyonia 239376
Pretty Maids 349419
Pretty Marshweed 230275
Pretty Meconopsis 247193
Pretty Michelia 252835
Pretty Millettia 254810
Pretty Moraea 258606
Pretty Mountain 273800
Pretty Muntries 218395
Pretty Nancy 349044,374916
Pretty Newlitse 264088
Pretty Oneflowerprimrose 270730
Pretty Onion 15112
Pretty Parodia 284665
Pretty Pricklythrift 2277
Pretty Prickly-thrift 2277
Pretty Primrose 314262
Pretty Protea 315742
Pretty Satinflower 365696
Pretty Schismatoglottis 351155
Pretty Sedge 76761
Pretty Senna 360485
Pretty Small Reed 127284
Pretty Smallreed 127284
Pretty Sneezeweed 188417
Pretty Spapweed 204809
Pretty Spikerush 143064
Pretty St. John's Wort 201775
Pretty Star-of-bethlehem 274786
Pretty Touch-me-not 204809
Pretty Trisetum 398526
Pretty Wild Buckwheat 151860
Pretty Wildbasil 97025
Pretty Willie 127607
Pretty-and-little 243209
Prettybean 67771
Prettybract Tanoak 233127
PrettyFlower Sealavender 230561
Prettyfruit Ninenode 319461
Prettyfruit Psychotria 319461
Pretty-fruited Psychotria 319461
Prettyleaf Dragonhead 137567
Prettyleaf Mucuna 259488
Prettyleaf Silkvine 291046
Prettyleaf Tanoak 233128
Pretty-leaf Winchia 18052
Pretty-leaved Alstonia 18052
Pretty-leaved Silkvine 291046
Pretty-leaved Tanoak 233128
Pretty-leaved Winchia 18052
Prettynerved Persimmon 132082
Prettynerved Slugwood 50498
Pretty-nerved Slugwood 50498
Prettynerved Storax 379457
Pretty-scaled Tanoak 233127
Prettyspike Knotweed 308939
Prettytooth Linden 391680
Prety Paphiopedilum 282791
Pre-vernal Bamboo 297483
Price Privet 229587
Price's Aster 380969
Prick Devil 350425
Prick Hollin 203545
Prick Holly 203545
Prick Madam 357114
Prick Needle 11947
Prick Willow 196757
Prick-bush 203545
Prick-devil 350425
Pricket 356468
Prick-hollin 203545
Prickle Euonymus 157283
Prickle Grass 113305

Prickle Meconopsis 247104
Prickle Pine 300168
Prickle Weed 126176
Prickleback 325612
Prickle-cone Pine 300098
Pricklefruit Licorice 177932
Pricklegrass 113305
Prickle-grass 113305,113306
Pricklepoppy 32411
Prickless Firethorn 322470
Prickley Cardinal 154747
Prickly Aapplecactus 185914
Prickly Alyxia 18522
Prickly Amaranth 18822
Prickly Ash 30760,274259,417132,417157,417231
Prickly Bark 155762
Prickly Barnyardgrass 140483
Prickly Beehive 133478
Prickly Beehive Cactus 107022
Prickly Bog Sedge 73782,73784
Prickly Bottlebrush 67252
Prickly Box 64064
Prickly Brazilwood 65002
Prickly Bridelia 60189
Prickly Broom 172905,401388
Prickly Caterpillar 18822
Prickly Cedar 213841
Prickly Chaff Flower 4259
Prickly Clover 396885
Prickly Coat 92485
Prickly Coats 92485
Prickly Comfrey 381026
Prickly Cucumber 114221,140527
Prickly Currant 334058
Prickly Cypress 213775
Prickly Dewberry 338399
Prickly Elder 30760
Prickly Euonymus 157282
Prickly Fuchsia Bush 180277
Prickly Ghost 401388
Prickly Glasswort 344585
Prickly Goldenfleece 402663
Prickly Gooseberry 333945
Prickly Gousblom 52715
Prickly Groundberry 338839
Prickly Heath 172115,291367
Prickly Hornwort 83558
Prickly Hound's Tongue 117970
Prickly Ivy 366341
Prickly Juniper 213841
Prickly Knotweed 309527
Prickly Lettuce 219497,219507
Prickly Macartney Rose 336406
Prickly Mallow 362662
Prickly Medick 247381
Prickly Moses 1321,1686
Prickly Myrtle 329166
Prickly Nightshade 367567
Prickly Oplopanax 272707
Prickly Palm 46379
Prickly Paperbark 248119
Prickly Parsnip 140640
Prickly Pear 272775,272856,272891,272915,273085,273088
Prickly Pear Cactus 272832
Prickly Pet Tigrue 340990
Prickly Pettigree 340990
Prickly Pettigrue 340990
Prickly Phlox 175676
Prickly Poison 171972
Prickly Poppy 32411,32415,32419,32422,32423,32429,282515
Prickly Rose 336320,336322
Prickly Russian Thistle 344743
Prickly Saltwort 344585
Prickly Sandwort 31731
Prickly Scorpion's-tail 354761
Prickly Sedge 74423,75451,76328
Prickly Sida 362662
Prickly Smartweed 291707,308922
Prickly Sow Thistle 368649
Prickly Sowthistle 368649
Prickly Sow-thistle 368649
Prickly Spider Flower 180609
Prickly Spineflower 89110
Prickly Tea-tree 226459
Prickly Thrift 2208,2237
Prickly Water Lily 160637
Prickly Wattle 1321,1647
Prickly Wild Coffee 95936
Prickly Wild Gooseberry 333945
Pricklyash 417132,417330
Prickly-ash 30760,417132
Prickly-ash-leaved Ash 168137
Pricklybox 238996
Pricklyburr 123061
Prickly-calyx Raspberry 338079,339015
Pricklyfruit Bedstraw 170352
Pricklyfruit Euonymus 157269
Pricklyfruit Jurinea 214060
Prickly-fruited Macaranga 240255
Pricklyleaf Dogweed 391036
Prickly-leaf Paperbark 248109,248119
Prickly-leaved Starwort 375078
Prickly-leaved Tea Tree 248119
Prickly-leaved Wattle 1686
Pricklypear 272775
Prickly-pole 46377
Prickly-poppy 32411
Pricklyscale Alfredia 14594
Prickly-seeded Spinach 371713
Prickly-sepaled Begonia 49799
Pricklythrift 2208
Prickly-thrift 2208
Prick-madam 356468,356503,356537,357114,358062
Prick-my-nose 266225
Prick-needle 11947
Pricksong-wort 238371
Prick-timber 157429,380499
Prick-tree 380499
Prickwind 366240
Prickwood 157429
Prick-wood 157429,380499
Pricky Thistle 91770
Prickyback 133478
Pridda 367696
Pride 188908
Pride of Barbados 65055
Pride of Bolivia 392342,392345
Pride of Burma 19462,19463
Pride of California 222842
Pride of California Peavine 222842
Pride of China 217613,248895
Pride of De Kaap 49094
Pride of India 217626,219966,248895
Pride of Indian 45908,219966
Pride of Madeira Viper's Bugloss 141164
Pride of Maderira 141129
Pride of Persia 248895
Pride of Sussex 298065
Pride of the Evening 236022
Pride of the Thames 64184
Pride of the Woods 145487
Pride Paphiopedilum 282901
Pride-o'-London 349044
Pride-of-barbados 65055
Pride-of-Bolivia 392345
Pride-of-Burma 19463
Pride-of-India 217610,217626,219966,248895,392345
Pride-of-India Family 346316
Pride-of-madeira 141164
Pride-of-ohio 135165
Pride-of-texas 295267
Pride-of-the-evening 236022
Prideweed 103446
Prie 229662
Prieldy Salad Burnet 345881
Priendly Lily 229723
Priest 273531
Priest Valley Spineflower 89127
Priest's Crown 384714
Priest's Hood 37015
Priest's Pilly 37015
Priest's Pintel 37015,273531,329935
Priest's Pintle 5442,37015,273531,329935
Priestand-pulpit 37015
Priest-and-pulpit 37015
Priesties 37015

Priest-in-the-pulpit 37015
Prieur's Umbrellagrass 146010
Pright Pellitory 284167
Prikly Bridelia 60162
Prim 229428,229662
Prima Vera 115784
Primarose 315101
Primaticfruit Epidendrum 146465
Primavera 115784
Primet 229662
Primmily 314143
Primmy Rose 315101
Primoglauous Strong-sepal Barberry 52299
Primp 229662
Primp Rint 229662
Primrose 50825,229662,314083,314085,314189,314342,314407,314684,315082,315101
Primrose Dendrobium 125317
Primrose Didymocarpus 129967
Primrose Family 315138
Primrose Gladiolus 176151
Primrose Jasmine 211905
Primrose Meconopsis 247164
Primrose Pearls 262438
Primrose Peerless 262417
Primrose Peerless Narcissus 262405
Primrose Raspberry 339096
Primrose Rhodiola 329926
Primrose Roscoea 337077
Primrose Rose 336872
Primrose Soldiers 30081
Primrose Tree 220228
Primrose Willow 214210,238150,238183,238188
Primroseflower Azalea 331539
Primrose-flower Rhododendron 331539
Primrose-flowered Rhododendron 331539
Primrose-leaf Ardisia 31565
Primroseleaf Sonerila 368897
Primroselike Coelogyne 98723
Primrose-like Lily 229999
Primrose-like Oreocereus 273878
Primrose-like Rhodiola 329926
Primroselike Sonerila 368897
Primrose-peerless 262417
Primrose-yellow Sunrose 188763
Primula Family 315138
Primula Gladiolus 176151
Primulaflower Gentian 173737
Primulaflower Isometrum 210190
Primulaleaf Ardisia 31565
Primulaleaf Isometrum 210190
Primula-leaved Ardisia 31565
Primula-leaved Violet 410360
Primule Rose 336872
Primulina 315146,315148
Primwort 229662
Prince Albert Vygie 54363
Prince Albert Yew 349016,349018
Prince Albert's Feather 18804
Prince Albert's Feather 18752
Prince Albert's Fir 399905
Prince Albert's Yew 349016,349018
Prince Alberts Yew 349018
Prince Cattleya 79554
Prince Edward Island Blackberry 339145
Prince-of-Wales'-feather 18752
Prince of Wales Juniper 213800
Prince of Wales' Feathers 81768,382329
Prince of Wales' Feathers Tree 58709
Prince Protea 315761
Prince Rupprecht's Larch 221939
Prince Wood 104185
Prince's Feather 18734,18752,309494
Prince's Feathers 18810,182942,227533,271280,309494,312360,316127,349044,382329,401711
Prince's Flower 68737
Prince's Pea 319114
Prince's Pine 87480,87498
Prince's Plume 309494,373709
Prince's Rock Cress 30415
Prince's Rock-cress 30415
Prince's-feather 18752,309494
Prince's-feather Amaranth 18752
Prince's-pine 87498
Prince's-plume 309494
Prince-of-wales Feather 18752
Princeps Begonia 50190
Princes Chalote's Passion-flower 285695
Princes Protea 315810
Princespine 87498
Princess Bean 319114
Princess Flower 391577
Princess Leaf 255694
Princess of the Night 357748
Princess Palm 129649,129679,129684,321164
Princess Pine 299818
Princess Tree 285981
Princess' Robe 262441
Princess-flower 391577
Princess-of-the-night 357748
Princesstree 285981
Princess-tree 285981
Princewood 104185,161825
Pringeleaf Ruellia 339694
Pringle Pincution 244197
Pringle Pine 300159
Pringle's Aster 380955
Pringle's Catchfly 364031
Pringle's Fleabane 150889
Pringle's Hawthorn 109945
Pringle's Pine 300159
Pringle's Rabbit-tobacco 317766
Pringle's Woolly Sunflower 152831
Pringler Yam 131780
Prinos-like Salacia 342709
Prinsepia 315171,315177,365003
Print 229662
Print Pinafore 174877,309570
Prinus Salacia 342709
Priotropis 315261
Pripet 229662
Prism Cactus 227757
Prismatic Manchurian Linden 391762
Prismatomeris 315318,315332
Pristimera 315352
Pristly Spiny-club Palm 46378
Pritchard Cyrtandra 120507
Pritchardia 315376
Pritchardia Palm 315376
Privellike Honeysuckle 235916
Privet 229428,229529,229558,229617,229662
Privet Honeysuckle 236035
Privet-leaf Memecylon 249996
Privet-leaved Honeysuckle 235916
Privet-leaved Memecylon 249996
Privet-like Jasmine 211893
Privet-like Premna 313673
Privy 229662
Privy Saugh 229662
Prncumbent Garden Speedwell 406973
Proboscidea 315430,315434
Proboscis Flower 315434
Probst's Goosefoot 87137
Procession Flower 308118,308454
Prococious Weigela 413621
Procris 315461,315476
Procumbent Asarina 37554
Procumbent Aster 41306
Procumbent Bastardtoadflax 389825
Procumbent Buckthorn 328831
Procumbent Coldenia 99361
Procumbent Crazyweed 278894
Procumbent Cuphea 114618
Procumbent Currant 334157
Procumbent Dragon Juniper 213645
Procumbent Embelia 144787
Procumbent Flemingia 166889
Procumbent Hedgehog Cactus 140288
Procumbent Honeysuckle 235713
Procumbent Indian Mallow 402267
Procumbent Juniper 213864
Procumbent Leucas 227691
Procumbent Marshwort 29341
Procumbent Mazus 247020
Procumbent Medick 247438
Procumbent Motherweed 231559

Procumbent Neurada 265715
Procumbent Paramicrorhynchus 283506
Procumbent Pearlwort 342279
Procumbent Sibbaldia 362361
Procumbent Skullcap 355683
Procumbent Spapweed 205252
Procumbent Tridax 396697
Procumbent Triumfetta 399303
Procumbent Water Parsnip 365861
Procumbent Wildberry 362361
Procumbent Yellow Sorrel 277747
Procumbent Yellow-sorrel 277747
Procumbrnt Falsepimpernel 231559
Procupine Barberry 52169
Proeumbent Meadow-grass 321378
Prof. Beijerinck's Rhododendron 330221
Professor-weed 169935
Profolium 34536
Profund Cacalia 283861
Profuse-flowering Senna 78301
Profusely Flowering Primrose 314405
Projectingvein Qinggang 116088
Proliferating Bulrush 210056
Proliferous Beautyberry 66901
Proliferous Loosestrife 239796
Proliferous Pink 292664,292665
Proliferous Primrose 314843
Proliferous Purplepearl 66901
Proliferous Wild Buckwheat 152499
Prolific Barberry 52065
Prominent Bamboo 47416,297400
Prominent Bambusa 47416
Prominent Monkshood 5497
Prominentlynerveid Bluegrass 305770
Promotion-nut 21195
Pronaya 315517
Prone Ironweed 406320
Prong 83944
Prong Ceriops 83946
Propeller Flower 17583
Propeller Plant 109264
Propellers 2837,3462
Propellor Trillium 397617
Prophet Bower 34635
Prophet Flower 34629,34635,241386
Prophet's Flower 34635
Prophet-flower 34635
Propinquity Bamboo 297401
Proskowetz Barley 198368
Proso 281916
Proso Millet 281916
Prostrate Abelia 132
Prostrate Acacia 1522
Prostrate Amaranth 18668
Prostrate Amaranthus 18668
Prostrate Armodorum 30534
Prostrate Axyris 45856
Prostrate Azalea 331551
Prostrate Begonia 49899
Prostrate Box 64341
Prostrate Broomsedge 217353
Prostrate Buckthorn 328833
Prostrate Bugle 13174
Prostrate Bulrush 353845
Prostrate Buttercup 326208
Prostrate Celery 29319
Prostrate Centaurea 81034
Prostrate Chamomile 26881
Prostrate Chinese Juniper 213696
Prostrate Coleus 303539
Prostrate Crabdaisy 413549
Prostrate Crazyweed 279087
Prostrate Cudweed 178463
Prostrate Cyathula 115738
Prostrate Euphorbia 159634
Prostrate False Pimpernel 231559
Prostrate Fargesia 162686
Prostrate Fescuegrass 163817
Prostrate Fraser Balsam Fir 378
Prostrate Geniosporum 172871
Prostrate Goosefoot 87058
Prostrate Heath Aster 380879
Prostrate Honeysuckle 236045
Prostrate Japanese Plum Yew 82527
Prostrate Knotweed 308816,309836
Prostrate Leek 15632
Prostrate Marshwort 29344
Prostrate Myoporum 260720
Prostrate Nepeta 265071
Prostrate Nujiang Azalea 331722
Prostrate Onion 15632
Prostrate Panisea 282413
Prostrate Pernettya 291371
Prostrate Pigweed 18655,18668
Prostrate Polycarpon 307748
Prostrate Pulicaria 321595
Prostrate Rhododendron 331551
Prostrate Rosemary 337190
Prostrate Ruellia 339792
Prostrate Salwen Rhododendron 331722
Prostrate Screwtree 190109
Prostrate Seepweed 379584
Prostrate Skullcap 355698
Prostrate Snowberry 172110
Prostrate Sophora 369111
Prostrate Speedwell 407299
Prostrate Spineflower 89100
Prostrate Spiraea 372049
Prostrate Spurge 85770,159286,159634
Prostrate Summer Cypress 217353
Prostrate Summercypress 217353
Prostrate Tick-trefoil 126586
Prostrate Toadflax 231140
Prostrate Velvetplant 183122
Prostrate Verbena 405875,405893
Prostrate Vervain 405815
Prostrate Wedelia 413549,413559
Prostrate Whitepearl 172136
Prostrate Wild Buckwheat 152407
Prostrate Wild Petunia 339792
Prostrate Willowherb 146644
Prostrate Woodbetony 287792
Prostrate Wrinkledleaf Box 64341
Prostrate Yellowcress 336265
Prostratehair Japanese Buttercup 325989
Prostratehair Kusnezoff Monkshood 5338
Prostratehair Ramie 56325
Prostratehair Windflower 23906
Prostrte Tadehagi 383007
Protea 315702,315949
Protea Family 316007
Protean Knapweed 80893
Protected Rockvine 387822
Proteoid Rhododendron 331556
Protium 316018
Protolaciniate Lilac 382256
Protracte Rattlebox 112562
Protracted Anemone 23906
Protruded Fruit Broadleaf Maple 2784
Protruding Eucalyptus 155578
Protruding Wild Buckwheat 152323
Proud Carpenter 316127
Proud Datura 123065
Provence Orchid 273596
Provence Reed 37467
Provevce Rose 336474
Provicar Rabdosia 324851
Provided Asparagus 39102
Provincial Barberry 52066
Provincial Orchis 273596
Provins Rose 336581
Provision Tree 279381
Provot Curvedtube Woodbetony 287133
Pruina-shoot Willow 344022
Pruinate Begonia 50194
Pruinose Acacia 1499
Pruinose Barberry 52069
Pruinose Pea-shrub 72317
Pruinose Sedge 75882
Pruinose-fruit Barberry 52068
Pruinous-shooted Willow 344022
Pruit's Candytuft 203189
Prune 316166
Prune Leaved Service-berry 19286
Prune Leaved Shadberry 19286
Prune Tree 316382
Prune-leaved Juneberry 19286
Prunell 316127
Pruniputamen Hackberry 80717
Prushia 364557
Pruss Hawkweed 195888

Prussian Asparagus 274748
Prussian French Asparagus 274748
Przevalski Dandelion 384759
Przevalsky Littledalea 234125
Przevalsky Small Reed 127317
Przewalsk 173778
Przewalsk Ajania 13023
Przewalsk Azalea 331561
Przewalsk Cattail 401132
Przewalsk Crazyweed 279089
Przewalsk Ephedra 146238
Przewalsk Euonymus 157809
Przewalsk Fritillary 168523
Przewalsk Glacial Knotweed 309144
Przewalsk Globethistle 140769
Przewalsk Gymnocarpos 182498
Przewalsk Incarvillea 205584
Przewalsk Juniper 213871
Przewalsk Larkspur 124020
Przewalsk Leek 15633
Przewalsk Milkvetch 42919
Przewalsk Onion 15633
Przewalsk Poplar 311434
Przewalsk Rhododendron 331561
Przewalsk Rockfoil 349808
Przewalsk Rush 213400
Przewalsk Sage 345327
Przewalsk Sagebrush 36110
Przewalsk Sandwort 32151
Przewalsk Sedge 75887
Przewalsk Skullcap 355699
Przewalsk Spindle-tree 157809
Przewalsk St. John's Wort 202104
Przewalsk Stinkinggrass 249075
Przewalsk Stonecrop 357049
Przewalsk Windhairdaisy 348675
Przewalsk Woodbetony 287544
Przewalski 388627
Przewalski Goldenray 229151
Przewalski Peashrub 72333
Przewalski Pea-shrub 72333
Przewalski Seepweed 379589
Przewalski's Leopard Plant 229151
Przewalskia 316953
Przewalsky Asparagus-fren 39155
Przewalsky Feathergrass 376887
Przewalsky Needlegrass 376887
Psammochloa 317000
Psammophil Willow 343914
Psammosilene 317035
Psathyrostachys 317073
Pseudaechmanthera 317135,317137
Pseudanthistria 317192
Pseudelephantopus 317231
Pseuderanthemum 317245
Pseudobanks Rose 336875
Pseudobartsia 317399
Pseudocerastium 317528,317529
Pseudochinolaena 317222
Pseudo-ginseng 280771
Pseudohirsute Crazyweed 279099
Pseudolephantopus 317231
Pseudopogonatherum 318147
Pseudopunctulate Rockfoil 349829
Pseudopyxis 318183
Pseudoraphis 318193
Pseudosasa 318277
Pseudosmilax 194127,318448,318450
Pseudosowthisitle Tephroseris 385916
Pseudostellaria 318513
Pseudostrobus Pine 300162
Pseudoswampy Bluegrass 305901
Pseuduvaria 318661
Psilate Primrose 314435
Psilateleaf Arrowwood 407915
Psilopeganum 318945
Psilotrichum 319000
Psilotum-like Bastardtoadflax 389829
Psilurus 319062
Psychotria 319392
Psychrogeton 319925
Psyllium 301864
Psyllium Seed 301864
Ptarmiganberry 31311
Ptarmigan-berry 31311
Pteridophytaleaf Woodbetony 287565
Pternandra 320198
Pterocaulon 320389
Pteroceras 320446
Pterocypsela 320508
Pterodiscus 320527,320541
Pterolobium 320594
Pteroptychia 320791
Pterospermous Chinabell 16304
Pterospermum 320827
Pteroxygonum 320911
Pterygiella 320914,320918
Pterygocalyx 320924
Pterygopleurum 320975
Pterygota 320985,320987
Ptilagrostis 321004
Ptilotrichum 321087,321089
Ptychococcus 321128
Ptychosperma 321161
Pu'er Lily 230005
Pua 276395
Puahilahila 255101
Puapilo 71871
Puberulate Ampelopsis 20444
Puberulent Azalea 331540
Puberulent Barberry 51924
Puberulent Chastetree 411422
Puberulent Crazyweed 279102
Puberulent Dinochloa 131284
Puberulent Eargrass 187628
Puberulent Fiveleaf Chastetree 411422
Puberulent Listera 264727
Puberulent Monkshood 5057
Puberulent Oligostachyum 270443
Puberulent Premna 313717
Puberulent Rhododendron 331540
Puberulent Swollennoded Cane 87600
Puberulent Vinebamboo 131284
Puberulous Cliffliving Eurya 160598
Puberulous Elatostema 142705,142758
Puberulous Glochidion 177170
Puberulous Hornbeam 77348
Puberulous Javan Meadowrue 388542
Puberulous Ocotea 268724
Puberulous Stylos Staff-tree 80331
Puberulous Sweetleaf 381452
Puberulous Tickseed 104834
Puberulous Tsai Monkshood 5646
Pubescence Gentian 173767
Pubescens Chestnut Sage 344943
Pubescent Acute Fieldcitron 400519
Pubescent Acute Turpinia 400519
Pubescent Ailanthus-like Pricklyash 417146
Pubescent Angelica 24441
Pubescent Anisadenia 25273
Pubescent Aphananthe 29005
Pubescent Apiculate Rockvine 387747
Pubescent Ash 168074
Pubescent Atylosia 65153
Pubescent Azalea 331575
Pubescent Barberry 52086
Pubescent Barrenwort 147039
Pubescent Betony 373386
Pubescent Bigflower Greenvine 204628
Pubescent Bigleaf Laurel Cherry 316950
Pubescent Big-leaved Prickly-ash 417279
Pubescent Birch 53563,53590
Pubescent Bird Cherry 280007
Pubescent Bitterbamboo 304083
Pubescent Blueflower Skullcap 355442
Pubescent Blumea 55788
Pubescent Bowedsepal Spiraea 371911
Pubescent Box 64324
Pubescent Brachyactis 58278
Pubescent Bridelia 60211
Pubescent Brownbract Yam 131762
Pubescent Bumalda Bladder-nut 374095
Pubescent Bushclover 226940
Pubescent Bush-clover 226977
Pubescent Butterflypea 81876
Pubescent Cajan 65153
Pubescent Cane 37286,318342
Pubescent Canebrake 37286
Pubescent Cardaria 73098
Pubescent Centrosema 81876
Pubescent Ceropegia 84211

Pubescent Chinese Altingia 18193
Pubescent Chinese Leptopus 226320
Pubescent Chinese Redbud 83774
Pubescent Christina Loosestrife 239584
Pubescent Cinchona 91086
Pubescent Cloudy Buttercup 326074
Pubescent Coccineus Colquhounia 100032
Pubescent Colquhounia 100032
Pubescent Colubrina 100070
Pubescent Coral Holly 203656
Pubescent Corydalis 106106
Pubescent Cowparsnip 192343
Pubescent Cudrania 240829
Pubescent Currant 334163
Pubescent Decumbent Mountainash 369500
Pubescent Dissected Violet 409928
Pubescent Drooping Razorsedge 354176
Pubescent Dzungaria Monkshood 5588
Pubescent Emei Larkspur 124430
Pubescent Emptygrass 98802
Pubescent Epimedium 147039
Pubescent Euonymus 157310
Pubescent Euscaphis 160960
Pubescent Forsythia 167460
Pubescent Germander 388126
Pubescent Glochidion 177170
Pubescent Gomphandra 178921
Pubescent Groundcherry 297720
Pubescent Gynostemma 183026
Pubescent Hackberry 80719
Pubescent Hairyleaf Lyonia 239408
Pubescent Hangzhou Rhododendron 330822
Pubescent Head Rattlebox 112384
Pubescent Holly 204184,204188
Pubescent Hornbeam 77373
Pubescent Hubei Catchfly 363560
Pubescent Japanese Pagodatree 369049
Pubescent Japanese Privet 229506
Pubescent Japanese Spiraea 371962
Pubescent Kangding Willow 343836
Pubescent Keteleeria 216150
Pubescent Knotweed 309644
Pubescent Lady's Slipper 120314
Pubescent Lavender 223321
Pubescent Leptospermum 226476
Pubescent Leucas 227692
Pubescent Leymus 228386
Pubescent Longleaf Raspberry 338337
Pubescent Longstalk Corydalis 106106
Pubescent Long-stalk Creeper 285109
Pubescent Longstamen Litse 233985
Pubescent Lupine 238482
Pubescent Meyer Currant 334095
Pubescent Mezoneuron 252752
Pubescent Microtoena 254258
Pubescent Mockorange 294517
Pubescent Mongolian Spiraea 372019
Pubescent Mysorethorn 64993
Pubescent Narrowleaf Eurya 160603
Pubescent Oak 324314
Pubescent Oat 45558
Pubescent Ormosia 274430
Pubescent Pachygone 279559
Pubescent Pachystoma 279859
Pubescent Paris 284347
Pubescent Pendent Inflorescence Indigo 206382
Pubescent Pendentlamp 84211
Pubescent Pepper 300415
Pubescent Periwinkle 409345
Pubescent Pertybush 292065
Pubescent Pimpinella 299502
Pubescent Pinnatenerve Christmasbush 14214
Pubescent Polystachya 310560
Pubescent Pricklyash 417273
Pubescent Prickly-ash 417273
Pubescent Purpleflower Violet 410033
Pubescent Qiaomaidi Sage 345144
Pubescent Racemose Supplejack 52456
Pubescent Rattlesnake 179680
Pubescent Redbark Litse 234032
Pubescent Redbud 83804
Pubescent Redflower Blueberry 404034
Pubescent Resupinate Woodbetony 287600
Pubescent Rhododendron 331575
Pubescent Ribbed Birch 53391
Pubescent Rockvine 387841
Pubescent Roscoea 337078
Pubescent Roughleaftree 29005
Pubescent Scabrous Aphananthe 29006
Pubescent Seashore Vetchling 222738
Pubescent Sedge 74810
Pubescent Siberian Apricot 34474
Pubescent Silvervine Kiwifrui 6684
Pubescent Slenderstyle Acanthopanax 2406
Pubescent Smoketree 107312
Pubescent Snakevine 100070
Pubescent Spapweed 205260
Pubescent Spathoglottis 370402
Pubescent Spiraea 372067
Pubescent Squarebamboo 87602
Pubescent Sterculia 376167
Pubescent Stonecrop 356766
Pubescent Strawberryleaf Raspberry 338431
Pubescent Swertia 380321
Pubescent Tea 69525
Pubescent Thaumastochloa 388914
Pubescent Three-forked Evodia 161383
Pubescent Toona 392833
Pubescent Toothed Waterlily 267701
Pubescent Tulip 400132
Pubescent Twining Monkshood 5675
Pubescent Umbellate Spicebush 231468
Pubescent Vanillagrass 196153
Pubescent Vetch 408271
Pubescent Violet 409975,410466
Pubescent Water Lily 267701
Pubescent Whing-hackberry 320412
Pubescent Willow 343865
Pubescent Winged Euonymus 157310
Pubescent Winged Spindle-tree 157310
Pubescent Woolly Buttercup 326074
Pubescent Wushan Newlitse 264115
Pubescent Xizang Luculia 238107
Pubescent Yam 131762
Pubescent Yanyuan Maple 3582
Pubescent Yunnanclave 238107
Pubescent Zanonia 417082
Pubescentbranch Box 64336
Pubescent-follicled Peony 280154
Pubescentfruit Codariocalyx 98156
Pubescentfruit Pearlsedge 354141
Pubescentfruit Razorsedge 354141
Pubescentleaf Holly 204189
Pubescentleaf Persimmon 132303
Pubescent-leaf Slenderstyle Acanthopanax 2406
Pubescent-leaved Rhododendron 331266
Pubescentpedicel Henry Acanthopanax 2414, 143612
Pubescent-veined Staff-tree 80267
Pubescest Litseleaf Tanoak 233296
Pubescet Bigleaf Pricklyash 417279
Pubescet Starbush 400598
Pubescet Star-bush 400598
Pubescet-branchlet Birch 53590
Pubescnt Waxmallow 243947
Pubicalyx Lilac 382390
Pubicostate Tutcheria 400713
Pubigenous Raspberry 338492
Pubigerous-flower Latua 222888
Publicans 68189
Puccoon 200175,233692,233698,233707, 233731,233749,233753,345803,345805
Pucelage 409335
Pucellage 409339
Puck Needle 11947,350425
Pucker Needle 350425
Pucte 61909
Pudding 248411
Pudding Pipe Tree 78300
Pudding-berry 85581
Pudding-grass 250432
Pudding-herb 250432
Pudding-pine Tree 78300
Pudgy Munronia 259881
Puding Childvine 400975
Pueblo Mountains Wild Buckwheat 152408
Pueple Diapensia 127921
Pueple Milkvetch 42955
Pueple-flower Milkvetch 42955

Puer Chirita 87949
Puer Gingerlily 187464
Puer Premna 313719
Puer Tea 69644
Pueraria 321419
Puerto Rica 44972
Puerto Rican Royal Palm 337875
Puerto Rico Acrocomia 6133
Puerto Rico Hat Palm 341407
Puerto Rico Royalpalm 337875
Puff Clocks 384714
Puff Clover 396908
Puff-sheath Drop-seed 372782
Puge Azalea 331579
Puge Monkshood 5507
Puge Orchis 310928
Puge Rhododendron 331579
Puget Sound Agoseris 11457
Pug-in-a-pinner 315106
Pugionium 321478
Pugley's Marsh Orchid 273681
Puiruellaine 311899
Puka 181267,250949
Pukamole 146633,240057
Pukang Brome 60898
Pukanui 250949
Pukapuka 58477
Pukeen Monkshood 5507
Pukeweed 234547
Pukotea 223012
Pulai Rattan Palm 65773
Pulan Barberry 52087
Pulan Buttercup 325639
Pulan Catchfly 363946
Pulan Goosecomb 335365
Pulan Kobresia 217130
Pulan Larkspur 124534
Pulan Roegneria 335365
Pulan Sandwort 32168
Pulan Swertia 380159
Pulan Windhairdaisy 348096
Pulasan 265137,265138
Pulas-tree 64148
Puliall Mountain 391397
Puliall-mountain 391365
Puliall-royal 250432
Pulicaria 321509
Pull Poker 217055,302034,401112
Pull-and-hold-back 300944
Pull-dailies 273469
Pullen Paraserianthes 283889
Pull-poker 401112
Pulplish Doubleflower Shrubalthea 195294
Pulque Agave 10933
Pulsatilla 321653
Pulsatille 23969,321708
Pulse Family 224085
Pulszky Pleurospermum 304838
Pulszky Ribseedcelery 304838
Pulu 342134
Pulverulent Primrose 314863
Pulvinate Cudweed 178366
Pulvinate Pussytoes 26583
Pulvinate Sandwort 32155
Pulvinate Spindle-tree 157814
Pulvinate Sumac 332787
Pulvinate Supine Chickweed 374831
Pulvinate Windhairdaisy 348696
Pulza 212127
Pumice Alpinegold 199225
Pumila Norway Spruce 298208
Pummelo 93579
Pummelo Orange 93717
Pumpernel 21339
Pumpion 114245
Pumpkin 114271,114273,114288,114292,114300
Pumpkin Ash 168073,168120
Pumpkin Pine 300211
Pumpkin Winter Squash 114292
Pumpwood 80003
Punah 387305
Punaluu White Hibiscus 194722
Punctate Ardisia 31502
Punctate Ash 168077
Punctate Aspidistra 39571
Punctate Bauhinia 49094
Punctate Bladderwort 403295
Punctate Bothriochloa 57549
Punctate Bunge Pricklyash 417182
Punctate Camellia 69531
Punctate Combretum 100728
Punctate Corydalis 106366
Punctate Cyrtopodium 120715
Punctate Dendrotrophe 125791
Punctate Dipodium 133399
Punctate Gentian 173776
Punctate Goniostemma 179401
Punctate Holly 204190
Punctate Maxillaria 246720
Punctate Monomeria 257711
Punctate Nervilia 265424
Punctate Notylia 267218
Punctate Orchid 273602
Punctate Orchis 273602
Punctate Parasiticvine 125791
Punctate Persimmon 132364
Punctate Pruinose Barberry 52069
Punctate Rabbitbrush 150362
Punctate Rockfoil 349691
Punctate Setose Sedge 76242
Punctate Staff-tree 80285
Punctate Stinkroot Holeglumegrass 57549
Punctate Tea 69531
Punctate Temple-bells 366718
Punctate Wampee 94207
Punctate Yushan Gentian 173875
Punctated Landpick 308635
Punctateleaf Slugwood 50597
Punctate-leaved Cheirostylis 86690
Punctate-leaved Slugwood 50597
Punctatelimb Holly 204190
Punctatelip Woodbetony 287372
Punctuate Hawthorn 109970
Punctulate Ceriman 258178
Punctulate Mosla 259323
Punctulate Rockfoil 349827
Punctulate Sweetleaf 381423
Punctulate Tailgrape 35048
Punctulate Yushania 416810
Puncture Vine 395082,395146
Puncturebract 278661
Puncturevine 395146
Puncture-vine 395146
Puncturevine Caltrop 395146
Pune-tree 248104
Punge Knotwood 44271
Pungent Crossandra 111755
Pungent Litse 234045
Pungent Oak 324316
Pungent Oryzopsis 276055
Pungent Rabbitbrush 150341
Pungent Raspberry 339130
Pungent Sensitivebrier 352636
Punjab Fig 165421
Punjab Sumac 332789
Punk Needle 11947,153767,350425
Punk Tree 248104
Punktree 248114
Punnai-nutus 67860
Punting Pole Bamboo 47408,47494
Punting Pole Bambusa 47408
Puntingpole Bamboo 47408,47494
Puntong Bambusa 47408
Puny Triumfetta 399197
Puppy Dog's Mouth 231183
Puqi Fritillary 168491
Pur Apple 300223
Puran Sandwort 32168
Pur-apple 300223
Purau 195311
Purchury Bean 263059
Purdie Podocarpus 306515
Purdie Yaccatree 306515
Purdom Barberry 52089
Purdom Butterflybush 62149
Purdom Butterfly-bush 62149
Purdom Gentian 173778
Purdom Greenorchid 137640
Purdom Leptodermis 226127
Purdom Peashrub 72321

Purdom Pea-shrub 72321
Purdom Poplar 311444
Purdom Primrose 314867
Purdom Rhododendron 331588
Purdom Stonecrop 357054
Purdom Wildclove 226127
Purdy Iris 208783
Purdy's Brodiaea 60519
Purdy's Fritillary 168541
Purdy's Iris 208783
Pure Red Rose 336918
Pure White Azalea 332094
Pure White Mockorange 294565
Pure White Rhododendron 332094
Pure White Scabrous Deutzia 127078
Purecolor Azalea 331933
Purecolor Butterflygrass 392895
Purecolor Vanda 404669
Purered Azalea 331866
Purestwhite Lycaste 238752
Pure-white Butterfly-bush 61989
Purewhite Iris Moraea 258531
Pure-whiteflower Martagon Lily 229925
Pureyellow Azalea 330396
Pureyellow Rhododendron 330396
Pure-yellow Rhododendron 330396
Purge Nut 212127
Purging Buckthorn 328642
Purging Cassia 78300
Purging Cassia Bark 78300
Purging Croton 113039
Purging Fistula 78300
Purging Flax 231884
Purging Nut 212127
Purging Root 158692
Purging Thorn 328642
Purgingnut 212127
Purification Flower 169719
Puriri 411329
Purloined Slipper 120375
Purple 177906
Purple African Nightshade 367362
Purple Ajania 13024
Purple Allamanda 14872,113823
Purple Amaranth 18670,18687,18696,18788
Purple Amole 88481
Purple Amur Arisaema 33246
Purple Anemone 23890
Purple Anhui-Hubei Sage 345289
Purple Anise 204517
Purple Apple 54436,123077
Purple Apple Berry 54436
Purple Apricot 34431
Purple Aster 41342,107161,240377,380947
Purple Avens 175444
Purple Awngrass 255873
Purple Azalea 331352
Purple Baby Toes 168330
Purple Balkan Maple 3014
Purple Bamboo 297367
Purple Barberry 52226
Purple Bauhinia 49211
Purple Beauty Berry 65670
Purple Beauty Bush 65670
Purple Beautyberry 65670
Purple Beech 162404,162415,162418
Purple Begonia 50205
Purple Bell Vine 329987
Purple Beneath Synotis 381948
Purple Bergamot 257169,257178
Purple Bergenia 52533
Purple Berry Barberry 51303
Purple Bindweed 208120
Purple Biond's Magnolia 242010
Purple Birdlingbran 87260
Purple Bladderpod 227023
Purple Bladderwort 403296
Purple Bletia 55549
Purple Blow Maple 3714
Purple Bluegrass 306143
Purple Boneset 158264
Purple Bract Larkspur 124539
Purple Broad Vetch 408505
Purple Broom 85401
Purple Broomrape 275173
Purple Butterbur 292372,292386
Purple Buttons 216753
Purple Cabbage 258793
Purple Calyx Milkvetch 42904
Purple Camel's-foot Tree 49211
Purple Camellia 69535
Purple Camomile 8332,398134
Purple Cane Raspberry 338970
Purple Cestrum 84406,84425
Purple Cherry Plum 316302
Purple Chinese Houses 99838
Purple Chokeberry 34869,323170
Purple Cinnabar-red Rhododendron 330415
Purple Cinquefoil 100257
Purple Clematis 95168,95449
Purple Clover 397019,397037
Purple Cock's Head 42262
Purple Cockle 11947
Purple Colt's-foot 197943,197944
Purple Cone Flower 140081
Purple Coneflower 140066,140067,140081
Purple Cone-flower 140081
Purple Coral Pea 185764,215981
Purple Corydalis 106282
Purple Corydalls 106453
Purple Cow-wheat 248161
Purple Crab 243545,243555,243667,323097
Purple Crane's-bill 174720,174852
Purple Cranesbill 174852
Purple Crazyweed 278875,279185
Purple Cress 72740
Purple Crocus 99297
Purple Crowberry 145061
Purple Crown Vetch 105324
Purple Crown-vetch 105324
Purple Cudweed 170926,178367
Purple Cupdaisy 115641
Purple Curvedstamencress 72945,238020
Purple Cushion Wild Buckwheat 152364
Purple Custardapple 25880
Purple Cyathocline 115641
Purple Dawyck 162408
Purple Dayflowerlike Thorowax 63608
Purple Dead Nettle 220416
Purple Deadnettle 220416
Purple Dead-nettle 220416
Purple Dewdrop 234707
Purple Dew-plant 134395,340548,340601
Purple Drop 129308
Purple Duckweed 224380
Purple Early Lilac 382230
Purple Ebei Fritillary 168344
Purple Echinacea 140081
Purple Eight Trasure 200812
Purple Eulophia 156951,157020
Purple European Beech 162404
Purple Eyebright 160252
Purple False Brome 58578
Purple False Foxglove 10153
Purple False Holly 276318
Purple False Oats 398499
Purple Falsebrome 58578
Purple Fan 350324
Purple Fescue 164243
Purple Fingers 130383
Purple Fishscale Bamboo 297298
Purple Flag 208583
Purple Fleabane 150915
Purple Fleablane 150419
Purple Flower Bean 241261
Purple Flower Meconopsis 247194
Purple Flower Onion 15780
Purple Flowering Raspberry 338917
Purple Fountain Grass 289248
Purple Foxglove 130383,130386
Purple Fringe 107300
Purple Fringed Orchid 302491
Purple Fringeless Orchid 302482
Purple Fringeless Orchis 302482
Purple Funiushan Magnolia 242120
Purple Galangal 17745
Purple Gariand-lily 68033
Purple Gentian 157020,173781,173834
Purple Geranium 174485
Purple Gerardia 10153
Purple Giant Hyssop 10420

Purple Giant-hyssop 10420
Purple Ginger 418017
Purple Glabrous Stringbush 414182
Purple Glands Azalea 330673
Purple Glands Ovary Rhododendron 330673
Purple Glasswort 342893
Purple Globe-tulip 67562
Purple Glycine 185764
Purple Goat's Beard 394327
Purple Goatsbeard 394327
Purple Goose Grass 362097
Purple Granadilla 285637,285639
Purple Grape-hyacinth 260301
Purple Gromwell 233772
Purple Grondbox Milkwort 307984
Purple Ground-chen-y 297693
Purple Groundcherry 324643
Purple Groundsel 358788
Purple Guava 318737
Purple Hair-grass 259688
Purple Hakea 184631
Purple Haricot 218721
Purple Hawkweed 348121
Purple Heart 288870,361985,361987,394057
Purple Heath 149138,149955
Purple Heather 149205
Purple Hedge Nettle 220416
Purple Hellebore 190955
Purple Helleborine 147112
Purple Hoarhound 245764
Purple Hoarypea 386257
Purple Horned Poppy 335638
Purple Hyacinth 273531
Purple Hypoestes 202604
Purple Ice Plant 220651
Purple Indian Dropseedgrass 372666
Purple Iris 208924
Purple Jacobea 358788
Purple Jimsonweed 123077
Purple Joe-pye-weed 158264
Purple Lantana 221284
Purple Largeflower Sobralia 366788
Purple Largewingbean 241261
Purple Larkspur 124539
Purple Lasianthus 222248
Purple Leaf Plum 316294
Purple Leaf Sand Cherry 316175
Purple Leaved Lesser Celandine 325820
Purple Lettuce 313866
Purple Lily 230035
Purple Liparis 232253
Purple Loco 42739,278940,279174
Purple Locoweed 278940
Purple Loose Strife 240068
Purple Loosestrife 239836,240025,240068
Purple Loosestrife Family 240020
Purple Love Grass 147971
Purple Lymegrass 144182
Purple Lythrum 240068
Purple Magnolia 416707
Purple Mallow 67128
Purple Manyleaf Woodbetony 287448
Purple Marsh Orchid 273594
Purple Mat 262121
Purple Meadow Rue 388469
Purple Meadow-parsnip 388899,388902
Purple Meadow-rue 388469,388638
Purple Meconopsis 247194
Purple Medick 247456
Purple Mexican Aster 107161
Purple Michelia 252858
Purple Milk Vetch 42262
Purple Milkvetch 42262
Purple Milk-vetch 42262
Purple Milkweed 38077
Purple Milkwort 308275,308276,308331,
308449
Purple Mint 290940
Purple Mint Bush 315605
Purple Mistress 258793
Purple Mombin 372479
Purple Monkshood 5177
Purple Moor Grass 256601
Purple Moorgrass 256601
Purple Moor-grass 256597,256601
Purple Mosquitotrap 117660
Purple Mountain Avens 175444
Purple Mountain Milk Vetch 278875
Purple Mountain Saxifrage 349719
Purple Mountainavens 175444
Purple Muhly 259695
Purple Mullein 405748
Purple Nan 295403
Purple Nanmu 295403
Purple Narenga 262552
Purple Needle Grass 262620
Purple Needlegrass 398358
Purple Needle-grass 376891
Purple Newlitse 264089
Purple Nightshade 367120,367753
Purple Nut 106760
Purple Nutgrass 119503
Purple Nut-sedge 119503
Purple Orchid Tree 49211
Purple Orchid-tree 49247
Purple Orchis 273604
Purple Origano 274223
Purple Orris-root 208583
Purple Osier 343937
Purple Osier Willow 343938,343939
Purple Osier-willow 343937
Purple Owl's Clover 79124
Purple Oxytropis 278875
Purple Paintbrush 79133
Purple Pampas Grass 105453
Purple Panicgrass 58128
Purple Pappus Galatella 169770
Purple Passion Flower 285654
Purple Pea 198762,198763
Purple Perilla 290968
Purple Phacelia 293154
Purple Phaius 293523
Purple Phyllary Fleabane 150885
Purple Pitcher Plant 347156
Purple Pitcher-plant 347156,347158
Purple Plantainlily 198592
Purple Pleat-leaf 17580
Purple Poppy Mallow 67125,67128
Purple Prairie Clover 121902
Purple Prairie-clover 121902
Purple Prickly Pear 272970
Purple Pricklypear 272970,273042
Purple Purplepearl 65670
Purple Queen Tradescantia 394057
Purple Ragwort 211295,358788,358996
Purple Ramping-fumitory 169151
Purple Raspberry 338646,338873
Purple Rattlebox 112474,361422
Purple Rattle-box 361422
Purple Rattlesnake Root 313866
Purple Rattlesnakeroot 313871
Purple Rattlesnake-root 313866,313871
Purple Red Silkyarn 238965
Purple Reedbentgrass 65471
Purple Rhododendron 330331
Purple Robe 367753
Purple Robe Locust 334948
Purple Rockcress 44868
Purple Rock-cress 44867,44868
Purple Rocket 85875,207353
Purple Rockrose 93199
Purple Roscoea 337099
Purple Rose 336495
Purple Rose-root 329937
Purple Rugose Rose 336910
Purple Sage 344838,345011,345169
Purple Salsify 394327
Purple Sand Grass 397939
Purple Sand Verbena 691
Purple Sandspurry 370690
Purple Sand-spurry 370696
Purple Sanicle 345943
Purple Saxifrage 349719
Purple Schizachne 351254
Purple Sea Spurge 159557
Purple Sesban 361422
Purple Sesbane 361422
Purple Shiningfruit Dragon Mallow 367498
Purple Side-saddle Flower 347156
Purple Silvergrass 255873
Purple Small Bugbane 91024

Purple Small-reed 65313
Purple Smoke Bush 107306,107313
Purple Smoke Indigo 47855
Purple Smoketree 107313
Purple Spapweed 205265
Purple Speargrass 300770
Purple Spike Bristlegrass 361959
Purple Spike Roegneria 335478
Purple Spikerush 143053
Purple Spiradiclis 371777
Purple Spots Rhododendron 331543
Purple Spring-cress 72740
Purple Sprouting 59545
Purple Spurge 159557
Purple Spurless Spapweed 205123
Purple Spurless Touch-me-not 205123
Purple Square-bamboo 87603
Purple Star-thistle 80981
Purple Stauntonvine 374436
Purple Stinkweed 123077
Purple Stokes Aster 377293
Purple Stonecrop 200812,357203
Purple Strife Loose 240068
Purple Sweet Sultan 18983
Purple Tansy Aster 240386
Purple Taro 99936
Purple Teasel 133459
Purple Tephrosia 386257
Purple Thistle 91843
Purple Thorn-apple 123077
Purple Threeawn 34002
Purple Toadflax 231084
Purple Toadshade 397551
Purple Toothwort 222644
Purple Touch-me-not 205265
Purple Trachydium 304826
Purple Trailing Lantana 221284
Purple Trillium 397556,397592,397599
Purple Turtlehead 86795,86796
Purple Tussock Grass 262620
Purple Twayblade 232225
Purple Twining Pea 185764
Purple Veined Willowherb 146670
Purple Velvet Flower 18687
Purple Velvet Plant 183056
Purple Verbena 405885
Purple Vermilion Azalea 330415
Purple Vetch 408257,408305
Purple Viper's Bugloss 141258
Purple Viper's Grass 354925
Purple Virgin's-bower 44226
Purple Wake Robin 397599
Purple Water Flag 208924
Purple Water Iris 208924
Purple Weeping Beech 162414
Purple Wen-dock 59148
Purple Wild Sage 345455
Purple Wildberry 362369
Purple Wildginger 37703
Purple Wildryegrass 144182
Purple Wildsenna 389514
Purple Willow 343937,344277
Purple Willowherb 240068
Purple Wing Pod Pea 237771
Purple Wing-pod Pea 237771
Purple Winter Creeper 157473
Purple Wood Spurge 158434
Purple Woodsorrel 278039,278147
Purple Wreath 292517,292518
Purple Wreath Petrea 292518
Purple Wreath Vine 292518
Purple Xinjiang opium 335638
Purple Yulan 416707
Purple Zinger 418017
Purpleanise 204517
Purpleanther Field Pepperweed 225371
Purpleanther Sandwort 31963
Purpleawn Lymegrass 144436
Purpleawn Wildryegrass 144436
Purpleback Addermonth Orchid 243116
Purpleback Azalea 330724
Purpleback Bogorchis 243116
Purpleback Buckthorn 328867
Purpleback Cacalia 283830
Purpleback Clearweed 299029
Purpleback Coldwaterflower 299029
Purpleback Guizhou Sage 344946
Purpleback Murdannia 260093
Purpleback Pyrola 322798
Purpleback Snakegourd 396250
Purple-back Strigose Hydrangea 200116
Purpleback Tailanther 381948
Purpleback Velvetplant 183124
Purpleback Waterbamboo 260093
Purple-backed Buckthorn 328867
Purplebeak Sedge 76233
Purplebell Cobaea 97757
Purple-bell Cobaea 97757
Purple-berry 345631
Purpleberry Barberry 51303
Purple-berry Barberry 51303
Purple-berry Cotoneaster 107337
Purplebloom Maple 3479
Purpleblotch Gladiolus 176431
Purpleblow Maple 3714
Purpleblue Azalea 331713
Purple-blue Cape-cowslip 218928
Purpleblue Rhododendron 331713
Purple-blue Rhododendron 331713
Purplebract Argyreia 32651
Purplebract Didymocarpus 129951
Purplebract Everlasting 21664
Purplebract Gentian 150885
Purplebract Iris 208797
Purplebract Pothos 313195
Purplebract Saussurea 348395
Purplebract Swordflag 208797
Purplebract Windhairdaisy 348395,348699
Purple-bracted Plantain Lily 198633
Purplebranch Willow 343477
Purple-branched Willow 343477
Purpleburnet 174935
Purple-bush 386964
Purplecalyx Butterflygrass 392948
Purplecalyx Chalkplant 183229
Purplecalyx Chirita 87822
Purplecalyx Cranesbill 174868
Purplecalyx Milkvetch 42904
Purplecalyx Swertia 380211
Purplecalyx Torenia 392948
Purplecone Fir 455
Purplecone Norway Spruce 298215
Purple-cone Spruce 298417
Purple-coned Spruce 298417
Purpleconed Sugar Pine 300011
Purplecup Mockorange 294457
Purple-cupped Mockorange 294457
Purpledaisy 267166
Purpledisc Sunflower 188926
Purpledot Paphiopedilum 282827
Purpledoubleflower Shrubalthea 195299
Purplefloer Baldstem Eriocycla 151750
Purpleflower Amplexicaul Swallowwort 117368
Purpleflower Angelica 24336
Purpleflower Asiabell 98405
Purpleflower Azalea 330081
Purpleflower Bifrenaria 54244
Purpleflower Bigleaf Thorowax 63737
Purpleflower Blue Orchid 193164
Purpleflower Boeica 56368
Purpleflower Brownhair Clematis 94938
Purpleflower Butterflybush 62054
Purpleflower Chesneya 87260
Purpleflower Childvine 400898
Purpleflower Chinese Sage 345393
Purpleflower Clethra 96544
Purple-flower Cliffbean 254737
Purpleflower Clubfruitcress 273969
Purpleflower Columbine 30078
Purple-flower Common Lilac 382372
Purpleflower Currant 334078
Purpleflower Daphne 122590
Purpleflower Dayflowerlike Thorowax 63608
Purpleflower Deadnettle 220401
Purpleflower Deerdrug 242698
Purple-flower Deutzia 127054
Purpleflower Divaricate Serpentroot 354846,354850
Purpleflower Eight Trasure 200790
Purpleflower Elsholtzia 143962

Purpleflower Erysimum 154457
Purpleflower Euonymus 157800
Purpleflower False Solomonseal 242698
Purpleflower Feathergrass 376893
Purpleflower Fewtwig Asparagus 39127
Purpleflower Goldenray 229023
Purpleflower Hairleaf Nervilia 265420
Purpleflower Hairleaf Taroorchis 265420
Purpleflower High Monkshood 5563
Purpleflower Hirsute Rockcress 30304
Purpleflower Isometrum 210183
Purple-flower Laelia 219690
Purpleflower Lecanorchis 223658
Purpleflower Licorice 177939
Purpleflower Lily 230035
Purple-flower Longleaf Deutzia 126994
Purpleflower Lycianthes 238965
Purpleflower Michelia 252858
Purple-flower Millettia 254737
Purpleflower Mosquitotrap 117368
Purpleflower Needlegrass 376893
Purpleflower New Eargrass 262958
Purpleflower Oreocharis 273834
Purple-flower Ormosia 274431
Purple-flower Pagodatree 369144
Purple-flower Porana 131254
Purpleflower Primrose 314117
Purpleflower Rattlebox 112474
Purpleflower Resurrectionlily 215004
Purpleflower Rockfoil 349098
Purpleflower Rockjasmine 23290
Purple-flower Rosewood 121623
Purpleflower Roundbract Sage 344992
Purpleflower Rugged Corydalis 106409
Purple-flower Sabia 341550
Purpleflower Scabrate Corydalis 106409
Purpleflower Scalecalyx Crazyweed 279177
Purpleflower Sibbaldia 362369
Purple-flower Sideritis 362746
Purpleflower Siebold Elder 345696
Purpleflower Silvery Cinquefoil 312393
Purpleflower Sophora 369113
Purple-flower Sophora 369144
Purpleflower Speargrass 4157
Purpleflower Spiraea 372069
Purpleflower Stanavoi Clematis 94938
Purpleflower Starjasmine 393625
Purpleflower Sterigmostemum 273969
Purpleflower Stonecrop 200790
Purpleflower Sugarmustard 154457
Purpleflower Summerlilic 62054
Purpleflower Swallowwort 117660
Purpleflower Syncalathium 381651
Purpleflower Taiwan Toadlily 396614
Purpleflower Thermopsis 389514
Purpleflower Thinstem Goldenray 229223
Purpleflower Toadflex 230925
Purpleflower Tsai Monkshood 5643
Purpleflower Twayblade 232253
Purpleflower Twocolored Monkshood 5027
Purpleflower Violet 410030,410412
Purpleflower Yushanbamboo 416822
Purple-flowered Butterfly-bush 62054
Purple-flowered Clethra 96544
Purple-flowered Cold Orchis 116928
Purple-flowered Cotoneaster 107352
Purple-flowered Datura 123077
Purple-flowered Deutzia 127054
Purple-flowered Elsholtzia 143962
Purple-flowered Elsholtzia-like Keiskea 215808
Purple-flowered Euonymus 157800
Purple-flowered Fritillary 168542
Purple-flowered Holly 203625
Purple-flowered Jacaranda 211238
Purple-flowered Maple 3479
Purple-flowered Michelia 252858
Purple-flowered Ormosia 274431
Purple-flowered Raspberry 338917
Purple-flowered Sabia 341550
Purple-flowered Sandwort 32169
Purple-flowered Spindle-tree 157800
Purple-flowered Spiraea 372069
Purple-flowered Star-jasmine 393625
Purple-flowered Sweetvetch 187792
Purple-flowered Tickclover 126551
Purple-flowered Yushania 416822
Purple-flowering Raspberry 338917
Purplefruit Actinidia 6525
Purplefruit Dandelion 384825
Purplefruit Holly 204354
Purple-fruit Japanese Raspberry 338995
Purplefruit Kiwifruit 6525
Purplefruit Manyflower Cotoneaster 107584
Purple-fruit Sophora 368958
Purplefruit Spikesedge 143053
Purplefruit Tea 69535
Purple-fruited Holly 204354
Purple-fruited Prickly Pear 273003
Purple-globe Clover 396818
Purple-grass 240068,247246
Purplegreen Rhodiola 329928
Purplehair Bistamengrass 127481
Purplehair Croton 112999
Purplehair Gentian 174058
Purplehair Groundsel 365079
Purplehair Melastoma 248767
Purplehair Pocktorchid 282914
Purplehair Rabdosia 209659
Purplehair Taiwan Plumegrass 148873
Purple-haired Melastoma 248767
Purple-haired Rabdosia 209659
Purple-hairy Melastoma 248767
Purplehammer 313552
Purplehead Rattlesnakeroot 313866
Purplehead Serpentroot 354925
Purplehead Sneezeweed 188419
Purple-headed Sneezeweed 188419,188429
Purpleheart 288863
Purpleheart Skullcap 355702
Purpleleaf 180270
Purple-leaf 200145
Purple-leaf Begonia 50394
Purple-leaf Birch 53565
Purpleleaf Canna 71199
Purpleleaf Cherry Plum 316302
Purple-leaf Clover 397044
Purpleleaf Cremanthodium 110450
Purpleleaf European Grape 411981
Purple-leaf Filbert 106760
Purpleleaf Hedge Maple 2841
Purpleleaf Lagotis 220192
Purpleleaf Malpighia 243526
Purpleleaf Maximowicz Maple 3138
Purpleleaf Metabriggsia 252388
Purpleleaf Nutantdaisy 110450
Purple-leaf Sand Cherry 316175
Purpleleaf Slugwood 50598
Purpleleaf Tylophora 400952
Purpleleaf Violet 410059,410060
Purpleleaf Willowweed 146670
Purpleleaf Wintercreeper 157509
Purple-leaved Black Elderberry 345643
Purple-leaved Common Barberry 52343
Purple-leaved Elderberry 345635
Purple-leaved Geranium 174833
Purple-leaved Ivy 187228
Purple-leaved Ninebark 297844
Purple-leaved Ottawa Barberry 51992
Purple-leaved Planetree Maple 3474
Purple-leaved Plantain 302069
Purple-leaved Plum 316302
Purple-leaved Sage 345273
Purple-leaved Sand Cherry 316175
Purple-leaved Slugwood 50598
Purple-leaved Smokebush 107304
Purple-leaved Spiderwort 394076
Purple-leaved Tylophora 400952
Purple-leaved Willow Herb 146670
Purplelines Pocktorchid 282889
Purplenerved Hornbeam 77379
Purplenet Toadflax 231106
Purplenymph Rhododendron 330452
Purpleosier 343937
Purpleosier Willow 343937
Purplepearl 66727,66743
Purplepearl Clubfilment 179177
Purplepearlleaf Meliosma 249369
Purple-phyllaries Pearleverlasting 21664
Purpleprickle Euonymus 157321
Purplequeen Azalea 330714

Purplequeen Rhododendron 330714
Purplered Addermonth Orchid 110652
Purple-red Amomum 19907
Purplered Bogorchis 110652
Purple-red Epidendrum 146459
Purplered Gentianella 174115
Purplered Lily 229725
Purplered Middayflower 289684
Purplered Mullein 405748
Purplered Pentapetes 289684
Purplered Raspberry 339320
Purple-red Raspberry 339320
Purplered Spapweed 205221
Purple-red Stelis 374714
Purplered Touch-me-not 205221
Purpleridge Lily 229891
Purples 273531
Purple-scaled Sedge 75936
Purplesepal Rabdosia 209675
Purple-sepal Touch-me-not 205229
Purple-sepaled Falsecuckoo 48313
Purplesheath Seseli 361558
Purple-sheathed Bambusa 47421
Purple-sheathed Graceful Sedge 74694
Purple-sheathed Sedge 75933
Purple-shelled Fargesia 162737
Purple-spike Galingale 119382
Purplespike Goosecomb 335478
Purplespike Vernalgrass 27959
Purple-spiked Loosestrife 240068
Purplespine Euonymus 157321
Purplespot Alplily 234224
Purplespot Azalea 331895
Purplespot Butterflygrass 392904
Purplespot Habenaria 184002
Purplespot Lady's Slipper 120357
Purplespot Ladyslipper 120357
Purplespot Orchis 121382,121389
Purple-spot Parrot-lily 18069
Purplespot Xiangcheng Lily 230078
Purple-spoted Rhododendron 331543
Purple-spotted Scarletball Rhododendron 331895
Purplespur Corydalis 106342
Purplespur Epimedium 146989
Purplessel Goldenray 229135
Purple-stained Toloache 123049
Purple-stalked Basil 268614
Purplestamen Pulsatilla 321688
Purple-stamened Rhododendron 330452
Purplestem 376428
Purplestem Aster 380970
Purple-stem Aster 380970
Purple-stem Beggar-ticks 53845
Purplestem Beijing Skullcap 355672
Purple-stem Cat's-tail 294989
Purplestem Cremanthodium 110459
Purplestem Dock 339922
Purplestem Fleabane 150914
Purplestem Gentian 150914
Purplestem Hogfennel 293064
Purplestem Nutantdaisy 110459
Purplestem Sinocarum 364878
Purplestem Taro 415202
Purple-stemmed Angelica 24303,24304
Purple-stemmed Tickseed 53845
Purplestreak Alstroemeria 18073
Purplestriate Galangal 17666
Purplestriped Chirita 87947
Purpletassel Milkaster 169770
Purple-tongue Eulophia 156769
Purpletop 396706
Purple-top 396706,405812
Purpletop Vervain 405812,405813
Purpletwig Ainsliaea 12728
Purpletwig Rabbiten-wind 12728
Purplevein Ardisia 31628
Purplevein Knotweed 309676
Purplevein Ophiorrhiza 272285
Purple-veined Ardisia 31570,31628
Purple-veined Mitreola 256217
Purplevine 414541,414576
Purple-washed Helleborine 147212
Purplewinged Russianthistle 344434
Purplish Croton 112999
Purplish Larkspur 124272
Purplish Lightviolet Nepeta 265107
Purplish Newlitse 264089
Purplish Red Mock Azalea 250532
Purplr Spathoglottis 370401
Purpule Japanese Maple 3302
Purpurascent Biond's Magnolia 242011
Purpurascent Whytockia 414005
Purpus Privet 229593
Purpus Wild Buckwheat 152190
Purpus' Birch 53310,53598
Purpus' Larkspur 124540
Purret 15621
Purrut 93492
Purse 329521
Purse Flower 72038
Purse Tassels 260301
Pursh Plantain 302122,302144
Pursh Ribgrass 302144
Pursh's Bulrush 352252,352253
Pursh's Crowfoot 326269
Pursh's Seepweed 379503
Pursh's Seep-weed 379503
Purslane 94292,311817,311890,311915
Purslane Family 311954
Purslane Speedwell 407275,407283
Purslane Tree 311956
Pursley 311890,311899
Purtorican Palmetto 341407
Puschkinia 321892
Pusley 311890,311899,311921
Pussies 401112
Pussley 311890,311899
Pussy Cat's Tail 106703
Pussy Cat's Tails 106703
Pussy Ear 215277
Pussy Ears 67631,115587,215277
Pussy Face 410677
Pussy Foot 397043
Pussy Tail 321102
Pussy Toe 26385,26490
Pussy Toes 26299,26490,26530,26551
Pussy Willow 343151,343302
Pussy's Tail 401112
Pussy's Tails 401112
Pussy's Toes 26299,26385,26536
Pussy's-toes 26299
Pussy-cats 106703
Pussypaws 93087
Pussy-paws 68607
Pussytoes 26299,26385,26490,26530,26551
Pussy-toes 26596
Pussywillow 343302
Pustulate Macaranga 240312
Putaputaweta 77483
Putchock 208640
Puto Hornbeam 77380
Putty Root 29365
Puttyroot 29365
Putty-root 29359,29365
Putuo Hornbeam 77380
Putuo Sedge 75940
Putuo Southstar 33476
Puwen Nanmu 295401
Puwen Phoebe 295401
Puya 321936,321938
Puzzl Gentian 173384
Puzzle Primrose 314497
Puzzled Mallotus 243352
Puzzle-Monkey 30832
Pyanot 280231
Pycnarrhena 321977,321981
Pycnoplinthus 322060
Pycnospora 322076
Pycreus 322179
Pygeum 322386,322412,322465
Pygmaeopremna 322421
Pygmean Milkvetch 42768
Pygmy Alpinegold 199229
Pygmy Aster 380974
Pygmy Bamboo 347279
Pygmy Barrel Cactus 140618
Pygmy Bluehearts 61847
Pygmy Buttercup 326259,326273
Pygmy Date Palm 295475,295484
Pygmy Datepalm 295484

Pygmy Date-palm 295484
Pygmy Deer-weed 237636
Pygmy Euonymus 157816
Pygmy Fleabane 150919,151036
Pygmy Galangal 17743
Pygmy Mahonia 242622
Pygmy Pine 225602
Pygmy Pink 346439
Pygmy Pussytoes 26474
Pygmy Sage 36131
Pygmy Serpentweed 392781
Pygmy Smartweed 309395
Pygmy Spindle-tree 157816
Pygmy Starwort 375043
Pygmy Water Lily 267767,267770
Pygmy Waterlily 267767,267770
Pygmy Water-lily 267695,267767
Pygmy Willow 343476
Pygmy Woodbetony 287570
Pygmy Yunnan Pine 300306
Pygmy-flowered Vetch 408492
Pygmy-like Buttercup 326259
Pygmy-poppy 71079
Pygmyweed 108775,391918,391923
Pygmy-weed 391930
Pygny Little Fragrant Bamboo 87635
Pyinkado 415693
Pyinma 219966
Pyjama Lily 111212
Pylzow Cranesbill 174855
Pyracantha 322454
Pyramid Blechum 55207
Pyramid Chinese Juniper 213658
Pyramid Hybrid Yew 385389
Pyramid Japanese Yew 385357
Pyramid Magnolia 242267
Pyramid Orchid 21176
Pyramid Rattlesnakeroot 313895
Pyramid Silver Maple 3537
Pyramid Tree 220228
Pyramidal Ajuga 13167
Pyramidal American Linden 391651
Pyramidal Asparagus 38987
Pyramidal Billbergia 54456
Pyramidal Bugle 13167
Pyramidal Epimedium 147029
Pyramidal Grevill 180639
Pyramidal Italian Cypress 114765
Pyramidal Japanese Spindle-tree 157620
Pyramidal Juniper 341779
Pyramidal Largeleaved Linden 391821
Pyramidal Larkspur 124548
Pyramidal Lobelia 234715
Pyramidal Loosestrife 239807
Pyramidal Machilus 240685
Pyramidal Magnolia 242064
Pyramidal Microtropis 254320
Pyramidal Mullein 405756
Pyramidal Orchid 21162,21176
Pyramidal Poplar 311226
Pyramidal Premna 313722
Pyramidal Saxifrage 349216
Pyramidal Siberian Elm 401607
Pyramidal Star-of-bethlehem 274746
Pyramidal Watson Flower 413391
Pyramidal Yellow Cedar 85303
Pyramidal-bud Microtropis 254320
Pyramidate Galanga Galangal 17679
Pyramidate Melochia 249638
Pyramidical Koeleria 217550
Pyrenacantha 322517,322562
Pyrenaria 322564
Pyrencan Honeysuckle 236051
Pyrene Oil 270099
Pyrenean Asphodel 39454,39455
Pyrenean Aster 41133
Pyrenean Avens 175441
Pyrenean Bellflower 70309
Pyrenean Broom 120995
Pyrenean Buttercup 325570,326273
Pyrenean Columbine 30070
Pyrenean Crane's-bill 174856
Pyrenean Cranesbill 174856
Pyrenean Gentian 173785
Pyrenean Hemp Nettle 170074
Pyrenean Hempnettle 170074
Pyrenean Hyacinth 60339
Pyrenean Oak 324319
Pyrenean Pheasant's Eye 8377
Pyrenean Pine 300112
Pyrenean Pink 127806
Pyrenean Poppy 282737
Pyrenean Saxifrage 349562,350011
Pyrenean Scurvy-grass 98027
Pyrenean Snake's Head 168543
Pyrenean Snake's-head 168543
Pyrenean Snakeshead 168543
Pyrenean Squill 352944
Pyrenean Thistle 92335
Pyrenean Toadflax 231088
Pyrenean Valerian 404337
Pyrenean-violet 325223,325225
Pyrenean-violet Family 175306
Pyrenees Aster 41133
Pyrenees Fritillary 168543
Pyrenees Geranium 174856
Pyrenees Lily 230015
Pyrenees Monk's-hood 5039
Pyrenees Monkshood 5039
Pyrenees Oak 324319
Pyrenees Poppy 282674
Pyrenees Soapwort 346420
Pyrenees Star of Bethlehem 274748
Pyrenees Star-of-bethlehem 274746,274748
Pyreness Honeysuckle 236051
Pyrenocarpa 322632
Pyrethrum 322640,322665
Pyrethrum Flower 322660
Pyrethum Daisy 322665
Pyrgophyllum 322761,322762
Pyriferous Gentian 173239
Pyriform Siberian Crabapple 243562
Pyrola 322775,322801
Pyrola Family 322935
Pyrolaeleaf Valerian 404338
Pyrolaformed Speedwell 407291
Pyrolaleaf Agapetes 10360
Pyrolaleaf Whitepearl 172142
Pyrolaleaf Willow 343968
Pyrolaleaf Wintergreen 172142
Pyrola-leaved Agapetes 10360
Pyrola-leaved Willow 343968
Pyrola-leaved Wintergreen 172142
Pyrostegia 322948
Pyrrie 323133
Pythagorean Bean 263272
Pyxie 323366
Pyysic-nut 212098

Q

Qat 79395
Qi Shan St. John's Wort 201837
Qian Beech 162366
Qian-Gui Machilus 240569
Qiangui Maclure Tolypanthus 392713
Qiangui Ramie 56358
Qian-Gui Tolypanthus 392713
Qianlingshan Holly 204199
Qiannan Bauhinia 49214
Qianning Monkshood 5121
Qianning Pennisetum 289222
Qianning Wolftailgrass 289222
Qianning Yinshancress 416347
Qianning Yinshania 416347
Qianshan Mockorange 294561
Qianshan Sagebrush 35328
Qianshan Tsang Raspberry 339412
Qianshan Vetch 408338

Qianyang Azalea 331591
Qianyang Rhododendron 331591
Qiaojia Barberry 52091
Qiaojia Protected Rockvine 387831
Qiaojia Spapweed 205428
Qiaojia Yushanbamboo 416811
Qiaojia Yushania 416811
Qiaomaidi Sage 345143
Qilianshan Crazyweed 279111
Qilianshan Cremanthodium 110385
Qilianshan Kobresia 217221
Qilianshan Milkvetch 42176
Qilianshan Monkshood 5123
Qilianshan Nutantdaisy 110385
Qilianshan Savin 213871
Qilianshan Sedge 73638
Qilianshan Stonecrop 356708
Qilianshan Swertia 380318
Qilin Larkspur 124099
Qimen Fritillary 168491
Qimen Loosestrife 239808
Qimen Sage 345336
Qimen Skullcap 355409
Qingcheng Clematis 95270
Qingcheng Deutzia 126978
Qingcheng Mountain Greenbrier 366611
Qingcheng Primrose 314236
Qingcheng Smilax 366611
Qingchengshan Bashanbamboo 37293
Qingchengshan Bashania 37293
Qingchengshan Greenbrier 366611
Qingchuan Arrowbamboo 162743
Qingchuan Fritillary 168563
Qingdao Bulrush 353901
Qingdao Cranesbill 174982
Qingdao Lily 230065
Qingdao Sedge 75947
Qinggang 116044,116100
Qinghai Angelica 24419
Qinghai Asparagus 39163
Qinghai Azalea 331592
Qinghai Bicolored Everlasting 21526
Qinghai Bicolored Pearleverlasting 21526
Qinghai Braya 59745
Qinghai Corydalis 106346
Qinghai Draba 137200
Qinghai Dwarfnettle 262226
Qinghai Eritrichium 153480
Qinghai Gooseberry 334162
Qinghai Goosecomb 335386
Qinghai Iris 208784
Qinghai Kangding Larkspur 124632
Qinghai Kengyilia 215932
Qinghai Larkspur 124549
Qinghai Leymus 228389
Qinghai Mandrake 244344
Qinghai Milklettuce 259763
Qinghai Monkshood 5519
Qinghai Morina 258849
Qinghai Nannoglottis 262226
Qinghai Parnassia 284609
Qinghai Peashrub 72206
Qinghai Pea-shrub 72206
Qinghai Pleurospermum 304851
Qinghai Primrose 314874
Qinghai Raspberry 338678
Qinghai Reedbentgrass 65386
Qinghai Rhododendron 331592
Qinghai Ribseedcelery 304851
Qinghai Rockfoil 349835
Qinghai Roegneria 335386
Qinghai Russianthistle 344493
Qinghai Sagebrush 35422
Qinghai Saussurea 348702
Qinghai Sedge 74871,75948
Qinghai Small Reed 127253
Qinghai Smallreed 127253
Qinghai Soroseris 381653
Qinghai Speargrass 4162
Qinghai Spikesedge 143315
Qinghai Stonecrop 357255
Qinghai Swordflag 208784
Qinghai Syncalathium 381653
Qinghai Thermopsis 389546
Qinghai Tickseed 104799
Qinghai Whitlowgrass 137200
Qinghai Wildsenna 389546
Qinghai Willow 343972
Qinghai-Gansu Woodbetony 287304
Qinghai Medic 247251
Qinghe Buttercup 325719
Qinghe Jerusalemsage 295087
Qinghe Milkvetch 42960
Qinghe Sweetvetch 188066
Qingliu Azalea 331442
Qingliu Rhododendron 331442
Qinglong Raspberry 339143
Qingmingflower 49355
Qingshui Akebia 13211,13240
Qingshui Buckthorn 328652
Qingshui Loosestrife 239581
Qingshui Orchis 273355
Qingshui St. John's-wort 202037
Qingshuihe Torularia 264630
Qingshuishan Barberry 51811
Qingshuishan Platanthera 302275
Qingshuishan Solomonseal 308520
Qingyang Sedge 75949
Qingyuan Eargrass 187509
Qingyuan Hedyotis 187509
Qingyuan Holly 204200
Qingyuan Sasa 347284
Qing-Zang Milkvetch 42864
Qingzang Sagebrush 35455
Qinhai Sandwort 32172
Qinjiu 173615
Qinlangdang Elatostema 142871
Qinling Angelica 24495
Qinling Arrowbamboo 162739
Qinling Asiabell 98422
Qinling Azalea 332024
Qinling Bergenia 52539
Qinling Cacalia 283879
Qinling Clematis 95166
Qinling Columbine 30042
Qinling Crazyweed 278783
Qinling Edelweiss 224846
Qinling Fargesia 162739
Qinling Figwort 355068
Qinling Fir 311
Qinling Gentian 174019
Qinling Germander 388296
Qinling Goldsaxifrage 90441
Qinling Goldwaist 90441
Qinling Ladybell 7761
Qinling Larkspur 124228
Qinling Litse 234083
Qinling Maple 3724
Qinling Monkshood 5370
Qinling Mountainash 369547
Qinling Mountain-ash 369547
Qinling Oak 324501
Qinling Pecteilis 285985
Qinling Petrocosmea 292577
Qinling Piratebush 61932
Qinling Pittosporum 301379
Qinling Rhododendron 332024
Qinling Rhubarb 329381
Qinling Rockfoil 349382
Qinling Rose 337000
Qinling Sage 345303
Qinling Sagebrush 36132
Qinling Sedge 74335
Qinling Simon Poplar 311507
Qinling Spapweed 205095
Qinling Speedwell 407422
Qinling Stonebutterfly 292577
Qinling Stonecrop 357002
Qinling Swallowwort 117400
Qinling Thorowax 63714
Qinling Touch-me-not 205095
Qinling Trigonotis 397415
Qinling Willow 343974
Qinling Windhairdaisy 348887
Qinzang Small Reed 127248
Qinzang Smallreed 127248
Qinzhou Garcinia 171178
Qinzhou Tanoak 233353
Qiongdao Persimmon 132279
Qiongdao Poplar 311454
Qionghai Falsenettle 56194

Qionghai Ramie 56194
Qiongnan Decaspermum 123445
Qiongpalm 90619,90620
Qiongya Lasianthus 222288
Qiongya Ophiorrhiza 272181
Qiongya Roughleaf 222288
Qiongzhong Tanoak 233142
Qiongzhu 87622,323381
Qishan St. John's Wort 201837
Qitai Kneejujube 67038
Qiu Peony 280269
Qiubei Camellia 68983
Qiubei Clematis 94824
Qiubei Cymbidium 117025
Qiubei Orchis 117025
Qiubei Rattlebox 112592
Qiujiang Barberry 51767
Qiujiang Cayratia 79866
Qiujiang Corydalis 106040
Qiujiang Cyclobalanopsis 116123
Qiujiang Fleabane 150729
Qiujiang Gentian 150729,173786
Qiujiang Maple 3063
Qiujiang Milkvetch 42183
Qiujiang Mountainash 369439
Qiujiang Oak 116123
Qiujiang Qinggang 116123
Qiujiang Raisin Tree 198768
Qiujiang Rose 336985
Qiujiang Sweetspire 210390
Qiujiang Turnjujube 198768
Qiujiang Waxpetal 106686
Qiujiang Winterhazel 106686
Qiwu Pleione 304225
Qiyang Wildginger 37678
Qiyun Sedge 75950
Qiyunshan Catchfly 363949
Q-tips 253845
Qu Bulrush 353308
Quack 144661
Quack Grass 144661
Quack Salver's Turbitit 158857
Quackgrass 144661
Quacksalver's Spurge 158857
Quacksalver's Turbith 158857
Quadrangular Combretum 100735
Quadrangular Eurya 160617
Quadrangular Exacum 161585
Quadrangular Olive 270171
Quadrangular Sauropus 348059
Quadrangular Stem Maxillaria 246722
Quadrangular Windmill 100735
Quadrinerved Leucosyke 228147
Quadripetalous Amoora 19970
Quail Brush 44495
Quail Bush 44490
Quail Grass 80381
Quail Plant 190597
Quaint Forsythia 167451
Quake Grass 60372,60382
Quakegrass 60372
Quake-grass 60653
Quaker Buttons 378836
Quaker Grass 60382
Quaker Ladies 187526,198708
Quaker Lady 187526,198708,371795
Quaker-Ladies 187526,198708
Quakers 37015,60382
Quakin Esp 311537
Quaking Ash 311537
Quaking Aspen 311537,311547
Quaking Asr 311547
Quaking Grass 60372,60382
Quaking Oats 60379
Quaking-grass 60372,60382
Quakiu Asp 311537
Qualulo Bell 299312
Qualup Bells 299312
Quamash 68789,68794,68803
Quandong 346209
Quandong Nut 155793
Quante Akebia 13225
Quanzhou Hackberry 80615
Quart Gladiolus 176487
Quarter Vine 54303
Quarters 389935
Quartervine 54303
Quarter-vine 54303
Quash 114189
Quassia 298506,298510,298516,323542
Quassia Family 364387
Quassia Wood 298506,298510,364382
Quassiawood 298506
Quassia-wood 298506
Quater 409335
Quat-vessel 92485
Quebec Berry 19295
Quebec Birch 53338
Quebec Hawthorn 110060
Quebec Linden 391786
Quebec Oak 323625
Quebracha 39702
Quebracho 323563,350973
Queen Aechmea 8573
Queen Agave 10955
Queen Alps 154280
Queen Anne's Flower 262441
Queen Anne's Irish Jonquil 262354
Queen Anne's Jonquil 262411
Queen Anne's Lace 28044,123141,123215,393883
Queen Anne's Lace Handkerchief 28044,392992
Queen Anne's Needlework 81768,174997
Queen Anne's Plume 182942
Queen Anne's Plumes 182942
Queen Anne's Pocket Melon 114150,114201
Queen Anne's Thimble Flower 175677
Queen Anne's Thimble-flower 175677
Queen Anne's Thistle 73430,91842
Queen Anne's-lace 123141
Queen Anthurium 28140
Queen Bird-of-paradise Flower 377576
Queen Crape Myrtle 219966
Queen Crapemyrtle 219966
Queen Crape-myrtle 219966
Queen Elizabeth in Her Bath 220729
Queen Flower 382329
Queen Lady's Slipper 120438
Queen Lady's-slipper 120438
Queen Lagerstroemia 219966
Queen Lily 114874,293252
Queen Mary's Thistle 271666
Queen of Flowering Tree 19462
Queen of Hearts 201787
Queen of Orchid 180126
Queen of the Alps 154280
Queen of the Bromeliads 8554
Queen of the Fields 166125
Queen of the Lilies 229730
Queen of the Marshes 208771
Queen of the Meadow 158219,158264,166125,166129
Queen of the Mist 349044
Queen of the Namib 198073
Queen of the Night 83874,357743
Queen of the Prairie 166122
Queen of the River 267294
Queen Ofthe Wood 203545
Queen Orchid 180116
Queen Palm 31702,31705,380553
Queen Protea 315876
Queen Sago 115812
Queen Stock 246478
Queen Victiria Centuryplant 10955
Queen Victoria Agave 10955
Queen Wattle 1255
Queen's Button 325518
Queen's Crape Myrtle 219966
Queen's Crepe-myrtle 219966
Queen's Cushion 349496,356468,407989
Queen's Delight 376645
Queen's Feather 166125,349044,382329
Queen's Feathers 349044,382329
Queen's Fingers 273531,273545
Queen's Flower 219966
Queen's Gilliflower 193417
Queen's Gillyflower 193417
Queen's Pincushion 407989
Queen's Rogues 193417
Queen's Root 376645

Queen's Tears 54454
Queen's Wreath 28422,292517,292518
Queen's-cup 97098
Queen's-Jewels 28422
Queen-cup 97098
Queen-devil 195643
Queen-devil Hawkweed 195629
Queen-flower 219966
Queen-of the-night 200713
Queen-of-flowers 219966
Queen-of-night 83874,357744
Queen-of-the-alps 154280
Queen-of-the-meadow 166125
Queen-of-the-meadows 166125
Queen-of-the-night 83874,185924
Queen-of-the-prairie 166122
Queen-of-the-summer 229
Queensiland Red Beech 130914
Queenslamd Cypress-pine 67407
Queensland Acacia 1181
Queensland Arrowroot 71169,71181
Queensland Asthma Weed 159069
Queensland Aussiepoplar 197625,270563
Queensland Bean 145899
Queensland Blue Gum 155756
Queensland Bottle Tree 58348
Queensland Bottle-tree 58348
Queensland Camphor 91388
Queensland Elemi Tree 71008
Queensland Fire-wheel Tree 375512
Queensland Golden Myrtle 252624
Queensland Grass Tree 415177
Queensland Gum 155532
Queensland Hemp 362473,362617
Queensland Kauri 10505
Queensland Kauri Pine 10505
Queensland Kurrajong 58348
Queensland Laurel 301207
Queensland Maple 166976
Queensland Nut 240207,240209,240210,
240212,241448
Queensland Peppermint 155578
Queensland Pittosporum 301390
Queensland Pyramidal Tree 220228
Queensland Raspberry 339100
Queensland Sedge 323569
Queensland Silver Wattle 1488
Queensland Stringybark 155694
Queensland Umbrella Tree 350648
Queensland Walnut 50636,145320
Queensland Wattle 1488
Queensland Wild Lime 253290
Queensland-nut 240210
Queensland-nut Tree 240210
Queenslang Maple 166981
Queenssland Firewheel Tree 375512
Queenwood 123267
Queer Windhairdaisy 348312
Quelite 44531
Quelite Cenizo 44517
Quelpaert Barberry 52092
Quenepa 249126
Quenoa 309988
Quer Citron Oak 109857
Quercitron Oak 144661,324539
Queshan Vetch 408446
Quetschen Plum 316382
Quetscheplum 316382
Quick 109790,109857
Quick Set Thorn 109790
Quick Wood 109857
Quickbane 369339
Quickbeam 311537,369322,369339
Quicken 109857
Quicken-berry 369339
Quicken-grass 144661
Quicken-tree 369339,401512
Quickenwood 369339
Quickgrass 144661
Quick-in-the-hand 205175
Quickly Spreading Tickclover 126603
Quickset 109857
Quicksilver Weed 388486
Quicksilver-weed 388486
Quickstick 176972
Quickthorn 109857
Quickthorn Quickset 109857
Quick-tree 369339
Quickweed 170137,170142,170146
Quick-weed 170137,170146
Quiet Neighbours 81768
Quietvein 244984
Quilete 18687
Quill Fleabane 150672
Quill Sedge 76508,76509
Quill Spike-rush 143237
Quillai 324624
Quillet 397043
Quillings 91446
Quill-leaved Sagittaria 342420
Quinault Fawn-lily 154942
Quince 116532,116546,317602
Quince-leaved Medlar 107319
Quindio Barberry 52094
Quindiu Nucleonwood 291485
Quindiu Perrottetia 291485
Quinice Bush 171552
Quinine 91055,91086
Quinine Bush 292294,292296
Quinine Cherry 316411
Quinine Plant 364450
Quinine Tree 18036,91055,91082,292294,
292296
Quinine-bush 171550
Quinoa 87145
Quinquefoliate Raspberry 339149
Quinquefoliolate Clematis 95271
Quinquelocular Eriolaena 152692
Quinsey-berry 334117
Quinsey-wort 39328
Quinsy Berry 334117
Quinsyberry 334117
Quinsywort 39328
Quintinia 324660
Quinua 87145
Quisco Del Desierto 103771
Quisco Peludo 153357
Quisqualis 324669
Quitch 144661
Quitch-grass 144661
Quito Orange 367543
Quitte 116546
Quiver Tree 16768
Quiver-leaf 311547
Quiver-tree 16768
Quiver-tree Aloe 16768
Quixote Plant 193534
Qujiang Milkwort 308138
Qunaxina Linden 391840
Quxian Bamboo 297453
Quxian Bitter Bamboo 304045
Quxian Bitterbamboo 304045
Quxiang Woodbetony 287572

R

R. B. Cooke's Rhododendron 331828
R. C. Grierson's Rhododendron 330793
Rabbie-rinnie-hedge 170193
Rabbit 28617,28631,231183
Rabbit Brush 54268,90484,90522
Rabbit Bush 19162
Rabbit Flower 28631,116736,130383,
220729,231183
Rabbit Meat 18101,192358
Rabbit Milkvetch 42570
Rabbit Milkweed 210525
Rabbit Pea 386365

Rabbit Thorn 239092
Rabbit Tobacco 178323
Rabbit Tracks 245022,245026
Rabbit's Chops 231183
Rabbit's Ear 373166
Rabbit's Ears 373166
Rabbit's Food 277648
Rabbit's Meat 192358,220416,277648,368771,384714
Rabbit's Mouth 28617,116736,176824,231183
Rabbit's Victuals 368771
Rabbit's-foot Clover 396828
Rabbitberry 362090
Rabbit-berry 362091
Rabbitbrush 54268,90484,90522,150326,150332
Rabbitbush 236495
Rabbitbush Chamisa 150332
Rabbitear Iris 208672
Rabbit-ear Iris 208672
Rabbitear Orchis 116938
Rabbit-ears 389259
Rabbiten-wind 12599
Rabbit-eye Blueberry 403730,404050
Rabbitfoot Beardgrass 310125
Rabbitfoot Clover 396828
Rabbit-foot Clover 396828
Rabbitfoot Grass 310125
Rabbit-foot Grass 310125
Rabbitfoot Polypogon 310125
Rabbitleaf Speedwell 317927
Rabbit-meat 18101,28044,192358,220416
Rabbit-pea 386365
Rabbit-root 30712
Rabbit-Tail Grass 220252
Rabbit-tail-grass 220252
Rabbit-thorn 239092
Rabbit-tobacco 127927,178318,317770
Rabbit-weed 54268
Rabdosia 209610,324712
Rabittail Milkvetch 42570
Rabone 326616
Racadal 34590
Racalus 314143
Racconals 315082
Raccoon 411631
Raccoon Grape 20334
Raccoon-berry 306636
Racem Goldenray 228995
Raceme Acrachne 5871
Raceme Bluebell 115340
Raceme Catnip 265034
Raceme Clearweed 299031
Raceme Coldwaterflower 299031
Raceme Corydalis 106350
Raceme Crazyweed 279114
Raceme Cremanthodium 110351
Raceme Flowers Rhododendron 331596
Raceme Globba 176995
Raceme Grepe-hyacinth 260333
Raceme Hartia 376447
Raceme Meconopsis 247177
Raceme Nutantdaisy 110351
Raceme Pencilwood 139657
Raceme Przewalsk Sedge 75888
Raceme Pussytoes 26572
Raceme Rattlesnakeroot 313895
Raceme Redbud 83807
Raceme Schefflera 350772
Raceme Sweetleaf 381423
Raceme Swertia 380333
Raceme Touch-me-not 205269
Racemed Milkwort 308275,308276
Racemiform Pearweed 239812
Racemiformis Spicegrass 239577
Racemobambos 324932,324933
Racemoce Bluegrass 305380
Racemoce Shrubcress 368570
Racemosa Cherry 280016
Racemosa Littledalea 234126
Racemose Adina 364768
Racemose Altai Draba 136926
Racemose Altai Whitlowgrass 136926
Racemose Ardisia 31369
Racemose Arrowbamboo 162740
Racemose Barringtonia 48518
Racemose Bauhinia 49215
Racemose Brandisia 59130
Racemose Brome 60912
Racemose Bugbane 91034
Racemose Caper 71694
Racemose Clearweed 299032
Racemose Corydalis 106350
Racemose Cyclea 116032
Racemose Dinetus 131254
Racemose Distylium 134946
Racemose Drimycarpus 138045
Racemose Fig 165541
Racemose Germander 388015
Racemose Goldenrod 368406
Racemose Incarvillea 205578
Racemose Inula 207206
Racemose Jurinea 207206
Racemose Leymus 228379
Racemose Lumnitzera 238354
Racemose Marshweed 230318
Racemose Masdevallia 246118
Racemose Mosquitoman 134946
Racemose Muehlenbergia 259689
Racemose Nepeta 264869
Racemose Neuropeltis 265794
Racemose Orixa 274280
Racemose Oryzopsis 276058
Racemose Osmanther 276390
Racemose Osmanthus 276390
Racemose Passionflower 285695
Racemose Pencilwood 139657
Racemose Pencil-wood 139657
Racemose Pentapanax 289663
Racemose Perrottetia 291486
Racemose Porana 131254
Racemose Premna 313726
Racemose Rabdosia 209804
Racemose Rattlesnakeroot 313895
Racemose Rhododendron 331596
Racemose Ringvine 116032
Racemose Sinoadina 364768
Racemose Spapweed 205269
Racemose Supplejack 52418,52454
Racemose Sweetleaf 381379
Racemose Thistle 92341
Racemose White-lettuce 313871
Racemulose Barberry 52095
Raceptacle Clearweed 299033
Raceptacle Coldwaterflower 299033
Racers 408571
Racheoled Sedge 75361
Rack 144661
Rackeler Rackliss 314143
Radde Anemone 24013
Radde Corydalis 106355
Radde Enemion 145498
Radde Sedge 75957
Radde Violet 410478
Radde Windflower 24013
Radde's Willow 343975
Raddeana Skullcap 355709
Radial Sawgrass 245518
Radiant Epidendrum 146469
Radiant Galingale 245518
Radiant Pecteilis 286835
Radiant Red Water Lily 267635
Radiant Sawgrass 245518
Radiata Pine 300173
Radiata White Pine 300218
Radiate Acronema 6221
Radiate Beggarticks 54081
Radiate Eucalyptus 155716
Radiate Gentian 173802
Radiate Hibiscus 195132
Radiate Spapweed 205273
Radiateleaf Beesian Larkspur 124063
Radiating Sedge 75959
Radiation Chickweed 375071
Radiative Calanthe 65866
Radiator Plant 290286,290380
Radicat Corydalis 106356
Radiccio Radicchio 90901
Radiciferous Japanese Spindle-tree 157637
Radiocarpellate Euonymus 157281

Radish 326578,326616
Radishleaf Groundsel 359863
Radish-leaved Bittercress 72948
Radishroot Gentian 173657
Radishroot Nepeta 265035
Radishweed 326609
Raffia 326638,326639,326649
Raffia Palm 326638,326639,326649
Rafflesia 325061
Rafflesia Family 325065
Rafinesqui's Chicory 325068
Rafort 326616
Rafted Hair Grass 126039
Rafting Fargesia 162640
Rag Gourd 238257
Rag Jack 86901
Rag Paper 405788
Rag-a-tag 363461
Ragee 143522
Ragged Evening Primrose 269456
Ragged Evening-primrose 269456
Ragged Fringed Orchid 302382
Ragged Jack 359158,363461
Ragged Lady 266225
Ragged Orchid 302382
Ragged Orchis 302382
Ragged Pink 127819
Ragged Robin 146724,174877,240068, 359158,363461,364193
Ragged Rock Flower 111817
Ragged Shirt 102933
Ragged Urchin 363461
Ragged Willie 363461
Ragged-robin 363461
Raggee 143522
Ragi 143522
Ragi Millet 143522
Ragimillet 143512,143522
Rag-Jack 86901
Ragonet 31680
Rag-paper 405788
Rags 106703,234269
Rags-and-tatters 30081,243840
Ragweed 19140,19145,19150,19191, 35084,359158
Ragwort 273531,279885,358159,359158, 360328
Ragwort Groundsel 359158
Rai 59438
Raillardella Scabrida Eastwood 25371
Railroad Canyon Wild Buckwheat 152468
Railroadfence 49194
Railway Beggarticks 54048
Railway Creeper 207934
Railway Daisy 54048
Railway Fence Bauhinia 49194
Railway Poplar 311146,311157
Raimond Puya 321942
Rain Daisy 131186
Rain Lily 103671,103674,184249,417606, 417608,417613,417624
Rain Orchid 302241
Rain Tree 61293,61315,235521,345517, 345525
Rainberry Thorn 328642
Rainberry-thorn 328642
Rainbow Cactus 140282,140313
Rainbow Candle Cactus 140285
Rainbow Hedgehog Cactus 140313
Rainbow Hills Rabbitbrush 150344
Rainbow Lily 229871
Rainbow Pink 127654
Rainbow Plant 198745
Rainbow Shower 78336
Rainbow Star 113380
Rainbow's End Spruce 298292
Rainlily 417608
Rain-lily 184249,417606
Raintree 61293,345517,345525
Rain-tree 61293,345517,345525
Raisehead Azalea 331557
Raisespur Calanthe 65887
Raisin 411979
Raisin Barberry 51333
Raisin Tree 198766,198767,198769,334179
Raisin-tree 198766,198767,198769
Rait 325580
Rajah Begonia 50220
Rajah Cane 156115
Rajah Sealing-wax Palm 120780
Rakkyo 15185
Rakum 342579
Ram Gall 250502
Ram Habenaria 183425
Ram's Claws 325518,326287,374968, 400675
Ram's Foot 175454,325580
Ram's Glass 325518
Ram's Head Lady's Slipper 120299
Ram's Horn 15861,37015,273531,273545, 315434
Ram's Horns 15861,37015,273531,273545
Ram's-head 120299
Ram's-head Lady's Slipper 120299
Ram's-head Lady's-slipper 120299
Rama 195196
Ramaine Lettuce 219492
Ramanas Rose 336901
Rambai 46193,46198
Rambler Rose 336783
Rambling Bellflower 70210
Rambling Dewberry 339438
Rambling Fleabane 151050
Rambling Milkweed 347013
Rambling Sailor 116736,239755,349936
Rambotang 265129
Rambustan 265129
Rambutan 265124,265129
Rambutang 265129
Ramenas 16991
Ramentaceous Maesa 241824
Ram-gall 250502
Ramie 56094,56229,56233,56236
Ramielike Stairweed 142623
Ramified Skullcap 355710
Ramified Woodbetony 287574
Ramiflorous Baccaurea 46198
Ramiflorus Cleistogenes 94620
Ramiflorus Hideseedgrass 94620
Ramin 179532,179533,179535
Ramipilose Rhododendron 331604
Rami-spined Bambusa 47425
Ramispiny Bamboo 47425
Ramlose Indigo 206466
Ramona Spineflower 135145
Ramonda 325223
Ramondia 325223
Ramontchi 166761,166773,166786
Ramos Luisia 238319
Ramose Greyawngrass 372425
Ramose Involucred Cudweed 178234
Ramose Lettuce 219493
Ramose Loosestrife 239707,239814
Ramose Monkshood 5526
Ramose Muhly 259691
Ramose Sharpdissected Monkshood 5325
Ramose Trikeraia 397507
Ramose Woodbetony 287574
Ramosest Tamarisk 383595
Ramp 15831,15832
Rampant Fumitory 169030
Rampe 37015
Ramping Fumitory 169030
Ramping-fumitory 169030
Rampion 70254,70257,298054,298065
Rampion Bellflower 70254,70255
Rampion Mignonette 327904
Ramps 15831,15861,70254
Ramrod Bamboo 318303
Rams 15861,99297
Ramsden 15861
Ramsden's Rhododendron 331606
Ramsey 15861,271563
Ramshorn Willow 343071
Ramsies 15861
Ramsins 15861
Ramsons 15861,37015,271563
Ramstead 231183
Ramsthorn 328642
Ramtil 181873
Ramtilla 181873

Ranalisma 325268
Rancheria Grass 228346
Rand's Goldenrod 368399,368403
Randa Blueberry 403976
Randa Cinnamon 91429
Randa Illicium 204598
Randa Sageretia 342191
Randaishan Cinnamon Tree 91429
Randaishan Tanoak 79064
Randia 12519,325278
Randramboay 159268
Ranevea 325466
Range Ratany 218078
Range Ritatany 218082
Rangiora 58477
Rangoon 385531
Rangoon Bean 294010
Rangoon Creeper 324677
Rangooncreeper 324677
Rangpur Lime 93546
Rangtang Larkspur 124551
Rankan Hornbeam 77381
Rankkopieva 62390
Rannd-leaved Fluella 216292
Ranner Pepper 300504
Rannoch-rush 350859,350861
Rannoch-rush Family 350864
Rantabun Dishspurorchis 171884
Rantabun Gastrochilus 171884
Ranting Widow 85875
Rantipole 123141
Rantle-tree 369339
Ran-tree 369339
Rantry 369339
Ranty-berry 369339
Rantytanty 340151
Ranunculifoliate Monkshood 5528
Ranunculifoliose Larkspur 124287
Ranunculus 325498
Ranunkel 325912
Raoping Photinia 295761
Raoping Raspberry 339154
Raoul 266876,266881
Raouli 266876
Raoulia 326513,326514
Rapanea 326523
Rape 59278,59493,59524,59603,364557
Rape Crowfoot 325666
Rape Kale 59493,59524
Rape Mustard 59575
Rape Radice 326616
Rape Violet 115949
Rapeform Mustard 59461
Rapeseed 59493,59603
Raphanistrum-like Radish 326629
Raphia 326638,326639
Raphia Palm 326638
Raphiolepis 329056
Raphistemma 326798
Rapistrum 326812
Rapper Dandy 31142
Rapper-dandy 31142
Rappers 130383
Rapunzel Plant 346995
Rara Dishspurorchis 171885
Rara Gastrochilus 171885
Rare Spring-sedge 74461
Rare-branched Yushania 416808
Rare-flowered Barberry 52096
Rare-spiked Bashania 37314
Rarshape Rockfoil 349089
Rasamala 18197
Rasamala Altingia 18197
Rasp 338557,339240
Rasp Pod 393872
Raspberries-and-cream 158062
Raspberry 334179,338059,338557
Raspberry Acacia 1030
Raspberry Jam 1030
Raspberry Jam Tree 1030
Raspberryleaf Clematis 95289
Raspberry-leaved Clematis 95289
Rasp-pod 393872
Rassels 271563
Rastewort 292957
Rat Tail 134230
Rat Tail Cactus 134230
Rat's Bane 28044
Rat's Ear 101023
Rat's Foot 176824
Rat's Mouth 176824,220346,220416
Rat's Tail 11549,302034,302068
Rat's Tail Cactus 29715
Rat's Tails 11549,302034,302068
Rat's-tail Cactus 29715
Rat's-tail Fescue 412465
Rata 171092,252625
Rata Tree 252607
Rata Vine 252610
Ratan 65782
Ratany 218083
Rata-tree 252607
Rataugueressa 166777
Ratdesnake Weed 158411
Ratdesnake-root 313901
Rathairdaisy 146565
Rathertomentose Spapweed 205388
Rathertomentose Touch-me-not 205388
Rati 765
Ratibida 326957
Ratpalm Cranesbill 174906
Rats-and-mice 118008
Rattail Cactus 29712,29715,244196
Rat-tail Cactus 29715
Rattail Crassula 109180
Rattail Elsholtzia 144062
Rattail Fescue 412465
Rat-tail Fescue 412465
Rattail Fescuegrass 412465
Rat-tail Grass 372579
Rat-tail Nipple Cactus 244196
Rat-tail Plant 290315
Rat-tail Plantain 302068
Rat-tailed Radish 326621
Rattan 52466,65637,65782,121486,121514,
217918
Rattan Bambusa 47321
Rattan Cane 65782
Rattan Palm 65637,65782,65800,121488
Rattan Palms 65637
Rattan Vine 52466
Rattanpalm 65637
Rattan-palm 65637
Ratten Tails 302068
Ratten-tail 302068
Rattesnakerooot-like Paraprenanthes 283693
Rattle Box 361422
Rattle Jack 329521
Rattle Penny 329521
Rattle Weed 42137
Rattle-bags 329521,364193
Rattle-basket 60382,287721,329521
Rattlebox 111851,112636,238160,329504,
329527,361378,361422
Rattle-box 111851,112636,329521,361378
Rattlebrush 361378
Rattlebush 47880,361422
Rattle-caps 329521
Rattlegrass 60653
Rattle-grass 60382,287494,329521
Rattle-Jack 329521
Rattle-penny 329521
Rattlepod 111851,287721
Rattle-pod Grevillea 180650
Rattle-pouch 72038
Rattle-root 91034
Rattlesnake Brome 60653
Rattlesnake Chess 60653
Rattlesnake Flower 59812
Rattlesnake Grass 60379,177577
Rattlesnake Manna Grass 177577
Rattlesnake Mannagrass 177577
Rattlesnake Master 154276,154284,154355,
244377
Rattlesnake Orchid 295510
Rattlesnake Orchis 179571
Rattlesnake Plant 66165,66166
Rattlesnake Plantain 179571,179685
Rattlesnake Root 313778,313787,313788,
313796,313809,313871
Rattlesnake Tail 108846

Rattlesnake Weed 196077
Rattlesnake-master 154284,154355
Rattlesnake-plantain 179571
Rattlesnakeroot 313778
Rattle-snakeroot 91034
Rattlesnake-root 85505,91034,308347,313787
Rattlesnake-weed 123215,140067,158411,159491,196077,239224
Rattlesnak-master 228519
Rattle-traps 329521
Rattleweed 47880,91034,287494,329504,364193
Rattlewort 329521
Rattling Asp 311537
Ratwife Lovegrass 147509
Rau Ram 309482
Raukumara 58476
Rauli 266876,266881
Rauntree 369339
Raupo 401094
Rautini 58473
Rauvolfia 326995
Rauwolfia 326995
Raven's Claws 326287
Raven's-foot Sedge 74219
Ravenala 327095
Ravenna Grass 148906
Ravennagrass 148906
Ravenna-grass 148906
Raven-tree 369339
Ravine Rhododendron 331906
Rawbone 326616
Rawheads 325580
Rawntree 369339
Rawp 59493
Raxen 64184
Ray 11947
Ray's Knotgrass 309503
Ray's Knot-grass 309503
Ray's Knotweed 309504
Rayado Bundle Bundleflower 126187
Rayado Bundleflower 126187
Rayed Tansy 383794
Ray-flowered Protea 315795
Ray-grass 235334,235373
Rayhair Ampelocalamus 20193
Ray-haired Ampelocalamus 20193
Rayless Alkali Aster 380850
Rayless Annual Aster 380850
Rayless Arnica 34705
Rayless Aster 40154
Rayless Chamomile 246375
Rayless Golden Rod 185489
Rayless Marigold 383088
Rayless Mayweed 246375
Rayless Mock Goldenrod 60344
Rayless Ragwort 279918
Rayless Rock Daisy 291297
Rayless Sunflower 189041,189044
Rayless-goldenrod 54257
Raymond Stonecrop 357075
Raywood Ash 168043
Razorsedge 353998
Razor-sedge 353998
Reactiongrass 54532,54555
Reaf-a-robber 31051
Real Mastic Tree 301005
Real Yellowwood 306450
Real Yellow-wood 306450
Realseed Leek 15268
Realseed Onion 15268
Realtooth Sandwort 31871
Reate 325580
Reathery Pincution 244192
Reaumuria 327194
Reautiful Lagochilus 220088
Rebizond Dates 141932
Rebol 19228
Rechinger Barberry 52097
Reckless 314143
Reclining Bedstraw 170193
Reclining Date Palm 295479
Rectanglesickle Medic 247251
Recti-cuneate Neosinocalamus 264531
Rectispine Randia 278321
Rectivein Fig 165527
Recurvate Ginger 418018
Recurvate Hakea 184632
Recurvate Leptospermum 226477
Recurvate Nolina 49343
Recurvate Tetracme 386955
Recurve Gladiolus 176494
Recurve Woodbetony 287575
Recurvebeak Spapweed 205279
Recurvebeak Touch-me-not 205279
Recurved Aechmea 8590
Recurved Blackberry 339158
Recurved Brownleea 61191
Recurved Larkspur 124554
Recurved Minuartia 255564
Recurved Muntries 218396
Recurved Woodbetony 287575
Recurved Yushania 416776
Recurvedsegment Fritillary 168548
Red Abyssinian Banana 145846
Red Aechmea 8577
Red African Clover 396811
Red Alder 16429,16437
Red Almond 17613,17614
Red Alpine Barrenwort 146962
Red Amaranth 18696,18788
Red American Osier 104949
Red Amitostigma 19536
Red Anemone 23891
Red Angel's Trumpet 61240
Red Anise 204517
Red Aphania 225678
Red Apple 4884,29855
Red Archangel 220416,373455
Red Arctous 31322
Red Argyroderma 32750
Red Ascocentrum 38193
Red Ash 17613,168057,168073
Red Astyleorchis 19536
Red Awamp Banksia 47656
Red Awngrass 255872
Red Axillary Flowered Rhododendron 332062
Red Azalea 332062
Red Balan 362207,362208
Red Bamboo 297448
Red Banana 145846,260266
Red Baneberry 6441
Red Bane-berry 6441
Red Barberry 51694,52227
Red Bark 91090
Red Bark Oak 116098
Red Bark Slugwood 50507
Red Bark Tree 91090
Red Baron Blood 205497
Red Barrenwort 147047
Red Bartsia 48626,268983,268986,268991,268999
Red Bauhinia 49017,49094
Red Bay 291491,291505
Red Bay Persea 291505
Red Bayberry 261212
Red Bead Tree 7190
Red Beadvine 765
Red Beakrush 333672
Red Bean 139665,369116,408839
Red Bean-tree 139665
Red Bearberry 31142
Red Bee Nettle 220416
Red Beech 162375,166971,266870
Red Beet 53259,53269
Red Bells 145673
Red Bentgrass 12280
Red Bergamot 257161
Red Berlinia 52850
Red Berry 105023,336419
Red Betty 234351
Red Bigbloom Magnoliavine 351095
Red Bilberry 403956,404051
Red Birch 53536,53540,53549,53590
Red Bird 287846,287853
Red Bird of Paradise 65055
Red Bird's Eye 21339,174877,248411
Red Bird's Eyes 21339,174877,248312
Red Bishop's Hat 147047
Red Bistort 308769

Red Bloodwood 106805,106820
Red Bobby's Eye 174877
Red Bobby's Eyes 174877
Red Bogles 102791
Red Boloflower 205557
Red Bopple Nut 195375
Red Bopplenut 195375
Red Boronia 57265
Red Box 155700,236462
Red Bractquitch 201563
Red Brome 60943
Red Bromegrass 60943
Red Broodhen Bamboo 297328
Red Broomrape 274934
Red Brush 105193
Red Bryony 61497
Red Buckeye 9720
Red Buckwheat 152107,152110
Red Bud 83752,83757
Red Bud Mallee 155681
Red Buffalo-bur 367617
Red Bugle 55109
Red Bugles 102791
Red Bulrush 55895
Red Butcher 248411,273531
Red Buttons Opuntia 273030
Red Cabbage 59534
Red Cabbage Rose 336481
Red Cactus 107004
Red Calla 417118
Red Callalily 417118
Red Calyx Rhododendron 330640
Red Camomile 8332
Red Campion 248411,363149,363411
Red Canella 91250
Red Canyon Sage 345213
Red Cap Gum 155574
Red Cape Lily 184345
Red Cardinal 154668
Red Carpet 109287
Red Cassia Tree 36922
Red Cat's Tail 1970
Red Catchfly 363411,410957
Red Cat-tails 1970
Red Cedar 5983,54620,213853,213922,213984,228639,390567,390645,392829,401233
Red Celandine 176738
Red Cestrum 84400,84409
Red Chickweed 21339
Red Chief 84555
Red Chile Bells 221341
Red Chile-bells 221341
Red Chinese Waxmyrtle 261218
Red Chinkapin 78955
Red Chokeberry 34854,34855,295760,323090
Red Chokecherry 316918
Red Cinchona 91090
Red Clematis 20361
Red Climonia 97090
Red Clover 397019
Red Cob Cactus 234933
Red Cock's Comb 18752
Red Cockle 363411
Red Cockscomb 18752
Red Coelogyne 98697
Red Cogon 205497
Red Coldwaterflower 299038
Red Cole 34590,59532,68737
Red Coliseum Maple 2851
Red Columbine 30007,30034
Red Com-lily 210723
Red Copperleaf 2011
Red Corn Rose 282685
Red Cornflower 11947,282685
Red Cow Basil 81768
Red Cowpea 408839
Red Cranesbill 174891
Red Creek Howell's-buckwheat 212399
Red Crown Cactus 327283
Red Cryptochilus 113515
Red Cuphea 114604
Red Currant 334179
Red Cushion 397019
Red Cypress 385262
Red Cypress Pine 67417
Red Dahlia 121561
Red Daisy 195478
Red Dali Rhododendron 330265
Red Darnel 235334
Red Dead Nettle 220416
Red Deadnettle 220416
Red Dead-nettle 220401,220416
Red Deal 300223
Red Dendrobium 125253
Red Desert Wild Buckwheat 151881
Red Deutzia 127071
Red Devil Maple 2928
Red Dhup 284421
Red Diapensia 127921
Red Dock 339961,339989,340082,340249
Red Dogtongueweed 385917
Red Dogwood 104949,380499
Red Dolly 282685
Red Dracaena 104367
Red Ebony 328908
Red Eightangle 204507
Red Elder 345594,345660,407989
Red Elderberry 345579,345660,345677
Red Elm 401430,401509,401593,401620,401626
Red Enkianthus 145718
Red Epimedium 146962,147047
Red Escallonia 155149
Red Eulalia 156500
Red Eulophia 156974
Red Euphorbia 158700
Red Evergreen Chinkapin 78955
Red Eyebright 268991
Red Fairy Duster 66665
Red False Beardgrass 90123
Red False Buck's-beard 41789
Red False Mallow 243892
Red False Yucca 193229
Red Fescue 163816,163819,164243,164272
Red Fescuegrass 164243
Red Fescue-grass 163816
Red Fetchling 271280
Red Fig-nut Flower 212144
Red Fingers 396934
Red Fir 272,409,453,318580
Red Fire Thorn 322454
Red Fire-thorn 322454
Red Fitchling 271280
Red Five-corners 379277
Red Flax 231907
Red Flower Parodia 284693
Red Flowered Camel's Foot 49017
Red Flowering Gum 106800
Red Four o'clock 255691
Red Frangipani 305225
Red Fruit Faber Maple 2956
Red Fruit Fig 165841
Red Fruit Fig-tree 165721
Red Funnel Flower 130229
Red Garden Beet 53259
Red Garden-currant 334179
Red Gaura 172191
Red Gentian 81065
Red German Catchfly 410957
Red Gilia 175704
Red Ginger 17745
Red Ginger-lily 187428
Red Gleadovia 176799
Red Goldenearrring 37702
Red Goldquitch 156500
Red Gooseberry 334179
Red Goosefoot 87147
Red Grain 65146
Red Granadilla 285632
Red Granfer-gregor 273531
Red Granfer-greygles 248312
Red Grape 411834,411896
Red Grevillea 180601
Red Groundsel 385917
Red Gum 103696,106800,155506,155514,155517,155520,155644,232565
Red Gum Fiver 155517
Red Gum Murray 155517
Red Hairy Scorpiothyrsus 354744

Red Haw 109754,110060,110065
Red Hawksbeard 110985
Red Hawthorn 109850,109936,110014
Red Head 37888,345881,392992
Red Heath 67456
Red Heather 297018
Red Hedge Nettle 170058
Red Helleborine 82070
Red Hemp Nettle 170058,170066
Red Hempnettle 170066,170070
Red Hemp-nettle 170058,170066
Red Hesperaloe 193229
Red Hibiscus 194804
Red Hickory 77894,77897,77912
Red Hills Soap Plant 88473
Red Hoarypea 386000
Red Hogfennel 292999
Red Holly Grape 51694
Red Homalomena 197796
Red Honeysuckle 47667,187842,235760,397019
Red Hoppel Nut 195375
Red Horned Poppy 176728
Red Horned-poppy 176728
Red Horse Chestnut 9680
Red Horsechestnut 9680,9720
Red Horse-chestnut 9680,9720,9726
Red Hot Cattail 1894
Red Hot Copper-leaf 1894
Red Hot Poker 216923,217055
Red Hot Poker Tree 154620
Red Hove 176824
Red Huckleberry 403956,404051
Red Huntsman 282685
Red Hybrid Lilac 382335
Red Hyparrhenia 201563
Red Incarvillea 205557
Red Indian Bean-tree 79243
Red Indian Water-lily 267756
Red Ink Berry 298094
Red Ink Plant 298094
Red Inkberry 298094
Red Inkplant 298116
Red Iris 208578
Red Ironbark 155545,155743,155744
Red Ironbark Gum 155741
Red Ivory 328908
Red Ivory Wood 328908
Red Ivy 20361,191488
Red Ixora 211067
Red Jack 248312
Red Jane 248312
Red Japanese Maple 3300
Red Jasmine 305225
Red Javatea 275759
Red Joints 309570
Red Juniper 213922,213984
Red Jurinea 211749
Red Justicia 214403,214410
Red Kamala 243427
Red Kex 392992
Red Khaya 216205
Red Knees 309199,309570
Red Knob 345881
Red Knot-grass 309531
Red Knotweed 309494
Red Kowhai 96639
Red Laceleaf Japanese Maple 3342
Red Lace-leaf Maple 3325
Red Laelia 219674
Red Larkspur 124410
Red Latan Palm 222626
Red Lauan 362186,362212,362217,362226
Red Lead 273469,273531,273545
Red Leaf 107309
Red Leaf Loropetalum 237195
Red Leaf Plantain 302069
Red Leaf-flower 296751
Red Legs 308816,308893,309570
Red Lily 229984
Red Lime 391691,391817,391829
Red Linden 391829
Red Ling 67456
Red Loosestrife 240068
Red Louro 268729
Red Love Grass 147962
Red Ludwigia 238221
Red Lumnitzera 238350
Red Lungwort 321646
Red Madder 338038
Red Mahoany 155689
Red Mahoganies 155720
Red Mahogany 10127,155720,216205,216211
Red Maids 65835
Red Mailkes 282685
Red Malabar Nightshade 48689
Red Malaysian Guava 318742
Red Mallee 155676
Red Mangrove 329761
Red Maple 3505
Red Margin Calathea 66200
Red Margins Dracaena 137447
Red Mathes 8332
Red Mathet 8332
Red Maydweed 8332
Red Maythe 8332
Red Meconopsis 247171
Red Melastome 248778
Red Meranti 362187
Red Mexican Thorn 110047
Red Milkweed 38099,159046
Red Milkwood 255356
Red Milkwort 308415
Red Millet 130612
Red Mint 250291,250362,250368,257161
Red Mintdrop 248411
Red Mirbelia 255785
Red Molly 217327
Red Mombin 372477,372479
Red Money 81768
Red Moneywort 18279
Red Monkey Flower 255197
Red Monkswort 266629
Red Moons 97894
Red Morning Glory 323459
Red Morocco 8332
Red Mountain Catchfly 363284
Red Mountain Juniper 213922
Red Mountain Leech 126588
Red mountain Spinach 44468
Red Mountain Wild Buckwheat 152183
Red Mountain-heather 297024
Red Mountain-spinach 44469
Red Mulberry 259186
Red Mustard 59515
Red Nanmu 240707
Red Nap 282685
Red Nerine 265294
Red Nidularium 266157
Red Night-blooming Jasmine 84412
Red Nightshade 297643
Red Ninenode 319810
Red None Pepper 72074
Red Norway Maple 3441
Red Oak 116150,320278,323894,324158,324232,324344
Red Oat 45424
Red Olive Plum 142434
Red Onion 15160,15682
Red Orache 44468,44629
Red Orange 93721
Red Orchid Cactus 266648
Red Osier 343937
Red Osier Dogwood 105193
Red Oxalis 278140
Red Paper Tree 13654
Red Parabarium 283067
Red Parrot-beak 96639
Red Passion Flower 285695
Red Passionflower 285631,285632,285678,285695
Red Passion-flower 285678
Red Passion-vine 285678
Red Pea 222694
Red Pear 354618
Red Pearweed 239718
Red Pear-wood 354618
Red Peashrub 72334
Red Pencil Tree 159975
Red Pepper 72068,72070,72075

Red Persimmon 132377
Red Petti Coats 282685
Red Pigweed 87147
Red Pimpernel 21339
Red Pine 121094,299890,300180,318580
Red Pineapple 21475
Red Pitcher Plant 347163
Red Plum 34431,316206
Red Poker 184593
Red Pokers 184593
Red Pondweed 312037
Red Poppy 282685
Red Pouzolzia 313471
Red Powderpuff 66673
Red Prince Weigela 413591
Red Princess Palm 129688
Red Privet 229608
Red Psychotria 319810
Red Puccoon 345805
Red Quakegrass 60387
Red Quebracho 323561,350973
Red Quina 91086,91090
Red Rags 282685
Red Ramie 56317
Red Raspberry 338557,338600,339310
Red Rattle 268991,287494
Red Rattle-grass 287494
Red Rattlesnake Plantain 179623
Red Ray 235334
Red Reekkudzu 313471
Red Restharrow 271347
Red Rhatany 218085
Red Ribbons 94135
Red Rice 275926,275957
Red Riding Hood 244329,248411
Red Robin 174877,248411,273531,308816,363461
Red Robin Hood 248411
Red Rocket 193417,228100
Red Rockfoil 349872
Red Root 233707,233717
Red Rose 336581,336607
Red Rose-of-lancaster 336593
Red Rot 138348,200388,299759
Red Royal Oak 407907
Red Ruellia 339717
Red Rugosa Rose 336909
Red Sage 217327,221238,344980,345271
Red Sally 240068
Red Salvia 345405
Red Sand Spurrey 370615
Red Sand Spurry 370696
Red Sand Verbena 704
Red Sandal Wood 7190
Red Sandalwood 7190,320327
Red Sandal-wood 320327
Red Sandal-wood Tree 7190
Red Sanders 320327
Red Sand-spurrey 370696
Red Sandspurry 370696
Red Sand-verbena 704
Red Sandwort 255573,370696
Red Sassafras 347407
Red Satin Flower 187842
Red Saunders 320327
Red Scale 44629
Red Sedum 357107
Red Sepal Evening Primrose 269443
Red Shanks 8033
Red Shoot Persimmon 132140
Red Shrubby Pemtemon 289334
Red Side-saddle Flower 347163
Red Silk Oak 180570
Red Silkcotton Tree 56784
Red Silk-cotton Tree 56784,56802
Red Silkyarn 238939
Red Silkyoak 375511
Red Silky-oak 273804
Red Silver Fir 280
Red Silvergrass 255872
Red Siris 283890
Red Skullcap 355785
Red Sliky Oak 375511
Red Snakebark Maple 2846
Red Soldiers 248411,282685
Red Sorrel 195149,195196,339897
Red Sour Leek 339887
Red Spatholobus 370419
Red Spider Flower 180638,180648
Red Spider Lily 239266
Red Spider-lily 196439
Red Spiderling 56413,56425
Red Spiderwort 394069
Red Spike 82282
Red Spike Ice Plant 82276,82279,82381
Red Spine Palm 2553
Red Spiradiclis 371750
Red Spire Tea Tree 226472
Red Splitlemma 351307
Red Spot Honeysuckle 235860
Red Spotleaf-orchis 179623
Red Sprangletop 226021
Red Sprangle-top 225989
Red Spruce 298423
Red Spurrey 370696
Red Squill 137958,352960
Red Star Grass 332185
Red Star Thistle 80981
Red Star-thistle 80981
Red Stelis 374716
Red Stem 153767
Red Sterculia 376095
Red Stickyleaf Rhododendron 330776
Red Stinkwood 316203,322389
Red Stonecrop 357107
Red Streptanthera 377628
Red Stringybark 155636
Red Striped Rhododendron 330501
Red Sugarbush 315810
Red Tangle 115008
Red Tassel Flower 66689
Red Tasselflower 144975
Red Tatarian Maple 3658
Red Tea 195149
Red Tea Rose 336816
Red Temple-bells 366715
Red Thistle 92285
Red Thorn 1339
Red Thorowax 63813
Red Three-awn 33915
Red Tingle 155612
Red Tiny Whortleberry 403987
Red Tip Photinia 295690
Red Tobacco 266042
Red Toona 392829
Red Top 11972,12118
Red Trefoil 397019
Red Tremacron 394678
Red Triangles 81900
Red Trillium 397556,397599,397609
Red Tritonia 399063
Red Turtlehead 86795,86796
Red Turtle-head 86795
Red Twig 228181
Red Twig Dogwood 105193
Red Valerian 81743,81768
Red Velvet Shamrock 277749
Red Vinespinach 48689
Red Violet 147393
Red Viper's Bugloss 141350
Red Water Lily 267651
Red Water Tree 154973
Red Waterlily 267651
Red Whortleberry 403956,404051
Red Wild Buckwheat 151843
Red Wild Purslane 311915
Red Willow 104949,105193,343360,343590
Red Willow Weed 85884
Red Willowleaf Achyranthes 4305
Red Wingseedtree 320847
Red Withyherb 146724
Red Woleberry 380749
Red Wolf 248411
Red Wood 65060,360565
Red Woodbetony 287645
Red Woodlotus 244449
Red Woodnettle 273917
Red Woodsorrel 277776
Red Worts 404051
Red Xinjiang Opium 335650
Red Yucca 193229,193232

Red Zinger 418019
Red-and-green Kangaroo Paw 25220
Red-and-green Kangaroo-paw 25220
Red-and-yellow Garden Raspberry 338557
Redback Azalea 331700
Redback Bauhinia 49222
Redback Christmasbush 14217
Redback Christmas-bush 14217
Redback Conehead 290921
Redback Evergreenchinkapin 78992
Redback Osmanthus 161647
Red-back Rhododendron 331700
Redback Xmas Bush 14217
Red-backed Bauhinia 49222
Red-backed Evergreen Chinkapin 78992
Redball Ginger 418031
Redball Snakegourd 396279
Redball Zinger 418031
Redband Lily 229735
Redbark Camellia 69015
Redbark Cinchona 91090
Red-bark Cinchona 91086
Red-bark Cyclobalanopsis 116098
Redbark Fewnerve Linden 391808
Redbark Litse 234031
Redbark Oak 116098
Redbark Qinggang 116098
Redbark Slugwood 50507
Redbark Stewartia 376471
Redbark Wendlandia 413836
Red-barked Apple Tree 24606
Red-barked Calligonum 67075
Red-barked Camellia 69015
Red-barked Cinchona 91090
Red-barked Dogwood 104920
Red-barked Litse 234031
Red-barked Slugwood 50507
Red-barked Stewartia 376471
Red-barked Willow 344121
Redbay 291505
Red-bay Persea 291505
Redbay Swampbay 291505
Redbead Barberry 51311
Red-bead Barberry 51311
Redbead Cotoneaster 107637
Redbell Azalea 331816
Red-berried Bamboo 366621
Red-berried Bryony 61497
Red-berried Elder 345660,345677
Red-berried Greenbrier 366621
Red-berried Juniper 213841
Red-berried Moonseed 97900
Red-berrier Elder 345660
Red-berriered Elder 345660
Redberry 105023,280799,328630,328674
Red-berry Bryony 61497
Redberry Buckthorn 328674
Red-berry Elder 345579
Redberry Juniper 213764,213827,213853
Red-berry Juniper 213853
Red-berry Laurel 122515
Redbird Flower 287853,287855
Redblood Gladiolus 176140
Redbox Cotoneaster 107659
Red-box Cotoneaster 107659
Redbox Gum 155700
Redbract Agapetes 10363
Redbract Cautleya 79756
Redbract Hemiboea 191381
Redbract Philodendron 294805
Red-bracted Agapetes 10363
Redbracted Lysidice 239527
Red-bracted Lysidice 239527
Redbranch Lanceolate Elaeagnus 142047
Red-branch Rockvine 387769
Redbranch Wolfberry 239031
Red-branched Ameliančhier 19290
Redbranched Barberry 51593
Red-branched Barberry 51593
Red-branched Largeleaved Linden 391824
Red-branched Striped Maple 3394
Red-branched White Linden 391729
Redbreast 174877
Redbrown Azalea 331683
Redbrown Przewalsk Sage 345332
Red-brown Sedge 75887,76080
Redbrown Woodrush 238577
Redbrownbristle Cottonsedge 152772
Redbrush 232492
Redbud 3505,83752,83757,83778,83798
Redbud Maple 3699
Redbud Pearlbush 161744
Red-buded Pearl-bush 161744
Red-bugle Vine 9475
Redcalyx Azalea 330640,331209
Redcalyx Catchfly 364001
Redcalyx Cliffbean 254690
Redcalyx Forkliporchis 269016
Redcalyx Garcinia 171098
Redcalyx Glorybower 96083
Red-calyx Millettia 254690
Redcalyx Saurauia 347983
Red-calyx Violet 410491
Red-calyxed Garcinia 171098
Red-calyxed Millettia 254690
Red-calyxed Rhododendron 330640,331209
Redcandle Balanophora 46863
Redcandle Snakemushroom 46863
Red-cap Gum 155574
Redcarweed 288705,288762
Redcarweed Coldwaterflower 299000
Redcaudex Inula 138899
Redcedar 213606
Red-cedar 213984,390567,392826
Redcenter Morning-glory 207579
Red-centered Hibiscus 18252
Redclawed Habenaria 184025
Redclaws 155141
Redco 34590
Redcrown Dolomiaea 135733
Redcup Redcap 282685
Redcurrant 334179,334223
Redden Ophiorrhiza 272298
Redden Woodrush 238690
Reddening Circaea 91545
Reddening Dewdrograss 91545
Reddening Maple 3209
Redding's Wild Buckwheat 152481
Reddish Arrowwood 407820
Reddish Bamboo 297449
Reddish Beautyberry 66913
Reddish Begonia 50257
Reddish Discocleidion 134146
Reddish Emei Bergenia 52521
Reddish Evergreen Chinkapin 79014
Reddish Jurinea 212013
Reddish Litse 234051
Reddish Mayten 182772
Reddish Obtuseleaf Camellia 69408
Reddish Oriental Blueberry 403746
Reddish Orostachys 218358
Reddish Pondweed 312037
Reddish Punctulation Rhododendron 330901
Reddish Purplepearl 66913
Reddish Raspberry 339218
Reddish Sandwort 255573
Reddish Stonecrop 356535
Reddish Tufted Vetch 408305
Reddish Uraria 402146
Reddish Viburnum 407820
Reddish Wallflower 154532
Reddish Woodnettle 273917
Reddish-brown Clovershrub 70809
Reddish-brown Rhododendron 331683
Reddishbrown Sedge 75724
Reddishflower Lily 229746
Reddish-flower Rhododendron 331647
Reddishmargin Bamboo 297449
Reddish-margined Bamboo 297449
Reddish-orange Bamboo 297453
Reddish-orange Gelidocalamus 172746
Reddish-pediseled Phoebe 295402
Reddish-punctulate Rhododendron 330901
Reddish-purple Iris 208453
Reddish-tomentose Honeysuckle 235796
Red-disk Pond-lily 267324
Reddrop Barberry 51548
Red-drop Barberry 51548
Reddy Bauhinia 49213
Redfibre Shrimpalga 297182
Redflame 191501

Redfleshy Yangtao Actinidia 6566
Redflower Abutilon 984
Redflower Agapetes 10315
Redflower Amomum 19914
Red-flower Anneslea 25777
Redflower Azalea 331863
Redflower Bedstraw 170240
Red-flower Black Locust 334987
Redflower Blueberry 404033
Redflower Braya 59758
Redflower Calanthe 66051
Redflower Camellia 68980
Redflower Canada Lily 229782
Redflower Canscora 71283
Redflower Cephalanthera 82070
Redflower Ceratostylis 83656
Redflower Chesneya 87248
Redflower China Houndstongue 117911
Red-flower Clearweed 299038
Redflower Cleisostoma 94483
Redflower Cleistes 94549
Redflower Closedspurorchis 94483
Redflower Corydalis 106088,106340
Redflower Cotoneaster 107661
Redflower Cremanthodium 110388
Red-flower Deutzia 127103
Redflower Dew Azalea 330941
Redflower Edelweiss 224929
Redflower Fairybells 134496
Red-flower Fake Woollypistil Willow 343920
Redflower Fistulous Marshmarigold 68196
Redflower Gentian 173811
Redflower Greenvine 204648
Redflower Gum 106800
Red-flower Hankow Willow 343674
Redflower Hawksbeard 110880
Red-flower Honeysuckle 236090
Redflower Houndstongue 118008
Redflower Illigera 204648
Redflower Iris 208712
Redflower Kalanchoe 215147
Redflower Kopsia 217869
Redflower Laelia 219691
Redflower Loropetal 237195
Redflower Loropetalum 237195
Redflower Maack Honeysuckle 235931
Redflower Madder 337964,338017
Redflower Magnoliavine 351095
Red-flower Manglietia 244449
Redflower Mayten 246876
Redflower Meconopsis 247171
Redflower Merremia 250836
Redflower Neillia 263159
Redflower Nutantdaisy 110388
Red-flower Oxalis 278064
Redflower Parodia 284667
Redflower Peashrub 72334
Redflower Phalaenopsis 293600
Redflower Pleurothallis 304926
Redflower Pyrenees Lily 230018
Redflower Pyrola 322790
Redflower Ragleaf 108736
Redflower Raspberry 338637
Red-flower Rhamnoneuron 328587
Redflower Rockvine 387858
Redflower Salsify 394337
Redflower Salver-shaped Cherry 83167
Redflower Sandwort 32177
Redflower Sanicle 345998
Redflower Schefflera 350776
Redflower Scurrula 355317
Red-flower Seve-petal Camellia 69622
Redflower Slenderstem Marshmarigold 68224
Redflower Snakegourd 396262
Redflower Sobralia 366791
Redflower Sophora 369014,369114
Redflower Spapweed 204940
Redflower Stimpsonia 376684
Redflower Sweetvetch 188016
Redflower Swordflag 208712
Redflower Toona 392839
Redflower Touch-me-not 204940
Redflower Tuber Ullucus 401414
Red-flower Wendlandia 413774
Red-flower Woodsorrel 277702
Red-flowered Anneslea 25777
Red-flowered Basketvine 9463
Red-flowered Bauhinia 49017
Redflowered Blueberry 404033
Red-flowered Blueberry 404033
Red-flowered Camellia 68980
Red-flowered Checkered-like Fritillary 168469
Red-flowered Cotoneaster 107661
Red-flowered Currant 334189,334192,334194
Red-flowered False Bracteate Chelonopsis 86833
Red-flowered Garden Nicotiana Tobacco Plant 266054
Red-flowered Illigera 204648
Red-flowered Magnolia-vine 351095
Red-flowered Mallee 155575
Red-flowered Manglietia 244449
Red-flowered Neillia 263159
Red-flowered Olive 270163
Red-flowered Pacific Island Wild Buckwheat 152110
Red-flowered Pea-shrub 72334
Red-flowered Persimmon 132383
Red-flowered Pussy Willow 343186
Red-flowered Raspberry 338637
Red-flowered Schefflera 350776
Red-flowered Scurrula 355317
Red-flowered Wendlandia 413774
Red-flowered Wood 106800
Redflowering Currant 334189
Red-flowering Gum 106800
Red-flowering Silk-cotton Tree 56788
Redflush 220504
Redfoot Sagebrush 36162
Redfoot Wormwood 36162
Redfruit Actinodaphne 6778
Redfruit Baneberry 6422
Redfruit Bladderwort 403854
Redfruit Dandelion 384540
Redfruit Devilpepper 327076
Redfruit Dragon Mallow 367726
Redfruit Elatostema 142607
Redfruit Elm 401634
Redfruit Glycosmis 177833
Redfruit Greenbrier 366526
Redfruit Kneejujube 67075
Red-fruit Mahonia 242548
Red-fruit Money-leaved Barberry 51969
Redfruit Nertera 265358
Redfruit North Pole Fruit 31322
Red-fruit Oppositeleaf Fig 165128
Red-fruit Pencilwood 139640
Redfruit Pepper 300501
Redfruit Primrose 314350
Redfruit Ptarmiganberry 31322
Redfruit Raspberry 338361
Redfruit Rosewood 121847
Redfruit Roureopsis 337772
Redfruit Saurauia 347956
Redfruit Smilax 366526
Redfruit Spicebush 231334
Red-fruit Sycamore Maple 3478
Redfruit Youngia 416418
Red-fruited Actinodaphne 6778
Redfruited Axial Paris 284294
Red-fruited Barberry 51694
Red-fruited Dandelion 384614
Red-fruited Elm 401634
Red-fruited Greenbrier 366526
Red-fruited Nightshade 367726
Red-fruited Pencil-wood 139640
Red-fruited Pepper 300501
Red-fruited Ptarmigan-berry 31322
Red-fruited Raspberry 338361
Red-fruited Roureopsis 337772
Red-fruited Saurauia 347956
Red-fruited Spice-bush 231334
Redgland Honeysuckle 235860
Redgland Premna 313733
Redgland Purplepearl 66776
Red-glanded Premna 313733
Red-glandular Purplepearl 66776
Redglandular Raspberry 339335
Red-glandular Raspberry 339335

Redglandule Loosestrife 239830
Redgold Honeysuckle 236166
Redgum 155514,232565
Redhair Actinidia 6698
Redhair Ainsliaea 12631
Redhair Bauhinia 49213
Redhair Beccarinda 49396
Redhair Cacalia 283867
Redhair Cactus 140883
Redhair Jadeleaf and Goldenflower 260430
Redhair Kiwifruit 6698
Redhair Loosestrife 239832
Redhair Mountainash 369509
Redhair Ophiorrhiza 272291
Redhair Plumegrass 148915
Redhair Rockfoil 349861
Redhair Scorpiothyrsus 354744
Redhair Slugwood 50602
Redhair Woodbetony 287628
Red-haired Actinidia 6698
Red-haired Beccarinda 49396
Red-haired Evergreen Chinkapin 79016
Red-haired Mussaenda 260430
Red-haired Rhododendron 331689,331703
Red-haired Yunnan Azalea 331703
Redhairgrass 332996
Redhairs Caryodaphnopsis 77967
Red-hairy Rhododendron 331689
Red-handed Orchid 182230
Redhaw Hawthorn 110014
Red-Haw Hawthorn 110014
Red-head Beautyberry 66831
Redhead Cremanthodium 110454
Redhead Nutantdaisy 110454
Redhead Vine 765
Red-headed Calliandra 66673
Red-headed Vine 765
Red-head-grass 312220
Redheart Bulbophyllum 63054
Redheart Ceanothus 79972
Redheart Stonebean-orchis 63054
Redheart Tanoak 233130
Redhillall Azalea 331203
Red-hirtellous Slugwood 50602
Redhorn Dandelion 384647
Red-hot Cat Tail 1894
Red-hot Cat's Tail 1894
Redhot Cat's Tails 1894
Red-hot Cat's-tail 1894
Red-hot Cattail 1894
Redhot Cat-tail 1894
Red-hot Cat-tail 1894
Redhot Poker 217055
Red-hot Poker 216923,217055
Red-hot Poker Tree 154620
Redhot-pokerplant 216923
Redink Plant 298094
Red-ink Plant 298094
Red-ink-plant 298094
Red-ironbark 155741
Redish Azalea 331705
Redish Ginger 418019
Redish Meadowrue 388650
Redish Powder 309716
Redish Rhynchosia 333392
Redish-tomentose Cinnamon 91422
Redivy 191488
Red-knees 309199
Redleaf Ainsliaea 12723
Red-leaf Ampelopsis 20444
Redleaf Anacampseros 21136
Redleaf Anemone 23808
Redleaf Begonia 50238
Redleaf Cherry Plum 316308
Redleaf Cranesbill 174888
Redleaf Grape 411664
Red-leaf Japanese Barberry 52225
Redleaf Japanese Maple 3345
Red-leaf Leafflower 296751
Redleaf Litse 234051
Redleaf Medinilla 247558
Red-leaf Mitragyna 256123
Redleaf Mountainash 369541
Redleaf Rabbiten-wind 12723
Redleaf Rose 336557,336611
Red-leaf Rose 336611
Redleaf Rourea 337733
Redleaf Salem-rose 339205
Redleaf Speedwell 407329
Redleaf Spiradiclis 371778
Red-leaf Underleaf Pearl 296751
Redleaf Wildtung Lushan Wildtung 243424
Redleaf Windhairdaisy 348634
Red-leafed Palm 234185
Red-leaved Grape 411664
Red-leaved Helicia 189918
Red-leaved Japanese Barberry 52250
Red-leaved Litse 234051
Red-leaved Medinilla 247558
Red-leaved Rose 336611
Red-leaved Rourea 337733
Red-leaved Switch Grass 282368
Red-leaved Wattle 1544
Redleg 291840
Red-ligulate Sinobambusa 364811
Red-ligulate Tephroseris 385917
Red-ligule Sinobambusa 364811
Red-linear Mekong River Rhododendron 331219
Red-lined Rhododendron 331219
Redlip Bulbophyllum 62536
Redlip Calanthe 66053
Redlip Cymbidiella 116766
Redlip Irisorchis 267987
Redlip Oberonia 267987
Redlip Sage 344980
Redlip Stonebean-orchis 62536
Redmahogany 155720
Redmargin Calathea 66200
Red-margin Phyllostachys 297449
Red-margin Sanguineous-scale Pycreus 322354
Redmarginate Stahlianthus 373636
Red-narrowleaf Raspberry 338922
Redneck Palm 139366
Rednerve Ainsliaea 12724
Rednerve Maple 3504
Rednerved Chaydaia 86318
Red-nerved Chaydaia 86318
Red-nerved Lyonia 239399
Red-nerved Reevesia 327377
Red-onion 143570
Red-osier Dogwood 105193
Redoul 104700
Redowsk Therorhodion 389577
Redowsk's Rhododendron 389577
Red-paper Tree 13654
Redpappo Aster 40537
Redpappo Gerbera 224122
Redpappo Goldenray 229165
Redpappo Tailanther 381935
Red-pappus Aster 40537
Redpappus Cissampelopsis 92519
Red-pappus Synotis 381935
Red-pedicel Barberry 52130
Redpedisel Nanmu 295402
Redpepper 72068,72070
Red-pepper 72068
Redpetal Rockfoil 349574
Redpilose Mountainash 369509
Red-pilose Rhododendron 331689
Red-pilosed Mountainash 369509
Red-pulp Australian Finger-lime 253286
Redpunctate Ophiorrhiza 272292
Redpunjab Sumac 332791
Redpurple Larkspur 124537
Redpurple Premna 313720
Redpurple Ragwort 358788
Redray Alpinegold 199218
Redroot 18787,18810,71218,79905,79940,79957,219026,219030,345805
Red-root 79905
Redroot Amaranth 18810
Redroot Amaranthus 18810
Redroot Arisaema 33288
Red-root Buckwheat 152420
Redroot Flatsedge 118820
Redroot Gromwell 233731
Redroot Loosestrife 239645
Redroot Pigweed 18810
Redroot Sage 345324

Redroot Southstar 33288
Red-root Wild Buckwheat 152420
Red-rooted Flatsedge 118820
Red-rooted Sedge 118820
Redsandplant 327194
Redscale 44629
Redscale Fimbristylis 166298
Redscale Flatsedge 322342
Redscale Fluttergrass 166298
Redscale Triangular Bulrush 352230
Red-scaled Flat-sedge 322342
Red-seed Dandelion 384614
Redseed Stonecrop 356715
Red-seeded Dandelion 384540,384614
Red-seeded Plantain 302158
Redseedtree 155032
Red-seepal Gooseberry 334178
Redsepal Buttercup 326317
Redsepal Currant 334178
Redsepal Evening-primrose 269443
Red-sepal Evening-primrose 269443
Redsepaled Garcinia 171098
Red-sepaled Saurauia 347983
Redshank 291840,309570
Redshanks 8033,79905,174877,308739,309199,309570,339887,340151
Redsheath Arrowbamboo 162737
Redsheath Bamboo 297327,297328
Redsheath Coldbamboo 172746
Red-sheath Sasa 347286
Red-sheathed Bulrush 353773
Red-sheathed Green Sedge 76563
Red-shelled Gelidocalamus 172746
Red-shelled Sasa 347286
Redshoot Gooseberry 334207
Redspike Copperleaf 1894
Redspike Mexican-hat 326960
Red-spiked Lovegrass 147947
Redspine Raspberry 339214
Red-spined Fishhook Cactus 354311
Redspiny Aechmea 8581
Redspot Azalea 331565
Redspot Beautyberry 66776
Redspot Begonia 50264
Redspot Dishspurorchis 171851
Redspot Gastrochilus 171851
Redspot Swertia 380205
Redspoted Bulbophyllum 62900
Red-spots Rhododendron 331565
Red-spotted Galangal 17748
Red-spotted Gum 155643
Red-spotted Rhododendron 330251,330901,331696
Redspray Ruellia 339731
Red-spurred Valerian 81768
Redstalk Cryptocarya 113490
Redstalk Machilus 240690
Red-stalk Phoebe 295402
Redstalk Thickshellcassia 113490
Red-stalked Aster 380970
Red-stalked Machilus 240690
Red-stalked Plantain 302159
Redstar 323459
Redstar Zinnia 418058
Red-star Zinnia 418058
Red-stem Acacia 1544
Redstem Actinidia 6692
Redstem Catchfly 363998
Redstem Ceanothus 79971
Red-stem Dogwood 105193
Red-stem Fig 165594
Redstem Filaree 153767
Red-stem Filaree 153767
Redstem Kiwifruit 6692
Redstem Skullcap 355858
Redstem Stork's Bill 153767,153768,153835
Red-stem Stork's-bill 153767
Redstem Woodbetony 287739
Redstem Wormwood 36232
Red-stemmed Acacia 1544
Red-stemmed Actinidia 6692
Red-stemmed Aster 380970
Red-stemmed Filaree 153767
Red-stemmed Gentian 173836
Redstemmed Lady's Mantle 14022
Red-stemmed Plantain 302159
Redstipe Bladderwort 403729
Redstipe Windhairdaisy 348305
Redstriate Spapweed 205296
Redstriate Touch-me-not 205296
Redstriate Woodbetony 287705
Redstriatelip Calanthe 66068
Redstring 345405
Red-stripe Rhododendron 330722
Red-striped Rhododendron 330722
Redstyle Flower 144847
Red-tail Fescue 412465
Red-tinged Rhododendron 330070
Red-tinted Pussy Toes 26386
Red-tip Rabbit-tobacco 317748
Red-tip Spines Denmoza 125818
Red-tipped Cudweed 165990
Red-tipped Fluffweed 165990
Red-tomentose Cinnamon 91422
Redtomentose Evergreenchinkapin 79016
Redtongue Tangbamboo 364811
Redtooth Catchfly 363888
Redtop 11972,12118,12323
Red-top 11972
Redtop Panic Grass 282173
Red-top Panic Grass 282173
Red-top Panicum 282173
Red-topped Sage 345474
Redtwig Dogwood 105193
Red-twig Leucothoe 228181
Redtwig Slugwood 50543
Red-twigged Lime 391817,391824
Red-twigged Slugwood 50543
Reduce Motherwort 224978
Reduced Sandwort 32176
Redvein Abutilon 997
Redvein Dianxiong 297962
Redvein Dock 340249
Redvein Enkianthus 145677
Redvein Honeysuckle 235981
Redvein Lovegrass 147950
Redvein Lyonia 239399
Redvein Maple 3529
Red-vein Maple 3529
Redvein Ophiorrhiza 272289
Redvein Physospermopsis 297962
Redvein Rabbiten-wind 12724
Redvein Reevesia 327377
Redvein Rhubarb 329344
Redvein Spicebush 231432
Redvein Suoluo 327377
Redvein Tongoloa 392796
Redvein Trachydium 393857
Red-veined Dock 340151,340249
Red-veined Honeysuckle 235981
Red-veined Maple 3529
Red-veined Spice-bush 231432
Red-vien Dock 340249
Red-villose Bauhinia 49222
Red-villose Hydrangeavine 351819
Red-water Tree 154973
Redweed 21339,174877,282685,298094,308816,309199,309570
Redwing Cypressgrass 119342
Redwing Galingale 119342
Redwing Russianthistle 344580
Redwithe 100309
Redwood 8186,79905,300223,320330,360561,360565
Red-wood 7190
Redwood Banksia 47667
Redwood Family 114645,385250
Redwood Inside-out Flower 404619
Redwood Sorrel 277644,277994,278389
Red-wood Wendlandia 413779
Red-wood Willow 343396
Redwood-ivy 404619
Redwort 59532
Red-yellow Hairs Rhododendron 330041
Reed 295881,295888
Reed Bamboo 47190,268113
Reed Canary Grass 293709
Reed Canarygrass 293709
Reed Canary-grass 293709
Reed Cinna 91236
Reed Fescue 163819

Reed Foxtail 17512
Reed Mace 401085
Reed Manna Grass 177610,177629
Reed Mannagrass 177629
Reed Meadow Grass 177629
Reed Meadow-grass 177629
Reed Palm 65637,85437
Reed Phalaris 293709
Reed Stem Epidendrum 146428
Reed Sweet-grass 177629
Reed Themeda 389314
Reed's Green Condalia 101743
Reedbentgrass 65254
Reedbind 102933
Reedflower Wobamboo 362129
Reed-grass 177629,293709,293773,295888
Reedlike Alopecurus 17512
Reedlike Plumegrass 341837
Reedlike Sugarcane 341837
Reedlike Sweetcane 341837
Reedmace 401085,401112
Reed-mace 401085
Reed-stemmed Mahonia 242502
Reef Pemphis 288927
Reefers 71218
Reefort 326616
Reekkudzu 313411,313483
Reel Grass 19766
Reeve's Spirea 371868
Reeves Leptospermum 226461
Reeves Skimmia 365974
Reeves Spiraea 371868,371894
Reeves' Meadowsweet 371868
Reeves' Spiraea 371868
Reevesia 327357
Reflex Bethlehemsage 283642
Reflex Calanthe 66046
Reflex Cranesbill 174870
Reflex Deutzia 127057
Reflex Eritrichium 184298
Reflex Paraphlomis 283642
Reflexbract Windhairdaisy 348706
Reflexed Appendicula 29819
Reflexed Ardisia 31589
Reflexed Dodder 115123
Reflexed Komarov Lilac 382190
Reflexed Meadow-grass 321249
Reflexed Roegneria 335480
Reflexed Saxifrage 349841
Reflexed Sepal Trillium 397599
Reflexed Spiderwort 394053
Reflexed Spines Chinquapin 78868
Reflexed Stonecrop 357114
Reflexed Trillium 397599
Reflexed Twayblade 232304
Reflexed-sepal Pittosporum 301383
Reflexed-sepal Seatung 301383
Reflexedtwig Rockjasmine 23260
Reflexflower Gentian 173340
Reflexhair Cranesbill 174937
Reflexive-spined Evergreen Chinkapin 78927
Reflexpetal Calanthe 66046
Reflexpetal Pearweed 239820
Refracte Bastardtoadflax 389844
Refracte Conehead 290927
Refracte Hundredstamen 389844
Refracted Asparagus 39171
Refracted Hairgrass 126102
Refracted Monkshood 5529
Refracted Raspberry 339167
Refractedcalyx Anemone 24019
Refractedcalyx Windflower 24019
Refractedfruit Monkshood 5529
Refract-petal Cranesbill 174870
Refrected Agapetes 10362
Refrectpetal Agapetes 10362
Refuge Orach 44568
Regal Geranium 288206
Regal Lily 230019
Regal Maximiliana 246735
Regal Milicia 254494
Regal Pelargonium 288206,288264
Regallily 230019
Regals 273531
Regel Barberry 52100
Regel Ephedra 146243
Regel Eyebright 160254
Regel Feathergrass 376899
Regel Iljinia 204435
Regel Mouseear 82997
Regel Sympegma 380682
Regel Three-winged Nut 398318
Regel Threewingnut 398318
Regel Thundergodvine 398318
Regel's Mouse-ear Chickweed 82997
Regel's Rush 213416
Regenerata Poplar 311157
Reggedbranch St. John's Wort 202147
Regler Barberry 52101
Regnell Wickerware Cactus 329708
Regnier Liparis 232305
Regnier Twayblade 232305
Rehder Acanthopanax 143650
Rehder Barberry 52102
Rehder Clematis 95286
Rehder Euonymus 157832
Rehder Hophornbeam 276816
Rehder Hop-hornbeam 276816
Rehder Ironwood 276816
Rehder Machilus 240686
Rehder Micrechites 203336
Rehder Mountainash 369501
Rehder Mountain-ash 369501
Rehder Oak 324322
Rehder Pittosporum 301384
Rehder Seatung 301384
Rehder Silk Mockorange 294544
Rehder Sinojackia 364957
Rehder Skullcap 355713
Rehder Spindle-tree 157832
Rehder Syzygium 382654
Rehder Tree 327410
Rehder Willow 343984
Rehder Wing Nut 320371
Rehder's Deutzia 127059
Rehder's Jasmine 211979
Rehder's Leptodermis 226128
Rehder's Taiwan Hydrangea 199947
Rehder's Wildclove 226128
Rehder's Wingnut 320371
Rehdertree 327410
Rehder-tree 327410
Rehe Alkaligrass 321332
Rehman's Halfchaff Sedge 232421
Rehmannia 327429
Reho Alkaligrass 321332
Reiche Barberry 52103
Reichenbach Iris 208787
Reichenbach Pleione 304291
Reichenheim Pansy Orchid 254935
Rein Orchid 98586,183389,302318
Rein Orchis 98586,182213,183389
Reina De La Noche 288986,288988
Reine Claude 316333,316391
Reineckea 327525
Reineckia 327524
Reinhardt Palm 327538
Reinhold's Bee-orchid 272477
Reinorchis 182213,182230,182244
Reinweisse Lycaste 238736
Reis Dwarf 85328
Reitz Aloe 17220
Rejoua 327575,327577,382727
Related Cousinia 108221
Related Neillia 263143
Related Vinegentian 108221
Relict Trillium 397603
Remarkable Barberry 51766
Remarkable Cymbidium 116916
Remarkable Goatsbeard 394319
Remarkable Holly 203906
Remarkable Macleania 240755
Remarkable Nemophila 263505
Remarkable Rhododendron 330932
Remcope 379660
Remember-me 407072
Remirea 327663,327664
Remote Lemongrass 117162
Remote Mountains Meadowrue 388703
Remote Rhubarb 329384
Remote Sedge 75993

Remote-denticulate Evergreen Chinkapin 79010
Remoteflower Beauty-berry 66907
Remoteflower Draba 137203
Remoteflower Whitlowgrass 137203
Remote-flowered Sea Lavender 230643
Remote-leaved Thistle 92342
Remotelobated Hawthorn 109994
Remote-lobated Hawthorn 109994
Remote-lobed Hawthorn 109994
Remoteserrate Beauty-berry 66909
Remote-serrated Gugertree 350940
Remote-serrulate Beautyberry 66909
Remotespike Sedge 76005
Remotetooth Lycianthes 238974
Remotetooth Red Silkyarn 238974
Remote-toothed Croton 112940
Remotidenticulate Evergreenchinkapin 79010
Remotiflower Bamboo 47427
Remotiflower Bambusa 47427
Remotiflowered Bambusa 47427
Remotiserrulate Beautyberry 66909
Ren Knotweed 309691
Renanthera 327681
Renchang Childvine 400954
Renchang Kadsura 214982
Renchang Tylophora 400954
Rendle's Meadow Foxtail 17564
Rengarenga 36835
Rengas 177542,373628
Rengench Eurya 160591
Reniform Cremanthodium 110451
Reniform Nutantdaisy 110451
Reniformleaf Acianthus 4515
Reniform-leaf Epipactis 147149
Reniformleaf Habenaria 184019
Reniformleaf Wildginger 37709
Reniformlip Calanthe Calanthe 65895
Renosterveld Bulbine 62337
Repand Indian Mallow 402271
Repand Magnoliavine 351093
Repand Sauropus 348062
Repand Saxifrage 349845
Repanda Juniper 213715
Repandens Yew 385319
Repand-leaved Allomorphia 16133
Repandous Blumea 55813
Repandous Sauropus 348062
Repandus Podocarpium 200747
Repeatedsplit Corydalis 105804
Repent Alloplectus 16180
Repent Bittercress 72942
Repent Cupflower 266206
Repent Honeygrass 249323
Repent Melinis 249323
Repent Skullcap 355714
Repent Spapweed 205281
Repent Touch-me-not 205281
Repent Wildbasil 97044
Repestrine Leptospermum 226480
Replicate Ardisia 31588
Replicate Knotwood 44273
Reps 334179
Reptant Annual Bluegrass 305348
Reptant Primrose 314893
Reptant Sedge 76008
Reptant Woodbetony 287583
Reptent Lilyturf 272128
Reriifoliate Camellia 69549
Rescue Brome 61008
Rescue Grass 60661,61008,83470
Rescuegrass 61008
Reseda 327793
Resedaleaf Bittercress 72717
Resembling Hippophae Rhododendron 330871
Resembling Ledum Rhododendron 332000
Resembling Protea Rhododendron 331556
Resembling R. Glischrum Rhododendron 330774
Resembling R. Lanatum Rhododendron 331044
Resembling R. Pulcherum Rhododendron 331580
Resembling R. Recurvum Rhododendron 331612
Resembling R. Semnum Rhododendron 331793
Resembling Sedge 76283
Resembling Sorghum 369689
Reshes 213165
Resin Birch 53454,53533
Resin Hogfennel 293033
Resin Tree 316018
Resinbush 404608
Resindot Sunflower 189043
Resinose Milkwort 308303
Resintree 316018
Resin-weed 181092
Respis 338557
Ressurection Lily 239280
Rest Harrow 271563
Rest Haven 269404
Restharrow 271290,271333,271347,271563
Resupinate Bladderwort 403311
Resupinate Twayblade 232307
Resupinate Woodbetony 287584
Resurrection Flower 21805
Resurrection Lily 214995,215011
Resurrection Plant 21805
Resurrectionlily 214995
Retam 328232,328243
Retama 284479
Retamo 63408
Retamo Wax 63408
Reticular-veined Cinnamon Tree 91420
Reticulate Bamboo 297435
Reticulate Bauhinia 49217
Reticulate Camellia 69552
Reticulate Drypetes 138679
Reticulate Embelia 144802
Reticulate Epimedium 147044
Reticulate Helicia 189949
Reticulate Holboellia 197249
Reticulate Indigo 206476
Reticulate Iris 208788
Reticulate Japanese Maple 3343
Reticulate Leafflower 296734
Reticulate Leaf-flower 296734
Reticulate Mahonia 242631
Reticulate Mallotus 243442
Reticulate Meadowrue 388636
Reticulate Microstegium 254042
Reticulate Milkvetch 42974
Reticulate Mitreola 256218
Reticulate Nerve Daphniphyllum 122721
Reticulate Oreocharis 273843
Reticulate Osmanthus 276393
Reticulate Primrose 314894
Reticulate Raspberry 339172
Reticulate Rhubarb 329385
Reticulate Saffron 111582
Reticulate Sichuan Woodbetony 287725
Reticulate Skullcap 355715
Reticulate Snakegourd 396256
Reticulate Spapweed 205282
Reticulate Tickclover 126580
Reticulate Touch-me-not 205282
Reticulate Underleaf Pearl 296734
Reticulate Willow 344003
Reticulate Wingseedtree 320846
Reticulated Nut-rush 354225
Reticulatefruit Larkspur 124181
Reticulate-nerve Machilus 240687
Reticulate-nerved Maesa 241826
Reticulate-veined Camphortree 91420
Reticulate-veined Cinnamon Tree 91420
Reticulinerve Barberry 52110
Retinerve Maesa 241826
Retinerve Marking Nut 357883
Retinerve Markingnut 357883
Retinerved Blueberry 403792
Retinerved Embelia 144802
Retinerved Marking-nut 357883
Retinerved Persimmon 132369
Reting Woodbetony 287604
Retinospora 85345
Retivein Markingnut 357883
Retrofracted Pepper 300495
Retropilose Shrubcress 368578
Retrorse Sedge 76014

Retrorsesetose Spikesedge 143404
Retrosehair Elatostema 142819
Retrose-hairy Willow 343898
Retroserrate Asiabell 98406
Retrospiral Cryptocoryne 113540
Retrostyle Elaeagnus 142182
Retro-styled Elaeagnus 142182
Retunse-leaf Grewia 180938
Retuse African Myrsine 261610
Retuse Ash 167994
Retuse Bauhinia 49219
Retuse Breynia 60078
Retuse Conehead 290928
Retuse Cup-calyx Honeysuckle 236064
Retuse Gurjun 133577
Retuse Haworthia 186679
Retuse Nitraria 266362
Retuse Obovateleaf Woodnettle 273915
Retuse Privet 229599
Retuse Rhododendron 330516
Retuse Rhynchostylis 333731
Retuse Stachyurus 373568
Retuse Whitethorn 266362
Retuse Wild Arrowhead 342435
Retuseleaf Daphne 122594
Retuse-leaf Holly 204208
Retuse-leaved Daphne 122594
Retuselobe Adenosma 7991
Retusifoliolate Ash 168084
Reunion Bamboo 262758
Reunion Isle Bamboo 262758
Revelute Biondia 54528
Revelute Mountainash 369431
Revelute Mountain-ash 369431
Reverchon Hawthorn 109995
Reversed Clover 397050
Reversepanicled Gentian 173674
Reversetooth Windhairdaisy 348709
Revolute Cupule Oak 116167
Revolute Holly 204329
Revolute Lilyturf 272130
Revolute Meadowrue 388638
Revolute Waxplant 198892
Revolvebill Woodbetony 287271
Rewa Rewa 216922
Rewarewa 216922
Rewa-rewa 216922
Rex Begonia 50232,50235
Rex Begonia Vine 92777
Rex Woodbetony 287605
Rex-begonia Vine 92692
Reynoutria 328328
Rhabdothamnopsis 328439
Rhabdothamus 328445,328446
Rhambic-nut Tanoak 233383
Rhamnella 328544,328550
Rhamnoneuron 328586
Rhamnusleaf Holly 204210
Rhaphidophora 328990
Rhapis 329173
Rhapis Palm 329184
Rhapontic 329197
Rhas Indofevillea 206848
Rhatany 218074,218076,218085
Rhatany Root 218074
Rhazya 329261
Rheed Cansjera 71292
Rheedia 329268
Rheingold Arborvitae 390618
Rhektophyllum 83713,329282
Rhenoster 326104
Rheumatism Root 212299
Rheumatism Weed 29473,87498
Rheumatism Wood 327022
Rheumatism-root 212299
Rhiannon's Aster 380978
Rhinacanthus 329454
Rhine Berry 328642
Rhineberry 328642
Rhinoceros Cactus 107016
Rhinocerous Cactus 107022
Rhipsalidopsis 329679
Rhipsalis 329683
Rhizoma Peanut 30497
Rhizomate Congested Sedge 73615
Rhizomatic Banana 260237
Rhizome Arisaema 33473
Rhizome Eria 148646
Rhizome Hairorchis 148646
Rhizome Lilyturf 272131
Rhizome Rush 212834
Rhizome Waterbamboo 260095
Rhizome Wheatgrass 11793
Rhizome Woodbetony 287627
Rhizomeless Spikesedge 143058
Rhizophor Pelliciera 288699
Rhodamnia 329793,329794
Rhode Island Bent 12025,12034
Rhode Island Bent Grass 12034
Rhodes Grass 88352
Rhodes' Violet 356368
Rhodesia Mahogany 216211
Rhodesian Bloodwood 320278
Rhodesian Chestnut 46573
Rhodesian Currant 332545
Rhodesian Holly 319301
Rhodesian Ironwood 103716
Rhodesian Mahogany 10127
Rhodesian Rubber 133167
Rhodesian Teak 46573
Rhodesian Timothy 361902
Rhodesian Tree Hibiscus 389935
Rhodesian Wattle 288884
Rhodesian Wistaria 56735
Rhodeslan Ash 63931
Rhodiola 329826
Rhodocoma 330014
Rhododaphne 265327
Rhododendron 330017,331520
Rhododendron-inhabiting Scurrula 385221
Rhododendronleaf Buckthorn 328841
Rhododendron-leaf Fig 165270
Rhododendron-leaved Buckthorn 328841
Rhododendron-leaved Willow 344006
Rhododendron-like Willow 344265
Rhodohypoxis 332184
Rhodoleia 332204,332205
Rhodophanes 265327
Rhodora 330321
Rhodostachys 162859,162861
Rhodostachys Family 60557
Rhoeo 332293
Rhoiptelea 332328
Rhoiptelea Family 332330
Rhomb-branched Birch 53604
Rhombic Copperleaf 1971
Rhombic Copper-leaf 1971
Rhombic Leaved Grewia 180939
Rhombic Leaves Rhododendron 331650
Rhombic Lip Galangal 17712
Rhombic-foliate Grewia 180939
Rhombicleaf Ancylostemon 22172
Rhombicleaf Ascitesgrass 407479
Rhombic-leaf Burretiodendron 161625
Rhombicleaf Daisy 124846
Rhombic-leaf Euonymus 157912
Rhombic-leaf Fig 165736
Rhombicleaf Goosefoot 86977
Rhombicleaf Grape 411885
Rhombicleaf Greenbrier 366361
Rhombic-leaf Grewia 180939
Rhombicleaf Passionflower 285697
Rhombicleaf Pimpinella 299512
Rhombic-leaf Poplar 311504
Rhombicleaf Pricklyash 417323
Rhombicleaf Purpledaisy 267178
Rhombicleaf Rockvine 387861
Rhombicleaf Spapweed 205283
Rhombic-leaf Spicebush 231444
Rhombic-leaf Spindle-tree 157912
Rhombicleaf Stairweed 142820
Rhombicleaf Touch-me-not 205283
Rhombic-leaf Xianmu 161625
Rhombic-leaved Cayratia 79886
Rhombicleaved Elatostema 142820
Rhombic-leaved Euonymus 157912
Rhombic-leaved Fig 165736
Rhombic-leaved Hsienmu 161625
Rhombic-leaved Manyflower Ladybell 7777
Rhombic-leaved Persimmon 132371
Rhombic-leaved Prickly-ash 417323

Rhombic-leaved Rhododendron 331650
Rhombic-leaved Rockvine 387861
Rhombic-leaved Spice-bush 231444
Rhombic-leaved Sunflower 189066
Rhombiclesf Monkshood 5532
Rhombiclip Dendrobium 125218
Rhombiclip Eria 148760
Rhombiclip Galangal 17712
Rhombiclip Hairorchis 148760
Rhombifolious Mountainash 369505
Rhombipetal Parnassia 284610
Rhombleaf Azalea 331650
Rhombleaf Copperleaf 1983
Rhombleaf Euonymus 157912
Rhomboid Bushmint 203062
Rhomboid Skullcap 355716
Rhomboidal Mahonia 242544
Rhomboidbract Birch 53604
Rhomboidleaf Alajja 13336
Rhomboidleaf Ivy 187318
Rhomboidleaf Monkshood 5532
Rhomboidleaf Veronicastrum 407479
Rhomboid-leaved Ivy 187325
Rhomboid-leaved Triumphweed 399312
Rhomicleaf Dolichos 135600
Rhomicleaf Sicklehairicot 135600
Rhopaloblaste 332375
Rhopalocnemis 332409,332411
Rhopalostylis 332434
Rhubarb 329306,329308,329309,329372,329386,329388,338447,339914
Rhubarb-weed 292353
Rhum Palm 57117
Rhus Tree 393479
Rhynchelytrum 332996
Rhynchodia 333076
Rhynchoglossum 333083
Rhynchosia 333131,333456
Rhynchosinapis wrightii 99094
Rhynchospermum 333471
Rhynchospore 333477
Rhynchostylis 333725
Rhynchotechum 333733
Rhytidophyllum 333861
Rib 262722
Rib Eargrass 187558
Rib Grass 302034,302068
Rib Plantain 150901,302034
Ribbeak Buttercup 326470
Ribbed Birch 53389
Ribbed Camellia 69012
Ribbed Hedeoma 187178
Ribbed Hedyotis 187558
Ribbed Lycaste 238738
Ribbed Melilot 249232
Ribbed Murainagrass 209356
Ribbed Paspalum 285462
Ribbed Vetch 408350
Ribbedculm Spikesedge 143217
Ribbed-pod Pigeonwings 97193
Ribbon Bush 8033,197769,197772,202506
Ribbon Cactus 287853,287855
Ribbon Eucalyptus 155772
Ribbon Fan Palm 234175
Ribbon Grass 293705,293709,293710
Ribbon Gum 155725,155772
Ribbon Plant 88553,137487
Ribbon Pocktorchid 282883
Ribbonbush 197769,197772
Ribboned Maranta 66213
Ribbon-grass 293709,293710,293720
Ribbonleaf Pondweed 312096
Ribbon-leaf Pondweed 312096
Ribbon-leaved Pondweed 312096,312265
Ribbon-leaved Water-plantain 14734
Ribbon-tree 53563
Ribbonwood 8033,197154,197158,301610,301612,301613
Ribcolumn Rush 213389
Ribedge Buttercup 326405
Riberry 382604
Ribfruit Himalayan Trevesia 394760
Ribfruit Linden 391762
Ribfruit Memecylon 250036
Ribfruit Moxanettle 221557
Ribfruit Sinojackia 364953
Ribfruit Tanoak 233400
Ribgrass 302034
Rib-grass 302034
Rible Beech 266879
Ribleaf Leek 15194
Riblines Yushanbamboo 416779
Ribpod Butterfly Pea 97193
Ribpod Ormosia 274424
Ribseed Evodia 161380
Rib-seed Sand-mat 85753,158963
Ribseedcelery 304752
Rib-seeded Sand Mat 85753
Rib-seeded Sand-mat 158963
Ribsilique 102808,102815
Ribspine Angulate-fruited Evergreen Chinkapin 78897
Rib-striped Yushania 416779
Ribwort 105352,301832,302034,302068
Ribwort Plantain 150901,302034
Ribworth 302034
Ricason Podranea 306706
Rice 170335,275909,275958
Rice Barnyardgrass 140477
Rice Bean 409085
Rice Cut Grass 223992
Rice Cutgrass 223992
Rice Cut-grass 223992
Rice Cypressgrass 119041
Rice Flower 279345,299303,299309,350125
Rice Galingale 119041
Rice Grass 275991,285507,370155,370182
Rice Grass Mau'u-Laiki 285507
Rice Paper 387432
Rice Paper Plant 162879,387432
Rice Tree 162876
Ricebag 181626
Ricebag Crazyweed 278866
Rice-button Aster 380866,380869
Ricefield Bulrush 352223
Ricefield Flatsedge 119041
Ricefield Waternymph 262040
Rice-flower 299303
Ricegrass 275991,285507
Rice-grass 275991
Rice-like Razorsedge 354199
Ricepaper Plant 387432
Rice-paper Plant 387428,387432
Ricepaperplant 387428,387432
Ricepaper-plant 387428,387432
Rice-root 168444
Riceshape Pearlsedge 354199
Ricesieve Bamboo 47402
Ricesteamer Qinggang 116092
Rich Wolfberry 239104
Richardii Rose 336883
Richardson Fringed Brome 60599
Richardson Monkshood 5534
Richardson Willow 344010
Richardson's Alumroot 194433
Richardson's Bitterweed 201325
Richardson's Geranium 174876
Richardson's Muehlenbergia 259694
Richardson's Pondweed 312253
Richardson's Sagewort 35213
Richardson's Sedge 76033
Richardson's Willow 344009
Richella 334358
Richleaf 99844
Richmond River Pine 30841
Richtube Woodbetony 286998
Richweed 91034,99844,158354,299024
Rich-woods Sedge 75589
Riddell's Goldenrod 368353
Riddell's Groundsel 359897
Riddell's Lazydaisy 29105
Ridge Beggartick 53920
Ridge White Oak 323625
Ridgecalyx Gentian 173731
Ridged Goosefoot 86986
Ridged Podograss 397168
Ridged Sedge 74722
Ridge-fruited Mallee 155544
Ridgelip Spotleaf-orchis 179621
Ridge-seeded Spurge 85753,158963
Ridgybranch Elaeocarpus 142322

Ridgyfruit Memecylon 250036
Ridgyshoot Agapetes 10287
Ridley Arecapalm 31694
Ridley Licuala Palm 228763
Ridley Licualapalm 228763
Ridley Peony 280270
Ridley Rattan Palm 65780
Ridong Sedge 76035
Riely 235373
Riengalily 36835
Rifle-the-Ladies'-purse 72038
Rifoliate Jewel Vine 126007
Riga Fir 300223
Riga Pea 222832
Riga Pine 300223
Rigescent Gentian 173814
Rigescent Hawksbeard 110979
Rigescent Sedge 74407
Rigglers 314143
Right Dittany 274210
Right Service 369381
Rightangle Arrowwood 407850
Rightangle Viburnum 407850
Rightangule Spapweed 205277
Rightangule Touch-me-not 205277
Rightbeak Spapweed 205278
Rightbeak Touch-me-not 205278
Rigid Ant Palm 217924
Rigid Arisaema 33476
Rigid Artocarpus 36945
Rigid Azalea 331652
Rigid Bamboo 47430
Rigid Bambusa 47430
Rigid Barbate Deadnettle 220359
Rigid Barberry 52115
Rigid Begonia 50244
Rigid Bottlebrush 67291
Rigid Brome 60935
Rigid Bromegrass 60935
Rigid Chung Sedge 74131
Rigid Coelogyne 98737
Rigid Cowparsnip 192346
Rigid Cymbidium 116822
Rigid Draba 137212
Rigid Dragonhead 137642
Rigid Epidendrum 146473
Rigid Goldenrod 368354,368356
Rigid Greenbrier 366475
Rigid Hair Branches Blueberry 404023
Rigid Heterocaryum 193744
Rigid Hornwort 83545
Rigid Inule 207087
Rigid Kengyilia 215954
Rigid Large Woodreed 65420
Rigid Manyflower Woodrush 238650
Rigid Milkvetch 42982
Rigid Nectandra 263062
Rigid Onosma 271820
Rigid Roegneria 335483
Rigid Sedge 76532
Rigid Seepweed 379594
Rigid Skullcap 355758,355804
Rigid Smilax 366475
Rigid Sweetvetch 188108
Rigid Swollennoded Cane 87609
Rigid Verbena 405885
Rigid Whitetop Aster 380977
Rigid Whitlowgrass 137212
Rigid Woodbetony 287633
Rigid Yellow Bamboo 47430
Rigidbranch Rockjasmine 23264
Rigidde Rhododendron 331652
Rigidform Woodbetony 287634
Rigidleaf Batrachium 48919
Rigidleaf Cymbidium 117008
Rigidleaf Globethistle 140688
Rigidleaf Indigo 206488
Rigidleaf Sedge 75724
Rigidleaf Silverspikegrass 227956
Rigidleaf Syzygium 382668
Rigidleaf Windhairdaisy 348903
Rigidleaf-like Silverspikegrass 227955
Rigid-leaved Barberry 52116
Rigid-leaved Crowfoot 48919
Rigid-leaved Indigo 206488
Rigid-leaved Syzygium 382668
Rigidwing Craneknee 221739
Rigley Acriopsis 5952
Rikeze Sagebrush 36558
Rim Ash 80698
Rimann Sunipia 379781
Rimose Primrose 314901
Rimose Rosewood 121805
Rimose-vaginate Ferrocalamus 163568
Rimu 121094
Rincon Bitterweed 201324
Rincon Rubberweed 201324
Rindera 334540,334555
Ring Cupped Oak 116100
Ring Fig 164633
Ring Fingers 273531
Ring Grass 259711
Ring Muhly 259711
Ring O' Bells 145487
Ringbract Macaranga 240229
Ring-calyxed Agapetes 10291
Ring-cupped Oak 116100
Ringe Heather 150131
Ringed Biflorgrass 128568
Ringed Calyx Agapetes 10291
Ringed Coldbamboo 172738
Ringed Dichanthium 128568
Ringed-stem Bastard-sage 152670
Ringflower 21230,21249
Ringfruit Mucuna 259501
Ring-leaved Willow 343027
Ring-o'-bells 145487
Ringpod Milkvetch 42218
Ringstem 28797
Ringvine 116008
Ringwood 46335
Ringworm Cassia 360409
Ringworm Cassia Bark 360409
Ringworm Root 329471
Ringworm Senna 360409
Ringworm Shrub 360409
Rinorea 334562
Rio Arrowroot 244507
Rio Bravo 227922
Rio Grande Bugheal 362160
Rio Grande Cottonwood 311324
Rio Grande Fire Pink 363891
Rio Grande Fleabane 151002
Rio Grande Globe-amaranth 179239
Rio Grande Nailwort 284826
Rio Grande Palmetto 341418
Rio Grande Pearlhead 209552
Rio Grande Wild Buckwheat 152297
Rio Nufiez Coffee 98941,99008
Rio Nunez Coffee 98941,99008
Rio Rosewood 121770
Ri-oak Casuarina 79157
Riparial Silverreed 255861
Riparian Awngrass 255861
Riparian Greenbrier 366548
Riparian Homonoia 197963
Riparian Silvergrass 255861
Riparian Sonerila 368898
Riparian Sweetspire 210415
Ripgut 60935
Ripgut Brome 60696,60700,60935
Ripgut Grass 60696,60700
Rip-gut Sedge 75034
Ripley's Wild Buckwheat 152431
Ripple Grass 302034,302068
Ripple-margined Plantain Lily 198598
Rippleseed Plantain 302068
Ripple-seed Plantain 302068
Rippling Grass 302034
Ririe's Rhododendron 331660
Rising Sun 227533
Risleya 334781,334782
Risso's pummelo 93751
Ritozan Scurrula 385208
Ritter Aztekium 45994
Ritter Matucana 246601
Riukiu Cowpea 409036
Rium 121084
Rivalling Serruria 361158
Rivebanic Sweetspire 210415
Rivebank Cragcress 98028

Rivebank Scurvyweed 98028
River Alder 16464
River Bamboo 297446
River Bank Grape 411887
River Banksia 47649
River Bells 296112
River Birch 53536,53540
River Black Box 155499
River Black Oak 79185
River Bottlebrush 67297
River Bulrush 56636,353422
River Bushwillow 100454
River Cooba Wattle 1623
River Crowfoot 48918
River Cyperus 119497
River Eupatorium 11171
River Feathergrass 376848
River Grape 411887
River Grass 354579
River Hawthorn 109684,109996
River Juniper 213922
River Lomatia 235410
River Loosestrife 239679
River Mangrove 8646
River Maple 3532
River Oak 79157
River Oats 86140
River Peppermint 155573
River Plum 316206
River Poison 161632
River Poplar 311292
River Protea 315761
River Red Gum 155517
River Redgum 155517
River Rose 48977
River Sallow 344016
River Sar Iris 208811
River She Oak 79157,79165
River Sheoak 79157
River She-oak 79157
River Side Azalea 330718
River Side Rhododendron 331661
River Tea Tree 248104
River Tea-tree 248077
River Water Dropwort 269320
River Water-crowfoot 48918
River Water-dropwort 269320
River Wattle 1139
River White-gum 155573
River Wild Rice 418096
River Willow 343392,343726
Riverbank Anemone 24021
Riverbank Dicliptera 129309
Riverbank Grape 411887
Riverbank Rattan Palm 65781
Riverbank Skullcap 355391
Riverbank Spapweed 205285
Riverbank Swallowwort 117666
Riverbank Sweetspire 210415
Riverbank Touch-me-not 205285
Riverbank Wild-rye 144449
River-beach Sweetvetch 187872
Rivers' Chinese Flowering Crabapple 243707
Rivershore Hilliella 196299
River-sida Sonerila 368898
Riverside Bogorchid 11171
River-side Flemingia 166862
Riverside Holly 204047
River-side Holly 203981
Riverside Homonoia 197963
Riverside Psychotria 319548
Riverside Rhododendron 330718
River-side Rhododendron 331661
Riverside Willow 344017
Rivervalley Stephania 375868
Riverweed 306694,306695
Riverweed Family 306688
Rivery 235373
Rives Pelatantheria 288604
Rivet Wheat 398986
Rivier Giantarum 20132
Riviera Camomile 26836
Rivin 334892,334897
Rivini Violet 410493
Rivularislike Waterdropwort 269337
Rivulet Violet 409958
Rix 295888
Rixford's Rockwort 354710
Riyuetan Liparis 232190
Riyuetan Twayblade 232190
Rizards 334179
Rizhao Woodbetony 287635
Rizhaoshan Rockfoil 349852
Rizzer-berries 334179
Rizzles 334179
Road Kill Cactus 102825
Road-grass 143063
Roadside Agrimony 11598
Roadside Bittercress 72749
Roadside Brome 60982
Roadside Bushclover 227001
Roadside Bush-clover 227001
Roadside Croton 113046
Roadside False Madwort 53039
Roadside Hawksbeard 110808
Roadside Pennycress 390205
Roadside Pepperweed 225447
Roadside Pepper-weed 225447
Roadside Rush 213507
Roadside Sand Spurry 370696
Roadside Thistle 91730
Roadside Toadflax 230884
Road-to-heaven 307197
Roan Mountain Goldenrod 368366
Roan Mountain Rattlesnakeroot 313877
Roan Mountain Sedge 76048
Roanoke Bells 250914
Roan-thee 369339
Roast Beef 208566
Roast Beef Plant 208566
Roastbeef Plant 208566
Roast-beef Plant 208566
Rob Roy 248411
Robb's Euphorbia 159730
Robber's Lanterns 9701
Robbin's Cory Cactus 155218
Robbins' Pondweed 312254
Robbins' Senecio 359904
Robbins' Spike-rush 143329
Robble 266879
Robert 174877
Robert Cranesbill 174877
Robert Geranium 174877
Robert's Geranium 174877
Robeson Hawthorn 110001
Robie 302863
Robillo 79258
Robin 174877,248411
Robin Flower 174877,248411
Robin Hood 11947,23767,174877,363461
Robin Hood's Feather 95029
Robin Hood's Feathers 95029
Robin Hood's Fetter 95029,95431
Robin Red Breast Bush 248103
Robin Red Breast Melaleuca 248103
Robin Redbreast 174877,248411
Robin Redbreast's Cushion 336419
Robin Redshanks 174877
Robin Run-p-the-hedge 170193
Robin Run-the-dyke 102933
Robin Tree Cactus 299264
Robin's Eye 174877,248411,260760,308454
Robin's Eyes 174877,248312,260760,
308454
Robin's Flower 174877,248411
Robin's Pillow 336419
Robin's Pincush 336419
Robin's Pincushion 216753,336419
Robin's Plantain 150901
Robin's-plantain 150862
Robin-catch-the-hedge 170193
Robinet Rose 363411
Robin-I'-the-hedge 174877
Robinia 334946
Robin-in-the-hedge 248411
Robin-in-the-house 248411
Robin-plantain 150901
Robin-redbreast Bush 248103
Robin-round-the-hedge 170193
Robin-run-by The-grass 170193
Robin-run-in-the-field 102933

Robin-run-in-tile-grass 170193
Robin-run-the-hedge 68713,170193,170743,176824,248411,367120
Robinson Ceriops 83946
Robinson Chessbean 116606
Robinson Cylindrokelupha 116606
Robinson Drooping Clinacanthus 96905
Robinson Macrosolen 241321
Robinson Sheathflower 241321
Robinson's Catchfly 364021
Robinson's Starwort 318510
Robin-under-the-hedge 82758
Robiquetia 335035
Roble 266879
Roble Amarillo 382711
Roble de Chile 156063
Roble de Olor 79258
Roblum 316294
Robomwsk Sage 345347
Roborovsk Calligonum 67074
Roborovsk Orchis 169896
Roborovsky Dandelion 384779
Roborovsky Pea-shrub 72333
Roborowsk 32179
Roborowsk Nitraria 266365
Roborowsk Prickly-thrift 2281
Roborowsk Sealavender 230653
Roborowsk Stonecrop 357084
Roborowsk Woodbetony 287636
Roborowski Cacalia 283864
Roborowsky Feathergrass 376903
Rob-run-up-dyke 170193,176824
Robur Oak 324335
Robust Acaulescent Pegaeophyton 287959
Robust Aloe 16817
Robust Anisochilus 25390
Robust Ascitesgrass 407480
Robust Asparagus-fren 39151
Robust Aucuba 44948
Robust Begonia 50246
Robust Bhesa 53679
Robust Blue Cohosh 79732
Robust Bouquet Larkspur 124257
Robust Ceratozamia 83688
Robust China Cane 329188
Robust Clinopodium 97023
Robust Coffee 98886
Robust Columnea 100120
Robust Corneredcalyx Larkspur 124120
Robust Dunn Wampee 94187
Robust Eria 148763
Robust Evax 193377
Robust Evodia 161369
Robust Fargesia 162742
Robust Fedde Elsholtzia 144019
Robust Fieldcitron 400542
Robust Fishvine 125994
Robust Gelonium-leaved Glyptic-petal Bush 177992
Robust Hedgehog 140256
Robust Holly 204213
Robust Hydrangea 200070
Robust Imbricateleaf Stairweed 142826
Robust Javatea 275755
Robust Jewelvine 125994
Robust Kobresia 217263
Robust Ladypalm 329188
Robust Ladypalms 329188
Robust Largeflower Larkspur 124257
Robust Larkspur 124557
Robust Leea 223954
Robust Manaraceme Sweetvetch 188052
Robust Marsh Orchid 121373
Robust Microtoena 254265
Robust Milkvetch 42493
Robust Monoflorcress 287959
Robust Neoporteria 198499
Robust Pipewort 151452
Robust Poplar 311158
Robust Privet 229601
Robust Rattanpalm 65699
Robust Rhodiola 329933
Robust Rockjasmine 23266
Robust Roegneria 335281
Robust Rungia 340373
Robust Scabrousleaf Elatostema 142826
Robust Silk Grevillea 180642
Robust Silk Oak 180642
Robust Silver-oak 180642
Robust Slugwood 50599
Robust Small Reed 127209
Robust Small Swordflag 208769
Robust Spapweed 205287
Robust Spineflower 89112
Robust Spiraea 372071
Robust Sweetvetch 188052
Robust Triflower Holly 204344
Robust Turpinia 400542
Robust Veronicastrum 407480
Robust Walsura 413129
Robust Wild Smallreed 127209
Robust Wildbasil 97023
Robust Wildginger 37678
Robust Windflower 24030
Robust Woodbetony 287637
Robust Zephyrlily 184255
Robusta China Weasel-snout 170022
Robusta Coffee 98872
Robuste Aucuba 44948
Robuste Eucalyptus 155722
Robuste Hartia 185964
Robuste Simon Poplar 311505
Robuste-branch Bamboo 297447
Robuste-toothed Eurya 160448
Robuste-vein Holly 204215
Robust-styled Anemone 24031
Robustus Holly 204213
Robyns' Aster 380979
Rocambole 15698,15726
Roccombole 15726
Rochebrun Sedge 76051
Rochel Hornrim Saxifrage 349609
Rochelia 335103
Rochlis 329521
Rock Alyssum 45192
Rock Amoora 19955
Rock Anemone 24032
Rock Apple 243687
Rock Aster 40016,41161
Rock Ballota 47021
Rock Balsam 97260,290286
Rock Barberry 52117
Rock Beauty 292527,292530
Rock Bells 30007
Rock Birch 53530
Rock Birthwort 34132
Rock Bittercress 72960
Rock Buckthorn 328628,328849
Rock Cacalia 283865
Rock Campion 364002
Rock Candytuft 203236
Rock Catchfly 364002
Rock Cedar 213620
Rock Cherry 316710
Rock Chestnut Oak 324166,324201
Rock Childvine 400956
Rock Cinnamon 91424
Rock Clovershrub 70882
Rock Columbine 30071
Rock Coriaceousleaf Lagotis 220159
Rock Cotoneaster 107486
Rock Crane's-bill 174715
Rock Cress 30160,30186,30358,30451,400638
Rock Currant 334149
Rock Daisy 291296
Rock Dandelion 384614
Rock Daphne 122416
Rock Draba 137213
Rock Elatostema 142733
Rock Elm 88506,254493,401509,401549,401618,401638
Rock Euonymus 157869
Rock Evergreen Chinkapin 79013
Rock Evergreenchinkapin 79013
Rock Fig 165423,165455,165471
Rock Five-finger 312962
Rock Fleabane 150946,150956
Rock Fringed Cress 30292
Rock Fumewort 106117
Rock Gaultheria 172147

Rock Golden Rod 368013
Rock Goldenrod 114516,292500,292503,368377
Rock Grape 411900
Rock Grevillea 180601,180662
Rock Groundsel 92521
Rock Hakea 184609
Rock Halfribpurse 191533
Rock Harlequin 105557,106430
Rock Hemilophia 191533
Rock Holly 204216,204236
Rock Honeysuckle 236073
Rock Hygrophila 200655
Rock Jasmine 23107,23139,23211
Rock Knotweed 309001
Rock Ligusticum 340463
Rock Lily 36835,125365,230020
Rock Live-forever 138839,138840
Rock Madwort 45192
Rock Mallow 243877
Rock Maple 2991,3236,3549
Rock Mat 292645
Rock Maurandya 246657
Rock Maxburretia 246698
Rock Mazus 247031
Rock Melon 114199,114214
Rock Microula 254365
Rock Midget 255241
Rock Milkvetch 42624
Rock Mint 388275
Rock Monkshood 5538
Rock Mosquitotrap 117667
Rock Muhly 259700
Rock Oak 324166,324201,324204
Rock Orchid 125365
Rock Palm 59095,59098
Rock Pennyroyal 257198
Rock Penstemon 289323
Rock Pimpinella 299520
Rock Pine 299822
Rock Pink 383304,383327
Rock Pink Spotted Rhododendron 330894
Rock Pittosporum 301291,301394
Rock Plant 356468
Rock Plumegrass 148913
Rock Poison 171970
Rock Primrose 314903
Rock Pueslane 65827
Rock Purdom Poplar 311445
Rock Purslane 65853
Rock Ragwort 359938
Rock Redcurrant 334149
Rock Rex Woodbetony 287615
Rock Rhododendron 330695
Rock Rose 34536,93120,93124,93137,93141,188583,188757
Rock Rose Family 93046
Rock Rose Mallow 194831
Rock Rosemallow 286649
Rock Rue 160236
Rock Sage 221275,345305
Rock Sallow 343876
Rock Samphire 111343,111347,342859
Rock Sampier 111347
Rock Sand Spurrey 370705
Rock Sandwort 32180,32242,255524,255582
Rock Satin Grass 259700
Rock Scurvyweed 98030
Rock Sea Lavender 230545
Rock Sea-lavender 230545
Rock Sea-spurrey 370705
Rock Seatung 301291
Rock Sedge 76094
Rock Skullcap 355448
Rock Snapdragon 28654
Rock Soapwort 346433
Rock Speedwell 407129,407130,407313
Rock Spicegrass 239839
Rock Stitchwort 32245,255458
Rock Stonecrop 356756,357114
Rock Syzygium 382655
Rock Tailsacgrass 402653
Rock Thorowax 63803
Rock Thryptomene 390550
Rock Tickclover 126584
Rock Tongoloa 392795
Rock Trachydium 393854
Rock Trumpet 241310
Rock Vallecula Rhododendron 332088
Rock Violet 410495
Rock Walnut 212646
Rock Whitepearl 172147
Rock Whitlowgrass 137213
Rock Whitlow-grass 136936,137147
Rock Wild Buckwheat 152483
Rock Willow 344020
Rock Windflower 24032
Rock Windhairdaisy 348729
Rock Wormwood 36162,36166
Rock's Peony 280272
Rockaster 218242
Rockbell 412568
Rockberry 31142
Rock-bush 300382
Rockcastle Aster 160678
Rockcolous Holly 204170
Rockcress 30160,30164
Rock-cress 30160,30186,44867
Rockcress Draba 136911
Rock-cress Draba 136911
Rockcress Whitlowgrass 136911
Rock-crop 356468
Rockdwelling Cinnamon 91424
Rockd-welling Cinnamon 91424
Rocked Pincushion 228100
Rockelm 254493
Rockery Blistercress 154525
Rockery Willowherb 146644
Rocket 94233,154019,154390,193387,365386
Rocket Candytuft 203184,203208,203239
Rocket Consolida 102833
Rocket Cress 47964
Rocket Galant 47964
Rocket Gentle 154019
Rocket Larkspur 102833,102834,102841
Rocket Salad 154019
Rocket Water Cress 262722
Rocket Weed 154073,154091
Rocketsalad 153998,154019
Rocket-salad 154019
Rockface Primrose 314798
Rockfoil 349032
Rockgap Corydalis 106401
Rockgap Figwort 355085
Rockgarden Spinepink 2565
Rock-growing Ormosia 274432
Rockhill Croton 112891
Rockhill Microtoena 254255
Rockhouse White Snakeroot 11167
Rockjasmine 23107
Rock-jasmine 23243
Rockjasmine Sandwort 31741
Rockjasmine Stringbush 414200
Rock-jasmine Wild Buckwheat 151814
Rockjasmineleaf Wikstroemia 414200
Rocklettuce 299686
Rockliving Beancaper 418753
Rockliving Begonia 50269
Rockliving Bluegrass 305668
Rockliving Buckeye 9737
Rockliving Chirita 87957
Rock-living Cinnamon 91397
Rockliving Clematis 95100
Rockliving Corydalis 106405
Rockliving Currant 334149
Rockliving Dendrobium 125348
Rockliving Gesneria 175303
Rockliving Hornbeam 77383
Rockliving Lacquer-tree 393446
Rockliving Larkspur 124166
Rock-living Milkvetch 43011
Rockliving Pearlwood 296458
Rockliving Phyllanthodendron 296458
Rock-living Tansy 383726
Rockliving Woodbetony 287650
Rocklotus Metalleaf 296380
Rock-loving Bamboo 297334
Rock-loving Milkvetch 42624,42878
Rock-loving Moltkia 256752

Rock-loving Straight-raceme Barberry 51988
Rock-mat 292645
Rock-moss 357053
Rockpearl 94129,94130
Rockphilous Evodia 161316
Rock-poppy 86755
Rock-portulaca 294334
Rockpueslane 65827
Rockroot 10973
Rockrose 93120,93161,93199,93212,188583,188605,188610,188757
Rock-rose 188757
Rockrose Family 93046
Rock-rose Family 93046
Rockslide Fleabane 150749
Rockspray 107549
Rockspray Cotoneaster 107486,107549
Rock-spray Cotoneaster 107549
Rocktaro 327670
Rockvine 387740
Rock-violet 394718
Rockweed 189052,298974
Rockwood 181496
Rocky Arundinella 37426
Rocky Aster 40972
Rocky Barberry 52028
Rocky Begonia 50224
Rocky Bentgrass 12281
Rocky Bergenia 52532
Rocky Bittercress 72712
Rocky Bladderwort 403983
Rocky Camellia 69260
Rocky Candytuft 203236
Rocky Cephalorrhynchus 82421
Rocky Cinquefoil 312946
Rocky Corydalis 106270
Rocky Craneknee 221737
Rocky E. H. Wilson Magnolia 242357
Rocky Euonymus 157871
Rocky Eustachys 160991
Rocky Lychnis 292544,292545
Rocky Montain Locoweed 278679
Rocky Mountain Alpine Fir 402
Rocky Mountain Alpine Fleabane 150674
Rocky Mountain Aspen 311547
Rocky Mountain Bee Plant 95791
Rocky Mountain Bee-plant 95791
Rocky Mountain Beeweed 95791
Rocky Mountain Birch 53437
Rocky Mountain Bristlecone Pine 299795
Rocky Mountain Bulrush 352260,353786
Rocky Mountain Cedar 213922
Rocky Mountain Columbine 30051
Rocky Mountain Cutleaf Coneflower 339577
Rocky Mountain Daisy 393400
Rocky Mountain Douglas Fir 318574
Rocky Mountain Douglas-fir 318585
Rocky Mountain Fescue 164301
Rocky Mountain Fir 402
Rocky Mountain Flax 231917
Rocky Mountain Fringed Thistle 91882
Rocky Mountain Garland 94139
Rocky Mountain Goldenrod 368258
Rocky Mountain Grape 242478
Rocky Mountain Groundsel 279950
Rocky Mountain Iris 208718
Rocky Mountain Juniper 213834,213922
Rocky Mountain Larkspur 124568
Rocky Mountain Lily 229984
Rocky Mountain Lodgepole Pine 299874
Rocky Mountain Maple 2991
Rocky Mountain Nailwort 284888
Rocky Mountain Oak 324515
Rocky Mountain Parnassia 284538
Rocky Mountain Penstemon 289378
Rocky Mountain Pinyon 299926
Rocky Mountain Ponderosa Pine 300157
Rocky Mountain Pussytoes 26462
Rocky Mountain Raspberry 338318
Rocky Mountain Redcedar 213922
Rocky Mountain Sage 36400,345340
Rocky Mountain Sandwort 255445
Rocky Mountain Sedge 73846,76157
Rocky Mountain Spring Bbeauty 94355
Rocky Mountain Starwort 375026
Rocky Mountain Subalpine Fir 402
Rocky Mountain White Fir 317
Rocky Mountain White Oak 323927
Rocky Mountain White Pine 299942
Rocky Mountain Whortleberry 403930
Rocky Mountain Whortleberry Whun 401388
Rocky Mountain Yellow Pine 299822
Rocky Mountain Zinnia 418046
Rocky Mountains Canada Goldenrod 368032
Rocky Mts. Dwarf Mahonia 242596
Rocky Pennywort 200271
Rocky Pepper 300491
Rocky Persimmon 132384
Rocky Pink 127792
Rocky Pleione 304302
Rocky Point Ice Plant 243262
Rocky Pyrethrum 322728
Rocky Raspberry 339240
Rocky Ribseedcelery 304844
Rocky Sedge 76155
Rocky Spindle-tree 157869
Rocky Star Wort 375136
Rocky Syzygium 382662
Rocky Tansy 383846
Rocky Typhonium 401154
Rocky Violet 409975,410536
Rocky Wall-cress 30164
Rocky Wang Buckeye 9737
Rocky Yarrow 4019
Rockycabbage 394835
Rocky-valley Rhododendron 332088
Rodden 369339
Roddin 369339
Roddon 369339
Roden 369339
Roden-quicken-rowan 369339
Rodewort 66445
Rodger's Bronze-leaf 335151
Rodgersflower 335141
Rodgersia 41870,335141,335153
Rodicled Spapweed 205274
Rodicled Touch-me-not 205274
Rodie Ocotea 268728
Rodin 369339
Rodon 369339
Rodowsk Craneknee 221727
Rodowsk Stickseed 221727
Rodriguezia 335161
Rods Gold 66445
Rodsgold 66445
Rod-shaped Rabdosia 209819
Rodway Ozothamnus 279350
Roe Briar 336419
Roe Chloranthus 88272
Roebelen Date 295484
Roebelen's Palm 295484
Roe-briar 336419
Roebuck 339240
Roebuck Berry 339240
Roebuck-berry 338243,339240
Roegneria 335183
Roemer Acacia 1538
Roemer Catclaw 1538
Roemer's Acacia 1538
Roezl Begonia 50248
Roffia Palm 326639,326649
Rogation Flower 308454
Rogation-flower 308454
Rogeria 335742,335758
Rogers Firethorn 322480
Rogue's Gilliflower 193417,193418
Roguery 81768
Rohan Soymida 369934
Rohutu 236405
Rok Begonia 50247
Rokfa 386656
Rokugatsu 93752
Roland's Sea-blite 379596
Rolfe's Raspberry 339188
Rolledge Milkaster 169795
Rolledleaf Rhododendron 331675
Rolledsepal Paphiopedilum 282775
Rollhair Sweetleaf 381446
Rollhairy Tanoak 233216
Roll-hairy Tanoak 233216
Rollinghair Dogwood 380510

Rollsepal Pocktorchid 282775
Rollstyle Elaeagnus 142182
Romaine Lettuce 219493
Roman Camomile 85526
Roman Candle 416607
Roman Candles 9701,217055
Roman Chamomile 85526
Roman Coriander 266241
Roman Cypress 114753
Roman Fennel 167146
Roman Hyacinth 50770,199598
Roman Jasmine 53257,294426
Roman Jessamine 53257,294426
Roman Kale 53257
Roman Laurel 223203
Roman Myrtle 261739
Roman Nettle 402992
Roman Pine 300151
Roman Plant 86970,261585
Roman Ragweed 19145
Roman Rocket 36088,154019
Roman Willow 382329
Roman Wormwood 19145,36088,36902
Romanet Grape 411890
Romanian Beech 162400
Romanzoff Syagrus 31705
Romanzofi's Ladies'-tresses 372242
Romanzov's Syagrus 31705
Romatic Aster 40948
Romero-macho 286894
Rommy 15861
Roms 15861
Rondeletia 336096,336104
Rone 369339
Rongan Ancylostemon 22173
Rongan Chirita 87954
Rongcheng Bamboo 297244
Rongcheng Bulrush 353765
Rongchu Hill Spindle-tree 157377
Rongjiang Begonia 50250
Rongjiang Camellia 69781
Rongjiang Tea 69781
Rongshui Kiwifruit 6690
Rongxia Rabdosia 324789
Ronnochs 144661
Rood-mark Woodbetony 287657
Roof Brome Grass 60989
Roof Houseleek 358062
Roof Iris 208875
Roof-iris 208875
Roofwing Corydalis 105562
Rooibos Tea 38645
Rooigras 389361
Rooiwortel 62404
Rook's Flower 145487
Roosevelt Dam Rock Daisy 291332
Roosevelt Weed 46236
Rooster Flower 34130
Roostertree 68095
Roosvygie 220660
Root Beet 53249
Root Crabgrass 130730
Root Spine Palm 113290,113293
Rooted Pellionia 288762
Rooted Trigonotis 397459
Rooted Water Hyacinth 141807
Rooted Water-hyacinth 141806
Roothead Inula 207209
Roothead-like Inula 207210
Rooting 70512
Rooting Branches Rhododendron 331001
Rooting Corydalis 106356
Rooting Nippon Cinquefoil 312482
Rooting-fortune Spindle-tree 157515
Rootless Duckweed 414649,414653
Rootless Vine 78726
Rootless Wolffia 414653
Rootlet Woodbetony 287382
Rootspine Palm 113290
Rootstock Southstar 33473
Rooty Milkvetch 42958
Rope Dodder 115038
Rope Grass 327963
Rope Tickseed 104765
Ropebark 133656
Rope-bark 133656
Rope-grass 327963
Ropegrass Family 328218
Ropevine 95223
Ropewind 102933
Roquette 153998,154019,193417
Rorgai Buttercup 326048
Rosa Mundi 336585
Rosamund 15861
Rosary Pea 738,765
Rosary Plant 109349
Rosary Root 29298
Rosary Vine 84305
Rosarypea 738,765
Rosary-pea 738,765
Rosarypea Tree 765
Rosary-pea Tree 765
Roscher Palm 337060
Roscoea 337065
Rose 336283,336692,336870,336901,
336944
Rose Acacia 334965
Rose Anise 204484
Rose Anise-tree 204484
Rose Apple 382582,382606,382616,382660
Rose Bay 146614,265327,330017,331207
Rose Beauty Bush 217789
Rose Beet 53267
Rose Black Wood 121653
Rose Blackberry 339192
Rose Bladdernut 374105
Rose Bladderpod 227023
Rose Box 107549
Rose Cactus 290708
Rose Callalily 417118
Rose Campion 11947,363141,363312,
363370,363477
Rose Catchfly 363982
Rose Chestnut 252370
Rose China Redbud 83776
Rose Chinese Waxmyrtle 261216
Rose Cistus 93154
Rose Clarkia 94138
Rose Clover 396926
Rose Cone Flower 210240
Rose Coneflower 210240
Rose Coreopsis 104577
Rose Cowwheat 248195
Rose Crazyweed 279130
Rose Cremanthodium 110386
Rose Crested Orchid 306883
Rose Daphne 122416
Rose David Vetchling 222709
Rose Dendrobium 125082
Rose Deutzia 127061
Rose Dwarf Tulip 400216
Rose Edelweiss 224929
Rose Elder 407989
Rose Elf 94373
Rose Eria 113514
Rose Eveningprimrose 269498
Rose Everlasting 190819
Rose Family 337050
Rose Flowering Dogwood 105030
Rose Forrest Sandwort 31913
Rose Fuchsia 168783
Rose Garlic 15666
Rose Gentian 341463,341466
Rose Geranium 288268
Rose Glorybower 95978
Rose Glow Japanese Barberry 52239
Rose Gum 155598
Rose Hairorchis 113514
Rose Heath 84933
Rose Hubei Crabapple 243634
Rose Irisorchis 267986
Rose Jackbean 71072
Rose Juteleaf Raspberry 338282
Rose Largeflower Larkspur 124240
Rose Laurel 265327
Rose Leek 15149
Rose Linearleaf Cremanthodium 110409
Rose Linearleaf Nutantdaisy 110409
Rose Lupin 238442
Rose Lycoris 239270
Rose Makino's Rhododendron 331192

Rose Mallee 155721
Rose Mallow 13934,194680,194955,194963,195032,195083,195149,402245
Rose Mandarin 377934
Rose Margine Japanese Maple 3344
Rose Milkywhite Pearleverlasting 21587
Rose Monkswort 266624
Rose Moss 311852
Rose Musciform Woodbetony 287440
Rose Myrtle 332218,332221
Rose Natal Grass 249323
Rose Noble 118008,355061,355202
Rose Nut 195375
Rose Oberonia 267986
Rose of Castile 336515
Rose of China 195149
Rose of Heaven 11947,363331
Rose of Jericho 21804,21805
Rose of Melaxo 336514
Rose of Provins 336581
Rose of Sharon 195269,195284,201732,201787,229789,262381,400114
Rose of the Virgin 21805
Rose of Venezuela 61155,61158
Rose Ophiorrhiza 272290
Rose Oxalis 278062
Rose Parsley 23846
Rose Pearlwood 296460
Rose Pelargonium 288268
Rose Pennywort 350032
Rose Periwinkle 79418
Rose Phyllanthodendron 296460
Rose Pincushion 244265
Rose Pink 329681,341463,341466
Rose Pink Zephyr Lily 417613
Rose Plumbago 305185
Rose Pogonia 306883
Rose Primrose 314904
Rose Rockcress 30200
Rose Rockfoil 349856
Rose Root 329935,357087
Rose Russianthistle 344686
Rose Sage 345281
Rose Sandwort 32181
Rose Scented Geranium 288133
Rose Silene 363331
Rose Spatholobus 370418
Rose Stonecrop 357088
Rose Stonegarlic 239270
Rose Syncalathium 381654
Rose Thistle 91739
Rose Tree 265327
Rose Tree of China 20970
Rose Turtlehead 86795
Rose Turtle-head 86795
Rose Verbena 405820
Rose Vervain 405820
Rose Vetchling 222851
Rose Watson Flower 413398
Rose Willow 104949
Rose Willow Weed 146859
Rose Wood 320301,389946
Rose Yarrow 3980
Rosea Ice Plant 138171,138183
Rosea Iceplant 138171
Rose-acacia 32988,334965,335013
Roseal Milkvetch 42985
Rosealind Deutzia 126835
Rose-among-the-thorns 266225
Rose-and-white Wild Buckwheat 152105
Roseapple 382522,382582
Rose-apple 382522,382582,382616
Rose-a-ruby 8332
Roseate Chelonopsis 86835
Rosebaby Rhododendron 331207
Rosebay 85875,265327,330017,331207
Rose-bay 85875
Rosebay Willowherb 85875,146586
Rose-bay Willow-herb 146614
Rose-box 107319
Rosebud Cherry 83334
Rosebud Orchis 94547
Rosecampi 337180,363370,368635,400489
Rose-colored Larkspur 124540
Rose-de-Meuse 336481
Rose-flower Bloodwood 155582
Roseflower Brandisia 59131
Roseflower Camellia 69572
Rose-flower Indianmulberry 258915
Rose-flower Olive 270163
Roseflower Snakegourd 396276
Roseflower Specious Lily 230039
Roseflower Storax 379439
Roseflower Tea 69572
Rose-flowered Brandisia 59131
Rose-flowered Storax 379439
Rose-flowered Yumin Fritillary 168647
Rose-fruit Rocky Mountain Maple 2994
Rose-gentian 341435,341463,341466
Rosegold Pussy Willow 343444
Rose-heath 84933
Rosehell Azalea 331546
Rose-leaf Begonia 50063
Roseleaf Raspberry 339194,339195
Roseleaf Sage 345105
Rose-leaved Raspberry 339194
Roselike Rhododendron 331667
Rose-like Rhododendron 331665
Rose-lilac Young Epimedium 147079
Rosella 195149,195196
Rosella Spotleaf-orchis 179586
Roselle 195149,195196
Rosemallow 194680
Rose-mallow 194680,195149
Rosemary 337174,337180,337185
Rosemary Barberry 52169,52184
Rosemary Bog 22445
Rosemary Grevill 180644
Rosemary Grevillea 180644
Rosemary Mint 307281,307282,307284
Rosemary Pine 300128,300264
Rosemary Willow 343335
Rosemary-leaf Grevillea 180644
Rosemaryleaf Leucas 227699
Rosemaryleaf Willow 344031
Rosemary-leaved Willow 344031
Rosemary-like Edelweiss 224930
Rose-moss 311852
Rosems 15861
Rosemyrtle 332218,332221
Rose-myrtle 332218
Rosen Iris 208792
Rosenbach Onion 15665
Rose-of-China 195149
Rose-of-heaven 363331
Rose-of-Jericho 21805
Rose-of-sharon 195269,195271,195276,195278,195281,195288,201787,262457
Rose-of-the-mount 280253
Rose-of-Venezuela 61158
Rosepanicle Deutzia 127061
Rose-pemlate Euonymus 157838
Rose-perulate Spindle-tree 157838
Rose-pink 341466
Rosepink Zephyrlily 417613
Rosepurple Herbclamworm 398112
Rose-purple Violet 409949
Rosepurpreum Onion 15082
Rosered Chelonopsis 86835
Rose-red Chelonopsis 86835
Rosered Lily 229725
Rosered Masdevallia 246119
Rosered Ruellia 339803
Rose-red Small Reed 127290
Rosering Gaillardia 169603
Rose-ring Gaillardia 169603
Roseroot 329935,357088
Rose-root 329935
Roseroot Stonecrop 329935
Rose-root Stonecrop 329935
Roserush 239323
Rose-scented Geranium 288133,288268
Rose-scented Pelargonium 288133
Rose-spined Spindle-tree 157834
Rosestripe Leek 15704
Rosestripe Onion 15704
Rosette Aloe 17186
Rosette Asiabell 98408
Rosette Beadcress 264632
Rosette Christolea 89261
Rosette Mullein 325225

Rosette Oxalis 278124
Rosette Plateaucress 89261
Rosette Satinflower 365743
Rosette Sisyrinchium 365743
Rosette Thistle 91958,92364
Rosette Torularia 264632
Rosette Valerian 404259
Rosetterfly Petrocosmea 292578
Rosetterfly Stonebutterfly 292578
Roseus Milkvetch 42985
Rosewater Jamb 156134
Rosewood 121607,121750,121770,360481
Rosewood Wendlandia 413779
Rosha 117204
Roshvits Ptilagrostis 321012
Rosilla 188434
Rosin Plant 364255
Rosin Rose 201787,202086
Rosin-brush 46255
Rosinweed 68295,364255,364285,364290
Rosinweed Sunflower 189057
Rosita 81490
Rosmaryleaf Daphne 122596
Rosmary-leaved Daphne 122596
Rosrate Cleisostoma 94464
Ross Anemone 23713
Ross Violet 410498
Ross Windflower 23713
Ross' Sandwort 255569
Rossberg Bluegrass 306053
Rossia Globethistle 140779
Rostellularia 337232
Rostewort 292957
Rosthorn Bigginseng 241176
Rosthorn Big-ginseng 241176
Rosthorn Bittersweet 80295
Rosthorn Bodinier Beautyberry 66748
Rosthorn Briggsia 60283
Rosthorn Bulrush 353767
Rosthorn Camellia 69573
Rosthorn Currant 334176
Rosthorn Curvedflower Corydalis 105785
Rosthorn Cystopetal 320238
Rosthorn Hydrangea 200071
Rosthorn Lily 230022
Rosthorn Primrose 314906
Rosthorn Rabdosia 209809
Rosthorn Sagebrush 36148
Rosthorn Snakegourd 396257
Rosthorn Spiraea 372072
Rosthorn Spirea 372072
Rosthorn Staff-tree 80295
Rosthorn Stonecrop 357101
Rosthorn Tanoak 233359,233378
Rosthorn Waterdropwort 269329
Rosthorn Willow 344039
Rosthorn Yam 131736
Rosthorns Buckthorn 328842
Rostkov Eyebright 160258
Rostrate Chinese Purplestem 376484
Rostrate Chinese Stewartia 376484
Rostrate Liparis 232312
Rostrate Osbeckia 276152
Rostrate Poison-ivy 393477
Rostrate Stewartia 376469
Rostrate Tutcheria 400717
Rostrate Viet Nam Xantolis 415236
Rostratefruit Blackhatflower 126727
Rostratefruit Breynia 60080
Rostratefruit Daphne 122595
Rostratefruit Dasymaschalon 126727
Rostratefruit Tailgrape 35052
Rostrate-fruited Dasymaschalon 126727
Rostrat-fruited Euonymus 157843
Rostrat-fruited Spindle-tree 157843
Rostratum Nightshade 367567
Rostrinucula 337289
Rosula Ardisia 31565
Rosula Habenaria 183977
Rosulaleaf Gentian 173345
Rosularia 337292
Rosulate Asiabell 98408
Rosulate Chinese Groundsel 365076
Rosulate Mimulicalyx 255150
Rosulate Sinosenecio 365076
Rosulate Swertia 380340
Rosy Arrowroot 66200
Rosy Baeckea 46437
Rosy Camphorweed 305077
Rosy Crazyweed 279130
Rosy Cress 30229
Rosy Cudweed 317769
Rosy Daun' Kalanchoe 215142
Rosy Dew-plant 220660
Rosy Dipelta 132601
Rosy Falsetamarisk 261291
Rosy False-tamarisk 261291
Rosy Garlic 15666
Rosy Heart 220729
Rosy Heath Myrtle 46437
Rosy Heath-myrtle 46437
Rosy Jasmine 211749
Rosy Largeleaf Hydrangea 199983
Rosy Lily 230023
Rosy Malanga 415200
Rosy Mockorange 294525
Rosy Morn 237539
Rosy Onion 15666
Rosy Pussytoes 26579
Rosy Rabbit-tobacco 317769
Rosy Rhododendron 331667
Rosy Rice Flower 299307
Rosy Sandcrocus 336030,336031
Rosy Sedge 76056
Rosy Smallreed 127290
Rosy Snowflake 227867
Rosy Sunray 329799
Rosy Treelike Rhododendron 330113
Rosy Waterhyacinth 141810
Rosy Weigela 413588
Rosy Woodsorrel 278062
Rosybetls 377934
Rosydandrum 331520
Rosy-flowered Camellia 69572
Rosy-flowered Leadwort 305185
Rosy-red Bracted Bougainvillea 57857
Rotala 337320
Rotan Ajer 65724
Rotan Lilin 65713
Rotan Segar 65656
Rotan Semaboe 65787
Rotang Palm 303099
Rotang Rattan Palm 65782
Rotang Rattan-palm 65782
Rotate Jerusalemsage 220336
Rotcoll 34590
Rot-grass 200388,299759
Roth Hawkweed 195924
Roth Rhynchosia 333389
Rothrock Currant 334269
Rothrock Sagebrush 36149
Rothrock's Basketflower 303085
Rothrock's Knapweed 303085
Rothrock's Snakeroot 11172
Rothrock's Thistle 91760
Rothschild Glorylily 177244
Rothschild Paphiopedilum 282893
Rothschild Rhododendron 331672
Rothschild's Glorylily 177244
Rothschild Cirrhopetalum 63048
Rotkiefer 300223
Rottle-penny 329521
Rottnest Daisy 393883
Rottnest Island Pine 67433
Rotula 337651
Rotunbase Tea 69575
Rotund Boesenbergia 56537
Rotund Honeysuckle 236132
Rotund Milkvetch 42830
Rotundate Camellia 69575
Rotundate Japanese Staunton-vine 374414
Rotundate Trigonotis 397465
Rotundifolious Tylophora 400957
Rotundifolious Violet 410509
Rotundileaf Changed Sagebrush 35353
Rotundilobe Pennsylvania Bittercress 72818
Rotundilobed Bittercress 72818
Rotundity Milkvetch 42830
Rotundleaf Aralia 30600
Rotundleaf Eight Trasure 200788
Rotundleaf Grewia 180944

Rotundleaf Hydrangea 200073
Rotundleaf Monkshood 5543
Rotundlobe Corydalis 105602
Rotundtooth White Bittercress 72859
Roudfruit Rush 212836
Roudleaf Cotyledon 107976
Rouen Lilac 382132,382153,382241
Rouen Pansy 410508
Rouge Daffodil 262438
Rouge Narcissus 262438
Rouge Plant 77748,334892,334897
Rougeflower Primrose 314624
Rougeplant 334892,334897
Rougewood 414799
Rough Agalinis 10139
Rough Agave 10935
Rough Alumroot 194425
Rough Amaranth 18810
Rough Asper 125465
Rough Aster 41137
Rough Avens 175423
Rough Azalea 330665,330776,331759
Rough Bamboo 125465
Rough Barked Apple 24607
Rough Barnyard Grass 140451,140453
Rough Barnyardgrass 140451
Rough Basil 4666
Rough Bedstraw 170193,170228
Rough Bent 12285
Rough Bent Grass 12140
Rough Blackfoot 248147
Rough Blazing Star 228439
Rough Blazing-star 228439
Rough Bluegrass 306098
Rough Boissiera 56563
Rough Boneset 158259
Rough Branches Rapanea 261662
Rough Branches Rhododendron 330153
Rough Bristle-grass 361930
Rough Brome 60918,60977
Rough Buckwheat 162335
Rough Bugleweed 239193
Rough Buttercup 325942,325943,325948
Rough Buttonweed 131351,131358
Rough Chervil 28023,84768,392992
Rough Chive 15717
Rough Cinquefoil 312784,312828
Rough Clcely 392992
Rough Clover 397068
Rough Cocklebur 415057
Rough Comfrey 381026
Rough Coneflower 339552
Rough Corn Bedstraw 170694
Rough Dandelion 384714
Rough Dashel 271666
Rough Dichaea 128400
Rough Dog's-tail 118319
Rough Dogstail 118319
Rough Dogwood 104959
Rough Dropseed 372629,372630
Rough Drop-seed 372625,372629,372630
Rough Ephedra 146230
Rough False Foxglove 10139
Rough False Pennyroyal 187181
Rough Fangcheng Rhododendron 330684
Rough Fleabane 150980
Rough Gayfeather 228439
Rough Gentian 173847
Rough Gerardia 10139
Rough Goldenrod 368345,368373
Rough Green Amaranth 18810
Rough Hairsedge 63280
Rough Halgania 184761
Rough Hawk's-beard 110752
Rough Hawkbit 224679,224698
Rough Hawksbeard 110752
Rough Hawkweed 195948
Rough Hawthorn 110021
Rough Hedge-nettle 373126
Rough Hedgeparsley 393004
Rough Heliopsis 190525
Rough Hydrangea 199817
Rough Kex 192358
Rough Knotweed 309442
Rough Leaf Actinidia 6695
Rough Leaved Agave 10803
Rough Leaved Goldenrod 368373
Rough Lemon 93505,93539
Rough Leucas 227554
Rough Licorice 177877
Rough Long-beaded Poppy 282515
Rough Mallow 18166
Rough Manystem Stonecrop 356954
Rough Marshmallow 18166
Rough Meadow Grass 306098
Rough Meadowgrass 306098
Rough Meadow-grass 306098
Rough Meadowrue 388651
Rough Melic 249085
Rough Menodora 250267
Rough Nut 78811
Rough Oryzopsis 275998
Rough Ox-eye 190525
Rough Parsnip 192358
Rough Pea Vine 222726
Rough Pellionia 288769
Rough Pennyroyal 187181
Rough Pigweed 18810
Rough Pipewort 151464
Rough Poppy 282574
Rough Potato 252514
Rough Prickly Poppy 32441
Rough Pricklyash 417200
Rough Rabbitbrush 150366
Rough Rattlesnakeroot 313796
Rough Rattlesnake-root 313796
Rough Rhododendron 330665,330776
Rough Robin 363461
Rough Rose 336900,336901
Rough Rush Grass 372625
Rough Sage 344865
Rough Sand Sedge 76182
Rough Saxifrage 349078
Rough Schismatoglottis 351147
Rough Sedge 75451,76160
Rough Skullcap 355378
Rough Spear-grass 376913
Rough Star Thistle 80939
Rough Star-thistle 80939
Rough Stinkinggrass 249085
Rough stone-break 349078
Rough Streblus 377538
Rough Sunfllower 188958
Rough Sunflower 188980,189063
Rough Thistle 92358
Rough Thorough-wort 158360
Rough Threeawngrass 34015
Rough Trefoil 396828,397068
Rough Triawn 34015
Rough Water Milfoil 261357
Rough Water-horehound 239193
Rough White Lettuce 313796
Rough White-lettuce 313796
Rough Zexmenia 413559,417826
Rough-bark 114652
Roughbark Eucalyptus 155689
Roughbark Lignumvitae 181506
Rough-barked Apple 24607
Rough-barked Arizona Cypress 114652
Rough-barked Robbon Gum 155608
Rough-brached Mexican Pine 300081
Rough-branch Machilus 240664
Rough-branched Rhododendron 330153
Roughear Tangbamboo 364812
Roughflower Arrowbamboo 162745
Roughflower Poorspikebamboo 270446
Rough-flowered Arundinaria 270446
Rough-flowered Cane-bamboo 270446
Rough-fruit Amaranth 18847
Roughfruit Buttercup 326440
Roughfruit Corn Bedstraw 170694
Roughfruit Corydalis 106541
Rough-fruit Hakea 184600
Rough-fruit Pittosporum 301388
Rough-fruited Amaranth 18844
Rough-fruited Buttercup 326108
Rough-fruited Cinquefoil 312912
Rough-fruited Corn Bedstraw 170694
Rough-fruited Corn-bedstraw 170694
Rough-fruited Fairy-bells 315531
Rough-fruited Mandarin 315531

Rough-fruited Sedge 76087
Rough-fruited Tanoak 233396
Rough-fruited Water-hemp 18844,18847
Roughfruitparsley 393937
Roughhair Azalea 331992
Roughhair Coneflower 339556
Roughhair Fig 165111
Roughhair Knotweed 309831
Rough-hair Raspberry 338864
Rough-haired Bauhinia 49122
Rough-haired Blueberry 403854
Rough-haired Eurya 160606
Rough-haired Fig 165111
Rough-haired Glochidion 177141
Roughhaired Holly 203578
Rough-haired Holly 203578
Rough-haired Rhododendron 330805
Rough-haired Saurauia 347990
Roughhairflower Storax 379331
Roughhairy Coneflower 339556
Roughhairy Indigo 206081
Roughhairy Yam 131474
Rough-headed Bastard Poppy 282515
Roughish Seseli 361467
Roughleaf 222081,222113
Roughleaf Aster 160677
Roughleaf Bedstraw 170218
Roughleaf Catchfly 364133
Roughleaf Coneflower 339606
Roughleaf Dogwood 104959,105017
Rough-leaf Earanther 276914
Roughleaf Falmeflower 295028
Roughleaf Fescuegrass 163850
Rough-leaf Fig 165166
Roughleaf Gentian 173917
Roughleaf Groundsel 358312
Roughleaf Ironweed 406119
Roughleaf Knotwood 44256
Roughleaf Largeleaf Falsenettle 56212
Rough-leaf Leptodermis 226129
Roughleaf Leucas 227660
Roughleaf Narrowhead Goldenray 229206
Roughleaf Pearleverlasting 21578
Roughleaf Peristrophe 291180
Roughleaf Phlogacanthus 295028
Roughleaf Pricklyash 417200
Roughleaf Raspberry 338097
Roughleaf Ricegrass 275998
Roughleaf Rosinweed 364307
Roughleaf Syzygium 382595
Rough-leaf Wildclove 226129
Roughleaf Yellow Bedstraw 170765
Roughleaftree 29004,29005
Rough-leaved Actinidia 6695
Rough-leaved Aphananthe 29005
Rough-leaved Aster 41137
Rough-leaved Begonia 49635
Rough-leaved Caspian Sea Inula 207071
Rough-leaved Dogwood 105017
Rough-leaved Earanther 276914
Rough-leaved Goldenrod 368304,368373
Rough-leaved Pavetta 286452
Rough-leaved Raspberry 338097
Rough-leaved Rice Grass 275998
Rough-leaved Sunflower 188997,189063
Rough-leaves Rhododendron 331756
Roughness Kneejujube 67080
Rough-node Bastard-sage 152675
Roughpod Copperleaf 1945
Roughpod Crazyweed 279028
Rough-podded Tine Tare 408423
Rough-podded Tine-tare 408423
Rough-podded Yellow Vetch 408469
Roughroot Mouseear Cress 113150
Rough-seed Bulrush 352223
Rough-seed Clammy-weed 307141
Rough-seeded Fameflower 294331
Rough-seeded Fame-flower 294331
Rough-skinned Lime 93492
Rough-skinned Plum 284215
Roughstalk Bluegrass 306098
Rough-stalk Bluegrass 306098
Roughstalk Greenbrier 366605
Rough-stalk Greenbrier 366605
Rough-stalked Blue-grass 306098
Rough-stalked Meadow-grass 306098
Roughstem Aster 380973
Roughstem Fishvine 125997
Roughstem Greenbrier 366243
Rough-stem Greenbrier 366243
Roughstem Jewelvine 125997
Roughstem Smilax 366243
Rough-stemmed Gerardia 10144
Rough-stemmed Goldenrod 368371,368373
Rough-stemmed Jewelvine 125997
Rough-stemmed Spindle 157952
Roughstraw 39298,39299
Rough-style Rhododendron 331842
Rough-styled Honeysuckle 235749
Roughtooth Cherry-laurel 316686
Rough-tooth Cherry-laurel 316686
Roughtooth Clearweed 299046
Roughtooth Coldwaterflower 299046
Roughweed 373346
Roug-leaved Holly 203578
Roulett China Rose 336490
Roulett Rose 336490
Roulette Chinese Rose 336490
Roullet Keteleeria 216151
Round Arborvitae 302728
Round Base Leaves Leucothoe 228184
Round Boxleaf Distylium 134920
Round Campion 364193
Round Cardamom 19836
Round Currant 334250
Round Dock 243840
Round Fruit Licorice 177944
Round Kumquat 167503
Round Leaf Actinidia 6596
Round Leaved Orchis 302472
Round Leaves Lyonia 239378
Round Manateagrass 117308
Round Monkeyflower 255233
Round Netleaf Willow 343809
Round Netleaf-willow 343809
Round Pimpernel 345733
Round Pricklyhead Poppy 282574
Round Prickly-headed Poppy 282574
Round Robin 174877,248411
Round Tower 240068
Round Towers 240068
Round Wingfruit Cyclocarya 116240
Round Woolly Marbles 318805
Round Zedoary 114884
Roundauricle Cacalia 283801
Roundbase Knotweed 309348
Roundbase Woodnettle 125541
Round-base-leaves Leucothoe 228184
Roundbeak Woodbetony 287271
Roundberry 369339
Round-bodied Oak Sedge 74887
Roundbract Cowwheat 248175
Roundbract Sage 344991
Round-bracted Tricklover 297013
Roundbud Arrowbamboo 137866
Round-bud Evergreen Chinkapin 78944
Roundcalyx Chickweed 375106
Roundcalyx Gentian 173941
Round-calyx Hartia 376437
Round-cephaloid Wormwood 36307
Round-crown Elm 401499
Roundear Aster 41296
Roundear False Rattlesnakeroot 283681
Round-ear Willow 343063
Round-eared Sallow 343063
Rounded Fargesia 162724
Rounded-buds Rhododendron 331870
Roundedcrown Elm 401499
Roundfan Eight Trasure 200798
Round-flowered Rhododendron 330094
Round-foliated Greenbrier 366307
Roundfoliolate Yam 131816
Roundfolious Smilax 366307
Roundfruit Blyxa 55913
Roundfruit Clearweed 299037
Roundfruit Coldwaterflower 299037
Roundfruit Dyetree 302676
Round-fruit Hemsleia 191964
Roundfruit Licorice 177944
Roundfruit Litse 234065
Roundfruit Microtropis 254327

Roundfruit Mountainash 369399
Roundfruit Rush 213008
Roundfruit Spikesedge 143200
Round-fruit Stewartia 376452
Round-fruit Tutcheria 322612
Roundfruit Violet 410594
Roundfruit Yam 131551
Round-fruited Cockspur-thorn 110065
Round-fruited Dye-tree 302676
Round-fruited Hedge Hyssop 180304
Round-fruited Litse 234065
Round-fruited Mountain-ash 369399
Round-fruited Nitraria 266380
Round-fruited Rush 213008,213138
Round-fruited St. John's-wort 202161
Round-fruited Yellow Wild Indigo 47878
Roundglandula Swertia 380344
Roundhead Azalea 331793
Roundhead Bulrush 353174
Round-head Conebush 210243
Roundhead Cryptomeria 113708
Roundhead Mosquitoman 134920
Roundhead Wormwood 36307
Round-headed Bush Clover 226729
Round-headed Bush-clover 226729
Round-headed Club-rush 353170,353174
Round-headed Garlic 15048,15752
Round-headed Leek 15752
Round-headed Lespedeza 226729
Round-headed Mealy Primrose 314212
Round-headed Poppy 282574
Round-headed Prairie Clover 121895
Round-headed Rampion 298065
Round-headed Rush 213438
Round-headed Trefoil 396917
Roundiflowered Holly 203851
Roundifruited Glochidion 177178
Roundleaf Actinidia 6619
Roundleaf Alumroot 194417
Round-leaf Ash 168053,168093
Roundleaf Azalea 331400
Roundleaf Begonia 49837,50255
Roundleaf Bittersweet 80221
Roundleaf Bladderwort 403351
Round-leaf Blue Gum 155555
Roundleaf Bluebell 70271
Roundleaf Blueberry 403767
Roundleaf Bonyberry 276550
Roundleaf Bredia 59874
Roundleaf Buckthorn 328713
Roundleaf Bugle 13154
Roundleaf Bulbophyllum 62696
Roundleaf Buttercup 325975
Roundleaf Cancerwort 216292
Round-leaf Candyleaf 376410
Roundleaf Chastetree 411430
Roundleaf Cherry 83253
Roundleaf Childvine 400957
Round-leaf Chinese Aspen 311202
Round-leaf Chinese Groundsel 365071
Roundleaf Chirita 87851,87956
Roundleaf Codariocalyx 98156
Roundleaf Combbush 292624
Roundleaf Conehead 320136
Roundleaf Cotoneaster 107659
Roundleaf Cranesbill 174886
Roundleaf Creeping Mahonia 242628
Roundleaf Danceweed 98156
Roundleaf Dogwood 105171
Roundleaf Draws ear grass 403351
Roundleaf Dumasia 138975
Roundleaf Dunbaria 138975
Roundleaf Eupatorium 158290
Roundleaf Fieldhairicot 138975
Round-leaf Fountain Palm 234190
Roundleaf Geranium 174886
Round-leaf Glomerate Arrowwood 407872
Round-leaf Glorybower 96310
Roundleaf Goatwheat 44280
Roundleaf Gooseberry 334177
Roundleaf Grape 411723
Roundleaf Grayback Grape 411617
Roundleaf Greenbrier 366254
Round-leaf Greenbrier 366254
Roundleaf Groundsel 279930
Roundleaf Gum 155555
Roundleaf Hawthorn 109598
Roundleaf Hirsute Spiraea 371934
Round-leaf Honeysuckle 235979
Roundleaf Hydrangeavine 351787
Roundleaf Juneberry 19290
Roundleaf Kalidium 215330
Roundleaf Kiwifruit 6619
Roundleaf Knotweed 309257
Round-leaf Largeleaf Falsenettle 56211
Roundleaf Litse 234047
Roundleaf Luisia 238310
Roundleaf Lycianthes 238966
Roundleaf Lyonia 239378
Roundleaf Maesa 241841
Roundleaf Magnolia 279250
Round-leaf Mallotus 243443
Roundleaf Mayten 182751
Roundleaf Meadowrue 388648
Roundleaf Milkvetch 42829
Roundleaf Mitragyna 256122
Roundleaf Monkey Flower 255203
Roundleaf Morning glory 208120
Round-leaf Mulberry 259174
Roundleaf Orchis 273614
Roundleaf Oreocharis 273880
Round-leaf Pearlwood 296457
Round-leaf Pentapanax 289628
Round-leaf Phyllanthodendron 296457
Roundleaf Pilea 299005
Roundleaf Polyscias 310237
Round-leaf Poplar 311459
Roundleaf Premna 313758
Round-leaf Privet 229513
Round-leaf Puncturebract 278670
Round-leaf Rabbitbrush 150379
Roundleaf Rattlebox 112238
Roundleaf Red Silkyarn 238966
Roundleaf Reevesia 327370,327376
Roundleaf Rhodiola 329942
Roundleaf Rhubarb 329406
Roundleaf Rockcress 30286
Roundleaf Rostellularia 337261
Roundleaf Rotala 337391
Roundleaf Sandwort 32115
Roundleaf Sapium 346404
Roundleaf Schoenorchis 352311
Roundleaf Serviceberry 19260,19290
Roundleaf Sida 362498
Round-leaf Simon Poplar 311506
Roundleaf Sinosenecio 365071
Roundleaf Sloanea 366073
Round-leaf Snakeweed 182112
Round-leaf Star Orchid 146392
Round-leaf Stephania 375896
Roundleaf Stonebean-orchis 62696
Roundleaf Sundew 138348
Roundleaf Suoluo 327370
Roundleaf Syncalathium 381649
Roundleaf Tallowtree 346404
Roundleaf Thoroughwort 158290
Roundleaf Thorowax 63805
Roundleaf Treebine 92934
Roundleaf Trigonotis 397465
Round-leaf Viburnum 407872
Roundleaf Vineclethra 95495
Roundleaf Violet 410604
Round-leaf Vitex 411430
Round-leaf Wattle 1523
Roundleaf Waxpetal 106674
Round-leaf Wild Buckwheat 152439,152550
Roundleaf Wildtung 243443
Round-leaf Willow 344045
Roundleaf Windhairdaisy 348738
Roundleaf Winterhazel 106674
Roundleaf Woodbetony 287639
Roundleaf Woollystalk Wormwood 35470
Roundleaf Yam 131551
Round-leafed Dogwood 105171
Round-leafed Gum 155555
Round-leafed Mallee 155678
Round-leafed Moort 155699
Round-leafed Pyrola 322872
Round-leaved Anemone 23740
Round-leaved Barberry 52118
Round-leaved Bellflower 70271

Round-leaved Blueberry 403767
Round-leaved Boneset 158290
Round-leaved Bonyberry 276550
Round-leaved Bridelia 60210
Round-leaved Buchu 10575
Round-leaved Cancerwort 216292
Round-leaved Catchfly 363990
Round-leaved Cotoneaster 107496
Round-leaved Crane's-bill 174886
Round-leaved Cranesbill 174886
Round-leaved Crowfoot 326147
Round-leaved Danceweed 98156
Round-leaved Dog-rose 336419
Round-leaved Egyptian Melon 114136
Round-leaved Elder 345650
Roundleaved Felwort 235456
Round-leaved Fluellen 216292
Round-leaved Fluellin 216292
Round-leaved Goldenrod 368304
Round-leaved Goosefoot 86893,87130
Round-leaved Greenbrier 366291,366559
Round-leaved Grewia 180944
Round-leaved Hare's-ear 63805
Round-leaved Hawthorn 109598,109810
Round-leaved Henbane 201372
Round-leaved Hooker Maple 3022
Round-leaved Illigera 204640
Round-leaved Juneberry 19290
Round-leaved Kalidium 215330
Round-leaved Lasianthus 222095
Round-leaved Lavender 223323
Round-leaved Leptospermum 226479
Round-leaved Litse 234047
Round-leaved Lyonia 239378
Round-leaved Maesa 241841
Round-leaved Mallow 243823
Round-leaved Maple 2893
Round-leaved Mint 250439,250457
Round-leaved Mint Bush 315606
Round-leaved Mint-bush 315606
Round-leaved Monkey-flower 255203,255206
Round-leaved Moort 155699
Round-leaved Navelwort 107976
Round-leaved Navel-wort 107976
Round-leaved Oak 324343
Round-leaved Orchid 19380
Round-leaved Orchis 19380,273614
Round-leaved Pea Vine 222831
Round-leaved Pentapanax 289628
Round-leaved Poplar 311459
Round-leaved Premna 313758
Round-leaved Primrose 314909
Round-leaved Reevesia 327370,327376
Round-leaved Rest Harrow 271571
Roundleaved Rockjasmine 23270
Round-leaved Saxifrage 349859
Round-leaved Serviceberry 19290
Round-leaved Shin-leaf 322872,322875
Round-leaved Sloanea 366073
Round-leaved Snow Gum 155691
Round-leaved Speedwell 407124
Round-leaved Spurge 159823
Round-leaved Squaw-weed 359606
Round-leaved St. John's-wort 202051
Round-leaved Staff Tree 80260
Round-leaved Sundew 138348
Round-leaved Tallow-tree 346404
Round-leaved Thoroughwax 63805
Round-leaved Trillium 397592
Round-leaved Triodanis 397759
Round-leaved Viburnum 407865
Round-leaved Violet 410509
Round-leaved Water Pimpernel 345733
Round-leaved Willow 344045
Round-leaved Wintergreen 322872
Round-leaved Winter-green 322872
Round-leaved Winterhazel 106674
Roundlip Calanthe 66028
Roundlip Liparis 232091
Roundlip Twayblade 232091
Roundlobe Buttercup 325789
Roundlobe Hepatica 192135
Round-lobed Hepatica 23703,192126,192135
Round-lobed Sichuan Peony 280181
Roundpetal Coldwaterflower 298854
Roundpetal Loosestrife 239764
Roundpetal Rockfoil 349860
Roundpod Jute 104072
Roundpod St. John's-wort 201814
Round-podded Jute 104059,104072
Round-rooted Crowfoot 325666
Round-rooted Galingale 215035
Roundscaled Keteleeria 216143
Roundscape Sealavender 230794
Roundseed Snakegourd 396200
Round-seeded St. John's-wort 202161
Roundsepal Hartia 376437
Round-sepaled Hartia 185965
Roundspike Knotweed 309360
Roundspike Lagotis 220193
Roundspike Sedge 73690
Round-stem Foxglove 10144
Round-stemmed False Foxglove 10144
Roundtooth Catnip 265106
Roundtooth Clearweed 298889
Roundtooth Coldwaterflower 298889
Roundtooth Eryuan Cystopetal 320232
Roundtooth Euonymus 157395
Roundtooth Fleshspike 346939
Roundtooth Isometrum 210177
Roundtoothed Cinquefoil 312475
Round-toothed Ookow 128879
Roundtoothleaf Droundsel 290821
Roundtooth-truncare Gentian 173360
Roundtop Blueberry 403764
Roundtop Rhododendron 331793
Roundtop Syzygium 382649
Round-topped Rhododendron 331793
Round-tree-leaved Cayratia 79897
Roundwarm Spunsilksage 360825
Roundwing Begonia 49993
Roundwing Twayblade 232118
Roundwingfruit Cyclocarya 116240
Round-wing-fruited Cyclocarya 116240
Roundwood 323083,369313
Roung-leaved Holly 203860
Roun-tree 369339
Roupala 337676
Rourea 337690,337733
Roureopsis 337770,337772
Rover Bellflower 70255
Roving Jenny 116736,349936
Roving Sailor 246649
Row Dashel 271666
Row Dashle 271666
Row Twisted-stalk 377934
Rowan 369290,369339
Rowan Tree 369290,369339
Rowberry 61497,383675
Rowntree 369339
Roxburgh Aglaia 11309
Roxburgh Amaranth 18814
Roxburgh Amaranthus 18814
Roxburgh Anoectochilus 26003
Roxburgh Argyreia 32653
Roxburgh Begonia 50256
Roxburgh Blackalga 200192
Roxburgh Ceriops 83944
Roxburgh Combretum 100758
Roxburgh Engelhardia 145517
Roxburgh Evodia 161350
Roxburgh Fig 164661,165121
Roxburgh Glaucous Oil-wood 142452
Roxburgh Goldlineorchis 26003
Roxburgh Mahonia 242632
Roxburgh Maizailan 11309
Roxburgh Mallotus 243443
Roxburgh Marshweed 230320
Roxburgh Milkwood 255368
Roxburgh Nitta-tree 284473
Roxburgh Peristrophe 291138
Roxburgh Pine 300186
Roxburgh Rose 336885
Roxburgh Roughfruitparsley 393949
Roxburgh Sagebrush 36153
Roxburgh Sesbania 361427
Roxburgh Slugwood 50601
Roxburgh Spatholobus 370419
Roxburgh Sumac 332513
Roxburgh Windmill 100758
Roxburgh's Kydia 218444

Roxburgh's Wormwood 36153
Roxburghia 337810,375336
Roxburth Starapple 90090
Roxie's Rhododendron 331675
Royal Agave 10955
Royal Azalea 331768
Royal Beard Tongue 289377
Royal Bellflower 70120
Royal Bluebell 412689
Royal Catchfly 363957
Royal Cudweed 317740
Royal Dendrobium 125334
Royal Dewflower 138233
Royal Goldfields 222597
Royal Grevillea 180659
Royal Hakea 184651
Royal Helleborine 147112
Royal Jasmine 211848
Royal Jewel Orchid 25999
Royal Knight's-spur 102841
Royal Lady's-slipper 120438
Royal Larkspur 124670
Royal Lily 230019
Royal Mallow 223393
Royal Paintbrush 350321
Royal Palm 337873,337885
Royal Palm-of-porto Rico 337875
Royal Paulownia 285981
Royal Peavine 222842
Royal Penny 401711
Royal Poinciana 123811
Royal Purple Lily 267783
Royal Red Bugler 9475
Royal Rhododendron 330205,331768
Royal Staff 39328
Royal Walnut 212636
Royal Water Lily 408733,408734
Royalfernleaf Meadowrue 388610
Royalpalm 337873,337885
Royal-red Bugler 9475
Royl Epipactis 147218
Royl Euphorbia 159739
Royl Zeuxine 417780
Royle Barberry 52119
Royle Brachyactis 58280
Royle Everlasting 21676
Royle Flatstalkgrass 220286
Royle Germander 388262
Royle Helictotrichon 190211
Royle Lallemantia 220286
Royle Orthoraphium 275614
Royle Spapweed 205291
Royle Touch-me-not 205291
Royle Trachydium 393856
Royle Woodbetony 287640
Royle's Balsam 204982,205291
Royle's Groundsel 359925
Royle's Mayten 182768
Royle's Pearleverlasting 21676
Royle's Snapweed 204982,205291
Royles Euphorbia 159739
Rozelle 195196
Rreflexisquamose Begonia 50227
Rrice-cooking Bamboo 82452
Rsimmon Pe 132466
Rubber 194453
Rubber Creeper 139959
Rubber Dandelion 384606
Rubber Euphorbia 159975
Rubber Hedge 159975
Rubber Plant 164925,164934
Rubber Rabbit Brush 90502
Rubber Rabbitbrush 90502,150332
Rubber Rabbitbush 90556
Rubber Tree 164925,169229,194451,350648
Rubber Vine 113820,113823,113825
Rubbertree 194451
Rubber-tree 194451
Rubberweed 201293
Rubbery Sage 345057
Rubble Dock 340254
Rubblefruit Ophiorrhiza 272212
Rubellum Lily 230023
Rubescent Cacalia 283866
Rubescent Rabdosia 209811
Rubescent Spindle-tree 157845
Rubgrass 398138
Rubicon White Cedar 85379
Rubiginose Eurya 160592
Rubimuricate Camellia 69577
Rubituberculate Camellia 69579
Rubrifolius Windflower 23808
Rubtzov Bigpetalcelery 357911
Rubwort 174877
Ruby Dock 339961
Ruby Giangt Crocus 111620
Ruby Grass 249323
Ruby Mountains Wild Buckwheat 152195
Ruby Protea 315761
Ruby Salt Bush 145268
Ruby Saltbush 145268
Ruby Sheepbush 145268
Ruby Spring Heath 149162
Ruby-grass 333024
Rubywood 320327
Ruchang Holly 203983
Rucheng Camellia 69525
Ruckerleaf Rhododendron 330613
Rud 66445
Rudbeckia 339512
Ruddes 66445
Ruddles 66445
Ruddy Woodsorrel 278140
Rude Knotweed 309420
Rue 341039,341064
Rue Anemone 24133
Rue Family 341086
Rue Lemongrass 117162
Rue Singlebamboo 56952
Rue-anemone 24133,388417
Ruecker Tulip Orchid 25138
Rueleaf Meadowrue 388653
Rueleaf Sagebrush 36173
Rue-leaved Saxifrage 349999
Ruellia 339658,339684,339709
Ruellia Pink 339684
Rue-weed 388506
Rufescent Evergreenchinkapin 79014
Rufescent Launaea 222941
Ruffet 401388
Ruffle Palm 12771,12772
Ruffled Sword Plant 140550
Rufous Bulrush 353701
Rufous Butterfly Orchid 145295
Rufous Mayten 182772
Rufous Privet 229608
Rugby Football Plant 290298,290299
Rugel Sugar Maple 3574
Rugel's Nailwort 284892
Rugel's Pawpaw 123576
Rugel's Plantain 302159
Rugel's Ragwort 339849
Rugged Corydalis 106407
Rugged Dogwood 380501
Rugged Dolomiaea 135738
Rugged Goldenrod 368435
Rugged Philodendron 294795
Rugged Seseli 361580
Rugged Touch-me-not 205303
Ruggedbud Schefflera 350726
Ruggedleaf Aralia 30751
Ruggedshoot Machilus 240664
Ruggeed Centaurea 80907
Rugolose Consolida 102842
Rugose Balanophora 46869
Rugose Dendrolobium 125579
Rugose Dysophylla 139582
Rugose Elaeocarpus 142381
Rugose Evodia 161373
Rugose Fortune Spindle-tree 157505
Rugose Glaucous-leaved Staff-tree 80285
Rugose Jerusalemsage 283623
Rugose Ophiorrhiza 272293
Rugose Paris 284397
Rugose Plectranthus 209815
Rugose Rapistrum 326833
Rugose Rattan Palm 65784
Rugose Rose 336901
Rugose Smallligulate Aster 40012
Rugose Willow 344243
Rugous Bitter-bamboo 304094

Rugulose Cenesolida 102842
Rugulose Elsholtzia 144086
Rugulose Grewia 180948
Rui 61254
Ruil 266862
Ruili Ardisia 31598
Ruili Arrowwood 408130
Ruili Azalea 331821
Ruili Bigumbrella 59240
Ruili Bladdernut 374108
Ruili Bladder-nut 374108
Ruili Brassaiopsis 59240
Ruili Helicia 189951
Ruili Holly 204264
Ruili Machilus 240697
Ruili Photinia 295767
Ruili Pseuderanthemum 317279
Ruili Randia 12539
Ruili Rhododendron 331821
Ruili Schefflera 350781
Ruili Skullcap 355763
Ruili Spapweed 205297
Ruili Spine olive 415240
Ruili Stonebean-orchis 63087
Ruili Storax 379466
Ruili Tupelo 267863
Ruili Viburnum 408130
Ruili Xantolis 415240
Ruizia 339857
Rukam 166777,166786,166788
Rukam Ramontchi 166788
Rum Cherry 83311,316787,316918
Rum-cherry 83311
Rumpet Creeper 70503
Rumph Cycas 115888
Rumph Greenstar 307524
Rumph Salacca 342576
Rumph Sweet Flag 5820
Rumphius Sago Palm 252634
Runaway Jack 176824
Runaway-robin 176824
Run-by-tile-ground 250432
Runch 326609,364545,364557
Runch-balls 364557
Runchie 364557
Runchik 364557
Runcinate Knotweed 309711
Runcinate Primrose 314918
Runcinate Saussurea 348742
Runcinate Windhairdaisy 348742
Rund-leaved Grewia 180768
Rungia 340336
Rungy 364557
Runner Bean 293985
Runner Oak 324315
Runner Violet 410511
Runnery Cinquefoil 312540
Runnidyke 176824
Running Bamboo 297203
Running Buckwheat 162519
Running Buffalo Clover 397093
Running Burning Bush 157750
Running Euonymus 157750
Running Five-fingers 312435
Running Fleabane 151019
Running Jacob 399601
Running Jenny 239755
Running Mallow 243823
Running Marsh Sedge 76147
Running Mountaingrass 272625
Running Myrtle 409335,409339
Running Postman 215982
Running Rockcress 30402
Running Savanna Sedge 76261
Running Shadbush 19295
Running Strawberry Bush 157750
Running Strawberry-bush 157750
Running Thyme 391397
Running-myrtle 409339
Runningpop 285639
Runningwater Nematanthus 263323
Runyon's Huaco 244370
Runyon's Sunflower 189036
Ruoergai Woodbetony 287649
Ruogh Syzygium 382657
Ruoqiang Aster 41173
Ruoqiang Calligonum 67036
Ruoqiang Kneejujube 67036
Ruoqiang Leymus 228382
Ruoqiang Windhairdaisy 348747
Rupestrine Dragonhead 137647
Rupestrine Eritrichium 153518
Rupestrine Geniostoma 172885
Rupestrine Greenorchid 137647
Rupestrine Hornbeam 77383
Rupestrine Rockvine 387844
Rupicolous Winkledleaf Box 64342
Ruppia 340470
Rupple-grass 302034
Ruprecht Honeysuckle 236076
Ruprecht Larkspur 124560
Ruprecht Sainfoin 271259
Ruprecht Silene 364003
Rupreeht's Silene 364002
Ruptile Jerusalemsage 295184
Ruptofoliate Purplepearl 66924
Rupture-grass 192941
Rupturewort 192926,192965
Rupture-wort 192941
Rusby Barberry 52125
Rusby's Bitterweed 201327
Rusby's Chinchweed 286902
Rusby's Jimmyweed 209583
Rusby's Milkweed 38100
Rusby's Rubberweed 201327
Rusby's Sedge 119611
Ruscusleaf Hakea 184633
Ruscus-leaved Barberry 52126
Ruseleaf Meliosma 249445
Rush 37469,212797,212911,212970,
213060,213066,213479,213531,213562
Rush Aster 40147,40923,380842
Rush Bebbia 49384
Rush Bushclover 226860
Rush Bush-clover 226860
Rush Daffodil 262429,262475
Rush Family 212752
Rush Fuirena 168883
Rush Garlic 15709
Rush Grass 372574
Rush Leek 15876
Rush Lily 365678
Rush Milkweed 38129
Rush Nul 118837
Rush Nut 118822
Rush Pea 64999,65024
Rush Pink 239322
Rush Pussytoes 26451
Rush Rabbitbrush 150345
Rush Sandwort 31967
Rush Skeletonplant 239334
Rush Skeletonweed 88776
Rush Wheatgrass 144297
Rusha Grass 117204
Rushes 212797
Rush-featherling 303955
Rushfoil 112963,113058
Rushleaf Jonquil 262353
Rushleaf Schoenorchis 352306
Rush-leaved Fescue 163816
Rush-leaved Narcissus 262410
Rushleek 15709,15876
Rush-like Bulrush 352200
Rushlike Dopatricum 136214
Rushlike Elytrigia 144654
Rushlike Moraea 258534
Rush-like Plantain 302091
Rush-like Psathyrostachys 317081
Rush-like Rattlebox 112269
Rush-lily 352147
Rusot 51333
Russell Lupin 238484
Russet Bud Rhododendron 331048
Russet Buffaloberry 362091
Russet Buffalo-berry 362091
Russet Cotton-grass 152749
Russet Sedge 76153
Russet-budded Rhododendron 331048
Russett Cotton-grass 152749
Russia Almond 20931
Russia Bedstraw 170603

Russia Broom 121000
Russia Cistancheherb 57437
Russia Dock 340116
Russia Elaeagnus 141932
Russia Elm 401549
Russia Euphorbia 159733
Russia Fritillary 168556
Russia Hogfennel 293002
Russia Salsify 394339
Russia Sedge 76096
Russia Thorowax 63804
Russia Torularia 264631
Russia Vetch 408693
Russia Wild Rye 144654
Russian Aconite 5461
Russian Almond 20931
Russian Arborvitae 253164
Russian Atriplex 44523
Russian Boschniakia 57437
Russian Box-thorn 239106
Russian Boxthoru 239106
Russian Carraway 266241
Russian Centaurea 81112
Russian Cinquefoil 312683
Russian Clover 105324
Russian Comfrey 381038,381042
Russian Currant 334114
Russian Cypress 253163,253164
Russian Dandelion 384472,384606
Russian Dock 339987
Russian Dwarf Almond 316859
Russian Elaeagnus 141932,141936
Russian Elm 401549
Russian Fenugreek 247452
Russian Fritillary 168556
Russian Globe Thistle 140704
Russian Hemp 71218
Russian Henbane 201372
Russian Iris 208797
Russian Knapweed 6272,6278
Russian Knotgrass 309094
Russian Leafy Spurge 158863
Russian Liquorice 177897
Russian Maple 2984
Russian Milkvetch 42355
Russian Millet 281916
Russian Mulberry 259067,259092
Russian Mustard 365649
Russian Olive 141929,141930,141932,
142152
Russian Pea Shrub 72233,72235
Russian Peashrub 72233
Russian Pea-tree 72233
Russian Pigweed 45843,45844
Russian Poplar 311375
Russian Purple Crab 243659
Russian Rhubarb 329366
Russian Rock Birch 53411
Russian Rocket 365644
Russian Sage 291427,291430
Russian Tamarisk 383440
Russian Tarragon 35407,35416
Russian Thistle 344425,344463,344496,
344571,344585,344589,344743
Russian Tumbleweed 344585,344743
Russian Turnip 59507
Russian Vetch 408693
Russian Vine 162509,162512
Russian Wild Rye 317081
Russian Wildrye 317081
Russian Willow 344037
Russian Wormwood 36177
Russian-olive 141930,141932
Russianthistle 344425,344743
Russian-thistle 344425
Russianthistle-like Sagebrush 36211
Russianthistle-like Wormwood 36211
Russles 334179
Russowia 341026
Rust Azalea 330265
Rust Liparis 232157
Rust Paraboea 283125
Rust Purplebract Saussurea 348396
Rust Purplestem 376471
Rustback Mananthes 244287
Rustbract Sagebrush 35626
Rustburn 271563
Rustcolor Rattlebox 112138
Rustcolored Rattlebox 112138
Rustcoloured Erioglossum 151783
Rust-coloured Fimbristylis 166313
Rustcoloured Loosestrife 239827
Rust-coloured Rhododendron 330069
Rustcoloured Scurrula 355302
Rusthair Agapetes 10288
Rusthair Ash 167974
Rusthair Biformbean 20576
Rusthair Boeica 56366
Rusthair Chessbean 116589
Rusthair Corallodiscus 103950
Rusthair Dinggongvine 154239
Rusthair Glycosmis 177835
Rusthair Hydrangea 199947
Rusthair Jurinea 207142
Rusthair Oreocharis 273850
Rusthair Pagodatree 369104
Rusthair Pentapanax 289636
Rusthair Scurrula 355302
Rusthair Tailanther 381940
Rusthair Wildtung 243319
Rusthair Woodlotus 244466
Rustleaf Azalea 331826
Rustpod Bauhinia 49075
Rust-red Eria 148683
Rustscale Fluttergrass 166313
Rustscale Olive 270089
Rustshoot Ormosia 274390
Rust-spoted Hypecoum 201607
Ruststem Elatostema 142657
Ruststem Spiradiclis 371757
Ruststem Stairweed 142657
Rustvein Meadowsweet 166131
Rusty Albizzia 13554
Rusty Ash 167974
Rusty Astronia 43462
Rusty Azalea 330449
Rusty Blackhaw 408095
Rusty Bog Rush 352385
Rusty Bog-rush 352385
Rusty Coated Leaves Rhododendron 331826
Rusty Cotton-grass 152749,152796
Rusty Fig 164904,165586
Rusty firmiana 155000
Rusty Flat Sedge 393205
Rusty Flatsedge 393205
Rusty Foxglove 130358
Rusty Glycosmis 177835
Rusty Iguanura 203507
Rusty Indigo 205806
Rusty Leaf 250513
Rusty Licuala Palm 228737
Rusty Licualapalm 228737
Rusty Mountainash 369387
Rusty Mountain-ash 369387
Rusty Olive 270089
Rusty Ormosa 274390
Rusty Podocarp 316086
Rusty Psilotrichum 319022
Rusty Rhododendron 331683
Rusty Rosewood 121678
Rusty Scurrula 355302
Rusty Shield-bearer 288897
Rusty Siris 13554
Rusty Spider Flower 180597
Rusty Wood Rush 238690
Rusty-branch Ormosa 274390
Rusty-colored Rhododendron 330695
Rusty-coloured Dendrobenthamia 105060
Rusty-coloured Erioglossum 151783
Rusty-coloured Glycosmis 177835
Rusty-coloured Lyonia 239379
Rusty-hair Anzacwood 310770
Rustyhair Boea 283125
Rustyhair Bowfruitvine 393530
Rustyhair Brier Grape 411649
Rustyhair Clematis 95077
Rustyhair Creeper 285161
Rustyhair Crucianvine 356285
Rustyhair Diversifolius Ampelopsis 20394
Rustyhair Elaeocarpus 142337
Rustyhair Erycibe 154239

Rustyhair Fishvine 125960
Rusty-hair Flemingia 166861
Rusty-hair Himalayan Rockvine 387843
Rustyhair Holly 203808
Rustyhair Honeysuckle 235790
Rustyhair Jewelvine 125960
Rusty-hair Kangding Acanthopanax 2429
Rustyhair Loosestrife 239620
Rusty-hair Manglietia 244466
Rustyhair Millettia 66647
Rustyhair Raphiolepis 329061
Rustyhair Raspberry 339162
Rusty-hair Rehder Mountainash 369502
Rustyhair Secamone 356285
Rustyhair Sophora 369104
Rustyhair Taxillus 385207
Rustyhair Toxocarpus 393530
Rustyhair Wampee 94194
Rustyhair Windhairdaisy 348228
Rusty-haired Agapetes 10288
Rusty-haired Ash 167974
Rusty-haired Brassaiopsis 59201
Rusty-haired Championella 85943
Rusty-haired Clematis 95077
Rusty-haired Elaeocarpus 142337
Rusty-haired Erycibe 154239
Rusty-haired Holly 203808
Rusty-haired Honeysuckle 235790
Rusty-haired Japanese Cayratia 79855
Rusty-haired Jewelvine 125960
Rusty-haired Longpetiole Hydrangea 199947
Rusty-haired Millettia 66647
Rusty-haired Raphiolepis 329061
Rusty-haired Raspberry 339162
Rusty-haired Rhododendron 330445
Rusty-haired Secamone 356285
Rusty-haired Taxillus 385207
Rusty-haired Toxocarpus 393530
Rusty-hairs Dogwood 105060
Rusty-hairs Rhododendron 330445
Rustyhairy Boea 283125
Rustyhairy Brassaiopsis 59201
Rustyhairy Euaraliopsis 59201
Rustyhairy Litse 233999
Rusty-hairy Litse 233999
Rusty-hairy Mallotus 243319
Rustyhairy Palmleaftree 59201
Rustylea Kiwifruit 6695
Rustyleaf 250513
Rustyleaf Astronia 43462
Rusty-leaf Begonia 49831
Rusty-leaf Fig 165586
Rustyleaf Mock Azalea 250513
Rusty-leaf Mucuna 259535
Rustyleaf Newlitse 264034
Rustyleaf Slugwood 50575
Rustyleaf Twayblade 232157
Rusty-leaved Astronia 43462
Rusty-leaved Beech 162375
Rusty-leaved Fig 165586
Rusty-leaved Newlitse 264034
Rusty-leaved Rhododendron 331826
Rusty-podded Bauhinia 49075
Rusty-vein Anzacwood 310771
Rusy Gum 24606
Rutabaga 59493,59507,59603
Ruth's Grass-leaved Goldenaster 301513
Ruthenia Medic 247452
Ruthenian Centaurea 81345
Rutland Beauty 68713
Ruvo Kale 59619
Ruyang Camellia 69785
Ruyschia 341172
Ruyschianum Dragon's-head 137651
Ruyschianum Dragonhead 137551
Ruyuan Azalea 331651
Ruyuan Fig 165607
Ruyuan Grape 411902
Ruyuan Manglietia 244479
Ruyuan Rhododendron 331651
Ruyuan Woodlotus 244479
Rydberg's Arnica 34765
Rydberg's Fleabane 150938
Rydberg's Goldenbush 150357
Rydberg's Orach 44322
Rydberg's Penstemon 289374
Rydberg's Poison-Ivy 393478
Rydberg's Spring Beauty 94338
Rydberg's Sunflower 189015
Rydberg's Thistle 92355
Rydberg's Vervain 405887
Rye 356232,356237
Rye Brome 60963
Rye Grass 163819,164219,235300,235354
Ryegrass 235300
Rye-grass 235300,235334
Rye-grass Sedge 75185
Ryegrass-like Eragrostiella 147465
Ryelike Brome 60963
Rye-like Brome 60963
Ryfart 326616
Ryssopterys 341223,341227
Ryukyu Viburnum 408151

S

S. Africa Withania 414601
S. Africa Yelloweyegrass 416020
S. America Feijoa 163117
S. America Heliotrope 190559
S. America Mountain Leech 126649
S. Anhui Fritillary 168586
S. Asia Blastus 55141
S. Asia Michelia 252860
S. Asia Panicgrass 281733
S. Asia Pine 300030
S. Burma Clovershrub 70861
S. Caucasia Sainfoin 271277
S. Changjiang Pipewort 151302
S. China Amomum 19827
S. China Aspidistra 39522
S. China Barthea 48552
S. China Bigleaf Thorowax 63729
S. China Birch 53354
S. China Chloranthus 88300
S. China Cinnamon 91270
S. China Clethra 96496
S. China Cycas 115888
S. China Dogwood 104961
S. China Dutchmanspipe 34111
S. China Elaeocarpus 142283
S. China Euphorbia 158534
S. China Evergreen Chinkapin 78899
S. China Evodia 161309
S. China Gentian 173606,173848
S. China Hemiboea 191356
S. China Holly 204276
S. China Honeylocust 176869
S. China Jurinea 193930,211775
S. China Keiskea 215806
S. China Kiwifruit 6616
S. China Knotweed 309189
S. China Litse 233934
S. China Lovableflower 148061
S. China Mahonia 242563
S. China Mananthes 244283
S. China Meliosma 249402
S. China Milkwort 308081
S. China Olax 269649
S. China Pecteilis 285953
S. China Pepper 300346
S. China Persimmon 132444
S. China Pipewort 151487
S. China Privet 229617
S. China Qinggang 116086
S. China Rabbiten-wind 12743
S. China Redcarweed 288727

S. China Roughleaf 222096
S. China Saddletree 240118
S. China Sedge 73837
S. China Slantspike 301722
S. China Slatepentree 400687
S. China Spikesedge 143213
S. China Spurge 158534
S. China Stairweed 142616
S. China Sweetleaf 381121
S. China Syzygium 382483
S. China Tarenna 384920
S. China Telosma 385759
S. China Xylosma 415877
S. Europe Ephedra 146221
S. Florida Marsh Pine 299930
S. Fujian Hydrangea 200026
S. Gansu Fritillary 168523
S. Gansu Stonecrop 357259
S. Guangdong Rosewood 121756
S. Guangdong Skullcap 355849
S. Guangxi Machilus 240667
S. Guangxi Mananthes 244280
S. Guangxi Rattanpalm 65642
S. Guangxi Tanoak 233345
S. Guangxi Woodlotus 244427
S. Guizhou Bauhinia 49214
S. Guizhou Cryptocarya 113431
S. Guizhou Indigo 205962
S. Guizhou Photinia 295657
S. Guizhou Thickshellcassia 113431
S. Guizhou Willow 343555
S. Hainan Decaspermum 123445
S. Hainan Litse 234103
S. Japan Barnyardgrass 140384
S. M. Weld Rhododendron 331707
S. Mongol Tamarisk 383449
S. Shaanxi Gentian 173724
S. Sichuan Azalea 331864
S. Sichuan Comanthosphace 100236
S. Sichuan Dutchmanspipe 34112
S. Sichuan Goldenray 229116
S. Sichuan Hogfennel 292837
S. Sichuan Ironweed 406153
S. Sichuan Linden 391784
S. Sichuan Loosestrife 239745
S. Sichuan Sicklebamboo 20207
S. Sichuan Tanoak 233359,233378
S. Sichuan Wildclove 226090,226097
S. Sichuan Willow 344278
S. Taiwan Begonia 49645
S. Taiwan Kalanchoe 215149
S. Taiwan Skullcap 355384
S. Xingjiang Bellflower 69913
S. Xinjiang Catnip 264921
S. Xinjiang Corydalis 106049
S. Xinjiang Ziziphora 418131
S. Xizang Azalea 331543
S. Xizang Bastardtoadflax 389680
S. Xizang Bellflower 70191
S. Xizang Chickweed 375156
S. Xizang Corydalis 106024
S. Xizang Cotoneaster 107702
S. Xizang Dendrobium 125263
S. Xizang Didymocarpus 129946
S. Xizang Dolomiaea 135741
S. Xizang Euonymus 157330
S. Xizang Hundredstamen 389680
S. Xizang Lilac 382300
S. Xizang Meconopsis 247200
S. Xizang Platanthera 302289
S. Xizang Ploughpoint 401151
S. Xizang Raspberry 338171
S. Xizang Rockfoil 349294
S. Xizang Spiraea 371821
S. Xizang Touch-me-not 205313
S. Xizang Typhonium 401151
S. Xizang Willow 343068
S. Yunnan Arisaema 33273
S. Yunnan Basketvine 9425
S. Yunnan Bauhinia 49136
S. Yunnan Beautyleaf 67866
S. Yunnan Birthwort 34295
S. Yunnan Bulbophyllum 63012
S. Yunnan Cinnamon 91271
S. Yunnan Cliffbean 254616
S. Yunnan Dishspurorchis 171880
S. Yunnan Eightangle 204564
S. Yunnan Elaeocarpus 142285
S. Yunnan Eria 299648
S. Yunnan Eurya 160428
S. Yunnan Forkliporchis 25966
S. Yunnan Glinus 176948
S. Yunnan Habenaria 183848
S. Yunnan Hairorchis 299648
S. Yunnan Hetaeria 193586
S. Yunnan Hibiscus 194729
S. Yunnan Hopea 198131
S. Yunnan Horsfieldia 198523
S. Yunnan Irisorchis 267929
S. Yunnan Linden 391771
S. Yunnan Liparis 232329
S. Yunnan Litse 233990
S. Yunnan Maire Camellia 69344
S. Yunnan Mayten 246812
S. Yunnan Meadowrue 388725
S. Yunnan Monkshood 5046
S. Yunnan Mountain Leech 126462
S. Yunnan Neonauclea 264226
S. Yunnan Oberonia 267929
S. Yunnan Ophiorrhiza 272182
S. Yunnan Pere Valentin's Rhododendron 332049
S. Yunnan Poisonnut 378833
S. Yunnan Primrose 314459
S. Yunnan Rattanpalm 65706
S. Yunnan Redcarweed 288752
S. Yunnan Rosewood 121721
S. Yunnan Sagebrush 35192
S. Yunnan Sicklehairicot 135508
S. Yunnan Snowbell 379311
S. Yunnan Southstar 33273
S. Yunnan Spapweed 204792
S. Yunnan St. John's Wort 201768
S. Yunnan Stonebean-orchis 63012
S. Yunnan Stranvaesia 377466
S. Yunnan Sweetleaf 381240
S. Yunnan Syzygium 382484
S. Yunnan Tailanther 381924
S. Yunnan Tainia 383159
S. Yunnan Tanoak 233340
S. Yunnan Touch-me-not 204915
S. Yunnan Twayblade 232329
S. Yunnan Uraria 402130
S. Zhejiang Sedge 73838
Sabadilla 352120
Sabah Climbing Bamboo 131286
Sabah Scrambling Bamboo 131286,216392
Sabal 172149
Sabal Palm 341422
Sabbatia 341466
Sabi Star 7343
Sabia 341475
Sabia Family 341582
Sabialeaf Actinidia 6700
Sabialeaf Kiwifruit 6700
Sabia-leaved Actinidia 6700
Sabicu 239538
Sabigrass 402533
Sabin 213902
Sabina 213827,341696
Sabine Pine 300189
Sabine's Pine 300189
Sabino 385281
Sabline 255435
Saboo Leard 375896
Sabra Spike Sage 344985
Sabre Bean 71042
Sabulicolous Barberry Barberry 52127
Sabut Devil Rattan 121524
Sacahuista 266504
Sacasil 140293
Sacasile 26269
Sacbearing Gagea 169499
Saccate Calyx Milkvetch 42994
Saccate Dishspurorchis 171893
Saccate Fruit Milkvetch 42991
Saccate Gastrochilus 171893
Saccate Honeysuckle 236078
Saccate Lily 230026
Saccate-calyx Crazyweed 279139
Saccate-calyx Milkvetch 42255

Saccate-calyx Peashrub 72257
Saccate-calyxed Milkvetch 42255
Saccateflower Honeysuckle 236078
Saccatefruit Seepweed 379582
Saccatepetal Phaeanthus 293241
Saccatespur Larkspur 124076
Saccolabium 342015
Sacculate Calanthe 66056
Sacfruit Corydalis 106387
Sachalin Angelica 24293
Sachalin Cherry 83300
Sachalin Clinopodium 97031
Sachalin Corktree 294251
Sachalin Draba 137224
Sachalin Elder 345690
Sachalin Eupatorium 158305
Sachalin Fir 460,463
Sachalin Honeysuckle 236085
Sachalin Meadowrue 388654
Sachalin Milkvetch 42995
Sachalin Mint 250443
Sachalin Monkshood 5546
Sachalin Mountainash 369367
Sachalin Raspberry 339227
Sachalin Rhodiola 329944
Sachalin Sedge 74029
Sachalin Thorowax 63807
Sachalin Violet 410517
Sachalin Windflower 23917
Sacharosa 290713
Sachu Tamarisk 383601
Sack Tree 28248
Sackysac 206934
Sacpetal Corydalis 106402
Sacramento Mountain Fleabane 150937
Sacramento Mountains Thistle 92474
Sacrau Poplar 79 311160
Sacred Bamboo 262189
Sacred Bean 263267,263272
Sacred Datura 123071,123092
Sacred Fig 165553
Sacred Fig Tree 165553
Sacred Fir 458
Sacred Flower of the Incas 71544
Sacred Lily 263272
Sacred Lily-of-China 335760
Sacred Lotus 263272
Sacred Place Rhododendron 331953
Sacred Spindle-tree 157310
Sacret Flower of the Incas 71547
Sacseed 390967
Sacsepal Crazyweed 279139
Sacymore 3462
Sad Melandrium 248440
Sad Stock 246461
Sad Tree 267430
Sadaphal 93763
Saddle Bluestem 22982
Saddleorchis 2071
Saddletree 240112
Saddle-tree 232609
Sad-flowered Iris 208870
Sadler's Willow 344056
Saffen 213902
Saffern 213902
Safflor 77748
Safflower 77684,77748
Safflowerlike Swisscentaury 375268
Safforne 111589
Saffron 111480,111566,111589,213902
Saffron Crocus 111589
Saffron Lily 229766
Saffron Ragwort 279898
Saffron Spike 29127
Saffron Thistle 77716,77748
Saffron Tree 347407
Saffron Tritonia 399062
Saffron Wild Buckwheat 151968
Saffron's Spike 29127
Saffron-wood 78555
Saffron-yellow Tritonia 399062
Safistan 104218
Safronyellow Cinquefoil 312457
Saf-saf 311308
Sagapenum 163574
Sage 35084,36474,44609,295049,344821,
345049,345080,345108,345271
Sage Brush 36400
Sage Colorado 35505
Sage Garlick 388275
Sage of Bethlehem 250450
Sage Plant 36400
Sage Rose 93202,400493
Sage Willow 343142,343506
Sagebrush 35084,36166,36400,44334,44609
Sagebrush False Dandelion 266829
Sagebrushleaf Russianthistle 344426
Sageflower Didymocarpus 129955
Sageflower Woodbetony 287655
Sageleaf Catnip 265042
Sageleaf Fleaweed 321598
Sageleaf Nepeta 265042
Sageleaf Pulicaria 321598
Sage-leaf Rockrose 93202
Sage-leaved Cistus 93202
Sage-leaved Germander 388275
Sage-leaved Rock Rose 93202
Sage-leaved Rockrose 93202
Sage-leaved Willow 343142
Sageretia 342155
Sagewort 35837
Sage-wort 35090,36474
Sagewort Wormwood 35237
Saggin Saggan 208771
Saghalien Fir 298298
Saghalin Spruce 298298
Sagisi Palm 194134,194135
Sagitate Knotweed 309742
Sagittate Barrenwort 147048
Sagittate Blumea 55817
Sagittate Parabaena 283060
Sagittate Rockcress 30424
Sagittate Saltbush 44375
Sagittate Tinospora 392269
Sagittate Violet 410527
Sagittate-leaf Kumquat 93492
Sagittateleaf Paraprenanthes 283694
Sagittateleaf Spodiopogon 372427
Sagittate-leaved Epimedium 147048
Sagittatepetal Stonecrop 357121
Saglionis Chin Cactus 182476
Sago 115884
Sago Cycad 115812,115884
Sago Cycas 115794,115884
Sago Palm 78056,98549,115794,115812,
115884,145217,156116,252631,252634,
252636
Sago Plum 115884
Sago Pondweed 378991
Sago-palm Family 417024
Sago-plum 115884
Sago-pondweed 378991
Saguaragy Bark 100068
Saguared Cactus 77077
Saguaro 77075,77077
Saguira 375529
Sahmarsh Water Hemp 18683
Saho 398826
Saigon Cinnamon 91366
Sailor Buttons 216753,248411,374916,
379660
Sailor's Buttons 216753,248411,374916,
379660
Sailor's Knot 174877
Sailor's Tobacco 36474
Sailor's-tobacco 35084
Sail-pod Dalea 121886
Saim Rhododendron 331910
Sain 386656
Sainfoin 188028,247366,247456,271166,
271280
Saint Augustine Grass 375774
Saint Barnaby's Thistle 81382
Saint Bernard's Lily 27190
Saint Bruno's Lily 283300
Saint Catherine's Lace 152086
Saint Dabeoc's Heath 121066
Saint Ignatius' Bean 378761
Saint John's Wort 201705
Saint Lucie Cherry 83253
Saint Patrick's Cabbage 349918

Saint Vincent Arrow-root 245014
Saintpaulia 342482
Saints Fleabane 150945
Sairam Thistle 92356
Sairim Thistle 92356
Sajia Peashrub 72173
Sajia Pea-shrub 72173
Sakaguchi Microtropis 254304
Sakaki 96587,96617
Sakden Large-sepal Barberry 51899
Sake 275958
Sakebed 97697,97719
Sakhalian Knotweed 309723
Sakhalin Corktree 294251
Sakhalin Fir 460,463
Sakhalin Spindle-tree 157848
Sakhalin Spruce 298298
Sakhaline Spruce 298298
Sakhu 362220
Sakiki Palm 315386
Sal 362220
Sal Mistletoe 355317
Sal Shorea 362220
Sal Tree 362220
Salab-misri 121358,156521,273317
Salacca 342570
Salacia 342581,342709
Salad 172135
Salad Burnet 345860,345866,345881
Salad Chervil 28024
Salad Rocket 154019
Salad-chervil 28024
Saladine 86755
Salai Tree 57539
Salak 342570,342580
Salak Palm 342570
Salamander Tree 28316
Salamander-tree 28316
Salatao Crazyweed 279003
Salebrosa Goldenrod 368032
Salem-rose 339194,339195
Salep 121358,156521,273317
Salicornia 342847
Saligna Gum 155732
Saligo 68189
Saligot 394500
Saligot Nut 394500
Salimu Tulip 400243
Salinas Valley Goldfields 222609
Saline Cistanche 93075
Saline Plantain 301967
Saline Sakebed 97727
Saline Saussurea 348763
Saline Seepweed 379598
Saline Smithia 366695
Saline Sphaerophysa 371161
Saline Swainsona 371161
Saline Tamarisk 383526
Saline Windhairdaisy 348763
Salisbury's Lily 229811
Salish Fleabane 150940
Sallet 339887
Sallit 219485
Sallleaf Chamabainia 85110
Sallleaf Ministylenettle 85110
Sallow 342923,343151
Sallow Thorn 196757
Sallstyle 63204
Sally 343151
Sally Wattle 1083,1528
Sally Withy 343151
Sally-my-handsome 77430,211295,279768
Sally-wood 343151
Salmon Barberry 51284
Salmon Berry 338985
Salmon Bloodlily 350321
Salmon Blood-lily 184428,350312
Salmon Bush Monkey Flower 255217
Salmon Gum 155733
Salmon Mountains Wake-robin 397587
Salmon River Fleabane 150941
Salmon River Rabbitbrush 150374
Salmonberry 339285
Salmon-berry 339285
Salmwood 104153,104185
Salomon Sago Palm 252637
Salomonia 344345
Salouen Fir 347
Salpiglossis 344381
Salry 29327,339887
Salsify 394257,394327,394333
Salsilla 56758
Salsola-like Inula 207224
Salt Birch 53468
Salt Bush 344804
Salt Cedar 383412,383469,383491,383591,383595
Salt Cellar 277648
Salt Crazyweed 279141
Salt Globepea 371161
Salt Grass 134842
Salt Marsh Cord Grass 370171
Salt Marsh Elder 46227
Salt Plantain 302092
Salt Rheum Weed 86783
Salt Rheumweed 86783
Salt River Rock Daisy 291313
Salt Rose 34536
Salt Sandspurry 370661
Salt Tree 184832,184838
Salt-and-pepper 11549
Salt-and-pepper Plant 302122
Saltbane 184810
Saltbeantree 184832,184838
Saltbush 44298,44317,44334,44446,44689,346610
Salt-Bush 44298
Saltbushleaf Nightshade 367416
Saltbushleaf Perovskia 291430
Saltbushleaf Wolfberry 239055
Saltcedar 383412,383469,383595
Salt-cellar 277648
Saltclaw 215322,215327
Saltcress 389197,389200
Saltgrass 184967,184979
Saltlake Fenugreek 397208
Saltlivedgrass 184947
Saltliving Anabasis 21058
Saltliving Nitraria 266367
Saltliving Sagebrush 35595
Saltliving Whitethorn 266367
Saltliving Wormwood 35595
Saltlover 184951
Salt-loving Birch 53468
Saltloving Goosefoot 87082
Salt-loving Iris 208606
Salt-loving Milkvetch 43000
Salt-loving Swordflag 208606
Saltmarsh Aster 41317,380996,380999
Saltmarsh Baccharis 46221
Saltmarsh Bulrush 352166
Salt-marsh Bulrush 56641,353759
Salt-marsh Cockspur Grass 140509
Saltmarsh Dilophia 130965
Saltmarsh Flat-sedge Red Clubrush 55895
Saltmarsh Goosefoot 86990
Saltmarsh Mallow 217944
Salt-marsh Mallow 217944
Saltmarsh Rush 213127,353587
Salt-Marsh Rush 353587
Saltmarsh Sand Spurrey 370661
Salt-Marsh Sand Spurry 370661
Salt-marsh Sand-spurrey 370706
Saltmarsh Sand-spurry 370661
Saltmarsh Sealavender 230700
Saltmarsh Sedge 76135
Salt-marsh Starwort 374919
Salt-marsh Water-hemp 18683
Saltmeadow Rush 213127
Salt-meadow Rush 213127
Saltnodetree 184930,184933
Salton Saltbush 44343
Saltplace Halerpestes 184713
Saltshrimp Flower 406667
Saltspike 185034
Salttree 184832
Salt-tree 184832,184838
Saltwater False Willow 46212
Saltwater Paperbark 248088
Saltweed 55873
Saltwleed 212929

Saltwort 48880,48885,176775,342847,342859,344425,344571,344585,344718,346761
Salt-wort 344425
Saltwort Wild Buckwheat 152446
Salvador Douglas Fir 318599
Salvador Henequen 10883
Salvador Leadtree 227431
Salvador Sisal 10883
Salvadora 344798
Salvadora Family 344816
Salvat Hackberry 80733
Salvation Jane 141258
Salver Qinggang 116172
Salver-shaped Cherry 83165
Salvia 344821,344824,345308,345405
Salvia Cistus 93202
Salvia Rockrose 93202
Salvialeaf Alangium 13384
Salween Clockvine 390871
Salween Mahonia 242597
Salween Waxplant 198893
Salweenia 345495
Salwen Rhododendron 331720
Salwin Aster 41195
Salwin Dolomiaea 135735
Salwin Eria 148768
Salwin Loquat 151171
Salwin Pennywort 200360
Salwin Syzygium 382659
Salwin Windhairdaisy 348765
Salzburg Saxifrage 349225
Salzmann Crocus 111594,111596
Salzmann's Mille Graines 269983
Salzmann's Restharrow 271338
Sama 282286
Samall Polemonium 307215
Saman 345517,345525
Saman Rain Tree 345525
Samaradaisy 320508,320515
Samarang Rose-apple 382660
Samaranga Syzygium 382660
Sambucus Baboon Flower 46122
Sambucus Iris 208805
Samecolor Pearleverlasting 21528
Samecolor Pocktorchid 282806
Samecolor Rush 213018
Samecolor Thrixsperm 390532
Samecolor Thrixspermum 390532
Samecolor Vanda 404632
Sameleaf Snakegourd 396197
Samfer 111347
Samite Azalea 331581
Samll Erysimum 154544
Samll Evolvulus 161418
Samll Privet 229617
Samll Storkbill 288268
Samll Sugarmustard 154544
Samllball Crazyweed 279016
Samllflower Bentgrass 12201
Samll-flowerbuds Biond's Magnolia 242007
Samll-head Milkvetch 42713
Samll-leafed Privet 229617
Sammy Gussets 273531
Samphire 111343,111347,342859,346761
Samphire Leafed Geranium 288183
Samphireleaf Grevill 180583
Sampion 342859
Sampkins 111347
Sampson Aster 41196
Sampson Chastetree 411441
Sampson Chaste-tree 411441
Sampson Dysophylla 139584
Sampson Helixanthera 190855
Sampson Jurinea 190855
Sampson Macaranga 240320
Sampson Peristylus 291189
Sampson Perotis 291189
Sampson Scurrula 190855
Sampson Snakeweed 173675
Sampson St. John's Wort 202146
Sampson Thyrocarpos 391421
Sampson Watercandle 139584
Sampson's Macaranga 240320
Sampson's Snakeroot 174059,273237,319223
Samuyao 93635,93869
Samyda Family 345760
San Angelo Yucca 416638
San Benito Fritillary 168622
San Benito Pentachaeta 289424
San Bernardino Aster 380860
San Bernardino Ragwort 279889
San Bernardino Wild Buckwheat 152273
San Carlos Wild Buckwheat 151897
San Clemente Island Bristleweed 186883
San Clemente Island Brodiaea 60487
San Clemente Island Larkspur 124672
San Clemente Island Triteleia 398784
San Clemente Island's St. Catherine's-lace 152090
San Diego Barrel Cactus 163479
San Diego Ceanothus 79922
San Diego County Viguiera 46550
San Diego Golden Star 55647
San Diego Rabbit-tobacco 317759
San Diego Silverpuffs 253922
San Diego Sunflower 46550
San Diego Wirelettuce 375961
San Domingo Box 297536
San Domingo Boxwood 297536
San Felipe Dyssodia 139717
San Fernando Spineflower 89095
San Francisco Campion 364162
San Francisco Lessingia 227101
San Francisco Nailwort 284851
San Francisco Peaks Ragwort 279911
San Francisco Spineflower 89070
San Gabriel Boykinia 58024
San Gorgonio Wild Buckwheat 152186
San Hemp 112269
San Jacinto Valley Crownscale 44362
San Joaquin Adobe Sunburst 317365
San Joaquin Orach 44477
San Joaquin Silverpuffs 253917
San Joaquin Snakeweed 182096
San Jose Hesper Palm 59097
San Juan Wild Buckwheat 152226
San Luis Mariposa-lily 67615
San Luis Obispo Mariposa-lily 67626
San Marcos Hibiscus 179899
San Matco Woolly Sunflower 152824
San Pedro Cactus 140877
San Pedro Macho 140880
San Pedro Matchweed 415085
San Pedro River Wild Buckwheat 152515
San Rafael Cactus 287891
San Rafael Fleabane 150567
San Simeon 46246
Sanbiling Milkvetch 43003
Sanbokan 93820
Sancarl Cactus 244119
Sanchezia 345769
Sanchi 280771,280778
Sanctuary 81065
Sand 163328
Sand Amaranth 18661
Sand Ammania 19554
Sand Apple 284200,284248
Sand Bitter-cress 72920,72922
Sand Blackberry 338307
Sand Bottlebrush 49350
Sand Bracted Sedge 75435,75436
Sand Brier 367014
Sand Bur 367567
Sand buttons 84503
Sand Cat's Tail 294966
Sand Cat's-tail 294966
Sand Catchfly 363363
Sand Cedar 213917
Sand Cherry 83157,316735,316739,316740
Sand Chickasaw Plum 316220
Sand Club-cholla 181458
Sand Coprosma 103783
Sand Coreopsis 104531,104532
Sand Cress 30342
Sand Crocus 335879,335926
Sand Croton 112895,112896
Sand Cyperus 119139,119140
Sand Cypress Pine 67414
Sand Dogwood 380434

Sand Dollar Cactus 43493
Sand Dropseed 372642
Sand Drop-seed 372642,372644
Sand Dune Chinchweed 286888
Sand Dune Thistle 92307
Sand Dune Willow 343343
Sand Evening Primrose 269424
Sand Evening-primrose 269422
Sand Fame-flower 294331
Sand Fleabane 150494
Sand Flower 34536
Sand Grape 411900
Sand Grass 397939
Sand Groundsel 359897
Sand Heath 83397
Sand Hickory 77926
Sand Jointweed 308669
Sand Juniper 213902
Sand Leaf-flower 296480
Sand Leek 15113,15726
Sand Lily 148263,148264,250488
Sand Live Oak 323934
Sand Love Grass 147468,148019
Sand Lucerne 247507
Sand Mat 85753
Sand Milkweed 37827
Sand Mustard 133286
Sand Myrtle 224196
Sand Palm 234182
Sand Pansy 410683
Sand Parsley 19786
Sand Pavetta 286095
Sand Pea 222735
Sand Pear 323268,323330
Sand Pear Tree 323268
Sand Penstemon 289319
Sand Phlox 295252
Sand Pine 299864
Sand Pink 127598
Sand Plantain 301864
Sand Plum 316219,316543
Sand Post Oak 324153,324443
Sand Pricklypear 273015
Sand Primrose 269497
Sand Rabbitbrush 150335
Sand Rock-cress 30186
Sand Rocket 133286
Sand Rye Grass 228346
Sand Ryegrass 228346
Sand Sacahuista 266483
Sand Sage 35488
Sand Sagebrush 35488
Sand Sedge 73747,75435,75436,119139
Sand Slimpod 20862
Sand Spurrey 370609,370696,379553
Sand Spurry 370609
Sand Stars 20862
Sand Systemon 382011
Sand Three-awn 33827
Sand Tickseed 104532
Sand Tickseed Lance Coreopsis 104531
Sand Timothy 294966
Sand Toadflax 230895
Sand Underleaf Pearl 296480
Sand Verbena 687
Sand Vine 117552
Sand Violet 409648,409650,409654,
409698,410528
Sand Wash Groundsel 358757
Sand Wild Buckwheat 152225
Sand Willow 343046,343193,343914
Sand Wormwood 35237
Sandal 346208,346210
Sandal Bead Tree 7190
Sandal Beadtree 7190
Sandal Bean Tree 7190
Sandal Wood 346208,346210
Sandal-bead Tree 7190
Sandaltree 346210
Sandal-tree 346210
Sandalwood 320269,346208,346210,389946
Sandalwood Family 346181
Sandalwood Padauk 320327
Sandal-wood Padauk 320327
Sandalwood Tree 7190
Sandankwa Viburnum 408151
Sandarac 386942
Sandaster 104646
Sandbank Hakea 184628
Sandbar Love Grass 147686
Sandbar Willow 343360,343361,343392,
343491,343622,344099
Sand-bar Willow 343622
Sandbell 211672
Sandberg's Birch 53311,53607
Sandberg's Sulphur Flower 152601
Sandberry 31142
Sandbog Death Camas 417903
Sandbox Tree 199465
Sandbox-tree 199465
Sandbur 80803,80813,80830,80855
Sandbur Pricklypear 273028
Sandcarpet 73172
Sand-cedar 213939
Sand-corn 417921
Sandcress 321478
Sand-dollar 43493
Sand-dune Milkvetch 42203
Sand-dune Thistle 92307
Sand-dune Wallflower 154419
Sandeman Primrose 314926
Sander Mandevilla 244328
Sander Ptychosperma 321174
Sander Tabacco 266054
Sander's Billbergia 54461
Sander's Dracaena 137487
Sandera Coelogyne 98740
Sanders 320327,346210
Sandersonia 345782
Sanderswood 320327
Sandhill Blackbuut 106805
Sandhill Jointweed 308676
Sandhill Magnolia 241979
Sand-hill Muhly 259688
Sandhill Rosemary 83398
Sandhill Sage 36127
Sandhill Spider Flower 180650
Sand-hill Thistle 92347
Sandhill Tylophora 400852
Sandhill Wild Buckwheat 152527
Sandhill Wormwood 36127
Sand-hills Amaranth 18661
Sandhills Gayfeather 228451
Sandhills Pigweed 18661
Sandholly 19776,19777
Sandlace 308680
Sandland Sedge 76150
Sand-lily 227818
Sandlive Willow 343914
Sandliving Crazyweed 278858
Sandliving Fimbristylis 166446
Sandliving Fluttergrass 166446
Sand-living Milkvetch 41975
Sandliving Sedge 75857
Sandliving Sophora 369078
Sandloving Ceratocarpus 83419
Sand-loving Milkvetch 41976
Sand-loving Willow 343914
Sandoricum 345785
Sandpaper Bush 141602
Sandpaper Fig 164853
Sandpaper Leaf Fig 164967
Sandpaper Tree 80004,104225,164650
Sandpaper Verbena 405885
Sandpaper Vine 292518
Sandpear Fishwood 110224
Sandpot Greenstar 307517
Sand-puffs 398234
Sandrach 386942
Sand-reed Grass 65633
Sandspike 148466
Sand-spurrey 370696
Sandspurry 370609
Sand-spurry 370609
Sandt Milkvetch 41975
Sandthorn 196743,196757
Sandtorch 228496
Sandu Machilus 240611
Sand-verbena 687
Sandweed 370522
Sandwhip 317000

Sandwort 31727,32095,32120,32212,255435,256441
Sandy Bristlegrass 361700
Sandy Bulbostyl 63216
Sandy Chamaerhodos 85657
Sandy Chickweed 374755
Sandy Craneknee 221655
Sandy Dragonhead 137637
Sandy Dropseedgrass 372591
Sandy Ephedra 146209
Sandy Feathergrass 376785
Sandy Fescue 164222
Sandy Fescuegrass 164222
Sandy Gentian 174164
Sandy Giantfennel 163604
Sandy Greenorchid 137637
Sandy Heteropappus 193925
Sandy Iris 208770
Sandy Jurinea 193925
Sandy Knotwood 44251
Sandy Madder 337949
Sandy Milkvetch 42262
Sandy Minorrose 85657
Sandy Needlegrass 376785
Sandy Pagodatree 369078
Sandy Pear 323268
Sandy Peashrub 72216
Sandy Plumegrass 148906
Sandy Purpleflower Feathergrass 376897
Sandy Purslane 311920
Sandy Rockaster 218245
Sandy Sedge 76098
Sandy Skeletonweed 88761
Sandy Wheatgrass 11647
Sandy Windhairdaisy 348154
Sandybeach Skullcap 355778
Sandy-woods Chaffhead 77226
Sanfendan Childvine 400856
Sang Rose Apple 382660
Sanglamu Woodbetony 287463
Sanguinarea 18101
Sanguinaria 345803,345805
Sanguinary 3978,72038
Sanguine Begonia 50273
Sanguine Cotoneaster 107676
Sanguine Japanese Maple 3346
Sanguine Purple Coneflower 140087
Sanguine Schizachyrium 351307
Sanguineous Scale Pycreus 322342
Sanguineous-scale Galingale 322342
Sanguineous-scale Pycreus 322342
Sanguinespikelet Spikesedge 143302
Sangzhi Lysionotus 239980
Sangzhi Oblongleaf Linden 391791
Sanhe Vetch 408276
Sanicle 13174,345940,345957,345980
Sanideleaf Anemone 23714
Saniker 345957
Sanke's Beard 272049
Sanming Bitter Bamboo 304096
Sanming Bitterbamboo 304096
Sanming Bitter-bamboo 304096
Sann Hemp 112269
Sanqi 280771
Sansevieria 346047
Sansho 417132
Sanson 59575
Sant Pods 1427
Santa Barbara Ceanothus 79924,79943
Santa Barbara Daisy 150724
Santa Barbara Island's St. Catherine's-lace 152089
Santa Barbara Poppy 199407,199408
Santa Catalina Island's St. Catherine's-lace 152086
Santa Catalina Prairie Clover 121901
Santa Clara County Lessingia 227124
Santa Clara Sulphur Flower 152570
Santa Cruz Beehive 107059
Santa Cruz Beehive Cactus 107059
Santa Cruz Cypress 114698
Santa Cruz Desertdandelion 242963
Santa Cruz Island Buckwheat 151834
Santa Cruz Island Desertdandelion 242946
Santa Cruz Island Ironwood 239412
Santa Cruz Island Pine 300098
Santa Cruz Water Lily 408735
Santa Cruz Water-lily 408735
Santa Cruz Water-platter 408735
Santa Lucia Fir 507
Santa Maria 67851,67855,285086
Santa Maria Feverfew 285086
Santa Marta 134232
Santa Rita Mountain Aster 380958
Santa Rita Prickly Pear 273042
Santa Rita Snakeroot 11169
Santa Rosa Plum 316761
Santa Ynez Groundstar 22029
Santa Ynez Wild Buckwheat 151928
Santalucia Fir 507
Santiria 346225
Santol 345785,345786,345788
Santolina 346243,346250
Santolina Pincushion 84522
Santonica 35333,35868
Santwood 1427
Sanvitalia 346300
Sanwa Millet 140367,140423,140503
Sanza Minik 132382
Sanzao Bambusa 47434
Sao 198145
Sap Gum 155542,232565
Sap Pine 300181
Sap Tree 369339
Sapdilla Family 346460
Sapele 145948,145959
Sapello Canyon Larkspur 124562
Sapgreen 328642
Sapgreen Rhododendron 330237
Sapgum 232565
Sapieangy 382737
Sapindus-leaf Millettia 254830
Sapium 346361
Sapodilla 67524,244605
Sapodilla Chicle Tree 244605
Sapodilla Family 346460
Sapodilla Plum 244605
Saposhinikov Sagebrush 36221
Saposhnicov Crazyweed 279143
Saposhnikov Willow 344065
Saposhnikovia 346447
Sapote 67524,238115,244605,313399
Sapote Borracho 238117
Sapote Family 346460
Sapoton 279387
Sappan 65060
Sappan Caesalpinia 65060
Sappan Wood 65060
Sapphire Berry 381341
Sapphire Dragon Tree 285966
Sapphireberry 381341
Sapphire-berry 381341
Sapphireberry Sweetleaf 381341
Sappire Flower 61131
Sapree Wood 61339
Sapria 346469
Saprophylic Cymbidium 116975
Sapu 252841
Sapucaia Nut 223764,223778
Sapucaia Nut Tree 223768
Saraca 346497,346504
Saracen Birthwort 34149,34150
Saracen Corn 162312
Saracen's Birthwort 34149
Saracen's Consound 358902
Saracen's Woundwort 358902
Sarah's Wild Buckwheat 152199
Sarana Lily 230005,230009
Sarawak Bean 408917
Sarawak Thin-walled Bamboo 216392
Sarbati 93757
Sarcandra 346525
Sarcocaulon 346642
Sarcochilus 346686
Sarcochlamys 346720,346721
Sarcococca 346723,346733
Sarcocolla Tree 344786
Sarcoculla 41885
Sarcodum 96633
Sarcoleaf Sagebrush 36350
Sardian Nut 78811

Sardinia Glory-of-the-blue 87770
Sarepta Mustard 59438
Sareptan Euphorbia 159770
Sargent Arrowwood 408009
Sargent Ash 168097
Sargent Barberry 52131
Sargent Buckthorn 328848
Sargent Cherry 83301
Sargent Cottonwood 311465
Sargent Crabapple 243690
Sargent Crandberrybush 408009
Sargent Cypress 114751
Sargent Grand Torreya 393065
Sargent Hydrangea 200076
Sargent Juniper 213696
Sargent Lily 230029
Sargent Magnolia 416717
Sargent Mountainash 369515
Sargent Mountain-ash 369515
Sargent Savin 213696
Sargent Spiraea 372078
Sargent Spruce 298258
Sargent Viburnum 408009,408098,408101
Sargent Vineclethra 95559
Sargent's Barberry 52131
Sargent's Catchfly 364011
Sargent's Cherry Palm 347076
Sargent's Chinese Juniper 213696
Sargent's Crabapple 243690
Sargent's Glory Vine 347077,347078
Sargent's Juniper 213696
Sargent's Rhododendron 331749
Sargent's Rowan 369515
Sargent's Spruce 298258
Sargentglorivine 347077,347078
Sargent-glory-vine 347077,347078
Sargent-glory-vine Family 347083
Sargentodoxa Family 347083
Saribu Palm 234190
Sarkand Crazyweed 279144
Sarmentose Azima 45974
Sarmentose Elaeagnus 142188
Sarmentose Fig 165619
Sarmentose Neoregelia 264425
Sarmentose Pegia 287989
Sarmentose Pepper 300504
Sarmienta 347103
Sarncen's Comfrey 358902
Saro 398826
Sarock 339887
Sarococca 346727
Sarracenia 347144
Sarrat 173082
Sarsaparilla 97904,366230,366341,366494,
366542
Sarsaparilla Vine 366536
Sartwell's Sedge 76147
Saruma 347185
Saruwatar Thrixsperm 390527
Saruwatar Thrixspermum 390527
Sarvis 19243
Sarviss Berry 19243
Sarviss Tree 19243
Sasa 347191
Sasa Lily 229872
Sasaki Actinodaphne 233840
Sasaki Fagraea 162346
Sasaki Sweetleaf 381423
Sasaki's Litse 233840
Sasamorpha 347377
Sasanqua 69594
Sasanqua Camellia 69594
Sasanqua Oil Camellia 69594
Sasanqua Tea 69594
Saskatoon 19241
Saskatoon Juneberry 19241
Saskatoon Serviceberry 19241
Saskatoon-berry 19241
Sassafras 136766,136768,347402,347407,
347413
Sassafras Tree 347407
Sassifrax 349419
Sasswood 154973
Sassy Bark 154983
Sassy Jack 159222
Sassybark 154973
Satan's Apple 244346
Satan's Bread 28044
Satan's Cherry 44708
Sates 109857
Satin 238371
Satin Bauhinia 49135
Satin Brome 60867
Satin Flower 94133,238369,238371,
365678,374916,374968
Satin Grass 259660
Satin Leaf 90082
Satin Oak 16251
Satin Philodendron 294810
Satin Poppy 247157
Satin Tail 205470
Satin Walnut 232565
Satin Wood 88663
Satin-balls 67456
Satine 61062
Satinflower 238371,365678
Satinpod 238371
Satintail 205470
Satin-tail 205470
Satinwood 80080,88663,386563
Satiny Willow 343855
Sat-kara 93430
Sato Milkvetch 43008
Satsuma 93717
Satsuma Mandarin 93736
Satsuma Mock Orange 294530
Satsuma Mock-orange 294530
Satsuma Orange 93717
Satsuma-kikoku 93647
Satsuman Mockorange 294530
Sattmarsh Club-flower 104327
Saturate Rose 336918
Saturday Night's Pepper 159027
Saturday's Pepper 159027
Satyrion Royal 273531
Satyrium 347690
Sau 162473
Sauce Sour 339887
Sauce-alone 14953,14964,192358
Saucer Magnolia 416682
Saucer Peach 20952
Saucer Plant 9072,9074,9075
Saucer-plant 9074
Saucershaped Flowers Rhododendron 330018
Saucerwood 253409,253413
Sauch Saugh 342973,343151
Sauchweed 309570
Saucy Alice 309570
Saucy Bet 81768
Saucy Jack 81198
Sauerkraut 59520
Sauf 343151
Saugh-tree 342973,343151
Saul Tree 362220
Saul-tree 362220
Saunders 320327
Saunders Cinquefoil 312956
Saunderswood 320327
Saundra Gladiolus 176524
Saur Crazyweed 279145
Saurauia 347952,347994
Saurauia Family 348003
Sauromatum 348022
Sauropus 348038
Sausage Tree 216313,216324
Sausage Vine 197206,197223,374381
Sausagetree 216313,216324
Sausage-tree 216313,216324
Saussurea 348089
Saussurea-like Blumea 55819
Saussurea-like Groundsel 359974
Saute 343151
Savanilla Rhatany 218079
Savanna Aster 380848
Savanna Flower 141051
Savanna Gayfeather 228516
Savanna Iris 208616
Savanna Sneezeweed 188441
Savannah Iris 208906
Savannah Meadow-beauty 329420
Savatier Monochasma 257542

Savatier Violet 410531
Save 345271
Savennah Palm 341416
Saver-margined Holly 203549
Save-whallop 336419
Savin 35868,213902,341696
Savin Juniper 213785,213902
Savin Savin 213902
Savine Juniper 213902
Saving-tree 213902
Savinier 213785
Saviour's Blanket 373166
Savonette 235570,347506
Savory 65528,347446,347560
Savoryfruit 78172,78175
Savory-leaved Aster 40767
Savoury Rhododendron 330092
Savoury-nut 53057
Savoy 59530
Savoy Cabbage 59530,59532
Savoy Hawkweed 195931
Savoyana 103850
Saw Banksia 47667
Saw Cabbage Palm 4951
Saw Grass 93901,245332
Saw Greenbrier 366264
Saw Palm 360645
Saw Palmetto 360645,360646
Saw Palmetto Palm 360646
Sawai Grass 156513
Sawara Chamaecyparis 85345
Sawara Cypress 85345
Sawara False Cypress 85345
Sawara Falsecypress 85345
Sawara Nut 77946
Sawari Nut 77946
Sawblade 139256
Sawbrier 366341
Sawfly-orchid 272494
Sawg 343151
Sawgrass 93901,245332,245556,266504
Saw-grass 93901,93913
Sawleaf banksia 47667
Sawleaf Groundsel 381944
Saw-leaf Mugwort 36277
Sawleaf Zelkova 417534,417558
Saw-leaved Speedwell 407198
Sawpetal Waterhyacinth 141807
Sawtooth Blackberry 338933
Saw-tooth Bristleweed 186887
Saw-tooth Candyleaf 376417
Sawtooth Elatostema 142642
Sawtooth Euphorbia 158534
Sawtooth Gentian 173889
Sawtooth Oak 323611
Sawtooth Stairweed 142642
Sawtooth Sunflower 188974
Saw-tooth Sunflower 188974
Sawtooth Wirelettuce 375981
Saw-tooth Wormwood 36277
Saw-toothed Sagebrush 36277
Sawuer Thermopsis 389543
Sawur Larkspur 124575
Sawwort 360982
Saw-wort 348089,360982,361139
Sawwortlike Thistle 92380
Saxatile Blueberry 403983
Saxatile Buckthorn 328849
Saxatile Cnesmone 97505
Saxatile Currant 334203
Saxatile Holly 204236
Saxatile Medick 247467
Saxatile Milkwort 308337
Saxatile Pittosporum 301394
Saxatile Raspberry 339240
Saxatile Rhododendron 331755
Saxatile Syzygium 382662
Saxaul 185062
Saxicolous Agapetes 10356
Saxicolous Anisadenia 25274
Saxicolous Azalea 332088
Saxicolous Beancaper 418753
Saxicolous Blueberry 403983
Saxicolous Casearia 78153
Saxicolous Catchfly 364027
Saxicolous Chickweed 375051
Saxicolous Childseng 318513
Saxicolous Cissampelopsis 92521
Saxicolous Corydalis 106270
Saxicolous Drepanostachyum 20207
Saxicolous Eurya 160597
Saxicolous Evodia 161316
Saxicolous Gerbera 175215
Saxicolous Hornbeam 77383
Saxicolous Milkvetch 43011
Saxicolous Ophiorrhiza 272278
Saxicolous Ormosia 274432
Saxicolous Pholidota 295536
Saxicolous Pleione 304301
Saxicolous Primrose 314930
Saxicolous Rhododendron 331755
Saxicolous Seatung 301394
Saxicolous Stairweed 142733
Saxifraga 349032
Saxifrage 52503,288840,349032,349989
Saxifrage Catchfly 364013
Saxifrage Family 350063
Saxifrage Pimpinella 299523
Saxifrage Pink 292668
Saxifrage Silene 364013
Saxifrage Tunicflower 292668
Saxitil Monocladus 56955
Saxitile Ormosia 274432
Saxoul 185064
Say Rhatany 218080
Sayanuka Cutgrass 224002
Scab Flower 28044
Scab Gowks 273531
Scab Rockfoil 349674
Scabbit Dock 130383
Scabbit-dock 130383
Scabby Begonia 50004
Scabby Hands 28044,101852,102727,
192358
Scabby Heads 28044,392992
Scaber Altai Heteropappus 193922
Scabgowk 273531
Scabiosa 350099
Scabious 103446,350086
Scabious Jasione 211670
Scabiousflower Figwort 355235
Scabland Sagebrush 36146
Scabland Wild Buckwheat 152487
Scabrate Corydalis 106407
Scabria Sinobambusa 364812
Scabrid Begonia 50277
Scabrid-flowered Fargesia 162745
Scabridge 14964
Scabridous Spapweed 205303
Scabridulous Windflower 24054
Scabriflower Fescuegrass 164307
Scabrifolious Meadowrue 388657
Scabril 14964
Scabrin 190502
Scabriosa Centaurea 81356
Scabripilose Creeping Pink 127810
Scabrous Agrostis 12285
Scabrous Alfredia 14595
Scabrous Anisochilus 25392
Scabrous Anthosachne Roegneria 144169
Scabrous Aphananthe 29005
Scabrous Arisaema 33266
Scabrous Betony 373427
Scabrous Burvine 64471
Scabrous Byttneria 64466
Scabrous Centaurea 80907
Scabrous Cleistogenes 94632
Scabrous Cowparsnip 192348
Scabrous Dendrocalamus 125465
Scabrous Deutzia 126843,127072,127075
Scabrous Doellingeria 135264
Scabrous Dolomiaea 135738
Scabrous Elephantfoot 143464
Scabrous Falsenettle 56212
Scabrous Fescue-grass 163802
Scabrous Globethistle 140786
Scabrous Goosecomb 335489
Scabrous Hedyotis 187694
Scabrous Hideseedgrass 94632
Scabrous Hydrangea 199817
Scabrous Indigo 206510

Scabrous Longflower Lily 229905
Scabrous Longnode Eargrass 187694
Scabrous Mosla 259323
Scabrous Patrinia 285855
Scabrous Petrosimonia 292727
Scabrous Rattan Palm 65776
Scabrous Rhodiola 329853
Scabrous Roegneria 335489
Scabrous Roughfruitparsley 393951
Scabrous Sinobambusa 364812
Scabrous Small Reed 127292
Scabrous Smallreed 127292
Scabrous Southstar 33266
Scabrous Spapweed 205304
Scabrous Spiderwort 394082
Scabrous Stem St. John's Wort 202147
Scabrous Threevein Aster 39989
Scabrous Touch-me-not 205304
Scabrous Wendlandia 413817
Scabrousbeak Sedge 76167
Scabrouseed Snakegourd 396260
Scabrousflower Cane 270446
Scabrousflower Canebrake 270446
Scabrous-flowered Oligostachyum 270446
Scabrous-hair Milkvetch 43013
Scabroushair Primrose 314172
Scabrous-haired Leptopus 226321
Scabrousleaf Anemone 24054
Scabrousleaf Aralia 30751
Scabrousleaf Borreria 57296
Scabrousleaf Buddhamallow 402250
Scabrousleaf Cadillo 402250
Scabrousleaf Elatostema 142823
Scabrousleaf Meadowrue 388657
Scabrous-leaf Milkvetch 43012
Scabrousleaf Pavetta 286452
Scabrousleaf Pricklyash 417200
Scabrousleaf Rhododendron 331756
Scabrousleaf Rose Mallow 402250
Scabrousleaf Sedge 76163
Scabrous-leaved Aralia 30751
Scabrous-leaved Prickly-ash 417200
Scabrous-leaved Rhododendron 331756
Scabrousscale Spikesedge 143298
Scabrousscape Leek 15065
Scabrousscape Onion 15065
Scabrous-stem Greenbrier 366566
Scabrousstem Smilax 366566
Scabrous-utricle Sedge 74600
Scabwort 207135
Scad 316470
Scaddie 402886
Scad-tree 316470
Scaevola 350323,350344
Scag 282717
Scalar Mountainash 369516
Scalar Mountain-ash 369516
Scalariform-nerved Ardisia 31594
Scalar-nerved Blueberry 404018
Scalarvein Ardisia 31594
Scald 114947,115008
Scaldberry 338447
Scalded Apple 248312
Scald-head 118008,338447
Scaldrick 364557
Scaldweed 115008
Scald-weed 115040
Scale Bird Cherry 280042
Scale Birdcherry 280042
Scale Debregeasia 123337
Scale Waternettle 123337
Scalebract Milkaster 169780
Scale-bract Wild Buckwheat 152259
Scalebud 25400
Scalecalyx Crazyweed 279175
Scaled Alphonsea 17631
Scaled Falsetamarisk 261293
Scale-flower Puberulent Azalea 331541
Scale-flower Puberulent Rhododendron 331541
Scale-fruit Euonymus 157534
Scalefruit Sanicle 345973
Scalefruit Tanoak 233287
Scalefruit Tickseed 104798
Scalehair Blastus 55181
Scaleleaf Aster 380834
Scaleleaf Corydalis 105670
Scaleleaf Cryptomeria 113693
Scaleleaf Pyrola 322915
Scaleleaf Tanoak 233280
Scaleless Catchfly 363443
Scaleless Luohanfruit 365326
Scaleless Rhododendron 330576
Scaleless Siraitia 365326
Scaleless Tanoak 233292
Scales Honeysuckle 236033
Scale-seed 370893,370895
Scaleseed Sedge 225567
Scaleseedsedge 225567
Scaletail 225644
Scaletepal Kobresia 217209
Scalies 364557
Scaligeria 350360,350377
Scallion 15170,15621
Scallions 15876
Scallop Gourd 114301
Scalloped Leaf Phacelia 293161
Scalloped Phacelia 293169
Scalloped Salpiglossis 344383
Scalloped Summer Squash 114300
Scallops 184599
Scalopendriform Sedge 76206
Scaly Apple 25898
Scaly Azalea 331089,331483
Scaly Blastus 55181
Scaly Blazing Star 228534
Scaly Buds Rhododendron 331240
Scaly Fig 165685
Scaly Grindelia 181174
Scaly Hawkbit 187709,187718
Scaly Kyllingia 218628
Scaly Liatris 228534
Scaly Phebalium 294107
Scaly Rhododendron 331483
Scaly Sedge 75668
Scaly Styles Rhododendron 331088
Scaly Windhairdaisy 348619
Scaly Zamia 225624
Scaly-bark Beefwood 79165
Scalybark Hickory 77920
Scalybark Spruce 298421
Scaly-barked Spruce 298421
Scalycalyx Crazyweed 279175
Scaly-fruit Tanoak 233287
Scalyfruit Tickseed 104798
Scaly-fruited Tanoak 233287
Scalyhair Deutzia 127104
Scalyhair Linden 391755
Scaly-haired Lime 391755
Scaly-haired Linden 391755
Scaly-leaved Barberry 51848
Scaly-leaved Nepal Juniper 213943
Scaly-leaved Nepal Savin 213943
Scalyseed Ephedra 146244
Scalyseeded Ephedra 146244
Scaly-styled Rhododendron 331088
Scambling Nightshade 367661
Scamman's Claytonia 94360
Scammony 103255,208037
Scammony Bindweed 103255
Scammony Glorybind 103255
Scandent Acanthopanax 143654
Scandent Asparagus 39185
Scandent Bauhinia 49226
Scandent Bredia 59855
Scandent Fieldorange 249674
Scandent Hookvine 401778
Scandent Melodinus 249674
Scandent Premna 313735
Scandent Rosewood 121703
Scandent Schefflera 350654
Scandent Skullcap 355727
Scandent Spiny-fruited Spindle-tree 157276
Scandinavian Small-reed 65471
Scant Ash 167951
Scant Randia 162194
Scape Chorispora 89006
Scape Groundsel 359991
Scape Javatea 275768
Scape Monkshood 5549
Scape Swertia 380348

Scapebearing Mischobulbum 255941
Scapeflower Loosestrife 239841
Scapeflower Spapweed 205305
Scapeflower Touch-me-not 205305
Scapelike Ligusticum 229383
Scape-like Sage 345365
Scaphoidhelmet Monkshood 5452
Scapose Carpesium 77186
Scapose Hawksbeard 110990
Scapose Larkspur 124583
Scapose Marshmarigold 68217
Scapose Primrose 314932
Scapose Sedge 76169
Scaraba Cypress 85272
Scarborough Anemone 24116
Scarborough Lily 120550,404575,404581
Scarborough-lily 404575,404581
Scarce Fiddleneck 20837
Scarce Londonpride 349473
Scarce Marsh Helleborine 147121
Scarce Serapias 360623
Scarce Water-figwort 355261
Scaredy Cat Plant 99538
Scargrass 203312,203317
Scaribeus 355202
Scariola 350504
Scariose Lettuce 219497
Scarious Iris 208813
Scarious Lettuce 219497
Scarious Swordflag 208813
Scarious-sepaled Camellia 69610
Scarlet Avens 175396,175400,175442
Scarlet Azalea 330741
Scarlet Ball Cactus 59164
Scarlet Ball-cactus 59164
Scarlet Banana 260217
Scarlet Banksia 47636
Scarlet Barrenwort 146971
Scarlet Basketvine 9475
Scarlet Bauhinia 49053
Scarlet Bee Balm 257161
Scarlet Beebalm 257161
Scarlet Bee-blossom 172191
Scarlet Begonia 49729
Scarlet Bennet 175400
Scarlet Blood Lily 184372
Scarlet Bloodlily 184372
Scarlet Bottlebrush 67262,67292
Scarlet Bottle-brush 67262
Scarlet Bourvardia 57957
Scarlet Bouvardia 57954,57958
Scarlet Buckeye 9720
Scarlet Bugler 94556,289324,289331
Scarlet Bush 185169
Scarlet Canada Maple 3505
Scarlet Canadian Maple 3505
Scarlet Catchfly 364175
Scarlet Clematis 94845,95371
Scarlet Clockvine 390742
Scarlet Clover 396934
Scarlet Cob Cactus 140857
Scarlet Columnea 100115
Scarlet Cordia 104245
Scarlet Cordia-tree 104245
Scarlet Cross 363312
Scarlet Crowfoot 325614
Scarlet Crowncactus 327274
Scarlet Curls Willow 343669
Scarlet Cynomorium 118133
Scarlet Echeveria 139982
Scarlet Elder 345660
Scarlet Elderberry 345660,345677
Scarlet Epidendrum 146397
Scarlet Epimedium 146971
Scarlet Eucalyptus 106800
Scarlet Fire Thorn 322454
Scarlet Firethorn 322454
Scarlet Flame Bean 61158
Scarlet Flame 269100
Scarlet Flax 231907
Scarlet Flowering Gum 106800
Scarlet Four-o'clock 255691
Scarlet Fritillary 168383,168548,168549
Scarlet Fuchsia Bush 180271
Scarlet Gaura 172191
Scarlet Geranium 288300,288310
Scarlet Gilia 71547,175671,175673,175704,175708,208318
Scarlet Gingerlily 187428
Scarlet Ginger-lily 187428
Scarlet Globe Mallow 370938,370939,370942
Scarlet Globemallow 370942
Scarlet Glochidion 177115
Scarlet Glorybower 96257
Scarlet Gum 106800
Scarlet Haw 109609,109818
Scarlet Hawkweed 195478
Scarlet Hawthorn 109609,109688,109913,109918
Scarlet Hedge Nettle 373179
Scarlet Hedgehog Cactus 140231
Scarlet Hedgenettle 373179
Scarlet Helixanthera 190832
Scarlet Hibiscus 194804
Scarlet Holm Oak 324027
Scarlet Honey Myrtle 248095
Scarlet Honey-myrtle 248095
Scarlet Honeysuckle 236089
Scarlet Ipomoea 323459
Scarlet Ixora 211068
Scarlet Jacobinia 279768
Scarlet Jungleflame 211068
Scarlet Kadsura 214959,214971
Scarlet Kaffir Lily 97218
Scarlet Kaffirlily 97218
Scarlet Kaffir-lily 97218
Scarlet Kleinia 216668
Scarlet Knotweed 308986
Scarlet Kunzea 218389
Scarlet Larkspur 124103,124410
Scarlet Lightning 81768,363312
Scarlet Lychnis 363312
Scarlet Maple 3505
Scarlet Martagon 229806
Scarlet Masdevallia 246105
Scarlet Maxillaria 246708
Scarlet Mint Bush 315596
Scarlet Morning-glory 323459
Scarlet Muskflower 267423
Scarlet Musk-flower 267423
Scarlet Oak 285233,323780,324027
Scarlet Painted-cup 79120
Scarlet Passionflower 285632
Scarlet Pear Gum 155751
Scarlet Pepper 72070
Scarlet Pimpernel 21339
Scarlet Plume 158927,159675
Scarlet Pompona Lily 229995
Scarlet Premna 313720
Scarlet Renanthera 327685
Scarlet Root-blosson 10191
Scarlet Rose Mallow 194804
Scarlet Rose-mallow 194804
Scarlet Runner 293985,409085
Scarlet Runner Bean 293985,409085
Scarlet Sage 344980,345405,373179
Scarlet Sophronitis 369180
Scarlet Spiderling 56413
Scarlet Sterculia 376130
Scarlet Strawberry 167663
Scarlet Sumac 332608
Scarlet Sumach 332608
Scarlet Swertia 380324
Scarlet Tofieldia 392603
Scarlet Trompetilla 57958
Scarlet Trumpent Honeysuckle 235684
Scarlet Trumpet 175673
Scarlet Trumpet Vine 134877
Scarlet Turk's Cap Lily 229806
Scarlet Turk's-cap Lily 229806
Scarlet Turkscap Lily 229806
Scarlet Turn-again-gentleman 229806
Scarlet Violet 147393
Scarlet Wild Buckwheat 152283
Scarlet Willow 342975
Scarlet Windflower 23823
Scarlet Wistaria Tree 361422
Scarlet Wisteria Tree 361384,361449
Scarlet Wrightia 414801
Scarletball Rhododendron 331894

Scarlet-balled Rhododendron 331894
Scarletbead Cotoneaster 107718
Scarlet-beaded Cotoneaster 107718
Scarlet-berried Elder 345660
Scarlet-bugler 94556
Scarletflower Pyrethrum 322665
Scarlet-flowered Camellia 69532
Scarlet-flowered Gum 106800
Scarlet-flowered Rhododendron 332054
Scarlet-flowering Gum 106800
Scarlet-flowering Rose Geranium 288459
Scarlet-flowers Rhododendron 332054
Scarletfruit Passionflower 285643
Scarlet-fruited Horse-gentian 397830
Scarlet-seeded Iris 208566
Scarred Machilus 240573
Scatter Skullcap 355564
Scatterbranch Russianthistle 344474
Scattered Groundsel 358847
Scattered Paspalum 285418
Scattered Primrose 314266
Scattered Yam 131474
Scatteredflower Ladybell 7798
Scattered-flower Tansy 383842
Scatteredhair Scorpiothyrsus 354747
Scatteredleaf Martagon Lily 229926
Scatterflor Spiradiclis 371762
Scatterflower Ash 167951
Scatterflower Balanophora 46859
Scatterflower Dinggongvine 154253
Scatterflower Eritrichium 153474
Scatterflower Forget-me-not 260880
Scatterflower Gynostemma 183011
Scatterflower Olive 270140
Scatterflower Snakemushroom 46859
Scatterflower Wendlandia 413792
Scatterflower Woodbetony 287347
Scatterfruit Barringtonia 48515
Scatterfruit Stoneclave 263959
Scatterhair Azalea 330576
Scatterhair Cherry 83272
Scatterhair Ribseedcelery 304830
Scatterleaf Pfeiffer Root 276458
Scatterlflower Primrose 314332
Scatternode Pagodatree 369101
Scatterspike Woodbetony 287349
Scatterspine Eritrichium 153496
Scattertongue Goldenray 229125
Scattertooth Purplepearl 66909
Scattertooth Skullcap 355632
Scattredtooth Gugertree 350940
Scatybark Hickory 77934
Scaw 345631
Scaybril Scabs 14964
Sceau-de-Salomon 308493
Scendess Mayweed 397964
Scent Bottle 302103
Scent-bottle 183571,230348
Scented Bells 337515
Scented Boronia 57267
Scented Bouvardia 57955
Scented Bramble 338917
Scented Broom 77058
Scented Buttercup 312360
Scented Cranesbill 174715
Scented Daisy 383874
Scented Fern 383874
Scented Goosefoot 87095
Scented Grass 90107
Scented Lippia 232492
Scented Mignonette 327896
Scented Oak 288459
Scented Or Aromatic Pussytoes 26339
Scented Orchid 182230
Scented Paperbark 248118
Scented Paper-bark 248118
Scented Poplar 311237
Scented Rosewood 121782
Scented Solomon's Seal 308616
Scented Stanhopea 373692
Scented Vanda 404641
Scented-grass 90107
Scentless Camomile 397970
Scentless Chamomile 397964,397970
Scentless False Chamomile 246384,397964
Scentless False Mayweed 397976
Scentless Feverfew 397970
Scentless Geranium 288308
Scentless Honeysuckle 235871
Scentless Matricary 397970
Scentless Maydweed 397970
Scentless Mayweed 397964
Scentless Mock Orange 294479
Scentless Mock-orange 294479
Scentless Rosewood 381965
Scentless Ruse 336650
Scepter Banksia 47666
Scepter Begonia 50281
Scepter-bearing Fleabane 150950
Sceptre Woodbetony 287657
Schaffner's Bahia 46541
Schaffner's Spikerush 143340
Schafta Campion 364023
Schangin Corydalis 106414
Scharff Begonia 50283
Scharnhorst Ajania 13032
Scheelea 350635
Scheffer Garcinia 171187
Schefflera 350646,350648,350706
Scheidecker Apple 243693
Schelle Helictotrichon 190191
Scheneider Cherry 83306
Schepherospurse 72038
Scheuchzer's Bellflower 70294
Scheuchzeria 350859,350861
Scheuchzeria Family 350864
Schidiger Agave 10936
Schiede Columnea 100126
Schiede Mahonia 242634
Schilinky Frailea 167704
Schiller Phalaenopsis 293633
Schima 350945
Schimper Embelia 144808
Schimper Harpachne 185833
Schinz's Pepperweed 225451
Schisandra Family 351121
Schischkin Medick 247468
Schischkin Minuartia 255575
Schischkinia 351124
Schismatoglottis 351146
Schismus 351162
Schizachyrium 351261
Schizocalyx Strawberry 167613
Schizo-calyxed Rhododendron 331765
Schizocapsa 351380
Schizoleaf Woollyfruit Cinquefoil 312525
Schizomussaenda 351740
Schizostachyum 351844
Schlaffes Blumenrohr 71171
Schlechter Argyroderma 32753
Schleicher Fumitory 169168
Schlim Alloplectus 16182
Schmid Helictotrichon 190192
Schmidt Aralia 30753
Schmidt Begonia 50287
Schmidt Birch 53610
Schmidt Giant Orchid 180124
Schmidt Goldenray 229175
Schmidt Sedge 76178
Schmidt Spreading Cyrtococcum 120634
Schmidt's Birch 53610
Schnabelia 352059,352060
Schneick Sugar Maple 3576
Schneider Barberry 52355
Schneider Bentgrass 12292
Schneider Buckthorn 328851
Schneider Cathay Poplar 311270
Schneider Condorvine 245844
Schneider Copperleaf 1975
Schneider Densespike Woodbetony 287157
Schneider Deutzia 127088
Schneider Honeysuckle 236086
Schneider Leptodermis 226130
Schneider Maple 3581
Schneider Milkgreens 245844
Schneider Oatgrass 122286
Schneider Photinia 295771
Schneider Poranopsis 311655
Schneider Sedge 76179
Schneider Spiraea 372079
Schneider Wildclove 226130

Schneider Zelkova 417552
Schneidler Falsehellebore 405633
Schneilder Erysimum 154535
Schneilder Sugarmustard 154535
Schoberia-like Asparagus 39187
Schoch Craspedolobium 108700
Schoch Crotonvine 108700
Schoch Mahonia 242607
Schoch Spiraea 372081
Schoenland Kungia 218358
Schoenland Orostachys 218358
Schoenoplecte 352158
Schoenorchis 352298
Schoenus-flower Lemongrass 117245
Schoenus-like Bluegrass 305957
Schoepfia 352430
Scholar Tree 369037
Scholars' Tree 369037
Scholar-tree 369037
Schomburgk Globba 176996
School Bells 70271
School of Christ 202086
Schoolboy's Clock 384714
Schoolcraft's Wild Buckwheat 152284
Schott Yucca 416647
Schott's Agave 10937
Schott's Century Plant 10937
Schott's Yucca 416647
Schottky Cyclobalanopsis 116109
Schottky Oak 116109
Schrader Hawthorn 110032
Schreber's Aster 41217,160679
Schrenk Crazyweed 279151
Schrenk Goosecomb 335491
Schrenk Lily 228596
Schrenk Mock Orange 294536
Schrenk Mockorange 294536
Schrenk Mock-orange 294536
Schrenk Roegneria 335491
Schrenk Spruce 298426
Schrenk Tulip 400222
Schrenk's Spruce 298426
Schubert Onion 15724
Schuette's Hawthorn 110033
Schultes Coelogyne 98741
Schultes' Bedstraw 170616
Schultzia 352688
Schulz Begonia 50289
Schuman Combretum 100770
Schuman Windmill 100770
Schumann's Abelia 200
Schumannia 352702
Schuster Corydalis 106421
Schwarz Widdringtonia 414065
Schweinfurth Cherry Orange 93280
Schweinitz's Cyperus 119549
Schweinitz's Flat Sedge 119549
Schweinitz's Flatsedge 119549
Schweinitz's Ragwort 279948
Schweinitz's Sedge 76182
Schweinitz's Sunflower 189054
Schwetzov Groundsel 360007
Sciaphila 352863
Scilla 352880
Scilla Campanulata 199553
Scilly Buttercup 326108
Scilly Pygmyweed 108957
Scimitar Shrub 58678
Scimitar-podded Kidney Bean 294010, 297698
Scimitar-shaped Dimeria 131007
Sciney 193417
Scirpes 352158
Scirpoid Sedge 76185
Scirpus 353179
Scirpuslike Rrush 213438
Scissor Plant 208524
Scissorsvine 197475
Sclerachne 353969
Sclero-fruited Glyptic-petal Bush 178004
Sclero-fruited Glyptopetalum 178004
Scleronut Persian Peach 20958
Sclerophyllous Willow 344072
Sclerophyllous-like Willow 344075
Scleropyrum 354500
Scobe 121001
Scoke 298094
Scolochloa 354579
Scolopendrifolious Cleisostoma 288605
Scolopia 354596
Scootberry 377902
Scopol Figwort 355236
Scopol Serpentroot 354948
Scopolia 25442,354678
Scopulous Bladderwort 403988
Scorb 369290
Scorched Alpine-sedge 73814
Scordion 388272,388275
Scorpioid Malcolmia 243230
Scorpion Broom 173056
Scorpion Grass 260737,260905
Scorpion Iris 208437
Scorpion Orchid 30526,30532
Scorpion Senna 105281,105324
Scorpion's Senna 105269
Scorpion's Tail 190622
Scorpion's Tails 190622
Scorpion's-tail 354755
Scorpionbush 77067
Scorpiongrass 175865,175879
Scorpion-grass 260737,260833,301625
Scorpionlike Malcolmia 243230
Scorpionsenna 105269
Scorpionsenna Coronilla 105281
Scorpion-senna Coronilla 105281
Scorpionstail Desmodium 126601
Scorpionstail Mountain Leech 126601
Scorpion-vetch 105269
Scorpionweed 293151,293173
Scorpion-weed 293173
Scorpiothyrsus 354743
Scorsonera 354870
Scortechin Ant Palm 217926
Scortechin Calospatha 68016
Scortechin Devil Rattan 121525
Scortechin Licuala Palm 228765
Scortechin Licualapalm 228765
Scortechin Pinanga 299674
Scortechin Salacca 342577
Scortechini Yam 131836
Scorzoner 354873
Scorzonera 354870
Scotch Asphodel 392629
Scotch Bean 408393
Scotch Briar 336851
Scotch Broom 121001,121005,172922
Scotch Camomile 85526
Scotch Cotton Thistle 271666
Scotch Cottonthistle 271666
Scotch Cotton-thistle 271666
Scotch Crocus 111498,111504
Scotch Elm 401512
Scotch False Asphodel 392630
Scotch Fir 300223
Scotch Flame Flower 399608
Scotch Gale 261162
Scotch Geranium 174877
Scotch Granfer-Griggles 316127
Scotch Heath 149205
Scotch Heather 67456,149138,149205
Scotch Laburnum 218757
Scotch Larch 221855
Scotch Ligusticum 229384
Scotch Lovage 229384
Scotch Mahogany 16352
Scotch Marigold 66427,66445
Scotch Mercury 130383
Scotch Mint 250362,250368
Scotch Moss 32301,255598
Scotch Parsley 229384
Scotch Pine 300223
Scotch Pink 127793
Scotch Rose 336851,336969
Scotch Thistle 73430,91713,92485,271663, 271666
Scotch-mist 170654
Scotino 107300
Scots Elm 401512
Scots Fir 300223
Scots Lovage 229283
Scots Pine 300223

Scots Plane 3462
Scots Rose 336851
Scots Whitebeam 369432
Scott Azalea 331427
Scott Jugflower 7201
Scott Mountain Sandwort 255580
Scott Mountain Sulphur Flower 152588
Scott Rhododendron 331427
Scott Valley Spineflower 89113
Scott's Clematis 95018
Scottis Asphodel 392630
Scottish Asphodel 392598,392629,392630
Scottish Bluebell 70271
Scottish Dock 339925
Scottish Gentian 174153
Scottish Heather 67456,149205
Scottish Lupin 238472
Scottish Maple 3462
Scottish Moss 342298
Scottish Pearlwort 342275
Scottish Sandwort 32103
Scottish Scurvy-grass 68737,97996,98014
Scottish Thistle 92485,271666
Scottish Wormwood 35993
Scouler Penstemon 289375
Scouler Willow 344077
Scouler's Catchfly 364028
Scouler's Popcorn-flower 301627,301628
Scouler's Surf-grass 297185
Scoulfer Surfgrass 297185
Scourwort 346434
Scoutch 144661
Scow 345631
Scrab Tree 243711
Scrabs 243711
Scrag 109857
Scrambled Eggs 105631
Scrambling Rocket 365564
Scrape-clean 359158
Scratch Bur 325612
Scratch Coco 99910
Scratch Daisy 185467
Scratch Grass 259634
Scratch-bur 325612
Scratchdaisy 111681
Scratchgrass 170193,259634
Scratchweed 170193
Scree Saxifrage 349060
Screw Auger 372192
Screw Augers 372192
Screw Bean Mesquite 315569
Screw Mesquite 315569
Screw Pine 168238,280968,299869
Screw Tree 190092
Screwbean 315569
Screwbean Mesquite 315569
Screwpine 280968,281138,281166
Screw-pine 280968
Screwpine Family 280963
Screw-pine Family 280963
Screwpine Rattan Palm 65763
Screwpineleaf Thoracostachyum 390378
Screw-pod Wattle 1302
Screwstem 48592
Screw-stem 48591
Screwtree 190092
Screw-tree 190092
Scribbly Gum 155603
Scribner's Panic Grass 128537
Scribner's Rosette Grass 128537
Scrntchbush 402281
Scrobiculate Dragonhead 137655
Scrofella 355033
Scrofula-plant 355202
Scrog 109857,243711,316819
Scrog-bush 109857
Scrooby Grass 98020
Scrub Beefwood 375511
Scrub Beef-wood 180569
Scrub Blood Wood 47042
Scrub Bloodwood 47042
Scrub Bottle-tree 58339
Scrub Chestnut Oak 324304
Scrub Cypress-pine 67434
Scrub Hickory 77893
Scrub Kurrajong 194919
Scrub Mugo Pine 300086
Scrub Oak 323849,323850,323982,324039,324087,324172,324210,324304,324316,324509,324515
Scrub Palmetto 341411,360646
Scrub Pine 299786,299818,300296
Scrub She-oak 15947
Scrub Wild Buckwheat 152242
Scrubby Grass 98020
Scrubhoneysuckle Banksia 47654
Scrubland Goldenaster 90267
Scruby Grass 98020
Scrump 243675
Scryle 144661
Scttish Bluebell 70271
Scub Pine 299864
Scuch Grass 144661
Scullion 15165
Sculptured Rhododendron 330931
Scuppernong 411893
Scurf Pea 114427,319116,319134
Scurfpea 114427,319116
Scurfy Fruit Syndiclis 381793
Scurfy Oilfruitcamphor 381793
Scurfy Pea 319116,319134
Scurfy Stonecrop 357069
Scurfymallow 243957
Scurfy-pea 319116,319125
Scurrula 236507,355287,355317
Scurrula Family 236503
Scurrula Mistletoe 411052
Scurve Three-awn 34006
Scurvy Bean 250502
Scurvy Grass 98020
Scurvy Sorrel 277822
Scurvy Weed 97967
Scurvy-berries 242675,366198
Scurvy-grass 97967,98020,170193,277822,374916
Scurvyweed 97967,98020
Scutate-dentate Bittercress 72971
Scutia 355868
Scyphellandra 355905
Scyphiphora 355908
Sczukin Monkshood 5558
Sdadlman Woodbetony 287694
Se Bolt 1465
SE. China Creeper 416526
SE. China Didymocarpus 129910
SE. China Grape 411613
SE. China Ophiorrhiza 272254
SE. China Raspberry 339413
SE. China Sinosenecio 365062
SE. Xizang Rockfoil 349967
Sea Almond 386500
Sea Alyssum 235090
Sea Apple 156235
Sea Arrow-grass 397159
Sea Asparagus 39120
Sea Aster 398134
Sea Barley 198319
Sea Bean 145875,145899,259479
Sea Beet 53231,53254
Sea Bells 68737
Sea Bent 19766
Sea Bindweed 68737
Sea Bittersweet 367127
Sea Blite 379503,379553
Sea Box 18495,18530
Sea Buckthorn 196743,196757
Sea Bulrush 353587
Sea Cabbage 108627,405788
Sea Campion 364146
Sea Carrot 123141,123149
Sea Catchfly 364146
Sea Cawle 68737
Sea Chickweed 198010
Sea Clover 397087,397089
Sea Club-rush 353587
Sea Coconut 235159
Sea Colewort 59532,68737,108627
Sea Coreopsis 104545
Sea Couch 144634
Sea Cushion 34536
Sea Daffodil 200906,200945,280900

Sea Dahlia 104429,104545
Sea Daisy 34536
Sea Dallisgrass 285534
Sea Dew 337180
Sea Fennel 111347
Sea Fern-grass 79305
Sea Fig 77431,165721,251199,251265,260253
Sea Foalfoot 68737
Sea Gilliflower 34536
Sea Grae 97865
Sea Grape 97861,97865,146154,146253,342859,344585
Sea Grape Tree 97865
Sea Green 377485
Sea Hard-grass 283659
Sea Hearse 192913
Sea Heart 145899
Sea Heath 167765,167802
Sea Hibiscus 195311
Sea Hog's Fennel 292957
Sea Holly 2672,154276,154281,154327,193839
Sea Holme 154327
Sea Horned Poppy 176738
Sea Hulver 154327
Sea Island Cotton 179884
Sea Isle Japanese Sedge 75003
Sea Kale 108627
Sea Keele 108627
Sea Kemps 302091
Sea Knifebean 71072
Sea Knotgrass 309376
Sea Knot-grass 309376
Sea Knotweed 309149,309393
Sea Lacquer 161639
Sea Lavender 230508,230655,230712,230761,230815
Sea Leek 352960,397159
Sea Lettuce 350344
Sea Lily 280877,280900
Sea Lime-grass 228346
Sea Lovage 229384
Sea Lungwort 250894,321637
Sea Lyme Grass 228346
Sea Lyme-grass 228346
Sea Mallow 223372
Sea Manna-grass 321336
Sea Marsh Bugloss 230815
Sea Matting 275958
Sea Mayweed 397970
Sea Meadow-grass 321336
Sea Medick 247378
Sea Milkwort 176772,176775
Sea Mouse-ear 82793
Sea Mugwort 35868
Sea Myrtle 46227
Sea Nut Palm 244495
Sea Oats 86139,401859
Sea of Flowers 81710
Sea Onion 57984,137958,145487,352960,353099,402314
Sea Oxeye Daisy 57383
Sea Parsley 229384
Sea Paspalum 285534
Sea Pea 222735
Sea Pearlwort 342258
Sea Pink 34488,34536,230508,230815,341471
Sea Plantain 302091,370661
Sea Poppy 176724,176738
Sea Purslane 44446,44468,44611,198003,198010,394903
Sea Putat 48510
Sea Radish 326601
Sea Ragwort 358395
Sea Reed 19766
Sea Rocket 65164,65185
Sea Rose 34536
Sea Rush 213263,213272
Sea Samphire 111347
Sea Sandwort 198010
Sea Scaevola 350344
Sea Scirpus 353587
Sea Scurvy-grass 97976
Sea Seepweed 379553
Sea Shore Bluegrass 305883
Sea Shore Spergularia 370661
Sea Spinach 53254,70119,86901,86970,250600
Sea Spray 337180
Sea Spurge 159526
Sea Spurrey 370661
Sea Squill 352960
Sea Starwort 398134
Sea Stock 246547
Sea Storksbill 153869
Sea Stork's-bill 153859
Sea Sulphurwort 292957
Sea Thongs 344585
Sea Thrift 34488,34536,230815
Sea Tomato 336901
Sea Tree Mallow 223355
Sea Turf 34536
Sea Urchin 184615
Sea Urchin Cactus 43493,140842
Sea Urchin Hakea 184625
Sea Wartwort 159557
Sea Wormwood 35868,360825
Sea Wrack 344585
Seabeach Amaranth 18808
Seabeach Dock 340167
Seabeach Evening-primrose 269451
Seabeach Knotweed 309149
Seabeach Orach 44520
Seabeach Sandwort 31999,32126
Seabeach Sedge 76275
Seabeach Senecio 359811
Seaberry 184995,196757
Seaberry Family 184993
Seablite 379499,379553
Sea-blite 379486
Seablite Glasswort 342889
Seabuckthorn 196743
Sea-buckthorn 196743,196757
Sea-buckthorn Family 141928
Sea-buckthorn Leaf Tanoak 116118
Sea-buckthorn Rhododendron 330871
Seabuckthorn-like Rhododendron 330871
Sea-carrot Broomrape 275131
Seacliff Wild Buckwheat 152373
Sea-coast Angelica 98778
Seacoast Bulrush 56652
Sea-coast Cutandia 115258
Seacoast Dutchmanspipe 34355
Seacoast Flax 231920
Sea-daffodil 200906,200955,280900
Seadaffodil Pancratium 280900
Sea-fennel 111347
Sea-fig 77431
Seaflag 145642,145643
Seafoam Statice 230712
Seagrape 97861
Sea-grape 97861,97865,146154
Seagrass 34536,342859
Sea-grass 340476,418392
Sea-heath 167765,167802
Sea-heath Family 167827
Sea-holly 154276,154327
Sea-holly Eryngo 154327
Seahore Jackbean 71072
Sea-Island Cotton 179884
Sea-kale 53257,108591,108627
Seakale Beet 53257
Sea-kale Beet 53249
Seal Flower 220729
Sealavender 230508
Sea-lavender 230508,230566,230761,230804,230815
Sealavenderleaf Krylovia 218246
Sealavenderleaf Rockaster 218246
Seale Tree 342982,343151
Sealing Wax Palm 120782
Sealing-wax Palm 120779,120780,120782
Sealwort 308607
Sealy Custard-apple 25898
Seamantree 379812,379813
Seamilkwort 176772,176775
Seamulberry 368910
Seamulberry Family 368926
Sea-myrtle 46227

Sea-onion 352960,402314
Sea-pine 300146
Sea-pink 230508
Sea-poppy 176724,176738
Seaport Goosefoot 87112
Seapurslane 361654,361667
Sea-purslane 44611,198010,361654,361667
Searchlight 231183
Searell Aralia 30755
Searocket 243209
Sea-rocket 65182
Sears Rhododendron 331776
Sea-sandwort 198003,198010
Sea-shore Aeluropus 8862
Seashore Alkali Grass 321336
Seashore Alkaligrass 321336
Seashore Alkali-grass 321336
Sea-shore Allophylus 16155
Seashore Barberry 51869
Seashore Beardgrass 310121
Seashore Betony 373303
Seashore Birthwort 34355
Seashore Catapodium 79305
Seashore Chamomile 397973
Seashore Cowpea 408957
Seashore Dropseed 372880
Seashore Dropseedgrass 372880
Seashore Glorybind 68737
Seashore Hairgrass 126088
Seashore Indigo 206183
Seashore Iris 208740,208854
Seashore Ironweed 406566
Seashore Mallow 217958
Seashore Mayweed 397970
Seashore Orach 44602
Seashore Orache 44510
Seashore Paspalum 285534
Seashore Remirea 327664
Seashore Sagebrush 35831
Seashore Seepweed 379553
Sea-shore Silene 364146
Seashore Suriana 379813
Sea-shore Suriana 379813
Seashore Toadflex 231015
Seashore Vervain 405865
Seashore Vetchling 222735
Seashore Woad 209210
Seaside Agoseris 11412
Seaside Agrostis 12323,18844
Seaside Alder 16404
Seaside Alkaligrass 321336
Seaside Angelica 98778
Seaside Arrow-grass 397159
Seaside Aster 41293
Seaside Balsam 112884
Seaside Barley 198319
Seaside Brookweed 345731,345733
Seaside Buttonweed 131357
Seaside Centaury 81506
Sea-side Clerodendrum 96140
Seaside Crowfoot 184717,325749
Seaside Daisy 150665
Seaside Delosperma 123915
Seaside Dock 340167
Seaside Eryngo 154327
Seaside Evax 193380
Seaside Fleabane 150665
Seaside Gerardia 10147
Seaside Golden Rod 368386
Seaside Goldenrod 368386
Seaside Goldfields 222613
Seaside Grape 97865
Seaside Heliotrope 190597
Seaside Joyweed 18117
Seaside Lavender 230566
Seaside Lobularia 235090
Seaside Mahoe 389946
Seaside Milkweed 176775
Sea-side Pancratium Lily 280900
Seaside Pansy 410683
Seaside Pea 222735
Sea-side Persimmon 132291
Seaside Pimpernel 198010
Seaside Plantain 302091
Seaside Plum 415558
Seaside Rush 213263
Seaside Sage 112842
Seaside Sandwort 198003
Seaside Smooth Gromwell 250894
Seaside Speedwell 317927
Seaside Spurge 85783,159351
Seaside Sword Bean 71066
Seaside Thistle 73504,92437
Seaside Three-awn 34062
Seaside Thrift 230566
Seaside Water Crowfoot 325644
Seaside Wild Buckwheat 152206
Seaside Woolly Sunflower 152832
Seaside Yam 208067,208069
Seaside-balsam 98177
Season Vine 92967
Sea-spurrey 370609
Seastarwort 398126,398134
Seattle Dock 359158
Seatung 301207,301433
Seatung Family 301203
Seatung Nightshade 367503
Seatungleaf Litse 234037
Seatunglike Spicegrass 239790
Sea-urchin Cactus 43493,107022,140842
Sea-urchin Dryandra 138440
Seaurchin Hakea 184615
Sea-urchin Tree 184615
Seaves 213021
Seawfell Pink 34536
Sea-wrack 418392
Sebaea 356034
Sebastan Plum Cordia 104175
Sebastiania 356196
Sebastiao-de-arruda 121663
Sebestan Plum 104218
Secale-like Brome 60963
Secamone 356259
Secamone-like Tylophora 400958
Seccomaria 138470
Sechsblattrige Staunton-vine 374409
Second Lemongrass 117218
Secud-flower Sophora 369116
Secund Macrozamia 241464
Secund Phreatia 295961
Secund Rodriguezia 335168
Secundflower Chirita 87959
Secundflower Melic 249090
Secundflower Rabdosia 209824
Secundflower Rattle Snake Plantain 179694
Secundflower Skullcap 355800
Secundiflorous Rabdosia 209824
Secundiflower Monkshood 5561
Secundspike Elsholtzia 143965
Securidaca 356361
Securinega 356392
Secymore 3462
Seda Milkvetch 43031
Sedge 73543, 73582, 73617, 73621, 73626, 73630, 73632, 73633, 73675, 73703, 73732, 73765, 73782, 73869, 73887, 73933, 73942, 73976, 73977, 74053, 74078, 74155, 74157, 74160, 74179, 74180, 74195, 74196, 74209, 74219, 74265, 74274, 74312, 74385, 74421, 74452, 74503, 74544, 74774, 74810, 74816, 74855, 74862, 74886, 74894, 74935, 75034, 75037, 75043, 75081, 75095, 75120, 75238, 75319, 75360, 75399, 75436, 75513, 75527, 75589, 75652, 75740, 75744, 75830, 75865, 75959, 76006, 76011, 76056, 76147, 76160, 76182, 76207, 76256, 76295, 76307, 76335, 76368, 76375, 76388, 76420, 76463, 76464, 76508, 76532, 76561, 76564, 76576, 76577, 76582, 76623, 76630, 76697, 93913, 118428, 118476, 208771
Sedge Family 118403
Sedge Kobresia 217138
Sedge Rush 213438
Sedgeleaf Orchis 116813
Sediformis Rockfoil 349883
Sedirea 356456
Sedocke 2695
Sedra 418169

Sedum 356467
Sedumlike Sealavender 230580
Seebright 345379
Seedbox 238150,238160
Seed-box 238160
Seedcake 77784
Seed-cake Thyme 391201
Seed-eake Thyme 391201
Seed-edible Dioon 131429
Seeder 199384
Seedless Common Barberry 52338
Seedless Sweetgum 232573
Seedling 235090
Seeds 235334
Seedy Resurrectionlily 214979
Seedy Salacia 342705
Seedy Sandwort 32144
Seedy Tarenna 385015
Seedybean 322076
Secly's Catchfly 364045
Seely's Silene 364045
Seemann Acuminate Balaca 46744
Seemann's Sunbonnet 224117
See-me-contract 154316
Seep Willow 46225,46251
Seep-spring Monkeyflower 255210
Seepweed 379486,379553,379573
Seepwillow 46251
Seepwood 346610
Seersucker Begonia 50321
Seersucker Plant 128990,174333
Seg 208771
Segawa Lady's Slipper 120446
Segawa Ladyslipper 120446
Segg 208771
Seggins 208771
Seggrum 359158
Seggy 3462,359158
Sego-lily 67611
Seguier Euphorbia 159808
Seguin Argyreia 32657
Seguin Chinkapin 78823
Seguin Colquhounia 100044
Seguin Iodes 207371
Seguin Jasmine 211996
Seguin Lobelia 234759
Seguin Loquat 151172
Seguin Melodinus 249676
Seguin Torchflower 100044
Segumber 3462
Sehima 357361
Sehwerin Barberry 52137
Sehwerin Corydalis 106425
Sehwerin Maple 3586
Seifriz's Chamaedorea 85437
Seikoo Spotleaf-orchis 179700
Seinghku Rhododendron 331778
Seje 269374
Sejila Trilobe Buttercup 325940
Sekitei Stonecrop 357138
Selangan 362186,362207
Selangor Licuala Palm 228740
Selangor Licualapalm 228740
Select Purple Smoketree 107307
Seleng Sagebrush 36241
Seleng Wormwood 36241
Selenipedium 357751
Self Heal 316127
Self-fruitful Bittrswt 80295
Selfheal 316095,316127
Self-heal 13174,299523,316095,316127,316159,345957
Selgreen 358062
Selinum 357780,357794
Selkirk's Violet 410547
Scll Green 358062
Seller Tassels 60382
Sello Basil 268622
Sellon Pampas Grass 105456
Sellow Barberry 52138
Sellow Feijoa 163117
Selly 343151
Selon's Rose 265327
Selsk Stonecrop 294126
Selu 104218
Selver Clary 344855
Semaphore Cactus 102821
Semaphore Grass 304695
Semaphore Plant 98159
Semaphore Thoroughwort 158236
Semaphore-grass 304695
Semeiandra 357886
Semenov Amur Maple 3587
Semenov Corydalis 106427
Semenov Crazyweed 279154
Semenov Euonymus 157876
Semenov Fir 471
Semenov Leek 15732
Semenov Onion 15732
Semenov Rhodiola 329950
Semenov Sealavender 230581
Semenov Spindle-tree 157876
Semenov Sweetvetch 188094
Semenov Thistle 92375
Semenov Woodbetony 287668
Semenov's Honeysuckle 236088
Semiaesert Spunsilksage 360837
Semiamplectant Speedwell 407352
Semi-amplectant Windhairdaisy 348774
Semiamplexicaul Speedwell 407352
Semiaquilegia 357918
Semiarundinaria 357932,357939
Semi-circular Milkvetch 43033
Semicostate Roegneria 335495
Semiexserted Spindle-tree 157877
Semifalcate Longauricle Indocalamus 206803
Semiglabrous Tea 69673
Semihaired Euphorbia 159811
Semiliquidambar 357968
Semi-lyrate Windhairdaisy 348776
Seminole Pumpkin 114292
Seminole Tea 38334
Semiorbicular Fargesia 137866
Semipaniculate Microtropis 254324
Semiprostrate Heteropappus 193982
Semiprostrate Jurinea 193982
Semirosulate Cyathula 115745
Semiserrate Camellia 69613
Semiserrate Oil Camellia 69613
Semitooth Cyclobalanopsis 116191
Semitooth Oak 116191
Scmnostachya 357999
Semolina 398900
Semper 111347,342859
Semperflorens Rose 336496
Sempervirent Evergreenchinkapin 79019
Sempervivum 358015
Sempilor 121095
Sen Cion 360328
Sen Tree 215442
Sendgreen 358062
Sene 100153
Seneca Grass 27967
Seneca Snakeroot 308347,308348
Senecio 358159
Senega Root 308347
Senega Snakeroot 308347
Senegal Coralbean 154718
Senegal Coral-tree 154718
Senegal Custardapple 25885
Senegal Date 295479
Senegal Date Palm 295479
Senegal Date-palm 295479
Senegal Ebony 121750
Senegal Gum 1572
Senegal Khaya 216214
Senegal Lilac 235570
Senegal Mahogany 216214
Senegal Mayten 246927
Senegal Nitraria 266374
Senegal Rosewood 320293
Senegal Senna 78436
Senegal Soapberry 225678
Senegal Sphaeranthus 371020
Senegal Tea 182537
Senegal Whitethorn 266374
Seneka 308347
Senencent Draba 137229
Senencent Whitlowgrass 137229
Sengreen 349419,356468,409339
Seniavin Rhododendron 331795

Seniawin St. John's Wort 202151
Senile Mammillopsis 243977
Senita 279492
Senna 78204,78227,78284,100153,360444
Senna Family 65084
Senna-pods 81065
Senno Campion 363265
Sensespine Hedge Gooseberry 333900
Senshon 360328
Sension 360328
Sensitive Brier 255102,265228
Sensitive Indigo 206534
Sensitive Jointvetch 9641
Sensitive Onosma 271785
Sensitive Partridge Pea 78421
Sensitive Pea 78421
Sensitive Plant 255098
Sensitive Smithia 366698
Sensitive Stonecrop 357006
Sensitivebriar 255074
Sensitivebrier 352629
Sensitive-brier 255082
Sensitive-plant 255082
Sensitiveplant-like Senna 85232
Sensly Hairy Mouthed Rhododendron 332001
Sentinel Silver Fir 276
Sention 360328
Sentry in the Box 13756
Sentry Palm 198804,198806,198808
Sentry Sugar Maple 3572
Sentry-boxes 13729
Sentul 345786,345788
Senvie 59515,364545,364557
Senvy 59515,364545,364557
Seoul Catchfly 364048
Seoul Siebold Wildginger 37729
Sepal Devil Rattan 121526
Separated Sedge 74360
Separated Xylopia 415778
Separateleaf Chickensage 371147
Separatestyle Acanthopanax 143592
Separatestyle Camellia 69272
Separatestyle Schefflera 350718
Separate-styled Acanthopanax 143592
Separate-styled Schefflera 350718
Separating-patal Hibiscus 195216
Separating-pataled Hibiscus 195216
Sepi 305085
Sepiaria 8783,8787
Septeless Erysimum 154448
September Bells 337500
September Clematis 94866
September Elm 401626
September Lily 97218
Septemlobate Kalopanax 215442
Septfoil 312520
Septicidal Fig 165658
Septifoliate Yam 131580
Sequier's Buttercup 326350
Sequin Myrsine 261655
Sequoia 360561,360565
Serai 117153
Serapia's Turbith 398134
Serapias 360582
Serate Bigginseng 241177
Serav Mountain Fleabane 150959
Serb 369339,369541
Serb Apple 369541
Serbian Bellflower 70229
Serbian Spruce 298387
Serge 401112
Sericea 226742
Sericea Lespedeza 226742
Sericeous Bird Cherry 280061
Sericeous Catchfly 364050
Sericeous Catnip 264963
Sericeous Chalkplant 183244
Sericeous Crabgrass 130763
Sericeous Craneknee 221744
Sericeous Dogwood 105193
Sericeous Fimbristylis 166473
Sericeous Fluttergrass 166473
Sericeous Glochidion 177175
Sericeous Hanging-down Milkvetch 42287
Sericeous Heteroplexis 194013
Sericeous Hiptage 196848
Sericeous Hydrangea 199883
Sericeous Indian Gugertree 350921
Sericeous Jurinea 207229
Sericeous Kitefruit 196848
Sericeous Larkspur 124195
Sericeous Leptospermum 226501
Sericeous Litse 234061
Sericeous Magnolia 232592
Sericeous Meadowrue 388439
Sericeous Mockorange 294540
Sericeous Mongolian Willow 343714
Sericeous Nepeta 264963
Sericeous Newlitse 264090
Sericeous Oreocharis 273881
Sericeous Peashrub 72244
Sericeous Pea-shrub 72244
Sericeous Petal Crazyweed 279157
Sericeous Sagebrush 36262
Sericeous Spiraea 372083
Sericeous Tall Larkspur 124195
Sericeous Tibetan Hydrangea 199803
Sericeous Vetch 408272
Sericeous Windhairdaisy 348777
Sericeous Youngia 416476
Sericeousflor Porterandia 311797
Sericeousflower Meranti 362222
Sericeousfruit Willow 344088
Sericeousfruited Willow 344088
Sericeous-leaf Litse 234061
Sericeous-leaf Rosewood 121822
Sericeousleaf Snakegourd 396270
Sericeus Bluebell 115420
Sericeus Bluebellflower 115420
Sericfous Elatostema 142711
Sericocalyx 360698
Seringa 194453
Seriseous Betony 373431
Serissa 360926,360935
Serisse 360923
Sermountain 222028
Serotina Poplar 311163
Serpent Agapetes 289750
Serpent Apple 25876
Serpent Draba 137230
Serpent Gourd 396151
Serpent Groundsel 360027
Serpent Melon 114209
Serpent Root 354797
Serpent Whitlowgrass 137230
Serpent's Tongue 154899
Serpent's Tongue Spearwort 326148
Serpentagourd 396150
Serpentaria 34325,91034
Serpentary 34325,91034
Serpentgourd 396151
Serpentina Manzanita 31129
Serpentine 137747
Serpentine Arnica 34685
Serpentine Aster 380861
Serpentine Bristleweed 186890
Serpentine Fleabane 150960
Serpentine Garlic 15861
Serpentine Goldenbush 150358
Serpentine Indian Pink 364052
Serpentine Sandwort 255520
Serpentine Sunflower 188965
Serpentroot 354797,354813
Serpent-root 354797
Serpentweed 392771
Serpolet 391365
Serpolet Oil 391327
Serradella 274895
Serralate Osmanthus 276396
Serrate Abelia 167
Serrate Amur Arisaema 33253
Serrate Amur Southstar 33253
Serrate Arisaema 33496
Serrate Barberry 52140
Serrate Chloranthus 88297
Serrate Distylium 134943
Serrate Drymonia 138527
Serrate Engelhardia 145520
Serrate Glorybower 96339
Serrate Gomphia 70748
Serrate Hedge Prinsepia 315178

Serrate Holly 204239
Serrate Hydrangea 200085
Serrate Japanese Ash 168014
Serrate Leaf Mayten 246930
Serrate Loquat 151173
Serrate Mosquitoman 134943
Serrate Mulberry 259191
Serrate Oak 324384
Serrate Oncidium 270840
Serrate Osmanthus 276396
Serrate Paniculate Staff-tree 80276
Serrate Pimpinella 299534
Serrate Plagiopetalum 301677
Serrate Poplar 311474
Serrate Protium 316031
Serrate Rabdosia 209826
Serrate Rhodiola 329951
Serrate Rockcress 30435
Serrate Rustyhair Raphiolepis 329062
Serrate Sikoku Arisaema 33509
Serrate Sikoku Southstar 33509
Serrate Slantpetal 301677
Serrate Southstar 33496
Serrate Spapweed 205313
Serrate Spurge 159828
Serrate Triratna Tree 397372
Serrate Twocolor Maple 2802
Serrate Violet 410452
Serrated Synotis 381944
Serrated Tailanther 381944
Serrated Tussock 262622
Serrated Wintergreen 275480,275486
Serrateleaf Banksia 47661
Serrateleaf Carallia 72394
Serrateleaf Emei Skullcap 355635
Serrate-leaf Myrsine 261650
Serrateleaf Pearlbush 161755
Serrateleaf Raspberry 339481
Serrateleaf Sanicle 346001
Serrateleaf Sinocrassula 364923
Serrate-leaf Storax 379449
Serrateleaf Willow 343180
Serrateleaf Windhairdaisy 348554,348775
Serrate-leaved Carallia 72394
Serrate-leaved Dense-flower Barberry 51545
Serrate-leaved Mayten 246930
Serrate-leaved Myrsine 261650
Serrate-leaved Pearlbush 161755
Serrate-leaved Pruinose Barberry 52076
Serrate-leaved Raspberry 339481
Serrate-leaved Sage 36277
Serrate-leaved Storax 379449
Serrate-leaved Willow 344096
Serrate-sepaled Neillia 263160
Serrate-toothed Barberry 52141
Serratifoliate Premna 313740
Serratisepal Neillia 263160
Serrete Ladybell 7855
Serrulate Altingia 18201
Serrulate Aucuba 44904
Serrulate Blueberry 403993
Serrulate Cherry 83312
Serrulate Denseleaved Maple 2900
Serrulate Enkianthus 145720
Serrulate Eryuan Barberry 51856
Serrulate Evergreenchinkapin 79010
Serrulate Hornbeam 77346
Serrulate Leaves Enkianthus 145720
Serrulate Linden 391680
Serrulate Pear 323294
Serrulate Pendent-bell 145720
Serrulate Rockvine 387815,387845
Serrulate Sikkim Maple 3613
Serrulate Spurge 158534
Serrulate Twocolor Maple 2803
Serrulate Ward Wintergreen 172175
Serrulate Woodnettle 273920
Serrulateleaf Willow 344096
Serrulate-leaved Willow 344096
Serruria 361154
Servian Ramonda 325228
Service Berry 19240,19243
Service Tree 19276,369381,369541
Service Tree of Fontainebleau 369429
Service Viburnum 408202
Serviceberry 19240,19251,19260,19262,
19279,19283,19288,19290,19298
Service-berry 19240,19241,19276,19286,
369322
Service-tree 369381
Service-tree Mountain Ash 369381
Service-tree-of-fontainebleau 369446
Servoile 236022
Sesame 361294,361317
Sesame Milkvetch 43039
Sesamum 361317
Sesban 361430
Sesbania 361350,361370,361449
Sesbanie 361350
Seseli 361456
Seselike Lily 228597
Seselopsis 361605,361606
Sesleria 361611
Sessban 361380
Sessile Alternanthera 18147
Sessile Azalea 332097
Sessile Bellwort 403675
Sessile Biophytum 54548
Sessile Blazing-star 228529
Sessile Blumea 55823
Sessile Bredia 59880
Sessile Catnip 265047
Sessile Chartaceous Fig 164797
Sessile Date 295453
Sessile Deutzia 126934
Sessile Didissandra 129832
Sessile Embelia 144810
Sessile Exacum 161577
Sessile Galangal 17753
Sessile Jadeleaf and Goldenflower 260492
Sessile Joyweed 18147
Sessile Ladybell 7835
Sessile Little Man Orchid 178896
Sessile Lobelia 234766
Sessile Marshweed 230323
Sessile Microtropis 254325
Sessile Neonauclea 264224
Sessile Nepeta 265047
Sessile Oak 324278
Sessile Paintbrush 79135
Sessile Pretty Fig 164834
Sessile Primrose 314308
Sessile Qinggang 116193
Sessile Reactionggrass 54548
Sessile Shiquan Willow 344107
Sessile Skullcap 355748
Sessile Sobralia 366792
Sessile Speedwell 317945
Sessile Spindle-tree 157899
Sessile Stemona 375351
Sessile Syrenia 382105
Sessile Syzygium 382490
Sessile Terniopsis 121979
Sessile Thistle 92361
Sessile Tonkin Spicebush 231455
Sessile Toothcup 19621
Sessile Tooth-cup 19621
Sessile Trillium 397609
Sessile Tutcheria 400726
Sessile Twistedstalk 377934
Sessile Vinegentian 110329
Sessileflower Acanthopanax 143665
Sessileflower Goldenaster 194246
Sessileflower Hoarypea 386308
Sessileflower Monkshood 5564
Sessileflower Rattlebox 112667
Sessileflower Salacia 342724
Sessileflower Willow 343057
Sessile-flowered Acanthopanax 143665
Sessile-flowered Cress 336268
Sessile-flowered Microtropis 254325
Sessile-flowered Salacia 342724
Sessile-flowered Tutcheria 400719
Sessile-fruited Arrowhead 342397
Sessile-headed Blazing-star 228529
Sessile-leaf Bellwort 403675
Sessile-leaf Blueberry 403853,404011
Sessileleaf False Oxtongue 55823
Sessileleaf Grewia 180965
Sessileleaf Japanese Maple 3364
Sessile-leaf Oak 116193

Sessile-leaved Bellwort 403675
Sessile-leaved Caper 71893
Sessile-leaved Grewia 180965
Sessile-leaved Lysionotus 239983
Sessile-leaved Mussaenda 260492
Sessile-leaved Water Horehound 239189
Sessilflower Seseli 361569
Sessilflower Slatepentree 400726
Sessilfruit Chinaure 56392
Sessilleaf Lysionotus 239983
Seta Daisy 124817
Seta Dendranthema 124817
Setacalyx Spinestemon 306985
Setaceous Greenbrier 366408
Setaceous Haworthia 186728
Setaceous Mitriform Rhododendron 331255
Setaceous Scaligeria 350377
Setaceous Yarrow 4027
Setaceous-barct Fimbristylis 166290
Setaceous-bract Fluttergrass 166290
Setafruit 84852
Setchuan Sweetleaf 381390
Setfoil 8826,312520
Setiferous Azalea 331808
Setiferous Sweetvetch 188099
Setiform Square-bamboo 87612
Setigrass 361993,361994
Setiperulate Camellia 69627
Setistyled Southern Rhododendron 331224
Seton-grass 190936
Setose Abelmoschus 233
Setose Allomorphia 16030
Setose Arundinella 37429
Setose Asparagus 39195
Setose Azalea 331424,331809
Setose Bamboo 297460
Setose Begonia 50303
Setose Blackberry 339262
Setose Blueberry 403812
Setose Blue-flag 208819
Setose Cephalanoplos 82024
Setose Deutzia 127097
Setose Draba 137231
Setose Eggleaf Rhododendron 331424
Setose Falsepimpernel 231575
Setose Gaultheria 172077
Setose Hawksbeard 111017
Setose Honeysuckle 236097
Setose Iris 208819
Setose Jerusalemsage 295191,295192
Setose Leea 223959
Setose Mahonia 242636
Setose Maoutia 244895
Setose Motherweed 231575
Setose Oak 324409
Setose Onosma 271824
Setose Rhododendron 331809
Setose Sandwort 32225
Setose Sedge 76240
Setose Smallfruit Microula 254364
Setose Spapweed 205315
Setose Sweetvetch 188100
Setose Thintube Woodbetony 287256
Setose Thistle 92384
Setose Touch-me-not 205315
Setose Vineclethra 95543
Setose Wallich Lasianthus 222311
Setose Wallich Roughleaf 222311
Setose Whitepearl 172077
Setose Whitlowgrass 137231
Setoseconnective Metalleaf 296412
Setose-connective Metalleaf 296412
Setoseleaf Begonia 50303
Setoseleaf Koeleria 217563
Setoseleaf Leek 15740
Setoseleaf Onion 15740
Setose-leaved Large-sepal Barberry 51900
Setter 190936,190959
Settergrass 190936,190959
Setterwort 190936,190959
Setulose Acanthopanax 143676
Setulose Actinidia 6567
Setulose Bethlehemsage 283650
Setulose Blastus 55175
Setulose Cherry 83322
Setulose Kiwifruit 6567
Setulose Lobelia 234379
Setulose Paraphlomis 283650
Setulose Pennywort 200364
Setulose Rockfoil 349411
Setuloso-peltate Begonia 50304
Setwall 404316,404337
Sevadilla 352124
Sevanlobed Nightshade 367604
Sevelobes Tenuity-leaved Maple 3675
Seven River Hills Wild Buckwheat 152115
Seven Sisters 159027,159557
Seven Sisters Rose 336470,336789
Seven Son Flower 192168
Seven Stars 33217
Seven Years'love 3978,4005
Seven-angle Nervilia 265408
Seven-angle Pipewort 151232
Seven-angle Taroorchis 265408
Sevenangular Pipewort 151481
Sevenbark 199806
Seven-bark 199806,200063
Sevenclaw Japanese Maple 3355
Sevendara 407585
Sevendara Grass 407585
Sevenhead False Rattlesnakeroot 283685
Seven-leaf 312520
Sevenleaf Arisaema 33507
Sevenleaf Fishvine 125972
Sevenleaf Southstar 33507
Sevenleaf Tubergourd 390148
Seven-leaved Cinquefoil 312658
Sevenleaves Cinquefoil 312658
Sevenleaves Gentian 173255
Seven-leaves Milkvetch 42485
Sevenleaves Tubergourd 390148
Sevenlobate Primrose 314938
Sevenlobe Japanese Maple 3363
Sevenlobe Maple 3017
Seven-lobe Maple 3017
Sevenlobed Maple 3163
Sevenlobed Sinosenecio 365074
Sevenlobed Yam 131840
Sevenlobes Chinese Groundsel 365074
Sevenpetal Michelia 252953
Sevenpetal Tea 69620
Sevenriver Rush 213150
Sevenrivulet Milkvetch 42485
Sevenseedflower 192166
Seven-sons Plant 192166,192168
Sevenstar Lotus 409898
Seven-stars Living-rock Cactus 33217
Seven-up Plant 373179
Sevenvein Slantpetal 301669
Seven-veins Esquirol Plagiopetalum 301669
Seven-years'-love 3978,4005
Severinias 362032
Severzov Crazyweed 279159
Seville Orange 93332,93368,93408,93414
Sewartia 376452
Sewertzov Brome 60968
Sewerzov Corydalis 106431
Sewerzow Fritillary 168557
Seychelles Coconut 235159
Seychelles Nut 235159
Seyny-tree 218761
Sfrican Yellowwood 306425
Sgeach 109857
Sgreat Tarwort 374916
Shaanxi Arrowwood 408118
Shaanxi Azalea 331588,332155
Shaanxi Barberry 52128,52145
Shaanxi Butterflybush 62164
Shaanxi Camellia 69628
Shaanxi Clematis 95297
Shaanxi Corydalis 106436
Shaanxi Cranesbill 174899
Shaanxi Douglas Fir 318589
Shaanxi Ellipticfruit Corydalis 105856
Shaanxi Euonymus 157874
Shaanxi Fir 306,311
Shaanxi Grape 411917
Shaanxi Hawthorn 110040
Shaanxi Honeysuckle 236046
Shaanxi Hornbeam 77388
Shaanxi Hydrangea 199905

Shaanxi Lacquertree 393492
Shaanxi Lycoris 239277
Shaanxi Maple 3596
Shaanxi Meadowrue 388664
Shaanxi Monkshood 5566
Shaanxi Paulownia 285975
Shaanxi Peashrub 72339
Shaanxi Pea-shrub 72339
Shaanxi Pennisetum 289257
Shaanxi Poplar 311479
Shaanxi Primrose 314450
Shaanxi Purplestem 376482
Shaanxi Raspberry 339052
Shaanxi Rhododendron 331409,332155
Shaanxi Sabia 341565
Shaanxi Sedge 76244
Shaanxi Seleng Wormwood 36251
Shaanxi Spiraea 372135
Shaanxi Stewartia 376482
Shaanxi Stonegarlic 239277
Shaanxi Toona 392846
Shaanxi Viburnum 408118
Shaanxi Violet 410539
Shaanxi Wolftailgrass 289257
Shaanxi-Gansu Maple 3595
Shaaxi Sabia 341565
Shaba Fig 164795
Shaba Holboellia 197221
Shaba Holly 203617
Shaba Michelia 252844
Shaba Tanoak 233190
Shabdar 397050
Shabub 238371
Shachai Riparial Silverreed 394940
Shackle Basket 60382
Shackle Box 60382
Shackle-backles 364193
Shackle-bag 329521
Shackle-basket 60382,329521
Shackle-box 60382,287721
Shackle-cap 329521
Shackle-grass 60382
Shacklers 2837,167955,329521
Shackles 329521
Shackshack Cratalaria 112238
Shad 19240
Shad Blow 19241
Shad Bush 19240
Shadblow 19240,19243
Shadblow Serviceberry 19251,19279
Shad-blow Servlceberry 19251
Shadbuah 19240
Shadbush 19241,19243,19276,19290
Shadbush Serviceberry 19251
Shaddock 93579
Shaddock Pummelo 93579
Shade Clearweed 299075
Shade Fescue 164006
Shade Knotweed 309918
Shade Meadowrue 388705
Shade Ophiorrhiza 272302
Shade Pine 300011
Shadecelery 248542,248543
Shaded Parsnip 285738
Shadegrass 352863,352875
Shadeloving Bugle 13183
Shadeloving Calamintha 65620
Shadeloving Eremotropa 258082
Shadeloving Newlitse 264107
Shadeloving Skullcap 355732
Shades of Evening 363673
Shades-of-evening 363673
Shadiloving Sporoxeia 372898
Shadily Raspberry 338808
Shad-loving Sporoxeia 372898
Shadow 60382
Shadow Grass 238709
Shadow Palm 174341
Shadow Witch 311023
Shadowy Goldenrod 368383
Shadscale 44355
Shadscale Wingscale 44334
Shady Anemone 24105
Shady Bladderwort 403986
Shady Chirita 87984
Shady Coldwaterflower 299075
Shady Coneflower 339548
Shady Dogwood 380429
Shady Greenbrier 366612
Shady Groundsel 359565
Shady Larkspur 124662
Shady Lilyturf 272154
Shady Marshmarigold 68212
Shady Milkvetch 43022
Shady Night 367120
Shady Orchis 121422
Shady Platanthera 302558
Shady Ramie 56350
Shady Raspberry 338808
Shady Sage 345450
Shady Stauranthera 374449
Shady Violet 410784
Shady Windflower 24105
Shady-humid Barberry 51744
Shaeklle Box 60382,287721
Shaftal Clover 397050
Shag Azalea 330805
Shag Camellia 69627
Shag Duckbeakgrass 209261
Shag Milkvetch 43013
Shag Mucuna 259558
Shag Peashrub 72215
Shag Slatepentree 322585
Shagbark Hickory 77920,77934
Shaggy Basketvine 9468
Shaggy Dryandra 138444
Shaggy Fleabane 150906
Shaggy Galinsoga 170146
Shaggy Hawkweed 196081
Shaggy Hydrangea 199846
Shaggy Jack 363461
Shaggy Mouse-ear Hawkweed 299226
Shaggy Oxhneedaisy 170146
Shaggy Pea 278494
Shaggy Prairie-turnip 287932
Shaggy Raspberry 338727
Shaggy Rosinweed 364296
Shaggy Soldier 170146
Shaggy Soldiers 170146
Shaggy-bark Manzanita 31140
Shaggy-footed Rhododendron 330876
Shaggy-soldier 170146
Shaggy-style Linden 391701
Shagvine 97499
Shaining Deutzia 127028
Shakaro Privet 229614
Shakers 60382
Shaking Asp 311537
Shaking Grass 60372,60382
Shaking Shadow 60382,238709
Shaky Grass 60382
Shalder 208771
Shaldon 208771
Shale Barren Wild Buckwheat 151810
Shale Groundsel 92521
Shale-barren Goldenrod 367970
Shalebarren Pussytoes 26618
Shalebarren Ragwort 279887
Shallon 172149
Shallon Bush 172149
Shallon Lemonleaf 172149
Shallot 15167,15170
Shallotlike Rush 212820
Shallow Sedge 75233,75238
Shallowbract Goldenray 229012
Shallow-cup Mongolian Oak 324190
Shallowdivided Larkspur 124675
Shallowfid Kobresia 217212
Shallowtooth Goldenray 229147
Sham Bendtooth Milkvetch 42132
Sham Borodin Milkvetch 42921
Sham Broom Milkvetch 42933
Sham Calabash 383007
Sham Calmshell Pricklyash 417250
Sham Chaff Camellia 69430
Sham Glacial Larkspur 124521
Sham Honey-flower 21176
Sham Horsebean Caltrop 418646
Sham Medic 112396
Sham Milkvetch 43123
Sham Milkvetch Sweetvetch 188062

Sham Roughleaf Milkvetch 42930
Sham Shootkeel Milkvetch 42923
Sham Skyblue Larkspur 124516
Sham Tongol Larkspur 124530
Sham Variant Milkvetch 42938
Sham Wampee 94191
Shama Millet 140360
Shame Face 410677
Shame Plant 255101
Shame-bush 255098
Shame-face 174717,410677
Shame-faced Maiden 23925,274823
Shamel Ash 168124
Shameweed 255098
Shamgoldquitch 318147
Shammanytooth Camellia 68908
Shampoo Orchid 372227
Shamrock 247366,262722,277644,277648,
396882,397019,397043,407072
Shamrock Pea 284647,284650
Shamrockpea 284647,284650
Shamrock-pea 284647,284650
Shamrue 185615
Shamseparatestyle Camellia 69274
Shamsher 111347
Shan Azalea 331811
Shan Bulbophyllum 63086
Shan Stonebean-orchis 63086
Shan's Rhododendron 331811
Shanche Tamarisk 383601
Shandan Sedge 76245
Shandan Willow 344101
Shander-grass 60382
Shandong Anemone 24058
Shandong Bentgrass 12301
Shandong Borreria 370841
Shandong Fairybells 134486
Shandong Fimbriate Orostachys 275367
Shandong Galingale 119578
Shandong Hawthorn 110039
Shandong Madder 338040
Shandong Pink 127675
Shandong Roegneria 335501
Shandong Seseli 293071
Shandong Stemona 375352
Shandong Willow 343578
Shandong Windflower 24058
Shangan Mountainash 369443
Shangcheng Rose 336945
Shangcheng Sedge 76246
Shangcheng Willow 344102
Shanghai Rosewood 121815
Shanghai Sedge 76247
Shanghang Azalea 331705
Shanghang Evergreen Chinkapin 78971
Shanghang Evergreenchinkapin 78971
Shanghang Rhododendron 331705
Shang-hang Rhododendron 331705
Shanghang Sedge 76248
Shanglin Elatostema 142837
Shanglin Furfuraceousfruit Camellia 69088
Shanglin Rockvine 387852
Shanglin Sonerila 368900
Shanglin Stairweed 142837
Shangnan Sagebrush 36279
Shangshi Beautyberry 66930
Shangshi Purplepearl 66930
Shangsi Bennettiodendron 51029
Shangsi Distinchous Rattan Palm 65674
Shangsi Eargrass 187615
Shangsi Fissistigma 166687
Shangsi Hedyotis 187615
Shangsi Holly 204154
Shangsi Hydrangea 200103
Shangsi Ixora 211196
Shangsi Maple 3593
Shangsi Oak 116080
Shangsi Reevesia 327378
Shangsi Scorpiothyrsus 354749
Shangsi Slugwood 50608
Shangsi Suoluo 327378
Shangsi Tanoak 233345
Shangsi Tupelo 267862
Shanhaiguan Sharphead Sagebrush 36037
Shaniodendron 284957,284964
Shannan Hawthorn 109656
Shansi Chicory 90913
Shansi Figwort 355190
Shansi Monkshood 5582
Shansi Tubergourd 390127
Shansi Woodbetony 287670
Shantou Box 64249
Shantou Capitate Box 64249
Shantou Pavetta 286500
Shantung Cabbage 59595
Shantung Maple 3714
Shanwamillet 140351,140360
Shanxi Anemone 23812
Shanxi Bluegrass 305968
Shanxi Cacalia 283802
Shanxi Chicory 90913
Shanxi Cowparsnip 192350
Shanxi Craneknee 221746
Shanxi Dandelion 384632
Shanxi Dimorphostemon 131144
Shanxi Figwort 355190
Shanxi Fritillary 168655
Shanxi Honeysuckle 236184
Shanxi Lady's Slipper 120447
Shanxi Ladyslipper 120447
Shanxi Monkshood 5582
Shanxi Pyrola 322909
Shanxi Sagebrush 35471
Shanxi Seseli 361562
Shanxi Skullcap 355759
Shanxi Stickseed 221746
Shanxi Woodbetony 287670
Shan-you-ma 394635
Shaoyao 280213
Shaped Sedge 73589
Shappetal Leadwort 305204
Share 3462
Share-like Violet 410108
Sharewort 201732,201787,262381,398134
Sharon Tulip 400223
Sharp Apices Rhododendron 330452
Sharp Calyx Himalaya Honey Suckle 228326
Sharp Cedar 213841
Sharp Club-rush 352249
Sharp Concavepetal Parnassia 284577
Sharp Cupgrass 151679
Sharp Dock 339887,339989
Sharp Edge Sedge 118982,119220
Sharp Gayfeather 228435
Sharp Goatwheat 44271
Sharp Hawksbeard 110717
Sharp Nepeta 265032
Sharp Riparian Greenbrier 366553
Sharp Rush 212808,212809
Sharp Spapweed 204782
Sharp Spines Chinkapin 79030
Sharp Tailanther 381919
Sharp Thistle 91770
Sharp Tree-nerved Neolitsea 264015
Sharpangle Anisodus 25439
Sharpangular Sea-urchin Cactus 140876
Sharpappendage Skullcap 355652
Sharpbeak Sedge 75101
Sharpbeak Slatepentree 400717
Sharpbill Crazyweed 278803
Sharpbill Heronbill 153883
Sharpbill Milkvetch 42834
Sharpbind 366240
Sharpbract Phrynium 296049
Sharpcalyx Ash 168038
Sharpcalyx Azalea 331422
Sharpcalyx Bladderwort 403324
Sharpcalyx Camellia 68882
Sharpcalyx Purplepearl 66838
Sharpcalyx Randia 12530
Sharpcalyx St. John's Wort 201710
Sharpcalyx Syzygium 382535
Sharpcleft Denseleaf Larkspur 124320
Sharpdent Oak 116165
Sharpdent Qinggang 116165
Sharp-dented Cyclobalanopsis 116165
Sharp-dented Oak 116165
Sharpdissected Monkshood 5323
Sharpe Arrowbamboo 162636
Sharped Ash 168028
Sharped-fruit Rattanpalm 65755

Sharped-fruited Rattan Palm 65755
Sharp-flowered Rush 212805
Sharp-flowered Signal-grass 402543
Sharpfruit Alplily 234228
Sharpfruit Aphragmus 29169
Sharpfruit Ash 168043
Sharpfruit Beancaper 418727
Sharpfruit Calathodes 66220
Sharpfruit Cogflower 154185,154192,382822
Sharpfruit Elaeagnus 141932,142140
Sharpfruit Falsepimpernel 231531
Sharpfruit Motherweed 231531
Sharpfruit Snakegourd 396274
Sharp-fruited Elaeagnus 142140
Sharp-fruited Rush 212800
Sharpfuit Woodbetony 287490
Sharp-glume Bluegrass 305275
Sharpglume Hymenacue 200824
Sharpglume Water Hymenacue 200824
Sharphead Sagebrush 36032
Sharpleaf 410450
Sharpleaf Acmena 4879
Sharpleaf Actinidia 6531
Sharp-leaf Adinandra 8210
Sharpleaf Agapetes 10307
Sharpleaf Ash 167942,168116
Sharpleaf Atalantia 43719
Sharpleaf Baby's-breath 183158
Sharp-leaf Baby's-breath 183158
Sharp-leaf Beautiful Azalea 330286
Sharp-leaf Beautiful Rhododendron 330286
Sharpleaf Begonia 50129
Sharpleaf Bigumbrella 59188
Sharpleaf Bluebell Comastoma 100269
Sharpleaf Bluebell Throathair 100269
Sharpleaf Bulbophyllum 62621
Sharpleaf Bushclover 226860
Sharpleaf Cancerwort 216262
Sharpleaf Cayratia 79860
Sharpleaf Cotoneaster 107322
Sharp-leaf Cotoneaster 107322
Sharpleaf Cowparsnip 192301
Sharp-leaf Currant 334087
Sharpleaf Cyrtococcum 120630
Sharpleaf Cystopetal 320249
Sharpleaf Euonymus 157761
Sharpleaf Eurya 160416
Sharpleaf Everlasting 21507,21658
Sharpleaf Fig 165097
Sharpleaf Fighazel 134907
Sharpleaf Four-involucre 124882
Sharpleaf Galangal 17736
Sharpleaf Gambirplant 401773
Sharpleaf Garcinia 171203
Sharpleaf Gooseberry 334228
Sharpleaf Ground Cherry 297641
Sharpleaf Hookvine 401773
Sharp-leaf Jacaranda 211235,211239
Sharpleaf Kiwifruit 6531
Sharpleaf Lacquertree 393443
Sharpleaf Ligusticum 229284
Sharpleaf Lilyturf 272051
Sharpleaf Mangrove 329745
Sharpleaf Meadowrue 388401
Sharpleaf Microula 254338
Sharpleaf Newlitse 264018
Sharpleaf Oak 324251
Sharpleaf Ophiorrhiza 272214
Sharpleaf Oxtail Greenbrier 366553
Sharpleaf Pearleverlasting 21507
Sharp-leaf Pearleverlasting 21658
Sharpleaf Pimpinella 299346
Sharpleaf Pinnate Groundsel 263513
Sharpleaf Pondweed 312206
Sharpleaf Premna 313596
Sharpleaf Pricklyash 417294
Sharpleaf Privet 229522
Sharpleaf Purple Willow 343579
Sharpleaf Sabia 341570
Sharpleaf Sedge 74922
Sharp-leaf Senna 78227
Sharpleaf Spiradiclis 371747
Sharpleaf Sporoxeia 372895
Sharpleaf Stonebean-orchis 62621
Sharpleaf Stonecrop 356733
Sharpleaf Swallowwort 117690
Sharpleaf Tanoak 233115
Sharpleaf Valerian 404216
Sharpleaf Vineclethra 95489
Sharp-leaf Willow 342961
Sharpleaf Winter Grape 411635
Sharp-leafed Willow 342961
Sharpleaf-like Eurya 160417
Sharp-leaved Acacia 1107
Sharp-leaved Acmena 4879
Sharp-leaved Agapetes 10307
Sharp-leaved Ash 168116
Sharp-leaved Asparagus 38908
Sharp-leaved Aster 268682
Sharp-leaved Barberry 51532
Sharp-leaved Beardtongue 289316
Sharp-leaved Blueberry 403792
Sharp-leaved Cancerwort 216262
Sharp-leaved Cayratia 79860
Sharp-leaved Currant 334228
Sharp-leaved Eurya 160416
Sharp-leaved Fighazel 134907
Sharp-leaved Fluella 216262
Sharp-leaved Fluellin 216262
Sharp-leaved Gambir-plant 401773
Sharp-leaved Goldenrod 367965
Sharp-leaved Gooseberry 334228
Sharp-leaved Hiptage 196822
Sharp-leaved Jacaranda 211226,211239,
211241
Sharp-leaved Lacquer-tree 393443
Sharp-leaved Mangrove 329745
Sharp-leaved Oak 324251
Sharp-leaved Pondweed 312036
Sharp-leaved Premna 313596
Sharp-leaved Prickly-ash 417294
Sharp-leaved Privet 229522
Sharp-leaved Sabia 341570
Sharp-leaved Spindle-tree 157761
Sharp-leaved Tanoak 233115
Sharpleaved Violet 409643
Sharp-leaved Yellow-wood 306397
Sharplemma Barley 198366
Sharplemma Paspalidium 285387
Sharplobe Briggsia 60252
Sharplobe Hepatica 192137
Sharplobed Adjoin Clearweed 298865
Sharplobed Daphne 122361
Sharp-lobed Daphne 122361
Sharp-lobed Hepatica 23698,192123,192137
Sharp-lobed Liverleaf 192123
Sharplobed Yellow Toadflex 230870
Sharpperianth Lily 229906
Sharpperula Camellia 68907
Sharppetal Begonia 49599
Sharp-petal Bredia 59846
Sharppetal Gansu Sandwort 31974
Sharppetal Indigo 205619
Sharppetal Loosestrife 239629
Sharppetal Pimpinella 299396
Sharppetal Sinocarum 364883
Sharp-petal Tiger Jaws 162899
Sharp-petaled Bletia 55549
Sharppinnate Groundsel 358184
Sharp-point Cleistogenes 94615
Sharp-pointed Fluellin 216262
Sharp-pointed Pernettya 291367
Sharp-pointed Rabdosia 209780
Sharp-scale Goldenweed 150320
Sharpscale Sedge 75652
Sharp-scaled Oak Sedge 73627
Sharpsepal Falsehellebore 405627
Sharpsepal Pittosporum 301397
Sharpsepal Seatung 301397
Sharpsepal Spapweed 205308
Sharpsepal Tarenna 384910
Sharpsepal Ternstroemia 386718
Sharp-sepaled Pittosporum 301397
Sharp-sepaled Tarenna 384910
Sharp-sepaled Ternstroemia 386718
Sharpserrateleaf Eurya 160579
Sharp-sheath Pseudosasa 318279
Sharpspike Bluntleaf Grass 375773
Sharpspine Herelip 220090
Sharpspine Lagochilus 220090
Sharpspine Peashrub 72249

Sharpspine Rose 336835
Sharp-spined Pea-shrub 72249
Sharpspiny Rose 336835
Sharpsplit Slimtop Meadowrue 388668
Sharpspur Corydalis 106432
Sharpspur Larkspur 124440
Sharpspurred Larkspur 124440
Sharptail Clematis 94808
Sharptail Copperleaf 1811
Sharp-tipped Aphananthe 29009
Sharp-tipped Mountain-ash 369375
Sharptongue Milkvetch 42843
Sharptooth 314118
Sharptooth Blumea 55800
Sharptooth Buckthorn 328609
Sharp-tooth Buckthorn 328609
Sharptooth Camellia 68888
Sharptooth Catnip 265094
Sharptooth Cherry 223119
Sharptooth Cherry-laurel 223119
Sharptooth Forrest Chirita 87867
Sharptooth Hornsage 205548
Sharptooth Incarvillea 205548
Sharptooth Indigo 205673
Sharptooth Milkvetch 42844
Sharptooth Mountainash 369321,369413
Sharptooth Nepeta 265094
Sharptooth Oak 323635
Sharptooth Photinia 295625
Sharptooth Pimpinella 299353
Sharptooth Premna 313596
Sharptooth Redcarweed 288708
Sharptooth Seseli 361506
Sharptooth Shiny Cinquefoil 312628
Sharptooth Slatepentree 400685
Sharptooth Slender Heterolamium 193818
Sharptooth Touch-me-not 204782
Sharptooth Veinyroot 237485
Sharptooth Wet Clearweed 298868
Sharp-toothed Buckthorn 328609
Sharptoothed Cinquefoil 312628
Sharp-toothed David Milkvetch 42270
Sharp-toothed Eurya 160579
Sharp-toothed Indigo 205673
Sharp-toothed Mahonia 242485
Sharptoothed Maple 2793
Sharp-toothed Mint 250385
Sharp-toothed Mountain-ash 369321
Sharp-toothed Photinia 295625
Sharp-toothed Willow 343812
Sharp-toothed Woolly Sedge 74806
Sharptoothleaf Willow 343812
Sharpvein Litse 233830
Sharp-veined Litse 233830
Sharpwing Monkey Flower 255187
Sharp-winged Monkeyflower 255187
Sharp-winged Monkey-flower 255187
Sharsmith Pincushion 84479
Shartspike Pair Vetch 408296
Shaslagh 19766
Shasta County Arnica 34783
Shasta Daisy 89402,227467,227512
Shasta Fir 409,412
Shasta Pincushion 84527
Shasta Red Fir 409,412
Shasta Star-tulip 67608
Shasta Valley Thistle 92365
Shasta Wild Buckwheat 152417
Shattercane 369707
Shattercane Gaoliang 369600
Shatterwood 46339
Shaved Sedge 76561
Shaving Brush 81237
Shaving Brush Palm 332438
Shaving Brush Plant 184347
Shaving Brush Tree 279381,317435
Shavings Stonecrop 357069
Shaw's Royle Woodbetony 287642
Shawan Fritillary 168624
Shawan Hawksbeard 111018
Shawan Kengyilia 215963
Shawan Spunsilksage 360871
Shawnee Wood 79260
Shawnee-salad 200488
Shaxian Masson Pine 300060
She Balsam 377
She Broom 173082
She Bulkishawn 383874
She Heather 149205,150131
She Holly 203545
She Oak 15953,79155,79164
She Pine 306424
Shea Butter 64207,64223
Shea Butter Tree 64223
Shea Butternut 64223
Shea Butter-seed 64223
Shea Butter-tree 64223
Shea Nut 64207,64223
Sheabutter Tree 64223
Shea-nut Tree 64207
Shearer Corydalis 106432
Shearer Monochasma 257543
Shearer Phoebe 295403
Sheareria 362082
Shear-grass 196818
Shearthed Monochoria 257583
Sheath Addermonth Orchid 243084
Sheath Chainpodpea 18289
Sheath Hemlockparsley 101839
Sheathbearing Arrowbamboo 162660
Sheathbract Netseed 260124
Sheathed Cholla 116674
Sheathed Cotton Sedge 152791
Sheathed Cottonsedge 152791
Sheathed Death Camas 417927
Sheathed Dropseed 372870
Sheathed Drop-seed 372870
SHeathed Greenbrier 366583
Sheathed Monkshood 5556
Sheathed Pondweed 312285,378994
Sheathed Primrose 315075
Sheathed Sainfoin 271278
Sheathed Sedge 76656
Sheathedleaf Gagea 169470
Sheathflower 241313,241317
Sheathing Bluegrass 306123
Sheathing Larkspur 124138
Sheathingbase Windhairdaisy 348220
Sheathless Goldenray 229055
Sheathpalm Primrose 315075
Sheathstalk Larkspur 124138
Sheathstipe Globeflower 399548
Sheathstipe Greenbrier 366583
Sheathstipe Smilax 366583
Sheath-stiped Greenbrier 366583
Shebalsam 377
She-balsam 377
She-barfoot 190959
She-broom 173082
She-bulkishawn 383874
Shed Arrowbamboo 162650
Shedand Mouse-ear Hawkweed 232483
Shedhair Azalea 330300
Sheegie Thimbles 130383
Sheep 300223
Sheep Bells 70271
Sheep Burr 1751
Sheep Bush 172478
Sheep Fat 44355
Sheep Fescue 163850,164126,164364
Sheep Fescuegrass 164126
Sheep Foot 237539
Sheep Laurel 215387,215391
Sheep Rot 200388
Sheep Scabious 211670
Sheep Soórag 277648
Sheep Sorrel 277644,339897
Sheep Sorrel Dock 339887,339897
Sheep Souce 339887
Sheep's Beard-like Sedge 74524
Sheep's Bit 211639
Sheep's Bit Scabious 211672
Sheep's Brisken 373346
Sheep's Cheese 144661
Sheep's Ear 190557,373166
Sheep's Ears 190557,373166
Sheep's Fescue 164126
Sheep's Gowan 397043
Sheep's Herb 302091
Sheep's Knapperty 312520
Sheep's Laurel 215387
Sheep's Parsley 28044,84768

Sheep's Sorrel 277648,277688,339897
Sheep's Sorrell 339887
Sheep's Sourack 339887
Sheep's Sourock 339897
Sheep's Tails 106703
Sheep's Thistle 91770
Sheep's Thyme 391365
Sheep's-beard 402660
Sheep's-bit 211639,211672
Sheep's-bit Jasione 211672
Sheep's-bit Scabious 211672
Sheep's-bit Sheep's Bit 211670
Sheep's-rescue 164126
Sheep's-scabious 211672
Sheepbane 200388
Sheepberry 407917
Sheepbine 102933
Sheepbit 211672
Sheep-bur 1739
Sheepbush 290222
Sheepear Jurinea 138888
Sheepeye Jurinea 207209
Sheepeye-like Jurinea 207210
Sheep-fat 44355
Sheepforage Giantfennel 163689
Sheephorn Aloe 16681
Sheephorn Maple 3758
Sheephorn Sedge 74051
Sheepkill 215387
Sheep-killing Penny-grass 200388
Sheep-laurel 215386,215387
Sheeploitered Azalea 331257
Sheep-root 299759
Sheep-rot 200388
Sheep-shearing Rose 280231
Sheep-Soórag 277648
Sheepsorrel 277644
Sheep-sorrel Dock 339897
Sheep-souce 339887
Sheepweed 182103
Sheer Burr 1744
Sheesham 121826
Sheet Paleface Rosemallow 194831
Shefherd's Needle 261585
She-heather 145073,149205,150131
She-holly 203545
Shejila Rockfoil 349894
Shekel Basket 60382,329521
Shekel Box 60382,329521
Sheldon's Pincushion 244226
Shell Begonia 49732
Shell Flower 17774,86783,256761,391627
Shell Ginger 17639,17774
Shellbark Hickory 77905,77920,77934
Shellfish Pricklyash 417216
Shellfish Prickly-ash 417216
Shell-fish-like Prickly-ash 417250
Shellflower 17774,86781,256761
Shell-flower 17774,256761,263307,391627
Shellflower Bush 57985
Shell-leaf Blueberry 403777
Shell-leaf Penstemon 289349
Shell-leaved Blueberry 403777
Shellplant 17774
Shellseed 98100
Shell-shaped Epidendrum 146399
Shemis 229384
Shen's Mahonia 242637
Shengnongjia Holly 204262
Shengtangshan Rhododendron 330494
Shenlong Corydalis 106435
Shennong Arrowbamboo 162717
Shennong Corydalis 106435
Shennong Cyclobalanopsis 116195
Shennong Oak 116195
Shennong Qinggang 116195
Shennongjia Chirita 87980
Shennongjia Clematis 95296
Shennongjia Fargesia 162717
Shennongjia Sagebrush 36281
Shennongjia Umbrella Bamboo 162717
Shensi Clematis 95297
Shensi Fir 306
Shensi Hawthorn 110040
Shensi Hornbeam 77388
Shensi Meadowrue 388664
Shensi Monkshood 5566
Shensi Rhubarb 329366
Sheo Kan 93734
She-oak 15947,15953,79155,79164
She-oak Family 79193
Shepher's Daisy 50825
Shepherd's Bag 72038
Shepherd's Bags 72038
Shepherd's Barometer 21339
Shepherd's Bedstraw 39328
Shepherd's Blue Thyme 307969
Shepherd's Bodkin 350425
Shepherd's Calendar 21339
Shepherd's Clock 21339,384714,394333
Shepherd's Club 405788
Shepherd's Comb 350425
Shepherd's Cress 385549,385555
Shepherd's Daisy 50825
Shepherd's Daylight 21339
Shepherd's Delight 21339,95431,407907
Shepherd's Dial 21339
Shepherd's Dock 21339
Shepherd's Flock 30220
Shepherd's Friend 369339
Shepherd's Glass 21339
Shepherd's Joy 21339
Shepherd's Knot 312520
Shepherd's Love 307969
Shepherd's Myrtle 340990
Shepherd's Needle 261585,350425
Shepherd's Pansy 410187
Shepherd's Pedlar 72038
Shepherd's Pocket 72038
Shepherd's Pouch 72038
Shepherd's Pounce 72038
Shepherd's Purse 72037,72038,220790,237539,238371,308454,329521
Shepherd's Rod 133478,133502
Shepherd's Scabious 211670
Shepherd's Scrip 72038
Shepherd's Staff 133478,405788
Shepherd's Sundial 21339
Shepherd's Thyme 307969,308454,391365
Shepherd's Warning 21339
Shepherd's Watch 21339
Shepherd's Weatherglass 21339,374916
Shepherd's Weather-glass 21339
Shepherd's Yard 133478
Shepherd's-needle 350402,350425
Shepherd's-purse 72037,72038
Shepherd's-scabious 211672
Shepherd's Pouch 275135
Shepherds Purse 72038
Shepherdspurse 72037
She-pine 306424
Sherard's Downy-rose 336946
Shergaon Remarkable Barberry 51770
Sheridan Mahonia 242638
Sherman Pass Sulphur Flower 152571
Sherriff Albizia 13671
Sherriff Barberry 52147
Sherriff Bluebell 115421
Sherriff Bluebellflower 115421
Sherriff Corydalis 106437
Sherriff Cotoneaster 107684
Sherriff Larkspur 124578
Sherriff Primrose 314944
Sherriff Woodbetony 287672
Sherriff's Rhododendron 331816
Sherves 369541
Sherwin Grade Sulphur Flower 152572
Shes Butter Seed 64223
Shetland Mouse-ear 82673
Shetland Pondweed 312259
Shevock's Goldenaster 194253
Shevock's Wild Buckwheat 151868
Shibata Bamboo 362121
Shibataea 362121
Shibataea-like Indosasa 206898
Shickle Shacklers 60382
Shickle-shacklers 60382
Shick-shack Tree 324335
Shield Aralia 310237
Shield Clover 396869
Shield Cress 225428
Shield Dock 340254

Shield Fern 138575
Shield Floatingheart 267825
Shield Ivy 187285
Shieldfruit 391419,391421
Shieldleaf Macaranga 240227
Shield-leaved Duck's Foot 306636
Shieldstyle 304892
Shieldwort 288818
Shifang Goldenray 229177
Shifun Grape 411918
Shiggy 109857
Shigucao Swertia 380353
Shiha Raspberry 339266
Shihezi Licorice 177943
Shihtsuan Willow 344104
Shihwe Maple 3598
Shiikuwasha 93448
Shikikitsu 93566
Shilin Clearweed 298911
Shilin Coldwaterflower 298911
Shilling-grass 200388
Shilling-rot 200388
Shillings 238371
Shilly Thim Bles 130383
Shilly Thimbles 130383
Shilu Michelia 252955
Shimada Aster 215366
Shimada Cowlily 267327
Shimada Fairybells 134484
Shimada Formosa Fig 164995
Shimada Gardneria 171455
Shimada Hedge Bamboo 47370
Shimada Lespedeza 226751
Shimada Tutcher Maple 3731
Shimada's Fig 164995
Shimei Holly 204263
Shimen Azalea 331817
Shimen Barberry 51976
Shimen Hornbeam 77390
Shimen Rhododendron 331817
Shimian Arisaema 33503
Shimian Azalea 331818
Shimian Chloranthus 88285
Shimian Corydalis 106438
Shimian Loosestrife 239845
Shimian Monkshood 5568
Shimian Primrose 314945
Shimian Rhododendron 331818
Shimian Southstar 33503
Shimian Trinerve Poplar 311556
Shimmies 68713,374916
Shimmies-and-shirt 68713
Shimmies-and-shirts 68713,374916
Shimmy-and-buttons 68713
Shimmy-shirt 68713
Shimmy-shirts 68713
Shin Oak 323927,324172
Shinano Bluegrass 305970
Shindagger 10881,10937
Shine Coelogyne 98701
Shine Gaoliang 369677
Shine Ungeargrass 36690
Shiners 37015
Shineseed 337802
Shingan Erysimum 154451
Shingle Oak 79161,324041
Shingle Plant 258165,258168
Shingle Tree 5983
Shingled Pricklypear 272924
Shinglewood 390645
Shingle-wood 386666
Shining Ajania 13015
Shining Artocarpus 36937
Shining Aster 40797,40911,380890
Shining Bedstraw 170325
Shining Begonia 49860
Shining Bird of Paradise 190031
Shining Blue Star 20854
Shining Bur Sedge 74904
Shining Camellia 69480
Shining Cock's-comb 194523
Shining Cotoneaster 107593
Shining Covering Rhododendron 330040
Shining Cranesbill 174711
Shining Crazyweed 279037
Shining Damascisa 176959
Shining Devilpepper 327041
Shining Dumasia 138938
Shining Eucalyptus 155668
Shining Euphorbia 159270
Shining Eurya 160553
Shining Everlasting 21596
Shining Fargesia 162718
Shining Figwort 355175
Shining Flat Sedge 118565
Shining Fleabane 150491
Shining Geranium 174711
Shining Gum 155668
Shining Gymnanthera 182326
Shining Habenaria 183820
Shining Hypserpa 203023
Shining Islandvine 182326
Shining Ixora 211091
Shining Jasmine 211927
Shining Ladies'-tresses 372227
Shining Lady's Tresses 372227
Shining Leaf Birch 53510
Shining Leaves Jewelvine 125964
Shining Luvunga 238534
Shining Mahonia 242607
Shining Manglietia 244453
Shining Meadowrue 388556
Shining Parakmeria 283421
Shining Pegia 287988
Shining Peppermint 155668
Shining Poisonnut 378833
Shining Pondweed 312168
Shining Raspberry 338768
Shining Rhododendron 331001
Shining Rose 336807
Shining Sage 345166
Shining Scabious 350191
Shining Sedge 75233
Shining Smythea 366749
Shining Sorghum 369677
Shining Spapweed 205102
Shining Speargrass 4163
Shining Spiculate Bluegrass 305780
Shining Spike-rush 143237
Shining Spurge 159270
Shining Starwort 375021
Shining Sterculia 376112
Shining Sumac 332529
Shining Sumach 332529
Shining Sweetvetch 188129
Shining Touch-me-not 205102
Shining Valerian 404314
Shining Wedge Grass 371492
Shining Willow 343633
Shining Wintersweet 87521
Shining Yam 131741
Shining-bark Birch 53510
Shining-branch Hawthorn 109628
Shining-calyx Milkvetch 42641
Shiningfruit Dragon Mallow 366934
Shiningfruit Nightshade 366934
Shininggreenutricle Sedge 74540
Shiningleaf Adinandra 8261
Shiningleaf Appendicula 29817
Shining-leaf Beech 162390
Shiningleaf Birch 53510
Shiningleaf Buckthorn 328754
Shiningleaf Cotoneaster 107594
Shiningleaf Distylium 134940
Shiningleaf Elaeocarpus 142363
Shining-leaf Falsenettle 56192
Shiningleaf Millettia 66637
Shiningleaf Mosquitoman 134940
Shiningleaf Rose 336714
Shining-leaved Aster 40488
Shining-leaved Birch 53510
Shining-leaved Cotoneaster 107594
Shining-leaved Millettia 66637
Shining-leaved Rhododendron 330741
Shining-leaved Violet 410186
Shining-lepidote Rhododendron 330841
Shining-nuted Tanoak 233328
Shiningpurple Skullcap 355554
Shining-white Milkvetch 42140
Shinko Gordonia 322612
Shinleaf 322775,322823

Shin-leaf 275486,322804,322872
Shinleaf Yellow-horn 415096
Shinlock 154019
Shinner's Sunflower 189018
Shinners' Gayfeather 228545
Shinners' Three-awn 33830
Shinners' Three-awned Grass 33830
Shinnery Shin 323998
Shinscale Azalea 330841
Shiny Albizzia 13609
Shiny Barleria 48256
Shiny Bugleweed 239215
Shiny Bugseed 104816
Shiny Calathea 66180
Shiny Camwood 47788
Shiny Cinquefoil 312627
Shiny Coneflower 339593
Shiny Cotoneaster 107531
Shiny Falsecuckoo 48256
Shiny Fig 165379
Shiny Glasswort 342878
Shiny Goldenrod 368284
Shiny Indianmulberry 258900
Shiny Leaf Prickly-ash 417282
Shiny Paperelder 290401
Shiny Peperomia 290401
Shiny Rhododendron 332055
Shiny Strychnos 378833
Shiny Willow 343633
Shiny Xylosma 415869
Shiny-bract Rabbitbrush 150350
Shinygreen Creeper 285130
Shiny-green Creeper 285130
Shiny-green Lime 391753
Shinygreen Linden 391753
Shiny-green Linden 391753
Shinyleaf Argyreia 32660
Shinyleaf Aster 40911
Shinyleaf Blueberry 403884
Shinyleaf Buckthorn 328722
Shinyleaf Camellia 69480
Shinyleaf Ford Fishvine 125964
Shinyleaf Ford Jewelvine 125964
Shinyleaf Galangal 17650
Shinyleaf Glycosmis 177839
Shinyleaf Heteropanax 193912
Shinyleaf Holly 204088
Shinyleaf Ixora 211091
Shiny-leaf Magnolia 283421
Shinyleaf Michelia 252879
Shinyleaf Mitreola 256214
Shinyleaf Oak 116177
Shiny-leaf Parakmeria 283421
Shinyleaf Pricklyash 417282
Shinyleaf Rose 336714
Shinyleaf Ternstroemia 386722
Shinyleaf Willow 343968
Shinyleaf Yellowhorn 415096
Shiny-leaf Yellowhorn 415096
Shiny-leaved Adinandra 8261
Shiny-leaved Argyreia 32660
Shiny-leaved Aster 380890
Shiny-leaved Blueberry 403884
Shiny-leaved Buckthorn 328754
Shiny-leaved Elaeocarpus 142363
Shiny-leaved Holly 204088
Shiny-leaved Magnolia 283421
Shiny-leaved Michelia 252879
Shiny-leaved Oak 116177
Shiny-leaved Prickly-ash 417282
Shiny-leaved Ternstroemia 386722
Shiny-leaved Willow 342939
Shiny-leaved Yellow-horn 415096
Shiping Tanoak 233385
Shipmast Locust 334976
Shiquan Willow 344104
Shir 369541
Shirasawa Maple 3599
Shirley Meadows Star-tulip 67648
Shirley Poppy 282685
Shirley's Poppy 282685
Shirt 59493,102933,364557
Shirt Buttons 4005,363673
Shirts-and-shimmies 102933
Shisham 121826
Shisham Wood 121826
Shishan Aspidistra 39574
Shishan Elaeagnus 141950
Shishan Newlitse 264033
Shiso 290940
Shit-a-bed 384714
Shittim Bark 328838
Shittim Wood 1518,1588
Shiuying Oligostachyum 270449
Shive 15709
Shiver-grass 60382
Shivering Jimmy 60382
Shiver-shakes 60382
Shiver-tree 311398,311537
Shiwandashan Azalea 331819
Shiwandashan Machilus 240696
Shiwandashan Rhododendron 331819
Shixing Rattlesnake Plantain 179702
Shixing Spotleaf-orchis 179702
Shme Plant 255098
Shoals Spider-lily 200921
Shoalweed 184938
Shockley's Wild Buckwheat 152459
Shoddy Ragwort 359830
Shoe Black 195149
Shoe Button Spiraea 372050
Shoe Flower 195149
Shoe Nut 53057
Shoeblackplant 195149
Shoebutton 31437
Shoebutton Ardisia 31437,31607
Shoe-button Ardisia 31607
Shoe-buttons 381831
Shoeflower 195149
Shoe-flower 195149
Shoemaker's Heel 86970
Shoes 174877
Shoes-and-slippers 23925
Shoes-and-socks 30081,237539
Shoes-and-stocking 30081,37015,237539
Shoes-and-stockings 30081,72934,220346,222784,231183,237539,273531,302103,410493,410677
Shoestring Acacia 1623
Shoestring Smartweed 308739
Shoestrings 70512
Shoe-strings 19997
Shola Pith 9499,9560
Sholgirse 4005
Shoo Fly 265944
Shoofly 64990
Shoo-fly 265944
Shoofly Plant 265944
Shoo-fly Plant 265944
Shortleaf White Fir 323
Shoot Everlasting 21705
Shoot Pearleverlasting 21705
Shooter Yew 385302
Shooting Star 115953,135153,135158,135165
Shooting Stars 135165
Shooting-star 135153
Shootspike Rhubarb 329389
Shop-consound 381035
Shore Aster 381003
Shore Banksia 47649
Shore Buttercup 325749
Shore Dock 340232
Shore Drugs Quill 352960
Shore Eurya 160458
Shore Gentian 173821
Shore Gumweed 181161
Shore Juniper 213739
Shore Orach 44510
Shore Orache 44510
Shore Pine 299869
Shore Pod Grass 397159
Shore Pod-grass 397159
Shore Podograss 397159
Shore Ribbonwood 301612
Shore Rush 213254
Shore Sagebrush 35517
Shore Sand Spurrey 370661
Shore Sea Onion 352960
Shore Sea-onion 352960
Shore Sedge 75106

Shore Shadbush 19290
Shore Weed 234150
Shorea 362186
Shorebay 291505
Shoredt Schizachyrium 351266
Shore-grape 97865
Shoregrass 288847,288848
Shoreline Figwort 355061
Shoreline Sea-purslane 361667
Shorepine 299869
Shoreweed 234146,234150,234151
Shorn Hairs Rhododendron 330560
Shorstyle Parnassia 284511
Short Aglaia 11276
Short Beak Huernia 199076
Short Bud Scales Rhododendron 330255
Short Caudate Wildginger 37584
Short Cluster Plantain Lily 198646
Short Corneredcalyx Larkspur 124117
Short Corydalis 106345
Short Drooping Stringbush 414225
Short Dune-grass 281317
Short Green Milkweed 38168
Short Greenhoods 320874
Short Honeysuckle 235953
Short Inflorescence Falserobust Vetch 408555
Short Inflorescens Vetch 408252
Short Lanceolate Draba 137072
Short Lanceolate Whitlowgrass 137072
Short Larkspur 124536
Short Leaf Aloe 16656
Short Oat 45402
Short Panicle Slugwood 50481
Short Pappus Synotis 381926
Short Petiole Rhodiola 329846
Short Petiole St. John's Wort 202107
Short Petioles Azalea 330358
Short Petioles Rhododendron 330358
Short Primrose 314483
Short Ragweed 19145
Short Rhubarb 329361
Short Sagebrush 35788
Short Shavegrass 196818
Short Spike Tamarisk 383535
Short Spikelet Bamboo 58698
Short Spine Chinkapin 79027
Short Stamen Rhododendron 331280
Short Sweetvetch 188099
Short Threeawngrass 33783
Short Thyrse Slugwood 50478
Short Tinctorial Marsdenia 245855
Short Tinctorial Milkgreens 245855
Short Triawn 33783
Short Tube Lycoris 239266
Short Whorled Milkwort 308440
Short Wildrice 418082
Short Woollyheads 318802
Short's Aster 380983
Short's Goldenrod 368397
Short's Rock-cress 30451
Short's Sedge 76256
Shortacuminate Elatostema 142626
Short-ament Simon Poplar 311492
Shortangle Pellionia 288714
Shortangle Redcarweed 288714
Shortanther Felwort 235433
Shortanther Sonerila 368902
Shortanther Syzygium 382491
Shortanther Tea 68943
Short-anthered Staunton-vine 374384
Short-anthered Syzygium 382491
Shortaristate Jijigrass 4118
Shortauricle Stroganowia 378290
Shortawl Yushanbamboo 416755
Short-awn Foxtail 17498
Shortawn Glabrous-spiked Lymegrass 335235
Shortawn Goldencattail Narenga 262550
Shortawn Hairy-callus Roegneria 335225
Shortawn Japan Barnyardgrass 140388
Shortawn Lymegrass 144546
Shortawn Roegneria 335235
Shortawn Sedge 73911
Shortawn Speargrass 4118
Shortawn Ungeargrass 36618
Shortawn Wildryegrass 144546
Short-awned Ciliate Roegneria 335270
Shortawned Confuse Roegneria 144259
Short-awned Foxtail 17498
Shortawned Roegneria 144259
Shortaxis Rattanpalm 65663
Shortbeak Buttercup 326078
Shortbeak Clearweed 299036
Shortbeak Coldwaterflower 299036
Shortbeak Dandelion 384482
Shortbeak Lettuce 219238
Short-beaked Arrowhead 342319
Short-beaked Brassica 59380
Short-beaked Salsify 394265
Short-beaked Skeletonweed 88766
Short-beaked Smalscale Sedge 74643
Short-beaked Touch-me-not 205288
Shortbentspur Monkshood 5077
Shortbill Rush 213191
Shortbract Arisaema 33286
Shortbract Bastardtoadflax 389625
Shortbract Hazelnut 106717
Shortbract Hibiscus 195300
Shortbract Honeysuckle 236086
Shortbract Hundredstamen 389625
Shortbract Maple 2808
Shortbract Petrosimonia 292725
Shortbract Redroot Amaranth 18811
Shortbract Redroot Amaranthus 18811
Shortbract Roscoea 337071
Shortbract Shrubalthea 195300
Shortbract Thorowax 63572
Shortbract Thrixspermum 390495
Shortbract Windhairdaisy 348173
Shortbranch Fishvine 125944
Short-branch Sasa 351849
Shortbranch Spicegrass 239555
Shortbranch Willow 343119
Short-branched Jewelvine 125944
Short-branched Schizostachyum 351849
Shortbristled Needlegrass 4118
Shortcalyx Craneknee 221748
Short-calyx Euchresta 155904
Shortcalyx Hartia 376454
Shortcalyx Lysidice 239526
Shortcalyx Nabilous Azalea 330353
Shortcalyx Pittosporum 301230
Short-calyx Pittosporum 301230
Shortcalyx Purplestem 376480
Shortcalyx Pyrenaria 322566
Shortcalyx Seatung 301230
Shortcalyx Stewartia 376480
Short-calyx Winged-stem Sonerila 368872
Shortcalyx Wood Didymocarpus 129906
Shortcalyx Yaoshan Pipewort 151551
Short-calyxed Honeysuckle 235683
Short-calyxed Jasmine 211760
Short-calyxed Lysidice 239526
Short-caudate Rhododendron 330253
Short-caudate Rush 212924
Shortcaudate Tanoak 233122
Shortcaudated Tanoak 233122
Shortclustered Plantainlily 198646
Short-column Camellia 68946
Shortconnective Sage 344915
Shortcorn Wet Clearweed 298869
Shortcorolla Ajania 12993
Shortcorolla Pogostemon 306963
Shortcrown Mosquitotrap 117411
Shortcuspidate Sedge 73930
Shortcyme Chirita 87850
Shortdiscid Ladybell 7613
Shortdisck Ladybell 7613
Shortear Bulbophyllum 62665
Shortear Irisorchis 267947
Shortear Oberonia 267947
Shortear Stonebean-orchis 62665
Shortened Aglaia 11276
Shortened Chelonopsis 86806
Shortenpanicle Maizailan 11276
Shortenspike Chelonopsis 86806
Shortest Fargesia 162648
Shortfilament Bugle 13058
Shortfilament Flower 221353
Shortfilament Tulip 400136
Shortflor Azalea 330237
Shortfloss Soja 177792

Shortflower Acuteangular Anisodus 25440
Short-flower Bambusa 47220
Shortflower Dragonhead 137563
Shortflower Feathergrass 376720
Short-flower Formosa Wendlandia 413785
Shortflower Greenorchid 137563
Shortflower Jerusalemsage 295076
Shortflower Leptospermum 226454
Shortflower Needlegrass 376720
Shortflower Pearweed 239563
Shortflower Rattlebox 111975
Shortflower Rhododendron 330237
Short-flower Stauntonvine 374384
Short-flower Sweetleaf 381148
Short-flower Tongol Milkvetch 43165
Shortflower Triadenum 394809
Shortflower Wendlandia 413828
Shortflower Whitebracteole Bugle 13135
Short-flower Wild Buckwheat 151864
Shortflower Woodbetony 287058
Short-flowered Rhododendron 330237
Shortfoot Peashrub 72195
Shortfoot Rose 336418
Short-footed Rose 336418
Short-fringed Knapweed 81243
Shortfruit Bauhinia 49022
Shortfruit Bird's Foot Trefoil 237725
Shortfruit Bugbane 91002
Shortfruit Ema Azalea 331385
Shortfruit Erect Ephedra 146137
Shortfruit Evening Primrose 269417
Shortfruit Glomerate Mouseear 82852
Shortfruit Hedgemustard 365584
Shortfruit Hooktea 52407
Shortfruit Loesel Sisymbrium 365530
Shortfruit Microstigma 254084
Short-fruit Oma Rhododendron 331385
Shortfruit Pantling Woodbetony 287499
Shortfruit Pimpinella 299365
Shortfruit Rush 213143
Shortfruit Slatepentree 400689
Shortfruit Spiny Naiad 262077
Shortfruit Stork's Bill 153743
Shortfruit Styltcress 254084
Shortfruit Tanoak 233211
Shortfruit Tutcheria 400689
Shortfruit Veinyroot 237725
Short-fruited Arabis 30207
Short-fruited Dysentery-herb 258106
Shortfruited Rhododendron 330240
Short-fruited Sisymbrium 365414
Shortfruited Supplejack 52407
Short-fruited Supple-jack 52407
Short-fruited Tutcheria 400689
Short-fruited Willowherb 146805
Shortglandular Meadowrue 388634
Shortglume Armgrass 58168
Shortglume Crabgrass 130656
Shortglume Cunealglume 29450
Shortglume Falsebrome 58622
Shortglume Goosecomb 335238
Shortglume Maned Crabgrass 130617
Shortglume Roegneria 335238
Shorthair Aster 40160
Shorthair Awngrass 255829
Shorthair Bistamengrass 127469
Shorthair Calcicole Nutantdaisy 110360
Shorthair California Buckthorn 328637
Shorthair Campanulate Cremanthodium 110360
Shorthair Chirita 87830
Shorthair Clematis 95264
Shorthair Cowparsnip 192312
Short-hair Goldenrod 368026
Shorthair Linden 391689
Shorthair Metalleaf 296373
Shorthair Onespikegrass 310722
Shorthair Purple Bergenia 52538
Shorthair Silvergrass 255829
Short-hair Simon Poplar 311483
Shorthair Small Reed 127200
Shorthair Thyme 391164
Shorthair Vanilla 404982
Shorthair Wild Smallreed 127200
Shorthaired Bluebell 115418
Short-haired Caladium 127197
Short-haired Linden 391689
Short-haired Stipule Elatostema 142847
Shorthairy Antenoron 26626
Shorthairy Metalleaf 296373
Short-hairy Rose 336991
Short-headed Bracted Sedge 74078
Short-headed Rush 212912
Short-headed Sedge 73933
Shorthelmet Woodbetony 287054
Shorthook Fishhook Cactus 354289
Shorthookedhair Tickclover 126663
Shorthorn Barrenwort 146966
Shorthorn Calanthe 65895
Shorthorn Zexmenia 212295
Shorthorned Epimedium 146966
Short-horned Rice Grass 276055
Shorthusk 58447
Shortia 362242,362246
Shortihook-haired Tickclover 126663
Shortin Florescens Butterflybush 61987
Shortinflorescence Machilus 240562
Shortinflorescence Snakegourd 396155
Short-inflorescence Tischler Barberry 52259
Shortinflorescence Urariopsis 402151
Shortinflorescenced Beijing Mockorange 294413
Short-inflorescenced Machilus 240562
Short-inflorescenced Mahonia 242499
Short-inflorescenced Phoebe 295350
Shortinflorescens Summerlilic 61987
Shortinflorescens Vetch 408296
Shortinvolucre Goldenyellow Thorowax 63572
Shortispined Evergreen Chinkapin 78924
Shortkeel Milkvetch 42853
Shortleaf Aloe 16656
Shortleaf Alpinegold 199212
Shortleaf Anabasis 21034
Shortleaf Arcuate Calanthe 65883
Shortleaf Baccharis or False Willow 46214
Shortleaf Beauverd Photinia 295631
Shortleaf Bethlehemsage 283612
Shortleaf Bushclover 226899
Shortleaf Cedar 80085
Shortleaf Clovershrub 70792
Shortleaf Common Amentotaxus 19358
Shortleaf Cryptomeria 113687
Shortleaf Diarthron 128028
Shortleaf Douglas Fir 318564
Shortleaf Drupetea 322579
Shortleaf Dyckia 139256
Shortleaf Erysimum 154411
Shortleaf Eulalia 156447
Shortleaf Fescuegrass 163847
Shortleaf Galingale 119162
Shortleaf Gayfeather 228542
Shortleaf Gentian 173261
Shortleaf Goldquitch 156447
Short-leaf Hainan Elaeocarpus 142329
Short-leaf Hartia 185972
Shortleaf Holly 203595
Shortleaf Hyacinthus 260299
Short-leaf Jiangxi Barberry 51796
Shortleaf Juniper 213624
Shortleaf Kyllingia 218480
Shortleaf Malaca Cypressgrass 119162
Shortleaf Nepeta 264891
Short-leaf Odontoglossum 269058
Shortleaf Panicgrass 281408
Shortleaf Paraphlomis 283612
Shortleaf Peashrub 72196
Shortleaf Photinia 295642
Shortleaf Pindrow Fir 440
Shortleaf Pine 299925,300264
Short-leaf Pine 299925,300264
Shortleaf Podocarpus 306407
Shortleaf Pyrenaria 322579
Shortleaf Rabdosia 209637
Shortleaf Rattan Palm 65675
Shortleaf Rattanpalm 65675
Shortleaf Redcarweed 288715
Shortleaf Rockfoil 349106
Shortleaf Sagebrush 35221
Shortleaf Sansevieria 346163
Shortleaf Senna 78366

Short-leaf Shrubby Baeckea 46431
Short-leaf Slugwood 50480
Shortleaf Sneezeweed 188412
Shortleaf Splitlemma 351266
Shortleaf Spunsilksage 360822
Shortleaf Stopper 156133
Shortleaf Tainan Greyawngrass 372436
Shortleaf Veratrilla 405568
Shortleaf Water-centipede 218480
Shortleaf Willow Weed 146639
Shortleaf Willowweed 146639
Shortleaf Yaccatree 306407
Shortleaf Yellowish Woodbetony 287387
Shortleaf Yew 385342
Short-leafed Westringia 413912
Short-leaved Barberry 51384
Short-leaved Clovershrub 70792
Short-leaved Douglas Fir 318564
Short-leaved Evax 193380
Short-leaved Kyllinga 218480
Short-leaved Leucothoe 228183
Short-leaved Nervedflower Phoebe 295389
Short-leaved Pea-shrub 72196
Shortleaved Pine 299925
Short-leaved Podocarpus 306407
Short-leaved Pruinose Barberry 52071
Short-leaved Pyrenaria 322579
Short-leaved Rabdosia 209637
Short-leaved Slugwood 50480
Short-leaved Smooth Barberry 51856
Short-leaved Water-starwort 67396
Shortleaves Cassiope 78623
Shortligule Bluegrass 305412
Shortligule Pyrethrum 322724
Short-liguled Ammophila 19773
Shortlimb Monkshood 5081
Shortlimbate Arisaema 33284
Shortlimbate Isometrum 210182
Shortlimbate Southstar 33284
Shortlip Neottia 264663
Shortlip Nestorchid 264663
Shortlip Paphiopedilum 282793
Shortlip Sage 344916
Shortlip Tulip Orchid 25135
Shortlip Woodbetony 287060
Shortlived Beardgrass 310109
Shortlobe Aganosma 10235
Shortlobe Ajania 12994
Shortlobe Jadeleaf and Goldenflower 260390
Shortlobe Ligusticum 229296
Shortlobe Malouetia 243507
Shortlobed Deutzia 126853
Shortlobed Dualbutterfly 398256
Shortlobed Fimbriate Tupistra 70607
Short-lobed Marsdenia 245788
Short-lobed Mussaenda 260390
Short-lobed Oplopanax 272708
Shortlobed Sowthistle 368842
Shortlobed Wormwood 35220
Short-needle Pindrow Fir 440
Shortneedle Pine 299925
Shortnode Bitterbamboo 304012
Shortnode Square Bamboo 87553
Shortnode Thyme 391264
Short-noded Square-bamboo 87553
Shortooth Jerusalemsage 295075
Shortpanicle Maesa 241748
Shortpanicle Rabdosia 324732
Short-panicled Agapetes 10329
Short-paniculate Barberry 51284
Shortpaniculate Stringbush 414216
Short-paniculate Wendlandia 413775
Short-paniculate Yushania 416755
Shortpappo Sage 344914
Shortpappo Tailanther 381926
Shortpappus Goldenray 229022
Shortpedicel Asparagus 39075
Shortpedicel Blueberry 403748
Short-pedicel Elegant Barberry 51484
Shortpedicel Elliott Monkshood 5173
Shortpedicel Glomerulate Brassaiopsis 59210
Short-pedicel Green Holly 204389
Shortpedicel Gugertree 350909
Short-pedicel Heteropanax 193898
Shortpedicel Homocodon 197895
Shortpedicel Indianmulberry 258911
Shortpedicel Newlitse 264030
Shortpedicel Nutmug 261431
Shortpedicel Paris 284402
Short-pedicel Pruinose Barberry 52069
Short-pedicel Slightly Wing-branched Barberry 52208
Short-pedicel Stellate Hairs Rhododendron 330156
Short-pedicel Wisteria 414551
Short-pediceled Blueberry 403748
Short-pediceled Cassiope 78623
Short-pediceled Heteropanax 193898
Short-pediceled Newlitse 264030
Short-pediceled Schefflera 350670
Short-pediceled Sterculia 376088
Shortpedicle Paris 284402
Short-peduncle Blood-red Spindle-tree 157857
Shortpeduncle Bluebeard 78040
Shortpeduncle Corydalis 105665
Shortpeduncle Crazyweed 278755
Short-peduncled Stephania 375832
Short-pedunculate Hubei Spindle-tree 157588
Short-perula Rhododendron 330255
Short-petal Aganosma 10235
Shortpetal Camellia 69109
Shortpetal Chickweed 374777
Shortpetal Daphne 122431
Shortpetal Globeflower 399515
Shortpetal Melandrium 248298
Shortpetal Monkshood 5082
Shortpetal Monomeria 257709
Shortpetal Orchis 257708,257709
Short-petal Osmanthus 276270
Shortpetal Paris 284366
Shortpetal Parnassia 284563
Shortpetal Sandwort 31777
Shortpetal Sisymbrium 365487
Shortpetal Yarrow 4014
Short-petaled Daphne 122431
Shortpetalflower 58965,58966
Short-petalled Campion 363578
Shortpetiolate Loxostigma 238033
Shortpetiole Acanthopanax 143579
Shortpetiole Arisaema 33285
Shortpetiole Azalea 330256
Shortpetiole Camellia 68949
Shortpetiole Cleidion 94406
Short-petiole Cleidion 94406
Short-petiole Enkianthus 145727
Shortpetiole Oak 323942
Short-petiole Pendent-bell 145727
Shortpetiole Raspberry 338204,339486
Shortpetiole Rhodiola 329846
Shortpetiole Serrate Oak 323942
Shortpetiole Southstar 33285
Short-petioled Bennettiodendron 51025
Short-petioled Camellia 69552
Short-petioled Raspberry 338204
Shortpistil Azalea 331240
Shortpistil Camellia 68945
Shortplume Clematis 94778
Shortpod Draba 136954
Shortpod Korshinsk Peashrub 72260
Shortpod Korshinsk Pea-shrub 72260
Shortpod Mustard 196952
Short-podium Rhododendron 330247
Shortpoined Honeysuckle 235973
Shortraceme Abacubeadbean 402151
Shortraceme Arrowwood 407706
Short-raceme Barberry 51373
Shortraceme Blueberry 403736
Shortraceme Clovershrub 70792
Shortraceme Homalium 197646
Short-raceme Mahonia 242499
Shortraceme Mockorange 294413
Short-raceme Neillia 263146
Shortraceme Stauntonvine 374387
Shortraceme Stringbush 414148
Shortraceme Viburnum 407706
Shortraceme Vineyam 26268
Shortraceme Wild Quince 374387
Short-racemed Arrowwood 407706
Short-racemed Blueberry 403736
Short-racemed Butterfly-bush 61987

Short-racemed Staunton-vine 374387
Shortradial Edelweiss 224806
Shortradiate Bigleaf Thorowax 63735
Short-radiate Cyclobalanopsis 116066
Short-ray Fleabane 150763
Shortray Prairie Coneflower 326966
Short-ray Rock Daisy 291325
Shortray Zinnia 418041
Short-rayed Alkali Aster 380898
Short-rostrate Longbeak Eucalyptus 155519
Shortrstar Edelweiss 224806
Short-scale Dropseed 372698
Shortscap Bulbophyllum 62845
Shortscap Curculigo 114818
Shortscap Stonebean-orchis 62845
Shortscape Arachnis 30529
Short-scape Barberry 51388
Shortscape Braya 59780
Shortscape Coelogyne 98617
Shortscape Corydalis 106319
Shortscape Fleabane 150513
Shortscape Gentian 150513
Shortscape Leek 15507
Shortscape Northern Rockjasmine 23296
Shortscape Onion 15507
Shortscape Sedge 73937
Shortscape Stapfiophyton 167292
Shortscape Taeniophyllum 383057
Shortseed Waterwort 142563
Short-sepal Barberry 51811
Shortsepal Drupetea 322566
Shortsepal Goldthread 103830
Short-sepal Homalium 197647
Short-sepal Honeysuckle 235683
Shortsepal Jurinea 211760
Short-sepal Leptodermis 226077
Short-sepal Mycetia 260617
Shortsepal Porana 131245
Shortsepal Pratia 234331
Short-sepal Wildclove 226077
Short-sepaled Pyrenaria 322566
Shortseta Shortawn Ungeargrass 36621
Shortsetose Spikesedge 143348
Short-shoots Jewelvine 125944
Shortspadix Rattan Palm 65652
Shortspik Devil Rattan 121489
Shortspike Alopecurus 17515
Shortspike Bladderwort 403736
Short-spike Branchy Pycreus 322319
Shortspike Brome 60649
Shortspike Bulbophyllum 62601
Shortspike Burreed 370066
Shortspike Canary Grass 293722
Shortspike Canarygrass 293722
Shortspike Cattail 401109
Short-spike Clethra 96478
Shortspike Craneknee 221638
Shortspike Elaeocarpus 142294
Short-spike Eugeissoma 156114
Shortspike Faber Rattanpalm 65684
Shortspike Fishtailpalm 78047
Shortspike Fox Tail 17515
Shortspike Galangal 17741
Shortspike Goatgrass 8728
Shortspike Hairliporchis 395868
Shortspike Lagotis 220162
Shortspike Lepironia 225666
Shortspike Luisia 238309
Shortspike Pogostemon 306965
Shortspike Rattanpalm 65684
Shortspike Sedge 73907
Shortspike Spinestemon 306965
Shortspike Stonebean-orchis 62601
Shortspike Sweetleaf 381148
Shortspike Syzygium 382493
Shortspike Tamarisk 383535
Shortspike Tanoak 233121
Shortspike Tropidia 399641
Short-spike Water-milfoil 261360
Shortspike Whitepearl 172126
Shortspike Wintergreen 172126
Shortspike Woodrush 238672
Shortspike Yinshancress 416351
Shortspike Yinshania 416351
Shortspikebamboo 58697,58698
Short-spiked Barberry 51553
Short-spiked Elaeocarpus 142294
Short-spiked Homalium 197646
Short-spiked Syzygium 382493
Short-spiked Tamarisk 383535
Shortspiked Tanoak 233121
Short-spiked Wintergreen 172126
Short-spikeed Clethra 96478
Shortspikilet Bamboo 58697,58698
Short-spikileted Bamboo 58697
Shortspine Begonia 49682
Shortspine Carles Evergreenchinkapin 78880
Shortspine Chinquapin 78924
Shortspine Damnacanthus 122034
Shortspine Evergreenchinkapin 78924
Short-spine Fishhook 354289
Shortspine Fishhook Cactus 354291
Shortspine Herelip 220068
Shortspine Lagochilus 220068
Short-spine Medick 247382
Shortspine Sanicle 345989
Shortspine Spikesedge 143068
Shortspine Tigerthorn 122034
Short-spined Damnacanthus 122034
Shortspiny Barberry 51372
Short-spiny Cutleaf Barberry 51470
Shortspoke Bracketplant 88541
Short-spur Brachycorythis 58389
Short-spur Columbine 30004
Shortspur Corydalis 105832
Shortspur Hemipilia 191612
Shortspur Holcoglossum 197258
Shortspur Monkshood 5075
Shortspur Neofinetia 263868
Shortspur Orchis 310864
Shortspur Platanthera 302265
Shortspur Rabdosia 209635
Shortspur Reinorchis 182235
Shortspur Spapweed 204825
Shortspur Touch-me-not 204825
Shortspur Violet 409762
Shortspur Windorchis 263868
Short-spurred Fragrant Orchid 182261
Short-stachys Chinese Stachyurus 373531
Shortstachys Forkstyleflower 374473
Shortstachys Heartleaf Hornbeam 77277
Short-stachys Songjiang Willow 344164
Shortstal Arisaema 33527
Shortstalk Acanthopanax 143579
Shortstalk Actephila 6481
Shortstalk Agapetes 10289
Shortstalk Aralia 30778
Shortstalk Ardisia 31602
Short-stalk Arrowbamboo 162647
Shortstalk Arrowwood 407710
Shortstalk Azalea 330247
Shortstalk Balanse Michelia 252820
Short-stalk Barberry 51374
Short-stalk Bennettiodendron 51025
Shortstalk Bentgrass 12020,12219
Shortstalk Bird Cherry 280009
Shortstalk Birdcherry 280009
Shortstalk Bittersweet 80295
Shortstalk Bladderwort 403748
Shortstalk Blastus 55142
Short-stalk Blueberry 403788
Shortstalk Bractquitch 201489
Short-stalk Bredia 59880
Shortstalk Broadfruit Shrubcress 368552
Shortstalk Broadfruit Solms-Laubachia 368552
Shortstalk Bushclover 226746
Shortstalk Bushmint 203042
Shortstalk Camellia 68949
Short-stalk Cane Acanthopanax 2438
Shortstalk Cherry 83193
Short-stalk Chickweed 82750
Short-stalk Clematis 94788
Shortstalk Clethra 96476
Shortstalk Cointree 280552
Short-stalk Copper-leaf 1876
Shortstalk Craneknee 221760
Shortstalk Crazyweed 278755,279190
Shortstalk Daphne 122572
Shortstalk Dewberry 338204
Short-stalk Dewberry 338309

Shortstalk Diels Cherry 83193
Shortstalk Draws Ear Grass 403132
Shortstalk Duthiea 139122
Short-stalk E. H. Wilson Poplar 311569
Short-stalk Echinodorus 140542
Shortstalk Elatostema 142628
Short-stalk Faber Clethra 96497
Shortstalk False Bindweed 68732
Shortstalk Falsepimpernel 231492
Shortstalk Fumitory 169183
Shortstalk Gentian 173929
Shortstalk Greenbrier 366566
Shortstalk Grewia 180707
Shortstalk Guizhou Hazelnut 106750
Short-stalk Hairy Metalleaf 296399
Shortstalk Hemsleia 191913
Shortstalk Hyparrhenia 201489
Shortstalk Iguanura 203504
Shortstalk Indianmulberry 258878
Short-stalk Japanese Camellia 69207
Shortstalk Kobresia 217147
Shortstalk Leathery Holboellia 197225
Shortstalk Leucothoe 228183
Shortstalk Lizardtaro 348025
Shortstalk Lobelia 234360
Short-stalk Loxostigma 238033
Short-stalk Lysionotus 239983
Shortstalk Malcolmia 243195
Short-stalk Melastoma-like Metalleaf 296399
Shortstalk Michelia 252830
Shortstalk Monkshood 5069
Shortstalk Motherweed 231492
Shortstalk Mycetia 260616
Short-stalk Needletooth Myrsine 261651
Shortstalk Nettleleaf Microtoena 254273
Shortstalk Nospurflower 167292
Shortstalk Oak 323944
Shortstalk Olive 270068
Shortstalk Oreorchis 274036
Short-stalk Ovalleaf Camellia 69445
Shortstalk Paliurus 280552
Shortstalk Pendentseed Jointfir 178558
Shortstalk Pommereschea 310830
Shortstalk Primrose 314184
Shortstalk Pruinose Barberry 52069
Shortstalk Purplevine 414551
Shortstalk Raspberry 339486
Shortstalk Redstem Skullcap 355862
Shortstalk Rhododendron 330256
Shortstalk Rinorea 334684
Shortstalk Rockcress 30222
Shortstalk Rockfoil 349107
Shortstalk Roegneria 335240
Short-stalk Rosemaryleaf Willow 344034
Shortstalk Sauromatum 348025
Shortstalk Slimtop Meadowrue 388669
Shortstalk Stairweed 142628
Shortstalk Stephania 375832
Shortstalk Sterculia 376088
Shortstalk Sugerok Holly 204299
Shortstalk Syzygium 382487
Shortstalk Taxillus 385246
Shortstalk Tipularia 392351
Shortstalk Tubergourd 390178
Shortstalk Tuberous Violet 410696
Shortstalk Turnflowerbean 98066
Short-stalk Venulose Holly 204372
Shortstalk Viburnum 407710
Shortstalk Wildorchis 274036
Shortstalk Wildtung 243352
Short-stalk Zhejiang Persimmon 132177
Short-stalked Acanthopanax 143579
Shortstalked Actephila 6481
Short-stalked Agapetes 10289
Short-stalked Ardisia 31602
Short-stalked Barberry 51374
Short-stalked Bedstraw 170284
Short-stalked Bird Cherry 280009
Short-stalked Blastus 55142
Short-stalked Buckthorn 328625
Short-stalked Bush-clover 226746
Short-stalked Camellia 68952
Short-stalked Cayratia 79837
Short-stalked Clethra 96476
Short-stalked Grewia 180707
Short-stalked Indian-mulberry 258878
Short-stalked Mouse-ear Chickweed 82750
Short-stalked Oak 324278
Short-stalked Paliurus 280552
Short-stalked Peashrub 72195
Short-stalked Pea-shrub 72195
Short-stalked Pepper 300515
Short-stalked Puge Monkshood 5509
Short-stalked Rhododendron 330256
Short-stalked Spapweed 204828
Short-stalked St. John's-wort 202173
Short-stalked Stink Corydalis 105908
Short-stalked Taxillus 385246
Short-stalked Viburnum 407710
Short-stamen Blueberry 403735
Short-stamen Camellia 68943
Shortstamen Fairybells 134420
Shortstamen Gentian 173608
Shortstamen Glorybower 95967
Shortstamen Greenvine 204613
Shortstamen Lycoris 239258
Short-stamen Manyprickle Acanthopanax 2471
Short-stamen Pagodatree 368979
Shortstamen Sibbaldia 362359
Short-stamen Sophora 368979
Shortstamen Spicegrass 239561
Shortstamen Stonegarlic 239258
Shortstamen Wildberry 362359
Short-stamened Blueberry 403735
Short-stamened Camellia 68943
Short-stamened Glorybower 95967
Short-stamened Illigera 204613
Short-stamened Sophora 368979
Short-starhair Oak 116066
Short-starhair Qinggang 116066
Shortstem Archangelica 30930
Shortstem Ardisia 31371
Shortstem Aster 40170
Shortstem Begonia 50311
Shortstem Bugle 13070
Shortstem Chinese Groundsel 365069
Shortstem Didymocarpus 129924
Short-stem Encephalartos 145236
Shortstem Epimedium 146965
Shortstem Fordiophyton 167291
Shortstem Gingerlily 187424
Shortstem Hemiboea 191384
Shortstem Holcoglossum 38191
Shortstem Indian Skullcap 355514
Shortstem Jute 104060
Shortstem Milkvetch 42676
Shortstem Primrose 314255
Shortstem Rush 213372
Shortstem Sedirea 356459
Shortstem Sinosenecio 365069
Shortstem Thorowax 63792
Shortstem Threefork Gentian 174002
Short-stem Wild Buckwheat 151869
Shortstem Woodbetony 287018
Shortstem Woolly Draba 137134
Shortstemen Holboellia 197219
Short-stemmed Ardisia 31371
Short-stemmed Bastard-sage 152674
Short-stemmed Iris 208471
Short-stemmed Stephania 375831
Shortstigma Chirita 87829
Shortstigma Dodder 115124
Shortstigma Pearweed 239635
Shortstipe Beautyberry 66755
Shortstipe Cassiope 78623
Shortstipe Clethra 96476
Shortstipe Duthiea 139122
Shortstipe Elsholtzia 143976
Shortstipe Gentian 173929
Shortstipe Giantfennel 163637
Shortstipe Grewia 180707
Shortstipe Leucothoe 228183
Shortstipe Mycetia 260616
Shortstipe Pepper 300515
Shortstipe Purplepearl 66755
Shortstipe St. John's Wort 202107
Shortstipe Sterculia 376088
Short-stiped Beautyberry 66755
Shortstlak Gugertree 350909
Shortstraw Pine 299925

Shortstyle Adonis 8353
Shortstyle Anemone 23735
Shortstyle Azalea 330841
Shortstyle Bowfruitvine 393546
Shortstyle Camellia 68953
Short-style Camellia 68953
Shortstyle Cratoxylum 110251
Shortstyle Dendropanax 125594
Shortstyle Devilpepper 326999
Shortstyle Eightangle 204488
Shortstyle Eurya 160432
Shortstyle Flax 231941
Shortstyle Forkliporchis 269012
Shortstyle Heterosmilax 194129
Short-style Holly 204488
Shortstyle Korean Willow 343576
Shortstyle Listera 264705
Short-style Narrowsepal Xantolis 415242
Shortstyle Pimpinella 299366
Short-style Rhododendron 330841
Shortstyle Rush 212917
Shortstyle Sichuan Fritillary 168655
Shortstyle Speedwell 406981
Shortstyle St. John's Wort 201743
Shortstyle Starjasmine 393630
Short-style Star-jasmine 393630
Short-style Thistle 91822
Shortstyle Treerenshen 125594
Shortstyle Villous Toxocarpus 393546
Shortstyle Wampee 94175
Shortstyle Waxpetal 106630
Shortstyle Wildginger 37572
Shortstyle Winterhazel 106630
Short-styled Camellia 68953
Short-styled Dendropanax 125594
Short-styled Eurya 160432
Short-styled Field-rose 336974
Short-styled Grevillea 180576
Short-styled Holly 204488
Shortstyled Honeysuckle 235795
Short-styled Rhododendron 330841
Short-styled Snakeroot 345945
Short-styled Winterhazel 106630
Shortstyly Craneknee 221698
Short-stypes Oak 323944
Shortsubulate Barley 198267
Shorttail Azalea 330253
Shorttail Bladderwort 403760
Shorttail Hornbeam 77335
Shorttail Sawtooth Elatostema 142643
Shorttail Tanoak 233122
Shorttail Wildginger 37584
Short-tailed Leaf Tanoak 233122
Shorttassell Cremanthodium 110352
Shorttassell Nutantdaisy 110352
Short-tassled Bamboo 58698
Shorttaxis Microstegium 254012
Short-thicked Camellia 69428
Short-thready-atalk Haines Barberry 51697
Short-thyrse Cinnamon 91279
Shortthyrse Cryptocarya 113433
Shortthyrse Nanmu 295350
Short-thyrse Rabdosia 324732
Short-thyrse Shortraceme 295350
Short-thyrse Slugwood 50478
Shortthyrse Thickshellcassia 113433
Short-thyrsed Cryptocarya 113433
Shorttine Azalea 330930
Shorttine Corydalis 106170
Shorttine Stairweed 142626
Short-tomentose Soybean 177792
Shorttongue Indosasa 206874
Shorttongue Poorspikebamboo 270447
Shorttonguedaisy 58038,58041
Shorttooth Blackcalyx Crazyweed 279167
Shorttooth Broomrape 275102
Shorttooth Bulbophyllum 62763
Shorttooth Elatostema 142624
Shorttooth Garlic 15134
Shorttooth Leek 15235
Shorttooth Onion 15235
Shorttooth Ophiorrhiza 272184
Shorttooth Stairweed 142624
Shorttooth Stonebean-orchis 62763
Shorttooth Whitehairy Bethlehemsage 283606
Short-toothed Jasmine 211760
Short-toothed Mountain Mint 321958
Short-toothed Purplevine 414550
Short-toothed Wisteria 414550
Shorttube Arrowwood 407712
Short-tube Daphne 122392
Shorttube Gentian 173894
Shorttube Heraldtrumpet 49356
Shorttube Hippeastrum 196452
Shorttube Knight's Star 196452
Shorttube Lagotis 220163
Shorttube Landpick 308498
Shorttube Lycoris 239266
Shorttube Primrose 314261
Shorttube Purplebract Iris 208799
Shorttube Purplebract Swordflag 208799
Shorttube Qingmingflower 49356
Shorttube Solomonseal 308498
Shorttube Swertia 380165
Shorttube Viburnum 407712
Shorttube Watson Flower 413326
Shorttube Wendlandia 413776
Short-tubed Herald-trumpet 49356
Short-tubed Rhododendron 330248
Short-tubed Viburnum 407712
Shorttwig Schizostachyum 351849
Shortvein Azalea 330254
Shortvein Rhododendron 330254
Shortveined Rhododendron 330254
Shortvelvet Serpentroot 354922
Shortwing Anhui Maple 2792
Shortwing Coliseum Maple 2861
Shortwing Euonymus 157832
Shortwing Milkvetch 42110
Shortwing Sweetvetch 187806
Shortwoolled Goldsaxifrage 90393
Shortwoolled Goldwaist 90393
Shortwoolled Photinia 295720
Short-woolled Photinia 295720
Shosen Out Rhododendron 330609
Shotbeak Corydalis 105666
Shotbush 30712
Shoter 385302
Shotlobed Bipinnate Dragonhead 137560
Shotlobed Bipinnate Greenorchid 137560
Shouan Spurge 159159
Shoucheng Chirita 87963
Shouliang Roegneria 335502
Shoups 336419
Shouyang Sanicle 345969
Shouzhu BambooMi 297230
Shovelweed 72038
Show Geranium 288206
Show Pea 222789
Show Sour-bamboo 4618
Show Washington Geranium 288206
Showcalyx Gentian 173416
Showeli Xantolis 415240
Shower Tree 78204
Showvein Bigginseng 241161
Showvein Rhodoleia 332207
Showvein Skullcap 355715
Showvein Wildberry 362360
Showy Antennaria 26428
Showy Aster 41293
Showy Attalea 44810
Showy Baby's-breath 183191
Showy Babysbreath 183191
Showy Balloonvine 73201
Showy Banksia 47669
Showy Blackberry 338346
Showy Blazing-star 228487
Showy Bleeding Heart 220729
Showy Bleedingheart 220729
Showy Blue Lettuce 219458
Showy Bossiaea 57491
Showy Bottle-brush 67298
Showy Bugloss 21925
Showy Bush Honeysuckle 235674
Showy Buttercup 325518
Showy Centaury 81517
Showy Chloris 88421
Showy Clover 396930
Showy Coneflower 339515,339539,339614
Showy Cotoneaster 107690
Showy Crab Apple 243617

Showy Crabapple 243617
Showy Crotalaria 112682
Showy Curvedstamencress 72944,238017
Showy Daisy 150972
Showy Dandelion 384813
Showy Dewflower 138171
Showy Dicentra 220729
Showy Dryandra 138431
Showy Dustymaidens 84511
Showy Eremostachys 148508
Showy Eriolaena 152693
Showy Evening Primrose 269504
Showy Evening-primrose 269446,269504
Showy Firerope 152693
Showy Fleabane 150972
Showy Fly Honeysuckle 235674
Showy Forsythia 167435
Showy Germander 388246
Showy Goat's-beard 394333
Showy goldeneye 190253
Showy Goldenrod 368422,368430
Showy Halfribpurse 191527
Showy Hebe 186973
Showy Hemilophia 191527
Showy Himalaya Honeysuckle 228321
Showy Himalaya-honeysuckle 228321
Showy Honey Myrtle 248108
Showy Indocalamus 206779
Showy Jasmine 211821
Showy Jerusalemsage 295165,295197
Showy Lady's Slipper 120438
Showy Lady's Slipper Orchid 120438
Showy Lady's-slipper 120438
Showy Larkspur 124595
Showy Lily 230036
Showy Marshweed 230314
Showy Meconopsis 247190
Showy Menodora 250266
Showy Milkweed 38119
Showy Monsonia 258155
Showy Mountain Ash 369377
Showy Mountainash 369377
Showy Mountain-ash 369377
Showy Mullein 405779
Showy Myrtle Milkwort 308210
Showy Navajo Tea 389154
Showy Onion 15665
Showy Or Handsome Pussytoes 26562
Showy Orchid 169899
Showy Orchis 169884,169899
Showy Oxeye 63531
Showy Pancratium 280909
Showy Partridge Pea 78284
Showy Patrinia 285872
Showy Penstemon 289377
Showy Pigeonpea 65156
Showy Rattlebox 112682
Showy Sandspike 148508
Showy Sandwort 31934
Showy Sedge 119602
Showy Skullcap 355747
Showy Spapweed 205262
Showy Stonecrop 200802
Showy Sun Rose 188863
Showy Sunflower 188987
Showy Thisde 92290
Showy Tick Trefoil 126278
Showy Tickclover 126296
Showy Tick-trefoil 126278
Showy Touch-me-not 205262
Showy Vetch 408419,410016
Showy Waterhyacinth 141808
Showy Wattle 1167
Showy White Thoroughwort 158257
Showy Windmill Grass 88417
Showy Woodbetony 287034,287794
Showyflower Embelia 144789
Showy-flowered Embelia 144789
Showyfruit Devil Rattan 121490
Shred-leaf Staghorn Sumac 332919
Shreve Iris 208826
Shreve's Iris 208933
Shrimp Begonia 49877,50012
Shrimp Plant 66725,214379,279769
Shrimpalga 297181
Shrimpfeelergrass 362082,362083
Shrimpfeelerswood 413828
Shrimpflower 414716
Shrimpgrass 263018,263020
Shrimpplant 214379
Shrimp-plant 66725
Shrimptail-orchis 283704,283706
Shrine Jimmyweed 209584
Shrink Forkliporchis 332337
Shrpleaf Alyxia 18509
Shrpleaf Dischidia 134029
Shrub Althea 195269,195271,195276,195278,195284,195288
Shrub Angelica 143677
Shrub Aster 41752
Shrub Azalea 330603
Shrub Bush Clover 226969
Shrub Bushclover 226698
Shrub Chastetree 411464
Shrub Chaste-tree 411464
Shrub Goldilocks 89863,89871
Shrub Harebell 210313
Shrub Lespedeza 226698
Shrub Live Oak 324509
Shrub Maple 3490
Shrub Milkvetch 42313
Shrub Morning-glory 207792
Shrub Mountain-ash 369377
Shrub Nypa 267838
Shrub of Beltaine 68189
Shrub Peashrub 72237
Shrub Primrose 314328
Shrub Ragwort 58468
Shrub Schizostachyum 351853
Shrub Senna 360426
Shrub Sweetvetch 187881
Shrub Thoroughwax 63668
Shrub Verbena 221229,221238
Shrub Watercoconut 267838
Shrub Yellow Root 415164,417124
Shrub Yellowroot 415164,415166,417124
Shrub Yellow-root 415164,417124
Shrubalthea 195269
Shrubb Vinca 217869
Shrubbamboo 388747
Shrubberry Rhodiola 329866
Shrubby 306469
Shrubby Ajania 12998
Shrubby Alkali Tansy-aster 33037
Shrubby Althaea 195269
Shrubby Althea 195269
Shrubby Amorpha 20005
Shrubby Anisochilus 25395
Shrubby Anzacwood 310769
Shrubby Argyranthemum 32563
Shrubby Aster 253455
Shrubby Asterbush 41752
Shrubby Asterothamnus 41752
Shrubby Atraphaxis 44260
Shrubby Baeckea 46430
Shrubby Basil 268518
Shrubby Begonia 49857
Shrubby Bindweed 102998
Shrubby Buckwheat 44270
Shrubby Bullseye 178750
Shrubby Bush-clover 226698,226969
Shrubby Calendula 66467
Shrubby Camphor-weed 305115
Shrubby Cananga 70961
Shrubby Cinquefoil 100259,289709,289713,312542,312574,312577,312580,312581
Shrubby Clematis 94920
Shrubby Coldenia 392358
Shrubby Daisybush 276608
Shrubby Decaspermum 123448
Shrubby Deervetch 237741
Shrubby Dendrotrophe 125794
Shrubby Dillenia 130926
Shrubby Elaeagnus 141995
Shrubby Elsholtzia 144023
Shrubby Everlasting 189814
Shrubby False Buttonweed 370857
Shrubby Falsenettle 56145
Shrubby False-nettle 56145
Shrubby Five-fingers 289709,289713,362389
Shrubby Flemingia 166865

Shrubby Flueggea 167092
Shrubby Fragrant-bamboo 87634
Shrubby Fuchsia 168779
Shrubby Germander 388085
Shrubby Goatwheat 44260
Shrubby Hare's Ear 63668
Shrubby Hare's-ear 63668
Shrubby Hawkweed 195491
Shrubby Honeysuckle 236208
Shrubby Hypericum 201886
Shrubby Ichnocarpus 203325
Shrubby Jerusalem Sage 295111
Shrubby Jointvetch 9535
Shrubby Kopsia 217869
Shrubby Leucas 227715
Shrubby Marguerite 32563
Shrubby Marsh Cinquefoil 100259
Shrubby Mayweed 270996
Shrubby Meliosma 249385
Shrubby Milkweed 179032
Shrubby Milkwort 307983
Shrubby Musk 255192
Shrubby Nightshade 367564
Shrubby Orache 44446
Shrubby Oreocnide 273903
Shrubby Osier 343398
Shrubby Parasiticvine 125794
Shrubby Peashrub 72233
Shrubby Pea-shrub 72237
Shrubby Pencilflower 379243
Shrubby Penstemon 289323
Shrubby Pimpernel 21383
Shrubby Pinanga 299661
Shrubby Plantain 302170
Shrubby Potentilla 289713
Shrubby Primrose Willow 238229
Shrubby Restharrow 271395
Shrubby Rhodamnia 329794
Shrubby Russian Thistle 344758
Shrubby Rusty Petals 222451
Shrubby Sabia 341551
Shrubby Saltwort 379616
Shrubby Scorpion-vetch 105317
Shrubby Seablite 379616
Shrubby Sea-blite 379616
Shrubby Sealavender 230782
Shrubby Securinega 167092
Shrubby Senna 360434,360497
Shrubby Shrimpflower 414718
Shrubby Simpoh 130926
Shrubby Spapweed 204966
Shrubby Speedwell 407129
Shrubby Spicebush 231401
Shrubby Spice-bush 231401
Shrubby St. John's-wort 202103,202160
Shrubby St. Johnswort 201886,202103
Shrubby Sundrops 269438
Shrubby Sweetvetch 187881
Shrubby Syzygium 382538
Shrubby Tiquilia 392358
Shrubby Touch-me-not 204966
Shrubby Trefoil 320071
Shrubby Umbrella Thoroughwort 217110
Shrubby Verbena 232483
Shrubby Veronica 186925
Shrubby Violet 409692
Shrubby White Cinquefoil 312442
Shrubby Whitevein 345772
Shrubby Woodbetony 287758
Shrubby Woodfordia 414718
Shrubby Woodnettle 273903
Shrubby Wormwood 35146
Shrubby Yew Podocarpus 306457,306469
Shrubby Yew Pine 306457
Shrubby Zinnia 418036
Shrubcress 368546
Shrub-harebell 210313,367861
Shrublover Chirita 87869
Shrun Catchfly 363431
Shuajingsi Monkshood 5183
Shuangbai Sedge 76259
Shuanghu Alkaligrass 321384
Shuanghu Beadcress 365382
Shuanghu Torularia 365382
Shuanghua Indianmulberry 258918
Shuangjiang Mayten 246933
Shuangliu Bamboo 47433,47440
Shuangliu Bambusa 47433,47440
Shuangpai Hilliella 196305
Shucheng Knotweed 309790
Shucheng Sedge 76260
Shui Elatostema 142838
Shui's Chirita 87964
Shuicheng Buttercup 326362
Shuicheng Epimedium 147060
Shuicheng Larkspur 124579
Shuiding Milkvetch 43111
Shui-Hsa 252540
Shuishan Small Reed 127299
Shuishan Smallreed 127299
Shuishe Glochidion 177181
Shuliang Yam 131522
Shumard Oak 324415
Shumard Red Oak 324415
Shumard's Oak 324415
Shumard's Red Oak 324415
Shun Oak 323982
Shunas 229384
Shunchang Paraphlomis 283651
Shungiku 89481
Shunkokan 93640,93764
Shunning Hackberry 80737
Shunning Lycianthes 238976
Shunning Red Silkyarn 238976
Shunning Tanoak 233338,233363
Shuteria 362292
Shuttle-cock Oak 116174
Shuttleleaffruit 139781
Shuttleworth Combbush 292627
Shuttleworth's Ginger 37719
Shuzhi Elatostema 142839
Shuzhi Stairweed 142839
Shwelien Bladdernut 374108
Shwelien Bladder-nut 374108
Shwelien Helicia 189951
Shweli-river Rhododendron 331821
Shy Wallflower 154483
Shy Widow 168476
Shy Widows 168476
Shyness Azalea 331578
Siam 316819
Siam Acuminate Glochidion 177186
Siam Aglaonema 11361
Siam Alyxia 18527
Siam Anchorstyleorchis 130034
Siam Bractquitch 201588
Siam Copperleaf 1983
Siam Corydalis 106439
Siam Crapemyrtle 219965
Siam Crape-myrtle 219965
Siam Cycas 115897
Siam Disperis 134326
Siam Drymoda 138518
Siam Eulophia 157008
Siam Falsenettle 56312
Siam Garrettia 171541
Siam Glorybower 96096
Siam Gurjun 133553
Siam Hyparrhenia 201588
Siam Luohanfruit 365333
Siam Mahonia 242517
Siam Ninenode 319837,319898
Siam Passionflower 285702
Siam Psychotria 319898
Siam Rattan Palm 65789
Siam Reevesia 327373
Siam Rosewood 121653
Siam Senna 360481
Siam Siraitia 365333
Siam Suoluo 327373
Siam Thyrsostachys 391460
Siam Wallichia 413099
Siam Weed 89299
Siamese Cycas 115897
Siamese Ginger 17677
Siamese Gurjun 133553
Siamese Senna 360481
Siamweed 89299
Sib Magnoliavine 351078
Siba Milkvetch 42590
Sibbaldia 362339

Siberia Adonis 8381
Siberia Alder 16362
Siberia Bentgrass 12323
Siberia Bluebells 250911
Siberia Bluegrass 305973
Siberia Bromegrass 60969
Siberia Burnet 345917
Siberia Catchfly 363691
Siberia Catnip 265048
Siberia Chorispora 89013
Siberia Clematis 95302
Siberia Cocklebur 415046
Siberia Columbine 30074
Siberia Columbine Meadowrue 388428
Siberia Cowparsnip 192352
Siberia Crees 47956
Siberia Currant 333955
Siberia Edelweiss 224893
Siberia Elder 345692
Siberia Elm 401602
Siberia Falseoat 398528
Siberia Filbert 106736
Siberia Flax 231942
Siberia Fritillary 168506
Siberia Gentian 154946
Siberia Globeflower 399494
Siberia Goldenray 229179
Siberia Hawksbeard 111019
Siberia Juniper 213934
Siberia Knotweed 309792
Siberia Landpick 308641
Siberia Larch 221944
Siberia Lettuce 219806
Siberia Linedaisy 166066
Siberia Madwort 18469
Siberia Merremia 250831
Siberia Milklettuce 219806
Siberia Mosquitotrap 117346
Siberia Motherwort 225009
Siberia Mountainash 369521
Siberia Patrinia 285870
Siberia Petrosimonia 292726
Siberia Phlox 295318
Siberia Pine 300197
Siberia Rocket 193444
Siberia Rockfoil 349895
Siberia Salsify 394345
Siberia Saltbush 44646
Siberia Stone Pine 300197
Siberia Swallowwort 117346
Siberia Sweetclover 249213
Siberia Sweetgrass 196135
Siberia Sweetvetch 188104
Siberia Swordflag 208829
Siberia Thorowax 63827
Siberia Thyme 391378
Siberia Tickseed 104844
Siberia Tournefortia 393269
Siberia Waldsteinia 413034
Siberia Waterchestnut 394542
Siberia Wheatgrass 11735
Siberia Whitlowgrass 137237
Siberia Windflower 23919
Siberia Yarrow 4028
Siberian Alder 16362,16473
Siberian Altelnateleaf Goldsaxifrage 90331
Siberian Anemone 23919
Siberian Apricot 34470
Siberian Aster 41257
Siberian Barberry 52149
Siberian Barley 198360
Siberian Bellflower 70305
Siberian Bluegrass 305973
Siberian Brome 60969
Siberian Buckwheat 162335
Siberian Bugloss 61366
Siberian Bugseed 104824
Siberian Burnet 345917
Siberian Campion 363691
Siberian Carpet Cypress 253164
Siberian Carpet Grass 253164
Siberian Catchfly 364062
Siberian Cedar 299842,300197
Siberian Centaurea 81378
Siberian Chickweed 375094
Siberian Chives 15709
Siberian Chorispora 89013
Siberian Clematis 95302
Siberian Cocklebur 415046
Siberian Columbine 30038,30074
Siberian Corydalis 106453
Siberian Cow Parsnip 192352
Siberian Cow-parsley 192352
Siberian Crab 243555
Siberian Crab Apple 243555
Siberian Crabapple 243555,243558
Siberian Crab-apple 243555
Siberian Crane's-bill 174906
Siberian Cranesbill 174906
Siberian Currant 333955,334203
Siberian Cypress 253163,253164
Siberian Dock 340256
Siberian Dogwood 104920,104928,105207
Siberian Draba 137237
Siberian Elder 345692
Siberian Elm 401602,417540
Siberian Falseoat 398528
Siberian Falsespiraea 369277
Siberian Fawnlily 154946
Siberian Feathergrass 4160
Siberian Filbert 106736
Siberian Filifolium 166066
Siberian Fir 476
Siberian Flag 208829
Siberian Fritillary 168506
Siberian Geranium 174906
Siberian Ginseng 143657
Siberian Globeflower 399494
Siberian Globethistle 140722
Siberian Goldenray 229179
Siberian Golumbine Meadowrue 388428
Siberian Graybeard 372428
Siberian Hazel 106736
Siberian Hazelnut 106736
Siberian Hemarthria 191251
Siberian Iris 208806,208829
Siberian Jellygrass 393269
Siberian Jijigrass 4160
Siberian Juniper 213934
Siberian Knotweed 309792
Siberian Larch 221944
Siberian Larkspur 124237
Siberian Leucanthemum 227528
Siberian Lily 210999
Siberian Linden 391832
Siberian Lymegrass 144466
Siberian Madwort 18469
Siberian Marshmarigold 68210
Siberian Melic 248951
Siberian Melicgrass 248951
Siberian Melick 248951
Siberian Merremia 250831
Siberian Messerschmidia 393269
Siberian Milkwort 308359
Siberian Millet 361794
Siberian Miners' Lettuce 94362
Siberian Monkshood 5055
Siberian Morningglory 250831
Siberian Motherwort 225009
Siberian Mountain Ash 369512
Siberian narrow-leaved Claytonia 94294
Siberian Nepeta 265048
Siberian Nitraria 266377
Siberian Orange Lily 229828
Siberian Patrinia 285870
Siberian Pea Shrub 72180
Siberian Pea Tree 72180
Siberian Peashrub 72180,72340
Siberian Pea-shrub 72180,72340
Siberian Pea-tree 72180
Siberian Pepperweed 225455
Siberian Pine 300197
Siberian Polemonium 307243
Siberian Primrose 314948
Siberian Pygmyweed 109404
Siberian Rocket 193444
Siberian Rockfoil 349895
Siberian Salt Tree 184838
Siberian Saltbush 44646
Siberian Salttree 184838
Siberian Salt-tree 184838

Siberian Saxifrage 52503,52532
Siberian Scilla 353065
Siberian Sea Lavender 230801
Siberian Sea Rosemary 190652
Siberian Sealavender 230631
Siberian Sea-lavender 230801
Siberian Skullcap 355799
Siberian Solomonseal 308641
Siberian Speargrass 4160
Siberian Spodiopogon 372428
Siberian Spruce 298382
Siberian Squill 353065
Siberian St-John's-wort 201840
Siberian Stone Pine 300197
Siberian Sweetclover 249213
Siberian Thistle 92268
Siberian Thorowax 63827
Siberian Veronicastrum 407485
Siberian Vetch 408693
Siberian Violet-willow 342961
Siberian Wallflower 14964,154369,154467,154505
Siberian Wheatgrass 11735
Siberian White Fir 427
Siberian White Pine 300197
Siberian Whitethorn 266377
Siberian Whitlowgrass 137237
Siberian Wildrye 144466
Siberian Willow Dock 340256
Siberian Winteraconite 148111
Siberian Woodrush 238694
Siberian Wormwood 35101
Siberian Yellow Pine 300197
Siberian Zigadenus 417925
Sibhald's Potentilla 362361
Sibiraea 362392
Sibirea 362392
Sibirian Sainfoin 271267
Sibthorp Sage 345389
Sibthorp's Sage 345389
Sibthorpia Europaea 401711
Sibuyan Fig 165663
Siccocolous Chelonopsis 86836
Sichuan Acanthopanax 143634
Sichuan Acronema 6223
Sichuan Acuminate Eurya 160417
Sichuan Adonis 8382
Sichuan Ailanthus 12563
Sichuan Ainsliaea 12641
Sichuan Alpine Pine Dwarf-mistletoe 30918
Sichuan Alpinepine Dwarfmistletoe 30918
Sichuan Angelica 24465
Sichuan Arborvitae 390661
Sichuan Ash 168097
Sichuan Asiabell 98417
Sichuan Asparagus 39198
Sichuan Aspidistra 39575
Sichuan Aster 41244
Sichuan Azalea 331911
Sichuan Barberry 52150
Sichuan Barrenwort 147067
Sichuan Bastard Cress 390232
Sichuan Bedstraw 170359
Sichuan Bentgrass 12061
Sichuan Bitterbamboo 304098
Sichuan Bittercress 72829
Sichuan Bladderwort 403770,403832
Sichuan Blueberry 403832
Sichuan Bluegrass 306050
Sichuan Boloflower 205552
Sichuan Bushbeech 178042
Sichuan Bush-beech 178042
Sichuan Caesalpinia 64983
Sichuan Calanthe 66122
Sichuan Camellia 69684
Sichuan Carpesium 77187
Sichuan Catchfly 364094
Sichuan Ceropegia 84089
Sichuan Cherry 83340
Sichuan Chinaberry 248925
Sichuan China-berry 248925
Sichuan Chirita 87965
Sichuan Chloranthus 88299
Sichuan Cinnamon 91449
Sichuan Citrus 93462
Sichuan Clovershrub 70855
Sichuan Corydalis 105688
Sichuan Crazyweed 279164
Sichuan Currant 334206
Sichuan Cycas 115909
Sichuan Cyclea 116036
Sichuan Cymbidium 116814
Sichuan Cypressgrass 119657
Sichuan Deerdrug 242703
Sichuan Deutzia 127093
Sichuan Dogwood 380510
Sichuan Ducksmeat 372304
Sichuan Dutchmanspipe 34147
Sichuan Dwarfmistletoe 30918
Sichuan Eightangle 204515
Sichuan Elm 401455,401553,401634
Sichuan Enkianthus 145724
Sichuan Epimedium 147067
Sichuan Eria 299643
Sichuan Euonymus 157906
Sichuan Eurya 160417
Sichuan Everlasting 21707
Sichuan Evodia 161381
Sichuan Eyebright 160240
Sichuan False Solomonseal 242703
Sichuan Fieldorange 249665
Sichuan Filbert 106740
Sichuan Forkstyleflower 374494
Sichuan Fritillary 168563
Sichuan Galangal 17756
Sichuan Galeobdolon 170026
Sichuan Galingale 119657
Sichuan Gentian 173948
Sichuan Goldenpoppy 379233
Sichuan Goldenray 229049
Sichuan Gordonia 179731
Sichuan Gymnaster 256310
Sichuan Habenaria 184114
Sichuan Hairorchis 299643
Sichuan Hartia 376476
Sichuan Headflower Woodbetony 287070
Sichuan Helwingia 191190
Sichuan Hemsleia 191967
Sichuan Himalayahoneysuckle 228325
Sichuan Holly 204311
Sichuan Honeysuckle 236138
Sichuan Hydrangea 200152
Sichuan Illicium 204515
Sichuan Incarvillea 205552
Sichuan Indigo 206637
Sichuan Iris 208834
Sichuan Isometrum 210191
Sichuan Jasmine 212052
Sichuan Jerusalemsage 295202
Sichuan Kadsura 214960
Sichuan Keiskea 215818
Sichuan Kobresia 217280
Sichuan Leek 15794
Sichuan Lilac 382298
Sichuan Lily 229829
Sichuan Lilyturf 272150
Sichuan Litse 234006
Sichuan Lobelia 234407
Sichuan Machilus 240699
Sichuan Maddencherry 241503
Sichuan Madden-cherry 241503
Sichuan Mahonia 242534
Sichuan Manglietia 244471
Sichuan Maple 3648
Sichuan Marsdenia 245844
Sichuan Melia 248925
Sichuan Michelia 252975
Sichuan Microtropis 254332
Sichuan Milkvetch 43044
Sichuan Mockorange 294521
Sichuan Monkeyflower 255244
Sichuan Monstruocalamus 258184
Sichuan Mosquitotrap 117713
Sichuan Mountainash 369520
Sichuan Mouseear 83042
Sichuan Nakeaster 256310
Sichuan Newlitse 264101
Sichuan Onion 15794
Sichuan Ophiorrhiza 272296
Sichuan Orchid 110654
Sichuan Orchis 116814,310933

Sichuan Ostryopsis 276848
Sichuan Parnassia 284518
Sichuan Paulownia 285959
Sichuan Pear 323251
Sichuan Pearleverlasting 21707
Sichuan Peashrub 72251
Sichuan Pea-shrub 72251
Sichuan Pecteilis 285959
Sichuan Pendent-bell 145724
Sichuan Pennisetum 289258
Sichuan Pennycress 390232
Sichuan Peony 280180
Sichuan Pepper 417301
Sichuan Persimmon 132423
Sichuan Petrocosmea 292580
Sichuan Pink 127880
Sichuan Pipewort 151221
Sichuan Pleione 304276
Sichuan Podocarpium 200745
Sichuan Pollia 307333
Sichuan Poplar 311514
Sichuan Pricklyash 417348
Sichuan Primrose 315028
Sichuan Pyrola 322919
Sichuan Queen Lily 114878
Sichuan Rabbiten-wind 12641
Sichuan Rabdosia 209828
Sichuan Raspberry 339259
Sichuan Reedbentgrass 65488
Sichuan Rhododendron 331911
Sichuan Rhombic-leaf Spicebush 231447
Sichuan Ringvine 116036
Sichuan Rockfoil 349956
Sichuan Rockjasmine 23309
Sichuan Roscoea 337104
Sichuan Sabia 341556
Sichuan Sagebrush 36283
Sichuan Sandwort 32267
Sichuan Secamone 356303
Sichuan Sedge 76461
Sichuan Sichuania 362423
Sichuan Sida 362672
Sichuan Silvergrass 255913
Sichuan Sinobambusa 258184
Sichuan Small Reed 127239
Sichuan Smallreed 127239
Sichuan Smoketree 107316
Sichuan Smoke-tree 107316
Sichuan Speedwell 407398,407400
Sichuan Spicebush 231435
Sichuan Spice-bush 231435
Sichuan Spiny-fruited Spindle-tree 157278
Sichuan Splitmelon 351767
Sichuan Square Bamboo 87616
Sichuan Square-bamboo 87616
Sichuan Stachyurus 373576
Sichuan Stephania 375903
Sichuan Stinkcherry 241503
Sichuan Stonebutterfly 292580
Sichuan Stylophorum 379233
Sichuan Sugarmustard 154398
Sichuan Sumac 332969
Sichuan Swallowwort 117713
Sichuan Sweetleaf 381390
Sichuan Swordflag 208834
Sichuan Synotis 381952
Sichuan Syzygium 382670
Sichuan Tailanther 381952
Sichuan Taxillus 385234
Sichuan Tea 69266
Sichuan Ternstroemia 386730
Sichuan Thistle 92304
Sichuan Tipularia 392353
Sichuan Tubergourd 390122
Sichuan Valerian 404351
Sichuan Vineclethra 95548
Sichuan Violet 410622
Sichuan Weasel-snout 170026
Sichuan White Birch 53572
Sichuan Whitebractcelery 266976
Sichuan Whitepearl 172066
Sichuan Willow 343512
Sichuan Windhairdaisy 348835
Sichuan Wintergreen 172066
Sichuan Wolftailgrass 289258
Sichuan Woodbetony 287723
Sichuan Woodlotus 244471
Sichuan Woodreed 65488
Sichuan Woodrush 238695
Sichuang Dichocarpum 128941
Sichuang Forkfruit 128941
Sichuan-Guizhou Glorybower 96014
Sichuan-Guizhou Machilus 240572
Sichuan-Hubei Tea 69573
Sichuania 362422
Sichuan-Shaanxi Globeflower 399497
Sichuan-Xizang Rabdosia 209794
Sichuan-Yunnan Alder 16340
Sichuan-Yunnan Chesneya 87257
Sichuan-Yunnan Cinquefoil 312533
Sichuan-yunnan Groundsel 358381
Sichuan-Yunnan St. John's Wort 202106
Sichuan-Yunnan St. John's-wort 202106
Sicilian Beet 53257
Sicilian Chamomile 26860
Sicilian Fennel 167170
Sicilian Fir 426
Sicilian Melilot 249225
Sicilian Star-thistle 81410
Sicilian Sumac 332533
Sicilian Sumach 332533
Sicily 172901
Sicily Calendula 66434
Sickle Alfalfa 247457
Sickle Awlquitch 131008
Sickle Elatostema 142856
Sickle Euphorbia 158876
Sickle Grass 309877
Sickle Hare's Ear 63625
Sickle Medic 247457
Sickle Nailorchis 9286
Sickle Neofinetia 263867
Sickle Plant 109264
Sickle Pod 129136
Sickle Senna 360493
Sickle Spurge 158895
Sickle Stairweed 142856
Sickle Windorchis 263867
Sicklebamboo 137841,137849
Sicklecalyx Calanthe 66037
Sicklecalyx Comastoma 100271
Sicklecalyx Crazyweed 278825
Sicklecalyx Spapweed 204914
Sicklecalyx Throathair 100271
Sicklecalyx Touch-me-not 204914
Sicklefruit Azalea 330749
Sicklefruit Fenugreek 397229
Sicklefruit Hypecoum 201611
Sickle-fruited Fenugreek 397211
Sickle-grass 283656,295555
Sicklehairicot 135399,135646
Sickleleaf Aster 40416
Sickleleaf Bladderwort 404019
Sickleleaf Catchfly 363566
Sickleleaf Coldwaterflower 299043
Sickleleaf Dacrydium 121098
Sickleleaf Dishspurorchis 171828
Sickleleaf Eucalyptus 155566
Sickleleaf Gastrochilus 171828
Sickleleaf Goldaster 90193
Sickleleaf Grewia 180773
Sickleleaf Mouseear 82808
Sickleleaf Passionflower 285719
Sickleleaf Peashrub 72187
Sickleleaf Primrose 314360
Sickleleaf Seepweed 379513
Sickleleaf Spinestemon 306973
Sickleleaf Tickseed 104776
Sickleleaf Yaccatree 306426
Sickle-leaved Albizia 162473
Sickle-leaved Dew-plant 220549
Sickle-leaved Hare's-ear 63625
Sicklelobe 139537,139539
Sicklepod 360461
Sickle-pod 30214,360461,360493
Sicklepod Milkvetch 42021
Sickle-shap Azalea 330678
Sickle-shap Leaves Rhododendron 330678
Sickle-shaped Helicia 189930
Sickleweed 162465
Sickle-weed 162465

Sicklewort 13073,13174
Sicktepod 30214
Sida 362473
Side Oat 45533
Sideballs Pyrola 275486
Sidebark Pencil-flower 379239
Side-bells Pyrola 275486
Side-cluster Milkweed 37983
Side-flowering Aster 40704,40706,40708,40709
Side-flowering Skullcap 355563
Sideoats Grama 57922,57924
Side-oats Grama 57924
Sideritis 362728
Siderodendron 211029
Sidesaddle Flower 347156
Side-saddle Flower 347144
Side-saddle Plant 347156
Sidewinder Vriesea 412360
Sids 235334
Siebeld Aspen 311480
Sieber Crocus 111597
Sieber Statice 230758
Sieber's Crocus 111597
Siebold Acanthopanax 143677
Siebold Alder 16458
Siebold Ardisia 31599
Siebold Arrowwood 408131
Siebold Ash 168100
Siebold Barberry 52151
Siebold Beech 162368
Siebold Blueberry 403998
Siebold Buttercup 326365
Siebold Calanthe 66068
Siebold Campion 364064
Siebold Chinese Arborvitae 302734
Siebold Crabapple 243717
Siebold Cranesbill 174684
Siebold Deutzia 127099
Siebold Elder 345679
Siebold Euphorbia 159841
Siebold Filbert 106777
Siebold Greenbrier 366573
Siebold Hemlock 399923
Siebold Maple 3604
Siebold Pipewort 151268
Siebold Primrose 314952
Siebold Ramie 56140
Siebold Rust-coloured Fimbristylis 166313
Siebold Serrate Holly 204255
Siebold Smilax 366573
Siebold Spindle-tree 157879
Siebold Spurge 159841
Siebold Stonecrop 200798
Siebold Viburnum 408131,408132
Siebold Walnut 212626
Siebold Wildginger 37722
Siebold Willow 344112
Siebold's Arrowwood 408131
Siebold's Barberry 52151
Siebold's Lily 229775
Siebold's Magnolia 279249
Siebold's Maple 3604
Siebold's Plantainlily 198646
Siebold's Primrose 314952
Siebold's Raspberry 339270
Sieglingia 122180
Siekle Medick 247457
Sie-la Rhododendron 331779
Sierra Alder 16435
Sierra Ancha Fleabane 150458
Sierra Arnica 34754
Sierra Blanca Aankle-aster 207396
Sierra Boykinia 58023
Sierra Chinkapin 79019,89973
Sierra Foothills Fawn-lily 154934
Sierra Foothills Silverpuffs 253910
Sierra Gooseberry 334175
Sierra iris 208608
Sierra Juniper 213834,300125
Sierra Laurel 228154,228161
Sierra Leone Butter 289476
Sierra Leone Copal 181770
Sierra Leone Copal Tree 103696
Sierra Leone Copal-tree 103696
Sierra Leone Gum Copal 103696
Sierra Lessingia 227121
Sierra Lily 229981
Sierra Live Oak 324556
Sierra Madre Acacia 1473
Sierra Madre Larkspur 124368
Sierra Madre St. Johnswort 202067
Sierra Madre Yucca 416622
Sierra Negra 121888
Sierra Nevada Claytonia 94339
Sierra Nevada Fawn-lily 154940
Sierra Nevada Gooseberry 334175
Sierra Nevada Pine 300301
Sierra Nevada Sulphur Flower 152583
Sierra Palm 172224,172227
Sierra Pincushion 84513
Sierra Plum 316833
Sierra Pussytoes 26561
Sierra Redwood 360561,360576
Sierra Sedum 356972
Sierra Starwort 318514
Sierra Sundrop 68506
Sierra Sweet-bay 261168
Sierra Wax Myrtle 261168
Sierra White Fir 407
Sierra-brodiaea 398797
Sierran Cluster-lily 60519
Sierran Crest Wild Buckwheat 152329
Sierran Cushion Wild Buckwheat 152360
Sierra-redwood 360576
Sieva Bean 294010
Sievabean 294010
Sieve Bean 294010
Sieve Sedge 75003
Sieve-like Willow 343228
Sievers Apple 243698
Sievers Wormwood 36286
Sieversia 363057
Sieves 213021
Sifton Bush 78608
Sigapore Holly 243523
Sigesbeckia 363090
Sigillate Walnut 212652
Siglaton 76817
Sigmoid Woodbetony 287675
Signal Grass 58088
Signal-grass 58052,402497
Signdaisy 143894
Signet Marigold 383107
Sigrim 359158
Sikang Small Reed 127295
Sikayo Sageretia 342193
Sikayotaizan Rhododendron 331830
Sikiyou Cypress 114662
Sikkim Ash 168103
Sikkim Aster 41264
Sikkim Barberry 52153
Sikkim Begonia 50305
Sikkim Bentgrass 12306
Sikkim Blueberry 404000
Sikkim Bluegrass 305980
Sikkim Clematis 95298
Sikkim Coriaria 104706
Sikkim Cowslip 314961
Sikkim Cucumber 114250
Sikkim Dandelion 384806
Sikkim Dendrocalamus 125508
Sikkim Draba 137238
Sikkim Dragonbamboo 125508
Sikkim Ephedra 146250
Sikkim Euphorbia 159845
Sikkim Felwort 235466
Sikkim Fir 341
Sikkim Gentian 173895
Sikkim Holly 204270
Sikkim Iris 208836
Sikkim Larch 221881
Sikkim Lasianthus 222269
Sikkim Leek 15744
Sikkim Lycianthes 238971
Sikkim Mahonia 242597
Sikkim Maple 3610
Sikkim Microula 254366
Sikkim Onion 15744
Sikkim Osbeckia 276166
Sikkim Platanthera 302509

Sikkim Primrose 314961
Sikkim Pyrola 322910
Sikkim Raspberry 339272
Sikkim Red Silkyarn 238971
Sikkim Roughleaf 222269
Sikkim Roughleaf Bedstraw 170222
Sikkim Rush 213450
Sikkim Sage 345390
Sikkim Siraitia 365334
Sikkim Spruce 298437
Sikkim Stonecrop 356763
Sikkim Sweetvetch 188105
Sikkim Vernalgrass 28005
Sikkim Violet 410580
Sikkim Whitlowgrass 137238
Sikkim Willow 344113
Sikkim Willowweed 146881
Sikkim Wrightia 414817
Sikledcalyx Gentian 173430
Sikleleaf Hogfennel 292850
Sikok Loosestrife 239847
Sikoku Arisaema 33505
Sikoku Arrowwood 407804
Sikoku Birch 53614
Sikoku Bladderwort 403996
Sikoku Fir 502
Sikoku Linden 391743
Sikoku Mockorange 294532
Sikoku Pipewort 151380
Sikoku Purplepearl 66927
Sikoku Rose 336954
Sikoku Sachalin Fir 467
Sikoku Southstar 33505
Sikoku Stachyurus 373547
Sikoku Stringbush 414257
Sikoku Yellowwood 94032
Sikta Burnet 345911
Sikta Columbine 30034
Sikta Vetch 408407
Siku Aster 41265
Silaus 363123
Silberpappel 311208
Silene 363141
Siler Fishhook Cactus 354318
Siler Pincushion Cactus 287908
Siler's Pincushion Cactus 287908
Silesian Willow 344115
Silgreen 358062
Silinderplakkie 108904
Siliquamomum 364245
Siliquamon 364245
Silique Braya 59770
Silique Parrya 285030
Silk Bay 291540
Silk Cogongrass 205517
Silk Cotton 80120
Silk Cotton Tree 56770,56784,56802,80120,98113
Silk Cottontree 98116
Silk Cotton-tree 80120
Silk Fig 260253
Silk Floss Tree 88973
Silk Flower 235
Silk Grass 416649
Silk Leaf Inula 207229
Silk Meadowrue 388509
Silk Mockorange 294540
Silk Oak 73530,180561,180642
Silk Oak Grevillea 180642
Silk Reed 265923
Silk Rubber 169229
Silk Satintail 205517
Silk Spindle Tree 157345
Silk Tassel 171545,171546
Silk Tassel Bush 171546
Silk Tree 13474,13578,13595,180561
Silk Vine 291031,291063
Silk Weed 38135
Silk-ball Bamboo 125441
Silkcalyx Gentian 173443
Silk-Cotton 80114
Silk-Cotton Family 56761
Silk-cotton Tree 56770,56802,80120,98113
Silken Cicely 38135
Silkfabric Bauhinia 49135
Silk-floss Tree 88973
Silkfruit Willow 344088
Silkglume Needlegrass 376730
Silkgrass 416584
Silk-grass Goldenaster 194215
Silkhair Dallisgrass 285533
Silkhair Globethistle 140753
Silkhair Paspalum 285533
Silkhairy Willow 343639
Silk-leaf 219795
Silkleaf Corydalis 105873
Silk-leaf Heteroplexis 194013
Silkleaf Iris 208558
Silkleaf Lagascea 219799
Silkleaf Ligusticum 229325
Silkleaf Onion 15282
Silkleaf Pipewort 151485
Silkleaf Pondweed 378986
Silkleaf Pyrethrum 322641
Silkleaf Serpentroot 354841
Silkleawn Needlegrass 376737
Silk-like Machilus 240588
Silklike Rush 213114
Silknode Rush 212999
Silkpalm 413289,413298
Silkpetal Gentian 173429
Silkpetal Habenaria 183937
Silk-plant 56229,56233
Silkreed 265923
Silkrubber 169229
Silk-Rubber 169226
Silk-Rubber Tree 169229
Silks And SATINS 238371
Silks-and-satins 238371
Silkspike Chloranthus 88276
Silksplit Habenaria 183983
Silkstalk Elatostema 142663
Silkstalk Speedwell 407125
Silkstalk Stairweed 142663
Silkstalk Triratna Tree 397361
Silkstem Bellflower 69962
Silkstem Knotweed 309421
Silkstem Milkvetch 42366
Silkstyle Gentian 173444
Silk-tassel Bush 171546,171550
Silk-tassel Tree 171545,171546
Silktree 13578
Silk-tree 13578
Silk-tree Albizia 13578
Silktree Albizzia 13578
Silk-tree Albizzia 13578
Silktree Siris 13578
Silkvine 291031,291063
Silk-vine 291031,291063
Silkvineleaf Wissadula 414527
Silk-vine-leaved Wissadula 414527
Silkweed 37811,38135
Silkwood 166976,166981
Silkworm Mulberry 259067
Silkwormbean 408393
Silky Acacia 1525
Silky Anisochilus 25393
Silky Apera 28972
Silky Apricot 34436
Silky Aster 41230
Silky Barberry 51368
Silky Beach Pea 222760
Silky Bent Grass 28969,28971,118307
Silky Bent-grass 28972
Silky Bluestem 128586
Silky Blumea 55820
Silky Bush-clover 226742
Silky Camellia 376452
Silky Cinquefoil 312969
Silky Corallodiscus 103952
Silky Cornel 104949
Silky Cottonseed 253846
Silky Creeping Cinquefoil 312931
Silky Daisy Bush 270189
Silky Dogwood 104949,104951,105133
Silky Dwarf Morning Glory 161450
Silky Flossy 344383
Silky Frosty Cinquefoil 312632
Silky Grevillea 180646
Silky Guinea-flower 194670
Silky Hair Maple 3590

Silky Indigo 206297
Silky Jackbean 71074
Silky Kangaroo Grass 389339
Silky Ladypalms 329178
Silky Lupin 238485
Silky Lupine 238485
Silky Machilus 240588
Silky Naroowleaf Peashrub 72354
Silky Naroowleaf Pea-shrub 72354
Silky Net Bush 68068
Silky Oak 73530,166982,180642
Silky Osier 344244
Silky Peashrub 72244
Silky Perennial Sagebr 35505
Silky Phacelia 293175
Silky Pincution 244015
Silky Prairie-clover 121908
Silky Purple Flag 285797
Silky Raspberry 338751
Silky Rock Sallow 344049
Silky Rose 336930
Silky Rosewood 121822
Silky Sibbaldia 362373
Silky Silverweed Cinquefoil 312372
Silky Sophora 306285
Silky Spicebush 231433
Silky Taxillus 385233
Silky Thatching Grass 201493
Silky Tickclover 126588
Silky Wildberry 362373
Silky Wild-rye 144524
Silky Willow 342986,343762,344085,344122
Silky Wistaria 414584
Silky Woadwaxen 173030
Silky Wormwood 35411
Silky-bent 28969
Silkybract Ophiorrhiza 272207
Silky-camellia 376452
Silky-dogwood 104949
Silky-fruited Willow 344088
Silkyhair Chessbean 116608
Silkyhair Newlitse 264090
Silkyhair Petrocosmea 292579
Silkyhair Sophora 369088
Silkyhair Stonebutterfly 292579
Silky-hairy Willow 343639
Silky-heads 117225
Silky-leaf Osier 344126
Silkyleaf Woadwaxen 173030
Silky-leaf Woadwaxen 173030
Silky-leaved Berry 338751
Silky-shining Ormosia 274436
Silkyspike Melic 248975
Silkystipe Paraboea 283116
Sillcotton Tree 80120
Siller Tassel 60382
Silly Green 358062
Silly Lovers 37015
Silphium Laciniatum 219507
Siltbush 418476
Silva Oligostachyum 270439
Silvaberry 338059
Silvedeaf Scurfpea 319134
Silvemargin Spurge 159313
Silver Acacia 1165
Silver Apple Mint 250439
Silver Ash 166975,166984
Silver Ash Northern 166982
Silver Ball 407989
Silver Ball Cactus 267051
Silver Ballcactus 267051
Silver Ball-cactus 267051
Silver Banksia 47650
Silver Beads 108960
Silver Beardgrass 57572
Silver Beech 266874
Silver Beet 53257
Silver Bell 184726,184731,184746
Silver Bells 23925,407989
Silver Berry 141941,141981
Silver Birch 53338,53549,53563,53590
Silver Bladder-pod 227022
Silver Blue-beard 78012
Silver Bluestem 22948,57572,57589
Silver Broom 7379,32682,32683
Silver Buckthorn 362986
Silver Buffalo Berry 362090
Silver Buffaloberry 362090
Silver Buffalo-berry 362090
Silver Bush 28200,95431
Silver Bush Morning Glory 102998
Silver Cap 204841
Silver Cassia 360419,360474
Silver Cephalanthera 82044
Silver Chain 125531,334976
Silver Chickensage 371143
Silver Chinese Poplar 311530
Silver Cholla 116654
Silver Cineraria 358395
Silver Cinquefoil 312389
Silver Cluster Cactus 244198
Silver Cock's-comb 80381
Silver Cockscomb 80381
Silver Crab-grass 143530
Silver Crown 108075
Silver Cup Vine 366866
Silver Cushion 28200
Silver Dalea 121879,121880
Silver Date Palm 295487
Silver Dichondra 128956
Silver Dock 308893
Silver Dollar 238371,273035
Silver Dollar Gum 155700
Silver Dollar Plant 108818,415509
Silver Dollar Tree 155529
Silver Dragon 232640
Silver Elaeagnus 141941
Silver Falls dichondra 128964
Silver Feather 312360
Silver Feathers 312360
Silver Fern 312360
Silver Fir 269,272,280,317
Silver Fleabane 150473
Silver Flower Vineclethra 95521
Silver Ginglers 60382
Silver Glorybind 102998
Silver Grass 255827,312360
Silver Grevill 180642
Silver Groundsel 358395
Silver Gum 155538,155547,155710
Silver Hair Grass 12794
Silver Hairgrass 12794
Silver Hair-grass 12794
Silver Hawthorn 109788
Silver Heart 290385
Silver Hedgehog Holly 203551
Silver Hoarypea 386129
Silver Honeysuckle 180651
Silver Horehound 245733
Silver Horse Nettle 367139
Silver Inch Plant 394093
Silver Inch-plant 394093
Silver Jade Plant 108818
Silver Joey Palm 212403
Silver July Croton 112838
Silver King Artemisia 35839
Silver Lace Cob Cactus 155202
Silver Lace Vine 162509,308690
Silver Lady's-mantle 13998
Silver Largeleaf Hydrangea 200010
Silver Latania 222634
Silver Lavender 346250
Silver Lawson Cypress 85273,85295
Silver Leaf Cassia 360474
Silver Leaf Poplar 311208
Silver Leaf Vineclethra 95481
Silver Leaves 312360
Silver Leucadendron 227233
Silver Lime 391836
Silver Linden 391836
Silver Mallee 155548
Silver Manzanita 31109
Silver Maple 3532
Silver Mist Shore Juniper 213743
Silver Moons 238371
Silver Morning-glory 32660
Silver Moss 83060
Silver Mound Artemisia 36225
Silver Mountain Gum 155710
Silver Nerve Plant 166709

Silver Net-leaf 166709
Silver Oak 180630
Silver Orache 44317
Silver Palm 4951,97883,97885
Silver Pear 323281
Silver Pendent Linden 391647,391816
Silver Pennies 50825
Silver Penny 238371
Silver Peperomia 290298
Silver Peppermint 155755
Silver Pine 244664,244665,300082
Silver Plate 238371
Silver Plectranthus 303150
Silver Plume Grass 148857
Silver Poplar 311208
Silver Puffs 253933
Silver Pussytoes 26332
Silver Queen 11329
Silver Queen Euonymus 157491
Silver rabbit-tobacco 127930
Silver Ragwort 358394,358395
Silver Red Gum 155517
Silver Rod 39328,367986
Silver Rose 312360
Silver Sage 35488,35837,291430,344855
Silver Sagebrush 35176,35488
Silver Saw Palm 4951
Silver Saw Palmetto 4951
Silver Scale 44680
Silver Scale Saltbrush 44680
Silver Senna 360419
Silver Shackle 60382
Silver Shackles 60382
Silver Shakers 60382
Silver Shekels 60382
Silver Shield 197799
Silver Shilling Flower 238371
Silver Silkwood 166972
Silver Slippers 266225
Silver Speedwell 317912
Silver Spreader 35314
Silver Spreader Juniper 213992
Silver Spruce 298273,298401,298435
Silver Squill 223819
Silver Sword 32960
Silver Thicket 159883
Silver Thistle 76990,271666
Silver Top 155668
Silver Topped Gimlet 155523
Silver Torch 94566
Silver Torch Cactus 94566
Silver Tree 227224,227233
Silver Vase 8558
Silver Vase Plant 8558
Silver Vase-plant 8558
Silver Vine 6682,147347,353132
Silver Wattle 1165,1422
Silver Weed 32600,312360
Silver Weeping Pear 323282
Silver Willow 342979,342986
Silver Wormwood 35268,35837
Silverback Artocarpus 36922
Silverback Saussurea 348585
Silverback Windhairdaisy 348585
Silver-backed Artocarpus 36922
Silverballi 263044
Silverbell 184726,184731
Silver-bell 184726
Silverbell Tree 184726,184731
Silver-bell Tree 184726
Silverberry 141941,142152,362090
Silverbush 32964,102998,369134
Silvercalyx Gentian 173196
Silver-crown 108075
Silver-dollar 43493
Silver-dollar Gum 155700
Silverdust Primrose 314863
Silveredge Polyscias 310204
Silverflower Globeamaranth 179227
Silver-flowered Jasmine 211926
Silver-frosted Plant 358395
Silvergrass 255827
Silver-grass 209259,255827
Silvergray Poplar 311261
Silvergreen-wattle Acacia 1165
Silver-green-wattle Acacia 1165
Silvergrey Data 386478
Silver-grey Rhododendron 331824
Silverhair Achyranthes 4262
Silver-hair Cassiope 78624
Silverhair Messerschmidia 393242
Silver-haired Cassiope 78624
Silver-haired Cyclobalanopsis 116055
Silverhairleaf Wildjute 394655
Silverhairy Trema 394655
Silver-hairy Trema 394655
Silver-hairyfruit Willow 343056
Silver-hairy-fruited Willow 343056
Silverhead 55871
Silverhook 256221
Silverlace Vine 162509
Silver-lavis Pearleverlasting 21554
Silverleaf 32769,227916,227919,227926,238371,312360,373166
Silver-leaf 17614,178358,372106,376645
Silverleaf Amomum 19916
Silver-leaf Atylosia 65157
Silverleaf Azalea 330134
Silverleaf Cinquefoil 312392,312714
Silverleaf Clovershrub 70785
Silverleaf Cotoneaster 107615
Silver-leaf Cotoneaster 107615
Silverleaf Daisy 124771
Silverleaf Dendranthema 124771
Silverleaf Evergreenchinkapin 78859
Silverleaf Gentian 112853
Silver-leaf Grape 411532,411551
Silverleaf Greenbrier 366299
Silverleaf Hydrangea 200068
Silverleaf Japanese Buttercup 325984
Silverleaf Magnolia 242193
Silverleaf Maple 3532
Silverleaf Mountain Gum 155710
Silver-leaf Mountain Gum 155710
Silverleaf Nightshade 367139
Silverleaf Oak 324021,324023
Silver-leaf Peperomia 290354
Silver-leaf Philodendron 294842
Silverleaf Poplar 311208
Silver-leaf Poplar 311208
Silverleaf Red Fir 410
Silverleaf Reekkudzu 313416
Silverleaf Rhododendron 330134
Silverleaf Sagebrush 35174
Silverleaf Scurf Pea 287930
Silverleaf Smilax 366299
Silverleaf Sunflower 188922
Silver-leaf Sunflower 188922
Silverleaf Terminalia 386460
Silverleaf Willow 343203
Silver-leafed Broom 172995
Silver-leaved Cineraria 358395
Silver-leaved Cotoneaster 107615
Silver-leaved Cranesbill 174475
Silver-leaved Evergreen Chinkapin 78859
Silver-leaved Ironbark 155647
Silver-leaved Poplar 311208
Silver-leaved Rhododendron 330134
Silver-leaved Terminalia 386460
Silver-leaved Tree 53563,311208
Silver-leaved Vineclethra 95481
Silver-leaved Willow 343203
Silverlight Willow 343055
Silver-lighted Willow 343055
Silverline Calanthe 65885
Silverling 46224,284774
Silverlraf Grape 411551
Silver-margined Fortune Spindle-tree 157508
Silver-margined Japanese Barberry 52242
Silver-margined Redvein Maple 3531
Silver-nerve Fitonia 166709
Silvernet Plant 166706,166709
Silveroak 180642
Silver-oak 180561
Silverpennies 50825
Silverpuff 85998
Silverpuffs 253909
Silver-puffs 402584
Silver-qeen Japanese Spindle-tree 157612
Silverreed 394926
Silverrod 367986

Silver-rod 367986
Silver-ruffles 108075
Silvers Thistle 92396
Silversandtree 19728,19729
Silverscale 44317
Silver-scale 44317
Silver-scale Saltbush 44317
Silverseed Gourd 114273
Silver-sericeous Sasa 347194
Silversheath Knotweed 308791
Silver-sheathed Knotweed 308791
Silversilk Eulalia 156474
Silversilk Goldquitch 156474
Silverspikegrass 227946,227948
Silverspot Begonia 49603
Silversquill 223819
Silverstar Begonia 49629
Silverstripe Hedge Bamboo 47350
Silverstripe-leaved Bamboo 347195
Silverstripe-leaved Japanese Bamboo 347195
Silversword 32959,32960
Silverthorn 142152
Silver-thread Peperomia 290300
Silvertip 409
Silvertop Gimlet 155523
Silvertop Stringybark 155617
Silver-topped Gimlet 155523
Silver-torch 94566
Silvertree 192477
Silver-tree 227233
Silver-variegared Platanus-leaved Maple 3456
Silvervein Creeper 285110
Silver-veined Creeper 285110
Silver-vine 6682
Silvervine Actinidia 6682
Silver-vine Actinidia 6682
Silvervine Kiwifruit 6682
Silver-vined Actinidia 6682
Silverweed 32600,55873,312360
Silver-weed 312360
Silverweed Cinquefoil 312360
Silverwood Cinnamon 91425
Silvery Argyrolobium 32783
Silvery Aster 40076
Silvery Bindweed 102998
Silvery Brown Pussytoes 26451
Silvery Cassia 360474
Silvery Cinquefoil 312360,312392
Silvery Creeper 285110
Silvery Crocus 111498
Silvery Cudweed 170920
Silvery Emu-bush 148371
Silvery Five-fingers 312389
Silvery Gugertree 350905
Silvery Guihaia 181803
Silvery Krameria 218076
Silvery Luina 238296
Silvery Lupin 238432
Silvery Lupine 238432
Silvery Messerschmidia 393242
Silvery Milfoil 3942
Silvery Nailwort 284774
Silvery Psoralea 287930
Silvery Scurf-pea 287930
Silvery Sedge 74004,74007
Silvery Sweetvetch 187781
Silvery Tournefortia 393242
Silvery White-striped Sasa 347195
Silvery Wormwood 35488
Silvery Yarrow 3942
Silvery-flowered Sedge 73752
Silveryleaf Cassia 91372
Silveryleaf Cinnamon 91372
Silveryleaf Clovershrub 70785
Silveryleaf Maesa 241739
Silveryleaf Pouzolzia 313416
Silvery-leafed Grevillea 180565
Silvery-leaved Cinnamon 91372
Silvery-leaved Cinquefoil 312389
Silvery-leaved Clovershrub 70785
Silvery-leaved Maesa 241739
Silvery-leaved Sunflower 188922
Silvery-nerved Caladium 16506
Silveryscale Spikesedge 143051
Silvestri Cotoneaster 107686
Silvestri Loosestrife 239851
Silvianthus 364338,364339
Sima Woodbetony 287677
Simal 56784,56802
Simao Balanophora 46872
Simao Camellia 69685
Simao Clematis 95261
Simao Cliffbean 254752
Simao Cowparsnip 192271
Simao Cupscale 341987
Simao Cyclobalanopsis 116213
Simao Dysophylla 139593
Simao Eriosolena 153103
Simao Evergreen Chinkapin 78934
Simao Evergreenchinkapin 78934
Simao Fimbristylis 166488
Simao Fluttergrass 166488
Simao Ginger 418021
Simao Gingerlily 187466
Simao Indigo 206151
Simao Litse 234074
Simao Meadowrue 388666
Simao Milkwort 308140
Simao Oak 116213
Simao Parnassia 284616
Simao Pepper 300526
Simao Pimpinella 299583
Simao Pine 299999
Simao Premna 313754
Simao Pseudoraphis 318200
Simao Qinggang 116213
Simao Rattlebox 112729
Simao Rattle-box 112729
Simao Rosewood 121837
Simao Sanqi 215036
Simao Snakemushroom 46872
Simao Sterculia 376187
Simao Sweetleaf 381341
Simao Syzygium 382671
Simao Tea 69685
Simao Teasel 133510
Simao Ternstroemia 386731
Simao Tetrathyrium 387935
Simao Watercandle 139593
Simao Wendlandia 413771
Simao Zinger 418021
Simaovine 146558
Simarouba 364380
Simaruba 364380,364382
Simaruba Family 364387
Similar Barberry 52131
Similar Brightleaf Gentian 173633
Similar Cystacanthus 120796
Similar Diploclisia 132929
Similar Eargrass 187581
Similar Elaeocarpus 142278
Similar Evil Windhairdaisy 348681
Similar Glabrate Rockfoil 349688
Similar Iris 208623
Similar Japan Maesa 241756
Similar Jiala Rockfoil 349560
Similar Merremia 250836
Similar Microtoena 254242
Similar Neosinocalamus 47264
Similar Rapanea 261600
Similar Rattlebox 112671
Similar Sagebrush 36294,36351
Similar Thyme 391334
Similar Yam 131844
Similarest Barberry 51494
Similarleather Bethlehemsage 283652
Similar-longania Eargrass 187613
Similar-longania Hedyotis 187613
Similarpalmvein Sage 345415
Simizu's Lasianthus 222294
Simizu's Roughleaf 222294
Simla Box-thorn Barberry 51892
Simlar Mitchella Ophiorrhiza 272254
Simle Leaf Sophora 369146
Simllex Common Pentapanax 289655
Simmeren 315101
Simmerin 315101
Simmon 132466
Simmonds' Aster 380985
Simmonos-tree 132466
Simon Bamboo 304098

Simon Bitter Bamboo 304098
Simon Bitter-bamboo 304098
Simon Cane-bamboo 304098
Simon Plum 316813
Simon Poplar 311482
Simond Cleisostoma 94467
Simond Closedspurorchis 94467
Simons Anise-tree 204590
Simons Barberry 52159
Simons Cotoneaster 107687
Simons Illicium 204590
Simons Mahonia 242641
Simons' Cotoneaster 107687
Simons-like Illicium 204579
Simple Cape-cowslip 218951
Simple Coelachne 98457
Simple Glorybower 96273
Simple Leafflower Simple 296809
Simple Petal Camellia 69552
Simple Raspberry 339273
Simple Rattan Palm 65790
Simple Sinorchis 365029
Simple Stem Vriesea 412369
Simple Style St. John's Wort 201894
Simple Undulate Bigginseng 241182
Simplebeak Ironwort 362864
Simpledental Hornbeam 77393
Simpleflower Bridalwreath Spiraea 372062
Simple-hair Anzacwood 310773
Simpleleaf Allomorphia 16133
Simpleleaf Barrenwort 147061
Simple-leaf Chaste Tree 411464
Simpleleaf Chastetree 411208,411464,
411471
Simpleleaf Epimedium 147061
Simple-leaf Evodia 161378
Simpleleaf Gynostemma 183029
Simple-leaf Indigo 206556
Simpleleaf Local Milkvetch 266465
Simpleleaf Meconopsis 247180
Simpleleaf Meliosma 249452
Simpleleaf Milkvetch 42319
Simpleleaf Munronia 259893
Simple-leaf Ormosia 274437
Simpleleaf Panisea 282413
Simpleleaf Pinanga 299675
Simpleleaf Sargentglorivine 347078
Simpleleaf Separatestyle
Acanthopanax 143593
Simpleleaf Shrub Chastetree 411430
Simple-leaf Tickclover 126679
Simpleleaf Trachydium 393859
Simple-leaf Trefoil 126679
Simpleleaf Vinelime 283509
Simple-leaf Wartcress 225360
Simpleleaf Wildginger 37642
Simple-leaved Evodia 161378
Simple-leaved Meliosma 249452
Simple-leaved Ormosia 274437
Simple-leaved Sargent-glory-vine 347078
Simple-leaved Tickclover 126679
Simplelobe Jadeleaf and
Goldenflower 260494
Simple-lobed Mussaenda 260494
Simplepinnata Chamaesciadium 85701
Simpler's Joy 405872
Simpler's-joy 405844
Simplestem Bittercress 72989
Simple-stem Cudweed 170930
Simple-stem Everlasting 170930
Simplestem Rockfoil 349949
Simpletooth Photinia 295792
Simpletoothed Hornbeam 77393
Simple-trnk Aloe 16592
Simplex Germander 388279
Simplex Murdannia 260113
Simplex Sedge 75846
Simplex Thinfruit Kobresia 217292
Simplex Waterbamboo 260113
Simpson 360328
Simpson Eugenia 261045
Simpson Milkvetch 43049
Simpson's Applecactus 185924
Simpson's Footcactus 287909
Simpson's Hedegehog Cactus 287909
Simpson's Prickly Apple 185924
Simpson's Wild Buckwheat 152285
Sims Azalea 331839
Sims Crean Clematis 94915
Sims Philodendron 294841
Sims' Azalea 331839
Sims' Indian Azalea 331839
Sims' Rhododendron 331839
Simsim 361317
Simul 56784
Sin Erycibe 154262
Sin Hartia 376485
Sin Meliosma 249391
Sin Petrosavia 292674
Sin Pinangapalm 299676
Sin Pinanga-palm 299676
Sin's Clematis 95314
Sinacalia 364517
Sinadoxa 364523,364524
Sinaloa Sage 345392
Sinchiang Monkshood 5572
Sinclair Acianthus 4516
Sindechites 364697
Sindora 364703
Siney 193417
Sinfull 358062
Singalang Schefflera 350782
Singapore Cedar 392829
Singapore Daisy 371233,413559
Singapore Mahogany 248529
Singapore Oak 323785
Singapore Plumeria 305221
Singhara Nut 394436
Singharanut 394436
Sing-kwa 238258
Sing-kwa of China 238258
Singkwa Towelgourd 238258
Sing-kwa Towelgourd 238258
Single Awlquitch 131031
Single Bamboo 47230
Single Castle 273531,273545
Single Delight 257374
Single Dimeria 131031
Single Fishwood 110235
Single Ghost 273531
Single Gusses 273531
Single Gussies 145487
Single Head Chinese Groundsel 365055
Single Leaflet Threefork 396795
Single Roxburgh Rose 336888
Single Spike Enteropogon 146006
Single Yushanbamboo 416819
Singleanther Papposedge 375806
Singlebamboo 56951,56955,257595
Singleberry Cotoneaster 107715
Single-berry Cotoneaster 107715
Singlecolor Gentian 173644
Single-ear Spatholobus 370423
Singleflower 381339
Singleflower Amitostigma 19520
Singleflower Antlerpilose Grass 257541
Singleflower Barberry 52292
Singleflower Chirita 87929
Singleflower Cotoneaster 107715
Singleflower Cranesbill 174923
Singleflower Dendrobium 125398
Singleflower Disa 133975
Singleflower Eulophia 156868
Singleflower Gentian 156868,173946
Singleflower Gueldenstaedtia 391539
Singleflower Honeysuckle 236130
Singleflower Knapweed 81443,81444
Singleflower Lily 230042
Singleflower Milkvetch 42742
Singleflower Panisea 282418
Singleflower Primrose 314123,314229
Singleflower Rush 213371
Singleflower Spapweed 205422
Singleflower St. John's Wort 202217
Singleflower Stonecrop 356635
Singleflower Touch-me-not 205422
Singleflower Tulip 400248
Singleflower Tulip Orchid 25139
Single-flowered Honeysuckle 236130
Single-flowered Mariposa-lily 67605
Single-flowered Pure White

Mockorange 294581
Single-flowered Rose 337004
Single-flowered Sweetleaf 381339
Singlefruit Craneknee 221708
Singlefruit Lycianthes 238977
Singlefruit Pondweed 312036
Singlefruit Privetleaf Memecylon 249990
Singlefruit Red Silkyarn 238977
Single-fruited Tanoak 233350
Single-grooved Holly 204363
Singlehead Edelweiss 224909
Singlehead Goldenbush 150378
Singlehead Groundsel 365055
Singlehead Jerusalemsage 295222
Singlehead Milklettuce 259760
Singlehead Nepal Pearleverlasting 21645
Singlehead Pussytoes 26596
Singlehead Pyrethrum 322734
Singlehead Sinosenecio 365055
Singlehead Thistle 92205
Single-leaf Ash 167922
Single-leaf Black Locust 334986
Single-leaf Corydalis 106114
Single-leaf Gymnadenia 264775
Singleleaf Hoodorchis 264775
Singleleaf Milkvetch 42319
Singleleaf Mombin 372469
Singleleaf Pine 300079
Single-leaf Pine 300079
Singleleaf Pinyon 299854,300079
Singleleaf Pyrola 322857
Singleleaf Rattanpalm 65791
Singleleaf Rush 213560
Singleleaf Sedge 76642
Singleleaf Sevanlobed Nightshade 367609
Singleleaf Threefork 396795
Singleleaf Tumorfruitparsley 393859
Singleleaf Yushanbamboo 416801
Singleleaf Yushania 416801
Singleleaflet Threefork 396795
Single-leaved Ash 167922
Single-leaved Bog Orchid 243084
Single-leaved Mombin 372469
Single-leaved Rockvine 387812
Single-leaved Threefork 396795
Singlelip Coelogyne 98688
Singlelip Leaflessorchis 29220
Singlepetal Yellow Rose 337034
Singlepistil Milkvetch 42740
Single-ramaled Yushania 416819
Single-scaled Spike-rush 143397
Singlescape Dendrobium 125320
Singleseed Argyreia 32640
Singleseed Bittersweet 257504
Singleseed Chickweed 375000
Single-seed Hawthorn 109857
Singleseed Juniper 213943
Single-seed Juniper 213943
Singleseed Ploughpoint 401154
Singleseed Savin 213943
Single-seeded Juniper 213943
Single-seeded Rockvine 387769
Singleshoot Rush 213388
Singlespike Fimbristylis 166411
Singlespike Fishtailpalm 78048
Singlespike Fluttergrass 166411
Singlespike Phacelurus 293219
Singlespike Weakchloa 209403
Single-spiked Fishtail Palm 78048
Singlespikelet Reedbentgrass 65325
Singlespikelet Woodreed 65325
Singlestalk Privet 229587
Singlestem Involucred Cudweed 178235
Singlestem Pleurospermum 304846
Singlestem Ribseedcelery 304846
Single-stem Wild Buckwheat 151800
Single-stemmed Bog 160664
Singlestyle Forkliporchis 269047
Singletary Pea 222726,222825
Singletary Vetchling 222825
Singletongue Goldenray 229164
Singnalgrass 58052,58088
Singreen 358062
Sinia 364732,364733
Sinicuichi 188264
Sinjiang Peony 280154
Sinkfield 312925
Sinkiang Aellenia 8850
Sinkiang Asparagus 39111
Sinkiang Bastard Cress 390231
Sinkiang Bird's Foot Trefoil 237609
Sinkiang Bugle 13190
Sinkiang Calophaca 67786
Sinkiang Fritillary 168624
Sinkiang Gagea 169470
Sinkiang Gentian 174066
Sinkiang Juniper 213873
Sinkiang Linosyris 231838
Sinkiang Madwort 18427
Sinkiang Onion 12997,15662
Sinkiang Pear 323306
Sinkiang Peashrub 72200
Sinkiang Pea-shrub 72200
Sinkiang Pyrethrum 322644
Sinkiang Rockcress 30205
Sinkiang Russianthistle 344716
Sinkiang Solomonseal 308639
Sinkiang Thermopsis 389557
Sinkiang Thorowax 63623
Sinkiang Tulip 400224
Sinko Raspberry 339268
Sinnabar Mosquitotrap 117637
Sinnegar 246478
Sinningia 364737
Sinoadina 364767
Sinobacopa 364774
Sinobambusa 364783
Sino-Buema Woodlotus 244448
Sino-Burma Eightangle 204490
Sino-Burma Euonymus 157669
Sino-Burma Habenaria 184060
Sino-Burma Mountainash 369535
Sino-Burma Tea 69757
Sinocarum 364872
Sinochasea 364898
Sinocrassula 364913
Sinodielsia 247711
Sinofranchetia 364945
Sinohaircot 364943,364944
Sinojackia 364951
Sinojohnstonia 364961,364966
Sinolimptichtia 364969
Sinolister Primrose 314977
Sinomarsdenia 364983,364984
Sinomontanous Mayten 182723
Sinopanax 364996
Sinopurple Primrose 314988
Sinorchis 365028
Sino-Russia Pipewort 151263
Sinosassafras 365031,365034
Sinosenecio 365035
Sinosideroxylon 365082
Sino-tailand Kadsura 214948
Sino-Thailand Habenaria 184061
Sino-Vietnam Goldlineorchis 25969
Sino-Vietnam Tea 69148
Sino-wilson Longtooth Rose 336693
Sinqua 238258
Sinqua Melon 238258
Sinslon 360328
Sintan Straight-raceme Barberry 51989
Sintenis Nepeta 265050
Sinuate Acacia 1142
Sinuate Arisaema 33411
Sinuate Bamboo 297281
Sinuate Bambusa 47275
Sinuate Casearia 78118
Sinuate Cragcress 98038
Sinuate Erysimum 154500
Sinuate Fivelobe Cacalia 283863
Sinuate Hilliella 196307
Sinuate Horsegentian 397849
Sinuate Leaf-flower 296565
Sinuate Mountain Leech 126603
Sinuate Primrose 314990
Sinuate Pseuderanthemum 317280
Sinuate Rabdosia 324879
Sinuate Rhodiola 329956
Sinuate Scurvyweed 98038
Sinuate Southstar 33411
Sinuate Sugarmustard 154500

Sinuate Sweetleaf 381390
Sinuate Tickclover 126603
Sinuate Windhairdaisy 348788
Sinuate Woodnettle 125548
Sinuolate Rabdosia 324879
Sinuoustooth Rabdosia 324879
Sinvey 364557
Sion 29341
Siphocranion 365192
Siphonia 194451
Siphonoglossa 365282
Siphonostegia 365295
Sip-sap 369339
Sir Lowry's Pass 315860
Sir. J. Brooke's Rhododendron 330261
Siraitia 365325
Siris 13474,13595
Siris Tree 13595
Siris-acacia 13595
Sisal 10940,169245
Sisal Agave 10940
Sisal Hemp 10940
Sisal Plant 10787
Sishan Cinquefoil 313000
Siskinling 315261,315263
Siskiyou Aster 155809
Siskiyou False Hellebore 405599
Siskiyou Fleabane 150542
Siskiyou Hawksbeard 110900
Siskiyou Inside-out Flower 404612
Siskiyou Iris 208470
Siskiyou Mariposa-lily 67620
Siskiyou Mountain Ragwort 279921
Siskiyou Rue-anemone 145500
Siskiyou Silverpuffs 253930
Sisso 121826
Sisso Rosewood 121826
Sissoo 121826
Sister Violet 410587
Sister-in-law 410677
Sisu 121826
Sisymbrium 365386,365529
Sisymbriumleaf Nightshade 367617
Sisyrinchium 365678
Sitapai Crazyweed 279166
Sitchen Brome 60974
Sitchen Bromegrass 60974
Sitfast 271563,326287
Sithes 15709
Sitka Alder 16459,16472
Sitka Cypress 85301
Sitka Mountain Ash 369525
Sitka Mountainash 369525
Sitka Spruce 298435
Sitka Starwort 374769
Sitka Willow 344122
Sit-siccar 325518,325666,326287
Sit-sicker 325518,325666,326287
Siunas 229384
Siuzev Willow 344124
Siver Crazyweed 278719
Siver Ivy-arum 147347
Siver Trumpet Tree 382706
Siverback Asiabell 98279
Siverback Cinquefoil 312389
Siverhead Jurinea 214041
Siverleaf Snowbell 379308
Siverleaf Storax 379308
Siver-leaved Storax 379308
Sivervein Gentian 173257
Siverweed Yarrow 3942
Sivery Willow 343052
Siveryfruit Elaeagnus 142077
Sivery-fruited Elaeagnus 142077
Sivery-grey Glorybind 102926
Sives 15709
Sivestre Indigo 206636
Sivinski's Fleabane 150966
Sivven 338557
Six o'clock 274823
Six o'clock Flower 274823
Six Weeks' Grass 305334
Six-angle Spurge 159056,159057
Six-angled Bruguiera 61267
Sixangled Iris 208614
Six-angular Clearweed 298934
Sixangular Dysosma 139623
Sixangular Pipewort 151487
Sixanther Microtropis 254298
Sixbract Elatostema 142732
Sixbract Mabian Stairweed 142732
Six-flowered Persimmon 132197
Sixleaf Gentian 173514
Sixleaf Rhodiola Rhodiola 329952
Sixleaves Rhodiola 329952
Sixleaves Stauntonvine 374409
Sixleves Bedstraw 170225
Sixloculed Tutcheria 400695
Sixlow Pinangapalm 299663
Sixpetal Eagleclaw 35016
Sixpetal Primrose Willow 238176
Sixpetal Tailgrape 35016
Sixroom Slatepentree 400695
Sixrow Barley 198376
Six-rowed Barley 198376
Six-scab Rockfoil 349431
Sixstamen Balata 244551
Sixstamen Microtropis 254298
Six-stamen Microtropis 254298
Sixstamen Waterwort 142568
Six-stamened Balata 244551
Six-stamened Waterwort 142568
Sixtooth Mouseear 82772
Sixweeks Fescue 412474
Six-weeks Fescue 412474,412475
Sixweeks Three-awn 33737
Sixweeks Threeawngrass 33737
Sixweeks Triawn 33737
Siziwang Crazyweed 279168
Sjambok Bush 216679
Sjolgirse 4005
Skally 144661
Skapanthus 365904
Skaw 345631
Skayug 109857
Skeat Legs 273531
Skeat LegsSkeat-legs 273531
Skeat-legs 273531
Skedge 229662
Skedgewith 229662
Skedlock 359158,364557
Skeeog 109857
Skeet 366743
Skeets 192358,366743
Skeg 109857,208771,316470,316819
Skeldick 326609
Skeldock 326609
Skeleton Plant 239334,239340
Skeleton Weed 88776,151981
Skeleton-leaf Bur Ragweed 19190
Skeleton-leaf Bur-sage 19190
Skeletonleaf Goldeneye 409169
Skeletonplant 239322,375964,375987
Skeletonweed 84794,88759,375955
Skellies 364557
Skellock 326609,364557
Skerret 365871
Skevish 150862
Skew 345631
Skewerwood 157429,380499
Skeyblue Onion 15194
Skidgy 229662
Skilful Habenaria 183581
Skilful Liparis 232133
Skilful Platanthera 302378
Skillock 326609,364557
Skillog 364557
Skimmia 365916,365974
Skinned Calystegia 68701
Skinner Columbine 30075
Skinner Hybrid 382171
Skinner Michelia 252958
Skinner's False Foxglove 10156
Skinner's Lycaste 238749
Skipping Rope 95431
Skirret 365835,365871
Skirret Water Parsnip 365871
Skirret Water-parsnip 365871
Skirt Buttons 374916,374968
Skirwits 365871
Skirwort 365871

Skit 366743
Skiver 157429,380499
Skiver-timber 157429
Skiver-tree 157429,380499
Skiver-wood 157429,380499
Skiwer 380499
Skiwet 285741
Skoke 298094
Skorniakov Milkvetch 43058
Skull Cap 355340
Skull Orchid 82028,82113
Skullcap 30081,355340,355387,355396,355449
Skullcap Speedwell 407349
Skullcaplike Falsepimpernel 231572
Skullcaplike Motherweed 231572
Skunk 71218
Skunk Bugbane 91011
Skunk Bush 171552,332908
Skunk Cabbage 239516,239517,381070,381073,381082,405581
Skunk Currant 333997,334159
Skunk Goldenrod 368123,368434
Skunk Grape 411764
Skunk Meadow-rue 388638
Skunk Spruce 298286
Skunk Sumash 332908
Skunk Tree 1541
Skunk Vine 280078
Skunk Weed 113037,262921
Skunk-bumash Sumac 332481
Skunkbush 250505,332908
Skunkbush Sumac 332908
Skunkcabbage 381070,381073
Skunk-cabbage 239516,239517,381070,381073,405570
Skunkflower 208318
Skunkweed 95791,95793,381073
Skunk-weed 113037
Skurwemannetjie 109283
Skvortsov Willow 344125
Sky Blue Lily 210999
Sky Flower 139062
Sky Lupine 238469
Sky Plant 392020
Sky Scale 200707,200713
Sky Tree 12585
Sky Vine 390787
Skyb Leek 15220
Skyblue Adenosma 7979
Skyblue Adonis 8344
Skyblue Amethystea 19459
Skyblue Aster 380943
Sky-blue Aster 380943
Skyblue Awlbillorchis 333727
Skyblue Bogorchid 101955
Skyblue Broomrape 275006,275010
Sky-blue Brownleea 61174
Skyblue Burmannia 63964
Skyblue Crazyweed 278757
Skyblue Gentian 173311
Sky-blue Hairgrass 256601
Skyblue Ladybell 7620
Skyblue Larkspur 124083
Sky-blue Memecylon 249923
Skyblue Oxypetalum 278547
Sky-blue Phyllodoce 297019
Sky-blue Sage 345449
Skyblue Sesleria 361615
Skyblue Teasel 133461
Skyblue Vanda 404630
Skybluewing Passionflower 285612
Skyflower 139060,139062,390822
Sky-flower 139060,139065
Skymallow 357918,357919
Skyrocket 71547,175671,175673,175708,208318
Skyrocket Pincushion 228100
Skyscraper 188908
Skytes 24483,192358
Sky-vine 390787
Slaa Thorn 316819
Slaa-thorn 316819
Slaa-tree 316819
Slacen-bush 316819
Sladenia 366015
Sladenia Family 366019
Slae 316819
Slag 316819
Slaigh 316819
Slanderpedicel Hooktea 52437
Slanderpedicel Supplejack 52437
Slanlus 302068
Slant Stephania 375850
Slantdaisy 301619
Slanting Arrowbamboo 162723
Slanting Herelip 220063
Slanting Hetaeria 193616
Slanting Sawwort 361110
Slantingawn Splitlemma 351277
Slantingleaf Rosewood 121793
Slantingsepal Jerusalemsage 295118
Slantingstem Swertia 380298
Slantleaf Pepper 300507
Slantpetal 301663
Slantspike 301721
Slantwing 301688
Slantwing Family 301687
Slantwing Gentian 173770
Slap 315082
Slash Pine 299836,299928
Slashed Philodendron 294830
Slashed-serrate Maddencherry 241504
Slashed-serrate Madden-cherry 241504
Slatepentree 400684
Slater's Crimson China Rose 336496
Slateyhide Eucalyptus 155689
Slath 316470
Slaty Gum 155756
Slaty-purple Crocus 111504
Slaun-bush 316819
Slaun-tree 316819
Slavin's Dogwood 104920
Slavonian Ash 167955
Slavonian Oak 324278,324335
Slaw 316819
Slawnes 316819
Slea 316819
Sledge's Pitcher-plant 347164
Sleep-at-noon 394333
Sleeping Beauty 277648,277747
Sleeping Clover 277648
Sleeping Hibiscus 243951
Sleeping Maggie 397019
Sleeping Nightshade 44708
Sleeping Waxmallow 243951
Sleepingeeg 414586,414601
Sleepinglily Azalea 331364
Sleeping-plant 78284
Sleepwort 219485,219609,260596
Sleepy Catchfly 363181
Sleepy Dick 274823
Sleepy Grass 376905
Sleepy Hibiscus 243932,243949,286663
Sleepy Mallow 243929
Sleepy Silene 363181
Sleepy-daisy 414971
Sleepydick 274823
Sleepy-dick 274823
Sleepydose 359158
Sleepyhead 282685,394333
Sleeve Palm 244495
Slendent Bladderwort 403321
Slender Acanthopanax 143642
Slender Amaranth 18734,18848
Slender Amitostigma 19512
Slender Anisachne 25266
Slender Apios 29305
Slender Arrowhead 342420
Slender Aster 160654
Slender Bamboo 47482
Slender Barrel Cactus 163445
Slender Beard-tongue 289346
Slender Bedstraw 170289,170579
Slender Begonia 49885
Slender Bethlehemsage 283616
Slender Bigumbrella 59216
Slender Bird's Foot Trefoil 237768
Slender Bird's-foot Trefoil 237485
Slender Bird's-foot-trefoil 237485
Slender Bittercress 72784

Slender Bladderwort 403356
Slender Blue Flag 208767
Slender Blue Iris 208767
Slender Blue-eyed Grass 365785
Slender Bluegrass 305564,305888
Slender Bluestem 128587
Slender Bog Arrow-grass 397165
Slender Borage 57099
Slender Branch St. John's Wort 201978
Slender Brassaiopsis 59216
Slender Bristlethistle 73504
Slender Broadpetal Epimedium 147037
Slender Brome 60811
Slender Bugseed 104816
Slender Bulrush 352192,353457
Slender Bunch-flower 248657
Slender Bush Clover 227012
Slender Bush-clover 227012
Slender Cactus 244070
Slender Cairo Morningglory 207660
Slender Calamintha 65573
Slender Calanthe 82088
Slender Carpetweed 256705
Slender Carpet-weed 256695
Slender Catkin Willow 343040
Slender Cautleya 79753
Slender Centaury 81539
Slender Centranthera 81734
Slender Chickweed 82861
Slender Childvine 400893
Slender China Cane 329182
Slender Cicendia 90792
Slender Cinquefoil 312649
Slender Cleistogenes 94599
Slender Clinopodium 96999
Slender Club-rush 353290
Slender Cock's-foot 121264
Slender Codonanthe 98234
Slender Corydalis 106149
Slender Cotoneaster 107703
Slender Cotton Grass 152759
Slender Cotton-grass 152759
Slender Cottonheads 263285
Slender Cottonsedge 152759
Slender Cottonseed 253845
Slender Cottonweed 168691
Slender Cotton-weed 168691
Slender Crab Grass 130554
Slender Crookstem Japanese Bamboo 297223
Slender Cudweed 178176,235267
Slender Cyclea 116017
Slender Daphne 122452
Slender Date Palm 295478
Slender Decaspermum 123451
Slender Dendropanax 125609
Slender Deutzia 126946
Slender Dropseed 372641
Slender Dutchmans Pipe 34162
Slender Dutchmanspipe 34162
Slender Dwarf Morning Glory 161418
Slender Dwarf Morning-glory 161421
Slender Dysophylla 139561
Slender Echinodorus 140563
Slender Enteropogon 146015
Slender Epidendrum 146424
Slender Eritrichium 153456
Slender Fairy Lantern 67595
Slender False Brome 58619
Slender False Garlic 266965
Slender Falsebrome 58625
Slender False-brome 58572,58619
Slender Falsenettle 56318
Slender Fimbristylis 166343
Slender Fimbry 166190
Slender Five-finger 312647
Slender Flat Sedge 118565
Slender Fleabane 151003
Slender Flowered Thistle 92437
Slender Fluffweed 235267
Slender Fluttergrass 166343
Slender Foxtail 17548
Slender Fringe-rush 166190
Slender Fritillary 168585
Slender Fumewort 106149
Slender Gayfeather 228471
Slender Ghostfluting 228330
Slender Glasswort 342859
Slender Goldenbush 185482
Slender Goldenrod 368095
Slender Goldentop 161100
Slender Goldenweed 414975
Slender Goldshower 390542
Slender Goldwaist 90366
Slender Goose Grass 312647
Slender Gueldenstaedtia 181659
Slender Gyperus 119139
Slender Hardhead 81230
Slender Hare's Ear 63855
Slender Hare's-ear 63855
Slender Hartia 376447
Slender Hawksbeard 110739,110767
Slender Heath 149516
Slender Hemsle Dogwood 380454
Slender Heterolamium 193816
Slender Hideseedgrass 94599
Slender Himalayahoneysuckle 228330
Slender Himalaya-honeysuckle 228330
Slender Hinoki Cypress 85318
Slender Honey Myrtle 248097
Slender Honeysuckle 236189
Slender Hop-bush 135197
Slender Hornwort 83578
Slender Iris 208595
Slender Janusia 211594
Slender Kalanchoe 61564
Slender Knapweed 81237
Slender Knotgrass 309395
Slender Knot-grass 309395
Slender Knotweed 309866
Slender Kurz Casearia 78134
Slender Ladies' Tresses 372214
Slender Ladies'-tresses 372214
Slender Lady Palm 329176,329184
Slender Lady's Tresses 372214
Slender Ladymantle 14042
Slender Ladypalm 329184
Slender Lady-palm 329184
Slender Lampranthus 220697
Slender Leaf Blysmus 55899
Slender Leaf Rattlebox 112475
Slender Leptocodon 226062
Slender Lespedeza 227012
Slender Lettuce 219480
Slender Liberbaileya 228621
Slender Litwinowia 234158
Slender Loose-flower Sedge 74687
Slender Loosestrife 240100
Slender Machilus 240593
Slender Maesa 241844
Slender Mahonia 242546
Slender Mariposa-lily 67572
Slender Marsh Bedstraw 170345
Slender Marsh-bedstraw 170345
Slender Marsh-pink 341467
Slender Meadow Foxtail 17548
Slender Melandrium 248435
Slender Metalleaf 296387
Slender Milkvetch 42723
Slender Milkwort 308118
Slender Monkeyflower 255208
Slender Mountain Mint 321961
Slender Mountain Sandwort 31787
Slender Mountainash 369401
Slender Mugwort 35598
Slender Muhly 259708
Slender Naiad 262028,267294
Slender Najad 262039
Slender Neillia 263150
Slender Nettle 402923
Slender Oak 116112
Slender Oat 45389
Slender One-seed Juniper 213827
Slender Orach 44697
Slender Palm Lily 104366
Slender Panic Grass 281618
Slender Paraphlomis 283616
Slender Parnassia 284622
Slender Parsley Piert 29025
Slender Parsley-piert 29025
Slender Passion-flower 285645
Slender Phlox 254080

Slender Pigweed 18734
Slender Pinanga 299662
Slender Pinanga-palm 299662
Slender Pincution 244085
Slender Pitred-seed 57689
Slender Plantain 302152
Slender Podocarpium 200732
Slender Pondweed 312236,312241
Slender Pride of Rochester 126946
Slender Primrose 314441,315045
Slender Pseudosasa 318296
Slender Qinggang 116112
Slender Raspberry 338546
Slender Rattlesnakeroot 313798
Slender Reddish Arrowwood 407826
Slender Reddish Viburnum 407826
Slender Rice Flower 299309
Slender Rice-flower 299309
Slender Ricegrass 300714
Slender Riparial Silverreed 394937
Slender Rock Daisy 291314
Slender Rockjasmine 23184
Slender Rosette Grass 128562
Slender Rosinweed 364308
Slender Rotang Palm 121487
Slender Rush 213138,213507
Slender Russian Thistle 344496
Slender Russian-thistle 344496
Slender Salsify 394296
Slender Sand Sedge 119140
Slender Sandwort 32120,255585,255586
Slender Satin Grass 259708
Slender Scratchdaisy 111683
Slender Sea Knotgrass 309503
Slender Sealavender 230791
Slender Sea-purslane 361660
Slender Sedge 74687,75065,75120
Slender Serpentroot 354926
Slender Sheep's Sorrel 340278
Slender Shortstalk Agapetes 10290
Slender Siberian Crabapple 243569
Slender Skullcap 355806
Slender Snake Cotton 168691
Slender Snake-cotton 168691
Slender Soft Brome 60811
Slender Soja 177722
Slender Sowthistle 368821
Slender Sow-thistle 368821
Slender Soybean 177722
Slender Speedwell 407124
Slender Spicebush 231360
Slender Spice-bush 231360
Slender Spike Rush 143022
Slender Spike-rush 143237,143365,143369,143397
Slender Spikesedge 143022,143365
Slender Spindle-tree 157546
Slender St. John's Wort 202117
Slender St. John's-wort 202117
Slender Starwort 374731
Slender Stelis 374706
Slender Stem Fleabane 151004
Slender Stem St. John's Wort 201877
Slender Stem Starjasmine 393616
Slender Stitchwort 255585
Slender Strawberry 167617
Slender Stringbush 414185
Slender Stylosanthes 379245
Slender Sunflower 188973
Slender Sweet-flag 5803
Slender Tamarisk 383516,383622
Slender Tare 408531,408633,408638
Slender Thistle 92437
Slender Three-awn 33914,33915,34006
Slender Three-seeded-mercury 1876
Slender Thryallis 390542
Slender Tick Clover 126291
Slender Tongoloa 392790
Slender Toothwort 125844
Slender Trailliaedoxa 394401
Slender Trefoil 396977
Slender Trillium 397569
Slender Tufted-sedge 73589
Slender Twigs Rhododendron 330075
Slender Tylophora 400893
Slender Umbrella Galingale 118481
Slender Vervain 405885
Slender Vetch 408418,408458,408638
Slender Wart Cress 105340
Slender Wart-cress 105340
Slender Water-milfoil 261374
Slender Water-nymph 262039
Slender Waterweed 143931,200190
Slender Weavers' Bamboo 47482
Slender Wedge Grass 371489
Slender Western Rosemary 413913
Slender Westringia 413913
Slender Wheat Grass 144498,144501
Slender Wheat-grass 144498
Slender White Aster 380842
Slender White Prairie-clover 121882,121883
Slender Wild Basil 96999
Slender Wildberry 362380
Slender Wildginger 37627
Slender Willow 343360,343868
Slender Wirelettuce 375984
Slender Wireweed 308678
Slender Wistaria Currant 334074
Slender Wood Sedge 74687
Slender Woodbetony 287246
Slender Woodland Sedge 74312
Slender Woolly Marbles 318810
Slender Woolly Wild Buckwheat 152100
Slender Yellow Trefoil 396977
Slender Yellow-eyed-grass 416173
Slender Youngia 416484
Slender Yunnan Hartia 376447
Slender Zeuxine 417739
Slender-andrus Sabia 341520
Slender-brabch Holly 203863
Slenderbract Honeysuckle 236104
Slenderbract Pepper 300379
Slender-bracted Honeysuckle 236104
Slender-bracted Pepper 300379
Slender-bracted Silverpuffs 253929
Slenderbranch Asparagus 39238
Slender-branch Caper 71925
Slenderbranch Cherry Elaeagnus 142112
Slender-branch Dracaena 137419
Slender-branch Dragonbood 137419
Slenderbranch Eurya 160535
Slender-branch Holly 204351
Slender-branch Hydrangea 199885
Slenderbranch Kalidium 215329
Slenderbranch Pearleverlasting 21577
Slender-branch Sweetvetch 188088
Slender-branched Eurya 160535
Slender-branched Gragontree 137419
Slender-branched Kalidium 215329
Slender-branched St. John's-wort 201978
Slender-branched Sweetvetch 188088
Slender-branched Willow 343440
Slenderculm Eulalia 156471
Slender-culm Rush-like Bulrush 352200
Slenderest Thorowax 63680
Slenderflower Aganosma 10237
Slenderflower Cherry 83292
Slenderflower Hemsleia 191926
Slender-flower Ixora 211100
Slenderflower Nepeta 264939
Slender-flower Rose 336615
Slenderflower Snowgall 191926
Slender-flowered Cherry 83292
Slender-flowered Holly 203861
Slender-flowered Malaxis 242994
Slender-flowered Rose 336615
Slender-flowered Thistle 73504
Slenderfruit Didymocarpus 129973
Slender-fruit Yu Maple 3762
Slender-fruited Anemone 23776
Slenderhair Fescuegrass 164050
Slenderhair Rabdosia 209698
Slender-haired Rabdosia 209698
Slender-hiared Machilus 240706
Slender-leaf Anneslea 25779
Slenderleaf Betony 373459
Slender-leaf Cretalaria 112248
Slenderleaf Falsepimpernel 231587
Slender-leaf Fleabane 151005
Slenderleaf Glochidion 177174
Slenderleaf Iceplant 251753

Slenderleaf Iris 208595,208882
Slenderleaf Ixeris 210539
Slenderleaf Ligusticum 229404
Slenderleaf Melic 249081
Slender-leaf Milkvetch 42697
Slenderleaf Monkshood 5052
Slenderleaf Motherweed 231587
Slenderleaf Mountain Leech 126368
Slenderleaf Needlegrass 376836
Slenderleaf Oxygraphis 278480
Slenderleaf Pholidota 295516
Slenderleaf Pulsatilla 321716
Slenderleaf Rabdosia 209842
Slenderleaf Sage 345167
Slenderleaf Saltbush 44508
Slenderleaf Sedge 76354
Slenderleaf Stonecrop 356696
Slenderleaf Swordflag 208882
Slenderleaf Tickclover 126368
Slenderleaf Tongoloa 392801
Slenderleaf Vetch 408624
Slender-leaf Wild Buckwheat 152227
Slender-leaved Bluets 187617
Slender-leaved Camellia 69694
Slender-leaved Cyclobalanopsis 116112
Slender-leaved Flax 231992
Slender-leaved Glochidion 177174
Slender-leaved Iceplant 251753
Slender-leaved Lily 230005,230009
Slender-leaved Naiad 262089
Slender-leaved Panic Grass 128525
Slender-leaved Peony 280318
Slender-leaved Pinweed 223706
Slender-leaved Pondweed 378986
Slender-leaved Rabdosia 209842
Slender-leaved Sundew 138304
Slender-leaved Valdiv Barberry 52296
Slender-leaved Venus' Looking Glass 397758
Slender-leaves Lampranthus 220697
Slender-lipped Helleborine 147168
Slenderlobe Maple 3637
Slender-lobed Maple 3637
Slenderlongstipe Primrose 314442
Slender-nerved Osmanthus 276309
Slenderpedicel Bushclover 227009
Slenderpedicel Meadowrue 388697
Slender-pedicel Pruinose Barberry 52070
Slender-pedicel Tephrosia 386070
Slender-pedicelled Barberry 51684
Slenderpedicelled Purpledaisy 267171
Slenderpedisel Sedge 76501
Slenderpeduncle Spapweed 205081
Slender-peduncle Spicebush 231360
Slenderpeduncle Touch-me-not 205081
Slender-peduncled Spice-bush 231360
Slenderprickle Holly 203901
Slender-prickled Holly 203901
Slenderrhizome Landpick 308552
Slenderrhizome Pearlsedge 354203
Slenderrhizome Solomonseal 308552
Slenderspicate Goosefoot 87037
Slender-spike Gouania 179947
Slenderspike Jointfir 178544
Slender-spiked Gouania 179947
Slender-spiked Willow 343040
Slenderspur Spapweed 205259
Slenderspur Touch-me-not 205259
Slenderstalk Acanthopanax 2391
Slenderstalk Acronema 6210
Slender-stalk Adinandra 8220
Slenderstalk Alseodaphne 17799
Slenderstalk Altingia 18198
Slenderstalk Arrowwood 407749
Slenderstalk Barberry 52222
Slenderstalk Dendropanax 125609
Slenderstalk Deutzia 126946
Slenderstalk Dicranostigma 129571
Slenderstalk Dogwood 104974
Slender-stalk Glyptic-petal Bush 177993
Slenderstalk Helwingia 191161
Slenderstalk Jointfir 178537
Slenderstalk Maesa 241798
Slender-stalk Mahonia 242542
Slender-stalk Microtropis 254295
Slenderstalk Milkvetch 42761
Slenderstalk Murdannia 260124
Slenderstalk Panicgrass 282286
Slenderstalk Privet 229646
Slenderstalk Starjasmine 393616
Slenderstalk Supplejack 52437
Slenderstalk Swallowwort 117492
Slenderstalk Thorowax 63679
Slenderstalk Trigonotis 397416
Slenderstalk Viburnum 407749
Slenderstalk Waterbamboo 260124
Slenderstalk Woodnettle 273905
Slenderstalk Yunnan Draba 137299
Slenderstalk Yunnan Whitlowgrass 137299
Slender-stalked Adinandra 8220
Slender-stalked Alseodaphne 17799
Slender-stalked Altingia 18198
Slender-stalked Ardisia 31617
Slender-stalked Jointfir 178537
Slender-stalked Mahonia 242542
Slender-stalked Maple 2846
Slender-stalked Microtropis 254295
Slender-stalked Oligostachyum 270431
Slenderstem Aster 40526
Slenderstem Cowpea 408900
Slenderstem Falsepimpernel 231562
Slenderstem Gentian 151004
Slenderstem Giantfennel 163623
Slenderstem Lessingia 227126
Slenderstem Marshmarigold 68222
Slenderstem Motherweed 231562
Slender-stem Pea-vine 222800
Slenderstem Rush 213137
Slenderstem Thorowax 63783
Slenderstem Turnflowerbean 98065
Slenderstem Woodbetony 287751
Slender-stemmed Pricklypear 272948
Slenderstyle Acanthopanax 143642
Slenderstyle Willow 343444
Slender-styled Acanthopanax 143642
Slendertail Elatostema 142869
Slender-Thoroughwort 166836
Slendertooth Clubfilment 179190
Slendertooth Gomphostemma 179190
Slender-tooth Milkvetch 42437
Slender-tooth Screwtree 190102
Slender-tooth Screw-tree 190102
Slendertoothed Cherry 83312
Slender-toothed Gomphostemma 179190
Slender-toothed Holly 203633
Slendertube Lagarosolen 219789
Slendertube Shortfilament Flower 221377
Slendertwig Clovershrub 70899
Slendertwig Cotoneaster 107468
Slender-twig Holly 204351
Slender-twigged Cotoneaster 107468
Slender-twigged Leaf-flower 296633
Slender-twigged Rhododendron 330075
Slender-twigs Blueberry 404050
Slenser Spiraea 371924
Slexder Bambusa 47483
Slice Qinggang 116130
Slick Greens 59532
Slickrock Wild Buckwheat 151836
Slickseed Bean 378504
Slick-seed Fuzzy Bean 378504
Slide Mountain Cushion Wild Buckwheat 152356
Slight-fragrant Holly 204287
Slight-lobed Largeleaved Linden 391825
Slightly Downy Flower Heath 149878
Slightly Small Iguanura 203510
Slightly Wing-branched Barberry 51683
Slightlyglabrous Pendentfruit Rockcress 30389
Slightly-hairy Polystachya 310559
Slightlyhard Greenorchid 137642
Slightlyobtuse Physospermopsis 297959
Slightlyrough Ehretia 141606
Slightly-scabrous Rattan Palm 65785
Slightyellow Sweetvetch 187875
Slim Amaranth 18734
Slim Tridens 396708
Slimflower Lovegrass 147692
Slimflower Scurfpea 319270
Slimfoot Century Plant 10863
Slimjim Flatsedge 118962
Slimleaf Flax 231992

Slimleaf Panicgrass 128525
Slimleaf Rosewood 405225
Slimleaf Sneezeweed 188427
Slimleaf Wallrocket 133394
Slim-leaved Flax 231992
Slim-leaved Wall Rocket 133313
Slim-leaved Wall-rocket 133313
Slimlobe Globeberry 203251,203254
Slim-lobe Rock Daisy 291308
Slimpod Rush 213049
Slimspike Three-awn 33914
Slim-spike Three-awn Grass 33914,33915
Slim-stem Reed Grass 65438,65490,65491,127276
Slim-stem Reed-grass 127276
Slim-stem Small Reed Grass 65490
Slimsten Lily 229775
Slim-top Meadow Rue 388667
Slimtop Meadowrue 388667
Slim-top Meadow-rue 388667
Slimy Rabdosia 209686
Slimyhair Sida 362590
Slimy-haired Sida 362590
Slink Pod 354573
Slinking Camomile 26768
Slinking Cedar 12559,85301
Slinking Christopher 355202
Slink-lily 354573
Slinkweed 182103
Sliperorchids 282768
Slipper Flower 66252,66263,66271,66273,287853,287855
Slipper Orchid 120287,282768,282824,282830,282839
Slipper Plant 287850
Slipperplant 287846
Slipperweed 204841
Slipperwort 66252,66261,66263,66273
Slippery Elm 401509,401620
Slippery Root 381035
Slite 115949
Slitlydivided Anemone 24069
Slitlydivided Windflower 24069
Slnnegar 246478
Sloanea 366034
Sloanes 316819
Sloe 83331,316206,316819,316890
Sloe Plum 316204
Sloe-bush 316819
Sloethorn 316819
Slogwood 50465
Slon 316819
Slon-bush 316819
Slone 316819
Slon-tree 316819
Sloon 316819
Slop 315082
Slope Azalea 330523
Slope Euonymus 157376
Slope Willow 343744
Slopeginseng Habenaria 183801
Slough Grass 49532,370173,370174
Slough Sedge 73782
Slough Thistle 91898
Sloughgrass 49532
Slough-grass 370173
Slough-heal 316127
Sloughwort 316127
Sloumi 122700
Slovenwood 35088
Slow 316819
Slue 316819
Slugwood 50465
Slugwoodleaf Litse 233853
Slugwood-leaved Litse 233853
Smack 282717
Smaillwort 325820
Smair Dock 340151
Smalach 29327
Smali-leaved Elm 401500
Small Aeluropus 8878
Small Agriophyllum 11616
Small Allson 18327
Small Aloe 17127
Small Androcorys 22326
Small Anemone 23811
Small Anise Tree 204570
Small Anisetree 204570
Small Arisaema 33449
Small Aromatic Bamboo 87634
Small Arrowhead 342394
Small Ash-leaf Raspberry 338999
Small Asian Toddalia 392560
Small Aulacolepis 25333
Small Awlquitch 131029
Small Azalea 331243
Small Balsam 205212
Small Bambooleaf Thorowax 63747
Small Basketvine 9437
Small Batrachium 48915
Small Beadcress 264627
Small Bedstraw 170696
Small Begonia 50143
Small Bellwort 403675
Small Bent 12139
Small Bindweed 102933
Small Bishop's-cap 256032
Small Bittersweet 80168
Small Black Poplar 311191
Small Bladderwort 403252
Small Bletilla 55563
Small Bluebasin 350209
Small Bluebellflower 115365
Small Blueberry 404003
Small Bluets 187559
Small Bogorchis 268004
Small Boniodendron 56971
Small Borreria 57329,370847
Small Bracteate Gingerlily 187462
Small Briggsia 60268
Small Bugbane 91024
Small Bugloss 21920
Small Bur 31056
Small Bur Clover 247381
Small Bur Marigold 53816
Small Burdock 31056,415057
Small Burnet 345860,345863,345866
Small Bur-parsley 79664
Small Burreed 370096
Small Bur-reed 370088,370096
Small Bush Violet 48323
Small Bushy Rhododendron 331477
Small Butterflyorchis 293623
Small Calabash 219854
Small Caltrops 395146
Small Camas 68800
Small Cancer-root 275234
Small Carpesium 77181
Small Carpgrass 36647,36659
Small Catchfly 363509
Small Cattleya 79566
Small Centaurium 81506
Small Centipeda 81687
Small Chamaesium 85721
Small Chickweed 375070
Small Chinese Brome 60973
Small Chirita 87972
Small Clover 247366
Small Clusters Rhododendron 332154
Small Common Alplily 234238
Small Cordgrass 370166
Small Cord-grass 370166
Small Cotton-weed 168691
Small Cowpea 408969
Small Cow-wheat 248211
Small Cowwheat-like Loosestrife 239734
Small Crab-grass 130612
Small Cranberry 278366,403894
Small Cranesbill 174854
Small Cremanthodium 110416
Small Crown Wirelettuce 375966
Small Crumbweed 87142
Small Cudweed 235267
Small Cup Spider-lily 200950
Small Cutfruit Tanoak 233402
Small Daffodil 262433
Small Daisy 50825
Small Dandelion 384614
Small Date Plum 132264
Small Datura 123049
Small Daylily 191312

Small Dendrocalamus 125495
Small Didissandra 87972
Small Didymocarpus 87972
Small Dimeria 131029
Small Diuranthera 135060
Small Dogwood 380486
Small Dragon 37015
Small Dragonbamboo 125495
Small Drop-seed 372782
Small Duckbeakgrass 209362
Small Duckweed 224375,224400
Small Dysosma 139612
Small Egretgrass 135060
Small Elaeagnus 142190
Small Elwes Woodbetony 287187
Small Enchanter's-nightshade 91508
Small Ephedra 146215
Small Euonymus 157737
Small European Cranberry 278366
Small Evening-primrose 269486
Small Fangcheng Rhododendron 330683
Small Fennel 266241
Small Fig-tree 165513
Small Fimbristylis 166183
Small Fingershapesplit Sagebrush 36399
Small Flatawndaisy 14916
Small Fleabane 321619
Small Floatingheart 267807
Small Flote Grass 177585
Small Flower Fishhook Cactus 354307
Small Flowered Agrimony 11571
Small Flowered Kudzuvine 321464
Small Flowered Rhododendron 331239
Small Flower-of-an-hour 383327
Small Flowers Rhododendron 330717
Small Footcatkin Willow 343658
Small Forget-me-not 260813
Small Forkleaf Maple 3499
Small Forkliporchis 269018
Small Fortune Spindle-tree 157503
Small Fragrantbamboo 87634
Small Fragrant-bamboo 87634
Small Fringed Gentian 174228
Small Fruit Bottle Gourd 219854
Small Fruit Coriaceousleaf Maple 2907
Small Fruit Crab Apple 243657
Small Furrowlemma 25333
Small Furze 401395
Small Gagea 169420,169446
Small Galangal 17686,17733
Small Garden Spurge 159557
Small Gastrodia 171937
Small Gentian 173666,173708
Small Gentianopsis 174224
Small Geophila 174366
Small Geranium 174854
Small Gingersage 253635
Small Glandhair Didymocarpus 129906
Small Globe Thistle 140776
Small Globeflower 399524
Small Globethistle 140776
Small Globe-thistle 140776
Small Globular Spike Pycreus 322242
Small Goose Grass 170694
Small Goosecomb 335435
Small Goosefoot 87158
Small Green Fringed Orchid 302287
Small Green Wood Orchid 302287
Small Green Wood-orchis 302287
Small Greenspine 71916
Small Guangxi Loosestrife 239548
Small Habenaria 183392
Small Hair-grass 12839
Small Hairycalyx Swertia 380229
Small Halophila 184977
Small Hare's-ear 63573
Small Head Edelweiss 224904
Small Head Heteroplexis 194013
Small Heartwort 34149
Small Heath 67456
Small Hemarthria 191250
Small Herbclamworm 398108
Small Heterolobe Pellionia 288730
Small Hiptage 196842
Small Hollowbamboo 82450
Small Hollow-bamboo 82450
Small Honesty 127793
Small Hooded Willowherb 355609
Small Hookleafvine 303094
Small Hop Clover 396882
Small Horaninowia 198243
Small Horse Bean 408395
Small Houseleek 356503
Small Indocalamus 206824
Small Iris 208768
Small Jack-in-the-pulpit 33544
Small Jadeleaf and Goldenflower 260472
Small Java Paraphlomis 283625
Small Jijigrass 4123
Small Kaffir Orange 378688
Small Kidney Bean 294056
Small Kingidium 293639
Small Kneejujube 67070
Small Knifebean 71040
Small Knotweed 309859
Small Lagenophora 219898
Small Lamb's-succory 34816
Small Landpick 308562
Small Lanyu Butterflyorchis 293600
Small Lateflowered Willow 343810
Small Leaf Edelweiss 224907
Small Leaf Joint-fir 178555
Small Leaved Lime 391691
Small Liparis 232156
Small Love Grass 147871
Small Lovegrass 147815
Small Love-grass 147815
Small Lupine 238483
Small Mahogany 400589
Small Malcolmia 243230
Small Manyleaf Paris 284370
Small Meadow Rue 388572
Small Meadowrue 388506
Small Medick 247381
Small Melilot 249218
Small Microcarp 253200
Small Microcarpaea 253200
Small Micromeria 253635
Small Microula 254377
Small Mistletoe 30917
Small Moonflower 208264
Small Morot Stonecrop 356946
Small Mouse-ear Chickweed 83020
Small Naiad 262089
Small Nettle 403039
Small Nightshade 367689
Small Northern Bog Orchid 302455
Small Northern Bog-orchis 302455
Small Nujiang Camellia 69589
Small Num-num 76894
Small Nutantdaisy 110416
Small Obcordatelimb Larkspur 124419
Small Ochotsk Corydalis 106355
Small Oncidium 270839
Small Oneflowerprimrose 270736
Small Orostachys 275386
Small Osmanther 276373
Small Ovary Rhododendron 331240
Small Oxwhipgrass 191250
Small Parnassia 284607
Small Pasqueflower 23998
Small Pasque-flower 23998
Small Patulous Anemone 23991
Small Pellionia 288749
Small Penny Pie 362406
Small Pepper 411189
Small Peppergrass 225340
Small Periwinkle 409339
Small Philippine Acacia 1145
Small Philippine Wattle 1145
Small Pink Lady's-slipper 120290
Small Pinweed 223701
Small Pipewort 151356
Small Plantain 302001,302152
Small Pogonia 306879
Small Pond-lily 267309
Small Pondweed 312236,312241
Small Pulicaria 321619
Small Purple Fringed Orchid 302491
Small Purpleflower Goldenray 229024
Small Pussy-toes 26428

Small Qinghai Peashrub 72207
Small Qinghai Pea-shrub 72207
Small Quaking-grass 60383
Small Radish 326616
Small Ragweed 19145
Small Ramie 56318
Small Rattanpalm 65644
Small Rattlebox 112447
Small Razorsedge 354185
Small Red Goosefoot 86990
Small Red Ramie 56318
Small Red Trubba 367157
Small Redcarweed 288749
Small Reddish Arrowwood 407831
Small Reddish Viburnum 407831
Small Redroot 79957
Small Reed 127192,127196,127243
Small Rest Harrow 271563
Small Restharrow 271540,271552
Small Rhododendron 331243
Small Robin's Eye 174877
Small Robin's Eyes 174877
Small Rockfoil 349791
Small rocklettuce 299687
Small Roegneria 335435
Small Round-leaved Orchid 273614
Small Round-leaved Orchis 273614
Small Rourea 337733
Small Rush 212929
Small Rush Sandwort 31968
Small Saltbush 44382
Small Saltcress 389198
Small Saltgrass 184977
Small Saltmarsh Aster 41317,256310
Small Scabious 350125
Small Scabrous Smallreed 127294
Small Screwpine 281039
Small Seaside Lavender 190597
Small Shell Ginger 17723
Small Shield Rhododendron 331446
Small Shin-leaf 322853
Small Short-flower Stauntonvine 374385
Small Skullcap 355383,355565,355661
Small Skullcaplike Coleus 99712
Small Smock 23925
Small Snapdragon 28631,84596
Small Soap Plant 88478
Small Soapweed 416603
Small Soaring Poplar 311192
Small Solomon's Seal 308508,308633
Small Solomonseal 308562
Small Sonerila 368887
Small Southstar 33449
Small Spearwort 325864
Small Spike Rush 143289
Small Spikenard 30712
Small Spike-rush 143289
Small Spine Bamboo 47275
Small Spiny-club Palm 46381
Small Spotleaf-orchis 179685
Small Spring Knotweed 309577
Small Spurge 158869
Small Squarebamboo 87557
Small St. John's-wort 201906
Small Starwort 375043
Small Staurochilus 374459
Small Straggly Rhododendron 331241
Small Strobilus Elsholtzia 144100
Small Sundrops 269486
Small Sword Jackbean 71040
Small Swordflag 208768
Small Swwet Indosasa 206902
Small Teasel 133502
Small Thellungiella 389198
Small Thomson Azalea 331972
Small Thorowax 63850
Small Thyrse Corydalis 106526
Small Toadflax 84596
Small Tongue Orchid 360629
Small Tooth Leaves Rhododendron 331807
Small Torularia 264627
Small Tree Willow 343041
Small Tripogon 398082
Small Tumbleweed Mustard 365529
Small Ungeargrass 36692
Small Vernalgrass 28001
Small Villous Gingerlily 187481
Small Waldheimia 14916
Small Wampee 94188
Small Warts Rhododendron 332057
Small Water Dock 340116
Small Water-pepper 309395
Small Waterplantain 14751
Small Water-plantain 14785
Small Waterwort 142572
Small Wax Lip 177315
Small Waxberry 261202
Small Waxmallow 243932
Small Welsh Stonecrop 356756
Small White Aster 40479,380945,380975
Small White Clover 396985
Small White Fawn-lily 154896
Small White Lady's Slipper 120320
Small White Lady's-slipper 120320
Small White Morning Glory 207921
Small White Morning-glory 207921
Small White Orchid 182216,318209,318210
Small White Violet 410189,410190
Small White Water-lily 267695,267767
Small Whorled Pogonia 210335
Small Whorled Pogonia Orchid 210335
Small Wild Banana 260199
Small Wild Buckwheat 162319
Small Wild Carrot 123215
Small Wild Clary 190552
Small Wild Gardenia 171250
Small Willow 343690
Small Windflower 23811
Small Windhairdaisy 348534
Small Wirelettuce 375964
Small Wood Sunflower 188999
Small Worm-seed Mustard 154483
Small Xizang Sedge 76539
Small Yellow Daylily 191312
Small Yellow Flax 231924
Small Yellow Foxglove 130371
Small Yellow Gentian 174076
Small Yellow Lady's Slipper 120310, 120312,120421
Small Yellow Lady's-slipper 120421,120423
Small Yellow Pond-lily 267309
Small Yellow Sedge 74223
Small Yellow Toadshade 397555
Small Yellow Trefoil 396882
Small Yellow Water-crowfoot 325901
Small Yucca 416603
Small Ziziphora 418140
Small's Goldenrod 368317
Small's Ragwort 279886
Small's Sulphur Flower 152602
Smallache 29327
Smalladge 29327
Smallage 29327
Small-and-pretty 243209
Small-anther Alkaligrass 321338
Smallanther Tea 69355
Smallbeak Habenaria 184038
Smallbeak Meadowrue 388647
Smallbell Ladybell 7673
Smallbetony Lopseed 295984
Smallbox Tanoak 233112
Smallbract Buttercup 326090
Smallbract Gingerlily 187462
Smallbract Hemiboea 191378
Smallbract Ligusticum 229301
Smallbract Milkvetch 42909
Smallbract Monkeyflower 255194
Smallbract Muli Larkspur 124395
Smallbract Orach 44618
Smallbract Spapweed 204858
Smallbristle Hound's Tongue 118002
Small-bud Neolitsea 264081
Smallbud Newlitse 264081
Smallbud Rockfoil 349375
Small-buded Newlitse 264081
Smallbulb Bulbophyllum 63100
Smallbulb Stonebean-orchis 63100
Smallcalyx Belltree 325000
Smallcalyx Fissistigma 166672,166682
Smallcalyx Jurinea 211907
Smallcalyx Lysionotus 239977

Smallcalyx Porana 131243
Smallcalyx Woodbetony 287151,287428
Small-calyxed Bell Tree 325000
Small-calyxed Fissistigma 166672
Small-calyxed Jasmine 211907
Smallcapitate Eargrass 187538
Smallcapitate Hedyotis 187538
Smallcarp Bellflower 69972
Smallcatkin Willow 343461
Smallclump Crazyweed 278762
Smallcluster Rockjasmine 23230
Smallcrenate Woodbetony 287125
Smallcristate Woodbetony 287127
Smallcup Rhodiola 329953
Small-cushion Milkvetch 42953
Small-cuspidate Camellia 69435
Smallcuspidate Tea 69435
Small-denticulated Willow 343840
Small-eared Indosasa 206897
Smalledge 29327
Smallegg Tea 69451
Smaller Bladderwort 403220
Smaller Burdock 45982
Smaller Cat's Tail 401094
Smaller Cat's Tails 401094
Smaller Cat's-tail 294968
Smaller Forget-me-not 260813
Smaller Fringed Gentian 173739
Smaller Hemlock 9832
Smaller Purple-fringed Orchis 302491
Smaller Pussytoes 26427
Smaller Rhododendron 331248
Smaller Spreading Pogonia 94546
Smaller Tree-mallow 223364
Smaller White Snakeroot 158062
Smallest Flowers Rhododendron 331249
Smallest Hare's Ear 63855
Smallest Redpepper 72113
Smallflower Adonis 8374
Smallflower Agapetes 10311
Smallflower Ajania 13021
Smallflower Alaishan Milkvetch 43005
Smallflower Alashan Panzerina 282479
Smallflower Alpine Rockcress 30174
Small-flower Anise Tree 204570
Smallflower Antlevine 88899
Smallflower Armenia Pocktorchid 282784
Smallflower Arundinella 37417
Smallflower Ashycoloured Ironweed 406228
Smallflower Asiabell 98373
Smallflower Aspidistra 39560
Smallflower Beggarticks 54042
Smallflower Bethlehemsage 283638
Smallflower Bigginseng 241179
Small-flower Biondia 54526
Smallflower Bittercress 72920
Smallflower Bitterorange 93635
Smallflower Boniodendron 56972
Smallflower Bracketplant 88589
Smallflower Broadleaf Adenostemma 8016
Smallflower Brooklet Anemone 24028
Smallflower Brooklet Windflower 24028
Smallflower Bruguiera 61266
Smallflower Bugle 13164
Smallflower Bunge Batrachium 48909
Smallflower Burnet 345921
Smallflower Buttercup 326194
Smallflower Butterflygrass 392927
Small-flower Camellia 69355
Smallflower Campanumoea 70408
Smallflower Camphor Tree 91378
Smallflower Capillary Feathergrass 376731
Smallflower Capillipedium 71608
Smallflower Catchfly 363259
Smallflower Catnip 265078
Smallflower Centaurea 81398
Smallflower Century Plant 10922
Smallflower Chastetree 411396
Smallflower Chaste-tree 411396
Smallflower Chee Woodreed 65337
Smallflower Chickweed 374986
Smallflower China Onosma 271759
Smallflower Chonemorpha 88899
Smallflower Chorispora 89008
Smallflower Cinnamon 91378
Smallflower Cinquefoil 312762
Small-flower Clovershrub 70797
Smallflower Clubfilment 179200
Smallflower Columbine 30068
Smallflower Crapemyrtle 219959
Smallflower Crazyweed 278842
Smallflower Dandelion 384732
Small-flower Death Camas 417897,417915
Smallflower Deutzia 127032
Smallflower Dichroa 129039
Smallflower Dogwood 380484
Smallflower Dragonflyoechis 302549
Smallflower Dragonhead 137611
Smallflower Dry Catchfly 363609
Smallflower Dutchmanspipe 34164
Small-flower Dwarf Peashrub 72330
Small-flower Dwarf Pea-shrub 72330
Smallflower Edelweiss 224825,224904
Smallflower Eightangle 204558
Small-flower Elaeocarpus 142357
Smallflower Embelia 144778
Smallflower Emei Larkspur 124429
Smallflower Epaulettetree 320885
Smallflower Epipactis 147149
Smallflower Falsehellebore 405615
Smallflower Falsetamarisk 261296
Smallflower Fieldjasmine 199451
Small-flower Ford Metalleaf 296385
Smallflower Forrest Sandwort 31911
Smallflower Foxglove 130374
Smallflower Fuchsia 168780
Smallflower Funnel Flower 130222
Smallflower Galangal 17653
Smallflower Galinsoga 170142
Small-flower Ginger 155398
Smallflower Gomphostemma 179200
Smallflower Greenorchid 137611
Smallflower Greenvine 204642
Smallflower Grewia 180703
Smallflower Gromwell 233760
Smallflower Hairy Willowherb 146834
Smallflower Hairyfruitgrass 222354
Smallflower Hartia 376455
Smallflower Hawksbeard 110896,110965
Small-flower Hawksbeard 110861
Smallflower Heliotrope 190676
Smallflower Hemiboea 191379
Smallflower Heteroplexis 194013
Small-flower Honeysuckle 235954
Smallflower Hornfennel 201616
Smallflower Houndstongue 117986
Smallflower Huodendron 199451
Smallflower Illigera 204642
Smallflower Iris 208853
Smallflower Irisorchis 267970
Smallflower Japan Pieris 298751
Smallflower Japanese Buckthorn 328747
Smallflower Junegrass 217494
Smallflower Ladyslipper 120414
Small-flower Lasianthus 222229
Smallflower Leopard 70408
Smallflower Linearstripe Rabdosia 209759
Smallflower Liparis 232290
Smallflower Listera 264746
Smallflower Lovegrass 147864
Smallflower Magnoliavine 351069
Small-flower Mastixia 246216
Smallflower Meadow Windhairdaisy 348135
Smallflower Milkwort 308398
Smallflower Milkwort Corydalis 106279
Smallflower Monkshood 5498
Smallflower Morina 258865
Smallflower Mountainash 369462
Smallflower Muping Corydalis 105888
Small-flower Mussaenda 260472
Small-flower Narrowleaf Peashrub 72352
Small-flower Narrowleaf Pea-shrub 72352
Smallflower Narrowwing Valerian 404360
Smallflower Nepeta 265022
Smallflower Oberonia 267970
Smallflower Oncidium 270835
Smallflower Ophiorrhiza 272272
Smallflower Opithandra 272590
Smallflower Oxhneedaisy 170142
Smallflower Oxtongue 298633
Smallflower Panzerina 282487

Smallflower Paphiopedilum 282867
Small-flower Parabarium 402194
Smallflower Paraboea 283135
Smallflower Peachgrass 172203
Smallflower Peliosanthes 288646
Smallflower Pellitory 284160
Smallflower Peristylus 291189
Smallflower Perotis 291189
Smallflower Plantain 302196
Smallflower Platanthera 302441
Smallflower Pricklyash 417271
Small-flower Quisqualis 324671
Smallflower Rabbit Milkvetch 42571
Smallflower Rabbittail Milkvetch 42571
Smallflower Rattlesnakeroot 313844
Smallflower Reedbentgrass 65337
Smallflower Resurrectionlily 215028
Smallflower Ricegrass 300704
Small-flower Rose 336741
Smallflower Rosewood 121788
Small-flower Roughleaf 222229
Smallflower Rush 212866
Smallflower Sabia 341539
Smallflower Sagebrush 36059
Smallflower Saltbush 44523
Smallflower Sarcochilus 346713
Smallflower Satyrium 347869
Smallflower Saussurea 348626
Smallflower Schoenorchis 352307
Smallflower Seaberry 179438
Smallflower Seatung 301359
Smallflower Showy Jerusalemsage 295166
Smallflower Sibbaldia 362356
Smallflower Silene 363259
Smallflower Silkstyle Gentian 173445
Smallflower Singlebamboo 56954
Smallflower Skullcap 355808
Small-flower Smoothleaf Kiwifruit 6645
Smallflower Snakegourd 396245
Smallflower Sobralia 366790
Smallflower Spapweed 205216
Smallflower Spiny Yam 131837
Smallflower Squarebamboo 87587
Smallflower Stephania 375890
Smallflower Sterculia 376142,376155
Smallflower Sugarpalm 32340
Smallflower Sundrops 269481
Smallflower Swallowwort 117454
Smallflower Sweetbrier 336741
Smallflower Swordflag 208853
Smallflower Tailleaftree 402635
Smallflower Tamarisk 383581
Small-flower Tamarisk 383581
Small-flower Tansy-aster 33038
Small-flower Tatarian Honeysuckle 236158
Smallflower Thoroughwort 158312
Smallflower Threevein Aster 39970
Smallflower Thymeleaf Eritrichium 153549
Smallflower Tibet Microula 254372
Smallflower Torenia 392927
Smallflower Touch-me-not 205212,205216
Smallflower Tuoli Fritillary 168591
Smallflower Twayblade 232290
Smallflower Vanda 404659
Smallflower Wallflower 154483
Smallflower Wallgrass 284160
Smallflower Ward Gentian 174069
Smallflower Wendlandia 413808
Smallflower Willow Weed 146834
Smallflower Willowweed 146834
Smallflower Windhairdaisy 348626
Smallflower Wingstalk Hilliella 196299
Smallflower Woodbetony 287427
Smallflower Woodrush 238678
Smallflower Wood-stem Edelweiss 224946
Smallflower Xizang Microula 254372
Smallflower Yam 131846
Smallflower Yarrow 3976
Smallflower Yellow Laelia 219678
Small-flower Yunnan Dutchmanspipe 34377
Small-flowered Acanthus 2672
Small-flowered Agapetes 10311
Small-flowered Androstephium 23380
Small-flowered Anemone 23980
Small-flowered Anise-tree 204558
Small-flowered Baliospermum 46969
Smallflowered Balsam 205212
Small-flowered Balsam 205212
Small-flowered Begonia 50079
Small-flowered Big-ginseng 241173
Small-flowered Bitter Cress 72920
Small-flowered Bitter-cress 72920,72922
Small-flowered Bitterorange 93634
Small-flowered Brookweed 345731
Small-flowered Buttercup 325506,326194
Small-flowered Caltrop 215384
Small-flowered Camellia 69437
Small-flowered Catchfly 363477
Small-flowered Chonemorpha 88899
Small-flowered Cinnamon 91378
Small-flowered Clovershrub 70797
Small-flowered Coralroot 104009
Small-flowered Coral-root 104009
Small-flowered Corydalis 106149
Small-flowered Crane's-bill 174854
Small-flowered Cranesbill 174805
Small-flowered Crowfoot 325506,326194
Small-flowered Deutzia 127032
Small-flowered Dogwood 380484
Small-flowered Dragonhead 297981
Small-flowered Elaeagnus 142081
Small-flowered Embelia 144778
Small-flowered Eucrypta 156074
Small-flowered Evening Primrose 269481
Small-flowered Evening-primrose 269418, 269481
Small-flowered False Foxglove 10148,10149
Smallflowered Felwort 235455
Small-flowered Forget-me-not 260885
Small-flowered Gaura 172203
Small-flowered Gerardla 10148
Small-flowered Glycosmis 177845
Small-flowered Grass-of-parnassus 284603
Small-flowered Greenstar 307497
Small-flowered Gypsum-loving Larkspur 124269
Small-flowered Hartia 376455
Small-flowered Hawkweed 202404
Small-flowered Hedgehog Cactus 140343
Small-flowered Helleborine 147112
Small-flowered Hemicarpha 232404
Small-flowered Heterosmilax 194123
Small-flowered Holboellia 197247
Small-flowered Illigera 204642
Smallflowered Japan Pine 300130
Small-flowered Japanese Abelia 184
Small-flowered Land Cress 47959
Small-flowered Lasianthus 222229
Small-flowered Leafcup 310023
Small-flowered Mallow 243810
Small-flowered Melilot 249218
Small-flowered Meliosma 249429
Small-flowered Microtropis 254312
Smallflowered Milkvetch 42810
Small-flowered Mussaenda 260472
Small-flowered Parnassia 284603
Small-flowered Pawpaw 38330
Small-flowered Phacelia 293164
Small-flowered Phoebe 295384
Small-flowered Photinia 295772
Small-flowered Prickly-ash 417271
Small-flowered Rhododendron 330716
Small-flowered Sabia 341539
Small-flowered Snapweed 205212
Small-flowered Square-bamboo 87587
Small-flowered St. John's-wort 202029
Small-flowered Sterculia 376142
Small-flowered Stringbush 414235
Small-flowered Sugarpalm 32340
Small-flowered Sweet Spire 210407
Small-flowered Sweetbrier 336741
Small-flowered Sweetspire 210407
Smallflowered Swertia 380217
Small-flowered Tick Clover 126514
Small-flowered Tickle Grass 126039
Small-flowered Trillium 397590
Small-flowered Twayblade 232275
Smallflowered Vancouveria 404619
Small-flowered Wild Bean 378504
Smallflowered Willow Herb 146834
Small-flowered Willowherb 146834

Small-flowered Willowweed 146834
Small-flowered Winter Cress 47959
Small-flowered Wood Rush 238678
Small-flowered Wood-rush 238678
Small-flowered Yellow Rocket 47959
Small-flowering Evening Primrose 269425
Smallflower-like Dendropanax 125628
Smallflower-like Treerenshen 125628
Smallflowerr Nielamu Gentian 173672
Small-flowred Barberry 51926,51928
Smallfruit Altai Draba 136921
Smallfruit Altai Whitlowgrass 136921
Smallfruit Amomum 19896
Smallfruit Aphragmus 29172
Smallfruit Bittersweet 80206
Small-fruit Blueberry 278360
Smallfruit Bottle Gourd 219854
Smallfruit Burreed 370080
Smallfruit Camelina 68850
Smallfruit Camellia 69006,69743
Small-fruit Chinese Aspen 311196
Smallfruit Coriaceousleaf Maple 2907
Smallfruit Corydalis 106067
Smallfruit Cragcress 98015
Small-fruit Cranberry 278360
Smallfruit Craneknee 221705
Smallfruit Dandelion 384636
Smallfruit Dock 340135
Smallfruit Elm 401569
Smallfruit Eritrichium 153535
Smallfruit Euonymus 157721
Smallfruit Falseflax 68850
Small-fruit Fig 165307,165320
Smallfruit Greenbrier 366309
Smallfruit Greenvine 204626
Smallfruit Gynostemma 183033
Small-fruit Hawthorn 110054
Smallfruit Heartseed 73208
Smallfruit Helicia 189918
Smallfruit Herb Seatung 350337
Smallfruit Holly 204050,204217
Small-fruit Honeylocust 176869
Smallfruit Hornyfruit Seepweed 379510
Smallfruit Iodes 207377
Smallfruit Japan Purplepearl 66826
Smallfruit Meadowrue 388569
Smallfruit Microula 254363
Smallfruit Milkvetch 42893,43199
Small-fruit Nearsmooth Barberry 52205
Smallfruit Nettle 402855
Smallfruit Nettletree 80610
Smallfruit Pagodatree 369066
Smallfruit Plumepoppy 240769
Small-fruit Queensland Nut 240210
Small-fruit Rehdertree 327421
Smallfruit Rockfoil 349642
Smallfruit Rose 336509
Smallfruit Scurvyweed 98015
Small-fruit Sessileflower Acanthopanax 143667
Smallfruit Sharpfruit Aphragmus 29172
Smallfruit Slatepentree 322596
Small-fruit Songar Cotoneaster 107689
Smallfruit Spicegrass 239735
Smallfruit Spiradiclis 371771
Smallfruit Tanoak 233317
Smallfruit Thickshellcassia 113471
Small-fruit Toona 392837
Smallfruit Waxpetal 106658
Smallfruit Wildtung 243391
Smallfruit Windhairdaisy 348786
Small-fruit Wingfruit Beancaper 418741
Smallfruit Winterhazel 106658
Smallfruit Woad 209215
Smallfruit Yinshancress 416320
Smallfruit Yinshania 416320
Small-fruited Blueberry 278360
Small-fruited Cactus 244152
Small-fruited Camellia 69006,69743
Small-fruited Chinese Holly 204217
Small-fruited Elm 401569
Small-fruited Euonymus 157721
Small-fruited False Flax 68850
Small-fruited Fig 165307,165320
Small-fruited Grape 411568
Small-fruited Gray Gum 155706
Small-fruited Holly 204050
Small-fruited Hooker Barberry 51736
Small-fruited Iodes 207377
Small-fruited Jujube 418197
Small-fruited Mallotus 243369
Small-fruited Mallow 243810
Small-fruited Marking-nut 357882
Small-fruited Pawpaw 38330
Small-fruited Rose 336509
Small-fruited Spindle-tree 157721
Small-fruited Toona 392837
Small-fruited Wildtung 243391
Small-fruited Winterhazel 106658
Smallgarliccress 253949
Smallglume Fescue 164198
Smallglume Fescuegrass 164198
Smallglume Goosecomb 335456
Smallglume Roegneria 335456
Smallgrain Rice 275946
Smallhead Adenosma 7989
Smallhead Arnica 34718
Smallhead Aster 380945
Smallhead Catsear 202426,202427
Smallhead Doll's-daisy 56713
Smallhead Gayfeather 228488
Smallhead Goldenray 229108
Smallhead Heathgoldenrod 236498
Smallhead Knotweed 309389
Smallhead Lakemelongrass 232423
Smallhead Lipocarpha 232423
Smallhead Nepeta 264996
Smallhead Serpentroot 354913
Small-head Snakeweed 182103
Smallhead Sneezeweed 188428
Smallhead Star-thistle 80982
Smallhead Thistle 91843
Smallhead Windhairdaisy 348239
Small-headed Bog Sedge 76513
Small-headed Rush 212912,212914
Small-headed Rush Pussytoes 26452
Small-headed Sweetshell Rhododendron 330527
Smallheartleaved Morningglory 208023
Smallhole Ardisia 31564
Smallhooked Hedyotis 187691
Smallhorn Iguanura 203505
Smallkatkin Willow 344195
Small-leaf Acaena 1751
Small-leaf African Myrsine 261608
Smallleaf Ainsliaea 12693
Smallleaf Anemone Clematis 95143
Smallleaf Arrowwood 408019
Small-leaf Arthraxon 36692
Small-leaf Balm 249507
Small-leaf Bamboo 297385
Smallleaf Beancaper 418703
Smallleaf Bedstraw 170696
Smallleaf Birch 53525
Smallleaf Bittercress 72897
Smallleaf Bladderwort 403252
Small-leaf Bluebeard 78009
Smallleaf Bluebell 115395
Smallleaf Bluebellflower 115395
Small-leaf Blueberry 403956
Smallleaf Bredia 59862
Smallleaf Briggsia 60280
Smallleaf Buchanania 61672
Smallleaf Caesalpinia 65038
Smallleaf Camelina 68854
Smallleaf Campanumoea 116257
Smallleaf Carmona 77067
Smallleaf Caudate Wildginger 37587
Smallleaf Chalkplant 183212
Small-leaf Childvine 400887
Smallleaf China Bonybery 276546
Small-leaf Chinese Aspen 311198
Smallleaf Chinese Bonyberry 276546
Smallleaf Chirita 87938
Small-leaf Cinnamon Tree 91421
Small-leaf Cinnamonleaf Maple 2892
Smallleaf Cinquefoil 312764
Small-leaf Cinquefoil 289718
Small-leaf Clematis 95151
Small-leaf Cleyera 96627
Small-leaf Cliffbean 254814

Smallleaf Climbing Fig 165520
Small-leaf Clovershrub 70839
Smallleaf Cockbone Elsholtzia 144030
Small-leaf Coinleaf Whitepearl 172129
Smallleaf Colicweed 128314
Small-leaf Columnea 100122
Smallleaf Crazyweed 279014
Small-leaf Crazyweed 278846
Smallleaf Cremanthodium 110414
Small-leaf Crystal Tea 223911
Smallleaf Cyrtococcum 120638
Small-leaf Desmodium 126465
Smallleaf Dischidia 134040
Smallleaf Dragonbamboo 125467
Smallleaf Draws Ear Grass 403252
Smallleaf Dualbutterfly 398289
Smallleaf Dunbaria 138973
Small-leaf Elaeocarpus 142373
Smallleaf Elatostema 142784
Smallleaf Elliptic Oreocharis 273855
Smallleaf Emarginate Eurya 160460
Smallleaf Epimedium 147031
Smallleaf Epipactis 147183
Smallleaf Eria 396476
Smallleaf Falseflax 68854
Smallleaf Fig 164688
Smallleaf Flatsheath Fimbristylis 166228
Small-leaf Flemingia 166876
Smallleaf Fleshspike 346957
Small-leaf Fordia 167286
Smallleaf Fullmoon Maple 3037
Small-leaf Gironniera 29009
Small-leaf Glaucousleaf Cotoneaster 107464
Smallleaf Goldenray 229130
Small-leaf Grape 411929
Small-leaf Grewia 180702
Small-leaf Gum 155685
Small-leaf Hackberry 80695
Small-leaf Helwingia 191167
Small-leaf Heritiera 192497
Smallleaf Hillcelery 276775
Small-leaf Himalayan Stachyurus 373540
Smallleaf Holly 204418
Small-leaf Honeylocust 176897
Small-leaf Honeysuckle 236751
Smallleaf Hornbeam 77345
Smallleaf Indian Skullcap 355510
Smallleaf Japan Eurya 160508
Smallleaf Japan Purplepearl 66827
Smallleaf Japanese Germander 388118
Smallleaf Jewelvine 125987
Small-leaf Jointfir 178555
Smallleaf Knotweed 309042
Small-leaf Kunming Mockorange 294487
Small-leaf Lasianthus 222231
Small-leaf Leafflower 296701
Smallleaf Leopard 116257
Small-leaf Leptodermis 226116
Smallleaf Longflower Willow 343620
Smallleaf Longspurlike Larkspur 124189
Smallleaf Meadowrue 388492
Smallleaf Meadowsweet 166073
Smallleaf Meconopsis 247200
Small-leaf Merrill Wendlandia 413801
Smallleaf Milkvetch 42505
Smallleaf Milkwort 308419
Small-leaf Millettia 254814
Smallleaf Mitragyna 256121
Smallleaf Mitreola 256215
Smallleaf Mockstrawberry 138801
Smallleaf Mosquitotrap 117373
Smallleaf Mountain Leech 126465
Smallleaf Myoporum 260720
Smallleaf Nanmu 295383
Small-leaf Neomicrocalamus 324933
Smallleaf North Pole Fruit 31321
Small-leaf Nummularylike Wintergreen 172129
Smallleaf Nutantdaisy 110414
Small-leaf Olive 270156
Smallleaf Orange Fig 164656
Small-leaf Oriental Grape 411686
Smallleaf Ormosia 274412
Smallleaf Pellionia 288715
Smallleaf Pepper 300533
Small-leaf Platea 302604
Smallleaf Pleione 304283
Smallleaf Premna 313710
Small-leaf Premna 313689
Smallleaf Privet 229525
Small-leaf Ptarmiganberry 31321
Smallleaf Pubescent Willow 343926
Small-leaf Pussytoes 26536
Small-leaf Raspberry 339349
Smallleaf Rose-root 329939
Small-leaf Roughleaf 222231
Smallleaf Roundscape Sealavender 230796
Smallleaf Schefflera 350757
Small-leaf Scrub Persimmon 132135
Small-leaf Scurrula 355315
Smallleaf Seepweed 379566
Smallleaf Selinum 357834
Small-leaf Shortfruit Bauhinia 49027
Small-leaf Shortstyle Camellia 69363
Small-leaf Shrubby Elsholtzia 144030
Small-leaf Siberian Elm 401604
Smallleaf Siebold Elder 345695
Small-leaf Silvertree 192497
Smallleaf Siphocranion 365193
Small-leaf Sophora 369106
Small-leaf Soulie Rose 336966
Smallleaf Spapweed 205216
Small-leaf Spiderwort 394021
Smallleaf Stephanotis 245849
Small-leaf Sterculia 376155
Smallleaf Stink New Eargrass 262967
Smallleaf Swallowwort 117373
Smallleaf Sweet Briar 336741
Small-leaf Sweet Briar 336741
Small-leaf Tamarind 132991
Smallleaf Tigerthorn 122041
Smallleaf Touch-me-not 205216
Smallleaf Tripterospermum 398289
Small-leaf Tylophora 400887
Small-leaf Underleaf Pearl 296701
Smallleaf Undulateleaf Oplismenus 272696
Small-leaf Villaintree 29009
Small-leaf Wendlandia 413793
Smallleaf Wildginger 37648
Small-leaf Winged Spindle-tree 157294
Small-leaf Winterhazel 106664,106680
Small-leaf Witchhazel 284964
Smallleaf Woodbetony 287104
Small-leaf Yellow-white Elsholtzia 144065
Smallleaf Yellowwood 94029
Small-leafed Fuchsia 168778
Small-leafed Lillypilly 382604
Small-leafed Lime 391691
Small-leafed Montia 258263
Small-leafed Moreton Bay Fig 165471
Small-leafed Peppermint 155666
Small-leaved Bamboo 297385
Small-leaved Barberry 51923
Smallleaved Barringtonia 48518
Small-leaved Barringtonia 48518
Small-leaved Birch 53525
Small-leaved Bittercress 72920
Small-leaved Blinks 258263
Small-leaved Box 64285
Small-leaved Breynia 60083
Small-leaved Buchanania 61672
Small-leaved Buckthorn 328816
Small-leaved Caesalpinia 65038
Small-leaved Camellia 69363,69443,69705
Small-leaved Canary Tree 71014
Small-leaved Canthium 71455
Small-leaved Carmona 77067
Small-leaved Chalkplant 183218
Small-leaved Cinnamon 91280
Small-leaved Clematis 95151
Small-leaved Cleyera 96627
Small-leaved Cotoneaster 107549
Small-leaved Currant 334017
Small-leaved Distylium 134932
Small-leaved Drepanostachyum 137861
Small-leaved Elaeocarpus 142373
Smallleaved European Linden 391691
Small-leaved European Linden 391691
Small-leaved Fargesia 162731
Small-leaved Fig 164833
Small-leaved Fordia 167286

Small-leaved Fortune Spindle-tree 157512
Small-leaved Grape 411929
Small-leaved Grewia 180925
Small-leaved Gum 155685
Small-leaved Heath 149205
Small-leaved Helleborine 147183
Small-leaved Heritiera 192497
Small-leaved Honeylocust 176897
Small-leaved Honeysuckle 235952
Small-leaved Hornbeam 77345
Small-leaved Indosasa 206894
Small-leaved Japanese Spindle-tree 157610
Small-leaved Jasmin-orange 260171
Small-leaved Jointfir 178555
Small-leaved Leaf-flower 296701
Small-leaved Lime 391691
Small-leaved Limetree 391691
Small-leaved Manglietia 244425
Small-leaved Maple 2837
Small-leaved Megacarpaea 247778
Small-leaved Mockorange 294501
Small-leaved Mrs. E. H. Wilson's Barberry 52371
Small-leaved Olive 270156
Small-leaved Ormosia 274412
Small-leaved Osmanther 276373
Small-leaved Pea-shrub 72285
Small-leaved Pepper 300338
Small-leaved Persimmon 132135
Small-leaved Phoebe 295383
Small-leaved Platea 302604
Small-leaved Psychotria 319880
Small-leaved Ptarmigan-berry 31321
Small-leaved Raspberry 338985
Small-leaved Rock Daisy 291324
Small-leaved Rockfoil 349653
Small-leaved Rubber Plant 164688
Small-leaved Scurrula 355315
Small-leaved Supple-jack 52450
Small-leaved Sweet-briar 336650
Smallleaved Viburnum 408019
Small-leaved Viburnum 408019
Small-leaved Wendlandia 413793
Small-leaved White Snakeroot 11159
Small-leaved Willow 343516
Small-leaved Wormseed 139689
Small-leaved Yellowwood 94029
Small-leaved Yunnan Holly 204418
Smallleved Apple 24604
Smallligulate Aster 40002
Smallligulatecorolla Aster 40002
Small-ligulated Aster 40002
Smalllip Paphiopedilum 282877
Smalllobed Clematis 95201
Small-marked Japanese Spindle-tree 157609
Smallmuricate Camellia 69447
Smallmuricatefruit Camellia 69447
Small-muricate-fruited Camellia 69447
Smallnerved Osmanther 276309
Small-nut 240210
Small-ovared Rhododendron 331240
Small-ovate-leaf Camellia 69451
Small-paper Mulberry 61101
Small-petal Barberry 51798
Smallpetal Camellia 69452
Smallpetal Draba 137111
Smallpetal Larkspur 124383
Smallpetal Sibbaldia 362352
Smallpetal Tea 69452
Smallpetal Wildberry 362352
Small-petaled Mouseear 82952
Small-petaled Pipewort 151401
Smallpiiar Raspberry 338269
Small-plate Dendrocalamus 125502
Smallpox Tanoak 233411
Smallprickle Holly 203540
Small-prickled Holly 203540
Smallpurple Skullcap 355608
Small-pyrene Holly 204058
Small-ray Goldfields 222614
Smallreed 127192
Small-reed 65254
Smallrhombicleaf Pimpinella 299513
Small-round-leaved Holly 204090
Smallscale Blotch Arrowwood 408075
Small-seed Bush Grape 411785
Smallseed Goldsaxifrage 90412
Smallseed Goldwaist 90412
Smallseed Greygreen Corydalis 105573
Smallseed Gynostemma 183014
Smallseed Jaeschkea 211404
Smallseed Rush 213212
Small-seeded Ephedra 146212
Small-seeded Sand-mat 159613
Smallsepal Camellia 69456
Smallsepal Tea 69456
Small-sepaled Camellia 69456
Smallserrated Willow 343840
Small-shielded Rhododendron 331446
Smallshrub Eritrichium 153453
Smallshrub Rockcress 30270
Smallshrublike Dragonhead 137587
Smallshrublike Greenorchid 137587
Small-spike Barnyardgrass 140450
Smallspike Birch 53409
Smallspike Enteropogon 146015
Small-spike False Nettle 56119
Smallspike Hairgrass 126045
Smallspike Lemongrass 117214
Smallspike Mannagrass 177661
Smallspike Sand Willow 343198
Smallspike Sawgrass 245560
Small-spike Schizachyrium 351294
Small-spiked Pondweed 312179
Small-spiked Spapweed 205145
Small-spiked Square-bamboo 87587
Small-spiked Willow 343690
Smallspine Caper 71803
Smallspine Duplicatefruit Craneknee 221661
Small-spined Caper 71803
Smallsping Skullcap 355604
Smallspur Corydalis 106039
Smallspur Hoodorchis 264770
Smallspur Thinnest Corydalis 105948
Smallspur Violet 410647
Smallsquarose Rockjasmine 23301
Smallstachys Awnless Hairgrass 126045
Smallstar Sedge 76342
Smallstar Shortleaf Kyllingia 218491
Smallstipe Ardisia 31617
Smallstipe Maesa 241798
Smallstipe Thorowax 63679
Smallstipe Vinegentian 110310
Small-stone Holly 204058
Smallstonefruit Camellia 69441
Smallstonefruit Tea 69441
Smalltail Tea 69433
Smallthorn Acanthopanax 143676
Smalltooth Basketvine 9433
Smalltooth Bladderwort 403993
Smalltooth Clematis 95401
Smalltooth Clubfilment 179195
Smalltooth Gomphostemma 179195
Smalltooth Mananthes 244293
Smalltoothed Elatostema 142742
Smalltoothed Osmanther 276396
Smalltoothed Primrose 314310
Smalltoothed Violet 410239,410570
Smalltoth Clearweed 299059
Smalltoth Coldwaterflower 299059
Smalltube Asiabell 98374
Smalltwig Arrowbamboo 162753
Smallumbel Chickweed 375044
Smallumbrella Primrose 314943
Smallviolet Skullcap 355608
Small-whorled Pogonia 210335
Smallwhorlleaf Blueberry 404036
Smallwoodyfruit Camellia 69435
Small-woody-fruited Camellia 69441
Smallyellow Swordflag 208715
Smalscale Sedge 74638
Smaoke Redcarweed 288767
Smaollseed Coral Peadtree 7187
Smara 397043
Smartarse 309199
Smartweed 291651,308690,308893,308960,309043,309199,309298,309570,309624,309654
Smartweed Amaranth 18801
Smartweed Dodder 115099,115116
Smear Docken 86970

Smearwort 34149,34312,86970
Smeli-foxes 23925,174877
Smell Badger 174877
Smell Foxes 23925,174877
Smell Smock 72934
Smelling-wood 35088
Smell-smock 72934
Smelly Geranium 174877
Smelly Wallflower 154513
Smelowskia 366115
Smere 397019
Smerewort 34149,250614
Smgon Cinnamon 91366
Smick Smock 72934
Smick-smock 72934
Smilax 38938,366230,366341,366358,366364,366580
Smilax Asparagus 38938
Smilax Family 366141
Smilax of Florists 38938
Smilax-leaf Clematis 95315
Smilaxleaf Snakegourd 396273
Smile Azalea 330901
Smilo Grass 300728
Smilograss 300728
Smilo-grass 275991,300728
Smirnov Japanese Buttercup 326375
Smirnov Kobresia 217286
Smirnow Leontice 224642
Smith Anemone 24063
Smith Arisaema 33513
Smith Aster 41280
Smith Barberry 52166
Smith Beakchervil 28066
Smith Bedstraw 170634
Smith Begonia 50316
Smith Bigbillorchis 346809
Smith Bittercress 72992
Smith Buckthorn 328860
Smith Cinquefoil 313002
Smith Corydalis 106452
Smith Cowparsnip 192309
Smith Curvedstamencress 72778,238023
Smith Edelweiss 224941
Smith Elytrigia 144675
Smith Eucalyptus 155745
Smith Fir 479
Smith Listera 264739
Smith Meadowrue 388672
Smith Meconopsis 247188
Smith Milkvetch 43065
Smith Neottia 264737
Smith Orchardgrass 121267
Smith Primrose 314991
Smith Purple Willow 343956
Smith Rabdosia 209836
Smith Rhodiola 329957
Smith Rhododendron 331859
Smith Rockfoil 349917
Smith Sage 345396
Smith Sagebrush 36298
Smith Southstar 33513
Smith Trigonotis 397418
Smith Willow 344126
Smith Windflower 24063
Smith Woodbetony 287682
Smith's Bulrush 352266,352269
Smith's Club-rush 353806
Smith's Cress 225371
Smith's Melic Grass 249094
Smith's Mint 250449
Smith's Pepper Cress 225371
Smith's Pepperwort 225371
Smith's Sunflower 189059
Smith's Wild Buckwheat 152467
Smith's Willow 344126
Smith's Wormwood 36298
Smithia 366632
Smithorchis 366724,366725
Smitinandia 366730,366731
Smock 68713,374916
Smock Frock 374916
Smoke Bush 102741,102743,102744,102745,107297,107300
Smoke Plant 107300
Smoke Tree 107297,107300,319359
Smokebush 62118,107300
Smokecolour Rattlesnake Plantain 179620
Smoke-coloured Rhododendron 330843
Smoketree 107297,107300
Smoke-tree 107297,107300,121904
Smoketreeleaf Arrowwood 407767
Smoketreeleaf Viburnum 407767
Smoke-tree-leaved Viburnum 407767
Smokeweed 54772
Smokewood 95431
Smoke-wood 107297
Smoking Cane 95431
Smoky Dragonhead 137590
Smoky Shining-lepidote Rhododendron 330843
Smoky Spotleaf-orchis 179620
Smollscape Daisy 124838
Smollscape Dendranthema 124838
Smoora 397043
Smooth Aeluropus 8860
Smooth Agalinis 10153
Smooth Alder 16437,16440,16445
Smooth Alplily 234219
Smooth Amaranth 18734
Smooth Annular Willow 343030
Smooth Aphragmus 29171
Smooth Arabis 30332
Smooth Arizona Cypress 114692
Smooth Arrow-leaf Wild Buckwheat 151938
Smooth Arrowwood 408081
Smooth Asiatic Star-jasmine 393619
Smooth Aster 380907
Smooth Azalea 330110
Smooth Baccharis 46246
Smooth Bank-cress 30332
Smooth Barberry 51856
Smooth Barley 198328
Smooth Beardtongue 289358
Smooth Beard-tongue 289337
Smooth Bees Barberry 51354
Smooth Beggar-ticks 53976
Smooth Black Sedge 75508
Smooth Blackberry 338225
Smooth Black-haw 408061
Smooth Blue Aster 380907
Smooth Bogrush 352351
Smooth Broadleaf Peppergrass 225403
Smooth Brome 60706,60760,60766,60912
Smooth Brome Grass 60760
Smooth Buttonweed 370782
Smooth Calyx Bird Cherry 280018
Smooth Carlina 76990
Smooth Carolina Allspice 68323
Smooth Cat's Ear 202404
Smooth Cat's-ear 202404
Smooth Catchfly 363385
Smooth Catnip 264973
Smooth Catsear 202404
Smooth Ceratolobus 83521
Smooth Chain Fruit Cholla 116657
Smooth Chastetree 411279
Smooth Chess 60912
Smooth China Weasel-snout 170023
Smooth Chysis 90714
Smooth Cicely 261585
Smooth Cord Grass 370156
Smooth Cordgrass 370156
Smooth Cord-grass 370156
Smooth Cotoneaster 107460
Smooth County Lessingia 227124
Smooth Crab Grass 130612
Smooth Crabgrass 130612,130825
Smooth Crab-grass 130612
Smooth Cranberrybush 408009
Smooth Cypress 114692
Smooth Dandelion 384614
Smooth Darling Pea 380067
Smooth Decaspermum 123450
Smooth Desertdandelion 242942
Smooth Distaff Thistle 77703
Smooth Dock 339916
Smooth Ehretia 141655
Smooth Elaeagnus 141999
Smooth Elderleaf Rodgersflower 335154
Smooth Elm-leaf Goldenrod 368077

Smooth Faberia 161890
Smooth False Foxglove 10143,10146,10148,10149,45167,45173,45175
Smooth Fanbract Corydalis 106389
Smooth Fescuegrass 164008
Smooth Fig 165225
Smooth Finger-grass 130612
Smooth Forked Nail-wort 284782
Smooth Franchet Monkshood 5210
Smooth Galinsoga 170142
Smooth Gambirplant 401765
Smooth Gambir-plant 401765
Smooth Geniculate Holly 203841
Smooth Giantgrass 125465
Smooth Goldbachia 178821
Smooth Goldenrod 368106
Smooth Goldenweed 323031
Smooth Goldfields 222602
Smooth Gooseberry 334030,334137
Smooth Greenbrier 366308,366338
Smooth Greenskin Bamboo 47482
Smooth Grevillea 180618
Smooth Ground Cherry 297697
Smooth Ground-cherry 297697
Smooth Hairy Stranvaesia 377438
Smooth Hakea 184611
Smooth Halgania 184762
Smooth Hawk's-beard 110767
Smooth Hawkbit 224655
Smooth Hawksbeard 110767,110993
Smooth Hawkweed 195857
Smooth Hawthorn 109790
Smooth Hedge-nettle 373459
Smooth Honeytree 78141
Smooth Hydrangea 199806,199807
Smooth Iris 208672
Smooth Ironweed 406150,406325
Smooth Ixeris 210543
Smooth Japanese Abelia 181
Smooth Joyweed 18123,18124
Smooth Juneberry 19276
Smooth Kangaroo-vine 92861
Smooth Kangding Buttercup 325775
Smooth Kesh 24483
Smooth Knotweed 291766
Smooth Lance-leaf Buckthorn 328756
Smooth Lanceolate Draba 137077
Smooth Lanceolate Whitlowgrass 137077
Smooth Lasianthus 222224
Smooth Long-headed Poppy 282546
Smooth Long-leaved Ground-cherry 297694
Smooth Loosestrife 239809
Smooth Lungwort 250883,250894
Smooth Lycianthes 238955
Smooth Mayten 246867
Smooth Meadow-grass 305855
Smooth Meadow-parsnip 388899,388902
Smooth Mesquite 315565
Smooth Milkweed 38132
Smooth Mnesithea 256337
Smooth Mojave Woody-aster 415855
Smooth Monocladus 56953
Smooth Nakestem Cystopetal 320237
Smooth Narrowfruit Draba 137243
Smooth Narrowfruit Whitlowgrass 137243
Smooth Nepeta 264973
Smooth Oblongleaf Dogwood 380476
Smooth Onosmodium 271854,271861
Smooth Oxeye 190504,190505
Smooth Ox-eye 190505
Smooth Pansy Orchid 254933
Smooth Parsonsia 285056
Smooth Penstemon 289358
Smooth Peruvian-daisy 170142
Smooth Pfeiffer Cicely 276472
Smooth Phlox 295275,295276
Smooth Pigweed 18734
Smooth Pittosporum 301366
Smooth Prickly Pear 272937
Smooth Prickly-headed Poppy 282546
Smooth Purple Coneflower 140073
Smooth Pussytoes 26530
Smooth Qiongzhu 87570
Smooth Rambutan 14374
Smooth Rattlebox 112491,112492
Smooth Rock Cress 30332
Smooth Rock-cress 30332
Smooth Rose 336389
Smooth Rose Mallow 194955
Smooth Roughleaf 222224
Smooth Round-beaded Poppy 282685
Smooth Ruellia 339819
Smooth Rupturewort 192965
Smooth Rupture-wort 192965
Smooth Sago Palm 252636
Smooth Saltbush 44484
Smooth Sawgrass 93918
Smooth Scabrousleaf Meadowrue 388553
Smooth Sedge 73887
Smooth Senna 360489
Smooth Serviceberry 19276
Smooth Shadbush 19276
Smooth Sheath Bamboo 297499
Smooth Shrubby Sweetvetch 187967
Smooth Sibiraea 362396
Smooth Singlebamboo 56953
Smooth Small Blueberry 404005
Smooth Small Skullcap 355667
Smooth Smallligulate Aster 40006
Smooth Smartweed 291766
Smooth Solomon's Seal 308508,308527
Smooth Solomon's-seal 308508
Smooth Sow Thistle 368771
Smooth Sow-thistle 368771
Smooth Spiderwort 394053
Smooth Spindle-tree 157545
Smooth Sterculia 376128
Smooth Stylose Staff-tree 80329
Smooth Sumac 332608
Smooth Sumach 332608
Smooth Sunflower 188991
Smooth Sunkenvein Eurya 160435
Smooth Sweet Shrub 68310
Smooth Sweetvetch 187967
Smooth Tanoak 233292
Smooth Tare 408638
Smooth Tasmanian Cedar 44047
Smooth Tea 69097
Smooth Three-ribbed Golden Rod 368106
Smooth Tick Clover 126417
Smooth Tower Mustard 400644
Smooth Toxocarpus 393534
Smooth Trailing Lespedeza 226943
Smooth Treebineleaf Acanthopanax 143583
Smooth Two-whorl Buckwheat 375584
Smooth Vaniot Staff-tree 80342
Smooth Viburnum 408081
Smooth Wampee 94208
Smooth Water Hyssop 46362
Smooth White Aster 380957
Smooth White Oldfield Aster 380975
Smooth White Violet 410189
Smooth White-rod 407970
Smooth Whitlow-grass 136998
Smooth Winterberry 203956
Smooth Withered 407970
Smooth Woody-aster 415849
Smooth Wrightia 414809
Smooth Yellow False Foxglove 45167
Smooth Yellow Vetch 408469
Smooth Yellow Violet 410466,410472
Smooth Yushanbamboo 416809
Smooth Zamia 417013
Smoothangular Crisped Corydalis 105763
Smoothaxis Pennisetum 289153
Smooth-bark Arizona Cypress 114692
Smoothbark Hickory 77894
Smooth-bark Hickory 77894
Smoothbark Himalayan Fir 481
Smooth-bark Kauri 10505
Smoothbark Oak 324539
Smooth-barked African Mahogany 216196
Smooth-barked Apple 24606
Smooth-barked Mexican Pine 300162
Smooth-barked White Mahogany 216196
Smoothbrached Hooktea 52451
Smoothbrached Supplejack 52451
Smooth-bracted Willow 344191
Smoothbranch Glomerate Falsenettle 56216
Smooth-branch Smallmuricatefruit Camellia 69449

Smoothbranched Largeleaf Falsenettle 56216
Smoothcalyx Asiabell 98352
Smooth-calyx Bird Cherry 280018
Smoothcalyx Bluebell 115361
Smoothcalyx Bluebellflower 115361,115386
Smooth-calyx Chinese Privet 229626
Smoothcalyx Chirita 87820
Smoothcalyx Clematis 95260
Smoothcalyx Fan-shaped Corallodiscus 103946
Smoothcalyx Flamenettle 99530
Smooth-calyx Ford's Fordiophyton 167304
Smooth-calyx Hawthorn 109663
Smooth-calyx Leptodermis 226092
Smoothcalyx Pricklythrift 2251
Smoothcalyx Primrose 314576
Smooth-calyx Silk Mockorange 294543
Smooth-calyx Taiwan Adinandra 8221
Smooth-calyx Wildclove 226092
Smooth-calyxed Ardisia 31546
Smoothcarp Blue Milkvetch 42127
Smoothcarp Branchy Milkvetch 42894
Smooth-cone Sedge 75037
Smoothest Sikkim Barberry 52153
Smoothflower Awnedglume Roegneria 335217
Smoothflower Bristlegrass 361852
Smooth-flowered Glaucous-dorsal Honeysuckle 235861
Smooth-flowered Kengyilia 215939
Smooth-forked Nailwort 284782
Smoothfruit Acomastylis 4991
Smooth-fruit Agapetes 10325
Smoothfruit Bedstraw 170205
Smoothfruit Bouquet Larkspur 124250
Smoothfruit Cherokee Rose 336678
Smooth-fruit Clovershrub 70871
Smoothfruit Denseleaf Larkspur 124322
Smoothfruit Erect Craneknee 221759
Smoothfruit Floccose Willow 343386
Smoothfruit Goniothalamus 179414
Smoothfruit Guinan Willow 343556
Smoothfruit Hairyleaf Draba 137085
Smoothfruit Hairyleaf Whitlowgrass 137085
Smoothfruit Hedge Gooseberry 333903
Smoothfruit Hedge Maple 2845
Smoothfruit Ironweed 405447
Smoothfruit Largeflower Larkspur 124250
Smoothfruit Lepironia 225664
Smoothfruit Linearleaf Solms-Laubachia 368563
Smoothfruit Mercury 250610
Smoothfruit Miaotai Maple 3146
Smoothfruit Microula 254351
Smoothfruit Mountain-loving Draba 137167
Smoothfruit Mountain-loving Whitlowgrass 137167
Smoothfruit Northern Bedstraw 170262
Smoothfruit Pearlsedge 354219
Smoothfruit Pinnaflower 4991
Smooth-fruit Piptanthus 300697
Smoothfruit Poppy 282633
Smooth-fruit Rabbitbrush 150347,150352
Smoothfruit Razorsedge 354219
Smoothfruit Rhomboidleaf Monkshood 5533
Smoothfruit Rochelia 335114
Smoothfruit Sageflower Woodbetony 287656
Smoothfruit Stenobracteoid Tickseed 104852
Smoothfruit Style-like Draba 137249
Smoothfruit Style-like Whitlowgrass 137249
Smooth-fruit Tanoak 233328
Smoothfruit Tauscheria 385134
Smoothfruit Tibet Microula 254371
Smoothfruit Tickseed 104836
Smoothfruit Trichophore Larkspur 124649
Smoothfruit Ventilago 405447
Smooth-fruit Wilhelms Willow 344275
Smoothfruit Woolly Draba 137136
Smoothfruit Xizang Microula 254371
Smoothfruit Yellow Bedstraw 170760
Smooth-fruited Agapetes 10325
Smooth-fruited Bureja Gooseberry 333930
Smooth-fruited Crazyweed 278874
Smooth-fruited Goniothalamus 179414
Smooth-fruited Oak Sedge 76561
Smooth-fruited Orange 93530
Smooth-fruited Piptanthus 300697
Smooth-fruited Ventilago 405447
Smooth-fruited Viburnum 407915
Smooth-headed Bastard Poppy 282546
Smoothish Salacca 342575
Smoothish Yellow Violet 409963
Smoothleaf Actinidia 6644
Smoothleaf Alpine Oak 324322
Smoothleaf Badong Loosestrife 239777
Smooth-leaf Begonia 50137
Smoothleaf Bournea 57900
Smoothleaf Chessbean 116596
Smoothleaf Chinacane 364654
Smoothleaf Chirita 87897
Smoothleaf Cylindrokelupha 116596
Smoothleaf Didymostigma 130070
Smoothleaf Elm 401468
Smooth-leaf Elm 401468
Smoothleaf Falserambutan 283534
Smoothleaf Forrest Larkspur 124342
Smoothleaf Galangal 17650
Smooth-leaf Hydrangea 199877
Smoothleaf Indigo 206295
Smoothleaf Kiwifruit 6644
Smoothleaf Longawn Roegneria 335293
Smooth-leaf Loxostigma 238037
Smoothleaf Machilus 240600
Smooth-leaf Manyhairs Indocalamus 206790
Smoothleaf Maple 3075
Smooth-leaf Mayten 182683
Smooth-leaf Michelia 252901
Smooth-leaf Mimosa 254997
Smoothleaf Mountain-loving Draba 137165
Smoothleaf Mountain-loving Whitlowgrass 137165
Smoothleaf Oneflowerprimrose 270731
Smoothleaf Oreocharis 57900
Smoothleaf Patrinia 285831
Smoothleaf Pine 300033
Smooth-leaf Pine 300033
Smoothleaf Raisin Tree 198789
Smoothleaf Scorpiothyrsus 354745
Smooth-leaf Shrubby Elsholtzia 144029
Smooth-leaf Sichuan Sumac 332970
Smoothleaf Spunsilksage 360816
Smoothleaf Swallowwort Heshouwu 162524
Smoothleaf Sweetleaf 381272
Smoothleaf Tanoak 233228
Smoothleaf Threevein Aster 39968
Smoothleaf Turnjujube 198789
Smoothleaf Yunnan Holly 204417
Smooth-leaved Actinidia 6644
Smooth-leaved Barberry 51867
Smooth-leaved Camellia 69705,69706
Smooth-leaved Climbing Violet 409911
Smooth-leaved Croton 112931
Smooth-leaved Elm 401468,401508,401593
Smooth-leaved Epimedium 147048
Smooth-leaved Honeysuckle 235760
Smooth-leaved Indigo 206295
Smooth-leaved Maple 3075
Smooth-leaved Nitta-tree 284464
Smooth-leaved Sotol 122905
Smooth-leaved Sweetleaf 381272
Smooth-leaved Tanoak 233228
Smoothlemma Bluegrass 305361
Smoothlip Cymbidium 116926
Smoothpedicel Elliott Monkshood 5174
Smoothpedicel Glandpetal Rockfoil 350036
Smoothpit Peach 20926
Smooth-podded Mimosa 254997
Smooth-podded Tine Tare 408638
Smooth-podded Tine-tare 408638
Smoothseed Euphorbia 159092
Smooth-seeded Rush 213207
Smoothsepal Martin Petrocosmea 292569
Smoothsheath Bamboo 47285,297499
Smooth-sheath Flower Bamboo 297360
Smooth-sheathed Fox Sedge 75043
Smooth-sheathed Sedge 75043
Smooth-shelled Macadamia Mut 240209
Smooth-shoot Gigantochloa 175600
Smoothspike Itchgrass 337583
Smoothspine Herelip 220077
Smoothspine Lagochilus 220077

Smooth-stalk Hydrangea 199879
Smoothstalk Madder 338028
Smooth-stalk Metalleaf 296400
Smoothstalk Small Reed 127259
Smoothstalk Smallreed 127259
Smooth-stalked Bluegrass 305653
Smooth-stalked Meadow-grass 305855
Smooth-stalked Sedge 75039
Smooth-stem Actinidia 6696
Smoothstem Bittercress 72862
Smoothstem Forrest Larkspur 124221
Smooth-stem Kiwifruit 6696
Smoothstem Obtuseleaf Elatostema 142765
Smoothstem Rockfoil 349385
Smooth-stemmed Bloodwood 106790
Smooth-stone Holly 203788
Smoothstyle Bewitch Azalea 330047
Smoothstyle Charming Rhododendron 330047
Smoothstyle Longscale Azalea 331127
Smooth-tube Milkvetch 42615
Smootjfruit Bentbranch Tickseed 104779
Smootjlraf Pyrethrum 322650
Smotherweed 48769
Smotherwood 36474
Smoth-leaf Anisetree 204545
Smothleaf Eightangle 204545
Smut Grass 372666
Smutgrass 372666
Smuts 238583
Smythea 366748
Snaffles 329521
Snag-bush 316819
Snaggs 364193
Snags 316819
Snail Clover 247456
Snail Flower 408862
Snail Medick 247469
Snail Plant 408862
Snail Saxifrage 349190
Snail Vine 293985,408862
Snailseed 97894,97933
Snail-seed 97894
Snail-seed Pondweed 312062
Snakcwood-tree 80003
Snake 381035
Snake Bark 100059
Snake Bean 409096
Snake Bean Tree 380083
Snake Bugloss 141347
Snake Bush 139211
Snake Cactus 258721,258722,267605,
357748
Snake Cholla 116650
Snake Cucumber 114209
Snake Flower 15861
Snake Gourd 396150,396151,396166
Snake Horemint 137747
Snake Lily 33234,128886,154932,208924
Snake Melon 114209
Snake Mockstrawberry 138793
Snake Mouth 306883
Snake Nut 272019
Snake Palm 20132
Snake Plant 15861,37015,346158
Snake Plantain 302034
Snake Range Thistle 91936
Snake Range Wild Buckwheat 152148
Snake River Goldenweed 323050
Snake Sansevieria 346158
Snake Vine 191332,194668
Snake Violet 410483,410493
Snake Weed 182107
Snake Wood 61062,327058,378816,378836
Snake's Berry 208566
Snake's Bugloss 141347
Snake's Cherries 380499
Snake's Cherry 380499
Snake's Fiddle 208566
Snake's Flower 15861,23925,68713,99297,
141347,145487,168476,174877,216753,
220346,250614,273531,321636,363461,
363673,374897,374916,405788
Snake's Food 15861,37015,174877,208566,
231183,250614,292372,308893,367120,
381965,383675
Snake's Grass 3978,260760
Snake's Head 168332,168476,192904,
242932,312520,405788
Snake's Head Desertdandelion 242932
Snake's Head Iris 192904
Snake's Head Lily 168476
Snake's Meat 37015,192358,208566,
250614,316127,367120,383675
Snake's Plant 15861
Snake's Poison 208566
Snake's Poison-food 367120
Snake's Rhubarb 31051,292372,332086
Snake's Skin Willow 344211
Snake's Tongue 154899,325864
Snake's Victuals 37015,250614
Snake's Violet 410483
Snake's-beard 272049
Snake's-head 168476,192904
Snake's-head Fritillary 168476
Snake's-head Iris 192898
Snake's-head Lily 168476
Snake's-skin Willow 344211
Snakebark Maple 3209,3529
Snake-bark Maple 2846,2920,3393,3529
Snakebeard 272086
Snakeberry 6405,61497
Snakebite 345805
Snakebubble Raspberry 338265
Snakebush 191332
Snakecactus 52571
Snake-cotton 168678
Snake-eyes 294092,294093
Snakeflower 15861,72038,168476,174877,
273531
Snakeflower Tylophora 400856
Snakefruit 342570
Snakefruit Corydalis 106214
Snakegoll Childvine 400958
Snakegourd 396150,396210
Snakegrape 20280
Snake-grass 374916
Snakehair Clearweed 299062
Snakehead 86783
Snake-head 86783,242932
Snakehead Dipoma 133411
Snake-head Pleurothallis 304922
Snakeheadcress 133410,133411
Snake-like Calligonum 67011
Snake-lily 128886
Snakelotus 191894,191964
Snake-mouth 306883
Snakemouth Orchid 306883
Snakemushroom 46810
Snakemushroom Family 46888
Snake-palm 20037
Snakepod Milkvetch 42826
Snakeroot 6433,11152,90924,90994,91034,
158354,228529,308893
Snake-root 34097
Snakeroot Eryngo 154301
Snakeroot Ononis 271464
Snakes-and-adders 23925,272398
Snakes-and-bee Orchis 272398
Snakeshcad Iris 192904
Snakeshead 86783
Snakes-head Iris 192898,192904
Snakeshead Lily 168476
Snake-skin Maple 3004
Snaketailgrass 272345
Snaketailgrass Centipedegrass 148254
Snaketongue Crowfoot 326148
Snake-tongue Crowfoot 326148
Snakevine 100058
Snakeweed 90924,101852,159069,162312,
170193,182092,182100,182107,182113,
250614,302068,308816,308893,374916,
383675,415079,415088
Snakewood 61059,61062,378836,378948
Snakewood Tree 80003,80008
Snap Bean 294056
Snap Jack 374916
Snap Willow 343396
Snapcracker 374916
Snapdragon 28576,28617,28631,30081,
116736,130383,169126,230866,231183
Snapdragon Vine 246649

Snapjack 21339,28617,130383,174877,231183,363673,374916
Snap-lion 28617
Snapper-flower 374916
Snappers 364193,374897
Snapping Hazelnut 185141
Snappish Nettle 403028
Snappy Gum 155623
Snaps 130383,231183,374916
Snapsen 311537
Snapstalks 374916
Snapweed 204841,205444
Snapwillow 343396
Snapwort 374916
Snares 170193
Snatberry 385302
Snauper 130383
Sneezeweed 4005,81681,188393,188402,188419,188424
Sneezeweedlike Thistle 92012
Sneezewood 320032,320033
Sneezewort 3978,4005,188393,284152,405571
Sneezewort Aster 368324
Sneezewort Yarrow 4005
Sneezlngs 3978
Sneez-wort 4005
Snegs 316819
Snitchback 115949
Snoder Gill 385302
Snod-goggles 385302
Snodgollion 292025
Snot Berry 385302
Snot-goblin 292025
Snots 385302
Snotter Gall 385302
Snotterberries 380756,385302
Snotty Gog 385302
Snottygobble 292025
Snottygobbles 385302
Snout Bean 333216,333288
Snout Woodbetony 287541
Snow 374916
Snow Argostemma 32527
Snow Arnica 34720,34721,34740
Snow Azalea 331284
Snow Bush 60057,60070
Snow Buttercup 325898,326128
Snow Camellia 69133
Snow Cinquefoil 312818
Snow Crazyweed 278784
Snow Creeper 311657
Snow Daisy 80367
Snow Daisy Bush 270207
Snow Draba 137146
Snow Drop Tree 184743
Snow Edelweiss 224914
Snow Falsecypress 85351
Snow Flower 370338
Snow Flower Rhododendron 330385
Snow Flurry Aster 380872
Snow Gentian 173642,173666
Snow Gooseberry 334126
Snow Gum 155667,155688
Snow Heath 149147
Snow in Summer 83060,248075,248105
Snow Lotus 348392
Snow Milkvetch 42796
Snow Mountain Wild Buckwheat 152303
Snow Myrtle 68756
Snow of June 360935
Snow of Kilimanjaro 159240
Snow on the Mountain 159313
Snow Orchid 82032
Snow Pea 301080
Snow Pear 323246
Snow Pearlwort 342268
Snow Phyllophyton 245695
Snow Piercer 169719
Snow Plant 346784,346785
Snow Poppy 146053,146054
Snow Queen 382053
Snow Rhododendron 331284
Snow Rose 360923
Snow Silvery Aster 40077
Snow Spirea 371890
Snow Sprite Deodara Cedar 80090
Snow Thoroughwort 158354
Snow Toss 407989
Snow Trillium 397570,397584
Snow White Cinquefoil 312818
Snow Whitlowgrass 137146
Snow Wild Buckwheat 152306
Snow Wood 283729
Snow Wreath 265856,265857
Snow Yunnan Draba 137304
Snow Yunnan Whitlowgrass 137304
Snowball 380756,384714,407645,407989
Snowball Arrowwood 408027
Snowball Bush 407989,407997
Snowball Cactus 155300,244012,244270,287895
Snowball Pincution 244018
Snowball Rockfoil 349700
Snowball Rose 407989
Snowball Saxifrage 349700
Snowball Tree 407989,407995,407997
Snowball-bush 407989
Snowbaw 407989
Snowbell 367762,379300,379304,379357,379378
Snowbell Tree 379374
Snowbellleaf Actinidia 6705
Snowbellleaf Kiwifruit 6705
Snowbell-leaf Tickclover 126623
Snowbell-leaved Actinidia 6705
Snowbells 15839,169719,379419
Snowberry 87673,172047,172057,380734,380736,380738,380756
Snowberry Mountain Ash 369379
Snowberry Mountainash 369379
Snowberry Mountain-ash 369379
Snowblossom Primrose 314237
Snowbrush Ceanothus 79978
Snowbush 60057,60070
Snow-bush 60070
Snowcreeper Porana 131254
Snowdon Lily 234214,234234
Snowdon Rose 329935
Snow-drift 235090
Snowdrop 168476,169705,169719,274823
Snowdrop Anemone 24061,24071
Snowdrop Bush 379300
Snowdrop Tree 184726,184731
Snowdrop Wind Flower 24071
Snowdrop Windflower 24061,24071
Snowdrop-anemone 24071
Snowdrop-bush 379419
Snowdropper 169719
Snowdrops 169705
Snowdroptree 184689
Snowdrop-tree 87736,184726,184731,184740,184746
Snowdrop-windflower 24071
Snowfield Sagebrush 36308
Snowflake 32518,109857,220346,227854,227875,267818,274823,374916,407989
Snowflake Loddon Lily 227856
Snowflake Rabdosia 324741
Snowflakelotus 227854
Snowflower 87736,169719,370338
Snowflower Rabdosia 324741
Snow-flowered Rabdosia 324741
Snow-flowered Rhododendron 330385
Snowgall 191894
Snow-gum 155667
Snow-in-harvest 83060,95431
Snow-in-summer 30220,82629,82739,83060,159313,248105,279351
Snowlayer Azalea 331346
Snowleaf Crazyweed 278785
Snowlike Rhododendron 331349
Snowline Pyrola 322853
Snowlotus 348089,348392
Snowloving Rockfoil 349175
Snowmound Spirea 372032
Snowmountain Bittersweet 367733
Snowmountain Eggplant 367733
Snowmountain Nightshade 367733
Snow-mountain Nightshade 367733
Snowmountain Primrose 314697

Snowmountain Sage 345019
Snowmountain Sagebrush 36427
Snowmountain Violet 410691
Snow-on-the-mountain 8826,30164,159313,235090,349419,374916
Snow-on-the-Mountain 83060
Snow-on-the-mountain Euphorbia 159313
Snow-peak Raspberry 338886
Snowpeaks Raspberry 338886
Snow-plant 83060
Snowpoppy 146053,146054
Snowrabbiten 348089,348344
Snowrange Draba 137174
Snowrange Whitlowgrass 137174
Snowrose 360926
Snowwhite Butterflybush 62131
Snow-white Butterfly-bush 62131
Snow-white Cajan 65154
Snowwhite Cinquefoil 312818
Snowwhite Cottondaisy 293445
Snowwhite Cypressgrass 119270
Snow-white False-nettle 56236
Snowwhite Galingale 119270
Snow-white Paphiopedilum 282882
Snowwhite Phagnalon 293445
Snow-white Platanthera 302453
Snow-white Slender Raspberry 339019
Snowwhite Summerlilic 62131
Snowwhite Waterlily 267664
Snow-white Winterhazel 106663
Snow-white Wood-rush 238668
Snowwhitehair Reekkudzu 313456
Snowwillow 167230,167233
Snow-wreath 265856
Snowy Bauhinia 48993
Snowy Cactus 244171
Snowy Campion 363814
Snowy Cinquefoil 312818
Snowy Fleece Tree 248089
Snowy Gomphostemma 179196
Snowy Lophanthus 236277
Snowy Mermaid 228628
Snowy Mespilus 19240,19243,19276,19279,19288
Snowy Mint Bush 315604
Snowy Mountain Tingitingi 155587
Snowy Nepeta 245695
Snowy Orchis 183903
Snowy Pouzolzia 313456
Snowy Rhododendron 330040,331346
Snowy River Wattle 1088
Snowy Thistle 92253
Snowy Wood-rush 238668
Snoxums 130383
Snuffbox Bean 145899
Snuff-candle 170025
Snut 1427
Soap Aloe 17002,17243
Soap Mallee 155563
Soap Nut 346323
Soap Nut Tree 346338
Soap Nuts 346323
Soap Plant 86982,88469,244371
Soap Pod 1142
Soap Tree 17613,182526
Soap Weed 416578
Soap-bark 1308
Soapbark Tree 301126,324623,324624
Soap-bark Tree 324624
Soap-bark Vine 145899
Soapberry 346323,346336,346351,362091
Soapberry Family 346316
Soapberry Tree 46768
Soapberryleaf Cliffbean 254830
Soapbush 181504,266532
Soapnut Tree 346338
Soap-nut Tree 346338,346351,346356
Soap-nut-tree 346338
Soap-plant 88476
Soappod 182524,182526
Soaproot 88476,416603
Soaptree 346323
Soaptree Yucca 416578
Soap-tree Yucca 416578
Soapweed 344585,416552,416603
Soapweed Yucca 416603
Soap-weed Yucca 416578
Soapwood 77944
Soapwort 346418,346420,346434
Soapwort Gentian 173840,346434
Soaring Poplar 311400
Soboliferous Cranesbill 174919
Sobralia 366776
Society Garlic 400095
Socotra Begonia 50317
Socotrine Aloe 17320
Sod Apple 146724
Soda Apple 366906
Soda Buttercup 184702
Soda Plant 344585
Sodaapple 366906
Soda-apple Nightshade 367735
Sodium Russianthistle 344644
Sodom apple 366910
Sodom Apple 367305
Sofar Iris 208846
Soft Acanthus 2695
Soft Agrimony 11595
Soft Albizzia 13621
Soft Aloe 17165
Soft Alphonsea 17627
Soft Aster 380928
Soft Bambusa 47313,47319,47344
Soft Bellflower 70174
Soft Bentgrass 12104
Soft Boronia 57271
Soft Brome 60745,60748,60853
Soft Chess 60745,60853
Soft Clover 397097
Soft Comfrey 381037
Soft Corn 417414
Soft Downy-rose 336762
Soft Elm 401431,401620
Soft Feather Pappusgrass 145763
Soft Flexuose Bittercress 72761
Soft Fox Sedge 74179
Soft Gentian 173733
Soft Goldenaster 59056
Soft Goldenrod 368250
Soft Hair Leptodermis 226113
Soft Hair Qiongzhu 87600
Soft Haired Tricalysia 133187
Soft Hairs Rhododendron 330805
Soft Hairy Tricalysia 133187
Soft Hawksbeard 110904
Soft Hornwort 83578
Soft Hydrangea 200029
Soft Knotted Clover 397097
Soft Knotweed 309418
Soft Leather Leaves Rhododendron 330068
Soft Legume Ormosia 274433
Soft Leymus 228369
Soft Lumbang 14549
Soft Maize 417414
Soft Maple 3505,3532
Soft Needleleaf 307832
Soft Persimmon 132305
Soft Petrocosmea 292555
Soft Pine 299942
Soft Pubescentfruit Pearlsedge 354132
Soft Pubescentfruit Razorsedge 354132
Soft Rockjasmine 23232
Soft Rose 336762
Soft Rush 213066,213507
Soft Sandwort 31848
Soft Skullcap 355806
Soft Sloanea 366064
Soft Snapdragon 28628
Soft Sophora 369070
Soft Stonebutterfly 292555
Soft Stork's-bill 153839
Soft Sunflower 189001
Soft Thermopsis 389542
Soft Thistle 91843
Soft Wallaba 146111
Soft Wild Oat 45457
Soft Yushanbamboo 416800
Soft Yushania 416800
Softbean 386403,386410
Soft-bearded Oat Grass 405429
Softcolor Hawksbeard 110904

Softflower Pondweed 312161
Soft-fruited Ormosia 274433
Soft-grass 197278
Softhair Ainsliaea 12684
Softhair Anna 25752
Softhair Aster 40872
Soft-hair Aucuba 44951
Softhair Azalea 331266
Softhair Bamboo 47344
Softhair Betony 373441
Softhair Brucea 61211
Softhair Coneflower 339589
Softhair Crazyweed 279020
Soft-hair Deutzia 127015
Softhair Didymocarpus 129936
Softhair Dumasia 138943
Softhair Dutchmanspipe 34273
Softhair Greenbrier 366295
Softhair Gugertree 350948
Softhair Guiyang Betony 373279
Softhair Hemiboea 191376
Softhair Hornbeam 77348
Softhair Isometrum 210192
Softhair Knotweed 309810
Softhair Leucas 226940
Softhair Magnoliavine 351110
Soft-hair Marble-seed 271855,271861
Softhair Milkvetch 42477
Softhair Mouseear Cress 113151
Softhair Pricklyash 417273
Softhair Primrose 314664
Softhair Rabbiten-wind 12684
Softhair Sinosenecio 365079
Softhair Twoflower Raspberry 338186
Softhair Wildryegrass 228369
Softhair Yam 131700
Soft-haired Leaves Rhododendron 331266
Softhaired Oak 324150
Soft-haired Oak 324150
Soft-haired Sweetleaf 381272
Softhairleaf Primrose 314615
Softhairleaf Raspberry 338492
Soft-hairy Box 64324
Softhairy Indigo 206399
Softhairy Larkspur 124389
Soft-hairy Primrose 314664
Softhairy Raspberry 338492
Softhairy Sweetleaf 381272
Softleaf Ash 168022
Softleaf Asterbush 41756
Softleaf Asterothamnus 41756
Soft-leaf Aucuba 44942
Softleaf Chirita 87928
Softleaf Dallisgrass 285462
Softleaf Datepalm 295484
Softleaf Jurinea 214086
Soft-leaf Muhly 259694
Softleaf Paspalum 285462
Softleaf Passionflower 285672
Softleaf Passion-flower 285672
Soft-leaf Sedge 74342
Soft-leaf Willow 344099
Soft-leather-leaved Rhododendron 330068
Soft-leaved Ash 168022
Softleaved Bear's Breech 2695
Soft-leaved Larkspur 124388
Soft-leaved Sedge 75409
Softlegume Ormosia 274433
Soft-legumed Ormosia 274433
Softpilose Torularia 264624
Softpod Ormosia 274433
Softsetose Lycianthes 238970
Softsetose Red Silkyarn 238970
Softstalk Trigonotis 397406
Softstem Bulrush 352278
Soft-stem Bulrush 352278
Softstem Metalleaf 301682
Softstem Spapweed 205367
Softstem Touch-me-not 205367
Soga 288897
Sogd Ash 168105
Sogd Russowia 341028
Sogdian Iris 208607
Sogot Milkvetch 43066
Soh-long 93631
Soh-phlong 166889
Soja 177689,177750
Soja Bean 177750
Sojabean 177689
Sokitsu Mandarin 93731
Sola Pith 9499
Solander Banksia 47668
Solander's Geranium 174922
Solandine 86755
Solan-like Nemosenecio 263516
Solanum 366902
Soldanellalike Primrose 314993
Soldier Boys 81768
Soldier Buttons 23925,30081,31051,68189,
170193,174877,175444,188757,216753,
248411,267294,267648,316127,325518,
325666,326287,336851,374916
Soldier Orchid 272460,273541,417795
Soldier Orchis 273541
Soldier Rush 213291
Soldier's Arrow 377485
Soldier's Buttons 23925,30081,31051,
68189,170193,174877,175444,188757,
216753,248411,267294,267648,316127,
325518,325666,326287,336851,374916
Soldier's Cap 5442,5676,273531
Soldier's Cross 86427
Soldier's Cullions 273541
Soldier's Drinking Cup 347156
Soldier's Feathers 18687,382329
Soldier's Jacket 273531
Soldier's Orchid 417795
Soldier's Pride 81768
Soldier's Ribwort 302034
Soldier's Tap Pie 302034
Soldier's Tappie 302034
Soldier's Tears 405788
Soldier's Woundwort 3978
Soldier's Yarrow 377485
Soldier-and-his-wife 321643
Soldier-in-the-box 13729
Soldiers 23925,130383,174877,217055,
240068,272398,273531,282685,302034,
336419,339887,396934,405788
Soldiers' Jackets 273531
Soldiers-and-angels 37015
Soldiers-and-sailors 37015,141347,208771,
321643
Soldiers-of-the-queen 170142
Soldiers-sailors-tinkers-tai Lors 235334
Soleded Pine 300288
Solemn Bells of Sodom 168476
Solena 367773,367775
Solenanthus 367804
Solflower 188757
Sol-flower 188757
Solid Bamboo 125510
Solid Bitter Bamboo 304103
Solid Bitterbamboo 304103
Solid Bitter-bamboo 304103
Solid Cliffbean 66627
Solid Coldbamboo 172747
Solid Fargesia 162687
Solid Fishscale Bamboo 297300
Solid Gelidocalamus 172747
Solid Millettia 66627
Solidago 367939
Solidleaf Bluegrass 306025
Solidstem Burnet Saxifrage 299523,299524
Solid-stem Burnet-saxifrage 299523
Solisia 368524
Solitaire Palm 321161,321164
Solitary Clematis 95029
Solitary Fishtail Palm 78056
Solitary Harebell 70233
Solitary Leaf-flower 296675
Solitary Palm 321164
Solitary Rose 336382
Solitary-flowered Reddrop Barberry 51548
Solitary-flowered Rhododendron 332039
Soliva 368527
Sollandine 86755
Sollar 343151
Sollery 29327
Sollop 339887
Solms-Laubachia 368546

Solomon Islands Ivyarum 147338
Solomon Nut Oil 71000
Solomon's Giant Seal 308527
Solomon's Lily 37024
Solomon's Puzzle 357203
Solomon's Sale 308607
Solomon's Seal 134400,201787,308493,
308508,308527,308607
Solomon's-plume 242699,366142
Solomon's-Seal 308493
Solomon's-seal 308508,308564,308607
Solomonplume 366142
Solomonseal 308493,366142
Solut-flowered Barberry 52167
Solwherf 4056
Soma 244349
Soma Dendrobium 125361
Somabutta 346995
Somalen Sophora 369120
Somali Kalanchoe 215263
Somali Lavender 223330
Somali Tea 79395
Somalia 57537
Someiyoshine 83365
Somen Stonecrop 357160
Somerset Hair-grass 217577
Somerset Rush 213492
Somerset Skullcap 355359
Somerset-grass 217577
Somes Bar Campion 363728
Somewhat Tawny Rhododendron 330742
Somewhat Yellow Rhododendron 330704
Somewhat-white Magnolia 198698
Somewhot Falcate Crazyweed 278795
Somniferous Poppy 282717
Sompong 387297
Son Brome 60853
Son Coya 25880
Son-before-the-father 99297,146724,292372,
400675
Soncoya 25880
Soneratia Family 368926
Sonerila 368869
Song of India 137478
Songar Clematis 95319
Songar Cotoneaster 107688
Songar Elaeagnus 142195
Songar Iris 208848
Songar Ixiolirion 210998
Songar Peashrub 72345
Songar Pepperweed 225460
Songar Sunrose 188846
Songar Willow 344127
Songar Woodbetony 287684
Songar Wormwood 36304
Songaria Buttercup 326365
Songaria Cynomorium 118133
Songaria Sedge 76299
Songaria Variousforms Sisymbrium 365589
Songarian Cleistogenes 94631
Songarian Eremosparton 148447
Songarian Sunrose 188846
Songarian Sun-rose 188846
Songjiang Waterchestnut 394476
Songjiang Willow 344163
Songming Camellia 69678
Songming Elaeagnus 141931
Songor Hawthorn 110048
Songorian Sweetvetch 188116
Songory Buckthorn 328861
Songory Reaumuria 327226
Songpan Angelica 24479
Songpan Bedstraw 170653
Songpan Catnip 265069
Songpan Chamaesium 85723
Songpan Chinese Groundsel 365077
Songpan Corydalis 106064
Songpan Hogfennel 293017
Songpan Hornbeam 77395
Songpan Hydrangea 200120
Songpan Larkspur 124438,124610
Songpan Leek 15750
Songpan Milkvetch 43113
Songpan Monkshood 5612
Songpan Nepeta 265069
Songpan Onion 15750
Songpan Savin Juniper 213911
Songpan Savin Savin 213911
Songpan Sinosenecio 365077
Songpan Yinshancress 416367
Songpan Yinshania 416367
Songshan Bluegrass 305434
Sonne's Mahonia 242642
Sonneratia 368910
Sonoma Spineflower 89126
Sonomen Sage 345399
Sonora Beeweed 95793
Sonora Stink Flower 95793
Sonoran Blue Oak 324235
Sonoran Chinchweed 286882
Sonoran Desert California Buckwheat 152059
Sonoran Desertdandelion 242962
Sonoran Globe-amaranth 179247
Sonoran Neststraw 379166
Sonoran Palmetto 341431
Sonoran Palo Verde 83749
Sonoran Rainbow Cactus 140324
Sonoran Scrub Oak 324509
Sonoran Slender-Thoroughwort 166839
Sonoran Tree Catclaw 1447
Soógar Calophaca 67786
Soógorian Pea-shrub 72345
Soong Primrose 314997
Soóngar Chorispora 89016
Soongarian Sophora 369122
Soór Dockin 339887
Soóse-toothed Croton 112940
Soót Dockin 339887
Soótep Dasymaschalon 126729
Soótep Gardenia 171393
Soótep Leaf-flower 296768
Soótep Pendentlamp 84259
Soótep Scurrula 355332
Soóty Beak-rush 333576
Soóty Groundsel 358925
Soóty-glandular Medinilla 247566
Sophia Daphne 122604
Sophia Sisymbrium 126127
Sophia Tansymustard 126127
Sophia-like Tansymustard 126131
Sophiopsis 368944
Sophora 368955
Sophronitis 369178
Sopor 123077
Sops-in-wine 127635
Sopubia 369185
Soranthus 369254,369256
Sorb Apple 369381
Sorbaria 369279
Sorbet 105111
Sorb-tree 369381
Sorcerer's Garlic 15480
Sorcerer's Violet 409339
Sore Eyes 407072
Sorgho 369600,369700
Sorghum 369590,369600,369720
Soroseris 369849
Sorrel 195196,277648,339880,339887,
339897,340075,340084,340254
Sorrel Dock 339887
Sorrel Family 277643
Sorrel Rhubarb 329372
Sorrel Tree 278389
Sorrel Wild Buckwheat 152399
Sorrell 195196
Sorrel-tree 278388
Sorrelvine 93007
Sorrow 339887
Sorrowless Tree 346504
Sort Woodbetony 287208
Sorva 108150
Soryokok 19228
Sosatieplakkie 109266
Sosaties 109349
Sost Hair Chinese Groundsel 365079
Sotol 122898,122912
Sotre-sotry 61564
Sotvine 194155
Sot-weed 266060
Souari-nut 77947
Souari-nut Tree 77946

Soucique 66445
Soukie Clover 397019
Souks Soukie Clover 397019
Soulie Arisaema 33516
Soulie Aster 41285
Soulie Barberry 52169
Soulie Cacalia 283870
Soulie Chelonopsis 86837
Soulie Dolomiaea 135739
Soulie Edelweiss 224942
Soulie Elsholtzia 144091
Soulie Everlasting 21699
Soulie Herminium 192876
Soulie Indigo 206564
Soulie Larkspur 124593
Soulie Milkvetch 43073
Soulie Monkshood 5589
Soulie Primrose 314998
Soulie Rhododendron 331861
Soulie Rose 336965
Soulie Southstar 33516
Soulie Syncalathium 381655
Soulie Thistle 92405
Soulie Willow 344128
Soulie Woodbetony 287686
Soulie's Currant 334211
Souliea 369894
Sound Milkvetch 42478
Soundbean 100149,100153,100185
Souphlong 166889
Sour 15165
Sour Bamboo 4590,4595
Sour Bean 383404,383407
Sour Cherry 83361
Sour Clover 249218,277648
Sour Creeper 346995,402201
Sour Dallisgrass 285417
Sour Dock 277648,339887,339897,339994,
340075,340151
Sour Duck 339887
Sour Eelgrass 418175
Sour Gog 339887
Sour Gogs 339887
Sour Gourd 7022
Sour Grabs 243711,339887
Sour Grass 277648,285417,339887,339897
Sour Greens 340304
Sour Gum 267867
Sour Leaves 339887
Sour Leek 339887,339897
Sour Mandarin 93717
Sour Orange 93332,93368,93408,93414,
93579,93748
Sour Oxalis 277664
Sour Paspalum 285417
Sour Plum 277638,415560
Sour Sab 277648,339887
Sour Sabs 277648,339887
Sour Sally 277648
Sour Salves 339887
Sour Sap 277648
Sour Sauce 339887
Sour Sobs 278008
Sour Sodge 339887
Sour Sog 339887
Sour Sogs 339887
Sour Sop 25868
Sour Sops 25868,339887
Sour Sower 15165
Sour Suds 277648,339887
Sour Top 403758
Sour Top Blueberry 403915
Sour Trefoil 277648
Sour Tupelo 267859
Sour Wood 278389
Sour Woodsorrel 277664
Sourack 339887,339897
Sourbamboo 4590
Sour-bamboo 4590,4595
Sourbean 383404,383407
Sourberry 332658
Sourbush 305085
Sour-clover 249218
Sourdock 339887
Sour-fig 77430
Sourgrass 130605,277877,277973
Sour-gum 267867
Sour-gum Family 267881
Sourock 277648,309570,339887
Soursob 278008
Soursop 25828,25868
Soursop Custard-apple 25868
Soursop-tree 25868
Sour-sour 195196
Sourthern Woodrush 238619
Sourtop Blueberry 172259,403915
Sour-top Blueberry 403916
Sourweed 339897
Sourwood 278388,278389,278559
Sour-wood 278388
Soutern Willow 343172
South Africac Sophora 369035
South African Aloe 16598
South African Asparagus 39058
South African Blackberry 339058
South African Coral Tree 154636
South African Coral-tree 154636
South African Daisy 276608
South African Hoarypea 386199
South African Honeysuckle 400592
South African Ironwood 254711
South African Jasmine 211931
South African Lovegrass 147893
South African Oatgrass 215602
South African Olive 270100
South African Orchid Bush 49094
South African Periwinkle 79418
South African Plum 185867
South African Pomegranate 63903,63904
South African Sagewood 62160
South African Spider Orchid 48566
South African Sumac 332685
South African Wild Almond 58032,58033
South African Wild Pear 135975
South African Wild Plum 185867
South African Wisteria 56735
South African Yellowwood 306425
South American Agave 10787
South American Air Plant 61570
South American Air-plant 61570
South American Bitterwood 364382
South American Bottle Tree 80118
South American Cedar 80030
South American Crowberry 145072
South American Diplachne 226008
South American Elodea 141568
South American Frailea 167690
South American Jelly Palm 64161
South American Lupine 238468
South American Mexican Clover 334325
South American Mock Vervain 176694
South American Rubber Plant 188424
South American Rubberweed 201304
South American Sage 344980
South American Saltbush 44486
South American Sassafras 268708
South American Skullcap 355708
South American Tobacco 266059
South American Vervain 405812
South American Walnut 104190
South American Waterweed 143918
South American Waxmallow 243929
South American Willow 343505
South Ameriean Cedar 80020
South Asia Michelia 252860
South Australia Blue Gum 155624
South Australia White Gum 155624
South Bigleaf Thorowax 63729
South Candle 403738
South China Astilbe 41812
South China Barthea 48552
South China Birch 53354
South China Blueberry 403899
South China Camellia 68928
South China Chinquapin 78899
South China Chloranthus 88300
South China Cinnamon 91270
South China Cyclobalanopsis 116086
South China Dogwood 104961
South China Elaeocarpus 142283
South China Elatostema 142616

South China Eranthemum 148061
South China Evergreenchinkapin 78899
South China Evodia 161309
South China Grevill 161309
South China Hemiboea 191356
South China Holly 203585,204276
South China Honeylocust 176869
South China Jasmine 211775
South China Keiskea 215806
South China Mandarin Blueberry 403900
South China Milkwort 308081
South China Oak 116086
South China Olax 269649
South China Paulownia 285953
South China Pellionia 288727
South China Pepper 300346
South China Persimmon 132444
South China Plagiostachys 301722
South China Pricklyash 417170
South China Prickly-ash 417170
South China Rattan Palm 65778
South China Rattanpalm 65778
South China Rosewood 121629
South China Screwpine 280986
South China Spikesedge 143213
South China St. John's Wort 202146
South China Sweetleaf 381121
South China Syzygium 382483
South China Tarenna 384920
South China Telosma 385759
South China Tutcheria 400687
South China Xylosma 415877
South Circaea 91575
South Coast Ranges pincushion 84503
South Esk Pine 67426
South European Hackberry 80577
South European Rocky Lychnis 292546
South European Tree Mallow 223364
South Florida Slash Pine 299930
South Gansu Birch 53620
South Guangxi Machilus 240667
South Guangxi Rattan Palm 65642
South Guangxi Syzygium 382574
South Guizhou Buckthorn 328651
South Guizhou Cryptocarya 113431
South Guizhou Machilus 240556
South Guizhou Photinia 295657
South Hainan Creeper 92919,285152
South Hainan Decaspermum 123445
South Hainan Litse 234103
South Hamilton Spindle-tree 157562
South Honeylocust 176869
South Island Edelweiss 227836
South Jiangsu Ladybell 7854
South Kansu Stonecrop 357259
South Kiangsu Ladybell 7854
South Kwangtung Skullcap 355849
South Mongolia Tamarisk 383449
South Mongolian Tamarisk 383449
South Mulberry 259097
South Paper-mulberry 61102
South Pricklypear 272802
South Queensland Kauri 10505
South Queensland Kauri Pine 10505
South Sea Arrowroot 382923
South Sea Arrowroot-plant 382923
South Sea Ironwood 79161
South Sea Laurel 98190
South Sea Persimmon 132300
South Sea Rose 265327
South Sichuan Leptodermis 226090,226097
South Sichuan Rhododendron 331864
South Sichuan Willow 344278
South Sinkiang Ziziphora 418131
South Snowy Rhododendron 331347
South Szechwan Willow 344278
South Taiwan Kalanchoe 215149
South Texas Gayfeather 228448
South Texas Rabbit-tobacco 317731
South Tibet Asiabell 98415
South Tibet Bellflower 70191
South Tibet Corydalis 106024
South Tibet Raspberry 338171
South Tibet Willow 343068
South Wales Wattle 1463
South Xinjiang Calligonum 67074
South Xinjiang Nepeta 264921
South Xinjiang Ziziphora 418131
South Xizang Raspberry 338171
South Xizang Spindle-tree 157330
South Xizang Willow 343068
South Yunnan Acanthopanax 2393
South Yunnan Adenia 7293
South Yunnan Allophylus 16065
South Yunnan Anise-tree 204564
South Yunnan Asparagus 39223
South Yunnan Basketvine 9425
South Yunnan Bauhinia 49136
South Yunnan Beauty-leaf 67866
South Yunnan Box 64239
South Yunnan Camellia 68929
South Yunnan Clematis 94930
South Yunnan Cyclobalanopsis 116060
South Yunnan Dolichos 135508
South Yunnan Dutchman's-pipe 34295
South Yunnan Dutchmanspipe 34295
South Yunnan Elaeocarpus 142285
South Yunnan Eurya 160428
South Yunnan Hibiscus 194729
South Yunnan Lime 391771
South Yunnan Linden 391771
South Yunnan Mayten 246812
South Yunnan Millettia 254616
South Yunnan Monkshood 5046
South Yunnan Oak 116060
South Yunnan Pere Valentin's Rhododendron 332049
South Yunnan Poplar 311461
South Yunnan Qinggang 116060
South Yunnan Storax 379311
South Yunnan Strychnos 378833
South Yunnan Synotis 381924
South Yunnan Syzygium 382484
South Yunnan Tanoak 233340
South Yunnan Treebine 92620
South Yunnan Uraria 402130
South Zhejiang Greenbrier 366248
South Zhejiang Smilax 366248
South-african Olive 270100
South-american Prickly Barberry 51739
South-asian Panicgrass 281733
Southasiatic Witchgrass 281733
South-Burma Clovershrub 70861
South-China Beautiful Grape 411574
South-China Holly 203585
South-China Lasianthus 222096
South-China Sinocalamus 125441
South-China Tutcheria 400687
South-China Urceola 402198
Southeast China Rhododendron 331202
Southeast Raspberry 339413
Southeastern Annual Saltmarsh Aster 380998
Southeastern China Creeper 416526
Southeastern Coral-bean 154668
Southeastern Cutleaf Coneflower 339582
Southeastern Sneezeweed 188432
Southeastern Sunflower 188903
Southeasthern China Grape 411613
Southern 88693
Southern Abelia 122
Southern Africcan Dogwood 328830
Southern Agrimony 11571
Southern Amaranth 18666
Southern Annual Saltmarsh Aster 380996
Southern Arrow Wood 407778
Southern Arrowwood 407778,407852
Southern Arrow-wood 407778,407782
Southern Ash 167910
Southern Asian Meliosma 249360
Southern Australian Mahogany 155505
Southern Balsam Fir 377
Southern Balsampear 256786
Southern Bayberry 261137
Southern Beech 266861,266864,266876,266879,266881
Southern Black Haw 408095
Southern Black-Haw 408095
Southern Blackhaw Viburnum 408095
Southern Blue Flag 208932,208933
Southern Blue Toadflax 267349
Southern Blue-gum 155589

Southern Brome 60982
Southern Buckthorn 362957
Southern Bulrush 352169,353274
Southern Bush Honeysuckle 130254
Southern Bush-honeysuckle 130254
Southern Calanthe 66001
Southern California Grape 411710
Southern California Walnut 212592
Southern Catalpa 79243
Southern Cattail 401105
Southern Cat-tail 401105
Southern Cherry Laurel 223097
Southern Cherry-laurel 223097
Southern Choerospondias 88691
Southern Chrysanthemum 89704
Southern Cottonwood 311292
Southern Crab Apple 243550
Southern Crab Grass 130489
Southern Crabapple 243550
Southern Crab-apple 243550
Southern Crapemyrtle 219970
Southern Crape-myrtle 219970
Southern Cross 415189
Southern Cucumbertree 242267
Southern Cut Grass 223984
Southern Cypress 385262
Southern Daisy 50831
Southern Dewberry 338357,339407
Southern Dewdrograss 91575
Southern Dodder 114972
Southern Epaltes 146098
Southern Europe Strawberry Tree 30880
Southern Fieldorange 249650
Southern Fir 377
Southern Flame Azalea 330190
Southern Flannel Bush 168218
Southern Flatseed Sunflower 405959
Southern Fleabane 150922
Southern Fox Grape 411893
Southern Gayfeather 228541
Southern Gentian 173206
Southern Globethistle 140776,140778
Southern Globe-thistle 140778
Southern Goldenrod 368052
Southern Gooseberry 404012
Southern Grass Tree 415174
Southern Grevillea 180568
Southern Hackberry 80673
Southern Hair Grass 12139,12140
Southern Harebell 70002
Southern Hawthorn 110107
Southern Heath 149046
Southern Hemlock 399874
Southern Indica Hybrid Azalea 330917
Southern Japanese Hemlock 399923
Southern Jimmyweed 209582
Southern Jointweed 308668
Southern Kingia 216415
Southern Ladiesstresses 372247
Southern Likiang Spruce 298331
Southern Live Oak 324544
Southern Magnolia 242135
Southern Mahogany 155505
Southern Marara 407570,407571
Southern Marigold 383100
Southern Marsh Orchid 273594
Southern Medick 247506
Southern Melodinus 249650
Southern Mint 250325
Southern Mosquitotrap 117722
Southern mountain Wild Buckwheat 152187
Southern Naiad 262044,262047,262048
Southern Needleleaf 392042
Southern Nettle-tree 80577
Southern Nodding Trillium 397608
Southern Oat 45513
Southern Oxeyedaisy 89704
Southern Pea 409086
Southern Pincushion Sierra 84479
Southern Pine 299928,300128,300264
Southern Pitch-pine 300128
Southern Plains Banksia 47651
Southern Pokeweed 298099
Southern Pond-iffy 267276
Southern Pondweed 312200
Southern Poplar 311237
Southern Prairie Aster 160664
Southern Prickly Ash 417199
Southern Queen Palm 31707
Southern Racemose Goldenrod 367964
Southern Ragweed 19150
Southern Rain-orchis 302326
Southern Rata 252628
Southern Red Cedar 213939
Southern Red Lily 229797
Southern Red Oak 323894,324344
Southern Red Trillium 397619
Southern Redcedar 213917,213939,213998
Southern Reed 295888
Southern Rhododendron 331222
Southern Rock Elm 401486
Southern Rockbell 412750
Southern Rosinweed 364259
Southern Sagebrush 36454
Southern Sallow 343064
Southern Sandbur 80821
Southern Sassafras 43972
Southern Sea Catchbird Tree 300955
Southern Sea-blite 379546
Southern Seepweed 379499
Southern Silky Oak 274259
Southern Single-headed Pussy-toes 26596
Southern Small Skullcap 355383
Southern Sneezeweed 188419
Southern Soapberry 346351
Southern Spikesedge 143209
Southern Star 400750
Southern Star Jasmine 211846
Southern Succisa 379657
Southern Succisella 379663
Southern Sugar Maple 2797
Southern Swallowwort 117722
Southern Swamp Aster 160673
Southern Swamp Crinum 111158
Southern Swamp Lily 111158,229938
Southern Sweet Bay 242341
Southern Syneilesis 381816
Southern Tainia 383169
Southern Taiwan Holly 204365
Southern Tall Flat-topped 135266
Southern Thistle 91751,92442
Southern Threecornerjack 144877
Southern Three-lobed Bedstraw 170668
Southern Twayblade 232894
Southern Viburnum 407852
Southern Warty-cabbage 63450
Southern Water Plantain 14785
Southern Water-hemp 18666
Southern Water-nymph 262044,262047,262048
Southern Water-plantain 14785
Southern Wax Myrtle 261137
Southern Waxmyrtle 261137
Southern Wax-myrtle 261137
Southern White Cedar 85375
Southern White-cedar 85375
Southern Whitetop Aster 135266
Southern Wild Buckwheat 152008
Southern Wild Ginger 37754
Southern Wild Indigo 47859
Southern Wild Rice 418080,418103
Southern Wild Senna 360453
Southern Wildjujube 88690
Southern Willowherb 146905
Southern Witchhazel 185127
Southern Witch-hazel 185141
Southern Wood-rush 238619
Southern Wood-violet 410064
Southern Yarrow 3970
Southern Yellow Orchis 183728
Southern Yellow Pine 299925,300128,300181
Southern Yellow Wood-sorrel 277803
Southern Yellow-cress 336268
Southern Yew 306457,385407
Southern Yunnan Mayten 246812
Southernwood 35088,35868
Southernwood Geranium 288046
South-Guizhou Photinia 295657
Southland Beech 266874
South-siberian Milkvetch 42051

Southstar 33234
Southward Habenaria 183844
Southwest Camellia 69722
Southwest China St. John's Wort 201919
South-west Chinese Groundsel 365036
Southwest Cyclea 116040
Southwest Poplar 311468
Southwest Rockfoil 349906
Southwestern Annual Saltmarsh Aster 380997
Southwestern Bedstraw 170767
Southwestern Big-tooth Maple 3001
Southwestern China Clematis 95258
Southwestern Chokecherry 316791
Southwestern Condalia 101735
Southwestern Coral Bean 154658
Southwestern Cup Grass 151675
South-western Cupgrass 151675
Southwestern Ehretia 141618
South-western Ehretia 141618
South-western Groundsel 359816
Southwestern Gymnadenia 182262
Southwestern Helictotrichon 190211
Southwestern Locust 334974
Southwestern Pricklypoppy 32444
Southwestern Rabbitbrush 236500
Southwestern Raspberry 338158
South-western Ringstem 28802
Southwestern Rose 336800
South-western Vetch 408512
Southwestern White Honeysuckle 235650
Southwestern White Pine 300210
South-westeru Thorn-apple 123092
Southwick's Papeda 93423
South-Yunnan Cassia 91271
South-Yunnan Horsfieldia 198523
Souvari Nut 77946
Sovereign Flower 216085
Sow Bread 115953,368771
Sow Dingle 368771
Sow Fennel 292957
Sow Flower 368771
Sow Foot 400675
Sow Grass 105352
Sow Sorrel 339887
Sow Thistle 368625,368635,368771
Sow Thristle 368771
Sow's Tits 308607
Sowbane 87048,87096,87147
Sowberry 52322
Sowbread 115931,115949,115953,115965, 368771
Sowd-wort 344585
Sower 15165
Sowkie 397019
Sowkie Soó 397019
Sowlers 45448
Sow-teat Blackberry 338105
Sow-teat Strawberry 167658
Sowthinstlelike Samaradaisy 320522
Sowthistle 368625
Sow-thistle 90835,368625
Sow-thistle Desertdandelion 242961
Sowthistle Tasselflower 144975
Sowthistle-leaf Ixeris 210548
Sowthistleleaf Primrose 314994
Sowthistleleaf Sage 345398
Sowthistleleaf Tasselflower 144975
Sowthristle 368771
Soy Bean 177750
Soya 177689,177750
Soya Bean 177750
Soya-bean 177689,177750
Soya-bean Goosefoot 86962
Soybean 177689,177750,177777
Soybeantree 13649
Spadeleaf 81570
Spadeleaf Cremanthodium 110458
Spadeleaf Nutantdaisy 110458
Spade-leaf Philodendron 294802
Spadepetal Stonecrop 356971
Spades 162519
Spadic Bush 155067
Spadix Bulbophyllum 63094
Spadix Stonebean-orchis 63094
Spaeth Ash 168068
Spaeth Barberry 52172
Spaeth Maple 3463
Spafling Poppy 364193
Spage Laurel 122359
Spaghetti 398900
Spain Barberry 51726
Spain Fir 442
Spainish Heath 149046
Spainish Tree Heath 149046
Spalding's Campion 364074
Spalding's Catchfly 364074
Spangled Bean 251265
Spangled Beau 251265
Spaniard 4748
Spanish Ash 382329
Spanish Bayonet 416535,416538,416556, 416583,416584,416603,416649
Spanish Bayonet Yucca 416538
Spanish Beet 53257
Spanish Blue Bell 199553
Spanish Bluebell 145478,145487,199553
Spanish Blue-bells 199553
Spanish Box 64240
Spanish Brome 60820
Spanish Broom 172982,370191,370202
Spanish Bugloss 14841,21959
Spanish Buttons 81237
Spanish Camomile 21266
Spanish Campion 363855
Spanish Cane 37467
Spanish Catch Fly 363855
Spanish Catchfly 363855
Spanish Cedar 80030
Spanish Cherry 255312
Spanish Chervil 261585
Spanish Chestnut 78766,78811,394500
Spanish Clover 126411,126596,237483, 237733,396909
Spanish Daffodil 262444
Spanish Dagger 416535,416538,416540, 416556,416583,416607,416646,416657, 416659
Spanish Dagger Yucca 416607
Spanish Daggers 416535
Spanish Daisy 50809
Spanish Elm 104185
Spanish False Fleabane 321589
Spanish Fir 442
Spanish Flag 255398
Spanish Furze 172982
Spanish Garlic 15726
Spanish Gold 361422
Spanish Gorse 172982
Spanish Gourd 114288
Spanish Grape 411575
Spanish Guava 79376
Spanish Heath 149703
Spanish Hyacinth 145487,199553
Spanish Iris 208946
Spanish Jasmine 211848,305226
Spanish Juniper 213969,386942
Spanish Lavender 223334
Spanish Lettuce 94349
Spanish Lime 249126
Spanish Liquorice 177893,177905
Spanish Mahogany 380528
Spanish Marjoram 402994
Spanish Melon 114212
Spanish Mint 250437
Spanish Moss 327751,392052
Spanish Mustard 59567
Spanish Needles 53797,54048,280416
Spanish Nut 365747
Spanish Oak 206934,323894,324010, 324011,324027,324262,324344
Spanish Onion 15289
Spanish Oregano 390987
Spanish Oyster 354651
Spanish Oyster Plant 354651
Spanish Oysterplant 354651
Spanish Pea 90801
Spanish Pendulous Fir 449
Spanish Pepper 72070,72100
Spanish Plane Tree 302582
Spanish Plum 88691,372477,372479
Spanish Poppy 282706

Spanish Potato 207623
Spanish Psyllium 301864
Spanish Reed 37467
Spanish Rhubarb Dock 339886
Spanish Root 271563
Spanish Saffron 77748
Spanish Sainfoin 187842
Spanish Saisify 354870
Spanish Salsify 354651
Spanish Sea Purslane 44446
Spanish Senna 78436,360444
Spanish Shawl 193747,351401
Spanish Shawl Flower 81709
Spanish Silver Spruce 442
Spanish Stone Crop 356802
Spanish Stonecrop 356802
Spanish Succory 88759
Spanish Tea 139678
Spanish Thistle 81112
Spanish Thyme 391402
Spanish Tree-heath 149046
Spanish Tuft 388423
Spanish Türpeth Root 388870
Spanish Violet 238463
Spanishash 382329
Spanish-bayonet 416535
Spanish-clover 237733,237781
Spanish-dagger 416538,416607
Spanish-hyacinth 60339
Spanish-moss 392052
Spanishneedles 53797
Spapweed 204756,204773,205071,205212
Sparage 39120
Sparagus 39120
Spared Bugloss 321643
Spareflower Primrose 314564
Sparked Grass 293709
Sparked Laurel 44915
Sparkleberry 403724
Sparkler Sedge 75775
Sparkling Rhododendron 331512
Sparmannia 370130
Sparra Grace 39120
Sparra Grass 39120
Sparraw Herb 230275
Sparrow False Pimpernel 231483
Sparrow Vetch 408638
Sparrow's Tongue 308816
Sparrow's-egg Lady's-slipper 120426
Sparrow-birds 174877
Sparrow-grass 39120
Sparrow-weed 326034
Sparrowwort 390994
Sparrow-wort 149847,285547
Sparseflower Alloplectus 16183
Sparseflower Arnebia 34639
Sparseflower Chirita 87896
Sparseflower Sageretia 342178
Sparseflower Touch-me-not 205074
Sparseflower Wildbasil 97011
Sparse-flowered Sageretia 342178
Sparse-flowered Sedge 76513
Sparsepilum Begonia 50320
Sparsepinna Ajania 13028
Sparse-veined Sageretia 342185
Sparsiflorous Neillia 263171
Sparte Rose 336309
Spartina 370155
Spathaceous Fargesia 162752
Spathaceous Moraea 258650
Spathe Flower 370332,370338,370340
Spathe Helicia 189947
Spathebract Fimbristylis 166493
Spathiphyllum 370332,370345
Spathodea 370351
Spathoglottis 370393
Spatholobus 370411,370422
Spathose Chirita 87970
Spathose Dolichandrone 135327
Spathulate Corydalis 106465
Spathulate Eightangle 204591
Spathulate Eritrichium 153536
Spathulate Fleawort 360085
Spathulate Illicium 204591
Spathulate Monkshood 5592
Spathulate Statice 230772
Spathulate Sundew 138354
Spathulate-sepal St. John's Wort 202204
Spathulate-sepaled St. John's-wort 202204
Spatling Poppy 364193
Spatlum 228298
Spatted Thistle 73318
Spatter Dock 267274,267645
Spatterdock 267274,267276,267294,267333
Spatulaleaf Loosestrife 240065
Spatulate Aeonium 9068
Spatulate Fluffweed 166025
Spatulate Glossostigma 177409
Spatulate Groundsel 360085
Spatulate Nepeta 265053
Spatulate Pilea 299006
Spatulate Rattan Palm 65794
Spatulate Robiquetia 335039
Spatulateleaf Sauropus 348066
Spatulate-leaved Heliotrope 190597
Spatulate-leaved Sauropus 348066
Spatulate-leaved Stringbush 414159
Spatulate-leaved Sundew 138299,138303
Spatulate-leaved Willow 344129
Speak 337180
Spear 182942,295888
Spear Crowfoot 325864,326034
Spear Grass 376687
Spear Head 359233
Spear Mint 250450
Spear Orach 44576
Spear Plume Thistle 92485
Spear Saltbush 44576
Spear Thistle 92485
Spearflower 31347
Speargrass 4113,4748,325864
Spear-grass 4748,144661,205497,305274
Spearleaf Agoseris 11503
Spearleaf Arnica 34738
Spearleaf Azalea 332155
Spearleaf Rabbitbrush 236497
Spearleaf Swamp Mallow 286634
Spearleaf Swampmallow 286634
Spear-leaved Dock 340075
Spear-leaved Fat Hen Saltbush 44621
Spear-leaved Fat-hen Saltbush 44621
Spear-leaved Geranium 288253
Spear-leaved Orache 44621
Spear-leaved Willow 343467
Spear-leaved Willowherb 146760
Spear-lily 136729
Spearmint 250287,250450
Spear-plume Thistle 92485
Spearscale 44576
Spear-scale 44576,44621
Spearwood 1292
Spear-wood 1184
Spearwort 207135,325864,326023,326034
Spearwort Buttercup 325864,325870,326303
Specious Dragonhead 137691
Specious Lily 230036
Specious Mirbelia 255786
Specious Pancratium 280909
Specious Wendlandia 413823
Speck 238371
Speckled Alder 16368,16375,16440,16445,16464
Speckled Dendrotrophe 125791
Speckled False Dandelion 266828
Speckled Jewels 204841
Speckled Loco 42599
Speckled Violet 410587
Speckled Wood Lily 97096
Speckled Wood-lily 97096
Spectacle Pod 131106,135008
Spectacle-pod 135008
Spectacular Ixora 211173
Spectacular Lampranthus 220674
Speedwell 317927,317964,406966,407018,407109,407256,407271,407309
Speedwell-creeping 407274
Speedwell-leaf Blumea 55839
Speedwellleaf Woodbetony 287812
Speedy Jenny 394021
Speek 337180
Speirantha 370488

Spekboom 311956
Speknel 252662
Spelt 398970
Spelt Wheat 398970
Spenceria 370501,370503
Sperage 39120,274748
Speranskia 370511,370515
Spergularia 370609
Sphacelate Rush 213460
Sphaeranthus 370956
Sphaere Dolichothele 135691
Sphaere-inflation Milkvetch 43076
Sphaere-sac Milkvetch 43075
Sphaerical Bulbophyllum 63098,63154
Sphaerical Curlylip-orchis 63098
Sphaerical Stonebean-orchis 63154
Sphaerical Turkestan Barberry 51719
Sphaerocaryum 371087
Sphaeroidalflower Woodbetony 287688
Sphaeroidalfruit Cardaria 73092
Sphaeroidflower Woodbetony 287688
Sphaerophysa 371160
Sphagnum Blackberry 339289
Sphallerocarpus 371237
Sphenoclea 371406
Sphenodesma 371416,371421
Sphereseed Millettia 66651
Sphere-seed Wheat 398972
Sphere-seeded Millettia 66651
Sphinctacanthus 371553,371554
Spicate Addermonth Orchid 243130
Spicate Centaurium 81536
Spicate Clerodendranthus 95862
Spicate Codonacanthus 226516
Spicate Currant 334214
Spicate Elsholtzia 144095
Spicate Elytrophorus 144701
Spicate Garcinia 171197
Spicate Nepeta 264973
Spicate Pencilwood 139674
Spicate Pencil-wood 139674
Spicate Pseudelephantopus 317233
Spicate Pseudolephantopus 317233
Spicate Rhubarb 329397
Spicate Screwtree 190112
Spicate Soft Hair Leptodermis 226119
Spicate Soft Hair Wildclove 226119
Spicate Woodbetony 287689
Spicate Yaccatree 306518
Spica-venti Apera 28972
Spice Bush 68329,231297,231306
Spice Cactus 186156
Spice Cliffbean 66624
Spice Litse 233907
Spice Ox in Night 115595
Spice Snowball 235065,235090
Spice Tree 401672
Spicebark Meliosma 249390
Spiceberry 31347,31396,31408
Spicebush 68306,68313,231297,231306
Spice-bush 231297
Spicedaisy 85510,85526
Spicegrass 239543
Spiceleaf Tree 231324
Spice-leaf Tree 231324
Spice-leaved Tree 231324
Spice-lily 244371
Spicknel 252660,252662
Spiculate Blueberry 404010
Spiculiflower Yam 131849
Spicy Jatropha 212160
Spicy Mandarin 93732
Spicy Wintergreen 172135
Spicy-leaf Tree 231324
Spider Aloe 16894
Spider Azalea 331886
Spider Cactus 182441
Spider Chin Cactus 182441
Spider Dendrobium 124998
Spider Flower 95606,95796,180561,180570
Spider Flower Cleome 95796
Spider Herb 95606
Spider Japanese Lily 265259
Spider Lily 111152,111210,191261,200906,200917,200922,200942,200946,239257,239266,239280,394088
Spider Milkweed 37837,38171
Spider Ophrys 272490
Spider Orchid 48564,59265,59270,272490
Spider Plant 26950,88546,88553,88557,95606,95791,95796,349936
Spider Smokebush 102747
Spider Thrixspermum 390491
Spider Tree 110216,110227
Spider Valerian 404285
Spider's Web 266225
Spider's Web Anacampseros 21080,21095
Spiderflower 95606,100892,182915
Spider-flower 95796
Spider-in-his-web 266225
Spiderlike Cryptostylis 113848
Spiderlily 200906
Spider-lily 200906,200940,200955,393987
Spiderling 56397,56437
Spidernet Grevill 180655
Spider-net Grevillea 180655
Spider-orchid 59265,272395,272490
Spiderorchis 30526
Spiderweb Blueeargrass 115526
Spiderweb Cyanotis 115526
Spiderweb Houseleek 358021
Spider-web Houseleek 358021
Spiderwisp 182915
Spiderwort 266225,283300,393987,394019,394053,394082
SpiderWort 394088
Spider-wort 100892
Spiderwort Family 101200
Spidery Groundsel 358284
Spiffy Azalea 330283
Spiffy Bamboo 297338
Spiffy Bushclover 226977
Spiffy Crazyweed 278738
Spiffy Herminium 192873
Spiffy Indocalamus 206779
Spiffy Knotweed 309647
Spiffy Phyllodium 297013
Spiffy Pleione 304290
Spiffy Pocktorchid 282911
Spiffy Sainfoin 271255
Spiffy Spatholobus 370417
Spiffy Spiderflower 95835
Spiffy Spotleaf-orchis 179714
Spiffy Windhairdaisy 348695
Spiffyflower Eulophia 156947
Spiffyflower Gentian 156947
Spiffyflower Mountain Leech 126329
Spiffyflower Orchis 116916
Spiffyflower Uraria 402144
Spiffyleaf Bogorchis 110640
Spigelia 371612
Spignel 167156,252660,252662
Spigneleaf Corydalis 106141
Spignel-meu 252662
Spiguel 361526
Spikate Corydalis 105824
Spikate Figwort 355246
Spikate Kohleria 217748
Spike Ainsliaea 12730
Spike Aletris 14529
Spike Austrotaxus 45318
Spike Balanophora 46859
Spike Bent 12096
Spike Broom 120983
Spike Broomrape 275143
Spike Bulbophyllum 62806
Spike Burgrass 394380
Spike Burreed 370042
Spike Bushmint 203064
Spike Curlylip-orchis 62806
Spike Eria 299639
Spike Falseoat 398531
Spike Gayfeather 228529
Spike Grass 86141,134842
Spike Hairgrass 126086
Spike Heath 61221,61223
Spike Indigo 206573
Spike Lavender 223251,223298
Spike Lilyturf 232640,272140
Spike Microtoena 254267
Spike Muhly 259717

Spike Oil-lavender 223294
Spike Pilocarpus 299172
Spike Primrose 56554,315006
Spike Rabbiten-wind 12730
Spike Rush 143019,143053,143143,143222,143412
Spike Sedge 143019
Spike Snakemushroom 46859
Spike Speedwell 317964
Spike Trisetum 398531
Spike Vinegentian 110330
Spike Wattle 1455
Spike Winter Hazel 106682
Spike Winter-hazel 106682
Spike Woodrush 238697
Spiked Beaksedge 333518
Spiked Blazing Star 228529
Spiked Bracted Sedge 76328
Spiked Capillipedium 71604
Spiked Centaury 81536
Spiked Cucumber 93292
Spiked Earthgallgrass 143472
Spiked Gayfeather 228529
Spiked Gingerlily 187468
Spiked Hyacinth Orchid 34899
Spiked Lobelia 234785,234787
Spiked Loosestrife 240068
Spiked Milfoil 261364
Spiked Millet 289116
Spiked Mixed-flower 298069
Spiked Monkshood 5593
Spiked Muhly 259667
Spiked Pentaphragma 289697
Spiked Rampion 298069
Spiked Rhododendron 331872
Spiked Speedwell 317964
Spiked Star of Bethlehem 274748
Spiked Star-of-bethlehem 274708,274748
Spiked Watermilfoil 261364
Spiked Water-milfoil 261364
Spiked Willow 85875
Spiked Willowherb 85875
Spiked Wood Rush 238697
Spiked Woodrush 238697
Spikedflower Pomatocalpa 310797
Spikeflower Axispalm 228738
Spikeflower Cousinia 108283
Spikeflower Figwort 355246
Spikeflower Germander 388114
Spikeflower Vinegentian 108283
Spike-grass 134842
Spike-heath 61223
Spikeled Toad Lily 396604
Spikeled Toadlily 396604
Spikenard 27969,30748,81065,242701,262497,365362
Spikenard Ampelopsis 20420
Spikenel 252662
Spike-rush 143019
Spikesedge 143019,143022,143141,143146,143385,218457
Spike-sedge 143019
Spikeweed 81825
Spiky-flowers 72802
Spinach 18708,371710,371713
Spinach Beet 53249,53257
Spinach Chard 53249
Spinach Dock 340178
Spinach Joint Fir 178535
Spinach Jointfir 178535,178536
Spinach Joint-fir 178535
Spinage 371710,371713
Spinate Catnip 265032
Spinatehair Jerusalemsage 295191
Spindle 157268,157429
Spindle Family 80127
Spindle Tree 157268,157429,157601
Spindleberry 157429
Spindle-fruited Barringtonia 48515
Spindlepalm 246028
Spindleroot Catchfly 363798
Spindle-tree 157268,157327,157429,157714
Spindlewood 157429
Spine Ailanthus 12588
Spine Aralia 30590
Spine Axispalm 228766
Spine Bambusa 47425,47448
Spine Birdlingbran 87263
Spine Bittergrass 404562
Spine Chrysopogon 90120
Spine Corydalis 106469
Spine Crazyweed 279172
Spine Date 418175
Spine Euonymus 157332
Spine Knotweed 309772
Spine Knotwood 44276
Spine Olive 415234
Spine Palm 2550,12772
Spine Pelargonium 288209
Spine Pricklyash 417134
Spine Randia 79595
Spine Shortawn Ungeargrass 36623
Spine Squarebamboo 87603
Spine Square-bamboo 87597
Spine Three-leaves Pricklyash 417360
Spine Wild Pomegranate 79595
Spine-Areca 2550
Spinebract Ironweed 406828
Spinebract Nightshade 366967
Spinebranch Wildclove 226114
Spinebranchlet Rhododendron 330210
Spinecalyx Bethlehemsage 283649
Spinecalyx Elegant Raspberry 338122
Spinecalyx Eremostachys 148467
Spinecalyx Paraphlomis 283649
Spine-elm 191626,191629
Spineflower 89051,222584
Spinefruit Beakchervil 28030
Spinefruit Buttercup 326108
Spinefruit Chervil 28030
Spinefruit Coldwaterflower 299049
Spinefruit Jurinea 214060
Spine-fruited Medick 247435
Spinegreens 92384
Spinehairy Blueberry 404023
Spineleaf Alpine Oak 324436
Spineleaf Calvepetal 177985
Spineleaf Milkvetch 42827
Spineleaf Oak 324436
Spineleaf Shamgoldquitch 318150
Spine-leaved Oak 324436
Spineless Acacia 1642
Spineless Acaena 1751
Spineless Arundinella 37430
Spineless Black Locust 334982
Spineless Brazil Mimosa 255056
Spineless Butcher's-broom 340994
Spineless Cactus 272891
Spineless Caper 71871
Spineless Common Jujube 418173
Spineless Eelgrass 418173
Spineless Five-angle Raspberry 339032
Spineless Gongshan Raspberry 338479
Spineless Hornwort 83578
Spineless Horsebrush 387075
Spineless Indian Bamboo 47493
Spineless Oilnut 323070
Spineless Oil-nut 323070
Spineless Rose 336649
Spineless Salem-rose 339202
Spineless Saltwort 344743
Spineless Teasel 133487
Spineless Waterchestnut 394529
Spineless Wenxian Peashrub 72377
Spineless Wenxian Pea-shrub 72377
Spineless Yellow Locust 334982
Spineless Yucca 416581
Spineless-stem Nettle 402884
Spinelike Raspberry 339294
Spine-like Raspberry 339294
Spine-pataled Meconopsis 247178
Spinepink 2555
Spinescent Grease-bush 177365
Spineseed Floatingheart 267818
Spinesheath Tanoak 233192
Spinest Thistle 92408
Spine-stalk Milkvetch 42827
Spinestemon 306956
Spinetaro 222038,222042
Spinetooth Barberry 51327
Spinetooth Chirita 87974

Spinetooth Dragonhead 137627
Spinetooth Greenorchid 137627
Spinetooth Indigo 205788
Spinetooth Woodbetony 287016
Spinetwig Azalea 330210
Spinevine Sageretia 342180
Sping Dragonhead 137665
Spingel 167156
Spinifex 303112,371723,397762
Spink 72934
Spinks 72934
Spinnage 371713
Spinning Gum 155691
Spinning Jenny 2837
Spinose Acanthopanax 143682
Spinose Orostachys 275394
Spinosy Honeysuckle 236109
Spinous Meyna 252707
Spinule Euonymus 157283
Spinule Podocarpus 306519
Spinulose Barberry 52176
Spinulose Leaf Cherry 223131,316824
Spinulose Raspberry 339294
Spinulose Statice 230760
Spinuloseleaf Cherry 223131,316824
Spinulose-leaved Cherry 223131
Spinulose-leaved Cherry Laurel 223131
Spiny Acanthus 2711
Spiny Alyssum 18474
Spiny Amaranth 18822
Spiny Ant Palm 217920
Spiny Aster 41297,88253,88254
Spiny Bamboo 47190,47448
Spiny Barrel Cactus 163418,163452
Spiny Bear's Breech 2711
Spiny Bear's-breech 2711
Spiny Bergmann Barberry 51361
Spiny Bittergrass 404544
Spiny Burweed 415053
Spiny Carissa 76962
Spiny Chesneya 87263
Spiny Clematis 94877
Spiny Clotbur 415053
Spiny Clot-burr 415053
Spiny Club Palm 46374
Spiny Club-palm 46374
Spiny Cocklebur 415053,415057
Spiny Crazyweed 279172
Spiny Currant 333961
Spiny Diversifolious Pricklyash 417292
Spiny Edible Yam 131579
Spiny Emex 144877,144880
Spiny Euphorbia 159081
Spiny Fake Needlegrass 318202
Spiny Fiber Palm 398825
Spiny Fibre-palm 398825
Spiny Glorybind 103338
Spiny Goatwheat 44276
Spiny Goldenweed 414980
Spiny Greenbrier 366332
Spiny Hackberry 80708
Spiny Hedgehog Cactus 140236
Spiny Holdback 65068
Spiny Honeysuckle 236109
Spiny Hornwort 83558
Spiny Horsebrush 387084
Spiny Lasia 222042
Spiny Leaf Sow-thistle 368649
Spiny Licuala Palm 228766
Spiny Licualapalm 228766
Spiny Meconopsis 247131
Spiny Metalleaf 296412
Spiny Milk Thistle 368649
Spiny Milkwort 308393
Spiny Naiad 262067
Spiny Nopal 273022
Spiny Oncosperma 271001
Spiny Peashrub 72348
Spiny Pea-shrub 72348
Spiny Persimmon 132051
Spiny Pigweed 18822
Spiny Plum 316819,415558
Spiny Plumeless Thistle 73281
Spiny Poisonnut 378676
Spiny Pricklyash 417134
Spiny Prickly-ash 417134
Spiny Pricklythrift 2229
Spiny Pseudopogonatherum 318150
Spiny Pseudoraphis 318202
Spiny Randia 79595
Spiny Raspberry 339130
Spiny Rest Harrow 271347
Spiny Restharrow 271347
Spiny Russian-olive 141937
Spiny Sago 252634
Spiny Sago Palm 252634
Spiny Saltbush 44355,44654,328535
Spiny Sandbur 80821
Spiny Sesbania 361367
Spiny Sida 362662
Spiny Smilax 366332
Spiny Solanum 367541
Spiny Sow Thistle 368649
Spiny Sowthistle 368649
Spiny Sow-thistle 368649
Spiny Spiderflower 95796
Spiny Spurge 159864
Spiny Star 155228
Spiny Sterculia 376077
Spiny Stickseed 221711,221755
Spiny Strychnos 378676
Spiny Swamp Currant 334058
Spiny Threecornerjack 144880
Spiny Thrift 34515
Spiny Yellow Spindle-tree 157280
Spinybract Bigumbrella 59237
Spinybract Brassaiopsis 59237
Spinybract Chinquapin 78955
Spinybract Evergreenchinkapin 78955
Spinybract Pertybush 292066
Spiny-bract Pouzolzia 313478
Spiny-bracted Evergreen Chinkapin 78955
Spiny-bracted Sedge 73634
Spinybranch Greenbrier 366392
Spinycalyx Raspberry 338101
Spiny-calyx Raspberry 338101
Spinycalyx Redflower Raspberry 338638
Spiny-calyx Twoflower Raspberry 338188
Spiny-club Palm 46374
Spinycupule Tanoak 233192
Spiny-cupuled Tanoak 233192
Spinyelephanetsear 222038
Spinyflower Alternanthera 18135
Spinyflower Raspberry 338079
Spinyflower Strophanthus 378401
Spiny-flowered Raspberry 338079
Spiny-flowered Strophanthus 378401
Spinyfruit Bedstraw 170199
Spinyfruit Caesalpinia 64971
Spiny-fruit Clearweed 299049
Spinyfruit Pricklyash 417221
Spiny-fruit Spindle-tree 157932
Spinyfruit Uvaria 403438
Spiny-fruited Aspidistra 39576
Spiny-fruited Euonymus 157269
Spiny-fruited Leaf-flower 296521
Spiny-fruited Prickly-ash 417221
Spiny-fruited Spindle-tree 157269
Spiny-fruited Uvaria 403438
Spinyhair Blastus 55175
Spinyhair Pennywort 200364
Spinyhair Seepweed 379489
Spiny-haired Blueberry 404023
Spiny-headed Mat-rush 235386
Spinyherb 135144
Spinyleaf Aralia 30759
Spinyleaf Cherry 223131,316824
Spinyleaf Cherrylaurel 223131,316824
Spinyleaf Cherry-laurel 223131
Spiny-leaf Crazyweed 278680
Spinyleaf Euonymus 177995
Spinyleaf Peashrub 72170
Spinyleaf Photinia 295754
Spinyleaf Rockjasmine 23300
Spiny-leaf Sow-thistle 368649
Spiny-leaved Aralia 30759
Spiny-leaved Barberry 52033
Spiny-leaved Crazyweed 278680
Spiny-leaved Holly 203590
Spiny-leaved Pea-shrub 72170
Spiny-leaved Photinia 295754

Spiny-leaved Sedge 73597
Spiny-leaved Sow Thistle 368649
Spiny-leaved Sow-thistle 368649
Spinymargine Holly 203590
Spinyseed Bedstraw 170199
Spinystem Aralia 30632
Spiny-stemmed Aralia 30632
Spiny-stemmed Tongue Flower 177365
Spiny-tooth Gum-weed 181132
Spiny-toothed Chalk-leaved Barberry 51557
Spiny-toothed Gumweed 181132
Spiradiclis 371742,371746
Spiraea 371785,371902,372000
Spiraea Meadowsweet 166110
Spiral Aloe 17182
Spiral Ditch-grass 340472
Spiral Eucalypt 155529
Spiral Flag 107221
Spiral Ginger 107221,107227,107254,
107263,107265
Spiral Macrozamia 241465
Spiral Marestail 196814
Spiral Spapweed 205329
Spiral Tall Dock 340268
Spiral Tasselweed 340472
Spiral Touch-me-not 205329
Spiral Wild Celery 404563
Spiralflag 107221
Spiral-fruit Gum 1615
Spiral-fruited Pondweed 312267
Spiralpetal Monkshood 5594
Spiralscale Spikesedge 143355
Spiralspurred Larkspur 124598
Spiranthes 372176
Spire Lily 170828
Spire Medinilla 247537
Spire Mint 250450
Spire Parabarium 402194
Spire Pepper 300512
Spire Urceola 402194
Spire's Ironweed 406827
Spirea 371785
Spires 293709
Spirewort 325864
Spiripetalous Monkshood 5594
Spirit Lily 280896
Spirit Plant 219026,219030
Spiritweed 154316
Spironema 67145
Spirorrhynchus 372333
Spitular Bridelia 60225
Spitulate Bridelia 60225
Spitzel's Orchid 273643
Splaycalyx St. John's Wort 201981
Splaying St. John's Wort 202070
Spleen Amaranth 18708
Spleenwortleaf Smelowskia 366119
Splended Bredia 59823
Splendid Bambusa 47462
Splendid Laelia 219694
Splendid mariposa Lily 67627
Splendid Mint Bush 315603
Splendid Oncidium 270842
Splendid Wildginger 37732
Splendida Water Lily 267636
Split Calyx Rhododendron 331765
Split Hamilton Spindle-tree 157563
Split Leaf Philodendron 258168,294839
Split Nematanthus 263322
Split Rock 304372
Splitbeard Bluestem 23052
Splitbract Copperleaf 1806
Splitbract Germander 388299
Splitbract Rattanpalm 65740
Splitbract Snakegourd 396184
Splitbract Trachydium 393840
Splitbract Tumorfruitparsley 393840
Splitcalyx Azalea 331765
Splitcalyx Sage 345375
Splitcalyx Spapweed 205064
Splitcalyx Touch-me-not 205064
Splitcalyx Vinegentian 110303
Splited Bambusa 47234
Splitedbract Germander 388299
Splitedge Gentian 173570
Splitehead Elatostema 142828
Splitehead Stairweed 142828
Splithair Azalea 331856
Spliting Living-rock Cactus 33209,337154
Splitleaf Groundsel 279903
Splitleaf Ladybell 7686
Splitleaf Lagotis 220191
Splitleaf Luohanfruit 365330
Splitleaf Primrose 314238
Splitleaf Snowbell 379464
Splitleaf Violet 409914,409925
Split-leaved Violet 409925
Splitlemma 351261
Splitlip Epipogium 147307
Splitlip Hemipilia 191609
Splitlip Hempnettle 170060
Splitlip Irisorchis 267983
Splitlip Liparis 232160
Splitlip Oberonia 267983
Splitlip Peristylus 291234
Splitlip Perotis 291234
Splitlip Zeuxine 417764
Splitmelon 351759,351762
Splitpetal Habenaria 183962
Splitpetal Liparis 416521
Splitpetal Sage 345376
Splitpetal Sinocarum 364893
Splitpetal Twayblade 416521
Splitpetal Wilford Cranesbill 175013
Splittedcalyx Sage 345375
Splittedpetal Sage 345376
Splittinglip Twayblade 232160
Splittongue Goldenray 229209
Splitwing Milkvetch 42566
Splny-tipped Tongue Flower 177364
Spodiopogon 372398,372428
Spogel Seed 302119
Spoil 144661
Spokeflower Azalea 330194
Spokewood 157429
Sponge Azalea 331493
Sponge Gourd 238261
Sponge Poorspikebamboo 270450
Sponge Rhododendron 331493
Sponge Tree 1219
Sponge-Berry Tree 199740
Spongetree 1219
Spongy Indocalamus 206901
Spongy Indosasa 206901
Spongy Oligostachyum 270450
Spongy Umbrella Bamboo 162692
Spongy Yam 131851
Spongystylebase Spikesedge 143303
Spontaneous China Rose 336496
Spoon Bottle Gourd 219847
Spoon Bush 114566
Spoonbract Corydalis 106465
Spoonbract Larkspur 124608
Spoonbract Monkshood 5592
Spoonbract Zinger 417972
Spooncalyx Bulbophyllum 63096
Spooncalyx Curlylip-orchis 63096
Spooncalyx Gentian 173931
Spooncalyx St. John's Wort 202204
Spooncress 63443,63446
Spoonflower 288805
Spoongrass 222871
Spoonlaef Aster 41289
Spoonleaf Anemone 24091
Spoonleaf Annular Willow 343031
Spoonleaf Barberry 52306
Spoonleaf Batang Aster 40124
Spoon-leaf Bouchea 57837
Spoonleaf Cottonweed 166025
Spoonleaf Cremanthodium 110460
Spoonleaf Cudweed 178342
Spoon-leaf Cudweed 170926
Spoonleaf Eritrichium 153536
Spoonleaf Gardenia 171245
Spoonleaf Gentian 173913
Spoonleaf Groundsel 360085
Spoonleaf Kalanchoe 215174
Spoonleaf Memecylon 249989
Spoonleaf Microula 254367
Spoonleaf Mitreola 256219
Spoonleaf Nardostachys 262497

Spoonleaf Nutantdaisy 110460
Spoonleaf Oak 323842
Spoonleaf Primrose 314581,314832
Spoonleaf Purple Everlasting 178367
Spoonleaf Rockfoil 349920
Spoonleaf Rockjasmine 23204
Spoonleaf Spapweed 205328
Spoonleaf Spiradiclis 371780
Spoonleaf Stringbush 414159
Spoon-leaf Sundew 138299
Spoonleaf Touch-me-not 205328
Spoonleaf Waxplant 198891
Spoon-leaf Wild Buckwheat 152470
Spoonleaf Willow 344129
Spoonleaf Windflower 24091
Spoonleaf Windhairdaisy 348218
Spoon-leaved Barberry 52306
Spoon-leaved Oak 323842
Spoon-leaved Saxifrage 349227
Spoon-leaved Soapwort 346419
Spoon-leaved Waxplant 198891
Spoonpetal Calanthe 66072
Spoon-sepal Spineflower 89089
Spoonshaped Ginger 417972
Spoonshapeleaf Oak 323842
Spoontree 189972
Spoonwood 215395
Spoonwort 97967,98020
Spoots 24483
Spoptflower Corydalis 105753
Sporoxeia 372889,372898
Spot Flower 4866,371627
Spot Gentian 173494
Spot Giant Timber Bamboo 297221
Spot Glabrous-fruit Viburnum 407916
Spot Psilateleaf Arrowwood 407916
Spot Yushanbamboo 416810
Spot-branch Camellia 69668
Spoted Basketvine 9456
Spoted Fruit Cryptocarya 113468
Spoted Scales Rhododendron 331250
Spotflower 4858,371627
Spot-leaf Aucuba 44888
Spotleaf Clearweed 298885
Spotleaf Coldwaterflower 298885
Spotleaf Cremastra 110516
Spotleaf Cuckoo-orchis 110516
Spotleaf Metalleaf 296411
Spotleaf Orchis 273524,273531
Spotleaf Peperomia 290380
Spot-leaf Privet 229590
Spot-leaf Southstar 33358
Spotleaf-orchis 179571,179694
Spot-leaved Aucuba 44888
Spot-leaved Rhododendron 331585
Spot-leaves Rhododendron 331585
Spotless Arrowbamboo 162684
Spotless Galangal 17670
Spotless Gansu Fritillary 168523
Spotless Japan Gentian 154929
Spotless Parnassia 284530
Spotless Rockfoil 349713
Spotlessflower Baker Lily 229747
Spotlip Curlylip-orchis 62984
Spotlip Orchis 169907
Spotseed 57675
Spotsheath Pseudosasa 318332
Spotstem Rhubarb 329354
Spotted African Cornlily 210843
Spotted Aglaonema 11332
Spotted Alder 185141
Spotted Alstroemeria 18079
Spotted Ardisia 31511
Spotted Arum Lily 417097
Spotted Ash 168077
Spotted Asian Poppy 335650
Spotted Bee Balm 257184
Spotted Beebalm 257183
Spotted Begonia 49632,50051
Spotted Bellflower 70234
Spotted Bitter Bamboo 304065
Spotted Bladderwort 403296
Spotted Bur Clover 247246
Spotted Burr Medic 247246
Spotted Calla Lily 417097
Spotted Callalily 417097
Spotted Cat's Ear 202424,202432
Spotted Cat's-car 202424
Spotted Cat's-ear 202432
Spotted Centaurea 81184
Spotted Chalcedonian Lily 229807
Spotted Clover 247246
Spotted Comfrey 321643
Spotted Coralroot 103990,104000
Spotted coral-root 103990
Spotted Cowbane 90924,101852
Spotted Crane's-bill 174717
Spotted Cranesbill 174717
Spotted Dead Nettle 220401
Spotted Deadnettle 220401
Spotted Dead-nettle 220401
Spotted Disporum 134450
Spotted Dog 273531,321643
Spotted Dumbcane 130113
Spotted Emu Bush 148367
Spotted Emu-bush 148367
Spotted Euphorbia 159286
Spotted Evergreen 11340
Spotted Fig 165841
Spotted Flowers Rhododendron 331975
Spotted Gasteria 171702
Spotted Geranium 174717
Spotted Gilia 175701
Spotted Goblin Orchid 258998
Spotted Goldenthistle 354655
Spotted Gongora 179293
Spotted Grass 247246
Spotted Gum 106809
Spotted Hawkweed 195780
Spotted Hemlock 90924,101852
Spotted Henbit 220401
Spotted Horse Mint 257184
Spotted Ixia 210843
Spotted Joe-pye Weed 158219
Spotted Joepyeweed 158219,161172
Spotted Joe-pye-weed 158219
Spotted Joe-Pye-weed 158221
Spotted Knapweed 80968,81184,81405,
81406
Spotted Lady's Thumb 291840
Spotted Lady's-slipper 120357
Spotted Lady's-thumb 309570
Spotted Ladysthumb 309570
Spotted Langloisia 221036
Spotted Laurel 44915
Spotted Loosestrife 239804
Spotted Mary 321643
Spotted Medic 247246
Spotted Medick 247246
Spotted Metalleaf 296411,296412
Spotted Morningstar Lily 229818
Spotted Morning-star Lily 229820
Spotted Mountain Gum 155596
Spotted Nemophila 263504
Spotted Nepal Fleshspike 346956
Spotted Nettle 220401
Spotted Notweed 309570
Spotted Oak 324225,324415
Spotted Odontoglossum 269077
Spotted Oncidium 270828
Spotted Orchis 273524,273531
Spotted Parsley 90924,101852
Spotted Persicaria 309468,309570
Spotted Phlox 295282
Spotted Pleione 304278
Spotted Pondweed 312235
Spotted Rhododendron 331176
Spotted Rock Rose 399957
Spotted Rockrose 399940,399957
Spotted Rodriguezia 335166
Spotted Sand-mat 85770,159286
Spotted Snapweed 204799
Spotted Snap-weed 204799
Spotted Spanish Oysterplant 354655
Spotted Spapweed 205112
Spotted Spider Orchid 59274
Spotted Spurge 158385,159286
Spotted St. John's-wort 202107,202118
Spotted Staunton-vine 374423
Spotted Sterculia 376118
Spotted Sun Orchid 389260

Spotted Swan Orchid 116521
Spotted Touch-me-not 204838,204841,205112
Spotted Trillium 397580
Spotted Tubergourd 390161
Spotted Tuftroot 130121
Spotted Virgin 321643
Spotted Water Hemlock 90924
Spotted Water-hemlock 90924,90925
Spotted Weavers' Bamboo 47476
Spotted Wild Buckwheat 152250
Spotted Wintergreen 87488
Spotted Yushania 416795
Spottedflower Patrinia 285849
Spottedfruit Thickshellcassia 113468
Spotted-laurel 44887,44915
Spottedleaf Chirita 87950
Spottedleaf Euphorbia 159286
Spottedleaf Spurge 159286
Spotted-leaves Rhododendron 331348
Spottedlip Corallorhiza 103990
Spotted-seed Tallow-tree 346370
Spotte-leaved Privet 229590
Spottongue Cymbidium 117081
Spottongue Orchis 117081
Spottwig Camellia 69668
Spotty Bitter-bamboo 304065
Spout Tanoak 233405
Spponleaf Sandwort 32234
Sprague Barberry 52178
Sprague Spindle-tree 157888
Sprangletop 225983
Sprangle-top 225983
Sprangletup 354583
Sprawl Crazyweed 278894
Sprawling Euonymus 157943
Sprawling Saltbush 44661
Sprawling Schizostachyum 351852
Sprawling Wirevine 259589
Spread Atylosia 65156
Spread Batweed 89207
Spread Bowfruitvine 393541
Spread Bushclover 226912
Spread Gentian 173234,173577
Spread Hairy Lilac 382268
Spread Ironweed 406320
Spread Pondweed 312214
Spread Swertia 380299
Spread Thomson Ligusticum 229406
Spread Wildcoral 170048
Spread-haired Cherry 83272
Spreading Amaranth 18692
Spreading Anemone 321721
Spreading Aster 380947
Spreading Bedstraw 170420
Spreading Bellflower 70210
Spreading Bladderpod 227021
Spreading Blue-eyed Grass 365767
Spreading Bluegrass 305605,305815
Spreading Bur Parsley 392968
Spreading Bushclover 226912
Spreading Bush-clover 226912
Spreading Buttercup 325779
Spreading Chervil 84759
Spreading Chinchweed 286899
Spreading Corydalis 105716
Spreading Cotoneaster 107435
Spreading Cottonrose 235259
Spreading Creosote Bush 221978
Spreading Creosote-bush 221978
Spreading Cyrtococcum 120631
Spreading Dalea 121890
Spreading Dogbane 29468
Spreading Elm 401501,401549,401618
Spreading English Yew 385319
Spreading Erysimum 154530
Spreading Fanpetals 362474
Spreading Fescue 163902
Spreading Fleabane 103580,150594
Spreading Gayfeather 228499
Spreading Globeflower 399512,399522
Spreading Globe-flower 399512
Spreading Hawthorn 109674
Spreading Hax-lily 127516
Spreading Hedge Parsley 392968
Spreading Hedgeparsley 392968,392977
Spreading Hedge-parsley 392968
Spreading Hedyotis 187565
Spreading Hogweed 56425
Spreading Indocalamus 206895
Spreading Jacob's-ladder 307241
Spreading Knapweed 81034
Spreading Lazy Daisy 84931
Spreading Liverseed Grass 58078
Spreading Lupine 238443
Spreading Mananthes 244295
Spreading Marigold 383103
Spreading Meadow-grass 305588
Spreading Melicope 249161
Spreading Nailwort 284830
Spreading Orache 44576
Spreading Oval Sedge 75527
Spreading Panicgrass 285379
Spreading Pasqueflower 321721
Spreading Pellitory 284152
Spreading Pigweed 18668
Spreading Pinanga 299670
Spreading Pine 300140
Spreading Pogonia 94545,94547
Spreading Pulsatilla 321703
Spreading Rattlebox 112449
Spreading Rhododendron 331457
Spreading Roman Cypress 114760
Spreading Rush 213360
Spreading Sand-mat 85764
Spreading Sandwort 31999
Spreading Sedge 75081,75879
Spreading Snakeroot 11171
Spreading Snake-root 11171
Spreading Sneezeweed 81687
Spreading Spartina 370171
Spreading Spiderflower 95662
Spreading Sprangletop 225996
Spreading St. John's Wort 202070
Spreading St. John's-wort 202070
Spreading St. -John's-wort 201930
Spreading Stringbush 414162
Spreading Sulphur Flower 152605
Spreading Toxocarpus 393541
Spreading Wallflower 154530
Spreading Wild Buckwheat 152015
Spreading Windmill Grass 88339
Spreading Wintergreen 172135
Spreading Yellow Cress 336272
Spreading Yellow-cress 336272
Spreadingbranch Ladybell 7639
Spreading-flowered Staff-tree 80279
Spreading-leaf Pine 300140
Spreading-leafed Pine 300140
Spreading-leaved Pine 300140
Spreading-pod Rock-cress 30246
Spreadleaf Spapweed 204941
Spreadspike Bluegrass 306038
Spreadspike Hymenacue 200839
Spreadspike Sawgrass 245558
Spreadspike Water Hymenacue 200839
Spread-styled Camellia 69272
Spread-style-like Camellia 69274
Spreadvein Melicope 249161
Spreesprinkle 273531
Sprengel Bladderwort 404011
Sprengel's Sedge 76333
Sprenger Asparagus 38985,39195
Sprenger Asparagus-fern 38985
Sprenger Lycoris 239278
Sprenger Magnolia 416721
Sprenger Magnolia's 416721
Sprenger Stonegarlic 239278
Sprenger Wildginger 37733
Sprenger's Asparagus Fern 38983
Spreusidany 292957
Spriding Euonymus 157651
Spring Adonis 8387
Spring Anemone 24109
Spring Avens 175457
Spring Beauty 94292,94305,94323,94349,94373
Spring Bedstraw 170737
Spring Bells 365726
Spring Bittercress 72829
Spring Blue-eyed-mary 99840

Spring Bouquet Viburnum 408172
Spring Cactus 351999
Spring Carrot 312360
Spring Cherry 83334
Spring Cinquefoil 312812,312813
Spring Cleavers 170193
Spring Colchicum 62525,99352
Spring Crees 47963
Spring Cress 72708
Spring Crocus 111625
Spring Cudweed 26551
Spring Cyclamen 115942,115974
Spring Cymbidium 116880
Spring Draba 153964
Spring Elm 401490
Spring Eupatorium 158362
Spring Fellwort 174049
Spring Felwort 174049
Spring Figwort 355265
Spring Forget-me-not 260905
Spring Fumewort 106453
Spring Gastrodia 171936
Spring Gentian 174049
Spring Grass 27930,27969
Spring Green Violet 199709
Spring Groundsel 360306
Spring Heath 149147,149401
Spring Iris 208922
Spring Knotweed 309570
Spring Ladies'-tresses 372277
Spring Lady's-tresses 372277
Spring Larkspur 124650
Spring Lessingia 227136
Spring Lily 154932
Spring Meadow Saffron 62525
Spring Meadow-saffron 99352
Spring Messenger 325820
Spring Milletgrass 254542
Spring Minuartwort 255598
Spring Mountain Ankle-aster 207395
Spring Mountains Thistle 91930
Spring Mouseear 82818
Spring Onion 15165,15289
Spring Orchis 116880
Spring Pea 222860
Spring Peony 280332
Spring Pheasant's Eye 8387
Spring Pleione 304272
Spring Puccoon 233707
Spring Pulsatilla 24109
Spring Rabbitbrush 387079
Spring Rabbit-tobacco 127934
Spring Saffron 111625
Spring Sandwort 255598
Spring Savory 96964
Spring Sedge 74059
Spring Skullcap 355836
Spring Sneezeweed 188441
Spring Snowflake 227875
Spring Snowflakelotus 227875
Spring Speedwell 407432
Spring Spurry 370604
Spring Squill 353099
Spring Starflower 207489
Spring Star-flower 60526,207489
Spring Vetch 408452,408571,408578
Spring Vetchling 222860
Spring Violet 174049
Spring Waterstarwort 67401
Spring Whitlow Grass 153964
Spring Whitlowgrass 153964
Spring Whitlow-grass 153964
Spring-beauty 94292,94305,94373
Springcress 72708
Spring-cress 72708
Spring-cyclamen 115942
Springdale Rock Daisy 291336
Spring-flower 174877,315106
Spring-flowering Goldenrod 368475
Spring-flowering Sulphur Flower 152609
Spring-flowering Tamarisk 383581
Spring-grass 27969
Spring-run Spider-lily 200952
Springstar 207489
Spring-twist Cactus 243996
Springwort 159222
Sprit 198376
Sprouting Broccoli 59545
Sprouting Spikerush 143416
Sprouting Spike-rush 143412
Sprowled Woodbetony 287168
Spruce 298189,298191
Spruce Barberry 52180
Spruce Beer 298191
Spruce Cone Cholla 385874
Spruce Fir 298191
Spruce Pine 299864,299954,399867
Spruce Stonecrop 357114
Spruced-leaved Stonecrop 357114
Spruce-fir Fleabane 150628
Spud 367696
Spunish-lime 249125
Spunsilksage 360807
Spur Cephalanthera 82035
Spur Flower 303319
Spur Gentian 184689
Spur Pepper 72070
Spur Pugionium 321479
Spur Sandcress 321479
Spur Skullcap 355398
Spur Valerian 81768
Spured Gentian 184691
Spurflower 303125,303224
Spurflower Honeysuckle 235708
Spurge 85764,158385,159286,159599,159856,279743
Spurge Family 160110
Spurge Flax 122416,122448,122515
Spurge Laurel 122492,122515
Spurge Laurel Daphne 122492
Spurge Nettle 97741,212228
Spurge Oil 122515
Spurge Olive 97491,122492,122515
Spurge Thyme 159557
Spurge Tree 159975
Spurgelaurel 122492
Spurge-laurel 122492
Spurgentian 184689
Spurgeolive 122492,122515
Spurge-olive 97491,97493
Spurgewort 208566
Spurless Baoshan Monkshood 5426
Spurless Barrenwort 146984
Spurless Calanthe 66106
Spurless Columbine 30024
Spurless Corydalis 105838
Spurless Ducloux Monkshood 5170
Spurless Epimedium 146984
Spurless Herminium 192841
Spurless Monkshood 5426
Spurless Paoshan Monkshood 5426
Spurless Spapweed 205121
Spurless Touch-me-not 205121
Spurred Appendicula 29809
Spurred Bellflower 69885
Spurred Butterfly Pea 81878
Spurred Coral Root 147307
Spurred Coral-root 147307
Spurred Coral-root Orchid 147307
Spurred Gentian 184694
Spurred Neottia 144123
Spurred Ophiorrhiza 272185
Spurred Snapdragon 231084
Spurred-flowered Honeysuckle 235708
Spurred-gentian 184694
Spurredpetal Urophysa 402653
Spurrey 370521,370522,370609
Spurry 370521,370522,370715
Spurry Knotweed 309813
Spurry Wild Buckwheat 152478
Spursepal Loosestrife 239604
Spurstandard 81872,81876
Spur-Valarian 81760
Spurwing Wattle 1669
Spur-wing Wattle 1669
Spurwood 325864
Spurwort 362097
Spyny Amaranth 18822
Spyny Amaranthus 18822
Squalid Iris 208860
Squam Carpet 79966

Squama Clearweed 299051
Squama Coldwaterflower 299051
Squamata Juniper 213943
Squamate Corydalis 106471
Squamate Falsetamarisk 261293
Squamate False-tamarisk 261293
Squamate Pine 300208
Squamose Ochna 268266
Squamulate Water-centipede 218628
Squamulose Licorice 177944
Squamulose Viburnum 408141
Square Bamboo 87547,87607
Square Parsley 63475
Square Rattlesnake Plantain 179707
Square Rush 143316
Square St. John's Grass 202179
Square St. John's Wort 202179
Square Tree 157429
Squareanfle Hedyotis 187686
Squarebamboo 87547
Square-bamboo 87547,87607
Squarebranch Olive 270171
Squarebranch Syzygium 382674
Square-branched Syzygium 382674
Squareculm Spikesedge 143316
Square-culmed Bamboo 87607
Squared Indocalamus 206825
Squareflower 284842
Squaregrass 226367
Square-leaved China-laurel 28331
Squarelip Liparis 232175
Squarelip Twayblade 232175
Squarengle St. John's-wort 202168
Square-pod Deerveteh 237771
Square-pod Water-primrose 238160
Squareseed Snakegourd 396277
Squarestalk Geranium 288546
Square-stalked Passion Fruit 285694
Square-stalked PassionflowerMaracuya 285694
Square-stalked St. John's-wort 202179
Square-stalked Willowherb 146595,146905
Squarestem 90556
Square-stem Bamboo 87607
Squarestem Butterflybush 62157
Squarestem Eargrass 187686
Squarestem Greenbrier 366539
Square-stem Greenbrier 366539
Square-stem Spike-rush 143316
Squarestem St. John's Wort 202168
Square-stemmed Bamboo 87547,87607
Square-stemmed Monkey Flower 255237
Square-stemmed Monkeyflower 255237
Squarestemmed Sandwort 32277
Square-stemmed Spike-rush 143316
Square-stemmed St. -John's-wort 201713
Squaretwig Skullcap 355426
Square-valved She-oak 79178
Squarevein Indocalamus 206825
Square-veined Indocalamus 206825
Squarrib 355202
Squarrosa Falsecypress 85354
Squarrose Agriophyllum 11619
Squarrose Calligonum 67080
Squarrose Knapweed 81453
Squarrose knapweed 81454
Squarrose Pseudoraphis 318202
Squarrose Sedge 76335
Squarrose White Aster 380872
Squarrose-leaved Gentian 173917
Squarrous Ironweed 406828
Squash 114189,114271,114292,114300
Squashberry 407799
Squash-berry 407799
Squat 176738
Squatmore 176738
Squaw 404012
Squaw Bush 332908
Squaw Currant 333936,334042
Squaw False Willow 46257
Squaw Huckleberry 403753,404012
Squaw Root 79733,102003
Squaw Tea 146262
Squaw Thorn 239123
Squaw Vine 256002
Squaw Weed 279888,279930,279939
Squaw's Carpet 79966
Squawberry 332716
Squaw-berry 332477,332908
Squawbusb 105193
Squawbush 238996,332908
Squaw-bush 105193,332477,332483,332908
Squawbush Sumac 332908
Squaw-grass 415441
Squawroot 77809,79733,91034,102003
Squaw-root 79730,79733,102003,397142,397556
Squaw-thorn 239123
Squaw-weed 358159
Squeeze Jaws 231183
Squeeze-jaws 231183
Squill 137958,352880,352886,352960,353065
Squill Onion 352960
Squinancy 334117
Squinancy Woodruff 39328
Squinancy-berry 334117
Squinancywort 39328
Squinaney 39328
Squinaney Woodrush 39328
Squinter-pip 174877
Squirrel Corn 128291
Squirrel Tail Small Reed 127211
Squirrel's Corn 128291
Squirrel's Tail 214359
Squirrel-corn 128291
Squirrel-taft Fescue 412407
Squirreltail 144379,198311
Squirrel-tail Grass 198311,198319
Squirreltail-grass 198311
Squirters 3505,380756
Squirtgourd 139851,139852
Squirting Cucumber 139851,139852
Squitch 144661
Squll Onion 352886
Squwreleaf Cremanthodium 110445
Squwreleaf Nutantdaisy 110445
Srgunen Bluegrass 305368
Sri Lank Doplopod 148258
Sri Lanka Lovegrass 147553
Sri Lanka Reed Bamboo 268120
Sriated Membranaceous Dendrocalamus 125493
Sriated Yellow Dragonbamboo 125493
Srilanka Glochidion 177192
Srilanka Hunteria 199440
Srilanka Hydrolea 200430
Srilanka Lobelia 234884
Srilanka Olax 269696
Ssetulose Oak 324409
Ssior Bird Cherry 280053
Ssior Birdcherry 280053
Sston Wire Shrub 259586
St Anthony's Lily 229789
St Catherine's Lily 229789
St Columcille's Plant 239750
St Francis' Wood 235982
St Joseph's Lily 229789
St. Agnes' Flower 227875
St. Andréw's Cross 38285,201933
St. Andréw's-cross 38285
St. Anne's Needlework 349044
St. Anthony's Nut 102727,374107
St. Anthony's Rape 325666
St. Anthony's Rope 325666
St. Anthony's Turnip 325666
St. Audre's Lace 71218
St. Augustine Grass 375774
St. Barbara's Cress 47964
St. Barbara's Herb 47964
St. Barnaby's Thistle 81382
St. Benedict's Herb 175454
St. Benedict's Thistle 73318
St. Bennet's Herb 101852,404316
St. Bernard Lily 27190
St. Bernard's Lily 26950,27190
St. -Bernard's-lily 27190
St. Bernard-lily 27190
St. Bride's Comb 53304
St. Bride's Forerunner 384714
St. Bridget's Wort 301910

St. Bruno's Lily 27195,283295,283300
St. -Bruno's-lily 283295
St. Catherine's Lace 152086
St. Catherine's-lace 152086
St. Cundida's Eyes 409339
St. Dabeoc's Heath 121060,121066
St. David's Plant 15048
St. David's Rose 336851
St. Domingo Apricot 243982
St. Domingo Mahogany 380528
St. Elmo's-Feather 19468
St. Foyne 271280
St. Francis' Thorn 154327
St. George's Bells 145487
St. George's Flower 145487
St. George's Herb 404316
St. Giles' Orpine 357203
St. Ignatius Poison Nut 378761
St. Ignatius Poison-nut 378761
St. James Lily 372912,372913
St. James' Lily 372913
St. James' Ragwort 359158
St. James' Wort 72038
St. John Rose 336883
St. John's Bloom 89704
St. John's Bread 83527
St. John's Bush 319707
St. John's Grass 202086
St. John's Herb 36474,158062
St. John's Lily 97218
St. John's River Oxytropls 278914
St. John's Rose 336883
St. John's Susan 339593
St. John's Sweet Bread 83527
St. John's Wort 36474,86755,201705,
201718,201733,201787,202070,202086,
202103
St. John's Wortleaf Spiraea 371938
St. John's Wort-leaved Spiraea 371938
St. John's-bread 83527
St. John's-wort 201787,201993,202086
St. John's-wort Family 97266
St. -John's-Wort Family 201681
St. Johns Lily 111171
St. Johns Wort 201705
St. Johnsedleaf Skullcap 355484
St. Johnswort 201886,202086
St. John's-wort 201705
St. Johnswortleaf Marshweed 230289
St. Johnswort-leaved Barberry 51748
St. Johnswort-like Loosestrife 239680
St. Joseph's Flower 407072
St. Joseph's Lily 229789
St. Joseph's Staff 262441,368480
St. Joseph's Wort 268438,268566
St. Leger-gordon Pixylily 374916
St. Lucie Cherry 83253
St. Lucie's Cherry 83253
St. Lucy Cherry 83253
St. Mark's Marsh Spider-lily 200931
St. Mark's Rose 336308,337007
St. Martin's Buttercup 326061
St. Martin's Herb 348975
St. Mary 383712
St. Mary Thistle 364360
St. Mary's Seal 308607
St. Mary's Thistle 364360
St. Mawe's Clover 247246
St. Nicholas Wild Buckwheat 152111
St. Olaf's Candlesticks 257374
St. Patrick's Cabbage 349918,350011,358062
St. Patrick's Spit 299759
St. Patrick's Staff 299759
St. Patrick's-cabbage 349918
St. Paul's-wort 363073
St. Paulswort 363073
St. Peter Plant 114604
St. Peter Port Daisy 150792
St. Peter's Bell 262441
St. Peter's Bells 262441
St. Peter's Cabbage 350032
St. Peter's Flower 329521
St. Peter's Herb 315082
St. Peter's Keys 315082
St. Peter's Palm 416593
St. Peter's Wort 201743,201847,202179,
322724,380734,380756
St. Peter's-wort 38291
St. Thomas Bauhinia 49241
St. Thomas Bean 145899
St. Thomas Lidpod 103362,103379
St. Thomas Tree 49241
St. Thomas' Onion 15165,15167,15289
St. Vincent Arrowroot 245014
Stable-broom Sedge 74866
Stabwort 277648
Stachelige Euryale 160637
Stachys 373085,373166
Stachyurus 373516,373556
Stachyurus Family 373515
Staellatehair Draba 137046
Staellatehair Whitlowgrass 137046
Staff Tree 80129
Staff Vine 80129,80260,80309
Stafftree 80309
Staff-tree 80129,80260,80309
Stafftree Family 80127
Staff-Tree Family 80127
Stag's Horn Plantain 301910
Stag's Horn Sumac 332916
Stag's Horn Sumach 332916
Stag's-horn Plantain 301910
Stag's-horn Sumach 332452,332916
Stagberry 380749
Stagbush 408061
Stagger Bush 239388
Staggerbush 239388
Staggergrass 19468
Staggers Weed 246379
Staggerweed 128291,373125
Stagger-weed 373125
Staggerwort 359158
Staggwort 359158
Staghorn Cholla 116675,273095
Staghorn Cinquefoil 312797
Staghorn Sumac 332639,332916
Staghorn Sumach 332639,332916
Staghorn Velvet Sumac 332639
Staghorn Velvet Sumach 332639
Stagshorn Sumach 332639
Stahl Stone Crop 357170
Stahlianthus 373633
Stainch 271563
Stained Crimson Rhododendron 331875
Staining Euonymus 157926
Stainless Bay 223203
Stainton's Cotoneaster 107691
Staintoniella 373642
Stairweed 142594,142694
Stalegrass 182620
Stalewort 35088
Stalinsky Milkvetch 43082
Stalked Alseodaphne 17809
Stalked Bulbine 62390
Stalked Bur Grass 394390
Stalked Burgrass 394390
Stalked Bur-grass 394390
Stalked Devil Rattan 121527
Stalked Fleabane 150442
Stalked Orach 44601,44633
Stalked Rhodiola 329920
Stalked Rockfoil 349051
Stalked Scurvy-grass 97981
Stalked Speedwell 407029
Stalked Wool-grass 353699
Stalkedfruit Falsenettle 56360
Stalkedfruit Pittosporum 301372
Stalkedfruit Pricklyash 417340
Stalkedfruit Rockvine 387776
Stalkedfruit Seatung 301372
Stalkedspike Nepeta 265027
Stalkfruit Buttercup 326240
Stalk-fruited Pittosporum 301372
Stalk-fruited Prickly-ash 417340
Stalkfruited Tanoak 233300
Stalkleaf Murainagrass 209321
Stalkleaf Primrose 314357
Stalkless Bladderwort 403851
Stalkless Butterflybush 62162
Stalkless Eritrichium 153533
Stalkless Honeysuckle 235669

Stalkless Mussaenda 260492
Stalkless Pendentseed Jointfir 178559
Stalkless Persimmon 132085
Stalkless Roscoea 337102
Stalkless Tutcheria 400726
Stalkless Windhairdaisy 348152
Stalkless Yellow-cress 336268
Stalklip Ginger 418022
Stalky Berula 53157
Stallions 37015
Stallions-and-mares 37015
Stallkless Caper 71893
Stamen Scarlet Root-blosson 10192
Stamenlike Brome 60982
Stamenlike Dragonhead 163089
Stamen-like Javatea 275642
Staminate Deutzia 127106
Stammerwort 359158
Stanavoi Clematis 94933
Stanche 72038
Stand Speedwell 407014
Stand Stemona 375351
Standard Crested Wheatgrass 11710
Standard Crested Wheat-grass 11710
Standard Wheatgrass 11696
Standelwelks 273531
Stander-grass 273531,372259
Standerwort 273531
Standflower Betony 373449
Standhair Skullcap 355642
Standing Bugle 13104
Standing Cypress 175704,217324
Standing Gusses 37015
Standing Gussets 273531
Standing Lobelia 234442
Standing Motherweed 231578
Standing Tallow-tree 346388
Standing-cypress 175704
Standingfruit Fenugreek 397261
Standish Arbor-vitae 390660
Standish Honeysuckle 235798
Standish's Honeysuckle 235798
Standle-grass 273531
Standlewort 273531
Standley Ceriman 258179
Standley Pincution 244236
Standley's Bloodleaf 208351
Standley's Goosefoot 87164
Standtooth Catnip 265016
Stanford Hesperantha 193343
Stanford Manzanita 31139
Stanhopea 373681
Stankewicz Pine 300209
Stanly's Club Cholla 181454
Stanmarch 366743
Stannen-grass 273531
Stansbury Cliffrose 108497
Stansbury's Rock Daisy 291334
Stapelia 373719
Stapeliaform Groundsel 360102
Stapf Barberry 52371
Stapf Cranesbill 174930
Stapf Fescuegrass 164330
Stapf Illicium 204593
Stapf Monkshood 5598
Stapf Rhodiola 329960
Stapf Sonerila 368874
Stapf Sweetleaf 381381
Stapfiophyton 374070
Stapflike Monkshood 5504
Star Acacia 1686
Star Anise 204517,204570,204575,204603
Star Anise Tree 204603
Star Apple 90007,90008,90015,90032,
90085
Star Astilbe 41851
Star Begonia 49915,50243
Star Brodiaea 60522
Star Bromelia 113365
Star Bur 80981
Star Bush 400592
Star Cactus 43492,43493,43526
Star Chickweed 374968,375067
Star Clover 397095
Star Cluster 289831
Star Clusters 289784,289831
Star Cucumber 362461
Star Daisy 231606,231607
Star Duckweed 224396
Star Flower 68755,396780
Star Fruit 45725
Star Gentian 173364,174049
Star Glory 323450,323465
Star Gooseberry 296466
Star Grass 14471,117852,117859,202796
Star Hyacinth 353099
Star Hyacinth Squill 352883
Star Ipomoea 323459
Star Jasmine 211912,393613,393657
Star Lily 155858,227818,229811
Star Lotus 267730
Star Magnolia 242331
Star Naked Boys 99297
Star Naked Ladies 99297
Star Nut Palm 43372
Star of Arabia 274513
Star of Bethlehem 21339,23925,70093,
169421,201787,210318,274495,274510,
274513,274708,274810,274823,285623,
374916,394333
Star of Bethlehem Orchid 25063
Star of Bethlehen 155858
Star of Hungary 274823
Star of Jerusalem 394333
Star of Persia 15188
Star of the Earth 105352,175454,301910
Star of the Lundi 279690
Star of the Night 97262
Star of the North 175454
Star of the Sabi 7340
Star of the Veldt 131150,131191
Star of the Wood 374916
Star Phlox 295272
Star Pine 300146
Star Pink 88022
Star Plum 90032
Star Rockfoil 349932
Star Sansevieria 346105
Star Saxifrage 349932
Star Sedge 74423,75711
Star Squill 352883
Star Thistle 80892,80981,81035
Star Tickseed 104572
Star Toadflax 100229
Star Tree 43472
Star Tulip 67554,67622
Star Violet 187559,198714
Star Window Plant 186368
Star Wort 374916
Star-and-garter 274823
Staranise 204603
Star-anise 204474,204603
Star-anise Family 204471
Starapple 90007,90032
Star-apple 90007,90032
Star-bearing Leucas 227709
Starbract Orach 44634
Starbur 2608
Star-bur 80981
Starbush 204517,400554,400598
Star-bush 400554
Star-cactus 43492
Starcalyx Gentian 173266
Starcalyx Swertia 380119
Starch Grape Hyacinth 260324,260333
Starch Grape-hyacinth 260324,260333
Starch Hyacinth 260322,260333
Starchcontaining Sage 345023
Starchey Sagebrush 36325
Starchey's Wormwood 36325
Starchwort 37015
Star-cluster 289784,289831
Star-clusters 289784
Star-cucumber 362461
Star-cudweed 155875
Star-daisy 231607
Starfish 113364
Starfish Flower 374013
Starfish plant 113365
Starflower 60431,366142,396780,396785,
398658,398777

Star-flower 15861,57101,239750,274823,312520,325820,356468,374013,374916,396780,396785,396790
Star-flower Cluster-lily 60522
Starflower Garlic 15226
Starflower Oil 57101
Starflower Onion 15226
Starflower Pincushions 350240
Starflower Rush 213045
Starflower Solomon's-seal 242701
Starflower Solomonseal 366211
Star-flower Ver Vine 292518
Starflower Vine 292518
Star-flowered False Solomon's-seal 366211
Star-flowered Lily of the Valley 366211
Star-flowered Lily-of-the-valley 366198,366211
Star-flowered Magnolia 242331
Starfruit 41732,41735,45725,121993,121995
Star-Fruit 121993
Star-fruit 121995
Starfruit False Solomonseal 366211
Starfruit Stonecrop 356473
Starfruitestraw 41732,41735
Star-glory 323450,323459
Stargrass 14471,117878,202796
Star-grass 14471,14486,67376,170524,202796,374916
Stargrass Family 202786
Starhaiey Bird Cherry 280054
Starhair Amoora 19967
Starhair Ardisia 31538
Starhair Azalea 331032
Star-hair Bamboo 303995
Star-hair Bitterbamboo 303995
Starhair Eustigma 161024
Starhair Handbamboo 20193
Starhair Kiwifruit 6703
Starhair Meadowrue 388449
Starhair Shorttonguedaisy 58047
Starhair Tanoak 233344
Starhair Tickseed 104773
Star-haired Schefflera 350747
Starhairy Birdcherry 280054
Starhead Elatostema 142606
Starhead Stairweed 142606
Star-headed Sedge 76342
Star-hyacinth 352883
Star-hyacinth Squill 352883
Starjasmine 393613
Star-jasmine 393613,393657
Starke Willow 344148
Starleaf Begonia 49915
Star-leaf Begonia 49915
Star-leaf Grevillea 180567
Starleaf Gum 232565
Starlight 174739
Starlike Gentian 173925
Star-lily 227818
Starlite Spineflower 89122
Star-mustard 99089,99090
Star-naked Boys 99297
Star-nut Plam 43372
Star-of-bethlehem 169375,274495
Star-of-bethlehem 274495,274823
Star-of-bethlehem Orchid 25063
Star-of-jerusalem 394333
Star-of-texas 414985
Star-of-the North 175454
Star-of-the-earth 175454
Star-of-the-veldt 131147,131150
Star-peyote 43493
Star-pink 88022
Star-rush 333524
Starry Campion 364076
Starry Cerastium 82680
Starry Clover 397089
Starry Eyes 274823
Starry False Solomon's Seal 242701
Starry False Solomon's-seal 242701
Starry Figmarigold 252044
Starry Gardenia 171408
Starry Grasswort 82680,82688
Starry Hare's-ear 63838
Starry Magnolia 242331
Starry Puncturebract 362995
Starry Rosinweed 364259
Starry Saussurea 348812
Starry Saxifrage 349932
Starry Sulphur Flower 152582
Starry Windhairdaisy 348812
Starryflower Aletris 14498
Starryleaf Acronema 6198
Starshape Sweetleaf 381412
Star-shape Sweetleaf 381412
Starshaped Hair Ural Falsespiraea 369282
Star-sky Alternanthera 18147
Star-spined Evergreen Chinkapin 79027
Starstyle Treerenshen 125641
Star-thistle 80981
Starthorn 200589
Star-thorn 200589
Startongue Aster 40091
Star-tree 43472
Starvation Prickly Pear 273014
Starvation Pricklypear 273014
Starve-acre 325612
Starved Aster 380861,380922
Starved Fleabane 150780
Starved Panic Grass 128502
Starved Panic-grass 128502
Starved Sedge 74478
Starved Wood-sedge 74291
Star-vine 351011,351040
Star-vine Family 351121
Starweed 374968
Starwert Chickweed 82772
Starwoetleaf Stonecrop 357174
Starwort 39910,40817,67336,67393,228506,370522,374724,374897,374916,374968,380930,398134
Starwort Gentian 173207
Starwort Mouse Ear 82772
Starwort Mouse-ear 82772
Starwort Mouse-ear Chickweed 82772
Starwort-leaf Rockfoil 349931
Stary Calyx Rhododendron 332093
Staryhair Schefflera 350747
Stately Bismarckia 54739
Statice 230508,230659,230712,230761,319996,320002
Staticelike Knotweed 162333
Staudre's Lace 71218
Staudt Lepidobotrys 225481
Staunch 28200,72038
Staunch-girs 3978
Staunch-grass 3978
Staunton Tickseed 104849
Staunton's Microcos 253391
Stauranthera 374445
Staurochilus 374457,374458
Stauton Elsholtzia 144096
Stave Oak 323625
Staverwort 359158
Stavesacre 124604
Stave-wood 364382
Stawberry Clover 397050
Stay-plough 271563
Steamboat Springs Wild Buckwheat 152366
Stearn Barberry 52182
Stebbins' Desertdandelion 242964
Stebbinsia 374537,374539
Stedfast 334435
Steel Begonia 50076
Steel Globe Thistle 140776
Steele's Dewberry 339297
Steens Mountain Thistle 91935
Steep Roegneria 335411
Steep Woodbetony 287538
Steepgrass 299759
Steeple 11549
Steeple Bellflower 70247
Steeple Bells 70247
Steeple Bush 372106
Steeplebush 371902,372106,372107
Steeplewort 11549
Steepweed 299759
Steepwort 299759
Steepy Rhododendron 330523
Steer's-head 128326

Steinberg Milkvetch 43083
Steininger Woodbetony 287696
Steinmetz's Bulrush 352270
Stelis 374699
Stellar Ardisia 31538
Stellar Brassaiopsis 59244
Stellar Hybrids Dogwood 105067
Stellar-calyxed St. John's-wort 202164
Stellaria-leaf Sandwort 32235
Stellarialike Loosestrife 239861
Stellate Actinidia 6703
Stellate Aglaia 19967
Stellate Bird Cherry 280054
Stellate Craibiodendron 108585
Stellate Dysophylla 139585
Stellate Elaeagnus 142199
Stellate Figmarigold 252044
Stellate Gelidocalamus 172748
Stellate Goldleaf 108585
Stellate Hairs Rhododendron 330155
Stellate Japanese Spiraea 371981
Stellate Parashorea 283904
Stellate Sedge 76056
Stellate Watercandle 139585
Stellate Winteraconite 148112
Stellate-calyx St. John's Wort 202164
Stellateflower Seepweed 379604
Stellate-frited Star-apple 90061
Stellatefruit Star Apple 90062
Stellatefruit Starapple 90062
Stellatehair Bigumbrella 59244
Stellatehair Brachanthemum 58047
Stellatehair Brassaiopsis 59244
Stellatehair Carpetweed 176949
Stellatehair Clethra 96532
Stellatehair Eriolaena 152684
Stellatehair Erysimum 154513
Stellatehair Firerope 152684
Stellatehair Osbeckia 276170
Stellatehair Sugarmustard 154513
Stellatehair Vatica 405178
Stellatehair Waxpetal 106683
Stellatehair Winterhazel 106683
Stellate-haired Actinidia 6703
Stellate-haired Eriolaena 152684
Stellate-haired Rhododendron 330155
Stellate-haired Winterhazel 106683
Stellate-scaled Amoora 19968
Stellatestyle Dendropanax 125641
Stellate-styled Dendropanax 125641
Stellato-spiny Evergreenchinkapin 79027
Steller Lagotis 220196
Steller's Wormwood 35779
Stellera 375179,375187
Stellteflower Ophiorrhiza 272179
Stellulate Barrenwort 147065
Stellulate Epimedium 147065
Stellulate Sandwort 32074
Stelmatocrypton 375234
Stem Lettuce 219485
Stem-clasping Butterflybush 62010
Stemflower Brandisia 59122
Stem-flower Eightangle 204494
Stem-flower Glochidion 177172
Stemflower Helicia 189913
Stem-flower Helicia 189913
Stem-flower Illicium 204494
Stem-flower Lepisanthes 225671
Stem-flower Sauropus 348045
Stemflower Spicegrass 239574
Stemless Aeginetia 8757
Stemless Arctic Bramble 338078
Stemless Arctotis 31196
Stemless Birdlingbran 87233
Stemless Carline Thistle 76990
Stemless Carpetweed 256711
Stemless Chesneya 87233
Stemless Cinquefoil 312325
Stemless Date 295453
Stemless Datepalm 295453
Stemless Dishspurorchis 171875
Stemless Emei Parnassia 284533
Stemless Evax 193375
Stemless Evening Primrose 269520
Stemless Forget Me Not 260897
Stemless Forgetmenot 260897
Stemless Forrest Wallflower 154465
Stemless Gastrochilus 171875
Stemless Gentian 173184,173778
Stemless Gesneria 175297
Stemless Goldenweed 375788
Stemless Gomphostemma 179175
Stemless Green Everlasting 21727
Stemless Green Pearleverlasting 21727
Stemless Horsebrush 387075
Stemless Lady's Slipper 120290
Stemless lady's-slipper 120290
Stemless Launaea 222909
Stemless Loco 278940,279174
Stemless Milkvetch 41912
Stemless Oberonia 267923
Stemless Parrya 224212
Stemless Pinanga 299650
Stemless Sawwort 361076
Stemless Selinum 357793
Stemless Spapweed 204760
Stemless Thistle 91713,92361
Stemless Touch-me-not 204760
Stemless Trumpet Gentian 173344
Stemless Youngia 416479
Stemmed Eelgrass 418384
Stemmy Vetch 408502
Stemona 375336
Stemona Family 375359
Stemonurus 375372
Stemroot Lycianthes 238959
Stemroot Red Silkyarn 238959
Stenanthium 375451
Stenless Cryptanthus 113365
Stenoauriculate Dendrocalamopsis 47451
Stenobracteoid Tickseed 104850
Stenocoelium 375551
Stenolobed Middling Agapetes 10320
Stenomesson 375607
Stenopetal Coelogyne 98728
Stenopetallous Crape-myrtle 219968
Stenopetalous Rhodoleia 332211
Stenophtllous Photinia 295794
Stenophyllous Acanthopanax 2505
Stenophyllous Gardenia 171401
Stenophyllous Metalleaf 296415
Stenophyllous Pea-shrub 72351
Stenophyllous Rosewood 121829
Stenophyllous Stringbush 414261
Stenophyllous Sweetspire 210370
Stenosepallous Xantolis 415241
Stenosepalous Ardisia 31613
Stenoseris 375710
Stenosolenium 375725
Stenostachyous Veronicastrum 407498
Stenotaphrum 375764
Step Palm 31014
Stepdaughters 410677
Stephan Rhodiola 329961
Stephanachne 375805
Stephanandra 375811
Stephania 375827,375839
Stephanotis 211693,211695,245803,376022
Stephen's Stork's Bill 153914
Stepmother 374916,410677
Stepmother's Blessing 28044
Stepmother's Son 410677
Steppe Agoseris 11495
Steppe Bluegrass 306139
Steppe Cabbage 326832
Steppe Draba 98764
Steppe Premna 313745
Steppeliving Premna 313745
Steppers Leek 15523
Steppers Onion 15523
Sterculia 376063,376144
Sterculia Family 376218
Sterculia Maple 3641
Sterculia Sloanea 366078
Sterculia-like Sloanea 366078
Stereo-leaves Rhododendron 331939
Stereosandra 376243
Sterigmostemum 376314
Sterile Oat 45586
Sterile Sedge 76368
Steriled Umbellate Hydrangea 200128

Stern's Cotoneaster 107692
Sternberg Pink 127842
Sternbergia 376350
Stertion 262722
Steudner Peristylus 291270
Steudnera 376380,376382
Steven Colewort 108637
Steven Gypsophila 183253
Steven Hawthorn 110056
Steven Heronbill 153916
Steven Maple 3645
Steven Mouseear 83032
Steven Rocket 193449
Steven Skullcap 355776
Steven Stonecrop 294128
Steven's Buttercup 326397
Steven's Gypsopbila 183253
Stevenia 376387
Stevenson Palm 376400
Stevenson Rosewood 121830
Stevensonia 295439
Stevevson Palm 295439
Stevia 300842,376405,376415
Steward Cyclobalanopsis 116201
Steward Holly 204278
Steward Oak 116201,324452
Steward Tongoloa 392799
Steward Woodbetony 287703
Stewardson's Jack-in-the-pulpit 33522
Stewart Catnip 265060
Stewart Cudweed 178444
Stewart Jerusalemsage 295200
Stewart Lawson Cypress 85297
Stewart Lily 230042
Stewart Nepeta 265060
Stewart Rhododendron 331890
Stewart Thorowax 63841
Stewarta 376428
Stewartia 376428
Sticadore 223334
Sticadove 223334
Stick Arrowbamboo 162742
Stick Buttons 31051,170193
Stick Coelodiscus 98558
Stick Donkey 170193
Stick Rattan Palm 65787
Stick Rattanpalm 65778,65787
Stick Robin 170193
Stick Touch-me-not 204858
Stickaback 170193
Stickbush 96009
Stick-button 31051,170193
Stick-donkey 170193
Stickers 31051
Stickfruit Sida 362628
Stickinghair Azalea 330775
Stickleaf 250483,250487,250489
Stickleaf Irisorchis 267936
Stickleaf Syzygium 382512
Stickleback 170193
Sticklewort 11549
Stick-robin 170193
Stickseed 184320,221627,221727,221755
Stickspur Astyleorchis 19503
Stickspur Calanthe 65906
Stickspur Habenaria 183839
Stickspur Platanthera 302500
Stickstyle Azalea 330468
Sticktight 53755,53797,53816,53913,54182
Stick-Tights 53755
Stickweed 375955
Stickwort 11549,370522
Sticky Adenosma 7982
Sticky Alder 16352
Sticky Arrowbamboo 196345
Sticky Aster 240386
Sticky Ball 31051,170193
Sticky Bobs 170193
Sticky Boronia 57259
Sticky Buttons 31051,170193
Sticky Catchfly 363824,364179,410957
Sticky Chickweed 82849
Sticky Cinquefoil 312734
Sticky Cockle 363815
Sticky Currant 334265
Sticky desert-sunflower 174429
Sticky False Asphodel 392608
Sticky Flower 275415,275416
Sticky Gayfeather 228470
Sticky Goldenrod 368398,368402
Sticky Goldenweed 323037
Sticky Goosefoot 139682
Sticky Grass 93913,170490
Sticky Groundsel 360321
Sticky Gypsophila 183270
Sticky Hawkweed 195948
Sticky Head 181174
Sticky Hedge Hyssop 180338
Sticky Hop-bush 135215
Sticky Jack 31051
Sticky Leaf Arnica 34755
Sticky Lovegrass 147700
Sticky Mouse Ear 82849
Sticky Mouse-ear 82849
Sticky Mouse-ear Chickweed 82849,82991
Sticky Nightshade 367617
Sticky Oxeye 193675
Sticky Rabbit-tobacco 317775
Sticky Ragwort 360321
Sticky Rehmannia 327435
Sticky Rhynchosia 333452
Sticky Rice 275966
Sticky Rosinweed 364281
Sticky Sage 345057
Sticky Sagebrush 35270,36149,36471
Sticky Sand-spurrey 370643
Sticky saw-wort 348147
Sticky Snakeroot 11153
Sticky Spiderwort 394001
Sticky Starwort 318489
Sticky Stork's-bill 153835
Sticky Storksbill 153767
Sticky tall Fleabane 151065
Sticky Tick Clover 126366
Sticky Wattle 1299
Sticky Waxweed 114603
Sticky White Catchfly 363583
Sticky Wild Buckwheat 151955,152657
Sticky Willie 170193
Sticky Willow 31051
Sticky Willow Weed 146587
Sticky Willowweed 146587
Sticky Willy 170193
Sticky-back 31051,138348,170193
Sticky-balls 31051,170193
Sticky-buds 9701,118008
Sticky-buttons 31051,170193
Sticky-fruited Fig 165711
Stickyhair Everlasting 21533
Stickyhair Groundsel 360321
Stickyhair Pearleverlasting 21533
Stickyhair Sage 345347
Sticky-head 181174
Sticky-leaf Rabbitbrush 90565
Stickyleaf Rhododendron 330775
Sticky-leaved Rhododendron 330775
Stickypappo 261084
Stickyseed 55226
Sticky-weed 170193
Sticky-willy 170193
Stictocardia 376524,376536
Stiebritz Barberry 51319
Stiff Arrowhead 342397
Stiff Aster 40767
Stiff Beargrass 266545
Stiff Bedstraw 170668
Stiff Bottle Brush 67291
Stiff Bottlebrush 67291
Stiff Bristless Rhododendron 331894
Stiff Brome 58578
Stiff Catapodium 354487
Stiff Cowbane 278559
Stiff Dogwood 105039
Stiff Eyebright 160268
Stiff Gentian 174148,174149
Stiff Goldenrod 368354,368356
Stiff Grass 354477
Stiff Hedgenettle 373389
Stiff Onosma 271820
Stiff Pondweed 312270
Stiff Reedbentgrass 65420

Stiff Rhododendron 330442
Stiff Sand-grass 79305
Stiff Sandwort 32242
Stiff Sedge 73875,74435
Stiff Sunflower 189028,189030,189044
Stiff Tickseed 104566
Stiff Yellow Flax 231923,231924
Stiff-branched Wirelettuce 375976
Stiffdrupe 354500
Stiffgrass 354400
Stiff-hair Raspberry 338513
Stiff-hair Wheat Grass 144680
Stiff-hair Wheat-grass 144680
Stiff-haired Lotus 237760
Stiffleaf Cheesewood 301248
Stiffleaf Goldenaster 194254
Stiffleaf Grevill 180574
Stiffleaf Juniper 213896
Stiffleaf Meliosma 249449
Stiff-leaf Scratchdaisy 111690
Stiff-leaved Aster 40767
Stiff-leaved Goldenrod 368354
Stiff-leaved Juniper 213896
Stiff-leaved Meliosma 249449
Stiff-leaved Panic Grass 128543,128544,
128545
Stiffpoint Jacob's-Ladder 307219
Stiffstem Flax 231952
Stiff-stem Yellow Flax 231952
Stigmatiferous Elatostema 142845
Stigmatiferous Stairweed 142845
Stigmose Begonia 50328
Stike-pile 153767
Stiking Yew 393047
Stilbanthus 376601
Stillman's Daisy 104593
Stilpnolepis 376650,376651
Stilt Palm 208366
Stilt-palm 208366
Stimpsonia 376682
Stimulant Stonecrop 357180
Stimulose Bamboo 297460
Stimulose Stonecrop 357180
Sting Fishtailpalm 78056
Sting Nettle 402886
Stingdng Nettle 402999
Stinger 125540,402886
Sting-haired Cnestis 97552
Stingiess Nettle 56119
Stinging Cevallia 84434
Stinging Cherry 243529
Stinging Nettle 170078,402886,402890,
403039
Stinging Nettle Tree 268020
Stinging Sinobambusa 364826
Stinging Woodnettle 125553
Stinginghair Cnestis 97552
Stingless Nettle 298848,299024
Sting-scaleless Shortleaf Kyllingia 218488
Stingy Nettle 402886
Stingy-wingies 170025
Stink Arrowwood 407844
Stink Asiabell 98316
Stink Celtis 80767
Stink Corydalis 105907
Stink Crazyweed 278780
Stink Daisy 322724
Stink Davie 359158,384714
Stink Dragon 137744
Stink Fleaweed 321570
Stink Flower 95791,174877
Stink Giantfennel 163731
Stink Grass 147593,147797
Stink Groundpine 70460
Stink Gymnema 182366
Stink Hawksbeard 110808
Stink Lily 168430
Stink Mayweed 26768
Stink New Eargrass 262966
Stink Plant 15861
Stink Tree 12559,407989
Stinkbells 168336
Stinkbush 204517
Stinkcherry 241498,241502
Stinker Bob 174877
Stinkgrass 147593
Stink-grass 147593
Stinking Airach 87200
Stinking Alisander 359158
Stinking Alisanders 359158
Stinking Arag 87200
Stinking Arrach 87200
Stinking Ash 320071
Stinking Benjamin 397556
Stinking Billy 359158
Stinking Blite 87200
Stinking Bob 174877
Stinking Buckeye 9694
Stinking Bush 360463
Stinking Camphorweed 305093
Stinking Cedar 393057,393089
Stinking Chamomile 26768
Stinking Christopher 355061,355202
Stinking Clover 95791,95793,249218
Stinking Cotton 360321
Stinking Cranesbill 174877
Stinking Davies 359158
Stinking Elder 345631,345660,345677
Stinking Elderberry 345660
Stinking Elshander 383874
Stinking Elshinder 359158
Stinking Elshinders 359158
Stinking Fleabane 135050,305093
Stinking Flueggea 167096
Stinking Gladdon 208566
Stinking Gladwin 208566
Stinking Gladwyn 208566
Stinking Goosefoot 87200
Stinking Groundsel 360321
Stinking Hawk's-beard 110808
Stinking Hawksbeard 110808
Stinking Hedge Mustard 14964
Stinking Hellebore 190936
Stinking Iris 208566
Stinking Jenny 15861,174877
Stinking Juniper 213774
Stinking Lily 15861
Stinking Madder 321909
Stinking Mathes 26768
Stinking Mayweed 26768
Stinking Motherwort 87200
Stinking Nancy 379660
Stinking Nanny 26768
Stinking Nightshade 201389
Stinking Onion 15839,15861
Stinking Orach 44644,87200
Stinking Orache 87200
Stinking Passion Flower 285639
Stinking Pea 360463
Stinking Pepperweed 225447
Stinking Pepper-weed 225447
Stinking Rabbitbrush 90502
Stinking Robert 174877
Stinking Roger 47013,174877,201389,
276579,328603,355061,355202,383100
Stinking Segg 208566
Stinking Shumac 12559
Stinking St. John's Wort 201924
Stinking Strawflower 189351
Stinking Tam 271563
Stinking Tarweed 241533
Stinking Tommy 271563
Stinking Trillium 397564
Stinking Tutsan 201924
Stinking Wall Rocket 133286
Stinking Wall-rocket 133286
Stinking Weed 220416,359158,360463
Stinking Willie 359158,383874,397556
Stinking Willow 19994
Stinking Winter Grape 411633
Stinking Yew 393057,393089
Stinkingash 320071
Stinking-ash 320071
Stinking-benjamin 397556
Stinking-Cedar 393047
Stinking-cedar 393089
Stinking-clover 95791
Stinking-cotula 26768
Stinkinggrass 248949
Stinking-marigold 139710
Stinkkest Bladderwort 403828

Stinknet 270993
Stinkroot 123077
Stinkroot Holeglumegrass 57548
Stinkvine 280078
Stinkweed 12559,95791,123077,133286,305083,360463,390213
Stinkwood 103786,182027,211279,268691,417867
Stink-wood 211279
Stinkwort 123077,135050,190936
Stipe Gentian 173723
Stipebract Corydalis 105649
Striped St. John's Wort 202082
Stipedflower Pennywort 200346
Stipefruit Mistletoe 411065
Stipefruit Pepper 300459
Stipe-fruited Mistletoe 411065
Stipeless Rinorea 334684
Stipeless St. John's Wort 202171
Stipespike Leymus 228371
Stipitate Giantarum 20139
Stipitate Goldenray 229134
Stipitate Peashrub 72356
Stipitate Pea-shrub 72356
Stipitate Pricklyash 417346
Stipitategland Senna 78461
Stipulaceous Begonia 50329
Stipulaceous Cherry 83332
Stipulata Dolichandrone 245629
Stipulate Cherry 83332
Stipulate Elatostema 142757
Stipulate Garcinia 171199
Stipulate Ghostfluting 228334
Stipulate Raspberry 338419
Stipulate Rosewood 121831
Stipulate Stairweed 142757
Stipule Clearweed 298935
Stipule Himalaya-honeysuckle 228334
Stipule Raspberry 338419,339307
Stipule Rosewood 121831
Stipule Sedge 76375
Stipule Willow 344151
Stipule-bearing Grease-bush 177366
Stipuled Raspberry 339307
Stipulete Albizzia 13515
Stipulete Caesalpinia 64971
Stipulose Fig 165698
Stirton Primrose 315011
Stitch Grass 374724
Stitch Hyssop 172905
Stitchwort 374724,374916
Stitchwort Sandwort 255520
Stitson 201732
Stixis 377121,377125
Stjame's Wort 359158
Stobar Butterflyorchis 293635
Stobwort 277648
Stochs Iris 208861
Stock 246438,246478,399504
Stock Gilliflower 246478
Stock Gillover 246478
Stock Nut 106703
Stock Rose 370136
Stockings-and-shoes 30081,37015
Stockorange 310848
Stocks Dysophylla 139592
Stocks Eriolaena 152695
Stock-violet-odor Epidendrum 146431
Stoddart Stock 246554
Stoke's Aster 377287
Stokes Aster 377286,377288
Stokes' Aster 77748,377286,377288
Stokesia 377286,377288
Stokoe Leucadendron 227394
Stoks Spapweed 205336
Stoks Touch-me-not 205336
Stoliczka Windhairdaisy 348817
Stoll Norway Maple 3448
Stolon Fluttergrass 166515
Stolon Sandwort 255580
Stolon Stonecrop 357181
Stolonbearing Anemone 24066
Stolonbearing Bentgrass 11972
Stolon-bearing Cypressgrass 119619
Stolon-bearing Fimbristylis 166515
Stolon-bearing Galingale 119619
Stolon-bearing Morningglory 207884
Stolon-bearing Myrsine 261657
Stolonbearing Sagebrush 36321
Stolonbearing Windflower 24066
Stolonbearing Wormwood 36321
Stolonifer Conehead 378229
Stolonifer Pimpinella 299400
Stolonifer Rungia 340376
Stoloniferous Barberry 52191
Stoloniferous Chinese Groundsel 365050
Stoloniferous Dogtongueweed 385919
Stoloniferous Edelweiss 224947
Stoloniferous Germander 388303
Stoloniferous Groundsel 358875
Stoloniferous Pussytoes 26385,26408
Stoloniferous Sanicle 345993
Stoloniferous Shadbush 19295
Stoloniferous Tephroseris 385919
Stoloniferous Thorowax 63613
Stoloniform Firethorn 322482
Stolon-rootstock Iris 208862
Stomach Motherwort 224976
Stone Bamboo 47412,297194
Stone Basil 4666,97060
Stone Beech 162375
Stone Bramble 339240
Stone Clover 396828
Stone Corallodiscus 103944
Stone Cress 9785
Stone Crop 356467,356468
Stone Drepanostachyum 20207
Stone Elsholtzia 259282
Stone Fig 164609
Stone Flower 103940
Stone Leek 15159,15289
Stone Millet 233760
Stone Mint 114520
Stone Minuartwort 255506
Stone Oak 323625
Stone Onion 15159
Stone Orpine 357114
Stone Osier 343937
Stone Parsley 365357,365362
Stone Pine 300151
Stone Plant 233459
Stone Plum 109967
Stone Root 99844
Stone Switch 233760
Stone Tanoak 233343
Stone Windmill 100850
Stonebean Eria 70384
Stonebean Hairorchis 70384
Stonebean-orchis 62530
Stoneberry 339240
Stoneblood 393657
Stonebreak 349419
Stone-break 349419
Stonebutterfly 292547
Stonecayenne 327548,327553
Stonecentiped Skullcap 355748
Stonechestnut 14538,14544
Stonecicada Paperelder 290305
Stoneclave 263958,263960
Stonecress 9816,9823
Stone-Cress 9785
Stonecrop 109421,200775,200789,200802,200804,275347,294117,294124,337292,356467,356468,356537,356633,356989,357114,357169
Stonecrop Family 109505
Stoneface 233459,233673
Stoneface Primrose 314341
Stonefruit Camellia 69260
Stonefruit Craneknee 221751
Stonefruit Pearlsedge 354143
Stonefruit Razorsedge 354143
Stonefruit Stickseed 221751
Stonegarlic 239253,239266
Stonehore 356468
Stone-hore 357114
Stone-hot 357114
Stonejujube 157855
Stonelilac 263958
Stoneliving Javatea 275720
Stoneliving Sage 345362

Stonelotus 364913,364919
Stone-mint 114520
Stonemuscle Coldwaterflower 299017
Stone-ornaments Azalea 331474
Stonepeak Azalea 331772
Stonepeak Rhododendron 331772
Stone-peak Rhododendron 331772
Stoneroot 99844
Stone-root 99844
Stone-rush 354267
Stone-seed Coralbean 154682
Stoneseed Gronwell 233707
Stoneseed Pinyon 299851
Stonespider 299720
Stone-switch 233760
Stonewood 192477
Stonewort 357114
Stonor Stonnord 357114
Stony Bambusa 47318
Stony Giantfennel 163654
Stony Living Aster 40972
Stony Place Dwellers Rhododendron 331710
Stony-hard 233760
Stony-in-the-well 72038
Stool Iris 208447
Stool Wood 18034
Stopper 156118
Storax 232562,379300,379304,379357,
379419
Storax Family 379296
Storehousebush 240842
Storie Renanthera 327697
Stork's Bill 174877,288045
Stork's-bill 153711,153767,288045
Storkbill 288045
Storks Storkbill 174877
Storksbill 153711,153767,174717,174877
Storksbill Wind Flower 23707
Storksbill Windflower 23707
Storshin 399601
Storshiner 399601
Story-of-tile-cross 285623
Story-tellers 313025
Stout Blue-eyed Grass 365745
Stout Blue-eyed-grass 365685,365747
Stout Camphor Tree 91378
Stout Eryngo 154317
Stout Goldenrod 368435
Stout Jointweed 308686
Stout Morningstar Lily 229818
Stout Rush 213323
Stout Smartweed 291924
Stout Woodreed 91236
Stout Wood-reed 91236
Stoutbranch Russianthistle 344728
Stoutroot Leek 15273
Stoutroot Onion 15273
Stover Nut 78811
Stover-nut 78811
Strachey Bergenia 52541
Strachey Corydalis 106477
Strachey Plectranthus 209861
Strachey Winterbottom Draba 137296
Strachey Winterbottom Whitlowgrass 137296
Stracheya 377409
Strachy Edelweiss 224948
Straddleleaf Wallichpalm 413093
Straggler Daisy 68553
Straggling Mariposa 67587
Straggly Corkbark 184607
Straggly Currant 333959
Straggly Gooseberry 333959
Straghtscape Bulbophyllum 63106
Straghtscape Stonebean-orchis 63106
Straight Arrowbamboo 162754
Straight Bluegrass 306009
Straight Carpetweed 256719
Straight Cinquefoil 312912
Straight Cockclawthorn 278321
Straight Coralbean 154722
Straight Coral-bean 154722
Straight Milkvetch 43098
Straight Spapweed 205255
Straight Spine Randia 278321
Straight Teasel 133487
Straight Touch-me-not 205255
Straight Twigs Rhododendron 331414
Straight-cuneate Neomicocalamus 264181
Straightfruit Anemone 23977
Straight-leaf Rush 213344
Straight-leaved Pondweed 312270
Straight-leaved Tanoak 233264
Straightlip Bulbophyllum 62682
Straightlip Cirrhopetalum 62682
Straightlip Curlylip-orchis 62682
Straight-nerve Cherry-leaf
Elaeocarpus 142375
Straightpod Milkvetch 42840
Straight-raceme Barberry 51983
Straightspine Evergreenchinkapin 78999
Straightspine Vinelime 283512
Straight-spined Evergreen Chinkapin 78999
Straightspiny Paramignya 283512
Straight-spiny Vine-lime 283512
Straightspur Larkspur 124438
Straightstalk Alpine Meadowrue 388406
Straightstem Rhodiola 329932
Straight-styled Wood Sedge 75959
Straight-tail Elatostema 142814
Straighttail Stairweed 142814
Straight-tooth Nepeta 265016
Straighttwig Azalea 331414
Straight-twigged Rhododendron 331414
Straight-veined Barberry 52098
Straighty-lobe-form Milkvetch 42840
Strainer Vine 238261
Straithelmet Woodbetony 287489
Straitjackets 253810
Stramineous Monkshood 5602
Stramineous Yushania 416817
Stramonium 123077
Stramony 123077
Strand Cabbage 108627
Strang Fig-tree 164688
Strange Iris 208750
Strange Onion 15585
Strange Violet 410246
Strangle Tare 115008,115031,408423
Strangler Fig 164659,165388,165841
Strangler Vetch 408423
Strangleweed 68713,115008,275135
Stranglewort 117345
Strangling Fig 164659,165762
Stranvaesia 295609,377433,377444
Strap Flower 28093
Strape Cactus 147283
Strap-grass 144661
Strapleaf Bergenia 52532
Strap-leaf Caladium 65237
Strapleaf Photinia 295728
Strap-leaf Sagittaria 342358
Strap-leaved Photinia 295728
Strap-leaved Violet 410166
Strapwort 105410,105416
Strasburg Turpentine 272
Strathmore Weed 299313
Straus Neoporteria 264371
Straw Bonnets 30081
Straw Foxglove 130371
Straw Sedge 76388
Straw Sweetdaisy 71090
Strawa Lycoris 239281
Strawbcrryshrub Family 68300
Strawbed 170743
Strawberry 167592,167593,167597
Strawberry Begonia 349032,349936
Strawberry Bladderwort 403714
Strawberry Blite 86984,87016
Strawberry Bramble 339018
Strawberry Bush 30888,157268,157315,
157429
Strawberry Cactus 140248,140258,140332,
243994,244054,244198
Strawberry Cinquefoil 312561
Strawberry Clover 396882,396905
Strawberry Fields Globe Amaranth 179239
Strawberry Geranium 288503,349419,349936
Strawberry Goosefoot 87016
Straw-berry Grevillea 180582
Strawberry Ground Cherry 297643,297645
Strawberry Groundcherry 297643

Strawberry Ground-cherry 297643,297645
Strawberry Guava 318737
Strawberry Hedgehog Cactus 140209,140242
Strawberry Madrone 30888
Strawberry Myrtle 401346
Strawberry Plant 313025,349936
Strawberry Raspberry 338429,338608
Strawberry Saxifrage 349936
Strawberry Shrub 68306,68313
Strawberry Shrub Family 68300
Strawberry Snowball Tree 135798
Strawberry Spapweed 204964
Strawberry Spinach 86984
Strawberry Tomato 297643,297658,297676,
297712,297716,297720
Strawberry Touch-me-not 204964
Strawberry Tree 30877,30888
Strawberry tree 261212
Strawberry Trefoil 396905
Strawberry-blight 86984
Strawberry-blite 86984,87016
Strawberry-bush 157315
Strawberryflower Azalea 330736
Strawberry-flowered Rhododendron 330736
Strawberryleaf Raspberry 338429
Strawberry-leaved Geranium 349936
Strawberry-leaved Potentil 312549
Strawberry-leaved Potentilla 313025
Strawberrylike Sage 345039
Strawberrylike Woodbetony 287221
Strawberry-raspberry 338608
Strawberryshrub 68306
Strawberry-shrub 68306
Strawberry-Shrub Family 68300
Strawberry-stone-break 349936
Strawberry-tomato 297643,297645,297711,
297716
Strawberrytree 259910
Strawberry-tree 30877
Strawberry-weed 312828
Straw-colooured Gentian 173932
Straw-colored Cyperus 119624
Straw-colored Hairy Bead Grass 285525
Straw-colored Hedgehog 140327
Strawcoloured Corydalis 106482
Straw-coloured Premna 313747
Straw-coloured Setaceous-barct
Fimbristylis 166291
Strawflower 189093,329797,415386
Straw-flower 189814,415386
Strawflower Wild Buckwheat 152135
Straw-lily 403675
Strawsandal Macaranga 240268
Straw-stem Beggar-ticks 53842
Straw-yellow Gentian 173932
Strawyellow Premna 313747
Strawyellow Stonegarlic 239281
Streaked Cranesbill 174997
Streaked Mint Bush 315608
Streakleaf Leek 15779
Streakleaf Onion 15779
Streak-leaved Garlic 266959
Streaky-leaved Garlic 15779
Stream Arundinella 37374
Stream Orchid 147142
Stream Shore Arundinella 37374
Stream Stone Side Rhododendron 331303
Stream Violet 409958,410008
Stream Water-crowfoot 326216
Streambank Bulrush 352284
Stream-bank Sedge 76044
Streambank Wheat Grass 144641
Streambank Wild Rye 144449
Streambank Wild-rye 144449
Streamer Bulbophyllum 62769
Streamer Stonebean-orchis 62769
Streamside Birch 53437
Streamside Corydalis 105957
Streamside Fleabane 150655,150658
Stream-side Rhododendron 331303
Streblus 377537
Street Watling Thistle 154301
Strelitzia 377550,377576
Strelitzia Family 377580
Strength Banksia 47665
Strengthvine 242974,242975
Strength-vine 242974,242975
Streptanthera 377625
Streptocalyx 377646
Streptocaulon 377844
Streptolirion 377877
Stretched Brassaiopsis 59237
Striate Acanthophippium 2080
Striate Agrimony 11598
Striate Butterflybush 62171
Striate Cleisostoma 94472
Striate Cleistogenes 94635
Striate Closedspurorchis 94472
Striate Emptygrass 98800
Striate Itchgrass 98800
Striate Kummerowia 218347
Striate Longpetal Parnassia 284564
Striate Loosestrife 239661
Striate Milkvetch 43096
Striate Mnesithea 98800
Striate Nothoscordum 266959
Striate Ormosia 274438
Striate Saddleorchis 2080
Striate Shining Umbrella Bamboo 162722
Striate Sinobambusa 364815
Striate Skullcap 355777
Striate Treebine 92977
Striate Violet 410602
Striated Broom 121015
Striated Gentian 173935
Striateflower Sterculia 376191
Striatefruit Corydalis 106487
Striatefruit Maesa 241839
Striate-fruited Maesa 241839
Striatepetal Knight's Star 196458
Striate-sheath Tangbamboo 364815
Striatod Catchfly 363363
Strickland's Gum 155752
Strict Beefwood 15953
Strict Blue-eyed Grass 365782,365784
Strict Borreria 370847
Strict Canarium 71023
Strict Conyza 103612
Strict Corydalis 106488
Strict Creeping Woodsorrel 278099
Strict Diptychocarpus 133641
Strict Eulophia 157035
Strict Evening Primrose 269509
Strict Eyebright 160268
Strict Falsepimpernel 231578
Strict Ferrocalamus 163570
Strict Fescuegrass 164333
Strict Fig 165703
Strict Flemingia 166895
Strict Fordiophyton 167317
Strict Forget-me-not 260885
Strict Garnotia 171520
Strict Globular Spike Pycreus 322244
Strict Goniolimon 179381
Strict Goosecomb 144482
Strict Goosefoot 87171
Strict Iron Bamboo 163570
Strict Ixora 211067
Strict Kudzuvine 321464
Strict Larkspur 124182
Strict Lavender 223340
Strict Lettuce 219547
Strict Leucas 227713
Strict Milkvetch 43098
Strict Mouseear Cress 30141
Strict Platanthera 302527
Strict Roegneria 144482
Strict Slender Woodbetony 287249
Strict Sopubia 369240
Strict Thorowax 63718
Strict Wild Buckwheat 152490
Strict Windhairdaisy 348818
Strictflower Betony 373449
Strictgalea Woodbetony 287489
Striga 377969
Strigila Snowbell 379450
Strigillose Argyreia 32661
Strigillose Begonia 50331
Strigillose Eurya 160606
Strigillose Padritree 376291
Strigillose Stairweed 142851

Strigilose Holly 204279
Strigose Acanthonema 2325
Strigose Actinidia 6704
Strigose Alangium 13355
Strigose Anemone Clematis 95128
Strigose Aralia 30793
Strigose Argyreia 32661
Strigose Arrowbamboo 162755
Strigose Differentfruit Tickclover 126395
Strigose Drymonia 138529
Strigose Euchresta 155892
Strigose Forkstyleflower 374497
Strigose Heterocarpus Mountain Leech 126395
Strigose Hydrangea 200108
Strigose Jerusalemsage 295201
Strigose Kiwifruit 6704
Strigose Onosma 271833
Strigose Poppy 282734
Strigose Ramie 56238
Strigose Raspberry 339310
Strigose Rhododendron 331896
Strigose Rockfoil 349947
Strigose Rockjasmine 23305
Strigose Sicklegrass 283659
Strigose Skullcap 355778
Strigose Small Bugbane 91024
Strigose Triangular Buttercup 326471
Strigose Vineclethra 95550
Strigosecalyx Padritree 376291
Strigose-flower Falsenettle 56325
Strigulose Nematanthus 263329
Strike 231183
Strike-fire 407072
Strindberg Knotweed 309836
String Bean 294056
String of Buttons 109266
String of Hearts 84305
String of Pearls 216637,359924
String Sedge 74118
Stringbark 155468
Stringbush 414118
Stringbushlike Rabdosia 209868
Stringbush-like Rabdosia 209868
String-lily 111152
String-of-beads Plant 359924
String-of-beads Senecio 359924
String-of-hearts 84305
Strings-of-sovereigns 239755
Stringy Gum 155719
Stringy Stonecrop 357123
Stringybark 155468,155475
Stringy-bark 155672,155761
Stringybark Cypress Pine 67422
Striolate Ginger 418023
Striped Abutilon 997
Striped Adobe-lily 168568
Striped Bamboo 303994
Striped Barbados Lily 196456
Striped Bulrush 352279
Striped Coralroot 104016,104019
Striped Coral-root 104016
Striped Corn 417417
Striped Corn Catchfly 363363
Striped Crotalaria 112491
Striped Dogwood 3393
Striped Dracaena 137367
Striped Gentian 174059
Striped Gomphia 70750
Striped Goosefoot 87171
Striped Hawksbeard 111096
Striped Hemlock 256741
Striped Himalayan Bamboo 196347
Striped Inch Plant 67144
Striped Japanese Silver-grass 255891
Striped Maple 3393,3549
Striped Mexican Marigold 383107
Striped Milk Thistle 364360
Striped Milky Thistle 364360
Striped Pipsissewa 87488
Striped Squill 321894,321896
Striped Toadflax 231105
Striped Torch 182177
Striped Violet 410602
Striped White Violet 410602
Striped Wintergreen 87488
Striped Yam 131908
Striped-edge Agave 10788
Stripedleaf Baboon Flower 46148
Stripedleaf Stringbush 414210
Stripe-leaved Bamboo 175611
Stripestem Fernleaf Bamboo 47351
Strip-for-strip 90901
Strip-jack-naked 99297
Stripped Fargesia 162677
Stripped Fieldjasmine 199453
Strirking Wallrocket 133286
Strobile Crazyweed 278954
Strobile Geniosporum 172875
Strobile Meriandra 250660
Strobiloides Bulrush 353828
Strobilus Elsholtzia 144099
Stroganowia 378287
Stroll 144661
Stromhosia 378319
Strong Angelica 24502
Strong Aromatic Azalea 330092
Strong Azalea 331182
Strong Barberry 52297
Strong Bulrush 352278
Strong Fordiophyton 167317
Strong Fragrance Rhododendron 330092
Strong Fragrant Rhododendron 330092
Strong Gentian 173822
Strong Greenbamboo 125459
Strong Machilus 240689
Strong Meadowrue 388644
Strong Monbeig Dogwood 380468
Strong Pearweed 239823
Strong Small Iris 208769
Strong Swertia 380230
Strong White Fir 365
Strong Windhairdaisy 348728
Strong Woodbetony 287637
Strongbark 57903
Strong-branched Barberry 52297
Strongfragrant Spicegrass 239640
Strong-scented Cranesbill 174877
Strongscented Garlic 266965
Strong-scented Glorybower 95978
Strong-scented Lettuce 219609
Strong-sepal Barberry 52298
Strong-sepaled Barberry 52298
Strong-smelling Casearia 78123
Strong-smelling Currant 334006
Strong-spined Medick 247493
Stroped Hakea 184645
Strophanthus 378363
Strophioblachia 378473
Stroyl 144661
Strumose Sunflower 189063
Strut Knotweed 309841
Strychnine 378836
Strychnine Tree 378836
Strychnos 378633,378836
Stuart Camellia 69669
Stuart Phalaenopsis 293636
Stuart Primrose 315019
Stub Apple 243711
Stub-apple 243711
Stubbleberry 367416
Stubwort 277648
Stuck Hairs Rhododendron 331479
Stuebel Barberry 52192
Stuffed Flower Bamboo 297361
Stuhlmann Lannea 221209
Stunted Sedge 75281
Sturdy 235373
Sturdy Cuttonguetree 413129
Sturdy Cypress-pine 67433
Sturdy Lowries 122492
Sturdy Sagebrush 36147
Sturt Desert Pea 380066
Sturt Desert Rose 179929
Sturt's Desert Pea 96637,380066
Sturt's Desert Rose 179929
Sturt's Desert-pea 96637
Sturt's Desert-rose 179929
Sturt's Pigface 181970
Stwnostachyous Barberry 52187
Stycky Swinglea 380539

Styckyfruit Conehead 320113
Stylate Elaeagnus 142200
Style Ash 168112
Style Benchfruit 279748
Style Pachysandra 279748
Stylecorolla Seseli 361475
Styled Bittersweet 80328
Stylefruit Clematis 95396
Stylefruit Meconopsis 247155
Stylegrass 379076
Stylegrass Family 379071
Styleless Burreed 370071
Styleless Mahonia 242608
Style-like Draba 137248
Style-like Whitlowgrass 137248
Stylepersistent Ash 168112
Stylewort 379076
Stylidium 379076
Stylidium Family 379071
Stylisma 379090,379091
Stylo 379237,379245
Stylophorum 379219
Stylosanthes 379237
Stylose Mangrove 329771
Stylose Oblongleaf Eurya 160563
Stylose Staff-tree 80328
Styloselike Monkshood 5604
Styltcress 254082
Styptic Weed 360463
Styrax 379300,379419
Styrophyton 379480
Suakwa Towelgourd 238261
Suakwa Vegetable Sponge 238261
Suakwa Vegetablesponge 238261
Suari 77942
Suaveolent Mint 250457
Suaveolent Paracryphia 283251
Subacuminate Barberry 52193
Subacuminate Evergreen Chinkapin 79032
Subacuminate Evergreenchinkapin 79032
Subaequifoliate Rockfoil 349952
Subakh 386448
Subal Tern's Butter 291494
Subalpine Arrowwood 408146
Subalpine Aster 160671
Subalpine Breaks Goldenbush 150382
Subalpine Buttercup 325800
Subalpine Coldwaterflower 299032
Subalpine Dapaoshan Stonecrop 356709
Subalpine Fir 288,402
Sub-alpine Fir 402
Subalpine Fleabane 150662
Subalpine Larch 221913
Subalpine Larkspur 124054
Subalpine Sulphur Flower 152591
Subalpine Viburnum 408146
Subaltern's Butter 291494
Subamplexicauline Rockfoil 349954
Subangular Momordica 256884
Subantarctic Barberry 52193,52194
Subarctic Dock 340275
Subarctic Willow Dock 340275
Subaristate Bentgrass 12356
Subarticulate Persimmon 132422
Subcaespitose Skullcap 355782
Subcalvous Yam 131857
Subcapitate Amomum 19921
Subcapitate Aralia 30768
Subcapitate Crazyweed 279184
Subcapitate Draba 137255
Subcapitate Dragonhead 137670
Subcapitate Premna 313749
Subcaulialate Mrs. E. H. Wilson's Barberry 52371
Subclimbing Premna 313751
Subclover 397106
Subcompressed Orange 93731
Subconcavepetal Parnassia 284620
Subconvex Broadflower Dendrocalamus 125484
Subcordate Cordia 104251
Subcordate Ermans Birch 53426
Subcordate Eurya 160607
Subcordate Sida 362668
Subcordate Spapweed 205340
Subcordate Touch-me-not 205340
Subcordate Water-plantain 14785
Subcordateleaf Euonymus 157898
Subcordateleaf Heteropanax 193905
Subcoriaceous Barberry 52029
Subcoriaceous Chinese Groundsel 365075
Subcoriaceous Embelia 144812
Subcoriaceous Holly 204283
Subcoriaceous Paraphlomis 283652
Subcoriaceous-leaf Embelia 144812
Subcostate Crapemyrtle 219970
Subcostate Crape-myrtle 219970
Subcrenate Gugertree 350935
Subcrenate Holly 204284
Subdichotomous Iris 208866
Subdigitate Cinquefoil 313034
Subdigitate Wormwood 36340
Subei Milkvetch 42668
Subentire Camellia 69675
Subentire Eurya 160608
Subentire Pinanga 299677
Subentire Serralateleaf Willow 344098
Subentire Skullcap 355784
Subentire-leaf Thinleaf Clethra 96485
Subentire-leaved Barberry 52199
Subequal Gironniera 175920
Subequal Honeysuckle 236128
Subequel Bamboo 47462
Suberect Barberry 52198
Suberect Spatholobus 370422
Suberleaf Kiwifruit 6706
Suberose Creeper 285151
Suberose Elm 401633
Suberous Greenstar 307531
Suberous Mundulea 259854
Subfalcate Blueberry 404019
Subfalcate Clematis 95332
Subfalcateleaf Clematis 95332
Subfalcate-leaved Clematis 95332
Subfalcate-leaved Garcinia 171203
Sub-fern-like Kobresia 217167
Subflatted Haworthia 186747
Subfragrant Holly 204287
Subgibbosous Neoporteria 264372
Subglabros Bitter Bamboo 304039
Subglabrous Cacalia 283871
Subglabrous Camellia 69673
Subglabrous Longestapex Beautyberry 66854
Subglabrous Longestapex Purplepearl 66854
Subglabrous Pearlbloomtree 307292
Subglabrous Reddish Beautyberry 66922
Subglabrous Reddish Purplepearl 66922
Subglabrous Uvaria 403587
Subglabrousentire Woodnettle 273909
Subglabrousfruit Trichophore Larkspur 124649
Subglobose Asiabell 98413
Subhastate Nepeta 265063
Subhispid Honeysuckle 236130
Subinbricated Mahonia 242644
Subkorean Raspberry 339317
Sublabiate Honeysuckle 236131
Sublaceoleaf Wight Fig 165844
Subladder-veins Blueberry 404018
Subleather-leaved Elatostema 142854
Sublethorleaf Orixa 274282
Sublobate Hewittia 194467
Sublobed Hewittia 194467
Sublobed Spiraea 372093
Sublongicaudate Holly 204286
Sublongtail Holly 204286
Submarginate Buttercup 326405
Submembranaceous Microtropis 254328
Submerged Ottelia 277386
Subnaviculare Monkshood 5212
Suboblong Smalllobed Clematis 95210
Subochre-yellow Mountain-ash 369527
Subopposite Milkwort 308392
Suborbicular Chiogenes 87679
Subpalatenerve Sage 345415
Subpeltate Stephania 375901
Subpeltate-leaf Pepper 313178
Subpetiolar Sedge 75055
Subpinnate Anemone 24070
Subplicate Twayblade 232342
Subprostrate Pagodatree 369141

Subprostrate Sophora 369141
Subpsilate Barberry 52201
Subpubescent Yam 131858
Subrepent Adenosma 7992
Subrhombicleaf Chirita 87976
Subrotate Aspidistra 39577
Subroundleaf Ternstroemia 386712
Subrugged Holly 204290
Subruminate Pinanga 299678
Subrusty Creeper 92919,285152
Subscapes Parnassia 284621
Subscapose Asiabell 98414
Subscapose Glorybower 96370
Subscapose Knotweed 309839
Subserrate Chinese Sweetspire 210374
Subsessile Euonymus 157899
Subsessile Ixora 211181
Subsessile Monkshood 5564
Subsessile St. John's Wort 202171
Subsessile St. John's-wort 202171
Subsessile-flower Feathergrass 376924
Subsessileflower Needlegrass 376924
Subsessile-flowered Barberry 52210
Subsessileleaf Turpinia 400545
Subsessile-leaved Turpinia 400545
Subsessiliflorous Rockfoil 349960
Subshrub Dragon Mallow 367651
Subshrub Indigo 206626
Subshrub Nightshade 367651
Subshrubby Edelweiss 224958
Subshrubby Peony 280286
Subsimplex Sage 345094
Subsoja 272360
Subsolid Coldbamboo 172749
Subsolid Gelidocalamus 172749
Subsolid Pseudosasa 318344
Subspike Goldenray 229212
Subspurless Spapweed 205341
Subspurless Touch-me-not 205341
Subsulcate Hakea 184643
Subterminal Club-rush 353837
Subternate Rockfoil 349966
Subterranean Clover 397106
Subtle Orach 44569
Subtomentose Cowparsnip 192383
Subtrichotomous Stairweed 142859
Subtrigonosperma Evodia 161380
Subtrinerved Mahonia 242542
Subtrinerved Spindle-tree 157904
Subtriplinerved Mahonia 242542
Subtrisect Threeleaf Chastetree 411471
Subtropical Lady's Slipper 120457
Subtropical Ladyslipper 120457
Subtruncate Bamboo 47463
Subtruncate Stone Bamboo 47463
Subturbinate Eritrichium 153507
Subularia 379650
Subulate Canarytree 71024
Subulate Cherry 83175,316346
Subulate Evergreenchinkapin 79034
Subulate False Incised Monkshood 5578
Subulate Gland Cherry 83175,316346
Subulate Mudweed 230833
Subulate Thistle 92415
Subulatetooth Woodbetony 287717
Subumbel Chickweed 375107
Subumbellate Bauhinia 49189
Subumbellate Narrow-petal Hydrangea 199939
Subundulate Pellionia 288771
Subverticillate Daphniphyllum 122726
Subverticillate Tigernanmu 122726
Subvesper Iris 208866
Subvillose Begonia 50340
Subvillous Panic-grass 128478
Subvirescent Box-thorn Barberry 51894
Succamore 3462
Succatespur Corydalis 105647
Succisella 379662
Succor Creek Mugwort 36043
Succory 90889,90901
Succulent Hawthorn 110065
Succulent Spiderwort 394013
Succulent Stalk Slugwood 50555
Succulentleaf Chirita 87836
Succulentleaf Garcinia 171204
Succulentleaf Honeysuckle 235714
Succulentleaf Windhairdaisy 348857
Succulentstem Pepper 300485
Suchow Mosla 259324
Suchow Willow 344161
Suchur Cinquefoil 312964
Suck Bottle 397019
Suck-bottle 220346,397019
Sucker Flax 231923
Sucker Windmillpalm 393810
Suckering Australian Pine 79165
Suckering Australian-pine 79165
Suckers 381035
Suckery 90901
Suckle-bush 236022
Sucklers 397019,397043
Suckles 236022,287494,397019
Suckley's Orach 44666
Suckling Clover 396882
Sucklings 236022,397019,397043
Sucks 397019
Sucky Calves 37015
Sucky Sue 220346
Sucupira 57968,133358,380101
Sud Grass 297181
Sudan Cola Nut 99158
Sudan Cola-nut 99158
Sudan Crowfoot Grass 121285
Sudan Ebony 121750
Sudan Gaoliang 369707
Sudan Grass 369600,369707
Sudan Gum Arabic Tree 1572
Sudan Gum-arabic 1572
Sudan Teak 104148,104218
Sudanese Tea 195149,195196
Sudangrass 369707
Sudan-grass 369707
Sudden Corydalis 106014
Sudetic Pedicularis 287718
Sudetic Wood Rush 238706
Suds 346434
Suffrutescent Everlasting 21704
Suffrutescent Pearleverlasting 21704
Suffrutescent Sandwort 32264
Suffrutescent Securinega 167092
Suffruticosa Moltkia 256753
Suffruticosa Peony 280286
Suffruticose Senna 360490
Suffurutescent Serpentroot 354906
Sugar Aapple 25882
Sugar Apple 25828,25898
Sugar Basin 374916
Sugar Basins 374916
Sugar Bean 294010
Sugar Beet 53249,53255,53266
Sugar Berry 80569
Sugar Bush 315702,315718,315937,332751
Sugar Candy 336419
Sugar Cane 341887
Sugar Codlins 146724
Sugar Corn 417433
Sugar Date 295487
Sugar Date-palm 295487
Sugar Grass 90107,156442,177599,307355
Sugar Gum 155531,155542
Sugar Hackberry 80673
Sugar Hawthorn 109573,110050
Sugar Maple 3549,3576
Sugar Palm 32330,32343
Sugar Pea 301070,301080
Sugar Pine 300011
Sugar Plant 376415
Sugar Plum 19240,19243,397019,401233
Sugar Protea 315937
Sugar Sorghum 369700
Sugar Sumac 332751
Sugar Sumach 332751
Sugarapple 25898
Sugar-apple 25882,25898
Sugarberry 80569,80673,80698,80728
Sugar-bosses 397019
Sugarbush 315937,332751
Sugar-busses 397019
Sugarcane 341829,341887
Sugarcane Plume Grass 148877

Sugarcane Plumegrass 148877
Sugarcaneflower Silverreed 394943
Sugarcane-yellow Azalea 331862
Sugared-almond Plum 279639
Sugar-grass 307355
Sugargum 155531
Sugarmustard 154363
Sugarpalm 32330,32343
Sugarplum Tree 220226
Sugarplum-tree 220226
Sugar-wood 260721
Sugerok Holly 204292,204299
Sugi 113685
Sugong Elatostema 142862
Suhui Orange 93722,93815
Suhulate Deutzia 127112
Suichang Bamboo 297398
Suichang Holly 204303
Suichang Spapweed 205342
Suichang Touch-me-not 205342
Suichuan Indocalamus 206832
Suiding Jurinea 214182
Suijiang Barberry 51773
Suijiang Spapweed 205343
Suijiang Yushanbamboo 416818
Suijiang Yushania 416818
Suisha Panicgrass 281533
Suisun Goldenbush 209567
Suisun Marsh Aster 380925
Suisun Thistle 92042
Suitable Rhododendron 331944
Suiyang Arisaema 33474
Sukaczew Birch 53619
Sukhaku Paphiopedilum 282900
Suksdorf Sagewort 36352
Suksdorf's Catchfly 364090
Sulcate Begonia 50344
Sulcate Clovershrub 70898
Sulcate Dendrobium 125375
Sulcate Fescuegrass Rocky 164295
Sulcate Sheep-fescuegrass 164144
Sulcate Sinobambusa 364817
Sulcate Spapweed 205344
Sulcate Tea Arrowwood 408128
Sulcate Tea Viburnum 408128
Sulcate Touch-me-not 205344
Sulcatefruit Flatsedge 322371
Sulcatefruit Galingale 322371
Sulcate-fruit Pycreus 322371
Sulcatestem Clovershrub 70898
Sulcate-stem Clovershrub 70898
Sulcatestem Clubfilment 179206
Sulcate-stem Common Barberry 52344
Sulcatestem Gomphostemma 179206
Sulcatestem Milkvetch 43112
Sulfur Crazyweed 279192
Sulfur Paintbrush 79136
Sulfur Sweetvetch 188137
Suling Armodorum 34578
Sulla 187842
Sulla Sweet Vetch 187842
Sulla Sweet-Vetch 187842
Sullen Lady 168476
Sullen Milkvetch 43063
Sullivant's Coneflower 339547
Sullivant's Cool-wort 379733
Sullivant's Milkweed 38132
Sullivantia 379733
Sulphucoloured Spapweed 205376
Sulphucoloured Touch-me-not 205376
Sulphur Azalea 331908
Sulphur Bamboo 297464
Sulphur Cinquefoil 312912
Sulphur Clover 396989
Sulphur Cosmos 107172
Sulphur Five-fingers 312912
Sulphur Flower 152550
Sulphur Hot Springs Wild Buckwheat 151840
Sulphur Knapweed 81410
Sulphur Lily 230043
Sulphur Plant 152550
Sulphur Rhododendron 331908
Sulphur Root 292957
Sulphur Rose 336623
Sulphur Striga 378042
Sulphur-colored Epimedium 147074
Sulphur-colored Sicilian Thistle 81410
Sulphur-coloured Rhododendron 331908
Sulphur-flower Buckwheat 152550
Sulphurweed 292957
Sulphurwort 269358,292957,363123
Sultan Flower 81116
Sultan Spapweed 205346,205444
Sultan Touch-me-not 205444
Sumac 332452,332477,332533,332646,
332916
Sumac Family 21190
Sumach 315101,332533
Sumach Family 21190
Sumac-leaved Meliosma 249445
Sumark 315101,397019
Sumatra Azalea 330159
Sumatra Camphor 138548
Sumatra Cleistanthus 94539
Sumatra Conyza 103617
Sumatra Cratoxylum 110283
Sumatra Crinum 111157
Sumatra Devilpepper 327061
Sumatra Devil-pepper 327061
Sumatra Fleabane 103617
Sumatra Mahonia 242647
Sumatra Mezoneuron 252758
Sumatra Morningglory 208224
Sumatra Oxwood 110283
Sumatra Pearlsedge 354253
Sumatra Pentastemona 290061
Sumatra Phalaenopsis 293637
Sumatra Pine 300068
Sumatra Rhododendron 330159
Sumatra Sealing-wax Palm 120782
Sumatra Siraitia 365335
Sumatra Snowbell 379311,379312
Sumatra Storax 379423
Sumatra Sugarpalm 32342
Sumatra Vanda 404674
Sumatra Yellow-wood Barberry 52387
Sumatra-dragonsblood 121493
Sumatran Eightangle 204595
Sumatran Illicium 204595
Sumatran Pearlsedge 354253
Sumbaviopsis 379747
Sumbul 136257,136259,163681
Sumch 332452
Summa Sage 345419
Summer Adonis 8325
Summer and Autumn Pumpkins 114300
Summer Chrysanthemum 89481
Summer Citrus 93577
Summer Coralroot 103990,104000
Summer Cypres 217361
Summer Cypress 48785,217324,217361
Summer Daisy 188757,227533
Summer Fimbristylis 166164
Summer Fimbry 166164
Summer Fluttergrass 166164
Summer Forget-me-not 21933
Summer Gastrodia 171934
Summer Gayfeather 228437
Summer Glow Tamarix 383598
Summer Grape 411531,411532,411551,
411578
Summer Hat 410677
Summer Hats 410677
Summer Haw 109850
Summer Hawthorn 109709
Summer Heather 67455,67499
Summer Holly 31113,100250
Summer Hyacinth 170827,170828
Summer Jasmine 211940
Summer Lady's Tresses 372178
Summer Lilac 61953,62019,62035,193417
Summer Lilic 62000,62019
Summer Orange 93340
Summer Perennial Phlox 295282,295288
Summer Pheasant's Eye 8325
Summer Pheasant's Eyes 8325
Summer Phlox 295288
Summer Poinsettia 18836
Summer Poppy 215381,282685
Summer Radish 326616
Summer Ragwort 229016

Summer Red Azalea 389577
Summer Rose 216085
Summer Saucers 363673
Summer Savory 347506,347560,347638
Summer Scented Wattle 1541
Summer Sedge 73610
Summer Snapdragon 24522
Summer Snow 305194
Summer Snowdrop 227856
Summer Snowflake 227856,227857
Summer Snowflakelotus 227856
Summer Squash 114300,114303
Summer Starwort 151102
Summer Sweet 96454,96457
Summer Torch 54456
Summer Tree 3462
Summer Vetch 408284
Summer Violet 410493
Summer's Bride 66445
Summer's Darling 94133
Summer's Farewell 166125,359158,398134
Summer's Goodbye 284591
Summer-cypres 217361
Summercypress 217324
Summer-cypress 48753,48767,217324,
217373
Summer-flower Caesalpinia 64966
Summergrass 16222
Summer-hyacinth 170827,170828
Summer-lilac 61953,62019
Summerlilacleaf Arrowwood 407713
Summerlilaclike Honeysuckle 235688
Summerlilic 61953
Summersweet 96454,96457
Summer-sweet 96454,96459,96460,96461,
96462
Summersweet Clethra 96457
Summit Cedar 44048,44049
Sumntia Sweetleaf 381423
Sump Weed 210462,210468
Sumpweed 210462,210478
Sumushan Polemonium 307244
Sun Berry 297703,338059,367558
Sun Cactus 190239
Sun Cress 190261
Sun Daisy 188757,227533
Sun Dew 138261,138329
Sun Dial Lupine 238473
Sun Drops 68505
Sun Euphorbia 159027
Sun Flower 188902,188908
Sun Flower Flat Wild Buckwheat 152007
Sun Hemp 112269
Sun Melocactus 249599
Sun Plant 311852
Sun Rose 188583,188757
Sun Sedge 74887
Sun Spurge 159027
Sun Star 274595
Sun's Bride 66445
Sun's Eye 188908
Sun's Eyes 188908
Sun's Herb 66445
Sun's Ray 188908
Sun's Rays 188908
Sun's Spapweed 205348
Sunberry 297703
Sunbonnet 85989,262441
Sunbright 294330
Sunburst 317361
Suncup 269414,269447
Sunda Mischocarpus 255954
Sunday Whites 374916
Sundcorns 349419
Sundermann Pink 127851
Sundew 138261,138328,138329,138348,
337217
Sundew Family 138369
Sundial Lupine 238473
Sundrop 68506
Sundrops 269394,269438,269461,269479,
269486,269489,269519
Sunfacing Coneflower 339554
Sunflower 21339,21427,66445,89704,
188589,188757,188902,188908,189017,
189057,207135,274823
Sunflower Artichoke 189073
Sunflower Cremanthodium 110397
Sunflower Everlasting 190502
Sunflower Goldeneye 409155
Sunflower Heliopsis 190505
Sunflower Nutantdaisy 110397
Sunflower-everlasting 190505,190525
Sunflowerweed 392428
Sun-following Spurge 159027
Sungari Clematis 95319
Sungari Redbead Cotoneaster 107688
Sungjiang Willow 344163
Sunglory 190502
Sungming Camellia 69678
Sungpan Milkvetch 43113
Sungpan Monkshood 5612
Sungpanensis Soulie Rose 336967
Sungreen 358062
Sun-hemp 112269
Sunipia 379768,379784
Sunken Sedge 74882
Sunkenvein Ardisia 31374
Sunkenvein Aucuba 44891
Sunkenvein Blueberry 403861
Sunkenvein Camellia 69147
Sunkenvein Ching Arrowwood 407746
Sunkenvein Ching Viburnum 407746
Sunkenvein Erycibe 154237
Sunkenvein Eurya 160434
Sunkenvein Greenbrier 366417
Sunkenvein Maesa 241752
Sunkenvein Schefflera 350774
Sunkenvein Sterculia 376125
Sunken-veined Ardisia 31374
Sunken-veined Camellia 69147
Sunken-veined Eurya 160434
Sunken-veined Schefflera 350774
Sunken-veined Sterculia 376125
Sunki 93717
Sun-lotus 267714
Sun-loving Tanoak 233111
Sunmoon Panicgrass 281533
Sunn Crotalaria 112269
Sunn Hemp 112269
Sunn Rattlebox 112269
Sunnan-corn 233760
Sunningclothes Tangbamboo 364800
Sunny Catchfly 248269
Sunny Melandrium 248269
Sunny Tanoak 233111
Sunnybells 352151
Sunpitcher 188554
Sunray 145204,145209,190791
Sunrise Abelia 133
Sunrose 93120,188583,188757
Sun-Rose 188583
Sunroseleaf Leucas 227608
Sunscape Daisy 276599
Sunset 223355
Sunset Abelmoschus 229
Sunset Hibiscus 229
Sunset showtime 219673
Sunset-tree 56788
Sunshade 102933,188908,401711
Sunshine Rose 336361
Sunshine Tree 115784
Sunshine Wattle 1458,1646
Sunshine-tree 360490
Suntara 93717
Sunward Gutzlaffia 182126
Sunward Melandrium 248269
Sunwheel 63524,385616
Sunyi Maple 3647
Sunyi Premna 313753
Suobaila Rockfoil 349291
Suolun Vetch 408406
Suoluo 327357
Suoxian Corydalis 106485
Suoxiyu David Stranvaesia 377450
Supa Storax 379464
Superb Beardtongue 289379
Superb Cape-cowslip 218922
Superb Eremostachys 148495
Superb Fairy Lantern 67630
Superb Lily 230044

Superb Penstemon 289379
Superb Pink 127852
Superb Statice 230784
Superior Pink Deutzia 127066
Superior Rockfoil 349286
Super-spathe Fargesia 162734
Superstition Mallow 967
Supine Bulrush 353845
Supine Chickweed 374826
Supine Skullcap 355786
Supine Vervain 405893
Supple Fleabane 150999
Supple Jack 52436,329677
Supple Jack Vine 329677
Supple-hair Actinidia 6685
Supplejack 52400,52466,360936
Supple-jack 52400,52436
Supple-Jack 52466
Suppy Orange 93732
Supraglandulose Slugwood 50612
Supra-vein Leaves Blueberry 404020
Surat Hibiscus 195257
Surat Senna 360490
Surattense Nightshade 367735
Suregada 379792
Suren Toona 392829,392847
Surette 64425
Surface Twitch 308816
Surf-grass 297188
Surfgrass Surf-grass 297181
Suriana 379812
Surinam Calliandra 66686
Surinam Cherry 156385
Surinam Dwarf Bamboo 47309
Surinam Nutmeg 261461
Surinam Ouassia 364382
Surinam Quassia Wood 364382
Surinamcherry 156118
Surinam-cherry 156385
Surinamese Stickpea 66686
Suronded Devil Rattan 121521
Surprise Corydalis 106160
Surrounded Swertia 380160
Surry 369541
Suruga-yuko 93531
Susann Pecteilis 286836
Susanna Stonecrop 357194
Susin Cymbidium 116843
Susquehana Sand Cherry 316740
Susquehanna Viburnum 408101
Sussex Weed 324335
Sussex Yellow-sorrel 277803,278099
Susumber 367682
Sutchuen Ailanthus 12563
Sutchuen Fir 356
Suterberry 417157
Sutherland Begonia 50347
Sutherland's Larkspur 124612
Suthywood 35088
Sutured Pittosporum 301367
Sutured Seatung 301367
Suwanee Carolina Silverbell 184735
Suworow Statice 320002
Suzanne's Spurge 159916
Suzhou Mosla 259324
Suzuk Holly 204306
Suzuki Acuminate Eurya 160535
Suzuki Raspberry 338433
Suzuki Thistle 92419
Suzuki Toadlily 396617
Suzukia 380044
Svenhedin Lipfernleaf Woodbetony 287086
SW. China Amitostigma 19533
SW. China Astyleorchis 19533
SW. China Bladderwort 403880
SW. China Buttercup 325832
SW. China Calanthe 65977
SW. China Camellia 69494
SW. China Cattailtree 245629
SW. China Clematis 95258
SW. China Cupgrass 151675
SW. China Daylily 191283
SW. China Diphylax 132671
SW. China Dualsheathorchis 85451
SW. China Ehretia 141618
SW. China Elsholtzia 144091
SW. China Groundsel 359816
SW. China Honeysuckle 235681
SW. China Indianmulberry 258917
SW. China Indigo 206270
SW. China Milkwort 308012
SW. China Oreorchis 274034
SW. China Reinorchis 182262
SW. China Rockfoil 349906
SW. China Rush 213166
SW. China Sagebrush 36297
SW. China Sedge 73836
SW. China Sinosenecio 365036
SW. China Strawberry 167629
SW. China Ungeargrass 36730
SW. China Vetch 408512
SW. China Wildorchis 274034
SW. Sichuan Rockfoil 349957
Swainson Pea 380065
Swainsona 371160
Swainsonpea 371161
Swainson-pea 380065
Swallow Dendrobium 125121
Swallow Grundy 360328
Swallow Pear 369541
Swallow Swordflag 208672
Swallow Wort 37888,86755,117334
Swallow-grass 86755
Swallow-herb 86755
Swallow-pear 369541
Swallowtail Pinangapalm 299676
Swallow-tailed Willow 342973
Swallowwort 117334,117351,117398,117399,117411,117415,117476,117488,117597,117690,117745
Swallow-wort 37888,38135,38147,68095,86755,117334,117597,117743,325820
Swallow-wort Gentian 173264
Swallowwort Heshouwu 162523
Swamp Agrimony 11571
Swamp Alder 16375
Swamp Ash 168023,168032,168057
Swamp Aster 380970
Swamp Azalea 332086
Swamp Baeckea 46436
Swamp Banksia 47649,47665
Swamp Bay 242341
Swamp Bayberry 261169
Swamp beaked orchid 375684
Swamp Bedstraw 170284,170717
Swamp Beggar Ticks 54158
Swamp Beggarticks 53845
Swamp Beggar-ticks 53858,53879
Swamp Betony 287335
Swamp Birch 53338,53594
Swamp Black Gum 267872
Swamp Blackberry 338190,338528,339248
Swamp Bladderwort 403377
Swamp Bloodwood 106820
Swamp Bluebell 69899
Swamp Blueberry 403780
Swamp Bottlebrush 49349
Swamp Bottle-brush 49349
Swamp Box 236465
Swamp Buttercup 325942,326351
Swamp Candle 352151
Swamp Candleberry 261202
Swamp Candles 239879
Swamp Cedar 85375
Swamp Chestnut Oak 324166
Swamp Chestnut-oak 324166
Swamp Coneflower 339585
Swamp Cottonwood 311347
Swamp Crinum 111158
Swamp Cudweed 178090,178481
Swamp Currant 334058
Swamp Cypress 385253,385254,385262
Swamp Cypress Family 385250
Swamp Cyrilla 120488
Swamp Dewberry 338528
Swamp Dock 340305
Swamp Dodder 115040
Swamp Dog Wood 380499
Swamp Dogwood 104949,105039,105201,320071
Swamp Dragon 238167

Swamp Ebony 132299
Swamp Elm 401431
Swamp False Solomon's-seal 242706
Swamp Fleabane 150611
Swamp Fly Honeysuckle 235996
Swamp Fly-honeysuckle 235996
Swamp Foxtail Grass 289015
Swamp Goldenrod 368304,368454
Swamp Gooseberry 334030,334058
Swamp Grass-of-parnassus 284591,284601
Swamp Gum 155680,155719,155722,155727
Swamp Haw 407737
Swamp Heath 146074,146075
Swamp Hellebore 405643
Swamp Hibiscus 194837,195032
Swamp Hickory 77881,77889,77907,77910
Swamp Honey Suckle 332086
Swamp Honeysuckle 235874,332086
Swamp Isotoma 210317,223053
Swamp Larkspur 124660
Swamp Laurel 215387,215391,215395,215405,242341
Swamp Laurel Oak 324095,324134
Swamp Leather Flower 94870
Swamp Leather-flower 94870
Swamp Lily 111210,111244,111250,230044,348085,417606
Swamp Locust 176864
Swamp Loosestrife 123526,123527,239879,239881
Swamp Lousewort 287335
Swamp Magnolia 242341
Swamp Mahogany 88506,155492,155722
Swamp Mallet 155746
Swamp Mallow 195032
Swamp Maple 3505
Swamp Marigold 53773,53842
Swamp Mazus 247025
Swamp Meadow-grass 305804
Swamp Messmate 155722
Swamp Milkvetch 43204
Swamp Milkweed 37967
Swamp Morningglory 207590
Swamp Morning-glory 207590
Swamp Oak 15953,79155,79165,323707,324137,324262
Swamp Oat 371498
Swamp Oats 371498
Swamp Oval Sedge 75466
Swamp Palmetto 341421
Swamp Paperbark 248093
Swamp Pedicularis 287494
Swamp Persicaria 309428
Swamp Pickle Grass 113316
Swamp Pine 299928,300128
Swamp Pink 32373,191046
Swamp Poplar 311347
Swamp Post Oak 324137,324422
Swamp Potato 342400
Swamp Pricklegrass 113316
Swamp Privet 167322
Swamp Rattlebox 112773
Swamp Red Bay 291587
Swamp Red Currant 334240
Swamp Red Oak 323894,324350,324415
Swamp Redbay 291587
Swamp Red-currant 334240
Swamp Rose 195032,336839
Swamp Rosemallow 194901
Swamp Rose-mallow 195083
Swamp Saltbush 44312
Swamp Sassafras 242341
Swamp Saxifrage 349771
Swamp Shadbush 19251
Swamp She Oak 79161
Swamp She-oak 79161,79165
Swamp Smartweed 291792,308739,308986,309428
Swamp Spike-rush 143271
Swamp Spruce 298349
Swamp Squaw-weed 358336
Swamp Star-anise 204570
Swamp Sumac 393491
Swamp Sunflower 188402,188906,188968
Swamp Sweet Pea 222800
Swamp Taro 120774
Swamp Tea Tree 248104
Swamp Thistle 91770,92213
Swamp Thistle Thistle 73420
Swamp Tickseed 53842
Swamp Tupelo 267850,267867,267872
Swamp Turpentine 236465
Swamp Valerian 404383
Swamp Verbena 405844
Swamp Wallaby Grass 20551
Swamp Wallaby-grass 20549
Swamp Wattle 1199
Swamp White Oak 323707
Swamp Wild Rose 336839
Swamp Willow 343172,343767
Swamp Willowherb 146812
Swamp Willow-weed 309039
Swamp Wood Betony 287335
Swamp's Companion 72934
Swampbay 242341
Swamp-bay 242341
Swamp-candles 239879
Swamp-cypress 385253
Swamphaw 407970
Swamp-laurel 215405
Swamp-lily 111152
Swamp-lousewort 287335
Swampmahogany 155722
Swamp-mahogany 155722
Swamp-pink 32373,191046,191050
Swamp-privet 167322
Swamps Rose Mallow 195032
Swampy Cyrtopodium 120710
Swampy Dipodium 133396
Swampy Gentianopsis 174225
Swampy Sagebrush 36048
Swampy Scheuchzeria 350861
Swampy Wormwood 36048
Swan Bill 208771
Swan Flower 34198
Swan Neck 116516
Swan Neck Orchid 116516
Swan Orchid 116516
Swan River Cypress 6933
Swan River Daisy 58669,58671,58672
Swan River Everlasting 329801
Swan River Everlasting Flower 190791
Swan River Myrtle 202377
Swan River Pae 58677
Swan River Pea Shrub 58678
Swan River Peabush 58678
Swan's Sedge 76463
Swan-among-the-flowers 267648
Swan-bill 208771
Swane's Golden Pencil-pine 114756
Swanflower 34174
Swan-orchid 116516
Swan-plant 116516
Swanriver Daisy 58669
Swan-river Daisy 58669,58671
Swanriver Hakea 184613
Swardleaf Elatostema 142690
Swardleaf Stairweed 142690
Swarms 14964
Swarri Nut 77942
Swart Hoat 1384
Swartz's Threeawn 34049
Swasey Mahonia 242649
Swaying Bulrush 352271
Sweating Plant 158250
Sweatroot 307241
Swede 59493,59507
Swede Turnip 59507
Sweden Currant 334179
Swedish Aspen 311537
Swedish Balsam Poplar 311237
Swedish Birch 53563,53564
Swedish Clover 396930
Swedish Coffee 42086
Swedish Goosefoot 87179
Swedish Hemp 402886
Swedish Ivy 303158,303539
Swedish Juniper 213738,213960
Swedish Mountain Ash 369432
Swedish Pondweed 312276

Swedish Service-tree 369429
Swedish Tummit 59575
Swedish Turnip 59493,59507
Swedish White Beam 369432
Swedish Whitebeam 369432
Swedish-coffee 42086
Sweellfruit Mayten 246862
Sweep 50825,81237,201787,238583
Sweep's Brush 133478,238583,302034,
302103,400675
Sweepina Przewalsk Juniper 341751
Sweepina Przewalsk Savin 213864
Sweeping Pond Cypress 385254
Sweet Abelia 152
Sweet Acacia 1219,1630
Sweet Alice 30164,235090,299351
Sweet Alison 235065,235090
Sweet Almond 20890,20897
Sweet Alyson 235065
Sweet Alyssum 235065,235090
Sweet Angelica 261585
Sweet Anise 167146,167156
Sweet Annie 35132
Sweet Archangel 220416
Sweet Arrowbamboo 162725
Sweet Arrowwood 407917,407977
Sweet Ash 28044
Sweet Autumn Clematis 95108,95358
Sweet Autumn Virginsbower 95358
Sweet Azalea 330110
Sweet Ballocks 372259
Sweet Balls 228180
Sweet Balm 249504
Sweet Bamboo 125482,297270
Sweet Basil 268438
Sweet Bay 223163,223203,242341
Sweet Bay Magnolia 242341
Sweet Bay Willow 343858
Sweet Bean 83527
Sweet Bedstraw 170708
Sweet bells Leucothoe 228180
Sweet Bent 238583
Sweet Bergamot 257161
Sweet Berry Joint-fir 178549
Sweet Beth 397628
Sweet Betsy 68313,81768,220729,397551
Sweet Betty 81768,346434
Sweet Billers 192358
Sweet Birch 53437,53505
Sweet Black-eyed Susan 339617
Sweet Bluebasin 350099
Sweet Blueberry 403798,403957
Sweet Bouvardia 57950
Sweet Box 346723,346735,346743
Sweet Bracken 261585
Sweet Brahea Palm 59098
Sweet Breath of Spring 235796
Sweet Briar 1219,336741,336892
Sweet Briar Rose 336892
Sweet brier 336892
Sweet Broom 121013,340990
Sweet Broomwort 354660
Sweet Bubby Bush 68313
Sweet Buckeye 9692
Sweet Bugle 239248
Sweet Bursaria 64064
Sweet Bush 68313,96457
Sweet Calabash 285668,285672
Sweet Calamus 117204
Sweet Calomel 5793
Sweet Cane 5793,341829
Sweet Cassava 244506
Sweet Cassawa 244512
Sweet Chamomile 85526
Sweet Cherry 83155
Sweet Chervil 261585
Sweet Chestnut 78766,78790,78811
Sweet Cicely 261576,261585
Sweet Cicely of Europe 261585
Sweet Cis 261585
Sweet Cistus 93161
Sweet Clethra 96457
Sweet Clockvine 390767
Sweet Clover 249197,249203,249218
Sweet Cods 372259
Sweet Colsfoot 292353
Sweet Coltsfoot 292353,292403
Sweet Common Barberry 52336
Sweet Coneflower 339617
Sweet Corn 417417,417433
Sweet Covey 153869
Sweet Crab 243583
Sweet Crab Apple 243583
Sweet Crabapple 243583
Sweet Crowfoot 325628
Sweet Cullins 372259
Sweet Cup 285637,285660,285668
Sweet Cymbidium 117067
Sweet Cypress 119125
Sweet Daphne 122532
Sweet Devil's-claws 315436
Sweet Dock 308893
Sweet Elder 345580
Sweet Elderberry 345580
Sweet Elm 401509
Sweet Everlasting 178318,178358,317761
Sweet False Chamomile 246396
Sweet Fennel 24213,167146,167156
Sweet Fern 101664,101667,261126,261585
Sweet Fern Shrub 101664
Sweet Flag 5788,5793,5794
Sweet Four-o'clock 255727
Sweet Fruits Blueberry 403738
Sweet Gale 261120,261126,261162
Sweet Gale Family 261243
Sweet Galingale 119125
Sweet Gallberry 203660
Sweet Gaoliang 369633
Sweet Garlic 400095
Sweet Goldenrod 368290
Sweet Granadilla 285662
Sweet Grass 27967,170524,177556,177599,
177648,187179,196116,196137
Sweet Green Basil 268438
Sweet Gum 232550,232565
Sweet Gum Tree 232550
Sweet Hairhoof 170524
Sweet Hakea 184602,184641
Sweet Hay 166125
Sweet Hearts 170193
Sweet hearts 374916
Sweet Herb 347506
Sweet Holly Rose 93161
Sweet Honey Leaf 376415
Sweet Honeysuckle 235713
Sweet Horsechestnut 9692
Sweet Humlick 261585
Sweet Hurts 403716
Sweet Indian-plantain 186137
Sweet Indosasa 206872
Sweet Iris 208744
Sweet Jasmine 211938
Sweet Joe-pye Weed 158264
Sweet Joepyeweed 161175
Sweet John 127607
Sweet Jute 104059
Sweet Kabious 350099
Sweet Lavender 223289
Sweet Leaf 201732
Sweet Lemon 93534
Sweet Lime 93534
Sweet Locust 176901,176903
Sweet Lucerne 249232
Sweet Lupin 238463
Sweet Machilus 240700
Sweet Magnolia 242341
Sweet Majoram 274224
Sweet Manna Grass 177556
Sweet Maple 3549
Sweet Marigold 383097
Sweet Marjoram 274218
Sweet Mary 81768
Sweet Maudlin 3978
Sweet Melon 114189
Sweet Michelia 252860
Sweet Mignonette 327896
Sweet Milk Vetch 42417
Sweet Mock Orange 294426
Sweet Mockorange 294426
Sweet Mock-orange 294426
Sweet Mountain Grape 411805

Sweet Myrtle 5793
Sweet Nance 374916
Sweet Nancy 262438,262441,374916
Sweet Nancy Lily 262414
Sweet Noor Euphorbia 158666
Sweet Nut 3978
Sweet Nuts 3978
Sweet Ombrocharis 270588
Sweet Orange 93765
Sweet Orchid 192862
Sweet Passionflower 285662
Sweet Pea 222671,222750,222789
Sweet Pea Bush 306249
Sweet Pea Shrub 308016
Sweet Pecan 77902
Sweet Pepper 72068,72070,72075,96457
Sweet Pepper Bush 96457
Sweet Pepperbush 96457
Sweet Pepper-bush 96457
Sweet Pignum 77912
Sweet Pignut 77894
Sweet Pigweed 139678
Sweet Pinesap 258091
Sweet Pink 127607
Sweet Pitcher Plant 347147,347163
Sweet Pitcher-plant 347163
Sweet Pittosporum 301433
Sweet Plantain 302103
Sweet Poplar 311509
Sweet Potato 207623
Sweet Potato Vine 323457
Sweet Quandong 346209
Sweet Resin Bush 160837
Sweet Ridge Wild Buckwheat 152189
Sweet Rocket 193417
Sweet Root-corn 66158
Sweet Rush 5793
Sweet Sage 307282
Sweet Sagebrush 35132
Sweet Sagewort 35132
Sweet Sansevieria 346155
Sweet Sansevierin 346158
Sweet Sarsaparilla 366348
Sweet Scabious 150464,350086,350099
Sweet Scented Geranium 288268
Sweet Scented Pincushion 184602
Sweet Sea Grass 418392
Sweet Sedge 5793
Sweet Seg 5793
Sweet Shade 201238,201240
Sweet Shrub 68306,68313
Sweet Sies 261585
Sweet Signalgrass 58088
Sweet Slumber 345805
Sweet Sop 25828,25898
Sweet Sops 25828
Sweet Sorghum 369633,369700
Sweet Spire 210359,210421
Sweet Spurge 158805
Sweet Stem 17602
Sweet Suckers 381035
Sweet Suckle 236022
Sweet Sultan 18946,18966,18983,81116,81219,81408
Sweet Sumach 332477
Sweet Tanglehead 194040
Sweet Thorn 1324
Sweet Tobacco 266035
Sweet Trillium 397628
Sweet Verbena Tree 46337
Sweet Vernal Grass 27969
Sweet Vernalgrass 27930,27969
Sweet Vernal-grass 27969
Sweet Vetch 187753,187802
Sweet Viburnum 407917,407977
Sweet Violet 410320
Sweet Virgin's Bower 94901
Sweet Vivian Rose 336310
Sweet Wattle 1395,1630
Sweet White Trillium 397615
Sweet White Violet 409751,409752,410189
Sweet William 86427,127607,349044
Sweet William Catchfly 363214
Sweet William Phlox 295257,295282
Sweet William Silene 363214
Sweet Willie 248411,273469,343858
Sweet Willow 261162,343858
Sweet Willy 343858
Sweet Winter Grape 411615
Sweet Withy 261162,261585
Sweet Wood Ruff 170524
Sweet Wood-reed 91236
Sweet Woodruff 170175,170524
Sweet Woodrush 170524
Sweet Wormwood 35132
Sweet Yarrow 3917
Sweet Yellow Bells 192619
Sweetalyssum 235065
Sweet-alyssum 235090
Sweet-amber 201732
Sweetbay 242341
Sweetbay Magnolia 242341
Sweetberry 381982
Sweet-berry Honeysuckle 235691
Sweetberry Jointfir 178549
Sweetbox 346733
Sweetbread 312360
Sweet-briar 336892
Sweetbrier 336538,336741
Sweetbrier Rose 336538,336892
Sweetbush 49382
Sweetcane 341829
Sweetcaneflower Silvergrass 255873,394943
Sweetch Sorrel 135215
Sweetclover 249197
Sweetcloverlike Milkvetch 42696
Sweet-colt's-foot 292409
Sweetdaisy 71085,71090
Sweetest Jasmine 211938
Sweetflag 5793
Sweet-flag 5788,5789,5793,5794
Sweetflaglike Enhalus 145643
Sweetflower Rockjasmine 23144
Sweet-fruited Blueberry 403738
Sweetgrass 5793,27967,177556,196116
Sweet-grass 27967,177556,196116
Sweetgum 232550,232565
Sweetgum-leaved Tonkin Maple 3697
Sweetgum-loving Scurrula 385211
Sweetgumshoot Mistletoe 411048
Sweet-gum-shooted Mistletoe 411048
Sweethaw 408061
Sweetheart Geranium 288209
Sweetheart Plant 294838
Sweetheart Tree 160954
Sweethearts 11549,31051,37015,53797,54048,170193,170524,232950,405788
Sweetie 93534
Sweetleaf 381090,381423,381441
Sweetleaf Corydalis 105938
Sweetleaf Family 381067
Sweetlime 93536
Sweet-lime-of-India 93536
Sweet-nancy 3917
Sweetpalm 57116,57122
Sweetpea 222789
Sweet-pea 222789
Sweet-pea Bush 306236,306249
Sweetpotato 207623
Sweetscent 305115
Sweet-scented Asparagus 39009
Sweetscented Basil 268518,268523
Sweet-scented Basil 268518
Sweetscented Bedstraw 170524
Sweet-scented Bedstraw 170524,170708
Sweet-scented Birthwort 34286
Sweet-scented Bush 68306
Sweetscented Candytuft 203229
Sweet-scented Coltsfoot 292372
Sweet-scented Geranium 288268
Sweet-scented Grass 27967
Sweet-scented Hakea 184602
Sweet-scented Heliotrope 190592
Sweet-scented Indian Plantain 64731
Sweetscented Joepyeweed 161175
Sweet-scented Life Everlasting 178358
Sweetscented Marigold 383097
Sweet-scented Marigold 383097
Sweet-scented Mexican Marigold 383097
Sweet-scented Mignonette 327896
Sweet-scented Narcissus 262429

Sweetscented Oleander 265327
Sweet-scented Oleander 265327
Sweet-scented Orchid 182230
Sweet-scented Pink 127713
Sweet-scented Shrub 68306
Sweetscented Snowbell 379417
Sweetscented Squinancy 170524
Sweetscented Storax 379417
Sweet-scented Storax 379417
Sweet-scented Waterlily 267730
Sweet-scented Water-lily 267730
Sweet-scented Wattle 1630
Sweet-scented-grass 27967
Sweet-scentedJoe-pye Weed 158264
Sweetshell Azalea 330524
Sweetshell Rhododendron 330524
Sweet-shelled Rhododendron 330524
Sweetshoot Bamboo 297263
Sweet-shooted Bamboo 297263
Sweetshrub 68306,68310,68313,68329
Sweetsop 25898
Sweetspire 210359,210364,210421
Sweet-spire 210421
Sweetspire Barberry 51778
Sweetspire Family 155155
Sweetspire-leaf Bodinier Beautyberry 66745
Sweetspireleaf Tanoak 233261,233263
Sweetspire-leaved Barberry 51778
Sweetspireleaved Tanoak 233263
Sweetsultan 18966
Sweet-sultan 18946,18966,81219
Sweettea Raspberry 338252
Sweettree Nectandra 263063
Sweetwilliam 127607
Sweet-william 127607
Sweet-william Campion 363214
Sweet-william Catchfly 363214
Sweet-william Silene 363214
Sweetwood 263056
Sweet-wood 112884
Sweginzow Rose 336981
Sweli Storax 379466
Swellen Chloris 88320
Swellenfruit Celery 295035,295038
Swellennode Poorspikebamboo 270440
Swellen-node Swellnode Oligostachyum 270440
Swelling Euonymus 157539
Swelling Ormosia 274404
Swelling Wheat 398986
Swellingcalyx Woodbetony 287517
Swellingpedicelled Groundsel 359811
Swellpod 27464
Swellstipe Elaeocarpus 142330
Swertia 380105,380131
Sweth 15709
Swichen 360328
Swift 34536
Swigoid-leaved Holly 203791
Swimming Plantain 312190
Swine Arnut 373346
Swine Carse 308816
Swine Cress 105338,105340,105352
Swine Grass 308816
Swine Succory 34816
Swine Thistle 368635,368771
Swine Wartcress 105340
Swine Weed 192358
Swine's Beads 312360,353099,373346
Swine's Cress 29341,105340,105352, 221787,308816,359158
Swine's Grass 308816,359158
Swine's Grease 308816
Swine's Maskert 373346
Swine's Mosscorts 373346
Swine's Murrill 353099,373346
Swine's Murrills 353099,373346
Swine's Skir 308816
Swine's Snout 384714
Swine's Succory 34816,202387
Swine's Tusker 308816
Swine's-cress 105338,105340,105352
Swinebread 102727
Swinecress 105352
Swine-cress 105338
Swinesnap 381035
Swingle Brandisia 59134
Swingle Chirita 87978
Swingle's Kumquat 93492,167513,167515
Swinglea 380538
Swinhoe Boea 283134
Swinhoe Fig 165766
Swinhoe Sabia 341570
Swinhoe's Raspberry 339338
Swinies 368771
Swiss Catchfly 364156
Swiss Centaury 329197,375263
Swiss Chard 53249,53257
Swiss Cheese Plant 258168
Swiss Forget-me-not 260794
Swiss Melilot 249208
Swiss Mountain Pine 300086
Swiss Mountain-pine 300086
Swiss Rock Jasmine 23190
Swiss Rock-jasmine 23190
Swiss Ryegrass 235354
Swiss Stone Pine 299842
Swiss Stone-pine 299842
Swiss Treacle Mustard 154466
Swiss Wild Pine 300086
Swiss Willow 343474
Swisscentaury 375263,375274
Swiss-cheese Plant 258168
Switch 261585
Switch Cane 37191
Switch Elm 401512
Switch Grass 281274,282366,282369
Switch Ivy 228162
Switchcane 37191
Switchgrass 282366
Switch-grass 282366
Switiks 24483
Swizzlestick Tree 327080
Swoed Lily 208771
Swollen Bamboo 47508
Swollen Buttercup 403214
Swollen Cylindrokelupha 116610
Swollen Duckweed 224366
Swollen Haworthia 186801
Swollen Rattan Palm 65808
Swollen Sedge 74904
Swollennoded Cane 87575,87622,323381
Swollennoded Cane Mountain 87588
Swollennoded Indosasa 206877
Swollen-noded Indosasa 206877
Swollen-spurred Bladderwort 403191
Swollen-stemmed Bamboo 47498
Swollen-thorn Acacia 1150
Sword Bean 71042,71050,145875
Sword Flag 208771
Sword Grass 208771,255827
Sword Jack Bean 71050
Sword Jackbean 71050,71053
Sword Jack-bean 71042
Sword Leaf Eulophia 156687
Sword Lily 4534,175990,176260,176600, 208771
Sword Orchis 116829
Sword Plant 140536,415990
Sword Plantainlily 198600
Sword Rush 225574
Sword Sedge 225574
Sword Tupistra 70605
Sword-and-spear 302034
Swordbean 71050
Swordflag 208433,208875
Swordflag Family 208377
Swordform Jackbean 71042
Swordgrass 255827
Sword-grass 255827
Swordleaf Apostasia 29793
Swordleaf Ardisia 31439
Swordleaf Calanthe 65915
Swordleaf Cladium 93908
Swordleaf Cymbidium 116829
Swordleaf Dendrobium 124958
Swordleaf Dianella 127507
Swordleaf Dracaena 137359
Swordleaf Dragonbood 137359
Swordleaf Eargrass 187598
Swordleaf Eulophia 157013

Swordleaf Gentian 157013
Swordleaf Greenstar 307505
Swordleaf Habenaria 183945
Swordleaf Hedyotis 187598
Swordleaf Holly 203960
Swordleaf Inula 207105
Sword-leaf Inula 207105
Swordleaf Irisorchis 267944
Swordleaf Oberonia 267944
Sword-leaf Rush 213093
Swordleaf Sedge 73875
Swordleaf Sterculia 376106
Swordleaf Suoluo 327364
Swordleaf Triratna Tree 397374
Swordleaf Tupistra 70605
Swordleaf Twigrush 93908
Sword-leaved Ardisia 31439
Sword-leaved Dracaena 137359
Sword-leaved Helleborine 82060
Sword-leaved Holly 203960
Sword-leaved Inula 207105
Sword-leaved Trigonostemon 397374
Swordlike Atractylodes 44208
Sword-like Iris 208543
Sword-like Plantainlily 198600
Swordlip Androcorys 22325
Swords 208771
Swords-and-spears 302034
Swordshape Oncidium 270807
Swordshape Rush 213571
Swort Beak Huernia 199043
Syagrus 380553
Sybie 15289
Sybow 15289
Sycamore 3462,165726,302574,302588
Sycamore Fig 165726
Sycamore Maple 3462
Sycamoreleaf Snowbell 379431
Sydney Acacia 1358
Sydney Blue Gum 155730,155732
Sydney Bluegum 155732
Sydney Christmas Bush 83535
Sydney Eucalyptus 155730
Sydney Golden Wattle 1358
Sydney Peppermint 155697
Sydney Red Gum 24606
Sydney Rock-rose 57277
Sydney Wattle 1358
Sydny Boronia 57266
Sykes Coral Tree 154618
Sylvan Blueberry 403986
Sylvan Bluegrass 306048
Sylvan Goat's-beard 37060
Sylvan Goatsbeard 37085
Sylvan Scorzonella 253942
Sylvan Violet 410483
Sylvan Waxplant 198896
Sylvestral Elaeocarpus 142396
Sylvestris Fargesia 162757
Sylvestris Persimmon 132230
Sympegma 380681,380682
Sympetalous Deerdrug 242707
Sympetalous False Solomonseal 242707
Symphonia 380696
Symphorema 380712
Symphyllocarpus 380791
Symplocarpus 381070
Symplocos 381090
Symplocos Family 381067
Symplocosleaf Holly 204307
Symplocosleaf Ninenode 319864
Symplocosleaf Psychotria 319864
Symplocos-leaved Psychotria 319864
Sympodial Arrowwood 408152
Sympodial Nucleonwood 291489
Sympodial Perrottetia 291489
Sympodial Viburnum 408152
Symyl 315082
Syncalathium 381644
Syncalathiumlike Hippolytia 196709
Syndery Green Wattle 1168
Syndiclis 381788
Syndow 14065
Synechanthus 381806
Synedrella 381808,381809
Syneilesis 381813
Syngonium 381855
Synonyms 75346
Synotis 381918
Synpyrene Holly 204308
Synstemon 382004,382011
Synstone Holly 204308
Synurus 382067
Syphelt 358062
Syreitschikovia 382096
Syreitschikow Bedstraw 170657
Syrenia 382101
Syria Ixiolirion 210999
Syria Juniper 213763
Syrian Alder 16431
Syrian Ash 168114
Syrian Beancaper 418641
Syrian Bean-caper 418641
Syrian Bindweed 68686,103255
Syrian Cephalaria 82173
Syrian Christ-thorn 418216
Syrian Hawberry 110077
Syrian Hibiscus 195269
Syrian Hyssop 274226
Syrian Ixiolirion 210999
Syrian Juniper 213763
Syrian Katmia 195269
Syrian Knapweed 80954
Syrian Maple 3651
Syrian Marjoram 274226
Syrian Mesquite 315554
Syrian Mustard 155984,155985
Syrian Oregano 274226
Syrian Rue 287978
Syrian Scabious 82173
Syrian-privet 167235
Syrian-rose 195269
Syringa 294407,294426,382114
Syringa-like Cliff Honeysuckle 236073
Syrup Jubaea 212557
Syrup Palm 212557
Syves 15709
Syzigium-leaved Raspberry 338650
Syzygium 382465
Syzygium-leaf Holly 204310
Syzygium-leaved Holly 204310
Szechuan Asiabell 98417
Szechuan Bastard Cress 390232
Szechuan Birch 53572
Szechuan Bramble 339259
Szechuan Cherry 83340
Szechuan Clovershrub 70855
Szechuan Corydalis 105688
Szechuan Cycas 115909
Szechuan Cyclea 116036
Szechuan Cymbidium 116814
Szechuan Epimedium 147067
Szechuan Galeobdolon 170026
Szechuan Habenaria 184114
Szechuan Holly 204311
Szechuan Jerusalemsage 295202
Szechuan Juniper 213972
Szechuan Manyflowered May-apple 139628
Szechuan Poplar 311514
Szechuan Primrose 315028
Szechuan Secamone 356303
Szechuan Swallowwort 117713
Szechuan Tipularia 392353
Szechuan White Birch 53572
Szechuan Woodbetony 287723
Szechwan Acanthopanax 143634
Szechwan Adonis 8382
Szechwan Arborvitae 390661
Szechwan Arbor-vitae 390661
Szechwan Aspen 311514
Szechwan Blueberry 403832
Szechwan Bluegrass 306050
Szechwan Cherry 83340
Szechwan Chinaberry 248925
Szechwan Cyclea 116036
Szechwan Deutzia 127093
Szechwan Dichocarpum 128941
Szechwan Dogwood 380510
Szechwan Eria 299643
Szechwan False Solomonseal 242703
Szechwan Filbert 106740

Szechwan Hemsleia 191967
Szechwan Indigo 206637
Szechwan Keiskea 215818
Szechwan Larkspur 124610
Szechwan Lilac 382298
Szechwan Lilyturf 272150
Szechwan Manglietia 244471
Szechwan Maple 3648
Szechwan Michelia 252975
Szechwan Monkeyflower 255244
Szechwan Mountain-ash 369520
Szechwan Onion 15794
Szechwan Raspberry 339259
Szechwan Rhododendron 331911
Szechwan Sabia 341556
Szechwan Sida 362672
Szechwan Speedwell 407400
Szechwan Square Bamboo 87616
Szechwan Taxillus 385234
Szechwan Willow 343512
Szechwan-Kweichow Glorybower 96014
Szechwan-shensi Globeflower 399497
Szechwan-tibet Rabdosia 209794
Szechwan-tibet Rockjasmine 23129
Szechwan-Yunnan Alder 16340
Szechwan-Yunnan Sanicle 345941
Szemao Dysophylla 139593
Szemao Milkwort 308140
Szemao Syzygium 382671
Szovits Heteraeia 193659
Szowits' Madwort 18480
Szu-ui-kan 93815

T

T. Nuttall's Rhododendron 331361
T. Thomson's Rhododendron 331971
Tab Mallow 243840
Tabacco 266031,266060
Tabacco Plant 266031
Tabaccoodored Primulina 315148
Tabaccooloured Devil Rattan 121528
Tabaek 219914
Tabalba 159207
Taban Merah 280441
Tabardillo 300842
Tabasco 72100
Tabasco Mahogany 380527
Tabasco Pepper 72070
Tabeau Peak Wild Buckwheat 152133
Tabernaemontana 382727
Tabernaemontana Divaricate 154162
Tabernaemontanus Bulrush 352278
Tabernemontan Cinquefoil 312812
Table Beet 53269
Table Dogwood 57695
Table Mountain Orchis 133684
Table Mountain Pine 300168
Table Tanoak 233382
Table Top Juniper 213930
Tables-and-chairs 64345
Tabletforme Aeonium 9072
Tabletop Dogwood 57695
Tabletop Scotch Elm 401515
Tab-mawn 34536
Tabog 84982
Tabor Barley 198365
Tabu Woodbetony 287739
Tabua Palm 298051
Tabular Tanoak 233382
Tabularformed Pine 300243
Tabular-formed Pine 300243
Tabu-no-ki Tree 240707
Tacaco 307129
Tacamahac 67860,67871,311237
Tacay 77971
Tacca 382908,382923
Tacca Family 382934
Tacheng Fritillary 168642
Tacheng Larkspur 124012
Tacheng Nepeta 265073
Tacheng Tulip 400239
Tacheng Willow 344189
Tachibana 93823,167499
Tacker-grass 308816
Tacker-weed 72038
Tackstem 68444,68445,68446
Tack-stem 68444
Tacky Goldenweed 323026
Tadehagi 383006
Taemon Woodsorrel 277659
Taeniophyllum 383054,383064
Tafel Koeleria 217489
Tafell Mountain-loving Draba 137170
Taffeta Plant 197103
Tag Alder 16375,16437,16440,16445
Tagasaste 85394
Tagawa Willow 344167
Tagetes 383087
Tagg's Rhododendron 331914
Taggar 91455
Tagua Nut 298051
Tagua Passionflower 285639
Tahai Woodbetony 287738
Tahan Fan Palm 234195
Tahan Fanpalm 234195
Tahini 361317
Tahiti Arrowroot 382923
Tahiti Chestnut 56033,207000
Tahiti Lime 93528
Tahitian Bridal Veil 175489
Tahitian Bridal-veil 175489
Tahitian Gardenia 171404
Tahitian Gooseberry Tree 296466
Tahitian Lime 93528
Tahitian Quince 372467
Tahitian Screwpine 281138
Tahiti-lime 93330
Tahl Gum 1588
Tahoka Daisy 240458
Tahoka-daisy 240458
Tahotta Daisy 240458
Taibai Anemone 24073
Taibai Buttercup 326223
Taibai Deutzia 127113
Taibai Fleabane 151000
Taibai Fritillary 168575
Taibai Gentian 151000
Taibai Larkspur 124617
Taibai Maple 2827
Taibai Monkshood 5616
Taibai Mountainash 369531
Taibai Orychophragmus 275874
Taibai Rhododendron 331588,331915
Taibai Rockfoil 349350
Taibai Sweetvetch 188138
Taibai Vetch 408622
Taibai Willow 344169
Taibai Windflower 24073
Taibai Yellowish Everlasting 21572
Taibai Yellowish Pearleverlasting 21572
Taibaishan Acanthopanax 2505
Taibaishan Cacalia 283859
Taibaishan Callianthemum 66718
Taibaishan Chinese Dipteronia 133601
Taibaishan Clovershrub 70842
Taibaishan Coinmaple 133601
Taibaishan Corydalis 106495
Taibaishan Ghstplant Sagebrush 35775
Taibaishan Goldsaxifrage 90459
Taibaishan Goldwaist 90459
Taibaishan Honeysuckle 236140
Taibaishan Milkvetch 43121
Taibaishan Mountainash 369531
Taibaishan Mountain-ash 369531
Taibaishan Rock's Peony 280279
Taibaishan Sagebrush 36360
Taibaishan Sandwort 32269

Taibaishan Sedge 76476
Taibaishan Willow 344169
Taibamboo 391457,391460
Taibei Arisaema 33476
Taibei Azalea 330990
Taibei Cleyera 96602
Taibei Dualbutterfly 398253
Taibei Heteropappus 193977
Taibei Jurinea 193977
Taibei Randia 12517
Taibei Raspberry 338433
Taibei Rhododendron 330990
Taibei Southstar 33476
Taidong Altai Heteropappus 193923
Taidong Arrowwood 408155
Taidong Corydalis 105693
Taidong Cycas 115910
Taidong Gentian 173967
Taidong Groundsel 360177
Taidong Lasianthus 222242
Taidong Nervilia 265439
Taidong Orchis 273653
Taidong Photinia 295775
Taidong Roughleaf 222242
Taidong Spingflower Raspberry 339344
Taidong Tanoak 233383
Taidong Viburnum 408155
Taihang Blushred Rabdosia 324864
Taihang Bulrush 353789
Taihang Flower 383115,383117
Taihang Gueldenstaedtia 181689
Taihang Mosquitotrap 117715
Taihang Mountain Bulrush 353789
Taihang Ricebag 181689
Taihang Swallowwort 117715
Taihangdaisy 272585,272588
Taihangia 383115,383117
Taihangshan Clematis 95051
Taihangshan Elm 401635
Taihangshan Figwort 355254
Taihangshan Giantfennel 163660
Taihangshan Pear 323312
Taihu Sedge 76475
Tail Fig 165231
Tail Flower 28071,28084
Tail Grape 34995
Tail Microtropis 254287
Tail Pholidota 295534
Tail Plant 28119
Tail Rhamnell 328546
Tail Themeda 389318
Tail Waterbolt 55918
Tail Windhairdaisy 348762
Tailand Mango 244409
Tailand Sunipia 379786
Tailand Swallowwort 117716
Tailand Wallichpalm 413099
Tailand Yellowleaftree 415147
Tailand Yellow-leaved Tree 415147
Tailanther 381918
Tailcalyx Bulbophyllum 62626
Tailcalyx Curlylip-orchis 62626
Tailcalyx Leaflessorchis 29214
Tailders 350425
Tailed Croton 112856
Tailed Pepper 300377
Tailed Spicebush 231313
Tailed Spice-bush 231313
Tailed-leaf Indigo 205784
Tailed-leaf Sweetleaf 381423
Tailflower 28071,28084,28119
Tailgrape 34995,35016
Tail-grape 34995,35016
Tailgrass 402497
Tailing Monkshood 5066
Tailleaf Acanthopanax 2370
Tailleaf Agapetes 10312
Tailleaf Allomorphia 16032
Tailleaf Azalea 332041
Tailleaf Banfenghe 357974
Tailleaf Basketvine 9481
Tail-leaf Blueberry 403818
Tail-leaf Camellia 68970
Tailleaf Copperleaf 1771
Tailleaf Crapemyrtle 219915
Tailleaf Eargrass 187540
Tailleaf Fishvine 125947
Tailleaf Indigo 205784
Tailleaf Jurinea 212052
Tailleaf Macaranga 240278
Tailleaf Milkwort 307982
Tailleaf Mountainash 369527
Tailleaf Olive 270075
Tailleaf Osmanther 276272
Tail-leaf Pericome 290866
Tailleaf Pimpinella 299380
Tailleaf Rabdosia 209665
Tailleaf Raspberry 338238
Tailleaf Rhamnella 328546
Tailleaf Schefflera 350768
Tailleaf Skullcap 355402
Tailleaf Tanoak 233136
Tailleaf Vetchling 222692
Tailleaf Violet 410708
Tailleaf Windhairdaisy 348191
Tailleaftree 402605
Tail-leaved Indigo 205784
Tail-leaved Knotweed 162336
Tail-leaved Olive 270075
Tail-leaved Osmanthus 276270,276272
Taillemma Armgrass 58196
Tailless Gentian 173414
Tailless Waterbolt 55913
Taillikeleaf Rabdosia 209665
Tailor Tree 123448
Tailor's Needle 350425,370522
Tailsacgrass 402651
Tailsepal Tupistra 70626
Tailshapeleaf Begonia 50383
Tailshapeleaf Schefflera 350768
Tailstachys Kobresia 217139
Tailuger Oak 324467
Tailuko Hornbeam 77302
Tailwort 57101
Tainan Greyawngrass 372434
Tainan Spodiopogon 372434
Tainan Tanoak 233360
Tainia 383119
Taining Sagebrush 36361
Taipai Anemone 24073
Taipai Larkspur 124617
Taipai Monkshood 5616
Taipai Mountain Fleabane 151000
Taipai Sweetvetch 188138
Taipaishan Milkvetch 43121
Taipei Fritillary 168575
Taipei Heteropappus 193977
Taipei Honeysuckle 236140
Taipei Randia 12517
Tai-pin Cleyera 96606
Taiping Cleyera 96606
Taiping Eranthemum 148096
Tai-ping Raspberry 339067
Taipingshan Epimedium 147029
Taipingshan Plum 316549
Taipingshan Pseuderanthemum 317282
Taipingshan Viburnum 407850
Taishan Amaranth 18801
Taishan Amaranthus 18801
Taishan Falsepimpernel 231586
Taishan Hogfennel 293071
Taishan Leek 15798
Taishan Linden 391834
Taishan Motherweed 231586
Taishan Mountainash 369529
Taishan Mountain-ash 369529
Taishan Onion 15798
Taishan Pipewort 151517
Taishan Procumbent Willow 344170
Taishan Sargent Arrowwood 408111
Taishan Sumac 332881
Taishan Willow 344170
Taishun Azalea 331917
Taisyleaf Cinquefoil 313054
Taitung Copper-leaf 1997
Taitung Firethorn 322471
Taitung Lasianthus 222242
Taitung Raspberry 339344
Taitung Viburnum 408155
Taiug Acalypha 1997
Taiwan Abelia 113

Taiwan Abutilon 905
Taiwan Acacia 1145
Taiwan Actinidia 6530,6567
Taiwan Actinodaphne 233945,234075
Taiwan Adanson Caper 110216
Taiwan Adenia 7245
Taiwan Adina 364768
Taiwan Adinandra 8221
Taiwan Aglaia 11293
Taiwan Agrostophyllum 12449
Taiwan Alder 16345
Taiwan Aletris 14491
Taiwan Alpine Rhododendron 331565
Taiwan Alpine Willow 344173
Taiwan Alyxia 18531
Taiwan Amentotaxus 19363
Taiwan Anemone 24074
Taiwan Angelica 24326,24331
Taiwan Angelica-tree 30593
Taiwan Anisetree 204484
Taiwan Anise-tree 204484
Taiwan Antlerorchis 310798
Taiwan Aphanamixis 28997
Taiwan Apios 29307
Taiwan Appendicula 29814
Taiwan Apple 243610
Taiwan Aralia 30593
Taiwan Arisaema 33332
Taiwan Arrowwood 407853,408156
Taiwan Ascitesgrass 407461
Taiwan Aspidistra 39534
Taiwan Aster 40474,41335,193977
Taiwan Astronia 43462
Taiwan Astyleorchis 19499
Taiwan Aulacolepis 25333
Taiwan Awlquitch 131009
Taiwan Azalea 330722
Taiwan Balanophora 46824,364778
Taiwan Bamboo 297337
Taiwan Barberry 51811
Taiwan Barthea 48552
Taiwan Basil 268642
Taiwan Batweed 89202
Taiwan Beadruby 242683
Taiwan Beautyberry 66779
Taiwan Bedstraw 170352,170354,170658
Taiwan Beech 162380
Taiwan Begonia 49848,50352
Taiwan Bentgrass 12029
Taiwan Black Bamboo 297375
Taiwan Bluebasin 350185
Taiwan Blueberry 403796,404061
Taiwan Bluegrass 305617,306056
Taiwan Blumea 55738,55769
Taiwan Bogorchid 158063
Taiwan Brome 60724
Taiwan Bromegrass 60724
Taiwan Buckthorn 328700
Taiwan Buerger Maple 2815
Taiwan Bugle 13166
Taiwan Bugleweed 239226
Taiwan Bulbophyllum 62571,63122
Taiwan Bulrush 353896,396025
Taiwan Butterbur 292352
Taiwan Buttercup 326420
Taiwan Butterflybush 62015
Taiwan Butterfly-bush 62015
Taiwan Calanthe 65952
Taiwan Camellia 69216
Taiwan Camphor Tree 91420
Taiwan Caper 71751
Taiwan Cardiandra 73134
Taiwan Catchfly 248269
Taiwan Cephalanthera 82075
Taiwan Champereia 85935
Taiwan Cherry 83158
Taiwan Chilosehista 87463
Taiwan Chinabells 16304
Taiwan Chinese Sumac 12564
Taiwan Chinkapin 78942
Taiwan Chloranthus 88295
Taiwan Chloris 88351
Taiwan Christia 89202
Taiwan Christmasbush 14219
Taiwan Cinnamon 91351
Taiwan Cinquefoil 313086
Taiwan Cissampelos 116029
Taiwan Citrullus 93289
Taiwan Claoxylon 94055
Taiwan Clematis 94916,95042
Taiwan Cliffbean 254854
Taiwan Cockscomb 80422,80473
Taiwan Cogflower 154192,382822
Taiwan Coldbamboo 172741
Taiwan Collabium 99775
Taiwan Columbine 30048
Taiwan Composite Oplismenus 272629
Taiwan Condorvine 245805
Taiwan Copperleaf 1781
Taiwan Copper-leaf 1781
Taiwan Cordate-leafed Deutzia 127114
Taiwan Cordia 104203
Taiwan Coriaria 104694
Taiwan Cork-tree 294256
Taiwan Corybas 105532
Taiwan Corydalis 105639,106508
Taiwan Cotoneaster 107575
Taiwan Cowlily 267327
Taiwan Cowpea 408973
Taiwan Cow-plant 182384
Taiwan Cow-tail Fir 216125
Taiwan Crabapple 243610
Taiwan Cragcress 97983
Taiwan Crateva 110216
Taiwan Crepidiastrum 110627
Taiwan Crossandra 111774
Taiwan Cryptomeria 383189
Taiwan Cryptostylis 113863
Taiwan Curlylip-orchis 63122
Taiwan Currant 333978
Taiwan Cycas 115912
Taiwan Cyclea 116029
Taiwan Cypress 85268
Taiwan Dallisgrass 285448
Taiwan Damnacanthus 122028
Taiwan Daphne 122377
Taiwan Daphniphyllum 122708
Taiwan David Gentian 173379
Taiwan Deerdrug 242691
Taiwan Dendranthema 124772
Taiwan Dendrobium 125378
Taiwan Dendropanax 125604
Taiwan Deutzia 127114
Taiwan Devilpepper 327064,327069
Taiwan Dichocarpum 128923
Taiwan Dimeria 131009
Taiwan Dinggongvine 154245
Taiwan Dischidia 134037
Taiwan Dishspurorchis 171847
Taiwan Distylium 134932
Taiwan Dodder 115053
Taiwan Dogtongueweed 385921
Taiwan Douglas Fir 318599
Taiwan Draba 137228
Taiwan Dragonflyoechis 302307
Taiwan Dregea 137799
Taiwan Drypetes 138611
Taiwan Dualbutterfly 398295
Taiwan Dutchmanspipe 34203
Taiwan Earanther 276914
Taiwan Ear-anther 276914
Taiwan Eargrass 187525
Taiwan Ebony 132351
Taiwan Eccoilopus 372412
Taiwan Ehretia 141691
Taiwan Eightangle 204484
Taiwan Elaeagnus 141991
Taiwan Elder 345586
Taiwan Elm 401642
Taiwan Elsholtzia 144067
Taiwan Enkianthus 145710
Taiwan Entada 145904
Taiwan Epigeneium 146543
Taiwan Epipactis 147190
Taiwan Epithema 147431
Taiwan Eria 148684
Taiwan Ervatamia 154192,382822
Taiwan Erycibe 154245
Taiwan Euchresta 155890
Taiwan Eugenia 382535
Taiwan Eulophia 156572,156660

Taiwan Euonymus 157675
Taiwan Eupatorium 158063
Taiwan Euphorbia 159159
Taiwan Eurya 160553
Taiwan Euscaphis 160963
Taiwan Evergreen Chinkapin 78942
Taiwan Evergreenchinkapin 78933,78942,79035
Taiwan Excoecaria 161657
Taiwan Eyebright 160286
Taiwan Fairybells 134441
Taiwan False Cypress 85268,85338
Taiwan False Loosestrife 238230
Taiwan False Solomonseal 242691
Taiwan Falsebrome 58588
Taiwan Falsehellebore 405593
Taiwan Falsenettle 56140,56195
Taiwan Fescuegrass 163969
Taiwan Fieldcitron 400522
Taiwan Fig 164992
Taiwan Fig-tree 164992
Taiwan Filbert 106735
Taiwan Fingergrass 88351
Taiwan Fir 396
Taiwan Firethorn 322471
Taiwan Fishwood 110216
Taiwan Five-leaf Akebia 13219
Taiwan Fleabane 150789
Taiwan Fleshyorchis 346900
Taiwan Flowering Cherry 83158
Taiwan Forkliporchis 269026
Taiwan Freycinetia 168245
Taiwan Galangal 17674
Taiwan Gambirplant 401754,401761
Taiwan Gambir-plant 401754
Taiwan Gastrochilus 171847
Taiwan Gaultheria 172059
Taiwan Gelidocalamus 172741
Taiwan Geniostoma 172885
Taiwan Gentian 150789,156660,174080
Taiwan Giant Bamboo 125482
Taiwan Giantarum 20088
Taiwan Gianthyssop 10408
Taiwan Giantreed 37477
Taiwan Gingersage 253661
Taiwan Globeflower 399543
Taiwan Glochidion 177148
Taiwan Glutinousmass 179494
Taiwan Goldenmelon 182575
Taiwan Goldenray 229078
Taiwan Goldraintree 217620
Taiwan Goldwaist 90396
Taiwan Goniothalamus 179405
Taiwan Gonocaryum 179442
Taiwan Gonostegia 179494
Taiwan Goosecomb 335312
Taiwan Goosefoot 87019
Taiwan Gordonia 179745
Taiwan Grassleaf orchis 12449
Taiwan Greenbrier 366324,366414
Taiwan Greenstar 307508
Taiwan Greenvine 204634
Taiwan Guger-tree 350946
Taiwan Hackberry 80764
Taiwan Hair Bamboo 297282
Taiwan Hairorchis 148684
Taiwan Hedyotis 187525
Taiwan Helicia 189931
Taiwan Helictotrichon 190130
Taiwan Heliotrope 190630
Taiwan Helmet Orchid 105532
Taiwan Hemlock 399901
Taiwan Hemlockparsley 101831
Taiwan Hetaeria 193592
Taiwan Heterosmilax 194127,318450
Taiwan Heterostemma 194157
Taiwan Hibiscus 195305
Taiwan Hilliella 196299
Taiwan Hinoki Falsecypress 85338
Taiwan Hoarypea 386257
Taiwan Hogfennel 292857
Taiwan Holcoglossum 197265
Taiwan Holly 203821,204299
Taiwan Homalium 197659
Taiwan Hooktea 52423
Taiwan Hornbeam 77386
Taiwan Horseorchis 215366
Taiwan Hydrangea 199947
Taiwan Hypolytrum 202727
Taiwan Illicium 204484
Taiwan Incense Cedar 67530
Taiwan Incense-cedar 67530
Taiwan Indianpine 258041
Taiwan Indigo 206639
Taiwan Iris 208574
Taiwan Ironweed 406392
Taiwan Ivy 187325
Taiwan Jadeleaf and Goldenflower 260499
Taiwan Jadeolive 179442
Taiwan Java Skullcap 355535
Taiwan Jobstears 99115
Taiwan Juniper 213775
Taiwan Kalimeris 215366
Taiwan Keteleeria 216125
Taiwan Kudzubean 321469
Taiwan Kylinleaf 147342
Taiwan Lady's Slipper 120353
Taiwan Ladybell 7702
Taiwan Ladyslipper 120353
Taiwan Lasianthus 222138,222232
Taiwan Lecanorchis 223658
Taiwan Leea 223937
Taiwan Lemon 93823
Taiwan Lily 229845
Taiwan Liparis 232333
Taiwan Listera 264707
Taiwan Litse 233832,233884
Taiwan Littleleaf Box 64373
Taiwan Loosestrife 239553
Taiwan Lophotocarpus 236482
Taiwan Loquat 151144
Taiwan Luisia 238316
Taiwan Luohanfruit 365336
Taiwan Lycianthes 238962
Taiwan Maackia 240124
Taiwan Macaranga 240322
Taiwan Maesa 241844
Taiwan Magnific Deutzia 127008
Taiwan Magnolia 283418
Taiwan Magnoliavine 351012
Taiwan Magnolia-vine 351012
Taiwan Maizailan 11293
Taiwan Mallotus 243453
Taiwan Maple 2815,3654
Taiwan Mayten 182687
Taiwan Mazus 246978
Taiwan Meadowrue 388710
Taiwan Meadowsweet 166092
Taiwan Medinilla 247564
Taiwan Melicop 161353
Taiwan Melothria 417492
Taiwan Meytea 28349
Taiwan Michelia 252849
Taiwan Micromeria 253661
Taiwan Milkbean 169650
Taiwan Milkgreens 245805
Taiwan Milkpea 169650
Taiwan Milkwort 307919
Taiwan Millettia 254854
Taiwan Mistletoe 410972
Taiwan Miterwort 256019
Taiwan Mitrastemon 256178
Taiwan Monkshood 5204
Taiwan Montane Azalea 330901
Taiwan Montane Buttercup 325879
Taiwan Mosla 259284
Taiwan Mosquitoman 134932
Taiwan Mosquitotrap 117469
Taiwan Mountain Longan 189931
Taiwan Mountainash 369498
Taiwan Mountain-ash 369498
Taiwan Mouseear 82924
Taiwan Mulberry 259097
Taiwan Mussaenda 260499
Taiwan Myriactis 261069
Taiwan Nanmu 295361
Taiwan Narrowleaf Qinggang 116200
Taiwan Natotree 280440
Taiwan Nemosenecio 263512
Taiwan Neonauclea 264222
Taiwan Nepeta 264926

Taiwan Nervilia 265440
Taiwan Nettle 403026
Taiwan Nettletree 80764
Taiwan New Eargrass 262959
Taiwan Newlitse 264015
Taiwan Nightshade 367208
Taiwan Nothaphoebe 266792
Taiwan Notoseris 267168
Taiwan Nucleonwood 291482
Taiwan Oak 116200,324467
Taiwan Oberonia 267958
Taiwan Oilawn 372412
Taiwan Oliver Maple 3278
Taiwan Orange 93341
Taiwan Orchis 310939
Taiwan Oreomyrrhis 273979
Taiwan Ormosia 274392
Taiwan Osmanther 276370
Taiwan Osmanthus 276286,276370
Taiwan Pachycentria 279479
Taiwan Palaktree 280440
Taiwan Paspalum 285448
Taiwan Patrinia 285826
Taiwan Paulownia 285966
Taiwan Pear 323207
Taiwan Pecteilis 285966,285978
Taiwan Pencilwood 139655
Taiwan Pendent-bell 145710
Taiwan Pennsylvania Bittercress 72809
Taiwan Pentapanax 289629
Taiwan Pepper 300528,300540
Taiwan Pericampylus 290841
Taiwan Peristylus 291216
Taiwan Perotis 291216
Taiwan Perrottetia 291482
Taiwan Persimmon 132351
Taiwan Pertybush 292077
Taiwan Phoebe 295361
Taiwan Photinia 295729,295749
Taiwan Phreatia 295962
Taiwan Pieris 298734
Taiwan Pimpinella 299480
Taiwan Pine 300269
Taiwan PinnateGroundsel 263512
Taiwan Pittosporum 301363
Taiwan Platanthera 302406,302531
Taiwan Pleione 304218
Taiwan Plumegrass 341858
Taiwan Plumyew 82550
Taiwan Podocarpus 306431
Taiwan Pogostemon 306974
Taiwan Porkkidneybean 254796
Taiwan Prickly-ash 417330
Taiwan Privet 229432
Taiwan Pterocypsela 320513
Taiwan Purpledaisy 267168
Taiwan Purplepearl 66779
Taiwan Pyrola 322858
Taiwan Qinggang 116150
Taiwan Rabdosia 209625
Taiwan Ramie 56140
Taiwan Raspberry 338425,339347,339349
Taiwan Rattanpalm 65697
Taiwan Red Cypress 85268
Taiwan Red False Cypress 85268
Taiwan Red Lily 229845
Taiwan Red Pine 300269
Taiwan Red Silkyarn 238962
Taiwan Redhair Azalea 331689
Taiwan Reevesia 327362
Taiwan Rhododendron 330617,330722
Taiwan Rhynchotechum 333742
Taiwan Rockcress 30268
Taiwan Rocktaro 327672
Taiwan Rockvine 387774
Taiwan Roegneria 335312
Taiwan Rose 336938,336983
Taiwan Roughleaf 222138
Taiwan Rush 213341
Taiwan Sabia 341570
Taiwan Sage 345254
Taiwan Sagebrush 36302
Taiwan Sakebed 97720
Taiwan Samaradaisy 320513
Taiwan Sandwort 32136
Taiwan Sanicle 345996
Taiwan Sapium 346379
Taiwan Sassafras 347411
Taiwan Saurauia 347996
Taiwan Scabious 350185
Taiwan Schefflera 350786
Taiwan Schoenorchis 352312
Taiwan Scolopia 354621
Taiwan Scorpiongrass 175877
Taiwan Screwpine 281169
Taiwan Scurvyweed 97983
Taiwan Seatung 301363
Taiwan Sedge 73877,74596,76477
Taiwan Seedbox 238230
Taiwan Serrate Chloranthus 88298
Taiwan Shortia 362251
Taiwan Short-leaf Pine 300083
Taiwan Silverlineorchis 25980
Taiwan Sinopanax 364997
Taiwan Siraitia 365336
Taiwan Skimmia 365924
Taiwan Skullcap 355789
Taiwan Slatepentree 400730
Taiwan Slugwood 50514
Taiwan Small Reed 127217
Taiwan Smallreed 127217
Taiwan Snakemushroom 46824
Taiwan Snowbell 379347
Taiwan Southstar 33332
Taiwan Spapweed 205352
Taiwan Speedwell 407403
Taiwan Spicy Litse 233884
Taiwan Spinestemon 306974
Taiwan Spiraea 371912
Taiwan Spodiopogon 372412
Taiwan Spruce 298375
Taiwan Spurge 158906
Taiwan St. John's Wort 201880
Taiwan St. John's-wort 201880
Taiwan Stairweed 142642
Taiwan Starjasmine 393616
Taiwan Stauntonvine 374425
Taiwan Staunton-vine 374400
Taiwan Stellera 375198
Taiwan Stephania 375897
Taiwan Sterculia 376092
Taiwan Stickflower Syzygium 382672
Taiwan Stolon Willow 344176
Taiwan Stonebean-orchis 62571
Taiwan Storax 379347
Taiwan Stranvaesia 377446
Taiwan Strengthvine 242976
Taiwan Strigillose Eurya 160606
Taiwan Stringbush 414265
Taiwan Summerlilic 62015
Taiwan Suoluo 327362
Taiwan Supplejack 52423
Taiwan Supple-jack 52423
Taiwan Suregada 379794
Taiwan Suzukia 380046
Taiwan Swallowwort 117469
Taiwan Sweetleaf 381104,381310
Taiwan Sweetspire 210402
Taiwan Swordflag 208574
Taiwan Syneilesis 381818
Taiwan Syzygium 382535,382672
Taiwan Taiwan Tinospore 392253
Taiwan Tallowtree 346379
Taiwan Tallow-tree 346379
Taiwan Tanoak 233219
Taiwan Taro 99918
Taiwan Taroorchis 265440
Taiwan Taxillus 385237
Taiwan Tea 69636
Taiwan Telosma 385757
Taiwan Tephroseris 385921
Taiwan Theligon 389185
Taiwan Theligonum 389185
Taiwan Thistle 91982,92066
Taiwan Thrixsperm 390503
Taiwan Thrixspermum 390503
Taiwan Tigerthorn 122028
Taiwan Tinospora 392253
Taiwan Tipularia 392352
Taiwan Toadlily 396588
Taiwan Tongavine 147342

Taiwan Touch-me-not 205352
Taiwan Tournefortia 393267
Taiwan Towelgourdflower 95148
Taiwan Treerenshen 125604
Taiwan Trichodesma 395736
Taiwan Trident Maple 2815
Taiwan Trigonotis 397410
Taiwan Trillium 397620
Taiwan Tripterospermum 398295
Taiwan Tubelolabium 399976
Taiwan Tubergourd 390175
Taiwan Turpinia 400522
Taiwan Tutcheria 322612,400730
Taiwan Twayblade 232164,232333
Taiwan Tylophora 400945
Taiwan Vanilla 405028
Taiwan Velvetplant 183084
Taiwan Vernalgrass 27947
Taiwan Veronicastrum 407461
Taiwan Viburnum 408156
Taiwan Violet 409993,410412
Taiwan Walnut 212595
Taiwan Waterchestnut 394433
Taiwan Wendlandia 413784,413827
Taiwan White Pine 300083
Taiwan Whitepearl 172059
Taiwan Whiter Daphne 122484
Taiwan Whitetree 379794
Taiwan Whitetung 94055
Taiwan Whitfordiodendron 254796
Taiwan Whitlowgrass 137228
Taiwan Whytockia 414006
Taiwan Wild Quince 374425
Taiwan Wildginger 37615
Taiwan Wildmedia 28997
Taiwan Willow 343321,344176
Taiwan Willowweed 146901
Taiwan Windflower 24074
Taiwan Windhairdaisy 348256,348419
Taiwan Wingdrupe 405445
Taiwan Wingseedtree 320843
Taiwan Wing-seed-tree 320843
Taiwan Wintercress 47962
Taiwan Wintergreen 172059
Taiwan Winterhazel 106656,106684
Taiwan Wood Sorrel 277650
Taiwan Woodbetony 287766
Taiwan Woodrush 238712
Taiwan Wright Blueberry 404061
Taiwan Wurrus 166897
Taiwan Yaccatree 306431,306502
Taiwan Yam 131593
Taiwan Yelloweyegrass 416069
Taiwan Yew 385407
Taiwan Youngia 416437
Taiwan Zehneria 417492
Taiwan Zelkova 417547
Taiwanese Cheesewood 301363
Taiwanese Creeping Rubus 339188
Taiwanese Miscanthus 255920
Taiwanese Photinia 295773
Taiwania 383188,383189
Taiyuan Hackberry 80754
Taiyuan Milkvetch 43122
Taizhong Arrowwood 407853
Taizhong Buckthorn 328793
Taizhong Cheirostylis 86720
Taizhong Clematis 95392
Taizhong Viburnum 407853
Takahas Pine 300272
Takasago Fescuegrass 164349
Takase Nipplewort 221802
Takelagan Kengyilia 215966
Takelamagan Tamarisk 383618
Takeo Tainan Greyawngrass 372437
Takhian 198176
Takil Palm 393815
Takil Windmill-palm 393815
Taklamakan Tamarisk 383618
Takoru Palm 413091
Takuma-sudachi 93833
Talamisan 93553
Talang Tanoak 233385
Talas 99910
Talauma 383236
Tale Mint 250331
Talewort 57101
Tali Asparagus 39225
Tali Betony 373456
Tali Camellia 69691
Tali Corydalis 106500
Tali Falsehellebore 405638
Tali Helmet Orchid 105533
Tali Larkspur 124618
Tali Lily 230048
Tali Rabdosia 324890
Tali Woodbetony 287741
Taliangshan Aster 41338
Taliesin's Cress 357203
Taliew Syrenia 382108
Talimu Kneejujube 67074
Talimu Tamarisk 383619
Talinum 383293
Talipot Palm 106984,106999
Tali-range Rhododendron 331921
Talispot Rhododendron 331707
Tall 407254
Tall Acomastylis 4988
Tall Agrimony 11558
Tall Albizia 13649
Tall Albizzia 13649
Tall Alphitonia 17614
Tall Amaranth 18804
Tall Anabasis 21037
Tall Anemone 24112,24114
Tall Attalea 44805
Tall Baby's-breath 183225
Tall Baeckea 46439
Tall Beard-tongue 289337
Tall Beggarticks 54182
Tall Beggar-ticks 54182
Tall Bellflower 69892
Tall Bilberry 403904
Tall Birch 53338
Tall Bitterweed 201308
Tall Blue Lettuce 219229,219532
Tall Blueberry 403780
Tall Bluegrass 305886
Tall Blumea 55813
Tall Bog-sedge 75281
Tall Boloflower 205547
Tall Boneset 158037
Tall Boronia 57272
Tall Boykinia 58020
Tall Broomrape 275044,275128
Tall Buckwheat 162309
Tall Bugbane 91009
Tall Bulbophyllum 62706
Tall Burhead 140539
Tall Bur-head 140539
Tall Bush Clover 226965,226969
Tall Bush-pea 321727
Tall Buttercup 325518
Tall Camomile 26743
Tall Camomileleaf Woodbetony 287010
Tall Chalkplant 183170
Tall Chinchweed 286890
Tall Chloranthus 88272
Tall Cinquefoil 312391,312406
Tall Cluster Plantain Lily 198612
Tall Coelogyne 98645
Tall Common Corkwood 101396
Tall Coneflower 339574,339585
Tall Coreopsis 104611
Tall Corydalis 105850
Tall Cosmos 107161
Tall Cotton-grass 152739,152769,152798
Tall Crowfoot 325518
Tall Cupflower 266197
Tall Cup-flower 266197
Tall Curlylip-orchis 62706
Tall Currant 333913
Tall Dacrydium 121095
Tall Deutzia 127005
Tall Diffusive Woodbetony 287169
Tall Dock 339916
Tall Dove Orchid 291129
Tall Draba 136986
Tall Drop-seed 372629,372630
Tall Dyckia 139255
Tall Eria 396479

Tall Eupatorium 158037
Tall Euryodendron 160742
Tall False Buck's-beard 41841
Tall False Foxglove 10139
Tall Falsepimpernel 231521
Tall Fescue 163819,163993,164219,350629
Tall Fescuegrass 163806
Tall Fescue-grass 163819
Tall Field Buttercup 325518
Tall Firethorn Corkwood 101396
Tall Fiveangular Fimbristylis 166457
Tall Flat-topped Aster 135253
Tall Flat-topped White Aster 41448,135268
Tall Fleabane 150464,150610
Tall Forked Chickweed 284782
Tall Fothergilla 167541
Tall Gansu Barberry 51524
Tall Gastrodia 171918
Tall Gay-feather 228439
Tall Gentian 173209
Tall Germander 388076
Tall Globethistle 140704
Tall Globe-thistle 140704
Tall Glyceria 177610
Tall Golden Rod 368013
Tall Goldenrod 367950,368013,368035,368106
Tall Green Milkweed 37964
Tall Greyleaf Dogwood 380493
Tall Ground-cherry 297694,297697
Tall Hairorchis 396479
Tall Hairy Agrimony 11558
Tall Hawkweed 195857
Tall Hedge Mustard 365529
Tall Hedge-mustard 365529
Tall Helictotrichon 190132
Tall Hemarthria 191227
Tall Henry Mazus 246983
Tall Hibiscus 194850
Tall Holly 203792
Tall Hymenodictyon 201080
Tall Hypogomphia 202650
Tall Incarvillea 205547
Tall Ladybell 7643
Tall Lagotis 220195
Tall Larkspur 124054,124194,124209,124233
Tall Lespedeza 226965,226969
Tall Lettuce 219249
Tall Ligusticum 229319
Tall Liparis 232252
Tall Little Sophora 369004
Tall Love Grass 148019
Tall Lovegrass 147486
Tall Lungwort 250902
Tall Manna Grass 177610,177629
Tall Marigold 383090
Tall Marshmarigold 68169
Tall Meadow Rue 388469
Tall Meadow-rue 388469,388621,388628
Tall Meconopsis 247191
Tall Melic Grass 249053,249054
Tall Melick 248951
Tall Melilot 249205
Tall Micrandra 253041
Tall Milicia 254493
Tall Milkweed 37914
Tall Mint 250449
Tall Monkshood 5574
Tall Morning-glory 208120
Tall Mountain Thistle 91933
Tall Mouse-ear Hawkweed 195873
Tall Nettle 402886,402999
Tall Nightshade 367028
Tall Northern Bog Orchid 302361
Tall Northern Green Orchid 302361
Tall Nothern Bog Orchid 302360
Tall Nut Grass 354267
Tall Nut-rush 354267
Tall Oat Grass 34930
Tall Oatgrass 34930,34935
Tall Oat-grass 34930
Tall Ocotillo 167558
Tall Oncidium 270789
Tall Onion 15260
Tall Oplopanax 272706
Tall Oregon Grape 242478
Tall Oxeye 63542
Tall Oxwhipgrass 191227
Tall Pencilwood 139670
Tall Pepper-grass 225475
Tall Pepperwort 225369
Tall Pink 127707
Tall Pinnaflower 4988
Tall Porkbloodtree 160742
Tall Potentilla 312391
Tall Pricklypear 272868
Tall Pterocypsela 320511
Tall Purple Moor Grass 256605,256606
Tall Pussytoes 26321
Tall Ramping-fumitory 169013
Tall Rattanpalm 65711
Tall Rattlesnake Plantain 179679
Tall Rattlesnakeroot 313788
Tall Rattlesnake-root 313901
Tall Redtop 396706
Tall Rehmannia 327434
Tall Rice Flower 299308
Tall Rock-cress 30186
Tall Rocket 365398
Tall Roegneria 335202
Tall Rorippa 336196
Tall Rye Grass 163819
Tall Samaradaisy 320511
Tall Satintail 205514
Tall Saw-wort 348129
Tall Saxifrage 349056
Tall Scurrula 355300
Tall Sea-blite 379546
Tall Sedge 79747
Tall Sisymbrium 365398
Tall Skeleton Weed 151989
Tall Skullcap 355359
Tall Skyblue Adonis 8363
Tall Spotleaf-orchis 179679
Tall St. John's Wort 201937
Tall Stewartia 376457
Tall Stokes Aster 377292
Tall Sunflower 188968
Tall Swamp Marigold 53858
Tall Swertia 380191
Tall Switch Grass 282367
Tall Thimbleweed 24112,24114
Tall Thistle 91730
Tall Thoroughwort 158037
Tall Tick Clover 126363,126507,126517
Tall Tickseed 104611
Tall Torreya 393061
Tall Tubergourd 390170
Tall Tumble Mustard 365398
Tall Tumblemustard 365398
Tall Tumble-mustard 365398
Tall Tutsan 201937
Tall Twayblade 232146
Tall Verschaffeltia 407521
Tall Vervain 405812
Tall Violet 409947
Tall Water Hemp 18844
Tall Water-hemp 18844,18847
Tall Western Meadow-rue 388620
Tall Wheatgrass 144297,144643
Tall Wheat-grass 144643
Tall White Aster 380912
Tall White Beard-tongue 289337
Tall White Bog Orchid 183571
Tall White Bog Orchis 183571
Tall White Lettuce 313788
Tall White Violet 409798
Tall White-lettuce 313788
Tall Whitlowgrass 136986
Tall Whorled Milkwort 308437
Tall Wild Buckwheat 152033,152507
Tall Wild Lettuce 219249
Tall Windmill Grass 88342
Tall Woodbetony 287182
Tall Wood-sorrel 278099
Tall Woolly Wild Buckwheat 152037
Tall Woollyheads 318806
Tall Wormwood 35315
Tall Yarrow 3946
Tall Yellow Sweetclover 249205

Tall Yellow Sweet-clover 249205
Tall Yellowcress 336196
Tall Yellow-eyed-grass 416045
Tall Yucca 416578
Tall-cluster Plantainlily 198608
Tallculm Cypressgrass 118841
Tallculm Galingale 118841
Taller Discoid Groundsel 279918
Tallerack 155759
Taller-ligular Bitter-bamboo 303997
Tallest Goosecomb 335202
Tallest Sainfoin 271170
Tallest Skullcap 355359
Tallhelmet Monkshood 5374
Tallicona 72640
Tallnode Bamboo 297400
Tallor Wild Sunflower 188968
Tallow Bayberry 261202
Tallow Nut 415558
Tallow Shrub 261137
Tallow Tree 126805,289476,346361
Tallow Wood 155654
Tallow Wood Gum 155654
Tallowshrub 261202
Tallowtree 261202,346361,346408
Tallow-tree 126805,289476,346361,346408
Tallowwood 155654,415555
Tallow-wood 155654,415555,415558
Tallowwood Gum 155654
Tallstem Meconopsis 247191
Talong Spindle-tree 157640
Talss Crazyweed 279196
Talus Fritillary 168400
Tam Furze 401395
Tam Fuzz 401390
Tam Juniper 213902
Tama Bamboo 125479
Tamala Cassia 91432
Tamalpais Lessingia 227122
Tamanian Leatherwood 156066
Tamaquar Malouetia 243512
Tamarack 221904
Tamarack Larch 221904
Tamarack Pine 299869
Tamarilla 119974
Tamarillo 119973,119974
Tamarind 383404,383407
Tamarind Date 383407
Tamarind Tree 383407
Tamarindo 383407
Tamarind-tree 383407
Tamarisk 383412,383433,383437,383469,383491,383595
Tamarisk Family 383397
Tamarisk Galls 383437,383491
Tamarisk Salt Cedar 383412
Tamarisk-leaf Russianthistle 344735
Tamarix 383412,383595
Tamarix Family 383397
Tamarix-flowering Mimosa 255026
Tambookie Thorn 154622
Tambootie 372350
Tame 401390
Tame Cornel 105111
Tame Withy 85875
Tamil Nadu Reed Bamboo 268118,268119,268122
Tampala 18836
Tampico Fibre 10778
Tampico Sea-blite 379607
Tamsui Lily 229845
Tamura Clematis 95344
Tan Oak 233099,233176,233228,285203,323811
Tana River Poplar 311354
Tanais Sainfoin 271273
Tanaka Ailanthus 12564
Tanaka Cystopetal 320241
Tanaka Rattan Palm 65797
Tanaka Scurrula 237077
Tanaka Spicegrass 239876
Tanaka Stephanandra 375821
Tanaka Sweetleaf 381429
Tanaka Thistle 92425
Tanaka Tylophora 400972
Tanaka's Rattlesnakeroot 313894
Tanakaea 383887
Tanariver Poplar 311354
Tanbark Oak 233099,233176
Tan-bark Oak 233176
Tanbark-oak 233176
Tang Bluegrass 306061
Tang Honeysuckle 236084
Tang Monkshood 5620
Tang Sedge 76484
Tang Willow 344178
Tang's Zhejiang Camellia 68981
Tanga-tanga 13554
Tangbamboo 258184,364783,364820
Tangchi Eight Trasure 200806
Tangchi Stonecrop 200806
Tangerian Iris 208899
Tangerine 93446,93717,93721,93837
Tangerine Orange 93446,93837
Tanghin 83706
Tangier Iris 208899
Tangier Pea 222724,222847,222851
Tangier Pea Vine 222724
Tangile 362217
Tangin Tree 383907
Tanging Nettle 402886
Tangjiaqi Spapweed 205358
Tangjiaqi Touch-me-not 205358
Tangkula Eritrichium 153545
Tangkula Larkspur 124625
Tangleberry 172257
Tangled Rhododendron 330913
Tanglefoot Beech 266872
Tangle-grass 326287
Tanglehead 194015,194020
Tanglehead Grass 194020
Tangle-legs 407674
Tango-plant 89134,89135
Tangor 93649
Tangshen 98395
Tangtsinia 383909,383911
Tanguile 362217
Tangul Larkspur 124625
Tangula Range Speedwell 407431
Tangula Rockfoil 349446
Tangula Rockjasmine 23312
Tangula Sedge 76486
Tangula Speedwell 407431
Tanguru 270185
Tangut Almond 20968
Tangut Anisodus 25457
Tangut Barberry 51524
Tangut Bittercress 73001
Tangut Bluebeard 78037
Tangut Blue-beard 78037
Tangut Buckthorn 328868
Tangut Cavea 79786
Tangut Clematis 95345
Tangut Corydalis 106506
Tangut Currant 334231
Tangut Daphne 122614
Tangut Dragonhead 137672
Tangut Dwarf Globeflower 399535
Tangut Globeflower 399535
Tangut Greenorchid 137672
Tangut Honeysuckle 236144
Tangut Horsebladder 316956
Tangut Leek 15801
Tangut Lymegrass 144489
Tangut Milkvetch 43125
Tangut Monkshood 5621
Tangut Nitraria 266381
Tangut Olgaea 270247
Tangut Onion 15801
Tangut Peach 20968
Tangut Peashrub 72358
Tangut Pea-shrub 72358
Tangut Primrose 315033
Tangut Przewalskia 316956
Tangut Rhodiola 329963
Tangut Rockfoil 349976
Tangut Rush 213501
Tangut Sagebrush 36365
Tangut Sweetvetch 188139
Tangut Valerian 404367
Tangut Whitethorn 266381

Tangut Whorledleaf Woodbetony 287807
Tangut Wildryegrass 144489
Tangut Willowweed 146902
Tangut Windhairdaisy 348845
Tangut Wormwood 36365
Tania 415201
Tanias 415201
Tanier 415190
Tanjong Tree 255312
Tankan 93734
Tankan Mandarine 93734
Tanket 104350
Tanner Grass 58039,58058
Tanner's Apron 314143
Tanner's Barberry 51327
Tanner's Cassia 78236
Tanner's Cassia Bark 78236
Tanner's Dock 340084
Tanner's Oak 323625
Tanner's Sumac 332533
Tanner's Tree 104701
Tannia 415201
Tanning Sumach 332533
Tanno Fig 165734
Tanoak 233099,233112,233176,233228
Tan-oak 233099,285203
Tansy 3978,89402,89704,312360,359158,
383690,383874
Tansy Aster 240377,240451,240458
Tansy Chrysanthemum 89599
Tansy Leaf Aster 240458
Tansy Mustard 126112,126120,126127,
365500
Tansy Phacelia 293176
Tansy Ragwort 359158
Tansyaster 240376
Tansy-aster 240458
Tansyleaf Aster 240458
Tansyleaf Sagebrush 36363
Tansyleaf Wormwood 36363
Tansy-leafed Hawthorn 110077
Tansy-leafed Thorn 110077
Tansy-leaved Evening-primrose 269517
Tansy-leaved Rocket 365630
Tansymustard 126112
Tansy-mustard 126112
Tantalus White Hibiscus 194722
Tantoon Tea-tree 226472
Tanya 99910
Taohe Barberry 51703
Taohe Crazyweed 279197
Taohe Cutleaf Barberry 51471
Taohe Rhodiola 329883
Taohe Willow 344181
Taohe Windhairdaisy 348678
Taoho Crazyweed 279197
Taoho Rhodiola 329883
Taonan Rush 213502
Tao-river Cutleaf Barberry 51471
Taoyuan Aster 41341
Taoyuan Manyflowered Rose 336798
Taoyuan Marshweed 230324
Taozhou Angelica 24506
Tap Aloe 16817
Tapa Cloth 61107
Tapa Cloth Tree 61107
Tapa Mountain Prickly-ash 417345
Tapao Woodbetony 287744
Tapa-tree 61107
Tape Grass 404536
Tapegrass 404531
Tape-grass 404531,404563,418377
Tape-grass Family 200235
Tapeleaf Galingale 118442
Taper 405788
Taper Leaf 290866
Taper Tanoak 233341
Taper Tarenna 384917
Tapered Cudweed 317730
Tapered Jasmine 211739
Tapered Tarenna 384917
Tapering Pfeiffer-root 276463
Taperleaf Camellia 69033
Taperleaf Japanese Spiraea 371964
Taperleaf Spiraea 371964
Taper-leaved Camellia 69033
Taper-leaved Garlic 15698
Taperstem Giantfennel 163598
Tapertip False Wheatgrass 148400
Tapertip Hawksbeard 110717
Taper-tip Rush 212800
Tapeweed 311969
Tapeworm 359712
Tapeworm Plant 197772
Taphropermum 383989
Tapingtze Skullcap 355791
Tapintze Sedge 76487
Tapioca 244507
Tapioca Plant 244507
Tappasha Raspberry 338967
Tapria Siris 13609
Taproot Dock 340279
Tap-rooted Valerian 404254,404255
Tapu Maple 3652
Taqiwan Swallowwort 117717
Taquet Skullcap 355778
Tar 408571
Tar Grass 408352
Tar Tree 357879
Tar Vetch 408571
Tara 65068
Tara Actinidia 6515
Tara Vine 6515
Tara Wingceltis 320412
Tarahumara Sage 345402
Taraike Willow 344184
Tarajo 203961
Tarajo Holly 203961
Tarata 301261
Taraw Palm 234191
Tarbagatai Pricklythrift 2291
Tarbagatai Prickly-thrift 2291
Tarbottle 81237
Tarbush 167054
Tare 68713,102933,222681,408251,
408423,408571,408611
Tare Fitch 408352,408423
Tare Vetch 408352,408423
Tare-fitch 408352,408423
Tare-grass 408352,408423
Tarenna 384909
Tarennoidea 385052,385055
Tares 235373
Tare-vetch 408423
Target 59144,59148
Tarhui 238468
Tarim Tamarisk 383619
Tark's-cap Lily 229922
Taro 16512,99899,99910
Tarogayo 93847
Taroko Angelica 24484
Taroko Bedstraw 170661
Taroko Begonia 50356
Taroko Elaeagnus 142204
Taroko Euphorbia 158857
Taroko Eyebright 160275
Taroko Gentian 173957
Taroko Groundsel 360191
Taroko Oak 324467
Taroko Small-leaved Box 64323
Taroko Spiraea 372094
Taro-leaf Steudnera 376382
Taroleaf Wildginger 37646
Taron Azalea 331934
Taron Litse 234077
Taron Mahonia 242650
Taron Monkshood 5625
Taron Rhododendron 331934
Taroorchis 265366
Taroucan Barberry 52156
Tarphochlamys 385065,385067
Tarragon 35411
Tarry Pea 222789
Tarry Tongue 262722
Tarry-tongue 262722
Tart Rhubarb 329388
Tartago 212202
Tartar Dogwood 104920
Tartar Gum 376093
Tartar Heteropappus 193983
Tartar Jurinea 193983

Tartarian Cephalaria 82139
Tartarian Oats 45448
Tartarian Statice 179368,179384
Tartarian Violet 409682
Tartar-root 280799
Tartary Buckwheat 162335
Tartogo 212202
Tar-vetch 408571
Tarweed 20837,20840,20842,20843, 181174,191719,241523,241528,241543
Tarweed Fiddle-neck 20837
Tarwi 238468
Tarwood 121088
Tarwort 167054
Tasajo 272793
Tashan Woodbetony 287737
Tashikuergan Hedinia 187367
Tashikuergan Microsisymbrium 253962
Tashiro Azalea 331936
Tashiro Bogorchid 158090
Tashiro Clematis 95350
Tashiro Eightangle 204598
Tashiro Indian Hawthorn 329075
Tashiro Kalanchoe 215272
Tashiro Maackia 240131
Tashiro Milkbean 169657
Tashiro Milkpea 169657
Tashiro Mucuna 259516
Tashiro Parachampionella 378231
Tashiro Rhododendron 331936
Tashiro Saddletree 240131
Tashiro Skullcap 355494
Tashiro's Eupatorium 158090
Tashkent Goatweed 8828
Tasmania Oak 155558
Tasmania Pencil Pine 44047
Tasmanian Beech 266867
Tasmanian Blue Eucalyptus 155589
Tasmanian Blue Gum 155589
Tasmanian Bluegum 155589,155640
Tasmanian Cedar 44046,44049,44050
Tasmanian Cider Tree 106805
Tasmanian Cypress Pine 67426
Tasmanian Cypress-pine 67426
Tasmanian Daisy Bush 270207
Tasmanian Fitzruya 166725
Tasmanian Honey Myrtle 248072
Tasmanian Laurel 26183,26184
Tasmanian Myrtle 266867
Tasmanian Ngaio 260717
Tasmanian Oak 155558,155672
Tasmanian Peppermint 155478
Tasmanian Podocarp 306399
Tasmanian Podocarpus 306399
Tasmanian Sassafras 43972,44883
Tasmanian Snow Gum 95431,155533, 235090
Tasmanian Teak 121100
Tasmanian Wallaby Grass 341252
Tasmanian Waratah 385743
Tasmanian Yellow Gum 155613
Tassel 81237,167955
Tassel Amaranth 18788
Tassel Bushy Silk 171546
Tassel Calanthe 65871
Tassel Coelogyne 98651
Tassel Cord Rush 328196
Tassel Dendrobium 125135
Tassel Euonymus 157539
Tassel Flickingeria 166961
Tassel Flower 18687,60126,144884,144903, 144976,388506,397964
Tassel Fragrantbamboo 87636
Tassel Gentian 173702
Tassel Grape Hyacinth 260308
Tassel Grape-hyacinth 260308
Tassel Hyacinth 260301,260308
Tassel Nervilia 265386
Tassel Pondweed 340476
Tassel Rags 343257
Tassel Rockfoil 350033
Tassel Rue 394584
Tassel Sedge 74489
Tassel Sunipia 379785
Tassel Taroorchis 265386
Tasselcalyx Bladderwort 403827
Tasseled Grape Hyacinth 260308
Tasseleflower Thrumwort 18687
Tasselflower 144884
Tasselleaf Euonymus 157466
Tasselleaf Stonecrop 356572
Tasselled Grape-hyacinth 260308
Tasselled Primrose 314259
Tasselless Goldenray 228992
Tasselpeta Corydalis 105875
Tasselpetal Dendrobium 125177
Tassel-rags 343151
Tassel-rue 394583,394584
Tasselscale Windhairdaisy 348499
Tasseltree 87694
Tasselweed 340470
Tassel-white 210421
Tasteless Stonecrop 357147
Tasteless Yellow Stonecrop 357147
Tasua 19957
Tatajuba 46514,46515
Tatak Gentian 173958
Tatar Catchfly 364100
Tatar Colewort 108641
Tatar Crazyweed 279200
Tatar Dogwood 105207
Tatar Eyebright 160239
Tatar Giantfennel 163729
Tatar Heronbill 153920
Tatar Honeysuckle 236146
Tatar Ixiolirion 210999
Tatar Lettuce 259772
Tatar Madder 338033
Tatar Maple 3656
Tatar Motherwort 225015
Tatar Mulberry 259193
Tatar Reedbentgrass 65362
Tatar Saltbush 44668
Tatar Stock 246556
Tatarian Aster 41342
Tatarian Bread 108641
Tatarian Buckwheat 162335
Tatarian Cephalaria 82151
Tatarian Dogwood 104920,104921,104923
Tatarian Honeysuckle 236146,236150
Tatarian Ixiolirion 210999
Tatarian Lettuce 259772
Tatarian Maple 3656
Tatarian Milkvetch 42893
Tatarian Motherwort 225015
Tatarian Orach 44668
Tatarian Orache 44668
Tatarian Pricklythrift 2292
Tatarian Prickly-thrift 2292
Tatarian Saltbush 44668
Tatarian Sea Lavender 179384
Tatarian Sea-lavender 179384
Tatarian Silene 364100
Tatarian Statice 230787
Tatarinow Catchfly 364103
Tatarinow Clematis 95351
Tatarinow Honeysuckle 236161
Tatarinow Melandrium 364103
Tatarinow Milkwort 308397
Tatarinow Rattlesnakeroot 313895
Tatarinow Stonecrop 200807
Tatarinow Woodbetony 287745
Tatartan Catchfly 364100
Tatary Knotweed 162335
Taters 367696
Tatewaki Cheirostylis 86725
Tateys 367696
Tatian Skullcap 355798
Tatntalobeak Woodbetony 287743
Tatsien Woodbetony 287746
Tatsienlu Rhododendron 331939
Tattered Medick 247337
Tatung Larkspur 124545
Tatunsan Rhododendron 331305
Tau Foo 177750
Taupata 103798
Taur Candytuft 203248
Taur Gagea 169509
Taur Serpentroot 354959
Taur Shamrue 185672
Taur Woad 209228

Tauri Wormwood 36367
Taurian Bellflower 70255
Taurian Thistle 271705
Taurus Candytuft 203248
Taurus Cerastium 82739
Taurus Cotton Thistle 271705
Taurus Cottonthistle 271705
Taurus Desertcandle 148559
Taurus Hedge Maple 2842
Tau-saghyz 354961
Tau-saghyz Rubber 354961
Tausch Goatgrass 8727
Tausch Hawkweed 196028
Tausch's Goatgrass 8727
Tauscheria 385129
Tavaresia 385146
Tavoy Phoebe 295410
Tawa 50636
Tawada Lasianthus 222281
Tawang Buchanan Barberry 51397
Tawhai 266870
Tawhiwhi Kohuhu 301398
Tawnish Hydrangea 199947
Tawny Cotton-grass 152791,152796
Tawny Day Lily 191284
Tawny Daylily 191284
Tawny Day-lily 191284
Tawny Fig 165023
Tawny Pea 301064
Tawny Sedge 74830
Tawnycolour Horn Moth Orchid 293595
Tawnyhair Honeysuckle 235801
Taxan Oak 324479
Taxan Walnut 212627
Taxas Ebony 301136
Taxas Oak 324479
Taxas Walnut 212627
Taxi Fuzz 401395
Taxillus 385186,385242
Taxkorgan Nepeta 265075
Taxodium Family 385250
Taxol 385301
Taxus Family 385175
Taybercy 338059
Taygetus Saxifrage 349982
Taylor Barberry 51691
Taylor Cotoneaster 107702
Taylor Duabanga 138726
Taylor Germander 388291
Taylor Primrose 315043
Taylor Skullcap 355800
Taylor Woodbetony 287747
Taylor's Arctic Campion 363581
Taylor's Fawn-lily 154947
Taylor's Parches 109097
Tazzle 92022
Tchirisch 39328
Tchitola 278631
Tea 68877,69634,339887
Tea Arrowwood 408127
Tea Balm 249504
Tea Bush 68877,69634,268671
Tea Camellia 69634
Tea Crabapple 243630
Tea Family 389046
Tea Flower 146782,166125,345631
Tea Grass 391397
Tea of Heaven 200085
Tea Olive 276291
Tea Plant 11549,68877,69634,69644,
239055,362617
Tea Rose 336813
Tea Senna 85232
Tea Sweetleaf 381437
Tea Tree 226450,226479,226481,239021,
248075,345631,380756
Tea Viburnum 408127
Tea Wood 299835
Teabalm 249504
Teaberry 172089,172135
Tea-berry 407737
Teab-your-mother's-eyes-out 407072
Tea-coloured Sweetleaf 381437
Tea-coured Spindle-tree 157917
Teacups 325518,325666,326287
Teaflower Azalea 330302
Tea-girse 391397
Teak 385529,385531
Teal Love Grass 147733
Tealeaf Agapetes 10297
Tealeaf Euonymus 157921
Tealeaf Sageretia 342162
Tealeaf Sweetleaf 381437
Tealeaf Willow 343886
Tea-leaved Agapetes 10297
Tea-leaved Euonymus 157921
Tea-leaved Sageretia 342162
Tea-leaved Spindle-tree 157921
Tea-leaved Willow 343484,343886,343890,
343894
Teamster's Tea 146224,146262
Tea-oil Camellia 69411
Tea-oil Plant 69411
Tea-plant 238996
Tear Thumb 309796
Tear-blanket 1264,30760
Tearthumb 308690
Tear-thumb 309796
Teasel 133451,133454,133478,133505,
158250
Teasel Clover 397002,397056
Teasel Family 133444
Teasel Nut Sedge 119476
Teasel sedge 118784
Teasel-flatsedge 118763
Teaser 73430
Teastick Bamboo 318283
Teatfruit Rhodiola 329918
Tea-tree 226450,226481,239021,248075
Teatshaped Ardisia 31518
Teat-shaped Euphorbia 159305
Teazel 133451
Teazle 133451,133505
Tecamulose Spapweed 205272
Tecamulose Touch-me-not 205272
Tecate Cypress 114702,114703
Techin Aster 41351
Techin Sinocrassula 364934
Tecoma 385461
Tecomaria 385505
Teddy Bear Cholla 116648
Teddy Bear Palm 139366
Teddy Buttons 216753
Teddy-bear Banksia 47631
Teddybear Cholla 272805,272814
Teddy-bear Cholla 116648,272814
Teddy-bear Plant 115070
Teddy-bear Vine 115556
Teesdale Sandwort 255582
Teesdale Violet 410513
Teethear Chinese Groundsel 365040
Teff 147468,147994
Tegmentose Maple 3665
Tehachapi Ragwort 279919
Tehkeh Dwarf Globeflower 399536
Teil 391706
Teinturier 411981
Teinturier Grape 411981
Tejocote 109548
Tekes Milkvetch 43142
Teke-saghyz Rubber 354798
Tekesi Goldenray 229077
Tekesi Peashrub 72361
Tekesi Pea-shrub 72361
Telegraph Plant 98152,98159
Telegraph Tick Clover 98159
Telegraph-plant 194264
Telegraph-weed 194215
Telescope Peak Wild Buckwheat 152043
Tellichery Bark 197199
Tell-time 384714
Telosma 385749,385752
Tembusu 162348
Temiche Cap 244497
Temola Barberry 52220
Temple Bamboo 357939
Temple Bells 366714,366715
Temple Flower 305225
Temple Juniper 213896
Temple Orange 93848
Temple Tree 305206,305213,305225,305226

Temple-bells 366714
Templetree 305225
Temu 238337
Ten Camellia 69692
Ten Epipactis 147235
Ten Neottia 264747
Ten Nestorchid 264747
Ten o'clock 274823
Ten o'clock Flower 311921
Ten Pendentlamp 84276
Ten Spiraea 372095
Ten Wattle 1642
Ten Weeks Stock 246478
Tenacious Condorvine 245852
Tenacious Marsdenia 245852
Tenacious Milkgreens 245852
Tenaciouspod Ormosia 274403
Tenacular Woodbetony 287750
Tenagocharis 64177
Tenangle Pipewort 151286
Tenasserium Pine 300068
Tenaza Huajillo 186238
Tenby Daffodil 262427,262441
Tenchweed 312190
Tender Bedstraw 170208
Tender Begonia 50360
Tender Bluegrass 305760
Tender Bothriospermum 57689
Tender Catchweed Bedstraw 170208
Tender Chorispora 89020
Tender Comastoma 100284
Tender Cystopetal 320218
Tender Gagea 169510
Tender Haworthia 186753
Tender Meadowrue 388688
Tender Monkeyflower 255246
Tender Parnassia 284622
Tender Pimpinella 299418
Tender Poppy 282740
Tender Silverpuffs 253923
Tender Spotseed 57689
Tender Throathair 100284
Tenderflower Eargrass 187681
Tender-flowered Hedyotis 187681
Tendergreens 59567
Tenderstem Rabdosia 209673
Tendertwig Sandwort 32174
Tendril Childseng 318493
Tendriled Iodes 207357
Tendrilleaf Fritillary 168563
Tendrilleaf Landpick 308523
Tendrilleaf Solomonseal 308523
Tendrilled Meadowrue 388449
Tendrilled Odontoglossum 269061
Tendrillous Yam 131872
Tenerife Broom 121017
Teneriffe Viper's Bugloss 141177
Tenfeed Arundinella 37368
Teng Tongue 262722
Teng Tylophora 400975
Tengchong Allomorphia 16023
Tengchong Anemone 24077
Tengchong Aralia 30775
Tengchong Arisaema 33534
Tengchong Arrowbamboo 162750
Tengchong Arrowhead 342419
Tengchong Arundinella 37431
Tengchong Azalea 330591
Tengchong Basketvine 9483
Tengchong Begonia 49728
Tengchong Clearweed 298939
Tengchong Cowparsnip 192380
Tengchong Euonymus 157913
Tengchong Evergreenchinkapin 79067
Tengchong Gingerlily 187473
Tengchong Indigo 206101
Tengchong Loosestrife 239878
Tengchong Premna 313738
Tengchong Rhododendron 330591
Tengchong Southstar 33534
Tengchong Willow 344193
Tengchong Windflower 24077
Tengchong Yam 131507
Tengchong Yushanbamboo 416770
Tengia 385836,385838
Teng-tongue 262722
Tengu 93466,93849
Tengyue Arrowwood 408160
Tengyue Aster 41161
Tengyue Loquat 151174
Tengyue Viburnum 408160
Tenjiku Sedge 75775
Ten-months Yam 131458
Ten-mouths Yam 131458
Tennessee Chickweed 374806
Tennessee Pondweed 312277
Tennessee Purple Coneflower 140090
Tenous Feathergrass 376836
Tenpetal Anemone 23720
Ten-point Phlox 295252
Ten-rayed Sunflower 188953
Tenrows Onosma 271750
Tens-o'-thousands 243209
Tenstamen Bigginseng 241163
Tenstamen Maple 2926
Ten-stamen Maple 2926
Tenstamened Begonia 49767
Ten-stamened Big-ginseng 241163
Tentacle Catchfly 364219
Tentacle-bearing Yam 131872
Tentaculate Lasianthus 222283
Tentaculate Roughleaf 222283
Tentes 212636
Tenuicauliferous Monkshood 5628
Tenuidivided Woodbetony 287754
Tenuifolious Bittercress 73018
Tenui-inflorescence Willow 344195
Tenuis Garnotia 171528
Tenuis Stenoseris 375715
Tenuis Tamarisk 383622
Tenuistyle Thick Hairs Rhododendron 331432
Tenuity Bluegrass 306067
Tenuity-leaved Maple 3674
Tenuous Bigbeak Monkshood 5404
Tenuous Clovershrub 70899
Tenuous Eutrema 161149
Tenuous Fruit Milkvetch 43199
Tenuous Leaf Galingale 119682
Tenuous Lipocarpha 232426
Tenuous Sibbaldia 362380
Tenuous Twinballgrass 209130
Tenuous Vine Actinidia 6622
Tenuous Yunnan Eutrema 161159
Tenuous-branch Clovershrub 70899
Tenuousculm Galingale 119674
Tenuous-flower Barley 198307
Tenuousflower Ervatamia 154197,382766
Tenuousflower Nepeta 265078
Tenuousflower Skullcap 355808
Tenuous-fruited Milkvetch 43199
Tenuousleaf Actinidia 6572
Tenuousleaf Kiwifruit 6572
Tenuousspanicle Schefflera 350787
Tenuous-stem Milkvetch 43143
Tenuousstem Monkshood 5628
Tenuousstem Woodbetony 287753
Tenuous-vined Actinidia 6622
Ten-week Stock 246478
Tenweeks Stock 246478
Teosinte 155876,155878
Tepa 223014
Tepary Bean 293969,293970,408829
Tepejilote Palm 85438
Tepescohuite 255126
Tephroseris 385884
Tephrosia 385926,386090
Tequila 10778
Terai Bamboo 249617
Teramnus 386403
Terap 36940
Tere 408571
Terebinth 301002
Terebinth Pistache 301002
Terebinth Tree 301002
Terete Glabripetal Elaeocarpus 142324
Terete Leucas 227718
Terete Stock Hedyotis 187557
Teretedleaf Gladiolus 176601
Teretelaef Luisia 238324
Terete-leaf Dendrobium 125383
Terete-leaf Epidendrum 146490

Teretepetiolet Poplar 311144
Teretiraceme Clovershrub 70900
Terminal Camellia 69492
Terminal Dracaena 104367
Terminal Heliciopsis 189976
Terminal Moxanettle 221538
Terminal Woodnettle 221538
Terminal-flower Hedyotis 187685
Terminalflower Pachysandra 279757
Terminal-flower Pepper 300529
Terminalia 386448,386504
Terminate Cranesbill 174959
Terminate Dendrobium 125384
Ternata Henry Clematis 94994
Ternate Carpet Cinquefoil 313047
Ternate Clematis 95066
Ternate Crabgrass 130797
Ternate Dyeingtree 346495
Ternate Dyeing-tree 346495
Ternate Larkspur 124637
Ternate Metalleaf 296420
Ternate Mikania 254448
Ternate Oldham Sping Raspberry 339137
Ternate One-leaf Dyetree 302681
Ternate Pinellia 299724
Ternate Waldsteinia 413034
Ternate Wild Buckwheat 152513
Ternateleaf Corydalis 106521
Ternateleaf Plectranthus 209844
Ternateleaf Rabdosia 209844
Ternate-leaved Cinquefoil 312828
Ternate-leaved Euonymus 157915
Terniopsis 121978
Ternstroemia 386689,386696
Ternstroemia-like Anise Tree 204600
Ternstroemia-like Anise-tree 204600
Ternstroemialike Eightangle 204600
Ternstroemia-like Illicium 204600
Ternstroemia-like Pittosporum 301385
Ternstroemia-like Seatung 301385
Terrateleaf Australnut 240210
Terrateleaf Macadamia 240210
Terrell Grass 144528
Terrestrial Cowhorn Orchid 120714
Terrestrial Helixanthera 190863
Terrestrial Loosestrife 239879
Terrestrial Starwort 67395
Terrestrial Water-starwort 67395
Terrydevil 162519,367120
Terrydiddle 162519,367120
Tesota 270527
Tessariumfruit Begonia 50364
Tesselated Rattlesnake-plantain 179707
Tessellate Fritillary 168531
Tessellate Gelidocalamus 172750
Tessellatted Haworthia 186756
Tesselpetal Primrose 314554
Tessllateleaf Landpick 308651
Tessllateleaf Solomonseal 308651
Testicular Aglaia 11313
Testiculate Aglaia 11313
Tether Devil 162519,367120
Tether Toad 326287
Tether-devil 162519,367120
Tether-grass 170193
Tether-toad 326287
Tetra Mad 44708
Tetracentron 386850,386851
Tetracentron Family 386847
Tetracera 386853
Tetracme 386946
Tetracoccous Camellia 69686
Tetracoccous Mallotus 243452
Tetradoxa 387068,387069
Tetradrous Hackberry 80764
Tetraena 387093,387116
Tetrafloret Bluegrass 306069
Tetrafloriferous Litse 233907
Tetragoga 387149
Tetragon Cassiope 78648
Tetragonal Fimbristylis 166530
Tetragonal Fluttergrass 166530
Tetragonal Sweetleaf 381291
Tetragonal Waterchestnut 394477,394535
Tetragonia 387156,387221
Tetragonous Hinoki False Cypress 85330
Tetragonousseed Snakegourd 396277
Tetra-leaved Queensland Nut 240212
Tetrameles 387293
Tetramerous Jasminorange 260178
Tetramerous Whitepearl 172162
Tetramerous Wintergreen 172162
Tetrandous Stephania 375904
Tetrandrous Wildberry 138411
Tetraplasia 387484
Tetrasepaled Nightshade 367632
Tetrasepalous Actinidia 6708
Tetrastamen Maple 3677
Tetrathyrium 387933
Tetsan 201732
Tetter 86755,325820
Tetter-berries 61497,383675
Tetterberry 61497
Tetterbush 298770
Tetterwort 86755,345805,383675
Tetty 367696
Teuscher Begonia 50366
Texan Phacelia 293169
Texan Pride 295267
Texan Redbud 83795
Texan Rhatany 218081
Texan Sumac 332686
Texas Amorpha 20029
Texas Ash 168119
Texas Aster 380865
Texas Barrel Cactus 163447
Texas Bean 293969
Texas Bear Grass 266509
Texas Betony 373179
Texas Black Walnut 212646
Texas Blue Grass 305356
Texas Bluebell 161027
Texas Bluebonnet 238488
Texas Blueweed 188933
Texas Boxelder Maple 3230
Texas Broomweed 20485
Texas Buckeye 9677
Texas Buckthorn 418196
Texas Candyleaf 376411
Texas Catclaw 1707
Texas Centaury 81543
Texas Chinchweed 286878
Texas Colubrina 100078
Texas Cone Cactus 264121
Texas Coneflower 339621
Texas Croton 113037
Texas Desert Goldenrod 264234
Texas Dutchman's-pipe 34307
Texas Dwarf Dandelion 218207
Texas False Agave 187137
Texas Firecracker Bush 185169
Texas Goldentop 161086
Texas Goldenweed 264234
Texas Groundsel 358250
Texas Hawthorn 110086
Texas Kidneywood 161864
Texas Knotweed 309827
Texas Lantana 221313
Texas Live Oak 323926
Texas Madrone 30894
Texas Mesquite 315557
Texas Millet 282304,402567
Texas Mimosa 1264
Texas Moonseed 250221
Texas Mountain Laurel 369116
Texas Mountain-laurel 369116
Texas Mulberry 259164,259165
Texas Nipple Cactus 244202
Texas Olive 104159
Texas Panic Grass 402567
Texas Persimmon 132431
Texas Pistache 301004
Texas Plume 49161
Texas Poison-ivy 393474
Texas Prickly Pear 272876
Texas Pricklypear 272876
Texas Prickly-pear 272825
Texas Pride 389205
Texas Primrose 68504
Texas Purple Thistle 92442
Texas Purple-spike 194527

Texas Rabbitbrush 150355
Texas Rainbow 140236
Texas Rainbow Cactus 140236
Texas Rainbow Hedgehog 140236
Texas Ranger 227919
Texas Red Oak 323725,324415
Texas Redbud 83764
Texas Rose 355785
Texas Rush 213524
Texas Sacahuiste 266509
Texas Sage 227919,344885,344980
Texas Sea-purslane 361676
Texas Shrub 208353
Texas Silverleaf 227919,227926
Texas Skeletonplant 239340
Texas Sleepy-daisy 414985
Texas Snake-cotton 168694
Texas Snakeroot 34307
Texas Snakeweed 182113
Texas Snowbell 379469
Texas Sophora 368958
Texas Sotol 122910
Texas Star 231607
Texas Stork's Bill 153921
Texas Sunflower 189034
Texas Thistle 92442,367567
Texas Tuberose 10892,244371
Texas Umbrella Tree 248893,248909
Texas Violet 345023
Texas Walnut 212646
Texas Western-daisy 43302
Texcalamate 165455
Tex-mex Tobacco 266051
Texoka Buffalograss 61724
Texsel Greens 59348,59438
Teyl 391706
Teyleria 388335
Teysmann Palm 212402
Thai Bai Makrut 93492
Thai Basil 268644
Thailand Cogflower 154192,382822
Thailand Disperis 134326
Thailand Ebony 132305
Thailand Loosestrife 239846
Thailand Papeda 93523
Thailand Ramie 56312
Thalassia 388357,388360,388362
Thale Cress 30105,30146
Thale-cress 30146
Thalia 388380,388385
Thalia Lovegrass 147509
Thalia-like Dracena 137342,137514
Thames Yellow-cress 336168
Thamnocharis 388765,388766
Thanaka 260173
Thanet Cress 73090
Thanet Weed 73090
Thape 334250
Thatch Leaf Palm 198808
Thatch Palm 97883,97887,198808,341422,
390451,390462,390463,390467
Thatch Screw Pine 281138
Thatch Screwpine 281138
Thatch Screw-pine 281138
Thatching Grass 201514
Thatch-leaf Palm 198808
Thaumastochloa 388906
Thayer Azalea 331968
Thayer's Rhododendron 331968
Thcknode Arrowbamboo 162665
Thck-noded Fargesia 162665
The Bride Pearlbush 161742
The Longer-the-dearer 410677
The Wig Japanese Maple 3315
Theabberry 334250
Theafruit Holly 204341
Thea-fruited Holly 204341
Thea-leaved Sweetleaf Bogoti Tea 381437
Theetsee 177545
Thelasis 389122
Theligon 389183
Theligon Family 389181
Theligona Family 389181
Theligonum 389183
Thellung Pimpinella 299555
Thellung Sagebrush 36377
Thellungiella 389197
Thelma Rose 336312
Thelocactus 389203
Thelygonum 389183
Thelymitra 389258
Themeda 389307
Thenardia 389396
Therds-grassop 11972
Thermopsis 389505,389558
Theropogon 389567,389568
Therorhodion 389570
Thesionlike Mosquitotrap 117721
Thespis 389961
Thetch 408571,408611
Theves Poplar 311398
Thevetia 389967
Thew Roxburghia 337810
Thick Anemone 24030
Thick Cajan 65147
Thick Goatgrass 8671
Thick Hairs Rhododendron 331430
Thick Hedgehogcactus 140874
Thick Ladypalms 329188
Thick Premna 313621
Thick Primrose 314766
Thick Privet 229601
Thick Rachis Camellia 69016
Thick Redpepper 72075
Thick Sepal Gastrochilus 171915
Thick Shellbark Hickory 77920
Thick Stem Batrachium 48928
Thick Styles Rhododendron 330468
Thick Tooth Actinidia 6626
Thick Tooth Kiwifruit 6626
Thick Umbrella Bamboo 162696
Thickanther Tea 69428
Thick-bark Harbin Poplar 311275
Thickbark Sweetleaf 381291
Thick-barked Sweetleaf 381291
Thickbill Tickseed 104770
Thickbracted Goldenbush 150360
Thick-branch Arrowbamboo 162726
Thickbranch Glaudular Eurya 160477
Thick-branch Machilus 240665
Thickbranch Willow 343283,344264
Thick-branched Amoora 19958
Thick-branched Barberry 51523
Thick-branched Willow 343283
Thickcalyx Clematis 95457
Thickcalyx Crabapple 243586
Thickcalyx Purplepearl 66800
Thick-collared Yushania 416766
Thickcolumn Tea 69016
Thick-columned Camellia 69016
Thickconered Everlasting 21659
Thickculm Dallisgrass 285536
Thickculm Paspalum 285536
Thickculm Spikesedge 143312
Thickcupule Tanoak 233198
Thick-cupuled Tanoak 233198
Thickdge Qinggang 116210
Thicked Onosma 242397
Thicked Woodbetony 287758
Thickedge Gentian 173899
Thickening Grass 299759
Thickening-hairs Pearleverlasting 21659
Thicket Creeper 285127
Thicket Dewberry 339037
Thicket Hawthorn 109609
Thicket Juneberry 19251
Thicket Rattlebox 112799
Thicket Sage 345014
Thicket Sedge 75833
Thicket Serviceberry 19251
Thicket Wintergreen 172069
Thick-fibred Guihaia 181805
Thickfilmtree 163396
Thick-flower Masdevallia 246116
Thickflower Monkshood 5150
Thick-flower Waterlily 267771
Thickflower Waxplant 198838
Thickflowered Bluegrass 305799
Thick-flowered Rhododendron 330901
Thick-flowers Rhododendron 330901
Thick-footed Manglietia 244430

Thick-footed Rhododendron 331427
Thickfruit Cliffbean 254796
Thickfruit Linearleaf Alyssum Alyssum 18404
Thickfruit Madwort 18370
Thick-fruit Mayten 246812
Thick-fruit Michelia 252941
Thick-fruit Millettia 254796
Thickfruit Snakehead Dipoma 133413
Thick-fruit Sophora 369095
Thickfruit Vetch 408371
Thick-fruitea Sloanea 366047
Thick-fruited Mayten 246903
Thick-fruited Millettia 254796
Thickhair Aralia 30755
Thickhair Grewia 180813
Thickhair Primrose 315065
Thickhair Wendlandia 413833
Thickhair Willow 343283
Thick-haired Rhododendron 331430
Thickhairy Waldheimia 413023
Thickhead 108736
Thickhorn Disa 133754
Thickhorn Stairweed 142773
Thickhull Camellia 69579
Thickleaf Actinidia 6610,6618
Thickleaf Ainsliaea 12626
Thickleaf Alangium 13372
Thick-leaf Alpine Holly 203829
Thickleaf Appendicula 29812
Thickleaf Banfenghe 357969
Thickleaf Beauverd Photinia 295635
Thickleaf Bedstraw 170333
Thickleaf Begonia 49791
Thick-leaf Biondia 54517
Thickleaf Boea 283114
Thickleaf Camellia 69023
Thickleaf Chickweed 374807
Thickleaf Citrus 93643
Thickleaf Clearweed 299044
Thickleaf Clematis 94866
Thickleaf Coldwaterflower 299044
Thickleaf Crabapple 243589
Thickleaf Cratoxylum 110256
Thickleaf Croton 112871
Thickleaf Curculigo 114821
Thickleaf Daphne 122562
Thick-leaf Deutzia 126887
Thickleaf Dolomiaea 135722
Thickleaf Dontostemon 136162
Thickleaf Drymary 138493
Thickleaf Eargrass 187697
Thickleaf Eightangle 204569
Thickleaf Elatostema 142634
Thickleaf Eria 299626
Thick-leaf Eurya 160447,160474
Thickleaf Evergreenchinkapin 78901
Thickleaf Fox-tail Orchid 9280
Thickleaf Gentian 173965
Thickleaf Globethistle 140688
Thickleaf Hairfruit Azalea 331796
Thickleaf Hairorchis 299626
Thickleaf Hemlock 399892
Thickleaf Holly 203786
Thickleaf Hydrangea 200068
Thick-leaf Hydrangeavine 351784
Thick-leaf Illicium 204569
Thick-leaf Jasmine 211959
Thickleaf Kiwifruit 6610,6618
Thickleaf Kumquat 167493
Thickleaf Lagotis 220165
Thickleaf Larkspur 124151
Thick-leaf Leptopus 226336
Thickleaf Lilyturf 272071
Thickleaf Maple 2909
Thickleaf Maxillaria 246710
Thickleaf Mayten 182751
Thickleaf Mazus 246972
Thickleaf Milkvetch 42233
Thickleaf Monkshood 5429
Thickleaf Mume 34458
Thickleaf Nothapodytes 266801
Thickleaf Oblongleaf Maple 3267
Thickleaf Ophiorrhiza 272194
Thickleaf Orach 44379
Thickleaf Orange 93643
Thickleaf Osmanther 276368
Thickleaf Paraboea 283114
Thickleaf Parnassia 284521,284605
Thickleaf Peashrub 72215
Thickleaf Pepperweed 225336
Thickleaf Photinia 295661
Thickleaf Primrose 314201,314245
Thick-leaf Pubescent Hornbeam 77296
Thickleaf Rabbiten-wind 12626
Thickleaf Raspberry 338299
Thickleaf Rockvine 387835
Thickleaf Slugwood 50590
Thickleaf Spicegrass 239602
Thick-leaf Spruce 298272
Thickleaf Stairweed 142634
Thick-leaf Stonecrop 356660
Thick-leaf Storax 379361
Thickleaf Sweetleaf 381163
Thickleaf Sweetspire 210378
Thickleaf Tanoak 233337
Thickleaf Ternstroemia 386712
Thickleaf Thrixsperm 390531
Thickleaf Thrixspermum 390531
Thickleaf Trigonotis 397448
Thickleaf Vatica 405182
Thickleaf Vineclethra 95530
Thickleaf Violet 410087
Thickleaf Vladimiria 135722
Thickleaf Whiteflower Embelia 144794
Thickleaf Woodlotus 244463
Thick-leafed 184597
Thickleaf-like Begonia 50197
Thick-leaved Acacia 1291
Thick-leaved Anise-tree 204569
Thick-leaved Croton 112871
Thick-leaved Deutzia 126887
Thick-leaved Erycibe 154236
Thick-leaved Evergreen Chinkapin 78901
Thick-leaved Glochidion 177141
Thick-leaved Holly 203786
Thick-leaved Honeysuckle 235745
Thick-leaved Hydrangea-vine 351784
Thick-leaved Jasmine 211959
Thick-leaved Manglietia 244463
Thick-leaved Maple 2909
Thick-leaved Photinia 295661
Thick-leaved Pittosporum 301248
Thick-leaved Raspberry 338299
Thick-leaved Rhododendron 331426
Thick-leaved Rockvine 387835
Thick-leaved Sallow 343252
Thick-leaved Saxifrage 349147,349216
Thick-leaved Slugwood 50590
Thick-leaved Spruce 298272
Thick-leaved Starwort 374807
Thick-leaved Stone Crop 356660
Thick-leaved Stonecrop 356660
Thick-leaved Sweetleaf 381163
Thick-leaved Tanoak 233162,233337
Thick-leaved Ternstroemia 386712
Thick-leaved Vineclethra 95530
Thick-leaved Wild Strawberry 167663
Thick-leaved Yong Shuh 165310
Thick-leaves Rhododendron 331426
Thick-limb Barberry 51509
Thicklip Herminium 192833
Thick-lobbed Millettia 254797
Thicklobe Millettia 254797
Thickly Leaved Rhododendron 330208
Thickly Woolly Aster 88160
Thickly-leaved Rhododendron 330208
Thick-nerve Anthurium 28091
Thick-nerve Pearlwood 296454
Thick-nerve Phyllanthodendron 296454
Thicknerve Sedge 76090
Thick-nerved Ardisia 31391
Thicknerved Cinnamon 91444
Thick-nerved Cinnamon 91444
Thick-nerved Holly 204215
Thick-nerves Ardisia 31391
Thick-pedicel Barberry 51767
Thick-pedicel Nanmu 295354,295355
Thick-pedicel Phoebe 295354,295355
Thick-pediceled Barberry 51767
Thick-pediceled Phoebe 295354
Thick-pedicelled Ardisia 31392

Thickpedicelled Corydalis 106224
Thick-pedicled Baneberry 6433
Thickpericarp Atalantia 43723
Thick-pericarped Atalantia 43723
Thickpeta Tea 69020
Thickpetal Camellia 69020
Thickpetal Habenaria 183553
Thick-petaled Camellia 69020
Thickpod Ormosia 274386
Thick-pod Ormosia 274422
Thickpod Pagodatree 369095
Thick-podded Ormosia 274422
Thickrachis Oberonia 267978
Thickrachis Stringbush 414230
Thick-rachised Stringbush 414230
Thick-ramel Bamboo 297447
Thickribcelery 279644
Thick-root Buttercup 325812
Thickroot Corydalis 105761
Thickroot Euphorbia 159285
Thickroot Gentian 173314
Thickroot Iris 208897
Thickroot Nettle 402958
Thickroot Saxifrage 349682
Thickroot Sedge 75660
Thickroot Swordflag 208897
Thick-rooted Ardisia 31394
Thickrostrate Begonia 49744
Thickrostrum Tickseed 104770
Thickscale Tanoak 233334
Thick-scaled Tanoak 233334
Thickseed Spapweed 204890
Thickseed Touch-me-not 204890
Thickshellcassia 113422
Thick-shelled Camellia 69023
Thick-spike Blazing-star 228511
Thick-spike Dendrocalamus 125500
Thick-spike Dragonbamboo 125500
Thick-spike Gay-feather 228511
Thickspike Indosasa 206884
Thickspike Leymus 228359
Thickspike Pepper 300537
Thickspike Sedge 75489
Thick-spike Square Bamboo 87597
Thick-spike Wheat Grass 144640,144641
Thick-spiked Indosasa 206884
Thick-spiny Barberry 51996
Thickspur Corydalis 105864,106223
Thickspur Dali Larkspur 124622
Thickspur Hemipilia 191597
Thickspur Larkspur 124441
Thickspur Trichophore Larkspur 124648
Thickspur Wright Larkspur 124700
Thickspurred Larkspur 124441
Thickspurred Tali Larkspur 124622
Thickstalk Ardisia 31392
Thickstalk Aucuba 44948
Thickstalk Azalea 331427
Thickstalk Corydalis 106224
Thickstalk Dyeingtree 346480
Thick-stalk Lasianthus 222103
Thickstalk Madder 337947
Thick-stalk Manglietia 244430
Thickstalk Oriental Clematis 95186
Thick-stalk Roughleaf 222103
Thickstalk Tanoak 233247
Thickstalk Tea 69019
Thickstalk Woodlotus 244430
Thickstalk Yushanbamboo 416766
Thick-stalked Camellia 69019
Thick-stamened Camellia 69428
Thickstem Ardisia 31422
Thickstem Aster 160667
Thickstem Beancaper 418695,418769
Thickstem Dysophylla 139552
Thickstem Epimedium 147045
Thickstem Fritillary 168385
Thickstem Goldenray 229041
Thickstem Hairorchis 299603
Thickstem Monkshood 5148
Thickstem Pepper 300449
Thickstem Pleurospermum 304858
Thickstem Resupinate Woodbetony 287586
Thickstem Rhaphidophora 328996
Thickstem Ribseedcelery 304858
Thickstem Spapweed 204876
Thickstem Touch-me-not 204876
Thickstemen Gentian 173355
Thick-stemmed Ardisia 31422
Thick-stemmed Colquhounia 100034
Thickstipe Clematis 94867
Thickstipe Desertcandle 148544
Thick-stiped Clematis 94867
Thick-stiped Staunton-vine 374394
Thick-stslk Stauntonvine 374394
Thick-stslk Wild Quince 374394
Thickstyle Azalea 330469
Thickstyle Cherry-laurel 223158
Thickstyle Oncidium 270805
Thickstyle Rush 213031
Thickstyle Xmas Bush 14199
Thick-styled Rhododendron 330469
Thick-tooth Clematis 94964
Thicktooth Guiyang Betony 373276
Thicktooth Spiny Naiad 262080
Thick-toothed Deutzia 126886
Thicktube Lagotis 220181
Thicktwig Amoora 19958
Thicktwig Azalea 330205
Thicktwig Yushanbamboo 416806
Thick-twigged Yushania 416806
Thickvein Ardisia 31391
Thickvein Azalea 330441
Thickvein Pholidota 295532
Thick-veined Rhododendron 330441
Thick-velutinous Grewia 180815
Thickvelutinous Japan Edelweiss 224886
Thickvelvety Grewia 180815
Thickwall Begonia 50306
Thickwing Corydalis 105804
Thickwing Ormosia 274424
Thick-winged Ormosia 274424
Thief 338447
Thief Palm 295439
Thikleaf Cleyera 96625
Thik-leaf Oak 116167
Thik-leaved Cleyera 96625
Thik-leaved Cyclobalanopsis 116167
Thikspine Lanceleaf Arthraxon 36678
Thill Bamboo 297334
Thimble Cactus 244085
Thimble Lily 229750
Thimbleberry 338808,338909,338917,
338970,338972,338985,339194
Thimble-berry 338909,338917,338970
Thimble-flower 130383
Thimbles 30081,70271,81020,130383,
237539,364146
Thimbleweed 23696,23776,24021,24109,
24112,339574
Thin Abelia 127
Thin Ainsliaea 12645
Thin Angularfruit Iris 208589
Thin Angularfruit Swordflag 208589
Thin Awlquitch 131011
Thin Basketvine 9437
Thin Bedstraw 170662
Thin Bluegrass 305649
Thin Bractquitch 201505
Thin Brassbuttons 107837
Thin Brome 60735
Thin Caraway 77785
Thin Catchfly 363508
Thin Chinese Shibataea 362126
Thin Cinnamon 91439
Thin Cogflower 154197,382766
Thin Coldwaterflower 298926
Thin Deutzia 126946
Thin Devil Rattan 121510
Thin Dimeria 131011
Thin Eurysolen 160917
Thin Evodia 161350
Thin Faber Maple 2954
Thin Flaccid Aster 40441
Thin Forkstyleflower 374492
Thin Fruit-cup Litse 234089
Thin Gentian 173944
Thin Glands Rhododendron 331094
Thin Goosefeather Wobamboo 362126
Thin Gorgeous Gentian 173904
Thin Grass 12240

Thin Hedgehogcactus 140872
Thin Hemiboea 191359
Thin Henry Clematis 94990
Thin Holly 204331
Thin Holow Bamboo 82452
Thin Hornfennel 201612
Thin Hydrangea 199885
Thin Hyparrhenia 201505
Thin Indian Horseorchis 215346
Thin Ladypalm 329182
Thin Ladypalms 329182
Thin Laminae Rhododendron 331965
Thin Lampranthus 220697
Thin Leaf Blastus 55182
Thin Leaf Sunflower 188953
Thin Leathery Leaves Blueberry 403837
Thin Leymus 228390
Thin Linden 391722
Thin Losefruit 371240
Thin Lysionotus 239944
Thin Madder 338035
Thin Mazus 246981
Thin Mycetia 260624
Thin Naiad 262039
Thin Paspalum 285522,285524
Thin Pinangapalm 299662
Thin Privet 229460
Thin Purple Willow 343949
Thin Rabbiten-wind 12645
Thin Raspberry 338776
Thin Rattan Palm 65681,65700
Thin Rhododendron 331963
Thin Riparial Silverreed 394936
Thin Rockfoil 349816
Thin Roegneria 335330
Thin Sageretia 342169
Thin Schefflera 350787
Thin Small Reed 127264
Thin Smallreed 127264
Thin Sphallerocarpus 371240
Thin Spunsilksage 360835
Thin Stenoseris 375712
Thin Swertia 380372
Thin Tailleaf Aucuba 44907
Thin Tamarisk 383516
Thin Tonkin Pseudosasa 318287
Thin Twinballgrass 209053
Thin Wildginger 37608
Thin Willow 343880
Thin Woodbetony 287245,287400
Thin Yellow Spapweed 205455
Thin Yellow Touch-me-not 205455
Thin Yellowcrown Sealavender 230577
Thinanther Corydalis 105720
Thinaxis Syzygium 382673
Thin-bark Siberian Elm 401603
Thinbrabch Cotoneaster 107703
Thinbract Argyreia 32653
Thinbract Spapweed 205368
Thinbract Touch-me-not 205368
Thinbract Windhairdaisy 348486
Thinbrahch Amaranth 18724
Thinbrahch Amaranthus 18724
Thinbranch Aphanopleura 29087
Thinbranch Comastoma 100282
Thin-branch Elm 401613
Thin-branch Leafflower 296633
Thin-branch Leptodermis 226088
Thin-branch Spicebush 231363
Thinbranch St. John's Wort 201978
Thin-branch Syzygium 382667
Thinbranch Throathair 100282
Thinbranch Tianshan Craneknee 221766
Thin-branch Underleaf Pearl 296633
Thin-branch Wildclove 226088
Thin-branched Fargesia 162753
Thin-branched Syzygium 382667
Thin-branched Youngia 416489
Thinbristle Oak 324409
Thincalyx Barbed Gentianopsis 174211
Thin-calyx Lysionotus 239948,239977
Thincapsulewort 226615,226621
Thinculm Lakemelongrass 232426
Thinculm Qiongzhu 87564
Thinculm Sugarcane 341840
Thincupula Oak 116206
Thincupula Qinggang 116206
Thin-cupule Cyclobalanopsis 116206
Thinest Hollow Bamboo 82452
Thinest Hollowbamboo 82452
Thinest Hollow-bamboo 82452
Thinestwalled Bamboo 318487
Thinestwalled Bamboon 318485
Thinest-walled Bamboon 318485,318487
Thin-flower Aganosma 10237
Thin-flower Agapetes 10311
Thinflower Ardisia 31458
Thin-flower Bauhinia 49237
Thinflower Cherry 83292
Thin-flower Cherry 83292
Thinflower Colquhounia 100039
Thinflower Corydalis 106182
Thin-flower Daphne 122619
Thinflower Felwort 235449
Thinflower Habenaria 183820
Thinflower Holly 203861
Thinflower Horse Mint 250388
Thinflower Linearstripe Rabdosia 209757
Thinflower Microtoena 254269
Thinflower Purplepearl 66871
Thin-flower Stephania 375857
Thinflower Valerian 404307
Thin-flowered Ardisia 31458
Thin-flowered Daphne 122619
Thin-flowered Sedge 76513
Thinfruit Hypecoum 201612
Thinfruit Kobresia 217289
Thinfruit Sedge 74555
Thinfruit Sloanea 366062
Thin-fruit Yu Maple 3762
Thin-fruited Oleander-leaved Fig 165362
Thin-fruited Sloanea 366062
Thinfruitgrass 225917
Thingadu 283904
Thingan 198176
Thingan Merawan 198176
Thin-glanded Rhododendron 331094
Thinhair Balansa Michelia 252818
Thinhair Bedstraw 170583
Thinhair Cinnamon 91439
Thinhair Indianmulberry 258912
Thinhair Machilus 240706
Thinhair Microula 254341
Thinhair Thick-pod Ormosia 274423
Thin-haired Cinnamon 91439
Thin-headed Hollow Bamboo 82452
Thinhorn Elatostema 142872
Thinhorn Spapweed 205082
Thinhorn Stairweed 142872
Thinhorn Touch-me-not 205082
Thinhull Camellia 69695
Thin-ieaf Barberry 51811
Thininflorescence Falsenettle 56165
Think Skullcap 355428
Think Wildbasil 96999
Thin-laminated Rhododendron 331965
Thin-leaf Actinidia 6653
Thinleaf Adina 8199
Thinleaf Ajania 13037
Thinleaf Alder 16464
Thinleaf Altingia 18206
Thinleaf Alyssum 18482
Thinleaf Amaranth 18829
Thinleaf Apea-earring 301164
Thinleaf Asiabell 98294
Thinleaf Aspidistra 39520
Thinleaf Azalea 331097
Thin-leaf Barberry 51363
Thinleaf Bauhinia 49105
Thinleaf Beggarticks 53982
Thinleaf Bentgrass 12413
Thinleaf Betony 373459
Thinleaf Bittersweet 80210
Thinleaf Blastus 55182
Thinleaf Bluegrass 305328
Thinleaf Buckthorn 328760
Thinleaf Camellia 69704
Thinleaf Camptotheca 70577
Thinleaf Caper 71899
Thinleaf Carpesium 77169
Thinleaf Casearia 78137

Thinleaf Celery 116384
Thinleaf Chickweed 374882
Thinleaf Childseng 318515
Thin-leaf Chinese Sinobambusa 364824
Thin-leaf Chinese Tangbamboo 364824
Thinleaf Chirita 87853,87979
Thinleaf Cinquefoil 312455
Thinleaf Clearweed 299048
Thinleaf Clethra 96483
Thinleaf Cliffbean 254808
Thinleaf Coiledleaf Pearleverlasting 21552
Thinleaf Coldwaterflower 299048
Thinleaf Crazyweed 278958
Thinleaf Creeping Catchfly 363963
Thinleaf Cypressgrass 119682
Thinleaf Cystopetal 320226
Thinleaf Dali Jerusalemsage 295108
Thinleaf Eargrass 187512
Thin-leaf Eightangle 204599
Thinleaf Elaeagnus 142207
Thinleaf Elatostema 142873
Thinleaf Erysimum 154492
Thinleaf Eurya 160527
Thinleaf Evergreenchinkapin 79038
Thinleaf False Brome 58601
Thin-leaf Falsepanax 266926
Thinleaf Fieldorange 249680
Thinleaf Fig 165364
Thinleaf Flatspike 55899
Thinleaf Franchet Terminalia 386542
Thinleaf Globethistle 140824
Thinleaf Grassleaf Sweetflag 5809
Thinleaf Habranthus 184253
Thinleaf Heliotrope 190738
Thinleaf Hillcelery 276778
Thinleaf Hiptage 196835
Thin-leaf Holly 203975
Thinleaf Honeysuckle 235717,236116
Thin-leaf Honeysuckle 235955
Thinleaf Huckleberry 403904
Thin-leaf Huckleberry 403904
Thin-leaf Illicium 204599
Thinleaf Impressed Sarmentose Fig 165626
Thinleaf Iris 208687
Thinleaf Ixora 211088
Thinleaf Jasmine 211732
Thinleaf Jointfir 178564
Thinleaf Kitefruit 196835
Thin-leaf Kiwifruit 6653
Thinleaf Knotweed 309853
Thinleaf Lampranthus 220697
Thin-leaf Late Purple Aster 380952
Thinleaf Lawngrass 418453
Thinleaf Leafspike Rabdosia 209798
Thin-leaf Ledum 223909
Thinleaf Leek 15813
Thinleaf Leptodermis 226138
Thinleaf Leptopus 226318
Thinleaf Liriope 232619
Thinleaf Machilus 240615
Thinleaf Maesa 241799
Thinleaf Mallotus 243372
Thin-leaf Mallotus 243372
Thinleaf Manglietia 244472
Thinleaf Maple 3099,3674
Thinleaf Melodinus 249680
Thinleaf Milkwort 308403
Thinleaf Millettia 254808
Thinleaf Mock Orange 294556
Thinleaf Mockorange 294556
Thinleaf Mouseear 83036
Thinleaf New Eargrass 262960
Thinleaf Nothapodytes 266808
Thinleaf Notopterygium 267158
Thinleaf Oak 116112,116189,324073
Thinleaf Oldham Elaeagnus 142133
Thinleaf Oleander-leaved Fig 165364
Thinleaf Onion 15813
Thinleaf Orach 44621
Thin-leaf Orach 44621
Thin-leaf Osyris 276893
Thinleaf Parnassia 284558
Thinleaf Peony 280318
Thinleaf Pine 300278
Thin-leaf Pine 300065
Thinleaf Pittosporum 301410
Thinleaf Poisoncelery 90940
Thinleaf Primrose 314569
Thinleaf Ptilotrichum 321098
Thinleaf Qinggang 116112,116189
Thinleaf Rattlebox 112509
Thinleaf Rhododendron 331097,331964
Thinleaf Sandwort 31965
Thin-leaf Sauropus 348052
Thinleaf Sawwort 361079
Thinleaf Selinum 357834
Thinleaf seudostellaria 318515
Thin-leaf Sichuan Dogwood 380513
Thinleaf Slatepentree 400732
Thinleaf Sloanea 366047
Thinleaf Snakegourd 396287
Thinleaf Sneezeweed 188437
Thinleaf Stairweed 142873
Thinleaf Stonecrop 356876
Thinleaf Stringbush 414203
Thinleaf Sunflowe 188953
Thin-leaf Sweetleaf 381104
Thinleaf Swertia 380391
Thinleaf Swordflag 208687
Thinleaf Syreitschikovia 382098
Thinleaf Tanoak 233387
Thinleaf Tarenna 384950
Thinleaf Tea 69267,69704
Thinleaf Thistle 92438
Thinleaf Tupelo 267857
Thinleaf Tutcheria 400732
Thinleaf Veinyroot 237768
Thinleaf Vervain 405896
Thin-leaf Wampee 94174
Thinleaf Waterdropwort 269312
Thinleaf Wendlandia 413774
Thinleaf Wildclove 226138
Thinleaf Youngia 416490
Thinleaf Yucca 416595
Thinleaf Yunnan Pine 300308
Thin-leathered Leaves Blueberry 403837
Thin-leaved Actinidia 6653
Thin-leaved Adina 8199
Thin-leaved Altingia 18206
Thin-leaved Apea-earring 301164
Thin-leaved Bamboo 37158
Thin-leaved Blastus 55182
Thin-leaved Bottlebrush 67250
Thin-leaved Buckthorn 328760
Thin-leaved Camellia 69267
Thin-leaved Caper 71782,71899
Thin-leaved Coneflower 339622
Thin-leaved Eurya 160527
Thin-leaved Evergreen Chinkapin 79038
Thin-leaved Flax 231992
Thin-leaved Ixora 211088
Thin-leaved Lovegrass 147847
Thin-leaved Machilus 240615
Thin-leaved Mahonia 242651
Thin-leaved Maple 3099,3674
Thin-leaved Mazus 246980
Thin-leaved Melodinus 249680
Thin-leaved Millettia 254808
Thin-leaved Mockorange 294556
Thin-leaved Oak 324073
Thin-leaved Rhododendron 331097,331964
Thin-leaved Sloanea 366047
Thin-leaved Snowberry 380736
Thin-leaved Sunflower 188953,189006
Thin-leaved Sweetleaf 381104
Thin-leaved Tanoak 233387
Thin-leaved Tupelo 267857
Thin-leaved Tutcheria 400732
Thin-leaved Virginia Copperleaf 1794
Thin-leaved Wild Strawberry 167653,167658
Thinlip Cryptostylis 113857
Thinlobate Shanxi Monkshood 5583
Thinlobe Sealavender 230657
Thinlobed Catclaw Buttercup 326432
Thinlobed Chie Acronema 6203
Thinlobed Ghstplant Sagebrush 35774
Thinlobed Jehol Ligusticum 229345
Thinlobed Sagebrush 35560
Thinlobed Schizonepeta 264869
Thinner Willow 343440
Thin-nerve Osmanthus 276309

Thinnest Corydalis 105947
Thinnest Draba 137009
Thinnest Whitlowgrass 137009
Thinnest Yam 131607
Thinpanicle Schefflera 350787
Thinpedicel Clovershrub 70793
Thin-pedicel Milkvetch 42761
Thinpedicel Mycetia 260624
Thinpedicel Pimpinella 299417
Thinpeduncle Eargrass 187684
Thinpeduncle Hedyotis 187684
Thinpeel Crazyweed 279072
Thin-pericarped Atalantia 43727
Thinpetal Raspberry 338305
Thinpetal Rotala 337378
Thinpetiole Spapweed 205081
Thin-pored Ardisia 31564
Thin-pyrene Holly 204331
Thin-racemose Millettia 254752
Thin-racemosed Millettia 254752
Thin-rachis Syzygium 382673
Thin-rachised Syzygium 382673
Thinrecepatacle Elatostema 142874
Thin-reticulate-veined Ardisia 31558
Thinreticulateveins Ardisia 31558
Thinrhizome Epimedium 147016
Thinrhizomous Barrenwort 147016
Thinroot Ginger 417995
Thin-rooted Sweetgrass 177621
Thin-sepal Camellia 69727
Thinsepal Clematis 95267
Thinsepal Ladybell 7617
Thin-sepal Pittosporum 301318
Thinsepal Seatung 301318
Thin-sepaled Pittosporum 301318
Thinshoot Syzygium 382476
Thinsilk Stairweed 142711
Thin-snake Rattail Cactus 29717
Thinspiculate Sedge 74915
Thinspike Denseflower Elsholtzia 144005
Thinspike Tamarisk 383542
Thin-spiked Sedge 76522
Thin-spiked Tamarisk 383542
Thin-spiked Willow 344195
Thin-spiked Wood-sedge 76410
Thinspine Craneknee 221761
Thinsplit Bittercress 72715
Thinsplit Habenaria 183778
Thinsplit Hogfennel 292920
Thinsplit Ligusticum 229403
Thinsplite Catchfly 364083
Thinspur Columbine 30051
Thinspur Habenaria 183896
Thinspur Platanthera 302260
Thinstachys Conehead 178863
Thin-stachysed Sedge 76519
Thin-stalk Ardisia 31617
Thinstalk Buckwheat 162316
Thinstalk Bulbophyllum 63104
Thinstalk Chickweed 375050
Thin-stalk Clematis 95355
Thinstalk Curvebeak Monkshood 5098
Thinstalk Hoarypea 386070
Thinstalk Knotweed 162319
Thinstalk Meconopsis 247125
Thin-stalk Milkvetch 42438
Thinstalk Mosquitotrap 117492
Thinstalk Oblongleaf Betony 373324
Thinstalk Ormosia 274409
Thinstalk Privet 229646
Thinstalk Sagebrush 35365,35856
Thinstalk Sandwort 31899
Thinstalk Smalllobed Clematis 95211
Thinstalk Spapweed 204991
Thinstalk Stonebean-orchis 63104
Thin-stalk Trigonostemon 397361
Thinstalk Trigonotis 397416
Thinstalk Wedgebract Elatostema 142639
Thinstalk Wedgebract Stairweed 142639
Thin-stalk Wedgeleaf Fig 165793
Thinstalk Youngia 416425
Thin-stalked Fig 165169
Thinstem Ainsliaea 12734
Thinstem Beancaper 418607
Thinstem Burreed 370107
Thinstem Dendrobium 125257
Thinstem Dishspurorchis 171862
Thinstem Dualbutterfly 398270
Thinstem Eria 116622
Thinstem Erysimum 154493
Thinstem Gastrochilus 171862
Thinstem Gentian 173966
Thinstem Goldenray 229222
Thinstem Hairorchis 116622
Thinstem Knotweed 309106
Thinstem Larkspur 124408
Thinstem Liparis 232122
Thinstem Milkvetch 43143
Thinstem Onion 15023
Thinstem Orchis 273421
Thinstem Rabbiten-wind 12734
Thinstem Sandwort 32015
Thinstem Speedwell 407409
Thinstem St. John's Wort 201877
Thinstem Strachy Edelweiss 224951
Thinstem Tripterospermum 398270
Thinstem Twayblade 232122
Thinstem Woodbetony 287751
Thin-stemmed Privet 229646
Thinstipe Goldenray 229225
Thinstipe Metalleaf 296387
Thinstipe Milkvetch 42438
Thinstipe Spicegrass 239637
Thinstiped Yam 131873
Thinstrip Yellow Bamboo 47234
Thinstype Treerenshen 125609
Thintongue Shorthair Aster 40165
Thintooth Bird Cherry 280013
Thintooth Clearweed 298940,299040
Thintooth Coldwaterflower 299040
Thintooth Guiyang Betony 373277
Thintooth Howell Coldwaterflower 298940
Thintooth Slender Heterolamium 193817
Thintooth Woodnettle 273920
Thin-tooth Xizang Fieldjasmine 199454
Thin-tooth Xizang Huodendron 199454
Thin-toothed Mahonia 242576
Thintube Chirita 87988
Thintube Skullcap 355567
Thintube Woodbetony 287255,287354
Thintwig Azalea 330075
Thin-twig Holly 203863
Thintwig Knotwood 44259
Thin-valvated Pittosporum 301410
Thinvalve Camellia 69695
Thinvein Syzygium 382533
Thin-veine Holly 204370
Thin-veined Eucalyptus 155622
Thinwall Arrowbamboo 162758
Thin-walled Bamboo 318487
Thin-walled Bambusa 47404
Thin-walled Fargesia 162758
Thinwin 310962
Thinwinged Russianthistle 344662
Thiny Evodia 161350
Thinyellow Tea 69071
Thirst 177545
Thirsty Heath 150057
Thirty Thorn 1588
Thistelike Windhairdaisy 348187
Thistle 31051,73278,76988,81112,91697,91843,368771
Thistle Broomrape 275195
Thistle Hemp 71218
Thistle Sage 344939
Thistleleaf Aster 160659
Thistle-upon-thistle 271666
Thitefruit Moorberry 404028
Thitka 289407,289408
Thitmin 306506
Thnderstruck Fieldorange 249682
Thochwood Family 64088
Thomas' Bird's-eye Bush 268273
Thomas' Wild Buckwheat 152517
Thompson Yucca 416654
Thompson's Curse 73090
Thompson's Pincushion 84533
Thompson's Wild Buckwheat 152518
Thompson's Yucca 416654
Thompsongrass 285426
Thompson's Sandwort 31917

Thoms Euphorbia 159967
Thomson Aralia 30777
Thomson Arnebia 34641
Thomson Azalea 331971
Thomson Barberry 52224
Thomson Catnip 265081
Thomson Cousinia 108425
Thomson Creeper 416529
Thomson Elsholtzia 144103
Thomson Falselily 266902
Thomson Felwort 235470
Thomson Fimbristylis 166531
Thomson Fluttergrass 166531
Thomson Glorybower 96387
Thomson Goldenray 229226
Thomson Kudzuvine 321466
Thomson Ligusticum 229405
Thomson Maple 3690
Thomson Meliosma 249469
Thomson Mountainash 369535
Thomson Mountain-ash 369535
Thomson Olgaea 270248
Thomson Pleurospermum 304809
Thomson Ribseedcelery 304809
Thomson Rush 213525
Thomson Siberian Knotweed 309795
Thomson Spicebush 231448
Thomson Spice-bush 231448
Thomson Swertia 380386
Thomson Tanoak 233395
Thomson Vinegentian 108425
Thomson Violet 410660
Thomson Waterdropwort 269368
Thomson Yua 416529
Thomson's Glochidion 177183
Thomson's Rhododendron 331971
Thomson's Waxplant 198897
Thonmson Spapweed 205379
Thor's Beard 358062
Thor's Mantle 31051,130383,312520
Thora Buttercup 326439
Thoracostachyum 390376
Thore's But Tercup 326439
Thore's Buttercup 326439
Thorel Colona 99968
Thorel Cyclobalanopsis 116210
Thorel Elaeocarpus 142411
Thorel Meliosma 249472
Thorel Mitrephora 256235
Thorel Oak 116210
Thorel Silverhook 256235
Thorel Sweetspire 210419
Thorel Tinospora 392252
Thorel Villous Bowfruitvine 393547
Thorel Villous Toxocarpus 393547
Thorn 109509
Thorn Apple 123036,123065,123071,
123077
Thorn Apple Datur 61235
Thorn Broom 172905,401388
Thorn Dock 340116
Thorn Lily 79376
Thorn Pear 354605
Thorn Plum 316206
Thorn Rockfoil 349117
Thorn Tree 1272
Thorn Wattle 1149
Thornapple 32429,109509
Thorn-apple 123036,123077
Thorn-apple-leaved Goosefoot 87048
Thornbearing Peashrub 72346
Thorn-bearing Pea-shrub 72346
Thornber Cholla 116642
Thornber Yucca 416558
Thornber's Nipple Cactus 244245
Thornberries 109857
Thornbract Reekkudzu 313478
Thornbush Cactus 244057
Thorncrest Century Plant 10888
Thorn-crested Agave 10888
Thorne's Larkspur 124673
Thorne's Royal Larkspur 124673
Thorne's Thistle 92366
Thorne's Wild Buckwheat 152522
Thorng Bamboo 47213
Thornless Blackberry 338225
Thornless Cockspur Hawthorn 109636
Thornless Common 176904
Thornless Dewberry 339432
Thornless Honey Locust 176903
Thornless Thistle 303082
Thorntwig Honeysuckle 236109
Thorny Acacia 1427
Thorny Amaranth 18822
Thorny Amaranthus 18822
Thorny Bamboo 47190,47448,47516
Thorny Bambusa 47190
Thorny Bridelia 60223
Thorny Broom 68469,68471
Thorny Bur 31051
Thorny Burnet 313160
Thorny Burr 31051
Thorny Burweed 415053
Thorny Catalpa 215442
Thorny Elaeagnus 142152
Thorny Gymnosporia 182683
Thorny Locust 176903
Thorny Mallow 195196
Thorny Olive 142152
Thorny Ononis 271347
Thorny Peashrub 72348
Thorny Pigweed 18822
Thorny Pittosporum 301352
Thorny Rest Harrow 271347
Thorny Skeletonweed 303961,303966
Thorny Wingnut 280555
Thorny-fruit Euonymus 157418
Thorny-locust 176903
Thorny-olive 142152
Thorold Kengyilia 215967
Thorold Lymegrass 335521
Thorold Roegneria 335521
Thoroldgrass Goosecomb 335521
Thorough-stem 158250
Thoroughwax 63553,63805
Thorough-wax 63805,158250
Thoroughword 158250
Thoroughwort 158021,158149,158157,
158312,159911
Thorough-wort 158037,158062,158250
Thoroughwort Pennycress 390249
Thorowax 63553,63805
Thorowaxleaf Cremanthodium 110355
Thorowaxleaf Euphorbia 158586
Thorowaxleaf Nutantdaisy 110355
Thorowax-like Hymenolyma 199647
Thorowax-like Rhodiola 329848
Thorowort Pondweed 312220
Thorow-wax 63805
Thottea 390413
Thouars Cycas 115915
Thouin Statice 230798
Thouinia 390424
Thousand Holes 201924
Thousand Jacket 172245
Thousand Mothers 392668
Thousand Stars 360926
Thousand-flower 116736
Thousandhead Goldenray 229114
Thousand-holes 201924
Thousand-leaf 3978
Thousand-leaf Grass 3978
Thousand-leaved Clover 3978
Thousandleaves Crazyweed 278756
Thousand-seal 3978
Thousand-weed 3978
Thread Agave 10852
Thread Bamboo 297278
Thread Clover 396896
Thread Flower 158062
Thread Green Violet 199695
Thread Love Grass 148019
Thread Palm 413307
Thread Plant 263292
Thread Rush 213114
Thread Violet 410014
Threadflower Gilia 175694
Threadleaf 135079
Threadleaf Agapetes 10328
Thread-leaf Agave 10852
Threadleaf Arborvitae 390601

Threadleaf Blue Star 20853
Threadleaf Coreopsis 104616
Threadleaf Eulalia 318151
Thread-leaf Fleabane 150631
Threadleaf Giant Hyssop 10419
Threadleaf Groundsel 358757,358881,358882
Thread-leaf Groundsel 359381
Threadleaf Knotweed 309526
Threadleaf Milkweed 37991
Threadleaf Phacelia 293170
Thread-leaf Pondweed 378986,378988,378989
Thread-leaved Beak-seed 63229
Thread-leaved Crowfoot 48933
Thread-leaved Elderberry 345637
Thread-leaved Pondweed 378986,378988,378989
Thread-leaved Sundew 138289
Thread-leaved Sundrops 269461
Thread-leaved Water-crowfoot 48933
Threadlike Crazyweed 278829
Thread-like-leaf Milkvetch 42784
Threadlobe Ladybell 7616
Thread-of-life 116736,349936
Thread-pedicel Clovershrub 70916
Threadpendulous Rattan Palm 65689
Threadsected Anemone 23816
Thread-shaped Taeniophyllum 383062
Threadshapedstem Skullcap 355438
Threadstalk Bulrush 353411
Threadstalk Chinchweed 286885
Threadstalk Rose 336558
Threadstalk Small Reed 127233
Threadstalk Smallreed 127233
Threadstalk Speedwell 407124
Thread-stalked Lilyturf 272076
Thread-stalked Primrose 314391
Thread-stalked Trigonostemon 397361
Threadstem Carpetweed 256695
Thread-stem Carpet-weed 256695
Thready Herbclamworm 398078
Thready Tripogon 398078
Threadyleaf Meadowrue 388509
Threadyleaf Serpentroot 354841
Threadylobe Ajania 13013
Thready-stalk Black Barberry Barberry 51643
Threawing Thelasis 389128
Threc-leaved Liverwort 192130
Thred Cypress 85348
Thredbranch Cypress 85348
Three Basinerved Aster 41404
Three Bird Orchid 397892
Three Birds 397885,397892
Three Birds Toadflax 231154
Three Faces-under-a-hood 410677
Three Flowered Rhododendron 332002
Three Goldenpoint 322078
Three Head Aster 41396
Three Leaflets Bittercress 73023
Three Leaved Lantana 221307
Three Leaved Sage 221307
Three Onion 15166
Three Pee-abed 384714
Three Teethed-leaf Vetch 408644
Three Tongues Synotis 381956
Three Two One 384714
Three Veined Chinese Groundsel 365078
Threeangle Oak 167343,167344
Threeangle Sonerila 368872
Threeanglesbranch Clovershrub 70904
Threeangular Iris 208767
Threeangularfruit CamelliaTriangleaf Tea 69717
Threeanguledleaf Violet 410672
Three-anthers Krameria 218085
Threearista Stonecrop 357236
Three-awn 33730
Three-awned Grass 33730
Three-awned-grass 33730
Threeawngrass 33730
Three-banded Passionflower 285709
Threebasevein Aster 41404
Threebibachene 397948
Threebineshaped Gomphogyne 179156
Three-birds Orchid 397876,397877,397885,397892
Threebract Chirita 87983
Threebract Loosestrife 240094
Threebract Small Thorowax 63854
Threebract Thorowax 63854
Three-bracted Paraboea 283137
Threebristle Garnotia 171531
Threebristle Threeawngrass 34059
Threecolor Rosewood 121845
Threecolor Wand-flower 370009
Threecolored Gentian 174003
Threecolored Pepper 300536
Three-coloroured Lily 229736
Three-coloured Amaranth 18836
Three-coloured Amaranthus 18836
Three-coloured Bindweed 102884,103340
Three-coloured Hibiscus 195326
Three-coloured Violet 410677
Threecorner Leek 15839
Three-cornered Garlic 15839
Three-cornered Leek 15839
Three-cornered Palm 139336
Threecusp Moraea 258680
Threecusp Sage 345439
Threecuspidate Snakegourd 396280
Threedleaf Burreed 370030
Three-faces-in-a-hood 410677
Threefaces-under-a-hood 410677
Threefinger Windhairdaisy 348870
Three-fingered Jack 349999
Threeflower Azalea 331411
Threeflower Bedstraw 170708
Threeflower Blueberry 404025
Threeflower Clematis 95358
Threeflower Gagea 234244
Threeflower Gentian 174004
Threeflower Hawthorn 110092
Threeflower Hypoestes 202632
Threeflower Maple 3710
Three-flower Microtropis 254331
Threeflower Pancratium 280912
Threeflower Purpledaisy 267179
Threeflower Raspberry 339390
Threeflower Rhododendron 331411
Three-flower snakeweed 390940
Threeflower Stenoseris 375716
Threeflower Tickclover 126651
Three-flowered Beggarweed 126651
Threeflowered Bluebeard 78039
Three-flowered Campion 364073
Three-flowered Gladiolus 176429
Threeflowered Maple 3710
Three-flowered Melic Grass 249053,249054
Three-flowered Nightshade 367689
Three-flowered Rush 213545
Threefold 250502,399370,399394
Three-foliate Akebia 13240
Threefoliolate Larkspur 124651
Threefork 396794
Threefork Gentian 173998
Threefork Hakea 184647
Threefork Saxifrage 350002
Three-forked Evodia 161382
Threefruit Datong Larkspur 124546
Threefruit Tanoak 233389
Threefruit Tatung Larkspur 124546
Three-fruited Sedge 76589
Three-fruited Spindle-tree 157930
Threeglume Rush 213545
Threeglume Wildryegrass 144415
Threehair Thrixsperm 390518
Threehair Thrixspermum 390518
Three-hairs Small Reed 127320
Threehead Aster 41396
Threehead Kyllingia 218643
Threehead Rush 212990
Threehead Sedge 76581
Threehead Water-centipede 218643
Threehorned Bedstraw 170694
Threeleaf Agelaea 11058
Threeleaf Akebia 13238
Threeleaf Arrowwood 408162
Threeleaf Ash 168123
Threeleaf Blackberry 338985
Threeleaf Chastetree 411464

Threeleaf Cinquefoil 312565,313047
Threeleaf Coreopsis 104611
Three-leaf Corydalis 105969
Threeleaf Cranesbill 174975
Threeleaf Crazyweed 279220
Three-leaf Creeper 285144
Threeleaf Ehretia 141708
Three-leaf Evodia 161384
Threeleaf Fieldcitron 400546
Threeleaf Fuchsia 168789
Threeleaf Gentian 173969
Threeleaf Ginsen 280817
Three-leaf Goldthread 103850
Threeleaf Illigera 204656
Threeleaf Indigo 206684
Threeleaf Japanese Buttercup 325991
Three-leaf Jewel-vine 126007
Threeleaf Lacquertree 386676,386678
Threeleaf Licuala Palm 228769
Threeleaf Licualapalm 228769
Threeleaf Melicope 249168
Three-leaf Millettia 254866
Threeleaf Patrinia 285874
Threeleaf Ploughpoint 401175
Threeleaf Prickly-ash 417290
Threeleaf Sage 345439
Threeleaf Slenderstyle Acanthopanax 2408
Threeleaf Soapberry 346356
Three-leaf Solomon's-plume 242706
Three-leaf Solomon's-seal 242706
Threeleaf Spicegrass 239683
Threeleaf Stonecrop 356617
Three-leaf Sumach 332908
Threeleaf Tubergourd 390149
Threeleaf Turpinia 400546
Threeleaf Viburnum 408162
Threeleaf Woodbetony 287757
Threeleaflet Ash 168123
Three-leaved Acanthopanax 143694
Three-leaved False Solomon's Seal 242706
Three-leaved Gold-thread 103850
Three-leaved Grass 397019
Three-leaved Hop Tree 320071
Three-leaved Laverock 277648
Three-leaved Nightshade 397556
Three-leaved Pair Vetch 408664
Threeleaved Rattlesnakeroot 313901
Three-leaved Rush 213537
Three-leaved Stonecrop 357224
Three-leaved Syzygium 382548
Three-leaved Toothwort 72895
Threeleaved Turpinia 400546
Three-leaved White-lettuce 313901
Three-leaved Yam 131567
Threeleaves Curvedstamencress 72786, 238011
Three-leaves Pricklyash 417358
Three-leaves Taiwan Crateva 110233
Threeleaves Tubergourd 390149
Threeleaves Yam 131471
Threeleft Spunsilksage 360845
Threeleves Buttercup 326443
Threeleves Trachydium 393866
Threelobe Buttercup 326472
Threelobe Cremanthodium 110466
Threelobe False Mallow 243893
Threelobe False-mallow 243893
Threelobe Nutantdaisy 110466
Threelobe Primrose 315059
Threelobe Spirea 372117
Three-lobe Spirea 372117,372118,372119
Threelobed Ajania 13040
Three-lobed Arum 401172
Threelobed Central-south Raspberry 338489
Three-lobed Coneflower 339622
Three-lobed Crowfoot 326475
Three-lobed Habenaria 184149
Three-lobed Malope 243494
Threelobed Paraprenanthes 283688
Three-lobed Rock Daisy 291314
Three-lobed Sage 345440
Three-lobed Sagebrush 36416
Threelobed Spiraea 372117
Three-lobed Spirea 372117
Three-lobed Starry Puncturebract 362998
Three-lobed Stork's-bill 153754
Three-lobed Violet 410340
Three-lobed Water Crowfoot 326475
Three-lobed Wormwood 36416
Three-lobes Dolichos 135646
Three-lobes Sumac 332908
Three-men-in-a-boat 394076
Three-monthly Lavatera 223393
Threenerve Fighazel 380604
Threenerve Greenbrier 366607
Three-nerve Greenbrier 366607
Threenerve Parnassia 284624
Threenerve Smilax 366607
Threenerved Dualbutterfly 398306
Three-nerved Duckweed 224360
Three-nerved Goldenrod 368470
Three-nerved Pericampylus 290847
Three-nerved Sandwort 256441,256456
Three-nerved Sedge 217163
Threenerved Tripterospermum 398306
Threepairleaf Raspberry 339398
Threepenny-bit Herb 213902
Threepenny-bit Rose 336542,336551
Threepetal Iris 208906
Threepetal Metalleaf 296420
Threepetal Spapweed 205405
Three-rayed Chinchweed 286882
Threerib Flickingeria 166968
Threeribfruit 397948
Threesect Bulrush 353901
Threeseed Mercury 1769
Three-seed Mercury 1971
Three-seeded Bog Sedge 76589
Three-seeded Mercury 1971
Three-seeded Sedge 76589
Three-seeded-mercury 1971
Three-sharp-leaves Rattlebox 112405
Threespike Eulalia 156502
Threespike Goldquitch 156502
Threespike Goosegrass 143553
Threespike Sedge 76591
Three-spikegrass 310713
Threespine Barberry 52265
Threespine Triawn 34059
Three-spiny Barberry 52265
Three-square Bulrush 352249
Threestamen Orchis 265836,265840
Threestamen Sonerila 368902
Threestamen Waterwort 142574
Threestamen Willow 344211
Three-stamened Waterwort 142574
Threestamen-like Willow 344220
Threestarfruit 398681,398684
Threeteeth Merremia 415299
Threeteeth Rockfoil 350001
Three-thorned Acacia 176903
Threethorned Barberry 52265
Three-threaded Willow 344211
Threetine Vetch 408635
Three-tip Sagebrush 36423
Three-tipped Sagebrush 36423
Threetongue Tailanther 381956
Threetooth Obtuseleaf Elatostema 142766
Threetooth Primrose 315058
Threetooth Ragwort 279957
Three-toothed Cinquefoil 313085,362389
Three-toothed Nitraria 266362
Three-toothed Sagebrush 36400
Three-toothed Whitethorn 266362
Threevein 256456
Threevein Aster 39928
Threevein Buerger Maple 2822
Threevein Caruncle grass 256456
Three-vein Clearweed 298971
Threevein Eargrass 187688
Threevein Euonymus 157904
Threevein Hedyotis 187688
Threevein Kobresia 217163
Threevein Poplar 311555
Threevein Rabbiten-wind 12739
Threevein Trident Maple 2822
Threevein Vetchling 222747
Threevein Woad 209184
Threeveined Arrowwood 408193
Three-veined Sandwort 256456
Threeveins Sinosenecio 365078

Threeway Sedge 138919
Three-way Sedge 138919
Three-wing Triangular Bulrush 352233
Three-winged Nut 398310
Threewinged Sawgrass 245572
Threewingedcalyx 224044
Threewingedculm Sawgrass 245572
Threewingedculm Spikesedge 143385
Threewingnut 398310
Thrift 34488,34536,127573,230508,
295324,357114
Thrift Family 305165
Thriftlike Catchfly 363214
Thrifty Goldenweed 375791
Thrixsperm 390486
Thrixspermum 390486
Throat Root 175459
Throat Spot Rhododendron 330692
Throathair 100263
Throat-spoted Rhododendron 330692
Throatwort 70331,130383,355202,393604,
393608
Throat-wort 70331
Throttlewort 130383
Throughhill Yam 131734
Throughstem Clearweed 299024
Throughstem Coldwaterflower 299024
Throughwort 101958
Throw-wax 63805
Thrum Liparis 232307
Thrumwort 18687,121993,121995
Thryallis 390538
Thryptomene 390548
Thsmanian Snowgum 155533
Thteabe 334250
Thua Khao Rui 61254
Thuarea 390551
Thuja 390567
Thujopsis 390673
Thumble 70271,81020,81237
Thumbs-and-fingers 237539,401388
Thunberg Astilbe 41852
Thunberg Chinese Knotweed 308974
Thunberg Cranesbill 174966
Thunberg Day Lily 191318
Thunberg Daylily 191318
Thunberg Elaeagnus 142207
Thunberg Evergreenchinkapin 78917
Thunberg Filbert 106736
Thunberg Fritillary 168586
Thunberg Hydrangea 200125
Thunberg Iris 208894
Thunberg Japanese Maple 3366
Thunberg Knotweed 309877
Thunberg Kudzu Bean 321441
Thunberg Kudzu Vine 321441
Thunberg Kudzu-bean 321441
Thunberg Kudzu-vine 321441
Thunberg Leek 15822
Thunberg Lespedeza 226969,226971,226977
Thunberg Nanmu 240707
Thunberg Nepal Cranesbill 174966
Thunberg Onion 15822
Thunberg Osmanther 276260
Thunberg Pfeiffer Osmanther 276297
Thunberg Pfeiffer Osmanthus 276297
Thunberg Pine 300281
Thunberg Sarmentose Fig 165630
Thunberg Sedge 76545
Thunberg Spicebush 231334
Thunberg Spiraea 371815
Thunberg Spirea 372101
Thunberg's Amaranthus 18833
Thunberg's Barberry 52225
Thunberg's Bush-clover 226969
Thunberg's Day-lily 191318
Thunberg's Geranium 174966
Thunberg's Lespedeza 226969
Thunberg's Pigweed 18833
Thunder Clover 13174
Thunder Daisy 227533
Thunder Flower 102933,227533,282685,
363461,363673,374916
Thunder God Hornbeam 77420
Thunder Plant 358062
Thunder Vine 176824
Thunder-and-lightning 13174,321643
Thunder-ball 282685
Thunderbolt Iris 208946
Thunderbolt Plant 361317
Thunderbolts 282685,363673,364193,
374916,407072
Thunder-cup 282685
Thundergod Caper 71802
Thundergodvine 398310,398322
Thundergot Qinggang 116114
Thunderhead Pine 300279
Thundermelia 327556
Thunia 390916,390917
Thurber's Beardtongue 289380
Thurber's Catchfly 364127
Thurber's Desert Honeysuckle 25262
Thurber's Penstemon 289380
Thurber's Sneezeweed 188440
Thurber's Wild Buckwheat 152523
Thurber's Wirelettuce 375987
Thuringian Mallow 223392
Thuringian Tree Mallow 223392
Thuringian Tree-mallow 223392
Thurlow Weeping Willow 343341
Thurlow's Weeping Willow 343341
Thuya 386942,390567
Thwaites Dutchmanspipe 34355
Thwaites Sterculia 376195
Thyme 391061,391365,391397
Thyme Broomrape 274934
Thyme Honey Myrtle 248123
Thyme Honeymyrtle 248123
Thyme Leaves Azalea 331305
Thyme Pennyroyal 187188
Thyme-flowered Dragon's-head 137677
Thyme-flowered Dragonhead 137677
Thymeleaf Azalea 331975
Thymeleaf Bluet 187627
Thymeleaf Dragonhead 137677
Thymeleaf Eritrichium 153546
Thymeleaf Loosestrife 240090
Thyme-leaf Melaleuca 248123
Thymeleaf Penstemon 289334
Thyme-leaf Pinweed 223701
Thyme-leaf Rhododendron 331975
Thymeleaf Sandwort 32212
Thymeleaf Speedwell 407358
Thyme-leaf Wild Buckwheat 152524
Thyme-leafed Willow 344094
Thyme-leaved Chickweed 32212
Thyme-leaved Cotoneaster 107549,107558
Thyme-leaved Flaxseed 325041
Thyme-leaved Rhododendron 331975
Thyme-leaved Sandwort 32212
Thyme-leaved Speedwell 407358,407362
Thyme-leaved Spurge 85792,159826
Thyme-leaves Rhododendron 331975
Thymifolious Euphorbia 159971
Thymifolious Spurge 159971
Thyrocarpos 391419
Thyrse Corydalis 106524
Thyrse Dichorisandra 128997
Thyrse Loosestrife 239881
Thyrse Microtropis 254330
Thyrse Saurauia 347992
Thyrse Tuber Sweet vetch 187765
Thyrseflower Dendrobium 125387
Thyrse-flower Fishvine 126003
Thyrse-flower Jewelvine 126003
Thyrseflower Mayten 246938
Thyrseflower Orostachys 275396
Thyrse-flowered Mayten 246938
Thyrse-flowered Microtropis 254330
Thyrse-flowerod Loosestrife 239881
Thyrselike Negundo Chastetree 411381
Thyrselike Triratna Tree 397373
Thyrsia 391440
Thyrsiferous Argyreia 32649
Thyrsostachys 391457
Thyrsus Orostachys 275396
Thyrsus Sansevieria 346155
Thysanospermum 103862
Ti 104350
Ti Kouka 104339
Ti Plant 104334,104350,104367

Ti Tree 104350
Tian'e Aspidistra 39524
Tian'e Spicegrass 239852
Tian'e Tylophora 400893
Tianbaoshan Barberry 51943
Tianlin Amesiodendron 19421
Tianlin Azalea 331977
Tianlin Elatostema 142877
Tianlin Gingerlily 187477
Tianlin Rhododendron 331977
Tianmor Fritillary 168491
Tianmu Arrowwood 408009
Tianmu Bamboo 297475
Tianmu Litse 233847
Tianmu Magnolia 416684
Tianmu Maple 3628
Tianmu Mountain Litse 233847
Tianmu Mountain Magnolia 416684
Tianmu Mountain Stewartia 376442
Tianmu Pearweed 239884
Tianmu Rehmannia 327433
Tianmu Sanicle 346005
Tianmu Stewartia 376442
Tianmu Teasel 133518
Tianmu Viburnum 408009
Tianmushan Angelica 24490
Tianmushan Bamboo 297475
Tianmushan Cacalia 283843
Tianmushan Chloranthus 88302
Tianmushan Cleistogenes 94621
Tianmushan Fritillary 168491
Tianmushan Hemsleia 191927
Tianmushan Hideseedgrass 94621
Tianmushan Litse 233847
Tianmushan Magnolia 416684
Tianmushan Maple 3628
Tianmushan Nettletree 80612
Tianmushan Peristrophe 291181
Tianmushan Purplestem 376442
Tianmushan Sedge 76549
Tianmushan Spapweed 205382
Tianmushan Stonecrop 357228
Tianmushan Teasel 133518
Tianmushan Thistle 92444
Tianmushan Touch-me-not 205382
Tianquan Aster 41376
Tianquan Barrenwort 146992
Tianquan Bergenia 52542
Tianquan Calanthe 65941
Tianquan Currant 334236
Tianquan Cuspidateleaf Acanthopanax 2372
Tianquan Dandelion 384446
Tianquan Epimedium 146992
Tianquan Maple 3649
Tianquan Skullcap 355810
Tianquan Spapweed 205381
Tianquan Spicebush 231453
Tianquan Spice-bush 231453
Tianquan Touch-me-not 205381
Tianquan Villous Hydrangea 200138
Tianquan Windflower 23990
Tianschan Mountain Groundsel 360214
Tianshan Adonis 8383
Tianshan Alkaligrass 321404
Tianshan Arnebia 34647
Tianshan Bedstraw 170667
Tianshan Birch 53625
Tianshan Bluegrass 305664
Tianshan Buttercup 326248
Tianshan Calophaca 67787
Tianshan Cancrinia 71105
Tianshan Catchfly 364128
Tianshan Centaurea 81161
Tianshan Cherry 83345
Tianshan Chorispora 89021
Tianshan Columbine 30076
Tianshan Cousinia 108426
Tianshan Craneknee 221764
Tianshan Crazyweed 279208
Tianshan Cudweed 178240
Tianshan Cuscuta 115143
Tianshan Dandelion 384832
Tianshan Dock 340285
Tianshan Dogtongueweed 385924
Tianshan Doronicum 136378
Tianshan Draba 137107
Tianshan Eremostachys 148512
Tianshan Euphorbia 159967
Tianshan Fakestellera 375215
Tianshan Feathergrass 376936
Tianshan Fescuegrass 163791
Tianshan Fleabane 151018
Tianshan Gentian 151018,173982
Tianshan Gianthyssop 236274
Tianshan Globethistle 140820
Tianshan Goldenray 229117,229228
Tianshan Goldsaxifrage 90461
Tianshan Goosecomb 335526
Tianshan Groundsel 360214
Tianshan Hawksbeard 111059
Tianshan Helictotrichon 190199
Tianshan Honeysuckle 236174
Tianshan Hornpoppy 176734
Tianshan Houndstongue 118022
Tianshan Hyssop 203105
Tianshan Iris 208690
Tianshan Knotweed 309889
Tianshan Lagochilus 220094
Tianshan Larkspur 124643
Tianshan Leek 15807
Tianshan Leymus 228392
Tianshan Listera 264748
Tianshan Lophanthus 236274
Tianshan Lymegrass 335526
Tianshan Maple 3587
Tianshan Milkaster 169803
Tianshan Milkvetch 42600
Tianshan Mountain Adonis 8383
Tianshan Mountain Mountain-ash 369536
Tianshan Mountain Onion 15807
Tianshan Mountain Spruce 298429
Tianshan Mountain Thorowax 63859
Tianshan Mountain Tulip 400242
Tianshan Mountainash 369536
Tianshan Mountain-ash 369536
Tianshan Mouseear 83059
Tianshan Needlegrass 376936
Tianshan Onion 15823
Tianshan Parrya 284974
Tianshan Pearlbush 161757
Tianshan Poppy 282743
Tianshan Pricklythrift 2294
Tianshan Prickly-thrift 2294
Tianshan Primrose 314713
Tianshan Pyrethrum 322750
Tianshan Reedbentgrass 127314
Tianshan Rhubarb 329415
Tianshan Ribseedcelery 304817
Tianshan Ricegrass 276062
Tianshan Rindera 334556
Tianshan Rockjasmine 23250
Tianshan Roegneria 335526
Tianshan Russianthistle 344583
Tianshan Schrenk Spruce 298429
Tianshan Sedge 76550
Tianshan Selinum 357828
Tianshan Serpentroot 354962,354966
Tianshan Small Reed 127314
Tianshan Smallreed 127314
Tianshan Sow-thistle 90871
Tianshan Spiraea 372104
Tianshan Sweetdaisy 71105
Tianshan Sweetvetch 188094
Tianshan Swordflag 208690
Tianshan Thistle 91724
Tianshan Thorowax 63859
Tianshan Thyme 391231
Tianshan Trachydium 393864
Tianshan Trianglegrass 397508
Tianshan Trikeraia 397508
Tianshan Tulip 400242
Tianshan Tumorfruitparsley 393864
Tianshan Violet 410662
Tianshan Waxdaisy 189859
Tianshan Whitlowgrass 137107
Tianshan Willow 344203
Tianshan Willowweed 146919
Tianshan Windflower 23921
Tianshan Windhairdaisy 348474
Tianshan Ziziphora 418141
Tianshanyarrow 185259

Tianshui Albizzia 13581
Tianshui Barberry 52253
Tianshui Windhairdaisy 348859
Tiantai Broadleaf Maple 2786
Tiantai Clematis 95222
Tiantai Hornbeam 77399
Tiantai Maple 2786
Tiantai Spicebush 231298
Tiantaishan Beech 162437
Tiantaishan Hornbeam 77399
Tiantang Fissistigma 166689
Tiantangshan Azalea 331978
Tiantangshan Holly 203941
Tiantong Acute Maple 2773
Tiantong Maple 2773
Tianyang Giantarum 20144
Tianyang Hiptage 196849
Tianyang Kitefruit 196849
Tianyang Spicegrass 239883
Tianzhu Corydalis 106529
Tianzhushan Cowwheat 248159
Tianzhushan Hogfennel 292768
Tiara 332766
Tiare 171404
Tiare Tahite 171404
Tibet Ajania 13038
Tibet Alplily 234241
Tibet Anemone 24088
Tibet Aromatic Bamboo 87646
Tibet Asparagus 39234
Tibet Barberry 52223
Tibet Bastard Cress 390210
Tibet Berneuxia 52929
Tibet Berneuxine 52929
Tibet Betony 373466
Tibet Blueberry 403979
Tibet Bluegrass 306074
Tibet Braya 59778
Tibet Buckthorn 328872
Tibet Catalpa 79265
Tibet Chickweed 375114
Tibet Chinquapin 79043
Tibet Clematis 95376
Tibet Corydalis 106535
Tibet Crazyweed 278780
Tibet Cymbidium 117090
Tibet Dandelion 384833
Tibet Dewberry 339370
Tibet Dinetus 131239
Tibet Diploknema 133007
Tibet Dipoma 383997
Tibet Draba 137266
Tibet Dutchmanspipe 34200
Tibet Dwarfmistletoe 30919
Tibet Dwarf-mistletoe 30919
Tibet Eutrema 161152
Tibet Evergreen Chinkapin 79043
Tibet Evergreenchinkapin 79043
Tibet Falsespiraea 369288
Tibet Filbert 106734
Tibet Flatawndaisy 413009
Tibet Floweringquince 84591
Tibet Flowering-quince 84591
Tibet Fragrant Bamboo 87646
Tibet Gentian 173988
Tibet Goldenray 229166
Tibet Greenbrier 366344
Tibet Habenaria 184136
Tibet Hardleaf Willow 344074
Tibet Hedinia 187373
Tibet Helicia 189955
Tibet Helictotrichon 190204
Tibet Heron's Bill 153923
Tibet Himalayahoneysuckle 228335
Tibet Honeysuckle 236070
Tibet Hornstedtia 198476
Tibet Intermediate Ephedra 146197
Tibet Iris 208492
Tibet Jerusalemsage 295208
Tibet Juniper 213972
Tibet Keeledsiliclecress 302793
Tibet Knotweed 309891
Tibet Kobresia 217299
Tibet Lancea 220796
Tibet Larch 221881
Tibet Larkspur 124639
Tibet Leptodermis 226142
Tibet Ligusticum 229415
Tibet Lilac 382300
Tibet Lily 229977
Tibet Listera 264722
Tibet Litse 234078
Tibet Lobelia 234827
Tibet Lophanthus 236278
Tibet Lyonia 239391
Tibet Machilus 240725
Tibet Madder 338037
Tibet Manglietia 244425
Tibet Meconopsis 247122
Tibet Melic 249102
Tibet Merrilliopanax 250874
Tibet Michelia 252899
Tibet Microgynoecium 253488
Tibet Microula 254370
Tibet Milkvetch 43154
Tibet Milkwort 308012
Tibet Monkeyflower 255256
Tibet Mountainash 369533
Tibet Mountain-ash 369533
Tibet Mouseear Cress 30472
Tibet Neillia 263172
Tibet Neoalsomitra 263568
Tibet Neopallasia 264250
Tibet Nepeta 265084
Tibet Nervilia 265442
Tibet Newlitse 264070
Tibet Oak 324119
Tibet Onewayflor 275483
Tibet Onosma 271846
Tibet Parapteropyrum 283708
Tibet Paris 284402
Tibet Parnassia 284623
Tibet Pearweed 239889
Tibet Peashrub 72362
Tibet Pea-shrub 72362
Tibet Pepper 300450
Tibet Peristylus 291206
Tibet Phyllophyton 245699
Tibet Plantain 302199
Tibet Platycraspedum 302793
Tibet Pratia 234721
Tibet Pricklyash 417294
Tibet Primrose 315049
Tibet Ptilotrichum 321099
Tibet Purple-branched Willow 343831
Tibet Pyrenaria 322624
Tibet Pyrola 322802
Tibet Raspberry 339370
Tibet Rhodiola 329967
Tibet Rhododendron 331349
Tibet Ricepaperplant 387435
Tibet Rockfoil 349990
Tibet Rockjasmine 23224
Tibet Roscoea 337109
Tibet Rose 336990
Tibet Savin 213972
Tibet Seabuckthorn 196780
Tibet Sea-buckthorn 196780
Tibet Sedge 76538
Tibet Seepweed 379511
Tibet Seseli 361545
Tibet Shortspur Corydalis 105833
Tibet Small Reed 127316
Tibet Speedwell 407416
Tibet Spicebush 231396
Tibet Spiraea 372136
Tibet Spruce 298437
Tibet St. John's Wort 201922
Tibet Stracheya 377411
Tibet Strawberry 167640
Tibet Swallowwort 117675
Tibet Swertia 380387
Tibet Thickribcelery 279653
Tibet Tickseed 104856
Tibet Toadflex 231148
Tibet Tofieldia 392640
Tibet Trachydium 393865
Tibet Tricholepis 395967
Tibet Trigonotis 397473
Tibet Trillium 397568
Tibet Vetch 408641

Tibet Viburnum 408165
Tibet Vicatia 408247
Tibet Walsura 177858
Tibet Wang Willow 344266
Tibet Watercress 262751
Tibet Wendlandia 413787
Tibet Whitebractcelery 266977
Tibet Whitlowgrass 137266
Tibet Willow 343371,344294,344306
Tibet Windhairdaisy 348861
Tibet Woodbetony 287761
Tibetan Aralia 30783
Tibetan Barberry 51691
Tibetan Cherry 83312
Tibetan Cotoneaster 107396
Tibetan Cowslip 314407
Tibetan Crabapple 243723
Tibetan Filbert 106734
Tibetan Hazel 106733
Tibetan Helleborus 190956
Tibetan Hydrangea 199800
Tibetan Manyflowered May-apple 139625
Tibetan Maple 3692
Tibetan Mazus 247043
Tibetan Princess Bamboo 196344
Tibetan Staff Tree 80203
Tibetan Taxillus 385238
Tibetblood Rhododendron 331726
Tibetia 391535,391548
Tibetian Thistle 92473
Tibetic Ricegrass 300752
Tibetica Medinilla 247616
Tibet-sikkim Bluegrass 305296
Tibinagua 152318
Tibourbou 28963
Tiburon Wild Buckwheat 152247
Tick Bush 218388
Tick Clover 126242,126482,126486,
126487,126605,126669
Tick Quackgrass 390081
Tick Sunflower 53858
Tick Trefoil 126242
Tickberry 221238
Tick-bush 218388
Tickclover 126242
Tick-clover 126242
Tickle Grass 12139,12140
Tickseed 53755,53797,53959,104429,
104509,104598,104750,334435
Tick-seed 104750
Tickseed Coreopsis 104531
Tickseed Sunflower 53773,53858,54155
Tickseedlike Trigonotis 397402
Tickseed-sunflower 54155
Tick-trefoil 126242,126278
Tickweed 104429,104509,405973
Tidal Arrowhead 342384
Tidalmarsh Flatsedge 212778
Tidal-marsh Water-hemp 18683
Tiddy 367696
Tideland Spruce 298435
Tide-marsh Water Hemp 18683
Tidewater Red Cypress 385262
Tidy Tips 223497
Tidytips 223476,223497
Tie Palm 104359
Tieghem Rhodiola 329968
Tiehm's Blazing Star 250490
Tiehm's Dwarf Rush 213528
Tiehm's Wild Buckwheat 152526
Tiekling Tommies 336419
Tien Shad Spruce 298426
Tien Shanzspruce 298426
Tienchuan Aster 41376
Tienchuan Maple 3649
Tienchuan Skullcap 355810
Tienlin Gingerlily 187477
Tienmu Sanicle 346005
Tienshan Fleabane 151018
Tientai Clematis 95222
Ti-es 238117
Tie-tongue 97862
Tifton Burclover 247443
Tigasco Oil 70475
Tigen Stick 328345
Tiger Aloe 17136
Tiger Cacao 389403
Tiger Cocoa 389403
Tiger Flower 391622,391627
Tiger Grass 391477,391487
Tiger Jaws 162898,162951
Tiger Lily 229797,229810,229978,230058
Tiger Nut 118822,118837
Tiger Orchid 269069
Tiger Orchis 116905
Tiger Paphiopedilum 282902
Tiger Pear 272799
Tiger Plant 29127
Tiger Pocktorchid 282902
Tiger Stanhopea 373703
Tiger Stripe Bamboo 47518
Tiger Wood 43475,237927
Tiger's Claw 154647,154734
Tiger's Claws 154734
Tiger's Jaw 162951
Tiger's Mouth 28617,130383
Tiger's-jaws 162898,162951
Tigerbeard 400852
Tigerflower 391622,391627
Tiger-flower 391622,391627
Tigergrass 391477,391487
Tiger-grass 391477
Tigerheadthistle Schmalhausenia 352004
Tiger-iris 391622
Tiger-jaws 162951
Tiger-like Rockfoil 349991
Tigernanmu 122661,122705
Tigernut 118822
Tigerpalm 299721
Tigerpalm Morningglory 208077
Tigerstick 328328,328345
Tigertail Grass 239558
Tigertail Spruce 298442
Tiger-tail Spruce 298442
Tigertaillily 346047,346158
Tigerthorn 122027,320594
Tigerwood 237927
Tight-flower Swampmallow 286606
Tight-flowered Scrambling Bamboo 131280
Tigridiopalma 391631
Tigrine Goblin Orchid 258999
Tigrine Oncidium 270843
Tigwer Lily 230058
Tihi-tihi 93676
Tikoki Alectryon 14362
Tikor 114858
Tikouka 104339
Tilandsialike Aechmea 8595
Tile 391706
Tile Sideritis 362807
Tiles Sagebrush 36383
Tilet-tree 391706
Tili Herelip 220070
Tilia-leaf Birch 53400
Tilia-leaf Mulberry 259195
Tillaea 391918
Tillaelike Marshweed 230327
Tillandsia 391966
Tiller Violet 410323
Tillet 391706
Tillettree 391706
Tills 224473
Tilt-head Aloe 17285
Timber Bamboo 297215
Timber Milk-vetch 42731
Timberline Larch 221913
Timbo 235513,235558
Timbre 1051
Time Flower 384714
Time Teller 384714
Timite Palm 244496
Tim-kat 93677
Timonius 392086
Timopheev Wheat 398981
Timor Allophylus 16155
Timor Black Bamboo 47317,175595
Timor Hackberry 80767
Timor Nettletree 80767
Timor Nittatree 284473
Timor Ryssopterys 79894,341227
Timor Screwpine 281122

Timor White Gum 155472
Timothy 294960,294992
Timothy Grass 294992
Timothygrass 294992
Timothy-grass 294992
Tinaroo Bottlebrush 67290
Tinctiria Money Maple 133598
Tinctor Mulberry 259197
Tinctorial Condorvine 245854
Tinctorial Fig 165761
Tinctorial Lachnanthes 219030
Tinctorial Marsdenia 245854
Tinctorial Milkgreens 245854
Tinctorial Onosma 271837
Tinctorial Wendlandia 413829,413836
Tine 408352,408423
Tine Tare 408423
Tineanther 5884
Tinebract Pipewort 151298
Tinebract Windhairdaisy 348827
Tinecalyx Camellia 69056
Tinefruit Knotweed 309696
Tine-grass 408352,408423
Tinehorn Dandelion 384740
Tineleaf Epipactis 147238
Tineleaf Pholidota 295530
Tinelip Nestorchid 264651
Tinelobe Windhairdaisy 348386
Tineo 413706
Tinepetal Corydalis 106221
Tinesepal Androcorys 22324
Tinesepal Knifebean 71053
Tinesepal Maddertree 12530
Tinesepal Tarenna 384910
Tine-tare 408423
Tinetooth Windhairdaisy 348306
Tine-weed 408352,408423
Tingbang Agapetes 10357
Tinged Oval Sedge 76551
Tinged Sedge 76551
Tinged Spindle-tree 157926
Tinghu Clematis 95380
Tingiring-gum Eucalyptus 155587
Tingiringi Gum 155587
Tingletongue 249139
Tingo Fibre 313454
Tingri Milkvetch 43159
Tinifoliate Raspberry 339378
Tinker's Weed 397844
Tinker's-weed 397844
Tinker-tailor Grass 235334,302034
Tinnevelly Senna 78227
Tinomiscium 392222
Tinospora 392243
Tinospore 392243,392269
Tinted Spurge 158672
Tinted Wood Spurge 158672
Tintern Spurge 159892
Tinwa Bamboo 82452
Tiny Ardisia 31571
Tiny Attenuated Haworthia 186304
Tiny Crocus 111557
Tiny Edelweiss 224928
Tiny Hawksbeard 110916
Tiny Henbane 201401
Tiny Lazy Daisy 84923
Tiny Mouse's-tail 260940
Tiny Mousetail 260940
Tiny Pentachaeta 289420
Tiny Tim 174285
Tiny Vetch 408423
Tinyhair Elatostema 142743
Tinyhair Stairweed 142743
Tinyleaf Saxifrage 349065
Tinypetal Cuphea 114611
Tinyscale Elatostema 142744
Tinyscale Stairweed 142744
Tinytail Stairweed 142869
Tinytooth Stairweed 142742
Tioman Licuala Palm 228768
Tioman Licualapalm 228768
Tipa 392345
Tipa Tree 392345
Tipo Ideale Rose 336491
Tipsen 201732
Tipsy-leaves 201732
Tipsy-wood 169928
Tipton-weed 202086
Tipu 392345
Tipu Tree 392342,392345
Tipularia 392346
Tirite 209410,209411
Tirpitzia 392368
Tirucalli 159975
Tirucalli Rubber 159975
Tischler Barberry 52258
Tisso Flowers 64148
Tisty-tosty 216085,315082,407989
Tit Crazyweed 278729
Titan Arum 20145
Titan Giant Arum 20145
Titanotrichum 392423
Titatch 271280,408352,408571
Tithonia 392428
Titi 96881,120488
Titoki 14362
Titov Brachanthemum 58048
Titov Skullcap 355811
Titsum 201732
Titsy-leaf 201732
Titters 408423
Tittle-my-fancy 410677
Titty Bottle 336419
Titty-bottles 336419
Titzen 201732
Tivers 170193
Tizon 93687
Tjempaka Petih 252907
Tlanochtle 238973
Tlme Table 384714
Tnorold Orinus 274254
Toad Flower 374013
Toad Lily 396583,396588,396598,397609
Toad Rush 212929,212955
Toad Tether 326287
Toad Tree 101980,382780
Toad Trillium 397609
Toad's Brass 370522
Toad's Head 168476
Toad's Meat 37015
Toad's Mouth 28617,28631,168476
Toadcolour Gongora 179291
Toadflax 84594,100229,230866,230931,231084,231106,231108,231183,267347,370522
Toadflax-leaved St. John's-wort 201992
Toadflaxlike Corydalis 106083
Toadflex 230866
Toadflexleaf Milkwort 308164
Toad-lilies 396583
Toadlily 396583,396598,396604
Toad-lily 396583
Toadroot 6433,6448
Toadshade Trillium 397609
Toadwort 398134
Tobacco Brush 79978
Tobacco Flower 266035
Tobacco Plant 266047
Tobacco Sumac 332956
Tobacco Tree 367146
Tobacco Weed 44297
Tobacco-brush 79978
Tobaccoleaf Chirita 87882
Tobacco-root 228298,404255
Tobacco-weed 44296
Tobaco Soybean 177789
Tobago Cane 46377
Tobguelikeleaf Sterculia 376135
Tobira 301413
Tobira Family 301203
Tobira Pittosporum 301413
Tobira Seatung 301413
Tobosa 196199
Tobosa Grass 196199
Tobosa-grass 196199
Tobusch Fishook Cactus 354289
Tocalote 81198
Tocusso 143522
Today and Tomorrow 61296
Today-and-tomorrow 141347,321643
Toddalia 392556

Toddling Grass 60382
Toddy Fishtail Palm 78056
Toddy Fishtail-palm 78056
Toddy Palm 57122,78056
Todomatsu Fir 414
Tofieldia 392598
Tofu 177750
Togo Tangle 84190
Toi 104359
Token Blackberry 338209
Tokin Aspidopterys 39680
Tokinkan 93566
Toksun Milkvetch 43161
Tokudama Plantain Lily 198650
Tokyo Azalea 332144
Tokyo Cherry 83365
Tolda 158460
Tollon 193839,193841
Tolmeiner 127607
Tolmeneer 127607
Tolmiea 392667
Toloache 123071
Tolu Balsam 261554,261559
Tolu Balsam Balm Tree 261561
Tolu Balsam Tree 261561
Tolu Balsam-balm Tree 261561
Tolu Balsam-tree 261561
Toluca Mahonia 242653
Tolypanthus 392710,392715
Tom Bacca 95431
Tom Paine 324335
Tom Pimpernel 21339
Tom Thumb 222820,237539,273531,
396854,399602
Tom Thumb Cactus 284678
Tom Thumb's Fingers-and-thumbs 237539
Tom Thumb's Honeysuckle 237539
Tom Thumb's Thousand Fingers 339887
Tomasinini Crocus 111617
Tomathlo 239092,297686
Tomatillo 239092,297650,297686,297712
Tomato 239154,239157
Tomato Egg-plant 366920
Tomatofruited Eggplant 367673
Tomcat Clover 397121
Tomentellate Honeysuckle 236176
Tomentosa Pummelo 93477
Tomentosa Sasa 347316
Tomentose Ampelosis 20453
Tomentose Anemone 24090
Tomentose Barbed Cinquefoil 312355
Tomentose Barberry 52262
Tomentose Bauhinia 49241
Tomentose Beadcress 264624
Tomentose Beautyleaf 67873
Tomentose Bedstraw 170763
Tomentose Begonia 50370
Tomentose Burdock 31068
Tomentose Bushmint 203068
Tomentose Calaba 67873
Tomentose Callery Pear 323119
Tomentose Capitate Paraphlomis 283653
Tomentose Carpesium 77195
Tomentose Cavalerie Blastus 55145
Tomentose Chestnut Sage 344944
Tomentose Cleistanthus 94540
Tomentose Cliffbean 254816
Tomentose Clovershrub 70873,70902
Tomentose Coelogyne 98749
Tomentose Condorvine 245858,245860
Tomentose Congea 101778
Tomentose Corchoropsis 104046
Tomentose Crapemyrtle 219976
Tomentose Crape-myrtle 219976
Tomentose Cyathula 115749
Tomentose David Poplar 311291
Tomentose Delavay Ampelopsis 20341
Tomentose Dendrocalamus 125514
Tomentose Diverse Sagebrush 35138
Tomentose Diverse Wormwood 35138
Tomentose Dysophylla 139596
Tomentose Elephantfoot 143474
Tomentose Entire Elatostema 142693
Tomentose Entireleaf Stairweed 142693
Tomentose Falsenettle 56339
Tomentose Fevervine 280106
Tomentose Fieldjasmine 199456
Tomentose Flatawndaisy 413018
Tomentose Flemingia 166867
Tomentose Fluttergrass 166537
Tomentose Germander 388294
Tomentose Glorybower 96200
Tomentose Grape 411882
Tomentose Guazuma 181618
Tomentose Hairycalyx Elsholtzia 144010
Tomentose Henry Pentapanax 289642
Tomentose Huodendron 199456
Tomentose Italian Maple 3290
Tomentose Japanese Viburnum 408034
Tomentose Javatea 275794
Tomentose Jurinea 211862
Tomentose Kalanchoe 215277
Tomentose Khasia Bauhinia 49148
Tomentose Kidney-shaped
Whitfordiodendron 9854
Tomentose Leptodermis 226124
Tomentose Lilac 382302
Tomentose Littlefruit Grape 411570
Tomentose Longspike Willow 343980
Tomentose Machilus 240711
Tomentose Magnolia 242331
Tomentose Magnoliavine 351110
Tomentose Marsdenia 245860
Tomentose Meadowsweet 372106
Tomentose Medicine Terminalia 386506
Tomentose Melhania 248887
Tomentose Millettia 254816
Tomentose Mockorange 294560
Tomentose Mongolian Spiraea 372109
Tomentose Montanoa 258211
Tomentose Nepeta 264964
Tomentose Newlitse 264106
Tomentose Nothapodytes 266811
Tomentose Oreocharis 273867
Tomentose Orientvine 364991
Tomentose Ormosia 274413
Tomentose Ovateleaf Croton 112857
Tomentose Pagodatree 369134
Tomentose Pavetta 286520
Tomentose Pearleaf Raspberry 339070
Tomentose Peashrub 72293
Tomentose Pea-shrub 72293
Tomentose Pedicel Clovershrub 70902
Tomentose Pentapanax 289665
Tomentose Peony 280319
Tomentose Pileostegia 299112
Tomentose Piptanthus 300699
Tomentose Pometia 310826
Tomentose Porkkidneybean 9854
Tomentose Qiaomaidi Sage 345145
Tomentose Raspberry 339383
Tomentose Rattan Palm 65805
Tomentose Reevesia 327384
Tomentose Rhododendron 331070
Tomentose Rose 336993
Tomentose Rough-leaved Begonia 49636
Tomentose Sainfoin 271274
Tomentose Serrate Oak 324401
Tomentose Shinyleaf Pricklyash 417285
Tomentose Sibiraea 362399
Tomentose Sichuan Sagebrush 36284
Tomentose Sloanea 366080
Tomentose Soapberry 346354
Tomentose Sophora 368965,369134
Tomentose Spapweed 205389
Tomentose Spathoglottis 370403
Tomentose Spicebush 231398
Tomentose Spice-bush 231398
Tomentose Spinyfruit Pricklyash 417222
Tomentose Spiraea 372109
Tomentose Srilanka Glochidion 177194
Tomentose Stranvaesia 377470
Tomentose Suoluo 327384
Tomentose Sweet Gale 261164
Tomentose Tangut Sagebrush 36366
Tomentose Tangut Wormwood 36366
Tomentose Tianshan Mountainash 369538
Tomentose Tinctorial Marsdenia 245858
Tomentose Tinctorial Milkgreens 245858
Tomentose Touch-me-not 205389
Tomentose Trema 394664

Tomentose Tubergourd 390121
Tomentose Viburnum 408096
Tomentose Waldheimia 413018
Tomentose Wildclove 226124
Tomentose Willowleaf Agapetes 10324
Tomentose Windflower 24090
Tomentose Windhairdaisy 348864
Tomentose Woodbetony 287762
Tomentose Wrightia 414799
Tomentose Yellowgland Everlasting 21522
Tomentose Yellowgland Pearleverlasting 21522
Tomentoseanther Clematis 95245
Tomentosecalyx Elsholtzia 143998
Tomentosefruit Tanoak 233116
Tomentose-fruited Tanoak 233116
Tomentoseleaf Aralia 30606
Tomentose-leaved Eria 148790
Tomentose-nerve Grape 411788
Tomentulose Barberry 52263
Tomentulose Pricklyash 417356
Tomentulose Prickly-ash 417356
Tommy Tottles 237539
Toms Herb 136417
Tonga-creeper 147349
Tongan Oil 67860
Tongavine 147336
Tongbiguan Spapweed 205390
Tongchuan Euphorbia 159984
Tongchun Indocalamus 206834
Tongga Monkshood 5273
Tonggu Maple 2957
Tongjiang Wildginger 37747
Tongking Rubber 377544
Tongling Fritillary 168491
Tongmai Lemongrass 117267
Tongmai Oak 324505
Tongo Corydalis 106539
Tongol Aster 41380
Tongol Bastardtoadflax 389899
Tongol Corydalis 106539
Tongol Desert Sagebrush 35392
Tongol Desert Wormwood 35392
Tongol Gentian 173992
Tongol Goldenray 229230
Tongol Hundredstamen 389899
Tongol Larkspur 124644
Tongol Milkvetch 43164
Tongol Primrose 315050
Tongol Windhairdaisy 348617
Tongol Woodbetony 287763
Tongoloa 392782
Tongs Screwpine 281029
Tongshan Giantfennel 163661
Tongtianhe Sweetvetch 187916
Tongue 40817
Tongue Buttercup 326034
Tongue Chirita 87900
Tongue Habenaria 183801
Tongue Leaf 177465
Tongue Orchid 125219,360582,360610
Tongue Saxifrage 52532,349147
Tongue-bleed 170193
Tongue-flowered Orchid 360582
Tongue-flowered Orchis 360610
Tongue-grass 225450,262722,374968
Tongueleaf 177434
Tongueleaf Aster 40779
Tongueleaf Cremanthodium 110410
Tongueleaf Flower 177465
Tongueleaf Nutantdaisy 110410
Tongue-leaved Crowfoot 326034
Tongueless Cremanthodium 110408
Tongueless Groundsel 358747
Tongueless Jurinea 193937
Tongueless Nutantdaisy 110408
Tongueless Pyrethrum 322749
Tongueless Threebibachene 397962
Tongueless Threeribfruit 397962
Tongue-like Aster 40779
Tongueshaped Guzmania 182172
Tongue-under-tongue 42262
Tongziguo Ginkgo 175825
Tonka Bean 133626,133629
Tonka-bean 133626,133629
Tonkin Acacia 1651
Tonkin Acroceras 263955
Tonkin Artocarpus 36947
Tonkin Bamboo 318283
Tonkin Bauhinia tonkinensis 49054
Tonkin Beccarinda 49401
Tonkin Buckthorn 328875
Tonkin Camellia 69704
Tonkin Canary Tree 71027
Tonkin Canarytree 71027
Tonkin Cane 318283
Tonkin Canebrake 318283
Tonkin Caryodaphnopsis 77969
Tonkin Cenocentrum 80872
Tonkin Chaydaia 86320
Tonkin Cinnamon 91440
Tonkin Cladogynos 93953
Tonkin Cleistanthus 94541
Tonkin Clethra 96496
Tonkin Cnesmone 97505
Tonkin Creeper 385752
Tonkin Croton 112920
Tonkin Cyclea 116038
Tonkin Cylindrokelupha 116608
Tonkin Dendrobenthamia 105063
Tonkin Deutzianthus 127130
Tonkin Dogwood 105063
Tonkin Eberhardtia 139784
Tonkin Elm 401552
Tonkin Euonymus 157928
Tonkin Evergreen Chinkapin 79045
Tonkin Evergreenchinkapin 79045
Tonkin Excentrodendron 161627
Tonkin Falsenettle 56340
Tonkin Falsenettleleaf Pepper 300359
Tonkin Fishvine 126005
Tonkin Fissistigma 166690
Tonkin Four-involucre 105063
Tonkin Galangal 17769
Tonkin Giantarum 20146
Tonkin Habenaria 184141
Tonkin Hartia 185970
Tonkin Hickory 77935
Tonkin Hiptage 196825
Tonkin Hsienmu 161627
Tonkin Jasmine 212040
Tonkin Jewelvine 126005
Tonkin Keenan 215751
Tonkin Kudzuvine 321469
Tonkin Leucothoe 228184
Tonkin Lilyturf 272152
Tonkin Maple 3694
Tonkin Myrioneuron 261331
Tonkin Neohusnotia 263955
Tonkin Parasite 293196
Tonkin Peltophorum 288901
Tonkin Peltostyle 288901
Tonkin Pepper 300359
Tonkin Phacellaria 293196
Tonkin Phrynium 296056
Tonkin Pilular Adina 8193
Tonkin Pipewort 151524
Tonkin Pittosporum 301423
Tonkin Pseudosasa 318283
Tonkin Ringvine 116038
Tonkin Rockvine 387860
Tonkin Rosewood 121841
Tonkin Saurauia 347961
Tonkin Screwpine 281157
Tonkin Seatung 301423
Tonkin Sedge 76492
Tonkin Shuttleleaffruit 139784
Tonkin Siliquamomum 364247
Tonkin Siliquamon 364247
Tonkin Silvianthus 364344
Tonkin Sindora 364715
Tonkin Snowbell 379471
Tonkin Sophora 369141
Tonkin Spicebush 231454
Tonkin Spice-bush 231454
Tonkin Spindle-tree 157928
Tonkin Spiralflag 107281
Tonkin Staff-tree 80336
Tonkin Sterculia 376198
Tonkin Storax 379471
Tonkin Streblus 377544

Tonkin Teonongia 377544
Tonkin Tinomiscium 392223
Tonkin Tubergourd 390121
Tonkin Uvaria 403585
Tonkin Whiteflower Neillia 263179
Tonkin Wingnut 320384
Tonkin Wing-nut 320384
Tonkin Zelkova 417565
Tonroku Galangal 17770
Tonth Violet 125836
Tooart Tree 155594
Toog 54620
Toolur 155598
Toon 392829
Toona 392822
Tooth Ammi 19672
Tooth Catnip 264910
Tooth Cress 125836
Tooth Gavelstyle Orchis 243289
Tooth Primrose 314747
Tooth Sweetclover 249212
Tooth Violet 125836
Toothache Grass 113955
Toothache Plant 4870,53905,405936
Toothache Tree 30760,417157,417199
Toothbract Begonia 49772
Toothbract Bladderwort 403826
Tooth-bract Corydalis 105991
Toothbract Whitebracteole Bugle 13136
Toothbrush Tree 344804
Toothcalyx Ash 168038
Toothcalyx Azalea 330554
Toothcalyx Draws Ear Grass 403377
Toothclaw Dualsheathorchis 269040
Toothcleft Sishan Cinquefoil 313001
Toothcleft Xishan Cinquefoil 313001
Toothcup 19575,337384
Tooth-cup 19621,337384
Toothear Sinosenecio 365040
Toothed Beautyberry 66767
Toothed Broadleaf Nettle 402949
Toothed Burr Medic 247425
Toothed Chinese Sinobambusa 364821
Toothed Cress 30451
Toothed Cyperus 118723
Toothed Diverse-leaved Hemiphragma 191575
Toothed Dock 340019
Toothed Dodder 115005
Toothed Dolomiaea 135728
Toothed Euphorbia 159828
Toothed Evening Primrose 68508
Toothed Habenaria 183559
Toothed Lancewood 318067
Toothed Lavender 223284
Toothed Medick 247425
Toothed Orchid 273374,273686
Toothed Sage 36277
Toothed Spider-flower 95791
Toothed Spurge 158758,158761
Toothed Sweetclover 249212
Toothed Tangbamboo 364821
Toothed Tinospore 392253
Toothed Waterlily 267698
Toothed White-topped Aster 360704
Toothed Wintergreen 275486
Toothed Woodbetony 287467
Toothed-bract Willow 343029
Toothedcalyx Primrose 314748
Toothedcalyx Raspberry 338223
Toothedcrown Redflower Corydalis 106089
Toothedfruit Dock 340019
Toothedleaf Japanese Windhairdaisy 348409
Toothedleaf Primrose 314941
Toothedleaf Pterygiella 415621
Toothedleaf Stonecrop 294125
Toothedleaf Yarrow 3914
Toothedpetal Corydalis 106564
Toothed-stipuled Wild Violet 409630
Toothfruit Amomum 19901
Toothfruit Dock 340019
Toothleaf 409155
Tooth-leaf 376641
Tooth-leaf Button Flower 194665
Toothleaf Coldwaterflower 299010
Toothleaf Dwarf Clearweed 299010
Toothleaf Elaeocarpus 142308
Toothleaf Engelhardia 145520
Toothleaf Goldenray 229016
Toothleaf Gomphia 70748
Toothleaf Hemiboea 191353
Toothleaf Lepidagathis 225158
Tooth-leaf Lysionotus 239981
Toothleaf Pterygiella 415621
Toothleaf Rockfoil 349474
Toothleaf Snowbell 379449
Toothleaf Stephania 375838
Toothleaf Stonecrop 294125
Toothleaf SW. China Cattailtree 135332
Toothleaf Tanoak 233269
Toothleaf Tubergourd 390123
Toothleaf Wampee 94185
Toothleaf Woodbetony 415621
Toothleaf Yunnan Pennycress 390272
Toothleaf Yushan Groundsel 359568
Tooth-leaved Croton 112895,112896
Tooth-leaved Gilia 175689
Tooth-leaved Spapweed 205190
Toothless Argy's Wormwood 35170
Toothless Camellia 69054
Toothless Guangzhou Salomonia 344349
Toothless Irisorchis 267939
Toothless Oberonia 267939
Toothless Qinggang 116191
Toothless Sagebrush 35170
Toothless Sotol 122908
Toothless Tea 69054
Toothless Vetch 408384
Toothlip Liparis 232114
Toothlip Suzukia 380045
Toothlip Twayblade 232114
Toothlip Woodbetony 287466
Toothpetal Catchfly 363564
Toothpetal Irisorchis 267949
Toothpetal Oberonia 267949
Toothpetal Platanthera 302477
Tooth-petaled Onion 15902
Tooth-petaled Spapweed 205189
Toothpick Ammi 19672
Toothpick Cactus 376364
Toothpick-plant 19672
Toothpickweed 19672
Toothplait Gentian 173419
Toothscale Windhairdaisy 348604
Toothsepal Chirita 87987
Toothsepal Loxostigma 238035
Toothsepal Spapweed 204903
Toothsepal Touch-me-not 204903
Tooth-sepaled Ash 168038
Toothshape Cenolophium 80879
Toothwing Broomsedge 217349
Toothwing Heshouwu 162525
Toothwing Summercypress 217349
Toothwing Sweetvetch 187852
Toothwort 72038,72723,125831,125836,125841,125846,222641,222644,222651
Toowoomba Canary Grass 293708,293764,293768
Top Knot 81237
Top Onion 15169
Top Primrose 314717
Topal Holly 203529
Topee-tamlbo 66158
Topeka Purple Coneflower 140070
Topeng Pygeum 322412
Topeng Rambutan 265139
Toper's Plant 345881
Topflower Anplectrum 247537
Topflower Eargrass 187685
Topflower Hedyotis 187685
Topflower Indianmulberry 258895
Topflower Lobelia 234821
Topflower Maesa 241743
Topflower Medinilla 247537
Topflower Pepper 300529
Top-flowered Maesa 241743
Top-flowered Pepper 300529
Topfruit 249557,249560
Tophair Epilasia 146567
Topinambour 66158
Topi-tamboo 66158
Top-knot 81237

Topola Canadsca 311146
Top-pod Water-primrose 238210
Toquilla 77043
Torcbwood 21003
Torch 405788
Torch Azalea 330966
Torch Cactus 83911,140888
Torch Flame Blanket Flower 169603
Torch Ginger 19884,155398
Torch Lily 216923,217055,350312
Torch Lobelia 234592
Torch Pine 300264
Torch Plant 16605,405788
Torch Rhododendron 331877
Torch Tree 211148,211150
Torch Wood Ixora 211148
Torchflower 100030
Torch-ginger 155398
Torch-lily 216923,217055
Torchorchis 181304,181305
Torchwood 21005
Torchwood Ixora 211148
Torchwort 405788
Torenia 392880
Tor-grass 58604
Toringa Crabapple 243735
Toringo Crab Apple 243717
Toringo Crabapple 243717
Toringo Crab-apple 243717
Tormentil 220416,312520
Tormentilla 312520
Tormentilla Cinquefoil 312520
Tormenting Root 312520
Tormerik 312520
Torn Leptodermis 226132
Tornsole 159027
Torote 64080
Torote Blanco 64081,279505
Torote Colorado 64080
Torote Prieto 64077,64086
Torpalm 181802,181803
Torpedo Grass 282162
Torpedograss 282162
Torree Saltbush 44675
Torree's Amaranth 18835
Torree's Saltbush 44675
Torrey Mormon Tea 146260
Torrey Pine 300288
Torrey Rush 213531
Torrey Surfgrass 297187
Torrey Three-square 352281
Torrey Yucca 416657
Torrey's Bulrush 352281
Torrey's Desertdandelion 242966
Torrey's Mountain-mint 321963
Torrey's Rush 213531
Torrey's Sedge 76563
Torrey's Surf-grass 297187
Torrey's Three-square Bulrush 352281
Torrey's Yucca 416657
Torreya 393047
Torricellia 393106
Torricellia Family 393113
Torsional Woodbetony 287765
Torsionsepal Spapweed 205392
Torsionsepal Touch-me-not 205392
Tortedfruit Screwtree 190105
Tortile-fruited Screw-tree 190105
Tortoise Plant 131451,131573
Tortoise Shell Bamboo 297266
Tortoiseshell Bamboo 297301
Tortuous Branch Rosewood 121647
Tortuous Caesalpinia 65073
Tortuous Figmarigold 252112
Tortuous Hankow Willow 343670
Tortuous Harrisia 185925
Tortuous Madwort 18483
Tortuous Mulberry 259069
Tortuous Weeping Willow 343074
Tortuousfruit Caesalpinia 65073
Tortuousfruit Draba 137278
Tortuousfruit Whitlowgrass 137278
Tortured Willow 343668
Toru Persoonia 292034
Torularia 264603
Torulinium 393185,393194
Torulosous Dactylicapnos 128323
Torus Herb 136470
Tory-top 300223
Tosa Azalea 331985
Tosa Dendrobium 125390
Tosa-asahi 93850
Tossa Jute 104103
Tossy Ball 407989
Tossy Balls 407989
Tosty 315082
Totai Palm 6137
Totara 306438,306522
Totara Pine 306522
Totara Podocarpus 306522
Totara Yaccatree 306522
Totora 353274
Totsan 201732
Totter Grass 60382
Totter-grass 60382
Tottering Grass 60382
Totty Grass 60382
Touch-and-heal 201732,202086,316127
Touched-leaf 201732
Touchen-leaf 201732
Touch-leaf 201732
Touch-me-not 30146,31051,72802,204756,204799,204841,205175,205405,255098
Touch-me-not Balsam 205175
Tou-fou 177750
Tough Bully 362986
Tough Bumelia 362986
Tough Flax-lily 295600
Tough Rice Flower 299306
Toughbark Star-jasmine 393642
Toughhair Corydalis 105864
Tough-leaf Iris 208879
Tough-leaved Iris 208879
Tough-leaved Sedge 74403
Toumey Oak 324493
Toumey's Century Plant 10946
Toumey's Groundsel 279929
Tournefort Skullcap 355813
Tournefort Speedwell 407287
Tournefort's Speedwell 407287
Tournefortia 393240
Tournesol 89330
Tournier Parabarium 402203
Tourpin 358062
Tous-les-mois 71169,71181
Tow Tree 9701
Towai Bark 413702
Towel Gourd 238257,238261
Towelgourd 238257
Tower 273531
Tower Cress 30477,400644
Tower Mustard 30278,30477,400644
Tower Mustard Rock-cress 400644
Tower of Babel 405788
Tower of Jewels 141350
Tower Poplar 311261
Tower Rock Cress 30477
Tower Rock-cress 30278,30477,400644
Tower Tree 351729,351732
Tower's Treacle 400644
Towercress 400638
Tower-mustard Rock Cross 400644
Towers Mustard 400644
Town Clock 8399
Town Cress 225450
Town-cress 225450
Town-hall Clock 8399
Townhan Clock 8399
Townsend's Cord-grass 370182
Townsville Stylo 379248
Town-weed 250614
Towri 64997
Toxinbean 218754,218761
Toxocarpus 393525
Toyo Nishiki 84555
Toyon 193839,193841
Toy-wort 72038
Tozzia 393568
Trabeculate Alder 16470
Trabeculate Winterhazel 106686
Trabeculose Viburnum 408187

Trac 121653
Trachelium 393608
Trachydium 393818
Tracy Orchis 117090
Tracy's Beakrush 333709
Tracy's Larkspur 124169
Tracy's Thistle 92450
Tradescant Aster 41387
Tradescant's Aster 381003
Tradescantia 393987,394076
Tragacanth 43173
Tragacanth Milk-vetch 42453
Tragacanthus-like Haplophyllum 185675
Tragaeanth Milkvetch 42453
Tragasol 83527
Trailing Abutilon 957
Trailing African Daisy 276608
Trailing Allionia 14992
Trailing Arbatus 146522
Trailing Arbutus 31142,146522,146526
Trailing Azalea 235279,235283
Trailing Begonia 92777
Trailing Bellflower 70229
Trailing Bell-flower 115338
Trailing Bindweed 207590
Trailing Bush Clover 226936
Trailing Bush-clover 226936
Trailing Chinkapin 78807
Trailing Coleus 99711
Trailing Crabgrass 130730
Trailing Currant 334064
Trailing Dewberry 338648
Trailing Dusty Miller 189659,189661
Trailing Fenugreek 397272
Trailing Fleabane 150635
Trailing Four-o'clock 14976,14992
Trailing Fuchsia 168781
Trailing Fuzzy Bean 378502
Trailing Gazania 172304
Trailing Guinea-flower 194665
Trailing Hollyhock 195326
Trailing Ice Plant 89959,220674
Trailing Indigo 206573
Trailing Indigo Bush 121890
Trailing Juniper 213785
Trailing Krameria 218081
Trailing Lantana 221284
Trailing Lespedeza 226936
Trailing Mallow 256432
Trailing Myrtle 409339
Trailing Pearlwort 342245,342279
Trailing Pelargonium 288430
Trailing Petunia 66535
Trailing Phlox 295286
Trailing Raspberry 338985
Trailing Rose 336364
Trailing Sallow 343063
Trailing Shrubverbena 221284
Trailing Snake Herb 139491
Trailing Snapdragon 37548,37554
Trailing St. John's Wort 201930
Trailing St. John's-wort 201930
Trailing St. -John's-wort 201930
Trailing St. Johnswort 201930
Trailing Strawberry Tree 31142
Trailing Tansy 312360
Trailing Tormentil 312357
Trailing Velvet Plant 339762
Trailing Watermelon Begonia 288767
Trailing Wild Bean 378502
Trailing Windmills 14992
Trailing-arbutus 146526
Traill Azalea 331987
Traill Rhododendron 331987
Trailliaedoxa 394400
Traling Queen 168781
Tramman 345631
Tranquil Goldenweed 323017
Trans Pecos Goldenshrub 181933
Trans Pecos Mouse-ear Chickweed 82727
Transala Craneknee 221767
Transarisan Sabia 341575
Transbaical Bluegrass 306082
Transcaucasia Nepeta 265088
Transcaucasian Birch 53523
Transcucent Haworthia 186385
Transe Ili Milkvetch 43175
Transeverse Cranesbill 174972
Transmorrison Bentgrass 12147
Transmorrison Silvergrass 255920
Transnoko Everlasting 21714
Transnoko Pearleverlasting 21714
Transparent Agapetes 10316
Transparent Spikesedge 143298
Trans-pecos Amaranth 18782
Transsect Monkshood 5635
Transsecte Dutchmanspipe 34357
Transsected Monkshood 5635
Transsylvania Melic 249104
Transvaal Beech 162999
Transvaal Candelabra Tree 158688
Transvaal Daisy 175107,175172
Transvaal Gardenia 171424
Transvaal Hard Pear 270470
Transvaal Kaffirboom 154687
Transvaal Millet 282209
Transvaal Privet 170822
Transvaal Red Milkwood 255392
Transvaal Teak 320278
Transvasl Strophanthus 378405
Transvers Dutchman's-pipe 34357
Transwall Statice 230803
Transylvanian Melick 249104
Tranzschell Waterchestnut 394551
Trapaleaf Spiraea 372126
Trapella 394559
Trappers' Tea 223894
Traslucence Haworthia 186780
Traunsteiner Orchis 273681
Trautvetter Buttercup 326442
Trautvetter Sagebrush 36396
Traveler's Fan 327095
Traveler's Joy 95431
Traveler's Palm 327095
Traveler's Tree 327095
Traveler's-joy 95431
Traveler's-tree 327093,327095
Travelerstree 327093,327095
Traveller Palm 327095
Traveller Tree 327095
Traveller's Comfort 170193
Traveller's Ease 3978,170193,312360
Traveller's Foot 302068
Traveller's Leaf 312360
Traveller's Palm 327095
Traveller's-joy 94676
Treacle Erysimum 154424
Treacle Hare's-ear 102811
Treacle Mustard 102811,154424,154530,
225315,390213
Treacle Sugarmustard 154530
Treacle Wormseed 154424
Treacle-berries 366198
Treacle-leaf 201732
Treadsoftly 97739
Treasure Flower 172348
Treasure Maesa 241817
Treasureflower 172276,172309,172321,
172348
Treasure-flower 172276
Treat Orchid 235131
Trebizond Dale 141932
Trebizond Date 141936
Treculia 394601
Tree Aeonium 9023
Tree Alfalfa 247249
Tree Aloe 16592,16617,16627
Tree Amelianchier 19243
Tree Anemone 77141
Tree Anise 204517
Tree Aralia 30760,215442
Tree Aster 270207
Tree Avens 110251
Tree Azalea 330111
Tree Beargrass 266502
Tree Beautyberry 66738
Tree Bedstraw 103798
Tree Blow-pipe 382329
Tree Boxwood 64346
Tree Buckthorn 17613
Tree Buekthorn 17614

Tree Cactus 299238
Tree Calabash 111104
Tree Ceanothus 79911
Tree Celandine 56027,240765
Tree Chinquapin 78807
Tree Cholla 116661,272915,272924
Tree Clethra 96466
Tree Clover 396865
Tree Correa 105393
Tree Cotoneaster 107454
Tree Cotton 179876,179884
Tree Crinum 111167
Tree Currant 334104
Tree Dahlia 121553
Tree Daisy 258202,270184
Tree Euphorbia 158982,159140
Tree Eurotia 218121
Tree Falsespiraea 369265
Tree False-spiraea 369265
Tree Fuchsia 168740,168746,184875,
352493
Tree Gardenia 337500
Tree Germander 388085
Tree Glxinia 217738
Tree Glycosmis 177819
Tree Grape 120108
Tree Groundsel 46210,46227,125716,
360100
Tree Hakea 184605
Tree Hazelnut 106724
Tree Heath 149017,149138,149703,334355
Tree Heather 149017
Tree Hibiscus 195311
Tree Hollyhock 195269
Tree Honeysuckle 47647
Tree Huckleberry 403724
Tree Hydrangea 199806,199870
Tree Indigo 205859,205962
Tree Ipomoea 207792
Tree Ironweed 406109
Tree Lavatera 223378,223392
Tree Limonium 230518
Tree Lupin 238430
Tree Lupine 238430
Tree Mallow 223350,223355,223357,
223372,223378,223392,243850
Tree Marigold 392432
Tree Mayten 246804
Tree Medic 247249
Tree Medick 247249
Tree Melon 76813
Tree Mignonette 158730
Tree Milkvetch 41997
Tree Morning Glory 207593
Tree Morningglory 207593
Tree Morning-glory 207792
Tree Nettle 402914
Tree Nightshade 367528
Tree Ocotillo 167557
Tree of Fountainbleu 369450
Tree of God 85268
Tree of Gold 382706
Tree of Heaven 12558,12559
Tree of Heaven Ailanthus 12559
Tree of Sadness 267429,267430
Tree of the Gods 12559
Tree Onion 15166,15169
Tree Parsnip 192394
Tree Peony 280223,280286
Tree Phillyrea 294772
Tree Phillyroa 294772
Tree Philodendron 294796
Tree Pink 127596
Tree Pop 335870
Tree Poppy 125583,125585,335869,335870
Tree Primrose 269404
Tree Privet 229529
Tree Purplepearl 66738
Tree Purslane 44446,311956
Tree Rhododendron 330110,330111
Tree Sacahuista 266505
Tree Scarlet Elder 345565
Tree Scarlet-elder 345565
Tree Skimmia 365918
Tree Sorrel 45724,278389
Tree Sow Thistle 368635,368781
Tree Spinach 87025
Tree Spurge 158754
Tree St. John's-wort 202134
Tree Strwberry 82096
Tree Tabacco 266043
Tree Tea 68911
Tree Timonius 392088
Tree Tobaco 367364
Tree Tomato 119973,119974
Tree Trefoil 121001
Tree Tutu 104693
Tree Waratah 273804
Tree Waxmallow 243929
Tree Wormwood 35146
Tree Yucca 416563,416593,416640
Tree-attached Rhododendron 330550
Tree-basil 268518
Treebeard Tillandsia 392052
Treebine 92586
Treebine Leaf Maple 2896
Treebineleaf Acanthopanax 143582
Treebineleaf Maple 2896
Treebine-leaved Acanthopanax 143582
Tree-celandine 56027,240763
Treeclimbing Centipede 329003
Treeclimbing Rhaphidophora 328997
Treed Medinilla 247535
Treedaisy 258202
Tree-fern Family 130928
Tree-fern Tree 351732
Tree-horned Acacia 1572
Tree-in-a-hurry 410878
Tree-inhabiting Viburnum 407679
Treeleaf Cayratia 79896
Tree-leaved Cayratia 79896
Tree-leaved Mahonia 242525
Tree-like Aloe 16592
Tree-like Bamboo 125461
Tree-like Blueberry 403714
Treelike Fleshseed Tree 346980
Treelike Fleshspike Tree 346980
Tree-like Ironweed 406109
Tree-like Medic 247249
Tree-like Milkvetch 42277
Treelike Peashrub 72180
Treelike Rhododendron 330111
Tree-like Rhododendron 330111
Treelived Rhododendron 330552
Tree-lived Rhododendron 330552
Treemallow 223350
Tree-mallow 223350,223355
Tree-of-chastity 411189
Tree-of-heaven 12558,12559
Tree-of-heaven Ailanthus 12559
Tree-of-heaven Family 364387
Tree-of-life 246672
Tree-of-sadness 267429
Treephilous Azalea 330552
Treerenshen 125589,125597,125604
Treetomato 119973,119974
Tree-tomato 119973,119974
Tree-wisteria 254711
Trefoil 237539,396806
Trefoil Cress 73019
Trefoil Milkmaid 73019
Trefold 250502
Trefoy 397043
Trelease's Beavertail Pricklypear 272810
Trelease's Larkspur 124645
Trellis Coldbamboo 172750
Trema 394626
Tremacron 394672
Tremble 311537
Trembling Aspen 311537,311547
Trembling Grass 60382
Trembling Heath 150233
Trembling Jockies 60382,102727
Trembling Jocks 60382
Trembling Poplar 311537,311547
Trembling Shadow 60382,238709
Tremblingbill Woodbetony 287743
Trencilla 418331
Tresco Rhodostachys 268103,268104
Treutler Raspberry 339389
Trevesia 394755

Trewia 394780
Triadenum 394807
Triander Sonerila 368902
Triander Themeda 389361
Triandrous Glochidion 177184
Triandrous Willow 344211
Triandrous-like Willow 344220
Triangle Clearweed 299061
Triangle Coldwaterflower 299061
Triangle Leaf Bursage 19162
Triangle Listera 232910
Triangle Oak 167343,167344
Triangle Onion 15839
Triangle Orache 44621
Triangle Palm 139336
Triangle Sweetvetch 188153
TriangleFig 164900
Trianglefruit Tanoak 233400
Trianglegrass 397505,397506
Triangleleaf Beesia 49579
Triangleleaf Bursage 19162
Triangleleaf Gentian 173391
Triangleleaf Nettle 403032
Triangle-leaf Windhairdaisy 348256
Triangleoak 167343,167344
Trianglespine Eritrichium 153444
Trianglestalk Bulrush 353605
Triangletooth Clubfilment 179181
Triangletooth Metalleaf 296379
Triangul Lettuce 320523
Triangular Alangium 13355,13356
Triangular Bulrush 353889,353899
Triangular Buttercup 326470
Triangular Calanthe 66099
Triangular Clubrush 352284
Triangular Club-rush 352284
Triangular Dendrolobium 125580
Triangular Dutchmanspipe 34359
Triangular Euphorbia 158456
Triangular Gasteria 171763
Triangular Living-rock Cactus 33221
Triangular Metalleaf 296379
Triangular Pittosporum 301424
Triangular Pterocypsela 320523
Triangular Scirpus 352284,353899
Triangular Violet 410672
Triangularfruit Seatung 301424
Triangular-fruited Pittosporum 301424
Triangularleaf Woodbetony 287155
Triangularpod Sweetvetch 188153
Triangularsepal Stonecrop 357238
Triangularteeth Woodbetony 287767
Triangulartooth Gomphostemma 179181
Triangular-valved Dock 340244
Triangulastalk Bulrush 353605
Triangulate Daphniphyllum 122693
Triangulate Indocalamus 206905
Triangulate Tigernanmu 122693
Triangule Groundsel 360244
Triangleleaf Loosestrife 239615
Triangleleaf Woodbetony 287155
Triangletooth Woodbetony 287767
Trianther Areca 31696
Trianther Tea 69708
Triawn 33730
Tricalysia 395164
Tricalysia-like Honeysuckle 236182
Tricalyxed Osmanthus 276400
Tricarinate Calanthe 66099
Tricarpelema 395370
Tricarpweed 395370
Tricaudate Coliseum Maple 2867
Tricaudate Maple 3701
Tricaudated Maple 3701
Trichantha 395427
Trichiate Barberry 52266
Trichilia 194638,395466
Trichocereus 395630
Trichodesma 395708
Tricholepis 395964
Trichomanefolious Cystopetal 320245
Trichophore 396003
Trichophore Larkspur 124646
Trichostomous Cherry 83350
Trichotomous Elatostema 142859
Trichotomous Evodia 161382
Trichotomous Feathergrass 376950
Trichotomous Metadina 252396
Trichurus 396491,396492
Trick 235131
Trick-madam 357114
Tricolor Amaranth 18836
Tricolor Cape Cowslip 218948
Tricolor Cape-cowslip 218948
Tricolor Chrysanthemum 89466
Tricolor Coliseum Maple 2853
Tricolor Daisy 89466
Tricolor Flowering Dogwood 105032
Tricolor Hypericum 201706
Tricolor Lycaste 238751
Tricolor Macrosolen 241323
Tricolor Raspberry 339392
Tricolor Sheathflower 241323
Tricolor Tillandsia 392050
Tricolor Vanda 404680
Tricolor Willow 343222
Tricolor Woodbetony 287774
Tricolour Pepper 300536
Tricorner Saxifrage 349846
Tricuspid Cudrania 240842
Tricuspid Exbucklandia 161612
Tricuspidate Falsenettle 56343
Tricuspidate Halerpestes 184719
Tricuspidate Snakegourd 396280
Tricuspidate Soda Buttercup 184719
Tridax 396695
Tridens Corchorus 104133
Trident Amitostigma 19538
Trident Astyleorchis 19538
Trident Maple 2811,2820
Tridentate Paphiopedilum 282910
Tridentate Snakegourd 396282
Tridenwort 394835
Trientalis 396785
Trifarious Calligonum 67083
Triffid Weed 89299
Trifid Bedstraw 170696
Trifid Bur Marigold 54158
Trifid Bur-marigold 54158
Trifid Calanthe 66100
Trifid Chamaerhodos 85660
Trifid Corallorhiza 104019
Trifid Hemp Agrimony 54158
Trifid Hippolytia 196711
Trifid Minorrose 85660
Trifid Miterwort 256040
Trifid Sopubia 369241
Trifid Speedwell 407421
Triflor Azalea 332002
Triflorous Ainsliaea 12669
Triflorous Blue-beard 78039
Triflorous Blueberry 404025
Triflorous Camellia 69708
Triflorous Clematis 95358
Triflorous Conehead 320108
Triflorous Elaeagnus 142210
Triflorous Hydrocera 200217
Triflorous Raspberry 339390
Triflorous Rhododendron 332002
Triflorous Rush 213541
Triflower Camellia 69708
Triflower Holly 204341
Triflower Hydrocera 200217
Triflower Maple 3710
Triflower Microtropis 254331
Triflower Rockjasmine 23321
Triflower St. John's Wort 202632
Triflowered Holly 204341
Triflowered Maple 3710
Trifolialate Mahonia 242654
Trifoliate Acanthopanax 143694
Trifoliate Akebia 13238
Trifoliate Ash 168123
Trifoliate Calabash-tree 111101
Trifoliate Citrus 310850
Trifoliate Clearweed 299068
Trifoliate Cliffbean 254866
Trifoliate Coldwaterflower 299068
Trifoliate Crateva 110233
Trifoliate Euchresta 155897
Trifoliate Fishvine 126007

Trifoliate Greenvine 204656
Trifoliate Illigera 204656
Trifoliate Jewelvine 126007
Trifoliate Melicope 249168
Trifoliate Orange 310848,310850
Trifoliate Stockorange 310850
Trifoliate Viburnum 408162
Trifoliate Wild Quince 374390
Trifoliate Yam 131471
Trifoliate-orange 310848,310850
Trifoliolate Angelica 24489
Trifoliolate Bigumbrella 59250
Trifoliolate Bittercress 73023
Trifoliolate Borthwickia 57390
Trifoliolate Brassaiopsis 59250
Trifoliolate Corydalis 106547
Trifoliolate Dichocarpum 128943
Trifoliolate Forkfruit 128943
Trifoliolate Indigo 206684
Trifolious Curvedstamencress 72786,238011
Trifolium 396806
Trifulcate Evodia 161350
Trifurcate Barley 198396
Trifurcate Japanese Mahonia 242494
Trifurcate Leek 15837
Trifurcate Onion 15837
Trigger Plant 379076,379079
Trigger Stylewort 379079
Triggerplant 379076
Triglandulose Dichromatic Willow 343298
Triglume Bluegrass 306092
Triglume Rush 213545
Trigona Cucumber 114261
Trigonel 397259
Trigonella 397186
Trigonostemon 397358
Trigonotis 397380
Trigonous Barberry 52270
Trigonous Clovershrub 70904
Trigonous-branch Dendrolobium 125580
Trigonous-fruited Camellia 69715
Trigynous Mercury 250648
Trijugate Raspberry 339398
Trikeraia 397505
Trileaf Butterfly Pea 97195
Trileaf Corydalis 106547
Trileaf Philodendron 294847
Trileaf Rocky Mountain Maple 2995
Trileaf Southstar 33544
Trileaved Lacquer-tree 386676
Tri-leaved Leea 223963
Trileaved Turpinia 400546
Trillium 397540,397556
Trillium Family 397533
Trilobate Brassaiopsis 59249
Trilobate Cowpea 409080
Trilobate Ketmia 195326
Trilobate Leaf Cowpea 409080
Trilobateleaf Morningglory 208255
Trilobatepetal Corydalis 106550
Trilobe Bigumbrella 59249
Trilobe Biswarea 54835
Trilobe Brassaiopsis 59249
Trilobe Buttercup 325933
Trilobe Pinellia 299728
Trilobe Toadlily 396698
Trilobe Typhonium 401175
Trilobed Liparis 232244
Trilobed Mono Maple 3204
Trilobed Mulberry 259200
Trilobed Oblongleaf Maple 3268
Trilobed Red Maple 3528
Trilobed Rocky Mountain Maple 2992
Trilobed Spapweed 205404
Trilobed Touch-me-not 205404
Trilobed Twayblade 232244
Trilobedleaf Kudzuvine 321460
Trilobedleaf Leea 223929
Trilobeleaf Kudzuvine 321460
Trilobular Silver Maple 3543
Trilower Notoseris 267179
Trimerous Splended Bredia 59829
Trimo Barberry 51684
Trin Bentgrass 12413
Trincomali Wood 52957
Trinervate Obtuseleaf Skullcap 355629
Trinervate Sauropus 348069
Trinervate Violet 410690
Trinerve Agelaea 11059
Trinerve Pink 127680
Trinerve Poplar 311555
Trinerved Agelaea 11059
Trinerved Everlasting 21716
Trinerved Fighazel 380604
Trinerves Sauropus 348069
Trinerves Stauntonvine 374439
Trinerves Viburnum 408193
Trinervious Staunton-vine 374439
Trinervose Rockfoil 349489
Trinia 397729
Trinidad Tournefortia 393255
Trinided Begonia 49937
Trinity 394088
Trinity Flower 394088,397540,397556, 397570,410677
Trinity Violet 410677
Trinity Wild Buckwheat 151811
Triodia 397762
Tripartite Bogorchid 158347
Tripartite Bur Marigold 54158
Tripartite Japanese Eupatorium 158347
Tripartite Monkshood 5107
Tripartite Philodendron 294847
Tripetaled Metalleaf 296420
Tripinate Chastetree 411477
Tripinate Chaste-tree 411477
Tripinnate Eugenia 382585
Tripinnatisect Ajania 13041
Triple Moles Rhododendron 330070
Triplenerved Pearleverlasting 21716
Triplet Lily 60447,398806
Triplet-lily 398777,398806
Triplex Crabgrass 130797
Triplicate Calanthe 66101
Triplinerved Viburnum 408193
Triplospermum 397964
Triplostegia 398022
Trip-madam 357114
Tripod Wild Buckwheat 152543
Tripogon 398064,398110,398112
Tripoli Aster 398134
Tripoli Senna 78436
Tripterospermum 398251
Tripterygium 398310
Triquetrous Garlic 15839
Triquetrous Oncidium 270844
Triquetrous Tadehagi 383009
Triquetrous Tanoak 233400
Triradiate Thorowax 63864
Triratna Tree 397358,397359
Tririb Clovershrub 70904
Triribmelon 141461
Trirow Kneejujube 67083
Trisect Corydalis 106551
Trisect Larkspur 124652
Trisect Monkshood 5499
Trisected Ward Monkshood 5684
Trisepaled Touch-me-not 205403
Trisepalum 398385,398386
Trisetum 398437
Trispike Sedge 76591
Trispring Falsetamarisk 261284
Tristamen Willow 344220
Tristania 398666
Triste Black-galingale 169561
Tristellateia 398681,398684
Tristram's Knot 71218
Tristylous Dichroa 200145
Triteleia 398777
Triternate Grevill 180657
Triternate Larkspur 124067
Triternate Peony 280320
Triternateleaf Corydalis 106553
Trithrinax Palm 398824,398826
Triticum 398834
Tritoma 216923
Tritonia 399043,399062
Triumfetta 399190
Triumphal Odontoglossum 269088
Triuris Family 399365
Trivalved Onion 15831

Trixis 399397
Trncate-leaf Actinidia 6713
Trncate-leaf Kiwifruit 6713
Trnvener's Rest 383874
Troll Barberry 52271
Troll Oreocereus 273826
Troll-flower 399504
Trollius 399484
Trompenburg Maple 3336
Tronadora 266043
Tropaeolum 399597
Tropic Ageratum 11199
Tropic Croton 112895
Tropic Daisy 141578
Tropical Almond 386500
Tropical Amaranth 18801
Tropical Ash 168124
Tropical Birch 53536
Tropical Black Bamboo 47317,175596
Tropical Blue Bamboo 37172
Tropical Burnweed 148164
Tropical Bushmint 203056
Tropical Cowpea 408927
Tropical Croton 112896
Tropical Cup Grass 151672
Tropical Cupgrass 151672,151693
Tropical Daisy 141579
Tropical Dock 340150
Tropical Duchweed 301021
Tropical False Goldeneye 190258
Tropical Fanleaf 166604
Tropical Finger-grass 130489
Tropical Green Amaranth 18848
Tropical Guava 318742
Tropical Hibiscus 195149
Tropical Kudzu 321460
Tropical Kudzubean 321460
Tropical Leaf-flower 296725
Tropical Milkweed 37888
Tropical Needlegrass 262617
Tropical Pine 300289
Tropical Pokeweed 298112
Tropical Rose Mallow 195341
Tropical Sage 344980
Tropical Seapurslane 361656
Tropical Sensitive Pea 78210
Tropical Signalgrass 402512
Tropical Soda Apple 367722
Tropical Sprangletop 226036
Tropical Spreading Amaranth 18692
Tropical Threefold 399398
Tropical Ticktrefoil 126276
Tropical Walnut 212599
Tropical White Gum 155607
Tropical Whiteweed 11199
Tropical Wild Petunia 339683
Tropical Woodsorrel 277889
Tropidia 399633
Tropillo 367139
Trossachs Dock 339925
Troublesome Sagebrush 36464
Troublesome Sedge 75399
Trout Begonia 49629
Trout Lily 154895,154899,154909
Trout-leaf Begonia 49629
Trout-lily 154895
Trowelpetal Stonecrop 357247
Trowie Girse 367157
Trowie Gliv 367157
Troy Iris 208910
Trucatecalyx Dragonhead 137678
Trucatecalyx Greenorchid 137678
Trucatelobed Bittercress 73024
Truckee Orach 44606
Truckee Rabbitbrush 90497
Truckleberry 403916
Truckles of Cheese 243840
Truckles-of-cheese 243840
True Aloe 17381
True Anise 204603
True Asphodel of the Ancients 39438
True Bay 223203
True Buckwheat 162301
True Camphortree 91287
True Cedar 80074
True Corydalis 105699
True Cypress 114649
True Daisy 50825
True Forget-me-not 260842,260868
True Fox-sedge 76716
True Frost Grape 411631
True Indigo 206669
True Jalap Plant 208117
True Jasmine 211938,211940
True Kino 155720
True Lacquertree 393491
True Lacquer-tree 393491
True Laurel 223203
True Lavender 223251
True Love 238371
True Lovers' Knot 284387
True Mahogany 380527
True Monkshood 5442
True Myrtle 261739
True Pitch Pine 300128
True Senna 78227
True Service 369381
True Service-tree 369381
True Star-anise Tree 204603
True Sterculia 376166
True Tarragon 35411
True Virginia Creeper 285136
True Wood Sorrel 277648
True Yellow Jasmine 211938
True-love 284387
True-star Anise Tree 204603
Truestar Anisetree 204603
True-star Anise-tree 204603
Truestar Illicium 204603
Truffle 102727
Truffle Oak 324335
Trulung Redbranched Barberry 51593
Trumpet 68713,347144,347150,347156,399601
Trumpet Achimenes 4082
Trumpet Bitter Bamboo 304005
Trumpet Bush 385493
Trumpet Climber 70512
Trumpet Creeper 70502,70512
Trumpet Creeper Family 54357
Trumpet Cup 255210
Trumpet Daffodil 262441
Trumpet Flower 54303,68713,205557,205571,236022,275401,275403,317477,366863,385461,385493
Trumpet Gentian 173184,173344
Trumpet Gilia 175672
Trumpet Gooseberry 334065
Trumpet Gourd 219843
Trumpet Honey Suckle 236089
Trumpet Honeysuckle 70512,236089
Trumpet Keck 24483
Trumpet Leaf 347150,347156
Trumpet Lily 68877,229900,334310,417092,417093,417118
Trumpet Narcissus 262441
Trumpet Pitcherplant 347147
Trumpet Tree 80003,80007,80008,80009,382704
Trumpet Vine 70502,70507,70512,97468
Trumpet Weed 158264
Trumpetbush 385461
Trumpet-climber 70512
Trumpetcreeper 70502,70512,385461,385493
Trumpet-creeper 70502,70512
Trumpet-creeper Family 54357
Trumpet-creeper 54292
Trumpetflower 275401
Trumpet-flower 54303,70512,389978
Trumpet-honeysuckle 70512
Trumpetilla 57957
Trumpets 4777
Trumpet-tree 80003
Trumpetvine 97468,97471
Trumpet-vine 70512,97468,97471
Trumpetweed 158113,161171
Truncate Ajania 13042
Truncate Begonia 50373
Truncate Bitterbamboo 304106
Truncate Camellia 69719

Truncate Centipedegrass 148256
Truncate Conehead 378240
Truncate Dumasia 138942
Truncate Epimedium 147071
Truncate Gongora 179297
Truncate Maple 3714
Truncate Newlitse 264222
Truncate Orange 93857
Truncate Pipewort 151532
Truncate Pittosporum 301426
Truncate Pseudosasa 318347
Truncate Seatung 301426
Truncate Separatestyle Schefflera 350720
Truncate Snakegourd 396284
Truncate Spapweed 205407
Truncate Tea 69719
Truncate Touch-me-not 205407
Truncate Twinballgrass 209133
Truncate Yalu Monkshood 5297
Truncate Yulongshan Anemone 24130
Truncate Yulongshan Windflower 24130
Truncatebase Hemsley Monkshood 5249
Truncatebract Willow 344001
Truncate-bracted Willow 344001
Truncate-calyx Blueberry 404026
Truncatecalyx Honeysuckle 235663
Truncatecalyx Lycianthes 238974
Truncatecalyx Red Silkyarn 238974
Truncatecalyx Wolfberry 239126
Truncate-calyxed Blueberry 404026
Truncate-calyxed Honeysuckle 235663
Truncate-calyxed Lycianthes 238974
Truncate-calyxed Wolfberry 239126
Truncated Haworthia 186792
Truncatefruit Tanoak 233401
Truncate-fruited Tanoak 233401
Truncateglume Sedge 76601
Truncateleaf Buttercup 326441
Truncate-leaf Chinese White Poplar 311536
Truncateleaf Flowering Plum 20974
Truncateleaf Raspberry 339378
Truncateleaf Rockfoil 349188
Truncate-leaved Maple 3714
Truncatelobed Wingseedtree 320850
Truncatelobed Wing-seed-tree 320850
Truncatesepal Bellsepal Sage 344928
Truncatestigma Stonecrop 357251
Truncatirostris Sedge 76603
Truncatum Ginger 418006
Trunk Pennilabium 289008
Trunk Rosewood 121760
Trunkorchis 266856,266857
Truxillo Barberry 52272
Truxillo Leaf 155114
Tsai Anise-tree 204601
Tsai Blastus 55186
Tsai Camellia 69721
Tsai Damnacanthus 122084
Tsai Eurya 160620
Tsai Hornbeam 77400
Tsai Milkgreens 245821
Tsai Monkshood 5642
Tsai Pearweed 239888
Tsai Photinia 295799
Tsai Sedge 76607
Tsai Tailleaftree 402646
Tsai Taxillus 385244
Tsai Tea 69721
Tsai Woodbetony 287782
Tsai's Actinodaphne 6821
Tsai's Clematis 95390
Tsai's Rhododendron 332009
Tsama Melon 93304
Tsang Aglaia 19972
Tsang Amoora 19972
Tsang Cinnamon 91442
Tsang Cryptocarya 113490
Tsang Ehretia 141717
Tsang Eightangle 204602
Tsang Faberdaisy 161891
Tsang Holly 204351
Tsang Illicium 204602
Tsang Lasianthus 222270,222296
Tsang Raspberry 339409
Tsang Roughleaf 222270
Tsang Slugwood 50619
Tsang Smallflower Eightangle 204602
Tsang Tanoak 233153
Tsangpo Barberry 52274
Tsangpo River Rhododendron 330370
Tsangshan Goldenray 229234
Tsangshan Wintergreen 172061
Tsangshan Woodbetony 287783
Tsangyuan Alphonsea 17632
Tsaoko Amomum 19927
Tsaoko-like Amomum 19903
Tsari Barberry 52275
Tsari Rhododendron 332016
Tsarong Barberry 52281
Tsarong Larkspur 124134
Tsarung Aster 41439
Tsarung Woodbetony 287784
Tsata Milkvetch 42496
Tsawa Meadowrue 388702
Tsayu Onosma 271849
Tsayu Speedwell 407076
Tsayul Maple 3692
Tsayul Rockjasmine 23347
Tschangbai Monkshood 5649
Tscherniev Sagebrush 36426
Tschernjaew Anemone 24100
Tschimgan Skullcap 355818
Tschonosk Hornbeam 77401
Tschonosk Trillium 397622
Tschonoski Azalea 332018
Tschonoski Barberry 52282
Tschonoski Maple 3718
Tschonosky Privet 229650
Tschuili Seseli 361589
Tsekou Woodbetony 287785
Tseku Pleurospermum 304855
Tsiang Buckeye 9731
Tsiang Clearweed 299074
Tsiang Dendrocalamus 125515
Tsiang Distylium 134953
Tsiang Dysophylla 139597
Tsiang Fig 165796
Tsiang Holly 204353
Tsiang Mosquitoman 134953
Tsiang Onosma 271841
Tsiang Persimmon 132442
Tsiang Primrose 315064
Tsiang Sauropus 348070
Tsiang Sedge 76609
Tsiang Stonecrop 357253
Tsiang Watercandle 139597
Tsiang Woodbetony 287786
Tsiang's Childvine 400981
Tsiang's Silkvine 291089
Tsiang's Tylophora 400981
Tsiangia 399832,399833
Tsien's Barberry 52283
Tsin Bean 49077
Tsingling Mountain Clematis 95166
Tsingpien Camellia 69724
Tsingpien Eurya 160621
Tsingtao Lily 230065
Tsingtau Lily 230065
Tsinling Asiabell 98422
Tsinling Germander 388296
Tsinling Mountain Thorowax 63714
Tsinling Paulownia 285985
Tsinling Spapweed 205095
Tsinyun Skullcap 355820
Tso Azalea 332025
Tso Cassia 91443
Tso Cinnamon 91443
Tso Holly 204354
Tso Michelia 252844
Tso Rosewood 121847
Tso Sedge 76610
Tso's Nitid Spindle-tree 157746
Tso's Rhododendron 332025
Tsofu Camellia 69727
Tsofu Tea 69727
Tsona Stonecrop 357256
Tsoong Begonia 50377
Tsoong Caesalpinia 252754
Tsoong Heterostemma 194167
Tsoong Sotvine 194167
Tsoong Syzygium 382679

Tsoong Willow 343333
Tsoong's Tree 399855,399856
Tsu Millettia 66653
Tsukaoshan Holly 204356
Tsunghua Eurya 160546
Tsunya 290701
Tualang 217852
Tuan Lime 391843
Tuan Linden 391843
Tuart 155594
Tuba 125958
Tuba Root 125958
Tuba-root 125958
Tubarroot 125958
Tubarroot Jewelvine 125958
Tubar-rootted Jewelvine 125958
Tube Beard-tongue 289381
Tube Clematis 94995
Tube Gilia 175682
Tube Penstemon 289381
Tube Tongue 214712
Tubecalyx Euchresta 155903
Tube-calyx Lasianthus 222171,222298
Tube-calyx Leptodermis 226135
Tube-calyx Roughleaf 222171,222298
Tube-calyx Wildclove 226135
Tubecell Oreocharis 273886
Tube-flower 96139
Tubeflower Arrowwood 407769,408196
Tubeflower Ascitesgrass 407502
Tube-flower Basketvine 9485
Tubeflower Chive 15749
Tubeflower Dutchmanspipe 34364
Tubeflower Elaeagnus 142212
Tubeflower Gentian 173906,174022
Tube-flower Honeysuckle 236197
Tubeflower Kalanchoe 215278
Tubeflower Kohleria 217750
Tubeflower Laelia 219696
Tubeflower Liparis 232326
Tubeflower Morningglory 208289
Tubeflower Onion 15749
Tubeflower Oreocharis 273887
Tube-flower Pittosporum 301430
Tubeflower Sea-urchin Cactus 140891
Tubeflower Tupistra 70602
Tubeflower Twayblade 232326
Tubeflower Veronicastrum 407502
Tubeflower Viburnum 407769
Tube-flower Viburnum 408196
Tubeflower Woodbetony 287678
Tube-flowered Hairy-stomatic Rhododendron 332000
Tube-flowered Pittosporum 301430
Tube-flowered Viburnum 407769
Tube-form Rhododendron 332030
Tubefruit Osmanther 276279
Tubeleaf Ceratostylis 83657
Tubeleaf Hornstyle 83657
Tubeleaf Kalanchoe 215129
Tubeorchis 106871
Tubepetalorchis 27655,27657
Tuber Anemone 24101
Tuber Camellia 69733
Tuber Dayflower 101178
Tuber Deadnettle 220422
Tuber Fescuegrass 164369
Tuber Fleeceflower 162542
Tuber Flower 95934,365217
Tuber Galeobdolon 220422
Tuber Gourd 390107
Tuber Hanceola 185253
Tuber Jerusalemsage 295212
Tuber Nasturtium 399611
Tuber Oat Grass 34935
Tuber Sedge 76543
Tuber Sweetvetch 187763
Tuber Ullucu 401412
Tuber Ullucus 401412
Tuber Vine 325185
Tuber Weasel-snout 220422
Tuber-bearing Ardisia 31390
Tuber-bearing Blue Berry 403821
Tubercle Waterchestnut 394552
Tubercled Orchid 302326,302328
Tubercled Orchis 302326
Tubercled-fruit Sanicle 346008
Tubercledseed Crateva 110224
Tubercorm Cranesbill 174972
Tuberculate Amomum 19928
Tuberculate Bicolor Magnoliavine 351018
Tuberculate Bredia 59884
Tuberculate Camellia 69730
Tuberculate Childvine 400982
Tuberculate Gastrodia 171957
Tuberculate Guiyang Betony 373278
Tuberculate Gurjun 133587
Tuberculate Manchurian Linden 391765
Tuberculate Spapweed 205409
Tuberculate Square Bamboo 87620
Tuberculate Touch-me-not 205409
Tuberculate Tylophora 400982
Tuberculate-anthered Bredia 59884
Tuberculate-branched Grape 411648
Tuberculate-fruited Camellia 69385
Tuberculate-seeded Crateva 110224
Tuberflower 95934,365217
Tubergen Iris 208911
Tubergourd 390107
Tuber-gourd 390107
Tuberless Whiteflower Patrinia 285882
Tubernode Spapweed 205227
Tubernode Touch-me-not 205227
Tuberoculate Holeglumegrass 57593
Tuberolabium 399974
Tube-root 99297
Tuberose 10924,307265,307269
Tuberose Corydalis 106560
Tuberose Lettuce 219585
Tuberose Serpentroot 354968
Tuberose Valerian 404377
Tuberose Velvetplant 183124
Tuberous Basketvine 9485
Tuberous Begonia 50298,50378
Tuberous Birthwort 34363
Tuberous Bulrush 56633,352187
Tuberous Chervil 84731
Tuberous Comfrey 381040
Tuberous Crane's-bill 174986
Tuberous Cypressgrass 119718
Tuberous Galingale 119718
Tuberous Grass Pink 67894
Tuberous Hawk's-beard 9771,9773
Tuberous Indian Plantain 64749
Tuberous Jerusalem Sage 295212
Tuberous Merremia 250796
Tuberous Nasturtium 399611
Tuberous Nightshade 367696
Tuberous Passionflower 285713
Tuberous Pea 222851
Tuberous Plume Thistle 92454
Tuberous Skullcap 355822
Tuberous Sweet Pea 222851
Tuberous Thistle 92454
Tuberous Vervain 405885
Tuberous Vetchling 222851
Tuberous Violet 410694
Tuberous Water Lily 267777
Tuberous Water-lily 267777
Tuberousroot Jerusalemsage 295212
Tuberous-root Peavine 222851
Tuberousroot Peony 280204
Tuberous-rooted Begonia 50378
Tuber-root 38147
Tubeshaped Cistanche 93082
Tubeshaped Flower Cistanche 93082
Tubesheath Angelica 24439
Tubespike Water-centipede 218518
Tubespur Platanthera 302536
Tubestem Loosestrife 239638
Tubestem Spapweed 205410
Tubestem Touch-me-not 205410
Tube-stem Windhairdaisy 348317
Tube-Tongue 344381
Tubiflorous Elaeagnus 142212
Tubiflorous Honeysuckle 236197
Tubiflorous Phacellanthus 293186
Tubiflorous Seatung 301430
Tubilarflower Mananthes 244297
Tubinate Bottle Gourd 219856
Tubinate Microula 254376

Tubocapsicum 399995
Tubular Corn-lily 210871
Tubular Euchresta 155903
Tubular Flowers Blueberry 403793
Tubular Tanoak 233405
Tubular Water Dropwort 269319
Tubular Water-dropwort 269319
Tubular-bract Rattan Palm 65792
Tubularflower Asiabell 98423
Tubularflower Cherry Elaeagnus 142110
Tubularflower Leek 15846
Tubularflower Onion 15846
Tubularflower Zephyrlily 417637
Tubulate Enkianthus 145735
Tubulate Pendent-bell 145735
Tubulate Sterculia 376204
Tubule Tanoak 233405
Tubur Onion 15843
Tucerne 85394
Tucher Azalea 332032
Tucher Rhododendron 332032
Tuckahoe 288808
Tucker Oak 324067
Tuckerman's Panic-grass 282333
Tuckerman's Pondweed 312076
Tuckerman's Sedge 76613
Tuckeroo 114585
Tucum Palm 43372,43378
Tucuma 43372,43373
Tucuman Cactus 140892
Tucuman Euphorbia 158991
Tudnoore 176824
Tuemur Mouseear Cress 264609
Tuffybells 412725
Tufted Bamboo 82442,125479
Tufted Bauhinia 49056
Tufted Bracketplant 88553
Tufted Bugle 13104
Tufted Bulrush 396010
Tufted Catchfly 364013
Tufted Cerastium 82758
Tufted Christolea 89256,126160
Tufted Cleistogenes 94587
Tufted Club-rush 353272,396010
Tufted Columbine 388423
Tufted Evening Primrose 269417
Tufted Fishtail Palm 78047
Tufted Fishtailpalm 78047
Tufted Fleabane 150522
Tufted Forget Me Not 260772
Tufted Forgetmenot 260772
Tufted Forget-me-not 260772,260813,
260892
Tufted Gasteria 171635
Tufted Globe Amaranth 179226
Tufted Globe-amaranth 179226
Tufted Grass 74435
Tufted Hair Grass 126039,126041
Tufted Hair Linden 391786
Tufted Hairgrass 126039
Tufted Hair-grass 126039
Tufted Hideseedgrass 94587
Tufted Lake Sedge 76677
Tufted Lily 122898
Tufted Loosestrife 239881
Tufted Love Grass 147871
Tufted Marshmarigold 68167
Tufted Pansy 409858
Tufted Phlox 295254
Tufted Plateaucress 89256,126160
Tufted Rush 212800
Tufted Sandwort 32301,255598
Tufted Saunders Cinquefoil 312957
Tufted Saxifrage 349145
Tufted Sea-blite 379507
Tufted Sedge 73987,74435,75106
Tufted Smallleaf Cinquefoil 312767
Tufted Vetch 408251,408352
Tuftedhair Cottonsedge 152753
Tuftedhairy Javatea 275662
Tuftroot 130083,130093
Tufty Bells 412568
Tufuling 366338
Tugarinovia 400035
Tugua Palm 298051
Tuguancun Monkshood 5650
Tuibaghia 400050
Tuju Mandarin 93428
Tukou Cycas 115870
Tula Ixtle 10881
Tulan Willow 344230
Tulare Orach 44683
Tuld Bamboo 47493
Tule 352169
Tule Mint 250331
Tule Potato 342363
Tulepotato 342363
Tuleroot 342315
Tulip 178524,232609,400110,400162
Tulip Gentian 161027
Tulip Lancewood 185900
Tulip Magnolia 416682
Tulip Oak 192477
Tulip Orchid 25134,79532
Tulip Poplar 232609
Tulip Poppy 282566
Tulip Tree 3462,232602,232609,370358,
389946
Tulip Wood 121650,185893
Tulipan Del Monte 195000
Tulip-poplar 232609
Tuliptree 232602,232609
Tulip-tree 232602,232609
Tulipwood 121650,121663,185893,185895,
185900
Tulip-wood Tree 185894
Tull Anemone 24109
Tull Germander 388085
Tulotis 400280
Tulufan Goldenray 229235
Tulufan Peashrub 72370
Tulufan Pea-shrub 72370
Tulung Maple 3655
Tumaorhair Swertia 380319
Tumbatsi Milkvetch 43195
Tumble Garliccress 365398
Tumble Grass 147971,350613
Tumble Knapweed 81034
Tumble Mustard 365398,365564
Tumble Pigweed 18655
Tumble Ringwing 116286
Tumble Weed 18687,344585
Tumbledown Eucalyptus 155554
Tumble-down Gum 155554
Tumble-down Red Gum 155554
Tumblegrass 350613
Tumble-grass 350613
Tumbleweed 18655,18668,18687,344496,
344571,344585,344743
Tumbleweed Amaranth 18655
Tumbling Mustard 365398
Tumbling Orach 44629
Tumbling Saltweed 44629
Tumbling Ted 346433
Tumbling-ted 346433
Tumen Sedge 76618
Tumen Skullcap 355824
Tumengela Sandwort 32289
Tumidenode Cranesbill 174684
Tumin Skullcap 355824
Tumitory 168964
Tummelberry 338059
Tumor Themeda 389309
Tumorbase Honeysuckle 235999
Tumorbranch Rapanea 261662
Tumorcomb 119968,119969
Tumorfruit Bladderwort 403954
Tumorfruit Mistletoe 411085
Tumorfruit Pimpinella 299565
Tumorfruit Ribseedcelery 304860
Tumorfruit Tanoak 233240
Tumorfruit Uvaria 403503
Tumorfruitparsley 393856
Tumorhull Clearweed 298909
Tumorhull Coldwaterflower 298909
Tumorleaf Tea 69385
Tumorless Camellia 68925
Tumorous Primrose 315016
Tumorsepal Angelica 24425
Tumorstem Stairweed 142754
Tumote Buckthorn 328817

Tun Oil Tree 406019
Tuna 272775,273085,273088
Tuna Cactus 272891
Tuna Colorada 273059
Tundra Azalea 331449
Tundra Fleabane 150700
Tundra Mountaincrown 274123
Tundra Rhododendron 331449
Tun-foot 176824
Tung Nut 406018
Tung Oil Tree 14544,406018
Tung Tree 14544,406018
Tungchuan Milkvetch 43196
Tungfang Firmiana 166692
Tungleaf Stephania 375861
Tungoil Tree 406018
Tung-oil Tree 406014,406018
Tungoiltree 406014,406018
Tung-oil-tree 14538,406014,406018
Tung-tree 14538
Tunhoof 176824
Tun-hoof 176824,400675
Tunic Flower 292656,292668
Tunica-like Psammosilene 317036
Tunicate Bluegrass 306107
Tunicate Corn 417434
Tunicflower 292656
Tunic-flower 292656,292668
Tuniclike Golden Ironlock 317036
Tunisian Gum Acacia 1653
Tunka 50998
Tunsing-wort 405571
Tuoli Fritillary 168591
Tuoli Giantfennel 163651
Tuoli Windhairdaisy 348888
Tuolumne Fawn-lily 154948
Tuomuer Poplar 311526
Tuomur Crazyweed 278777
Tupelo 267849,267850,267867
Tupelo Family 267881
Tupelo Gum 267849,267850,267867
Tupelo-gum 267850
Tupelo-gum Family 267881
Tupidanthus 400376
Tupistra 70593,400379
Turban Bell 266225
Turban Buttercup 325614
Turban Lily 229922
Turban Squash 114288
Turbinate Aster 41441
Turbinate Dillenia 130927
Turbinate Hillbubble Crazyweed 278959
Turbinate Lagerstroemia 219978
Turbinate Snowgall 191971
Turbinella Oak 324509
Turcoman Barberry 52284
Turcoman Wolfberry 239128
Turczaninov Depressed Plantain 301956
Turczaninov Draba 137282
Turczaninov Goosecomb 335547
Turczaninov Melic 249105
Turczaninov Pepperweed 225472
Turczaninov Reedbentgrass 65507
Turczaninov Roegneria 335547
Turczaninov Valerian 404379
Turczaninov Willow 344231
Turczaninovia 400466
Turczaninow Euphorbia 160019
Turczaninow Hornbeam 77411
Turczaninow Rush 213552
Turczaninow's Tephroseris 385924
Turf Spiraea 371853
Turfan Peashrub 72370
Turfing Daisy 246407
Turgai Feathergrass 376695
Turgai Needlegrass 376695
Turgai Skullcap 355825
Turgenia 400471
Turgenleaf Hogfennel 293053
Turgid Chessbean 116610
Turgid Fruit Milkvetch 43197
Turgid-calyx Milkvetch 42322
Turgidstipe Windhairdaisy 348224
Turion Duckweed 224399
Turk Terebinth Pistaehe 300998
Turk Terebinth-pistache 300998
Turk's Cap 5442,5676,229922,243929,
243951,249578,249590
Turk's Cap Cactus 249578,249581,249590
Turk's Cap Lily 229922,229939,230044
Turk's Caplily 230044
Turk's Head 140133,168476,230058
Turk's Head Cactus 163447
Turk's Turban 96139,243929
Turk's Turbin 96139
Turk's-cap Lily 229939
Turkenstan Draba 137253
Turkenstan Whitlowgrass 137253
Turkestan Adonis 8385
Turkestan Allium 15389
Turkestan Anemone 23921
Turkestan Ash 168071
Turkestan Barberry 51717
Turkestan Barley 198373
Turkestan Bentgrass 12403
Turkestan Bluestem 57568
Turkestan Brome 61003
Turkestan Burning Bush 157740
Turkestan Desertcandle 148560
Turkestan Dwarf Euonymus 157742
Turkestan Globeflower 399503
Turkestan Hawthorn 110096
Turkestan Juniper 213877
Turkestan Motherwort 225019
Turkestan Panderia 281175
Turkestan Pearlbush 161743
Turkestan Reedbentgrass 65509
Turkestan Rose 336685,336901,337002
Turkestan Rush 213554
Turkestan Salsify 394354
Turkestan Seabuckthorn 196783
Turkestan Sedge 76622
Turkestan Shrub Maple 3587
Turkestan Thermopsis 389557
Turkestan Tulip 400246
Turkestan Wildsenna 389557
Turkestanicus Pink 127892
Turkevicz Hoarhound 245769
Turkey Almond 20890
Turkey Apple 109850
Turkey Beard 415440
Turkey Berry 367682
Turkey Box 64345
Turkey Boxwood 64345
Turkey Bush 68759
Turkey Cap 168476
Turkey Corn 128291,128296,128297,417417
Turkey Eggs 168476
Turkey Fig 164763
Turkey Foot 22681
Turkey Gilliflower 383090
Turkey Grape 411782
Turkey Mountainash 369548
Turkey Mullein 148238
Turkey Oak 323745,324087
Turkey Pea 128291,128297,346009,386365
Turkey Pine 299827
Turkey Red 258882
Turkey Rhubarb 31051,292372,329366
Turkey Tangle 296121
Turkey Tangle Fogfruit 296121
Turkey's Food 170193
Turkey's Snout 18687
Turkeyalmond 20890
Turkey-beard 415440
Turkey-berry 256002
Turkey-cap 168476
Turkey-claw 22681
Turkey-corn 128291,128296,128297
Turkey-dish 250432
Turkey-eggs 168476
Turkeyfoot 22681
Turkey-foot 22681
Turkey-hen Flower 168476
Turkey-pod 365564
Turkish Baby's-breath 183233
Turkish Beech 162395
Turkish Comfrey 381037
Turkish Corn 417417
Turkish Corn-flag 176122
Turkish Cottonthistle 271702

Turkish Filbert 106720,106724
Turkish Galls 324314
Turkish Hawksbeard 110929
Turkish Hazel 106720,106724
Turkish Hazelnut 106724
Turkish Hornbeam 77359
Turkish Iris 208740
Turkish Liquidambar 232562
Turkish Millet 417417
Turkish Oak 324146
Turkish Pine 299827
Turkish Rhododendron 331858
Turkish Rhubarb 329372
Turkish Rocket 63452
Turkish Rugging 89116
Turkish Sage 295185
Turkish Tobacco 266053
Turkish Tulip 400112
Turkish Wartycabbage 63452
Turkish Warty-cabbage 63452
Turkman Larkspur 124659
Turkscap Lily 230044
Turmentill 312520
Turmentyne 312520
Turmeric 114852,114871
Turmeric Root 200175
Turmet 59575
Turmit 59575
Turn-again Gentleman 229806,229922
Turncap 229922
Turnera 400486,400493
Turnflowerbean 98064
Turn-in-the Wind 243420
Turn-in-the-wind 243427
Turnip 59493,59575,364557
Turnip Garden Parsley 292701
Turnip Neoporteria 264366
Turnip Weed 326833
Turnip Wood 13202
Turnip-cabbage 59541
Turnipleaf Blumea 55792
Turnip-rape 59603
Turnip-reoted Parsley 292701
Turniproot Cranesbill 174752
Turniprooted Celery 29329
Turnip-rooted Celery 29329
Turnip-rooted Chervil 84731,84768
Turnip-rooted Parsley 292695,292705
Turnip-shaped Hawksbeard 110920
Turnip-weed 326833
Turnipwood 13202
Turnjujube 198766,198767
Turn-Merick 114871
Turnopshape Neoporteria 264359
Turnsole 89320,89328,90901,113042,159027,188908,190622,190651
Turnsole Crozophore 89320
Turpentine 381692
Turpentine Broom 388857
Turpentine Bursh 53669,150327
Turpentine Pine 300128
Turpentine Tree 103716,301002,301003,381692
Turpentine Weed 182103,182107
Turpentine Wood 381692
Turpentine-brush 150327
Turpinia 400515
Turquoise-berry 20348
Turr 401388
Turraie 180651
Turtle Doves 5442,130383
Turtle Grass 388357,388360,388362
Turtleback 317093
Turtleback Bamboo 297301
Turtleback Plant 131573
Turtle-grass 388357,388360,388362
Turtlehead 86781,86783,86793
Turtle-head 86781,86783,86795
Turtschaninov Corydalis 106564
Turu Palm Jatrorrhiza 212346
Turulus Torularia 264639
Turwad Bark 78236
Tuscarora Rice 418080
Tushan Loosestrife 239625
Tushan Photinia 295800
Tushy-lueky Gowan 400675
Tusilla 136470
Tussac Grass 126039
Tussac-grass 305529
Tussock Bellflower 69942
Tussock Bluebell 69942
Tussock Bulrush 396010
Tussock Cotton-grass 152791
Tussock Dropseed 372650
Tussock Grass 126039
Tussock Paspalum 285494
Tussock Sedge 74435,76399
Tussock-grass 126039
Tutcher Fighazel 134913
Tutcher Holly 204357
Tutcher Maple 3727
Tutcher Ninenode 319880
Tutcher Psychotria 319880
Tutcheria 400684,400722
Tutsan 201732,201733,201743,202086,409339
Tutterflybuch-like Abelia 105
Tutterflybuch-like Honeysuckle 235688
Tutterflybuch-like Viburnum 407713
Tutties 83361
Tutty Pea 222789
Tuven Skullcap 355826
Tuxiandan Melia 248925
Tuzzy-muzzy 31051,95431
Twadger 408571
Twayblade 232072,232215,232890,232950
Tway-blade 232072
Twayblade Orchid 232890
Tweaksheath Lemongrass 117185
Tweed Calliandra 66673,66689
Tweedia 278547
Tweedy's Catsear 202394
Tweedy's Fleabane 151027
Tweeny-legs 284098
Twelve Apostles 321643
Twelve Disciples 50825,285623
Twelve-month Yam 131516
Twelve-o'clock 845,21339,274823,394333
Twelvestamen Melastoma 248748
Twelve-stamened Melastoma 248748
Twice Forked Vetch 408321
Twice-writhen 308893
Twickband 369339
Twig Bean 369339
Twig Rush 93901,245332
Twig Withy 344244
Twig-bean 369339
Twiggy Barberry 52318
Twiggy Buckthorn 328892
Twiggy Cinquefoil 313104
Twiggy Glasswort 342893
Twiggy Mullein 405796
Twiggy Spurge 159663
Twiggy Statice 230814
Twig-hanging Embelia 144753
Twigrush 93901
Twig-rush 93901,93918
Twig-withy 344244
Twike 144661
Twin Arnica 34770
Twin Berry 256002
Twin Cryptodiscus 113555
Twin Flower 231676,231679
Twin Flowered Cassia 78245
Twin Flowered Violet 409729
Twin Goosecomb 335318
Twin Lovebamboo 264531
Twin Merremia 250781
Twin Woodbetony 287046
Twinball Grass 209028
Twinballgrass 209028
Twinberry 235874,256002
Twinberry Eugenia 261045
Twin-berry Honeysuckle 235874
Twinbill Rockfoil 349236
Twine Dactylicapnos 128318
Twine Draws Ear Grass 403321
Twine Mosquitotrap 117745
Twinebark 301610
Twinebranch Billardiera 54437
Twine-grass 408352

Twiner 121623
Twinflower 87488,231676,231679,231682
Twinflower Abelia 416841
Twinflower Cinquefoil 312409
Twinflower Crabdaisy 413506
Twinflower Groundsel 381935
Twin-flower Sandwort 255548
Twinflower Skullcap 355733
Twinflower Violet 409729
Twin-flowered Clover 396844
Twin-flowered Marsh Marigold 68163
Twin-flowered Violet 409729
Twin-fruited Holly 203762
Twinginseng 398022,398024
Twin-headed Clover 396843
Twining Atylosia 65153
Twining Bluehood 278575
Twining Brodiaea 128886
Twining Guinea Flower 194665
Twining Milkweed 347008
Twining Monkshood 5674
Twining Pterygocalyx 320927
Twining Pyrenacantha 322562
Twining Rhynchosia 333456
Twining Rosewood 121856
Twining Screw-stem 48591
Twining Snapdragon 28598,246649
Twining Streptolirion 377884
Twinkle Star 374916
Twinkle-star 374916
Twinleaf 212296,212299,418585
Twin-leaf 212296,212299
Twinleaf Cassia 360425
Twinleaf Leek 15872
Twinleaf Onion 15872
Twinleaf Rockfoil 349719
Twinleaf Saxifrage 349719
Twinleaf Squill 352886
Twinleaf Zornia 418331
Twin-leaved Squill 352886
Twinlipped Roscoea 337077
Twinlobe Combbush 292622
Twins 28200
Twins Magnolia 416682
Twinscar Greenbamboo 47211
Twinseeded Garcinia 171207
Twinspike Ephedra 146154
Twinspot Corydalis 105651
Twinspur 128029,128032,128064
Twin-stemmed Bladderwort 403190
Twintail Pelatantheria 288599
Twintongue Catchfly 363250
Twiny Legs 48626,268983
Twiny-legs 268991,284098
Twisselmann's Wild Buckwheat 152549
Twist Hair Longstalk Maple 3120
Twisted Acacia 1560
Twisted Anther Flower 377625
Twisted Arisaema 33540
Twisted Babylon Weeping Willow 343074
Twisted Desert Wattle 1053
Twisted Draba 137039
Twisted Hazel 106705
Twisted Heath 149205,149398,149571
Twisted Ladies' Tresses 372277
Twisted Lemongrass 117185
Twisted Milkvetch 42218
Twisted Myrtle 261740
Twisted Parrot-pea 130937
Twisted Rhododendron 331984
Twisted Sedge 76564
Twisted Tickseed 104840
Twisted Trillium 397617
Twisted Whitlow Grass 137039
Twisted Whitlowgrass 137039
Twisted Woodbetony 287765
Twisted-arum 189982,189983
Twistedawn Goatgrass 8669
Twistedbeak Woodbetony 287704
Twisted-branch Yunnan Spiraea 372139
Twistedfruit Tickseed 104840
Twistedleaf Leek 15525
Twistedleaf Onion 15525
Twisted-leaf Pine 300277
Twisted-leaf Yucca 416642
Twisted-petal Eulophia 157029
Twisted-podded Whitlow Grass 137039
Twisted-rib Cactus 185155,185156
Twistedspadix Southstar 33540
Twistedstalk 377902,377930
Twisted-stalk 377902,377905,377920,377923,377930
Twistfruit Dactylicapnos 128323
Twisthair Longstalk Maple 3120
Twisting Asiabell 98403
Twisting Blueflowervine 292518
Twisting Dregea 137799
Twisting Petrea 292518
Twist-leaf Goldenrod 368451
Twistleaf Nordmann Fir 432
Twistsepal Cymbidium 117082
Twistwood 407907
Twist-wood 407907
Twisty Everlasting 21554
Twistybill Woodbetony 287704
Twistysplit Cheirostylis 86729
Twitch 144661
Twitch-grass 144661
Twlckbine 369339
Two Leaves Bergenia 52508
Two Ray-florets Sinacalia 364519
Twoanther Dropseedgrass 372650
Twoanther Mosla 259284
Twoanther Sandspurry 370624
Two-auricled Arisaema 33278
Two-auricled Spatholobus 370412
Two-awn Goatgrass 8663
Twoawn Goldenhairgrass 306830
Twoawn Microstegium 253985
Twoawn Pogonatherum 306830
Two-awned Microstegium 253985
Two-ball Nitta Tree 284473
Twoblotch Doplopod 148250
Twobract Landpick 308567
Twobract Solomonseal 308567
Two-bristle Rock Daisy 291301
Twocalyx Peristrophe 291139
Twocalyx Stonecrop 356671
Twocolor Arrowroot 245016
Twocolor Azalea 330565
Two-color Crazyweed 278740
Twocolor Flickingeria 166957
Twocolor Maple 2801
Twocolor Sealavender 230544
Twocolor Snowflake 32520
Twocolor Sunipia 379773
Twocolor Woodbetony 287041
Twocolored Acanthophippium 2072
Twocolored Ardisia 31403
Twocolored Asiabell 98281
Two-colored Maple 2801
Twocolored Monkshood 5021
Two-colored Rhododendron 330565
Twocolored Ribseedcelery 304769
Twocolour Acidanthera 4534
Twocolour Azalea 330223
Twocolour Baboon Flower 46149
Two-colour Cattleya 79529
Two-colour Leaf Mahonia 242608
Two-colour Rhododendron 330223
Two-coloured Begonia 49775
Two-coloured Cactus 244010
Two-coloured Catasetum 79332
Two-coloured Indigo 205902
Two-coloured Peashrub 72190
Two-coloured Pea-shrub 72190
Twocoloured Pleurospermum 304769
Two-coloured Rhododendron 330223
Twocoloured Sesbania 361432
Two-coloured Sorghum 369600
Two-colours Cactus 244055
Two-cusp Milkvetch 42078
Twodentate Germander 388012
Twoear Addermonth Orchid 110638
Twoear Bogorchis 110638
Two-ear Peashrub 72252
Two-ear Pea-shrub 72252
Two-edged Begonia 49611
Twoedged Loosestrife 240030
Two-eyed Berry 256002
Two-faces-in-a-hood 410677

Two-faces-under-a-hat 30081
Two-faces-under-one-hat 410677
Two-faces-under-the-sun 410677
Twoflower Bipinnate Dragonhead 137558
Twoflower Bipinnate Greenorchid 137558
Twoflower Burgrass 394381
Twoflower Burmannia 63956
Twoflower Butterflygrass 392891
Twoflower Cherry 83186
Twoflower Cinquefoil 312409
Twoflower Dogtoothgrass 117862
Twoflower Eria 148637
Twoflower Gingersage 253637
Twoflower Habenaria 183580
Twoflower Hedyotis 187514
Twoflower Jerusalemcherry 367528
Twoflower Leucas 227558
Twoflower Litosanthus 233811
Twoflower Loosestrife 239559
Twoflower Lycianthes 238942
Twoflower Micromeria 253637
Twoflower Microtropis 254284
Twoflower Monkshood 5065
Twoflower Pancratium 280880
Twoflower Passionflower 285620
Twoflower Pearlsedge 354022
Twoflower Plectocomiopsis 303099
Twoflower Raspberry 338183
Twoflower Rattlebox 111955
Twoflower Rattlesnake Plantain 179580
Twoflower Razorsedge 354022
Twoflower Red Silkyarn 238942
Twoflower Saxifrage 349100
Twoflower Small Reed 127215
Twoflower Smallreed 127215
Twoflower Spicegrass 239559
Twoflower Torenia 392891
Twoflower Tulip 400132
Twoflower Wedelia 413506
Twoflowered Cathayanthe 79442
Twoflowered Crazyweed 278744
Two-flowered Everlasting-pea 222724
Two-flowered Narcissus 262417
Two-flowered Passion Vine 285620
Two-flowered Pea 222724
Two-flowered Rush 212905
Two-flowered Touch-me-not 204817
Two-flowers Aniseia 25316
Twoflowers Crazyweed 278814
Twoflowers Trigonella 397254
Twoflowers Vetch 408406
Twoflower-type Tulip 400134
Two-fold Rhododendron 331742
Twofork Falseoat 398448
Twofork Rejoua 327577
Twoforked Celerycress 366121
Twoform Crazyweed 278816
Two-form Holly 203763
Two-formed Microstegium 253987
Two-formes Caladenia 65207
Twoformflower 128477
Twoformflower Tubergourd 390127
Twofruit Dichocarpum 128930
Two-glandules Bredia 59831
Twoglobular Nittatree 284473
Two-grain Wheat 398890
Two-grooved Milk-vetch 42082
Twohead Woodbetony 287165
Twohorn Dandelion 384472
Two-horn Rhododendron 330225
Twohorned Iguanura 203503
Two-horned Stock 246443
Two-horns Azalea 330225
Two-horns Rhododendron 330225
Twohorny Touch-me-not 204816
Two-in-a-purse Pennies 238371
Two-jointed Dendrolobium 125576
Twolayer Oreorchis 274035
Twolayer Wildorchis 274035
Twoleaf Beadruby 242672
Twoleaf Cassia 360477
Twoleaf Epigeneium 146544
Twoleaf Epimedium 146980
Twoleaf Hoodorchis 264765
Twoleaf Hoodshaped Orchid 264765
Two-leaf Miterwort 256015
Twoleaf Nightshade 367107
Two-leaf Nightshade 367107
Twoleaf Nut Pine 299926
Twoleaf Orchis 169897
Twoleaf Pine 299926
Two-leaf Pinion 299926
Twoleaf Pinyon 299926
Twoleaf Platanthera 302280
Twoleaf Pleione 304302
Twoleaf Squill 352886
Twoleaf Tulip 400156
Two-leaf Water-milfoil 261348
Twoleaf Zornia 418331
Twoleaveas Amitostigma 19503
Two-leaved Bead-ruby 242672
Two-leaved False Solomon's Seal 242675
Two-leaved Miterwort 256015
Two-leaved Solomon's-seal 242675
Twoleaved Swertia 380126
Two-leaved Toothwort 125841
Two-leaved Vetch 408648
Two-leaves Morningglory 208067
Twoleaves Oncidium 270798
Twoleaves Satyrium 347717
Twolinehairy Eyebright 160286
Twolobe Dutchman's Pipe 34122
Twolobe Pipewort 151250
Twolobe Speedwell 407040
Two-lobe Speedwell 407040
Two-lobe Spineflower 89055
Two-lobed Fruit Grewia 180700
Two-lobed Leschenaultia 226670
Twolobed Officinal Magnolia 242234
Twolobed Vineclethra 95487
Twolobed Woodbetony 287045
Twolow Wallichpalm 413093
Twonectary Lepistemon 225702
Two-needle Pinyon 299926
Twonerve Statice 230545
Twony-flowered Rhododendron 330749
Twooawn Goatgrass 8665
Two-pennies-in-a-purse 238371
Two-petal Ash 167953
Two-petaled Begonia 49783
Twopistil Sarcococca 346732
Twoply Liparis 232116
Tworow Barley 198288
Two-row Calanthe 65952
Two-row Stickseed 221711,221755
Tworow Stonecrop 357169
Two-row Stonecrop 357169
Tworow Thorowax 63618
Two-rowed Barley 198288
Two-rowed Stonecrop 357169
Tworows Rattan Palm 65672
Tworowsleaves Eustachys 160985
Twoscale Saltbush 44523
Twoseed Dendrolobium 125576
Two-seed Dendrolobium 125576
Twoseed Fake Woodbean 125576
Two-seed Garcinia 171207
Twoseed Jasmine 211795
Twoseed Palm 130057
Twoseed Phrynium 296043
Twoseed Pristimera 315356
Two-seeded Bog Sedge 74342
Two-seeded Orach 44523
Two-seeded Sedge 74342
Twosexflower Cunealglume 29453
Twospike Cupgrass 151670
Twospike Pennisetum 289083
Twospikelet Fimbristylis 166518
Twospikelet Fluttergrass 166518
Twospine Pincution 244081
Two-spined Acaena 1753
Twospined Herelip 220063
Twospined Lagochilus 220063
Twospot Swertia 380131
Twostamen Fiveangular Fimbristylis 166456
Twostamen Orchid 132807,132808
Two-stamened Cyperus 118734
Twostamens Brome 60696
Two-style Eurya 160430
Two-styled Begonia 49782
Two-swelling Gastrochilus 171834

Twoteeth Stelis 374701
Twoteeth Woodbetony 287043
Twotooth Achyranthes 4273
Twotooth Dalechampia 121914
Two-toothed Barberry 51364
Two-toothed Wild Buckwheat 152655
Twotoothedbeak Sedge 74685
Twovalve Microtropis 254285
Two-velate Rhododendron 330228
Two-velates Rhododendron 330228
Twovittae Cryptanthus 113374
Two-whorl Buckwheat 375581
Twowing Abelia 147
Two-wing Hemsleia 191914
Two-wing Silverbell 184740
Two-winged Abelia 147
Two-winged Silverbell 184740
Twyblade 232950
Tydaea 217738
Tyle-berry 212186
Tylophora 400844
Typhonium 401148,401172
Typic Twig-hanging Embelia 144752
Typical Canada Lily 229783
Tyrol Knapweed 81243
Tzekwei Buckthorn 328880
Tzetsou Primrose 315068

U

Uala-kahiki 367696
Uapaca 401210
Uauassu Palm 44811
Uba Cane 341887
Ubame Oak 324283
Ubim Palm 201435
Ucahuba 261461
Udo 30619
Udo Salad Plant 30619
Uebelmannia 401332
Uek Neillia 263183
Ugam Goosecomb 335558
Ugame Roegneria 335558
Uganda Clover 396850
Uganda Grass 117869,117886
Uganda Mahogany 216196
Ugly Willow 343524
Ugni Shrub 261765
Uhde Ash 168124
Uinta Basin Hookless Cactus 354295
Uinta Fleabane 151028
Uinta Mountain Fleabane 150667
Ujukitsu 93860
Ukraine Chalkplant 183265
Ukraine Dock 340298
Ukraine Gagea 169520
Ukraine Hawthorn 110097
Ukraine Hedgeparsley 393007
Ukraine Salsify 394355
Ukraine Scabious 350279
Ukraine Sweetvetch 188162
Ukrainian Catnip 265094
Ukurundu Maple 3732
Ulé Rubber 79110
Uler Otaheite Potato 16352
Ulex Ceratostigma 83649
Ulex-like Barberry 52287
Uli 79112
Uline Yam 131889
Ulla Grass 389333
Ullimer 35090
Ulluco 401412
Ullucu 401412
Ullucus 401411
Ullum 401593
Ulmer Pipes 2837
Ulm-leaved Falsemallow 243893
Ulmo 156063
Ulotricha Sweetleaf 381446
Ulotrichy Footcatkin Willow 343660
Ulva Marina 418392
Umbaba 216211
Umbei Tansy 383857
Umbel Chickweed 375128
Umbel Clerodendrum 96419
Umbel Cranesbill 174992
Umbel Dendrotrophe 125792
Umbel Eargrass 187690
Umbel Goatgrass 8737
Umbel Hedyotis 187690
Umbel Spiradiclis 371782
Umbel Taxillus 385245
Umbel Yarrow 4059
Umbelflower Pollia 307341
Umbell Bulbophyllum 63159
Umbell Curlylip-orchis 63159
Umbell Mulgedium 259776
Umbellaflower Abelia 198
Umbellaflower Strychnos 378922
Umbella-flowered Abelia 198
Umbellaflowered Poisonnut 378922
Umbella-flowered Poisonnut 378922
Umbellare Japanese Pine 299897
Umbellate Alangium 13373
Umbellate Aster 41448
Umbellate Barberry 52289
Umbellate Blachia 54880
Umbellate Bulbophyllum 63087
Umbellate Candytuft 203249
Umbellate Centaurium 81544
Umbellate Chickweed 197491
Umbellate Chinese Knotweed 308976
Umbellate Dendrolobium 125582
Umbellate Dendrotrophe 125792
Umbellate Fake Woodbean 125582
Umbellate Fluttergrass 166550
Umbellate Fuirena 168903
Umbellate Hawkweed 196057
Umbellate Heloniopsis 191068
Umbellate Hydrangea 200126
Umbellate Leptodermis 226136
Umbellate Merremia 250853
Umbellate Parasiticvine 125792
Umbellate Rockjasmine 23323
Umbellate Rust Paraboea 283126
Umbellate Spapweed 205418
Umbellate Spicebush 231461
Umbellate Star Wort 375128
Umbellate Starwort 375128
Umbellate Touch-me-not 205418
Umbellate Water Pennywort 200383
Umbellate Wildclove 226136
Umbellate Wintergreen 87498
Umbellateflower Litse 234085
Umbellate-flowered Basketvine 9455
Umbellateleaf Loosestrife 239773
Umbelliferous Pleasing Barberry 51297
Umbelliform Hankow Willow 343676
Umbellike Actinidia 6715
Umbel-like Sedge 76630
Umbellike Woodbetony 287789
Umbellulate Holly 204358
Umbellulate Rockfoil 350007
Umbels Heath 150187
Umbelspike Goatgrass 8737
Umbelstyle Tupistra 400394
Umber Pussytoes 26615
Umbilicate Milkwort 308428
Umbinza 184875
Umbonate Milkwort 308428
Umbra Tree 298109
Umbraticolor Barberry 52291
Umbraticous Barberry 52291
Umbrella 14760,68713,401711,409339
Umbrella Bamboo 162635,162711,162717,162752,388747,388761,391457,391460
Umbrella Cheese Tree 177182
Umbrella Dracaena 127507
Umbrella Eastern Arborvitae 390624

Umbrella Flat Sedge 118476,118477,118734
Umbrella Flatsedge 118476
Umbrella Flat-sedge 118476
Umbrella Galingale 118476
Umbrella Grass 168885
Umbrella Larkspur 124661
Umbrella Leaf 132677,132678
Umbrella Lily 61390
Umbrella Locust 334992
Umbrella Magnolia 242333
Umbrella Merremia 250853
Umbrella Milkvetch 43022
Umbrella Palm 118476,188190,188191
Umbrella Pine 300151,352854,352856
Umbrella Plant 118476,122766,122779,152056,152240,181958,292372,349770
Umbrella Sedge 118428,118442,118476,118734,118890,119041,119624,245351
Umbrella Thorn 1602,1653
Umbrella Thoroughwort 217105
Umbrella Tree 115190,115217,242193,242333,260287,310169,310227,350648
Umbrella-bamboo 388747
Umbrellaflower Agapetes 10309
Umbrellaflower Basketvine 9455
Umbrellaflower Kiwifruit 6715
Umbrella-flower Litsea 234085
Umbrellaflower Loosestrife 239842
Umbrella-flowered Cyrtandra 120508
Umbrella-flowered Litse 234085
Umbrella-grass 118476,168819,168876
Umbrellaleaf 132677
Umbrella-leaf 132677,132678
Umbrellaleaf Loosestrife 239843
Umbrella-leaves 292372
Umbrella-plant 118476,118477
Umbrella-sedge 118428
Umbrellastalk Arrowbamboo 162760
Umbrellate Deodar Cedar 80091
Umbrellate Soroseris 374539
Umbrellateflower Taxillus 385245
Umbrella-tree 241896,242110,242193,242333,260288
Umbrellawort 255672
Umbrella-wort 14976,14992
Umbu 372482
Umburana 19229
Umiry Balsam 199297
Umkokolo 136839
Ummersweet 96457
Umnonono 378746
Umplescrump 192358
Umpqua Chief Nandina 262201
Umpqua Mariposa-lily 67633
Umu Palm 263552
Umzimbeet 254711
Unarmed Bamboo 47516
Unarmed Glorybower 96140
Unarmed Hornwort 83578
Unarmed Lambertia 220310
Unarmed Magellan Barberry 51403
Unarmed Ramontchi 166776
Unarmed Sainfoin 271222
Unarmed Variable Mayten 246949
Unarmed Wild Buckwheat 152157
Unbranched Bur-reed 370044
Uncertain Mahonia 242562
Uncifera 401795
Unciform Calathodes 66225
Uncinate Childvine 400984
Uncinate Solomonseal 308656
Uncinate Spapweed 205420
Uncinate Touch-me-not 205420
Uncinate Tylophora 400984
Uncommon Barberry 51773
Undate Fescuegrass 164372
Under Pine Tree Rhododendron 331491
Underbloue Magnoliavine 351054
Underbrown Cherry 83314
Underbrown Japanese Cherry 83234
Underground Ivy 116736,176824
Underground Nut 102727
Underground Shepherd 273531
Underleaf Pearl 296465,296562,296565,296801,296809
Underpine Indianpine 202767
Undersized Fig 165515
Underwood's Trillium 397626
Undulate Aeonium 9074
Undulate Alumroot 194438
Undulate Aralia 30788
Undulate Arisaema 33553
Undulate Begonia 50382
Undulate Bigginseng 241179
Undulate Big-ginseng 241179
Undulate Bird Cherry 280016
Undulate Butterflybush 62137
Undulate Cherrylaurel 223133,316922
Undulate Chinense Maple 3626
Undulate Creeping Stem 256005
Undulate Embelia 144817
Undulate Fig 165807
Undulate Indianmulberry 258927
Undulate Laelia 219697
Undulate Lettuce 219589
Undulate Mahonia 242469
Undulate Paris 284404
Undulate Pearleverlasting 21530
Undulate Poplar 311558
Undulate Serpentroot 354878
Undulate Southstar 33553
Undulate Spapweed 205421
Undulate Speedwell 407430
Undulate Stranvaesia 377452
Undulate Summerlilic 62137
Undulate Touch-me-not 205421
Undulate Vandopsis 404754
Undulate Windhairdaisy 348893
Undulateleaf Aralia 30788
Undulate-leaf Ardisia 31604
Undulateleaf Evergreen Chinkapin 79063
Undulateleaf Oplismenus 272684
Undulateleaf Pittosporum 301432
Undulateleaf Pricklyash 417366
Undulateleaf Seatung 301432
Undulate-leaved Aralia 30788
Undulate-leaved Ardisia 31351
Undulate-leaved Pittosporum 301432
Undulate-leaved Prickly-ash 417366
Undulate-leaved Thyrsia 391443
Unequal Crabgrass 130592
Unequal Eurya 160502
Unequal Lepidagathis 225178
Unequal Size Rhododendron 330916
Unequalbase Cocklebur 415020
Unequalglume Bentgrass 12142
Unequalglume Jijigrass 4141
Unequalglume Speargrass 4141
Unequal-glumed Wild-rye 144292
Unequalhair Baoshan Monkshood 5429
Unequalhair Paoshan Monkshood 5429
Unequalleaf Clearweed 298858
Unequalleaf Hemlock 399905
Unequalleaf Metalleaf 296366
Unequal-leaved Metalleaf 296366
Unequallip Rabdosia 324728
Unequallobed Woodbetony 287289
Unequal-perianth Woodrush 238627
Unequalsepal Jerusalemsage 295118
Unexpected Corydalis 106014
Unexpected Larkspur 124306
Unexpected Rhododendron 330930,331444
Unexpected Viburnum 407894
Unfortunate Glorybower 96141
Ungeargrass 36612,36647
Ungern Rhododendron 332038
Ungle-pigle 374916
Unguentine Cactus 17381
Unguicular Geniculatepetal Monkshood 5223
Unguiculate Hemsley Monkshood 5265
Ungulate Fargesia 162759
Unharmed Greyhair Raspberry 338642
Unhlamalala 378699
Unialate Laborde Begonia 49972
Unialate Rough-leaved Begonia 49637
Unibract Fritillary 168605
Unibract Iris 208445
Unibract Swordflag 208445
Unibranch Yushania 416819
Unicanaliculat Holly 204363
Unicanaliculate China Fir 114555

Unicom Root 14486,85505
Unicorn Plant 315430,315434
Unicorn Root 14486
Unicornplant 245977,245979
Unicorn-plant 245977,245979,315431
Unicornplant Family 245991
Unicorn-plant Family 245991
Unicorn-root 14486
Unida-like Pycreus 322376
Uniflor Azalea 332039
Uniflored Melic 249106
Uniflorous Currant 334162
Uniflorous Lambertia 220312
Uniflorous Litse 135137
Uniflorous Lycianthes 238957
Uniflorous Meconopsis 247142
Uniflorous Primrose 315072
Uniflorous Rhododendron 331270,332039
Uniflorous Rose 337004
Uniflorous Touch-me-not 205422
Uniflower Cinquefoil 313091
Uniflower Garcinia 171157
Uniflower Iris 208917
Uniflower Leek 15486
Uniflower Littlefruit Rockfoil 349643
Uniflower Onion 15486
Uniflower Orchid 85973,85974
Uniflower Swordflag 208917
Uniflowered Eriolaena 152697
Uniflowered Firerope 152697
Uniflowered Garcinia 171157
Uniflowered Goldsaxifrage 90468
Uniflowered Goldwaist 90468
Uniflowered Japanese Barberry 52251
Uniflowerorchid 85973,85974
Unifoliate Swollennoded Cane 87624
Unifolious Buttercup 326093
Unifolious Rattan Palm 65791
Uniform Bramble 339432
Uniglume Spikesedge 143397
Unijugate Engelhardia 145526
Unijugate Millettia 254866
Unijugous Milkvetch 43207
Unileaf Sage 344947
Unilocular Crateva 110235
Uninectary Stringbush 414222
Uniola-like Galingale 322376
Uniseminal Persimmon 132446
Unisexual Sedge 76644
Unispike Kyllingia 218571
Unispike Water-centipede 218571
Unistyle Camellia 69007
Unistyle Dodder 115080
Unitedsepal Felwort 235448
United-sepal Lysionotus 239942
United-style St. John's Wort 201818
Universe Plant 31142
Univerve Redhair Rockfoil 349863
Unlovely Willow 343524
Unmaculate Fargesia 162684
Unpleasant Dracaena 137367
Unpleasant Dragonbood 137367
Unpleasant Mosquitotrap 117523
Unpleasant Swallowwort 117523
Unsavoury Marjoram 316127
Unshiu Mandarin 93736
Unshoe-the-horse 196630,238371
Unspotted Lungwort 321641
Unspottedflower Canada Lily 229786
Unstriped Holly 203788
Unstyle Eutrema 161152
Untrod Den-to-pieces 308816
Untrodden-to-death 308816
Unusual Azalea 331987
Unusual Milkvetch 42913
Unusual Rhododendron 331987
Upas 28248
Upas Climber 378761
Upas Tree 28248
Upas-tree 28248
Upland Bent 12240
Upland Bent Grass 12240
Upland Boneset 158037,158315,158316
Upland Burnet 345881
Upland Coffee 99008
Upland Cotton 179906,179923
Upland Cranberry 31142
Upland Cress 47931,47963,47964
Upland Enchanter's-nightshade 91553
Upland Geranium 174548
Upland Helixanthera 190863
Upland Hickory 77920
Upland Sumach 332608
Upland Tupelo 267867
Upland Violet Iris 208923
Upland White Aster 41086,368324
Upland White Goldenrod 368324
Upland Wild-timothy 259689
Upland Willow 343506
Upland Wintercress 47964
Uponnaked Begonia 49814
Upper Burma Red Rhododendron 331866
Upright Axyris 45844
Upright Baboon Flower 46148
Upright Bamboo 47246
Upright Bambusa 47246
Upright Bindweed 68742
Upright Brome 60707
Upright Brome Grass 60707
Upright Brome-grass 60707
Upright Bur-head 140539
Upright Carrion-flower 366317
Upright Chickweed 256462,256470
Upright Clover 397097
Upright European Mountain-ash 369342
Upright Fig 164671
Upright Goosefoot 87186
Upright Heath 150127
Upright Hedge Bedstraw 170372
Upright Hedge Nettle 373389
Upright Hedge Parsley 392992
Upright Hedge-nettle 373389
Upright Hedge-parsley 392992
Upright Hyptianthera 203031
Upright Italian Celery 29327
Upright Japanese Plum-yew 82525
Upright Laceleaf Maple 3331
Upright Ladybell 7830
Upright Meadow 325518
Upright Meadow Crowfoot 325518
Upright Mignonette 327795
Upright Myrtle Spurge 159726
Upright Pearlwort 256470
Upright Prairie Coneflower 326960
Upright Rattan Palm 65677
Upright Redcurrant 334149
Upright Red-leaved Barberry 52236
Upright Roman Cypress 114768
Upright Speedwell 317964
Upright Spotted Spurge 159286
Upright Spurge 158534,159892
Upright St. John's Wort 202117
Upright Stewarta 376469
Upright Velvetflower 383090
Upright Vervain 405889
Upright Vetch 408522
Upright Virgin's Bower 94901
Upright Yellow Sorrel 278099
Upright Yellowwood 306450
Upright Yellow-wood 306450
Upright Yew 385313
Uprighthair Skullcap 355642
Upright-spiked Cypressgrass 119319
Upright-spiked Galingale 119319
Upstart 99297
Uptight Sedge 74435,76399
Uptight Yellow-sorrel 278099
Urai Evergreen Chinkapin 79064
Urai Evergreenchinkapin 79064
Urai Holly 204365
Urai Tanoak 79064
Urai Wikstroemia 414222
Uraku Camellia 69739
Ural Anemone 24107
Ural Bellflower 70366
Ural Cephalaria 82183
Ural Chalkplant 183266
Ural Crazyweed 279228
Ural Euphorbia 159663
Ural False Spiraea 369279
Ural Falsespiraea 369279

Ural False-spiraea 369279
Ural Globeflower 399547
Ural Larkspur 124669
Ural Licorice 177947
Ural Peony 280154
Ural Pleurospermum 304766
Ural Ribseedcelery 304766
Ural Roegneria 335560
Ural Willowweed 146930
Uraria 402106,402132
Urariopsis 402149
Urat Spiraea 372123
Uratu Plum 20979
Urban Spurge 158407
Urceola 402187
Urceolar Arrowwood 408198
Urceolar Sweetleaf 381423
Urceolar-flowered Agapetes 10342
Urceolar-flowered Osmanthus 276402
Urceolate Aspidistra 39581
Urceolate Himalayan Currant 333965
Urceolate Ladybell 7617
Urceolate Viburnum 408198
Urceolate Woodbetony 287791
Urceolatue Osmanthus 276402
Urchin Crowfoot 325612
Urchin Dryandra 138440
Urchins 167955
Urd 408982
Urd Bean 408982
Urjanchaic Bluegrass 306113
Urles 408571
Urn Gum 155770
Urn Orchid 55575
Urn Plant 8558,302652
Urn-fruited Gum 155770
Urn-like Flowers Blueberry 404033
Urnshaped Diphylax 132672
Urnshaped Dischidanthus 134026
Urnshapedfruit Sweetleaf 381423
Urn-tree Hawthorn 109573
Urnu Gentian 174032
Urobotrya 402471,402477
Urochloa 402497
Urophyllous Caper 71916
Urophyllous Clematis 95401
Urophyllous Jasmine 212052
Urophyllum 402605
Urophysa 402651
Ursine Garlic 15861
Ursinia 402743
Urts 403916
Urucury Wax 380556
Uruguayan Fountaingrass 289143
Uruguayan Needlegrass 262618
Uruguayan Pampas Grass 105456
Uruguayan Vervain 405869
Urumchi Sweetvetch 188119
Urumqi Milkvetch 43254
Urunday 43480
Usambara Violet 342493
Usawa Pseudosasa 318350
Useful Arisaema 33557
Useful Bamboo 47501
Useful Bambusa 47501
Useful Dinochloa 131288
Useful Fargesia 162760
Useful Pricklyash 417369
Useful Southstar 33557
Useful Umbrella Bamboo 162760
Useful Viburnum 408202
Useful Vinebamboo 131288
Useless 309570
Ushy Milkvetch 42313
Ussur Neottia 264750
Ussuri Asiabell 98424
Ussuri Bluegrass 306119
Ussuri Buckthorn 328882
Ussuri Buttercup 326487
Ussuri Cinquefoil 312944
Ussuri Currant 334249
Ussuri Dandelion 384846
Ussuri Dock 340270
Ussuri Draba 137286
Ussuri Flueggea 167094
Ussuri Fritillary 168612
Ussuri Germander 388298
Ussuri Greenbrier 366548
Ussuri Iris 208700
Ussuri Leaf-flower 296803
Ussuri Maple 2798
Ussuri Parrotfeather 261377
Ussuri Pear 323330
Ussuri Peashrub 72373
Ussuri Pea-shrub 72373
Ussuri Pipewort 151536
Ussuri Plum 316900
Ussuri Poplar 311562
Ussuri Rose 337005
Ussuri Saussurea 348902
Ussuri Sedge 76651
Ussuri Skullcap 355675
Ussuri Soja 177777
Ussuri Spikesedge 143201
Ussuri Swordflag 208700
Ussuri Whitlowgrass 137286
Ussurian Pear 323330
Ussurian Plum 316900
Uster Barberry 51781
Usual Knotweed 309602
Utah Agave 10948
Utah Breadroot 319206
Utah Cedar 213840
Utah Firecracker 289382
Utah Fleabane 151047
Utah Juniper 213625,213840
Utah Knotweed 309930
Utah Mortonia 259056
Utah Oak 324525
Utah Uniper 213980
Utah White Oak 323927
Utah Willow Dock 340303
Ute Anemone 24103
Ute Windflower 24103
Utility Oak 324528
Utricularflower Dutchmanspipe 34365
Utriculateflower Birthwort 34365
Uunequal-sepaled Mycetia 260609
Uva Grass 182946
Uvalde Big-tooth Maple 3002
Uvaria 403413,403521
Uvariafolious Tanoak 233409
Uvarialeaf Azalea 332042
Uvarialeaf Rhododendron 332042
Uvarialeaf Tanoak 233409
Uvarialeaf Wendlandia 413842
Uvaria-leaved Rhododendron 332042
Uvaria-leaved Wendlandia 413842
Uvaria-like Rhododendron 332042
Uva-ursi 31142
Uvilla 97862,313263
Uvula-wort 70331
Uwa-pomelo 93709
Uygur Perilla 290952
Uzon-kunnebu 93704

V

V. Naray Tanoak 233366
Vaal Populier 311261
Vaccinium Fig 165814
Vaccinium-like Blueberry 404036
Vadlapudi Orange 93565
Vagabond's Friend 308607
Vagabondage Microstegium 254004
Vagetable Ivary 298050
Vagetable Ivary Palm 298050

Vaginate Omphalotrigonotis 270762
Vaginate Sainfoin 271278
Vaginate Violet 410710
Vagrant Spindle-tree 157943
Vaillant Bedstraw 170727
Vaillant Hawkweed 196073
Vaillant Tillaea 391959
Vakhan Nepeta 265098
Valaeria 404316
Valais Catchfly 364156
Valameurto 360471
Valamuerto 360467,360469
Valara 404316
Valdiv Barberry 52295
Valencia Orange 93785
Valentin's Rhododendron 332047
Valeri Begonia 50386
Valerian 81743,120290,307197,404213,404316
Valerian Family 404392
Valid Euphorbia 160039
Valid Roundtooth Clearweed 298890
Vallaris 404508
Vallesia 404526
Vallesian Koeleria 217577
Valley Arisaema 33464
Valley Crownscale 44363
Valley Grape 411710
Valley Knotweed 309338
Valley Larkspur 124554
Valley Lily 102856,102867
Valley Mayweed 246381
Valley Oak 324114
Valley Pimpinella 299575
Valley Southstar 33464
Valley Spinycape 34088
Valley White Oak 324114
Valley Willow 343491
Valley Windhairdaisy 348292,348920
Valleys 102867
Vallonea Oak 324059
Valonea 324146
Valonia Oak 324146
Valuable Spiny Pincution 244233
Valvate Actinidia 6717
Valvate Kiwifruit 6717
Valveless Loosestrife 239633
Van Fleet Barberry 52300
Van Houtte Plant 265204
Van Houtte Spiraea 372126
Van Houtte Spirea 372126
Van Houtte's Spiraea 372126
Van Volxem's Maple 3578
Vancouver Fir 384
Vancouveria 404610
Vanda 404620
Vandell Lafoensia 219731
Vandenhecke Calathea 66196
Vandervoet Hackberry 80777
Vanderwolf Pine 299942
Vandopsis 404745,404747
Vandopsis Gigantic 404747
Vang Litse 234089
Vanheurck Thorowax 63598
Vanhout Spirea 401286
Vanhoutte Spiraea 372126
Vanikoro Kauri 10500
Vanilla 404971,405017
Vanilla Bran 405017
Vanilla Grass 27967,196116,196137
Vanilla Leaf 4105
Vanilla Orchid 266260,405017
Vanilla Trumpet Vine 134878
Vanillagrass 27967
Vanilla-grass 27967
Vanillaleaf 77230
Vanilla-leaf 4103,4105
Vaniot Bird Cherry 280040
Vaniot Clematis 95405
Vaniot Paris 284405
Vaniot Staff-tree 80341
Vaniot Tailanther 381957
Vaniot Vetchling 222852
Vaniot's Bittersweet 80341
Vanishing Wild Buckwheat 152052
Vanvolsem Velvet Maple 3738
Vapour 169126
Vara Dulce 161864
Vardar Valley Boxwood 64357
Varenut 102727
Varges 243711
Variabile Glaucous Bamboo 297288
Variable Bambusa 47376
Variable Bigwhite Pelargonium 288264
Variable Blackberry 338857
Variable Bluegrass 306141
Variable Bossiaea 57494
Variable Cereus 83913
Variable Coloured Rhododendron 332095
Variable Dendrobium 125275
Variable Fig 165819
Variable Flatfruit Camellia 69004
Variable Flatsedge 118744
Variable Giantarum 20147
Variable Goosecomb 335562
Variable Hirsute Roegneria 335345
Variable Indocalamus 206835
Variable Kengyilia 215927
Variable Litse 234091
Variable Mayten 182800
Variable Mountainash 369336
Variable Mountain-ash 369336
Variable Pitard Camellia 69503
Variable Rhododendron 331779
Variable Roegneria 335562
Variable Spatholobus 370424
Variable Tango-plant 89139
Variable Tuftroot 130121
Variable Vineclethra 95557
Variable-auricled Bamboo 297476
Variable-coloured Rhododendron 331838
Variable-coloured Spatholobus 370413
Variable-flowered Barberry 52301
Variablele Pepper 300461
Variableleaf Bushbean 241262
Variable-leaf Mitragyna 256115
Variableleaf Pepper 300461
Variable-leaf Ticktrefoil 126396
Variableleaf Yellowcress 336211,336215
Variable-leaved Buckthorn 328726
Variable-leaved Mayten 182683
Variable-leaved Pondweed 312126
Variable-leaved Prickly-ash 417290
Variable-leaved Wing-seed-tree 320845
Variable-lobe Roughleaf Raspberry 338099
Variable-striate Dendrocalamopsis 47505
Variant Milkvetch 43214
Variantcolor Gentian 150444
Variant-coloured Spatholobus 370413
Variantleaf Cremanthodium 110467
Variantleaf Nutantdaisy 110467
Variantleaf Wingseedtree 320845
Varicolor Chirita 87986
Varicose Epidendrum 146499
Varieble Neolitsea 264016
Varieble-leaved Tickclover 126396
Varied Azalea 331779
Varied Brome 60680
Varied Fescuegrass 164393
Varied Hedgehog 140342
Varied Leaf Fig 165828
Varied Litse 234091
Varied Milkvetch 43214
Varied Rhododendron 331779
Varied Tanoak 233411
Variedawn Roegneria 335541
Variedcolor Peashrub 72374
Varied-leaf Cinquefoil 312975
Variedleaf Fig 165828
Variedleaf Hakea 184650
Variedleaf Newlitse 264016
Variegaled Laurel 98190
Variegaled Leaf Croton 98190
Variegata Giant Dogwood 105003
Variegata Gypsy Rose 216089
Variegata Hardy Fuchsia 168771
Variegata Italian Buckth 328596
Variegata Japan Nakeaster 182351
Variegata Norway Maple 3430
Variegata Pfeiffer Olive 276316,276320
Variegata Porcelain Vine 20299

Variegata Queen of Meadowsweet 166128
Variegata Rose of Sharon 195287
Variegata Spearmint 250421
Variegata Star Jasmine 393658
Variegata Sweet Daphne 122534
Variegata Sycamore Maple 3464,3465
Variegata Waxleaf Privet 229489
Variegata Wayfaringtree 407910
Variegate Azalea 330609
Variegate Carrionflower 273206
Variegate Dandelion 384848
Variegate Ginkgo 175827
Variegate Grassleaf Sweetflag 5805
Variegate Lady's Slipper 120411
Variegate Ladyslipper 120411
Variegate Leopardflower 273206
Variegate M. Degron's Rhododendron 330537
Variegate Primrose 314109
Variegated Aloe 17136,17243
Variegated Apple Mint 250458
Variegated Bermuda Arrowroot 245015
Variegated Bishop's Weed 8827
Variegated Bitter Bamboo 304109
Variegated Black Elder 345634,345639
Variegated Black Elderberry 345638
Variegated Box Elder 3218
Variegated Box Leaf 157601
Variegated Boxelder Maple 3232
Variegated Boxwood 64348,64350
Variegated Broadleaf Liriope 232637
Variegated Buckwheat 162301
Variegated Butterfly Bush 62025
Variegated Castillon Bamboo 297218
Variegated Ceanothus 79935
Variegated Chinese Lantern 975
Variegated Confederate-jasmine 393659
Variegated Corn Plant 137397
Variegated Creeping Charlie 116734
Variegated Creeping Fig 165515
Variegated Creeping Soft-grass 197308
Variegated Dead-nettle 220401
Variegated Dwarf Lilyturf 272097
Variegated Dwarf Myrtle 202376
Variegated English Ivy 203548
Variegated Feather Reed Grass 65262
Variegated Fig 165819
Variegated Figwort 355039,355264
Variegated Flowering Maple 913,987
Variegated Gout Weed 8827
Variegated Goutweed 8827
Variegated Grape 411982
Variegated Ground-ivy 176825
Variegated Hebe 186975
Variegated Hedge Maple 2838
Variegated Horsechestn 9704
Variegated Horseradish 34591
Variegated Iris 208919
Variegated Irish Yew 385323
Variegated Japanese Aralia 30638
Variegated Japanese Sedge 75421
Variegated Japanese Silver Grass 255891
Variegated Leafcroton 98190
Variegated Leaf-croton 98190
Variegated Liriope 232637
Variegated Maiden Grass 255888
Variegated Manioc 244508
Variegated Manna Grass 177630
Variegated Meadow Saffron 99351
Variegated Milkweed 38157
Variegated Miscanthus 255891
Variegated Monkshood 5659
Variegated Moor Grass 256597
Variegated Nettle 220401
Variegated New Zealand Christmas Tree 252613
Variegated Oncidium 270849
Variegated Pagoda Tree 104948
Variegated Pedilanthus 287855
Variegated Planetree Maple 3476
Variegated Plantainlily 198647
Variegated Pterolobium 320607
Variegated Purple Moor-grass 256602
Variegated Rush 49284
Variegated Sand Rose 21136
Variegated Shore Juniper 213745
Variegated Silverberry 142157
Variegated Solomonseal 308635
Variegated St. Augustine Grass 375775
Variegated Star Jasmine 393659
Variegated Stonecrop 356885
Variegated Sweet Flag 5794
Variegated Sweetheart Hoya 198850
Variegated Tartarian Dogwood 104921
Variegated Tatarian Dogwood 104932
Variegated Thistle 364360
Variegated Tigerthorn 320607
Variegated Tuftroot 130121
Variegated Water Snowflake 267813
Variegated Wax Vine 359416
Variegated Wayfaring Tree 407910
Variegated White Cedar 85384
Variegated Willow 343530,344240
Variegated Winter Daphne 122532
Variegated Woodbetony 287793
Variegatedleaf Begonia 49701
Variegated-leaf Chirita 87950
Variegatedleaf Loosestrife 239806
Variegatedleaf Periwinkle 409336
Variegatedleaf Tiger Lily 230062
Variegatedleaf Viburnum 408074
Variegatedleaf Violet 410721
Variegated-leaved Viburnum 408074
Variegateleaf Ajania 13044
Variegateleaf Aucuba 44888
Variegatestalk Primrose 314230
Varigated Chastetree 411476
Varigated Cockspur Grass 140481
Varioauriculate Bamboo 297476
Varioeal Side-saddle Flower 347153
Variole Tanoak 233411
Various Colours Rhododendron 330592
Various Iris 208919
Various Vineclethra 95557
Variousforms Sisymbrium 365585
Variousleaf Fescue 164006
Various-leaved Canary -Grass 293710
Various-leaved Crowfoot 325929
Various-leaved Fescue 164006
Various-leaved Fig 165828
Various-leaved Hawthorn 109752
Various-leaved Pondweed 312126,312138
Various-leaved Water-milfoil 261348
Various-leaved Water-starwort 67389
Variously Coloured Peashrub 72374
Variously Slightly Wing-branched Barberry 51683
Various-spiny Barberry 51714
Varnish Tree 12559,14544,217626,393491
Varnish Tree-leaved Meliosma 249445
Varnished Dickasonia 129192
Varnish-tree 12559,217626,393491
Varnish-tree-leaved Meliosma 249445
Varrea 413243
Vartan Iris 208920
Varvine 405872
Varyleaf Mayten 182683
Vase Flower 95016
Vases 174739
Vasey Grass 285533
Vasey Oak 324537
Vasey's Bitterweed 201336
Vasey's Grass 285533
Vasey's Pondweed 312288
Vasey's Rabbitbrush 90564
Vasey's Rock Daisy 291337
Vasey's Rubberweed 201336
Vasey's Rush 213566
Vasey's Thistle 92043
Vasey's Trillium 397628
Vasey-grass 285533
Vassilczenko Medic 247509
Vatch 408571
Vateria 405153
Vatica 405158,405178
Vaudois Sallow 344241
Vaupel Brain Cactus 140580
Vavilov Goatgrass 8745
Vavilov Wheat 399004
Vcnus' Navelwort 270677
Veech Palm 405253
Vegetable Antimony 158250

Vegetable Calomel 5793
Vegetable Dolomiaea 135730
Vegetable Down 179032
Vegetable Gold 103850
Vegetable Hummingbird 361384
Vegetable Ivory 280434,298051
Vegetable Ivory Palm 298051
Vegetable Ivory Substitute 202321
Vegetable Marrow 114300,114304
Vegetable Mercury 61304,119974
Vegetable Orange 114200
Vegetable Oyster 394327
Vegetable Oyster Salsify 394327
Vegetable Pear 356352
Vegetable Rennet 414590
Vegetable Sheep 319938,319941,326513,326515
Vegetable Spaghetti 114300
Vegetable Sponge 238257,238261
Vegetable Tallow 346408
Vegetable Tallow Tree 346408
Vegetable Wild-rice 418095
Vegetable-hummingbird 361385
Vegetable-oyster 394327
Vegetable-oyster Salsify 394327
Vegetablepalm 341401,341422
Vegetable-sponge 238257
Vegtable Ivory Palm 202300
Veincalyx Primrose 314693
Veined Anzacwood 310753
Veined Argyreia 32642
Veined China-laurel 28407
Veined Dock 340304
Veined Fig 165370
Veined Holly 204370
Veined Holly Grape 242513
Veined Illigera 204638
Veined Indigo 206714
Veined Inula 138897
Veined Jasmine 211918
Veined Jurinea 138897
Veined Meadow-rue 388712
Veined Osmanthus 276403
Veined Rabdosia 209784
Veined Skullcap 355619
Veined Vervain 405885
Veined Vetch 408672
Veined Viburnum 407964
Veined Windhairdaisy 348203
Veined Yellow-eyed Grass 365771
Veined Zeuxine 417768
Veined-leaf Blueberry 404036
Veinfruit Sandthorn 196755
Veinfruit Seabuckthorn 196755
Veing Pepper 300518
Vein-leaved Indigo 206714
Veinless Barberry 51967
Veinless Cockclawthorn 278318
Veinless Sedge 74453
Veinless Sheathflower 241315
Veinrise Coldwaterflower 298960
Veiny Actinidia 6720
Veiny Begonia 50390
Veiny Blueberry 404044
Veiny Boeica 56369
Veiny Catchfly 364193
Veiny Clovershrub 70790
Veiny Dock 340304
Veiny Elm 401476
Veiny Geranium 174997
Veiny Jadeleaf and Goldenflower 260465
Veiny Kiwifruit 6720
Veiny Lined Aster 380959,380962,380964
Veiny Lizardtaro 348036
Veiny meadow-rue 388712
Veiny Michelia 252948
Veiny Pea 222854,222860
Veiny Pea-vine 222854,222860
Veiny Qinggang 116152
Veiny Sauromatum 348036
Veiny Skullcap 355619
Veiny Stairweed 142801
Veiny Tanoak 233218
Veinyroot 237476,237539
Veitch Barberry 52302
Veitch Begonia 50251,50388
Veitch Calathea 66211
Veitch Catnip 265100
Veitch Ceanothus 79977
Veitch Didymocarpus 129951
Veitch Dysosma 139628
Veitch Fir 496
Veitch Gentian 174041
Veitch Glabrous Cinquefoil 312638
Veitch Goldenray 229241
Veitch Hawthornleaf Maple 2913
Veitch Hogfennel 293061
Veitch Holly 204167
Veitch Mahonia 242620
Veitch Maple 3735
Veitch Meliosma 249476
Veitch Nepeta 265100
Veitch Peony 280155
Veitch Primrose 315080
Veitch Sasa 347324
Veitch Screwpine 281169
Veitch Screw-pine 281169
Veitch Spiraea 372127
Veitch Ural Peony 280155
Veitch Viburnum 408210
Veitch Waxpetal 106687
Veitch Windhairdaisy 348915
Veitch Winterhazel 106687
Veitch's Anthurium 28138
Veitch's Bamboo 297477,347324
Veitch's Litse 234095
Veitch's Rhododendron 332052
Veitch's Screwpine 281169
Veitch's Silver Fir 496
Veitch's Winter Hazel 106687
Veitch's Yunnan Crabapple 243733
Veitchberry 338059
Veitchia 405253
Veitchia Family's Rhododendron 332061
Veitch's Screwpine 281141
Veivetypetal Indigo 206046
Velanidi Oak 324146
Veld Grape 92910
Veld Lily 111152
Veldt Daisy 175172
Veldt Grass 141743
Velentin Primrose 315078
Velety Metalleaf 296422
Velezia 405269,405271
Vellayim 197451,197454
Vellvetgrass 418453
Veltheimia 405356,405358
Veltman's Listera 232892,232975
Velutinous Anzacwood 310775
Velutinous Aster 41473
Velutinous Carpesium 77195
Velutinous Cinquefoil 312653
Velutinous Indigo 206583
Velutinous Machilus 240711
Velutinous Meliosma 249477
Velutinous Michelia 252972
Velutinous Moquinia 258368
Velutinous Rattlesnake Plantain 179708
Velutinous Soapberry 346345
Velutinous Spiraea 372128
Velutinous Spotleaf-orchis 179708
Velutinous Tickclover 126663
Velutinous Waxpetal 106688
Velutinousstipe Yam 131893
Veluty Clovershrub 200732
Velvent Sagebrush 36111
Velvet Apple 132030,132351
Velvet Ash 168127
Velvet Atylosia 64145
Velvet Banana 260279
Velvet Bean 259479,259559
Velvet Bells 373608
Velvet Bent 12025
Velvet Bent Grass 12025
Velvet Bentgrass 12025
Velvet Bent-grass 12025
Velvet Butea 64145
Velvet Crabgrass 130817
Velvet Dock 207135,405788
Velvet Edelweiss 224960
Velvet Flower 18687,18752,344383,370009

Velvet Flower-de-Luce 192904
Velvet Glochidion 177188
Velvet Goldenrod 368250
Velvet Grass 197278,197301,418453
Velvet Hookvine 401765
Velvet Leaf 1000,223355,405788
Velvet Leaf Bamboo 125469
Velvet Leaf Philodendron 294850
Velvet Leaf Senna 360451
Velvet Maple 3736
Velvet Mesquite 315557,315575
Velvet Mullein 405788
Velvet Nerisyrenia 265308
Velvet Nettle 183056
Velvet Ophiorrhiza 272236
Velvet Plant 108723,183051,183056,405788
Velvet Pod Mimosa 255026
Velvet Poppy 405788
Velvet Prickly Pear 273077
Velvet Rattlebox 112757
Velvet Ricebag 181670
Velvet Sage 345167
Velvet Sumac 332639,332916
Velvet Sumach 332639,332916
Velvet Tamarind 127396,127397
Velvet Touch-me-not 205260
Velvet Tree 253007
Velvet Tree Mallow 223355
Velvet Treemallow 223355
Velvet Tree-mallow 223355
Velvet Trumpet Flower 344383
Velvet Weed 1000
Velvet William 127607
Velvet Willow 344099,344122
Velvet Woodbetony 287434
Velvet-bush 219795
Velvetcalyx Milkvetch 42641
Velvetfruit Euonymus 157811
Velvetfruit Tanoak 233116
Velvetgrass 197278,197301
Velvet-grass 197278,197301
Velvetleaf 837,1000,230244,403915
Velvet-leaf 1000
Velvet-leaf Blueberry 403915
Velvetleaf Glorybower 96196
Velvet-leaf Huckleberry 403915
Velvetleaf Indian Mallow 317128
Velvetleaf Mayten 182683
Velvetleaf Rhynchosia 333426
Velvetleaf Soldierbush 393242
Velvet-like Argyreia 32665
Velvet-like Ash 168136
Velvet-like Buckthorn 328889
Velvet-like Casearia 78156
Velvet-like Eurya 160623
Velvet-like Millettia 254876
Velvet-like Premna 313761
Velvetplant 183051,183056
Velvetseed 181728,181743
Velvet-seed 181728,181743
Velvetseed Waterbolt 55924
Velvetsheath Bitterbamboo 304039
Velvet-tree 183056
Velvettube Milkvetch 42615
Velvet-weed 172203
Velvety Astridia 43342
Velvety Bird Cherry 280056
Velvety Birdcherry 280056
Velvety Casearia 78156
Velvety Cinquefoil 312653
Velvety Clovershrub 70863
Velvety Coldbamboo 172751
Velvety Dragonhead 137688
Velvety Epimedium 147039
Velvety Evodia 161387
Velvety Gaura 172203
Velvety Gelidocalamus 172751
Velvety Glochidion 177188
Velvety Goldenrod 368250,368470
Velvety Hawthorn 110060
Velvety Honeylocust 176887
Velvety Jurinea 214115
Velvety Lilac 382268
Velvety Longleaf Privet 229449
Velvety Maire Camellia 69344
Velvety Maple 3736
Velvety Melochia 249642
Velvety Metalleaf 296422
Velvety Millettia 254876
Velvety Motherwort 225021
Velvety Nepeta 265101
Velvety Premna 313761
Velvety Primrose 314980
Velvety Saussurea 348916
Velvety Sophora 369149
Velvety Spindle-tree 157946
Velvety Spiraea 372128
Velvety Swellenfruit celery 295041
Velvety Tickclover 126663
Velvety Wild Buckwheat 151964
Velvety Windhairdaisy 348916
Velvety Winterhazel 106688
Velvetyleaf Bugbane 91021
Velvetymouth Hornbeam 77302
Velvetyoid Machilus 240713
Vened Eranthemum 148087
Venenate Excoecaria 161680
Venetian Sumach 107300
Venezuela Heliconia 190031
Venezuela Zephyrlily 417638
Venezuelan Box 78145
Venezuelan Mahogany 380524
Venezuelan Panicgrass 282366
Venezuelan Treebine 92934
Vengai Padauk 320307
Venice Mallow 195326
Venice Turpentine 221855
Venns'-navelwort 401711
Venose 267923
Venose Ainsliaea 12687
Venose Euonymus 157947
Venose Kumquat 167516
Venose Leontice 224644
Venose Rabbiten-wind 12687
Venose Spindle-tree 157947
Venosus Fendler Ceanothus 79927
Vente Conmigo 112895,112896
Ventenata Grass 405429
Ventilago 405438
Ventricose Gendarussa 172831
Ventricose Goatgrass 8746
Ventricose Heath 150214
Venulose Holly 204371
Venulose Rockvine 387865
Venus Cup Teasel 133517
Venus Fly Trap 131397
Venus Flytrap 131397
Venus Fly-trap 131397
Venus Pride Angel-eyes 198708
Venus Thistle 92257
Venus Tree 3978
Venus' Basin 133478
Venus' Bath 133478
Venus' Chariot 5442
Venus' Comb 350425
Venus' Flytrap 131397
Venus' Looking Glass 224067,238371,397755,397756,397757,397759
Venus' Looking-glass 224058,224067,224073
Venus' Navelwort 270705,401711
Venus' Needle 350425
Venus' Sumach 107300
Venus'-chariot 5442
Venus'-comb 350425
Venus'-flytrap 131397
Venus'-looking-glass 224058
Venus-cup Teasel 133478,133517
Venus-in-her-car 355061
Venuskamm 350425
Venuslooking-glass 224067
Vera Cruz Pepper 300344,300345
Veratrilla 405564,405567
Veratrin 352124
Verawood 63406
Verbascum 405657
Verbena 176683,405801,405852,405868,405872,405881,405885
Verbena Family 405907
Verbena Sage 345458
Verbenifolius Woodbetony 287796

Verdant Bamboo 47494
Verdant Primrose 314927
Verdigris Sweetleaf 381100
Veriegated Coral Tree 154734
Veriegated Coralbean 154734
Veriegated Coral-bean 154734
Verlot's Mugwort 36454
Vermeil Azalea 330793
Vermicelli 398900
Vermifuge Ironweed 406102
Vermilion Azalea 330409
Vermilion Passionflower 285631
Vermilion Rhododendron 330409
Vermilion Wood 320269
Vermont Blackberry 339443
Verna Barberry 52306
Vernal Anemone 24109
Vernal Barberry 52306,52307
Vernal Caesalpinia 65077
Vernal Clematis 94835
Vernal Ferdinandi-coburg Barberry 52307
Vernal Grass 27930
Vernal Sandwort 32193,255598
Vernal Sedge 74059,75859
Vernal Speedwell 407432
Vernal Stitchwort 255598
Vernal Water Starwort 67376
Vernal Water-starwort 67376
Vernal Whitlow Grass 153964
Vernal Whitlow-grass 153964
Vernal Witch Hazel 185138
Vernal Witchhazel 185138
Vernal Witch-hazel 185138
Vernalgrass 27930
Vernal-grass 27930
Vernicose Rhododendron 332055
Veronica 186967,186968,406966,407274
Veronicastrum 407449
Verruca-like Spindle-tree 157948
Verrucato-glandurose Syzygium 382680
Verrucolose Fimbristylis 166557
Verrucolose Fluttergrass 166557
Verrucolose Merremia 250858
Verrucose Amomum 19929
Verrucose Greenbrier 366243
Verrucose India Locust 200848
Verrucose Lasianthus 222301
Verrucose Rattlebox 112792
Verrucose Roughleaf 222301
Verrucose Spindle-tree 157952
Verrucosewing Melandrium 248444
Verrucosohispid Argyreia 32666
Verrucousbract Saltbush 44688
Verrucousstem Elatostema 142754
Verruculose Barberry 52308
Verruculose Himalayan Currant 334029
Verruculose Machilus 240714
Verruculose Rapanea 261662
Verruculose Swollennoded Cane 87626
Verruculose Tupistra 70628
Versailles Laurel 223116
Versatilihirte Dinggongvine 154264
Versatilihirte Erycibe 154264
Verschaffelt Barberry 52309
Verschaffelt Palm 407520,407521
Verschaffeltia 407520
Versicolor Birthwort 34367
Versicolor Caper 71919
Versicolor Craniotome 108669
Versicolor Dutchmanspipe 34367
Versicolor Epimedium 147072
Versicolor Japanese Maple 3347
Versicolor Rose 336585
Versicolor Sterculia 376207
Versicolorline Greenbamboo 47505
Versicolorous Crocus 111634
Versicolorous Pea-shrub 72374
Versicolour Heath 150225
Versicoloured Rhododendron 331838
Versicolous Caper 71919
Versicolous Dutchman's-pipe 34367
Vertical Leaf 358633
Verticellate-sepaled Lasianthus 222302
Verticilate Carpetweed 256727
Verticilate Heath 150230
Verticileaf Mouseear 83077
Verticillaster Acanthopanax 143698
Verticillate Acanthopanax 143698
Verticillate Anisochilus 25396
Verticillate Barberry 52310
Verticillate Bedstraw 170742
Verticillate Bentgrass 12410
Verticillate Bristlegrass 361930
Verticillate Coreopsis 104616
Verticillate Devil Rattan 121529
Verticillate Dysophylla 139585
Verticillate Foxtail 361930
Verticillate Hydrilla 200190
Verticillate Kalanchoe 215288
Verticillate Mosquitotrap 117735
Verticillate Paris 284406
Verticillate Parrotfeather 261379
Verticillate Pentapanax 289671
Verticillate Rhynchospermum 333475
Verticillate Snowflake 32527
Verticillate Spapweed 205431
Verticillate Swallowwort 117735
Verticillate Touch-me-not 205431
Verticillate Wildsenna 389527
Verticillate-flowered Indigo 206624
Verticillate-flowered Touch-me-not 205273
Verticillateleaf Basketvine 9421
Verticillateleaf Rockfoil 349431
Verticillateleaf Swertia 380401
Verucatespot Euonymus 157948
Verucose Clearweed 299086
Verucosetwig Euonymus 157952
Vervain 405801,405819,405835,405840,405872,405880,405887
Vervain Family 405907
Vervain Mallow 243746
Vervain Sage 345458
Vervain Thoroughwort 158360
Vervain Thorough-wort 158360
Vervose Ungeargrass 36704
Very Branched Rhododendron 331348
Very Foft-hairy Montanoa 258206
Very Fragrant Rhododendron 330916
Very Large Rhododendron 331207
Very Pleasing Rhododendron 330547
Verylittle-flower Camellia 69368
Verylong Goatgrass 8688
Very-slender Iris 208886
Verysmall Licuala Palm 228761
Verysmall Licualapalm 228761
Verysmall Twayblade 232282
Verythin Violet 410654
Veryweak Woodbetony 287150
Verywhite Milkvetch 42140
Vesica Sedge 76687
Vesicular-haired Rhododendron 332059
Vesper Flower 193417
Vesper Iris 208524
Vesper Swordflag 208524
Vespers Primrose 269475
Vest Greenbrier 366414
Vesta Fairy Lantern 67644
Vesture Willow 344240,344243
Vetch 222671,408251,408611
Vetch Grass 222784
Vetchingleaf Corydalis 106056
Vetchleaf Mutisia 260554
Vetchleaf Pagodatree 368994
Vetchleaf Sophora 368994
Vetchling 222671,222800
Vether Vaw 322724
Vether-vo 322724
Vethervow 322724
Veticillate Nowallcress 373643
Veticillate Staintoniella 373643
Vetiver 407579,407585
Vetivergrass 407585
Vevine 405872
Vew 385302
Vewe 385302
Vexillary Flower Bamboo 297365
Vial Primrose 315089
Vial Rhododendron 332062
Vial Woodbetony 287813
Vi-apple 372467
Viatnam 112920,127130,242477,330667,

347962,378476,387860,415235
Viatnam Waterchestnut 394430
Viburnum 407645,408203
Viburnum Leaf Raspberry 339447
Viburnum-leaf Caper 71920
Viburnum-leaf Pittosporum 301438
Viburnumleaf Raspberry 339447
Viburnumleaf Seatung 301438
Viburnum-leaved Caper 71920
Viburnum-leaved Pittosporum 301438
Viburnum-leaved Raspberry 339447
Viburnum-leaved Rhododendron 331839
Viburnum-like Euonymus 157960
Viburnum-like Spindle-tree 157960
Vicar's Tresses 68713
Vicary Eremostachys 148516
Vicary Privet 229661
Vicatia 408238,408244
Viciu Cracca 408423
Vicksburg Blackberry 339440
Victirious Hakea 184651
Victor Nut 106703
Victor's Laurel 223203
Victoria 408733
Victoria Bloodwood 106805
Victoria Dendrobium 125407
Victoria-maria Paphiopedilum 282913
Victorian Ash 155719
Victorian Box 301433
Victorian Christmas-bush 315601
Victorian Common Heath 146071
Victorian Mint Bush 315601
Victorin's Evening-primrose 269521
Victorin's Gentian 174057
Victor-nut 106703
Victory Indocalamus 206837
Victory Pyracantha 322474
Viejo 181452
Viejos 299265
Viellard Ochrosia 268362
Vierhapper Fescuegrass 164395
Viet Nam Bauhinia 49244
Viet Nam Camellia 68902,69741
Viet Nam China-laurel 28319
Viet Nam Cratoxylum 110256
Viet Nam Croton 112920
Viet Nam Elaeagnus 142209
Viet Nam Enkianthus 145718
Viet Nam Fissistigma 166690
Viet Nam Goniothalamus 179408
Viet Nam Gouania 179949
Viet Nam Keteleeria 216130
Viet Nam Leaf-flower 296522
Viet Nam Litse 234034
Viet Nam Melodinus 249658
Viet Nam Rhododendron 330086
Viet Nam White-flower Hiptage 196827
Viet Nam Xantolis 415235
Vietnam Acacia 1691
Vietnam Allocheilos 16008
Vietnam Arachnis 30527
Vietnam Aralia 30793
Vietnam Ardisia 31642
Vietnam Azalea 330086
Vietnam Beccarinda 49401
Vietnam Buckthorn 328875
Vietnam Bushbeech 178035
Vietnam Camellia 69741
Vietnam Cassia 91366
Vietnam Chastetree 411477
Vietnam Chaulmoogra-tree 199743
Vietnam Chaydaia 86320
Vietnam Chinagreen 11363
Vietnam Cladogynos 93953
Vietnam Coelogyne 98604
Vietnam Cordia 104164
Vietnam Cratoxylum 110256
Vietnam Croton 112920
Vietnam Cudrania 240813
Vietnam Cuttonguetree 413125
Vietnam Densevein 261331
Vietnam Eargrass 187542
Vietnam Elaeagnus 142209
Vietnam Elm 401552
Vietnam Euonymus 157539
Vietnam Fakesnaketailgrass 193989
Vietnam Falsestairweed 223682
Vietnam Fieldcitron 400521
Vietnam Fir 355
Vietnam Garcinia 171187
Vietnam Gouania 179949
Vietnam Gurjun 133577
Vietnam Hedyotis 187542
Vietnam Hickory 77935
Vietnam Holly 203642
Vietnam Hooktea 52402
Vietnam Hopea 198135
Vietnam Keteleeria 216130
Vietnam Kitefruit 196825
Vietnam Kudzuvine 321453
Vietnam Litse 234034
Vietnam Mint 309482
Vietnam Mosquitotrap 117730
Vietnam Olive 71027
Vietnam Pagodatree 369141
Vietnam Paris 284412
Vietnam Pavieasia 286574
Vietnam Pendent-bell 145718
Vietnam Pine 300007
Vietnam Qinggang 116059
Vietnam Ramie 56340
Vietnam Rehdertree 327415
Vietnam Rhamnoneuron 328587
Vietnam Rosewood 121841
Vietnam Sour Bamboo 4593
Vietnam Swallowwort 117730
Vietnam Sweetleaf 381143
Vietnam Taller Rhododendron 330667
Vietnam Underleaf Pearl 296522
Vietnam Waterelm 417565
Vietnam White Kitefruit 196827
Vietnam Wingnut 320384
Vietnamese Canary Tree 71027
Vietnamese Cleistanthus 94541
Vietnamese Cuttonguetree 413125
Vietnamese Glycosmis 177822
Vietnamese Ixonanthes 211023
Vietnamese Maple 3694
Vietnamese Turpinia 400521
Vietnamse Pleurostylia 304898
View 385302
Vigate Sagebrush 36232
Vigated Wormwood 36232
Vigongo 131567
Vigorous Fortune Spindle-tree 157517
Vigorous Mahobarnerry 242452
Vilip 410320
Village-oak Geranium 288459
Villaintree 175912,175920
Villaresia 409251
Villera 404316
Villform Tea 69743
Villi-transverse Sedge 76573
Villose Adonis 8389
Villose Aporosa 29778
Villose Ardisia 31630
Villose Awlfruit 179158
Villose Babylon Weeping Willow 343076
Villose Bamboo 297479
Villose Bikou Willow 343108
Villose Bouquet Larkspur 124264
Villose Broadstamen Burnet 345833
Villose Brome 61019
Villose Bureja Gooseberry 333931
Villose Buttercup 326494
Villose Camellia 69745
Villose Caudate-leaf Leptopus 226329
Villose Cinnamon 91447
Villose Common Pentapanax 289656
Villose Crapemyrtle 219981
Villose Crape-myrtle 219981
Villose Currant 334260
Villose Edelweiss 224960
Villose Eritrichium 153508,153555
Villose European Aspen 311545
Villose Eveningprimrose 269522
Villose Flaxaster 231842
Villose Franchet Monkshood 5213
Villose FritschSpiraea 371920
Villose Gansu Woodbetony 287305
Villose Gomphogyne 179158

Villose Gugertree 350948
Villose Harlequin Glorybower 96398
Villose Hawkweed 196081
Villose Holly 204005
Villose Hydrangeavine 351815
Villose Indianmulberry 258928
Villose Largeflower Larkspur 124264
Villose Linosyris 231842
Villose Machilus 240590,240716
Villose Manschurian Currant 334086
Villose Manynode Capillipedium 71613
Villose Milkbean 169661
Villose Mitracarpus 256104
Villose Motherwort 225022
Villose Neoporteria 264376
Villose Oldham Sping Raspberry 339138
Villose Pennisetum 289303
Villose Peony 280216
Villose Plateaucress 89268
Villose Rangoon Creepet 324679
Villose Red Fescuegrass 164266
Villose Reddish Beautyberry 66917
Villose Rockjasmine 23326
Villose Rorippa 336281
Villose Sand Willow 343199
Villose Shortspikilet Bamboo 58700
Villose Siberian Crabapple 243563
Villose Sichuan Holly 204315
Villose Skullcap 355840
Villose Smallpiiar Raspberry 338271
Villose Snakegourd 396286
Villose South China Milkwort 308083
Villose Spineflower 89071
Villose Spiny Pricklyash 417137
Villose Stairweed 142888
Villose Taibai Rhododendron 331589
Villose Tang Willow 344179
Villose Themeda 389339
Villose Tubergourd 390190
Villose Vetch 408693
Villose Wendlandia 413850
Villose Willow 343898
Villose Woodnettle 273928
Villose Xizang Willow 343372
Villose Yang Monkshood 5705
Villose Yellowcress 336281
Villose Yunnan Chickweed 375153
Villose Yunnan Fescuegrass 164403
Villoseleaf Slenderstyle Acanthopanax 2410
Villosulous Ascitesgrass 407503
Villosulous Veronicastrum 407503
Villosus Branchy Elatostema 142813
Villosus Elatostema 142614
Villous Alumroot 194439
Villous Amomum 19930
Villous Armgrass 58199
Villous Bowfruitvine 393545
Villous Casearia 78157
Villous Chervil 84773
Villous Christolea 89268
Villous Dumasia 138943
Villous Dunbaria 138978
Villous Fatoua 162872
Villous Fortune Spindle-tree 157519
Villous Gingerlily 187480
Villous Glorybower 96434
Villous Gongbu Monkshood 5330
Villous Groundcherry 297655
Villous Gugertree 350948
Villous Halimocnemis 184829
Villous Hartia 376487
Villous Henry Monkshood 5271
Villous Heterostemma 194169
Villous Hydrangea 200134
Villous Lilac 382312
Villous Moraea 258608
Villous Motherwort 225021
Villous Purplevine 414585
Villous Rhododendron 331991
Villous Saltbane 184829
Villous Singnalgrass 58199
Villous Sotvine 194169
Villous Stargrass 202987
Villous Sterculia 376210
Villous Summercypress 217356
Villous Toxocarpus 393545
Villous Veined Jasmine 211921
Villous Watersnake 162872
Villous Waxplant 198900
Villous Wisteria 414585
Villousleaf Begonia 50394
Villouspetal Hemsleia 191972
Villouspetal Snowgall 191972
Villus Alyxia 18532
Vilmorin Ailanthus 12588
Vilmorin Barberry 52311
Vilmorin Cryptomeria 113706
Vilmorin Currant 334262
Vilmorin Deutzia 127121
Vilmorin Monkshood 5667
Vilmorin Mountainash 369558
Vilmorin Mountain-ash 369558
Vilmorin Rhododendron 330165
Vilmorin Rockfoil 350028
Vilose Paphiopedilum 282914
Vimineous Hornbeam 77420
Vimineous Lettuce 219600
Vimineous Microstegium 254046
Vinca 79418
Vincetoxicum 117743
Vine 61497,411521,411979
Vine Cactus 167562
Vine Catchfly 363958
Vine Corydalis 105666
Vine Grape 411979
Vine Hill Manzanita 31112
Vine Leaf Passion Flower 285717
Vine Lilac 185764
Vine Maple 2893,250221
Vine Mesquite 281995
Vine Mesquite Grass 281995
Vine Panicgrass 281984
Vine Penstemon 289333
Vine Redcarweed 288769
Vine Rosewood 121703
Vine Spinach 48688,48689
Vine Yellowwood 94031
Vinebamboo 131277
Vinebamboo Panicgrass 282200
Vineclethra 95471
Vinegar Plant 332916
Vinegar Tree 236462,332608
Vinegar Weed 111146
Vinegar-tree 332916
Vinegentian 108217,110292
Vinegreens 48688
Vine-leaf Maple 2896
Vine-like Mayten 246952
Vinelike Milkwort 308415
Vine-like Milkwort 308415
Vinelike moonlight cactus 357749
Vinelime 283508
Vine-lime 283508
Vinenettle 315461,315476
Vine-of-sodom 93297
Vineorange 238530,238536
Vinepeony 132874
Vinepepper 313172
Vinescrewpine 168238
Vinespinach 48688
Vine-spinach 48688,48689
Vinetwig Bamboo 47321
Vine-yaed Bentgrass 12413
Vineyam 26264,26265
Vineyard Stinkweed 133324
Vinhalico 302612
Vining Velvet 339762
Viny-coloured Hydrangea 200145
Viola 410749
Violaceous Iris 208930
Violet 409339,409608,409656,409715,
410027,410320,410434,410442,410749
Violet Ajania 13024
Violet Amur Arisaema 33248
Violet Awlbillorchis 333732
Violet Barley 198375
Violet Bird's Nest Orchid 230378
Violet Bird's-nest Orchid 230378
Violet Bittercress 73034
Violet Bluegrass 305359
Violet Boxelder Maple 3233

Violet Bush　207340
Violet Bush-clover　227005
Violet Butterwort　299739,299759
Violet Caesalpinia　65079
Violet Cattleya　79576
Violet Churur　207342
Violet Collinsia　99841
Violet Crabgrass　130825
Violet Cress　212474,212476
Violet Cuphea　114601
Violet Deinostemma　123730
Violet Dendrobium　125001
Violet Dock　340314
Violet Dog-tooth　154899
Violet Draba　137293
Violet Dwarf Tulip　400217
Violet Easter Lily Cactus　140875
Violet Figmarigold　252187
Violet Fnmily　410792
Violet Fruit Finger-grass　130825
Violet Gentian　173949
Violet Grass　354871
Violet Helleborine　147212
Violet Horned Poppy　335638
Violet Iris　208922
Violet Ivy　97757
Violet Korea Monkshood　5143
Violet Kunzea　218393
Violet Large-flowered Epimedium　147006
Violet Larkspur　124484
Violet Lespedeza　227005
Violet Ligusticum　229329
Violet Limodore　230378
Violet Nightshade　367498
Violet Orchid　207448
Violet Orchids　207446
Violet Orychophragmus　275876
Violet Passionflower　285714
Violet Pearweed　239893
Violet Petunia　292754
Violet Poplar　311564
Violet Prairie-clover　121902
Violet Pulsatilla　321720
Violet Ribseedcelery　304760
Violet Rosewood　121855
Violet Sage　345246
Violet Short-stemmed Ardisia　31638
Violet Silverleaf　227926
Violet Skullcap　355842
Violet Slipper Gloxinia　364752,364754
Violet Sobralia　366793
Violet Tea　410320
Violet Texas Ranger　227926
Violet Tooth Cress　125836
Violet Trumpet Vine　97471
Violet Westringia　413915
Violet Whitlowgrass　137293
Violet Wildryegrass　144280
Violet Willow　343280
Violet Wood　1292
Violet Wood Sorrel　278147
Violet Woodbetony　287819
Violet Wood-sorrel　278147
Violet Woodsorrel Oxalis　277776
Violetbead Barberry　51750
Violetbloom Rhododendron　331343
Violet-bloom Rhododendron　331343
Violetbract Corydalis　106580
Violetchaeta Goldenray　229059
Violet-colored Rhododendron　331348
Violetcress　264164,264167
Violetflower Japanese Pagodatree　369041
Violetflower Petunia　292749
Violetflower Phalaenopsis　293643
Violetflower Primrose　314988
Violetflower Spapweed　205434
Violetflower Touch-me-not　205434
Violet-flowered Petunia　292754
Violetish Poplar　311564
Violetleaf Bittercress　73036
Violetleaf Childvine　400952
Violetleaf Mockorange　294556
Violetleaf Neomartinella　264167
Violetleaf Paris　284351
Violet-like Platystemma　302960
Violetlike Primrose　315096
Violetpetal Pimpinella　299503
Violet-purple Rhododendron　331343
Violet-scented Orchid　269055
Violetspike Barnyardgrass　140503
Violetspike Primrose　315093
Violet-stalked Poplar　311564
Violettongue Aster　40635
Violetvein Loosestrife　239829
Violet-vein Viper's Bugloss　141217
Violin Strings　302034
Violineleaf Mullein　405685
Violinleaf Mazus　246971
Violinleaf Nanmu　295397
Violinleaf Rockjasmine　23280
Violin-leaved Wallichpalm　413086
Viper Gourd　396151
Viper Grass　131224
Viper's Bowstring Hemp　346158
Viper's Bugloss　141087,141347
Viper's Grass　141347,354873
Viper's Herb　141347
Viper's Victuals　37015
Viper's-bugloss　141347
Viper's-grass　354797,354870,354873
Vipertail　402215
Vippe　300223
Viraru　340507
Virescent Anzacwood　310755
Virescent Anzac-wood　310755
Virescent Anzae Wood　310755
Virescent Barberry　52313
Virescent Bauhinia　49255
Virescent Fairybells　134504
Virescent Fig　165841
Virescent Pseudosasa　318353
Virescent Rhododendron　332077
Virgate Epidendrum　146502
Virgate Everlasting　21724
Virgate Leptodermis　226140
Virgate Lespedeza　227009
Virgate Meadowrue　388716
Virgate Melic　249110
Virgate Nepeta　265103
Virgate Pearleverlasting　21724
Virgate Serpentroot　354918
Virgate Wildclove　226140
Virgate Wirelettuce　375989
Virgdnia Mountain-mint　321964
Virghl's-bower　95431
Virgilia　94022,410869
Virgin Mary　158062,321643
Virgin Mary's Candle　405788
Virgin Mary's Candles　405788
Virgin Mary's Cowslip　321643
Virgin Mary's Honeysuckle　321643
Virgin Mary's Milkdrops　321643
Virgin Mary's Nipple　158433,159027
Virgin Mary's Nipples　158433,159027
Virgin Mary's Tears　321643
Virgin Mary's Thistle　73318,364360
Virgin Oil　270099
Virgin Wild Buckwheat　152519
Virgin's Bower　94676,95085,95426,95431
Virgin's Bower Clematis　95426
Virgin's Fingers　130383
Virgin's Hair　60382
Virgin's Milk　364360,368771
Virgin's Palm　131429
Virgin's Pinch　309570
Virgin's Robe　57101
Virgin's Tears　102867
Virgin's Thistle　364360
Virgin's-bower　94676,94807,94835,94901,95085,95426
Virginal　294408
Virginal Mockorange　294565
Virginalis Water Lily　267637
Virgineous Hydrangea　199851
Virginia Ash　168079
Virginia Bartonia　48592
Virginia Birch　53630
Virginia Bird-cherry　316918
Virginia Bluebells　250914
Virginia Brome　60867
Virginia Bugleweed　239248

Virginia Bunch-flower 248677
Virginia Buttonweed 131361
Virginia Copperleaf 1790,2011
Virginia Copper-leaf 2008
Virginia Cotton-grass 152796
Virginia Cottonsedge 152796
Virginia Cowslip 250914
Virginia Creeper 285097,285136,285157
Virginia Dayflower 101193
Virginia Dwarfdandelion 218213
Virginia Dwarf-dandelion 218213
Virginia False Dragonhead 297984
Virginia False-dragonhead 297984
Virginia Germander 388033
Virginia Ground Cherry 297740
Virginia Ground-cherry 297740
Virginia Hackelia 184320
Virginia Horehound 239248
Virginia Hound's-tongue 118031
Virginia Iris 208932,208933
Virginia Knotweed 309940,393363
Virginia Lespedeza 227012
Virginia Lime-grass 144528
Virginia Lionsheart 297984
Virginia Marsh St. John's-wort 394814
Virginia Meadow Beauty 329427
Virginia Meadow-beauty 329427
Virginia Mountain Mint 321964
Virginia Oak 324544
Virginia Pepper Grass 225475
Virginia Peppergrass 225475
Virginia Pepper-grass 225475
Virginia Pepperweed 225475
Virginia Pepper-weed 225475
Virginia Pine 300296
Virginia Plantain 302209
Virginia Rose 238463,337007
Virginia Roundleaf Birch 53630
Virginia Savin 213984
Virginia Saxifrage 350032
Virginia Snake Root 34325
Virginia Sneezeweed 188442
Virginia Spiderwort 393987,394088
Virginia Spirea 372134
Virginia Spring Beauty 94373
Virginia Spring-beauty 94373
Virginia St. John's-wort 202207
Virginia Stewartia 376452
Virginia Stickseed 184320
Virginia Stock 243156,243209
Virginia Strawberry 167663
Virginia Sumach 332916
Virginia Sweet Spire 210421
Virginia Sweetspire 210421
Virginia Tephrosia 386365
Virginia Thistle 92476
Virginia Three-seeded-mercury 2008
Virginia Virgin's-bower 95426
Virginia Water Horehound 239248
Virginia Water Lily 267638
Virginia Water-horehound 239248
Virginia Water-leaf 200488
Virginia Wild Ginger 37754
Virginia Wild Rye 144528
Virginia Wild-rye 144528
Virginia Willow 210421
Virginia Witch Hazel 185141
Virginiahorehound 239248
Virginia-lily 417608
Virginian Agave 244377
Virginian Angelica Tree 30760
Virginian Bird Cherry 316918
Virginian Boxwood 105023
Virginian Cedar 213984
Virginian Chestnut 78807
Virginian Cowslip 135165,321643
Virginian Daffodil 417608
Virginian Datelum 132466
Virginian Dogwood 105023
Virginian Hemp 18683
Virginian Juniper 213984
Virginian Pencil Cedar 213984
Virginian Poke 298094
Virginian Pokeweed 298094
Virginian Raspberry 338909
Virginian Rose 337007
Virginian Silk 38135
Virginian Skullcap 355563
Virginian Snakeroot 34325
Virginian Stock 243209
Virginian Strawberry 167663
Virginian Sumac 332916
Virginian Waterleaf 200488
Virginian Willow 210421
Virginian Winterberry 204376
Virginian Witch Hazel 185141
Virginiawillow 210421
Viriculis Guangxi Triratna Tree 397367
Viridistriatus Tsiang Dendrocalamus 125516
Viridleaved Falsenettle 56241
Virmorin Dovetree 123247
Viscaria 363331
Viscid Acacia 1421
Viscid Aster 240377
Viscid Bartsia 284098
Viscid Camphor-daisy 327153
Viscid Candyleaf 376421
Viscid Euthamia 161086
Viscid Germander 388303
Viscid Grass-leaved Goldenrod 161086
Viscid Melandrium 248453
Viscid Nightshade 367583
Viscid Pink 127901
Viscid Rhododendron 332079
Viscid-buds Rhododendron 332080
Viscidhair Screwtree 190117
Viscidhair Skullcap 355846
Viscid-haired Screw-tree 190117
Viscidhairy Falsepimpernel 231594
Viscidhairy Larkspur 124687
Viscidhairy Motherweed 231594
Viscidity Azalea 332079
Viscidity Knotweed 309946
Visco Acacia 1697
Viscoid Leaves Rhododendron 332078
Viscoid-leaved Rhododendron 332078
Viscose Haworthia 186840
Viscose-budded Rhododendron 332080
Viscosus Rhododendron 332079
Viscous Campion 363583
Viscous Catchfly 363583
Viscous Meadowrue 388718
Viscous Melandrium 248453
Viscous Rhynchosia 333452
Visher's Wild Buckwheat 152658
Vismia 411140
Visnaga 19672,140111
Vitelline Barberry 52321
Vites 411362
Vitex 411237
Vitifolious Larkspur 124584
Vitis-cell Clematis 95431
Vittate Calathea 66213
Vittate Eria 148796
Vittate Hairorchis 148796
Vittate Spapweed 205439
Vittate Touch-me-not 205439
Vivax Bamboo 297499
Viviparous Bistort 309954
Viviparous Bistort Serpentgrass 309954
Viviparous Bluegrass 305989
Viviparous Corydalis 106592
Viviparous Eight Trasure 200820
Viviparous Fescue 163806,164396
Viviparous Rocktaro 327676
Viviparous Sheep's-fescue 164396
Viviparous Spike-rush 143412
Viviparous Stonecrop 200820
Vivipary Corydalis 106592
Vivvervaw 322724
Vivvyvew 322724
Vixivixio 273027
Vladimiria 135721
Vlassoviana Thistle 92479
Vlix 232001
Voa Vanga 404833
Voavanga 404833
Vochin Knapweed 81243
Vodka 367696
Vogel Hoarypea 386368
Vogel Tephrosia 386368

Vogel's Fig 165263
Volador 397931
Volantines Preciosos 95796
Volcan Mahonia 242659
Volcanic Daisy 150614
Volcanica Loasa 234252
Volga Adonis 8390
Volga Betony 373480
Volga Catchfly 364216
Volga Cousinia 108239
Volga Euphorbia 160067
Volga Fescue 164382
Volga Garliccress 365649
Volga Hawkweed 196091
Volga Hawthorn 110110
Volga Milkvetch 43251
Volga Salsify 394358
Volga Sisymbrium 365649
Volga Sweetclover 249263
Volga Vinegentian 108239
Volkamer Lemon 93866
Volkens Adenia 7323
Volkens Garcinia 171217
Voloschilov Cowparsnip 192395
Voluble Cissampelopsis 92522
Voluble Microglossa 253479
Vomit Nut 212127
Voodoo Lily 348036
Vorsicolorous Mosquitotrap 117734
Vorsicolorous Swallowwort 117734
Vortriede's Spinyherb 382451
Voss Calceolaria 66271
Voss' Laburnum 218756
Voveolate Maple 3745
Vriesea 412337
Vrydagzynea 412385
Vulcan Rabbitbrush 150375
Vulvet Ash 168136
Vuz 401388
Vuzz 401388
Vuzzen 401388

W

W. Africa Siris 13703
W. Africa Sterculia 376150
W. African Padauk 320330
W. Asia Cedar 213767
W. Asia Lemongrass 117192
W. Asian Maple 2890
W. B. Hemsley's Rhododendron 330849
W. China Fritillary 168563
W. China Honeysuckle 236113
W. China Monkshood 5352
W. China Orchis 310914
W. China Raspberry 339304
W. China Rose 336777
W. China Sagebrush 36017
W. China Spiraea 371990
W. China Willow 343792
W. Cuffe's Rhododendron 330483
W. Guangxi Indocalamus 206797
W. Hubei Arisaema 33510
W. Hubei Atractylodes 44194
W. Hubei Azalea 331533
W. Hubei Buckthorn 328880
W. Hubei Comastoma 100272
W. Hubei Mouseear 83108
W. Hubei Pennywort 200392
W. Hubei Rockfoil 350020
W. Hubei Sedge 75296
W. Hubei Southstar 33510
W. Hubei Spapweed 204939
W. Hubei Spicegrass 239798
W. Hubei Spiraea 372127
W. Hubei Swertia 380295
W. Hubei Throathair 100272
W. Hubei Violet 409876
W. Hubei Yinshancress 416330
W. Hubei Yinshania 416330
W. Hubei Yushanbamboo 416765
W. India Mahogany 380528
W. J. Bean's Rhododendron 330210
W. Sichuan Bentgrass 12137
W. Sichuan Bluebell 115348
W. Sichuan Bluebellflower 115348
W. Sichuan Corydalis 106599
W. Sichuan Cranesbill 174784
W. Sichuan Dandelion 384497
W. Sichuan Dodder 115052
W. Sichuan Dwarfnettle 262229
W. Sichuan Gentian 173392
W. Sichuan Hoodorchis 264764
W. Sichuan Indigo 205902
W. Sichuan Lilac 382139
W. Sichuan Linden 391738
W. Sichuan Loosestrife 239802
W. Sichuan Meconopsis 247128
W. Sichuan Nannoglottis 262229
W. Sichuan Nettletree 80777
W. Sichuan Oak 323936
W. Sichuan Onosma 271803
W. Sichuan Pagodatree 368995
W. Sichuan Peashrub 72225
W. Sichuan Peristylus 291252
W. Sichuan Perotis 291252
W. Sichuan Pyrethrum 322748
W. Sichuan Rhodiola 329833
W. Sichuan Rockliving Woodbetony 287651
W. Sichuan Sagebrush 36016
W. Sichuan Speedwell 407398
W. Sichuan Sterculia 376129
W. Sichuan Stringbush 122437
W. Sichuan Sweetvetch 187985
W. Sichuan Swertia 380285
W. Sichuan Tailanther 381954
W. Sichuan Touch-me-not 204779
W. Sichuan Willowweed 146695
W. Sichuan Windhairdaisy 348280
W. Taibai Crazyweed 279166
W. Wild Grape 411594
W. Xizang Honeysuckle 236088
W. Xizang Kobresia 217154
W. Xizang Sandwort 32241
W. Xizang Smallreed 127325
W. Xizang Sweetvetch 187868
W. Xizang Touch-me-not 205379
W. Yunnan Arundinella 37409
W. Yunnan Azalea 330085,330650
W. Yunnan Bastardtoadflax 389839
W. Yunnan China Houndstongue 117912
W. Yunnan Coelogyne 98622
W. Yunnan Cyclobalanopsis 116136
W. Yunnan Eightangle 204556
W. Yunnan Euonymus 157771
W. Yunnan Ghost lampstand 335143
W. Yunnan Hawthorn 109881
W. Yunnan Hogfennel 292829
W. Yunnan Holcoglossum 197266
W. Yunnan Honeysuckle 235687
W. Yunnan Hundredstamen 389839
W. Yunnan Ironweed 406338
W. Yunnan Lilyturf 272160
W. Yunnan Lysionotus 239941
W. Yunnan Meconopsis 247138
W. Yunnan Monkshood 5094
W. Yunnan Nomocharis 266555
W. Yunnan Platanthera 302511
W. Yunnan Qinggang 116136
W. Yunnan Rodgersflower 335143
W. Yunnan Sedge 75427
W. Yunnan Snakefruit 342578
W. Yunnan Spapweed 204963
W. Yunnan Touch-me-not 204963
W. Yunnan Waterparsnip 365844
Wa Arrowbamboo 162744
Wabran-leaf 302068
Wabret-leaf 302068
Wabu Fritillary 168385

Wachendorfia 412519
Wachin Delavy Barberry 52029
Wada Lily 230024
Wadalee-gum 1120
Wadaleetree 1120
Wadd 209229
Waddell Onosma 271844
Waddell Primrose 315107
Waddy Wood 1485
Wade 209229
Waeteplace Peppergrass 225447
Waeteplace Pepperweed 225447
Wafer Ash 320065
Waftwort 393240,407520
Wag Wafers 60382
Wag Wantons 60382
Waggering Grass 60382
Wagner Barberry 51700
Wagner Mahonia 242660,242661
Wagner Windmillpalm 393816
Wagtails 60382
Wag-wafers 60382
Wag-wams 60382
Wag-wanting 60382
Wag-wantons 60382
Wagwants 60382
Wahlenberg Wood Rush 238717
Wahlenberg's Wood Rush 238717
Wahlenberg's Wood-rush 238717
Wahoo 157268,157285,157315,157327,
401430
Wahoo Elm 401430
Wahoo-tree 227434
Waianae White Hibiscus 194722
Waifa 369037
Wailong Rattan Palm 65813
Wailong Rattanpalm 65813
Wainscot Oak 323745
Waistbone Vine 203321,203325
Wait-a-bit Thorn 1091,1386
Wait-a-minute 1264
Wait-a-minute Bush 254980,254981
Waiteweed 384714
Waiuatua 328446
Wake Pintel 37015
Wake Robin 37015,248411,397540,397556,
397570,397609
Wake Robine 37015
Wakerobin 397540,397556,397622
Wake-robin 37015,248411,273531,397540,
397556,397570,397609
Waldersii Maple 3445
Waldheim Spapweed 205442
Waldheimia 413007
Waldmeister tea 170524
Waldo Wild Buckwheat 152384
Waldstein Onion 15885
Waldsteinia 413030,413031
Walewort 345594
Walf Azalea 331495
Walf Rhododendron 331495
Walker Rattan Palm 65814
Walker Spapweed 205443
Walker Touch-me-not 205443
Walker's Ainsliaea 12743
Walking Iris 264158,264163
Walking Kalanchoe 215269,215270
Walking Sedge 143331
Walking-stick Bamboo 162742
Walkingstick Cactus 116672
Walking-stick Camwood 47804
Walking-stick Ebony 132307
Walkingstick Palm 231799,231803
Walking-stick Plant 233772
Wall Barley 198327,235334,235373
Wall Bedstraw 170536
Wall Bugloss 21920
Wall Bur Cucumber 362461
Wall Bur-cucumber 362461
Wall Cotoneaster 107486
Wall Cress 30160,30220,44870,133286
Wall Draba 137129
Wall Flag 208875
Wall Flower 86409,86427
Wall Germander 388042
Wall Gilliflower 86427
Wall Ginger 356468
Wall Grass 356468
Wall Hawkweed 195803
Wall Heshouwu 162528
Wall Ink 407029
Wall Iris 208875
Wall July Flower 86427
Wall July-flower 86427
Wall Lettuce 260594,260596
Wall Lilac 81768
Wall Moss 356468
Wall Mustard 133286
Wall Paper 356468
Wall Pellitory 284152,284167
Wall Pennyroyal 401711
Wall Pennywort 401711
Wall Pepper 356468
Wall Rock Cress 30220
Wall Rockcress 30220
Wall Rock-cress 30220
Wall Rocket 133226,133286,133313
Wall Sage 284152
Wall Speedwell 407014
Wall Spray 107486
Wall Stonecrop 356468
Wall Toadflax 116736
Wall Whitlow Grass 137129
Wall Whitlow-grass 137129
Wallaba 146107
Wallaba Tree 146107
Wallaba-tree 146107
Wallaby Grass 122180,341250
Wallaby Weed 270216
Wallaby-grass 341243,341251
Wallace's Woolly Daisy 152836
Wallangara Wattle 1034
Wallangarra White Gum 155737
Wallapata 183294
Wallaston Pratia 313575
Wallenia 413059
Waller Spapweed 205444
Waller Touch-me-not 205444
Wallfern Windhairdaisy 348661
Wallfernleaf Milklettuce 283692
Wallflower 86409,86427,154363,154390,
154445,154497,188908
Wallflower Cabbage 99090
Wallflower Mustard 154424
Wallgrass 284131
Wallich Achyrospermum 4489
Wallich Apostasia 29793
Wallich Argyreia 32668
Wallich Begonia 50399
Wallich Brachytome 59006
Wallich Bulbophyllum 63182
Wallich Bulrush 352293
Wallich Burmannia 63998
Wallich Catch 1125
Wallich Cirrhopetalum 63182
Wallich Cleyera 96604
Wallich Combretum 100850
Wallich Cowparsnip 192396
Wallich Crestpetal-tree 236420
Wallich Curlylip-orchis 63182
Wallich Debregeasia 123339
Wallich Dragonhead 137691
Wallich Eriophyton 152839
Wallich Euphorbia 160074
Wallich Fescuegrass 164398
Wallich Fivestamen Sphenodesma 371422
Wallich Flemingia 166905
Wallich Glorybower 96342,96439
Wallich Greenorchid 137691
Wallich Gugertree 350949
Wallich Heterostemma 194172
Wallich Houndstongue 118034
Wallich Italian Jasmine 211868
Wallich Ixora 211203
Wallich Juniper 213807
Wallich Knotweed 309616
Wallich Kudzuvine 321473
Wallich Lacquertree 393495
Wallich Lacquer-tree 393495
Wallich Lasianthus 222305
Wallich Laurel Cherry 223133,316922

Wallich Leek 15886
Wallich Ligusticum 229410
Wallich Madder 338043
Wallich Mountainash 369391
Wallich Mountain-ash 369391
Wallich Mouseear Cress 30155
Wallich Myriactis 261082
Wallich Onion 15886
Wallich Ophiorrhiza 272304
Wallich Palm 413085,413093
Wallich Paris 284381
Wallich Pepper 300548
Wallich Podocarpus 306526
Wallich Poisonnut 378948
Wallich Randia 385055
Wallich Rhodiola 329973
Wallich Rhododendron 332091
Wallich Roughleaf 222305
Wallich Savin 213807
Wallich Scleropyrum 354502
Wallich Snakegourd 396287
Wallich Sotvine 194172
Wallich Stiffdrupe 354502
Wallich Strophanthus 378468
Wallich Strychnos 378948
Wallich Swallowwort 117747
Wallich Tomentose Reevesia 327385
Wallich Violet 410757
Wallich Waternettle 123339
Wallich Willow 344261
Wallich Willowweed 146937
Wallich Woodbetony 287823
Wallich Yaccatree 306526
Wallich Yam 131909
Wallich Zehneria 417517
Wallich's Bindweed 103387
Wallich's Clematis 95452
Wallich's Glorybower 96439
Wallich's Juniper 213807
Wallich's Lily 230068
Wallich's Oak 324550
Wallich's Wedelia 413565
Wallichia 413085
Wallichpalm 413085,413090
Wall-lettuce 260596
Wallowa 1116
Wallpepper 356468
Wall-pepper 356468
Wallrocket 133226
Wallspray 107549
Wall-spray 107486
Wallum 47628
Wallum Banksia 47628
Wallum Bottlebrush 67281
Wallwort 86427,284152,345594,356468, 401711
Walnut 212582,212631,212636
Walnut Bean 145320
Walnut Family 212581
Walnut-leaved Mahonia 242483
Walnut-leaved White Ash 167908
Walsh Nut 212636
Walsh Primrose 315108
Walsura 413124
Walter Barberry 52354
Walter Dogwood 380514
Walter Pine 299954
Walter's Aster 381010
Walter's Millet 140509
Walter's Pine 299954,300237
Waltheria 413142
Walton Crisped Corydalis 105763
Walton Cyclorhiza 116373
Walton Primrose 315109
Walton Wormwood 36548
Waltzing Matilda 155655
Wamin Bamboo 47516
Wampee 94169,94177,94207,310993
Wampi 94207
Wand Blackroot 320400
Wand Fleabane 150837
Wand Flower 130187,130229,172201, 369975
Wand Goldenrod 368438
Wand Lespedeza 227009
Wand Lessingia 227138
Wand Lythrum 240100
Wand Mullein 405796
Wand Riverhemp 361454
Wand Sage 345452,345473
Wand Wild Buckwheat 152436
Wand Wirelettuce 375990
Wandering Fleabane 150851
Wandering Heath 150201
Wandering Jack 116736
Wandering Jenny 239755
Wandering Jew 116736,176824,349936, 361985,393987,394012,394021,394026, 394057,394088,394093
Wandering Rhododendron 331495
Wandering Sailor 116736,239755,349936
Wan-dering Sailor 116736
Wandering Tinker 239755
Wandering Vetch 408537
Wandering Willy 102933,174877
Wanderingjew 393987,394093
Wandering-jew 393987,394093
Wanderingjew Zebrina 394093
Wanderingvine Euonymus 157943
Wanderingvine-like Euonymus 157941
Wandflower 130187,130229,169819, 369975,370009
Wand-flower 130229,369975
Wanding Ginger 418027
Wand-like Bushclover 226846
Wandlike Goldenrod 368438
Wand-like Goldenrod 368438
Wandoo 155718,155774
Wang Adinandra 8277
Wang Ambiguous Cacalia 283790
Wang Barberry 52355
Wang Begonia 50400
Wang Black Holly 203584
Wang Bowfruitvine 393548
Wang Buckeye 9736
Wang Buttercup 326497
Wang Cystopetal 320253
Wang Daphne 122460
Wang Filbert 106784
Wang Gambirplant 401786
Wang Hackberry 80779
Wang Hedyotis 187700
Wang Holly 204395
Wang Jasmine 212018
Wang Meadowrue 388719
Wang Mitrephora 256236
Wang Morningglory 208295
Wang Mosquitotrap 117748
Wang Mucuna 259578
Wang Pine 300298
Wang Primrose 315111
Wang Raspberry 339469
Wang Slugwood 50627
Wang Stonecrop 357294
Wang Swallowwort 117748
Wang Tarenna 385047
Wang Topfruit 249560
Wang Toxocarpus 393548
Wang Willow 344265
Wang's Chirita 87990
Wang's Japanese Mazus 247024
Wang's Oak 324551
Wangchi Maple 3746
Wangee 297367
Wangmo Amoora 19962
Wangrangkura 143665
Wangyedian Monkshood 5679
Wanhuashan Barberry 52356
Wannan Stonecrop 357295
Wannianchun 228614
Wanning Syzygium 382572
Wanning Tanoak 233201
Wanong Azalea 332092
Wanong Rhododendron 332092
Wantian Camellia 69513
Wantian Kiwifruit 6725
Wantien Camellia 69513
Wantien-like Camellia 69432
Waoriki 326313
Wapato 342308,342325,342363
Wapatoo 342363,342400

Wara 68088
Waratah 385734,385738
Waratah Tree 16252
Warba-leaves 302068
Warburg Holly 204396
Warburg Pothos 313210
Warburg Willow 344268
Ward Anise-tree 204605
Ward Arisaema 33564
Ward Azalea 332093
Ward Barberry 52357
Ward Basketvine 9487
Ward Bluegrass 306151
Ward Butterfly-bush 62203
Ward Calophanoides 67832
Ward Camellia 69757
Ward Cassiope 78649
Ward Corydalis 106596
Ward Cotoneaster 107726
Ward Deutzia 127123
Ward Diapensia 127925
Ward Eightangle 204605
Ward Falsetamarisk 261296
Ward False-tamarisk 261296
Ward Holly 204398
Ward Illicium 204605
Ward Lady's Slipper 120470
Ward Ladyslipper 120470
Ward Lagotis 220202
Ward Larkspur 124692
Ward Lasianthus 222313
Ward Lily 230069
Ward Lysionotus 239979
Ward Maple 3748
Ward Micromeria 253734
Ward Monkshood 5680
Ward Onosma 271847
Ward Paphiopedilum 282922
Ward Pocktorchid 282922
Ward Rabdosia 209862
Ward Raspberry 339470
Ward Rockjasmine 23342
Ward Roscoea 337114
Ward Roughleaf 222313
Ward Salweenia 345496
Ward Schefflera 350803
Ward Southstar 33564
Ward Stanhopea 373704
Ward Summerlilic 62203
Ward Willow 343172
Ward Windhairdaisy 348932
Ward Wintergreen 172173
Ward Woodbetony 287826
Ward Xizang Jerusalemsage 295209
Ward's Butterflybush 62203
Ward's Cold Spindle-tree 157526
Ward's Dolomiaea 135741
Ward's Goldenweed 271917
Ward's Rhododendron 332093
Ward's Willow 343172
Wardseed 72038
Ware Desertcandle 148561
Ware-moth 35090
Warence 338038
Warker Rattlebox 112808
Warley Iris 208937
Warlock 364557
Warming Sinningia 364766
Warminster Broom 120991
Warmot 35090
Warner Mountains Sulphur Flower 152584
Warner Springs Lessingia 227115
Warnock's Javelina Bush 101744
Warnock's Rock Daisy 291340
Warnut 212636
Warocque Vandopsis 404755
Warpbill Woodbetony 287426
Warratau 385738
Warrigal 387221
Warrigal Cabbage 387221
Warscewicz Canna 71199
Warszewicz Dicyrta 129724
Warszewicz Pansy Orchid 254939
Wart Cress 105338,105352
Wart Spurge 159027
Wartberry Falrybell 134502
Wartcress 105338,105340,105352
Wart-cress 105340,105352
Wart-curer 86755
Warted Bamboo 297478
Warted Barberry 52308
Warted Bunias 63452
Wart-flower 86755,282685
Wart-grass 159027,159557
Wartleaf Elatostema 142878
Wartleaf Stairweed 142878
Wart-plant 86755
Wartpull Coldwaterflower 299086
Wartremoving Herb 260102
Wart-removing Herb 260102
Warts Rhododendron 332056
Wart-spurge 159027
Wartstalk Azalea 332056
Wartweed 86755,159027,159557
Wart-weed 85770,159027,159286
Wartwort 86755,159027,159557,178481,
407516
Warty Barberry 52308
Warty Bedstraw 170738
Warty Birch 53563
Warty Cabbage 63452
Warty Caltrop 215384
Warty Carrion Flower 374021
Warty Carrion-flower 374021
Warty Epidendrum 146500
Warty Gasteria 171767
Warty Mock Orange 294564
Warty Mock-orange 294564
Warty Panicum 282351
Warty Rattlebox 112792
Warty Rhododendron 332056
Warty Spurge 159856
Warty Trachydium 393862
Warty Twig Machilus Machilus 240714
Warty Yate 155646
Wartybark Euonymus 157952
Warty-barked Euonymus 157952
Wartybark-like Euonymus 157948
Warty-cabbage 63443,63452
Wartyfruit Amomum 19899
Warty-girse 159027
Wartyleaf Greenstar 307534
Warty-leaved Greenstar 307534
Wartylip Oncidium 270836
Warwickshire Weed 401593
Warytooth Loosestrife 239603
Wasabi 161154
Wasatch Aster 193132
Wasatch Fleabane 150471
Wasatch Plateau Wild Buckwheat 151873
Wase-satsuma Mandarin 93736
Wase-unshiu 93736
Washan Evergreen Chinkapin 78884
Washan Evergreenchinkapin 78884
Washan Gentian 174070
Washan Leptodermis 226112
Washan Sage 345082
Washan Snowbell 379427
Washan Storax 379427
Washan Wildclove 226112
Washiba Wood 382724
Washington Coneflower 339515
Washington Fan Palm 413307
Washington Fanwort 64496
Washington Grass 64495,64496
Washington Hawthorn 109931
Washington Lily 230070
Washington Lupine 238481
Washington Navel Orange 93797
Washington Palm 413289
Washington Plant 64496
Washington Thorn 109931
Washington's Hawthorn 109931
Washington's-thorn Washington-thorn 109931
Washington-grass 64496
Washingtonia 360561,413289
Washingtonia Palm 413298
Washoe Pine 300301
Washrag Gourd 238261
Washrag Sponge 238261
Washstand Tree 18052,414458

Wason Rhododendron 332095
Wasp Orchid 272399
Wasp-orchid 272399
Waster Ledges 308893
Waszewicz Barberry 52358
Watch Chain 109180
Watch Chain Crassula 109180
Watch Guards 218761
Watch Wheels 325612
Watchchain 218761
Watches 347156,374916
Watches-and-clocks 384714
Wate Violet 198680
Water Agrimony 53816,54158,158062
Water Aloe 377485
Water Anemone 325580,325923
Water Apple 382474
Water Archer 342400
Water Arum 66593,66600
Water Ash 167923,167934,168032,168057,
320071
Water Avens 175444
Water Awlwort 379650
Water Babies 68189
Water Baby 68189
Water Balsam 204982,205326
Water Bamboo 297296
Water Banyan 94576
Water Bedstraw 170529
Water Beech 77263,302588
Water Beggar-ticks 247882
Water Bells 267648
Water Bent 12297
Water Bentgrass 310134
Water Berry 156187,382519,382567
Water Betony 355061,373262
Water Birch 53437,53536,53540
Water Bird's Eye 407029
Water Bird's Eyes 407029
Water Bitny 355061
Water Bleb 68189
Water Blinks 258248
Water Blob 68189,267294,267648,326340
Water Blubbers 68189
Water Boats 68189,397019
Water Bubbles 68189
Water Bug Trap 14273
Water Bugle 239248
Water Bulrush 352271
Water Burreed 370062
Water Bush 260719
Water Buttercup 68189,325864
Water Buttons 107766
Water Caltrop 68189,394500
Water Caltrops 394436,394500
Water Can 267294,267648
Water Canna 388385
Water Carpet 90335
Water Case 29341
Water Cattail 401094
Water Celery 269326,326340
Water Chestnut 394412,394426,394436,
394500
Water Chestnut Family 394554
Water Chickweed 258237,258248,260920,
260922,374749
Water Chinkapin 263270
Water Chinquapin 263270
Water Chive 352278
Water Club-rush 352271
Water Collard 267274
Water Coltsfoot 267294
Water Cotoneaster 107583
Water Crash 262722
Water Crease 262722
Water Cress 72679,262722,336166,336242
Water Crowfoot 48918,325580
Water Cuckoo 72934
Water Cup 200388,267294,347156
Water Dionaea 14273
Water Dock 31051,339916,339925,340082,
340305
Water Docken 292745
Water Dragon 348085
Water Dragons 66611
Water Dropwort 269292,269299,269307
Water Elder 345594,407989
Water Elm 301801,301804,401431,417534
Water Feather 261342
Water Featherfoil 198680
Water Feathers 198680
Water Fennel 67376,269299
Water Figwort 355061,355062,355261
Water Finger-grass 285426
Water Fir 252540
Water Flag 208771,208924
Water Flower 175444
Water Flower De Luce 208771
Water Flower-de-Luce 208771
Water Forget-me-not 260842,260868
Water Fox Tail 17527
Water Foxtail 17527
Water Fringe 267825
Water Fucksia 205326
Water Garliccress 365503
Water Gentian 173242
Water Germander 388272
Water Gilliflower 198680
Water Gladiole 64184,234437
Water Gladiolus 64184
Water Goggles 68189
Water Golland 68189,267294
Water Gowan 68189
Water Gowland 68189
Water Grass 238556,262722,281934
Water Gress 262722
Water Gum 267850,398669,398675
Water Hair 79198
Water Hawthorn 29325,29643,29660,29678
Water Heart's-ease 308739,308750,308758
Water Heath 149309
Water Hedge 290480
Water Hemlock 90918,90924,90932,
192358,269299,269307,365851
Water Hemp 18815,18844,54158,158062
Water Hickory 77881
Water Honeysuckle 235749
Water Horehound 239186,239202,239241
Water Houseleek 377485
Water Hyacinth 141805,141808
Water-hyacinth 141808
Water Hymenacue 200822,200840
Water Hyssop 46348,46352,46362,46371,
203102
Water Jasmine 414815
Water Jobstears 99107
Water Kesh 24483
Water Knotweed 309199,309654
Water Lady's-thumb 308739
Water Larch 252540
Water Leek 15861
Water Lemon 285660
Water Lentil 224375
Water Lettuce 234379,301021,301025
Water Lily 68189,72934,208771,263267,
267627,267730,325580,417093
Water Lily Tulip 400181
Water Liverwort 325580
Water Lobelia 234437
Water Locust 176864
Water Loosestrife 239881,306960
Water Lovage 269307,269319
Water Maize 408734
Water Mallow 18173
Water Manna Grass 177599
Water Mannagrass 177599
Water Manna-grass 177599
Water Maple 3505
Water Marigold 53816,247882
Water Meal 414649,414659
Water Medick 247366
Water Melon 93277,93304
Water Milfoil 198680,261337,261348,
261379
Water Milfoil Family 184993
Water Millet 418103
Water Mint 250291,250385
Water Mosquitotrap 117519
Water Mudwort 230831
Water Murdannia 260102
Water Musk 255210

Water Nemony 325580
Water Nightshade 367682
Water Nut 68189
Water Nymph 267627
Water Nymph Flower 262381
Water Oak 324225,324415
Water Oat 418079
Water Oats 418095
Water Orchid 184021
Water Pachira 279381
Water palm 267838
Water Parsley 29327,269356,365872
Water Parsnip 53157,365835,365861,365872,377485
Water Paspalum 285423,285450,285496
Water Pear 204060
Water Pennywort 200253,200257,200385,200388
Water Pepper 142554,142568,142570,309199
Water Pimpernel 345729,345731,345733,406992,407029
Water Pine 377485
Water Plantain 14719,14760,14785,326023
Water Platter 408733
Water Pod 200427
Water Poplar 311398
Water Poppy 200241,200244,248411
Water Ppepper 291958
Water Primrose 214210,238150
Water Pumpy 407029
Water Purple 407029
Water Purslane 238172,238185,238192,238210,238221,240065,290476,290480,290489
Water Ragwort 358280
Water Ramie 56206
Water Rice 418095
Water Rocket 336277
Water Rose 267294,267648
Water Rose Apple 156134,382474
Water Rot 200388
Water Rotula 337652
Water Sapwort 269307
Water Saxifrage 349066
Water Scorpion-grass 260868
Water Sedge 73732,73739
Water Seedbox 238192
Water Seg 208771
Water Sengren 377485
Water Shield 59144,59148
Water Single Bambusa 47405
Water Sisymbrium 365503
Water Skeg 208771
Water Skirret 365861
Water Smartweed 308739,308750,308758,308986,309199,309654
Water Snow-cup 325580
Water Snow-cups 325580
Water Snowflake 267819
Water Socks 267648
Water Soldier 377482,377485
Water Soldiers 377485
Water Sorrel 340082
Water Spapweed 204780,205416
Water Spearwort 326023
Water Speedwell 406992,407029,407064
Water Spike 312028,312190
Water Spinach 207590,207623
Water Squirt 24483
Water Star-grass 193678
Water Starwort 67336,67376,67393
Water Starwort Family 67334
Water Stitchwort 260922
Water Strawcoat 200589,200652
Water Sweetgrass 177629
Water Target 59148
Water Thistle 92285
Water Thyme 143920
Water Touch-me-not 204780,205416
Water Tree 204060
Water Trefoil 250502
Water Tupelo 267850
Water Uintje 29660
Water Vine 92764
Water Violet 198679
Water Wally 46225,46251
Water Wattle 1528
Water Whorl-grass 79198
Water Willow 46225,123526,214320,344268
Water Wingnut 320372
Water Wisteria 200609
Water Wort 142554
Water Yam 131458
Water Yarrow 198680,261364
Water-apple 156134
Water-arum 66600
Water-ash 320071
Waterbamboo 260086,260121
Waterbamboo Syzygium 382534
Water-beech 77263
Waterbleb 68189
Waterbolt 55908,55922
Water-bug-trap 14273
Waterbuttons 107766
Watercandle 139544,139602
Water-celery 404536
Water-centipede 218457
Waterchestnulleaf Dendranthema 124846
Waterchestnut 143122,394412,394436
Waterchestnut Family 394554
Waterchestnut Tanoak 233383
Waterchestnutleaf Notoseris 267178
Water-chinquapin 263270
Watercoconut 267837
Water-convolvulus 207590
Watercress 72679,225450,262645,262707,262722,336232,336242
Water-cress 262722,336166
Watercup 200388
Waterdropwort 269292
Water-dropwort 269292,269299,269307,269319
Waterelm 417534,417558
Water-elm 301801,301804,417534,417540
Waterer's Cotoneaster 107727
Waterer's Gold Holly 203564
Waterfig 94571
Water-filter Nut 378860
Water-flag 208771
Waterghostbanana 200906,200941
Watergrass 63216
Water-gum 398675
Waterhawthorn 29643,29673
Water-hawthorn 29643,29660
Waterhawthorn Family 29696
Waterhemlock 90918
Water-hemlock 90926
Waterhemlockleaf Primrose 314247
Water-hemp Pigweed 18683
Waterhissop 46348
Waterhyacinth 141805
Water-hyacinth 141805
Water-hyssop 46348
Watering Currant 334047
Waterleaf 200482,200488,383345
Waterleaf Family 200476
Waterleaf Scorpion-weed 293167
Water-lemon 285660
Water-lemon Jamaica Honey Suckle 285660
Waterlettuce 301021,301025
Water-lettuce 301021,301025
Waterlily 267627
Water-lily 267274,267627,267648
Waterlily Family 267787
Water-lily Family 267787
Waterlily Tulip 400181
Water-lily Tulip 400181
Waterlilyleaf Begonia 50114
Waterlilyleaf Goldenray 229118
Waterlily-leaved Rhododendron 331364
Water-Locust 176860
Waterloving Swallowwort 117519
Water-marigold 247882
Watermeal 414656,414659
Water-meal 414649
Watermelon 93286,93304
Watermelon Begonia 288767
Watermelon Peperomia 290299
Water-melon Peperomia 290299

Watermelon Plant 290299
Water-milfoil 261337
Watermilfoil Family 184993
Water-millet 140509
Watermoss 200201
Watermotie 46225
Watern Wallflower 154390
Waternettle 123316,123332
Waternut 143122
Water-nymph 262015
Water-nymph Family 262013
Waterparsnip 365835
Water-parsnip 365835,365851,365872
Waterpear 382541
Water-pepper 309199
Water-pepper Smartweed 309199
Waterpine 178014,178019
Waterplantain 14719,14754
Water-plantain 14719,14760,14785
Water-plantain Family 14790
Waterplantain Ottelia 277369
Water-plantain Spearwort 325567
Waterplantainleaf Calanthe 65869
Water-platter 408733
Waterpod 143886
Water-pod 143886
Waterpoppy 200241
Water-poppy 200241,200244
Water-poppy Family 230253
Water-primrose 214210,238150
Water-purslane 142574,238192,240065,
290476,290480
Water-ribbons 397166
Water-root 167134
Watersal 343151
Watershield 59144,59148
Water-shield 59144,59148,200448,200453
Water-shield Family 64503
Watersnake Hemp 162868
Water-speedwell 406992
Water-spider Orchid 184021
Waterspinach 207590
Water-spinach 207590
Waterspring Currant 333977
Waterstar Wort 260920,260922
Waterstarwort 67336
Water-starwort 67336,67393
Waterstarwort Family 67334
Water-starwort Family 67334
Waterthread Pondweed 312091
Water-thread Pondweed 312091
Waterthyme 200190
Watertowerflower 54439,54456
Water-tree 184616
Water-violet 198676,198680
Waterwampee 310950,310962
Waterweed 143914,143931
Water-weed 141569,143914
Waterwheel Plant 14273
Waterwillow 197960,197963
Water-willow 123526,123527
Waterwood 204060
Waterwort 67376,142554,142568,142574
Waterwort Family 142552
Waterwortlike Rockfoil 349290
Watery Chickweed 374926
Watery Speedwell 406992
Watling Street Thistle 154301
Watlington Barberry 52359
Watson Caladium 16532
Watson Flower 413315
Watson Gladiolus 176661
Watson Primrose 315115
Watson Rhododendron 332097
Watson's Amaranth 18850
Watson's Fleabane 151068
Watson's Goldenbush 150381
Watson's Magnolia 242352
Watson's Orach 44695
Watson's Puncturebract 278674
Watson's Spineflower 89130
Watson's Wild Buckwheat 152660
Watson's Yellow Bloodwood 155775
Watsonia 413328
Watt Cyclea 116040
Watt Hawthorn 110111
Watt Oreosolen 274113
Watt Ringvine 116040
Watt Tupistra 70635
Watt's Holly 204399
Watter Milkwort 308457
Watters Clearweed 299093
Watters Coldwaterflower 299093
Wattery Drum 360328
Wattery Drums 360328
Wattle 1024,1375,1380,1529
Wave Cactus 375495
Waveedge Clearweed 298888
Waveedge Coldwaterflower 298888
Waveedge Redcarweed 288771
Waveleaf Allophylus 16059
Waveleaf Ardisia 31604
Waveleaf Buddhamallow 402271
Waveleaf Pricklyash 417366
Waveleaf Rhubarb 329386
Waveleaf Tinospora 392252
Waveleaf Tinospore 392252
Wave-leaved Bridelia 60198
Wave-leaved Ixora 211198
Wavemargin Dayflower 101185
Wavemargin Murdannia 260123
Wavemargin Waterbamboo 260123
Wavepetal Gladiolus 176620
Wavepetal Pocktorchid 282845
Waveran-leaf 302068
Waveribgourd 193057
Wavestalk Yushanbamboo 416767
Wavewind 68713
Wavewine 68713
Wavy Bittercress 72749
Wavy Bitter-cress 72749
Wavy Fig 165807
Wavy Hair Grass 126074
Wavy Hairgrass 126074
Wavy Hair-grass 101667,126074
Wavy Heliotrope 190718
Wavy Meadow-grass 305569
Wavy Mint 250385
Wavy Newlitse 264108
Wavy Rhubarb 329386
Wavy St. John's Wort 202202
Wavy St. John's-wort 202202
Wavyleaf Alder 16459
Wavyleaf Aster 381006
Wavyleaf Ceanothus 79928
Wavyleaf Honeysuckle 235781
Wavyleaf Mullein 405774
Wavyleaf Newlitse 264108
Wavyleaf Oak 324515
Wavyleaf Plantainlily 198651
Wavy-leaf Purple Coneflower 140088
Wavyleaf Sealavender 230761
Wavyleaf Silk-tassel 171546
Wavyleaf Thistle 92461
Wavy-leaf Wattle 1523
Wavy-leaved Gaura 172210
Wavy-leaved Litse 264108
Wavy-leaved Monkey Orchid 273476
Wavy-leaved Newlitse 264108
Wavy-leaved Plantain Lily 198651
Wavy-leaved Thistle 92461
Wavylobulate Bolbostemma 56658
Wavy-margined Antelope Sage 152176
Wavyrib Dianxiong 297959
Wawa 398004
Wawashan Lateflowered Dwarf
Willow 343808
Wawra Barberry 52360
Wawre Ladybell 7884
Wawu Sedge 76737
Wawushan Raspberry 339471
Wax Bean 294056
Wax Begonia 49755,50298
Wax Currant 333936
Wax Dolls 169126
Wax Flower 85852,153325,153330,153331,
198821,245803,294867,294868
Wax Goldenweed 181154
Wax Gourd 50994,50998
Wax Jambo 156372
Wax Jambu 382660

Wax Mallow 243929
Wax Maple 2883
Wax Myrtle 261120,261132,261137,261202
Wax Myrtle Family 261243
Wax Palm 84326,84328,84329,103731
Wax Plant 50298,153325,153328,153330,153332,198821,198827
Wax Plants 153325
Wax Privet 229558,290349
Wax Tree 229529,393479
Wax Trillium 397578
Wax Vine 359415
Wax-apple 382660
Waxberry 172083,261137,261176,261202,261208,380736,380749
Waxberry Gaultheria 172083
Waxberry Whitepearl 172083
Waxcolor Dove Orchid 291128
Waxdaisy 189093,415386
Waxflower 39328,87480,153325,198827,211550,211551
Wax-flower 382763
Wax-flower Shin-leaf 322823
Waxgourd 50994,50998
Waxgourd Poplar 311444
Waxleaf Acacia 1153
Waxleaf Azalea 331152
Waxleaf Cottonwood 311142
Wax-leaf Meadow-rue 388638
Waxleaf Privet 229529
Wax-leaf Privet 229485
Wax-leaved Meadow Rue 388638
Wax-leaved Meadow-rue 388638
Wax-lip 177313
Wax-lip Orchid 177313
Waxmallow 243927,243929
Waxmyrtle 261120
Wax-myrtle 261120,261168,261169
Waxmyrtle Family 261243
Waxpetal 106628
Waxplant 198821,198827
Waxtree 261176
Wax-tree 261137,393479
Waxworks 308454
Waxwort 80309
Wax-wort 80309
Waxworts 80309
Waxy Bedstraw 170289
Waxy Coneflower 339550
Waxy Heath 149352
Waxy Mannagrass 177585
Waxy Meadow Rue 388638
Waxy Meadow-rue 388638
Waxy Pink Rhododendron 331928
Waxy Saltbush 44688
Waxy Saurauia 347955
Waxy Yellow Rhododendron 331898
Waxyellow Lecanorchis 223636
Waxyellow Primrose 314227
Waxy-fruit Thorn 109949
Waxyleaf Nightshade 367178
Waxyleaf Privet 229593
Waxy-leaf Privet 229593
Waxy-powdery Peashrub 72317
Waxy-yellow Rhododendron 331898
Way Bennett 175454,235373
Way Thistle 91770
Waya Bamboo 125490
Wayaka Yam Bean 279729
Wayaka Yambean 279732
Waybent 198327
Wayberry 302068
Wayborn 302068
Waybread 302068
Waybroad 302068
Waybroadleaf 302068
Waybrow 302068
Wayburn-leaf 302068
Wayfaring Tree 407907,407908
Wayfaring Tree Viburnum 407909
Wayfaring Viburnum 407907
Wayfaringtree 407907
Wayfaring-tree 407907
Wayfaringtree Viburnum 407907
Wayforn 302068
Wayfron 302068
Waygrass 308816
Wayside Aster 155822
Wayside Beauty 316819
Wayside Bread 302068
Wayside Cerastium 83020
Wayside Cudweed 178481
Wayside Mallow 243823
Wayside Mouse Ear Chickweed 83064
Wayside Mouse-ear Chickweed 83064
Wayside Speedwell 407292
Waythorn 328642
Waywind 68713,102933
Waywort 21339
Wazaristan Barberry 51417
Weak Anemone 23780
Weak Azalea 331963
Weak Bentgrass 12034
Weak Bluegrass 305599
Weak Bothriochloa 57557
Weak Bulrush 352252,352253
Weak Buttercup 326106
Weak Corydalis 106515
Weak Flexuose Bittercress 72761
Weak Gentian 173427
Weak Goldsaxifrage 90397
Weak Groundsel 279902
Weak Holeglumegrass 57557
Weak Jumby Pepper 254204
Weak Kiwifruit 6622
Weak Loosestrife 239606
Weak Maesa 241844
Weak Mahobarnerry 242451
Weak Manna Grass 393101
Weak Meadow-grass 305471
Weak Microstegium 254001
Weak Milkvetch 42723
Weak Motherweed 231507
Weak Nettle 402875
Weak Ophiorrhiza 272208
Weak Philodendron 294814
Weak Plantain 301949
Weak Primrose 314399
Weak Ricebag 181659
Weak Roegneria 335288
Weak Rush 213035
Weak Sedge 74265,74271,76508
Weak Signalgrass 402536
Weak Spapweed 205365
Weak St. John's-wort 202029
Weak Stellate Sedge 76229
Weak Touch-me-not 205365
Weak Trigonotis 397472
Weak Twinballgrass 209098
Weak Willow 343880
Weak Windhairdaisy 348318
Weak Woodbetony 287149,287291
Weak Yellowflower Betony 373483
Weak Yushanbamboo 416771
Weakchloa 209401
Weakleaf Stringbush 122520
Weakleak Yucca 416595
Weakly-hair Aster 40872
Weak-stalk Bulrush 352252,352253
Weak-stalk Club-rush 352252
Weakstem Peristylus 291245
Weakstem Perotis 291245
Weak-stemmed Wood Sedge 75081
Weasel Flower 345437
Weasel's Nose 170025
Weasel's Snout 28631,170025,231183
Weasel's-snout 28631,255983
Weaselsnout 170025
Weasel-snout 170015,170025
Weather Clock 384714
Weather Fair Thistle 76990
Weather Flower 21339
Weather Plant 765
Weather Teller 21339
Weather Thistle 77022
Weatherglass 21339
Weatherplant 765
Weathervine 765
Weatherwind 68713
Weaver's Bamboo 47475
Weaver's Broom 370191,370202

Weaver's-broom 370191,370202
Weaversbroom 370191,370202
Webb Honeysuckle 236214
Webb Rose 337012
Webbract Stairweed 142601
Weber Agave 10960
Weber Blue Agave 10945
Weber's Century Plant 10960
Weberbauer Barberry 52362
Webseedvine 235202
Webstalk Knotweed 309807
Webstem Allocheilos 278610
Webstem Clearweed 299058
Webstem Coldwaterflower 299058
Webstem Pellionia 288716
Webstem Redcarweed 288716
Webster Azalea 332098
Webster Rabdosia 209864
Webster Rhododendron 332098
Weddel Palm 253347,253348
Weddell Barberry 52363
Weddell Palm 380561
Wedding Bush 334403,334404
Wedding Flower 167735,258636
Wedding-cake Tree 57695
Wedding-cake-tree 105003
Wedding-flower 374916
Wedelia 371233,413490
Wede-wixen 173082
Wedge Glaudular Eurya 160476
Wedge Grass 371488,371493
Wedge Orach 44680
Wedge Pea 179162,179163
Wedgebract Elatostema 142638
Wedgebract Stairweed 142638
Wedge-fruited Oval Sedge 76420
Wedgeleaf Calotis 68082
Wedge-leaf Candollea 194664
Wedgeleaf Chinese Aspen 311194
Wedgeleaf Cystopetal 320213
Wedgeleaf David Poplar 311283
Wedgeleaf Dianxiong 297946
Wedgeleaf Dock 340008,340059
Wedgeleaf Draba 136972
Wedge-leaf Dregea 137787
Wedge-leaf Eurya 160449
Wedgeleaf Fig 165791
Wedgeleaf Jilin Poplar 311331
Wedgeleaf Premna 313669
Wedgeleaf Saxifrage 349227
Wedge-leafed Fan Palm 228762
Wedge-leaved Barberry 51517
Wedge-leaved Dregea 137787
Wedge-leaved Hawthorn 110052
Wedge-leaved Komarov Currant 334056
Wedge-leaved Raspberry 338307
Wedge-leaved Rhododendron 330485
Wedge-leaved Saxifrage 349227
Wedge-leaved Wattle 1494
Wedgescale 44361,44680
Wedge-shaped Leaves Rhododendron 330485
Weebow 359158,383874
Weeby 359158
Weed Passionflower 285641
Weed Silene 363368
Weed Sunflower 188908
Weedbind 68713
Weedwind 102933,162519
Weedy 360404
Weedy Brome 60621
Weedy Dogfennel 85526
Weedy Hawksbeard 111073
Weedy Knotweed 309094
Weedy Rattle-box 112636
Weeping Acacia 1465
Weeping Alaska Cedar 85306
Weeping Alkali Grass 321248,321249
Weeping Alkaligrass 321248
Weeping Ash 167965
Weeping Aspen 311539
Weeping Beech 162413
Weeping Birch 53563,53590
Weeping Blue Atlas Cedar 80082
Weeping Boer-bean 352493
Weeping Boobialla 260716
Weeping Boree 1689
Weeping Bottlebrush 67302
Weeping Bottlebush 248122
Weeping Broom 88917
Weeping Buddleja 337290
Weeping Bulrush 353290
Weeping Cabbage Palm 234175
Weeping Cherry 316169
Weeping Chinese Banyan 164688
Weeping Common Linden 391711
Weeping Crack-willow 342942
Weeping Cypress 114690
Weeping Elm 401516
Weeping European Beech 162413
Weeping Fig 164682,164688
Weeping Fingergrass 160985
Weeping Flowering Dogwood 105028
Weeping Forsythia 167456,167468
Weeping Giant Sequoia 360577
Weeping Grass 141784
Weeping Gum 155740
Weeping Hardy Privet 229665
Weeping Hemlock 399873
Weeping Hers Lilac 382186
Weeping Himalayan Cedar 80089
Weeping Japanese Red Pine 299895
Weeping Juniper 213773
Weeping Lantana 221284
Weeping Lillypilly 413312
Weeping Limber Pine 299943
Weeping Love Grass 147612
Weeping Lovegrass 147612
Weeping Love-grass 147612,147864
Weeping Mulberry 259067
Weeping Myall 1465
Weeping Oak 324339
Weeping Pagoda Tree 369038
Weeping Paperbark 248104
Weeping Pine 300140
Weeping Pinyon 300148
Weeping Pittosporum 301369
Weeping Purple Beech 162414
Weeping Pussy Willow 343153
Weeping Sage 61982
Weeping Sally 155657
Weeping Satinash 413312
Weeping Sawara False Cypress 85357
Weeping Serbian Spruce 298390
Weeping Silver Fir 275
Weeping Silver Pear 323282
Weeping Silver-lime 391647
Weeping Snowbell 379375
Weeping Spanish Fir 449
Weeping Spring Cherry 316836
Weeping Spruce 298265
Weeping Tree Myrtle 156390
Weeping Wattle 1465,1553
Weeping Waxmallow 243951
Weeping White Fir 327
Weeping White Linden 391647
Weeping White Pine 300216
Weeping Widow 168476
Weeping Willow 218761,342942,342944,342978,342982,342985,343070,344084,350980,350990
Weeping Wych-elm 401516
Weeping Yaupon Holly 204393
Weeping Yellow Cedar 85306
Weeping Yucca 416610
Weeping-grass 141736,141784
Weevil Plant 114819
Wegwants 60382
Weidenblatt Callistemon 67294
Weigela 130242,413570,413591,413610
Weigold Corydalis 106327
Weigttree 364951,364959
Weihai Sage 345481
Weihe Yuhuangshan Willow 344302
Weilbach Aechmea 8598
Weiling Xian 94814
Weining Barberry 52364
Weining Camellia 69589
Weining Larkspur 124693
Weinmannia 413682
Weiser Wild Buckwheat 152485
Weishan Germander 388144

Weishan Pimpinella 299578
Weishan Skullcap 355847
Weisi Burmaflame Rhododendron 331867
Weisi False Incised Monkshood 5579
Weisi Rabdosia 209866
Weiss Sandwort 32312
Weisstanne 272
Weixi Barberry 52355,52365
Weixi Bittercress 73041
Weixi Ciliate Poplar 311279
Weixi Corydalis 106600
Weixi E. Xizang Rockfoil 349501
Weixi False Incised Monkshood 5579
Weixi Hydrangea-vine 351785
Weixi Longleaf Willow 343882
Weixi Longstalk Maple 3123
Weixi Milkvetch 43240
Weixi Monkshood 5687
Weixi Rabdosia 209866
Weixi Rose 337013
Weixi Spapweed 205446
Weixi Touch-me-not 205446
Weixi Violet 410760
Weixi Willow 344269
Weixi Windhairdaisy 348809
Weixi Woodbetony 287827
Weixi Yushania 416823
Weixin Barberry 52366
Welcome-home-husband 158730
Welcome-home-husbandthough-never-so-drunk 358062
Welcome-home-husbandthough-never-so-late 358062
Welcome-to-our-house 158730
Weld 327879
Weld Mignonette 327879
Welgers 344244
Well Girse 262722
Well Karse 262722
Well Kerse 262722
Wellby Sagebrush 36550
Wellby Windhairdaisy 348934
Wellflavoured Sterculia 376108
Well-flavoured Sterculia 376108
Well-girse 262722
Well-grass 262722
Wellingtonia 360561,360575,360576
Wellink 407029
Well-karse 262722
Welsh Corn 417417
Welsh Gentian 174164
Welsh Groundsel 358485
Welsh Mudwort 230833
Welsh Nut 212636
Welsh Onion 15289
Welsh Parsley 71218
Welsh Poppy 247099,247113
Welsh Ragwort 385903
Welsh Vine 345631
Welsh's Aster 381011
Welsh's Bugseed 104862
Welsh's Saltbush 44412
Welted Thistle 73337
Welwitsch Antiaris 28257
Welwitsch Lannea 221215
Welwitschia 413741,413748
Wenatchee Larkspur 124686
Wenatchee Mountain Coneflower 339515
Wenatchee Thistle 91942
Wenatchee Wild Buckwheat 151937
Wenchang Evergreen Chinkapin 79068
Wenchang Evergreenchinkapin 79068
Wenchang Machilus 240720
Wenchengia 413758
Wenchuan Azalea 332096
Wenchuan Bittercress 72748
Wenchuan Cowparsnip 192397
Wenchuan Curvedstamencress 238027
Wenchuan Larkspur 124694
Wenchuan Onion 15891
Wenchuan Purpleflower Isometrum 210184
Wenchuan Saxifrage 350038
Wenchuan Smith Curvedstamencress 238027
Wenchuan Starhair Azalea 330155
Wenchuan Stonecrop 357297
Wenchuan Thorowax 63868
Wenchuan Variable Coloured Rhododendron 332096
Wenchuan Willow 343793
Wendlandia 413765
Wendt's Water Trumpet 113545
Wenge 254747
Weniger's Hedgehog Cactus 140286
Wenquan Milkvetch 43241
Wensai Larkspur 124695
Wenshan Begonia 50405
Wenshan Briggsia 60285
Wenshan Camellia 69760
Wenshan Caper 71744
Wenshan Clematis 95454
Wenshan Clovershrub 70910
Wenshan Cymbidium 117100
Wenshan Elaeagnus 142244
Wenshan Eurya 160626
Wenshan Hemsleia 191966
Wenshan Lasianthus 346477
Wenshan Lily 230073
Wenshan Machilus 240722
Wenshan Metalleaf 296423
Wenshan Ophiorrhiza 272305
Wenshan Orchis 117100
Wenshan Phaius 293547
Wenshan Pholidota 295543
Wenshan Primrose 315117
Wenshan Rhododendron 332101
Wenshan Roughleaf 346477
Wenshan Sabia 341485
Wenshan Schefflera 350693
Wenshan Sedge 76751
Wenshan Skullcap 355848
Wenshan Snowgall 191966
Wenshan Spapweed 205447
Wenshan Sweetleaf 381221
Wenshan Syzygium 382682
Wenshan Tea 69760
Wenshan Treebine 93038
Wenshan Tupelo 267876
Wenshui Leptodermis 226082
Wenshui Wildclove 226082
Wensu Milkvetch 43242
Wentsai Chirita 87991
Wenxian Elatostema 142894
Wenxian Euonymus 157964
Wenxian Milkvetch 43244
Wenxian Milkwort 308462
Wenxian Oblongleaf Maple 3269
Wenxian Paris 284415
Wenxian Peashrub 72375
Wenxian Pea-shrub 72375
Wenxian Pittosporum 301287
Wenxian Seatung 301287
Wenxian Stairweed 142894
Wenxian Tanoak 233418
Wenxian Yinshancress 416365
Wenxian Yinshania 416365
Wenzhou Bamboo 47530
Wenzhou Bambusa 47530
Wenzhou Didymocarpus 129887
Wenzhou Grape 412002
Wenzhou Holly 204400
Wenzhou Migan 93736
Werdermann Cactus 244098
Wermout 35090
Wermud 35090
Weschniakow Leek 15893
Weschniakow Onion 15893
Wesffelton Yew 385307
Weskit Bamboo 47493
West Africa Afrormosia 290876
West African Albizzia 13554,13703
West African Black Pepper 300374,300397
West African Ebony 132128,132299
West African Kino 320293
West African Mahogany 216214
West African Mulberry 88506
West African Nutmeg 257653
West African Pennisetum 289057
West African Plum 411258
West African Ratbane 128835
West African Rosewood 320293
West African Rubber Tree 169229

West Asian Cedar 213767
West Asian Maple 2890
West Buckthorn 328880
West China Barberry 52156
West China Indigo 206283
West China Raspberry 339304
West China Spiraea 371990
West China Willow 343792
West Coast Goldenrod 368091
West Coast Hemlock 399905
West Coast Rhododendron 330277
West Himalayan Birch 53481
West Himalayan Elm 401643
West Himalayan Fir 439
West Himalayan Silver Fir 439
West Himalayan Spruce 298436
West Hubei Sabia 341492
West India Elemi 121116
West India Elm 181620
West Indian Aloe 17381
West Indian Arrrowroot 245014
West Indian Bay Tree 299328
West Indian Birch 64085
West Indian Blackthorn 1219
West Indian Box 78145
West Indian Boxwood 78145
West Indian Cedar 80030
West Indian Cherry 243526
West Indian cockscomb 80443
West Indian Cocoa Palm 89816
West Indian Dropseed 372666
West Indian Ebony 61427
West Indian Elder 345703
West Indian Gherkin 114122,114123
West Indian Gooseberry 290701
West Indian Greenheart 100066
West Indian Holly 223929,400493
West Indian Indigo 206626
West Indian Ipecacuanha 37888
West Indian Jasmine 211029,305225
West Indian Laurel Fig 164629
West Indian Locust 200844
West Indian Locust-bean Tree 284473
West Indian Locust-tree 200844
West Indian Mahogany 380528
West Indian Marsh Grass 200827
West Indian Mignonette 158730,223454
West Indian Mistletoe 296076
West Indian Raspberry 339194
West Indian Rattlebox 112757
West Indian Red Cedar 213623
West Indian Redwood 181559
West Indian Sandalwood 21005
West Indian Satin Wood 417230
West Indian Spigelia 371614
West Indian Sumach 61283
West Indian trema 394648
West Indian Tufted Airplant 182177
West Indian Vanilla 405021
West Indian Whitewood 71132
West Indies Balsa 268344
West Indies Ebony 61427
West Indies Smutgrass 372650
West Java Clumping Bamboo 262762
West Java Pipe-bamboo 175609
West Ochroma 268344
West Pear 323133
West Sichuan Broadpetal Winterhazel 106671
West Sichuan Cherry 83350
West Sichuan Edelweiss 224961
West Sichuan Oak 323936
West Sichuan Onosma 271803
West Sichuan Peashrub 72225
West Sichuan Pea-shrub 72225
West Sichuan Point Leaves Rhododendron 330286
West Sichuan Rose 336952
West Sichuan Softhairleaf Raspberry 338493
West Sichuan Sophora 368995
West Sichuan Webb Honeysuckle 236216
West Szechuan Bluebellflower 115348
West Szechuan Cherry 83350
West Szechuan Pyrethrum 322748
West Szechuan Rhodiola 329833
West Szechuan Saussurea 348280
West Szechwan Cherry 83350
West Szechwan Oak 323936
West Tibet Corydalis 106165
West Tibet Willow 343528
West Willow 343930
West Xizang Willow 343528
West Yunnan Abutilon 896
West Yunnan Anise-tree 204556
West Yunnan Azalea 331427
West Yunnan Ehretia 141617
West Yunnan Hawthorn 109881
West Yunnan Helixanthera 190856
West Yunnan Holly 203824
West Yunnan Meconopsis 247138
West Yunnan Monkshood 5094
West Yunnan Nomocharis 266555
West Yunnan Pittosporum 301302
West Yunnan Rhododendron 331427
West Yunnan Spindle-tree 157771
West-bengal Bamboo 249617
West-China Himalayan Birch 53635
West-China Honeysuckle 236113
West-coast Violet 410008
Westen Yellow Pine 299992
Western Alder 16437
Western Androsace 23243
Western Arborvitae 390645
Western Arbor-vitae 390645
Western Arrowwood 407958
Western Aster 380839
Western Australian Christmas-tree 267409
Western Australian Christnas Tree 267409
Western Australian Floodedgum 155727
Western Australian Golden Wattle 1553
Western Australian Grass Tree 415178
Western Azalea 331383
Western Balsam Fir 317
Western Balsam Poplar 311371,311554
Western Baneberry 6412
Western Banksia 47656
Western Birch 53540
Western Bistort 54772
Western Bitterweed 201309,201323
Western Black Wattle 1271
Western Black Willow 343596
Western Bleeding Heart 128297
Western Bleedingheart 128297
Western Bleeding-heart 128297
Western Blue Flag 208718
Western Blue Flax 231917,231942
Western Blue Virgin's-bower 95168
Western Blueberry 403924
Western Bog Aster 70949,380991
Western Bog Birch 53594
Western Boobialla 260719
Western Brome 60840
Western Bruning Bush 157751
Western Butterwort 299751
Western Canada Goldenrod 368211
Western Catalpa 79260
Western Catawba 79260
Western Catchfly 363836
Western Chinkapin 89968
Western Choke Cherry 83188
Western Chokecherry 316918
Western Clover 396988
Western Columbine 30034
Western Coneflower 339595
Western Coral Bean 154658
Western Coral-root 104002
Western Cork Tree 184618
Western Cottonwood 311324
Western Crane's-bill 174581
Western Creeper 285127
Western Cudweed 317753
Western Dayllower 100989
Western Dock 340157
Western Dog Violet 409648
Western Dwarfdandelion 218212
Western False Foxglove 45169
Western False Gromwell 271854,271857,271861
Western Fescue 164112
Western field Mouse-ear Chickweed 83081
Western Figwort 355261
Western Flowering Dogwood 105130

Western Fumitory 169125
Western Germander 388032
Western Glasswort 342894
Western Golden Ragwort 279939,279941
Western Golden Wattle 1167
Western Goldenrod 368289
Western Goldentop 161096
Western Gorse 401390
Western Great Angelica 24304
Western Hackberry 80728
Western Hawksbeard 110934
Western Heart-leaved Groundsel 279939
Western Heart-leaved Twayblade 232909
Western Heath 149703
Western Heath Aster 380883
Western Hemlock 399905
Western Hemlock-spruce 399905
Western Himalayan Pine 300297
Western Hophornbeam 276811
Western Hop-hornbeam 276811
Western Horse Nettle 367102
Western Hubei Rhododendron 331533
Western Hupeh Atractylodes 44194
Western Jacob's-ladder 307232
Western Jimson 123071
Western Juniper 213834,213840,213922
Western Kapok Bush 98105
Western Labrador Tea 223894
Western Larch 221917
Western Larkspur 124233,124280
Western Laurel 215402
Western Lined Aster 380916
Western Maple 3127
Western Marsh Cudweed 178332
Western Marsh Orchid 273529
Western Marsh Spider-lily 200940
Western Meadow Aster 380845
Western Meadow-rue 388602
Western Minniebush 250513
Western Mockorange 294493
Western Mojave Wild Buckwheat 152287
Western Monk's-hood 5136
Western Monkshood 5136
Western Mountain Ash 369519
Western Mountain Aster 380989
Western Mountainash 369519
Western Mountain-ash 369519
Western Mugwort 35404,35837
Western Ninebark 297838
Western Oat 45531
Western Orange-cup Lily 230066
Western Panic Grass 128481
Western Paper Birch 53549
Western Pasque Flower 23971
Western Pasqueflower 23969
Western Pearly Everlasting 21596
Western Peony 280166
Western Pepper-grass 225415
Western Pink 127723
Western Plane 302588
Western Poison Oak 332570
Western Poison-Ivy 393478
Western Poison-oak 332570
Western Polemonium 307197,307232
Western Poppy 282524
Western Prairie Fringed Orchid 302487
Western Prickly Moses 1511
Western Pricklypear 272972
Western Ragweed 19179
Western Ragwort 279916
Western Ramping-fumitory 169125
Western Raspberry 338741
Western Rattlesnake-plantain 179664
Western Rattlesnakeroot 313785
Western Red Cedar 213922
Western Red Ceder 390645
Western Redbud 83753,83798
Western Redcedar 390645
Western Rhododendron 331383
Western Rock Jasmine 23243
Western Rock-cress 30299
Western Rose 337026
Western Rosewood 14369
Western Rough Goldenrod 368345
Western Rue-anemone 145497
Western Rush 213337
Western Sage 35837
Western Sagewort 35237,35246
Western Salsify 394281
Western Sand Cherry 83157
Western Seepweed 379576
Western Serviceberry 19241
Western Service-berry 19241
Western Shadbush 19241
Western Shooting-star 135178
Western Showy Aster 160655
Western Shrubby Woodnettle 273906
Western Sichuan Dutchmanspipe 34227
Western Sichuan Gentian 173392
Western Silvery Aster 41230,380980
Western Snakeroot 11168
Western Snowberry 380738,380748
Western Soapberry 346336
Western Spiderwort 394050
Western Spiraea 371902
Western Spring Beauty 94323
Western Spruce 298435
Western Squaw Lettuce 200487
Western St. Paul's Wort 363096
Western Starflower 396790
Western Stickseed 221727
Western Stork's-bill 153796
Western Sunflower 189017
Western Swamp Laurel 215404
Western Sweet Coltsfoot 292359
Western Sycamore 302595
Western Szechwan Spapweed 204779
Western Tamarack 221917
Western Tansy Mustard 126120,126121,
365500
Western Tea-myrtle 248108
Western Thistle 91770,92249
Western Thoroughwort 158240
Western Thorough-wort 158240
Western Thuja 390645
Western Tickseed 53773
Western Turk's Cap Lily 229939
Western Turk's-cap Lily 229939
Western Wallflower 154390,154419
Western Water Hemlock 90923
Western Water-milfoil 261349
Western Waxmyrtle 261132
Western Wheat Grass 144675
Western Wheatgrass 144478
Western Wheat-grass 144675
Western White Anemone 23784
Western White Cedar 390587,390645
Western White Lettuce 313785
Western White Pine 300082
Western White Spruce 298286
Western White Trillium 397585
Western Wild Daisy 215350
Western Wild Ginger 37582
Western Wild Iris 208718
Western Wild Lettuce 219403
Western Wild Rose 337026
Western Willow 343596
Western Willow Dock 340079
Western Windflower 23833
Western Wolfberry 380748
Western Wood Anemone 23882,23976
Western Woody-pear 415729
Western Yarrow 3957
Western Yellow Pine 300151,300153
Western Yellow-cress 336272
Western Yew 385342
Western-daisy 43295
Western-nettle 193470
Western-purslane 361677
Western-Sichuan Pearleverlasting 21699
Westfeltan Yew 385305
Westfelton Yew 385305
West-himalayan Bluegrass 305927
Westindian Birch 64085
West-Java Climbing Bamboo 131285
West-Java Hollow-culmed Bamboo 175609
Westland Azalea 331280
Westland Rhododendron 331280
Weston's Wild Buckwheat 152331
Westonbirt Dogwood 104930
Westringia 413863,413911,413914

Westwater Wild Buckwheat 152454
Wet Chickweed 375120
Wet Clearweed 298866
Wet Coldwaterflower 298866
Wet Cuckoo 72934
Wet Cudweed 178475
Wet Dogtongueweed 385908
Wet Habenaria 183709
Wet Peristylus 168326
Wet Perotis 168326
Wet-a-bed 384714
Wetfoot Corydalis 105993
Wet-forest Bamboo 262758
Wetherill's Wild Buckwheat 152661
Wetland Anemone 24036
Wetland Cudweed 178475
Wet-land Deer Vetch 237713
Wetland Deervetch 237713
Wetland Holly 204374
Wetland Night Shade 367661
Wetland Nightshade 367661
Wetland Osbeckia 276138
Wetland Sagebrush 36394
Wetland Sunflower 188979
Wetland Windflower 24036
Wetland Woodbetony 287788
Wetland Wormwood 36394
Wets 45566
Wet-the-bed 384714
Wettstein Barberry 52367
Wettstein Nematanthus 263331
Wetweed 159027,384714
Weybred 302068
Weyer Hybrid Butterfly Bush 61957
Weyeriana Butterfly Bush 61957
Weymouth Pine 300211
Whakataka 105238,105241
Whalebone Tree 377539
Whangee 297367
Wharangi 249167
Wharre 243711
What o'clock 384714
What's-your-sweetheart 235334
Whau 145990,145991
Wheat 398834
Wheat Barley 198376
Wheat Thief 233700
Wheat Vodka 398839
Wheat Yellowquitch 194044
Wheatbine 102933
Wheatgrass 11628,11696
Wheat-grass 11628
Wheatley Elm 401575,401632
Wheat-thief 233700
Wheel Lily 229934
Wheel Tree 399432,399433
Wheeler Sotol 122912
Wheeler's Blackberry 339472
Wheeler's Spineflower 89131
Wheeler's Thistle 92493
Wheelscale 44386
Wheelscale Orach 44383
Wheelstamen Tree 399432,399433
Wheel-stamen Tree 399432,399433
Wheelstamentree 399432,399433
Wheel-wort Tooth-cup 337384
Wherry Foam Flower 391526
Wherry's Pink 363302
Whi 187432
Whicken 144661
Whicks 109857,144661
Whicky 369339
Whidow-wort 204469
Whie Sandalwood 346209
Whieflower Changbai Crazyweed 278713
Whieflower Przewalsk Sage 345328
Whiflowwort 204469
While Frills 50825
While Rhatany 218078
While Tansy 4005,11549
Whilowort-like Knotweed 309526
Whim Broomweed 285086
Whimberry 403916,404051
Whin 271563,401373,401388,401395
Whinberry 403916
Whinberry Whortleberry 403916
Whin-cammock 271563
Whinghack Berry 320411,320412
Whinghackberry 320411,320412
Whing-hackberry 320411,320412
Whinghead 320421
Whinghead Flower 320421
Whinshag 167955
Whip Nut-rush 354267
Whip Pussytoes 26408
Whip Tongue 170193,170490
Whipash Orange Magnoliavine 351066
Whipbeam 369322
Whipcord Hebe 186946
Whipcrop 369322,407907
Whipforme Rattail Cactus 29716
Whipformed False Manyroot 283766
Whipformed Pararuellia 283766
Whipformed Ploughpoint 401160
Whipgrass 354267
Whip-grass 354267
Whiplash Rose 336471
Whiplash Star-of-bethlehem 274495,274556
Whip-like Staff-tree 80189
Whipple Cholla 116676
Whipple Fishhook Cactus 354327
Whipple's Yucca 193538
Whip-poor-will Flower 397547,397551
Whip-poor-Will's Shoe 120438
Whip-poor-Will's Shoes 120438
Whip-stick Ash 155662
Whipstyle Meadowrue 388672
Whiptongue 170193,170490
Whiptop 407907
Whiptree 238242
Whiptwig Bittercress 72960
Whirl Mint 250432
Whirl-mint 250432
Whiskbranch Gentian 173535
Whisker Cactus Old-man-cactus 279492
Whiskered Brome 60739
Whiskers 8399
Whiskey 198376
Whisky 198376
Whispering Bells 145021
Whistle Tree 3462
Whistlewood 2837,3462,16352,369339
Whistling Jacks 175990,176122
Whistling Pine 79161
Whistling Thorn 1186
Whistling Tree 1588
Whit Aller 345631
Whit Sunday 262441
Whitbin Pear 369322
Whitcbeam Mountain Ash 369322
White 146958,310996
White Abele 311208
White Adder's Tongue 154896
White Adder's-mouth 243011,243084, 243085
White Adder's-tongue 154896
White Afara 386652
White Affodill 39328
White Alder 16368,16435,96454,96456, 96457
White Alder Family 96561
White Alisson 30164
White Allium 15509
White Alpine Speedwell 406980
White Althaea 18174
White Amaranth 18655
White Amaranthus 18655
White Amaryllis 417612
White Amberboa 18954
White Angel Orchid 302280
White Anisomeles 25468
White Anona 25850
White Apple 90015
White Archangel 220346
White Ardisia 31522
White Argyreia 32657
White Aril Yew 318538,318539
White Arrow Arum 288805
White Arrowbamboo 162746
White Arrowleaf Aster 381007
White Arum 417093

White Arum Lily 417093
White Ash 8826,155585,167897,167942,168032,168057,369339
White Asp 311208
White Asparagus 38921
White Aspen 311208
White Asphodel 39328,39449,39454
White Aster 40753,380872
White Astridia 43321
White Avens 175393,175459
White Azalea 330879,331284,332086
White Bachelor's Buttons 325516,363673
White Backed Hosta 198617
White Ball Mustard 66506
White Ballmustard 66506
White Balsam 178358
White Balsam Fir 317
White Baneberry 6433
White Bark Fig-tree 164875
White Basswood 391727
White Bauhinia 49247
White Bay 242341
White Beak-rush 333480
White Beak-sedge 333480
White Beam 369322
White Beamtree 369322
White Bean 12587,294010
White Bear Sedge 73633
White Beardtongue 289337
White Bedstraw 170186,170490
White Bee Nettle 220346
White Beech 77256,162375,178036,311208
White Beer 398839
White Beet 53257
White Behen 364193
White Bellflower 68713
White Bells 102867,169719,374916
White Ben 363671
White Beneath Aster 40599
White Beneath Leaves Heath 149709
White Bentgrass 12323
White Berry Nandina 262190
White Bigflower Morningglory 208197
White Birch 53549,53563,53572,53586,53590
White Bird's Eye 374916,374968
White Bird's Eyes 374916,374968
White Bird-of-paradise 377554
White Bird-of-paradise Flower 377554
White Bittercress 72858
White Bladder Flower 30862
White Bladderflower 30862
White Bladder-flower 30862
White Blazing Star 228532
White Blood Lily 184347
White Bloodlily 184347
White Blood-lily 184347
White Blow 349999
White Bluebell 15839
White Blueberry 403711
White Blue-eyed-grass White Blue-eyed Grass 365680
White Bobby's Eye 374916
White Bobby's Eyes 374916
White Bog Orchid 183571,230348
White Bog-orchis 183571
White Bolly Gum 264044
White Boltonia 56706
White Bombway 386613
White Bothen 227533
White Bottle 364193
White Bottlebrush 67293
White Box 155473,155605
White Brain Cactus 140569
White Breath of Heaven 99458,131966
White Brodiaea 398794
White Broodhen Bamboo 297263
White Broom 57329,120909,120979,328243
White Broomrape 274934
White Bryonia 61471
White Bryony 61464,61471,61497
White Bryony Family 114313
White Buckeye 9677,9719
White Bulbophyllum 62824,63044
White Bumet 345834
White Bur Sage 19163
White Bursage 19163
White Bush 335869
White Bush Clover 226972
White Butterbur 292374
White Buttercup 284591,325516
White Butterfly Orchid 125694
White Butterflybush 61989
White Butterfly-bush 61975
White Buttons 151232
White Cabbage 59533
White Caladenia 65203
White Callalily 417097
White Camas 417890,417891
White Camomile 85526
White Campion 363671,363673,363814
White Canary Tree 70989
White Canarytree 70989
White Candles 413951
White Canker Weed 313787
White Cap 372106
White Caroba 211244
White Casimiroa 78175
White Catchfly 363673
White Caterpillar 18848
White Cat-tail 401086,401108
White Cedar 85271,85375,90605,228639,248895,321973,390567,390587
White Cedar False Cypress 85375
White Celerycress 366116
White Chamaecyparis 85375
White Charlock 326609
White Cheesewood 18055
White Chelonopsis 86807
White Chervil 113872
White Chicory 90894
White Chile Bells 221342
White Chinaure 56382
White Chinese Birch 53335
White Chinese Flowering Crabapple 243704
White Chinese Olive 70989
White Chinese Waxmyrtle 261213
White Chinese Wisteria 414577
White Chuglam 386478
White Chuglam Wood 386478
White Cinquefoil 312337,312667,312946
White Clary 190552
White Clintonia 97096
White Cloud Tree 248089
White Clover 249203,397043
White Clustered Bellflower 70055
White Cock Robin 364193
White Cockle 363671,363673,364193
White Coelogyne 98686
White Cogongrass 205497
White Cohosh 6433
White Colic-root 14486,14518
White Colorado Fir 317
White Comfrey 381037
White Common Crapemyrtle 219934
White Confetti Bush 131966
White Copernicia 103732
White Coralberry 380736
White Corallita 311621
White Cordia 104151
White Cotton Bush 299311
White Cotton-grass 152781
White Cottony Nepeta 264980
White Craneknee 221700
White Cranesbill 174454
White Creeping Mazus 247030
White Crest Iris 208504
White Crested Dwarf Iris 208504
White Crowfoot 325580
White Crownbeard 405973
White Crownvetch 356385
White Cudweed 405304
White Cup 169719
White Curldoddy 397043
White Curlylip-orchis 62824
White Currant 334201
White Curvedstamencress 72778,238023
White Cushion Fleabane 150590,150934
White Cutch 401757
White Cypress 85375
White Cypress Pine 67414,67418

White Cypress-pine 67414,67418
White Daffodil 262438
White Daisy 227533
White Dalea 121877
White Dammar 405175
White Dandelion 299688,384628
White Daphne 122567
White Dayflower 100964
White Dead Nettle 220346
White Dead-netde 220346
White Deadnettle 220346
White Dead-nettle 220346
White Deal 298191
White Dendrobium 125033
White Desert Plume 373708
White Deutzia 126839,126861,127075
White Dicyrta 129722
White Dilly 262438
White Dittany 129618
White Diuris 135064
White Dock 340167,340244
White Dogtooth Violet 154896,154932
White Dog-tooth Violet 154896
White Dogwood 104920,279345,407989
White Doll's Daisy 56706
White Doll's-daisy 56704,56705,56706
White Doubleflower Shrubalthea 195270
White Doubleflowered Manyflowered Rose 336787
White Dungbeetle Cajan 65157
White Durra 369601
White Dutch Clover 397043
White Dwarf Tulip 400214
White Eardrops 128307
White Earliporchis 277240
White Early Lilac 382225,382227
White Easterbonnets 152823
White Ebony 132466
White Egyptian Lotus 267698
White Egyptianlotus 267698
White Eight Trasure 200793
White Eightangle 204575
White Elaeagnus 142142
White Elder 96496,345580,407989
White Elm 401431,401549
White Elsholtzia 144105
White Endive 90894
White Epidendrum 146379
White Erect Trillium 397557
White Escallonia 155134
White Evening Primrose 269415,269417,269504
White Evening-primrose 269471,269504
White Everlasting 178286
White False Asphodel 392607
White False Cypress 85351
White False Hellebore 405571
White False Helleborine 405571
White False Indigo 47856
White Falsehellebore 405571
White False-hellebore 405571
White Falsenettle 56110
White Fan Flower 350323
White Fargesia 162746
White Fawn Lily 154896
White Fawnlily 154896
White Featherling 392607
White Felt Thorn 387076
White Felt-thorn 387076
White Fig-tree 165830
White Fingers 65206
White Fir 280,317,384,114690
White Fishook Cactus 354298
White Flat-top Goldenrod 368324
White Flax 231862
White Floss Silk Tree 88971
White Flower Agapetes 10334
White Flower Amur Vetch 408276
White Flower Anise 204598
White Flower Scurrula 355324
White Flower Triadenum 394809
White Flowering Dogwood 105026
White Flowers Azalea 330057
White Flowers Rhododendron 330057
White Fluff 250502
White Forget-me-not 260905,301623,301627
White Forkliporchis 25967
White Forsythia 210
White Foster Tulip 400160
White Four-o'clock 255674
White Foxglove 70119
White Fragrant Orchid 318210
White Frangipani 305213,305221
White Fringe Tree 87736
White Fringed Iris 208642
White Fringed Orchid 302491
White Fringed Orchis 302261
White Fringetree 87736
White Fritillary 168454
White Frog Orchid 318210
White Fumitory 169030
White Gambirplant 401779
White Gambir-plant 401779
White Gardenia 171408
White Garland-lily 187432
White Garlic 15509
White Gastrodia 171919
White Gaura 172201
White Gentian 173193
White Giantarum 20040
White Gilia 175693
White Ginger 187432
White Ginger Lily 187432
White Ginger-lily 187432
White Globe-flower 399490
White Globe-lily 67555
White Globe-tulip 67555
White Glochidion 177100
White Gold 227533
White Golden Rod 367986
White Goldenrod 367986
White Goosefoot 86901
White Gourd 50998
White Gowan 227533
White Grass 224007
White Great Lobelia 234777
White Grevillea 180630
White Ground Ivy 176813
White Groundcherry 227930
White Groundsel 283839
White Guihaia 181803
White Guinea Yam 131814
White Gull 227533
White Gum 155472,155680,155772
White Hair Good Report Rhododendron 330662
White Hair Leibo Maple 3098
White Hairy Lilac 382263
White Hardy Begonia 49887
White Hardy Fuchsia 168776
White Hawkweed 195451
White Heath 380872
White Heath Aster 380872,380953
White Heather 67456,78637
White Heliotrope 190554
White Hellebore 405571,405643
White Helleborine 82038
White Hemlock 399867
White Henbane 201372
White Hibiscus 194722
White Hickory 77934
White Hoarhound 245770
White Hoarypea 385985
White Holcoglossum 197269
White Holly 204114
White Honeysuckle 47647,397043
White Hood 364193
White Hookvine 401779
White Horehound 245723,245770
White Horned Violet 409859
White Horse Nettle 367139
White Horsenettle 367139
White Hubei Anemone 23853
White Hubei Windflower 23853
White Hyssop 203099
White Ifafa Lily 120570
White Indian Jasmine 305213
White Iris 208744
White Iron Bark 155624
White Ironbark 155624
White Isometrum 210186

White Ixora 211105,211148
White Jacaranda 211244
White Japan Floweringquince 84559
White Japan Gentian 154928
White Japanese Spirea 372102
White Jasmine 211940
White Jerusalem Artichoke 18066
White Jew's-mallow 332284
White Jungleflame 211190
White Kerria 332284
White Khaya 216196
White Kitefruit 196826
White Knapweed 81034
White Lacey 30164,356503
White Lady 169719,215275
White Lady Banks Rose 336367,336368
White Lady's Petticoats 374916
White Lady's-slipper 120320
White Largeflower Sobralia 366787
White Larkspur 124106
White Lauan 283900,362201,362226
White Laurel 331207
White Lead Tree 227430
White Leaf Poplar 311208
White Leaf-back Mallotus 243320
White leafy Rock Daisy 291331
White Lettuce 313785,313787
White Lily 68713,229789,262438
White Lilyturf 272086
White Lily-turf 272086
White Lime 391671
White Linden 391727,391836
White Liparis 232075
White Liverwort 284591
White Loosestrife 239594,239712
White Lotus 267698
White Lotus of Egypt 267698
White Lower Surface Rhododendron 332157
White Lupin 238421
White Lupine 238421
White Maass Parodia 284675
White Magnolia 241969,416694
White Mahogany 155722,155769,216196,
395488
White Mahogany Gum 155766
White Mahogyany 155470
White Malabar Nightshade 48689
White Malabarnightshade 48689
White Mallee 155568
White Mallow 18173
White Man's Foot 302068
White Man's Footprints 302068
White Mandarin 377905,377930
White Mandragora 244349
White Mangrove 45746,45750,220233
White Maple 3127,3532
White Marginate Resurrectionlily 215022
White Marguerite 32563
White Mariposa Lily 67635
White Mariposa-lily 67635
White May 30164
White Meadowsweet 371795
White Meat 28044
White Meconopsis 247106
White Melastoma 248732
White Meliot 249203
White Meranti 362200,362226
White Meranti Phayom 362227
White Michelia 252809
White Microtoena 254243
White Mignonette 327795
White Milkweed 37822,38058,38157
White Milkwood 382732
White Milkwort 307898
White Millet 140423
White Mint 250422,250429
White Mintdrop 364193
White Mistletoe 410960
White Mockernut 77934
White Moho 190231
White Money 235090
White Moneywort 18289
White Morningglory 207570
White Morning-glory 207921
White Mountain Avens 138461
White Mountain Larkspur 124409
White Mountain Mint 321953
White Mountain Orchid 318210
White Mountain Ragwort 279901
White Mountain-heather 78637
White Mountains Hairy Goldenaster 194275
White Mountains Wild Buckwheat 152104
White Mucuna 259486
White Mugwort 35570,35770
White Mulberry 259067
White Mullein 405729,405788
White Musky Saxifrage 349303
White Mustard 59284,364545
White Myrtle 202375
White Nan 295388
White Nancy 262438
White Nanmu 295388
White Neesewort 405571
White Nettle 220346
White Nonesuch 235334
White Oak 180571,220228,232609,323625,
323859,323881,323931,324114,324278,
324314
White Ocotillo 167559
White Oldfield Aster 380953
White Old-field Aster 380872,380953,
380955
White Oleander 265327
White Olive 70989,107300
White Onosma 271730
White Orchis 310948
White Orris-root 208565
White Osyris 276870
White Oxalis 278139
White Paintbrush 184347
White Pampas Grass 105456
White Panicle Aster 380912,380918,380919,
380921,380986
White Panicled Aster 380912
White Pansy Orchid 254927
White Paphiopedilum 282817
White Papilioorchis 282931
White Paraboea 283117
White Pareira Rroot 831
White Parkinson Crocus 111502
White Passionflower 285704
White Patrinia 285880
White Pea 222787,222832
White Peachgrass 172201
White Pear 29587,323114
White Pearweed 239566
White Pea-vine 222832
White Peony 280285
White Pepper 300464
White Peppermint 250431
White Peppermint-gum 155709
White Petunia 292751
White Pigeonpea 65154
White Pigweed 18655,86901
White Pincushion 84484
White Pine 299785,299942,300211,306395,
306420
White Pinkshell Azalea 332051
White Pitard Camellia 69497
White Plantain 26551
White Plantain-lily 198639
White Pleione 304209
White Pleurospermum 304755
White Plume Grevillea 180614
White Pocktorchid 282817
White Pomegranate 321765
White Pond Lily 267730
White Popinac 227430
White Poplar 232609,311208,311547
White Poppy 282717
White Porana 131243
White Portugal Broom 120909
White Potato 367696
White Potherb 404439
White Powderpuff 66682
White Prairie Aster 40239,380872,380883,
380887
White Prairie Clover 121882
White Prairie Mallow 362703
White Prairie-clover 121882,121883
White Prairiemallow 362703

White Prickly Poppy 32415
White Primrose 315121
White Pumpkin 219843
White Purplebract Iris 208798
White Purplebract Swordflag 208798
White Purslane 158692
White pygmy-poppy 71081
White Pyrethrum 322751
White Quebracho 39693
White Queen 169719
White Rabbitbrush 150342
White Rabbit-tobacco 317747
White Radish 326620
White Ragweed 19190
White Raintree 61315
White Ramping Fumitory 169030
White Ramping-fumitory 169030
White Ratany 218078
White Rattanpalm 65800
White Rattle 329521
White Rattlesnakeroot 313787
White Ray-floret Fleabane 150754
White Ribseedcelery 304755
White Rice 369322
White Riding Hood 364193
White Rim Yucca 416575
White Robin 363673
White Robin Hood 363673,364193
White Robin's Eye 374916
White Robin's Eyes 374916
White Rock 30164
White Rock Rose 188590
White Rocket 133268,193417
White Rocklettuce 299688
White Rockrose 93124,93153,188590
White Rollingweed 18655
White Roof Iris 208876
White Root 313570
White Roscoea 337082
White Rose 336343
White Rose of York 336343
White Roseau 182946
White Rosebay 85550
White Rose-bay 330057
White Rose-of-York 336343
White Rosinweed 364256
White Rot 200388,299759
White Rugose Rose 336902
White Runch 326609
White Sade Orchid Tree 252809
White Sage 35837,218132
White Sagebrush 35837
White Sails 370344,370345
White Salix 342973
White Sallee 155688
White Sallow 1358
White Sallow Wattle 1528
White Sally 155667,155688,156068
White Sand Verbena 700
White Sandal Wood 346210
White Sandalwood 346210
White Sands Spider-lily 200923
White Sapote 78175
White Sassafras 347407
White Satin 238371
White Saxifrage 349419
White Saxifrage Tunicflower 400356
White Saxoul 185072
White Scale Beakrush 333480
White Sea-blite 379553
White Sedge 73628,74235
White Sepals Pieris 298729
White Seraya 283891,283902
White Shaddock 93469
White Shaving Brush Tree 56789
White Shield Rhododendron 331099
White Shirt 68713
White Shooting Stars 266059
White Silk Cotton 98113
White Silk Cotton Tree 80120
White Silk Floss Tree 88971
White Silkcotton 98113
White Silky Oak 180602
White Silky Wisteria 414584
White Silverballi Louro 268692
White Sincle 299759
White Siris 12587,13595,13649
White Skapanthus 365906
White Smelowskia 366116
White Smock 68713,102933
White Snake Root 158305
White Snakeroot 11152,11154,158295,
158354,313787
White Snapjack 364146
White Soaproot 183174
White Soldiers 23925
White Spanish Broom 120909,120979
White Speedwell 317912
White Spider Wort 394021
White Spiked Rhododendron 331873
White Spiny-club Palm 46382
White Spread Atylosia 65157
White Sprite Cherry-plum 316294
White Spruce 298286
White Squill 352960,352969
White Star 153333
White Star-apple 90015
White Stargrass 14486
White Stem Pine 299786
White Sticky Catchfly 363583
White Sting Nettle 220346
White Stinkweed 123077
White Stinkwood 80571
White Stokes Aster 377289,377294
White Stokes' Aster 377288
White Stoncbreak 349419
White Stone Crop 356503
White Stonebean-orchis 63044
White Stone-break 349419
White Stonecrop 356503
White Stringbush 414271
White Stringybark 155588
White Sumach 332477
White Sunbonnet 85990
White Sunday 374897,374916
White Sundrops 269504
White Sunipia 379776
White Sunrose 188762
White Suregada 379794
White Swallowwort 117743
White Swallow-wort 117743
White Swamp Azalea 332086
White Sweet Clover 249203,249232
White Sweet Pea 222787
White Sweetclover 249203
White Sweet-clover 249203
White Swertia 380111
White Swordflag 208665
White Tack-stem 68446
White Taiwan Beautyberry 66780
White Tamarisk 383431
White Tanoak 233174
White Tansy 312360
White Tassel Flower 66682
White Tatarian Statice 230788
White Tea Tree 248104
White Tephrosia 385985
White Thistle 44490,92030
White Thorn 1148,1290,109790,109857
White Thoroughwort 158031
White Tip Willow 343529
White Titi 120488
White Toadshade 397542
White Top 73090,73098,150464
White Torch Cactus 140888
White Trailing Iceplant 123915
White Tree 248104
White Trefoil 250502,397043
White Trigonotis 397446
White Trillium 397563,397570,397585
White Triratna Tree 397368
White Trout Lily 154896,154932
White Trout-lily 154896
White Trumpet Lily 229900
White Turtlehead 86783
White Turtle-head 86783
White Twayblade 232075
White Twinevine 347006
White Upland Aster 41086,368324
White Upright Mignonette 327795
White Vandopsis 404754

White Variegated Hakone Grass 184657
White Veind Arrowhead Vine 381869
White Velvet 394075
White Vervain 405903,405904
White Vine 61497
White Vine Spinach 48689
White Vinespinach 48689
White Vine-spinach 48689
White Violet 409661,409665,410165,410484
White Wake Robin 397563
White Wakerobin 397557
White Wake-robin 397570
White Wall Rocket 133268
White Wallflower 246478
White Wallrocket 133268
White Walnut 212599
White Wand Beard-tongue 289381
White Ward's Rhododendron 332094
White Watch-and-chain 334976
White Water Crowfoot 325580,325589,326045
White Water Lily 267648,267730
White Water Mint 250291
White Water-buttercup 48933,326045
White Waterlily 267767
White Water-lily 267627,267648,267730,267777
White Wax-tree 229529
White Western Laurel 215403
White Whitlow Wort 137205
White Wicky 215392
White Wild Indigo 47856,47857,47871
White Wild Vine 61497
White Willow 342973,342979,342986
White Willow Dock 340244
White Windhairdaisy 348462
White Winter 149565
White Winteraconite 148102
White Wisteria 414577
White Witch Hazel 237190,237191
White Woadwaxen 172903
White Wombgrass 365906
White Wood 43740,232602,232609,398004
White Wood Aster 40322,160657
White Wood Lily 397540
White Wood Sorrel 277648
White Woodland Aster 380922
White Wood-rush 238640
White Woolly daisy 152823
White Woolly Mint 250385
White Wormwood 35770,35839
White Yam 131458,131814
White Yerbadetajo 141374
White Yiel-yiel 180602
White Yulan 416694
White Yule 50825
White Zeuxine 417774
White Zinnia Southern Zinnia 418036
White-Alder 96454
White-and-red 37015
White-and-yellow-flower Cornlily 210880
Whitearil Yew 318538,318539
White-aril Yew 318538,318539
White-awhile Vine 65637
Whiteback 311208
Whiteback Actinidia 6633
Whiteback Alfredia 14599
Whiteback Aralia 30613,30634
Whiteback Aster 40599
Whiteback Azalea 330137,331099
Whiteback Bay-leaf Willow 343859
Whiteback Beautyberry 66900
Whiteback Cinquefoil 312671
Whiteback Deutzia 126976
Whiteback Erythrospermum 155034
Whiteback Gerbera 175199
Whiteback Groundsel 365062
White-back Honeysuckle 235862
Whiteback Hydrangeavine 351801
Whiteback India Pentanema 289569
White-back Japanese Helwingia 191184
Whiteback Kiwifruit 6633
White-back Leaf Mallotus 243320
Whiteback Magnolia 198698
Whiteback Purplepearl 66900
Whiteback Redseedtree 155034
Whiteback Saussurea 348160
Whiteback Schefflera 350716
Whiteback Sida 362617
Whiteback Thundergodvine 398317
Whiteback Windhairdaisy 348160
Whiteback Yushanbamboo 416778
White-backed Aster 40599
White-backed Beautyberry 66900
White-backed Deutzia 126976
White-backed Hydrangea-vine 351801
White-backed Mahonia 242559
White-backed Rhododendron 331099
White-backed Schefflera 350716
Whitebackleaf Wildtung 243320
White-back-leaved Tree 243320
White-baked Hydrangea 199904
Whiteball Acacia 1051
Whiteband Yam 131556
Whitebark 53635
Whitebark Acacia 1351
Whitebark Calligonum 67043
Whitebark Jurinea 211979
White-bark Maple 3100
Whitebark Peashrub 72270
Whitebark Pea-shrub 72270
Whitebark Pine 299786
White-bark Pine 299786
White-bark Poplar 311257
Whitebark Spruce 298233
Whitebark Tanoak 233174
Whitebark Tarenna 384936
Whitebark Willow 343890
White-barked Calligonum 67043
Whitebarked Himalayan Birch 53481
White-barked Himalayan Birch 53481
White-barked Hydrangea 199932
Whitebarked Maple 3100
White-barked Tanoak 233174
White-barked Tarenna 384936
Whitebeak Beakrush 333495
Whitebeam 369290,369322
Whitebeam Mountain Ash 369322
White-bear Sedge 73633
Whitebell Azalea 332016
White-bell Honeysuckle 235674
Whiteblow Grass 153964
Whiteblue Gentian 173580
Whitebowl Azalea 331861
Whitebox 155605
White-braced Stick-leaf 250484
Whitebract Arisaema 33289
Whitebract Gymnotheca 182849
Whitebractcelery 266974
White-bracted Thoroughwort 158198
Whitebracteole Bugle 13133
White-branch Linociera 87713
Whitebranch Liolive 87713
White-branch Milkvetch 42613
Whitebranch Oak 116050
Whitebranch Russianthistle 344453
WhitebranchAzalea 330476
White-branched Cyclobalanopsis 116050
White-branched Linociera 87713
White-branched Milkvetch 42613
White-branched Rhododendron 330476
White-branched Russianthistle 344453
White-branched Tanoak 233289
White-branches Tanoak 233289
Whitebristle Cottonsedge 152791
Whitebrush 17600,232499
Whitebugle Azalea 331914
Whitebutton Azalea 330338
Whitebutton Rhododendron 330338
White-button Rhododendron 330338
Whitebuttons 21276
Whitecalyx Cinquefoil 312408
Whitecalyx Dragonhead 137603
Whitecalyx Greenorchid 137603
Whitecalyx Hairycalyx Blueberry 403975
White-calyx Jasmine 211721
Whitecalyx Kiwifruit 6514
White-calyx Leafflower 296637
White-calyx Underleaf Pearl 296637
White-calyxed Jasmine 211721

White-cedar 85263
Whitecephalous Leek 15428
Whitecephalous Onion 15428
White-cloaked Cudweed 170922
Whitecloud Hundredstamen 389829
White-column Foxtail Cactus 107072
Whitecotton Didymocarpus 129939
Whitecottony Catnip 264980
Whitecreeper Euonymus 157473
White-creeper Euonymus 157473
White-cup 266206
Whitedaisy 227467
Whitedaisy Tidy Tips 223485
White-doesal Honeysuckle 235862
Whitedragon Ligusticum 229354
Whitedragon Rabdosia 324851
White-dragon Storax 379336
Whitedrumnail 307642,307656
Whiteear Leadwort 305175
Whiteedge Dandelion 384749
Whiteedge Gentian 173197
Whiteedge Pleurospermum 304754
Whiteedge Ribseedcelery 304754
White-edge Sedge 74265,74271
White-eye Rose 336506
White-eyed Grass 365715
Whiteface Azalea 332157
Whiteface Lupine 238439
Whitefeather Calathea 66180
Whitefelt Leucas 227654
Whitefloer Multijugate Sweetvetch 188018
Whiteflower 374916
Whiteflower Agapetes 10334
Whiteflower Aging Onion 15735
Whiteflower Albizia 13663
White-flower Albizzia 13526
Whiteflower Amberboa 18954
Whiteflower Angraecum 24823
Whiteflower Anna 25753
Whiteflower Arnold Milkvetch 42018
White-flower Asiaglory 32657
Whiteflower Autumn-crocus 99300
Whiteflower Baikal Skullcap 355388
White-flower Bauhinia 49248
Whiteflower Betonicalike Jerusalemsage 295065
Whiteflower Bicolor Bushclover 226706
Whiteflower Bird Vetch 408358
Whiteflower Bloodred Iris 208807
Whiteflower Bloodred Swordflag 208807
White-flower Bluebell Falsecuckoo 48141
Whiteflower Blueberry 403711
Whiteflower Bouquet Larkspur 124239
Whiteflower Broad-claw Vetch 408457
Whiteflower Broadleaf Vetch 408264
Whiteflower Bushclover 226843
Whiteflower Butterflybush 61968
Whiteflower Cacalia 65040
Whiteflower Cactus 244000
Whiteflower Canscora 71273
Whiteflower Cape-cowslip 218883
Whiteflower Cephalorrhynchu 82412
Whiteflower Chelonopsis 86807
Whiteflower China Houndstongue 117910
Whiteflower Chinese Betony 373174
Whiteflower Chinese Cymbidium 117051
Whiteflower Chinese Pulsatilla 321673
Whiteflower Clematis 95101
Whiteflower Clover 396959
White-flower Common Lilac 382368
Whiteflower Common Pyrola 322817
Whiteflower Conyza 103516
Whiteflower Corydalis 106068
Whiteflower Crazyweed 279187
White-flower Creeping Skyflower 139068
Whiteflower Dahurian Milkvetch 42259
Whiteflower Dahurian Pulsatilla 321677
Whiteflower Dahurian Thyme 391168
Whiteflower Dandelion 384628
Whiteflower Diapensia 127922
Whiteflower Doublelip Dendrobium 125185
Whiteflower Dragonhead 137596
Whiteflower Eagleclawbean 370204
Whiteflower Earthgallgrass 143474
Whiteflower Earth-silkworm Betony 373223
Whiteflower Echeveria 139975
Whiteflower Elaeagnus 142142
White-flower Elegant Tickclover 126334
Whiteflower Embelia 144793
Whiteflower Entireleaf Dragonhead 137602
Whiteflower Entireleaf Greenorchid 137602
Whiteflower Eritrichium 153475
Whiteflower Erysimum 154494
Whiteflower False Largeflower Motherwort 225006
Whiteflower Falserobust Vetch 408553
Whiteflower Falsestonecrop 318390
White-flower Farges Catalpa 79251
Whiteflower Fascicular Gentian 173978
Whiteflower Fine-leaved Vetch 408632
Whiteflower Fishvine 125940
Whiteflower Floatingheart 267802
Whiteflower Gansu Woodbetony 287303
Whiteflower Goldband Lily 229737
Whiteflower Goldenbush 150325
Whiteflower Gorgeous Gentian 173901
Whiteflower Greenorchid 137596
Whiteflower Greyblue Sophora 369152
Whiteflower Greyhair Pagodatree 369152
Whiteflower Gueldenstaedtia 181640
White-flower Hall Crabapple 243710
Whiteflower Hemsley Monkshood 5259
White-flower Hiptage 196826
Whiteflower Hollowstem Iris 208475
Whiteflower Hollowstem Swordflag 208475
Whiteflower Hoodshaped Orchid 264766
Whiteflower Huashan Milkvetch 42475
Whiteflower India Sundew 138296,138322
Whiteflower Indian Nightshade 367243
White-flower Indian Rhododendron 331841
Whiteflower Iris 208665
Whiteflower Ixora 211105
Whiteflower Japanese Lily 229874
Whiteflower Japanese Spiraea 371957
Whiteflower Jewelvine 125940
Whiteflower Knotweed 309015
Whiteflower Kurrajong 376102
Whiteflower Lagopsis 220126
Whiteflower Leadwort 305202
Whiteflower Leonotis 224588
Whiteflower Lily 229805,229890,229924
Whiteflower Lobate Wildcotfon 402248
Whiteflower Longpetal Parnassia 284562
Whiteflower Loosestrife 239676
Whiteflower Lovely-shaped Rhododendron 330280
Whiteflower Lycaste 238736
Whiteflower Lycoris 239255
Whiteflower Maire Camellia 69341
Whiteflower Makino's Rhododendron 331188
Whiteflower Manyleaf Paris 284361
Whiteflower Maries' Azalea 331204
Whiteflower Maxillaria 246706
Whiteflower Mazus 246990
Whiteflower Meconopsis 247106,247172
Whiteflower Melia 248896
Whiteflower Michelia 252927
Whiteflower Microtoena 254243
White-flower Milkvetch 42612
Whiteflower Mongolian Thyme 391287
Whiteflower Morina 2111,258862
Whiteflower Mucuna 259486
Whiteflower Nakedstamen Monkshood 5238
Whiteflower Navelwort 270705
Whiteflower NE. China Violet 410202
Whiteflower Neillia 263178
Whiteflower Neoregelia 264407
Whiteflower Nepal Iris 208518
Whiteflower Nepal Swordflag 208518
Whiteflower Northeast Crazyweed 278993
Whiteflower Oleander 265328
Whiteflower Olgaea 270236
Whiteflower Onion 15899
Whiteflower Onosma 271730
Whiteflower Ovateleaf Azalea 330280
Whiteflower Pagodatree 368959
Whiteflower Pair Vetch 408653
White-flower Paniculate Glorybower 96258
Whiteflower Paperbush 141467
Whiteflower Paraphlomis 283607
Whiteflower Parnassia 284613

Whiteflower Patrinia 285880
Whiteflower Peacock Poppy 282660
Whiteflower Peony 280147,280285
Whiteflower Periwinkle 79419
Whiteflower Persian Lilac 382242
Whiteflower Pilose Vetch 408542
Whiteflower Pomatocalpa 310797
Whiteflower Pricklepoppy 32415
White-flower Prickly Rose 336324
Whiteflower Purplevine 414577
White-flower Purplevine 414584
Whiteflower Pyrethrum 322751
Whiteflower Rabbitbrush 150307
Whiteflower Raspberry 338737
Whiteflower Resupinate Woodbetony 287596
Whiteflower Resurrectionlily 214999
Whiteflower Ricebag 181640
Whiteflower Rockjasmine 23202
Whiteflower Rockvine 387745
Whiteflower Rodriguezia 335163
Whiteflower Rose Sandwort 32182
Whiteflower Roughleaf Gentian 173918
Whiteflower Roundleaf Bugle 13155
Whiteflower Rush 213220
Whiteflower Salttree 184840
Whiteflower Sarcochilus 346712
Whiteflower Schefflera 350731
Whiteflower Schultzia 352689
Whiteflower Selfheal 316148
Whiteflower Shawur Larkspur 124576
Whiteflower Siberia Burnet 345921
Whiteflower Silver Maple 3533
Whiteflower Skullcap 355773
White-flower Skyrocket 208320
Whiteflower Slenderleaf Pulsatilla 321717
Whiteflower Small Bletilla 55566
Whiteflower Smooth Sweetvetch 187968
Whiteflower Songpan Milkvetch 43114
Whiteflower Songpan Monkshood 5613
Whiteflower Sophora 368959
Whiteflower Spapweed 205449
Whiteflower Specious Lily 230037
Whiteflower Spured Gentian 184693
Whiteflower Stanhopea 373689
Whiteflower Star-of-bethlehem 274810
White-flower Stauntonvine 374420
Whiteflower Stringbush 414271
Whiteflower Suborbicular Chiogenes 87681
Whiteflower Sungpan Monkshood 5613
Whiteflower Tangut Milkvetch 43126
Whiteflower Tarenna 384988
Whiteflower Tenuidivided Woodbetony 287755
Whiteflower Thorowax 63596
Whiteflower Thrixspermum 390488
Whiteflower Touch-me-not 205087,205449
White-flower Tree Azalea 330062
White-flower Tree Rhododendron 330062
White-flower Trigonostemon 397368
Whiteflower Trigonotis 397426
Whiteflower Tuber Ullucus 401413
Whiteflower Tuoli Fritillary 168591
Whiteflower Twocolored Monkshood 5024
White-flower Velvety Sophora 368959
Whiteflower Vetch 408254
Whiteflower Villose Vetch 408694
Whiteflower Violet 410360
White-flower Weaversbroom 370204
Whiteflower Whytockia 414007
White-flower Wild Quince 374420
Whiteflower Wild Soja 177781
Whiteflower Wild Soybean 177781
White-flower Wisteria 414584
Whiteflower Wood Elsholtzia 144097
Whiteflower Wormwoodlike Motherwort 224991
Whiteflower Yellow Bedstraw 170757
Whiteflower Yinshania 416317
Whiteflower Zhejiang Hilliella 196312
Whiteflower Zhongdian Swordflag 208867
White-flowered Agapetes 10334
White-flowered Albizia 13526
White-flowered Anise-tree 204575
Whiteflowered Bittercress 72858
White-flowered Bush-clover 226844
White-flowered Butterfly-bush 61968
White-flowered Cacalia 65040
White-flowered Caesalpinia 65040
White-flowered Calotrope 68095
White-flowered Chaste-tree 411365
White-flowered Chelonopsis 86807
White-flowered Chinese Lilac 382133
White-flowered Currant 334198,334200
White-flowered Elaeagnus 142142
White-flowered Embelia 144793
White-flowered Glorybower 96083
White-flowered Gourd 219827,219843
White-flowered Grass 374916
White-flowered Ground-cherry 227932
White-flowered Hiptage 196826
White-flowered Ixora 211105
White-flowered Jewelvine 125940
White-flowered Leadwort 305202
White-flowered Leaf-cup 310023
White-flowered Lily 229890
White-flowered Menzies' Larkspur 124380
White-flowered Michelia 252927
White-flowered Milkweed 38157
White-flowered Mucuna 259486
White-flowered Neillia 263178
White-flowered Parsley 99047
White-flowered Paulownia 285960
White-flowered Peony 280147
White-flowered Pyrenaria 322565
White-flowered Raspberry 338737
White-flowered Redbud 83771
White-flowered Schefflera 350731
White-flowered Semiserrate Camellia 69616
White-flowered Serissa 360935
White-flowered Sophora 368959
White-flowered Staunton-vine 374420
White-flowered Stringbush 414271
White-flowered Sweetvetch 187815
White-flowered Tarenna 384988
White-flowered Wandering Jew 394021
White-flowered Wisteria 414584
White-flowered Young Epimedium 147077
White-flowered Yumin Fritillary 168644
Whitefoot Portiatree 389941
White-fruied Common Barberry 52328
White-fruit Agapetes 10327
Whitefruit Alleizettella 14928
Whitefruit Amomum 19870
White-fruit Blueberry 403887
White-fruit China Whitepearl 172153
White-fruit Chinese Magnoliavine 351023
White-fruit Chinese Wintergreen 172153
White-fruit Gaultheria 172097
Whitefruit Japan Eurya 160506
Whitefruit Japan Purplepearl 66809
Whitefruit Mistletoe 410960
White-fruit Paper-mulberry 61108
Whitefruit Stonecrop 356877
Whitefruit Violet 410403
Whitefruit Whitepearl 172097
White-fruited Agapetes 10327
White-fruited Blueberry 403887
Whitefruited Grape 79850
White-fruited James Barberry 51782
White-fruited Mulberry 259070
White-fruited Randia 14928
Whitefruited Redflower Salsify 394338
White-fruited Skimmia 365931
White-fruited Wintergreen 172097
Whitefuit Heavenly Bamboo 262200
Whitefuit Nandina 262200
White-giant Camellia 68894
Whiteglabrous Mustard 59526
Whitegrass 223973,224007,289096
White-grass 224007,261120
Whitegreen Sedge 75131
Whitegreen Squill 353061
White-grey Edelweiss 224770
White-greyleaf Blueberry 403841
Whitehair Atylosia 65154
Whitehair Azalea 332053
Whitehair Bulbophyllum 62541
White-hair Bush Cinquefoil 289717
Whitehair Camellia 68968
Whitehair Chickweed 375045
Whitehair Cinquefoil 312721,312727

Whitehair Colquhounia 100048
Whitehair Conehead 320130
Whitehair Cotoneaster 107726
White-hair Cotoneaster 107557
Whitehair Crazyweed 279048
Whitehair Curlylip-orchis 62541
Whitehair Decaspermum 123444
Whitehair Dogwood 124930
Whitehair Dontostemon 136191
Whitehair Four-involucre 124930
White-hair Giantgrass 175589
White-hair Glabrous Indocalamus 206882
Whitehair Glochidion 177103
White-hair Hydrangea 200029
Whitehair Indosasa 206882
Whitehair Leibo Maple 3098
Whitehair Mountainash 369301
Whitehair Negundo Chastetree 411362
Whitehair Peashrub 72273
White-hair Raspberry 338617
Whitehair Smallleaf Cinquefoil 312858
Whitehair Torchflower 100048
Whitehair Xizang Taxillus 385239
White-haired Balfour's Rhododendron 330199
White-haired Beautyberry 66761
Whitehaired Camellia 68968
White-haired Decaspermum 123444
White-haired Giant-grass 175589
White-haired Goldenrod 367945
White-haired Leather-flower 94709
White-hairs Camellia 68898
Whitehairy Beautyberry 66761
Whitehairy Bethlehemsage 283605
Whitehairy Chastetree 411362
Whitehairy Nanchuan Clethra 96538
Whitehairy Paraphlomis 283605
Whitehairy Purplepearl 66761
Whitehairy Rockfoil 349654
Whitehart Mockernut 77934
Whitehaw 109970
Whitehead 285086,401112
Whitehead Cacalia 283839
Whitehead Elatostema 142714
Whitehead Milkvetch 42612
Whitehead Sedge 218593
Whitehead Spikesedge 218579
Whitehead Stairweed 142714
Whiteheadtree 171561
White-heart Hickory 77934
White-hill Azalea 331239
White-horny-leaf Iris 208494
White-hypophyllous Litsea 234048
Whitelady 390767
Whitelanose Eria 125571
Whitelanose Hairorchis 125571
White-large Camellia 68894
Whiteleaf 372106
Whiteleaf Crystal Tea 223912
Whiteleaf Crystal Tea Ledum 223912
Whiteleaf Currant 334044
Whiteleaf Everlasting 189458
Whiteleaf Jadeleaf and Goldenflower 260479
Whiteleaf Japanese Magnolia 198698
White-leaf Japanese Magnolia 198698
Whiteleaf Ledum 223912
Whiteleaf Ligusticum 229331
Whiteleaf Magnolia 198698
Whiteleaf Manzanita 31123,31159
Whiteleaf Mountainash 369375
Whiteleaf Mussaenda 260479
Whiteleaf Nepal Osbeckia 276132
Whiteleaf Nepeta 264981
Whiteleaf Oak 324021
Whiteleaf Olgaea 270234
Whiteleaf Orach 44504
Whiteleaf Rabdosia 209742
Whiteleaf Raspberry 338628
White-leaf Rock-rose 93124
Whiteleaf Sagebrush 35816
Whiteleaf Sunflower 188972
Whiteleaf Tree 369322
Whiteleaf Windhairdaisy 348488
White-leaved Marlock 155759
White-leaved Rabdosia 209742
White-leaved Raspberry 338628
White-leaved Rock Rose 93154
White-leaved Rockrose 93124
White-leaved Willow 343527
White-lettuce 313787
Whiteligulate Aster 40105
Whiteligulatecorolla Aster 40105
White-line Bamboo 47178
White-lineate Bambusa 47178
Whitelip Lady's Slipper 120332
Whitelip Ladyslipper 120332
Whitelip Oncidium 270819
White-longspur Calanthe 65867
Whitemaculate Boesenbergia 56533
Whitemaculate Wildginger 37674
Whitemargin Calathea 66164
Whitemargin Plantainlily 198592
Whitemargin Pussytoes 26460
White-margin Sasa 347324
White-margin Starry Puncturebract 362997
White-marginata Japanese Spindle-tree 157614
White-margined Agave 10782
White-margined Barberry 51907
White-margined Cluster Plantainlily 198598
White-margined Spurge 159313
White-margined Wax-plant 178010
White-mealy Metasasa 252534
White-moon Petunia 292751
White-moss Cabbage Rose 336478
Whitemouth Dayflower 101008
White-mouth Dayflower 101008
Whiten Eriocycla 151747
Whitened Barberry 51528
Whitened Cowparsnip 192249
Whitenerved Kudoacanthus 218273
Whitenetvein Pyrola 322779
White-netvein Spotleaf-orchis 179625
Whitening Iris 208440
Whitenode Bambusa 47243
Whitenode Spine Bambusa 47243
Whitepearl 172047,172067,172133
Whitepetal Rockfoil 349276
White-petiole Sophora 368961
Whitephloem Acacia 1351
Whitepilose Rabdosia 209621
Whitepine Pagodatree 368994
White-plume Grevillea 180614
Whitepoint Loco 278940
Whitepopinac Leadtree 227430
White-popinaced Leadtree 227430
Whitepurple Azalea 330913
Whitepurple Groundsel 358223
White-purple Maple 2775
Whiter Daphne 122532
White-race Poppy 282722
Whiteraceme Goldenray 228984
White-rayed Pentachaeta 289422
Whiterib Calathea 66178
White-ring Metasasa 252534
White-rod 407737
Whiteroot 299759,308607
Whiteroot Acanthopanax 143628
White-rooted Acanthopanax 143628
Whitescale Beakrush 333480
Whitescale Cypressgrass 119262
Whitescale Galingale 119262
Whitescale Litse 233840
White-scale Saltbush 44383
Whitescale Sedge 75792
Whitescale Woodsorrel 277932
White-scaled Sedge 73622
Whitescale-like Sedge 75852
White-sericeous Camellia 68896
Whitesh Cactus 243997
Whitesheep Holeglumegrass 57567
Whiteshing Blueberry 403759
Whitesilk Camellia 68896
Whitesilkgrass 87775
Whitesimpleflower Shrubalthea 195298
White-snaketotue-grass 187565
Whitespathe Southstar 33289
Whitespike 370488,370490
Whitespike Acanthus 2690
Whitespike Tanoak 233161
White-spiked Acanthus 2690
Whitespine Peashrub 72271

White-spine Pea-shrub 72271
Whitespine Schischkinia 351125
White-spine Thistle 91930
Whitespinedaisy 351124,351125
Whitespire Japanese White Birch 53575
Whitespot 373455
Whitespot Betony 373455
Whitespot Giant Arum 20057,20125
Whitespot Giant-arum 20142
Whitespot Haworthia 186644
Whitespot Taiwan Begonia 49849
White-spotted Scabrous Deutzia 127083
Whitestachys Corydalis 106543
Whitestamen Indianmulberry 258884
Whitestamen Manglietia 244421
Whitestar 207921
Whitestem Distaff Thistle 77726
White-stem Distaff Thistle 77726
Whitestem Globethistle 140654
Whitestem Goldenbush 150320
Whitestem Gooseberry 334043
White-stem Gooseberry 334043
Whitestem Meadowrue 388552
Whitestem Milkvetch 41948
Whitestem Pondweed 312232
White-stem Pondweed 312232
Whitestem Serpentroot 354801
Whitestem Spunsilksage 360879
Whitestem Undulate Lettuce 219590
White-stemmed Bramble 338267
White-stemmed Filaree 153869
Whitestemmed Gooseberry 334043
Whitestemmed Heronbill 153869
White-stemmed Milkweed 37822
White-stemmed Pondweed 312232
Whitestriate Gentian 173307
Whitestripe Bamboo 304007
White-striped Square-bamboo 87565
White-striped-leaf Pleioblastus 304110
Whitestyle Vanda 404627
Whitethorn 1148,109790,109857,266355
Whitethorn Acacia 1148
Whitethread Yam 375831
Whitethroat Monkshood 5359
White-throat Rhododendron 331944
White-tinge Sedge 73626
Whitetomentose Ainsliaea 12698
White-tomentose Eria 148622
Whitetongue Gentian 150754
Whitetongue Shortscape Fleabane 150514
Whitetop 73085,73090,150925,201203,
225279
White-top 73090,155558
Whitetop Fleabane 150464,150980
Whitetop Sedge 333477
White-topped Aster 360702
Whitetopped Bluegrass 305278
White-topped Sedge 333524
Whitetree 379792
Whitetree Calvepetal 177991
Whitetung 94051
Whitetwig Milkvetch 42613
Whitetwig Qinggang 116050
Whitetwig Tanoak 233289
White-variegated Common Barberry 52329
White-variegated Eastern Arborvitae 390592
White-variegated Ivy 187249
Whitevein Gynostemma 183015
Whitevein Hetaeria 193592
Whitevein Onion 15560
Whitevein Ploughpoint 401149
Whitevein Typhonium 401149
Whitevein Zeuxine 417738
White-veined Shinleaf 322869
Whitevelvety Rabdosia 209621
Whitevillose Camellia 68898
White-villosed Camellia 68898
Whitevillous Monkshood 5663
Whiteviolet Groundsel 358223
Whitewater Spineflower 89133
Whiteweed 4005,28044,227533
Whitewing Grevill 180614
Whitewing Hakea 184616
Whitewood 269,272,43740,71132,204060,
232609,249175,298191,311208,391646,
391649,391706,407907
White-wood Ailanthus 12560
Whitewoolly Rose 336993
Whitewort 26746,308607,322724
Whitey Wood 5877
White-yellow Barbed Gentianopsis 174206
Whiteyellow Dischidia 134047
Whiteyellow Iris 208734
Whiteyellow Sobralia 366784
Whiteywood 249175
Whit-flower Cuspidata Hyssop 203085
Whitflower M. Degron's
Rhododendron 330534
Whitfordiodendron 413975
Whithe Adnate Elder 345563
Whithin Pear 369322
Whitish Adnate Elder 345564
Whitish Blueberry 403711
Whitish Elytranthe 144595
Whitish Indumentum Rhododendron 330040
Whitish Laelia 219669
Whitish Meliosma 249357
Whitish Notylia 267214
Whitish Onion 15032
Whitish Rattlebox 111879
Whitish Rhododendron 330040
Whitish Sedge 74004
Whitish Skullcap 355353,355583
Whitish Water-milfoil 261360
Whitlavia 293179
Whitle Hairyflower Actinidia 6589
Whitle Hairyflower Kiwifruit 6589
Whitlow Grass 136899,136954,136972,
137126,153964,159027,349999
Whitlow Pepperwort 73090
Whitlow Wort 137126
Whitlowgrass 136899
Whitlow-grass 136899,153949,153964
Whitlowwort 136899,284735
Whitlow-wort 136899,284735
Whitney 407907
Whitney's bristleweed 186891
Whitsun Ball 280231,407989
Whitsun Balls 407989
Whitsun Boss 407989
Whitsun Flower 277648,407989
Whitsun Gillies 193417
Whitsun Gilliflower 193417
Whitsun Tassel 407989
Whitsun Tassels 407989
Whitsunfide Boss 407995
Whitsuntide 382329
Whitsuntide Gillif Ower 72934
Whitten 369322
Whittenbeam 407907
Whitten-tree 369339,407907,407989
Whitty 369339
Whitty Pear 369381,369541
Whitty-bush 2837,369541
Whitty-pear 369381,369541
Whit-veined Mananthes 244282
Whltty-tree 369339,407907
Whole Woolly Aster 40587
Wholebeak Woodbetony 287020
Wholebill Woodbetony 287020
Wholeglabrous Eurya 160474
Whole-gland Machilus 240598
Wholeleaf Rosinweed 364285
Whole-leaf Rosinweed 364285
Wholelipflower 197386
Wholen-leaf Hawthorn 329086
Whored Cinquefoil 313098
Whoreman's Permacetty 72038
Whoreman's Permacity 72038
Whorl Flower 258831
Whorl Leaved Lily 284358
Whorled Aster 39920,268682
Whorled Bristlegrass 361930
Whorled Carraway 77851
Whorled Clary 345471
Whorled Coreopsis 104616
Whorled Dropseed 372813
Whorled Giant-grass 175611
Whorled Hedyotis 187697
Whorled Knotweed 204469
Whorled Loosestrife 123526,123527,

239804,239810
Whorled Mallow 243862
Whorled Milkweed 38080,38160
Whorled Milkwort 308437,308438,308442
Whorled Mint 250465
Whorled Nut Grass 354274
Whorled Pedicularis 287802
Whorled Plantain 301858,301864,305185
Whorled Plectranthus 303748
Whorled Pogonia 210334,210336
Whorled Pogonia Orchid 210334
Whorled Rosinweed 364264,364321
Whorled Sage 345471
Whorled Silkweed 38160
Whorled Solomon's Seal 308659
Whorled Solomon's-seal 308659
Whorled Stonecrop 357224
Whorled Sunflower 189078
Whorled Water-milfoil 261379
Whorled Wood Aster 39920,268682
Whorled Yellow Loosestrife 239810
Whorledleaf Landpick 308659
Whorledleaf Solomonseal 308659
Whorledleaf Woodbetony 287802
Whorled-leaved Syzygium 382548
Whorlflower 258831,258853
Whorl-flower 258853
Whorlflower Spicegrass 239868
Whorl-grass 79194,79198
Whorlleaf Ardisia 31547
Whorlleaf Banksia 47673
Whorlleaf Basketvine 9421
Whorlleaf Blueberry 404036
Whorlleaf Cinquefoil 313098
Whorlleaf Eight Trasure 200816
Whorlleaf Elangate Litse 233903
Whorlleaf Litse 234098
Whorlleaf Loosestrife 239700
Whorlleaf Rotala 337368
Whorlleaf Sandwort 31920
Whorlleaf Stonecrop 200816
Whorlleaf Stringbush 414261
Whorlleaf Syzygium 382548
Whorl-leaved Acacia 1686
Whorl-leaved Ardisia 31547,31571
Whorl-leaved Blueberry 404044
Whorl-leaved Litse 234098
Whortleberry 403710,403915,403916
Whortleberry Cactus 261706
Whortleberry Willow 343745
Whortleberry-like Willow 343745
Whortle-bilberry 403916
Whortle-leaved Willow 343743
Whorts 403916
Who-stole-the-donkey 170193
Whurt 403916
Whuttle-grass 249232
Whya-tree 334976
Whytockia 413997,413999
WI Locust 200844
Wibrow-wobrow 109857,302068
Wichita Juniper 213922
Wichura Rose 337015
Wichura Spikesedge 143414
Wicked Herb 166038
Wicked Tree 115008
Wicken 144661
Wicken Tree 369339
Wickenwood 369339
Wicker-stem Wild Buckwheat 152629
Wickerware Cactus 329683
Wickham's Grevillea 180661
Wickup 85875,146614
Wicky 215387,215394
Wicopy 85875,133656
Widbin 236022,380499
Widbin Pear 369322
Widbin Pear Tree 369322
Widdringtonia 414058
Widdy 344244
Widdy-wine 102933
Wide Calyx Rhododendron 330591
Wide Jaeschkea 211400
Wide Wing Synotis 381920
Wide-calyxed Rhododendron 330591
Widefilament Swertia 380296
Wide-globe Eastern Arborvitae 390626
Widehead Groundsel 279906
Wideleaf Bamboo 125482
Wideleaf Osbeckia 276098
Wideleaf Razorsedge 354255
Wideleaf Rockfoil 349584
Wideleaf Sealavender 230655
Wideleaf Wood Lily 229985
Wideleaved Cinna 91239
Wide-leaved Ladies'-tresses 372227
Wide-leaved Lady's Tresses 372227
Wide-leaved Osbeckia 276098
Wide-leaved Sea Lavender 230655
Wide-leaved Sea-lavender 230655
Wide-leaved Spiderwort 394082
Wide-leaved Spring Beauty 94305
Widelip Orchid 232146
Widewind 236022
Wideword Parnassia 284591
Wide-world Parnassia 284591
Widgeon Grass 340470
Widgeon Weed 340476
Widgeongrass 340476
Widgeonweed 340470,340476
Widgeonweed Family 340502
Widow Iris 192904
Widow Veil 168476
Widow Wail 122515,168476
Widow's Cross 357053
Widow's Flower 350099
Widow's Frill 364076
Widow's Tears 394088
Widow's Weeds 30081
Widow's Willow 343396
Widow's-cross 357053
Widow's-frill 364076
Widow's-tears 100892,392133,394088
Widow-veil 168476
Widow-wail 168476
Widow-wisse 173082
Widtsoe Wild Buckwheat 151838
Widwind 102933
Wiebel Cinquefoil 313113
Wiegand's Lime-grass 144541
Wiegand's Wild-rye 144541
Wiers Weeping Maple 3544
Wiesner's Meranti 362238
Wig Knapweed 81277
Wig Tree 107300
Wig Wagons 60382
Wigandla 414092
Wigeon-grass 418392
Wiggan 369339
Wiggen 369339
Wiggers 384714
Wiggin 369339
Wiggle Waggles 60382
Wiggle-waggle Grass 60382
Wiggle-waggle Wantons 60382
Wigglewaggles 60382
Wiggle-waggle-wantons 60382
Wiggle-wants 60382
Wiggy 369339
Wight Andrographis 22413
Wight Anisochilus 25397
Wight Anotis 262971
Wight Azalea 332109
Wight Barberry 52368
Wight Chaulmoogratree 199758
Wight Clearweed 299097
Wight Coldwaterflower 299097
Wight Crestpetal-tree 236422
Wight Dioecious Olive 270096
Wight Eulalia 156507
Wight Fig 165841
Wight Goldquitch 156507
Wight Groundsel 360344
Wight Hackberry 80780
Wight Ironolive 365086
Wight Leucas 227737
Wight Nakedanther Ternstroemia 386700
Wight New Eargrass 262971
Wight Olax 269663
Wight Osyris 276887
Wight Parnassia 284628

Wight Rhododendron 332109
Wight Sinosideroxylon 365086
Wight Sophora 369161
Wight Spapweed 205448
Wight Touch-me-not 205448
Wight Toxocarpus 393549
Wight Vinenettle 315465
Wight's St. John's Wort 202217
Wightia 414109
Wigtree 107300
Wig-tree 107300
Wigwag Wantons 60382
Wig-wagons 60382
Wigwag-wantons 60382
Wigwams 60382
Wigwants 60382
Wikkowleaf Randia 12538
Wikstroemia 414118
Wilcox's Fishhook Cactus 244263
Wilcox's Panic Grass 128561
Wild Ageratum 101955
Wild Agrimony 312360
Wild Allspice 231297,231306
Wild Almond 58032,58033,316413,386500
Wild Amaranth 18810
Wild Ananas 21485
Wild Angelica 24483
Wild Anise 77817,261585
Wild Apple 158692,243711,256797
Wild Apple Tree 243711
Wild Apricot 34477,243982
Wild Arrach 44576,87200
Wild Arrowhead 342421
Wild Artocarpus 36910
Wild Arum 37015
Wild Ash 369339
Wild Asparagus 39124,86970,274748
Wild Aster 163200,216753
Wild Azalea 235283,330322
Wild Bachelor's Buttons 308173
Wild Balsam 204841,204982,205175
Wild Balsam Apple 140527,256797
Wild Balsam-apple 140526,256797
Wild Bamboo 366245
Wild Banana 260206
Wild Banks Rose 336378
Wild Banyan Tree 164813
Wild Barley 198365
Wild Basil 4666,97060
Wild Basil Savory 97060
Wild Basil-savory 97060
Wild Bay 408167
Wild Bean 29298,238473,294034,378497,
378502,378506
Wild Bee Balm 257159
Wild Beet 18810,349771
Wild Beggarman 363461
Wild Begonia 340304
Wild Bergamot 257169
Wild Betony 138461
Wild Bishop 54239
Wild Bittertea 243405
Wild Bittervetch 408392
Wild Black Cherry 83311,316918
Wild Black Currant 333916
Wild Black Hellebore 190959
Wild Blackcurrant 333916
Wild Bleeding Heart 128296
Wild Bleeding-heart 128296
Wild Blue Flax 231942
Wild Blue Larkspur 124104
Wild Blue Phlox 295257,295258
Wild Buckthorn 328900
Wild Buckwheat 151792,152056,152240,
152246,152261,162519
Wild Bugloss 21920,21959,141347,239167
Wild Bushbean 241263
Wild Buttercup 326104
Wild Cabbage 59520,79696
Wild Calla 66593,66600
Wild Camelina 68871
Wild Camomile 26768,246396
Wild Candytuft 203184
Wild Cane 37467
Wild Caper 159222
Wild Caprifig 164774
Wild Carraway 28044
Wild Carrot 90924,123141,123215
Wild Cassada 212144
Wild Celandine 204841
Wild Celery 29327,366743,404531
Wild Century Plant 10911
Wild Chamomile 85526,246307,397973
Wild Charlock 364557
Wild Chastetree 411420
Wild Cherry 83155,83311,316411
Wild Chervil 28044,84759,84766,113872,
350425
Wild Chestnut 58032,58033,67667,279383
Wild Chicory 90889,90901
Wild Chier 86427
Wild China Tree 346336
Wild Chinatree 346336
Wild Chinese Arrowwood 407942
Wild Chinese Viburnum 407942
Wild Chinese Waxmyrtle 261217
Wild Chive 15189
Wild Chives 15709,15717
Wild Christmas Rose 190959
Wild Chrysanthemum 359158
Wild Cicely 28044
Wild Cineraria 358788
Wild Cinnamon 71132,91276
Wild Citron flower 71699
Wild Clary 190552,190651,345321,345458,
345471
Wild Clear-eye 345458
Wild Clematis 95426,95431
Wild Clover 277648,396956
Wild Cluster Cherry 280003
Wild Clusterberry 280003
Wild Coco 156540
Wild Coffee 310204,319392,319467,
319810,360463,397844
Wild Coffee Tree 310204
Wild Coldwaterflower 298945
Wild Cole 364557
Wild Columbine 30007
Wild Comfrey 117926,118031,184320,
321643
Wild Cornel 380499
Wild Cotton 29473,179931,195032,207571
Wild Crab 243550,243583,243636
Wild Crabapple 243711
Wild Crabdaisy 413565
Wild Cranberry 278366
Wild Cranesbill 174717
Wild Cress 225475,390213
Wild Crocus 394040
Wild Cucumber 61497,114122,139852,
140526,140527,256797
Wild Cucumber Vine 140527
Wild Currant 334042,334179
Wild Custard 25838
Wild Custard-apple 25838,25885
Wild Cyclamen 135165
Wild Daffodil 262441
Wild Dagga 224585
Wild Damson 316470
Wild Date 416556
Wild Date Palm 295479,295487
Wild Dicoccum Wheat 398886
Wild Eightangle 204590
Wild Elder 8826,267370,345594
Wild Embelia 144802
Wild False Indigo 47871
Wild Falseflax 68871
Wild Fennel 266225
Wild Fetch 408352
Wild Field Onion 274823
Wild Fig 164751,165471,165630
Wild Fig Tree 165726
Wild Fig-tree 3462
Wild Fire 4005
Wild Fitch 408423,408571
Wild Flag 208924
Wild Flaskform Barley 198314
Wild Flax 115007,231183,231861,231961,
231963,232014
Wild Floating Rice 200675,200677
Wild Four-o'clock 255723,255735

Wild Foxglove 83672,289349
Wild French Guava 360463
Wild Garden Parsnip 285744
Wild Gardenia 171425
Wild Garden-parsnip 285744
Wild Garlic 15149,15254,15861,15876,400065,400095
Wild Gentian 88029,150732
Wild Geranium 18173,174717,174877,248312
Wild Germander 388032,407072
Wild Gilliflower 127635
Wild Ginger 36967,37548,37556,37561,37574,37582,418031
Wild Gladiolus 176262
Wild Gold Chain 249205
Wild Golden-glow 339574
Wild Goose Plum 316444,316583
Wild Gooseberry 297658,297703,333945,334042,334099
Wild Gourd 93297
Wild Granaat 63904
Wild Granadilla 7238
Wild Grape 411590,411887
Wild Grass Nettle 373455
Wild Green Bean 294019,408975
Wild Guava 76799,318744
Wild Hazel 364450
Wild Heart Cherry 83155
Wild Hedge Maple 2844
Wild Heliotrope 293162
Wild Hemp 71228,170078
Wild Holly 263498,263499
Wild Honeysuckle 235742,235760,235762,287494,331546
Wild Hop 53304,61497,162519,199384,383675
Wild Horehound 158360
Wild Hyacinth 60431,68789,68790,68791,68794,68806,128870,128879,145487,352880,398777,398789,398794
Wild Hydrangea 199806,340304
Wild Hyssop 405844
Wild Ice-leaf 405788
Wild Indigo 47854,47859,47880,386257
Wild Ipecacuanha 37888,159150
Wild Iris 130296,208771,208924
Wild Irishman 134021
Wild Isaac 158250
Wild Ixora 209472
Wild Jalap 208050
Wild Jasmine 23925,211834
Wild Jessamine 23925,172781
Wild Jonquil 262410,262441
Wild Kale 364557
Wild Kale Sea Cabbage 59520
Wild Laburnum 249205,249232
Wild Lady's Slipper 204841
Wild Lavender 190597
Wild Leek 15048,15684,15709,15831,15832,15861,15876
Wild Lemon 93539,306636
Wild Lentil 42193
Wild Lepisanthes 225675
Wild Lettuce 94349,219206,219249,219507,219609,320515
Wild Licorice 170316,170317,170457,177922
Wild Lilac 79902
Wild Lily 37015,68713,112667
Wild Lily of the Valley 322872
Wild Lily of the Velley 322823
Wild Lily-of-the-valley 242675,242676,242680,322872
Wild Lilyturf 272132
Wild Lime 415558,417228
Wild Liquorice 765,30712,42417,82098
Wild Live-forever 357202
Wild Lobelia 308454
Wild London Pride 91557,345957
Wild Loquat 401238
Wild Lupin 238472,238473
Wild Lupine 238472,238473,238474
Wild Lychee 233079
Wild Madder 170490,170520,170668,337906,338007
Wild Mallow 243840
Wild Mandrake 306636
Wild Mango-tree 208990
Wild Marigold 66395,89704,321554
Wild Marjoram 274224,274237
Wild Marjorana 274197
Wild Marsh Beet 230815
Wild Marshmarigold 68221
Wild Masterwort 8826
Wild Maws 282685
Wild Medlar 404924
Wild Mercury 130383
Wild Michelia 252958
Wild Mignonette 327866
Wild Mint 13174,250291,250294,250331,250362,250370,250385,250439
Wild Monk's-hood 5654
Wild Monkshood 5654
Wild Morning Glory 68713
Wild Mouseear 82994
Wild Musk 153767
Wild Mustard 34406,59575,326609,364557
Wild Myrobalan Plum 316294
Wild Myrtle 261162,340990
Wild Nancy 262438
Wild Nanmu 295353
Wild Nard 37616
Wild Nasturtium 336277
Wild Navew 61497
Wild Nep 61497,383675
Wild Nepeta 265040
Wild Nept 61497
Wild Nettle 402992
Wild Nigella 11947
Wild Nightshade 367044
Wild Nodding Onion 15173
Wild Nutmeg 321973
Wild Oat 45448
Wild Oatgrass 122180
Wild Oats 45448,45566,86140,403675
Wild Oats Bellwort 403665
Wild Ochre 242866
Wild Okra 195341,242866
Wild Oleander 8186
Wild Olive 142382,184731,267850,270100,270110,276250,415558
Wild Olive Tree 270150
Wild Onion 15020,15125,15149,15173,15660,15684,15765,15861,15876
Wild Orache 86901
Wild Orande 316285
Wild Orange 71808,378688
Wild Orange Lily 229984
Wild Ornithoboea 274473
Wild Palm 393813
Wild Panicgrass 281917,282256
Wild Pansy 409703,409728,410480,410677
Wild Parsley 28044,235412,350425,365362,366743
Wild Parsnip 90924,192278,285741,285744
Wild Passionflower 285639,285654
Wild Passion-flower 285654
Wild Pea 222671,222844,408251
Wild Peach 155794,216354,223094,316285
Wild Pear 323133,323267,369322
Wild Pear Tree 267867
Wild Pellitory 4005
Wild Pennycress 390265
Wild Peony 280182
Wild Pepper 3978,122515,300408,411189
Wild Peppergrass 225475
Wild Pepper-grass 225340,225475
Wild Peppermint 250291
Wild Peristrophe 291151
Wild Petunia 339658,339736,339783,339819
Wild Phlox 146724
Wild Pinangapalm 299652
Wild Pine 300223,392001
Wild Pineapple 21475,392001
Wild Pink 127698,127816,174877,363300,363957
Wild Pink Clover 396811
Wild Pistachio 301004
Wild Pistacia 374107

Wild Plantain 189992,190005
Wild Plum 185867,282958,306423,313344,316206,316382,316470,316553
Wild Poinsettia 158727,413269
Wild Poisonnut 378834
Wild Pomegranate 79590,321764
Wild Poplar 311281
Wild Potato 88476,208050,287932,342315
Wild Potato Flower 367120
Wild Potato Vine 208050
Wild Prairie Rose 336361
Wild Pricklyash 417340
Wild Privet 229662
Wild Proso Millet 281916
Wild Pumpkin 114278
Wild Purplehammer 313562
Wild Purplepearl 66829,66856
Wild Purse 72038
Wild Purslane 159557
Wild Quince 374392
Wild Quinine 285088
Wild Radish 326609,326616
Wild Raisin 407737,407917,408095
Wild Rape 326833,326835,364557
Wild Raspberry 338557
Wild Red Cherry 83274
Wild Red Raspberry 338557,338600
Wild Rhubarb 31051,292372,340084,340151,400675
Wild Rice 275939,275957,418079,418080,418095,418096
Wild Rocambole 15726
Wild Rocket 72934,133286,327879
Wild Rose 336389,336783,336892
Wild Rose Campion 248411
Wild Rosemary 22445,170743,223901
Wild Rubber Lagos 169229
Wild Rue 269307,287978
Wild Rum-cherry 83311
Wild Rye 144144,144295,144297,175454,228346,356253
Wild Rye-grass 235373
Wild Safflower 77734
Wild Saffron 77748,99297
Wild Sage 35570,221238,344838,345246,345453,345458,388275
Wild Sago 302034,302068
Wild Salsify 394333
Wild Saltbush 44400
Wild Sandheath 81061
Wild Sarsaparilla 30712,185760,366237,366341
Wild Sarsaparilla Aralia 30712
Wild Savager 11947
Wild Scarlet Lily 229811
Wild Selfheal 316097
Wild Senna 360437,360453
Wild Sensitive Plant 78421
Wild Sensitive-plant 78421
Wild Service 369339,369541
Wild Service Tree 369541
Wild Service-tree 369541
Wild Shamrock 8399,277648
Wild Shower 360453
Wild Sierra Currant 334116
Wild Siris 13586
Wild Sixrow Barley 198262
Wild Skirret 312360
Wild Smallreed 127197
Wild Snapdragon 85875,231183
Wild Snap-dragon 231183
Wild Snowball 79905
Wild Soja 177777
Wild Sorb 369339
Wild Sorrel 278008,339887,340075
Wild Sour Sop 25838
Wild Soybean 177777
Wild Spider 242699
Wild Spikenard 203066,242699
Wild Spinach 86901,86970
Wild Squill 352894,352975
Wild Stonecrop 356988,357224
Wild Strawberry 167653,167663
Wild Succory 90901
Wild Sugar Cane 341912
Wild Sugarcane 341912
Wild Sugarpalm 32336
Wild Sumach 261162
Wild Sunflower 188908,207135
Wild Sweet Crab Apple 243583
Wild Sweet Crabapple 243583
Wild Sweet Crab-apple 243583
Wild Sweet Pea 271563
Wild Sweet Potato 298109
Wild Sweet William 295257,295282,346434,363461
Wild Sweetcane 341912
Wild SweetPea 187994
Wild Sweet-william 295257,295258,295282
Wild Swordflag 208819
Wild Syringa 63931
Wild Syzygium 382599
Wild Tallowtree 346379
Wild Tamarind 78307
Wild Tansy 11549,312360
Wild Tare 408352,408611
Wild Tarragon 35411
Wild Teasel 133478,133482,133517
Wild Teazle 133478
Wild Thetch-grass 408352,408423
Wild Thistle 368771
Wild Thyme 237539,391346,391365,391397
Wild Tiger Lily 230044
Wild Tobacco 4941,234547,266053,305085
Wild Tobacco 367364
Wild Tom Thumb 387251
Wild Tomato 367416,367690
Wild Tulip 168476,400229
Wild Turnip 33544,59603,59651,326833,364557
Wild Turnjujube 307290,307291
Wild Valerian 404316
Wild Vervain 405844
Wild Vetch 408262,408352
Wild Vetchling 222844
Wild Vine 61497,95431,383675
Wild Vinescrewpine 168245
Wild Violet 409884,410770
Wild Violet Tree 356368
Wild Wallflower 154390
Wild Walnut 212595
Wild Water Lemon 285639
Wild Water Pepper 291792
Wild Water Plum 414815
Wild Waxplant 198896
Wild Wheat 398886
Wild White Clover 397043
Wild White Indigo 47856
Wild White Violet 410189,410190
Wild Wildginger 37713
Wild William 384714
Wild Willow 146724,343726,344159,344279
Wild Winter Pea 222726
Wild Wistaria 56735,356368
Wild Woad 173082,327879
Wild Wormwood 35097,35237
Wild Yam 131592,131798,131896
Wild Yamroot 131896
Wild Yellow Flax 232014
Wild Yellow Lily 229779
Wild Zinger 418023
Wild-bamboo 366245
Wildbasil 96960
Wild-beet Amaranth 18810
Wild-begonia 340304
Wildberry 362339
Wild-celery 404531
Wild-cinnamon 71129
Wild-cinnamon Family 71138
Wildclove 226075,226113
Wild-coffee 319392,397844
Wildcoral 170031
Wild-cucumber 140527
Wild-dragon Dendrocalamus 125507
Wildeggplant 58299,58300
Wildenow's Leopard's-bane 136337
Wilderness Violet 410547
Wild-fire 4005
Wildfire Barberry 51665
Wildfireball Trefoil 396956

Wildginger 37556
Wild-ginger 37556,37574
Wildginger Metalleaf 296368
Wildgingerleaf Metalleaf 296368
Wild-gingerleaf Primrose 314131
Wildgingerleaf Swertia 380118
Wildgoose Plum 316583
Wild-goose Plum 316444
Wildhops 166897
Wilding Tree 243711
Wild-ixora 209472
Wildjute 394626,394664
Wildlaurel 51024,51039
Wild-lettuce 222946
Wild-lime 417228
Wild-mandrake 306636
Wildmangrove 288682
Wild-marjoram 274237
Wildmedia 28993
Wild-oat 45448
Wild-oats 403675
Wildolive 301760
Wild-olive 276250
Wildorchis 274033
Wild-rhubarb 340084
Wildrice 418079
Wild-rice 418080
Wildrye 144144
Wild-rye 144144,144295
Wildryegrass 144144
Wildsenna 389505,389540
Wildstickiness 50471
Wildthyme Azalea 331305
Wild-tuberose 244371
Wildtung 243315,243372
Wildwood Buttercup 326118
Wilf 343151
Wilfire 325864
Wilfoed Swallowwort 117751
Wilford Bushclover 227013
Wilford Bush-clover 227013
Wilford Campion 364213
Wilford Cranesbill 175006
Wilford Mosquitotrap 117751
Wilford Three-winged Nut 398322
Wilford Three-wing-nut 398322
Wilga 172478
Wilgers 344244
Wilgers Welgers 344244
Wilhelm's Wattle 1704
Wilhelms Willow 344273
Wilken's Fleabane 151070
Wilkes Copperleaf 2011
Wilkes' Acalypha 1780
Wilkinson's Nailwort 284908
Will Fleming Yaupon 204394
Will Kale 364557
Willamette Fleabane 150578
Willamette Rue-anemone 145495
Willard's Acacia 1704,1705
Willd Bellcalyxwort 231267
Willd Cranesbill 174524
Willdenow Flueggea 167067,167084
Willdenow's sedge 76756
William Pine 44050
William Sweet Catchfly 363214
William's Campanilla 14876
William-and-Mary 87488,243209,321643
Williams Alumroot 194441
Williams Azalea 332110
Williams Elder 345708
Williams Kobresia 217178
Williams Podocarpium 200748
Williams Rhododendron 332110
Williams Vinescrewpine 168251
Williams Waterwort 142580
Williams Willow Weed 146942
Williams' Catchfly 364215
Williamson's Spruce 399916
Willkomm Eyebright 160292
Willmott Bluesnow 83650
Willmott Ceratostigma 83650
Willmott Iris 208940
Willmott Leadwort 83650
Willmott Rose 337022
Willmott Waxpetal 106689
Willmott Winter Hazel 106689
Willmott Winterhazel 106689
Willmott Winter-hazel 106689
Willmott's Lily 229831
Willow 342923,343111,343343,343414,343937,344046
Willow Acacia 1552
Willow Amsonia 20860
Willow Aster 380959,380962,380964
Willow Baccharis 46253
Willow Bay 343858
Willow Bottlebrush 67293
Willow Dock 340237,340244
Willow Family 342841
Willow Fig 164688
Willow Fircwheel Tree 375511
Willow Fire-wheel Tree 375511
Willow Flower 146724
Willow Gentian 173264
Willow Glowweed 236501
Willow Gum 155737
Willow Hakea 184634
Willow Hawthorn 110013
Willow Hedge Bamboo 47352
Willow Herb 85875,146586
Willow Leaf Foxglove 130376
Willow Leaf Groundsel 359959
Willow Lettuce 219480
Willow Machilus 240693
Willow Myrtle 11397,11399
Willow Oak 116187,324282
Willow Peppermint 155666
Willow Persicaria 309298
Willow Pittosporum 301369
Willow Podocarp 306517
Willow Poplar 311398
Willow Ragwort 48074
Willow Rhus 332685
Willow Scurrula 385194
Willow Spiraea 372075
Willow Thorn 196757
Willow Toothwort 222644
Willow Wattle 1552
Willow Weed 146586
Willow-baccharis 46253
Willow-bay 343858
Willow-blossom 295257
Willowcelery 121044,121045
Willow-grass 308739,308816
Willowherb 146586,146614,239898
Willow-herb 146586
Willowherb Family 270773
Willowherb Skullcap 355435
Willowleaf Achyranthes 4304
Willowleaf Actinodaphne 6789
Willowleaf Agapetes 10322
Willow-leaf Agapetes 10365
Willowleaf Ajania 13030
Willowleaf Anodendron 25946
Willowleaf Ardisia 31474
Willowleaf Aster 41190,380959
Willow-leaf Bay 223204
Willowleaf Beautyberry 66925
Willowleaf Beggarticks 53816
Willowleaf Begonia 50271
Willowleaf Bentham Photinia 295639
Willowleaf Blueberry 403863
Willowleaf Bottlebrush 67293
Willowleaf Buckthorn 328694
Willowleaf Bugle 13181
Willowleaf Camellia 69586
Willow-leaf Candyleaf 376416
Willowleaf Catchfly 364005
Willowleaf Ceropegia 84234
Willowleaf Chickweed 375083
Willowleaf Cotoneaster 107668,107671
Willow-leaf Cotoneaster 107668
Willowleaf Cyclobalanopsis 116187
Willowleaf Damnacanthus 122060
Willow-leaf Daphniphyllum 122712
Willowleaf Debregeasia 123333
Willow-leaf Drypetes 138690
Willowleaf Dysophylla 139583
Willowleaf Eelvine 25946
Willowleaf Eucalyptus 155730

Willow-leaf Eugenia 4879
Willowleaf Euphorbia 159759
Willowleaf Fighazel 134911
Willowleaf Frostweed 188829
Willowleaf Gardneria 171451
Willow-leaf Goldenrod 368438
Willow-leaf Hakea 184634
Willowleaf Hawkweed 195696
Willowleaf Heimia 188265
Willowleaf Holly 204234
Willowleaf Homalium 197717
Willowleaf Honeysuckle 235910
Willowleaf Hygrophila 200652
Willowleaf Inula 207219
Willowleaf Ironweed 406772
Willowleaf Jurinea 207219
Willowleaf Knotweed 308922,309318
Willowleaf Lettuce 219480
Willowleaf Lobelia 234555
Willow-leaf Maack Euonymus 157702
Willowleaf Machilus 240691
Willowleaf Madder 338026
Willowleaf Maesa 241833
Willow-leaf Magnolia 242274
Willowleaf Meadowsweet 372075
Willowleaf Michelia 252894
Willowleaf Monkeyflower 255195
Willowleaf Mosquitotrap 117692
Willowleaf Multiradiate Fleabane 150805
Willowleaf Onion 15888
Willowleaf Ophiorrhiza 272294
Willowleaf Oxeye 63539
Willow-leaf Oxeye 63539
Willowleaf Oxeyedaisy 63539
Willowleaf Pagodatree 369002
Willowleaf Pear 323281
Willow-leaf Peppermint 155666
Willowleaf Podocarp 306517
Willowleaf Podocarpus 306517
Willowleaf Prismatomeris 122060
Willowleaf Prive 229610
Willowleaf Purplepearl 66925
Willowleaf Qinggang 116187
Willowleaf Raphiolepis 329105
Willow-leaf Red Quebracho 350974
Willowleaf Redstem Skullcap 355860
Willowleaf Rhododendron 331089
Willowleaf Sagebrush 35648
Willowleaf Salacia 342609
Willowleaf Sandthorn 196773
Willowleaf Sarcococca 346747
Willowleaf Scurrula 385194
Willowleaf Seabuckthorn 196773
Willow-leaf Sharptooth Photinia 295627
Willowleaf Smallligulate Aster 40013
Willowleaf Spapweed 205300
Willow-leaf Spindle-tree 157551
Willowleaf Spiradiclis 371748
Willowleaf Spiraea 372075
Willowleaf Spurge 158837
Willowleaf Stachyurus 373569
Willowleaf Stranvaesia 377446
Willowleaf Stringbush 414253
Willowleaf Sunflower 189052
Willow-leaf Sunflower 189052
Willow-leaf Sunwheel 63539
Willowleaf Swallowwort 117692
Willowleaf Tanoak 233181
Willowleaf Tea 69586
Willowleaf Thistle 92357
Willow-leaf Tigernanmu 122712
Willowleaf Tigerthorn 122060
Willowleaf Touch-me-not 205300
Willowleaf Vetch 408672
Willowleaf Vlassoviana Thistle 92357
Willowleaf Waternettle 123333
Willowleaf Wendlandia 413816
Willowleaf Windhairdaisy 348757
Willowleaf Wintersweet 87542
Willowleaf Woodbetony 287654
Willowleaf Wormwood 35648
Willowleaf Yaccatree 306517
Willowleaf Yam 131683
Willowleaf Yarrow 4021
Willowleaf Yellowhead 207219
Willow-leafed Oak 324282
Willow-leafed Pear 323281
Willow-leaved Actinodaphne 6789
Willow-leaved Agapetes 10365
Willow-leaved Ajania 13030
Willow-leaved Amsonia 20860
Willow-leaved Ardisia 31474
Willow-leaved Aster 41190,380959,380962,380964
Willow-leaved Barberry 52128
Willow-leaved Beautyberry 66925
Willow-leaved Blueberry 403863
Willow-leaved Bottlebrush 67293
Willow-leaved Buckthorn 328694
Willow-leaved Camellia 69586
Willow-leaved Cotoneaster 107668
Willow-leaved Damnacanthus 122060
Willow-leaved Dock 340128,340135,340237,340244
Willow-leaved Drypetes 138690
Willow-leaved Eucalyptus 155730
Willow-leaved Fighazel 134911
Willow-leaved Gentian 173264
Willow-leaved Heimia 188265
Willow-leaved Holly 204234
Willow-leaved Homalium 197717
Willow-leaved Honeysuckle 235910
Willow-leaved Inula 207219
Willow-leaved Jessamine 84419
Willow-leaved Lawson's Spindle-tree 157852
Willow-leaved Lettuce 219480
Willow-leaved Machilus 240691
Willow-leaved Maesa 241833
Willow-leaved Magnolia 242274
Willowleaved Pendentlamp 84234
Willow-leaved Pondweed 312260
Willow-leaved Prismatomeris 122060
Willow-leaved Randia 325377
Willow-leaved Raphiolepis 329105
Willow-leaved Rhododendron 330943,331089
Willow-leaved Salacia 342609
Willow-leaved Sarcococca 346747
Willow-leaved Sea-buckthorn 196773
Willowleaved Spiraea 372075
Willow-leaved Spiraea 372075
Willow-leaved Stachyurus 373569
Willow-leaved Sunflower 189052
Willow-leaved Wendlandia 413816
Willow-leaved Wild Sunflower 189052
Willow-leaved Wintersweet 87542
Willow-leaved Yellow-head 207219
Willowleaved-jassamine 84419
Willow-like Machilus 240693
Willow-like Oak 116187
Willowlike Rabbitbrush 150353
Willowmore Cedar 414065
Willowmyrtle 11399
Willowshoot Azalea 332071
Willowshoot Rhododendron 332071
Willow-shooted Rhododendron 332071
Willow-strife 240068
Willow-thorn 196757
Willowtwig Rattanpalm 65809
Willowweed 85870,85875,146586
Willow-weed 308739,309570
Willowweed Sonerila 368880
Willow-weed Sonerila 368880
Willowweedleaf Saussurea 348292
Willowweedleaf Windhairdaisy 348292
Willowweedlike Spapweed 204930
Willowweedlike Touch-me-not 204930
Willow-wind 95431,102933,162312
Willow-wort 239898
Willowy Twigs Rhododendron 332071
Willy-run-the-hedge 170193
Willywind 102933
Wilman Lovegrass 147990
Wilowleaf Alyxia 18525
Wilslizen Barrel Cactus 163483
Wilslizen Poplar 311567
Wilson Acanthopanax 143701
Wilson Barberry 52371
Wilson Berchemiella 52481
Wilson Bird Cherry 280061
Wilson Buckeye 9738
Wilson Buckthorn 328900

Wilson Cinnamon 91449
Wilson Clovershrub 70911
Wilson Deutzia 127124
Wilson Dogwood 380520
Wilson Elaeagnus 142207
Wilson Euonymus 157966
Wilson Fir 464,510
Wilson Glochidion 177190
Wilson Grape 412004
Wilson Grevill 180663
Wilson Hairystyle Mockorange 294553
Wilson Hawthorn 110114
Wilson Holly 204401
Wilson Indigo 206744
Wilson Indocalamus 206839
Wilson Juniper 213857
Wilson Litse 234108
Wilson Lysionotus 239991
Wilson Maddenia 241507
Wilson Magnolia 279251
Wilson Magnolia-vine 351118
Wilson Maple 3749
Wilson Michelia 252974
Wilson Mountain-ash 369565
Wilson Pearlbush 161745
Wilson Pentapanax 289672
Wilson Poplar 311568
Wilson Pygeum 322413
Wilson Sachalin Fir 464
Wilson Sakhalin Fir 510
Wilson Sophora 369162
Wilson Spiraea 372135
Wilson Spirea 372135
Wilson Spruce 298449
Wilson St. John's-wort 202220
Wilson Sumac 332969
Wilson Willow 344277
Wilson Woodbetony 287830
Wilson Yellowwood 94035
Wilson's Barberry 52371
Wilson's Clusterpea 131328
Wilson's Grevillea 180663
Wilson's Honeysuckle 235984
Wilson's Magnolia 279251
Wilson's Spruce 298449
Wilson's Yarrow 4062
Wilsontree 365092
Wilson-tree 365092
Wilton Azalea 332113
Wilton Rhododendron 332113
Wiltshire Weed 401593
Wimble-straw 118315
Wimmera Ryegrass 235354
Wimote 18173
Winberry 403916
Winchia 18027,18052,414458
Wincopipe 21339
Wind Bent Grass 28972
Wind Bent-grass 28972
Wind Flower 23696,23925,24004,321721
Wind Grass 28972
Wind Mill Palm 266677,266678
Wind Plant 23925
Wind Root 38147
Wind Rose 282515
Windamere Palm 393812
Windberry 403916
Windflower 23696,23724,23767,23891,24004,24133,321653,321721,417606
Wind-flower 24133
Windflowers 23767
Wind-grass 28972
Windhairdaisy 348089
Windhairdaisylike Groundsel 359974
Windhairlike Blumea 55819
Winding Lily 68713
Windles 302034
Wind-loving Wild Buckwheat 151815
Windmill 100309
Windmill Cypressgrass 118477
Windmill Grass 88316
Windmill Orchid 63030
Windmill Palm 393803,393809
Windmillgrass 88316
Windmill-grass 88316,88414
Windmillpalm 393803
Windmill-palm 393803
Windmilltung 147353
Windorchis 263865
Window Bearing Orchid 113738
Window Box Oxalis 278064
Window Orchid 113738
Window Plant 163329,360027
Window's Weed 208870
Windowbox Woodsorrel 278064
Windowleaf 258162
Windpipe 21339
Windsor Bean 408393
Wine Cup 67128
Wine Cups 67128
Wine Grape 411979
Wine Mauritia 246674
Wine Palm 57122,78045,78056,212556,212557,246671,326649
Wine Plant 329388
Wine Raspberry 339047
Wine See 411979
Wine-bearing Barberry 52312
Wineberry 34394,34400,334117,334179,334250,339047
Wineberryleaf Honeysuckle 235978
Wineberry-leaved Honeysuckle 235978
Wineberry-leaved Willow 343982
Winecups 46119
Wineglass Comastoma 100285
Wineglass Throathair 100285
Wineglasses 70164
Winejar Spapweed 204768
Winejar Touch-me-not 204768
Wine-plant 329386
Wing Cactus 320258
Wing Celtis 320411
Wing Mosquitotrap 117356
Wing Nut 320346
Wing Ophiorrhiza 272176
Wing Pipewort 151555
Wing Pod Pea 387239
Wing Scale 44334
Wing Snowgall 191914
Wing Thistle 92485
Wingbract Elatostema 142601
Wingbranch Summerlilic 61967
Wing-cactus 320411
Wingcalyx Dendrobium 125037
Wingcalyx Gentian 173759
Wingcalyx Loosestrife 239799
Wingceltis 320411,320412
Wingcurved Mono Maple 3190
Wingdrupe 405438,405447
Wing-dstalked Prickly-ash 417330
Winged Acacia 1494
Winged Bean 319114
Winged Branch Treebine 92907
Winged Broom 173051
Winged Burning Bush 157285
Winged Burning-bush 157285
Winged Butterflybush 61967
Winged Butterfly-bush 61967
Winged Citron 93318
Winged Clematis 95246
Winged Corolla Corydalis 106052
Winged Cragcress 97971
Winged Dock 340304
Winged Elaeocarpus 142322
Winged Elm 401430
Winged Epidendrum 146375
Winged Euonymus 157285,157290
Winged Eurya 160419
Winged Everlasting 19691
Winged Gladiolus 176003
Winged Grape 411868
Winged Greenweed 173051
Winged Hackberry 320411
Winged Heterostemma 194156
Winged Highstem Sedge 73660
Winged Laggera 219996
Winged Loosestrife 240025,240027,240028
Winged Lophira 236336
Winged Lotus 237771
Winged Lythrum 240027
Winged Monkeyflower 255187

Winged Orchid 273604
Winged Pararuellia 283761
Winged Pauldopia 285917
Winged Pea 237771,319114,387251
Winged Petal Corydalis 106331
Winged Petiole Synotis 381921
Winged Pigweed 116284,116286
Winged Plumeless Thistle 73504
Winged Pterygota 320987
Winged Rockvine 387743
Winged Scurvyweed 97971
Winged Sea Lavender 230761
Winged Seed Sterculia 376070
Winged Senna 360409
Winged Sotvine 194156
Winged Spindle 157285
Winged Spindle Tree 157285
Winged Spindle-tree 157285
Winged Spine Emei Rose 336828
Winged Sterculia 376070
Winged Sumac 332529
Winged Tabacco 266035
Winged Thistle 91723
Winged Water-starwort 67370
Winged Wild Buckwheat 151804
Winged Yam 131458
Wingedbeak Woodbetony 287426
Winged-everlasting 19691
Winged-flowered-like Loosestrife 239802
Winged-fruit Coralbean 154726
Winged-fruit Sophora 369070
Wingedfruit Swallowwort 117356
Wingedleaf Hainan Marsdenia 245816
Wingedleaf Hainan Milkgreens 245816
Wingedpetiole Purplestem 376467
Wingedpetiole Sage 344840
Wingedpetiole Stewartia 376467
Winged-seed China-bell 16304
Wingedseed Rockcress 30406
Winged-spindle Tree 157285
Wingedspine Silky Rose 336936
Wingedstalk Pluchea 305123
Wingedstalk Pricklyash 417330
Wingedstem Allomorphia 278610
Wingedstem Inula 138898
Wingedstem Jurinea 138898
Wingedstem Lysionotus 239982
Winged-stem Sonerila 368870
Winged-stem Treebine 92747
Winged-stemmed Sonerila 368870
Wingedtooth Laggera 220032
Wingespine Pricklyash 417317
Wingflower Ophiorrhiza 272177
Wingfruit Beancaper 418740
Wingfruit Coral-bean 154726
Wingfruit Eargrass 187649
Wingfruit Elaeagnus 142082
Wingfruit Hedyotis 187649
Wingfruit Lindelofia 231253
Wingfruit Pagodatree 369070
Wingfruit Peltophorum 288897
Wingfruit Peltostyle 288897
Wingfruit Sedge 75500
Wingfruit Thorowax 63559
Wing-fruited Elaeagnus 142082
Wing-fruited Pterocarpous
Peltophorum 288897
Wing-fruited Sophora 369070
Wingfruit-knotweed 283707,283708
Wingfruitvine 261387
Wing-fruit-vine 261387
Wingknotweed 320911,320912
Wingleaf Butterfly Flower 351345
Wingleaf Euaraliopsis 59199
Wingleaf Jasminorange 260160
Wingleaf Palmleaftree 59199
Wingleaf Passion Flower 285619
Wingleaf Passionflower 285619
Wingleaf Pleurospermum 304787
Wingleaf Pricklyash 417161
Wingleaf Ribseedcelery 304787
Wingleaf Windhairdaisy 348685
Wing-leaved Brassaiopsis 59199
Wing-leaved Jasmin-orange 260160
Wing-leaved Primrose-willow 238166
Wing-leaved Rattlesnakeroot 313785
Wingless Ainsliaea 12613
Wingless Begonia 49594
Wingless Fruit Sophora 369073
Wingless Marsh Peavine 222802
Wingless Rabbiten-wind 12613
Wingless Russianthistle 344443
Wingless Sedge 75837
Wingless Synotis 381958
Wingless Tailanther 381958
Wingless Willowcelery 121048
Wingless Yam 131581
Wingnut 320346,320356
Wingpetal Corydalis 106331
Wingpetal Dandelion 384428
Wingpod Bird's Foot Trefoil 237771
Wingpod Purslane 311945
Wingpod Senna 360409
Wingpod Veinyroot 237771
Wing-podded Senna 360409
Wingridge Elatostema 142603
Wingridge Stairweed 142603
Wings 3462,167955
Wingseed 320071
Wingseed Tree 320827
Wing-seeded Spurry 370568
Wingseedtree 320827
Wing-seed-tree 320827
Wingsepal Spapweed 205259
Wingsepal Touch-me-not 205259
Wing-spined Prickly-ash 417317
Wingstalk Azalea 330708
Wingstalk Hilliella 196299
Wingstalk Sage 344840
Wingstalk Windhairdaisy 348116
Wingstem 405915
Wing-stem 405915
Wing-stem Camphorweed 305130
Wingstem Cousinia 108224
Wingstem Figwort 355261
Wing-stem Holboellia 197248
Wingstem Horseegg 183003
Wingstem Madder 338020
Wing-stem Meadow-pitchers 329427
Wingstem Passionflower 285611
Wingstem Pluchea 305130
Wingstem Rattlebox 111951
Wingstem Rush 212816
Wingstem Spurry 370568
Wingstem Vinegentian 108224
Wingstem Waterdropwort 269351
Wingstem Windhairdaisy 348113,348192,
348456
Wingstem Woodbetony 287458
Wing-stemmed Allomorphia 278610
Wingstipe Aster 40000
Wingstipe Beancaper 418695
Wingstipe Chirita 87948
Wingstipe Dendrobium 125396
Wingstipe Dubyaea 138778
Wingstipe Goldenray 228975
Wingstipe Laurel 122488
Wingstipe Samaradaisy 320523
Wingstipe Tailanther 381921
Wingstipule Rattlebox 111874
Wingtooth Gerbera 224110
Wing-wangs 60382
Wingystem Ascitesgrass 407460
Wingystem Veronicastrum 407460
Wink-and-peep 21339
Wink-a-peep 21339
Winkle Knotweed 309624
Winkled Eurya 160594
Winkled Marshweed 230322
Winkled Sawwort 361119
Winkledleaf Camellia 69027
Winkledleaf Eurya 160594
Winkledleaf Horny Tanoak 233159
Winkler Onion 15895
Winkler's Cactus 287885
Winkler's Footcactus 287926
Winkler's Groundsel 365047
Winkler's Pincushion Cactus 287926
Winkler's White Blanketflower 169571
Winter Aconite 148101,148104,148112
Winter Arrowbamboo 162699

Winter Barberry 51394
Winter Bean 222750
Winter Beech 162375
Winter Bell 55109
Winter Bent Grass 12139
Winter Blooming Bergenia 52514
Winter Buddleja 62160
Winter Camellia 69133
Winter Candytuft 203249
Winter Cassia 360426
Winter Cattleya 79555
Winter Cherry 73207,83112,297640,297643,297645,297711,297720,367522,367528
Winter Clover 256002
Winter Creeper 157473,157515
Winter Creeper Spindle Tree 157473
Winter Cress 47931,47947,47963,47964,154019
Winter Crocus 111531
Winter Crookneck Squash 114292
Winter Currant 334189
Winter Cymbidium 116926
Winter Daffodil 376350,376356
Winter Daphne 122532
Winter Elaeocarpus 142278
Winter Fat 160387,218116,218122
Winter Forget-me-not 270723
Winter Gastrodia 171949
Winter Gilliflower 86427,169719,193417
Winter Gladiolus 351884
Winter Grape 411575,411631,411887
Winter Green 53505,172047,322775
Winter Greens 59387
Winter Hazel 106628,106669
Winter Heath 149147,149398
Winter Hedge Mustard 47964
Winter Heliotriope 292353
Winter Heliotrope 292342
Winter Hellebore 148104,190944
Winter Honeysuckle 235796,236049
Winter Hyacinth 352885
Winter Iris 208863,208914
Winter Jasmine 211931,211963
Winter Kecksies 316819
Winter Mango 244396
Winter Marguerite 104545
Winter Marjoram 274218,274231
Winter Melon 114212
Winter Mint 250468
Winter Nerine 265299
Winter Oak 324027
Winter Onion 15165
Winter Pink 39871
Winter Poker 216960
Winter Pot Kalanchoe 215102
Winter Primrose 314821
Winter Purslane 94349
Winter Rape 59493
Winter Red Winterberry 204380
Winter Rocket 47964
Winter Rose 190944
Winter Sage 218122
Winter Savory 347446,347597
Winter Spinach 371713
Winter Squash 114271,114288,114292
Winter Strawberry 30888
Winter Sweet 87525
Winter Thorn 1035
Winter Thyme 391366
Winter Vetch 408666,408693,408700
Winter Well 339849
Winter Wild-oat 45586
Winter Wolf's Bane 148104
Winter Yellow Jasmine 211931
Winter's Bark 138058
Winter's Bark Drimys 138058
Winter's Cinnamon 138058
Winter's Rose Camellia 68879
Winteraconite 148101,148112
Winter-aconite 148101,148104
Winterbeam 369322
Winterberry 172135,203752,204376
Winterberry Currant 333970
Winterberry Euonymus 157345
Winter-berry Euonymus 157345
Winterberry Holly 204378
Winterberry-like Currant 334162
Winterbottom Draba 137295
Winterbottom Whitlowgrass 137295
Wintercherry 297643
Wintercreeper 157473
Winter-creeper 157473
Wintercreeper Euonymus 157515
Wintercress 47931,47964
Winter-cress 47964
Winterfat 218116,218132
Winter-flowering Begonia 49921
Winter-flowering Cherry 83112,83334
Winter-flowering Heather 149147
Winter-flowering Jasmine 211931
Winterflowering Renanthera 327693
Winter-flowering Wallflower 66427
Wintergreen 87480,172047,172087,172135,322775,322853
Winter-green 322775
Wintergreen Barberry 51802
Wintergreen Family 322935
Winter-greenleaf Blueberry 403837
Winterhazel 106628
Winter-hazel 106628
Winter-heliotriope 292353
Winterling 148104
Winterpicks 316819,338447
Wintersweet 4963,4965,87510,87525,274218
Winter-sweet 4954,4965,87525
Winterthorn 1035
Winterweed 374968,406973,407144
Winton Barberry 52378
Wintsun Rose 407989
Winuner's Osier 343134
Wippul-squip 192358
Wire Crabgrass 130635
Wire Grass 33772,143530
Wire Ling 145073,150131
Wire Netting Bush 105240
Wire Plant 259591
Wire Plants 259584
Wire Rush 212844,212852,213165
Wire Sage 344848
Wire Shrub 259584
Wire Thorn 385302
Wire Vine 259584,259591
Wiregrass 75065,143530,308816
Wirelettuce 375955
Wire-netting Bush 105240
Wireplant 259584,259591
Wirestem Muhly 259660,259679
Wire-stem Muhly 259660,259679
Wire-stem Wild Buckwheat 152387
Wirevine 259584,259596
Wireweed 68713,166125,308667,308816
Wirilda 1528
Wirtgen's Bedstraw 170766
Wirtgen's Water-crowfoot 325499
Wiry Jack 365564
Wiry Knotweed 309366
Wiry Lotus 237741
Wiry Panic Grass 281618
Wiry Panic-grass 281618
Wiry Witch Grass 281618
Wiry Witchgrass 281618
Wisconsin Beard-tongue 289347
Wisconsin Blackberry 339476
Wisconsin Penstemon 289347
Wisconsin Weeping Willow 342942,343112
Wisconsin Willow-herb 146943
Wisdom of Surgeons 126127
Wisdom Tree 259180
Wise Tree 259180
Wishbone Bush 255681
Wishbone Flower 392884,392905
Wishes 384714
Wishing-flower 227533
Wish-me-well 407072
Wisley Barberry 52379
Wisley Coral-red-fruited Barberry 52120
Wislizenus Dalea 121907
Wissadula 414522
Wissel Cypress 85300

Wissmann Clematis 95457
Wissmann Inula 138901
Wistaria 414541,414554
Wistaria Currant 334072
Wister's Coralroot 104023
Wister's coral-root 104023
Wisteria 414541,414554,414556,414565,414576
Witan Elm 401512
Witch 401512
Witch Alder 167540
Witch Bells 70271,81020,130383
Witch Elm 369339,401512
Witch Gowan 384714,399504
Witch Grass 144661,281274,281438
Witch Halse 106703,401512
Witch Hazel 77256,185092,185104,185130,185141,284957,369339,401512
Witch Hazel Family 185089
Witch Hobble 407674
Witch Tree 345631,401512
Witch Weed 377973
Witch Wicken 369339
Witch Wiggin 369339
Witch's Bells 81020
Witch's Milk 196818
Witch's Moneybags 200812
Witch's Tongue 96339
Witch's-moneybags 357203
Witch-alder 167537
Witchbeam 369339
Witchelower 91557,367120
Witchen 369339
Witchen-tree 369339
Witches Thimble 364146
Witches' Arms 170078
Witches' Blood 173082
Witches' Cap 188908
Witches' Flower 86755
Witches' Gloves 130383
Witches' Milk 196818
Witches' Needle 350425
Witches' Pouch 72038
Witches' Thimbles 70271,81020,130383,364146,412697
Witches' Tongue 96339
Witches'blood 173082
Witches'thimbles 412698
Witchetty Bush 1327
Witchgeass 281274
Witchgrass 226229
Witch-grass 281274,281438
Witchhazel 185092,185141
Witch-hazel 185092,185141
Witchhazel Family 185089
Witchhobble 407674
Witch-hobble 407674
Witchin Tree 369339
Witchwood 157429,323083,369339,401512
Witchwort 91557
Witeh-Hazel Family 185089
With Ywinny 383675
Withania 414586
Withen 369339
Wither Arrowbamboo 162691
Witherileaved Fargesia 162691
Witherod 407737
Withe-rod 407737,407971
Witherod Arrowwood 407737
Witherod Viburnum 407737
Witherspail 170193
Witherwine 68713
Withe-tree 343063
Withewind 68713
With-vine 68713,102933
Withwind 68713,102933
Withwine 102933,344244,369339
Withy Pear 369381
Withybind 102933
Withyvine 95431,102933
Withyweed 102933
Withywind 68713,95431,102933,236022,261162
Withywine 102933
Withywing 102933
Withywiny 383675
Witloof 90901
Witloof Chicory 90901
Wittall Tulip 400252
Witten Pear 369381
Witten Pear Tree 369381
Wittily Mulberry 259202
Witty 369339
Wizzard 144661
Wlassow Cranesbill 175018
Wlbrow 109857,302068
Wlcklow Marsh Orchid 273681
Wltchbane 369339
Woad 173082,209167,209229,327879
Woadmesh 173082
Woadwax 173082
Woadwaxen 172898,173082
Woad-waxen 172898,173082
Woadwex 173082
Woad-wise 173082
Wobamboo 362121,362131
Wocomahi Agave 10962
Wod Sndwort 256444
Wode Whistle 101852
Wode-whistle 101852
Wodystyle Bonvalot Larkspur 124204
Woila Gum 155677
Wold 327879
Wolewort 146834
Wolf Cholla 116680
Wolf Lilac 382383
Wolf Velvet Maple 3739
Wolf Willow 141941
Wolf's Bane 5014,5442,5676
Wolf's Bluegrass 306152
Wolf's Comb 133478
Wolf's Eye 21920
Wolf's Eyes 21920
Wolf's Milk 158730,158857,159027,159526
Wolf's Orach 44697
Wolf's Spike Rush 143418
Wolf's Spike-rush 143418
Wolf's Teasel 133478
Wolf's-bane 5676
Wolf's-milk 158857
Wolfbean 238432
Wolfberry 238996,239052,380748
Wolff Acronema 364894
Wolff Cystopetal 320255
Wolff Physospermopsis 297964
Wolff Trachydium 393869
Wolffia 414649
Wolffiella 414684
Wolfgang Milkwort 308468
Wolfgang Pondweed 312290
Wolfsbane 5014,5389,5676
Wolfsbane Monkshood 5676
Wolftail Sedge 74095
Wolftailgrass 289011
Wolf-willow 141941
Wollaston Primrose 315124
Wollemi Pine 414712,414713
Wolong Azalea 332115
Wolong Chinese Litse 233881
Wolong Cowparsnip 192400
Wolong Habenaria 184205
Wolong Milkvetch 43252
Wolong Rattlesnake Plantain 179715
Wolong Rhododendron 332115
Wolong Spotleaf-orchis 179715
Woman's Nightcap 277648
Woman's Tobacco 26551
Woman's Tongue 13595,60382,201732,311537
Woman's Tongue Tree 13595
Woman's-tobacco 26551
Woman-drake 61497
Wombat Berry 161031,161032
Wombatberry 161031
Wombat-berry 161031,161032
Wombgrass 365904
Wondeful Primrose 314348
Wonder Bean 71042
Wonder Tree 203417,203422,334435
Wonder Violet 410246
Wonder Willow 343057

Wonderbean 71042
Wonderberry 367256,367558
Wonderflower Willow 343057
Wonderful Forsythia 167451
Wonderful Haworthia 186550
Wonderful Larkspur 124386
Wonderful Licuala Palm 228750
Wonderful Licualapalm 228750
Wonderful Rattan Palm 65796
Wonderful Rhektophyllum 329286
Wonderful Saussurea 348312
Wonderful Seepweed 379580
Wonderful Spapweed 205210
Wonderful Spikesedge 143285
Wonderful Touch-me-not 205210
Wonderfully Bristled Turk's-cap Cactus 249588
Wondering Cowpea 409060
Wonder-tree 203422
Wong Rhododendron 332116
Wonga-wonga Vine 281227
Wongderful Woodbetony 287008
Wongdful Paraphlomis 283637
Wongoola 155472
Wood 184586,209229
Wood Anemone 23925,24004,24133
Wood Angelica 24503
Wood Apple 163499,163502
Wood Ash 277648
Wood Aster 40256,40906
Wood Avens 175454
Wood Bamboo 47433
Wood Barley 198258,198259
Wood Bedstraw 170631
Wood Bells 145487
Wood Betony 13073,13174,53304,286972,287063
Wood Bine 236022
Wood Bittercress 72749
Wood Bitter-vetch 408522
Wood Blades 238709
Wood Bluegrass 305743
Wood Blue-grass 305743
Wood Boneset 158250
Wood Broney 167955
Wood Burdock 31061
Wood Calamint 96966,97027
Wood Ceropegia 84305
Wood Chickweed 375008
Wood Club-rush 353179,353856
Wood Cotton-tree 56802
Wood Cow-wheat 248211
Wood Crane's-bill 174951
Wood Cranesbill 174951
Wood Crowfoot 23925,325628
Wood Cudweed 178465
Wood Datepalm 295487
Wood Dendrobium 125087
Wood Didymocarpus 129961
Wood Dock 339887,340249
Wood Dog Violet 410483
Wood Draba 137133
Wood Elder 345957
Wood Elsholtzia 144096
Wood Eupatorium 158090
Wood Fern 138575
Wood Fescue 163806
Wood Figwort 355202
Wood Flowers 295575
Wood Forget-me-not 260892
Wood Garlic 15509,15861
Wood Germander 388275
Wood Goldenrod 368282
Wood Goldilocks 325628
Wood Gossip Caesalpinia 64983
Wood Groundsel 360172
Wood Hyacinth 145487
Wood Hyacinths 199553
Wood Laurel 122492
Wood Leek 15832
Wood Lettuce 219507
Wood Lily 82038,97086,102867,229984,397540,397570
Wood Loosestrife 239750
Wood Marche 345957
Wood Meadow Grass 305743
Wood Meadow-grass 305743
Wood Melick 249106
Wood Milkvetch 42277
Wood Millet 254515
Wood Mint 55522,55524
Wood Moneywort 239750
Wood Nettle 221529,221547
Wood Nightshade 367120
Wood Nymph 257374
Wood Oil 406018
Wood Oil Tree 406019
Wood Oil-tree 406019
Wood Pea 222758,222844
Wood Pennyroyal 407256
Wood Pimpernel 239750,239898
Wood Pink 127678,127874
Wood Poppy 379220
Wood Rasp 338557
Wood Reed 91235,91236
Wood Reed Grass 91236
Wood Rose 32642,121192,121193,250850
Wood Rush 238563,238583,238712
Wood Sage 316127,387989,388026,388032,388033,388166,388275
Wood Sandwort 32004
Wood Sanical 345957
Wood Sanicle 345957
Wood Saxifrage 349227,350011
Wood Scabious 216764
Wood Scorpion-grass 260892
Wood Sedge 73887,75084,76465,76656
Wood Small-reed 65330
Wood Sorrel 277644,277648,277747,278008
Wood Speedwell 407238
Wood Spurge 158433,158672
Wood Starwort 375008
Wood Stitchwort 375008
Wood Strawberry 167601,167653
Wood Thistle 91730
Wood Tickseed 104542
Wood Vamp 123543
Wood Vetch 408336,408435,408620
Wood Vetchling 222844
Wood Vine 61497
Wood Violet 410306,410340,410483,410493,410530
Wood Water-cress 336277
Wood Waxen 173082
Wood Willowherb 146782
Wood Woundwort 373455
Wood's Aster 380866
Wood's Bunch-flower 248679
Wood's Stiff Sedge 76761
Wood's Willow 344279
Wood-alone 8399
Woodapple 163499,163502
Wood-apple 163499,163502
Woodas 173082
Wood-bank Sedge 74078
Woodberry 405571
Woodbetony 286972
Wood-betony 287063
Woodbind 187222,236022
Woodbine 20361,68713,95271,95426,172781,235630,236022,285097,285136,285164,367120
Woodbine Honeysuckle 236022
Wood-broney 167955
Woodcock Orchid 272481
Woodcock-orchid 272481
Wooded Chroesthes 89287
Wooded Lilyturf 272149
Wooden Rose 32642,121193,250850
Woodenbroom Cotoneaster 107424
Woodford Phaius 293494
Woodfordia 414716
Woodfruit Camellia 69260,69773
Woodfruit Euonymus 157539
Woodfruit Pittosporum 301454
Woodfruit Tanoak 233421
Woodgossip Caesalpinia 64983
Woodland 200946
Woodland Agrimony 11597,11598
Woodland Angelica 24483
Woodland Arctic-cudweed 270581

Woodland Beakchervil 28044
Woodland Beaked Chervil 28044
Woodland Bittercress 72749
Woodland Bluegrass 306048
Woodland Bluets 198717
Woodland Boneset 158315,158316
Woodland Bulrush 353856
Woodland Burdock 31069
Woodland Calamint 96966
Woodland Chee Woodreed 65338
Woodland Chickweed 374781
Woodland Crocus 111617
Woodland Cudweed 178465
Woodland Draba 137133,137136
Woodland Drop-seed 259707,259708
Woodland European Grape 411983
Woodland Forget Me Not 260892
Woodland Forgetmenot 260892
Woodland Forget-me-not 260892
Woodland Germander 388275
Woodland Globethistle 140814
Woodland Goldenrod 368003
Woodland Goosefoot 87164
Woodland Greenstar 307515
Woodland Groundsel 360172
Woodland Hawthorn 109790
Woodland Helicia 189952
Woodland Hogfennel 292815
Woodland Jurinea 190863
Woodland Knotweed 309940
Woodland Ladybell 7821
Woodland Lettuce 219229,219321
Woodland Lilyturf 272149
Woodland Mazus 247035
Woodland Mint 250385
Woodland Monkshood 5453
Woodland Orchid 302287
Woodland Paraprenanthes 283697
Woodland Passionflower 285679
Woodland Phlox 295257,295258
Woodland Pimpinella 299547
Woodland Pinedrops 320853
Woodland Pink 127874
Woodland Pinkroot 371616
Woodland Rabdosia 209834
Woodland Ragwort 360172
Woodland Sage 345246
Woodland Sagebrush 36354
Woodland Satin Grass 259707
Woodland Silverpuffs 253942
Woodland Spotted Orchid 121382
Woodland Spurge 159488
Woodland Star 233456
Woodland Strawberry 167653,167658
Woodland Sunflower 188958,188999,
189063,189073
Woodland Tanoak 233365
Wood-land Tanoak 233365
Woodland Thistle 92367
Woodland Threadstem 320855
Woodland Violet 410483
Woodland Whitlow-grass 137133,137136
Woodland Windhairdaisy 348818,348836
Woodland Wormwood 36354
Woodland Yam 131863
Wood-legume Ormosia 274440
Woodlot Poorspikebamboo 270439
Woodlotus 244420,244440
Woodnettle 125540,125553,221529,273899,
273903
Wood-nettle 221529,221547,273899
Woodnettle Like Elatostema 142768
Woodnettlelike Stairweed 142768
Wood-nut 106703
Woodnymph One-flowered Shinleaf 257374
Wood-oil Tree 406019
Wood-oilnut Tree 334411
Woodorchis 200770
Woodpit Sweetleaf 381460
Woodpod Ormosia 274440
Wood-poppy 379220
Woodreed 65254,91235
Woodreeve 170524
Woodrep 170524
Woodrip 170524
Woodroof 170524
Woodroot Sawwort 361134
Woodrose 170524
Wood-rose 250850
Woodrow 170524
Wood-rowe 170524
Wood-rowell 170524
Woodruff 39300,39328,170524
Woodrush 238563,238583
Wood-rush 238563
Woods Anemone 23925
Woods Bedstraw 170316
Woods Bluegrass 305743
Woods Buttercup 326118
Woods Ceropegia 84305
Woods Crazyweed 279195
Wood's Dwarf Nandina 262202
Woods Epidendrum 146447
Woods Falsebrome 58619
Woods Greenstar 307515
Woods Hypolytrum 202727
Woods Lacquertree 393485
Woods Parsnip 285744
Woods Reedbentgrass 65338
Woods Rhododendron 331944
Woods Rorippa 336277
Woods Tanoak 233365
Woods Tylophora 400961
Woods' Rose 337026
Wood-sage Speedwell 407019
Wood-sedge 76465
Woodside Wild Buckwheat 152548
Woodsmint 250459
Woodsore 52322
Woodsorrel 277644,277648
Wood-sorrel 277644,277648,277688,277973
Woodsorrel Family 277643
Wood-sorrel Family 277643
Wood-sorrel Oxalis 277648
Woodsour 52322,278161
Woodsow 277648
Woodsower 277648
Wood-stalk Calophanoides 67836
Wood-stem Edelweiss 224944
Woodt Stem Edelweiss 224944
Woodvine Heshouwu 162509
Woodward 170524
Wood-ward Peony 280330
Woodward Primrose 315125
Woodward's Blackbutt 155776
Woodwax 121001,172905,173082
Woodwaxa 173082
Woodwaxen 173082
Woodwesh 173082
Woodwex 173082
Woodwind 236022
Woody Aloe 16592
Woody Angelica 24483
Woody Bambusa 47433
Woody Bristlegrass 361728
Woody Clubfilment 179176
Woody Elder 345631
Woody Fleabane 135051
Woody Goldenrod 89978,89981
Woody Gomphostemma 179176
Woody Goosecomb 144488
Woody Green Violet 199705
Woody Hawksbeard 110886
Woody Ironweed 406854
Woody Jointweed 308680
Woody Lacquer-tree 393485
Woody Leaf Flower 296554
Woody Medic 247249
Woody Milkvetch 41997
Woody Nightshade 367120,367416
Woody Ophiorrhiza 272240
Woody Pear 323346,415728,415730
Woody Roegneria 144488
Woody Russianthistle 344452
Woody Sweetvetch 187981
Woody Tarenna 384917
Woody Tylophora 400961
Woody Windhairdaisy 348208
Woody-aster 415842
Woodycupule Tanoak 233421
Woody-cupuled Tanoak 233421

Woodyfruit Afzelia 10129
Woodyfruit Melliodendron 249560
Woodyfruit Seatung 301454
Woody-fruit Stereospermum 376299
Woody-fruited Afzelia 10129
Woody-fruited Melliodendron 249560
Woody-fruited Oil-wood 142476
Woody-fruited Pittosporum 301454
Woody-legumed Ormosia 274440
Woody-pear 415728
Woodypetal Fissistigma 166695
Woody-petaled Fissistigma 166695
Woody-putamened Sweetleaf 381460
Woody-rhizomed Fimbristylis 166462
Woody-rhizomed Fluttergrass 166462
Woodyroot Everlasting 21729
Woodyroot Lilyturf 272159
Woodyroot Pearleverlasting 21729
Woody-ruffee 170524
Wool Azalea 331195
Wool Grass 353179,353342
Wool Mullein 405694,405788
Wool Thistle 91954
Wool-bearing Blueberry 403885
Woolflower 80378,80476
Wool-grass 353342
Woolgrass Bulrush 353342
Woollen 405788
Woollen Breeches 200483
Woollstamen Japanese Cranberry 403871
Woolly 47631
Woolly Actinidia 6607
Woolly Aechmanthera 8539
Woolly Asiaglory 32642
Woolly Aster 40700
Woolly Beach-heather 198955
Woolly Beautyberry 66773
Woolly Begonia 49842
Woolly Bellflower 70113
Woolly Bentgrass 12425
Woolly Betony 53304,373166
Woolly Birthwort 34276
Woolly Blackberry 339383,339384
Woolly Blue Curls 396439
Woolly Blue Violet 410587
Woolly Blueberry 403885
Woolly Bluestar 20862
Woolly Buckthorn 362953
Woolly Buckwheat 152663
Woolly Bur Sage 19166
Woolly Burdock 31068
Woolly Bur-sage 19166
Woolly Bush 7197
Woolly Bush-clover 226989
Woolly Buttercup 326018,326072
Woolly Chaffhead 77235
Woolly Childvine 400934
Woolly Chinese Everlasting 21691
Woolly Chinese Pearleverlasting 21691
Woolly Christolea 160692
Woolly Clary 344836
Woolly Clover 397121
Woolly Cottonflower 179852
Woolly Crazyweed 278941
Woolly Croton 112852
Woolly Cup Grass 151702
Woolly Daisy 152809,152836
Woolly Desertdandelion 242935
Woolly Distaff Thistle 77716
Woolly Draba 136984,137133
Woolly Dutchman's-pipe 34275,34356
Woolly Dutchmanspipe 34275
Woolly Easterbonnets 152836
Woolly Fishhooks 22028
Woolly Fleabane 150740,150743
Woolly Foxglove 130369
Woolly Fruit Jewelvine 125959
Woolly Glorybower 96268
Woolly Goatsbeard 394286
Woolly Goldenweed 375794
Woolly Grevillea 180610
Woolly Groundsel 279894,359377,359381
Woolly Hardhead 379660
Woolly Hawkweed 195715
Woolly Hawthorn 109797
Woolly Head 23925
Woolly Heads 23925
Woolly Hedeg-nettle 373166
Woolly Hedgenettle 373166
Woolly Indian Wheat 302144
Woolly Kohleria 217745
Woolly Lavender 223297
Woolly Leptodermis 226095
Woolly Lespedeza 226989
Woolly Loco 42739
Woolly Locoweed 42739
Woolly Machilus 240711
Woolly Marbles 318801
Woolly Milkvetch 42580
Woolly Milkweed 37983
Woolly Mint 250439,250457,250468
Woolly Morning Glory 32642
Woolly Motherwort 225004
Woolly Mullein 405788
Woolly Net Bush 68068
Woolly Netbush 68063,68068
Woolly Nightshade 367364
Woolly Ovary Rhododendron 330675
Woolly Paintbrush 79126
Woolly Panicgrass 128478
Woolly Panic-grass 128478
Woolly Paperflower 318981
Woolly Pea-shrub 72300
Woolly Pfeiffer Cicely 276466
Woolly Photinia 295720
Woolly Pipe-vine 34356
Woolly Planetree Maple 3475
Woolly Plantain 302122
Woolly Plateaucress 160692
Woolly Pomaderris 310765
Woolly Psathyrostachys 317084
Woolly Pussytoes 26441,26586
Woolly Pyrul 408982
Woolly Ragwort 279956
Woolly Rambutan 14375
Woolly Raspberry 338719
Woolly Rhododendron 330710,331195
Woolly Sea-blite 379608
Woolly Sedge 75059
Woolly Senna 78507
Woolly Shrubcress 368560
Woolly Solms-Laubachia 368560
Woolly Solomonseal 308580
Woolly Speedwell 317912
Woolly Sterculia 376156
Woolly Sumpweed 227763
Woolly Sunbonnet 85999
Woolly Sunflower 152799,152809
Woolly Tea-tree 226462
Woolly Terminalia 386656
Woolly Thistle 91954
Woolly Thorn 109850
Woolly Three-awn 33907
Woolly Thyme 391340,391365
Woolly Torch 299262
Woolly Trefoil 397116
Woolly Vetch 408693
Woolly Vineclethra 95512
Woolly Wattle 1340
Woolly Wildclove 226095
Woolly Willow 343283,343592,343594
Woolly Woundwort 373166
Woolly Yarrow 3957,3990,4056
Woolly Yellowish Everlasting 21569
Woolly Yellowish Pearleverlasting 21569
Woollybranch Willow 343344
Woolly-branched Barberry 51832
Woollybutt 155558,155628
Woolly-butt 155628
Woollychin Woodbetony 287340
Woolly-cup Wild Buckwheat 152198
Woolly-flower Kohleria 217744
Woollyflower Oat 45446
Woollyflower Summercypress 217344
Woolly-flowered Oat 45446
Woolly-flowered Persimmon 132145
Woolly-flowered Rinorea 334618
Woolly-footed Rhododendron 331665
Woollyfruit Cinquefoil 312522
Woollyfruit Lily 228581
Woollyfruit Pubescent Spiraea 372068

Woollyfruit Roquette 154031
Woolly-fruit Sedge 75065,75066
Woollyfruit Stylophorum 379228
Woolly-fruited Grewia 180768
Woollyhair Rockjasmine 23308
Woollyhead Fanbract 379164
Woollyhead Lessingia 227116
Woolly-headed Clover 396891
Woolly-headed Thistle 91954
Woollyheads 318801
Woollyjoint Pricklypear 273077
Woolly-leaf Anthurium 28108
Woollyleaf Buttercup 325777
Woolly-leaf Clematis 95067
Woolly-leaf Mock Orange 294560
Woolly-leaf Mock-orange 294560
Woolly-leaved Alder 16400
Woolly-leaved Clematis 95067
Woolly-leaved Lupin 238460
Woolly-leaved Nepal Maple 3740
Woolly-leaved Pea-shrub 72215
Woolly-leaved Rhododendron 331045
Woollypetiole Willow 343598
Woolly-petioled Willow 343598
Woolly-pod Milkweed 37910
Woolly-pod Vetch 408371,408693
Woolly-podded Broom 120958
Woolly-silke Daphne 122458
Woolly-silky Daphne 122458
Woollyspike Lagopsis 220121
Woollystalk Anabasis 21038
Woollystalk Draba 136990
Woollystalk Sagebrush 35466
Woollystalk Whitlowgrass 136990
Woollystalk Wormwood 35466
Woollystem-ciliate Roegneria 335260
Woollystyle Raspberry 338719
Woollystyle Rhododendron 331062
Woolly-styled Adinandra 8247
Woolly-styled Raspberry 338719
Woolly-styled Rhododendron 331062
Woolly-tube Camellia 69257
Woollytwig Sageretia 342189
Woolly-twig Sageretia 342189
Woollytwig Tuan Linden 391850
Woolly-twigged Willow 343283
Wooly Butterfly Bush 62123
Wooly Dalea 121892,121905
Wooly Jelly Palm 64162
Wooly Panicum 282201
Wooly Spine Palm 2551
Wooly-fruited Gomphostemma 179182
Wooly-leaved Caper 71727
Woolyleaved Ciliate Roegneria 335267
Woomung Barberry 52380
Woonyoung Primrose 315126
Wooton Loco 43253
Wooton's Larkspur 124698
Wooton's Wild Buckwheat 152662
Worcesterberry 333959
Worldwise 345733
Worle Sycamore Maple 3469
Worm Vine 404982
Wormgrass 371616
Worm-grass 356503
Wormit 35090
Wormod 35090
Wormseed 35868,86934,139680,139681,
365564
Worm-seed 139678
Wormseed Erysimum 154424
Wormseed Goose foot 139678
Wormseed Goosefoot 139678
Wormseed Mustard 154424
Worm-seed Mustard 154424
Worm-seed Sand-mat 160047
Wormseed Sugarmustard 154424
Wormseed Tea 139678
Wormseed Wallflower 154424
Worm-seed Wallflower 154424
Wormseed Weed 139678
Wormweed 139678
Wormwood 35084,35090,35237,35307,
35534,35837,36474,139678,221787,
285086
Wormwood Cassia 360419
Wormwoodleaf Edelweiss 224802
Wormwoodleaf Woodbetony 286973
Wormwoodlike Motherwort 224989
Woronow Orchardgrass 121273
Woronow Skullcap 355850
Worral 47013
Worts 403916
Wothflower Raspberry 339096
Wo-tone Sage 345381
Woudwix 173082
Wound Rocket 47964
Wound Root 383675
Woundweed 368480
Woundwort 3978,28200,53293,222020,
227533,358902,368480,373085,373346,
373365,373455,381035
Woundwort Laser 222020
Wrapfruit Tanoak 233148
Wrappers 130383
Wray 235373
Wray Mischobulbum 255942
Wray Pinanga 299681
Wray Plectocomiopsis 303100
Wrayi Twayblade 232094
Wreath Aster 380872
Wreath Flower 409335
Wreath Golden Rod 368003
Wreath Goldenrod 368003
Wren 174877
Wren-flower 174877
Wrest Harrow 271563
Wrest-harrow 271563
Wretched Sedge 75389
Wretweed 86755
Wriggle Azalea 330264
Wright Acacia 1707
Wright Anisacanthus 25261
Wright Apocopis 29457
Wright Arrowwood 408224
Wright Balanophora 46882
Wright Begonia 50411
Wright Blue Berry 404060
Wright Blueberry 404060
Wright Catclaw 1707
Wright Cunealglume 29457
Wright Glochidion 177191
Wright Larkspur 124699
Wright Lotus 237787
Wright Morningglory 208305
Wright Nightshade 367749
Wright Philoxerus 294882
Wright Pleurospermum 304860
Wright Pycreus 322378
Wright Sageretia 342217
Wright Silktassel 171557
Wright Verbena 176696
Wright Viburnum 408224
Wright's Amaranth 18851
Wright's Baccharis 46272
Wright's Bastard-sage 152663
Wright's Buckwheat 152663
Wright's Cactus 354333
Wright's Catchfly 364217
Wright's Catkin Mistletoe 28307
Wright's Dutchman's-pipe 34374
Wright's Dwarfdandelion 218214
Wright's Fishhook Cactus 244262,354333
Wright's Giant Hyssop 10423
Wright's Goldenrod 368517
Wright's Jimson Weed 123092
Wright's Marsh Thistle 92496
Wright's Milkpea 169668
Wright's Morning-glory 208305
Wright's Nutrush 354129
Wright's Nut-rush 354129
Wright's Orach 44699
Wright's Penstemon 289385
Wright's Plantain 302214
Wright's Prairie Clover 121909
Wright's Rabbit-tobacco 317735
Wright's Sealavender 230816
Wright's Snakeroot 11175
Wright's Snakeweed 182116
Wright's Waternymph 262115
Wright's Wild Buckwheat 152663

Wrightia 414794,414813
Wringchain Azalea 331984
Wringleaf Pondweed 312151
Wrinkl 354838
Wrinkl Indianmulberry 258916
Wrinkl Serpentroot 354835
Wrinkle Agapetes 10317
Wrinkle Bitterbamboo 304094
Wrinkle Floweringquince 84573
Wrinklecalyx Syzygium 382657
Wrinkled Alseodaphne 17810
Wrinkled Alysicarpus 18279
Wrinkled Anzacwood 310756
Wrinkled Bamboo 297222
Wrinkled Bitter Bamboo 304094
Wrinkled Christmas-bush 14213
Wrinkled Duckbeakgrass 209356
Wrinkled Fruit Millettia 66641
Wrinkled Gianthyssop 10414
Wrinkled Goldenrod 368373
Wrinkled Honeysuckle 236066
Wrinkled Jujube 418207
Wrinkled Medick 247450
Wrinkled Spineflower 89069
Wrinkled Syzygium 382657
Wrinkled Viburnum 408089
Wrinkled Yushania 416816
Wrinkledfruit Amaranth 18848
Wrinkledfruit Amaranthus 18848
Wrinkledfruit Camellia 69569
Wrinkledfruit Mockstrawberry 138793
Wrinkledfruit Woodnettle 221543
Wrinkled-fruited Millettia 66641
Wrinkledleaf Agapetes 10317
Wrinkledleaf Bennettiodendron 51044
Wrinkledleaf Bittersweet 80304
Wrinkledleaf Box 64338
Wrinkledleaf Bristlegrass 361868
Wrinkledleaf Buckthorn 328843
Wrinkledleaf Lettuce 219491
Wrinkledleaf Maesa 241830
Wrinkledleaf Mint 250349
Wrinkledleaf Oreocharis 273879
Wrinkledleaf Privet 229630
Wrinkledleaf Pyrola 322902
Wrinkledleaf Rabdosia 209815
Wrinkledleaf Rhododendron 330556
Wrinkledleaf Sageretia 342192
Wrinkledleaf Storax 379442
Wrinkledleaf-like Rabdosia 209813
Wrinkled-leaved Agapetes 10317
Wrinkled-leaved Willow 344003
Wrinkled-seeded Oak Sedge 76562
Wrinkleedge Comastoma 100277
Wrinkleedge Throathair 100277
Wrinklefruit Bittersweet 80336
Wrinklefruit Pleurospermum 304827
Wrinklefruit Ribseedcelery 304827
Wrinklefruit Windhairdaisy 348710
Wrinkle-fruited Camellia 69569
Wrinkle-fruited Leafflower 296801
Wrinkleleaf Calvepetal 178001
Wrinkleleaf Cinquefoil 312351
Wrinkleleaf Clematis 95398
Wrinkleleaf Dock 339994
Wrinkleleaf Dolomiaea 135726
Wrinkleleaf Everlasting 21579
Wrinkle-leaf Goldenrod 368373
Wrinkleleaf Greenorchid 137564
Wrinkleleaf Honeysuckle 236066
Wrinkleleaf Magnolia 242169
Wrinkleleaf Opithandra 272595
Wrinkleleaf Parsley 292694
Wrinkleleaf Pearleverlasting 21579
Wrinkleleaf Pimpinella 299388
Wrinkleleaf Primrose 314917
Wrinkle-leaf Rockrose 93141
Wrinkleleaf Sanicle 345999
Wrinkleleaf Serpentroot 354878
Wrinkleleaf Twinballgrass 209136
Wrinkleleaf Yushanbamboo 416816
Wrinkle-leaved Box 64338
Wrinkle-leaved Buckthorn 328843
Wrinkle-leaved Goldenrod 368373
Wrinkle-leaved Honeysuckle 236066
Wrinkle-leaved Maesa 241830
Wrinkle-leaved Rabdosia 209815
Wrinkle-leaved Sageretia 342192
Wrinkle-leaved Storax 379442
Wrinkle-leaved Willow 344003
Wrinklelemma Dallisgrass 285481
Wrinklepod Cliffbean 66641
Wrinkleseed Snakegourd 396266
Wrinkle-seed Spiderflower 95781
Wrinklesepal Snakegourd 396164
Wrinkle-sheathed Spike-rush 143154
Wrinklespadix Arisaema 33293
Wrinklespadix Southstar 33293
Wrinklewave Corydalis 105763
Wrinklsheath Arrowbamboo 162735
Wrinkly Willowleaf Cotoneaster 107675
Wrinkly-leaf Glyptopetalum 178001
Wrinkly-leaved Glyptic-petal Bush 178001
Wryleaf Coldwaterflower 298897
Wu Holly 204403
Wu Microtropis 254333
Wu Ophiorrhiza 272306
Wu Poplar 311575
Wu Rattlesnake Plantain 179716
Wu Sedge 76764
Wu Spotleaf-orchis 179716
Wu Tutcheria 322628
Wu Waterchestnut 394426
Wu's Caper 71932
Wu's Copperleaf 2021
Wu's Euonymus 157967
Wuchagou Monkshood 5697
Wud 209229
Wudan Wormwood 36553
Wudangshan Greenbrier 366502
Wudangshan Smilax 366502
Wudu Crazyweed 278887
Wudu Sedge 76766
Wudu Stringbush 414188
Wudwise 173082
Wugang Chickweed 375147
Wugang Magnolia 242361
Wugang Oak 324283
Wugongshan Cragcress 98002
Wugongshan Fimbristylis 166564
Wugongshan Fluttergrass 166564
Wugongshan Hilliella 196289
Wugongshan Holly 204404
Wugongshan Scurvyweed 98002
Wuhan Grape 412005
Wuk 324335
Wukungshan Fimbristylis 166564
Wulai Holly 204365
Wulate Feathergrass 376791
Wulate Milkvetch 42493
Wulfenia 414829
Wulff Broom 121030
Wulfsberg's Holarrhena 197203
Wulian Poplar 311576
Wuliangshan Arrowbamboo 162764
Wuliangshan Barberry 52381
Wuliangshan Fargesia 162764
Wuliangshan Sweetleaf 381100
Wuliangshan Umbrella Bamboo 162764
Wuling Hazelnut 106785
Wuling Lemperg's Barberry 51847
Wuling Lilac 382386
Wuling Monkshood 5472
Wuling Mountains Ladybell 7894
Wuling Onion 15769
Wuling Panicled Monkshood 5472
Wuling Pine 299982
Wuling Rocket 193426
Wuling Thorowax 63828
Wuling Wildginger 37758
Wulingshan Ladybell 7894
Wulingshan Larch 221951
Wulingshan Sessile Skullcap 355757
Wullow 16352
Wullow Wallow 16352
Wulong Hogfennel 293078
Wulong Wildginger 37759
Wulongshan Goosecomb 335350
Wulumuqi Milkvetch 43254
Wulumuqi Sweetvetch 188119
Wumeng Barberry 52380

Wumeng Broadlobe 106062
Wumeng Corydalis 106606
Wumeng Lady's Slipper 120472
Wumeng Ladyslipper 120472
Wumeng Meconopsis 247199
Wumeng Pimpinella 299574
Wumeng Primrose 315098
Wumeng Rounded-buds Rhododendron 331871
Wumengshan Rhododendron 332118
Wuming Azalea 332119
Wuming Buckthorn 328902
Wuming Rhododendron 332119
Wumung Meconopsis 247199
Wuning Indocalamus 206907
Wunu 225623
Wuqia Fritillary 168401
Wuqia Hawksbeard 110872
Wuqia Larkspur 124701
Wuqia Milkvetch 43059
Wurmplakkie 108846
Wurral 47013
Wushan Azalea 331681
Wushan Barrenwort 147075
Wushan Chickweed 375145
Wushan Elaeagnus 142247
Wushan Epimedium 147075
Wushan Newlitse 264114
Wushan Pertybush 292082
Wushan Raspberry 339480
Wushan Rhododendron 331681
Wushan Sedge 76765
Wushan Snakemushroom 43256
Wushan Vetch 408710
Wushan Violet 410060
Wushe Machilus 240728
Wusheh Cherry 316852
Wushleen 401711
Wusu Fritillary 168591
Wusuli Dragonflyoechis 302549
Wusuli Peashrub 72373
Wusuli Windhairdaisy 348902
Wutai Bluegrass 305763
Wutai Gentian 174074
Wutai Goldsaxifrage 90448
Wutai Goldwaist 90448
Wutai Pricklyash 417372
Wutai Prickly-ash 417372
Wutai Shortstem Woodbetony 287019
Wutaishan Corydalis 105992
Wutaishan Crazyweed 279237
Wutaishan Motherwort 225023
Wutaishan Oak 324559
Wutaishan Sedge 75415
Wutang Mountain Greenbrier 366502
Wutu Crazyweed 278887
Wuwei Begonia 49686
Wuweishan Loquat 151145
Wuweishan Newlitse 264031
Wuwishan Neolitsea 264031
Wuxi Corydalis 105671
Wuxi Indocalamus 206840
Wuxi Meliosma 249465
Wuxi Windflower 24034
Wuxing Clematis 95022
Wuxinyinxing Ginkgo 175828
Wuxuhai Willow 344285
Wuyang Fritillary 168344
Wuyi Barberry 52382
Wuyi Chinese Groundsel 365081
Wuyi Mountain Meadowrue 388720
Wuyisha Pseudosasa 318356
Wuyishan Azalea 332120
Wuyishan Bitter Bamboo 304117
Wuyishan Bitterbamboo 304117
Wuyishan Bitter-bamboo 304117
Wuyishan Clethra 96555
Wuyishan Dendrobenthamia 124887
Wuyishan Four-involucre 124887
Wuyishan Illicium 204606
Wuyishan Maple 3755
Wuyishan Meadowrue 388720
Wuyishan Mountainash 369310
Wuyishan Mountain-ash 369310
Wuyishan Photinia 295818
Wuyishan Poorspikebamboo 270452
Wuyishan Pseudosasa 318356
Wuyishan Salem-rose 339206
Wuyishan Sedge 76767
Wuyishan Sinosenecio 365081
Wuyishan Square Bamboo 87612
Wuyishan Yushanbamboo 416824
Wuyishan Yushania 416824
Wuyuan Maple 3756
Wuyuan Sage 344954
Wuyuan Snowbell 379476
Wuyuan Spapweed 205453
Wuyuan Storax 379476
Wuyuan Touch-me-not 205453
Wuyuzhen Goldsaxifrage 90472
Wuyuzhen Goldwaist 90472
Wuzhishan Persimmon 132422
Wyalong Wattle 1119
Wych Elm 401512
Wych Halse 401512
Wych Hazel 77256,401512
Wych Tree 401512
Wychen 369339
Wychwood 401512
Wylliespoort Aloe 16576
Wyman Creek Wild Buckwheat 152443
Wymote 18173
Wyoming Goldenbush 150330
Wyoming Sagebrush 36409
Wyoming Thistle 92326

X

Xalapen Strawberry Tree 30893
Xanthopappus 415130
Xanthophyllum 415143
Xanthophyllum Family 415139
Xanthophytum Rockyellowtree 415154
Xantolis 415234
Xantus Pincushion 84535
Xeranthemumlike Visnaga 140175
Xeric Arisaema 33263
Xeric Bromegrass 60989
Xeric Rockjasmine 23213
Xeric Southstar 33263
Xerophillous Rabdosia 209870
Xerophilous Calophanoides 67835
Xerophilous Keteleeria 216155
Xerophilous Willow 344286
Xerospermum 415519,415520
Xiahe Kobresia 217288
Xiahe Valerian 404390
Xiamen Acanthus 2673
Xianbeiflower 362392,362396
Xiangcheng Arisaema 33567
Xiangcheng Jujube 418231
Xiangcheng Leek 15896
Xiangcheng Lily 230077
Xiangcheng Monkshood 5700
Xiangcheng Onion 15896
Xiangcheng Poplar 311578
Xiangcheng Southstar 33567
Xiangcheng Sweetvetch 188110
Xiangcheng Woodbetony 287832
Xianggui Newlitse 264057
Xianggui Oreocharis 273888
Xianggui Persimmon 132476
Xiangxi Cyclobalanopsis 116214
Xiangxi Qinggang 116214
Xiangxi Sedge 76772
Xiangyang Cherry 83184
Xiangye Sweet Flag 5822

Xiangyuan 93869
Xianju Bitterbamboo 304037
Xianmu 161621
Xiaofoshow Ginkgo 175829
Xiaoguangshan Willow 344293
Xiaoguongshan Willow 344293
Xiaohei Poplar 311191
Xiaojin Cowparsnip 192401
Xiaojin Crabapple 243731
Xiaojin Holly 204406
Xiaojin Milkvetch 43258
Xiaojin Superior Rockfoil 349288
Xiaokuai Lilyturf 272157
Xiaowutaishan Barberry 52149
Xiaoxidong Azalea 332126
Xiaoxidong Rhododendron 332126
Xiaozuan Poplar 311192
Xiashan Bamboo 47532
Xiashan Bambusa 47532
Xiashan Mud Bamboo 47532
Xiazang Achyrospermum 4489
Xiberian Seabuckthorn 196777
Xichang Azalea 332127
Xichang Douglas Fir 318602
Xichang Larkspur 124702
Xichang Lemongrass 117274
Xichang Mint 250333
Xichang Neofinetia 263869
Xichang Pondweed 312292
Xichang Rhododendron 332127
Xichang Sida 362698
Xichou Alseodaphne 17811
Xichou Cyclobalanopsis 116198
Xichou Cylindrokelupha 30978
Xichou Daphne 122634
Xichou Dendrobium 125419
Xichou Elaeagnus 142248
Xichou Elatostema 142896
Xichou Euonymus 157791
Xichou Evergreen Chinkapin 79069
Xichou Evergreenchinkapin 79069
Xichou Fig 165871
Xichou Hornbeam 77275
Xichou Machilus 240698
Xichou Maple 3603
Xichou Meliosma 249481
Xichou Michelia 252857
Xichou Oak 116198
Xichou Oilfruitcamphor 381799
Xichou Paris 284308
Xichou Persimmon 132397
Xichou Pricklyash 417374
Xichou Prickly-ash 417374
Xichou Qinggang 116198
Xichou Raspberry 339488
Xichou Rockvine 387853
Xichou Sedge 76262
Xichou Skullcap 355764
Xichou Sloanea 366083
Xichou Slugwood 50609
Xichou Stairweed 142896
Xichou Syndiclis 381799
Xichou Trigonotis 397425
Xichuan Leek 15897
Xichuan Onion 15897
Xide Privet 229536
Xigu Aster 41265
Xigu Azalea 332129
Xigu Barberry 51326
Xigu Rhododendron 332129
Xikang Azalea 331828
Xikang Cotoneaster 107685
Xikang Deutzia 127101
Xikang Goldsaxifrage 90451
Xikang Goldwaist 90451
Xikang Ligusticum 229387
Xikang Mountainash 369490
Xikang Rhododendron 331828
Xikang Rose 336952
Xikang Smallreed 127295
Xikang Spruce 298248
Xilan Lasianthus 222289
Xilan Roughleaf 222289
Xilanshan Drypetes 138627
Xilanshan Roughleaf 222289
Xilin Aspidistra 39582
Xilin Vine Catchfly 363977
Xilinhaote Speedwell 407439
Ximalaya Alkaligrass 321297
Ximalaya Pygeum 322403
Ximeng Giantarum 20101,20151
Ximeng Psilotrichum 319023
Ximeng Raspberry 339085
Xinba Milkvetch 42501
Xincheng Aspidistra 39518
Xincheng Hornbeam 77302
Xindian Swertia 380354
Xinduqiao Monkshood 5633
Xing'an Azalea 330495,332130
Xing'an Barberry 52388
Xing'an Bushclover 226751
Xing'an Chickweed 374794
Xing'an Cowparsnip 192255
Xing'an Erysimum 154451
Xing'an Hogfennel 292780
Xing'an Ladybell 7744
Xing'an Larkspur 124296
Xing'an Maple 2871
Xing'an Milkaster 169773
Xing'an Monkshood 5032
Xing'an Parnassia 284634
Xing'an Pine 300197
Xing'an Pink 127682
Xing'an Pondweed 312293
Xing'an Poplar 311350
Xing'an Sedge 74101
Xing'an Small Reed 127321
Xing'an Smallreed 127321
Xing'an Stonecrop 294121
Xing'an Sugarmustard 154451
Xing'an Tickseed 104758
Xing'an Violet 410010
Xing'an Willow 343497
Xingjiang Mustard 364557
Xingjiang Nettle 402893
Xingjiang Peashrub 72200
Xingjiang Pricklythrift 2273
Xingkaihu Sharphead Sagebrush 36039
Xinglong Crazyweed 279239
Xinglong Petrov Synstemon 382015
Xinglong Stonecrop 200778
Xinglong Synstemon 382015
Xinglongshan Buttercup 326426
Xingren Briggsia 60286
Xingren Forkliporchis 26011
Xingren Privet 229667
Xingshan Barberry 52157
Xingshan Indigo 205879
Xingshan Magnoliavine 351063
Xingshan Meadowrue 388721
Xingshan Spiraea 371932
Xingshan Viburnum 408056
Xingshan Willow 343695
Xingwen Barberry 52389
Xingwen Elaeagnus 142249
Xingyi Begonia 50414
Xingyi Nan 295390
Xingyi Nanmu 295390
Xingyi Phoebe 295390
Xingyi Spicegrass 239738
Xinhu Schizostachyum 351865
Xinhua Bulrush 353655
Xinjiang Aellenia 8850
Xinjiang Aeluropus 8870
Xinjiang Ajania 12997
Xinjiang Amberboa 18983
Xinjiang Ammopiptanthus 19778
Xinjiang Asiabell 98312
Xinjiang Asparagus 39111
Xinjiang Bastard Cress 390231
Xinjiang Beancaper 418760
Xinjiang Bedstraw 170777
Xinjiang Bird's Foot Trefoil 237609
Xinjiang Bluebasin 350288
Xinjiang Bugle 13190
Xinjiang Calophaca 67786
Xinjiang Carlina 76994
Xinjiang Carunclegrass 256459
Xinjiang Chervil 84758
Xinjiang Chive 15662
Xinjiang Craneknee 221773

Xinjiang Crazyweed 279165
Xinjiang Epipactis 147195
Xinjiang Eritrichium 153542
Xinjiang Feathergrass 376908
Xinjiang Figwort 355137
Xinjiang Fir 476
Xinjiang Flaxaster 231838
Xinjiang Fleabane 150968
Xinjiang Fritillary 168624
Xinjiang Gagea 169470
Xinjiang Gentian 173747,174066
Xinjiang Gentianella 174163
Xinjiang Giantfennel 163711
Xinjiang Goldenray 229252
Xinjiang Hawkweed 195692
Xinjiang Hawthorn 110048
Xinjiang Heliotrope 190562
Xinjiang Herelip 220096
Xinjiang Hippolytia 196698
Xinjiang Hornpoppy 176754
Xinjiang Juniper 213873
Xinjiang Knotweed 309766
Xinjiang Knotwood 44264
Xinjiang Lagochilus 220096
Xinjiang Landpick 308639
Xinjiang Leek 15296
Xinjiang Lily 229929
Xinjiang Linosyris 231838
Xinjiang Lymegrass 144474
Xinjiang Madwort 18427
Xinjiang Medic 247468
Xinjiang Milkvetch 42580
Xinjiang Milkwort 308116
Xinjiang Minuartwort 255504
Xinjiang Moehringia 256459
Xinjiang Monkshood 5572
Xinjiang Needlegrass 376908
Xinjiang Olgaea 270243
Xinjiang Onion 15662
Xinjiang opium 335624
Xinjiang Oxtongue 298640
Xinjiang Parnassia 284556
Xinjiang Peach 20914
Xinjiang Pear 323306
Xinjiang Peashrub 72371
Xinjiang Pea-shrub 72371
Xinjiang Pennycress 390231
Xinjiang Peony 280154
Xinjiang Prettybean 67786
Xinjiang Pyrethrum 322644
Xinjiang Pyrola 322932
Xinjiang Ricegrass 300747
Xinjiang Rockcress 30205
Xinjiang Roegneria 335510
Xinjiang Russianthistle 344716
Xinjiang Savin 213873
Xinjiang Sawwort 361119
Xinjiang Sedge 76622
Xinjiang Silverspikegrass 227953
Xinjiang Skullcap 355765
Xinjiang Solomonseal 308639
Xinjiang Spruce 298382
Xinjiang Statice 230737
Xinjiang Teasel 133519
Xinjiang Thorowax 63623
Xinjiang Threeribfruit 397964
Xinjiang Toadflex 230870
Xinjiang Tulip 400224
Xinjiang Valerian 404267
Xinjiang Veinyroot 237609
Xinjiang Vetch 408350
Xinjiang Vetchling 222720
Xinjiang Violet 335042,335043
Xinjiang Wheatgrass 11876
Xinjiang Wild Apple 243698
Xinjiang Willowweed 146609
Xinjiang Windhairdaisy 348118
Xinjiang Wolfberry 239030
Xinjiangdaisy 382096,382098
Xinjiang-Xizang Arnebia 34615
Xinjiashan Cacalia 283881
Xinjiashan Sakebed 97731
Xinning Buttercup 326498
Xinning Chirita 87992
Xinning Elatostema 142897
Xinning Newlitse 264097
Xinning St. John's Wort 201918
Xinning Stairweed 142897
Xinyi Bambusa 47463
Xinyi Begonia 49899
Xinyi Eurya 160623
Xinyi Gugertree 350955
Xinyi Machilus 240719
Xinyi Maple 3647
Xinyi Metalleaf 296426
Xinyi Persimmon 132421
Xinyi Poorspikebamboo 270432
Xinyi Sterculia 376193
Xinyuan Dandelion 384860
Xinyuan Fritillary 168624
Xinzhu Leafflower 296695
Xiongyue Apricot 34480
Xiphobracteate Shorttail Hornbeam 77338
Xiqingshan Milkvetch 43261
Xiqingshan Woodbetony 287833
Xique-Xique 299255
Xisha Hoarypea 386167
Xishan Cinquefoil 313000
Xishi Azalea 330617
Xishuaagbanna Persimmon 132477
Xishuanbanna Jasmine 211746
Xishuangbana Ixora 211147
Xishuangbanna Arisaema 33275
Xishuangbanna Eria 61448
Xishuangbanna Garcinia 171222
Xishuangbanna Ginger 418028
Xishuangbanna Hairorchis 61448
Xishuangbanna Lasianthus 222262
Xishuangbanna Rattanpalm 65746
Xishuangbanna Rockvine 387868
Xishuangbanna Roughleaf 222262
Xishuangbanna Vatica 405184
Xishui Begonia 50416
Xishui Primrose 314590
Xita Wild Smallreed 127211
Xitaibai Milkvetch 43263
Xitou Begonia 49716
Xitou Bulbophyllum 62630
Xitou Stonebean-orchis 62630
Xiuning Chiritopsis 88006
Xiuning Mazus 247051
Xiuning Stringbush 414221
Xiuren Dendrobium 125420
Xiushan Goldencalyx Rhododendron 330395
Xiushan Rhododendron 330395
Xiushan Sour Bamboo 4619
Xiuying Gentian 157582
Xiuying Poorspikebamboo 270449
Xizang Acanthopanax 2521
Xizang Acronema 6226
Xizang Actinocarya 6743
Xizang Agapetes 10379
Xizang Alashan Woodbetony 286981
Xizang Albizzia 13671
Xizang Alp Corydalis 106535
Xizang Alplily 234241
Xizang Amitostigma 19534
Xizang Anemone 24088
Xizang Aphragmus 29177
Xizang Aralia 30783
Xizang Arisaema 33467
Xizang Arrowbamboo 162708
Xizang Arrowwood 408165
Xizang Asiabell 98431
Xizang Asparagus 39234
Xizang Astyleorchis 19534
Xizang Azalea 330876
Xizang Baliosperm 46965
Xizang Baliospermum 46965
Xizang Barberry 52223,52255
Xizang Beadcress 264637
Xizang Betony 373466
Xizang Bigumbrella 59247
Xizang Birthwort 34200
Xizang Bistamengrass 127483
Xizang Blueberry 403979
Xizang Bluegrass 306074,306077
Xizang Bluestem 22833
Xizang Boloflower 205591
Xizang Brassaiopsis 59247
Xizang Braya 59778

Xizang Broadflower Corydalis 106059
Xizang Broadlobe Corydalis 106063
Xizang Broomrape 275002
Xizang Bubbleweed 297883
Xizang Buckthorn 328872,328903
Xizang Cacalia 283799
Xizang Calanthe 66061
Xizang Catalpa 79265
Xizang Catchfly 364085
Xizang Catnip 264947
Xizang Chickweed 375114
Xizang Childseng 318517
Xizang Cinquefoil 313114
Xizang Clabgourd 263568
Xizang Climbing Apple Bamboo 249610
Xizang Climbing-bamboo 249610
Xizang Common White Jasmine 211952
Xizang Corallodiscus 103960
Xizang Corydalis 106531
Xizang Cotoneaster 107707
Xizang Currant 334270
Xizang Cuttonguetree 177858
Xizang Cyclobalanopsis 116216
Xizang Cymbidium 117090
Xizang Dandelion 384833
Xizang Dendrocalamus 125513
Xizang Diploknema 133007
Xizang Dipoma 383997
Xizang Doronicum 136376
Xizang Draba 137266
Xizang Dragonbamboo 125513
Xizang Drake Corydalis 105833
Xizang Drupetea 322624
Xizang Duck Woodbetony 287000
Xizang Dutchmanspipe 34200
Xizang Dysosma 139625
Xizang Eightangle 204524
Xizang Elaeagnus 142250
Xizang Ephedra 146197
Xizang Euonymus 157925
Xizang Everlasting 21711
Xizang Falsenettle 56336
Xizang Falseoat 398565
Xizang Fargesia 162708
Xizang Fieldjasmine 199455
Xizang Flatawndaisy 413009
Xizang Floweringquince 84591
Xizang Fragrant Bamboo 87646
Xizang Fragrant-bamboo 87646
Xizang Fritillary 168640
Xizang Gaultheria 172173
Xizang Gentian 150515,159974,173988
Xizang Geraniumleaf Primrose 314432
Xizang Ghostfluting 228335
Xizang Gianthyssop 236278
Xizang Glorybower 96392
Xizang Goosecomb 335529
Xizang Goosefoot 87182
Xizang Greenbrier 366344
Xizang Groundsel 360220
Xizang Habenaria 184136
Xizang Hackberry 80786
Xizang Hairvein Azalea 331349
Xizang Halogeton 184952
Xizang Haplosphaera 185704
Xizang Helicia 189955
Xizang Helictotrichon 190204
Xizang Hemsley Monkshood 5267
Xizang Heronbill 153923
Xizang Himalayahoneysuckle 228335
Xizang Himalaya-honeysuckle 228335
Xizang Holly 204407
Xizang Hornstedtia 198476
Xizang Houndstongue 118017
Xizang Huodendron 199455
Xizang Incarvillea 205591
Xizang Iris 208492
Xizang Ixora 211192
Xizang Jasmine 212071
Xizang Jerusalemsage 295208
Xizang Jijigrass 4131
Xizang Juniper 213972
Xizang Jurinea 207113
Xizang Knotweed 309891
Xizang Kobresia 217299
Xizang Lady's Slipper 120461
Xizang Ladyslipper 120461
Xizang Larch 221881
Xizang Large Rhinanthus-like
Woodbetony 287620
Xizang Largeleaf Newlitse 264070
Xizang Launaea 222996
Xizang Lemongrass 117258
Xizang Leptodermis 226142
Xizang Ligusticum 229415
Xizang Lilac 382300
Xizang Lily 229977
Xizang Linearleaf Kobresia 217299
Xizang Listera 264722
Xizang Litse 234078
Xizang Lobelia 234827
Xizang Luculia 238106
Xizang Machilus 240725
Xizang Madder 338037
Xizang Manglietia 244425
Xizang Maple 3692
Xizang Mazus 247043
Xizang Meconopsis 247122
Xizang Michelia 252899
Xizang Micromelum 253734
Xizang Micromeria 253734
Xizang Microula 254370
Xizang Middle Ephedra 146197
Xizang Milkvetch 43154
Xizang Monkeyflower 255256
Xizang Mosquitotrap 117515
Xizang Mouseear 83058
Xizang Mouseear Cress 30472
Xizang Mugwort Wormwood 36544
Xizang Muskroot 8405
Xizang Needlebractdaisy 395967
Xizang Neoalsomitra 263568
Xizang Neopallasia 264250
Xizang Nepeta 265084
Xizang Nervilia 265442
Xizang Nettle 403031
Xizang Ninenode 319532
Xizang Oak 116216,324119
Xizang Oneflowerprimrose 270740
Xizang Onosma 271846
Xizang Oppositeleaf Corydalis 106536
Xizang Orinus 274255
Xizang Parasite 30919
Xizang Paris 284402
Xizang Parnassia 284537,284623
Xizang Pearbamboovine 249610
Xizang Pearleverlasting 21711
Xizang Pearweed 239889
Xizang Peashrub 72346
Xizang Pennisetum 289141
Xizang Pennycress 390210
Xizang Pepper 300450
Xizang Perotis 291206
Xizang Physochlaina 297883
Xizang Pimpinella 299585
Xizang Plantain 302199
Xizang Pricklyash 417294
Xizang Prickly-ash 417294
Xizang Primrose 315049
Xizang Pseudostellaria 318517
Xizang Psychotria 319532
Xizang Psychrogeton 319930
Xizang Ptilotrichum 321099
Xizang Purplebranch Willow 343831
Xizang Purplehammer 234721
Xizang Pyrenaria 322624
Xizang Pyrethrum 322653
Xizang Pyrola 322802
Xizang Qinggang 116216
Xizang Rabdosia 209862
Xizang Raspberry 339370
Xizang Rhamnella 328552
Xizang Rhodiola 329967
Xizang Rhododendron 330876
Xizang Rhubarb 329407
Xizang Ricepaperplant 387435
Xizang Rockfoil 349990
Xizang Rockjasmine 23224
Xizang Rockvine 387869
Xizang Roegneria 335529
Xizang Roscoea 337109

Xizang Rose 336990
Xizang Rush 213527
Xizang Sage 345480
Xizang Saltlivedgrass 184952
Xizang Sandthorn 196780
Xizang Seabuckthorn 196780
Xizang Sedge 76149,76538
Xizang Seepweed 379511
Xizang Seseli 361545
Xizang Shortscape Fleabane 150515
Xizang Shrubby Elsholtzia 144026
Xizang Sichuan Poplar 311516
Xizang Silverspikegrass 227951
Xizang Slugwood 50629
Xizang Smallreed 127316
Xizang Smallspar Violet 410757
Xizang Snakeheadcress 383997
Xizang Solomonseal 308663
Xizang Southstar 33467
Xizang Sow-thistle 90864
Xizang Spapweed 204881
Xizang Speedwell 407416
Xizang Spicebush 231396
Xizang Spiradiclis 371784
Xizang Spiraea 372136
Xizang Splitmelon 351774
Xizang Spunsilksage 360882
Xizang Staff-tree 80203
Xizang Stephania 375854
Xizang Stinkinggrass 249102
Xizang Strawberry 167640
Xizang Swallowwort 117515
Xizang Sweetvetch 188187
Xizang Swertia 380387
Xizang Swordflag 208492
Xizang Syzygium 382684
Xizang Tanoak 233420
Xizang Taxillus 385238
Xizang Thickribcelery 279653
Xizang Tibet 416825
Xizang Tickseed 104856
Xizang Tigernanmu 122691
Xizang Tofieldia 392640
Xizang Torularia 264637
Xizang Touch-me-not 204881
Xizang Trachydium 393865
Xizang Tricarpelema 395373
Xizang Tricarpweed 395373
Xizang Trigonotis 397473
Xizang Trisetum 398565
Xizang Truncate-bract Willow 344002
Xizang Truncate-bracted Willow 344002
Xizang Tumorfruitparsley 393865
Xizang Vetch 408641
Xizang Viburnum 408165
Xizang Vicatia 408247
Xizang Vinegentian 110333
Xizang Wakerobin 397568
Xizang Waldheimia 413009
Xizang Walsura 177858
Xizang Watercress 262751
Xizang Waxplant 198862
Xizang Wendlandia 413787
Xizang White Sagebrush 36561
Xizang Whitebractcelery 266977
Xizang Whitepearl 172173
Xizang Whitlowgrass 137266
Xizang Wildclove 226142
Xizang Willow 343371,344294
Xizang Windflower 24088
Xizang Windhairdaisy 348861
Xizang Wintergreen 172173
Xizang Wolftailgrass 289141
Xizang Woodbetony 287761
Xizang Woodlotus 244425
Xizang Yam 131913
Xizang Yew 385404
Xizang Yushanbamboo 416825
Xizanggrass 293365,293370
Xizangia 415619,415622
Xizang-sikkim Bluegrass 305296
Xiznag Diapensia 127925
Xmas Bush 14177
Xoconochtli 375533
Xoconoxtle 290719
Xu Changqing 117643
Xuan'en Dishspurorchis 171899
Xuan'en Gastrochilus 171899
Xuancheng Sedge 76774
Xuanen Marsdenia 245865
Xuanen Milkgreens 245865
Xuedan Hemsleia 191903
Xuedan Snowgall 191903
Xuefeng Calanthe 65967
Xuefengshan Reevesia 327387
Xuefengshan Suoluo 327387
Xuelihong Leaf-mustard 59458
Xupu Silky Apricot 34438
Xylobium 415704
Xylocarp Euonymus 157539
Xylocarpous Camellia 69773
Xylocarpous Sinojackia 364959
Xylocarpus 415714
Xylopia 415760,415826
Xylopyrene Sweetleaf 381460
Xylosma 415863,415869
Xylosmae-leaved Holly 204408
Xylosma-leaf Holly 204408

Y

Ya'an Conehead 378177
Ya'an Cranesbill 175020
Ya'an Cryptocarya 113496
Ya'an Thickshellcassia 113496
Yaan Aletris 14536
Yabe Columbine 30090
Yabulai Windhairdaisy 348941
Yabunikki 91351
Yacca 306524,415172
Yacca Podocarpus 306418
Yaccatree 306395,306437,306457,306467,306468,306485,306519
Yaccatree Family 306345
Yachan 80118
Yachiang Monkshood 5701
Yachiang Rockjasmine 23344
Yackrod 359158
Yackyar 359158
Yacon Strawberry 310029
Yadong Acronema 6227
Yadong Alpbean 391547
Yadong Barberry 51728
Yadong Cremanthodium 110470
Yadong Dandelion 384677
Yadong Kobresia 217310
Yadong Milkvetch 42303
Yadong Monkshood 5593
Yadong Mouseear Cress 30157,30406
Yadong Nutantdaisy 110470
Yadong Paratetradgland Willow 343839
Yadong Poplar 311579
Yadong Sagebrush 36560
Yadong Tibetia 391547
Yadong Willow 344295
Yadong Yushanbamboo 416826
Yadong Yushania 416826
Yadorik Scurrula 355337
Yagara Bulrush 56637
Yahusuo 106609
Yaihsien Rattlebox 112819
Yajiang Clovershrub 70911
Yajiang Corydalis 106611
Yajiang Gansu Woodbetony 287306
Yajiang Larkspur 124703
Yajiang Leptodermis 226079
Yajiang Monkshood 5701
Yajiang Onosma 271848
Yajiang Pleurospermum 304763

Yajiang Primrose 314504
Yajiang Ribseedcelery 304763
Yajiang Rockjasmine 23344
Yajiang Sedge 76776
Yajiang Stinkinggrass 249111
Yajiang Wildclove 226079
Yak 324335
Yake Synotis 381960
Yakibaran Cymbidium 116845
Yakkron 324335
Yakou Tailanther 381960
Yaku Fescue 164023
Yakushima Gentian 174080
Yakushima Spindle-tree 157970
Yakusima Rhododendron 331237
Yaller 359158
Yallow 3978
Yalluc 381035
Yalong Hemsleia 191912
Yalong Snowgall 191912
Yalongjiang Windhairdaisy 348377
Yalu Monkshood 5292
Yalu River Primrose 314514
Yalu Sedge 74934
Yalung Hemsleia 191912
Yaluzangbu Agapetes 10356
Yaluzangbu Rockfoil 350042
Yam 131451,131501,131772,131896,
207623
Yam Arum 36985
Yam Bean 279728,279732,279737,371540
Yam Family 131918
Yam Root 131896
Yama Gumi 141999
Yamabuki Mandarin 93871
Yamada Drypetes 138649
Yamamikan 93501
Yamamoto Mitrastemon 256180
Yamamoto Monkshood 5702
Yambean 279728
Yamleaf Bauhinia 49069
Yam-leaved Clematis 95358
Yamp 77809
Yampa 77809
Yampee 131880
Yampi 131880
Yampi Yam 131880
Yamroot Corydalis 106066
Yanagi 123322
Yanan Peony 280145
Yanbian Asparagus 39258
Yanbian Trefoil 396920
Yanbian Willow 344296
Yandang Buerger Maple 2824
Yandang Machilus 240648
Yandang Maple 2824
Yandang Trident Maple 2824
Yandee 155631
Yanfen Acacia 1642
Yanfeng Cacalia 283875
Yang Goosecomb 335566
Yang Khao 133588
Yang Kobresia 217311
Yang Monkshood 5704
Yang Pearleverlasting 21730
Yang Roegneria 335566
Yang Snowbell 379477
Yang's Larkspur 124704
Yangbi Bentgrass 12214
Yangbi Elatostema 142899
Yangbi Marsdenia 245867
Yangbi Milkgreens 245867
Yangbi Stairweed 142899
Yangchun Camellia 68933
Yangchun Chiritopsis 88005
Yangchun Cleyera 96630
Yangchun Eargrass 187704
Yangchun Galangal 17760
Yangchun Hedyotis 187704
Yangchun Helicia 189961
Yangchun Holly 204410
Yangchun Rattan Palm 65815
Yangchun Rattanpalm 65815
Yangchun Sweetspire 210422
Yanggona 300458
Yangjue Maple 3758
Yangmingshan Azalea 332138
Yangmingshan Rhododendron 332138
Yangona 300458
Yangping Persimmon 132442
Yangping Spapweed 205088
Yangquan Hackberry 80788
Yangshan Peony 280258
Yangsho Galeobdolon 170029
Yangshuo Galeobdolon 170029
Yangshuo Loosestrife 239573
Yangshuo Sedge 76779
Yangshuo Weasel-snout 170029
Yangtao 6553,6558
Yangtao Actinidia 6553
Yangtao Kiwifruit 6553
Yangtze Milkvetch 43269
Yangxian Windhairdaisy 348444
Yangzi Milkvetch 43269
Yangzijiang Cystacanthus 120798
Yanhui Croton 113060
Yanhui Sauropus 348071
Yanhusuo 105557
Yanjin Euphorbia 160096
Yanjing Delavay Taxillus 385196
Yankee Blackberry 338444,338843
Yankeeweed 158097
Yanling Holly 204411
Yanquapin 263270
Yanshan Buttercup 326499
Yanshan Eria 148799
Yanshan Grape 411546
Yanshan Hairorchis 148799
Yantai Fimbristylis 166511
Yantai Fluttergrass 166511
Yantai Larkspur 124124
Yantai Thorowax 63598
Yanwa Larkspur 124705
Yanyuan Alplily 234246
Yanyuan Asparagus 39259
Yanyuan Butterbur 292408
Yanyuan Ligusticum 229416
Yanyuan Maple 3581
Yanyuan Monkshood 5706
Yanyuan Parnassia 284636
Yanyuan Reedbentgrass 65524
Yanyuan Tickclover 126338
Yanyuan Woodbetony 287834
Yanyuan Woodreed 65524
Yao Buttercup 326500
Yao's Cherry 83364
Yaogangxiang Rhododendron 332141
Yaoluo Pinellia 299733
Yaoshan Arisaema 33512
Yaoshan Arrowwood 408141
Yaoshan Azalea 332142
Yaoshan Blueberry 404063
Yaoshan Bostrychanthera 57505
Yaoshan Cicadawingvine 356375
Yaoshan Corallodiscus 103943
Yaoshan Corydalis 106018
Yaoshan Dichroa 129050
Yaoshan Dinggongvine 154262
Yaoshan Dumasia 138938
Yaoshan Elatostema 142900
Yaoshan Erycibe 154262
Yaoshan Falsepimpernel 231600
Yaoshan Holly 204352
Yaoshan Litse 234110
Yaoshan Maple 3759
Yaoshan Motherweed 231600
Yaoshan Pauciannulated Camellia 69461
Yaoshan Pipewort 151549
Yaoshan Rattan Palm 65736
Yaoshan Rattanpalm 65736
Yaoshan Rhododendron 332142
Yaoshan Sandwort 31962
Yaoshan Securidaca 356375
Yaoshan Southstar 33512
Yaoshan Stairweed 142900
Yaoshan Stauntonvine 374440
Yaoshan Staunton-vine 374440
Yaoshan Sweetleaf 381463
Yaoshan Viburnum 408141
Yap Mouth 28617
Yapana 158349

Yap-mouth 28617
Yar 79161
Yaran 1292
Yard Daisy 322724
Yard Dock 340109
Yard Grass 143530
Yard Long Cowpea 409096
Yardgrass 143512
Yard-grass 143512,143530
Yardgrass Galingale 118798
Yardgrass-like Cypressgrass 118798
Yard-long Bean 409086,409096
Yardlong Cowpea 409096
Yard-long Cow-pea 409096
Yarey 103733
Yarey Hembra 103733
Yarkrod 359158
Yarlungzangbo Rockfoil 350042
Yarnut 102727
Yarr 370522
Yarra Gum 155777
Yarra-grass 3978
Yarran 1292
Yarrel 3978
Yarreyon 103733
Yarroconalli 386312
Yarrow 3913
Yarrow Broomrape 275173
Yarrow Gilia 175669
Yarroway 3978
Yarrowleaf Pimpinella 299345
Yarrowleaf Woodbetony 286974
Yarrow-like Ajania 12989
Yashino-zakura 83365
Yatabe Blueberry 404066
Yatay Palm 64159,64163
Yate 155540
Yate Tree 155540
Yatsushiro 93873
Yatung Milkvetch 42303
Yaupon 204078,204391
Yaupon Holly 204391
Yaupon Tea 204391
Yautia 65230,415190
Yautia Amarilla 415191
Yavering Bells 275486
Yawa 409086,409096
Yaweiyinxing Ginkgo 175830
Yawl 144661
Yawroot 376645
Yaxian Leaf-flower 296476
Yaxian Nanmu 295415
Yaxian Phoebe 295415
Yaxian Rattlebox 112819
Yaxian Windmill 100674
Yazhou Blachia 54881
Yazhou Hedyotis 187705
Yazhou Munronia 259893
Yazhou Rattlebox 112819
Yazhou Tarenna 384977
Ye'eb Nut 104144
Yeaker 324335
Yecheng Buttercup 326501
Yecheng Larkspur 124706
Yecheng Microsisymbrium 365384
Yecheng Smallgarliccress 365384
Yeddo Euonymus 157559
Yeddo Hawthorn 329068
Yeddo Hornbeam 77401
Yeddo Raphiolepis 329114
Yeddo Spruce 298307
Yeddo-hawthorn 329114
Yedo Euonymus 157570
Yedo Spindle-tree 157570
Yedwark 282685
Yeenga 174305
Yegoma Oil 290940
Yeheh Nut 104144
Yeilow Pansy Violet 410394
Yeilowpurple Odontoglossum 269076
Yek 324335
Yellow 364557
Yellow ' Loosestrife' 188263
Yellow Aconite 5676,148104
Yellow Adder's Tongue 154899,154945
Yellow Adder's-tongue 154899
Yellow Adonis 8387
Yellow Agapetes 10308
Yellow Ageratum 235503
Yellow Agrimony 11549
Yellow Alfalfa 247457
Yellow Alicoche 140280
Yellow Alkanet 21957
Yellow Allemande 14873
Yellow Alopecurus 17525
Yellow Alpbean 391543
Yellow Alpine Milk Vetch 42384
Yellow Alpine Skullcap 355580
Yellow Alplily 234217
Yellow Alstroemeria 18062
Yellow Alyssum 18327,45192
Yellow Amitostigma 19530
Yellow and Blue Scorpion-grass 260791
Yellow Anemone 24020
Yellow Angel's Trumpet 61235
Yellow Angledtwig Magnoliavine 351081
Yellow Anise Tree 204570
Yellow Anthers Rhododendron 330703
Yellow Arborvitae 390610
Yellow Archangel 170025,220326,220401
Yellow Areca Palm 89360
Yellow Argyreia 32626
Yellow Arnebia 34624
Yellow Arum Lily 417104
Yellow Ash 94026,145028,155632,166985
Yellow Asia Osmanther 276259
Yellow Asphodel 39438,262574
Yellow Astyleorchis 19530
Yellow Avens 175378,175446,175452,
175454
Yellow Azalea 330276,330706,331159,
331161
Yellow Bachelor's Button 81183
Yellow Bachelor's Buttons 308173,325518
Yellow Bachelor's Cornflower 81183
Yellow Bachelor's-buttons 325519
Yellow Ball Cactus 267035
Yellow Ball Mustard 265553
Yellow Balm 249499
Yellow Balsam 112842
Yellow Bamboo 47518,297203
Yellow Banks Rose 336368
Yellow Banksian Rose 336368
Yellow Barberry 51889
Yellow Bark 91059
Yellow Bark Oak 324539
Yellow Bartsia 284098
Yellow Basket Willow 145517
Yellow Basket-willow 145517
Yellow Bauhinia 49241
Yellow Bean 389547
Yellow Beavertail 272800
Yellow Bedstraw 170335,170737,170743
Yellow Bee Plant 95727
Yellow Bee-orchid 272455
Yellow Beeplant 95727
Yellow Beet 53261
Yellow Begonia 49839,49987
Yellow Bell 168538
Yellow Bell Bauhinia 49241
Yellow Bell Rhododendron 332123
Yellow Bellflower 70328
Yellow Bells 262441,385493
Yellow Berries 328737
Yellow Berry 328737,328849
Yellow Betony 373104
Yellow Bigbractginger 79727
Yellow Biond's Magnolia 241999
Yellow Birch 53338,53432
Yellow Bird of Paradise 65014
Yellow Bird's Nest 202767
Yellow Bird's-nest 202767,258006
Yellow Bird-lime Flower 210843
Yellow Bird's-foot 274880,274890
Yellow Birthwort 106117
Yellow Blackhatflower 126729
Yellow Bladderwort 403108
Yellow Blanket 18385
Yellow Bletilla 55572
Yellow Blob 68189
Yellow Bloodwood 106798

Yellow Blue-bead-lily 97091
Yellow Bluebell 115350
Yellow Bluebellflower 115350
Yellow Bluegrass 305531
Yellow Bluestem 57567,57568
Yellow Blumeopsis 55862
Yellow Bonnet 30054
Yellow Boots 68189
Yellow Bottle 89704
Yellow Box 155649,155658
Yellow Boys 359158
Yellow Bozzom 89704
Yellow Branchlet Keteleeria 216122
Yellow Bristlegrass 361856,361877
Yellow Bristle-grass 361877
Yellow Broomrape 275177
Yellow Bubbleleaf Azalea 331778
Yellow Buckeye 9692
Yellow Buckthorn 328641,328706
Yellow Bugle 13073
Yellow Bulbophyllum 62949
Yellow Bunge Batrachium 48908
Yellow Bunny-ears 272980
Yellow Bussell 89704
Yellow Buttercup 325785
Yellow Butterfly Bush 61957
Yellow Butterfly Palm 89360
Yellow Butterwort 299752
Yellow Buttons 383874
Yellow Cactus 395630
Yellow Calla 36967
Yellow Camellia 69070
Yellow Camomile 26900
Yellow Campanulate Cremanthodium 110361
Yellow Canada Lily 229785
Yellow Caralluma 72530
Yellow Catalpa 79257
Yellow Cattleya 79528
Yellow Caul 325518,325666,326287
Yellow Cedar 85263,85301,390587
Yellow Centaurea 81382
Yellow Centaury 90790,90792
Yellow Cephalaria 82116
Yellow Chamomile 26900,107295
Yellow Chestnut Oak 324204
Yellow Chinese Poppy 247139
Yellow Cinnamon 91392
Yellow Clintonia 97091
Yellow Clover 237539,247366,396806,
396854,396882,396989,397096
Yellow Clovershrub 70809
Yellow Cob Cactus 234926
Yellow Colic-root 14509
Yellow Columbine 30017
Yellow Coneflower 140079,326964
Yellow Corallodiscus 103948
Yellow Coralroot 104019
Yellow Coris 201824
Yellow Corn Flower 89704
Yellow Corydalis 106117
Yellow Cosmos 107172
Yellow Cottonwood 311292
Yellow Cotula 26900
Yellow Crabgrass 130567
Yellow Crain 325820,325864
Yellow Crane 325820,325864
Yellow Cranesbill 174948
Yellow Craw 325666,360409
Yellow Crazy 68189
Yellow Creams 325518,325666
Yellow Crees 325666,326287
Yellow Crepis 110741
Yellow Cress 47964,336268
Yellow Crimson Ixora 211074
Yellow Crocus 111535,111558,417616
Yellow Crossandra 111713
Yellow Crow Bells 262441
Yellow Crow Foot 28200
Yellow Crow's Foot 28200
Yellow Crownbeard 405941
Yellow Crownvetch 105310
Yellow Cucumber Tree 242043
Yellow Cup 325518,325612,325666,326287
Yellow Curlylip-orchis 62949
Yellow Cypress 85301
Yellow Daffodil 262441
Yellow Daisy 339556,359158
Yellow Dalea 121894
Yellow Day Lily 191284
Yellow Dayflowerlike Thorowax 63609
Yellow Daylily 191304
Yellow Day-lily 191284,191304
Yellow Dead Nettle 170025
Yellow Dead-nettle 170025
Yellow Deal 300223
Yellow Delavay Peony 280192
Yellow Dendrobium 125115
Yellow Devil 208771
Yellow Dianthus 127747
Yellow Diapensia 127912
Yellow Dishspurorchis 171845
Yellow Dock 339994
Yellow Dodder 114986
Yellow Dog-fennel 188397
Yellow Dogtooth Violet 154899,154945
Yellow Dog-tooth Violet 154899
Yellow Dots 413559
Yellow Dragonbamboo 125490
Yellow Drott 231183
Yellow Dwarf Lily 229956
Yellow Edge 198610
Yellow Elder 385493
Yellow Elsholtzia 144021
Yellow Enkianthus 145677
Yellow Epimedium 146954
Yellow Evening Primrose 269464
Yellow Evening-primrose 68508
Yellow Everlasting 189147
Yellow Evodia 161393
Yellow Eye 415990
Yellow False Foxglove 10143
Yellow False Garlic 266959
Yellow False Indigo 47878
Yellow False Mallow 243898
Yellow False Yucca 193229
Yellow Falsecuckoo 48295
Yellow Falsegrniculate Monkshood 5501
Yellow Field Cress 336277
Yellow Field Pumpkin 114288
Yellow Fieldcress 336277
Yellow Field-cress 336277
Yellow Figwort 355265
Yellow Fingers-and-thumbs 28200
Yellow Fir 318580
Yellow Flag 208771
Yellow Flag Iris 208771
Yellow Flamboyant 288897
Yellow Flame 288897
Yellow Flame Grevillea 180595
Yellow Flame Tree 288897
Yellow Flamenettle 99748
Yellow Flame-tree 288897
Yellow Flax 231882,231947,327548,327553
Yellow Flax Bush 327553
Yellow Fleabane 321554
Yellow Floating Heart 267825
Yellow Floatingheart 267825
Yellow Flour-de-Luce 208771
Yellow flower Chirita 87912
Yellow Flower De Luce 208771
Yellow Flower Erysimum 154416
Yellow Flowered Rhododendron 330706
Yellow Forest Violet 410466,410472
Yellow Forget-me-not 20840
Yellow Fox-and-cubs 299220
Yellow Foxglove 130365
Yellow Foxtail 289116,361762,361856,
361877
Yellow Fox-tail 361877
Yellow Fringed Orchid 183501,302285
Yellow Fritillary 168538
Yellow Fumewort 169065
Yellow Fumrrory 106117
Yellow Gagea 169460
Yellow Garden-hawkweed 392672
Yellow Gariand-lily 68032
Yellow Garlic 15480
Yellow Gastrochilus 171845
Yellow Gastrodia 171921
Yellow Gentian 54896,156708,173193,
173610,173778,174221

Yellow Gentianopsis 174221
Yellow Ghostfluting 228320
Yellow Giant Fox Tail Grass 361877
Yellow Giant Hyssop 10410
Yellow Giant-hyssop 10410
Yellow Gingerlily 187447
Yellow Ginseng Papoose Root 79733
Yellow Glandweed 284098
Yellow Glasswort 342864
Yellow Goat's Beard 394281
Yellow Goat's-beard Buck's Beard 394333
Yellow Goatsbeard 394281,394333
Yellow Gold 89704
Yellow Gold Mahur 288897
Yellow Gollan 325518,325666,326287
Yellow Gowan 89704,325518,384714
Yellow Gowlan 68189
Yellow Granadilla 285660
Yellow Groove Bamboo 297204
Yellow Guava 318737
Yellow Guinea Flower 194663
Yellow Guinea Yam 131516
Yellow Gum 103696,155552,155624,
232565,267867,415172,415181
Yellow Habenaria 183393
Yellow Hairgrass 12839
Yellow Hairy Knotweed 308943
Yellow Harlequin 169065
Yellow Haw 109709
Yellow Hawkweed 195506,195511,392672
Yellow Hawthorn 109709
Yellow Head 360328
Yellow Hedgehyssop 180306
Yellow Hedge-hyssop 180297
Yellow Hemiboea 191351
Yellow Henbane 201385
Yellow Himalayan Raspberry 338347,338354
Yellow Hoarypea 386353
Yellow Hollyhock 32429
Yellow Honeysuckle 235791,236066,330276
Yellow Hop Clover 397096
Yellow Hoptree 320068
Yellow Horn 415094
Yellow Horn Poppy 176738
Yellow Horned-poppy 176738
Yellow Hornpoppy 176738
Yellow Horn-poppy 176738
Yellow Horse Daisy 89704
Yellow Houseteek 358066
Yellow Ice Plant 123932
Yellow Indian Blanket 169601
Yellow Indian Grass 369582
Yellow Indiangrass 369582
Yellow Indigo 205927
Yellow Ipomoea 268229
Yellow Iris 208560,208771
Yellow Ironweed 405915,406402
Yellow Ixora 211125
Yellow Jack 262429
Yellow Jade Orchid Tree 252841
Yellow Japanese Azalea 331261
Yellow Jasmine 211821,211834,211863,
211864,211868,211905
Yellow Jasmine-root 172781
Yellow Jessamine 172781
Yellow Jewelweed 205204
Yellow Joyweed 18108
Yellow Jugflower 7201
Yellow Kangaroo-paw 25218
Yellow Kapok 98105
Yellow Kawakami Azalea 330994
Yellow Kawakami Rhododendron 330994
Yellow Keenan 215750
Yellow King-devil 195506
Yellow Knapweed 81352
Yellow Korean Lily 229724
Yellow Kowhai 369130
Yellow Lady Banks Rose 336368
Yellow Lady Slipper 120310
Yellow Lady's Slipper 120310,120421
Yellow Lady's Slipper Orchid 120310
Yellow Lady's-slipper 120310,120421
Yellow Ladyslipper 120352
Yellow Laelia 219677
Yellow Lagopsis 220124
Yellow Lamb's Head 399601
Yellow Lanterns 24167,216473,216474
Yellow Largeleaf Primrose 314606
Yellow Lark's Heel 399601
Yellow Larkspur 124359,124712
Yellow Latan 222636
Yellow Latan Palm 222636
Yellow Leaf Tree 415143
Yellow leafy Rock Daisy 291330
Yellow Licorice 177888
Yellow Licorice Weed 354661
Yellow Lily 262441
Yellow Lily-of-incans 18062
Yellow Liparis 232236
Yellow Lobelia 257704
Yellow Locust 334976
Yellow Longtube Lycoris 239265
Yellow Longtube Stonegarlic 239265
Yellow Loosestrife 239543,239804,239879,
239898
Yellow Lotus 263270
Yellow Lupine 238463
Yellow Lysionotus 239985
Yellow Mahonia 242517
Yellow Maiden 262441
Yellow Mallow 286672
Yellow Mandarin 134442,315525
Yellow Mangosteen 171221
Yellow Margineted Rubber Tree 164940
Yellow Mariposa Lily 67570,67595
Yellow Mariposa Tulip 67595
Yellow Marsh Marigold 68189
Yellow Marsh Saxifrage 349452
Yellow Marsh-marigold 68189
Yellow Meadow Rue 388506
Yellow Meadow Vetchling 222820
Yellow Meadow-rue 388506
Yellow Meconopsis 247124
Yellow Medick 247457
Yellow Melancholy Thistle 91955
Yellow Melilot 249232
Yellow Melliot 249205
Yellow Menodora 250267
Yellow Messmate 155578
Yellow Mexican Water Lily 267714
Yellow Mexican Water-lily 267714
Yellow Mexico Waterlily 267714
Yellow Michelia 252871
Yellow Mignonette 327866
Yellow Milk Vetch 278764
Yellow Milkvetch 42648
Yellow Milkwort 308173
Yellow Mombin 88691,372477
Yellow Mombin Tree 372477
Yellow Monk's-hood 5039
Yellow Monkey Flower 255203,255217,
255218
Yellow Monkeyflower 255196,255218
Yellow Monkshood 5200
Yellow Monkswort 266608
Yellow Moraea 258415
Yellow Morning Glory 250758,250850
Yellow Mountain Saxifrage 349052
Yellow Mountain-heather 297025
Yellow Mouse Ear 195858
Yellow Musk 255218
Yellow Mustard 59348,59438,59603,364545
Yellow Myrtle 239755
Yellow Nailwort 284905
Yellow Narcissus 262441
Yellow Nicker Bean 64971
Yellow Nightshade 367311,367726
Yellow Nut Grass 118822
Yellow Nut Sedge 118822,118829
Yellow Nutsedge 118822,118837
Yellow Oak 324204,324539
Yellow Oat Grass 398460,398518
Yellow Oat-grass 398437,398518
Yellow Oleander 389978
Yellow Ononis 271464
Yellow Ophiorrhiza 272268
Yellow Orchid Vine 246022
Yellow Orchis 310871
Yellow Oreocharis 273837
Yellow Ormocarpum 274345
Yellow Ox Eyes 89704

Yellow Oxeye 63524,63539,385614,385616
Yellow Ox-eye 89704
Yellow Oxytropis 278764
Yellow Paintbrush 79132
Yellow Palm 89360
Yellow Panic-grass 282393
Yellow Pansy Violet 410394
Yellow Pareirs Root 198800
Yellow Parilla 250221
Yellow Passion Fruit 285638
Yellow Passionflower 285666
Yellow Patrinia 285859
Yellow Pea 130935,222681,222820,321731
Yellow Pea Vine 222762
Yellow Pendent-bell 145677
Yellow Peony 280192,280223
Yellow Pepper-grass 225365
Yellow Pereira 198800
Yellow Periwinkle 79420
Yellow Phaius 293494
Yellow Pheasant's Eye 8387
Yellow Pheasant's Eyes 8387
Yellow Pheasant's-eye 8387
Yellow Phlox 154390
Yellow Pimpernel 239750,383043
Yellow Pincushion 84498,84504
Yellow Pincushion Flower 350207
Yellow Pine 121090,184915,299925,
300033,300153,300211
Yellow Pine Flax 231935
Yellow Pitaya 140236,357745
Yellow Pitcher Plant 347150
Yellow Pitcherplant 347150
Yellow Pittosporum 301388
Yellow Pleander 389978
Yellow Pleione 304248
Yellow Plum 316206
Yellow Plumegrass 148915
Yellow Pmemone 24018
Yellow Poinciana 288897
Yellow Pomegranate 321767
Yellow Pond Lily 267274,267276,267294
Yellow Pond-lily 267274,267276,267294,
267309,267333
Yellow Poplar 232602,232609
Yellow Poppy 32429,176738,282676
Yellow Poui 382724
Yellow Prickly Apple 185914
Yellow Primrose 314242
Yellow Princess Palm 129685
Yellow Provence Rose 336623
Yellow Puccoon 233707,233749
Yellow Pulsatilla 321707,321714
Yellow Pygmy-poppy 71080
Yellow Rabbitbrush 90565
Yellow Rabdosia 209822
Yellow Ragwort 359158
Yellow Raspberry 338773
Yellow Ratde 329543
Yellow Ratdebox 329521
Yellow Rattan Palm 121514
Yellow Rattle 329504,329521,329543
Yellow Resin 415181
Yellow Restharrow 271464
Yellow Rhododendron 331161
Yellow Rhynchosia 333298
Yellow River Tickseed 104785
Yellow Rocket 47947,47964,228100,239898
Yellow Rod 231183
Yellow Root 200175
Yellow Rose 216085,336718,337029
Yellow Russian Boschniakia 57438
Yellow Sage 221238
Yellow Saggan 208771
Yellow Salsify 394281,394333
Yellow Sanctuary 54896
Yellow Sand Paspalum 285525
Yellow Sand Verbena 702
Yellow Sandalwood 346210
Yellow Sand-verbena 702
Yellow Sanide 345971
Yellow Sarsapar Illa 250221
Yellow Satintail 205516
Yellow Saugh 239898
Yellow Saxifrage 349052,349068,349904
Yellow Scabious 82139,350207
Yellow Screw-stem 48592
Yellow Scurf Pea 319199
Yellow Sedge 74557,75744,76701,208771
Yellow Sedum 356468
Yellow Seraya 362186
Yellow Serradella 274880
Yellow Shellflower Bush 57987
Yellow Shinders Ell 16352,345631,359158
Yellow Shrimp Plant 66725,214379
Yellow Sickle Medic 247457
Yellow Side-saddle Flower 347150
Yellow Silver Currant 333916
Yellow Silver Pine 121101,225599,225601
Yellow Skunk-cabbage 239517
Yellow Slash Pine 299928
Yellow Slender Deutzia 126952
Yellow Smallreed 127235
Yellow Smithia 366643
Yellow Snakeroot 103850
Yellow Snowdrop 154899
Yellow Soapwort 346429
Yellow Spanish Plum 372477
Yellow Spapweed 204958,205454
Yellow Spathoglottis 370394
Yellow Spiderflower 34406
Yellow Spiderwort 95727
Yellow Spike Orchid 310365
Yellow Spike-rush 143154
Yellow Spinecape 179569
Yellow Spiny Daisy 185566,240419
Yellow Spit 86755
Yellow Spotted Ardisia 31514
Yellow Spotty Azalea 330235
Yellow Spring Bedstraw 170743
Yellow Spruce 298423,298435
Yellow Stamen Gordonia 179751
Yellow Star 148104,400675
Yellow Star Flower 376356
Yellow Star Grass 202870
Yellow Star Jasmine 393616
Yellow Star of Bethlehem 169460
Yellow Star Thistle 81382
Yellow Star-grass 202870
Yellow Star-of-bethlehem 169460
Yellow Star-thistle 81382
Yellow Star-tulip 67566,67606
Yellow Sterculia 376150
Yellow Stiff Aster 367984
Yellow Stingbush 156000
Yellow Stock Gilliflower 86427
Yellow Stokes Aster 377291
Yellow Stonecrop 356468,357114,357123
Yellow Strawberry 175454,413034
Yellow Strawberry Guava 318739
Yellow Strawberry-flower 312925
Yellow Streamers 416910
Yellow Striga 377973
Yellow Stringybark 155470,155661
Yellow Succory 298589
Yellow Sundrops 68508
Yellow Sunipia 379771
Yellow Sweet Clover 249232
Yellow Sweetclover 249232
Yellow Sweet-clover 249232
Yellow Sweetvetch 187834
Yellow Swertia 380244
Yellow Swordflag 208771
Yellow Syntrichopappus 382061
Yellow Syringa 84419
Yellow tack-stem 68445
Yellow Tangut Anisodus 25459
Yellow Tare-tine 222820
Yellow Tar-fitch 222820
Yellow Tea 69765
Yellow Tea-tree 226472
Yellow Terminalia 386563
Yellow Thatch 222820
Yellow Thistle 32411,32429,91955,92033,
92222
Yellow Throat 231183
Yellow Tiger-lily 398791
Yellow Toadflex 231183
Yellow Toadshade 397578
Yellow Top 364557
Yellow Touch-me-not 204958,205204,

205317,205454
Yellow Tree Lupin 238430
Yellow Tree Peony 280223
Yellow Trefoil 247366,397096
Yellow Trillium 397578,397579
Yellow Triratna Tree 397371
Yellow Trisetum 398518
Yellow Triteleia 398785
Yellow Trout Lily 154899,154945
Yellow Trout-lily 154899
Yellow Trumpet 70513,262441,347164,
400675
Yellow Trumpet Bush 385493
Yellow Trumpet Flower 385493
Yellow Trumpet Tree 385493
Yellow Trumpet Vine 24173,240368
Yellow Trumpets 347147
Yellow Tuberose 191304
Yellow Tulip 247113,400146
Yellow Tulipwood 138586
Yellow Turban 152416
Yellow Turbans 152416
Yellow Turk's Cap Lily 230015
Yellow Turk's Caplily 230015
Yellow Turk's-cap Lily 230015
Yellow Turkscap Lily 230015
Yellow Turnip 59507
Yellow Twig Dogwood 105194
Yellow Twigged Magnolia-vine 351078
Yellow Twining Snapdragon 28598
Yellow Unicorn Plant 203260
Yellow Unicorn-plant 203260
Yellow Uvaria 403502
Yellow Vancouveria 404612
Yellow Velvetleaf 230250
Yellow Vetch 408469
Yellow Vetchling 222681,222820
Yellow Violet 86427,410187,410466,
410472
Yellow Viscidhairy Larkspur 124688
Yellow Wall Bedstraw 170507
Yellow Wallflower 86427
Yellow Walnut 50474
Yellow Water Buttercup 325838
Yellow Water Cress 336250
Yellow Water Crowfoot 325838
Yellow Water Dropwort 269307
Yellow Water Flag 208771
Yellow Water Flower De Luce 208771
Yellow Water Flower-de-Luce 208771
Yellow Water Fringe 267825
Yellow Water Iris 208771
Yellow Water Lily 267294,267714
Yellow Water-buttercup 325838
Yellow Water-crowfoot 325764
Yellow Waterhyacinth 141809
Yellow Waterlily 267645
Yellow Water-lily 267274,267276,267294,
267320,267714
Yellow Wax Bells 216474
Yellow Waxbells 216473
Yellow Waxdaisy 189147
Yellow Weed 327879
Yellow White Swordflag 208669
Yellow Whitlow Grass 136901
Yellow Whitlow-grass 136901
Yellow Whitlowwort 204570
Yellow Whitlow-wort 204570
Yellow Wild Bastard Poppy 247113
Yellow Wild Indigo 47878,47880
Yellow Willow 342982,343438,343596
Yellow Willow-weed 239898
Yellow Wintergreen 322804
Yellow Wolfberry 239012
Yellow Wood 94014,94026,166971,
254406,268351,306395,417132
Yellow Wood Anemone 24018
Yellow Wood Poppy 379220
Yellow Wood Sorrel 277849,278099
Yellow Wood Violet 409729
Yellow Wood-anemone 24018
Yellow Woodbetony 287209
Yellow Woodsorrel 278008
Yellow Wood-sorrel 278099
Yellow Yam 131516,131754
Yellow Yarrow 3948,4056,152803
Yellow-and-blue Forget-me-not 260791,
260905
Yellow-androus Gordonia 179751
Yellow-anise 204570
Yellowanther Azalea 330703
Yellowanther Gordontea 179751
Yellow-anther Hairy Lilac 382269
Yellowanther Michelia 252976
Yellow-anthered Rhododendron 330703
Yellowarrow Dendrobium 125058
Yellow-aster 139749
Yellow-back Argyreia 32626
Yellow-back Barberry 51753
Yellowback Eurya 160555
Yellowback Oak 324264
Yellowback Qinggang 116180
Yellowback Themeda 389355
Yellow-backed Barberry 51753
Yellow-backed Oak 324264
Yellowbackleaf Eurya 160555
Yellowband Iris 208734,208740
Yellow-bark 91059,294231
Yellow-bark Barberry 52384
Yellowbark Oak 324539
Yellow-bark Oak 324539
Yellowbark Willow 343171
Yellow-barked Barberry 52384
Yellowbell Azalea 331045,332123
Yellowbells 385493,385505
Yellow-berried Nightshade 367311,367726
Yellow-berried Yew 385315
Yellowblack Epipactis 147255
Yellow-blossom 325690
Yellow-blue Skullcap 355582
Yellowbow Dendrobium 125058
Yellow-box 155649
Yellowbox Eucalyptus 155649
Yellow-boy 359158
Yellow-bract Pachystachys 279769
Yellowbract Thistle 91877
Yellow-branch Machilus 240715
Yellow-branch Oak 323924
Yellowbranch Premna 313747
Yellow-branched Barberry 52383
Yellow-branched Cyclobalanopsis 323924
Yellow-branched Grodtmann Barberry 51687
Yellow-branched Largeleaved Linden 391822
Yellow-branched Machilus 240715
Yellow-branchlet Keteleeria 216122
Yellow-brodiaea 398796
Yellow-brown Arrowwood 407795
Yellow-brown Azalea 331250
Yellow-brown Habenaria 183642
Yellowbrown Ophiorrhiza 272248
Yellow-brown Pearleverlasting 21612
Yellowbrown Rush 213338
Yellowbubble Raspberry 339014
Yellow-bud Hickory 77889
Yellowby 89704
Yellowcalyx Rhodiola 329905
Yellowcodded Rhododendron 332123
Yellowcone Spruce 298336
Yellowcoralla Rhododendron 332125
Yellow-corallate Rhododendron 332125
Yellowcotton 252395,252396
Yellowcress 336166
Yellow-cress 262746,336166,336187
Yellowcrown Sealavender 230575
Yellow-cupule Yushania 416817
Yellowcyme Argyreia 32624
Yellow-cymose Argyreia 32624
Yellow-cypress 85301
Yellow-devil Hawkweed 195603
Yellowdicks 188397
Yellow-dicks 188397
Yellowdot Glorybower 96191
Yellowdotted Beautyberry 66861
Yellow-dotted Beautyberry 66861
Yellow-dotted Glorybower 96191
Yellowdotted Purplepearl 66861
Yellowed Iris 208561
Yellow-edge Elaeagnus 142158
Yellower Cherryred Cotoneaster 107475
Yellowerflower Huernia 199053
Yellow-eyed Grass 365714,415990,416088,

416173
Yellow-eyed-grass Family 415975
Yellow-eyed-grasses 415990
Yelloweyegrass 415990,416082
Yelloweyegrass Family 415975
Yellowflag Iris 208771
Yellow-flame 288897
Yellowflax 327548
Yellow-flax 327548
Yellowfloer Bigbractginger 79727
Yellowflorrush 230244,230250
Yellowfloss Hairorchis 125573
Yellow-flower 59493,364557
Yellowflower Acineta 4650
Yellow-flower Agapetes 10308
Yellowflower Ajania 13016
Yellowflower Amberboa 18983
Yellowflower Ancylostemon 22165
Yellowflower Ardisia 31447
Yellowflower Arisaema 33331
Yellowflower Aspidistra 39541
Yellowflower Barleria 48295
Yellowflower Batrachium 48908
Yellowflower Betony 373482
Yellowflower Bladderwort 403108
Yellowflower Bletilla 55572
Yellowflower Briggsia 60256
Yellowflower Broadleaf Poplar 311427
Yellowflower Broomrape 275177
Yellowflower Butterflygrass 392903
Yellowflower Calathodes 66221
Yellowflower Camellia 69329
Yellow-flower Camellia 69765
Yellowflower Canna 71182
Yellowflower Cattleya 79547
Yellowflower Championella 85953
Yellowflower Changruicaoia 85977
Yellowflower Chinese Iris 208669
Yellowflower Chinese Stellera 375189
Yellowflower Cinquefoil 312457
Yellowflower Clockvine 390828
Yellowflower Cockclawflower 66221
Yellowflower Coleus 99748
Yellowflower Corydalis 106227
Yellowflower Crazyweed 279053
Yellowflower Daisy 124789
Yellowflower Dasymaschalon 126729
Yellowflower Dayflowerlike Thorowax 63609
Yellowflower Daylily 191304
Yellowflower Delavay Microtoena 254249
Yellowflower Dendranthema 124789
Yellowflower Dendrobium 125390
Yellowflower Diapensia 127912
Yellow-flower Didymocarpus 22159
Yellowflower Diuris 135065
Yellowflower Doublefurrow 23130
Yellowflower Eulophia 156708
Yellowflower False Loosestrife 238230
Yellowflower Farinose Dendrocalamus 125474
Yellowflower Fleawee 321535
Yellowflower Fritillary 168614
Yellowflower Gentian 173451
Yellowflower Gloriosa 177253
Yellowflower Gueldenstaedtia 181644
Yellow-flower Himalayah 228320
Yellowflower Hisute Crazyweed 278884
Yellow-flower Honeysuckle 236091
Yellowflower Hornsage 205566
Yellowflower Houlletia 198686
Yellowflower Incarvillea 205566
Yellow-flower Indigo 205927
Yellowflower Japane Abelia 180
Yellowflower Korean Pulsatilla 321668
Yellowflower Lady's Slipper 120352
Yellowflower Lagopsis 220124
Yellowflower Lecanorchis 223636
Yellow-flower Long-stem Clearweed 298963
Yellow-flower Long-stem Coldwaterflower 298963
Yellow-flower Lysionotus 239985
Yellowflower Mananthes 244296
Yellowflower Martagon Lily 229927
Yellowflower Maximowicz Fritillary 168467
Yellowflower Maximowicz Primrose 314628
Yellowflower Meadowrue 388506
Yellowflower Meconopsis 247124
Yellowflower Michelia 252976
Yellow-flower Milkvetch 42648
Yellowflower Milkwort 307923,308050
Yellowflower Mockstrawberry 138793
Yellowflower Mussot Swertia 380286
Yellowflower Nepeta 265012
Yellowflower Oncidium 270792
Yellowflower Onion 15189
Yellowflower Onosma 271827
Yellowflower Ophrys 272455
Yellowflower Oreocharis 273859
Yellowflower Pagodatree 369163
Yellowflower Pepper 300392
Yellowflower Phaius 293494
Yellowflower Photinia 295707
Yellowflower Primrose 314400
Yellowflower Pulicaria 321535
Yellowflower Pulsatilla 321714
Yellowflower Rabdosia 209822
Yellowflower Reaumuria 327230
Yellowflower Redsandplant 327230
Yellowflower Rhynchosia 333298
Yellowflower Rockfoil 349177
Yellow-flower Sacred Place Rhododendron 331962
Yellowflower Sage 345029
Yellowflower Silkvine 291053
Yellowflower Sinocrassula 364921
Yellowflower Sisymbrium 365532
Yellowflower Small Reed 127235
Yellowflower Smallreed 127235
Yellowflower Sophora 369163
Yellowflower Spapweed 205188
Yellowflower Spiderflower 34406
Yellowflower Star-of-bethlehem 274692
Yellowflower Sweetvetch 187834
Yellowflower Syncalathium 381646
Yellowflower Tibetia 391543
Yellowflower Tigwer Lily 230060
Yellowflower Toadlily 396611
Yellowflower Torenia 392903
Yellowflower Tricalysia 133185
Yellow-flower Trigonostemon 397371
Yellowflower Trillium 397579
Yellowflower Tuoli Fritillary 168591
Yellowflower Turmeric 114867
Yellowflower Twayblade 232236
Yellowflower Waxplant 198844
Yellowflower Willow 343151,344128
Yellowflower Woodbetony 287209
Yellow-flower Wright's Sealavender 230821
Yellowflower Yabe Columbine 30091
Yellowflower Yinshan Bushclover 226844
Yellow-flower Yun-Zang Azalea 331962
Yellowflower Zeuxine 417731
Yellow-flowered Agapetes 10308
Yellow-flowered Alicoche 140280
Yellow-flowered Camellia 69765
Yellow-flowered Coleus 99748
Yellow-flowered Columbine 30017
Yellow-flowered Crossandra 111713
Yellow-flowered Currant 333919
Yellow-flowered Gourd 114300
Yellow-flowered Gourds 114310
Yellow-flowered Gum 155776
Yellow-flowered Himalaya-honeysuckle 228320
Yellow-flowered Honeysuckle 236024
Yellow-flowered Hornpoppy 176738
Yellow-flowered Horse Gentian 397829
Yellow-flowered Indigo 205927
Yellow-flowered Leaf Cup 276724
Yellow-flowered Milkwort 307923
Yellow-flowered Nutmeg 257633
Yellow-flowered Onion 15206
Yellow-flowered Pepper 300392
Yellow-flowered Reaumuria 327230
Yellow-flowered Restharcow 271464
Yellow-flowered Rhododendron 330704
Yellow-flowered Roseflower Brandisia 59132
Yellow-flowered Smithia 366643
Yellowflowered Soapwort 346429
Yellow-flowered Sophora 369163
Yellow-flowered Sterigmostemum 376327

Yellow-flowered Strawberry 138790,138796
Yellow-flowered Teasel 133513
Yellow-flowered Tricalysia 395277
Yellow-flowered Violet 410763
Yellow-flowering Rattlebox 112615
Yellow-fringed Orchis 183501
Yellow-fruit Artocarpus 36948
Yellow-fruit Creeping Mallotus 243439
Yellowfruit Crowncactus 327312
Yellowfruit Cryptocarya 113438
Yellow-fruit Fig 164686
Yellowfruit Hackberry 80670
Yellow-fruit Hubei Hawthorn 109761
Yellowfruit Integral Holly 203912
Yellowfruit Nightshade 367735
Yellowfruit Raspberry 339483
Yellowfruit Rush 212999
Yellowfruit Siebold Elder 345697
Yellowfruit Snakegourd 396203
Yellowfruit Snowbell 379326
Yellowfruit Storax 379326
Yellowfruit Thickshellcassia 113438
Yellow-fruited Common Barberry 52341
Yellow-fruited Cranberry Bush 407998
Yellow-fruited Cryptocarya 113438
Yellow-fruited European Cranberry Bush 407998
Yellow-fruited Hawthorn 109709
Yellow-fruited Nippon Hawthorn 109651
Yellow-fruited Sargent Viburnum 408099
Yellow-fruited Sikkim Coriaria 104707
Yellow-fruited Storax 379326
Yellow-fruited Wolfberry 239012
Yellowgland Everlasting 21516
Yellow-globular Barberry 51466
Yellow-grass 262583
Yellowgreen Baker Lily 229745
Yellowgreen Catchfly 363315,363458
Yellow-green Coelogyne 98725
Yellow-green Corydalis 106510
Yellowgreen Everlasting 21723
Yellowgreen Milkvetch 42369
Yellowgreen Pearleverlasting 21723
Yellowgreen Sagebrush 36556
Yellow-green Silene 363312,363315
Yellowgreenbract Windhairdaisy 348319
Yellowgroove Arcane Bamboo 297196
Yellow-groove Bamboo 297204
Yellowhair Ainsliaea 12639
Yellow-hair Anzacwood 310762
Yellowhair Aralia 30628
Yellowhair Azalea 331707
Yellowhair Birthwort 34189
Yellow-hair Bois' Merremia 250761
Yellowhair Chastetree 411486
Yellowhair Cinquefoil 312738
Yellowhair Clockvine 390816
Yellowhair Crazyweed 279043
Yellowhair Dutchmanspipe 34189
Yellowhair Fieldhaircot 138968
Yellow-hair Hartia 376485
Yellowhair Jokul Azalea 330041
Yellowhair Lacquertree 393454
Yellow-hair Lasianthus 222190
Yellowhair Linden 391719
Yellowhair Machilus 240571
Yellowhair Maple 2977
Yellowhair Minjiang Azalea 330894
Yellowhair Monbeig Dogwood 380472
Yellowhair Monkshood 5131
Yellowhair Mucuna 259487
Yellowhair Qinggang 116079
Yellowhair Rabbiten-wind 12639
Yellow-hair Rhododendron 331707
Yellowhair Roughleaf 222190
Yellow-hair Roughleaf 222190
Yellowhair Sagebrush 36447
Yellowhair Sandwort 31766
Yellowhair Tephrosia 386353
Yellow-haired Actinidia 6599
Yellow-haired Aralia 30628
Yellow-haired Chaste-tree 411486
Yellow-haired Dutchman's-pipe 34189
Yellow-haired Machilus 240571
Yellow-haired Maple 2977
Yellow-haired Pink Spotted Rhododendron 330894
Yellow-haired Premna 313635
Yellow-haired Rhododendron 331707
Yellowhairs Argyreia 32665
Yellow-hairs Gansu Butterflybush 62151
Yellowhairy Actinidia 6599
Yellow-hairy Calyx Kudzuvine 321426
Yellowhairy Chastetree 411486
Yellow-hairy China-laurel 28329
Yellowhairy Clematis 94982
Yellow-hairy Clematis 94982
Yellowhairy Goldenray 229251
Yellowhairy Goldenrod Goldenray 229249
Yellowhairy Kiwifruit 6599
Yellow-hairy Lacquer-tree 393454
Yellowhairy Premna 313635
Yellowhairy Sibbaldia 362350
Yellowhairy Strawberry 167632
Yellowhairy Wildberry 362350
Yellowhead 396124,396125
Yellowhead Spapweed 205456
Yellowhead Sweetdaisy 71088
Yellowhead Touch-me-not 205456
Yellow-headed Fox Sedge 73703,73704
Yellowheart Euonymus 157705
Yellow-heart Prickly Ash 417230
Yellowhiry Larkspur 124132
Yellowhorn 415094,415096
Yellow-horn 415094
Yellowish Arrowwood 407841
Yellowish Banks Rose 336372
Yellowish Camellia 69071
Yellowish Cattleya 79563
Yellowish Chrysalidocarpus 89360
Yellowish Colored Mistletoe 410994
Yellowish Corymborkis 106876
Yellowish Edelweiss 224916
Yellowish Everlasting 21568
Yellowish Flower Dendrobium 125240
Yellowish Gentian 173193
Yellowish Glochidion 177155
Yellowish Hanging-down Milkvetch 42285
Yellowish Lady's Slipper 120352
Yellowish Lily 230043
Yellowish Narrosepal Pearweed 239863
Yellowish Onion 15296
Yellowish Pansy Orchid 254931
Yellowish Paspalidium 285380
Yellowish Pearleverlasting 21568
Yellowish Plicate Woodbetony 287523
Yellowish Polystachya 310492
Yellowish Premna 313629
Yellowish Rabdosia 209669
Yellowish Scarlet Swertia 380326
Yellowish Showy Hemilophia 191528
Yellowish Sweetvetch 187875
Yellowish Twocolor Crazyweed 278742
Yellowish Viburnum 407841
Yellowish Woodbetony 287386
Yellowish-green Cactus 244075
Yellowish-green California Buckthorn 328636
Yellowishgreen Spapweed 204852
Yellowishgreen Touch-me-not 204852
Yellowish-greenflower Eria 299603
Yellowishpink Oreocharis 273846
Yellowish-white Larkspur 124067
Yellowish-white Maxillaria 246715
Yellowish-white Rattlebox 112475
Yellowjuice Garcinia 171221
Yellow-juice Garcinia 171221
Yellow-kudzu Fig 165841
Yellowleaf Aucuba 44919
Yellowleaf Eargrass 187703
Yellowleaf Elatostema 142895
Yellowleaf Flowering Dogwood 105029
Yellowleaf Hedyotis 187703
Yellowleaf Iris 208488
Yellow-leaf Iris 208488
Yellow-leaf Silk Tassel 171550
Yellowleaf Stairweed 142895
Yellowleaf Stephania 375907
Yellowleafback Oak 324264
Yellowleaftree 415143
Yellow-leaved Cornelissen Dogw 105112
Yellow-leaved Elderberry 345633

Yellow-leaved Iris 208488
Yellow-leaved Ivy 187230
Yellow-leaved Mahonia 242517
Yellow-leaved Queen of Meadowsweet 166126
Yellow-leaved Tree 415143
Yellowlinear Willow 343111
Yellowlip Eria 148797
Yellowlip Zeuxine 417768,417786
Yellow-lipped Ladies' Tresses 372227
Yellow-margin Agave 10788
Yellow-margined Madonna Lily 229790
Yellowmouth Dutchmanspipe 34203
Yellownappy Premna 313761
Yellow-nerve Kurz Lasianthus 222196
Yellow-nerve Kurz Roughleaf 222196
Yellownerve Ninenode 319850
Yellownerve Psychotria 319850
Yellownerve Raspberry 339485
Yellow-nerved Psychotria 319850
Yellow-nerved Raspberry 339485
Yellowpetal Parnassia 284569
Yellowpilose Acutatesepal Monkshood 5017
Yellow-pimpernel 383043
Yellowpistil Camellia 69722
Yellow-poplar 232609
Yellow-poppy 379220
Yellowpowder 314921
Yellowpubescent Hornbeam 77348
Yellow-pubescent Hornbeam 77348
Yellow-puccoon 200173,200175
Yellowpurple Bluebellflower 115386
Yellow-pyramid Eastern Arborvitae 390611
Yellow-qeen Japanese Spindle-tree 157613
Yellowquitch 194015
Yellowraceme Raspberry 338256
Yellow-racemed Raspberry 338256
Yellowray Fremont's-gold 382061
Yellow-ray Goldfields 222603
Yellow-rayed Goldenweed 271916
Yellow-red Turmeric 114881
Yellowrim 360926
Yellowrocket 47964
Yellow-rocket 47964
Yellowroot 415164,415166,417124
Yellowroot Muping Corydalis 105896
Yellow-rooted Water Dock 339957
Yellow-seed False Pimpernel 231476,231514
Yellowseed Sophora 369117
Yellowsheath Yushanbamboo 416817
Yellowshrub Statice 230726
Yellow-sorrel 277747
Yellowspathe Southstar 33331
Yellowspike Barberry 51374
Yellowspike Cogongrass 205516
Yellowspike Melic 249096
Yellow-spiked Sedge 74122
Yellowspine Barrel Cactus 163427
Yellowspine Crowncactus 327262
Yellowspine Thistle 92259
Yellow-spiny Barberry 51465
Yellow-spirted Thistle 92259
Yellowspot Dockleaved Knotweed 309322
Yellow-spot Ginger 417983
Yellowspot Scorpiothyrsus 354751
Yellowspot Spread Gentian 173235
Yellowspotted Chirita 87862
Yellowspotted Scabrous Deutzia 127081
Yellowstem Indosasa 206885
Yellowstem Koehne Barberry 51822
Yellowstem Sandwort 31764
Yellow-stem White Willow 342982
Yellow-stemmed Bamboo 47516
Yellowstone Sulphur Flower 152574
Yellowstriate Gentian 173475
Yellowstripe Hedge Bamboo 47353
Yellowtop 359158,364557
Yellow-top Ash 155632
Yellow-topped Mallee-ash 155632
Yellowtrumpet 385461
Yellowtube Azalea 330992
Yellowtube Gentian 173676
Yellow-tube Rhododendron 330992
Yellowtuft 18444
Yellow-tuft 18332
Yellowtung 145413
Yellowtwig Dogwood 105193
Yellowtwig Qinggang 323924
Yellow-twig Spruce 298233
Yellowvariegate Pearleverlasting 21516
Yellow-variegated Sycamore Maple 3472
Yellowvein Hydrangea 200148
Yellowvein Mountainash 369567
Yellowvein Spicebush 365034
Yellow-veined Hydrangea 200148
Yellow-veined Mountain-ash 369567
Yellow-veined Pittosporum 301350
Yellow-veined Seatung 301350
Yellow-veined Spice-bush 365034
Yellow-veins Cinnamon 91451
Yellow-vetch 408469
Yellow-villosed Argyreia 32626
Yellowvine 121486,121514
Yellow-weed 182103,257943,359158,364557
Yellowwhire Cudweed 178265
Yellow-white Calophanoides 67833
Yellow-white Elsholtzia 144064
Yellow-white Gingerlily 187427
Yellow-white Indosasa 206869
Yellow-white Milkvetch 41949
Yellow-white Sobralia 366795
Yellow-white Synotis 381959
Yellowwish Tailanther 381959
Yellowwish-white Diastema 128146
Yellowwood 94014,94022,94026,107315,306395,306450
Yellow-wood 51640,94014,94026,306395,306435,386638,417157
Yellow-wood Barberry 52385
Yellow-wood Family 306345
Yellow-wood Trigonostemon 397365
Yellowwort 88242
Yellow-wort 54891,54896
Yelowerflower Duck Woodbetony 287002
Yemem Iris 208565
Yemtani 178025
Yenissei River Pondweed 312275
Yenow Lupin 238463
Yercum 68088
Yernut Yennut 102727
Yes Smart 309199,309570
Yeso Spruce 298307
Yes-or-no 235334
Yesso Hackberry 80600
Yesterday 61296
Yesterday Today and Tomorrow 61297,61310
Yesterday-today-and-tomorrow 61310
Yesterday-today-tomorrow 61310
Yeth 67456
Yethnut 102727
Yeugh 385302
Yew 385301,385302,385342,385384
Yew Brimmle 336419
Yew Family 385175
Yew Pine 306395,306457
Yew Plum Pine 306457,306469
Yew Podocarpus 306457
Yew Willow 344190
Yew-brimmle 336419
Yewleaf Willow 344190
Yew-leaf Willow 344190
Yew-leaved Torreya 393089
Yew-like Phyllochlamys 377543
Yew-like Streblus 377543
Yew-pine 306457,306522
Yezhi Rockfoil 350043
Yezo Dyer's Woad 209238
Yezo Figwort 355129
Yezo Laurel 122471
Yezo Linden 391768
Yezo Madder 337970
Yezo Meadowrue 388722
Yezo Needlejade 297021
Yezo Nettle 402996
Yezo Primrose 314280
Yezo Raspberry 338583
Yezo Rosewood 121718
Yezo Sharp-leaved Spindle-tree 157767
Yezo Skullcap 355856

Yezo Spruce 298307
Yichang Arrowwood 407802
Yichang Bitterorange 93495
Yichang Box 64264
Yichang Childvine 400847
Yichang Clovershrub 70833
Yichang Elatostema 142689
Yichang Fimbristylis 166348
Yichang Indigo 205881
Yichang Japanese Pagodatree 369051
Yichang Lemon 93495
Yichang Litse 233947
Yichang Loosestrife 239668
Yichang Machilus 240599
Yichang Orange 93495
Yichang Privet 229634
Yichang Raspberry 338554
Yichang Sedge 73774
Yichang Stairweed 142689
Yichang Tylophora 400847
Yichang Viburnum 407802
Yielyiel Grevill 180602
Yigisar Kneejujube 67086
Yik 324335
Yilan Raspberry 338756
Yilan Southstar 33356
Yili Barberry 51757
Yili Beancaper 418677
Yili Bubbleweed 297876
Yili Clematis 95026
Yili Dicranostigma 129568
Yili Favusheadflower 129568
Yili Gagea 169450
Yili Globethistle 140816
Yili Iris 208636
Yili Larkspur 124302
Yili Lily 228583
Yili Milkvetch 42516
Yili Monkshood 5619
Yili Ononis 271320
Yili Physochlaina 297876
Yili Poplar 311355
Yili Restharrow 271320
Yili Rose 336956
Yili Serpentroot 354877
Yili Spunsilksage 360883
Yili Sweetvetch 187940
Yili Tulip 400178
Yili Willow 343523
Yiliang Cystopetal 320256
Yiliang Gentian 174083
Yiliang Parnassia 284637
Yiliang Poplar 311374
Yilicelery 383233
Yilicelery Talassia 383235
Yilin Youngia 416496
Yimen Fig 165142
Yimen Mayten 246955
Yimen Onosma 271750
Yin-Chen Wormwood 35282
Yinchow Thorowax 63872
Ying'ao Grape 411590
Yingde Champion Bauhinia 49049
Yingde Liriope 232650
Yingde Loosestrife 239906
Yingde Skullcap 355857
Yingde Spapweed 204932
Yingjiang Amomum 19937
Yingjiang Aspidistra 39583
Yingjiang Begonia 50417
Yingjiang Cyclobalanopsis 116218
Yingjiang Falsenettle 56179
Yingjiang Featherlip Orchis 274491
Yingjiang Ginger 418029
Yingjiang Gingerlily 187483
Yingjiang Grewia 181023
Yingjiang Jasmine 211817
Yingjiang Meranti 362194
Yingjiang Oak 116218
Yingjiang Pepper 300552
Yingjiang Qinggang 116218
Yingjiang Ramie 56179
Yingjiang Rattanpalm 65747
Yingjiang Southstar 33357
Yingjiang Tanoak 233266
Yingjiang Touch-me-not 205457
Yingjiang Yushanbamboo 416777
Yingjiang Yushania 416777
Yingjiang Zinger 418029
Yingkiang Pepper 300552
Yingkili Ermans Birch 53429
Yingtak Skullcap 355857
Yining Sevenriver Rush 213151
Yining Whitlowgrass 137248
Yining Windhairdaisy 348185
Yinkeng Hackberry 80791
Yinkun Maple 3760
Yinma 90605
Yinping Suffruticosa Peony 280303
Yinquania 416312,416313
Yinshan Bushclover 226843
Yinshan Buttercup 326502
Yinshan Crazyweed 278905
Yinshan Dandelion 384865
Yinshan Euphorbia 160098
Yinshan Ladybell 7891
Yinshan Monkshood 5708
Yinshan Woodbetony 287373
Yinshancress 416316,416317
Yinshania 416316,416317
Yinzhou Thorowax 63872
Yirnin-girse 299759
Yishan Begonia 50418
Yiwu Ghostfluting 228394
Yiwu Oak 324566
Yiwu Raspberry 339498
Yiwu Rockvine 387870
Yiwu Tanoak 233208
Yixian Hilliella 196318
Yixing Bitter Bamboo 304119
Yixing Bitterbamboo 304119
Yixing Bitter-bamboo 304119
Yixing Bloodred Iris 208810
Yixing Sinobambusa 364827
Yixing Snakegourd 396288
Yixing Swordflag 208810
Yiyang Indocalamus 206804
Yizhang Azalea 332147
Yizhang Mahonia 242505
Yizhang Sweetleaf 381221
Ylang Ylang 70960
Ylang-ylang 70960
Ylan-ylan 70960
Yne-brimmel 338447
Yoania 416378
Yoco 285935
Yodogawa Azalea 331530
Yoe Brimmle 336419
Yoe-brimmle 336419,338447
Yohimbe Bark 286054
Yoke Cactus 418532
Yoke Elm 77256
Yoke Passionflower 285657
Yoke Waterbamboo 260105
Yoke-wood Tree 110251,110256
Yokohama Bean 259520
Yokusa Cinquefoil 312942
Yolkcolour Pomatocalpa 310800
Yolk-of-egg Willow 342982
Yomhin 90605
Yomitwort 234547
Yon 26022
Yong'an Cyclobalanopsis 116219
Yong'an Oak 116219
Yong'an Qinggang 116219
Yongchun Yellowwood 94036
Yongdeng Leek 15901
Yongdeng Onion 15901
Yongfu Cymbidium 117106
Yongfu Dendrobium 125421
Yongfu Orchis 117106
Yongfu Tanoak 233423
Yongning Azalea 332154
Yongning Cowparsnip 192402
Yongning Dock 340319
Yongning Groundsel 360360
Yongning Larkspur 124707
Yongning Rhododendron 332154
Yongning Rhubarb 329417
Yongning Sedge 76786
Yongning Spicegrass 239670

Yongshan Ampelocalamus 20211
Yongshan Spapweed 205458
Yongshan Squarebamboo 87620
Yongshan Square-bamboo 87620
Yongsheng Rabdosia 209828
Yongshuh 165307
Yongshun Bittercress 73051
Yongshun Elatostema 142904
Yongshun Neomartinella 264168
Yongshun Stairweed 142904
Yongtai Skullcap 355517
Yongxiu Twinballgrass 209074
York And Lancaster Rose 336517
York Gum 155631
York Road Poison 171971
York-and-lancaster Rose 336517
Yorkshire Fog 197301
Yorkshire Milkwort 307905
Yorkshire Sandwort 31931
Yorkshire Sanicle 299759
Yornut 102727
Yorrell 155598
Yoruba Ebony 132307
Yoruba Wild Indigo 235527
Yosemite Woolly Sunflower 152828
Yoshimura Figwort 355272
Yoshino Buckthorn 328904
Yoshino Cherry 83365
Yoshino Cranesbill 175034
Youe 385302
Youle Giantarum 20152
Youman's Sringybark 155778
Young Epimedium 147076
Young Fustic 107300
Young Man's Death 102933
Young Palm Orchid 399653
Young Sagebrush 36562
Young's Golden Juniper 213635
Young's Helleborine 147256
Young's Weeping Birch 53566
Youngberry 338059,338837
Younger Primrose 315131
Younghusband Jerusalemsage 295227
Youngia 416390
Youth On Age 392668
Youth-and-old-age 418035,418043
Youthful Goatgrass 8683
Youth-on-age 392668
Youthwort 138348
Youxian Camellia 69778
Youxiu Indocalamus 206842
Youyang Elatostema 142902
Youyang Stairweed 142902
Yowe Yornut 102727
Yowe-yornut 102727
Yowie Yorlin 102727
Yowie-yorlin 102727
Yu Acanthopanax 2522
Yu Asparagus 39260
Yu Baliospermum 46971
Yu Begonia 50420
Yu Corydalis 106612
Yu Elatostema 142903
Yu Hackberry 80790
Yu Holly 204412
Yu Maple 3761
Yu Mountainash 369569
Yu Parnassia 284639
Yu Poplar 311583
Yu Stairweed 142903
Yu Tailanther 381961
Yu Waxpetal 106694
Yu Winterhazel 106694
Yu Woodbetony 287839
Yu's Barberry 52390
Yu's Clematis 95462
Yu's Goldenray 229258
Yu's Leptodermis 226143
Yu's Spapweed 205459
Yu's Synotis 381961
Yu's Wildclove 226143
Yua 416525
Yuan Amitostigma 19540
Yuan Astyleorchis 19540
Yuan Calanthe 66123
Yuan Habenaria 184211
Yuan Larkspur 124708
Yuan Leek 15902
Yuanbao Mountain Fir 513
Yuanbaoshan Fir 513
Yuandifoshow Ginkgo 175831
Yuanjiang Arrowbamboo 162765
Yuanjiang Bauhinia 49079
Yuanjiang Bladdernut 374111
Yuanjiang Camellia 69776
Yuanjiang Caper 71932
Yuanjiang Clematis 95461
Yuanjiang Fargesia 162765
Yuanjiang Jasmine 212074
Yuanjiang Pricklyash 417376
Yuanjiang Prickly-ash 417376
Yuanjiang Rattlebox 112821
Yuanjiang Scurrula 355332
Yuanjiang Sesbania 361432
Yuanjiang Sterculia 376212
Yuanjiang Tea 69776
Yuanjiang Umbrella Bamboo 162765
Yuanjiang Windmill 100862
Yuanling Bitterbamboo 303999
Yuanling Didymocarpus 129990
Yuanling Grape 412008
Yuanmou Hirsute Abutilon 912
Yuanmou Tailgrass 402527
Yuanmou Themeda 389380
Yuanmou Urochloa 402527
Yuanqu Gymnaster 182355
Yuanqu Nakeaster 182355
Yuanyang Bulbophyllum 63195
Yuanyang Stonebean-orchis 63195
Yubaicai Pakchoi 59358
Yuca 244507
Yucatan 10854
Yucatan Camphorweed 305144
Yucatan Elemi 21011
Yucatan Plum 372477
Yucatan Tithonia 392432
Yucca 416535,416538,416540,416584,416595
Yucca Wild Buckwheat 152393
Yuchuan Onion 15903
Yuefeng Rhododendron 332150
Yue-Gui Skullcap 355784
Yuelu Aster 41278
Yuelushan Pseudosasa 318358
Yuexi Cranesbill 175035
Yuexi Eutrema 161161
Yuexi Greyawngrass 372447
Yuexi Sedge 76784
Yuexi Spodiopogon 372447
Yuge-hyokan 93874
Yugh 385302
Yugoslav Globe Thistle 140660
Yugoslav Globe-thistle 140660
Yugoslavian Mallow 216532
Yuhuangshan Willow 344301
Yujiecai Leaf-mustard 59438
Yuko 93875
Yuko Citrus 93875
Yukon Aster 381012
Yukon Fleabane 151072
Yukon Goldenweed 265639
Yukon Stitchwort 255610
Yukon Wild Buckwheat 152073
Yulan 241951,416694
Yulan Magnolia 416694
Yulan Tree 416694
Yule Girse 166125
Yuli Raspberry 339501
Yulin Rhododendron 332151
Yulin Savin Juniper 213915
Yulin Savin Savin 213915
Yulong Kobresia 217295,217303
Yulong Ligusticum 229380
Yulong Nannoglottis 262222
Yulong Pleurospermum 304861
Yulong Sedge 76785
Yulong Teasel 133521
Yulong Webster Rhododendron 332099
Yulongshan Anemone 24129
Yulongshan Arrowbamboo 162766
Yulongshan Barberry 51926

Yulongshan Fargesia 162766
Yulongshan Larkspur 124710
Yulongshan Parnassia 284640
Yulongshan Pipewort 151456
Yulongshan Sandwort 31919
Yulongshan Umbrella Bamboo 162766
Yulongshan Windflower 24129
Yumen Milkvetch 43274
Yumen Rockjasmine 23131
Yumen Willow 344304
Yumin Fritillary 168566,168642
Yun Garcinia 171086
Yun Muxiang 44881
Yunan Cypress 114775
Yunchang Rhododendron 332153
Yuncon Hackberry 80791
Yungfu Chirita 87993
Yungkiang Camellia 69012
Yungning Groundsel 360360
Yungning Rhododendron 332154
Yun-Gui Peristylus 291254
Yun-Gui Perotis 291254
Yun-Gui Pipewort 151470
Yungui Privet 229671
Yun-Gui Stairweed 142800
Yunhe Bamboo 297506
Yunhe Broodhen Bamboo 297506
Yunhe Newlitse 264026
Yunhe Poorspikebamboo 270436
Yunho Bamboo 297506
Yunjian Woodrush 238717
Yunling Monkshood 5709
Yunling Rockfoil 350045
Yunling Sedge 76788
Yunlong Arrowbamboo 162730
Yunlong Umbrella Bamboo 162730
Yunnan 93627
Yunnan Acacia 1711
Yunnan Achasma 155408
Yunnan Actinostemma 6908
Yunnan Addermonth Orchid 110636
Yunnan Aganosma 10220
Yunnan Aglaia 19974
Yunnan Ainsliaea 12744
Yunnan Alangium 13393
Yunnan Albizia 13526
Yunnan Albizzia 13526
Yunnan Allomorphia 16095
Yunnan Alpbean 391548
Yunnan Alplily 234248
Yunnan Alseodaphne 17812
Yunnan Alstonia 18059
Yunnan Altingia 18207
Yunnan Alysicarpus 18302
Yunnan Amalocalyx 18562
Yunnan Amentotaxus 19365
Yunnan Amomum 19911,19938
Yunnan Amoora 19974
Yunnan Anemone 23795
Yunnan Aphananthe 29009
Yunnan Apios 29303
Yunnan Aporosa 29780
Yunnan Aralia 30798
Yunnan Ardisia 31644
Yunnan Argostemma 32528
Yunnan Arisaema 33574
Yunnan Arrowbamboo 162767
Yunnan Arrowwood 408235
Yunnan Artocarpus 36928
Yunnan Arundinella 37442
Yunnan Ascitesgrass 407510
Yunnan Ash 168016
Yunnan Aspen 311584
Yunnan Aster 41520
Yunnan Aucuba 44949
Yunnan Azalea 332155
Yunnan Balanophora 364781
Yunnan Baliosperm 46968
Yunnan Balm 249513
Yunnan Bamboo 47535
Yunnan Banana Shrub 252979
Yunnan Barberry 52391
Yunnan Barringtonia 48515
Yunnan Basketvine 9425
Yunnan Bastard Cress 390271
Yunnan Bauhinia 49268
Yunnan Bean 294059
Yunnan Beautyberry 66948
Yunnan Beautyleaf 67866
Yunnan Bedstraw 170779
Yunnan Begonia 50087
Yunnan Bell Tree 325007
Yunnan Bellflower 70368
Yunnan Belltree 325007
Yunnan Belostemma 50936
Yunnan Berchemiella 52484
Yunnan Berneuxia 52929
Yunnan Berneuxine 52929
Yunnan Bigbractginger 322762
Yunnan Bigstylrvine 247969
Yunnan Biondia 54530
Yunnan Birdfootorchis 347942
Yunnan Birthwort 34375
Yunnan Bistamengrass 127486
Yunnan Bittercress 73048
Yunnan Blastus 55186
Yunnan Bletilla 55563
Yunnan Blooddisperser 297636
Yunnan Blueberry 403810
Yunnan Bluestem 23086
Yunnan Boeica 56373
Yunnan Bogorchis 110636
Yunnan Bothriochloa 57595
Yunnan Bowfruitvine 393527
Yunnan Brachytome 59004
Yunnan Bredia 59889
Yunnan Bristlegrass 361734,361971
Yunnan Broomrape 275262
Yunnan Buchanania 61675
Yunnan Buckeye 9736
Yunnan Buddhamallow 402260
Yunnan Bugbane 91042
Yunnan Bulbophyllum 62961,63197
Yunnan Bulleyia 63389
Yunnan Bulrush 353794
Yunnan Burreed 370109
Yunnan Buttercup 326503
Yunnan Butterflybush 62206
Yunnan Butterfly-bush 62206
Yunnan Cadillo 402260
Yunnan Calliandra 66690
Yunnan Calophanoides 67837
Yunnan Camelina 68873
Yunnan Camellia 69782
Yunnan Caper 71933
Yunnan Casearia 78161
Yunnan Catchfly 364222
Yunnan Centrostemma 81914
Yunnan Chainpodpea 18302
Yunnan Chastetree 411498
Yunnan Chaste-tree 411498
Yunnan Cheirostylis 86730
Yunnan Cherry 83368
Yunnan Cherry Laurel 223094
Yunnan Cherrylaurel 223094
Yunnan Chesneya 87266
Yunnan Chessbean 116613
Yunnan Chickweed 375151
Yunnan Childgrass 340378
Yunnan Childvine 400987
Yunnan Chilosehista 87469
Yunnan Ching Scurrula 355299
Yunnan Clematis 95463
Yunnan Cliffbean 254819
Yunnan Clovershrub 70914
Yunnan Coelogyne 98753
Yunnan Cogflower 154201,382766
Yunnan Coinmaple 133598
Yunnan Combretum 100501
Yunnan Condorvine 245870
Yunnan Conehead 320143
Yunnan Connarus 101913
Yunnan Coralbean 154746
Yunnan Coral-bean 154746
Yunnan Coraltree 407986
Yunnan Corydalis 106613
Yunnan Cowparsnip 192403
Yunnan Cowpea 409124
Yunnan Crab Apple 243732
Yunnan Crabapple 243732
Yunnan Craibiodendron 108587

Yunnan Craigia 108590
Yunnan Cranesbill 174676,175037
Yunnan Crapemyrtle 219952
Yunnan Crape-myrtle 219952
Yunnan Crawfurdia 110298
Yunnan Crazyweed 279243
Yunnan Croton 113061
Yunnan Cryptocarya 113497
Yunnan Cryptocoryne 113546
Yunnan Cryptomeria 113721
Yunnan Cuttonguetree 413136
Yunnan Cyathostemma 115692
Yunnan Cyclea 116024
Yunnan Cylindrokelupha 116613
Yunnan Cypress 114680
Yunnan Cystacanthus 120799
Yunnan Daimyo Oak 324572
Yunnan Dalechampia 121915
Yunnan Daphne 122635
Yunnan Daphniphyllum 122730
Yunnan Delavaya 123766
Yunnan Dendrocalamus 125519
Yunnan Dendropanax 125596
Yunnan Desmos 126730
Yunnan Deutzia 127126
Yunnan Devilpepper 327082,327093
Yunnan Dichroa 129051
Yunnan Dicliptera 129310
Yunnan Didymocarpus 129992
Yunnan Dipelta 132605
Yunnan Diploknema 133008
Yunnan Dipteronia 133598
Yunnan Dischidia 134033
Yunnan Dishspurorchis 171900
Yunnan Docynia 135114
Yunnan Dodder 115123
Yunnan Dogbonbavin 133187
Yunnan Douglas Fir 318571
Yunnan Draba 137298
Yunnan Dragonbamboo 125519
Yunnan Dregea 137802
Yunnan Drupetea 322616
Yunnan Dumasia 138944
Yunnan Dutchman's-pipe 34375
Yunnan Dutchmanspipe 34375
Yunnan Dwarfnettle 262232
Yunnan Dysosma 139607
Yunnan Eaglewood 29985
Yunnan Eelgrass 418184
Yunnan Ehretia 141617
Yunnan Elaeagnus 142253
Yunnan Elaeocarpus 142345,142412
Yunnan Engelhardia 145522
Yunnan Entada 145929
Yunnan Ephedra 146248
Yunnan Epipactis 147258
Yunnan Eranthemum 148093
Yunnan Eriolaena 152698
Yunnan Eritrichium 153449
Yunnan Ervatamia 154201,382766
Yunnan Erysimum 154565
Yunnan Eulalia 156510
Yunnan Eulophia 157105
Yunnan Euonymus 157974
Yunnan Eurya 160630
Yunnan Eustigma 161022
Yunnan Eutrema 161158
Yunnan Everlasting 21733
Yunnan Evodia 161307
Yunnan Evolvulus 161447
Yunnan Exacum 161584
Yunnan False Rattlesnakeroot 283700
Yunnan Falseflax 68873
Yunnan Falsepistache 384375
Yunnan False-pistache 384375
Yunnan Falserambutan 283532
Yunnan Falsestairweed 223684
Yunnan Fargesia 162744,162767
Yunnan Felwort 235445
Yunnan Fescue 164401
Yunnan Fescuegrass 164401
Yunnan Fevervine 280121
Yunnan Fewnerve Linden 391813
Yunnan Fibrous Crabgrass 130551
Yunnan Fig 165875
Yunnan Fighazel 134915
Yunnan Figwort 355273
Yunnan Filbert 106786
Yunnan Fimbristylis 166565
Yunnan Firerope 152698
Yunnan Firethorn 322465
Yunnan Fishvine 126012
Yunnan Flabellateleaf Maple 2965
Yunnan Flameorchis 327688
Yunnan Flemingia 166905
Yunnan Flickingeria 166955
Yunnan Floscopa 167046
Yunnan Fluttergrass 166565
Yunnan Forangledbranch Eurya 160576
Yunnan Forkstamenflower 416510
Yunnan Forkstyleflower 374498
Yunnan Galangal 17649
Yunnan Gambirplant 401788
Yunnan Gambir-plant 401788
Yunnan Garcinia 171086,171223
Yunnan Gastrochilus 171900
Yunnan Gentian 157105,174088
Yunnan Giantarum 20153
Yunnan Giantgrass 175611
Yunnan Giant-grass 175598
Yunnan Ginger 418030
Yunnan Gingerlily 187484
Yunnan Gironniera 29009
Yunnan Gleadovia 176800
Yunnan Globeflower 399549
Yunnan Glorybower 96449
Yunnan Glycosmis 177839
Yunnan Goldenray 229259
Yunnan Goldleaf 108587
Yunnan Goldquitch 156510
Yunnan Goldthread 103846
Yunnan Gonatanthus 179264
Yunnan Goniothalamus 179420
Yunnan Grape 412009
Yunnan Greenbrier 366623
Yunnan Greenstar 307500
Yunnan Grewia 181024
Yunnan Greyawngrass 372411
Yunnan Gromwell 233743
Yunnan Grumewood 211023
Yunnan Gueldenstaedtia 391548
Yunnan Gutzlaffia 182137
Yunnan Gymnema 182395
Yunnan Hairy Ailanthus 12573
Yunnan Hairyfruitgrass 222355
Yunnan Hairy-veined Rhododendron 331576
Yunnan Hairy-veins Rhododendron 331576
Yunnan Haplanthoides 185338
Yunnan Harrysmithia 185938
Yunnan Hartia 376433
Yunnan Hawthorn 110030
Yunnan Helictotrichon 190146
Yunnan Hemlock 399896
Yunnan Hemslet Buckthorn 328724
Yunnan Heraldtrumpet 49362
Yunnan Herald-trumpet 49362
Yunnan Herbclamworm 398121
Yunnan Herminium 192889
Yunnan Heteropanax 193913
Yunnan Heterosmilax 194132
Yunnan Hibiscus 195359
Yunnan Himalayan Speedwell 407152
Yunnan Hippolytia 196712
Yunnan Hiptage 196851
Yunnan Hogfennel 293080
Yunnan Holeglumegrass 57595
Yunnan Holly 204413
Yunnan Honeysuckle 236229,236230
Yunnan Hooffruit 316035
Yunnan Hookettea 52484
Yunnan Hooktea 52474
Yunnan Hookvine 401788
Yunnan Hop 199394
Yunnan Hopea 198200
Yunnan Hophornbeam 276826
Yunnan Hop-hornbeam 276826
Yunnan Hornbeam 77350
Yunnan Hydrangea 200157
Yunnan Inflated 125647
Yunnan Ironwood 276826
Yunnan Ixeris 210552

Yunnan Ixora 211204
Yunnan Jasminanthes 211698
Yunnan Jasmine 211987
Yunnan Jewelvine 126012
Yunnan Jujube 418184
Yunnan Keteleeria 216134
Yunnan Kibatalia 216232
Yunnan Kitefruit 196851
Yunnan Knotweed 309807
Yunnan Kobresia 217312
Yunnan Kudzuvine 321459
Yunnan Kyllingia 218492
Yunnan Lacquertree 393497
Yunnan Lacquer-tree 393497
Yunnan Lady's Slipper 120474
Yunnan Ladyslipper 120474
Yunnan Lagotis 220204
Yunnan Landpick 308572
Yunnan Larkspur 124711
Yunnan Laureleaf Currant 334063
Yunnan Lecanthus 223684
Yunnan Leek 15453
Yunnan Leptopus 226337
Yunnan Licorice 177950
Yunnan Ligusticum 229417
Yunnan Lilac 382388
Yunnan Lilyturf 272151
Yunnan Lime 391856
Yunnan Linden 391856
Yunnan Linociera 270155
Yunnan Liolive 270155
Yunnan Listera 264754
Yunnan Litse 234111
Yunnan Loeseneriella 235231
Yunnan Longan 131064
Yunnan Longleaf Deutzia 127003
Yunnan Loosespike Oplismenus 272679
Yunnan Loosestrife 239546
Yunnan Lovegrass 147674
Yunnan Luculia 238108
Yunnan Luohanfruit 365331
Yunnan Lycianthes 238984
Yunnan Machilus 240723
Yunnan Madder 338044
Yunnan Maddertree 12544
Yunnan Magnoliavine 351056
Yunnan Mallotus 243457
Yunnan Mammea 243990
Yunnan Mananthes 244298
Yunnan Manglietia 244478
Yunnan Manyflowered May-apple 139607
Yunnan Manyleaf Paris 284382
Yunnan Marsdenia 245821,245870
Yunnan Meadowrue 388724
Yunnan Medinilla 247640
Yunnan Meliosma 249482
Yunnan Melodinus 249682
Yunnan Memecylon 250044
Yunnan Meranti 362190
Yunnan Merremia 250862
Yunnan Michelia 252979
Yunnan Micrechites 203336
Yunnan Microstegium 254057
Yunnan Microtropis 254334
Yunnan Miliusa 254559
Yunnan Milkgreens 245821,245870
Yunnan Milkvetch 43275
Yunnan Millettia 254819
Yunnan Mistletoe 411127
Yunnan Mitrephora 256236
Yunnan Momordica 256815
Yunnan Mountain Leech 126675
Yunnan Mulberry 259178
Yunnan Munronia 259875
Yunnan Murdannia 260125
Yunnan Mycetia 260656
Yunnan Nanmu 295364
Yunnan Nannoglottis 262232
Yunnan Nemosenecio 263517
Yunnan Neomicocalamus 264182
Yunnan Neosinocalamus 47533
Yunnan Nerium-leaf Rapanea 261648
Yunnan Nettle 402960
Yunnan Ninenode 319914
Yunnan Notoseris 267181
Yunnan Nutmeg 261463
Yunnan Nutmeg Yew 393090
Yunnan Oak 116109,324572
Yunnan Obviousvein Tea 69062
Yunnan Ochrocarpus 268335
Yunnan Olive 71023,270172
Yunnan Ormosia 274442
Yunnan Orophea 275337
Yunnan Osmanther 276405
Yunnan Osmanthus 276405
Yunnan Ostodes 276787
Yunnan Ostryopsis 276849
Yunnan Ottelia 277418
Yunnan Oxyspora 278613
Yunnan Pachygone 279562
Yunnan Pagodatree 369164
Yunnan Palegreenvine 279562
Yunnan Palmcalyx 317400
Yunnan Panisea 282420
Yunnan Paperbush 141472
Yunnan Papery Daphne 122572
Yunnan Papilliferous Barberry 52002
Yunnan Papillose Spiraea 372046
Yunnan Paraboea 283121
Yunnan Parakmeria 283423
Yunnan Paranephelium 283532
Yunnan Paraprenanthes 283700
Yunnan Paravallaris 216232
Yunnan Parnassia 284641
Yunnan Pavieasia 286577
Yunnan Pear 323265
Yunnan Pearleverlasting 21733
Yunnan Pearlwood 296462
Yunnan Peashrub 72227
Yunnan Peliosanthes 288662
Yunnan Pellacalyx 288684
Yunnan Pellionia 288784
Yunnan Pennilabium 289009
Yunnan Pennycress 390271
Yunnan Pentapanax 289674
Yunnan Pepper 300554
Yunnan Peristrophe 291187
Yunnan Persimmon 132478
Yunnan Phoebe 295364,295386
Yunnan Phoenix Tree 166620
Yunnan Phoenix-tree 166620
Yunnan Pholidota 295545
Yunnan Phrynium 296056
Yunnan Phyllanthodendron 296462
Yunnan Physaliastrum 297636
Yunnan Pimpinella 299586
Yunnan Pine 300305
Yunnan Pineapple Oak 324572
Yunnan Pinnate Groundsel 263517
Yunnan Pipewort 151253
Yunnan Piptanthus 300672
Yunnan Pistache 301006
Yunnan Pitard Camellia 69504
Yunnan Pittosporum 301427
Yunnan Plantain 301879
Yunnan Pleione 304313
Yunnan Pleurospermum 304862
Yunnan Plumegrass 148913
Yunnan Podocarpium 200734
Yunnan Pogonia 306913
Yunnan Poplar 311584
Yunnan Potatobean 29303
Yunnan Premna 313772
Yunnan Pricklyash 417248
Yunnan Primrose 315134
Yunnan Protium 316035
Yunnan Pseudobartsia 317400
Yunnan Psychotria 319914
Yunnan Purpledaisy 267181
Yunnan Purplepearl 66948
Yunnan Purplestem 376433
Yunnan Pygeum 322397
Yunnan Pygmyweed 391920
Yunnan Pyrenaria 322616
Yunnan Qinggang 116109
Yunnan Qingmingflower 49362
Yunnan Rabbiten-wind 12744
Yunnan Rabdosia 209872
Yunnan Racemobambos 324935
Yunnan Ramontchi 166777
Yunnan Randia 12544

Yunnan Raspberry 338369,339502
Yunnan Rattan Palm 65816
Yunnan Rattanpalm 65816
Yunnan Rattlebox 112823
Yunnan Rattlesnake Plantain 179722,347942
Yunnan Rattlesnakeroot 313908
Yunnan Red Azalea 331703
Yunnan Red Silkyarn 238984
Yunnan Redbud 83812
Yunnan Redcarweed 288784
Yunnan Redleaf Litse 234053
Yunnan Rehdertree 327415
Yunnan Renanthera 327688
Yunnan Rhodiola 329975
Yunnan Rhododendron 332155
Yunnan Rhubarb 329418
Yunnan Rhynchosia 333461
Yunnan Ribseedcelery 304862
Yunnan Ricebag 391548
Yunnan Ringvine 116024
Yunnan Rockfoil 349627
Yunnan Rockvine 387871
Yunnan Roscoea 337115
Yunnan Rose Mallow 402260
Yunnan Rose Rhododendron 330560
Yunnan Rosewood 121859
Yunnan Roughleaf Bedstraw 170223
Yunnan Roughleaftree 29009
Yunnan Rungia 340378
Yunnan Rush 213579
Yunnan Sabia 341579
Yunnan Sage 345485
Yunnan Sagebrush 36564
Yunnan Sandthorn 196770
Yunnan Sandwort 32319
Yunnan Sanicle 345978
Yunnan Santiria 346232
Yunnan Sapium 346382
Yunnan Saraca 346503
Yunnan Sarcococca 346750
Yunnan Saurauia 348001
Yunnan Sawwort 361052
Yunnan Schefflera 350806
Yunnan Seabuckthorn 196770
Yunnan Sedge 76789
Yunnan Shinyleaf Buckthorn 328724
Yunnan Sida 362693
Yunnan Silverhook 256236
Yunnan Sinocrassula 364935
Yunnan Sinosideroxylon 365087
Yunnan Siraitia 365331
Yunnan Skullcap 355363,355858
Yunnan Slatepentree 322616
Yunnan Slugwood 50633
Yunnan Snakegourd 396249
Yunnan Snowflake 32528
Yunnan Sonerila 368905
Yunnan Sophora 369164
Yunnan Sotvine 194173
Yunnan Soulie Rose 336968
Yunnan Southstar 33574
Yunnan Spapweed 205460
Yunnan Spatholobus 370424
Yunnan Speedwell 407441
Yunnan Speranskia 370516
Yunnan Spiderflower 95835
Yunnan Spikesedge 143424
Yunnan Spindle-tree 157974
Yunnan Spiny Jujube 418184
Yunnan Spiraea 372138
Yunnan Spotleaf-orchis 179722
Yunnan Spotseed 57680
Yunnan Square Bamboo 87627
Yunnan Square-bamboo 87627
Yunnan St. John's Wort 201829,202094
Yunnan Stachyurus 373578
Yunnan Starjasmine 393626
Yunnan Stemona 375346
Yunnan Stephania 375908
Yunnan Stephanotis 211698
Yunnan Stinkwood 322397
Yunnan Stonebean-orchis 63197
Yunnan Stonelotus 364935
Yunnan Storax 379311
Yunnan Strict Slender Woodbetony 287254
Yunnan Sugarmustard 154565
Yunnan Sumac 332883
Yunnan Summerlilic 62206
Yunnan Supplejack 52474
Yunnan Supple-jack 52474
Yunnan Sweet Viburnum 407986
Yunnan Sweetleaf 381465
Yunnan Sweetspire 210423
Yunnan Sweetvetch 187984
Yunnan Swertia 380418
Yunnan Swordflag 208575
Yunnan Syzygium 382685
Yunnan Tallow-tree 346382
Yunnan Tarenna 385048
Yunnan Tephrosia 386269
Yunnan Ternstroemia 386734
Yunnan Themeda 389382
Yunnan Thickshellcassia 113497
Yunnan Thorowax 63873
Yunnan Three Flowered Rhododendron 332003
Yunnan Tickclover 126675
Yunnan Tigernanmu 122730
Yunnan Tillaea 391920
Yunnan Tinospora 392273
Yunnan Tinospore 392273
Yunnan Toadflex 231194
Yunnan Tongoloa 392791
Yunnan Toona 392835
Yunnan Torreya 393090
Yunnan Toxocarpus 393527
Yunnan Triflor Azalea 332003
Yunnan Tripogon 398121
Yunnan Tubergourd 390177
Yunnan Tumorfruitparsley 393841
Yunnan Tupelo 267879
Yunnan Tupistra 70637
Yunnan Turmeric 114882
Yunnan Tutcheria 322616
Yunnan Tylophora 400987
Yunnan Umbrella Bamboo 162767
Yunnan Underleaf Sand pearl 296481
Yunnan Veronicastrum 407510
Yunnan Viburnum 408235
Yunnan Villaintree 29009
Yunnan Vinegentian 110298
Yunnan Violet 410789
Yunnan Wallichia 413094
Yunnan Wallichpalm 413094
Yunnan Walsura 413136
Yunnan Wampee 94225
Yunnan Waterbamboo 260125
Yunnan Waxpetal 106695
Yunnan Webseedvine 235231
Yunnan White Beautyberry 65669
Yunnan Whitepearl 172099
Yunnan Whitlowgrass 137298
Yunnan Wildmangrove 288684
Yunnan Wildtung 243457
Yunnan Willow 343182
Yunnan Willowleaf Honeysuckle 235919
Yunnan Windflower 23795
Yunnan Windhairdaisy 348524,348959
Yunnan Windmill 100501
Yunnan Wingseedtree 320851
Yunnan Wing-seed-tree 320851
Yunnan Wintergreen 172099
Yunnan Winterhazel 106695
Yunnan Wolfberry 239134
Yunnan Woodbetony 287841
Yunnan Woodlotus 244478
Yunnan Yam 131915
Yunnan Yellowflower Photinia 295710
Yunnan Yellowflower Sisymbrium 365534
Yunnan Yellowleaftree 415148
Yunnan Yellow-leaved Tree 415148
Yunnan Yew 385409
Yunnan Zangolive 133008
Yunnan Zehneria 354668
Yunnan Zinger 418030
Yunnan-bamboo 278645
Yunnan-burma Asiabell 98288
Yunnan-Burma Azalea 330440
Yunnan-Burma Camellia 69153
Yunnan-Burma Ironweed 406662
Yunnan-Burma Rattanpalm 65678

Yunnanclave 238100,238106
Yunnan-Guangxi Elatostema 142800
Yunnan-Guizhou Fevervine 280108
Yunnan-Guizhou Holly 203761
Yunnan-Guizhou Millettia 66633
Yunnan-Guizhou Pipewort 151470
Yunnan-Guizhou Privet 229671
Yunnan-Kuichou Falsenettle 56297
Yunnan-Kweichow Millettia 66633
Yunnanopilia 416737,416738
Yunnan-Xizang Raspberry 338550
Yunshan Cyclobalanopsis 116193
Yunshan E. H. Wilson Holly 204402
Yunshan Lime 391793
Yunshan Linden 391793
Yunshan Oak 116193
Yuntai Southstar 33316
Yunxiaoshan Poplar 311589
Yun-Zang Azalea 331953
Yunzhu 297286
Yuquan Larkspur 124709
Yushan Actinodaphne 234000
Yushan Angelica 24413
Yushan Arrowwood 407934
Yushan Azalea 331565
Yushan Barberry 51940
Yushan Bentgrass 12145
Yushan Brome 60860
Yushan Cacalia 283846,283847
Yushan Cane 416804
Yushan Catchfly 363774
Yushan Daphne 122525
Yushan Dewberry 339188
Yushan Elaeagnus 142207
Yushan Euonymus 157728
Yushan Everlasting 21630
Yushan Gentian 150788,173868
Yushan Groundsel 359516
Yushan Honeysuckle 235999
Yushan Listera 264704
Yushan Litse 234000
Yushan Mahonia 242608
Yushan Mouseear 82924
Yushan Mout Pearleverlasting 21630
Yushan Neolitsea 264015
Yushan Pink 127802
Yushan Primrose 314655
Yushan Privet 229553
Yushan Raspberry 339188
Yushan Red Fescuegrass 164290
Yushan Rhododendron 331565
Yushan Rockcress 30345
Yushan Rose 336938
Yushan Sagebrush 35981
Yushan Spindle-tree 157728
Yushan Stonecrop 356947
Yushan Viburnum 407696
Yushan Willow 343721
Yushanbamboo 416742,416804
Yushania 416742,416783,416803,416825
Yushu Azalea 331564,332156
Yushu Goosecomb 335568
Yushu Kobresia 217314
Yushu Parnassia 284513
Yushu Rhododendron 331564,332156
Yushu Rockfoil 350048
Yushu Roegneria 335568
Yushu Sagebrush 36549
Yushu Sedge 76790
Yushu Thermopsis 389559
Yushu Wildsenna 389559
Yushu Windhairdaisy 348963
Yuzhong Fritillary 168655
Yuzu 93515
Yves Stonecrop 357326

Z

Zabala Fruit 221816
Zabei Spirea 372141
Zabel Barberry 52393
Zabel's Cherry Laurel 316508
Zabon 93579
Zacate 266499
Zaduo Corydalis 106616
Zaduo Rockjasmine 23118
Zaduo Sandwort 32326
Zag Willow 343670
Zahlbruckner Mountainash 369571
Zahlbruckner Mountain-ash 369571
Zahlbruckner Peashrub 72380
Zahlbruckner Pea-shrub 72380
Zahlbruckner Glomerate Barberry 51674
Zaidam Calligonum 67087
Zaidam Russianthistle 344777
Zaissan Milkvetch 43286
Zalil 124572
Zaman 345525
Zambak 211990
Zambesi Coffee 49200
Zambesi Redwood 46573
Zamia 417002
Zamia Palm 115852,241458,241463
Zang Corydalis 106554
Zang Willow 344306
Zangbean 377409,377411
Zangbu Barberry 52274
Zangbu Elegent Azalea 330370
Zangbu Elegent Rhododendron 330370
Zangbu Threeawngrass 34061
Zangbu Triawn 34061
Zangnan Rockfoil 349294
Zangolive 133006,133007
Zangxi Small Reed 127325
Zanlanscian Barberry 52396
Zanonia 417077,417081
Zante Fustic 107300
Zante Wood 107300
Zanthoxylum 417132
Zanzibar Aloe 17381
Zanzibar Copal 200848
Zanzibar Ebony 132299
Zanzibar Oil Vine 385655
Zanzibar Rattlebox 112757
Zanzibar Redheads 382477
Zanzibar Waterlily 267783
Zanzibar Yam 131824
Zaoju Mandarin 93731
Zapeteru 78145
Zapieandi 382737
Zapote Jnjerto 313408
Zapote Negro 132132
Zarzaparrilla 366264
Zawa 236334
Zawadsk Daisy 124862
Zawadsk's Dendranthema 124862
Zayu Barberry 52397
Zayu Fargesia 162768
Zayu Machilus 240567
Zayu Milkvetch 43288
Zayu Willow 344307
Zea Thyrsia 391445
Zebra Aloe 17002,17420
Zebra Basket-plant 9458
Zebra Caladium 16535
Zebra Calathea 66215
Zebra Grass 255886,255893
Zebra Haworthia 186422
Zebra Plant 29127,66215,186422
Zebra Rush 352206,352279
Zebra Wood 101873,181743
Zebrano 253158
Zebraplant 66215
Zebra-striped Rush 255886
Zebra-striped Temple-bells 366719
Zebrawood 43475,81821,132030,253158
Zebrina 393987
Zedang Elagant Falsetamarisk 261261
Zedary 114884

Zedoary 114884
Zehneria 417452
Zeko Rockfoil 350054
Zeku Azalea 332161
Zeku Crazyweed 279244
Zeku Pleurospermum 304855
Zeku Rhododendron 332161
Zeku Ribseedcelery 304855
Zeku Rockfoil 350054
Zeku Sedge 76791
Zela Remarkable Barberry 51772
Zelenetzky Thyrocarpos 391399
Zelkova 417534
Zen Magnolia 416727
Zenbean 417573,417574
Zenia 417573,417574
Zenobia 417590
Zenry 364557
Zenvy 364557
Zeodary 114884
Zephyr Flower 417612
Zephyr Lily 417606,417612
Zephyr-flower 23925,417606
Zephyrlily 417606,417612
Zephyr-lily 417606
Zeshan Gentian 150959
Zeuxine 417698,417795
Zew Zealand Tea-tree 226481
Zeyher Mock Orange 294590
Zeyher Mock-orange 294590
Zeyland Centipedegrass 148258
Zhacai Leaf-mustard 59469
Zhada Catnip 265109
Zhada Clematis 95467
Zhada Milkvetch 42496
Zhada Nepeta 265109
Zhaisang Crazyweed 279118
Zhaisang Milkvetch 43286
Zhang Arundinella 37368
Zhang's Actinidia 6549
Zhang's Kiwifruit 6549
Zhang's Larkspur 124713
Zhangbei Cinquefoil 313119
Zhanghe Hankow Willow 343672
Zhanglangxiang Arisaema 33575
Zhangmu Fargesia 137842
Zhangmu Milkvetch 42173
Zhangping Goldenhairy Tanoak 233145
Zhangxian Fritillary 168524
Zhaojue Monkshood 5710
Zhaojun Dennes Dendrobium 125007
Zhaoling Milkvetch 43289
Zhaoqian Synotis 381930
Zhaoqian Tailanther 381930
Zhaosu Catchfly 363933
Zhaosu Kengyilia 215972
Zhaosu Milkaster 169792
Zhaotong Actinidia 6694
Zhaotong Azalea 332009
Zhaotong Chinese Fir 114550
Zhaotong Kiwifruit 6694
Zhaotong Onosma 271747
Zhatong Meadowrue 388520
Zhayou Cotoneaster 107731
Zhegu Azalea 332160
Zhegu Rhododendron 332160
Zhegushan Cystopetal 320246
Zhegushan Willow 344308
Zhejiang Acanthopanax 2531
Zhejiang Actinidia 6726
Zhejiang Adelostemma 54525
Zhejiang Arrowwood 408120
Zhejiang Barberry 52320
Zhejiang Bittercress 73052
Zhejiang Buckeye 9684
Zhejiang Buckthorn 328844
Zhejiang Bulbophyllum 63019
Zhejiang Bushcherry 316488
Zhejiang Camellia 68980
Zhejiang Chelonopsis 86810
Zhejiang Clematis 94811
Zhejiang Corydalis 106240
Zhejiang Cragcress 98041
Zhejiang Curlylip-orchis 63019
Zhejiang Damnacanthus 122064
Zhejiang Flatsedge 322196
Zhejiang Fritillary 168586
Zhejiang Glorybower 96157
Zhejiang Goldenray 229003
Zhejiang Goldlineorchis 26015
Zhejiang Grape 412010
Zhejiang Hackberry 80612
Zhejiang Helwingia 191192
Zhejiang Hemlock 399925
Zhejiang Hemsleia 191977
Zhejiang Hilliella 196312
Zhejiang Holly 204421
Zhejiang Indigo 206350
Zhejiang Kiwifruit 6726
Zhejiang Landpick 308666
Zhejiang Leafflower 296517
Zhejiang Leaf-flower 296517
Zhejiang Loosestrife 239578
Zhejiang Maackia 240120
Zhejiang Machilus 240568
Zhejiang Mockorange 294591
Zhejiang Mosquitotrap 117424
Zhejiang Nanmu 295352
Zhejiang Newlitse 264024
Zhejiang Persimmon 132175
Zhejiang Phoebe 295352
Zhejiang Photinia 295820
Zhejiang Pycreus 322196
Zhejiang Rockfoil 350055
Zhejiang Saddletree 240120
Zhejiang Scorpiongrass 175868
Zhejiang Scurvyweed 98041
Zhejiang Sinojohnstonia 364962
Zhejiang Skullcap 355406
Zhejiang Snowbell 379478
Zhejiang Snowgall 191977
Zhejiang Solomonseal 308666
Zhejiang Spapweed 204846
Zhejiang Stonecrop 357233
Zhejiang Storax 379478
Zhejiang Swallowwort 117424
Zhejiang Swertia 380225
Zhejiang Taperleaf Camellia 69034
Zhejiang Tigerthorn 122064
Zhejiang Touch-me-not 204846
Zhejiang Turbinate Aster 41442
Zhejiang Underleaf Pearl 296517
Zhejiang Viburnum 408120
Zhejiang Willow 343200
Zhejiang Wintersweet 87545
Zhenbian Eurya 160621
Zhenduo Youngia 416499
Zhenfeng Meliosma 249473
Zhenfeng Persimmon 132480
Zhenghe Apricot 34482
Zhengyi Begonia 50431
Zhengyi Corydalis 106607
Zhengyi Euonymus 157967
Zhenjiang Mosquitotrap 117698
Zhenkang Bigumbrella 59196
Zhenkang Brassaiopsis 59196
Zhenkang Coelogyne 98761
Zhenkang Cotoneaster 107387
Zhenkang Deutzia 126844
Zhenkang Didymocarpus 129993
Zhenkang Griffith Lacquertree 393458
Zhenkang Onosma 242400
Zhenkang Sandwort 32101
Zhenkang Sedge 76793
Zhenkang Windflower 23945
Zhennan 295417
Zhenping Barrenwort 147010
Zhewan Hydrangea 200159
Zhidan Apricot 34481
Zhiduo Rockfoil 350057
Zhijiang Wingnut 320383
Zhijin Camellia 69738
Zhongba Bluegrass 306161
Zhongba Catchfly 364227
Zhongba Larkspur 124136
Zhongdian Azalea 332162
Zhongdian Bluebell 115343
Zhongdian Bluebellflower 115343
Zhongdian Buttercup 326504
Zhongdian Catchfly 363321
Zhongdian Clovershrub 70918

Zhongdian Corydalis 106620
Zhongdian Cotoneaster 107525
Zhongdian Cowparsnip 192265
Zhongdian Cremanthodium 110363
Zhongdian Deerdrug 242698
Zhongdian Deutzia 127128
Zhongdian Draba 137183
Zhongdian Dutchmanspipe 34381
Zhongdian Everlasting 21540
Zhongdian False Solomonseal 242698
Zhongdian Felwort 235471
Zhongdian Gentian 173341
Zhongdian Groundsel 358550
Zhongdian Hawthorn 109606
Zhongdian Lady's Slipper 120303
Zhongdian Ladyslipper 120303
Zhongdian Larkspur 124708
Zhongdian Longfruit Speedwell 407081
Zhongdian Mahonia 242497
Zhongdian Milkvetch 42382
Zhongdian Monkshood 5486
Zhongdian Nutantdaisy 110363
Zhongdian Oneflowerprimrose 270734
Zhongdian Pearleverlasting 21540
Zhongdian Pearweed 239588
Zhongdian Pimpinella 299382
Zhongdian Primrose 314246
Zhongdian Rhododendron 332162
Zhongdian Rockfoil 349278
Zhongdian Rose 336868,337048
Zhongdian Sagebrush 36570
Zhongdian Sandwort 32328
Zhongdian Skullcap 355411
Zhongdian Spapweed 204854
Zhongdian Sweetvetch 188143
Zhongdian Swordflag 208866
Zhongdian Tongoloa 392803
Zhongdian Touch-me-not 204854
Zhongdian Whitlowgrass 137230
Zhongdian Windhairdaisy 348274
Zhongdian Woodbetony 287844
Zhonghai Sedge 76794
Zhongnan Hornbeam 77422
Zhongning Milkvetch 42813
Zhongshangrass 292485,292487
Zhongtiao Maple 3763
Zhongtiaoshan Maple 3763
Zhou Catchfly 364228
Zhou's Bittercress 72720
Zhouqu Carpesium 77196
Zhouqu Fritillary 168563
Zhouqu Goldenray 229260
Zhouqu Milkvetch 43291
Zhouqu Striate Butterflybush 62173
Zhouzhi Willow 344178
Zhufeng Eritrichium 153513
Zhufeng Fritillary 168563
Zhumu Willow 343218
Zhuoba Lily 230069
Zhuoni Rhododendron 330964
Zhushan Barrenwort 147080
Zhushan Epimedium 147080
Zicaitai Mustard 59342
Zieria 417866
Zigadenus 417880
Zigzag 2837,3462
Zigzag Aster 380966
Zigzag Bamboo 297281
Zig-zag Bamboo 297281
Zigzag Begonia 50138
Zigzag Clover 396967
Zig-zag Clover 396967
Zigzag Golden Rod 368100
Zigzag Goldenrod 368100
Zig-zag Goldenrod 368100
Zigzag Grass 281561
Zigzag Iris 208471
Zigzag Jujube 418170
Zimapan Mahonia 242663
Zimbabwe Aloe 16810
Zimbabwe Creeper 306706
Zinegar 246478
Zingana 253158,253160
Zinger 417964,418010
Zinger Broom 121031
Zinger Milkvetch 43292
Zinnia 418035,418038,418043,418050
Zion Chickensage 371150
Zion Fleabane 150964
Zion Goldenaster 194278
Zion Jimmyweed 209574
Zion Tansy 371150
Zion Wild Buckwheat 152677
Zippel Yerbadetajo 141378
Zippelia 418072
Zirieote 104177
Zitchiku Bamboo 297233
Zitherform Leaf Phoebe 295397
Zit-Kawa 50998
Zit-Kwa 50998
Zixi Spapweed 205463
Ziyang Whorlleaf Stringbush 414262
Ziyuan Azalea 332164
Ziyuan Fir 294
Ziyuan Rhododendron 332164
Ziyun Barberry 52398
Ziziphora 418119
Zocoxochiti 140293
Zodoary 114884
Zodoary Turmeric 114884
Zollinger Birthwort 34382
Zollinger Cypressgrass 119769
Zollinger Dutchmanspipe 34382
Zollinger Eulophia 157110
Zollinger Falsenettle 56357
Zollinger False-nettle 56357
Zollinger Galingale 119769
Zollinger Gentian 174095
Zollinger Gromwell 233786
Zollinger Indigo 206761
Zonal Geranium 288297,288594
Zonal Pelargonium 288594
Zonate Cryptanthus 113387
Zonate Neoregelia 264427
Zoned Pink 127628
Zongbaye Aspidistra 39584
Zornia 418320,418331
Zostera Family 418405
Zostera-like Pondweed 312299
Zoysia Grass 418424
Zschack's Goosefoot 86966
Zuccarini Pincution 244268
Zuiho Machilus 240726
Zulu Fescue 166590
Zulu Fig 165263
Zulu Giant 373821
Zulu Nul 118837
Zulu Nut 118822
Zum Crabapple 243735
Zuni Fleabane 150933
Zunyi Hornbeam 77408
Zunyi Sedge 76797
Zuogong Rockfoil 350059
Zuolinhe Water Lobelia 234381
Zygopetalum 418573

日文序号—名称索引

Japanese

91 ツクバネウツギ属
113 シナツクバネウツギ,タイワンツクバネウツギ,タイワンバネウツギ,テイワントウツクバネウツギ
114 タイワンツクバネウツギ,タイワンバネウツギ,テイワントウツクバネウツギ
117 カラツクバネウツキ
135 ハナゾノツクバネウツギ,ハナツクハネ
167 キバナツクバネウツギ,コツクバネウツギ
168 ヒロハコツクバネウツギ
169 ホソバコツクバネウツギ
170 ベニバナコツクバネウツギ
176 オニツクバネウツギ
177 コツクバネ,ツクバネウツギ,ツクバネタニウツギ
178 タキネツクバネウツギ
179 ヤエノツクバネウツギ
180 ウスギツクバネウツキ,キバナノツクバネウツキ
181 アツバツクバネウツギ,テリハツクバネウツギ
182 ケツクバネウツギ
184 コバナツクバネウツギ
185 ベニツクバネ,ベニバナノツクバネウツギ
186 ウゴツクバネウツギ
191 オオツクバネウツギ
209 ウチワノキ属
210 ウチワノキ
211 ウチワノキ
212 アベルモスクス属,トロロアオイ属,トロロアフヒ属
219 アメリカネリ,オクオ,オクラ,オクヲ
225 ノリアサ
228 オガサハライチビ,シマイチビ,タカサゴイチビ,ヲガサハライチビ
229 オウスゲ,クサダモ,トロロ,トロロアオイ,トロロアフヒ,ネリ
232 センカクトロロアオイ
235 トロロアオイモドキ,リウキウトロロアフヒ,リュウキュウトロロアオイ
236 センカクトロロアオイ
243 シャウジャウクワ,シャウジャウボク
269 モミ属
272 オウシウモミ,シラモミ,ヨーロッパモミ
282 カナダバルサムノキ,バルサムフヤー,バルサムモミ,ベルサムモミ
317 コロラドモミ,ベイモミ,ホワイトフヤー
365 アオモリトドマツ,サナギ,タウモミ,モミ,モミソ,モミノキ
366 シダレモミ
369 ウンナンモミ
377 フラセリーモミ
384 アメリカオオモミ
388 アオトド,チョウセンモミ,テウセンモミ
389 アオボウモミ,アヲボウモミ,ウラジロモミ,ダケモミ,ニックワウモミ,ニッコウモミ,ニレモミ
396 ニイタカトドマツ,ニヒタカトドマッ
414 アオモミトドマッ,アオモリトドマツ,アヲモミトドマッ,オオシマビソ,オオシラビソ,オオシラベ,オオジラベ,オオリュウセン,オホシマビソ,オホシラベ,オホリウセン,カラツガ,ツガ,モロビ
427 アカトド,タウシラベ,テウセンシラベ,トウシラベ
429 クリミャモミ
453 ノーブルモミ
460 アカトド,アカトドマツ,トドマツ
463 アオトド,アオトドマツ,アカトド,アカトドマツ,オニトド,ネムロトド,ネムロトドマツ
464 エゾシラビソ
467 シュクシラベ
476 シベミアモミ
480 ヒマラャモミ
495 ミツミネモミ
496 アオヒ,コリュウセン,シコクシラベ,シラツガ,シラビソ,シラベ,トラノオモミ,リュウセン
497 アオシラベ,アヲシラベ
498 コマガタケシラベ
502 シコクシラベ
504 アオシコクシラベ
510 エゾシラビソ,カマフテシラビソ
675 トゲアオイモドキ
687 アブロニア属,ハイビジョザクラ属,バイビジョザクラ属
702 キバナハイビジョザクラ,キバナビジョザクラ
710 アブロニア,バイビジョザクラ
738 タウアヅキ属,トウアヅキ属
765 タウアヅキ,トウアヅキ,ナンバンアヅキ
837 アブーチロン属,イチビ属
905 タイワンイチビ,ヒメイチビ
913 フイリアブチロン
923 ゴールデンーフリース
928 サングラン
934 オガサハライチビ,シマイチビ,タカサゴイチビ,ヲガサハライチビ
935 サキシマイチビ
957 ウキツリボク
989 タウイチビ
997 シャウジャウクワ,シャウジャウボク,ショウジョウカ
998 キフアブチロン
1000 イチビ,キリアサ,キリアザ,クサギリ,ゴサイバ,ヒナハギリ
1018 アカカリス属
1024 アカシア属
1060 ハリアカシア
1069 ギンヨウアカシア
1120 アカシアアセンヤク,アセンヤク,アセンヤクノキ,ガイジチャ,カラキュ,ジチャ,ペグアセンヤク,ペグノキ
1142 オキナハネム,ニンニンバ,ネムカヅラ
1145 サウシジュ,ソウシジュ,タイワンアカシア,タイワンガフクワン,タイワンゴウカン
1156 ウロコアカシア,サンカクバアカシア
1165 ギンワットル,ハナアカシア,フサアカシア
1168 アサアカシア,シュウアカシア,ミモサアカシア,モリシマアカシア
1219 キンガフクワン,キンコウカン
1308 ツルアカシア
1358 ナガバアカシア
1380 クロワットル,モリシマアカシア
1384 ブラックウッドアカシア,メラノクシロンアカシア
1401 モリシマアカシア
1422 ヤナギバアカシア
1427 アラビヤゴムノキ,アラビヤゴムモドキ
1466 トゲガフクワン
1488 ムクゲアカシア
1516 ゴールデンワットル,ピクナンサアカシア
1572 アフリカゴムノキ,アラビアゴム,アラビヤコムノキ
1613 アリアカシア
1686 スキバアカシア
1739 アケーナ属
1769 アカリーファ属,エノキグサ属
1775 アコウカミフア
1790 アミガサソウ,エノキグサ,ナガバエノキグサ,ビロオドエノキグサ
1794 ホソバエノキグサ,ホソバノエノキグサ
1869 アカリファ,キフクリンアカリーファ
1876 ヒメアミガサソウ
1894 ナガボアミガサノキ,ベニヒモノキ
1900 キダチアミガサソウ,キダチアミガサノキ
1923 ホソバアミガサノキ
2011 サンシキアカリフア
2014 フクリンアカリフア
2019 ニシキアカリフア
2032 アカンペ属
2060 キツネノマゴ科
2071 アカンソフィッピューム属,アカンテフィピウム属,アカントフィピウム属,タイワンショウキラウ属
2080 タイワンアオイラン

2081 エンレイショウキラン,タイワンショウキラウ
2084 タイワンアオイラン
2096 アカリトガリキウム属
2101 ホウカンマル
2103 オウシュンマル
2105 カカンマル
2107 シセイマル
2137 アカントスレウス属
2208 アガントリ－モン属
2313 トウショクマル
2421 タカノツメ
2502 フイリウコギ
2550 トゲノャシ属
2551 オニトゲノャシ
2553 アカトゲノャシ,アレカ－ルブラ
2601 アカントシキオス属
2612 アメリカトゲミギク
2623 アカントスタキス属
2625 マッカサアナナス
2657 アカンサス属,ハアザミ属
2684 ムラサキミズヒイラギ
2691 ナガバナアザミ
2695 アカンサス,ハアザミ
2711 トゲハアザミ
2769 カエデ属,カヘデ属
2775 クスノハカエデ,ナガバカエデ
2776 アマミカジカエデ
2778 オオモミジ
2780 フカギレオオモミジ
2781 ヤマモミジ
2782 ナンブコハモミジ
2793 アサノハカエデ,ミヤマモミジ,ミヤマモミヂ
2796 ナンゴクミネカエデ
2797 ウスゲカエデ,チョウセンアサノハカエデ
2798 ウスゲカヘデ,テウセンアサノハカヘデ
2811 トウカエデ
2812 ヒトッバトウカエデ
2815 タイワントウカエデ
2837 カムベストルカエデ,コブカエデ,コルクカエデ
2846 アシボソウリノキ,オオカラバナ,オソエウリハダ,オホカラバナ,ホソエカエデ,ヤケマオナガカエデ
2847 ヤクミマオナガカエテ
2868 アラハゴ,シラシデ,タニアサ,チドリノキ,ヤマシバカエデ
2869 オオバチドリノキ
2872 オナガカエデ,カハカエデ,トガリバカエデ,ヲナガカエデ
2891 アッバクズノハカエデ
2896 アマクキ,アマコギ,アンマッコカエデ,ミッデカエデ,ミッデモミジ
2910 ウリカエデ,コウリカエデ,シラハシノキ,ホンウリ,メウリノキ,ヤマカエデ
2912 オオバノウリカエデ
2913 ウリカエデ,フイリウリカエデ,フイリコウリカエデ,メウリノキ
2928 オニモミジ,オニモミヂ,カジカエデ,カヂカエデ,キリハカエデ
2939 ヒトツバカエデ,マルバカエデ
2942 シマヤマモミヂ,タイワンヤマモミヂ,タカサゴモミヂ
2984 カラコギカエデ,カラコギカヘデ,チョウセンカラコギカエデ,マイラチャ
2987 カラコギカエデ
3003 アカハダメグスリノキ
3026 ウラジロカエデ,バタアンカエデ
3031 シマウリカエデ,シマウリハダカエデ
3034 アカバナウチワカエデ,オグラヤマ,キハウチワカエデ,ハウチワカエデ,メイケッカエデ,メイゲッカエデ
3035 マイタジャク
3036 キハウチハカエデ
3037 エゾメイゲッカエデ
3038 マイクジャク
3043 ケハウチノカエデ,ケハウチハカエデ,ケハウチワカエデ
3048 カサド
3050 オオバケハウチワカエデ
3052 サヨシグレ
3058 オナガカエデ,カハカエデ,トガリバカエデ,ヲナガカエデ
3064 チョウセンミネカエデ
3087 ウラジロイタヤ
3105 アオガシカエデ,アヲガシカエデ,カゴノキカエデ
3127 オレゴソメ－ブル,ヒロハノカエデ
3129 マンシウカエデ,マンシウカヘデ
3140 アカイタヤ,ベニイタヤ
3147 コバナカエデ,コミネカエデ
3149 エゾイタヤ,クロビイタヤ
3150 シバタカエデ
3154 アサヒカエデ,イタヤ,イタヤカエデ,エンコウカエデ,ツタモミジ,トキワカエデ
3160 オニイタヤ
3174 オニイタヤ
3179 エンコウカエデ
3198 アカイタヤ,ベニイタヤ
3205 エゾハウチワカエデ
3208 ヤクシマオナガカエデ
3209 シマウリバカエデ,タイワンウリハダカエデ,タカサゴウリハダカエデ
3218 サトウカエデ,トネリコバノカエデ,ネグンドカエデ
3236 クロカエデ
3243 オオミッデカエデ,コチョウノキ,コテフノキ,チャウヂャノキ,チョウジャノキ,チョオジャノキ,ミツバナ,メグズリノキ,メグロ
3248 テツカエデ,テツノキ
3249 キタノテツカエデ
3250 コウシンテツカエデ
3251 ナンゴクテツカエデ
3255 クスノハカエデ,ナガバカエデ
3256 クスノハカエデ
3258 クスノハカエデ
3278 イトマキシマモミヂ
3300 イロハカエデ,イロハモミジ,イロハモミヂ,カエデ,カヘルデノキ,タヵオカエデ,タヵオカヘデ,タカオモミジ,タヵヲカヘデ,ハイトベニ,モミジ,モミヂ
3309 サニシサ,チリメンカエデ
3356 シメノウチ
3398 イタヤカエデ,オニイタヤ,フイリオニイタヤ
3400 オニイタヤ
3404 イタヤカエデ,エンコウカエデ
3405 ウラゲエンコウカエデ
3406 ケエンコウカエデ
3407 ケウラゲエンコウカエデ
3408 ウラジロイタヤ
3411 オオエゾイタヤ
3412 イトマキイタヤ
3413 タイシャクイタヤ
3425 ノルウエ－カエデ
3458 イロハモミジ
3462 セイヨウカジガエデ
3479 タウハウチハカヘデ,チョウセンハウチワカエデ,トウハウチワカエデ
3492 アカモミジ,ハナカエデ,ハナノキ,メグスリノキ
3505 アカカエデ,アメリカハナノキ,ベニカエデ
3529 ウリノキ,ウリハダカエデ,オオウリカエデ,オオミネカエデ,オホウリカエデ,オホミネカエデ,クシノキ,コウモリカエデ,コンジノキ
3530 ホソバウリハダカエデ
3531 ハツュキカエデ,フイリウリハダカエデ
3532 ウラジロサトウカエデ,ギンカエデ,ギンヨウカエデ,サッカリヌムカヘデ,サトウカヘデ
3549 サトウカエデ
3599 オオイタヤメイゲッ,オホイタヤメイゲッ
3604 イタヤメイゲッ,コハウチハカエデ,コハウチワカエデ,コバナウチワカエデ
3605 フカギレイタヤメイゲッ
3607 カサトリャマ
3608 コハウチハカエデ,コハウチワカエデ,ヒメウチワカエデ
3609 コハイタヤメイゲッ
3665 ケナシウリハダカヘデ,マンシウウリハダ,マンシウウリハダカエデ
3676 ヒナウチワカエデ

3710 オニメグスリ
3714 マンシウイタヤ
3718 ハクサンモミジ,ハクサンモミチ,バンダイカエデ,ヒメオガラバナ,ヒメヲガラバナ,ミネカエデ
3722 オオバミネカエデ
3732 オガラバナ,ホザキカエデ,ホザキカヘデ,ヲガラバナ
3733 ウスゲオガラバナ
3765 カエデ科,カヘデ科
3818 アセリフィラム属
3913 ノコギリサウ属,ノコギリソウ属
3914 カラノコギリソウ
3917 エダウチノコギリソウ
3918 イチゲノコギリソウ
3921 イッジゴウ,シロバナノコギリサウ,ノコギリサウ,ノコギリソウ,ハゴロモサウ,ハゴロモソウ,ヤマノコギリサウ
3922 シュムシュノコギリソウ
3923 キタノコギリソウ,ホロマンノコギリソウ
3924 アカバナエゾノコギリソウ
3925 アソノコギリソウ
3930 ヤマノコギリソウ
3942 ギンバノコギリソウ
3948 キバナノギコリソウ,キバナノコギリソウ,キバナノノコギリサウ,キバナノノコギリソウ
3978 セイヤウノコギリサウ,セイヤウノコギリソウ,セイヨウノコギリソウ
3995 イワノココギリソウ,ジャカウノコギリサウ,ジャコウノコギリソウ
3997 ケノコギリサウ,チャボノコギリサウ,チャボノコギリソウ
4005 エゾノコギリサウ,エゾノコギリソウ,オオバナノコギリサウ,オオバナノコギリソウ,オホバナノコギリサウ,オホバナノコギリソウ
4008 エゾノコギリソウ
4009 ホソバエゾノコギリソウ
4014 シロバナノノコギリサウ,ヤマノコギリソウ
4028 シロバナノコギリサウ,シロバナノコギリソウ,ノコギリサウ,ノコギリソウ,ハゴロモサウ,ハゴロモソウ,ヤマノコギリサウ,ヤマノコギリソウ
4052 ノコギリソウモドキ
4056 ヒメノコギリソウ
4070 アキメネス属,ハナギリソウ属
4076 バニバナギリソウ
4078 サビハナギリソウ,サビバハナギリソウ
4082 ハナギリソウ
4103 ナンブソウ属
4104 ナンブソウ
4113 ハネガヤ属
4134 ハネガヤ
4148 マンシウハネガヤ
4150 ハネガヤ
4160 オオハネガヤ,シベリアハネガヤ
4163 ラクダガヤ
4249 イノコズチ属,イノコヅチ属,ヰノコヅチ属
4259 イノコヅチ,ケイノコヅチ,ケヰノコヅチ,シマイノコヅチ,ナガバノシマイノコヅチ,ナガバノシマヰノコヅチ,モンパイノコヅチ,モンパヰノコヅチ
4263 ケイノコヅチ,ケヰノコヅチ,シマイノコヅチ,シロイノコヅチ,シロヰノコヅチ
4269 ムラサキイノコヅチ
4273 イノコズチ,イノコヅチ,モンパイノコヅチ,ヰノコヅチ
4274 ハチジョウイノコヅチ
4275 イノコズチ,イノコヅチ,コマノヒザ,ヒカゲイノコズチ,ヒカゲヰノコヅチ,フシダカ,ヰノコヅチ
4286 ヒナタイノコズチ,ヒナタイノコヅチ,ヒナタヰノコヅチ
4287 マルバイノコヅチ
4304 ドゴシッ,ナガバイノコヅチ,ナガバヰノコヅチ,ヤナギイノコズチ,ヤナギイノコヅチ,ヤナギヰノコヅチ
4319 ナンテンイノコヅチ,ナンテンヰノコヅチ,ハヒイノコヅチ,ハヒヰノコヅチ
4523 アシダンテラ属
4648 アシネタ属
4863 ヌマツルギクモドキ
4866 オトウガラシ,オランダセンニチ,キバナオランダセンニチ,センニチギク,ハトラガラシ
4868 ヌマツルギク
4869 ヌマツルギク
4870 センニチコモドキ,センニチモドキ
4876 ヒメセンニチモドキ
4951 ライトヤシ
4954 アコカンテーラ属,サンダンカモドキ属
4965 サンダンカモドキ
5014 トリカブト属
5021 レイジンカズラ
5043 セイタカブシ
5052 ホソバキエボシソウ
5055 キエボシソウ
5057 オオキエボシソウ
5100 カブトギク
5108 カブトギク,トクリシトリカブト,トリカブト,ハナトリカブト
5144 キバナトリカブト
5185 カブトギク,トリカブト,ハナトリカブト
5193 オオブシ,オホブシ,セロハノカラフトブシ,ブジ,ブス,ヤマトリカブト
5214 フクトメブシ
5231 エゾノレイジンソウ,オオレイジンサウ,オホレイジンサウ
5275 ミタケウヅ
5292 アレナレブシ
5294 マンシウトリカブト
5298 オクトリカブト,ヤマトリカブト
5302 ムレイウズ
5313 チシマトリカブト
5318 ヒロハキエボシソゥ
5322 ウンゼントリカブト
5331 キバナトリカブト
5335 ホザキブシ
5363 カブトギク,トリカブト,リョウトウブシ
5374 シラチャレイジンソウ
5403 イトウウズ,ホソバハナカズラ
5413 オオチシマトリカブト,オホチシマトリカブト
5421 ヒメトリカブト
5442 アカニット,ホソバノトリカブト,ヨゥシュトリカブト
5466 オオレイジンサウ,オホレイジンサウ,タチキエボシソウ
5469 マンシウヤチマタブシ
5472 オオムレイウズ,トキウズ
5525 オオバナハナカズラ
5546 カラフトブシ,ホソバトリカブト
5558 ヒロハハナカズラ
5639 マンシウミツバトリカブト
5663 シラゲトリカブト
5674 ツルブシ,ハナカヅラ,ホソバヅルウズ
5675 タチハナカズラ,ハナカズラ,ハナヅル
5707 エゾトリカブト
5771 ショウブ科
5788 シャウブ属,ショウブ属
5793 シャウブ,ショウブ
5794 フイリショウブ
5798 アヤメグサ,オニセキシャウ,シャウブ,ショウブ
5803 セキシャウ,セキショウ
5809 アリスガハゼキシャウ,アリスガワセキショウ,カウライゼキシャウ,ビロウドゼキシャウ
5947 アクリオプシス属
5989 タマザキニガクサ属
6039 タマザキニガクサ
6128 アグロコーミア属,オニトゲココヤシ属
6129 アメリカアブラヤシ
6130 グムグムャシ
6134 アメリカアブラヤシ
6233 オホバグッケイ属
6251 オホバグッケイ
6405 ルイヨウショウマ属,ルヰエフショウマ属

6408 ルイヨウショウマ
6414 ルイヨウショウマ,ルヰエフショウマ
6422 アカシノルイヨウショウマ,アカシノルヰエフショウマ
6513 サルナシ属,マタタビ属
6515 オオミノサルナシ,コクワ,サルナシ,シラクチズル,シラクチヅル,ヤブナシ
6516 コクワ
6517 サビサルナシ
6518 サルナシ
6520 ウラジロマタタビ
6523 コクワ,テハノマタタビ
6530 アリサンサルナシ,タイワンサルナシ,ランカンサルナシ,レモガンサルナシ
6553 オニマタタビ,キーウイ,キウイフルーツ,シナサルナシ,シナマタタビ
6633 ウラジロマタタビ,ウラヅロマタタビ
6639 ウスバシラクチズル,ウスバシラクチヅル,ミヤママタタビ
6648 ウスバサルナシ,タカサゴサルナシ
6682 ナシウメ,マタタビ,ワタタビ
6697 シマサルナシ,ナシカズラ,ナシカヅラ,ホンバサルナシ
6728 マタタビ科
6761 アオカゴノキ属,アゴノキ属,アヲカゴノキ属
6762 アオカゴノキ,アヲカゴノキ,カウボク,ナントウカゴノキ,ナントウクロモジ,ナントウダモ,バリバリノキ
6771 シマクロモジ,タイワンヤマクロモジ
6800 ムシャカゴノキ,ムシャダモ
6826 アミダケンチャ属
6827 ヨッアミダケンチャ
6849 アクティノフレウズ属
6861 カラッパヤシ属
6862 カラッパヤシ
6874 オオサンカクイ
6888 ゴキヅル属
6907 ゴキヅル,モミジバゴキヅル
6988 アーダ属,アダ属
7018 アダンソニア属,バオバブノキ属
7022 バオバブ,バオバブノキ
7113 アデナンドラ属
7178 アデナンテーラ属,ナンバンアカアヅキ属
7190 ゼンダン,ナンバンアカアズキ,ナンバンアカアヅキ
7220 アデーニア属
7333 アデニウム属,アデニューム属
7343 アデニウム,アデニューム,オベスム
7427 ノブキ属
7433 ノブキ
7569 アデノンコス属
7596 ツリガネニンジン属
7618 ウスイロシャジン
7639 ヒロバニンジン,フクシマシャジン,フクシマヤジン
7640 シロバナフクシマシャジン
7643 ムレイシャジン
7649 ナカバシャジン,ヤナギバシャジン
7659 キキョウシャジン
7660 ツクシイハシャジン,ツクシイワシャジン
7678 ミヤマシャジン
7695 ヒナシャジン
7702 ニヒタカシャジン
7706 ヒメシャジン,ミヤマシャジン
7707 マリシャジン
7708 ケヒメシャジン
7712 ヒメイワシャジン
7713 ミョウギシャジン
7715 シライワシャジン
7728 ヌマシャジン,ヤチシャジン
7729 シロバナヤチシャジン
7739 シラトリシャジン,ノコギリシャジン,マンシウツリガネニンジン,マンシュウツリガネニンジン,モイハシャジン,モイワシャジン
7765 マツバニンジン
7768 ハナシャジン,フウリンハナシャジン
7798 ソバナ,マルバシャジン
7799 ホソバソバナ
7800 ケソバナ
7801 シロソバナ
7816 オカシャジン,ナカバモウコシャジン
7824 オカシャジン
7826 ヒロマモウコシャジン
7828 イトバシャジン,ケナシイトバシャジン
7830 タウシャジン,トウシャジン,トオシャジン,マルバニンジン,マルバノニンジン
7831 シロバナトウシャジン
7836 ナガバノニンジン
7843 イワシャジン
7844 ホウオウシャジン
7845 シマシャジン
7847 ケシライワシャジン
7850 サイヨウシャジン,ニオイシャジン
7853 アツバソバナ
7855 コウアンシャジン
7857 シロバナリュウキュウシャジン
7863 ツリガネニンジン
7864 シロバナツリガネニンジン
7865 シラゲシャジン
7867 ハイツリガネニンジン
7868 マルバノハマシャジン
7869 ハクサンシャジン
7871 オトメシャジン
7878 ウリュウシャジン,シラトリシャジン
7879 ニオイシャジン
7884 エダウチシャジン
7888 ケエダウチシャジン
7894 クルマバムレイシャジン
8004 ヌマダイコン属
8012 ヌマダイコン
8016 コバナヌマダイコン
8018 オカダイコン
8170 タニワタリノキ属
8192 タニワタリノキ
8199 シマタニワタリノキ
8205 ナガエサカキ属
8221 ウラキサカキ,キイバサカキ,ナガエサカキ
8235 ハイナンサカキ
8247 アリサンサカキ
8256 キールンサカキ,サカキヒサカキ,シマサカキ
8270 リュウキュウナガエサカキ
8278 ケナガエサカキ
8299 ツルコマクサ
8302 モモチドリ
8324 フクジュサウ属,フクジュソウ属
8325 ナツザキフクジュソウ
8331 グワンジッサウ,シカギク,フクジュサウ,フクジュソウ
8332 アキザキフクジュソウ
8378 フクジュソウ
8381 シベリアフクジュソウ
8387 ヨウシュフクジュソウ
8396 レンプクサウ属,レンプクソウ属
8399 レンプクサウ,レンプクソウ
8402 シマレンプクソウ
8406 レンプクサウ科,レンプクソウ科
8417 アドロミスクス属
8544 エクメア属,サンゴアナシス属,ツブアナナス属
8549 ウスバサンゴアナナス
8553 フイリサンゴアナナス
8554 テャンティニイ,トラフサンゴアナナス
8555 シロツブアナナス
8558 シマサンゴアナナス
8562 サンゴアナナス
8563 ウラベニサンゴアナナス,プラジ
8573 メオトアナナス
8576 メキシコアナナス
8578 ヒメベニサンゴアナナス
8582 オオシマサンゴアナナス
8589 オオサンゴアナナス
8593 タマサンゴアナナス
8597 ナガボサンゴアナナス
8599 ショウジョウアナナス
8672 ヤギムギ
8724 クサビコムギ
8732 タルホコムギ
8756 ナンバンギセル属
8760 オホナンバンギセル,オモヒグサ,キセルサウ,ナンバンギセル,ヒメナンバンギセル,ヤマナンバンギセル

8767 オオナンバンギセル,オホナンバンギセル,ヤマナンバンギセル
8768 オオナンバンギセル,シロバナオオナンバンギセル
8810 エゾバウフウ属,エゾボウフウ属
8811 エゾバウフウ,エゾボウフウ
8812 イヌエゾボウフウ
8818 ホソバエゾボウフウ
8826 イハミツバ,イワミツバ
8862 シオギリソウ,ツルオニシバ
8884 シオギリソウ
9021 エオニューム属
9051 クンビメイ
9058 カガミジシ
9091 エランギス属
9191 エランテス属
9271 エリデス属,ナコラン属,ナゴラン属
9297 フィリピンナコラン
9417 エスキナンツス属,エスキナントゥス属,ハナツルグサ属
9419 ナガミカツラ,ハゼカツラ,ハヅカシバナ
9475 ハナツルクサ
9489 クサネム属
9495 エダウチクサネム
9560 クサネム
9665 アメリカクサネム
9675 トチノキ属
9680 バチバナトチノキ,ベニバナトチノキ
9683 シナトチノキ,トチノキ,ナガミトチノキ
9690 エゾトチノキ
9692 キバナトチノキ
9701 ウマグリ,セイヨウトチノキ,マロニエ,ヨウシュトチノキ
9706 インドトチノキ
9720 アカバナアメリカトチノキ,アカバナトチノキ
9732 ウラケトチノキ,ケトチノキ,トチノキ
9733 ウラゲトチノキ,ケトチノキ
9779 エチフィルム属
9785 エチオネーマ属
9798 タイリンミヤコナズナ
9832 イヌニンジン
10145 アメリカウンランモドキ
10244 アガパンサス属,アガパンドゥス属,ムラサキクンシラン属
10245 ムラサキクンシラン
10284 アガペーテス属
10402 カハミドリ属,カワミドリ属
10407 シロバナカワミドリ
10408 タイワンカハミドリ
10414 カハミドリ,カワミドリ
10415 シロカワミドリ
10492 アガシス属,インヨウヌギ属
10494 カウリマツ,ナギモドギ,ニューシーランドマツ
10496 ダンマルジュ
10775 リュウゼツラン科
10778 アガーベ属,リュウゼツラン属
10787 アオノリュウゼツラン,アガベ,アヲノリュウゼツラン,マンネンラン
10788 リュウゼッサウ,リュウゼツラン
10794 シマリュウゼッ
10801 ニッセンロウ
10806 ハツミドリ
10829 キシモジン
10850 ササフブキ
10852 ミダレユキ
10854 ヘネケン
10855 フランツォシニイ
10874 シロフャリンリュウゼッ
10887 ハクジロ
10892 センイチヤ
10918 ライコウ
10925 ライジン
10926 フウライジン
10933 サケリョウゼッ
10940 サイザル,シサルアサ,シサルヘンペ
10942 フキアゲ
10943 フキアゲ
10945 テキーラーリュウゼツラン
10948 セイジロ
10955 ササノユキ
11154 マルバフジバカマ
11171 アメリカフジバカマ
11194 カッコウアザミ属,クワクカウアザミ属
11199 カゲラータム,カッコアザミ,カッコウアザミ,カッコオアザミ,クワクカウアザミ,ムラサキクワクカウアザミ
11206 オオカッコウアザミ,ムラサキクワクカウアザミ
11273 モラン属
11291 オオバジュラン
11293 グミトベラ
11294 ゴルテイア
11300 ジュラン,モラン
11329 アグラオネマ属,リョクチク属
11332 シラフイモ
11340 アグラオネマ,セスジゲサ
11341 ヒトスジグサ
11342 セスジグサ
11346 シモフリチク
11355 シラフチク,ナガバセスジゲサ
11359 アサヒリョクチク
11360 マルバセスジグサ
11362 リョクチク
11364 ハクチク
11397 アゴニス属
11542 キンミズヒキ属,キンミヅヒキ属
11543 アイノコキンミズヒキ
11546 チョウセンキンミズヒキ
11549 キンミズヒキ,キンミヅビキ,セイヨウキンミズヒキ
11566 ヒメキンミズヒキ,ヒメキンミヅヒキ
11572 キンミズヒキ,キンミヅヒキ,シベリアキンミズヒキ,ダウリアキンミズヒキ,ナンマンキンミズヒキ
11580 キンミズヒキ
11584 ウスゲキンミズヒキ
11587 ケキンミズヒキ
11591 ダルマキンミズヒキ
11609 サバクソウ属
11619 サバクソウ
11628 カモジグサ属
11630 ザラゲカモジ
11696 コムギダマシ
11710 ニセコムギダマシ
11938 ムギセンノウ属,ムギナデシコ属
11947 ムギセンノウ,ムギナデシコ
11968 コヌカグサ属,ヌカボ属
11970 バケヌカボ
11972 コヌカゲサ
12000 アリサンヌカボ
12008 ナンカイヌカボ
12025 タカネヌカボ,ヒメヌカボ
12034 イトコヌカグサ
12038 イベリアヌカボ
12053 ヤマヌカボ
12080 カスミヌカボ
12096 チシマヌカボ
12104 ヒメコメススキ,ミヤマヌカボ
12105 コモチミヤマヌカボ
12118 クロコヌカグサ,コヌカグサ
12127 ユキクラヌカボ
12139 エゾヌカボ,フユヌカボ
12145 ニヒタカヌカボ,ミヤオヌカボ,ミヤヌカボ
12147 サウザンヌカボ
12164 アフリカヌカボ
12186 ヌカボ,ヤマヌカボ
12196 コミヤマヌカボ
12217 カスミヌカボ
12221 クロコヌカグサ
12227 キタヤマヌカボ
12240 ヤマヌカボ
12285 エゾヌカボ
12323 ハイコヌカグサ
12369 タテヤマヌカボ
12407 ヒメコヌカグサ
12413 クロヌカボ,タカネヌカボ,チシマヌカボ
12442 アグロストフィルルム属,ヌカボラン属
12449 ヌカボラン
12517 シマミサオノキ,シマミサヲノキ,タイワンミサオノキ,タイワンミサヲノキ,ダシチャウ

12519 ミサオノキ,ミサヲノキ
12521 ミサオノキ
12558 シンジュ属,ニハウルシ属,ニワウルシ属
12559 シンジュ,ニハウルシ,ニワウルシ
12564 タイワンシンシュノキ,タイワンニハウルシ
12587 キカラパ
12599 モミジハグマ属,モミヂハグマ属
12600 モミジハグマ,モミヂハグマ
12602 オクモミジハグマ
12608 キッコウハグマ
12610 マルバキッコウハグマ
12611 リュウキュウハグマ
12623 テイショウソウ,テイショオソウ
12625 ヒロハテイショウソウ
12627 エンシュウハグマ
12628 マルバエンシュウハグマ
12629 ムラサキエンシュウハグマ
12633 ホソバハグマ
12635 マルバテイショウソウ
12636 マルバテイショオソウ
12656 アイノコハグマ
12676 アリサンハグマ,ナカハラハグマ
12678 オキナワハグマ
12679 ナカハラハグマ
12689 オオシマハグマ,ナガバハグマ
12694 ノウカウハグマ
12741 クサヤツデ
12771 ハリクジャクソン属,ハリクジャグヤシ属
12772 オビレハリクジャグヤシ
12773 アンチルハリクジャクヤシ
12774 トゲハリクジャクヤシ
12776 アイ-ラ属,ヌカススキ属
12789 アイグラス,マカススキ
12794 アイグラス,マカススキ
12795 ヌカススキ
12816 ハナヌカススキ
12818 ヒメヌカススキ
12911 サクロサウ科,ツルナ科,ハマミズナ科
13018 オオイワインチン
13051 キランサウ属,キランソウ属
13057 シマカコソウ,シマキランソウ
13063 ヤヘヤマジフニヒトヘ
13079 カイジンドウ,カヒジンドウ,キクバジフニヒトヘ
13088 カイジンドウ
13091 キランサウ,キランソウ,ジゴクノカマノフタ,ヂゴクノカマノフタ
13093 シロバナキランソウ
13094 モモイロキランソウ
13097 オニキランソウ,ケナシツルカコサウ,ケナシツルカコソウ
13104 ジフニヒトヘ
13114 ヒイラギソウ
13115 シロバナヒイラギソウ
13116 ベニバナヒイラギソウ
13126 オウギカズラ,オオギカズラ
13127 シロバナオウギカズラ
13131 ホリバルリカコソウ
13137 タイワンキランサウ,タイワンキランソウ
13140 タチキランソウ
13142 ジュウニキランソウ
13143 ルリカコソウ
13145 アキノルリカコソウ
13146 ジフニヒトヘ,ジュウニヒトエ
13147 シロバナジュウニヒトエ
13166 ヒメキランサウ,ヒメキランソウ
13174 セイヨウジュウニヒトエ
13184 ツルカコソウ
13185 ケブカツルカコソウ
13187 ヤエヤマキランソウ,ヤエヤマジュウニヒトエ
13191 ニシキゴロモ
13192 シロバナニシキゴロモ
13193 ツクバキンモンソウ
13194 ウドキランソウ
13208 アケビ属
13219 ホザキアケビ
13221 ゴエフアケビ,ゴヨウアケビ
13222 クワヅメアケビ
13225 アオアケビ,アケビ,アケビカヅラ,アヲアケビ,ハダッカヅラ,ハンダッカヅラ
13226 フタエアケビ
13231 シロバナアケビ
13238 マルハミツバアケビ,ミツバアケビ
13241 ホザキアケビ
13343 ウリノキ科
13345 ウリノキ属
13376 ウリノキ,モミジウリノキ
13378 ウリノキ
13379 ビロ-ドウリノキ
13380 ウリノキ
13381 シマウリノキ
13474 ネムノキ属
13565 ヒロハネム,ヒロハネムノキ
13578 カオカ,ゴウカンボク,ネブカ,ネブタ,ネブノキ,ネブリ,ネブリノキ,ネムノキ
13580 シロバナネムノキ
13586 オオバネムノキ
13595 オオバネム,オホバネム,ビルマガフクワン,ビルマゴウカン,ビルマネム,ビルマネムノキ
13611 タイワンネム
13635 カン
13649 タイワンネムノキ
13653 ヤエヤマネムノキ
13729 アルブ-カ属
13911 タチアオイ属
13934 オホアフヒ,カラアフヒ,タチアオイ,タチアフヒ,ツュアフヒ,ハナアオイ,ハナアフヒ
13951 アルケミラ属,ハゴロモグサ属
13966 ノミノハゴロモグサ
14065 セイヨウハゴロモグサ,ハゴロモグサ
14166 セイヨウハゴロモグサ
14177 アミガサギリ属,オオバベニガシワ属
14184 オオバベニガシワ
14199 キ-ルンアミガサギリ
14204 アミガサギリ
14217 アミガサギリ,オオバアカメガシワ,オオバベニガシワ
14271 ムジナモ属
14273 ムジナモ
14471 ソクシンラン属
14487 ネバリノギラン
14491 タイワンノギラン
14510 ノギラン
14529 ソクシンラン
14538 アブラギリ属
14544 ククイノキ,ケミリ
14631 マンナ
14719 オモダカ属,サジオモダカ属,ヘラオモダカ属
14725 ヘラオモダカ
14726 アズミノヘラオモダカ
14727 ホソバヘラオモダカ,ホソパヘラオモダカ
14734 ヒメオモダカ
14754 サジオモダカ
14760 サジオモダカ,トウゴクヘラオモダカ
14779 トウゴクヘラオモダカ
14790 オモダカ科
14862 イヌパココヤシ属
14871 アラマンダ属,アリアケカズラ属,アリアゲキヅラ属,ヘンデルギ属
14873 アリアケカズラ,アリアゲキズラ,アリアゲキヅラ,オオバナアリアゲキズラ,オホバナアリアゲキズラ,オホバナアリアゲキヅラ,ヘンデルギ
14874 アリアゲキヅラ,オオバナアリアゲキズラ,オオバナアリアゲキヅラ,ヘンデルギ
14876 コバナアリアゲキズラ
14881 ヒメアリアケカズラ,ヒメアリアゲキズラ
15018 アリュ-ム属,アルリウム属,ネギ属
15064 ヒメラッキョウ
15105 ニヒタカラッキョウ
15113 スナジニラ
15165 タマネギ,タマブギ
15170 ワケギ
15185 オオニラ,サトニラ,シナラッキョ

ウ,ラッキヤウ,ラッキョウ,ラムキョ
15198 キバナラッキョウ
15220 ルリネキ
15289 キ,ネギ,ネブカ,ヒトモジ
15291 フユキ,フユネギ,ワケギ
15292 イッボンネギ,オホネギ,オホネブカ,シモニタネギ
15293 オランダネギ,サンガイネギ,タウネギ,マンネンネギ,ヤブラネギ
15373 ステゴビル
15381 ヤマラッキヤウ,ヤマラッキョウ
15420 アサツキ,センブキ,センボンワケギ
15428 サキュウニラ
15432 カラフトラッキョウ
15445 アリウム,ギガンチウム
15450 チョウセンノビル,ノビル
15467 シベリアラッキョウ,シロウマアサツキ
15480 キバナノギャウジャニンニク,キバナノギョウジャニンニク
15486 ヒメアマナ,ヒメニラ,ヒメビル,メメニラ
15534 ニラ
15550 ベニバナニラ
15621 タマブキ,ニラネギ,ホロ,リーキ,リーク
15632 モウコラッキョウ
15690 チョウセンヤマラッキョウ,テマリラッキョウ
15698 オオニンニク,オオビル,ニンニク
15709 アサツキ,エゾネギ
15715 シロウマアサツキ
15720 ヒメエゾネギ
15721 シロバナヒメエゾネギ
15726 ニンニク
15734 セッカヤマネギ
15760 ミヤマラッキョウ
15779 キョウ,ミヤマラ
15813 チゴラッキョウ,ヒメラッギョウ
15822 サンカクニラ,タマムラサキ,ムラサキビル,ヤマラッキヤウ,ヤマラッキョウ,ヤラニラ
15843 エゾラッキョウ,ニラ,ミラ
15868 キトビル,ギャウジャニンニク,ギョウジャニンニク,ゼンジャウ,ヤマビル
15874 ギョウジャニンニク
15953 オガサハラマツ,オガサワラマツ,モクマオウ,ヲガサハラマツ
16006 アロモルフィア属
16039 ムラサキフロックス
16043 アカギモドキ属
16155 アカギモドキ
16174 アロプレクタス属,ビロトイワギリ属,ビロードイワギリ属
16184 ビロードイワギリ
16231 ハネキビ
16291 エゴハンノキ属
16295 シマエゴノキ,ハイナンエゴノキ,ハンノキバエゴノキ,ハンノハエゴノキ,フオリエゴノキ
16304 エゴハンノキ,ハンノキエゴノキ
16309 ハンノキ属
16310 アイノコヤシャブシ
16311 イワキハンノキ
16312 モリオカハンノキ
16313 タルミヤシャブシ
16314 ウラジロカワラハンノキ
16328 ミヤマハンノキ
16329 オオバミヤマハンノキ
16330 カラフトミヤマハンノキ
16339 ミヤマカワラハンノキ
16341 ヤシャブシ
16342 ミヤマヤシャブシ
16345 タイワンハンノキ
16348 オクヤマハンノキ,マンシュウハンノキ
16352 オウシュウクロハンノキ,セイヨウヤマハンノキ,ヨーロッパハンノキ
16359 ケシベリヤハンノキ,ケヤマハンノキ,ヤマハンノキ
16362 シベリアハンノキ,ヤマハンノキ
16384 タニガワハンノキ
16386 ハリノキ,ハンノキ,ヤチハンノキ
16389 エゾハンノキ,ハリノキ,ハンノキ,ヤチハンノキ
16393 ケハンノキ
16401 マンシウハンノキ
16412 ヤハズハンノキ
16415 ウスゲヒロハハンノキ
16416 ヒロハハンノキ
16421 ネパールハンノキ
16433 ガケシバリ,ツチシバリ,ハゲシバリ,ヒメハンノキ,ヒメヤシャブシ,ヤシャ,ヤマシバリ
16447 カワックナベ,カワハラハンノキ,カワラハンノキ,ネバリハンノキ,メハリノキ
16448 ケカワラハンノキ
16458 オオバヤシャブシ,オホバヤシャブシ
16466 エゾヤマハンノキ,ケハリノキ,ケヤマハンノキ,ヤマハンノキ
16469 マルバハンノキ,ヤマハリノキ,ヤマハンノキ
16470 サクラバハンノキ
16479 キョクホクミヤマハンノキ
16484 アロガーシア属,クハズイモ属,クワズイモ属
16489 アイノコクワズイモ
16492 ヤエヤマクワズイモ
16494 アマゾンダコ
16495 シマクハズイモ,シマクワズイモ,タイワンクハズイモ,タイワンクワズイモ
16496 キッコウダコ
16501 ハスイモ
16507 シロスジクワズイモ
16508 カブトダコ
16509 キングハニシキ,テイオウハイモ,ナガバクワズイモ
16510 オオバテテイオウハイモ
16511 シロフイリクヮズイモ
16512 イシイモ,インドクワズイモ,クハズイモ,クワズイモ,ドクイモ,バシガイモ,マンシウイモ
16518 クワズイモ
16525 コウライダコ
16526 シュスジコウライダコ
16535 トラフイモ
16540 アロエ属,ロクワイ属
16592 キダチアロエ,キダチロクワイ
16598 アロエ,キダチロカイ
16605 アヤニシギ
16617 タイザンニシキ
16657 フシチョウ
16687 ニンギョウニシキ
16768 オオジニシキ
16817 アオワニ,ロクワイ
16894 ミカドニシキ
17017 オニキリマル
17136 チヨダニシキ,フイリロクワイ
17163 ソコトラアロエ
17175 ゴジュウノトウ
17289 シッポウニシキ
17381 バルバドスアロエ
17383 クサアタン,タウアダン,タウロクワイ,ロクワイ
17424 アロエ科
17433 アロイノプシス属
17445 テンニョショウ
17447 テンニョウン
17452 ハナニシキ
17459 テンニョノマイ
17475 アロンソア属
17476 タイリンヒメジギタリス,ヒメジギタリス
17479 タイリンヒメジギタリス
17481 ホソバベニゴチョウ
17490 ベニコチョウ
17497 スズメノテッパウ属,スズメノテッポウ属
17498 スズメノテッポウ,ノハラスズメノテッポウ
17501 スズメノテツバウ,スズメノテッポウ
17512 マンシウスズメノテッポウ
17515 コウアンスズメノテッポウ
17521 ヒメスズメノテッポウ
17525 スズメノテッパウ,スズメノヤリ,ヒヨヒヨグサ,ヤリクサ,ヤリンボ
17537 セトガヤ
17548 ノスズメノテッポウ

17553 オオスズメノテッポウ,オホススズメノテッパウ
17621 カラビオブ
17639 アルピニア属,ハナミョウガ属,ハナメウガ属
17640 ツクシハナミョウガ
17641 クマタケゲットウ
17656 アオノクマタケラン,アヲノクマタケラン
17666 ナガボノゲッタウ
17671 イリオモテクマタケラン
17674 クマタケラン,コウシュンゲッタウ,タイワンゲッタウ,タイワンゲットウ,ホクトクマタケラン
17675 フイリクマタケラン
17677 サンキョウ,ナンキャウサウ
17693 アオノクマタケラン,キールンクマタケラン,ヒロハクマタケラン
17695 ハナミョウガ,ハナメウガ
17696 キミノハナミョウガ
17703 カハカミゲッタウ
17711 フイリクマタゲラン
17712 グッシャクゲツタウ
17722 カッパンクマタクラン
17724 イオウクマタケラン
17728 チクリンカ
17733 カウリャウキャウ,カウリョウキョウ,コウリョウキョウ
17736 ヤクチ
17741 ブイサンハナメウガ,プライスハナメウガ
17751 タグスイゲッタウ,タグスヰゲッタウ
17754 シチセイハナメウガ
17766 イシガキクマタケラン
17771 セイタカフイリゲットウ
17772 タイリンゲッタウ,タイリンゲットウ
17774 オホクマタケラン,ゲッタウ,ゲットウ,サニン
18060 アルストレメ－リア属,エリズイセン属
18062 キバナユリズイセイ
18074 アルストロメリア ペレグリナ
18076 ユリズイセン
18089 ツルノゲイトウ属
18095 モヨウビユ
18124 ケツルノゲイトウ
18128 ナガエツルノゲイトウ
18135 マルバツルノゲイトウ
18147 ツルノゲイトウ,ホシノゲイトウ,ホシバツルノゲイトウ,ホソバツルノゲイトウ
18155 タチアオイ属,タチアフヒ属,ビロードアオイ属
18157 シチゴサンアオイ
18173 アルテヤ,ウスベニアルテア,ウスベニタチアフヒ,タチアオイ,ビロードアオイ
18254 ブル－ハイビスカス
18257 ササハギ属
18259 オホバタケハギ,ナガバタケハギ
18273 フシナシササハギ
18289 ササハギ,マルバタケハギ
18313 イワナズナ属,ニハナヅナ属,ニワナズナ属
18317 ミヤマナズナ,ミヤマニワナズナ
18327 アレチナズナ
18357 ステップナズナ
18405 モウコナズナ
18433 ヤマナズナ,ヤマニワナズナ
18469 シベリアナズナ
18504 シマテイカカズラ
18568 アマナ属
18569 アマナ属,ムギグワイ,ムギグワヰ
18646 ヒユ科
18652 ヒユ属
18655 シロビユ,ヒメシロビユ
18661 ヒメアオゲイトウ
18668 アメリカビユ,イヌヒメシロビユ
18670 イヌビユ,ノビユ,ハゲイトウ
18687 センニンコク,ヒモゲイトウ,ヒモズイトウ
18692 サジビユ
18701 ハイビユ,ハヒビユ
18734 ホソアオゲイトウ
18752 シロミセンニンコク
18765 イヌビユ
18776 ヒユ,ヒユナ
18779 ハゲイトウ
18786 アレチアオゲイトウ
18787 オオホナガアオゲイトウ
18788 スギモリゲイトウ,フヂゲイトウ
18795 ホソアオゲイトウ,ホソアヲゲイトウ
18804 イガホビユ
18810 アオゲイトウ,アオビユ,アヲビユ
18822 ハリビユ,ヒユナ
18836 ハゲイトウ,ヒユ
18847 ヒユモドキ
18848 アオビユ,アヲビユ,ホナガイヌビユ
18861 ヒガンバナ科
18862 アマリリス属
18865 シンセイアマリリス,ホンアマリリス
18908 アムソニア属
18946 アンベルボア属
18966 サルタン,スイ－ト,ニオイヤグルマギク
19045 ヒメルリザクラ
19140 ブタクサ属,ブタグサ属
19145 ブタクサ
19179 ブタクサモドキ
19191 オオブタクサ,クワモドキ
19240 ザイフリボク属
19241 アメリカザイフリボク
19248 ウネモジリ,ザイフリボク,シデザクラ,シデヤナキ,ニレザクラ,ヤグラバナ,ヤマムロ
19357 ウラジロイヌガヤ,ウラジロマキ
19363 タイワンウラジロイヌガヤ
19413 アメシエラ属
19458 ルリハクカ属,ルリハッカ属
19459 ルリジソ,ルリハクカ,ルリハッカ
19462 ヨウラクボク属
19463 ヨウラクボク
19491 ヤンバルミョウガ
19493 ヤンバルミョウガ
19498 アミトスチグマ属,グンダイモチヅミ属,ヒナラン属
19512 クモラン,ヒナラン,ヒメイハラン,ヒメイワラン
19513 シロヒナラン
19515 イワチドリ
19516 コアニチドリ
19518 オキナワチドリ
19550 ヒメミソハギ属
19554 アメリカミソハギ,ナンゴクヒメミソハギ
19566 シマミソハギ,ナガトミソハギ
19575 ホソバヒメミソハギ
19602 ヒメミソハギ
19659 ドクゼリモドキ属
19664 ドクゼリモドキ
19690 カイザイク属
19691 カイザイク
19773 オオハマガヤ
19817 アモマム属
19839 サウグワ
19933 シュクシャミツ
19952 アモラギ属
19989 クロバナエンジュ属
20005 イタチハギ,イヌアイュンジュ,イヌアヰュンジュ,クロバナエンジュ,クロバナクララ,ロシャハギ
20037 コンニャク属,ヒドロスメ属
20080 オニコンニャク
20088 ヘンリ－イモ,ヤマコンニャク
20090 ケコンニャク
20098 ヤマコンニャク
20125 イシウスイモ
20132 コニャク,コンニャク,コンニャクイモ,ヘビイモ
20145 ショクダイオオコンニャク
20280 ノブダウ属,ノブドウ属
20284 オフクカズラ,オフクカヅラ
20321 テリハノブドウ,テリハノブドオ,ノブドオ
20327 ウドカズラ,ウドカヅラ,シマウドカヅラ,タイワンウドカヅラ
20329 ウドカズラ

20348 ザトウエビ,ノブドウ,マンシウノブドウ
20352 テリハノブドウ
20354 ウマブダウ,ノブダウ,ノブドウ,ノブドオ
20400 アッバノブドウ,テリハノブドウ,ノブドウ
20402 ミッバノブドウ
20408 カガミグサ,ビヤクレン,ヤマカガミ
20558 ヤブマメ属
20566 ヤブマメ
20570 ヤブマメ
20837 ワルタビラコ
20843 ハリゲタビラコ
20844 チャウジサウ属,チョウジソウ属
20846 ホソバチョウジソウ
20850 チャウジサウ,チョウジソウ
20860 ヤナギチョウジソウ,ヤナギバチョウジソウ
20866 エビブレムノブシス属
20890 アメンダウ,アメンドウ,アーモンド,ヘントウ
20892 ビター・アーモンド
20897 スイート・アーモンド,ヘントウ
20908 サントウ,ノモモ,ハヤザキモモ
20935 モモ
20952 バントウ
20955 ズバイモモ,スバシモモ,ツバシモモ,ネクタリン
20970 オヒョウモモ,オヒョモモ,コオヒョウモモ,ユエフバイ,ユヨウバイ
20974 オヒョウモモ
21068 アナカンプセロス属
21097 ハナフブキ
21117 キンショウジョ
21124 グンサン
21136 フブキノマツ
21190 ウルシ科
21191 アナカルディウム属
21195 カシュ-,カシュウナット,カシュ-ナットノキ
21335 ルリハコベ属
21339 アカバナルリハコベ,ルリハコベ
21340 アカバナルリハコベ,ルリハコベ
21472 アナナス属,パイナップル属
21479 アナナス,パイナップル,パイン,パインアップル
21506 ヤマハハコ属,ヤマホオコ属
21580 シナタカネヤハズハハコ
21586 タカネウスユキソウ,タカネヤハズハハコ
21596 ヤマハウコ,ヤマハハコ,ヤマホオコ
21607 ホソバノヤマハハコ
21616 ホソバノヤマハウコ,ホソバノヤマハハコ,ホソバノヤマホウコ,ホソバノヤマホオコ
21630 ニイタカヤマハハコ,ニヒタカヤマハハコ,ホソバノヤマハハコモドキ
21643 コダマギグ,ニヒタカウスユキ
21645 コダマギク
21676 ノウコウウスユキソウ
21682 ヤハズハハコ,ヤバネホオコ
21683 ニオイヤハズハハコ
21692 タンナヤハズハハコ
21697 クリヤマハハコ
21698 ヤクシマウスユキソウ
21731 アラレギク,カハラハウコ,カハラハハコ,カワラホオコ
21804 アンザンジュ属
21805 アンザンジュ,テマリグサ
21914 アンク-サ属,アンチュ-サ属,ウシノシタグサ属
21920 アラゲムラサキ,アレチウシノシタグサ
21933 アフリカワスレナグサ
21949 アンチュ-サ,ウシノシタグサ
21959 アルカネット
22017 アンシストロカクタス属
22037 サワルリソウ属
22038 サワルリソウ
22039 シロバナサワルリソウ
22040 アンキストロキラス属,アンシストロキルス属
22047 アンシストロクラヅス科
22320 ミスズラン属
22326 ミスズラン
22418 ヒメシャクナゲ属
22445 ニッコウシャクナゲ,ヒメシャクナゲ
22450 カラフトヒメシャクナゲ
22451 シロバナノヒメシャクナゲ
22476 ウシクサ属,ヒメアブラススキ属,モロコシ属
22836 カウスイガヤ,カウスヰガヤ,シトロネグラス,シトロネブラス
23077 メリケンカルカヤ
23107 アンドロサセ属,チシマザクラ属,トチナイソウ属
23139 イワハナガタ
23170 コンガコザクラ,サカコザクラ
23182 ヒナコザクラ
23202 サイトウソウ
23209 ユキハナガタ
23211 ギンハナガタ
23213 チシマコザクラ,チシマザクラ,トチナイサウ,トチナイソウ
23219 ホソバコザクラ
23281 ツルハナガタ
23294 ハリコザクラ
23323 トチナイソウ,リウキウコザクラ,リュウキュウコザクラ
23669 ハナスゲ属
23670 カラスノススキ,チモ,ハナスゲ,ヤマシ,ヤマトコロ
23696 アネモネ属,イチゲソウ属,イチリンソウ属,オキナグサ属,ニリンソウ属
23701 カザフキサウ,キクザキイチゲサウ,キクザキイチゲソウ,キクザキイチリンサウ,キクザキイチリンソウ,ルリイチゲサウ,ルリイチゲソウ
23705 ヒロハヒメイチゲ,ヤチイチゲ,ヤブイチゲ
23709 ヒロハイチゲ
23713 コウライニリンソウ,マンシウイチゲ
23724 ハナアネモネ
23745 アメリカオキナグサ
23767 アネモネ,ハナイチゲ,ベニバナオキナグサ,ボタンイチゲ
23773 ダウリアイチゲ
23780 ヒメイチゲ,ヒメイチゲソウ
23796 アウシキナ,オウシキナ,オホエゾイチゲ,フタマタイチゲ
23817 ガジャウサウ,ガショウソウ,クワジャウイチゲ,ニリンサウ,ニリンソウ,フクベラ,フクベライチゲ
23853 コタイワンシウメイギク
23854 アキボタン,キブネギク,シウメイギク,シュウメイギク
23868 イチゲソウ,ユキワリイチゲ,ルリイチゲ
23901 トウハクサンイチゲ,ハクサンイチゲ
23917 エゾノハクサンイチゲ,カラフトセンクワサウ
23925 ヤブイチゲ
23936 イチゲソウ,イチリンソウ
24013 アズマイチゲ,アヅマイチゲ,ウラベニイチゲ
24024 サワイチゲ
24066 サンリンサウ,サンリンソウ,トキハイチゲ
24071 オオバナイチゲ,バィカイチグ,マツユキオキナグサ
24103 バイカイチゲ
24105 モリイチゲ
24109 ハルオキナグサ
24116 シマキブネギク,タイワンシウメイギク,タイワンシュメイギク,タカサゴイチゲ
24128 エゾイチゲ,ヒロハノイチゲ
24166 レンゲショウマ属
24167 キレンゲショウマ
24177 レンゲショウマ属
24208 イノンド属
24213 イノンド,ヒメウイキャウ,ヒメウイキョウ
24281 シシウド属,シラネセンキウ属
24282 タウキ,トウキ
24283 ツクバトウキ
24284 ミヤマトウキ

24290 ホッカイトウキ
24293 アメナレノダケ,エゾオオヨロイグサ,エゾオホヨロヒグサ,エゾノヨロイグサ,エゾノヨロヒグサ
24307 キビノノダケ
24313 ヒメノダケ
24314 ヒメノダケ
24315 コウライヒメノダケ
24322 ミヤマノダケ
24323 ツクシミヤマノダケ
24325 エゾセンキウ,オオシシウド,オホシシウド,ヨロイグサ
24326 タイワンオニウド,タイワンシシウド
24336 ウダケ,コマゼリ,ゼンゴ,ノゼリ,ノダケ
24337 シロバナノダケ
24339 ウスイロノダケ
24340 シロバナノダケ
24341 ホソバノダケ
24345 アマニュウ,マルバエゾニュウ
24355 エゾオオバセンキュウ,エゾオホバセンキク,オオバセンキュウ,オホバセンキク
24357 ホソバエゾセンキュウ
24358 オニノダケ
24366 イワニンジン
24372 ナンゴクハマウド
24376 ハナビゼリ
24380 オニウド,ハマウド
24381 ムニンハマウド
24384 アシタバ,ハチジョウソウ
24396 ツクシゼリ
24397 ヒナボウフウ
24407 ヒュウガセンキュウ
24411 モリゼンゴ
24413 ニチタカシシウド
24432 シシウド
24433 シラネセンキュウ,シラペセンキウ,スズカゼリ,ヤマセンキウ
24434 ホソバシラネセンキュウ
24436 コバノヨロイグサ
24440 ツクシトウキ
24441 ウドダラシ,シシウド,タカオキョウカツ,タカヲキャウクワツ
24445 ミヤマシシウド
24453 エゾノヨロイグサ
24455 ケエゾノヨロイグサ
24456 ミチノクヨロイグサ
24457 ホソバノヨロイグサ
24461 イシヅチボウフウ
24466 イヌトウキ
24468 クマノダケ
24480 ホソバトウキ
24481 トカチトウキ
24484 タロコシシウド,タロコタウキ
24485 カワゼンゴ
24486 ヒュウガトウキ
24497 ウバタケニンジン
24498 オオウバタケニンジン
24500 エゾニュウ
24508 ヤクシマノダケ
24509 トサボウフウ
24519 アンゲロンソウ属
24522 ホソバアンゲロンソウ
24523 アンゲロンソウ
24525 ヤナギバアンゲロン
24687 アングレカム属,フウラン属
25134 アングロア属,アンゲローア属
25216 アニゴザンッス属
25316 ネコアサガオ
25323 ナガバアサガオ
25331 ヒロハノコヌカグサ属
25343 ココメイチゴツナギ,ヒロハノコヌカグサ
25346 ヒロハノコヌカグサ
25439 サンブンサン
25466 ブソロイバナ属
25771 ナガバモクコク属
25779 ナガバモクコク
25828 バンレイシ属
25836 チエリモヤ
25850 イラマ
25866 ホシババンレイシ,マウンティン－サワ－ソップ
25868 オランダドリアン,シャウシャップ,トゲバンレイシ
25880 ソウコヤ,ソウゴヤ
25882 ギウシンリ,ギュウシンリ
25898 バンレイシ
25909 バンレイシ科
25922 ヤノネアオイ属
25923 ニシキアオイ
25924 ニシキアオイ
25926 ヤノネアオイ
25927 サカキカズラ属,サカキカヅラ属
25928 クチナシカズラ,サカキカズラ,サカキカヅラ,ニシキカズラ,ニシキラン
25932 オオサカキカズラ,オオサカキカヅラ
25933 オホサカキカヅラ
25959 アネクトキールス属,キバナシュスラン属,タイワンシュスラン属
25980 キバナシュスラン
25986 コウシュンシュスラン,コウシュンスラン
26003 キバナシュスラン
26264 アンレデラ属
26265 アカザカズラ,マデイラカズラ
26273 アンセリァ属,ジョン－アンセル属
26293 アンテギッバエウム属
26299 エゾノチチコグサ属
26385 エゾノチチコグサ
26393 ヒツナヴ
26579 モモイロマットギク
26624 ミズヒキ,ミヅヒキ
26626 シンミズヒキ,シンミヅヒキ
26738 アンセミス属,ロウマカミツレ属,ローマカミツレ属
26746 キゾメカミツレ,キソメカミルレ,ノカミツレ
26768 カミツレモドキ
26900 カウヤカミツレ,コウヤカミツレ,コウヤカミルレ
26950 アンスリュム属,アンセリカム属,アンテリクム属,ベニウチワ属
27664 アントリ－ザ属
27930 ハルガヤ属
27934 ヒメハルガヤ
27942 ヒメコウボウ
27947 タイワンハルガヤ
27949 タカネコウボウ
27963 コウボウ,タカネコウボウ,ミヤマカウバウ,ミヤマコウボウ
27967 カウバウ,コウボウ,セイヨウコウボウ
27969 ハルガヤ,ミヤマハルガヤ
27971 ケナシハルガヤ
28013 アンスリスクス属,シャク属
28019 ケジャク
28023 ノハラジャク
28030 オニジク,オニジャク,コシャク,ジャク
28044 コジャク,シャク,ヤマニンジン
28046 オニジャク,ケジャク
28071 アンスリュ－ム属,ベニウチハ属,ベニウチワ属
28084 アンスリユム,オオベニウチワ,オホベニウチハ
28093 シロシマウチワ
28103 ハランウチハ,ハランウチワ
28107 ホソバハランウチワ
28108 ビロ－ドウチワ
28113 ヤツデウチワ
28119 ベニウチハ,ベニウチワ,モモイロフイリイモ
28130 ホソバウチワ
28132 ミミガタウチワ
28133 フナガタウチワ
28134 ズブシグナ－ツム
28138 ビロ－ドオオウチワ
28140 ナガバオオウチワ,ナガバオオベニウチワ,フイリイモ
28200 クマノアシツメクサ
28248 ウパス,ウバスノキ
28309 ヤマヒハツ属
28316 サラマンド
28341 ウゲヨシ,コウシュンヤマハヅ,シマヤマヒハツ,ヤマッバ,ヤマヒハツ
28343 ナントウヤマハヅ
28349 コウトウセイシボク,コウトウヤマ

イハヅノキ
28378 コウトウヤマヒハツ
28381 カイガンセイシボク,コウトウヤマヒハツ,コオトオヤマヒハツ,ヤマハヅモドキ
28418 アサヒカズラ属,ニトベカヅラ属
28422 アサヒカズラ,アサヒカヅラ,ニトベカズラ,ニトベカヅラ
28470 ヤヨイ
28576 キンギョサウ属,キンギョソウ属
28617 キンギョサウ,キンギョソウ
28631 アレチキンギョソウ
28932 アパテシア属
28971 ホソセイヨウヌカボ
28972 セイヨウヌカボ
28997 アモラギ
29004 ムクノキ属
29005 オムク,ムクエノキ,ムクノキ
29121 アフェランドラ属,キンヨウボク属
29125 キンヨウボク
29212 タネガシマムエフラン属,タネガシマムヨウラン属
29216 タネガシマムエフラン,タネガシマムヨウラン
29230 セリ科
29295 ホドイモ属,ホド属
29298 アメリカホド,アメリカホドイモ
29304 カナホド,ツチグリ,フド,ホド,ホドイモ
29307 タイワンホドイモ
29312 アーピウム属,アピューム属,オランダミツバ属,オラン属,マツバゼリ属
29325 リズサンザシ
29326 レースソウ
29327 オランダミツバ,キョマサニンジン,セリニンジン,セルリー,セレリー,セロリ
29328 セロリ
29329 セルリアック
29422 ヒメガルカヤ属
29430 オキナハガルカヤ,オキナワカルカヤ,ヒメガルカヤ
29462 キョウチクトウ科,ケフチクタウ科
29465 バシクルモン属
29546 エキサイゼリ
29643 アホノゲトン属,アポノゲトン属
29660 ケープヒルムシロ,ミズサンザシ
29663 レースソウ
29673 ウリカワヒルムシロ
29676 レースソウ
29696 レースソウ科
29712 アポロカクタス属,ヒモサホテン属
29715 キンヒモ,ヒモサボテン
29716 キンヒモ
29787 ヤクシマラン属
29794 ヤクシマラン
29796 ヤクシマラン科
29807 アッペンディキュラ属,ヒメタケラン属
29814 コウトウタケラン,ヒメタケラン
29853 アプテニア属,ハナズルソウ属,ハナヅルソウ属
29855 ハナズルソウ,ハナツルクサ,ハナヅルサウ,ハナヅルソウ
29968 モチノキ科
29973 キャラ,ジンコウ
29977 マレージンコウ
29994 アキレジア属,オダマキ属,ヲダマキ属
29997 アキレギアーアルピナ
30005 ヤマオダマキ
30007 アメリカオダマキ,アメリカヲダマキ
30017 キバナォダマキ,キバナヲダマキ
30031 イトクリ,イトクリサウ,オダマキ,ヒメオダマキ,ミヤマオタマキ,ムラサキオダマキ,ムラサキヲダマキ,ヲダマキ
30032 ホウカゾウ
30043 ミヤマォダマキ,ミヤマヲダマキ
30061 エゾノヤマオダマキ,オオヤラオダマキ
30062 キバナオオヤラオダマキ
30068 カラフトヒナオダマキ
30077 ウスキオダマキ
30078 クロバナオダマキ
30081 アメリカオダマキ,アメリカヲダマキ,セイヨウオダマキ
30090 トウヤマオダマキ
30105 シロイヌナズナ属,シロイヌナヅナ属
30111 ハクサンハタザオ
30121 ミヤマハタザオ
30146 シロイヌナズナ,シロイヌナヅナ
30160 ハタザオ属,ハタザホ属
30164 ニイタカハタザオ,ニヒタカハタザオ,ニヒタカハタザホ,ユキハタザオ
30220 ニワハタザオ
30229 ウスベニハタザオ
30263 スズシロサウ,スズシロソウ
30264 ケスズシロソウ
30265 カワチスズシロソウ
30272 ツルタガラシ,ハクサンハタザオ,ハクサンハタザヲ
30275 イブキハタザオ
30286 ハクサンハタザオ,ハクサンハタザホ,マルバハタザオ,マルバハタザホ
30292 ノハタザホ,ヤマハタザオ,ヤマハタザホ
30337 ヘラハタザオ
30342 ミヤマシロイヌナヅナ,ミヤマハタザオ,ミヤマハタザホ
30345 ナヅナハタザオ,ナヅナハタザホ,ミヤマハタザオ
30363 ウスベニハタザオ
30387 エゾハタザオ,エゾハタザホ
30435 フジハタザオ,フジハタザヲ
30437 イハハタザホ,イワハタザオ,エゾノイハハタザホ,エゾノイワハタザオ,コイハハタザホ
30439 イワハタザオ
30441 イワテハタザオ
30442 ケナシイワハタザオ
30444 ウメハタザオ
30447 シコクハタザオ
30456 ハマハタザオ
30458 キールンハタザホ,ハマハタザオ,ハマハタザヲ
30461 ケナシハマハタザオ
30462 ベニバナハマハタザオ
30467 クモマナズナ
30491 サトイモ科,テンナンショウ科
30494 ナンキンマメ属
30498 オニマメ,ダウジンマメ,タウマメ,ナンキンマメ,ラッカセイ
30526 アラクナンテ属,アラクニス属
30569 アレオコックス属
30587 タラノキ属
30593 ウラジロタラノキ
30618 マンセンウド,メダウ
30619 ウド
30620 オオバウド
30622 カラフトウド
30628 タイワンタラノキ
30634 ウドモドキ,オニダラ,タラノキ,マンシウダラ,マンシュウダラ
30638 フクリンタラノキ
30640 キモンタラノキ
30643 メダラ
30655 メダラ
30675 ミヤマウド
30749 リュウキュウタラノキ
30750 シチトウタラノキ
30753 カラフトウド
30760 タラノキ
30800 ウコギ科
30830 ナンヤウスギ属,ナンヨウスギ属
30831 バラナマツ,パラナマツ,フサナンヨウスギ,ブラジルアラウカリア,ブラジルマツ
30832 アメリカウロコモミ,チリーマツ,チリマツ,モリアラウカリア
30835 カウエフザヰバノウロコモミ,カウエフザンバノウロコモミ,ヒモハノナンヨウスギ,ヒロハノナンヤウスギ,ヒロハノナンヨウスギ
30839 シマナンヤウスギ
30841 スギバノウロコモミ,ナンヤウスギ,ナンヨウスギ
30846 シマナンヨウスギ
30848 コバノナンヤウスキ,コバノナンンヨ

ウスキ,コバノナンヨウスギ,シマナンヤウスギ,シマナンヨウスキ,ノーフオークマツ,ノルホオクアラウカリア
30852 クリンキパイン
30857 ナンヨウスギ科
30860 チョウトリカズラ属
30862 チョウトリカズラ
30877 アルブス属,アルブッス属,アルブートゥス属
30888 アルブスウネドウ,イチゴノキ
30932 アンゼリカ
31012 ユスラヤシ属
31013 ユスラヤシ
31015 ユスラヤシドキ
31038 コメバツガザクラ,ハマザクラ
31044 ゴバウ属,ゴボウ属
31051 ゴバウ,ゴボウ
31079 シバヨナメ属
31080 シバヨナメ
31142 クマコケモモ
31162 アフリカヒマワリ,ワタゲハナグルマ
31169 ワタゲツルハナグルマ
31176 アークトチス属,アークトティス属,ハゴロモギク属
31178 ヒメアフリカギク
31294 ァークトチス,アフリカギヶ,ハゴロモギヶ
31311 ウラシマツツジ属
31314 ウラシマツツジ
31316 ウラシマツツジ,クマコケモモ
31322 アカミノウラシマツツジ,アカミノクマコケモモ
31347 マンリャウ属,マンリョウ属,ヤブカウジ属,ヤブコウジ属
31380 シナタチバナ,シナヤブカウジ,シナヤブコウジ,シナヤブコオジ
31387 アリサンカウジ,アリサンマンリャウ,オニタチバナ,クラルカウジ
31388 コバノマンリャウ,ニヒタカタチバナ,ニヒタカマンリャウ
31396 オオミマンリョウ,ハナタチバナ,マンリャウ,マンリョウ
31400 シロミノマンリョウ
31402 キミノマンリョウ
31408 カラタチバナ,コウジ,タチバナ,マンリャウ,マンリョウ
31410 シロミタチバナ
31411 キミタチバナ
31414 コバタチバナ
31471 カラタチバナ
31477 アカダマノキ,ヤブカウジ,ヤブコウジ,ヤブタチバナ,ヤマタチバナ
31479 シラタマコウジ
31480 ホソバヤブコウジ
31489 オホバマンリャウ,クスクスカウジ
31502 カラタチバナ,コウジ,タチバナ,タチバナマンリョウ
31527 オオツルコウジ,オオツルコオジ
31571 ツルカウジ,ツルコウジ
31574 リュウキュウツルコウジ
31578 シシアクチ,ミヤマアクチ,ミヤマシシアクチ
31599 アクチノキ,モクタチバナ
31600 クロミノモクタチバナ
31601 アカミノモクタチバナ
31613 アリサンカウジ,アリサンマンリャウ,オニタチバナ,クラルカウジ
31630 ツルカウジ
31677 アレカ属,ビンラウジ属,ビンロウジュ属
31680 アレカヤシ,ビンォウジ,ビンロウ,ビンロウジ,ビンロウジュ,ビンヲウジ
31702 ジョオウヤシ属
31705 ギリバヤシ,ジョオウヤシ
31707 ピンドジョオウヤシ
31727 ノミノツヅリ属
31742 エゾタカネツメクサ
31744 タカネツメクサ
31745 レブンタカネツメクサ
31767 ハイユキソウ
31787 カラフトツメクサ,ヌンシュウツメクサ
31795 ハリツメクサ
31934 オオユキソウ
31967 イトフスマ
31968 ヒメイトフスマ
31969 ケナシイトフスマ
31977 カトウハコベ
31978 アポイツメクサ
32004 バイカツメクサ
32015 ノミノツヅリ
32056 ミヤマツメクサ
32057 エゾミヤマツメクサ
32070 メアカンフスマ
32085 ハナフスマ
32136 シマノミノツヅリ,タイワンノミノツヅリ
32212 ノミノツヅリ
32222 ネバリノミノツヅリ
32301 コケツメクサ,ヤマハルユキソウ
32306 ホソバツメクサ
32330 クロツグ属
32336 クロツグ,ツグ,マアニ,ヤマシュロ
32343 サゴバイ,サトウマツ,サトウヤシ
32346 コダネクロツグ
32350 アレニフェラ属
32357 アレクイパ属
32361 サイオウギョク
32362 スイビギュク
32366 スイユウギュク
32371 アレッーサ属
32411 アザミゲシ属
32423 オオバナアザミゲシ,オホバナアザミゲシ
32424 シロアザミゲシ
32429 アザミゲシ,デロジョウ
32525 イリオモテソウ
32563 キダチカミツレ,マーガレット,モクシュンギク,モクシュンク
32600 オオバアサガオ属,オホバアサガホ属
32642 オホバアサガホ,ギンヨウアサガオ
32689 アルギロデルマ属
32691 ヘキレイ
32696 ヘキカンギョク
32712 ソウビギョク,ホウツイギョク
32723 シンショウギョク
32733 ギンレイ
32743 シュゥビギョク
32750 アカバナキンレイ
32959 ギンケンソウ属
32960 ギンケンソウ
32988 アズキナシ,アヅキナシ,カタスギ,ハカリノメ
33008 ウラジロノキ
33010 キミノウラジロノキ
33011 コモノウラジロノキ
33197 アリクリヤシ属
33200 アリクリヤシ
33206 アリオカルプス属
33209 キッコウボタン
33221 サンカクボタン
33234 テンナンシャウ属,テンナンショウ属
33236 ツルギテンナンショウ
33237 ヒガンマムシグサ
33245 アムールテンナンショウ
33248 ムラサキアムールテンナンショウ
33258 ホソバテンナンシャウ,ホソバテンナンショウ
33260 カウライテンナンシャウ,コウライテンナンショウ
33264 アリサンムサシアブミ
33295 キールンテンナンシャウ
33304 ホロテンナンショウ
33331 キバナテンナンショウ
33332 アリサンテンナンシャウ,ホソバテンナンシャウ
33337 ゴエフテンナンシャウ,フデボテンナンシャウ
33343 ハチジョウテンナンショウ
33345 アマミテンナンショウ
33346 オオアマミテンナンショウ
33347 オキナワテンナンショウ
33349 マイズルテンナンショウ,マイヅルテンナンショウ,マヒヅルテンナンシャウ
33361 イシヅチテンナンショウ
33362 カミコウチテンナンショウ
33363 ハリノキテンナンショウ

33365 オモゴウテンナンショウ
33366 シコクテンナンショウ
33369 テンナンシャウ,マムシグサ
33377 トクノシマテンナンショウ
33381 キシダマムシグサ
33383 ヒメウラシマソウ
33385 アマギテンナンショウ
33389 ミミガタテンナンショウ
33390 キイロミミガタテンナンショウ
33400 シコクヒロハテンナンショウ
33401 ヤクシマヒロハテンナンショウ
33406 マツダテンナンシャウ
33407 ツクシマムシグサ
33408 ツクシヒトツバテンナンショウ
33413 ヒュウガヒロハテンナンショウ
33414 ハリママムシグサ
33416 ヒトツバテンナンショウ
33417 アキタテンナンショウ
33418 クロハシテンナンショウ
33425 タカハシテンナンショウ
33429 シマテンナンショウ
33431 ユモトマムシグサ
33432 クボタテンナンショウ
33433 オオミネテンナンショウ
33437 ヤマナシテンナンショウ
33440 オガタテンナンショウ
33444 アシウテンナンショウ
33445 イナヒロハテンナンショウ
33446 ヒロハテンナンショウ
33459 ミクニテンナンショウ
33476 ムサシアブミ
33483 カラフトヒロハテンナンショウ
33486 キリシマテンナンショウ
33487 ミドリテンナンショウ
33495 セッピコテンナンショウ
33496 オオマムシグサ,カルイザワテンナンショウ,カントウマムシグサ,コウライテンナンショウ,ヘビノダイハチ,ホソバテンナンショウ,マムシグサ,マムシサウ,ムシグサマムシグサ,ムシグサマムシサウ,ムラサキマムシゲサ,ヤマゴンニャク,ヤマジノテンナンショウ,ヤマトテンナンショウ
33500 ヒトヨシテンナンショウ
33501 ヤマグチテンナンショウ
33505 ユキモチサウ,ユキモチソウ
33526 ナガヒゲウラシマソウ
33535 ミツバテンナンショウ
33536 ナンゴクウラシマソウ
33538 ウラシマソウ
33543 アオテンナンショウ
33547 ナガバマムシグサ
33549 ウワジマテンナンショウ
33554 ウンゼンマムシグサ
33569 ムロウテンナンショウ
33570 スルガテンナンショウ
33730 マツバシバ属
33737 ノゲエノコロ
33775 マツバシバ
33797 タイワンマツバシバ
33817 マツバシバ
33914 ヒメマツバシバ
34050 オオマツバシバ
34097 ウマノスズクサ属
34154 コウマノスズクサ,マルバノウマノスズクサ
34157 コロバウマノスズクサ
34162 ウマノスズカケ,ウマノスズクサ,オハグロバナ
34171 サラサバナ,バイブカズラ,バイブバナ
34198 ペリカンバナ
34203 タイワンウマノスズクサ
34219 オオバウマノスズクサ,オオバノウマノスズクサ,オホバウマノスズクサ
34223 ナガバノウマノスズクサ
34224 ナガバウマノスズクサ
34229 ホソバウマノスズクサ
34230 タンザワウマノスズクサ
34246 サラサバナ,パイプカズラ,パイプバナ
34247 リュウキュウウマノスズクサ
34268 キダチウマノスズクサ,ヒロハウマノスズクサ,マンシウウマノスズクサ,マンシュウウマノスズクサ
34275 ビロ-ドウマノスズクサ
34332 アリマウマノスズクサ
34382 コウシュンウマノスズクサ
34384 ウマノスズクサ科
34394 アリストテ-リア属
34406 ヒメフウチョウソウ
34419 アルマトセレウス属
34420 テッカン
34421 カガイチュウ
34422 マテンロウ
34431 ニグラスモモ
34443 マンシウアンズ,マンシュウアンズ
34448 ウメ
34455 アオジク,アオヂク,アヲジク,アヲヂク,チャセイバイ,リョクガクバイ
34470 モウコアンズ
34475 アンズ,カラモモ,セイヨウアンズ
34477 アンズ,カラモモ
34488 アルメリア属,ハマカンザシ属
34502 チシマハマカンザシ
34522 ヒメカンザシ
34536 オオハマカンザシ,ハマカンザシ
34550 オオハマカンザシ
34560 オオハマカンザシ
34582 セイヨウワサビ属
34590 セイヨウワサビ,ワサビダイコン
34591 セイヨウワサビ
34600 アルネ-ビア属
34655 アルニカ属,ウサギギク属
34746 クマギク,チョウジギク
34752 アルニカ
34766 オオウサギギク
34780 ウサギギク,エゾウサギギク,キンクルマ
34781 ウサギギク
34782 ハンヤエウサギギク
34845 アロ-ニア属
34895 アーポフィラム属,アルポフィムム属
34920 オオカニツリ属,オホカニツリ属
34930 オオカニツリ,オホカニツリ
34934 フタヒゲオオカニツリ
34935 チョロギガヤ
34936 リボンガヤ
34966 アロヤドラ属
34995 アウソウクワ属,オウソウカ属
35016 オウソウカ
35084 ヨモギ属
35088 オキナヨモギ,カハオニンジン,カハヲニンジン
35090 アブシント,アリセム,アルセム,ニガヨモギ
35095 アダムスヨモギ
35119 ステップヨモギ
35130 イトヨモギ
35132 クソニンジン,ホソバニンジン
35136 リトウザンヨモギ
35153 サマニヨモギ
35157 サマニヨモギ,チシマヨモギ
35158 シロサマニヨモギ
35167 チョウセンヨモギ
35171 アイノコヨモギ,シロチョウセンヨモギ
35187 キンキョギ,キンヨモギ
35211 アライトヨモギ,イソヨモギ,エトロフヨモギ,キタヨモギ
35220 マリヨモギ
35221 タカネヒトツバヨモギ
35237 ニイタカヨモギ,ノヨモギ,リュキュヨモギ
35282 カハラヨモギ,カワラヨモギ,シロカワラヨモギ,ネズミヨモギ,フナバハギ
35308 カハラニンジン,カワラニンジン,カワラヨモギ,ノニンジン
35356 オニオトコヨモギ
35373 カブヨモギ,サバクオトコヨモギ
35411 タンゴン
35416 ホソバアオヨモギ
35430 ケショウヨモギ
35466 ナンマンオトコヨモギ
35481 イヌフクド
35483 ヒメヨモギ
35503 ヒメイワヨモギ
35505 キタノアサギリサウ,キタノアサギリソウ,マンシウアサギリサウ,マンシウアサギリソウ

35517 ハマヨモギ,フクド
35520 エゾハハコヨモギ
35527 オオワタヨモギ,オホヨモギ,オホワタヨモギ,ワタヨモギ
35550 チシマハハコヨモギ,ハハコヨモギ
35560 イワヨモギ,ウラジロヒメイワヨモギ
35595 サバクヨモギ
35622 エダウチヒトツバヨモギホクチヨモギ
35630 ニシヨモギ
35634 ニシヨモギ,ヨモギ
35638 カズサキヨモギ,ヨモギ
35648 コウアンホソバヨモギ,ヒトツバヨモギ,マンシウヒトツバヨモギ,マンシュウヒトツバヨモギ
35674 オトコヨモギ,マンシウオトコヨモギ,ヲトコヨモギ
35732 カハカミヨモギ
35733 イヌヨモギ
35737 キタダケヨモギ
35742 ヒロハウラジロヨモギ
35746 オオバヨモギ
35753 クラムヨモギ
35759 キクヨモギ,シコタンヨモギ
35770 ヨモギナ
35779 イハキヌヨモギ,イワキヌヨモギ,シロヨモギ,タカネキヌヨモギ
35788 ヒメヨモギ
35789 ヒロハキクヨモギ
35794 ケショウヨモギ,ヒメヨモギ
35816 ウスユキョモギ,シラゲヨモギ
35831 ハマオトコヨモギ
35868 ミビヨロギ,ミブヨモギ
35915 ユキヨモギ
35916 ホソバヨモギモウゴホソバヨモギ
35945 ヒトツバヨモギ
35948 エゾヨモギ,オオヨモギ,ヤマヨモギ
35950 エゾノユキヨモギ
35964 ニイタカヨモギ,ニヒタカヨモギ
35972 イヌフクド,ヒメハマヨモギ
35981 キダチイトヨモギ,ゴトウサウ,ゴトウヨモギ
36020 イトヨモギ,ニトベヨモギ
36032 スナジヨモギ
36048 ノジヨモギ,ノヂヨモギ,モリヨモギ
36070 ミヤマオトコヨモギ
36097 カズザキヨモギ,モチグサ,ヨモギ
36111 タチスナジヨモギ
36162 ヤブヨモギ
36164 オオヤブヨモギ
36177 イワヨモギ,カムイヨモギ,マンシウイワヨモギ
36179 イワヨモギ
36225 アサギリソウ,ハクサンヨモギ
36232 ハマヨモギ,ヒロハハマヨモギ
36234 キヌゲハマヨモギ
36241 タカヨモギ,ヒトツバタカヨモギ
36262 チシマアサギリサウ
36286 スズフリヨモギ,ハイイロヨモギ
36295 タカネヨモギ
36302 サウマヨモギ
36303 タロコヨモギ,バタカンヨモギ
36321 ヒロハノヒトツバヨモギ,ヒロハヒトツバヨモギ,ヒロハヤマヨモギ
36340 アオヨモギ
36348 マキノハヨモギ
36354 モリヨモギ
36363 キクヨモギ,シコタンヨモギ,ミヤマキクヨモギ
36428 マシュウヨモギ
36440 チシマヨモギ
36453 コバノヨモギ,リョゥトゥヨモギ
36468 ペキンヨモギ
36469 コウライヒトツバヨモギ
36474 エゾオオヨモギ,オホヨモギ,ハタヨモギ,モチグサ,ヨモギ
36612 コブナグサ属
36647 カイナグサ,カリヤス,コブナグサ
36648 アイダコブナグサ
36650 ホンコブナグサ
36673 オニコブナグサ
36710 ヒメコブナ
36739 アルスロセレウス属
36902 アルトカルプス属,パンノキ属
36913 パンノキ,マルミパンノキ
36920 ナガミパンノキ,パラミツ
36924 コパラミツ
36928 モンキ-ジャック
36967 アラム属
37005 モエギアルム
37015 マムシアルム
37024 クロバナアルム
37027 マルバアルム
37053 ヤマブキショウマ属
37060 ヤマブキショウマ
37063 タンナショウマ
37064 ミヤマヤマブキショウマ
37065 シマヤマブキショウマ
37066 エゾヤマブキショウマ
37067 キレハヤマブキショウマ
37069 ヤマブキショウマ
37073 アポイヤマブキショウマ
37085 ヤマブキショウマ
37089 ミヤマブキショウマ
37109 アルンディナ属,ナリヤラン属
37115 ナリヤラン
37127 アズマザサ属,メダケ属
37196 ヒメスズタケ
37305 ウラゲサドザサ
37350 トダシバ属
37352 ウスゲトダシバ,トダシバ
37379 アラゲトダシバ,ケトダシバ,コウリュウトダシバ,トダシバ,バレンジバ
37389 ウスゲトダシバ,トダシバ,バレンシバ
37392 シロトダシバ
37398 オニトダシバ
37421 シバガヤ
37424 ミギワトダシバ
37429 ヒガヤ
37450 ダンチク属
37467 ダンチク,ヨシタケ
37469 シマダンチク,セイヤウダンチク,フイリダンチク,フイリノセイヨウダンチク
37470 ムラサキダンチク
37471 クカサゴチク,タカサゴダンチク
37477 タイワンアシ,ヒナヨシ
37548 キリカヅラ属
37553 ツタキンギョソウ
37554 ツルキンギョソウ
37556 カンアオイ属,カンアフヒ属,サイシン属,フタバアオイ属
37566 タイリンアオイ
37568 ミヤコアオイ
37571 カンアフヒ,ランヨウアオイ
37574 アメリカカンアオイ,アメリカカンアフヒ,カナダサイシン
37582 セイガンサイシン
37585 ウスバカンアフヒ,オナガサイシン,ケカンアフヒ,サンカクカンアフヒ
37590 カモアオイ,カモアフヒ,フタバアオイ,フタバアフヒ
37591 シロガネアオイ
37595 ミヤビカンアオイ
37600 ツクバネアオイ
37602 トサノアオイ
37604 ナンゴクアオイ
37606 カギガタアオイ
37612 クロフネサイシン
37613 オモロカンアオイ
37614 ウハミカンアフヒ
37616 オウシュウサイシン
37619 ミチノクサイシン
37620 ミヤマアオイ
37621 ソノウサイシン
37625 エクボサイシン
37630 イロバカンアフヒ
37632 ハッシマカンアオイ
37633 オオカンアオイ
37635 オクエゾサイシン
37637 オクエゾサイシン,ケイリンサイシン
37638 ケイリンサイシン
37639 サンヨウアオイ
37640 シジキカンアオイ
37646 オオカンアオイ,シタミカンアフヒ
37649 ユキグニカンアオイ
37650 アラカワカンアオイ
37654 ツクシアオイ

37655 キキョウカンアオイ
37656 アケボノアオイ
37657 コウヤカンアオイ
37658 スズカカンアオイ
37659 カンアオイ
37662 クワイバカンアオイ
37663 ムラクモアオイ
37665 イワタカンアオイ
37668 タニムラアオイ
37673 オオバカンアオイ
37674 シロフカンアフヒ,タイトンカンアフヒ,ホウライアオイ,ホウライアフヒ,ムラサキバノカンアフヒ,ヤンバルサイシン
37681 コシノカンアオイ
37684 オナガカンアオイ
37686 フクエジマカンアオイ
37687 モノドラカンアオイ
37688 アマギカンアオイ
37689 シモダカンアオイ
37692 ナンカイアオイ
37696 ノマダケカンアオイ
37697 コツブアオイ
37698 ヒナカンアオイ
37699 コバナカンアオイ
37700 トリガミネカンアオイ
37701 キンチャクアオイ
37705 ズソウカンアオイ
37710 アツミカンアオイ
37714 サカワサイシン,サツマアオイ
37715 サツマアオイ
37716 イセノカンアオイ
37717 オトメアオイ
37718 センカクアオイ
37722 ウスバサイシン,サイシン,ニッポンサイシン
37723 アツバサイシン
37729 ウスゲサイシン
37730 トクノシマカンアオイ
37734 ホシザキカンアオイ
37735 ツルダシアオイ
37736 マルミカンアオイ
37741 ヒメカンアオイ
37742 スエヒロアオイ
37743 ゼニバサイシン
37744 タマノカンアオイ
37746 トカラカンアオイ
37748 サンコカンアオイ
37749 カケロマカンアオイ
37751 ウンゼンカンアオイ
37752 モエギウンゼンカンアオイ
37753 コバノカンアオイ
37756 モエギカンアオイ
37760 ヤエヤマカンアオイ
37761 ヤクシマアオイ
37762 ヒゴカンアオイ
37763 クロヒメカンアオイ
37806 ガガイモ科,タウワタ科
37811 タウワタ属,トウワタ属
37888 タウワタ,ツルワタ,トウワタ
37889 キバナトウワタ
38135 オオトウワタ,オホタウワタ
38147 ヤナギトウワタ
38187 アスコセントウム属,アスコセントラム属
38194 フヂイラン,フヂヰラン
38312 モエギオクエゾサイシン
38316 オオバナサイシン
38319 フイリウスバサイシン
38322 アシミナ属
38338 パポー,ポーポー,ポポー
38904 アスパラガス属,クサスギカズラ属,クサスギカヅラ属
38938 クサナギカヅラ
38950 トウキジカクシ,ハマキジカクシ
38960 クサスギカズラ,ズマイラックス,テンモンドウ
38974 ヤナギボウキ
38977 トウキジカクシ,ヤマキジカクシ
38983 ホウキテンモンドウ
38985 シダレキジカクシ,スギノハカズラ
39009 マキバアスパラガス,ヤナギバテンモンドウ
39014 ホウライチク
39051 ハマタマボウキ
39069 クサスギカズラ,テンモンドウ
39079 ナギボウキ
39105 タチボウキ
39120 アスパラガス,オランダキジカクシ,コバノタマボウキ,マツバウド
39126 タマボウキ,ツクシタマバハキ,ツクシタマボウキ
39150 シノブボウキ
39161 タチテンモウドウ
39185 ギョリュウカズラ
39187 キジカクシ,サウチク
39195 アスパラガス・プルモーサス,オオミドリボウキ,シノブバハキ
39251 クルマバテンモンドウ
39267 アスパーシア属,アスパシア属
39277 アズパヅマ属
39299 トゲムラサキ
39300 クルマバサウ属,クルマバソウ属
39328 ツルボラン
39352 カスミムグラ
39383 タマクルマバサウ,タマクルマバソウ
39413 アカゾメムグラ,アカネムグラ,アカペムグラ
39435 アスフォデリーネ属
39449 アスフォデルス属,ツルボラン属
39463 ハナツルボラン
39516 ハラン属
39520 アリサンハラン,アリサンバラン
39529 ダイブハラン
39532 ハラン,バラン,バレン
39533 フイリハラン
39536 ホシハラン
39557 ハラン,バラン,バレン
39563 ムシャバラン
39910 アスター属,シオン属,シヲン属
39915 コシキギク
39928 シロヨナメ,チョウセンノコンギク,ツルヤマシロギク,ノヤマシロギク,ヤマシロギク
39930 タマバシロヨメナ
39931 サガミギク
39943 コマチギク
39965 ケシロヨメナ
39966 ケタカサゴヤマシロギク,ビロウドヤマシロギク,ビロードヤマシロギク
39968 シロヨメナ
39972 キントキシロヨメナ
39980 ハルノコンギク
39993 ナガバシロヨメナ
40016 イワアズマギク,タイリクアズマギク,ミヤマアヅマギク
40081 ヨナクニイソノギク
40105 シマコンギク,タイワンコンギク
40309 タテヤマギク
40311 モモイロタテヤマギク
40334 アズマギク
40474 タイワンシラヤマギク
40518 エゾゴマナ
40519 ゴマナ
40577 ブゼンノギク
40609 ユウガギク
40610 トキワギク
40621 コヨメナ
40652 カワラノギク,ヤマジノキク
40657 アイノコヨメナ
40663 コモノギク
40664 チョウセンシオン,チョウセンヨレナ
40805 ハコネギク,ヒゴシオン,ヒゴシヲン
40817 アメリカシオン,アメリカシヲン,ヒロハアメリカシオン
40835 ハマベヨメナ
40845 センボンギク
40846 ホソバコンギク
40847 ハマコンギク
40848 コンギク
40849 ノコンギク
40850 タニガワコンギク
40851 シロバナチョクザキヨメナ
40852 チョクザキヨメナ
40854 エゾノコンギク
40860 オオバヨメナ
40867 オキナワギク
40878 ニヒタカヤマヂノギク,ヒメタカサゴヤマシロギク

40923 ニュ-ヨ-クシオン,ユウゼンギク
40976 シンチクシラヤマギク
41084 イソカンギク,カンヨメナ
41086 テリアッバキク
41159 オオユウガギク
41171 サワシロギク
41172 シブカワシロギク
41198 サツマシロギク
41201 ノシュンギク,ミヤコワスレ,ミヤマヨメナ
41202 シンジュギク
41225 ナガバシラヤマギク
41226 ヤマシロギク
41257 シベリヤヨメナ,タカスギク
41282 ホソバノギク
41289 ダルマギク
41290 オオダルマギク
41317 ハハキギク,ホウキギク
41322 オオホウキギク
41328 ヒロハホウキギク
41329 アキバギク
41335 ツルギバノギク
41336 テリハノギク
41337 ナンコシラヤマギク
41342 シオン,シヲン
41366 クルマギク
41398 アレノノギク
41404 コンギク,ノコンギク,ハハキギク
41432 ウラギク
41434 シロバナウラギク
41444 オオイソノギク
41495 ハコネギク
41496 タカネコンギク
41511 ヤクシマノギク,ヤクシマノコンギク
41516 ヨメナ
41517 カントウヨメナ
41518 シコクシロギク
41647 ホシヤシ属
41786 アスティルベ属,チダケサシ属
41787 チダケトリアシ
41788 テリハチダケサシ
41789 アスチルベ
41795 アワユキサウ,アワユキソウ,オオチダケサシ,オホチダケサシ
41806 オホチダケサシ
41808 ハナチダケサシ
41810 ヤクシマショウマ
41811 コヤクシマショウマ
41812 タリノホチダケサシ
41813 ハチジョウショウマ
41817 アワモリサウ,アワモリショウマ,アワモリソウ
41818 ベニバナアワモリショウマ
41822 タリホノチダケサシ
41825 シマアハモリ,タイワンアハモリ
41827 アリサンアハモリ
41828 チダケサシ
41829 キレバチダケサシ
41832 トリアシショウマ
41833 ウスベニトリアシショウマ
41834 バンダイショウマ
41835 ホソバトリアシショウマ
41836 ミカワショウマ
41839 モミジバショウマ
41847 オオチダケサシ
41848 タンナトリアシ
41851 ヒトツバショウマ
41852 アカショウマ,トリアショウマ
41853 ウスベニアカショウマ
41861 フジアカショウマ
41863 テリハアカショウマ
41864 ツクシアカショウマ
41866 シコクトリアシショウマ
41867 ヒメアカショウマ
41872 フキモドキ
41906 ゲンケ属,ゲンゲ属
41918 シロバナムラサキモメンヅル
42145 マルバノコゴメオウギ
42177 キバナノモメンズル
42193 シロバナモメンヅル
42208 ツルゲンゲ
42258 ムラサキオウギ
42262 クロガクモメンズル
42264 ムラサキワウギ
42385 リシリオウギ
42391 チヤボゲンゲ
42417 モメンヅル,ヤハラグサ
42537 エゾモメンヅル,チシマモメンヅル
42546 カワカミモメンヅル
42594 ムラサキモメンズル,ムラサキモメンヅル
42696 シナガワハギモドキ
42697 ヒメシナガワハギモドキ
42699 キバナオウギ,タイツリオウギ
42704 モウコモメンヅル
42705 タイツリオウギ
42723 ヒナモメンズル
42969 モメンヅル
42995 シュミットザウ
43008 コゴメオウギ
43012 ハイゲンゲ
43018 カラフトモメンヅル
43024 ネッカオウギ
43030 リシリオウギ
43042 シロウマオウギ
43048 ナルトウオウギ
43053 ゲンゲ,ゲンゲバナ,ホウゾウバナ,レンゲ,レンゲサウ,レンゲソウ,レンゲバナ
43054 シロバナゲンゲ
43160 トカチオウギ
43204 ヤチオウギ
43265 カリバオウギ
43284 ダッタンモメンズル
43309 アストランチア
43320 アストリディア属
43329 アスホ
43330 アスハル
43334 アスヒ
43342 アスベニ
43372 ホシダネヤシ属
43428 アストロロバ属
43457 オホノボタンノキ属
43462 オホノボタンノキ,キダチノボタン,シマノボタン
43492 アストロフィツム属
43493 カブトマル
43495 グンボウギョク,ズイホウギョク
43496 オウホウギョク
43497 タイホウギョク
43498 ホウオウギョク
43499 グンホウギョク
43504 ハクズイホウギョク
43509 ランポウギョク
43510 ランポウカク
43512 ヘキルリランポウカク
43524 ハクランポウ
43526 ハンニャ
43527 キトゲハンニャ
43529 ハタカハンニャ
43580 シデシャジン
43718 ツホバゲッケイ属
43721 ツゲカウジ
44046 タスマニアスギ属
44050 タスマニアスギ
44192 ウケラ属,オケラ属,ヲケラ属
44200 ナンマンオケラ
44202 チョウセンオケラ
44205 オケラ,ビャクジュツ,マルバオケラ,マルバヲケラ
44208 ウケラ,オケラ,サウジュツ,シナオケラ,ノオケラ,ホソバオケラ,ホソバノヲケラ,ヲケラ
44210 シナオケラ
44218 オオバナオケラ,ビャクジュッ,マルバオケラ
44269 タデノキ
44298 アトリプレックス属,ハマアカザ属
44347 エダハリハマアカザ
44400 モウコハマアカザ
44430 ホソバノハマアカザ
44468 ヤマホウレンソウ
44520 サキシマハマアカザ,シマハマアカザ,ミヤコジマハマアカザ
44542 シマハマアカザ,タンワンハマアカザ
44554 セイヨウハマアカザ
44575 ホソバハマアカザ
44576 エゾハマアカザ,コハマアカザ,セイヨウハマアカザ

44621 ホコガタアカザ
44646 ウラジロハマアカザ
44659 ハマアカザ
44661 ミナミハマアカザ
44668 ハマアカザ
44704 アトロパ属,オオカミナスビ属,ベラドンナ属
44708 オオカミナスビ,オオハシリドコロ,オホハシリドコロ,セイヤウハシリドコロ,セイヨウハシリドコロ,ベラドンナ
44801 アッタレア属
44807 ブラジルゾウゲヤシ
44824 ビロウドヒメクヅ属
44867 ムラサキナズナ属
44868 ムラサキナズナ
44881 モッコウ
44887 アオキ属,アヲキ属
44893 シナアオキ,タイワンアオキ,タイワンアヲキ
44911 ヒマラヤアオキ
44915 アオキ,アオキバ,アヲキ,アヲキバ
44919 フイリアオキ
44920 アマノガワ
44921 ヒロハノアオキ
44922 クロエアオキ
44923 シロミノアオキ
44924 ホソバアオキ
44925 キミノアオキ
44926 タガヤサン
44927 アオバナアオキ,シロミアオキ
44932 ヒメアオキ
44933 ウチダシヒメアオキ
44937 ナンゴクアオキ,ヒゴアオキ
45192 イワナズナ
45231 アウストロカクタス属,オーストロカクタス属
45232 ロゥソゥギョク
45233 ヒョゥソゥギョク
45236 ユゥソゥギョク
45241 アウストロケファロセレウス属
45245 ハクレイオク
45254 オーストロシリンドロプンティア属,オストロシリンドロプンティア属
45260 サンゴジュ
45265 オキナウチワ
45363 カラスムギ属
45389 ミナトカラスムギ
45448 カラスムギ,ススメムギ,チャヒキ
45454 コカラスムギ
45495 オニカラスムギ
45566 エンバク,オートエンバク,オートムギ,カラスムギ,マカラスムギ
45602 セイヤウチャヒキ,セイヨウチャヒキ
45721 ゴレンシ属
45724 ビリンビ
45725 カランボー,カランボク,ゴゴレンシ,ゴレンジ,ヨウトウ,レンシャウタウ
45740 ヒルギダマシ属
45746 ヒルギダマシ,ヒルギモドキ
45821 ホソバツルメヒシバ
45827 ツルメヒシバ
45833 ホソバツルメヒシバ
45843 イヌハハキギ属
45844 イヌハハキギ,イヌホウキギ
45853 イヌハハキギ
45879 アイロステラ属
45886 ユウホウマル
45887 ユウシウマル
45888 エンレイマル
45892 コウショウマル
45893 シュウテンマル
45895 ジュウホウマル
45907 カュバワソ
45908 インドセンダン
45952 アサーラ属,アザラ属
45979 アカウキクサ属
45993 アズテキウム属,アズテキューム属
45994 ハナカゴ
46008 アズレオセレウス属
46022 バビアーナ属,ホザキアヤメ属
46148 ホザキアャメ
46176 パッコーメア属
46193 ランバイ,レンバイ
46348 オトメアゼナ属
46362 オトメアゼナ
46366 キバナオトメアゼナ
46371 ウキアゼナ
46374 ステッキヤシ属,モモミヤシ属,ユーユーヤシ属
46743 バラカヤシ属
46744 シーマンバラカヤシ
46810 ツチトリモチ属
46824 タイワンツチトリモチ
46826 タイワンツチモチ,リュウキュウツチモチ
46848 シャウジャウタケ,ツチトリモチ,ツチモチ,ツチャマモチ,ミヤマツチトリモチ,ヤマトリモチ
46859 シマトリモチ,ホザキツチトリモチ
46863 フデガタツチトリモチ
46882 キイレツチトリモチ,トベラニンギョウ
46885 ヤクシマツチトリモチ
46888 ツチトリモチ科
47077 ツリフネソウ科,ホウセンクワ科
47174 バンブーサ属,ホウワウチク属
47178 ヒフキダケ
47213 シチク
47220 コバナダケ
47250 チャウシチク,チョウシチク
47254 ヒフキダケ
47345 ホウビチク,ホウライチク,ホウワウチク,ヨウタケ
47346 スハウチク,スホウチク
47347 ホウオウチク
47350 フイリホウワウチク,ホウショウチク
47351 ヒメホウワウチク,ベニホウワウチク
47370 セキカクチク
47402 パチナダケ
47494 トウドウチク
47508 ダイフクチク
47516 ダイサンチク,マーテコダケ
47518 キンシチク,リョクチク
47626 バンクシア属
47845 ヤマアイ,ヤマアヰ,リウキウア,リウキウアイ,リウキウアヰ,リュウキュウアイ
47854 ムラサキセンダイハギ属
47859 ムラサキセンダイハギ
47931 ヤマガラシ属
47949 シベリアマガラシ,ミヤマガラシ,ヤマガラシ
47956 シベリヤヤマガラシ
47962 ニイタカガラシ
47963 キバナクレス
47964 ハルザキヤマガラシ
48039 バーケリア属
48080 バーレーリア属,バーレリア属
48140 バルレリア
48241 マツカサバーレリア
48295 トゲバーレリア
48508 ゴバンノアシ属,サガリバナ属
48510 ゴバンノアシ
48518 アマキ,サカミハナ,サガリバナ,サハフヂ
48524 サガリバナ科
48551 ミヤマノボタン属
48552 ミヤマノボタン
48688 ツルムラサキ属
48689 シンツルムラサキ,ツルムラサキ
48698 ツルムラサキ科
48762 ムヒョウソウ
48829 バテマニヤ属
48907 トウウメバチソウ
48922 オオウメバチモ
48975 エリカモドキ属,バウエラ属
48977 エリカモドキ
48990 ハカマカズラ属,ハカマカヅラ属
48993 キワンジュ,ソシンカ,ソシンクワ,モクワンジュ
49046 キククワボク,キッカボク,ヤハズカツラ
49140 ハカマカズラ,ハカマカヅラ,ワンジュ
49173 ナツザキソシンカ
49211 ムラサキソシンカ,ムラサキソシンクワ,ムラサキモクワンジュ
49241 キバナモクワンジュ

49247 アカバナハカマノキ,コチョオボク,ソシンカ,フイリソシンカ
49248 シロバナソシンカ
49343 トックリノキ,トックリラン
49347 ベアウフォルティア属
49359 オオバナカズラ
49532 カズノコグサ属,ミノゴメ属
49540 エッタムギ,カズノコグサ,ミノゴメ,ムギカラグサ
49544 ケミノゴメ
49587 シウカイダウ属,シュウカイドウ属,ベゴニア属
49603 ギンボシベゴニア
49617 ビロ-ドベゴニア
49626 マルミシウカイダウ
49629 アマノガワ
49711 クリスマスバゴニヤ,クリスマスベゴニア,ハナバゴニヤ
49729 ベニバナベコニア
49827 ウトウシュウカイドウ,コウトウシウカイダウ,コウトウシュウカイドウ,コオトオシュウカイドオ,シマシウカイダウ
49844 コバベゴニア
49845 コバベゴニア,ベニバナベゴニヤ
49848 マルヤマシウカイダウ,マルヤマシュウカイドウ,マルヤマシュウカイドオ
49877 シダレベゴニア
49885 シュウカイドウ
49886 シウカイダウ,シュウカイドウ,ヨウラクサウ,ヨウラクソウ
49887 シロバナシュウカイドウ
49890 シナシュウカイドウ,モミヂバシウカイダウ
49915 ヤツデベゴニア
49942 モエギベゴニア
49943 ワタゲベゴニア
49974 シマミヤマシウカイダウ,タイワンミヤマシウカイダウ,ランダイシュウカイドウ
50012 シダレベゴニア
50051 シラホシベゴニア,ホシベゴニヤ
50076 ケテリハナゴニア,メタルベゴニア
50111 テリハベゴニア
50232 オオバベゴニヤ,オホバベゴニヤ,タイエフベゴニヤ,タイヨウベゴニア,ベゴニヤ
50235 レックスベゴニア
50273 ウラベゴニア
50284 キダチベゴニア
50287 ヒメベゴニア
50298 シキザキシウカイダウ,シキザキベゴニア,シロバナベゴニア
50317 ハスノハベゴニア
50329 ビロ-ドベゴニア
50352 タイワンシウカイダウ
50378 キウコンベゴニヤ,キュウコンベゴニア,バナベゴニヤ
50433 シウカイダウ科,シュウカイドウ科
50465 アカハダクスノキ属,アカハダノキ属
50507 アカハダグス,アカハダクスノキ,アカハダノキ
50667 ヒアフギ属,ヒオウギ属
50669 カラスオウギ,ヒアフギ,ヒオウギ,ヒオオギ
50717 ヒサウチソウ
50808 ヒナギク属
50809 ヒナギクモドキ
50825 エンメイギク,チョウメイギク,デージ-,トキシラズ,ヒナギク
50841 ベリュ-ム属
50922 コエビソウ属
50939 コウシュンツュクサ
50994 トウガン属,トウグワ属
50998 カモウリ,タウグワ,トウガ,トウガン,トウグワ
51140 ハナヤマボウシ
51141 ベニヤマボウシ
51142 アオヤマボウシ
51163 ベンティンクヤシ属
51164 ニコバルヤシ
51165 ベンティンキンブシス属
51261 メギ科
51268 メギ属
51301 オホトリトマラズ,オホバメギ,ヒトハリヘビノボラズ,ヒロハノヘビノボラズ,ヒロハヘビノボラズ
51303 アカジクヘビノボラズ
51331 アミバヘビノボラズ
51422 キヤナダメギ
51733 フッカ-メギ
51811 クロミノヘビノボラズ,タイワンヘビノボラズ
51926 ウスバヘビノボラズ
51940 ニヒタカヘビノボラズ
52049 タウメギ,トウメギ,ホソバメギ
52069 テンガイメギ,テンジクメギ
52100 オオトリトマラズ,オオバノヘビノボラズ,オオバヘビノボラズ,ナガミノノヘビノボラズ,ヒトハリヘビノボラズ,ヒロハノヘビノボラズ,ヒロハヘビノボラズ
52149 シベリャメギ
52151 コガネェンジュ,コヘビノボラズ,トリトマラズ,ヘビノボラズ,マルミノヘビノボラズ
52225 コトリトマラズ,メギ,ヨコイドオシ,ヨコイドホシ,ヨロイドオシ
52246 コトリトマラス,ナギ,ヨロヒドホシ
52282 オオバメギ,ミヤマヘビノボラズ,ミヤマメギ
52322 ヘビノボラズ
52339 オホトリトマラズ,オホバヘビノボラズ,ヒロハノヘビノボラズ
52400 クマヤナギ属
52418 クマヤナギ,ナンゴククマヤナギ
52423 ウスバクマヤナギ,タイワンクマヤナギ
52436 ヒメクマヤナギ,マコウギ
52439 ホナガクマヤナギ
52448 ミヤマクマヤナギ
52454 クマヤナギ
52456 ケオオクマヤナギ,ケクマヤナギ
52457 ナガミクマヤナギ
52459 ナンゴククマヤナギ
52461 オオクマヤナギ,オホクマヤナギ,ケヤマヤナギ
52464 ウスゲクマヤナギ
52478 エイノキ,ヨコグラノキ
52503 チョウセンイワウチワ属,チゥウセンイワウチワ属
52514 アルタイユキノシタ,シベリアユキノシタ,ナガバユキノシタ
52532 カガミユキノシタ
52541 オオイワウチワ,ヒマラヤユキノシタ
52548 ベルゲランツス属
52556 セナス
52561 テルナミ
52563 スイホコ
52565 ダイグンポ
52571 ベルゲロカケタス属,ベルゲロセレウス属
52587 シマバラサウ属,シマバラソウ属
52592 シマバラサウ,ヤンバルミゾハコベ
52617 シマバラソウ
52947 ベリスフォルディア属
53038 ウスユキナズナ
53045 ハナナズナ属
53047 ハナナズナ
53055 ブラジルナットノキ属
53057 ブラジルナット,ブラジルナットノキ
53130 ヒメノボタン属,ベルトロ-ニア属
53136 ヒメニシキノボタン
53138 ヒメニシキノボタン,ヒメノボタン
53213 ベッセラ属,ボッセラ属
53224 タウヂサ属,トウヂサ属
53249 インゲンナ,ウヅダイコン,シャムロダイコン,テンサイ,ハマアカザ,フダンサウ,フダンソウ,ヰンゲンナ
53254 ハマフダンソウ
53257 イッモヂシャ,トウヂシャ,フダンソウ
53259 テ-ブルビ-ト
53266 カエンサイ,サトウダイコン,テンサイ
53269 インゲンナ,ウズマキダイコン,サンゴジュナ,ニシキダイコン
53304 カッコウソウ,カッコウチョロギ

53309 カバノキ属,カンバ属,シラカンバ属
53338 アメリカミネバリ,キハダカンバ
53353 アポイカンバ,アポイカンボ
53355 オクエゾシラカンバ
53378 チチブミネバリ
53379 シナカンバ,タウカンバ,トウカンバ
53380 ソナカンバ
53388 ウラジロ,ウラジロカンバ,ネコシデ
53389 チョウセンミネバリ,テウセンミネバリ,ヤエガワカンバ
53400 ウタイカンバ,ウダイカンバ,クロカンバ,コオノオレ,コノオレ,コヲノヲレ,サイハダカンバ,マカンバ,ミヤマカンバ,ヤエガワカンバ,ヤヘガハカンバ
53402 ヒダカヤエガワ
53411 エゾノタケカンバ,カラフトダケカンバ,ソウシカンバ,タケカンバ,ダケカンバ,ドスカンバ
53413 アツハダカンバ
53420 キレハダケカンバ
53421 ナガバシラカンバ,ナガバダケカンバ,ナガバノシラカンバ,ナガバノダケカンバ,マカンバ
53425 コバノダケカンバ
53426 アカカンバ
53427 マルミノダケカンバ
53434 ヒメカンバ
53444 コウアンヒメオノオレ,チャボカンバ,チャボヲノオレ,ャボオノオレ
53452 オオダケカンバ
53454 ヒメカンバ
53460 イヌブシ,ジゾウカンバ,ヂザウカンバ
53465 アジサ,アズサ,イタヤミネバリ,カハラブナ,コッパダミネバリ,ハンサ,ミズメ,ミヅネ,ミヅメ,モーカザクラ,ヨグソミネバリ
53483 カニワ,カバノキ,カンバ,ガンピ,クサザクマ,シラカバ,シラカンバ,シロザクラ
53503 オボバシラカンバ,カニワ,カバノキ,カマフトシラカンバ,カンバ,カンビ,クサザクラ,コバノシラカンバ,シラカバ,シラカンバ,マンシウシラカンバ
53505 オウシュウカンバ
53520 ウタイカンバ,サイハダカンバ,マカバ
53527 ポロナイカンバ
53544 ヒメオノオレ,ヤチカンバ
53549 アメリカシラカンバ
53563 シダレカンバ
53572 コウアンシラカンバ,シラカバ,シラカンバ,マンシウシラカンバ
53574 ホソバシラカンバ
53578 キレハシラカンバ
53580 エゾノシラカンバ
53581 カラフトシラカンバ,マンシウシラカバ
53584 エゾノオオシラカンバ
53610 アズサミネバリ,オノオノカンバ,オノオレ,オノオレカンバ,オンノレ,ホンアズサ,ミネバリ,モウカ,ヲノオレ,ヲノオレカンバ,ヲノヲレ,ヲノヲレカンバ
53612 ナカバオノオレ,ホソバオノオレ
53614 シコクダケカンバ
53631 アヅサ,オホバミネバリ,ヨグソアヅサ,ヨグソカンバ,ヨグンミネバリ
53651 カバノキ科,ハンノキ科
53755 センダングサ属,タウコギ属
53773 オトメセンダングサ
53786 キンバイタウコギ
53797 キツネバリ,キバナノセンダングサ,コバナノセンダングサ,コバノセンダングサ,センダングサ
53801 センダングサ
53805 マルバタウコギ
53816 ナガバタウコギ,ヤナギタウコギ
53913 アメリカセンダングサ,セイタカウコギ
53976 キクザキセンダングサ
54001 エゾノタウコギ,キクバタウコギ,セイタカタウコギ
54042 ホソバセンダングサ,ホソバノセンダングサ
54048 コセンダングサ,センダングサ
54054 コシロノセンダングサ,シロノセンダングサ,シロバナセンダングサ
54059 オオバナノセンダングサ
54060 ハイシロノセンダングサ
54061 マルバアアワユキセンダングサ
54065 タホウタウコギ
54081 エゾノタウコギ
54158 シラシグサ,タウコギ,タウマギ,トッカンバラ,ヒッツキ
54168 ハイタウコギ
54240 フランスゼリ
54243 ビフレナーリア属,ビフレナリア属
54292 ツリガネカズラ属
54303 ツリガネカズラ,ビグノニア・カプレオラータ
54357 ノイゼンカツラ科,ノウゼンカズラ科
54362 ビイリア属
54432 ツツアナナス属
54439 アカバナハナオンライ属,ツツアナナス属
54451 コバナツツアナナス,ベニガクツツアナナス
54452 ヒロハツツアナナス
54453 モーレルツツアナナス
54454 ヨウラクツツアナナス
54456 ベニアヂツツアナナス
54461 サンダーツツアナナス
54464 アカバナハナオンライ
54467 トラフツツアナナス
54532 オサバフウロ属,ビオフィタム属
54555 オサバフウロ
54618 アカギ属
54620 アカギ,アタン,カタンノキ,ギンツンガソ
54738 ビスマルクヤシ属
54763 アブクマトラノオ
54781 ナンブトラノオ
54799 イブキトラノオ
54800 エゾイブキトラノオ
54818 ケクリンユキフデ
54821 ハルトラノオ
54822 オオハルトラノオ
54826 ウラゲムカゴトラノオ
54834 イブキトラノオ属
54859 ベニノキ属
54862 アケノキ,ベニノキ
54863 ベニノキ科
54874 アカギ属
55135 ミヤマハシカンボク属
55147 ミヤマハシカンボク
55212 ホソバヤロオド,ホソバヤロード
55529 ブレッティア属
55559 シラン属,ブレティラ属
55563 アマナラン,ニヒタカアマナラン
55575 シケイ,シラン,ビャクギフ,ベニラン
55577 フクリンシラン
55579 シロバナジマン
55581 コウトウアマナラン
55644 ブローメリア属
55656 ブロスフェルディア属
55657 ヨウムギョク
55658 ショウロギョク
55669 ツルハグマ属
55693 カイナウカウ,カイノウコウ,カイパウカウ,ダイワウサウ,タカサゴギク
55738 ウラジロワタナ
55754 ダカサゴカウゾリナ,タカサゴコオゾリナ,ヤヘヤマコウリンクワ
55766 ヤエヤマコウゾリナ,ヤヘヤマカウゾリナ,ヤマコオゾリナ
55768 オキナハカウゾリナ,ザケバカウゾリナ,サケバコウゾリナ,ヒラギギク
55769 オホキバナムカシヨモギ,カズザキコウゾリナ,タイワンハグマ,ツルヤブタバコ,ツルヤブタビラコ
55783 ツルハグマ
55784 キバナノムカショモキ,キバナノヤブタバコ
55796 タイワンコウゾリナ
55820 シマヂワウギク,タイワンヂワウギク
55851 ブルーメンバッキア属
55908 スブタ属

55911 セトヤナギスブタ
55913 オオスブタ,オホスブタ,マルミスブタ
55918 コスブタ,スブタ,ナガヒグミスブタ
55922 ヤナギスブタ
55924 ミカハスブタ,ミカワスブタ
56023 ボッコニア属
56031 タイヘイヨウクルミ属,ボトァ属
56033 タイヘイヨウクルミ
56044 ホクチグサ属
56050 コヨリソウ
56063 スミレイワギリ
56094 カラムシ属,マオ属,マヲ属
56096 ハマヤブマオ
56100 ラセイタソウ
56105 オガサワラモクマオ
56112 ヒラバヒメマオ
56125 モクマオ,モクマヲ,ヤナギバモクマオ,ヤナギバヤブマオ,ヤナギバヤブマヲ
56134 カタバヤブマオ
56135 シマナガバヤブマオ
56140 タイワントリアシ,ナガバヤブマオ,ナガバヤブマヲ
56145 カラムシ,コロモグサ,シロウ,ヒウジ,マオ,マヲ
56166 ササグリヤブマオ
56169 ケナガバヤブマオ
56180 オオシマヤブマオ
56181 キアカソ,コアカゾ,トガリバヤブマオ,ニオウヤブマオ,ヤブマオ,ヤブマヲ
56189 ツクシヤブマオ
56190 キヨズミヤブマオ
56195 タイワンコアカソ,ヤブマオ,ヤブマヲ
56221 コヤブマオ
56222 ゲンカイヤブマオ
56229 カラムシ,チョマ,ナンバンカラムシ,ナンベンカラムシ,マオ,マヲ,ラミー
56233 ラミー
56238 カラムシ
56240 アオカラムシ,アヲカラムシ,クサマオ,クサマヲ
56247 フトボヤブマオ
56254 モクマオ,モクマヲ,ヤナギバヤブマオ,ヤナギバヤブマヲ
56257 ヒロバヒメマオ
56260 メヤブマオ,メヤブマヲ
56300 タンナヤブマオ
56302 マルバヤブマオ
56317 アカソ
56318 キアカソ,クサコアカソ,コアカソ
56322 コバノコアカソ
56338 ムラダチヤブマオ
56342 リュウノヤブマオ
56343 アカソ,メヤブマオ,メヤブマヲ
56353 クマヤブマオ
56356 ヤエヤマラセイタソウ
56357 イトザキヒメマオ,イトザキヒメマヲ,ナガバヒメマオ,ナガバヒメマヲ
56360 ナガバヒメマオ
56380 マツカゼサウ属,マツカゼソウ属
56382 ケマツカゼサウ,ケマツカゼソウ,マツカゼソウ
56383 マツカゼソウ
56386 マツカゼソウ
56392 マツカゼサウ,マツカゼサウダ,マツカゼソウ
56397 ナハカノコサウ属,ナハカノコソウ属
56420 タイトウカノコサウ
56425 ナハカノコサウ,ナハカノコソウ
56437 タチナハカノコソウ
56526 コウトウヤツデ属
56637 ウキヤガラ,ヤガラ
56688 ボルレア属
56700 アメリカギク属
56704 オオアメリカギク
56706 アメリカギク
56755 ヅルユリズイセン属,ボマーレア属
56759 ヅルユリズイセン
56761 パンヤ科
56770 キワタノキ属,キワタ属
56784 インドワタノキ
56802 キワタ,ハンシジュ,バンヤ,パンヤノキ,ワタノキ
57084 ムラサキ科
57095 ルリヂサ属,ルリヂシャ属
57101 ルリジサ
57116 アウギヤシ属,アフギヤシ属,オアギヤシ属,オウギヤシ属
57117 エチオピアオウギヤシ,エチオピオアギヤシ
57122 アフギヤシ,オウギヤシ,バイタラ
57126 ハイネオアギヤシ,ハイネオウギヤシ
57257 ボロ－ニア属
57296 ハリフタバ
57333 ヒロハフタバムグラ
57389 オニク属
57390 オニク,キムラタケ
57429 オニク属
57434 シマオニク
57437 オカサダケ,オニク,キムラタケ,キムラタゲ,ヲカサダケ
57545 ボスリオキラス属,モンツキガヤ属
57548 モンツキガヤ
57553 モンツキガヤ
57567 カモノハシガヤ,チョウゼンカモノハシガヤ
57568 マンシウカモノハシガヤ
57579 オオモンツキガヤ
57675 ハナイバナ属
57681 トウハナイバナ
57683 オオハナイバナ
57685 ハナイバナ
57689 タカサゴハナイバナ,ハナイバナ
57695 クルマミズキ,クロミズキ,ダンゴノキ,ハシカノキ,ミズキ,ミズクサ,ミヅキ
57847 プルマ・マンゴ－
57852 イカダカズラ属,イカダカヅラ属,ブ－ケンビレア属
57857 テリハイカダカズラ,テリハイカダカヅラ
57864 ブゲンカズラ
57868 イカダカズラ,イカダカヅラ,ココノエカズラ,ココノヘカツラ,ブ－ゲンヴヰレヤ,ブ－ゲンビレア
57924 アゼガヤモドキ
57948 カンチョウジ属,ブバルディア属
57954 カンチョウジ
57955 シロカンチョウジ,ナガバナカンチョウジ
57957 ミツバカンチョウジ
57958 ミツバカンチョウジ
57972 ボエ－ニア属
57977 タマツルグサ属,ボ－ウィエア属
58022 アラシグサ
58052 ニクキビ属
58088 ヒメスズメノヒエ
58128 パラグラス
58139 ニクキビモドキ
58155 ケニクキビ
58183 ニクキビ
58199 ビロ－ドキビ
58201 ラシャキビ
58261 アレチシオン属
58265 アレチシオン,モウコムカシヨモギ
58275 ホシバムカシヨモギ
58299 クロキソウ属
58300 クロキソウ
58301 ブラキカリキウム属
58303 シンセガイ
58320 ブラキセレウス属
58334 ツボノキ属
58335 ゴウシュウアオギリ
58348 カエデゴウシュウアオギリ
58389 マンキンヤマサギサウ,ランタイヤマサギサウ
58447 カウヤザサ属
58462 カウヤザサ,コウヤザサ
58572 ヤマカモジグサ属
58578 セイヨウヤマカモジ
58588 タイワンヤマカモジグサ
58592 イハセサウ,イワイソウ
58596 ヒロハヤマカモジグサ
58604 コウアンヤマカモジグサ
58619 エゾヤマカモジグサ,ヤマカヅラ,ヤマカモジグサ

58629 ヤマカモジグサ
58669 ヒメコスモス属
58671 ヒメコスモス
58813 ブラキステルマ属
59095 ハクセンヤシ属,ブラヘアヤシ属,ブラヘア属
59096 トゲハクセンヤシ
59097 ブランデゲーハクセンヤシ
59098 オオミブラヘア
59099 メキシコハクセンヤシ
59144 ジュンサイ属
59148 ジュンサイ,ヌナワ,ネヌナワ
59158 ブラシリカクタス属
59163 キセッコウ
59164 セッコウ
59167 ブラシリセレウス属
59254 ブラッサボラ属
59265 ブラッシア属
59278 アブラナ属
59348 アビシニアガラシ
59358 アブラナ,ツケナ,ナタネナ
59434 ミブナ
59435 イセナ
59436 スイグキナ,スヰグキナ
59438 オホガラシ,オホバガラシ,カキナ,カラシナ,キガラシ,キャウナ,キョウナ,セリフォン,ダイシンサイ,タカナ,ナガラシ,ニンスーカ,ハガラシ,ミズナ,ミヅナ
59458 セツリコ
59461 ネカラシナ
59469 ザーサイ
59493 セイヨウアブラナ
59507 スウェーデンカブ
59512 ヒサゴナ
59515 クロガラシ
59520 カンラン,キャベツ,タマナ,ハボタン,ボタンナ,ヤセイカンラン
59524 ケイル,ケール,コラード,ハゴロモカンラン,ハボタン,リョクヨウカンラン
59525 ハボタン
59526 カランチョウ
59529 カリフラワー,コーリフラワー,ハナヤサイ
59532 カンラン,キャベツ,タマナ,ハボタン
59538 ハゴロモカンラン
59539 コモチカンラン,コモチタマナ,コモチハボタン,メキャベツ
59541 カブカンラン,カブラカンラン,カブラタマネ,キュウケイカンラン,コールラビ
59545 イタリアン・ブロッコリー,スプラウチング・ブロッコリー,ブロッコリー,ミドリハナヤサイ,メハナヤサイ
59556 シロクキタイサイ
59575 ウグイスナ,カブ,カブナ,カブラ,カブラナ,コマッナ,フユナ
59595 オタマナ,シャクシナ,タイサイ,タイナ,ホテイナ
59598 アキザキナタネ
59600 ハクサイ
59603 アブラナ,カブラ,シュンフラン,シラクキナ,スイグキナ,ムラサキナ
59604 サントウサイ
59607 コマツナ
59651 ハリゲナタネ
59714 ベニバナゲッカコウ属
59715 ベニバナゲッカコウ
59822 ハシカンボク属,ハシカン属,ブレディア属
59851 ハシカン,ハシカンボク
59855 ツルハシカンボク
59863 コバノミヤマノボタン
59864 ヒナノボタン
59888 ヤエヤマノボタン
59889 ヤエヤマノボタン
59982 ボレボールティア属
60049 オオシマコバンノキ属,タカサゴコバンノキ属
60053 ヒメコバンノキ
60061 タイワンヒメコバナノキ
60073 タカサゴコバンノキ
60083 オオシマカンコノキ,オオシマコバンノキ,オホシマコバンノキ,タイワンヒメコバンノキ,タカサゴコバンノキ,ヒメコバンノキ
60162 アルヤマカンコノキ属,カンコモドキ属
60189 マルヤマカンコノキ
60230 カンコモドキ
60295 ブリルランタイシア属
60372 コバンサウ属,コバンソウ属
60379 オホスズガヤ,オホユレケサ,コバンサウ,コバンソウ,タハラムギ,タワラムギ,ムギラン
60382 チュウコバンソウ
60383 スズガヤ,ヒメコバンサウ,ヒメコバンソウ
60431 ハナニラ属
60447 ムラサキハナニラ
60526 セイヨウアマナ,ハナニラ
60544 ブロメーリア属
60555 フイリブロメリア
60557 アナナス科,パイナップル科
60599 クシロチャヒキ
60612 スズメノチャヒキ属
60631 チシマチャヒキ
60653 ニセコバンソウ
60654 クシロチャヒキ
60659 ヤクナガイヌムギ
60667 ミヤマチャヒキ
60680 ムクゲチャヒキ
60692 オニチャヒキ
60696 ヒゲナガスズメノチャヒキ
60724 タカサゴチャヒキ
60745 ハマチャヒキ
60748 ハトノチャヒキ
60760 コスズメノチャヒキ
60777 スズメノチャヒキ
60797 オオチャヒキ,オホチャヒキ
60820 マドリードチャヒキ
60853 ハマチャヒキ
60860 ニヒタカキツネガヤ
60928 キツネガヤ
60935 オオキツネガヤ,オホキツネガヤ,ヒゲナガスズメノチャヒキ
60943 チャボチャヒキ
60963 カラスノチャヒキ
60974 ノゲイヌムギ
60985 アレチノチャヒキ
60989 ウマノチャヒキ
60992 メウマノチャヒキ
61008 イヌムギ
61087 ブラウトーニア属
61096 カウゾ属,コウゾ属
61101 ツルカウゾ,ツルコウゾ,ハラミツ
61102 ツルコウゾ,ナンゴクコウゾ,ムクミカズラ,ムクミコウゾ
61103 カウゾ,カズ,カゾ,カミキ,カミギ,カミノギ,コウゾ,コーズ,コーゾ,コゾノキ,ヒメコウゾ,ヤマカジ
61107 オウコウズ,オブチ,カジノキ,カヂノキ,カミノキ,マムシカヅ,マムシラヅ
61118 アガリバナ属,ルリマガリバナ属
61119 ルリマガリバナ
61126 マガリナデシコ
61131 タイリンマガリバナ,タイリンルリマガリバナ,ブロワリアースベシオーサ
61132 ネバリマガリバナ,ネバリルリマガリバナ
61135 ブラウナンッス属
61154 ブラウーネア属
61157 ホウカンボク
61165 ブロウニンギア属
61166 グンダチュウ,セイドウリュウ
61167 ブットウ
61199 ニガキモドキ属
61208 ニガキモドキ
61231 コダチチョウセンアサガオ,ピンクダチュラ
61235 コダチチョウセンアサガオ
61240 ベニバナチョウセンアサガオ
61242 オオバナチョオセンアサガオ,カシワバチョオセンアサガオ,キダチチョウセンアサガオ
61250 オヒルギ属,ヲヒルギ属
61258 アカバナヒムギ
61263 アカバナヒルギ,オヒルギ,タンガラ,ベニガクヒルギ,ヲトコヒルギ,ヲヒ

ルギ
61293 バンマツリ属
61295 アメリカバンマツリ
61297 オオバンマツリ,オホバナマウリンクワ
61304 バンソケイ,バンマツリ
61306 ニオイバンマツリ
61310 オオバンマツリ
61316 バンマツリ
61381 ブルーノニア属,ブルノニア属
61384 ブルーノニア科,ブルノニア科
61390 ブルンズウィギア属
61430 チシマツガザクラ属
61432 チシマツガザクラ
61441 ホガエリガヤ属,ホガヘリガヤ属
61442 ハジガヘリ,ホガエリガヤ,ホガヘリガヤ
61549 ブリオポプシア属
61561 トウロウサウ属,トウロウソウ属
61568 コダカラベンケイ,シコロベンケイ
61572 トウロウソウ
61665 ヤマンワヤ属
61666 ウミンヤ,ヤマンワヤ
61929 ツクバネ属
61935 コギノコ,ツクバネ,ハゴノキ,マメギ
61936 タニガワツクバネ
61953 フジウツギ属,フヂウツギ属
61954 ハナフジウツギ
61955 ニワフジウツギ
61975 タイワンフジウツギ,タイワンフヂウツギ,タカサゴフジウツギ,タカサゴフヂウツギ,ニオイフジウツギ
62000 バニバナフジウツギ,バニバナフッドレア
62015 ウラジロブジウツギ,コフジウツギ
62016 ウラジロフジウツギ
62019 チチブフジウツギ,フサフジウツギ
62064 タマフジウツギ
62093 フジウツギ,フヂウツギ
62099 アキウツギ,ニッコウウツギ,フジウツギ,フヂウツギ
62102 シロバナフジウツギ
62110 シマフジウツギ,タウフヂウツギ,トウフジウツギ,トオフジウツギ,リュウキュウフジウツギ,リュウキュウフヂウツギ
62118 アフリカフジウツギ,アフリカフヂウツギ
62208 フジウツギ科
62334 ブルビーネ属
62471 ブルビネラ属
62518 ブルボコジューム属
62530 バルボフィラム属,バルボフィルム属,マメヅタラン属
62536 クスクスラン
62571 ヒメマメラン
62591 オガサワラシコウラン
62696 サウマラン,マメヅタラン,マメラン
62697 ベニマメヅタラン
62799 イボラン,ムギラン
62806 ホザキクシノハラン
62813 ミヤマムギラン
62814 キバナミヤマムギラン
62873 ウライラン,シコウラン
62875 タネガシマシコウラン
62900 タシロクシノハラン
62958 トサカクシノハラン
62985 ユリラン
63035 キバナクシノハラン
63159 ケイタオクシノハラン
63204 ハタガヤ属
63216 ハタガヤ
63253 イトハナビテンツキ
63258 イトテンツキ
63327 オオハタガヤ
63443 テンシンナズナ属
63446 テンシンナズナ
63475 アレチウイキョウ
63524 タウカセン属,トウカセン属,ブプタルムム属
63539 タウカセン,トウカセン
63542 キハマギク
63553 ミシマサイコ属
63558 レブンサイコ
63562 ホソバノミシマサイコ
63571 コガネサイコ
63594 ヒロハミシマサイコ,マンシウミシマサイコ
63622 タイゲキサイコ
63625 カマクラサイコ,ミシマサイコ
63665 クルマバサイコ
63699 ニセツキヌキサイコ
63728 オオホタルサイコ,ダイサイコ,ホタルサイコ,ホタルソウ
63735 ホタルサイコ
63736 ホタルサイコ
63739 オオハクサンサイコ
63740 エゾホタルサイコ
63741 コガネサイコ
63764 ハクサンサイコ
63765 シナノサイコ
63766 エゾサイコ
63805 ツキヌキサイコ
63807 ホタルサイコ,ホタルサウ,マルバサイコ
63813 ホソバミシマサイコ,ミシマサイコ
63820 イキノサイコ
63821 ミシマサイコ
63827 ダウリアザイコ
63864 レブンサイコ
63903 ブルケルリア属
63949 ヒナノシャクジョウ属,ヒナノシャクヂャウ属
63960 ヒナノシャクジョウ,ヒナノシャクヂャウ
63964 ミドリシャクジョウ
63966 シロシャクジョウ,シロシャクヂャウ
63977 ヤヘヤマシャクヂャウ,ルリシャクジョウ,ルリシャクヂャウ
63981 キリシマシャクジョウ
63987 キリシマサウ,キリシマシャクヂャウ
64000 ヒナノシャクジョウ科,ヒナノシャクヂャウ科
64088 カンラン科
64148 ハナモツヤク,ハナモツヤクノキ
64159 ブティア属
64161 ブティア
64173 ハナイ科,ハナヰ科
64180 ハナイ属,ハナヰ属
64184 ハナイ,ハナサワイ,ハナサワヰ,ハナヰ
64224 ツゲ科
64235 ツゲ属,ブクスス属
64273 オキナワツゲ
64285 アサマツゲ,クサツゲ,ツゲ,ニハツゲ,ニワツゲ,ヒメツゲ
64301 チョウセンヒメツゲ
64303 アサマツゲ,コツゲ,ツゲ,ベンテンツゲ,ホンツゲ
64304 アリマツゲ
64305 コメツゲ
64306 サンゴジュツゲ
64312 ベンテンツゲ
64316 コツゲ
64345 アサマツゲ,セイヨウツゲ,ツゲ,ホンツゲ
64369 タイワンアサマツゲ,テウセンヒメツゲ
64371 タイワンアサマツゲ
64373 オキナハツケ
64491 カボンバ属,ハゴロモモ属,フサジュンサイ属
64496 ハゴロモモ,フサジュンサイ
64503 ハゴロモモ科
64822 サボテン科
64965 ケザルピニア属,ジャケツイバラ属
64971 シロツブ,ハスノミカヅラ
64973 シロツブ
64975 ブラジルジャケツイバラ
64982 ケザルピニア,ユリアリア
64983 ナンテンカズラ,ナンテンカヅラ
64990 カハラフヂ,カワラフジ,キサイカチ,サルカケイバラ,ジヤケツイバラ,ハナサイカチ,ハマゴケ,ハマササゲ,マメイバラ
64991 ジャケツイバラ
64993 ジヤケツイバラ
65002 ベルナンブコ,ペルナンブコ
65014 ジリエホウオウボク,ホウオウボク

モドキ
65034 ハスノミカズラ
65055 オウゴチョウ,ワウゴテフ
65060 スオウ,スハウ,スホウ
65084 ジャケツイバラ科
65128 カヨフォラ属
65143 カヤ－ヌス属,キマメ属,リュウキュウマメ属
65146 キマ,キマメ,ジュトウ,リヴキウマメ,リュウキュウマメ
65156 ビロウドヒメクヅ,ビロ－ドヒメクズ
65185 オニハマダイコン
65202 カラデニア属
65214 カラジュ－ム属,ニシキイモ属,ハイモ属
65218 カラジュ－ム,カラヂュ－ム,ニシキイモ,ハイモ,ハニシキ
65229 ヒメハイモ,ヒメハニシキ
65230 シロスジカイウ
65237 アオハイモ,アヲハイモ,チリメンハイモ
65239 ソテツバカイウ
65254 ノガリヤス属
65255 シコクガリヤス
65256 ゴヨウザンガリヤス
65257 ムラマツノガリヤス
65258 ヤマアワモドキ
65265 コバナノガリヤス
65270 ハクトウイワガリヤス
65277 マンシウノガリヤス
65291 キリシマノガリヤス
65292 シマノガリヤス
65294 クジュウノガリヤス
65299 タイワンサイトウガヤ
65300 サイシュウノガリヤス
65330 ヤマアハ,ヤマアワ
65333 ホソヤマアハ
65343 カニツリノガリヤス
65354 オニノガリヤス
65359 オオヒゲガリヤス
65363 チョウセンノガリヤス
65413 ヒゲノガリヤス
65423 ヤクシマノガリヤス
65425 ムツノガリヤス
65429 モコキノガリヤス
65434 ヒナガリヤス
65435 ザラッキヒナガリヤス
65436 オオミネヒナノガリヤス
65438 チシマガリヤス
65450 オニビトノガリヤス
65464 トシマガヤ,ホッスガヤ
65471 イハガリヤス,イハノガリヤス,エゾノガリヤス,ネムロガヤ
65473 イワノガリヤス
65477 タカネノガリヤス
65486 ミヤマノガリヤス
65487 ミヤマノガリヤス
65499 タシロノガリヤス
65500 シコクノガリヤス
65507 キタノガリヤス
65626 カラモフィルム属
65637 カラムス属,トウ属
65670 コシキブ,コムラサキ,コムラサキシキブ,ミムラサキ
65697 シマトウヅル,タイワントウ,ミズトウズル,ミヅトウヅル
65782 ロタントウ
65827 マツゲボタン属
65835 マツゲボタン
65843 タイリンマツゲボタン
65855 ハイマツゲボタン
65862 エビネ属
65863 ユウヅルエビネ
65865 リュウキュウエビネ
65869 ダルマエビネ,ヒロハノカラン
65871 キソエビネ,コラン
65874 アマミエビネ
65881 トガリバナエビネ
65886 アリサンエビネ,ササキエビネ
65887 エリッブエビネ,キリシマエビネ,ヒロハノエビネ,ライシャエビネ
65915 タガネラン,マツダエビネ
65920 タマザキエビネ
65921 エビネ
65923 アカエビネ
65924 ダイダイエビネ
65926 キヌタエビネ
65927 アカクラエビネ
65928 ヤブエビネ
65930 タカネ
65931 ハノジエビネ
65934 カツウダケエビネ
65952 シマエビネ,タイワンエビネ,ホソバナエビネ
65966 ホソバナエビネ
65975 アサヒエビネ
65980 オオキミシマエビネ,オオキリシマエビネ,ニオイエビネ
66001 リュウキュウエビネ,レンギョウエビネ
66005 サクラジマエビネ
66008 シロバナオナガエビネ
66020 キンセイエビネ,キンセイラン
66034 オクシリエビネ
66035 ナツエビネ
66037 ナツエビネ
66046 ナツエビネ
66063 キソエビネ
66068 カハカミエビネ,キエビネ
66074 シマエビネ,タイワンエビネ,ホソバナエビネ
66086 オナガエビネ
66095 トクノシマエビネ
66099 エゾエビネ,サルメンエビネ,ヒダブチエビネ
66101 カラン,クワラン,シンナシエビネ,ツルラン,リウキウエビネ,リュウキュウエビネ
66157 カラテ－ア属,テブラサウ属,テブラソウ属
66159 シロバヤバネバショウ
66160 バッベムヒメバショウ
66165 ヤバネシハイヒメバショウ
66175 カラテア,ゴシキヤバネバショウ
66181 フイリヒメバショウ
66185 ベニスジヒメバショウ
66190 ヤマズヒメバショウ
66191 シハイヒメバショウ
66198 ナガジクヤバネバショウ
66201 チャボベニスジヒメバショウ
66205 オオヤバネバショウ
66211 オオゴシキヤバネバショウ
66215 テブラサウ,トラフヒレバショウ
66252 キンチャクソウ属
66262 キンチャクソウ
66271 キンチャクソウ
66273 チリメンキンチャク,チリメンキンチャクソウ
66284 ハネバキンチャクソウ
66287 アツバキンチャクソウ,ダキバキンチャクソウ
66337 マルバオモダカ属
66343 マルバオモダカ
66344 ウキマルバオモダカ
66380 キンセンカ属,キンセンクワ属,ホンキンセンカ属
66395 キンセンカ,キンセンクワ,トウキンセン,トウキンセンカ,ヒメキンセンカ,ホンキンセンカ
66445 キンセンカ,タウキンセン,トウキンセン,トウキンセンカ
66449 キンセンカ
66593 カラ－属,ガリアンドラ属,ヒメカイウ属,ミズイモ属,ミヅイモ属
66600 ヒメカイウ,ミズイモ,ミズザゼン,ミヅイモ,ミヅザゼン
66637 シマダフヂ
66643 サクカウ,サッコウフジ,ムラサキナツフジ,ムラサキナツフヂ
66660 カリアンドラ属,ベニガフクワン属,ベニゴウカン属
66669 ヒゴウカン,ヒネム,ベニガフクワン,ベニゴウカン
66673 アカバナブラッシマメ,オオベニゴウカン
66697 ウメザキサバノオ属,カリアンセマム属,キタダケソウ属,ヒダカサウ属,ヒダカソウ属
66725 コエビソウ
66727 ムラサキシキブ属

66728 イヌムラサキシキブ
66733 アメリカムラサキ,アメリカムラサキシキブ
66741 オオムラサキシキブ
66769 シラタマコシキブ,シロミノコムラサキ
66779 ホウライムラサキ,ホオライムラサキ
66794 シマムラサキ
66808 オオムラサキシキブ,オホムラサキシキブ,カナモドリ,コメゴメ,コメノキ,シロシキブ,ナンメイラ,ハシノキ,ミムラサキ,ムラサキシキブ,ヤマムラサキ
66809 シロシキブ,ヤマシロ
66810 シロバナムラサキシキブ
66817 ナガバムラサキシキブ
66824 オオムラサキシキブ,オホムラサキシキブ
66825 オオシロシキブ
66826 コミムラサキシキブ
66827 コバムラサキシキブ
66829 オニヤブムラサキ,カラチムラサキ,ケムラサキ,コウヂムラサキ,コメゴメ,シマムラサキ,タイワンシロシキブ,ビロウドムラサキ,ビロオドムラサキ,ビロ-ドムラサキ,ヤマムラサキ
66831 アンタオムラサキ,コウトウムラサキ,シマムラサキ
66840 ナガバムラサキ
66853 タカクマムラサキ
66856 オニヤブムラサキ,ケムラサキ,コメゴメ,タイワンシロシキブ,ヤマムラサキ
66873 ケムラサキ,コメゴメ,ヤブムラサキ,ヤマムラサキ
66874 シロミノヤブムラサキ
66875 ナガバヤブムラサキ
66882 オオシマムラサキ,オホシマムラサキ
66883 イリオモテムラサキ
66884 オキナワヤブムラサキ
66888 ウラジロコムラサキ
66897 アオゲムラサキ,アヲゲムラサキ,ホソバムラサキ
66904 ランダイムラサキ
66909 コウシュンムラサキ,タイワンムラサキ
66913 ナンバンムラサキ,ネンバレムラサキ
66927 タカクマムラサキ,トサムラサキ,ヤクシマコムラサキ
66928 シロバナトサムラサキ
66931 オオバシマムラサキ,ムニンムラサキ
66943 オオバムラサキ,オホバムラサキ
66975 カリコマ属
66976 カリコマ
67124 カリロエ属,ケシバナアオイ属
67128 ケシバナアオイ,フウロアオイ,ボタンフウロ
67142 カリシア属
67145 オリヅルツユクサ,シタレツユクサ
67248 アキバブラシノキ属,カリステモン属,ブラッシノキ属,マキバブラッシノキ属
67253 ハナマキ
67262 サメカハブラッシノキ,サメカワブラッシノキ
67275 キンポウジュ,ハナマキ,ハマキ
67278 ホソバブラッシノキ
67286 マルバブラッシノキ
67291 マキバブラッシノキ
67293 シロバナブラシノキ,シロバナマキ
67294 シロバナマキ
67298 ブラシノキ,ブラッシノキ
67302 シダレハナマキ
67312 エゾギク属,サツマギク属
67314 アスタ-,エゾギク,サツマギク,サツマコンギク,テウセンギク,マンザイギク
67334 アワゴケ科,ミヅハコベ科
67336 アワゴケ属,ミヅハコベ属
67361 チシマミズハコベ
67367 アハゴケ,アワゴケ
67376 テウセンアハゴケ,ナガバアワゴケ,ミギハハコベ,ミズハコベ
67382 ミズハコベ
67393 イケノミズハコベ,ミヅハコベ
67395 アメリカアワゴケ
67405 カリトリス属,マオウヒバ属
67455 カルナ属,ギョリュウモドキ属
67456 カル-ナ,ギョリュウモドキ,ギョリュウロドキ,ナツザキエリカ
67526 オニヒバ属,ショウナンバク属
67529 シマヒノキ,ショウナンバク,ショウナンボク,セウナンボク,ワウニクジュ
67530 オウニクジュ,シマヒノキ,ショウナンボク
67532 カロセファラス属
67554 カロコルタス属
67635 カロコルタス-ベヌスタス
67734 ヨルガオ属,ヨルガホ属
67736 シロバナユウカオ,ヤカイソウ,ユウカオ,ヨルガオ,ヨルガホ
67846 テリハボク属
67860 タマナ,タマワ,テラハボク,テリハボク,ヒイタマナ,ヤラブ,ヤラボ
67882 カロポゴン属
67898 クズモドキ
68082 イガギク
68088 アコン,カイガンタバコ
68157 リウキンクワ属,リュウキンカ属
68186 ヒメエンコウソウ
68189 シベリアリュウキンカ,リウキンクワ,リュウキンカ
68196 エゾノリウキンクワ,エゾノリュウキンカ,ヤチブキ
68201 リュウキンカ
68206 コバノリウキンクワ
68210 エンコウサウ
68300 ラフバイ科,ロウバイ科
68306 クロバナラフバイ属,クロバナロウバイ属
68310 アメリカロウバイ
68313 クロバナラフバイ,クロバナロウバイ,ニオイロウバイ,フロリダロウバイ
68321 アメリカラフバイ,クロバナロウバイ
68531 ホテイラン属
68538 ヒメホテイラン,ホテイラン
68539 シロバナヒメホテイラン
68542 ツリフネラン,ホテイラン
68544 ホテイラン
68614 カリブトロカリテクス
68617 ホバナヘナヘナヤシ
68639 マナックヤシ属
68671 ヒルガオ属,ヒルガホ属
68678 ヒルガオ
68686 コヒルガオ,コヒルガホ
68688 フギレヒルガオ
68689 シロバナコヒルガオ
68692 ヒルガオ,ヒルガホ
68701 タチヒルガオ,マンシウヒルガオ
68705 ヒルガオ
68706 テンシボタン
68707 シロバナヒルガオ
68708 ヒルガオ
68713 ヒロガホ,ヒロハヒルガオ,ヒロハヒルガホ
68716 ヒロハヒルガオ
68721 ヒロハヒルガオ
68724 ヒルガオ
68737 アフヒカヅラ,ハマヒルガオ,ハマヒルガホ
68739 シロバナハマヒルガオ
68740 ベニバナハマヒルガオ
68789 カマシア属
68797 コヒナユリ
68800 オオヒナユリ
68803 ヒナユリ
68839 アマナヅナ属
68841 アマナズナ
68850 ヒメアマナズナ
68860 アマナヅナ,タマナヅナ,ナガミノアマナズナ
68877 チャノキ属,ツバキ属
68920 ユカリッバキ
68939 オホシマサザンクワ,フタバナサザンクワ
68953 シマサザンクワ,ミヤマサザンクワ

68970 シマヒメツバキ,タイワンヒメツバキ,トガリバヒメツバキ
69015 クラフネリアーナ
69033 クスビダータ
69064 サカキモドキ,ヒサカキモドキ
69082 シラバトツバキ,フラテルナ
69108 グランサムツバキ
69133 カンツバキ,シシガシラ
69135 ホンコンツバキ
69153 イラワジエンシス
69156 カタシ,ツバキ,ヤブツバキ,ヤマツバキ
69193 ロクベンヤブツバキ
69194 ホソバヒラギツバキ
69195 オトメツバキ
69196 ナガバヤマツバキ,ヤブツバキ
69198 シロバナヤブツバキ
69199 ユリツバキ,ユリバツバキ
69201 コバナヤブツバキ,ヤブツバキ
69203 チリツバキ
69204 キンギョツバキ,キンギョバツバキ
69207 サルイワツバキ,ユキツバキ
69210 シロバナユキツバキ
69211 ヤエノユキツバキ
69213 ノザワツバキ
69214 ユキツバキ
69218 ヤクシマツバキ
69235 トガラバサザンカァ,ヒマラヤサザンカ
69327 ヒメサザンカ,リュウキュウツバキ
69348 テマリツバキ
69370 オキナハサザンカ,オキナハサザンクワ,オキナワサザンカ
69398 ウスバヒメツバキ
69411 アブラツバキ,オオサザンカ,オオシマサザンカ,オホシマサザンクワ,フタバナサザンクワ,ユチャ
69480 キンカチャ
69494 ビタールツバキ・ピタルデイー
69504 ピタルツバキ・ユンナン
69552 カラツバキ,タウツバキ,チョウセンツバキ,テウセンツバキ,トウツバキ,ナンキンツバキ,ヤマトウツバキ
69572 ローセーフローラ
69586 キンモウサカキ,ヤナギバサザンカ,ヤナギバツバキ
69589 サルウィンツバキ
69594 コカタシ,コツバキ,サザンカ,サザンクワ,ヒメツバキ
69606 カンツバキ
69634 サウジンボク,サウニンボク,ソウジンボク,ソウニンボク,チャ,チャノキ,メザマシグサ
69641 ベニバナチャ
69644 アッサムチャ,ホソバチャ
69651 タウチャ,トウチャ,ニガチャ
69691 タリエンシス
69694 ワンンサザンカ
69705 タイワンヒメサザンカ,タイワンヒメサザンクワ,テリハヒメサザンクワ
69706 テリバヒメザザンクワ
69721 ツアイ
69739 ウラク,ウラクツバキ
69740 ハルサザンカ
69753 ベニワビスケ
69754 コチョウワビスケ
69830 カモエンシア属
69870 カンパニュラ属,ホタルブクロ属
69942 ニワギキョウ
69956 チシマギキョウ
69957 シロバナチシマギキョウ
70005 ヒメギキョウ
70008 ホシザキキキョウ
70054 ハナヤッシロソウ,ヤツシロサウ,ヤツシロソウ
70058 ビロードヤッシロソウ,ヤツシロソウ
70061 ヤツシロソウ
70062 シロバナヤッシロソウ
70117 イワギキョウ
70118 シロバナイワギキョウ
70164 フウリンサウ,フウリンソウ
70170 シマホタルブクロ
70214 モモノハギキョウ
70228 オトメギキョウ
70234 ホタルブクロ
70235 フナシホタルブクロ
70236 テリハホタルブクロ
70237 ムラサキホタルブクロ
70241 ヤマホタルブクロ
70242 シロバナヤマホタルブクロ
70255 ハタザオギキョウ
70271 イトシャジン
70279 ホソバイワギキョウ
70372 キキヤウ科,キキョウ科
70386 ツルギキヤウ属,ツルギキョウ属
70398 ツルギキヤウ,ツルギキョウ
70403 シマギキョウ,タイワンツルギキョウ,タンゲブ
70502 ノウゼンカズラ属,ノウゼンカツラ属
70503 アイノコノウゼンカズラ
70507 ノウゼン,ノウセンカズラ,ノウゼンカズラ,ノウゼンカツラ,ノウゼンカヅラ,ノゼウ,マカヤキ,メショウ
70509 カバイロノウゼンカズラ
70512 アメリカノウセンカズラ
70575 カンレンボク
70833 ハナハギ
70847 トンボハギ
70954 イランイランノキ属
70960 イランイランノキ
70971 カナリア属
70988 カンラン属
70989 ウオノホネヌキ,カンラン,ハクラン,リョクラン
70996 カナリカンカン,カナリヤノキ
71004 マニラエレミ
71013 ピリナッツツリ
71015 ウラン
71027 ケドンドン
71036 ナタマメ属
71040 タカナタマメ
71042 タチナタマメ,ツルナシナタマメ,ナタマメ
71050 タチハキ,タテハキ,トウズ,ナタマメ
71052 シロナタマメ
71055 ハマナタマメ
71056 シロバナハマナタマメ
71072 ナガミハマナタマメ
71129 カネラ属
71138 カネラ科
71149 カニストルム属
71156 カンナ属,ダンドク属
71166 アカバナカンナ,アカバナダンドク
71169 ショクヨウカンナ
71171 キバナカンナ,キバナノダンドク
71173 カンナ,ハナカンナ
71179 シロバナカンナ
71181 ダンドク
71182 キバナダンドク
71186 ダンドク
71189 タイリンカンナ,タイリンダンドク
71193 イタリアンカンナ
71196 オランダダンドク,ホソバダンドク
71199 アメリカダンドク
71202 アサ科
71209 アサ属
71218 アサ,タイマ
71220 インドアサ
71235 カンナ科,ダンドク科
71544 カンツア属
71561 カパネミア属
71607 カショウアブラススキ
71608 ヒメアブラススキ
71610 ムラサキヒメアブラススキ
71611 コモチヒメアブラススキ
71613 チョウセンヒメアブラススキ,ムラサキヒメアブラススキ
71614 リュウキュウヒメアブラススキ
71671 フウチョウソウ科,フウテウサウ科
71676 カッパリス属,フウチウボク属,フウチョウボク属
71679 ウスバフウテウボク
71751 オホバフウテウカヅラ,オホバフウテウボク
71760 フウテウボク,フウテフボク
71762 トゲフウチョウボク
71775 ヒメフウテウボク,ホソバフウテウボク

71782 ウスバフウテウボク,コフウテウボク
71871 トゲフウチョウボク
71955 スイカズラ科,スヒカヅラ科
72037 ナズナ属,ナヅナ属
72038 ナズナ,ナヅナ,バチグサ,ペンペングサ,ホソミナズナ
72047 ナズナ
72059 ルベラナズナ
72068 タウガラシ属,トウガラシ属
72070 ウワムキトウガラシ,コシキトウガラシ,テンヂクマモリ,トウガラシ
72071 トウガラシ
72072 ゴシキトウガラシ,ホシトウガラシ
72073 ゴシキトウガラシ
72074 テンジクマモリ,テンヂクマモリ,ヤツブサ,ヤツブサ
72075 シシタウガラシ,ピーマン
72077 ナガミトウガラシ
72081 シシトウガラシ
72100 キダチタウガラシ,キダチトウガラシ,シマタウガラシ
72113 ヒメタウガラシ
72168 ムラスズメ属,ムレスズメ属
72180 オオムレスズメ,オホムレスズメ
72233 コムレスズメ
72235 コムレスズメ
72279 マンシュウムレスズメ
72285 アオムレスズメ,コバノムレスズメ
72288 コバノムレスズメ
72322 ヒメムレスズメ,ホソバムレスズメ
72334 コムレスズメ
72342 ムレスズメ
72351 ヒメムレスズメ,ホソバムレスズメ
72402 カラルマ属
72425 リュウカク
72674 タネツケバナ属
72678 ツルワサビ
72680 ミツバコンロンソウ
72681 オオミツバコンロンソウ
72682 ヒトツバコンロンソウ
72685 ヒロハコンロンソウ
72686 オオマルバコンロンソウ
72689 アフヒバタネツケバナ,アリサンナヅナ
72717 オクヤマナズナ
72721 シマユリワサビ,タイワンユリワサビ,ナカイサウ,ナカヰサウ
72735 オオケタネツケバナ
72737 ニシノオオタネツケバナ
72744 タチタネツケバナ
72747 エゾワサビ
72749 アフヒバタネツケバナ,アリサンナヅナ,イイラギ,タガラシ,タチタネツケバナ,タネツケバナ,タビラキ
72764 タチタネツケバナ
72802 ミチタネツケバナ
72829 ジャニンジン
72835 ケジャニンジン
72838 ナガエジャニンジン
72851 サジガラシ
72858 コンロンソウ
72861 ハダカコンロンソウ
72873 ミズタガラシ,ミヅタガラシ
72876 コンロンサウ,コンロンソウ,ムラサキコンロンソウ
72909 コシジタネツケバナ
72910 ミネガラシ,ミヤマタネツケバナ
72920 コタネツケバナ,コバナノタネツケバナ,ヒメタネツケバナ
72924 コタネツケバナ
72934 カルダミネ-プラテンシス,セキサウ,セキソウ,ハナタネツケバナ
72942 ハイタネツケバナ,ハヒタネツケバナ
72965 エゾノジャニンジン
72966 ケエゾノジャニンジン
72971 オオバタネツケバナ,タネツケバナ,マンシュウタネツケバナ,ヤマタネツケバナ
72979 ミズタネツケバナ
73000 マルバコンロンソウ
73013 オクヤマガラシ
73018 ミヤウチソウ
73044 アイヌワサビ
73045 タカチホガラシ
73090 アコウグンバイ
73119 クサアジサイ属,クサアヂサヰ属
73120 クサアジサイ
73121 ハナクサアジサイ
73122 オオバナノクサガク
73123 ヨウラククサアジサイ
73124 ミヤマクサアジサイ
73128 ハコネクサアジサイ
73131 アマミクサアジサイ
73134 オオクサアジサイ,オホクサアヂサヰ,シマクサアジサイ,シマクサアヂサヰ
73136 オオクサアジサイ
73154 ウバユリ属
73158 ウバユリ,カバユリ,ネズミユリ
73159 ヒマラヤウバユリ
73161 ウンバイロ,エゾウバユリ,オオウバユリ
73194 フウセンカズラ属,フウセンカヅラ属
73201 シュッコンフウセンカズラ
73207 フウセンカズラ,フウセンカヅラ
73208 コフウセンカズラ,コフウセンカヅラ
73278 ヒレアザミ属,ヤハズアザミ属
73281 オオシュウヒラアザミ,オオヒレアザミ
73329 ガンクビヒレアザミ
73337 オニノマユハキ,ヒレアザミ,ヤハズアザミ
73339 ヒレアザミ
73340 シロバナヒレアザミ
73377 セイヨウフジアザミ
73467 ヒメヒレアザミ
73504 イヌヒレアザミ
73543 スゲ属
73544 カヅノスゲ
73545 フトボタニガワスゲ
73546 オオタヌキラン
73547 ヒラギシスゲモドキ
73548 ミカワオオイトスゲ
73549 エンシュウカワラスゲ
73551 ハシナガアワボスゲ
73552 ゴウソモドキ
73553 ゴヨウザンスゲ
73554 ネムロゴウソ
73555 ホカリヒゴクサ
73556 タヌキナルコ
73557 イナテキリスゲ
73558 ノトロスゲ
73559 オニアゼスゲ
73560 エゾシロハリスゲ
73561 アイノコナルコ
73562 モリヨシスゲ
73563 タカオスゲ
73564 シラホウマスゲ
73565 オクシリカンスゲ
73566 ヤシマスゲ
73567 アイノコシラスゲ
73568 サドスゲモドキ
73569 フイリミチノクアゼスゲ
73570 クロシマシバスゲ
73571 タニガワタヌキラン
73573 スミカワスゲ
73574 ムジナクグ
73575 オクタヌキラン
73576 オゼクロスゲ
73589 アゼスゲ
73608 アハスゲ,アワスゲ,トダスゲ
73619 アキアオスゲ
73636 リュウキュウスゲリウギウスゲ
73641 カハカミスゲ,シラスゲ
73646 シラスゲ
73663 オオイトスゲ,オホイトスゲ
73666 キイトスゲ
73673 アムグンスゲ
73696 キタノカワズスゲ
73699 ヤマタヌキラン
73703 アメリカミコシガヤ
73714 カウライアゼスゲ,シユミットスゲ
73718 エナシヒゴクサ
73719 タテヤマスゲ
73724 アポイタヌキラン
73732 クロアゼスゲ
73738 クロアゼスゲ

73748 クロカワズスゲ
73751 ワンドスゲ
73755 アリサンタマツリスゲ
73767 アオナルコスゲ,アルネルスゲ,オクエゾアイスゲ
73768 アイズスゲ
73774 タイワンスゲ
73782 ヒメヤガミスゲ
73788 クロボスゲ
73824 ウミノサチスゲ
73826 エゾアゼスゲ,ヒラギシスゲ,ヤチバウズ
73828 シャリスゲ
73839 オオキリスゲ,オオナキリスゲ,オホキリスゲ
73842 ヒエボスゲ
73875 オハグロスゲ
73877 ヒラミスゲ
73884 ビッチュウヒカゲスゲ
73885 マツバスゲ
73888 ショウジョウスゲ
73891 ナガミショウジョウスゲ
73894 カヤツリスゲ
73906 イハスゲ,イマジスゲ,タカネスゲ,ヤマジスゲ,ヤマヂスゲ
73913 アオスゲ,アヲスゲ,イトアオスゲ,イワアオスゲ,ヒゲアオスゲ
73925 イソスゲ,ハマアオスゲ,ハマアヲスゲ
73928 オオアオスゲ
73930 オニカンスゲ
73933 ヒレミヤガミスゲ
73937 オキナワスゲ
73945 アハボスゲ,アワボスゲ
73952 ココメスゲ,ナキリスゲ
73959 ヒメカハヅスゲ,ヒメカワズスゲ,ヒメカワヅスゲ
73961 ヒメカワズスゲ
73987 カブスゲ,クロオスゲ,クロヲスゲ
73992 イトヒカゲスゲ,ホソバイトヒカゲスゲ,ホソバヒカゲスゲ
74004 ハクサンスゲ
74023 ハリガネスゲ,ハリガネッゲ
74029 ゴンゲンスゲ,ミチノクハリスゲ
74032 タカネシバスゲ
74051 ジャウブスゲ,ジョウロウスゲ
74085 カブスゲ
74086 クロメスゲ
74098 クミアヒスゲ,シナスゲ
74102 タテシナヒメスゲ
74111 タカネシバスゲ
74121 チョウセンハリスゲ
74122 キバナスゲ,コイハカンスゲ,コイワカンスゲ
74124 カンサイイワスゲ
74125 ミヤマイワスゲ
74133 ケタガネソウ
74135 ヌマアゼスゲ
74141 ヤマオオイトスゲ
74149 リュウキュウヒエスゲ
74171 ミヤマシラスゲ
74174 ヒメカンスゲ
74176 ウジカンスゲ
74177 トカラカンスゲ
74187 タワラスゲ
74196 クシロヤガミスゲ
74211 ジフモンジスゲ,ハナビスゲ
74221 カラフトヤラメスゲ,ヤラメスゲ
74224 カクレボスゲ
74230 ホソボノクロカハヅスゲ
74235 ハクサンスゲ
74241 ナルコスゲ
74257 ダイセンスゲ
74297 ホスゲ
74299 ホスゲ
74304 クリイロスゲ
74309 オニスゲ,ミクリスゲ
74327 アゼナルコ,アゼナルコスゲ
74333 マスノスゲ
74339 カサスゲ,スゲ,ミノスゲ
74342 ホソスゲ
74343 ミヤマジュズスゲ
74354 アサマスゲ
74368 コタヌキラン
74369 シマタヌキラン
74374 ナガボスゲ
74377 オキナワナガボスゲ
74383 シラスゲ
74392 アカンカサスゲ,エゾカサスゲ,モリカサスゲ
74394 アカンカサスゲ
74406 イトノヤマスゲ,ノヤマスゲ
74407 ノヤマスゲ
74408 イトノヤマスゲ
74415 ケスゲ
74423 キタノカワズスゲ
74431 タマツリスゲ,ハネスゲ,ムギスゲ
74438 ヒメアゼスゲ
74453 ホソノヤマスゲ
74460 モウコスグ
74483 ケヤリスゲ,サヤスゲ
74497 イトスゲ
74513 イヌハナビスゲ,イヌヒナビスゲ,ハナスゲ
74514 イヌハナビスゲ
74524 タマツリスゲ
74529 ヒロハノオオタマツリスゲ
74531 オクタマツリスゲ
74534 オオタマツリスゲ
74537 シナツリスゲ
74539 ヒメジュズスゲ
74544 オオアメリカミコシガヤ
74546 ヤマテキリスゲ
74569 ミヤマクロスゲ
74570 チャイロタヌキラン
74579 タマノヤガミスゲ
74584 オクノカンスゲ
74586 ウスイロオクノカンスゲ
74592 タニガハスゲ,タニガワスゲ
74594 オオタニガワスゲ
74596 オオミヤマカンスゲ,オホミヤマカンスゲ,ゲンカイモエギスゲ,タイワンスゲ
74601 エノコロスゲ
74608 ニッコウハリスゲ
74610 チャイロスゲ
74611 オンブスゲ
74636 ゲイホクスゲ
74638 アマミナキリスゲ
74643 ナカハラスゲ
74644 コマトメ,マスクサ,マスクサスゲ
74645 クロヒナスゲ
74646 アカヒナスゲ
74650 イヌカサスゲ,スナジスゲ
74664 コノッブスゲ
74669 コソッブスゲ
74672 タマツヅリスゲ,トナカイスゲ
74676 ネムロスゲ
74685 コウライヤワラスゲ
74692 ナゴスゲ
74697 ヒナスゲ
74700 サナギスゲ
74709 サヤシロスゲ
74722 カンチスゲ
74724 ハチジョウカンスゲ
74733 イトキンスゲ
74736 ハコネイトスゲ
74737 コハリスゲ
74752 クロカワスゲ
74764 サヤマスゲ
74768 ムニンナキリスゲ
74797 ヤマアゼスゲ
74801 ブンゲスゲ
74824 ホウザンスゲ,ホウワウスゲ
74839 フトボアゼスゲ
74842 ヒナタスゲ
74847 イトヒカゲスゲ
74849 ヒメヒカゲスゲ,ホソバヒカゲスゲ
74851 ヒナタスゲ
74857 ヤマクボスゲ
74866 ウマスゲ
74869 ヒルゼンスゲ
74872 カワラスゲ,タニスゲ
74888 ヒロバスゲ
74890 アオバスゲ
74891 アオヒエスゲ
74913 ダゲスゲ
74915 ジュズスゲ,ヒロバノヤハラスゲ
74916 オキナワジュズスゲ
74919 ハガクレスゲ
74920 ケハガクレスゲ

74934 チョウセンカサスゲ
74941 ヒゴクサ,ヒゴスゲ
74942 シラスゲ
74966 サンインヒエスゲ
74969 ヤリスゲ
74970 アキイトスゲ
74976 イセアオスゲ
74982 コツブスゲ
74997 テキリスゲ
74998 ホクリョウスゲ,ムシャスグ
75003 コウボウムギ,コウボフムギ,フデクサ
75012 ホソバスゲ
75021 クジュウツリスゲ
75030 タカネヤガミスゲ
75042 ヒメミコシガヤ
75043 セイタカカワズスゲ
75048 ヒカゲスゲ
75051 フイリヒカゲスゲ
75055 マンシウヒカゲスゲ
75065 ムジナスゲ
75073 アズマスゲ
75074 オオアズマスゲ
75076 オオムギスゲ,オホムギスゲ
75079 ハタベスゲ
75080 イトナスゲ,イトナルコスゲ
75095 リ－ベンボルシ－スゲ
75099 センジャウスゲ,センジョウスゲ,センヂャウスゲ
75101 ヤマミコシガヤ
75103 ナキリスゲ
75104 センダイスゲモドキ
75131 アオスゲ
75135 メアオスゲ
75136 ヒメアオスゲ
75139 ミセンアオスゲ
75140 オオアオスゲ
75144 セトウチコアオスゲ
75149 サツマスゲ
75161 サツマスゲ
75162 ウスイロサツマスゲ
75165 カラヒトヤチスゲ,ヤチスゲ
75178 アサマスゲ,タイリクカワズスゲ
75184 ムセンスゲ
75185 アカンスゲ
75189 キタヒエスゲ,ヒエスゲ,マツマエスゲ,マツマヘスゲ
75192 ウスイロヒエスゲ,チュウゼンジスゲ,ヒエスゲ
75252 ヤガミスゲ
75253 カラフトスゲ,ノルゲスゲ
75258 カタスゲ
75259 エゾノコウボウムギ,エゾノコウボフムギ,カラバウムギ,コウボフスゲ,ハマムギ,フデクサ
75279 タチスゲ
75280 リュウキュウタチスゲ
75281 ダケスゲ
75282 ダケスゲ
75287 イハカンスゲ,イワカンスゲ,バケイスゲ
75310 キノクニスゲ
75314 ガウソ,ゴウソ,タイツリスゲ
75316 ホシナシゴウソ
75317 チョウセンゴウソ
75318 ケヒエスゲ
75330 タカネヒメスゲ
75339 イソアオスゲ
75343 キンチャクスゲ
75347 シラボスゲ,フサスゲ
75348 オホミヤマスゲ,シラカハスゲ,シラカワスゲ,ヌマクロネスゲ,ヌマクロボスゲ
75350 ミタケスゲ
75357 ココメアゼスゲ,ヌマアゼスゲ
75362 コキンスゲ
75364 クロヤガミスゲ
75369 チャシバスゲ
75370 クロスゲ,トマリスゲ,ホロムイスゲ
75378 クロメスゲ
75384 サワヒメスゲ
75387 ヒメタヌキラン
75391 ヌカスゲ
75393 ノゲヌカスゲ
75394 ビンゴヌカスゲ
75396 ビモ－ドスゲ,ビロ－ドスゲ
75400 ヒメシラスゲ
75403 フクロスゲ
75417 モリスゲ
75418 ニヒタカスゲ
75421 カンスゲ
75423 シマカンスゲ,フイリカンスゲ
75425 ヤクシマカンスゲ
75443 ミヤマカンスゲ
75444 ケナシミヤマカンスゲ
75445 ヤワラミヤマカンスゲ
75446 アオミヤマカンスゲ
75447 ツルミヤマカンスゲ
75448 コミヤマカンスゲ
75471 キシウナキリスゲ,キシュウナキリスゲ
75488 アキカサスゲ,ネブラスカスゲ
75490 ホソバオゼヌマスゲ
75496 シバスゲ
75500 ミコシガヤ,ミコミスゲ
75544 ニヒタカカハヅスゲ
75547 ミノボロスゲ
75550 ツクシミノボロスゲ
75592 ホロムイクグ
75596 ミヤマシラスゲ
75604 ヤチカハッスゲ,ヤチカワズスゲ
75606 カワズスゲ
75607 チャボカワズスゲ
75609 スルガスゲ
75610 ハリスゲ,ヒカゲハリスゲ
75627 ノゲスゲ
75632 アリサンモエギスゲ
75635 オオシマカンスゲ
75637 フイリオオシマカンスゲ
75638 オタルスゲ,ヒメテキリスゲ,ヲタルスゲ
75639 ミカワナルコ
75640 ナガエスゲ
75649 ヒメスゲ
75650 ナガミヒメスゲ
75658 ササノハスゲ
75675 ウスイロスゲ,エゾカハヅスゲ,エゾカワズスゲ
75690 マキバクロカワズスゲ
75692 エゾツリスゲ
75698 グレ－ンスゲ
75699 コジュズスゲ
75700 ムギスゲ
75701 ヒロハノコジュズスゲ
75702 ナガボノコジュズスゲ
75711 タカネハリスゲ,ミガエリスゲ,ミガヘリスゲ
75720 キビノミノボロスゲ
75724 タカヒカゲスゲ
75737 マンシウクロカワスゲ,マンシュウクロカワスゲ
75740 ヒメビロ－ドスゲ
75756 ダンスゲ
75759 クロボスゲ
75761 キンキカサスゲ
75762 フクイカサスゲ
75769 アオゴウソ,カスガスゲ,ヒメゴウソ,ヒメナルコスゲ
75771 ハシナガカンスゲ
75775 テンジクスゲ
75785 ケスゲ,サッポロスゲ,ハナマガリスゲ,ハナマルコスゲ,ミミスゲ
75786 サッポロスゲ
75792 イトスゲ,シロホンモンジスゲ,ホンモンジスゲ,ワタリスゲ
75826 タカネマスクサ
75827 ヒカゲシラスゲ
75828 チチブシラスゲ
75841 アカネスゲ
75843 タヌキラン
75882 ガウソ
75894 ツルカハヅスゲ,ツルカワズスゲ,ツルスゲ
75895 クグスゲ
75910 ヒロハイッポンスゲ
75926 マメスゲ
75930 コウボウシバ,コウボフシバ,ハマムギ
75941 ホソバノクロカハヅスゲ
75942 キンスゲ
75954 アカスゲ

75957 コノゲスゲ
75970 ウシオスゲ
75980 チシマスゲ
75992 コカンスゲ
75993 ショタイサウ,ヤブスゲ
76005 イトヒキスゲ
76015 オオヒカゲスゲ
76022 シラコスゲ
76029 オオカサスゲ,オホカサスゲ
76051 マバマスゲ,ヤブスゲ
76068 カラフトカサスゲ
76070 ヌマスゲ
76073 コヌマスゲ
76086 クサスゲ
76090 オオクグ,オホクグ
76094 カラフトイハスゲ,カラフトイワスゲ
76100 カミカハスゲ,カミカワスゲ,チョウセンアカスゲ
76103 ツルカミカワスゲ
76114 チャイトスゲ
76117 クジュウスゲ
76119 キイトスゲ
76120 ゴンゲンスゲ
76121 ミヤマアオスゲ
76122 クラサワスゲ
76123 マツカゼスゲ
76126 ベニイトスゲ
76130 ジングウスゲ
76131 オキナワヒメナキリ
76132 サドスゲ
76150 アブラシバ,アブラスゲ
76163 シオクグ,シホクグ,ハマクグ
76178 カウライアセスゲ,コウライアセスゲ,シュミットスゲ
76195 ミヤマアシボソスゲ
76200 ダイセンアシボソスゲ
76201 リシリスゲ
76202 シコタンスゲ
76203 アシボソスゲ
76204 オオタヌキラン
76207 アメリカヤガミスゲ
76219 ヒメオノエスゲ,ミヤマアオスゲ,ミヤマアヌスゲ
76221 ユキグニハリスゲ
76226 センダイスゲ
76230 エゾサハスゲ
76254 アズマナルコ
76264 ギャウジャサウ,ササスゲ,タガネサウ,タガネソウ,ヤマオホバコ
76266 フイリタガネソウ
76268 ケタガネソウ
76271 ホソバタガネソウ
76289 タカネナルコ
76296 クミアイスゲ,クミアヒスゲ,タシロスゲ
76312 ケヤリスゲ,サヤスゲ
76338 クロアゼスゲ
76342 カハヅスゲ,キタノカワズスゲ
76348 イワスゲ
76349 タイセツイワスゲ
76365 ニシノホンモンジスゲ
76366 ミチノクホンモンジスゲ
76367 コシノホンモンジスゲ
76375 オオカワズスゲ,オホカハヅスゲ
76413 ラウススゲ
76416 ツクシナルコ,ツクシナルゴスゲ
76418 ツルカミカハスゲ,ツルカミカワスゲ
76431 ヒカゲスゲ
76437 ヒメアシホスゲ,ヒメウシオスゲ,ヒメウシホスゲ
76446 ミヤケスゲ
76447 クモマシバスゲ
76473 タイホクスゲ
76477 ヒメミヤマジュズスゲ
76479 タイワンヒエスゲ
76490 タルマイスゲ
76491 ノスゲ
76492 ダンスゲ
76497 タイホクスゲ,タッタカスゲ
76501 フサナキリスゲ
76503 ホソバカンスゲ
76513 イッボンスゲ,イッポンスゲ,シロハリスゲ
76515 オノエスゲ,オノヘスゲ,テンセンスゲ,ヲノヘスゲ
76516 ケオノエスゲ
76517 チャイトスゲ
76518 コバケイスゲ
76545 アゼスゲ
76547 オオアゼスゲ,コアゼスゲ
76553 フサカンスゲ
76568 セキモンスゲ
76570 ヒロハオゼヌマスゲ
76571 ダイブスゲ,ミヤマスゲ
76573 ヤハラスゲ,ヤワラスゲ
76585 タカネヤガミスゲ
76591 モエギスゲ
76593 コップモエギスゲ,ヒメモエギスゲ
76612 ツシマスゲ
76617 イワヤスゲ
76618 ハハキカサスゲ,ホウキカサスゲ,ミツマタスゲ
76626 エゾハリスゲ,オオハリスゲ,オホハリスゲ
76628 マンシゥクロヒナスゲ
76644 カタガワヤガミスゲ
76651 ココメスゲ,ヒロハヒエスゲ
76656 ケヤリスゲ,サヤスゲ
76658 サヤスゲ
76665 シロウマヒメスゲ,ヌイオスゲ,ヌイヲスゲ
76677 オニナルコスゲ
76683 ジュズナルコスゲ
76687 オニナルコスゲ,ホソバナルコスゲ
76701 エゾサワスゲ
76735 ナガバアメリカミコシガヤ
76737 ヒゲスゲ
76740 ヒゲスゲ
76773 ヒロバヒエスゲ
76777 ヤクシマハマスゲ
76778 ヒメアサマスゲ
76808 パパイヤ属,バンクワジュ属
76813 チチウラ,チチウリ,パパイヤ,パパヤ,バンクワジュ,マンジュカ,マンジュクワ,モククロ,モッカ
76820 パパイア科,バンカジュ科,バンクワジュ科
76854 カリッサ属
76873 カリッサ
76926 オオバナカリッサ
76986 サントウゼリ属
76987 イワボウフウ,サントウゼリ
76988 チャボアザミ属
76990 チャボアザミ
77038 バナマサウ属,バナマソウ属,パナマソウ属
77043 バナマサウ,パナマソウ,バナマハットサウ,バナマハットソウ
77050 ニュ-ジ-ランドイチビ属
77067 フクマンギ
77075 カ-ネキエカ属
77077 ベンケイチュウ
77131 カルパンテア属
77145 ヤブタバコ属
77146 ヤブタバコ
77149 ガンクビサウ,ガンクビソウ,コヤブタバコ,サジガンクビサウ
77156 ガンクビサウ,ガンクビソウ,キバナガンクビソウ
77157 ホソバガンクビソウ
77158 ノッポロガンクビソウ
77160 コバナガンクビソウ,バンジンガンクビソウ
77162 サジガンクビソウ
77178 オオガンクビソウ,オホガンクビサウ
77181 バンジンガンクビソウ
77185 ヒメガンクビソウ
77190 カンクビヤブタバコ,ミヤマガンクビサウ,ミヤマガンクビソウ,ミヤマヤブタバコ
77191 マンシウガンクビソウ
77201 イヌノグサ
77252 イヌシデ属,クマシデ属,ケマシテ属,シデゾク属
77256 オウシュウシデ
77268 オナガクマシデ,ヲナガクマシデ
77276 イシゾネ,サハシデ,サハシバ,サハマキ,サワシデ,サワシバ,シバシデ,ブ

ナゾロ
77278 ビロ－ドサワシバ
77281 オオサワシバ
77283 コシデ
77288 チョウセンアカシデ
77295 サイシュウイヌシデ
77312 イシソネ,イシゾノ,オオソネ,カタシデ,カナシデ,クマシデ,クロソネ,シデ
77313 オオクマシデ,オホクマシデ
77314 アリサンシデ,シマシデ,ホウゴウシデ
77323 アカシデ,カナシデ,コシデ,コソネ,シデノキ,ソノ,ソヤ,ソロ,ソロノキ,ホンソネ
77325 シダレアカシデ
77346 タンダイシデ,ホンバシデ
77381 ランカンシデ
77401 アカシデ,イヌシデ,シロシデ,シロソネ,ソネ,ソロ
77402 シダレイヌシデ
77411 イハスデ,イワシデ,コバノイヌシデ,ヤマシデコイヌシデ
77429 カルポプロ－ツス属
77441 ハクヤギク
77639 カンムリナズナ
77656 カルアンツス属
77684 ベニバナ属
77716 アレチベニバナ
77748 ウレツムハパ,クレナイ,クレナヰ,クレノアイ,クレノアヰ,スエツムハナ,ベニ,ベニバナ
77749 ベニバナ
77766 イブキゼリ属,カルム属,シムラニンジン属
77777 ノハラニンジン
77784 ヒメウイキョウ,ヒメウヰゼリ
77874 ペカン属
77902 ベカン,ペカン
77910 ナットメッグ
77920 ヒッコリ－
77934 ヒッコリ－
77940 カリオカル属
77947 カリオカル
77980 セキチク科,ナデシコ科,ニデシコ科
77992 カリガネサワ属,カリガネソウ属
78002 カリガネサウ,カリガネソウ,ホカケサウ,ホカケソウ,ヤマドリサウ
78012 ダンギク,ランギク
78013 シロバナダンギク
78014 モモイロダンギク
78045 クジャクヤシ属
78046 クジャクヤシ
78047 カブダチクジャクヤシ,コモチクジャクヤシ
78056 クジヤクヤシ
78065 カリオトフォラ属
78095 イヌカンコノキ属
78137 イヌカンコノキ
78172 カシミロ－ア属
78204 カハラケツメイ属,カワラケツメイ属
78212 センナ
78227 アレクサンドリアセンナ,センナ,ホソバセンナ
78300 ナンバンサイカチ
78301 オオバノハブソウ,オホバノハブサウ,キダチハブソウ
78303 ガランビネムチャ
78308 コバノセンナ,モクセンナ
78316 ウマセンナ,モモイロナンバンサイカテ
78336 コチョウセンナ,ジャワセンナ
78338 バライロモクセンナ
78366 タイワンカワラケツメイ
78429 ガハラケツメイ,カワラケツメイ,キジマメ,コウボフチャ,ネムチャ,ノマメ
78622 イハヒゲ属,イワヒゲ属
78627 カラフトイワヒゲ
78631 イハヒゲ,イワヒゲ
78632 タマザキイワヒゲ
78646 ジムカデ,ヂムガデ
78726 スナヅル属
78731 シマネナシカッラ,スナヅル
78733 イトスナヅル
78738 ケスナヅル
78766 クリ属
78777 クリ
78778 ヤッブサグリ
78779 タンバグリ
78780 ハゼグリ
78781 シダレグリ
78782 ハコグリ
78783 ハナグリ
78784 トゲナシグリ
78790 アメリカグリ
78795 ヘンリ－グリ
78802 アマグリ,ウラジログリ,シナアマグリ,シナグリ,タイワングリ,チュウゴクグリ
78807 チンカピン,チンカピングリ
78811 クリ,スペイングリ,セイヨウクリ,ヨケロッパグリ,ヨ－ロッパグリ
78823 モ－パングリ
78848 クラガシ属,クリガシ属,シイノキ属,シイ属,シヒノキ属
78857 サニンテン
78877 シマシラカシ,タカサゴジイ,ノコギリジヒ
78903 コジイ,シイノキ,タイコジイ,ツブラジイ,ツブラジヒ
78907 シダレジイ
78913 オキナワジイ
78914 シダレュジイ
78916 イタジイ,シイ,シイガシ,シイノキ,スダジイ,ナガジイ
78932 イラガシ,クサノクリガシ,シマクリガシ,タイワンクリガシ,ビロウドソバグリ
78933 アカテッガシ,クリガシ,テツガシ,ヒシグリ
78942 シマヒシグリ,タイワンヒシグリ
78966 オホクリガシ,クリガシハ
78977 ナガバシイ
79022 シイ,スダジイ
79023 オキナワジイ
79027 セイジヤウクリガシ,ハリジヒ,ヒシガタクリガシ,ミヤマクリガシ
79032 ヒ－ランクリガシ,ホソバクリガシ
79035 アカテッガシ,クリガシ,テツガシ,ヒシグリ
79054 ヒシグリ
79064 ウライガシ,ランダイガシ
79096 カステラノシア属
79108 カスチア属,カスチロア属
79130 カステリソウ,タイマツグサ,ミヤマガラガラ
79155 モクマオウ属,モクマワウ属
79157 カンニングハムモクマワウ,トキハギョりウ,トキハギョリュウ,モクマオウ
79161 トキハギョりウ,トキワギョリュウ,トクサバモクマオウ,モクマオウ
79164 フラセリアナモクマオウ,フラセリアモクマオウ
79165 グラウカモクマオウ
79166 ヒュ－ゲリモクマオウ
79178 クアドソバミスモクマオウ
79182 モクマオウ
79185 スベロサモクマオウ
79193 モクマオウ科,モクマワウ科
79242 キササゲ属
79243 アメリカキササゲ,ヒナカキササゲ
79247 シナキササゲ,タウキササゲ,トウキササゲ
79250 コバノキササゲ
79257 アズサ,アヅサ,カハラヒサギ,カミナリササゲ,カワギリ,カワラササギ,カワラヒサギ,キササギ,キササゲ,ハブテコブラ
79260 オオアメリカキササゲ,ハナキササゲ
79271 ルリニカナ属,ルリニガナ属
79276 コヤグルマギク,ルリニガナ
79295 ルリニカナ属
79329 カタセ－ツム属
79410 ニチニチソウ属
79418 ニチニチカ,ニチニチクワ,ニチニチソウ
79424 ヒメニチニチソウ
79438 ギンサン

79500 エゾハタザオ
79526 カトレヤ属
79546 ヒノデラン
79555 ハツヒノデラン
79583 カトレイオパシス属
79595 ハリクチナシ,ハリザクロ
79664 ハナヤブジラミ
79730 ハイヨウボタン属,ルイヨウボタン属,ル井エフボタン属
79732 ボタンサウ,ルイヨウボタン,ル井エフボタン
79778 カバニレシア属
79788 キャベンディシア属
79831 ヤブカラシ属,ヤブガラシ属
79841 アカミノヤブカラシ
79850 シロミノブダウ,ビンボウガズラ,ビンボフガヅラ,ヤブガラシ,ヤブガラツ
79888 ヒイラギヤブカラシ
79901 アカミノヤブカラシ
79902 ソリチャ属
79905 ソリチャ
79916 メキシコソリチャ
80015 チャンチン属
80030 セドロ
80074 ヒマラヤスギ属
80080 アトラス・シーダー,アトラスシーダー
80087 インドスギ,カラマツモミ,デオダラモミ,ネズモミ,ヒマラヤ・シーダー,ヒマラヤシーダー,ヒマラヤスキ,ヒマラヤスギ
80100 レバノン・ジーダー,レバノンジーダー
80114 ケイバ属
80120 インドワタノキ,カボック,パンヤノキ
80127 ニシキギ科
80129 ツルウメモドキ属
80168 コツルウメモドキ
80189 イハウメヅル,イワウメズル,イワウメヅル,トリモチカズラ,トリモチカヅラ
80203 アミサンツルウメモドキ,ダイブウメモドキ,ダイブツルウメモドキ,ホザキツルウメモドキ
80221 オホバツルウメモドキ,リュウキュウツルウメモドキ
80260 ツルウメモドキ,ツルマユミ,ツルモドキ,ヤマガキ
80261 キミツルウメモドキ
80262 ナガミノツルウメモドキ
80263 イヌツルウメモドキ,イプッルウメモドキ
80265 コツルウメモドキ
80267 オオツルウメモドキ,オホツルウメモドキ
80273 アミバマユミ
80279 ムシャツルウメモドキ
80285 アリサンツルウメモドキ,ウチダシツルウメモドキ,コハノツルウメモドキ,テリハツルウメモドキ,テリミノツルウメモドキ
80324 オオツルウメモドキ,シタキツルウメモドキ
80327 オニツルウメモドキ
80378 ケイトウ属
80381 ノゲイトウ,ノケイトウ
80395 カラアイ,カライ,カラヰ,ケイトウ,トサカケイトウ
80396 ジャワケイトウ,スギナリゲイトウ,ヤリゲイトウ
80399 チャボゲイトウ
80422 タイトウノゲイトウ
80469 イトゲイトウ
80473 タイトウノゲイトウ
80569 エノキ属
80580 カウライエノキ,コバノチウセンエノキ,コバノチョウセンエノキ,チュウゴクエノキ
80586 チュウゴク,チュウゴクエノキ
80590 クワノハエノキ
80596 エゾエノキ,オクノエノキ,コバノエゾエノキ
80600 エゾエノキ,オクエノキ,オクジリエノキ
80639 タイワンエノキ
80659 ナガバエゾエノキ
80660 カンサイエノキ
80663 オガサワラエノキ
80664 オヒョウエノキ,オホバエノキ,テウセンエノキノ
80695 コバノエノキ
80698 アメリカエノキ
80709 コウトウエノキ
80739 エ,エノキ,エノキエ,コバノエノキ,メムクノキ
80740 シダレエノキ
80741 エ,エノキ
80764 タイワンエノキ,ナンバンエノキ
80811 クリノイガ
80815 ヒゲクリノイガ
80821 シンクリノイガ
80826 コウベクリノイガ
80830 ヒメクリノイガ
80855 オオクリノイガ
80877 チシマゼリ属
80892 セントーレア属,ヤグルマギク属
80981 ムラサキイガヤグルマギク
81004 シラタエヤグルマギク,シロタエギク
81020 マギク,ヤグルマギク,ヤグルマソウ
81024 ウラジロヤグルマギク
81033 シュッコンヤグルマギク
81150 ヤグルマアザミ
81183 オウゴンヤグルマソウ,キバナヤグルマギク
81198 ヒレハリギク
81214 ヤマヤグルマソウ
81219 スイート・サルタン,ニオイヤグルマギク,ヤマヤグルマギク
81237 クロアザミ
81356 ヒメタマバハキ,ヒメタマボウキ
81382 イガヤグルマギク
81408 イエローサルタン
81485 シマセンブリ属,セントーリューム属
81495 ベニバナセンブリ
81501 アメリカホウライセンブリ
81502 シマセンブリ,ホオライセンブリ
81520 コゴメセンブリ
81539 ハナハマセンブリ
81564 ツボクサ属
81570 クックサ,ツボクサ
81681 トキンサウ属,トキンソウ属
81687 トキンサウ,トキンソウ,ハナヒリグサ
81692 ラッパグサ属
81694 ラッパグサ
81714 ゴマクサ属
81718 ゴマクサ
81719 ゴマクサ
81720 シロバナゴマクサ
81743 セントランサス属,ベニカノコソウ属
81760 ウスベニカノコソウ
81768 ベニカノコソウ
81787 ルリアザミ
81811 カツマダソウ科
81814 カツマダソウ
81872 チョゥマメモドキ属
81876 ムラサキチョウマメモドキ
81878 チョゥマメモドキ
81933 ウドノキ,オキナハアオキ,オキナハアヲキ,オホクサボク
81945 トコン属
81970 トコン
82022 アレチアザミ
82024 エゾノキツネアザミ
82028 キンラン属,ハクリン属
82044 ギンラン,ハクサンラン,ハクリン
82046 エゾギンラン
82047 クゲヌマラン
82048 ユウシュンラン
82051 アサマサウ,アサマソウ,アリマサウ,アリマソウ,オウラン,キサンラン,キンラン,ワウラン
82052 シロバナキンラン
82058 ササバギンラン
82059 ニシダケササバギンラン
82088 クサイロエビネ,トキサラン,トクサラン,ヒメクワラン,ホソバエビネ

82091 ヤマタマガサ属
82098 アメリカヤマタマガサ,タマガサノキ
82107 カギカヅラモドキ,ヤマタマガサ
82113 キバナノマツムシソウ属,キンラン属,セファラリ－ア属,ハクリン属
82166 キバナマツムシソウ
82204 ケファロセレウス属
82211 オキナマル
82276 セファロフィルム属
82282 アサヒミネ
82287 カクレイ
82295 リョウメイ
82300 ケイレイ
82307 ショウエン
82312 シンレイ
82320 テイオウカ
82327 ヨウエン
82332 ショウエン
82345 ロウエン
82355 ザイエン
82365 スイエン
82366 ソウレイ
82372 ソウエン
82380 リュウエン
82385 ショウレイ
82387 テンレイ
82390 トリコウ
82399 ヨウレイ
82490 フクロユキノシタ科
82493 イヌガヤ科
82496 イヌガヤ属
82499 アブラガヤ,イヌガヤ,ヒョウビ,ベベガヤ,ヘボガヤ
82523 アブラガヤ,イヌガヤ,デバガヤ,ヒビガヤ,ヒヨウビ,ヘッタマ,ベベガヤ,ヘボガヤ,メガヤ
82524 イヌガヤ
82525 チョウセンガヤ,チョウセンマキ,トウガヤ
82528 ホソハィヌカャ
82532 ハイイヌガヤ
82550 タイワンイヌガヤ
82560 セファロタス属,フクロユキノシタ属
82561 フクロユキノシタ
82629 ミミナグサ属
82680 セイヤウミミナグサ,セイヨウミミナグサ
82692 エダウチミミナグサ
82705 ミッモリミミナグサ
82732 タカネミミナグサ
82758 オオミミナグサ
82764 ミミナグサ
82772 コケミミナグサ
82813 オオバナノミミナグサ,オホバナミミナグサ
82817 ゲンカイミミナグサ
82819 タイリンミミナグサ
82824 ミミナグサ
82826 オオミミナグサ
82835 コバノミミナグサ
82849 オランダミミナグサ
82924 ニシウチサウ
82955 タガソデソウ
82956 タガソデソウ
82958 タガソデソウ
83007 タカネミミナグサ
83009 キクザキタカネミミナグサ
83013 ミヤマミミナグサ
83014 クモマミミナグサ
83036 テラモトハコベ
83060 シロミミナグサ,ナツユキソウ
83112 シキザクラ,ジュウガツザクラ
83114 モリオカシダレ
83115 チチブザクラ
83116 アイズシダレ
83117 オグラヤマザクラ
83118 アカツキザクラ
83119 ヤママメザクラ
83120 イブキザクラ,キンキヤマメザクラ
83122 ヤブザクラ
83123 マツマエベニソメイ
83124 タカネオオヤマザクラ
83125 チョウジマメザクラ
83126 ヤツガタケザクラ
83127 ホウキザクラ
83128 モチヅキザクラ
83129 モニワザクラ
83130 オネヤマザクラ
83131 オオミネザクラ
83132 カッテザクラ
83133 ショウドウザクラ
83135 タキノザクラ
83136 カスミオクチョウジザクラ
83137 ニッコウザクラ
83138 ナルサワザクラ
83139 ソメイヨシノ
83140 フジカスミザクラ
83141 ブコウカスミザクラ
83145 アカカバ,カバザクラ,カンバザクラ,コメザクラ,タニノゾキ,チョウジザクラ,メジロザクラ,ヤマカバ
83147 ミヤマチョウジザクラ
83148 オクチョウジザクラ
83149 キクザキオクチョウジザクラ
83155 サクランボ,セイヤウミザクラ,セイヨウミザクラ
83158 カンヒザクラ,グワンジッザクラ,タイワンザクラ,ヒカンザクラ,ヒザクラ
83162 シマニハウメ,タイワンニハウメ,タカサゴニハウメ
83165 ヒマラヤザクラ
83167 ヒマラヤヒザクラ
83208 セトエノニワザクラ,セトヘノニハザクラ,ニワザクラ
83220 コニハザクラ
83224 オマキザクラ,カバザクラ,フジザクラ,マメザクラ,ミヅザクラ,ヲマキザクラ
83225 フジキクザクラ
83226 アメダマザクラ
83227 オオバナマメザクラ
83228 ミドリザクラ,リョクガクザクラ
83229 ブコウマメザクラ
83231 キンキマメザクラ
83232 ヤエノキンキマメザクラ
83234 シロヤマザクラ,ヤマザクラ
83235 ワカキノサクラ
83236 ウスゲヤマザクラ
83237 ツクシヤマザクラ
83238 コウメ,コンメ,ニハウメ,ニハザクラ,ニワウメ,リンショウバイ
83241 チョウセンニワウメ
83250 キリフリザクラ
83255 シロザクラ,ミヤマザクラ,メザクラ
83266 ミドリヨシノ
83267 オホマメザクラ,タカネザクラミネザクラ
83268 クモイザクラ
83269 チシマザクラ
83271 コバザクラ
83284 カラノミザクラ,カラミザクラ,シナアウタウ,シナオウトウ,シナノザクラ,シロバナカラミザクラ,ミザクラ
83301 エゾザクラ,エゾヤマザクラ,オオヤマザクラ,オホヤマザクラ,ベニバナザクラ,ベニヤマザクラ
83302 ハツユキザクラ
83303 シダレオオヤマザクラ
83304 ケエゾヤマザクラ
83305 キリタチヤマザクラ
83311 ブラックチェリ－
83314 ヤマザクラ
83316 オホシマザクラ,サトザクラ
83319 オクヤマザクラ,カスミザクラ,ケヤマザクラ
83324 イシヅチザクラ
83326 ナデン
83327 イトザクラ
83328 アヅマヒガン,エドヒガン,シロヒガン,タチザクラ,タチヒガン
83330 コシノヒガンザクラ
83331 オオシマザクラ
83334 アケボノヒガン,コヒガンザクラ,ヒガンザクラ
83337 イトザクラ,クサイザクラ,シダレザクラ,シダレヒガン
83342 ホシザクラ
83347 ユスラウメ

83361 スミセイヤウミザクラ,スミセイヨウミザクラ,スミミザクラ
83365 ソメイヨシノ,ヤマトザクラ,ヨシノザクラ
83523 イナゴマメ属
83527 イナゴマメ
83533 ケラトペタ-ルム属
83539 キンギョモ科,マツモ科
83540 キンギョモ属,マツモ属
83545 キンギョモ,キンチャウモ,マツモ
83568 マンシウマツモ
83575 ゴハリキンギョモ,ヨツバリキンギョモ
83577 ゴハリキンギョモ
83578 トゲナシキンギョモ
83641 ルリマツリモドキ属
83646 ルリマツリモドキ
83651 ケラトスティリス属
83680 ツノミザミア属
83682 クエスタ-ツノミザミア
83683 ヒロハツノミザミア
83685 ナガバツノザミア,ナガハツノミザミア
83690 ミフクラギ属
83696 グレイミルクウット
83698 ミフクラギ
83699 オトメオキナワキョウチクトウ
83700 アマミケフチクタウ,オキナハケフチクタウ,ミフクラギ
83732 カツラ科
83734 カツラ属
83736 アカキ,オカツラ,カツラ,カモカツラ,シロカツラ
83737 シダレカツラ
83743 ウチハカッラ,ヒロハカッラ
83744 シダレカッラ
83752 ハナズオウ属,ハナズハウ属
83757 アメリカハナズオウ,アメリカハナズハウ
83769 オナズオウ,スオウギ,スオウバナ,スオウバナノキ,スハウバナ,スハウバナノキ,ハナスオウ,ハナズオウ,ハナズハウ,ハナズホウ,ハナムラサキ
83771 シロバナハナズオウ,スハウバナ,スハウバナノキ,ハナズハウ
83809 セイヤウズハウ,セイヨウズオウ,セイヨウズハウ,セイヨウハナズオウ
83855 セレウス属,ヒモサボテン属
83874 タイリンバシロ
83890 キメンカク,ロクカクサボテン
83908 レンジョウカク
83912 ハリヅス
83913 ジンダイ
83918 キバナルリソウ属
83923 キバナルリソウ
83939 タカオコヒルギ属,タカヲコヒルギ属
83967 ケロクラミス属
83969 ギョクサイリン
83975 セロプジア属
84305 ハ-トカズラ
84326 アンデスロウヤシ属
84400 キチャウジ属,キチョウジ属
84404 キチャウジ,キチョウジ
84417 ヤクワウクワ,ヤクワウボク,ヤコウカ,ヤコウボク
84425 ベニチョウジ
84549 ボケ属
84552 ヒボケ
84553 マボケ
84554 シロボケ
84556 クサボケ,コボケ,シドミ,シドメ,ジナシ,チナシ,ノボケ,ボケ
84559 シロバナクサボケ
84569 ウンリュウボケ
84573 カイドウボケ,カラボケ,シロボケ,チョウシュンボケ,ボケ,モケ,ヨドボケ
84596 ヒナウンラン
84723 ケロフィルム属
84731 カブラゼリ
84773 ウスゲヤマニンジン
84958 コヘラナレン
85054 イスパニヤガヤ属
85058 イスパニヤガヤ
85106 モリサウ属,モリソウ属
85107 ッルイラクサ,ヒメモリサウ,モリサウ
85163 カマエセレウス属,カメエケレウス属
85165 ビャクダン
85232 カハラケツメイ
85263 ヒノキ属
85268 タイワンサワラ,タイワンヒノキ,ベニヒ
85271 グランドヒノキ,ロ-ソンヒノキ
85301 アメリカヒノキ,アラスカヒノキ,ベイヒバ
85310 ヒノキ,ヒバ,ホンヒ
85311 オウコンヒバ
85312 カマクマヒバ,カマクラヒノキ,チャボヒバ
85313 カマクラヒノキ,カマクラヒバ,チャボヒバ
85316 アオノクジャクヒバ,クジャクヒバ
85317 オウゴンクジャクヒバ
85321 シャモヒバ
85331 クリハダヒノキ
85338 タイワンヒノキ
85339 ツノミノヒノキ
85345 アスナロ,サワラ,ヌカビ
85346 シノブヒバ
85347 ブル-バ-ル
85348 イトヒバ,ヒヨクヒバ
85349 オウゴンシノブヒバ
85350 シノブヒバ
85352 アヤスギ,シモフリヒバ,ヒムロ,ヒメムロ
85353 アカスギ,シモフリヒバ,ヒムロ,ヒメムロ
85364 アッカワサワラ
85375 ヌマヒノキ,ヌマヒバ
85401 ベニバナエニシダ
85412 チヤツツジ属
85414 ホロムイツツジ,ヤチツツジ
85420 テ-ブルヤシ属,バ-ラ-ヤシ属,モレノヤシ属
85428 キレバテ-ブルヤシ
85432 ヒロバモレノヤシ
85437 キレバテ-ブルヤ
85455 ヒメノヤガラ
85478 セントウソウ
85479 イブキセントウソウ
85480 オオギバセントウソウ
85482 ミヤマセントウソウ
85483 ヒナセントウソウ
85485 ヤクシマセントウソウ
85526 ロウマカミツレ,ロ-マカミツレ,ロ-マカミルレ,ロ-ムセカミツレ
85579 ゴゼンタチバナ属
85581 ゴゼンタチバナ
85646 インチンロウゲ属
85648 ハナインチンロウゲ
85651 インチンロウゲ
85660 ヒメインチンロウゲ
85665 チャボトウジュロ属
85671 チャボトウジュロ
85768 リュウキュウタイゲキ
85779 オオニシキソウ
85842 イヌケンチャ属
85875 ヤナギサウ,ヤナギソウ,ヤナギラン
85880 ウスゲヤナギラン,ケヤナギラン
85884 キタダケヤナギラン,ヒメヤナギラン
85932 カナビキボク属
85935 カナビキボク
85944 イセハナビ
85949 スズムシサウ,スズムシソウ,スズムシバナ,スズルシサウ,スズルシソウ
86025 ヒメアフリカギク属
86027 アフリカヨメナ
86144 カスマトフィルム属
86409 ニオイアラセイトウ属,ニホヒアラセイトウ属
86427 キアラセイトウ,キバナアラセイトウ,ケイランサウ,ケイランソウ,ニオイアラセイトウ,ニホヒアラセイトウ
86480 ケイリドブシス属
86481 キョウギョ
86490 シュカイギョク
86506 ジコウニシキ
86510 ダイソウケン

86511 サカホキ
86517 バンコウ
86521 ヒョウレイ
86530 ギボウシュ
86537 ヤヨイ
86544 コウギョ
86551 ロウゲツ
86572 ショウホウ
86575 シンプウギョク
86585 リョウウン
86587 シュンイギョク
86589 シヨウギョク
86610 アヤホ
86612 ユウヒギョク
86620 レイギョク
86624 シンホウギョク
86667 カイロラン属
86670 タカオラン,タカヲラン
86698 アカバシュスラン,タネガシマカイロラン
86707 ガイロラン
86725 アリサンムエフラン,アリサンムシャラン,アリサンムヨウラン
86733 クサノオウ属,クサノワウ属
86755 クサノオウ,クサノワウ,コクサノオウ,タムシグサ,ヨウシュクサノオウ
86757 クサノオウ
86761 クサノオウ
86781 ジャコウソウモドキ属
86793 ジャコウソウモドキ
86805 ジャカウサウ属,ジャコウソウ属
86818 タニジャコウソウ,タニジャコオソウ
86819 シロバナタニジャコウソウ
86821 ジャコウソウ,ジャコオソウ
86838 アシタカジャコウソウ,アシタカジャコオソウ
86855 アブラギ,イモギ,コシアブラ,ゴンゼツ,ヤマウコギ
86873 アカザ科
86891 アカザ属
86892 カハラアカザ,カワラアカザ,キバナアカザ,マルバアカザ
86893 カハラアカザ,カワラアカザ,ホソバアカザ,マルバアカザ
86901 アカザ,ギンザ,シロアカザ,シロザ
86913 アカザ
86927 ホソバアカザ
86934 アリタソウ,ルウダソウ
86941 アメリカアリタソウ,アリタサウ,アリタソウ
86970 キクバアカザ
86977 コアカザ,ミドリアカザ
87007 コアカザ
87029 ウラジロアカザ
87037 イワアカザ,ヤマアカザ
87048 アオアカザ,アヲアカザ,ウスバアカザ,オホバアカザ
87070 ヒメハマアカザ
87073 ホソバヒメハマアカザ
87096 ミナトアカザ
87112 ヒロハアカザ
87133 ヒロハヒメハマアカザ
87136 アイノコアカザ
87142 ゴウシュウアリタソウ
87144 ムラサキアカザ
87145 キノア
87147 イヌアリタサウ
87186 テリハアカザ
87204 ノハラアカザ
87352 ツクシガヤ属
87353 ツクシガヤ
87354 イリオモテガヤ
87480 ウメガササウ属,ウメガサソウ属
87485 ウメガササウ,ウメガサソウ,キヌガササウ
87490 タイワンウメガサソウ
87491 タイワンウメガサソウ
87498 オオウメガサソウ,オホウメガササウ
87510 ラフバイ属
87525 カラウメ,カラムメ,トウウメ,ナンキンウメ,ラフバイ,ロウバイ
87526 シロバナノラフバイ,ソシンラフバイ,ソシンロウバイ
87528 シンノラフバイ,タウラフバイ,ダンカウバイ,トウロウバイ
87532 カカバイ,カクワバイ
87547 カンチク属
87581 カンチク
87582 チゴカンチク
87607 イボダケ,シカクダケ,シハウチク,シホウチク
87668 ハリガネカズラ属,ハリガネカヅラ属
87694 ヒトツバタゴ属
87729 アンニヤモンニヤ,ナンジヤモンジヤ,ヒトツバタゴ,ロクダウボク
87757 チオノドクサ属
87759 ユキゲユリ
87765 ユキゲユリ
87775 シライトサウ属,シライトソウ属
87777 シライトサウ,シライトソウ
87815 イワギリソウ属,キリタ属,ツノギリサウ属
87895 ゾウイロッノギリソウ
88111 クスクスジュラン属
88215 クリダンツス属
88263 センリヤウ科,センリョウ科,チャラン科
88264 センリヤウ属,センリョウ属,チャラン属
88276 キゼセトリシズカ
88289 ヒトリシズカ,ヒトリシヅカ,マユハキサウ,ヨシノシズカ
88295 タイワンフタリシヅカ
88297 フタリシズカ,フタリシヅカ
88301 チャラン
88316 オヒゲシバ属,ヒゲシバ属
88320 シマヒゲシバ,ムラサヒゲシバ
88339 ヒメヒゲシバ
88340 ヒメヒゲシバ
88344 セイヨウヒゲシバ
88351 シマヒゲシバ,タイワンヒゲシバ
88352 アフリカヒゲシバ
88383 クシヒゲシバ
88396 カセンガヤ,コウセンガヤ,ヒゲシバ
88414 チャボヒゲシバ
88421 オヒゲシバ
88527 オリヅルラン属
88537 シャムオリヅルラン
88553 オリヅルラン,チョウラン,フウチョウラン
88565 ヒロハオリヅルラン
88567 フイリヒロハオリヅルラン
88690 チャンチンモドキ属
88691 カナメ,チャンチンモドキ
88696 ケイシヤ属,コイシヤ属
88854 コンドモミンカ属
88970 コリシア属,トッケリキワタ属
88973 トッケリキワタ
88977 イハイテフ,ハマイチョウ,ハマニガナ
89020 ツノミナズナ
89134 コリゼマ属,ヒイラギハギ属
89135 ハナヒイラギ,ヒイラギマメ
89136 ヒイラギハギ,ホソバヒイラギマメ
89139 カワリハナヒイラギ
89159 ケショウヤナキ属
89160 カラフトクロヤナギ,クロヤナギ,ケシャウヤナギ,ケショウヤナギ
89201 ホオズキハギ属
89207 ホオズキハギ
89299 ヒマワリヒヨドリ
89345 アレカヤシ属,タケヤシ属
89360 アレカヤシ,コガネタケヤシ,トガネタケヤシ
89402 キク属
89404 ワジキギク
89405 トガクシギク
89406 シロバナアブラギク
89407 ハナイソギク
89408 ミヤトジマギク
89409 ヒノミサキギク
89410 ニジガハマギク
89424 アキノコハマギク,チシマコハマギク
89428 コハマギグ,チシマコハマギク
89429 コハマキク
89466 クリサンテマム,サンシキカミツレ,ハナワギク

89472 コ－カシアムシヨケギク
89481 オランダギク,カウライギク,キクナ,コウライギク,シュンギク,フダンギク,ムジンサウ,ムジンソウ,リュゥキュウギク
89499 オオシマノジギク
89536 シロシマカンギク
89541 シロバナハマカンギク
89546 イヨアブラギク
89547 シロイヨアブラギク
89550 オッタチカンギク
89551 ハイシマカンギク
89552 ツルギカンギク
89556 ノシギク
89557 アシズリノジギク
89567 キイシオギク,キノクニシオギク
89603 リュノウギク
89619 キク
89628 クリサンセラメ,ムルティカウレ
89646 ハマギク
89652 サツマノギク
89657 トカラノギク
89659 イソギク
89698 イワインチン
89704 アラゲシュンギク,カモメギク,クヅャクギク,リウキウシュンギク,リュウキュウシュンギク
89709 キクタニギク
89715 シオギク
89766 ワカサハマギク
89768 ピレオギク
89773 ナカガワノギク
89786 オグラギク
89812 イガコ属,クリソバラヌス属
89816 イカコ
89863 グリゾゴ－マ属
89936 ヒメクリソラン
89938 クリソラン属
89948 クリソラン
89952 クリソゴナム属
89959 キンセイギク
90007 オ－ガストノキ属
90032 カイニット
90108 オキナワミチシバ
90317 ネコノメサウ属,ネコノメソウ属
90321 シロバナネコノメソウ
90322 キバナハナネコノメ
90323 キイハナネコノメ
90324 ハナネコノメ
90325 カラフトネコノメサウ,カラフトネコノメソウ,ヤマネコノメサウ,ヤマネコノメソウ
90331 エゾネコノメソウ
90356 イワネコノメソウ
90358 ホクリクネコノメ
90359 サンインネコノメ
90362 コバノネコノメサウ,ツルネコノメサウ,ツルネコノメソウ,ヒメネコノメサウ
90370 ネコノメソウ
90383 ヤマネコノメサウ,ヤマネコノメソウ
90384 ヨツシベヤマネコノメ
90387 チシマネコノメソウ
90388 トビシマネコノメソウ
90389 ミチノクネコノメソウ
90390 キンシベボタンネコノメ
90391 ボタンネコノメソウ
90400 ヒカゲネコノメソウ,マンシウネコノメソウ
90406 イワボタン
90407 ヨゴレネコノメ
90408 キシュウネコノメ
90409 ニッコウネコノメ
90410 サツマネコノメ
90411 ムカゴネコノメソウ
90413 ヒダボタン
90414 ヒメヒダボタン
90415 アカヒダボタン
90429 ケネコノメソウ
90430 オオコガネネコノメソウ
90432 コガネネコノメソウ
90434 オオイワボタン
90436 ヒメオオイワボタン
90437 トウノウネコノメ
90438 ヤマシロネコノメ
90442 クモノスネコノメソウ,マルバネコノメ,マルバネコノメソウ
90445 ツクシネコノメソウ
90446 トゲミツクシネコノメ
90448 エゾネコノメソウ
90464 タチネコノメソウ
90465 オオイワボタン
90711 チシス属
90767 キトログッサ属
90797 ヒヨコマメ属
90801 ヒヨコマメ
90802 シロヒヨコマメ
90889 キクヂシャ属
90894 アサギク,エンダイブ,オランダヂサ,オランダヂシャ,キクヂサ,キクヂシャ,チリメンヂシャ,ニガチシャ,ハナヂシャ,メリケンサラダ
90901 キクニガナ
90918 ドクゼリ属
90932 オホゼリ,ドクゼリ,ハナワラビ,ヤナギバドクゼリ
90935 ヒメドクゼリ
90937 ドクゼリ,ヒロハオホゼリ,ヒロハドクゼリ
90940 ホソハドクゼリ
90994 サラシナショウマ属
91008 フブキショウマ
91011 サラシナショウマ
91023 オオバショウマ,オオミツバショウマ
91024 アハボ,アワボ,イヌショウマ,オオバショウマ,オホバショウマ,ミツバショウマ,ミツフデ
91038 サラシナショウマ,ヤサイショウマ
91055 キナノキゾク属,キナノキ属
91075 キナノキ,バリビヤキナノキ,ボリビヤキナ,ボリビヤキナノキ,レドゲリアナ
91082 オヒチナリス,キナノキ
91090 アカキナノキ,キナ
91235 フサガヤ属
91239 フサガヤ
91252 クスノキ属
91255 ヒロハヤブニッケイ
91256 シバヤブニッケイ
91280 マルバヤブニッケイ
91287 クス,クスノキ,ナンジヤモンジヤ,ニッケイ
91289 ホウショウ
91295 クスノキダマシ,ハウシャウ
91302 ケイ,トンキンニッケイ
91318 カウチニクケイ,コウチニッケイ,ツンナメ,マルバニクケイ,マルバニッケイ,ミサキシバ
91322 シバニッケイ
91323 ケシバニッケイ
91349 ウスゲヤブニッケイ
91351 アブラダモ,クスタブ,クロコガ,クロダモ,コガノキ,タイワンニクケイ,ホソバグス,ホソバクスノキ,マツラニッケイ,ミヤマクスノキ,ヤブニクケイ,ヤブニッケイ
91353 ジャワニッケイ
91366 ケイヒジュ,ニクケイ,ニッケイ
91368 ギウシャウ,シャウギウ,シャウジュ,タイナングス,タイナンニクケイ
91369 タイナングス,タイナンニクケイ
91378 アッパクスノキ,オホバグス,オホバグスノキ,ギウシャウ,シャウギウ,シャウジュ
91389 ニクケイモドキ
91392 インドグス
91405 ウスバクスノキ,マルバグス,ミヤマニクケイ
91420 コマルバクスノキ,ハマグス
91426 ニッケイ
91429 シマニクケイ,ナガミグス,マルバグス,マルバクスノキ,ランタイグス,ランダイグス,ランタイニクケイ,ランダイニクケイ
91432 タマラニッケイ
91437 ヤブニッケイ
91438 オガサハラヤブニクケイ,オガサワラヤブニッケイ,コヤブニクケイ,ヤブニッケイ,ヲガサハラヤブニクケイ

91446 カュマニス,セイロンニクケイ,セイロンニッケイ
91479 アメリカニガキ属
91499 ミズタマソウ属,ミヅタマサウ属
91501 タニタデモドキ
91504 マルヤマタニタデ
91506 ハヤチネミズタマソウ
91508 ミヤマタニタデ
91511 ケミヤマタニタデ
91512 タイワンミヤマタニタデ,タッタカタニタデ
91520 オオミヤマタニタデ
91524 エゾミズタマソウ,ミヅタマサウ,ヤマタニタデ
91537 ウシタキサウ,ウシタキソウ,オランダゴシツ,タニタデ,マンシュウウシタキソウ
91542 オオタニタデ
91543 ケタニタデ
91545 タニタデ
91553 ヤマタニタデ
91557 エゾミズタマソウ,ヤマタニタデ
91575 ナガバノウシタキサウ,ミズタマソウ
91576 ミヤマミズタマソウ
91580 ヒロハノミズタマソウ
91611 キレア属
91697 アザミ属
91698 センダイキツネアザミ
91699 ハダカアザミ
91700 イブリアザミ
91701 オバケアザミ
91702 ミサワアザミ
91703 エダハリアザミ
91704 マヨワセアザミ
91705 ケマアザミ
91706 スギモトアザミ
91708 アブクマアザミ
91722 アイズヒメアザミ
91729 ミネアザミ
91736 ダキバヒメアザミ
91737 キンカアザミ
91741 アオモリアザミ
91742 シロバナオオノアザミ
91743 ウニアザミ
91745 アポイアザミ
91753 アリサンアザミ
91770 エゾノキツネアザミ,セイヨウトゲアザミ
91771 シロバナセイヨウトゲアザミ
91785 アシノクラアザミ
91791 ダイニチアザミ
91799 ビッチュウアザミ
91800 マニサンアザミ
91806 オガサワラアザミ
91808 オニアザミ
91816 シマアザミ
91817 シロバナシマアザミ
91826 シロバナヒメヤマアザミ
91862 ノマアザミ
91863 シロバナノマアザミ
91864 アキョシアザミ,カラアザミ,ヤナギアザミ
91873 チョウカイアザミ
91892 コイブキアザミ
91895 ヒッツキアザミ
91912 アキヨシアザミ
91913 キクゴボウ,ゴボウアザミ,モリアザミ,ヤマゴボウ
91958 レンザアザミ
91966 キソアザミ
91967 ツワモノアザミ
91986 ハリカガノアザミ
91996 リョウノウアザミ
91997 ホウキアザミ
91998 ミヤマホソエノアザミ
91999 マルバヒレアザミ
92002 ハチジョウアザミ
92003 ハチマンタイアザミ
92007 ハナマキアザミ
92010 ハッポウアザミ
92011 エゾヤマアザミ
92022 ノアザミ
92026 ヒダカアザミ
92029 オゼヌマアザミ
92032 オガアザミ
92037 モモヤマアザミ
92040 オオヤナギアザミ
92055 トゲナシアザミ
92058 タチアザミ
92065 イシヅチウスバアザミ
92066 コアザミ,ドイツアザミ,ノアザミ
92067 シロバナアザミ
92070 タカサゴアザミ
92072 オニオオノアザミ
92074 トゲアザミ
92075 ミヤマコアザミ
92078 オキノアザミ
92079 シロバナタカサゴアザミ
92083 ケショウアザミ
92084 シロバナケショウアザミ
92085 ビャッコウアザミ
92087 カガノアザミ
92088 シロバナカガノアザミ
92089 チシマアザミ
92090 ビロードエゾアザミ
92094 コバナアザミ
92096 ニヒタカアザミ
92127 アラムシャアザミ
92132 ヤナギアザミ
92161 ナガエノアザミ
92163 テリハアザミ
92164 ヘイケモリアザミ
92165 カッラカワアザミ
92167 カラノアザミ
92168 シロバナカラノアザミ
92182 イナベアザミ
92186 ゴバウアザミ,ハマアザミ,ハマゴバウ
92187 シロバナハマアザミ
92189 ムラクモアザミ
92190 ハクサンアザミ
92191 ホッコクアザミ
92198 アズマヤマアザミ
92199 ネバリアズマヤマアザミ
92200 オハラメアザミ
92201 エチゼンアザミ
92210 シマアザミモドキ
92217 ナンブタカネアザミ
92226 オニアザミ,オニノアザミ,ナンブアザミ,ニッポウアザミ
92228 シロバナオニアザミ
92230 ウラゲヒメアザミ
92234 トネアザミ
92236 サドアザミ
92238 シコクアザミ
92239 シロウマアザミ
92241 ヨシノアザミ
92242 ノリクラアザミ
92258 エチゼンオニアザミ
92266 ジョウシュウオニアザミ
92267 シロバナジョウシュウオニアザミ
92268 アザミゴボウ
92269 ノハラアザミ
92270 シロバナノハラアザミ
92274 ニッコウアザミ
92280 タテヤマアザミ
92281 オクヤマアザミ
92287 エゾノサワアザミ
92293 エゾノサワアザミ
92296 タカアザミ
92297 シロバナタカアザミ
92332 フジアザミ
92333 シロバナフジアザミ
92357 ナガバヤナギアザミ
92367 オクエゾアザミ,テウセンキセルアザミ
92374 シロバナアレチアザミ
92378 センジョウアザミ
92384 アレチアザミ,エゾノキツネアザミ
92385 シロエゾノキツネアザミ
92387 シドキヤマアザミ
92388 ヤチアザミ
92390 キセルアザミ,サツマアザミ,マアザミ
92391 シロバナキセルアザミ
92407 オニアザミ,ツクシヤマアザミ,ヤマアザミ
92410 オイランアザミ
92416 ツクシアザミ
92417 シロバナツクシアザミ

92418 スズカアザミ
92419 スズキアザミ
92425 クルマアザミ,ゴバラアザミ,ノハラアザミ
92430 タネガシマアザミ
92431 ワタムキアザミ
92432 ヒダアザミ
92435 ウスバアザミ
92439 ホソエノアザミ
92440 サンベサワアザミ
92449 トヨシマアザミ
92458 ウゴアザミ
92468 ウゼンアザミ
92479 ウラジロアザミ,ウラジロヤナギアザミ
92485 アメリカオニアザミ
92486 エチゼンヒメアザミ
92498 ヤクシマアザミ
92499 ヤツタカネアザミ
92501 サワアザミ
92503 ザオウアザミ
92523 ミャコジマツヅラフジ属,ミャコジマツヅラフヂ属
92586 キススス属,シッサス属,ヤブガラシ属,リュゥキュウヤブカラシ属
92777 セイシカズラ
92907 ヒレブダウ
92920 ハヒカヅラ
93046 ニチバナ科,ハンニチバナ科
93054 ホンオニク
93120 キストゥス属,コシアオイ属,ゴジアフイ属,シスタス属
93124 ゴジアオイ,ゴジアフイ
93219 ムラサキゴジアオイ
93269 シトロプシス属
93286 スイカ属,スヰクワ属
93289 スイカ
93297 コロシントウリ
93304 スイカ,スヰクワ
93314 カンキツ属,ミカン属
93318 ウイングドシトロン
93322 ジェロウクリモー
93323 ヒョウカン
93324 シレンボー
93326 アサヒカン,アシヒカン
93327 アダジャミー
93328 オウゴンキツ
93329 ライム
93330 タヒチライム
93332 ダイダイ
93333 ヒヨンカン
93340 ナツダイダイ,ナツミカン
93341 ナンショウダイダイ
93368 ダイダイ
93403 アダムリング
93404 カワバタ
93406 バリンコロング
93407 パロチンベルガモット,ベルガモット
93408 ベルガモット
93414 サワーオレンジ
93415 カンシ
93417 キクダイダイ
93422 セレベスパペダ
93423 サウスウイックパペダ
93429 クレメンティンマンダリン
93430 アンナンパペダ,サットカラ
93435 ケオンラ
93439 カルビー
93446 キシュウミカン,チチュウカイマンダリン
93448 シイカアシャア,シイクワシャー,シーカーシャー,シークワシャー,ヒラミレモン
93451 ハシュクリー
93456 レモンリアル
93457 マエダカン
93460 フクライミカン
93461 フナドコ
93463 ゲンショカン
93464 キヌカワ,コウジロキツ
93466 テング
93468 ブンタン
93470 アンセイカン
93471 バンオウカン
93476 タニカワブンタン
93480 ハイナンワイルドカット
93482 ハッサク
93483 ヒメキツ
93485 ヒロシマナツザボン
93492 カフィールライム,プルット,モーリシャスパペダ
93495 イーチャンジェンシス,イーチャンパペダ
93499 モチユ
93500 カーフクルー
93501 ヤマミカン
93502 フサラ
93503 イワイカン
93504 イヨ,イヨカン
93505 ラフレモン
93513 キンカン,ジェローくハンジ
93515 ユ,ユズ
93517 カルナカッタ
93519 ケラジ
93520 カーブチ
93521 ウンゾキ
93523 タイランドパペダ,パペダ
93526 アカツナ,コトウカン,コトーカン
93528 タヒチライム
93529 カシーパペダ,ラテイベス
93530 コウジ
93531 スルガユコウ
93534 スィートレモン,リメッタ,リメッターオーディネール
93536 スイートライム,ミタニムプ
93539 レモン
93546 カントンレモン,ヒメレモン,リモニア
93550 バジュウラシトロン,マルブッシュカン
93552 アッサムレモン
93553 タラミサン
93554 ルミー
93555 デーデー
93556 コクニー
93557 ガルガルラージ
93558 アレモウ
93559 カブヤオ,メラネシアパペダ
93560 アンナンパペダ
93565 キチリー
93566 カラモンジン,シキキツ,トウキンカン
93579 ウチムラサキ,ザボン,ジヤボン,パメロ,ブンタン
93603 シトロン,ブシュカン,マルブシュカン
93604 テブシュカン,ブシュカン
93628 アミルベッド
93630 メラローサ
93631 グラントレモン,ソーロング,マイヤーレモン
93633 ミアライ
93634 ビアソング
93635 サムヤオ
93636 シキキツ,シキナリキンカン,トウキンカン
93639 カラモンジン,サイセイキツ,シキキツ,シキサリミガシ,タウキンカン,トウキンカン
93640 ミツハル
93641 ビロロ
93643 チノット
93645 ナヤチャングニイ
93647 サツマキコク
93648 コウライタチバナ,コオライタチバナ,コーライタチバナ
93649 キングマンダリン,クネンボ
93660 クネンボ
93671 ダエンカン
93672 キンコウジ
93674 ジェローくバリック
93676 タイヒタイヒ
93677 ユヒキツ
93678 オオミカントウ
93679 オオタチバナ
93680 オートー
93684 パニュバン
93685 ジェローくパパヤ
93687 チゾン
93690 グレープフルーツ

93692 ビューティオブグレンリトリート,ラドー
93693 ガジャニマ
93694 ベレッタ
93695 ビンキツ
93698 ポンキー,ポンキツ
93703 ヘンカミカン
93704 ウゾンクネボ,ザックネボ
93705 シシユ,ジャガタラユ
93706 ヒルレモン
93707 フォールズレモン
93708 バランガオレンジ
93709 ウワポメロ
93710 フィリピンスンキー
93712 ポンデローザ,ロチンベルガモット
93716 クレオバトラ
93717 サンキ,サンキー,スンキ,ポンカン,マンダリン,ミカン
93721 コベニミカン
93724 キシュウ,キシュウミカン,コミカン,ホンミカン
93728 ポンカン
93731 ソウキツ
93732 ジミカン,ホンジソ
93734 タンカン
93736 ウンシウミカン
93737 ベニコウジ
93748 サンキツ,リソーポメロ
93752 ロクガツミカン
93753 アタニー
93754 ウチムラサキ,ザボン,ジヤボン
93757 サーバテイ
93763 サダファル
93764 シュンコウカン
93765 アマダイダイ,キンクネンボ,スイートオレンジ,タウミカン,ネーブル
93785 バレンシア
93797 ネーブルオレンジ
93801 キンクネンボ,セッカン
93803 オートーミカン
93805 カボス
93806 オウカン
93814 スダチ
93815 シカイカン
93818 スイザボン
93820 サンボウ,サンボウカン,サンポウカン
93823 タチバナ,ニッポンタチバナ
93829 タニブータ
93833 ナオシチ
93834 コナツミカン,シンナッダイダイ,タルラミカン,ニューサンマー,ヒュウガナツ,ヒュウガナツミカン,ョヒュウガナツミカン
93837 オオヘニミカン,オーベニミカン
93844 マンキツ
93846 ギリミカン
93847 タロガヨ
93848 テンプル
93849 テング
93850 トサアサヒカン
93857 カイコウカン,ジャガタラミカン
93859 フクレミカン
93860 ウジュキツ,ホウライカン,ヤマブキ
93865 モリクリクリ
93868 マウンテンライム
93869 イーチャンレモン,サマヤオ,ハナユ
93871 ヤマブキミカン
93872 ユークニブ
93873 ヤツシロ
93874 ユゲヒョウカン
93875 ユコウ
93882 クラダンサス属
93901 ヒトモトススキ属
93913 シシキリガヤ,ヒトモトススキ
93915 ヒトモトススキ
93927 アンペライ,アンペライネビキグサ,アンペラヰネビキグサ
93968 カワゴケソウ属
93969 タシロカワゴケソウ
93970 トキワカワゴケソウ
93972 マノセカワゴケソウ
93975 カワゴケソウ
94014 フジキ属,フヂキ属
94026 アメリカユクノキ,オオバユク,オホバユク
94030 フジキ,フヂキ,ヤマユンジュ
94032 ココメバナ,ミヤマフジキ,ミヤマフヂキ,ユクノキ
94051 アカリフヤモドキ属
94055 アカリフアモドキ,アカリフャモドキ
94057 セキモンノキ
94068 ワンビ
94132 サンジサウ属,サンジソウ属
94133 イロマツヨイ,タイリンマツヨイグサ
94138 イヌサンジソウ,サンジサウ
94140 サンジソウ,ネソバノサンジサウ,ホソバノサンジソウ
94169 ワンビ属
94191 スメルノキ,ヒメワンピ
94207 ワンピ
94241 オオハナハタザオ
94292 クレイトーニア属
94373 ハルヒメソウ
94404 エノキフヂ属
94425 クレイソストーマ属,ニウメンラン属,ムカデラン属
94458 オホムカデラン,タイワンクレソラン,タイワンムカデラン
94478 ウライムカデラン
94545 クレイステス属
94551 クレイスタントセレウス属,クレイストカクタス属,セチセレゥス属,ボラビセレウス属,ロクサントセレウス属
94552 ジャケイチュウ
94553 アミヂチュウ,カスミガキ,カスミガセキ
94554 コウサイチュウ
94555 ミダレフブキ
94556 バウマンチュウ,リョウウンカク
94557 オオシチュウ,ジャモンチュウ
94559 ヤマフブキ
94560 ショウジョウフブキ
94561 シンノウ
94562 ユウフブキチュウ
94563 ミロクチュウ
94564 キッショウテン
94566 フブキチュウ
94567 タニマノトモシヒ,ハクセン,ハクセンチュウ
94568 ベナフブキ
94585 テウセンガリヤス属
94591 ノゲヤシチョウセンガリヤス
94601 チョウセンガリヤス,テウセンガリヤス
94604 ヒロハチョウセンガリヤス
94605 トウガリヤス,ムラサキチョウセンガリヤス
94608 リョウヶンガリヤス
94632 シナガリヤス,ヒゲナガシナガリヤス
94676 クレマチス属,センニンサウ属,センニンソウ属
94704 カラクサハンショウズル
94707 ヒイランボタンヅル
94708 アコウセンニンサウ,アコウセンニンソウ,オホワタリボタンヅル
94714 ミヤマハンショウヅル
94740 カラクサ,ボタンヅル,ワクヅル,ワクノテ
94778 コボタンヅル,マンシウボタンヅル,メボタンヅル
94814 サキシマボタンヅル,シナセンニンサウ,シナボタンヅル,ミヤコジマボタンヅル
94818 フジセンニンサウ,フジセンニンソウ
94853 シャクヂヤウヅル
94866 ヤマハンシャウヅル,ヤマハンショウヅル
94910 テッセン
94916 オホバボタンヅル
94933 キンチャクズル,クロバナハンショウヅル,シラゲキンチャクズル,チシマハンショウヅル,ムクゲキンチャクズル
94953 ホウライボタンヅル
94969 ケボタンヅル,シナボタンヅル
94995 オオクサボタン,トウクサボタン,ムリクサボタン

95000 ホソバハンシャウヅル
95031 ヒメワキノテ
95037 ハンショウヅル
95042 タイワンボタンヅル
95061 ミツバハンショウズル
95068 タカネハンショウヅル,タカネヒンシャウヅル,ヒメハンシヨウヅル
95077 ビロウボタンヅル,ビロ－ドボタンヅル
95100 キクザキハンショウズル
95113 オホバセンニンサウ,テリハノセンニンサウ,テリハノセンニンソウ,ヤンバルセンニンサウ,ヤンバルセンニンソウ
95148 モリハンジヤウヅル
95194 キイセンニンサウ,キノクニセンニンサウ,タニモダマ
95201 シマコバノボタンヅル
95216 カザグルマ
95218 ヤエカザクルマ
95219 フジボタン,ルリコシ
95221 ヤエカザクルマ
95295 オオワクノテ
95302 シベリアハンショウズル
95304 ェゾミヤマハンショウズル,ミヤマハンショウズル,ミヤマハンショウヅル,ルリハンショウズル
95350 コウデンセンニンサウ,トウサンセンニンサウ,ヤヘヤマセンニンサウ
95358 センニンサウ,センニンソウ,タカタデ,ハコボレ,ハボロシ,フックサ
95361 ガランビボタンヅル
95366 タチセンニンソウ,ミョウトウセンニンソウ
95371 ベニバナハンショウヅル
95396 サンエフボタンヅル
95606 グレオメ属,セイヤウフウチウサウ属,セイヨウフウチョウソウ属
95687 ミツバセイヨウソウフウチョウソウ,ミツバフウチョウソウ
95759 フウチョウソウ
95781 アフリカフウチョウソウ
95796 クレオメサウ,クレオメソウ,クレヲメサウ,セイヤウフウテウサウ,セイヨウフウチョウソウ,ノボリバナ,ハリフウテウサウ
95807 ミツバフウチョウソウ
95862 クミスクチク,クミスクチン,ネコノヒゲソウ
95934 クサギ属,クレロデンドロングレオメ属
95978 ボタンクサギ
95984 シラゲクサギ
96009 ヤヘザキクサギ
96028 マキバクサギ
96044 マレ－クサギ
96078 シナクサギ,ベニバナクサギ
96139 クルマバジヤウザン,クルマバジョウザン,クルマバタガヤサン
96140 イボタクサギ,ガジヤンギ
96146 シマクサギ
96147 タウギリ,トウギリ,ヒギリ
96180 リンドレイクサギ
96257 シマヒギリ,リュウセンクワ
96356 ジヤワヒギリ
96387 ゲンペイカズラ,ゲンペイクサギ,ハリガネカズラ,ハリガネカツラ
96398 ギナ,クサギ,クサギナ,クサギラ,トウノキ,ビロウドクサギ,ビロ－ドクサギ,ヤマコウズ
96399 シロミクサギ
96400 ビロ－ドクサギ
96401 ジウロクサギナ,ショウロウクサギ,ショウロクサギ
96403 アマクサギ
96408 アマクサギ
96439 クラリンドウ,タガヤサン
96454 リャウブ属,リョウブ属
96457 アメリカリャウブ,アメリカリョウブ
96466 サルダメシ,ナタツモリ,ハタツマリ,ハタツモリ,ミヤマリョウブ,リャウブ,リョウブ,リョウボウ,リョオブ,レウボフ
96561 リャウブ科,リョウブ科
96576 サカキ属
96587 サカキ
96588 フクリンサカキ
96598 マルバサカキ,マルバノサカキ
96617 サカギ
96633 クリアンサス属
96650 クリデミア属
96960 タフバナ属,トウバナ属
96970 オキナワクルマバナ,クルマバナ
96972 マンシウクルマバナ,マンシュウクルマバナ
96973 ニッコウクルマバナ
96974 シロバナクルマバナ
96979 エゾクルマバナ,ヤマクルマバナ
96981 タウバナ,トウバナ
96993 ジャカゴトウバナ
96999 タフバナ,トウバナ,トオバナ
97008 ヒロハヤマトオバナ
97011 ニヒタカタフバナ
97015 ミヤマクルマバナ
97029 イヌトウバナ
97030 シロバナイヌトウバナ
97031 ミヤマトウバナ
97034 ヤマトウバナ
97035 ヒロハヤマトウバナ
97037 コケトオバナ,ヤクシマトウバナ
97056 クルマバナ
97072 ノヤシ属,マガクチヤシ属
97074 オトコヤシ
97076 ノヤシ
97086 ツバメオモト属
97093 ササニンドウ,タウチサウ,ツバメオモト,トウチソウ
97184 チョウマメ属,テフマメ属
97203 キリトリア,コテフマメ,チョウマメ,テフマメ
97213 クンシラン属
97218 アバナクンシラン,ウケザキクンシラン,オオバナクンシラン,バナラン
97223 クンシラン
97260 オトギリソウ属
97266 オトギリサウ科,オトギリソウ科
97468 クリトストマ属
97491 クネオルム属
97581 サントリサウ属,サントリソウ属
97599 キバナアザミ,サントリサウ,サントリソウ
97697 センキウ属,センキュウ属,ハマゼリ属
97703 ダウリアゼリ
97711 ハマゼリ,ハマニンジン
97719 オカゼリ,ジャショウシ
97720 カギノニンジン
97727 モウコゼリ
97752 コ－ベア属
97757 ツルコ－ベア,ツルコベア
97861 コッコロバ属
97865 ハマベブドウ
97883 ホソエクマデヤシモドキ属
97894 アオツヅラフジ属,アヲツヅラフヂ属
97919 イソヤマアオキ,イソヤマアヲキ,イソヤマダケ,カウシウウヤク,コウシュウウヤク,コメゴアジン,コメゴメジン
97933 アオツヅラフジ,ホウザンツヅラフヂ,ホソバツヅラフヂ
97938 ホウザンツヅラフジ
97947 アオツズラフジ,アオツヅラフジ,アヲツヅラフヂ,カミエビ,チンチンカヅラ,ツヅラカヅラ,ツヅラフヂ,ピンピカヅラ
97967 コクレア－リア属,トモシリサウ属,トモシリソウ属
97983 ユリワサビモドキ
98020 トモシリサウ,トモシリソウ
98022 トモシリソウ
98085 コクリオ－ダ属
98116 バタ－カップノキ
98134 コ－コス属,ココス属,ココヤシ属,ヤシ属
98136 ケラパ,ココヤシ,ヤギ,ヤシ,ヤシホ
98156 マヒハギモドキ
98157 ヒメノハギ
98159 デンシンサウ,マイクサ,マイグサ,マイハギ,マヒエフハギ,マヒグサ,マヒハギ,ユレハギ,ユレバハギ

98177 クロトンノキ属,ヘンエフボク属
98190 クロトンノキ,ヘンエフボク,ヘンヨウボク
98191 クロトンノキ,ヘンヨウボク
98220 アリモリサウ属,アリモリソウ属
98223 アリモリサウ,アリモリソウ,ツノツクバネ
98273 ツルニンジン属
98341 ヒメツルニンジン
98343 ツルニンジン
98345 ミドリツルニンジン
98346 シブカワニンジン
98395 ヒカゲツルニンジン
98417 トウサン
98424 バアソブ
98425 ミドリバアソブ
98446 ヒナザサ属
98451 ヒナザサ
98586 アオチドリ,チシマアオチドリ
98591 タカネアオチドリ
98593 アオチドリ
98601 キンヤウラク属,セロヂネ属
98634 マシロラン
98640 キンヨウラク
98653 フラシダ
98743 ガンショウラン,チャイロヤウラクラン
98772 エゾノシシウド属,ミヤマセンコ属
98777 ウシウド,エゾノシシウド,エゾノハマウド,ハマシシウド
98781 ミヤマゼンゴ
98784 クモイノダケ
98785 エゾヤマゼンゴ
98786 タカネシシウド
98844 コーヒーノキ属,コーヒー属
98857 コーヒーノキ
98866 ベンガルコーヒー,ベンガルコーヒーノキ
98872 ロブスタコーヒーノキ
98886 ロブスタコーヒーノキ
98896 コンゴコーヒーノキ
98941 リベリアコーヒーノキ,リベリャコーヒー
99008 ステノフイラコーヒー
99089 キバナスズシロモドキ
99090 キバナスズシロモドキ
99105 ジュズダマ属
99124 ジュズダマ,ズズコ,チョウセンムギ,トウムギ,ハトムギ
99131 オニジュズダマ
99134 ハトムギ
99138 ナガミノジュズダマ
99157 コラナットノキ属,コラノキ属,コラ属
99158 コラ,コラナットノキ,ヒメコラノキ
99240 コラノキ
99293 イヌサフラン属,コルチカム属
99297 イヌサフラン
99358 ホウザンカラクサ属
99361 ホウザンカラクサ
99411 コヌカシバ属
99413 コヌカシバ
99506 コレウス属,サヤバナ属
99519 ナンヨウサヤバナ
99523 ナンヨウサヤバナ
99573 ヒメサヤバナ
99581 ケサヤバナ
99711 キランジソ,キンランジソ,ケサヤバナ,コモウソウ,サヤバナ,ニシキジソ,ハゲイトウモドギ,ヒメコレウス,ヒメニシキジソ
99712 ヒメニシキジソ
99770 コラビラン属
99775 コラビラン
99833 コリンシア属
99836 オオバナコリンソウ
99838 コリンソウ,フタイロコリンシア
99850 コロミア属
99854 ヤマギバナシノブ
99855 ヤナギハナシノブ
99858 タイリンハナシノブ,タイリンヒメハナシノブ
99859 ホソバヤナギハナシノブ
99899 サトイモ属
99910 イヘツイモ,イモ,サトイモ,タイモ,ハタクイモ,ヤマサトイモ
99913 ズイキ
99918 コニシイモ,サトイモモドキ
99919 シロイモ,ハスイモ
99928 コウトウイモ
99936 トウノイモ
100027 キューバヤシ属
100028 キューバヤシ
100058 ヤヘヤマハマナツメ属
100061 ヤエヤマナツメ,ヤエヤマハマナツメ,ヤヘヤマハマナツメ
100110 コルムネア属
100149 ギリシア属,バウクワウマメ属,ボウコウマメ属
100153 バウクワウマメ,ボウコウマメ
100232 テンニンサウ属,テンニンソウ属
100233 タカサゴテンニンソウ
100254 クロバナラフゲ属,クロバナロウゲ属
100257 クロバナラフゲ,クロバナロウゲ,ヌマラフゲ
100263 サンプクリンドウ属
100277 サンプクリンドウ
100278 サンプクリンドウ
100280 サンプクリンドウ
100281 シロバナサンプクリンドウ
100296 シクンシ科
100892 ツユクサ属
100930 ホウライツユクサ
100931 シロバナホウライツユクサ
100940 マルバツユクサ,マルバノバツユクサ
100961 アイバナ,アオバナ,アヰバナ,カマツカ,ツキクサ,ツユクサ,バツユクサ,ハナダグサ,ボウシバナ
100964 シロバナツユクサ
100966 ケツユクサ
100967 ウサギツユクサ
100970 オオボウシバナ
100971 シロバナオオボウシバナ
100972 ホソバツユクサ
100982 ヒメオオボウシバナ
100990 シマツユクサ
101112 ナンバンツユクサ
101178 タマツュクサ
101185 ナミツユクサ
101200 ツユクサ科
101285 コンミフォラ属,ミルラノキ属
101626 シンパレッティア属
101642 キク科
101680 イハタバコ属,イワタバコ属
101681 イハタバコ,イワタバコ
101682 シロバナイワタバコ
101684 ケイワタバコ,フイリケイワタバコ
101686 イリオモテイワタバコ,タイワンイワタバコ
101798 コニコシア属
101820 ミヤマセンキウ属,ミヤマセンキュウ属
101823 カラフトニンジン
101825 ミヤマセンキュウ
101827 カラフトニンジン
101831 ニヒタカセンキウ
101837 チシマニンジン,テウカイゼリ,ミヤマセンキウ,ミヤマセンキュウ
101845 ドクニンジン属
101852 シキュータ,ドクニンジン
101858 コウトウマメ科,コンナルス科,ヒルガホ科
102017 コノフィルム属
102037 コノフィツム属
102086 ショウショウ
102090 シキテン
102122 ウンエイギョク
102136 セソン
102150 コブエ
102163 シチセイザ
102209 シュウソウ
102215 タマヒコ
102227 ジャクコウ
102272 セイエン
102324 ツバキヒメ
102377 シンレイ
102392 キョヒメ
102457 オウキュウデン
102460 セイシュンギョク

102508 チュウナゴン
102510 スイコウギョク
102542 グンケイ
102550 チョウウギョク
102613 サクラガイ
102628 スイセイ
102652 テンシ
102657 コウスイギョク
102658 コウスイギョク
102668 シュンジギョク
102692 ヒナバト
102699 メイソウギョク
102710 コヅチ
102811 ナタネハタザオ
102827 ヒエンソウ属
102833 チドリソウ,ヒエンソウ,ルリヒエンサウ,ルリヒエンソウ
102856 キミカゲサウ属,スズラン属,ヨキミカゲソウ属
102863 キミカゲサウ,キミカゲソウ,スズラン
102867 キミカゲサウ,キミカゲソウ,スズラン,タニマノヒメユリ,ドイツスズラン
102884 アサガオバナ,サンシキアサガオ
102892 ビルガオ科
102903 サンシキヒルガオ属,サンシキヒルガホ属
102914 アオイヒルガオ
102926 ヒナヒルガオ,ヒナヒルガホ
102933 セイヤウヒルガホ,セイヨウヒルガオ,ヤマブヒルガオ
102980 ムラダチヒルガオ
103048 ハナヒルガオ
103194 ヒメムラダチヒルガオ
103340 アサガオバナ,サンシキアサガオ,サンシキヒルガホ
103362 フウセンアサガオ
103402 イズハハコ属,ヤマヂワウギク属
103407 キクバイヅハハコ,キクバハハコ
103438 アレチノギク
103446 ケナシヒメムカシヨモギ,ゴイッシングサ,テツダウグサ,ヒメムカシヨモギ,メイヂサウ
103449 ケナシヒメムカシヨモギ
103509 イズハハコ,イズホオコ,イヅハウコ,イヅハハコ,ヤマジオウギク,ヤマヂワウギク,ワタナ
103516 ネバリイヅハハコ
103554 ケナシヒメムカシヨモギ
103617 オオアレチノギク
103671 クーペーリア属
103731 フトエウチワヤシ属
103737 ブラジルロウヤシ
103738 トレーフトエウチワヤシ
103739 コピアポア属
103740 ブリュウマル
103741 テイリュウカン
103742 コクセンギョク
103743 カナルギョク
103744 リュウガギョク
103745 コクオウマル
103746 コリュウマルォ
103747 シロトゲコクオウマル
103749 リュウソウキョク
103750 オゲンギョク,オゴンギョク
103751 クンコウマル
103752 コクシカン
103753 ドウラマル
103754 リュウマギョク
103755 モウコギョク
103756 ライケツマル
103757 シュウソウギョク
103758 ヒヨウゼンギョク
103759 コウシマル
103760 キオウギョク
103761 メイオウマル
103762 キシンリュウ
103763 ゴウソソウマル
103764 リュウリンギョク
103765 コゼンギョク
103766 ショウフウギョク
103767 カイリュウギョク
103769 セッキギョク
103770 コクランジョウ
103771 コクオウデン
103772 ギョリンギョク
103824 オウレン属,ワウレン属
103835 オウレン,オホバワウレン,キクバワウレン,セリバワウレン,ナガバワウレン,ホソバワウレン,ワウレン
103843 ゴカエフワウレン,ゴカヨウオウレン,バイカオウレン,バイクワワウレン
103850 カタバミオウレン,カタバミワウレン,ミツバオウレン,ミツバワウレン
103853 ミツバノバイカオウレン
103862 サルトカヅラ,シマヒョオタンボク,シマヘウタンボク,ニンドウモドキ,ヒョオタンカズラ,ヘウタンカヅラ,ヘウタンボク
104019 チョウセンラン
104039 カラスノゴマ属
104046 カラスノゴマ
104047 アオカラスノゴマ
104051 チョウセンカラスノゴマ
104056 ツナソ属
104057 ツナソモドキ
104059 シマツナソ
104072 インドアサ,カナビキオ,カナビキヲ,コウマ,ジュート,ツナソ
104103 シマツナソ,タイワンツナソ
104147 イヌヂシャ属
104175 イヌジシャ,カキバチシャノキ
104203 トゲミイヌジシャ,トゲミチシャノキ,トゲミノイヌヂシャ
104218 アドダン,イヌジシャ,カギバチシャノキ,ワレナ
104334 センネンボク属
104339 センネンボクラン,ニオイシュロラン,ニホヒシュロラン
104344 アカスジアツバセンネンボク
104350 センネンボク
104357 ハーゲセンネンボク
104359 アツバセンネンボク,アツバセンネンボクラン
104367 コウチク,センネンソゥ,センネンボク
104415 イヌヤマブキソウ
104429 ハルシャギク属
104448 キンケイギク
104482 キンケイギク
104509 ホソバハルシャギク
104531 オオキンケイギク,オホキンケイギク
104545 ハマベハルシャギク
104598 クジャクサウ,クジャクソウ,ジャノメサウ,ジャノメソウ,ソメモノギク,タリヤス,ハルシャギク
104604 クロバナハルシャギク
104616 イトバハルシャギク
104638 ケモウコオウギ
104640 モウコオウギ
104687 コエンドロ属
104690 エンスイ,コエンドロ,コニシ
104691 ドクウツギ属
104694 キンバンドクウツギ,タイワンドクウツギ
104696 イチロベゴロシ,カワラウツギ,シマウツギ,ドクウツギ,ナベワリ
104700 セイヨウドクウツギ
104701 ヒマラヤドクウツギ
104710 ドクウツギ科
104750 カハラヒジキ属,カワラヒジキ属
104755 トウカワラヒジキ,マツカサヒジキ
104758 コウアンスナヂヒジキ
104765 ヒモヒジキ
104767 ウスイロマツカサヒジキ
104771 カハラヒジキ,カワラヒジキ
104772 ヒロハカワラヒジキ
104786 アレチヒジキ
104831 ウロコヒジキ
104834 スナヂヒジキ
104836 エリミノスナヂヒジキ
104849 アオヒジキ
104850 ホソバウロコヒジキ
104880 ミズキ科,ミヅキ科
104917 コルヌス属,ズミノキ属,ミズキ属,ミヅキ属
104920 シラタマミズキ,シラタマミヅキ,シロミノミズキ
104949 アメリカミズキ

104966 クマノミズキ,サワミズキ
105003 フイリミズキ
105023 アメリカヤマボウシ,ハナミズキ
105111 セイヤウサンジュユ,セイヨウサンシュユ
105146 アキサンゴ,サハグミ,サワグミ,サンシュユ,ハルコガネバナ,ヤマグミ
105147 マルバサンシュユ
105202 カラフトミズキ
105204 エゾゴゼンタチバナ,ユゾゴゼンタチバナ
105207 シラタマミズキ
105269 オウゴンハギ属,ワウゴンハギ属
105284 オウゴンハギ
105310 ツリシャクジョウ
105324 タマザキクサフジ
105340 カラクサナズナ
105434 コリヨカクタス属
105436 キョウリュウカク
105437 シンリョクカク
105438 ヒャクマントウ
105439 ハテンコウ
105440 カンノンチュウ
105450 シロガネヨシ属
105456 シロガネヨシ,パンパス・グラス
105488 サクラサウモドキ属,サクラソウモドキ属
105496 サクラサウモドキ,セイヨウサクラサウモドキ
105499 サクラソウモドキ
105501 サクラソウモドキ,センバサクラサウモドキ
105518 ゴリアーンテス属
105557 キケマン属,ムラサキケマン属,ヤブケマン属
105594 エゾエンゴサク,デハエンゴサク,ヤマエンゴサク
105596 ササバエンゴサク,シロバナシラゲエンゴサク
105599 ホソバェンゴサク
105602 マルバェンゴサク
105607 ホソバエンゴサク
105639 シマキケマン
105680 イヌキグマン
105682 マンシゥエンゴサク
105693 タイトウケマン
105721 エンゴサク,オランダエンゴサク
105776 エゾオオケマン
105810 シラウバウエンゴサク,ジロボウエンゴサク,ビッチリ,ヤブエンゴサク
105812 シロバナジロボウエンゴサク
105874 タケシマエンゴサク
105915 オトメエンゴサク
105917 ヤチマタエンゴサク
105918 エゾエンゴサク
105929 カラフトオオケマン,カラフトゲマン
105976 キケマン,ツクシキケマン,ハマキケマン
105977 ムニンキケマン
105978 キケマン
106004 シロヤブケマン,ムラサキケマン,ヤブケマン
106005 シナノサラサケマン
106006 ユキヤブケマン
106021 ヤマキケマン
106051 チドリケマン
106076 ササバエンゴサク,ヤマエンゴサク
106078 ヒメエンゴサク
106197 オホバナミヤマキケマン,ツルケマン
106214 ヤマキケマン
106217 ミチノクエンゴサク
106218 タイワンケマン,ナンゴクキケマン
106227 フウロケマン,ミヤマキケマン
106235 タカネキケマン
106238 ミヤマキケマン
106246 キンキエンゴサク
106350 キケマン,ホザキキケマン
106355 ナガミノツルキケマン
106370 ササバエンゴサク
106380 ヒナエンゴサク
106385 ヒナエンゴサク
106453 オランダエンゴサク
106466 エゾキケマン,エゾケマン
106508 シマキケマン
106519 エンゴサク,オオバエンゴサク,オホバノエンゴサク,タンエフエンゴサク,タンヨウエンゴサク
106564 チョウセンエンゴサク
106568 ホソバノチョウセンエンゴサク
106571 リョウトウエンゴサク
106609 エンゴサク
106628 トサミズキ属,トサミヅキ属
106636 ショウコウミズキ
106637 キリシマミズキ
106644 カウヤミズキ,コウヤミズキ,ミヤマトサミズキ,ミヤマトサミヅキ
106647 ヒゴミズキ
106656 シマミヅキ
106669 イヨミズキ,イヨミヅキ,ヒウガミヅキ,ヒュウガミズキ,ヒュウガミヅキ
106675 シナミズキ
106682 トサミズキ
106684 イトウミヅキ,タイワントサミヅキ
106696 ハシバミ属
106698 アメリカハシバミ
106703 セイヨウハシバミ
106724 カナダハシバミ
106736 オオハシバミ,オヒヤウハシバミ,オヒョウハシバミ,オヒョウバハシバミ,オホハシバミ,ハシバミ
106741 ハシバミ
106743 エゾハシバミ
106751 オオツノハシバミ,オホッハシバミ
106777 ツノハシバミ,ナガハシバミ
106778 コツノハシバミ
106779 トックリハシバミ
106781 オオツノハシバミ
106793 レモンユウカリ,レモンユーカリ
106880 チクセツラン
106883 バイケイラン
106957 コリノプンテイア属
106958 マメキリン
106961 ムシャウチワ
106970 ヒメムシャ
106984 コウリバヤシ属
106999 コウリバヤシ,タリポットヤシ
107004 コリファンタ属
107016 シシフンジン
107021 アカメジシ,ハリネズミマル
107023 ゾウゲマル
107026 ヨウキヒ
107053 ダイショウカン,ホウジョ
107059 レイヨウマル
107076 ホッギョクマル,ヨコヅナ
107158 オオバルシャ属,オホバルシャ属,コスモス属
107160 ベニコスモス
107161 アキザケラ,アキノハルシャギク,オオバルシャ,オオバルシャギク,オホバルシャ,オホバルシャギク,コスモス
107168 ダリアコスモス
107172 キバナコスモス
107189 オオホザキアヤメ科
107221 オオホザキアヤメ属,オホホザキアヤメ属,コスタス属
107244 ベニバナフクジンソウ
107271 オオホザキアヤメ,オホホザキアヤメ,フクジンソウ
107297 コチヌス属
107300 オウロ,ケムリノキ,コチナス,ハゲマノキ
107312 コッギグリア
107319 コトネアスター属,シャリンタウ属,シャリントウ属,ベニシタン属
107322 シタンシュ
107325 コニシカマツカ
107336 シセンシャリントウ
107370 ツケバシャリントウ
107454 ヒマラヤリントウ
107486 コトネアスター,ベニシタン
107504 マルバシヤリントウ
107516 コニシカマツカ
107543 クロミシャリントウ
107549 コバノシャリントウ,ヒメシタン,ヒメシャリントウ
107575 コケモモカマシカ,ロクジャウコケモモカマツカ
107583 シロバナシャリントウ
107615 ギンヨウシャリントウ

107637 フサザキシャリントウ
107649 チリメンシャリントウ
107668 ヤナギバシャリントウ
107686 コホクシャリントウ
107728 タケシマシャリントウ
107747 タカサゴトキンサウ属,タカサゴトキンソウ属
107752 タカサゴトキンサウ,タカサゴトキンソウ
107755 マメカミツレ
107766 ウシオシカギク
107840 コチレドン属
107852 ハクチョウ
108045 サオヒメ
108075 ギンバニシキ
108193 ホウガンノキ属
108195 ホウガンノキ,ホウガンボク
108591 ハマナ属
108627 ハマナ
108736 ベニバナボロギク
108775 クラッスラ属
108828 リュウグラジョウ
108844 ゲッコウ
108846 タマツバキ
108905 レイジン
108918 オトヒメ
108921 シロタエ
108952 チコスガタ
109060 トモエ
109097 ラクトウ
109151 ギンセン
109174 イソベノマツ,ナルト
109180 セイサリュウ
109229 フチベニベンケイ
109264 シントウ,ホウトウ
109266 ホシオトメ
109293 リョクトウ
109374 ツクバネ
109446 サヨギネ
109452 トウゲンキョウ
109463 コケマンネングサ
109505 ベンケイサウ科,ベンケイソウ科
109509 サンザシ属
109597 クロミサンザシ
109650 サモモ,サンザシ
109651 キミサンザシ,キミノサンザシ
109780 エゾオオバサンザシ,エゾサンザシ,エゾノオホサンザシ,ヤチザクラ
109790 セイヤウサンザシ,セイヨウサンザシ
109839 アラゲアカサンザシ,オオバサンザシ,オホバサンザン,カラフトオオサンザシ,カラフトオホサンザシ
109857 アカバナヤエサンザシ,ヒトシベサンザシ
109885 アカバナサンザシ
109924 オオウサンザシ,オオサンザシ,オホサンザシ,タウサンザシ,トウサンザシ,リュウキュウサンザシ
109933 オオサンザシ,オホサンザシ,ミサンザシ
109936 オオサンザジ,オオミサンザジ,ヒロハサンザシ
109937 ホソバサンザシ
110014 アカサンザシ,オオバサンザシ,ベニサンザシ
110205 ギョボク属
110216 ギョボク
110227 アマギ,ギョボク
110246 ゲロンガン
110251 オハグロノキ
110292 ツルリンダウ属
110500 サイハイラン属
110501 モイワラン
110502 サイハイラン,タイワンサイハイラン,ハックリ
110505 アオサイハイラン
110506 サイハイラン
110516 トケンラン
110602 アゼタウナ属,アゼトウナ属
110603 ヤクシアゼトウナ
110604 ユズリハワダン
110605 クサノオウバノギク
110606 ヤクシソウ
110607 ウスイロヤクシソウ
110608 ハナヤクシソウ
110610 コヘラナレン
110611 アゼトウナ
110612 ソテツバアゼトウナ
110615 ソテツナ,ハマナレン,ホソバワダン
110621 ダイトウワダン
110622 ヘラナレン,モクワダン
110623 ワダン
110626 イヌヤクシソウ
110627 アシブトワダン
110628 ナガバヤクシソウ
110637 イリオモテヒメラン
110648 マツダヒメラン
110710 オニタビラコ属,クレビス属,クレピス属,フタマタタンポポ属
110767 セイヨウニガナ
110785 シラゲタンポポ
110835 エゾタカネニガナ
110841 フタマタタンポポ
110985 センボンタンポポ,モモイロタンポポ
111017 アレチニガナ
111019 シベリアニガナ
111050 ヤネタビラコ
111100 フクベノキ属
111104 フクベノキ
111145 トリクスピダーリア属
111152 クリナム属,ハマオモト属
111157 エンレイハマオモト
111167 オオハマオモト,ハマオモト
111169 ハマオモト
111170 ハマオモト,ハマモメン,ハマユウ
111171 ダイワンハマオモト
111180 ナガバハマオモト
111200 オガサワラハマユウ
111214 インドハマユウ,クリナム
111230 ジュウガツハマオモト,モモイロハマオモト
111236 ナンヨウハマオモト
111343 クリスマム属
111455 クロコスミア属,ヒメトウショウブ属
111458 ヒオウギズイセン
111464 ヒメヒオウギズイセン
111480 クロッカス属,サフラン属
111558 イノゲラン,キバナサフラン
111589 サフラン,バンサンジコ
111625 クロッカス,ハナサフラン,ハルサフラン,ムラサキサフラン
111674 ナベワリ属
111676 ナベワリ
111677 ヒメナベワリ
111693 クロッサンドラ属,ヘリドリオシベ属,ヘリドリヲシベ属
111724 キツネノヒガサ,ジョゥゴバナ,ヘリドリオシベ
111741 シラゲキツネノヒガサ
111774 ハリドリオシベ,ハリドリヲシベ
111819 クロッソソマ科
111823 モクビヤクコウ属
111826 モクビヤクコウ,モクビャッコ,モクビャッコウ
111851 タヌキマメ属
111856 ヒメタヌキマメ
111874 ヤハズマメ
111879 シロタヌキマメ
111915 コガネタヌキマメ
111951 ハネタヌキマメ
111987 ガクタヌキマメ
112013 アコウタヌキマメ,シマタヌキマメ,タイワンタヌキマメ
112108 ダエンタヌキマメ
112138 オホバタヌキマメ,スズナリヤー,ナンバンタヌキマメ
112238 クロタラミア,コウシュンタヌキマメ
112269 サンヘンプ
112340 ホソバタヌキマメ,ヤヘヤマタヌキマメ
112405 アメリカタヌキマメ
112432 ヤエヤマタヌキマメ
112433 ヤエヤマタヌキマメ
112486 ヒレタヌキマメ
112491 オオミツバタヌキマメ
112492 オオミツバタヌキマメ

112601 ハウチワタヌキマメ
112615 コガネタヌキマメ
112667 タヌキマメ
112671 ガランビタヌキマメ,ガランビマメ
112757 アフリカタヌキマメ
112776 エダウチタヌキマメ
112792 オホバタヌキマメ
112829 グミモドキ属,クロトン属,チャンカニ属,ハズ属
112853 グミモドキ,チャンカニイ
112884 エルテリア
113039 ハズ,ハヅ
113102 クロ-ウェア属
113110 クルシアネラ属
113144 アブラナ科
113290 ハリネヤシ属
113293 バナマハリネヤシ
113305 トキンガヤ属
113306 トキンガヤ
113316 ホガクレシバ
113364 クリプタンサス属
113365 ヒメアナナス
113366 ムラサキヒメアナナス
113372 ヘラヒメアナナス
113374 ビロ-ドヒメアナナス
113380 ナガバ,ナガバヒメアナナス
113381 フイリナガバヒメアナナス
113387 トラフヒメアナナス,フヒメアナナス
113397 クスモドキ属
113422 クスモドキ属,シナクスモドキ属
113436 シナクスモドキ,マルバダモ
113438 コニシグス,コニシグスモドキ
113507 クリプトセレウス属
113511 クリプトキ-ルス属
113527 クリプトコリ-ネ属
113566 マツムラカヅラ属
113633 スギ属
113685 サンブスギ,スギ,マキ
113686 メジロスギ
113687 アヤスギ,エンコウスギ
113690 アヤスギ,エンコウスギ
113692 セッカスギ
113693 イトスギ
113696 クロベスギ,タウスギ,トウスギ,ヒメスギ,ヒラスギ,ベニスギ,ヤハラスギ,ヤワラスギ
113699 イカリスギ
113700 チャボスギ,バンタイスギ,ビロウドスギ,マンネンスギ
113702 セッカンスギ
113703 ヨレスギ
113705 エイザンスギ
113710 マンキチスギ,ムレスギ
113711 ミタマスギ
113717 チャボスギ,バンダイスギ,ビロウドスギ,マンネンスギ
113720 アシウスギ,アシオスギ
113721 カワイスギ
113738 クリプトフォランツス属
113760 クリプト-プス属
113823 オオバナアサガオ
113846 オホスズムシラン属
113848 オオスズムシラン
113863 タカオオオスズムシラン
113868 ミツバゼリ属,ミツバ属
113872 ミツバ,ミツバゼリ
113879 ウシミツバ,ハニヤミツバ,ミツバ,ミツバゼリ
113881 ムラサキミツバ
113883 ウシミツバ
113944 クテナンテ属
114057 ナンバンハコベ属
114060 ナンバンハコベ
114066 ツルセンノウ,ナンバンハコベ
114108 キウリ属,キュウリ属
114122 ニシインドコキュウリ
114150 ボケットメロン
114189 アミメロン,ジャコウウリ,マスクメロン,メロン
114194 ザッソウメロン
114197 ナシウミ
114199 カンタロ-ブ
114200 マンゴ-メロン
114201 シロウリ,ツケウリ
114208 キンウリ
114212 トウメロン
114213 ナシウリ,マクワウリ
114214 ツナメロン
114231 トゲスズメウリ
114245 カラウリ,キウリ,キュウリ,ソバウリ
114252 キュウリ
114261 キンカクウリ
114271 カボチャ属,タウナス属
114277 クロダネカボチャ
114288 セイヨウカボチャ,タウナス,ナタウリ,ポンキン
114292 アコダ,カボチャ,トウナス,ニホンカボチャ,ボウブラ
114293 ツルクビカボチャ,ヘチマカボチャ
114294 キクザカボチャ,キクザノタウナス,サツマ,ナンクワ,ボウブナ,ボウブラ,ワセボウブラ
114295 カボチャ,タウナス,ナンキンボウフラ,ニホンカボチャ,バンナンクワ,ヒウガウリ
114300 セイヤウカボチャ,セイヨウカボチャ,ナタウリ,ペポカボチャ,ボウフラ
114301 ウリカボチャ,ズッキ-ニ
114305 キントウガ,キントオガ
114310 オモチャカボチャ,カザリカボチャ,コナタウリ
114313 ウリ科
114319 ハリグハ属,ハリグワ属
114325 カカツガユ,ソンノイゲ,ヤマミカン
114427 オランダビュ,ハシコ
114502 クミヌム属,ヒメウイキョウ属
114503 クミン,ヒメウイキョウ
114532 クワウエフザン属,コウヨウザン属
114539 オランダモミ,カントンスギ,ギョウジヤモミ,クワウエフザン,コウヨウザン,ランダイスギ,リウヒガヤ,リュウキュウスギ,リュウキュウモミ,リュウヒ
114548 カハリクワウエフザン,ダイスギ,ランダイスギ
114563 クノニア属
114567 クノニア科
114595 クフエア属,クフエヤ属,タバコソウ属
114599 ネバリミソハギ
114601 ヒメホクシャ
114605 メキシコハナヤナギ
114606 タバコソウ,ベニチョウジ
114611 ハナヤナギ
114613 ヒメハナヤナギ
114614 ネバリクフュア
114616 ムラサキミソハギ
114618 クサミソハギ
114621 ネバリハナヤナギ
114645 ヒノキ科
114649 イトスギ属,クプレッズスス属,シダレイトスギ属
114652 アリゾナイトスギ
114668 カシミ-ルイトスキ
114690 イトヒバ,シダレイトスギ
114698 カシュウイトスギ
114714 クプレサス
114721 マクナブイトスギ
114723 モントレイイトスギ,モントレ-サイプレス,モントレ-サイプレス
114751 サ-ジェントイトスギ
114753 イタリアサイプレス,セイヤウヒノキ,セイヨウヒノキ,ホソイトスギ,ホソイトヒバ
114760 ニオイヒバ
114763 ピオイヒバ
114768 イタリアホソヒバ
114775 オオイイトスギ,オオイトスギ,オホイイトスギ
114814 キンバイザサ属
114819 オオギンバイザサ,オオセンボウ,オオバセンボウ,キンバイザサ
114838 キンバイザサ
114852 ウコン属
114859 キャウワウ,キョウオウ,キョウワウ,ノルウコン,ハルウコン
114871 ウコン,キゾメグサ,キヤウワウ
114874 クルクマ
114884 ウスグロ,ガジュツ,シロウコン

114947 ネナシカズラ属,ネナシカヅラ属
114972 カハカミネナシカヅラ,マメダオシ
114986 アメリカネナシカズラ
114994 ハマネナシカズラ,マメダフシ
115007 アマダオシ
115008 ツメクサダオシ
115031 クシロネナシカズラ,クシロネナシカヅラ
115050 ネナシカズラ,ネナシカツラ
115051 ミドリネナシカズラ
115053 タイワンネナシカヅラ
115072 ハマネナシカヅラ
115099 アメリカネナシカズラ
115155 ネナシカズラ科
115239 アフリカヤツデ
115426 キアナストルム科
115429 キアナストルム属
115521 アラゲツユクサ属,シアノチス属
115526 アラゲツユクサ
115587 アラゲツユクサ
115595 ムラサキムカシヨモギ,ヤンバルノキク,ヤンバルヒゴタイ
115697 イノコヅチモドキ属,ヰノコヅチモドキ属
115738 イノコヅチモドキ,ヰノコヅチモドキ
115784 キンレイジュ
115793 ソテツ科
115794 ソテツ属
115812 インドソテツ,ナンヨウソテツ,モルッカソテツ
115884 ソテツ
115888 ナンヨウソテツ,ルンフソテツ
115910 タイワンソテツ
115912 タイワンソテツ
115931 シクラメン属,ブタノマンヂュウ属
115949 アキザキシクラメン,カガリビバナ,ザイクラメン,シクラメン,ネアポリタヌム,ブタノマンヂュウ,マルバシクラメン
115953 アキザキシクラメン,ハミズシクラメン
115965 カガミビバナ,カガリビバナ,シクラメン,ツタバカガリビバナ,ヒロハブタノマンヂャウ,ブタノマンジュウ
115974 ツタバシクラメン
115989 パナマサウ科,パナマソウ科
115990 シクランテラ属,バクダンウリ属
115991 バクダンウリ
116020 ミャコジマツヅラフジ,ミャコジマツヅラフヂ
116029 フサザキツヅラフヂ
116044 アカガシ属
116098 イチイ,イチイガシ,イチガシ,イチヰガシ,イヌガシ
116100 アラガシ,アランボガシシ,クロガシ,ツボガシ,ナラバガシ,ハトガシ
116106 シマガシ,ヨコメガシ
116118 モンパガシ
116139 ナガバシラカシ,ナガミシラカシ,ホソバシラカシ
116143 ホソバシラカシ
116150 タイワンアカガシ,モラガシ
116153 クロガシ,ササガシ,シラカシ,ホソバガシ
116156 クロガシ
116167 ナガエガシ
116174 ツクバネガシ
116187 ウラジロガシ,シラカシ,ホソバウジロガシ,ホンバガシ,ヤナギガシ
116193 カワガシ,センバカジ,ツクバネガシ,ハイギ,ハヒギ,メンガシ
116200 タイワンウラジロガシ,リサンアラカシ
116258 タンゲブ
116286 ホシサンゴ
116384 マツバコエンドロ,マツバゼリ
116516 キクノケス属
116532 クワリン属,シドニア属
116546 オニメロ,カマクラカイダウ,カマクラカイドウ,クワリン,マルメ,マルメル,マルメロ
116623 キリンドロフィルム属
116636 キリンドロブンティア属
116648 マツアラシ
116652 コブサンゴ
116657 ウロコウチワ
116661 キシカク
116667 クレタケ
116732 シンバラリア属
116736 ウンランカズラ,ツタガラクサ,ツタバウンラン,マルバノウンラン
116747 ウスギヌソウ,シンバリソウ
116755 シンビディエラ属
116769 ガンラン属,キペロルキス属,キンビディウム属,シュンラン属,シンビジューム属
116773 カンポクラン,ホウラン
116774 シュンポクラン,ホウラン
116779 ココンリン
116781 ケンラン,ダイケンラン
116815 カンポウラン,ヘツカラン,ホウラン,リ-チラン
116829 アカメソシン,オラン,ケイラン,コンリンザイ,シマホウザイ,ジラン,スルガラン,ヂラン,ママン,ヤキバラン
116832 メラン
116851 イッケイキウクワ,イッケイキュウカ,オビワケラン
116863 キンリュウヘン,キンリョウヘン,タイワンキンリョヘン,チャウジュラン,チョウジュラン
116880 イトラン,エクリ,ジイババ,シュンラン,タイワンシュンラン,ハックリ,ホクリ,ホクロ
116887 ホソバシュンラン
116900 ギョクチンラン
116916 シンゼジューム
116926 カンラン,シカンラン,セイバンソシン,ヨゴレ
116927 サラサカンラン
116928 シカンラン
116930 セイカンラン
116933 オオバカンラン
116936 カンラン,コラン,シマサキラン
116938 アキザキナギラン,ナキラン,ナギラン
116941 アキザキナギラン
116975 マヤラン
116976 サガミランモドキ
117000 ギョッカラン
117008 ダイケンラン
117022 キンリュウヘン,キンリョウヘン,チョウジュラン
117042 シロバナホウサイ,シロバナホウザイ,タイワンホウサイ,ハックワホウサイ,フチベニホウサイラン,ホウサイ,ホウサイラン
117045 ダイシンラン,ハクラン
117051 シロバナホウサイ
117090 コドウラン
117104 コンリンザイ,ヤキバラン
117136 オガルカヤ属,ヲガルカヤ属
117153 レモングラス
117181 イヌガルカヤ,オガルカヤ,カルカヤ,ヲガルカヤ
117185 オガルカヤ,シマオガルカヤ,シマヲガルカヤ,ハマオガルカヤ,ハマヲガルカヤ
117204 パルマロサグラス
117218 コウスイガヤ,シトロネラグラス
117245 カウスイガヤ,カウスヰガヤ
117296 シホニラ属
117308 ベニアマモ
117309 リュウキュウアマモ
117334 イケマ属,カモメヅル属,シナンクム属
117339 クサタチバナ
117364 クロバナロクオンソウ,ヒゴヒャクゼン,ロクオンサウ,ロクオンソウ
117368 クロバナロクオンサウ
117385 クサダチイケマ,クロベンケイ,フナバラサウ,フナバラソウ,ホソバノロクオンソウ,ロクオンソウ
117387 ホソバノロクオンソウ,ヤナギフナバラソウ
117388 アオフナバラソウ
117408 アマミイケマ
117412 ヤハズイグマ,ヤハズイケマ
117423 タンザワイケマ
117425 ヒメイグマ,ヒメイケマ

117469 ヒメガガイモ,ホウライイケマ
117488 タチカモメヅル
117497 ツクシガシワ,ツルガシワ
117499 ツルガシワ
117523 エゾノクサタチバナ
117532 シロバナクサタチバナ
117535 クロバナイヨカズラ
117537 イヨカズラ,イヨカヅラ,カラズノヒルヅル,スズメノオゴケ,スズメノヲゴケ
117538 スズメノオゴケ,スズメノヲゴケ
117571 リュウキュウガシワ
117597 イケマ,ケイマ,コサ,ヤマコガメ
117643 スズサイコ
117645 ヒロハスズサイコ
117660 ムラサキカモソツル
117698 コバノカモメツル
117700 アズマカモメヅル,トキワカモメヅル
117703 ジョウシュウカモメヅル
117705 キノクニカモメヅル
117706 シロバナカモメヅル
117721 イトカモメズル,ヒメイヨカズラ,ヒロバノヤナギカモメズル
117722 ホリバイヨカズラ
117734 シロバナオオカモメズル
117745 チョセンイケマ
117751 コイケマ
117765 チョウセンアザミ属,テウセンアザミ属
117770 カルドン
117787 アーチチョーク,チョウセンアザミ,テウセンアザミ
117852 ギャウギシバ属,ギョウギシバ属
117859 ギャウギシバ,ギョウギシバ
117867 オオギョウギシバ
117874 ティフトン
117878 オニギョウギシバ
117900 オオルリソウ属,オホルリサウ属,シノグロッサム属
117904 タカネルリソウ
117908 シナワスレナグサ,ホソバルリソウ
117956 エグウチルリソウ
117961 シマスナビキソウ,シマルリソウ,タイワンルリソウ
117964 シロバナシマルリソウ
117965 インドルリソウ,ウスバルリサウ,ウスバルリソウ,オオルリサウ,オオルリソウ
117966 オオルリソウ
117986 シマスナビキサウ,シマルリサウ,シマルリソウ,タイワンルリサウ,タイワンルリソウ
118038 オオルリソウ
118040 シロバナオオルリソウ
118047 ナムナム属
118060 ナムナム
118147 キノルキス属
118307 クシガヤ属
118315 クシガヤ
118319 ヒゲガヤ
118393 シペラ属
118398 アカバナサカズキアヤメ,トラユリモドキ
118403 カヤツリグサ科
118428 カヤツリグサ属,シペラス属
118430 アイノコガヤツリ
118431 オニチャガヤツリ
118432 コウヤガヤツリ
118433 フサガヤツリ
118434 カズサガヤツリ
118436 ヒラボガヤツリ
118437 チャハマスゲ
118438 コオニガヤツリ
118458 アレチクグ
118473 オキナワオオガヤツリ
118476 カラカサガヤツリ,シュロガヤツリ
118477 シュロガヤツリ
118478 フイリシュロガヤツリ
118481 コシュロガヤツリ
118491 キガヤツリ,チャガヤツリ
118493 コチャガヤツリ
118520 フトイガヤツリ
118642 クグガヤツリ
118646 ユメノシマガヤツリ
118737 タチガヤツリ,ヒトリカヤツリ
118744 タマガヤツリ
118749 オホノシスゲ
118756 オオホウキガヤツリ,オホハハキガヤツリ
118765 ハハキガヤツリ,ホウキガヤツリ
118784 ミクリガヤツリ
118798 ヒメハハキガヤツリ
118803 ホソミキンガヤツリ
118822 ショクヨウガヤツリ
118841 カンエンガヤツリ
118845 カンエンガヤツリ
118867 ヒメムツオレガヤツリ
118889 ヒナガヤツリ
118928 クロガヤツリ
118957 ヌマガヤツリ
118982 コアゼガヤツリ,ミヅハナビ
118987 ツルナシコアゼガヤツリ
118990 コアゼガヤツリ
119016 オオガヤツリ
119020 オホコゴメガヤ
119041 コゴメガヤツリ
119125 セイタカハマスゲ
119139 アレチハマスゲ
119160 オオシチトウ,シチタウ,シチタウヰ,シチトウ,ホンシチタウイ,ホンシチタウヰ,リウキウヰ,リュウキュウイ
119162 シチタウ,シチトウ,シチトウイ,リウキウヰ,リュウキュウイ
119196 コシロガヤツリ
119208 カヤツリグサ,キガヤツリ,コガヤツリ,マスクサ
119259 ニイガタガヤツリ
119262 アオガヤツリ,アヲガヤツリ,オオタマガヤツリ,オホタマガヤツリ
119265 オオシロガヤツリ
119282 ウナヅキガヤツリ
119284 ヒメホウキガヤツリ
119316 ツクシオオガヤツリ
119319 ウシクグ,コウシクグ
119321 ウシクグ
119337 シロガヤツリ
119347 カミイ,カミガヤツリ
119378 オニガヤツリ
119390 ウキガヤツリ
119410 イガカヤツリ
119419 ホクトガヤツリ
119422 カラカサガヤツリ
119461 ヒメアオガヤツリ
119503 カウブシ,コウブシ,ハマスゲ
119518 トサノハマスゲ
119535 シデガヤツリ
119607 ゴマフガヤツリ
119619 スナハマスゲ
119624 コガネガヤツリ
119687 アゼガヤツリ,ヒメガヤツリ,ミヅハナビ
119693 コシュロガヤツリ
119718 タカオガヤツリ,タカヲガヤツリ
119769 ザウリガヤツリ
119973 コダチトマト属
119974 コダチトマト
119988 ヘソノヤシ属
120287 アツモリサウ属,アツモリソウ属,シプリペジューム属
120310 オオキバナアツモリ,オオキバナノアツモリ,オホキバナアツモリ,カラフトアヅモリサウ,カラフトアヅモリソウ
120341 コアツモリ,コアツモリサウ,コアツモリソウ
120353 タイワンクマガイソウ,タイワンマガイサ
120357 エゾノクマガイサウ,エゾノクマガエソウ,チョウセンキバナアツモリソウ,チョウセンキバナアヅロリ,テウセンキバナアツモリ,ミヤマアツモリサウ,ミヤマアツモリソウ
120371 クマガイサウ,クマガイソウ,クマガエソウ,クマガユサウ,クマガユソウ,ホテイソウ,ホライサウ
120372 キバナクマガイソウ
120374 ヒタチクマガイソウ
120396 アツモリサウ,アツモリソウ,シロバナアツモリソウ,ホテイアツモリ,ホテイアツモリソウ
120397 シロバナアツモリソウ

120400 レブンアツモリソウ
120402 アツモリソウ
120407 ホテイアツモリ
120438 アメリカアツモリ
120446 タイワンキバナアツモリソウ
120464 ホチイアツモリ
120473 キバナノアツモリソウ
120501 ミズビワソウ属
120509 ミズビワソウ
120532 キルタンサス属
120631 ヒメチゴザサ
120632 ヒロハヒメチゴザサ
120705 キルトポジューム属
120723 キルトルキス属
120764 ツチアケビ
120768 シルトスペルマ属
120774 キルトスペルマ
120779 ショウジョウヤシ属
120780 ヒメショウジョウヤシ
120782 ショウジョウヤシ
120903 エニシダ属
120909 シロエニシダ
120979 シロバナエニシダ
121001 エニシダ,エニスダ
121005 ホオベニエニシダ
121013 ヒメエニシダ
121045 コバノノダケ
121060 ゾボエシア属
121084 リムノキ属
121226 カモガヤ属,ダクティリス属
121233 カモガヤ
121281 タツノツメガヤ属
121285 タツノツメガヤ,ミヤコジマオヒシバ,ミヤコジマヲヒシバ,ヤンバルオヒシバ,ヤンバルヲヒシバ,リュウノツメガヤ,リュウノツメビエ
121333 ダクチロプシス属
121358 ハクサンチドリ属
121359 ハクサンチドリ
121361 シロバナハクサンチドリ
121363 ウズラバハクサンチドリ
121435 イチヨウラン
121436 ヒメウズラヒトハラン
121486 ヒメトウ属
121487 ホソボヒメトウ
121493 キリンケットウ
121495 ワレバヒメトウ
121497 オオバヒメトウ
121500 オオバヒメトウ
121506 カーツヒメトウ
121507 カーツヒメトウ
121514 トウ,マーガレットヒメトウ
121539 ダリア属,テンジクボタン属,テンヂクボタン属
121541 ダリア
121547 ヒグルマダリア,ヒグルマテンジクボタン
121556 コダチダリア,タラノハダリア
121557 カクタスダリア
121559 フジイロテンジクボタン
121561 ダリア,テンジクボタン,テンヂクボタン,ヒグル,ヒグルマダリア,ヒグルマテンジクボタン
121574 ダイス属
121607 ツルサイカチ属
121621 アリサンコセウノギ
121632 カムライ
121634 ツルサイカチ
121647 ヒルギカズラ
121653 クラニュソ,シタン
121714 オウダソ
121718 ナニハヅ
121720 カラフトナニハヅ,テウセンオニシバリ,ナニハヅ
121730 ヒロハノハネミノキ
121810 ツルサイカチ,ハネノミカヅラ,ハネノミマメ
121826 シッソノキ
121871 ダレーア属
121912 ケショウボク属,ダレカンピア属
121933 ケショウボク,フウチョウガシワ
122027 アリドオシ属,アリドホシ属
122028 タイワンアリドホシ,トゲナシアリドホシ,ホンバルリダマノキ
122031 リュウキュウアリドオシ
122034 ナガバシュズネノキ
122040 アリドオシ,アリドホシ,コトリトマラズ,テクサワバナ,ネズミノハナドオシ,ヒイラギ
122041 ヒメアリドオシ
122046 ナガバジュズネノキ
122048 オオシマアリドオシ,ビシンジュズネノキ
122049 ホソバオオアリドオシ,ホソバニセジュズネノキ
122054 コバンバニセジュズネノキ
122056 タマゴバアリドオシ,マルバアリドオシ
122064 オオシマアリドオシ,オオバジュズネノキ,オオバノアリトオシ,オホバジュズネノキ,ジュズネノキ
122067 オオアリドオシ,ジュズネノキ,ニセジュズネノキ
122078 ヤンバルアリドオシ,ヤンバルジュズネノキ,リュウキュウジュズネノキ
122121 ダナーエ属
122124 ナギイカダ,ルスガス
122359 ジンチョウゲ属,ヂンチャウゲ属,ヂンチョウゲ属
122377 アリサンコセウノキ
122393 チェリーピンク
122416 クネオルム,ダフネ
122438 サツマフジ,シゲンジ,チャウジザクラ,チョウジザクラ,フジモドキ,フヂモドキ
122439 タイトウフジモドキ
122471 エゾナニワズ,ナニワズ,レブンナニワズ
122477 カラフトナニハズ,カラフトナニワズ,チョウセンオニシバリ,テウセンオニシバリ
122482 カラスシキミ,コショウノキ,コセウノキ,ハナチャウジ,ハナチョウジ,ヤマシンチョウ,ヤマリンチャウ,ヤマリンチョウ
122486 チョウセンオニシバリ,チョウセンナニワズ
122515 セイヤウオニシバリ,セイヨウオニシバリ,ヨウシュジンチョウゲ
122519 カラスシキミ
122532 ジンチョウゲ,ジンチョオゲ,チャウジグサ,チョウジグサ,ヂンチャウゲ,ハナゴショウ,ハナゴセウ,リンチャウ,リンチョウ
122534 フクリンジンチョウゲ,フクリンヂンチャウ
122538 シロバナジンチョウゲ
122540 シロバナジンチョウゲ
122543 ウスイロジンチョウ,ウスイロジンチョウゲ
122586 オニシバリ,ナツバウズ
122587 ムラサキオニシバリ
122654 ツチビノキ
122655 シャクナンガンピ
122656 ユズリハ科
122661 ユヅリハ属
122679 ヒメユヅリハ
122695 エゾユズリハ
122700 ユズリハ,ユヅリハ
122701 フイリユズリハ
122702 アオジクユズリハ
122708 ウスバユヅリハ,オホバヒメユヅリハ
122710 イヅシベユヅリハ
122713 イッシベユヅリハ,シマユズリハ,シマユヅリハ
122727 ヒメユズリハ
122729 スルガヒメユズリハ
122742 サンアソウ
122758 ダーリングトニア属,ランチュウソウ属
122759 ランチュウソウ
122779 ダーウィニア属
122786 ワックスフラワー
122898 ダシリリオン属
123028 ナギナタソウ
123033 ダティスカ科,ナギナタソウ科
123036 チウセンアサガホ属,チョウセンアサガオ属
123054 ツノミチョウセンアサガオ
123065 ケチョウセンアサガオ,ケテウセン

アサガホ,チョウセンアサガオ,ヤエチョウセンアサガオ
123066 ヤエチョウセンアサガオ
123071 アメリカアサガオ,アメリカチョウセンアサガオ
123077 シロバナチョウセンアサガオ,シロバナヤウシュテウセンアサガホ,シロバナヨウシュチョウセンアカガオ,ヤウシュテウセンアサガホ,ヨウシュチョウセンアサガオ
123081 トゲナシチョウセンアサガオ
123092 ケチョウセンアサガオ
123105 ダウベニア属
123135 ニンジン属
123141 セリニンジン,ナニンジン,ニンジン,ハタニンジン
123164 ニンジン
123180 ゴウシュウヤブジラミ
123244 シノブ属,ダビディア属
123245 ハトノキ,ハンカチツリ-,ハンカチノキ
123253 ダビデイア科,ハンカチノキ科
123307 ディミア属
123316 ヤナギイチゴ属
123322 カワイチゴ,スズメノコウメ,トウイチゴ,ヤナギイチゴ
123332 ヤナギイチゴ
123367 デカネマ属
123401 デカネマ属
123425 デカリア属
123443 コウシュンツゲ属
123448 モチアデク
123451 コウシュンツゲ
123456 コウジュンツゲ,モチアデク
123497 トグノヤシモドキ属
123558 インドヒモカズラ属
123569 インドヒモカズラ
123585 デゲネリア
123587 デゲネリア科
123654 ギンバイサウ属,ギンバイソウ属
123655 ギンガサウ,ギンガソウ,ギンバイサウ,ギンバイソウ,タヅイモ
123657 マルバギンバイソウ
123658 ムラサキギンバイソウ
123727 サワトウガラシ属
123729 マルバノサワトウガラシ
123730 サワトウガラシ
123750 ツタギク
123801 ホウオウボク属
123811 ホウオウボク
123819 デロスペルマ属
123821 ハナアスカ
123823 ハナウジ
123843 コウリンギク
123854 ハナランザン
123867 ハナガサ
123904 ハナコセ
123909 ユウナミ
123920 ハナダイゴ
123959 ハナカタノ
123963 ハナカンゼ
123969 ハナゴショ
123974 ササラン
124008 デルフィニューム属,ヒエンサウ属,ヒエンソウ属
124039 セリバヒエンサウ,セリバヒエンソウ
124067 シロヒエンサウ
124141 ルリヒエンソウ
124151 ホザキヒエンソウ
124237 オオバナヒエンソウ,オオヒエンソウ,オホバナヒエンサウ,オホヒエンサウ,シベリアヒエンソウ,タカネヒエンソウ
124301 ラークスパー
124360 ブジバヒエンソウ,ブッバヒエンソウ
124712 キバナヒエンソウ
124772 アリサンアブラギク
124775 アブラギク,アワコガネギク,キクタニギク
124778 チョウセンイワギク
124780 ワジキギク
124785 キク
124790 アブラギク,シマカンギク,タイワンカンギク,ハマカンギク
124793 ホソバカンギク
124806 アブラギク,ビロ-トガンギク,ホソバアブラギク
124817 アブラギク,アワコガネギク,キクタニギク
124821 シロバナアブラギク
124826 アマギク,イエギク,キク,ショウヨウギク,ショクヨウギク,リョウリギク
124829 モリギク
124832 チョウセンノギク
124836 ヒノミサキギク
124837 オキノアブラギク
124857 ウラゲノギク
124862 イワギク,ピレオギク,ホソバチョウセンギク
124864 チョウセンイワギク
124891 ナガバクマノミヅキ,ヒマラヤヤマボウシ
124923 ヤマクワ,ヤマボウシ
124925 イッキ,カラグハ,ヤエヤマヤマボウシ,ヤマクハ,ヤマグハ,ヤマバウシ,ヤマボウシ
124957 セキコク属,セッコク属,デンドロビウム属,デンドロビューム属
125003 キバナセキコク
125044 ナガツメセキコク,ランダイセキコク
125087 オホバセキコク,カショウセッコク,クワセウセキコク
125104 シラタエセッコク,シロタエセッコク
125121 ツバナセキコク,ツバメセッコク
125135 デンドロビューム,フィンブリアーム,フチトリセッコク
125138 デンドロビューム
125158 マタザキセキコク
125166 ゴールドセキコク
125173 ベニバナセキコク
125218 イトウセキコク
125221 サクラセキコク,タイワンセキコク
125253 タイワンベニバナセッコク,ミヤケセキコク,ミヤケセッコク
125257 セキコク,セッコク
125279 オホバナセキコク,カウキセキコク,シマセキコク,タイヘイラク,タイワンセキコク,ニオイセッコク,モクセキコク
125288 ホンセッコク
125291 オキナワセッコク
125304 コチョウセッコク,コテフセキコク
125361 コバナタマザキセキコク
125365 タイミンッコク
125383 ボウセキコク
125386 シカクセッコク
125390 キバナノセキコク,キバナノセッコク
125433 ワダンノキ
125447 コダイサンチク
125453 ウヤクチク,リョクチク
125455 リョクチク
125461 マチク属
125477 ダイマチク
125482 マチク
125522 テンドロセレウス属
125523 ジュモクチュウ
125526 タイワンムカゴサウ属,タイワンムカゴソウ属,デンドロキールム属
125535 タイワンムカゴサウ,タイワンムカゴソウ,ホザキヒトッバラン
125544 コウトウイラクサ,コウトウイラノキ
125545 イラクシノキ,イラノキ,モクイラクサ
125582 ナハキハギ
125583 デンドロメーコン属
125589 カクレミノ属
125604 タイワンカクレミノ
125620 イリオモテカクレミノ
125626 チョウセンカクレミノ
125644 カクレミノ,カラミツデ,ミソブタ,ミツデ,ミツナガシハ,ミツナガシワ
125645 フイリカクレミノ
125749 デンドロシキオス属
125817 デンモーザ属,デンモザ属
125818 カエンリュウ

125863 ホソバコンロンソウ,ホソバタネツケバナ
125868 タイワンミゾハゴヘ属
125872 コケムグラ,シマミゾハゴヘ,タイワンミゾハゴヘ,ヒメフタバムグラ,ミチハコベ
125937 デリス属,ドクフヂ属
125958 デリス,ドクフジ,ハイトバ,ビヨウリッフヂ
125960 インドトバ
125978 ヒロハシヒノキカヅラ
125983 タチトバ
126007 ケンカヅラ,シイノキカズラ,シヒノキカヅラ
126025 コメススキ属
126031 タカネコメススキ
126032 タカネコメススキ
126039 ヒロハノコメススキ,ミヤマコメススキ
126055 コミヤマコメススキ,ヒロハノコメススキ
126061 ユウバリカニツリ
126064 オニコメススキ
126070 カシュウコメススキ
126074 コメススキ
126112 クジラグサ属
126120 ヒメクジラグサ
126127 クジラグサ
126168 アメリカガフクワン属
126171 アメリカガフクワン
126176 ハイクサネム
126187 タチクサネム
126242 デスモジューム属,ヌスビトハギ属
126278 アメリカハギ,メリケンハギ
126290 シラゲマメハギ
126329 ハナヌスビトハギ,ハマバハギ
126359 タマツナギ
126368 ヒメコハギ
126389 クサハギ,シバハギ
126396 オホハヒマキエハギ,カハリバマキエハギ,カワリバマキエハギ,シバハギ,ハヒマキエハギ
126407 イリノイヌスビトハギ
126408 タチシバハギ
126411 フジボツルハギ
126424 ホソミハギ
126465 コバノハギ,ヒメノハギ
126471 フヂバナマメ
126486 アメリカヌスビトハギ
126501 ヤブハギ
126507 アレチヌスビトハギ
126527 シロバナケヤブハギ,ヤブハギ
126528 シロバナヌスビトハギ
126529 オキチハギ
126532 シロバナヤブハギ
126596 ハナタチシバハギ
126601 アコウマイハギ
126603 オホバマヒハギ
126649 ムラサキヌスビトハギ
126651 ハイマキエハギ,ハヒマキエハギ
126688 コモチトゲココヤシ属
126689 ハリコモチトゲココヤシ
126832 ウツギ属
126836 ヒメマルバウツギ
126841 アム-ルウツギ
126855 マルバコウツギ
126891 ウツギ,ウノハナ,クヂベニウツギ
126892 シロバナヤエウツギ
126893 オオミウツギ
126894 サラサウツギ
126895 ジクゲウツギ
126896 ムラサキウツギ
126898 ビロ-ドウツギ
126904 ウラジロウツギ
126929 コウツギ
126933 チョウセンウツギ
126946 ウメウツギ,ヒメウツギ
126947 ヒロハヒメウツギ
126948 ハナヒメウツギ
126949 アオヒメウツギ
126954 ナチウツギ
126956 イハウツギ,ウラジロウメウツギ,ウラジロヒメウツギ
126963 イハウツギ,イワウツギ,ウメウツギ
126967 コミノヒメウツギ
126976 ウラジロウツギ
127012 ウラジロウツギ
127023 オオシマウツギ
127024 オキナワヒメウツギ
127031 アオコウツギ
127032 タウウツギ,テマリウツギ,トウウツギ
127049 オオバウツギ,オニウツギ,オホバウツギ,タイワンウツギ,マルバオニウツギ
127060 アツアルバウツギ
127072 ウツギ,ウノハナ,コウツギ,セノハナ,マルバウツギ
127076 フイリマルバウツギ
127085 ヒナウツギ
127099 ツクシウツギ,マルバウツギ
127114 キ-ルンウツギ,シマヒメウツギ,ダイワンウツギ,タイワンヒメウツギ
127118 ウメウツギ,ニックワウウツギ,ニッコウウツギ,ミヤマウツギ
127125 ヤエヤマウツギ
127127 ブンゴウツギ
127197 ノガリヤス,マンシュウノガリヤス
127200 サイトウガヤ,ノガリヤス
127202 タイワンサイトウガヤ
127209 オホサイトウガヤ,ノガリヤス
127211 サイトウガヤ
127217 タイワンノガリヤス,タイワンヒゲノガリヤス,ノガリヤス
127243 ヒメノガリヤス
127257 コガリヤス
127268 ムツノガリヤス
127276 チシマガリヤス
127299 スイザンカニツリ,スヰザンカニツリ
127460 ディアモルファ属
127504 キキャウラン属,キキョウラン属
127507 キキャウラン,キキョウラン
127573 カハラナデシコ属,ナデシコ属
127574 イセナデシコ
127575 カラスナデシコ
127600 ノハラナデシコ
127607 アメリカナデシコ,ヒゲナデシコ,ビジュナデシコ,ビジョナデシコ,フヂナデシコ
127608 ホソバヒゲナデシコ
127621 カンナデシコ
127634 ホソバナデシコ,ホソバハマナデシコ
127635 アンジャベル,オランダセキチク,オランダナデシコ,カ-ネ-ション,ジャカウナデシコ
127654 カラナデシコ,セキチク,チャボナデシコ
127656 シロバナセキチク,シロバナモウコナデシコ
127657 ヒオドシナデシコ
127667 イセナデシコ,オオサカナデシコ,サツマナデシコ
127670 ミヤマカラナデシコ
127674 トコナツ
127676 モウコナデシコ
127682 コウアンナデシコ
127700 ヒメナデシコ
127728 シバナデシコ
127739 ハマナデシコ,フジナデシコ,フヂナデシコ
127740 シロバナハマナデシコ
127746 ヒメバナデシコ,ヒメハマナデシコ
127747 ホタルナデシコ
127756 カワラナデシコ,ナデシコ
127793 タツタナデシコ,トコナデシコ
127802 エゾカハラナデシコ,ニヒタカセキチク
127833 シナノナデシコ,ミロマナデシコ
127834 ミヤマナデシコ
127852 エゾカワラナデシコ,カハラナデシコ,カワラナデシコ,ナデツコ
127853 シロバナタカネナデツコ
127854 シロバナタカネナデシコ
127855 ヒロハカワラナデシコ
127856 シロバナエゾカワラナデシコ
127861 クモイナデシコ
127862 ニイタカセキチク
127863 シロバナカワラナデシコ

127864 サンシキナデシコ
127869 イワテナデシコ
127870 タカネナデシコ
127910 イハウメ属,イワウメ属
127916 イハウメ,イワウメ,フキズツメソウ,フキズメサウ,フキヅメサウ,ホソバイワウメ
127918 ベニバナイワウメ
127919 イワウメ,フキズツメソウ,フキズメサウ,フキヅメサウ
127926 イハウメ科,イワウメ科
127936 ディアファナンテ属
128011 タツノヒゲ属
128015 ヒロハヌマガヤ
128017 タツノヒゲ
128019 マンシウタツノヒゲ,マンシュウタツノヒゲ
128022 オオバタツノヒゲ
128023 コゴメアマ属
128026 コゴメアマ
128029 ディアスキア属
128032 ニカクソウ
128115 ユキヤッデ
128288 コマクサ属
128291 カナダケマンソウ
128297 ハナケマンソウ
128310 コマクサ
128312 シロバナコマクサ
128314 キンギンサウ,コマクサ
128321 ケマンソウ
128397 ディケア属
128478 ニコゲヌカキビ
128483 ホソヌカキビ
128516 ケヌカキビ
128568 ヒメオニササガヤ
128571 オニササガヤ
128586 シラゲオニササガヤ
128955 アオイゴケ属,アフヒゴケ属,ダイコンドラ属
128964 アオイゴケ,アフヒゴケ
128972 カロリナアオイゴケ,ダイコンドラ
128987 ディコリサンドラ属
128997 オオタチカラクサ
129019 シロガヤツリ
129030 ジヤウザン属,ディクロア属
129033 ジヤウザン,ジョウザン,ジョウザンアジサイ
129062 ブクリャウサイ属,ブクリュウサイ属,ブクリョウサイ属
129068 ブクリャウサイ,ブクリュウサイ,ブクリョウサイ
129080 ブクリョオサイ
129220 ハグロサウ属,ハグロソウ属
129243 ヤンバルハグロサウ,ヤンバルハグロソウ
129273 シロバナハグロソウ
129615 ハクセン属
129618 サンショグサ,ハクセン,ヨウシュハクセン
129629 ハクセン
129679 アミダネヤシ属
129684 シロアミダネヤシ
129685 キイアミダネヤシ
129688 アカアミダネヤシ
129768 ヒトツボクロモドキ属
129772 ディディエーア属
129781 ディディエーア科
130034 コカゲラン
130037 ヒメヤッシロラン属,ヤッシロラン属
130041 ヒメヤッシロラン
130043 イウレイン,ヤッシロラン,ユウレイラン
130057 フタッブダネヤシ属
130065 ヒメフタッブダネヤシ
130067 ムラサキフタッブダネヤシ
130083 シロガスリソウ属,ディーフュンバッキア属
130096 フクリンカスリソウ
130097 ヒロハカスリソウ
130120 ナガバカスリソウ
130128 カスリソウ
130135 シロジクカスリソウ
130139 シロカスリソウ
130185 ホザキヒメラン
130187 ディエラーマ属
130344 キツネノテブクロ属,ジギタリス属,ヂギタリス属
130358 サビイロジギタリス
130365 オオバナジギタリス
130369 ケジギタリス
130371 キバナジギタリス,キバナヂギタリス
130383 キツネノテブクロ,ジギタリス,ヂギタリス
130404 メヒシバ属
130481 コメヒシバ
130489 ジシバリ,メシバ,メヒシバ
130590 ヘンリーメヒシバ
130592 フタマタメヒシバ
130605 ススキメヒシバ
130612 アキメヒジハ,キタメヒシバ
130627 イトメヒシバ,タイワンイトメヒシバ
130635 チャボメヒシバ
130656 イヌメヒシバ
130666 ビロードメヒシバ
130703 イトメヒシバ
130709 シマギョウギシバ
130721 ヒトタバメヒシバ
130730 コメヒシバ
130731 アラゲコメヒシバ
130745 オニメヒシバ,メヒジハ
130763 フタマタメヒシバ
130768 イヌメヒシバ
130825 アキメヒシバ
130826 アラゲメヒシバ
130913 ディレーニア属,ビハモドキ属,ビワモドキ属
130915 シンボー
130919 ビハモドキ,ビワモドキ,ホンダパラ
130928 サルナシ科,ディレニア科,ビワモドキ科
131006 カリマタガヤ属
131008 ケカリマタガヤ
131018 カリマタガヤ
131026 カリマタガヤ
131028 ヒメカリマタガヤ
131061 リュウガン
131124 ディモルフォルキス属
131147 アフリカキンセンカ属,ディモルフォセカ属
131191 アフリカキンセンカ
131224 ハキダメガヤ
131300 ディンテラ属
131306 ディンテランッス属
131358 オオフタバムグラ
131361 メリケンムグラ
131397 ハエジゴク属,ハエトリグサ属,ハエトリソウ属
131427 ディオオン属
131451 ヤマノイモ属
131458 グンバイドコロ,タイジョ
131473 ツクシタチドコロ
131485 タイセイイモ
131501 カシュウイモ,カショウイモ,ケイモ,ゼンプ,ナリイモ,ニガガシュウ,ベンケイイモ,マルドコロ,マルバドコロ
131516 キイロギニアヤム
131522 ソメモノイモ
131529 コウトウイモ
131531 イズドコロ
131573 ツルカメソウ
131577 ハリイモ
131579 トゲイモ
131592 ナガイモ
131593 ワーブルクドコロ
131605 タチドコロ
131607 タチドコロ
131630 ウチハバドコロ,オホニガイモ,コバンドコロ,ミツバドコロ
131645 ジネンジャウ,ジネンジョウ,ヤマノイモ
131691 ルゾンヤマノイモ
131734 ウチハドコロ,ウチワドコロ,カウモリドコロ,コウモリドコロ
131736 ケナシウチワドコロ
131759 アケビドコロ
131761 ライシャイモ
131772 タカサゴドコロ,ナガイモ

131784 キ-ルンヤマノイモ
131804 カエデドコロ,カヘデドコロ
131814 シロギニアヤム
131840 キクバドコロ,モミジドコロ,モミヂドコロ
131842 コシジドコロ
131873 エドドコロ,ヒメドコロ
131877 オニドコロ,トコロ,ナガドコロ
131918 ヤマノイモ科
132030 カキノキ属,カキ属
132077 アオコクダン
132093 セレベスコロマンデル
132137 エボニ-,コクタン
132139 ヤエヤマコクタン,リュウキュウコクタン
132145 ヤハラゲガキ,ヤワラケガキ
132153 エボニ-,ヤエヤマコクタン
132216 シナノガキ,リュウキュウマメガキ
132217 ウスゲシナノガキ
132219 カキ,カキノキ
132230 アマメ,グロガキ,ビング,ヤマガキ
132244 フイリコクタン
132245 クラルガキ,クロシブガキ,サノガキ
132264 シナノガキ,マメガキ
132267 シナノガキ
132291 クサノガキ,クロボウ,リウキウガキ,リュウキュウガキ,リュキュガキ
132298 コロマンテル
132309 クロカキ,タイワンマメガキ,トキハガキ,トキワガキ,トキワマメガキ
132320 トキハガキ,ヤマガキ
132335 オルドガキ,シマガキ
132339 アブラガキ
132351 カムゴソ,クロガキ,ケガキ,タンワンコクタン
132383 アカバナガキ
132466 アメリカガキ
132677 サンカエフ属,サンカヨウ属
132678 アメリカサンカヨウ
132682 サンカエフ,サンカヨウ
132685 サンカヨウ
132720 テウセンガリヤズ属,ハマガヤ属
132787 カガシラ属
132789 カガシラ,ヒメシンジュガヤ
132947 ティプロシアサ属
132949 ティプロシアサ
132957 オキナワスズメウリ
133143 サガリラン属
133145 クスクスサガリラン,サガリラン
133177 ディプモソマ属
133181 シロミミズ属
133185 シロミミズ
133226 エダウチナヅナ属
133313 ロボウガラシ
133324 エダウチナズナ
133444 マツムシサウ科,マツムシソウ科
133451 ナベナ属
133463 シナナベナ
133478 オニナベナ,チイゼル,チ-ゼル,ラシャカキグサ
133486 グメリニイナベナ
133487 トゲナシナベナ
133490 ナベナ
133517 オニナベナ
133551 フタバガキ科
133552 フタバガキ属
133553 ヤン
133558 グルイン
133563 アピトン
133587 イン,エン,ス-アソ
133599 キンセンセキ
133684 ディ-サ属
134009 ディサンツス属,ベニマンサク属,マルバノキ属
134010 クロジシャ,ヒトッバ,ベニマンサク,マルバノキ,ユバズサ,ラフバイモドキ,ロウバイモドキ
134027 ディスキ-ディア属,マメヅタカヅラ属
134037 フクロネカヅラ,マメズタカズラ,マメヅタカズラ,マメヅタカヅラ
134038 アケビカズラ,アケビモドキ
134040 キカズラ
134044 フクロカズラ
134115 ディスコカクタス属
134147 エノキフジ,エノキフヂ
134228 アカバナクジャク
134326 ジョウロウラン
134394 ディスフィマ属
134400 ハウチャケモドキ属
134405 ウスバハウチャクモドキ,ハウチャクモドキ
134417 チゴユリ属
134425 トウチクラン
134441 タイワンチゴザサ
134467 コガネホウチャクソウ,ハウチャクサウ,ホウチャクソウ
134477 コホウチャクソウ
134478 ホソバホウチャクソウ
134484 キバナタウチクラン
134486 チゴユリ
134503 キバナホウチャクソウ
134504 オオチゴユリ,オホチゴユリ
134918 イスノキ属,イス属
134922 イズ
134932 タイワンイス
134934 シマイスノキ
134946 イス,イスノキ,サルフエ,シダレイス,ヒョソノキ,フクベノキ,ユスノキ
134947 ホソバイスノキ
134948 シダレイスノキ,シライス
135079 ディジゴテ-カ属
135153 ドデカテオン属
135165 カタクリモドキ
135192 ドドヌア属,ドドネア属,ハウチハノキ属
135215 シマアハブキ,ハウチハノキ,ハウチワノキ
135264 シラヤマギク
135399 ド-リコス属,フジマメ属,フヂマメ属
135646 ミヅバフヂマメ
135649 コウシュンフジマメ
136072 コウトウケマタケラン属
136081 コウトウケマタケラン,ドナクス
136156 ハナハタザオ属,ハナハタザホ属
136163 ネバリハナハタザオ,ネバリハナハタザホ,ハナハタザオ,ハナハタザホ
136173 ニオイハタザオ
136179 コバナハタザオ
136185 クシバハタザオ
136209 アブノメ属
136214 アブノメ,パチパチグサ
136312 ドリティス属
136333 ドロニカム属,ドロニクム属
136391 ドロテアトサス属,ドロテアンッス属
136394 ヘラマツバギク
136417 ドルステ-ニア属
136470 アメリカドルステニア
136784 ドシニァ属
136832 ドビア-リス属
136846 セイロン・グ-スベリ-,セイロンスグリ
136877 ドクサンタ属
136899 イヌナズナ属,イヌナヅナ属
136901 ハリイヌナズナ
136911 タカネナズナ
136951 エゾイヌナズナ,シロバナノイヌナズナ
136952 エゾイヌナズナ,シロバナノイヌナズナ
136996 モイワナズナ
137010 イシノナズナ
137052 ナンブイヌナズナ
137056 キタダケナズナ
137131 ヨウシュイヌナズナ
137133 イヌナズナ,イヌナヅナ
137136 メイヌナズナ
137143 クモマナズナ,クモマナヅナ
137144 モノガタリナズナ
137145 ケナシクモマナズナ
137151 ヒナナズナ
137154 ヒメハリイヌナズナ
137222 モイワナズナ
137223 カブダチナズナ
137224 モイハナズナ,モイハナヅナ,モイワナズナ
137225 トガクシナズナ
137236 シロウマナズナ

137286 オクエゾナズナ
137330 ドラセナ属,ミュウケッジュ属,リュウケッジュ属
137337 ホソバセンネンボク
137342 ナカエセンネンボク,パカエセンネンボク
137360 ベニフクリンセンネンボク
137367 オオシロシマセンネンボク
137368 オオシロシマセンネンボク
137380 シロシマセンネンボク
137382 リュウケツジュ,リョウケツジュ
137387 モンセンネンボク
137397 ニオイセンネンボク
137398 ウスイモフクリンセンネンボク
137401 シマセンネンショウ
137403 クリンセンネンボク
137409 ホシセンネンボク
137416 トラフセンネンボク
137487 キンヨウセンネンショウ,キンヨウセンネンボク
137514 ナカエセンネンボク,パカエセンネンボク
137545 ムシャリンダウ属,ムシャリンドウ属
137551 セイラン,ムシャリンダウ,ムシャリンドウ
137553 シロバナムシャリンドウ
137613 タチムシャリンダウ,タチムシャリンドウ,ホザキムシャリンダウ,ホザキムシャリンドウ
137617 コウアンムシャリンドウ
137624 コバナムシャリンドウ
137647 ラショウモンソウ
137651 オクムシャリンドウ
137723 ドラコフィラス属
137744 ドゥパクンクルス属,ドゥラクンクルス属
137774 タシロカヅラ属
137799 タシロカヅラ
138006 ドリミオプシス属
138131 ドロサンテモプシス属
138135 ドロサンテムム属
138149 ハナキョウソウ
138171 ビコウ
138183 ハナヤヨイ
138205 ハナキユウ
138220 ハナホウショウ
138229 ギョクキ
138233 ハナショウジョウ
138237 ハナコゾメ
138261 マウセンゴケ属,モウセンゴケ属
138267 ナガバノモウセンゴケ
138270 サスマタモウセンゴケ,フタマタモウセンゴケ
138273 イッシベマウセンゴケ,クルマバモウセンゴケ
138275 アフリカナガバモウセンゴケ
138289 ィトバモウセンゴケ,インモウセンゴケ
138295 ナガバノイシモチサウ,ナガバノイシモチソウ
138296 シロバナナガバノイシモチソウ
138299 パガエノモウセンゴケ
138307 チシママウセンゴケ,ナガエノモウセンゴケ,ナガバノマウセンゴケ
138322 サジバモウセンゴケ
138326 ヨツマタモウセンゴケ
138328 イシモチソウ
138331 イシモチサウ,イシモチソウ
138338 イシモチソウ
138340 ビグミ－モウセンゴケ
138348 ウシノハヘトリ,マウセングサ,マウセンゴケ,マゴノテ,モウセンゴケ
138350 コマウセンゴケ,ミッシベマウセンゴケ
138354 ゴウシュウコモウセンゴケ,コマウセンゴケ,コモウセンゴケ,ミッシベマウセンゴケ
138355 シロバナコモウセンゴケ
138357 コモウセンゴケ
138361 トウカイコモウセンゴケ
138362 ヒュウガモウセンゴケ
138363 アオイトバモウセンゴケ
138369 アシモチサウ科,マウセンゴケ科,モウセンゴケ科
138374 ドロソフィラム科
138375 ドロソフィラム属
138376 イヌイシモチソウ
138445 チャウノスケサウ属,チョウノスケソウ属
138450 キバナチョウノスケソウ
138460 チョウノスケサウ,チョウノスケソウ,ミヤマグルマ,ミヤマチングルマ
138461 キョクチチョウノスケソウ,チャウノスケサウ,チョウノスケソウ,ミヤマグルマ,ミヤマチングルマ
138470 ヤンバルハコベ属
138478 オムナグサ
138482 ネバリハコベ,ミヅクサ,ヤンバルハコベ
138532 ノコギリバケンチャヤシ属
138534 ベクインケンチャヤシ
138536 オリブ－ケンチャヤシ
138547 リュウノウジュ属
138548 リュウノウジュ
138634 ツゲモドキ
138635 ハツバキ
138649 コウシュンテッボク
138658 ツゲモデキ,ツゲモドキ
138790 ヘビイチゴ属
138792 アイノコヘビイチゴ
138793 ヘビイチゴ
138794 シロミノヘビイチゴ
138796 クチナハイチゴ,ヘビイチゴ,ヤブヘビイチゴ,ヤマヘビイチゴ
138797 シロミノヤブヘビイチゴ
138829 ダドレヤ属
138831 センニョハイ
138928 ノササゲ属
138929 シマノササゲ,タイワンノササゲ
138942 カラスマメ,キツネササゲ,ノササゲ
138966 ノアズキ属,ノアヅキ属
138974 カイナンノアズキ
138978 キツネアヅキ,ノアズキ,ノアヅキ,ヒメクズ
139060 ハリマツリ属
139062 タイワンレンギョウ,タイワンレンギョオ,ハリマツリ
139065 ジュランカヅラ,タイワンレンギョウ,タイワンレンゲウ,ハリマツリ
139068 シロバナタイワンレンギョウ,シロバナハリマツリ
139089 ヅリオ属
139092 ドリアン
139133 ヅバリア属
139134 カクレミノ,ツカサギュウカク,ツカュカク
139176 カクレミノ,カババハナギュウカク
139246 ディアキア属
139254 ディッキア属
139256 シマケンザン
139261 ホソバシマケンザン
139263 ケンザンノシマ
139300 ヒメタケヤシ属
139372 ロ－ベルヒメタケヤシ
139544 ミズトラノオ属,ミヅトラノヲ属
139585 オオバノミズトラノオ,オホバノミヅトラノヲ,ミズトラノオ,ミズネコノオ,ミヅネコノヲ
139602 セイスイソウ,セイス井サウ,ミズトラノオ,ミヅトラノヲ,ムラサキミズトラノオ
139606 ミヤオソウ属
139623 ハッカクレン,ミヤオサウ,ミヤオソウ,ミヤヲサウ
139629 キキウ,ハスノハグサ
139637 シマセンダン属
139641 シマセンダン
139655 クスクスジウマン
139678 アリタサウ,アリタソウ,ケアリタソウ,ナンバンルウダ,ルウダサウ,ルウダソウ
139681 ハリセンボン,ヒメアカザ
139693 キクバアリタソウ
139733 セリモドキ,タニセリモドキ
139734 タケシマシシウド
139763 カキノキ科
139785 エ－ベルランジア属,エベルランジア属
139831 エブラクテオ－ラ属

139851 テツパウウリ属,テツポウウリ属
139852 テツパウウリ,テツポウウリ
139905 アブラスズキ属
139911 タカサゴアブラススキ
139913 ダンチアブラススキ
139934 エックレモカクタス属
139935 エックレモカクタス属,チョウチンバナノノウゼンカズラ属
139936 チョウチンバナノノウゼンカズラ
139945 ゴムカヅラモドキ属
139959 ゴムカズラ,ゴムカヅラ
139970 エケベリア属
139974 シノノメ
139980 ギンメイショク
139987 ツキカゲ
139988 タカサキレンゲ
139993 ハクトジ
139998 ベニツカサ
140000 ヨウロウ
140005 ヤマトニシキ
140009 キンシコウ
140017 エキドノプシス属
140021 セイリュウカク
140066 エキナセア属
140081 ムラサキバレンギク
140111 エキノカクタス属,タマサボテン属
140116 ボウンカク
140127 キンシャウ,キンシャチ
140133 タイベイマル
140135 イワオ
140150 イワオ
140152 ダイリュウカン
140209 エキノセレウス属
140214 ハナサカズキ
140236 スハタ
140239 オキナニシキ
140242 ブユウマル
140282 サンコウマル
140288 ヒカカク
140296 アカシマル
140320 ソウモクカク
140327 ブユウマル
140332 カガリビ
140342 アオバナエビ
140351 ヒエ属
140356 イヌビエ,ミヅビエ
140360 ワセビエ
140367 イヌビエ,サルビエ,タイヌビエ,ヒエ,ヒメイヌビエ,ヒメタイヌビエ
140369 アオビエ
140382 ケイヌビエ
140383 ウスゲケイヌビエ
140384 イヌビエ
140408 ノビエ
140421 ヒエ
140423 インドビエ,ヒエ
140430 ヒメタイヌビエ
140438 タイヌビエ,タビエ
140462 タイヌビエ
140466 ノゲタイヌビエ
140477 ノゲタイヌビエ
140503 ヒエ
140526 エキノシスチス属
140536 エキノドルス属
140543 シャゼンオモダカ
140545 ヒロハシャゼンオモダカ
140566 エキノフォッシュロカクタス属
140577 タリョウギョク
140579 タチアラシ
140583 チヂミダマ
140612 エキノマスツス属
140616 サクラマル
140618 エイカン
140620 タイハクマル
140622 シホウギョク
140652 エキノプス属,ヒゴタイ属,ルリヒゴタイ属
140697 オニルリヒゴタイ
140712 コルリヒゴタイ,ルリヒゴタイ
140716 タカサゴルリヒゴタイ
140732 オクルリヒゴタイ
140770 トウルリヒゴタイ
140776 ルリタマアザミ
140786 ヒゴタイ,ルリヒゴタイ
140789 セイタカヒゴタイ
140842 ウニサボテン属,エキノプシア属,ヘリアントセレアス属
140843 ハクセイマル
140844 ホウシュンマル
140846 コウコウマル
140847 オウショウマル
140848 ビショウマル
140849 ゴウショウマル
140850 ボタンマル
140852 キンセイマル
140857 キッコウラム
140858 コンポル
140860 タンゲマル
140861 キョウフウマル
140862 ホウレイマル
140863 シュンホウマル
140865 コウレイマル
140866 ハクシュマル
140867 マケンマル
140868 トウコウマル
140869 オウシュマル
140870 キホウマル
140872 カブト,チョウセイマル
140875 ケンボウマル
140876 オウセイマル
140878 ズイコウマル
140879 セイギョク
140881 ケイレイマル
140882 マオウマル
140883 ニオウマル
140885 セイボンギョク
140886 トウウンマル
140887 ギンレイギョク
140888 キダイモンジ
140889 ライホウマル
140891 カセイマル
140894 マスアラマル
140981 エキヌス属
141087 エキウム属,エキューム属,シャゼンムラサキ属
141258 シャゼンムラサキ
141347 シベナガムラサキ
141365 タカサブラウ属,タカサブロウ属
141366 アメリカタカサブロウ
141374 タカサブラウ,タカサブロウ,タカサブロオ
141439 エクトトロピス属
141466 ミツマタ属
141470 ミツマタ,ミツマタヤナギ,ムスビギ
141472 チェンチエシャン,ミツマタ
141484 エヂィトコレァ属
141568 オオカナグモ属
141569 オオカナダモ
141592 チシャノキ属
141595 カキノキダマシ,チシャノキ,ヤメラガシハ
141597 タイワンチシャノキ,タカサゴチシャノキ,チシャノキ
141611 フクマンギ
141629 オオバチシャノキ,オホバチシャノキ,マルバチシャノキ
141633 マルバチシャノキ
141662 ナガバチシャノキ
141683 カキノキダマシ,チシャノキ,トウビワ,ヤラメガシワ
141687 リュウキュウチシャノキ
141691 ヤニチシャノキ
141805 ホテイアオイ属,ホテイアフヒ属
141808 ホテイアオイ,ホテイアフヒ,ホテイサウ,ホテイソウ
141928 グミ科
141929 グミ属
141930 ホソグミ
141932 ホソグミ,ホソバグミ,ヤナギバグミ
141940 タンゴグミ
141941 ギンヨウグミ
141943 アサカワグミ
141965 アキグミ
141981 フユグミ
141987 キリシマグミ,クマヤマグミ
141991 コウトウナハシログミ,タイワンナハシログミ,ノウカウグミ
141992 タイワンマルバグミ
141999 オヒワケグミ,サガリグミ,ダイブ

グリ,ツルグミ,ツルグミモドキ,ナガバニヒタカグミ,ノコバツルグミ,ブイサングミ
142001　ヒロバツルグミ
142007　コバノツルグミ
142008　ホソバツルグミ
142013　ナガバツルグミ
142056　リュウキュウツルグミ
142075　オオバグミ,オオバツルグミ,オオバグミ,コウトウナハシログミ,マルバグミ
142078　アカバグミ,オオバツルグミ
142079　ハコネグミ
142083　マメグミ
142084　ツクバグミ
142088　ナツグミ,ホソバナツグミ
142092　カンドウナツグミ,サツキグミ,タウエグミ,ナツグミ,マルバナツグミ,ヤマグミ
142100　ダイオウグミ
142102　トウグミ
142103　イスズグミ,サイグミ
142114　アリマグミ
142115　ナガバアリマグミ
142117　ハナツルグミ
142122　コウヤグミ
142132　タカサゴグミ,タカサゴグミモドキ
142151　ワセアキグミ
142152　タワラグミ,ナハシログミ,ナワシログミ,ハルグミ
142153　キフクリンナワシログミ
142158　キンギンボク,フイリナワシログミ,フクリンナワシログミ
142164　センベイグミ
142181　ウラギンツルグミ,マルバツルグミ
142184　オガサワラグミ
142202　カツラギグミ
142207　オヒワケグミ,ダイブグミ,タイワンアキグミ,ニヒタカグミ,ノコバツルグミ
142214　アキグミ,カワラグミ,タカサゴグミモドキ,ツルグミ,ナツグミ,フユグミ
142215　シロアキグミ
142217　キミノアキグミ
142218　オオアキグミ
142219　ヒロハアキグミ
142225　タテミゾアキグミ
142230　カラアキグミ,ミチノクアキグミ
142238　マルバアキグミ
142251　ヤクシマグミ
142252　サツキアサドリ,ナツアサドリ,ババノシリ,バンバ
142255　アブラヤシ属,アフリカアブラヤシ属
142257　アブラヤシ,ギネヤアブラヤシ,キネロアブラヤシ
142267　ホルトノキ科
142269　エレオカルプス属,ホルトノキ属
142272　インドジュズノキ,ジェゼボグイジュ
142303　ホルトノキ,モガシ
142340　コバンモチ,ヅキ
142351　ホンバコバンモチ
142362　ナガバコバンモチ
142372　シマホルトノキ
142382　セイロンオリ－ブ
142396　アケイノギ,イヌヤマモモ,シイトギ,シラキ,ズクノキ,ヅクノキ,ハボソ,ハボソノキ,ホルトノギ,モガシ
142399　シラキ,ヅクノキ,ハボソノキ,ホルトノキ,モガシ
142404　チギ
142552　ミゾハコベ科
142554　ミゾハコベ属
142574　イヌミゾハコベ,タイワンミゾハコベ,ナンゴクミゾハコベ,ミヅハコベ
142578　ミゾハコベ,ミヅタハコベ
142579　ミゾハコベ
142594　ウハバミサウ属,ウワバミサウ属,ウワバミソウ属
142642　クサダチキミヅ
142647　トキホコリ
142654　ヒロハノキミヅ
142658　ヒコキミヅ
142694　ウワバミソウ,ミズ,ミズナ
142696　ヒメウワバミソウ
142697　ウワバミソウ
142703　ヤマトキホコリ
142704　ヤマトキホコリ
142719　ホソバノキミズ
142741　タマザキウハバミサウ
142759　トキホコリ
142766　ヒメミヅ
142771　アマミサンショウソウ
142793　ランダイミズ
142794　ランダイキミヅ
142867　クニガミサンショウヅル
142898　ヒメトキホコリ
142901　ヨナクニトキホコリ
143019　エレオカリス属,ハリイ属,ハリ井属
143022　オゴケ,コウゲ,コゲ,チシママツバイ,マツバイ,マツバ井,ヲゴケ
143026　マツバイ
143029　マツバイ,マツバ井
143042　ミスミイ
143053　クロミノハリイ
143054　クロミノハリイ
143055　オオハリイ,セイタカハリイ
143056　チョウセンハリイ
143092　カヤツリマツバイ
143100　オオハリイ,ハリイ
143101　オオハリイ
143102　アサヒラン,エゾハリイ,サワラン,ハリイ,ハリ井
143107　ヤリハリイ
143108　エゾハリイ
143122　イヌクログワイ,イヌクログワ井,オオクログワイ,シナクログワイ
143123　イヌクログワイ,シナクログワイ
143132　シバヤマハリイ
143139　スジヌマハリイ
143143　オウギシマヒメハリイ
143149　タウユン井,ミスミイ,ミスミ井
143158　タマハリイ
143168　クロヌマハリイ,クロヌマハリ井,ヌマハリイ
143169　クロヌマハリイ
143178　クロハリイ,ヒメハリイ,ヒメハリ井
143179　クロハリイ
143198　オオヌマハリイ,フトヌマハリイ
143200　オオヌマハリイ,ヌマハリイ
143201　オホヌマハリイ,オホヌマハリ井,ヌマハリイ,ヌマハリ井
143205　シロミノハリイ
143257　トクサイ
143261　タイリクハリイ,マルホハリイ
143271　クロヌマハリイ,ヌマハリイ,ヌマハリ井
143287　コツブヌマハリイ
143289　チャボイ,チャボ井
143298　オホハリイ,オホハリ井,ハリイ,ハリ井
143300　ハリイ
143312　クログワイ
143318　キタヌマハリイ
143326　カヤツリマツバイ
143327　カヤツリマツバイ
143358　オキナワハリイ
143365　イトハリイ
143374　シマシカクイ,シマシカク井,マシカクイ,マシカク井
143376　カドハリイ
143397　ヒメヌマハリイ,ヒメヌマハリ井
143401　スジヌマハリイ,スヂヌマハリ井
143414　シカクイ,シカク井
143415　ミツカドシカクイ
143416　コモチシカクイ
143417　オキナワイヌシカクイ
143421　ヒメシカクイ
143438　サワレン属
143440　アサヒレン,サワレン
143441　シロバナサワラン
143442　キリガミネアサヒラン
143453　ミスミギク属
143460　シロバナイガカウゾリナ
143464　イガカウゾリナ,ミスミギク,ミスミグサ
143472　ホザキイガカウゾリナ
143499　エレッタ－リア属

143512 オヒシバ属,ヲヒシバ属
143522 カモマタキビ,カモマタビエ,カラビエ,コウボウビエ,コウボフビエ,シコクビエ,ノラビエ
143530 オヒシバ,オヒジハ,チカラグサ,ヲヒシバ
143553 ズングリオヒシバ
143570 アカネズイセン属,アカネズヰセン属,エレウテリ－ネ属
143573 アカネズイセン,アカネズヰセン,ビッキ,ボスメン
143575 ウコギ属
143588 アブラコ,オニウコギ,ケヤマウコギ,ノウコギ
143591 ケヤマウコギ,トゲナシエゾウコギ,トゲナシオニウコギ
143613 ヒゴウコギ
143617 ウラジロウコギ,タカノツメ,ミサヤマコシアブラ,ミヤマコシアブラ
143657 エゾウコギ,ハリウコギ
143665 マンシウウコギ
143677 ウコギ,ヒメウコギ
143679 フイリウコギ
143682 ウコギ,オニウコギ,ヤマウコギ
143684 トゲナシウコギ
143685 オカウコギ,マルハウコギ,ヤマウコギ,ヲカウコギ
143686 クロバナヤマウコギ
143687 ツクシウコギ
143688 ウラゲウコギ
143693 ミヤマウコギ
143694 ミツバウコギ
143707 チャボテ－ブルヤシ属
143712 ケオノマテ－ブルヤシ
143809 エリセナ属
143841 エルレアントウス属
143894 キクガラクサ属
143896 キクガラクサ,ホロギク
143897 キクガラクサ
143914 エロデア属,カナダモ属
143920 カナダモ
143958 ナギナタカウジユ属,ナギナタコウジュ属
143974 ナギナタカウジュ,ナギナタコウジュ
143975 シロバナナギナタコウジュ
143993 ナギナタカウジュ
144002 ミヤマコウジュ
144063 フトボナギナタコウジュ
144067 タイワンシモバシラ
144079 ハナナギナタコウジュ
144083 ニシキコウジュ
144088 イワナギナタコウジュ
144093 ニシキコウジュ,ホソバナギナタコウジュ
144096 キダチナギナタコウジュ
144144 エゾムギ属,エリムス属,ハマムギ属
144228 イブキカモジグサ
144239 ウスゲカモジグサ
144266 コハマムギ
144271 ハマムギ
144279 ヤマムギ
144303 オオハマムギ
144320 イヌカモジグサ,コウアンカモジグサ
144327 イヌカモジグサ
144336 ミズタカモジ
144365 コウリョウカモジグサ
144412 エゾカモジグサ
144439 アオカモジグサ
144440 タチカモジ
144466 エゾムギ,ホソテンツキ
144661 シバムギ,ヒメカモジグサ
144667 ノゲシバムギ
144716 サンショオモドキ属,サンセウモドキ属
144880 イヌスイバ
144884 ウスベニニガナ属,ベニニガナ属
144903 ベニニガナ
144916 ナンカイウスベニニガナ
144975 ウスベニニガナ
144976 ウスベニニガナ
145052 ガンカウラン科,ガンコウラン科
145054 エンペトルム属,ガンカウラン属,ガンコウラン属
145069 ガンコウラン
145071 シロミノガンコウラン
145073 イワモミ,ガンカウラン,ガンコウラン,ガンコオラン,コケノミ,シロミノガンコウラン,セイヨウガンコウラン
145158 エナルガンテ属
145217 オニソテツ属
145220 アルテンスタインオニソテツ
145221 スナオニソテツ
145224 トゲオニソテツ
145226 ゲリンクオニソテツ
145229 ヒメオニソテツ
145233 ツヤオニソテツ
145234 カフェルオピソテツ
145235 ツヤオニソテツ,ヒロバオニソテツ
145238 ナガバオニソテツ
145241 トノサマオニソテツ
145244 ナガゲオニソテツ
145245 ウッドオニソテツ
145247 エンセファロカルプス属
145248 ショウキュウギョク
145418 センドク
145420 グバス
145498 チチブシロカネソウ
145504 フヂバシデ属
145517 セオ,フヂバシデ
145642 ウミシャウブ属
145643 ウミシャウブ,ウミショウブ
145673 ドウダンツツジ属
145677 カイナンサラサドウダン,サラサドウダン,フウリンツツジ,ヨウラクツツジ
145679 シロバナフウリンツツジ
145680 キバナフウリンツツジ
145682 ミヤマドウダン
145683 ツクシドウダン
145684 ベニサラサドウダン
145689 シロドウダン
145690 チチブドオダン,ベニドウダン
145708 コアブラツツジ
145710 ツクモドウダン,ドウダンツツジ,ドオダンツツジ,ヒロハドウダンツツジ
145711 ヒロハドウダンツツジ
145727 アブラツツジ,ホウキヤシオ,ヤマドウダン
145730 ホソバアブラツツジ
145766 エノコロモトキ
145846 アビシニアバショウ
145857 モダマ属
145894 コウシュンモダマ
145899 モダマ,モタマズル,モタマヅル
145908 タイワンモダマ
145992 サマン属
145999 ヒトモトメヒシバ属
146002 ムラサキヒゲシバ
146015 ヒトモトメヒシバ
146054 シラユキゲシ
146065 エパクリス科
146067 エパクリス属
146096 オホトキンサウ属,オホトキンソウ属
146098 オホトキンサウ
146122 マオウ属,マワウ属
146125 ヒレマオウ
146128 セイタカマオウ
146154 フタタマオウ
146155 キダチマオウ
146183 マワウ
146253 アマナ,カツネグサ,シナマオウ,マオウ,マワウ
146270 マオウ科
146305 コイテョウラン属
146307 ハコネラン
146308 コイチヨウラン
146371 エピデンドルム属
146522 イワナシ属
146523 イワナシ
146528 エピゲネイウム属
146543 サンセイセキコク,ナカハラセキコク
146586 アカバナ属,エピロ－ビウム属
146587 カラフトアカバナ
146599 アシボソアカバナ,ニヒタカアカバナ
146605 アム－ルアカバナ,ケゴンアカバ

ナ,ナガバアカバナ,ホソアカバナ
146606 イハアカバナ,イワアカバナ,オオアカバナ,マンシウアカバナ
146607 シロバナイワアカバナ
146609 アシボソアカバナ,ヤナギサウ,ヤナギラン
146614 ヤナギサウ,ヤナギソウ,ヤナギラン
146663 オオチシマアカバナ,カラフトアカバナ
146696 エダウチアカバナ
146697 ヒメアカバナ
146709 カラフトアカバナ
146724 オオアカバナ
146734 ミヤマアカバナ
146755 シロウマアカバナ
146775 カラフトアカバナ
146782 エゾアカバナ
146787 タイリンアカバナ
146801 コバンゴケ
146812 ホソバアカバナ,ヤナギアカバナ
146834 ススヤアカバナ
146840 イダイアカバナ,サイヨウアカバナ,トダイアカバナ
146842 モウコアカバナ
146849 アカバナ,ムツアカバナ
146859 シマアカバナ,タイワンアカバナ
146879 タラオアカバナ
146953 イカリサウ属,イカリソウ属
146955 スズフリイカリソウ
146956 オオバイカイカリソウ
146959 ウメザキイカリソウ
146972 クモイイカリソウ
146980 バイカイカリソウ,バイクワイカリサウ
146981 サイコクイカリソウ
146995 イカリサウ,イカリソウ,シロイカリサウ,シロバナイカリサウ,ヤチマタイカリソウ
147003 イカリソウ
147004 シロイカリソウ
147005 ジンバイカリソウ
147013 キバナイカリソウ,シロバナイカリソウ,チョウセンイカリソウ,ニッコウイカリソウ
147021 イカリサウ,イカリソウ
147048 クハナ,ホザキイカリソウ,ホザキノイカリサウ,ホザキノイカリソウ,マラタケリグサ,ヤマトリサウ
147050 クハナ,ホザキノイカリサウ,ホザキノイカリソウ,マラタケリグサ,ヤマトリクサ
147054 トキワイカリソウ
147055 ウラジロイカリソウ
147059 オオバイカイカリソウ
147068 ヒメイカリソウ
147069 ウスベニヒメイカリソウ
147070 シオミイカリソウ
147076 ウメザキイカリソウ
147084 ブゾロイバナ
147106 カキラン属,スズラン属
147149 カラフトスズラン
147196 アオスズラン,エゾスズラン,オスズラン
147197 ハマカキラン
147198 マルバハマカキラン
147238 カキラン,スズラン
147240 キバナカキラン
147241 イソマカキラン
147279 エビフィロブシス属
147283 エピフィルム属,カニサボテン属
147291 ゲッカビジン
147303 トラキチラン属
147307 トラキチラン
147314 アオキラン,アヲキラン
147323 クシロラン,クスクスムエフラン
147336 エピプレムヌム属,ハブカツラ属
147338 オウゴンカズラ,ポトス
147342 ミヤマハブカツラ
147347 オオシラフカズラ
147349 ハブカズラ,ハブカヅラ
147384 エニスシア属,エピスキア属,エピスシア属
147386 ハイベニギリ,ベニギリソウ,ベニハエギリ
147422 エピテランサ属,エピテランタ属
147425 ツキセカイ
147468 エラグロスティス属,カゼクサ属,スズメガヤ属
147496 ヌマカゼクサ
147509 イトスズメガヤ,クロカゼクサ,ナガスズメガヤ,ホソバノスズメガヤ
147553 イトスズメガヤ
147593 オオスズメガヤ,スズメガヤ
147594 アメリカカゼクサ
147609 イトスズメガャ,ナガスズメガャ,ホソバノスズメガャ
147612 シナダレスズメガヤ
147671 カゼクサ,ミチシバ
147672 アオカゼクサ
147699 タチホスズメガヤ
147737 ノハラカゼクサ
147746 コゴメカゼクサ
147797 スズメガヤ
147815 コスズメガヤ
147823 ニハホコリ,ニワホコリ
147883 オオニワホコリ,オホニハホコリ,ニハホコリ,ムラサキニワホコリ
147891 コスズメガヤ,ヒメスズメガヤ
147969 シロカゼクサ
147990 コバンソウモドキ
147994 テフ
147995 ヌカカゼクサ
148019 スナジカゼクサ
148028 ベニスズメガヤ
148038 アンデスカゼクサ
148053 エランテムム属,ルリハナガヤ属
148087 ルリハナガサ
148101 エランティス属,セッブンサウ属,セッブンソウ属
148104 キバナセッブンソウ
148112 チョウセンセッブンソウ
148134 エルシア属
148141 オウセンリュウ,メイロギョク
148143 シセンリュウ
148144 ギョウセンリュウ,スクワギョク
148153 ダンドボロギク
148154 ウシノタケダグサ
148164 シマボロギク,タケダグサ
148249 ヂャボウシノシッペイ属
148254 ヂャボウシノシッペイ,ナガバヂャボウシノシッペイ
148261 エレモシトラス属
148262 エレモシトラス,オーストラリアデザートライム
148531 エレムールス属
148562 エレプシア属
148619 エリア属,オサラン属
148656 オオオサラン,オホバヲサラン,ホザキヲサマン
148684 タイワンオサラン,タイワンヲサラン
148712 リュウキュウセッコク
148754 アリサンオサラン,アリサンヲサラン,オサラン,バッコクラン,マッダオサラン,マッダヲサラン,ヲサラン
148790 チャイロオサラン,チャイロヲサラン,ヒダイリオサラン,ヒダイリヲサラン
148810 イゼナガヤ
148855 エリアンツス属,タカオススキ属,タカヲススキ属
148873 ムラサキタカオススキ
148941 エリカ属
149017 エイジュ
149138 アフリカエリカ,エリカ,クロジベエリカ,ジャノメエリカ
149195 ケエリカ
149398 エイカン
149750 アフリカエリカ,クロシベエリカ,ジャノメエリカ
150027 ザンセツ
150099 ジラユキエリカ
150286 シャクナケ科,ツツジ科
150414 エリゲロン属,ヒメジョオン属,ヒメジヨヲン属,ムカシヨモギ属
150419 エゾムカシヨモギ,ムカシヨモギ
150421 ヒロハムカシヨモギ
150422 ムカシヨモギ,ヤナギヨモギ
150423 ホソバムカシヨモギ
150446 エゾアズマギク,タカネアズマギ

ク,ミヤマアズマギク
150449 アノイアズマギク,アホイアズマギク
150454 ヨウシュタカネアズマギク
150464 イヌヨメナ,ヒメジョオン,ヒメジョヲン,ボウズヒメジョオン,ヤナギバヒメギク
150599 アズマギク
150635 オオツルギク
150724 ペラペラヨメナ
150732 ホソバムカシヨモギ,マンシウムカシヨモギ,ミヤマアズマギク
150781 ミヤマノギク
150788 ニイタカアズマギク,メキシコヒナギク
150792 ニヒタカアヅマギク,メキシコヒメギク
150851 チシマアズマギク
150862 ハルジオン,ハルジョオン
150894 ヤナギバヒメジョオン
150953 シコタンアズマギク
150972 ヒロハヒメジョオン
150980 ヘラバヒメジョオン
151008 アズマギク
151009 シロバナアズマギク
151010 ミヤマアズマギク
151011 シロバナミヤマアズマギク
151012 ユウバリアズマギク
151013 アポイアズマギク,シロバナアポイアズマギク
151014 シロバナアポイアズマギク
151015 ジョウシュウアズマギク
151098 エリヌス属
151102 イワカラクサ
151130 エリオボトリア属,ビハ属,ビワ属
151144 シマビハ,タイワンビハ,タイワンビワ,ヤマビハ,ヤンバルビハ
151145 ブイサンヤマビハ
151162 ビハ,ビワ
151178 エリオカクタス属
151206 ホシクサ科
151208 ホシクサ属
151219 ヒロハノイヌノヒゲ
151222 アマノホシクサ
151237 クロイヌノヒゲモドキ
151238 クロイヌノヒゲ
151243 ホシケサ
151257 オオホシクサ,オホホシクサ,タイワンヒロハノイヌノヒケ,ホシクサ,ミズタマソウ
151261 タカノホシクサ
151268 タイワンホシクサ,タカサゴホシクサ,ホシクサ,ミヅタマサウ
151288 イトイヌノヒゲ,コイヌノヒゲ
151295 ユキイヌノヒゲ
151321 コシガヤホシクサ
151336 ヤマトホシクサ
151341 ツクシクロイヌノヒゲ
151356 シマイヌノヒゲ,タイワンイヌノヒゲ
151369 ルゾンホシクサ
151377 ミカワイヌノヒゲ
151378 アズミイヌノヒゲ
151380 イヌノヒゲ,オホイヌノヒゲ,シロイヌノヒゲ
151381 ムツイヌノヒゲ
151382 タカユイヌノヒゲ
151383 オキナワホシクサ
151384 マツムライヌノヒゲ
151387 エゾホシクサ
151394 ミヤマヒナホシクサ
151395 シロバナミヤマヒナホシクサ
151398 ノソリホシクサ
151399 コヌマイヌノヒゲ
151400 ナントウホシクサ
151404 サツマホシクサ
151416 コンペイトウゲサ,シラタマホシクサ,チクトウササウ
151419 オオムラホシクサ
151421 ハライヌノヒゲ
151423 シロエゾホシクサ
151428 クロホシクサ
151430 エゾイヌノヒゲ
151452 オオミズタマソウ,ヒロハイヌノヒゲ,ヒロハノイヌノヒゲ
151461 カラフトホシクサ
151463 コケヌマイヌノヒゲ
151475 イヌノヒゲモドキ
151476 ヤシュウイヌノヒゲ
151479 ゴマシオホシクサ
151486 ヒュウガホシクサ
151487 イヌノヒゲモドキ,オオシラタマホシクサ,オホシラタマホシクサ,ミヅタマサウ
151500 ナガトホシクサ
151519 アズマホシクサ
151520 ニッポンイヌノヒゲ
151532 スイシャホシクサ,スヰシャホシクサ
151535 ガリメギイヌノヒゲ
151536 マンシウイヌノヒゲ
151556 イズノシマホシクサ
151630 エリオセレウス属
151655 ナルコビエ属
151668 アメリカノキビ
151675 ホソナルコビエ
151693 ノキビ,ムラサキノキビ
151702 スズメノアハ,スズメノアワ,ナルコビエ
151703 ホソバナナルコビエ
152734 サギスゲ属,ワタスゲ属
152759 サギスゲ,マユハキグサ,ワセワタスゲ
152765 チシマワタスゲ,チョウセンサギスゲ,ヒロハサギスゲ
152769 シムシュワタスゲ,テウセンサギスゲ,ホソバワタスゲ
152772 キツネスゲ
152781 エゾワタスゲ
152782 エゾワタスゲ
152791 スズメノケヤリ,タテヤマワタ,ワタスゲ
152792 マユハキグサ,ワタスゲ
152844 エリオプシス属
153325 エリオステモン属
153351 エリオシケ属
153353 キョッコウマル
153354 イオヅギョク
153420 ミヤマムラサキ属
153491 ミヤマムラサキ
153492 エゾルリムラサキ,シロバナエゾルリムラサキ
153493 エゾルリムラサキ
153502 ルリザクラ
153518 カブムラサギ,ホソバルリザクラ
153534 ルリザクラ
153555 チシマルリザクラ
153711 エロジューム属,オランダフウロ属
153738 ツノミオランダフウロ
153767 オランダフウロ
153769 ヒロハオランダフウロ
153788 ミツバオランダフウロ
153869 ジャコウオランダフウロ
153914 キクバフウロ,セリバフウロ
153964 ヒメナズナ,ヒメナヅナ
153998 エルーカ属,キバナスズシロ属
154019 キバナスズシロ,ヒメキバナスズシロ
154091 オハツキガラシ
154162 サンユウカ
154192 マニラサンユウカ
154233 ホルトカズラ属,ホルトカヅラ属
154245 ホルトカズラ,ホルトカヅラ
154252 チョウコゥトウ
154266 エリキーナ属
154276 エリンジューム属,ヒゴタイサイコ属
154280 ミヤマヒゴタイサイコ
154281 オオヒゴタイサイコ
154284 ナガベエリンジューム
154317 ナガベエリンジューム,ヒゴタイサイコ
154327 ヒイラギサイコ
154337 マルバノヒゴタイサイコ,マルバヒゴタイサイコ
154363 エゾスズシロ属,エリシマム属
154369 チェランサス
154370 チェランサス
154380 シベリアスズシロ
154414 オオスズシロソウ
154416 キバナオオスズシロ

154419 キバナアラセイトウ
154424 エゾスズシロ,マンシゥスズシロ
154452 モウコスズシロ
154466 ヒメコガネナズナ
154485 エゾスズシロ,キタミハタザオ
154500 トウスズシロ
154507 ムラサキナズナ
154530 エゾスズシロモドキ
154533 コガネナズナ
154571 ハクセンヤシ属
154617 エリスリナ属,デイコ属
154622 トゲミデイゴ
154632 サンゴシトウ,ヒシバデイコ
154643 サンゴシドウ
154647 アメリカデイコ,カイコウズ,カイコウヅ
154648 アメリカデイコ,マルバデイコ
154734 カイコウジ,シトウ,シトウヅ,デイグ,デイゴ,ドイツハギ,ハリギリ
154735 シロバナデイゴ
154871 ホソフデラン属
154872 ホソフデラン
154895 エリスロニューム属,カタクリ属
154899 エリスロニウム
154909 カタクリ
154926 カガユリ,カタカコ,カタクリ,カタコ,カタユリ,タコユリ,ハヅユリ,ブンダイユリ
154952 コカノキ科
154962 アカバノキ属
154971 アカハダノキ,アカバノキ
155012 オホテウビツルラン,タカツルラン,ツルツチアケビ
155017 エリスロリブサリス属
155048 コカノキ科,コカ科
155051 コカ属
155065 アマゾンコカ
155067 コカ,コカノキ,ナガバコカノキ
155130 エスカロニア属
155170 キンエイカ属,キンエイクワ属,ハナシソウ属
155171 ヒメハナビシソウ
155173 キンエイカ,キンエイクワ,ハナビシサウ,ハナビシソウ
155201 エスコバリア属
155220 ルロウマル
155225 ショウキュウマル
155246 エスコントリア属
155277 エスペレチア属
155296 エスポストア属
155299 エイラグキュウ,ギンガラク
155300 オイラク
155302 エテンラク
155303 ハクラクオウ,ロウジュラク
155304 ギョウウンカク
155306 タイヘイラク
155354 ナガサハサウ属,ナガサハソウ属
155365 ナガサハサウ,ナガサハソウ
155398 ケットウ,サニン
155432 ユ-アンテ属
155434 ダイオウラン
155468 ユウカリノキ属,ユ-カリノキ属
155472 ガム
155478 セイタカユ-カリ,ナガバユ-カリ
155514 カロフイラユ-カリ
155517 セキザイユ-カリ,ロストラ-タユ-カリ
155529 キンマルバユ-カリ
155542 サトウゴムノキ
155545 クレブラユ-カリ
155558 デレゲ-トユ-カリ
155562 カリ-
155589 ユウカリ,ユウカリジュ,ユウカリノキ,ユ-カリ,ユ-カリジュ,ユ-カリノキ
155594 ゴムフオセフアラユ-カリ
155624 ヤナギユウカリ,ヤナギユ-カリ
155628 ナカバュ-カリ
155691 ツキヌキユ-カリ
155710 コマルバユ-カリ,マルバユウカリ
155719 セイタカユ-カリ
155720 キノユ-カリ
155722 オオバユ-カリ,ロブスタユ-カリ
155743 アカゴムノキ
155857 ユ-チャリス属
155858 アマゾンユリ,ギボウシズイセン
155876 ブタモロコシ属
155878 テオシント,ブタモロコシ
155888 ミヤマトベラ属
155890 タイワンミヤマトベラ,リウキウミヤマトベラ
155897 イシャダオシ,ミヤマトベラ,ヤマニガキ
156011 ユ-コミス属
156040 トチュウ属
156041 トチュウ
156042 トチユウ科
156058 エウクリフィア属
156072 エウクリフィア科
156113 チリメンウロコヤシ属
156116 イチゴチリメンウロコヤシ
156117 クイミチリメンウロコヤシ
156118 フトモモ属
156134 ミズレンブ
156338 サラル
156385 タイバナアデク,タチバナアデク,ピタンガ
156439 ウンヌケ属
156450 オオササガヤ
156471 コウシュンウンヌケ
156490 ウンヌケモドキ,コカリヤス,チャヒキカリヤス
156499 ウンヌケ
156513 ワタガヤ
156521 イモラン属,ユ-ロフィア属,リッソキ-ルス属
156572 シルトラン,タイワンヤガラ,ヒロハノヤガラ
156660 タカサゴヤガラ
156723 エタウチヤガラ
156732 イモラン
157110 イモネヤガラ
157111 ミドリイモネヤガラ
157145 ユ-ロフィエラ属
157163 エウリクニア属,ユ-リクニア属
157164 タカノスマル
157168 ニオウモン
157169 ハクゲンカク
157171 センオウマル
157268 ニシキギ属,マユミ属
157285 ケコマユミ,ケニシキギ,コマユミ,ニシキギ,ヤハズニシキギ
157293 コマユミ
157294 コバノコマュミ
157310 オホバニシキギ,ケニシキギ
157312 オオコマュミ
157343 ヒメマサキ
157345 ヒメマユミ
157355 アッバマュミ,クロトチウ,クロトチュウ,コクタンノキ,コクテンギ,バタカンマュミ
157365 ヒゼンマユミ
157416 チャウチンマユミ
157418 クラルマユミ,トゲマユミ
157429 セイヨウマュミ
157473 ツルマサキ,マサキカヅラ
157495 シロミノツルマサキ
157496 ツルマサキ,ナカバツルマサキ
157499 ヒロハツルマサキ
157501 ツルノキンマサキ
157503 コツルマサキ,コバノツルマサキ
157504 キンミャクツルマサキ
157505 ウチダシツルマサキ
157508 シロフクリンツルマサキ
157509 ムラサキツルマサキ
157511 ウズマサキ
157512 コバノツルマサキ
157515 オオツルマサキ,ツルオオマサキ,ツルマサキ
157517 ツルマサキ,マルバツルマサキ
157519 ケツルマサキ
157539 ミヤケマユミ
157559 マユミ
157570 オホバマユミ
157601 シタワレ,マサキ
157614 ギンマサキ
157615 キフクリンマサキ,フイリマサキ
157616 キンマサギ,フイリマサキ
157617 オオバマサキ,オホバマサキ,マサキ
157620 チャボマサキ

157621 ウチダシマサキ
157625 ギソフイリマサキ
157631 ナガバマサキ,マサキ
157637 オオツルマサキ,オホツルマサキ,ツルオオバマサキ
157647 カハチマサキ,カワチマサキ,マサキ
157651 サントウマサギ
157663 ムラサキマユミ
157698 リウキウマユミ,リュウキュウマユミ
157699 チョウセンマユミ,ホソバマユミ,モウコマユミ
157705 ヒロハツリバナ,ヒロハノツリバナ
157709 ナンバンマユミ
157713 アオジクマユミ,アヲヂクマユミ,サハダツ,サワタチ,サワダツ
157745 カラマユミ
157748 ナガバヒゼンマユミ
157753 アンドンマユミ
157761 ツリバナ,ツリバナマユミ
157763 タンザワツリバナ
157764 ニッコウマユミ
157765 エゾツリバナ
157767 エゾツリバナ
157769 オホミマユミ,ニシガキマユミ
157777 アンドンマユミ,イトツリバナ,イトマユミ
157784 タイワンアヅサ
157797 オオツリバナ,オホツリバナ,ミドリツリバナ
157848 カラフトツリバナ,ムラサキツリバナ
157879 オオコマユミ,オニマユミ,カントウマユミ,マユミ,ヤマニシキギ
157882 カントウマユミ,ユモトマユミ
157883 チョウチンマユミ
157884 ホソバマユミ
157888 アリサンツルマユミ,クラルマユミ,シマママユミ,トゲミマユミ
157910 コクテンギ
157912 チャカンマユミ,ヒシガタマユミ,ヒツガタマユミ,ヤンバルマユミ
157930 クロツリバナ,ムラサキツリバナ
157932 ニイタカマユミ,ニヒタカマユミ,ハリミマユミ
157970 アオツリバナ,アヲツリバナ
158021 ヒヨドリバナ属,フジバカマ属
158022 サワフジバカマ
158023 サワシマフジバカマ
158038 ミカエリヒヨドリ
158043 アメリカフジバカマ
158063 タイワンヒヨドリ,タイワンヒヨドリバナ,タイワンヒヨドリバナモドキ
158066 イトバヒヨドリ
158070 シナヒヨドリ,フヂバカマ
158090 タシロヒヨドリ
158118 フジバカマ
158136 クルマバヒヨドリ,ハコネヒヨドリ,ホソバヨツバヒヨドリ,ヨツバヒヨドリ
158139 クスクスバカマ
158161 ヒヨドリバナ,フジバカマ,フヂバカマ
158182 キ-ルンフジバカマ,シマフジバカマ
158187 サケバヒヨドリ
158200 サハヒヨドリ,サワヒヨドリ
158202 ホシナシサハヒヨドリ,ホシナシサワヒヨドリ
158207 ミツバサワヒヨドリ
158210 ハマサワヒヨドリ
158214 キイルンフジバカマ,シマフジバカマ
158230 キクバヒヨドリ,ニオイヒヨドリ,ニバイタイ,ヒヨドリバナ
158232 オオヒヨドリバナ,バイスタイ
158250 ツキヌキヒヨドリ
158263 ルリアザミ
158264 ルリアザミ
158305 クルマバヒヨドリ,ヨツバヒヨドリバナ
158333 サワシマフジバカマ
158347 ホソバノミツバヒヨドリ,ミツバヒヨドリ,ミツバヒヨドリバナ
158354 マルバフジバカマ
158358 ヤマヒヨドリ,ヤマヒヨドリバナ
158373 ヤクシマヒヨドリ
158385 タカトウダイ属,トウダイグサ属,ユ-ホルビア属
158386 ランガク
158397 ノウルシ
158404 ベニキリン
158456 サイウンカク,サボテンタイゲキ,フクロ,フクロギ,フクロノキ
158490 スナダイゲキ,ハマダイゲキ
158534 アハユキニシキサウ,アワユキニシキソウ,ミヤコジマニシキサウ,ミヤコジマニシキソウ
158586 テッコウマル
158614 テンコウリュウ
158649 ゲキリンリュウ
158653 ヒトウバン
158665 タイホウカ
158700 ケツヨウボク
158727 ショオジョオソウ
158730 マツバトウダイ
158758 ニシキダイゲキ,ノコギリダイゲキ
158809 ベニタイゲキ,マルミノウルシ,リョジョ
158829 コウサイカク
158857 イトバトウダイ,セイヨウハギクソウ,タロコタカトウダイ,ハギクサウ,ハギクソウ,マツバトウダイ,マンシウダイゲキ
158895 ヒロハタイゲキ,ヒロハタカトウダイ
158906 キダチダイゲキ,タイワンタカトウダイ
158926 シコウカク
158935 ガランビニシキサウ,ニシキサウダイゲキ
158958 タマリンボウ
159027 スズフリバナ,トウダイゲサ,トオダイグサ
159046 クサシャウジャウボク,シャウジャウサウ,ショウジュウソウ,ショウジョウソウモ,ショウジョウソウモドキ
159069 シマニシキソウ,タイワンニシキサウ,タイワンニシキソウ,ツマニシキサウ
159071 テリハニシキソウ
159081 カイイギョク
159092 チチグサ,ニシキサウ,ニシキソウ,ノコバニシキサウ
159102 ウンリンニシキサウ,オトギリバニシキソウ,オホニシキサウ
159109 セイタカオオニシキソウ
159140 チュウテンカク
159159 アカメダイゲキ,イヒダイゲキ,イワダイゲキ,キダチダイゲキ,タイワンタカトウダイ
159204 ミカドニシキ
159215 タカトウダイ
159217 ハマタカトウダイ
159218 イブキタイゲキ
159222 クサホルト,コハズ,ゾクズヰシ,ハンシレン,ホルトサウ,ホルトソウ
159242 ハクギンサンゴ
159245 キリンカク
159272 チョウセントウダイクサ
159286 イリオモテニシキサウ,イリオモテニシキソウ,オオニシキソウ,コニシキサウ,コニシキソウ
159299 コバノニシキサウ
159305 リンボウ
159313 ハツユキソウ,フクリンダイゲキ
159363 ハナキリン
159366 ハナキリン
159457 フクロギ
159492 ハギクソウ
159540 イブキタイゲキ,タカトウダイ,トウタカトウダイ
159543 アソタイゲキ
159544 ミヤマタイゲキ
159557 キダチニシキサウ,チャボダイゲキ
159617 ホウリンギョク
159634 ハイニシキソウ
159641 ハルゴマ
159675 シャウジヤウボク,ショウジョウボク,ポインセチヤ

159705 ゴサイカク
159710 シャボテンダイゲキ
159785 ハナグシ
159791 トウギュウカク
159813 センダイタイゲキ,ムサシタイゲキ
159828 セイロンオリ-ブ
159841 イズナットオダイ,オオバカンスイ,ナットウダイ,ナンゴクナットオダイ
159844 ナンゴクナットウダイ
159847 エチゴタイゲキ,シナノタイゲキ
159854 オホアカリニシキサウ,スナダイゲキ,ボロヂノニシキサウ
159880 グンセイカン
159883 ギンツノサンゴ
159916 ルリコウ
159971 イリオモテニシキソウ
159975 アオサンゴ,アヲサンゴ,ミドリサンゴ,レダマキリン
159980 オゼヌマタイゲキ,ハクサンタイゲキ,ミヤマノウルシ
159997 オオマトイ
160007 ヒメタイゲキ,ヒメナットウダイ
160012 ヨウギョク
160064 ハチダイリュオウ,ヤドクキリン
160076 フジタイゲキ
160077 ヒュウガタイゲキ
160110 トウダイグザ科
160128 コゴメグサ属
160148 ネバリコゴメグサ
160168 ハチジョウコゴメグサ
160176 ミヤマコゴメグサ
160177 イブキコゴメグサ
160178 イズコゴメグサ
160179 キュウシュウコゴメグサ
160180 トサコゴメグサ
160182 ダイセンコゴメグサ
160183 ホソバコゴメグサ
160184 マルバコゴメグサ
160185 オオミコゴメグサ
160186 マツラコゴメグサ
160187 トガクシコゴメグサ
160199 コケコゴメグサ
160210 ノウコウコゴメグサ
160212 コバノコゴメグサ,ヒメコゴメグサ
160213 センナシヒメコゴメグサ
160214 タチコゴメグサ
160216 シライワコゴメグサ,ミチノクコゴメグサ
160218 エゾコゴメグサ
160221 ナヨナヨコゴメグサ
160224 チシマコゴメグサ
160225 カラフトコゴメグサ
160226 ツクシコゴメグサ
160227 ツクシコゴメグサ
160228 イナコゴメグサ
160229 クモイコゴメグサ
160239 オクエゾコゴメグサ
160241 タチコゴメグサ
160242 エゾノダッタンコゴメグサ
160275 タロココゴメグサ
160277 ダッタンコゴメグサ
160286 ニヒタカコゴメグサ
160294 ヒナコゴメグサ
160325 オオコゴメスナビキソウ
160338 フサザクラ属
160347 カウヤマンサク,コウヤマンサク,サハグハ,サワグワ,タニグワ,タニダイ,ナシマンサク,フサザクラ,ヤマグハ,ヤマグワ
160348 ウラジロフサザクラ
160349 フサザクラ科
160405 サカキ属,ヒサカキ属
160406 トガリバヒサカキ,ナンゴクヒサカキ
160458 イソチジミ,イソヒサカキ,ハマヒサカキ
160460 ヒメハマヒサカキ
160462 ケナシハマヒサカキ
160464 マメヒサカキ
160474 アッバヒサカキ
160503 アクシバ,ヒサカキ,ビシャコ
160504 ホソバヒサカキ
160506 シロシヒサカキ
160507 ヤナギバヒサカキ
160508 コヒメヒサカキ,ツゲバヒサカキ
160510 ムニンヒサカキ
160513 ケヒサカキ
160518 ケンロクヒサカキ
160527 ニヒタカヒサカキ
160535 アリサンヒサカキ,フトミノヒサカキ
160553 モチバヒサカキ
160571 アマミヒサカキ
160572 オキナワヒサカキ
160595 テリバヒサカキ
160596 サキシマヒサカキ
160606 ミマヒサカキ,ミヤマヒサカキ,ラゲサカキ
160628 ヤエヤマヒサカキ
160629 ヒメヒサカキ
160631 クニガミヒサカキ
160635 オニバス属,ユ-リアレ属
160637 イバラバス,オニバス,ミズブキ,ミヅブキ
160707 エウリコ-ネ属
160928 ユリスチグマ属,ユ-リスティグマ属
160946 ゴンズイ属
160954 キツネノチャブクロ,クロクサギ,ゴンズイ,ハゼナ
160955 シロゴンズイ
160956 タネガシマゴンズイ
160963 コニシゴソズイ
161020 ナガバマンサク属
161023 ナガバマンサク
161026 トルコギキョウ属
161027 トルコギキョウ
161053 キャベツヤシ属
161056 ラングロイスキャベツヤシ
161057 ワカバキャベツヤシ
161059 ワインニセダイオウヤシ
161120 ワサビ属
161137 カラフトワサビ
161149 ユリワサビ
161154 オオユリワサビ,ワサビ
161302 ゴシュユ属
161323 イヌゴシュユ,シュユ
161335 ハマセンダン
161336 ウラジロゴシュユ,シマクロギ,ハマセンダン
161342 ケハマセンダン
161350 オホバアハダン,オホバアワセンダン
161356 ハマセンダン
161373 イタチキ,カラハジカミ,カワハジカミ,ゴシュユ,ニセゴシュユ,ハハジカミ,ホンゴシュユ
161376 ホンゴシュユ
161384 アハダン,ドモノキ
161416 アサガホカラクサ属
161418 アサガオカラクサ,アサガホカラクサ,ヤヘヤマカラクサ
161422 アサガオガラクサ
161426 マルバアサガオガラクサ,マルバアサガホカラクサ
161430 シロガネガラクサ
161528 エキサカム属
161531 ベニヒメリンドウ
161618 ホソバヤロオド,ホソバヤロ-ド
161628 サイシボク属,セイシボク属
161632 シマシラキ
161639 オキナワジンコオ,キヤラ,シマシラキ
161643 セイシボク
161647 セイシボク
161656 ダイトウセイシボク
161657 シマセイシボク,タイワンセイシボク
161666 コウトウシラキ,コウトウユヅリハ
161741 エキソコルダ属,ヤナギザクラ属
161749 ウメザキウツギ,バイカシモツケ,マルバヤナギザクラ,リキュウバイ
161755 ヤナギザクラ
161896 ファビアナ属
161919 キヌゲチチコグサ
161921 ファケイロア属
161922 ヒカチュウ
162033 ブナ科
162301 ソバムギ属,ソバ属,ソマムギ属,ファゴピラム属

162309 シャクチリソバ
162312 ソバ
162335 インドソバ,ダッタンソバ,ニガソバ
162338 クロムギ,ソバ,ソバムギ
162343 ゴムミカズラ属,ゴムミカヅラ属
162346 キベンケイ,ゴムミカヅラ,モクベンケイ
162359 ブナノキ属,ブナ属
162368 イボブナ,クロブナ,シロブナ,ソバグリ,ソバグルミ,ブナ,ブナノキ,ホンブナ
162370 オオバブナノキ,オホバブナノキ
162372 エングラ-ブナ
162375 アメリカブナ
162380 タイワンブナ,タイワンブナノキ
162382 イヌブナ,イボブナ,クロブナ
162386 ナガエブナ,ブナ
162390 テリハブナ
162395 オリエントブナ
162400 セイヨウブナノキ,ヨ-ロッパブナ
162473 ジエウジソ,モルッカネム
162505 ツバカズラ属
162508 アイイタドリ
162509 ナツユキカズラ
162516 チョウセンツルドクダミ
162519 ソバカズラ,ソバカヅラ,ツルイタドリ
162525 オオツルイタドリ
162528 ツルイタドリ,ツルタデ
162530 サナエツルイタドリ
162531 カライタドリ
162532 イタドリ
162536 メイゲツソウ
162538 ハチジョウイタドリ
162539 ケイタドリ
162542 カシウ,ツルドクダミ
162546 タイワンツルドクダミ
162549 オオイタドリ
162550 エゾイタドリ
162612 ツハブキ属,ツワブキ属
162615 カンツワブキ
162616 ツハブキ,ツワブキ
162617 キモンツワブキ,キンモンツワブキ
162619 カワリツワブキ
162621 ウコンツワブキ
162622 ヤエツワブキ
162624 タイワンツハブキ
162626 オオツワブキ,オオバノツワブキ,トオツワブキ
162627 リュウキュウツワブキ,リュキュツワブキ
162868 クハクサ属,クワクサ属
162871 クハクサ,テンデククハクサ
162872 クハクサ,クワクサ
162876 ヤツデ属
162879 テングノハウチワ,ヤツデ
162883 シロブチャツデ
162884 フクリンヤツデ
162885 キモンヤツデ
162886 ヤグルマヤツデ
162887 チヂミバヤツデ
162889 リュウキュウヤツデ
162891 ムニンヤツデ
162895 タイワンヤツデ
162898 ファウカリア属,フォ-カリア属
162902 カタオナミ
162908 シラナミ
162934 サカナミ
162951 シカイナミ
162953 アラナミ
163116 フェイジョア属
163117 アナナスガヤバ,フェイジョア
163121 フエリシア属,ルリヒナギク属
163131 ルリヒナギク
163326 フェネストラリア属
163328 グンギョク
163329 イスズギョク
163417 フエロカクタス属
163418 シャチガシラ
163422 イソウギョク
163423 センプウギョク
163425 ルリマル
163427 キンカンリュウ
163429 エモリ
163435 シキンジョウ
163438 リュウコ
163440 シンリョクギョク
163442 リュウホウギョク
163443 オウカンリュウ
163445 カリホギョク
163446 シンセンウギョク
163451 シュンロウ
163452 キョシュウギョク
163453 シンチョウマル
163455 ハクテイジョウ
163456 セキリュウマル,ヒノデマル
163458 アカギ
163462 ハントウギョク
163463 イカンリュウ
163464 セキホウ
163471 シンジュ
163472 ユウソウマル
163477 セキホウ
163479 リュウガン
163480 コウショウリュウ
163481 シホウリュウ
163483 キンセキリュウ
163499 フェロ-ニア属
163502 ウッドアップル
163510 フェラ-リア属
163516 チリメンショウブ
163574 オオウイキョウ属,オホウイキャウ属,オホウヰキャウ属
163580 アギ
163584 エダウチゼリ,エダハリゼリ
163590 オオウイキョウ
163709 アギ
163778 ウシノケグサ属,フェスツ-カ属
163817 ミヤマウシノケグサ
163818 アリゾナウシノケグサ
163819 オニウシノケグサ,ヒロハノウシノケグサ
163836 ミヤマオオウシノケグサ
163864 チイサンウチノケグサ
163950 オオトボシガラ,オニトボシガラ,オホトボシガラ,タウトボシガラ,トウトボシガラ
163969 タロコガヤ
163993 オウシュウトボシガラ,オニトボシガラ
164006 ハガワリトボシガラ
164008 タロコガヤ
164025 ココメヌカボ,ヤマトボシガラ
164126 ウシノケグサ,ギンシンサウ,ギンシンソウ,シンウシノケグサ
164131 アオウシノケグサ
164139 ケウシノケグサ,ヤマオオウシノケグサ
164141 ミヤマウシノケグサ
164153 チイサンウチノケグサ
164159 コウライウシノケグサ
164182 タカネウシノケグサ
164198 トボシガラ
164200 イブキトボシガラ
164219 ヒロハノウシノケグサ
164243 オオウシノケグサ,オホウシノケグサ
164260 ヒロハノオオウシノケグサ
164272 イトウシノケグサ
164286 ハマオオウシノケグサ
164287 アサカワソウ
164351 タカネソモソモ
164608 イチジク属,イチヂク属,イヌビワ属,フィカス属
164630 ホソバムクイヌビワ
164635 コウトウイチジク
164661 オオバイチジク
164671 ケイヌビハ,ケイヌビワ
164684 バンヤンジュ,ベンカルボダイジュ
164685 クリシュナボダイジュ
164686 アカメイヌビワ
164688 シダレガジュマル
164711 トキワイヌビワ
164763 イチジク,イチヂク,ウドンゲ,タウガキ,トウガキ,ナンバンガキ
164783 オオバアコウ
164875 シロカエマル,シロカジュマル
164900 コバンボダイジュ
164925 イントゴムノキ,インドゴムノキ,オオバゴムノキ,ゴムイチジク,ゴムノ

キ,ゴムビハ,ダンセイゴムノキ
164933 マルバインドゴムノキ
164940 フイリインドゴムノキ
164947 イタビ,イタブ,イチヂク,イヌビハ,イヌビワ,ウシノヒタイ,コイチジク,コイチヂク,チョウセンイチジク,ヤブビワ
164956 キッコウボク,ホソバイスビハ,ホソバイスビワ
164985 オホバケイヌビハ,ハルライヌビハ
164992 タイワンイヌビハ,タカサゴイヌビハ,モモノミイヌビハ
165002 イタビカヅラ
165086 ワセオホイタビ
165140 オオヤマイチジク
165166 ムクイヌビワ
165203 コウシュンイヌビハ
165210 クリスナボダイジュ
165274 オオバゴムノキ,オオバゴムビワ,オホバゴムノキ,オホバゴムビハ
165307 ガジュマル,ガヅマル,タイワンマツ
165370 カシイヌビハ,ナガバアコウ
165378 オオトキワイヌビワ
165390 オホケイヌビハ
165426 カシワバゴムノキ
165429 キンダイビハ
165444 イハイヌビハ
165515 オオイタビ,オホイタビ,コヅタ,ヒメイタビ
165518 アイギョクシ,アイギョクシイタビ,カンテンイタビ
165527 モモノミイヌビハ
165541 ウドンゲ
165553 インドボダイジュ,テンジクボダイジュ,テンヂクボダイジュ,ボダイジュ
165586 コバノゴムビワ
165623 アリサンイタビカツラ
165628 イタツラ,イタビカズラ,イタビカヅラ,ツルイチジク
165630 クライタボ,ヒゴヅタ,ヒメイタビ,ヒメクライタボ,ヒメビタイ
165658 オオバイヌビワ
165717 アカウ,アコウ,アコノキ
165726 イチジクグワ
165734 ツルイヌビハ
165761 ガランビイヌビハ,シマイヌビハ
165762 オホイヌビハ,キンダイヌビハ,ホコバイヌビハ,ムクイヌビハ
165814 コケモモイタビ,テリハイヌビハ
165819 ギランイヌビワ
165826 ギランイヌビワ
165828 ギランイヌビワ,トキハイヌビハ
165830 ハマイヌビハ
165841 アカウ,アコウ,アコギ,アコノキ,アコミズギ,ウスク,オホギ
165847 ハマイヌビワ
166038 キヨミギク
166056 シダノキ属
166059 シダノキ
166066 キバナイトヨモギ
166071 シモツケサウ属,シモツケソウ属
166073 ホソバシモツケソウ
166076 コシジシモツケソウ
166077 シロバナコシジシモツケソウ
166083 シロバナエゾノシモツケソウ
166089 オニシモツケ
166092 タイワンシモツケサウ
166093 コウライカノコソウ
166097 クサシモツケ,シモツケサウ,シモツケソウ
166098 シロバナシモツケソウ
166099 ハコネシモツケソウ
166101 モミジシモツケ
166105 アカバナシモツケソウ
166106 ウスイロシモツケソウ
166109 オレゴンシモツケソウ
166110 ウラジモシモツケサウ,ウラジロシモツケソウ,チシマシモツケ,チシマシモツケソウ
166117 エゾノシモツケソウ,キャウガノコ,キョウガノコ
166119 ナツユキソウ
166122 アメリカシモツケソウ
166124 シコクシモツケソウ
166125 セイヤウナツユキサウ,セイヨウナツユキソウ
166127 ヤエセイヨウナツユキソウ
166132 ヤウシュシモツケ,ヨウシュシモツケ,ロクベンシモツケ,ロクベンシモツケソウ
166156 テンツキ属
166157 インバテンツキ
166158 ナガボトネテンツキ
166164 コアゼテンツキ
166165 エチゴテンツキ
166190 ヒメヒラテンツキ
166223 オオヒラテンツキ,オホヒラテンツキ,ノテンツキ,ヒラテンツキ
166224 ノテンツキ
166226 ノテンツキ,ヒラテンツキ
166228 クサテンツキ,ヒメテンツキ,ヒメヒラテンツキ
166236 シオカゼテンツキ,シバテンツキ
166238 クジュウクリテンツキ
166242 タマテンツキ
166248 テンツキ
166251 ホソバテンツキ
166258 ケックシテンツキ
166259 ツクシテンツキ
166265 クグテンツキ
166266 ケテンツキ
166268 アカンテンツキ
166271 テンツキ
166272 ホソバテンツキ
166290 クロテンツキ
166293 アオテンツキ
166304 カゼクサテンツキ
166313 イソヤマテンツキ,シマテンツキ,スナテンツキ,ハマテンツキ
166315 アンピンテンツキ,イソヤマテンツキ,シマテンツキ
166324 トモエバテンツキ
166329 オノエテンツキ,オノヘテンツキ,ノヤマテンツキ,ヲノヘテンツキ
166367 イッスンテンツキ
166379 チャイロテンツキ
166380 ヒデリコ
166383 タイワンヒデリコ
166386 オオテンツキ,オホテンツキ,ナガボテンツキ
166387 ムニンテンツキ
166388 ハハジマテンツキ
166393 メアゼテンツキ
166402 ヒデリコ,ヤリテンツキ
166411 ヤリテンツキ
166421 ウナヅキテンツキ
166428 ヤリテンツキ
166431 イソテンツキ
166434 イシガキイトテンツキ
166435 ノハラテンツキ,ブゼンテンツキ
166444 スギゴケテンツキ
166454 イヌヒデリコ
166471 ショウノテンツキ,ヒメヤマイ,ヒメヤマヰ
166473 ビロウドテンツキ,ビロ-ドテンツキ
166482 イソヤマテンツキ
166493 シオカゼテンツキ,シバテンツキ,シホカゼテンツキ
166499 アゼテンツキ
166500 エゾアゼテンツキ
166501 メアゼテンツキ
166511 ハタケテンツキ
166513 トネテンツキ
166518 タマイ,タマヰ,ヤマイ,ヤマヰ
166530 シンチクテンツキ
166531 オニテンツキ
166550 ハナシテンツキ
166555 メアゼテンツキ
166557 アオテンツキ,アヲテンツキ
166612 アオギリ属,アヲギリ属
166627 アオギリ,アヲギリ
166628 フイリアオギリ
166706 アミメグサ属,フィット-ニア属,フィト-ニア属
166709 ベニアミメグサ
166710 シロアミメグサ
166761 フラク-ルティア属
166773 インドルカム
166776 オオミイヌカンコ,ロビロビ

166777 ナンヨウイヌカンコ
166788 ジャワルカム
166793 アイギリ科,イイギリ科,イヒギリ科
166795 タウツルモドキ属,トウツルモドキ属,フラゲラリア属
166797 サントウ,タウツルモドキ,トウツルモドキ
166801 タウツルモドキ科,トウツルモドキ科
166816 キアレチギク
166818 カツマタギク
166824 フラベリギク
166850 ソロハギ属
166888 エノキマメ
166897 ソロハギ,ソロフヂ
166959 クスクスセキコク
166964 バレンセキコク
167010 ツルヤブメウガ属
167040 ツルヤブミョウガ,ツルヤブメウガ
167067 シマヒトッバハギ属
167092 アツバコバンノキ,アマミヒトツバハギ,イチヨウシュウ,ヒトツバハギ
167096 シマヒトッバハギ,タイワンヒトッバハギ
167122 フォッケア属
167124 ウチュウセン,キョウマイギ
167141 ウイキャウ属,ウイキョウ属,ウヰキャウ属
167146 イタリアウイキョウ,イタリ-ウイキョウ
167156 ウイキャウ,ウイキョウ,ウヰキャウ,クレノオモ,クワイカウ
167201 フッケンヒバ
167230 コバタゴ属,フォンタネ-シア属
167233 ガラユキヤナギ,コバタゴ
167235 カラュキヤナギ,コバタゴ
167427 レンギョウ属,レンゲウ属
167444 ヤマトレンギョウ,ヤマトレンギョオ,ヤマトレンゲウ
167446 イワレンギョウ
167450 マンシウレンギョウ
167452 ヒロハレンギョウ
167456 イタグサ,イタチグサ,イタチハゼ,レンギヤウ,レンギョウ,レンギョウウツギ,レンゲウ
167470 ショウドシマレンギョウ,ショオドシマレンギョオ
167471 シナレンギョウ,シナレンゲウ,テウセンレンゲウ
167473 チョウセンレンギョウ
167489 キンカン属
167493 キソズ,タイワンキンカン,ニンポウキンカン,ネイハキンカン,ネンパキンカン,ネンボキンカン,メイワキンカン
167499 キンズ,マメキンカン
167503 キンカン,ネイハ,マルキンカン,マルミキンカン
167506 ナガキンカン,ナガミキンカン
167511 チョウジュキンカン,フクシュウキンカン
167513 ナガハキンカン,マライキンカン
167515 スウィングルキンカン,ナガバキンカン
167516 キンズ,マメキンカン
167529 フォステレラ属
167537 シロバナマンサク属,フォサ-ギラ属
167556 フキエ-ラ属
167592 イチゴ属,オランダイチゴ属
167597 オランダイチゴ,ストロベリ
167604 チリイチゴ
167620 シマシロヘビイチゴ,タイワンシロヘビイチゴ
167622 ノウゴウイチゴ
167627 シロバナヘバイチゴ
167635 エゾノクサイチゴ,シロバナノヘビイチゴ,モリイチゴ
167636 ベニバナノヘビイチゴ
167637 キミノモリイチゴ
167641 マンシウクサイチゴ
167653 エゾヘビイチゴ,オヲンダイチゴ,ミヤマクサイチゴ
167654 シロミノベスカイチゴ
167674 コオランダイチゴ
167689 フレ-レア属
167690 シドウ
167691 ヒレイガグリギョグ
167692 テンケイマル
167693 クマノル
167695 シウンデン
167696 シウンマル
167698 ハツヒメマル
167699 タメキノコ
167700 ヒメノコ
167702 トラノコ
167703 ヒョウノコ
167704 コジシマル
167733 フランコ-ア属
167765 フランケ-ニア属
167893 トネリコ属
167897 アッジュ,アメリカトネリコ,テリハトネリコ
167923 コバシジノキ,ナガバアヲダモ,ミヤラアオダモ
167931 トネリコ,ヒメトネリコ
167940 アオダモ,トネリコ
167955 セイヤウトネリコ,セイヨウトネリコ
167970 ヤチダモ
167982 シマトネリコ,タイワンシホヂ,タイワントネリコ
167984 シマトネリコ
167994 シマタゴ,タイワンタゴ
167999 サトトネリコ,タゴノキ,トスベリ,トヌリキ,トネリコ
168002 ナガミノトネリコ
168003 シオジノキ,デワノトネリコ
168013 ビロ-ドアオダモ
168014 アオダモ,アラゲアヲダゴ,コバノトネリコ
168018 コバノトネリコ,ヤマトアオダモ,ヤマトアヲダモ
168019 ツクシトネリコ
168023 オオバトネリコ,オクエゾヤチダモ,オホバトネリコ,タモノキ,ナンブタモ,ヤチダモ
168025 ヤチダモ
168041 マンナノキ
168065 ソウマシオジ
168068 シオジ,シホジ,シホジノキ,シヲジ,デハトネリコ
168076 ツクシトネリコ,トネリコ
168086 オオトネリコ,チョウセントネリコ,テウセントネリコ
168100 アオダモ,コバノトネリコ,ホソバアヲダモ,マルバアオダモ,ミルバアオダモ,ミルバアヲダモ
168106 シオジ
168109 ヤマシオジ
168153 アサギズイセン属,フリ-ジア属
168178 アヤメズイセン
168183 アサギズイセン,フリ-ジア
168189 アヤメズイセン
168215 フレモンティアデンドロン属,フレモンティア属
168221 フレ-レア属
168238 ツルアダン属,フレイシネチヤ属
168243 タコヅル
168245 ツルアダン
168251 ヒメツルアダン
168328 フリシア属
168332 バイモ属
168361 クロユリ
168430 ヨウラクユリ
168434 コバイモ,テンガイユリ
168467 イチリンバイモ
168586 アミガサユリ,バイモ,ハツユリ,ハハクリ,ハルユリ
168612 チョウセンバイモ
168691 ハマデラソウ
168735 フクシア属,ホクシャ属
168748 ショウジョウフクシア
168750 タイリンフクシア,ツリウキソウ,フクシア,ホクシャ
168767 ツリウキサウ,ツリウキソウ,フクシャ,ホクシャ
168774 タマメツリウキソウ
168780 ツッバナホクシャ
168819 クロタマガヤツリ属

168831 クロタマガヤツリ
168903 ヒロハノクロタマガヤツリ,ヤヘヤマススキ
168964 カラクサケマン属
169011 セイヨウムラサキケマン
169111 セイヨウエンゴサク
169114 セイヨウムラサキケマン
169126 カラクサケマン
169194 ケマンソウ科
169242 マンネンラン属
169245 オオマンネンラン
169375 キバナノアマナ属
169446 ヒメアマナ
169452 ヒメアマナ
169460 キバナノアマナ,コウライアマナ
169470 エゾヒメアマナ
169548 クロガヤ属
169550 ムニンクロガヤ
169561 クロガヤ
169568 テンニンギク属
169575 オオテンニンギク,オホテンニンギク
169589 ホソテンニンギク
169603 テンニンギク
169608 テンニンギク
169645 ハギカズラ属,ハギカヅラ属
169650 ウスバノハギカズラ,ウスバノハギカヅラ
169657 ハギカズラ,ハギカヅラ
169658 ヤエヤマハギカズラ
169660 コバナハギカヅラ
169705 ガランサス属,マツユキソウ属
169710 オオユキノハナ
169719 スノ-ドロップ,マツユキソウ,ユキノハナ
169773 ダウリアヨナメ
169865 ガレアンドラ属
169887 カモメラン
169906 カモメラン
169919 ガレ-ガ属
169953 ガレニア属
170031 ツチアケビ属
170044 オホタカツルラン
170056 チシマオドリコソウ属,チシマオドリコ属,チシマヲドリコ属
170060 イタチジソ,キバナノクルマバナ,チシマオドリコ,チシマオドリコソウ
170078 タヌキジソ,チシマオドリコソウ
170137 ハキダメギク属
170142 コゴメギク
170146 ハキダメギク
170175 ヤエムグラ属,ヤヘムグラ属
170193 トゲナシヤエムグラ,ヤエムグラ,ヤヘムグラ
170199 ヤエムグラ
170205 トゲナシヤエムグラ
170246 エゾキヌタソウ,シベリアキヌタソウ,ホソバノキヌタサウ
170259 エゾキヌタサウ,エゾキヌタソウ
170289 コバノヨツバムグラ,トウヨツバムグラ,ヒメヨツバムグラ
170294 ヤマムグラ
170340 エゾムグラ
170342 ハナムグラ,ハムグラ
170350 コメツブヤエムグラ
170352 ゴエフムグラ,シマヨツバムグラ,タイワンヨツバムグラ
170407 リュウキュウヨツバムグラ
170430 ホソバノクルマムグラ
170434 エゾノヨツバムグラ
170436 ケナシエゾノヨツバムグラ
170437 オオバノヨツバムグラ
170438 ヤクシマムグラ
170444 キクムグラ
170445 キヌタサウ,キヌタソウ,クルマムグラ
170446 オトギリキヌタソウ
170447 バライロキヌタソウ
170448 アオキヌタソウ
170461 イトムグラ
170474 マンシウムグラ
170477 グアマバモドキ
170490 トゲナシムグラ
170503 モリムグラ
170511 ミヤマキヌタソウ
170517 ヤフムグラ
170524 クルマバサウ,クルマバソウ
170526 ヒメヤエムグラ
170534 ミヤマムグラ
170535 ミヤマムグラ
170545 ヒメヤエムグラ
170558 キヌタモドキ
170560 ケナシヤマムグラ
170563 オオヤマムグラ
170564 ヤクシマヤマムグラ
170570 オオバノヤエムグラ,オホバノヤヘムグラ
170571 ビンゴムグラ
170574 エゾムグラ
170589 ニヒタカムグラ
170603 カワラマツバ
170627 ウスユキムグラ
170661 タロコヨツバムグラ,モリムグラモドキ
170678 ヨツバムグラ
170686 ケナシヨツバムグラ
170694 ミナトムグラ
170696 イトバヨツバムグラ,ホソバノヨツバムグラ
170697 ホソバノヨツバムグラ
170705 オククルマバムグラ,オククルマムグラ,クルマムグラ
170708 ヤツガタケムグラ
170743 カハラマツバ,カワラマツバ,キバナカワラマツバ
170746 シナノカワラマツバ
170747 チョウセンカワラマツバ
170751 カワラマツバ,キバナカワラマツバ
170757 カハラマツバ
170761 カワラマツバ
170763 エゾノケカワラマツバ
170764 エゾカワラマツバ
170827 ガルトニア属,ツソガネオモト属
170828 ツリガネオモト
170902 イモノキ,タカノツメ
170923 ウラジロチチコグサ
171046 フクギ属,マンゴスチン属
171092 オオバノマンゴスチン
171115 ハママンゴスチン
171136 マンゴスタン,マンゴスタンノキ,マンゴズチウ,マンゴスチン
171144 ガンボウジノキ,シオウフクギ,シワウ
171145 タイワンフクギ,マサキボク
171197 フクギ
171201 フクギ
171221 タマゴノキ
171238 クチナシ属
171246 タイワンクチナシ
171253 ォガサハラクチナシ,カラクチナシ,クチナシ,ケンサキ,コクチナシ,コリンクチナシ,テウセンクチナシ,ハナクチナシ,ヒトエノコクチナシ,ヒトヘノコクチナシ,フイリケンサキ,フクリンクチナシ,フクリンサンヒチ,ホソバクチナシ,マルバクチナシ,ヤエクチナシ,ヤヘクチナシ
171266 オガサワラクチナシ
171307 フクリンクチナシ
171308 ヤエクチナシ
171310 マルバクチナシ
171312 フイリクチナシ
171313 マルミノクチナシ
171333 クチナシ,センブク
171339 ヤエクチナシ,ヤヘクチナシ
171374 カラクチナシ,コクチナシ,チョウセンクチナシ,テウセンクチナシ,ハナクチナシ,ヤエクチナシ,ヤヘクチナシ
171378 フイリケンサキ
171439 ホウライカズラ属,ホウライカヅラ属
171449 エイシュウカズラ
171455 タイワンチトセガヅラ,チトセカズラ,チトセカヅラ
171456 セイリュウカズラ,チラヨウカズラ,ニシキラン,ホウライカズラ,ホウライカヅラ,ホオライカズラ,リュウキュウホウライカズラ
171487 アオシバ
171607 ガステリア属
171622 ガギュウ

171624 シュンオウテン
171654 セイリュウトウ
171673 トラノマキ
171702 スミホコ
171738 ハクコウリュウ
171764 ギュウゼッ
171767 ハクセイリュウ
171797 ガストルキス属
171825 カシノキラン属,ガストロキ-ルア属
171836 ムツゲカヤラン
171864 カシノキラン
171869 ベニカヤラン,マツラン
171870 マツラン
171897 モミラン
171902 グルグルヤシ
171905 オニノヤガラ属
171911 ムニンヤツシロラン
171912 フサザキムニンヤツシロラン
171913 アキザキヤツシロラン,ヤツシロラン
171918 オニノヤガラ,テンマ,ヌスビトノアシ
171925 シロテンマ
171928 アオテンマ,アヲテンマ
171930 ヒメテンマ
171937 シロテンマ,ナガイモヤガラ,ナヨチンム,ナヨテンマ,ヒメテンマ
171939 コンジキヤガラ
171940 ヒスイヤガラ
171944 ハルザキヤツシロラン
171949 クロヤツシロラン
171952 ナンゴクヤツシロラン
171958 アキザキヤツシロラン
172047 シラタマノキ属
172052 アカダマノギ,アカモノ,イワハゼ,イワバゼ
172053 シロイワハゼ
172054 ベニバナイワハゼ
172059 ニイタカシラタマ,ニヒタカシラタマ
172092 ハリガネカズラ,ハリカネカヅラ
172100 シマシラタマ,タイワンシラタマ
172111 シラタマノキ,シロモノ
172135 ヒメコウジ
172142 シラタマノキ
172185 ガウ-ラ属,ヤマモモサウ属,ヤマモモソウ属
172201 セイヤウフウテウサウ,セイヤウフウテウソウ,ハクチョウソウ,ヤマモモサウ,ヤマモモソウ
172203 イヌヤマモモサウ,イヌヤマモモソウ,バナヤマモモソウ
172224 オヤマヤシ属
172225 カヤバオヤマヤシ
172227 トノサマオヤマヤシ
172248 ゲイル-サキア属
172276 ガザニア属,クンショウギク属
172335 ハネバクンショウギク
172348 ジャノメクンショウギク
172357 クンショウギク
172741 タイワンヤダケ
172754 オホバツゲ属
172778 ゲルセミウム属
172781 カロライナ,カロライナジャスミン,ジャスミン
172832 キダチキツネノマゴ,ケンダルサウ
172881 オガサハラモクレイシ属,オガサワラモクレイシ属,ヲガサハラモクレイシ属
172885 オガサハラモクレイシ,オガサワラモクレイシ,ヲガサハラモクレイシ
172888 ゲンパ属
172898 ゲニスタ属,ヒトッバエニシダ属
173082 ヒトッバエニシダ
173181 リンダウ属,リンドウ属
173182 イセリンドウ
173184 チャボリンドウ
173198 トウヤオリンドウ,トウヤクリンドウ,トヤクリンドオ
173199 クモイリンドウ
173242 ヒナリンドウ,ヒナリンドオ
173260 ミヤマコケリンダウ,ミヤマコケリンドウ
173379 サクマリンダウ,サクマリンドウ,ネバリリンダウ,ネバリリンドウ
173387 ハイリンドウ
173451 アサギリンダウ,ニヒモトリンダウ
173455 サクマリンダウ,サクマリンドウ,ナカハラリンダウ,ナカハラリンドウ,ホウライリンダウ,ホウライリンドウ
173478 ヨコヤマリンドウ,ヨコヤマリンドオ
173540 クモマリンドウ,クモマリンドオ,リシリリンドウ
173541 シロバナリシリリンドウ
173573 コヒナリンドウ,コヒナリンドオ
173598 オクヤマリンダウ,オクヤマリンドウ
173610 ゲンチアナ
173615 オオバリンドウ
173622 シロバナオヤマリンドウ
173623 ホソバオヤマリンドウ
173624 オヤマリンドウ
173625 マンシウリンドウ
173662 ミヤマリンドウ,ミヤマリンドオ
173663 シロバナミヤマリンドウ
173665 イイデリンドウ
173751 ヤチリンドウ
173844 リュウキュウコケリンドウ
173847 チョウセンリンドウ,トウリンドウ,トオリンドオ,リンダウ,リンドウ
173852 エヤミグサ,オコリンダウ,オコリンドウ,ササリンダウ,ササリンドウ,タツノイグサ,ホソバリンドウ,リンダウ,リンドウ
173853 シロバナリンドウ
173854 アケボノリンドウ
173856 キリシマリンドウ
173857 クマガワリンドウ
173858 ホソバリンドウ
173861 ツクシリンダウ
173863 キタダケリンドウ
173868 タイワンコケリンダウ,タイワンコケリンドウ,ニヒタカリンダウ,ニヒタカリンドウ,ニヒモトリンダウ,ニヒモトリンドウ
173886 ナシリンダウ,ナツリンドウ
173896 アサマリンドウ,アサマリンドオ
173897 シロバナアサマリンドウ
173917 コケリンダウ,コケリンドウ,コケリンドオ
173955 ミヤコケリンドウ
173967 タイトウリンダウ,タイトウリンドウ,ホソリンダウ,ホソリンドウ
173974 ハルリンダウ,ハルリンドウ,ハルリンドオ
173975 シロバナハルリンドウ
173978 コミヤマリンダウ,コミヤマリンドウ,タテヤマリンダウ,タテヤマリンドウ,タテヤマリンドオ
173979 シロバナタテヤマリンドウ
174004 エゾリンドウ,ホソバエゾリンダウ,ホソバエゾリンドウ
174007 エゾリンドウ,エゾリンドオ
174008 シロバナエゾリンドウ
174009 ハマエゾリンドウ
174010 ホロムイリンドウ,ホロムイリンドオ
174011 エゾオヤマリンドウ,エゾオヤマリンドオ
174012 タマザキエゾリンドウ,タマザキエゾリンドオ
174029 ナガバオオリンドウ
174079 ヤクシマコケリンドウ,ヤクシマコケリンドオ
174080 ヤクシマリンダウ,ヤクシマリンドウ,ヤクシマリンドオ
174095 フデ,フデリンドウ,フデリンドオ
174096 シロバナフデリンドウ
174097 トキイロフデリンドウ
174099 ミドリヤクシマコケリンドウ
174100 リンダウ科,リンドウ科
174101 オノエリンドウ属,チシマリンドウ属
174108 オノエリンドウ
174109 シロバナオノエリンドウ
174110 エゾオノエリンドオ,ユウバリリンドウ,ユウパリリンドオ
174111 シロバナユウバリリンドウ
174118 チシマリンドウ

174119 シロバナチシマリンドウ
174156 ヤチリンドウ
174204 シロウマリンドウ属,タカネリンドウ属
174205 ヒゲリンドウ
174212 チチブリンドウ,チチブリンドオ,ヒロハヒゲリンドウ
174234 シロウマリンドウ,タカネリンドオ
174235 ムラサキシロウマリンドウ
174236 アカイシリンドウ,アカイシリンドオ
174237 シロバナアカイシリンドウ
174299 ゲオドルム属
174305 トサカメオトラン
174341 ウスバヒメヤツ属
174355 アフヒモドキ属
174431 フウロサウ科,フウロソウ科
174440 フウロサウ属,フウロソウ属
174524 アメリカフウロソウ
174559 ダウリアフウロ
174569 オトメフウロ
174583 オホフウロ,チシマフウロ
174584 シロバナノチシマフウロ
174585 トカチフウロ
174586 レブンフウロ
174587 ホソバノチシマフウロ
174589 ゲンナイフウロ
174592 ウラシロフウロ
174593 ハナフウロ
174595 グンナイフウロ
174645 ニイタカフウロ,ニヒタカフウロ
174671 タチフウロ
174682 チョウセンフウロ
174684 ゲイナイフウロ,タチフウロ,ミツバフウロ
174685 フシゲタチフウロ
174729 カントウフウロ,モモイロフウロ
174739 ヤワゲフウロ
174755 ゲンノショウコ,タチマチグサ,ツルウメサウ,フウロサウ,ミコシグサ
174779 グンナイフウロ,タカネグンナイフウロ
174780 シロバナグンナイフウロ
174781 エゾグンナイフウロ
174832 ノハラフウロ
174854 チゴフウロ
174856 ピレネ-フウロ
174877 イョフウロ,シオヤキソウ,シオヤキフウロ,ヒメフウロ
174900 イヨフウロ
174901 カイフウロ
174902 シロバナカイフウロ
174903 ヤエザキカイフウロ
174904 ヤマトフウロ
174905 ヤクシマフウロ
174906 イチゲフウロ,ヒトリフウロ
174919 アサマフウロ
174920 アサマフウロ
174921 ツクシフウロ
174966 ゲンノショウコ,フウロソウ
174967 シロバナゲンノショウコ
174968 ヤエザキゲンノショウコ
174978 コフウロ
174979 ケナシコフウロ
174980 ホコガタフウロ
175000 ビロ-ドフウロ
175006 コフウロ,フシダカフウロ,ミツバフウロ
175007 ブコウミツバフウロ
175015 エゾミツバフウロ
175018 ビロウドフウロ
175023 イブキフウロ,エゾフウロ,フウロソウ
175024 シロバナエゾフウロ
175027 ヒダフウロ
175028 イブキフウロ
175029 ハクサンフウロ
175030 シロバナハクサンフウロ
175031 ハマフウロ
175032 オガフウロ
175033 シロバナハマフウロ
175034 キビフウロ,ビッチュウフウロ,ヒッチュウフウロサウ
175107 ガ-ベラ属,センボンヤリ属
175129 アカバナセンボンヤリ
175172 アフリカセンボンヤリ,オオセンボンヤリ,ハナグルマ
175295 ゲスネリア属
175306 アワタバコ科,イハタバコ科
175375 ダイコンサウ属,ダイコンソウ属
175378 オオダイコンソウ,オホダイコンサウ
175382 ヤエオオダイコンソウ
175390 ミヤマダイコンソウ
175392 ミヤマダイコンソウ
175396 チリ-ダイコンソウ
175400 ベニバナダイコンソウ
175410 カラフトダイコンソウ
175417 ダイコンサウ,ダイコンソウ
175418 コダイコンソウ
175419 ヤエザキダイコンソウ
175426 カラフトダイコンソウ,チシマダイコンサウ,チシマダイコンソウ
175431 カラフトダイコンソウ
175433 イワダインソウ
175489 ブライダルベ-ル
175494 ギッバエウム属,ギッベウム属
175496 ハクマ
175504 ムヒギョク
175506 ヘキギョク
175508 アオジュズダマ
175509 アオザメ
175512 ナミマクラ
175524 スイレイ
175526 モレイギョク
175527 ハツザメ
175530 シュンキンギョク
175531 スイテキギョク
175533 タチサメ
175539 オオザメ
175668 ギリア属
175669 ホソバギリア
175671 ホソベンギリア
175672 ラッパギリア
175677 タマザキギリア,タマザキヒメハナシノブ
175685 タイリンギリア
175691 アママツバ
175692 マツバギリア
175694 ツツナガギリア
175704 アカバナギリア
175707 イガギリア,イガサクラ
175708 アメリカハナシノブ,サンシキギリア,ヒメハナシノブ
175760 ミツバシモツケソウ属
175766 ミツバシモツケソウ
175812 イチャウ属,イチョウ属,イテフ属
175813 イチウ,イチャウ,イチョウ,イテフ,ギンナン
175824 シダレイチョウ
175827 フイリイチョウ,フイリイテフ
175834 キレハイチョウ
175839 アチョウ科,イチャウ科
175865 オニイラクサ属,セイバンイラクサ属
175877 セイバンイラクサ
175990 アシダンテラ属,グラジオラス属,タウシャウブ属,トウショウブ属
176121 グラジオラス
176151 キバナトウショウブ
176222 グラジオラス,タウシャウブ,トウショウブ,ナ-ガルブル-ム
176260 グラジオラス
176697 グランジュリカクタス属
176722 シラネアオイ属
176723 シラネアオイ
176724 ツノゲシ属
176728 ベニバナツノゲシ
176738 ツノゲシ
176772 ウミミドリ属,シホマツバ属
176775 ウミミドリ,シオマツバ,シホマツバ,シエハコベ,ヒメハマボッス
176776 ウミミドリ,シオマツバ,シホマツバ
176812 カキドオシ属
176821 カキドオシ
176824 カキドホシ,セイヨウカキドオシ
176828 フイリカキドオシ
176829 シロバナカキドオシ
176839 コウライカキドオシ
176860 サイカチ属

176870 シマサイカチ,タイワンサイカチ
176881 カワラフジノキ,サイカチ,テウセンサイカチ
176882 トゲナシサイカチ
176885 チョウセンサイカチ
176886 ヒメサイカチ
176900 タイワンサイカチ
176901 サウキヤウ,トウサイカチ
176903 アメリカサイカチ
176921 ハマボウフウ属
176923 イセボウフウ,ハマボウフウ,ヤオヤボウフウ
176955 ホソバモンパミミナグサ
176985 グロッバ属
177022 グロブラ-リア属
177054 グロブラリア科
177096 カンコノキ属
177097 ウラジロカンコノキ,コバノカンコノキ
177100 ケカンコノキ
177114 ハヤタカンコノキ
177123 アカゲカンコノキ
177141 アカカンコ,オホバノケカンコノキ
177148 クスクスカンコノキ
177149 キイルンカンコノキ,キ-ルンカンコノキ,コウトウカンコノキ
177150 キ-ルンカンコノキ,コウトウカンコノキ
177153 リウキウカンコノキ
177162 カンコノキ
177170 オホミノカンコノキ,タカサゴカンコノキ,ツシマカンコノキ,ヒラミカンコノキ,マルミカンコノキ
177174 ヒラミカンコノキ
177178 ナガバカンコノキ
177181 スイシャカンコノキ,スヰシャカンコノキ
177184 ウラジロカンコノキ,コバノカンコノキ,タカサゴケカンコノキ
177192 カキバカンコノキ,ホンコンカンコノキ
177194 オオバケカンコノキ,ケカンコノキ
177230 キツネユリ属,グロリオ-サ属,ユリグルマ属
177251 クロリオサ・スーパーパ,クロリオサリリ,ユリグルマ
177285 セリバノセンダングサ
177434 グロッチフィルム属,グロッティフィルム属
177449 カミホコ
177465 ホウロク
177475 サオトメ
177515 グロキシニア属
177556 ドジョウツナギ属,ドゼウツナギ属
177557 マンゴクドジョウツナギ
177559 アメリカミノゴメ,ムツオレグサ
177560 タムギ,ミノゴメ,ムツオレグサ,ムツヲレグサ
177562 ミヤマドジョウツナギ
177574 チシマイチゴツナギ,ミヤマイチゴツナギ,ミヤマドゼウツナギ
177586 ヒメウキガヤ
177587 ウキガヤ
177612 ドジョウツナギ
177618 ヒロハノドジョウツナギ,ヒロハノドゼウツナギ
177621 ウキガヤ
177626 カラフトドジャウツナギ,カラフトドジョウツナギ
177629 ヒロハノドゼウツナギ
177639 セイヨウウキガヤ
177648 ヒロハウキガヤ
177661 ヌマドジョウツナギ
177668 オホドゼウツナギ,ドジャウツナギ,ドゼウツナギ
177672 クロバナドジョウツナギ
177673 ヒロハドジョウツナギ
177679 ハイドジョウツナギ
177689 ダイズ属,ダイヅ属
177714 ナガミツルマメ
177722 ヒロハツルマメ
177739 ミヤコジマツルマメ
177750 オオマメ,ダイズ,ダイヅ,ミソマメ
177751 ホソバツルマメ
177777 オホマメ,ダイヅ,ツルマメ,ノマメ,マメ,ミソマメ
177789 タバノマメ,ナガバャブマメ,バウコャブマメ,ボウコツルマメ
177792 コバノタバノマメ,ヒロハヤブマメ
177817 ハナシンボウギ属
177822 ハナシンボウギ
177833 コミノキンカン
177845 ハナシンボウギ
177846 ハナシンボウギ
177873 カンザウ属,カンゾウ属
177885 カンザウ,カンゾウ
177893 アマキ,アラクサ,カンザウ,カンゾウ
177897 アマキ,アマクサ,カンゾウ
177932 イヌカンゾウ,ノカンザウ
177947 ウラルカンゾウ
178014 イヌスギ属,スイショウ属
178019 イヌスキ,スイショウ,スヰシャウ,スヰショウ
178024 キバナエウラク属,キバナヨウラク属,グメリ-ナ属
178039 キバナエウラク,キバナヨウラク
178056 ハハコグサ属,ホオコグサ属
178061 タイワンハハコグサ
178062 ハハコグサ
178101 ホソバノチチコグサモドキ
178218 アキノハウコグサ,アキノハハコグサ,アキノホオコグサ
178220 アキノハハコグサモドキ
178232 ニイタカチチコグサ
178237 チチコグサ
178265 シロバナハハコグサ,セイタカハハコグサ
178297 ゴギョウ,ハハコクサ,ホウコクサ
178342 チチコグサモドキ
178367 ウスベニチチコグサ
178430 ウラジロチチコグサ
178465 エダウチチチコグサ
178481 エゾノハハコグサ,チチコグサ,ヒメチチコグサ
178522 グネツム科
178524 グネツム属
178535 グネツム,グネモソ,グネモンノキ
178541 インドグネツム
178564 ウスバノゲネツム
178851 タイワンヤマアイ,タイワンヤマアヰ,ヤマアイモドキ,ヤマアヰモドキ
178893 ゴメサ属
178974 フウセントウワタ属
179032 フウセンダマノキ,フウセントウワタ
179096 フウセントウワタ
179221 センニチコウ属,センニチサウ属,センニチソウ属
179227 センニチノゲイトウ
179236 センニチコウ,センニチサウ,センニチソウ
179238 アメリカセンニチコウ
179239 アメリカセンニチコウ,アメリカセンニチサウ,アメリカセンニチソウ
179288 ゴンゴ-ラ属
179380 トゲハマサジ
179384 ヨレハナビ
179405 キダチアウソウクワ
179438 アリノタフグサ,アリノトウグサ,ヨノミトリグサ
179441 クラルガキ属
179485 ツルマオ属,ツルマヲ属
179488 ツルマオ,ツルマヲ
179495 オトギリマオ
179500 オトギリマオ
179563 クサトベラ科
179571 シュスラン属
179577 アリサンウヅラ
179580 ベニシュスラン
179582 ナガバウヅラ
179585 ムニンシュスラン
179610 タカネシュスラン,ツユクサシュスラン
179612 シロバナアケボノシュスラン
179613 アケボノシュスラン
179620 タカサゴキンギンサウ,ヤブミョウガラン
179623 ナンバンキンギンザウ,ナンバンキンギンソウ,ヒゲナガキンギンザウ
179625 ハチジョウシュスラン,ハチヂャウ

シュスラン
179626 オオシマシュスラン
179627 シライトシュスラン
179630 ヤクシマシュスラン
179631 アケボノシュスラン
179644 ベニシュスラン
179650 カゴメラン,フイリウヅラ,リウキウシュスラン
179653 アケボノシュスラン
179672 ツリシュスラン
179673 ヒロハツリシュスラン
179679 キンギンサウ,キンギンソウ
179685 ヒメミヤマウズラ,ヒメミヤマウヅラ,ホウライウヅラ
179694 カモメラン,トヨシマラン,フナシミヤマウズラ,ミヤマウズラ,ミヤマウヅラ
179708 シュスラン,ビロウドラン,ビロ-ドラン,ミヤマシュスラン
179709 オホシュスラン,シマシュスラン,ナガバナウジラ
179729 ゴルドニア属,タイワンツバキ属
179745 タイワンツバキ,ナンバンツバキ
179865 ワタ属
179876 キダチワタ
179878 キワタ,ナンキンワタ,ワタ
179879 アカバナワタ
179884 カイトウメン,ベニバナワタ
179890 アメリカワタ
179900 シロバナワタ,ワタ
179906 キヌワタ
180078 イネ科
180090 クラマンギス属
180116 グラマトフィルム属
180126 ホウオウラン
180163 タカサゴハナヒリグサ属
180177 タカサゴハナヒリグサ
180265 グラプトペタルム属
180269 オボロヅキ,ハッシモ
180270 グラプトフィルム属
180292 オオアブノメ属,オホアブノメ属
180304 カミガモソウ
180308 オオアブノメ,オホアブノメ,ミヅサハタウカラシ
180504 グリ-ノビア属
180535 グレイギア属
180561 グレビレア属,シノブノキ属
180642 キヌガシハ,キヌガシワ,サラマンダ,シノブノキ,ハゴロモノキ
180655 ホソバハゴロモノキ
180665 ウオトリギ属,ウヲトリギ属
180682 インドウオトリギ
180700 アツバウオトリギ
180703 ウオトリギ,ウオノキ,ウヲトリキ,ウヲノキ,エノキウツギ
180925 ヒメウオトリキ,ヒメウヲトリキ
180939 ヒシバウオトリギ
180995 オホバウオトリギ,オホバウヲトリギ
181092 グリンデ-リア属
181174 ネバリオグルマ
181260 グリセリ-ニア属
181282 グロビア属
181295 キリハヤシ属
181296 コップキリハヤシ
181339 スグリ科
181450 グルソニア属
181452 ハクホウ
181496 グアイクウット属,ユソウボク属
181505 グアイクウット,ユソウボク,リグナムバイタ
181687 ホソバヒナゲンゲ
181693 ヒナゲンゲ
181695 イヌゲンゲ
181728 ハテルマギリ属
181743 シマハビロ,ハテルマギリ
181821 モモミヤシ属
181873 キバナタカサブロウ,キバナタカサブロオ
181915 マルミケンチャ属
181917 バラオキリハヤシ
181946 グンネ-ラ属,コウモリガサソウ属
181951 オニブキ
181958 コウモリガサソウ
181960 グンネラ科
182149 グズマニア属
182213 チドリサウ属,チドリソウ属,テガタチドリ属
182225 ノヒネチドリ
182230 チドリサウ,チドリソウ,テガタチドリ
182254 ノビネチドリ
182361 ホウライアオカヅラ属,ホウライアヲカヅラ属
182384 シルベストルアオカズラ,ホウライアオカズラ,ホオライアオカズラ
182406 ギムノカクタス属
182410 ハクロゥギョク
182414 ビシンギョク
182419 ギムノカリキウム属
182421 ギンセイギョウ
182422 オウダマル
182423 スイコウカン
182424 ホウガシラ
182427 コクチョウギョク
182428 リュウラン
182429 キタンマル
182430 ラセイマル
182433 カセイマル
182434 カラコマル
182435 トラカンギョク
182436 コウリンギョク
182437 ケンマギョク
182438 キュウテンマル
182439 レイダマル
182440 ゲッカンマル
182441 テンオウマル
182443 ジャモンギョク
182444 クモンリュウ
182445 キョリンギョク
182446 キョウリュウマル
182447 ヘキガンギョク
182448 セッカンギョク
182449 メイホウギョク
182451 セックンマル
182452 チリュウマル
182453 レイフギョク
182454 アヤッヅミ
182455 マテンリュウ
182456 ホウランギョク
182458 ジャハンリュウ
182459 ズィランマル
182462 ヒボタン
182463 ボタンギョク
182464 ランリュウ
182465 コウダマル
182466 タカダマ
182467 モウシュウギョク
182468 クンソウギョク
182469 ジュンヒギョク
182470 クンカンギョク
182471 テンシギョク
182472 コウリュウマル
182473 タツガシラ
182476 シンテンチ
182477 ハコウリュウ
182478 セイカンギョク
182479 テンペイマル
182480 シュデンギョク
182481 マンジュギョク
182482 オウカンギョク
182484 ジュウジグン
182485 ヘキバンギョク
182487 セイシマル
182488 カッカザン
182489 ハコウリョウ
182511 キムノセレウス属
182516 ビスイチュウ
182526 ヒサウキヤウ
182537 ミズヒマワリ
182574 アンナンカラスウミ属
182575 アンナンカラスウミ
182634 イヌエンゴサク,ヒメルイヨウボタン
182683 グンバイウメヅル,トゲマサキ,ハリウメヅル,ハリツルマサキ
182687 コウトウマサキ
182836 デブランチェアモクマオ
182915 フウチョウソウ
182961 タイフウシ属
182962 タイフウシ

182998 アマチャヅル属
183023 アマクサ,アマチャズル,アマチャヅル,ツルアマチャ
183025 ソナレアマチャヅル
183051 サンシチサウ属,サンシチソウ属,スイゼンジナ属
183056 ビロ-ドサンシチ
183066 スイゼンジソウ,スイゼンジナ,スヰゼンジサウ,スヰゼンジナ,ハルタマ
183084 タカサゴサンシチソウ
183086 オホバノスヰゼンジナ,コウトウスイゼンジナ,コウトウスヰゼンジナ,シマスヰゼンジナ
183097 オランダグサ,サンシチ,サンシチソウ,サンシンサウ,チドメ,フサナリサンシチ
183134 ツリダチスイゼンジナ,ツルビロ-ドサンシチ
183157 イトナデシコ属,カスミソウ属,ジプソフィラ属
183183 オノエマンテマ
183185 ノハラコゴメナデシコ
183191 カスミソウ,ハナイトナデシコ,フタマタコゴメナデシコ,ムレナデシコ
183192 アカバナカスミソウ
183221 ヌカイトナデシコ
183222 イワコゴメナデシコ,コゴメナデシコ
183223 イトナデシコ
183225 コゴメナデシコ,シュッコンカスミソウ
183231 イトナデシコ
183366 ハ-ゲオセレウス属
183367 キンコウチュウ
183372 ケゴン,サイコウチュウ
183376 キンエンチュウ
183377 キンギンカク,キンヨゥチュウ
183378 モウキンシュウ
183379 タイオウカヶ,チャバシラ
183380 サイカカク
183389 サギサウ属,サギソウ属,ミズトンボ属
183503 ネバリサギサウ
183559 シマサギサウ,セイタカサギサウ,ダイサギサウ,ダイサギソウ,トウホサギサウ,ニヒタカトンボ
183625 ムカゴトンボ
183629 ヒゲナガトンボ
183736 イヨトンボ
183765 タコガタサギソウ
183797 アオサギソウ,オオミズトンボ,オホミヅトンボ,サハトンボ,サワトンボ,ミズトンボ
183798 チョウセンミズトンボ
183799 ヒメミズトンボ
183921 アオサギサウ,アオトンボ,アヲサギサウ,アヲトンボ,ミヅトンボ
183937 ナメラサギサウ,ナメラサギソウ
183983 イトヒキサギサウ,イトヒキサギソウ,ヒゲサギサウ
184085 ナガバサギサウ,ナガバサギソウ
184222 ハベルレア属
184249 ハブランサス属
184298 イワムラサキ,オカムラサキ
184323 ヤエガヤ
184345 ハエマンサス属,ホテイラン属,マユハケオモト属
184347 マユハケオモト
184372 ホテイラン
184428 センコウハナギ
184513 ヘマトキシルム属
184518 ロッグウッド
184524 ハエモドルム科
184581 ハリノホ
184586 ハケア属
184655 ウラハグサ属
184656 ウラハグサ,チフウソウ,フウチソウ
184657 ギンウラハグサ
184658 キンウラハグサ
184659 シラキンウラハグサ
184689 ハナイカリ属
184691 ハナイカリ
184692 ムラサキハナイカリ
184711 オオバコキンポウゲ
184717 カエデキンポウゲ
184726 アメリカアサガラ属,ハレ-シア属
184731 アメリカアサガラ
184776 ハリミウム属
184832 ハリモデンドロン属
184838 シオノキ
184940 マツバウミジグサ
184943 ウミジグサ
184967 ウミヒルモ属
184977 ヒメウミヒルモ
184979 ウミヒルモ,オオウミヒルモ,オホウミヒルモ
184993 アリノタフグサ科,アリノトウグサ科
184995 アリノタフグサ属,アリノトウグサ属
184997 ナガバアリノトウグサ,ホソバアリノトオグサ
185014 タネガシマアリノトオグサ,ナガバアリノトオグサ,ホソバアリノトウグサ
185089 マンサク科
185092 マンサク属
185104 アオモミ,ウメズエ,カタソゲ,マンサク
185108 ウラジロマルバマンサク
185109 ニシキマンサク
185110 アカバナマンサク
185111 シダレマンサク
185114 アテツマンサク
185115 ウラジロマルバマンサク
185116 コガネマンサク
185117 ニシキマンサク
185118 アカバナマンサク,マルバマンサク
185119 ウラジロマンサク
185120 オオバマンサク
185130 シナマンサク
185141 アメリカマンサク
185155 ハマトカクタス属
185157 オオニジ
185158 リュウオウマル
185200 ヤチラン
185226 ハナブサソウ
185256 ヒメクリソラン
185635 クサヘンルウダ
185753 ニオイラン属,ニボヒラン属
185755 オホバカヤラン,ニオイラン,ニホヒラン,ニボヒラン
185758 ハ-デンベルギア属,ヒトツバマメ属
185764 ツルヒトツバ,ヒトツバマメ,ムラサキツルマメ
185912 ハリシア属
185919 ビケイチュウ
186105 ハセルトニア属
186107 ウウンリュウ
186143 ハスティングシア属
186151 ハティオラ属
186251 ハウォルティア属
186422 ジュウニノマキ
186444 チョウドウ
186545 マンゾウ
186546 ヤチヨニシキ
186644 タカノツメ
186655 クミントウ
186728 アヤギヌエマキ
186776 コテング
186792 タマオウキ
186925 ヘ-ベ属
186973 トラノオノキ
187129 ヘヒティア属
187156 ヘッケリア属
187195 キヅタ属
187200 カナリ-キヅタ
187222 キズタ,セイヨウキズタ,セイロウキズタ
187227 フイリセイヨウキヅタ
187318 アイビ,オニヅタ,キズタ,キヅタ,ダイワンキヅタ,フユズタ,フユヅタ
187319 フクリンキズタ,フクリンキヅタ
187321 ナガバキヅタ
187322 ナガボキヅタ
187324 シロバキヅタ
187325 タイワンヤッデ
187417 ガランガ属,シュクシャ属
187425 ニクイロシュクシャ
187427 シュクシャ,ジンジャ,ジンジャ-

187428 ガランガ,ベニバナシュクシャ
187432 シュクシャ,ホザキシャウガ
187451 キバナシュクシャ
187468 サンナ
187497 ニホヒグサ属,フタバムグラ属
187510 ヤエヤマハシカグサ
187514 シマソナレムグラ
187515 ソナレムグラ
187523 フタバムグラ
187526 トキワナッナ,ヒナソウ
187544 コバンムグラ
187549 ソナレムグラ
187555 タマザキフタバムグラ
187565 フタバムグラ
187592 マルバシマザクラ
187602 シマザクラ
187630 ナントウニホヒグサ
187638 アツバシマザクラ
187639 シマソナレムグラ
187674 ソナレムグラ
187676 オオソナレムグラ
187677 ソナレムグラ
187681 ケニオイグサ,ケニホヒグサ,タイワンフタバムグラ
187682 ナガエケニオイグサ
187691 ニオイグサ,ニホヒグサ
187697 ヒロハケニオイグサ
187753 イバワウギ属,イワオウギ属,ヘディサルム属
187769 ムラサキイワオウギ
187842 アカバナオウギ,フランスオウギ
187881 モウコオウギ
188023 チシマゲンゲ
188028 カラフトゲンゲ,チウセンイハワウギ,チウセンイワオウギ,チシマゲンゲ,チョウセンイワオウギ
188099 シラゲオウヂ
188177 イハワウギ,イバワウギ,イワオウギ,タテヤマオウギ,タテヤマワウギ
188178 イワオウギ
188179 ケイワオウギ
188190 マルジクホエア属
188191 カンタベリ-マルジクホエア
188263 キバナノミソハギ属,キバナミソハギ属
188264 キバナノミソハギ,キバナミソハギ
188365 ヘルキア属
188393 ダンゴギク属,ヘレニュ-ム属,マツバハルシャギク属
188402 ダンゴギク
188410 ヤハズギク,ヤハズダンゴ,ヤバネダンゴギク
188437 イトギク,マツバハルシャギク
188583 ハンニチバナ属
188586 タカネハンニチバナ
188590 ハンニチバナ
188846 ハンニチバナ
188902 ヒマハリ属,ヒマワリ属
188908 テンガイバナ,ニチリンサウ,ニチリンソウ,ヒグルマ,ヒフガアフヒ,ヒマハリ,ヒマワリ,ヒュウガアオイ
188909 ヤエヒマワリ
188922 シラヤマブキ,シロタエヒマワリ,シロタヘヒマハリ,ハクマウヒマハリ,ハクモウヒマワリ
188936 コキクイモ,ヒメヒマハリ,ヒメヒマワリ
188941 コキクイモ,ヒメヒマハリ,ヒメヒマワリ
188953 コヒマワリ,ノヒマハリ,ノヒマワリ
188991 ヤナギヒマワリ
189006 コヒマワリ
189052 ヤナギバヒマワリ
189063 イヌキクイモ
189073 キクイモ
189093 ハナカンザシ属,ムギワラギク属
189910 ヤマモガシ属
189918 カマウド,カマノキ,ヤマモガシ
189931 タイワンヤマモガシ,ヤマリウガン
189982 ヘリコディケロス属
189986 ヘリコニア属
190038 ヒメゴクラクチョウカ
190073 オウムバナ科
190092 ヤンバルコマ属
190094 ナツメヤシ,ヤンバルコマ
190129 ミサヤマチャヒキ属
190130 ホウライチャヒキ
190145 ミヤマエンバク
190155 ミサヤマチャヒキ
190191 オロシヤエンバク
190206 コウアンチャヒキ
190237 ヘリオセレウス属
190261 ヘリオフィラ属
190502 キクイモモドキ属,ヒメキクイモ属,ヘリオプシス属
190505 キクイモモドキ,ヒメキクイモ
190540 キダチルリサウ属,キダチルリソウ属,ヘリオトロピュ-ム属
190559 キダチルリサウ,キダチルリソウ,コウスイソウ,コウスイボク,ニオイムラザキ,ヘリオトロ-ブ
190597 アレチムラサキ
190622 ヨウシュキダチルリソウ
190630 オオココメスナビキソウ
190651 ナンバンルリサウ,ナンバンルリソウ
190652 スナビキソウ,ハマムラサキ
190702 キダチルリソウ,コウスイソウ,コウスイボク,ヘリオトロブ
190708 オオココメスナビキソウ
190738 ココメスナビキサウ
190791 ハナカンザシ属
190819 ハナカンザシ
190925 クリスマス・ロ-ズ属,クリスマスロ-ズ属
190936 コダチクリスマス・ロ-ズ
190944 クリスマス・ロ-ズ,クリスマスロ-ズ
191059 シャウジャウバカマ属,ショウジョウバカマ属
191065 シャウジャウバカマ,ショウジョウバカマ,シロバナシャウジャウバカマ
191068 シマシャウジャウバカマ,タイワンシャウジャウバカマ,ヒメシャウジャウバカマ
191139 ハナイカダ属
191157 ヒマラヤハナイカダ
191173 コバノハナイカダ,ツカデノキ,ツキデノキ,ハナイカダ,ママコノキ,ママッコ
191174 ホソバノハナイカダ
191185 リュウキュウハナイカダ
191192 タイワンハナイカダ
191225 ウシノシッペイ属
191228 コバノウシノシッペイ
191234 ウシノシッペイ
191251 ウシノシッベイ,ウシノシッペイ,バリン
191261 キスゲ属,ヘメロカリス属,ワスレグサ属
191263 カンゾウ,クワンザウ,トウカンゾウ,ナンバンカンゾウ,ワスレグサ
191268 チョウセンキスゲ
191272 ヒメカンゾウ,ヒメクワンザウ
191278 アサマクワンザウ,エゾゼンテイカ,センテンカ,ニッコウキスゲ
191284 オニカンゾウ,シナクワンザウ,ホンクワンザウ,ホンワスレグサ,マンシウキスゲ,ヤブカンゾウ,ヤブクワンザウ
191286 ワスレグサ
191289 ノカンゾウ
191291 オニカンゾウ,オニクワンザウ,スジカンゾウ,ヤブカンゾウ,ヤブクワンザウ,ヤヘクワンザウ,ワスレグサ
191296 ヒメカンゾウ
191298 アキノワスレグサ
191301 ハクウンキスゲ
191304 クワンザウ
191309 エゾゼンテイカ,エゾゼンテイクワ,セッテイクワ,ゼンテイクワ,ゾクワンザウ,ニックワウキスゲ,ニッコウキスゲ
191312 キスゲ,ホソバキスゲ,ユフスゲ,ヨシノキスゲ
191317 ニュ-サイラン,マオラン
191318 キスゲ,ユウスゲ
191319 エゾキスゲ
191325 ヒヤハサギゴケ属
191347 ツノギリソウ属

191349　ツノギリソウ
191486　ヒロハサギゴケ属
191501　ヒロハサギゴケ,ミヤコジマソウ
191573　サクマサウ属,サクマソウ属
191574　サクマサウ,サクマソウ
191575　サクマソウ
191582　ニヒタカヒトツバラン属,ムミピリア属
191591　ニイタカヒトツバラン,ニヒタカヒトツバラン
191598　ニイタカヒトツバラン
191626　ハリゲヤキ属
191629　ハリゲヤキ
191665　キツネアザミ属
191666　キツネアザミ
191671　キツネアザミ
191673　シロバナキツネアザミ
192120　スハマソウ属,ユキワリサウ属,ユキワリソウ属
192138　マンセンスハマソウ
192139　ミスミソウ,ユキワリソウ
192140　スハマソウ
192227　ハナウド属
192255　アラゲハナウド
192259　オオハナウド
192278　オオハナウド,クハズウド,サガウド,ゾウジャウジビャクシ,ゾウジョウジビャクシ,ハナウド,ヤブウド
192282　キレハオオハナウド
192283　ベニバナオオハナウド
192284　ホソバハナウド
192312　ハナウド
192320　ホソバノマンシウハナウド
192375　ハナウド
192379　ツルギハナウド
192405　チリ－アヤメ属
192414　チリ－アヤメ
192427　ヘレロア属
192436　シュラハッコウ
192438　コウリュウ
192439　カンリュウ
192442　ホンリュゥ
192445　ユウギリ
192448　センリュウ
192452　ホウリュウ
192454　レイオン
192477　サキシマスハウノキ属
192494　サキシマスオウノキ,サキシマスホウノキ,スオウギ,スハウギ,ハマグルミ
192809　ムカゴサウ属,ムカゴソウ属
192851　アカヌマラン,ナガバノムカゴサウ,ヌガゴサウ,ムカゴサウ,ムカゴソウ
192862　クシロチドリ
192898　クロバナイリス属
192904　クロバナイリス
192909　ハスノハギリ属
192913　ハスノハギリ,ハマギリ
192921　ハスノハギリ科
192965　ココメビユ
193109　ヘッレアンツス属
193117　ヘルレリア属
193119　ヘルレリア科
193134　ヘルシエリア属
193202　ヘルチア属
193232　ヘスペランサ属,ヘスペランタ属
193387　キバナノハタザオ属,キバナノハタザホ属,ハナダイコン属
193417　ハナスズシロ,ハナダイコン
193426　ムレイソウ
193581　ヒメノヤガラ属
193592　シロスジカゲロウラン,ヤクシマアカシュスラン
193617　オオカゲロウラン
193683　アメリカコナギ
193746　メキシコノボタン属
193749　メキシコノボタン,メキシコヒメノボタン
193914　ハマベノギク属
193918　アルタイノギク,ステップノギク,ヤマヂノギク
193923　アルタイノギク
193925　ハマベノギク
193930　イソノギク
193939　アレノノギク,ハマベノギク,ヤマジノギク,ヤマヂノキク
193941　シロバナアレノノギク
193946　ヤナギノギク
193954　ソナレノギク
193973　コウライヤマジノギク
193977　シマノギク,シロブナノヤマヂノギク,タイワンノギク
193989　ヒメウシノシッペイ
194015　アカヒゲガヤ属,ダイワンアカヒゲガヤ属
194020　アカヒゲガヤ,ダイワンアカヒゲガヤ
194044　ウスアカヒゲガヤ
194106　カラスギバサンキライ属,カラスバサンキライ属
194120　イヌサンキライ,カラスギバサンキライ,ナンバンサンキライ,ヒラエサンキライ
194127　シホデモドキ,ホウゴウシホデモドキ
194134　イヌヘラヤシ属
194135　セタカイヌヘラヤシ
194155　ブラオンカヅラ属
194157　ブラオンカヅラ,ホソバキジュラン
194256　アレチオグルマ
194318　カンアオイ属
194325　ツクバネアオイ
194334　ツチグリカンアオイ
194345　ジュロウカンアオイ
194358　サヌキカンアオイ
194369　モモイロカンアオイ
194371　ホシザキカンアオイ
194379　オキノシマカンアオイ
194383　キナンカンアオイ
194411　ツボサンゴ属
194413　アメリカツボサンゴ
194437　サンゴバナ,ツボサンゴ
194451　パラゴムノキ属
194453　パラゴムノキ,ブラジルゴムノキ,ヘベアゴムノキ,ヘベヤゴムノキ
194459　ツリガネヒルガホ属
194467　アサギヒルガネ,キバナヒルガネ,ツリガネヒルガホ
194609　ホシダネヤシモドキ属
194610　メキシコホシダネヤシモドキ
194661　ヒッベルティア属
194680　フヨウ属
194774　カメロンムクゲ
194779　アオイツナソ,アンバリ麻,ガンボ麻,ケナフ,ボンベイ麻
194804　モミジアオイ,モミヂアオイ,モミヂアフヒ
194830　デニソンムクゲ
194895　オガサワラハマボウ,テリハノハマボウ,テリハハマボオ,モンテンボク
194911　ハマボ,ハマボウ,ハマボオ
194912　シロバナハマボウ
194955　ソコベニアオイ
194971　ユリザキムクゲ
194989　サキシマフヨウ
195032　アメリカフヨゥ
195040　フヤウ,フヨウ,モクフヨウ
195044　スイフヨウ
195052　ヤエザキフヨウ
195077　イオウトウフヨウ
195119　ハリアオイ
195149　ブッサウゲ,ブッソウゲ,リウキウムクゲ,リュウキュウムクゲ
195179　ユリザキムクゲ
195196　ロ－ゼリサウ,ロ－ゼリソウ,ロ－ゼル,ロ－ゼルソウ
195216　フウリンブッサウゲ,フウリンブッソウゲ
195257　マンキンアフヒ
195269　キハチス,ハチス,フヨウ,ホコ,ムクゲ,モクゲ
195305　ヤマフヨウ
195311　オオハマボウ,オオハマボオ,オホハマボウ,シマハマボウ,ハブ,ハマイチビ,ハマボウ,ヤマアサ
195326　ギンセンカ,ギンセンクワ,チウロサウ,チョウロソウ
195437　ミヤマコウゾリナ属
195478　コウリンタンポポ,コオリンタンポポ
195548　ヒロハコウゾリナ
195655　マンシュウスイラン,ヤナギスイラ

ン
195672 ミヤマカウゾリナ,ミヤマコウゾリナ,ミヤマコオゾリナ
195696 スイラン,スヰラン
195780 ウズラバタンポポ
195797 タイワンミヤマカウゾリナ
196027 ツイミソウ
196043 ヒメコウゾリナ
196057 アザミタンポポ,キタンポポ,ヤナギタンポポ
196086 イヌコウゾリナ
196116 カウバウ属,コウボウ属
196121 オオミヤマコウボウ
196122 ワシベツミヤマコウボウ
196134 コウボウ
196153 カウバウ,コウボウ
196155 コウボウコウシュ
196159 エゾカウバウ
196161 エゾコウボウ
196163 エゾヤマコウボウ
196439 アマリリス属,ジャガタラズイセン属,ジャガタラズヰセン属,ヒッペアストラム属
196443 インチアアセン,キンサンジュ,ナルシス
196446 アマリリス
196450 キンサンジゴ
196452 ジャガタラズイセン,ジャガタラズヰセン
196454 アミメアマリリス
196456 ヒイロサンジコ
196458 ベニスヂサンジュ
196494 トチノキ科
196616 ヒポクラテア科
196743 ヒッポフェ属
196809 スギナモ科
196810 スギナモ属
196816 ヒロハスギナモ
196818 スギナモ
196821 ホザキサルノオ属,ホザキサルノヲ属
196824 ウスバサルノオ,ホザキサルノオ,ホザキサルノヲ
196835 ウスバサルノオ,ウスバサルノヲ
196952 アレチガラシ
197101 ホフマンニア属
197109 ホフマンセジア属
197143 ホ-エンベルギア属
197253 ホルコグロッサム属,マツノハラン属
197265 マツノハラン
197278 シラゲガヤ属
197301 シラゲガヤ
197306 ニセシラゲガヤ
197361 ホルムスキオ-ルディア属
197491 カギザケハコベ
197633 タカサゴノキ属
197659 タカサゴノキ
197761 ボマロケファラ属
197762 アヤナミ
197769 カンキチク属
197772 カンキチク,カンメイチク
197785 セントンイモ属
197793 セントンイモ
197796 アカジクセントンイモ,ホマロメナ・ルベッセンス
197799 ハルユキソウ
197823 ホメ-リア属
197960 ナンバンヤナギ属
197963 ナンバンヤナギ
198010 ハマハコベ
198035 フ-ディア属
198040 ンポイカク
198044 マハイカク
198053 レイハイカク
198079 フ-ディアプシス属
198080 マセイカク
198176 タキエン
198260 オオムギ属,オホムギ属
198266 チシマムギクサ
198267 チシマムギクサ,ライムギモドキ
198288 ヤバネオオムギ
198293 ハダカエンバク,ハダカムギ
198297 ヒメムギクサ
198311 ホソノゲムギ,ムギクサ
198316 オオムギクサ
198319 ハマムギクサ
198327 ムギクサ
198337 ホソムギクサ
198344 ミナトムギクサ
198354 ライムギモドキ
198376 オオムギ,オホムギ,カチカタ,シジョウオオムギ,ハダカムギ,フトムギ
198391 ハダカムギ
198480 ミヤマカラサナズナ
198491 ホリドハクタス属
198493 アガンギョク
198494 ソウソウギョク
198495 トウヨウマル
198496 コクライギョク
198497 キシュウギョク
198498 コクテンギョク
198499 ロウシン
198500 カイソウギョク
198501 バンレイギョク
198574 クロタキカズラ属,クロタキカヅラ属
198575 クロタキカズラ,クロタキカヅラ
198576 トウクロタキカズラ
198589 ギバウシ属,ギボウシ属
198590 オヒガンギボウシ
198592 コバギバウシ,コバギボウシ
198593 カンザシギボウシ
198594 アキカゼギボウシ
198598 サザナミギバウシ,サザナミギボウシ
198599 オタフクギバウシ,オタフクギボウシ
198600 サクハナギボウシ,ツルギギボウシ
198608 レンゲギボウシ
198612 オオバギボウシ
198615 ヒュウガギボウシ
198618 コバギバウシ,コバギボウシ
198619 ヒメイヮギボウシ,ヒュウガギボウシ
198620 ケヤリギボウシ
198621 トサノギボウシ
198622 ヤクシマギボウシ
198623 キヨスミギボウシ,ハヤサキギボウシ
198624 ミズギボウシ
198628 コバギボウシ
198633 イワギボウシ
198634 ミズギボウシ
198635 ケイリンギボウシ
198637 クロギボウシ
198638 ニシキギボウシ
198639 タマノカンザシ,マルバタマノカンザシ
198643 タチギボウシ
198644 オモトギボウシ
198645 ハチジョウギボウシ
198646 オオバギボウシ,オホギバウシ,タウギバウシ,トウギボウシ
198648 アキギボウシ
198649 ナガサキギボウシ
198650 トクダマ
198651 ギボウシ,スジギボウシ,スヂギバウシ
198652 ムラサキギボウシ
198655 オトメギボウシ
198684 ウレティア属
198698 ホウノキ,ホオガシワノキ,ホオノキ,ホホガシハ,ホホガシハノキ,ホホノキ
198699 コウボク
198703 トキワナズナ属
198708 トキワナズナ,ヒナソゥ
198743 ドクダミ属
198745 ジフヤク,ジュウヤク,ダクダミ,ドクダミ,ドクダメ
198746 フイリドクダミ
198747 ヤエドクダミ
198753 ミドリドクダミ
198766 ケンポナシ属
198769 ケンポナシ
198770 オウゴンケンポナシ
198784 ケケンポナシ
198786 ケケンポナシ
198789 ケケンポナシ
198804 ケンチャヤシ属,ケンチャ属,ホエ

ア属
198806 カレ-ヤシ,ケンチャヤシ,ベルモアホエア
198808 フォ-スタ-ホエア
198821 サクララン属,ホヤ属
198827 サクララン
198831 フイリサクララン
198850 シャムサクララン
198969 フエルニア属
199020 リュウショウカク
199033 シカノアタマ
199049 アシュラ
199053 リュウオウカク
199057 バンビカク
199076 ガカク
199252 ユンベールティア属
199379 カナムグラ属,カラハナサウ属,カラハナソウ属
199382 カナムグラ,カラムグオ,カラムグヲ
199384 カラハナサウ,カラハナソウ,セイヨウカラハナソウ,ホップ
199386 カラハナソウ,ホップ
199392 カナムグラ
199407 ハンネマンニア属
199408 カラクサゲシ
199443 フントレア属
199480 フッチンシア属
199553 ツリガネズイセン,ツリガネズヰセン
199565 ヒアシンス属
199583 ヒアシンス,ヒアシント,ヒヤシンス
199740 ダイフウジノキ属
199744 ダイフウジノキ
199787 アジサイ属,アヂサイ属,アヂサヰ属
199788 アマギコアジサイ
199789 ミズシマアジサイ
199792 キダチノコガク,トカラアジサイ,ヤクシマアジサイ
199800 タイワンツアヂサイ,タイワンツアヂサヰ,タイワンツルアジサイ
199806 アメリカノリノキ
199817 シマイハガラミ,タイワンゴトウヅル
199846 トウアジサイ
199853 カラコンテリギ,シマコンテリギ
199856 ヤエヤマコンテリギ
199877 コバノアマチャ
199897 コアジサイ
199898 シロバナコアジサイ
199899 ミドリコアジサイ
199910 アコウバアヂサイ,アコウバアヂサヰ,クスノハアヂサイ,クスノハアヂサヰ,タイワンアヂサイ,タイワンアヂサヰ,マルバアヂサイ,マルバアヂサヰ,ユヅヒハアヂサイ,ユヅヒハアヂサヰ
199911 キアジサイ,キアヂサヰ,ギョクダンカ,ギョクダンクワ,ギンガサウ,ギンガソウ,サハフサギ,サワフサギ,タマアジサイ,タマアヂサヰ,ヤマタバコ
199912 ギョクダンカ
199914 シロバナタマアジサイ
199915 ヨウラクタマアジサイ
199916 ココノエタマアジサイ
199917 テマリタマアジサイ
199919 ラセイタタマアジサイ
199921 トカラタマアジサイ
199926 トカラアジサイ
199927 ヤクシマアジサイ
199942 リュウキュウコンテリギ
199945 ナガバアジサイ,ナガバアヂサイ,ナガバアヂサヰ
199950 コガクウツギ
199952 ヤクシマガクウツギ
199953 アジサイ
199956 アジサイ,アマチャ,ガク,ガクアジサイ,ガクアヂサヰ,ハマアジサイ
199986 セイヨウアジサイ
199990 フチナシガクアジサイ
199991 アジサイ,テマリアジサイ,テマリカ,テマルバナ
199994 ガクソウ,ベニガク
200014 ガクアジサイ
200019 ナガバコンテリギ
200031 ムラサキアヂサイ,ムラサキアヂサヰ
200038 キネリ,サビタ,トロロノキ,ニベノキ,ノリウツギ,ノリノキ,ヤマウツギ
200051 ヒダカノリウツギ
200053 ビロ-ドノリウツギ
200055 ミナヅキ
200056 エゾノリウッギ,エゾノリノギ
200059 ゴトウズル,ゴトウヅル,ツルアジサイ,ツルアヂサヰ,ツルデマリ
200063 カシワバアジサイ
200077 ガクウヅギ,コンテリギ,ズイボウ,ヤマドウシン
200079 ベニガクウツギ
200085 ヤマアジサイ
200089 ガクバナ,サハアヂサイ,サハアヂサヰ,サワアジサイ,ヤマアジサイ,ヤマアヂサイ,ヤマアヂサヰ
200090 マイコアジサイ
200091 シチダンカ
200092 イハガク
200093 ベニガク
200096 アマギアマチャ
200097 ナンゴクヤマアジサイ
200099 ヒュウガアジサイ
200100 アマチャ,コアマチャ
200101 ニワアジサイ
200102 エゾアジサイ
200105 ヤハズアジサイ
200125 アマチャノキ
200160 アジサイ科
200173 ヒドラスチス属
200180 イズミケンチャ属
200182 ホソバイズミケンチャ
200184 クロモ属
200190 クロモ,コカナダモ
200192 エビモ,クロモ
200201 ウスカハゴロモ属,カワゴロモ属
200202 ウスカワゴロモ
200204 カワゴロモ
200205 オオヨドカワゴロモ
200206 ヤクシマカワゴロモ
200225 トチカガミ属,ヒドロカリス属
200228 カエルノエンザ,スッポンノカガミ,ドウガメバス,トチカガミ,ドチモ
200235 トウカガミ科,トチカガミ科
200241 ヒドロクレイス属
200244 ミズヒナゲシ
200253 チドメグサ属
200261 モミジチドメ
200275 シマチドメグサ,タイワンチドメグサ
200281 ケチドメグサ
200312 オオバチドメ,オオバチドメグサ,オホバチドメグサ,ミヤマチドメグサ
200321 アマゾンチドメグサ
200329 ノチドメ
200349 タイワンチドメグサ
200350 オオチドメ,オホチドメ,ヤマチドメ
200354 ブラジルチドメグサ
200364 アリサンチドメグサ,ハリチドメ
200366 チドメグサ
200370 オキナワチドメグサ
200386 ウチワゼニクサ
200390 ノチドメ
200392 オオチドメ
200393 ヒメチドメ
200394 ミヤマチドメ
200411 セイロンハコベ属
200430 セイロンハコベ
200476 ハゼリサウ科,ハゼリソウ科
200589 オギノツメ属,ヲギノツメ属
200641 ケヤナギハグロ
200652 オギノツメ,ヤナギハグロ,ヲギノツメ
200677 ヒルムシロシバ
200707 ヒロセレウズ属
200709 サンカクチュウ
200712 サンカクチュウ
200727 オオバヌスビトハギ,ザイコクトキワヤブハギ,リュウキュウヌスビトハギ
200728 オオバヌズビトハギ,ザイコクトキワヤブハギ
200732 オオバヌスビトハギ,オホバヌビト

ハギ,トキハヤブハギ,トキワブハギ
200736 カンザウダマシ,ヌスビトノアシ,フジカサ,フジカンザウ,フジカンゾウ,ヤブカンサウ
200739 マルバヌスヒトハギ
200740 ケヤブハギ,ヤブハギ
200742 ヌスヒトハギ
200753 ヤマブキサウ属,ヤマブキソウ属
200754 クサヤマブキ,タイリクヤマブキソウ,ヤマブキサウ,ヤマブキソウ
200756 ウスイロヤマブキソウ
200758 セリバヤマブキサウ,セリバヤマブキソウ
200759 ホソバヤマブキソウ
200760 ホソバヤマブキソウ
200777 ミセバヤベンケイ
200783 ヒダカミセバヤ
200784 ベンケイソウ
200788 アルタイミセバヤ,ヒマラヤミセバヤ
200793 シロベンケイソウ,セイタカベンケイソウ,ムラサキベンケイソウ
200794 カラフトミセバヤ
200798 タマノオ,タマノヲ,チチッパベンケイ,ミセバヤ
200799 エッチュウミセバヤ
200800 チチッパベンケイ
200801 オオチチッパベンケイ
200802 オオベンケイソウ
200804 ホソバノオオベンケイ
200805 タマザキマンネングサ
200812 エゾベンケイ,ムラサキベンケイサウ,ムラサキベンケイソウ
200814 ウスリ-ミセバヤ
200815 ツガルミセバヤ
200816 ミツバベンケイサウ,ミツバベンケイソウ
200817 コモチミツバベンケイソウ
200818 ショウドシマベンケイソウ
200819 アオベンケイ
200820 コモチベンケイソウ
200827 シマエノコロ,ミズエノコロ,ミヅエノコロ
200840 ミズエノコロ
200906 ヒメノカリス属
200913 ヒメノカリス
200941 ササガニユリ
200955 ナガエササガニユリ
201093 ヒメノギネ属
201356 トックリヤシ属,ヒダカトックリヤシ属
201359 トックリヤシ,ブ-ルボンセダカトックリヤシ
201360 トックリヤシ
201363 モ-リシャセダカトックリヤシ
201370 ヒヨス属
201372 シロバナヒヨス
201389 ヒヨス
201392 シナヒヨス,ラウトウ,ロウトウ
201433 テ-ブルヤシモドキ属
201435 ヒメテ-ブルヤシモドキ
201563 ヒッパリガヤ
201597 ケシモドキ属
201605 ケシモドキ
201647 ゲンパ属
201705 オトギリサウ属,オトギリソウ属
201708 クモイオトギリ
201710 タカサゴキンシバイ
201732 コボウズオトギリ
201741 ダイセンオトギリ
201742 シロウマオトギリ
201743 オホオトギリ,クサビョウ,ススヤトモユ,トモエサウ,トモエソウ,ヒメトモエサウ,ヒメトモエソウ,ビヨウサウ,ビヨウソウ
201745 コトモエソウ,ヒメトモエソウ
201751 オオトモエソウ
201761 シナオトギリ
201787 セイヨウキンシバイ
201853 オトギリサウ,オトギリソウ
201859 メイテンオトギリ
201865 フジオトギリ
201880 タイワンキンシバイ,タカサゴキンシバイ,ニリンオトギリ
201885 トモエオトギリ
201888 オソバキンシバイ
201892 タカサゴキンシバイ,ニリンオトギリ
201894 ノウカウキンシバイ,ノウコウキンシバイ
201899 マルミノキンシバイ
201902 オクヤマオトギリ
201903 ホソバヒメオトギリ
201907 ハチジョウオトギリ
201908 コオトギリ,コオトギリソウ
201909 クロテンコオトギリ
201911 ニッコウオトギリ
201912 アカテンオトギリ
201925 ヒマラヤキンシバイ,ボウズオトギリ
201942 コケオトギリ,ヒメオトギリ,ミヤマオトギリ
201954 ハイオトギリ
201956 ウックシオトギリ
201957 チシマオトギリ
201959 シナノオトギリ
201961 カワラオトギリ
201964 ミヤコオトギリ
201967 ヒカゲオトギリ
201968 ユフダケオトギリ
201970 キタミオトギリ
201971 ナガサキオトギリ
201972 ヤクシマコオトギリ
201976 エゾヤマオトギリ
202010 オオカナダオトギリ
202013 トウゲオトギリ
202014 セイタカオトギリ
202021 ビヤウヤナギ,ビヨウヤナギ
202029 トミサトオトギリ
202035 ニヒタカオトギリ
202047 ニッコウオトギリ
202059 アゼオトギリ
202060 フデオトギリ
202062 オオシナノオトギリ
202063 トガクシオトギリ
202070 キンシバイ
202086 セイヨウオトギリ
202087 コゴメバオトギリ
202097 オオバオトギリ
202099 ハナオトギリ
202107 サワオトギリ
202132 ハリガネオトギリ
202145 サマニオトギリ
202146 ツキヌキオトギリ
202150 イワオトギリ
202152 センカクオトギリ
202154 タカネオトギリ
202168 タイワンオトギリ
202178 シラトリオトギリ
202179 シカクオトギリ
202185 ビロ-ドオトギリ
202192 トサオトギリ
202204 ヒマラヤキンシバイ
202213 オシマオトギリ
202216 クロテンシラトリオトギリ
202224 マシケオトギリ
202226 タニマノオトギリ
202227 ミネオトギリ
202228 センゲンオトギリ
202230 エゾオトギリ
202231 ダイセツヒナオトギリ
202256 ド-ムヤシ属
202268 クロミド-ムヤシ
202321 テベスド-ムヤシ,ド-ムヤシ
202387 オウゴンソウ属,ワウゴンサウ属
202400 オウゴンソウ,オオゴンソウ,ワウゴンサウ
202403 エゾコウゾリナ
202404 ヒメブタナ
202432 ブタナ
202450 ヒポキルタ属
202495 シタイシャウ属,ヒポエステス属
202538 シタイシャウ
202604 ムラサキシタイシャウ
202702 スゲガヤ属
202724 スゲガヤ
202727 スゲガヤ
202767 シャクジョウソウ,シャクジョウバナ,シャクヂャウサウ,シャクヂャウバナ
202786 コギンバイザサ科
202796 コキンバイザサ属

202812 コキンバイザサ
203036 イガニガクサ属
203042 ナントウイガニガクサ
203044 イガニガクサ,タマザキウラガヘリバナ
203062 イガニガクサ
203064 オザキイガニガクサ
203066 ニオイニガクサ,ニホヒニガクサ
203079 ヤナギハクカ属,ヤナギハッカ属
203095 ヒソップ,ヤナギハクカ,ヤナギハッカ
203099 シロバナヤナギハッカ
203100 タイリンヤナギハッカ
203102 アカバナヤナギハッカ
203126 コウライハマムギ
203128 イワタケソウ
203129 アズマガヤ
203136 アイヌムギ
203175 イベルウィルレア属
203181 イベリス属,マガミバナ属,マガリバナ属
203184 キャンディタフト,ナヅナザクラ,マガミバナ,マガリバナ
203209 オニマガリバナ
203229 ニオイナズナ,ニオイマガリバナ
203235 ハタザオナズナ
203239 トキワマガリバナ
203249 イロマガリバナ
203260 キバナノツノゴマ
203303 クロタキカズラ科,クロタキカヅラ科
203312 タイワンササキビ属
203317 タイワンササキビ
203318 タイワンササキビ
203417 イイギリ属,イヒギリ属
203422 イイギリ,イヌギリ,イヒギリ,ケラノキ,タウセンダン,トウセンダン,ナンデンギリ
203423 シロミイイギリ
203444 イドリア属
203445 カンポウギョク
203501 マラヤヒメヤシ属
203513 ワリチャマラヤヒメヤシ
203527 モチノキ属
203530 オオツルツゲ
203538 マルバタラヨウ
203545 イングリッシュ・ホーリー,セイヤウヒヒラギ,セイヨウヒイラギ,ヒヒラギモチ,ヒラギモチ
203571 サカキバイヌツゲ,マンリャウイヌツケ
203575 ウスバアリサンソヨゴ
203578 シマウメモドキ,タイワンウメモドキ,ヤンハルウメモドキ
203590 ヒイラギソヨゴ,ヒラギソヨゴ
203600 シイモチ,シチモチ,シヒモチ,ハクサンモチ,ヒゼンモチ,ビゼンモチ
203625 ナナミノキ,ナナメノキ
203626 ホソバナナメノキ
203636 オホシヒバモチ,ムッチャギ
203660 チャイニーズ・ホーリ,ヒイラチモチ,ヤバネヒイラギモチ
203668 マルバヒイラギモチ
203670 イヌツゲ,キッコウツゲ,コバモチ,シダレイヌツゲ,チグロ,ツクシイヌツゲ,ハハキイヌツゲ,ビンカ,マメツゲ,マンキチイヌツゲ,ヤマッゲ
203672 イヌマメツゲ,マメイヌツゲ,マメツゲ
203690 ホソバイヌツゲ
203692 ナガエイヌツゲ
203694 フイリイヌツゲ
203696 コバノイヌツゲ
203704 キミイヌツゲ
203715 ハチジョウイヌツゲ
203721 オオバイヌツゲ
203734 ハイイヌツゲ,ヤチイヌツゲ
203763 アマミヒイラギモチ,アラミヒイラギモチ,ヒイラギモチ,ヒヒラギモチ
203821 シマナナメノキ,タイワンナナメノキ
203840 フウリンウメモドキ
203841 オクノフウリンウメモドキ
203851 アリサンソヨゴ,タマザキモチノキ
203860 オキナワソヨゴ,オキナワモチ,ツゲモチ,リュウキュウソヨゴ
203871 オキナハソヨゴ,オキナハモチ,ツゲモチ,リウキウソヨゴ
203908 トリモチノキ,モチノキ
203909 イヌモチ
203911 キミノモチ,キミノモチノキ
203912 キミノモチノキ
203914 エナシモチノキ
203916 イヌモチ
203946 ナリヒラモチ
203952 イヌソヨゴ,クサノソヨゴ,タイワンイヌツゲ
203961 オオモチ,タウエフ,タラヨウ,タラヨオ,ノコギリバ,モンッキシバ
203964 フイリタラヨウ
203977 ヒメモチ
203992 ニンドウモチノキ
203995 ケナシニンドウモチノキ
204001 ヒロハタマミズキ
204008 アオハダ,アヲハダ,イヌゲヤキ,ケナシアオハダ,コウシュウブナ,コウボウチャ,ホソバアオハダ,マルバウメモドキ,マルバノウメモドキ
204010 ホソバアオハダ
204023 ムニンイヌツゲ
204025 カネヒライヌツゲ,シマイヌツゲ,ナガバイヌツゲ,ムッチャガラ
204042 アッバモチノキ,シマモチ,ムニンモチ
204044 ムニンモチ
204050 アカミヅキ,タマミズキ,タマミヅキ
204051 キミノタマミズキ
204071 シマイヌツゲ,ムッチャガラ
204081 オオバウメモドキ,オホバウメモドキ
204087 ホソバウメモドキ,ミヤマウメモドキ
204114 アメリカヒイラギ,アメリカヒイラギモチ,アメリカ・ホーリー
204126 キミノアメリカヒイラギ
204135 パラグアイチャ,パラグワイチャ,マテ,マテチャ
204138 サヤゴ,スズガシ,ソヨゴ,タカネソヨゴ,ナガエソヨゴ,フクラシバ,フクラモチ
204140 キミソヨゴ
204144 フイリソヨゴ
204159 アツバモチ
204162 ペルニーヒイラキ,ペルピーヒイラキ
204172 サワフサギ
204184 ケイヌツゲ
204217 クロガネモチ,フクラシバ
204220 キミノクロガネモチ
204226 エゾツルツゲ,ツルツゲ,マルバツルツゲ
204230 ホソバツルツゲ
204239 イヌウメモドキ,ウメモドキ,オオバウメモドキ
204240 シロイヌウメモドキ
204242 イヌウメモドキ
204245 フジウメモドキ
204248 イヌウメモドキ,シロウメモドキ
204251 キミノウメモドキ
204255 ウメモドキ
204276 ホソバツルツゲ
204292 アカツゲ,アカミノイヌツゲ,アブラギ,ウシカバ,クロソヨゴ,ミヤマクロソヨゴ
204299 アカツゲ,アカマノイヌツゲ,アカミノイヌツゲ,コウトウツゲ,タイワンイヌツゲ,ミヤマクロソヨゴ
204301 アブラギ,ウシカバ,クロギ,クロソヨゴ
204344 カネヒライヌツゲ
204365 メルテンモチ,リュウキュウモチ
204396 オオシイバモチ,ワルブルギモチ
204418 ニヒタカイヌツゲ,ニヒタカツゲ,ミヤマヒメツゲ
204471 シキミ科
204474 シキミ属
204479 ウスベニシキミ
204482 オキナワシキミ
204484 アカバナシキミ,マルバアカバナシキミ

204533 シキミ,ランダイシキミ
204575 アリサンシロバナシキミ,オキナハシキミ,シキミ,ランダイシキミ
204583 カウシバ,コウシバ,シキミ,ハナシバ,ハナノキ
204598 ヤエヤマシキミ,ヤヘヤマシキミ,リウキウシキミ
204603 ダイウイキョウ,トウシキミ,ハッカクウイキョウ,ハッカクウヰキヤウ
204609 テンダノハナ属
204634 テンダノハナ
204752 イミタリア属
204756 ツリフネソウ属,ホウセンカ属,ホウセンクワ属
204797 ハナツリフネソウ
204799 ツマクレナイ,ツマクレナヰ,ツマベニ,ホウセンカ,ホウセンクワ,ホネヌキ
204838 アカボシツリフネソウ
204841 アカボシツリフネソウ
204969 ヤマツリフネソウ
204982 ツリフネ,ロイルツリフネソウ
205025 ハガクレツリフネ
205027 エンシュウツリフネソウ
205175 キツリフネ,ホラガヒサウ
205176 コウサカツリフネ
205177 オクエゾキツリフネ
205178 ウスキツリフネ
205181 コバナキツリフネ
205346 アフリカホウセンカ,インパチエンス
205361 センザンツリフネ
205369 シロツリフネ,ツリフネサウ,ツリフネソウ,ムラサキツリフネ
205370 クロバナツリフネ
205372 ナメラツリフネソウ
205422 イチゲツリフネ,ニヒタカツリフネ
205444 アフリカホウセンカ,インパチエンス
205470 チガヤ属
205473 チガヤ
205497 カワラチガヤ,ケナシチガヤ,チガヤ
205506 チガヤ
205517 チ,チガヤ,ツバナ,フシゲチガヤ
205546 インカ-ビレア属,インカルビレア属,ツノシホガマ属
205557 インカ-ビレア・デラバ-イ,ウンナンハナゴマ
205571 オオバナハナゴマ
205580 ハナゴマ
205585 ツノシオガマ
205611 コマツナギ属
205698 オオコマツナギ,オホコマツナギ
205752 トウコマツナギ
205876 イハフジ,イハフヂ,イワフジ,ニハフヂ,ニワフジ
205877 シロバナニワフジ
206002 マンスウコマツナギ
206005 コダチノニワフジ
206018 コウシュンコマツナギ
206081 タヌキコマツナギ
206140 カウライニハフヂ,コウライニワフヂ,チウセンニハフヂ,チョウセンニワフジ
206181 ヒメコマツナギ
206445 コマツナギ
206447 シロバナコマツナギ
206573 アフリカコマツナギ
206626 アメリカアイ,アメリカアヰ,インドアイ,インドアヰ,キアイ,キアヰ,ナンバンコマツナギ,ナンバンタイセイ,モラン
206669 キアイ,キアヰ,シマコマツナギ,タイワンコマツナギ,ナンハンアイ,ナンハンアヰ,マメアイ,マメアヰ,モクラン
206684 ミツバノコマツナギ
206714 コニハフヂ,シマクサハギ,タイワンクサハギ
206761 リュウキュウコマツナギ
206765 ニヒタカヤダケ属
206833 オオバヤダケ
206998 イノカルプス属
207000 タイヘイヨウグルミ
207015 タシロマメ
207025 イヌラ属,オグルマ属,ヲグルマ属
207046 オグルマ,カラフトオグルマ,ツイミオグルマ,ヨウシュオグルマ,ヲグルマ
207051 マルザキオグルマ
207052 ヤエオグルマ
207053 エダウチオグルマ
207059 オグルマ
207066 カラフトオグルマ
207068 エゾオグルマ
207082 ミズギク
207083 オゼミズギク
207084 オクノミズギク
207135 オオグルマ,オホグルマ,ドモッゴウ
207151 オグルマ,サクラオグルマ,ヲグルマ
207165 ホソバオグルマ,ホソバヲグルマ
207219 オヤマオグルマ,オヤマヲグリマ,カセンサウ,カセンソウ,ミセンギク,ヲヤマヲグリマ
207222 カセンソウ
207224 スナジオグルマ
207446 イオノプシス属
207546 イポメア属,サツマイモ属
207570 ヨルガオ
207590 アサガホナ,カンコン,ヤウサイ,ヨウサイ
207623 アメリカイモ,カライモ,カンショ,サツマイモ,ベンリイモ,リュゥキュウイモ
207659 モミジヒルガオ
207780 コバナミミアサガオ
207792 キアサガホ,コダチアサガオ,ゴヨウアサガオ
207839 アメリカアサガオ,マルバアメリカアサガオ
207845 ツタノハルコウ
207873 ゴヨウアサガオ
207884 アツバアサガオ
207891 ノアサガオ
207892 シロバナノアサガオ
207921 マメアサガオ
207922 ベニバナマメアサガオ
207940 コハマアサガオ
207953 ソコベニヒルガオ
207971 キバナハマヒルガオ
207981 キバナハマヒルガオ
207988 タイワンアサガホ,テガタアサガホ,モミチヒルガホ,ヤツデアサガオ,ヤツデアサガホ
208016 アサガオ,アサガホ,アメリカアサガオ,ケニゴシ
208023 ヒメアサガホ,ヒメノアサガオ,ヒメノアサガホ
208050 イモネアサガオ
208051 ヤツデアサガオ
208067 グンバイヒルガオ
208069 グンバイヒルガオ
208077 キクザアサガオ,キクサアサガホ
208092 タマザキアサガオ
208104 カワリバアサガオ
208117 ヤラッパ
208120 マルバアサガオ
208249 イモネノホシアサガオ
208250 アメリカソライロアサガオ
208253 メキシコアサガオ
208255 ホシアサガオ
208264 タウアサガホ,タウナスゼ,チャウジナスゼ,ハリアサガオ,ハリアサガホ
208289 キバナハマヒルガオ
208305 フウリンアサガオ
208323 イプセア属
208344 イレシネ属,マルバビユ属
208349 イレシネ,ケシャウビユ,ケショウビユ,センニチサウ,マルバビユ
208350 キフマルバビユ
208354 ナガバケショウビユ
208366 タケウマヤシ属
208367 サンカクタケウマヤシ
208368 トックリタケウマヤシ
208372 ヒメタケウマヤシ属
208373 トビヒメタケウマヤシ
208377 アヤメ科
208433 アヤメ属,イリス属
208466 オオコガネアヤメ

208524 ヒアフギモドキ,ヒオギモドキ
208527 ヒオウギ
208543 ドンドバナ,ネジアヤメ,ネヂアヤメ,ノハナショウブ,ハナショウブ
208547 ハナシャウブ,ハナショウブ
208552 ノハナショウブ
208562 コガネアヤメ
208565 シオバナイリス,シモバナイリス,シロバナイリス,ニオイアヤメ,ニオイイリス,ニホヒイリス,ニホヒシャウブ
208574 タイワンシャガ
208578 チャショウブ
208583 ジャーマン・アイリス,ジャーマンアイリス,ドイツアヤメ,ムラサキイリス
208595 ヒメシャガ
208606 ロクカクアヤメ
208626 オランダアヤメ
208639 カリフォルニア・アヤメ
208640 シャガ
208650 カマヤマショウブ
208657 ヒメネジアヤメ
208665 ネジアヤメ
208666 モウコネジアヤメ
208667 ネジアヤメ,ネヂアヤメ,ネヂバリン,バリン,バレン
208672 ガオバナ,カオヨグサ,カキツバタ
208679 イギリスアヤメ
208700 カラフトキシャウブ
208706 マンシウカキツバタ
208715 キンカキツバダ
208744 シボリイリス
208771 キシャウブ,キショウブ
208793 エヒメアヤメ,タレユエサウ,タレユエソウ,ヒメアヤメ
208795 シロバナエヒメアヤメ
208797 コカキツバタ,マンシュウアヤメ
208801 コカキツバタ
208805 アカネイリス
208806 アヤメ,ハナアヤメ
208807 シロアヤメ
208809 カマヤマショウブ
208819 ヒアブギアヤメ,ヒオウギアヤメ
208823 キリガミネヒオウギアヤメ
208824 ナスノヒオウギアヤメ
208825 シガアヤメ
208829 コアヤメ
208830 シロバナコアヤメ
208870 グロアヤメ
208874 ナンキンアヤメ
208875 イチハツ,コヤスグサ
208876 シロバナイチハツ
208882 イトバアヤメ
208894 カマヤマショウブ
208897 アンザンアヤメ,フトネアヤメ
208899 モロッコアヤメ
208913 ホクリョウアヤメ
208914 カンザキアヤメ
208917 ヒロハコアヤメ
208918 コアヤメ
208921 シヘイネヅアヤメ,フクロネジアヤメ
208944 イギリスアヤメ,イングリッシュ・アリイス
208946 スペインアヤメ,ラッキョウアヤメ
209024 イサベリア属,ネオラウケア属
209025 イサベリア
209028 チゴザサ属
209030 アリサンチゴザサ
209048 タイワンチゴザサ
209054 ナンテンチゴザサ
209059 チゴザサ,ヤナギバザサ
209060 コツブチゴザサ
209081 アツバハイチゴザサ,アツバハヒチゴザサ
209087 ケナシハイチゴザサ
209098 ヒメチゴザサ
209105 ダイトンチゴザサ,ハイチゴザサ,ハヒチゴザサ
209107 ヒメハイチゴザサ
209121 アツバハイチゴザサ
209128 オオチゴザサ
209167 タイセイ属
209201 タイセイ
209217 タイセイ,ハトクサ
209229 タイセイ,ハマタイセイ,ホソバタイセイ,マタイセイ
209232 タイセイ
209238 エゾタイセイ,ハマタイセイ
209249 カモノハシ属
209251 コタイワンアイアシ
209255 ケカモノハシ
209259 タイワンカモノハシ
209261 コブカモノハシ
209266 カモノハシ
209267 カモノハシ
209276 ハナカモノハシ
209279 コブカモノハシ
209288 カモノハシ,ハサミガヤ
209321 チモルカモノハシ,ヒメカモノハシ
209327 シマカモノハシ
209345 タイワンオニシバ,ヤエヤマカモノハシ
209356 タイワンアイアシ
209362 コハナカモノハシ
209503 イスラヤ属
209505 カイジンテッドウ
209506 フクメンギョク
209507 ハナワオウジ
209508 フラビマル
209509 ハッキンギョク
209510 モウコギョク
209511 ボウシギョク
209512 コクガリュウ
209557 イソキルス属
209610 ヤマハッカ属
209612 セキヤノヒキオコシ
209613 クロバナアキチョウジ
209614 クロバナヤマハッカ
209615 カメバヤマハッカ
209616 コシジヒキオコシ
209631 ヤマアキチョウジ,ヤマアキチョオジ
209658 シロバナセキヤノアキチョウジ
209665 カメバヒキオコシ,マンシュウカメバヒキオコシ
209667 シロチョウセンカメバヒキオコシ
209702 ヤマハッカ
209704 シロバナヤマハッカ
209705 メヤマハッカ
209706 オオヤマハッカ
209713 エンメイサウ,エンメイソウ,ヒキオコシ
209715 シロヒキオコシ
209717 ヒキオコシ,マンシウヒキオコシ,マンシュウヒキオコシ
209735 タイワンヒキオコシ
209748 アキチャウジ,アキチョウジ,キリツボ
209750 シロバナノアキチョウジ
209826 オオヒキオコシ,オホヒキオコシ,ムラサキヒキオコシ
209830 ミヤマヒキオコシ
209831 タカクマヒキオコシ
209832 サンインヒキオコシ
209833 シロバナサンインヒキオコシ
209846 クロバナヒキオコシ
209847 アカバナヒキオコシ
209848 ノハラヒキオコシ
209851 イヌヤマハッカ
209852 シロバナノイヌヤマハッカ
209853 メイヌヤマハッカ
209854 タイリンヤマハッカ
209855 ハクサンカメバヒキオコシ
209857 コウシンヤマハッカ
209858 シロバナコウシンヤマハッカ
209859 カメバヒキオコシ
209860 シロバナカメバヒキオコシ
209983 ビャッコイ
210251 シロカネサウ属,シロカネソウ属
210281 マンシウシロカネソウ
210313 イソトマ属
210359 ズイナ属,ズヰナ属
210364 シマズイナ,シマズヰナ
210374 ノコギリズイナ,ノコギリズヰナ,ヒヒラギズイナ,ヒヒラギズヰナ
210389 ズイナ,ズヰナ,ヨメナノキ
210402 シマズイナ,ヒイラギズイナ
210407 アリサンヒラギズイナ,アリサンヒラギズヰナ,カシノハズイナ,カシノハズヰナ,ヒメズイナ,ヒメズヰナ

210421　コバノズイナ,コバノズヰナ
210485　フナバシソウ
210521　タカネニガナ
210522　ホソバニガナ
210525　ウサギソウ,タカサゴサウ,タカサゴソウ
210527　タカネニガナ,ニガナ,ハナニガナ
210529　クモマニガナ
210530　ドロニガナ
210531　イソニガナ
210532　ハナニガナ
210534　オゼニガナ
210535　シラネニガナ
210536　シロバナニガナ
210537　ハイニガナ
210539　ヒメヂシバリ
210541　イトバニガナ
210543　アツバニガナ,オホバニガナ,ニガナモドキ,ヤナギニガナ
210546　ヤクシマニガナ
210548　イヌヤクシソウ
210549　タカサゴソウ
210551　コスギニガナ
210553　ニガナ属
210567　マンシウタカサゴソウ,マンシュウタカサゴソウ
210572　ホソバタカサゴソウ
210592　イソニガナ
210594　オバナニガナ,シロバナニガナ,ハナニガナ
210616　ノニガナ
210623　オオジシバリ,ジシバリ,ツルニカナ,ホソバチシバリ
210645　ツルワダン
210648　タカサゴニガナ
210649　ミヤコジシバリ
210650　ツルカワラニガナ
210654　ノニガナ
210666　ノヂシバリ
210673　イハニガナ,イワニガナ,シシバリ,ジシバリ,ヂシバリ,ハイジシバリ,ヒメヂシバリ
210674　シロバナイワニガナ
210678　ミヤマイワニガナ
210679　タカサゴソウ
210681　カハラニガナ,カワラニガナ,タイトウタカサゴサウ
210689　イキシア属,ヤリズイセン属
210843　モンイキシア,ヤリズイセン
210988　イギシオリリオン属
211029　サンタンカ属,サンタンクワ属
211037　テマリサンタンカ
211067　サンタンカ,サンダンカ,サンタンクワ,センタンクワ
211096　コバナサンタンクワ
211113　ジャワサンタンカ
211125　キバナサンタンカ
211143　ニオイサンタンカ
211148　シロバナサンタンカ,シロバナサンタンクワ
211225　ジャカランダ属
211235　オオバジャカランダ,オオバジャカランダノキ
211239　キリモドキ
211241　ブラジルジャカランダ
211338　サンゴバナ属,ジャコビニア属,ヤコビニア属
211344　キイロサンゴバナ
211350　ヤコブセニア属
211602　オセソウ属
211603　オセソウ
211639　ヤシオネ属
211693　シタキサウ属,シタキソウ属,ステファノ－ティス属
211695　オキナハシタキヅル,オキナワシタキヅル,シタキサウ,シタキソウ
211697　シタキソウ
211700　ジャスミノセレウズ属
211701　ジャコウチュウ
211715　オオバイ属,ソケイ属,ワウバイ属
211749　ベニバナソケイ
211821　モクカウクワ,モツコウカ,リウキウワウバイ,リュウキュウオウバイ
211846　ジャバソケイ,ボルネオソケイ
211848　オオバナソケイ,ソケイ,タイワンソケイ,ツルソケイ
211863　ヒマラヤソケイ,ヒメラヤソケイ
211864　キソケイ
211868　ウンナンソケイ
211869　ウンナンソケイ
211905　ウンナンオウバイ,オウバイモドキ
211912　ケソケイ,ボルネオソケイ
211918　イヌシロソケイ
211931　オウバイ,オオシュクバイ,オオバイ,キンバイ,ワウシュクバイ,ワウバイ
211938　キソケイ,シンソケイ
211940　ソケイ,ツルマツリ
211956　ヒメオウバイ
211963　ジャスミナム・ポリアンツム,ホシソケイ
211990　マツリ,マツリカ,マツリクワ,マリ,マリカ,モウリンカ,モウリンクワ,モリカ,モリクワ,モリンカ
212004　オキナハソケイ,オキナワソケイ
212024　オキナワソケイ
212043　シロソケイ,シロバナツルソケイ
212098　タイワンアブラキリ属,タイワンアブラ属
212127　シマアブラギリ,タイワンアブラキリ,ナンヤウアブラギリ,ナンヨウアブラギリ,マアブラキリ
212186　モミジバアブラギリ
212202　サンゴアブラギリ,トックリアブラギリ
212296　タッタソウ属
212299　アメリカタツタソウ
212323　イェンゼノボトリア属
212343　サケミヤシモドキ属
212346　オニサケミヤシモドキ
212401　ヒトツバクマデヤシ属
212402　セダカヒトツバクマデヤシ
212474　イオノプシジューム属
212541　アヤザクラ
212553　ユアヌルロア属
212556　チリ－ヤシ属,チリヤシ属
212557　チリ－ヤシ,チリヤシ
212560　アフリカチリ－ヤシ属,アフリカチリヤシ属
212561　アフリカチリ－ヤシ,アフリカチリヤシ
212581　クルミ科
212582　クルミ属
212595　ネッカグルミ
212599　バタグルミ
212608　タイワングルミ,タカサゴグルミ
212621　クルミ,ナガグルミ,マンシウグルミ,マンシュウグルミ,ヤマグルミ
212622　ミゾナシオタフクグルミ
212624　ヒメグルミ
212626　オグルミ,オニグルミ,クルミ,ヲグルミ
212631　ウオルナット,ニグラゲルミ,ブラックウオルナット
212636　アシカワグルミ,オウシュウグロミ,カシグルミ,クルミ,クワシグルミ,チョウセングルミ,テウセングルミ,テウチグルミ,トウクルミ,ハグルミ,ペルシアグルミ,ロニングルミ
212640　ロニングルミ
212641　クマオングルミ
212643　テウチグルミ
212655　オタフクグルミ,ヒメグルミ
212656　ハリサキタフクグルミ
212695　ユメルレア属,ユメ－レア属
212752　イグサ科,ヰ科
212753　シバナ科,ヰ科
212774　シオガヤツリ
212778　オホガヤツリ,ミズガヤツリ,ミヅガヤツリ
212783　ヤブガヤツリ
212797　イグサ属,イ属,ヰ属
212816　ハナビゼキシャウ,ハナビゼキショウ,ヒロハノコウガイゼキショウ
212866　カラフトハナビゼキショウ
212890　イヌヰ,ネヂヰ,ヒラヰ
212898　ミヤマイ
212905　ホクトイ
212915　コウライイ
212929　ヒメカウガイゼキシャウ,ヒメコウガイゼキショウ
212990　クロカウガイゼキシャウ,クロコウ

ガイゼキショウ,チョウセンクロコウガイゼキショウ
213028 セキショウイ
213036 イ,イグサ,トウシンサウ,トウシンソウ,ヰ,ヰグサ
213037 ラセンイ
213038 コヒゲ
213040 オキナワイ
213041 タマイ
213042 ヒメイ
213045 ヒロハノカウガイゼキシャウ,ヒロバノカウガイゼキシャウ,ヒロハノコウガイゼキショウ
213046 タマコウガイゼキショウ
213066 コヒゲ
213067 ラセンイ
213093 ミクリゼキショウ
213110 イヌイ
213111 ホソコウガイゼキショウ
213114 エゾホソイ,エゾホソヰ,カラフトホソイ,カラフトホソヰ,リシリイ,リシリヰ
213138 トロイ,ドロイ,トロヰ,ミズイ,ミヅヰ
213145 ハマイ
213173 トミサトクサイ
213182 ミヤマホソコウガイゼキショウ
213191 タチガウガイゼキシャウ,タチコウガイゼキショウ
213213 カウガイゼキシャウ,コウガイゼキショウ,ヒラカウガイゼキシャウ,ヒラコウガイゼキショウ
213272 イトイ,イトヰ
213281 エゾノミクリゼキショウ
213297 ニヒタカイトヰ
213320 ニッコウコウガイゼキショウ
213355 アオカウガイ,アオカウガイゼキシャウ,アオコウガイゼキショウ,アヲカウガイ,アヲカウガイゼキシャウ,ホソバノカウガイゼキシャウ,ホソバノコウガイゼキショウ
213362 ホソヰ
213388 エゾイトイ,エゾイトヰ
213448 ホソイ,ホソヰ
213476 ヤチイ
213502 マンシウクサイ
213507 アメリカクサイ,アリサンクサヰ,クサイ,クサヰ,シラネイ,シラネヰ,ヰゼキシャウ
213530 ホロムイコウガイ
213531 オテダマゼキショウ
213545 シロウマゼキシャウ,シロウマゼキショウ,タカネイ,タカネヰ
213552 コウアンゼキショウ
213553 ネッカハナビゼキショウ
213569 ナガミタチコウガイゼキショウ,ハリカウガイゼキシャウ,ハリコウガイゼキショウ
213606 ネズミサシ属,ビャクシン属
213615 オキアガリネズ
213634 イブキ,イブキビャクシン,カマクライブキ,カマクラビャクシン,ビャクシン,ビャクダン
213642 タマイブキ
213647 カイズカイブキ,カイヅカイブキ,ヤクシン
213696 シンパク,ミヤマビャクシン
213702 セイヨウネズ,セイヨウビャクシン,トショウ,ヒメネズ,ヨウシュネズ
213712 リシリビャクシン
213728 ホンドミヤマネズ
213735 シンシュウネズ,ミヤマネズ
213739 ハイスギ,ハイネズ,ハイムロ,ハト,ハヒネズ
213752 ダウリャビヤクシン,ダブリャビヤクシン
213775 シマビャクシン,タイワンビヤクシン,ハマネズ
213776 ハマネズ
213785 アメリカハイネズ
213864 ソナレ,ハイビャクシン,ハヒビャクシ
213896 ネズ,ネズサシ,ネズミサシ,ムロ
213934 ハシリビャクシン,ミヤマトショウ,リシリビャクシン
213943 ニイタカビャクシン,ニヒタカビャクシン
213962 シマムロ,ヒゴ,ヒデ
213963 オオシマハイネズ,オキナハハヒネズ,オキナワハイネズ,ハマハイネズ
213984 エンピツノキ,エンピツビャクシン
214304 キツネノマゴ属
214308 アドハトダ
214379 コエビソウ
214410 サンゴバナ
214512 キツネノメマゴ
214734 シロバナキツネノマゴ
214735 キツネノマゴ
214739 キツネノヒマゴ
214911 ユッタディンテリア属
214947 カズラ属,サネカズラ属,サネカヅラ属
214971 アンダカジャー,サネカズラ,サネカヅラ,ドロリカヅラ,トロロカヅラ,ビナンカズラ,ビナンカヅラ,ビンツケズル,フノリカズラ,ヤマオウレン
214995 バンウコン属
215011 バンウコン
215035 サンナ,バンジュツ
215086 カランコエ属,リウキウベンケイ属,リュウキュウベンケイ属
215097 テンニンノマイ
215102 カランコエ,ベニバナベンケイ,ベニベンケイ
215160 ヒメトウロウサウ
215174 ヘラバトウロウサウ,リウキウベンケイ,リュウキュウベンケイ
215198 ミョウギ
215206 エドムラサキ
215220 テンニンノマイ
215272 コウトウベンケイ
215277 ゲットジ,ツキトジ
215329 ギョリウダマシ
215334 ヨメナ属
215339 オオユウガギク,コウアンヨメナ,チョウセンヨメナ
215343 コヨメナ
215350 ホソバヨナメ
215352 キタヨナメ,キレハヨナメ,ヤマヨメナ
215356 オオバヨメナ,モウコヨナメ
215386 カルミア属
215387 ナガバハナガサシャクナケ,ホソバアメリカシャクナゲ
215395 アメリカシャクナゲ,ハナガサシクナゲ
215405 ホソバハナガサシャクナゲ
215411 カルミオプシス属
215420 ハリギリ属
215442 アクダラ,イヌダラ,センノキ,ハリギリ,ボウダラ,ミャコダラ,ヤマギリ,リュウギュウハリギリ
215443 キレハハリギリ,ケハリギリ
215445 ミヤコダラ,リュウキュウハリギリ
215508 メヒルギ属
215510 オホヒルギ,タカツク,ヒルギ,メヒルギ,リウキウカウガイ,リュウキュウコオガイ
215511 メヒルギ
215805 シモバシラ属
215811 シモバシラ,ホソバシモバシラ,ユキョセサウ,ユキョセソウ
215814 ウスベニシモバシラ
215993 ケンシティア属
216005 ケンチヤモドキ属
216084 ヤマブキ属
216085 ゴエフイチゴ,ゴヨウイチゴ,トゲゴエフイチゴ,ヤマブキ
216088 ヤエヤマブキ
216090 シロバナヤマブキ
216095 キクザキヤマブキ
216114 アブラスギ属,シマモミ属,ユサン属
216120 アブラスギ,カタモミ,シマモミ,タイワンモミ,テッケンユサン,ユサン
216125 アブラスギ,シマモミ,タイワンモミ,タイワンユサン,ユサン
216142 フッケンユサン
216181 カディア属
216214 アフリカンマホガニー
216262 ヒメツルウンラン

216313 ソ-セ-ジノキ属
216324 ソ-セ-ジノキ
216464 キヌガサソウ属
216473 キレンゲショウマ属
216474 キレンゲショウマ
216637 アマクサ,ゲンゲツ,タゴトノツキ,テンソウ
216642 フウセンアカメガシワ
216654 シッボウジュ,ブンパイクライニヤ
216923 クニフォフィア属
217094 シソノミグサ
217101 シソノミグサ
217121 ヒゲハリスゲ
217234 ヒゲハリスゲ
217324 ハハキギ属,ホウキギ属
217345 イソホウキギ
217353 イトバハハキギ
217361 ニハクサ,ニワクサ,ネンドウ,ハハキギ,ハハキグサ,ホウキギ,マキグサ
217372 シラゲハハキギ,シラゲホウキギ
217373 ハナホウキギ
217416 ミノボロ属
217441 ミノボロ
217494 ミノボロ
217518 ミノボロモドキ
217593 ケルレンステイニア属
217610 モクゲンジ属
217613 オオモクゲンジ,フクワバモクゲンジ
217620 タイワンセンダンボダイジュ,タイワンモクゲンジ,モクレンジ
217626 センダンバノボダイジュ,モクゲンジ,モクレンジ
217640 チシマミチャナギ属
217648 チシマミチャナギ
217738 コ-レリア属,ゴ-レリア属
217739 ナキリ,ベニギリソウ
217785 コルクウィツィア属
217864 コプシア属
217869 ビルマコプシア
217901 ヒノキバヤドリギ属
217906 ヒノキバヤドリギ
217918 トウサゴヤシ属
217920 ノキトウサゴヤシ
217922 テイヌマントウサゴヤシ
217925 フナトウサゴヤシ
218121 ワタフキノキ
218294 ハクウンラン属
218296 ヤクシマヒメアリドオシラン
218344 ヤマズソウ属
218345 マルバヤマズソウ
218347 ヤマズソウ
218457 ヒメクグ属
218480 アイダクグ,ヒメクグ
218482 オニヒメクグ
218518 タイトウクグ
218556 タチヒメクグ
218571 オオヒメクグ,オヒメクグ,シロヒメクグ,ヲヒメクグ
218697 ミソ科
218715 フジマメ属
218721 アジマメ,アヂマメ,インゲンマメ,シャクヂャウマメ,センゴクマメ,テンジクマメ,フジマメ,フヂマメ
218722 シロフジマメ
218754 キングサリ属,キンレンクワ属
218761 キバナフジ,キバナフヂ,キングサリ,キンレンカ,キンレンクワ,ツェテセス
218824 ラケナリア属
218920 ラケナリアペンジユラ
219206 アキノノゲシ属,ニガナ属
219307 リュウゼッサイ
219458 アメリカニガナ
219485 カキヂサ,チサ,チシャ,チシャナ
219487 カキチシャ
219488 カキチシャ,クキチシャ,チシャトウ
219489 タマチシャ,チシャ
219490 チリメンチシャ
219492 タチチシャ
219497 トゲチシャ
219508 マルバトゲチシャ
219513 エゾムラサキニガナ
219522 シロバナムラサキニガナ
219609 トゲハニガナ
219668 レ-リア属
219729 サルスベリ属
219806 エゾムラサキニガナ,ムラサキノゲシ
219821 ユウガオ属,ユフガオ属,ユフガホ属
219827 ユウガオ
219843 ヒョウタン
219844 ヒョウタン,ヘウタン
219847 シャクヘウタン
219848 フクベ
219851 ユウガオ
219854 センナリヒョウタン,センナリヘウタン
219892 コケセンボンギク属
219898 コケセンボンギク
219900 コケセンボンギク
219901 コケセンボンギク
219908 サルスベリ属
219909 コサルスベリ,ムラサキサルスベリ
219922 ヤクシマサルスベリ
219933 サルスベリ,ヒャクジッコウ
219934 シロバナサルスベリ
219966 オオバナサルスベリ,ジャワザクラ
219970 アカブリ,シマサルスベリ,タイワンサルスベリ
219972 ヤクシマサルスベリ
219976 サラオ
219990 ヒレギク属
219996 ヒレギク
220126 コゴメオドリコソウ,シロバナノホトケノザ
220155 ウルップサウ属,ウルップソウ属,ハマレンゲ属
220167 ウルップサウ,ウルップソウ,ハマレンゲ
220169 シロバナウルップソウ
220196 ホソバウルップソウ
220200 ユウバリソウ
220203 ホソバウルップソウ
220226 ラグナリア属
220228 ラグナリア
220252 ラグルス属
220254 ウサギノオ
220295 ノレンガヤ属
220298 ノレンガヤ
220345 オドリコソウ属,ラミュ-ム属,ヲドリコサウ属
220346 エゾオドリコソウ,オドリコサウ,オドリコソウ,タイリクオドリコソウ
220348 キタダケオドリコソウ
220355 カスミグサ,サンガイグサ,トンビグサ,ホトケノザ,ホトケノツヅレ
220356 シロバナホトケノザ
220359 オドリコソウ
220389 ヤマジオウ
220390 シロバナヤマジオウ
220392 モミジバヒメオドリコソウ
220416 ヒメオドリコソウ
220417 シロバナヒメオドリコソウ
220422 ウライオドリコサウ,ウライヲドリコサウ,キ-ルンオドリコサウ,キ-ルンヲドリコサウ,ヒメキサワタ
220464 オスキュ-リア属,オスクラリア属,マツバギク属,ランプランサス属
220494 ハクコウ
220523 コウリンギク
220674 キクボタン,サボテンギク,マツバギク
220697 ヒメマツバギク
220729 キンチャクボタン,クマンボタン,ケマンサウ,ケマンソウ,タイツリソウ,フジボタン,フヂボタン,ヤウラクボタン,ヨウラクボタン
221007 シマウキクサ,タイワンウキクサ,ヒメウキクサ
221224 ランサ属,ランシウム属
221226 ランサ,ランザット
221229 コウオウカ属,コウワウクワ属,ランタナ属
221238 コウオウカ,コウワウクワ,サンダンサウ,シチヘンゲ,シャヘンゲ,セイヤウサンダンクワ,セイヨウサンダンカ,トゲナシランタナ,ランタナ
221252 タチゲランタナ

221266 ムラサキランタナ
221272 キバナランタナ
221284 コバノコウオウカ,コバノシチベンゲ,コバノランタナ,コバノランタナー
221340 ラパゲーリア属
221341 ツバキカズラ
221515 ラピダリア属
221529 ムカゴイラクサ属
221538 ギョクシュクサウ,ムカゴイラクサ
221539 コモチミヤマイラクサ
221552 オイラクサ,コモチミヤマイラクサ,ミヤマイラクサ
221627 ノムラサキ属
221685 ノムラサキ
221711 ノムラサキ
221727 モウコノムラサキ
221780 ヤブタビラコ属
221784 カハラケナ,カワラケナ,コオニタビラコ,コニタビラコ,タビラコ
221787 ナタネタビラコ
221792 ヤブタビラコ
221802 ヤノネジシバリ
221815 ラルディザバラ属
221818 アケビ科
221829 カラマツ属
221855 オウシュウカラマツ,ヨウシュカラマツ
221866 グイマッ,グイマツ,グヒマツ,シコタンマツ,ダウリヤカラマッ,ダフリャカラマツ,チシマカラマッ
221878 グイマツ,シコタンマツ
221894 カラマッ,ニッコウマツ,フジマツ,ラクエフショウ,ラクヨウショウ
221895 シダレカラマツ
221897 アカミカラマツ
221904 アメリカカラマッ
221919 チョウセンカラマツ,トオホクカラマツ,マンシウカラマツ,マンシュウカラマッ
221939 ホクシカラマツ,ムレイカラマツ
221944 シベリアカラマツ,シベリヤラマツ
222081 ルリミノキ属
222095 マルバルリミノキ
222115 オホバルリミノキ
222123 ケシンテンルリミノキ,ケハダルリミノキ
222124 オホバノヤマルリミノキ,シマルリミノキ,タイワンルリミノキ
222133 タシロルリミノキ,ミヤマルリミノキ,リウキウルリミノキ,リュウキュウルリミノキ
222138 ケハダルリミノキ,ケミヤマルリミノキ,シンテンルリミノキ,タイワンルリミノキ
222158 タイワンルリミノキ
222172 ホルリミノキ,ルリダマノキ,ルリミノキ
222173 サツマルリミノキ
222231 ウスバルリミノキ
222237 オオバルリミノキ
222281 シンテンルリミノキ
222302 オオバルリミノキ
222305 マルバルリミノキ
222625 ニオウギヤシ属,ベニオウギヤシ属
222626 ブールボンベニオウギヤシ,ベニラタンヤシ
222634 アオラタンヤシ,ロディゲスベニオウギヤシ
222636 キイラタンヤシ,キラタンヤシ,バーシャフェルトベニオウギヤシ
222641 ヤマウッボ属
222645 ケヤマウッボ,ヤマウッボ
222647 シロケヤマウッボ
222671 レンリサウ属,レンリソウ属
222681 タクエフレンリサウ,タクヨウレンリソウ
222699 オトメレンリソウ
222707 イタチササゲ,エンドウサウ,エンドウソウ
222708 ヒゲナシイタチササゲ
222724 オオレンリソウ,ハットクマメ
222727 コエンドウ
222731 スズメノレンリソウ
222735 ハマエンドウ
222736 シロバナハマエンドウ
222737 ユキイロハマエンドウ
222738 ケハマエンドウ
222743 ハマエンドウ
222747 タチレンリソウ
222750 ヒロハノレンサウ,ヒロハノレンソウ
222782 セイヤウエビラフヂ,セイヨウエビラフジ
222788 ヒゲレンリソウ
222789 ジャカウエンドウ,ジャコウエンドウ,スイートピー,スヰートピー,ニオイエンドウ,ニホヒエンドウ
222800 セイヨウレンリソウ,レンリサウ,レンリソウ
222803 エゾノレンリサウ,エゾノレンリソウ
222805 ヤチエンドウ
222807 カマキリサウ,レンリサウ
222819 セイヤウエビラフヂ
222820 キバナノレンリサウ,キバナノレンリソウ
222828 レンリソウ
222829 シロバナレンリソウ
222844 ヤナギバレンリソウ
222851 キュウコンエンドウ
222852 チョウセンヤマエンドウ
222860 ツルナシレンリソウ
223006 クスノキ科
223101 セイヨウバクチノキ
223116 セイヨウタデキ,セイヨウバクチノキ,ロウレルザクラ
223117 クロボシイヌザクラ,クロボシザクラ,タカサゴイヌザクラ
223131 アズサ,カタザクラ,タテギ,ヒイラギガシ,ヒラギガン,ホカノキ,ミズアオイ,メヒイラギ,メヒラギ,リンボク
223150 ケイマ,ゴイノキ,サルコカシ,バクチノキ,ハダカノキ,ビラン,ビランジュ
223163 ゲッケイジュ属
223203 ゲッケイジュ,ロウレルノキ,ローレル
223248 ラバンジュラ属,ラワンデル属
223251 ラベンダー
223298 ヒロハラベンダー
223350 ハナアオイ属,ハナアフヒ属
223355 モクアオイ,モクアフヒ
223393 ハナアオイ,ハナアフヒ
223451 シカウクワ属,シコウカ属
223454 シカウクワ,シカフクワ,シコウカ,ツマクレナイノキ,ツマクレナヰノキ
223634 ムエフラン属,ムヨフラン属
223636 ロウバイスケロクラン
223643 サキシマスケロクラン
223644 シラヒゲムヨウラン
223645 ホクリクムヨウラン
223646 キイムヨウラン
223647 ムエフラン,ムヨフラン
223652 ヤエヤマスケロクラン
223654 ウスギムヨウラン
223658 アワムヨフラン,クロムヨウラン,クロムヨフラン
223659 ヤクムヨウラン
223663 ムラサキムヨウラン
223665 エンシュウムヨウラン
223666 キバナエンシュウムヨウラン
223669 アワムヨウラン
223670 オキナワムヨウラン
223671 ミドリムヨウラン
223674 シマミヅ属
223679 チョクザキミズ,チョクザキミヅ
223762 サガリバナ科
223776 パラダイスナットノキ
223888 イソツツジ属,レヅム属
223894 ミドリイソツツジ
223901 イソツツジ,カラフトイソツツジ
223904 イソツツジ,カラフトイソッッジ
223905 イソツツジ
223906 エゾイソツツジ
223909 ホソバイソツツジ
223911 ヒメイソツツジ,ホソバイソツツジ
223912 カバフトイソツツジ,カラフトイソツツジ
223913 イソツツジ,カラフトイソツツジ,ニクケイソウ
223919 ウホウドカズラ属,オオウドノキ属,オホウドカズラ属

223937 オオウドカズラ,オオウドノキ
223943 ウホウドカズラ,ウホウドノキ,サンゴジュウドノキ
223967 ウドノキ科
223973 サヤヌカグサ属
223984 タイワンアシカキ
223987 アシカキ
223992 エゾサヤヌカグサ,エゾノサヤヌカグサ
224002 サヤヌカグサ
224003 ヒロハノサヤヌカグサ
224073 オオミゾカクシ
224085 マメ科
224106 センボンヤリ属
224110 ガフクワンセンボンヤリ,センボニャリ,センボンヤリ,ムラサキタンボ,ムラサキタンポポ
224112 シロバナセンボンヤリ
224244 レイポルツティア属
224335 レマイレオセレウス属
224336 ベニモンジ
224352 オドントグロッサム
224359 アオウキクサ属,アヲウキクサ属
224360 ナンゴクアオウキクサ
224362 ホクリクアオウキクサ
224366 イボウキクサ
224369 ムラサキコウキクサ
224375 コウキクサ
224378 ヒナウキクサ
224385 アオウキクサ,チビウキクサ
224396 サンカクナ,ヒンジモ
224399 キタグニコウキクサ
224400 チリウキクサ
224403 ウキクサ科
224471 レンズ属
224473 ヒラマメ,レンズマメ
224499 タヌキモ科
224521 レオセレウス属
224558 カエンキセワタ属,レイポチス属,レオノチス属
224585 カエンキセワタ
224600 ケカニア
224731 カワリミタンポポモドキ,タンポポモドキ
224767 ウスユキサウ属,ウスユキソウ属
224772 エ－デルワイス
224822 ハナウスユキソウ
224827 エゾウスユキソウ,レブンウスユキソウ
224833 ヒナウスユキソウ,ミヤマウスユキソウ
224834 ホソバノヒメウスユキソウ
224852 ハヤチネウスユキソウ
224854 オオヒラウスユキソウ
224872 ウスユキサウ,ウスユキソウ
224873 ハッポウウスユキソウ
224874 ヤマウスユキソウ
224881 カワラウスユキソウ
224882 ミネウスユキソウ
224883 コウスユキソウ
224890 チシマウスユキソウ
224893 ノウスユキソウ
224907 カハカミウスユキ,カワカミウスユキソウ
224908 オオヒラウスユキソウ
224932 コマウスユキソウ,ヒメウスユキソウ,ミヤマウスユキソウ
224969 メハジキ属,レオヌ－ルスゾク属
224976 モミジバキセワタ
224989 ニガヨモギ,メハジキ,ヤクモサウ
224990 シロバナメハジキ
224991 シロバナホソバメハジキ
224996 キセワタ
225009 ホソバ,メハジキ,ヤクモサウ,ヤクモソウ
225015 ダッタンメハジキ
225037 レオボルドヤシ属
225061 レパンテス属
225134 ウロコマリ属
225162 ウロコマリ
225178 リュウキュウウロコマリ
225224 ホソバウロコマリ,ヤナギバウロコマリ
225279 コショウソウ属,コセウサウ属,コセウソウ属,マメグンバイナズナ属
225283 ダイコクマメグンバイナズナ
225295 ヒメグンバイナズナ
225305 キレハマメグンバイナズナ
225315 ウロコナズナ,オニグンバイナズナ
225326 アツバナズナ
225340 ヒメグンバイナズナ
225360 ハマガラシ
225363 タマザキマメグンバイナズナ
225398 ベンケイナズナ
225428 コシミノナズナ
225433 ニセマメグンバイナズナ
225447 コバノコショウソウ
225450 コショウソウ,コセウサウ,セルデレ－
225475 カウベナヅナ,コウベナヅナ,セイヤウグンバイナヅナ,マメグンバイナズナ,マメグンバイナヅナ
225493 ウロコゴヘイヤシ属
225495 ヒメウロコゴヘイヤシ
225554 ウロコケンチヤ属
225557 モ－アウロコケンチヤ
225622 ウロコザミア属
225623 ヒロバウロコザミア
225624 ホソバウロコザミア
225659 アンペラ属
225661 アンペラ
225664 アンペラ,アンペラサウ,アンペラソウ,アンペラヰ
225668 シチモウゲ属,シチャウゲ属
225691 レピスミウム属
225694 アオアシ
225712 オオバケアサガオ,オバケアサガオ
225974 レプトセレウス属
225983 アゼガヤ属
225989 アゼガヤ
226005 タカオバレンガヤ,タカヲバレンガヤ,ハマガヤ,ミツバガヤ
226008 オニアゼガヤ
226009 ニセアゼガヤ
226016 ホウキアゼガヤ
226021 イトアゼガヤ
226075 イハハギ属,シチャウゲ属,シチョウゲ属
226106 イワシチョウゲ
226125 イワハギ,シチョウゲ,シチョオゲ,ムラサキチョウジ
226231 ニセクサキビ
226320 スズフリノキ
226339 ヒメウズサバノオ属
226341 ヒメウズサバノオ
226450 ネズモドキ属,レプトスペルマム属
226451 ネズモドキ
226481 ギョリュウバイ
226539 レプト－テス属
226585 ハイシバ属,ハリノホ属
226599 ハイシバ,ハヒシバ
226621 ハナハコベ
226669 レスケナウルティア属
226677 ハギ属
226678 ヤマツクシハギ
226679 オクタマハギ
226680 マルバツクシハギ
226682 ツルメドハギ
226683 シロヤマハギ
226684 オオマキエハギ
226685 ナガサキハギ
226698 エゾヤマハギ,ヤマハギ
226701 シロバナエゾヤマハギ
226708 シラハギ,ハギ,ヤマハギ
226712 チャボヤマハギ
226721 キハギ,ノハギ,ヤマハギ
226722 シロバナキハギ
226723 タチゲキハギ
226739 オホバメドハギ,タママキエハギ
226742 メドキ,メドハギ
226744 キバナメドハギ
226745 ハイメドハギ
226746 マルバハギ,ミヤマハギ
226747 カワチハギ
226751 オオバメドハギ,オオメドハギ,オホバメドハギ
226764 オクシモハギ
226793 タウクサハギ,トウクサハギ
226800 シロバナハギ
226801 テヤマハギ
226804 シロバナチョウセンヤマハギ

226805 ビッチュウヤマハギ
226806 サツマハギ
226836 サガミメドハギ
226838 ツクシハギ,ヤブハギ
226839 シロバナツクシハギ
226843 カラメドハギ
226849 シラハギ
226851 ヒメニシキハギ
226852 シロバナハギ,ニシキハギ
226853 ソメワケハギ
226858 チョウセンヤマハギ
226860 イヌメドハギ,カラメドハギ,シベリアメドハギ,メドハギ,モリメドハギ
226891 アヤコハギ
226892 チョウセンキハギ
226895 クロバナキハギ
226896 ベニクロバナキハギ
226912 ケハギ,ヤマミヤギノハギ
226913 タテヤマハギ
226914 ユキハギ
226916 ミャギノハギ
226919 ミャギノハギ
226924 ネコハギ
226925 ムラサキネコハギ
226926 タチネコハギ
226940 シマヤマハギ,タイワンハギ
226965 ビロードハギ
226969 ミヤギノハギ
226974 シラハギ,シロバナハギ
226977 タイワンハギ,ナツハギ,ビッチュウヤマハギ,ミヤキノハギ,ミヤギノハギ
226989 イヌハギ,シラハギ
227009 マキエハギ
227010 ムラサキマキエハギ
227013 リュウキュウハギ
227224 ギンヨウジュ属
227233 ギンノキ,ギンヨウジュ
227345 ホソバギンヨウジュ
227420 ギンガフクワン属,ギンゴウカン属
227430 ギンガフクワン,ギンゴウカン,ギンネム,タマザキセンナ
227453 ホソバノセイタカギク,ミコシギク
227512 シャスタ・デージー,シャスタデージー
227533 フランスギク
227537 ヤンバルツルハクカ属
227572 ヤンバルツルハッカ
227577 モンパヤンバルクルマバナ
227657 ヤンバルツルハクカ,ヤンバルツルマバナ
227697 リュウキュウクルマバナ
227755 リューヒテンベルギア属
227813 リュウココリネ属
227854 オオマツユキソウ属,スノーフレーク属
227856 オオマツユキソウ,スノーフレーク
227858 アキザキスノーフレーク
227946 コウボウモドキ属
227948 コウボウモドキ
227999 テンニンソウ属
228002 テンニンサウ,テンニンソウ,ムラサキテンニンサウ,ムラサキテンニンソウ
228004 シロバナテンニンソウ
228005 フジテンニンソウ
228011 ミカエリソウ
228012 シロバナミカエリソウ
228014 オオマルバノテンニンソウ,ツクシミカエリソウ,トサノミカエリソウ
228029 リューコスペルマム属
228134 リューコステレ属
228144 ウラジロイハガネ属
228147 ウラジロイハガネ,ウラジロイワガネ
228154 イハナンテン属,イワナンテン属
228159 アメリカイワナンテン
228166 ハナヒリノキ
228167 エゾウラジロハナヒリノキ,ヒロハハナヒリノキ
228169 ウラジロハナヒリノキ
228171 ヒメハナヒリノキ
228172 ウスユキハナヒリノキ
228173 ハコネハナヒリノキ
228178 イハナンテン,イワツバキ,イワナンテン
228244 レビスチカム属
228250 ロベージ
228256 レウィシア属
228319 レイケステーリア属
228346 テンキ
228356 シバムギモドキ
228369 アア,クサドウ,テンキ,テンキグサ,ハマニンニク,ムリッ
228384 ケハマムギ
228433 ユリアザミ属,リアトリス属
228487 タマザキキリンギク,タマザキリアトリス
228511 ヒメキリンギク,ユリアザミ
228519 マツガサギク
228529 キリンギク,ユリアザミ,リアトリス
228572 イブキボウフウ,ハマイブキボウフウ
228575 タカネイブキボウフウ
228597 アムールボゥフウ,マンシウボゥフウ
228599 シベリアボゥフウ
228609 ケタカネイブキボウフウ
228611 チシマイブキボウフウ
228625 イボクサアヤメ属
228635 オニヒバ属
228639 オニヒバ
228729 ウチワヤシ属,ゴヘイヤシ属
228732 ダイオウゴヘイヤシ
228736 チャボゴヘイヤシ
228741 マルハウチワヤシ
228756 タテバゴヘイヤシ
228758 サケバゴヘイヤシ
228759 チャボゴヘイヤシ
228764 ランフィーゴヘイヤシ
228766 ゴヘイヤシ,トゲゴヘイヤシ
228969 タガラカウ属,タカラコウ属,ツハブキ属,メタカラコウ属
228982 ヤマタバコ
228992 イトウソウ
228999 タカラカウ,タガラカウ
229005 マルバダケブキ
229016 マルバダケブキ,マルバノチョウリョウソウ
229034 ミチノクメタカラコオ,ミチノクヤマタバコ
229035 エゾオタカラコオ,オタカラコウ
229038 オニオタカラコウ
229049 エゾタガラカウ,エゾタガラコウ,エゾタカラコオ,オニタガラカウ,オニタガラコウ,カラフトトウゲブキ,タウゲブキ,タガラコウ,トウゲブキ,トオゲブキ
229052 カラフトトウゲブキ
229060 コウライメタカラコウ
229063 サンカクツワブキ,ジンユウツハブキ,チョウセンタカラコウ,テフセンタカラカウ
229065 ヤノネツハブキ,ヤノネツワブキ
229066 ハンカイソウ,ハンクワイサウ,ハンクワイソウ
229068 シマハンクワイサウ,タイワンハンクワイサウ
229070 カイタカラコウ
229078 ミヤマオタカラコウ
229109 ハマタバコ,モウコハマタバコ
229175 シカナ,チョウセンヤマタバコ,ヤマタバコ,リウキウモクカウ
229179 オタカラカウ,オタガラカウ,ダッタンタカラコウ,ヲタカラカウ,ヲタガラカウ
229202 メタカラカウ,メタガラカウ,メタカラコウ,メタカラコオ
229206 オニメタガラカウ
229208 オオメタカラコウ
229221 カニオタカラコウ
229257 ダケブキ,チョオリョオソウ
229283 タウキ属,トウキ属,マルバトウキ属
229343 イワテトウキ
229344 ムレイセンキュウ
229371 センキウ,センキュウ
229384 マルバトウキ
229385 マルバトウキ
229404 ニオイウイキョウ

229428 イボタノキ属,タウキ属
229432 オキナワイボタ,コバノタマツバキ
229472 イボタノキ,ケイボク,サイコクイボタ
229485 イヌツバキ,イハテタウキ,タニワタシ,タマツバキ,テラツバキ,ナンブタウキ,ネズミモチ,ミヤマタウキ
229489 キモンネズミモチ
229492 フクリンネズミモチ
229493 キミノネズミモチ
229499 イワキ
229503 キマダラネズミモチ
229506 ケネズミモチ
229512 ハイネズミモチ
229513 フクラモチ,フクロモチ,マルバネズミモチ
229514 イワキ
229522 ヤブイボタ
229524 オキナハイボタ,オキナワイボタ
229525 コバノオキナハイボタ
229529 タウネズミモチ,トウネズミモチ,トオネズミモチ
229533 キフクリントウネズミモチ
229546 ムニンネズミモチ
229547 ホソバムニンネズミモチ
229558 イボタ,イボタノキ,イボタロウ,イボトリ,イボノキ
229560 ビロ－ドイボタ
229561 コバイボタ,コバノイボタ
229562 カオリイボタ
229563 セッツイボタ
229565 オニイボタ
229566 ニガクイボタ
229569 イヌイボタ,オオバイボタ,オホバイボタ,オホバイボタノキ,ケオオバイボタ,コネズミモチ,ヨメノツジラ
229571 フイリオオバイボタ
229573 ケオオバイボタ
229578 オカイボタ
229579 ケオカイボタ
229580 ハチジョウイボタ
229587 アリサンイボタ,ケネズミモチ
229593 クロイゲイボタ
229608 ナガサギイボタ
229610 オニイボタ,ハナイボタ,ヤナギイボタ
229643 トゲイボタ
229650 ミヤマイボタ
229651 エゾイボタ
229652 ミウライボタ
229656 オオミイボタ
229658 オクノハマイボタ
229662 セイヨウイボタ,ヨウシュイボタ,ヨウシュイボタノキ
229700 ユリ科
229717 バイモ属,ユリ属
229723 コマユリ,シマユリ
229730 エイザンユリ,カマクラユリ,カントウユリ,キツネユリ,キンセンユリ,シロユリ,スジュリ,トウネミネユリ,ニオイユリ,ハコネユリ,ホウイジユリ,ホウランジュリ,ヤマユリ,ヨシノユリ,リョウリュリ,レウリユリ
229732 ヒロハヤマユリ
229733 クチベニ
229737 シロボシ
229758 サツマユリ,シハイユリ,ニオイハカタユリ,ハカタユリ
229775 スゲユリ,タイワンヒメユリ,ノヒメユリ
229776 キバナノヒメユリ
229789 トキワユリ,ニワシロユリ,フランスユリ,ホザキユリ,ヨレハユリ
229803 ゲンジユリ,コウライホソバユリ,チョウセンホソバユリ,マツバユリ,ムラサキホソバユリ
229805 シロマツバユリ
229811 ヒメユリ,ベニユリ
229813 トウヒメユリ,ヒメユリ
229815 キヒメユリ
229817 ヒメユリ
229818 コヒメユリ,トウヒメユリ,ヒメユリ
229822 ウバユリ
229828 エゾスカシユリ,エゾズカシユリ,エゾユリ,ミカドユリ
229834 チョウセンクルマユリ
229836 スカシユリ
229845 スジテッパウユリ,スヂテッパウユリ,タイワンユリ,タカサゴユリ,ホソバテッパウユリ,ホソバテッポウユリ
229857 オオクルマユリ,タケシマユリ,チョウセンクルマユリ
229862 キォノコユリ,キカノコユリ,キンコウデン
229872 ササユリ,サシキユリ,サユリ,ニオイユリ,ノユリ,ヒメユリ,ホンゴウロ,ヤマユリテングユリ
229873 ジンリョウユリ
229874 シロバナササユリ
229887 キバナノオニユリ,キヒラトユリ,コオニユリ
229888 アカヒラトユリ,アマユリ,コオニユリ,スゲユリ,ナツユリ,ノユリ
229900 イワユリ,サガリユリ,サシマユリ,ジャコウユリ,タカサゴユリ,タメトモユリ,ツツナガユリ,テッパウユリ,テッポウユリ,ニオイユリ,リウキウユリ,リュウキュウユリ
229915 アオガシマスカシユリ,イソユリ,イワトユリ,テンジョウユリ,ナツスカシユリ,ハマユリ
229919 ササユリ,サユリ
229934 カサユリ,クルマユリ,コメユリ
229947 チベットササユリ,モモハユリ
229958 ウコンユリ
229966 コウユリ,タモツユリ,タモトユリ,テモチユリ
229992 サクユリ,サックイネラ,タメトモユリ,ニオイユリ,ハチジョウユリ
230005 イトハユリ,イトユリ,スゲユリ,ホソバユリ
230009 イトハユリ,イトユリ,スゲユリ,ホソバユリ
230019 オウカンユリ,ホソバハカタユリ,リ－ガル・リリ－
230023 アイヅユリ,オトメユリ,コマチユリ,ハルユリ,ヒメサユリ
230024 シモバナオトメユリ
230036 カノコユリ,カラユリ,スズユリ,タキユリ,タナバタユリ,ドヨウユリ
230038 アキザキカノコユリ,カネヒラユリ,タイワンカノコユリ,ヌニシユリ
230041 シラタマユリ,シロカノコユリ
230058 アワユリ,オニユリ,サツマユリ,テンガイユリ,テンガユリ,ノユリ
230060 オウゴンオニユリ
230061 ヤエオニユリ
230065 チウセンカサユリ,チョウセンカサユリ
230183 リムナンテス科
230205 リムナンテス属
230238 アマゾントチカガミ
230244 リムノカリス属
230250 キバナオモタカ,キバナオモダカ
230253 キバナオモダカ科
230274 シソクサ属
230275 シソクサ
230297 エナシシソクサ
230307 コキクモ,タイワンキクモ
230320 ホウライシンクサ
230322 ホウライシソクサ,ホオライシソクサ
230323 キクモ
230328 コキクモ
230508 イソマツ属,リモニューム属
230524 キバナハマサジ
230544 イヌトウゴウソウ
230627 トウゴウイゾマツ,トウゴウソウ
230631 モウコハマサジ
230655 ニワハナビ,ヒロハノハマサジ
230712 リモニューム・ペレジ－
230759 カタバナハマサジ,センカクハマサジ,タイワンハマサジ,トウハマサジ
230761 ハナハマサジ
230762 スタ－チスシヌアタソビア
230787 ダッタンハマサジ,ヨレハナビ
230797 ハマサジ,ハマヂサ
230816 イソハナビ,イソマツ,ウコンイソマツ,ムラサキイソマツ
230817 シロバナイソマツ

230818 ウスジロイソマツ
230820 イソマツ
230828 キタミサウ属,キタミソウ属
230831 キタミサウ,キタミソウ
230854 アマ科
230866 ウンラン属
230889 ヒメウンラン
230911 ヒメキンギョソウ,ムラサキウンラン
230927 ヒメウンラン
230931 マツバウンラン
230978 キバナウンラン
231015 ウンラン
231084 ヒメキンギョソウ
231183 セイヤウウンラン,ホザキウンラン,ホソバウンラン
231184 シロバナセイヨウウンラン
231243 リンデローフィア属
231297 クロモジ属
231298 テンダイウヤク
231301 ナンバンクロモジ
231306 ニオイベンゾイン
231324 コウシュンカウバシ,コウジュンカウバシ,タイワンカウバシ,タイワンコウバシ,タイワンヤマカウバシ,ホソバヤマカウバシ
231327 オキナワコウバシ
231334 アサタ,アワガラ,カナクギノキ,カノコ,ヌカガラ,ヒエダンゴ
231355 アハキ,アワキ,カハカミカウバシ,シャウブノキ,ショウブノキ,セキショウボク,ソバノキ,タイワンヤマカウバシ,モチシバ,ヤマカウバシ,ヤマコウバシ,ヤマコショウ,ヤマトロロ
231379 ヒメクロモジ
231385 オホバカウバシ
231403 イハヅサ,イワヅサ,ウコンバナ,ウコンバメ,コウジバナ,シロジシャ,シロヂシャ,ダンカウバイ,ダンコウバイ,ヤマヅサ
231415 アブラチャン
231417 ケアブラチャン
231433 クロモジ,ケクロモジ
231434 ウスゲクロモジ
231456 シロモジ
231457 マルバシロモジ
231461 キリキシバ,クロトリノキ,クロモジ,タマワンサク
231466 オオバクロモジ
231467 キミノオオバクロモジ
231473 アゼタウガラシ属,アゼナ属
231476 ヒメアメリカアゼナ
231479 シマウリクサ,スズメノトウガラシ
231483 スズメノトウガラシ
231485 エダウチスズメノトウガラシ
231486 ヒロハスズメノトウガラシ
231496 スズメノトウガラシモドキ
231503 ウリクサ
231514 アメリカアゼナ,タケトアゼナ
231515 アメリカアゼナ
231544 アゼタウガラシ,アゼトウガラシ
231559 アゼナ
231568 クチバシグサ
231575 シソバウリクサ
231587 ヒメクチバシグサ
231594 ケウリクサ
231676 リンネーア属,リンネサウ属,リンネソウ属
231679 エゾアリドオシ,リンネサウ,リンネソウ
231680 エゾアリドホシ,メオトバナ,メヲトバナ,リンネサウ
231729 コウトウナタヲレ属
231739 コウトウナタオレ,コウトウナタヲレ
231856 アマ属,リナム属
231863 タカヌアマ
231880 ヒメアマ
231907 ニバナアマ,ベニアマナ,ベニバナアマ
231908 ベニバナアマ
231923 キバナノマツバニンジン
231942 アマ,シュクコンアマ,シュクコンヌメゴマ,シュッコンアマ,ヌメゴマ
231953 スギバアマ
231961 マツバナデシコ,マツバニンジン
231963 アレチアマ
232001 アカゴマ,アマ,イチネンアラ,ヌメゴマ
232072 クモキリサウ属,クモキリソウ属,スズムシラン属
232086 ギバウシラン,ギボウシラン,クモキリサウ,クモチリサウ,タカノハ
232106 ウチャマラン,キノエササラン,キノヘササラン,チクカクラン,チクカラン,チクケイラン,チケイラン
232114 クモキリソウ,クモチリソウ,タカノハ
232117 ヒメキノヘラン
232122 ホザキノキノヘラン,ラッキョウラン
232125 アリサンスズムシサウ,ミドリスズムシ
232139 スズナリラン,タイワンキノヘラン,ナカハララン
232151 コゴメキノヘラン
232164 イウコクラン,ユウコクラン,リウキウコクラン
232166 コガネユウコクラン
232168 シマササバラン
232169 フガクスズムシソウ
232188 ヘンリーコクラン
232193 シマクモキリソウ
232197 セイタカスズムシ,セイタカスズムシソウ
232200 フタハキノヘラン
232205 コウライスズムシソウ
232206 ジガバチサウ,ジガバチソウ
232207 フクリンジガバチソウ
232208 アオジガバチソウ
232211 ヒメジガバチソウ
232215 クモキリサウ,クモキリソウ
232225 スズムシサウ
232229 ホソバクモキリサウ
232252 コクラン
232253 タイリンコクラン,ンリーコクラン
232256 ヒメスズムシソウ
232261 ササバラン
232318 ホソバスズムシサウ
232322 スズムシソウ,スズムシラン,ミヤマコクラン
232333 ヘラキノヘラン
232359 クモイジガバチ
232364 コゴメキノエラン,ラッキョウラン
232380 ヒンジガヤツリ属
232391 オオヒンジガヤツリ
232405 ヒンジガヤツリ
232423 オオヒンジガヤツリ,ヒンジガヤツリ
232457 イワダレゾウ属,リッピア属
232483 コウスイボク,ボウシュウボク
232532 コウスイボク
232550 フウ属
232557 シャウ,タイワンフウ,フウ
232565 アメリカフウ,カエデバフウ,モミジバフウ,モミヂバフウ
232594 シラタマモクレン,トキハレンゲ,トキワレンゲ
232599 オオトキワレンゲ
232602 ウツコンカウジュ属,ユリノキ属
232603 シナユリノキ
232609 ウッコンカウジュ,ウッコンコウジュ,ウッコンジュ,チューリップノキ,ハンテンボク,ハンボク,ユリノキ,レンゲボク
232617 ヤブラン属
232623 ヤブラン
232628 コヤブラン
232630 ヒメヤブラン
232631 ホンヤブラン,ヤブラン
232640 コヤブラン,ヤブラン,リュウキュウヤブラン
232890 フタバラン属,リステラ属
232905 コフタバラン,フタバラン
232908 コフタバラン
232919 オフタバラン,ヒメフタバラン,ムラサキフタバラン
232921 フイリヒメフタバラン
232922 ナガバヒメフタバラン
232923 ミドリヒメフタバラン
232934 アオフタバラン

232946 ミヤマフタバラン
232947 フイリミヤマフタバラン
232948 ミドリミヤマフタバラン
233077 レイシ属
233078 ライチ,ライチイ,レイシ,レイチイ
233099 アリサンガシ属,マテバシイ属,マテバシヒ属,リトカルプス属
233104 アミガシ,オニガシ
233121 オニジヒ,コイガガシ,ハリジヒノキ
233122 オホミノアカガシ,セイシャウガシ,トガリバガシ
233132 オニガシ
233181 ヒラミガシ,レナギバガシ
233193 サシマジイ,サシマジヒ,サツマジイ,シブカガシ,マタジイ,マタジヒ,マテガシ,マテバシイ,マテバシヒ
233219 タイワンガシ,マルバノマテガシ
233228 シリブカ,シリブカガシ,ヨシガシ
233238 ナンバンガシ
233269 カハカミガシ
233278 コニシガシ
233287 オホアミガシ
233303 オキナハジヒ
233323 ナントウガシ
233360 シンズイユイガシ,シンス井ユイガシ
233383 コマミガシ,タイトウガシ,ヒシミガシ
233389 ナンバンガシ
233434 ミヤマホタルカズラ
233459 イシコロケサ属,イシコロマツバギク属,リトープス属
233464 ニチリンギョク
233468 コハクギョク
233471 フクジュギョク
233473 セキシュンギョク
233475 ザクロギョク,シチホウギョク
233483 タイコギョク
233490 デンボウギョク
233492 テンテキギョク
233498 ホウスイギョク
233500 レイコウギョク
233510 エイルギョク
233511 バリ-ギョク
233513 コテンギョク
233524 ビモンギョク
233526 ゲンジギョク
233531 アラタマ
233538 セイジギョク
233539 チョウセイギョク
233554 メイグンギョク
233557 ジュレイギョク
233558 フクライギョク
233564 カモンギョク
233570 ビクンギョク
233572 ハククンギョク
233574 エイギョク,サカエダマ
233584 シクン
233603 マルガタダマ
233606 ケンランギョク
233608 ジャクランギョク
233610 キクスイギョク
233617 オオウチギョク
233619 ベニオオウチギョク
233620 オオツエ
233624 マガタマ
233635 ザイコウギョク
233637 ハクロウセキ
233639 メノウ
233642 ベニダマ
233643 フクインギョク
233650 ルチョウギョク
233651 ミフジン
233654 ショウフクギョク
233658 ギョク,コウシャクギョク
233661 コクヨウギョク
233664 ダイコウシャク
233666 ヘキロウギョク
233667 スイガ
233673 ツユビギョク
233677 ヘキシギョク
233683 レイテンギョク
233684 ビミギョク
233692 ムラサキ属,リソスペルマム属
233700 イヌムラサキ
233731 ムラサキ
233760 セイヨウムラサキ,ムラサキ
233786 ホタルカズラ,ホタルカヅラ,ホタルソウ,ルリサウ,ルリソウ
233787 シロバナホタルカズラ
233791 リスレア属
233809 コバンバヤナギ属
233811 コバンバヤナギ
233827 ハマビハ属,ハマビワ属
233832 アコウクロダモ,アコウクロモジ,オホバクロダモ,コウトウクロダモ,ヒロハダモ
233876 カゴガシ,カゴノキ,カノコガ,コガタブ,ゴカノキ
233882 アオモジ
233920 オホクロモジ,ゲンペイダモ
233945 ウラチャクロダモ,シマカゴノキ,タイワンカゴノキ,ヤンバルカゴノキ,リウキウカゴノキ
233949 イソジラキ,イソビワ,ケイジュ,シバギ,シャクナンショ,ハマビハ,ハマビワ
233967 ホカゴノキ,ホゴノキ
234000 ニヒタカクロモジ
234061 クロモヅ
234075 タイワンカノキ
234130 リットーニア属
234162 ビラウ属,ビロウ属
234166 オーストラリアビロウ
234167 オガサワラビロウ
234170 タガヤサン,トウビロウ,ビラウ,ビレウ,ビロウ
234175 ビロウモドキ
234181 セルダンビロウ
234183 ヒマラアビロウ
234185 マリアビロウ
234187 メリルビロウ
234188 ミューラービロウ
234190 ジャワビロウ,マルバビロウ
234194 ビラウ,ビロウ,マルバビラウ,リウイストナ,ワビロウ
234214 チシマアマナ属
234234 チシマアマナ
234244 ホソバアマナ,ホソバノアマナ
234251 ロアサ属
234252 シロバナシレンゲ
234254 シレンクワ科,シレンゲ科,ロアサ科
234275 ミゾカクシ属,ロベリア属
234291 タチミゾカクシ
234303 サクラダソウ
234327 オオハマギキョウ,オオハマギキョオ
234351 ベニバナサワギキョウ
234360 タチミゾカクシ
234363 アセムシロ,ミゾカクシ,ルリチョウチョウ,ルリミゾカクシ
234365 シロバナミゾカクシ
234366 ヤエノミゾカクシ
234447 ルリミゾカクシ
234547 ロベリアソウ,ロベリヤ,ロベリヤソウ
234615 マルバハタケムシロ
234715 フヂサキャウ
234766 サハギキャウ,サハギキョウ,サワギキョウ,サワギキョオ,チャウジナ,チョウジナ
234767 シロバナサワギキョウ
234813 オオロベリアソウ
234884 マルバミゾカクシ
234890 ミゾカクシ科
234900 アカントロビビア属,ロビビア属
234901 ヒレイマル
234903 クントウマル
234908 タンレイマル
234909 トウレイマル
234910 タンリンマル
234911 ヨウレイマル
234913 ヨウレイマル
234914 トウセイマル
234917 オウカマル
234919 オウジュマル
234921 ヨウセイマル
234922 トウコウマル
234923 ハクレイマル

234924 マキエマル
234925 セツレンマル
234926 キボタンマル
234927 レンゲマル
234928 シュレイマル
234929 ヒセイマル
234930 トウエンマル
234932 ヒエンマル
234933 ニガサマル
234935 シュウレイマル
234936 シエンマル
234937 ハックンマル
234939 モモガサマル
234940 ヨシヒメマル
234941 フジムスメ
234943 シフンマル
234945 エンレンマル
234946 シュエンマル
234947 ヒサマル
234948 ニヒメマル
234949 ビヨウマル
234950 エンキマル
234951 カクヨウマル
234952 ウタヒメマル
234953 オウショクマル
234954 セイレイマル
234955 セキレイマル
234957 コウエンマル
234958 キレイマル
234959 トウエイマル
234960 カナガサマル
234962 カセンマル
234964 ムユウゲ
234965 トウエンマル
234966 シレイマル
234968 トウリンマル
235065 ニワナズナ属,ロブラ－リア属
235090 アリッスム,ゴバナレセダ,ニハナヅナ,ニワナズナ,ハマアリッスム
235131 ロックハ－ティア属
235151 オオミヤシ属
235159 ウミヤシ,オオミヤシ
235246 ロガニア属
235251 フジウツキ科,フヂウツキ科,マチン科
235279 ミネズオウ属,ロワズル－リア属
235283 ミネズオウ,ミネズオオ
235287 ベニバナミネズオウ
235288 ツマキミネズオウ
235300 ドクムギ属,ホソムギ属
235301 ネズミホソムギ
235303 ノゲナシドクムギ
235310 ボウムギ
235315 イタリアン・ライグラス,ネズミムギ
235334 ペレニアル・ライグラス,ホソムギ
235353 アマドクムギ
235354 ボウムギ
235373 ドクムギ
235431 ヒメセンブリ属
235435 ヒメセンブリ
235474 ロ－ナス属,ロマトフィム属
235503 ロマトフィルム属
235504 アフリカヒナギク
235630 スイカズラ属,スイカヅラ属,スヒカヅラ属
235633 アリサンニンドウ,ホソバスヒカヅラ
235639 ハマニンドウ,ハマニンドオ
235646 ケハマニンドウ
235684 キバナノツキヌキニンドウ
235691 ケヨノミ
235694 ナガヒゲヨノミ
235696 マルバヨノミ
235702 ケヨノミ
235705 ヒロハヨノミ
235717 ウスバヒョウタンボク,ウスバヒョオタンボク,ウスバヘウタンボク,トオゴクヒョオタンボク
235721 チシマヒョウタンボク,チシマヒョオタンボク
235722 シロバナチシマヒョウタンボク
235730 ネムロブシダマ,ヒメブシダマ
235733 ネムロブシダマ,ヒラエブシダマ
235734 ナガエブシダマ
235757 イボタヒョウタンボク
235758 キタカミヒョウタンボク
235781 タマヒョウタンボク,ホクシヒョウタンボク
235796 ツシマヒョウタンボク,ツシマヒョオタンボク
235798 アカバナスイカズラ
235811 エゾヒョウタンボク,エゾヘウタンボク
235812 アズキグミ,ウグイスカグラ,コジキグミ,ヤマウグイスカグラ
235813 ウグイスカグラ,ウグイスノキ,ウグヒスカグラ,ウグヒスノキ,ウグヒスホク,コジキグミ
235814 シロバナウグイスカグラ
235815 ミヤマウグイスカグラ
235860 キダチニンドウ,キダチニンドオ,タチニンドウ,チョウセンニンドウ,ドウニンドウ
235878 コバナスヒカヅラ,スイカズラ,スヒカヅラ,ニンドウ,ニンドオ,ベニバナスイカズラ,ベニバナスヒカヅラ
235883 テリハニンドウ
235885 ベニバナスイカズラ
235887 ヒメスイカズラ
235897 タイワンシテウゲ,ニヒタカハクテウゲ
235902 ココゴメヒョウタンボク,ココゴメヘウタンボク
235905 ムレイヒョウタンボク
235908 クロブシヒョウタンボク
235921 ヤブヒョウタンボク
235929 ハナヒョウタンボク,ハナヒョオタンボク,ハナヘウタンボク
235935 シマスヒカヅラ
235945 ニバナヘウタンボク,マンシウヒョウタンボク
235960 ニッコウヒョウタンボク
235962 アカイシヒョウタンボク
235963 ヤマヒョウタンボク
235970 キンギンボク,ヒョウタンボク,フタゴシバ,フタコロビ,ヘウタンボク,ヤエガワ,ヤヘカハ
235971 キミノヒョウタンボク
235999 オヒワケヘウタンボク
236022 ニオイニンドウ
236040 ハヤザキヒョウタンボク,ハヤザキヘウタンボク,ヒロハヒョウタンボク,ヒロハヘウタンボク
236041 ハヤザキヒョウタンボク
236053 コウグイスカグラ,チチブヒョオタンボク
236054 チチブヒョウタンボク
236058 キンキヒョウタンボク
236076 ヒロウドヘウタンボク,ヒロ－ドヘウタンボク
236085 アイナニ,アカバナヒョウタンボク,アカバナヘウタンボク,エノミタンネ,ニバナヘウタンボク,ベニバナヒョウタンボク,ポネチ
236089 ツキヌキニンドウ
236122 アラゲヒョウタンボク
236124 ナンブヒョウタンボク
236125 ホソバアラゲヒョウタンボク
236126 ダイセンヒョウタンボク
236130 グミヒョウタンボク
236146 モモイロヘウタンボク
236161 オオウラジロヒヨオタンボク,ヤナギヒョウタンボク
236196 オオヒョウタンボク
236199 ウゼンベニバナヒョウタンボク
236205 オニヒョウタンボク,オニヒョオタンボク,オニヘウタンボク
236211 エゾヒョウタンボク,スルガヘウタンボク,スンシウヘウタン
236281 ササクサ属
236284 ササクサ
236285 ムサシノササクサ
236295 タウササクサ,トウササクサ
236296 ケナシトウササクサ
236362 ロフォセレウス属
236423 ロッフォフォラ属,ロフォフォラ属
236425 ウバタマ
236461 ロフォステモン属
236462 セルンス－ル,トベラモドキ
236503 オオバヤドリギ科,ヤドリギ科

236507 マツグミ属
237077 ホザキヤドリギ
237190 トキハマンサク属,トキワマンサク属
237191 トキハマンサク,トキワマンサク
237476 ミヤコグサ属,ロ－タス属
237506 アマミエボシグサ,オキナハミヤコグサ,シロバナノミヤコグサ,ヤンバルエボシグサ
237509 ツルミヤコグサ
237539 セイヨウミヤコグサ,ミヤコグサ
237554 エボシバナ,エボスサウ,キメンゲ,コガネバナ,ミヤコグサ
237555 ニシキミヤコグサ
237617 ワタリミヤコグサ
237689 ワタゲミヤコグサ
237706 シロバナミヤコグサ
237713 ネビキミヤコグサ
237762 セイヨウヒメミヤコグサ
237957 マネキグサ属
237958 マネキグサ
237959 キレハマネキグサ
237971 アカメヤシ属
237972 アカメヤシ
238100 ルクレア属
238106 アッサムニオイザクラ
238110 ルクマ属
238133 ホンコンシュスラン属
238150 チャウジタデ属,チョウジタデ属,ミズユキノシタ属
238152 ケミズキンバイ,ミズキンバイ
238166 アメリカミズキンバイ,ヒレタゴボウ,ヒレタゴボオ
238167 タゴボオ,チョウジタデ,チョオジタデ
238168 ウスゲチョウジタデ
238178 タゴボウモドキ
238188 ウスゲキダチキンバイ,オキナハチャウジタデ,キダチキンバイ,ホソバノミヅキンバイ
238190 キダチキンバイ
238191 ミズユキノシタ,ミヅユキノシタ
238192 セイヨウミズユキノシタ
238200 コバノタゴボウ
238203 コバノミズキンバイ
238207 ホソバタゴボウ,ホソバタゴボオ
238211 タゴバウ,チャウジタデ,チョウジタデ
238221 アメリカミズユキノシタ
238230 ミズキンバイ
238257 ヘチマ属
238258 トカドヘチマ
238261 イトウリ,ナガウリ,ヘチマ
238268 ヘチマ
238301 ボウラン属
238307 ムニンボウラン
238316 タイワンボウラン
238322 ボウラン
238324 マリアナボウラン
238345 ヒルギモドキ属
238350 アカバナヒルギモドキ,ニバナヒルギ
238354 コバノヒルギモドキ,ヒルギモドキ
238369 キンセンソウ属,ゴウダソウ属,ルナア属,ルナリア属
238371 キンセンソウ,コウダソウ,ルナア
238419 ハウチハマメ属,ハウチワマメ属,ルピヌス属
238421 シロバナハウチワマメ
238430 キダチハウチワマメ
238444 アフリカハウチハマメ,アフリカハウチワマメ
238448 ニシキハウチワマメ
238450 カサバルピナス,ケノボリフジ
238457 オオハウチワマメ,オホハウチハマメ
238463 キバナノハウチハマメ,キバナノハウチワマメ,キバナハウチワマメ,キバナルピナス,ノボリフジ,ノボリフヂ,ルピナス
238467 カサバルピナス,ケノボリフジ,ケハウチワマメ,ヒメハウチハマメ,ヒメハウチワマメ
238468 ザッショクノボリフジ,ザッショクノボリフヂ,マダラハウチワマメ
238469 チャボハウチワマメ,ワイゼイノボリフジ,ワイゼイノボリフヂ
238473 ノボリフジ,ノボリフヂ,ハウチハマメ,ハウチワマメ
238481 タヨウハウチワマメ,ワシントンルピナス
238563 スズメノヒエ属,スズメノヤリ属
238573 クモマスズメノヒエ
238583 スズメノヒエ,スズメノヤリ
238600 スズメノヒエ,スズメノヤリ
238614 ニヒタカヌカボシ
238616 セイタカヌカボシソウ
238617 ナスヌカボシソウ
238632 ジンボソウ
238633 ミヤマヌカボシソウ
238637 チシマスズメノヒエ
238639 アサギスズメノヒエ
238647 ヤマスズメノヒエ,ヤマスズメノヤリ
238664 ミヤマスズメノヒエ
238671 タカネスズメノヒエ
238675 オカスズメノヒエ,ヲカスズメノヒエ
238677 オカスズメノヒエ
238678 コゴメヌカボシ
238684 コゴメヌカボシ
238685 シマヌカボシソウ,ヌカボシサウ,ヌカボシソウ
238686 クロボシソウ
238690 クロボシサウ,クロボシソウ,シャマクロボシサウ,ヌカボシサウ
238691 コウライヌカボシソウ
238697 シマスズメノヒエ,タイワンスズメノヒエ
238717 クモマスズメノヒエ
238732 リカステ属
238733 ニオイミツビシラン
238942 メジロホオズキ,メジロホホヅキ
238951 ムニンホオズキ
238956 ヤエヤマメジロホオズキ
238996 クコ属
239011 ナガバクコ
239021 クコ,トウグコ,ヒロハクコ
239108 アツバクコ,ハマクコ
239154 トマト属
239157 アカナス,トマト
239158 マメアカナス,マメトマト
239184 シロネ属
239194 イヌシロネ,コシロネ
239215 イプシロネ,シロネ
239222 ケシロネ,ヒメシロネ
239224 ヒメシロネ
239235 エゾシロネ
239238 コシロネ,ヒメサルダヒコ
239246 エゾシロネ
239253 ヒガンバイザサ属,ヒガンバナ属
239255 シロバナヒガンバナ,シロバナマンジュシャゲ
239257 シャウキラン,ショウキズイセン,ショウキラン
239266 シタマガリ,シビトバナ,テンガイバナ,ヒガンガナ,ヒガンバナ,マンジュシャゲ,リコリス
239267 ニシキヒガンバナ
239268 ワラベノカンザシ
239269 シナヒガンバナ
239271 キツネノカミソリ
239272 シロバナキツネノカミソリ
239273 ヤエキツネノカミソリ
239274 オオキツネノカミソリ
239275 シロバナオオキツネノカミソリ
239276 ムジナノカミソリ
239278 リコリス・スプリンゲリ
239280 ナツズイセン,ナツズヰセン,ワスレグサ
239282 ショウキズイセン
239369 ネジキ属,ネヂキ属
239391 カシオシミ,ネヂキ
239392 ウジコロシ,カシオシミ,カシオズミ,カシオスミノキ,カスノキ,ネジキ,ネヂキ,メシツブノキ
239395 ホロムイツツジ,ヤチツツジ
239516 ミズバショウ属
239517 アメリカミズバショウ
239518 ミズバショウ
239525 シタンノキ属

239527 シタンノキ
239543 オカトラノオ属,ヲカトラノヲ属
239544 ギンレイカ,ミヤマタゴボウ,ミヤマタゴボオ
239558 ノジトラノオ,ノヂトラノヲ
239566 トウサワトラノオ
239589 アメリカクサレダマ
239594 オカトラノオ,チカトラノオ,チカトラノヲ,トラノオ,トラノヲ,ヲカトラノヲ
239595 ヤエトラノオ
239608 シマギンメイソウ,シマギンレイカ
239645 ヌマトラノオ,ヌマトラノヲ
239687 コナスビ
239689 コナスビ
239693 ヒメコナスビ,ヤクシマコナスビ
239710 サワトラノオ
239719 ヒメミヤマコナスビ
239727 ハマボッス
239728 ベニバナハマボッス
239729 オオハマボッス
239755 コバンコナスビ,ヨウシュコニスビ
239761 ヘツカコナスビ
239781 バイカトラノオ
239789 イヌヌマトラノオ
239847 モロコシサウ,モロコシソウ,ヤマネクンボ
239876 ミヤマコナスビ
239877 オニコナスビ
239881 ヤナギトラノオ,ヤナギトラノヲ
239898 イワウサウ,セイヨウクサレダマ,ヒロハクサレダマ
239901 ヒロハクサレダマ
239902 イオウソウ,イワウサウ,クサレダマ,セイヨウクサレダマ
239926 シシンラン属
239967 シシンラン,タイワンシシンラン
240020 ミソハギ科
240025 ミソハギ属
240030 サウハギ,シャウリャウバナ,タマノヤグサ,ミソハギ,ミゾハギ
240041 アメリカミソハギ
240047 コメバミソハギ
240068 エゾミソハギ,ミソハギ
240072 ケナシエゾミソハギ
240112 イヌエンジュ属
240114 イヌエンジュ,オオエンジュ,カライヌエンジュ,クロユンジュ,ユンジュ
240124 ハネミイヌエンジュ
240130 タイワンエンジュ
240131 シマユンジュ
240207 マカダミア属
240209 マカダミア
240210 クインスランドナット,クインスランドナットノキ
240226 オホバギ属
240325 オホバギ
240368 ウングイス-カティ
240478 ヒラアンペライ
240481 ネビキグサ
240511 マカエロケレウス属,マセレウス属
240516 マカイロフィルム属
240550 タブノキ属
240559 オガサワラアオグス
240604 アオガシ,アヲガシ,ホソバタブ,ホタウガシ
240605 オホバタブ
240606 コブガシ
240625 ナガバイヌグス
240658 コウシュンクスノキ,タマゴバクスノキ
240680 タブガシ
240707 アリサンタブ,イヌグス,イブノキ,タブ,タブノキ,タマグス,タモ
240726 アオグスモドキ,アヲグスモドキ,ズイハウダモ,ズヰハウダモ,ニホヒタブ
240735 マカカヤ属
240763 タケニグサ属,タニケグサ属
240765 オホカミグサ,オホカメダフシ,ササヤキグサ,タケニグサ,チャンパギク
240766 ケナシチャンパギク
240767 マルバタケニグサ
240806 アメリカハリグワ属,オセ-ジオレンヂ属
240813 アマズラ,カカツガユ,カクワツガユ,ソンノイゲ,ヤマミカン
240828 アメリカハリグワ,オセ-ジ・オレンジ
240842 ドミヤ,ハリグハ,ハリグワ
240860 マコデス属
240862 ナンバンカモメラン
240901 マクラデニア属
240965 サンシュユ属
241047 ウコンウツギ
241144 マクロレリア属
241248 ハクセンナズナ属,ハクセンナヅナ属
241250 ハクセンナズナ,ハクセンナヅナ
241261 クロバナツルアズキ
241263 ナンバンアカバナアズキ
241448 オニザミア属
241449 ヤブオニザミア
241454 マタバオニザミア
241457 ホソバオニザミア
241458 ナガバオニザミア
241464 マガオニザミア
241465 ネジレオニザミア
241734 イジセンリョウ属,イズセンリョウ属,イヅセンリャウ属,イヅセンリョウ属
241774 ナガバイズセンリョウ
241781 イズセンリョウ,イヅセンリャウ,ウバガネモチ,カシラン,カシワラン,シロウメモドキ,ツルセンリョウ,ミカドガシワ
241808 シマイズセンリョウ
241817 カラウバガネ,シナセンリャウ,シマイヅセンリャウ
241818 シマイズセンリョウ,シマイズセンリョオ,タイワンセンリャウ
241844 カラウバガネ,シナセンリョウ,シマイズセンラリャウ,シマイズセンラリョウ,タイワンセンリョウ
241951 ホオノキ属,ホホノキ属,モクレン属
241955 シバコブシ
241962 キモクレン
242135 タイサンボク,ハクレンボク
242169 キタコブシ,コブシ,コブシハジカミ,ヒキザクラ,ヤマアララギ,ヤマモクレン
242171 エゾコブシ,キタコブシ,ヒキザクラ
242193 オオバタイサンボク
242260 コブシモドキ
242274 カムシバ,ゴマガラ,サタウシバ,サトウシバ,タムシバ,ニオイコブシ
242285 オオヤマレンゲ
242331 シデコブシ,ヒメコブシ
242341 ヒメタイサンボク
242352 ウケザキオオヤマレンゲ,ギョクスイ,ヨクセイ
242370 モクレン科
242458 ヒイラギナンテン属,ヒラギナンテン属
242478 ヒイラギメギ
242487 シナヒイラギナンテン
242534 オソバヒヒラギナンテン,フォーチュネイ-,フォーチュン,ホソバノヒイラギナンテン,ホソバヒイラギナンテン,ホソバヒヒラギナンテン,ホソバヒラギナンテン
242563 オホバヒラギナンテン,チクシヒラギナンテン,テウセンナンテン,トウナンテン,ヒイラギナンテン,ヒラギナンテン
242580 オイワケヒイラギナンテン,ニイタカヒイラギナンテン
242608 オヒワケヒラギナンテン,ニヒタカヒラギナンデン,モリソンヒラギナンデン
242668 マイヅルソウ属,マヒヅルサウ属,マヤンテムム属
242672 コマイヅルソウ,シラヤマアフヒ,ヒメマイヅルソウ,マイヅルソウ,マヒヅルサウ
242679 カラフトユキザサ
242680 オオマイヅルソウ,マイヅルソウ
242691 オホボユキザサ,タイワンユキザサ,ヒロバユキザサ,ユキザサ

242706 トナカイサウ,ミツバザサ
242717 マイフェニア属
242720 マイフェニオプシス属
242792 ハナハツカ属,マヨラ－ナ属
242882 キラギョク
242883 ボウホウギョク
242886 シュウケンマル
242887 キンオウマル
242890 シシギョク
242891 セイビマル
242893 ハッケンギョク
242974 クハイタビ属
242976 キダチネヂレバナ,クハイタビ,ネヂレギ
242993 ヤチラン属
243010 シマホザキラン
243042 ハハジマホザキラン
243084 アリサンホザキイチエフラン,ホザキイチョウラン
243156 ヒメアラセイトウ属
243209 ハマアラセイトウ,ヒメアラセイトウ
243252 マレフォーラ属
243253 ホウカ
243261 イワヤマブキ
243266 ハナゼノ
243315 アカメガシハ属,アカメガシワ属
243371 アカガシハ,アカカヂ,アカベ,アカメガシハ,アカメガシワ,アズサ,カシワギ,ゴサイバ,シャウグンボク,ヒサキ
243395 ヤンバルアカメガシハ
243420 アンナンアカメガシハ,ウラジロアカメガシハ,ウラジロアカメガシワ,ホザキアカメガシハ
243427 カマラ,クスノハガシハ,クスノハガシワ
243437 ツルアカメガシハ,ツルアカメガシワ
243453 シマアズサ,タイワンアカメガシハ
243472 マロ－ペ属
243519 バルバドスチエリ－属,ヒイラギトラノオ属,マルピギア属
243523 ヒイラギトラノオ
243526 バルバドスサクラ,バルバドスチユリ－
243530 キントウノオ科,キントウノヲ科,キントラノオ科
243543 リンゴ属
243551 クラブリンゴ,チョウセンリンゴ,ワリンゴ
243555 アラフトズミ,エゾズミ,エゾノコリンゴ,ガラフトカイドウ,サンナシ,シベリアリンゴ,マンシュウズミ
243580 エゾリンゴ,ヒメリンゴ
243610 タイワンリンゴ
243617 カイダウズミ,カイドウズミ
243623 カイダウ,カイドウ,スイシカイダウ,スイシカイドウ,ハナカイドウ
243624 ヤエカイダウ,ヤエカイドウ
243625 スイジカイドウ
243630 チャカイダウ,チャカイドウ,ツクシカイダウ,ツクシカイドウ
243649 エゾズミ,エゾノコリンゴ,カラフトカイダウ,カラフトズミ,サンナシ,ヒロハオオズミ,マンシウズミ
243657 カイダウ,カイドウ,ナガサキズミ,ナガサキリンゴ,ミカイダウ,ミカイドウ
243667 イヌリンゴ,カイダウリンゴ
243670 オオバイヌリンゴ
243672 マルバカイドウ
243675 オオリンゴ,セイヤウリンゴ,セイヨウリンゴ,リンゴ,ワリンゴ
243702 ホンカイドウ
243710 ノカイダウ,ノカイドウ,ヤマカイダウ
243717 エゾズミ,キリンボク,コナシ,コリンゴ,サナシ,ズミ,トシミカイドウ,ヒメカイドウ,ミツバカイダウ,ミツバカイドウ,ミナリカイドウ,ミヤマカイドウ,ヤブリンゴ,ヤマナシ
243729 オオウラジノキ
243735 オオズミ
243738 ゼニアオイ属,ゼニアフヒ属
243771 オカノリ
243797 ゼニアオイ
243803 シャコウアオイ,ジャコウアオイ,ジャコオアオイ
243810 ウサギアオイ
243823 ゼニバアオイ,ナガエアオイ,ハイアオイ,ハヒアフヒ
243840 ウスベニアオイ,ウスベニアフヒ,ゼニアオイ,セニアフヒ
243850 ゼニアオイ,セニアフヒ
243862 アオイ,アフイ,アフヒ,カンアオイ,カンアフヒ,フユアオイ,フュアフイ,フュアフヒ
243873 アオイ科,アフヒ科
243877 アオイモドキ属,アフヒモドキ属,エノキアオイ属,マルバストラム属
243887 ホシアオイ
243893 エノキアオイ
243927 ヒメフヨウ属
243929 タイリンヒメフヨウ,ヒメブッソウゲ,ヒメフヨウ
243932 ヒメフヨウ
243947 タイリンヒメフヨウ
243974 ゲッキュウデン属
243977 ゲッキュウデン
243979 マンメア属
243994 マル属
243999 ハクヨウマル
244005 マイボシ
244012 タカザゴ
244015 ホウメイマル
244017 コトイトマル
244018 セツパクマル
244024 キジンマル
244026 コンゴウマル
244031 セットウマル
244033 カンバク
244035 コウリンマル,ニコンゴウ
244037 ハクリュウマル
244047 ハクウンマル
244054 タンドクマル
244061 チッゲッガ
244081 ハクギョクト,ハクジュマル
244085 ギンテマリ,ギンモウマル
244090 クンレイマル,リキュウマル
244097 タマオカナ,タマオキナ
244104 ハクチョウ
244105 ミユキマル
244124 コウフクマル
244134 キンボシ
244137 ダイコウマル
244141 ムゲンジョウ
244152 ホウメイデン
244154 アサギリ
244164 ハクヒョウマル
244171 キンギンツカス
244182 ハクオウマル
244187 カゲロウ
244192 シロボシ
244198 マツガスミ
244210 アサヒマル
244224 カイジンマル
244234 ニシキマル
244250 コンゴウセキ
244254 ミヤコドリ
244258 ナナコマル
244259 ダイユウマル
244260 ムラサキマル
244320 チリ－ソケイ属,チリソケイ属
244327 チリ－ソケイ
244346 マンドラゴラ
244358 カエンソウ属,マネッチア属
244363 カエンソウ
244367 アラゲカエンソウ
244386 マンゴウノキ属,マンゴウ属,マンゴ－属
244390 ゼンゼイ,ビンゼイ
244397 マンゴ－,マンゴウ,マンゴウノキ
244495 フクロヤシ属
244496 チミテフクロヤシ
244501 イモノキ属,マニホット属
244506 アマタピオカノキ
244507 イモノキ,カッサバ,キャッサバ,タカノツメ,タピオカノキ,マニオク,マニホットノキ
244509 セアラゴムノキ,セ－ラゴムノキ,マニホットゴムノキ

244520 サポジラ属
244605 サノニサポタ,サポジラ,チューインガムノキ
244728 キダチキツネアザミ
244892 コウトウウラジロマオ属,コウトウウラジロマヲ属
244895 コウトウウラジロマオ,コウトウウラジロマヲ
245013 クズウコン属,マランタ属
245014 クズウコン
245015 フイリクズウコン
245016 フタイロマランタ
245026 モンヨウショウ
245027 ヒョウモンヨウショウ
245039 クズウコン科,クヅウコン科
245152 マレノプンティア属
245155 グンリュウ
245220 アカハダコバンノキ
245252 ハクウンカク
245332 クグ属
245351 オニクグ
245376 ビトウクグ
245392 シマクグ,タイワンクグ
245453 オニクグ
245518 シンチククグ
245556 イヌクグ,クグ
245723 ニガハクカ属,ニガハッカ属
245770 ニガハクカ,ニガハッカ,マルバハクカ
245777 キジョラン属
245803 アフリカシタキヅル,マダガスカルシタキソウ
245805 タイワンキジョラン
245854 アイカヅラ,アヰカヅラ,ソメモノカズラ,ソメモノカヅラ
245858 アイカズラ,ソメモノカズラ
245860 キジョラン,シマイケマ,フヨウラン
245977 マルチニア属
245991 ツノゴマ科
246026 トックリヤシ属
246028 トックリヤシモドキ
246101 マスデバリア属
246307 カミツレ属,カミルレ属,シカギク属
246315 シカギク
246375 オロシャギク,コシカギク
246396 カミツレ,カミルレ
246438 アラセイトウ属
246443 ヨルザキアラセイトウ
246478 アラセイトウ,ストック
246479 アラセイトウ,コアラセイトウ
246590 マツカナ属
246593 サイセンギョク
246594 トウセンギョク
246595 ハクセンギョク
246601 ウズシオ
246603 セイセンギョク
246649 ツルキンギョソウ
246651 キリカズラ属
246653 ツタバキリカズラ
246654 キリカズラ
246656 コキリカズラ
246665 ノースポールギク
246671 テングヤシ属
246672 オオミテングヤシ
246675 チャビテングヤシ属
246676 トゲチャビテングヤシ
246704 マキシラーリア属,ルルニティディウム属
246731 マキシミリヤンヤシ属,マクシミリァンヤシ属
246964 サギゴケ属,マズス属
246978 セイタカサギゴケ
246980 ヒメサギゴケ
246987 シロバナトキワハゼ
247008 サギゴケ,ムラサキサギゴケ
247009 シロバナサギゴケ
247010 ジャカゴソウ
247011 モモイロサギゴケ
247012 ヤマサギゴケ
247025 トキハハゼ,トキワハゼ,ナツハゼ
247039 タチサギゴケ
247099 メコノプシス属
247239 ウマゴヤシ属
247246 モンツキウマゴヤシ
247262 トゲミノウマゴヤシ
247337 キレハウマゴヤシ
247366 カラハナウマゴヤシ,コメツブウマゴヤシ
247381 コウマゴヤシ
247397 マルミウマゴヤシ
247425 ウマゴヤシ,ニセウマゴヤシ,マゴヤシ
247431 トゲナシウマゴヤシ
247452 マンシウウマゴヤシ
247456 アルフアルフア,ムラサキウマゴヤシ
247457 コガネウマゴヤシ
247469 ウズマキウマゴヤシ,カギュウソウ
247493 タルウマゴヤシ
247516 メディコスマ属
247526 ノボタンカヅラ属,メディニラ属
247564 ノボタンカズラ,ノボタンカツラ
247648 メディォカクタス属
247664 メディオロビビア属
247667 ギョウエイマル
247668 ビトウマル
247669 コクエイマル
247670 ハナチゴ
247671 リョクキマル
247672 コウシンマル
247674 カホウマル
247720 ラシャウモンカヅラ属,ラショウモンカズラ属
247731 オチフジ
247739 ラシャウモンカヅラ,ラショウモンカズラ,ラショオモンカズラ,ルリテフサウ
247741 シロバナラショウモンカズラ
247743 モモイロラショウモンカズラ
248072 コバノブラッシノキ属
248099 コバノブラッシノキ
248104 カュプッド,カュプテ,ゲラム
248155 ママコナ属
248161 タカネママコナ
248175 シコクママコナ,ミヤマママコナ
248177 ミヤジマママコナ
248178 シロバナシコクママコナ
248179 タカネママコナ
248181 ミヤマママコナ
248182 シロバナミヤマママコナ
248184 ヤクシマママコナ
248188 オオママコナ
248195 ツシマママコナ,ミシマママコナ
248196 シロバナツシマママコナ
248201 ママコナ
248202 シロバナママコナ
248203 ツシママ
248204 マルバママコナ
248205 シロバナマルバママコナ
248208 ホソバママコナ
248210 オオホソバママコナ,ヒカゲママコナ
248213 エゾママコナ
248235 フシグロ属
248255 タカネマンテマ
248269 ケフシグロ,ヒメケフシグロ,ヒメフシグロ,モリビランデ
248271 シロバナヒメケフシグロ
248382 ツキミセンノウ
248500 ヤンバルアカメガシワ
248718 メラスフェヶルラ属
248722 ウスキヒメショウブ
248726 ノボタン属
248727 コウトウノボタン,コウトウボタン
248732 コウトウノボタン,ノボタン
248734 シロバナノボタン
248735 イオウノボタン
248762 マラバルノボタン
248778 オオナンヨウノボタン
248786 ムニンノボタン
248787 ハハジマノボタン
248806 ノボタン科
248822 ノジアオイ属,ノヂアフヒ属
248893 センダン属
248895 アウチ,アウチノキ,アラノキ,シンラン,センダ,センダン,タイワンセンダン
248896 シロバナセンダン
248899 アイノコセンダン

248903 アフチ,アフチノキ,アラノキ,オオチ,センダ,センダン
248925 トウセンダン
248929 センダン科
248949 コメガヤ属
249015 アオコメガヤ,アヲコメガヤ,シロコメガヤ
249025 アオコメガヤ
249054 コメガヤ
249060 ハナビガヤ,ミチシバ
249061 ウスゲミチシバ
249081 イトバコメガヤ
249085 チョウセンミチシバ,フサコメガヤ
249105 オオコメガヤ
249110 ホナガコメガヤ
249124 メリコッカ属
249143 オオバシロテツ
249144 アツバシロテツ
249156 オガサワラゴシュユ,ムニンゴシュユ
249163 シロテツ
249168 アワダン,ミツバゴシュユ
249197 シナガハハギ属,シナガワハギ属
249203 ココメハギ,シロバナノシナガハハギ,シロバナノシナガワハギ
249205 セイタカコゴメハギ
249212 モウコエビラハギ
249218 コシナガワハギ
249232 エビラハギ,シナガハハギ,シナガワハギ,セイヤウエビラバギ,セイヨウエビラバギ,メリロート
249237 ヒシバシナガワハギ
249303 トミツグラス
249355 アハブキ属,アワブキ属
249369 アリサンアハブキ,ウスバアハブキ
249395 サクダモ,サクノキ
249414 アハブキ,アブクタラシ,アワキ,アワブキ
249415 シロミアワブキ
249424 ヌルデアワブキ,フシノハアワブキ,リュウキュウアワブキ
249445 ハビバアハブキ,ハビバアワブキ,フシノハアワブキ,ヤンバルアハブキ,リウキウアハブキ
249449 シマアワブキ,スグワ,スゴノキ,タイワンアハブキ,タイワンアワブキ,ヤマビハ,ヤマビワ
249463 ナンバンアハブキ,ナンバンアワブキ,リュウキュウアワブキ
249468 ミヤマハハソ
249491 セイヤウヤマハクカ属,セイヨウヤマハッカ属
249504 カウスイハクカ,カウスヰハクカ,コウスイハッカ,セイヤウヤマハクカ,セイヨウヤマハッカ,メリッササウ,メリッサソウ
249506 フイリセイヨウヤマハッカ
249578 メロカクタス属
249580 ソウウン
249586 ヒウン
249590 サイラン
249591 ロウウン
249592 カクウン
249597 ゲンウン
249601 カウン
249603 ショウウン,スイウン
249617 ナシタケ
249627 ノジアオイ属,ノヂアフヒ属
249631 キダチノジアオイ
249633 ノジアオイ,ノヂアフヒ,ヤマカハゴジクワ
249648 シマダカツラ属
249649 シマダカツラ,ムベカヅラ
249760 スズメウリ属
249945 ソメモノコメツブノボタン
250217 ツヅラフジ科,ツヅラフチ科
250218 カウモリカヅラ属,コウモリカズラ属
250228 カウモリカヅラ,ケナシコウモリカズラ,コウモリカズラ
250229 コウモリカズラ
250287 ハクカ属,ハッカ属,メンタ属
250291 ヌマハッカ
250294 ノハッカ,ヨウシュハッカ
250331 ハッカ
250341 ベルガモットハッカ,ルガモットハッカ
250353 ダウリアハッカ
250362 アメリカハッカ
250370 ハクカ,ハッカ
250376 ハッカ
250381 ヒメハッカ
250382 ハイヒメハッカ
250385 ナガバハッカ
250420 セイヨウハッカ
250432 ペニロイアルハッカ,メグサハッカ
250443 エゾハッカ,ハッカ
250450 オランダハッカ,ミドリハッカ
250457 マルバハッカ
250481 メンツェ-リア属
250486 ニオイシレンゲ
250492 ミツガシワ科
250493 ミヅガシハ属,ミツガシワ属
250502 ミズガシワ,ミヅガシハ,ミツガシワ
250505 ヨウラクツツジ属
250507 ウスギョウラク,サイリンヨウラク,ツリガネツツジ
250509 アキツリガネツツジ
250510 アラゲツリガネツツジ
250514 ゴヨウザンヨウラク
250515 ウラジロコヨウラク
250516 ホザキツリガネツツジ
250517 フタエツリガネツツジ
250518 ムラサキツリガネツツジ
250519 ハコネツリガネツツジ
250520 ウラジロコウラク,ウラジロコウラクツツジ,ウラジロヨウラク,ウラジロヨオラク
250521 アズマツリガネツツジ
250526 フジツリガネツツジ
250527 トネツリガネツツジ
250528 コヨウラクツツジ
250529 ツリガネツツジ
250532 コヨウラクツツジ,ヨウラクツツジ,ヨオラクツツジ
250533 ヤクシマヨウラクツツジ
250597 ヤマアイ属,ヤマアヰ属
250600 ゼイヤウヤマアイ,ゼイヤウヤマアヰ,セイヨウヤマアイ
250610 ヤマアイ,ヤマアヰ
250627 メレンデラ属
250749 コガネヒルガオ属,ツタノハヒルガホ属,フウセンアサガオ属
250779 ヒメコガネヒルガオ
250792 ツタノハヒルガオ
250812 ハスノハヒルガオ
250821 コバノモミジアサガオ
250831 モウコアサガオ
250850 ウッドローズ
250853 ミツバフサアサガホ
250854 ミミバフサアサガオ
250883 ハアベンケイソウ属,ハマバンケイサウ属,ハマベンケイソウ属,メルテンシア属
250895 ハマバンケイサウ,ハマベンケイソウ
250896 シロバナハマベンケイソウ
250905 エゾルリソウ,チシマルリソウ
250906 エゾルリソウ
250907 タカオカソウ
251025 オスクラリア属,マツバギク属
251194 コトツメギク
252241 センサウ属,センソウ属
252242 センサウ,ハヒメソナ,ヒメセンサウ
252277 メソスピニディア属
252291 セイヨウカリン属
252305 セイヨウカリン,メドラ-
252333 スナビキサウ属,スナビキソウ属
252361 メストクレマ属
252369 テツザイノキ属
252370 セイロンテッボク,セイロンテリハボク,タガヤサン,テッサイノキ,テッザイノキ
252508 ガガイモ属
252514 ガガイモ
252515 シロバナガガイモ
252538 メタセコイア属
252540 アケボノスギ,ヌマスギモドキ,メタセコイア

252577 チャボツメレンゲ属
252578 チャボツメレンゲ
252607 メトロシデ-ロス属
252609 オガサワラフトモモ,ムニンフトモモ
252631 サゴヤシ属,メトロキシロン属
252633 タイヘイヨウゾウゲヤシ
252634 サゴヤシ,トゲサゴヤシ
252636 サゴヤシ,マサゴヤシ
252697 マイエロフィツム属
252803 オガタマノキ属,ヲガタマノキ属
252809 ギョクラン,ギンコウボク
252817 ギオイ
252841 キンカウボク,キンコウボク
252849 オガタマ,オカタマノキ,オガタマノキ,トキハコブシ,トキワコブシ,ヲガタマ,ヲガタマノキ
252850 ホソバオガタマノキ
252851 タイワンオガタマノキ
252855 ヒロハオガタマノキ
252869 カラタネオガタマ,タウオガタマ,タウヲガタマ
252907 ギョクラン,ギンカウボク,ギンコウボク
253003 オオバノボタン属,ミコニア属
253092 ミクラントセレウス属
253093 レイソウオウ
253197 スズメノハコベ属
253200 スズメノハコベ
253284 ミクロシトラス属
253285 オ-ストラリアホソミライム,ミクロシトラス
253288 オ-ストラリアマルミライム
253289 ガロウエイオ-ストラリアワイルドライム
253291 マイデンオ-ストラリアワイルドライム
253293 ニュ-ギニアワイルドライム
253315 ミクロケリア属
253399 ミクロシカス属
253450 シマイヅハハコ属
253476 シマイズハハコ
253479 シマイヅハハコ,ツルイヅハハコ,ツルハハコ
253816 ミクロプテルム属
253832 ミクロプンティア属
253833 グンモウゾウ
253980 アシボソ属
253992 オオササガヤ,オホササガヤ,フササガヤ
253997 ヒメアシボソ
254005 フオ-リ-ササガヤ
254010 ススキササガヤ
254012 ヒメササガヤ
254020 ササガヤ
254022 キタササガヤ
254034 アシボソ
254036 アリサンササガヤ,ササガヤ,ミヤマササガヤ
254043 メンテンササガヤ
254046 アシボソ,ヒメアシボソ
254226 ニラハラン属,ニラバラン属
254238 ニラハラン,ニラバラン
254283 アリサンモクレイシ属,モクレイシ属
254293 アリサンモクレイシ,マツダモクレイシ
254303 コウトウモクレイシ,ハマモチ,フクボク,モクレイシ
254304 ホソバモクレイシ
254312 コバナモクレイシ
254414 ツルギク属
254424 ツルギク
254443 ツルギク,ツルヒヨドリ
254466 ミラ属
254467 グンコヅチ
254468 タカラコヅチ
254469 コトブキコヅチ
254496 イブキヌカボ属
254515 イブキヌカボ
254517 マツゲイブキヌカボ
254562 ミラ属
254564 ナガエアマナ
254596 ナツフジ属,ナツフヂ属
254796 ドクフジ
254803 クロヨナ
254854 アッカハマメ,アッカワマメ,カハマメカヅラ,カワマメカズラ,ギョトウ,ドクフヂ
254924 ミルト-ニア属
254927 トラフコチョウ
254979 オジギソウ属,ネムリグサ属
255098 オジギソウ,オヂキサウ,ネムリグサ
255134 シガフグワン
255141 ネムノキ科
255184 サワホオズキ属,ミゾホオズキ属,ミゾホホヅキ属
255186 ニシキミゾホオズキモドキ
255187 ヒレミゾホオズキ,ヤハズミゾホオズキ
255197 ニミゾホオズキ,ベニバナミゾホオズキ
255210 セイタカミゾホオズキ
255218 ニシキミゾホオツキ,ニシキミゾホホツキ
255221 アメリカミゾホオズキ
255223 アメリカミゾホオズキ,ジャコウミゾホオズキ,ニオイホオズキ,ニオイミゾホオズキ,ミムラス
255242 オオバミゾホオズキ
255246 ヒメミゾホオズキ
255253 ミゾホオズキ,ミゾホホヅキ
255435 タカネツメクサ属
255442 エゾタカネツメクサ
255447 ハイツメクサ
255506 タイリンツメクサ
255598 コケツメクサ,コバノツメクサ,ホソバツメクサ
255672 オシロイバナ属
255711 オシロイバナ,ユウゲショウ
255727 ナガバナオシロイ,ナガバナオシロイバナ
255729 ナガバナオシロイバナ
255735 イヌオシロイバナ
255790 オオモミジガサ,トサノモミジソウ
255827 ススキ属
255828 ムニンススキ
255838 ハチジョウススキ
255849 アリハラススキ,オニススキ,カンススキ,トキハススキ,トキワススキ
255852 オオヒゲナガカリヤスモドキ
255867 カリヤスモドキ
255868 ケカリヤスモドキ
255871 シナノカリヤスモドキ
255873 ウミガヤ,オギ,オギヨシ,ムラサキススキ,ヲギ,ヲギヨシ
255886 オバナ,カヤ,コガネススキ,ススキ,スズキ,タカネススキ,ハチジョウススキ,ホソススキ,ヲバナ
255887 イトススキ
255893 シマスズキ,タカノハススキ
255896 シマススキ
255918 オウミカリヤス,カリヤス,ヤマカリヤス
256001 ツルアリドオシ属
256005 ツルアリドオシ,ヒメツルアリドオシ
256010 チャルメルサウ属,チャルメルソウ属
256012 ミノチャルメルソウ
256018 ヒメチャルメルソウ
256019 タイワンチャルメルソウ
256020 ミカワチャルメルソウ
256021 チャルメルソウ
256022 エゾノチャルメルソウ
256024 オオチャルメルソウ,チャルメルソウ
256027 ツクシチャルメルソウ
256028 コシノチャルメルソウ
256032 マルバチャメルソウ,マルバチャルメルソウ
256036 コチャルメルソウ
256038 タキミチャルメルソウ
256039 シコクチャルメルソウ
256041 トサノチャルメルソウ
256147 アイナエ属,アヰナヘ属
256154 アイナエ,アヰナヘ,ヒメナエ
256161 アイナエ,アヰナヘ
256171 ヤッコソウ
256175 ヤッコサウ属,ヤッコソウ属

256178 タイワンヤッコサウ
256180 ヤッコサウ,ヤッコソウ
256182 ヒシガタヤッコサウ,ヒシガタヤッコソウ,ヤッコソウ
256186 ヤッコソウ科
256257 ミトロフィルム属
256260 ソウシチョウ
256264 ゲンソウチョウ
256268 フシチョウ
256270 メイソウチョウ
256271 カイキチョウ
256301 ジムナスタ－属,ミヤマヨメナ属
256429 キクノハアオイ
256441 オオヤマフスマ属,オホヤマフスマ属,タチハコベ属
256444 オオヤマフスマ,オホヤマハコベ,オホヤマフスマ,ヒメタガソデソウ
256456 エゾフスマ,タチハコベ
256582 ヒロハキンバイザサ
256597 ヌマガヤ属
256601 ヨウシュヌマガヤ
256632 ススキヨシ,ヌマガヤ,ヤチガヤ
256684 ザクロソウ科
256689 ザクロサウ属,ザクロソウ属
256711 ハナビザクロソウ
256719 ザクロサウ,ザクロソウ,マツバギク
256727 クルマバザクロサウ,クルマバザクロソウ,ヘラザクロサウ
256746 モルトキア属
256758 カイガラサルビア属
256761 カイガラサルビア,カイガラソウ
256780 ツルレイシ属,ニガウリ属,モモルディカ属
256797 ツルレイシ,ナガレイシ,ニガウリ
256804 ナンバンカラスウリ,モクベッシ
256966 モナデニウム属
257069 モナンセス属
257157 モナルダ属,ヤグリマクワクカウ属
257161 タイマツバナ
257169 モナルドバナ,ヤグリマクワクカウ,ヤグルマハッカ
257184 ケショウヤグルマハッカ
257187 サオトメハッカ
257371 イチゲイチヤクサウ属,イチゲイチヤクソウ属,タイワンウメガササウ属,タイワンウメガサソウ属
257374 イチゲイチャクサウ,イチゲイチヤクソウ,タイワンウメガササウ,タイワンウメガサソウ
257395 モニラリア属
257398 コウリンホウ
257403 ヘキコウカン
257405 ホウセキコウ
257407 キコウギョク
257411 トブヒノ
257412 カンコウホウ
257537 クチナシグサ属
257538 カガリビソウ,クチナシグサ
257542 ウスユキクチナシグサ
257543 カガリビソウ,クチナシグサ
257564 ミズアオイ属,ミヅアフヒ属
257572 ミズアオイ,ミヅアフヒ
257573 シロバナミズアオイ
257583 コナギ,ササナギ,ミヅナギ
257587 ホソバコナギ
257687 ヤリノホアカザ
257708 アクロケネ属,モノメリア属
257906 モノタグマ属
258006 ギンリャウサウ属,ギンリャウソウ属,ギンリョウサウ属,ギンリョウソウ属,シャクジョウソウ属
258019 ハダカシャクジョウバナ
258033 シャクジョウソウ
258044 アキノギンリョウソウ,アキノギンリョオソウ,イウレイサウズヰシャウレン,イウレイソウ,イウレイタケ,イウレイバナ,ギンリャウサウ,ギンリョウソウモトギ,スイシュウラン,ユウレイソウモトギ
258049 ギンリョウソウ科,シャクジョウソウ科
258162 ホウライショウ属,ホウライセウ属
258168 ホウライショウ,ホウライセウ,モンセテラ
258170 マドカズラ
258177 ヒメモンステラ
258237 モンチア属
258248 ヌマハコベ
258258 マキバヌマハコベ
258336 モンビレア属
258378 クワ科
258380 モレ－ア属
258791 モリカンディア属
258793 イタリアソウ
258871 ハナガサノキ属,ヤエヤマアオキ属,ヤヘヤマアヲキ属
258882 アカダマノキ,ヤエヤマアオキ,ヤヘヤマアオキ,ヤヘヤマアヲキ
258910 コバノアカダマカヅラ
258923 アカダマカヅラ,ハナガサノキ
258924 コハナガサノキ,ムニンハナガサノキ
258925 ハハジマハナガサノキ
258926 ハナガサノキ
258935 モリンガ属,ワサビノキ属
258945 モリンガ,ワサビノキ
258961 ワサビノキ科
258987 モルモ－デス属
259000 モルモリカ属
259065 クハ属,クワ属
259067 カラクワ,カラヤマグハ,カラヤマグワ,クハ,クワ,ササグワ,トウクワ,マグハ,マグワ,ヤマグハ
259068 シダレグワ
259085 マルグハ,モチグハ,ログハ,ロサウ,ロソウ
259094 アカキイチベイ,イチベイ,シマノウチ,ダテイチベイ,ネゴヤタカスケ
259097 クワ,ササクワ,シマグワ,タイワンクワ,ハチジョウグワ,ヤマグワ
259099 ホソバヤマグワ
259100 ハマグワ
259117 オナガグワ
259121 オガサワラグワ
259122 ケグワ
259152 マルグワ,モチゲワ,ログワ,ロソウ
259165 テウセングハ,モウコグハ,モウコグワ
259180 クロミグワ,クロミノクワ
259186 アカミグワ
259195 ケグワ,ノグワ
259278 イヌカウジュ属,イヌコウジュ属
259282 ホソバヤマジソ
259284 イヌジソ,タイワンヒメジソ,ヒカゲヒメジソ,ヒメジソ
259292 ヒメジソ,ミゾカウシュ,ミゾカウジュ
259299 シラゲヒメジソ,ヒカゲヒメジソ
259300 ヤマジソ
259302 オオヤマジソ,オホヤマジソ
259305 タンナヤマジソ
259306 アオヤマジソ,シロバナヤマジソ
259323 イヌカウジュ,イヌコウジュ,イヌコオジュ,シロバナイヌカウジュ,シロバナヤマジソ
259328 モッシア属
259479 スチゾロビュ－ム属,トビカズラ属,ハッショウマメ属
259512 ウジルカンダ,タイワンクズマメ,タカモダマ,モクワンジュ,ワニグモチダマ
259519 コバノワニグチ
259520 オシャラクマメ,ハッショウマメ
259535 ウジルカンダ,クズノハカヅラ,クズモダマ,ヒメワニクチ
259541 カショウクズマメ
259545 クワセウクズマメ,ヤブハッショウマメ
259559 ハッショウマメ
259566 アイラトビカズラ,トビカズラ
259630 ネズミガヤ属
259636 フサガヤ
259646 コシノネズミガヤ,ミヤマネズミガヤ
259649 ミヤマネズミガヤ
259669 タチネズミガヤ
259671 アリサンネズミガヤ,オオネズミガヤ,オホネズミガヤ,ネズミガヤ
259672 ウシガヤ,ネズミガヤ
259691 キダチノネズミガヤ

259696 コネズミガヤ
259725 ミュイリア属,ムイリア属
259726 ホウキギョク
259730 タンチョウソウ属
259731 イワヤツデ,タンチョウソウ
259739 ザラメキスズメウリ,サンゴジュススズメウリ
259746 ムラサキノゲシ属
259772 ムラサキハチジョウナ
259910 ナンヨウザクラ
260095 タカサゴイボクサ
260102 イボクサ
260105 シマイボクサ
260110 シマイボクサ,タイワンイボクサ,ヒメイボクサ,ヒメツユクサ
260113 シナイボクサ
260158 ゲッキツ属
260168 オオバゲッキツ,オホバゲッキツ
260173 ゲッキツ
260198 バショウ属,バセウ属
260199 マレ-ヤマバショウ
260206 リュウキュウバショウ
260208 バショウ,バセウ,バセヲ
260215 センナリバナナ
260217 ビジンショウ,ビジンセウ,ヒナバショウ,ヒナバセウ,ヒメバショウ,ヒメバセウ
260228 タイワンバショウ,タイワンバセウ,ヤマバショウ,ヤマバセウ
260236 コウトウアバカ,コウトウバショウ,コウトウバセウ,コウトウヤマバショウ,コウトウヤマバセウ
260241 イトバショウ,リュウキュウバショウ
260252 リンゴバショウ
260253 バナナ,ミバショウ,レウリバショウ,レウリバセウ
260258 ダイワンバショウ,ダイワンバセウ,ヤマバショウ,ヤマバセウ
260275 マニラアサ,マニライトバショウ,マニライトバセウ
260284 バショウ科,バセウ科
260293 ムスカリ属
260301 ルリムスカリ
260375 コンロンカ属,コンロンクワ属
260414 ヒゴロモコンロンカ
260418 ケコンロンカ
260451 シロバナコンロンカ
260472 コンロンカ,コンロンクワ
260475 ヤエヤマコンロンカ
260483 ケコンロンカ,ケコンロンクワ
260499 ヒロハコンロンカ,マルバコンロンクワ
260530 ムティシア属
260577 ハエトリナズナ
260700 ハマジンチョウ科,ハマヂンチャウ科
260706 ハマジンチョウ属,ハマヂンチャウ属
260710 コハマジンチョウ
260711 キンギョシバ,ハマジンチョウ,ハマジンチョオ,ハマヂンチャウ,ハマベンケイ,ハマリンチョウ,モクベンケイ
260720 コバノハマヂンチャウ
260734 ミオソティディウム属
260737 ワスレナグサ属,ワスレナサウ属,ワスレナソウ属
260747 ノハラワスレナグサ,ワスレナグサ
260760 ノハラムラサキ
260769 イヌハナイバナ
260772 カブムラサキ
260791 ハマワスレナグサ
260807 エゾムラサキ
260813 タビラコモドキ
260815 ナヨナヨワスレナグサ
260868 シンワスレナグサ,ワスルナグサ,ワスレナグサ,ワスレナサウ
260870 コリンワスレナグサ
260892 エゾムラサキ,ミヤマワスレナサウ,ミヤマワスレナゾウ
260922 ウシハコベ
261049 ジャボチカバ
261060 ミヤオギク属,ミヤヲギク属
261069 キクタビラコ,ミヤオギク,ミヤヲギク
261071 ヒメキクタビラコ
261090 ジャモントウ属
261091 ヒメジャモントウ
261120 ヤチャナキ属,ヤマモモ属
261122 コウシュンヤマモモ,シマヤマモモ
261137 シロコヤマモモ
261162 ゼイヨウヤチャナキ,ヤチャナキ
261164 エゾヤマモモ,ヤチヤナギ
261212 ヤマモモ
261213 シロモモ,シロヤマモモ,ダンゴモモ
261237 ヤチヤナギ
261243 ヤマモモ科
261299 ミリオカルパ属
261337 フサモ属
261342 オオフサモ
261355 オグラノフサモ,オグラフサモ
261364 キンギョモ,ホザキノフサモ,ボザキノフサモ
261370 トゲボザキノフサモ
261377 タチモ
261379 キツネノオ,フサモ
261399 ハグマノキ
261402 ニカヅク属,ニクズク属,ニクヅク属
261412 コウトウニクズク
261424 シシズク,ニカヅク,ニクズク,ニクズクキ,ニクヅク
261457 アンタオニクヅク
261464 ニクズク科,ニクヅク科
261467 アリドウシラン属,アリドオシラン属,アリドホシラン属
261470 アリサンリドホシラン,タイワンアリドホシラン
261474 アリドウシラン,アリドオシラン,アリドホシラン
261478 ツクシアリドオシラン
261482 アリノスダマ属
261549 ミロキシロン属
261559 ペル-バルサムノキ
261576 ミリス属
261596 ヤブカウジ科,ヤブコウジ科
261598 ツルアカミノキ属,ツルマンリョウ属,ミルシネ属
261632 シマタイミンタチバナ,マルバタイミンタチバナ
261648 ソゲキ,タイミンタチバナ,ヒチノキ
261657 ツルアカミノキ,ツルマンリャウ,ツルマンリョウ
261686 テンニンクワ科,フトモモ科
261705 ミルチロカクタス属
261706 リュウジンバク,リュウジンボク
261708 センニンカク
261726 ギンバイカ属,ギンバイクワ属
261739 イハヒノキ,イワイノキ,ギンバイカ,ギンバイクワ
261772 ミスタシジュ-ウル属
261926 コウライオオニガナ
261992 イバラモ科
262013 イバラモ科
262015 イバラモ属
262018 ムサシモ
262024 サガミトリゲモ
262035 サガミトリゲモ
262039 イトトリゲモ
262040 ホッスモ
262067 イバラモ
262089 トリゲモ
262096 オオトリゲモ
262098 オホトリゲモ,サガミトリゲモ
262108 ヒメイバラモ
262116 イトイバラモ
262127 ナマクァンテス属
262149 ナナンッス属
262188 ナンテン属
262189 ナンテン
262200 シロミナンテン
262234 チャボウチワヤシ属
262235 リッチ-チャボウチワヤシ
262243 カテンサウ属,カテンソウ属
262247 カテンサウ,カテンソウ,ヒシバカキドオシ,ヒシバカキドホシ
262249 シマカテンソウ,トウカテンソウ,ヤヘヤマカテンサウ
262348 スイセン属

262361 ナ-シッサス・ボルボコデイウム
262405 アサギノヒトヘズイセン,カンランズイセン
262410 キズイセン
262417 ウスギズイセン
262420 ツツザキズイセン
262429 カンランズイセン,キブサスイセン
262433 ホンシロズイセン
262438 クチベニズイセン
262441 ラッパズイセン
262442 ヤエラッパズイセン
262457 エダザキズイセン,スイセン,フサザキズイセン,マルザキズイセン
262458 ヤエズイセン
262468 スイセン,ズイセン,スヰセン,フサザキズイセン
262497 カンショウコウ
262552 シマオバナ,シマヲバナ,ムラサキワセオバナ,ムラサキワセヲバナ
262645 イヌガラシ属,オランダガラシ属
262722 オランダガラシ,ミヅタカラシ
262795 タニワタリノキ属
262823 ゲンポール,タニワタリノキ
262960 ハシカグサ
262961 マルミノハシカグサ
262962 オオハシカグサ
262964 ヤクシマハシカグサ
263183 スグリウツギ
263198 シマソケイ
263199 ヤロオド,ヤロ-ド
263267 ハス属
263270 キバナハス,キバナバス
263272 ハス,ハチス,レングエ
263281 ハス科
263320 ヒポキルタ属
263374 アフリカウンラン属,ネメ-シア属
263410 アフリカウンラン
263459 ウンランモドキ,フクロウンラン
263501 ネモフィラ属,ルクカラクサ属
263504 モンカラクサ
263505 コモンカラクサ,ルリカラクサ
263512 タイワンサワギク
263515 サハギク,サワギク,ボロギク
263549 ネンガヤシ属,ネンガ属
263552 タケネンガ
263559 ネオアボッチア属
263560 キョウボクチュウ
263641 ネオベンタミア属
263650 ネオビンガミア属
263652 カカンチュウ
263697 ハクオウチュウ
263732 ネオキレニア属
263737 ショクロウギョク
263738 ハクロギョク
263739 オウメイギョク
263740 ウンサイギョク
263741 スイエンギョク
263742 シュウセンギョク
263743 トウグンギョク
263744 ヤミウシ
263745 シシャチギョク
263746 レイムギョク
263747 ヒョウトウ
263748 ライトウギョク
263749 ロウトウギョク
263750 コクカンマル
263752 タルタルギョク
263800 ネオダウソニア属
263802 ハナオキナ
263836 ミツヤヤシ属
263843 ミエミツヤヤシ
263865 フウラン属
263867 フウラン
264009 シロバナノビネチドリ
264011 シロダモ属
264015 イヌガシ,マツテニッケイ
264021 キンショクダモ,キンマウダモ,コガネシロダモ,シロダモモドキ
264029 オガサワラシロダモ
264048 ナガバシロダモ
264090 ウラジロ,オホシバキ,シロダブ,シロダモ,シロッヅ,タマガラ,ツヅ,ハナゴ,マカゴ
264091 ホソバシロダモ
264093 キミノシロダモ
264094 ダイトウシロダモ
264120 ネオロイディア属
264121 ミヤコニシキ
264158 ネオマリカ属
264163 アメリカシャカ
264205 ネガモ-レア属
264222 ハビロモドキ,マルバハナタマ,ランジンモドキ
264236 ニコルソンヤシ属
264249 ニオイヨモギ
264332 ネオポルテリア属
264334 コン
264337 ケイセキマル
264338 マリュウギョク
264339 オウリュマル
264342 アンコクギョク
264343 リュセイギョク
264344 コイマギョク
264348 ハクオウギョク
264350 シリュウギョク
264356 リュウジュマル
264357 ミロケマル
264358 タザイギョク
264360 ギンオウギョク
264362 クンリュウギョク
264365 ゴウリュウギョク
264366 ラソッギョク
264368 タマヒメ
264369 コクロギョク
264372 ギャクリュウギョク,リトウテン
264376 コンランギョク
264386 ネオライモンディア属
264387 ドセイカン
264388 タイショッカン
264406 ネオレゲリア属
264421 ミスタマアナナス
264426 ツマベニアナナス
264461 アサギシザサ
264506 シラキ
264650 サカネラン属
264651 ヒメムョウラン
264685 カイサカネラン
264689 ツクシサカネラン
264704 ニヒタカフタバラン
264709 エゾサカネラン,サカネラン
264711 サカネラン
264722 タカネフタバラン
264727 タカネフタバラン
264757 ミヤマモジズリ属
264765 ウズラモヂズリ,シヤマモジズリ,シヤマモヂズリ
264769 フジチドリ
264798 フィジ-ノヤシモドキ属
264809 ネオウエルデルマンニア属
264811 ヨウレイ
264812 テンレイ
264813 グンレイ
264827 ウツボカズラ科,ウツボカツラ科
264829 ウツボカズラ属,ウツボカツラ属,ネペンテス属
264832 ツボウツボカズラ
264834 ヒョウモンウツボ
264837 コウツボカズラ
264840 フッカ-ウツボ
264841 シビンウツボ
264845 ムラサキウツボカズラ
264846 ウツボカズラ,ウツボカツラ,ナンヨウウツボカズラ
264849 ウツボカズラ
264850 ミドリウツボ
264857 キエリウツボ
264858 ビロ-ドウツボ
264859 イヌハッカ属,カキドオシ属,カキドホシ属
264897 イヌハクカ,イヌハッカ,チクマハクカ,チクマハッカ
264946 カキドオシ,カントリソウ
264958 アリダサウ,ケイガイ
264977 トウイヌハッカ
265060 チョウセンミソガワソウ
265061 オオムシヤリンドウ
265064 ミソガワソウ
265065 シロバナエゾミソガワソウ
265115 ネフェラフィルルム属
265124 ネフェリウム属,ネフェリュ-ム属
265129 ランブ-タン

265136 マレ-ルュウガン
265137 バラサン,ボルネオランブタン
265195 イワイチョウ属
265197 アメリカイワイチョウ,イワイチョウ
265198 イワイチョウ
265202 アカエトグノヤシ属,アカトゲノヤシ属
265204 バンホウテコブダネヤシ,バンホウテヤシ
265211 ネフティティス属
265228 ミズオジギソウ属
265233 カイジンソウ,ミズオジギソウ
265238 オカミズオジギソウ
265254 ネリネ属
265299 ヒメヒガンバナ
265310 キョウチクトウ属,ケフチクタウ属
265327 キョウチクトウ,ケフチクタウ,ケヨウチクトウ,ケワチクタウ,セイヤウケフチクタウ,セイヨウキョウチクトウ,タイミンカ
265335 ヤエキョウチクトウ
265339 キョウチクトウ
265353 アリサンアハゴケ属,ネルテ-ラ属
265358 コケサンゴ,タマツヅリ
265359 アリサンアハゴケ,アリサンサウ,タイワンアワゴケ
265364 アマミアワゴケ
265366 ムカゴサイシン属
265373 アフヒボクロ,ヤエヤマヒトツボクロ,ヤヘヤマクマガユサウ
265412 ムカゴサイシン
265418 ヒロハヒトツバラン
265420 ヒロハヒトツバラン
265439 タイトウアフヒラン
265550 タマガラシ属
265553 タマガラシ
265916 ヨシガヤ
265923 ヨシガヤ
265940 オオセンナリ属,オホセンナリ属,ニカンドラ属
265944 オオセンナリ,オホセンナリ
266031 タバコ属
266035 シュクコンタバコ
266043 キダチタバコ
266048 ナガバタバコ,ナガバナタバコ,ハナタバコ,ホウライアマ
266053 マルバタバコ
266054 ハナタバコ
266060 タバコ
266062 タバコオオ
266063 アレチタバコ
266148 ウラベニアナナス属,ニヅラリュ-ム属
266155 ウラベニアナナス
266194 アマダマシ属,アマモドキ属
266197 アマダマシ,アマモドキ
266198 ヒメアマダマシ,ヒメアマモドキ
266206 ギンサカズキ,ギンパイソウ
266215 クロタネサウ属,クロタネソウ属
266225 クロタネサウ,クロタネソウ
266319 ハマギク
266355 ソ-ダノキ属
266367 ウモコソ-ダノキ
266371 ソ-ダノキ
266470 ノラナ属
266474 ノラナ
266475 ハイナス
266476 ノラナ科
266482 トックリラン属
266532 トウリンボク属
266548 ノモキリス属
266608 キバナムラサキ
266637 ノパレア属
266641 コウマノケン
266642 メイケンシ
266647 ノバ-ルホッキア属
266648 アカバナクジャク,ベニクジャク
266677 ノルマンビ-属
266678 ノルマンビ-ヤシ
266787 コニシタブ属
266792 コニシイヌグス,コニシタブ
266800 ワダツミノキ
266807 クサミズキ
266858 ナンキョクブナ科
266864 ナンキョクブナ
266893 ノソリリオン属
266957 ニラモドキ属
266959 ニラモドキ
266965 ハタケニラ
266970 ハナビニラ
266974 カサモチ属
266975 カウボン,カサモチ
267008 シリアアザミ
267027 ノトカクタス属
267030 コウチマル
267034 カカンマル
267043 キウンマル
267044 シラギョク
267045 セイオウマル
267046 ヘホウセイ
267049 コウカンマル
267051 コマチ
267053 オウコンコマチ
267057 サイリュウギョク
267058 シシオウマル
267060 バンギョク
267213 ノティリア属
267274 カハホネ属,コウホネ属
267275 サイジョウコウホネ
267286 カウホネ,カハホネ,コウホネ
267287 ベニコウホネ
267290 ナガバコウホネ
267313 オグラコウホネ
267314 ベニオグラコウホネ
267320 エゾカハホネ,エゾコウホネ,ネムロカハホネ,ネムロカワホネ,ネムロコウホネ,マンシウカワホネ
267322 オゼコウホネ
267327 ダイワンカハホネ
267331 ヒメコウホネ
267347 マツバウンラン
267349 オオマツバウンラン
267421 オシロイバナ科
267429 ニクタンテス属
267430 ヨルソケイ
267603 ニクトセレウス属
267605 ダイモンジ
267627 スイレン属,ニンフェア属,ヒツジグサ属
267641 オトメスイレン
267648 セイヨウスイレン
267658 ルリスイレン
267660 アフリカスイレン
267665 ケ-プスイレン
267684 アサギスイレン
267698 ヨザキスイレン
267714 キバナスイレン
267723 ホシザキスイレン
267730 シハイスイレン,シハイスヰレン,ニオイスイレン,ニオイヒツジグサ,ニホヒヒツジグサ
267756 アカバナスイレン,アカバナヒツジグサ
267767 エゾノヒツジグサ,オオヒツジグサ,ヒツジグサ,ヒツヂグサ
267770 スイレン,ヒツジグサ
267771 オオヒツジグサ
267773 エゾベニヒツジグサ
267777 アメリカスイレン
267787 スイレン科,スヰレン科,ヒツジグサ科
267800 アサザ属,リムナンセマム属
267801 ハナガガブタ
267802 ヒメアサザ
267807 ヒメシロアサザ
267818 ヒメガガブタ
267819 ガガブタ
267825 アサザ
267837 ニッパヤシ属,ニ-パ属
267838 ニッパヤシ,ニ-パヤシ
267849 ヌマミズキ属
267856 キルソ
267864 ニッサボク
267867 ヌマミズキ
267922 ヤウラクラン属,ヨウラクラン属
267927 アリサンクスクスラン,アリサンヤウラクラン,クスクスヤウラクラン
267935 サケバナヤウラクラン
267958 タイワンヤウラクラン,タカサゴヤウラクラン,ヒオウギラン,モミジラン,

ヨウラクラン
267959 ベニバナヨウラクラン
268000 オオバヨウラクラン
268085 オブレゴニア属,テイカン属
268086 テイカン
268132 オクナ属
268261 オクナ・セルラ－タ
268277 オクナ科
268341 バルサ属
268344 バルサ
268377 カワグチヒメノボタン
268418 オシマム属,メバハキ属
268438 バシル,メバウキ,メバハキ,メボウキ
268518 インドメボウキ
268523 インドメボウキ
268526 シラゲメボウキ
268614 カミバハキ
268642 タイワンメバハキ
268644 カミメボウキ
268845 オクトポマ属
268957 マスノグサ属
268986 マスノグサ
268999 マスノグサ
269006 イナバラン属,タシロラン属
269025 ハツシマラン
269026 イナバシュスラン,イナバラン,コウトウオホギミラン
269033 タイワンシュスラン
269040 ツシマラン
269046 オオギミラン
269055 オドントグロッサム属
269094 ツツサンゴバナ属
269100 ツツサンゴバナ
269114 オトントフォルス属
269292 セリ属
269300 テンヂクゼリ
269326 サケバゼリ,セリ
269335 イトバゼリ
269373 サケミヤシ属
269374 バカバヤシ
269375 オニサケヤシモドキ
269394 マツヨイグサ属,マツヨヒグサ属
269395 タマザキマツヨイグサ
269396 チャボツキミソウ,ツキミタンポポ
269404 メマツヨイグサ,メマツヨヒグサ
269437 オオメマツヨイグサ
269438 キダチマツヨイグサ
269441 シモフリマツヨイグサ
269443 オオマツヨイグサ,ツキミソウ
269446 オオバナコマツヨイグサ
269452 ミナトマツヨイグサ
269454 オニマツヨイグサ
269456 コマツヨイグサ
269457 マルバコマツヨイグサ
269464 ミズ－リマツヨイグサ
269475 マツヨイグサ,マツヨヒグサ,ヤハズキンバイ
269481 アレチマツヨイグサ
269486 ケヒナマツヨイグサ,ヒナマツヨイグサ
269498 アカバナノユウゲシャウ,アカバナユウゲショウ,ユウゲシャウ,ユウゲショウ
269504 ヒルザキツキミソウ
269506 モモイロヒルザキツキミソウ
269509 マツヨイグサ
269519 ツキミグサ,ツキミサウ,ツキミソウ
269522 ノハラマツヨイグサ
269536 エオピア属
269550 エオニエルラ属
269613 ウジクサ,ウジコロシ,ミソクサ,ミソナオシ,ミソナヲシ
269640 ボロボロノキ科
270058 オリ－ブ属,オリブ属,オレイフ属
270099 オリ－ブ,オリブ,オルトノキ,オレイフ,オレイフノキ,ホルトノキ
270110 コバノオレイフ
270182 ヒヒラギ科,モクセイ科
270184 オレアリア属
270234 ウラジロヒレアザミ
270247 ウニヒレアザミ
270677 オンファロ－デス属,ヤマルリソウ属,ルリソウ属
270696 ヤマウタイス,ヤマルリソウ
270697 シロバナヤマルリソウ
270698 トゲヤマルリソウ
270699 ルリソウ
270700 シロバナルリソウ
270701 エチゴルリソウ
270702 シロバナエチゴルリソウ
270705 シロウメソウ
270714 ハイルリソウ
270723 ハナルリソウ
270757 ココメタツナミソウ属
270758 ココメタツナミソウ
270773 アカベナ科
270787 オンキディウム属,オンシジウム属,オンシジュ－ム属
270799 オンシジウムケイロフオルム
270809 キバナスズメラン
270832 ヒメスズメラン
270841 ムレスズメラン
270846 スズメラン
270997 コブダネヤシ属
270998 ヤブコブダネヤシ
271001 トゲコブダネヤシ
271166 イガマメ属,オノブリキス属
271280 イガマメ
271290 ハリモクシュク属
271347 ハリモクシュク
271663 オオヒレアザミ属,オホヒレアザミ属
271666 オホヒレアザミ,ゴロツキアザミ
271692 オニウロコアザミ
271919 オオフィツム属
271920 クルミギョク
272049 ジャノヒゲ属
272086 ノシラン
272090 ジャノヒゲ,タツノヒゲ,ハズミダマ,リュウノヒゲ
272122 オオバジャノヒゲ
272125 シロバナオオバジャノヒゲ
272173 キダチイナモリサウ属,キダチイナモリソウ属,キダチイナモリ属,サツマイナモリ属
272187 オオイナモリ
272221 イナモリソウ,オキナワイナモリ,キダチイナモリ,キダチイナモリサウ,キダチイナモリソウ,サツマイナモリ,ジャコンサウ
272222 ヤエキダチイナモリ
272224 アマミイナモリ
272227 ヤマイナモリ
272228 ナガバイナモリ
272241 オキナハイナモリ,オキナワイナモリ,リウキウイナモリ,リュウキュウイナモリ
272272 コウトウイナモリ,シマイナモリ
272282 シマイナモリ,タイワンイナモリ,チャボイナモリ,ニチタカイナモリ,ヤエヤマイナモリソウ
272345 ヒメウシノシッペイ属,ヒメウシノシャバイ属
272350 ヒメウシノシッペイ
272395 オフリス属
272510 オフタルモフィルム属
272516 フウリンギョク
272530 レイザン
272532 セイレイギョク
272535 シュウレイギョク
272540 フウキンギョク
272580 カナビキボク科
272589 イワギリソウ属
272598 イワギリソウ
272599 シロバナイワギリソウ
272611 オプリスメヌス属,チヂミザサ属
272612 タイトンコブナシバ,ダイトンチヂミザサ
272614 リボンザサ
272625 アラゲチヂミザサ,エダウチチヂミザサ,オホエダウチチヂミザサ,オホバチヂミザサ
272629 オホエダウチチヂミザサ
272632 アラゲチヂミザサ
272676 オオバチヂミザサ,オホバチヂミザサ
272684 ケチヂミザサ,チヂミザサ
272691 ホソバチヂミザサ
272693 コチヂミザサ,チヂミザサ

272696 チャボチヂミザサ
272704 ハリブキ属
272706 チョウセンハリブキ
272707 アメリカハリブキ
272708 ヒロハハリブキ
272712 クマダラ,ハリブキ
272756 マヤヤシ属
272775 ウチハ属,ウチワサボテン属,ウチワ属,オプンティア属
272799 カンジンチョウキ
272812 アカバナウチワ
272818 ブラジルウチワ
272830 コチニ-ルウチワ
272849 オロチ
272856 キンブセン,センニンサボテン
272867 オオウチワ
272868 ツヤハダウチワ
272891 イロイロ,ウチワサボテン,サボテン,サンボテ,サンボテイ,ニョロリ
272898 アサヒ
272915 エンブセン,チヂミウチワ
272928 チリウチワ
272938 ダイゴクデン
272950 ギンセカイ
272951 アカマルチワ,ルリキョウ
272960 カエンダイコ
272976 オオガタホウケン
272980 キンエボシ
272981 ハクトウセン
272997 シロタエ
273029 ノギツネ
273043 シッケンウチワ
273067 ショウグン
273070 イオウウチワ
273103 タンシウチワ,ヒラウチハ,ヒラウチハサボテン
273120 クワズヤシ属
273123 ナガエクワズヤシ
273124 フィリビンクワズヤシ
273241 オルビグニ-ア属,オルビグニ-ヤシ属
273244 コフネヤシ
273267 ラン科
273317 ハクサンチドリ属
273427 オノエラン
273481 ノウカウチドリ
273583 チョウセンチドリ
273809 オレオセレウス属,モラウェッチア属
273812 ハクリョウリュウ
273819 ハクムオウ
273820 セイウンリュウ
273821 キョオウニシキ
273822 ライオンニシキ
273823 ハクテンマル
273824 ソウウンリュウ
273826 ハクウンニシキ
273827 サイオウニシキ
273830 イハギリサウ属,イハギリソウ属
273878 イワギリソウ
273899 ハドノキ属
273903 イハガネ,イワガネ,コショウボク,ヤブマオ
273906 イハガネ,コセウボク
273908 ハドイワガネ
273916 カハシャシャブ,ハダラ,ハドノキ,ハドハノキ
273927 オホイハガネ
273976 イシダサウ属,イシダソウ属
273979 イシダサウ,イシダソウ
274033 コケイラン属
274037 タマササギキコケイラン
274047 コハクラン
274058 コケイラン,ササエビネ,ヒメケイラン,ヒメコケイラン
274155 イシワリソウ属
274156 イシワリソウ
274197 ハナハクカ属,ハナハッカ属
274224 マジョラム,マヨラナ
274237 タイワンハナハクカ,ニヒタカジャカウサウ,ハナハクカ,ハナハッカ
274276 コクサギ属
274277 コクサギ,サワウルシ
274337 ハマユンジュ属
274343 ハマセンナ
274375 オルモ-シア属,ホリシャアカマメ属
274392 ベニマメノキ,ホリシャアカマメ
274477 オルニソケファルス属
274495 オオアマナ属,オホアマナ属,オルニソガルム属
274513 クロボシオオアマナ
274556 コモチカイソウ
274799 ホソバオホハマニラ
274823 オオアマナ,オ-ニソガラム,オホアマナ
274895 ツノウマゴヤシ
274921 ハマウツボ科
274922 ハマウツボ属
274952 チョウセンハマウツボ
275010 オカウツボ,ハマウツボ
275011 シロバナハマウツボ
275135 ヤセウツボ
275137 キバナヤセウツボ
275177 キバナハマウツボ
275347 イワレンゲ属
275356 ツメレンゲ
275363 エダウチレンゲ
275370 レブンイワレンゲ
275374 イワレンゲ
275375 ツメレンゲ
275376 ヤツガシラ
275379 アオノイワレンゲ,ゲンカイイワレンゲ
275382 アオノイワレンゲ
275383 ウスベニレンゲ
275384 コモチレンゲ
275385 イワレンゲ
275386 コレンゲ
275394 タウツメレンゲ
275405 オロヤ属
275407 クゼンギョク
275408 キョクビマル
275410 ビゼンギョク,レイゼンギョク
275480 コイチャクソウ属
275482 ヒナイチャク
275486 コイチヤクサウ,コイチヤクソウ
275576 オルトフィッム属
275601 オルトプテルム属
275607 ヒロハノハネガヤ属
275624 ネコノヒゲソウ属
275639 クミスクチン,ネコノヒゲソウ
275865 オオアラセイトウ属,ムラサキハナナ属
275876 オオアラセイトウ,ショカッサイ,ハナタイコン
275882 ケショカッサイ
275909 イネ属
275926 アフリカイネ
275957 ノイネ
275958 イネ,ウルチ,オカボ,コメ,ノネ,モチゴメ,ヲカボ
275966 モチイネ,モチゴメ
276072 ヒメノボタン属
276090 クサノボタン,ササバノボタン,ヒメノボタン
276159 セイロンヒメノボタン
276242 モクセイ属
276257 ギンモクセイ,モクセイ
276259 キンモクセイ
276260 ウズギモクセイ
276268 リュウキュウモクセイ
276286 ナンゴクモクセイ
276287 ヒラギモクセイ
276290 ヒイラギモクセイ
276291 ウスギモクセイ,ギンモクセイ,シキザキモクセイ,モクセイ
276292 キンモクセイ
276294 シロモクセイ
276295 ギンモクセイ
276297 ウスギモクセイ
276311 シマモクセイ
276313 シマモグセイ,タイワンモグセイ,タイワンモグセイモドキ,ヒイラギ,ヒヒラギ,ヒラギ,マルバモグセイ
276327 ヒイラギ
276336 サツマモクセイ,サツマモクゼイ,シマモクゼイ,ナタオレノキ,ナタハジキ,ハチジョウモクセイ
276337 キンナタオレノキ
276338 ヤナギバモクセイ

276341 ヤエヤマヒイラギ
276344 トガリバモクセイ,ホソバモクセイ
276362 リュウキュウモクセイ
276370 オホバモクセイ,ナガバモクセイ
276394 オオモクセイ,オホモクセイ
276450 ヤブニンジン属
276455 オナガヤブニンジン,ナガジラミ,ミヤマヤブニンジン,ヤブニンジン,ヲナガヤブニンジン
276460 オナガヤブニンジン,ミヤマヤブニンジン
276486 コウトウヤッデ
276533 テンノウメ属
276534 テンノウメ
276543 シラゲテンノウメ
276544 タチテンノウメ
276550 イシマメ,イソザンショウ,イソザンセウ,イソノカリガネ,テンノウメ
276743 ヤマゼリ属
276747 ミヤマニンジン
276748 チョウセンノダケ,ニオイウド
276752 ホソバセンキウ,ホソバセンキュウ,ホソバノダケ
276755 コバノセンキュウ
276768 ヤマゼリ
276769 ケヤマゼリ
276770 ハナヤマゼリ
276780 ミドリノダケ
276795 オストロウスキア属
276801 アサダ属
276808 アサダ,シナアサダ
276809 コアサダ
276818 アメリカアサダ
276846 ハミバミモドキ
276914 ケヒメノボタン
277000 オトンナ属
277361 ミズオオバコ属,ミヅオオバコ属,ミヅオホバコ属
277369 オオミズオオバコ,オホミヅオホバコ,カハホホヅキ,タオホバコ,ミズオオバコ,ミヅアサガホ,ミヅオオバコ,ミヅオホバコ,ミヅホコリ
277397 ミヅオホバコ
277441 オットーソンデリア属
277643 カタバミ科
277644 オキザリス属,カタバミ属
277648 コミヤマカタバミ
277649 エゾミヤマカタバミ
277661 ヒョウノセンカタバミ
277688 イモカタバミ
277702 ハナカタバミ
277706 ベニカタバミ
277747 カタバミ,スイモノグサ,スグサ,スズメノハカマ
277749 ウスアカカタバミ
277751 ホシザキカタバミ
277752 アカカタバミ
277773 ケカタバミ
277776 キキョウカタバミ,タクワカタバミ,ムラサキカタバミ
277803 オッタチカタバミ
277832 アマミカタバミ
277878 アリサンカタバミ,エイザンカタバミ,ミヤマカタバミ,ヤマカタバミ
277880 ベニバナミヤマカタバミ
277881 カントウミヤマカタバミ
277889 キダチハナカタバミ
277985 オオヤマカタバミ
278008 オオキバナカタバミ
278019 ゴヨウカタバミ
278039 フヨウカタバミ
278099 エゾタチカタバミ
278124 モンカタバミ
278137 フヨウカタバミ
278142 シボリカタバミ
278147 キキャウカタバミ,ムラサキカタバミ
278323 シナミサオノキ,シナミサヲノキ,ヒジハリノキ,ヒヂハリノキ
278351 ツルコケモモ属
278360 チョウセンコケモモ,ヒメツルコケモモ
278363 シロバナヒメツルコケモモ
278366 ツルコケモモ
278378 シロバナツルコケモモ
278388 オキシデンドラム属
278483 ヒナノボンボリ
278484 ホシザキシャクジョウ
278546 ルリトウワタ属
278547 ルリトウワタ
278576 マルバギシギシ属
278577 ジンエフスイバ,ジンヨウスイバ,マルバギシギシ,ミヤマギシギシ,ミヤマスイバ
278679 オヤマノエンドウ属
278710 ミヤマゲンゲ
278722 ミヤマオウギ
278757 フジボゲンゲ
278765 リシリゲンゲ
278829 モウコゲンゲ
278862 オオゲンゲ
278883 ホザキゲンゲ
278910 オヤマノエンドウ
278911 シロバナオヤマノエンドウ
278912 エゾオヤマノエンドウ
278913 オヤマノエンドウ,シロバナエゾ
278936 ヒダカゲンゲ
278937 クナシリオヤマノエンドウ
278941 クルマバゲンゲ
278958 シベリアゲンゲ
279000 レブンソウ
279001 シロバナレブンソウ
279014 コゲンゲ
279029 スギナエンドウ
279043 キバナノフジボエンドウ
279060 イワゲンゲ
279114 キヌゲンゲ
279121 ヒダカミヤマノエンドウ
279122 オカダゲンゲ
279162 マシケゲンゲ
279163 ウスゲマシケゲンゲ
279195 ヒロハキゲンゲ
279249 オオバオオヤマレンゲ,オオヤマレンゲ,オホヤマレンゲ,ビャケレンゲ,ミヤマコブシ,ミヤマレンゲ
279380 パキラ属
279476 コノボタン属
279479 アカミノボタン,コノボタン
279482 パキセレウス属
279484 タンバタロウ
279488 ヒウンカク
279489 ドジンノクシバシラ
279490 ブリンチュウ
279493 ブエイチュウ
279504 パキコルムス属
279635 パキフィツム属
279639 ホシビジン
279666 パキポジューム属
279728 クズイモ属
279732 クズイモ
279743 フッキサウ属,フッキソウ属
279744 タイワンフッキサウ
279757 キチジサウ,キチジソウ,フッキサウ,フッキソウ
279758 フイリフッキソウ
279765 ニサンゴバナ属,ベニサンゴバナ属
279768 ベニサンゴバナ
279769 パキスタキス・ルテア
279852 シナヤガラ属
280003 エゾノウワミズザクラ
280007 エゾノウワミズザクラ,カラフトウハミヅザクラ,カラフトウワミヅザクラ,ケウワミズ,ケウワミズザクラ,チョウセンウワミズザクラ
280016 イヌザクラ,シタミズザクラ,シタミヅザクラ,シロザクラ
280023 アハカ,ウハミヅザクラ,ウワミズザクラ,ウワミヅザクラ,コンガウザクラ,コンゴウザクラ,ハハカ,メズラ
280028 ウラボシザクラ
280044 エゾノウワミズザクラ
280053 シウリザクラ
280063 ヘクソカズラ属,ヘクソカヅラ属
280078 ヘクソカズラ,ヤイトバナ
280097 サオトメバナ,ヘクソカズラ,ヘクソカヅラ,ヤイトバナ
280100 アケボノヤイトバナ
280102 ツツナガヤイトバナ
280104 ハマサオトメカズラ
280105 ホシザキハマサオトメカズラ
280107 ビロードヤイトバナ

280143 ボタン属
280147 エビスグサ,シベリアシャクヤク,シャクヤク
280149 マンシウシャクヤク
280207 クサボタン,ノシャクヤク,ボタングサ,ヤマシャクヂャウ,ヤマシャクヤク
280209 ケヤマシャクヤク
280213 シベリアシャクヤク,シャクヤク
280214 シャクヤク
280241 ニバナヤマシャクヤク,ベニバナヤマシャクヤク,ヤマシャクヤク
280242 ケナシシロシャクヤク
280248 ケナシベニバナヤマシャクヤク
280253 オランダシャクヤク,セイヨウシャクヤク
280286 ナトリグサ,ハッカグサ,フカミグサ,ボタン,ヤマタチバナ
280304 カンボタン
280318 ホソバシャクヤク
280350 オクトネホシクサ
280437 オオバアカテツ属,オホバアカテツ属
280440 オオバアカテツ,オホバアカテツ,タココン
280441 グッタペルカノキ,ゲタペルチャ
280539 ハマナツメ属
280555 サルカキイバラ,ハマナツメ
280608 シュロ科,ヤシ科
280655 パルンビ－ナ属
280712 チョウセンニンジン属,ニンジン属,パナックス属
280722 ヒマラヤニンジン
280741 オタネニンジン,コマニンジン,チョゥセンニンジン,テウゼンニンジン,ニンジン,ヤクヨウニンジン
280748 ソウシシヨウニンジン
280749 キレハトチバニンジン
280750 キミノトチバニンジン
280751 ホソバチクセツニンジン
280771 サンシチ
280778 サンシチニンジン
280793 オタネニンジン,キニンジン,キミノチクセツニンジン,タチセツニンジン,チョウセンニンジン,トチバニンジン,トチハラニンジン,ナカバチクセツニンジン,ニンジン,ノコバチクセツニンジン,ミツバチクセツニンジン,ヨシノニンジン
280799 アメリカニンジン,カントンニンジン,セイヤウニンジン
280877 パンクラチューム属
280960 タコノキ科
280963 アダン科,タコノキ科
280968 タコノキ属,パンダ－ヌス属
280974 ホソバタコノキ
280993 タコノキ
280997 フサナリタコノキ
281037 イトバタコノキ
281089 アダン
281092 トゲナシアダン
281114 ヒメタコノキ
281138 アダン,エラン,シマタコノキ,リントウ
281166 アガタコノキ,ビヨウタコノキ
281169 シマタコノキ,フイリタコノキ
281223 ソケイノウゼン属
281225 ソケイノウゼン,ダイソケイ,ナンテンソケイ,バンソケイ
281274 キビ属
281370 ケハヒキビ,ヒメビエ
281390 ヌカキビ
281408 クサキビ
281438 ハナクサキビ
281532 ヒゲメヒシバ
281533 スイシャセトカヤ,スヰシャセトカヤ
281561 オオクサキビ,オホクサキビ
281733 ヒメヌカキビ
281887 ギネアキビ
281916 キビ,キミ,コキビ
281917 イヌキビ
281984 シマキビ,ヤマキビ
282031 オオヌカキビ
282162 ノキビ,ハイキビ,ハヒキビ
282200 イヌヤマキビ,サルメンキビ,ナカハラキビ
282214 ホウキヌカキビ
282286 スマトラキビ
282496 ケシ属,パパ－パル属
282504 アライドヒナゲシ
282506 タカネヒナゲシ,ミヤマヒナゲシ
282522 ハカマオニゲシ,ボタンゲシ
282530 パパベル・コンシュ－タム,モンツキヒナゲシ
282546 ナガミヒナゲシ
282563 リシリヒナグシ
282566 チュ－リップゲシ
282574 トゲミゲシ
282612 チシマヒナゲシ
282617 アイスラント・ポピ－,シベリアヒナゲシ,シュクコンゲシ,チシマヒナゲシ
282622 シロバナヒナゲシ
282633 シロバナヒナゲシ
282647 オオゲシ,オニゲシ,オホゲシ
282655 ボタンゲシ
282683 ミヤマヒナゲシ
282685 グビジンソウ,ビジンサウ,ビジンソウ,ヒナゲシ
282713 アツミゲシ
282717 ケシ
282736 カラフトヒナゲシ
282751 ケシ科
282764 パフィニア属
282768 トキハラン属,パフイオペデイルム属
282791 アクラトキワラン
282797 ハンヨウトキワラン
282845 インドアツモリサウ,トキハラン,トキワラン
282889 パフイオペデイルム・ロスチャイルデイアナム
282895 オバケトキワラン
282914 パフイオペデイルム・ビロ－サム
282934 ハナボウラン
283092 シロモジ属
283134 チリノヤマビト,ホクチグサ
283209 アリサンアイ,アリサンアヰ,セイタカスズムシソウ,タイトウアイ,タイトウアヰ
283211 コバノアリサンアイ,コバノアリサンアヰ
283269 オニルリソウ
283282 コウトウフヂ
283295 パラディセア属
283386 クサノオウバノギク,クサノオオバノギク,クサノワウバノギク,クサノワウバノケマン,クサノワウバノヤクミサウ,クサノワウバノヤクミソウ
283388 ヤクシソウ
283400 アキノイヌヤクシソウ
283407 アンデスチリ－ヤシ属,アンデスチリヤシ属
283418 カチラチライノキ,キンカウボク
283421 オオノキ,ホオガシワ,ホオナラ
283461 ニセマガホヤシ属
283598 パラファレノブシス属
283615 オホジフニヒトヘ
283625 クラルオドリコソウ
283654 スズメノナギナタ属
283656 スズメノナギナタ
283689 ケムラサキニガナ
283695 ムラサキニガナ
283781 コウモリソウ属,ヤブレガサ属
283782 アブクマオオカニコウモリ
283783 ハクサンカニコウモリ
283784 アカイシコウモリ
283785 シロウマオオカニコウモリ
283786 カニコウモリ
283788 イズカニコウモリ
283792 カラフトコウモリソウ,カラフトミミコウモリ,ミミカウモリ
283793 コモチミミコウモリ
283794 コウモリソウ,ミミコウモリ
283798 コバナノコウモリソウ
283804 シドケ,モミジガサ
283805 オガサモミジガサ
283806 モミジガサ
283808 ウスゲタマブキ
283810 ミヤマコウモリソウ,モミジタマブ

キ
283811 タマブキ
283812 オニタイミンガサ
283816 ウラゲヨブスマソウ
283820 ヨブスマソウ
283821 ニッコウコウモリ
283822 オオバコウモリ
283824 イヌドウナ,イヌドオナ
283825 ヨブスマソウ
283827 ハヤチネコウモリ
283833 モミジコウモリ
283834 アカイシコウモリ
283844 コウモリソウ
283845 オクヤマコウモリ
283847 ニイタカコウモリ
283848 オオカニコウモリ
283849 ツクシコウモリソウ
283852 オガコウモリ
283856 タイミンガサ
283868 ヒメコウモリ
283874 テバコモミジガサ
283882 ヤクシマコウモリ
283883 ヤマタイミンガサ
283884 ニシノヤマタイミンガサ
284098 セイヨウヒキヨモギ
284131 ヒカゲミズ属,ヒカゲミヅ属
284152 カベイラクサ
284160 ヒカゲミズ,ヒカゲミヅ
284162 タチゲヒカゲミズ
284174 オオヒカゲミズ
284289 ツクバネサウ属,ツクバネソウ属,パリス属
284340 キヌガササウ,キヌガサソウ,ハナガササウ,ハナガサソウ
284343 ニヒタカツクバネ
284358 タイワンツクバネ
284378 オホツクバネサウ,ニフタカツクバネ
284401 ツクバネサウ,ツクバネソウ
284406 クルマバツクバネソウ,ムラサキツクバネソウ
284473 クパン
284490 ロウソクノキ属
284497 ロウソクノキ
284500 ウメバチサウ属,ウメバチソウ属
284503 タカネウメバチソウ,ヒメウメバチソウ
284539 シラヒゲソウ
284542 オオシラヒゲソウ
284591 ウメバチサウ,ウメバチソウ
284593 ベニシベウメバチソウ
284595 イズノシマウメバチソウ
284596 ウメバチサウ,ウメバチソウ
284601 コウメバチソウ
284602 ヤクシマウメバチソウ
284645 ウメバチソウ科
284655 パロディア属
284656 サカボコマル
284657 キンショウギョク
284659 ムシュギョク
284660 ホウシュギョク
284661 ホウシュウギョク
284662 ラシュウギョク
284663 キンオウギョク
284664 グンシンマル
284665 ビソウギョク
284666 チキュウマル
284667 サイシュウギョク
284668 オウショウギョク
284673 キンコウマル
284674 マジンマル
284675 ハクシマジンマル
284679 ブジンマル
284680 ホウギョク
284681 ギョウショウギョク
284682 ボンシュウギョク
284683 レイシュウギョク
284684 ギンショウギョク
284685 セイテンマル
284687 オウシュウギョク
284690 カショウギョク
284693 ヒシュウギョク
284695 セッシュウギョク
284696 コウシュウギョク
284699 キシュウギョク
284700 メイシュウギョク
284701 トウシュウギョク
284702 コクウンリュン
284706 ロウシウギョク
284969 グンジサウ属,グンジソウ属
285004 グンジソウ
285055 ホウライカガミ属
285056 ホウライカガミ
285086 アメリカブクリョウサイ
285097 ツタ属
285102 ツタ
285110 ヘンリ-ヅタ
285117 アマミナツズタ,アマミナツヅタ,コウヨウズタ,ツタ,ナツズタ
285130 テンモクヅタ
285136 アメリカズタ,アメリカヅタ
285157 アマヅラ,コウエフヅタ,コウヨウヅタ,ツタ,ナツズタ,ナツヅタ
285203 パサニア属,マテバシイ属
285233 コダイホガシ,ナカキガシ
285379 コゴメビエ
285381 スズメノヒエツナギ
285390 スズメノヒエ属
285417 オガサハラスズメノヒエ,オガサワラスズメノヒエ,スズメノナガビエ
285423 シマススズメノヒエ
285426 キシュウスズメノヒエ,サシュウスズメノヒエ,サハススズメノヒエ
285435 ハネスズメノヒエ
285448 ハヒスズメノヒエ
285460 ナガバススズメノヒエ
285463 コアメリカスズメノヒエ
285469 アメリカスズメノヒエ
285471 スズメノコビエ
285476 コゴメススズメノヒエ
285507 スズメノコビエ,マルミススズメノヒエ
285508 アコウスズメノヒエ
285520 ケマルミススズメノヒエ
285522 ヒゲススズメノヒエ
285530 スズメノヒエ
285533 タチススズメノヒエ
285534 サハススズメノヒエ,サワススズメノヒエ
285607 トケイサウ属,トケイソウ属
285623 トケイサウ,トケイソウ,ボロンカズラ,ボロンカヅラ
285626 シロバナヒメトケイソウ
285632 ベニバナトケイソウ
285637 クタモノトクイ,クダモノトケイソウ,ショクョウトケイサウ,ショクョウトケイソウ,バッション・フルート,マルミクダモノトケイ
285639 クサトケイソウ
285641 クサトケイサウ,クサトケイソウ
285645 ヒメミナリトケイソウ
285654 チャボトケイソウ
285660 ウォーター-レモン,キミトケイソウ,キミノトケイソウ,ミゼレモン
285671 ヒメトケイソウ
285694 オオナガミクダモノトケイ,オオミトケイソウ,オホナガミクダモノトケイ,オホミトケイサウ,オホミトケイソウ
285695 ホザキトケイソウ,ホザキノトケイサウ,ホザキノトケイソウ
285703 スズメノトケイソウ
285709 ムラサキフイリバトケイソウ
285725 トケイサウ科,トケイソウ科
285731 アメリカバウフウ属,アメリカボウフウ属
285741 アメリカニンジン,アメリカバウフウ,アメリカボウフウ,パ-スニップ
285814 オミナエシ属,ヲミナヘシ属
285821 ロウテッオミナエシ
285826 ヒトツバオトコエシ
285830 マルバキンレイカ
285834 ホクシオミナエシ
285838 オトコオミナエシ
285854 イワオミナエシ
285855 モウコオミナエシ
285859 オミナエシ,オミナヘシ,オミナベシ,オミナメシ,チメグサ,ヲミナヘシ,ヲミナベシ,ヲミナメシ
285860 ハマオミナエシ
285870 タカネオミナエシ,タカネヲミナヘ

シ,チシマキンレイカ,チシマキンレイタ
285874 ハクサンオミナエシ
285875 キンレイカ,コキンレイカ,ハクサンオミナエシ
285877 キンレイカ
285878 オオキンレイカ
285880 オトコエシ,オトコヘシ,オトコメシ,オヱツチ,ヲトコヘシ,ヲトコベシ,ヲトコメシ
285930 ガラナ
285951 キリ属
285959 キミソセヤマギリ,シナギリ,シナシロギリ,ラクタギリ
285960 ココノエギリ,ココノエノキリ,ココノヘノキリ,ミカドギリ
285966 タイワンギリ
285981 キリ,キリノキ,ハナギリ,ヒトハグサ,ヒトハグハ
286038 ライトヤシ属
286073 キダチハナカンザシ属
286275 キダチハナカンザシ
286580 ヤノネボンテンカ属
286634 ヤノネボンテンカ
286829 サギサウ属,サギソウ属
286835 サギサウ,サギソウ
286918 ゴマ科
286972 シオガマギク属,シホガマギク属
286977 カラフトシオガマ
287012 ミヤマシオガマ
287013 シロバナミヤマシオガマ
287067 タマザキシオガマ
287076 エゾヨツバシオガマ
287077 シロバナエゾヨツバシオガマ
287078 キタヨツバシオガマ
287079 ヨツバシオガマ
287081 クチバシシオガマ
287082 レブンシオガマ
287083 シロバナレブンシオガマ
287194 コゴメグサシホガマ,チシマシホガマ
287209 マウコシホガマ
287239 ハンカイアザミ,ハンカイシオガマ,ハンクワイアザミ,ハンクワイシホガマ
287240 シロバナノハンカイシオガマ
287259 オオシホガマギク,ハナシホガマ
287299 イワテシオガマ
287300 ヨツバシオガマ
287312 セリバシオガマ
287315 ベニシオガマ
287326 チシマシオガマ
287334 アイザワシオガマ,ワタシホガマ
287338 キバナヒメシホガマ
287411 マンシウシホガマ
287459 オニシオガマ
287460 シロバナオニシオガマ
287468 キバナシオガマ
287474 ウルップシオガマ,キバナシオガマ
287494 サワシホガマ
287577 ツクシシオガマ
287578 シロバナツクシシオガマ
287584 シオガマギク,シベリアシオガマ,シホガマギク,シホガマサウ
287589 シオガマギク
287590 シロシオガマ
287591 ウスベニシオガマ
287592 ミカワシオガマ
287593 ミカワシオガマ
287594 ビロオドシオガマ,ビロ－ドシオガマ
287595 トモエシオガマ
287600 ケシオガマ
287601 エタウチシオガマ
287645 コウアンソウ
287647 ホソバシオガマ
287659 ハタザオシホガマ
287664 カフカシオガマ
287689 ホザキシオガマ,ホザキシホガマ
287705 クミバシホガマ
287718 ヤチシオガマ
287766 ニヒタカシホガマ
287794 オトメシオガマ,オトメシホガマ
287802 タカネシオガマ,タカネシホガマ,ユキワリシホガマ
287805 シロバナタカネシオガマ
287837 エゾシオガマ
287838 ビロ－ドエゾシオガマ
287846 ペディランツス属
287881 ペヂトオカクタス属
287909 ゲッカギョク
288038 マ－ケサズヤシ属
288045 テンジクアオイ属,テンヂクアフヒ属,パラルゴニュ－ム属
288192 キレハテンジクアオイ
288206 ナツテンジクアオイ
288264 オオバナテンジクアオイ
288268 ニオイテンジクアオイ,ニホヒテンヂクアフヒ,モモイロテンジクアオイ
288297 ゼラニウム,ハナテンジクアオイ
288300 ツタバツルテンジクアオイ
288310 テンジクアオイ
288326 ツタバテンジクアオイ
288396 シロバナニイテンジクアオイ
288430 タテバテンジクアオイ,タテバテンヂクアフヒ,ツタバテンジクアオイ
288459 カシワバテンジクアオイ
288468 キクバテンジクアオイ,キクバテンヂクアフヒ,ホソバテンジクアオイ
288594 モンテンジクアオイ,モンテンヂクアフヒ
288605 ムカデラン
288615 ペレキフォラ属
288616 セイコウマル
288638 シマハラン属,シマバラン属
288639 アリサンヒメバラン
288658 アリサンヒメバラン,ヒメバラン
288705 サンショウソウ属,サンセウサウ属
288715 アラゲサンショウソウ
288729 キミズモドキ,ケイタオミズ
288749 サンショウソウ,サンセウサウ,サンセウヅル,ハヒミズ,ハヒミヅ
288762 アリサンサンショウソウ,アリサンサンセウヅル,オオサンショウソウ,オホサンセウサウ,ヒメアリサンサンセウヅル
288767 オランダミズ,ハナビソウ,モヨウガラクサ
288769 キミズ,キミヅ
288840 ペルティフィル－ム属
288847 ヤワタソウ属
288848 ヤワタソウ
288850 ワタナベソウ
288882 トゲナシジャケツ属
288897 トゲナシジャケツ
288926 ミヅガンピ属
288927 ミズガンピ,ミヅガンピ,ミヅハギ
288983 プニオセレウス属
289011 チカラシバ属
289015 チカラシバ,ベニチカラシバ
289018 アオチカラシバ
289067 キクユグラス
289096 イヌチガヤ
289116 タウジンキビ,トウジンキビ,トウジンヒエ,トウジンビエ
289135 チカラシバ
289143 ツリエノコロ
289185 エダウチチカラシバ
289187 エダウチチカラシバ
289205 ホソボチカラシバ
289208 ホソボチカラシバ
289218 ナピアグラス
289267 シマチカラシバ
289303 シロガネチカラシバ
289314 ペンステモン属,ペントステモン属
289318 ヒメツリガネヤナギ
289324 ヤナギチョウジ
289327 ツリガネヤナギ
289332 ウスムラサキツリガネヤナギ,ヒメツリガネヤナギ
289337 シロバナツリガネヤナギ
289341 イワブクロ,タルマイソウ
289342 シロバナイワブクロ
289350 リンドウツリガネヤナギ
289353 フウリンイワブクロ
289476 ニュウシジュ
289500 ペンタグロッティス属
289627 ヤドリタラノキ属
289629 ヤドリタラノキ
289682 ゴジカ属,ゴジクワ属
289684 ゴジカ,ゴジクワ,コムヤム

289713 キンロウバイ,キンロバイ
289784 ペンタス属
289831 クササンダンカ
290133 タコノアシ属
290134 サハシヲン,サワシオン,タコノアシ
290286 サダサウ属,サダソウ属,ペペロミア属
290305 ビロウドゴセウ,ビロ-ドゴショウ
290308 シマゴショウ
290315 シワアオイソウ,チヂミバシマアオイソウ
290331 オオバアオイソウ
290365 サダサウ,サダソウ,スナゴセウ
290386 サビアオイソウ
290389 ナカハラゴセウ
290397 オキナワスナゴショウ
290399 コシマゴショウ
290572 タニギキャウ属,タニギキョウ属
290573 タニギキャウ,タニギキョウ,ツクシタニギキョオ,ユキノシタ
290699 ペイレスキア属,モクキリン属
290701 ツルキリン,モクキリン
290708 サクラキリン
290717 ペイレスキオプシス属,ペレスキオプシス属
290818 フウキギク
290821 シネラミヤ,シネラリヤ,シュントウギク,フウキギク,フキザクラ,フキブキ
290840 ホウライツヅラフヂ属
290841 ホウライツヅラフジ,ミスヂツヅラフヂ
290921 ウラムラサキ,カザンラン
290932 シソ属,ペリラ属
290940 アオジソ,エゴマ,シソ,ジュウネン,チリメンジソ,トラノオジソ,レモンエゴマ
290952 イヌエ,オランダジソ,カウライジソ,カタメンジソ,シソ,チヂミジソ,チリメンジソ,テョウセンジソ,ヌカエ,ノラエ
290953 カタメンジソ
290954 チリメンアオジソ
290955 アカジソ
290956 マダラジソ
290957 アオジソ
290968 シソ,タイワンエゴマ
290972 トラノオジソ
290989 スズコウジュ
291031 クロバナカズラ属
291082 クロバナカズラ
291127 ペリステリア属
291134 ペリストロ-フェ属
291161 ハグロサウ,ハクロソウ,フチゲハグロサウ,フチゲハグロソウ,メゴシツ
291162 ハグロソウ
291188 アオチドリ属,アヲチドリ属,ムカゴトンボ属
291211 ムカゴトンボ
291216 タカサゴサギサウ,タカサゴサギソウ
291221 タイワンサギサウ
291224 コカゲトンボ
291229 ダケトンボ
291234 タカサゴサギソウ
291363 ペルネッチア属
291396 コササガヤ属
291405 コササガヤ
291416 オホコササガヤ
291481 ミヂンコザクラ属
291482 ミヂンコザクラ
291483 ミヂンコザクラ
291491 アボカド属,ペルセア属,ワニナシ属
291494 アボカド,ワニナシ
291651 サナエタデ属
291666 シラカワタデ
291705 ナガバノヤノネグサ
291707 ハリタデ
291729 シロバナハナサクラタデ
291732 ミヤマタニソバ
291744 ヒメタデ
291745 アオヒメタデ
291746 ホソバヌマタデ
291747 モリイヌタデ
291748 ホソバイヌタデ
291755 ミズヒキ
291757 ギンミズヒキ
291766 オオサクラタデ,オホサクラタデ,テリハオホイヌタデ
291775 ナガバノウナギツカミ,ナガバノウナギヅル
291785 アザブタデ
291786 イトタデ
291787 ナガボヤナギタデ
291789 サツマタデ
291790 アオホソバタデ
291818 サナエタデ
291822 イヌタデ
291825 シロバナイヌタデ
291827 マルバイヌタデ
291828 イザリタデ
291834 シロバナサデクサ
291835 サクラタデ
291840 ハルタデ
291853 コヤナギタデ
291868 イヌハナタデ
291872 ネバリタニソバ
291885 ヤマミゾソバ
291895 オトメサナエタデ
291915 オオバナボントクタデ
291916 ボントクタデ
291937 アキノウナギツカミ
291938 ウナギツカミ
291939 ウスゲアキノウナギツカミ
291940 ケアキノウナギツカミ
291941 シロバナアキノウナギツカミ
291954 シロバナトゲソバ
291955 トゲナシママコノシリヌグイ
291997 アオネバリタデ
291998 イヌネバリタデ
292040 カウヤバハキ属,コウヤボウキ属
292041 センダイハグマ
292042 イワキハグマ
292055 コウヤボウキ,コオヤボオキ,ナガバコウヤボウキ,ナガバノカウバハキ,ナガバノカウヤバウキ,ナガバノコウヤボウキ,ナガバノコオヤボオキ
292058 カコマハグマ
292059 センダイハグマ
292067 クルマバハグマ
292069 カシワバハグマ
292077 タイワンタマボウキ
292080 オヤリハグマ
292084 シマコウヤボウキ,シマコオヤボオキ
292138 ペスカトレア属
292342 フキ属
292352 タイワンブキ
292353 ニオイガントウ,ニホヒクワントウ
292374 フキ
292375 オカブキ
292376 ベニブキ
292377 アキタブキ
292394 トウブキ
292403 アイヌブキ,ホロナイブキ,ボロナイブキ
292404 タイワンブキ,モンジュブキ
292517 ペトラエア属,ヤモメカズラ属
292530 ムラサキイヌナズナ属
292531 ムラサキイヌナズナ
292656 ハリナデシコ属
292664 イヌコモチナデシコ
292665 コモチナデシコ
292668 ハリナデシコ
292671 サクライソウ属,サクラヰサウ属
292673 サクライサウ,サクライソウ
292685 オランダゼリ属,ペトロセリ-ヌム属
292694 オランダゼリ,パセリ
292741 ツクバネアサガオ属,ツクバネアサガホ属,ペチュニア属
292745 ツクバネアサガオ,ツクバネアサガホ
292753 ヒメツクバネアサガオ
292758 カワラボウフウ属,ノダケ属,ペウケダ-ヌム属
292780 コウアンニンジン
292821 タカネボウフウ
292843 ホソバボウフウ

292857 タイワンカハラボウフウ
292892 カワラボウフウ,ケヅリバウブウ,ケヅリボウフウ,ゴシャメンニンジン,ゴトウボウフウ,ゴメンニンジン,ショクヨウボウフウ,ヒラノニンジン,ボタンニンジン,ボタンバウブウ,ボタンボウフウ,ボタンボウブウ,ヤマナスビ
292893 ムラサキボタンボウフウ
292894 ナンゴクボタンボウフウ
292895 コダチボタンボウフウ
292942 ハクサンボウフウ
292943 キレハノハクサンボウフウ
292944 エゾノハクサンボウフウ
293033 カハラバウフウ,カワラボウフウ,シラカワボウフウ,ヤマニンジン
293034 モイワボウフウ
293038 シラカハバウフウ,シラカワボウフウ,ヤマニンジン
293091 ペウ－ムス属
293151 ハゼリソウ属
293160 タチホロギク
293171 ヒメタチホロギク
293176 ハゼリソウ
293178 ネバリホロギク
293181 キヨスミウツボ属
293186 キヨスミウツボ
293207 アイアシ属
293215 アイアシ
293216 ホゾバアイアシ
293252 フェドラナッサ属
293293 タキキビ属
293296 オホタツノヒゲ,カシマガヤ,タキキビ
293484 ガンゼキラン属,クワクラン属
293494 ガンゼキラン,キバナクワクラン,クンケイラン,ヒダベリクワクラン,ホシケイ,ホシケイラン
293495 ホシケイラン
293523 ヒメカクラン,ヒメクワクラン
293537 カクチョウラン,カグラン,クワクラン
293577 コチョウラン属,コテフラン属,ファレノプシス属
293578 マニラコチョウラン
293582 コチョウラン,コテフラン,チョゴテフ,テフラン,マニラコテフラン
293600 ヒメゴチョウ,ヒメコチョウラン
293633 アサヒコチョウラン,アサヒゴテフ,シルレルコテフラン
293643 ヒロバコチョウラン,ヒロバゴテフ
293705 クサヨシ属
293708 オニクサヨシ
293709 クサヨシ
293710 シマガヤ,シマヨシ,チグサ,リボングラス
293714 ホソボクサヨシ
293729 カナリアクサヨシ,カナリアサード,カナリ－クサヨシ,カナリヤクサヨシ,ヤリクサヨシ
293743 ヒメカナリ－クサヨシ
293750 セトガヤモドキ
293755 アレチクサヨシ
293861 アサガオ属,アサガホ属
293966 インゲンマメ属,インゲン属,ヰンゲンマメ属
293985 アカハナマメ,ハナササゲ,ハナマメ,バニバナインゲン,バニバナヰンゲン,ベニハナインゲン
293986 シロバナササゲ
293989 ヒメヤエナリ
293999 カハリバツルアヅキ,タイワンツルアヅキ,ホンバツルアヅキ
294010 アオイマメ,アフヒマメ,イチコクマメ,クワウテイマメ,ゴモンマメ
294030 ツルナシインゲン,ツルナシヰンゲン
294039 アヅキ
294040 カニメ,コメアヅキ,シャボンマメ,ツルアヅキ,ニラコ
294056 インゲンササゲ,インゲンマメ,ゴガッササゲ,ダウササゲ,ツルナシインゲンマメ,トウササゲ,ニドササゲ,ヰンゲンササゲ,ヰンゲンマメ
294114 トウキリンソウ,ナガバキリンサウ,ホソバノキリンサウ,ホソバノキリンソウ,ヤマキリンサウ
294115 キリンソウ
294117 オオバキリンソウ,ハコダテキリンソウ
294119 コバノキリンソウ
294123 エゾノキリンソウ,キリンサウ,キリンソウ
294124 ヒメキリンソウ
294126 ケキリンソウ
294127 ヒメキリンソウ
294230 キハダ属,フェロデンドロン属
294231 キハダ,キワダ,シコロ
294233 アシキハダ,オオバノキハダ,オオバノキワダ,オホバノキハダ
294234 ミヤマキハダ
294236 タイワンキハダ
294238 シナキハダ,ヒロハノキハダ
294240 タイワンキハダ
294251 カラフトキハダ,カラフトキワダ,キハダ,シコロ,ヒロハノキハダ
294256 タイワンキハダ
294273 フェロスペルマ属
294404 バイカウツギ科
294407 バイカウツギ属,バイクワウツギ属
294426 サツマウツギ
294463 セイヨオバイカウツギ,セイヨオバイカッキ
294509 ヒメバイクワウツギ,ペキンバイカウツギ,ペキンバイクワウツギ
294530 サツマウツギ,バイカウツギ,フスマウツギ
294531 ケバイカウツキ,ニッコウバイカウツギ
294532 シコクウツギ
294536 マンシウバイカウツギ,マンシウバイクワウツギ,マンシュウバイカウツギ
294539 オオバイカウツギ
294556 ウスババイクワウツギ,ウスベニバイカウツギ,テウセンバイクワウツギ
294785 フィロデンドロン属
294792 ビロ－ドカズラ
294797 ホテイカズラ
294807 ベニヤッコガズラ
294825 ヒメビロ－ドカズラ
294827 ヒメカズラ
294830 ヤッコカズラ
294839 ヒトデカズラ
294842 シロガネカズラ
294844 ワタネカズラ
294877 イソフサギ属
294882 イソフサギ,コケゲイトウ
294886 タスキアヤメ科
294892 タヌキアヤメ属
294895 タヌキアヤメ
294960 アハガヘリ属,アワガエリ属
294962 シャマアハガヘリ,シャマアワガエリ,ミヤマアワガエリ
294988 アハガヘリ,アワガエリ
294989 イヌアワガエリ
294992 オオアワガエリ,オホアハガヘリ,キヌイトサウ,コモチオオアワガエリ,チモシ－グラス
295041 タカスゼリ
295049 オオキセワタ属,オホキセワタ属,フロ－ミス属
295123 ネッカキセカタ,ネッカキセワタ
295142 オオキセワタ,オオバキセワタ,オホキセワタ,オホバキセワタ
295150 アラゲキセワタ,モウコキセワタ
295212 タマキセワタ
295215 ヒカゲキセワタ
295239 クサケフチクタウ属,フロックス属
295267 キキャウナデシコ,キキョウナデシコ,ヒメフロックス,ベチュニヤ
295272 ホシザキフロックス
295286 シロバナツメクサ
295288 クサキョウチクトウ,クサケフチクタウ
295324 シバサクラ,ハナツメクサ,モスフロックス
295344 タイワンイヌグス属
295361 ウラジロダモ,タイワンイヌグス,ホザキハマビハ
295451 ナツメヤシ属,フェニックス属
295453 チャボナツメヤシ,フエニックス
295459 カナリ－ヤシ,キヤナリ－ヤシ

295461 カラナツメ,ナツメジュロ,ナツメヤシ
295465 クワウラウ,クワラン,ケンラン,コンロン,ソテツジュロ
295468 ヒメナツメヤシ
295475 シンノウヤシ
295477 ウラジロナツメヤシ
295478 オオカミヤシ,ヒメナツメヤシモドキ
295479 カブダチソテツジュロ,セネガルヤシ
295483 イヌナツメヤシ
295484 シンノウヤシ
295485 イワヤマナツメヤシ
295487 サトウナツメヤシ
295490 セイロンナツメヤシ
295501 オオミクマデヤシ属
295510 タマラン属,フォリドータ属
295516 ウライタマラン
295518 タマラン
295594 ニュウサイラン属
295600 ニュウサイラン,ニュージーマンドアサ,マオラン
295604 フイリニュウサイラン
295609 アカメモチ属,カナメモチ属,カマツカ属
295635 セイバンカマツオ,ミヤマカナメモチ
295669 ニイタカカマツカ
295674 ニヒタカカマツカ,ホソバカマツカ
295691 アカメ,アカメモチ,カナメ,カナメガシ,カナメノキ,カナメモチ,ソバノキ
295729 タイワンカナメモチ,ヤンバルカナメモチ,ヤンバルカマツカ
295773 オオカナメモチ,オホカナメモチ,ナガバカナメモチ
295775 マンリャウカナメモチ
295778 ケバナカメモチ
295808 オオカマツカ,カマツカ,ワタゲカマツカ
295817 シマカナメモチ
295875 フラグミペティラム属,フラグモペティラム属
295881 ヨシ属
295888 アシ,キタヨシ,タニハグサ,ハマオギ,ハマヲギ,ヨシ
295889 ケヨシ
295907 シラゲヨシ
295911 ヂシバリ,ツルヨシ
295912 ヂシバリ,ツルヨシ
295914 ホソバヨシ
295916 ウドノノヨジ,セイコノヨシ,セイタカヨシ
295919 ヨシ
295952 フレラン属
295958 フレラン
295960 モリフレラン
295978 ハエドクソウ属,ハヘドクサウ属
295984 アメリカハエドクソウ,ハエドクソウ,ハヘドクサウ
295986 クロミノハエドクソウ,ナガバハエドクソウ,ナガバハヘドクサウ,ハエドクソウ
295988 チャボハエドクソウ
295999 ナガバハエドクソウ
296006 ハエドクソウ科,ハヘドクサウ科,ハマヂンチャウ科
296061 コシオガマ属,コシホガマ属
296068 コシオガマコシホガマ,コシホガマ
296069 シロバナコシオガマ
296091 アケボノムグラ
296108 フィグ-ミウス属
296110 ケ-プフクシア
296112 ケ-プフクシア
296121 イワダレソウ
296465 コミカンサウ属,コミカンソウ属
296471 キダチコミカンソウ
296530 オガサワラコミカンソウ
296553 ミナミコミカンソウ
296554 アンマロク,マラッカノキ,ユカン
296565 キハギ,コバンノキ,コミカンノキ,ヤマナンテン
296632 コミカンソウ
296641 ハナコミカンボク
296679 キダチコミカンサウ,キダチコミカンソウ
296695 コカエフコバンノキ,タカオコバンノキ,タカヲコバンノキ
296734 シマコバンノキ,タイワンコバンノキ
296779 タカオコバンノキ,タカヲコバンノキ
296785 ナガエコミカンソウ
296801 キツネノチャブクロ,コミカンサウ,コミカンソウ
296802 テリミコミカンソウ
296803 チョウセンミカンソウ,ヒメミカンサウ,ヒメミカンソウ
296809 シマコバンノキ,シマヒメミカンソウ,タイワンコバンノキ
296868 フィロボルス属
297013 ウチワツナギ
297015 ツガザクラ属
297016 アオノツガザクラ,アヲノツガザクラ,オオツガマツ,オホツガザクラ,オホツガマツ,ハクサンガヤ
297017 オオツガザクラ,コツガザクラ,シロウマツガザクラ
297019 エゾツガザクラ,エゾノツガザクラ
297021 コエゾツガザクラ
297022 ユウバリツガザクラ
297030 ツガザクラ,ツガマツ
297031 ナガバツガザクラ
297033 ハイツガザクラ,ハヒツガザクラ
297181 エビアラロ属,スガモ属
297182 スガモ,ハイスガモ,ハヒスガモ
297183 エビアマモ,エビアラロ
297185 ウミスゲ,ゴモクサ,スガモ,ハマクサ
297188 マダケ属,モウソウチク属
297203 ケレタケ,コサンチク,ゴサンチク,ホテイチク
297215 カラタケ,ニガタケ,マダケ
297222 シボチク
297239 キンメイチク,シマダケ,ヒョンチク
297266 マウソウチク
297282 ジンメンチク
297301 キッカフチク,キッコウチク,ブツメンチク,モウソウチク
297306 マウソウチク,モウソウチク
297307 キンメイモウソウチク
297334 セキチク,チャウチク
297337 タイワンマダケケイチク
297367 クロチク,シチク,モウソウチグ
297368 キンメイハチク
297370 サカサダケ
297373 アハタケ,アワタケ,オホタケ,カラタケ,クレタケ,ハチク
297375 タイワンクロチク
297435 カラタケ,ニガタケ,マダケ
297464 キンメイチク,ワウゴンチク
297624 イガホオズキ属,イガホホヅキ属
297625 ヤマホオズキ,ヤマホホヅキ
297626 イガホオズキ
297628 アオホオズキ,イガホオズキ,イガホホズキ,イガホホヅキ
297640 ホオズキ属,ホオヅキ属,ホホヅキ属
297641 ナガエノセンナリホオズキ
297643 ホオズキ,ホホヅキ,ヨウシュホオズキ
297645 タンバホホヅキ,ホオズキ,ホホズキ
297650 セッナリホオズキ,センナリホオヅキ,センナリホホヅキ,ヒロハフウリンホオズキ
297652 ホソバフウリンホオズキ
297653 アイフウリンホオズキ
297676 ショクヨウホオズキ
297678 ビロードホオズキ
297686 オオブドホオヅキ
297697 ウスゲホオズキ
297703 センナリホオズキ,ヒメセンナリホホヅキ
297711 ケホオズキ,ケホオヅキ,ケホホヅキ,シマホオズキ,シマホオヅキ,シマホホヅキ,ブダウホホヅキ,ブドウホオズキ,ブドウホオヅキ
297712 キバナホオズキ
297716 ショクヨウホオズキ,ショクヨウホ

オヅキ
297720 ヒメセンナリホオズキ
297833 テマリシモッケ属
297837 テマリシモッケ
297842 アメリカシモッケ,ケアメリカシモッケ
297845 コカネシモッケ
297875 フクロヒヨス属
297882 フクロヒヨス
297973 ハナトラノオ属
297984 カクトラ,ハナトラノオ
298050 アメリカゾウグヤシ属
298051 ツケアアメリカゾウグヤシ
298054 シデシャジン属,フィテウ-マ属
298088 ヤマゴバウ属,ヤマゴボウ属
298093 イヌゴボウ,トウゴボウ,ヤマゴバウ,ヤマゴボウ
298094 アメリカヤマゴバウ,アメリカヤマゴボウ,ヤウシュヤマゴバウ,ヨウシュヤマゴボウ
298114 マルミノヤマゴバウ,マルミノヤマゴボウ
298124 ヤマゴバウ科,ヤマゴボウ科
298132 ピアランツス属
298189 タラヒ属,トウヒ属,ハリモミ属
298191 オウシュウトウヒ,ドイットウヒ
298226 イラモミ,トラノオモミ,マツハダ
298227 ミドリイラモミ
298230 シラネマツハダ
298286 カナダ-スプル-ス,シロトウヒ,ホワイトスプル-ス
298298 アカエゾ,アカエゾマツ,シンコマツ,テシオマツ,テシホマツ,ヤチエゾ
298299 アオミノアカエゾマツ
298303 シモツガ,タウヒ,トウヒ,トラノオモミ,トラノヲモミ
298307 エゾマツ,クロエゾ,クロエゾマツ
298313 オゼトウヒ
298323 チョウセンハリモミ,テウセンハリモミ,ヤツガタケトウヒ,ヤツガタケトオヒ
298329 ヒメマツハダ
298349 クロトウヒ,ブラックスプル-ス
298354 コバラモミ,ヒメバラモミ
298358 ミモフリハリモミ,ムレイタウヒ
298375 ニイタカトウヒ,ニヒタカタウヒ
298382 シベリャトウヒ,テフセンタウヒ
298401 アオトウヒ,アメリカハリモミ,コロラドトウヒ,プンゲンストウヒ
298406 ギンヨウコロラドトウヒ
298423 アカトウヒ,レッドスプル-ス
298426 シュレンクタウヒ
298435 シトカスプル-ス,シトカトウヒ
298436 ヒマラヤトウヒ,ヒマラヤハリモミ
298442 シロモミ,トラノヲモミ,バラモミ,ハリモミ
298444 ベニバナハリモミ
298506 ニガキ属
298516 ニガキ
298517 ケニガキ
298519 ニガキ
298550 カウゾリナ属,コウゾリナ属
298584 ハリゲコウゾリナ
298589 カウゾリナ,コウゾリナ,セイヨウコウゾリナ
298593 ナメラコウゾリナ
298594 ハマコウゾリナ
298595 アカイシコウゾリナ
298596 ホソバコウゾリナ
298602 カンチコウゾリナ,タカネコウゾリナ
298618 カウゾリナ,コウアンコウゾリナ,コウゾリナ
298625 チョウセンコウゾリナ
298670 コオウレン
298703 アセビ属
298722 コウザンアセビ,ヒマラヤアセビ
298734 アセビ,アセボ,アセボノキ,ウマクスバ,ウマクワズ,シカクハヅ,シカクワズ,タイワンアセビ,ホザキアセビ,ヨナバ
298749 フクリンアセビ
298750 ホナガアセビ
298751 ヒメアセビ
298757 ヤクシマアセビ
298758 リュウキュウアセビ
298830 セダカウロコヤシ属
298836 イトセダカウロコヤシ
298848 ミズ属,ミヅ属
298854 ミヤマミズ,ミヤマミヅ
298857 ミヤマミズ
298858 アニソミズ,アニソミヅ
298869 アリサンミズ,アリサンミヅ,シマミズ
298878 ミヅバウシ
298885 アサバソウ
298919 フンキコミズ,フンキコミヅ
298945 ツルミツ,ヒメミズ,ヒメミヅ,ヤマミズ,ヤマミヅ
298946 オオヤマミズ
298947 アキヤマミズ
298955 ミヤコミズ
298968 マツダミズ,マツダミヅ
298971 アオミヅバヴシ,アヲミヅバヴシ,オホミズ,オホミヅ,オホミヅバヴシ,ミスヂミズ,ミスヂミヅ
298974 コゴメミズ,コゴメミヅ,コメバコケミズ
298980 ヒメアオミズ
298989 コミヤマミズ
299006 コケミズ,コケミヅ
299010 オオコケミズ,コケミズ
299017 ゼンエンミズ,ゼンエンミヅ,タイトウミズ,タイトウミヅ
299024 アオミズ,アヲミズ,アヲミヅ,ミズ,ミヅ
299026 ミズ,ミヅ
299037 ナガバオホミズ,ナガバオホミヅ,ファ-リ-ミズ,ファ-リ-ミヅ
299048 ホソバミズ,ホソバミヅ
299109 シマユキカヅラ属
299115 シマユキカズラ
299224 キバナコウリンタンポポ
299238 ピロソセレウス属
299240 ハクテンリュウ
299242 エイキリュウ
299244 ハクモウリュウ
299246 セイウンカク
299247 キンホウリュウ
299248 オウゴンリュウ
299249 コウリンリュウ
299250 スイセイチュウ
299255 ゴウソウリュウ
299259 オキナジシ
299260 ハクショウリュウ
299261 ベニフデ
299262 ハルゴロモ
299265 オオエンリュウ
299266 ロウオウ
299269 サルトリカク
299270 ダイガカク
299303 ピメレア属
299325 オ-ルスパイス
299342 ダケゼリ属,ヒカゲミツバ属,ミツバグサ属
299351 アニス
299357 タイワンゼリ
299365 ヤマカノツメソウ
299366 マンシウヒカゲミツバ
299373 カノツメサウ,カノツメソウ,ダケゼリ
299395 ミツバグサ
299448 ヒカゲミツバ
299480 フヂイゼリ,フヂヰゼリ
299481 ヒカゲミツバ
299534 キソムカゴ,キンミツバ,タニミツバ
299555 コウアンボウフウ
299594 マツ科
299625 フシナシオサラン,フシナシヲサラン,リウキウセキコク,リュウキュウセッコク
299649 ソアグヤシ属,ピナンガ属
299660 マタバピナンガ
299664 ク-リ-ピナンガ
299667 マタラピナンガ
299668 マライピナンガ
299669 イヌピナンガ
299674 スコ-テチニピナンガ
299679 ソアグヤシ
299717 ハンゲ属

299719 ニオイハンゲ
299724 カラスビシャク,ハンゲ,ヘソクリ
299725 シカハンゲ,シカハンゲハンゲ,ホソバハンゲ
299726 ムラサキ,ムラサキハンゲ
299727 ヤマハンゲ
299728 オオハンゲ,オホハンゲ
299729 ムラサキオオハンゲ
299736 ムシトリスミレ属
299740 アシナガムシトリスミレ
299743 マルバムシトリスミレ
299751 ヒメムシトリスミレ
299752 キバナムシトリスミレ
299754 コウシンソウ,コオシンソウ
299755 シロバナコウシンソウ
299757 カバフトムシトリスミレ,カラフトムシトリスミレ
299759 ネバリサウ,ムシトリスミレ
299763 イイタカムシトリスミレ
299764 ムシトリスミレ
299765 シロバナムシトリスミレ
299780 マツ属
299781 アイグロマツ
299789 アマミゴエフマツ,アマミゴヨオ,ヤクタネゴヨウ,ヤクタネゴヨオ
299799 カザンマツ,シマチョウセンマツ,シマテウセンマツ,タカネゴエフ,タカネゴヨウ,ヤクタネゴヨウ
299802 タカネゴヨウ
299807 ダイオウマツ
299818 シャックパイン,バンクスマツ
299830 シロマツ,ハクショウ
299836 カリビアマツ
299842 アローラマツ,ヨーロッパハイマツ
299851 メキシコマツ
299854 アメリカヒトツバマツ
299864 クラウサマツ
299869 コントルタマツ,ロッジポールペイン
299890 アカマツ,シナアカマツ,メマツ,メンマツ
299894 ジャノメアカマツ
299895 シダレマツ
299897 ウックシマツ,タギョウショウ
299925 エキナタマツ,ショートリーフパイン
299928 スラッシュパイン,スラッシュマツ
299942 フレキシマツ
299964 アレッポマツ
299986 カシアマツ
299992 ジエフリーマツ
299999 カシヤマツ
300006 チョウセンマツ,チョオセンゴヨオ,チョセンウゴヨウ,テウセンゴエフ,テウセンゴマツ,テウセンマツ
300011 サトウマツ,シュガーパイン,ナガミマツ
300044 オキナハマツ,リウキウマツ,リュウキュウマツ
300054 シナアカマツ,シナクロマツ,タイワンアカマツ,タイワンマツ
300068 スラール,メルクシマツ
300082 モンチコラマツ,モンティコーラマツ
300083 アリサンゴエフ,タイワンアカマツ,タイワンゴエフマツ,タイワンゴヨウ,タイワンゴヨウマツ
300086 ムゴマツ,モンタナマツ
300098 ビショップマツ
300117 オウシュクロマツ,オーストリアマツ,ヨーロッパクロマツ
300128 ダイオウショウ,ダイオウマツ,ダイワウマツ,ナガハマツ,ヒマラヤマツ,ロングリーフパイン
300130 ゴヨウマツ,ヒメコマツ,マルミノゴエフ
300137 キタゴヨウ,キタゴヨウマツ,ゴエフマツ,ヒメコマツ
300138 トドハダゴヨウ
300146 オニマツ,カイガンショウ,フッコクカイガンショウ,フランスカイガンショウ
300151 カサマツ
300153 ポンデローサパイン,ポンデローサマツ
300163 イワマツ,ジョウマツ,センジョウマツ,ハイマツ,ハッコオダゴヨオ,ハヒマツ
300168 ブンゲンスマツ
300173 ラジアータマツ,ラジアタマツ,ラデイアタマツ
300180 リギダマツ,レジノーザマツ
300181 チャンマツ,ミツバマツ,リギダマツ
300186 ナガバマツ,ヒマラヤマツ
300189 サビンマツ
300211 イースタンホワイトパイン,ストロウブマツ,ストローブゴヨウマツ,ストローブマツ,ストロープマツ
300223 オウシウアカキシ,オウシュウアカキシ,オウシュウアカマツ,スコッッパイン,ドイツアカマツ,ヨーロッパアカマツ
300240 モウコアカマツ,モウゴアカマツ
300243 マンシウアカマツ,マンシウクロマツ,マンシエウクロマツ,ムレイショウ
300251 マンシウクロマツ
300264 タエダマツ,テーダマツ,ロビロリーパイン,ロブロリーパイン
300269 ニイタカアカマツ,ニヒタカアカマツ,ニヒタカクロマツ,ミヤマアカマツ
300281 オトコマツ,オマツ,オンマツ,クロマツ,ヲマツ
300285 タギョウクロマツ
300288 トレイマツ
300296 バージニアマツ
300297 ヒマラヤゴヨウ
300323 コショウ属,コセウ属,フウトウカズラ属,フウトウカヅラ属
300354 キンマ
300374 ギアナ・ペッパー
300377 ヒッチョウカ
300408 ヒハツモドキ
300427 ツルゴセウ,フウトウカズラ,フウトウカヅラ
300428 オオバフウトウカズラ
300432 クラルフウトウカヅラ
300434 クワセウフウトウカヅラ
300446 インドナガコショウ,ヒハツ
300458 カバ
300464 コショウ,コセウ
300495 コセウモドキ,ヒハツモドキ
300540 ガランビゴセウ
300557 コショウ科,コセウ科,ニショウ科
300722 イネガヤ
300728 アレチイネガヤ
300833 ヌノマオ属,ヌノマヲ属
300835 オホイハガネ,ヌノマオ,ヌノマヲ
300840 ピクェーリア属
300842 シラユキギク
300941 ウドノキ属,ピゾーニア属
300944 オハグロカヅラ,トゲオホジラミ,トゲカズラ,トゲカヅラ,トリモチカヅラ
300955 トゲミウドノキ
300956 キャベツノキ
300974 トネリバハゼノキ属
300980 カイノキ,トネリバハゼノキ,ランシンボク
300994 マスティクス
301002 テレビンノキ,テレピンノキ,トクノウコウ
301005 ピスタシオノキ,ピスタチオ
301006 セイコウボク
301021 ボタンウキクサ属
301025 オホバモ,ボタンウキクサ,リウキウウキクサ
301053 エンドウ属,ピズム属
301055 アカエンドウ,エンドウ
301070 エンドウ,サアエンドウ,サヤエンドウ,シロエンドウ
301080 サヤエンドウ
301099 ケイビアナナス属,ピトカイルニア属
301102 ウラジロアナナス
301106 ハランアナナス
301111 ケイビアナナス
301117 キンキジュ属
301134 キンキジュ
301147 アカハダノキ,アカバノキ,クマザキガフクワン,タマネム

301203 トベラ科
301207 トベラ属
301221 タイワントベラ
301226 シロトベラ
301227 オオミノトベラ
301229 オキナワトベラ
301253 オホバトベラ,ヤドリオホバトベラ
301266 ヒマラヤトベラ
301294 コヤスノキ,トガリバトベラ,ヒメシキブ,ヒメシキミ,ヒメトベラ,リウキウトベラ
301357 コバトベラ
301358 ハハジマトベラ
301360 トビラギ,リウキウトベラ
301363 タイワントベラ
301398 クロベトベラ
301413 タカサゴトベラ,トビラギ,トビラノキ,トベラ,トベラノキ,ピトスポルム
301416 オオバトベラ,オオバトベラノキ,オホバトベラノキ
301433 シマ,シマトベラ,タウソヨゴ,トウソヨゴ
301438 オニトベラ,ガマズミトベラ
301626 イヌミヤマムラサキ
301627 ヒナムラサキ
301629 アメリカキュウリグサ
301692 タッタソウ属
301694 イトマキグサ,イトマキソウ,タッタソウ
301769 ムニンノキ
301776 ケヅ
301777 アカテツ,クロテツ
301778 コバノアカテツ
301821 オオバコ科,オホバコ科
301832 オオバコ属,オホバコ属
301855 ダキバオオバコ,ダキバオホバコ
301864 エダウチオオバコ,ホソバオオバコ,ホソバオホバコ
301868 アメリカオオバコ
301871 オオバコ,オホバコ,オンバコ,ハハキオオバコ,ハハキオホバコ,ヤグラオオバコ,ヤグラオホバコ
301872 イサワオオバコ
301873 ケバノオオバコ
301877 フイリオオバコ
301884 マルミオオバコ
301885 ヤクシマオオバコ
301894 エゾオオバコ,エゾオホバコ
301895 ケナシエゾオオバコ
301910 セリバオオバコ
301952 ムジナオオバコ
301997 ハクサンオオバコ,ハクサンオホバコ
301998 ケナシハクサンオオバコ
301999 シロバナハクサンオオバコ
302001 ニチナンオオバコ
302014 タウオホバコ,トウオオバコ
302034 ヘラオオバコ,ヘラオホバコ
302068 イヌオオバコ,オオバコ,オホバコ,セイヨウオオバコ,タイワンオオバコ,タイワンオホバコ
302091 ハマオオバコ,ハマオホバコ
302103 シロバナオオバコ,シロバナオホバコ
302178 トゲオオバコ
302181 ハイオオバコ
302209 タチオオバコ,ツボミオオバコ
302229 スズカケノキ科
302241 ツレサギサウ属,ツレサギソウ属
302242 ハチジョウアイノコチドリ
302243 キソチドリモドキ
302248 ヤクシマチドリ
302260 エゾチドリ,フタバツレサギ,フタブツレサギ
302262 シマツレサギソウ
302265 ツクシチドリ,ニイタカチドリ,ニヒタカチドリ
302266 ツクシチドリ
302280 エゾチドリ,エゾノジンバイサウ
302290 マンシュウヤマサギソウ
302329 ジンバイソウ
302331 コウライチドリ
302355 ジャカウサギ,ジャカウチドリ,ジャコウチドリ,ミズチドリ,ミヅチドリ
302357 オオバナオオヤマサギソウ
302361 シロウマチドリ
302370 イイヌマムカゴ
302377 ツレサギサウ,ツレサギソウ
302395 ナガヅメトンボサウ
302404 オオキソチドリ,キソチドリ,ハシナガサギソウ,ハシナガヤマサギソウ,マイサギソウ,ヤマサギサウ,ヤマサギソウ,リトウトンボサウ,リトウトンボソウ
302407 アマミトンボ
302408 ヤクシマトンボ
302409 ハチジョウチドリ
302411 タカネサギソウ
302413 ナガバノキソチドリ
302414 ヒトツバキソチドリ
302415 リウトウンボサウ,リトウトンボ
302439 オオバノトンボソウ,タカネサギサウ,ノヤマトンボ,ノヤマトンボソウ
302440 ミクラトンボソウ
302458 ハチジョウツレサギ
302502 オオヤマサギソウ,オホヤマサギサウ,ナガバチドリ
302512 クニガミトンボソウ
302513 ヒロバトンボソウ
302519 タイトントンボソウ,ランカンンボサウ
302520 イリオモテトンボソウ
302533 ミヤマチドリ
302534 ガッサンチドリ
302536 ホソバノキソチドリ
302537 ナガバトンボソウ
302538 コバノトンボソウ
302541 ホソバノキソチドリ
302549 トンボソウ
302574 スズカケノキ属,プラタナス属,プラタヌス属
302582 カエテバスズカケノキ,モミジバスズカケ,モミジバスズカケノキ,モミヂバスズカケノキ
302588 アメリカスズカケ,アメリカスズカケノキ,ボタンノキ
302592 スズカケノキ,ボタンノキ
302674 ノグルミ属
302676 トウノブノキ
302684 ドクグルミ,ノグルミ,ノブノキ,フデノキ,ヤマグルミ
302721 コノテガシハ,コノテガシワ,ソノテ,テガシハ,トノテガシワ,ハリギ,フタオモテ
302740 プラティクリニス属
302747 キキャウ属,キキョウ属
302753 キキャウ,キキョウ,キキョオ,フタエギキョウ
302754 フタエキキョウ
302755 シロギキョウ
302795 バイカアマチャ属,バイクワアマチャ属
302796 タマウツギ,バイカアマチャ,バイクワアマチャ,モッコウバナ
302884 シマウツボ
303022 プラティティラ属
303082 アザミヤグルマ,アザミヤグルマギク,セントーレア
303087 シロジクトウ属
303089 セダカシロジクトウ
303096 シロジクトウモドキ属
303099 ヤブシロジクトウモドキ
303125 ヤマハッカ属
303561 ベニオウギ
303780 プレクトメルミンツス属
303994 メダケ属
304007 オキナダケ,シマメダケ
304008 オキナダケ
304009 オロシマチク
304010 ネザサ
304014 アズマシノ,アズマネザサ,アヅマネザサ,オオシマダケ,シナガハダケ,シナガワダケ
304015 ジョウボウジダケ
304016 スダレヨシ
304018 ヤシバダケ
304032 ダイミョウダケ,タイミンチク,ツウシチク
304033 アラゲネザサ
304034 ヒゴメダケ
304040 アオネザサ

304047 コンゴウタケ
304048 オニネザサ
304054 ギャウエフチク,リウキウチク,リュウキュウチク
304055 フイリリュウキュウチク
304069 ヨコハマダケ
304071 ヒロウザサ
304072 ユゲネザサ
304073 エチゼンネザサ
304082 エチゴメダケ
304083 ケオロシマチク,ケネザサ
304097 ケネザサ
304098 オナガタケ,オナゴダケ,オンナダケ,カハタケ,カワタケ,シノダケ,ナヨタケ,ニガタケ,ニガダケ,メダケ,ヲナガタケ
304100 シロシマメダケ
304101 ハガワリメダケ
304109 シマザサ,チゴザサ
304114 カムロザサ
304115 オウゴンカムロザサ
304116 ヤマカムロザサ
304204 タイリントキサウ属,タイリントキソウ属,プレイオネ属
304218 タイリンカヤラン,タイリントキサウ,タイリントキソウ
304248 プレイオネ・フォレステイ
304336 カクイシソウ属,プレイオスピロズ属
304340 ホウラン,ホプマン
304351 ニョライ
304353 ランジョウ
304367 ホウヨク
304372 テイギョク
304381 セイラン
304752 オオカサモチ属,オホカサモチ属
304765 オニカサモチ,オホカサモチ
304766 オオカサモチ
304772 オオカサモチ,オホカサモチ
304904 プレウロタリス属
305072 ヒヒラギギク属
305100 ヒイラギギク,ヒヒラギギク
305115 タワダギク
305123 ホソバヒイラギギク
305165 イソマツ科
305167 プルムバーゴ属,ルリマツリ属
305172 アオマツリ,アギバナジラミ,アヲマツリ,ルリマツリ
305185 アカマツリ
305202 インドマツリ,シマナガジラミ,シマワシグサ,セイロンマツリ
305206 インドソケイ属
305213 シロバナインドソケイ
305219 タイリンインドソケイ
305225 インドソケイ,トガリバインドソケイ,ベニバナインドソケイ
305226 インドソケイ,ホウライオホケフチク
305274 イチゴツナギ属
305278 ミゾイチゴツナギ
305280 オキナワミゾイチゴツナギ
305282 タマミゾイチゴツナギ
305303 タカネイチゴツナギ,ミヤマナガハグサ
305328 ホソバノナガハグサ
305334 イチゴツナギ,スズメノカタビラ,ニラミグサ
305348 ツルスズメノカタビラ
305383 ダウリアイチゴツナギ
305419 チャボノカタビラ
305422 ムカゴイチゴツナギ
305451 コイチゴツナギ
305454 ツクシスズメノカタビラ
305506 オニイチゴツナギ
305520 アイヌソモソモ
305551 タカネタチイチゴツナギ
305558 キタダケイチゴツナギ
305572 ハクサンイチゴツナギ
305573 ナンブソモソモ
305584 フジイチゴツナギ,ヤマミゾイチゴツナギ
305588 ミスジナガハグサ
305605 ミスジナガハグサ
305606 イワテイチゴツナギ
305630 チョウセンタチイチゴツナギ
305686 カラフトイチゴツナギ,ザラバナソモソモ
305687 ワタゲソモソモ
305705 ムラサキソモソモ
305720 イトイチゴツナギ
305733 マンシウイチゴツナギ
305743 ウッド・メドウ・グラス,タチイチゴツナギ
305776 オオイチゴツナギ,オホイチゴツナギ,オホニラミグサ,カラスノカタビラ
305788 オガタチイチゴツナギ
305804 ヌマイチゴツナギ,ファウル・メドウ・グラス
305855 ケンタッキ・ブルー・グラス,ソモソモ,ナガハグサ
305858 ホクセンイチゴツナギ
305875 ヒロハノナガハグサ
305882 オオナガハグサ
305916 イブキソモソモ,チシマソモソモ
305942 ヒメカラフトイチゴツナギ
305944 ムラサキエゾソモソモ
305970 タカネイチゴツナギ,ミヤマイチゴツナギ
305993 アオイチゴツナギ
306003 イチゴツナギ,ザラツキイチゴツナギ,タイワンヒメイチゴツナギ,ヒメイチゴツナギ
306010 ヒロハイチゴツナギ
306033 ホソバノナガハグサ
306038 エダハリシバ
306062 ホソバナソモソモ
306098 オオスズメノカタビラ,オホスズメノカタビラ,コイチゴツナギ,ラフ・ストークド・メドウ・グラズ
306099 タマオオスズメノカタビラ
306106 ムカゴツヅリ
306133 イチゴツナギ
306146 アオイチゴツナギ,アヲイチゴツナギ
306156 タニイチゴツナギ
306307 ポダンギス属
306345 イヌマキ科,マキ科
306395 イヌマキ属,マキ属
306419 コウトウマキ
306424 ナンヤウマキ
306431 タイワンマキ
306450 オホバナギ,ナギマキ
306457 アスナロ,イヌマキ,クサマキ,ヒトツハ,ヒトマキ,ホンマキ,マキ,ラカンマキ
306458 カクバマキ
306469 マカンマキ,マキ,ラカンマキ
306493 センビキ,チカラシバ,ナギ
306502 トガリバマキ
306503 ナンコウナギ
306506 カュチイナ,ネリイマキ,ヤママキ
306512 ナンバンイヌマキ
306564 ポドレピス属
306607 カワゴケソウ科
306609 ポドフィラム属
306636 ポドヒルマ,ポドフィルム
306688 カハゴケサウ科,カワゴケソウ科,ポドステモン科
306828 イタチガヤ属
306832 イタチガヤ,ヒメテウセンガリヤス
306834 オオイタチガヤ
306847 トキサウ属,トキソウ属
306870 トキサウ,トキソウ
306871 シロバナトキソウ
306879 ヤマトキサウ,ヤマトキソウ
306880 シロバナヤマトキソウ
306956 ヒゲオシベ属
306960 オホネコノオ,オホネコノヲ,オホバノミヅトラノオ,オホバノミヅトラノヲ
306964 パチョリ
306974 ヒゲオシベ
307055 ホウオウボク属
307124 イヌタムラソウ属
307125 イヌタムラソウ
307169 ボラスキア属
307170 カクランボウ,ライジンカク
307186 ハイシノブ科,ハナシノブ科
307190 ハイシノブ属,ハナシノブ属
307195 ヒメハナシノブ
307197 ハナシノブ,ヨウシュハナシノブ

307198　キョクチハナシノブ
307201　カラフトハナシノブ
307202　シロバナカラフトハナシノブ
307203　レブンハナシノブ
307204　クシロハナシノブ
307207　ミヤマハナシノブ
307209　キョクチハナシノブ
307215　カラフトハナシノブ,ハナシノブ
307222　ヒメハイシノブ,ヒメハナシノブ
307250　エゾノハナシノブ,エゾハナシノブ
307265　ゲッカカウ属,ゲッカコウ属,チューベロース属,チュベロース属
307269　オランダズイセン,ゲッカカウ,ゲッカコウ,ジャガタラズイセン,チュベロース,ナクトール,ナフトール
307312　ヤブミョウガ属,ヤブメウガ属
307324　メウガサウ,ヤブミョウガ,ヤブメウガ
307333　コヤブミョウガ,コヤブメウガ
307337　コヤブミョウガ,ザルゾコミョウガ
307484　キダチアウソウクワ属
307508　クロボウモドキ
307538　ロウココヤシ属
307755　ヨツバハコベ
307859　ポリキクニス属
307891　ヒメハギ属
307907　ハリヒメハギ
307919　シマキンチャク,タイワンキンチャク
307927　シンチクヒメハギ
308081　ホソバヒメハギ
308123　ハヒヒメハギ,ヒメハギ
308124　ホソバヒメハギ
308125　マツゲヒメハギ
308126　セイカヒメハギ
308170　リュウキュウヒメハギ
308247　コバナヒメハギ
308299　カキノハグサ
308302　ナガバカキノハグサ
308331　カンザシヒメハギ
308347　セネガ
308348　ヒロハセネガ
308359　オオヒメハギ,ヒメハギ,ホソバヒメハギ
308397　ヒナノキンチャク
308403　イトヒメハギ,オンジ
308437　クルマバヒメハギ
308474　ヒメハギ科
308481　タデ科
308493　アマドコロ属
308495　ヒナヨウラク
308498　アリサンアマドコロ
308529　コナルコユリ
308538　ナルコユリ,ワウセイ
308562　ヒメアマドコロ,ヒメイズイ,ヒメヰズヰ
308565　ミドリヨウラク
308567　ワニグチサウ,ワニグチソウ
308582　ミヤマナルコユリ
308592　オオナルコユリ
308593　トウアマドコロ
308613　アマドコロ,イズイ,オホアマドコロ,カラスユリ,ユミグサ,ヰズヰ
308616　アマドコロ,イスイ,エミクサ,カラスユリ,ナガバアマドコロ
308641　カギクルマバナルコユリ
308647　クルマバナルコユリ
308690　タデ属
308717　ヒメイワタデ
308726　ナガバノイブキトラノオ
308730　コウアンイワタデ,マンシウイワタデ
308739　エゾノミズタデ,エゾノミヅタデ
308777　ヤナギイワタデ
308788　アキノミチヤナギ,スナジミチヤナギ,ハイミチヤナギ
308791　ヌカボミチヤナギ
308816　オオミチヤナギ,ニハヤナギ,ニワヤナギ,ミチヤナギ
308820　タチニハヤナギ
308832　オクミチヤナギ,ホソバミチヤナギ
308850　タチホソバミチヤナギ,マンシウミチヤナギ
308863　ミチヤナギ
308877　クタデ,ケタデ
308886　シマミゾソバ,ニヒタカミゾソバ
308893　イブキトラノオ,イブキトラノヲ
308907　アカノマンマ,イヌタデ,ハナタデ
308922　ハリタデ
308946　ヒメツルソバ
308965　タイワンツルソバ,ツルソバ
308974　ツルソバ
308976　ツルソバ
309006　サクラタデ
309051　ナツノウナギツカミ,リュウキュウヤノネグサ
309054　ココメタデ,ヌカボシタデ
309056　エゾタデ,コバノイハタデ,コバノイワタデ,シベリアイワタデ,ホソバオンタデ
309116　ボントクタデ
309121　マンシウヌカボタデ,マンシュウヌカボタデ
309123　サイコクヌカボ
309125　ヤナギヌカボ
309192　ヒナミチヤナギ
309199　タデ,ホンタデ,マタデ,ヤナギタデ
309201　ムラサキタデ
309203　イトタデ,ヤオゼンタデ
309262　サクラタデ,シロイヌタデ,シロバナサクラタデ,ハマタデ
309284　ホソバヌマタデ
309294　キヌタデ
309298　オオイヌタデ,サナエタデ,サナヘタデ,ナツタデ
309312　キヌタデ
309318　サナエタデ
309325　ホソバイブキトラノオ
309334　リョウトウミチヤナギ
309338　ヤチタデ
309345　イヌタデ,ハナタデ
309355　ヒメタニタデ
309356　サデクサ,ミゾサデクサ
309373　エダハリタデ,マンセンイブキトラノオ
309395　コヤナギタデ,ヤナギヌカボ
309442　ザラメキタデ,ヤノネグサ
309447　オヤマソバ
309459　タニソバ
309465　シロバナヤノネグサ
309468　オオイヌタデ,オホイヌタデ,シロイヌタデ
309472　ウラジロオホイヌタデ
309494　イヌタデ,オオケタデ,オオタデ,オオベニタデ,オホケタデ,オホタデ,ハコウ,ハブテコブラ
309505　エゾイブキトラノオ
309542　アメリカサナエタデ,ケオトメサナエタデ
309564　イシミカハ,イシミカワ,サデクサ
309570　アオナシタデ,アヲナシタデ,ハチノジタデ,ハルタデ
309602　ヒメミチャナギ,ヤンバルミチャナギ
309624　イヌタデ,ハナタデ
309629　ホソバノウナギツカミ
309644　ボントクタデ
309647　アラゲタデ
309683　ホザキニワヤナギ
309711　ニイタカタニソバ,ニヒタカタニソバ,ミヤマアカヅラ,ミヤマタニソバ
309723　オオイタドリ,オホイタドリ,ドングイ,ドンゴ
309772　トゲソバ,ママコノシリヌグイ,ママコノシリヌグヒ
309792　スナジタデ,スナヂタデ
309796　アキノウナギツカミ,アキノウナギツル,ウナギツカミ,ウナギヅル
309831　タイワンヤノネグサ
309841　クリンユキフデ
309853　コヌカボタデ,ヌカボタデ
309857　ナガバハマミチヤナギ
309858　ウシオミチヤナギ
309859　フトボノヌカボタデ
309863　シマヒメタデ
309870　ハルトラノオ
309877　ウシノヒタイ,オホミゾソバ,ニヒタカミゾソバ,ミゾソバ,ミマミゾソバ
309888　オオミゾソバ
309893　アイ,アヰ,タデアイ,タデアヰ
309898　アラゲタデ

309909 ホソバイヌタデ
309926 コゴメウナギッカミ
309946 オオネバリタデ,オホネバリタデ,ケネバリタデ,ネバリタデ
309951 ニオイタデ,ニオヒタデ,ニホヒタデ
309954 コモチトラノオ,コモチトラノヲ,ムカゴトラノオ,ムカゴトラノヲ
309966 ウラジロタデ
309969 オンタデ
310102 ヒエガエリ属,ヒエガヘリ属
310109 ヒエガエリ,ヒエガヘリ
310125 ハマヒエガエリ,ハマヒエガヘリ
310134 ウォーターベントグラス
310169 タイワンモミジ属,タイワンモミヂ属,ポリスキアス属
310199 タイワンモミジ,タイワンモミヂ,ホソバアラリア
310204 オオバアラリア
310208 キレバアラリア
310309 ボリズダーキア属
310798 フトバクレソラン
310818 シマリュウガン属
310823 カサイ,シマリュウガン,タイトウリュウガン,タウン,バンリュウガン,マトア,マルガイ
310848 カラタチ属
310850 カラタチ,キコク,トリフォリア
310852 ヒリュウ
310859 ポネロルキス属
310866 ヒナチドリ
310867 シロバナチドリ
310868 チャボチドリ
310891 ウチョウラン
310892 シロバナウチョウラン
310893 アワチドリ
310894 クロカミラン
310897 ニョホウチドリ
310943 タカサゴチドリ
310950 クロヨナ属
310962 アッカハマメ,クロヨナ,ゴエフフヂ
311019 ミズアオイ科,ミヅアフヒ科
311020 ポンチエバ属
311131 ハコヤナギ属,ヤマナラシ属
311142 アクミナタポプラ
311144 トウシドロ,ベルリンドロ,ロリネンシスポプラ
311146 カナダポプラ,ユールアメリカナポプラ
311148 バケリエリイポプラ
311149 エウゲニイポプラ
311150 ゲルリカポプラ
311155 マリランデイカポプラ
311157 レゲネラータポプラ
311158 ロプスタポプラ
311163 セロチナポプラ
311164 ヴエルニルベンスポプラ
311208 ウラジロハコヤナギ,ギンドロ,ハクヤウ,ハクヨウ
311226 ピラミダリスポプラ
311231 アングラタポプラ,カロライナポプラ,カロリナハコヤナギ
311233 アングスチフオリアポプラ
311237 バルサミベラポプラ,バルサムポプラ
311259 カンジカンスポプラ
311261 オウバギンドロ
311265 ナガバドロ
311276 キリアタポプラ
311281 エゾヤマナラシ,カラフトヤマナラシ,チョウセンヤマナラシ
311292 ヒロハハコヤナギ,ヒロハヤマナラシ
311308 エウフラテカポプラ,コトカケヤナギ
311324 フレモンテイポプラ
311335 オオバヤマナラシ,グランデイデンターポプラ
311361 エゾヤマナラシ,チョウセンヤマナラシ
311368 チリメンドロ,ニホヒドロ
311371 オウバヨウ
311386 ソキホウヤマナラシ
311389 キワタ,デロ,デロノキ,ドロノキ,ドロブ,ドロヤナギ,ホトケギ,ワタドロ
311398 アメリカヤマナラシ,クロヤマナラシ,ニグラヤマナラシ,ヨーロッパクロヤマナラシ
311400 イタリカポプラ,イタリャマナラシ,セイチョウハコヤナギ,セイヤウハコヤナギ,セイヤウヤマナラシ,セイヨウハコヤナギ,ポプラ
311414 ヨーロッパポプラ
311423 ケツメクサ
311436 オオテリハドロ
311480 アオドロ,アヲドロ,ツラフリ,ハコヤナギ,マルバヤナギ,ヤマナラシ,ヨメフリ
311482 シモニードロ,テリハドロ,テリハドロノキ
311504 コバノテリハドロ
311509 コウアンドロ,サウヴエオレンスポプラ,デロ,ドロノキ,ドロヤナギ
311514 スゼクアニカポプラ
311530 オニヤマナラシ,ケハクヨウ
311537 テウセンヤマナラシ,ヨーロッパヤマナラシ
311545 ヤマナラシ
311547 アメリカヤマナラシ,トレムロイデスポプラ
311562 ケドノロキ
311565 ヒロハハコヤナギ
311584 ユンナンポプラ
311780 ポルテア属
311817 スベリヒユ属
311852 ホロビンソウ,マツバボタン
311877 ヒロハノマツバボタン
311889 オキナワマツバボタン
311890 イハイズリ,イハイヅル,イワイズリ,スベリヒユ,トンボグサ,ミンブトキ
311899 タチスベリヒユ
311915 ケツメグサ,ヒメマツバボタン
311921 タイワンスベリヒユ
311954 スベリヒユ科
311955 ポーチュラカリア属
311956 ギンイチョウ
312028 ヒルムシロ属
312030 ツツヤナギモ
312031 アイノコヤナギモ
312032 ヒロハノセンニンモ
312033 アイノコヒルムシロ
312034 ササエビモ
312035 アイノコイトモ
312037 ホソバヒルムシロ
312045 オオササエビモ
312075 エゾヤナギモ
312079 エビモ
312083 コバノヒルムシロ
312090 ヒルムシロ
312123 フトヒルムシロ
312126 エゾノヒルムシロ,カラフトササモ
312138 エゾノヒルムシロ
312168 ガシャモク
312170 センニンモ
312171 ササバモ,サジバモ
312173 ナガバイトモ
312188 オヒルムシロ
312190 オオヒルムシロ,オヒルムシロ,オホヒルムシロ,メヒルムシロ,ヲヒルムシロ
312199 イヌイトモ
312200 ホソバミヅヒキモ
312203 イトモ,ホソバミズヒキモ,ミズヒキモ,ミヅヒキモ
312206 ササモ,ヤナギモ
312220 ヒロハノエビモ
312228 ヒルムシロ
312232 ナガバエビモ
312236 イトモ,ツツイトモ
312288 モヅヒキモ
312291 ササバモ
312300 ヒルムシロ科,ヒルルシロ科
312322 ギジムシロ属
312323 オオミツバッチグリ
312325 ビローキギジムシロ
312347 コバナギジムシロ
312351 アオナイワキンバイ,イハキンバイ,キンエビネ
312357 ハイキジムシロ
312360 エゾツルキンバイ,トウツルキンバ

イ,ヨウシュノツルキンバイ
312365 エゾツルキンバイ
312389 ウラジロロウゲ
312392 オオロウゲ,オホラフゲ
312393 ベニバナキジムシロ,ベニバナミツモトソウ
312399 アラゲキンバイ
312408 ウラギシキンバイ,サンヨウッチグリ
312412 クサキンロウバイ
312421 ヒノキンロウバイ
312446 カラヒメヘビイチゴ,ヒメヘビイチゴ
312447 ヒメヘビイチゴ
312449 トウヘビイチゴ
312450 カハラサイコ,カワラサイコ
312453 ホソバカハラサイコ
312457 タチヘビイチゴ
312467 シラゲカワラサイコ
312478 マルバミツモト,ミツモト,ミツモトサウ,ミツモトソウ,ミナモトサウ,ラフゲ
312480 ミツモトソウ
312498 イワキンバイ,キンエビネ
312500 マルバイワキンバイ
312501 タケシマイワキンバイ
312502 ツチグリ,ツチナ,ブクリュウサウ
312528 ウチワキジムシロ
312540 ハイオオヘビイチゴ,モミジキンバイ
312550 ヒメキジムシロ
312552 キジムシロ
312565 ミツバツチグリ
312566 ヤエノミツバツチグリ
312567 ヒトツバキジムシロ
312599 キンロバイ
312606 ギンロバイ,ハクロバイ
312659 コバナキジムシロ
312690 オヘビイチゴ,ヲヘビイチゴ
312714 オホツキンバイ,タカネラフゲ,フクトメキンバイ
312745 ミヤマキンバイ
312746 ケミヤマキンバイ
312747 オクミヤマキンバイ
312748 アポイキンバイ
312750 モリキンバイ
312751 ユウバリキンバイ
312758 チシマキンバイ
312782 メアカンキンバイ
312810 ニベナロウゲ
312811 ウラジロキンバイ
312815 ヒロハカハラサイコ,ヒロハノカワラサイコ
312818 ウラジロキンバイ
312828 エゾノミツモトサウ,エゾノミツモトソウ,カラフトミツモト
312834 アオカワラサイコ
312912 オオヘビイチゴ,オホヘビイチゴ,オホラフゲ,タチラフゲ,タチロウゲ
312935 テリハキンバイ
312942 ツルキンバイ
312944 トウイワキンバィ
312946 シロキジムシロ,シロバナラフゲ,シロバナラロウゲ
312955 オグヂガワラザイコ
313013 キジムシロ
313015 ヤエキジムシロ
313028 ツルキジムシロ
313039 オキジムシロ
313047 コバナキジムシロ
313054 スナジカワラサイコ
313076 エチゴキジムシロ
313080 エチゴツルキジムシロ
313098 クルマバカワラサイコ
313103 ネバリカワラサイコ
313115 ヒメツルキジムシロ
313169 オホキノボリカヅラ属
313170 オホキノボリカヅラ
313178 アフヒゴセウ,タイヨウフウトウカズラ
313186 ハズノハカヅラ属,ポトス属,ユズノハカヅラ属
313197 ハズノハカヅラ,ユズノハカズラ,ユズノハカヅラ
313266 カマツカ属
313316 キミノカマツカ
313318 ナガエカマツカ
313411 オオバヒメマオ属,コケツルマヲ属
313424 キタチマオ,キタチマヲ,コケツルマオ,コケツルマヲ,ツルイハガネ
313450 アリエヒメマオ,アリエヒメマヲ,オホバツルマオ,オホバツルマヲ
313483 オオバヒメマオ,ツルマオモドキ,ヤンバルツルマオ
313552 サクラダサウ属,サクラダソウ属
313564 サクラダサウ
313593 ハマクサギ属
313692 ウオクサギ,ウヲクサギ,キバナハマクサギ,ハマクサギ
313702 ケウオクサギ,シマウオクサギ,シマウヲクサギ,タイワンウオクサギ,タイワンウヲクサギ
313740 タイワンウオクサギ
313778 フクオウソウ属,フクワウサウ属
313781 フクオウソウ
313782 マルバフクオウソウ
313783 フクオウニガナ
313822 タイワンフクオウソウ
313849 キレハノフクオウモドキ
313894 オオニガナ
313895 フクオウモドキ,フクワウドキ
313909 プレニア属
314083 サクラソウ属,サケリサウ属
314084 キュ-コサクラ,ヤグラサクラ
314085 クリンザクラ
314143 アツバサクラソウ
314144 ナガバノサクラサウ,ナガバノミミサクラソウ
314157 シナサクラソウ,ムラサキクリンソウ
314212 テマリコザクラ,ヒマラヤサクラソウ
314271 サクラサウ
314280 エゾコザクラ
314281 シロバナエゾコザクラ
314284 ハクサンコザクラ
314285 シロバナハクサンコザクラ
314286 シロバナミチノクコザクラ
314310 カサザキサクラソウ,ダマザキサクラソウ
314333 セイタカサクラソウ,セイタカセイヨウサクラソウ
314366 セイヨウユギワリソウ,ユギワリソウ,ヨウシュギワリソウ
314370 ユキワリソウ
314372 シロバナユキワリコザクラ
314373 オオユキワリソウ
314375 シロバナレブンコザクラ
314376 レブンコザクラ
314377 ユキワリソウ
314378 シベリアユキワリソウ
314387 ユキワリコサクラ
314396 テマリコザクラ
314405 キヒメザクラ
314408 ハルコザクラ,ヒメザクラ
314461 イキサコザクラ,ミチノクコザクラ
314465 ヒダカイワザクラ
314466 カムイコザクラ
314508 クリンサウ,クリンソウ,サクラソウ
314511 シロバナクリンソウ
314514 オオサクラソウ,オホサクラサウ
314516 シロバナオオサクラソウ
314519 オオサクラソウ,ミヤマサクラソウ
314521 エゾオオサクラソウ,エゾノサクラソウ,サントウザクラ
314522 シロバナエゾオオサクラソウ
314523 ウスゲノエゾザクラ
314531 カワシマコザクラ
314537 カッコウウ,カッコソウ,キソザクラ,シコクカッコソウ
314573 ウラジロクリンザクラ
314604 ヒメコザクラ
314613 オトメザクラ,ケショウザクラ,ヒメサクラサウ
314624 リウクリンソウ
314655 ニヒタカクリンサウ
314696 ヒナザクラ
314713 ラシュワコザクラ
314717 シキザキサクラソウ,トキハザクラ,トキワザクラ

314820 クリンザクラ,プリムラ・ポリアンサ
314885 コイワザクラ
314886 シロバナコイワザクラ
314888 クモイコザクラ
314890 ナガバコイワザクラ
314891 チチブイワザクラ
314904 ヒメザクラ
314930 イワコザクラ,イワサクラソウ
314948 ラシュワコザクラ
314952 サクラソウ
314956 キレザキサクラソウ
314957 シロバナサクラソウ
314975 カンコザクラ,カンザクラ,カンザクラソウ,サンザクラサウ,サンザクラソウ,チュウカサクラソウ,チュウクワサクラサウ
315029 テシオコザグラ
315051 イワザクラ,トサザクラ
315052 シロバナイワザクラ
315054 カマナシコザクラ,シナノコザクラ
315082 キバナノクリンザクラ,セイヤウサクラサウ,セイヨウサクラソウ
315087 アラビアコザクラ
315101 イチゲコザクラ,イチゲサクラソウ,オホバナクリンザクラ,ブリムロース
315102 ヤエイチゲサクラ
315136 ユウバリコザクラ,ユウパリコザクラ
315138 サクラサウ科,サクラソウ科
315171 タカサゴグミモドキ属,プリンセーピア属
315174 サイカチモドキ,タカサゴグミモドキ
315176 グミモドキ
315376 フトエクマデヤシ属
315379 ベッカリフトエクマテヤシ
315386 フィジーフトエクマテヤシ
315387 ヒメフトエクマテヤシ
315388 トックリフトエクマテヤシ
315430 ツノゴマ属
315434 タビビトナカセ,ツノゴマ
315461 ウライサウ属,ウライソウ属
315464 セキモンウライソウ
315465 ウライソウ
315476 ウライサウ,ウライソウ
315509 プロメネーア属
315547 ケベヤ属
315560 ケベヤ
315702 プロチア属,プロテア属
315776 キングプロテア,プロチアキナロィデス
315810 ピンクミンク
315937 シュガーブッシュ
316007 ヤマモガシ科
316095 ウツボグサ属,プルネラ属
316097 ウツボグサ
316100 シロバナウツボグサ
316123 タテヤマウツボグサ
316124 シロバナタテヤマウツボ
316125 ウスイロハクサンウツボ
316127 ウツボグサ,セイヨウウツボグサ,ミヤマウツボグサ
316133 タイワンウツボグサ
316136 ミヤマウツボグサ
316150 ウツボグサ,ベニバナウツボグサ
316166 サクラ属
316188 センダイヨシノ
316206 アメリカスモモ,アメリカナスモモ
316219 アンゲスチフォリアスモモ
316285 アメリカタテギ
316294 ミロバランスモモ
316302 アカバサクラ,ベニバスモモ
316382 セイヨウスモモ,ドメスチカスモモ,ヨーロッパスモモ
316436 カキバイヌザクラ
316439 シモツキザクラ
316444 ホーチュラナスモモ
316466 ヤブザクラ
316470 インシチチアスモモ
316485 ニワザクラ,ハネズ
316551 カバザクラ
316570 コウシュウバイ,コウメ,シナノウメ
316577 オオバイ
316581 ザロンウメ,ザロンバイ,ヤツブサウメ
316583 マンソニアナスモモ
316636 イトザクラ,クサイザクラ,シダレザクラ,シダレヒカン
316761 イクリョネモモ,スモモ,トガリスモモ,ニホンスモモ,ハダンキョウ
316767 ケスモモ
316777 ケエゾヤマザクラ
316809 タカサゴ,チャワンザクラ,ナテン,ナデン,ムシャザクラ
316813 サイモンスモモ,シモンプラム
316819 カタザクラ,タテギ,ヒラギガシ,メヒラギ
316824 アズサ,カタザクラ,タデキ,ヒイラギガシ,ヒラギガン,ホカノキ,ミズアオイ,メヒイラギ,メヒラギ,リンボク
316874 アリサンヤマザクラ
316884 コヨウバイ,ヤヘオヒョウモモ
316913 カスミザクラ,ケヤマザクラ
316936 ウスゲオオシマ
316939 ベニヅルザクラ
317026 プサンモフォラ属
317174 プセウドアナナス属
317233 ナホザキイガコウゾリ,ホザキイガコウゾリナ
317245 プセウデランテムム属
317598 カリン属
317602 アンラクワ,アンランカ,アンランジュ,カラナシ,カリン,キボケ,クワリン
317636 タカネトンボ
317685 プセウドエスポストア属
317687 ゲンラク
317688 アカトゲゲンラク,コウジュラク,ムゲンラク
317689 ハクキュウデン
317821 イヌカラマツ属
317822 イヌカラマツ,キンショウ
317869 プセウドロビビア属
317904 ルリトラノオ属
317909 キクバトラノオ
317912 タウテイラン,テウセントラノオ,テウセントラノヲ,トウテイラン
317921 ツクシトラノオ,ツクシトラノヲ,ヒロハトラノオ,ヒロバトラノヲ
317924 ホソバヒメトラノオ
317925 オオホソバトラノオ
317926 カラフトルリトラノオ
317927 カラフトルリトラノオ,セイヨウトラノオ,ベロニカ,ヤマルミトラノヲ,ヤマルリトラノオ
317930 サンイントラノオ
317931 トウテイラン
317932 ヒロハヤマトラノオ
317933 ツクシトラノオ,ヒロハトラノオ
317934 キタダケトラノオ
317935 エチゴトラノオ
317936 シラゲエチゴトラノオ
317937 エゾルリトラノオ
317938 ビロードトラノオ
317939 ヤマルリトラノオ
317942 シロバナヒメトラノオ
317943 ヒメトラノオ
317945 トラノオ
317949 ヤマトラノオ
317951 キクバクワガタ
317953 シロバナキクバクワガタ
317954 シラゲキクバクワガタ
317956 ダイセンクワガタ
317957 ミチノククワガタ
317958 ミヤマクワガタ
317959 ミヤマクワガタ
317960 アポイクワガタ
317961 エゾミヤマクワガタ
317962 ケミヤマトラノオ
317963 ハマトラノオ
317966 ヒメトラノオ,ヒメトラノヲ,ヤマトラノオ,ヤマトラノヲ
317968 ルリトラノオ
317969 シロバナルリトラノオ
317970 イブキルリトラノオ
318118 ニセダイオウヤシ属
318120 チャボニセダイオウヤシ
318121 セタカニセダイオウヤシ

318183 イナモリサウ属,イナモリソウ属
318184 イナモリサウ,イナモリソウ,ヨツバハコベ
318185 ホシザキイナモリソウ
318186 フイリイナモリソウ
318187 シロイナモリサウ,シロイナモリソウ,シロバナイナモリサウ,シロバナイナモリソウ
318193 ウキシバ属
318201 ウキシバ
318202 オオウキシバ
318218 シュ-ドリプサリス属
318277 プセウドササ属,ヤダケ属
318303 カンザンチク,ダイミャウチキ
318307 シノベ,シノベチク,ヤジノ,ヤダケ
318308 キイイシマヤダケ,キシマヤダケ
318309 ラッキョウヤダケ
318313 フイリヤダケ
318325 シラシマメダケ
318336 ヤクシマダケ
318350 カワカムリャダケ
318448 シホデモドキ属
318450 シホデモドキ,ホウゴウシホデモドキ
318489 ワチガイサウ属,ワチガイソウ属
318493 ツルワチガイ,ツルワチガヒ
318496 ワチガイサウ,ワチガイソウ
318499 ヒナワチガイソウ
318501 ヨツバハコベ,ワダソウ,ワチソウ
318506 ナンブワチガイ,ナンブワチガイソウ
318511 ヒゲネワチガイソウ
318513 イワワチガイソウ
318515 クシロワチガイ,クシロワチガイソウ,ホソバワチガイソウ
318562 トガサハラ属,トガサワラ属
318575 カハキトガ,カワキトガ,ゴエフトガ,ゴヨウトガ,サハラトガ,サワラトガ,トガサハラ,トガサワラ
318580 アメリカトガサワラ,アメリカマツ,オメゴソパイソ,オレゴンバイン,ダグラスモミ,ドグラスモミ,ベイマツ,メリケンマツ
318599 シマトガサハラ,タイワントガサハラ,タイワントガサワラ
318733 バンジラウ属,バンジロウ属
318734 ブラジルバンジロウ
318737 キバンザクロ,テリハバンジロウ,テリハバンジロオ
318739 キミノバンジロウ
318740 コスタリカバンジロウ
318742 グアバ,バンジラウ,バンジロウ,バンジロオ
318749 キミノバンジロウ
318820 プシロカウロン属
319101 シカクマメ属,トウサイ属
319112 ハネミササゲ
319114 シカクマメ,トウサイ,ワラス
319116 オランダビュ属,ヌソラ-レア属
319392 ボチャウジ属,ボチョウジ属,リュウギュウアオギ属
319437 オオシラタマカズラ
319477 コウトウボチャウジ
319671 オガサワラボチョウジ,ナガミボチョウジ,ナガミボチョオジ
319810 ボチャウジ,ボチョウジ,リウキウアヲキ,リュウキュウアオキ
319833 イシカツラ,イハヅタヒ,イワズタイ,シラタマカズラ,シラタマカヅラ,ミサキシラタマ,ワラベナカセ
320065 ホップノキ属
320071 ホップノキ
320181 オサバグサ
320203 イワセントウソウ属
320241 イワセントウソウ
320258 プテロカクタス属
320259 ドコウリュウ
320260 コクリュウ
320269 シタン属
320301 インドシタン,カリン,シタン,ヤエヤマシタン
320306 カリン,ビルマシタン
320307 マルスピウムシタン
320327 コウキシタン,サンタルロンジ,シタン
320346 サハグルミ属,サワグルミ属
320372 カハグルミ,カルメ,カワグルミ,コグルミ,サハグルミ,サワグルミ,フジグルミ,ヤシ,ヤスノキ,ヤマギリ
320377 カンペイジュ,カンポウフウ,シナサハグルミ,シナサワグルミ
320511 ヤマニガナ
320513 シマノゲシ,タイワンニガナ,モリアキノノグシ
320515 アキノノゲシ
320518 アキノノゲシ
320520 チョウセンヤマニガナ,チョオセンヤマニガナ,ヤマニガナ
320523 ミヤマアキノノゲシ
320525 プテロディスクス属
320827 シマウラジロノキ属
320828 シロギリ
320843 シマウラジロノキ,タイワンウラジロノキ
320862 プチロズティリス属
320871 プテロスティリス属
320878 アサガラ属
320880 アサガラ
320885 オホバアサガラ,ケアサガラ
320886 オオバアサガラ,オホバアサガラ,ケアサガラ
320924 ホソバノツルリンドウ属
320927 ホソバノツルリンドウ
320975 シルラニンジン属
320976 シムラニンジン,シルラニンジン
321004 ヒゲナガコメススギ属
321016 ヒゲナガコメススギ
321128 オオミヤハズ属
321161 シ-フォ-シア属,ニコバルヤシ属,ヤハズ属
321164 ダイオウヤハズ
321165 ホシノヤシ
321170 コモチケンチャヤシ,シュロチクヤシ
321174 ホソバケンチャ,ホソバシュロチクヤシ
321220 チシマドジャウツナギ属,チシマドジョウツナギ属
321221 コドジョウツナギ
321236 ヌカボガヤ
321241 チョウセンドジョウツナギ,チョウセンヌカボ,モウコドジョウツナギ
321248 アレチタチドジョウツナギ
321296 アオヌカボガヤ
321305 ネッカヌカボガヤ
321324 チシマドジャウツナギ
321342 タチドジャウツナギ,タチドジョウツナギ,チシマドジョウツナギ
321397 コゴメヌカボガヤ
321419 クズ属
321441 クズ,マクズ
321443 トキイロクズ
321444 シロバナクズ
321445 シナクズ
321447 フシゲクズ
321453 タイワンクズ
321460 クズインゲン,クズヰンゲン
321509 カセンサウモドキ属
321595 カセンサウモドキ
321633 プルモナ-リア属
321636 ヒメムラサキ
321643 ハイゾウソウ
321653 オキナグサ属
321659 モウコオキナグサ
321667 オキナグサ,カザフキサウ,キクザキイチゲサウ,キクザキイチリンサウ,シャグマサイコ,チゴグサ,チョウセンオキナグサ,ルリイチゲサウ
321668 キバナオキナグサ
321672 ヒロハオキナグサ
321675 アイノコオキナグサ
321676 マンシウオキナグサ
321697 ツクモグサ
321703 キタノオキナグサ
321716 ホソバオキナグサ
321721 セイヨウオキナグサ
321763 ザクロ属
321764 ザクロ,ジャクロ
321771 テウセンザクロ,ヒメザクロ
321777 ザクロ科
321892 プ-シュキニア属

321936 プヤ属
321956 アワモリハッカ
322076 キンチャクハギ属
322078 キンチャクハギ
322210 タチガヤツリ
322233 アゼガヤツリ
322235 メアゼガヤツリ
322304 アゼガヤツリ
322342 カハラスガナ,カワラスガナ,メリケンガヤツリ
322376 ムギガラガヤツリ
322386 カキバイヌザクラ属
322448 トキハサンザシ属,トキワサンザシ属
322450 タチバナモドキ,トキハサンザシ,トキワサンザシ,ピラカンサ,ピラカンサス,ホソバトキワサンザシ
322454 トキワサンザシ,ピラカンサス
322458 カザンテマリ,ヒマラヤピラカンサ
322465 ビラカンサ
322471 コイツミカマツカ,タイトウカマツカ,タイトウサンザシ,タイワンカマツカ,ホソバノタイトウサンザシ
322517 ピレナカンタ属
322596 ヒサカキサザンカ
322660 シロバナムシヨケギク
322665 アカバナムシヨケギク
322724 ナツシロギク
322775 イチヤクサウ属,イチヤクソウ属
322779 アリサンイチャク
322780 コバノイチヤクソウ
322781 ベニバナコバノイチヤクソウ
322790 ニバナイチヤクソウ,ベニイチヤクサウ,ベニイチヤクソウ,ベニバナイチヤクサウ,ベニバナイチヤクソウ
322823 イチヤクサウ
322825 カラフトイチヤクソウ
322839 ムヨウイチヤクソウ
322842 イチヤクサウ,イチヤクソウ
322843 オオベニバナイチヤクソウ
322852 マルバノイヂャクサウ
322853 エゾイチヤクソウ
322858 ニヒタカイチャク
322859 マルバノイチヤクソウ
322860 ベニバナマルバノイチヤクソウ
322870 ジンエフイチヤクサウ,ジンショウイチヤクソウ,ジンヨウイチヤクソウ,ジンヨオイチヤクソウ
322872 チョウセンイチヤクソウ,マルバノイチヤクサウ,マルバノイチヤクソウ
322915 ヒトツバイチヤクソウ
322935 イチヤクサウ科,イチヤクソウ科
322948 ピロステギア属
322986 ピロカクタス属
322989 テッシンマル
322990 カイマギョク
322991 コウユマル
322992 カンキギョク
323076 ナシ属
323085 シブナシ
323094 イワテヤマナシ
323110 タウマメナシ,トウマメナシ,ホクシマメナシ,マンシュウマメナシ
323114 チュウゴクナシ
323116 イヌナシ,マメナシ
323133 セイヤウナシ,セイヨウナシ,セイロウナシ
323151 セイヨウナシ
323156 オショウナシ
323202 ウメバナシ
323207 セイバンズミ,タイワンイヌナシ
323210 コウノワタシ
323216 サブロウナシ
323217 チョウライナシ
323240 キシブナシ
323246 ユキナシ
323251 ヒマラヤナシ
323268 アリノミ,イシナシ,イヌナシ,ナシ,ニホンヤマナシ,ヤマナシ,ユデナシ
323272 アリナシ,ナシ
323330 イヌナシ,チョウセンヤマナシ,テウセンヤマナシ,ホクシヤマナシ,ミチノクナシ
323332 アオナシ
323347 オオミネイヌナシ,オホミネイヌナシ
323348 ジャウバウナシ,ショウウナシ
323450 ルカウサウ属,ルカウソウ属,ルコウソウ属
323457 ハゴロモルコウソウ,モミジバルコウソウ,モミジルコウ
323459 ウチハルカウ,ウチワルコウ,マルバルコウ,マルバルコウソウ
323465 カボチャアサガオ,ルコウソウ
323470 モミジルコウ
323542 カッシア属
323580 カシノキ属,カシ属,コナラ属
323581 カシワコナラ
323584 ミズコナラ
323588 チンゼイガシ
323590 オオバコナラ
323591 ホソバガシワ
323592 モンゴリコナラ
323594 オオツクバネガシ
323599 アカガシ,アラアカ,オオガシ,オオバガシ,オホガシ,オホバガシ,カシ
323601 ヤナギアカガシ
323602 ヒロハアカガシ
323611 オクヌギ,カタギ,クニギ,クヌギ,クノギ,ジザイガシ,ジンダンボウ,ドングリ,フジマキ
323625 シロナラ,ホワイトオーク
323630 オウゴンカシワ,カシハナラ,カシハハハソ,カシワナラ,カシワハハソ,ナラガシハ,ナラガシワ,ワウゴンガシハ
323635 ツクシオオナラ,ツクシオホナラ
323647 アオナラガシワ
323780 スカーレットナラ
323797 ミヤマナラ
323814 エドガシワ,オオカシワ,カシハ,カシワ,カシワギ,カシワノキ,コガシハ,ゴヘイナラ,モチガシハ,モチカシワ
323819 クジャクガシワ
323820 ハゴロモガシワ
323821 ホウオウガシワ
323881 カラカシワナラ,フアーベルガシ
323942 コナラ,ナラ,ハハソ
323944 タイワンコナラ
323952 ヨコメガシ
323957 ヒリュウガシ
324027 オウシュウウバメカシ,セイヨウヒイラギガシ
324100 ハゴロモナラ
324171 オキナワウラジロガシ
324173 オオバナラ,オホバナラ,カウライミヅナラ,カラフトガシハ,カラフトガシワ,コバガシハ,コバガシワ,マンシウミヅナラ,モウコガシワ,モウコナラ,モンゴリナラ
324178 ハゴロモミズナラ
324190 オオナラ,オホナラ,ナラマキ,ノイシナラ,マナラ,ミズナラ,ヤマホソ
324223 ナガバコナラ
324262 アメリカガシワ,パルストリス
324283 イマメガシ,ウバシバ,ウバメガシ,ウマメガシ,クマノガシ,バメ,バメガシ,マベシイ
324284 チリメンガシ
324285 フクレウバメガシ
324286 ケウバメガシ
324335 アカガシハ,イギリスナラ,オウシュウナラ,ヨーロッバナラ
324344 アカガシハ,アカガシワ
324358 ウラジロガシ
324359 ヒロハウラジロガシ
324384 イシナラ,クヌギ,コナラ,コマキ,スノキ,ナラ,ノホソ,ハサコ,ハハソ,ホウソ,ホサノキ,ホソ,マキ,マホソ
324385 アオナラ
324386 シダレコナラ
324387 テリハコナラ
324388 ナガミコナラ
324389 ハゴロモアオナラ
324390 タレハハゴロモアオナラ
324395 マルバコナラ
324404 モンゴリコナラ
324436 ヒヒラギガシ,ヒロハウバメガシ
324449 ウラジロガシ
324453 コルクガシ,コルクノキ
324464 コガシハ,コガシワ
324467 タロコガシ

324490 ヂンガサガシ,ビロウドワンガシ,モンパガシ
324532 アベクヌギ,アベヌキ,オクヌギ,クリガシワ,コルククヌギ,メクヌギ,ワタクヌギ,ワタヌキ
324577 ケスネリア属
324596 クィアベンティア属
324623 キラア属
324624 シャボンノキ
324669 シクンシ属
324677 インドシクンシ,シクンシ
324679 カラクチナシ,シクンシ
324712 ヤマハッカ属
324757 セキヤノアキチョウジ
324918 ラビエア属
324919 アサヒナミ
324922 シズナミ
324993 センダンキササゲ属
325002 カタミノカツラ,センダンキササゲ
325061 ラフレシア属
325062 ラフレシア
325065 ヤッコサウ科,ヤッコソウ科,ラフレシア科
325223 ラモンダ属
325225 ピレネ-イワタバコ
325278 ミサオノキ属,ミサヲノキ属
325301 シマミサオノキ
325309 ミサオノキ
325393 ミサオノキ
325494 ウマノアシガタ科,キンポウゲ科
325498 ウマノアシガタ属,キンポウゲ属
325518 ミヤマキンポウゲ
325562 イワンポウゲ
325571 ホソバキンポウゲ
325589 ウレバチモ
325612 イトキッネノボタン
325614 ハナキッネノボタン,ハナキンパウゲ,ハナキンポウゲ
325628 チツマキンパウゲ,チツマキンポウゲ
325697 ケキッネノボタン
325717 ツルヒキノカサ,テガタヒキノカサ,ヒメキンポウゲ
325718 コキッネノボタン
325743 ケキンポウゲ
325882 エゾキンパウゲ,エゾキンポウゲ
325901 カラクサキンポウゲ
325916 オオウマノアシガタ,オホウマノアシガタ
325981 ウマノアシガク,キッネノボタン,キンポウゲ,コマノアシガタ,ヒトデ
325989 ノハラキンポウゲ
326108 トゲミノキッネノボタン
326168 コウリンキンポウゲ
326273 クモマキンパウゲ,クモマキンポウゲ,ミヤマヒキノカサ
326276 キッネノボタン
326278 カラクサキンパウゲ,ホソバツルヒキノカサ,モクズキンポウゲ
326287 ハイキンポウゲ,ハヒキンポウゲ
326303 イトキンポウゲ,マツバキンポウゲ
326340 カヘルノキツケ,コセウナ,タガラシ,タタラビ,ドブゴセウ
326365 シマキツネノボタン,ヤヘヤマキツネノボタン
326375 コウアンキンポウゲ
326411 タカネキンポウゲ
326417 オトコゼリ,ヲトコゼリ
326419 タイサンキンパウゲ,ニヒタカキパウゲ
326431 コキンパウゲ,コキンポウゲ,タイワンキンパウゲ,ヒキノカサ
326493 キツネノポウゲ
326507 トガクシショウマ属,トガクシソウ属
326508 トガクシショウマ,トガクシソウ
326513 ザンセツソウ属
326523 タイミンタチバナ属
326578 ダイコン属
326609 セイヨウノダイコン
326616 ダイコン,ネリマダイコン,ハツカダイコン
326622 ダイコン
326626 クロダイコン
326629 ハマダイコン
326638 ラフィア属
326639 ウラジロラフィア
326649 アメリカゾウゲヤシ,サケラフィア
326833 ケミヤガラシ,ミヤガラシ
326842 ヒメミヤガラシ
326957 バレンギク属
326995 インドジャボク属,ホウライアオキ属,ホウライアヲキ属
327058 インドジャボク
327069 ホウライアオキ,ホウライアヲキ
327093 アフギバセウ属,オウギバショウ属,タビビトノキ属
327095 アフギバセウ,オウギバショウ,タビビトノキ
327115 ラベ-ニア属
327117 ラベニア属
327226 ココメギョリュウ
327250 レブティア属
327256 オウレイマル
327257 ウガマル
327259 ハクゾウマル
327262 キンホウマル
327264 レイセイマル
327274 イホウマル
327278 ヒホウマル
327281 キンサンマル
327283 ホウザン
327286 シホウマル
327294 ハクキュウマル
327297 オウホウマル
327298 キンチョウマル
327299 トウチョウマル
327300 カクホウマル
327301 ソウレイマル
327310 ケンポウマル
327311 ギンボウマル
327312 クンポウマル
327313 トウホウマル
327314 レイホウマル
327315 センポウマル
327316 キョウポウマル
327317 トウホウマル
327322 レックステイネリア属
327357 チャセンギリ属
327362 チャセンギリ
327429 ジオウ属,ヂワウ属,レ-マンニア属
327434 セダカセンリゴマ
327435 アカヤジオウ,サオヒメ,ジオウ,シロジオウ,センリゴマ,ヂオウ
327437 シロヤジオウ
327450 センリゴマ,ハナジオオ
327498 レイケオカクタス属
327503 クロナナコ
327524 キチジャウサウ属,キチジャウソウ属,キチジョウソウ属,ライネッキア属
327525 キチジャウサウ,キチジャウソウ,キチジョウソウ
327538 アメリカチャボシ属,アメリカチャボヤシ属
327540 ヤハズアメリカチャボヤシ
327541 ヒメアメリカチャボヤシ
327543 アメリカチャボヤシ,ヒトツバアメリカチャボヤシ
327548 キバナアマ属
327553 キアマ,キバナアマ
327577 コウトウサンイウクワ,コウトウサンユウカ,シロバナキョウチクトウ,シロバナケフチクタウ
327663 コウシュンスゲ属
327664 コウシュンスゲ
327670 タコイモ属
327672 タコイモ
327681 レナンセラ属
327793 モクセイサウ属,モクセイソウ属
327795 シノブモクセイサウ,シノブモクセイソウ
327866 キバナモクセイソウ,ホソバモクセイソウ
327879 ホザキモクセイソウ
327896 ニオイレセダ,ニホヒレセダ,モクセイサウ,モクセイソウ
327934 モクセイサウ科
328218 クロウメモドキ科,サンアソウ科
328219 レストレピア属
328328 イタドリ属

328345 イタドリ,サイタツマ,タチビ
328357 ベニバナケイタドリ
328445 ラブドタムヌス属
328446 ニュージーランド
328543 クロウメモドキ科
328544 ネコノチチ属
328550 ナガミノイソノキ,ネコノチチ
328551 ヤエヤマネコノチチ
328592 クロウメモドキ属
328609 ウメバクロッバラ
328628 クニガミクロウメモドキ
328660 クロカンバ,ナンブクロカンバ,ヤマナシクロカンバ
328665 イソノキ,ウバギ,ホソバクロウメ
328671 ホソバイソノキ
328680 クロツバラ,ダウリアクロツバラ,チョウセンクロツバラ
328684 ウシコロシ,オベカウジ,オホクロウレモドキ,クロッバラ,ナベコウジ,ナベノキ
328685 ケクロツバラ
328689 ヤマクロウメモドキ
328700 シマクロウメモドキ,タイワンクロウメモドキ,ナンキンクロウメモドキ
328701 セイヨウイソノキ
328740 エゾノクロウメモドキ,クロウメモドキ
328741 エゾノクロウメモドキ,ナガハノクロウメモドキ
328742 クロウメモドキ
328743 アオミノクロウメモドキ
328744 シナノクロウメモドキ
328745 コバノクロウメモドキ
328747 シナノクロウメモドキ
328750 ヒメクロウメモドキ
328752 チョウセンクロッバラ,テウセンクロッバラ,マルバクロウメモドキ
328767 リュウキュウクロウメモドキ
328793 ナカハラクロウメ,ナカハラクロウメモドキ
328794 ソメモノノキ
328816 イバクロウメモドキ,イワクロウメモドキ,コゴメバクロウメ
328851 マンシウクロウメモドキ,ヤブクロウメモドキ
328853 マンシウクロウメモドキ
328882 ウスリクロッバラ
328883 シイボルトノキ,シーボルトノキ
328904 アオウメモドキ,アヲウメモドキ,キビノクロウメモドキ,タイシャククロウメモドキ
328990 ラフィドフォラ属
328992 オウゴンカズラ
329004 ヒメハブカズラ
329056 シャリンバイ属
329068 シャシャリンバイ,モクコクモドキ
329073 ホソバシャリンバイ
329077 ヒメシャリンバイ
329086 アヅバシャリンバイ,シャリンバイ,ハマモクコケ,マルバノシャリンバイ
329114 シャリンバイ,タチシャリンバイ,テイチギ,テッギ,トカチギ,ハマモクコケ,ハマモッコク
329164 ラフィサムヌス属
329169 ハリヤシ属
329173 カンノンジュロ属,クワンオンチク属,シュロチク属,ハリヤシ属
329176 カンノンチク
329184 イヌシュロチク,シュロチク
329186 シュロチク
329306 カラダイオウ属,カラダイワウ属
329318 ヒロハダイオウ
329320 チョウセンダイオウ
329363 ノビレダイオウ
329372 モミジバダイオウ
329386 オオギシギシ,オホシ,カラダイオウ,カラダイワウ,マルバダイオウ
329388 ダイオウ,マルバダイオウ,マルバダイワウ,ルバーブ
329401 タングートダイオウ
329419 レキシア属
329504 オクエゾガラガラ属
329527 オクエゾガラガラ
329578 リネフィルム属
329579 チゴマツリ
329580 ヒコウギョク
329594 リネルミーザ属
329679 リプサリドプシス属
329683 リプサリス属
329689 イトアイ,イトアシ
329692 セイリュウ
329699 ハナヤナギ
329705 タマヤナギ
329744 シュドリプサリス属,ヤエヤマヒルギ属,ヤヘヤマヒルギ属
329765 オオバヒルギ,オホバノヒルギ,オホバヒルギ,シロバナヒルギ,マングローブ,ヤエヤマヒルギ,ヤヘヤマヒルギ,ヤマキマヒルギ
329771 オオバヒルギ
329776 ヒルギ科
329801 ヒロハハナカンザシ
329826 イワベンケイ属
329836 ヒメイワベンケイ
329866 ムレイキリンソウ
329888 ムラサキイワベンケイ
329889 ホソバイワベンケイ
329935 イワベンケイ,イワベンケイソウ
329944 ナガバノイワベンケイ
329986 ロドーキトン属
330017 アザレア属,シャクナゲ属,ツツジ属
330023 カラフトミヤマツツジ
330058 エゾムラサキツツジ,ミヤマツツジ,ムラサキヤシオ,ムラサキヤシオツツジ,ムラサキヤシホ,ムラサキヤシホツツジ
330059 シロバナムラサキヤシオツツジ
330060 ウラゲムラサキヤシオツツジ
330061 ウラジロミヤマツツジ
330071 アマギツツジ
330072 アマクサミツバツツジ
330076 サキシマツツジ
330182 キバナシャクナゲ
330184 ゴシキシャクナゲ
330185 ヤエキバナシャクナゲ
330233 オガサラツツジ,オガサワラツツジ,ムニンツツジ,ヲガサラツツジ
330240 ウスギシャクナゲ,ハクサンシャクナゲ
330241 ケナシハクサンシャクナゲ
330242 ネモトシャクナゲ
330245 ウラケハクサンシャクナゲ,エゾシャクナゲ,シロバナシャクナゲ,ニシャクナゲ
330255 ナンオウツツジ
330457 モウゼンツツジ
330495 エゾムラサキツツジ,ダウリャツツジ,トキハゲンカイ,トキハツツジ,トキワゲンカイ,トキワツツジ
330497 シロバナエゾムラサキツツジ
330511 エゾルラサキツツジ
330532 アズマシャクナゲ,シャクナゲ,シャクナン,シャクナンゲ
330534 ハクカシャクナゲ
330535 シロバナアズマシャクナゲ
330536 フチベニシャクナゲ
330537 フイリシャクナゲ
330538 ツクシシャクナゲ,ヘノリンク
330543 アマギシャクナゲ
330577 ミツバツツジ
330578 ウラゲミツバツツジ
330579 シロバナミツバツツジ
330580 ヒダカミツバツツジ
330581 トサノミツバツツジ
330586 アワノミツバツツジ
330588 ハヤトミツバツツジ
330617 アコウシャクナゲ,セイシカ,セイシクワ,タイトンシャクナゲ,ナンバンツツジ,ヤエヤマセイシカ,ヤヘヤマセイシクワ
330629 マルバサツキ
330630 センカクツツジ
330675 マルバサツキ
330695 アルペンローゼ
330722 タイワンシャクナゲ,タイワンツツジ
330863 アオヤシャクナゲ
330883 ウスエフ,ムラサキリウキウツツジ,ムラサキリュウキュウ

330901 アッバナシャクナゲ
330909 ヒュウガミツバツツジ
330917 サツギ,サツキツツジ,サツギツヅジ,シナノサツキ
330952 アズマシャクナゲ
330953 ホンシャクナゲ
330954 シロバナシャクナゲ
330966 アカツツジ,ツツジ,ヤマツツジ
330967 シロヤマツツジ
330968 ホソバヤマツツジ
330969 キンシベヤマツツジ
330970 キソシベ,キソシベヤマツツジ
330971 ヤエザキヤナギラン,ヤエヤマツツジ
330973 エゾヤマツツジ
330975 ミカワツツジ,ムラサキヤマツツジ
330978 ニシキヤマツツジ
330979 シキザキヤマツツジ
330980 ヒメヤマツツジ
330984 ミソメキリシマ
330986 サイカイツツジ
330990 カネヒラツツジ
330993 シマシャクナゲ,チャクセイシャクナゲ
330997 ウラジロヒカゲツツジ
330998 サハテラシ,サワテラシ,ヒカゲツツジ,ヒメシャクナゲ,メシャクナゲ,ヤクシマヒカゲツツジ
330999 ウラジロヒカゲツツジ
331000 ハイヒカゲツツジ
331009 キソミツバツツジ
331010 クルミツヅジ,ミヤマキリシマ
331011 シロバナミヤマキリシマ
331014 サンヨウツツジ
331015 アシタカツツジ
331022 アワミツバツツジ
331023 クロヒメシャクナゲ
331039 ダイセンミツバツツジ
331040 シロバナユキグニミツバツツジ
331057 サカイツツジ
331058 サカイツツジ
331062 ケシベツツジ,タカネツツジ
331065 セイシカ
331066 アマミセイシカ
331107 セイガイツツジ
331113 セイガイツツジ
331170 オオシマツツジ
331173 モチツツジ
331174 セイガイツツジ
331175 シロバナモチツツジ
331187 エンシウシャクナゲ,エンシュウシャクナゲ,ホソバシャクナゲ
331188 シロバナホソバシャクナゲ
331189 クルマザキホソバシャクナゲ
331190 ヤエホソバシャクナゲ
331191 トキイロホソバシャクナゲ
331192 モモイロホソバシャクナゲ
331227 オオシャクナゲ,シャクナゲ,ツクシシャクナゲ
331230 シロバナホンシャクナゲ
331237 ヤクシマシャクナゲ
331239 コゴメツツジ,ホザキツツジ
331257 トウレンゲツツジ,レンゲツツジ
331258 イヌツツジ,ウマツツジ,オニツツジ,レンゲツツジ
331260 ビロ－ドレンゲツツジ
331261 キ,キツツジ,キレンゲ,キレンゲツツジ,シシクハズ
331262 ウラジロレン,ウラジロレンゲツツジ
331263 レンゲボタン,レンゲボタンゲツツジ
331276 モリシャクナゲ
331284 シロリウキウツツジ,シロリュウキュウ,シロリュウキュウツツジ,リウキウツツジ,リュウキュウツツジ
331285 シロマンヨウ
331286 フジマンヨウ
331289 カラムラサキツツジ,シロバナゲンカイツツジ
331290 シロバナカラムラサキツツジ,シロバナゲンカイツツジ
331296 ゲンカイツツジ
331297 シロバナゲンカイツツジ
331305 ウンゼンツツジ,タイトンッッジ,タイワンヒメツツジ,ナカハラツツジ
331311 ナンコシャクナゲ,ナンコタイサンシャクナゲ
331335 ニッコウキバナシャクナゲ
331341 オオバツツジ,オホバツツジ
331354 サイコクミツバツツジ
331358 ヒメミツバツツジ
331370 キリシマ,キリシマツツジ,サタツツジ,ヒリュウ,ホンキリシマ
331371 シロキリシマ
331380 クルメツツジ
331391 キンマウツツジ,キンモウツツジ
331397 オオムラサキ
331417 ウラジロミツバツツジ
331420 イサンッッジ,ランダイヒカゲツツジ
331421 ランダイヒカゲツツジ,リバツツジ
331449 サカイツツジ,シロガネツツジ
331453 シロバナサカイツツジ
331462 アケボノツツジ,ツクシアケボノツツジ
331463 アカヤシオ
331464 ユキヤシオ
331465 アケボノツツジ
331530 チョウセンヤマツツジ
331565 アカボシツツジ,ニイタカシャクナゲ,ニヒタカシャクナゲ,モリシャクナゲ
331581 オオムラサキ,オホムラサキ,ヒラドツツジ
331582 オオムラサキ
331593 キキョウツツジ,クルマツツジ,ゴエフツツジ,ゴヨウツツジ,シロヤシオ,シロヤシボ,マツノキハダ
331595 カノコゴヨウツツジ
331596 チベットシャクナゲ
331615 クモマツツジ
331623 コバノミツバツツジ,ツリガネツツジ,ムラサキツツジ
331624 シロバナコバノミツバツツジ,シロバナミツバツツジ
331625 ムギコバノミツバツツジ
331626 アラゲミツバツツジ
331627 ツクシコバノミツバツツジ
331628 ニシキコバノミツバツツジ
331629 ゲンペイコバノミツバツツジ
331657 イソツツジ,カワツツジ,キシツツジ
331658 シロバナキシツツジ
331689 アカグツツジ
331723 ジングウツツジ
331724 シブカワツツジ
331725 シロバナシブカワツツジ
331753 サクラオンツツジ
331759 カザンジマ,クワザンジマ,ケラマツツジ,トウツツジ
331760 シロケラマ
331762 モモケラマ
331768 クロフネツツジ
331791 バイカツツジ,バイケワツツジ,ミヤマツツジ
331801 ウンゼンツツジ
331804 シロウンゼン
331805 ウンゼンツツジ,シロバナウンゼンツツジ
331839 シナノサツキ,シナヤマツツジ,ダイトンッッジ,タイワンヤマツツジ,トオヤマツツジ,ホソバキンマウツツジ,マルバサッキ,リュウキュウヤマツツジ
331843 フヨウホウ,マルバサツキ
331926 アリサンシックナゲ,オホバセイシクワ
331936 サクラツツジ
331937 シロサクラツツジ
331938 アラゲサクラツツジ
331943 ミヤコツツジ
331985 トサツツジ,フジツツジ,フヂツツジ,メンツツジ
331986 シロバナフジツツジ
331989 オオヤマツツジ
332018 コメツツジ,シロバナノコメツツジ
332019 ベニバナコメツツジ
332020 オオコメツツジ,シロバナコメツツジ
332021 チョウジコメツツジ
332026 ツルギミツバツツジ

332027 アカイシミツバツツジ
332045 トキワバイカツツジ
332082 タカクマミツバツツジ
332083 アマクサミツバツツジ
332089 トウゴクミツバツツジ,トオゴクミツバツツジ
332090 シロバナトウゴクミツバツツジ
332104 アカツツジ,オツツジ,オンツツジ,ツクシアカツツジ,ホンツツジ,ヲンツツジ
332105 シロバナオンツツジ
332106 ムラサキオンツツジ
332107 ジングツツジ,タンナアカツツジ
332133 ヤクシマヤマツツジ
332134 ヤクシマミツバツツジ
332137 オオヤクシマシャクナゲ
332144 ボタンツツジ,ヨドガワツツジ
332145 チョウセンヤマツツジ
332184 ロードヒポキシス属
332204 ロドレイア属
332205 シャクナゲモドキ
332218 テンニンカ属,テンニンクワ属
332221 テンニンカ,テンニンクワ
332224 ロドフィアラ属
332253 ロドスパサ属
332282 シロヤマブキ属
332284 シロヤマブキ
332293 ムラサキオモト属
332353 ロンボフィルム属
332356 ギンホコ
332359 カイトウランマ
332360 セイガイ
332375 モルッカヤシ属
332377 シダメモルッカヤシ
332434 ハケヤシ属
332436 バウエルハケヤシ
332438 ナガバハケヤシ
332452 ウルシ属
332477 ニオイウルシ,ニホヒウルシ
332502 ニホヒウルシ
332509 カシノキ,カツカド,カツノキ,タイワンフシノキ,ヌルデ,ノデボ,フシノキ
332513 タイワンフシノキ
332533 シシリ-シュマツク
332662 タイワンフシノキ,ヌルデ
332869 アンナンウルシ
332901 コバノヤマウルシ,コヤマウルシ
333024 ルビ-ガヤ
333083 ルリブクロ属
333086 ルリブクロ
333089 ルリブクロ
333101 リンコレリァ属
333131 タンキリマメ属
333133 オオバタンキリマメ,オホバタンキリマメ,トキリマメ,ベニカハ,ベニカワ
333323 ヒメノアズキ,ヒメノアヅキ
333324 マルバヒメノアズキ
333456 キツネマメ,キンチャクマメ,タンキリマメ
333471 シウブンサウ属,シウブンソウ属,シュウブンソウ属
333475 シウブンサウ,シウブンソウ,シュウブンソウ,タイワンシウブンサウ
333477 イヌノハナヒゲ属,ツカヅキグサ属
333478 ミカヅキグサモドキ
333480 ツカヅキグサ
333490 シマイガクサ
333495 トラノハナヒゲ
333518 イヌノハナヒゲ
333530 ヤエヤマアブラスゲ
333560 イトイヌノハナヒゲ,イトイヌハナヒゲ,ヒメイヌハナヒゲ
333561 チョウセンヒメイヌノハナヒゲ
333562 ネズミノハナヒゲ
333568 オオイヌノハナヒゲ
333575 コイヌノハナヒゲ
333576 イヌノハナヒゲ
333628 ミクリガヤ
333672 イガクサ,イヌノハナヒゲ
333675 イヌノハナヒゲ
333715 ミヤマイヌノハナヒゲ
333725 リンコスティリス属
333729 オオホザキマツラン
333731 ウライムヨウラン,リンコステイリス・プレイオネッサ
333733 ヤマビワソウ属
333735 ヤマビワソウ
333738 キレバヤマビワソウ
333840 アンチルココヤシ属
333861 リチドフィルム属
333893 スグリ属
333914 キンメ,テンノウメ,テンバイ,ヤシオ,ヤシャビシャク
333915 ケナシヤシャビシャク
333916 アメリカクロスグリ,アメリフサスグレ
333929 スグリ,ハリスグリ
333955 トゲスグリ
333970 オオバスグリ,オホバスグリ,カラノキ,キヒヨドリ,キヒヨドリジャウゴ,キヒヨドリジョウゴ,コマガタケスグリ,ハマナシ,ヤブサンザシ,ヤマサンザシ
333972 シナノヤブサンザシ,シナヤブサンザシ,トウヤブサンザシ
333978 ニヒタカスグリ
333988 シホカゼスグリ
334030 アメリカスグリ
334034 クロミノハリスグリ
334055 ホザキヤブサンザシ,ホザキヤブスグリ
334061 エゾスグリ,ヤマスグリ
334084 オオモミジスグリ,オホモミヂスグリ
334085 チョウセンモミジスグリ
334087 ザリコミ
334089 ザリグミ,ザリコミ,チョウセンザリコミ
334117 クロサスグリ,クロスグリ,フラック・カーラント
334144 マンシウハイスグリ
334145 シベリアイワスグリ
334147 マンシウクロスグリ
334149 イワスグリ
334157 イハスグリ,ハイスグリ,ハヒスグリ
334166 コバノトゲスグリ
334179 アカスグリ,アカフサスグリ,ガーラント,フサスグリ
334188 オオザリコミ,オホザリコミ,トガスグリ,トガスグリアカスグリ
334201 アカスグリ
334209 スグリ
334210 トゲナシスグリ
334240 チシマスグリ,トカチスグリ
334249 ウスリ-スグリ
334250 オオスグリ,グ-スベリ,グ-スベリ-,グスベリ-,セイヨウスグリ,マルスグリ
334320 ブラジルハシカグサモドキ
334338 ハシカグサモドキ
334432 タウゴマ属,トウゴマ属
334435 カラエ,タウゴマ,トウゴマ
334892 ジュズサンゴ属
334897 ジュズサンゴ
334946 ハリエンジュ属
334965 ハナアカシア,ハナエンジュ,バラアカシア
334976 アカシャ,イヌアカシャ,ギガフクワン,ギゴウカン,ニセアカシヤ,ハリエンジュ
334982 トゲナシニセアカシャ
334995 エイコクトゲナシ,ニセアカシャ
335013 モモイロハリエンジュ
335070 ロ-ケア属
335079 クレナイロケア
335141 ヤグルマサウ属,ヤグルマソウ属
335151 ヤグルマサウ,ヤグルマソウ
335161 ロドリゲュ-チア属
335183 カモジグサ属
335204 ウスゲカモジグサ,タチカモジ
335258 アオカモジグサ,ケカモジグサ
335267 アラゲカモジグサ
335312 タイワンカモジグサ
335370 タチカモジ,タチカモジグサ
335371 タチカモジ
335423 オホカモジグジ,オホタチカモジグサ
335461 マンシウカモジグサ
335480 アオカモジグサ
335495 カモジグサ

335536 オニカモジグサ
335544 カモジグサ
335547 イヌカモジグサ,コウアンカモジグサ
335758 オモト属
335760 オモト
335869 ロムネヤ属
335879 ロムレア属
336096 ベニマツリ属
336104 ベニマツリ
336149 ロリデュラ属
336166 イヌガラシ属,オランダガラシ属
336167 ヒメイヌガラシ
336172 サケバミミイヌガラシ
336182 ミミイヌガラシ
336183 ケタマイヌガラシ
336187 コイヌガラシ
336191 マガリミイヌガラシ
336193 ミチバタガラシ
336200 タマイヌガラシ
336211 イヌガラシ
336212 ナガミノイヌガラシ
336215 アオイヌガラシ
336232 コバノオランダガラシ
336246 ミギワガラシ
336250 スカシタゴボウ
336277 キレハイヌガラシ
336279 コゴメイヌガラシ
336283 イバラ属,バラ属
336313 ヤブテリハノイバラ
336316 オオサクラバラ
336320 オオタカネイバラ,オオタカネバラ,オホタカネイバラ
336348 セッザンバラ
336353 カラフトイバラ
336366 スダレバラ,モクカウイバラ,モクカウバラ,モッコウバラ
336405 カカヤンバラ,ヤエヤマノイバラ,ヤヘヤマノイバラ
336408 モスカ-タバラ
336419 カニナバラ
336462 ヒメボタンバラ
336474 セイヤウバラ,セイヨウバラ
336476 コケバラ,モッスロ-ス
336485 カウシンバラ,コウシンバラ,シキザイバラ,シキザキイバラ,チャウシュン,チョウシュン
336486 グラインロウス,グリ-シロウズ,グリ-ンロ-ス,セイカ
336490 ナナコイバラ,ナンキンイバラ,ヒメバラ
336509 タウュンイバラ
336514 ダマスクバラ
336522 カラフトイバラ,ヤマハマナス
336563 オ-ストリアソライヤ-,オ-ストリヤンブライエル,フォエティダバラ
336565 オ-ストリアンカッパ-ロ-ズ
336568 ペルシアナ
336578 オオサクラバラ,オオフジイバラ,フジイバラ
336581 ガリカバラ
336602 ギガンテアバラ
336631 サンショウバラ
336640 セイヨウバラ
336652 コハマナス
336660 コウライバラ
336673 ハトヤバラ
336675 ナニハイバラ,ナニワイバラ
336697 テリハノイバラ
336700 リュウキュウテリハノイバラ
336701 トゲナシテリハノイバラ
336705 タイワンハマイバラ
336730 マイカイ,ヤエハマナシ
336736 カラストバラ,カラフトイバラ,ポンチカプマウ,ポンチカマウ,ミヤマバラ,ヤマハマナス,ヤマヤマナスミヤマバラ
336738 ツルノイバラ
336783 イバラ,ノイバラ,ノバラ
336785 ウスアカノイバラ
336786 ツクシイバラ
336789 ゴヤバラ,サクライバラ,サクラバラ,ボサツバラ,ムラサキゴヤバラ
336791 ゴヤバラ
336805 タカネイバラ,タカネバラ,ミヤマハマナス
336813 ボタンバラ
336829 ヤブイバラ
336830 モリイバラ
336831 アズマイバラ
336832 アケボノオオフジイバラ
336842 ミヤコイバラ
336843 ウスアカミヤコイバラ
336851 ミヤマバラ
336871 ダヅダカイハマ,タヅタカイバラ,ダヅダカイバラ
336885 イザヨイバラ
336901 ハマナシ,ハマナス,バラ
336902 シロバナハマナス
336914 タカサゴャマイバラ,ヤマイバラ
336915 タカサゴヤマイバラ
336930 ニイタカイバラ
336938 ニヒタカイバラ,ピビダカイハマ
336954 カラフトイバラ,コヤマイバラ,ニホヒバラ,ヤブイバラ
336955 センノオバラ
336983 タイワンノイバラ
336995 ニヒタカモリイバラ
337006 ヒメハマナシ
337015 アズマイバラ,オオフジイバラ,オホサクライバラ,テリハノイバラ,ハイイバラ,ハヒイバラ,ハマイバラ,ヤマテリハノイバラ
337017 ヤエテリハノイバラ
337029 キバナハマナシ,キバナハマナス
337046 オオバフジイバラ
337050 アバラ科,バラ科
337059 ロッシャ-ヤシ属
337060 ロッシャ-ヤシ
337065 ロスコ-エア属
337137 ロ-センベルギア属
337153 ロセォカクタス属
337154 キッコウボタン
337156 コクボタン
337157 レンザン
337174 マンネンロウ属
337180 マンネンラウ,マンネンロウ,マンルサウ,マンルソウ
337253 カグラサウ,キツネノヒマゴ,キツネノマゴ,シマキツネノマゴ,タイワンキツネノマゴ,リウキウヒマゴ
337255 カハリキツネノメマゴ,キツネノメマゴ
337256 ケバノキツネノメマゴ,ケブカキツネノメマゴ
337257 ホソバキツネノメマゴ
337292 ロスラリア属
337320 キカシグサ属
337330 ホソバキカシグサ
337349 ミズスギナ
337351 キカシグサ,ヒメキカシグサ
337353 キカシグサ,ホソバキカシグサ
337368 ミズマツバ,ミヅキカシグサ
337378 ミズキカシグサ
337382 ミズマツバ
337384 アメリカキカシグサ
337389 ミズキカシグサ
337391 ホザキキカシグサ,マルバキカシグサ
337544 アイアシ属,ウシノシッペイ属
337555 ケアイアシ,ツノアイアシ
337690 コウトウマメ属
337873 ダイオウヤシ属
337875 ボリンクエンダイオウヤシ
337879 フロリダダイオウヤシ
337883 カリブダイオウヤシ,セダカダイオウヤシ
337885 キュ-バダイオウヤシ
337906 アカネ属
337910 ナガバアカネ
337912 アカネ
337916 オオキヌタソウ,オホキヌタサウ,マンセンオオキヌタソウ,マンセンキヌタソウ
337919 オオキヌタソウ
337921 オオキヌタソウ
337925 アカミノアカネ,オホクルマアカネ,クルマバアカネ,ツシマアカネ
337945 ナガバアカネ
337965 オオアカネ
337970 アカアカネ,アカネムグラ,オオア

カネ
338021 ビロ-ドアカネ
338032 モリアカネ
338038 セイヤウアカネ,セイヨウアカネ,ムッバアカネ
338045 アカネ科
338059 キイチゴ属
338060 ビロ-ドクサイチゴ
338061 ゴショモミジイチゴ
338062 オオトックリイチゴ
338063 キレハクマイチゴ
338064 ヒメカジイチゴ
338065 アイノコキイチゴ
338066 オオミネイチゴ
338068 エゾエビガライチゴ
338069 アイノコフユイチゴ
338070 ナガバナワシロイチゴ
338072 ソマイチゴ
338079 オヂギイチゴ,ミンゲツイチゴ
338097 ケオリイチゴ
338105 クロミキイチゴ
338124 アマミフユイチゴ
338125 コバノアマミフユイチゴ
338142 チシマイチゴ
338151 セイヨウヤブイチゴ
338196 イオウトウキイチゴ
338205 カンイチゴ,ツルカシワ,トキシラズ,フユイチゴ
338209 オオナワシロイチゴ,オホナハシロイチゴ
338243 クラウドベリ,ホロムイイチゴ
338250 ゴショイチゴ
338265 セイヤウヤブイチゴ,セイヨウヤマイチゴ
338281 アリサンイチゴ,ズイシャイチゴ,スヰシャイチゴ,ヒメタマイチゴ,ビラウドイチゴ,ビロウドイチゴ,ビロ-ドイチゴ,ホリシャイチゴ
338292 ウラジロトックリイチゴ,オオトックリイチゴ,トックリイチゴ,フクボンシ
338300 エゾクマイチゴ,エゾノクマイチゴ,クマイチゴ,クワイチゴ,サンザシバノイチゴ,タチイチゴ,ヤマイチゴ
338302 メクマイチゴ
338303 キミノクマイチゴ
338305 アリサンオホバライチゴ,エフミャクイチゴ,オオバライチゴ,シマヤブイチゴ,タイワンヤブイチゴ,ビロウトバライチゴ,リウキウバライチゴ
338307 サンドブラックベリ-
338334 ナガミイチゴ
338347 ゴ-ルデン・エバ-グリ-ン・ラズベリ
338399 ヤツデイチゴ
338425 タカザイチゴ,テガタイチゴ,ナントウイチゴ,ビケイイチゴ,ランダイイチゴ
338433 カラピイチゴ,コウトウイチゴ
338435 コウトウイチゴ,トゲナシトネリコバノイチゴ,トネリコバノイチゴ
338447 セイヨウヤブイチゴ
338471 シチセイイチゴ
338487 シマアハイチゴ,シマアワイチゴ,リウキウイチゴ,リュウキュウイチゴ
338488 トゲリュウキュウイチゴ
338500 ツクシアキツルイチゴ
338508 ヒメフユイチゴ
338516 クサイチゴ,ナベイチゴ,ハンドイチゴ,マルバイチゴ,ヤブイチゴ,ワセイチゴ
338517 ヤエザキクサイチゴ
338518 マルバクサイチゴ
338519 キミノクサイチゴ
338557 アヂイチゴ,エゾイチゴ,チョウセンキイチゴ,ヨ-ロッパキイチゴ
338566 エゾイチゴ
338568 カナヤマイチゴ
338572 ミヤマウラジロイチゴ
338573 シナノキイチゴ
338574 ミヤマウラジロイチゴ
338576 イシヅチイチゴ
338583 エゾイチゴ
338606 ゴヨウイチゴ
338608 バライチゴ,ミヤマイチゴ
338610 ヤクシマバライチゴ
338664 タカサゴクハノハイチゴ
338678 ミヤマニガイチゴ
338679 カナヤマトチゴ,ナカナヤマトチゴ
338698 シマバライチゴ,フユノヤンバルイチゴ,ミヤマイチゴ,ミヤマフユイチゴ
338808 アリサンミヤマウラジロイチゴ,クロイチゴ,クロミキイチゴ
338810 ベニバナノクロイチゴ
338811 シモキタイチゴ
338822 ニガイチゴ
338824 ヤエザキニガイチゴ
338850 チャンカンイチゴ
338867 チチジマキイチゴ
338885 ハチジョウクサイチゴ
338886 アズキイチゴ,ウシイチゴ,キイチゴ,ゴガツイチゴ,コバノニガイチゴ,サツキイチゴ,シトウイチゴ,ニガイチゴ,フチグモイチゴ
338895 ノリクライチゴ
338909 クロミキイチゴ
338922 オキナワバライチゴ
338933 オニクロイチゴ
338942 イオウトウキイチゴ,イワウトウキイチゴ
338945 アハイチゴ,アワイチゴ,キイチゴ,ナガバキイチゴ,ナガバモミジイチゴ,ナガバモミヂイチゴ,ミヤマニガイチゴ,モミジイチゴ
338946 アハイチゴ,サガリイチゴ,モミジイチゴ,モミヂイチゴ
338947 ヤエノモミジイチゴ
338948 トゲナシモミジイチゴ
338949 キソイチゴ
338950 ヤクシマキイチゴ
338967 タウバイチゴ,タラバイチゴ
338985 アシクダシ,ウシイチゴ,ウラジロイチゴ,サツキイチゴ,サルイチゴ,ナハシロイチゴ,ナワシロイチゴ,ミツバイチゴ,ワセイチゴ
338987 アオナワシロイチゴ
338988 キミノナワシロイチゴ
338989 シロバナナワシロイチゴ
338991 タンヨウナワシロイチゴ
338993 カラナワシロイチゴ
338999 クワレンコウイチゴ
339014 コバノフユイチゴ,マルバイチゴ,マルバフユイチゴ
339015 ミツマタフユイチゴ
339018 コガネイチゴ
339024 ハスイチゴ,ハスノハイチゴ
339047 ウラジロイチゴ,エビガライチゴ,サルイチゴ,タタミイチゴ,ミヤマアシクダシ,ミヤマナワシロイチゴ,モリイチゴ,ワインベリ-
339048 キミノエビガライチゴ
339067 タイヘイイチゴ
339103 デュ-ベリ-
339111 ミヤマモミジイチゴ
339115 ヒメゴヨウイチゴ
339130 サナギイチゴ
339135 アリサンサナギイチゴ,ケサナギイチゴ,サナギイチゴ,シマバライチゴ
339136 ベニバナサナギイチゴ
339188 オホマルバフユイチゴ,ニヒタカイチゴ,ヒメフユイチゴ,ミヤマカヂイチゴ
339194 コバライチゴ,サクマイチゴ,バライチゴ,ヒメバライチゴ
339195 トキンイバラ
339208 アリサンバライチゴ
339227 エゾイチゴ,チョウセンウラジロイチコ
339240 キタイチゴ
339270 オニイチゴ,カシワイチゴ,タグリイチゴ,ナベイチゴ,ハウロクイチゴ,ホウロクイチゴ
339285 サ-モンベリ-
339290 マルミイチゴ
339310 アメリカアカミキイチゴ
339318 ミヤマニガイチゴ
339335 コジキイチゴ
339338 シマウラジロイチゴ,タイワンウラジロイチゴ
339344 オホビロウドイチゴ,タイトウイチゴ

339347 タイワンイチゴ,タカサゴクサイチゴ
339381 トキンイバラ
339390 タカサゴニガイチコ,フオーリーイチゴ,ファーリーイチゴ
339394 カジイチゴ
339395 ヤエザキカジイチゴ
339445 ベニバナイチゴ
339468 オニイチゴ,ヒヒランイチゴ
339499 キビナワシロイチゴ
339512 オオハンゴンソウ属,オホハンゴンサウ属,ルドベッキア属
339518 ダキバハンゴンサウ,ダキバハンゴンソウ
339522 テンニンギクモドキ
339539 トウゴウギク
339556 アラゲハンゴンサウ,アラゲハンゴンソウ,キヌガサギク
339569 アラゲハンゴンソウ
339574 オオハンゴンソウ,オホハンゴンサウ,ルドベキア
339581 ハナガサギク,ヤヘジキオホハンゴンサウ,ヤヘジキオホハンゴンソウ
339585 オオバナハンゴンソウ
339622 ミツバオオハンゴンソウ
339658 リュエリア属,ルイラサウ属,ルイラソウ属,ルエルリア属
339684 ヤナギバルイラソウ
339803 バライロルエリア
339816 ケブカルイラソウ
339828 ムラサキルエリア
339868 リュウリンジュ属
339869 チリメンバコトネアスタ,リュウリンジュ
339880 ギシギシ属
339881 アレチエゾノギシギシ
339883 アレチコギシギシ
339884 ノハラダイオウ
339887 スイバ,スカンポ,スシ,ダヤス
339897 ヒメスイバ
339899 ヒメスイバ
339918 アムールギシギシ
339925 ヌマギシギシ,ヌマダイオウ,ヌマダイワン
339940 タカネスイバ
339960 カギミギシギシ
339979 マンシウギシギシ
339989 アレチギシギシ
339994 ギシギシ,シブクサ,チヂミスイバ,ナガバギシギシ
340000 ギシギシ
340019 コギシギシ,ココメギシギシ,ノコギリスイバ
340023 コギシギシ
340064 カラズトノダイワウ,カラフトダイオウ,カラフトノダイオウ
340075 ハネミヒメスイバ
340082 ミゾダイオウ,ミゾダイワウ
340089 ウマノスカナ,ギシギシ,シブクサ,ノミノアッネ,ノミノフネ
340103 タカネスイバ
340109 ノダイオウ,ノダイワウ,ノダイヲウ
340114 マダイオウ
340116 コガネギシギシ,コゴメギシギシ,ハマギシギシ,ハマスイバ
340136 オオバスイハ,タカネスイバ
340141 キブネダイオウ
340142 キブネダイオウ
340151 エゾノギシギシ,ヒロバギシギシ
340171 パミールギシギシ
340178 コウライギシギシ,ワセスイバ
340214 ヒョウタンギシギシ
340249 ニセアレチギシギシ
340254 テガタスイバ,マルバスイバ
340269 ホソバギシギシ
340270 ホソバギシギシ
340293 ニセコガネギシギシ
340336 シロハグロ属
340344 ホダチハグロサウ
340367 シロハグロ
340463 オオイワウイキョウ,ミヤマウイキョウ
340470 カハツルモ属,カワツルモ属
340472 ネジリカワツルモ,ネテリカハツルモ
340476 カハツルモ,カワツルモ,ネジリカワツルモ
340491 ヤハズカワツルモ
340502 カワツルモ科
340541 ナギイカダ科
340548 ルシア属
340595 ハナシキギク
340626 ギンクジャク
340644 コウボウ
340655 カタシキギク
340659 ヒスイホコ
340673 スズカゴ
340687 サイリュウ
340757 グンケン
340781 ムレホコ
340786 シホウ
340835 オウシュンギョク
340844 コトジギク
340860 ハクテンシ
340918 アサジギク
340922 ケンリュウ
340954 ノボリリュウ
340989 ナギイカダ属
340990 ナギイカダ,ルスガス
341016 ハナチャウジ属,ハナチョウジ属
341020 ハナチャウジ,ハナチョウジ
341039 ヘンルウダ属,ヘンルーダ属
341047 コヘンルーダ
341052 ウンコウ,コヘンーダ,コヘンルーダ
341064 ヘンルウダ,ヘンルーダ
341086 ヘンルウダ科,ミカン科
341223 ササキカヅラ属
341227 ササキカズラ,ササキカヅラ
341401 クマデヤシ属,サバル属
341406 ブラックバーンサバル
341418 メキシコサバル
341421 チャボサバル
341422 アメリカパルメット
341423 キューバサバル
341430 テキサスビル
341475 アオカズラ属,アワブキ属,アヲカヅラ属
341513 アオカズラ,アオカヅラ,アヲカヅラ
341570 シマアオカヅラ,シマアヲカヅラ,タイワンアオカズラ,タイワンアオカヅラ,タイワンアヲカヅラ
341575 アリサンアオカヅラ,アリサンアヲカヅラ
341582 アオカヅラ科,アワブキ科,アヲカヅラ科
341696 イブキ属
341829 サタウキビ属,サトウキビ属
341830 ムラサキオバナ
341837 ヨシスス,ヨシススキ
341858 シロシマススキ,シロタカオススキ,シロタカヲススキ
341887 カンショ,カンショウ,ササウグサ,ササウサ,サタウキビ,サタウグケ,サタウノキ,サタタケ,サタノキ,サトウキビ
341909 カラサトウキビ
341912 トウエゾガヤ,ナンゴクワセオバナ,ハマススキ,ワセオバナ,ワセヲバナ
341915 ハマススキ,ワセオバナ
341917 カンショガヤ
341937 ヌメリグサ属
341960 ヌメリグサ,ハイヌメリ,ハイヌメリグサ,ハヒヌメリ
341962 ホソハイヌメリ
341977 ハラヌメリ
342015 カシノキラン属,ガストロキールア属,マツラン属
342022 マツゲカヤラン
342155 クロイゲ属
342170 ハマヲクロイゲ
342191 ヒロハクロイゲ
342198 クロイゲ
342206 ケクロイゲ
342221 ツメクサ属
342226 イトツメクサ
342251 タカノツメ,ツメクサ
342259 スズメグサ,タカノツメ,ツメクサ,ハマツメクサ
342261 エゾハマツメクサ

342279 アライドツメクサ
342290 チシマツメクサ
342308 オモダカ属,クワイ属,クワヰ属
342310 アギナシ
342338 ナガバオモダカ
342381 タイリンオモダカ
342385 ウキオモダカ,カラフトグワイ,カラフトグワヰ
342396 ウリカハ,ウリカワ,オオボシソウ,オホボシサウ
342400 オモダカ,セイヨウオモダカ
342409 ホソバオモダカ
342417 アメリカウリカワ
342421 オモダカ,ハナグワイ,ハナグワヰ,ホソバオモダカ
342424 クワイ,クワヰ,シナクワイ,シナクワヰ,シロクワイ,シロクワヰ,ハククワイ,ハククワヰ
342429 ヒトツバオモダカ
342482 アフリカスミレ属,セントポーリア属
342493 アフリカスミレ
342570 ザラッカ属
342571 ザラッカモドキ
342572 オオバザラッカ
342580 アマミザラッカ
342841 ヤナギ科
342847 アッケシサウ属,アッケシソウ属
342859 アッケシサウ,アッケシソウ,ハママツ,ヤチサンゴ
342900 カブダチアッケシソウ
342923 ヤナギ属
342924 ナガバノネコヤナギ
342926 コガネシダレ
342928 ロッコウヤナギ
342929 チクゼンヤナギ
342930 リュウゾウジヤナギ
342931 フジヤナギ
342933 ミョウジンヤナギ
342934 トヨハラヤナギ
342935 シロシダレヤナギ
342938 オクヤマサルコ
342940 ワケノカワヤナギ
342941 ナスノイワヤナギ
342943 コンゴウバッコヤナギ
342947 コセキヤナギ
342949 スミヨシヤナギ
342950 ツガルヤナギ
342952 ヤマトヤナギ
342954 ユウキシダレ
342961 オケシャウヤナギ,カスピエゾヤナギ
342967 ペルシャバッコヤナギ
342973 セイヨウシロヤナギ
343013 アメリカマルバヤナギ
343027 マガタマヤナギ
343045 チシマヤナギ
343063 ユスラバヤナギ
343070 イトヤナギ,オオシタレ,オホシダレ,シダレヤナギ,ヤナギ,ロクカクヤナギ,ロッカクヤナギ
343101 メギヤナギ
343106 タカネヤナギ
343119 ヌマキヌヤナギ
343151 オホサルコヤナギ,カウライバッコヤナギ,サルヤナギ,バッコヤナギ,ヤマネコヤナギ,ヤマヤナギ
343168 トカチヤナギ
343170 オオバヤナギ
343185 アカヂクマルヤナギ,アカメヤナギ,ウラシロヤナギ,マルバヤナギ
343221 ハヒイロヤナギ
343280 セイヨウエゾヤナギ
343321 シマオホネコヤナギ,トウザンヤナギ,トクンヤナギ,ドヰヤナギ,モリヤナギ
343342 オオシロヤナギ,コゴメヤナギ,ジャヤナギ
343388 チョウセンキツネヤナギ,テウセンキツネヤナギ
343396 サワヤナギ,ポッキリヤナギ
343400 チャイロヤナギ,トウサンヤナギ
343409 ミヤマヤチヤナギ
343410 アカゲオオキツネヤナギ
343414 カハヤナギ,カワヤナギ,ナガバカワヤナギ,ホソバコラヤナギ
343416 セイヨウミヤマヤナギ
343440 ホソバモウコヤナギ
343444 エノコロヤナギ,オオバカワヤナギ,オホバカハヤナギ,カワヤナギ,サルコヤナギ,タニガハヤナギ,チョウセンネコヤナギ,ネコヤナギ
343445 クロヤナギ
343448 シダレネコヤナギ
343449 フイリネコヤナギ
343467 タライカヤナギ
343529 イヌコリヤナギ,オオバコリヤナギ,オホバコリヤナギ,コブヤナギ,ヒロハコリヤナギ
343531 フイリイヌコリヤナギ
343532 シダレイヌコリヤナギ
343546 シバキツネヤナギ,シバヤナギ
343549 シロヤナギ
343560 コウカイヤナギ,ホウオウヤナギ
343567 キヌヤナギ
343569 マンシウコリヤナギ
343574 オオタチヤナギ,カウライヤナギ,コウライヤナギ
343578 シロヤナギ
343579 コリヤナギ
343587 ズイシャヤナギ,スヰシャヤナギ,ドイヤナギ,ドヰヤナギ
343602 フリソデヤナギ
343668 ネッカヤナギ,ペキンヤナギ
343670 ウンリュウヤナギ,ワンリュウヤナギ
343680 ヒロハタチャナギ
343691 キヌグコリヤナギ,キヌグフリヤナギ
343702 エゾカワヤナギ,エゾノカワヤナギ
343704 シダレカワヤナギ
343705 カワヤナギ
343706 モウコヤナギ
343745 コウアンヌマヤナギ,コバヤナギ
343747 ヌマヤナギ
343749 レンゲイワヤナギ
343750 ケタカネイワヤナギ
343752 ヒダカミネヤナギ
343753 エゾノタカネヤナギ
343754 イヌマルバヤナギ
343767 アメリカポッキリヤナギ
343780 エゾマメヤナギ
343784 エゾマメヤナギ
343795 オオシタシ,オオシタレヤナギ,シダレヤナギ
343858 セイヨウテリハヤナギ,テリハヤナギ
343859 エトロフヤナギ,テリハヤナギ
343886 テヤボヤナギ
343890 オオタチヤナギ
343901 ハクトウヤナギ
343911 オクヤマヤナギ,ヤマヤナギ
343917 オオタチャナギ,オホタチャナギ,ジャヤナギ
343918 カウライシダレヤナギ,コウライシダレヤナギ
343923 オオミヤマヤチャナギ
343937 カハヤナギ,カラュリヤナギ,コリヤナギ,セイヨウコリヤナギ,ヒロハコリヤナギ
343989 ミヤマヤナギ
343991 キヌゲミヤマヤナギ
343992 シダレミネヤナギ
343994 ヌマキヌヤナギ,ノヤナギ
344022 エゾヤナギ
344023 シダレエゾヤナギ
344028 コエゾヤナギ
344031 ホソバヌマヤナギ
344034 ヌマキヌヤナギ
344045 マメヤナギ
344050 コマイワヤナギ
344051 ケコマイワヤナギ
344053 オノエヤナギ
344060 ヤマヤナギ
344071 エゾノキヌヤナギ
344082 センダイヤナギ
344090 ココメヤナギ
344092 シダレコゴメヤナギ
344108 シライヤナギ,シラヰヤナギ,チチブヤナギ
344109 チチブヤナギ

344112 イワヤナギ,ヤマヤナギ
344124 チウセンオノエヤナギ,チウセンヨノヘヤナギ
344155 ノヤナギ
344184 タライカヤナギ
344186 タライカヤナギ
344194 ホソバコリヤナギ
344199 テンジクヤナギ,ヨッシベヤナギ
344211 ウラジロヤナギ,セイヨウタチヤナギ,タチャナギ,マンシウタチャナギ
344213 タチヤナギ
344233 オノエヤナギ
344244 キヌヤナギ,セイヨウキヌヤナギ,ダイリクキヌヤナギ
344245 エゾノキヌヤナギ
344255 イワヤナギ,オオシボヤナギ,オホシボヤナギ,キツネヤナギ,キンモウヤナギ,ヘッピヤナギ
344256 サイコクキツネヤナギ
344257 カンサイキツネヤナギ
344259 オオキツネヤナギ
344260 オオネコケネヤナギ,オホネコヤナギ
344268 タイワンアカメヤナギ,タイワンヤナギ,タカサゴアカメヤナギ,ハンスヤナギ
344300 ヨシノヤナギ
344345 ヒナノカンザシ属
344351 ヒナノカンザシ
344369 バコベホオズキ属
344372 ハコベホオズキ
344381 サルピグロッシス属,サルメンバナ属
344383 アサガオタバコ,サルメンバナ
344425 オカヒジキ属,ヲカヒジキ属
344496 ハマヒジキ,ヤマヒジキ
344538 マツナモドキ
344585 ノハラヒジキ
344606 オカヒジキ,ミルナ
344718 オカヒジキ,オカミル,ミルナ,ヲカヒジキ,ヲカミル
344743 ハリヒジキ,ホソバオカヒジキ
344821 アオギリ属,アキギリ属,アキノタムラサウ属,アキノタムラソウ属,サルビア属
344823 サクキバナアキギリ
344855 ビロ-ドアサギリ
344885 ベルバナサルビア
344957 アキノタムラサウ,コマトドメ
344980 ニバナサルゼヤ,ベニバナサルビア
345023 ケショウサルビア,ブル-・サルビア
345026 ケショウサルビア
345043 メキシコサルビア
345050 アキギリ,オオアキギリ
345052 オオアキギリ
345053 シボリミヤマアキギリ
345054 ハイコトジソウ
345067 アキノベニサルビア
345076 アリサンタタムラサウ,ミヤマタムラサウ,ヤンバルタムラサウ,ヤンバルタムラソウ
345077 ケイタオタムラサウ
345083 イスパンサルビア
345107 シマジタムラソウ
345108 アキノタムラサウ,アキノタムラソウ
345110 シロバナアキノタムラソウ
345113 ケブカアキノタムラソウ
345148 シナノアキギリ
345178 ウスギナツノタムラソウ,キバナナツノタムラソウ
345181 ミヤマタムラソウ
345182 シロバナミヤマタムラソウ
345183 ナツノタムラソウ
345184 シロバナナツノタムラソウ
345185 ダンドタムラソウ
345214 タンジン
345244 カブラバサルビア,カブラバノサルビア
345252 アキギリ,キバナアキギリ,コトジソウ
345254 タイワンアキギリ
345256 キソキバナアキギリ
345258 ミツデコトジソウ
345267 オトメサルビア
345271 サルビヤ,セ-ジ,ゼ-ジ,ヤクヨウサルビヤ
345280 ハイタムラソウ
345292 ソウイロサルビア
345310 ミゾコウジュ,ミゾコオジュ,ユキミソウ
345311 シロバナミゾコウジュ
345315 イヌヒキオコシ,ヒキオコシダマシ
345334 ヒメタムラソウ
345335 アマミタムラソウ
345338 ハルノタムラソウ
345340 イヌヒメコヅチ
345365 タカサゴタムラサウ,タカサゴタムラソウ
345379 オニサルビア
345405 オホバナヘニサルヒヤ,サルビア,ヒゴロモサウ,ヒゴロモソウ
345450 マンシウアキギリ
345471 ミナトタムラソウ
345474 ムラサキサルビア
345507 サマイパティセレウス属
345517 アメリカネム属,サアマネ-ア属
345525 アメフリノキ,アメリカネム,アメリカネムノキ
345558 ニハトコ属,ニワトコ属
345580 アメリカニワトコ
345586 クサタズ,クサニワトコ,ソクズ,ソクヅ,ソクド,タイワンソクズ,ニワタズ
345588 オガサワラソクズ,タイワンソクズ
345592 ニワトココウライ
345620 ヒロハニワトコ
345624 マンシウニワトコ
345629 ミヤマニワトコ
345631 セイヨウニワトコ
345652 ヨウラクニワトコ
345653 リョウトウニワトコ
345660 セイヨウアカニワトコ,ニハトコ,ニワトコ
345667 コバノニワトコ
345668 クロバナニワトコ
345669 エゾニワトコ
345670 キミノエゾニワトコ
345671 ホソバエゾニワトコ
345672 ハゴロモニワトコ
345673 ビロ-ドエゾニワトコ
345674 ケナシエゾニワトコ
345675 ヒロハエゾニワトコ
345679 キタズ,コモウツギ,タズノキ,タズバ,ニワトコ
345680 ダイダイミノニワトコ
345681 サケバニワトコ
345682 ケナシニハトコ,ケナシニワトコ,タンナニハトコ
345683 オオバニワトコ
345684 キミノニワトコ
345685 マルバニワトコ
345686 ケニワトコ
345687 オオニワトコ
345695 コバノニハトコ
345696 クロハナニハトコ
345697 キミノニハトコ
345708 トウニワトコ,ニオイニワトコ
345709 エゾニワトコ
345728 ハイハマボッス属,ヒメボッス属
345731 ハイハマボッス
345769 サンケジア属
345782 サンデルソニア属
345788 サントール
345803 サンギナリア属
345816 ワレモカウ属,ワレモコウ属
345818 ポロシリトウウチソウ
345820 シロバナトウウチソウ
345834 トウウチソウ
345835 リシリトウウチソウ
345836 ケトウウチソウ
345837 タカネトウウチソウ
345850 カライトソウ,トウチウソウ
345851 コウライカライトソウ
345853 エゾトウウチソウ
345860 オランダワメモコウ
345880 ナンブトウウチソウ
345881 ワレモカウ,ワレモコウ
345884 ウラゲワレモコウ
345888 エゾワレモカウ,エゾワレモコウ,ワレモコウ

345890 ウラゲワレモコウ,ケワレモコウ
345894 ウスイロワレモコウ,オナガワレモコウ,タイワンワレモカウ,ヒメワレモコウ,ホクリョウワレモコウ,ミヤマワレモコウ,メワレモコウ
345911 タカネトウウチソウ
345917 コバナワレモコウ,ナガボノアカワレモコウ,ナガボノワレモコウ
345921 コバナノワレモコウ,シロワレモカウ,シロワレモコウ,ナガボノシロワレモカウ,ナガボノシロワレモコウ,ワレモコウ
345923 オオバナシロワレモコウ,オオバナノワレモコウ,チシマワレモコウ
345924 コバナノワレモコウ
345927 ケナガボノシロワレモコウ
345940 ウマノミツバ属,オニミツバ属
345948 ウマノミツバ
345949 サイシュウウマノミツバ
345957 ウマノミツバ
345978 ヒメウマノミツバ
345979 キイウマノミツバ
345996 ニヒタカミツバ
345998 クロバナウマノミツバ,マンジウミツバ
346008 フキヤミツバ
346047 サンセベリア属,チトセラン属
346070 ツッチトセラン,ボウチトセラン
346089 オオヒロハチトセラン
346101 ホソツッチトセラン
346120 チトセラン
346148 ツッチトセラン
346155 ヒロハチトセラン
346158 アツバチトセラン,サンセベリア
346162 フイリマルバチトセラン
346163 マルバチトセラン
346165 フクリンチトセラン
346173 チトセラン,トラノオラン,トラノヲラン,ホウコトウオモト
346181 カナビキサウ科,ビャクダン科
346208 ビャクダン属
346210 センダン,ビャクダン
346211 ムニンビャクダン
346243 サントリ-ナ属
346300 ジャノメギク属
346304 サンビタリア,ジャノメギク
346314 サフェシア属
346316 ムクロジ科
346323 ムクロジ属
346338 ツブ,ムクレンジ,ムクロジ
346361 シラキ属
346379 ナガバナンキンハゼ
346390 アブラミ,コクドノカシ,コクドノクワシ,シラキ,ナベコワシ,モエカラ
346408 カムテラギ,カンテラギ,タウハゼ,トウハゼ,ナンキンハゼ,リウオウハゼ,リュウキュウハゼ,ロウノキ
346418 サボンサウ属,サボンソウ属,シャボンソウ属
346421 ヒメサボンソウ
346429 キバナサボンソウ
346433 ツルコザクラ,ヒメサボンソウ
346434 サボンサウ,サボンソウ
346436 サボンソウ
346448 バウフウ,ボウフウ
346460 アカテツ科
346477 ニコゲルリミノキ
346497 アソカノキ属,ムユウジュ属
346504 アリカノキ,ムユウジュ
346527 センリヨウ
346528 キミノセンリョウ
346642 サルコカウロン属
346686 カヤラン属,サルコキラス属
346703 カヤラン
346723 コッカノキ属,サルココッカ属
346747 コッカノキ
346797 サルコグロッチス属
346899 サルコフィトン属
346900 タカサゴクレソラン
346935 タカサゴイナモリ属
346938 タカサゴイナモリ
346950 タカサゴイナモリ,ヒメノボタンサウ
346993 サルコステンマ属
347074 ニセダイオウヤシ属
347076 ヒメニセダイオウヤシ
347144 サラセニア属,ヘイシソウ属
347149 アミメヘイシソウ,シラフヘイシソウ
347150 キバナヘイシソウ
347153 コウツボソウ,コヘイシソウ
347155 ヒメヘイシソウ
347156 ムラサキヘイシソウ
347163 アカバナヘイシソウ,モミジヘイシソウ
347164 ウスギヘイシソウ
347166 サラセニア科
347191 クマザサ属,ササ属
347192 アキウネマガリ
347199 ジャウバウザサ,ショウボウザサ
347200 スズ,スズタケ,チダケ
347202 シャコタンチク
347203 センダイザサ
347205 ビロ-ドミヤコザサ
347206 ニッコウザサ
347207 アズマミヤコザサ
347213 タンガザサ
347214 シマザサ,シマダケ,チゴザサ,ヤナギバザサ
347216 フゲシザサ
347217 ウンゼンザサ
347221 ミヤマクマザサ
347222 シコクザサ
347223 クテガワザサ
347224 イヌクテガワザサ
347228 ミクラザサ
347229 カガミナンブスズ
347230 アリマコスズ,ビッチュウミヤコザサ
347233 ナスノユカワザサ
347234 チシマザサ
347238 サヤゲチシマザサ
347240 ナガバネマガリ
347242 マキヤマザサ
347243 ケマキヤマザサ
347244 イッショウチザサ
347245 セトウチコスズ
347250 オオバザサ
347251 キンタイオオバザサ
347252 ミアケザサ
347253 ミネザサ
347254 アワノミネザサ
347259 ミヤコザサ
347262 サイゴクザサ
347264 オオシダザサ
347265 ケナシカシダザサ
347268 チマキザサ
347270 シャコハンチク
347272 ルベシベザサ
347273 ヨサチマキ
347275 ケザサ
347276 オモエザサ
347277 ミカワザサ
347278 ウックシザサ
347279 オロシマチク,ケオロシマチク,ケネザサ
347287 アポイザサ
347291 イヌトクガワザサ
347292 クマイザサ
347293 キンタイザサ
347295 ミナカミザサ
347297 ミヤマザサ
347299 ウスバザサ
347300 ハコネナンブスズ
347301 カシダザサ
347303 ケスズ
347305 オクヤマザサ
347306 サイヨウザサ
347309 カワウチザサ
347310 キリシマザサ
347312 エゾミヤマザサ
347317 ツクバナンブスズ
347318 イナコスズ
347324 クマザサ,ヘリトリザサ,ヤキバザサ
347326 オオササ
347327 チュウゴクザサ
347328 ヤヒコザサ
347329 シコタンザサ
347330 オゼザサ
347332 イワテザサ

347334 アズマザサ属
347338 ジョウボウザサ
347340 グジョウシノ
347341 オニグジョウシノ
347342 オオサカザサ
347343 ヒシュウザサ
347347 ミヤギザサ
347349 カリワシノ
347350 コガシアズマザサ,コガシザサ
347352 ヒメシノ
347353 アリマシノ
347354 タンゴシノチク
347355 ケスエコザサ
347356 クリオザサ
347359 ヨモギダコチク
347360 ミドウシノ
347361 アズマザサ
347365 トミクサザサ
347366 オオバアズマザサ
347367 オニウラジロシノ
347368 サドザサ
347369 トウゲダケ
347370 ハコネシノ
347372 シオバラザサ
347373 エッサシノ
347374 タキナガワシノ
347377 スズダケ属
347380 ホソバスズタケ
347381 ウラゲスズタケ
347383 ハチジョウスズタケ
347402 ランダイカウバシ属,ランダイコウバシ属
347407 サッサフラス,サッサフラスノキ
347411 クサノオホカウバシ,タイワンサッサフラス,ランダイカウバシ
347421 ヤエヤマヤシ
347446 トウバナ属
347560 キダチハッカ
347952 タカサゴシラタマ属
347993 タカサゴシラタマ
347996 タカサゴシラタマ
348003 ハンゲシャウ科
348022 サウロマ－ツム属
348036 ヘビイモ
348077 ドクダミ科,ハンゲシャウ科
348082 ハンゲシャウ属,ハンゲショウ属
348085 アメリカハンゲショウ
348087 オシロイカケ,カタジロ,カタシログサ,サンパクソウ,ハンゲシャウ,ハンゲショウ,ミツジロ
348089 キツネアザミ属,タウヒレン属,トウヒレン属
348091 イワテヤマトウヒレン
348092 カルイザワトウヒレン
348093 オバケヒゴタイ
348094 ホクチキクアザミ
348095 シナノトウヒレン
348104 タカネアザミ,トナカイアザミ,トナカヒアザミ
348121 タカネヒゴタイ
348127 コウシュウヒゴタイ
348128 キレバコウシュウヒゴタイ
348129 スナジヒゴタイ,ヒレハナガサヒゴタイ
348132 スナジヒゴタイ
348141 キヌヒゴタイ,ヤナギヒゴタイ
348142 ススヤアザミ
348172 イワテヒゴタイ
348201 ミギハヒゴタイ,ヤナギヒゴタイ
348206 ユキバヒゴタイ
348207 シロバナユキバヒゴタイ
348243 ワタフキヒゴタイ
348256 タイワンアザミ,タイワンヒゴタイ
348313 フォ－リ－アザミ
348323 タイワンヒゴタイ
348327 ミヤマキタアザミ
348336 アブラトウヒレン
348344 ワタゲトウヒレン
348349 ホクチアザミ
348355 シラネアザミ,マンシウシラネアザミ
348373 タンザワヒゴタイ
348385 イナトウヒレン
348388 シマトウヒレン
348392 ニッコウトウヒレン
348395 シホウヒゴタイ
348403 ヒナヒゴタイ,ヒメヒゴタイ
348406 シロバナハナガサヒゴタイ,シロバナヒナヒゴタイ
348411 シマヒメヒゴタイ,タイワンヒメヒゴタイ
348430 タカサゴヒゴタイ
348442 ヒダカトウヒレン
348443 ウリュウトウヒレン
348505 ヒメシラネアザミ
348510 ヤノネアザミ
348515 ミヤコアザミ
348516 マルバミヤコアザミ
348536 ネコヤマヒゴタイ
348538 ネコヤマヒゴタイ,ネッカキクアザミ
348554 サワヒゴタイ
348559 アキノヤハズアザミ,シラネアザミ
348560 ニッコウトウヒレン
348561 クロトウヒレン
348564 オオダイトウヒレン
348569 アサマヒゴタイ
348570 ツクシトウヒレン
348571 オオトウヒレン
348572 ヤクシマヒゴタイ
348578 ホクロクトウヒレン
348583 トサトウヒレン
348585 ウラギンヒゴタイ,ホクシヒゴタイ
348590 タカスアザミ
348596 ボンボリトウヒレン
348604 キクバヒゴタイ
348609 キクバヒアザミ
348635 オオキクバヒゴタイ,オニキクバヒゴタイ
348645 ミヤマトウヒレン
348690 ウラシロヒメヒゴタイ,コオライヒメヒゴタイ,ナカハヒメヒゴタイ,ヒメヒゴタイ,ヒレヒメヒゴタイ
348691 シロバナヒメヒゴタイ
348706 ナガバヒゴタイ
348711 チシマキタアザミ
348715 ナガバキタアザミ
348716 シロバナキタアザミ
348717 ダイセツヒゴタイ
348718 エゾトウヒレン
348719 レブントウヒレン
348720 オクキタアザミ
348742 シボヒゴタイ
348749 カラオトアザミ,カラフトアザミ
348752 ヤハズトウヒレン
348753 シロバナヤハズトウヒレン
348754 チャボヤハズトウヒレン
348757 マキノハヒゴタイ
348768 キントキヒゴタイ
348769 キリシマヒゴタイ
348772 シシヒゴタイ
348778 ノコギリヒゴタヒ
348788 キクバモリアザミ,タニヒゴタイ,マンシウヒゴタイ,マンシウヒゴタヒ
348792 タカオヒゴタイ
348794 ノコギリヒゴタイ
348795 キントキヒゴタイ
348829 ナンブトウヒレン
348841 セイタカトウヒレン
348843 シロバナセイタカトウヒレン
348854 ウスバシラネアザミ
348864 ミヤマウラジロヒゴタイ
348865 オオヤノネアザミ
348872 トゲキクアザミ,ヤハズヒゴタイ
348878 シラネヒゴタイ
348881 ミクラジマトウヒレン
348882 タカネヒゴタイ,ミヤマヒゴタイ
348892 タニヒゴタイ,タニヒゴタヒ
348902 イタチアザミ,キクアザミ
348903 アツバヒゴタイ
348908 ウスユキキクアザミ
348930 ワカサトウヒレン
348945 ウスユキトウヒレン
348954 ユキバトウヒレン
349032 サキシフラガ属,ユキノシタ属
349046 エチゼンダイモンジソウ
349078 コケシコタンソウ
349113 キヨシソウ
349117 シコタンサウ,シコタンソウ,ナガバシコタンソウ
349118 チャボシコタンソウ

349119 シコタンソウ
349120 ハイシコタンソウ
349121 ヒメクモマグサ
349122 ユウバリクモマグサ
349162 ムカゴユキノシタ
349204 ジンジソウ,モミジバダイモンジソウ
349207 ムラサキジンジソウ
349208 マルバジンジソウ
349209 フギレジンジソウ
349211 フイリジンジソウ
349214 ツルジンジソウ
349216 イシッヅリ
349344 ダイモンジソウ,チョウセンダイモンジソウ
349346 ダイモンジソウ
349347 ウラベニダイモンジソウ
349350 イズノシマダイモンジソウ
349354 ウチワダイモンジソウ
349355 ヤクシマダイモンジソウ
349357 タケシマダイモンジソウ
349358 ナメラダイモンジソウ
349361 エゾクロクモソウ
349365 ナンゴククロクモソウ
349366 チシマクロクモソウ
349419 タマユキノシタ
349510 フキユキノシタ
349529 クモマユキノシタ,ヒメヤマハナソウ
349530 ツルクモマグサ
349604 キクバダイモンジサウ,ジンジサウ,ジンジソウ,モミジバダイモンジサウ,ヤッデユキノシタ
349607 シロバナクロクモソウ,チョウセンクロクモソウ
349631 チシマクモマグサ
349632 クモマグサ
349691 チョウセンイワブキ
349694 タテヤマイワブキ
349697 ハルユキノシタ
349698 ベニバナハルユキノシタ
349699 エゾノクモマグサ
349838 カラフトキヨシソウ
349841 カラフトキヨシソウ
349843 ケシマイワブキ,チシマイハブキ,チシマイワブキ,マイワブキ
349851 ヒメキヨシソウ
349856 ヨウシュクモマクサ
349869 イワユキソウ,ヤマハナソウ
349886 センダイソウ
349887 モミジバセンダイソウ
349895 シベリアユキノシタ,ベキンソウ
349936 イトバス,イハカヅラ,イハブキ,キジンサウ,ユキノシタ
349938 ホシザキユキノシタ
349940 ニシキユキノシタ
349941 アオユキノシタ
349972 ツルクモマグサ
350011 ハルサメソウ,ヒカゲユキノシタ
350063 ユキノシタ科
350086 マツムシサウ属,マツムシソウ属
350099 クロバナマツムシサウ,セイヤウマツムシサウ,セイヨウマツムシソウ
350119 コーカサスマツムシサウ,コーカサスマツムシソウ,スカビオサ・コーカシカ
350125 セイヨウイトバマツムシソウ
350129 ホソバマツムシソウ
350132 シラゲマツムシソウ
350156 ウスイロマツムシソウ
350175 マツムシサウ,マツムシソウ,リンボウキク
350176 シロバナマツムシソウ
350177 ソナレマツムシソウ
350180 エゾマツムシソウ
350183 タカネマツムシソウ
350184 シロバナタカネマツムシソウ
350185 ニイタカマツムシソウ,ニヒタカマツムシサウ
350265 トウマツムシソウ
350312 ハエマンサス・ムルテイフロルス
350323 クサトベラ属
350331 ヒメクサトベラ
350337 ナンヤウクサトベラ
350344 クサトベラ,クシトベラ,タバコノキ,テリハクサトベラ,ヤギサウ
350346 ケクサトベラ
350425 ナガミノセリモドキ
350479 スカフォセパラム属
350494 スカフィグロティス属,ヘクシセア属
350635 シェーレア属
350646 シェフレーラ属,フカノキ属
350654 ヤドリフカノキ
350706 トウシンギ,フカノキ
350859 ホロムイサウ属,ホロムイソウ属
350861 エゾセキシャウ,ホリサウ,ホロムイサウ,ホロムイソウ
350864 シバナ科,ホロムイサウ科,ホロムイソウ科
350904 ヒマツバキ属,ヒメツバキ属
350932 イジュ,ヒメツバキ
350945 イジュ
350946 マルバヒメツバキ
350980 コショウボク属
350990 コショウボク,モレノキ
350995 サンショウモドキ
351011 マッブサ属
351012 アリサンマッブサ
351021 サネカヅラ,チョウセンゴミシ
351073 ウシブドウ,マッブサ,マツフジ,ワタフジ
351093 マッブサ
351094 ウラジロマッブサ
351121 マッブサ科
351146 シスマトグロッティス属
351250 フオーリーガヤ属
351251 フオーリーガヤ
351254 フォーリーガヤ
351266 ウシクサ
351343 ムレゴチョウ属
351345 コチョウソウシザンサス
351439 シゾコドン属
351441 ヒメイワカガミ
351443 タカネヒメイワカガミ
351445 アカバナヒメイワカガミ
351446 ヤマイワカガミ
351447 ヒメコイワカガミ
351448 ナンカイヒメイワカガミ
351450 イワカガミ
351451 コイワカガミ
351452 シロバナイワカガミ
351454 シロバナナガバイワカガミ
351455 ナガバイワカガミ
351457 オオイワカガミ
351458 シロバナオオイワカガミ
351759 ミヤマニガウリ属
351762 ミヤマニガウリ
351780 イハガラミ属,イワガラミ属
351793 イワガラミ
351795 ハナイワガラミ
351796 ケイワガラミ
351797 イハガラミ,イワガラミ
351798 テリハイワガラミ
351844 ヒヒランチク属
351852 ヒヒランチク
351883 シゾスティリス属
351917 イトバギク
351935 シュレヒテランッス属
351993 シュルンベルゲラ属
351997 カニバサボテン
352000 カニサボテン,シャコバサボテン
352160 オグライ
352161 オソレヤマオトコイ
352162 シカクホタルイ
352163 アイノコカンガレイ
352187 オオサンカクイ
352193 ミヤマホタルイ
352195 イガホタルイ
352200 イヌホタルイ,ホタルイ,ホタル井
352202 ホタルイ
352205 コホタルイ,コホタル井,マンシウホタルイ
352206 オオフトイ
352212 ヒメホタルイ
352223 カンガレイ,カンガレ井,ヒメカンガレイ,ヒメカンガレ井
352228 ホソイリカンガレイ
352231 タタラカンガレイ
352232 チョビヒゲタタラカンガレイ
352235 ツクシカンガレイ

352238 シズイ
352242 ミチノクホタルイ
352278 オウイ,オオイ,フトイ,フトヰ,マルスゲ
352279 シマフトイ
352284 サギノシリサシ,サンカクイ,サンカクスゲ,サンカクヰ,タイカフヰ
352289 オホフトイ,オホフトヰ,フトイ
352293 タイワンヤマイ,タイワンヤマヰ
352298 ショエノルキス属
352348 ノグサ属
352351 ノグサ,ヒゲクサ
352353 ジョウイ
352357 イヘヤヒゴクサ
352382 オオヒゲクサ,オホヒゲクサ
352430 ボロボロノキ属
352438 シトタゴ,ボロボロノキ
352477 ションバーキア属,ションバーグキア属
352753 シュワンテシア属
352758 ユウコウギョク
352761 ギョウコゥギョク
352765 リュウギョク
352769 トリブネ
352851 コウヤマキ科
352854 カウヤマキ属,コウヤマキ属
352856 カウヤマキ,クサマキ,コウヤマキ,ホンマキ,マキ
352863 ホンガウサウ属,ホンゴウサウ属,ホンゴウソウ属
352869 ホンガウサウ,ホンゴウソウ
352871 スズフリホンゴウソウ
352873 ウエマッサウ,ウエマッソウ,トキヒササウ
352874 タカクマソウ
352880 シラ-属,ツルボ属
352960 カイソウ
353001 オオツルボ,オホツルボ,シラ-
353057 サンダイガサ,スルボ,ツルボ
353121 オウゴンカズラ属,スキンダプスス属
353132 オオシラフカズラ
353179 ホタルイ属,ホタルヰ属
353186 サンカクホタルイ
353226 エゾアブラガヤ
353272 ミネハリイ,ミネハリヰ
353290 カワベホタルイ,フサハリイ
353422 ウキヤガラ
353429 コマツカサススキ
353434 セフリアブラガヤ
353451 イワキアブラガヤ
353510 チョウセンマツカサススキ,ヒメマツカサススキ
353516 メホタルイ
353557 ヒメホタルイ,ヒメホタルヰ
353569 イヌフトイ
353576 アイバサウ,アイバソウ,アブラガヤ,エゾアブラガヤ,シテアブラガヤ,チュウゴクアブラガヤ
353587 ウキヤガラ,ヤガラ
353606 タカネクロスゲ,ナンブスゲ,ミヤマワタスゲ
353618 ヒメクロアブラガヤ
353624 マツカサススキ
353660 シズイ,シヅヰ,テガヌマイ
353676 クロアブラガヤ,ヤマアブラガヤ
353707 イセウキヤガラ,エゾウキヤガラ,コウキヤガラ,ヒメウキヤガラ
353750 ケナシアブラガヤ,ツルアブラガヤ
353767 ツクシアブラガヤ
353768 ツクシアブラガヤ
353828 コウキヤガラ
353840 イヌフトイ
353856 アオアブラガヤ,アヲアブラガヤ,クロアブラガヤ,ケナシアブラガヤ
353878 イスガヤ,オオアブラガヤ,オホアブラガヤ
353889 カンガレイ
353899 サギノシリサシ,サンカクイ,サンカクスゲ,サンカクヰ,タイカフヰ
353935 アブラガヤ
353936 シデアブラガヤ
353987 シバツメクサ
353994 アオバナツメクサ
353998 シンジュガヤ属,シンショウガ属
354022 ホソバシンジュガヤ
354052 オオシンジュガヤ
354141 シンジュガヤ
354143 タカオシンジュガヤ,タカヲシンジュガヤ
354159 ミカワシンジュガヤ
354175 カガシラ,ヒメシンジュガヤ,マネキシンジュガヤ
354185 コシンジュガヤ
354242 オホシンジュガヤ
354253 クロミノシンジュガヤ
354254 オオシンジュガヤ,ハネシンジュガヤ
354260 コシンジュガヤ
354287 スクレロカクタス属
354311 ハクコウザン
354487 カタボウシノケグサ
354579 ミヅガヤ属
354583 ミヅガヤ
354596 トゲイヌツゲ属
354621 トゲイヌツゲ
354648 キバナアザミ属
354651 キバナアザミ
354658 シマカナビキサウ属,シマカナビキソウ属
354660 シマカナビキサウ,シマカナビキソウ,セイタカカナビキサウ,セイタカカナビキソウ,セイタカカナビクソウ,タイナンカナビキサウ,タイナンカナビキソウ
354678 ハシリドコロ属
354691 オニミルクサ,オメキグサ,ハシリドコロ,ヲメキグサ
354692 キバナハシリドコロ
354755 シャトリムシマメ属
354761 シャトリムシマメ
354778 ウズムシマメ
354797 キバナバラモンジン属,スコルゾネラ属,フタナミサウ属,フタナミソウ属
354801 ダウリアバラモンジン,ヤナギバラモンジン
354813 コバラモンジン,ヒロハバラモンジン
354846 エダハリバラモンジン
354870 キクゴボウ,キバナバラモンジン
354897 ヤマバラモンジン
354904 イソバラモンジン
354930 オウゴンソウ,オオフタナミソウ,ハヒバラモンジン,フタナミサウ,フタナミソウ,ホソバフタナミソウ
354952 トウバラモンジン
355039 ゴマノハグサ属
355042 エゾヒナノウスツボ
355046 ィワゴマノハグサ
355067 ゴマノハグサ
355106 ヒナノウスツボ
355107 ナガバヒナノウスツボ
355129 エゾヒナノウスツボ
355132 ハマヒナノウスツボ
355145 アツバゴマノハグサ
355148 オオヒメノウスツボ,ナガガクヒナノウスツボ,ヒメノウスツボ
355150 ツシマヒナノウスツボ
355182 マンシウヒナノウスツボ
355190 ヒメゴマノハグサ
355195 サツキヒナノウスツボ
355201 ゲンジン
355202 オホヒナノウスツボ
355204 ゴマノハグサ
355211 ヒナノウスツボ
355261 エソヒナノウスツボ
355276 ゴマノハグサ科
355319 ニンドウバノヤドリギ
355322 アツバヤドリギ,シナヤドリギモドギ
355337 オオバヤドリギ,オホバヤドリギ,コガノヤドリギ
355340 タシミサウ属,タツナミソウ属
355359 セイヤウナミキサウ,セイヨウナミキソウ
355360 ヤマジノタツナミソウ
355387 オウゴン,コガネバナ,コガネヤナギ,ワウゴン
355391 セイタカナミキソウ,ヤンバルナミキサウ
355392 オカタツナミソウ

355393 シロバナオカタツナミソウ
355428 ヒメナミキ
355440 タイワンタシミサウ
355464 コナミキ,ヤンバルスズカウジュ
355494 セシロタツナミ,タツナミサウ,タツナミソウ
355495 シロバナタツナミソウ
355505 ウズタツナミ,シソバタツナミソウ
355510 コタツナミソウ,コバノタツナミ,コバノタツナミサウ,コバノタツナミソウ,ビロウドタツナミ,ビロ-ドタツナミ
355511 シロバナコバノタツナミ
355512 ウスベニコバノタツナミ
355518 ヒカゲナミキソウ
355544 ツクシタツナミソウ
355545 シロバナツクシタツナミ
355552 ヤクシマシソバタツナミ
355554 シソバタツナミ
355555 アオシソバタツナミ
355556 トウゴクシソバタツナミ
355558 イガタツナミ
355559 ホナガタツナミソウ
355578 ムニンタツナミソウ
355615 ジュズネタツミ,ジュズネナミキソウ
355616 デワノタツナミソウ
355669 アツバタツナミ
355674 ヤマタツナミソウ
355675 エゾタツナミソウ,ミヤマタツナミソウ
355683 ハイタツナミソウ,ヒメタツナミサウ
355711 ホソバナミキ
355722 アカボシタツナミソウ
355724 ヒメアカボシタツナミ
355733 イヌナミキソウ
355739 マンシウナミキソウ
355757 ムレイナミキソウ
355760 ミヤマナミキ
355762 ケミヤマナミキ
355778 ナミキソウ,ハマナミソウ
355779 シロバナナミキソウ
355780 ケナミキソウ
355794 タシロタツナミソウ
355821 アツバタツナミキソウ
355824 カントウナミキソウ
355846 キバナオウコン
355856 エゾナミキ
355882 スクティカリア属
355908 ミツバヒルギ属
356232 ライムギ属
356237 クロムギ,ナツコムギ,ライムギ
356350 ハヤトウリ属
356352 チャヨ-テ,ハヤトウリ
356392 ヒトツバハギ属
356456 セディレア属,ナゴラン属
356457 ナゴラン,フウラン
356467 キリンソウ属,ベンケイサウ属,ベンケイソウ属,マンネングサ属
356468 オウシュウマンネングサ,ヨ-ロッパマンネングサ
356473 ホシザキマンネングサ
356503 シロバナマンネングサ
356512 コナシマンネングサ,シナマンネングサ
356535 ヨロ-ッパミセバヤ
356590 コモチマンネングサ,ハナツヅキ
356598 アオバナイチネングサ
356607 ヒダカミセバヤ
356628 ベニケシヨウ
356649 アカアツバベンケイ
356660 ヒメホシビジン
356685 ケマンネングサ,ナナツガママンネングサ,ハコベマンネングサ
356715 アカダネマンネングサ
356751 タカサゴマンネングサ,ハママンネングサ
356759 キダチベンケイ
356776 ハイイロキダチベンケイ
356778 マツノハマンネングサ
356802 イソコマツ,ウスユキマンネングサ
356841 コマノツメ,ハナツヅキ,マンネングサ,メノマンネングサ
356843 ムニンタイトゴメ
356844 タイトゴメ
356845 オカタイトゴメ
356846 コゴメマンネングサ
356847 ミヤママンネングサ
356848 シロバナミヤママンネングサ
356855 カガノベンケイソウ
356857 ホソバエゾキリンソウ
356858 コモチキリンソウ
356884 オノマンネングサ,タカノツメ,マンネングサ,ヲノマンネングサ
356885 フクリンマンネングサ
356902 テリハアツベンケイ
356917 イハガネサウ,マメコケ,マルバマンネングサ,ミヅウルシ,ヤマヅタヒ
356928 ヨ-ロッパベンケイソウ
356934 メキシコマンネングサ
356937 ニヒタカヒメレンゲ
356944 タマツヅリ
356947 ニヒタカマンネングサ
356955 コマツミドリ
356957 ナガサキマンネングサ
356964 ノウカウマンネングサ
356970 グンハイキダチベンケイ
356993 ウスバキダチベンケイ
356994 アツバベンケイ
356999 ウスケシヨウ
357007 ニイタカナガバマンネングサ,ニヒタカナガバマンネングサ
357036 ウンゼンマンネングサ,ツクシマンネングサ
357037 セトウチマンネングサ
357038 ツシママンネングサ
357039 シベリアベンケイ
357080 ヤハズキダチベンケイ
357102 コウライコモチマンネングサ
357111 コウギョク,ニジノタマ
357114 サカサベンケイ,サカサンネングサ,サビバマンネングサ
357116 オオメノマンネングサ
357123 ツルマンネングサ
357126 サツママンネングサ
357138 ハヤタマンネングサ
357147 ロクジョウマンネングサ
357164 シラギグ
357169 キバナツメレンゲ,コ-カサスキリンソウ
357170 タマバ
357174 ハコベマンネングサ
357189 コマンネングサ,コマンネンソウ,ヒメレンゲ,ミヅウルシ
357203 ムラサキベンケイソウ
357232 ヤハズマンネングサ
357240 タカネマンネングサ
357241 アオタカネマンネングサ
357282 ケイチマンネングサ
357327 ヒメマンネングサ
357372 セイデンファデニア属
357735 セレニセレウス属
357748 ヨルノジョオウ
357751 セレニペジューム属
357877 タイトウウルミ属
357881 タイトウウルミ
357898 ツルナギイカダ属
357918 ヒメウズ属,ヒメウヅ属
357919 チンチンバナ,トンボグサ,トンボソウ,ヒメウズ,ヒメウヅ
357932 ナリヒラダケ属
357939 ダイミャウチク,ダイミョウチク,ナリヒラダケ,フグシュウダケ
357941 アオナリヒラ
357942 クマナリヒラ
357945 リクチュウダケ
357958 アオナリヒラ
357959 ヤシャダケ
358001 ナガボヤマアイ,ナガボヤマアヰ
358015 クモノスバンダイソウ属,センペルビブム属,トキワソウ属,バンダイソウ属
358021 クモノスバンダイソウ,クモノスベンダイソウ
358030 ホウシュン
358056 カミカゼ
358057 セイダイ
358062 ヘイワ,ヤネバンダイソウ
358159 キオン属,サハギク属,サハヲグルマ属,サワギク属,セネキオ属,セネシオ

属
358234 セイコウボク
358240 オオバナコウリンギク
358271 ホウビリュウ,ヤナギバシッボウジュ
358292 コウリンギク
358307 シッボウニシキ
358395 シロタエギク
358408 マツバサワギク
358500 ウラゲハンゴンソウ,ダウリアハンゴンソウ,ハンゴンサウ,ハンゴンソウ,ヒロハハンゴンサウ
358504 ヒトツバハンゴンサウ,ヒトツバハンゴンソウ
358561 ハクシュラク
358621 ミドリサンゴジュ
358633 ギョビカン,シキンショゥ,シリュウ
358637 タイキンギクモドキ
358788 ムラサキオグルマ
358872 オオガタマンボウ
358925 ハクギンリュウ
358997 ヒノカムリ,ヒノカンムリ,ベニタカ
359026 ギンゲツ
359037 ダイゲンゲッジョウ
359103 シンコウサワギク
359158 ヤコブボロギク
359232 テンリュウ
359363 イトオグルマ
359415 キョゼツキオン
359423 ナルトサワギク
359516 ニイタカボロギク,ニヒタカバロギク,モリサウ
359565 キオン,キヲン,ヒゴオミナエシ,ヒゴヲミナヘシ
359568 ニヒタカシオン,ニヒタカシヲン
359588 シロバナサワギク
359654 ナナツバ,ニヒタカキヲン,ハンゴンサウ,ハンゴンソウ,ヤマアサ
359712 ハツタカ
359811 エゾオグルマ,エゾヲグルマ
359880 マンボウ,マンポウ
359924 ミドリノスズ,ミドリノタマ
359980 タイキンギク,ユキミギク
359991 シンゲツ
360102 テッシャクジョウ
360172 アレチボロギク
360306 ハナノボロギク
360321 ネバリノボロギク
360328 ノボロギク
360409 ハネセンナ
360429 ハナセンナ
360431 フタホセンナ
360434 ヨツバセンナ
360439 ケセンナ
360453 ツリハブソウ
360461 エビスグサ
360463 クサセンナ,ハブサウ,ハブソウ
360467 コバノセンナ
360481 タガヤサン,タガヤサンノキ
360483 オオバノセンナ,オホバノセンナ,ホソバハブサウ,ホソバハブソウ
360490 キダチセンナ,センナ,モクセンナ
360493 エビスグサ,コビスグサ,ホソミエビスグサ,ロッカククソウ
360561 イチイモドキ属,イチヰモドキ属,セコイア属,センペルセコイア属
360565 アメリカスギ,イチイモドキ,イチヰモドキ,セカイヤア,セコイア,セコイアメスギ,センペルセコイア,レッドウッド
360575 ギガントセコイア属,セコイアデンドロン属
360576 ギガントセコイア,セカイヤオスキ,セカイヤオスギ,セコイアオスギ,セコイヤ,セコイヤデンドロン,マツモ－スツリ－
360582 セラプアズ属,リングア属
360645 ノコギリパルメット属
360646 ノコギリパルメット
360825 セメンシナ
360832 セメンシナモトキ
360923 ハクチョウゲ属,ハクテウゲ属
360933 コウチョウゲ,コウテウゲ,ダンテウケ,タンテウボク,ハクチョウゲ,ハクチョウボク,ハクチョオゲ,ハクテウク,ハクテウゲ,バンテイシ,ホウワウッゲ,リトウハクチョウゲ,リトウハクテウゲ
360934 フイリハクチョウゲ
360935 シナハクチョウゲ
360982 タムラサウ属,タムラソウ属
361011 キクボクチ,コウアンキクボクチ,コキクボクチ
361023 タマバウキ,タマバハキ,タマボウキ,タムラサウ,タムラソウ,マンシウタムラソウ,マンシュウタムラソウ
361024 タムラソウ
361025 シロバナタムラソウ
361029 ノコギリボクチ
361079 ウスバキクボクチ
361104 アラゲキクボクチ,モウコボクチ
361284 セサモタムヌス属
361294 ゴマ属
361317 ウゴマ,ゴマ
361350 ツノクサネム属
361367 トゲナシツノクサネム
361370 キバナツノクサネム
361380 アメリカツノクサネム
361384 アカゴチョウ,シロゴョウ
361385 シロゴチョウ
361430 キダチデンセイ
361456 イブキバウフウ属
361654 ハマミヅナ属
361667 イソミヅナ,ハマスベリヒユ,ハマミヅナ,ミルスベリヒユ
361668 オオミルスベリヒユ
361669 シロミルスベリヒユ
361680 エノコログサ属
361681 イヌエノコロ
361682 オオエノコロ
361683 アイノコエノコロ
361709 ヒメササキビ
361728 イヌアハ,イヌアワ
361730 ウスギキンエノコロ
361743 アキノエノコグサ,アキノエノコログサ
361753 コツブキンエノコロ,フシネキンエノコロ
361762 キンエノコロ
361773 フシネキンエノコロ
361794 アハ,アワ,オオアハ,オオアワ,オホアハ,コアハ
361795 コアワ
361798 コアハ,コアワ
361840 クロボキンエノコロ
361847 コツブキンエノコロ
361850 ササキビ
361868 コササキビ
361877 キンエノコロ,キンエノゴロ
361902 アフリカキンエノコロ
361930 ザラッキエノコログサ
361935 エノコログサ,ネコジャラシ
361936 カタバエノコログサ
361937 ムラサキエノコロ
361942 ハマエノコロ
361943 オオエノコロ
361946 ホソバノエノコログサ
361948 チヤボエノコロ
361958 ムラサキハマエノコロ
361959 ムラサキエノコロ
361978 セトクレアセア属
362002 セティエキノプシス属
362003 キソウマル
362032 セベリニア属
362097 ハナヤエムグラ
362121 オカメザサ属
362123 トウオカメザサ
362124 シマオカメザサ
362131 オカメザサ,カゲラザサ,ゴマイザサ,ブンゴザサ
362186 サラノキ属
362204 セラッカソバツ
362212 レッドラワン
362217 タンギ－リ
362220 サラソウジュ,サラノキ
362226 マヤビス
362242 イハカガミ属,イワウチソウ属,イワウチワ属
362251 カメヤマサウ,シマイワウチワ,ムナガタイハウチハ,ランダイイハウメ,リュウキュウイワウチワ

362252 アマミイワウチワ
362270 イワウチワ,オオイワウチワ,トクワカソウ
362271 シロバナイワウチワ,シロバナオオイワウチワ
362273 カントオイワウチワ,コイワウチワ
362274 ヤエイワウチワ
362275 イワウチワ,トクワカソウ
362339 タテヤマキンバイ属
362344 ケタテヤマキンバイ
362361 ケタテヤマキンバイ,タテヤマキンバイ
362460 シキオス属
362461 アレチウリ
362462 トゲナシアレチウリ
362473 キンゴジカ属,キンゴジクワ属
362478 ホソバキンゴジカ,ホソバキンゴジクワ
362490 ヤハズキンゴジカ
362523 シロバナゴジクワ,マルバキンゴジカ,マルバキンゴジクワ
362590 ウスバキンゴジクワ
362617 キンゴジカ,キンゴジクワ
362622 ハイキンゴジカ
362662 アメリカキンゴジカ
362663 アマミキンゴジカ
362670 ホザキキンゴジカ
362702 キンゴジカモドキ属
362703 キンゴジカモドキ
362706 オレゴンアオイ
363051 シ－ベキンギア属
363057 チングルマ属
363063 イワグルマ,チングルマ
363064 ヤエチングルマ
363073 メナモミ属
363080 コメナモミ
363090 コメナモミ,ツクシメナモミ
363093 メナモミ
363116 シグマトスタリックス属
363141 シレネ属,センノウ属,ビランジ属,ビランヂ属,マンテマ属
363144 コケマンテマ
363154 エジプトマンテマ
363158 タカネビランジ
363159 シロバナタカネビランジ
363167 ユキマンテマ
363181 ムシトリマンテマ
363191 アオモリマンテマ
363202 ハマフシグロ
363203 リュウキュウフシグロ
363214 ハエトリナデシコ,ハヘトリナデシコ,ヘルトカ－ルス,ムシトリナデシコ
363222 ヒロハムシトリナデシロ
363235 ガンピ,ガンビセンノウ,ガンピセンノウ
363265 コウバイグサ,センオウ,センオウゲ,センノウ,センノウゲ,センヲウ
363295 タマザキマンテマ
363312 アメリカセンノウ,ヤグルマセンノウ
363331 コムギセンノウ
363337 エゾェンビセンノウ,チョウセンマツモト
363363 ヒメシラタマソウ
363368 オオシラタマソウ,オホシラタマサウ
363370 スイセンノウ,スヰセンノウ,フラネルサウ,フラネルソウ
363409 マンテマモドキ
363411 アケボノセンノウ,ヒロハノマンテマ
363451 カハラケシ,カハラゴマ,ケフシグロ,サシマニンジン,フシグロ
363454 ケフシグロ
363461 カッコウセンノウ
363463 エゾマンテマ
363467 シマビランヂ,シママンテマ,ナガバナビランヂ
363472 エゾセンノウ,ケエンビセンノウ,フシグロセンノウ
363477 シロバナマンテマ
363479 イタリ－マンテマ
363483 マンテマ
363515 センジュガンピ
363516 ホソバマンテマ
363543 カムイビランジ
363606 アミボソマンテマ
363622 オオビランジ
363625 ビランジ
363626 シロバナビランジ
363627 ツルビランジ
363638 オオサカソウ,オグラセンノウ
363641 ネバリマンテマ
363671 シラタマソウ
363673 ヒロハノマンテマ,マツヨイセンノウ,マツヨヒセンノウ
363691 シベリヤセンノウ
363718 クルマセンノウ
363755 オオサカソウ,フシグロセンノウ
363757 シロガネセンノウ
363758 ザクロガンピ
363774 フクロガンピ,モリソンガンピ
363815 ツキミセンノウ
363817 ツキミマンテマ
363827 ヨルザキマンテマ
363882 オオマンテマ,オホマンテマ,サクラマンテマ,フクロナデシコ
363958 カラフトマンテマ,チシママンテマ
363963 モウコマンテマ
363967 アポイマンテマ
363971 チシママンテマ
364013 ムシトリマンテマ
364023 ヒメサクラマンテマ
364062 シベリアセンノウ
364064 マツモト,マツモトセンノウ
364067 クルマガンピ
364072 コウアンフシグロ
364079 スガワラビランジ
364103 トウエダナシマンテマ
364106 エダナシマンテマ
364130 トカチビランジ
364134 ニヒタカガンピ
364146 ハマベマンテマ
364149 タカネマンテマ
364156 スイスマンテマ
364193 シラタマサウ,シラタマソウ
364202 タカネマンテマ
364213 エンビセン,エンビセンノウ,ホウテンカ,ホウテンクワ
364218 テバコマンテマ
364255 ツキヌキオグルマ属
364301 ツキヌキオグルマ
364353 オオアザミ属,オホアザミ属
364360 オオアザミ,オホアザミ,マリヤマザミ
364382 クワッシア,シマルバ
364387 ニガキ科
364451 シムモンドシア科,ホホバ科
364544 シロガラシ属
364545 キクバガラシ,シロガラシ
364557 ノハラガラシ
364563 ケノハラガラシ
364704 クラカス
364713 ス－パ
364715 グ－マト
364737 オオイワギリソウ属,シンニンギア属
364738 グロキシニア
364754 オオイワギリソウ,グロキシニア
364768 オオタマガサ,オホタマガサ,ケナシタマガサ,ザブンギ,ハニガキ,ヘツカニガキ
364783 タウチク属,トウチク属,ビゼンナリヒラ属
364819 ホソバトウチク
364820 タウチク,トウチク,ビゼンナリヒラ
364913 シノクラッスラ属
364917 タツタホウ
364935 スマロ
364985 オホツヅラフヂ属,ツヅラフジ属,ツヅラフヂ属
364986 アオカヅラ,アオツヅチ,オオツヅラフジ,オホツヅラフヂ,ツタノハカヅラ,ツヅチ,ツヅラフジ,ツヅラフヂ
364987 ヒメツヅラフジ
364989 ウラジロオホ,ケオオツヅラフジ,ツヅラフヂ
364991 ケオオツヅラフジ
364997 ウラジモヤツデ
365002 グミモドキ

365009 ヒマラヤハッカクレン
365060 キクバキョン,キクバキヲン
365295 オオヒキヨモギ属,オホヒキヨモギ属,ヒキヨモギ属
365296 ヒキヨモギ
365298 オオヒキヨモギ,オホヒキヨモギ
365336 タイワンノウリ
365386 カキネガラシ属,クジラグサ属,ハナナヅナ属
365398 ハタザオガラシ
365503 ホソエガラシ
365515 オホカキネガラシ
365529 ホコバガラシ
365532 キバナノハタザオ,キバナハタザオ,ヘスペリソウ
365564 カキネガラシ,ケカキネガラシ
365565 ハマカキネガラシ
365568 イヌカキネガラシ
365585 ハタザオモドモ
365678 ニハゼキシャウ属,ニワゼキショウ属
365695 ニワゼキショウ
365743 キバナニワゼキショウ,ニワゼキショウ
365747 ニハゼキシャウ,ニワゼキショウ,ルリニワゼキショウ
365767 コバナアヤメ
365785 ヒトフサニワゼキショウ
365835 ムカゴニンジン属
365858 サワゼリ,ヌマゼリ
365871 ムカゴニンジン
365872 トウヌマゼリ,ホソバトウヌマゼリ
365877 ヒロハヌマゼリ
365910 スキアトフィツム属
365916 ミヤマシキミ属
365919 アリサンミヤマシキミ
365929 ミヤマシキミ
365936 ウチコミッルミヤマシキミ,カラフトシキミ
365937 シロミノツルシキミ
365938 ナガバツルシャマシキミ
365940 モンタチバナ
365942 ツルシキミ
365943 アケボノミヤマシキミ
365945 ウチダシミヤマシキミ
365947 ウチダシモンタチバナ,ツクバシキミ
365949 ウチダシミヤマシキミ
365951 ソナレミヤマシキミ
365954 ウチダシツルシキミ,ツルシキミ
365956 シロミノツルシキミ
365957 ナガバツルミヤマシキミ
365958 ナガミノツルシキミ
365959 ツルシキミ
365961 ノコバツルシキミ
365962 リュウキュウミヤマシキミ
365963 リュウキュウミヤマシキミ
365965 リュウキュウシャマシキミ
365969 テンジクミヤマシキミ
365974 スグダチミヤマシキミ,ミヤマシキミ
366034 ハリミコバンモチ属
366047 ハリミコバンモチ
366136 スミクロスティグマ属
366137 サクラリュウ
366141 サルトリイバラ科
366142 ユキザサ属
366230 シオデ属,シホデ属
366238 アリサンサンキライ
366243 アラガタオホサンキライ,アラガタサンキライ
366257 ヒメカカラ,ヒメサルトリ
366271 オボサンキライ,サツマサンキライ
366284 カカラ,カラダチ,ガンタチイバラ,サクトリイバラ,サルトリイバラ,サルマメ,サンキライ,フクダンバラ,ヤキモチバラ,ワガキ
366286 キミノサルトリイバラ
366290 コミノサルトリイバラ
366291 タイヘイサンキライ
366294 トキワサルトリイバラ
366324 ナガエサンキライ,ホソバサンキライ
366338 ドブクリョウ,ナメラサンキライ
366364 シホデ
366392 アラグサンキライ
366395 ウラジロサンキライ
366414 シマサンキライ,タイワンサンキライ
366460 ランダイサンキライ
366483 ササバサンキライ
366486 タチシオデ,タチシホデ
366487 オオバタチシオデ
366488 ホソバタチシオデ
366518 コモチサンキライ
366548 シオデ,タカオサンキライ,タカヲサンキライ
366549 トガリシオデ
366551 サドシオデ
366557 ホソバシオデ
366573 サイカチバラ,サイゴクバラ,シオデ,シホデ,タイワンヤマカシュウ,タチシホデ,ヤマカシウ,ヤマカシュウ
366574 トゲナシヤマカシュウ
366583 ウスバサンキライ,マムバサンキライ,マルバサンキライ
366607 サルマメ
366632 シバクサネム属,シバネム属
366654 シバネム,ヒゲネムリハギ
366670 シバクサネム,シバネム
366698 ネムリハギ
366714 ビロ－ドギリ属
366715 ビロ－ドギリ
366719 トラフビロ－ドギリ
366730 スミティナンディア属
366776 ソブラリア属
366786 チクヨウラン
366806 ニセタケゥマヤシ属
366820 ソエレンシス属
366824 ショウヨウマル
366825 レイシギョグ
366826 キョオウリュウ
366827 キョウマギョク
366853 ナス科
366863 ラッパバナ属
366866 ラッパバナ
366870 ナガラッパバナ
366902 ナス属
366910 サンゴナス,トゲハリナスビ
366920 カザリナス,ヒラナス
366934 アメリカイヌホオズキ,テリミノイヌホオズキ,テリミノイヌホオホロシ
367011 キンギンナスビ
367014 オニナスビ,ノハラナスビ,ワルナスビ
367015 シロバナワルナスビ
367042 ムラサキハリナス,ムラサキハリナスビ
367120 マルバノホロシ
367133 ヤマホロシ
367139 ラシャナス
367140 シロバナラシャナス
367146 タバコホホヅキ,ヤンバルナスビ
367178 スズガケヤナギ,チョウジカ,リュキュウヤナギ,ルリヤナギ
367241 タイワンナスビ,テンヂクナスビ,ヤンバルナセビ
367262 タカオホロシ,ホソバノホロシ,ヤマホロシ
367263 キミノヤマホロシ
367265 ツルハナナス
367273 オキナワヒヨドリジョウゴ
367282 ヒロハノホロシ
367295 オニハリナスビ
367305 キダチハリナスビ
367311 アカミノイヌホオズキ
367322 ツツラコ,ヒヨドリジャウコ,ヒヨドリジョウゴ,ヒヨドリジョオゴ,ホロシ
367323 シロバナヒヨドリジョウゴ
367325 ムラサキヒヨドリジョウゴ
367326 キミノヒヨドリジョウゴ
367328 イトホロシ
367330 ヒロハノヒヨドリジョウゴ
367334 ヤイマナスビ
367357 ツノナス
367365 マルバノホロシ,ヤママルバノホロシ
367367 オオマルバノホロシ
367368 オオマルバノホロシ
367370 ナス,ナビス

367375 ヒラナス
367392 イラブナスビ
367414 オオイヌホオズキ
367416 イヌホオズキ,イヌホホズキ,イヌホホヅキ,ウシホホヅキ,ヤマホホヅキ
367437 アカミノイヌホオズキ
367455 テリミノイヌホオズキ
367499 ヒメケイヌホオズキ
367522 タマサンゴ,フユサンゴ,リュウノタマ
367528 ヒメタマサンゴ
367535 アメリカイヌホオズキ
367567 トマトダマシ
367583 ケイヌホオズキ
367596 ルリイロツルナス
367604 フルカワナスビ
367617 ハリナスビ
367632 キダチイヌホオズキ
367651 ナンゴクイヌホオズキ
367682 スズメナスビ,セイバンナスビ
367689 ハゴロモイヌホオズキ
367696 キンカイモ,ゴショウイモ,ジャガイモ,ジャガタライモ,ハッショウイモ,バレイショ
367706 トゲハリナスビ
367726 アカミノイヌホオズキ
367733 テンジクナスビ
367735 キンギンナスゼ,ニシキハリナスビ
367745 ウェンドランドツルナス
367762 ソルダネラ属
367763 イワカガミダマシ
367775 テンダスズメウリ
367848 コケタンポポ
367939 アキノキリンサウ属,アキノキリンソウ属,ソリダ－ゴ属
368013 カナダアキノキリンサウ,カナダアキノキリンソウ,セイタカアキノキリンソウ,セイタカアワダチソウ,ソリダゴ
368026 ケカナダアキノキリンソウ
368073 イッシオウカ,コガネギク,ミヤマアキノキリンサウ,ミヤマアキノキリンソウ
368106 オオアワダチソウ,ハダカアキノキリンソウ
368134 イトバアワダチソウ
368188 ハヤザキアワダチソウ
368237 イッスンキンカ,ヤクシマアキノキリンソウ
368268 オオアワダチソウ,オホアハダチサウ
368289 ヤナギバキリンソウ
368386 トキワアワダチソウ
368387 ホソバトキワアワダチソウ
368480 アキノキリンサウ,アキノキリンソウ,アハダチサウ,アワダチソウ,キンクワ,チョウセンアキノキリンソウ,ヨウシュアキノキリンソウ
368483 アキノキリンソウ,アワタチソウ,キンカ
368484 シマコガネギク
368485 オオアキノキリンソウ
368486 コガネギク,ミヤマアキノキリンソウ
368490 キリガミネアキノキリンソウ
368491 ハマアキノキリンソウ
368494 ハチジョウアキノキリンソウ
368502 コウアンアキノキリンソウ
368520 アオヤギバナ
368524 ソリシア属
368526 シロナナコ
368527 シマトキンサウ属,シマトキンソウ属
368528 イガトキンソウ,シマトキンサウ,シマトキンソウ
368534 メリケントキンソウ
368536 メリケントキンソウ
368539 ソリア属
368625 ノゲシ属,ハチジョウナ属,ハチヂャウナ属
368635 タイワンハチジョウナ,タイワンハチジョオナ,タイワンハチヂャウナ,ハチジョウナ,ハチヂャウナ
368649 オニノゲシ,ケナシオニノゲシ
368650 ケオニノゲシ
368675 ハチジョウナ,ハチジョオナ
368771 ゲシアザミ,ノゲシ,ノシ,ハルノノゲシ
368842 アレチノゲシ
368910 マヤブシキ属
368913 ハマザクロ,マヤブシキ
368915 オオバナヒルギ
368926 ハマザクロ科,マヤプシキ科
368955 クララ属
369010 キツネノササゲ,クサエンジュ,クサジュ,クララ,シカクミクララ,ヒロバクララ,ホソバクララ
369012 ムラサキクララ
369018 チクシムレスズメ,ツクシムレスズメ
369037 エゾジュ,エニス,エンジュ,カタクミ,キフジ,キフヂ
369038 シダレエンジュ
369130 コ－ハイ,ハネミエンジュ
369134 イソフジ,イソフヂ,ケクララ
369175 ゾフロニテラ属
369178 ソフロニチス属,ソフロニティス属
369180 ショウジョウラン
369259 ホザキナナカマド属
369265 キダチシモツケ
369267 オオニワナナカマド
369270 オニナナカマド
369272 チンシバイ,ニワナナカマド
369279 キタナナカマド,ホザキナナカマド,ホザクナナカマド
369280 エゾノホザキナナカマド
369282 イワキキンネ,ホザキナナカマド,ホザキノナナカマド
369290 ナナカマド属
369296 カワシロナナカマド
369299 リクチュウナナカマド
369303 アズキナシ
369306 フギレアズキナシ
369307 オクシモアズキナシ
369308 マルバアズキナシ
369313 アメリカナナカマド
369339 オウシュウナナカマド,オヤマノサンセウ,セイヨウナナカマド,ナナカマド
369362 オヤマノサンショウ,ナナカマド
369366 サビバナナカマド
369367 エゾナナカマド,オオナナカマド,ナナカマド
369369 ツシマナナカマド
369401 コバノナナカマド,ナンキンナナカマド,ヤマザンショウ
369432 スウェ－デンナナマド
369454 ウラジロナナカマド
369455 ナンキンナナカマドモドキ
369462 コゴメナナカマド,コバナノナナカマド
369484 オオバトウナナカマド,オホバタウ,トウナナカマド,マンシウナナカマド
369488 ナナカマド
369498 セイバンナナカマド,タカサゴサビハナナカマド
369509 アカテツナナカマド,サビハナナカマド
369512 オオミヤマナナカマド,オホミヤマナナカマド,タカネナナカマド
369513 ミヤマナナカマド
369525 ウラジロナナカマド
369561 タチナンキンナナカマド
369590 モロコシ属
369593 ブラックソルガム
369600 タカキビ,ナミモロコシ,モロコシ,モロコシキビ
369633 カザリモロコシ
369640 アズキモロコシ
369652 セイバンモロコシ,ヒメモロコシ
369653 ヒメモロコシ
369671 コウリャン
369677 コモロコシガヤ,モロコシガヤ
369678 シラゲモロコシガヤ
369681 モロコシガヤ
369689 セイバンモロコシ
369700 サトウモロコシ,ロゾク
369707 ス－ダングラス
369720 モロコシ
369877 ソテロサンツス属
369975 スイセンアヤメ属,スパラックシス

属
370018 ミクリ科
370021 ミクリ属
370023 エゾミクリ
370030 ホソバウキミクリ
370044 エゾミクリ
370050 ミクリ
370057 オオミクリ
370061 ヤマトミクリ
370066 コミクリ,タマミクリ
370067 ホソバタマミクリ
370068 ウキミクリ,ナガハウキミクリ
370071 タカネミクリ,チシマミクリ
370072 ナガエミクリ
370075 ミクリ
370088 ヒナミクリ
370096 エゾミクリ
370100 ヒメミクリ,ボソパミクリ
370102 オオミクリ,オボミクリ,カドハリミクリ,サンリョウ,ミクリ
370106 ヒメミクリ
370130 スパルマンニア属
370191 レダマ属
370202 レダマ
370315 スパティカルパ属
370332 スパシフィルム属
370336 オオササウチワ
370337 ニオイササウチワ
370338 オカメウチワ
370344 ササウチワ,スパシフィルムパティニー
370351 カエンボク属,クワエンボク属
370358 カエンボク
370393 コウトウシラン属,スパソグロッチス属,スパトグロティス属
370401 コウトウシラン
370515 ダイダイグサ
370521 オオツメクサ属,オホツメクサ属
370522 オオツメクサ,オホツメクサ,ノハラツメクサ
370526 オオツメクサモドキ
370609 ウシオツメクサ属,シオツメクサ属
370615 ウシオハナツメクサ
370661 ウシオツメクサ,シオツメクサ
370675 ウシツメクサ
370696 ウスベニツメクサ
370706 ウシホツメクサ
370736 ハリフタバ属
370744 ハリフタバ
370749 ナガバハリフタバ
370782 アメリカハリフタバ
370802 ヒロハフタバムグラ
370829 マルバフタバムグラ
370852 アメリカムグラ
370956 タマバナサウ属,タマバナソウ属
370958 タマバナソウ
371087 オホウシクサ属
371090 オホウシクサ
371161 クサボウコウマメ
371240 ズウエソウ,ヅウエソウ
371406 ナガバノウルシ属
371409 ナガバノウルシ
371413 ナガボノウルシ科
371493 クサビガヤ
371627 オランダセンニチ属
371710 ハウレンサウ属,ハウレンソウ属,ホウレンソウ属
371713 ハウレン,ハウレンサウ,ハウレンソウ,ホウレンソウ
371723 ツキイゲ属
371724 ツキイゲ,ハリハナムギ
371785 シモツケ属
371789 ウラジロイワガサ
371793 エゾノマルバシモツケ
371807 カラマツシモツケ,コバノユキヤナギ
371823 イワデマリ,マルバシモツケ
371824 アポイシモツケ
371826 テシオシモツケ
371828 エゾノマルバシモツケ
371829 ケナシエゾノマルバシモツケ
371830 アカンマルバシモツケ
371836 イハガサ,イワガサ
371837 コゴメイワガサ
371844 ミツバイワガサ
371868 コデマリ,スズカケ
371871 ヤエコデマリ
371877 アイズシモツケ,アヒヅシモツケ,ケナシアイズシモツケ,ケナシアイヅシモツケ,ホソバアイズシモツケ
371881 アイヅシモツケ,オオバユキヤナギ,シロバナシ,ヒロハノユキヤナギ
371884 タウシモツケ,ホソバノイブキシモツケ
371897 イブキシモツケ,ケミツデイワガサ,マンシウシモツケ
371906 エゾノシジミバナ,オシマシモツケ,コシジミバナ
371912 タイワンシモツケ
371915 チョウセンマンシウ
371944 キシモツケ,シモツケ,ホソバシモツケ
371945 オヤマシモツケ
371947 コシモツケ
371957 シロバナシモツケ,シロバナモッケ
371960 ウラジロシモツケ
371961 ヒメシモツケ
371962 ケホソバシモツケ
371971 オホケシモツケ
371972 ケナシシモツケ
371975 シロバナシモツケ
371977 ケシモツケ
371979 ケホソバシモツケ
371980 ドロノシモツケ
372006 エゾシモツケ
372009 エゾシモツケ,コエゾシモツケ
372012 エゾノシロバナシモツケ,コエゾシモツケ
372021 ニヒタカシモツケ
372029 イハシモツケ,イワシモツケ,コシモツケ
372031 イワシモツケ
372036 キイシモツケ
372038 リョゥトウマンシウ
372050 コゴメバナ,シジミバナ,ハゼバナ
372051 シジミバナ
372053 タイワンシジミバナ,ヒトエシジミバナ
372062 ヒトエノシジミバナ
372067 ウスゲシモツケ
372075 アカヌマシモツケ,エゾハギ,ホザキシモツケ
372083 エゾシモツケ
372094 クロシモツケ
372101 イハヤナギ,イワヤナギ,コゴメバナ,ユキヤナギ
372112 トサシモツケ
372114 チョウセンコデマリ
372117 ミツデイハガサ,ミツデシモツケ,ミツバイハガサ,ミツバイワガサ
372176 スピランテス属,ネジバナ属,ネヂバナ属
372247 ナンゴクネヂバナ,ネジバナ,ネヂバナ,モジズリ
372250 ネジバナ
372252 シロバナモジズリ
372253 アキネジバナ
372254 ヤクシマネジバナ
372255 アオモジズリ
372295 ウキクサ属
372300 ウキクサ,ウラベニウキクサ,カガミグサ,タネナシ
372398 オオアブラススキ属,ミヤマアブラススキ属
372406 アブラススキ,アブラスズキ,ヂョロガヤ
372408 ミヤマアブラススキ
372412 タイワンアブラススキ,タイワンアブラスズキ,トウホアブラスズキ
372428 オオアブラススキ,オホアブラススキ
372431 ホソバノアブラススキ
372434 タイナンアブラススキ
372459 スポンディアス属
372467 スポンジアス
372478 ケドンドンアラス
372574 ネズミノオ属,ネズミノヲ属
372642 スズメヒゲシバ
372650 フタシベネズミノオ,フタシベネズミノヲ
372659 ネズミノオ,ネズミノヲ

372666 ネズミノオ,ネズミノヲ,フタシベネズミノオ,フタシベネズミノヲ,ムラサキネズミノオ
372696 ヒメネズミノオ
372737 ヒゲシバ
372803 ヒゲクサ,ヒゲシバ
372874 サヤヒゲシバ
372880 ソナレシバ
372912 ツバメズイセン属
372913 スプレケリア,ツバメズイセン
373085 イヌゴマ属,スタキス属
373110 セイヤウイヌゴマ
373125 ノチョロギ,ヤブチョロギ
373126 エゾイヌゴマ
373139 イヌゴマ,エゾイヌゴマ,チョロギダマシ
373142 ホソバイヌゴマ
373147 イヌゴマ,チョロギダマシ
373166 ワタイヌゴマ,ワタカッコウ,ワタクワクカウ,ワタチョロギ
373179 ベニイヌゴマ
373232 タイリンカッコウ
373262 ケナシイヌゴマ,チョウセンイヌゴマ
373321 ナガバヤブチョロギ,ビロウドチョロギ
373324 コナガバヤブチョロギ
373346 オトメイヌゴマ
373365 ケオトメイヌゴマ
373400 シロバナイヌゴマ
373439 ジイナモ,チョウロギ,チョウロク,チョナ,チョロギ
373490 ナガボソウ属,ホナガサウ属,ホナガソウ属
373495 ウスイロホナガソウ
373497 チリメンナガボソウ
373507 インドナガボソウ,フトボナガボソウ
373511 ナガボソウ,ホナガソウ
373515 キブシ科
373516 キブシ属
373540 ヒマラヤキブシ
373547 シコクキブシ
373556 キフジ,キブシ,キブヂ,ゴンゼ,ズイノキシバ,マメバス,マメフジ,マメブシ,マメフヂ,マメヤナギ
373559 ニシキキブシ
373560 フクリンキブシ
373561 コバノキブシ
373562 マルバキブシ
373563 ナンバンキブシ
373564 ケキブシ
373565 ナガバキブシ,ハザクラキブシ
373567 ヒメキブシ
373673 シダソテツ属
373674 オオバシダソテツ
373681 スタンホ-ペア属,スタンホペア属
373719 スタペリア属
373800 セイテンカク
373806 ヨウセイカク
373821 オウサイカク
373832 ギュウカクモモドキ
373836 サイカク
373874 シュロウカク
373908 テイオウサイカク
373917 シスイカク
373975 トラサイカク
373982 オウサイカク
374013 ギュウカク
374018 ヤギュウカク
374031 バンサイカク
374037 スタペリアンサス属
374040 ライカク
374041 ライカク
374044 スタペリオプシス属
374087 ミツバウツギ属
374089 コメゴメ,ナマイ,ミツバウツギ
374090 ミドリミツバウツギ
374091 ウスベニミツバウツギ
374093 シロバナミツバウツギ
374107 セイヨウミツバウツギ
374114 ミツバウツギ科
374381 ムベ属
374400 タイワンアケビ
374409 ウベ,トキハアケビ,トキワアケビ,ムベ
374417 ゲイタオトキハアケビ
374425 タイワンアケビ,マルバトキハアケビ
374428 タエフトキハアケビ
374457 タイワンニウメンラン属
374458 タイワンニウメンラン
374460 イリモテラン,ニウメンラン
374476 シマサギゴケ,タイワンサギゴケ
374699 ステリス属
374724 ハコベ属
374731 ノミノフスマ
374749 ウシハコベ
374757 アリサンハコベ
374758 オホバアリサンハコベ
374780 エゾノミヤマハコベ,オオハコベ,ミヤマハコベ
374781 オオハコベ
374783 カンチヤチハコベ
374794 アオツメクサ
374836 ヒロハフタマハコベ,フタマタハコベ
374848 エダウチヒメハコベ
374858 マンシウフスマ
374859 サハハコベ,サワハコベ,ツルハコベ
374860 ナガバノサワハコベ
374861 オオサワハコベ
374868 ヤクシマハコベ
374880 シラオイハコベ
374881 ケナシシラオイハコベ
374882 イトハコベ
374897 カラフトホソバハコベ
374916 アワユキハコベ
374919 エゾハコベ
374944 エゾノミノフスマ,ナガバツメクサ,ナガバハコベ
374968 アサシラゲ,コハコベ,ハコベ,ハコベラ
374970 ハダカハコベ
374986 ヒメコハコベ
375002 オオヤマハコベ
375007 ミドリハコベ
375018 イワツメクサ
375020 オオイワツメクサ
375033 イヌコハコベ
375036 コウアンフスマ
375041 オオヤマハコベ,オホヤマハコベ
375066 エゾイワツメクサ
375071 エゾオオヤマハコベ,エゾオホヤマハコベ
375074 アミバハコベ,ヒメコハコベ
375078 シコタンハコベ
375093 ミヤマハコベ
375117 ヤマハコベ
375119 アオハコベ
375126 ノミノフスマ
375136 ナガサワハコベ
375187 クサナニワズ
375274 オオバナアザミ,オホバナアザミ
375336 ビャクブ属
375343 ツルビャクブ,ビャクブ,ホドヅラ
375351 タチビャクブ,ビャクブ
375355 タマビャクブ
375359 ビャクブ科
375411 ステナンドリウム属
375510 ステノカルプス属
375516 ステノセレウス属,ラスブニア属
375558 ステノコリネ属
375573 ステノグロッチス属,ステノグロッティス属
375604 ステノメッソン属
375726 オニムラサキ
375764 イヌシバ属
375773 ツノキビ
375774 イヌシバ
375805 コゴメウツギ属
375811 コゴメウツギ属
375817 ウバスカシ,キタバヤマブキ,コゴメウツギ,シロウツギ
375820 シマコゴメウツギ
375821 オオバノコゴメウツギ,カナウツギ,ヤマアサ,ヤマドウシン
375827 ハスノハカズラ属,ハスノハカヅラ属
375833 タマザキツヅラフジ,タマザキツヅ

ラフヂ
375861 ハスノハカヅラ
375870 イヌカヅラ,イヌツヅラ,ケハスノハカヅラ,ハスノハカズラ,ハスノハカヅラ
375884 ハスノハカズラ
375897 コウトウツヅラフヂ
375904 カンバウイ,シマハスノハカズラ,シマハスノハカヅラ
375918 ステファノセレウセ属
375919 モウカンオウ
376022 シタキサウ属,シタキソウ属,ステファノ-ティス属
376063 ゴウショウアオギリ属,ステルキュリア属,ピンポン属
376092 コウトウピンポン,シマピンポン,タイワンピンポン,ルゾンピンポン
376102 マルバゴウシュウアオギリ
376110 ヤツデアオギリ,ヤツデアヲギリ
376144 ピンポン,ピンポンノキ
376150 ピンポン
376184 バクタイカイ
376218 アオギリ科,アヲギリ科
376228 ステレオキラス属
376244 イリオモテムヨウラン
376257 センダンキササゲ属
376350 キバナタマスダレ属,ステルンベルギア属
376356 キバナノタマスダレ
376363 ステトソニア属
376364 コノエ
376389 アケボノハタザオ,ウスエキナズナ,ウスユキナズナ
376400 キリンヤシ属,ステベンソンヤシ属
376405 ステ-ウィア属,ステビア属
376410 アワユキギク
376414 ムラサキステビア
376415 アマハステビア,ステビア
376428 ナツツバキ属
376457 アカキ,アカラギ,コシャラ,コナツツバキ,サルスベリ,サルタノキ,サルナメリ,ヒメシャラ,ヤマチャ
376458 トチュウシャラノキ
376466 サルナメ,シャ,シャラノキ,ナツツバキ,ヤマカリン
376473 トウゴクヒメシャラ,ヒコサンヒメシャラ
376524 オオバハマアサガオ属
376536 アツバアサガホ,オオバハマアサガオ,オホバハマアサガホ
376554 ツルキントラノオ属
376557 ツルキントラノオ
376576 コオロギラン属,コホロギラン属,スチグマトダクチラス属
376578 コオロギラン,コホロギラン
376682 ホザキザクラ属
376683 ナゼザクラ,ホザキザクラ,ホザキノコケザクラ
376687 スティパ属,ハネガヤ属
376690 ヒゲナガコメススキ
376710 ノゲナガハネガヤ
376723 ホクシハネガヤ
376754 オオハネガヤ,ヒロハノハネガヤ
376755 ヒロハノハネガヤ
376877 ナガホハネガヤ
376917 ヤマアラシガヤ
376931 エスパルト
377254 ステベリア属
377286 ストケシア属,ルリギク属
377288 ストケシア,ルリギク
377331 ストマティウム属
377342 ササブネギョク
377343 ウキシロ
377360 ウキブネギョク
377365 ナンシュウ
377433 ストランベ-シャ属
377446 ニイタカカマツカ
377457 ニイタカカマツカ
377550 ゴクラクチョウカ属
377554 オウギバショウモドキ
377563 ルリゴクラクチョウカ
377565 コバストレリッチア
377576 ゴクラクチョウカ
377580 ゴクラクチョウカ科,ストレリチア科
377625 ストレプタンセラ属
377646 ストレプトカリックス属
377652 ストレプトカルプス属
377654 オオヒメギリソウ
377727 ウシノシタモドキ,オオウシノシタ
377747 ゴダチヒメギリソウ
377804 ヒメギリソウ
377838 ウシノシタ
377877 アオイカズラ属,アフヒカヅラ属
377884 アオイカズラ,アフヒカヅラ,ツルツユクサ
377902 タケシマラン属
377919 オタケシマラン
377930 オオバタケシマラン,オホバタケシマラン
377931 コンゴウソウ
377969 マスリマヘガミ属
377973 ヒメノマヘガミ
378024 マスリマヘガミ
378079 イセハナビ属,ストロビランテス属
378166 ムラサキイセハナビ
378231 オキナワスズムシソウ,タシロアイ,タシロアヰ
378305 ウラベニショウ属,ストロマンテ属
378307 ウラベニショウ
378312 ストロンボカクタス属
378313 キクスイ
378347 ストロンギロドン属
378363 キンリュウカ属
378375 キンリュウカ
378386 キンリュウカ
378401 ストロファンタス
378487 ストロフォカクタス属
378488 ストロフォカクタス
378633 ストリキニ-ネノキ属,ストリクノス属
378836 ストリキニ-ネノキ,バンボクベツ,マチン
378991 リウノヒゲモ,リュウノヒゲモ
379071 スティリディウム科
379076 スティリディウム属
379296 エゴノキ科
379300 エゴノキ属
379304 アメリカエゴノキ
379312 アンソクカウ,アンソクコウノキ
379347 タイワンエゴノキ,ヘンリエゴノキ,シリエゴノキ
379348 コウラジロエゴノキ,ヒメウラジロエゴノキ
379357 コウトウエゴノキ
379374 イチシ,イッシ,イッチャ,エゴ,エゴノキ,クロジシャ,コハゼ,コメジシャ,コメミズ,コヤス,ジサノキ,シチャノキ,ズサ,チサノキ,チシャノキ,チナイ,チナシ,チョウメン,ヤマチサ,ロクモギ,ロクロギ
379380 ホソバエゴノキ
379381 オオバエゴノキ
379382 ヒメエゴノキ
379383 シダレエゴノキ
379384 ベニガクエゴノキ
379388 オオバケエゴノキ
379389 コウトウエゴノキ
379408 タカサゴエゴノキ
379414 オオバジシャ,オオバヂシャ,オナガシ,オホバチシャ,ハクウンボク,ハビロ,ヲナガシ
379419 セイヨウエゴノキ
379423 スマトラエゴノキ
379450 コハクウンボク
379451 ウラジロコハクウンボク
379457 ウラジロエゴノキ
379486 マツナ属
379499 シマハママツナ,ハママツナ,リウキウマツナ
379509 モウコマツナ
379531 マツナ
379542 サンゴジュマツナ,シチメンサウ,シチメンソウ,ミルマツナ,ムレ
379551 ヒロハマツナ
379553 ハママツナ,ヒジキアカザ
379598 ホソバハママツナ,ホソバママツナ
379634 スブマッカナ属
379638 オウセンギョク
379650 ハリナズナ
379715 スルコレブティア属

379766 キメラク
379771 ササキナゴラン
379794 オホバツゲ
380045 ヤエヤマスズコオジュ
380046 シキクンソウ
380066 デザートピー
380105 センブリ属
380117 アリサンセンブリ,シマセンブリ,セイバンセンブリ
380128 ホシナシアケボノソウ
380131 アケボノサウ,アケボノソウ,ヨシノシヅカ
380181 ツリハコベ
380184 サワセンブリ,ムラサキセンブリ
380205 アカボシアケボノソウ
380234 センブリ,タウヤク,トウヤク
380235 シロバナセンブリ
380236 ハマセンブリ
380237 ヤエセンブリ
380238 ヒロハセンブリ
380264 ニヒタカセンブリ
380267 シマアケボノソウ
380269 マンシウアケボノソウ
380303 ミヤマアケボノソウ
380319 ムラサキセンブリ
380354 シンテンアケボノサウ
380367 シノノメサウ,シノノメソウ
380368 ジュウジアケボノソウ,ヘツカリンドウ,リュウキュウアケボノソウ
380369 ホシナシヘツカリンドウ
380376 チシマセンブリ
380377 シロバナチシマセンブリ
380378 フイリチシマセンブリ
380379 タカネセンブリ
380380 シロバナタカネセンブリ
380381 ハッポウタカネセンブリ
380382 ヤケイシセンブリ
380383 エゾタカネセンブリ
380385 チシマセンブリ
380389 イヌセンブリ
380391 タカサゴセンブリ,トウサンセンブリ
380400 チョウセンァケホノソゥ
380411 オオバセンブリ
380434 トウミズキ
380439 タカネミズキ
380440 イシヅチミズキ
380444 テウセンミヅキ
380461 クマノミズキ,クマノミヅキ,サハミヅキ,サワミヅキ,ミズキ,ミズハシカ,ヤマトミズキ
380499 セイヨウミヅキ
380514 チョウセンミズキ
380522 マホガニー属
380526 メキシコマホガニー
380527 オオバマホガニー,ホンジュラスマホガニー
380528 マホガニー
380553 スジミココヤシ属
380556 コバナスシミココヤシ
380559 アリクリヤシ
380561 フラジルヒメヤシ
380641 シンベゴニア属
380734 シンフォリカルポス属
380736 セッコウボク
380769 シンフィアンドラ属
380772 ハナブサソウ
380781 シンフィグロッサム属
380791 イヌトキンソウ属
380792 イヌトキンソウ
380872 キダチコンギク
380924 ヤマヨナメ
380930 アメリカシオン,ネバリノギク
381024 ヒレハリサウ属,ヒレハリソウ属
381026 オオハリサウ,オオハリソウ
381035 コンフリー,ヒレハリサウ,ヒレハリソウ
381042 コンフリー
381067 ハイノキ科,ハヒノキ科
381070 ザゼンサウ属,ザゼンソウ属
381073 アメリカザゼンソウ,ザゼンサウ,ザゼンソウ,ダルマサウ,ダルマソウ
381078 ナベクラザゼンソウ
381079 ヒメザゼンソウ
381080 フイリヒメザゼンソウ
381081 ミドリヒメザゼンソウ
381082 ザゼンソウ,タルマソウ,フイリザゼンソウ
381090 ハイノキ属,ハヒノキ属
381104 キライシハヒノキ,ドイフタギ,ニイタカハイノキ,ニヒタカハヒノキ,リュウキュウハイノキ
381124 ムニンクロキ
381137 サワフダギ,ニシゴリ,ルリミノウシコロシ
381143 アオバノキ,クロキ,コウトウカンザブロウ,コウトウハイノキ,コニシハイノキ
381148 オホバカンザブラウ,カンザブラウノキ,カンザブロウノキ,コニシカンザブラウ,コニシハイノキ,コニシハヒノキ,ヒメカンザブラウノキ
381149 アオバノキ,アヲバノキ,コウドウカンザブラウ,コウトウハヒノキ
381154 オホババハヒノキ,クラルハヒノキ,クラルフタギ,ナガババハヒノキ,ランダイフタギ
381158 タンナサワフタギ
381206 アマシバ,オホバナシロバヒ,シマハヒノキ,ズイシャリャウシロバヒ,ズイシャリャウハヒノキ,スヰシャリャウシロバヒ,スヰシャリャウハヒノキ,タイワンハヒノキ
381217 ミミズノマクラ,ミミズバイ,ミミスベリ,ミミヅバヒ,メメズリ
381235 ヘイシャナハヒノキ,リセキシロバヒ
381252 クロキ
381256 ウチダシクロギ
381272 アリサンハヒノキ,ケハヒノキ,シロバイ,シロバヒ
381288 アオバナハヒノキ
381289 イリオモテハイノキ
381291 クロキ
381310 ホソエダハヒノキ,ヤマシマハヒノキ,ワタメフタギ
381323 ハイノキ
381325 オモゴハイノキ
381326 ヒロハハイノキ
381327 ナカハラクロキ,リュキュクロキ
381330 コメバカンサビラウ
381341 カラサワフタギ,クロミノニシゴリ,コサハフタギ,シロサハフタギ,シロミノサハフタギ,シロミノサワフタギ,タイワンサワフタギ,トゲハマサジ,ニシゴリ
381355 ミヤマシロバヒ
381356 チチジマクロキ
381371 ナガバクロバイ
381390 クロキ,タマザキハヒノキ,モチバハヒノキ,ヤマキ
381404 ドスン,ミヤマシロバイ,ユワンギ,ルスン
381412 ビハバハヒノキ,ビハバミミズバヒ,ヤンバルミミズバイ
381423 オホガクフタギ,クロバイ,クロバヒ,コバノシロバヒ,サウマハヒノキ,ソウザンハヒノキ,ソメシバ,チクトウハヒノキ,トモシバ,ヤエヤマクロバイ
381429 オニクロギ,ヒロハノミミズバイ
381437 クラルハヒノキ,クラルフタギ
381452 ガンピバハヒノキ,ガンピハヒノキ
381628 シナプトフィルム属
381806 ボラヤシ属
381807 パナボラヤシ
381808 フシザキサウ属,フシザキソウ属
381809 フシザキサウ,フシザキソウ
381813 ヤブレガサ属
381814 コヤブレガサ,ホソバヤブレガサ
381815 タンバヤブレガサ
381818 タカサゴヤブレガサ
381820 ヤブレガサ
381823 ヤブレガサモドキ
381824 ヒロハヤブレガサモドキ
381835 ハナホシクサ
381855 シンゴニューム属
382067 ヤマボクチゾク属,ヤマボクチ属
382069 オニヤマボクチ,チョウセンヤマボクチ,ヤマボクチ,ヤマボクチゾク
382071 カラフトヤマボクチ
382073 ハバヤマボクチ

382076 キクバヤマボクチ
382078 ヤマボクチ
382079 オヤマボクチ
382114 ハシドイ属
382132 コバノシナハシドイ
382153 ヒマラヤハシドイ
382220 オニハシドイ
382241 ペルシャハシドイ
382247 トネリバハシドイ
382257 コハシドイ,シナハシドイ,シマハシドイ
382266 コバノハシドイ,チャボハシドイ
382268 ウスゲハシドイ,オオバハシドイ,チョウセンハシドイ
382276 キンツクバネ,ドスナラ,ハシドイ,ヤチカバ
382277 マンシウハシドイ,マンシュウハシドイ,マンショウハシドイ
382278 ペキンハシドイ
382281 ケオオバハシドイ
382282 タチハシドイ
382308 タケイマハシドイ
382329 スオウボク,スワウボク,ハナハシドイ,ムラサキハシドイ,ライラック
382368 シロライラック
382371 ヤエライラック,ヤヘライラック
382383 ハナハシドイ
382388 ウンナンハシドイ
382420 シオニラ
382465 アデク属
382474 ミズレンブ
382477 クロ-プノキ,チヤウジ,チョウコウ,チョウジ,チョウジノキ
382499 アデク
382512 ヤマアデク
382514 ヒメフトモモ
382522 ムラサキフトモモ
382533 サイミャクアデク
382535 イヌアデク,タイワンアデク,ヤマツゲモドキ
382582 フ-トノキ,フトモモ,ホタウ,ホトウ
382589 クスクスアデク,ゴウシュンアデク
382601 ナガハアデク
382606 マレ-フトモモ
382649 アヅバヅゲトモモ,クワセウッゲトモモ,シマツゲトモモ
382660 オオフトモモ,オホフトモモ,ジャワフトモモ,レンブ
382672 コンボウアデク,スチバアデク
382704 タベブ-イア属
382727 サンイウクワ属,サンユウカ属
382822 マニラサンユウカ
382908 タシロイモ属
382912 クロバナタシロイモ
382914 オオタシロイモ
382923 タシロイモ
382934 タシロイモ科
382991 タキンガ属
382992 ロウエンダイ
383007 タデハギモドキ
383009 タデハギ
383027 オビバナヤシ属
383054 クモラン属,タエニオフィラム属
383064 クモラン
383087 センジュギク属,マンジュギク属
383090 センジュギク,テンリンクワ,マンジュギク
383097 ニオイマンジュギク
383100 シオザキソウ
383103 クジャクサウ,クジャクソウ,コウオウソウ,コウワウサウ,フジギク,ホウワウサウ,ヰンゲサウ
383107 ヒメクジャクソウ,ヒメコウオウソウ
383119 ヒメトケンラン属
383127 アオラン,アフヒラン
383132 ナガバアフヒラン
383153 セメトケンラン
383188 タイワンスギ属
383189 アサン,タイワンスギ
383210 オニヤマボクチ属
383213 オニヤマボクチ
383236 ネッタイモクレン属
383293 ハゼラン属
383324 シュッコンハゼラン
383345 ハゼラン
383397 ギョリュウ科
383404 タマリンヅス属,チョウセンモダマ属,テウセンモダマ属
383407 タマリンド,チョウセンモダマ,テウセンモダマ,ラボウシ
383412 ギョリウ属,ギョリュウ属
383437 アフリカギョリュウ
383469 ギョリウ,ギョリュウ,サッキギョリュウ
383553 マナギョリュウ
383595 ヒナギョリュウ
383622 ギョリュウ
383690 ヨモギギク属
383874 エゾヨモギギク,ヨモギギク
383876 エゾノヨモギギク,エゾヨモギギク
383887 イハユキノシタ属,イワユキノシタ属
383890 イハユキノシタ,イワユキノシタ,ヤマユキノシタ
383920 タンクヮナ属
384420 タンポポ属
384430 シロタンポポ,シロバナタンポポ
384434 ミヤマタンポポ
384435 シロウマタンポポ
384447 ヤマザトタンポポ
384453 アジアタンポポ,ホソバアジアタンポポ
384463 カバイロタンポポ
384473 シナタンポポ
384479 ナノハタンポポ
384492 ケンサキタンポポ
384505 ウスイロタンポポ,ケイリンシロタンポポ
384509 ホクシタンポポ
384520 オクウスギタンポポ
384530 セイタカタンポポ
384538 ァツバタンポポ
384548 コウアンヒメタンポポ
384554 タカサゴタンポポ
384572 カイゲンタンポポ,ミズタタンポポ
384573 キビシロタンポポ
384591 カンサイタンポポ,カンセイタンポポ,クワンセイタンポポ
384601 ケトイタンポポ
384605 コタンポポ
384606 ゴムタンポポ
384610 クザカイタンポポ
384614 アカミタンポポ,キレハアザミタンポポ
384618 アカジクタンポポ
384629 ミョウトウタンポポ
384663 オキタンポポ
384681 モウコタンポポ
384714 エゾタンポポ,クモマタンポポ,ショクヨウタンポポ,セイヤウタンポポ,セイヨウタンポポ,タカネタンポポ,タンポポ,ミヤマタンポポ
384718 オオヒラタンポポ
384719 コウライキバナタンポポ
384734 クシバタンポポ
384736 アライトタンポポ
384741 ウスジロタンポポ
384742 シナノタンポポ
384744 トウカイタンポポ
384746 アズマタンポポ,ガントウタンポポ,グジナ,タナ,タンポポ,フジナ,ムジナ
384749 オダサムタンポポ,ネッカタンポポ
384760 マンシウシロタンポポ
384802 シコタンタンポポ
384834 トガクシタンポポ
384847 キツネタンポポ
384848 ヨゴレタンポポ
384851 エゾタンポポ
384862 ヤツガタケタンポポ
384867 タカネタンポポ,ユバリタンポポ
384868 オオタカネタンポポ
384909 ギョクシンカ属,ギョクシンクワ属
384950 ギョクシンカ,コギョクシンクワ,ホソバギョクシンクワ
384951 ヤエヤマギョクシンカ
385028 シマギョクシンカ
385049 ギョクシンクワ,コウトウギョクシンクワ

385146 タバレーシア属
385151 レイショウカク
385175 アチイ科,イチイ科,イチヰ科
385186 マッグミ属
385189 タイワンマッグミ
385202 カラスノウエキ,マッグミ,マッボヤ
385204 マルバマッグミ
385208 リトウヤドリギ
385211 オホバフウジュヤドリギ
385221 アナシベヤドリギ,シャクナゲヤドリギ,ニンドウバノヤドリギ
385225 シナヤドリギモドキ
385230 リトウヤドリギ
385237 チャヤドリギ
385250 スギ科
385253 スイショウ属,スヰショウ属,ヌマスギ属,ラクウショウ属
385262 アメリカズイショウ,アメリカスヰショウ,ニレッバズイショウ,ニレッバスヰショウ,ヌマスギ,ラクウショウ,ラクエフスギ
385264 タチラクウショウ
385301 イチイ属,イチヰ属
385302 オウシュウイチイ,セイヨウイチイ,ゼイヨウイチイ,ヨーロッパイチイ
385355 アララギ,イチイ,イチヰ,オンコ,スハウノキ,スホウノキ,ミネスホウ
385356 オウゴンキャラ
385363 キミノオンコ
385370 キャラボク
385372 キャラボク
385405 タイワンイチイ,タイワンイチヰ
385407 タイワンイチイ
385461 テコマ属
385505 テコマリア属,ヒメノウゼンカズラ属
385508 テコマリア・カペンシス,ヒメノウゼンカズラ,ヒメノウゼンカゼラ
385520 テコフィレーア属
385529 チークノキ属,チーク属,ナナバケンシダ属
385531 ジャテイ,チーク,チークノキ
385616 オオトウカセン
385734 テロペア属
385738 テロペア
385752 ヤライコウ
385771 ハリセンボン
385870 テプロカクタス属
385872 バンショウデン
385876 ヒメムサシノ
385878 ハッコッジョウ
385880 タコッボ
385881 ダイシュウチョウ
385883 サイオウウチワ
385892 エフデギク,コウリンクワ,シラゲコオリンカ,タカネコウリンギク,ベニニガナ
385894 コウリンカ
385898 キバナコウリンカ
385902 ミヤマオグルマ
385903 オカオグルマ,オカヲグルマ,キタオグルマ,サハオグルマ,サハヲグルマ,ヲカヲグルマ
385908 ミズオグルマ
385913 イワオグルマ
385914 サワオグルマ
385921 シマコウリンカ,シマコウリンクワ,タイトウコウリンクワ,タカネコウリンカ
385926 ナンバンクサフジ属,ナンバンクサフヂ属
385985 キダチナンバンクサフジ,シロバナナンバンクサフジ
386210 タイワンクサフジ
386257 クロハナナンバンクサフジ,クロハナナンバンクサフヂ,ナンバンクサフジ,ナンバンクサフヂ
386410 ナンバンヤブマメ
386448 コバテイシ属,モモタマナ属
386500 コハテイシ,コバテイシ,シマボウ,モモタマナ
386504 ミロバラン,ミロバランノキ
386590 テリハモモタマナ
386689 モクコク属,モッコク属
386696 イクキ,モクコク,モッコク
386698 ヒメモッコク
386847 スイセイジュ科
386851 スイセイジュ
386992 フブキバナ属
387032 フブキバナ
387136 イセハナビ
387156 ツルナ属
387221 ツルナ,ハマヂシャ
387321 テトラミクラ属
387343 テトラネマ属,メキシコジギタリス属
387346 メキシコジキタリス
387428 カミヤツデ属,ツウダッボク属,テトラパナックス属
387432 カミヤツデ,ツウサウ,ツウソウ,ツウダッボク
387740 ミッバカシラ属
387766 ミッバカズラ
387769 ビヨウリッヤブガラシ,ミッバカシラ
387774 ドクブダウ,ミッバビンボウカズラ,ミッバビンボオズル,ミッバビンボフカヅラ
387802 オモロカズラ
387825 アリサンヤブガラシ,タイワンヤブガラシ
387853 ブドウカズラ
387989 ニガクサ属
388114 イヌチョモキ,ニガクサ
388298 チョウセンニガクサ,マンシウニガクサ
388299 エゾニガクサ,ヒメニガクサ
388300 イヌニガクサ
388303 コニガクサ
388307 ツルニガクサ
388357 リュウキュウスガモ属
388360 リュウキュウスガモ
388380 ターリア属,ミズカンナ属
388385 ミズカンナ
388397 カラマツサウ属,カラマツソウ属
388399 シギンカラマツ
388412 ヒメカラマツ
388417 ハイカカラマツソウ
388423 カラマツサウ,カラマツソウ,マンセンカラマツ
388432 ハルカラマツ
388463 ハスノハカラマツ
388503 ミヤマカラマツ
388509 イトバカラマツ
388510 ニオイカラマツ
388511 チャボカラマツ
388536 トウハスノカラマツ
388541 ニヒタカカラマツ
388546 ツケシカラマツ
388583 アキカラマツ,エゾアキカラマツ
388584 コカラマツ
388594 モリカラマツ
388614 ミユキカラマツ
388615 ハナカラマツ
388617 ヤチマタカラマツ
388645 シキンカラマツ
388654 エゾカラマツ,ミヤマアキカラマツ
388667 ノカラマツ
388668 キカラマツ,ノカラマツ
388669 ノカラマツ
388673 カラフトカラマツ,ツリフネカラマツ
388680 エダハリカラマツ
388703 ミヤマカラマツ
388710 タカサゴカラマツ,ナカハラカラマツ
388909 ヒメウシノシッペイ
389046 ツバキ科
389103 テコステレ属
389122 セラシス属
389181 ヤマトグサ科
389183 ヤマトグサ属
389185 シンヤマトグサ
389186 ヤマトグサ
389203 テロカクタス属
389205 ダイトウリョウ
389216 テンコウ
389218 ハクシギョク
389221 カクソウマル
389222 オムリジシ

389258 テリミ-トラ属
389307 メガルカヤ属
389318 シマシバ,タイワンメガルカヤ
389355 カルカヤ,メガルカヤ
389399 カカオノキ属,カカオ属
389404 カカオ,カカオノキ,ココアノキ,チョコレ-トノキ
389505 センダイハギ属
389518 クソエンドウ
389533 ホソバセンダイハギ
389540 センダイハギ
389547 アメリカセンダイハギ
389570 エゾツツジ属
389572 エゾツツジ
389573 シロバナエゾツツジ
389575 アラゲエゾツツジ
389577 クモマツツジ
389592 カナビキサウ属,カナビキソウ属
389638 カナビキサウ,カナビキソウ
389769 ナカバノカナビキサウ,ナカバノカナビキソウ
389844 カマヤリソウ
389848 カマヤリサウ,カマヤリソウ,ハヒカナビキサウ,ハヒカナビキソウ
389857 イワカナビキソウ
389928 タウユウナ属,トウユウナ属
389946 サキシマハマボウ,サキシマハマボオ,シマアオイ,シマアフヒ,タイワンアオイ,タイワンアフヒ,タウユウナ,トウユウナ
389967 キバナキョウチクトウ属,キバナケフチクタウ属
389978 オオバナキョウチクトウ,オホバナケフチクタウ,キバナキョウチクトウ,キバナケフチクタウ,メキシコキョウチクトウ
390097 タヌキノショクダイ
390101 キリシマタヌキノショクダイ
390107 オホスズメウリ属
390128 オオスズメウリ,オホスズメウリ,キバナカラスウリ,ノウリ,ヒメカラスウリ
390175 シマウリ,テンツキノウリ
390202 グンバイナズナ属,グンバイナヅナ属
390213 グンバイウチハ,グンバイナズナ,グンバイナヅナ
390223 カラフトグンバイ,シャセシナズナ
390237 タカネグンバイ
390238 タカネグンバイ
390325 トンプソネラ属
390451 ホソエクマデヤシ属
390460 コダネクマデヤシ
390462 モラスクマデヤシ
390463 コバナクマデヤシ
390479 トリクサントセレウス属
390481 ギンイチュウ
390482 セイリョウテン
390486 カヤラン属,トリクススペルムム属,トリックススパーマム属
390502 ハガクレナガミラン
390503 タイワンフウラン
390506 カヤラン
390518 クスクスフウラン
390522 オビガタセキコク
390527 ケイタオフウラン
390538 キントラノオ属
390542 キントラノオ
390548 トリプトメネ属
390551 クロイハザサ属
390553 クロイワザサ
390567 クロベ属,コノテガシハ属,ヒボ属
390579 ニオイネズコ,ニホヒネズコ
390587 ニオイヒバ,ニホヒヒバ
390588 オウゴンコノテ
390599 オウゴンニオイヒバ
390628 イトヒバ
390645 アメリカネズコ,ベイスギ
390660 クロビ,クロベ,ゴラウヒバ,ゴロウヒバ,ネズコ,ヒメアスナロ
390661 シセンネズコ
390673 アスナロ属
390680 アスナロ,アスヒ,アテ,アテアスビ,シロビ,タニヒノキ,ヒノキ,ヒバ
390681 ヒメアスナロ
390682 フイリアスナロ,フイリヒバ
390684 アスナロ,ヒノキアスナロ
390692 ツンベルギア属,ヤバズカズラ属,ヤバズカッラ属,ヤバネカズラ属
390699 ムラサキヤハズカズラ
390701 ヤハズカズラ,ヤバズカッラ,ヤバネカズラ
390759 キンギョボク,コダチヤハズカズラ,ツンベルギア・エレクタ
390767 カオリカズラ
390778 カバイロヤハズカズラ
390787 ウリバロ-レルカヅラ,ベンガルヤハズカズラ
390822 ゲッゲイカズラ,ゲッゲイカヅラ,ルリバナカヅラ,ロオレルカズラ,ロ-レルカズラ,ロ-レルカヅラ
390916 ツ-ニア属,ツニア属
390917 ホザキカクラン
391030 ジンチョウゲ科,ヂンチャウゲ科
391061 イブキジャカウサウ属,イブキジャカウソウ属,イブキジャコウソウ属
391167 スナジジャコウソウ
391225 イブキジャコウソウ
391264 ケジャコウソウ
391303 ツメジャコウソウ
391347 イハジャカウサウ,イブキジャカウサウ,イブキジャコウソウ
391348 シロバナイブキジャコウソウ
391349 ハマジャコウソウ
391351 イブキジャコウソウ
391352 ヒメヒャクリコウ
391353 イブキジャコウソウ
391356 イワジャコウソウ
391365 ヨウシュイブキジャコウソウ
391369 ホソバタイム
391397 キダチヒャクリコウ,キダモヒャクリコウ,タイム,タチジャコウソウ
391477 ヤダケガヤ属
391487 ヤダケガヤ
391520 ズダヤクシュ属,ヅダヤクシュ属
391524 ズダヤクシュ,ヅタヤクシュ
391566 チボウキナ属
391577 コバノシコンノボタン,シコンノボタン
391622 チグリジャ属,トウュリ属,トラフユリ属
391627 チグリジャ,チグリジャ・パボニア,トウフュリ,トウュリ,トラフュリ
391646 シナノキ属
391649 アメリカシナノキ
391661 アム-ルシナノキ
391691 オウシュウシナノキ,フユボダイジュ
391706 オウシウボダイジュ,セイヤウシナノキ,セイヨウシナノキ
391717 コバノシノキ
391739 シナノキ,マダノキ,モウダ
391740 グンバイシナノキ
391741 チャボシナノキ
391742 アラカワシナノキ
391743 ケナシシナノキ,シコクシナノキ
391749 ヘラノキ
391760 オホシナノキ,マンシウボダイジュ,マンシュウボダイジュ
391764 ツクシボダイジュ
391766 アオシナ,オオシナノキ,オオバシナノキ,オオバボダイジュ,オホバボダイジュ,ヤイニペシ
391768 モイワボダイジュ
391773 オオバシナノキ,ボダイジュ
391781 モウコシナノキ
391795 オリバ-シナノキ
391817 ナツボダイジュ
391832 シベリアシナノキ
391850 ヒエイボダイジュ
391860 シナノキ科
391904 シラネニンジン
391906 ヒロハシラネニンジン
391907 ホソバシラネニンジン
391908 ヒメシラネニンジン
391911 イブキゼリモドキ
391916 ツシマノダケ
391918 アズマツメクサ属,アヅマツメクサ属
391923 アズマツメクサ,アヅマツメクサ
391966 ティランジア属

391982 コビトアナナス
391985 ラッキョウアナナス
392005 イトバアナナス
392006 ウチワアナナス
392015 タチバナアナナス,ハナアナナス
392026 ハナアナナス
392039 モスボール
392052 サルオガセモドキ
392248 キンクオラン
392346 ティプラリア属,ヒトツボクロ属
392349 ヒトツボクロ
392350 ヒトツボクロモドキ
392380 ティシュレリア属
392381 ギンキョウ
392405 チタノプシス属
392406 テンニョ
392408 テンニョサン
392410 テンニョセン
392412 テンニョハクセン
392414 テンニョハイ
392416 テンニョエイ
392418 テンニョカン
392423 マツムラサウ属,マツムラソウ属
392424 タカサゴミヅヒキ,マツムラサウ,マツムラソウ
392428 チトニア属,ニトベギク属
392432 ニトベギク
392434 ヒロハヒマワリ,メキシコヒマワリ
392556 サルカケミカン属
392559 サルカケミカン
392598 チシマゼキショウ属,チャボゼキシャウ属,チャボゼキショウ属
392603 チシマゼキショウ,チャボゼキシャウ,チャボゼキショウ
392622 イワショウブ
392786 イトバボウフウ
392822 チャンチン属
392827 カランタス
392836 シアンモック
392841 キヤンチン,クモヤブリ,スリアン,チャンチン,ヒヤンチン,ライデンボク
392847 スレン
392880 ウリクサ属,ツルウリクサ属,トレニア属
392884 ゲンジバナ,コバナツルウリクサ,ツルウリクサ,ヒメトレニア
392895 ツルウリクサ
392903 キバナノウリクサ,ホクトウリクサ
392904 タイワンミゾホホヅキ
392905 トレニヤ,ハナウリグサ,ムラサキミゾホホヅキ
392948 コミゾホオズキ,コミゾホホヅキ
392959 ヤブジラミ属
392992 ヤブジラミ
392993 セイヨウヤブジラミ
392996 タマヤブジラミ
393004 オヤブジラミ
393047 カヤ属
393075 ヒダリマキガヤ
393077 カエ,カヘ,カヤ,シロガヤ,ホンガヤ
393079 コツブガヤ
393080 ヒダリマキガヤ
393081 ハダカガヤ
393087 チャボガヤ
393100 ホソバドジョウツナギ
393185 ヒネリガヤツリ属
393194 キンガヤツリ,ムツオレガヤツリ,ムツヲレガヤツリ
393205 イガガヤツリ,キンガヤツリ,ムツオレガヤツリ
393233 トウメア属
393235 ツキノドウ
393240 ハマムラサキ属
393242 シロマラフクラギ,ハマムラサキノキ,モンパノキ
393267 キダチスナビキ
393269 スナビキサウ,スナビキソウ,ハマムラサキ,ミヅナ
393387 ヂギク属
393402 ヂギク
393466 ウルシヅタ,カキウルシ,ツウルシル,ツタウルシ
393479 トウハゼ,ハジノキ,ハゼ,ハゼウルシ,ハゼノキ,ラフノキ,リウキウハゼ,リュウキュウハゼ,ロウノキ
393485 ハゼ,ハゼウルシ,ヤブウルシ,ヤマハゼ
393488 ヤマウルシ
393491 ウルシ,ウルシノキ
393604 ユウキリソウ属
393608 ユウギリソウ
393613 テイカカズラ属,テイカカヅラ属
393616 セキダカズラ,チョウジカズラ,テイカカズラ,テイカカヅラ,マサキズル,マサキノカズラ,ムニンテイカカヅラ
393619 チョウセンテイカカズラ
393621 セキダカヅラ,テイカカズラ,テイカカヅラ,マサキヅル
393623 チョウジカズラ
393650 ヒメテイカカズラ
393656 オキナワテイカカズラ,ムニンテイカカズラ,リュウキュウテイカカズラ
393657 ケテイカカズラ,ケテイカカヅラ,タイワンテイカカヅラ,タウケフチクタウ,トウキョウチクトウ,トウテイカカズラ
393664 ケテイカカズラ,ケテイカカヅラ
393666 ニシキケテイカカズラ
393677 チョウジカズラ
393803 シュロ属
393809 アカヅク,シュロ,シロ,スロ,スロノキ,タウジュロ,トウジュロ,リウキウジュロ,ワジュロ
393813 チュ-サンヤシ,マ-チュ-スジュロ
393815 タキルジュロ
393816 タウジュロ,トウジュロ,リュウキュウジュロ
393882 トラキメ-ネ属
393883 ソウイロレ-スソウ,ソライロゼリ,ソライロレ-スソウ,ディディスクス
393987 シマフムラサキツユクサ属,シマムラサキツユクサ属,ゼブリ-ナ属,ムラサキツュクサ属
393989 シラフオオトキワツユクサ
393999 シキンツユクサ,フイリシキンツユクサ
394007 シキンツユクサ
394021 オオトキワツユクサ,シロフハカタツユクサ,ノハカタカラクサ
394024 シロフハカタカラクサ
394047 カサネオウギ
394053 ムラサキツユクサ
394076 アカオモト,シキンラン,ムラサキオモト
394088 アンダ-ソンムラサキツユクサ,オオムラサキツユクサ
394093 シマフムラサキツユクサ,シマムラサキツユクサ,ハカタガラクサ,ハマタガツラ
394257 バラモンジン属
394321 トライバラモンジン
394327 カキナ,サルシフィ,バラモンジン,ムギナデシコ
394333 キバナザキバラモンジン,キバナバラモンジン,キバナムギナデシコ,バラモンギク
394380 イガボシバ
394390 シラミシバ
394412 ヒシ属
394433 タウビシ,ツノビシ
394436 トウビシ,ヒシ
394459 ヒメビシ
394463 オタフクヒシ,ビシ
394471 イボビシ
394485 シリブトビシ
394496 マンシウヒメビシ
394500 オニビシ
394513 オトコビシ,オニビシ,ヲトコビシ
394516 コオニビシ
394526 マンセンビシ
394554 ヒシ科
394559 ヒシモドキ属
394560 ヒシモドキ,ムシヅル
394583 モミジカラマツ属
394626 ウラジロエノキ属
394628 キリエノキ,コバフンギ,ホソバウラジロエノキ
394635 キリエノキ,コバフンギ,ホソバウ

ラジロ
394656 ウラジロエノキ
394664 ウラジロエノキ
394807 ミズオトギリ属,ミヅオトギリ属
394811 ミズオトギリ,ミヅオトギリ
394814 アメリカミズオトギリ
394903 スベリヒユモドキ
394943 オギ
394947 トリアス属
395064 ハマビシ属
395082 オオバナハマビシ,オホバナハマビシ
395146 ハマビシ
395164 シロミミズ属
395581 トリコカウロン属
395616 トリコケントルム属,トリコセントラム属
395630 トリコセレウス属
395634 テンシュカク
395636 キンコウミュウ
395637 キンジョウマル
395638 モウカンチュウ
395639 キンハイリュウ
395642 キョレイマル
395643 ショウナンマル
395644 リショウリュウ
395645 ダィリョウチュウ
395647 コウリョクチュウ
395648 コキンチュウ
395650 キタカ
395651 キンセン
395653 キンリョウ,ハクコウマル
395655 コウモウチュウ
395656 タカハネカク
395658 ホクトカク
395660 コクホウ
395661 ユウレツリュウ
395662 トゥリコケーロス属
395708 ルリホホヅキ属
395791 トリコディアデア属
395855 クラルクレソラン属,トリコグロッチス属,トリコグロッティス属
395865 イリオモラフン,ニュウメンラン,ニューメンラン
395868 クラルクレソラン
395872 トリコゴニア属
395876 トリコゴニオプシス属
395879 トリコジネ属
395897 ドリコレーナ属
396005 シヤマサギスゲ,ヒメワタスゲ
396010 ミネハリイ
396025 ニイタカハリイ,ニヒタカハリイ,ニヒタカハリキ
396150 カラスウリ属
396151 ケカラスウリ,ゴーダー・ビーソ,ヘビウリ
396170 カラスウリ,キツネノマクラ,グドウジ,タマズサ,タマヅサ,ムスビジャウ
396177 カラスウリ
396197 イモノハカラスウリ
396198 イシガキカラスウリ
396203 キカラスウリ,キガラスウリ
396210 カラスウリ,チョウセンカラスウリ
396213 オオカラスウリ,フサウケバカラスウリ
396224 リュウキュウカラスウリ
396226 モミヂカラスウリ
396238 ケカラスウリ
396240 ムニンカラスウリ
396254 コウトウカラスウリ,シマカラスウリ
396280 オオカラスウリ
396464 トリコトシア属
396583 ホトトギス属
396584 ヤマジノホトトギス
396588 タイワンホトトギス,ホソバホトトギス
396598 ハナホトトギス,ホトトギス
396599 ケタイワンホトトギス
396601 タマガハホトトギス,タマガワホトトギス,マンシウホトトギス
396602 ジョウロウホトトギス,トサノジョウロウホトトギス
396603 キイジョウロウホトトギス
396604 ヤマホトトギス
396607 チャボホトトギス
396608 タカクマホトトギス
396610 キバナノツキヌキホトトギス,ツキヌキホトトギス
396614 ツルタイワンホトトギス
396697 コトブキギク
396780 ツマトリサウ属,ツマトリソウ属
396786 ツマトリサウ,ツマトリソウ
396789 コツマトリソウ
396806 シャジクソウ属,シャヂクサウ属
396822 クシバツメクサ
396823 トガリバツメクサ
396828 シャグマハギ
396829 シロバナシャグマハギ
396854 クスダマツメクサ
396859 チゴツメクサ
396882 コメツブツメクサ
396905 ツメクサダマシ
396908 ヤマブキツメクサ
396917 ダンゴツメクサ
396926 ビロードアカツメクサ
396930 タチオランダゲンゲ
396934 ベニバナウマゴヤシ,ベニバナツメクサ
396956 アミダガサ,カタワグルマ,シャジクソウ,シャヂクサウ,ボサツソウ
396959 シロバナシャジクソウ
396967 オオバノアカツメクサ,オホバノアカツメクサ
397019 アカツメクサ,ムラサキツメクサ
397020 シロバナアカツメクサ
397043 オランダゲンゲ,オランダマゴヤシ,クローバー,シロツメクサ,ツメクサ
397045 オオシロツメクサ
397047 モモイロシロツメクサ
397050 ヒナツメクサ
397052 オオヒナツメクサ
397053 コバナヒナツメクサ
397096 テマリツメクサ
397097 ハクモウアカツメクサ
397106 ジモグリツメクサ
397116 フウセンツメクサ
397121 ツバナツメクサ
397130 トックリツメクサ
397142 シバナ属
397144 シバナ
397159 ウミゼキシャウ,ウミゼキショウ,ウミニラ,オオシバナ,シバナ,ハマシバ,ヒロハナミサキサウ,ヒロハナミサキソウ,マルミノシバナ,モシオグサ
397165 ホソバシバナ,ホソバノシバナ,ミサキサウ,ミサキソウ
397186 レイリョウカウ属,レイリョウコウ属
397208 レイリョウカウ,レイリョウコウ
397229 コロハ
397314 トゥリゴニディウム属,トリゴニジューム属
397380 タビラコ属,ミズタビラコ属
397386 ミズタビラコ
397387 コシジタビラコ
397410 アリサンタビラコ,シマタビラコ
397411 ウチダシタビラコ
397420 ツルカメバソウ
397441 ヤチムラサキ
397451 カハラケナ,キウリゲサ,キウリナ,キュウリゲサ,タビラコ
397452 トウワスレナグサ
397459 キヌゲカメバソウ,ケカメバソウ,ケルリソウ
397462 キヌゲカメバソウ,チョウセンカメバソウ,チョオセンカメバソウ
397533 エンレイサウ科,エンレイソウ科
397540 エンレイサウ属,エンレイソウ属
397570 ヨウシュエンレイソウ
397574 オオバナノエンレイソウ,オホバナノエンレイサウ
397616 エンレイサウ,エンレイソウ,タチアオイ,タチアフヒ
397622 シロバナエンレイサウ,シロバナエンレイソウ,タカサゴエンレイサウ,ミヤマエンレイサウ,ミヤマエンレイソウ
397755 ヒナキキョウソウ
397759 キキョウソウ
397760 シロバナキキョウソウ
397828 キキヌキサウ属,キキヌキソウ属,

ツキヌキソウ属
397847 ホザキツキヌキソウ
397849 キキヌキサウ,キキヌキソウ,ツキヌキソウ
397858 ホツツジ属
397860 ハコツツジ,ミヤマホツツジ
397862 ホツツジ
397864 シロバナホツツジ
397964 イヌカミツレ
397969 ミゾカミツレ
397976 イヌカミツレ
397986 シカギク
398022 ヒメカノコサウ属,ヒメカノコソウ属
398024 ニヒタカカノコサウ,ニヒタカカノコソウ,ヒメカノコサウ,ヒメカノコソウ
398064 トリコグサ属
398071 トリコグサ,ネズミシバ
398091 フクロダガヤ
398093 フクロダガヤ
398134 ウラギク
398139 ガマグラス
398251 ツルリンドウ属
398268 ハナヤマツルリンドウ,ハナヤマツルリンドオ
398273 ツルリンダウ,ツルリンドウ,ツルリンドオ
398276 シロバナツルリンドウ
398282 シロバナテングノコヅチ
398285 ニヒタカツルリンダウ
398289 ヒメツルリンダウ
398306 ツルリンドウ
398307 テングノコヅチ
398310 クロヅル属
398312 コバノクロヅル
398318 アカネカズラ,アカネカヅラ,クロズル,クロヅル
398319 ウラジロクロヅル
398322 タイワンクロヅル
398437 カニツリグサ属
398448 カニツリグサ
398449 フタヒゲカニツリ
398486 ミヤマカニツリ
398528 カニツリススキ,チシマカニツリ
398531 タカネカニツリ,リシリカニツリ
398540 キタダケカニツリ
398547 タイワンカニツリ,リシリカニツリ
398566 ヒカゲカニツリ
398666 トベラモドキ属
398681 コウシュンカズラ属,ビヨウカヅラ属
398684 コウシュンカズラ,コウシュンカヅラ,コオシュンカズラ,ビヨウカヅラ
398777 トリテレイア属
398824 ブラジルクマデヤシ属
398834 コムギ属
398839 コムギ
398869 ペルシャコムギ
398877 クラブコムギ
398886 ノハラフタツブコムギ
398890 エンマ-コムギ,ツブコムギ
398900 マカロニコムギ
398924 ヒトツブコムギ
398939 ポ-ランドコムギ
398970 スペルトコムギ
398986 リベットコムギ
399043 トリト-ニア属,ヒメタウシャウブ属,ヒメトウショウブ属
399062 アカバナヒメアヤメ
399094 スイセンアヤメ
399116 ヒメトウショウブ
399190 ラセンサウ属,ラセンソウ属,ラセントウ属
399202 カジノハラセンソウ
399261 ラセンサウ,ラセンソウ,ラセントウ
399296 ナガハラセンサウ,ナガハラセンソウ,ホソバラセンサウ,ホソバラセンソウ
399303 ハテルマカズラ
399304 ケナシハテルマカズラ
399309 コンペイトウグサ
399312 カジノハラセンソウ,カヂノハラセンサウ,カヂノハラセンソウ
399325 カラピンラセンサウ,カラピンラセンソウ
399365 ホンガウサウ科,ホンガウソウ科,ホンゴウソウ科
399432 ヤマグルマ属
399433 イワグルマ,イワモチ,オオモチノキ,オホモチノキ,トリモチノキ,ヤマグルマ
399435 ナガバノヤマグルマ
399484 キンバイサウ属,キンバイソウ属
399500 トウキンハイソウ
399504 セイヨウキンバイ
399509 キンカソウ,キンクワサウ,キンバイサウ,キンバイソウ
399510 エゾキンバイサウ,シナノキンバイ,シナノキンバイソウ
399515 イブキキンバイサウ,イブキキンバイソウ,オクキンハイソウ
399519 ボタンザキキンバイサウ,ボタンザキキンバイソウ
399522 シナノキンバイ
399523 ボタンキンバイ
399524 ヒメキンバイ
399596 ノウゼンハレン科
399597 キンレンカ属,ノウゼンハレン属
399601 キンレンカ,ノウゼンハレン
399603 フイリノウゼンハレン
399604 ヒメキンレンカ,ヒメノウゼンハレン
399606 カナリヤヅル
399611 タマノウゼンハレン
399633 ネッタイラン属
399634 アコウネッタイラン,ネッタイラン
399650 ヤクシマネッタイラン
399651 ハチジョウネッタイラン
399860 ツガ属
399867 カナダツガ
399874 カロライナツガ
399879 シナツガ,テ-シャン
399892 コメヅガ
399895 クロツガ,コメツガ,ヒメツガ,ベニヅガ
399901 タイワンツガ
399905 アメリカツガ,ウエスタンヘムロック
399916 メルテンスツガ
399923 クロツガ,ツガ,ツガマツ,トガ,トガマツ,ホンツガ
399930 ハコネコメツツジ属
399932 ハコネコメツツジ
399995 ハダカホオズキ属,ハダカホホヅキ属
399998 ハダカホオズキ,ハダカホホヅキ
400000 ムニンハダカホオズキ
400001 マルバハダカホオズキ
400050 ツルバギア属
400110 アマナ属,チュ-リップ属
400156 ヒロハアマナ,ヒロハナアマナ,ヒロハムギグワイ
400162 ウツコンカウ,チュウリップ,チュ-リップ
400280 トンボソウ属
400310 ツマモカ属
400376 インドヤツデ属,ツピダンサス属
400377 インドヤツデ
400458 ガジョウマル
400467 ヒメシイン,ヒメシオン
400486 トルネラ属
400493 トルネラ
400515 シュウベンノキ属,ショウベンノキ属,セウベンノキ属
400522 タイワンセウベンノキ,ミヤマセウベンノキ
400536 マルバセウベンノキ
400539 シヨウベンノキ
400546 ショウベンノキ,セウベンノキ,ヤマデキ
400644 ハタザオ,ハタザホ
400671 クワントウ属,クントウ属,ツスシラ-コ属,フキタンポポ属
400675 カントウ,クワントウ,フキタンポポ
400844 オオカモメズル属,オオカモメヅル属,オホカモメヅル属,カモメヅル属
400853 オオカモメヅル
400888 コカモメズル,コカモメヅル,トサ

カモメヅル
400927 ヒメイヨカズラ
400944 ホソバカモメヅル
400945 ブラオンカモメヅル
400969 タイワンカモメヅル
400972 ツルモウリンカ
400973 ケナシツルモウリンカ
401085 ガマ属
401087 アイノコガマ
401094 ヒメガマ,ホソバガマ
401103 モウコガマ
401105 ヒメガマ
401112 ガマ,ヒラガマ,ミスクサ
401120 モウコガマ
401128 イドバガマ,チャボガマ
401129 コガマ,コヒラガマ
401134 コガマ
401140 ガマ科
401148 リウキウハンゲ属,リュウキュウハンゲ属
401152 リュウキュウハンゲ
401156 オヱフネサウ,タウハンゲ,バンコクハンゲ,リウキウハンゲ,リュウキュウハンゲ
401332 ウエベルマンナア-ナ属
401373 ハリエニシダ属
401388 ハリエニシダ
401411 ウルクス属
401415 ニレ科
401425 ニレ属
401431 アメリカニレ
401468 ヨ-ロッパニレ
401489 アカダモ,タウニレ,トウニレ,ニレ,ネリ,ハルニレ
401490 アカダモ,コブニレ,テリハニレ,ニガニレ,ニレ,ネリ,ネレ,ハルニレ,ヤニレ
401492 テリハニレ
401512 エルム,オウシュゥハルニレ,セイヨウハルニレ
401542 アッシ,アツニ,ウバニレ,オヒョウ,オヒョウニレ,ネバリジナ,ヤジナ
401545 ニックワアツニ,ニッコウアツコ,ニッコウアッシ,ニッコウオヒョウ
401556 オウミノニレ,オホミノニレ,チョウセンニレ,テウセンニレ
401563 オウバニレ
401581 アキニレ,イシゲヤキ,カハラゲヤキ,カワラゲヤキ,ニレ
401593 オウシュウニレ,コブニレ,ヨ-ロッパニレ
401602 タウハルニレ,ノニレ,ヒメニレ
401642 アリサンニレ
401737 カギカズラ属,カギカヅラ属
401753 トゲソヘバ
401754 タイワンカギカヅラ
401757 ガンビ-ル,ガンビ-ルノキ
401761 タイワンカギカヅラ,ハナダマノキ
401773 カギカズラ,カギカヅラ,カラスノカギズル,カラスノカギヅル,タケカズラ,フジツリバナ
401776 アラゲカギカズラ
401852 ウニ-オラ属
402106 フヂボグサ属
402118 オホバノフヂボグサ,フジボグサ,フヂボグサ,フヂボハギ
402132 オオバフジボグサ,オホバフヂボグサ
402144 ホソバフジボグサ,ホソバフヂボグサ
402194 ゴムカズラ,ゴムカヅラ
402201 アカハゼ,ゴムカヅラモドキ
402213 ウルケオリナ属
402238 ボンテンカ属,ボンテンクワ属
402245 オオバボンテンカ
402250 アマミボンテンカ,アマミボンテンクワ,オオバボンテンカ,オホバボンテンクワ,オボンテンカ,オボンテンクワ,ヲボンテンクワ
402267 オダンカ,ボンデンカ,ボンデンクワ,ヲダンクワ
402314 ウルギネア属,カイソウ属
402512 コニクキビ
402543 ニクキビモドキ
402546 メリケンニクキビ
402553 ヒメキビ
402660 ウロスベルムム属
402663 オニコウゾリナ
402743 ウルシニア属
402840 イラクサ属,ウルチカ属
402847 ホソバイラクサ
402848 エゾイラクサ,ナガバイラクサ
402869 アサノハイラクサ
402886 セイヨウイラクサ
402946 コバノイラクサ
402996 エゾイラクサ,ナガバイラクサ
403028 イタイタグサ,イラクサ,マムシグサ
403039 ヒメイラクサ
403050 イラクサ科
403095 タヌキモ属
403098 ムラサキミミカキグサ
403108 ノタヌキモ
403109 モンナシノタヌキモ
403110 イヌタヌキモ
403119 ミミカキグサ
403132 ホザキノミミカキグサ
403135 シロバナホザキノミミカキグサ
403156 フサタヌキモ
403176 ナガレイトタヌキモ
403191 イトタヌキモ,オオバナイトタヌキモ,シマヒメタヌキモ,ミカワタヌキモ,ワスレタヌキモ
403199 アミメミミカキグサ
403214 エフクレタヌキモ
403220 コタヌキモ
403222 タヌキモ,チョウシタヌキモ
403252 ヒメタヌキモ
403257 ヒメミミカキグサ
403258 シロバナヒメミミカキグサ
403268 ヤチコタヌキモ
403299 ホザキノミミカキグサ
403351 タマバタヌキモ,マルバノミミカキグサ,マルバミミカキグサ
403377 ムラサキミミカキグサ
403379 シロバナミミカキグサ
403382 タヌキモ,マンシウタヌキモ
403385 オオタヌキモ
403665 ウブラ-リア属
403686 ダウクワンサウ属,ダウクワンソウ属,ダドウカンソウ属,ドウカンソウ属
403687 カサクサ,ススクサ,ダウクワンサウ,ダウクワンソウ,ドウカンソウ
403706 コケモモ科
403710 コケモモ属,スノキ属
403716 ナガバスノキ,ヤナギバスノキ
403734 ムニンシャシャンボ
403738 サシブノキ,シャシャンボ,シャシャンポ,ワクラハ
403741 ナガバシャシャンボ,ナガバシャシャンポ
403776 アラゲナツハゼ
403780 ヌマスノキ
403796 ニヒタカコケモモ
403816 クスノハシャシャンボ
403821 オオバコケモモ,オホバコケモモ,ヤドリコケモモ
403854 アカモヂ,ウスノキ,カクミノスノキ,コウスノキ,コバウスノキ
403857 ツクシウスノキ
403859 ウスノキ
403868 アクシバ
403869 ハゴロモアクシバ
403870 ケアクシバ
403878 チョウセンスノキ
403886 カクミケウスノキ,ケウスノキ
403894 オオミノツルコケモモ
403899 オホバシャシャンポ,オホバシャシャンポモドキ
403916 ビルベリ-
403917 ヒメウスノキ
403919 クロミウスノキ
403925 アキゴロモ,ナツハゼ
403927 ウラジロナツハゼ
403932 クロウスゴ
403933 ナガバクロウスゴ
403937 ミヤミクロウスゴ
403938 ケバノクロウスゴ
403939 ミヤマエゾクロウスゴ
403942 オククロウスゴ
403957 アメリカスノキ

403967 イハツツジ,イワツツジ
403976 ランダイシャシャンポ
403996 シコクウスゴ,マルバウスゴ
403998 ナガボナツハゼ,ホナガナツハゼ
404003 オオバスノキ,オホバスノキ,ケナシオオバスノキ,スノキ
404005 ケナシスノキ,スノキ
404007 カンサイスノキ
404027 アサマブドウ,クロマメノキ,タケグミ
404030 コバノクロマメノキ,ヒメクロマメノキ
404031 クロマメノキ
404047 アオバウスノキ,アヲバウスノキ,ウスイチゴ,クロミウスノキ,スノキ,ナツゴロウ
404051 イハナシ,イハモモ,イワナシ,イワモモ,オヤマリンゴ,カンロバイ,コケモモ,ハマナシ,フレップ
404060 ギイマ,キイモ,ギーマ,タロコシャシャンポ,ヒメシャシャンポ
404062 アクシバモドキ
404066 アオジクスノキ,アヲジクスノキ,ヒメウスノキ
404213 カノコソウ属,バレリアナ属,ヤマカノコサウ属,ヤマカノコソウ属
404261 カノコソウ
404263 シロバナカノコソウ
404264 カノコソウ
404269 サルカノコサウ,ツルカノコソウ,ヤマカノコサウ
404292 ミヤマカノコサウ
404316 カノコソウ,コウアンカノコソウ,セイヨウカノコソウ,モリカノコソウ,ヤウジュカノコサウ,ヨウジュカノコソウ
404325 カノコサウ,カノコソウ,ハルオミナヘシ,ハルヲミナヘシ
404392 オミナエシ科,ヲミナヘシ科
404394 ノヂシャ属
404408 モモイロノジシャ
404439 ノヂシャ
404454 ノヂシャ
404468 シロノヂシャ
404469 シロノヂシャ
404531 セキシャウモ属,セキショウモ属,ハリスネーリア属
404536 アメリカセキショウモ
404544 カウガイモ,コウガイモ
404546 オオセキショウモ
404555 イトモ,セキシャウモ,セキショウモ,ヘラモ
404556 ネジレモ
404557 ヒラモ
404563 セイヨウセキショウモ
404575 バロータ属
404610 アメリカイカリソウ属
404620 バンダ属,ヒスヰラン属
404629 バンダ・セルーレア,ヒスイラン,ヒスヰラン
404652 コウトウヒスイラン
404680 ヒョウモンラン,ヘウモンラン
404696 スズメノトウガラシ
404703 シロバナアゼトウガラシ
404712 シマウリクサ
404745 バンドプシス属
404749 イリオモララン,ニューメンラン
404942 ワイヘルディア属
404945 スイチョウ
404946 ロウギョク
404948 ヒショクギョク
404971 バニラ属
405010 バニラ
405017 バニラ
405021 ニシインドバニラ
405028 タイワンバニラ,ミソボシラン,リモガソバニラ,ロウノウバニラ
405030 タヒチバニラ
405059 バンジリア属
405163 クラマス
405182 シックリーフドナリブ
405191 バトリカニア属
405193 キンソウリュウ
405253 フィジーノヤシ属
405256 ニューカレドニアフィジーノヤシ
405257 ジョーンフィジーノヤシ
405258 マニラヤシ
405356 ベルタイーミア属
405429 ヒトツノコシカニツリ
405438 カザナビキ属
405445 カザナビキ
405447 テリミノカザナビキ
405570 シュロサウ属,シュロソウ属
405585 ケバイケイソウ
405598 バイケイサウ,バイケイソウ
405601 シュロサウ,シュロソウ
405606 ナガバシュロソウ,ホソバシュロサウ,ホソバシュロソウ
405607 シュロソウ
405618 オオシュロソウ,オホシュロサウ,テウゼンシュロサウ
405627 エゾバイケイソウ,カラバイケイソウ,ミドリバイケイソウ
405633 アオヤギサウ,アヲヤギサウ
405657 モウズイカ属,モウズイクワ属,モウズヰクワ属
405659 キバナヒメモウズイカ
405667 ニハタバコ,ニワタバコ,モウズイカ,モウズイクワ,モウズヰクワ
405670 シロバナモウズイカ
405691 ヒメモウズイカ
405737 クロバナモウズイ,クロバナモウズイカ,クロバナモウズヰ,クロモウズイカ,クロモウズイクワ,クロモウズヰクワ
405748 ムラサキモウズイカ
405788 ニハタバコ,ビロウドモウズイクワ,ビロウドモウズヰクワ,ビロオドモオズイカ,ビロードモウズイカ,モンパモウズイ,モンパモウズヰ
405789 シロバナビロードモウズイカ
405796 アレチモウズイカ
405801 クマツヅラ属,バーベナ属
405804 ダキバアレチハナガサ
405812 サンジャク,サンジャクバーベナ,バーベーナ,ヤナギハナガサ,ヤマギハナガサ
405815 ミナトクマツヅラ
405819 アレチハナガサ
405820 ハナガサソウ
405852 シキザクラ,ハナガサ,バーベーナ,ビジョサクラ,ビジョザクラ
405865 ハマクマツヅラ
405872 クマツズラ,クマツヅラ
405885 シュッコンバーバナ
405889 マルバクマツヅラ
405890 シロバナマルバクマツヅラ
405896 ネバリビジョザクラ,ヒナビジョザクラ,ヒメビジョザクラ
405897 カラクサハナガサ
405907 クマツヅラ科
405911 ベルベシーナ属
405915 ハネミギク
406016 アブラキリ,アブラギリ,イヌギリ,エギリ,ダイガン,ドクエ,ヤマギリ
406018 オオアブラギリ,カントンアブラギリ,シナアブラギリ
406019 カントンアブラギリ,モンタナアブラギリ
406040 シャウジャウハダマ属,ショウジョウハグマ属,ベルノニア属,ヤンバルヒゴタイ属
406228 コバナムラサキムカシヨモギ
406266 アメリカムカシヨモギ,ヤナギアザミ
406392 イヌカイノウコウ,シマムカシヨモギ,ショウジョウハグマ
406566 ハマシャギク,ハマムカシヨモギ
406631 ヤナギタムラソウ
406664 コバナムラサキムカシヨモギ
406667 ウラジロカッコウ
406966 クガイサウ属,クガイソウ属,クワガタソウ属
406968 ホナガカワヂシャ
406973 イヌノフグリ,ハタケクガタ
406979 ヒメルリトラノオ
406983 エゾノカワジシャ,エゾノカワヂシャ
406984 シロバナエゾノカワヂシャ
406992 オオカワジシャ,オオカワヂシャ,カワヂサ

407002 オオカワヂサ
407003 オトメカワヂシャ
407014 タチイヌノフグリ,タチイヌフグリ
407029 マルバカワヂシャ
407050 クハガタサウ,クワガタソウ
407054 コクワガタソウ
407064 カワヂシャモドキ
407072 カラフトビヨクサウ,カラフトヒヨクソウ
407096 ココメイヌノフグリ
407142 シュムシュクワガタ
407144 フラサバソウ
407162 ヤマクワガタ
407164 ハマクワガタ
407200 ヒヨクサウ,ヒヨクソウ
407232 クワガタソウ
407233 シロバナクワガタソウ
407234 コクワガタ
407241 ニヒタカクハガタ,ニヒタカクワガタ
407246 サンインクワガタ
407252 ヒメクハガタ,ヒメクワガタ
407253 シナノヒメクワガタ
407259 コバノクハガタ
407261 グンバイヅル
407262 アレチイヌノフグリ
407265 トウテイラン
407275 ケナシムシクサ,ケムシクサ,ムシクサ
407283 ケムシクサ
407287 オオイヌノフグリ,オホイヌフグリ,オホハタケクハガタ,ヘウタングサ
407292 イヌノフグリ
407294 イヌノフグリ
407299 ノボリクワガタ,ハイタワガタ
407349 ホソカワヂシャ
407358 コテングクワガタ,テングクハガタ
407362 テングクワガタ
407371 クガイソウ,シベリアクガイソウ
407387 チシマクワガタ,ミヤマクハガタ
407388 エゾヒメクワガタ,ハクトウクワガタ
407401 テングクハガタ
407421 ミツバイヌノフグリ
407424 ヤナギバトラノオ
407430 イヌカワジサ,イヌカワヂサ,カハサ,カハヂサ,カワジサ,カワジシャ,カワヂシャ
407434 クガイサウ,クガイソウ,クルマサンシチ,トラノオ,トラノヲ,ヤマツツミ
407449 クガイソウ属
407453 トラノオスズカケ
407461 タロコソウ
407462 クガイソウ
407463 シロバナクガイソウ
407465 ナンゴククガイソウ
407466 シロバナナンゴククガイソウ
407467 イブキクガイソウ
407471 リュウキュウスズカケ
407485 クガイソウ,シベリアクガイソウ
407488 エゾクガイソウ
407492 クガイソウ
407493 エゾクガイソウ
407496 シロバナエゾクガイソウ
407501 キノクニスズカケ
407503 スズカケソウ,チョウケンカズラ
407520 タケウマキリンヤシ属
407521 タケウマキリンヤシ
407585 ベチベルソウ
407606 オオハクウンラン
407612 ハクウンラン
407645 ガマズミ属
407656 ヒゼンガマズミ
407658 オニコバノガマズミ
407679 ヤドリサンゴジュ
407705 ケガマズミ,シマガマズミ,ニイジマガマズミ,ニヒジマガマズミ
407715 カラスガマズミ,ブレヤガマズミ
407727 オオチョウジガマズミ,オホチャウジガマズミ,チョウジガマズミ
407731 チュウゴクガマズミ,チョウジガマズミ
407759 コニシガマズミ
407785 アラゲガマスミ,ガマズミ,ソゾミ,ヨソゾメ,ヨツズミ,ヨツドゾメ
407790 ヘンヨウガマズミ
407791 ハコネガマズミ
407792 ニッコウガマズミ
407793 キミノガマズミ
407802 コバノガマズミ,タイワンコバガマズミ
407803 タカオコバノガマズミ
407804 シコクガマズミ,シュクガマズミ
407806 キミノコバノガマズミ
407819 サイゴクガマズミ
407850 シマヨウゾメ,タイワンヨウゾメ
407853 シマガマズミ,タカサゴガマズミ,ルゾンガマズミ
407865 オオカメノキ,オホカメノキ,カベノキ,ビラカ,ムシカリ
407867 オキナハハクサンボク,オキナワハクサンボク
407897 イセビ,イセブ,イヌデマリ,カゴビチ,ハクサンボク,フサボタン,ヤマテラシ
407898 シマハクサンボク,トキワガマズミ
407899 コハクサンボク
407905 ヒロハガマズミ
407930 アッゲタカサゴガマズミ,ムシャガマズミ
407954 イワガマズミ
407964 コニシガマズミ
407977 アオサンゴジュボク,アヲサンゴジュボク,キサンゴ,サンゴジュ,サンゴジュボク,タイワンサンゴジュ
407980 キサンゴ,サンゴジュ,サンゴジュボク,メタラヨウ,ヤブサンゴ
407989 セイヨウカンボク,ヨウシュカンボク
408009 カンボク,クサギ,クソクサギ,ケナシカンボク,メドノキ
408010 キミノカンボク
408011 テマリカンボク
408012 ケカンボク
408019 コミヤガマズミ,ヒメガマズミ
408024 オトコヨウゾメ,コネス,コネソ,ヲトコヨウゾメ
408025 キミノオトコヨウゾメ
408027 オオデマリ,テマリバナ,ヤブデマリ
408031 ケナシヤブデマリ
408034 ウシノヒタイ,ウシノヒタヒ,ゴネズ,ゴネヅ,ヘミノキ,ヤブデマリ,ヤマデマリ
408042 コヤブデマリ
408044 ヤエザキチャボヤブデマリ
408045 チャボヤブデマリ
408048 ナガバノヤブデマリ
408056 オヒワケガマズミ
408098 カンボク
408131 ゴマキ,ゴマギ,ゴマシオヤナギ,ゴマシホヤナギ,テンコウボク
408134 マルバゴマギ
408151 コウルメ,ゴモジュ
408155 タイトウガマズミ
408156 タイワンガマズミ
408159 オオシマガマズミ
408188 アメリカカンボク
408198 タイワンガマズミ,マルバミヤマシグレ,ミヤマシグレ,ヤマシグレ
408199 ミヤマシグレ
408224 シグレノキ,ヘミノキ,ミヤマガマズミ
408227 コミヤマガマズミ
408230 テリハミヤマガマズミ
408231 オオミヤマガマズミ
408251 ソラマメ属
408262 ケナシツルフジバカマ,ツルフジバカマ,ツルフヂバカマ
408263 シロツルフジバカマ
408269 ナガバツルフジバカマ
408272 スナジツルフジバカマ
408276 シロバナノハラクサフジ
408278 ノハラクサフジ,ノハラクサフヂ
408280 モモイロノハラクサフジ
408284 カラスノエンドウ,ホソバノヤハズエンドウ,ヤハズエンドウ
408288 ヤマズエンドウ
408289 シロバナヤハズエンドウ
408296 タマザキナンテンハギ
408327 トウエンドウ

408342 チョウセンエビラフジ
408352 ケサフジ,ケサフヂ
408358 シロバナクサフジ
408360 ケクサフジ,ケクサフヂ
408361 クサフヂホ
408374 エビラフジ,エビラフヂ,ヒロハノエビラフヂ
408393 シカツマメ,シグワツマメ,ソラマメ,タウマメ,トウマメ,ナツマメ,ヤマトマメ
408399 ツガルフジ
408419 キバナカラスノエンドウ
408423 スズメノエンドウ
408435 ハマクサフジ,ヒロハクサフジ,ヒロハクサフヂ
408436 シロバナヒロハクサフジ
408439 エゾヒロハクサフジ
408442 ナガボヒロハクサフジ
408452 ヒナカラスノエンドウ
408469 オニカラスノエンドウ
408493 アレチノエンドウ
408502 タチノエンドウ
408508 ヨツバハギ
408509 シロバナヨツバハギ
408510 ツルナシヨツバハギ
408511 エダウチヨツバハギ
408551 オオバクサフジ,オホバクサフヂ
408553 シロバナオオバクサフジ
408554 アケボノオオバクサフジ
408556 マルバオオバクサフジ
408571 アカエンドウ,アコ,オオヤハズエンドウ,カラスノエンドウ,ノエンドウ,ノラマメ,メホウエンドウ,ヤハズエンドウ
408578 ヤハズエンドウ
408579 ツルナシヤハズエンドウ
408580 ホソバヤハズエンドウ
408611 イブキエンドウ,イブキノエンドウ
408638 カスマグサ
408648 タニワタシ,ナンテンサウ,ナンテンソウ,ナンテンハギ,ニックワウハギ,ニックワタニワタシ,フタバハギ,ミヤマタニワタシ
408649 フジガエソウ
408650 ヒメナンテンハギ
408651 ミツバナンテンハギ
408652 フジガエソウ
408654 フジガエソウ
408658 クマガワナンテンハギ
408661 エダウチナンテンハギ
408666 ナヨクサフジ
408672 ホソバヨツバハギ
408674 シロウマエビラフジ
408675 ビワコエビラフジ
408677 オオバノヨツバハギ
408678 ミネヨツバハギ
408693 ビロードクサフジ
408733 オオオニバス属
408734 オオオニバス
408735 パラブアイオニバス
408825 ササゲ属
408831 タイワンハマササゲ
408832 コチョウインゲン
408839 アズキ,アヅキ,ショウズ
408841 ヤブツルアヅキ
408917 サラワクマメ
408950 コハマササゲ,ナカバハマササゲ
408957 ハマアズキ
408969 コハマアヅキ,ヒナアズキ,ヒメツルアズキ
408976 ヒナアズキ
408982 ケツルアズキ
409008 ケハマササゲ
409025 シマアヅキ,ブンドウ,ヤエナリ,ヤヘナリ,リョクトウ
409026 シマアヅキ
409028 オオヤブツルアズキ,サカサハマササゲ
409065 シマアヅキ
409080 シマアズキ,シマアヅキ,ヤブツアズキ
409085 カニメ,ツルアズキ,ツルアヅキ
409086 ササゲ
409092 ハタササゲ
409096 アヅキササゲ,ジュウロクササゲ,ツブササゲ,ハタササゲ,ミササゲ,ミトリササゲ
409113 アカササゲ,サクヤアカササゲ,フヂササゲ
409122 アカササゲ
409230 ビラディア属
409325 ツルニチニチサウ属,ツルニチニチソウ属
409335 ツルギキヤウ,ツルギキョウ,ツルニチニチサウ,ツルニチニチソウ
409336 フイリツルニチニチソウ
409339 ヒメツルニチニチソウ
409394 ビンセトキシカル属
409395 ムラサキスズメノオゴケ
409402 アオカモメヅル
409416 ナンゴクカモメヅル
409422 イシダテクサタチバナ
409433 サツマビャクゼン
409447 タチカモメヅル
409448 アオタチカモメヅル
409449 マルバカモメヅル
409463 ヤクヨウカモメヅル
409467 ホウヨカモメヅル
409474 イズカモメヅル
409477 マルバオゴケ
409480 クサナギオゴケ
409481 シロバナクサナギオゴケ
409483 マルバノフナバラソウ
409494 タチガシワ
409508 オオアオカモメヅル
409509 ナガバクロカモメヅル
409551 イワキカモメヅル
409557 ミウラスズメノオゴケ
409572 ヤマワキオゴケ
409573 ヨナクニカモメヅル
409608 スミレ属
409609 スワキクバスミレ
409610 ヒメキクバスミレ
409613 キソスミレ
409615 アスマスミレ
409616 カツラギスミレ
409618 オクタマスミレ
409619 フイリオクタマスミレ,フイリフギレシハイスミレ
409622 エドスミレ
409624 コチョウスミレ,サンシキスミレ,パンジー,ユウチョウカ
409630 イヌスミレ,エゾタチツボスミレ,エゾノタチツボスミレ
409631 シロバナエゾノタチツボスミレ
409633 ケナシエゾノタチツボスミレ
409656 ツボスミレ,ニョイスミレ
409665 コマスミレ,フギレサキガケスミレ
409680 ジンヨウキスミレ
409682 アルタイスミレ
409685 アマミスミレ
409687 アムールスミレ
409716 タイワンヤノネスミレ,ヤノネシロバナスミレ,リウキウシロスミレ
409723 アリアケスミレ,コシロバナスミレ
409725 エナガスミレ,リュウキュウシロスミレ
409729 オオタカネスミレ,キスミレ,キバナノコマノツメ
409730 ジョウエツキバナノコマノツメ
409733 アカイシキバナノコマノツメ
409746 ナカバノスミレサイシン
409747 シロバナナカバノスミレサイシン
409749 フイリナガバノスミレサイシン
409751 アメリカウスバスミレ,ウスバスミレ
409753 ウスバスミレ
409755 ヒメミヤマスミレ
409765 ミヤマキスミレ
409768 アラゲキスミレ
409770 エゾキスミレ
409771 フギレキスミレ
409772 ケエゾキスミレ
409773 コバナエゾキスミレ
409774 トカチキスミレ
409775 ダイセンキスミレ
409777 フチゲオオバキスミレ
409780 ナエバキスミレ
409781 フギレオオバキスミレ
409784 オオバキスミレ
409804 ドッグバイオレット

409816 エゾスミレ,ナンザンスミレ,ヒゴスミレ
409819 ヒゴスミレ
409834 エゾアフヒスミレ,エゾノアオイスミレ,ニホヒケスミレ,マルバケスミレ
409843 タイワンコスミレ,リウキウコスミレ
409858 タフテッドパンジー,ツノスミレ
409862 エゾタカネスミレ,クモマスミレ,タカネスミレ,ヤツガダケキスミレ
409863 クモマスミレ
409864 エゾタカネスミレ
409878 テガタスミレ
409895 フキスミレ
409898 ツクミスミレ,ハヒスミレ
409904 ツクシスミレ
409911 ケナシハヒスミレ,ツクシスミレ
409914 ケナシマンシウスミレ,マンシュウスミレ
409925 ソウサイスミレ
409942 エイザンスミレ,エゾスミレ
409943 シロバナエゾスミレ
409945 ナルカミスミレ
409946 ヒトツバエゾスミレ
409958 イハスミレ,オクヤマスミレ,カラフトスミレ,タニマスミレ
409962 タニマスミレ
409976 テリハタチツボスミレ
409977 シロバナテリハタチツボスミレ
409993 ウラジロスミレ,シマスミレ,タイワンスミレ,トウザンスミレ,ニヒタカスミレ
409995 カハカミスミレ,ナンバンスミレ
409997 カハカミスミレ,ナンバンスミレ
410010 コウアンスミレ
410018 イソスミレ,ケイソスミレ,セナミスミレ
410030 タチツボスミレ,ヤブスミレ
410032 シロバナケタチツボスミレ
410033 ケタチツボスミレ
410034 オトメスミレ
410035 サクラタチツボスミレ
410037 アカフタチツボスミレ
410042 コタチツボスミレ
410043 シロバナコタチツボスミレ
410044 シロバナツヤスミレ
410048 ツルタチツボスミレ
410049 ケイリュウタチツボスミレ
410073 オホミヤマスミレ,サクラスミレ
410074 ケナシサクラスミレ
410075 ワタゲスミレ
410076 ケントクスミレ
410077 チシオスミレ
410081 アオイスミレ
410090 チシマウスバスミレ
410104 ウスイロヒメスミレ
410105 ナンゴクスミレ,ヒメスミレ
410107 ヤクシマスミレ
410108 コスミレ
410109 シロバナツクシコスミレ
410110 ヒゲコスミレ
410111 フイリコスミレ
410131 ケマルバスミレ,マルバスミレ
410132 ヒゲケマルバスミレ
410143 シレトコスミレ
410154 オオタチツボスミレ
410155 シロバナオオタチツボスミレ
410156 ヒダカタチツボスミレ
410157 ケオオタチツボスミレ
410165 シロコスミレ
410170 オオバタチツボスミレ
410172 ケオオバタチツボスミレ
410173 タカネタチツボスミレ
410201 ケナシスミレ,シロガネスミレ,スミレ,ノコギリスミレ
410206 アナマスミレ
410214 オオバナスミレ
410215 コモロスミレ
410226 アツバスミレ
410234 コミヤマスミレ
410235 アカコミヤマスミレ
410246 イブキスミレ,マルバタチスミレ
410255 イブキスミレ
410257 シロバナイブキスミレ
410269 モウコスミレ
410289 タカサゴスミレ,ナガサハスミレ
410312 ニオイタチツボスミレ
410313 シロバナニオイタチツボスミレ
410314 オトメニオイタチツボスミレ
410317 ケナシニオイタチツボスミレ
410320 ニオイスミレ,ニホヒスミレ
410326 イチゲキスミレ,イチゲスミレ,キスミレ
410327 ノコギリバキスミレ
410330 ナガバタチツボスミレ
410331 シロバナナガバノタチツボスミレ
410332 アサギケナガバノタチツボスミレ
410333 ケナガバノタチツボスミレ
410334 マダラナガバノタチツボスミレ
410340 クワガタスミレ
410360 ケシロスミレ,シロスミレ,シロバナスミレ,スミレ
410368 トヨコロスミレ
410370 ホソバシロスミレ
410403 アカネスミレ
410404 シロバナマスノスミレ
410405 コボトケスミレ
410406 オカスミレ
410407 シロバナウスゲオカスミレ
410408 ナガワスミレ
410409 ウスゲオカスミレ
410412 タイワンコスミレ,ノジスミレ,マスノスミレ
410420 リュウキュウコスミレ
410452 サキガケスミレ
410466 ビロウドキスミレ
410474 フモトスミレ
410478 タチスミレ
410491 ベニガクスミレ
410498 アケボノスミレ
410499 クロバナアケボノスミレ
410500 シロバナアケボノスミレ
410502 テングスミレ,ナガオスミレ,ナガバシスミレ,ナガヲスミレ
410503 シラユキナガハシスミレ
410504 ミヤマナガハシスミレ
410505 ナガハシスミレ
410517 アイヌタチツボスミレ,カウライタチツボスミレ,コウライタチツボスミレ,ミチノクスミレ
410520 アポイタチツボスミレ
410525 イワマタチツボスミレ
410531 アザミスミレ,エゾノスミレ,エドスミレ,ケナシエドスミレ,ドラバスミレ,ノコギリスミレ
410547 ミヤマスミレ
410548 シロバナミヤマスミレ
410550 ハダカミヤマスミレ
410551 フイリミヤマスミレ
410559 センザンコマノツメ,センザンスミレ
410571 シコクスミレ,ハコネスミレ
410576 フモトクスミレ
410577 フイリフモトスミレ
410587 アメリカスミレサイシン
410600 オリヅルスミレ
410633 キクバコマスミレ
410641 エゾアオイスミレ,マルバケスミレ,ヤエヤマスミレ
410642 フイリヤエヤマスミレ
410643 イリオモテスミレ
410645 イシガキスミレ
410647 トウケンジスミレ
410656 シロバナエゾアオイスミレ
410664 フジスミレ
410665 ミドリフジスミレ
410666 シロバナフジスミレ
410667 ヒナスミレ
410668 シロバナヒナスミレ
410669 エゾヒナスミレ
410670 フイリヒナスミレ
410677 コテフスミレ,サンシキスミレ,パンジー,ミイロスミレ
410710 スミレサイシン
410712 ウスジロスミレサイシン
410713 シロバナスミレサイシン
410719 ゲンジスミレ
410721 ゲンジスミレ,フイリゲンジスミレ,ミドリゲンジスミレ
410730 アギスミレ,ツボスミレ,ニョイスミレ,ヒメアギスミレ

410732 シラユキスミレ
410735 サルクラスミレ
410736 ハイツボスミレ
410738 マダラツボスミレ
410740 ミヤマツボスミレ
410741 アギスミレ
410743 ヒメアギスミレ
410744 コケスミレ
410749 シハイスミレ
410750 シロバナシハイスミレ
410751 ミドリシハイスミレ
410752 コンピラスミレ
410753 フイリシハイスミレ
410754 マキノスミレ
410755 フイリマキノスミレ
410756 シナノスミレ
410758 コウライタデスミレ
410763 イチゲキスミレ,イチゲスミレ,キスミレ
410769 ヒメスミレサイシン
410770 オトコノジスミレ,ケナシノジスミレ,ノジスミレ,ノヂスミレ
410772 シロノジスミレ
410779 シロノジスミレ
410783 シロバナリュウキュウコスミレ
410784 エゾコスミレ,エゾヒカゲスミレ,シンスミレ,ヒカゲスミレ
410785 タカオスミレ
410788 シソバキスミレ
410792 スミレ科,ヤドリギ科
410934 ヤドリギ科
410959 ヤドリギ属
410960 セイヨウヤドリギ,トビヅタ,ホヤ,ヤドリギ
410972 タイワンヤドリギ
410992 タイワンヤドリギ,トビヅタ,ホヤ,ヤドリギ
410994 トビヅタ,ホイ,ホヤ,ヤドリギ
410995 アカミヤドリギ
411016 カキノキヤドリギ,ヤウラクヤドリギ
411048 フウジュヤドリギ,ボンガリャドリギ,モリガシャドリギ
411065 ナガバヤドリギ
411165 ブダウ科,ブドウ科
411187 ニンジンボク属,ハマゴウ属,ビテックス属
411189 イタリアニンジンボク,セイヤウニンジンボク,セイヨウニンジンボク,セイロウニンジンボク
411208 ヤエヤマハマゴウ
411362 タイワンニソジンボク,ニソジンボク
411373 ニンジンボク
411374 クサニンジンボク,コンシジボク
411375 クサニンジンボク
411420 オオニソジンボク,オホニソジンボク,カハリバニンジンボク
411430 ハウ,ハマカヅラ,ハマグミ,ハマグワ,ハマゴウ,ハマシキミ,ハマツバキ,ハマハイ,ハマハギ,ハマハヒ,ハマホウ,ハマボウ
411431 シロバナハマゴウ
411433 モモイロハマゴウ
411434 シロタエハマゴウ
411436 カワリバハマゴウ
411464 シラガドウ,シルホ-ギ,タチハマゴウ,ミツバハマゴウ
411521 ブダウ属,ブドウ属
411540 アム-ルブドウ,シラガドウ,シラガブドウ,シラガブドオ,チョウセンヤマブドウ,テウセンヤマブドウ
411590 コエビヅル
411623 オオエビヅル,オホエビヅル,ガネブ,ヤマエビ,ヤマブダウ,ヤマブドウ
411624 タケシマヤマブドウ,ヤマブドオ
411669 イヌエビ,イヌブドウ,エビカズラ,エビズル,エビヅル
411671 ケナシエビヅル
411672 キクバエビヅル
411675 シチトウエビヅル
411686 ギャウジャノミズ,ギョウジャノミズ,ギョオジャノミズ,サヤウジャノミヅ,サンカクズル,サンカクヅル
411700 ケサンカクズル,ケサンカクヅル
411701 ウスゲサンカクヅル
411735 キ-ルンブドウ,サナツラブダウ,リウキウガネブ,ワタエビ
411764 アメリカブダウ,アメリカブドウ
411890 クヌガワブドウ,クマガワブドウ
411903 アマズル,アマヅル,オトコブドウ,オトコブドオ,ヤマズル,ヤマヅル
411918 ミツバヅル
411979 ブダウ,ブドウ,ヨ-ロッパブドウ
412022 ビッタディニア属
412164 ヴォキシア科
412286 オオミタケヤシ属
412337 フリ-セア属
412343 インコアナナス
412371 トラフアナナス
412385 ミソボシラン属
412388 ミソボシラン
412407 イヌナギナタガヤ
412465 シッポガヤ,ナギナタガヤ,ネズミノシッポ
412470 オオナギナタガヤ
412474 ムラサキナギナタガヤ
412568 ヒナギキャウ属,ヒナギキョウ属,ワ-レンベルギア属
412697 ヨウシュヒナギキョウ
412750 ヒナギキャウ,ヒナギキョウ,ヒナギキョオ
413030 コキンバイ属
413034 エゾキンバイ,コキンバイ
413085 アッサムヤシ属
413091 ナカバアッサムヤシ
413093 フタバアッサムヤシ
413142 コバンバノキ属
413143 コバンバノキ
413149 コバンバノキ
413243 ワルレア属
413254 ワ-セウィッチェラ属
413261 ワ-セウィッチア属
413273 ワサビ属
413289 ワシントンヤシ属
413298 オキナワシントンヤシ,カリフォルニアビロウ,ワシントンヤシ
413307 オニジュロ,セダカワシントンヤシ,ワシントンヤシモドキ
413315 ヒオウギズイセン属,ワトソ-ニア属,ワトソニア属
413326 ヒオウギズイゼン,ワトソニア
413479 ウェ-ベロセレウス属
413480 ビオウレン
413490 ネコノシタ属,ハマグルマ属
413506 キダチハマグルマ,シマハマグルマ,トミワハマグルマ
413507 オオキダチハマグルマ,キダチハマグルマ,トキワハマグルマ
413514 クマノギク,ハマグルマ
413535 ホコガタギク
413549 ネコノシタ,ハマグルマ
413550 オオハマグルマ,オホハマグルマ
413559 アメリカハマグルマ
413570 タニウツギ属,ハコネウツギ属
413571 サウシキウツギ,フジサンシキウツギ
413572 クリ-ムウツギ
413573 ハコネニシキウツギ
413576 ハコネウツギ
413577 シロバナハコネウツギ
413578 ベニバナハコネウツギ
413579 ニオイウツギ
413580 ニシキウツギ
413581 シロバナニシキウツギ
413582 ベニバナニシキウツギ
413583 アマギベニウツギ
413584 アマギニシキウツギ
413585 アマギアオウツギ
413588 ヤブウツギ
413589 シロバナヤブウツギ
413590 チシオウツギ
413591 オオタニウツギ,オオベニウツギ,オホタニウツギ,オホベニウツギ,カラタニウツギ
413604 シロバナオオベニウツギ
413606 チシオウツギ
413608 タニウツギ
413609 シロバナウツギ
413611 ツクシャウツギ,ツクシャブウツギ,ツクショウツギ

413615 カリヨセウツギ
413616 キバナウツギ
413617 ウコンウツギ
413619 ニシキウツギ
413621 ハヤザキウツギ,ビロ－ドウツギ
413623 オケヤマウツギ
413624 ケウツギ,ビロ－ドウツギ
413625 シロバナケウツギ
413626 ビロ－ドウツギ
413671 ウエインガルティア属
413674 オウンギョク
413675 カショクギョク
413678 ハナガサマル
413679 カデンギョク
413722 ゴンカヤシ属
413741 ウェルウィッチア属
413748 ウェルウィッチア,キソウテンガイ
413750 ヴェルヴィチア科,ウェルウィッチア科
413765 アカミミズギ属
413784 アカミズキ,アカミミズギ
413883 ベルックゼオセレウス属
413924 ウェッチンヤシ属,オオミウェッチンヤシ属
414006 コルリブクロ
414058 ウィドリント－ニア属
414092 コダチハゼリソウ属
414095 キダチハゼリソウ,コダチハゼリソウ
414118 アオガンピ属,ガンピ属
414123 ミヤマガンピ
414143 サクラガンピ,ミヤマコガンピ
414177 コガンピ
414193 インドガンピ,シマガンピリョウカオウ,ヤンバルガンピ
414222 ヒメガンピ
414227 タカクマキガンピ
414237 サクラガンピ
414240 シマサクラガンピ
414243 オオシマガンピ
414248 ムニンアオガンピ
414250 アオガンピ,アヲガンピ,オキナハガンピ,マルバガンピ
414257 カミノキ,ガンピ
414271 アカヂクキコガンピ,キガンピ,キコガンピ
414293 ウイルコキシア属
414429 ウイルマッテア属
414541 フジ属,フヂ属
414546 ヤマフジ
414554 ノダフジ,ノダフヂ,フジ,フヂ,ムラサキフジ
414555 シロバナフジ
414556 ノダナガフジ
414562 アケボノフジ
414565 アメリカフジ,アメリカフヂ
414567 ドヨウフジ,ナツフジ
414569 アケボノナツフジ
414570 ヒメフジ,メクラフジ
414576 シナフジ,シラフジ,フジ,フヂ
414577 シラフジ,シロカピタン,ヤマフジ
414636 ウィトロキア属
414649 ミジンコウキクサ属,ミヂンコウキクサ属
414653 コツブウキクサ,コナウキクサ,ミジンコウキクサ,ミヂンコウキクサ
414664 ミジンコウキクサ
414713 ウォレミマツ
414971 キサンシスマ属
414990 オナモミ属,ヲナモミ属
415003 オホオナモミ,オホヲナモミ
415020 オナモミ
415023 イガオナモミ
415031 モウコオナモミ
415033 オオオナモミ
415046 オナモミ
415053 トゲオナモミ
415057 オナモミ,ヲナモミ
415094 ブンカンカ属
415096 ブンカンカ
415164 ザンドリ－サ属
415172 ススキノキ属
415174 ミナミススキノキ
415178 ススキノキ
415181 ススキノキ
415182 ススキノキ科
415190 クサントソ－マ属
415191 ミドリセンニンイモ
415201 ヤバネイモ
415202 ムラサキセンニンイモ
415299 ホソバアサガオ
415315 トキワバナ属
415317 トキワバナ,ヒガサギク
415386 ムギワラギク
415558 ハマナツメモドキ
415693 ピンカドウ,ピンガドウ
415704 キシロビューム属,クシロビューム属
415716 ホウガンヒルギ
415863 クスドイゲ属
415869 クスドイゲ,コバノクスドイゲ
415975 タウユンサウ科,トウエンソウ科
415990 タウユンサウ属,タウユンソウ属,トウエンソウ属
416069 タウユンサウ,タウユンソウ,トウエンソウ
416378 ショウキラン属,ヨ－アニア属
416379 キバナノショウキラン
416381 シナノショウキラン
416382 ショウキラン,ヤマタウガラン,ランデンマ
416390 オニタビラコ属
416437 オニタビラコ,タカオタビラコ,タカヲタビラコ
416440 タカオタビラコ
416490 クシバニガナ
416535 イトラン属,キミガヨウラン属,ユッカ属
416538 センジュラン,チモラン,リンポウラン
416563 ヨシュアノキ
416581 メキシコチモラン
416607 アツバキミガヨラン,アメリカキミガヨラン
416610 イトナシイトラン,キミガヨウラン,キミガヨラン,ネジイトラン,ネヂイトラン
416649 イトラン,ジュモウラン
416659 トレクレイトラン
416682 ソコベニハクモクレン,ニシキモクレイ,モクレン
416694 ハクモクレン,ハクレンゲ
416707 シモクレン,モクレン,モクレンゲ
416721 サラサモクレン,サラサレンゲ
416804 ニイタカヤダケ,ニヒタカメダケ,ニヒタカヤダケ
416841 イハックバネウツギ,トウツクバネウツギ,ミヤマツクバネウツギ
416853 イワツクバネウツギ,マルバツクバネウツギ
416994 サクラカラクサ
417002 ザミア属
417005 ウスバザミア,ヒシバザミア
417006 フロリダザミア,フロリダソテツ,ホソボザミア
417010 ヒロバザミア
417016 ナガバザミア
417021 タテジマザミア
417026 ザミオクルカス属
417045 イトクズモ属,イトクヅモ属
417053 イトクズモ
417060 イトクズモ,イトクヅモ,ミカヅキイトモ
417073 イトクズモ科,イトクヅモ科
417081 ガンドウカズラ
417092 オランダカイウ属
417093 オランダカイウ,カラ,カラ－,クハズイモ,バンカイウ
417097 シラボシカイウ
417098 オオキバナカイウ
417104 キバナカイウ
417118 ムラサキカイウ,モモイロカイウ
417132 サンショウ属,サンセウ属
417139 カラスザンショウ,カラスザンセウ,カラスノサンショウ
417140 トゲナシカラスザンショウ
417144 アコウザンショウ
417146 ケカラスザンショウ
417156 アマミザンショウ
417161 オニザンショウ,フダンサンショウ,フユザンショウ,フユザンセウ

417163 フユザンショウ
417176 イハザンセウ,イワザンショウ
417177 ヒレザンショウ
417180 カホクザンショウ
417229 コカラスザンショウ
417247 ネワタノキ,バロック
417282 クメザンセウ,テリハザンショウ,テリバザンセウ
417290 ヒメハゼザンセウ
417298 サンセウ
417301 サンショウ,サンセウ,ハジカミ
417302 ヤマアサクラザンショウ
417304 アラゲザンショウ
417305 アサクラザンショ
417306 リュウジンザンショウ
417307 シダレヤマアサクラザンショウ
417329 オホバツルザンセウ,ツルザンショウ,ツルザンセウ,ナガツルザンセウ,マギヅルザンセウ
417330 アリサンザンセウ,イヌザンショウ,イヌザンセウ,ミヤマザンセウ,ヤマザンセウ
417333 トゲナシイヌザンショウ
417335 シマイヌザンショウ
417340 オホザンセウ,カホクザンショウ,トウザンショウ,トゲザンセウ,ヒメザンセウ,フダンザンショウ,フユザンショウ
417375 ヤクシマカラスザンショウ
417405 カリフォルニア・ホクシャ属
417407 カリフォルニア・ホクシャ,カリフォルニアフクシア
417412 タウモロコシ属,トウモロコシ属
417417 カウライキビ,コウライキビ,タウモロコシ,タマキビ,トウキビ,トウモロコシ,ナンバン,ナンバンキビ
417420 フイリトウモロコシ
417472 スズメウリ,タイワンスズメウリ
417488 サツマスズメウリ,ホソガタスズメウリ
417492 クロミノオキナワスズメウリ
417520 ゼントネレラ属
417534 ケヤキ属
417547 タイワンケヤキ
417558 ケヤキ,タロコゲヤキ,ツキ,ツキゲヤキ
417559 メゲヤキ
417560 シダレケヤキ
417590 ゼノ-ビア属
417592 スズランノキ
417606 タマスダレ属
417612 タマスダレ
417613 サフランモドキ
417623 カバイロタマスダレ
417693 ツ-クトフィルム属
417698 キヌラン属,ゼウクシネ属,ホソバラン属
417700 アオジクキヌラン
417702 カゲロウラン
417713 ムニンキヌラン
417731 タイトウキヌラン
417764 タイワンキヌラン
417768 オオキヌラン
417772 ジャコウキヌラン
417779 アリサンキヌラン
417786 イシガキキヌラン
417795 キヌラン,チクシキヌラン,ホソバラン
417804 タビヤラン
417808 ヤンバルキヌラン
417880 リシリソウ属
417925 リシリソウ
417964 ショウガ属,メウガ属
417990 タイワンメウガ
418002 ミョウガ,メウガ,メカ
418003 フイリミョウガ
418010 クレノバジカミ,シャウガ,ショウガ,バジカミ
418031 ハナシャウガ,ハナショウガ
418034 シャウガ科,ショウガ科
418035 ジニア属,ヒャクニチサウ属,ヒャクニチソウ属
418038 メキシコ・ジニア,メキシコヒャクニチソウ
418043 ウラシマサウ,ヒャクニチサウ,ヒャクニチソウ
418046 シュッコンヒャクニチソウ
418052 ホソバヒャクニチソウ
418057 キバナノジニア,コバノジニア,ヒメヒャクニチソウ
418058 マンサクヒャクニチソウ
418079 マコモ属
418080 アメリカマコモ,マコモ
418095 カツミ,カツミグサ,コモ,ハナカツミ,フシシバ,マコモ
418144 ナツメ属
418169 イヌナツメ,サネブトナツメ,ナツメ
418172 ダルマナツメ
418173 ナツメ
418175 サネブトナツメ
418177 サンソ
418184 イヌナツメ
418320 スナジマメ属,スナヂマメ属
418331 スナヂマメ,スナマメ,ヒメハギモドキ
418335 スナジマメ,スナヂマメ
418377 アマモ属
418381 オオアマモ,オホアマモ
418382 スゲアマモ
418384 タチアマモ
418389 コアヂモ,コアマモ,コモ,ニラモ
418392 アヂモ,アマモ,スゲモ,モシオグサ,モシホホグサ
418396 コアマモ
418405 アマモ科
418424 シバ属
418425 スナシバ
418426 シバ,ノシバ,ヤマシバ
418428 アオシバ
418431 オニシバ,スナシバ
418432 コウシュンシバ,コウラインシバ,チュウシバ,トウキョウシバ,ハリシバ
418439 コウライシバ
418445 コオニシバ,ナガミノオニシバ
418450 ナガミノオニシバ
418453 イトシバ,カウライシバ,キヌシバ,コウライシバ,シバ,チョゥセンシバ,テウセンシバ,ヒメシバ,ビロ-トシバ
418529 ジゴカクタス属
418532 カニバサボテン
418573 ジゴペタラム属
418578 ムラサキウズラフン
418580 ハマビシ科
418788 ジゴシキオス属
418793 シゴスタテス属

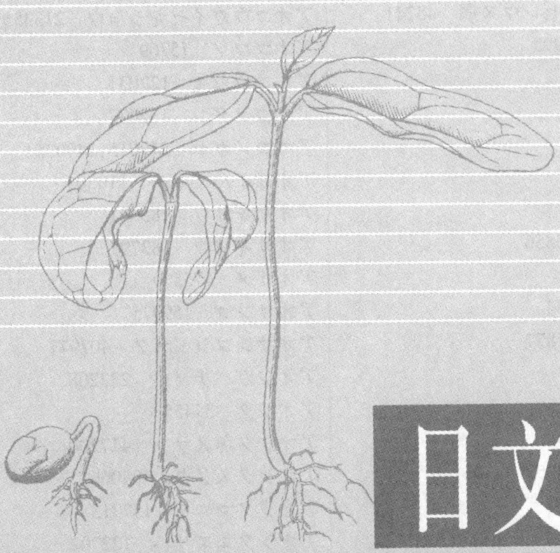

日文名称—序号索引

Japanese

A Dictionary
of Seed Plant Names
种子植物名称

アア 228369
アーポフィラム属 34895
アイ 309893
アイアシ 293215
アイアシ属 293207,337544
アイイタドリ 162508
アイカズラ 245858
アイカヅラ 245854
アイギョクシ 165518
アイギョクシイタビ 165518
アイギリ科 166793
アイグラス 12789,12794
アイグロマツ 299781
アイザワシオガマ 287334
アイズシダレ 83116
アイズシモツケ 371877
アイズスゲ 73768
アイズヒメアザミ 91722
アイスラント・ポビ－ 282617
アイダクグ 218480
アイダコブナグサ 36648
アイヅシモツケ 371881
アイヅユリ 230023
アイナエ 256154,256161
アイナエ属 256147
アイナニ 236085
アイヌソモソモ 305520
アイヌタチツボスミレ 410517
アイヌブキ 292403
アイヌムギ 203136
アイヌワサビ 73044
アイノコアカザ 87136
アイノコイトモ 312035
アイノコエノコロ 361683
アイノコオキナグサ 321675
アイノコガマ 401087
アイノコガヤツリ 118430
アイノコカンガレイ 352163
アイノコキイチゴ 338065
アイノコキンミズヒキ 11543
アイノコクワズイモ 16489
アイノコシラスゲ 73567
アイノコセンダン 248899
アイノコナルコ 73561
アイノコノウゼンカズラ 70503
アイノコハグマ 12656
アイノコヒルムシロ 312033
アイノコフユイチゴ 338069
アイノコヘビイチゴ 138792
アイノコヤシャブシ 16310
アイノコヤナギモ 312031
アイノコヨメナ 40657
アイノコヨモギ 35171
アイバサウ 353576
アイバソウ 353576
アイバナ 100961
アイビ 187318
アイフウリンホオズキ 297653
アイラトビカズラ 259566
アイ－ラ属 12776
アイロステラ属 45879
アウギヤシ属 57116
アウシキナ 23796
アウストロカクタス属 45231
アウストロケファロセレウス属 45241
アウソウクワ属 34995
アウチ 248895
アウチノキ 248895
アオアカザ 87048
アオアケビ 13225
アオアシ 225694
アオアブラガヤ 353856
アオイ 243862
アオイカズラ 377884
アオイカズラ属 377877
アオイゴケ 128964
アオイゴケ属 128955
アオイスミレ 410081
アオイチゴツナギ 305993,306146
アオイツナソ 194779
アオイトバモウセンゴケ 138363
アオイヌガラシ 336215
アオイヒルガオ 102914
アオイマメ 294010
アオイモドキ属 243877
アオイ科 243873
アオウキクサ 224385
アオウキクサ属 224359
アオウシノケグサ 164131
アオウメモドキ 328904
アオカウガイ 213355
アオカウガイゼキシャウ 213355
アオカゴノキ 6762
アオカゴノキ属 6761
アオガシ 240604
アオガシカエデ 3105
アオガシマスカシユリ 229915
アオカズラ 341513
アオカズラ属 341475
アオカゼクサ 147672
アオカヅラ 341513,364986
アオカヅラ科 341582
アオカモジグサ 144439,335258,335480
アオカモメヅル 409402
アオガヤツリ 119262
アオカラスノゴマ 104047
アオカラムシ 56240
アオカワラサイコ 312834
アオガンピ 414250
アオガンピ属 414118
アオキ 44915
アオキヌタソウ 170448
アオキバ 44915
アオキラン 147314
アオギリ 166627
アオギリ科 376218
アオギリ属 166612,344821
アオキ属 44887
アオグスモドキ 240726
アオゲイトウ 18810
アオゲムラサキ 66897
アオコウガイゼキショウ 213355
アオゴウソ 75769
アオコウツギ 127031
アオコクダン 132077
アオコメガヤ 249015,249025
アオサイハイラン 110505
アオサギサウ 183921
アオサギソウ 183797
アオザメ 175509
アオサンゴ 159975
アオサンゴジュボク 407977
アオジガバチソウ 232208
アオジク 34455
アオジクキヌラン 417700
アオジクスノキ 404066
アオジクマユミ 157713
アオジクユズリハ 122702
アオシコクシラベ 504
アオジソ 290940,290957
アオシソバタツナミ 355555
アオシナ 391766
アオシバ 171487,418428
アオジュズダマ 175508
アオシラベ 497
アオスゲ 73913,75131
アオスズラン 147196
アオタカネマンネングサ 357241
アオタチカモメヅル 409448
アオダモ 167940,168014,168100
アオチカラシバ 289018
アオヂク 34455
アオチドリ 98586,98593
アオチドリ属 291188
アオツズラフジ 97947
アオツヅチ 364986
アオツヅラフジ 97933,97947
アオツヅラフジ属 97894
アオツメクサ 374794
アオツリバナ 157970
アオテンツキ 166293,166557
アオテンナンショウ 33543
アオテンマ 171928
アオトウヒ 298401
アオトド 388,463
アオトドマツ 463
アオドロ 311480
アオトンボ 183921
アオナイワキンバイ 312351
アオナシ 323332
アオナシタデ 309570

アオナラ　324385
アオナラガシワ　323647
アオナリヒラ　357941,357958
アオナルコスゲ　73767
アオナワシロイチゴ　338987
アオヌカボガヤ　321296
アオネザサ　304040
アオネバリタデ　291997
アオノイワレンゲ　275379,275382
アオノクジャクヒバ　85316
アオノクマタケラン　17656,17693
アオノツガザクラ　297016
アオノリュウゼツラン　10787
アオハイモ　65237
アオバウスノキ　404047
アオハコベ　375119
アオバスゲ　74890
アオハダ　204008
アオバナ　100961
アオバナアオキ　44927
アオバナイチネングサ　356598
アオバナエビ　140342
アオバナツメクサ　353994
アオバナハヒノキ　381288
アオバノキ　381143,381149
アオヒ　496
アオビエ　140369
アオヒエスゲ　74891
アオヒジキ　104849
アオヒメウツギ　126949
アオヒメタデ　291745
アオビユ　18810,18848
アオフタバラン　232934
アオフナバラソウ　117388
アオベンケイ　200819
アオボウモミ　389
アオホオズキ　297628
アオホソバタデ　291790
アオマツリ　305172
アオミズ　299024
アオミヅバヴシ　298971
アオミノアカエゾマツ　298299
アオミノクロウメモドキ　328743
アオミヤマカンスゲ　75446
アオムレスズメ　72285
アオモジ　233882
アオモジズリ　372255
アオモミ　185104
アオモミトドマッ　414
アオモリアザミ　91741
アオモリトドマツ　365,414
アオモリマンテマ　363191
アオヤギサウ　405633
アオヤギバナ　368520
アオヤシャクナゲ　330863
アオヤマジソ　259306
アオヤマボウシ　51142
アオユキノシタ　349941
アオヨモギ　36340
アオラタンヤシ　222634
アオラン　383127
アオワニ　16817
アカアカネ　337970
アカアツバベンケイ　356649
アカアミダネヤシ　129688
アカイシキバナノコマノツメ　409733
アカイシコウゾリナ　298595
アカイシコウモリ　283784,283834
アカイシヒョウタンボク　235962
アカイシミツバツツジ　332027
アカイシリンドウ　174236
アカイシリンドオ　174236
アカイタヤ　3140,3198
アカウ　165717,165841
アカウキクサ属　45979
アカエゾ　298298
アカエゾマツ　298298
アカエトグノヤシ属　265202
アカエビネ　65923
アカエンドウ　301055,408571
アカオモト　394076
アカカエデ　3505
アカガシ　323599
アカガシハ　243371,324335,324344
アカガシワ　324344
アカガシ属　116044
アカカタバミ　277752
アカカヂ　243371
アカカバ　83145
アカカリス属　1018
アカカンコ　177141
アカカンバ　53426
アカキ　83736,376457
アカギ　54620,163458
アカキイチベイ　259094
アカキナノキ　91090
アカギモドキ　16155
アカギモドキ属　16043
アカギ属　54618,54874
アカグツツジ　331689
アカクラエビネ　65927
アカゲオオキツネヤナギ　343410
アカゲカンコノキ　177123
アカゴチョウ　361384
アカゴマ　232001
アカコミヤマスミレ　410235
アカゴムノキ　155743
アカザ　86901,86913
アカザカズラ　26265
アカササゲ　409113,409122
アカサンザシ　110014
アカザ科　86873
アカザ属　86891
アカシアアセンヤク　1120
アカシア属　1024
アカジクセントンイモ　197796
アカジクタンポポ　384618
アカジクヘビノボラズ　51303
アガシス属　10492
アカジソ　290955
アカシデ　77323,77401
アカシノルイヨウショウマ　6422
アカシノルヰエフショウマ　6422
アカシマル　140296
アカシャ　334976
アカショウマ　41852
アカスギ　85353
アカスグリ　334179,334201
アカスゲ　75954
アカスジアツバセンネンボク　104344
アカソ　56317,56343
アカゾメムグラ　39413
アガタコノキ　281166
アカダネマンネングサ　356715
アカダマカヅラ　258923
アカダマノキ　31477,258882
アカダマノギ　172052
アカダモ　401489,401490
アカヂクキコガンピ　414271
アカヂクマルヤナギ　343185
アカツキザクラ　83118
アカヅク　393809
アカツゲ　204292,204299
アカツツジ　330966,332104
アカツナ　93526
アカツメクサ　397019
アカテツ　301777
アカテツガシ　78933,79035
アカテツナナカマド　369509
アカテツ科　346460
アカテンオトギリ　201912
アカトウヒ　298423
アカトゲゲンラク　317688
アカトゲノャシ　2553
アカトゲノヤシ属　265202
アカトド　427,460,463
アカトドマツ　460,463
アカナス　239157
アカニット　5442
アカヌマシモツケ　372075
アカヌマラン　192851
アカネ　337912
アカネイリス　208805
アカネカズラ　398318
アカネカヅラ　398318
アカネズイセン　143573
アカネズイセン属　143570
アカネスゲ　75841
アカネスミレ　410403
アカネズヰセン　143573
アカネズヰセン属　143570

アカネムグラ 39413,337970
アカネ科 338045
アカネ属 337906
アカノマンマ 308907
アカバグミ 142078
アカバサクラ 316302
アカバシュスラン 86698
アカハゼ 402201
アカハダグス 50507
アカハダクスノキ 50507
アカハダクスノキ属 50465
アカハダコバンノキ 245220
アカハダノキ 50507,154971,301147
アカハダノキ属 50465
アカハダメグスリノキ 3003
アカバナ 146849
アカバナアメリカトチノキ 9720
アカバナウチワ 272812
アカバナウチワカエデ 3034
アカバナエゾノコギリソウ 3924
アカバナオウギ 187842
アカバナガキ 132383
アカバナカスミソウ 183192
アカバナカンナ 71166
アカバナギリア 175704
アカバナキンレイ 32750
アカバナクジャク 134228,266648
アカバナサカズキアヤメ 118398
アカバナサンザシ 109885
アカバナシキミ 204484
アカバナシモツケソウ 166105
アカバナスイカズラ 235798
アカバナスイレン 267756
アカバナセンボンヤリ 175129
アカバナダンドク 71166
アカバナトチノキ 9720
アカバナノユウゲシャウ 269498
アカバナハカマノキ 49247
アカバナハナオンライ 54464
アカバナハナオンライ属 54439
アカバナヒキオコシ 209847
アカバナヒツジグサ 267756
アカバナヒムギ 61258
アカバナヒメアヤメ 399062
アカバナヒメイワカガミ 351445
アカバナヒョウタンボク 236085
アカバナヒルギ 61263
アカバナヒルギモドキ 238350
アカバナブラッシマメ 66673
アカバナヘイシソウ 347163
アカバナヘウタンボク 236085
アカハナマメ 293985
アカバナマンサク 185110,185118
アカバナムシヨケギク 322665
アカバナヤエサンザシ 109857
アカバナヤナギハッカ 203102
アカバナユウゲショウ 269498
アカバナルリハコベ 21339,21340
アカバナワタ 179879
アカバナ属 146586
アカバノキ 154971,301147
アカバノキ属 154962
アガパンサス属 10244
アガパンドゥス属 10244
アカヒゲガヤ 194020
アカヒゲガヤ属 194015
アカヒダボタン 90415
アカヒナスゲ 74646
アカヒラトユリ 229888
アカフサスグリ 334179
アカフタチツボスミレ 410037
アカブリ 219970
アカベ 243371
アガベ 10787
アガベ－テス属 10284
アカベナ科 270773
アカペムグラ 39413
アガ－ベ属 10778
アカボシアケボノソウ 380205
アカボシタツナミソウ 355722
アカボシツツジ 331565
アカボシツリフネソウ 204838,204841
アカマツ 299890
アカマツリ 305185
アカマノイヌツゲ 204299
アカマルチワ 272951
アカミカラマツ 221897
アカミグワ 259186
アカミズキ 413784
アカミタンポポ 384614
アカミヅキ 204050
アカミノアカネ 337925
アカミノイヌツゲ 204292,204299
アカミノイヌホオズキ 367311,367437,367726
アカミノウラシマツツジ 31322
アカミノクマコケモモ 31322
アカミノボタン 279479
アカミノモクタチバナ 31601
アカミノヤブカラシ 79841,79901
アカミミズギ 413784
アカミミズギ属 413765
アカミヤドリギ 410995
アカメ 295691
アカメイヌビワ 164686
アカメガシハ 243371
アカメガシハ属 243315
アカメガシワ 243371
アカメガシワ属 243315
アカメジシ 107021
アカメソシン 116829
アカメダイゲキ 159159
アカメモチ 295691
アカメモチ属 295609
アカメヤシ 237972
アカメヤシ属 237971
アカメヤナギ 343185
アカモヂ 403854
アカモノ 172052
アカモミジ 3492
アカヤシオ 331463
アカヤジオウ 327435
アカラギ 376457
アカリトガリキウム属 2096
アガリバナ属 61118
アカリファ 1869
アカリフアモドキ 94055
アカリ－ファ属 1769
アカリファモドキ 94055
アカリフヤモドキ属 94051
アカンカサスゲ 74392,74394
アガンギョク 198493
アカンサス 2695
アカンサス属 2657
アカンスゲ 75185
アカンソフィツピュ－ム属 2071
アカンテフィピウム属 2071
アカンテンツキ 166268
アカントシキオス属 2601
アカントスタキス属 2623
アカントスレウス属 2137
アカントフィピウム属 2071
アガントリ－モン属 2208
アカントロビビア属 234900
アカンペ属 2032
アカンマルバシモツケ 371830
アギ 163580,163709
アキアオスゲ 73619
アキイトスゲ 74970
アキウツギ 62099
アキウネマガリ 347192
アキカサスゲ 75488
アキカゼギボウシ 198594
アキカラマツ 388583
アキギボウシ 198648
アキギリ 345050,345252
アキギリ属 344821
アキグミ 141965,142214
アキゴロモ 403925
アキザキカノコユリ 230038
アキザキシクラメン 115949,115953
アキザキスノ－フレ－ク 227858
アキザキナギラン 116938,116941
アキザキナタネ 59598
アキザキフクジュソウ 8332
アキザキヤツシロラン 171913,171958
アキザケラ 107161
アキサンゴ 105146
アギスミレ 410730,410741
アキタテンナンショウ 33417
アキタブキ 292377

アキチャウジ 209748
アキチョウジ 209748
アキツリガネツツジ 250509
アギナシ 342310
アキニレ 401581
アキネジバナ 372253
アキノイヌヤクシソウ 283400
アキノウナギツカミ 291937,309796
アキノウナギツル 309796
アキノエノコグサ 361743
アキノエノコログサ 361743
アキノキリンサウ 368480
アキノキリンサウ属 367939
アキノキリンソウ 368480,368483
アキノキリンソウ属 367939
アキノギンリョウソウ 258044
アキノギンリョオソウ 258044
アキノコハマギク 89424
アキノタムラサウ 344957,345108
アキノタムラサウ属 344821
アキノタムラソウ 345108
アキノタムラソウ属 344821
アキノノゲシ 320515,320518
アキノノゲシ属 219206
アキノハウコグサ 178218
アキノハハコグサ 178218
アキノハハコグサモドキ 178220
アキノハルシャギク 107161
アキノベニサルビア 345067
アキノホオコグサ 178218
アキノミチヤナギ 308788
アキノヤハズアザミ 348559
アキノルリカコソウ 13145
アキノワスレグサ 191298
アキバギク 41329
アギバナジラミ 305172
アキバブラシノキ属 67248
アキボタン 23854
アキメネス属 4070
アキメヒシバ 130825
アキメヒジハ 130612
アキヤマミズ 298947
アキョシアザミ 91864
アキヨシアザミ 91912
アキレギアーアルピナ 29997
アキレジア属 29994
アクシバ 160503,403868
アクシバモドキ 404062
アクダラ 215442
アクチノキ 31599
アクティノフレウズ属 6849
ァークトチス 31294
アークトチス属 31176
アークトティス属 31176
アクミナタポプラ 311142
アグラオネマ 11340
アグラオネマ属 11329
アクラトキワラン 282791
アクリオプシス属 5947
アクロケネ属 257708
アグロコーミア属 6128
アグロストフィルルム属 12442
アケイノギ 142396
アケーナ属 1739
アケノキ 54862
アケビ 13225
アケビカズラ 134038
アケビカヅラ 13225
アケビドコロ 131759
アケビモドキ 134038
アケビ科 221818
アケビ属 13208
アケボノアオイ 37656
アケボノオオバクサフジ 408554
アケボノオオフジイバラ 336832
アケボノサウ 380131
アケボノシュスラン 179613,179631,
179653
アケボノスギ 252540
アケボノスミレ 410498
アケボノセンノウ 363411
アケボノソウ 380131
アケボノツツジ 331462,331465
アケボノナツフジ 414569
アケボノハタザオ 376389
アケボノヒガン 83334
アケボノフジ 414562
アケボノミヤマシキミ 365943
アケボノムグラ 296091
アケボノヤイトバナ 280100
アケボノリンドウ 173854
アコ 408571
アコウ 165717,165841
アコウカミフア 1775
アコウクロダモ 233832
アコウクロモジ 233832
アコウグンバイ 73090
アコウザンショウ 417144
アコウシャクナゲ 330617
アコウスズメノヒエ 285508
アコウセンニンサウ 94708
アコウセンニンソウ 94708
アコウタヌキマメ 112013
アコウネッタイラン 399634
アコウバアヂサイ 199910
アコウバアヂサヰ 199910
アコウマイハギ 126601
アコカンテーラ属 4954
アコギ 165841
アコダ 114292
アゴニス属 11397
アコノキ 165717,165841
アゴノキ属 6761
アコミズギ 165841
アコン 68088
アサ 71218
アサアカシア 1168
アサガオ 208016
アサガオカラクサ 161418
アサガオガラクサ 161422
アサガオタバコ 344383
アサガオバナ 102884,103340
アサガオ属 293861
アサガホ 208016
アサガホカラクサ 161418
アサガホカラクサ属 161416
アサガホナ 207590
アサガホ属 293861
アサガラ 320880
アサガラ属 320878
アサカワグミ 141943
アサカワソウ 164287
アサギク 90894
アサギケナガバノタチツボスミレ 410332
アサギシザサ 264461
アサギズイセン 168183
アサギズイセン属 168153
アサギスイレン 267684
アサギスズメノヒエ 238639
アサギノヒトヘズイセン 262405
アサギヒルガネ 194467
アサギリ 244154
アサギリソウ 36225
アサギリンダウ 173451
アサクラザンショ 417305
アサザ 267825
アサザ属 267800
アサジギク 340918
アサシラゲ 374968
アサタ 231334
アサダ 276808
アサダ属 276801
アサツキ 15420,15709
アサノハイラクサ 402869
アサノハカエデ 2793
アサバソウ 298885
アサヒ 272898
アサヒエビネ 65975
アサヒカエデ 3154
アサヒカズラ 28422
アサヒカズラ属 28418
アサヒカヅラ 28422
アサヒカン 93326
アサヒコチョウラン 293633
アサヒゴテフ 293633
アサヒナミ 324919
アサヒマル 244210
アサヒミネ 82282
アサヒラン 143102
アサヒリョクチク 11359
アサヒレン 143440

アザブタデ 291785
アサマクワンザウ 191278
アサマサウ 82051
アサマスゲ 74354,75178
アサマソウ 82051
アサマツゲ 64285,64303,64345
アサマヒゴタイ 348569
アサマフウロ 174919,174920
アサマブドウ 404027
アサマリンドウ 173896
アサマリンドオ 173896
アザミゲシ 32429
アザミゲシ属 32411
アザミゴボウ 92268
アザミスミレ 410531
アザミタンポポ 196057
アザミヤグルマ 303082
アザミヤグルマギク 303082
アザミ属 91697
アサーラ属 45952
アザラ属 45952
アザレア属 330017
アサン 383189
アサ科 71202
アサ属 71209
アシ 295888
アジアタンポポ 384453
アシウスギ 113720
アシウテンナンショウ 33444
アシオスギ 113720
アシカキ 223987
アシカワグルミ 212636
アシキハダ 294233
アシクダシ 338985
アジサ 53465
アジサイ 199953,199956,199991
アジサイ科 200160
アジサイ属 199787
アシズリノジギク 89557
アシタカジャコウソウ 86838
アシタカジャコオソウ 86838
アシタカツツジ 331015
アシタバ 24384
アシダンテラ属 4523,175990
アシナガムシトリスミレ 299740
アシネタ属 4648
アシノクラアザミ 91785
アシヒカン 93326
アシブトワダン 110627
アシボソ 254034,254046
アシボソアカバナ 146599,146609
アシボソウリノキ 2846
アシボソスゲ 76203
アシボソ属 253980
アジマメ 218721
アシミナ属 38322
アシモチサウ科 138369
アシュラ 199049
アズキ 408839
アズキイチゴ 338886
アズキグミ 235812
アズキナシ 32988,369303
アズキモロコシ 369640
アスコセントウム属 38187
アスコセントラム属 38187
アズサ 53465,79257,223131,243371,316824
アズサミネバリ 53610
アスター 67314
アスター属 39910
アスチルベ 41789
アスティルベ属 41786
アズテキウム属 45993
アズテキューム属 45993
アストランチア 43309
アストリディア属 43320
アストロフィツム属 43492
アストロロバ属 43428
アスナロ 85345,306457,390680,390684
アスナロ属 390673
アスパーシア属 39267
アスパシア属 39267
アズパゾマ属 39277
アスパラガス 39120
アスパラガス・プルモーサス 39195
アスパラガス属 38904
アスハル 43330
アスヒ 43334,390680
アスフォデリーネ属 39435
アスフォデルス属 39449
アスベニ 43342
アスホ 43329
アズマイチゲ 24013
アズマイバラ 336831,337015
アズマカモメヅル 117700
アズマガヤ 203129
アズマギク 40334,150599,151008
アズマザサ 347361
アズマザサ属 37127,347334
アズマシノ 304014
アズマシャクナゲ 330532,330952
アズマスゲ 75073
アズマスミレ 409615
アズマタンポポ 384746
アズマツメクサ 391923
アズマツメクサ属 391918
アズマツリガネツツジ 250521
アズマナルコ 76254
アズマネザサ 304014
アズマホシクサ 151519
アズマミヤコザサ 347207
アズマヤマアザミ 92198
アズミイヌノヒゲ 151378
アズミノヘラオモダカ 14726
アズレオセレウス属 46008
アゼオトギリ 202059
アゼガヤ 225989
アゼガヤツリ 119687,322233,322304
アゼガヤモドキ 57924
アゼガヤ属 225983
アゼスゲ 73589,76545
アゼタウガラシ 231544
アゼタウガラシ属 231473
アゼタウナ属 110602
アゼテンツキ 166499
アゼトウガラシ 231544
アゼトウナ 110611
アゼトウナ属 110602
アゼナ 231559
アゼナルコ 74327
アゼナルコスゲ 74327
アゼナ属 231473
アセビ 298734
アセビ属 298703
アセボ 298734
アセボノキ 298734
アセムシロ 234363
アセリフィラム属 3818
アセンヤク 1120
アセンヤクノキ 1120
アソカノキ属 346497
アソタイゲキ 159543
アソノコギリソウ 3925
アダジャミー 93327
アタニー 93753
アダムスヨモギ 35095
アダムリング 93403
アタン 54620
アダン 281089,281138
アダンソニア属 7018
アダン科 280963
アーダ属 6988
アダ属 6988
アヂイチゴ 338557
アチイ科 385175
アヂサイ属 199787
アヂサヰ属 199787
アーチチョーク 117787
アヂマメ 218721
アヂモ 418392
アチョウ科 175839
アツアルバウツギ 127060
アッカハマメ 254854
アツカハマメ 310962
アツカワサワラ 85364
アッカワマメ 254854
アヅキ 294039,408839
アヅキササゲ 409096
アヅキナシ 32988
アッケシサウ 342859
アッケシサウ属 342847

アッケシソウ　342859
アッケシソウ属　342847
アッゲタカサゴガマズミ　407930
アヅサ　53631,79257
アッサムチャ　69644
アッサムニオイザクラ　238106
アッサムヤシ属　413085
アッサムレモン　93552
アッシ　401542
アッジュ　167897
アッタレア属　44801
アッニ　401542
アッバアサガオ　207884
アッバアサガホ　376536
アッバウオトリギ　180700
アッバキミガヨラン　416607
アッバキンチャクソウ　66287
アッバクコ　239108
アッバクスノキ　91378
アッバクズノハカエデ　2891
アッバコバンノキ　167092
アッバゴマノハグサ　355145
アッバサイシン　37723
アッバサクラソウ　314143
アッバシマザクラ　187638
アヅバシャリンバイ　329086
アッバシロテツ　249144
アッバスミレ　410226
アッバセンネンボク　104359
アッバセンネンボクラン　104359
アッバソバナ　7853
アッハダカンバ　53413
アッバタッナミ　355669
アッバタッナミキソウ　355821
ァッバタンポポ　384538
アッバチトセラン　346158
アッバックバネウッギ　181
アヅバヅゲトモモ　382649
アッバナシャクナゲ　330901
アッバナズナ　225326
アッバニガナ　210543
アッバノブドウ　20400
アッバハイチゴザサ　209081,209121
アッバハヒチゴザサ　209081
アッバヒゴタイ　348903
アッバヒサカキ　160474
アッバベンケイ　356994
アッバマュミ　157355
アッバモチ　204159
アッバモチノキ　204042
アッバヤドリギ　355322
アッペンディキュラ属　29807
アヅマイチゲ　24013
アヅマッメクサ　391923
アヅマッメクサ属　391918
アヅマネザサ　304014
アヅマヒガン　83328
アツミカンアオイ　37710
アツミゲシ　282713
アツモリサウ　120396
アツモリサウ属　120287
アツモリソウ　120396,120402
アツモリソウ属　120287
アテ　390680
アテアスヒ　390680
アデク　382499
アデク属　382465
アテツマンサク　185114
アデナンテ－ラ属　7178
アデナンドラ属　7113
アデ－ニア属　7220
アデニウム　7343
アデニウム属　7333
アデニューム　7343
アデニューム属　7333
アデノンコス属　7569
アドダン　104218
アドハトダ　214308
アトラス・シーダ－　80080
アトラスシーダ－　80080
アトリプレックス属　44298
アトロパ属　44704
アドロミスクス属　8417
アナカルディウム属　21191
アナカンプセロス属　21068
アナシベヤドリギ　385221
アナナス　21479
アナナスガヤバ　163117
アナナス科　60557
アナナス属　21472
アナマスミレ　410206
アニゴザンッス属　25216
アニス　299351
アニソミズ　298858
アニソミヅ　298858
アネクトキ－ルス属　25959
アネモネ　23767
アネモネ属　23696
アノイアズマギク　150449
アハ　361794
アハイチゴ　338945,338946
アハカ　280023
アハガヘリ　294988
アハガヘリ属　294960
アハキ　231355
アハゴケ　67367
アハスゲ　73608
アハタケ　297373
アハダチサウ　368480
アハダン　161384
アパテシア属　28932
アバナクンシラン　97218
アハブキ　249414
アハブキ属　249355
アハボ　91024
アハボスゲ　73945
アハユキニシキサウ　158534
アバラ科　337050
ア－ピウム属　29312
アビシニアガラシ　59348
アビシニアバショウ　145846
アヒヅシモッケ　371877
アピトン　133563
アピューム属　29312
アフイ　243862
アフェランドラ属　29121
アフギバセウ　327095
アフギバセウ属　327093
アフギヤシ　57122
アフギヤシ属　57116
アブクタラシ　249414
アブクマアザミ　91708
アブクマオオカニコウモリ　283782
アブクマトラノオ　54763
アブシント　35090
アフチ　248903
アフチノキ　248903
アブ－チロン属　837
アプテニア属　29853
アブノメ　136214
アブノメ属　136209
アフヒ　243862
アフヒカヅラ　68737,377884
アフヒカヅラ属　377877
アフヒゴケ　128964
アフヒゴケ属　128955
アフヒゴセウ　313178
アフヒバタネッケバナ　72689,72749
アフヒボクロ　265373
アフヒマメ　294010
アフヒモドキ属　174355,243877
アフヒラン　383127
アフヒ科　243873
アブラガキ　132339
アブラガヤ　82499,82523,353576,353935
アブラギ　86855,204292,204301
アブラギク　124775,124790,124806,124817
アブラキリ　406016
アブラギリ　406016
アブラギリ属　14538
アブラコ　143588
アブラシバ　76150
アブラスギ　216120,216125
アブラスギ属　216114
アブラスゲ　76150
アブラススキ　372406
アブラスズキ　372406
アブラスズキ属　139905
アブラダモ　91351
アブラチャン　231415
アブラッツジ　145727

アブラツバキ 69411
アブラトウヒレン 348336
アブラナ 59358,59603
アブラナ科 113144
アブラナ属 59278
アブラミ 346390
アブラヤシ 142257
アブラヤシ属 142255
アフリカアブラヤシ属 142255
アフリカイネ 275926
アフリカウンラン 263410
アフリカウンラン属 263374
アフリカエリカ 149138,149750
アフリカギヶ 31294
アフリカギョリュウ 383437
アフリカキンエノコロ 361902
アフリカキンセンカ 131191
アフリカキンセンカ属 131147
アフリカコマツナギ 206573
アフリカゴムノキ 1572
アフリカシタキヅル 245803
アフリカスイレン 267660
アフリカスミレ 342493
アフリカスミレ属 342482
アフリカセンボンヤリ 175172
アフリカタヌキマメ 112757
アフリカチリ-ヤシ 212561
アフリカチリヤシ 212561
アフリカチリ-ヤシ属 212560
アフリカチリヤシ属 212560
アフリカナガバモウセンゴケ 138275
アフリカヌカボ 12164
アフリカハウチハマメ 238444
アフリカハウチワマメ 238444
アフリカヒゲシバ 88352
アフリカヒナギク 235504
アフリカヒマワリ 31162
アフリカフウチョウソウ 95781
アフリカフジウツギ 62118
アフリカフヂウツギ 62118
アフリカホウセンカ 205346,205444
アフリカヤツデ 115239
アフリカヨメナ 86027
アフリカワスレナグサ 21933
アフリカンマホガニ- 216214
アブロニア 710
アブロニア属 687
アベクヌギ 324532
アベヌキ 324532
アベルモスクス属 212
アボイアザミ 91745
アホイアズマギク 150449
アポイアズマギク 151013
アポイカンバ 53353
アポイカンボ 53353
アポイキンバイ 312748
アポイクワガタ 317960
アポイザサ 347287
アポイシモツケ 371824
アポイタチツボスミレ 410520
アポイタヌキラン 73724
アポイツメクサ 31978
アポイマンテマ 363967
アポイヤマブキショウマ 37073
アボカド 291494
アボカド属 291491
アホノゲトン属 29643
アポノゲトン属 29643
アポロカクタス属 29712
アマ 231942,232001
アマキ 48518,177893,177897
アマギ 110227
アマギアオウツギ 413585
アマギアマチャ 200096
アマギカンアオイ 37688
アマギク 124826
アマギコアジサイ 199788
アマギシャクナゲ 330543
アマギツツジ 330071
アマギテンナンショウ 33385
アマギニシキウツギ 413584
アマギベニウツギ 413583
アマクキ 2896
アマクサ 177897,183023,216637
アマクサギ 96403,96408
アマクサミツバツツジ 330072,332083
アマグリ 78802
アマコギ 2896
アマシバ 381206
アマズラ 240813
アマズル 411903
アマゾンコカ 155065
アマゾンダコ 16494
アマゾンチドメグサ 200321
アマゾントチカガミ 230238
アマゾンユリ 155858
アマダイダイ 93765
アマダオシ 115007
アマタピオカノキ 244506
アマダマシ 266197
アマダマシ属 266194
アマチャ 199956,200100
アマチャズル 183023
アマチャヅル 183023
アマチャヅル属 182998
アマチャノキ 200125
アマヅラ 285157
アマヅル 411903
アマドクムギ 235353
アマドコロ 308613,308616
アマドコロ属 308493
アマナ 146253
アマナズナ 68841
アマナヅナ 68860
アマナヅナ属 68839
アマナラン 55563
アマナ属 18568,18569,400110
アマニュウ 24345
アマノガワ 44920,49629
アマノホシクサ 151222
アマハステビア 376415
アママツバ 175691
アマミアワゴケ 265364
アマミイケマ 117408
アマミイナモリ 272224
アマミイワウチワ 362252
アマミエビネ 65874
アマミエボシグサ 237506
アマミカジカエデ 2776
アマミカタバミ 277832
アマミキンゴジカ 362663
アマミクサアジサイ 73131
アマミケフチクタウ 83700
アマミゴエフマツ 299789
アマミゴヨオ 299789
アマミザラッカ 342580
アマミザンショウ 417156
アマミサンショウソウ 142771
アマミスミレ 409685
アマミセイシカ 331066
アマミタムラソウ 345335
アマミテンナンショウ 33345
アマミトンボ 302407
アマミナキリスゲ 74638
アマミナツズタ 285117
アマミナツヅタ 285117
アマミヒイラギモチ 203763
アマミヒサカキ 160571
アマミヒトツバハギ 167092
アマミフユイチゴ 338124
アマミボンテンカ 402250
アマミボンテンクワ 402250
アマメ 132230
アマモ 418392
アマモドキ 266197
アマモドキ属 266194
アマモ科 418405
アマモ属 418377
アマユリ 229888
アマリリス 196446
アマリリス属 18862,196439
アマ科 230854
アマ属 231856
アミガサギリ 14204,14217
アミガサギリ属 14177
アミガサソウ 1790
アミガサユリ 168586
アミガシ 233104
アミサンツルウメモドキ 80203
アミダガサ 396956
アミダケンチャ属 6826

アミダネヤシ属 129679
アミヂチュウ 94553
アミトスチグマ属 19498
アミバハコベ 375074
アミバヘビノボラズ 51331
アミバマユミ 80273
アミボソマンテマ 363606
アミメアマリリス 196454
アミメグサ属 166706
アミメヘイシソウ 347149
アミメミミカキグサ 403199
アミメロン 114189
アミルベッド 93628
アムグンスゲ 73673
アムソニア属 18908
アム-ルアカバナ 146605
アム-ルウツギ 126841
アム-ルギシギシ 339918
アム-ルシナノキ 391661
アム-ルスミレ 409687
アム-ルテンナンショウ 33245
アム-ルブドウ 411540
アム-ルボゥフウ 228597
アメシエラ属 19413
アメダマザクラ 83226
アメナレノダケ 24293
アメフリノキ 345525
アメリカアイ 206626
アメリカアカミキイチゴ 339310
アメリカアサガオ 123071,207839,208016
アメリカアサガラ 184731
アメリカアサガラ属 184726
アメリカアサダ 276818
アメリカアゼナ 231514,231515
アメリカアツモリ 120438
アメリカアブラヤシ 6129,6134
アメリカアリタソウ 86941
アメリカアワゴケ 67395
アメリカアヰ 206626
アメリカイカリソウ属 404610
アメリカイヌホオズキ 366934,367535
アメリカイモ 207623
アメリカイワイチョウ 265197
アメリカイワナンテン 228159
アメリカウスバスミレ 409751
アメリカウリカワ 342417
アメリカウロコモミ 30832
アメリカウンランモドキ 10145
アメリカエゴノキ 379304
アメリカエノキ 80698
アメリカオオバコ 301868
アメリカオオモミ 384
アメリカオキナグサ 23745
アメリカオダマキ 30007,30081
アメリカオニアザミ 92485
アメリカガキ 132466
アメリカガシワ 324262
アメリカカゼクサ 147594
アメリカガフクワン 126171
アメリカガフクワン属 126168
アメリカカラマッ 221904
アメリカカンアオイ 37574
アメリカカンアフヒ 37574
アメリカカンボク 408188
アメリカキカシグサ 337384
アメリカギク 56706
アメリカギク属 56700
アメリカキササゲ 79243
アメリカキミガヨラン 416607
アメリカキュウリグサ 301629
アメリカキンゴジカ 362662
アメリカクサイ 213507
アメリカクサネム 9665
アメリカクサレダマ 239589
アメリカグリ 78790
アメリカクロスグリ 333916
アメリカコナギ 193683
アメリカサイカチ 176903
アメリカザイフリボク 19241
アメリカザゼンソウ 381073
アメリカサナエタデ 309542
アメリカサンカヨウ 132678
アメリカシオン 40817,380930
アメリカシナノキ 391649
アメリカシモッケ 297842
アメリカシモッケソウ 166122
アメリカシャカ 264163
アメリカシャクナゲ 215395
アメリカシラカンバ 53549
アメリカシヲン 40817
アメリカズイショウ 385262
アメリカスイレン 267777
アメリカスギ 360565
アメリカスグリ 334030
アメリカスズカケ 302588
アメリカスズカケノキ 302588
アメリカスズメノヒエ 285469
アメリカズタ 285136
アメリカスノキ 403957
アメリカスミレサイシン 410587
アメリカスモモ 316206
アメリカスヰショウ 385262
アメリカセキショウモ 404536
アメリカセンダイハギ 389547
アメリカセンダングサ 53913
アメリカセンニチコウ 179238,179239
アメリカセンニチサウ 179239
アメリカセンニチソウ 179239
アメリカセンノウ 363312
アメリカゾウグヤシ属 298050
アメリカゾウゲヤシ 326649
アメリカソライロアサガオ 208250
アメリカタカサブロウ 141366
アメリカタッタソウ 212299
アメリカタテギ 316285
アメリカタヌキマメ 112405
アメリカダンドク 71199
アメリカチャボシ属 327538
アメリカチャボヤシ 327543
アメリカチャボヤシ属 327538
アメリカチョウセンアサガオ 123071
アメリカツガ 399905
アメリカヅタ 285136
アメリカツノクサネム 361380
アメリカツボサンゴ 194413
アメリカデイコ 154647,154648
アメリカトガサワラ 318580
アメリカトゲミギク 2612
アメリカトネリコ 167897
アメリカドルステニア 136470
アメリカナスモモ 316206
アメリカナデシコ 127607
アメリカナナカマド 369313
アメリカニガキ属 91479
アメリカニガナ 219458
アメリカニレ 401431
アメリカニワトコ 345580
アメリカニンジン 280799,285741
アメリカヌスビトハギ 126486
アメリカネズコ 390645
アメリカネナシカズラ 114986,115099
アメリカネム 345525
アメリカネムノキ 345525
アメリカネム属 345517
アメリカネリ 219
アメリカノウセンカズラ 70512
アメリカノキビ 151668
アメリカノリノキ 199806
アメリカハイネズ 213785
アメリカバウフウ 285741
アメリカバウフウ属 285731
アメリカハエドクソウ 295984
アメリカハギ 126278
アメリカハシバミ 106698
アメリカハッカ 250362
アメリカハナシノブ 175708
アメリカハナズオウ 83757
アメリカハナズハウ 83757
アメリカハナノキ 3505
アメリカハマグルマ 413559
アメリカハリグワ 240828
アメリカハリグワ属 240806
アメリカハリブキ 272707
アメリカハリフタバ 370782
アメリカハリモミ 298401
アメリカパルメット 341422
アメリカハンゲショウ 348085
アメリカバンマツリ 61295
アメリカヒイラギ 204114
アメリカヒイラギモチ 204114
アメリカヒトツバマツ 299854

アメリカヒノキ 85301
アメリカビユ 18668
アメリカフウ 232565
アメリカフウロソウ 174524
アメリカブクリョウサイ 285086
アメリカフジ 414565
アメリカフジバカマ 11171,158043
アメリカブダウ 411764
アメリカフヂ 414565
アメリカブドウ 411764
アメリカブナ 162375
アメリカフヨゥ 195032
アメリカボウフウ 285741
アメリカボウフウ属 285731
アメリカホウライセンブリ 81501
アメリカポッキリヤナギ 343767
アメリカホド 29298
アメリカホドイモ 29298
アメリカ・ホーリー 204114
アメリカマコモ 418080
アメリカマツ 318580
アメリカマルバヤナギ 343013
アメリカマンサク 185141
アメリカミコシガヤ 73703
アメリカミズオトギリ 394814
アメリカミズキ 104949
アメリカミズキンバイ 238166
アメリカミズバショウ 239517
アメリカミズユキノシタ 238221
アメリカミソハギ 19554,240041
アメリカミゾホオズキ 255221,255223
アメリカミネバリ 53338
アメリカミノゴメ 177559
アメリカムカシヨモギ 406266
アメリカムグラ 370852
アメリカムラサキ 66733
アメリカムラサキシキブ 66733
アメリカヤガミスゲ 76207
アメリカヤマゴバウ 298094
アメリカヤマゴボウ 298094
アメリカヤマタマガサ 82098
アメリカヤマナラシ 311398,311547
アメリカヤマボウシ 105023
アメリカユクノキ 94026
アメリカラフバイ 68321
アメリカリャウブ 96457
アメリカリョウブ 96457
アメリカロウバイ 68310
アメリカワタ 179890
アメリカヲダマキ 30007,30081
アメリフサスグレ 333916
アメンダウ 20890
アメンドウ 20890
アモマム属 19817
アモラギ 28997
アモラギ属 19952
アーモンド 20890
アヤギヌエマキ 186728
アヤコハギ 226891
アヤザクラ 212541
アヤスギ 85352,113687,113690
アヤツヅミ 182454
アヤナミ 197762
アヤニシギ 16605
アヤホ 86610
アヤメ 208806
アヤメグサ 5798
アヤメズイセン 168178,168189
アヤメ科 208377
アヤメ属 208433
アラアカ 323599
アライトタンポポ 384736
アライドツメクサ 342279
アライドヒナゲシ 282504
アライトヨモギ 35211
アラガシ 116100
アラガタオホサンキライ 366243
アラガタサンキライ 366243
アラカワカンアオイ 37650
アラカワシナノキ 391742
アラクサ 177893
アラグサンキライ 366392
アラクナンテ属 30526
アラクニス属 30526
アラゲアカサンザシ 109839
アラゲアヲダゴ 168014
アラゲエゾツツジ 389575
アラゲカエンソウ 244367
アラゲカギカズラ 401776
アラゲガマスミ 407785
アラゲカモジグサ 335267
アラゲキクボクチ 361104
アラゲキスミレ 409768
アラゲキセワタ 295150
アラゲキンバイ 312399
アラゲコメヒシバ 130731
アラゲサクラツツジ 331938
アラゲザンショウ 417304
アラゲサンショウソウ 288715
アラゲシュンギク 89704
アラゲタデ 309647,309898
アラゲチヂミザサ 272625,272632
アラゲツユクサ 115526,115587
アラゲツユクサ属 115521
アラゲツリガネツツジ 250510
アラゲトダシバ 37379
アラゲナツハゼ 403776
アラゲネザサ 304033
アラゲハナウド 192255
アラゲハンゴンサウ 339556
アラゲハンゴンソウ 339556,339569
アラゲヒョウタンボク 236122
アラゲミツバツツジ 331626
アラゲムラサキ 21920
アラゲメヒシバ 130826
アラシグサ 58022
アラスカヒノキ 85301
アラセイトウ 246478,246479
アラセイトウ属 246438
アラタマ 233531
アラナミ 162953
アラノキ 248895,248903
アラハゴ 2868
アラビアコザクラ 315087
アラビアゴム 1572
アラビヤコムノキ 1572
アラビヤゴムノキ 1427
アラビヤゴムモドキ 1427
アラフトズミ 243555
アラマンダ属 14871
アラミヒイラギモチ 203763
アラムシャアザミ 92127
アラム属 36967
アララギ 385355
アラレギク 21731
アランボガシシ 116100
アリアカシア 1613
アリアケカズラ 14873
アリアケカズラ属 14871
アリアゲキズラ 14873
アリアゲキヅラ 14873,14874
アリアゲキヅラ属 14871
アリアケスミレ 409723
アリウム 15445
アリエヒメマオ 313450
アリエヒメマヲ 313450
アリオカルプス属 33206
アリカノキ 346504
アリクリヤシ 33200,380559
アリクリヤシ属 33197
アリサンアイ 283209
アリサンアオカヅラ 341575
アリサンアザミ 91753
アリサンアハゴケ 265359
アリサンアハゴケ属 265353
アリサンアハブキ 249369
アリサンアハモリ 41827
アリサンアブラギク 124772
アリサンアマドコロ 308498
アリサンアヰ 283209
アリサンアヲカヅラ 341575
アリサンイタビカツラ 165623
アリサンイチゴ 338281
アリサンイチャク 322779
アリサンイボタ 229587
アリサンウヅラ 179577
アリサンエビネ 65886
アリサンオサラン 148754
アリサンオホバライチゴ 338305
アリサンカウジ 31387,31613
アリサンガシ属 233099

アリサンカタバミ 277878
アリサンキヌラン 417779
アリサンクサヰ 213507
アリサンクスクスラン 267927
アリサンゴエフ 300083
アリサンコセウノキ 122377
アリサンコセウノギ 121621
アリサンサウ 265359
アリサンサカキ 8247
アリサンササガヤ 254036
アリサンサナギイチゴ 339135
アリサンサルナシ 6530
アリサンサンキライ 366238
アリサンサンショウソウ 288762
アリサンザンセウ 417330
アリサンサンセウヅル 288762
アリサンシックナゲ 331926
アリサンシデ 77314
アリサンシロバナシキミ 204575
アリサンスズムシサウ 232125
アリサンセンブリ 380117
アリサンソヨゴ 203851
アリサンタタムラサウ 345076
アリサンタビラコ 397410
アリサンタブ 240707
アリサンタマツリスゲ 73755
アリサンチゴザサ 209030
アリサンチドメグサ 200364
アリサンツルウメモドキ 80285
アリサンツルマユミ 157888
アリサンテンナンシャウ 33332
アリサンナヅナ 72689,72749
アリサンニレ 401642
アリサンニンドウ 235633
アリサンヌカボ 12000
アリサンネズミガヤ 259671
アリサンハグマ 12676
アリサンハコベ 374757
アリサンハヒノキ 381272
アリサンバライチゴ 339208
アリサンハラン 39520
アリサンバラン 39520
アリサンヒサカキ 160535
アリサンヒメバラン 288639,288658
アリサンヒラギズイナ 210407
アリサンヒラギズヰナ 210407
アリサンホザキイチエフラン 243084
アリサンマップサ 351012
アリサンマンリャウ 31387,31613
アリサンミズ 298869
アリサンミヅ 298869
アリサンミヤマウラジロイチゴ 338808
アリサンミヤマシキミ 365919
アリサンムエフラン 86725
アリサンムサシアブミ 33264
アリサンムシャラン 86725
アリサンムヨウラン 86725
アリサンモエギスゲ 75632
アリサンモクレイシ 254293
アリサンモクレイシ属 254283
アリサンヤウラクラン 267927
アリサンヤブガラシ 387825
アリサンヤマザクラ 316874
アリサンリドホシラン 261470
アリサンヲサラン 148754
アリスガハゼキシャウ 5809
アリスガワセキショウ 5809
アリストテーリア属 34394
アリセム 35090
アリゾナイトスギ 114652
アリゾナウシノケグサ 163818
アリタサウ 86941,139678
アリダサウ 264958
アリタソウ 86934,86941,139678
アリッスム 235090
アリドウシラン 261474
アリドウシラン属 261467
アリドオシ 122040
アリドオシラン 261474
アリドオシラン属 261467
アリドオシ属 122027
アリドホシ 122040
アリドホシラン 261474
アリドホシラン属 261467
アリドホシ属 122027
アリナシ 323272
アリノスダマ属 261482
アリノタフグサ 179438
アリノタフグサ科 184993
アリノタフグサ属 184995
アリノトウグサ 179438
アリノトウグサ科 184993
アリノトウグサ属 184995
アリノミ 323268
アリハラススキ 255849
アリマウマノスズクサ 34332
アリマグミ 142114
アリマコスズ 347230
アリマサウ 82051
アリマシノ 347353
アリマソウ 82051
アリマツゲ 64304
アリモリサウ 98223
アリモリサウ属 98220
アリモリソウ 98223
アリモリソウ属 98220
アリューム属 15018
アルカネット 21959
アルギロデルマ属 32689
アルケミラ属 13951
アルストレメーリア属 18060
アルストロメリア ペレグリナ 18074
アルスロセレウス属 36739
アルセム 35090
アルタイスミレ 409682
アルタイノギク 193918,193923
アルタイミセバヤ 200788
アルタイユキノシタ 52514
アルテヤ 18173
アルテンスタインオニソテツ 145220
アルトカルプス属 36902
アルニカ 34752
アルニカ属 34655
アルネービア属 34600
アルネルスゲ 73767
アルピニア属 17639
アルファルファ 247456
アルブーカ属 13729
アルブスウネドウ 30888
アルブス属 30877
アルブッス属 30877
アルブートゥス属 30877
アルペンローゼ 330695
アルポフィムム属 34895
アルマトセレウス属 34419
アルメリア属 34488
アルヤマカンコノキ属 60162
アルリウム属 15018
アルンディナ属 37109
アレオコックス属 30569
アレカヤシ 31680,89360
アレカヤシ属 89345
アレカールブラ 2553
アレカ属 31677
アレクイパ属 32357
アレクサンドリアセンナ 78227
アレチアオゲイトウ 18786
アレチアザミ 82022,92384
アレチアマ 231963
アレチイヌノフグリ 407262
アレチイネガヤ 300728
アレチウイキョウ 63475
アレチウシノシタグサ 21920
アレチウリ 362461
アレチエゾノギシギシ 339881
アレチオグルマ 194256
アレチガラシ 196952
アレチギシギシ 339989
アレチキンギョソウ 28631
アレチクグ 118458
アレチクサヨシ 293755
アレチコギシギシ 339883
アレチシオン 58265
アレチシオン属 58261
アレチタチドジョウツナギ 321248
アレチタバコ 266063
アレチナズナ 18327
アレチニガナ 111017
アレチヌスビトハギ 126507
アレチノエンドウ 408493
アレチノギク 103438

アレチノゲシ 368842
アレチノチャヒキ 60985
アレチハナガサ 405819
アレチハマスゲ 119139
アレチヒジキ 104786
アレチベニバナ 77716
アレチボロギク 360172
アレチマツヨイグサ 269481
アレチムラサキ 190597
アレチモウズイカ 405796
アレッ-サ属 32371
アレッポマツ 299964
アレナレブシ 5292
アレニフェラ属 32350
アレノノギク 41398,193939
アレモウ 93558
アロイノプシス属 17433
アロエ 16598
アロエ科 17424
アロエ属 16540
アロガ-シア属 16484
アロ-ニア属 34845
アロプレクタス属 16174
アロモルフィア属 16006
アロヤドラ属 34966
アロ-ラマツ 299842
アロンソア属 17475
アワ 361794
アワイチゴ 338945
アワガエリ 294988
アワガエリ属 294960
アワガラ 231334
アワキ 231355,249414
アワコガネギク 124775,124817
アワゴケ 67367
アワゴケ科 67334
アワゴケ属 67336
アワスゲ 73608
アワタケ 297373
アワタチソウ 368483
アワダチソウ 368480
アワタバコ科 175306
アワダン 249168
アワチドリ 310893
アワノミツバツツジ 330586
アワノミネザサ 347254
アワブキ 249414
アワブキ科 341582
アワブキ属 249355,341475
アワボ 91024
アワボスゲ 73945
アワミツバツツジ 331022
アワムヨウラン 223669
アワムヨフラン 223658
アワモリサウ 41817
アワモリショウマ 41817
アワモリソウ 41817
アワモリハッカ 321956
アワユキギク 376410
アワユキサウ 41795
アワユキソウ 41795
アワユキニシキソウ 158534
アワユキハコベ 374916
アワユリ 230058
アヰ 309893
アヰカヅラ 245854
アヰナヘ 256154,256161
アヰナヘ属 256147
アヰバナ 100961
アヲアカザ 87048
アヲアケビ 13225
アヲアブラガヤ 353856
アヲイチゴツナギ 306146
アヲウキクサ属 224359
アヲウメモドキ 328904
アヲカウガイ 213355
アヲカウガイゼキシャウ 213355
アヲカゴノキ 6762
アヲカゴノキ属 6761
アヲガシ 240604
アヲガシカエデ 3105
アヲカヅラ 341513
アヲカヅラ科 341582
アヲカヅラ属 341475
アヲガヤツリ 119262
アヲカラムシ 56240
アヲガンピ 414250
アヲキ 44915
アヲキバ 44915
アヲキラン 147314
アヲギリ 166627
アヲギリ科 376218
アヲギリ属 166612
アヲキ属 44887
アヲグスモドキ 240726
アヲゲムラサキ 66897
アヲコメガヤ 249015
アヲサギサウ 183921
アヲサンゴ 159975
アヲサンゴジュボク 407977
アヲジク 34455
アヲジクスノキ 404066
アヲシラベ 497
アヲスゲ 73913
アヲヂク 34455
アヲヂクマユミ 157713
アヲチドリ属 291188
アヲツヅラフヂ 97947
アヲツヅラフヂ属 97894
アヲツリバナ 157970
アヲテンツキ 166557
アヲテンマ 171928
アヲドロ 311480
アヲトンボ 183921
アヲナシタデ 309570
アヲノクマタケラン 17656
アヲノツガザクラ 297016
アヲノリュウゼツラン 10787
アヲハイモ 65237
アヲバウスノキ 404047
アヲハダ 204008
アヲバノキ 381149
アヲビユ 18810,18848
アヲボウモミ 389
アヲマツリ 305172
アヲミズ 299024
アヲミヅ 299024
アヲミヅバヴシ 298971
アヲモミトドマツ 414
アヲヤギサウ 405633
アンキストロキラス属 22040
アンク-サ属 21914
アングスチフオリアポプラ 311233
アングラタポプラ 311231
アングレカム属 24687
アングロア属 25134
アンゲスチフォリアスモモ 316219
アンゲロ-ア属 25134
アンゲロンソウ 24523
アンゲロンソウ属 24519
アンコクギョク 264342
アンザンアヤメ 208897
アンザンジュ 21805
アンザンジュ属 21804
アンシストロカクタス属 22017
アンシストロキルス属 22040
アンシストロクラヅス科 22047
アンジャベル 127635
アンズ 34475,34477
アンスリスクス属 28013
アンスリユム 28084
アンスリュ-ム属 28071
アンスリュム属 26950
アンセイカン 93470
アンセミス属 26738
アンセリァ属 26273
アンゼリカ 30932
アンセリカム属 26950
アンソクカウ 379312
アンソクコウノキ 379312
アンタオニクヅク 261457
アンタオムラサキ 66831
アンダカジャ- 214971
アンダ-ソンムラサキツユクサ 394088
アンチュ-サ 21949
アンチュ-サ属 21914
アンチルココヤシ属 333840
アンチルハリクジャクヤシ 12773
アンテギッバエウム属 26293
アンデスカゼクサ 148038
アンデスチリ-ヤシ属 283407

アンデスチリヤシ属　283407
アンデスロウヤシ属　84326
アンテリクム属　26950
アントリ－ザ属　27664
アンドロサセ属　23107
アンドンマユミ　157777
アンドンマユミ　157753
アンナンアカメガシハ　243420
アンナンウルシ　332869
アンナンカラスウミ　182575
アンナンカラスウミ属　182574
アンナンパペダ　93430,93560
アンニヤモンニヤ　87729
アンバリ麻　194779
アンピンテンツキ　166315
アンペラ　225661,225664
アンペライ　93927
アンペライネビキグサ　93927
アンペラサウ　225664
アンペラソウ　225664
アンペラヰ　225664
アンペラヰネビキグサ　93927
アンペラ属　225659
アンベルボア属　18946
アンマッコカエデ　2896
アンマロク　296554
アンラクワ　317602
アンランカ　317602
アンランジュ　317602
アンレデラ属　26264
イ　213036
イイギリ　203422
イイギリ科　166793
イイギリ属　203417
イイタカムシトリスミレ　299763
イイデリンドウ　173665
イイヌマムカゴ　302370
イイラギ　72749
イウコクラン　232164
イウレイサウズヰシャウレン　258044
イウレイソウ　258044
イウレイタケ　258044
イウレイバナ　258044
イウレイン　130043
イエギク　124826
イエロ－サルタン　81408
イェンゼノボトリア属　212323
イオウウチワ　273070
イオウクマタケラン　17724
イオウソウ　239902
イオウトウキイチゴ　338196,338942
イオウトウフヨウ　195077
イオウノボタン　248735
イオヅギョク　153354
イオノプシジュ－ム属　212474
イオノプシス属　207446
イガオナモミ　415023
イガカウゾリナ　143464
イガカヤツリ　119410
イガガヤツリ　393205
イガギク　68082
イガギリア　175707
イガクサ　333672
イカコ　89816
イガコ属　89812
イガサクラ　175707
イカダカズラ　57868
イカダカズラ属　57852
イカダカヅラ　57868
イカダカヅラ属　57852
イガタツナミ　355558
イガトキンソウ　368528
イガニガクサ　203044,203062
イガニガクサ属　203036
イガホオズキ　297626,297628
イガホオズキ属　297624
イガボシバ　394380
イガホタルイ　352195
イガホビユ　18804
イガホホズキ　297628
イガホホヅキ　297628
イガホホヅキ属　297624
イガマメ　271280
イガマメ属　271166
イガヤグルマギク　81382
イカリサウ　146995,147021
イカリサウ属　146953
イカリスギ　113699
イカリソウ　146995,147003,147021
イカリソウ属　146953
イカンリュウ　163463
イキサコザクラ　314461
イキシア属　210689
イギシオリリオン属　210988
イキノサイコ　63820
イギリスアヤメ　208679,208944
イギリスナラ　324335
イクキ　386696
イグサ　213036
イグサ科　212752
イグサ属　212797
イクリョネモモ　316761
イケノミズハコベ　67393
イケマ　117597
イケマ属　117334
イサベリア　209025
イサベリア属　209024
イザヨイバラ　336885
イザリタデ　291828
イサワオオバコ　301872
イサンッッジ　331420
イシイモ　16512
イシウスイモ　20125
イシガキイトテンツキ　166434
イシガキカラスウリ　396198
イシガキキヌラン　417786
イシガキクマタケラン　17766
イシガキスミレ　410645
イシカッラ　319833
イシゲヤキ　401581
イシコロケサ属　233459
イシコロマツバギク属　233459
イジセンリョウ属　241734
イシソネ　77312
イシゾネ　77276
イシゾノ　77312
イシダサウ　273979
イシダサウ属　273976
イシダソウ　273979
イシダソウ属　273976
イシダテクサタチバナ　409422
イシヅチイチゴ　338576
イシヅチウスバアザミ　92065
イシヅチザクラ　83324
イシヅチテンナンショウ　33361
イシヅチボウフウ　24461
イシヅチミズキ　380440
イシッヅリ　349216
イシナシ　323268
イシナラ　324384
イシノナズナ　137010
イシマメ　276550
イシミカハ　309564
イシミカワ　309564
イシモチサウ　138331
イシモチソウ　138328,138331,138338
イシャダオシ　155897
イジュ　350932,350945
イシワリソウ　274156
イシワリソウ属　274155
イス　134946
イズ　134922
イスイ　308616
イズイ　308613
イズカニコウモリ　283788
イズカモメヅル　409474
イスガヤ　353878
イズコゴメグサ　160178
イスズギョク　163329
イスズグミ　142103
イズセンリョウ　241781
イズセンリョウ属　241734
イ－スタンホワイトパイン　300211
イズドコロ　131531
イズナットオダイ　159841
イスノキ　134946
イスノキ属　134918
イズノシマウメバチソウ　284595
イズノシマダイモンジソウ　349350
イズノシマホシクサ　151556
イスパニヤガヤ　85058

イスパニヤガヤ属 85054
イズハハコ 103509
イズハハコ属 103402
イスパンサルビア 345083
イズホオコ 103509
イズミケンチャ属 200180
イスラヤ属 209503
イス属 134918
イセアオスゲ 74976
イセウキヤガラ 353707
イセナ 59435
イゼナガヤ 148810
イセナデシコ 127574,127667
イセノカンアオイ 37716
イセハナビ 85944,387136
イセハナビ属 378079
イセビ 407897
イセブ 407897
イセボウフウ 176923
イセリンドウ 173182
イソアオスゲ 75339
イソウギョク 163422
イソカンギク 41084
イソギク 89659
イソキルス属 209557
イソコマツ 356802
イソザンショウ 276550
イソザンセウ 276550
イソジラキ 233949
イソスゲ 73925
イソスミレ 410018
イソチジミ 160458
イソツツジ 223901,223904,223905,223913,331657
イソツツジ属 223888
イソテンツキ 166431
イソトマ属 210313
イソニガナ 210531,210592
イソノカリガネ 276550
イソノキ 328665
イソノギク 193930
イソハナビ 230816
イソバラモンジン 354904
イソヒサカキ 160458
イソビワ 233949
イソフサギ 294882
イソフサギ属 294877
イソフジ 369134
イソフヂ 369134
イソベノマツ 109174
イソホウキギ 217345
イソマカキラン 147241
イソマツ 230816,230820
イソマツ科 305165
イソマツ属 230508
イソミヅナ 361667
イソヤマアオキ 97919
イソヤマアヲキ 97919
イソヤマダケ 97919
イソヤマテンツキ 166313,166315,166482
イソユリ 229915
イソヨモギ 35211
イダイアカバナ 146840
イタイタグサ 403028
イタグサ 167456
イタジイ 78916
イタチアザミ 348902
イタチガヤ 306832
イタチガヤ属 306828
イタチキ 161373
イタチグサ 167456
イタチササゲ 222707
イタチジソ 170060
イタチハギ 20005
イタチハゼ 167456
イタツラ 165628
イタドリ 162532,328345
イタドリ属 328328
イタビ 164947
イタビカズラ 165628
イタビカヅラ 165002,165628
イタブ 164947
イタヤ 3154
イタヤカエデ 3154,3398,3404
イタヤミネバリ 53465
イタヤメイゲツ 3604
イタリアウイキョウ 167146
イタリアサイプレス 114753
イタリアソウ 258793
イタリアニンジンボク 411189
イタリアホソヒバ 114768
イタリアン・ブロッコリ－ 59545
イタリアン・ライグラス 235315
イタリアンカンナ 71193
イタリ－ウイキョウ 167146
イタリカポプラ 311400
イタリ－マンテマ 363479
イタリヤマナラシ 311400
イチイ 116098,385355
イチイガシ 116098
イチイモドキ 360565
イチイモドキ属 360561
イチイ科 385175
イチイ属 385301
イチウ 175813
イチガシ 116098
イチゲイチャクサウ 257374
イチゲイチャクサウ属 257371
イチゲイチャクソウ 257374
イチゲイチャクソウ属 257371
イチゲキスミレ 410326,410763
イチゲコザクラ 315101
イチゲサクラソウ 315101
イチゲスミレ 410326,410763
イチゲソウ 23868,23936
イチゲソウ属 23696
イチゲツリフネ 205422
イチゲノコギリソウ 3918
イチゲフウロ 174906
イチコクマメ 294010
イチゴチリメンウロコヤシ 156116
イチゴツナギ 305334,306003,306133
イチゴツナギ属 305274
イチゴノキ 30888
イチゴ属 167592
イチシ 379374
イチジク 164763
イチジクグワ 165726
イチジク属 164608
イチヂク 164763,164947
イチヂク属 164608
イチネンアラ 232001
イチハツ 208875
イチビ 1000
イチビ属 837
イチベイ 259094
イチャウ 175813
イチャウ科 175839
イチャウ属 175812
イチヤクサウ 322823,322842
イチヤクサウ科 322935
イチヤクサウ属 322775
イチヤクソウ 322842
イチヤクソウ科 322935
イチヤクソウ属 322775
イ－チャンジェンシス 93495
イ－チャンパペダ 93495
イ－チャンレモン 93869
イチョウ 175813
イチヨウシュウ 167092
イチヨウラン 121435
イチョウ属 175812
イチリンソウ 23936
イチリンソウ属 23696
イチリンバイモ 168467
イチロベゴロシ 104696
イチヰ 385355
イチヰガシ 116098
イチヰモドキ 360565
イチヰモドキ属 360561
イチヰ科 385175
イチヰ属 385301
イツキ 124925
イッケイキウクワ 116851
イッケイキュウカ 116851
イツシ 379374
イッシオウカ 368073
イッジゴウ 3921
イツシベマウセンゴケ 138273
イツシベユヅリハ 122713
イヅシベユヅリハ 122710

イッショウチザサ　347244
イッスンキンカ　368237
イッスンテンツキ　166367
イヅセンリャウ　241781
イヅセンリャウ属　241734
イヅセンリョウ属　241734
イッチャ　379374
イヅハウコ　103509
イヅハハコ　103509
イッボンスゲ　76513
イッポンスゲ　76513
イツボンネギ　15292
イツモヂシャ　53257
イテフ　175813
イテフ属　175812
イトアイ　329689
イトアオスゲ　73913
イトアシ　329689
イトアゼガヤ　226021
イトイ　213272
イトイチゴツナギ　305720
イトイヌノハナヒゲ　333560
イトイヌノヒゲ　151288
イトイヌハナヒゲ　333560
イトイバラモ　262116
イトウウズ　5403
イトウシノケグサ　164272
イトウセキコク　125218
イトウソウ　228992
イトウミヅキ　106684
イトウリ　238261
イトオグルマ　359363
イトカモメズル　117721
イトギク　188437
イトキッネノボタン　325612
イトキンスゲ　74733
イトキンポウゲ　326303
イトクズモ　417053,417060
イトクズモ科　417073
イトクズモ属　417045
イトクヅモ　417060
イトクヅモ科　417073
イトクヅモ属　417045
イトクリ　30031
イトクリサウ　30031
イトゲイトウ　80469
イトコヌカグサ　12034
イトザキヒメマオ　56357
イトザキヒメマヲ　56357
イトザクラ　83327,83337,316636
イトシバ　418453
イトシャジン　70271
イトスギ　113693
イトスギ属　114649
イトスゲ　74497,75792
イトススキ　255887
イトスズメガャ　147609
イトスズメガヤ　147509,147553
イトスナヅル　78733
イトセダカウロコヤシ　298836
イトタデ　291786,309203
イトタヌキモ　403191
イトツメクサ　342226
イトツリバナ　157777
イトテンツキ　63258
イトトリゲモ　262039
イトナシイトラン　416610
イトナスゲ　75080
イトナデシコ　183223,183231
イトナデシコ属　183157
イトナルコスゲ　75080
イトノヤマスゲ　74406,74408
イトバアナナス　392005
イトバアヤメ　208882
イトバアワダチソウ　368134
イドバガマ　401128
イトバカラマツ　388509
イトバギク　351917
イトハコベ　374882
イトバコメガヤ　249081
イトバシャジン　7828
イトバショウ　260241
イトバス　349936
イトバゼリ　269335
イトバタコノキ　281037
イトバトウダイ　158857
イトハナビテンツキ　63253
イトバニガナ　210541
イトバハハキギ　217353
イトバハルシャギク　104616
イトバヒヨドリ　158066
イトバボウフウ　392786
ィトバモウセンゴケ　138289
イトハユリ　230005,230009
イトバヨツバムグラ　170696
イトハリイ　143365
イトヒカゲスゲ　73992,74847
イトヒキサギサウ　183983
イトヒキサギソウ　183983
イトヒキスゲ　76005
イトヒバ　85348,114690,390628
イトヒメハギ　308403
イトフスマ　31967
イトホロシ　367328
イトマキイタヤ　3412
イトマキグサ　301694
イトマキシマモミヂ　3278
イトマキソウ　301694
イトマユミ　157777
イトムグラ　170461
イトメヒシバ　130627,130703
イトモ　312203,312236,404555
イトヤナギ　343070
イトユリ　230005,230009
イトヨモギ　35130,36020
イトラン　116880,416649
イトラン属　416535
イドリア属　203444
イトヰ　213272
イナコゴメグサ　160228
イナコスズ　347318
イナゴマメ　83527
イナゴマメ属　83523
イナテキリスゲ　73557
イナトウヒレン　348385
イナバシュスラン　269026
イナバラン　269026
イナバラン属　269006
イナヒロハテンナンショウ　33445
イナベアザミ　92182
イナモリサウ　318184
イナモリサウ属　318183
イナモリソウ　272221,318184
イナモリソウ属　318183
イヌアイュンジュ　20005
イヌアカシャ　334976
イヌアデク　382535
イヌアハ　361728
イヌアリタサウ　87147
イヌアワ　361728
イヌアワガエリ　294989
イヌアヰュンジュ　20005
イヌイ　213110
イヌイシモチソウ　138376
イヌイトモ　312199
イヌイボタ　229569
イヌウメモドキ　204239,204242,204248
イヌエ　290952
イヌエゾボウフウ　8812
イヌエノコロ　361681
イヌエビ　411669
イヌエンゴサク　182634
イヌエンジュ　240114
イヌエンジュ属　240112
イヌオオバコ　302068
イヌオシロイバナ　255735
イヌカイノウコウ　406392
イヌカウジュ　259323
イヌカウジュ属　259278
イヌカキネガラシ　365568
イヌカサスゲ　74650
イヌガシ　116098,264015
イヌカヅラ　375870
イヌカミツレ　397964,397976
イヌカモジグサ　144320,144327,335547
イヌガヤ　82499,82523,82524
イヌガヤ科　82493
イヌガヤ属　82496
イヌガラシ　336211
イヌガラシ属　262645,336166
イヌカラマツ　317822

イヌカラマツ属 317821
イヌガルカヤ 117181
イヌカワジサ 407430
イヌカワヂサ 407430
イヌカンコノキ 78137
イヌカンコノキ属 78095
イヌカンゾウ 177932
イヌキクイモ 189063
イヌキグマン 105680
イヌキビ 281917
イヌギリ 203422,406016
イヌクグ 245556
イヌグス 240707
イヌクテガワザサ 347224
イヌクログワイ 143122,143123
イヌクログワヰ 143122
イヌゲヤキ 204008
イヌゲンゲ 181695
イヌケンチャ属 85842
イヌコウジュ 259323
イヌコウジュ属 259278
イヌコウゾリナ 196086
イヌコオジュ 259323
イヌゴシュユ 161323
イヌコハコベ 375033
イヌゴボウ 298093
イヌゴマ 373139,373147
イヌゴマ属 373085
イヌコモチナデシコ 292664
イヌコリヤナギ 343529
イヌザクラ 280016
イヌサフラン 99297
イヌサフラン属 99293
イヌサンキライ 194120
イヌサンジソウ 94138
イヌザンショウ 417330
イヌザンセウ 417330
イヌジシャ 104175,104218
イヌジソ 259284
イヌシデ 77401
イヌシデ属 77252
イヌシバ 375774
イヌシバ属 375764
イヌシュロチク 329184
イヌショウマ 91024
イヌシロソケイ 211918
イヌシロネ 239194
イヌスイバ 144880
イヌスキ 178019
イヌスギ属 178014
イヌスミレ 409630
イヌセンブリ 380389
イヌソヨゴ 203952
イヌタデ 291822,308907,309345,309494,309624
イヌタヌキモ 403110
イヌタムラソウ 307125
イヌタムラソウ属 307124
イヌダラ 215442
イヌチガヤ 289096
イヌヂシャ属 104147
イヌチョモキ 388114
イヌツゲ 203670
イヌツツジ 331258
イヌツヅラ 375870
イヌツバキ 229485
イヌツルウメモドキ 80263
イヌデマリ 407897
イヌトウキ 24466
イヌトウゴウソウ 230544
イヌドウナ 283824
イヌトウバナ 97029
イヌドオナ 283824
イヌトキンソウ 380792
イヌトキンソウ属 380791
イヌトクガワザサ 347291
イヌナギナタガヤ 412407
イヌナシ 323116,323268,323330
イヌナズナ 137133
イヌナズナ属 136899
イヌナヅナ 137133
イヌナヅナ属 136899
イヌナツメ 418169,418184
イヌナツメヤシ 295483
イヌナミキソウ 355733
イヌニガクサ 388300
イヌニンジン 9832
イヌヌマトラノオ 239789
イヌネバリタデ 291998
イヌノグサ 77201
イヌノハナヒゲ 333518,333576,333672,333675
イヌノハナヒゲ属 333477
イヌノヒゲ 151380
イヌノヒゲモドキ 151475,151487
イヌノフグリ 406973,407292,407294
イヌハギ 226989
イヌハクカ 264897
イヌパココヤシ属 14862
イヌハッカ 264897
イヌハッカ属 264859
イヌハナイバナ 260769
イヌハナタデ 291868
イヌハナビスゲ 74513,74514
イヌハハキギ 45844,45853
イヌハハキギ属 45843
イヌビエ 140356,140367,140384
イヌヒキオコシ 345315
イヌヒデリコ 166454
イヌヒナビスゲ 74513
イヌピナンガ 299669
イヌビハ 164947
イヌヒメコヅチ 345340
イヌヒメシロビユ 18668
イヌビユ 18670,18765
イヌヒレアザミ 73504
イヌビワ 164947
イヌビワ属 164608
イヌフクド 35481,35972
イヌブシ 53460
イヌフトイ 353569,353840
イヌブドウ 411669
イヌブナ 162382
イヌヘラヤシ属 194134
イヌホウキギ 45844
イヌホオズキ 367416
イヌホタルイ 352200
イヌホホズキ 367416
イヌホホヅキ 367416
イヌマキ 306457
イヌマキ科 306345
イヌマキ属 306395
イヌマメツゲ 203672
イヌマルバヤナギ 343754
イヌミゾハコベ 142574
イヌミヤマムラサキ 301626
イヌムギ 61008
イヌムラサキ 233700
イヌムラサキシキブ 66728
イヌメドハギ 226860
イヌメヒシバ 130656,130768
イヌモチ 203909,203916
イヌヤクシソウ 110626,210548
イヌヤマキビ 282200
イヌヤマハッカ 209851
イヌヤマブキソウ 104415
イヌヤマモモ 142396
イヌヤマモモサウ 172203
イヌヤマモモソウ 172203
イヌヨメナ 150464
イヌヨモギ 35733
イヌラ属 207025
イヌリンゴ 243667
イヌヰ 212890
イネ 275958
イネガヤ 300722
イネ科 180078
イネ属 275909
イノカルプス属 206998
イノゲラン 111558
イノコズチ 4273,4275
イノコズチ属 4249
イノコヅチ 4259,4273,4275
イノコヅチモドキ 115738
イノコヅチモドキ属 115697
イノコヅチ属 4249
イノンド 24213
イノンド属 24208
イハアカバナ 146606
イハイズリ 311890
イハイヅル 311890

イハイテフ　88977
イハイヌビハ　165444
イハウツギ　126956,126963
イハウメ　127916
イハウメヅル　80189
イハウメ科　127926
イハウメ属　127910
イハカガミ属　362242
イハガク　200092
イハガサ　371836
イハカヅラ　349936
イハガネ　273903,273906
イハガネサウ　356917
イハガラミ　351797
イハガラミ属　351780
イハガリヤス　65471
イハカンスゲ　75287
イハキヌヨモギ　35779
イハギリサウ属　273830
イハギリソウ属　273830
イハキンバイ　312351
イハクロウメモドキ　328816
イハザンセウ　417176
イハシモツケ　372029
イハジャカウサウ　391347
イハスグリ　334157
イハスゲ　73906
イハスデ　77411
イハスミレ　409958
イハセサウ　58592
イハタバコ　101681
イハタバコ科　175306
イハタバコ属　101680
イハックバネウツギ　416841
イハヅサ　231403
イハヅタヒ　319833
イハツツジ　403967
イハテタウキ　229485
イハナシ　404051
イハナンテン　228178
イハナンテン属　228154
イハニガナ　210673
イハノガリヤス　65471
イハハギ属　226075
イハハタザホ　30437
イハヒゲ　78631
イハヒゲ属　78622
イハヒノキ　261739
イハブキ　349936
イハフジ　205876
イハフヂ　205876
イハミツバ　8826
イハモモ　404051
イハヤナギ　372101
イハユキノシタ　383890
イハユキノシタ属　383887
イバラ　336783
イバラバス　160637
イバラモ　262067
イバラモ科　261992,262013
イバラモ属　262015
イバラ属　336283
イハワウギ　188177
イバワウギ　188177
イバワウギ属　187753
イヒギリ　203422
イヒギリ科　166793
イヒギリ属　203417
イヒダイゲキ　159159
イブキ　213634
イブキエンドウ　408611
イブキカモジグサ　144228
イブキキンバイサウ　399515
イブキキンバイソウ　399515
イブキクガイソウ　407467
イブキコゴメグサ　160177
イブキザクラ　83120
イブキシモツケ　371897
イブキジャカウサウ　391347
イブキジャカウサウ属　391061
イブキジャカウソウ属　391061
イブキジャコウソウ　391225,391347,391351,391353
イブキジャコウソウ属　391061
イブキスミレ　410246,410255
イブキゼリモドキ　391911
イブキゼリ属　77766
イブキセントウソウ　85479
イブキソモソモ　305916
イブキタイゲキ　159218,159540
イブキトボシガラ　164200
イブキトラノオ　54799,308893
イブキトラノオ属　54834
イブキトラノヲ　308893
イブキヌカボ　254515
イブキヌカボ属　254496
イブキノエンドウ　408611
イブキバウフウ属　361456
イブキハタザオ　30275
イブキビャクシン　213634
イブキフウロ　175023,175028
イブキボウフウ　228572
イブキルリトラノオ　317970
イブキ属　341696
イプシロネ　239215
イプセア属　208323
イプッルウメモドキ　80263
イブノキ　240707
イブリアザミ　91700
イヘツイモ　99910
イヘヤヒゴクサ　352357
イベリアヌカボ　12038
イベリス属　203181
イベルウィルレア属　203175
イボウキクサ　224366
イホウマル　327274
イボクサ　260102
イボクサアヤメ属　228625
イボタ　229558
イボタクサギ　96140
イボダケ　87607
イボタノキ　229472,229558
イボタノキ属　229428
イボタヒョウタンボク　235757
イボタロウ　229558
イボトリ　229558
イボノキ　229558
イボビシ　394471
イボブナ　162368,162382
イポメア属　207546
イボラン　62799
イマジスゲ　73906
イマメガシ　324283
イミタリア属　204752
イモ　99910
イモカタバミ　277688
イモギ　86855
イモネアサガオ　208050
イモネノホシアサガオ　208249
イモネヤガラ　157110
イモノキ　170902,244507
イモノキ属　244501
イモノハカラスウリ　396197
イモラン　156732
イモラン属　156521
イヨ　93504
イヨアブラギク　89546
イヨカズラ　117537
イヨカヅラ　117537
イヨカン　93504
イヨトンボ　183736
イョフウロ　174877
イヨフウロ　174900
イヨミズキ　106669
イヨミヅキ　106669
イラガシ　78932
イラクサ　403028
イラクサ科　403050
イラクサ属　402840
イラクシノキ　125545
イラノキ　125545
イラブナスビ　367392
イラマ　25850
イラモミ　298226
イラワジエンシス　69153
イランイランノキ　70960
イランイランノキ属　70954
イリオモテイワタバコ　101686
イリオモテカクレミノ　125620
イリオモテガヤ　87354
イリオモテクマタケラン　17671

イリオモテスミレ　410643
イリオモテソウ　32525
イリオモテトンボソウ　302520
イリオモテニシキサウ　159286
イリオモテニシキソウ　159286,159971
イリオモテハイノキ　381289
イリオモテヒメラン　110637
イリオモテムヨウラン　376244
イリオモテムラサキ　66883
イリオモラテン　395865,404749
イリス属　208433
イリノイヌスビトハギ　126407
イリモテラン　374460
イレシネ　208349
イレシネ属　208344
イロイロ　272891
イロハカエデ　3300
イロバカンアフヒ　37630
イロハモミジ　3300,3458
イロハモミヂ　3300
イロマガリバナ　203249
イロマツヨイ　94133
イワアオスゲ　73913
イワアカザ　87037
イワアカバナ　146606
イワアズマギク　40016
イワイカン　93503
イワイズリ　311890
イワイソウ　58592
イワイチョウ　265197,265198
イワイチョウ属　265195
イワイノキ　261739
イワインチン　89698
イワウサウ　239898,239902
イワウチソウ属　362242
イワウチワ　362270,362275
イワウチワ属　362242
イワウツギ　126963
イワウトウキイチゴ　338942
イワウメ　127916,127919
イワウメズル　80189
イワウメヅル　80189
イワウメ科　127926
イワウメ属　127910
イワオ　140135,140150
イワオウギ　188177,188178
イワオウギ属　187753
イワオグルマ　385913
イワオトギリ　202150
イワオミナエシ　285854
イワカガミ　351450
イワカガミダマシ　367763
イワガサ　371836
イワカナビキソウ　389857
イワガネ　273903
イワガマズミ　407954
イワカラクサ　151102
イワガラミ　351793,351797
イワガラミ属　351780
イワカンスゲ　75287
イワキ　229499,229514
イワキアブラガヤ　353451
イワキカモメヅル　409551
イワギキョウ　70117
イワキキンネ　369282
イワギク　124862
イワキヌヨモギ　35779
イワキハグマ　292042
イワキハンノキ　16311
イワギボウシ　198633
イワギリソウ　272598,273878
イワギリソウ属　87815,272589
イワキンバイ　312498
イワグルマ　363063,399433
イワクロウメモドキ　328816
イワゲンゲ　279060
イワコゴメナデシコ　183222
イワコザクラ　314930
ィワゴマノハグサ　355046
イワザクラ　315051
イワサクラソウ　314930
イワザンショウ　417176
イワシチョウゲ　226106
イワシデ　77411
イワシモツケ　372029,372031
イワジャコウソウ　391356
イワシャジン　7843
イワショウブ　392622
イワスグリ　334149
イワスゲ　76348
イワズタイ　319833
イワセントウソウ　320241
イワセントウソウ属　320203
イワダイゲキ　159159
イワダインソウ　175433
イワタカンアオイ　37665
イワタケソウ　203128
イワタバコ　101681
イワタバコ属　101680
イワダレソウ　296121
イワダレゾウ属　232457
イワチドリ　19515
イワツクバネウツギ　416853
イワヅサ　231403
イワツツジ　403967
イワツバキ　228178
イワツメクサ　375018
イワテイチゴツナギ　305606
イワテザサ　347332
イワテシオガマ　287299
イワテトウキ　229343
イワテナデシコ　127869
イワテハタザオ　30441
イワテヒゴタイ　348172
イワデマリ　371823
イワテヤマトウヒレン　348091
イワテヤマナシ　323094
イワトユリ　229915
イワナギナタコウジュ　144088
イワナシ　146523,404051
イワナシ属　146522
イワナズナ　45192
イワナズナ属　18313
イワナンテン　228178
イワナンテン属　228154
イワニガナ　210673
イワニンジン　24366
イワネコノメソウ　90356
イワノガリヤス　65473
イワノココギリソウ　3995
イワハギ　226125
イワハゼ　172052
イワバゼ　172052
イワハタザオ　30437,30439
イワハナガタ　23139
イワヒゲ　78631
イワヒゲ属　78622
イワブクロ　289341
イワフジ　205876
イワベンケイ　329935
イワベンケイソウ　329935
イワベンケイ属　329826
イワボウフウ　76987
イワボタン　90406
イワマタチツボスミレ　410525
イワマツ　300163
イワミツバ　8826
イワムラサキ　184298
イワモチ　399433
イワモミ　145073
イワモモ　404051
イワヤスゲ　76617
イワヤツデ　259731
イワヤナギ　344112,344255,372101
イワヤマナツメヤシ　295485
イワヤマブキ　243261
イワユキソウ　349869
イワユキノシタ　383890
イワユキノシタ属　383887
イワユリ　229900
イワヨモギ　35560,36177,36179
イワレンギョウ　167446
イワレンゲ　275374,275385
イワレンゲ属　275347
イワワチガイソウ　318513
イワンポウゲ　325562
イン　133587
インカ-ビレア・デラバ-イ　205557
インカ-ビレア属　205546
インカルビレア属　205546
イングリッシュ・ボ-リ-　203545

イングリッシュ・アリイス　208944
インゲンササゲ　294056
インゲンナ　53249,53269
インゲンマメ　218721,294056
インゲンマメ属　293966
インゲン属　293966
インコアナナス　412343
インシチチアスモモ　316470
インチアアセン　196443
インチンロウゲ　85651
インチンロウゲ属　85646
インドアイ　206626
インドアサ　71220,104072
インドアツモリサウ　282845
インドアヰ　206626
インドウオトリギ　180682
インドガンピ　414193
インドグス　91392
インドグネツム　178541
インドクワズイモ　16512
イントゴムノキ　164925
インドゴムノキ　164925
インドシクンシ　324677
インドシタン　320301
インドジャボク　327058
インドジャボク属　326995
インドジュズノキ　142272
インドスギ　80087
インドセンダン　45908
インドソケイ　305225,305226
インドソケイ属　305206
インドソテツ　115812
インドソバ　162335
インドトチノキ　9706
インドトバ　125960
インドナガコショウ　300446
インドナガボソウ　373507
インドハマユウ　111214
インドビエ　140423
インドヒモカズラ　123569
インドヒモカズラ属　123558
インドボダイジュ　165553
インドマツリ　305202
インドメボウキ　268518,268523
インドヤツデ　400377
インドヤツデ属　400376
インドルカム　166773
インドルリソウ　117965
インドワタノキ　56784,80120
インパチエンス　205346,205444
インバテンツキ　166157
インモウセンゴケ　138289
インヨウヌギ属　10492
イ属　212797
ウイキャウ　167156
ウイキャウ属　167141
ウイキョウ　167156
ウイキョウ属　167141
ウィドリントーニア属　414058
ウィトロキア属　414636
ウイルコキシア属　414293
ウイルマッテア属　414429
ウイングドシトロン　93318
ウウンリュウ　186107
ウエインガルティア属　413671
ウエスタンヘムロック　399905
ウェッチンヤシ属　413924
ウエベルマンナアーナ属　401332
ウェーベロセレウス属　413479
ウエマツサウ　352873
ウエマツソウ　352873
ヴェルヴィチア科　413750
ウェルウィッチア　413748
ウェルウィッチア科　413750
ウェルウィッチア属　413741
ヴエルニルベンスポプラ　311164
ウェンドランドツルナス　367745
ヴォキシア科　412164
ウオクサギ　313692
ウォーターベントグラス　310134
ウォーターレモン　285660
ウオトリギ　180703
ウオトリギ属　180665
ウオノキ　180703
ウオノホネヌキ　70989
ウオルナット　212631
ウォレミマツ　414713
ウガマル　327257
ウキアゼナ　46371
ウキオモダカ　342385
ウキガヤ　177587,177621
ウキガヤツリ　119390
ウキクサ　372300
ウキクサ科　224403
ウキクサ属　372295
ウキシバ　318201
ウキシバ属　318193
ウキシロ　377343
ウキツリボク　957
ウキブネギョク　377360
ウキマルバオモダカ　66344
ウキミクリ　370068
ウキヤガラ　56637,353422,353587
ウグイスカグラ　235812,235813
ウグイスナ　59575
ウグイスノキ　235813
ウグヒスカグラ　235813
ウグヒスノキ　235813
ウグヒスホク　235813
ウケザキオオヤマレンゲ　242352
ウケザキクンシラン　97218
ウゲヨシ　28341
ウケラ　44208
ウケラ属　44192
ウゴアザミ　92458
ウコギ　143677,143682
ウコギ科　30800
ウコギ属　143575
ウゴツクバネウツギ　186
ウゴマ　361317
ウコン　114871
ウコンイソマツ　230816
ウコンウツギ　241047,413617
ウコンツワブキ　162621
ウコンバナ　231403
ウコンバメ　231403
ウコンユリ　229958
ウコン属　114852
ウサギアオイ　243810
ウサギギク　34780,34781
ウサギギク属　34655
ウサギソウ　210525
ウサギツユクサ　100967
ウサギノオ　220254
ウシイチゴ　338886,338985
ウシウド　98777
ウシオシカギク　107766
ウシオスゲ　75970
ウシオツメクサ　370661
ウシオツメクサ属　370609
ウシオハナツメクサ　370615
ウシオミチヤナギ　309858
ウシカバ　204292,204301
ウシガヤ　259672
ウジカンスゲ　74176
ウシクグ　119319,119321
ウシクサ　351266
ウジクサ　269613
ウシクサ属　22476
ウシコロシ　328684
ウジコロシ　239392,269613
ウシタキサウ　91537
ウシタキソウ　91537
ウシツメクサ　370675
ウシノケグサ　164126
ウシノケグサ属　163778
ウシノシタ　377838
ウシノシタグサ　21949
ウシノシタグサ属　21914
ウシノシタモドキ　377727
ウシノシツペイ　191251
ウシノシッペイ　191234
ウシノシッペイ　191251
ウシノシッペイ属　191225,337544
ウシノタケダグサ　148154
ウシノハヘトリ　138348
ウシノヒタイ　164947,309877,408034
ウシノヒタヒ　408034
ウシハコベ　260922,374749
ウシブドウ　351073
ウシホツメクサ　370706

ウシホホヅキ 367416
ウシミツバ 113883
ウシミツバ 113879
ウジュキツ 93860
ウジルカンダ 259512,259535
ウスアカカタバミ 277749
ウスアカノイバラ 336785
ウスアカヒゲガヤ 194044
ウスアカミヤコイバラ 336843
ウスイチゴ 404047
ウスイモフクリンセンネンボク 137398
ウスイロオクノカンスゲ 74586
ウスイロサツマスゲ 75162
ウスイロシモツケソウ 166106
ウスイロシャジン 7618
ウスイロジンチョウ 122543
ウスイロジンチョウゲ 122543
ウスイロスゲ 75675
ウスイロタンポポ 384505
ウスイロノダケ 24339
ウスイロハクサンウツボ 316125
ウスイロヒエスゲ 75192
ウスイロヒメスミレ 410104
ウスイロホナガソウ 373495
ウスイロマツカサヒジキ 104767
ウスイロマツムシソウ 350156
ウスイロヤクシソウ 110607
ウスイロヤマブキソウ 200756
ウスイロワレモコウ 345894
ウスエキナズナ 376389
ウスエフ 330883
ウスカハゴロモ属 200201
ウスカワゴロモ 200202
ウスキオダマキ 30077
ウスギキンエノコロ 361730
ウスギシャクナゲ 330240
ウスギズイセン 262417
ウスギツクバネウツキ 180
ウスキツリフネ 205178
ウスギナツノタムラソウ 345178
ウスギヌソウ 116747
ウスキヒメショウブ 248722
ウスギヘイシソウ 347164
ウスギムヨウラン 223654
ウスギモクセイ 276291,276297
ウズギモクセイ 276260
ウスギョウラク 250507
ウスク 165841
ウスグロ 114884
ウスゲアキノウナギツカミ 291939
ウスゲオオシマ 316936
ウスゲオカスミレ 410409
ウスゲオガラバナ 3733
ウスゲカエデ 2797
ウスゲカヘデ 2798
ウスゲカモジグサ 144239,335204
ウスゲキダチキンバイ 238188
ウスゲキンミズヒキ 11584
ウスゲクマヤナギ 52464
ウスゲクロモジ 231434
ウスゲケイヌビエ 140383
ウスゲサイシン 37729
ウスゲサンカクヅル 411701
ウスゲシナノガキ 132217
ウスゲシモツケ 372067
ウスケシヨウ 356999
ウスゲタマブキ 283808
ウスゲチョウジタデ 238168
ウスゲトダシバ 37352,37389
ウスゲノエゾザクラ 314523
ウスゲハシドイ 382268
ウスゲヒロハハンノキ 16415
ウスゲホオズキ 297697
ウスゲマシケゲンゲ 279163
ウスゲミチシバ 249061
ウスゲヤナギラン 85880
ウスゲヤブニッケイ 91349
ウスゲヤマザクラ 83236
ウスゲヤマニンジン 84773
ウズシオ 246601
ウスジロイソマツ 230818
ウスジロスミレサイシン 410712
ウスジロタンポポ 384741
ウズタツナミ 355505
ウスノキ 403854,403859
ウスバアカザ 87048
ウスバアザミ 92435
ウスバアハブキ 249369
ウスバアリサンソヨゴ 203575
ウスバカンアフヒ 37585
ウスバキクボクチ 361079
ウスバキダチベンケイ 356993
ウスバキンゴジクワ 362590
ウスバクスノキ 91405
ウスバクマヤナギ 52423
ウスバサイシン 37722
ウスバザサ 347299
ウスバザミア 417005
ウスバサルナシ 6648
ウスバサルノオ 196824,196835
ウスバサルノヲ 196835
ウスバサンキライ 366583
ウスバサンゴアナナス 8549
ウスバシラクチズル 6639
ウスバシラクチヅル 6639
ウスバシラネアザミ 348854
ウスバスミレ 409751,409753
ウスバノゲネツム 178564
ウスバノハギカズラ 169650
ウスバノハギカヅラ 169650
ウスババイクワウツギ 294556
ウスバハウチャクモドキ 134405
ウスバヒメツバキ 69398
ウスバヒメヤツ属 174341
ウスバヒョウタンボク 235717
ウスバヒョオタンボク 235717
ウスバフウテウボク 71679,71782
ウスバヘウタンボク 235717
ウスバヘビノボラズ 51926
ウスバユヅリハ 122708
ウスバルリサウ 117965
ウスバルリソウ 117965
ウスバルリミノキ 222231
ウスベニアオイ 243840
ウスベニアカショウマ 41853
ウスベニアフヒ 243840
ウスベニアルテア 18173
ウスベニカノコソウ 81760
ウスベニコバノタツナミ 355512
ウスベニシオガマ 287591
ウスベニシキミ 204479
ウスベニシモバシラ 215814
ウスベニタチアフヒ 18173
ウスベニチチコグサ 178367
ウスベニツメクサ 370696
ウスベニトリアシショウマ 41833
ウスベニニガナ 144975,144976
ウスベニニガナ属 144884
ウスベニバイカウツギ 294556
ウスベニハタザオ 30229,30363
ウスベニヒメイカリソウ 147069
ウスベニミツバウツギ 374091
ウスベニレンゲ 275383
ウズマキウマゴヤシ 247469
ウズマキダイコン 53269
ウズマサキ 157511
ウズムシマメ 354778
ウスムラサキツリガネヤナギ 289332
ウスユキキクアザミ 348908
ウスユキクチナシグサ 257542
ウスユキサウ 224872
ウスユキサウ属 224767
ウスユキソウ 224872
ウスユキソウ属 224767
ウスユキトウヒレン 348945
ウスユキナズナ 53038,376389
ウスユキハナヒリノキ 228172
ウスユキマンネングサ 356802
ウスユキムグラ 170627
ウスユキヨモギ 35816
ウズラバタンポポ 195780
ウズラバハクサンチドリ 121363
ウズラモヂズリ 264765
ウスリクロッパラ 328882
ウスリ-スグリ 334249
ウスリ-ミセバヤ 200814
ウゼンアザミ 92468
ウゼンベニバナヒョウタンボク 236199
ウゾンクネボ 93704
ウタイカンバ 53400,53520
ウダイカンバ 53400

ウダケ 24336
ウタヒメマル 234952
ウチコミッルミヤマシキミ 365936
ウチダシクロギ 381256
ウチダシタビラコ 397411
ウチダシツルウメモドキ 80285
ウチダシツルシキミ 365954
ウチダシツルマサキ 157505
ウチダシヒメアオキ 44933
ウチダシマサキ 157621
ウチダシミヤマシキミ 365945,365949
ウチダシモンタチバナ 365947
ウチハカッラ 83743
ウチハドコロ 131734
ウチハバドコロ 131630
ウチハルカウ 323459
ウチハ属 272775
ウチムラサキ 93579,93754
ウチャマラン 232106
ウチュウセン 167124
ウチョウラン 310891
ウチワアナナス 392006
ウチワキジムシロ 312528
ウチワサボテン 272891
ウチワサボテン属 272775
ウチワゼニクサ 200386
ウチワダイモンジソウ 349354
ウチワツナギ 297013
ウチワドコロ 131734
ウチワノキ 210,211
ウチワノキ属 209
ウチワヤシ属 228729
ウチワルコウ 323459
ウチワ属 272775
ウツギ 126891,127072
ウツギ属 126832
ウツクシオトギリ 201956
ウツクシザサ 347278
ウツクシマツ 299897
ウツコンカウ 400162
ウツコンカウジュ 232609
ウツコンカウジュ属 232602
ウツコンコウジュ 232609
ウツコンジュ 232609
ウヅダイコン 53249
ウッド・メドウ・グラス 305743
ウッドアップル 163502
ウッドオニソテツ 145245
ウッドローズ 250850
ウツボカズラ 264846,264849
ウツボカズラ科 264827
ウツボカズラ属 264829
ウツボカツラ 264846
ウツボカツラ科 264827
ウツボカツラ属 264829
ウツボグサ 316097,316127,316150
ウツボグサ属 316095
ウド 30619
ウトウシュウカイドウ 49827
ウドカズラ 20327,20329
ウドカヅラ 20327
ウドキランソウ 13194
ウドダラシ 24441
ウドノキ 81933
ウドノキ科 223967
ウドノキ属 300941
ウドノノヨジ 295916
ウドモドキ 30634
ウドンゲ 164763,165541
ウナギツカミ 291938,309796
ウナギヅル 309796
ウナヅキガヤツリ 119282
ウナヅキテンツキ 166421
ウニアザミ 91743
ウニ－オラ属 401852
ウニサボテン属 140842
ウニヒレアザミ 270247
ウネモジリ 19248
ウノハナ 126891,127072
ウバガネモチ 241781
ウバギ 328665
ウバシバ 324283
ウパス 28248
ウバスカシ 375817
ウバスノキ 28248
ウバタケニンジン 24497
ウバタマ 236425
ウバニレ 401542
ウハバミサウ属 142594
ウハミカンアフヒ 37614
ウハミヅザクラ 280023
ウバメガシ 324283
ウバユリ 73158,229822
ウバユリ属 73154
ウブラ－リア属 403665
ウベ 374409
ウホウドカズラ 223943
ウホウドカズラ属 223919
ウホウドノキ 223943
ウマクスバ 298734
ウマグリ 9701
ウマクワズ 298734
ウマゴヤシ 247425
ウマゴヤシ属 247239
ウマスゲ 74866
ウマセンナ 78316
ウマツツジ 331258
ウマノアシガク 325981
ウマノアシガタ科 325494
ウマノアシガタ属 325498
ウマノスカナ 340089
ウマノスズカケ 34162
ウマノスズクサ 34162
ウマノスズクサ科 34384
ウマノスズクサ属 34097
ウマノチャヒキ 60989
ウマノミツバ 345948,345957
ウマノミツバ属 345940
ウマブダウ 20354
ウマメガシ 324283
ウミガヤ 255873
ウミジグサ 184943
ウミシャウブ 145643
ウミシャウブ属 145642
ウミショウブ 145643
ウミスゲ 297185
ウミゼキシャウ 397159
ウミゼキショウ 397159
ウミニラ 397159
ウミノサチスゲ 73824
ウミヒルモ 184979
ウミヒルモ属 184967
ウミミドリ 176775,176776
ウミミドリ属 176772
ウミヤシ 235159
ウミンヤ 61666
ウメ 34448
ウメウツギ 126946,126963,127118
ウメガササウ 87485
ウメガササウ属 87480
ウメガサソウ 87485
ウメガサソウ属 87480
ウメザキイカリソウ 146959,147076
ウメザキウツギ 161749
ウメザキサバノオ属 66697
ウメズエ 185104
ウメバクロッバラ 328609
ウメハタザオ 30444
ウメバチサウ 284591,284596
ウメバチサウ属 284500
ウメバチソウ 284591,284596
ウメバチソウ科 284645
ウメバチソウ属 284500
ウメバナシ 323202
ウメモドキ 204239,204255
ウモコソ－ダノキ 266367
ウヤクチク 125453
ウライオドリコサウ 220422
ウライガシ 79064
ウライサウ 315476
ウライサウ属 315461
ウライソウ 315465,315476
ウライソウ属 315461
ウライタマラン 295516
ウライムカデラン 94478
ウライムヨウラン 333731
ウライラン 62873
ウライヲドリコサウ 220422
ウラギク 41432,398134
ウラキサカキ 8221
ウラギシキンバイ 312408

ウラギンツルグミ 142181
ウラギンヒゴタイ 348585
ウラク 69739
ウラクツバキ 69739
ウラゲウコギ 143688
ウラゲエンコウカエデ 3405
ウラゲサドザサ 37305
ウラゲスズタケ 347381
ウラゲトチノキ 9732
ウラゲトチノキ 9733
ウラゲノギク 124857
ウラケハクサンシャクナゲ 330245
ウラゲハンゴンソウ 358500
ウラゲヒメアザミ 92230
ウラゲミツバツツジ 330578
ウラゲムカゴトラノオ 54826
ウラゲムラサキヤシオツツジ 330060
ウラゲヨブスマソウ 283816
ウラゲワレモコウ 345884,345890
ウラシマサウ 418043
ウラシマソウ 33538
ウラシマツツジ 31314,31316
ウラシマツツジ属 31311
ウラジモシモツケサウ 166110
ウラジモヤツデ 364997
ウラジロ 53388,264090
ウラジロアカザ 87029
ウラジロアカメガシハ 243420
ウラジロアカメガシワ 243420
ウラジロアザミ 92479
ウラジロアナナス 301102
ウラジロイカリソウ 147055
ウラジロイタヤ 3087,3408
ウラジロイチゴ 338985,339047
ウラジロイヌガヤ 19357
ウラジロイハガネ 228147
ウラジロイハガネ属 228144
ウラジロイワガサ 371789
ウラジロイワガネ 228147
ウラジロウコギ 143617
ウラジロウツギ 126904,126976,127012
ウラジロウメウツギ 126956
ウラジロエゴノキ 379457
ウラジロエノキ 394656,394664
ウラジロエノキ属 394626
ウラジロオホ 364989
ウラジロオホイヌタデ 309472
ウラジロカエデ 3026
ウラジロガシ 116187,324358,324449
ウラジロカッコウ 406667
ウラジロカワラハンノキ 16314
ウラジロカンコノキ 177097,177184
ウラジロカンバ 53388
ウラジロキンバイ 312811,312818
ウラジログリ 78802
ウラジロクリンザクラ 314573
ウラジロクロヅル 398319
ウラジロコウラク 250520
ウラジロコウラクツツジ 250520
ウラジロゴシュユ 161336
ウラジロコハクウンボク 379451
ウラジロコムラサキ 66888
ウラジロコヨウラク 250515
ウラジロサトウカエデ 3532
ウラジロサンキライ 366395
ウラジロシモツケ 371960
ウラジロシモツケソウ 166110
ウラジロスミレ 409993
ウラジロタデ 309966
ウラジロダモ 295361
ウラジロタラノキ 30593
ウラジロチチコグサ 170923,178430
ウラジロトックリイチゴ 338292
ウラジロナツハゼ 403927
ウラジロナツメヤシ 295477
ウラジロナナカマド 369454,369525
ウラジロノキ 33008
ウラジロハコヤナギ 311208
ウラジロハナヒリノキ 228169
ウラジロハマアカザ 44646
ウラジロヒカゲツツジ 330997,330999
ウラジロヒメイワヨモギ 35560
ウラジロヒメウツギ 126956
ウラシロヒメヒゴタイ 348690
ウラジロヒレアザミ 270234
ウラシロフウロ 174592
ウラジロフサザクラ 160348
ウラジロフジウツギ 62016
ウラジロブジウツギ 62015
ウラジロマキ 19357
ウラジロマタタビ 6520,6633
ウラジロマツブサ 351094
ウラジロマルバマンサク 185108,185115
ウラジロマンサク 185119
ウラジロミツバツツジ 331417
ウラジロミヤマツツジ 330061
ウラジロモミ 389
ウラジロヤグルマギク 81024
ウラシロヤナギ 343185
ウラジロヤナギ 344211
ウラジロヤナギアザミ 92479
ウラジロヨウラク 250520
ウラジロヨオラク 250520
ウラジロラフィア 326639
ウラジロレン 331262
ウラジロレンゲツツジ 331262
ウラジロロウゲ 312389
ウラジロワタナ 55738
ウラチャクロダモ 233945
ウラヅロマタタビ 6633
ウラハグサ 184656
ウラハグサ属 184655
ウラベゴニア 50273
ウラベニアナナス 266155
ウラベニアナナス属 266148
ウラベニイチゲ 24013
ウラベニウキクサ 372300
ウラベニサンゴアナナス 8563
ウラベニショウ 378307
ウラベニショウ属 378305
ウラベニダイモンジソウ 349347
ウラボシザクラ 280028
ウラムラサキ 290921
ウラルカンゾウ 177947
ウラン 71015
ウリカエデ 2910,2913
ウリカハ 342396
ウリカボチャ 114301
ウリカワ 342396
ウリカワヒルムシロ 29673
ウリクサ 231503
ウリクサ属 392880
ウリノキ 3529,13376,13378,13380
ウリノキ科 13343
ウリノキ属 13345
ウリハダカエデ 3529
ウリバロ－レルカヅラ 390787
ウリュウシャジン 7878
ウリュウトウヒレン 348443
ウリ科 114313
ウルギネア属 402314
ウルクス属 401411
ウルケオリナ属 402213
ウルシ 393491
ウルシヅタ 393466
ウルシニア属 402743
ウルシノキ 393491
ウルシ科 21190
ウルシ属 332452
ウルチ 275958
ウルチカ属 402840
ウルップサウ 220167
ウルップサウ属 220155
ウルップシオガマ 287474
ウルップソウ 220167
ウルップソウ属 220155
ウレツムハパ 77748
ウレティア属 198684
ウレバチモ 325589
ウロコアカシア 1156
ウロコウチワ 116657
ウロコケンチヤ属 225554
ウロコゴヘイヤシ属 225493
ウロコザミア属 225622
ウロコナズナ 225315
ウロコヒジキ 104831
ウロコマリ 225162
ウロコマリ属 225134
ウロスベルムム属 402660
ウワジマテンナンショウ 33549
ウワバミサウ属 142594

ウワバミソウ 142694,142697
ウワバミソウ属 142594
ウワポメロ 93709
ウワミズザクラ 280023
ウワミヅザクラ 280023
ウワムキトウガラシ 72070
ウヰキャウ 167156
ウヰキャウ属 167141
ウヲクサギ 313692
ウヲトリキ 180703
ウヲトリギ属 180665
ウヲノキ 180703
ウンエイギョク 102122
ウングイス-カティ 240368
ウンコウ 341052
ウンサイギョク 263740
ウンシウミカン 93736
ウンゼンカンアオイ 37751
ウンゼンザサ 347217
ウンゼンツツジ 331305,331801,331805
ウンゼントリカブト 5322
ウンゼンマムシグサ 33554
ウンゼンマンネングサ 357036
ウンゾキ 93521
ウンナンオウバイ 211905
ウンナンソケイ 211868,211869
ウンナンハシドイ 382388
ウンナンハナゴマ 205557
ウンナンモミ 369
ウンヌケ 156499
ウンヌケモドキ 156490
ウンヌケ属 156439
ウンバイロ 73161
ウンラン 231015
ウンランカズラ 116736
ウンランモドキ 263459
ウンラン属 230866
ウンリュウボケ 84569
ウンリュウヤナギ 343670
ウンリンニシキサウ 159102
エ 80739,80741
エイカン 140618,149398
エイギョク 233574
エイキリュウ 299242
エイコクトゲナシ 334995
エイザンカタバミ 277878
エイザンスギ 113705
エイザンスミレ 409942
エイザンユリ 229730
エイジュ 149017
エイシュウカズラ 171449
エイノキ 52478
エイラグキュウ 155299
エイルギョク 233510
エウクリフィア科 156072
エウクリフィア属 156058
エウゲニイポプラ 311149
エウフラテカポプラ 311308
エウリクニア属 157163
エウリコ-ネ属 160707
エオニエルラ属 269550
エオニューム属 9021
エオピア属 269536
エキウム属 141087
エキサイゼリ 29546
エキサカム属 161528
エキソコルダ属 161741
エキドノプシス属 140017
エキナセア属 140066
エキナタマツ 299925
エキヌス属 140981
エキノカクタス属 140111
エキノシスチス属 140526
エキノセレウス属 140209
エキノドルス属 140536
エキノフォッシュロカクタス属 140566
エキノプシア属 140842
エキノプス属 140652
エキノマスツス属 140612
エキューム属 141087
エギリ 406016
エグウチルリソウ 117956
エクトトロピス属 141439
エクボサイシン 37625
エクメア属 8544
エクリ 116880
エケベリア属 139970
エゴ 379374
エゴノキ 379374
エゴノキ科 379296
エゴノキ属 379300
エゴハンノキ 16304
エゴハンノキ属 16291
エゴマ 290940
エジプトマンテマ 363154
エスカロニア属 155130
エスキナンツス属 9417
エスキナントゥス属 9417
エスコバリア属 155201
エスコントリア属 155246
エスパルト 376931
エスペレチア属 155277
エスポストア属 155296
エゾアオイスミレ 410641
エゾアカバナ 146782
エゾアキカラマツ 388583
エゾアジサイ 200102
エゾアズマギク 150446
エゾアゼスゲ 73826
エゾアゼテンツキ 166500
エゾアフヒスミレ 409834
エゾアブラガヤ 353226,353576
エゾアリドオシ 231679
エゾアリドホシ 231680
エゾイソツツジ 223906
エゾイタドリ 162550
エゾイタヤ 3149
エゾイチゲ 24128
エゾイチゴ 338557,338566,338583,339227
エゾイチヤクソウ 322853
エゾイトイ 213388
エゾイトヰ 213388
エゾイヌゴマ 373126,373139
エゾイヌナズナ 136951,136952
エゾイヌノヒゲ 151430
エゾイブキトラノオ 54800,309505
エゾイボタ 229651
エゾイラクサ 402848,402996
エゾイワツメクサ 375066
エゾウキヤガラ 353707
エゾウコギ 143657
エゾウサギギク 34780
エゾウスユキソウ 224827
エゾウバユリ 73161
エゾウラジロハナヒリノキ 228167
エゾエノキ 80596,80600
エゾエビガライチゴ 338068
エゾエビネ 66099
エゾエンゴサク 105594,105918
エゾェンビセンノウ 363337
エゾオオケマン 105776
エゾオオサクラソウ 314521
エゾオオバコ 301894
エゾオオバサンザシ 109780
エゾオオバセンキュウ 24355
エゾオオヤマハコベ 375071
エゾオオヨモギ 36474
エゾオオヨロイグサ 24293
エゾオグルマ 207068,359811
エゾオタカラコオ 229035
エゾオトギリ 202230
エゾオドリコソウ 220346
エゾオノエリンドオ 174110
エゾオホバコ 301894
エゾオホバセンキク 24355
エゾオホヤマハコベ 375071
エゾオホヨロヒグサ 24293
エゾオヤマノエンドウ 278912
エゾオヤマリンドウ 174011
エゾオヤマリンドオ 174011
エゾカウバウ 196159
エゾカサスゲ 74392
エゾカハヅスゲ 75675
エゾカハホネ 267320
エゾカハラナデシコ 127802
エゾカモジグサ 144412
エゾカラマツ 388654
エゾカワズスゲ 75675
エゾカワヤナギ 343702
エゾカワラナデシコ 127852
エゾカワラマツバ 170764

エゾギク　67314
エゾギク属　67312
エゾキケマン　106466
エゾキスゲ　191319
エゾキスミレ　409770
エゾキヌタサウ　170259
エゾキヌタソウ　170246,170259
エゾキンバイ　413034
エゾキンバイサウ　399510
エゾキンパウゲ　325882
エゾキンポウゲ　325882
エゾギンラン　82046
エゾクガイソウ　407488,407493
エゾクマイチゴ　338300
エゾクルマバナ　96979
エゾクロクモソウ　349361
エゾグンナイフウロ　174781
エゾケマン　106466
エゾコウゾリナ　202403
エゾコウボウ　196161
エゾコウホネ　267320
エゾコゴメグサ　160218
エゾコザクラ　314280
エゾコスミレ　410784
エゾゴゼンタチバナ　105204
エゾコブシ　242171
エゾゴマナ　40518
エゾサイコ　63766
エゾサカネラン　264709
エゾザクラ　83301
エゾサハスゲ　76230
エゾサヤヌカグサ　223992
エゾサワスゲ　76701
エゾサンザシ　109780
エゾシオガマ　287837
エゾシモツケ　372006,372009,372083
エゾシャクナゲ　330245
エゾジュ　369037
エゾシラビソ　464,510
エゾシロネ　239235,239246
エゾシロハリスゲ　73560
エゾスカシユリ　229828
エゾズカシユリ　229828
エゾスグリ　334061
エゾスズシロ　154424,154485
エゾスズシロモドキ　154530
エゾスズシロ属　154363
エゾスズラン　147196
エゾズミ　243555,243649,243717
エゾスミレ　409816,409942
エゾセキシャウ　350861
エゾセンキウ　24325
エゾゼンテイカ　191278,191309
エゾゼンテイクワ　191309
エゾセンノウ　363472
エゾタイセイ　209238
エゾタカネスミレ　409862,409864
エゾタカネセンブリ　380383
エゾタカネツメクサ　31742,255442
エゾタカネニガナ　110835
エゾタガラカウ　229049
エゾタガラコウ　229049
エゾタカラコオ　229049
エゾタチカタバミ　278099
エゾタチツボスミレ　409630
エゾタツナミソウ　355675
エゾタデ　309056
エゾタンポポ　384714,384851
エゾチドリ　302260,302280
エゾツガザクラ　297019
エゾツツジ　389572
エゾツツジ属　389570
エゾツリスゲ　75692
エゾツリバナ　157765,157767
エゾツルキンバイ　312360,312365
エゾツルツゲ　204226
エゾトウウチソウ　345853
エゾトウヒレン　348718
エゾトチノキ　9690
エゾトリカブト　5707
エゾナナカマド　369367
エゾナニワズ　122471
エゾナミキ　355856
エゾニガクサ　388299
エゾニュウ　24500
エゾニワトコ　345669,345709
エゾヌカボ　12139,12285
エゾネギ　15709
エゾネコノメソウ　90331,90448
エゾノアオイスミレ　409834
エゾノイハハタザホ　30437
エゾノイワハタザオ　30437
エゾノウワミズザクラ　280003,280007,280044
エゾノオオシラカンバ　53584
エゾノオホサンザシ　109780
エゾノガリヤス　65471
エゾノカワジシャ　406983
エゾノカワヂシャ　406983
エゾノカワヤナギ　343702
エゾノギシギシ　340151
エゾノキツネアザミ　82024,91770,92384
エゾノキヌヤナギ　344071,344245
エゾノキリンソウ　294123
エゾノクサイチゴ　167635
エゾノクサタチバナ　117523
エゾノクマイチゴ　338300
エゾノクマガイサウ　120357
エゾノクマガエソウ　120357
エゾノクモマグサ　349699
エゾノクロウメモドキ　328740,328741
エゾノケカワラマツバ　170763
エゾノコウボウムギ　75259
エゾノコウボフムギ　75259
エゾノコギリサウ　4005
エゾノコギリソウ　4005,4008
エゾノコリンゴ　243555,243649
エゾノコンギク　40854
エゾノサクラソウ　314521
エゾノサヤヌカグサ　223992
エゾノサワアザミ　92287,92293
エゾノシシウド　98777
エゾノシシウド属　98772
エゾノシジミバナ　371906
エゾノシモツケソウ　166117
エゾノジャニンジン　72965
エゾノシラカンバ　53580
エゾノシロバナシモツケ　372012
エゾノジンバイサウ　302280
エゾノスミレ　410531
エゾノタウコギ　54001,54081
エゾノタカネヤナギ　343753
エゾノタケカンバ　53411
エゾノタチツボスミレ　409630
エゾノダッタンコゴメグサ　160242
エゾノチチコグサ　26385
エゾノチチコグサ属　26299
エゾノチャルメルソウ　256022
エゾノツガザクラ　297019
エゾノハクサンイチゲ　23917
エゾノハクサンボウフウ　292944
エゾノハナシノブ　307250
エゾノハハコグサ　178481
エゾノハマウド　98777
エゾノヒツジグサ　267767
エゾノヒルムシロ　312126,312138
エゾノホザキナナカマド　369280
エゾノマルバシモツケ　371793,371828
エゾノミクリゼキショウ　213281
エゾノミズタデ　308739
エゾノミヅタデ　308739
エゾノミツモトサウ　312828
エゾノミツモトソウ　312828
エゾノミノフスマ　374944
エゾノミヤマハコベ　374780
エゾノヤマオダマキ　30061
エゾノユキヨモギ　35950
エゾノヨツバムグラ　170434
エゾノヨモギギク　383876
エゾノヨロイグサ　24293,24453
エゾノヨロヒグサ　24293
エゾノリウキンクワ　68196
エゾノリウツギ　200056
エゾノリノギ　200056
エゾノリュウキンカ　68196
エゾノレイジンソウ　5231
エゾノレンリサウ　222803
エゾノレンリソウ　222803
エゾバイケイソウ　405627
エゾハウチワカエデ　3205
エゾバウフウ　8811

エゾバウフウ属　8810
エゾハギ　372075
エゾハコベ　374919
エゾハシバミ　106743
エゾハタザオ　30387,79500
エゾハタザホ　30387
エゾハッカ　250443
エゾハナシノブ　307250
エゾハハコヨモギ　35520
エゾハマアカザ　44576
エゾハマツメクサ　342261
エゾハリイ　143102,143108
エゾハリスゲ　76626
エゾハンノキ　16389
エゾヒカゲスミレ　410784
エゾヒナスミレ　410669
エソヒナノウスツボ　355261
エゾヒナノウスツボ　355042,355129
エゾヒメアマナ　169470
エゾヒメクワガタ　407388
エゾヒョウタンボク　235811,236211
エゾヒロハクサフジ　408439
エゾフウロ　175023
エゾフスマ　256456
エゾヘウタンボク　235811
エゾベニヒッジグサ　267773
エゾヘビイチゴ　167653
エゾベンケイ　200812
エゾボウフウ　8811
エゾボウフウ属　8810
エゾホシクサ　151387
エゾホソイ　213114
エゾホソヰ　213114
エゾホタルサイコ　63740
エゾマツ　298307
エゾマツムシソウ　350180
エゾママコナ　248213
エゾマメヤナギ　343780,343784
エゾマンテマ　363463
エゾミクリ　370023,370044,370096
エゾミズタマソウ　91524,91557
エゾミソハギ　240068
エゾミツバフウロ　175015
エゾミヤマカタバミ　277649
エゾミヤマクワガタ　317961
エゾミヤマザサ　347312
エゾミヤマツメクサ　32057
ェゾミヤマハンショウズル　95304
エゾムカシヨモギ　150419
エゾムギ　144466
エゾムギ属　144144
エゾムグラ　170340,170574
エゾムラサキ　260807,260892
エゾムラサキツツジ　330058,330495
エゾムラサキニガナ　219513,219806
エゾメイゲッカエデ　3037
エゾモメンヅル　42537
エゾヤナギ　344022
エゾヤナギモ　312075
エゾヤマアザミ　92011
エゾヤマオトギリ　201976
エゾヤマカモジグサ　58619
エゾヤマコウボウ　196163
エゾヤマザクラ　83301
エゾヤマゼンゴ　98785
エゾヤマツツジ　330973
エゾヤマナラシ　311281,311361
エゾヤマハギ　226698
エゾヤマハンノキ　16466
エゾヤマブキショウマ　37066
エゾヤマモモ　261164
エゾユズリハ　122695
エゾユリ　229828
エゾヨツバシオガマ　287076
エゾヨモギ　35948
エゾヨモギギク　383874,383876
エゾラッキョウ　15843
エゾリンゴ　243580
エゾリンドウ　174004,174007
エゾリンドオ　174007
エゾルラサキツツジ　330511
エゾルリソウ　250905,250906
エゾルリトラノオ　317937
エゾルリムラサキ　153492,153493
エゾワサビ　72747
エゾワタスゲ　152781,152782
エゾワレモカウ　345888
エゾワレモコウ　345888
エゾヲグルマ　359811
エダウチアカバナ　146696
エダウチオオバコ　301864
エダウチオグルマ　207053
エダウチクサネム　9495
エタウチシオガマ　287601
エダウチシャジン　7884
エダウチスズメノトウガラシ　231485
エダウチゼリ　163584
エダウチタヌキマメ　112776
エダウチチカラシバ　289185,289187
エダウチチチコグサ　178465
エダウチチヂミザサ　272625
エダウチナズナ　133324
エダウチナヅナ属　133226
エダウチナンテンハギ　408661
エダウチノコギリソウ　3917
エダウチヒトツバヨモギホクチヨモギ　35622
エダウチヒメハコベ　374848
エダウチミミナグサ　82692
エタウチヤガラ　156723
エダウチヨツバハギ　408511
エダウチレンゲ　275363
エダザキズイセン　262457
エダナシマンテマ　364106
エダハリアザミ　91703
エダハリカラマツ　388680
エダハリシバ　306038
エダハリゼリ　163584
エダハリタデ　309373
エダハリハマアカザ　44347
エダハリバラモンジン　354846
エディトコレァ属　141484
エチオネ-マ属　9785
エチオピアオウギヤシ　57117
エチオピオアギヤシ　57117
エチゴキジムシロ　313076
エチゴタイゲキ　159847
エチゴツルキジムシロ　313080
エチゴテンツキ　166165
エチゴトラノオ　317935
エチゴメダケ　304082
エチゴルリソウ　270701
エチゼンアザミ　92201
エチゼンオニアザミ　92258
エチゼンダイモンジソウ　349046
エチゼンネザサ　304073
エチゼンヒメアザミ　92486
エチフィルム属　9779
エックレモカクタス属　139935
エツクレモカクタス属　139934
エッサシノ　347373
エッタムギ　49540
エッチュウミセバヤ　200799
エ-デルワイス　224772
エテンラク　155302
エドガシワ　323814
エドスミレ　409622,410531
エドドコロ　131873
エドヒガン　83328
エドムラサキ　215206
エトロフヤナギ　343859
エトロフヨモギ　35211
エナガスミレ　409725
エナシシソクサ　230297
エナシヒゴクサ　73718
エナシモチノキ　203914
エナルガンテ属　145158
エニシダ　121001
エニシダ属　120903
エニス　369037
エニスシア属　147384
エニスダ　121001
エノキ　80739,80741
エノキアオイ　243893
エノキアオイ属　243877
エノキウツギ　180703
エノキエ　80739
エノキグサ　1790
エノキグサ属　1769
エノキフジ　134147
エノキフヂ　134147

エノキフヂ属 94404
エノキマメ 166888
エノキ属 80569
エノコログサ 361935
エノコログサ属 361680
エノコロスゲ 74601
エノコロモトキ 145766
エノコロヤナギ 343444
エノミタンネ 236085
エパクリス科 146065
エパクリス属 146067
エビアマモ 297183
エビアラロ 297183
エビアラロ属 297181
エビカズラ 411669
エビガライチゴ 339047
エピゲネイウム属 146528
エピスキア属 147384
エビスグサ 280147,360461,360493
エピスシア属 147384
エビズル 411669
エビヅル 411669
エピテランサ属 147422
エピテランタ属 147422
エピデンドルム属 146371
エビネ 65921
エビネ属 65862
エピフィルム属 147283
エピフィロプシス属 147279
エピプレムヌム属 147336
エピブレムノプシス属 20866
エヒメアヤメ 208793
エビモ 200192,312079
エビラハギ 249232
エビラフジ 408374
エビラフヂ 408374
エピロ－ビウム属 146586
エフクレタヌキモ 403214
エフデギク 385892
エフミャクイチゴ 338305
エブラクテオ－ラ属 139831
エ－ベルランジア属 139785
エベルランジア属 139785
エボシバナ 237554
エボスサウ 237554
エボニ－ 132137,132153
エミクサ 308616
エモリ 163429
エヤミグサ 173852
エラグロスティス属 147468
エラン 281138
エランギス属 9091
エランティス属 148101
エランテス属 9191
エランテムム属 148053
エリアンツス属 148855
エリア属 148619
エリオカクタス属 151178
エリオシケ属 153351
エリオステモン属 153325
エリオセレウス属 151630
エリオプシス属 152844
エリオボトリア属 151130
エリカ 149138
エリカモドキ 48977
エリカモドキ属 48975
エリカ属 148941
エリキ－ナ属 154266
エリゲロン属 150414
エリシマム属 154363
エリズイセン属 18060
エリスリナ属 154617
エリスロニウム 154899
エリスロニュ－ム属 154895
エリスロリブサリス属 155017
エリセナ属 143809
エリッブエビネ 65887
エリデス属 9271
エリヌス属 151098
エリミノスナヂヒジキ 104836
エリムス属 144144
エリンジュ－ム属 154276
エル－カ属 153998
エルシア属 148134
エルテリア 112884
エルム 401512
エルレアントウス属 143841
エレウテリ－ネ属 143570
エレオカリス属 143019
エレオカルプス属 142269
エレッタ－リア属 143499
エレプシア属 148562
エレム－ルス属 148531
エレモシトラス 148262
エレモシトラス属 148261
エロジュ－ム属 153711
エロデア属 143914
エン 133587
エンキマル 234950
エングラ－ブナ 162372
エンコウカエデ 3154,3179,3404
エンコウサウ 68210
エンコウスギ 113687,113690
エンゴサク 105721,106519,106609
エンシウシャクナゲ 331187
エンジュ 369037
エンシュウカワラスゲ 73549
エンシュウシャクナゲ 331187
エンシュウツリフネソウ 205027
エンシュウハグマ 12627
エンシュウムヨウラン 223665
エンスイ 104690
エンセファロカルプス属 145247
エンダイブ 90894
エンドウ 301055,301070
エンドウサウ 222707
エンドウソウ 222707
エンドウ属 301053
エンバク 45566
エンビセン 364213
エンビセンノウ 364213
エンピツノキ 213984
エンピツビャクシン 213984
エンブセン 272915
エンペトルム属 145054
エンマ－コムギ 398890
エンメイギク 50825
エンメイサウ 209713
エンメイソウ 209713
エンレイサウ 397616
エンレイサウ科 397533
エンレイサウ属 397540
エンレイショウキラン 2081
エンレイソウ 397616
エンレイソウ科 397533
エンレイソウ属 397540
エンレイハマオモト 111157
エンレイマル 45888
エンレンマル 234945
オアギヤシ属 57116
オイラク 155300
オイラクサ 221552
オイランアザミ 92410
オイワケヒイラギナンテン 242580
オウイ 352278
オウカマル 234917
オウカン 93806
オウカンギョク 182482
オウカンユリ 230019
オウカンリュウ 163443
オウギカズラ 13126
オウギシマヒメハリイ 143143
オウギバショウ 327095
オウギバショウモドキ 377554
オウギバショウ属 327093
オウギヤシ 57122
オウギヤシ属 57116
オウキュウデン 102457
オウコウズ 61107
オウゴチョウ 65055
オウゴン 355387
オウゴンオニユリ 230060
オウゴンカシワ 323630
オウゴンカズラ 147338,328992
オウゴンカズラ属 353121
オウゴンカムロザサ 304115
オウゴンキッ 93328
オウゴンキャラ 385356
オウゴンクジャクヒバ 85317
オウゴンケンポナシ 198770
オウゴンコノテ 390588

オウコンコマチ 267053
オウゴンシノブヒバ 85349
オウゴンソウ 202400,354930
オウゴンソウ属 202387
オウゴンニオイヒバ 390599
オウゴンハギ 105284
オウゴンハギ属 105269
オウコンヒバ 85311
オウゴンヤグルマソウ 81183
オウゴンリュウ 299248
オウサイカク 373821,373982
オウシウアカキシ 300223
オウシウボダイジュ 391706
オウシウモミ 272
オウシキナ 23796
オウシュウアカキシ 300223
オウシュウアカマツ 300223
オウシュウイチイ 385302
オウシュウウバメカシ 324027
オウシュウカラマツ 221855
オウシュウカンバ 53505
オウシュウギョク 284687
オウシュウクロハンノキ 16352
オウシュウグロミ 212636
オウシュウサイシン 37616
オウシュウシデ 77256
オウシュウシナノキ 391691
オウシュウトウヒ 298191
オウシュウトボシガラ 163993
オウシュウナナカマド 369339
オウシュウナラ 324335
オウシュウニレ 401593
オウシュゥハルニレ 401512
オウシュウマンネングサ 356468
オウシュクロマツ 300117
オウシュマル 140869
オウジュマル 234919
オウシュンギョク 340835
オウシュンマル 2103
オウショウギョク 284668
オウショウマル 140847
オウショクマル 234953
オウスゲ 229
オウセイマル 140876
オウセンギョク 379638
オウセンリュウ 148141
オウソウカ 35016
オウソウカ属 34995
オウダソ 121714
オウダマル 182422
オウニクジュ 67530
オウバイ 211931
オウバイモドキ 211905
オウバギンドロ 311261
オウバニレ 401563
オウバヨウ 311371
オウホウギョク 43496
オウホウマル 327297
オウミカリヤス 255918
オウミノニレ 401556
オウムバナ科 190073
オウメイギョク 263739
オウラン 82051
オウリュマル 264339
オウレイマル 327256
オウレン 103835
オウレン属 103824
オウロ 107300
オウンギョク 413674
オオアオカモメヅル 409508
オオアオスゲ 73928,75140
オオアカネ 337965,337970
オオアカバナ 146606,146724
オオアキギリ 345050,345052
オオアキグミ 142218
オオアキノキリンソウ 368485
オオアザミ 364360
オオアザミ属 364353
オオアズマスゲ 75074
オオアゼスゲ 76547
オオアハ 361794
オオアブノメ 180308
オオアブノメ属 180292
オオアブラガヤ 353878
オオアブラギリ 406018
オオアブラススキ 372428
オオアブラススキ属 372398
オオアマナ 274823
オオアマナ属 274495
オオアマミテンナンショウ 33346
オオアマモ 418381
オオアメリカギク 56704
オオアメリカキササゲ 79260
オオアメリカミコシガヤ 74544
オオアラセイトウ 275876
オオアラセイトウ属 275865
オオアリドオシ 122067
オオアレチノギク 103617
オオアワ 361794
オオアワガエリ 294992
オオアワダチソウ 368106,368268
オオイ 352278
オオイイトスギ 114775
オオイソノギク 41444
オオイタチガヤ 306834
オオイタドリ 162549,309723
オオイタビ 165515
オオイタヤメイゲツ 3599
オオイチゴツナギ 305776
オオイトスギ 114775
オオイトスゲ 73663
オオイナモリ 272187
オオイヌタデ 309298,309468
オオイヌノハナヒゲ 333568
オオイヌノフグリ 407287
オオイヌホオズキ 367414
オオイワインチン 13018
オオイワウイキョウ 340463
オオイワウチワ 52541,362270
オオイワカガミ 351457
オオイワギリソウ 364754
オオイワギリソウ属 364737
オオイワツメクサ 375020
オオイワボタン 90434,90465
オオウイキョウ 163590
オオウイキョウ属 163574
オオウキシバ 318202
オオウサギギク 34766
オオウサンザシ 109924
オオウシノケグサ 164243
オオウシノシタ 377727
オオウチギョク 233617
オオウチワ 272867
オオウドカズラ 223937
オオウドノキ 223937
オオウドノキ属 223919
オオウバタケニンジン 24498
オオウバユリ 73161
オオウマノアシガタ 325916
オオウミヒルモ 184979
オオウメガサソウ 87498
オオウメバチモ 48922
オオウラジノキ 243729
オオウラジロヒヨオタンボク 236161
オオウリカエデ 3529
オオエゾイタヤ 3411
オオエノコロ 361682,361943
オオエビヅル 411623
オオエンジュ 240114
オオエンリュウ 299265
オオオサラン 148656
オオオナモミ 415033
オオオニバス 408734
オオオニバス属 408733
オオカゲロウラン 193617
オオカサスゲ 76029
オオカサモチ 304766,304772
オオカサモチ属 304752
オオガシ 323599
オオカシワ 323814
オオガタホウケン 272976
オオガタマンボウ 358872
オオカッコウアザミ 11206
オオカナグモ属 141568
オオカナダオトギリ 202010
オオカナダモ 141569
オオカナメモチ 295773
オオカニコウモリ 283848
オオカニツリ 34930
オオカニツリ属 34920
オオカマツカ 295808

オオカミナスビ 44708
オオカミナスビ属 44704
オオカミヤシ 295478
オオカメノキ 407865
オオカモメズル属 400844
オオカモメヅル 400853
オオカモメヅル属 400844
オオガヤツリ 119016
オオカラスウリ 396213,396280
オオカラバナ 2846
オオカワジシャ 406992
オオカワズスゲ 76375
オオカワヂサ 407002
オオカワヂシャ 406992
オオカンアオイ 37633,37646
オオガンクビソウ 77178
オオキエボシソウ 5057
オオギカズラ 13126
オオキクバヒゴタイ 348635
オオギシギシ 329386
オオキセワタ 295142
オオキセワタ属 295049
オオキソチドリ 302404
オオキダチハマグルマ 413507
オオキツネガヤ 60935
オオキツネノカミソリ 239274
オオキツネヤナギ 344259
オオキヌタソウ 337916,337919,337921
オオキヌラン 417768
オオギバセントウソウ 85480
オオキバナアツモリ 120310
オオキバナカイウ 417098
オオキバナカタバミ 278008
オオキバナノアツモリ 120310
オオキミシマエビネ 65980
オオギミラン 269046
オオギョウギシバ 117867
オオキリシマエビネ 65980
オオキリスゲ 73839
オオキンケイギク 104531
オオギンバイザサ 114819
オオキンレイカ 285878
オオクグ 76090
オオクサアジサイ 73134,73136
オオクサキビ 281561
オオクサボタン 94995
オオクマシデ 77313
オオクマヤナギ 52461
オオクリノイガ 80855
オオグルマ 207135
オオクルマユリ 229857
オオクログワイ 143122
オオゲシ 282647
オオケタデ 309494
オオケタネッケバナ 72735
オオゲンゲ 278862
オオコガネアヤメ 208466
オオコガネネコノメソウ 90430
オオコケミズ 299010
オオコゴメスナビキソウ 160325,190630,190708
オオゴシキヤバネバショウ 66211
オオコマツナギ 205698
オオコマユミ 157879
オオコマユミ 157312
オオコメガヤ 249105
オオコメツツジ 332020
オオゴンソウ 202400
オオサカキカズラ 25932
オオサカキカヅラ 25932
オオサカザサ 347342
オオサカソウ 363638,363755
オオサカナデシコ 127667
オオサクラソウ 314514,314519
オオサクラタデ 291766
オオサクラバラ 336316,336578
オオササ 347326
オオササウチワ 370336
オオササエビモ 312045
オオササガヤ 253992
オオササガヤ 156450
オオサザンカ 69411
オオザメ 175539
オオザリコミ 334188
オオサワシバ 77281
オオサワハコベ 374861
オオサンカクイ 6874,352187
オオサンゴアナナス 8589
オオサンザシ 109924,109933
オオサンザジ 109936
オオサンショウソウ 288762
オオシイバモチ 204396
オオシシウド 24325
オオジシバリ 210623
オオシダザサ 347264
オオシタシ 343795
オオシタレ 343070
オオシタレヤナギ 343795
オオシチトウ 119160
オオシチュウ 94557
オオシナノオトギリ 202062
オオシナノキ 391766
オオジニシキ 16768
オオシバナ 397159
オオシホガマギク 287259
オオシボヤナギ 344255
オオシマアリドオシ 122048,122064
オオシマウツギ 127023
オオシマガマズミ 408159
オオシマカンコノキ 60083
オオシマカンスゲ 75635
オオシマガンピ 414243
オオシマコバンノキ 60083
オオシマコバンノキ属 60049
オオシマザクラ 83331
オオシマサザンカ 69411
オオシマサンゴアナナス 8582
オオシマシュスラン 179626
オオシマダケ 304014
オオシマツツジ 331170
オオシマノジギク 89499
オオシマハイネズ 213963
オオシマハグマ 12689
オオシマビソ 414
オオシマムラサキ 66882
オオシマヤブマオ 56180
オオシャクナゲ 331227
オオシュウヒラアザミ 73281
オオシュクバイ 211931
オオシュロソウ 405618
オオシラタマカズラ 319437
オオシラタマソウ 363368
オオシラタマホシクサ 151487
オオシラヒゲソウ 284542
オオシラビソ 414
オオシラフカズラ 147347,353132
オオシラベ 414
オオジラベ 414
オオシロガヤツリ 119265
オオシロシキブ 66825
オオシロシマセンネンボク 137367,137368
オオシロツメクサ 397045
オオシロヤナギ 343342
オオシンジュガヤ 354052,354254
オオスグリ 334250
オオスズシロソウ 154414
オオスズムシラン 113848
オオスズメウリ 390128
オオスズメガヤ 147593
オオスズメノカタビラ 306098
オオスズメノテッポウ 17553
オオスブタ 55913
オオズミ 243735
オオセキショウモ 404546
オオセンナリ 265944
オオセンナリ属 265940
オオセンボウ 114819
オオセンボンヤリ 175172
オオソナレムグラ 187676
オオソネ 77312
オオダイコンソウ 175378
オオダイトウヒレン 348564
オオタカネイバラ 336320
オオタカネスミレ 409729
オオタカネタンポポ 384868
オオタカネバラ 336320
オオダケカンバ 53452
オオタシロイモ 382914
オオタチカラクサ 128997
オオタチツボスミレ 410154
オオタチバナ 93679

オオタチャナギ　343917
オオタチヤナギ　343574,343890
オオタデ　309494
オオタニウツギ　413591
オオタニガワスゲ　74594
オオタニタデ　91542
オオタヌキモ　403385
オオタヌキラン　73546,76204
オオタマガサ　364768
オオタマガヤツリ　119262
オオタマツリスゲ　74534
オオダルマギク　41290
オオチ　248903
オオチゴザサ　209128
オオチゴユリ　134504
オオチシマアカバナ　146663
オオチシマトリカブト　5413
オオチダケサシ　41795,41847
オオチチッパベンケイ　200801
オオチドメ　200350,200392
オオチャヒキ　60797
オオチャルメルソウ　256024
オオチョウジガマズミ　407727
オオツエ　233620
オオツガザクラ　297017
オオツガマツ　297016
オオツクバネウツギ　191
オオツクバネガシ　323594
オオツヅラフジ　364986
オオツノハシバミ　106751,106781
オオツメクサ　370522
オオツメクサモドキ　370526
オオツメクサ属　370521
オオツリバナ　157797
オオツルイタドリ　162525
オオツルウメモドキ　80267,80324
オオツルギク　150635
オオツルコウジ　31527
オオツルコオジ　31527
オオツルツゲ　203530
オオツルボ　353001
オオツルマサキ　157515,157637
オオツワブキ　162626
オオデマリ　408027
オオテリハドロ　311436
オオテンツキ　166386
オオテンニンギク　169575
オオトウカセン　385616
オオトウヒレン　348571
オオトウワタ　38135
オオトキワイヌビワ　165378
オオトキワツユクサ　394021
オオトキワレンゲ　232599
オオトックリイチゴ　338062,338292
オオトネリコ　168086
オオトボシガラ　163950
オオトモエソウ　201751
オオトリゲモ　262096
オオトリトマラズ　52100
オオナガハグサ　305882
オオナガミクダモノトケイ　285694
オオナギナタガヤ　412470
オオナキリスゲ　73839
オオナナカマド　369367
オオナラ　324190
オオナルコユリ　308592
オオナワシロイチゴ　338209
オオナンバンギセル　8767,8768
オオナンヨウノボタン　248778
オオニガナ　313894
オオニジ　185157
オオニシキソウ　85779,159286
オオニソジンボク　411420
オオニラ　15185
オオニワトコ　345687
オオニワナナカマド　369267
オオニワホコリ　147883
オオニンニク　15698
オオヌカキビ　282031
オオヌマハリイ　143198,143200
オオネコケネヤナギ　344260
オオネズミガヤ　259671
オオネバリタデ　309946
オオノキ　283421
オオバアオイソウ　290331
オオバアカテツ　280440
オオバアカテツ属　280437
オオバアカメガシワ　14217
オオバアコウ　164783
オオバアサガオ属　32600
オオバアサガラ　320886
オオバアズマザサ　347366
オオバアラリア　310204
オオバイ　211931,316577
オオバイカイカリソウ　146956,147059
オオバイカウツギ　294539
オオバイチジク　164661
オオバイヌツゲ　203721
オオバイヌビワ　165658
オオバイヌリンゴ　243670
オオバイボタ　229569
オオバイ属　211715
オオハウチワマメ　238457
オオバウツギ　127049
オオバウド　30620
オオバウマノスズクサ　34219
オオバウメモドキ　204081,204239
オオバエゴノキ　379381
オオバエンゴサク　106519
オオバオオヤマレンゲ　279249
オオバオトギリ　202097
オオバガシ　323599
オオバカワヤナギ　343444
オオバカンアオイ　37673
オオバカンスイ　159841
オオバカンラン　116933
オオバキスミレ　409784
オオバキセワタ　295142
オオバギボウシ　198612,198646
オオバキリンソウ　294117
オオハクウンラン　407606
オオバクサフジ　408551
オオハクサンサイコ　63739
オオバグミ　142075
オオバクロモジ　231466
オオバケアサガオ　225712
オオバケエゴノキ　379388
オオバケカンコノキ　177194
オオバゲッキツ　260168
オオバケハウチワカエデ　3050
オオバコ　301871,302068
オオバコウモリ　283822
オオバコキンポウゲ　184711
オオバコケモモ　403821
オオバコナラ　323590
オオハコベ　374780,374781
オオバゴムノキ　164925,165274
オオバゴムビワ　165274
オオバコリヤナギ　343529
オオバコ科　301821
オオバコ属　301832
オオバザサ　347250
オオバザラッカ　342572
オオバサンザシ　109839,110014
オオハシカグサ　262962
オオバジシャ　379414
オオバシダソテツ　373674
オオバシナノキ　391766,391773
オオハシバミ　106736
オオバシマムラサキ　66931
オオバジャカランダ　211235
オオバジャカランダノキ　211235
オオバジャノヒゲ　272122
オオバジュズネノキ　122064
オオバジュラン　11291
オオバショウマ　91023,91024
オオハシリドコロ　44708
オオバシロテツ　249143
オオバスイハ　340136
オオバスグリ　333970
オオバスノキ　404003
オオバセンキュウ　24355
オオバセンブリ　380411
オオバセンボウ　114819
オオバタイサンボク　242193
オオハタガヤ　63327
オオバタケシマラン　377930
オオバタチシオデ　366487
オオバタチツボスミレ　410170
オオバタツノヒゲ　128022
オオバタネツケバナ　72971

オオバタンキリマメ　333133
オオバヂシャ　379414
オオバチシャノキ　141629
オオバチヂミザサ　272676
オオバチドメ　200312
オオバチドメグサ　200312
オオバチドリノキ　2869
オオバツツジ　331341
オオバツルグミ　142075,142078
オオバテテイオウハイモ　16510
オオバトウナナカマド　369484
オオバトネリコ　168023
オオバトベラ　301416
オオバトベラノキ　301416
オオバナアサガオ　113823
オオバナアザミ　375274
オオバナアザミゲシ　32423
オオバナアリアゲキズラ　14873,14874
オオバナアリアゲキヅラ　14874
オオバナイチゲ　24071
オオバナイトタヌキモ　403191
オオハナイバナ　57683
オオハナウド　192259,192278
オオバナオオヤマサギソウ　302357
オオバナオケラ　44218
オオバナカズラ　49359
オオバナカリッサ　76926
オオバナキョウチクトウ　389978
オオバナクンシラン　97218
オオバナコウリンギク　358240
オオバナコマツヨイグサ　269446
オオバナコリンソウ　99836
オオバナサイシン　38316
オオバナサルスベリ　219966
オオバナジギタリス　130365
オオバナシロワレモコウ　345923
オオバナスミレ　410214
オオバナソケイ　211848
オオバナチョオセンアサガオ　61242
オオバナテンジクアオイ　288264
オオバナノエンレイソウ　397574
オオバナノクサガク　73122
オオバナノコギリサウ　4005
オオバナノコギリソウ　4005
オオバナノセンダングサ　54059
オオバナノミミナグサ　82813
オオバナノワレモコウ　345923
オオハナハタザオ　94241
オオバナハナカズラ　5525
オオバナハナゴマ　205571
オオバナハマビシ　395082
オオバナハンゴンソウ　339585
オオバナヒエンソウ　124237
オオバナヒルギ　368915
オオバナボントクタデ　291915
オオバナマメザクラ　83227
オオバナラ　324173
オオバニワトコ　345683
オオバヌスビトハギ　200732
オオバヌスビトバギ　200727
オオバヌズビトハギ　200728
オオハネガヤ　4160,376754
オオバネム　13595
オオバネムノキ　13586
オオバノアカツメクサ　396967
オオバノアリトオシ　122064
オオバノウマノスズクサ　34219
オオバノウリカエデ　2912
オオバノキハダ　294233
オオバノキワダ　294233
オオバノコゴメウツギ　375821
オオバノセンナ　360483
オオバノツワブキ　162626
オオバノトンボソウ　302439
オオバノハブソウ　78301
オオバノヘビノボラズ　52100
オオバノボタン属　253003
オオバノマンゴスチン　171092
オオバノミズトラノオ　139585
オオバノヤエムグラ　170570
オオバノヨツバハギ　408677
オオバノヨツバムグラ　170437
オオバハシドイ　382268
オオバハマアサガオ　376536
オオバハマアサガオ属　376524
オオバヒメトウ　121497,121500
オオバヒメマオ　313483
オオバヒメマオ属　313411
オオバヒルギ　329765,329771
オオバフウトウカズラ　300428
オオバフジイバラ　337046
オオバフジボグサ　402132
オオバブナノキ　162370
オオバベゴニヤ　50232
オオバベニガシワ　14184,14217
オオバベニガシワ属　14177
オオバヘビノボラズ　52100
オオバボダイジュ　391766
オオバボンテンカ　402245,402250
オオハマオモト　111167
オオハマガヤ　19773
オオハマカンザシ　34536,34550,34560
オオハマギキョウ　234327
オオハマギキョオ　234327
オオハマグルマ　413550
オオバマサキ　157617
オオハマボウ　195311
オオハマボオ　195311
オオバマホガニ－　380527
オオハマボッス　239729
オオハマムギ　144303
オオバマンサク　185120
オオバミゾホオズキ　255242
オオバミネカエデ　3722
オオバミヤマハンノキ　16329
オオバムラサキ　66943
オオバメギ　52282
オオバメドハギ　226751
オオバヤシャブシ　16458
オオバヤダケ　206833
オオバヤドリギ　355337
オオバヤドリギ科　236503
オオバヤナギ　343170
オオバヤマナラシ　311335
オオバユ－カリ　155722
オオバユキヤナギ　371881
オオバユク　94026
オオバヨウラクラン　268000
オオバヨメナ　40860,215356
オオバヨモギ　35746
オオバライチゴ　338305
オオハリイ　143055,143100,143101
オオハリサウ　381026
オオハリスゲ　76626
オオハリソウ　381026
オオバリンドウ　173615
オオバルシャ　107161
オオバルシャギク　107161
オオバルシャ属　107158
オオハルトラノオ　54822
オオバルリミノキ　222237,222302
オオハンゲ　299728
オオハンゴンソウ　339574
オオハンゴンソウ属　339512
オオバンマツリ　61297,61310
オオヒエンソウ　124237
オオヒカゲスゲ　76015
オオヒカゲミズ　284174
オオヒキオコシ　209826
オオヒキヨモギ　365298
オオヒキヨモギ属　365295
オオヒゲガリヤス　65359
オオヒゲクサ　352382
オオヒゲナガカリヤスモドキ　255852
オオヒゴタイサイコ　154281
オオヒツジグサ　267767,267771
オオヒナツメクサ　397052
オオヒナユリ　68800
オオヒメギリソウ　377654
オオヒメクグ　218571
オオヒメノウスツボ　355148
オオヒメハギ　308359
オオヒョウタンボク　236196
オオヒヨドリバナ　158232
オオヒラウスユキソウ　224854,224908
オオヒラタンポポ　384718
オオヒラテンツキ　166223
オオビランジ　363622
オオビル　15698
オオヒルムシロ　312190
オオヒレアザミ　73281

オオヒレアザミ属　271663
オオヒロハチトセラン　346089
オオヒンジガヤツリ　232391,232423
オオフィツム属　271919
オオフサモ　261342
オオブシ　5193
オオフジイバラ　336578,337015
オオブタクサ　19191
オオフタナミソウ　354930
オオフタバムグラ　131358
オオフトイ　352206
オオブドホオヅキ　297686
オオフトモモ　382660
オオベニウチワ　28084
オオベニウツギ　413591
オオベニゴウカン　66673
オオベニタデ　309494
オオベニバナイチヤクソウ　322843
オオヘニミカン　93837
オオヘビイチゴ　312912
オオベンケイソウ　200802
オオホウキガヤツリ　118756
オオホウキギク　41322
オオボウシバナ　100970
オオホザキアヤメ　107271
オオホザキアヤメ科　107189
オオホザキアヤメ属　107221
オオホザキマツラン　333729
オオホシクサ　151257
オオボシソウ　342396
オオホソバトラノオ　317925
オオホソバママコナ　248210
オオホタルサイコ　63728
オオホナガアオゲイトウ　18787
オオマイヅルソウ　242680
オオマキエハギ　226684
オオマツバウンラン　267349
オオマツバシバ　34050
オオマツユキソウ　227856
オオマツユキソウ属　227854
オオマツヨイグサ　269443
オオマトイ　159997
オオママコナ　248188
オオマムシグサ　33496
オオマメ　177750
オオマルバコンロンソウ　72686
オオマルバノテンニンソウ　228014
オオマルバノホロシ　367367,367368
オオマンテマ　363882
オオマンネンラン　169245
オオミイヌカンコ　166776
オオミイボタ　229656
オオミウェッチンヤシ属　413924
オオミウツギ　126893
オオミカントウ　93678
オオミクマデヤシ属　295501
オオミクリ　370057,370102
オオミコゴメグサ　160185
オオミサンザシ　109936
オオミズオオバコ　277369
オオミズタマソウ　151452
オオミズトンボ　183797
オオミゾカクシ　224073
オオミゾソバ　309888
オオミタケヤシ属　412286
オオミチヤナギ　308816
オオミツデカエデ　3243
オオミツバコンロンソウ　72681
オオミツバショウマ　91023
オオミツバタヌキマメ　112491,112492
オオミツバッチグリ　312323
オオミテングヤシ　246672
オオミトケイソウ　285694
オオミドリボウキ　39195
オオミネイチゴ　338066
オオミネイヌナシ　323347
オオミネカエデ　3529
オオミネザクラ　83131
オオミネテンナンショウ　33433
オオミネヒナノガリヤス　65436
オオミノサルナシ　6515
オオミノツルコケモモ　403894
オオミノトベラ　301227
オオミブラヘア　59098
オオミマンリョウ　31396
オオミミナグサ　82758,82826
オオミヤシ　235159
オオミヤシ属　235151
オオミヤハズ属　321128
オオミヤマガマズミ　408231
オオミヤマカンスゲ　74596
オオミヤマコウボウ　196121
オオミヤマタニタデ　91520
オオミヤマナナカマド　369512
オオミヤマヤチャナギ　343923
オオミルスベリヒユ　361668
オオムギ　198376
オオムギクサ　198316
オオムギスゲ　75076
オオムギ属　198260
オオムシヤリンドウ　265061
オオムラサキ　331397,331581,331582
オオムラサキシキブ　66741,66808,66824
オオムラサキツユクサ　394088
オオムラホシクサ　151419
オオムレイウズ　5472
オオムレスズメ　72180
オオメタカラコウ　229208
オオメドハギ　226751
オオメノマンネングサ　357116
オオメマツヨイグサ　269437
オオモクゲンジ　217613
オオモクセイ　276394
オオモチ　203961
オオモチノキ　399433
オオモミジ　2778
オオモミジガサ　255790
オオモミジスグリ　334084
オオモンツキガヤ　57579
オオヤクシマシャクナゲ　332137
オオヤナギアザミ　92040
オオヤノネアザミ　348865
オオヤハズエンドウ　408571
オオヤバネバショウ　66205
オオヤブツルアズキ　409028
オオヤブヨモギ　36164
オオヤマイチジク　165140
オオヤマカタバミ　277985
オオヤマサギソウ　302502
オオヤマザクラ　83301
オオヤマジソ　259302
オオヤマツツジ　331989
オオヤマハコベ　375002,375041
オオヤマハッカ　209706
オオヤマフスマ　256444
オオヤマフスマ属　256441
オオヤマミズ　298946
オオヤマムグラ　170563
オオヤマレンゲ　242285,279249
オオヤラオダマキ　30061
オオユウガギク　41159,215339
オオユキソウ　31934
オオユキノハナ　169710
オオユキワリソウ　314373
オオユリワサビ　161154
オオヨドカワゴロモ　200205
オオヨモギ　35948
オオリュウセン　414
オオリンゴ　243675
オオルリサウ　117965
オオルリソウ　117965,117966,118038
オオルリソウ属　117900
オオレイジンサウ　5231,5466
オオレンリソウ　222724
オオロウゲ　312392
オオロベリアソウ　234813
オオワクノテ　95295
オオワタヨモギ　35527
オガアザミ　92032
オカイボタ　229578
オカウコギ　143685
オカウツボ　275010
オカオグルマ　385903
オガコウモリ　283852
オカサダケ　57437
オガサハライチビ　228,934
ォガサハラクチナシ　171253
オガサハラスズメノヒエ　285417
オガサハラマツ　15953
オガサハラモクレイシ　172885
オガサハラモクレイシ属　172881

オガサハラヤブニクケイ　91438
オガサモミジガサ　283805
オガサラツツジ　330233
オガサワラアオグス　240559
オガサワラアザミ　91806
オガサワラエノキ　80663
オガサワラクチナシ　171266
オガサワラグミ　142184
オガサワラグワ　259121
オガサワラゴシュユ　249156
オガサワラコミカンソウ　296530
オガサワラシコウラン　62591
オガサワラシロダモ　264029
オガサワラスズメノヒエ　285417
オガサワラソクズ　345588
オガサワラツツジ　330233
オガサワラハマボウ　194895
オガサワラハマユウ　111200
オガサワラビロウ　234167
オガサワラフトモモ　252609
オガサワラボチョウジ　319671
オガサワラマツ　15953
オガサワラモクマオ　56105
オガサワラモクレイシ　172885
オガサワラモクレイシ属　172881
オガサワラヤブニッケイ　91438
オカシャジン　7816,7824
オカスズメノヒエ　238675,238677
オ－ガストノキ属　90007
オカスミレ　410406
オカゼリ　97719
オカダイコン　8018
オカタイトゴメ　356845
オカダゲンゲ　279122
オガタチイチゴツナギ　305788
オカタツナミソウ　355392
オガタテンナンショウ　33440
オガタマ　252849
オカタマノキ　252849
オガタマノキ　252849
オガタマノキ属　252803
オカツラ　83736
オカトラノオ　239594
オカトラノオ属　239543
オカノリ　243771
オカヒジキ　344606,344718
オカヒジキ属　344425
オガフウロ　175032
オカブキ　292375
オカボ　275958
オカミズオジギソウ　265238
オカミル　344718
オカムラサキ　184298
オカメウチワ　370338
オカメザサ　362131
オカメザサ属　362121
オガラバナ　3732
オガルカヤ　117181,117185
オガルカヤ属　117136
オカヲグルマ　385903
オギ　255873,394943
オキアガリネズ　213615
オキザリス属　277644
オキシデンドラム属　278388
オキジムシロ　313039
オキタンポポ　384663
オキチハギ　126529
オキナウチワ　45265
オキナグサ　321667
オキナグサ属　23696,321653
オキナジシ　299259
オキナダケ　304007,304008
オキナニシキ　140239
オキナハアオキ　81933
オキナハアヲキ　81933
オキナハイナモリ　272241
オキナハイボタ　229524
オキナハカウゾリナ　55768
オキナハガルカヤ　29430
オキナハガンピ　414250
オキナハケフチクタウ　83700
オキナハサザンカ　69370
オキナハサザンクワ　69370
オキナハシキミ　204575
オキナハシタキヅル　211695
オキナハジヒ　233303
オキナハソケイ　212004
オキナハソヨゴ　203871
オキナハチャウジタデ　238188
オキナハツケ　64373
オキナハネム　1142
オキナハハクサンボク　407867
オキナハハヒネズ　213963
オキナハマツ　300044
オキナハミヤコグサ　237506
オキナハモチ　203871
オキナマル　82211
オキナヨモギ　35088
オキナワイ　213040
オキナワイナモリ　272221,272241
オキナワイヌシカクイ　143417
オキナワイボタ　229432,229524
オキナワウラジロガシ　324171
オキナワオオガヤツリ　118473
オキナワカルカヤ　29430
オキナワギク　40867
オキナワクルマバナ　96970
オキナワコウバシ　231327
オキナワサザンカ　69370
オキナワジイ　78913,79023
オキナワシキミ　204482
オキナワシタキヅル　211695
オキナワジュズスゲ　74916
オキナワジンコオ　161639
オキナワシントンヤシ　413298
オキナワスゲ　73937
オキナワスズムシソウ　378231
オキナワスズメウリ　132957
オキナワスナゴショウ　290397
オキナワセッコク　125291
オキナワソケイ　212004,212024
オキナワソヨゴ　203860
オキナワチドメグサ　200370
オキナワチドリ　19518
オキナワツゲ　64273
オキナワテイカカズラ　393656
オキナワテンナンショウ　33347
オキナワトベラ　301229
オキナワナガボスゲ　74377
オキナワハイネズ　213963
オキナワハクサンボク　407867
オキナワハグマ　12678
オキナワバライチゴ　338922
オキナワハリイ　143358
オキナワヒサカキ　160572
オキナワヒメウツギ　127024
オキナワヒメナキリ　76131
オキナワヒヨドリジョウゴ　367273
オキナワホシクサ　151383
オキナワマツバボタン　311889
オキナワミゾイチゴツナギ　305280
オキナワミチシバ　90108
オキナワムヨウラン　223670
オキナワモチ　203860
オキナワヤブムラサキ　66884
オキノアザミ　92078
オキノアブラギク　124837
オキノシマカンアオイ　194379
オギノツメ　200652
オギノツメ属　200589
オギヨシ　255873
オクウスギタンポポ　384520
オクエゾアイスゲ　73767
オクエゾアザミ　92367
オクエゾガラガラ　329527
オクエゾガラガラ属　329504
オクエゾキツリフネ　205177
オクエゾコゴメグサ　160239
オクエゾサイシン　37635,37637
オクエゾシラカンバ　53355
オクエゾナズナ　137286
オクエゾヤチダモ　168023
オクエノキ　80600
オクオ　219
オクキタアザミ　348720
オクキンハイソウ　399515
オククルマバムグラ　170705
オククルマムグラ　170705
オククロウスゴ　403942
オクシモアズキナシ　369307
オクシモハギ　226764

オクジリエノキ　80600
オクシリエビネ　66034
オクシリカンスゲ　73565
オクタヌキラン　73575
オクタマスミレ　409618
オクタマツリスゲ　74531
オクタマハギ　226679
オグヂガワラザイコ　312955
オクチョウジザクラ　83148
オクトネホシクサ　280350
オクトポマ属　268845
オクトリカブト　5298
オクナ・セルラータ　268261
オクナ科　268277
オクナ属　268132
オクヌギ　323611,324532
オクノエノキ　80596
オクノカンスゲ　74584
オクノハマイボタ　229658
オクノフウリンウメモドキ　203841
オクノミズギク　207084
オクミチヤナギ　308832
オクミヤマキンバイ　312747
オクムシャリンドウ　137651
オクモミジハグマ　12602
オクヤマアザミ　92281
オクヤマオトギリ　201902
オクヤマガラシ　73013
オクヤマコウモリ　283845
オクヤマザクラ　83319
オクヤマザサ　347305
オクヤマサルコ　342938
オクヤマスミレ　409958
オクヤマナズナ　72717
オクヤマハンノキ　16348
オクヤマヤナギ　343911
オクヤマリンダウ　173598
オクヤマリンドウ　173598
オクラ　219
オグライ　352160
オグラギク　89786
オグラコウホネ　267313
オグラセンノウ　363638
オグラノフサモ　261355
オグラフサモ　261355
オグラヤマ　3034
オグラヤマザクラ　83117
オグルマ　207046,207059,207151
オグルマ属　207025
オグルミ　212626
オクルリヒゴタイ　140732
オクヲ　219
オケシャウヤナギ　342961
オケヤマウツギ　413623
オケラ　44205,44208
オケラ属　44192
オゲンギョク　103750
オゴケ　143022
オコリンダウ　173852
オコリンドウ　173852
オゴンギョク　103750
オザキイガニガクサ　203064
オサバグサ　320181
オサバフウロ　54555
オサバフウロ属　54532
オサラン　148754
オサラン属　148619
オジギソウ　255098
オジキソウ属　254979
オシマオトギリ　202213
オシマシモツケ　371906
オシマム属　268418
オシャラクマメ　259520
オショウナシ　323156
オシロイカケ　348087
オシロイバナ　255711
オシロイバナ科　267421
オシロイバナ属　255672
オスキューリア属　220464
オスクラリア属　220464,251025
オスズラン　147196
オーストラリアデザートライム　148262
オーストラリアビロウ　234166
オーストラリアホソミライム　253285
オーストラリアマルミライム　253288
オーストリアソライヤー　336563
オーストリアマツ　300117
オーストリアンカッパーローズ　336565
オーストリヤンブライエル　336563
オストロウスキア属　276795
オーストロカクタス属　45231
オーストロシリンドロプンティア属　45254
オストロシリンドロプンティア属　45254
オゼクロスゲ　73576
オゼコウホネ　267322
オゼザサ　347330
オセージ・オレンジ　240828
オセージオレンヂ属　240806
オセソウ　211603
オセソウ属　211602
オゼトウヒ　298313
オゼニガナ　210534
オゼヌマアザミ　92029
オゼヌマタイゲキ　159980
オゼミズギク　207083
オソエウリハダ　2846
オソバキンシバイ　201888
オソバヒヒラギナンテン　242534
オソレヤマオトコイ　352161
オタカラカウ　229179
オタガラカウ　229179
オタカラコウ　229035
オタケシマラン　377919
オダサムタンポポ　384749
オタネニンジン　280741,280793
オタフクギバウシ　198599
オタフクギボウシ　198599
オタフクグルミ　212655
オタフクヒシ　394463
オダマキ　30031
オダマキ属　29994
オタマナ　59595
オタルスゲ　75638
オダンカ　402267
オヂギイチゴ　338079
オヂキサウ　255098
オチフジ　247731
オッタチカタバミ　277803
オッタチカンギク　89550
オツツジ　332104
オットーソンデリア属　277441
オテダマゼキショウ　213531
オートー　93680
オトウガラシ　4866
オートエンバク　45566
オトギリキヌタソウ　170446
オトギリサウ　201853
オトギリサウ科　97266
オトギリサウ属　201705
オトギリソウ　201853
オトギリソウ科　97266
オトギリソウ属　97260,201705
オトギリバニシキソウ　159102
オトギリマオ　179495,179500
オトコエシ　285880
オトコオミナエシ　285838
オトコゼリ　326417
オトコノジスミレ　410770
オトコビシ　394513
オトコブドウ　411903
オトコブドオ　411903
オトコヘシ　285880
オトコマツ　300281
オトコメシ　285880
オトコヤシ　97074
オトコヨウゾメ　408024
オトコヨモギ　35674
オトヒメ　108918
オートーミカン　93803
オートムギ　45566
オトメアオイ　37717
オトメアゼナ　46362
オトメアゼナ属　46348
オトメイヌゴマ　373346
オトメエンゴサク　105915
オトメオキナワキョウチクトウ　83699
オトメカワヂシャ　407003
オトメギキョウ　70228
オトメギボウシ　198655
オトメザクラ　314613
オトメサナエタデ　291895

オトメサルビア　345267
オトメシオガマ　287794
オトメシホガマ　287794
オトメシャジン　7871
オトメスイレン　267641
オトメスミレ　410034
オトメセンダングサ　53773
オトメツバキ　69195
オトメニオイタチツボスミレ　410314
オトメフウロ　174569
オトメユリ　230023
オトメレンリソウ　222699
オドリコサウ　220346
オドリコソウ　220346,220359
オドリコソウ属　220345
オドントグロッサム　224352
オドントグロッサム属　269055
オトントフォルス属　269114
オトンナ属　277000
オナガエビネ　66086
オナガカエデ　2872,3058
オナガカンアオイ　37684
オナガクマシデ　77268
オナガグワ　259117
オナガサイシン　37585
オナガシ　379414
オナガタケ　304098
オナガヤブニンジン　276455,276460
オナガワレモコウ　345894
オナゴダケ　304098
オナズオウ　83769
オナモミ　415020,415046,415057
オナモミ属　414990
オニアザミ　91808,92226,92407
オニアゼガヤ　226008
オニアゼスゲ　73559
オニイタヤ　3160,3174,3398,3400
オニイチゴ　339270,339468
オニイチゴツナギ　305506
オニイボタ　229565,229610
オニイラクサ属　175865
オニウコギ　143588,143682
オニウシノケグサ　163819
オニウッギ　127049
オニウド　24380
オニウラジロシノ　347367
オニウロコアザミ　271692
オニオオノアザミ　92072
オニオタカラコウ　229038
オニオトコヨモギ　35356
オニカサモチ　304765
オニガシ　233104,233132
オニカモジグサ　335536
オニガヤツリ　119378
オニカラスノエンドウ　408469
オニカラスムギ　45495
オニカンスゲ　73930
オニカンゾウ　191284,191291
オニキクバヒゴタイ　348635
オニギョウギシバ　117878
オニキランソウ　13097
オニキリマル　17017
オニク　57390,57437
オニクグ　245351,245453
オニクサヨシ　293708
オニグジョウシノ　347341
オニグルミ　212626
オニクロイチゴ　338933
オニクロギ　381429
オニクワンザウ　191291
オニグンバイナズナ　225315
オニク属　57389,57429
オニゲシ　282647
オニコウゾリナ　402663
オニコナスビ　239877
オニコバノガマズミ　407658
オニコブナグサ　36673
オニコメススキ　126064
オニコンニャク　20080
オニサケミヤシモドキ　212346
オニサケヤシモドキ　269375
オニササガヤ　128571
オニザミア属　241448
オニサルビア　345379
オニザンショウ　417161
オニシオガマ　287459
オニジク　28030
オニシバ　418431
オニシバリ　122586
オニジヒ　233121
オニシモツケ　166089
オニジャク　28030,28046
オニジュズダマ　99131
オニジュロ　413307
オニスゲ　74309
オニススキ　255849
オニセキシャウ　5798
オーニソガラム　274823
オニソテツ属　145217
オニタイミンガサ　283812
オニタガラカウ　229049
オニタガラコウ　229049
オニタチバナ　31387,31613
オニタビラコ　416437
オニタビラコ属　110710,416390
オニダラ　30634
オニチャガヤツリ　118431
オニチャヒキ　60692
オニツクバネウツギ　176
オニヅタ　187318
オニツツジ　331258
オニツルウメモドキ　80327
オニテンツキ　166531
オニトゲココヤシ属　6128
オニトゲノヤシ　2551
オニドコロ　131877
オニトダシバ　37398
オニトド　463
オニトベラ　301438
オニトボシガラ　163950,163993
オニナスビ　367014
オニナナカマド　369270
オニナベナ　133478,133517
オニナルコスゲ　76677,76687
オニネザサ　304048
オニノアザミ　92226
オニノガリヤス　65354
オニノゲシ　368649
オニノダケ　24358
オニノマユハキ　73337
オニノヤガラ　171918
オニノヤガラ属　171905
オニハシドイ　382220
オニバス　160637
オニバス属　160635
オニハマダイコン　65185
オニハリナスビ　367295
オニビシ　394500,394513
オニビトノガリヤス　65450
オニヒバ　228639
オニヒバ属　67526,228635
オニヒメクグ　218482
オニヒョウタンボク　236205
オニヒョオタンボク　236205
オニブキ　181951
オニヘウタンボク　236205
オニマガリバナ　203209
オニマタタビ　6553
オニマツ　300146
オニマツヨイグサ　269454
オニマメ　30498
オニマユミ　157879
オニミツバ属　345940
オニミルクサ　354691
オニムラサキ　375726
オニメグスリ　3710
オニメタガラカウ　229206
オニメヒシバ　130745
オニメロ　116546
オニモミジ　2928
オニモミヂ　2928
オニヤブムラサキ　66829,66856
オニヤマナラシ　311530
オニヤマボクチ　382069,383213
オニヤマボクチ属　383210
オニユリ　230058
オニルリソウ　283269
オニルリヒゴタイ　140697
オネヤマザクラ　83130
オノエスゲ　76515
オノエテンツキ　166329

オノエマンテマ 183183
オノエヤナギ 344053,344233
オノエラン 273427
オノエリンドウ 174108
オノエリンドウ属 174101
オノオノカンバ 53610
オノオレ 53610
オノオレカンバ 53610
オノブリキス属 271166
オノヘスゲ 76515
オノヘテンツキ 166329
オノマンネングサ 356884
オハグロカヅラ 300944
オハグロスゲ 73875
オハグロノキ 110251
オハグロバナ 34162
オバケアサガオ 225712
オバケアザミ 91701
オバケトキワラン 282895
オバケヒゴタイ 348093
オハツキガラシ 154091
オバナ 255886
オバナニガナ 210594
オハラメアザミ 92200
オビガタセキコク 390522
オヒガンギボウシ 198590
オヒゲシバ 88421
オヒゲシバ属 88316
オヒシバ 143530
オヒジハ 143530
オヒシバ属 143512
オヒチナリス 91082
オビバナヤシ属 383027
オヒメクグ 218571
オヒヤウハシバミ 106736
オヒョウ 401542
オヒョウエノキ 80664
オヒョウニレ 401542
オヒョウハシバミ 106736
オヒョウババハシバミ 106736
オヒョウモモ 20970
オヒョウモモ 20974
オヒョモモ 20970
オヒルギ 61263
オヒルギ属 61250
オヒルムシロ 312188,312190
オビレハリクジャグヤシ 12772
オヒワケガマズミ 408056
オヒワケグミ 141999,142207
オヒワケヒラギナンテン 242608
オヒワケヘウタンボク 235999
オビワケラン 116851
オフクカズラ 20284
オフクカヅラ 20284
オフタバラン 232919
オフタルモフィルム属 272510
オブチ 61107
オプリスメヌス属 272611
オフリス属 272395
オブレゴニア属 268085
オプンティア属 272775
オベカウジ 328684
オベスム 7343
オーベニミカン 93837
オヘビイチゴ 312690
オホアカリニシキサウ 159854
オホアザミ 364360
オホアザミ属 364353
オホアハ 361794
オホアハガヘリ 294992
オホアハダチサウ 368268
オホアブノメ 180308
オホアブノメ属 180292
オホアフヒ 13934
オホアブラガヤ 353878
オホアブラススキ 372428
オホアマドコロ 308613
オホアマナ 274823
オホアマナ属 274495
オホアマモ 418381
オホアミガシ 233287
オホイイトスギ 114775
オホイタドリ 309723
オホイタビ 165515
オホイタヤメイゲツ 3599
オホイチゴツナギ 305776
オホイトスゲ 73663
オホイヌタデ 309468
オホイヌノヒゲ 151380
オホイヌビハ 165762
オホイヌフグリ 407287
オホイハガネ 273927,300835
オホウイキャウ属 163574
オホウシクサ 371090
オホウシクサ属 371087
オホウシノケグサ 164243
オホウドカズラ属 223919
オホウマノアシガタ 325916
オホウミヒルモ 184979
オホウメガササウ 87498
オホウリカエデ 3529
オホウヰキャウ属 163574
オホエゾイチゲ 23796
オホエダウチヂミザサ 272625,272629
オホエビヅル 411623
オホオトギリ 201743
オホオナモミ 415003
オホカキネガラシ 365515
オホガクフタギ 381423
オホカサスゲ 76029
オホカサモチ 304765,304772
オホカサモチ属 304752
オホガシ 323599
オホカナメモチ 295773
オホカニツリ 34930
オホカニツリ属 34920
オホカハヅスゲ 76375
オホカミグサ 240765
オホカメダフシ 240765
オホカメノキ 407865
オホカモジグジ 335423
オホカモメヅル属 400844
オホガヤツリ 212778
オホガラシ 59438
オホカラバナ 2846
オホガンクビサウ 77178
オホギ 165841
オホキセワタ 295142
オホキセワタ属 295049
オホキツネガヤ 60935
オホキヌタサウ 337916
オホキノボリカヅラ 313170
オホキノボリカヅラ属 313169
オホギバウシ 198646
オホキバナアツモリ 120310
オホキバナムカシヨモギ 55769
オホキリスゲ 73839
オホキンケイギク 104531
オホクグ 76090
オホクサアヂザヰ 73134
オホクサキビ 281561
オホクサボク 81933
オホクマシデ 77313
オホクマタケラン 17774
オホクマヤナギ 52461
オホクリガシ 78966
オホグルマ 207135
オホクルマアカネ 337925
オホクロウレモドキ 328684
オホクロモジ 233920
オホケイヌビハ 165390
オホゲシ 282647
オホケシモツケ 371971
オホケタデ 309494
オホコゴメガヤ 119020
オホコササガヤ 291416
オホコマツナギ 205698
オホサイトウガヤ 127209
オホサカキカヅラ 25933
オホサクライバラ 337015
オホサクラサウ 314514
オホサクラタデ 291766
オホササガヤ 253992
オホザリコミ 334188
オホサルコヤナギ 343151
オボサンキライ 366271
オホサンザシ 109924,109933
オホザンセウ 417340
オホサンセウサウ 288762
オホシ 329386
オホシシウド 24325

オホシダレ　343070
オホシナノキ　391760
オホシバキ　264090
オホシヒバモチ　203636
オホジフニヒトヘ　283615
オホシボヤナギ　344255
オホシマコバンノキ　60083
オホシマザクラ　83316
オホシマサザンクワ　68939,69411
オホシマビソ　414
オホシマムラサキ　66882
オホシュスラン　179709
オホシュロサウ　405618
オホシラタマサウ　363368
オホシラタマホシクサ　151487
オホシラベ　414
オホシンジュガヤ　354242
オホスズガヤ　60379
オホスズムシラン属　113846
オホスズメウリ　390128
オホスズメウリ属　390107
オホスズメノカタビラ　306098
オホスズメノテッパウ　17553
オホスブタ　55913
オホゼリ　90932
オホセンナリ　265944
オホセンナリ属　265940
オホダイコンサウ　175378
オホタウワタ　38135
オホタカツルラン　170044
オホタカネイバラ　336320
オホタケ　297373
オホタチカモジグサ　335423
オホタチヤナギ　343917
オホタツノヒゲ　293296
オホタデ　309494
オホタニウツギ　413591
オホタマガサ　364768
オホタマガヤツリ　119262
オホチゴユリ　134504
オホチシマトリカブト　5413
オホチダケサシ　41795,41806
オホチドメ　200350
オホチャウジガマズミ　407727
オホチャヒキ　60797
オホッガザクラ　297016
オホツガマツ　297016
オホツキンバイ　312714
オホツクバネサウ　284378
オホツヅラフヂ　364986
オホツヅラフヂ属　364985
オホッハシバミ　106751
オホツメクサ　370522
オホツメクサ属　370521
オホツリバナ　157797
オホツルウメモドキ　80267
オホツルボ　353001
オホツルマサキ　157637
オホテウビツルラン　155012
オホテンツキ　166386
オホテンニンギク　169575
オホトキンサウ　146098
オホトキンサウ属　146096
オホトキンソウ属　146096
オホドゼウツナギ　177668
オホトボシガラ　163950
オホトリゲモ　262098
オホトリトマラズ　51301,52339
オホナガミクダモノトケイ　285694
オホナハシロイチゴ　338209
オホナラ　324190
オホナンバンギセル　8760,8767
オホニガイモ　131630
オホニシキサウ　159102
オホニソジンボク　411420
オホニハホコリ　147883
オホニラミグサ　305776
オホヌマハリイ　143201
オホヌマハリヰ　143201
オホネギ　15292
オホネコノオ　306960
オホネコノヲ　306960
オホネコヤナギ　344260
オホネズミガヤ　259671
オホネバリタデ　309946
オホネブカ　15292
オホノシスゲ　118749
オホノボタンノキ　43462
オホノボタンノキ属　43457
オホバアカザ　87048
オホバアカテツ　280440
オホバアカテツ属　280437
オホバアサガホ　32642
オホバアサガホ属　32600
オホバアサガラ　320885,320886
オホバアハダン　161350
オホバアリサンハコベ　374758
オホバアワセンダン　161350
オホバイボタ　229569
オホバイボタノキ　229569
オホバウオトリギ　180995
オホハウチハマメ　238457
オホバウツギ　127049
オホバウマノスズクサ　34219
オホバウメモドキ　204081
オホバウヲトリギ　180995
オホバエノキ　80664
オホバカウバシ　231385
オホバガシ　323599
オホバカハヤナギ　343444
オホバカヤラン　185755
オホバガラシ　59438
オホバカンザブラウ　381148
オホバギ　240325
オホバキセワタ　295142
オホバギ属　240226
オホバクサフヂ　408551
オホバグス　91378
オホバグスノキ　91378
オホバグッケイ　6251
オホバグッケイ属　6233
オホバグミ　142075
オホバクロダモ　233832
オホバケイヌビハ　164985
オホバケッキツ　260168
オホバコ　301871,302068
オホバコケモモ　403821
オホバゴムノキ　165274
オホバゴムビハ　165274
オホバコリヤナギ　343529
オホバコ科　301821
オホバコ属　301832
オホバサンザシ　109839
オホハシバミ　106736
オホバシャシャンポ　403899
オホバシャシャンポモドキ　403899
オホバジュズネノキ　122064
オホバショウマ　91024
オボバシラカンバ　53503
オホハシリドコロ　44708
オホバスグリ　333970
オホバスノキ　404003
オホバセイシクワ　331926
オホバセキコク　125087
オホバセンキク　24355
オホバセンニンサウ　95113
オホバタウ　369484
オホハタケクハガタ　407287
オホバタケシマラン　377930
オホバタケハギ　18259
オホバタヌキマメ　112138,112792
オホバタブ　240605
オホバタンキリマメ　333133
オホバチシャ　379414
オホバチシャノキ　141629
オホバチヂミザサ　272625,272676
オホバチドメグサ　200312
オホバツゲ　379794
オホバツゲ属　172754
オホバツツジ　331341
オホバツルウメモドキ　80221
オホバツルザンセウ　417329
オホバツルマオ　313450
オホバツルマヲ　313450
オホバトネリコ　168023
オホバトベラ　301253
オホバトベラノキ　301416
オホバナアザミ　375274
オホバナアザミゲシ　32423
オホバナアリアゲキズラ　14873
オホバナアリアゲキヅラ　14873

オホバナギ 306450
オホバナクリンザクラ 315101
オホバナケフチクタウ 389978
オホバナシロバヒ 381206
オホバナセキコク 125279
オホバナノエンレイサウ 397574
オホバナノコギリサウ 4005
オホバナノコギリソウ 4005
オホバナハマビシ 395082
オホバナヒエンサウ 124237
オホバナヘニサルヒヤ 345405
オホバナマウリンクワ 61297
オホバナミミナグサ 82813
オホバナミヤマキケマン 106197
オホバナラ 324173
オホバニガナ 210543
オホバニシキギ 157310
オホバヌビトハギ 200732
オホバネム 13595
オホバノアカツメクサ 396967
オホバノエンゴサク 106519
オホバノキハダ 294233
オホバノケカンコノキ 177141
オホバノスヰゼンジナ 183086
オホバノセンナ 360483
オホバノハブサウ 78301
オホバノヒルギ 329765
オホバノフヂボグサ 402118
オホバノミヅトラノオ 306960
オホバノミヅトラノヲ 139585,306960
オホバノヤヘムグラ 170570
オホバノヤマルリミノキ 222124
オホハハキガヤツリ 118756
オホバハヒノキ 381154
オホバハマアサガホ 376536
オホハヒマキエハギ 126396
オホバヒメユヅリハ 122708
オホバヒラギナンテン 242563
オホバヒルギ 329765
オホバフウジュヤドリギ 385211
オホバフウテウカヅラ 71751
オホバフウテウボク 71751
オホバフヂボグサ 402132
オホバブナノキ 162370
オホバベゴニヤ 50232
オホバヘビノボラズ 52339
オホバボダイジュ 391766
オホバボタンヅル 94916
オホバボンテンクワ 402250
オホハマグルマ 413550
オホバマサキ 157617
オホバマヒハギ 126603
オホハマボウ 195311
オホバマユミ 157570
オホバマンリャウ 31489
オホバミネバリ 53631
オホバムラサキ 66943
オホバメギ 51301
オホバメドハギ 226739,226751
オホバモ 301025
オホバモクセイ 276370
オホバヤシャブシ 16458
オホバヤドリギ 355337
オホバユク 94026
オホハリイ 143298
オホハリスゲ 76626
オホハリヰ 143298
オホバルシャ 107161
オホバルシャギク 107161
オホバルシャ属 107158
オホバルリミノキ 222115
オホバワウレン 103835
オホバヲサラン 148656
オホハンゲ 299728
オホハンゴンサウ 339574
オホハンゴンサウ属 339512
オホヒエンサウ 124237
オホヒキオコシ 209826
オホヒキヨモギ 365298
オホヒキヨモギ属 365295
オホヒゲクサ 352382
オホヒナノウスツボ 355202
オホヒラテンツキ 166223
オホヒルギ 215510
オホヒルムシロ 312190
オホヒレアザミ 271666
オホヒレアザミ属 271663
オホビロウドイチゴ 339344
オホフウロ 174583
オホブシ 5193
オホフトイ 352289
オホフトモモ 382660
オホフトヰ 352289
オホベニウチハ 28084
オホベニウツギ 413591
オホヘビイチゴ 312912
オホホザキアヤメ 107271
オホホザキアヤメ属 107221
オホホシクサ 151257
オホボシサウ 342396
オホボュキザサ 242691
オホマメ 177777
オホマメザクラ 83267
オホマルバフュイチゴ 339188
オホマンテマ 363882
オボミクリ 370102
オホミズ 298971
オホミゾソバ 309877
オホミヅ 298971
オホミヅオホバコ 277369
オホミヅトンボ 183797
オホミヅバウシ 298971
オホミトケイサウ 285694
オホミトケイソウ 285694
オホミネイヌナシ 323347
オホミネカエデ 3529
オホミノアカガシ 233122
オホミノカンコノキ 177170
オホミノニレ 401556
オホミマユミ 157769
オホミヤマカンスゲ 74596
オホミヤマスゲ 75348
オホミヤマスミレ 410073
オホミヤマナナカマド 369512
オホムカデラン 94458
オホムギ 198376
オホムギスゲ 75076
オホムギ属 198260
オホムラサキ 331581
オホムラサキシキブ 66808,66824
オホムレスズメ 72180
オホモクセイ 276394
オホモチノキ 399433
オホモミヂスグリ 334084
オホヤマサギサウ 302502
オホヤマザクラ 83301
オホヤマジソ 259302
オホヤマハコベ 256444,375041
オホヤマフスマ 256444
オホヤマフスマ属 256441
オホヤマレンゲ 279249
オホユレケサ 60379
オホヨモギ 35527,36474
オホラフゲ 312392,312912
オホリウセン 414
オホルリサウ属 117900
オホレイジンサウ 5231,5466
オボロヅキ 180269
オホワタヨモギ 35527
オホワタリボタンヅル 94708
オホヲナモミ 415003
オボンテンカ 402250
オボンテンクワ 402250
オマキザクラ 83224
オマツ 300281
オミナエシ 285859
オミナエシ科 404392
オミナエシ属 285814
オミナヘシ 285859
オミナベシ 285859
オミナメシ 285859
オムク 29005
オムナグサ 138478
オムリジシ 389222
オメキグサ 354691
オメゴソパイソ 318580
オモエザサ 347276
オモゴウテンナンショウ 33365
オモゴハイノキ 381325
オモダカ 342400,342421
オモダカ科 14790

オモダカ属 14719,342308
オモチャカボチャ 114310
オモト 335760
オモトギボウシ 198644
オモト属 335758
オモヒグサ 8760
オモロカズラ 387802
オモロカンアオイ 37613
オヤブジラミ 393004
オヤマオグルマ 207219
オヤマシモツケ 371945
オヤマソバ 309447
オヤマノエンドウ 278910,278913
オヤマノエンドウ属 278679
オヤマノサンショウ 369362
オヤマノサンセウ 369339
オヤマボクチ 382079
オヤマヤシ属 172224
オヤマリンゴ 404051
オヤマリンドウ 173624
オヤマヲグリマ 207219
オヤリハグマ 292080
オラン 116829
オランダアヤメ 208626
オランダイチゴ 167597
オランダイチゴ属 167592
オランダエンゴサク 105721,106453
オランダカイウ 417093
オランダカイウ属 417092
オランダガラシ 262722
オランダガラシ属 262645,336166
オランダギク 89481
オランダキジカクシ 39120
オランダグサ 183097
オランダゲンゲ 397043
オランダゴシツ 91537
オランダジソ 290952
オランダシャクヤク 280253
オランダズイセン 307269
オランダセキチク 127635
オランダゼリ 292694
オランダゼリ属 292685
オランダセンニチ 4866
オランダセンニチ属 371627
オランダダンドク 71196
オランダヂサ 90894
オランダヂシャ 90894
オランダドリアン 25868
オランダナデシコ 127635
オランダネギ 15293
オランダハッカ 250450
オランダビュ 114427
オランダビュ属 319116
オランダフウロ 153767
オランダフウロ属 153711
オランダマゴヤシ 397043
オランダミズ 288767
オランダミツバ 29327
オランダミツバ属 29312
オランダミミナグサ 82849
オランダモミ 114539
オランダワメモコウ 345860
オラン属 29312
オリエントブナ 162395
オリヅルスミレ 410600
オリヅルツユクサ 67145
オリヅルラン 88553
オリヅルラン属 88527
オリバ-シナノキ 391795
オリ-ブ 270099
オリブ 270099
オリブ-ケンチャヤシ 138536
オリ-ブ属 270058
オリブ属 270058
オ-ルスパイス 299325
オルドガキ 132335
オルトノキ 270099
オルトフィッム属 275576
オルトプテルム属 275601
オルニソガルム属 274495
オルニソケファルス属 274477
オルビグニ-ア属 273241
オルビグニ-ヤシ属 273241
オルモ-シア属 274375
オレアリア属 270184
オレイフ 270099
オレイフノキ 270099
オレイフ属 270058
オレオセレウス属 273809
オレゴソメ-ブル 3127
オレゴンアオイ 362706
オレゴンシモツケソウ 166109
オレゴンバイン 318580
オロシマチク 304009,347279
オロシヤエンバク 190191
オロシャギク 246375
オロチ 272849
オロヤ属 275405
オヱッチ 285880
オヱフネサウ 401156
オヲンダイチゴ 167653
オンキディウム属 270787
オンコ 385355
オンジ 308403
オンシジウムケイロフオルム 270799
オンシジウム属 270787
オンシジューム属 270787
オンタデ 309969
オンツツジ 332104
オンナダケ 304098
オンノレ 53610
オンバコ 301871
オンファロ-デス属 270677
オンブスゲ 74611
オンマツ 300281
カイイギョク 159081
カイガラサルビア 256761
カイガラサルビア属 256758
カイガラソウ 256761
カイガンショウ 300146
カイガンセイシボク 28381
カイガンタバコ 68088
カイキチョウ 256271
カイゲンタンポポ 384572
カイコウカン 93857
カイコウジ 154734
カイコウズ 154647
カイコウヅ 154647
カイザイク 19691
カイザイク属 19690
カイサカネラン 264685
ガイジチャ 1120
カイジンソウ 265233
カイジンテッドウ 209505
カイジンドウ 13079,13088
カイジンマル 244224
カイズカイブキ 213647
カイソウ 352960
カイソウギョク 198500
カイソウ属 402314
カイダウ 243623,243657
カイダウズミ 243617
カイダウリンゴ 243667
カイタカラコウ 229070
カイヅカイブキ 213647
カイドウ 243623,243657
カイドウズミ 243617
カイドウボケ 84573
カイトウメン 179884
カイトウランマ 332359
カイナウカウ 55693
カイナグサ 36647
カイナンサラサドウダン 145677
カイナンノアズキ 138974
カイニット 90032
カイノウコウ 55693
カイノキ 300980
カイパウカウ 55693
カイフウロ 174901
カイマギョク 322990
カイリュウギョク 103767
ガイロラン 86707
カイロラン属 86667
カウエフザヰバノウロコモミ 30835
カウエフザンバノウロコモミ 30835
カウガイゼキシャウ 213213
カウガイモ 404544
カウキセキコク 125279
カウシウウヤク 97919
カウシバ 204583
カウシンバラ 336485

カウスイガヤ 22836,117245
カウスイハクカ 249504
カウスヰガヤ 22836,117245
カウスヰハクカ 249504
カウゾ 61103
ガウソ 75314,75882
カウゾリナ 298589,298618
カウゾリナ属 298550
カウゾ属 61096
カウチニクケイ 91318
カウバウ 27967,196153
カウバウ属 196116
カウブシ 119503
カウベナヅナ 225475
カウボク 6762
カウホネ 267286
カウボン 266975
カウモリカヅラ 250228
カウモリカヅラ属 250218
カウモリドコロ 131734
カウヤカミツレ 26900
カウヤザサ 58462
カウヤザサ属 58447
カウヤバハキ属 292040
カウヤマキ 352856
カウヤマキ属 352854
カウヤマンサク 160347
カウヤミズキ 106644
カウライアセスゲ 76178
カウライアゼスゲ 73714
カウライエノキ 80580
カウライギク 89481
カウライキビ 417417
カウライジソ 290952
カウライシダレヤナギ 343918
カウライシバ 418453
カウライゼキシャウ 5809
カウライタチッボスミレ 410517
カウライテンナンシャウ 33260
カウライニハフヂ 206140
カウライバッコヤナギ 343151
カウライミヅナラ 324173
カウライヤナギ 343574
ガウ－ラ属 172185
カウリマツ 10494
カウリャウキャウ 17733
カウリョウキョウ 17733
カウン 249601
カエ 393077
カエデ 3300
カエデキンポウゲ 184717
カエデゴウシュウアオギリ 58348
カエデドコロ 131804
カエテバスズカケノキ 302582
カエデバフウ 232565
カエデ科 3765
カエデ属 2769
カエルノエンザ 200228
カエンキセワタ 224585
カエンキセワタ属 224558
カエンサイ 53266
カエンソウ 244363
カエンソウ属 244358
カエンダイコ 272960
カエンボク 370358
カエンボク属 370351
カエンリュウ 125818
カオカ 13578
ガオバナ 208672
カオヨグサ 208672
カオリイボタ 229562
カオリカズラ 390767
カガイチュウ 34421
ガガイモ 252514
ガガイモ科 37806
ガガイモ属 252508
カカオ 389404
カカオノキ 389404
カカオノキ属 389399
カカオ属 389399
ガカク 199076
カガシラ 132789,354175
カガシラ属 132787
カカツガユ 114325,240813
カガノアザミ 92087
カガノベンケイソウ 356855
カカバイ 87532
ガガブタ 267819
カガミグサ 20408,372300
カガミジシ 9058
カガミナンブスズ 347229
カガミビバナ 115965
カガミユキノシタ 52532
カカヤンバラ 336405
カガユリ 154926
カカラ 366284
カガリビ 140332
カガリビソウ 257538,257543
カガリビバナ 115949,115965
カカンチュウ 263652
カカンマル 2105,267034
カキ 132219
カキウルシ 393466
カギカズラ 401773
カギカズラ属 401737
カギガタアオイ 37606
カギカヅラ 401773
カギカヅラモドキ 82107
カギカヅラ属 401737
カギクルマバナルコユリ 308641
カギザケハコベ 197491
カキヂサ 219485
カキチシャ 219487,219488
カキツバタ 208672
カキドオシ 176821,264946
カキドオシ属 176812,264859
カキドホシ 176824
カキドホシ属 264859
カキナ 59438,394327
カキネガラシ 365564
カキネガラシ属 365386
カキノキ 132219
カキノキダマシ 141595,141683
カキノキヤドリギ 411016
カキノキ科 139763
カキノキ属 132030
カギノニンジン 97720
カキノハグサ 308299
カキバイヌザクラ 316436
カキバイヌザクラ属 322386
カキバカンコノキ 177192
カキバチシャノキ 104175
カギバチシャノキ 104218
カギミギシギシ 339960
ガギュウ 171622
カギュウソウ 247469
カキラン 147238
カキラン属 147106
カキ属 132030
ガク 199956
ガクアジサイ 199956,200014
ガクアヂサヰ 199956
カクイシソウ属 304336
ガクウツギ 200077
カクウン 249592
ガクソウ 199994
カクソウマル 389221
カクタスダリア 121557
ガクタヌキマメ 111987
カクチョウラン 293537
カクトラ 297984
ガクバナ 200089
カクバマキ 306458
カクホウマル 327300
カクミケウスノキ 403886
カクミノスノキ 403854
カクヨウマル 234951
カグラサウ 337253
カグラン 293537
カクランポウ 307170
カクレイ 82287
カクレボスゲ 74224
カクレミノ 125644,139134,139176
カクレミノ属 125589
カクワッガユ 240813
カクワバイ 87532
ガケシバリ 16433
カゲラザサ 362131
カゲラ－タム 11199
カゲロウ 244187
カゲロウラン 417702

カケロマカンアオイ 37749
カゴガシ 233876
カゴノキ 233876
カゴノキカエデ 3105
カゴビチ 407897
カコマハグマ 292058
カゴメラン 179650
カサイ 310823
カサクサ 403687
カザグルマ 95216
カサザキサクラソウ 314310
カサスゲ 74339
カサド 3048
カサトリャマ 3607
カザナビキ 405445
カザナビキ属 405438
ガザニア属 172276
カサネオウギ 394047
カサバルピナス 238450,238467
カザフキサウ 23701,321667
カサマツ 300151
カサモチ 266975
カサモチ属 266974
カサユリ 229934
カザリカボチャ 114310
カザリナス 366920
カザリモロコシ 369633
カザンジマ 331759
カザンテマリ 322458
カザンマツ 299799
カザンラン 290921
カシ 323599
カシアマツ 299986
カジイチゴ 339394
カシイヌビハ 165370
カシウ 162542
カシオシミ 239391,239392
カシオズミ 239392
カシオスミノキ 239392
カジカエデ 2928
カシグルミ 212636
カシダザサ 347301
カシノキ 332509
カジノキ 61107
カシノキラン 171864
カシノキラン属 171825,342015
カシノキ属 323580
カシノハズイナ 210407
カシノハズヰナ 210407
カジノハラセンソウ 399202,399312
カシハ 323814
カシハナラ 323630
カシハハハソ 323630
カシーパペダ 93529
カシマガヤ 293296
カシミールイトスキ 114668
カシミローア属 78172
ガジャウサウ 23817
ガジャニマ 93693
カシヤマツ 299999
ガシャモク 312168
ガジヤンギ 96140
カシュー 21195
カシュウイトスギ 114698
カシュウイモ 131501
カシュウコメススキ 126070
カシュウナット 21195
ガジュツ 114884
カシューナットノキ 21195
ガジュマル 165307
カショウアブラススキ 71607
カショウイモ 131501
カショウギョク 284690
カショウクズマメ 259541
カショウセッコク 125087
ガショウソウ 23817
ガジョウマル 400458
カショクギョク 413675
カシラン 241781
カシワ 323814
カシワイチゴ 339270
カシワギ 243371,323814
カシワコナラ 323581
カシワナラ 323630
カシワノキ 323814
カシワバアジサイ 200063
カシワバゴムノキ 165426
カシワバチョオセンアサガオ 61242
カシワバテンジクアオイ 288459
カシワバハグマ 292069
カシワハハソ 323630
カシワラン 241781
カシ属 323580
カズ 61103
カスガスゲ 75769
カズサガヤツリ 118434
カズザキコウゾリナ 55769
カズサキヨモギ 35638
カズザキヨモギ 36097
カスチア属 79108
カスチロア属 79108
カステラノシア属 79096
ガステリア属 171607
カステリソウ 79130
ガストルキス属 171797
ガストロキールア属 171825,342015
カスノキ 239392
カズノコグサ 49540
カズノコグサ属 49532
カスピエゾヤナギ 342961
カスマグサ 408638
カスマトフィルム属 86144
カスミオクチョウジザクラ 83136
カスミガキ 94553
カスミガセキ 94553
カスミグサ 220355
カスミザクラ 83319,316913
カスミソウ 183191
カスミソウ属 183157
カスミヌカボ 12080,12217
カスミムグラ 39352
カズラ属 214947
カスリソウ 130128
カセイマル 140891,182433
カゼクサ 147671
カゼクサテンツキ 166304
カゼクサ属 147468
カセンガヤ 88396
カセンサウ 207219
カセンサウモドキ 321595
カセンサウモドキ属 321509
カセンソウ 207219,207222
カセンマル 234962
カゾ 61103
カタオナミ 162902
カタカコ 154926
カタガワヤガミスゲ 76644
カタギ 323611
カタクミ 369037
カタクリ 154909,154926
カタクリモドキ 135165
カタクリ属 154895
カタコ 154926
カタザクラ 223131,316819,316824
カタシ 69156
カタシキギク 340655
カタシデ 77312
カタジロ 348087
カタシログサ 348087
カタスギ 32988
カタスゲ 75258
カタセーツム属 79329
カタソゲ 185104
カタバエノコログサ 361936
カタバナハマサジ 230759
カタバミ 277747
カタバミオウレン 103850
カタバミワウレン 103850
カタバミ科 277643
カタバミ属 277644
カタバヤブマオ 56134
カタボウシノケグサ 354487
カタミノカツラ 325002
カタメンジソ 290952,290953
カタモミ 216120
カタユリ 154926
カタワグルマ 396956
カタンノキ 54620
カヂカエデ 2928
カチカタ 198376
カヂノキ 61107

カヂノハラセンサウ　399312
カヂノハラセンソウ　399312
カチラチライノキ　283418
カツウダケエビネ　65934
カッカザン　182488
カツカド　332509
カッコアザミ　11199
カッコウアザミ　11199
カッコウアザミ属　11194
カッコウウ　314537
カッコウセンノウ　363461
カッコウソウ　53304
カッコウチョロギ　53304
カッコオアザミ　11199
カッコソウ　314537
カッサバ　244507
ガッサンチドリ　302534
カッシア属　323542
カッテザクラ　83132
カツネグサ　146253
カツノキ　332509
カヅノスゲ　73544
カッパリス属　71676
カッパンクマタクラン　17722
カ－ツヒメトウ　121506,121507
カツマタギク　166818
カツマダソウ　81814
カツマダソウ科　81811
ガヅマル　165307
カツミ　418095
カツミグサ　418095
カツラ　83736
カツラカワアザミ　92165
カツラギグミ　142202
カツラギスミレ　409616
カツラ科　83732
カツラ属　83734
カディア属　216181
カデンギョク　413679
カテンサウ　262247
カテンサウ属　262243
カテンソウ　262247
カテンソウ属　262243
カトウハコベ　31977
カドハリイ　143376
カドハリミクリ　370102
カトレイオパシス属　79583
カトレヤ属　79526
カナウツギ　375821
カナガサマル　234960
カナクギノキ　231334
カナシデ　77312,77323
カナダアキノキリンサウ　368013
カナダアキノキリンソウ　368013
カナダケマンソウ　128291
カナタサイシン　37574
カナダ－スプル－ス　298286
カナダツガ　399867
カナダハシバミ　106724
カナダバルサムノキ　282
カナダポプラ　311146
カナダモ　143920
カナダモ属　143914
カナビキオ　104072
カナビキサウ　389638
カナビキサウ科　346181
カナビキサウ属　389592
カナビキソウ　389638
カナビキソウ属　389592
カナビキボク　85935
カナビキボク科　272580
カナビキボク属　85932
カナビキヲ　104072
カナホド　29304
カナムグラ　199382,199392
カナムグラ属　199379
カナメ　88691,295691
カナメガシ　295691
カナメノキ　295691
カナメモチ　295691
カナメモチ属　295609
カナモドリ　66808
カナヤマイチゴ　338568
カナヤマトチゴ　338679
カナリアクサヨシ　293729
カナリアサ－ド　293729
カナリア属　70971
カナリカンカン　70996
カナリ－キヅタ　187200
カナリ－クサヨシ　293729
カナリヤクサヨシ　293729
カナリ－ヤシ　295459
カナリヤヅル　399606
カナリヤノキ　70996
カナルギョク　103743
カニオタカラコウ　229221
カニコウモリ　283786
カニサボテン　352000
カニサボテン属　147283
カニストルム属　71149
カニツリグサ　398448
カニツリグサ属　398437
カニツリススキ　398528
カニツリノガリヤス　65343
カニナバラ　336419
カニバサボテン　351997,418532
カニメ　294040,409085
カニワ　53483,53503
カ－ネキエカ属　77075
カ－ネ－ション　127635
カネヒライヌツゲ　204025,204344
カネヒラツツジ　330990
カネヒラユリ　230038
ガネブ　411623
カネラ科　71138
カネラ属　71129
カノコ　231334
カノコガ　233876
カノコゴヨウツツジ　331595
カノコサウ　404325
カノコソウ　404261,404264,404316,404325
カノコソウ属　404213
カノコユリ　230036
カノツメサウ　299373
カノツメソウ　299373
カバ　300458
カバイロタマスダレ　417623
カバイロタンポポ　384463
カバイロノウゼンカズラ　70509
カバイロヤハズカズラ　390778
カハオニンジン　35088
カハカエデ　2872,3058
カハカミウスユキ　224907
カハカミエビネ　66068
カハカミカウバシ　231355
カハカミガシ　233269
カハカミゲッタウ　17703
カハカミスゲ　73641
カハカミスミレ　409995,409997
カハカミネナシカヅラ　114972
カハカミヨモギ　35732
カハキトガ　318575
カハグルミ　320372
カハゴケサウ科　306688
カハサ　407430
カバザクラ　83145,83224,316551
カハシャシャブ　273916
カハタケ　304098
カハヂサ　407430
カハチマサキ　157647
カハヅスゲ　76342
カハツルモ　340476
カハツルモ属　340470
カバニレシア属　79778
カパネミア属　71561
カバノキ　53483,53503
カバノキ科　53651
カバノキ属　53309
カバハナギュウカク　139176
カバフトイソツツジ　223912
カバフトムシトリスミレ　299757
カハホネ　267286
カハホネ属　267274
カハホホヅキ　277369
カハマメカヅラ　254854
カハミドリ　10414
カハミドリ属　10402
カハヤナギ　343414,343937
カバユリ　73158
カハラアカザ　86892,86893
カハラケシ　363451

カハラケツメイ 85232
ガハラケツメイ 78429
カハラケツメイ属 78204
カハラケナ 221784,397451
カハラゲヤキ 401581
カハラゴマ 363451
カハラサイコ 312450
カハラスガナ 322342
カハラナデシコ 127852
カハラナデシコ属 127573
カハラニガナ 210681
カハラニンジン 35308
カハラハウコ 21731
カハラバウフウ 293033
カハラハハコ 21731
カハラヒサギ 79257
カハラヒジキ 104771
カハラヒジキ属 104750
カハラフヂ 64990
カハラブナ 53465
カハラマツバ 170743,170757
カハラヨモギ 35282
カハリキツネノメマゴ 337255
カハリクワウエフザン 114548
カハリバツルアヅキ 293999
カハリバニンジンボク 411420
カハリバマキエハギ 126396
カハヲニンジン 35088
カヒジンドウ 13079
カブ 59575
カフィールライム 93492
カフェルオピソテツ 145234
カフカシオガマ 287664
カブカンラン 59541
カーフクルー 93500
ガフクワンセンボンヤリ 224110
カブスゲ 73987,74085
カブダチアッケシソウ 342900
カブダチクジャクヤシ 78047
カブダチソテツジュロ 295479
カブダチナズナ 137223
カーブチ 93520
カブト 140872
カブトギク 5100,5108,5185,5363
カブトダコ 16508
カブトマル 43493
カブナ 59575
カブムラサキ 260772
カブムラサギ 153518
カブヤオ 93559
カブヨモギ 35373
カブラ 59575,59603
カブラカンラン 59541
カブラゼリ 84731
カブラタマネ 59541
カブラナ 59575
カブラバサルビア 345244
カブラバノサルビア 345244
カヘ 393077
カベイラクサ 284152
カヘデドコロ 131804
カヘデ科 3765
カヘデ属 2769
カベノキ 407865
ガーベラ属 175107
カヘルデノキ 3300
カヘルノキツケ 326340
カホウマル 247674
カホクザンショウ 417180,417340
カボス 93805
カボチャ 114292,114295
カボチャアサガオ 323465
カボチャ属 114271
カボック 80120
カボンバ属 64491
ガマ 401112
カマウド 189918
カマエセレウス属 85163
カマキリサウ 222807
カマクマヒバ 85312
カマクライブキ 213634
カマクラカイダウ 116546
カマクラカイドウ 116546
カマクラサイコ 63625
ガマグラス 398139
カマクラヒノキ 85312,85313
カマクラヒバ 85313
カマクラビャクシン 213634
カマクラユリ 229730
カマシア属 68789
ガマズミ 407785
ガマズミトベラ 301438
ガマズミ属 407645
カマツカ 100961,295808
カマツカ属 295609,313266
カマナシコザクラ 315054
カマノキ 189918
カマフテシラビソ 510
カマフトシラカンバ 53503
カマヤマショウブ 208650,208809,208894
カマヤリサウ 389848
カマヤリソウ 389844,389848
カマラ 243427
ガマ科 401140
ガマ属 401085
カミイ 119347
カミエビ 97947
カミカゼ 358056
カミカハスゲ 76100
カミガモソウ 180304
カミガヤツリ 119347
カミカワスゲ 76100
カミキ 61103
カミギ 61103
カミコウチテンナンショウ 33362
カミツレ 246396
カミツレモドキ 26768
カミツレ属 246307
カミナリササゲ 79257
カミノキ 61107,414257
カミノギ 61103
カミバハキ 268614
カミホコ 177449
カミメボウキ 268644
カミヤツデ 387432
カミヤツデ属 387428
カミルレ 246396
カミルレ属 246307
ガム 155472
カムイコザクラ 314466
カムイビランジ 363543
カムイヨモギ 36177
カムゴソ 132351
カムシバ 242274
カムテラギ 346408
カムベストルカエデ 2837
カムライ 121632
カムロザサ 304114
カメエケレウス属 85163
カメバヒキオコシ 209665,209859
カメバヤマハッカ 209615
カメヤマサウ 362251
カメロンムクゲ 194774
カモアオイ 37590
カモアフヒ 37590
カモウリ 50998
カモエンシア属 69830
カモカツラ 83736
カモガヤ 121233
カモガヤ属 121226
カモジグサ 335495,335544
カモジグサ属 11628,335183
カモノハシ 209266,209267,209288
カモノハシガヤ 57567
カモノハシ属 209249
カモマタキビ 143522
カモマタビエ 143522
カモメギク 89704
カモメヅル属 117334,400844
カモメラン 169887,169906,179694
カモンギョク 233564
カヤ 255886,393077
カヤツリグサ 119208
カヤツリグサ科 118403
カヤツリグサ属 118428
カヤツリスゲ 73894
カヤツリマツバイ 143092,143326,143327
カヤーヌス属 65143
カヤバオヤマヤシ 172225
カヤラン 346703,390506
カヤラン属 346686,390486

カヤ属　393047
カュチイナ　306506
カュバワソ　45907
カュプッド　248104
カュプテ　248104
カュマニス　91446
カヨフォラ属　65128
カラ　417093
カラー　417093
カラアイ　80395
カラアキグミ　142230
カラアザミ　91864
カラアフヒ　13934
カライ　80395
カライタドリ　162531
カライトソウ　345850
カライヌエンジュ　240114
カライモ　207623
カラウバガネ　241817,241844
カラウメ　87525
カラウリ　114245
カラエ　334435
カラオトアザミ　348749
カラカサガヤツリ　118476,119422
カラカシワナラ　323881
カラキュ　1120
カラクサ　94740
カラクサキンパウゲ　326278
カラクサキンポウゲ　325901
カラクサゲシ　199408
カラクサケマン　169126
カラクサケマン属　168964
カラクサナズナ　105340
カラクサハナガサ　405897
カラクサハンショウズル　94704
カラクチナシ　171253,171374,324679
カラグハ　124925
カラクワ　259067
カラコギカエデ　2984,2987
カラコギカヘデ　2984
カラコマル　182434
カラコンテリギ　199853
カラサトウキビ　341909
カラサワフタギ　381341
カラシナ　59438
カラジューム　65218
カラジューム属　65214
カラスウリ　396170,396177,396210
カラスウリ属　396150
カラスオウギ　50669
カラスガマズミ　407715
カラスギバサンキライ　194120
カラスギバサンキライ属　194106
カラスザンショウ　417139
カラスザンセウ　417139
カラスシキミ　122482,122519
カラズトノダイワウ　340064
カラストバラ　336736
カラスナデシコ　127575
カラスノウエキ　385202
カラスノエンドウ　408284,408571
カラスノカギズル　401773
カラスノカギヅル　401773
カラスノカタビラ　305776
カラスノゴマ　104046
カラスノゴマ属　104039
カラスノサンショウ　417139
カラスノススキ　23670
カラスノチャヒキ　60963
カラズノヒルヅル　117537
カラスバサンキライ属　194106
カラスビシャク　299724
カラスマメ　138942
カラスムギ　45448,45566
カラスムギ属　45363
カラスユリ　308613,308616
カラダイオウ　329386
カラダイオウ属　329306
カラダイワウ　329386
カラダイワウ属　329306
カラタケ　297215,297373,297435
カラタチ　310850
カラダチ　366284
カラタチバナ　31408,31471,31502
カラタチ属　310848
カラタニウツギ　413591
カラタネオガタマ　252869
カラチムラサキ　66829
カラヂューム　65218
カラツガ　414
カラックバネウツキ　117
カラツバキ　69552
カラッパヤシ　6862
カラッパヤシ属　6861
カラテア　66175
カラテーア属　66157
カラデニア属　65202
ガラナ　285930
カラナシ　317602
カラナツメ　295461
カラナデシコ　127654
カラナワシロイチゴ　338993
カラノアザミ　92167
カラノキ　333970
カラノコギリソウ　3914
カラノミザクラ　83284
カラバイケイソウ　405627
カラバウムギ　75259
カラハジカミ　161373
カラハナウマゴヤシ　247366
カラハナサウ　199384
カラハナサウ属　199379
カラハナソウ　199384,199386
カラハナソウ属　199379
カラピイチゴ　338433
カラビエ　143522
カラビオブ　17621
カラヒトヤチスゲ　75165
カラヒメヘビイチゴ　312446
カラピンラセンサウ　399325
カラピンラセンソウ　399325
カラフトアカバナ　146587,146663,146709,146775
カラフトアザミ　348749
カラフトアヅモリサウ　120310
カラフトアヅモリソウ　120310
カラフトイソッッジ　223904
カラフトイソツツジ　223901,223912,223913
カラフトイチゴツナギ　305686
カラフトイチヤクソウ　322825
カラフトイハスゲ　76094
カラフトイバラ　336353,336522,336736,336954
カラフトイワスゲ　76094
カラフトイワヒゲ　78627
カラフトウド　30622,30753
カラフトウハミヅザクラ　280007
カラフトウワミヅザクラ　280007
カラフトオオケマン　105929
カラフトオオサンザシ　109839
カラフトオグルマ　207046,207066
カラフトオホサンザシ　109839
カラフトカイダウ　243649
ガラフトカイドウ　243555
カラフトカサスゲ　76068
カラフトガシハ　324173
カラフトガシワ　324173
カラフトカラマツ　388673
カラフトキシャウブ　208700
カラフトキハダ　294251
カラフトキヨシソウ　349838,349841
カラフトキワダ　294251
カラフトクロヤナギ　89160
カラフトグワイ　342385
カラフトグワヰ　342385
カラフトグンバイ　390223
カラフトゲマン　105929
カラフトゲンゲ　188028
カラフトコウモリソウ　283792
カラフトコゴメグサ　160225
カラフトササモ　312126
カラフトシオガマ　286977
カラフトシキミ　365936
カラフトシラカンバ　53581
カラフトスゲ　75253
カラフトスズラン　147149
カラフトズミ　243649
カラフトスミレ　409958
カラフトセンクワサウ　23917
カラフトダイオウ　340064

カラフトダイコンソウ　175410,175426,175431
カラフトダケカンバ　53411
カラフトツメクサ　31787
カラフトツリバナ　157848
カラフトトウゲブキ　229049,229052
カラフトドジャウツナギ　177626
カラフトドジョウツナギ　177626
カラフトナニハズ　122477
カラフトナニハヅ　121720
カラフトナニワズ　122477
カラフトニンジン　101823,101827
カラフトネコノメサウ　90325
カラフトネコノメソウ　90325
カラフトノダイオウ　340064
カラフトハナシノブ　307201,307215
カラフトハナビゼキショウ　212866
カラフトヒナオダマキ　30068
カラフトヒナゲシ　282736
カラフトヒメシャクナゲ　22450
カラフトビヨクサウ　407072
カラフトヒヨクソウ　407072
カラフトヒロハテンナンショウ　33483
カラフトブシ　5546
カラフトホシクサ　151461
カラフトホソイ　213114
カラフトホソバハコベ　374897
カラフトホソヰ　213114
カラフトマンテマ　363958
カラフトミズキ　105202
カラフトミセバヤ　200794
カラフトミツモト　312828
カラフトミミコウモリ　283792
カラフトミヤマツツジ　330023
カラフトミヤマハンノキ　16330
カラフトムシトリスミレ　299757
カラフトモメンヅル　43018
カラフトヤマナラシ　311281
カラフトヤマボクチ　382071
カラフトヤラメスゲ　74221
カラフトュキザサ　242679
カラフトラッキョウ　15432
カラフトルリトラノオ　317926,317927
カラフトワサビ　161137
カラボケ　84573
カラマッ　221894
カラマツサウ　388423
カラマツサウ属　388397
カラマツシモツケ　371807
カラマツソウ　388423
カラマツソウ属　388397
カラマツモミ　80087
カラマッ属　221829
カラマユミ　157745
カラミザクラ　83284
カラミツデ　125644
カラムグオ　199382
カラムグヲ　199382
カラムシ　56145,56229,56238
カラムシ属　56094
カラムス属　65637
カラムメ　87525
カラムラサキツツジ　331289
カラメドハギ　226843,226860
カラモフィルム属　65626
カラモモ　34475,34477
カラモンジン　93566,93639
カラヤマグハ　259067
カラヤマグワ　259067
カラュキヤナギ　167235
ガラユキヤナギ　167233
カラユリ　230036
カラュリヤナギ　343937
カラルマ属　72402
カラヰ　80395
カラン　66101
ガランガ　187428
ガランガ属　187417
カランコエ　215102
カランコエ属　215086
ガランサス属　169705
カランタス　392827
カランチョウ　59526
ガ－ラント　334179
ガランビイヌビハ　165761
ガランビゴセウ　300540
ガランビタヌキマメ　112671
ガランビニシキサウ　158935
ガランビネムチャ　78303
ガランビボタンヅル　95361
ガランビマメ　112671
カランボ－　45725
カランボク　45725
カラ－属　66593
カリ－　155562
カリアンセマム属　66697
カリアンドラ属　66660
ガリアンドラ属　66593
カリオカル　77947
カリオカル属　77940
カリオトフォラ属　78065
カリガネサウ　78002
カリガネサワ属　77992
カリガネソウ　78002
カリガネソウ属　77992
ガリカバラ　336581
カリコマ　66976
カリコマ属　66975
カリシア属　67142
カリステモン属　67248
カリッサ　76873
カリッサ属　76854
カリトリス属　67405
カリバオウギ　43265
カリビアマツ　299836
カリフォルニア・アヤメ　208639
カリフォルニア・ホクシャ　417407
カリフォルニアビロウ　413298
カリフォルニアフクシア　417407
カリフォルニア・ホクシャ属　417405
カリブダイオウヤシ　337883
カリブトロカリテクス　68614
カリフラワ－　59529
カリホギョク　163445
カリマタガヤ　131018,131026
カリマタガヤ属　131006
ガリメギイヌノヒゲ　151535
カリヤス　36647,255918
カリヤスモドキ　255867
カリヨセウツギ　413615
カリロエ属　67124
カリワシノ　347349
カリン　317602,320301,320306
カリン属　317598
カルアンツス属　77656
カルイザワテンナンショウ　33496
カルイザワトウヒレン　348092
カルカヤ　117181,389355
ガルガルラ－ジ　93557
カルダミネ－プラテンシス　72934
ガルトニア属　170827
カルドン　117770
カル－ナ　67456
カルナカッタ　93517
カルナ属　67455
カルパンテア属　77131
カルビ－　93439
カルポプロ－ツス属　77429
カルミア属　215386
カルミオプシス属　215411
カルム属　77766
カルメ　320372
ガレアンドラ属　169865
ガレ－ガ属　169919
ガレニア属　169953
カレ－ヤシ　198806
ガロウエイオ－ストラリアワイルドライム　253289
カロコルタス－ベヌスタス　67635
カロコルタス属　67554
カロセファラス属　67532
カロフイラユ－カリ　155514
カロポゴン属　67882
カロライナ　172781
カロライナジャスミン　172781
カロライナツガ　399874
カロライナポプラ　311231
カロリナアオイゴケ　128972
カロリナハコヤナギ　311231
カワイスギ　113721
カワイチゴ　123322

カワウチザサ 347309
カワガシ 116193
カワカミウスユキソウ 224907
カワカミモメンヅル 42546
カワカムリャダケ 318350
カワキトガ 318575
カワギリ 79257
カワグチヒメノボタン 268377
カワグルミ 320372
カワゴケソウ 93975
カワゴケソウ科 306607,306688
カワゴケソウ属 93968
カワゴロモ 200204
カワゴロモ属 200201
カワジサ 407430
カワジシャ 407430
カワシマコザクラ 314531
カワシロナナカマド 369296
カワズスゲ 75606
カワゼンゴ 24485
カワタケ 304098
カワヂサ 406992
カワヂシャ 407430
カワヂシャモドキ 407064
カワチスズシロソウ 30265
カワチハギ 226747
カワチマサキ 157647
カワックナベ 16447
カワツツジ 331657
カワツルモ 340476
カワツルモ科 340502
カワツルモ属 340470
カワハジカミ 161373
カワバタ 93404
カワハラハンノキ 16447
カワベホタルイ 353290
カワマメカズラ 254854
カワミドリ 10414
カワミドリ属 10402
カワヤナギ 343414,343444,343705
カワラアカザ 86892,86893
カワラウスユキソウ 224881
カワラウツギ 104696
カワラオトギリ 201961
カワラグミ 142214
カワラケツメイ 78429
カワラケツメイ属 78204
カワラケナ 221784
カワラゲヤキ 401581
カワラサイコ 312450
カワラササギ 79257
カワラスガナ 322342
カワラスゲ 74872
カワラチガヤ 205497
カワラナデシコ 127756,127852
カワラニガナ 210681
カワラニンジン 35308
カワラノギク 40652
カワラハンノキ 16447
カワラヒサギ 79257
カワラヒジキ 104771
カワラヒジキ属 104750
カワラフジ 64990
カワラフジノキ 176881
カワラボウフウ 292892,293033
カワラボウフウ属 292758
カワラホオコ 21731
カワラマツバ 170603,170743,170751,
170761
カワラヨモギ 35282,35308
カワリツワブキ 162619
カワリバアサガオ 208104
カワリハナヒイラギ 89139
カワリバハマゴウ 411436
カワリバマキエハギ 126396
カワリミタンポポモドキ 224731
カン 13635
カンアオイ 37659,243862
カンアオイ属 37556,194318
カンアフヒ 37571,243862
カンアフヒ属 37556
カンイチゴ 338205
カンエンガヤツリ 118841,118845
ガンカウラン 145073
ガンカウラン科 145052
ガンカウラン属 145054
カンガレイ 352223,353889
カンガレヰ 352223
カンキギョク 322992
カンキチク 197772
カンキチク属 197769
カンキツ属 93314
ガンクビサウ 77149,77156
ガンクビソウ 77149,77156
ガンクビヒレアザミ 73329
カンクビヤブタバコ 77190
カンコウホウ 257412
ガンコウラン 145069,145073
ガンコウラン科 145052
ガンコウラン属 145054
ガンコオラン 145073
カンコザクラ 314975
カンコノキ 177162
カンコノキ属 177096
カンコモドキ 60230
カンコモドキ属 60162
カンコン 207590
カンサイイワスゲ 74124
カンサイエノキ 80660
カンサイキツネヤナギ 344257
カンサイスノキ 404007
カンサイタンポポ 384591
カンザウ 177885,177893
カンザウダマシ 200736
カンザウ属 177873
カンザキアヤメ 208914
カンザクラ 314975
カンザクラソウ 314975
カンザシギボウシ 198593
カンザシヒメハギ 308331
カンザブラウノキ 381148
カンザブロウノキ 381148
カンザンチク 318303
カンシ 93415
カンジカンスポプラ 311259
カンショ 207623,341887
カンショウ 341887
カンショウコウ 262497
ガンショウラン 98743
カンショガヤ 341917
カンジンチョウキ 272799
カンスゲ 75421
カンススキ 255849
カンセイタンポポ 384591
ガンゼキラン 293494
ガンゼキラン属 293484
カンゾウ 177885,177893,177897,191263
カンゾウ属 177873
ガンタチイバラ 366284
カンタベリ-マルジクホエア 188191
カンタロ-ブ 114199
カンチク 87581
カンチク属 87547
カンチコウゾリナ 298602
カンチスゲ 74722
カンチヤチハコベ 374783
カンチョウジ 57954
カンチョウジ属 57948
カンツア属 71544
カンツバキ 69133,69606
カンツワブキ 162615
カンテラギ 346408
カンテンイタビ 165518
カントウ 400675
ガンドウカズラ 417081
ガントウタンポポ 384746
カンドウナツグミ 142092
カントウナミキソウ 355824
カントウフウロ 174729
カントウマムシグサ 33496
カントウマユミ 157879
カントウマユミ 157882
カントウミヤマカタバミ 277881
カントウユリ 229730
カントウヨメナ 41517
カントオイワウチワ 362273
カントリソウ 264946
カントンアブラギリ 406018,406019
カントンスギ 114539
カントンニンジン 280799
カントンレモン 93546

カンナ 71173
カンナデシコ 127621
カンナ科 71235
カンナ属 71156
カンニングハムモクマワウ 79157
カンノンジュロ属 329173
カンノンチク 329176
カンノンチュウ 105440
カンバ 53483,53503
カンバウイ 375904
カンバク 244033
カンバザクラ 83145
カンパニュラ属 69870
カンバ属 53309
カンビ 53503
ガンピ 53483,363235,414257
カンヒザクラ 83158
ガンビセンノウ 363235
ガンピセンノウ 363235
ガンピバハヒノキ 381452
ガンピハヒノキ 381452
ガンビール 401757
ガンビールノキ 401757
ガンピ属 414118
カンペイジュ 320377
カンポウギョク 203445
ガンボウジノキ 171144
カンポウフウ 320377
カンポウラン 116815
カンボク 408009,408098
カンポクラン 116773
カンボタン 280304
ガンボ麻 194779
カンムリナズナ 77639
カンメイチク 197772
カンヨメナ 41084
カンラン 59520,59532,70989,116926,116936
カンランズイセン 262405,262429
カンラン科 64088
カンラン属 70988
ガンラン属 116769
カンリュウ 192439
カンレンボク 70575
カンロバイ 404051
キ 15289,331261
キアイ 206626,206669
キアカソ 56181,56318
キアサガホ 207792
キアジサイ 199911
キアヂサヰ 199911
ギアナ・ペッパー 300374
キアナストルム科 115426
キアナストルム属 115429
キアマ 327553
キアラセイトウ 86427
キアレチギク 166816
キアヰ 206626,206669
キイアミダネヤシ 129685
キイイシマヤダケ 318308
キイウマノミツバ 345979
キイシオギク 89567
キイシモツケ 372036
キイジョウロウホトトギス 396603
キイセンニンサウ 95194
キイチゴ 338886,338945
キイチゴ属 338059
キイトスゲ 73666,76119
キイバサカキ 8221
キイハナネコノメ 90323
ギイマ 404060
キイムヨウラン 223646
キイモ 404060
キイラタンヤシ 222636
キイルンカンコノキ 177149
キイルンフジバカマ 158214
キイレツチトリモチ 46882
キイロギニアヤム 131516
キイロサンゴバナ 211344
キイロミミガタテンナンショウ 33390
キーウイ 6553
キウイフルーツ 6553
キウコンベゴニヤ 50378
ギウシャウ 91368,91378
ギウシンリ 25882
キウリ 114245
キウリゲサ 397451
キウリナ 397451
キウリ属 114108
キウンマル 267043
キエビネ 66068
キエボシソウ 5055
キエリウツボ 264857
ギオイ 252817
キオウギョク 103760
キォノコユリ 229862
キオン 359565
キオン属 358159
キカシグサ 337351,337353
キカシグサ属 337320
キカズラ 134040
キカノコユリ 229862
ギガフクワン 334976
キガヤツリ 118491,119208
キガラシ 59438
キカラスウリ 396203
キガラスウリ 396203
キカラパ 12587
キカラマツ 388668
ギガンチウム 15445
ギガンテアバラ 336602
ギガントセコイア 360576
ギガントセコイア属 360575
キガンピ 414271
キキウ 139629
キキヌキサウ 397849
キキヌキサウ属 397828
キキヌキソウ 397849
キキヌキソウ属 397828
キキャウ 302753
キキャウカタバミ 278147
キキャウナデシコ 295267
キキャウラン 127507
キキャウラン属 127504
キキヤウ科 70372
キキャウ属 302747
キキョウ 302753
キキョウカタバミ 277776
キキョウカンアオイ 37655
キキョウシャジン 7659
キキョウソウ 397759
キキョウツツジ 331593
キキョウナデシコ 295267
キキョウラン 127507
キキョウラン属 127504
キキョウ科 70372
キキョウ属 302747
キキョオ 302753
キク 89619,124785,124826
キクアザミ 348902
キクイモ 189073
キクイモモドキ 190505
キクイモモドキ属 190502
キクガラクサ 143896,143897
キクガラクサ属 143894
キククワボク 49046
キクゴボウ 91913,354870
キクザアサガオ 208077
キクサアサガホ 208077
キクザカボチャ 114294
キクザキイチゲサウ 23701,321667
キクザキイチゲソウ 23701
キクザキイチリンサウ 23701,321667
キクザキイチリンソウ 23701
キクザキオクチョウジザクラ 83149
キクザキセンダングサ 53976
キクザキタカネミミナグサ 83009
キクザキハンショウヅル 95100
キクザキヤマブキ 216095
キクザノタウナス 114294
キクスイ 378313
キクスイギョク 233610
キクダイダイ 93417
キクタニギク 89709,124775,124817
キクタビラコ 261069
キクヂサ 90894
キクヂシャ 90894
キクヂシャ属 90889
キクナ 89481
キクニガナ 90901
キクノケス属 116516

キクノハアオイ 256429
キクバアカザ 86970
キクバアリタソウ 139693
キクバイヅハハコ 103407
キクバエビヅル 411672
キクバガラシ 364545
キクバキョン 365060
キクバキヲン 365060
キクバクワガタ 317951
キクバコマスミレ 410633
キクバジフニヒトへ 13079
キクバダイモンジサウ 349604
キクバタウコギ 54001
キクバテンジクアオイ 288468
キクバテンヂクアフヒ 288468
キクバドコロ 131840
キクバトラノオ 317909
キクバハハコ 103407
キクバヒアザミ 348609
キクバヒゴタイ 348604
キクバヒヨドリ 158230
キクバフウロ 153914
キクバモリアザミ 348788
キクバヤマボクチ 382076
キクバワウレン 103835
キクボクチ 361011
キクボタン 220674
キクムグラ 170444
キクモ 230323
キクユグラス 289067
キクヨモギ 35759,36363
キク科 101642
キク属 89402
キケマン 105976,105978,106350
キケマン属 105557
ギゴウカン 334976
キコウギョク 257407
キコガンピ 414271
キコク 310850
キサイカチ 64990
キササギ 79257
キササゲ 79257
キササゲ属 79242
キサンゴ 407977,407980
キサンシスマ属 414971
キサンラン 82051
キシウナキリスゲ 75471
キシカク 116661
キジカクシ 39187
ギシギシ 339994,340000,340089
ギシギシ属 339880
キシダマムシグサ 33381
キシツツジ 331657
キシブナシ 323240
キジマメ 78429
キシマヤダケ 318308
キジムシロ 312552,313013
ギジムシロ属 312322
キシモジン 10829
キシモツケ 371944
キシャウブ 208771
キシュウ 93724
キシュウギョク 198497,284699
キシュウスズメノヒエ 285426
キシュウナキリスゲ 75471
キシュウネコノメ 90408
キシュウミカン 93446,93724
キショウブ 208771
キジョラン 245860
キジョラン属 245777
キシロビューム属 415704
キジンサウ 349936
キジンマル 244024
キシンリュウ 103762
キズイセン 262410
キスゲ 191312,191318
キスゲ属 191261
キススス属 92586
キズタ 187222,187318
キストゥス属 93120
キスミレ 409729,410326,410763
キゼセトリシズカ 88276
キセッコウ 59163
キセルアザミ 92390
キセルサウ 8760
キセワタ 224996
キソアザミ 91966
キソイチゴ 338949
キソウテンガイ 413748
キソウマル 362003
キソエビネ 65871,66063
キソキバナアキギリ 345256
キソケイ 211864,211938
キソザクラ 314537
キソシベ 330970
キソシベヤマツツジ 330970
キソズ 167493
キソスミレ 409613
キソチドリ 302404
キソチドリモドキ 302243
ギソフイリマサキ 157625
キソミツバツツジ 331009
キソムカゴ 299534
キゾメカミツレ 26746
キソメカミルレ 26746
キゾメグサ 114871
キタイチゴ 339240
キダイモンジ 140888
キタオグルマ 385903
キタカ 395650
キタカミヒョウタンボク 235758
キタグニコウキクサ 224399
キタコブシ 242169,242171
キタゴヨウ 300137
キタゴヨウマツ 300137
キタササガヤ 254022
キタズ 345679
キタダケイチゴツナギ 305558
キタダケオドリコソウ 220348
キタダケカニツリ 398540
キタダケソウ属 66697
キタダケトラノオ 317934
キタダケナズナ 137056
キタダケヤナギラン 85884
キタダケヨモギ 35737
キタダケリンドウ 173863
キダチアウソウクワ 179405
キダチアウソウクワ属 307484
キダチアミガサソウ 1900
キダチアミガサノキ 1900
キダチアロエ 16592
キダチイトヨモギ 35981
キダチイナモリ 272221
キダチイナモリサウ 272221
キダチイナモリサウ属 272173
キダチイナモリソウ 272221
キダチイナモリソウ属 272173
キダチイナモリ属 272173
キダチイヌホオズキ 367632
キダチウマノスズクサ 34268
キダチカミツレ 32563
キダチキツネアザミ 244728
キダチキツネノマゴ 172832
キダチキンバイ 238188,238190
キダチコミカンサウ 296679
キダチコミカンソウ 296471,296679
キダチコンギク 380872
キダチシモツケ 369265
キダチスナビキ 393267
キダチセンナ 360490
キダチダイゲキ 158906,159159
キダチタウガラシ 72100
キダチタバコ 266043
キダチチョウセンアサガオ 61242
キダチデンセイ 361430
キダチトウガラシ 72100
キダチナギナタコウジュ 144096
キダチナンバンクサフジ 385985
キダチニシキサウ 159557
キダチニンドウ 235860
キダチニンドオ 235860
キダチネヂレバナ 242976
キダチノコガク 199792
キダチノジアオイ 249631
キダチノネズミガヤ 259691
キダチノボタン 43462
キダチハウチワマメ 238430
キダチハゼリソウ 414095
キダチハッカ 347560
キダチハナカタバミ 277889
キダチハナカンザシ 286275

キダチハナカンザシ属 286073
キダチハブソウ 78301
キダチハマグルマ 413506,413507
キダチハリナスビ 367305
キダチヒャクリコウ 391397
キダチベゴニア 50284
キダチベンケイ 356759
キタチマオ 313424
キダチマオウ 146155
キダチマツヨイグサ 269438
キタチマヲ 313424
キダチルリサウ 190559
キダチルリサウ属 190540
キダチルリソウ 190559,190702
キダチルリソウ属 190540
キダチロカイ 16598
キダチロクワイ 16592
キダチワタ 179876
キタナナカマド 369279
キタヌマハリイ 143318
キタノアサギリサウ 35505
キタノアサギリソウ 35505
キタノオキナグサ 321703
キタノガリヤス 65507
キタノカワズスゲ 73696,74423,76342
キタノコギリソウ 3923
キタノテツカエデ 3249
キタバヤマブキ 375817
キタヒエスゲ 75189
キタミオトギリ 201970
キタミサウ 230831
キタミサウ属 230828
キタミソウ 230831
キタミソウ属 230828
キタミハタザオ 154485
キタメヒシバ 130612
キダモヒャクリコウ 391397
キタヤマヌカボ 12227
キタヨシ 295888
キタヨツバシオガマ 287078
キタヨナメ 215352
キタヨモギ 35211
キタンポポ 196057
キタンマル 182429
キチジサウ 279757
キチジソウ 279757
キチジャウサウ 327525
キチジャウサウ属 327524
キチジャウソウ 327525
キチジャウソウ属 327524
キチジョウソウ 327525
キチジョウソウ属 327524
キチャウジ 84404
キチャウジ属 84400
キチョウジ 84404
キチョウジ属 84400
キチリ－ 93565
キッカフチク 297301
キッカボク 49046
キッコウダコ 16496
キッコウチク 297301
キッコウツゲ 203670
キッコウハグマ 12608
キッコウボク 164956
キッコウボタン 33209,337154
キッコウラム 140857
キッショウテン 94564
キヅタ 187318
キヅタ属 187195
キツツジ 331261
キツネアザミ 191666,191671
キツネアザミ属 191665,348089
キツネアヅキ 138978
キツネガヤ 60928
キツネササゲ 138942
キツネスゲ 152772
キツネタンポポ 384847
キツネノオ 261379
キツネノカミソリ 239271
キツネノササゲ 369010
キツネノチャブクロ 160954,296801
キツネノテブクロ 130383
キツネノテブクロ属 130344
キツネノヒガサ 111724
キツネノヒマゴ 214739,337253
キツネノポウゲ 326493
キツネノボタン 325981,326276
キツネノマクラ 396170
キツネノマゴ 214735,337253
キツネノマゴ科 2060
キツネノマゴ属 214304
キツネノメマゴ 214512,337255
キツネバリ 53797
キツネマメ 333456
キツネヤナギ 344255
キツネユリ 229730
キツネユリ属 177230
ギッパエウム属 175494
ギッベウム属 175494
キツリフネ 205175
キトゲハンニャ 43527
キトビル 15868
キトログッサ属 90767
キナ 91090
ギナ 96398
キナノキ 91075,91082
キナノキゾク属 91055
キナノキ属 91055
キナンカンアオイ 194383
キニンジン 280793
キヌイトサウ 294992
キヌガサギク 339556
キヌガササウ 87485,284340
キヌガサソウ 284340
キヌガサソウ属 216464
キヌガシハ 180642
キヌガシワ 180642
キヌカワ 93464
キヌグコリヤナギ 343691
キヌグフリヤナギ 343691
キヌゲカメバソウ 397459,397462
キヌゲチチコグサ 161919
キヌゲハマヨモギ 36234
キヌゲミヤマヤナギ 343991
キヌゲンゲ 279114
キヌシバ 418453
キヌタエビネ 65926
キヌタサウ 170445
キヌタソウ 170445
キヌタデ 309294,309312
キヌタモドキ 170558
キヌヒゴタイ 348141
キヌヤナギ 343567,344244
キヌラン 417795
キヌラン属 417698
キヌワタ 179906
ギネアキビ 281887
ギネヤアブラヤシ 142257
キネリ 200038
キネロアブラヤシ 142257
キノア 87145
キノエササラン 232106
キノクニカモメヅル 117705
キノクニシオギク 89567
キノクニスゲ 75310
キノクニスズカケ 407501
キノクニセンニンサウ 95194
キノヘササラン 232106
キノユ－カリ 155720
キノルキス属 118147
ギバウシラン 232086
ギバウシ属 198589
キハウチハカエデ 3036
キハウチワカエデ 3034
キハギ 226721,296565
キハダ 294231,294251
キハダカンバ 53338
キハダ属 294230
キハチス 195269
キバナアカザ 86892
キバナアキギリ 345252
キバナアザミ 97599,354651
キバナアザミ属 354648
キバナアマ 327553
キバナアマ属 327548
キバナアラセイトウ 86427,154419
キバナイカリソウ 147013
キバナイトヨモギ 166066
キバナウツギ 413616
キバナウンラン 230978
キバナエウラク 178039

キバナエウラク属　178024
キバナエンシュウムヨウラン　223666
キバナオウギ　42699
キバナオウコン　355846
キバナオオスズシロ　154416
キバナオオヤラオダマキ　30062
キバナオキナグサ　321668
キバナォダマキ　30017
キバナオトメアゼナ　46366
キバナオモタカ　230250
キバナオモダカ　230250
キバナオモダカ科　230253
キバナオランダセンニチ　4866
キバナカイウ　417104
キバナカキラン　147240
キバナカラスウリ　390128
キバナカラスノエンドウ　408419
キバナカワラマツバ　170743,170751
キバナガンクビソウ　77156
キバナカンナ　71171
キバナキョウチクトウ　389978
キバナキョウチクトウ属　389967
キバナクシノハラン　63035
キバナクマガイソウ　120372
キバナクレス　47963
キバナクワクラン　293494
キバナケフチクタウ　389978
キバナケフチクタウ属　389967
キバナコウリンカ　385898
キバナコウリンタンポポ　299224
キバナコスモス　107172
キバナザキバラモンジン　394333
キバナサフラン　111558
キバナサボンソウ　346429
キバナサンタンカ　211125
キバナシオガマ　287468,287474
キバナジギタリス　130371
キバナシャクナゲ　330182
キバナシュクシャ　187451
キバナシュスラン　25980,26003
キバナシュスラン属　25959
キバナスイレン　267714
キバナスゲ　74122
キバナスズシロ　154019
キバナスズシロモドキ　99089,99090
キバナスズシロ属　153998
キバナスズメラン　270809
キバナセキコク　125003
キバナセッブンソウ　148104
キバナタウチクラン　134484
キバナタカサブロウ　181873
キバナタカサブロオ　181873
キバナタマスダレ属　376350
キバナダンドク　71182
キバナヂギタリス　130371
キバナチョウノスケソウ　138450
キバナツクバネウツギ　167
キバナツノクサネム　361370
キバナツメレンゲ　357169
キバナテンナンショウ　33331
キバナトウショウブ　176151
キバナトウワタ　37889
キバナトチノキ　9692
キバナトリカブト　5144,5331
キバナナツノタムラソウ　345178
キバナニワゼキショウ　365743
キバナノアツモリソウ　120473
キバナノアマナ　169460
キバナノアマナ属　169375
キバナノウリクサ　392903
キバナノオニユリ　229887
キバナノギコリソウ　3948
キバナノギャウジャニンニク　15480
キバナノギョウジャニンニク　15480
キバナノクリンザクラ　315082
キバナノクルマバナ　170060
キバナノコギリソウ　3948
キバナノコマノツメ　409729
キバナノジニア　418057
キバナノショウキラン　416379
キバナノセキコク　125390
キバナノセッコク　125390
キバナノセンダングサ　53797
キバナノタマスダレ　376356
キバナノダンドク　71171
キバナノツキヌキニンドウ　235684
キバナノツキヌキホトトギス　396610
キバナノックバネウツキ　180
キバナノツノゴマ　203260
キバナノノコギリサウ　3948
キバナノノコギリソウ　3948
キバナノハウチハマメ　238463
キバナノハウチワマメ　238463
キバナノハタザオ　365532
キバナノハタザオ属　193387
キバナノハタザホ属　193387
キバナノヒメユリ　229776
キバナノフジボエンドウ　279043
キバナノマツバニンジン　231923
キバナノマツムシソウ属　82113
キバナノミソハギ　188264
キバナノミソハギ属　188263
キバナノムカショモキ　55784
キバナノモメンズル　42177
キバナノヤブタバコ　55784
キバナノレンリサウ　222820
キバナノレンリソウ　222820
キバナハイビジョザクラ　702
キバナハウチワマメ　238463
キバナハシリドコロ　354692
キバナハス　263270
キバナバス　263270
キバナハタザオ　365532
キバナハナネコノメ　90322
キバナハマウツボ　275177
キバナハマクサギ　313692
キバナハマサジ　230524
キバナハマナシ　337029
キバナハマナス　337029
キバナハマヒルガオ　207971,207981,
208289
キバナバラモンジン　354870,394333
キバナバラモンジン属　354797
キバナヒエンソウ　124712
キバナビジョザクラ　702
キバナヒメシホガマ　287338
キバナヒメモウズイカ　405659
キバナヒルガネ　194467
キバナフウリンツツジ　145680
キバナフジ　218761
キバナフヂ　218761
キバナヘイシソウ　347150
キバナホウチャクソウ　134503
キバナホオズキ　297712
キバナマツムシソウ　82166
キバナミソハギ　188264
キバナミソハギ属　188263
キバナミヤマムギラン　62814
キバナムギナデシコ　394333
キバナムシトリスミレ　299752
キバナムラサキ　266608
キバナメドハギ　226744
キバナモクセイソウ　327866
キバナモクワンジュ　49241
キバナヤグルマギク　81183
キバナヤセウツボ　275137
キバナユリズイセイ　18062
キバナヨウラク　178039
キバナヨウラク属　178024
キバナラッキョウ　15198
キバナランタナ　221272
キバナルピナス　238463
キバナルリソウ　83923
キバナルリソウ属　83918
キバナヲダマキ　30017
キハマギク　63542
キバンザクロ　318737
キビ　281916
キビシロタンポポ　384573
キビナワシロイチゴ　339499
キビノクロウメモドキ　328904
キビノノダケ　24307
キビノミノボロスゲ　75720
キビフウロ　175034
キヒメザクラ　314405
キヒメユリ　229815
キヒヨドリ　333970
キヒヨドリジャウゴ　333970
キヒヨドリジョウゴ　333970
キヒラトユリ　229887
キビ属　281274

キフアブチロン　998
キフクリンアカリ－ファ　1869
キフクリントウネズミモチ　229533
キフクリンナワシログミ　142153
キフクリンマサキ　157615
キブサスイセン　262429
キフジ　369037,373556
キブシ　373556
キブシ科　373515
キブシ属　373516
キフヂ　369037
キブヂ　373556
キブネギク　23854
キブネダイオウ　340141,340142
キフマルバビユ　208350
キペロルキス属　116769
キベンケイ　162346
ギボウシ　198651
ギボウシズイセン　155858
ギボウシュ　86530
ギボウシラン　232086
ギボウシ属　198589
キホウマル　140870
キボケ　317602
キボタンマル　234926
キマ　65146
ギ－マ　404060
キマダラネズミモチ　229503
キマメ　65146
キマメ属　65143
キミ　281916
キミイヌツゲ　203704
キミカゲサウ　102863,102867
キミカゲサウ属　102856
キミカゲソウ　102863,102867
キミガヨウラン　416610
キミガヨウラン属　416535
キミガヨラン　416610
キミサンザシ　109651
キミズ　288769
キミズモドキ　288729
キミソセヤマギリ　285959
キミソヨゴ　204140
キミタチバナ　31411
キミヅ　288769
キミツルウメモドキ　80261
キミトケイソウ　285660
キミノアオキ　44925
キミノアキグミ　142217
キミノアメリカヒイラギ　204126
キミノウメモドキ　204251
キミノウラジロノキ　33010
キミノエゾニワトコ　345670
キミノエビガライチゴ　339048
キミノオオバクロモジ　231467
キミノオトコヨウゾメ　408025
キミノオンコ　385363
キミノガマズミ　407793
キミノカマツカ　313316
キミノカンボク　408010
キミノクサイチゴ　338519
キミノクマイチゴ　338303
キミノクロガネモチ　204220
キミノコバノガマズミ　407806
キミノサルトリイバラ　366286
キミノサンザシ　109651
キミノシロダモ　264093
キミノセンリョウ　346528
キミノタマミズキ　204051
キミノチクセツニンジン　280793
キミノトケイソウ　285660
キミノトチバニンジン　280750
キミノナワシロイチゴ　338988
キミノニハトコ　345697
キミノニワトコ　345684
キミノネズミモチ　229493
キミノハナミョウガ　17696
キミノバンジロウ　318739,318749
キミノヒョウタンボク　235971
キミノヒヨドリジョウゴ　367326
キミノマンリョウ　31402
キミノモチ　203911
キミノモチノキ　203911,203912
キミノモリイチゴ　167637
キミノヤマホロシ　367263
ギムノカクタス属　182406
ギムノカリキウム属　182419
キムノセレウス属　182511
キムラタケ　57390,57437
キムラタゲ　57437
キメラク　379766
キメンカク　83890
キメンゲ　237554
キモクレン　241962
キモンタラノキ　30640
キモンツワブキ　162617
キモンネズミモチ　229489
キモンヤツデ　162885
ギヤウエフチク　304054
キヤウガノコ　166117
ギヤウギシバ　117859
ギヤウギシバ属　117852
ギヤウジヤサウ　76264
ギヤウジヤニンニク　15868
ギヤウジヤノミズ　411686
キヤウナ　59438
キヤウワウ　114859
キヤウワウ　114871
ギヤクリュウギョク　264372
キャッサバ　244507
キヤナダメギ　51422
キヤナリ－ヤシ　295459
キャベツ　59520,59532
キャベツノキ　300956
キャベツヤシ属　161053
キャベンディシア属　79788
キャラ　29973
キヤラ　161639
キャラボク　385370,385372
キヤンチン　392841
キャンディタフト　203184
ギュウカク　374013
ギュウカクモモドキ　373832
キュウケイカンラン　59541
キュウコンエンドウ　222851
キュウコンベゴニア　50378
キュウシュウココゴメグサ　160179
ギュウシンリ　25882
ギュウゼッ　171764
キュウテンマル　182438
キュウリ　114245,114252
キュウリゲサ　397451
キュウリ属　114108
キュ－コサクラ　314084
キュ－バサバル　341423
キュ－バダイオウヤシ　337885
キュ－バヤシ　100028
キュ－バヤシ属　100027
キョウ　15779
ギョウウンカク　155304
ギョウエイマル　247667
キョウオウ　114859
キョウガノコ　166117
ギョウギシバ　117859
ギョウギシバ属　117852
キョウギョ　86481
ギョウコゥギョク　352761
ギョウジャニンニク　15868,15874
ギョウジャノミズ　411686
ギョウジヤモミ　114539
ギョウショウギョク　284681
ギョウセンリュウ　148144
キョウチクトウ　265327,265339
キョウチクトウ科　29462
キョウチクトウ属　265310
キョウナ　59438
キョウフウマル　140861
キョウポウマル　327316
キョウボクチュウ　263560
キョウマイギ　167124
キョウマギョク　366827
キョウリュウカク　105436
キョウリュウマル　182446
キョウワウ　114859
キョオウニシキ　273821
キョオウリュウ　366826
ギョオジャノミズ　411686
ギョク　233658
ギョクキ　138229
ギョクサイリン　83969
ギョクシュクサウ　221538

ギョクシンカ　384950
ギョクシンカ属　384909
ギョクシンクワ　385049
ギョクシンクワ属　384909
ギョクスイ　242352
ギョクダンカ　199911,199912
ギョクダンクワ　199911
キョクチチョウノスケソウ　138461
キョクチハナシノブ　307198,307209
ギョクチンラン　116900
キョクビマル　275408
キョクホクミヤマハンノキ　16479
ギョクラン　252809,252907
キヨシソウ　349113
キョシュウギョク　163452
キヨスミウツボ　293186
キヨスミウツボ属　293181
キヨスミギボウシ　198623
キヨズミヤブマオ　56190
キョゼツキオン　359415
ギョッカラン　117000
キョッコウマル　153353
ギョトウ　254854
ギョビカン　358633
キョヒメ　102392
ギョボク　110216,110227
ギョボク属　110205
キョマサニンジン　29327
キヨミギク　166038
ギョリウ　383469
ギョリウダマシ　215329
ギョリウ属　383412
ギョリュウ　383469,383622
ギョリュウカズラ　39185
ギョリュウバイ　226481
ギョリュウモドキ　67456
ギョリュウモドキ属　67455
ギョリュウロドキ　67456
ギョリュウ科　383397
ギョリュウ属　383412
キョリンギョク　182445
ギョリンギョク　103772
キョレイマル　395642
キラア属　324623
キライシハヒノキ　381104
キラギョク　242882
キラタンヤシ　222636
ギランイヌビワ　165819,165826,165828
キランサウ　13091
キランサウ属　13051
キランジソ　99711
キランソウ　13091
キランソウ属　13051
キリ　285981
キリアサ　1000
キリアザ　1000
キリアタポプラ　311276
ギリア属　175668
キリエノキ　394628,394635
キリカズラ　246654
キリカズラ属　246651
キリカヅラ属　37548
キリガミネアキノキリンソウ　368490
キリガミネアサヒラン　143442
キリガミネヒオウギアヤメ　208823
キリキシバ　231461
ギリシア属　100149
キリシマ　331370
キリシマエビネ　65887
キリシマグミ　141987
キリシマサウ　63987
キリシマザサ　347310
キリシマシャクジョウ　63981
キリシマシャクヂャウ　63987
キリシマタヌキノショクダイ　390101
キリシマツツジ　331370
キリシマテンナンショウ　33486
キリシマノガリヤス　65291
キリシマヒゴタイ　348769
キリシマミズキ　106637
キリシマリンドウ　173856
キリタチヤマザクラ　83305
キリタ属　87815
キリツボ　209748
キリトリア　97203
キリノキ　285981
キリハカエデ　2928
ギリバヤシ　31705
キリハヤシ属　181295
キリフリザクラ　83250
ギリミカン　93846
キリモドキ　211239
キリンカク　159245
キリンギク　228529
キリンケツトウ　121493
キリンサウ　294123
キリンソウ　294115,294123
キリンソウ属　356467
キリンドロフィルム属　116623
キリンドロブンティア属　116636
キリンボク　243717
キリンヤシ属　376400
キリ属　285951
キルソ　267856
キルタンサス属　120532
キルトスペルマ　120774
キルトポジューム属　120705
キルトルキス属　120723
キールンアミガサギリ　14199
キールンウツギ　127114
キールンオドリコサウ　220422
キールンカンコノキ　177149,177150
キールンクマタケラン　17693
キールンサカキ　8256
キールンテンナンシャウ　33295
キールンハタザホ　30458
キールンフジバカマ　158182
キールンブドウ　411735
キールンヤマノイモ　131784
キールンヲドリコサウ　220422
キレア属　91611
キレイマル　234958
キレザキサクラソウ　314956
キレハアザミタンポポ　384614
キレバアラリア　310208
キレハイチョウ　175834
キレハイヌガラシ　336277
キレハウマゴヤシ　247337
キレハオオハナウド　192282
キレハクマイチゴ　338063
キレバコウシュウヒゴタイ　348128
キレハシラカンバ　53578
キレハダケカンバ　53420
キレバチダケサシ　41829
キレバテーブルヤ　85437
キレバテーブルヤシ　85428
キレハテンジクアオイ　288192
キレハトチバニンジン　280749
キレハノハクサンボウフウ　292943
キレハノフクオウモドキ　313849
キレハハリギリ　215443
キレハマネキグサ　237959
キレハマメグンバイナズナ　225305
キレバヤマビワソウ　333738
キレハヤマブキショウマ　37067
キレハヨナメ　215352
キレンゲ　331261
キレンゲショウマ　24167,216474
キレンゲショウマ属　216473
キレンゲツツジ　331261
キワタ　56802,179878,311389
キワダ　294231
キワタノキ属　56770
キワタ属　56770
キワンジュ　48993
キヲン　359565
ギンイチュウ　390481
ギンイチョウ　311956
キンウラハグサ　184658
ギンウラハグサ　184657
キンウリ　114208
キンエイカ　155173
キンエイカ属　155170
キンエイクワ　155173
キンエイクワ属　155170
キンエノコロ　361762,361877
キンエノゴロ　361877
キンエビネ　312351,312498
キンエボシ　272980
キンエンチュウ　183376
キンオウギョク　284663

ギンオウギョク 264360
キンオウマル 242887
キンカ 368483
キンカアザミ 91737
キンカイモ 367696
キンカウボク 252841,283418
ギンカウボク 252907
ギンカエデ 3532
キンカキツバタ 208715
キンカクウリ 114261
ギンガサウ 123655,199911
キンカソウ 399509
ギンガソウ 123655,199911
キンカチャ 69480
キンガフクワン 1219
ギンガフクワン 227430
ギンガフクワン属 227420
キンガヤツリ 393194,393205
ギンガラク 155299
キンカン 93513,167503
キンカンリュウ 163427
キンカン属 167489
キンキエンゴサク 106246
キンキカサスゲ 75761
キンキジュ 301134
キンキジュ属 301117
キンキヒョウタンボク 236058
キンキマメザクラ 83231
キンキヤママメザクラ 83120
ギンキョウ 392381
キンキョギ 35187
キンギョサウ 28617
キンギョサウ属 28576
キンギョシバ 260711
キンギョソウ 28617
キンギョソウ属 28576
キンギョツバキ 69204
キンギョバツバキ 69204
キンギョボク 390759
キンギョモ 83545,261364
キンギョモ科 83539
キンギョモ属 83540
キンギンカク 183377
キンギンサウ 128314,179679
キンギンソウ 179679
キンギンツカス 244171
キンギンナスゼ 367735
キンギンナスビ 367011
キンギンボク 142158,235970
キンクオラン 392248
キングサリ 218761
キングサリ属 218754
ギンクジャク 340626
キンクネンボ 93765,93801
キングハニシキ 16509
キングプロテア 315776
キングマンダリン 93649
キンクルマ 34780
キンクワ 368480
キンクワサウ 399509
キンケイギク 104448,104482
ギンゲツ 359026
ギンケンソウ 32960
ギンケンソウ属 32959
キンコウカン 1219
ギンゴウカン 227430
ギンゴウカン属 227420
キンコウジ 93672
キンコウチュウ 183367
キンコウデン 229862
キンコウボク 252841
ギンコウボク 252809,252907
キンコウマル 284673
キンコウミュウ 395636
キンゴジカ 362617
キンゴジカモドキ 362703
キンゴジカモドキ属 362702
キンゴジカ属 362473
キンゴジクワ 362617
キンゴジクワ属 362473
ギンザ 86901
ギンサカズキ 266206
ギンサン 79438
キンサンジゴ 196450
キンサンジュ 196443
キンサンマル 327281
キンシコウ 140009
キンシチク 47518
キンシバイ 202070
キンシベボタンネコノメ 90390
キンシベヤマツツジ 330969
キンシャウ 140127
キンシャチ 140127
キンショウ 317822
キンショウギョク 284657
ギンショウギョク 284684
キンショウジョ 21117
キンジョウマル 395637
キンショクダモ 264021
ギンシンサウ 164126
ギンシンソウ 164126
キンズ 167499,167516
キンスゲ 75942
キンセイエビネ 66020
キンセイギク 89959
ギンセイギョウ 182421
キンセイマル 140852
キンセイラン 66020
ギンセカイ 272950
キンセキリュウ 163483
キンセン 395651
ギンセン 109151
キンセンカ 66395,66445,66449
ギンセンカ 195326
キンセンカ属 66380
キンセンクワ 66395
ギンセンクワ 195326
キンセンクワ属 66380
キンセンセキ 133599
キンセンソウ 238371
キンセンソウ属 238369
キンセンユリ 229730
キンソウリュウ 405193
キンタイオオバザサ 347251
キンタイザサ 347293
キンダイヌビハ 165762
キンダイビハ 165429
キンチャウモ 83545
キンチャクアオイ 37701
キンチャクスゲ 75343
キンチャクズル 94933
キンチャクソウ 66262,66271
キンチャクソウ属 66252
キンチャクハギ 322078
キンチャクハギ属 322076
キンチャクボタン 220729
キンチャクマメ 333456
キンチョウマル 327298
キンツクバネ 382276
ギンツノサンゴ 159883
ギンツンガソ 54620
ギンテマリ 244085
キントウガ 114305
キントウノオ科 243530
キントウノヲ科 243530
キントオガ 114305
キントキシロヨメナ 39972
キントキヒゴタイ 348768,348795
キントラノオ 390542
キントラノオ科 243530
キントラノオ属 390538
ギンドロ 311208
キンナタオレノキ 276337
ギンナン 175813
ギンネム 227430
ギンノキ 227233
キンバイ 211931
ギンバイカ 261739
ギンバイカ属 261726
ギンバイクワ 261739
ギンバイクワ属 261726
キンバイサウ 399509
ギンバイサウ 123655
キンバイサウ属 399484
ギンバイサウ属 123654
キンバイザサ 114819,114838
キンバイザサ属 114814
キンバイソウ 399509
ギンバイソウ 123655
ギンパイソウ 266206
キンバイソウ属 399484

ギンバイソウ属　123654
キンバイタウコギ　53786
キンハイリュウ　395639
ギンハナガタ　23211
ギンバニシキ　108075
ギンバノコギリソウ　3942
キンバンドクウツギ　104694
キンビディウム属　116769
キンヒモ　29715,29716
キンブセン　272856
キンポウゲ　325981
キンポウゲ科　325494
キンポウゲ属　325498
キンポウジュ　67275
キンホウマル　327262
ギンボウマル　327311
キンホウリュウ　299247
ギンホコ　332356
キンボシ　244134
ギンボシベゴニア　49603
キンマ　300354
キンマウダモ　264021
キンマウツツジ　331391
キンマサギ　157616
ギンマサキ　157614
キンマルバユ－カリ　155529
キンミズヒキ　11549,11572,11580
ギンミズヒキ　291757
キンミズヒキ属　11542
キンミツバ　299534
キンミヅヒキ　11572
キンミヅビキ　11549
キンミヅヒキ属　11542
キンミャクツルマサキ　157504
キンメ　333914
ギンメイショク　139980
キンメイチク　297239,297464
キンメイハチク　297368
キンメイモウソウチク　297307
キンモウサカキ　69586
キンモウツツジ　331391
ギンモウマル　244085
キンモウヤナギ　344255
キンモクセイ　276259,276292
ギンモクセイ　276257,276291,276295
キンモンツワブキ　162617
キンヤウラク属　98601
ギンヨウアカシア　1069
ギンヨウアサガオ　32642
ギンヨウカエデ　3532
ギンヨウグミ　141941
ギンヨウコロラドトウヒ　298406
ギンヨウシャリントウ　107615
ギンヨウジュ　227233
ギンヨウジュ属　227224
キンヨウセンネンショウ　137487
キンヨウセンネンボク　137487
キンヨゥチユウ　183377
キンヨウボク　29125
キンヨウボク属　29121
キンヨウラク　98640
キンヨモギ　35187
キンラン　82051
ギンラン　82044
キンランジソ　99711
キンラン属　82028,82113
ギンリャウサウ　258044
ギンリャウサウ属　258006
ギンリャウソウ属　258006
キンリュウカ　378375,378386
キンリュウカ属　378363
キンリュウヘン　116863,117022
キンリョウ　395653
ギンリョウサウ属　258006
ギンリョウソウモトギ　258044
ギンリョウソウ科　258049
ギンリョウソウ属　258006
キンリョウヘン　116863,117022
ギンレイ　32733
キンレイカ　285875,285877
ギンレイカ　239544
ギンレイギョク　140887
キンレイジュ　115784
キンレンカ　218761,399601
キンレンカ属　399597
キンレンクワ　218761
キンレンクワ属　218754
キンロウバイ　289713
キンロバイ　289713,312599
ギンロバイ　312606
ギンワットル　1165
グアイクウット　181505
グアイクウット属　181496
クアドソバミスモクマオウ　79178
グアバ　318742
グアマバモドキ　170477
クィアベンティア属　324596
グイマッ　221866
グイマツ　221866,221878
クイミチリメンウロコヤシ　156117
クインスランドナット　240210
クインスランドナットノキ　240210
クエスタ－ツノミザミア　83682
クガイサウ　407434
クガイサウ属　406966
クガイソウ　407371,407434,407462,407485,407492
クガイソウ属　406966,407449
クカサゴチク　37471
クキチシャ　219488
クグ　245556
ククイノキ　14544
クグガヤツリ　118642
クグスゲ　75895
クグテンツキ　166265
クグ属　245332
クゲヌマラン　82047
クコ　239021
クコ属　238996
クサアジサイ　73120
クサアジサイ属　73119
クサアタン　17383
クサアヂサヰ属　73119
クサイ　213507
クサイザクラ　83337,316636
クサイチゴ　338516
クサイロエビネ　82088
クサエンジュ　369010
クザカイタンポポ　384610
クサギ　96398,408009
クサギナ　96398
クサキビ　281408
クサキョウチクトウ　295288
クサギラ　96398
クサギリ　1000
クサキンロウバイ　312412
クサギ属　95934
クサケフチクタウ　295288
クサケフチクタウ属　295239
クサコアカソ　56318
クサザクマ　53483
クサザクラ　53503
クササンダンカ　289831
クサシモツケ　166097
クサシャウジャウボク　159046
クサジュ　369010
クサスギカズラ　38960,39069
クサスギカズラ属　38904
クサスギカヅラ属　38904
クサスゲ　76086
クサセンナ　360463
クサタズ　345586
クサダチイケマ　117385
クサダチキミヅ　142642
クサタチバナ　117339
クサダモ　229
クサツゲ　64285
クサテンツキ　166228
クサドウ　228369
クサトケイサウ　285641
クサトケイソウ　285639,285641
クサトベラ　350344
クサトベラ科　179563
クサトベラ属　350323
クサナギオゴケ　409480
クサナギカヅラ　38938
クサナニワズ　375187
クサニワトコ　345586
クサニンジンボク　411374,411375
クサネム　9560
クサネム属　9489

クサノオウ 86755,86757,86761
クサノオウバノギク 110605,283386
クサノオウ属 86733
クサノオオバノギク 283386
クサノオホカウバシ 347411
クサノガキ 132291
クサノクリガシ 78932
クサノソヨゴ 203952
クサノボタン 276090
クサノワウ 86755
クサノワウバノギク 283386
クサノワウバノケマン 283386
クサノワウバノヤクミサウ 283386
クサノワウバノヤクミソウ 283386
クサノワウ属 86733
クサハギ 126389
クサビガヤ 371493
クサビコムギ 8724
クサビョウ 201743
クサフヂホ 408361
クサヘンルウダ 185635
クサボウコウマメ 371161
クサボケ 84556
クサボタン 280207
クサホルト 159222
クサマオ 56240
クサマキ 306457,352856
クサマヲ 56240
クサミズキ 266807
クサミソハギ 114618
クサヤツデ 12741
クサヤマブキ 200754
クサヨシ 293709
クサヨシ属 293705
クサレダマ 239902
クサヰ 213507
クサントソーマ属 415190
クシガヤ 118315
クシガヤ属 118307
クシトベラ 350344
グジナ 384746
クシノキ 3529
クシバタンポポ 384734
クシバツメクサ 396822
クシバニガナ 416490
クシバハタザオ 136185
クシヒゲシバ 88383
クジャクガシワ 323819
クジャクサウ 104598,383103
クジャクソウ 104598,383103
クジャクヒバ 85316
クジャクヤシ 78046
クジヤクヤシ 78056
クジャクヤシ属 78045
クジュウクリテンツキ 166238
クジュウスゲ 76117
クジュウツリスゲ 75021
クジュウノガリヤス 65294
グジョウシノ 347340
クジラグサ 126127
クジラグサ属 126112,365386
クシロチドリ 192862
クシロチャヒキ 60599,60654
クシロネナシカズラ 115031
クシロネナシカヅラ 115031
クシロハナシノブ 307204
クシロビューム属 415704
クシロヤガミスゲ 74196
クシロラン 147323
クシロワチガイ 318515
クシロワチガイソウ 318515
クス 91287
クズ 321441
クズイモ 279732
クズイモ属 279728
クズインゲン 321460
クズウコン 245014
クズウコン科 245039
クズウコン属 245013
クスクスアデク 382589
クスクスカウジ 31489
クスクスカンコノキ 177148
クスクスサガリラン 133145
クスクスジウマン 139655
クスクスジュラン属 88111
クスクスセキコク 166959
クスクスバカマ 158139
クスクスフウラン 390518
クスクスムエフラン 147323
クスクスヤウラクラン 267927
クスクスラン 62536
クスタブ 91351
クスダマツメクサ 396854
クスドイゲ 415869
クスドイゲ属 415863
クスノキ 91287
クスノキダマシ 91295
クスノキ科 223006
クスノキ属 91252
クスノハアヂサイ 199910
クスノハアヂサヰ 199910
クスノハカエデ 2775,3255,3256,3258
クスノハガシハ 243427
クスノハガシワ 243427
クズノハカヅラ 259535
クスノハシャシャンボ 403816
クスビダータ 69033
グースベリ 334250
グースベリー 334250
グスベリー 334250
グズマニア属 182149
クズモダマ 259535
クズモドキ 67898
クスモドキ属 113397,113422
クズヰンゲン 321460
クズ属 321419
クゼンギョク 275407
クソエンドウ 389518
クソクサギ 408009
クソニンジン 35132
クタデ 308877
クタモノトクイ 285637
クダモノトケイソウ 285637
クチナシ 171253,171333
クチナシカズラ 25928
クチナシグサ 257538,257543
クチナシグサ属 257537
クチナシ属 171238
クチナハイチゴ 138796
クチバシグサ 231568
クチバシシオガマ 287081
クチベニ 229733
クヂベニウツギ 126891
クチベニズイセン 262438
クヅウコン科 245039
クツクサ 81570
グツシャクゲツタウ 17712
グッタペルカノキ 280441
クヅャクギク 89704
クテガワザサ 347223
クテナンテ属 113944
グドウジ 396170
クナシリオヤマノエンドウ 278937
クニガミクロウメモドキ 328628
クニガミサンショウヅル 142867
クニガミトンボソウ 302512
クニガミヒサカキ 160631
クニギ 323611
クニフォフィア属 216923
クヌガワブドウ 411890
クヌギ 323611,324384
クネオルム 122416
クネオルム属 97491
グネツム 178535
グネツム科 178522
グネツム属 178524
グネモソ 178535
グネモンノキ 178535
クネンボ 93649,93660
クノギ 323611
クノニア科 114567
クノニア属 114563
クハ 259067
クハイタビ 242976
クハイタビ属 242974
クハガタサウ 407050
クハクサ 162871,162872
クハクサ属 162868
グバス 145420
クハズイモ 16512,417093
クハズイモ属 16484

クハズウド　192278
クハナ　147048,147050
クパン　284473
クハ属　259065
グビジンソウ　282685
グヒマツ　221866
クフエア属　114595
クフエヤ属　114595
クプレサス　114714
クプレッズスス属　114649
クーペーリア属　103671
クボタテンナンショウ　33432
クマイザサ　347292
クマイチゴ　338300
クマオングルミ　212641
クマガイサウ　120371
クマガイソウ　120371
クマガエソウ　120371
クマガユサウ　120371
クマガユソウ　120371
クマガワナンテンハギ　408658
クマガワブドウ　411890
クマガワリンドウ　173857
クマギク　34746
クマコケモモ　31142,31316
クマザキガフクワン　301147
クマザサ　347324
クマザサ属　347191
クマシデ　77312
クマシデ属　77252
クマタケゲットウ　17641
クマタケラン　17674
クマダラ　272712
クマツズラ　405872
クマツヅラ　405872
クマツヅラ科　405907
クマツヅラ属　405801
クマデヤシ属　341401
グーマト　364715
クマナリヒラ　357942
クマノアシツメクサ　28200
クマノガシ　324283
クマノギク　413514
クマノダケ　24468
クマノミズキ　104966,380461
クマノミヅキ　380461
クマノル　167693
クマヤナギ　52418,52454
クマヤナギ属　52400
クマヤブマオ　56353
クマヤマグミ　141987
クマンボタン　220729
クミアイスゲ　76296
クミアヒスゲ　74098,76296
クミスクチク　95862
クミスクチン　95862,275639
グミトベラ　11293
クミヌム属　114502
クミバシホガマ　287705
グミヒョウタンボク　236130
グミモドキ　112853,315176,365002
グミモドキ属　112829
クミン　114503
クミントウ　186655
グミ科　141928
グミ属　141929
グムグムャシ　6130
クメザンセウ　417282
グメリーナ属　178024
グメリニイナベナ　133486
クモイイカリソウ　146972
クモイオトギリ　201708
クモイコゴメグサ　160229
クモイコザクラ　314888
クモイザクラ　83268
クモイジガバチ　232359
クモイナデシコ　127861
クモイノダケ　98784
クモイリンドウ　173199
クモキリサウ　232086,232215
クモキリサウ属　232072
クモキリソウ　232114,232215
クモキリソウ属　232072
クモチリサウ　232086
クモチリソウ　232114
クモノスネコノメソウ　90442
クモノスバンダイソウ　358021
クモノスバンダイソウ属　358015
クモノスベンダイソウ　358021
クモマキンパウゲ　326273
クモマキンポウゲ　326273
クモマグサ　349632
クモマシバスゲ　76447
クモマスズメノヒエ　238573,238717
クモマスミレ　409862,409863
クモマタンポポ　384714
クモマツツジ　331615,389577
クモマナズナ　30467,137143
クモマナヅナ　137143
クモマニガナ　210529
クモマミミナグサ　83014
クモマユキノシタ　349529
クモマリンドウ　173540
クモマリンドオ　173540
クモヤブリ　392841
クモラン　19512,383064
クモラン属　383054
クモンリュウ　182444
クライタボ　165630
グラインロウス　336486
グラウカモクマオウ　79165
クラウサマツ　299864
クラウドベリ　338243
クラガシ属　78848
クラカス　364704
クラサワスゲ　76122
グラジオラス　176121,176222,176260
グラジオラス属　175990
クラダンサス属　93882
クラッスラ属　108775
クラニュソ　121653
クラブコムギ　398877
グラプトフィルム属　180270
グラプトペタルム属　180265
クラフネリアーナ　69015
クラブリンゴ　243551
クラマス　405163
グラマトフィルム属　180116
クラマンギス属　180090
クラムヨモギ　35753
クララ　369010
クララ属　368955
クラリンドウ　96439
クラルオドリコソウ　283625
クラルカウジ　31387,31613
クラルガキ　132245
クラルガキ属　179441
クラルクレソラン　395868
クラルクレソラン属　395855
クラルハヒノキ　381154,381437
クラルフウトウカヅラ　300432
クラルフタギ　381154,381437
クラルマユミ　157418,157888
グランサムツバキ　69108
グランジュリカクタス属　176697
グランデイデンタータポプラ　311335
グランドヒノキ　85271
グラントレモン　93631
クリ　78777,78811
クリアンサス属　96633
クリイロスゲ　74304
クリオザサ　347356
クリガシ　78933,79035
クリガシハ　78966
クリガシワ　324532
クリガシ属　78848
クリサンセラメ　89628
クリサンテマム　89466
クリシュナボダイジュ　164685
クリスナボダイジュ　165210
クリスマス・ローズ　190944
クリスマス・ローズ属　190925
クリスマスバゴニヤ　49711
クリスマスベゴニア　49711
クリスマスローズ　190944
クリスマスローズ属　190925
クリスマム属　111343
グリセリーニア属　181260
クリソゴナム属　89952
グリゾゴーマ属　89863
クリソバラヌス属　89812

クリソラン 89948
クリソラン属 89938
クリダンツス属 88215
クリデミア属 96650
クリトストマ属 97468
クリナム 111214
クリナム属 111152
クリノイガ 80811
グリ－ノビア属 180504
クリハダヒノキ 85331
ク－リ－ピナンガ 299664
クリプタンサス属 113364
クリプトキ－ルス属 113511
クリプトコリ－ネ属 113527
クリプトセレウス属 113507
クリプトフォランツス属 113738
クリプト－プス属 113760
クリミャモミ 429
クリ－ムウツギ 413572
クリヤマハハコ 21697
クリンキパイン 30852
クリンサウ 314508
クリンザクラ 314085,314820
クリンセンネンボク 137403
クリンソウ 314508
グリンデ－リア属 181092
クリンユキフデ 309841
グリ－ンロウズ 336486
グリ－ンロ－ス 336486
クリ属 78766
グルイン 133558
クルクマ 114874
グルグルヤシ 171902
クルシアネラ属 113110
グルソニア属 181450
クルマアザミ 92425
クルマガンピ 364067
クルマギク 41366
クルマザキホソバシャクナゲ 331189
クルマサンシチ 407434
クルマセンノウ 363718
クルマツツジ 331593
クルマバアカネ 337925
クルマバカワラサイコ 313098
クルマバゲンゲ 278941
クルマバサイコ 63665
クルマバサウ 170524
クルマバサウ属 39300
クルマバザクロサウ 256727
クルマバザクロソウ 256727
クルマバジヤウザン 96139
クルマバジョウザン 96139
クルマバソウ 170524
クルマバソウ属 39300
クルマバタガヤサン 96139
クルマバツクバネソウ 284406
クルマバテンモンドウ 39251
クルマバナ 96970,97056
クルマバナルコユリ 308647
クルマバハグマ 292067
クルマバヒメハギ 308437
クルマバヒヨドリ 158136,158305
クルマバムレイシャジン 7894
クルマバモウセンゴケ 138273
クルマミズキ 57695
クルマムグラ 170445,170705
クルマユリ 229934
クルミ 212621,212626,212636
クルミギョク 271920
クルミツヅジ 331010
クルミ科 212581
クルミ属 212582
クルメツツジ 331380
グレイギア属 180535
クレイスタントセレウス属 94551
クレイステス属 94545
クレイストカクタス属 94551
クレイソスト－マ属 94425
クレイト－ニア属 94292
グレイミルクウツト 83696
クレオバトラ 93716
クレオメサウ 95796
クレオメソウ 95796
グレオメ属 95606
クレタケ 116667,297373
クレナイ 77748
クレナイロケア 335079
クレナヰ 77748
クレノアイ 77748
クレノアヰ 77748
クレノオモ 167156
クレノバジカミ 418010
クレビス属 110710
クレピス属 110710
グレビレア属 180561
グレ－プフル－ツ 93690
クレブラユ－カリ 155545
クレマチス属 94676
クレメンティンマンダリン 93429
クレロデンドロングレオメ属 95934
クレヲメサウ 95796
グレ－ンスゲ 75698
クロアザミ 81237
クロアゼスゲ 73732,73738,76338
クロアブラガヤ 353676,353856
グロアヤメ 208870
クロイゲ 342198
クロイゲイボタ 229593
クロイゲ属 342155
クロイチゴ 338808
クロイヌノヒゲ 151238
クロイヌノヒゲモドキ 151237
クロイハザサ属 390551
クロイワザサ 390553
クロ－ウェア属 113102
クロウスゴ 403932
クロウメモドキ 328740,328742
クロウメモドキ科 328218,328543
クロウメモドキ属 328592
クロエアオキ 44922
クロエゾ 298307
クロエゾマツ 298307
クロオスゲ 73987
クロカウガイゼキシャウ 212990
クロカエデ 3236
クロカキ 132309
クロガキ 132351
グロガキ 132230
クロガクモメンズル 42262
クロガシ 116100,116153,116156
クロカゼクサ 147509
クロガネモチ 204217
クロカミラン 310894
クロガヤ 169561
クロガヤツリ 118928
クロガヤ属 169548
クロガラシ 59515
クロカワスゲ 74752
クロカワズスゲ 73748
クロカンバ 53400,328660
クロキ 381143,381252,381291,381390
クロギ 204301
グロキシニア 364738,364754
グロキシニア属 177515
クロキソウ 58300
クロキソウ属 58299
クロギボウシ 198637
クロクサギ 160954
クログワイ 143312
クロコウガイゼキショウ 212990
クロコガ 91351
クロコスミア属 111455
クロコヌカグサ 12118,12221
クロサスグリ 334117
クロジシャ 134010,379374
クロシブガキ 132245
クロシベエリカ 149750
クロジベエリカ 149138
クロシマシバスゲ 73570
クロシモツケ 372094
クロスグリ 334117
クロスゲ 75370
クロズル 398318
クロソネ 77312
クロソヨゴ 204292,204301
クロダイコン 326626
クロタキカズラ 198575
クロタキカズラ科 203303
クロタキカズラ属 198574
クロタキカヅラ 198575
クロタキカヅラ科 203303

クロタキカヅラ属 198574
クロダネカボチャ 114277
クロタネサウ 266225
クロタネサウ属 266215
クロタネソウ 266225
クロタネソウ属 266215
クロタマガヤツリ 168831
クロタマガヤツリ属 168819
クロダモ 91351
クロタラミア 112238
クロチク 297367
クロツガ 399895,399923
クロッカス 111625
クロッカス属 111480
クロツグ 32336
クロツグ属 32330
クロッサンドラ属 111693
クロッソソマ科 111819
グロッチフィルム属 177434
グロッティフィルム属 177434
クロッバラ 328684
クロツバラ 328680
グロッバ属 176985
クロツリバナ 157930
クロヅル 398318
クロヅル属 398310
クロテツ 301777
クロテンコオトギリ 201909
クロテンシラトリオトギリ 202216
クロテンツキ 166290
クロトウヒ 298349
クロトウヒレン 348561
クロトチウ 157355
クロトチュウ 157355
クロトリノキ 231461
クロトンノキ 98190,98191
クロトンノキ属 98177
クロトン属 112829
クロナナコ 327503
クロヌカボ 12413
クロヌマハリイ 143168,143169,143271
クロヌマハリヰ 143168
クローバー 397043
クロバイ 381423
クロハシテンナンショウ 33418
クロバナアキチョウジ 209613
クロバナアケボノスミレ 410499
クロバナアルム 37024
クロバナイヨカズラ 117535
クロバナイリス 192904
クロバナイリス属 192898
クロバナウマノミツバ 345998
クロバナエンジュ 20005
クロバナエンジュ属 19989
クロバナオダマキ 30078
クロバナカズラ 291082
クロバナカズラ属 291031
クロバナキハギ 226895
クロバナクララ 20005
クロバナタシロイモ 382912
クロバナツリフネ 205370
クロバナツルアズキ 241261
クロバナドジョウツナギ 177672
クロハナナンバンクサフジ 386257
クロハナナンバンクサフヂ 386257
クロハナニハトコ 345696
クロバナニワトコ 345668
クロバナハルシャギク 104604
クロバナハンショウヅル 94933
クロバナヒキオコシ 209846
クロバナマツムシサウ 350099
クロバナモウズイ 405737
クロバナモウズイカ 405737
クロバナモウズヰ 405737
クロバナヤマウコギ 143686
クロバナヤマハッカ 209614
クロバナラフゲ 100257
クロバナラフゲ属 100254
クロバナラフバイ 68313
クロバナラフバイ属 68306
クロバナロウゲ 100257
クロバナロウゲ属 100254
クロバナロウバイ 68313,68321
クロバナロウバイ属 68306
クロバナロクオンサウ 117368
クロバナロクオンソウ 117364
クロバヒ 381423
クロハリイ 143178,143179
クロビ 390660
グロビア属 181282
クロビイタヤ 3149
クロヒナスゲ 74645
クロヒメカンアオイ 37763
クロヒメシャクナゲ 331023
クロブシヒョウタンボク 235908
クロブナ 162368,162382
クロフネサイシン 37612
クロフネツツジ 331768
クロープノキ 382477
グロブラリア科 177054
グロブラーリア属 177022
クロベ 390660
クロベスギ 113696
クロベトベラ 301398
クロベンケイ 117385
クロベ属 390567
クロボウ 132291
クロボウモドキ 307508
クロボキンエノコロ 361840
クロボシイヌザクラ 223117
クロボシオオアマナ 274513
クロホシクサ 151428
クロボシサウ 238690
クロボシザクラ 223117
クロボシソウ 238686,238690
クロボスゲ 73788,75759
クロマツ 300281
クロマメノキ 404027,404031
クロミウスノキ 403919,404047
クロミキイチゴ 338105,338808,338909
クロミグワ 259180
クロミサンザシ 109597
クロミシャリントウ 107543
クロミズキ 57695
クロミドームヤシ 202268
クロミノオキナワスズメウリ 417492
クロミノクワ 259180
クロミノシンジュガヤ 354253
クロミノニシゴリ 381341
クロミノハエドクソウ 295986
クロミノハリイ 143053,143054
クロミノハリスグリ 334034
クロミノヘビノボラズ 51811
クロミノモクタチバナ 31600
クロムギ 162338,356237
クロムヨウラン 223658
クロムヨフラン 223658
クロメスゲ 74086,75378
クロモ 200190,200192
クロモウズイカ 405737
クロモウズイクワ 405737
クロモウズヰクワ 405737
クロモジ 231433,231461
クロモジ属 231297
クロモヅ 234061
クロモ属 200184
クロヤガミスゲ 75364
クロヤッシロラン 171949
クロヤナギ 89160,343445
クロヤマナラシ 311398
クロユリ 168361
クロユンジュ 240114
クロヨナ 254803,310962
クロヨナ属 310950
クロリオサ・スーパーパ 177251
クロリオサリリ 177251
グロリオーサ属 177230
クロワットル 1380
クロヲスゲ 73987
クワ 259067,259097
クワイ 342424
クワイカウ 167156
クワイチゴ 338300
クワイバカンアオイ 37662
クワイ属 342308
クワウエフザン 114539
クワウエフザン属 114532
クワウテイマメ 294010
クワウラウ 295465
クワエンボク属 370351
クワガタスミレ 410340

クワガタソウ　407050,407232
クワガタソウ属　406966
クワクカウアザミ　11199
クワクカウアザミ属　11194
クワクサ　162872
クワクサ属　162868
クワクラン　293537
クワクラン属　293484
クワザンジマ　331759
クワシグルミ　212636
クワジャウイチゲ　23817
クワズイモ　16512,16518
クワズイモ属　16484
クワズヤシ属　273120
クワセウクズマメ　259545
クワセウセキコク　125087
クワセウツゲトモモ　382649
クワセウフウトウカヅラ　300434
クワゾメアケビ　13222
クワッシア　364382
クワノハエノキ　80590
クワモドキ　19191
クワラン　66101,295465
クワリン　116546,317602
クワリン属　116532
クワレンコウイチゴ　338999
クワヰ　342424
クワヰ属　342308
クワンオンチク属　329173
クワンザウ　191263,191304
グワンジツサウ　8331
グワンジツザクラ　83158
クワンセイタンポポ　384591
クワントウ　400675
クワントウ属　400671
クワ科　258378
クワ属　259065
クンカンギョク　182470
グンギョク　163328
グンケイ　102542
クンケイラン　293494
グンケン　340757
クンコウマル　103751
グンコヅチ　254467
グンサン　21124
グンジサウ属　284969
グンジソウ　285004
グンジソウ属　284969
クンショウギク　172357
クンショウギク属　172276
クンシラン　97223
クンシラン属　97213
グンシンマル　284664
グンセイカン　159880
クンソウギョク　182468
グンダイモチヅミ属　19498
グンダチュウ　61166
クントウマル　234903
クントウ属　400671
グンナイフウロ　174595,174779
グンネラ科　181960
グンネーラ属　181946
グンバイウチハ　390213
グンバイウメヅル　182683
グンハイキダチベンケイ　356970
グンバイシナノキ　391740
グンバイヅル　407261
グンバイドコロ　131458
グンバイナズナ　390213
グンバイナズナ属　390202
グンバイナヅナ　390213
グンバイナヅナ属　390202
グンバイヒルガオ　208067,208069
クンビメイ　9051
グンホウギョク　43499
グンボウギョク　43495
クンポウマル　327312
グンモウゾウ　253833
グンリュウ　245155
クンリュウギョク　264362
グンレイ　264813
クンレイマル　244090
ケアイアシ　337555
ケアキノウナギツカミ　291940
ケアクシバ　403870
ケアサガラ　320885,320886
ケアブラチャン　231417
ケアメリカシモツケ　297842
ケアリタソウ　139678
ケイ　91302
ケイガイ　264958
ケイシヤ属　88696
ケイジュ　233949
ケイセキマル　264337
ケイソスミレ　410018
ケイタオクシノハラン　63159
ケイタオタムラサウ　345077
ケイタオトキハアケビ　374417
ケイタオフウラン　390527
ケイタオミズ　288729
ケイタドリ　162539
ケイチマンネングサ　357282
ケイトウ　80395
ケイトウ属　80378
ゲイナイフウロ　174684
ケイヌツゲ　204184
ケイヌビエ　140382
ケイヌビハ　164671
ケイヌビワ　164671
ケイヌホオズキ　367583
ケイノコヅチ　4259,4263
ケイバ属　80114
ケイビアナナス　301111
ケイビアナナス属　301099
ケイヒジュ　91366
ケイボク　229472
ゲイホクスゲ　74636
ケイマ　117597,223150
ケイモ　131501
ケイラン　116829
ケイランサウ　86427
ケイランソウ　86427
ケイリドブシス属　86480
ケイリュウタチツボスミレ　410049
ケイリンギボウシ　198635
ケイリンサイシン　37637,37638
ケイリンシロタンポポ　384505
ケイル　59524
ゲイル－サキア属　172248
ケイレイ　82300
ケイレイマル　140881
ケイワオウギ　188179
ケイワガラミ　351796
ケイワタバコ　101684
ケウオクサギ　313702
ケウシノケグサ　164139
ケウスノキ　403886
ケウツギ　413624
ケウバメガシ　324286
ケウラゲエンコウカエデ　3407
ケウリクサ　231594
ケウワミズ　280007
ケウワミズザクラ　280007
ケエゾキスミレ　409772
ケエゾノジャニンジン　72966
ケエゾノヨロイグサ　24455
ケエゾヤマザクラ　83304,316777
ケエダウチシャジン　7888
ケエリカ　149195
ケエンコウカエデ　3406
ケエンビセンノウ　363472
ケオオクマヤナギ　52456
ケオオタチツボスミレ　410157
ケオオツヅラフジ　364989,364991
ケオオバイボタ　229569,229573
ケオオバタチツボスミレ　410172
ケオオバハシドイ　382281
ケオカイボタ　229579
ケオトメイヌゴマ　373365
ケオトメサナエタデ　309542
ゲオドルム属　174299
ケオニノゲシ　368650
ケオノエスゲ　76516
ケオノマテ－ブルヤシ　143712
ケオリイチゴ　338097
ケオロシマチク　304083,347279
ケオンラ　93435
ケガキ　132351
ケカキネガラシ　365564
ケカタバミ　277773
ケカナダアキノキリンソウ　368026

ケカニア　224600
ケガマズミ　407705
ケカメバソウ　397459
ケカモジグサ　335258
ケカモノハシ　209255
ケカラスウリ　396151,396238
ケカラスザンショウ　417146
ケカリマタガヤ　131008
ケカリヤスモドキ　255868
ケカワラハンノキ　16448
ケカンアフヒ　37585
ケカンコノキ　177100,177194
ケカンボク　408012
ケキツネノボタン　325697
ケキブシ　373564
ケキリンソウ　294126
ゲキリンリュウ　158649
ケキンポウゲ　325743
ケキンミズヒキ　11587
ケクサトベラ　350346
ケクサフジ　408360
ケクサフヂ　408360
ケクマヤナギ　52456
ケクララ　369134
ケクリンユキフデ　54818
ケクロイゲ　342206
ケクロツバラ　328685
ケクロモジ　231433
ケグワ　259122,259195
ケケンポナシ　198784,198786,198789
ケコマイワヤナギ　344051
ケコマユミ　157285
ケゴン　183372
ケゴンアカバナ　146605
ケコンニャク　20090
ケコンロンカ　260418,260483
ケコンロンクワ　260483
ケザサ　347275
ケサナギイチゴ　339135
ケサフジ　408352
ケサフヂ　408352
ケサヤバナ　99581,99711
ケザルピニア　64982
ケザルピニア属　64965
ケサンカクズル　411700
ケサンカクヅル　411700
ケシ　282717
ゲシアザミ　368771
ケシオガマ　287600
ケジギタリス　130369
ケシバナアオイ　67128
ケシバナアオイ属　67124
ケシバニッケイ　91323
ケシベツツジ　331062
ケシベリヤハンノキ　16359
ケシマイワブキ　349843
ケシモツケ　371977
ケシモドキ　201605
ケシモドキ属　201597
ケシャウビユ　208349
ケシャウヤナギ　89160
ケジャク　28019,28046
ケジャコウソウ　391264
ケジャニンジン　72835
ケショウアザミ　92083
ケショウザクラ　314613
ケショウサルビア　345023,345026
ケショウビユ　208349
ケショウボク　121933
ケショウボク属　121912
ケショウヤグルマハッカ　257184
ケショウヤナギ　89160
ケショウヤナキ属　89159
ケショウヨモギ　35430,35794
ケショカツサイ　275882
ケシライワシャジン　7847
ケシロスミレ　410360
ケシロネ　239222
ケシロヨメナ　39965
ケシンテンルリミノキ　222123
ケシ科　282751
ケシ属　282496
ケスエコザサ　347355
ケスゲ　74415,75785
ケスズ　347303
ケスズシロソウ　30264
ケスナヅル　78738
ケスネリア属　324577
ゲスネリア属　175295
ケスモモ　316767
ケセンナ　360439
ケソケイ　211912
ケソバナ　7800
ケタイワンホトトギス　396599
ケタカサゴヤマシロギク　39966
ケタカネイブキボウフウ　228609
ケタカネイワヤナギ　343750
ケタガネソウ　74133,76268
ケタチツボスミレ　410033
ケタデ　308877
ケタテヤマキンバイ　362344,362361
ケタニタデ　91543
ゲタペルチャ　280441
ケタマイヌガラシ　336183
ケチヂミザサ　272684
ケチドメグサ　200281
ケチョウセンアサガオ　123065,123092
ケヅ　301776
ゲッカカウ　307269
ゲッカカウ属　307265
ゲッカギョク　287909
ゲッカコウ　307269
ゲッカコウ属　307265
ゲッカビジン　147291
ゲッカンマル　182440
ゲッキツ　260173
ゲッキツ属　260158
ゲッキュウデン　243977
ゲッキュウデン属　243974
ケツクシテンツキ　166258
ケックバネウツギ　182
ゲッゲイカズラ　390822
ゲッゲイカヅラ　390822
ゲッケイジュ　223203
ゲッケイジュ属　223163
ゲッコウ　108844
ゲッタウ　17774
ケットウ　155398
ゲットウ　17774
ゲットジ　215277
ケツメクサ　311423
ケツメグサ　311915
ケツユクサ　100966
ケツヨウボク　158700
ケヅリバウブウ　292892
ケヅリボウフウ　292892
ケツルアズキ　408982
ケツルノゲイトウ　18124
ケツルマサキ　157519
ケテイカカズラ　393657,393664
ケテイカカヅラ　393657,393664
ケテウセンアサガホ　123065
ケテリハナゴニア　50076
ケテンツキ　166266
ケトイタンポポ　384601
ケトウウチソウ　345836
ケトダシバ　37379
ケトチノキ　9732,9733
ケドノロキ　311562
ケドンドン　71027
ケドンドンアラス　372478
ケナガエサカキ　8278
ケナガバノタチツボスミレ　410333
ケナガバヤブマオ　56169
ケナガボノシロワレモコウ　345927
ケナシアイズシモツケ　371877
ケナシアイヅシモツケ　371877
ケナシアオハダ　204008
ケナシアブラガヤ　353750,353856
ケナシイトバシャジン　7828
ケナシイトフスマ　31969
ケナシイヌゴマ　373262
ケナシイワハタザオ　30442
ケナシウチワドコロ　131736
ケナシウリハダカヘデ　3665
ケナシエゾオオバコ　301895
ケナシエゾニワトコ　345674
ケナシエゾノタチツボスミレ　409633
ケナシエゾノマルバシモツケ　371829
ケナシエゾノヨツバムグラ　170436
ケナシエゾミソハギ　240072

ケナシエドスミレ　410531
ケナシエビヅル　411671
ケナシオオバスノキ　404003
ケナシオニノゲシ　368649
ケナシカシダザサ　347265
ケナシカンボク　408009
ケナシクモマナズナ　137145
ケナシコウモリカズラ　250228
ケナシコフウロ　174979
ケナシサクラスミレ　410074
ケナシシナノキ　391743
ケナシシモツケ　371972
ケナシシラオイハコベ　374881
ケナシシロシャクヤク　280242
ケナシスノキ　404005
ケナシスミレ　410201
ケナシタマガサ　364768
ケナシチガヤ　205497
ケナシチャンパギク　240766
ケナシツルカコサウ　13097
ケナシツルカコソウ　13097
ケナシツルフジバカマ　408262
ケナシツルモウリンカ　400973
ケナシトウササクサ　236296
ケナシニオイタチツボスミレ　410317
ケナシニハトコ　345682
ケナシニワトコ　345682
ケナシニンドウモチノキ　203995
ケナシノジスミレ　410770
ケナシハイチゴザサ　209087
ケナシハクサンオオバコ　301998
ケナシハクサンシャクナゲ　330241
ケナシハテルマカズラ　399304
ケナシハヒスミレ　409911
ケナシハマハタザオ　30461
ケナシハマヒサカキ　160462
ケナシハルガヤ　27971
ケナシヒメムカシヨモギ　103446,103449,103554
ケナシベニバナヤマシャクヤク　280248
ケナシマンシウスミレ　409914
ケナシミヤマカンスゲ　75444
ケナシムシクサ　407275
ケナシヤシャビシャク　333915
ケナシヤブデマリ　408031
ケナシヤマムグラ　170560
ケナシヨツバムグラ　170686
ケナフ　194779
ケナミキソウ　355780
ケニオイグサ　187681
ケニガキ　298517
ケニクキビ　58155
ケニゴシ　208016
ケニシキギ　157285,157310
ゲニスタ属　172898
ケニホヒグサ　187681
ケニワトコ　345686
ケヌカキビ　128516
ケネコノメソウ　90429
ケネザサ　304083,304097,347279
ケネズミモチ　229506,229587
ケネバリタデ　309946
ケノコギリサウ　3997
ケノハラガラシ　364563
ケノボリフジ　238450,238467
ケバイカウツキ　294531
ケバイケイソウ　405585
ケハウチノカエデ　3043
ケハウチハカエデ　3043
ケハウチワカエデ　3043
ケハウチワマメ　238467
ケハガクレスゲ　74920
ケハギ　226912
ケハクヨウ　311530
ケハスノハカヅラ　375870
ケハダルリミノキ　222123,222138
ケバナカメモチ　295778
ケバノオオバコ　301873
ケバノキツネノメマゴ　337256
ケバノクロウスゴ　403938
ケハヒキビ　281370
ケハヒノキ　381272
ケハマエンドウ　222738
ケハマササゲ　409008
ケハマセンダン　161342
ケハマニンドウ　235646
ケハマムギ　228384
ケハリギリ　215443
ケハリノキ　16466
ケハンノキ　16393
ケヒエスゲ　75318
ケヒサカキ　160513
ケヒナマツヨイグサ　269486
ケヒメシャジン　7708
ケヒメノボタン　276914
ケファロセレウス属　82204
ケブカアキノタムラソウ　345113
ケブカキツネノメマゴ　337256
ケブカツルカコソウ　13185
ケブカルイラソウ　339816
ケフシグロ　248269,363451,363454
ケープスイレン　267665
ケフチクタウ　265327
ケフチクタウ科　29462
ケフチクタウ属　265310
ケープヒルムシロ　29660
ケープフクシア　296110,296112
ケベヤ　315560
ケベヤ属　315547
ケホオズキ　297711
ケホオヅキ　297711
ケホソバシモツケ　371962,371979
ケボタンヅル　94969
ケホホヅキ　297711
ケマアザミ　91705
ケマキヤマザサ　347243
ケマシテ属　77252
ケマツカゼサウ　56382
ケマツカゼソウ　56382
ケマルバスミレ　410131
ケマルミスズメノヒエ　285520
ケマンサウ　220729
ケマンソウ　128321,220729
ケマンソウ科　169194
ケマンネングサ　356685
ケミズキンバイ　238152
ケミツデイワガサ　371897
ケミノゴメ　49544
ケミヤガラシ　326833
ケミヤマキンバイ　312746
ケミヤマタニタデ　91511
ケミヤマトラノオ　317962
ケミヤマナミキ　355762
ケミヤマルリミノキ　222138
ケミリ　14544
ケムシクサ　407275,407283
ケムラサキ　66829,66856,66873
ケムラサキニガナ　283689
ケムリノキ　107300
ケモウコオウギ　104638
ケヤキ　417558
ケヤキ属　417534
ケヤナギハグロ　200641
ケヤナギラン　85880
ケヤブハギ　200740
ケヤマウコギ　143588,143591
ケヤマウツボ　222645
ケヤマザクラ　83319,316913
ケヤマシャクヤク　280209
ケヤマゼリ　276769
ケヤマハンノキ　16359,16466
ケヤマヤナギ　52461
ケヤリギボウシ　198620
ケヤリスゲ　74483,76312,76656
ケヨウチクトウ　265327
ケヨシ　295889
ケヨノミ　235691,235702
ケラジ　93519
ケラトスティリス属　83651
ケラトペタ-ルム属　83533
ケラノキ　203422
ケラパ　98136
ケラマツツジ　331759
ゲラム　248104
ゲリンクオニソテツ　145226
ケール　59524
ゲルセミウム属　172778
ゲルリカポプラ　311150
ケルリソウ　397459
ケルレンステイニア属　217593
ケレタケ　297203

ケロクラミス属　83967
ケロフィルム属　84723
ゲロンガン　110246
ケワチクタウ　265327
ケワレモコウ　345890
ケヰノコヅチ　4259,4263
ゲンウン　249597
ゲンカイイワレンゲ　275379
ゲンカイツツジ　331296
ゲンカイミミナグサ　82817
ゲンカイモエギスゲ　74596
ゲンカイヤブマオ　56222
ケンカヅラ　126007
ゲンゲ　43053
ゲンゲツ　216637
ゲンゲバナ　43053
ゲンケ属　41906
ゲンゲ属　41906
ケンサキ　171253
ケンサキタンポポ　384492
ケンザンノシマ　139263
ゲンジギョク　233526
ゲンジスミレ　410719,410721
ケンシティア属　215993
ゲンジバナ　392884
ゲンジユリ　229803
ゲンショカン　93463
ゲンジン　355201
ゲンソウチョウ　256264
ケンタッキ・ブルー・グラス　305855
ケンダルサウ　172832
ゲンチアナ　173610
ケンチヤモドキ属　216005
ケンチャヤシ　198806
ケンチャヤシ属　198804
ケンチャ属　198804
ケントクスミレ　410076
ゲンナイフウロ　174589
ゲンノショウコ　174755,174966
ゲンパ属　172888,201647
ゲンペイカズラ　96387
ゲンペイクサギ　96387
ゲンペイコバノミッバッツジ　331629
ゲンペイダモ　233920
ケンボウマル　140875
ケンポウマル　327310
ケンポナシ　198769
ケンポナシ属　198766
ゲンポール　262823
ケンマギョク　182437
ゲンラク　317687
ケンラン　116781,295465
ケンランギョク　233606
ケンリュウ　340922
ケンロクヒサカキ　160518
コアカザ　86977,87007
コアカソ　56318
コアカゾ　56181
コアサダ　276809
コアザミ　92066
コアジサイ　199897
コアゼガヤツリ　118982,118990
コアゼスゲ　76547
コアゼテンツキ　166164
コアヂモ　418389
コアツモリ　120341
コアツモリサウ　120341
コアツモリソウ　120341
コアニチドリ　19516
コアハ　361794,361798
コアブラツツジ　145708
コアマチャ　200100
コアマモ　418389,418396
コアメリカスズメノヒエ　285463
コアヤメ　208829,208918
コアラセイトウ　246479
コアワ　361795,361798
コイガガシ　233121
コイケマ　117751
コイシヤ属　88696
コイチゴツナギ　305451,306098
コイチジク　164947
コイチヂク　164947
コイチヤクサウ　275486
コイチヤクソウ　275486
コイチャクソウ属　275480
コイチヨウラン　146308
ゴイッシングサ　103446
コイツミカマツカ　322471
コイテョウラン属　146305
コイヌガラシ　336187
コイヌノハナヒゲ　333575
コイヌノヒゲ　151288
ゴイノキ　223150
コイハカンスゲ　74122
コイハハタザホ　30437
コイブキアザミ　91892
コイマギョク　264344
コイワウチワ　362273
コイワカガミ　351451
コイワカンスゲ　74122
コイワザクラ　314885
コウアンアキノキリンソウ　368502
コウアンイワタデ　308730
コウアンカノコソウ　404316
コウアンカモジグサ　144320,335547
コウアンキクボクチ　361011
コウアンキンポウゲ　326375
コウアンコウゾリナ　298618
コウアンシャジン　7855
コウアンシラカンバ　53572
コウアンスズメノテッポウ　17515
コウアンスナヂヒジキ　104758
コウアンスミレ　410010
コウアンゼキショウ　213552
コウアンソウ　287645
コウアンチャヒキ　190206
コウアンドロ　311509
コウアンナデシコ　127682
コウアンニンジン　292780
コウアンヌマヤナギ　343745
コウアンヒメオノオレ　53444
コウアンヒメタンポポ　384548
コウアンフシグロ　364072
コウアンフスマ　375036
コウアンボウフウ　299555
コウアンホソバヨモギ　35648
コウアンムシャリンドウ　137617
コウアンヤマカモジグサ　58604
コウアンヨメナ　215339
コウエフヅタ　285157
コウエンマル　234957
コウオウカ　221238
コウオウカ属　221229
コウオウソウ　383103
コウガイゼキショウ　213213
コウガイモ　404544
コウカイヤナギ　343560
ゴウカンボク　13578
コウカンマル　267049
コウキクサ　224375
コウキシタン　320327
コウキヤガラ　353707,353828
コウギョ　86544
コウギョク　357111
コウグイスカグラ　236053
コウゲ　143022
コウコウマル　140846
コウサイカク　158829
コウサイチュウ　94554
コウサカツリフネ　205176
コウザンアセビ　298722
コウジ　31408,31502,93530
コウシクグ　119319
コウシバ　204583
コウジバナ　231403
コウシマル　103759
コウシャクギョク　233658
ゴウシュウアオギリ　58335
ゴウシュウアリタソウ　87142
コウシュウウヤク　97919
コウシュウギョク　284696
ゴウシュウコモウセンゴケ　138354
コウシュウバイ　316570
コウシュウヒゴタイ　348127
コウシュウブナ　204008
ゴウシュウヤブジラミ　123180
コウジュラク　317688
ゴウシュンアデク　382589
コウシュンイヌビハ　165203
コウシュンウマノスズクサ　34382

コウシュンウンヌケ　156471
コウシュンカウバシ　231324
コウジュンカウバシ　231324
コウシュンカズラ　398684
コウシュンカズラ属　398681
コウシュンカヅラ　398684
コウシュンクスノキ　240658
コウシュンゲッタウ　17674
コウシュンコマツナギ　206018
コウシュンシバ　418432
コウシュンシュスラン　25986
コウシュンスゲ　327664
コウシュンスゲ属　327663
コウシュンスラン　25986
コウシュンタヌキマメ　112238
コウシュンツゲ　123451
コウジュンツゲ　123456
コウシュンツゲ属　123443
コウシュンツュクサ　50939
コウシュンテッボク　138649
コウシュンフジマメ　135649
コウシュンムラサキ　66909
コウシュンモダマ　145894
コウシュンヤマハヅ　28341
コウシュンヤマモモ　261122
ゴウショウアオギリ属　376063
コウショウマル　45892
ゴウショウマル　140849
コウショウリュウ　163480
コウジロキツ　93464
コウシンソウ　299754
コウシンテツカエデ　3250
コウシンバラ　336485
コウシンマル　247672
コウシンヤマハッカ　209857
コウスイガヤ　117218
コウスイギョク　102657,102658
コウスイソウ　190559,190702
コウスイハッカ　249504
コウスイボク　190559,190702,232483,232532
コウスノキ　403854
コウスユキソウ　224883
コウセンガヤ　88396
コウゾ　61103
ゴウソ　75314
ゴウソウリュウ　299255
ゴウソウマル　103763
ゴウソモドキ　73552
コウゾリナ　298589,298618
コウゾリナ属　298550
コウゾ属　61096
コウダソウ　238371
ゴウダソウ属　238369
コウダマル　182465
コウチク　104367
コウチニッケイ　91318
コウチマル　267030
コウヂムラサキ　66829
コウチョウゲ　360933
コウツギ　126929,127072
コウツボカズラ　264837
コウツボソウ　347153
コウテウゲ　360933
コウデンセンニンサウ　95350
コウトウアバカ　260236
コウトウアマナラン　55581
コウトウイチゴ　338433,338435
コウトウイチジク　164635
コウトウイナモリ　272272
コウトウイモ　99928,131529
コウトウイラクサ　125544
コウトウイラノキ　125544
コウトウウラジロマオ　244895
コウトウウラジロマオ属　244892
コウトウウラジロマヲ　244895
コウトウウラジロマヲ属　244892
コウトウエゴノキ　379357,379389
コウトウエノキ　80709
コウトウオホギミラン　269026
コウトウカラスウリ　396254
コウトウカンコノキ　177149,177150
コウドウカンザブラウ　381149
コウトウカンザブロウ　381143
コウトウギョクシンクワ　385049
コウトウクロダモ　233832
コウトウケマタケラン　136081
コウトウケマタケラン属　136072
コウトウサンイウクワ　327577
コウトウサンユウカ　327577
コウトウシウカイダウ　49827
コウトウシュウカイドウ　49827
コウトウシラキ　161666
コウトウシラン　370401
コウトウシラン属　370393
コウトウスイゼンジナ　183086
コウトウスヰゼンジナ　183086
コウトウセイシボク　28349
コウトウタケラン　29814
コウトウツゲ　204299
コウトウツヅラフヂ　375897
コウトウナタオレ　231739
コウトウナタヲレ　231739
コウトウナタヲレ属　231729
コウトウナハシログミ　141991,142075
コウトウニクズク　261412
コウトウノボタン　248727,248732
コウトウハイノキ　381143
コウトウバショウ　260236
コウトウバセウ　260236
コウトウハヒノキ　381149
コウトウヒスイラン　404652
コウトウピンポン　376092
コウトウフヂ　283282
コウトウベンケイ　215272
コウトウボタン　248727
コウトウボチャウジ　319477
コウトウマキ　306419
コウトウマサキ　182687
コウトウマメ科　101858
コウトウマメ属　337690
コウトウムラサキ　66831
コウトウモクレイシ　254303
コウトウヤツデ　276486
コウトウヤツデ属　56526
コウトウヤマイハヅノキ　28349
コウトウヤマバショウ　260236
コウトウヤマバセウ　260236
コウトウヤマヒハツ　28378,28381
コウトウユヅリハ　161666
コウノワタシ　323210
コウバイグサ　363265
コウフクマル　244124
コウブシ　119503
コウベクリノイガ　80826
コウベナヅナ　225475
コウボウ　27963,27967,196134,196153,340644
コウボウコウシュ　196155
コウボウシバ　75930
コウボウチャ　204008
コウボウビエ　143522
コウボウムギ　75003
コウボウモドキ　227948
コウボウモドキ属　227946
コウボウ属　196116
コウボク　198699
コウホネ　267286
コウホネ属　267274
コウボフシバ　75930
コウボフスゲ　75259
コウボフチャ　78429
コウボフビエ　143522
コウボフムギ　75003
コウマ　104072
コウマゴヤシ　247381
コウマノケン　266641
コウマノスズクサ　34154
コウメ　83238,316570
コウメバチソウ　284601
コウモウチュウ　395655
コウモリカエデ　3529
コウモリガサソウ　181958
コウモリガサソウ属　181946
コウモリカズラ　250228,250229
コウモリカズラ属　250218
コウモリソウ　283794,283844
コウモリソウ属　283781
コウモリドコロ　131734
コウヤカミツレ　26900
コウヤカミルレ　26900

コウヤガヤツリ　118432
コウヤカンアオイ　37657
コウヤグミ　142122
コウヤザサ　58462
コウヤボウキ　292055
コウヤボウキ属　292040
コウヤマキ　352856
コウヤマキ科　352851
コウヤマキ属　352854
コウヤマンサク　160347
コウヤミズキ　106644
コウユマル　322991
コウユリ　229966
コウヨウザン　114539
コウヨウザン属　114532
コウヨウズタ　285117
コウヨウヅタ　285157
コウライアセスゲ　76178
コウライアマナ　169460
コウライイ　212915
コウライウシノケグサ　164159
コウライオオニガナ　261926
コウライカキドオシ　176839
コウライカノコソウ　166093
コウライカライトソウ　345851
コウライギク　89481
コウライギシギシ　340178
コウライキバナタンポポ　384719
コウライキビ　417417
コウライコモチマンネングサ　357102
コウライシダレヤナギ　343918
コウライシバ　418439,418453
コウライスズムシソウ　232205
コウライダコ　16525
コウライタチツボスミレ　410517
コウライタチバナ　93648
コウライタデスミレ　410758
コウライチドリ　302331
コウライテンナンショウ　33260,33496
コウライニリンソウ　23713
コウライニワフヂ　206140
コウライヌカボシソウ　238691
コウライハマムギ　203126
コウライバラ　336660
コウライヒトツバヨモギ　36469
コウライヒメノダケ　24315
コウライホソバユリ　229803
コウライメタカラコウ　229060
コウライヤナギ　343574
コウライヤマジノギク　193973
コウライヤワラスゲ　74685
コウラインシバ　418432
コウラジロエゴノキ　379348
コウリカエデ　2910
コウリバヤシ　106999
コウリバヤシ属　106984
コウリャン　369671
コウリュウ　192438
ゴウリュウギョク　264365
コウリュウトダシバ　37379
コウリュウマル　182472
コウリョウカモジグサ　144365
コウリョウキョウ　17733
コウリョクチュウ　395647
コウリンカ　385894
コウリンギク　123843,220523,358292
コウリンギョク　182436
コウリンキンポウゲ　326168
コウリンクワ　385892
コウリンタンポポ　195478
コウリンホウ　257398
コウリンマル　244035
コウリンリュウ　299249
コウルメ　408151
コウレイマル　140865
コウワウクワ　221238
コウワウクワ属　221229
コウワウサウ　383103
コエゾシモツケ　372009,372012
コエゾツガザクラ　297021
コエゾヤナギ　344028
コエビソウ　66725,214379
コエビソウ属　50922
コエビヅル　411590
ゴエフアケビ　13221
ゴエフイチゴ　216085
ゴエフツツジ　331593
ゴエフテンナンシャウ　33337
ゴエフトガ　318575
ゴエフフヂ　310962
ゴエフマツ　300137
ゴエフムグラ　170352
コエンドウ　222727
コエンドロ　104690
コエンドロ属　104687
コオウレン　298670
コオシュンカズラ　398684
コオシンソウ　299754
コオトオシュウカイドオ　49827
コオトオヤマヒハツ　28381
コオトギリ　201908
コオトギリソウ　201908
コオニガヤツリ　118438
コオニシバ　418445
コオニタビラコ　221784
コオニビシ　394516
コオニユリ　229887,229888
コオノオレ　53400
コオヒョウモモ　20970
コオヤボオキ　292055
コオライタチバナ　93648
コオライヒメヒゴタイ　348690
コオランダイチゴ　167674
コオリンタンポポ　195478
コオロギラン　376578
コオロギラン属　376576
コカ　155067
コカエフコバンノキ　296695
ゴカエフワウレン　103843
コカキツバタ　208797,208801
コガクウツギ　199950
コカゲトンボ　291224
コカゲラン　130034
コ－カサスキリンソウ　357169
コ－カサスマツムシサウ　350119
コ－カサスマツムシソウ　350119
コガシアズマザサ　347350
コ－カシアムシヨケギク　89472
コガシザサ　347350
コガシハ　323814,324464
コガシワ　324464
コカタシ　69594
コガタブ　233876
ゴガツイチゴ　338886
ゴガツササゲ　294056
コカナダモ　200190
コガネアヤメ　208562
コガネイチゴ　339018
コガネウマゴヤシ　247457
コガネエンジュ　52151
コガネガヤツリ　119624
コガネギク　368073,368486
コガネギシギシ　340116
コガネサイコ　63571,63741
コガネシダレ　342926
コカネシモッケ　297845
コガネシロダモ　264021
コガネススキ　255886
コガネタケヤシ　89360
コガネタヌキマメ　111915,112615
コガネナズナ　154533
コガネネコノメソウ　90432
コガネバナ　237554,355387
コガネヒルガオ属　250749
コガネホウチャクソウ　134467
コガネマンサク　185116
コガネヤナギ　355387
コガネユウコクラン　232166
コカノキ　155067
コガノキ　91351
ゴカノキ　233876
コカノキ科　154952,155048
コガノヤドリギ　355337
コガマ　401129,401134
コカモメズル　400888
コカモメヅル　400888
コガヤツリ　119208
ゴカヨウオウレン　103843
コカラスザンショウ　417229
コカラスムギ　45454
コカラマツ　388584

コカリヤス 156490
コガリヤス 127257
コカンスゲ 75992
コガンピ 414177
コカ科 155048
コカ属 155051
コキクイモ 188936,188941
コキクボクチ 361011
コキクモ 230307,230328
コギシギシ 340019,340023
コキツネノボタン 325718
ゴキヅル 6907
ゴキヅル属 6888
コギノコ 61935
コキビ 281916
ゴギョウ 178297
コギョクシンクワ 384950
コキリカズラ 246656
コキンスゲ 75362
コキンチュウ 395648
コキンバイ 413034
コキンバイザサ 202812
コギンバイザサ科 202786
コキンバイザサ属 202796
コキンバイ属 413030
コキンパウゲ 326431
コキンポウゲ 326431
コキンレイカ 285875
コクウンリュン 284702
コクエイマル 247669
コクオウデン 103771
コクオウマル 103745
コクガリュウ 209512
コクカンマル 263750
コクサギ 274277
コクサギ属 274276
コクサノオウ 86755
コクシカン 103752
コクセンギョク 103742
コクタン 132137
コクタンノキ 157355
コクチナシ 171253,171374
コクチョウギョク 182427
コクテンギ 157355,157910
コクテンギョク 198498
コクドノカシ 346390
コクドノクワシ 346390
コクニ- 93556
コクホウ 395660
コクボタン 337156
コクヨウギョク 233661
コクライギョク 198496
ゴクラクチョウカ 377576
ゴクラクチョウカ科 377580
ゴクラクチョウカ属 377550
コクラン 232252
コクランジョウ 103770
コクリオ-ダ属 98085
コクリュウ 320260
コグルミ 320372
コクレア-リア属 97967
コクロギョク 264369
コクワ 6515,6516,6523
コクワガタ 407234
コクワガタソウ 407054
コゲ 143022
コケイラン 274058
コケイラン属 274033
コケオトギリ 201942
コケゲイトウ 294882
コケコゴメグサ 160199
コケサンゴ 265358
コケシコタンソウ 349078
コケスミレ 410744
コケセンボンギク 219898,219900,219901
コケセンボンギク属 219892
コケタンポポ 367848
コケツメクサ 32301,255598
コケツルマオ 313424
コケツルマヲ 313424
コケツルマヲ属 313411
コケトオバナ 97037
コケヌマイヌノヒゲ 151463
コケノミ 145073
コケバラ 336476
コケマンテマ 363144
コケマンネングサ 109463
コケミズ 299006,299010
コケミヅ 299006
コケミミナグサ 82772
コケムグラ 125872
コケモモ 404051
コケモモイタビ 165814
コケモモカマシカ 107575
コケモモ科 403706
コケモモ属 403710
コケリンダウ 173917
コケリンドウ 173917
コケリンドオ 173917
コゲンゲ 279014
ココアノキ 389404
コ-コス属 98134
ココス属 98134
ココノエカズラ 57868
ココノエギリ 285960
ココノエタマアジサイ 199916
ココノエノキリ 285960
ココノヘカツラ 57868
ココノヘノキリ 285960
コゴメアゼスゲ 75357
コゴメアマ 128026
コゴメアマ属 128023
コゴメイチゴツナギ 25343
コゴメイヌガラシ 336279
コゴメイヌノフグリ 407096
コゴメイワガサ 371837
コゴメウツギ 375817
コゴメウツギ属 375805,375811
コゴメウナギツカミ 309926
コゴメオウギ 43008
コゴメオドリコソウ 220126
コゴメカゼクサ 147746
コゴメガヤツリ 119041
コゴメギク 170142
コゴメギシギシ 340019,340116
コゴメキノエラン 232364
コゴメキノヘラン 232151
コゴメギョリュウ 327226
コゴメグサシホガマ 287194
コゴメグサ属 160128
コゴメスゲ 73952,76651
コゴメスズメノヒエ 285476
コゴメスナビキサウ 190738
コゴメセンブリ 81520
コゴメタツナミソウ 270758
コゴメタツナミソウ属 270757
ココメタデ 309054
コゴメツツジ 331239
コゴメナデシコ 183222,183225
コゴメナナカマド 369462
コゴメヌカボ 164025
コゴメヌカボガヤ 321397
コゴメヌカボシ 238678,238684
コゴメバオトギリ 202087
コゴメハギ 249203
コゴメバクロウメ 328816
コゴメバナ 94032,372050,372101
コゴメビエ 285379
コゴメビユ 192965
コゴメヒョウタンボク 235902
コゴメヘウタンボク 235902
コゴメマンネングサ 356846
コゴメミズ 298974
コゴメミヅ 298974
コゴメヤナギ 343342,344090
ココヤシ 98136
ココヤシ属 98134
ゴゴレンシ 45725
ココンリン 116779
コサ 117597
ゴサイカク 159705
ゴサイバ 1000,243371
コササガヤ 291405
コササガヤ属 291396
コササキビ 361868
コサハフタギ 381341
コサルスベリ 219909
コサンチク 297203
ゴサンチク 297203
ゴジアオイ 93124
コシアオイ属 93120

ゴジアフイ　93124
ゴジアフイ属　93120
コシアブラ　86855
コジイ　78903
コシオガマコシホガマ　296068
コシオガマ属　296061
ゴジカ　289684
コシカギク　246375
コシガヤホシクサ　151321
ゴジカ属　289682
コジキイチゴ　339335
コシキギク　39915
コジキグミ　235812,235813
ゴシキシャクナゲ　330184
コシキトウガラシ　72070
ゴシキトウガラシ　72072,72073
コシキブ　65670
ゴシキヤバネバショウ　66175
ゴジクワ　289684
ゴジクワ属　289682
コシジシモツケソウ　166076
コシジタネツケバナ　72909
コシジタビラコ　397387
コシジドコロ　131842
コシジヒキオコシ　209616
コジシマル　167704
コシジミバナ　371906
コシデ　77283,77323
コシナガワハギ　249218
コシノカンアオイ　37681
コシノチャルメルソウ　256028
コシノネズミガヤ　259646
コシノヒガンザクラ　83330
コシノホンモンジスゲ　76367
コシホガマ　296068
コシホガマ属　296061
コシマゴショウ　290399
コシミノナズナ　225428
コシモツケ　371947,372029
コシャク　28030
コジャク　28044
ゴシャメンニンジン　292892
コシャラ　376457
ゴジュウノトウ　17175
コジュズスゲ　75699
ゴシュユ　161373
ゴシュユ属　161302
コシュロガヤツリ　118481,119693
ゴショイチゴ　338250
コショウ　300464
ゴショウイモ　367696
コショウソウ　225450
コショウソウ属　225279
コショウノキ　122482
コショウボク　273903,350990
コショウボク属　350980
コショウ科　300557
コショウ属　300323
ゴショモミジイチゴ　338061
コシロガヤツリ　119196
コシロネ　239194,239238
コシロノセンダングサ　54054
コシロバナスミレ　409723
コシンジュガヤ　354185,354260
コーズ　61103
コスギニガナ　210551
コスズメガヤ　147815,147891
コスズメノチャヒキ　60760
コスタス属　107221
コスタリカバンジロウ　318740
コスブタ　55918
コスミレ　410108
コスモス　107161
コスモス属　107158
コセウ　300464
コセウサウ　225450
コセウサウ属　225279
コセウソウ属　225279
コセウナ　326340
コセウノキ　122482
コセウボク　273906
コセウモドキ　300495
コセウ科　300557
コセウ属　300323
コセキヤナギ　342947
コゼンギョク　103765
ゴゼンタチバナ　85581
ゴゼンタチバナ属　85579
コセンダングサ　54048
コーゾ　61103
コソツブスゲ　74669
コソネ　77323
コゾノキ　61103
ゴーダー・ビーソ　396151
コダイコンソウ　175418
コダイサンチク　125447
コダイホガシ　285233
コタイワンアイアシ　209251
コタイワンシウメイギク　23853
コダカラベンケイ　61568
コダチアサガオ　207792
コダチクリスマス・ローズ　190936
コダチダリア　121556
コダチチョウセンアサガオ　61231,61235
コタチツボスミレ　410042
コダチトマト　119974
コダチトマト属　119973
コダチノニワフジ　206005
コダチハゼリソウ　414095
コダチハゼリソウ属　414092
ゴダチヒメギリソウ　377747
コダチボタンボウフウ　292895
コダチヤハズカズラ　390759
コタツナミソウ　355510
コタヌキモ　403220
コタヌキラン　74368
コダネクマデヤシ　390460
コダネクロッグ　32346
コタネツケバナ　72920,72924
コダマギク　21645
コダマギグ　21643
コタンポポ　384605
コチヂミザサ　272693
コチナス　107300
コチニールウチワ　272830
コチヌス属　107297
コチャガヤツリ　118493
コチャルメルソウ　256036
コチョウインゲン　408832
コチョウスミレ　409624
コチョウセッコク　125304
コチョウセンナ　78336
コチョウソウシザンサス　351345
コチョウノキ　3243
コチョウラン　293582
コチョウラン属　293577
コチョウワビスケ　69754
コチョオボク　49247
コチレドン属　107840
コツガザクラ　297017
コッカノキ　346747
コッカノキ属　346723
コツギグリア　107312
コツクバネ　177
コツクバネウツギ　167
コツゲ　64303,64316
コッコロバ属　97861
コヅタ　165515
コヅチ　102710
コツノハシバミ　106778
コツバキ　69594
コッパダミネバリ　53465
コツブアオイ　37697
コツブウキクサ　414653
コツブガヤ　393079
コツブキリハヤシ　181296
コツブキンエノコロ　361753,361847
コツブスゲ　74982
コツブチゴザサ　209060
コツブヌマハリイ　143287
コッブモエギスゲ　76593
コツマトリソウ　396789
コツルウメモドキ　80168,80265
コツルマサキ　157503
コテフスミレ　410677
コテフセキコク　125304
コテフノキ　3243
コテフマメ　97203
コテフラン　293582
コテフラン属　293577
コデマリ　371868

コテンギョク　233513
コテング　186776
コテングクワガタ　407358
コトイトマル　244017
コトウカン　93526
ゴトウサウ　35981
ゴトウズル　200059
ゴトウヅル　200059
ゴトウボウフウ　292892
ゴトウヨモギ　35981
コドウラン　117090
コトカケヤナギ　311308
コトーカン　93526
コトジギク　340844
コトジソウ　345252
コドジョウツナギ　321221
コトツメギク　251194
コトネアスター　107486
コトネアスター属　107319
コトブキギク　396697
コトブキコヅチ　254469
コトモエソウ　201745
コトリトマラス　52246
コトリトマラズ　52225,122040
コナウキクサ　414653
コナガバヤブチョロギ　373324
コナギ　257583
コナシ　243717
コナシマンネングサ　356512
コナスビ　239687,239689
コナタウリ　114310
コナツツバキ　376457
コナツミカン　93834
コナミキ　355464
コナラ　323942,324384
コナラ属　323580
コナルコユリ　308529
コニガクサ　388303
コニクキビ　402512
コニコシア属　101798
コニシ　104690
コニシイヌグス　266792
コニシイモ　99918
コニシガシ　233278
コニシガマズミ　407759,407964
コニシカマツカ　107325,107516
コニシカンザブラウ　381148
コニシキサウ　159286
コニシキソウ　159286
コニシグス　113438
コニシグスモドキ　113438
コニシゴソズイ　160963
コニシタブ　266792
コニシタブ属　266787
コニシハイノキ　381143,381148
コニシハヒノキ　381148
コニタビラコ　221784
コニハザクラ　83220
コニハフヂ　206714
コニャク　20132
コヌカグサ　12118
コヌカグサ属　11968
コヌカゲサ　11972
コヌカシバ　99413
コヌカシバ属　99411
コヌカボタデ　309853
コヌマイヌノヒゲ　151399
コヌマスゲ　76073
コネス　408024
ゴネズ　408034
コネズミガヤ　259696
コネズミモチ　229569
コネソ　408024
ゴネヅ　408034
コノエ　376364
コノオレ　53400
コノゲスゲ　75957
コノッブスゲ　74664
コノテガシハ　302721
コノテガシハ属　390567
コノテガシワ　302721
コノフィツム属　102037
コノフィルム属　102017
コノボタン　279479
コノボタン属　279476
コーハイ　369130
コハイタヤメイゲツ　3609
コバイボタ　229561
コバイモ　168434
ゴバウ　31051
ゴバウアザミ　92186
コバウスノキ　403854
コハウチハカエデ　3604,3608
コハウチワカエデ　3604,3608
ゴバウ属　31044
コバガシハ　324173
コバガシワ　324173
コバギバウシ　198592,198618
コバギボウシ　198592,198618,198628
コハクウンボク　379450
コハクギョク　233468
コハクサンボク　407899
コハクラン　274047
コバケイスゲ　76518
コハコベ　374968
コバザクラ　83271
コバシジノキ　167923
コハシドイ　382257
コハズ　159222
コバストレリッチア　377565
コハゼ　379374
コバタゴ　167233,167235
コバタゴ属　167230
コバタチバナ　31414
コハテイシ　386500
コバテイシ　386500
コバテイシ属　386448
コバトベラ　301357
コバナアザミ　92094
コバナアヤメ　365767
コバナアリアゲキズラ　14876
コバナウチワカエデ　3604
コバナエゾキスミレ　409773
コバナカエデ　3147
コハナガサノキ　258924
コハナカモノハシ　209362
コバナカンアオイ　37699
コバナガンクビソウ　77160
コバナキジムシロ　312659,313047
コバナギジムシロ　312347
コバナキツリフネ　205181
コバナクマデヤシ　390463
コバナサンタンクワ　211096
コバナシミココヤシ　380556
コバナスヒカヅラ　235878
コバナダケ　47220
コバナタマザキセキコク　125361
コバナックバネウツギ　184
コバナツツアナナス　54451
コバナツルウリクサ　392884
コバナヌマダイコン　8016
コバナノガリヤス　65265
コバナノコウモリソウ　283798
コバナノセンダングサ　53797
コバナノタネツケバナ　72920
コバナノナナカマド　369462
コバナノワレモコウ　345921,345924
コバナハギカヅラ　169660
コバナハタザオ　136179
コバナヒナツメクサ　397053
コバナヒメハギ　308247
コバナミミアサガオ　207780
コバナムシャリンドウ　137624
コバナムラサキムカシヨモギ　406228,406664
コバナモクレイシ　254312
コバナヤブツバキ　69201
ゴバナレセダ　235090
コバナワレモコウ　345917
コバノアカダマカヅラ　258910
コバノアカテツ　301778
コバノアマチャ　199877
コバノアマミフユイチゴ　338125
コバノアリサンアイ　283211
コバノアリサンアヰ　283211
コバノイチヤクソウ　322780
コバノイヌシデ　77411
コバノイヌツゲ　203696
コバノイハタデ　309056
コバノイボタ　229561
コバノイラクサ　402946

コバノイワタデ 309056
コバノウシノシッペイ 191228
コバノエゾエノキ 80596
コバノエノキ 80695,80739
コバノオキナハイボタ 229525
コバノオランダガラシ 336232
コバノオレイフ 270110
コバノガマズミ 407802
コバノカモメヅル 117698
コバノカンアオイ 37753
コバノカンコノキ 177097,177184
コバノキササゲ 79250
コバノキブシ 373561
コバノキリンソウ 294119
コバノクスドイゲ 415869
コバノクハガタ 407259
コバノクロウメモドキ 328745
コバノクロヅル 398312
コバノクロマメノキ 404030
コバノコアカソ 56322
コバノコウオウカ 221284
コバノコゴメグサ 160212
コバノコショウソウ 225447
コバノコマユミ 157294
コバノゴムビワ 165586
コバノシコンノボタン 391577
コバノシチベンゲ 221284
コバノシナハシドイ 382132
コバノジニア 418057
コバノシノキ 391717
コバノシャリントウ 107549
コバノシラカンバ 53503
コバノシロバヒ 381423
コバノズイナ 210421
コバノズヰナ 210421
コバノセンキュウ 276755
コバノセンダングサ 53797
コバノセンナ 78308,360467
コバノダケカンバ 53425
コバノタゴボウ 238200
コバノタツナミ 355510
コバノタツナミサウ 355510
コバノタツナミソウ 355510
コバノタバノマメ 177792
コバノタマツバキ 229432
コバノタマボウキ 39120
コバノチウセンエノキ 80580
コバノチョウセンエノキ 80580
コバノツメクサ 255598
コハノツルウメモドキ 80285
コバノツルグミ 142007
コバノツルマサキ 157503,157512
コバノテリハドロ 311504
コバノトゲスグリ 334166
コバノトネリコ 168014,168018,168100
コバノトンボソウ 302538
コバノナナカマド 369401
コバノナンヤウスキ 30848
コバノナンヨウスキ 30848
コバノナンヨウスギ 30848
コバノニガイチゴ 338886
コバノニシキサウ 159299
コバノニハトコ 345695
コバノニワトコ 345667
コバノネコノメサウ 90362
コバノノダケ 121045
コバノハギ 126465
コバノハシドイ 382266
コバノハナイカダ 191173
コバノハマヂンチャウ 260720
コバノヒルギモドキ 238354
コバノヒルムシロ 312083
コバノフュイチゴ 339014
コバノブラッシノキ 248099
コバノブラッシノキ属 248072
コバノマンリャウ 31388
コバノミズキンバイ 238203
コバノミツバツツジ 331623
コバノミミナグサ 82835
コバノミヤマノボタン 59863
コバノムレスズメ 72285,72288
コバノモミジアサガオ 250821
コバノヤマウルシ 332901
コバノユキヤナギ 371807
コバノヨツバムグラ 170289
コバノヨモギ 36453
コバノヨロイグサ 24436
コバノランタナ 221284
コバノランタナー 221284
コバノリウキンクワ 68206
コバノワニグチ 259519
コバフンギ 394628,394635
コバベゴニア 49844,49845
コハマアカザ 44576
コハマアサガオ 207940
コハマアヅキ 408969
コハマキク 89429
コハマギク 89428
コハマササゲ 408950
コハマジンチョウ 260710
コハマナス 336652
コハマムギ 144266
コバムラサキシキブ 66827
コバモチ 203670
コバヤナギ 343745
ゴバラアザミ 92425
コバライチゴ 339194
コパラミツ 36924
コバラモミ 298354
コバラモンジン 354813
ゴハリキンギョモ 83575,83577
コハリスゲ 74737
コバンゴケ 146801
コバンコナスビ 239755
コバンサウ 60379
コバンサウ属 60372
コバンソウ 60379
コバンソウモドキ 147990
コバンソウ属 60372
コバンドコロ 131630
ゴバンノアシ 48510
ゴバンノアシ属 48508
コバンノキ 296565
コバンバニセジュズネノキ 122054
コバンバノキ 413143,413149
コバンバノキ属 413142
コバンバヤナギ 233811
コバンバヤナギ属 233809
コバンボダイジュ 164900
コバンムグラ 187544
コバンモチ 142340
コピアポア属 103739
コヒガンザクラ 83334
コヒゲ 213038,213066
コビスグサ 360493
コビトアナナス 391982
コヒナユリ 68797
コヒナリンドウ 173573
コヒナリンドオ 173573
コーヒーノキ 98857
コーヒーノキ属 98844
コヒマワリ 188953,189006
コヒメヒサカキ 160508
コヒメユリ 229818
コヒラガマ 401129
コヒルガオ 68686
コヒルガホ 68686
コーヒー属 98844
コフウセンカズラ 73208
コフウセンカヅラ 73208
コフウテウボク 71782
コフウロ 174978,175006
コブエ 102150
コブカエデ 2837
コブガシ 240606
コブカモノハシ 209261,209279
コブサンゴ 116652
コブシ 242169
コプシア属 217864
コフジウツギ 62015
コブシハジカミ 242169
コブシモドキ 242260
コブダネヤシ属 270997
コフタバラン 232905,232908
コブナグサ 36647
コブナグサ属 36612
コブニレ 401490,401593
コフネヤシ 273244
コブヤナギ 343529
コーベア属 97752
コヘイシソウ 347153

ゴヘイナラ　323814
ゴヘイヤシ　228766
ゴヘイヤシ属　228729
コベニミカン　93721
コヘビノボラズ　52151
コヘラナレン　84958,110610
コヘンーダ　341052
コヘンルーダ　341047,341052
ゴボウ　31051
ゴボウアザミ　91913
コボウズオトギリ　201732
コホウチャクソウ　134477
ゴボウ属　31044
コホクシャリントウ　107686
コボケ　84556
コホタルイ　352205
コホタルヰ　352205
コボタンヅル　94778
コボトケスミレ　410405
コホロギラン　376578
コホロギラン属　376576
ゴマ　361317
ゴマイザサ　362131
コマイヅルソウ　242672
コマイワヤナギ　344050
コマウスユキソウ　224932
コマウセンゴケ　138350,138354
コマガタケシラベ　498
コマガタケスグリ　333970
ゴマガラ　242274
コマキ　324384
ゴマキ　408131
ゴマギ　408131
コマクサ　128310,128314
ゴマクサ　81718,81719
コマクサ属　128288
ゴマクサ属　81714
ゴマシオホシクサ　151479
ゴマシオヤナギ　408131
ゴマシホヤナギ　408131
コマスミレ　409665
コマゼリ　24336
コマチ　267051
コマチギク　39943
コマチユリ　230023
コマツカサススキ　353429
コマツナ　59575,59607
コマツナギ　206445
コマツナギ属　205611
コマツミドリ　356955
コマツヨイグサ　269456
コマトドメ　344957
コマトメ　74644
ゴマナ　40519
コマニンジン　280741
コマノアシガタ　325981
コマノツメ　356841
ゴマノハグサ　355067,355204
ゴマノハグサ科　355276
ゴマノハグサ属　355039
コマノヒザ　4275
ゴマフガヤツリ　119607
コマミガシ　233383
コマユミ　157285,157293
コマユリ　229723
コマルバクスノキ　91420
コマルバユーカリ　155710
コマンネングサ　357189
コマンネンソウ　357189
ゴマ科　286918
ゴマ属　361294
コミカン　93724
コミカンサウ　296801
コミカンサウ属　296465
コミカンソウ　296632,296801
コミカンソウ属　296465
コミカンノキ　296565
コミクリ　370066
コミゾホオズキ　392948
コミゾホホヅキ　392948
コミネカエデ　3147
コミノキンカン　177833
コミノサルトリイバラ　366290
コミノヒメウツギ　126967
コミムラサキシキブ　66826
コミャガマズミ　408019
コミヤマカタバミ　277648
コミヤマガマズミ　408227
コミヤマカンスゲ　75448
コミヤマコメススキ　126055
コミヤマスミレ　410234
コミヤマヌカボ　12196
コミヤマミズ　298989
コミヤマリンダウ　173978
コミヤマリンドウ　173978
ゴムイチジク　164925
ゴムカズラ　139959,402194
ゴムカヅラ　139959,402194
ゴムカヅラモドキ　402201
ゴムカヅラモドキ属　139945
コムギ　398839
コムギセンノウ　363331
コムギダマシ　11696
コムギ属　398834
ゴムタンポポ　384606
ゴムノキ　164925
ゴムビハ　164925
ゴムフオセフアラユーカリ　155594
ゴムミカズラ属　162343
ゴムミカヅラ　162346
ゴムミカヅラ属　162343
コムヤム　289684
コムラサキ　65670
コムラサキシキブ　65670
コムレスズメ　72233,72235,72334
コメ　275958
コメアヅキ　294040
コメガヤ　249054
コメガヤ属　248949
コメゴアジン　97919
コメゴメ　66808,66829,66856,66873,374089
コメゴメジン　97919
コメザクラ　83145
ゴメサ属　178893
コメジシャ　379374
コメススキ　126074
コメススキ属　126025
コメツガ　399895
コメヅガ　399892
コメツゲ　64305
コメツツジ　332018
コメツブウマゴヤシ　247366
コメツブツメクサ　396882
コメツブヤエムグラ　170350
コメナモミ　363080,363090
コメノキ　66808
コメバカンサビラウ　381330
コメバコケミズ　298974
コメバツガザクラ　31038
コメバミソハギ　240047
コメヒシバ　130481,130730
コメミズ　379374
コメユリ　229934
ゴメンニンジン　292892
コモ　418095,418389
コモウセンゴケ　138354,138357
コモウソウ　99711
コモウツギ　345679
ゴモクサ　297185
ゴモジュ　408151
コモチオオアワガエリ　294992
コモチカイソウ　274556
コモチカンラン　59539
コモチキリンソウ　356858
コモチクジャクヤシ　78047
コモチケンチャヤシ　321170
コモチサンキライ　366518
コモチシカクイ　143416
コモチタマナ　59539
コモチトゲココヤシ属　126688
コモチトラノオ　309954
コモチトラノヲ　309954
コモチナデシコ　292665
コモチハボタン　59539
コモチヒメアブラススキ　71611
コモチベンケイソウ　200820
コモチマンネングサ　356590
コモチミツバベンケイソウ　200817
コモチミミコウモリ　283793
コモチミヤマイラクサ　221539,221552

コモチミヤマヌカボ 12105
コモチレンゲ 275384
コモノウラジロノキ 33011
コモノギク 40663
コモロコシガヤ 369677
コモロスミレ 410215
コモンカラクサ 263505
ゴモンマメ 294010
コヤクシマショウマ 41811
コヤグルマギク 79276
コヤス 379374
コヤスグサ 208875
コヤスノキ 301294
コヤナギタデ 291853,309395
ゴヤバラ 336789,336791
コヤブタバコ 77149
コヤブデマリ 408042
コヤブニクケイ 91438
コヤブマオ 56221
コヤブミョウガ 307333,307337
コヤブメウガ 307333
コヤブラン 232628,232640
コヤブレガサ 381814
コヤマイバラ 336954
コヤマウルシ 332901
ゴヨウアケビ 13221
ゴヨウアサガオ 207792,207873
ゴヨウイチゴ 216085,338606
ゴヨウカタバミ 278019
ゴヨウザンガリヤス 65256
ゴヨウザンスゲ 73553
ゴヨウザンヨウラク 250514
ゴヨウツツジ 331593
ゴヨウトガ 318575
コヨウバイ 316884
ゴヨウマツ 300130
コヨウラクツツジ 250528,250532
コヨメナ 40621,215343
コヨリソウ 56050
コラ 99158
コーライタチバナ 93648
ゴラウヒバ 390660
コラード 59524
コラナットノキ 99158
コラナットノキ属 99157
コラノキ 99240
コラノキ属 99157
コラビラン 99775
コラビラン属 99770
コラン 65871,116936
コラ属 99157
ゴリアーンテス属 105518
コリシア属 88970
コリゼマ属 89134
コリノプンテイア属 106957
コリファンタ属 107004
コーリフラワー 59529
コリヤナギ 343579,343937
コリュウセン 496
コリュウマルォ 103746
コリヨカクタス属 105434
コリンクチナシ 171253
コリンゴ 243717
コリンシア属 99833
コリンソウ 99838
コリンワスレナグサ 260870
コルクウィツィア属 217785
コルクカエデ 2837
コルクガシ 324453
コルククヌギ 324532
コルクノキ 324453
コルチカム属 99293
ゴルテイア 11294
ゴールデン・エバーグリーン・ラズベリ 338347
ゴールデンーフリース 923
ゴールデンワットル 1516
ゴールドセキコク 125166
ゴルドニア属 179729
コルヌス属 104917
コルムネア属 100110
コールラビ 59541
コルリヒゴタイ 140712
コルリブクロ 414006
コレウス属 99506
コーレリア属 217738
ゴーレリア属 217738
コレンゲ 275386
ゴレンジ 45725
ゴレンシ属 45721
ゴロウヒバ 390660
コロシントウリ 93297
ゴロツキアザミ 271666
コロハ 397229
コロバウマノスズクサ 34157
コロマンテル 132298
コロミア属 99850
コロモグサ 56145
コロラドトウヒ 298401
コロラドモミ 317
コヲノヲレ 53400
コン 264334
コンガウザクラ 280023
コンガコザクラ 23170
ゴンカヤシ属 413722
コンギク 40848,41404
ゴンゲンスゲ 74029,76120
コンゴウザクラ 280023
コンゴウセキ 244250
コンゴウソウ 377931
コンゴウタケ 304047
コンゴウバッコヤナギ 342943
コンゴウマル 244026
コンゴコーヒーノキ 98896
ゴンゴーラ属 179288
コンジキヤガラ 171939
コンシジボク 411374
コンジノキ 3529
ゴンズイ 160954
ゴンズイ属 160946
ゴンゼ 373556
ゴンゼツ 86855
コンテリギ 200077
コンドモミンカ属 88854
コントルタマツ 299869
コンナルス科 101858
コンニャク 20132
コンニャクイモ 20132
コンニャク属 20037
コンピラスミレ 410752
コンフリー 381035,381042
コンペイトウグサ 399309
コンペイトウゲサ 151416
コンボウアデク 382672
コンポル 140858
コンミフォラ属 101285
コンメ 83238
コンランギョク 264376
コンリンザイ 116829,117104
コンロン 295465
コンロンカ 260472
コンロンカ属 260375
コンロンクワ 260472
コンロンクワ属 260375
コンロンサウ 72876
コンロンソウ 72858,72876
サアエンドウ 301070
サアマネーア属 345517
サイウンカク 158456
ザイエン 82355
サイオウウチワ 385883
サイオウギョク 32361
サイオウニシキ 273827
サイカイツツジ 330986
サイカカク 183380
サイカク 373836
サイカチ 176881
サイカチバラ 366573
サイカチモドキ 315174
サイカチ属 176860
サイグミ 142103
ザイクラメン 115949
ザイコウギョク 233635
サイコウチュウ 183372
サイコクイカリソウ 146981
サイコクイボタ 229472
サイゴクガマズミ 407819
サイコクキツネヤナギ 344256
サイゴクザサ 347262
ザイコクトキワヤブハギ 200727,200728
サイコクヌカボ 309123

サイゴクバラ　366573
サイコクミツバツツジ　331354
サイザル　10940
サイシボク属　161628
サイシュウイヌシデ　77295
サイシュウウマノミツバ　345949
サイシュウギョク　284667
サイシュウノガリヤス　65300
サイジョウコウホネ　267275
サイシン　37722
サイシン属　37556
サイセイキツ　93639
サイセンギョク　246593
サイタツマ　328345
サイトウガヤ　127200,127211
サイトウソウ　23202
サイハイラン　110502,110506
サイハイラン属　110500
サイハダカンバ　53400,53520
ザイフリボク　19248
ザイフリボク属　19240
サイミャクアデク　382533
サイモンスモモ　316813
サイヨウアカバナ　146840
サイヨウザサ　347306
サイヨウシャジン　7850
サイラン　249590
サイリュウ　340687
サイリュウギョク　267057
サイリンヨウラク　250507
サウヴエオレンスポプラ　311509
サウキヤウ　176901
サウグワ　19839
サウザンヌカボ　12147
サウシキウツギ　413571
サウシジュ　1145
サウジュツ　44208
サウジンボク　69634
サウスウイックパペダ　93423
サウチク　39187
サウニンボク　69634
サウハギ　240030
サウマハヒノキ　381423
サウマヨモギ　36302
サウマラン　62696
ザウリガヤツリ　119769
サウロマーツム属　348022
ザオウアザミ　92503
サオトメ　177475
サオトメハッカ　257187
サオトメバナ　280097
サオヒメ　108045,327435
サカイツツジ　331057,331058,331449
サガウド　192278
サカエダマ　233574
サカキ　96587
サカギ　96617
サカキカズラ　25928
サカキカズラ属　25927
サカキカヅラ　25928
サカキカヅラ属　25927
サカキバイヌツゲ　203571
サカキヒサカキ　8256
サカキモドキ　69064
サカキ属　96576,160405
サカコザクラ　23170
サカサダケ　297370
サカサハマササゲ　409028
サカサベンケイ　357114
サカサンネングサ　357114
サカナミ　162934
サカネラン　264709,264711
サカネラン属　264650
サカホキ　86511
サカボコマル　284656
サガミギク　39931
サガミトリゲモ　262024,262035,262098
サカミハナ　48518
サガミメドハギ　226836
サガミランモドキ　116976
サガリイチゴ　338946
サガリグミ　141999
サガリバナ　48518
サガリバナ科　48524,223762
サガリバナ属　48508
サガリユリ　229900
サガリラン　133145
サガリラン属　133143
サカワサイシン　37714
サキガケスミレ　410452
サギゴケ　247008
サギゴケ属　246964
サギサウ　286835
サギサウ属　183389,286829
サキシフラガ属　349032
サキシマイチビ　935
サキシマスオウノキ　192494
サキシマスケロクラン　223643
サキシマスハウノキ属　192477
サキシマスホウノキ　192494
サキシマツツジ　330076
サキシマハマアカザ　44520
サキシマハマボウ　389946
サキシマハマボオ　389946
サキシマヒサカキ　160596
サキシマフヨウ　194989
サキシマボタンヅル　94814
サギスゲ　152759
サギスゲ属　152734
サギソウ　286835
サギソウ属　183389,286829
サギノシリサシ　352284
サギノシリサン　353899
サキュウニラ　15428
サクカウ　66643
サクキバナアキギリ　344823
サクダモ　249395
サクトリイバラ　366284
サクノキ　249395
サクハナギボウシ　198600
サクマイチゴ　339194
サクマサウ　191574
サクマサウ属　191573
サクマソウ　191574,191575
サクマソウ属　191573
サクマリンダウ　173379,173455
サクマリンドウ　173379,173455
サクヤアカササゲ　409113
サクユリ　229992
サクライサウ　292673
サクライソウ　292673
サクライソウ属　292671
サクライバラ　336789
サクラオグルマ　207151
サクラオンツツジ　331753
サクラガイ　102613
サクラカラクサ　416994
サクラガンピ　414143,414237
サクラキリン　290708
サクラサウ　314271
サクラサウモドキ　105496
サクラサウモドキ属　105488
サクラサウ科　315138
サクラジマエビネ　66005
サクラスミレ　410073
サクラセキコク　125221
サクラソウ　314508,314952
サクラソウモドキ　105499,105501
サクラソウモドキ属　105488
サクラソウ科　315138
サクラソウ属　314083
サクラダサウ　313564
サクラダサウ属　313552
サクラダソウ　234303
サクラダソウ属　313552
サクラタチツボスミレ　410035
サクラタデ　291835,309006,309262
サクラツツジ　331936
サクラバハンノキ　16470
サクラバラ　336789
サクラマル　140616
サクラマンテマ　363882
サクラワン　198827
サクラワン属　198821
サクラリュウ　366137
サクラヰサウ属　292671
サクランボ　83155
サクラ属　316166
ザクロ　321764
ザクロガンピ　363758
ザクロギョク　233475

ザクロサウ　256719
サクロサウ科　12911
ザクロサウ属　256689
ザクロソウ　256719
ザクロソウ科　256684
ザクロソウ属　256689
ザクロ科　321777
ザクロ属　321763
ザケバカウゾリナ　55768
サケバコウゾリナ　55768
サケバゴヘイヤシ　228758
サケバゼリ　269326
サケバナヤウラクラン　267935
サケバニワトコ　345681
サケバヒヨドリ　158187
サケバミミイヌガラシ　336172
サケミヤシモドキ属　212343
サケミヤシ属　269373
サケラフィア　326649
サケリサウ属　314083
サケリョウゼッ　10933
サゴバイ　32343
サゴヤシ　252634,252636
サゴヤシ属　252631
ザ-サイ　59469
ササウグサ　341887
ササウサ　341887
ササウチワ　370344
ササエビネ　274058
ササエビモ　312034
ササガシ　116153
ササガニユリ　200941
ササガャ　254020
ササガヤ　254036
ササキエビネ　65886
ササキカズラ　341227
ササキカヅラ　341227
ササキカヅラ属　341223
ササキナゴラン　379771
ササキビ　361850
ササクサ　236284
ササクサ属　236281
ササグリヤブマオ　56166
ササクワ　259097
ササグワ　259067
ササゲ　409086
ササゲ属　408825
ササスゲ　76264
ササナギ　257583
サザナミギバウジ　198598
サザナミギボウシ　198598
ササニンドウ　97093
ササノハスゲ　75658
ササノユキ　10955
ササバエンゴサク　105596,106076,106370
ササハギ　18289
ササバギンラン　82058
ササハギ属　18257
ササバサンキライ　366483
ササバノボタン　276090
ササバモ　312171,312291
ササバラン　232261
ササブネギョク　377342
ササフブキ　10850
ササモ　312206
ササヤキグサ　240765
ササユリ　229872,229919
ササラン　123974
ササリンダウ　173852
ササリンドウ　173852
サザンカ　69594
サザンクワ　69594
ササ属　347191
サ-ジェントイトスギ　114751
サジオモダカ　14754,14760
サジオモダカ属　14719
サジガラシ　72851
サジガンクビサウ　77149
サジガンクビソウ　77162
サシキユリ　229872
サジバモ　312171
サジバモウセンゴケ　138322
サジビユ　18692
サシブノキ　403738
サシマジイ　233193
サシマジヒ　233193
サシマニンジン　363451
サシマユリ　229900
サシュウスズメノヒエ　285426
サスマタモウセンゴケ　138270
ザゼンサウ　381073
ザゼンサウ属　381070
ザゼンソウ　381073,381082
ザゼンソウ属　381070
サタウキビ　341887
サタウキビ属　341829
サタウグケ　341887
サタウシバ　242274
サタウノキ　341887
サダサウ　290365
サダサウ属　290286
サダソウ　290365
サダソウ属　290286
サタタケ　341887
サタツツジ　331370
サタノキ　341887
サダファル　93763
サッカリヌムカヘデ　3532
サツギ　330917
サツキアサドリ　142252
サツキイチゴ　338886,338985
サツキギョリュウ　383469
サツキグミ　142092
サツキツツジ　330917
サツギツヅジ　330917
サツキヒナノウスツボ　355195
サックイネラ　229992
ザックネボ　93704
サッコウフジ　66643
サッサフラス　347407
サッサフラスノキ　347407
ザッショクノボリフジ　238468
ザッショクノボリフヂ　238468
ザッソウメロン　114194
サットカラ　93430
サッボロスゲ　75785
サツボロスゲ　75786
サツマ　114294
サツマアオイ　37714,37715
サツマイナモリ　272221
サツマイナモリ属　272173
サツマイモ　207623
サツマイモ属　207546
サツマウツギ　294426,294530
サツマギク　67314
サツマギク属　67312
サツマキコク　93647
サツマコンギク　67314
サツマサンキライ　366271
サツマジイ　233193
サツマシロギク　41198
サツマスゲ　75149,75161
サツマスズメウリ　417488
サツマタデ　291789
サツマナデシコ　127667
サツマネコノメ　90410
サツマノギク　89652
サツマハギ　226806
サツマビャクゼン　409433
サツマフジ　122438
サツマホシクサ　151404
サツマママアザミ　92390
サツマママンネングサ　357126
サツマモクセイ　276336
サツマモクゼイ　276336
サツマユリ　229758,230058
サツマルリミノキ　222173
サデクサ　309356,309564
サドアザミ　92236
サトイモ　99910
サトイモモドキ　99918
サトイモ科　30491
サトイモ属　99899
ザトウエビ　20348
サトウカエデ　3218,3549
サトウカヘデ　3532
サトウキビ　341887
サトウキビ属　341829
サトウゴムノキ　155542
サトウシバ　242274
サトウダイコン　53266

サトウナツメヤシ 295487
サトウマツ 32343,300011
サトウモロコシ 369700
サトウヤシ 32343
サトザクラ 83316
サドザサ 347368
サドシオデ 366551
サドスゲ 76132
サドスゲモドキ 73568
サトトネリコ 167999
サトニラ 15185
サナエタデ 291818,309298,309318
サナエタデ属 291651
サナエツルイタドリ 162530
サナギ 365
サナギイチゴ 339130,339135
サナギスゲ 74700
サナシ 243717
サナツラブダウ 411735
サナヘタデ 309298
サニシサ 3309
サニン 17774,155398
サニンテン 78857
サヌキカンアオイ 194358
サネカズラ 214971
サネカズラ属 214947
サネカヅラ 214971,351021
サネカヅラ属 214947
サネブトナツメ 418169,418175
サノガキ 132245
サノニサポタ 244605
サハアヂサイ 200089
サハアヂサヰ 200089
サハオグルマ 385903
サハギキャウ 234766
サハギキョウ 234766
サハギク 263515
サハギク属 358159
サバクオトコヨモギ 35373
サバクソウ 11619
サバクソウ属 11609
サハグハ 160347
サハグミ 105146
サバクヨモギ 35595
サハグルミ 320372
サハグルミ属 320346
サハシデ 77276
サハシバ 77276
サハシヲン 290134
サハスズメノヒエ 285426,285534
サハダツ 157713
サーバテイ 93757
サハテラシ 330998
サハトンボ 183797
サハハコベ 374859
サハヒヨドリ 158200
サハフサギ 199911
サハフヂ 48518
サハマキ 77276
サハミヅキ 380461
サハラトガ 318575
サバル属 341401
サハヲグルマ 385903
サハヲグルマ属 358159
サビアオイソウ 290386
サビイロジギタリス 130358
サビサルナシ 6517
サビタ 200038
サビハナギリソウ 4078
サビハナナカマド 369509
サビバナナカマド 369366
サビバハナギリソウ 4078
サビバマンネングサ 357114
サビンマツ 300189
サフェシア属 346314
サフラン 111589
サフランモドキ 417613
サフラン属 111480
サブロウナシ 323216
ザブンギ 364768
サポジラ 244605
サポジラ属 244520
サボテン 272891
サボテンギク 220674
サボテンタイゲキ 158456
サボテン科 64822
ザボン 93579,93754
サボンサウ 346434
サボンサウ属 346418
サボンソウ 346434,346436
サボンソウ属 346418
サマイパティセレウス属 345507
サマニオトギリ 202145
サマニヨモギ 35153,35157
サマヤオ 93869
サマン属 145992
ザミア属 417002
ザミオクルカス属 417026
サムヤオ 93635
サメカハブラッシノキ 67262
サメカワブラッシノキ 67262
サモモ 109650
サーモンベリー 339285
サヤウジャノミヅ 411686
サヤエンドウ 301070
サヤエンドウ 301080
サヤゲチシマザサ 347238
サヤゴ 204138
サヤシロスゲ 74709
サヤスゲ 74483,76312,76656,76658
サヤヌカグサ 224002
サヤヌカグサ属 223973
サヤバナ 99711
サヤバナ属 99506
サヤヒゲシバ 372874
サヤマスゲ 74764
サユリ 229872,229919
サヨギネ 109446
サヨシグレ 3052
サラオ 219976
ザラゲカモジ 11630
サラサウツギ 126894
サラサカンラン 116927
サラサドウダン 145677
サラサバナ 34171,34246
サラサモクレン 416721
サラサレンゲ 416721
サラシナショウマ 91011,91038
サラシナショウマ属 90994
サラセニア科 347166
サラセニア属 347144
サラソウジュ 362220
ザラッカモドキ 342571
ザラッカ属 342570
ザラッキイチゴツナギ 306003
ザラッキエノコログサ 361930
ザラッキヒナガリヤス 65435
サラノキ 362220
サラノキ属 362186
ザラバナソモソモ 305686
サラマンダ 180642
サラマンド 28316
ザラメキスズメウリ 259739
ザラメキタデ 309442
サラル 156338
サラワクマメ 408917
ザリグミ 334089
ザリコミ 334087,334089
サルイチゴ 338985,339047
サルイワツバキ 69207
サルウィンツバキ 69589
サルオガセモドキ 392052
サルカキイバラ 280555
サルカケイバラ 64990
サルカケミカン 392559
サルカケミカン属 392556
サルカノコサウ 404269
サルクラスミレ 410735
サルコカウロン属 346642
サルコカシ 223150
サルコキラス属 346686
サルコグロッチス属 346797
サルコココッカ属 346723
サルコステンマ属 346993
サルコフィトン属 346899
サルコヤナギ 343444
サルシフィ 394327
サルスベリ 219933,376457
サルスベリ属 219729,219908
ザルゾコミョウガ 307337
サルタノキ 376457

サルダメシ　96466
サルタン　18966
サルトカヅラ　103862
サルトリイバラ　366284
サルトリイバラ科　366141
サルトリカク　299269
サルナシ　6515,6518
サルナシ科　130928
サルナシ属　6513
サルナメ　376466
サルナメリ　376457
サルビア　345405
サルビア属　344821
サルビエ　140367
サルピグロッシス属　344381
サルビヤ　345271
サルフエ　134946
サルマメ　366284,366607
サルメンエビネ　66099
サルメンキビ　282200
サルメンバナ　344383
サルメンバナ属　344381
サルヤナギ　343151
ザロンウメ　316581
ザロンバイ　316581
サワアザミ　92501
サワアジサイ　200089
サワイチゲ　24024
サワウルシ　274277
サワオグルマ　385914
サワオトギリ　202107
サワ－オレンジ　93414
サワギキョウ　234766
サワギキョオ　234766
サワギク　263515
サワギク属　358159
サワグミ　105146
サワグルミ　320372
サワグルミ属　320346
サワグワ　160347
サワシオン　290134
サワシデ　77276
サワシバ　77276
サワシホガマ　287494
サワシマフジバカマ　158023,158333
サワシロギク　41171
サワスズメノヒエ　285534
サワゼリ　365858
サワセンブリ　380184
サワタチ　157713
サワダツ　157713
サワテラシ　330998
サワトウガラシ　123730
サワトウガラシ属　123727
サワトラノオ　239710
サワトンボ　183797
サワハコベ　374859
サワヒゴタイ　348554
サワヒメスゲ　75384
サワヒヨドリ　158200
サワフサギ　199911,204172
サワフジバカマ　158022
サワフダギ　381137
サワホオズキ属　255184
サワミズキ　104966
サワミヅキ　380461
サワヤナギ　343396
サワラ　85345
サワラトガ　318575
サワラン　143102
サワルリソウ　22038
サワルリソウ属　22037
サワレン　143440
サワレン属　143438
サンアソウ　122742
サンアソウ科　328218
サンイウクワ属　382727
サンインクワガタ　407246
サンイントラノオ　317930
サンインネコノメ　90359
サンインヒエスゲ　74966
サンインヒキオコシ　209832
サンエフボタンヅル　95396
サンガイグサ　220355
サンガイネギ　15293
サンカエフ　132682
サンカエフ属　132677
サンカクイ　352284,353899
サンカクカンアフヒ　37585
サンカクスゲ　352284,353899
サンカクズル　411686
サンカクタケウマヤシ　208367
サンカクチュウ　200709,200712
サンカクヅル　411686
サンカクツワブキ　229063
サンカクナ　224396
サンカクニラ　15822
サンカクバアカシア　1156
サンカクホタルイ　353186
サンカクボタン　33221
サンカクヰ　352284,353899
サンカヨウ　132682,132685
サンカヨウ属　132677
サンキ　93717
サンキ－　93717
サンキツ　93748
サンギナリア属　345803
サンキョウ　17677
サンキライ　366284
サングラン　928
サンケジア属　345769
サンゴアナシス属　8544
サンゴアナナス　8562
サンゴアブラギリ　212202
サンコウマル　140282
サンコカンアオイ　37748
サンゴシトウ　154632
サンゴシドウ　154643
サンゴジュ　45260,407977,407980
サンゴジュウドノキ　223943
サンゴジュスズメウリ　259739
サンゴジュツゲ　64306
サンゴジュナ　53269
サンゴジュボク　407977,407980
サンゴジュマツナ　379542
サンゴナス　366910
サンゴバナ　194437,214410
サンゴバナ属　211338
サンザクラサウ　314975
サンザクラソウ　314975
サンザシ　109650
サンザシバノイチゴ　338300
サンザシ属　109509
サンシキアカリフア　2011
サンシキアサガオ　102884,103340
サンシキカミツレ　89466
サンシキギリア　175708
サンシキスミレ　409624,410677
サンシキナデシコ　127864
サンシキヒルガオ属　102903
サンシキヒルガホ　103340
サンシキヒルガホ属　102903
サンジサウ　94138
サンジサウ属　94132
サンジソウ　94140
サンジソウ属　94132
サンシチ　183097,280771
サンシチサウ属　183051
サンシチソウ　183097
サンシチソウ属　183051
サンシチニンジン　280778
サンジャク　405812
サンジャクバ－ベ－ナ　405812
サンシュユ　105146
サンシュユ属　240965
サンショウ　417301
サンショウソウ　288749
サンショウソウ属　288705
サンショウバラ　336631
サンショウモドキ　350995
サンショウ属　417132
サンショオモドキ属　144716
サンショグサ　129618
サンシンサウ　183097
サンセイセキコク　146543
サンセウ　417298,417301
サンセウサウ　288749
サンセウサウ属　288705
サンセウヅル　288749
サンセウモドキ属　144716
サンセウ属　417132

ザンセツ 150027
ザンセツソウ属 326513
サンセベリア 346158
サンセベリア属 346047
サンソ 418177
サンダイガサ 353057
サンダ-ツツアナナス 54461
サンタルロンジ 320327
サンタンカ 211067
サンダンカ 211067
サンダンカモドキ 4965
サンダンカモドキ属 4954
サンタンカ属 211029
サンタンクワ 211067
サンタンクワ属 211029
サンダンサウ 221238
サンデルソニア属 345782
サントウ 20908,166797
サントウサイ 59604
サントウザクラ 314521
サントウゼリ 76987
サントウゼリ属 76986
サントウマサギ 157651
サンドブラックベリ- 338307
サントリサウ 97599
サントリサウ属 97581
ザンドリ-サ属 415164
サントリソウ 97599
サントリソウ属 97581
サントリ-ナ属 346243
サント-ル 345788
サンナ 187468,215035
サンナシ 243555,243649
サンパクソウ 348087
サンビタリア 346304
サンプクリンドウ 100277,100278,100280
サンプクリンドウ属 100263
サンブスギ 113685
サンブンサン 25439
サンベサワアザミ 92440
サンヘンプ 112269
サンボウ 93820
サンボウカン 93820
サンポウカン 93820
サンボテ 272891
サンボテイ 272891
サンユウカ 154162
サンユウカ属 382727
サンヨウアオイ 37639
サンヨウッチグリ 312408
サンヨウツツジ 331014
サンリョウ 370102
サンリンサウ 24066
サンリンソウ 24066
シアノチス属 115521
シイ 78916,79022
シイカアシャア 93448
シイガシ 78916
シイクワシャ- 93448
シイトギ 142396
ジイナモ 373439
シイノキ 78903,78916
シイノキカズラ 126007
シイノキ属 78848
ジイババ 116880
シイボルトノキ 328883
シイモチ 203600
シイ属 78848
シウカイダウ 49886
シウカイダウ科 50433
シウカイダウ属 49587
シウブンサウ 333475
シウブンサウ属 333471
シウブンソウ 333475
シウブンソウ属 333471
シウメイギク 23854
シウリザクラ 280053
ジウロクサギナ 96401
シウンデン 167695
シウンマル 167696
ジエウジソ 162473
ジェゼボグイジュ 142272
ジエフリ-マツ 299992
シェフレ-ラ属 350646
シェ-レア属 350635
ジェロウクリモ- 93322
ジェロ-くパパヤ 93685
ジェロ-くバリック 93674
ジェロ-くハンジ 93513
シエンマル 234936
ジオウ 327435
シオウフクギ 171144
ジオウ属 327429
シオカゼテンツキ 166236,166493
シオガマギク 287584,287589
シオガマギク属 286972
シオガヤツリ 212774
シオギク 89715
シオギリソウ 8862,8884
シオクグ 76163
シオザキソウ 383100
シオジ 168068,168106
シオジノキ 168003
シオツメクサ 370661
シオツメクサ属 370609
シオデ 366548,366573
シオデ属 366230
シオニラ 382420
シオノキ 184838
シオバナイリス 208565
シオバラザサ 347372
シオマツバ 176775,176776
シオミイカリソウ 147070
シオヤキソウ 174877
シオヤキフウロ 174877
シオン 41342
シオン属 39910
シガアヤメ 208825
シカイカン 93815
シカイナミ 162951
シカウクワ 223454
シカウクワ属 223451
シカギク 8331,246315,397986
シカギク属 246307
シカクイ 143414
シカクオトギリ 202179
シカクセッコク 125386
シカクダケ 87607
シカクハヅ 298734
シカクホタルイ 352162
シカクマメ 319114
シカクマメ属 319101
シカクミクララ 369010
シカクワズ 298734
シカク井 143414
シ-カ-シャ- 93448
シカツマメ 408393
シカナ 229175
シカノアタマ 199033
ジガバチサウ 232206
ジガバチソウ 232206
シカハンゲ 299725
シカハンゲハンゲ 299725
シカフクワ 223454
シガフグワン 255134
シカンラン 116926,116928
シキオス属 362460
シキキツ 93566,93636,93639
シキクンソウ 380046
シキザイバラ 336485
シキザキイバラ 336485
シキザキサクラソウ 314717
シキザキシウカイダウ 50298
シキザキベゴニア 50298
シキザキモクセイ 276291
シキザキヤマツツジ 330979
シキザクラ 83112,405852
シキサリミガシ 93639
ジギタリス 130383
ジギタリス属 130344
シキテン 102090
シキナリキンカン 93636
シキミ 204533,204575,204583
シキミ科 204471
シキミ属 204474
シキュ-タ 101852
シキンカラマツ 388645
シギンカラマツ 388399
シキンショゥ 358633
シキンジョウ 163435
シキンツユクサ 393999,394007

シキンラン　394076
ジクゲウツギ　126895
シグマトスタリックス属　363116
シクラメン　115949,115965
シクラメン属　115931
シクランテラ属　115990
シグレノキ　408224
シークワシャー　93448
シグワツマメ　408393
シクン　233584
シクンシ　324677,324679
シクンシ科　100296
シクンシ属　324669
シケイ　55575
シゲンジ　122438
シコウカ　223454
シコウカ属　223451
ジコウニシキ　86506
シコウラン　62873
ジゴカクタス属　418529
シコクアザミ　92238
シコクウスゴ　403996
シコクウツギ　294532
シコクカッコソウ　314537
シコクガマズミ　407804
シコクガリヤス　65255
シコクキブシ　373547
シコクザサ　347222
シコクシナノキ　391743
シコクシモッケソウ　166124
シコクシラベ　496,502
シコクシロギク　41518
シコクスミレ　410571
シコクダケカンバ　53614
シコクチャルメルソウ　256039
シコクテンナンショウ　33366
シコクトリアシショウマ　41866
ジゴクノカマノフタ　13091
シコクノガリヤス　65500
シコクハタザオ　30447
シコクビエ　143522
シコクヒロハテンナンショウ　33400
シコクママコナ　248175
ジゴシキオス属　418788
シゴスタテス属　418793
シコタンアズマギク　150953
シコタンサウ　349117
シコタンザサ　347329
シコタンスゲ　76202
シコタンソウ　349117,349119
シコタンタンポポ　384802
シコタンハコベ　375078
シコタンマツ　221866,221878
シコタンヨモギ　35759,36363
ジゴペタラム属　418573
シコロ　294231,294251
シコロベンケイ　61568
シコンノボタン　391577
ジザイガシ　323611
ジサノキ　379374
シサルアサ　10940
シサルヘンペ　10940
シシアクチ　31578
シシウド　24432,24441
シシウド属　24281
シシオウマル　267058
シシガシラ　69133
シジキカンアオイ　37640
シシギョク　242890
シシキリガヤ　93913
シシクハズ　331261
シシズク　261424
シシタウガラシ　72075
シシトウガラシ　72081
シシバリ　210673
ジシバリ　130489,210623,210673
シシヒゴタイ　348772
シシフンジン　107016
シジミバナ　372050,372051
シシャチギョク　263745
シシユ　93705
シジョウオオムギ　198376
シシリ－シュマック　332533
シシンラン　239967
シシンラン属　239926
シズイ　352238,353660
シスイカク　373917
シスタス属　93120
シズナミ　324922
シスマトグロッティス属　351146
シセイマル　2107
シセンシャリントウ　107336
シセンネズコ　390661
シセンリュウ　148143
シソ　290940,290952,290968
ジゾウカンバ　53460
シソクサ　230275
シソクサ属　230274
シゾコドン属　351439
シゾスティリス属　351883
シソノミグサ　217094,217101
シソバウリクサ　231575
シソバキスミレ　410788
シソバタツナミ　355554
シソバタツナミソウ　355505
シソ属　290932
シタイシャウ　202538
シタイシャウ属　202495
シタキサウ　211695
シタキサウ属　211693,376022
シタキソウ　211695,211697
シタキソウ属　211693,376022
シタキツルウメモドキ　80324
シダソテツ属　373673
シダノキ　166059
シダノキ属　166056
シタマガリ　239266
シタミカンアフヒ　37646
シタミズザクラ　280016
シタミヅザクラ　280016
シダメモルッカヤシ　332377
シダレアカシデ　77325
シダレイス　134946
シダレイスノキ　134948
シダレイチョウ　175824
シダレイトスギ　114690
シダレイトスギ属　114649
シダレイヌコリヤナギ　343532
シダレイヌシデ　77402
シダレイヌツゲ　203670
シダレエゴノキ　379383
シダレエゾヤナギ　344023
シダレエノキ　80740
シダレエンジュ　369038
シダレオオヤマザクラ　83303
シダレガジュマル　164688
シダレカッラ　83744
シダレカツラ　83737
シダレカラマツ　221895
シダレカワヤナギ　343704
シダレカンバ　53563
シダレキジカクシ　38985
シダレグリ　78781
シダレグワ　259068
シダレケヤキ　417560
シダレコゴメヤナギ　344092
シダレコナラ　324386
シダレザクラ　83337,316636
シダレジイ　78907
シダレツユクサ　67145
シダレネコヤナギ　343448
シダレハナマキ　67302
シダレヒカン　316636
シダレヒガン　83337
シダレベゴニア　49877,50012
シダレマツ　299895
シダレマンサク　185111
シダレミネヤナギ　343992
シダレモミ　366
シダレヤナギ　343070,343795
シダレヤマアサクラザンショウ　417307
シダレユジイ　78914
シタワレ　157601
シタン　121653,320301,320327
シタンシュ　107322
シタンノキ　239527
シタンノキ属　239525
シタン属　320269
シチク　47213,297367
シチゴサンアオイ　18157
シチセイイチゴ　338471

シチセイザ 102163
シチセイハナメウガ 17754
シチタウ 119160,119162
シチタウヰ 119160
シチダンカ 200091
シチトウ 119160,119162
シチトウイ 119162
シチトウエビヅル 411675
シチトウタラノキ 30750
シチヘンゲ 221238
シチホウギョク 233475
シチメンサウ 379542
シチメンソウ 379542
シチモウゲ属 225668
シチモチ 203600
ジチャ 1120
シチャウゲ属 225668,226075
シチャノキ 379374
シチョウゲ 226125
シチョウゲ属 226075
シチョオゲ 226125
シックリ－フドナリブ 405182
シッケンウチワ 273043
シッサス属 92586
シッソノキ 121826
シッボウジュ 216654
シッボウニシキ 358307
シッポウニシキ 17289
シッポガヤ 412465
シヅヰ 353660
シデ 77312
シテアブラガヤ 353576
シデアブラガヤ 353936
シデガヤツリ 119535
シデコブシ 242331
シデザクラ 19248
シデシャジン 43580
シデシャジン属 298054
シデゾク属 77252
シデノキ 77323
シデヤナキ 19248
シトウ 154734
シドウ 167690
シトウイチゴ 338886
シトウヅ 154734
シトカスプルース 298435
シトカトウヒ 298435
シドキヤマアザミ 92387
シドケ 283804
シトタゴ 352438
シドニア属 116532
シドミ 84556
シドメ 84556
シトロネグラス 22836
シトロネブラス 22836
シトロネラグラス 117218
シトロプシス属 93269
シトロン 93603
シナアウタウ 83284
シナアオキ 44893
シナアカマツ 299890,300054
シナアサダ 276808
シナアブラギリ 406018
シナアマグリ 78802
シナイボクサ 260113
シナオウトウ 83284
シナオケラ 44208,44210
シナオトギリ 201761
シナガハダケ 304014
シナガハハギ 249232
シナガハハギ属 249197
シナガリヤス 94632
シナガワダケ 304014
シナガワハギ 249232
シナガワハギモドキ 42696
シナガワハギ属 249197
シナカンバ 53379
シナキササゲ 79247
シナキハダ 294238
シナギリ 285959
シナクサギ 96078
シナクズ 321445
シナクスモドキ 113436
シナクスモドキ属 113422
シナグリ 78802
シナクログワイ 143122,143123
シナクロマツ 300054
シナクワイ 342424
シナクワヰ 342424
シナクワンザウ 191284
シナサクラソウ 314157
シナサハグルミ 320377
シナサルナシ 6553
シナサワグルミ 320377
ジナシ 84556
シナシュウカイドウ 49890
シナシロギリ 285959
シナスゲ 74098
シナセンニンサウ 94814
シナセンリャウ 241817
シナセンリョウ 241844
シナタカネヤハズハハコ 21580
シナタチバナ 31380
シナダレスズメガヤ 147612
シナタンポポ 384473
シナツガ 399879
シナツクバネウツギ 113
シナツリスゲ 74537
シナトチノキ 9683
シナナベナ 133463
シナノアキギリ 345148
シナノウメ 316570
シナノオトギリ 201959
シナノガキ 132216,132264,132267
シナノカリヤスモドキ 255871
シナノカワラマツバ 170746
シナノキ 391739
シナノキイチゴ 338573
シナノキンバイ 399510,399522
シナノキンバイソウ 399510
シナノキ科 391860
シナノキ属 391646
シナノクロウメモドキ 328744,328747
シナノコザクラ 315054
シナノサイコ 63765
シナノザクラ 83284
シナノサツキ 330917,331839
シナノサラサケマン 106005
シナノショウキラン 416381
シナノスミレ 410756
シナノタイゲキ 159847
シナノタンポポ 384742
シナノトウヒレン 348095
シナノナデシコ 127833
シナノヒメクワガタ 407253
シナノヤブサンザシ 333972
シナハクチョウゲ 360935
シナハシドイ 382257
シナヒイラギナンテン 242487
シナヒガンバナ 239269
シナヒヨス 201392
シナヒヨドリ 158070
シナフジ 414576
シナプトフィルム属 381628
シナボタンヅル 94814,94969
シナマオウ 146253
シナマタタビ 6553
シナマンサク 185130
シナマンネングサ 356512
シナミサオノキ 278323
シナミサヲノキ 278323
シナミズキ 106675
シナヤガラ属 279852
シナヤドリギモドキ 385225
シナヤドリギモドギ 355322
シナヤブカウジ 31380
シナヤブコウジ 31380
シナヤブコオジ 31380
シナヤブサンザシ 333972
シナヤマツツジ 331839
シナユリノキ 232603
シナラッキョウ 15185
シナレンギョウ 167471
シナレンゲウ 167471
シナワスレナグサ 117908
シナンクム属 117334
ジニア属 418035
シネラミヤ 290821
シネラリヤ 290821
ジネンジャウ 131645
ジネンジョウ 131645

シノクラッスラ属　364913
シノグロッサム属　117900
シノダケ　304098
シノノメ　139974
シノノメサウ　380367
シノノメソウ　380367
シノブノキ　180642
シノブノキ属　180561
シノブバハキ　39195
シノブヒバ　85346,85350
シノブボウキ　39150
シノブモクセイサウ　327795
シノブモクセイソウ　327795
シノブ属　123244
シノベ　318307
シノベチク　318307
シバ　418426,418453
シハイスイレン　267730
シハイスミレ　410749
シハイス井レン　267730
シハイヒメバショウ　66191
シハイユリ　229758
シハウチク　87607
シバガヤ　37421
シバギ　233949
シバキツネヤナギ　343546
シバクサネム　366670
シバクサネム属　366632
シバコブシ　241955
シバサクラ　295324
シバシデ　77276
シバスゲ　75496
シバタカエデ　3150
シバツメクサ　353987
シバテンツキ　166236,166493
シバナ　397144,397159
シバナデシコ　127728
シバナ科　212753,350864
シバナ属　397142
シバニッケイ　91322
シバネム　366654,366670
シバネム属　366632
シバハギ　126389,126396
シバムギ　144661
シバムギモドキ　228356
シバヤナギ　343546
シバヤブニッケイ　91256
シバヤマハリイ　143132
シバヨナメ　31080
シバヨナメ属　31079
シバ属　418424
シビトバナ　239266
シヒノキカヅラ　126007
シヒノキ属　78848
シヒモチ　203600
シビンウツボ　264841
シ－フォ－シア属　321161
シブカガシ　233193
シブカワシロギク　41172
シブカワツツジ　331724
シブカワニンジン　98346
シブクサ　339994,340089
ジプソフィラ属　183157
シブナシ　323085
ジフニヒトヘ　13104,13146
ジフモンジスゲ　74211
ジフヤク　198745
シプリペジューム属　120287
シフンマル　234943
シヘイネヅアヤメ　208921
シ－ベキンギア属　363051
シベナガムラサキ　141347
シベミアモミ　476
シペラス属　118428
シペラ属　118393
シベリアイワスグリ　334145
シベリアイワタデ　309056
シベリアカラマツ　221944
シベリアキヌタソウ　170246
シベリアキンミズヒキ　11572
シベリアクガイソウ　407371,407485
シベリアゲンゲ　278958
シベリアシオガマ　287584
シベリアシナノキ　391832
シベリアシャクヤク　280147,280213
シベリアスズシロ　154380
シベリアセンノウ　364062
シベリアナズナ　18469
シベリアニガナ　111019
シベリアハネガヤ　4160
シベリアハンショウズル　95302
シベリアハンノキ　16362
シベリアヒエンソウ　124237
シベリアヒナゲシ　282617
シベリアフクジュソウ　8381
シベリアベンケイ　357039
シベリアボゥフウ　228599
シベリアマガラシ　47949
シベリアメドハギ　226860
シベリアユキノシタ　52514,349895
シベリアユキワリソウ　314378
シベリアラッキョウ　15467
シベリアリュウキンカ　68189
シベリアリンゴ　243555
シベリヤセンノウ　363691
シベリャトウヒ　298382
シベリャメギ　52149
シベリヤヤマガラシ　47956
シベリヤヨメナ　41257
シベリヤラマツ　221944
シホウ　340786
シホウギョク　140622
シホウチク　87607
シホウヒゴタイ　348395
シホウマル　327286
シホウリュウ　163481
シホカゼスグリ　333988
シホカゼテンツキ　166493
シホガマギク　287584
シホガマギク属　286972
シホガマサウ　287584
シホクグ　76163
シホジ　168068
シホジノキ　168068
シボチク　297222
シホデ　366364,366573
シホデモドキ　194127,318450
シホデモドキ属　318448
シホデ属　366230
シホニラ属　117296
シボヒゴタイ　348742
シホマツバ　176775,176776
シホマツバ属　176772
シボリイリス　208744
シボリカタバミ　278142
シボリミヤマアキギリ　345053
シ－ボルトノキ　328883
シマ　301433
シマアオイ　389946
シマアオカヅラ　341570
シマアカバナ　146859
シマアケボノソウ　380267
シマアザミ　91816
シマアザミモドキ　92210
シマアズキ　409080
シマアズサ　243453
シマアヅキ　409025,409026,409065,409080
シマアハイチゴ　338487
シマアハブキ　135215
シマアハモリ　41825
シマアフヒ　389946
シマアブラギリ　212127
シマアワイチゴ　338487
シマアワブキ　249449
シマアヲカヅラ　341570
シマイガクサ　333490
シマイケマ　245860
シマイズセンラリャウ　241844
シマイズセンラリョウ　241844
シマイズセンリョウ　241808,241818
シマイズセンリョオ　241818
シマイスノキ　134934
シマイズハハコ　253476
シマイチビ　228,934
シマイヅセンリャウ　241817
シマイヅハハコ　253479
シマイヅハハコ属　253450
シマイナモリ　272272,272282
シマイヌザンショウ　417335
シマイヌツゲ　204025,204071
シマイヌノヒゲ　151356

シマイヌビハ　165761
シマイノコヅチ　4259,4263
シマイハガラミ　199817
シマイボクサ　260105,260110
シマイワウチワ　362251
シマウオクサギ　313702
シマウキクサ　221007
シマウツギ　104696
シマウツボ　302884
シマウドカヅラ　20327
シマウメモドキ　203578
シマウラジロイチゴ　339338
シマウラジロノキ　320843
シマウラジロノキ属　320827
シマウリ　390175
シマウリカエデ　3031
シマウリクサ　231479,404712
シマウリノキ　13381
シマウリバカエデ　3209
シマウリハダカエデ　3031
シマウヲクサギ　313702
シマエゴノキ　16295
シマエノコロ　200827
シマエビネ　65952,66074
シマオカメザサ　362124
シマオガルカヤ　117185
シマオニク　57434
シマオバナ　262552
シマオホネコヤナギ　343321
シマガキ　132335
シマカコソウ　13057
シマカゴノキ　233945
シマガシ　116106
シマカテンソウ　262249
シマカナビキサウ　354660
シマカナビキサウ属　354658
シマカナビキソウ　354660
シマカナビキソウ属　354658
シマカナメモチ　295817
シマガマズミ　407705,407853
シマカモノハシ　209327
シマガヤ　293710
シマカラスウリ　396254
シマカンギク　124790
シマカンスゲ　75423
シマガンピリョウカオウ　414193
シマギキョウ　70403
シマキケマン　105639,106508
シマキツネノボタン　326365
シマキツネノマゴ　337253
シマキビ　281984
シマキブネギク　24116
シマギョウギシバ　130709
シマギョクシンカ　385028
シマキランソウ　13057
シマキンチャク　307919
シマギンメイソウ　239608
シマギンレイカ　239608
シマクグ　245392
シマクサアジサイ　73134
シマクサアヂサヰ　73134
シマクサギ　96146
シマクサハギ　206714
シマクハズイモ　16495
シマクモキリソウ　232193
シマクリガシ　78932
シマクロウメモドキ　328700
シマクロギ　161336
シマクロモジ　6771
シマグワ　259097
シマクワズイモ　16495
シマケンザン　139256
シマコウヤボウキ　292084
シマコウリンカ　385921
シマコウリンクワ　385921
シマコオヤボオキ　292084
シマコガネギク　368484
シマコゴメウツギ　375820
シマゴショウ　290308
シマコバノボタンヅル　95201
シマコバンノキ　296734,296809
シマコマツナギ　206669
シマコンギク　40105
シマコンテリギ　199853
シマサイカチ　176870
シマサカキ　8256
シマサギゴケ　374476
シマサギサウ　183559
シマサキラン　116936
シマザクラ　187602
シマサクラガンピ　414240
シマザサ　304109,347214
シマササバラン　232168
シマサザンクワ　68953
シマサルスベリ　219970
シマサルナシ　6697
シマサンキライ　366414
シマサンゴアナナス　8558
シマシウカイダウ　49827
シマシカクイ　143374
シマシカクヰ　143374
シマジタムラソウ　345107
シマシデ　77314
シマシバ　389318
シマシャウジャウバカマ　191068
シマシャクナゲ　330993
シマシャジン　7845
シマシュスラン　179709
シマシラカシ　78877
シマシラキ　161632,161639
シマシラタマ　172100
シマシロヘビイチゴ　167620
シマズイナ　210364,210402
シマススキ　255896
シマスズキ　255893
シマスズメノヒエ　238697,285423
シマスナビキサウ　117986
シマスナビキソウ　117961
シマスヒカヅラ　235935
シマスミレ　409993
シマスヰゼンジナ　183086
シマズヰナ　210364
シマセイシボク　161657
シマセキコク　125279
シマセンダン　139641
シマセンダン属　139637
シマセンネンショウ　137401
シマセンブリ　81502,380117
シマセンブリ属　81485
シマソケイ　263198
シマソナレムグラ　187514,187639
シマタイミンタチバナ　261632
シマタウガラシ　72100
シマダカツラ　249649
シマダカツラ属　249648
シマダケ　297239,347214
シマタゴ　167994
シマタコノキ　281138,281169
シマタニワタリノキ　8199
シマタヌキマメ　112013
シマタヌキラン　74369
シマタビラコ　397410
シマダフヂ　66637
シマダンチク　37469
シマチカラシバ　289267
シマチドメグサ　200275
シマチョウセンマツ　299799
シマヂワウギク　55820
シマツゲトモモ　382649
シマツナソ　104059,104103
シマツユクサ　100990
シマツレサギソウ　302262
シマテイカカズラ　18504
シマテウセンマツ　299799
シマテンツキ　166313,166315
シマテンナンショウ　33429
シマトウヅル　65697
シマトウヒレン　348388
シマトガサハラ　318599
シマトキンサウ　368528
シマトキンサウ属　368527
シマトキンソウ　368528
シマトキンソウ属　368527
シマトネリコ　167982,167984
シマトベラ　301433
シマトリモチ　46859
シマナガジラミ　305202
シマナガバヤブマオ　56135
シマナナメノキ　203821
シマナンヤウスギ　30839,30848
シマナンヨウスキ　30848

シマナンヨウスギ 30846
シマニクケイ 91429
シマニシキソウ 159069
シマニハウメ 83162
シマヌカボシソウ 238685
シマネナシカツラ 78731
シマノウチ 259094
シマノガリヤス 65292
シマノギク 193977
シマノゲシ 320513
シマノササゲ 138929
シマノボタン 43462
シマノミノツヅリ 32136
シマハクサンボク 407898
シマハシドイ 382257
シマハスノハカズラ 375904
シマハスノハカヅラ 375904
シマハヒノキ 381206
シマハビロ 181743
シマハマアカザ 44520,44542
シマハマグルマ 413506
シマハマボウ 195311
シマハマママツナ 379499
シマバライチゴ 338698,339135
シマバラサウ 52592
シマバラサウ属 52587
シマバラソウ 52617
シマバラソウ属 52587
シマハラン属 288638
シマバラン属 288638
シマハンクワイサウ 229068
シマヒギリ 96257
シマヒゲシバ 88320,88351
シマヒシグリ 78942
シマヒトッバハギ 167096
シマヒトッバハギ属 167067
シマヒノキ 67529,67530
シマビハ 151144
シマヒメウツギ 127114
シマヒメタデ 309863
シマヒメタヌキモ 403191
シマヒメツバキ 68970
シマヒメヒゴタイ 348411
シマヒメミカンソウ 296809
シマビャクシン 213775
シマヒョオタンボク 103862
シマビランヂ 363467
シマピンポン 376092
シマフジウツギ 62110
シマフジバカマ 158182,158214
シマフトイ 352279
シマフムラサキツユクサ 394093
シマフムラサキツユクサ属 393987
シマヘウタンボク 103862
シマボウ 386500
シマホウザイ 116829
シマホオズキ 297711
シマホオヅキ 297711
シマホザキラン 243010
シマホタルブクロ 70170
シマホホヅキ 297711
シマホルトノキ 142372
シマボロギク 148164
シマママユミ 157888
シマママンテマ 363467
シマミサオノキ 12517,325301
シマミサヲノキ 12517
シマミズ 298869
シマミゾソバ 308886
シマミソハギ 19566
シマミゾハゴヘ 125872
シマミヅキ 106656
シマミヅ属 223674
シマミヤマシウカイダウ 49974
シマムカシヨモギ 406392
シマムラサキ 66794,66829,66831
シマムラサキツユクサ 394093
シマムラサキツユクサ属 393987
シマムロ 213962
シマメダケ 304007
シマモクセイ 276311
シマモクゼイ 276336
シマモグセイ 276313
シマモチ 204042
シマモミ 216120,216125
シマモミ属 216114
シマヤブイチゴ 338305
シマヤマハギ 226940
シマヤマヒハツ 28341
シマヤマブキショウマ 37065
シマヤマモミヂ 2942
シマヤマモモ 261122
シマユキカズラ 299115
シマユキカヅラ属 299109
シマユズリハ 122713
シマユヅリハ 122713
シマユリ 229723
シマユリワサビ 72721
シマュンジュ 240131
シマヨウゾメ 407850
シマヨシ 293710
シマヨツバムグラ 170352
シマリュウガン 310823
シマリュウガン属 310818
シマリュウゼツ 10794
シマルバ 364382
シマルリサウ 117986
シマルリソウ 117961,117986
シマルリミノキ 222124
シマレンプクソウ 8402
シマワシグサ 305202
シマヲガルカヤ 117185
シマヲバナ 262552
シーマンバラカヤシ 46744
ジミカン 93732
ジムカデ 78646
シムシュワタスゲ 152769
ジムナスタ－属 256301
シムモンドシア科 364451
シムラニンジン 320976
シムラニンジン属 77766
シメノウチ 3356
シモキタイチゴ 338811
ジモグリッメクサ 397106
シモクレン 416707
シモダカンアオイ 37689
シモツガ 298303
シモツキザクラ 316439
シモツケ 371944
シモツケサウ 166097
シモツケサウ属 166071
シモツケソウ 166097
シモツケソウ属 166071
シモツケ属 371785
シモニタネギ 15292
シモニ－ドロ 311482
シモバシラ 215811
シモバシラ属 215805
シモバナイリス 208565
シモバナオトメユリ 230024
シモフリチク 11346
シモフリヒバ 85352,85353
シモフリマツヨイグサ 269441
シモンプラム 316813
シャ 376466
シャウ 232557
シャウガ 418010
シャウガ科 418034
シャウギウ 91368,91378
シャウキラン 239257
シャウグンボク 243371
ジヤウザン 129033
ジヤウザン属 129030
シャウジャウクワ 243,997
シャウジャウサウ 159046
シャウジャウタケ 46848
シャウジャウバカマ 191065
シャウジャウバカマ属 191059
シャウジャウハダマ属 406040
シャウジャウボク 243,997
シャウジヤウボク 159675
シャウシャップ 25868
シャウジュ 91368,91378
ジャウバウザサ 347199
ジャウバウナシ 323348
シャウブ 5793,5798
ジャウブスゲ 74051
シャウブノキ 231355
シャウブ属 5788
シャウリャウバナ 240030
シャガ 208640

ジャガイモ　367696
ジャカウエンドウ　222789
ジャカウサウ属　86805
ジャカウサギ　302355
ジャカウチドリ　302355
ジャカウナデシコ　127635
ジャカウノコギリサウ　3995
ジャカゴソウ　247010
ジャカゴトウバナ　96993
ジャガタライモ　367696
ジャガタラズイセン　196452,307269
ジャガタラズイセン属　196439
ジャガタラズヰセン　196452
ジャガタラズヰセン属　196439
ジャガタラミカン　93857
ジャガタラユ　93705
ジャカランダ属　211225
シャク　28044
ジャク　28030
ジャクコウ　102227
シャクシナ　59595
シャクジョウソウ　202767,258033
シャクジョウソウ科　258049
シャクジョウソウ属　258006
シャクジョウバナ　202767
シャクヂャウサウ　202767
シャクヂヤウヅル　94853
シャクヂャウバナ　202767
シャクヂャウマメ　218721
シャクチリソバ　162309
シャクナゲ　330532,331227
シャクナゲモドキ　332205
シャクナゲヤドリギ　385221
シャクナケ科　150286
シャクナゲ属　330017
シャクナン　330532
シャクナンガンピ　122655
シャクナンゲ　330532
シャクナンショ　233949
シャクヘウタン　219847
シャグマサイコ　321667
シャグマハギ　396828
シャクヤク　280147,280213,280214
ジャクランギョク　233608
ジャクロ　321764
シャク属　28013
ジャケイチュウ　94552
ジャケツイバラ　64991
ジヤケツイバラ　64990,64993
ジャケツイバラ科　65084
ジャケツイバラ属　64965
シャコウアオイ　243803
ジャコウアオイ　243803
ジャコウウリ　114189
ジャコウエンドウ　222789
ジャコウオランダフウロ　153869
ジャコウキヌラン　417772
ジャコウソウ　86821
ジャコウソウモドキ　86793
ジャコウソウモドキ属　86781
ジャコウソウ属　86805
ジャコウチドリ　302355
ジャコウチュウ　211701
ジャコウノコギリソウ　3995
ジャコウミゾホオズキ　255223
ジャコウユリ　229900
ジャコオアオイ　243803
ジャコオソウ　86821
シャコタンチク　347202
シャコバサボテン　352000
シャコハンチク　347270
ジャコビニア属　211338
ジャコンサウ　272221
シャジクソウ　396956
シャジクソウ属　396806
シャシャリンバイ　329068
シャシャンボ　403738
シャシャンポ　403738
ジャショウシ　97719
シャスタ・デージー　227512
シャスタデージー　227512
ジャスミナム・ポリアンツム　211963
ジャスミノセレウズ属　211700
ジャスミン　172781
シャセシナズナ　390223
シャゼンオモダカ　140543
シャゼンムラサキ　141258
シャゼンムラサキ属　141087
シャチガシラ　163418
シャヂクサウ　396956
シャヂクサウ属　396806
シャックパイン　299818
ジャテイ　385531
シャトリムシマメ　354761
シャトリムシマメ属　354755
ジャニンジン　72829
ジャノヒゲ　272090
ジャノヒゲ属　272049
ジャノメアカマツ　299894
ジャノメエリカ　149138,149750
ジャノメギク　346304
ジャノメギク属　346300
ジャノメクンショウギク　172348
ジャノメサウ　104598
ジャノメソウ　104598
ジャバソケイ　211846
ジャハンリュウ　182458
シャヘンゲ　221238
ジャボチカバ　261049
シャボテンダイゲキ　159710
ジヤボン　93579,93754
シャボンソウ属　346418
シャボンノキ　324624
シャボンマメ　294040
シャマアハガヘリ　294962
シャマアワガエリ　294962
シャマクロボシサウ　238690
シヤマサギスゲ　396005
シヤマモジズリ　264765
シヤマモヂズリ　264765
ジャーマン・アイリス　208583
ジャーマンアイリス　208583
シャムオリヅルラン　88537
シャムサクラララン　198850
シャムロダイコン　53249
シャモヒバ　85321
ジャモンギョク　182443
ジャモンチュウ　94557
ジャモントウ属　261090
ジャヤナギ　343342,343917
シャラノキ　376466
シャリスゲ　73828
シャリンタウ属　107319
シャリントウ属　107319
シャリンバイ　329086,329114
シャリンバイ属　329056
ジャワケイトウ　80396
ジャワザクラ　219966
ジャワサンタンカ　211113
ジャワセンナ　78336
ジャワニッケイ　91353
ジヤワヒギリ　96356
ジャワビロウ　234190
ジャワフトモモ　382660
ジャワルカム　166788
シュウアカシア　1168
シュウカイドウ　49885,49886
シュウカイドウ科　50433
シュウカイドウ属　49587
ジュウガツザクラ　83112
ジュウガツハマオモト　111230
シュウケンマル　242886
ジュウジアケボノソウ　380368
ジュウジグン　182484
シュウセンギョク　263742
シュウソウ　102209
シュウソウギョク　103757
シュウテンマル　45893
ジュウニキランソウ　13142
ジュウニノマキ　186422
ジュウニヒトエ　13146
ジュウネン　290940
シュゥビギョク　32743
シュウブンソウ　333475
シュウブンソウ属　333471
シュウベンノキ属　400515
ジュウホウマル　45895
シュウメイギク　23854
ジュウヤク　198745
シュウレイギョク　272535
シュウレイマル　234935

ジュウロクササゲ　409096
シュエンマル　234946
シュカイギョク　86490
シュガ-パイン　300011
シュガ-ブッシュ　315937
シュクガマズミ　407804
シュクコンアマ　231942
シュクコンゲシ　282617
シュクコンタバコ　266035
シュクコンヌメゴマ　231942
シュクシャ　187427,187432
シュクシャミツ　19933
シュクシャ属　187417
シュクシラベ　467
ジュズサンゴ　334897
ジュズサンゴ属　334892
シュスジコウライダコ　16526
ジュズスゲ　74915
ジュズダマ　99124
ジュズダマ属　99105
ジュズナルコスゲ　76683
ジュズネタツミ　355615
ジュズネナミキソウ　355615
ジュズネノキ　122064,122067
シュスラン　179708
シュスラン属　179571
シュッコンアマ　231942
シュッコンカスミソウ　183225
シュッコンハゼラン　383324
シュッコンバ-バナ　405885
シュッコンヒャクニチソウ　418046
シュッコンフウセンカズラ　73201
シュッコンヤグルマギク　81033
シュデンギョク　182480
ジュ-ト　104072
ジュトウ　65146
シュ-ドリプサリス属　318218
シュドリプサリス属　329744
シュミットザウ　42995
シュミットスゲ　76178
シユミツトスゲ　73714
シュムシュクワガタ　407142
シュムシュノコギリソウ　3922
ジュモウラン　416649
ジュモクチュウ　125523
シュユ　161323
シュラハッコウ　192436
ジュラン　11300
ジュランカヅラ　139065
シュルンベルゲラ属　351993
ジュレイギョク　233557
シュレイマル　234928
シュレヒテランッツ属　351935
シュレンクタウヒ　298426
シュロ　393809
シュロウカク　373874
ジュロウカンアオイ　194345
シュロガヤツリ　118476,118477
シュロサウ　405601
シュロサウ属　405570
シュロソウ　405601,405607
シュロソウ属　405570
シュロチク　329184,329186
シュロチクヤシ　321170
シュロチク属　329173
シュロ科　280608
シュロ属　393803
シュワンテシア属　352753
シュンイギョク　86587
シュンオウテン　171624
シュンギク　89481
シュンキンギョク　175530
シュンコウカン　93764
ジュンサイ　59148
ジュンサイ属　59144
シュンジギョク　102668
シュントウギク　290821
ジュンヒギョク　182469
シュンフラン　59603
シュンホウマル　140863
シュンポクラン　116774
シュンラン　116880
シュンラン属　116769
シュンロウ　163451
ジョウイ　352353
ショウウナシ　323348
ショウウン　249603
ジョウエッキバナノコマノツメ　409730
ショウエン　82307,82332
ショウガ　418010
ショウガ科　418034
ショウガ属　417964
ショウキズイセン　239257,239282
ショウキュウギョク　145248
ショウキュウマル　155225
シヨウギョク　86589
ショウキラン　239257,416382
ショウキラン属　416378
ショウグン　273067
ショウコウミズキ　106636
ジョゥゴバナ　111724
ジョウザン　129033
ジョウザンアジサイ　129033
ジョウシュウアズマギク　151015
ジョウシュウオニアザミ　92266
ジョウシュウカモメヅル　117703
ショウジュウソウ　159046
ショウショウ　102086
ショウジョウアナナス　8599
ショウジョウカ　997
ショウジョウスゲ　73888
ショウジョウソウモ　159046
ショウジョウソウモドキ　159046
ショウジョウバカマ　191065
ショウジョウバカマ属　191059
ショウジョウハグマ　406392
ショウジョウハグマ属　406040
ショウジョウフクシア　168748
ショウジョウフブキ　94560
ショウジョウボク　159675
ショウジョウヤシ　120782
ショウジョウヤシ属　120779
ショウジョウラン　369180
ショウズ　408839
ショウドウザクラ　83133
ショウドシマベンケイソウ　200818
ショウドシマレンギョウ　167470
ショウナンバク　67529
ショウナンバク属　67526
ショウナンボク　67529,67530
ショウナンマル　395643
ショウノテンツキ　166471
ショウブ　5793,5798
ショウフウギョク　103766
ショウフクギョク　233654
ショウブノキ　231355
ショウブ科　5771
ショウブ属　5788
ショウベンノキ　400546
シヨウベンノキ　400539
ショウベンノキ属　400515
ショウホウ　86572
ショウボウザサ　347199
ショウボウザサ　347338
ショウボウジダケ　304015
ジョウマツ　300163
ショウヨウギク　124826
ショウヨウマル　366824
ショウレイ　82385
ショウロウクサギ　96401
ジョウロウスゲ　74051
ジョウロウホトトギス　396602
ジョウロウラン　134326
ショウロギョク　55658
ショウロクサギ　96401
ショエノルキス属　352298
ジョオウヤシ　31705
ジョオウヤシ属　31702
ショオジョオソウ　158727
ショオドシマレンギョオ　167470
ショカッサイ　275876
ショクダイオオコンニャク　20145
ショクヨウガヤツリ　118822
ショクヨウカンナ　71169
ショクヨウギク　124826
ショクヨウタンポポ　384714
ショクョウトケイサウ　285637
ショクョウトケイソウ　285637
ショクヨウボウフウ　292892
ショクヨウホオズキ　297676,297716
ショクヨウホオヅキ　297716

ショクロウギョク　263737
ショタイサウ　75993
ショートリ－フパイン　299925
ジョン－アンセル属　26273
ションバーキア属　352477
ションバーグキア属　352477
ジョーンフィジ－ノヤシ　405257
シラ－　353001
シライス　134948
シライトサウ　87777
シライトサウ属　87775
シライトシュスラン　179627
シライトソウ　87777
シライトソウ属　87775
シライヤナギ　344108
シライワコゴメグサ　160216
シライワシャジン　7715
シラウバウエンゴサク　105810
シラオイハコベ　374880
シラカシ　116153,116187
シラガドウ　411464,411540
シラカバ　53483,53503,53572
シラカハスゲ　75348
シラカハバウフウ　293038
シラガブドウ　411540
シラガブドオ　411540
シラカワスゲ　75348
シラカワタデ　291666
シラカワボウフウ　293033,293038
シラカンバ　53483,53503,53572
シラカンバ属　53309
シラキ　142396,142399,264506,346390
シラギグ　357164
シラギョク　267044
シラキンウラハグサ　184659
シラキ属　346361
シラクキナ　59603
シラクチズル　6515
シラクチヅル　6515
シラゲエチゴトラノオ　317936
シラゲオウヂ　188099
シラゲオニササガヤ　128586
シラゲガヤ　197301
シラゲガヤ属　197278
シラゲカワラサイコ　312467
シラゲキクバクワガタ　317954
シラゲキツネノヒガサ　111741
シラゲキンチャクズル　94933
シラゲクサギ　95984
シラゲコオリンカ　385892
シラゲシャジン　7865
シラゲタンポポ　110785
シラゲテンノウメ　276543
シラゲトリカブト　5663
シラゲハハキギ　217372
シラゲヒメジソ　259299
シラゲホウキギ　217372
シラゲマツムシソウ　350132
シラゲマメハギ　126290
シラゲメボウキ　268526
シラゲモロコシガヤ　369678
シラゲヨシ　295907
シラゲヨモギ　35816
シラコスゲ　76022
シラシグサ　54158
シラシデ　2868
シラシマメダケ　318325
シラスゲ　73641,73646,74383,74942
シラタエセッコク　125104
シラタエヤグルマギク　81004
シラタマカズラ　319833
シラタマカヅラ　319833
シラタマコウジ　31479
シラタマコシキブ　66769
シラタマサウ　364193
シラタマソウ　363671,364193
シラタマノキ　172111,172142
シラタマノキ属　172047
シラタマホシクサ　151416
シラタマミズキ　104920,105207
シラタマミヅキ　104920
シラタマモクレン　232594
シラタマユリ　230041
シラチャレイジンソウ　5374
シラツガ　496
シラトリオトギリ　202178
シラトリシャジン　7739,7878
シラナミ　162908
シラネアオイ　176723
シラネアオイ属　176722
シラネアザミ　348355,348559
シラネイ　213507
シラネセンキウ属　24281
シラネセンキュウ　24433
シラネニガナ　210535
シラネニンジン　391904
シラネヒゴタイ　348878
シラネマツハダ　298230
シラネヰ　213507
シラハギ　226708,226849,226974,226989
シラハシノキ　2910
シラバトツバキ　69082
シラヒゲソウ　284539
シラヒゲムヨウラン　223644
シラビソ　496
シラフイモ　11332
シラフオオトキワツユクサ　393989
シラフジ　414576,414577
シラフチク　11355
シラフヘイシソウ　347149
シラベ　496
シラペセンキウ　24433
シラホウマスゲ　73564
シラボシカイウ　417097
シラホシベゴニア　50051
シラボスゲ　75347
シラミシバ　394390
シラモミ　272
シラヤマアフヒ　242672
シラヤマギク　135264
シラヤマブキ　188922
ジラユキエリカ　150099
シラユキギク　300842
シラユキゲシ　146054
シラユキスミレ　410732
シラユキナガハシスミレ　410503
シラヰヤナギ　344108
シラン　55575
ジラン　116829
シラン属　55559
シラ－属　352880
シリアアザミ　267008
ジリエホウオウボク　65014
シリブカ　233228
シリブカガシ　233228
シリブトビシ　394485
シリュウ　358633
シリュウギョク　264350
シルトスペルマ属　120768
シルトラン　156572
シルベストルアオカズラ　182384
シルホ－ギ　411464
シルラニンジン　320976
シルラニンジン属　320975
シルレルコテフラン　293633
シレイマル　234966
シレトコスミレ　410143
シレネ属　363141
シレンクワ科　234254
シレンゲ科　234254
シレンボ－　93324
シロ　393809
シロアカザ　86901
シロアキグミ　142215
シロアザミゲシ　32424
シロアミダネヤシ　129684
シロアミメグサ　166710
シロアヤメ　208807
シロイカリサウ　146995
シロイカリソウ　147004
シロイナモリサウ　318187
シロイナモリソウ　318187
シロイヌウメモドキ　204240
シロイヌタデ　309262,309468
シロイヌナズナ　30146
シロイヌナズナ属　30105
シロイヌナヅナ　30146
シロイヌナヅナ属　30105
シロイヌノヒゲ　151380
シロイノコヅチ　4263
シロイモ　99919

シロイヨアブラギク　89547
シロイワハゼ　172053
シロウ　56145
シロウコン　114884
シロウツギ　375817
シロウマアカバナ　146755
シロウマアサツキ　15467,15715
シロウマアザミ　92239
シロウマエビラフジ　408674
シロウマオウギ　43042
シロウマオオカニコウモリ　283785
シロウマオトギリ　201742
シロウマゼキシャウ　213545
シロウマゼキショウ　213545
シロウマタンポポ　384435
シロウマチドリ　302361
シロウマツガザクラ　297017
シロウマナズナ　137236
シロウマヒメスゲ　76665
シロウマリンドウ　174234
シロウマリンドウ属　174204
シロウメソウ　270705
シロウメモドキ　204248,241781
シロウリ　114201
シロウンゼン　331804
シロエゾノキツネアザミ　92385
シロエゾホシクサ　151423
シロエニシダ　120909
シロエンドウ　301070
シロカエマル　164875
シロカジュマル　164875
シロカスリソウ　130139
シロガスリソウ属　130083
シロカゼクサ　147969
シロカツラ　83736
シロガネアオイ　37591
シロガネカズラ　294842
シロガネガラクサ　161430
シロカネサウ属　210251
シロガネスミレ　410201
シロガネセンノウ　363757
シロカネソウ属　210251
シロガネチカラシバ　289303
シロガネツツジ　331449
シロガネヨシ　105456
シロガネヨシ属　105450
シロカノコユリ　230041
シロカピタン　414577
シロガヤ　393077
シロガヤツリ　119337,129019
シロガラシ　364545
シロガラシ属　364544
シロカワミドリ　10415
シロカワラヨモギ　35282
シロカンチョウジ　57955
シロギキョウ　302755
シロキジムシロ　312946
シロギニアヤム　131814
シロギリ　320828
シロキリシマ　331371
シロクキタイサイ　59556
シロクワイ　342424
シロクワヰ　342424
シロケヤマウツボ　222647
シロケラマ　331760
シロコスミレ　410165
シロゴチョウ　361385
シロコメガヤ　249015
シロコヤマモモ　261137
シロゴョウ　361384
シロゴンズイ　160955
シロザ　86901
シロザクラ　53483,83255,280016
シロサクラツツジ　331937
シロサハフタギ　381341
シロサマニヨモギ　35158
シロジオウ　327435
シロシオガマ　287590
シロシキブ　66808,66809
シロジクカスリソウ　130135
シロジクトウモドキ属　303096
シロジクトウ属　303087
シロジシャ　231403
シロシダレヤナギ　342935
シロシデ　77401
シロシヒサカキ　160506
シロシマウチワ　28093
シロシマカンギク　89536
シロシマススキ　341858
シロシマセンネンボク　137380
シロシマメダケ　304100
シロシャクジョウ　63966
シロシャクヂャウ　63966
シロスジカイウ　65230
シロスジカゲロウラン　193592
シロスジクワズイモ　16507
シロスミレ　410360
シロソケイ　212043
シロソネ　77401
シロソバナ　7801
シロタエ　108921,272997
シロタエギク　81004,358395
シロタエセッコク　125104
シロタエハマゴウ　411434
シロタエヒマワリ　188922
シロタカオススキ　341858
シロタカヲススキ　341858
シロタヌキマメ　111879
シロダブ　264090
シロタヘヒマハリ　188922
シロダモ　264090
シロダモモドキ　264021
シロダモ属　264011
シロタンポポ　384430
シロヂシャ　231403
シロチョウセンカメバヒキオコシ　209667
シロチョウセンヨモギ　35171
シロツヅ　264090
シロップ　64973
シロツブ　64971
シロツブアナナス　8555
シロツメクサ　397043
シロツリフネ　205369
シロツルフジバカマ　408263
シロテツ　249163
シロテンマ　171925,171937
シロドウダン　145689
シロトウヒ　298286
シロトゲコクオウマル　103747
シロトダシバ　37392
シロトベラ　301226
シロナタマメ　71052
シロナナコ　368526
シロナラ　323625
シロネ　239215
シロネ属　239184
シロノジスミレ　410772,410779
シロノセンダングサ　54054
シロノヂシャ　404468,404469
シロバイ　381272
シロバキヅタ　187324
シロハグロ　340367
シロハグロ属　340336
シロバナアカイシリンドウ　174237
シロバナアカツメクサ　397020
シロバナアキノウナギツカミ　291941
シロバナアキノタムラソウ　345110
シロバナアケビ　13231
シロバナアケボノシュスラン　179612
シロバナアケボノスミレ　410500
シロバナアサマリンドウ　173897
シロバナアザミ　92067
シロバナアズマギク　151009
シロバナアズマシャクナゲ　330535
シロバナアゼトウガラシ　404703
シロバナアツモリソウ　120396,120397
シロバナアブラギク　89406,124821
シロバナアポイアズマギク　151013,151014
シロバナアレチアザミ　92374
シロバナアレノノギク　193941
シロバナイガカウゾリナ　143460
シロバナイカリサウ　146995
シロバナイカリソウ　147013
シロバナイソマツ　230817
シロバナイチハツ　208876
シロバナイナモリサウ　318187
シロバナイナモリソウ　318187
シロバナイヌカウジュ　259323
シロバナイヌゴマ　373400
シロバナイヌタデ　291825
シロバナイヌトウバナ　97030

シロバナイブキジャコウソウ　391348
シロバナイブキスミレ　410257
シロバナイリス　208565
シロバナイワアカバナ　146607
シロバナイワウチワ　362271
シロバナイワカガミ　351452
シロバナイワギキョウ　70118
シロバナイワギリソウ　272599
シロバナイワザクラ　315052
シロバナイワタバコ　101682
シロバナイワニガナ　210674
シロバナイワブクロ　289342
シロバナインドソケイ　305213
シロバナウグイスカグラ　235814
シロバナウスゲオカスミレ　410407
シロバナウチョウラン　310892
シロバナウツギ　413609
シロバナウツボグサ　316100
シロバナウラギク　41434
シロバナウルップソウ　220169
シロバナウンゼンツツジ　331805
シロバナエゾ　278913
シロバナエゾアオイスミレ　410656
シロバナエゾオオサクラソウ　314522
シロバナエゾカワラナデシコ　127856
シロバナエゾクガイソウ　407496
シロバナエゾコザクラ　314281
シロバナエゾスミレ　409943
シロバナエゾツツジ　389573
シロバナエゾノカワヂシャ　406984
シロバナエゾノシモツケソウ　166083
シロバナエゾノタチツボスミレ　409631
シロバナエゾフウロ　175024
シロバナエゾミソガワソウ　265065
シロバナエゾムラサキツツジ　330497
シロバナエゾヤマハギ　226701
シロバナエゾヨツバシオガマ　287077
シロバナエゾリンドウ　174008
シロバナエゾルリムラサキ　153492
シロバナエチゴルリソウ　270702
シロバナエニシダ　120979
シロバナエヒメアヤメ　208795
シロバナエンレイサウ　397622
シロバナエンレイソウ　397622
シロバナオウギカズラ　13127
シロバナオオイワウチワ　362271
シロバナオオイワカガミ　351458
シロバナオオカモメズル　117734
シロバナオオキツネノカミソリ　239275
シロバナオオサクラソウ　314516
シロバナオオタチツボスミレ　410155
シロバナオオナンバンギセル　8768
シロバナオオノアザミ　91742
シロバナオオバクサフジ　408553
シロバナオオバコ　302103
シロバナオオバジャノヒゲ　272125
シロバナオオベニウツギ　413604
シロバナオオボウシバナ　100971
シロバナオオルリソウ　118040
シロバナオカタツナミソウ　355393
シロバナオナガエビネ　66008
シロバナオニアザミ　92228
シロバナオニシオガマ　287460
シロバナオノエリンドウ　174109
シロバナオホバコ　302103
シロバナオヤマノエンドウ　278911
シロバナオヤマリンドウ　173622
シロバナオンツツジ　332105
シロバナカイフウロ　174902
シロバナガガイモ　252515
シロバナカガノアザミ　92088
シロバナカキドオシ　176829
シロバナカノコソウ　404263
シロバナカメバヒキオコシ　209860
シロバナカモメヅル　117706
シロバナカラノアザミ　92168
シロバナカラフトハナシノブ　307202
シロバナカラミザクラ　83284
シロバナカラムラサキツツジ　331290
シロバナカワミドリ　10407
シロバナカワラナデシコ　127863
シロバナカンナ　71179
シロバナキキョウソウ　397760
シロバナキクバクワガタ　317953
シロバナキシツツジ　331658
シロバナキセルアザミ　92391
シロバナキタアザミ　348716
シロバナキツネアザミ　191673
シロバナキツネノカミソリ　239272
シロバナキツネノマゴ　214734
シロバナキハギ　226722
シロバナキョウチクトウ　327577
シロバナキランソウ　13093
シロバナキンラン　82052
シロバナクガイソウ　407463
シロバナクサタチバナ　117532
シロバナクサナギオゴケ　409481
シロバナクサフジ　408358
シロバナクサボケ　84559
シロバナクズ　321444
シロバナクリンソウ　314511
シロバナクルマバナ　96974
シロバナクロクモソウ　349607
シロバナクワガタソウ　407233
シロバナグンナイフウロ　174780
シロバナケウツギ　413625
シロバナケショウアザミ　92084
シロバナケタチツボスミレ　410032
シロバナケフチクタウ　327577
シロバナケヤブハギ　126527
シロバナゲンカイツツジ　331289,331290,331297
シロバナゲンゲ　43054
シロバナゲンノショウコ　174967
シロバナコアジサイ　199898
シロバナコアヤメ　208830
シロバナコイワザクラ　314886
シロバナコウシンソウ　299755
シロバナコウシンヤマハッカ　209858
シロバナコシオガマ　296069
シロバナゴジクワ　362523
シロバナコシジシモツケソウ　166077
シロバナコタチツボスミレ　410043
シロバナコバノタツナミ　355511
シロバナコバノミツバツツジ　331624
シロバナコヒルガオ　68689
シロバナコマクサ　128312
シロバナゴマクサ　81720
シロバナコマツナギ　206447
シロバナコメツツジ　332020
シロバナコモウセンゴケ　138355
シロバナコンロンカ　260451
シロバナサカイツツジ　331453
シロバナサギゴケ　247009
シロバナサクラソウ　314957
シロバナサクラタデ　309262
シロバナササゲ　293986
シロバナササユリ　229874
シロバナサデクサ　291834
シロバナサルスベリ　219934
シロバナサワギキョウ　234767
シロバナサワギク　359588
シロバナサワラン　143441
シロバナサワルリソウ　22039
シロバナサンインヒキオコシ　209833
シロバナサンタンカ　211148
シロバナサンタンクワ　211148
シロバナサンプクリンドウ　100281
シロバナシ　371881
シロバナシコクママコナ　248178
シロバナシハイスミレ　410750
シロバナシブカワツツジ　331725
シロバナシマアザミ　91817
シロバナシマルリソウ　117964
シロバナジマン　55579
シロバナシモツケ　371957,371975
シロバナシモツケソウ　166098
シロバナシャウジャウバカマ　191065
シロバナシャクナゲ　330245,330954
シロバナシャグマハギ　396829
シロバナシャジクソウ　396959
シロバナシャリントウ　107583
シロバナシュウカイドウ　49887
シロバナジュウニヒトエ　13147
シロバナジョウシュウオニアザミ　92267
シロバナシラゲエンゴサク　105596
シロバナシレンゲ　234252
シロバナジロボウエンゴサク　105812
シロバナジンチョウゲ　122538,122540
シロバナスミレ　410360
シロバナスミレサイシン　410713

シロバナセイタカトウヒレン 348843
シロバナセイヨウウンラン 231184
シロバナセイヨウトゲアザミ 91771
シロバナセキチク 127656
シロバナセキヤノアキチョウジ 209658
シロバナセンダン 248896
シロバナセンダングサ 54054
シロバナセンブリ 380235
シロバナセンボンヤリ 224112
シロバナソシンカ 49248
シロバナタイワンレンギョウ 139068
シロバナタカアザミ 92297
シロバナタカサゴアザミ 92079
シロバナタカネシオガマ 287805
シロバナタカネセンブリ 380380
シロバナタカネナデシコ 127854
シロバナタカネナデッコ 127853
シロバナタカネビランジ 363159
シロバナタカネマツムシソウ 350184
シロバナタツナミソウ 355495
シロバナタテヤマウツボ 316124
シロバナタテヤマリンドウ 173979
シロバナタニジャコウソウ 86819
シロバナタマアジサイ 199914
シロバナタムラソウ 361025
シロバナダンギク 78013
シロバナタンポポ 384430
シロバナチシマギキョウ 69957
シロバナチシマセンブリ 380377
シロバナチシマヒョウタンボク 235722
シロバナチシマリンドウ 174119
シロバナチドリ 310867
シロバナチョウセンアサガオ 123077
シロバナチョウセンヤマハギ 226804
シロバナチョクザキヨメナ 40851
シロバナツクシアザミ 92417
シロバナツクシコスミレ 410109
シロバナツクシシオガマ 287578
シロバナツクシタツナミ 355545
シロバナツクシハギ 226839
シロバナツシマママコナ 248196
シロバナツメクサ 295286
シロバナツヤスミレ 410044
シロバナツユクサ 100964
シロバナツリガネニンジン 7864
シロバナツリガネヤナギ 289337
シロバナツルコケモモ 278378
シロバナツルソケイ 212043
シロバナツルリンドウ 398276
シロバナデイゴ 154735
シロバナテリハタチツボスミレ 409977
シロバナテングノコヅチ 398282
シロバナテンニンソウ 228004
シロバナトウウチソウ 345820
シロバナトウゴクミツバツツジ 332090
シロバナトウシャジン 7831
シロバナトキソウ 306871
シロバナトキワハゼ 246987
シロバナトゲソバ 291954
シロバナトサムラサキ 66928
シロバナナガバイワカガミ 351454
シロバナナガバノイシモチソウ 138296
シロバナナカバノスミレサイシン 409747
シロバナナガバノタチツボスミレ 410331
シロバナナギナタコウジュ 143975
シロバナナツノタムラソウ 345184
シロバナナミキソウ 355779
シロバナナワシロイチゴ 338989
シロバナナンゴククガイソウ 407466
シロバナナンバンクサフジ 385985
シロバナニイテンジクアオイ 288396
シロバナニオイタチツボスミレ 410313
シロバナニガナ 210536,210594
シロバナニシキウツギ 413581
シロバナニシキゴロモ 13192
シロバナニワフジ 205877
シロバナヌスビトハギ 126528
シロバナネコノメソウ 90321
シロバナネムノキ 13580
シロバナノアキチョウジ 209750
シロバナノアサガオ 207892
シロバナノイヌナズナ 136951,136952
シロバナノイヌヤマハッカ 209852
シロバナノコギリサウ 3921,4028
シロバナノコギリソウ 4028
シロバナノコメツツジ 332018
シロバナノシナガハハギ 249203
シロバナノシナガワハギ 249203
シロバナノダケ 24337,24340
シロバナノチシマフウロ 174584
シロバナノノコギリサウ 4014
シロバナノハラアザミ 92270
シロバナノハラクサフジ 408276
シロバナノハンカイシオガマ 287240
シロバナノビネチドリ 264009
シロバナノヒメシャクナゲ 22451
シロバナノヘビイチゴ 167635
シロバナノボタン 248734
シロバナノホトケノザ 220126
シロバナノマアザミ 91863
シロバナノミヤコグサ 237506
シロバナノラフバイ 87526
シロバナハウチワマメ 238421
シロバナハギ 226800,226852,226974
シロバナハクサンオオバコ 301999
シロバナハクサンコザクラ 314285
シロバナハクサンチドリ 121361
シロバナハクサンフウロ 175030
シロバナハグロソウ 129273
シロバナハコネウツギ 413577
シロバナハナガサヒゴタイ 348406
シロバナハナサクラタデ 291729
シロバナハナズオウ 83771
シロバナハハコグサ 178265
シロバナハマアザミ 92187
シロバナハマウツボ 275011
シロバナハマエンドウ 222736
シロバナハマカンギク 89541
シロバナハマゴウ 411431
シロバナハマナス 336902
シロバナハマナタマメ 71056
シロバナハマナデシコ 127740
シロバナハマヒルガオ 68739
シロバナハマフウロ 175033
シロバナハマベンケイソウ 250896
シロバナハマボウ 194912
シロバナハリマツリ 139068
シロバナハルリンドウ 173975
シロバナヒイラギソウ 13115
シロバナヒガンバナ 239255
シロバナヒナゲシ 282622,282633
シロバナヒナスミレ 410668
シロバナヒナヒゴタイ 348406
シロバナヒメエゾネギ 15721
シロバナヒメオドリコソウ 220417
シロバナヒメケフシグロ 248271
シロバナヒメツルコケモモ 278363
シロバナヒメトケイソウ 285626
シロバナヒメトラノオ 317942
シロバナヒメヒゴタイ 348691
シロバナヒメホテイラン 68539
シロバナヒメミミカキグサ 403258
シロバナヒメヤマアザミ 91826
シロバナヒヨス 201372
シロバナヒヨドリジョウゴ 367323
シロバナビランジ 363626
シロバナヒルガオ 68707
シロバナヒルギ 329765
シロバナヒレアザミ 73340
シロバナビロードモウズイカ 405789
シロバナヒロハクサフジ 408436
シロバナフウリンツツジ 145679
シロバナフクシマシャジン 7640
シロバナフジ 414555
シロバナフジアザミ 92333
シロバナフジウツギ 62102
シロバナフジスミレ 410666
シロバナフジツツジ 331986
シロバナフデリンドウ 174096
シロバナブラシノキ 67293
シロバナベゴニア 50298
シロバナヘバイチゴ 167627
シロバナホウサイ 117042,117051
シロバナホウザイ 117042
シロバナホウライツユクサ 100931
シロバナホザキノミミカキグサ 403135
シロバナホソバシャクナゲ 331188
シロバナホソバメハジキ 224991
シロバナホタルカズラ 233787
シロバナホツツジ 397864
シロバナホトケノザ 220356

シロバナホンシャクナゲ　331230
シロバナマキ　67293,67294
シロバナマスノスミレ　410404
シロバナマツムシソウ　350176
シロバナママコナ　248202
シロバナマルバクマツヅラ　405890
シロバナマルバママコナ　248205
シロバナマンサク属　167537
シロバナマンジュシャゲ　239255
シロバナマンテマ　363477
シロバナマンネングサ　356503
シロバナミカエリソウ　228012
シロバナミズアオイ　257573
シロバナミゾカクシ　234365
シロバナミゾコウジュ　345311
シロバナミチノクコザクラ　314286
シロバナミツバウツギ　374093
シロバナミツバツツジ　330579,331624
シロバナミミカキグサ　403379
シロバナミヤコグサ　237706
シロバナミヤマアズマギク　151011
シロバナミヤマキリシマ　331011
シロバナミヤマシオガマ　287013
シロバナミヤマスミレ　410548
シロバナミヤマタムラソウ　345182
シロバナミヤマヒナホシクサ　151395
シロバナミヤマママコナ　248182
シロバナミヤママンネングサ　356848
シロバナミヤマリンドウ　173663
シロバナムシトリスミレ　299765
シロバナムシャリンドウ　137553
シロバナムショケギク　322660
シロバナムラサキシキブ　66810
シロバナムラサキニガナ　219522
シロバナムラサキモメンヅル　41918
シロバナムラサキヤシオツツジ　330059
シロバナメハジキ　224990
シロバナモウコナデシコ　127656
シロバナモウズイカ　405670
シロバナモジズリ　372252
シロバナモチツツジ　331175
シロバナモッケ　371957
シロバナモメンヅル　42193
シロバナヤウシュテウセンアサガホ　123077
シロバナヤエウツギ　126892
シロバナヤチシャジン　7729
シロバナヤツシロソウ　70062
シロバナヤナギハッカ　203099
シロバナヤノネグサ　309465
シロバナヤハズエンドウ　408289
シロバナヤハズトウヒレン　348753
シロバナヤブウツギ　413589
シロバナヤブツバキ　69198
シロバナヤブハギ　126532
シロバナヤマジオウ　220390
シロバナヤマジソ　259306,259323
シロバナヤマトキソウ　306880
シロバナヤマハッカ　209704
シロバナヤマブキ　216090
シロバナヤマホタルブクロ　70242
シロバナヤマルリソウ　270697
シロバナユウカオ　67736
シロバナユウバリリンドウ　174111
シロバナユキグニミツバツツジ　331040
シロバナユキツバキ　69210
シロバナユキバヒゴタイ　348207
シロバナユキワリコザクラ　314372
シロバナヨウシュチョウセンアカガオ　123077
シロバナヨツバハギ　408509
シロバナラシャナス　367140
シロバナラショウモンカズラ　247741
シロバナラフゲ　312946
シロバナラロウゲ　312946
シロバナリシリリンドウ　173541
シロバナリュウキュウコスミレ　410783
シロバナリュウキュウシャジン　7857
シロバナリンドウ　173853
シロバナルリソウ　270700
シロバナルリトラノオ　317969
シロバナレブンコザクラ　314375
シロバナレブンシオガマ　287083
シロバナレブンソウ　279001
シロバナレンリソウ　222829
シロバナワタ　179900
シロバナワルナスビ　367015
シロバヒ　381272
シロバヤバネバショウ　66159
シロハリスゲ　76513
シロビ　390680
シロヒエンサウ　124067
シロヒガン　83328
シロヒキオコシ　209715
シロヒナラン　19513
シロヒメクグ　218571
シロビユ　18655
シロヒヨコマメ　90802
シロフイリクヮズイモ　16511
シロフカンアフヒ　37674
シロフクリンツルマサキ　157508
シロフェリンリュウゼッ　10874
シロフジマメ　218722
シロブチャッデ　162883
シロブナ　162368
シロブナノヤマヂノギク　193977
シロフハカタカラクサ　394024
シロフハカタツユクサ　394021
シロベンケイソウ　200793
シロボウエンゴサク　105810
シロボケ　84554,84573
シロボシ　229737,244192
シロホンモンジスゲ　75792
シロマツ　299830
シロマツバユリ　229805
シロマラフクラギ　393242
シロマンヨウ　331285
シロミアオキ　44927
シロミアワブキ　249415
シロミイイギリ　203423
シロミクサギ　96399
シロミセンニンコク　18752
シロミタチバナ　31410
シロミナンテン　262200
シロミノアオキ　44923
シロミノガンコウラン　145071,145073
シロミノコムラサキ　66769
シロミノサハフタギ　381341
シロミノサワフタギ　381341
シロミノツルシキミ　365937,365956
シロミノツルマサキ　157495
シロミノハリイ　143205
シロミノブダウ　79850
シロミノベスカイチゴ　167654
シロミノヘビイチゴ　138794
シロミノマンリョウ　31400
シロミノミズキ　104920
シロミノヤブヘビイチゴ　138797
シロミノヤブムラサキ　66874
シロミミズ　133185
シロミミズ属　133181,395164
シロミミナグサ　83060
シロミルスベリヒユ　361669
シロモクセイ　276294
シロモジ　231456
シロモジ属　283092
シロモノ　172111
シロモミ　298442
シロモモ　261213
シロヤシオ　331593
シロヤジオウ　327437
シロヤシボ　331593
シロヤナギ　343549,343578
シロヤブケマン　106004
シロヤマザクラ　83234
シロヤマツツジ　330967
シロヤマハギ　226683
シロヤマブキ　332284
シロヤマブキ属　332282
シロヤマモモ　261213
シロユリ　229730
シロヨナメ　39928
シロヨメナ　39968
シロヨモギ　35779
シロライラック　382368
シロリウキウツツジ　331284
シロリュウキュウ　331284
シロリュウキュウツツジ　331284
シロワレモカウ　345921
シロワレモコウ　345921
シロヰノコヅチ　4263

シワアオイソウ 290315
シワウ 171144
シエハコベ 176775
シヲジ 168068
シヲン 41342
シヲン属 39910
シンウシノケグサ 164126
ジンエフイチヤクサウ 322870
ジンエフスイバ 278577
ジングウスゲ 76130
ジングウツツジ 331723
ジングツツジ 332107
シンクリノイガ 80821
シンゲツ 359991
ジンコウ 29973
シンコウサワギク 359103
シンゴニューム属 381855
シンコマツ 298298
ジンジサウ 349604
ジンジソウ 349204,349604
ジンジャ 187427
ジンジャー 187427
シンジュ 12559,163471
シンシュウネズ 213735
シンジュガヤ 354141
シンジュガヤ属 353998
シンジュギク 41202
シンジュ属 12558
ジンショウイチヤクソウ 322870
シンショウガ属 353998
シンショウギョク 32723
シンズイユイガシ 233360
シンスミレ 410784
シンスヰユイガシ 233360
シンセイアマリリス 18865
シンセガイ 58303
シンゼジューム 116916
シンセンウギョク 163446
シンソケイ 211938
ジンダイ 83913
ジンダンボウ 323611
シンチククグ 245518
シンチクシラヤマギク 40976
シンチクテンツキ 166530
シンチクヒメハギ 307927
ジンチョウゲ 122532
ジンチョウゲ科 391030
ジンチョウゲ属 122359
ジンチョオゲ 122532
シンツルムラサキ 48689
シンテンアケボノサウ 380354
シンテンチ 182476
シンテンルリミノキ 222138,222281
シントウ 109264
シンナシエビネ 66101
シンナツダイダイ 93834
シンニンギア属 364737
シンノウ 94561
シンノウヤシ 295475,295484
シンノラフバイ 87528
ジンバイカリソウ 147005
ジンバイソウ 302329
シンバク 213696
シンバラリア属 116732
シンバリソウ 116747
シンビジューム属 116769
シンビディエラ属 116755
シンフィアンドラ属 380769
シンフィグロッサム属 380781
シンプウギョク 86575
シンフォリカルポス属 380734
シンベゴニア属 380641
シンボー 130915
シンホウギョク 86624
ジンボソウ 238632
シンミズヒキ 26626
シンミヅヒキ 26626
ジンメンチク 297282
シンヤマトグサ 389185
ジンユウツハブキ 229063
ジンヨウイチヤクソウ 322870
ジンヨウキスミレ 409680
ジンヨウスイバ 278577
ジンヨオイチヤクソウ 322870
シンラン 248895
ジンリョウユリ 229873
シンリョクカク 105437
シンリョクギョク 163440
シンレイ 82312,102377
シンワスレナグサ 260868
スーアソ 133587
スイウン 249603
スイエン 82365
スイエンギョク 263741
スイカ 93289,93304
スイガ 233667
スイカズラ 235878
スイカズラ科 71955
スイカズラ属 235630
スイカヅラ属 235630
スイカ属 93286
ズイキ 99913
スイグキナ 59436,59603
スイコウカン 182423
スイコウギョク 102510
ズイコウマル 140878
スイザボン 93818
スイザンカニツリ 127299
スイシカイダウ 243623
スイシカイドウ 243623
スイジカイドウ 243625
ズイシャイチゴ 338281
スイシャカンコノキ 177181
スイシャセトカヤ 281533
スイシャホシクサ 151532
ズイシャヤナギ 343587
ズイシャリャウシロバヒ 381206
ズイシャリャウハヒノキ 381206
スイシュウラン 258044
スイショウ 178019
スイショウ属 178014,385253
スイスマンテマ 364156
スイセイ 102628
スイセイジュ 386851
スイセイジュ科 386847
スイセイチュウ 299250
スイセン 262457,262468
ズイセン 262468
スイセンアヤメ 399094
スイセンアヤメ属 369975
スイゼンジソウ 183066
スイゼンジナ 183066
スイゼンジナ属 183051
スイセンノウ 363370
スイセン属 262348
スイチョウ 404945
スイテキギョク 175531
スイート 18966
スイート・アーモンド 20897
スイートオレンジ 93765
スイート・サルタン 81219
スイートピー 222789
スイートライム 93536
スィートレモン 93534
ズイナ 210389
ズイナ属 210359
ズイノキシバ 373556
スイバ 339887
ズイハウダモ 240726
スイビギュク 32362
スイフヨウ 195044
ズイボウ 200077
ズイホウギョク 43495
スイホコ 52563
スイモノグサ 277747
スイユウギュク 32366
スイラン 195696
ズィランマル 182459
スイレイ 175524
スイレン 267770
スイレン科 267787
スイレン属 267627
スウィングルキンカン 167515
ズウエソウ 371240
スウェーデンカブ 59507
スウェーデンナナマド 369432
スエツムハナ 77748
スエヒロアオイ 37742
スオウ 65060
スオウギ 83769,192494
スオウバナ 83769

スオウバナノキ　83769
スオウボク　382329
スカシタゴボウ　336250
スカシユリ　229836
スカビオサ・コ-カシカ　350119
スカフィグロティス属　350494
スカフォセパラム属　350479
スガモ　297182,297185
スガモ属　297181
スカ-レットナラ　323780
スガワラビランジ　364079
スカンポ　339887
スギ　113685
スキアトフィツム属　365910
スギゴケテンツキ　166444
スギナエンドウ　279029
スギナモ　196818
スギナモ科　196809
スギナモ属　196810
スギナリゲイトウ　80396
スギノハカズラ　38985
スキバアカシア　1686
スギバアマ　231953
スギバノウロコモミ　30841
スギモトアザミ　91706
スギモリゲイトウ　18788
スキンダプスス属　353121
スギ科　385250
スギ属　113633
スグサ　277747
スグダチミヤマシキミ　365974
スクティカリア属　355882
ズクノキ　142396
スグリ　333929,334209
スグリウツギ　263183
スグリ科　181339
スグリ属　333893
スクレロカクタス属　354287
スグワ　249449
スクワギョク　148144
スゲ　74339
スゲアマモ　418382
スゲガヤ　202724,202727
スゲガヤ属　202702
スゲモ　418392
スゲユリ　229775,229888,230005,230009
スゲ属　73543
スコッツパイン　300223
スコ-テチニピナンガ　299674
スゴノキ　249449
スコルゾネラ属　354797
ズサ　379374
スシ　339887
スジカンゾウ　191291
スジギボウシ　198651
スジテッパウユリ　229845
スジヌマハリイ　143139,143401
スジミココヤシ属　380553
スジュリ　229730
スズ　347200
スズカアザミ　92418
スズカカンアオイ　37658
スズカケ　371868
スズカケソウ　407503
スズカケノキ　302592
スズカケノキ科　302229
スズカケノキ属　302574
スズガケヤナギ　367178
スズカゴ　340673
スズガシ　204138
スズカゼリ　24433
スズガヤ　60383
ススキ　255886
スズキ　255886
スズキアザミ　92419
ススキササガャ　254010
ススキノキ　415178,415181
ススキノキ科　415182
ススキノキ属　415172
ススキメヒシバ　130605
ススキョシ　256632
ススキ属　255827
ススクサ　403687
ズズコ　99124
スズコウジュ　290989
スズサイコ　117643
スズシロサウ　30263
スズシロソウ　30263
スズタケ　347200
スズダケ属　347377
スズナリヤ-　112138
スズナリラン　232139
スズフリイカリソウ　146955
スズフリノキ　226320
スズフリバナ　159027
スズフリホンゴウソウ　352871
スズフリヨモギ　36286
スズムシサウ　85949,232225
スズムシソウ　85949,232322
スズムシバナ　85949
スズムシラン　232322
スズムシラン属　232072
スズメウリ　417472
スズメウリ属　249760
スズメガヤ　147593,147797
スズメガヤ属　147468
スズメグサ　342259
スズメナスビ　367682
スズメノアハ　151702
スズメノアワ　151702
スズメノエンドウ　408423
スズメノオゴケ　117537,117538
スズメノカタビラ　305334
スズメノケヤリ　152791
スズメノコウメ　123322
スズメノコビエ　285471,285507
スズメノチャヒキ　60777
スズメノチャヒキ属　60612
スズメノテツバウ　17501
スズメノテッパウ　17525
スズメノテッパウ属　17497
スズメノテッポウ　17498,17501
スズメノテッポウ属　17497
スズメノトウガラシ　231479,231483,404696
スズメノトウガラシモドキ　231496
スズメノトケイソウ　285703
スズメノナガビエ　285417
スズメノナギナタ　283656
スズメノナギナタ属　283654
スズメノハカマ　277747
スズメノハコベ　253200
スズメノハコベ属　253197
スズメノヒエ　238583,238600,285530
スズメノヒエツナギ　285381
スズメノヒエ属　238563,285390
スズメノヤリ　17525,238583,238600
スズメノヤリ属　238563
スズメノレンリソウ　222731
スズメノヲゴケ　117537,117538
スズメヒゲシバ　372642
ススメムギ　45448
スズメラン　270846
ススヤアカバナ　146834
ススヤアザミ　348142
ススヤトモユ　201743
スズユリ　230036
スズラン　102863,102867,147238
スズランノキ　417592
スズラン属　102856,147106
スズルシサウ　85949
スズルシソウ　85949
スゼクアニカポプラ　311514
ズソウカンアオイ　37705
スタキス属　373085
スダジイ　78916,79022
スダチ　93814
スタ-チスシヌアタソビア　230762
スタペリアンサス属　374037
スタペリア属　373719
スタペリオプシス属　374044
ズダヤクシュ　391524
ズダヤクシュ属　391520
スダレバラ　336366
スダレヨシ　304016
ス-ダングラス　369707
スタンホ-ペア属　373681
スタンホペア属　373681
スチギバウシ　198651
スチグマトダクチラス属　376576
スチゾロビューム属　259479

スヂテッパウユリ　229845
スヂヌマハリヰ　143401
スヂバアデク　382672
ズッキ－ニ　114301
スッポンノカガミ　200228
スティパ属　376687
スティリディウム科　379071
スティリディウム属　379076
ステ－ウィア属　376405
ステゴビル　15373
ステッキヤシ属　46374
ステップナズナ　18357
ステップノギク　193918
ステップヨモギ　35119
ステトソニア属　376363
ステナンドリウム属　375411
ステノカルプス属　375510
ステノグロッチス属　375573
ステノグロッティス属　375573
ステノコリネ属　375558
ステノセレウス属　375516
ステノフイラコ－ヒ－　99008
ステノメッソン属　375604
ステビア　376415
ステビア属　376405
ステファノセレウセ属　375918
ステファノ－ティス属　211693,376022
ステベリア属　377254
ステベンソンヤシ属　376400
ステリス属　374699
ステルキュリア属　376063
ステルンベルギア属　376350
ステレオキラス属　376228
ストケシア　377288
ストケシア属　377286
ストック　246478
ストマティウム属　377331
ストランベ－シャ属　377433
ストリキニ－ネノキ　378836
ストリキニ－ネノキ属　378633
ストリクノス属　378633
ストレプタンセラ属　377625
ストレプトカリックス属　377646
ストレプトカルプス属　377652
ストレリチア科　377580
ストロウブマツ　300211
ストロビランテス属　378079
ストロファンタス　378401
ストロフォカクタス　378488
ストロフォカクタス属　378487
ストロ－ブゴヨウマツ　300211
ストロ－ブマツ　300211
ストロ－プマツ　300211
ストロベリ　167597
ストロマンテ属　378305
ストロンギロドン属　378347
ストロンボカクタス属　378312
スナオニソテツ　145221
スナゴセウ　290365
スナジオグルマ　207224
スナジカゼクサ　148019
スナジカワラサイコ　313054
スナジジャコウソウ　391167
スナジスゲ　74650
スナジタデ　309792
スナジツルフジバカマ　408272
スナジニラ　15113
スナシバ　418425,418431
スナジヒゴタイ　348129,348132
スナジマメ　418335
スナジマメ属　418320
スナジミチヤナギ　308788
スナジヨモギ　36032
スナダイゲキ　158490,159854
スナヂタデ　309792
スナヂヒジキ　104834
スナヂマメ　418331,418335
スナヂマメ属　418320
スナヅル　78731
スナヅル属　78726
スナテンツキ　166313
スナハマスゲ　119619
スナビキサウ　393269
スナビキサウ属　252333
スナビキソウ　190652,393269
スナビキソウ属　252333
スナマメ　418331
スノキ　324384,404003,404005,404047
スノキ属　403710
スノ－ドロップ　169719
スノ－フレ－ク　227856
スノ－フレ－ク属　227854
ス－パ　364713
ズバイモモ　20955
スハウ　65060
スハウギ　192494
スハウチク　47346
スハウノキ　385355
スハウバナ　83769,83771
スハウバナノキ　83769,83771
スパシフィルムパティニ－　370344
スパシフィルム属　370332
スバシモモ　20955
スパソグロッチス属　370393
スハタ　140236
スパティカルパ属　370315
スパトグロティス属　370393
スハマソウ　192140
スハマソウ属　192120
スパラックシス属　369975
スパルマンニア属　370130
スヒカヅラ　235878
スヒカヅラ科　71955
スヒカヅラ属　235630
スピランテス属　372176
ズブシグナ－ツム　28134
スブタ　55918
スブタ属　55908
スブマツカナ属　379634
スブラウチング・ブロッコリ－　59545
スプレケリア　372913
スペインアヤメ　208946
スペイングリ　78811
スベリヒユ　311890
スベリヒユモドキ　394903
スベリヒユ科　311954
スベリヒユ属　311817
スペルトコムギ　398970
スベロサモクマオウ　79185
スホウ　65060
スホウチク　47346
スホウノキ　385355
スポンジアス　372467
スポンディアス属　372459
ズマイラックス　38960
スマトラエゴノキ　379423
スマトラキビ　282286
スマロ　364935
ズミ　243717
スミカワスゲ　73573
スミクロスティグマ属　366136
スミセイヤウミザクラ　83361
スミセイヨウミザクラ　83361
スミティナンディア属　366730
ズミノキ属　104917
スミホコ　171702
スミミザクラ　83361
スミヨシヤナギ　342949
スミレ　410201,410360
スミレイワギリ　56063
スミレサイシン　410710
スミレ科　410792
スミレ属　409608
スメルノキ　94191
スモモ　316761
スラッシュパイン　299928
スラッシュマツ　299928
スラ－ル　300068
スリアン　392841
スルガスゲ　75609
スルガテンナンショウ　33570
スルガヒメユズリハ　122729
スルガヘウタンボク　236211
スルガユコウ　93531
スルガラン　116829
スルコレブティア属　379715
スルボ　353057
スレン　392847
スロ　393809
スロノキ　393809
スワウボク　382329

スワキクバスミレ　409609
スヰグキナ　59436
スヰクワ　93304
スヰクワ属　93286
スヰザンカニツリ　127299
スヰシャイチゴ　338281
スヰシャウ　178019
スヰシャカンコノキ　177181
スヰシャセトカヤ　281533
スヰシャホシクサ　151532
スヰシャヤナギ　343587
スヰシャリャウシロバヒ　381206
スヰシャリャウハヒノキ　381206
スヰショウ　178019
スヰショウ属　385253
スヰセン　262468
スヰゼンジサウ　183066
スヰゼンジナ　183066
スヰセンノウ　363370
スヰ-トピ-　222789
ズヰナ　210389
ズヰナ属　210359
ズヰハウダモ　240726
スヰラン　195696
スヰレン科　267787
スンキ　93717
ズングリオヒシバ　143553
スンシウヘウタン　236211
セアラゴムノキ　244509
セイウンカク　299246
セイウンリュウ　273820
セイエン　102272
セイオウマル　267045
セイカ　336486
セイガイ　332360
セイガイツツジ　331107,331113,331174
セイカヒメハギ　308126
セイカンギョク　182478
セイガンサイシン　37582
セイカンラン　116930
セイギョク　140879
セイコウボク　301006,358234
セイコウマル　288616
セイコノヨシ　295916
セイサリュウ　109180
セイシカ　330617,331065
セイシカズラ　92777
セイジギョク　233538
セイシクワ　330617
セイシボク　161643,161647
セイシボク属　161628
セイシマル　182487
セイシャウガシ　233122
セイジヤウクリガシ　79027
セイシュンギョク　102460
セイジロ　10948
セイスイソウ　139602
セイスヰサウ　139602
セイセンギョク　246603
セイダイ　358057
セイタカアキノキリンソウ　368013
セイタカアワダチソウ　368013
セイタカウコギ　53913
セイタカオオニシキソウ　159109
セイタカオトギリ　202014
セイタカカナビキサウ　354660
セイタカカナビキソウ　354660
セイタカカナビクソウ　354660
セイタカカワズスゲ　75043
セイタカコゴメハギ　249205
セイタカサギゴケ　246978
セイタカサギサウ　183559
セイタカサクラソウ　314333
セイタカスズムシ　232197
セイタカスズムシソウ　232197,283209
セイタカセイヨウサクラソウ　314333
セイタカタウコギ　54001
セイタカタンポポ　384530
セイタカトウヒレン　348841
セイタカナミキソウ　355391
セイタカヌカボシソウ　238616
セイタカハハコグサ　178265
セイタカハマスゲ　119125
セイタカハリイ　143055
セイタカヒゴタイ　140789
セイタカフイリゲットウ　17771
セイタカブシ　5043
セイタカベンケイソウ　200793
セイタカマオウ　146128
セイタカミゾホオズキ　255210
セイタカュ-カリ　155478
セイタカユ-カリ　155719
セイタカヨシ　295916
セイチョウハコヤナギ　311400
セイテンカク　373800
セイデンファデニア属　357372
セイテンマル　284685
セイドウリュウ　61166
セイバンイラクサ　175877
セイバンイラクサ属　175865
セイバンカマツオ　295635
セイバンズミ　323207
セイバンセンブリ　380117
セイバンソシン　116926
セイバンナスビ　367682
セイバンナナカマド　369498
セイバンモロコシ　369652,369689
セイビマル　242891
セイボンギョク　140885
セイヤウアカネ　338038
セイヤウイヌゴマ　373110
セイヤウウンラン　231183
セイヤウエビラバギ　249232
セイヤウエビラフヂ　222782,222819
セイヤウオニシバリ　122515
セイヤウカボチャ　114300
セイヤウグンバイナヅナ　225475
セイヤウケフチクタウ　265327
セイヤウサクラサウ　315082
セイヤウサンザシ　109790
セイヤウサンジュユ　105111
セイヤウサンダンクワ　221238
セイヤウシナノキ　391706
セイヤウズハウ　83809
セイヤウダンチク　37469
セイヤウタンポポ　384714
セイヤウチャヒキ　45602
セイヤウトネリコ　167955
セイヤウナシ　323133
セイヤウナツュキサウ　166125
セイヤウナミキサウ　355359
セイヤウニンジン　280799
セイヤウニンジンボク　411189
セイヤウノコギリサウ　3978
セイヤウノコギリソウ　3978
セイヤウハコヤナギ　311400
セイヤウハシリドコロ　44708
セイヤウバラ　336474
セイヤウヒノキ　114753
セイヤウヒヒラギ　203545
セイヤウヒルガホ　102933
セイヤウフウチウサウ属　95606
セイヤウフウテウサウ　95796,172201
セイヤウフウテウソウ　172201
セイヤウマツムシサウ　350099
セイヤウミザクラ　83155
セイヤウミミナグサ　82680
セイヤウヤブイチゴ　338265
ゼイヤウヤマアイ　250600
ゼイヤウヤマアヰ　250600
セイヤウヤマナラシ　311400
セイヤウヤマハクカ　249504
セイヤウヤマハクカ属　249491
セイヤウリンゴ　243675
セイヨウアカニワトコ　345660
セイヨウアカネ　338038
セイヨウアジサイ　199986
セイヨウアブラナ　59493
セイヨウアマナ　60526
セイヨウアンズ　34475
セイヨウイソノキ　328701
セイヨウイチイ　385302
ゼイヨウイチイ　385302
セイヨウイトバマツムシソウ　350125
セイヨウイボタ　229662
セイヨウイラクサ　402886
セイヨウウキガヤ　177639
セイヨウウツボグサ　316127
セイヨウエゴノキ　379419
セイヨウエゾヤナギ　343280
セイヨウエビラバギ　249232

セイヨウエビラフジ　222782
セイヨウエンゴサク　169111
セイヨウオオバコ　302068
セイヨウオキナグサ　321721
セイヨウオダマキ　30081
セイヨウオトギリ　202086
セイヨウオニシバリ　122515
セイヨウオモダカ　342400
セイヨウカキドオシ　176824
セイヨウカジガエデ　3462
セイヨウカノコソウ　404316
セイヨウカボチャ　114288,114300
セイヨウカラハナソウ　199384
セイヨウカリン　252305
セイヨウカリン属　252291
セイヨウガンコウラン　145073
セイヨウカンボク　407989
セイヨウキズタ　187222
セイヨウキヌヤナギ　344244
セイヨウキョウチクトウ　265327
セイヨウキンシバイ　201787
セイヨウキンバイ　399504
セイヨウキンミズヒキ　11549
セイヨウクサレダマ　239898,239902
セイヨウクリ　78811
セイヨウコウゾリナ　298589
セイヨウコウボウ　27967
セイヨウコリヤナギ　343937
セイヨウサクラサウモドキ　105496
セイヨウサクラソウ　315082
セイヨウサンザシ　109790
セイヨウサンシュユ　105111
セイヨウサンダンカ　221238
セイヨウシナノキ　391706
セイヨウシャクヤク　280253
セイヨウジュウニヒトエ　13174
セイヨウシロヤナギ　342973
セイヨウスイレン　267648
セイヨウズオウ　83809
セイヨウスグリ　334250
セイヨウズハウ　83809
セイヨウスモモ　316382
セイヨウセキショウモ　404563
セイヨウタチヤナギ　344211
セイヨウタデキ　223116
セイヨウタンポポ　384714
セイヨウチャヒキ　45602
セイヨウツゲ　64345
セイヨウテリハヤナギ　343858
セイヨウドクウツギ　104700
セイヨウトゲアザミ　91770
セイヨウトチノキ　9701
セイヨウトネリコ　167955
セイヨウトラノオ　317927
セイヨウナシ　323133,323151
セイヨウナツユキソウ　166125
セイヨウナナカマド　369339
セイヨウナミキソウ　355359
セイヨウニガナ　110767
セイヨウニワトコ　345631
セイヨウニンジンボク　411189
セイヨウヌカボ　28972
セイヨウネズ　213702
セイヨウノコギリソウ　3978
セイヨウノダイコン　326609
セイヨウハギクソウ　158857
セイヨウバクチノキ　223101,223116
セイヨウハコヤナギ　311400
セイヨウハゴロモグサ　14065,14166
セイヨウハシバミ　106703
セイヨウハシリドコロ　44708
セイヨウハッカ　250420
セイヨウハナズオウ　83809
セイヨウハマアカザ　44554,44576
セイヨウバラ　336474,336640
セイヨウハルニレ　401512
セイヨウヒイラギ　203545
セイヨウヒイラギガシ　324027
セイヨウヒキヨモギ　284098
セイヨウヒゲシバ　88344
セイヨウヒノキ　114753
セイヨウヒメミヤコグサ　237762
セイヨウビャクシン　213702
セイヨウヒルガオ　102933
セイヨウフウチョウソウ　95796
セイヨウフウチョウソウ属　95606
セイヨウフジアザミ　73377
セイヨウブナノキ　162400
セイヨウマツムシソウ　350099
セイヨウマユミ　157429
セイヨウミザクラ　83155
セイヨウミズユキノシタ　238192
セイヨウミヅキ　380499
セイヨウミツバウツギ　374107
セイヨウミミナグサ　82680
セイヨウミヤコグサ　237539
セイヨウミヤマヤナギ　343416
セイヨウムラサキ　233760
セイヨウムラサキケマン　169011,169114
ゼイヨウヤチャナキ　261162
セイヨウヤドリギ　410960
セイヨウヤブイチゴ　338151,338447
セイヨウヤブジラミ　392993
セイヨウヤマアイ　250600
セイヨウヤマイチゴ　338265
セイヨウヤマカモジ　58578
セイヨウヤマハッカ　249504
セイヨウヤマハッカ属　249491
セイヨウヤマハンノキ　16352
セイヨウユギワリソウ　314366
セイヨウリンゴ　243675
セイヨウレンリソウ　222800
セイヨウワサビ　34590,34591
セイヨウワサビ属　34582
セイヨオバイカウツギ　294463
セイヨオバイカッキ　294463
セイラン　137551,304381
セイリュウ　329692
セイリュウカク　140021
セイリュウカズラ　171456
セイリュウトウ　171654
セイリョウテン　390482
セイレイギョク　272532
セイレイマル　234954
セイロウキズタ　187222
セイロウナシ　323133
セイロウニンジンボク　411189
セイロンオリ－ブ　142382,159828
セイロン・グ－スベリ－　136846
セイロンスグリ　136846
セイロンテッボク　252370
セイロンテリハボク　252370
セイロンナツメヤシ　295490
セイロンニクケイ　91446
セイロンニッケイ　91446
セイロンハコベ　200430
セイロンハコベ属　200411
セイロンヒメノボタン　276159
セイロンマツリ　305202
ゼウクシネ属　417698
セウナンボク　67529
セウベンノキ　400546
セウベンノキ属　400515
セオ　145517
セカイヤア　360565
セカイヤオスキ　360576
セカイヤオスギ　360576
セキカクチク　47370
セキコク　125257
セキコク属　124957
セキザイユ－カリ　155517
セキサウ　72934
セキシャウ　5803
セキシャウモ　404555
セキシャウモ属　404531
セキシュンギョク　233473
セキショウ　5803
セキショウイ　213028
セキショウボク　231355
セキショウモ　404555
セキショウモ属　404531
セキソウ　72934
セキダカズラ　393616
セキダカヅラ　393621
セキチク　127654,297334
セキチク科　77980
セキホウ　163464,163477
セキモンウライソウ　315464
セキモンスゲ　76568
セキモンノキ　94057
セキヤノアキチョウジ　324757

セキヤノヒキオコシ 209612
セキリュウマル 163456
セキレイマル 234955
セコイア 360565
セコイアオスギ 360576
セコイアデンドロン属 360575
セコイアメスギ 360565
セコイア属 360561
セコイヤ 360576
セコイヤデンドロン 360576
セサモタムヌス属 361284
セージ 345271
ゼージ 345271
セシロタツナミ 355494
セスジグサ 11342
セスジゲサ 11340
セソン 102136
セタカイヌヘラヤシ 194135
セダカウロコヤシ属 298830
セダカシロジクトウ 303089
セダカセンリゴマ 327434
セダカダイオウヤシ 337883
セタカニセダイオウヤシ 318121
セダカヒトツバクマデヤシ 212402
セダカワシントンヤシ 413307
セチセレゥス属 94551
セッカスギ 113692
セッカヤマネギ 15734
セッカン 93801
セッカンギョク 182448
セッカンスギ 113702
セッキギョク 103769
セックンマル 182451
セッコウ 59164
セッコウボク 380736
セッコク 125257
セッコク属 124957
セツザンバラ 336348
セッシュウギョク 284695
セッツイボタ 229563
セッテイクワ 191309
セットウマル 244031
セッナリホオズキ 297650
セツパクマル 244018
セッピコテンナンショウ 33495
セップンサウ属 148101
セップンソウ属 148101
セツリコ 59458
セツレンマル 234925
セティエキノプシス属 362002
セディレア属 356456
セトウチコアオスゲ 75144
セトウチコスズ 347245
セトウチマンネングサ 357037
セトエノニワザクラ 83208
セトガヤ 17537
セトガヤモドキ 293750
セトクレアセア属 361978
セトヘノニハザクラ 83208
セトヤナギスブタ 55911
セドロ 80030
セナス 52556
セナミスミレ 410018
ゼニアオイ 243797,243840,243850
ゼニアオイ属 243738
セニアフヒ 243840,243850
ゼニアフヒ属 243738
ゼニバアオイ 243823
ゼニバサイシン 37743
セネガ 308347
セネガルヤシ 295479
セネキオ属 358159
セネシオ属 358159
セノハナ 127072
ゼノービア属 417590
セファラリーア属 82113
セファロタス属 82560
セファロフィルム属 82276
セフリアブラガヤ 353434
ゼブリーナ属 393987
セベリニア属 362032
セメトケンラン 383153
セメンシナ 360825
セメンシナモトキ 360832
セーラゴムノキ 244509
セラシス属 389122
セラッカソバッ 362204
ゼラニウム 288297
セラプアズ属 360582
セリ 269326
セリニンジン 29327,123141
セリバオオバコ 301910
セリバシオガマ 287312
セリバノセンダングサ 177285
セリバヒエンサウ 124039
セリバヒエンソウ 124039
セリバフウロ 153914
セリバヤマブキサウ 200758
セリバヤマブキソウ 200758
セリバワウレン 103835
セリフォン 59438
セリモドキ 139733
セリ科 29230
セリ属 269292
セルダンビロウ 234181
セルデレー 225450
セルリー 29327
セルリアック 29329
セルンスール 236462
セレウス属 83855
セレニセレウス属 357735
セレニペジューム属 357751
セレベスコロマンデル 132093
セレベスパペダ 93422
セレリー 29327
セロチナポプラ 311163
セロヂネ属 98601
セロハノカラフトブシ 5193
セロプジア属 83975
セロリ 29327,29328
センイチヤ 10892
ゼンエンミズ 299017
ゼンエンミヅ 299017
センオウ 363265
センオウゲ 363265
センオウマル 157171
センカクアオイ 37718
センカクオトギリ 202152
センカクツツジ 330630
センカクトロロアオイ 232,236
センカクハマサジ 230759
センキウ 229371
センキウ属 97697
センキュウ 229371
センキュウ属 97697
センゲンオトギリ 202228
ゼンゴ 24336
センコウハナギ 184428
センゴクマメ 218721
センサウ 252242
センサウ属 252241
センザンコマノツメ 410559
センザンスミレ 410559
センザンツリフネ 205361
ゼンジャウ 15868
センジャウスゲ 75099
センジュガンピ 363515
センジュギク 383090
センジュギク属 383087
センジュラン 416538
センジョウアザミ 92378
センジョウスゲ 75099
センジョウマツ 300163
ゼンゼイ 244390
センソウ属 252241
センダ 248895,248903
センダイキツネアザミ 91698
センダイザサ 347203
センダイスゲ 76226
センダイスゲモドキ 75104
センダイソウ 349886
センダイタイゲキ 159813
センダイハギ 389540
センダイハギ属 389505
センダイハグマ 292041,292059
センダイヤナギ 344082
センダイヨシノ 316188
センダン 248895,248903,346210
ゼンダン 7190
センダンキササゲ 325002
センダンキササゲ属 324993,376257

センダングサ　53797,53801,54048
センダングサ属　53755
センタンクワ　211067
センダンバノボダイジュ　217626
センダン科　248929
センダン属　248893
センヂャウスゲ　75099
ゼンテイクワ　191309
センテンカ　191278
セントウソウ　85478
センドク　145418
ゼントネレラ属　417520
セントポーリア属　342482
セントランサス属　81743
セントーリューム属　81485
セントーレア　303082
セントーレア属　80892
セントンイモ　197793
セントンイモ属　197785
センナ　78212,78227,360490
センナシヒメコゴメグサ　160213
センナリバナナ　260215
センナリヒョウタン　219854
センナリヘウタン　219854
センナリホオズキ　297703
センナリホオヅキ　297650
センナリホホヅキ　297650
センニチギク　4866
センニチコウ　179236
センニチコウ属　179221
センニチコモドキ　4870
センニチサウ　179236,208349
センニチサウ属　179221
センニチソウ　179236
センニチソウ属　179221
センニチノゲイトウ　179227
センニチモドキ　4870
センニョハイ　138831
センニンカク　261708
センニンコク　18687
センニンサウ　95358
センニンサウ属　94676
センニンサボテン　272856
センニンソウ　95358
センニンソウ属　94676
センニンモ　312170
センネンソゥ　104367
センネンボク　104350,104367
センネンボクラン　104339
センネンボク属　104334
センノウ　363265
センノウゲ　363265
センノウ属　363141
センノオバラ　336955
センノキ　215442
センバカジ　116193
センバサクラサウモドキ　105501
センビキ　306493
ゼンブ　131501
センプウギョク　163423
センブキ　15420
センブク　171333
センブリ　380234
センブリ属　380105
センベイグミ　142164
センペルセコイア　360565
センペルセコイア属　360561
センペルビブム属　358015
センポウマル　327315
センボニャリ　224110
センボンギク　40845
センボンタンポポ　110985
センボンヤリ　224110
センボンヤリ属　175107,224106
センボンワケギ　15420
センリゴマ　327435,327450
センリヤウ科　88263
センリヤウ属　88264
センリュウ　192448
センリヨウ　346527
センリョウ科　88263
センリョウ属　88264
センヲウ　363265
ソアグヤシ　299679
ソアグヤシ属　299649
ソウイロサルビア　345292
ゾウイロッノギリソウ　87895
ソウイロレースソウ　393883
ソウウン　249580
ソウウンリュウ　273824
ソウエン　82372
ソウキッ　93731
ゾウゲマル　107023
ソウコヤ　25880
ソウゴヤ　25880
ソウサイスミレ　409925
ソウザンハヒノキ　381423
ソウシカンバ　53411
ソウシジュ　1145
ソウシシヨウニンジン　280748
ソウシチョウ　256260
ゾウジャウジビャクシ　192278
ゾウジョウジビャクシ　192278
ソウジンボク　69634
ソウソウギョク　198494
ソウニンボク　69634
ソウビギョク　32712
ソウマシオジ　168065
ソウモクカク　140320
ソウレイ　82366
ソウレイマル　327301
ソエレンシス属　366820
ソキホウヤマナラシ　311386
ソクシンラン　14529
ソクシンラン属　14471
ソクズ　345586
ゾクズヰシ　159222
ソクヅ　345586
ソクド　345586
ゾクワンザウ　191309
ソケイ　211848,211940
ソケイノウゼン　281225
ソケイノウゼン属　281223
ソケイ属　211715
ソゲキ　261648
ソコトラアロエ　17163
ソコベニアオイ　194955
ソコベニハクモクレン　416682
ソコベニヒルガオ　207953
ソシンカ　48993,49247
ソシンクワ　48993
ソシンラフバイ　87526
ソシンロウバイ　87526
ソーセージノキ　216324
ソーセージノキ属　216313
ソゾミ　407785
ソーダノキ　266371
ソーダノキ属　266355
ソテツ　115884
ソテツジュロ　295465
ソテツナ　110615
ソテツバアゼトウナ　110612
ソテツバカイウ　65239
ソテツ科　115793
ソテツ属　115794
ソテロサンツス属　369877
ソナカンバ　53380
ソナレ　213864
ソナレアマチャヅル　183025
ソナレシバ　372880
ソナレノギク　193954
ソナレマツムシソウ　350177
ソナレミヤマシキミ　365951
ソナレムグラ　187515,187549,187674,187677
ソネ　77401
ソノ　77323
ソノウサイシン　37621
ソノテ　302721
ソバ　162312,162338
ソバウリ　114245
ソバカズラ　162519
ソバカヅラ　162519
ソバグリ　162368
ソバグルミ　162368
ソバナ　7798
ソバノキ　231355,295691
ソバムギ　162338
ソバムギ属　162301
ソバ属　162301
ソブラリア属　366776

ソフロニチス属 369178
ソフロニティス属 369178
ゾフロニテラ属 369175
ゾボエシア属 121060
ソマイチゴ 338072
ソマムギ属 162301
ソメイヨシノ 83139,83365
ソメシバ 381423
ソメモノイモ 131522
ソメモノカズラ 245854,245858
ソメモノカヅラ 245854
ソメモノギク 104598
ソメモノコメツブノボタン 249945
ソメモノノキ 328794
ソメワケハギ 226853
ソモソモ 305855
ソヤ 77323
ソヨゴ 204138
ソライロゼリ 393883
ソライロレ-スソウ 393883
ソラマメ 408393
ソラマメ属 408251
ソリア属 368539
ソリシア属 368524
ソリダゴ 368013
ソリダ-ゴ属 367939
ソリチャ 79905
ソリチャ属 79902
ソルダネラ属 367762
ソロ 77323,77401
ソロノキ 77323
ソロハギ 166897
ソロハギ属 166850
ソロフヂ 166897
ソ-ロング 93631
ソンノイゲ 114325,240813
ダイウイキョウ 204603
タイエフベゴニヤ 50232
ダイオウ 329388
タイオウカヶ 183379
ダイオウグミ 142100
ダイオウゴヘイヤシ 228732
ダイオウショウ 300128
ダイオウマツ 299807,300128
ダイオウヤシ属 337873
ダイオウヤハズ 321164
ダイオウラン 155434
ダイガカク 299270
タイカフヰ 352284,353899
ダイガン 406016
タイキンギク 359980
タイキンギクモドキ 358637
ダイグンポ 52565
タイゲキサイコ 63622
ダイゲンゲッジョウ 359037
ダイケンラン 116781,117008
ダイコウシャク 233664
ダイコウマル 244137
タイコギョク 233483
ダイゴクデン 272938
ダイコクマメグンバイナズナ 225283
タイコジイ 78903
ダイコン 326616,326622
ダイコンサウ 175417
ダイコンサウ属 175375
ダイコンソウ 175417
ダイコンソウ属 175375
ダイコンドラ 128972
ダイコンドラ属 128955
ダイコン属 326578
タイサイ 59595
ダイサイコ 63728
ダイサギサウ 183559
ダイサギソウ 183559
タイサンキンパウゲ 326419
ダイサンチク 47516
タイザンニシキ 16617
タイサンボク 242135
タイシャクイタヤ 3413
タイシャククロウメモドキ 328904
ダイシュウチョウ 385881
タイジョ 131458
ダイショウカン 107053
タイショッカン 264388
ダイシンサイ 59438
ダイシンラン 117045
ダイズ 177750
ダイスギ 114548
ダイス属 121574
ダイズ属 177689
タイセイ 209201,209217,209229,209232
タイセイイモ 131485
タイセイ属 209167
タイセツイワスゲ 76349
ダイセツヒゴタイ 348717
ダイセツヒナオトギリ 202231
ダイセンアシボソスゲ 76200
ダイセンオトギリ 201741
ダイセンキスミレ 409775
ダイセンクワガタ 317956
ダイセンココメグサ 160182
ダイセンスゲ 74257
ダイセンヒョウタンボク 236126
ダイセンミツバツツジ 331039
ダイソウケン 86510
ダイソケイ 281225
ダイダイ 93332,93368
ダイダイエビネ 65924
ダイダイグサ 370515
ダイダイミノニワトコ 345680
ダイヅ 177750,177777
タイツリオウギ 42699,42705
タイツリスゲ 75314
タイツリソウ 220729
ダイヅ属 177689
タイトウアイ 283209
タイトウアフヒラン 265439
タイトウアヰ 283209
タイトウイチゴ 339344
タイトウウルミ 357881
タイトウウルミ属 357877
タイトウガシ 233383
タイトウカノコサウ 56420
タイトウガマズミ 408155
タイトウカマツカ 322471
タイトウキヌラン 417731
タイトウクグ 218518
タイトウケマン 105693
タイトウコウリンクワ 385921
タイトウサンザシ 322471
ダイトウシロダモ 264094
ダイトウセイシボク 161656
タイトウタカサゴサウ 210681
タイトウノゲイトウ 80422,80473
タイトウフジモドキ 122439
タイトウミズ 299017
タイトウミヅ 299017
タイトウリュウガン 310823
ダイトウリョウ 389205
タイトウリンダウ 173967
タイトウリンドウ 173967
ダイトウワダン 110621
タイトゴメ 356844
タイトンカンアフヒ 37674
タイトンコブナシバ 272612
タイトンシャクナゲ 330617
ダイトンチゴザサ 209105
ダイトンチヂミザサ 272612
タイトンッツジ 331305
ダイトンッツジ 331839
タイトントンボソウ 302519
タイナ 59595
タイナンアブラススキ 372434
タイナンカナビキサウ 354660
タイナンカナビキソウ 354660
タイナングス 91368,91369
タイナンニクケイ 91368,91369
ダイニチアザミ 91791
タイヌビエ 140367,140438,140462
タイハクマル 140620
タイバナアデク 156385
タイヒタイヒ 93676
タイフウシ 182962
ダイフウジノキ 199744
ダイフウジノキ属 199740
タイフウシ属 182961
ダイブウメモドキ 80203
ダイフクチク 47508
ダイブグミ 142207
ダイブグリ 141999
ダイブスゲ 76571

ダイブツルウメモドキ　80203
ダイブハラン　39529
タイヘイイチゴ　339067
タイヘイサンキライ　366291
タイベイマル　140133
タイヘイヨウクルミ　56033
タイヘイヨウグルミ　207000
タイヘイヨウクルミ属　56031
タイヘイヨウゾウゲヤシ　252633
タイヘイラク　125279,155306
タイホウカ　158665
タイホウギョク　43497
タイホクスゲ　76473,76497
タイマ　71218
ダイマチク　125477
タイマツグサ　79130
タイマツバナ　257161
ダイミャウチキ　318303
ダイミャウチク　357939
ダイミョウダケ　304032
ダイミョウチク　357939
タイミンカ　265327
タイミンガサ　283856
タイミンタチバナ　261648
タイミンタチバナ属　326523
タイミンチク　304032
タイミンッコク　125365
タイム　391397
タイモ　99910
ダイモンジ　267605
ダイモンジソウ　349344,349346
ダイユウマル　244259
タイヨウフウトウカズラ　313178
タイヨウベゴニア　50232
タイランドパペダ　93523
タイリクアズマギク　40016
タイリクオドリコソウ　220346
タイリクカワズスゲ　75178
ダイリクキヌヤナギ　344244
タイリクハリイ　143261
タイリクヤマブキソウ　200754
ダイリュウカン　140152
ダィリョウチュウ　395645
タイリンアオイ　37566
タイリンアカバナ　146787
タイリンインドソケイ　305219
タイリンオモダカ　342381
タイリンカッコウ　373232
タイリンカヤラン　304218
タイリンカンナ　71189
タイリンギリア　175685
タイリンゲッタウ　17772
タイリンゲットウ　17772
タイリンコクラン　232253
タイリンダンドク　71189
タイリンツメクサ　255506
タイリントキサウ　304218
タイリントキサウ属　304204
タイリントキソウ　304218
タイリントキソウ属　304204
タイリンバシロ　83874
タイリンハナシノブ　99858
タイリンヒメジギタリス　17476,17479
タイリンヒメハナシノブ　99858
タイリンヒメフヨウ　243929,243947
タイリンフクシア　168750
タイリンマガリバナ　61131
タイリンマツゲボタン　65843
タイリンマツヨイグサ　94133
タイリンミミナグサ　82819
タイリンミヤコナズナ　9798
タイリンヤナギハッカ　203100
タイリンヤマハッカ　209854
タイリンルリマガリバナ　61131
ダイワウサウ　55693
ダイワウマツ　300128
タイワンアイアシ　209356
タイワンアオイ　389946
タイワンアオイラン　2080,2084
タイワンアオカズラ　341570
タイワンアオカヅラ　341570
タイワンアオキ　44893
タイワンアカガシ　116150
タイワンアカシア　1145
タイワンアカバナ　146859
ダイワンアカヒゲガヤ　194020
ダイワンアカヒゲガヤ属　194015
タイワンアカマツ　300054,300083
タイワンアカメガシハ　243453
タイワンアカメヤナギ　344268
タイワンアキギリ　345254
タイワンアキグミ　142207
タイワンアケビ　374400,374425
タイワンアサガホ　207988
タイワンアサマツゲ　64369,64371
タイワンアザミ　348256
タイワンアシ　37477
タイワンアシカキ　223984
タイワンアセビ　298734
タイワンアヂサイ　199910
タイワンアヂサヰ　199910
タイワンアヅサ　157784
タイワンアデク　382535
タイワンアハブキ　249449
タイワンアハモリ　41825
タイワンアフヒ　389946
タイワンアブラキリ　212127
タイワンアブラキリ属　212098
タイワンアブラススキ　372412
タイワンアブラスズキ　372412
タイワンアブラ属　212098
タイワンアリドホシ　122028
タイワンアリドホシラン　261470
タイワンアワゴケ　265359
タイワンアワブキ　249449
タイワンアヲカヅラ　341570
タイワンアヲキ　44893
タイワンイス　134932
タイワンイチイ　385405,385407
タイワンイチゴ　339347
タイワンイチビ　905
タイワンイチヰ　385405
タイワンイトメヒシバ　130627
タイワンイナモリ　272282
タイワンイヌガヤ　82550
タイワンイヌグス　295361
タイワンイヌグス属　295344
タイワンイヌツゲ　203952,204299
タイワンイヌナシ　323207
タイワンイヌノヒゲ　151356
タイワンイヌビハ　164992
タイワンイボクサ　260110
タイワンイワタバコ　101686
タイワンウオクサギ　313702,313740
タイワンウキクサ　221007
タイワンウツギ　127049
ダイワンウツギ　127114
タイワンウツボグサ　316133
タイワンウドカヅラ　20327
タイワンウマノスズクサ　34203
タイワンウメガササウ　257374
タイワンウメガササウ属　257371
タイワンウメガサソウ　87490,87491,
257374
タイワンウメガサソウ属　257371
タイワンウメモドキ　203578
タイワンウラジロイチゴ　339338
タイワンウラジロイヌガヤ　19363
タイワンウラジロガシ　116200
タイワンウラジロノキ　320843
タイワンウリハダカエデ　3209
タイワンウヲクサギ　313702
タイワンエゴノキ　379347
タイワンエゴマ　290968
タイワンエノキ　80639,80764
タイワンエビネ　65952,66074
タイワンエンジュ　240130
タイワンオオバコ　302068
タイワンオガタマノキ　252851
タイワンオサラン　148684
タイワンオトギリ　202168
タイワンオニウド　24326
タイワンオニシバ　209345
タイワンオホバコ　302068
タイワンカウバシ　231324
タイワンカギカヅラ　401754,401761
タイワンカクレミノ　125604
タイワンカゴノキ　233945
タイワンガシ　233219
タイワンカナメモチ　295729
タイワンカニツリ　398547

タイワンカノキ　234075
タイワンカノコユリ　230038
タイワンカハホネ　267327
タイワンカハミドリ　10408
タイワンカハラボウフウ　292857
タイワンガフクワン　1145
タイワンガマズミ　408156,408198
タイワンカマツカ　322471
タイワンカモジグサ　335312
タイワンカモノハシ　209259
タイワンカモメヅル　400969
タイワンカワラケツメイ　78366
タイワンカンギク　124790
タイワンキクモ　230307
タイワンキジョラン　245805
タイワンキヅタ　187318
タイワンキツネノマゴ　337253
タイワンキヌラン　417764
タイワンキノヘラン　232139
タイワンキハダ　294236,294240,294256
タイワンキバナアツモリソウ　120446
タイワンキランサウ　13137
タイワンキランソウ　13137
タイワンギリ　285966
タイワンキンカン　167493
タイワンキンシバイ　201880
タイワンキンチャク　307919
タイワンキンパウゲ　326431
タイワンキンリョヘン　116863
タイワンクグ　245392
タイワンクサハギ　206714
タイワンクサフジ　386210
タイワンクズ　321453
タイワンクズマメ　259512
タイワンクチナシ　171246
タイワンクハズイモ　16495
タイワンクマガイソウ　120353
タイワンクマヤナギ　52423
タイワングリ　78802
タイワンクリガシ　78932
タイワングルミ　212608
タイワンクレソラン　94458
タイワンクロウメモドキ　328700
タイワンクロチク　297375
タイワンクロヅル　398322
タイワンクワ　259097
タイワンクワズイモ　16495
タイワンゲッタウ　17674
タイワンゲットウ　17674
タイワンケマン　106218
タイワンケヤキ　417547
タイワンコアカソ　56195
タイワンゴウカン　1145
タイワンコウゾリナ　55796
タイワンコウバシ　231324
タイワンゴエフマツ　300083
タイワンコケリンダウ　173868
タイワンコケリンドウ　173868
タイワンコスミレ　409843,410412
タイワンゴトウヅル　199817
タイワンコナラ　323944
タイワンコバガマズミ　407802
タイワンコバンノキ　296734,296809
タイワンコマツナギ　206669
タイワンゴヨウ　300083
タイワンゴヨウマツ　300083
タイワンコンギク　40105
タイワンサイカチ　176870,176900
タイワンサイトウガヤ　65299,127202
タイワンサイハイラン　110502
タイワンサギゴケ　374476
タイワンサギサウ　291221
タイワンザクラ　83158
タイワンササキビ　203317,203318
タイワンササキビ属　203312
タイワンサッサフラス　347411
タイワンサルスベリ　219970
タイワンサルナシ　6530
タイワンサワギク　263512
タイワンサワフタギ　381341
タイワンサワラ　85268
タイワンサンキライ　366414
タイワンサンゴジュ　407977
タイワンシウカイダウ　50352
タイワンシウブンサウ　333475
タイワンシウメイギク　24116
タイワンシシウド　24326
タイワンシジミバナ　372053
タイワンシシンラン　239967
タイワンシテウゲ　235897
タイワンシホヂ　167982
タイワンシモツケ　371912
タイワンシモツケサウ　166092
タイワンシモバシラ　144067
タイワンシャウジャウバカマ　191068
タイワンシャガ　208574
タイワンシャクナゲ　330722
タイワンシュスラン　269033
タイワンシュスラン属　25959
タイワンシュメイギク　24116
タイワンシュンラン　116880
タイワンショウキラウ　2081
タイワンショウキラウ属　2071
タイワンシラタマ　172100
タイワンシラヤマギク　40474
タイワンシロシキブ　66829,66856
タイワンシロヘビイチゴ　167620
タイワンシンシュノキ　12564
タイワンスギ　383189
タイワンスギ属　383188
タイワンスゲ　73774,74596
タイワンスズメウリ　417472
タイワンスズメノヒエ　238697
タイワンスベリヒユ　311921
タイワンスミレ　409993
タイワンセイシボク　161657
タイワンセウベンノキ　400522
タイワンセキコク　125221,125279
タイワンゼリ　299357
タイワンセンダン　248895
タイワンセンダンボダイジュ　217620
タイワンセンリャウ　241818
タイワンセンリョウ　241844
タイワンソクズ　345586,345588
タイワンソケイ　211848
タイワンソテツ　115910,115912
タイワンタカトウダイ　158906,159159
タイワンタゴ　167994
タイワンタシミサウ　355440
タイワンタヌキマメ　112013
タイワンタマボウキ　292077
タイワンタラノキ　30628
タイワンチゴザサ　134441,209048
タイワンチシャノキ　141597
タイワンチトセガヅラ　171455
タイワンチドメグサ　200275,200349
タイワンチャルメルソウ　256019
タイワンヂワウギク　55820
タイワンツアヂサイ　199800
タイワンツアヂサヰ　199800
タイワンツガ　399901
タイワンツクバネ　284358
タイワンツクバネウツギ　113,114
タイワンツチトリモチ　46824
タイワンツチモチ　46826
タイワンツツジ　330722
タイワンツナソ　104103
タイワンツバキ　179745
タイワンツバキ属　179729
タイワンツハブキ　162624
タイワンツルアジサイ　199800
タイワンツルアヅキ　293999
タイワンツルギキョウ　70403
タイワンツルソバ　308965
タイワンツルドクダミ　162546
タイワンテイカカヅラ　393657
タイワントウ　65697
タイワントウカエデ　2815
タイワントガサハラ　318599
タイワントガサワラ　318599
タイワンドクウツギ　104694
タイワントサミヅキ　106684
タイワントネリコ　167982
タイワントベラ　301221,301363
タイワントリアシ　56140
タイワンナスビ　367241
タイワンナナメノキ　203821
タイワンナハシログミ　141991
タイワンニウメンラン　374458
タイワンニウメンラン属　374457
タイワンニガナ　320513

タイワンニクケイ 91351
タイワンニシキサウ 159069
タイワンニシキソウ 159069
タイワンニソジンボク 411362
タイワンニハウメ 83162
タイワンニハウルシ 12564
タイワンネナシカヅラ 115053
タイワンネム 13611
タイワンネムノキ 13649
タイワンノイバラ 336983
タイワンノウリ 365336
タイワンノガリヤス 127217
タイワンノギク 193977
タイワンノギラン 14491
タイワンノササゲ 138929
タイワンノミノツヅリ 32136
タイワンハギ 226940,226977
タイワンハグマ 55769
タイワンバショウ 260228
ダイワンバショウ 260258
タイワンバセウ 260228
ダイワンバセウ 260258
タイワンハチジョウナ 368635
タイワンハチジョオナ 368635
タイワンハチヂャウナ 368635
タイワンハナイカダ 191192
タイワンハナハクカ 274237
タイワンバニラ 405028
タイワンバネウツギ 113,114
タイワンハハコグサ 178061
タイワンハヒノキ 381206
タイワンハマイバラ 336705
ダイワンハマオモト 111171
タイワンハマササゲ 408831
タイワンハマサジ 230759
タイワンハルガヤ 27947
タイワンハンクワイサウ 229068
タイワンハンノキ 16345
タイワンヒエスゲ 76479
タイワンヒキオコシ 209735
タイワンヒゲシバ 88351
タイワンヒゲノガリヤス 127217
タイワンヒゴタイ 348256,348323
タイワンヒシグリ 78942
タイワンヒデリコ 166383
タイワンヒトツバハギ 167096
タイワンヒノキ 85268,85338
タイワンビハ 151144
タイワンヒメイチゴツナギ 306003
タイワンヒメウツギ 127114
タイワンヒメコバナノキ 60061
タイワンヒメコバンノキ 60083
タイワンヒメサザンカ 69705
タイワンヒメサザンクワ 69705
タイワンヒメジソ 259284
タイワンヒメツツジ 331305
タイワンヒメツバキ 68970
タイワンヒメヒゴタイ 348411
タイワンヒメユリ 229775
タイワンビヤクシン 213775
タイワンヒヨドリ 158063
タイワンヒヨドリバナ 158063
タイワンヒヨドリバナモドキ 158063
タイワンヒロハノイヌノヒケ 151257
タイワンビワ 151144
タイワンピンポン 376092
タイワンフウ 232557
タイワンフウラン 390503
タイワンブキ 292352,292404
タイワンフクオウソウ 313822
タイワンフクギ 171145
タイワンフジウツギ 61975
タイワンフシノキ 332509,332513,332662
タイワンフタバムグラ 187681
タイワンフタリシヅカ 88295
タイワンフヂウツギ 61975
タイワンフッキサウ 279744
タイワンブナ 162380
タイワンブナノキ 162380
タイワンベニバナセッコク 125253
タイワンヘビノボラズ 51811
タイワンホウサイ 117042
タイワンボウラン 238316
タイワンホシクサ 151268
タイワンボタンヅル 95042
タイワンホドイモ 29307
タイワンホトトギス 396588
タイワンマガイサ 120353
タイワンマキ 306431
タイワンマダケケイチク 297337
タイワンマツ 165307,300054
タイワンマツグミ 385189
タイワンマツバシバ 33797
タイワンマメガキ 132309
タイワンマルバグミ 141992
タイワンミサオノキ 12517
タイワンミサヲノキ 12517
タイワンミゾハコベ 142574
タイワンミゾハゴヘ 125872
タイワンミゾハゴヘ属 125868
タイワンミゾホホヅキ 392904
タイワンミヤマカウゾリナ 195797
タイワンミヤマシウカイダウ 49974
タイワンミヤマタニタデ 91512
タイワンミヤマトベラ 155890
タイワンムカゴサウ 125535
タイワンムカゴサウ属 125526
タイワンムカゴソウ 125535
タイワンムカゴソウ属 125526
タイワンムカデラン 94458
タイワンムラサキ 66909
タイワンメウガ 417990
タイワンメガルカヤ 389318
タイワンメバハキ 268642
タイワンモクゲンジ 217620
タイワンモグセイ 276313
タイワンモグセイモドキ 276313
タイワンモダマ 145908
タイワンモミ 216120,216125
タイワンモミジ 310199
タイワンモミジ属 310169
タイワンモミヂ 310199
タイワンモミヂ属 310169
タイワンヤウラクラン 267958
タイワンヤガラ 156572
タイワンヤダケ 172741
タイワンヤッコサウ 256178
タイワンヤツデ 162895,187325
タイワンヤドリギ 410972,410992
タイワンヤナギ 344268
タイワンヤノネグサ 309831
タイワンヤノネスミレ 409716
タイワンヤブイチゴ 338305
タイワンヤブガラシ 387825
タイワンヤマアイ 178851
タイワンヤマアヰ 178851
タイワンヤマイ 352293
タイワンヤマカウバシ 231324,231355
タイワンヤマカシュウ 366573
タイワンヤマカモジグサ 58588
タイワンヤマクロモジ 6771
タイワンヤマツツジ 331839
タイワンヤマモガシ 189931
タイワンヤマモミヂ 2942
タイワンヤマヰ 352293
タイワンュキザサ 242691
タイワンユサン 216125
タイワンユリ 229845
タイワンユリワサビ 72721
タイワンヨウゾメ 407850
タイワンヨツバムグラ 170352
タイワンリンゴ 243610
タイワンルリサウ 117986
タイワンルリソウ 117961,117986
タイワンルリミノキ 222124,222138,222158
タイワンレンギョウ 139062,139065
タイワンレンギョオ 139062
タイワンレンゲウ 139065
タイワンワレモカウ 345894
タイワンヲサラン 148684
タウアサガホ 208264
タウアダン 17383
タウアヅキ 765
タウアヅキ属 738
タウイチビ 989
ダ－ウィニア属 122779
タウウツギ 127032
タウエグミ 142092
タウエフ 203961
タウオガタマ 252869

タウオホバコ　302014
タウガキ　164763
タウカセン　63539
タウカセン属　63524
タウガラシ属　72068
タウカンバ　53379
タウキ　24282
タウキササゲ　79247
タウギバウシ　198646
タウギリ　96147
タウキンカン　93639
タウキンセン　66445
タウキ属　229283,229428
タウクサハギ　226793
タウグワ　50998
ダウクワンサウ　403687
ダウクワンサウ属　403686
ダウクワンソウ　403687
ダウクワンソウ属　403686
タウゲブキ　229049
タウケフチクタウ　393657
タウコギ　54158
タウコギ属　53755
タウゴマ　334435
タウゴマ属　334432
タウササクサ　236295
ダウササゲ　294056
タウサンザシ　109924
タウシモツケ　371884
タウシャウブ　176222
タウシャウブ属　175990
タウシャジン　7830
タウジュロ　393809,393816
タウシラベ　427
タウジンキビ　289116
ダウジンマメ　30498
タウスギ　113696
タウセンダン　203422
タウソヨゴ　301433
タウチク　364820
タウチク属　364783
タウチサウ　97093
タウヂサ属　53224
タウチャ　69651
タウツバキ　69552
タウツメレンゲ　275394
タウツルモドキ　166797
タウツルモドキ科　166801
タウツルモドキ属　166795
タウテイラン　317912
タウトボシガラ　163950
タウナス　114288,114295
タウナスゼ　208264
タウナス属　114271
タウニレ　401489
タウネギ　15293
タウネズミモチ　229529
タウバイチゴ　338967
タウハウチハカヘデ　3479
タウハゼ　346408
タウバナ　96981
タウハルニレ　401602
タウハンゲ　401156
タウヒ　298303
タウビシ　394433
タウヒレン属　348089
タウフヂウツギ　62110
ダウベニア属　123105
タウマギ　54158
タウマメ　30498,408393
タウマメナシ　323110
タウミカン　93765
タウメギ　52049
タウモミ　365
タウモロコシ　417417
タウモロコシ属　417412
タウヤク　380234
タウユウナ　389946
タウユウナ属　389928
タウュンイバラ　336509
タウユンサウ　416069
タウユンサウ科　415975
タウユンサウ属　415990
タウユンソウ　416069
タウユンソウ属　415990
タウユン井　143149
タウラフバイ　87528
ダウリアイチゲ　23773
ダウリアイチゴツナギ　305383
ダウリアキンミズヒキ　11572
ダウリアクロツバラ　328680
ダウリアザイコ　63827
ダウリアゼリ　97703
ダウリアハッカ　250353
ダウリアバラモンジン　354801
ダウリアハンゴンソウ　358500
ダウリアフウロ　174559
ダウリアヨナメ　169773
ダウリヤカラマツ　221866
ダウリヤツツジ　330495
ダウリヤビヤクシン　213752
タウロクワイ　17383
タウワタ　37888
タウワタ科　37806
タウワタ属　37811
タウヲガタマ　252869
タウン　310823
タエダマツ　300264
タエニオフィラム属　383054
タエフトキハアケビ　374428
ダエンカン　93671
ダエンタヌキマメ　112108
タオホバコ　277369
タカアザミ　92296
タカオオオスズムシラン　113863
タヵオカエデ　3300
タカオカソウ　250907
タヵオカヘデ　3300
タカオガヤツリ　119718
タカオキョウカツ　24441
タカオコバノガマズミ　407803
タカオコバンノキ　296695,296779
タカオコヒルギ属　83939
タカオサンキライ　366548
タカオシンジュガヤ　354143
タカオスゲ　73563
タカオススキ属　148855
タカオスミレ　410785
タカオタビラコ　416437,416440
タカオバレンガヤ　226005
タカオヒゴタイ　348792
タカオホロシ　367262
タカオモミジ　3300
タカオラン　86670
タカキビ　369600
タカクマキガンピ　414227
タカクマソウ　352874
タカクマヒキオコシ　209831
タカクマホトトギス　396608
タカクマミツバツツジ　332082
タカクマムラサキ　66853,66927
タカザイチゴ　338425
タカサキレンゲ　139988
タカサゴ　316809
タカザゴ　244012
タカサゴアカメヤナギ　344268
タカサゴアザミ　92070
タカサゴアブラススキ　139911
タカサゴイチゲ　24116
タカサゴイチビ　228,934
タカサゴイナモリ　346938,346950
タカサゴイナモリ属　346935
タカサゴイヌザクラ　223117
タカサゴイヌビハ　164992
タカサゴイボクサ　260095
タカサゴウリハダカエデ　3209
タカサゴエゴノキ　379408
タカサゴエンレイサウ　397622
ダカサゴカウゾリナ　55754
タカサゴガマズミ　407853
タカサゴカラマツ　388710
タカサゴカンコノキ　177170
タカサゴギク　55693
タカサゴキンギンサウ　179620
タカサゴキンシバイ　201710,201880,
201892
タカサゴクサイチゴ　339347
タカサゴクハノハイチゴ　338664
タカサゴグミ　142132
タカサゴグミモドキ　142132,142214,
315174

タカサゴグミモドキ属　315171
タカサゴグルミ　212608
タカサゴクレソラン　346900
タカサゴケカンコノキ　177184
タカサゴコオゾリナ　55754
タカサゴコバンノキ　60073,60083
タカサゴコバンノキ属　60049
タカサゴサウ　210525
タカサゴサギサウ　291216
タカサゴサギソウ　291216,291234
タカサゴサビハナナカマド　369498
タカサゴサルナシ　6648
タカサゴサンシチソウ　183084
タカサゴジイ　78877
タカサゴシラタマ　347993,347996
タカサゴシラタマ属　347952
タカサゴスミレ　410289
タカサゴセンブリ　380391
タカサゴソウ　210525,210549,210679
タカサゴタムラサウ　345365
タカサゴタムラソウ　345365
タカサゴダンチク　37471
タカサゴタンポポ　384554
タカサゴチシャノキ　141597
タカサゴチドリ　310943
タカサゴチャヒキ　60724
タカサゴテンニンソウ　100233
タカサゴトキンサウ　107752
タカサゴトキンサウ属　107747
タカサゴトキンソウ　107752
タカサゴトキンソウ属　107747
タカサゴドコロ　131772
タカサゴトベラ　301413
タカサゴニガイチコ　339390
タカサゴニガナ　210648
タカサゴニハウメ　83162
タカサゴノキ　197659
タカサゴノキ属　197633
タカサゴハナイバナ　57689
タカサゴハナヒリグサ　180177
タカサゴハナヒリグサ属　180163
タカサゴヒゴタイ　348430
タカサゴフジウツギ　61975
タカサゴフヂウツギ　61975
タカサゴホシクサ　151268
タカサゴマンネングサ　356751
タカサゴミヅヒキ　392424
タカサゴモミヂ　2942
タカサゴヤウラクラン　267958
タカサゴヤガラ　156660
タカサゴヤブレガサ　381818
タカサゴャマイバラ　336914
タカサゴヤマイバラ　336915
タカサゴユリ　229845,229900
タカサゴルリヒゴタイ　140716
タカサブラウ　141374
タカサブラウ属　141365
タカサブロウ　141374
タカサブロウ属　141365
タカサブロオ　141374
タカスアザミ　348590
タカスギク　41257
タカスゼリ　295041
タガソデソウ　82955,82956,82958
タカタデ　95358
タカダマ　182466
タカチホガラシ　73045
タカック　215510
タカツルラン　155012
タカトウダイ　159215,159540
タカトウダイ属　158385
タカナ　59438
タカナタマメ　71040
タカヌアマ　231863
タカネ　65930
タカネアオチドリ　98591
タカネアザミ　348104
タカネアズマギク　150446
タカネイ　213545
タカネイチゴツナギ　305303,305970
タカネイバラ　336805
タカネイブキボウフウ　228575
タカネウシノケグサ　164182
タカネウスユキソウ　21586
タカネウメバチソウ　284503
タカネオオヤマザクラ　83124
タカネオトギリ　202154
タカネオミナエシ　285870
タカネカニツリ　398531
タカネキケマン　106235
タカネキヌヨモギ　35779
タカネキンポウゲ　326411
タカネクロスゲ　353606
タカネグンナイフウロ　174779
タカネグンバイ　390237,390238
タカネコウゾリナ　298602
タカネコウボウ　27949,27963
タカネコウリンカ　385921
タカネコウリンギク　385892
タカネゴエフ　299799
タカネコメススキ　126031,126032
タカネゴヨウ　299799,299802
タカネコンギク　41496
タガネサウ　76264
タカネサギサウ　302439
タカネサギソウ　302411
タカネザクラミネザクラ　83267
タカネシオガマ　287802
タカネシシウド　98786
タカネシバスゲ　74032,74111
タカネシホガマ　287802
タカネシュスラン　179610
タカネスイバ　339940,340103,340136
タカネスゲ　73906
タカネススキ　255886
タカネスズメノヒエ　238671
タカネスミレ　409862
タカネセンブリ　380379
タガネソウ　76264
タカネソモソモ　164351
タカネソヨゴ　204138
タカネタチイチゴツナギ　305551
タカネタチツボスミレ　410173
タカネタンポポ　384714,384867
タカネツツジ　331062
タカネツメクサ　31744
タカネツメクサ属　255435
タカネトウウチソウ　345837,345911
タカネトンボ　317636
タカネナズナ　136911
タカネナデシコ　127870
タカネナナカマド　369512
タカネナルコ　76289
タカネニガナ　210521,210527
タカネヌカボ　12025,12413
タカネノガリヤス　65477
タカネバラ　336805
タカネハリスゲ　75711
タカネハンショウヅル　95068
タカネハンニチバナ　188586
タカネヒエンソウ　124237
タカネヒゴタイ　348121,348882
タカネヒトツバヨモギ　35221
タカネヒナゲシ　282506
タカネヒメイワカガミ　351443
タカネヒメスゲ　75330
タカネビランジ　363158
タカネヒンシャウヅル　95068
タカネフタバラン　264722,264727
タカネボウフウ　292821
タカネマスクサ　75826
タカネマツムシソウ　350183
タカネママコナ　248161,248179
タカネマンテマ　248255,364149,364202
タカネマンネングサ　357240
タカネミクリ　370071
タカネミズキ　380439
タカネミミナグサ　82732,83007
タカネヤガミスゲ　75030,76585
タカネヤナギ　343106
タカネヤハズハハコ　21586
タカネヨモギ　36295
タカネラフゲ　312714
タガネラン　65915
タカネリンドウ属　174204
タカネリンドオ　174234
タカネルリソウ　117904
タカネヰ　213545
タカネヲミナヘシ　285870
タカノスマル　157164
タカノツメ　2421,143617,170902,186644,

244507,342251,342259,356884
タカノハ　232086,232114
タカノハススキ　255893
タカノホシクサ　151261
タカハシテンナンショウ　33425
タカハネカク　395656
タカヒカゲスゲ　75724
タカモダマ　259512
タガヤサン　44926,96439,234170,252370,360481
タガヤサンノキ　360481
タカユイヌノヒゲ　151382
タカヨモギ　36241
タカラカウ　228999
タガラカウ　228999
タガラカウ属　228969
タガラコウ　229049
タカラコウ属　228969
タカラコヅチ　254468
タガラシ　72749,326340
タヵヲカヘデ　3300
タカヲガヤツリ　119718
タカヲキャウクワッ　24441
タカヲコバンノキ　296695,296779
タカヲコヒルギ属　83939
タカヲサンキライ　366548
タカヲシンジュガヤ　354143
タカヲススキ属　148855
タカヲタビラコ　416437
タカヲバレンガヤ　226005
タカヲラン　86670
タキエン　198176
タキキビ　293296
タキキビ属　293293
タキナガワシノ　347374
タキネックバネウッギ　178
タキノザクラ　83135
ダキバアレチハナガサ　405804
ダキバオオバコ　301855
ダキバオホバコ　301855
ダキバキンチャクソウ　66287
ダキバハンゴンサウ　339518
ダキバハンゴンソウ　339518
ダキバヒメアザミ　91736
タキミチャルメルソウ　256038
タキユリ　230036
タギョウクロマツ　300285
タギョウショウ　299897
タキルジュロ　393815
タキンガ属　382991
タクエフレンリサウ　222681
タグスイゲッタウ　17751
タグスヰゲッタウ　17751
ダクダミ　198745
ダクチロプシス属　121333
ダクティリス属　121226
タクヨウレンリソウ　222681
ダグラスモミ　318580
タグリイチゴ　339270
タクワカタバミ　277776
タケイマハシドイ　382308
タケウマキリンヤシ　407521
タケウマキリンヤシ属　407520
タケウマヤシ属　208366
タケカズラ　401773
タケカンバ　53411
ダケカンバ　53411
タケグミ　404027
タケシマイワキンバイ　312501
タケシマエンゴサク　105874
タケシマシシウド　139734
タケシマシャリントウ　107728
タケシマダイモンジソウ　349357
タケシマヤマブドウ　411624
タケシマユリ　229857
タケシマラン属　377902
ダケスゲ　75281,75282
ダゲスゲ　74913
ダケゼリ　299373
ダケゼリ属　299342
タケダグサ　148164
タケトアゼナ　231514
ダケトンボ　291229
タケニグサ　240765
タケニグサ属　240763
タケネンガ　263552
ダケブキ　229257
ダケモミ　389
タケヤシ属　89345
タコイモ　327672
タコイモ属　327670
タコガタサギソウ　183765
タココン　280440
タコツボ　385880
タコヅル　168243
タゴトノツキ　216637
タコノアシ　290134
タコノアシ属　290133
タコノキ　280993
タゴノキ　167999
タコノキ科　280960,280963
タコノキ属　280968
タゴバウ　238211
タゴボウモドキ　238178
タゴボオ　238167
タコユリ　154926
タザイギョク　264358
ダシチャウ　12517
タシミサウ属　355340
ダシリリオン属　122898
タシロアイ　378231
タシロアヰ　378231
タシロイモ　382923
タシロイモ科　382934
タシロイモ属　382908
タシロカヅラ　137799
タシロカヅラ属　137774
タシロカワゴケソウ　93969
タシロクシノハラン　62900
タシロスゲ　76296
タシロタツナミソウ　355794
タシロノガリヤス　65499
タシロヒヨドリ　158090
タシロマメ　207015
タシロラン属　269006
タシロルリミノキ　222133
タスキアヤメ科　294886
タズノキ　345679
タズバ　345679
タスマニアスギ　44050
タスマニアスギ属　44046
タタミイチゴ　339047
タタラカンガレイ　352231
タタラビ　326340
タチアオイ　13934,18173,397616
タチアオイ属　13911,18155
タチアザミ　92058
タチアフヒ　13934,397616
タチアフヒ属　18155
タチアマモ　418384
タチアラシ　140579
タチイチゴ　338300
タチイチゴツナギ　305743
タチイヌノフグリ　407014
タチイヌフグリ　407014
タチオオバコ　302209
タチオランダゲンゲ　396930
タチガウガイゼキシャウ　213191
タチガシワ　409494
タチカモジ　144440,335204,335370,335371
タチカモジグサ　335370
タチカモメヅル　117488,409447
タチガヤツリ　118737,322210
タチキエボシソウ　5466
タチギボウシ　198643
タチキランソウ　13140
タチクサネム　126187
タチゲキハギ　226723
タチゲヒカゲミズ　284162
タチゲランタナ　221252
タチコウガイゼキショウ　213191
タチコゴメグサ　160214,160241
タチサギゴケ　247039
タチザクラ　83328
タチサメ　175533
タチシオデ　366486
タチシバハギ　126408
タチシホデ　366486,366573
タチジャコウソウ　391397
タチシャリンバイ　329114
タチスゲ　75279

タチスズメノヒエ　285533
タチスナジヨモギ　36111
タチスベリヒユ　311899
タチスミレ　410478
タチセツニンジン　280793
タチセンニンソウ　95366
タチタネツケバナ　72744,72749,72764
タチチシャ　219492
タチツボスミレ　410030
タチテンノウメ　276544
タチテンモウドウ　39161
タチドコロ　131605,131607
タチドジャウツナギ　321342
タチドジョウツナギ　321342
タチトバ　125983
タチナタマメ　71042
タチナハカノコソウ　56437
タチナンキンナナカマド　369561
タチニハヤナギ　308820
タチニンドウ　235860
タチネコノメソウ　90464
タチネコハギ　226926
タチネズミガヤ　259669
タチノエンドウ　408502
タチハキ　71050
タチハコベ　256456
タチハコベ属　256441
タチハシドイ　382282
タチバナ　31408,31502,93823
タチバナアデク　156385
タチバナアナナス　392015
タチハナカズラ　5675
タチバナマンリョウ　31502
タチバナモドキ　322450
タチハマゴウ　411464
タチビ　328345
タチヒガン　83328
タチヒメクグ　218556
タチビャクブ　375351
タチヒルガオ　68701
タチフウロ　174671,174684
タチヘビイチゴ　312457
タチボウキ　39105
タチホスズメガヤ　147699
タチホソバミチヤナギ　308850
タチホロギク　293160
タチマチグサ　174755
タチミゾカクシ　234291,234360
タチムシャリンダウ　137613
タチムシャリンドウ　137613
タチモ　261377
タチャナギ　344211
タチヤナギ　344213
タチラクウショウ　385264
タチラフゲ　312912
タチレンリソウ　222747
タチロウゲ　312912
タヅイモ　123655
タツガシラ　182473
ダヅダカイハマ　336871
タヅタカイバラ　336871
ダヅダカイバラ　336871
タツタカスゲ　76497
タツタカタニタデ　91512
タツタソウ　301694
タツタソウ属　212296,301692
タツタナデシコ　127793
タツタホウ　364917
ダッタンコゴメグサ　160277
ダッタンソバ　162335
ダッタンタカラコウ　229179
ダッタンハマサジ　230787
ダッタンメハジキ　225015
ダッタンモメンズル　43284
タツナミサウ　355494
タツナミソウ　355494
タツナミソウ属　355340
タツノイグサ　173852
タツノツメガヤ　121285
タツノツメガヤ属　121281
タツノヒゲ　128017,272090
タツノヒゲ属　128011
タデ　309199
タデアイ　309893
タデアヰ　309893
ダティスカ科　123033
ダテイチベイ　259094
タテギ　223131,316819
タデキ　316824
タテシナヒメスゲ　74102
タテジマザミア　417021
タデノキ　44269
タテハキ　71050
タデハギ　383009
タデハギモドキ　383007
タテバゴヘイヤシ　228756
タテバテンジクアオイ　288430
タテバテンヂクアフヒ　288430
タテミゾアキグミ　142225
タテヤマアザミ　92280
タテヤマイワブキ　349694
タテヤマウツボグサ　316123
タテヤマオウギ　188177
タテヤマギク　40309
タテヤマキンバイ　362361
タテヤマキンバイ属　362339
タテヤマスゲ　73719
タテヤマヌカボ　12369
タテヤマハギ　226913
タテヤマリンダウ　173978
タテヤマリンドウ　173978
タテヤマリンドオ　173978
タテヤマワウギ　188177
タテヤマワタ　152791
タデ科　308481
タデ属　308690
ダドウカンソウ属　403686
ダドレヤ属　138829
タナ　384746
ダナ－エ属　122121
タナバタユリ　230036
タニアサ　2868
タニイチゴツナギ　306156
タニウツギ　413608
タニウツギ属　413570
タニガハスゲ　74592
タニガハヤナギ　343444
タニガワコンギク　40850
タニガワスゲ　74592
タニガワタヌキラン　73571
タニガワツクバネ　61936
タニガワハンノキ　16384
タニカワブンタン　93476
タニギキャウ　290573
タニギキャウ属　290572
タニギキョウ　290573
タニギキョウ属　290572
タニグワ　160347
タニケグサ属　240763
タニジャコウソウ　86818
タニジャコオソウ　86818
タニスゲ　74872
タニセリモドキ　139733
タニソバ　309459
タニダイ　160347
タニタデ　91537,91545
タニタデモドキ　91501
タニノゾキ　83145
タニハグサ　295888
タニヒゴタイ　348788,348892
タニヒゴタヒ　348892
タニヒノキ　390680
タニブ－タ　93829
タニマスミレ　409958,409962
タニマノオトギリ　202226
タニマノトモシヒ　94567
タニマノヒメユリ　102867
タニミツバ　299534
タニムラアオイ　37668
タニモダマ　95194
タニワタシ　229485,408648
タニワタリノキ　8192,262823
タニワタリノキ属　8170,262795
タヌキアヤメ　294895
タヌキアヤメ属　294892
タヌキコマツナギ　206081
タヌキジソ　170078
タヌキナルコ　73556
タヌキノショクダイ　390097
タヌキマメ　112667
タヌキマメ属　111851

タヌキモ　403222,403382
タヌキモ科　224499
タヌキモ属　403095
タヌキラン　75843
タネガシマアザミ　92430
タネガシマアリノトオグサ　185014
タネガシマカイロラン　86698
タネガシマゴンズイ　160956
タネガシマシコウラン　62875
タネガシマムエフラン　29216
タネガシマムエフラン属　29212
タネガシマムヨウラン　29216
タネガシマムヨウラン属　29212
タネツケバナ　72749,72971
タネツケバナ属　72674
タネナシ　372300
タバコ　266060
タバコオオ　266062
タバコソウ　114606
タバコソウ属　114595
タバコノキ　350344
タバコホホヅキ　367146
タバコ属　266031
タバノマメ　177789
タハラムギ　60379
タバレ-シア属　385146
タビエ　140438
タピオカノキ　244507
タヒチバニラ　405030
タヒチライム　93330,93528
ダビデイア科　123253
ダビディア属　123244
タビビトナカセ　315434
タビビトノキ　327095
タビビトノキ属　327093
タビヤラン　417804
タビラキ　72749
タビラコ　221784,397451
タビラコモドキ　260813
タビラコ属　397380
タブ　240707
タブガシ　240680
タフテッドパンジ-　409858
ダフネ　122416
タブノキ　240707
タブノキ属　240550
タフバナ　96999
タフバナ属　96960
ダフリャカラマツ　221866
ダブリャビヤクシン　213752
タベブ-イア属　382704
タホウタウコギ　54065
タマアジサイ　199911
タマアヂサヰ　199911
タマイ　166518,213041
タマイヌガラシ　336200
タマイブキ　213642
タマウツギ　302796
タマオウキ　186792
タマオオスズメノカタビラ　306099
タマオカナ　244097
タマオキナ　244097
タマガサノキ　82098
タマガハホトトギス　396601
タマガヤツリ　118744
タマガラ　264090
タマガラシ　265553
タマガラシ属　265550
タマガワホトトギス　396601
タマキセワタ　295212
タマキビ　417417
タマグス　240707
タマクルマバサウ　39383
タマクルマバソウ　39383
タマコウガイゼキショウ　213046
タマゴノキ　171221
タマゴバアリドオシ　122056
タマゴバクスノキ　240658
タマザキアサガオ　208092
タマザキイワヒゲ　78632
タマザキウハバミサウ　142741
タマザキウラガヘリバナ　203044
タマザキエゾリンドウ　174012
タマザキエゾリンドオ　174012
タマザキエビネ　65920
タマザキギリア　175677
タマザキキリンギク　228487
タマザキクサフジ　105324
ダマザキサクラソウ　314310
タマザキシオガマ　287067
タマザキセンナ　227430
タマザキツヅラフジ　375833
タマザキツヅラフヂ　375833
タマザキナンテンハギ　408296
タマザキニガクサ　6039
タマザキニガクサ属　5989
タマザキハヒノキ　381390
タマザキヒメハナシノブ　175677
タマザキフタバムグラ　187555
タマザキマツヨイグサ　269395
タマザキマメグンバイナズナ　225363
タマザキマンテマ　363295
タマザキマンネングサ　200805
タマザキモチノキ　203851
タマザキリアトリス　228487
タマサザキコケイラン　274037
タマサボテン属　140111
タマサンゴ　367522
タマサンゴアナナス　8593
ダマスクバラ　336514
タマズサ　396170
タマスダレ　417612
タマスダレ属　417606
タマチシャ　219489
タマヅサ　396170
タマツヅリ　265358,356944
タマツヅリスゲ　74672
タマツナギ　126359
タマツバキ　108846,229485
タマツュクサ　101178
タマツリスゲ　74431,74524
タマツルグサ属　57977
タマテンツキ　166242
タマナ　59520,59532,67860
タマナヅナ　68860
タマネギ　15165
タマネム　301147
タマノウゼンハレン　399611
タマノオ　200798
タマノカンアオイ　37744
タマノカンザシ　198639
タマノヤガミスゲ　74579
タマノヤグサ　240030
タマノヲ　200798
タマバ　357170
タマバウキ　361023
タマバシロヨメナ　39930
タマバタヌキモ　403351
タマバナサウ属　370956
タマバナソウ　370958
タマバナソウ属　370956
タマバハキ　361023
タマハリイ　143158
タマヒコ　102215
タマヒメ　264368
タマビャクブ　375355
タマヒョウタンボク　235781
タマブキ　15621,283811
タマブギ　15165
タマフジウツギ　62064
タマボウキ　39126,361023
タママキエハギ　226739
タママンサク　231461
タマミクリ　370066
タマミズキ　204050
タマミゾイチゴツナギ　305282
タマミヅキ　204050
タマムラサキ　15822
タマメツリウキソウ　168774
タマヤナギ　329705
タマヤブジラミ　392996
タマユキノシタ　349419
タマラニッケイ　91432
タマラン　295518
タマラン属　295510
タマリンヅス属　383404
タマリンド　383407
タマリンボウ　158958
タマワ　67860
タマヰ　166518
タムギ　177560

タムシグサ　86755
タムシバ　242274
タムラサウ　361023
タムラサウ属　360982
タムラソウ　361023,361024
タムラソウ属　360982
タメキノコ　167699
タメトモユリ　229900,229992
タモ　240707
タモツユリ　229966
タモトユリ　229966
タモノキ　168023
ダヤス　339887
タヨウハウチワマメ　238481
タライカヤナギ　343467,344184,344186
タラオアカバナ　146879
タラノキ　30634,30760
タラノキ属　30587
タラノハダリア　121556
タラバイチゴ　338967
タラヒ属　298189
タラミサン　93553
タラヨウ　203961
タラヨオ　203961
ダリア　121541,121561
ダリアコスモス　107168
ターリア属　388380
ダリア属　121539
タリエンシス　69691
タリノホチダケサシ　41812
タリポットヤシ　106999
タリホノチダケサシ　41822
タリヤス　104598
タリョウギョク　140577
ダーリングトニア属　122758
タルウマゴヤシ　247493
タルタルギョク　263752
タルホコムギ　8732
タルマイスゲ　76490
タルマイソウ　289341
ダルマエビネ　65869
ダルマギク　41289
ダルマキンミズヒキ　11591
ダルマサウ　381073
タルマソウ　381082
ダルマソウ　381073
ダルマナツメ　418172
タルミヤシャブシ　16313
タルラミカン　93834
ダレーア属　121871
ダレカンピア属　121912
タレハハゴロモアオナラ　324390
タレユエサウ　208793
タレユエソウ　208793
タロガヨ　93847
タロコガシ　324467
タロコガヤ　163969,164008
タロコゲヤキ　417558
タロコココゴメグサ　160275
タロコシシウド　24484
タロコシャシャンポ　404060
タロコソウ　407461
タロコタウキ　24484
タロコタカトウダイ　158857
タロコヨツバムグラ　170661
タロコヨモギ　36303
タワダギク　305115
タワラグミ　142152
タワラスゲ　74187
タワラムギ　60379
タンエフエンゴサク　106519
ダンカウバイ　87528,231403
タンガザサ　347213
タンガラ　61263
タンカン　93734
ダンギク　78012
タンギーリ　362217
タンキリマメ　333456
タンキリマメ属　333131
タングートダイオウ　329401
タンクヮナ属　383920
タンゲブ　70403,116258
タンゲマル　140860
ダンコウバイ　231403
ダンゴギク　188402
ダンゴギク属　188393
タンゴグミ　141940
タンゴシノチク　347354
ダンゴツメクサ　396917
ダンゴノキ　57695
ダンゴモモ　261213
タンゴン　35411
タンザワイケマ　117423
タンザワウマノスズクサ　34230
タンザワツリバナ　157763
タンザワヒゴタイ　348373
タンシウチワ　273103
タンジン　345214
ダンスゲ　75756,76492
ダンセイゴムノキ　164925
タンダイシデ　77346
ダンチアブラススキ　139913
ダンチク　37467
ダンチク属　37450
タンチョウソウ　259731
タンチョウソウ属　259730
ダンテウケ　360933
タンテウボク　360933
ダンドク　71181,71186
タンドクマル　244054
ダンドク科　71235
ダンドク属　71156
ダンドタムラソウ　345185
ダンドボロギク　148153
タンナアカツツジ　332107
タンナサワフタギ　381158
タンナショウマ　37063
タンナトリアシ　41848
タンナニハトコ　345682
タンナヤハズハハコ　21692
タンナヤブマオ　56300
タンナヤマジソ　259305
タンバグリ　78779
タンバタロウ　279484
タンバホホヅキ　297645
タンバヤブレガサ　381815
タンポポ　384714,384746
タンポポモドキ　224731
タンポポ属　384420
ダンマルジュ　10496
タンヨウエンゴサク　106519
タンヨウナワシロイチゴ　338991
タンリンマル　234910
タンレイマル　234908
タンワンコクタン　132351
タンワンハマアカザ　44542
チ　205517
チイサンウチノケグサ　163864,164153
チイゼル　133478
チウセンアサガホ属　123036
チウセンイハワウギ　188028
チウセンイワオウギ　188028
チウセンオノエヤナギ　344124
チウセンカサユリ　230065
チウセンニハフヂ　206140
チウセンヨノヘヤナギ　344124
チウロサウ　195326
チェランサス　154369,154370
チェリーピンク　122393
チエリモヤ　25836
チェンチエシャン　141472
ヂオウ　327435
チオノドクサ属　87757
チカトラノオ　239594
チカトラノヲ　239594
チガヤ　205473,205497,205506,205517
チガヤ属　205470
チカラグサ　143530
チカラシバ　289015,289135,306493
チカラシバ属　289011
チギ　142404
ヂギク　393402
ヂギク属　393387
ヂギタリス　130383
ヂギタリス属　130344
チキュウマル　284666
チーク　385531
チクカクラン　232106
チクカラン　232106
チクケイラン　232106
チグサ　293710

チクシキヌラン 417795
チクシヒラギナンテン 242563
チクシムレスズメ 369018
チクセツラン 106880
チクゼンヤナギ 342929
チクトウサウ 151416
チクトウハヒノキ 381423
チ－クノキ 385531
チ－クノキ属 385529
チクマハクカ 264897
チクマハッカ 264897
チクヨウラン 366786
チグリジャ 391627
チグリジャ・パボニア 391627
チグリジャ属 391622
チクリンカ 17728
チグロ 203670
チ－ク属 385529
チケイラン 232106
チゴカンチク 87582
チゴグサ 321667
ヂゴクノカマノフタ 13091
チゴザサ 209059,304109,347214
チゴザサ属 209028
チコスガタ 108952
チゴツメクサ 396859
チゴフウロ 174854
チゴマツリ 329579
チゴユリ 134486
チゴユリ属 134417
チゴラッキョウ 15813
チサ 219485
ヂザウカンバ 53460
チサノキ 379374
チシオウツギ 413590,413606
チシオスミレ 410077
チシス属 90711
ヂシバリ 210673,295911,295912
チシマアオチドリ 98586
チシマアサギリサウ 36262
チシマアザミ 92089
チシマアズマギク 150851
チシマアマナ 234234
チシマアマナ属 234214
チシマイチゴ 338142
チシマイチゴツナギ 177574
チシマイハブキ 349843
チシマイブキボウフウ 228611
チシマイワブキ 349843
チシマウスバスミレ 410090
チシマウスユキソウ 224890
チシマオトギリ 201957
チシマオドリコ 170060
チシマオドリコソウ 170060,170078
チシマオドリコソウ属 170056
チシマオドリコ属 170056
チシマカニツリ 398528
チシマカラマッ 221866
チシマガリヤス 65438,127276
チシマギキョウ 69956
チシマキタアザミ 348711
チシマキンバイ 312758
チシマキンレイカ 285870
チシマキンレイタ 285870
チシマクモマグサ 349631
チシマクロクモソウ 349366
チシマクワガタ 407387
チシマゲンゲ 188023,188028
チシマコゴメグサ 160224
チシマコザクラ 23213
チシマコハマギク 89424,89428
チシマザクラ 23213,83269
チシマザクラ属 23107
チシマザサ 347234
チシマシオガマ 287326
チシマシホガマ 287194
チシマシモツケ 166110
チシマシモツケソウ 166110
チシマスグリ 334240
チシマスゲ 75980
チシマスズメノヒエ 238637
チシマゼキショウ 392603
チシマゼキショウ属 392598
チシマゼリ属 80877
チシマセンブリ 380376,380385
チシマソモソモ 305916
チシマタイコンサウ 175426
チシマタイコンソウ 175426
チシマチャヒキ 60631
チシマツガザクラ 61432
チシマツガザクラ属 61430
チシマツメクサ 342290
チシマドジャウツナギ 321324
チシマドジャウツナギ属 321220
チシマドジョウツナギ 321342
チシマドジョウツナギ属 321220
チシマトリカブト 5313
チシマニンジン 101837
チシマヌカボ 12096,12413
チシマネコノメソウ 90387
チシマハハコヨモギ 35550
チシマハマカンザシ 34502
チシマハンショウヅル 94933
チシマヒナゲシ 282612,282617
チシマヒョウタンボク 235721
チシマヒョオタンボク 235721
チシマフウロ 174583
チシママウセンゴケ 138307
チシママツバイ 143022
チシママンテマ 363958,363971
チシマミクリ 370071
チシマミズハコベ 67361
チシマミチャナギ 217648
チシマミチャナギ属 217640
チシマムギクサ 198266,198267
チシマモメンヅル 42537
チシマヤナギ 343045
チシマヨモギ 35157,36440
チシマリンドウ 174118
チシマリンドウ属 174101
チシマルリザクラ 153555
チシマルリソウ 250905
チシマワタスゲ 152765
チシマワレモコウ 345923
チシマヲドリコ属 170056
チシャ 219485,219489
チシャトウ 219488
チシャナ 219485
チシャノキ 141595,141597,141683,379374
チシャノキ属 141592
チ－ゼル 133478
チゾン 93687
チダケ 347200
チダケサシ 41828
チダケサシ属 41786
チダケトリアシ 41787
チタノプシス属 392405
チチウラ 76813
チチウリ 76813
チチグサ 159092
チチコグサ 178237,178481
チチコグサモドキ 178342
チチジマキイチゴ 338867
チチジマクロキ 381356
チチッパベンケイ 200798,200800
チチブイワザクラ 314891
チチブザクラ 83115
チチブシラスゲ 75828
チチブシロカネソウ 145498
チチブドオダン 145690
チチブヒョウタンボク 236054
チチブヒョオタンボク 236053
チチブフジウツギ 62019
チチブミネバリ 53378
チチブヤナギ 344108,344109
チチブリンドウ 174212
チチブリンドオ 174212
チヂミウチワ 272915
チヂミザサ 272684,272693
チヂミザサ属 272611
チヂミジソ 290952
チヂミスイバ 339994
チヂミダマ 140583
チヂミバシマアオイソウ 290315
チヂミバヤッデ 162887
チチュウカイマンダリン 93446
チッゲッガ 244061
チツマキンパウゲ 325628
チツマキンポウゲ 325628
チトセカズラ 171455
チトセカヅラ 171455

チトセラン 346120,346173
チトセラン属 346047
チトニア属 392428
チドメ 183097
チドメグサ 200366
チドメグサ属 200253
チドリケマン 106051
チドリサウ 182230
チドリサウ属 182213
チドリソウ 102833,182230
チドリソウ属 182213
チドリノキ 2868
チナイ 379374
チナシ 84556,379374
チノット 93643
チビウキクサ 224385
チフウソウ 184656
チベットササユリ 229947
チベットシャクナゲ 331596
チボウキナ属 391566
チマキザサ 347268
チミテフクロヤシ 244496
ヂムガデ 78646
チメグサ 285859
チモ 23670
チモシ-グラス 294992
チモラン 416538
チモルカモノハシ 209321
チャ 69634
チャイトスゲ 76114,76517
チャイニーズ・ホーリ 203660
チャイロオサラン 148790
チャイロスゲ 74610
チャイロタヌキラン 74570
チャイロテンツキ 166379
チャイロヤウラクラン 98743
チャイロヤナギ 343400
チャイロヲサラン 148790
チヤウジ 382477
チャウジグサ 122532
チャウジサウ 20850
チャウジサウ属 20844
チャウジザクラ 122438
チャウジタデ 238211
チャウジタデ属 238150
チャウシチク 47250
チャウジナ 234766
チャウジナスゼ 208264
チャウジュラン 116863
チャウシュン 336485
チャウチク 297334
チャウヂャノキ 3243
チャウチンマユミ 157416
チャウノスケサウ 138461
チャウノスケサウ属 138445
チャカイダウ 243630
チャカイドウ 243630
チャガヤツリ 118491
チャカンマユミ 157912
チャクセイシャクナゲ 330993
チャシバスゲ 75369
チャショウブ 208578
チャセイバイ 34455
チャセンギリ 327362
チャセンギリ属 327357
チヤツツジ属 85412
チャノキ 69634
チャノキ属 68877
チャバシラ 183379
チャハマスゲ 118437
チャヒキ 45448
チャヒキカリヤス 156490
チャビテングヤシ属 246675
チャボアザミ 76990
チャボアザミ属 76988
チャボイ 143289
チャボイナモリ 272282
ヂャボウシノシッペイ 148254
ヂャボウシノシッペイ属 148249
チャボウチワヤシ属 262234
チヤボエノコロ 361948
チャボガマ 401128
チャボガヤ 393087
チャボカラマツ 388511
チャボカワズスゲ 75607
チャボカンバ 53444
チャボゲイトウ 80399
チヤボゲンゲ 42391
チャボゴヘイヤシ 228736,228759
チャボサバル 341421
チャボシコタンソウ 349118
チャボシナノキ 391741
チャボスギ 113700,113717
チャボゼキシャウ 392603
チャボゼキシャウ属 392598
チャボゼキショウ 392603
チャボゼキショウ属 392598
チャボダイゲキ 159557
チャボチヂミザサ 272696
チャボチドリ 310868
チャボチャヒキ 60943
チャボツキミソウ 269396
チャボツメレンゲ 252578
チャボツメレンゲ属 252577
チャボテ-ブルヤシ属 143707
チャボトウジュロ 85671
チャボトウジュロ属 85665
チャボトケイソウ 285654
チャボナツメヤシ 295453
チャボナデシコ 127654
チャボニセダイオウヤシ 318120
チャボノカタビラ 305419
チャボノコギリサウ 3997
チャボノコギリソウ 3997
チャボハウチワマメ 238469
チャボハエドクソウ 295988
チャボハシドイ 382266
チャボヒゲシバ 88414
チャボヒバ 85312,85313
チャボベニスジヒメバショウ 66201
チャボホトトギス 396607
チャボマサキ 157620
チャボメヒシバ 130635
チャボヤハズトウヒレン 348754
チャボヤブデマリ 408045
チャボヤマハギ 226712
チャボリンドウ 173184
チャボヰ 143289
チャボヲノオレ 53444
チャヤドリギ 385237
チャヨ-テ 356352
チャラン 88301
チャラン科 88263
チャラン属 88264
チャルメルサウ属 256010
チャルメルソウ 256021,256024
チャルメルソウ属 256010
チャワンザクラ 316809
チャンカニイ 112853
チャンカニ属 112829
チャンカンイチゴ 338850
チャンチン 392841
チャンチンモドキ 88691
チャンチンモドキ属 88690
チャンチン属 80015,392822
チャンパギク 240765
チャンマツ 300181
チュ-インガムノキ 244605
チュウカサクラソウ 314975
チュウクワサクラサウ 314975
チュウゴク 80586
チュウゴクアブラガヤ 353576
チュウゴクエノキ 80580,80586
チュウゴクガマズミ 407731
チュウゴクグリ 78802
チュウゴクザサ 347327
チュウゴクナシ 323114
チュウコバンソウ 60382
チュウシバ 418432
チュウゼンジスゲ 75192
チュウテンカク 159140
チュウナゴン 102508
チュウリップ 400162
チュ-サンヤシ 393813
チュベロ-ス 307269
チュ-ベロ-ス属 307265
チュベロ-ス属 307265
チュ-リップ 400162
チュ-リップゲシ 282566
チュ-リップノキ 232609
チュ-リップ属 400110

チョウウギョク　102550
チョウカイアザミ　91873
チョウケンカズラ　407503
チョウコウ　382477
ヂョウコゥトウ　154252
チョウジ　382477
チョウジカ　367178
チョウジカズラ　393616,393623,393677
チョウジガマズミ　407727,407731
チョウジギク　34746
チョウジグサ　122532
チョウジコメツツジ　332021
チョウジザクラ　83145,122438
チョウジソウ　20850
チョウジソウ属　20844
チョウジタデ　238167,238211
チョウジタデ属　238150
チョウシタヌキモ　403222
チョウシチク　47250
チョウジナ　234766
チョウジノキ　382477
チョウジマメザクラ　83125
チョウジャノキ　3243
チョウジュキンカン　167511
チョウジュラン　116863,117022
チョウシュン　336485
チョウシュンボケ　84573
チョウセイギョク　233539
チョウセイマル　140872
チョウセンアカシデ　77288
チョウセンアカスゲ　76100
チョウセンアキノキリンソウ　368480
チョウセンァケホノソゥ　380400
チョウセンアサガオ　123065
チョウセンアサガオ属　123036
チョウセンアサノハカエデ　2797
チョウセンアザミ　117787
チョウセンアザミ属　117765
チョウセンイカリソウ　147013
チョウセンイチジク　164947
チョウセンイチヤクソウ　322872
チョウセンイヌゴマ　373262
チョウセンイワウチワ属　52503
チョウセンイワオウギ　188028
チョウセンイワギク　124778,124864
チョウセンイワブキ　349691
チョウセンウツギ　126933
チョウセンウラジロイチコ　339227
チョウセンウワミズザクラ　280007
チョウセンエビラフジ　408342
チョウセンエンゴサク　106564
チョウセンオキナグサ　321667
チョウセンオケラ　44202
チョウセンオニシバリ　122477,122486
チョウセンカクレミノ　125626
チョウセンカサスゲ　74934
チョウセンカサユリ　230065
チョウセンカメバソウ　397462
チョウゼンカモノハシガヤ　57567
チョウセンガヤ　82525
チョウセンカラコギカエデ　2984
チョウセンカラスウリ　396210
チョウセンカラスノゴマ　104051
チョウセンカラマツ　221919
チョウセンガリヤス　94601
チョウセンカワラマツバ　170747
チョウセンキイチゴ　338557
チョウセンキスゲ　191268
チョウセンキツネヤナギ　343388
チョウセンキハギ　226892
チョウセンキバナアツモリソウ　120357
チョウセンキバナアヅロリ　120357
チョウセンキンミズヒキ　11546
チョウセンクチナシ　171374
チョウセンクルマユリ　229834,229857
チョウセングルミ　212636
チョウセンクロクモソウ　349607
チョウセンクロコウガイゼキショウ　212990
チョウセンクロツバラ　328680,328752
チョウセンゴウソ　75317
チョウセンコウゾリナ　298625
チョウセンコケモモ　278360
チョウセンコデマリ　372114
チョウセンゴミシ　351021
チョウセンサイカチ　176885
チョウセンサギスゲ　152765
チョウセンザリコミ　334089
チョウセンシオン　40664
チョゥセンシバ　418453
チョウセンスノキ　403878
チョウセンセップンソウ　148112
チョウセンダイオウ　329320
チョウセンダイモンジソウ　349344
チョウセンタカラコウ　229063
チョウセンタチイチゴツナギ　305630
チョウセンチドリ　273583
チョウセンツバキ　69552
チョウセンツルドクダミ　162516
チョウセンテイカカズラ　393619
チョウセントウダイクサ　159272
チョウセンドジョウツナギ　321241
チョウセントネリコ　168086
チョウセンナニワズ　122486
チョウセンニガクサ　388298
チョウセンニレ　401556
チョウセンニワウメ　83241
チョウセンニワフジ　206140
チョゥセンニンジン　280741
チョウセンニンジン　280793
チョウセンニンジン属　280712
チョウセンニンドウ　235860
チョウセンヌカボ　321241
チョウセンネコヤナギ　343444
チョウセンノガリヤス　65363
チョウセンノギク　124832
チョウセンノコンギク　39928
チョウセンノダケ　276748
チョウセンノビル　15450
チョウセンバイモ　168612
チョウセンハウチワカエデ　3479
チョウセンハシドイ　382268
チョウセンハマウツボ　274952
チョウセンハリイ　143056
チョウセンハリスゲ　74121
チョウセンハリブキ　272706
チョウセンハリモミ　298323
チョウセンヒメアブラススキ　71613
チョウセンヒメイヌノハナヒゲ　333561
チョウセンヒメツゲ　64301
チョウセンフウロ　174682
チョウセンホソバユリ　229803
チョウセンマキ　82525
チョウセンマツ　300006
チョウセンマツカサススキ　353510
チョウセンマツモト　363337
チョウセンマユミ　157699
チョウセンマンシウ　371915
チョウセンミカンソウ　296803
チョウセンミズキ　380514
チョウセンミズトンボ　183798
チョウセンミソガワソウ　265060
チョウセンミチシバ　249085
チョウセンミネカエデ　3064
チョウセンミネバリ　53389
チョウセンムギ　99124
チョウセンモダマ　383407
チョウセンモダマ属　383404
チョウセンモミ　388
チョウセンモミジスグリ　334085
チョウセンヤマエンドウ　222852
チョウセンヤマタバコ　229175
チョウセンヤマツツジ　331530,332145
チョウセンヤマナシ　323330
チョウセンヤマナラシ　311281,311361
チョウセンヤマニガナ　320520
チョウセンヤマハギ　226858
チョウセンヤマブドウ　411540
チョウセンヤマボクチ　382069
チョウセンヤマラッキョウ　15690
チョウセンヨメナ　215339
チョウセンヨモギ　35167
チョウセンヨレナ　40664
チョウセンラン　104019
チョウセンリンゴ　243551
チョウセンリンドウ　173847
チョウセンレンギョウ　167473
チョウチンバナノウゼンカズラ　139936
チョウチンバナノウゼンカズラ属　139935
チョウチンマユミ　157883

チョウドウ　186444
チョウトリカズラ　30862
チョウトリカズラ属　30860
チョウノスケサウ　138460
チョウノスケソウ　138460,138461
チョウノスケソウ属　138445
チョウマメ　97203
チョゥマメモドキ　81878
チョゥマメモドキ属　81872
チョウマメ属　97184
チョウメイギク　50825
チョウメン　379374
チョウライナシ　323217
チョウラン　88553
チョウロギ　373439
チョウロク　373439
チョウロソウ　195326
チョオジタデ　238167
チョオジャノキ　3243
チョオセンカメバソウ　397462
チョオセンゴヨオ　300006
チョオセンヤマニガナ　320520
チョオリョオソウ　229257
チョクザキミズ　223679
チョクザキミヅ　223679
チョクザキヨメナ　40852
チョゴテフ　293582
チョコレ－トノキ　389404
チョセンイケマ　117745
チョセンウゴヨウ　300006
チヨダニシキ　17136
チョナ　373439
チョビヒゲタタラカンガレイ　352232
チョマ　56229
ヂョロガヤ　372406
チョロギ　373439
チョロギガヤ　34935
チョロギダマシ　373139,373147
チラヨウカズラ　171456
ヂラン　116829
チリ－アヤメ　192414
チリ－アヤメ属　192405
チリイチゴ　167604
チリウキクサ　224400
チリウチワ　272928
チリ－ソケイ　244327
チリ－ソケイ属　244320
チリソケイ属　244320
チリ－ダイコンソウ　175396
チリツバキ　69203
チリノヤマビト　283134
チリ－マツ　30832
チリマツ　30832
チリメンアオジソ　290954
チリメンウロコヤシ属　156113
チリメンカエデ　3309
チリメンガシ　324284
チリメンキンチャク　66273
チリメンキンチャクソウ　66273
チリメンジソ　290940,290952
チリメンシャリントウ　107649
チリメンショウブ　163516
チリメンチシャ　219490
チリメンヂシャ　90894
チリメンドロ　311368
チリメンナガボソウ　373497
チリメンハイモ　65237
チリメンバコトネアスタ　339869
チリ－ヤシ　212557
チリヤシ　212557
チリ－ヤシ属　212556
チリヤシ属　212556
チリュウマル　182452
チヮウセンイワウチワ属　52503
ヂワウ属　327429
ヂンガサガシ　324490
チンカピン　78807
チンカピングリ　78807
チングルマ　363063
チングルマ属　363057
チンシバイ　369272
チンゼイガシ　323588
ヂンチャウゲ　122532
ヂンチャウゲ科　391030
ヂンチャウゲ属　122359
ヂンチョウゲ属　122359
チンチンカヅラ　97947
チンチンバナ　357919
ツアイ　69721
ツイミオグルマ　207046
ツイミソウ　196027
ヅウエソウ　371240
ツウサウ　387432
ツウシチク　304032
ツウソウ　387432
ツウダツボク　387432
ツウダッボク属　387428
ツウルシル　393466
ツェテセス　218761
ツガ　414,399923
ツカサギュウカク　139134
ツガザクラ　297030
ツガザクラ属　297015
ツカヅキグサ　333480
ツカヅキグサ属　333477
ツカデノキ　191173
ツガマツ　297030,399923
ツカュカク　139134
ツガルフジ　408399
ツガルミセバヤ　200815
ツガルヤナギ　342950
ツガ属　399860
ツキ　417558
ヅキ　142340
ツキイゲ　371724
ツキイゲ属　371723
ツキカゲ　139987
ツキクサ　100961
ツキゲヤキ　417558
ツキセカイ　147425
ツキデノキ　191173
ツキトジ　215277
ツキヌキオグルマ　364301
ツキヌキオグルマ属　364255
ツキヌキオトギリ　202146
ツキヌキサイコ　63805
ツキヌキソウ　397849
ツキヌキソウ属　397828
ツキヌキニンドウ　236089
ツキヌキヒヨドリ　158250
ツキヌキホトトギス　396610
ツキヌキユ－カリ　155691
ツキノドウ　393235
ツキミグサ　269519
ツキミサウ　269519
ツキミセンノウ　248382,363815
ツキミソウ　269443,269519
ツキミタンポポ　269396
ツキミマンテマ　363817
ツグ　32336
ツクシアオイ　37654
ツクシアカショウマ　41864
ツクシアカツツジ　332104
ツクシアキツルイチゴ　338500
ツクシアケボノツツジ　331462
ツクシアザミ　92416
ツクシアブラガヤ　353767,353768
ツクシアリドオシラン　261478
ツクシイヌツゲ　203670
ツクシイハシャジン　7660
ツクシイバラ　336786
ツクシイワシャジン　7660
ツクシウコギ　143687
ツクシウスノキ　403857
ツクシウツギ　127099
ツクシオオガヤツリ　119316
ツクシオオナラ　323635
ツクシオホナラ　323635
ツクシカイダウ　243630
ツクシカイドウ　243630
ツクシガシワ　117497
ツクシガヤ　87353
ツクシガヤ属　87352
ツクシカンガレイ　352235
ツクシキケマン　105976
ツクシクロイヌノヒゲ　151341
ツクシコウモリソウ　283849
ツクシコゴメグサ　160226,160227
ツクシコバノミツバツツジ　331627
ツクシサカネラン　264689
ツクシシオガマ　287577

ツクシシャクナゲ　330538,331227
ツクシスズメノカタビラ　305454
ツクシスミレ　409904,409911
ツクシゼリ　24396
ツクシタチドコロ　131473
ツクシタツナミソウ　355544
ツクシタニギキョオ　290573
ツクシタマバハキ　39126
ツクシタマボウキ　39126
ツクシチドリ　302265,302266
ツクシチャルメルソウ　256027
ツクシテンツキ　166259
ツクシトウキ　24440
ツクシドウダン　145683
ツクシトウヒレン　348570
ツクシトネリコ　168019,168076
ツクシトラノオ　317921,317933
ツクシトラノヲ　317921
ツクシナルコ　76416
ツクシナルゴスゲ　76416
ツクシネコノメソウ　90445
ツクシハギ　226838
ツクシハナミョウガ　17640
ツクシヒトツバテンナンショウ　33408
ツクシフウロ　174921
ツクシボダイジュ　391764
ツクシマムシグサ　33407
ツクシマンネングサ　357036
ツクシミカエリソウ　228014
ツクシミノボロスゲ　75550
ツクシミヤマノダケ　24323
ツクシムレスズメ　369018
ツクシメナモミ　363090
ツクシャウツギ　413611
ツクシャブウツギ　413611
ツクシヤブマオ　56189
ツクシヤマアザミ　92407
ツクシヤマザクラ　83237
ツクショウツギ　413611
ツクシリンダウ　173861
ツークトフィルム属　417693
ヅクノキ　142396,142399
ツクバキンモンソウ　13193
ツクバグミ　142084
ツクバシキミ　365947
ツクバトウキ　24283
ツクバナンブスズ　347317
ツクバネ　61935,109374
ツクバネアオイ　37600,194325
ツクバネアサガオ　292745
ツクバネアサガオ属　292741
ツクバネアサガホ　292745
ツクバネアサガホ属　292741
ツクバネウツギ　177
ツクバネウツギ属　91
ツクバネガシ　116174,116193
ツクバネサウ　284401
ツクバネサウ属　284289
ツクバネソウ　284401
ツクバネソウ属　284289
ツクバネタニウツギ　177
ツクバネ属　61929
ツクミスミレ　409898
ツクモグサ　321697
ツクモドウダン　145710
ツゲ　64285,64303,64345
ツケアアメリカゾウグヤシ　298051
ツケウリ　114201
ツゲカウジ　43721
ツケシカラマツ　388546
ツケナ　59358
ツケバシャリントウ　107370
ツゲバヒサカキ　160508
ツゲモチ　203860,203871
ツゲモデキ　138658
ツゲモドキ　138634,138658
ツゲ科　64224
ツゲ属　64235
ツシマアカネ　337925
ツシマカンコノキ　177170
ツシマスゲ　76612
ツシマナナカマド　369369
ツシマノダケ　391916
ツシマヒナノウスツボ　355150
ツシマヒョウタンボク　235796
ツシマヒョオタンボク　235796
ツシママ　248203
ツシマママコナ　248195
ツシマママンネングサ　357038
ツシマラン　269040
ツスシラーコ属　400671
ツソガネオモト属　170827
ツタ　285102,285117,285157
ツタウルシ　393466
ツタガラクサ　116736
ツタギク　123750
ツタキンギョソウ　37553
ツタノハカヅラ　364986
ツタノハヒルガオ　250792
ツタノハヒルガホ属　250749
ツタノハルコウ　207845
ツタバウンラン　116736
ツタバカガリビバナ　115965
ツタバキリカズラ　246653
ツタバシクラメン　115974
ツタバツルテンジクアオイ　288300
ツタバテンジクアオイ　288326,288430
ツタモミジ　3154
ヅタヤクシュ　391524
ヅダヤクシュ属　391520
ツタ属　285097
ツチアケビ　120764
ツチアケビ属　170031
ツチグリ　29304,312502
ツチグリカンアオイ　194334
ツチシバリ　16433
ツチトリモチ　46848
ツチトリモチ科　46888
ツチトリモチ属　46810
ツチナ　312502
ツチビノキ　122654
ツチモチ　46848
ツチャマモチ　46848
ツヅ　264090
ツツアナナス属　54432,54439
ツツイトモ　312236
ツツザキズイセン　262420
ツツサンゴバナ　269100
ツツサンゴバナ属　269094
ツツジ　330966
ツツジ科　150286
ツツジ属　330017
ツヅチ　364986
ツツチトセラン　346070,346148
ツツナガギリア　175694
ツツナガヤイトバナ　280102
ツツナガユリ　229900
ツツバナホクシャ　168780
ツツヤナギモ　312030
ツヅラカヅラ　97947
ツツラコ　367322
ツヅラフジ　364986
ツヅラフジ科　250217
ツヅラフジ属　364985
ツヅラフヂ　97947,364986,364989
ツヅラフチ科　250217
ツヅラフヂ属　364985
ツナソ　104072
ツナソモドキ　104057
ツナソ属　104056
ツナメロン　114214
ツーニア属　390916
ツニア属　390916
ツノアイアシ　337555
ツノウマゴヤシ　274895
ツノキビ　375773
ツノギリサウ属　87815
ツノギリソウ　191349
ツノギリソウ属　191347
ツノクサネム属　361350
ツノゲシ　176738
ツノゲシ属　176724
ツノゴマ　315434
ツノゴマ科　245991
ツノゴマ属　315430
ツノシオガマ　205585
ツノシホガマ属　205546
ツノスミレ　409858
ツノツクバネ　98223
ツノナス　367357
ツノハシバミ　106777

ツノビシ　394433
ツノミオランダフウロ　153738
ツノミザミア属　83680
ツノミチョウセンアサガオ　123054
ツノミナズナ　89020
ツノミノヒノキ　85339
ツバカズラ属　162505
ツバキ　69156
ツバキカズラ　221341
ツバキヒメ　102324
ツバキ科　389046
ツバキ属　68877
ツバシモモ　20955
ツバナ　205517
ツバナセキコク　125121
ツバナツメクサ　397121
ツハブキ　162616
ツハブキ属　162612,228969
ツバメオモト　97093
ツバメオモト属　97086
ツバメズイセン　372913
ツバメズイセン属　372912
ツバメセッコク　125121
ヅバリア属　139133
ツピダンサス属　400376
ツブ　346338
ツブアナナス属　8544
ツブコムギ　398890
ツブササゲ　409096
ツブラジイ　78903
ツブラジヒ　78903
ツボウツボカズラ　264832
ツボガシ　116100
ツボクサ　81570
ツボクサ属　81564
ツボサンゴ　194437
ツボサンゴ属　194411
ツボスミレ　409656,410730
ツボノキ属　58334
ツホバゲッケイ属　43718
ツボミオオバコ　302209
ツマキミネズオウ　235288
ツマクレナイ　204799
ツマクレナイノキ　223454
ツマクレナヰ　204799
ツマクレナヰノキ　223454
ツマトリサウ　396786
ツマトリサウ属　396780
ツマトリソウ　396786
ツマトリソウ属　396780
ツマニシキサウ　159069
ツマベニ　204799
ツマベニアナナス　264426
ツマモカ属　400310
ツメクサ　342251,342259,397043
ツメクサダオシ　115008
ツメクサダマシ　396905
ツメクサ属　342221
ツメジャコウソウ　391303
ツメレンゲ　275356,275375
ツヤオニソテツ　145233,145235
ツヤハダウチワ　272868
ツュアフヒ　13934
ツユクサ　100961
ツユクサシュスラン　179610
ツユクサ科　101200
ツユクサ属　100892
ツユビギョク　233673
ツラフリ　311480
ツリウキサウ　168767
ツリウキソウ　168750,168767
ツリエノコロ　289143
ヅリオ属　139089
ツリガネオモト　170828
ツリガネカズラ　54303
ツリガネカズラ属　54292
ツリガネズイセン　199553
ツリガネズヰセン　199553
ツリガネツツジ　250507,250529,331623
ツリガネニンジン　7863
ツリガネニンジン属　7596
ツリガネヒルガホ　194467
ツリガネヒルガホ属　194459
ツリガネヤナギ　289327
ツリシャクジョウ　105310
ツリシュスラン　179672
ツリダチスイゼンジナ　183134
ツリハコベ　380181
ツリバナ　157761
ツリバナマユミ　157761
ツリハブソウ　360453
ツリフネ　204982
ツリフネカラマツ　388673
ツリフネサウ　205369
ツリフネソウ　205369
ツリフネソウ科　47077
ツリフネソウ属　204756
ツリフネラン　68542
ツルアカシア　1308
ツルアカミノキ　261657
ツルアカミノキ属　261598
ツルアカメガシハ　243437
ツルアカメガシワ　243437
ツルアジサイ　200059
ツルアズキ　409085
ツルアダン　168245
ツルアダン属　168238
ツルアヂサヰ　200059
ツルアヅキ　294040,409085
ツルアブラガヤ　353750
ツルアマチャ　183023
ツルアリドオシ　256005
ツルアリドオシ属　256001
ツルイタドリ　162519,162528
ツルイチジク　165628
ツルイヅハハコ　253479
ツルイヌビハ　165734
ツルイハガネ　313424
ツルイラクサ　85107
ツルウメサウ　174755
ツルウメモドキ　80260
ツルウメモドキ属　80129
ツルウリクサ　392884,392895
ツルウリクサ属　392880
ツルオオバマサキ　157637
ツルオオマサキ　157515
ツルオニシバ　8862
ツルカウジ　31571,31630
ツルカウゾ　61101
ツルカコソウ　13184
ツルカシワ　338205
ツルガシワ　117497,117499
ツルカノコソウ　404269
ツルカハヅスゲ　75894
ツルカミカハスゲ　76418
ツルカミカワスゲ　76103,76418
ツルカメソウ　131573
ツルカメバソウ　397420
ツルカワズスゲ　75894
ツルカワラニガナ　210650
ツルギカンギク　89552
ツルギギボウシ　198600
ツルギキヤウ　70398,409335
ツルギキャウ属　70386
ツルギキョウ　70398,409335
ツルギキョウ属　70386
ツルギク　254424,254443
ツルギク属　254414
ツルキジムシロ　313028
ツルギテンナンショウ　33236
ツルギハナウド　192379
ツルギバノギク　41335
ツルギミツバツツジ　332026
ツルキリン　290701
ツルキンギョソウ　37554,246649
ツルキントラノオ　376557
ツルキントラノオ属　376554
ツルキンバイ　312942
ツルクビカボチャ　114293
ツルグミ　141999,142214
ツルグミモドキ　141999
ツルクモマグサ　349530,349972
ツルケマン　106197
ツルゲンゲ　42208
ツルコウジ　31571
ツルコウゾ　61101,61102
ツルコケモモ　278366
ツルコケモモ属　278351
ツルコザクラ　346433
ツルゴセウ　300427
ツルコ－ベア　97757

ツルコベア　97757
ツルコマクサ　8299
ツルサイカチ　121634,121810
ツルサイカチ属　121607
ツルザンショウ　417329
ツルザンセウ　417329
ツルシキミ　365942,365954,365959
ツルジンジソウ　349214
ツルスゲ　75894
ツルスズメノカタビラ　305348
ツルセンノウ　114066
ツルセンリョウ　241781
ツルソケイ　211848
ツルソバ　308965,308974,308976
ツルタイワンホトトギス　396614
ツルタガラシ　30272
ツルダシアオイ　37735
ツルタチツボスミレ　410048
ツルタデ　162528
ツルツゲ　204226
ツルツチアケビ　155012
ツルツユクサ　377884
ツルデマリ　200059
ツルドクダミ　162542
ツルナ　387221
ツルナギイカダ属　357898
ツルナシインゲン　294030
ツルナシインゲンマメ　294056
ツルナシコアゼガヤツリ　118987
ツルナシナタマメ　71042
ツルナシヤハズエンドウ　408579
ツルナシヨツバハギ　408510
ツルナシレンリソウ　222860
ツルナシヰンゲン　294030
ツルナ科　12911
ツルナ属　387156
ツルニガクサ　388307
ツルニカナ　210623
ツルニチニチサウ　409335
ツルニチニチサウ属　409325
ツルニチニチソウ　409335
ツルニチニチソウ属　409325
ツルニンジン　98343
ツルニンジン属　98273
ツルネコノメサウ　90362
ツルネコノメソウ　90362
ツルノイバラ　336738
ツルノキンマサキ　157501
ツルノゲイトウ　18147
ツルノゲイトウ属　18089
ツルバギア属　400050
ツルハグマ　55783
ツルハグマ属　55669
ツルハコベ　374859
ツルハシカンボク　59855
ツルハナガタ　23281
ツルハナナス　367265
ツルハハコ　253479
ツルヒキノカサ　325717
ツルヒトツバ　185764
ツルビャクブ　375343
ツルヒヨドリ　254443
ツルビランジ　363627
ツルビロ－ドサンシチ　183134
ツルブシ　5674
ツルフジバカマ　408262
ツルフヂバカマ　408262
ツルボ　353057
ツルボラン　39328
ツルボラン属　39449
ツルボ属　352880
ツルマオ　179488
ツルマオモドキ　313483
ツルマオ属　179485
ツルマサキ　157473,157496,157515,157517
ツルマツリ　211940
ツルマメ　177777
ツルマユミ　80260
ツルマヲ　179488
ツルマヲ属　179485
ツルマンネングサ　357123
ツルマンリャウ　261657
ツルマンリョウ　261657
ツルマンリョウ属　261598
ツルミツ　298945
ツルミヤコグサ　237509
ツルミヤマカンスゲ　75447
ツルムラサキ　48689
ツルムラサキ科　48698
ツルムラサキ属　48688
ツルメドハギ　226682
ツルメヒシバ　45827
ツルモウリンカ　400972
ツルモドキ　80260
ツルヤブタバコ　55769
ツルヤブタビラコ　55769
ツルヤブミョウガ　167040
ツルヤブメウガ　167040
ツルヤブメウガ属　167010
ツルヤマシロギク　39928
ヅルユリズイセン　56759
ヅルユリズイセン属　56755
ツルヨシ　295911,295912
ツルラン　66101
ツルリンダウ　398273
ツルリンダウ属　110292
ツルリンドウ　398273,398306
ツルリンドウ属　398251
ツルリンドオ　398273
ツルレイシ　256797
ツルレイシ属　256780
ツルワサビ　72678
ツルワタ　37888
ツルワダン　210645
ツルワチガイ　318493
ツルワチガヒ　318493
ツレサギサウ　302377
ツレサギサウ属　302241
ツレサギソウ　302377
ツレサギソウ属　302241
ツワブキ　162616
ツワブキ属　162612
ツワモノアザミ　91967
ツンナメ　91318
ツンベルギア・エレクタ　390759
ツンベルギア属　390692
ディアキア属　139246
ディアスキア属　128029
ディアファナンテ属　127936
ディアモルファ属　127460
ディエラ－マ属　130187
テイオウカ　82320
テイオウサイカク　373908
テイオウハイモ　16509
ディオオン属　131427
テイカカズラ　393616,393621
テイカカズラ属　393613
テイカカヅラ　393616,393621
テイカカヅラ属　393613
テイカン　268086
テイカン属　268085
テイギョク　304372
デイグ　154734
ディクロア属　129030
ディケア属　128397
デイゴ　154734
ディコリサンドラ属　128987
デイコ属　154617
ディサンツス属　134009
ディ－サ属　133684
ディジゴテ－カ属　135079
ティシュレリア属　392380
テイショウソウ　12623
テイショオソウ　12623
ディスキ－ディア属　134027
ディスコカクタス属　134115
ディスフィマ属　134394
テイチギ　329114
ディッキア属　139254
ディディエ－ア科　129781
ディディエ－ア属　129772
ディディスクス　393883
テイヌマントウサゴヤシ　217922
ティフトン　117874
ディプモソマ属　133177
ディ－フュンバッキア属　130083
ティプラリア属　392346
ティプロシアサ　132949
ティプロシアサ属　132947
ディミア属　123307
ディモルフォセカ属　131147

ディモルフォルキス属 131124
ティランジア属 391966
テイリュウカン 103741
ディレニア科 130928
ディレ－ニア属 130913
テイワントウックバネウツギ 113,114
ディンテランツス属 131306
ディンテラ属 131300
テウカイゼリ 101837
テウセンアサノハカヘデ 2798
テウセンアザミ 117787
テウセンアザミ属 117765
テウセンアハゴケ 67376
テウセンエノキノ 80664
テウセンオニシバリ 121720,122477
テウセンガリヤス 94601
テウセンガリヤス属 94585
テウセンガリヤズ属 132720
テウセンギク 67314
テウセンキセルアザミ 92367
テウセンキツネヤナギ 343388
テウセンキバナアツモリ 120357
テウセンクチナシ 171253,171374
テウセングハ 259165
テウセングルミ 212636
テウセンクロッバラ 328752
テウセンゴエフ 300006
テウセンゴマツ 300006
テウセンサイカチ 176881
テウセンサギスゲ 152769
テウセンザクロ 321771
テウセンシバ 418453
テウゼンシュロサウ 405618
テウセンシラベ 427
テウセンツバキ 69552
テウセントネリコ 168086
テウセントラノオ 317912
テウセントラノヲ 317912
テウセンナンテン 242563
テウセンニレ 401556
テウゼンニンジン 280741
テウセンバイクワウツギ 294556
テウセンハリモミ 298323
テウセンヒメツゲ 64369
テウセンマツ 300006
テウセンミヅキ 380444
テウセンミネバリ 53389
テウセンモダマ 383407
テウセンモダマ属 383404
テウセンモミ 388
テウセンヤマナシ 323330
テウセンヤマナラシ 311537
テウセンヤマブドウ 411540
テウセンレンゲウ 167471
テウチグルミ 212636,212643
テオシント 155878
デオダラモミ 80087
テガシハ 302721
テガタアサガホ 207988
テガタイチゴ 338425
テガタスイバ 340254
テガタスミレ 409878
テガタチドリ 182230
テガタチドリ属 182213
テガタヒキノカサ 325717
テガヌマイ 353660
デカネマ属 123367,123401
デカリア属 123425
テキサスビル 341430
テキ－ラ－リュウゼツラン 10945
テキリスゲ 74997
テクサワバナ 122040
デゲネリア 123585
デゲネリア科 123587
テコステレ属 389103
テコフィレ－ア属 385520
テコマリア・カペンシス 385508
テコマリア属 385505
テコマ属 385461
デザ－トピ－ 380066
デ－ジ－ 50825
テシオコザグラ 315029
テシオシモツケ 371826
テシオマツ 298298
テシホマツ 298298
テ－シャン 399879
デスモジュ－ム属 126242
テ－ダマツ 300264
テツカエデ 3248
テツガシ 78933,79035
テッカン 34420
テツギ 329114
テッケンユサン 216120
テッコウマル 158586
テッサイノキ 252370
テツザイノキ 252370
テツザイノキ属 252369
テツシャクジョウ 360102
テッシンマル 322989
テッセン 94910
テツダウグサ 103446
テツノキ 3248
テツパウウリ 139852
テツパウウリ属 139851
テツパウユリ 229900
テツポウウリ 139852
テツポウウリ属 139851
テツポウユリ 229900
デ－デ－ 93555
テトラネマ属 387343
テトラパナックス属 387428
テトラミクラ属 387321
デニソンムクゲ 194830
デハエンゴサク 105594
デバガヤ 82523
テバコマンテマ 364218
テバコモミジガサ 283874
デハトネリコ 168068
テハノマタタビ 6523
テフ 147994
テブシュカン 93604
テフセンタウヒ 298382
テフセンタカラカウ 229063
テフマメ 97203
テフマメ属 97184
テブラサウ 66215
テブラサウ属 66157
テブラソウ属 66157
テフラン 293582
デブランチェアモクマオ 182836
テ－ブルビ－ト 53259
テ－ブルヤシモドキ属 201433
テ－ブルヤシ属 85420
テプロカクタス属 385870
テベスド－ムヤシ 202321
テマリアジサイ 199991
テマリウツギ 127032
テマリカ 199991
テマリカンボク 408011
テマリグサ 21805
テマリコザクラ 314212,314396
テマリサンタンカ 211037
テマリシモツケ 297837
テマリシモツケ属 297833
テマリタマアジサイ 199917
テマリツバキ 69348
テマリツメクサ 397096
テマリバナ 408027
テマリラッキョウ 15690
テマルバナ 199991
テモチユリ 229966
テヤボヤナギ 343886
テヤマハギ 226801
テャンティニイ 8554
デュ－ベリ－ 339103
テョウセンジソ 290952
テラツバキ 229485
テラハボク 67860
テラモトハコベ 83036
テリアッバキク 41086
デリス 125958
デリス属 125937
テリハアカザ 87186
テリハアカショウマ 41863
テリハアザミ 92163
テリハアツベンケイ 356902
テリハイカダカズラ 57857
テリハイカダカヅラ 57857
テリハイヌビハ 165814
テリハイワガラミ 351798
テリハオホイヌタデ 291766

テリハキンバイ　312935
テリハクサトベラ　350344
テリハコナラ　324387
テリハザンショウ　417282
テリバザンセウ　417282
テリハタチツボスミレ　409976
テリハチダケサシ　41788
テリハックバネウツギ　181
テリハツルウメモドキ　80285
テリハトネリコ　167897
テリハドロ　311482
テリハドロノキ　311482
テリハニシキソウ　159071
テリハニレ　401490,401492
テリハニンドウ　235883
テリハノイバラ　336697,337015
テリハノギク　41336
テリハノセンニンサウ　95113
テリハノセンニンソウ　95113
テリハノハマボウ　194895
テリハノブドウ　20321,20352,20400
テリハノブドオ　20321
テリハハマボオ　194895
テリハバンジロウ　318737
テリハバンジロオ　318737
テリバヒサカキ　160595
テリハヒメサザンクワ　69705
テリバヒメザザンクワ　69706
テリハブナ　162390
テリハベゴニア　50111
テリハボク　67860
テリハボク属　67846
テリハホタルブクロ　70236
テリハミヤマガマズミ　408230
テリハモモタマナ　386590
テリハヤナギ　343858,343859
テリミコミカンソウ　296802
テリミ-トラ属　389258
テリミノイヌホオズキ　366934,367455
テリミノイヌホオホロシ　366934
テリミノカザナビキ　405447
テリミノツルウメモドキ　80285
テルナミ　52561
デルフィニュ-ム属　124008
デレゲ-トユ-カリ　155558
テレビンノキ　301002
テレピンノキ　301002
デロ　311389,311509
テロカクタス属　389203
デロジョウ　32429
デロスペルマ属　123819
デロノキ　311389
テロペア　385738
テロペア属　385734
デワノタッナミソウ　355616
デワノトネリコ　168003
テンオウマル　182441
テンガイバナ　188908,239266
テンガイメギ　52069
テンガイユリ　168434,230058
テンガユリ　230058
テンキ　228346,228369
テンキグサ　228369
テング　93466,93849
テングクハガタ　407358,407401
テングクワガタ　407362
テングスミレ　410502
テングノコヅチ　398307
テングノハウチワ　162879
テングヤシ属　246671
テンケイマル　167692
テンコウ　389216
テンコウボク　408131
テンコウリュウ　158614
テンサイ　53249,53266
テンシ　102652
テンシギョク　182471
テンジクアオイ　288310
テンジクアオイ属　288045
テンジクスゲ　75775
テンジクナスビ　367733
テンジクボダイジュ　165553
テンジクボタン　121561
テンジクボタン属　121539
テンジクマメ　218721
テンジクマモリ　72074
テンジクミヤマシキミ　365969
テンジクメギ　52069
テンジクヤナギ　344199
テンシボタン　68706
テンシュカク　395634
テンジョウユリ　229915
デンシンサウ　98159
テンシンナズナ　63446
テンシンナズナ属　63443
テンセンスゲ　76515
テンソウ　216637
テンダイウヤク　231298
テンダスズメウリ　367775
テンダノハナ　204634
テンダノハナ属　204609
テンヂクアフヒ属　288045
テンヂクゼリ　269300
テンヂクナスビ　367241
テンヂクボダイジュ　165553
テンヂクボタン　121561
テンヂクボタン属　121539
テンヂクマモリ　72070,72074
テンツキ　166248,166271
テンツキノウリ　390175
テンツキ属　166156
テンテキギョク　233492
テンデククハクサ　162871
デンドロキ-ルム属　125526
デンドロシキオス属　125749
テンドロセレウス属　125522
デンドロビウム属　124957
デンドロビュ-ム　125135,125138
デンドロビュ-ム属　124957
デンドロメ-コン属　125583
テンナンシャウ　33369
テンナンシャウ属　33234
テンナンショウ科　30491
テンナンショウ属　33234
テンニョ　392406
テンニョウン　17447
テンニョエイ　392416
テンニョカン　392418
テンニョサン　392408
テンニョショウ　17445
テンニョセン　392410
テンニョノマイ　17459
テンニョハイ　392414
テンニョハクセン　392412
テンニンカ　332221
テンニンカ属　332218
テンニンギク　169603,169608
テンニンギクモドキ　339522
テンニンギク属　169568
テンニンクワ　332221
テンニンクワ科　261686
テンニンクワ属　332218
テンニンサウ　228002
テンニンサウ属　100232
テンニンソウ　228002
テンニンソウ属　100232,227999
テンニンノマイ　215097,215220
テンノウメ　276534,276550,333914
テンノウメ属　276533
テンバイ　333914
テンプル　93848
テンペイマル　182479
デンボウギョク　233490
テンマ　171918
テンモクヅタ　285130
デンモ-ザ属　125817
デンモザ属　125817
テンモンドウ　38960,39069
テンリュウ　359232
テンリンクワ　383090
テンレイ　82387,264812
ドイツアカマツ　300223
ドイツアザミ　92066
ドイツアヤメ　208583
ドイツスズラン　102867
ドイットウヒ　298191
ドイツハギ　154734
ドイフタギ　381104
ドイヤナギ　343587
トウ　121514
トウアジサイ　199846

トウアヅキ 765
トウアヅキ属 738
トウアマドコロ 308593
トウイチゴ 123322
トウイヌハッカ 264977
トウイワキンバィ 312944
トウウチソウ 345834
トウウツギ 127032
トウウメ 87525
トウウメバチソウ 48907
トウウンマル 140886
トウエイマル 234959
トウエゾガヤ 341912
トウエダナシマンテマ 364103
トウエンソウ 416069
トウエンソウ科 415975
トウエンソウ属 415990
トウエンドウ 408327
トウエンマル 234930,234965
トウオオバコ 302014
トウオカメザサ 362123
トウガ 50998
トウカイコモウセンゴケ 138361
トウカイタンポポ 384744
トウカエデ 2811
トウカガミ科 200235
トウガキ 164763
トウカセン 63539
トウカセン属 63524
トウカテンソウ 262249
ドウガメバス 200228
トウガヤ 82525
トウガラシ 72070,72071
トウガラシ属 72068
トウガリヤス 94605
トウカワラヒジキ 104755
トウガン 50998
トウカンゾウ 191263
ドウカンソウ 403687
ドウカンソウ属 403686
トウカンバ 53379
トウガン属 50994
トウキ 24282
トウキササゲ 79247
トウキジカクシ 38950,38977
トウキビ 417417
トウギボウシ 198646
トウギュウカク 159791
トウキョウシバ 418432
トウキョウチクトウ 393657
トウギリ 96147
トウキリンソウ 294114
トウキンカン 93566,93636,93639
トウキンセン 66395,66445
トウキンセンカ 66395,66445
トウキンハイソウ 399500
トウキ属 229283
トウグコ 239021
トウクサハギ 226793
トウクサボタン 94995
トウグミ 142102
トウクルミ 212636
トウクロタキカズラ 198576
トウクワ 259067
トウグワ 50998
トウグワ属 50994
トウグンギョク 263743
トウゲオトギリ 202013
トウゲダケ 347369
トウゲブキ 229049
トウゲンキョウ 109452
トウケンジスミレ 410647
トウゴウイゾマツ 230627
トウゴウギク 339539
トウゴウソウ 230627
トウコウマル 140868,234922
トウゴクシソバタツナミ 355556
トウゴクヒメシャラ 376473
トウゴクヘラオモダヵ 14760
トウゴクヘラオモダカ 14779
トウゴクミツバツツジ 332089
トウゴボウ 298093
トウゴマ 334435
トウコマツナギ 205752
トウゴマ属 334432
トウサイ 319114
トウサイカチ 176901
トウサイ属 319101
トウサゴヤシ属 217918
トウササクサ 236295
トウササゲ 294056
トウサワトラノオ 239566
トウサン 98417
トウサンザシ 109924
トウザンショウ 417340
トウザンスミレ 409993
トウサンセンニンサウ 95350
トウサンセンブリ 380391
トウサンヤナギ 343400
トウザンヤナギ 343321
トウシキミ 204603
トウシドロ 311144
トウシャジン 7830
トウシュウギョク 284701
トウジュロ 393809,393816
トウショウブ 176222
トウショウブ属 175990
トウショクマル 2313
トウシラベ 427
トウシンギ 350706
トウジンキビ 289116
トウシンサウ 213036
トウシンソウ 213036
トウジンヒエ 289116
トウジンビエ 289116
トウズ 71050
トウスギ 113696
トウスズシロ 154500
トウセイマル 234914
トウセンギョク 246594
トウセンダン 203422,248925
トウソヨゴ 301433
トウダイグザ科 160110
トウダイグサ属 158385
トウダイゲサ 159027
トウタカトウダイ 159540
ドウダンツツジ 145710
ドウダンツツジ属 145673
トウチウソウ 345850
トウチク 364820
トウチクラン 134425
トウチク属 364783
トウヂサ属 53224
トウヂシャ 53257
トウチソウ 97093
トウチャ 69651
トウチョウマル 327299
トウツクバネウツギ 416841
トウツツジ 331759
トウツバキ 69552
トウツルキンバイ 312360
トウツルモドキ 166797
トウツルモドキ科 166801
トウツルモドキ属 166795
トウテイカカズラ 393657
トウテイラン 317912,317931,407265
トウドウチク 47494
トウトボシガラ 163950
トウナス 114292
トウナナカマド 369484
トウナンテン 242563
トウニレ 401489
トウニワトコ 345708
ドウニンドウ 235860
トウヌマゼリ 365872
トウネズミモチ 229529
トウネミネユリ 229730
トウノイモ 99936
トウノウネコノメ 90437
トウノキ 96398
トウノブノキ 302676
トウハウチワカエデ 3479
トウハクサンイチゲ 23901
ドゥパクンクルス属 137744
トウハスノカラマツ 388536
トウハゼ 346408,393479
トウバナ 96981,96999
トウハナイバナ 57681
トウバナ属 96960,347446
トウハマサジ 230759
トウバラモンジン 354952

トウヒ 298303
トウビシ 394436
トウヒメユリ 229813,229818
トウヒレン属 348089
トウビロウ 234170
トウビワ 141683
トウヒ属 298189
トウブキ 292394
トウフジウツギ 62110
トウフェリ 391627
トウヘビイチゴ 312449
トウホアブラスズキ 372412
トウホウマル 327313,327317
トウホサギサウ 183559
トウマツムシソウ 350265
トウマメ 408393
トウマメナシ 323110
トウミズキ 380434
トウムギ 99124
トウメア属 393233
トウメギ 52049
トウメロン 114212
トウモロコシ 417417
トウモロコシ属 417412
トウヤオリンドウ 173198
トウヤク 380234
トウヤクリンドウ 173198
トウヤブサンザシ 333972
トウヤマオダマキ 30090
トウユウナ 389946
トウユウナ属 389928
トウュリ 391627
トウュリ属 391622
トウヨウマル 198495
トウヨツバムグラ 170289
ドゥラクンクルス属 137744
ドウラマル 103753
トゥリコケ-ロス属 395662
トゥリゴニディウム属 397314
トウリンドウ 173847
トウリンボク属 266532
トウリンマル 234968
トウルリヒゴタイ 140770
トウレイマル 234909
トウレンゲツツジ 331257
トウロウサウ属 61561
トウロウソウ 61572
トウロウソウ属 61561
トウロウバイ 87528
トウワスレナグサ 397452
トウワタ 37888
トウワタ属 37811
トウ属 65637
トオゲブキ 229049
トオゴクヒョオタンボク 235717
トオゴクミツバツツジ 332089
トオシャジン 7830
トオダイグサ 159027
ドオダンツツジ 145710
トオツワブキ 162626
トオネズミモチ 229529
トオバナ 96999
トオフジウツギ 62110
トオホクカラマツ 221919
トオヤマツツジ 331839
トオリンドオ 173847
トガ 399923
トガクシオトギリ 202063
トガクシギク 89405
トガクシコゴメグサ 160187
トガクシショウマ 326508
トガクシショウマ属 326507
トガクシソウ 326508
トガクシソウ属 326507
トガクシタンポポ 384834
トガクシナズナ 137225
トガサハラ 318575
トガサハラ属 318562
トガサワラ 318575
トガサワラ属 318562
トガスグリ 334188
トガスグリアカスグリ 334188
トカチオウギ 43160
トカチギ 329114
トカチキスミレ 409774
トカチスグリ 334240
トカチトウキ 24481
トカチビランジ 364130
トカチフウロ 174585
トカチヤナギ 343168
トカドヘチマ 238258
トガネタケヤシ 89360
トガマツ 399923
トカラアジサイ 199792,199926
トカラカンアオイ 37746
トカラカンスゲ 74177
トカラタマアジサイ 199921
トカラノギク 89657
トガラバササザンカァ 69235
トガリシオデ 366549
トガリスモモ 316761
トガリバインドソケイ 305225
トガリバカエデ 2872,3058
トガリバガシ 233122
トガリバツメクサ 396823
トガリバトベラ 301294
トガリバナエビネ 65881
トガリバヒサカキ 160406
トガリバヒメツバキ 68970
トガリバマキ 306502
トガリバモクセイ 276344
トガリバヤブマオ 56181
トキイロクズ 321443
トキイロフデリンドウ 174097
トキイロホソバシャクナゲ 331191
トキウズ 5472
トキサウ 306870
トキサウ属 306847
トキサラン 82088
トキシラズ 50825,338205
トキソウ 306870
トキソウ属 306847
トキハアケビ 374409
トキハイチゲ 24066
トキハイヌビハ 165828
トキハガキ 132309,132320
トキハギョリウ 79157,79161
トキハギョリュウ 79157
トキハゲンカイ 330495
トキハコブシ 252849
トキハザクラ 314717
トキハサンザシ 322450
トキハサンザシ属 322448
トキハススキ 255849
トキハツツジ 330495
トキハハゼ 247025
トキハマンサク 237191
トキハマンサク属 237190
トキハヤブハギ 200732
トキハラン 282845
トキハラン属 282768
トキハレンゲ 232594
トキヒササウ 352873
トキホコリ 142647,142759
トキリマメ 333133
トキワアケビ 374409
トキワアワダチソウ 368386
トキワイカリソウ 147054
トキワイヌビワ 164711
トキワカエデ 3154
トキワガキ 132309
トキワガマズミ 407898
トキワカモメヅル 117700
トキワカワゴケソウ 93970
トキワギク 40610
トキワギョリュウ 79161
トキワゲンカイ 330495
トキワコブシ 252849
トキワザクラ 314717
トキワサルトリイバラ 366294
トキワサンザシ 322450,322454
トキワサンザシ属 322448
トキワススキ 255849
トキワソウ属 358015
トキワツツジ 330495
トキワナズナ 198708
トキワナズナ属 198703
トキワナツナ 187526
トキワバイカツツジ 332045
トキワハゼ 247025
トキワバナ 415317

トキワバナ属 415315
トキワハマグルマ 413507
トキワブハギ 200732
トキワマガリバナ 203239
トキワマメガキ 132309
トキワマンサク 237191
トキワマンサク属 237190
トキワユリ 229789
トキワラン 282845
トキワレンゲ 232594
トキンイバラ 339195,339381
トキンガヤ 113306
トキンガヤ属 113305
トキンサウ 81687
トキンサウ属 81681
トキンソウ 81687
トキンソウ属 81681
ドクイモ 16512
ドクウツギ 104696
ドクウツギ科 104710
ドクウツギ属 104691
ドクエ 406016
ドクグルミ 302684
トクサイ 143257
トクサバモクマオウ 79161
トクサラン 82088
ドクサンタ属 136877
ドクゼリ 90932,90937
ドクゼリモドキ 19664
ドクゼリモドキ属 19659
ドクゼリ属 90918
トクダマ 198650
ドクダミ 198745
ドクダミ科 348077
ドクダミ属 198743
ドクダメ 198745
ドクニンジン 101852
ドクニンジン属 101845
トクノウコウ 301002
トクノシマエビネ 66095
トクノシマカンアオイ 37730
トクノシマテンナンショウ 33377
トグノヤシモドキ属 123497
ドクフジ 125958,254796
ドクブダウ 387774
ドクフヂ 254854
ドクフヂ属 125937
ドクムギ 235373
ドクムギ属 235300
ドグラスモミ 318580
トクリジトリカブト 5108
トクワカソウ 362270,362275
トクンヤナギ 343321
トゲアオイモドキ 675
トゲアザミ 92074
トケイサウ 285623
トケイサウ科 285725
トケイサウ属 285607
トケイソウ 285623
トケイソウ科 285725
トケイソウ属 285607
トゲイヌツゲ 354621
トゲイヌツゲ属 354596
トゲイボタ 229643
トゲイモ 131579
トゲオオバコ 302178
トゲオナモミ 415053
トゲオニソテツ 145224
トゲオホジラミ 300944
トゲカズラ 300944
トゲカヅラ 300944
トゲガフクワン 1466
トゲキクアザミ 348872
トゲゴエフイチゴ 216085
トゲコブダネヤシ 271001
トゲゴヘイヤシ 228766
トゲサゴヤシ 252634
トゲザンセウ 417340
トゲスグリ 333955
トゲスズメウリ 114231
トゲソバ 309772
トゲソへバ 401753
トゲチシャ 219497
トゲチャビテングヤシ 246676
トゲナシアザミ 92055
トゲナシアダン 281092
トゲナシアリドホシ 122028
トゲナシアレチウリ 362462
トゲナシイヌザンショウ 417333
トゲナシウコギ 143684
トゲナシウマゴヤシ 247431
トゲナシエゾウコギ 143591
トゲナシオニウコギ 143591
トゲナシカラスザンショウ 417140
トゲナシキンギョモ 83578
トゲナシグリ 78784
トゲナシサイカチ 176882
トゲナシジャケツ 288897
トゲナシジャケツ属 288882
トゲナシスグリ 334210
トゲナシチョウセンアサガオ 123081
トゲナシツノクサネム 361367
トゲナシテリハノイバラ 336701
トゲナシトネリコバノイチゴ 338435
トゲナシナベナ 133487
トゲナシニセアカシャ 334982
トゲナシママコノシリヌグイ 291955
トゲナシムグラ 170490
トゲナシモミジイチゴ 338948
トゲナシヤエムグラ 170193,170205
トゲナシヤマカシュウ 366574
トゲナシランタナ 221238
トゲノャシ属 2550
トゲハアザミ 2711
トゲハクセンヤシ 59096
トゲハニガナ 219609
トゲハマサジ 179380,381341
トゲハリクジャクヤシ 12774
トゲハリナスビ 366910,367706
トゲバーレリア 48295
トゲバンレイシ 25868
トゲフウチョウボク 71762,71871
トゲボザキノフサモ 261370
トゲマサキ 182683
トゲマユミ 157418
トゲミイヌジシャ 104203
トゲミウドノキ 300955
トゲミゲシ 282574
トゲミチシャノキ 104203
トゲミツクシネコノメ 90446
トゲミデイゴ 154622
トゲミノイヌヂシャ 104203
トゲミノウマゴヤシ 247262
トゲミノキツネノボタン 326108
トゲミマユミ 157888
トゲムラサキ 39299
トゲヤマルリソウ 270698
トゲリュウキュウイチゴ 338488
トケンラン 110516
ドコウリュウ 320259
ドゴシツ 4304
トコナツ 127674
トコナデシコ 127793
トコロ 131877
トコン 81970
トコン属 81945
トサアサヒカン 93850
トサオトギリ 202192
トサカクシノハラン 62958
トサカケイトウ 80395
トサカメオトラン 174305
トサカモメヅル 400888
トサコゴメグサ 160180
トサザクラ 315051
トサシモツケ 372112
トサツツジ 331985
トサトウヒレン 348583
トサノアオイ 37602
トサノギボウシ 198621
トサノジョウロウホトトギス 396602
トサノチャルメルソウ 256041
トサノハマスゲ 119518
トサノミカエリソウ 228014
トサノミツバツツジ 330581
トサノモミジソウ 255790
トサボウフウ 24509
トサミズキ 106682
トサミズキ属 106628
トサミヅキ属 106628
トサムラサキ 66927
ドシニァ属 136784

トシマガヤ 65464
トシミカイドウ 243717
ドジャウツナギ 177668
トショウ 213702
ドジョウツナギ 177612
ドジョウツナギ属 177556
ドジンノクシバシラ 279489
ドスカンバ 53411
ドスナラ 382276
トスベリ 167999
ドスン 381404
ドセイカン 264387
ドゼウツナギ 177668
ドゼウツナギ属 177556
トダイアカバナ 146840
トダシバ 37352,37379,37389
トダシバ属 37350
トダスゲ 73608
トチカガミ 200228
トチカガミ科 200235
トチカガミ属 200225
トチナイサウ 23213
トチナイソウ 23213,23323
トチナイソウ属 23107
トチノキ 9683,9732
トチノキ科 196494
トチノキ属 9675
トチバニンジン 280793
トチハラニンジン 280793
ドチモ 200228
トチュウ 156041
トチュウシャラノキ 376458
トチユウ科 156042
トチュウ属 156040
トッカンバラ 54158
ドッグバイオレット 409804
トックリアブラギリ 212202
トックリイチゴ 338292
トツクリタケウマヤシ 208368
トックリツメクサ 397130
トックリノキ 49343
トックリハシバミ 106779
トックリフトエクマテヤシ 315388
トックリヤシ 201359,201360
トックリヤシモドキ 246028
トックリヤシ属 201356,246026
トックリラン 49343
トックリラン属 266482
トッケリキワタ 88973
トッケリキワタ属 88970
ドデカテオン属 135153
ドドヌア属 135192
ドドネア属 135192
トドハダゴヨウ 300138
トドマツ 460
トナカイアザミ 348104
トナカイサウ 242706
トナカイスゲ 74672
トナカヒアザミ 348104
ドナクス 136081
トヌリキ 167999
トネアザミ 92234
トネツリガネツツジ 250527
トネテンツキ 166513
トネリコ 167931,167940,167999,168076
トネリコバノイチゴ 338435
トネリコバノカエデ 3218
トネリコ属 167893
トネリバハシドイ 382247
トネリバハゼノキ 300980
トネリバハゼノキ属 300974
トノサマオニソテツ 145241
トノサマオヤマヤシ 172227
トノテガシワ 302721
ドビアーリス属 136832
トビカズラ 259566
トビカズラ属 259479
トビシマネコノメソウ 90388
トビヅタ 410960,410992,410994
トビヒメタケウマヤシ 208373
トビラギ 301360,301413
トビラノキ 301413
ドブクリョウ 366338
ドブゴセウ 326340
トブヒノ 257411
トベラ 301413
トベラニンギョウ 46882
トベラノキ 301413
トベラモドキ 236462
トベラモドキ属 398666
トベラ科 301203
トベラ属 301207
トボシガラ 164198
トマト 239157
トマトダマシ 367567
トマト属 239154
トマリスゲ 75370
トミクサザサ 347365
トミサトオトギリ 202029
トミサトクサイ 213173
トミツグラス 249303
ドミヤ 240842
トミワハマグルマ 413506
ドームヤシ 202321
ドームヤシ属 202256
ドメスチカスモモ 316382
トモエ 109060
トモエオトギリ 201885
トモエサウ 201743
トモエシオガマ 287595
トモエソウ 201743
トモエバテンツキ 166324
トモシバ 381423
トモシリサウ 98020
トモシリサウ属 97967
トモシリソウ 98020,98022
トモシリソウ属 97967
ドモッゴウ 207135
ドモノキ 161384
トヤクリンドオ 173198
ドヨウフジ 414567
ドヨウユリ 230036
トヨコロスミレ 410368
トヨシマアザミ 92449
トヨシマラン 179694
トヨハラヤナギ 342934
トライバラモンジン 394321
トラカンギョク 182435
トラキチラン 147307
トラキチラン属 147303
トラキメーネ属 393882
ドラコフィラス属 137723
トラサイカク 373975
ドラセナ属 137330
トラノオ 239594,317945,407434
トラノオジソ 290940,290972
トラノオスズカケ 407453
トラノオノキ 186973
トラノオモミ 496,298226,298303
トラノオラン 346173
トラノコ 167702
トラノハナヒゲ 333495
トラノマキ 171673
トラノヲ 239594,407434
トラノヲモミ 298303,298442
トラノヲラン 346173
ドラバスミレ 410531
トラフアナナス 412371
トラフイモ 16535
トラフコチョウ 254927
トラフサンゴアナナス 8554
トラフセンネンボク 137416
トラフツツアナナス 54467
トラフヒメアナナス 113387
トラフヒレバショウ 66215
トラフビロードギリ 366719
トラフユリ 391627
トラフユリ属 391622
トラユリモドキ 118398
トリアシショウマ 41832
トリアショウマ 41852
トリアス属 394947
ドリアン 139092
トリカブト 5108,5185,5363
トリカブト属 5014
トリガミネカンアオイ 37700
トリクサントセレウス属 390479
トリクススペルムム属 390486
トリクスピダーリア属 111145
トリゲモ 262089
トリコウ 82390

トリコカウロン属 395581
トリコグサ 398071
トリコグサ属 398064
トリコグロッチス属 395855
トリコグロッティス属 395855
トリコケントルム属 395616
トリコゴニア属 395872
トリコゴニオプシス属 395876
トリコジネ属 395879
ドーリコス属 135399
トリコセレウス属 395630
トリコセントラム属 395616
トリコディアデア属 395791
トリコトシア属 396464
トリゴニジューム属 397314
ドリコレーナ属 395897
トリックススパーマム属 390486
ドリティス属 136312
トリテレイア属 398777
トリトーニア属 399043
トリトマラズ 52151
トリフォリア 310850
トリプトメネ属 390548
トリブネ 352769
ドリミオプシス属 138006
トリモチカズラ 80189
トリモチカヅラ 80189,300944
トリモチノキ 203908,399433
トルコギキョウ 161027
トルコギキョウ属 161026
ドルステーニア属 136417
トルネラ 400493
トルネラ属 400486
トレイマツ 300288
トレクレイトラン 416659
トレニア属 392880
トレニヤ 392905
トレーフトエウチワヤシ 103738
トレムロイデスポプラ 311547
トロイ 213138
ドロイ 213138
ドロサンテムム属 138135
ドロサンテモプシス属 138131
ドロソフィラム科 138374
ドロソフィラム属 138375
ドロテアトサス属 136391
ドロテアンッス属 136391
ドロニガナ 210530
ドロニカム属 136333
ドロニクム属 136333
ドロノキ 311389,311509
ドロノシモツケ 371980
ドロブ 311389
ドロヤナギ 311389,311509
ドロリカヅラ 214971
トロロ 229
トロロアオイ 229
トロロアオイモドキ 235
トロロアオイ属 212
トロロアフヒ 229
トロロアフヒ属 212
トロロカヅラ 214971
トロロノキ 200038
トロヰ 213138
ドヰヤナギ 343321,343587
トンキンニッケイ 91302
ドングイ 309723
ドングリ 323611
ドンゴ 309723
ドンドバナ 208543
トンビグサ 220355
トンプソネラ属 390325
トンボグサ 311890,357919
トンボソウ 302549,357919
トンボソウ属 400280
トンボハギ 70847
ナエバキスミレ 409780
ナオシチ 93833
ナカイサウ 72721
ナガイモ 131592,131772
ナガイモヤガラ 171937
ナガウリ 238261
ナガエアオイ 243823
ナガエアマナ 254564
ナガエイヌツゲ 203692
ナガエガシ 116167
ナガエカマツカ 313318
ナガエクワズヤシ 273123
ナガエケニオイグサ 187682
ナガエコミカンソウ 296785
ナガエサカキ 8221
ナガエサカキ属 8205
ナガエササガニユリ 200955
ナガエサンキライ 366324
ナガエジャニンジン 72838
ナガエスゲ 75640
ナカエセンネンボク 137342,137514
ナガエソヨゴ 204138
ナガエツルノゲイトウ 18128
ナガエノアザミ 92161
ナガエノセンナリホオズキ 297641
ナガエノモウセンゴケ 138307
ナガエブシダマ 235734
ナガエブナ 162386
ナガエミクリ 370072
ナガオスミレ 410502
ナガガクヒナノウスツボ 355148
ナカガワノギク 89773
ナガキンカン 167506
ナガグルミ 212621
ナガゲオニソテツ 145244
ナガサギイボタ 229608
ナガサキオトギリ 201971
ナガサキギボウシ 198649
ナガサキズミ 243657
ナガサキハギ 226685
ナガサキマンネングサ 356957
ナガサキリンゴ 243657
ナガサハサウ 155365
ナガサハサウ属 155354
ナガサハスミレ 410289
ナガサハソウ 155365
ナガサハソウ属 155354
ナガサワハコベ 375136
ナガジイ 78916
ナガジクヤバネバショウ 66198
ナガジラミ 276455
ナガススズメガャ 147609
ナガスズメガヤ 147509
ナガツメセキコク 125044
ナガヅメトンボサウ 302395
ナガツルザンセウ 417329
ナガドコロ 131877
ナガトホシクサ 151500
ナガトミソハギ 19566
ナカナヤマトチゴ 338679
ナガバ 113380
ナガバアカシア 1358
ナガバアカネ 337910,337945
ナガバアカバナ 146605
ナガバアコウ 165370
ナガバアサガオ 25323
ナガバアジサイ 199945
ナガバアヂサイ 199945
ナガバアヂサヰ 199945
ナカバアッサムヤシ 413091
ナガハアデク 382601
ナガバアフヒラン 383132
ナガバアマドコロ 308616
ナガバアメリカミコシガヤ 76735
ナガバアリノトウグサ 184997
ナガバアリノトオグサ 185014
ナガバアリマグミ 142115
ナガバアワゴケ 67376
ナガバアヲダモ 167923
ナガバイズセンリョウ 241774
ナガバイトモ 312173
ナガバイナモリ 272228
ナガバイヌグス 240625
ナガバイヌツゲ 204025
ナガバイノコヅチ 4304
ナガバイラクサ 402848,402996
ナガバイワカガミ 351455
ナガハウキミクリ 370068
ナガバウヅラ 179582
ナガバウマノスズクサ 34224
ナガバエゾエノキ 80659
ナガバエノキグサ 1790
ナガバエビモ 312232
ナガバオオウチワ 28140
ナガバオオベニウチワ 28140

ナガバオオリンドウ　174029
ナガバオニザミア　241458
ナガバオニソテツ　145238
ナカバオノオレ　53612
ナガバオホミズ　299037
ナガバオホミヅ　299037
ナガバオモダカ　342338
ナガバカエデ　2775,3255
ナガバカキノハグサ　308302
ナガバカスリソウ　130120
ナガバカナメモチ　295773
ナガバカワヤナギ　343414
ナガバカンコノキ　177178
ナガバキイチゴ　338945
ナガバギシギシ　339994
ナガバキタアザミ　348715
ナガバキヅタ　187321
ナガバキブシ　373565
ナガバキリンサウ　294114
ナガハキンカン　167513
ナガバキンカン　167515
ナガバクコ　239011
ナガハグサ　305855
ナガバクマノミヅキ　124891
ナガバクロウスゴ　403933
ナガバクロカモメヅル　409509
ナガバクロバイ　381371
ナガバクワズイモ　16509
ナガバケショウビユ　208354
ナガバコイワザクラ　314890
ナガバコウホネ　267290
ナガバコウヤボウキ　292055
ナガバコカノキ　155067
ナガバコナラ　324223
ナガバコバンモチ　142362
ナガバコンテリギ　200019
ナガバサギサウ　184085
ナガバサギソウ　184085
ナガバザミア　417016
ナガバシイ　78977
ナガバシコタンソウ　349117
ナガハシスミレ　410505
ナガバシスミレ　410502
ナガハシバミ　106777
ナガバシャシャンボ　403741
ナガバシャシャンポ　403741
ナカバシャジン　7649
ナガバシュズネノキ　122034
ナガバジュズネノキ　122046
ナガバシュロソウ　405606
ナガバシラカシ　116139
ナガバシラカンバ　53421
ナガバシラヤマギク　41225
ナガバシロダモ　264048
ナガバシロヨメナ　39993
ナガバスズメノヒエ　285460
ナガバスノキ　403716
ナガバセスジゲサ　11355
ナガバタウコギ　53816
ナガバダケカンバ　53421
ナガバタケハギ　18259
ナガバタチツボスミレ　410330
ナガバタバコ　266048
ナカバチクセツニンジン　280793
ナガバチシャノキ　141662
ナガバチドリ　302502
ナガバヂャボウシノシッペイ　148254
ナガバツガザクラ　297031
ナガバツノザミア　83685
ナガハツノミザミア　83685
ナガバツメクサ　374944
ナガバツルグミ　142013
ナガバツルシャマシキミ　365938
ナガバツルフジバカマ　408269
ナカバツルマサキ　157496
ナガバツルミヤマシキミ　365957
ナガバドロ　311265
ナガバトンボソウ　302537
ナガバナアザミ　2691
ナガバナウジラ　179709
ナガバナオシロイ　255727
ナガバナオシロイバナ　255727,255729
ナガバナカンチョウジ　57955
ナガバナタバコ　266048
ナガバナビランヂ　363467
ナガバナワシロイチゴ　338070
ナガバナンキンハゼ　346379
ナガバニヒタカグミ　141999
ナガバネマガリ　347240
ナガバノイシモチサウ　138295
ナガバノイシモチソウ　138295
ナガバノイブキトラノオ　308726
ナガバノイワベンケイ　329944
ナガバノウシタキサウ　91575
ナガバノウナギツカミ　291775
ナガバノウナギヅル　291775
ナガバノウマノスズクサ　34223
ナガバノウルシ　371409
ナガバノウルシ属　371406
ナガバノカウバハキ　292055
ナガバノカウヤバウキ　292055
ナカバノカナビキサウ　389769
ナカバノカナビキソウ　389769
ナガバノキソチドリ　302413
ナガハノクロウメモドキ　328741
ナガバノコウヤボウキ　292055
ナガバノコオヤボオキ　292055
ナガバノサクラサウ　314144
ナガバノサワハコベ　374860
ナガバノシマイノコヅチ　4259
ナガバノシマヰノコヅチ　4259
ナガバノシラカンバ　53421
ナカバノスミレサイシン　409746
ナガバノダケカンバ　53421
ナガバノニンジン　7836
ナガバノネコヤナギ　342924
ナガバノマウセンゴケ　138307
ナガバノミミサクラソウ　314144
ナガバノムカゴサウ　192851
ナガバノモウセンゴケ　138267
ナガバノヤノネグサ　291705
ナガバノヤブデマリ　408048
ナガバノヤマグルマ　399435
ナガバハエドクソウ　295986,295999
ナガババハグマ　12689
ナガババハケヤシ　332438
ナガババハコベ　374944
ナガババハッカ　250385
ナガババハナガサシャクナゲ　215387
ナガババハヒノキ　381154
ナガバハヘドクサウ　295986
ナガバハマオモト　111180
ナカバハマササゲ　408950
ナガバハマミチヤナギ　309857
ナガバハリフタバ　370749
ナガバヒゴタイ　348706
ナガバヒゼンマユミ　157748
ナガバヒナノウスツボ　355107
ナガバヒメアナナス　113380
ナカハヒメヒゴタイ　348690
ナガバヒメフタバラン　232922
ナガバヒメマオ　56357,56360
ナガバヒメマヲ　56357
ナガバマサキ　157631
ナガハマツ　300128
ナガバマツ　300186
ナガバマムシグサ　33547
ナガバマンサク　161023
ナガバマンサク属　161020
ナガバムラサキ　66840
ナガバムラサキシキブ　66817
ナカバモウコシャジン　7816
ナガバモクコク　25779
ナガバモクコク属　25771
ナガバモクセイ　276370
ナガバモミジイチゴ　338945
ナガバモミヂイチゴ　338945
ナガバヤクシソウ　110628
ナガバヤドリギ　411065
ナガバヤナギアザミ　92357
ナガバヤブチョロギ　373321
ナガバヤブマオ　56140
ナガバャブマメ　177789
ナガバヤブマヲ　56140
ナガバヤブムラサキ　66875
ナガバヤマツバキ　69196
ナカバューカリ　155628
ナガバユーカリ　155478
ナガバユキノシタ　52514
ナカハラカラマツ　388710
ナカハラキビ　282200

ナカハラクロウメ 328793
ナカハラクロウメモドキ 328793
ナカハラクロキ 381327
ナカハラゴセウ 290389
ナカハラスゲ 74643
ナカハラセキコク 146543
ナガハラセンサウ 399296
ナガハラセンソウ 399296
ナカハラツツジ 331305
ナカハラハグマ 12676,12679
ナカハララン 232139
ナカハラリンダウ 173455
ナカハラリンドウ 173455
ナガバワウレン 103835
ナガバヰノコヅチ 4304
ナガヒグミスブタ 55918
ナガヒゲウラシマソウ 33526
ナガヒゲヨノミ 235694
ナガベエリンジューム 154284,154317
ナガボアミガサノキ 1894
ナガボキヅタ 187322
ナガボサンゴアナナス 8597
ナガボスゲ 74374
ナガボソウ 373511
ナガボソウ属 373490
ナガボテンツキ 166386
ナガボトネテンツキ 166158
ナガボナツハゼ 403998
ナガボノアカワレモコウ 345917
ナガボノウルシ科 371413
ナガボノゲッタウ 17666
ナガボノコジュズスゲ 75702
ナガボノシロワレモカウ 345921
ナガボノシロワレモコウ 345921
ナガボノワレモコウ 345917
ナガホハネガヤ 376877
ナガボヒロハクサフジ 408442
ナガボヤナギタデ 291787
ナガボヤマアイ 358001
ナガボヤマアヰ 358001
ナガミイチゴ 338334
ナガミカツラ 9419
ナガミキンカン 167506
ナガミグス 91429
ナガミクマヤナギ 52457
ナガミコナラ 324388
ナガミショウジョウスゲ 73891
ナガミシラカシ 116139
ナガミタチコウガイゼキショウ 213569
ナガミツルマメ 177714
ナガミトウガラシ 72077
ナガミトチノキ 9683
ナガミノアマナズナ 68860
ナガミノイソノキ 328550
ナガミノイヌガラシ 336212
ナガミノオニシバ 418445,418450
ナガミノジュズダマ 99138
ナガミノセリモドキ 350425
ナガミノッルウメモドキ 80262
ナガミノツルキケマン 106355
ナガミノツルシキミ 365958
ナガミノトネリコ 168002
ナガミノノヘビノボラズ 52100
ナガミハマナタマメ 71072
ナガミパンノキ 36920
ナガミヒナゲシ 282546
ナガミヒメスゲ 75650
ナガミボチョウジ 319671
ナガミボチョオジ 319671
ナガミマツ 300011
ナガラシ 59438
ナガラッパバナ 366870
ナ－ガルブル－ム 176222
ナガレイシ 256797
ナガレイトタヌキモ 403176
ナガワスミレ 410408
ナカヰガシ 285233
ナカヰサウ 72721
ナガヲスミレ 410502
ナギ 52246,306493
ナギイカダ 122124,340990
ナギイカダ科 340541
ナギイカダ属 340989
ナギナタカウジュ 143974,143993
ナギナタカウジユ属 143958
ナギナタガヤ 412465
ナギナタコウジュ 143974
ナギナタコウジュ属 143958
ナギナタソウ 123028
ナギナタソウ科 123033
ナギボウキ 39079
ナギマキ 306450
ナギモドギ 10494
ナキラン 116938
ナギラン 116938
ナキリ 217739
ナキリスゲ 73952,75103
ナクト－ル 307269
ナゴスゲ 74692
ナゴラン 356457
ナコラン属 9271
ナゴラン属 9271,356456
ナシ 323268,323272
ナシウミ 114197
ナシウメ 6682
ナシウリ 114213
ナシカズラ 6697
ナシカヅラ 6697
ナシタケ 249617
ナ－シッサス・ボルボコデイウム 262361
ナシマンサク 160347
ナシリンダウ 173886
ナシ属 323076
ナス 367370
ナズナ 72038,72047
ナズナ属 72037
ナスヌカボシソウ 238617
ナスノイワヤナギ 342941
ナスノヒオウギアヤメ 208824
ナスノユカワザサ 347233
ナス科 366853
ナス属 366902
ナゼザクラ 376683
ナタウリ 114288,114300
ナタオレノキ 276336
ナタツモリ 96466
ナタネタビラコ 221787
ナタネナ 59358
ナタネハタザオ 102811
ナタハジキ 276336
ナタマメ 71042,71050
ナタマメ属 71036
ナチウツギ 126954
ナツアサドリ 142252
ナツエビネ 66035,66037,66046
ナツグミ 142088,142092,142214
ナツコムギ 356237
ナツゴロウ 404047
ナツザキエリカ 67456
ナツザキソシンカ 49173
ナツザキフクジュソウ 8325
ナツシロギク 322724
ナツズイセン 239280
ナツスカシユリ 229915
ナツズタ 285117,285157
ナツズヰセン 239280
ナツダイダイ 93340
ナツタデ 309298
ナツヅタ 285157
ナツツバキ 376466
ナツツバキ属 376428
ナツテンジクアオイ 288206
ナツトウダイ 159841
ナツトメッグ 77910
ナヅナ 72038
ナヅナザクラ 203184
ナヅナハタザオ 30345
ナヅナハタザホ 30345
ナヅナ属 72037
ナツノウナギツカミ 309051
ナツノタムラソウ 345183
ナツバウズ 122586
ナツハギ 226977
ナツハゼ 247025,403925
ナツフジ 414567
ナツフジ属 254596
ナツフヂ属 254596
ナツボダイジュ 391817
ナツマメ 408393
ナツミカン 93340
ナツメ 418169,418173

ナツメジュロ　295461
ナツメヤシ　190094,295461
ナツメヤシ属　295451
ナツメ属　418144
ナツユキカズラ　162509
ナツユキソウ　83060,166119
ナツユリ　229888
ナツリンドウ　173886
ナデシコ　127756
ナデシコ科　77980
ナデシコ属　127573
ナデツコ　127852
ナテン　316809
ナデン　83326,316809
ナトリグサ　280286
ナナカマド　369339,369362,369367,369488
ナナカマド属　369290
ナナコイバラ　336490
ナナコマル　244258
ナナツガママンネングサ　356685
ナナツバ　359654
ナナバケンシダ属　385529
ナナミノキ　203625
ナナメノキ　203625
ナナンッス属　262149
ナニハイバラ　336675
ナニハヅ　121718,121720
ナニワイバラ　336675
ナニワズ　122471
ナニンジン　123141
ナノハタンポポ　384479
ナハカノコサウ　56425
ナハカノコサウ属　56397
ナハカノコソウ　56425
ナハカノコソウ属　56397
ナハキハギ　125582
ナハシロイチゴ　338985
ナハシログミ　142152
ナピアグラス　289218
ナビス　367370
ナフトール　307269
ナベイチゴ　338516,339270
ナベクラザゼンソウ　381078
ナベコウジ　328684
ナベコワシ　346390
ナベナ　133490
ナベナ属　133451
ナベノキ　328684
ナベワリ　104696,111676
ナベワリ属　111674
ナホザキイガコウゾリ　317233
ナマイ　374089
ナマクァンテス属　262127
ナミキソウ　355778
ナミツユクサ　101185
ナミマクラ　175512
ナミモロコシ　369600
ナムナム　118060
ナムナム属　118047
ナメラコウゾリナ　298593
ナメラサギサウ　183937
ナメラサギソウ　183937
ナメラサンキライ　366338
ナメラダイモンジソウ　349358
ナメラツリフネソウ　205372
ナヤチャングニイ　93645
ナヨクサフジ　408666
ナヨタケ　304098
ナヨチンム　171937
ナヨテンマ　171937
ナヨナヨココゴメグサ　160221
ナヨナヨワスレナグサ　260815
ナラ　323942,324384
ナラガシハ　323630
ナラガシワ　323630
ナラバガシ　116100
ナラマキ　324190
ナリイモ　131501
ナリヒラダケ　357939
ナリヒラダケ属　357932
ナリヒラモチ　203946
ナリヤラン　37115
ナリヤラン属　37109
ナルカミスミレ　409945
ナルコスゲ　74241
ナルコビエ　151702
ナルコビエ属　151655
ナルコユリ　308538
ナルサワザクラ　83138
ナルシス　196443
ナルト　109174
ナルトウオウギ　43048
ナルトサワギク　359423
ナワシロイチゴ　338985
ナワシログミ　142152
ナンオウツツジ　330255
ナンカイアオイ　37692
ナンカイウスベニニガナ　144916
ナンカイヌカボ　12008
ナンカイヒメイワカガミ　351448
ナンキャウサウ　17677
ナンキョクブナ　266864
ナンキョクブナ科　266858
ナンキンアヤメ　208874
ナンキンイバラ　336490
ナンキンウメ　87525
ナンキンクロウメモドキ　328700
ナンキンツバキ　69552
ナンキンナナカマド　369401
ナンキンナナカマドモドキ　369455
ナンキンハゼ　346408
ナンキンボウフラ　114295
ナンキンマメ　30498
ナンキンマメ属　30494
ナンキンワタ　179878
ナンクワ　114294
ナンコウナギ　306503
ナンゴクアオイ　37604
ナンゴクアオウキクサ　224360
ナンゴクアオキ　44937
ナンゴクイヌホオズキ　367651
ナンゴクウラシマソウ　33536
ナンゴクカモメヅル　409416
ナンゴクキケマン　106218
ナンゴククガイソウ　407465
ナンゴククマヤナギ　52418,52459
ナンゴククロクモソウ　349365
ナンゴクコウゾ　61102
ナンゴクスミレ　410105
ナンゴクテツカエデ　3251
ナンゴクナットウダイ　159844
ナンゴクナットオダイ　159841
ナンゴクネヂバナ　372247
ナンゴクハマウド　24372
ナンゴクヒサカキ　160406
ナンゴクヒメミソハギ　19554
ナンゴクボタンボウフウ　292894
ナンゴクミゾハコベ　142574
ナンゴクミネカエデ　2796
ナンゴクモクセイ　276286
ナンゴクヤッシロラン　171952
ナンゴクヤマアジサイ　200097
ナンゴクワセオバナ　341912
ナンコシャクナゲ　331311
ナンコシラヤマギク　41337
ナンコタイサンシャクナゲ　331311
ナンザンスミレ　409816
ナンジヤモンジヤ　87729,91287
ナンシュウ　377365
ナンショウダイダイ　93341
ナンテン　262189
ナンテンイノコヅチ　4319
ナンテンカズラ　64983
ナンテンカヅラ　64983
ナンデンギリ　203422
ナンテンサウ　408648
ナンテンソウ　408648
ナンテンソケイ　281225
ナンテンチゴザサ　209054
ナンテンハギ　408648
ナンテンキノコヅチ　4319
ナンテン属　262188
ナントウイガニガクサ　203042
ナントウイチゴ　338425
ナントウカゴノキ　6762
ナントウガシ　233323
ナントウクロモジ　6762
ナントウダモ　6762
ナントウニホヒグサ　187630
ナントウホシクサ　151400
ナントウヤマハヅ　28343

ナンバン 417417
ナンハンアイ 206669
ナンバンアカアズキ 7190
ナンバンアカアヅキ 7190
ナンバンアカアヅキ属 7178
ナンバンアカバナアズキ 241263
ナンバンアヅキ 765
ナンバンアハブキ 249463
ナンバンアワブキ 249463
ナンハンアヰ 206669
ナンバンイヌマキ 306512
ナンバンエノキ 80764
ナンバンガキ 164763
ナンバンガシ 233238,233389
ナンバンカモメラン 240862
ナンバンカラスウリ 256804
ナンバンカラムシ 56229
ナンバンカンゾウ 191263
ナンバンギセル 8760
ナンバンギセル属 8756
ナンバンキビ 417417
ナンバンキブシ 373563
ナンバンキンギンザウ 179623
ナンバンキンギンソウ 179623
ナンバンクサフジ 386257
ナンバンクサフジ属 385926
ナンバンクサフヂ 386257
ナンバンクサフヂ属 385926
ナンバンクロモジ 231301
ナンバンコマツナギ 206626
ナンバンサイカチ 78300
ナンバンサンキライ 194120
ナンバンスミレ 409995,409997
ナンバンタイセイ 206626
ナンバンタヌキマメ 112138
ナンバンツツジ 330617
ナンバンツバキ 179745
ナンバンツユクサ 101112
ナンバンハコベ 114060,114066
ナンバンハコベ属 114057
ナンバンマユミ 157709
ナンバンムラサキ 66913
ナンバンヤナギ 197963
ナンバンヤナギ属 197960
ナンバンヤブマメ 386410
ナンバンルウダ 139678
ナンバンルリサウ 190651
ナンバンルリソウ 190651
ナンブアザミ 92226
ナンブイヌナズナ 137052
ナンブクロカンバ 328660
ナンブコハモミジ 2782
ナンブスゲ 353606
ナンブソウ 4104
ナンブソウ属 4103
ナンブソモソモ 305573
ナンブタウキ 229485
ナンブタカネアザミ 92217
ナンブタモ 168023
ナンブトウウチソウ 345880
ナンブトウヒレン 348829
ナンブトラノオ 54781
ナンブヒョウタンボク 236124
ナンブワチガイ 318506
ナンブワチガイソウ 318506
ナンベンカラムシ 56229
ナンマンオケラ 44200
ナンマンオトコヨモギ 35466
ナンマンキンミズヒキ 11572
ナンメイラ 66808
ナンヤウアブラギリ 212127
ナンヤウクサトベラ 350337
ナンヤウスギ 30841
ナンヤウスギ属 30830
ナンヤウマキ 306424
ナンヨウアブラギリ 212127
ナンヨウイヌカンコ 166777
ナンヨウウツボカズラ 264846
ナンヨウザクラ 259910
ナンヨウサヤバナ 99519,99523
ナンヨウスギ 30841
ナンヨウスギ科 30857
ナンヨウスギ属 30830
ナンヨウソテツ 115812,115888
ナンヨウハマオモト 111236
ニイガタガヤツリ 119259
ニイジマガマズミ 407705
ニイタカアカマツ 300269
ニイタカアズマギク 150788
ニイタカイバラ 336930
ニイタカカマツカ 295669,377446,377457
ニイタカガラシ 47962
ニイタカコウモリ 283847
ニイタカシャクナゲ 331565
ニイタカシラタマ 172059
ニイタカセキチク 127862
ニイタカタニソバ 309711
ニイタカチチコグサ 178232
ニイタカチドリ 302265
ニイタカトウヒ 298375
ニイタカトドマツ 396
ニイタカナガバマンネングサ 357007
ニイタカハイノキ 381104
ニイタカハタザオ 30164
ニイタカハリイ 396025
ニイタカヒイラギナンテン 242580
ニイタカヒトツバラン 191591,191598
ニイタカビャクシン 213943
ニイタカフウロ 174645
ニイタカボロギク 359516
ニイタカマツムシソウ 350185
ニイタカマユミ 157932
ニイタカヤダケ 416804
ニイタカヤマハハコ 21630
ニイタカヨモギ 35237,35964
ニウメンラン 374460
ニウメンラン属 94425
ニオイアヤメ 208565
ニオイアラセイトウ 86427
ニオイアラセイトウ属 86409
ニオイイリス 208565
ニオイウイキョウ 229404
ニオイウツギ 413579
ニオイウド 276748
ニオイウルシ 332477
ニオイエビネ 65980
ニオイエンドウ 222789
ニオイカラマツ 388510
ニオイガントウ 292353
ニオイグサ 187691
ニオイコブシ 242274
ニオイササウチワ 370337
ニオイサンタンカ 211143
ニオイシャジン 7850,7879
ニオイシュロラン 104339
ニオイシレンゲ 250486
ニオイスイレン 267730
ニオイスミレ 410320
ニオイセッコク 125279
ニオイセンネンボク 137397
ニオイタチツボスミレ 410312
ニオイタデ 309951
ニオイテンジクアオイ 288268
ニオイナズナ 203229
ニオイニガクサ 203066
ニオイニワトコ 345708
ニオイニンドウ 236022
ニオイネズコ 390579
ニオイハカタユリ 229758
ニオイハタザオ 136173
ニオイハンゲ 299719
ニオイバンマツリ 61306
ニオイヒツジグサ 267730
ニオイヒバ 114760,390587
ニオイヒヨドリ 158230
ニオイフジウツギ 61975
ニオイベンゾイン 231306
ニオイホオズキ 255223
ニオイマガリバナ 203229
ニオイマンジュギク 383097
ニオイミゾホオズキ 255223
ニオイミツビシラン 238733
ニオイムラザキ 190559
ニオイヤグルマギク 18966,81219
ニオイヤハズハハコ 21683
ニオイユリ 229730,229872,229900,229992
ニオイヨモギ 264249
ニオイラン 185755
ニオイラン属 185753
ニオイレセダ 327896
ニオイロウバイ 68313

ニオウギヤシ属 222625
ニオウマル 140883
ニオウモン 157168
ニオウヤブマオ 56181
ニオヒタデ 309951
ニガイチゴ 338822,338886
ニガウリ 256797
ニガウリ属 256780
ニガガシュウ 131501
ニガキ 298516,298519
ニガキモドキ 61208
ニガキモドキ属 61199
ニガキ科 364387
ニガキ属 298506
ニガクイボタ 229566
ニガクサ 388114
ニガクサ属 387989
ニカクソウ 128032
ニガサマル 234933
ニガソバ 162335
ニガタケ 297215,297435,304098
ニガダケ 304098
ニガチシャ 90894
ニガチャ 69651
ニカヅク 261424
ニカヅク属 261402
ニガナ 210527
ニガナモドキ 210543
ニガナ属 210553,219206
ニガニレ 401490
ニガハクカ 245770
ニガハクカ属 245723
ニガハッカ 245770
ニガハッカ属 245723
ニガヨモギ 35090,224989
ニカンドラ属 265940
ニクイロシュクシャ 187425
ニクキビ 58183
ニクキビモドキ 58139,402543
ニクキビ属 58052
ニクケイ 91366
ニクケイソウ 223913
ニクケイモドキ 91389
ニクズク 261424
ニクズクキ 261424
ニクズク科 261464
ニクズク属 261402
ニクタンテス属 267429
ニクヅク 261424
ニクヅク科 261464
ニクヅク属 261402
ニクトセレウス属 267603
ニグラゲルミ 212631
ニグラスモモ 34431
ニグラヤマナラシ 311398
ニコゲヌカキビ 128478
ニコゲルリミノキ 346477
ニコバルヤシ 51164
ニコバルヤシ属 321161
ニコルソンヤシ属 264236
ニコンゴウ 244035
ニサンゴバナ属 279765
ニシインドコキュウリ 114122
ニシインドバニラ 405021
ニシウチサウ 82924
ニシガキマユミ 157769
ニジガハマギク 89410
ニシキアオイ 25923,25924
ニシキアカリフア 2019
ニシキイモ 65218
ニシキイモ属 65214
ニシキウツギ 413580,413619
ニシキカズラ 25928
ニシキギ 157285
ニシキキブシ 373559
ニシキギボウシ 198638
ニシキギ科 80127
ニシキギ属 157268
ニシキケテイカカズラ 393666
ニシキコウジュ 144083,144093
ニシキコバノミツバツツジ 331628
ニシキゴロモ 13191
ニシキサウ 159092
ニシキサウダイゲキ 158935
ニシキジソ 99711
ニシキソウ 159092
ニシキダイゲキ 158758
ニシキダイコン 53269
ニシキハウチワマメ 238448
ニシキハギ 226852
ニシキハリナスビ 367735
ニシキヒガンバナ 239267
ニシキマル 244234
ニシキマンサク 185109,185117
ニシキミゾホオズキモドキ 255186
ニシキミゾホオツキ 255218
ニシキミゾホホツキ 255218
ニシキミヤコグサ 237555
ニシキモクレイ 416682
ニシキヤマツツジ 330978
ニシキユキノシタ 349940
ニシキラン 25928,171456
ニシゴリ 381137,381341
ニシダケササバギンラン 82059
ニシノオオタネツケバナ 72737
ニジノタマ 357111
ニシノホンモンジスゲ 76365
ニシノヤマタイミンガサ 283884
ニシャクナゲ 330245
ニショウ科 300557
ニシヨモギ 35630,35634
ニセアカシャ 334995
ニセアカシヤ 334976
ニセアゼガヤ 226009
ニセアレチギシギシ 340249
ニセウマゴヤシ 247425
ニセクサキビ 226231
ニセコガネギシギシ 340293
ニセゴシュユ 161373
ニセコバンソウ 60653
ニセコムギダマシ 11710
ニセジュズネノキ 122067
ニセシラゲガヤ 197306
ニセダイオウヤシ属 318118,347074
ニセタケゥマヤシ属 366806
ニセツキヌキサイコ 63699
ニセマガホヤシ属 283461
ニセマメグンバイナズナ 225433
ニソジンボク 411362
ニチタカイナモリ 272282
ニチタカシシウド 24413
ニチナンオオバコ 302001
ニチニチカ 79418
ニチニチクワ 79418
ニチニチソウ 79418
ニチニチソウ属 79410
ニチバナ科 93046
ニチリンギョク 233464
ニチリンサウ 188908
ニチリンソウ 188908
ニックワアツニ 401545
ニックワウウツギ 127118
ニックワウキスゲ 191309
ニックワウハギ 408648
ニックワウモミ 389
ニックワタニワタシ 408648
ニッケイ 91287,91366,91426
ニッコウアザミ 92274
ニッコウアッコ 401545
ニッコウアッシ 401545
ニッコウイカリソウ 147013
ニッコウウツギ 62099,127118
ニッコウオトギリ 201911,202047
ニッコウオヒョウ 401545
ニッコウガマズミ 407792
ニッコウキスゲ 191278,191309
ニッコウキバナシャクナゲ 331335
ニッコウクルマバナ 96973
ニッコウコウガイゼキショウ 213320
ニッコウコウモリ 283821
ニッコウザクラ 83137
ニッコウザサ 347206
ニッコウシャクナゲ 22445
ニッコウトウヒレン 348392,348560
ニッコウネコノメ 90409
ニッコウバイカウツギ 294531
ニッコウハリスゲ 74608
ニッコウヒョウタンボク 235960
ニッコウマツ 221894
ニッコウマユミ 157764
ニッコウモミ 389

ニッサボク 267864
ニツセンロウ 10801
ニッパヤシ 267838
ニッパヤシ属 267837
ニッポウアザミ 92226
ニッポンイヌノヒゲ 151520
ニッポンサイシン 37722
ニッポンタチバナ 93823
ニヅラリューム属 266148
ニデシコ科 77980
ニドササゲ 294056
ニトベカズラ 28422
ニトベカヅラ 28422
ニトベカヅラ属 28418
ニトベギク 392432
ニトベギク属 392428
ニトベヨモギ 36020
ニバイタイ 158230
ニハウメ 83238
ニハウルシ 12559
ニハウルシ属 12558
ニハクサ 217361
ニハザクラ 83238
ニハゼキシャウ 365747
ニハゼキシャウ属 365678
ニハタバコ 405667,405788
ニハツゲ 64285
ニハトコ 345660
ニハトコ属 345558
ニバナアマ 231907
ニバナイチヤクソウ 322790
ニバナサルゼヤ 344980
ニハナヅナ 235090
ニハナヅナ属 18313
ニバナヒルギ 238350
ニバナヘウタンボク 235945,236085
ニバナヤマシャクヤク 280241
ニハフヂ 205876
ニハホコリ 147823,147883
ニーパヤシ 267838
ニハヤナギ 308816
ニーパ属 267837
ニヒジマガマズミ 407705
ニヒタカアカバナ 146599
ニヒタカアカマツ 300269
ニヒタカアザミ 92096
ニヒタカアヅマギク 150792
ニヒタカアマナラン 55563
ニヒタカイチゴ 339188
ニヒタカイチャク 322858
ニヒタカイトヰ 213297
ニヒタカイヌツゲ 204418
ニヒタカイバラ 336938
ニヒタカウスユキ 21643
ニヒタカオトギリ 202035
ニヒタカカノコサウ 398024
ニヒタカカノコソウ 398024
ニヒタカカハヅスゲ 75544
ニヒタカカマツカ 295674
ニヒタカカラマツ 388541
ニヒタカガンピ 364134
ニヒタカキツネガヤ 60860
ニヒタカキパウゲ 326419
ニヒタカキヲン 359654
ニヒタカクハガタ 407241
ニヒタカグミ 142207
ニヒタカクリンサウ 314655
ニヒタカクロマツ 300269
ニヒタカクロモジ 234000
ニヒタカクワガタ 407241
ニヒタカコケモモ 403796
ニヒタカコゴメグサ 160286
ニヒタカシオン 359568
ニヒタカシホガマ 287766
ニヒタカシモツケ 372021
ニヒタカジャカウサウ 274237
ニヒタカシャクナゲ 331565
ニヒタカシャジン 7702
ニヒタカシラタマ 172059
ニヒタカシヲン 359568
ニヒタカスグリ 333978
ニヒタカスゲ 75418
ニヒタカスミレ 409993
ニヒタカセキチク 127802
ニヒタカセンキウ 101831
ニヒタカセンブリ 380264
ニヒタカタウヒ 298375
ニヒタカタチバナ 31388
ニヒタカタニソバ 309711
ニヒタカタフバナ 97011
ニヒタカチドリ 302265
ニヒタカツクバネ 284343
ニヒタカツゲ 204418
ニヒタカツリフネ 205422
ニヒタカツルリンダウ 398285
ニヒタカトドマツ 396
ニヒタカトンボ 183559
ニヒタカナガバマンネングサ 357007
ニヒタカヌカボ 12145
ニヒタカヌカボシ 238614
ニヒタカハクテウゲ 235897
ニヒタカハタザオ 30164
ニヒタカハタザホ 30164
ニヒタカハヒノキ 381104
ニヒタカハリイ 396025
ニヒタカハリヰ 396025
ニヒタカバロギク 359516
ニヒタカヒサカキ 160527
ニヒタカヒトツバラン 191591
ニヒタカヒトツバラン属 191582
ニヒタカヒメレンゲ 356937
ニヒタカビャクシン 213943
ニヒタカヒラギナンデン 242608
ニヒタカフウロ 174645
ニヒタカフタバラン 264704
ニヒタカヘビノボラズ 51940
ニヒタカマツムシサウ 350185
ニヒタカマユミ 157932
ニヒタカマンネングサ 356947
ニヒタカマンりゃウ 31388
ニヒタカミゾソバ 308886,309877
ニヒタカミツバ 345996
ニヒタカムグラ 170589
ニヒタカメダケ 416804
ニヒタカモリイバラ 336995
ニヒタカヤダケ 416804
ニヒタカヤダケ属 206765
ニヒタカヤマヂノギク 40878
ニヒタカヤマハハコ 21630
ニヒタカヨモギ 35964
ニヒタカラッキョウ 15105
ニヒタカリンダウ 173868
ニヒタカリンドウ 173868
ニヒメマル 234948
ニヒモトリンダウ 173451,173868
ニヒモトリンドウ 173868
ニフタカツクバネ 284378
ニベナロウゲ 312810
ニベノキ 200038
ニホヒアラセイトウ 86427
ニホヒアラセイトウ属 86409
ニホヒイリス 208565
ニホヒウルシ 332477,332502
ニホヒエンドウ 222789
ニホヒグサ 187691
ニホヒグサ属 187497
ニホヒクワントウ 292353
ニホヒケスミレ 409834
ニホヒシャウブ 208565
ニホヒシュロラン 104339
ニホヒスミレ 410320
ニホヒタデ 309951
ニホヒタブ 240726
ニホヒテンヂクアフヒ 288268
ニホヒドロ 311368
ニホヒニガクサ 203066
ニホヒネズコ 390579
ニホヒバラ 336954
ニホヒヒツジグサ 267730
ニホヒヒバ 390587
ニホヒラン 185755
ニボヒラン 185755
ニボヒラン属 185753
ニホヒレセダ 327896
ニホンカボチャ 114292,114295
ニホンスモモ 316761
ニホンヤマナシ 323268
ニミゾホオズキ 255197
ニュウサイラン 295600
ニュウサイラン属 295594
ニュウシジュ 289476

ニュウメンラン　395865
ニューカレドニアフィジーノヤシ　405256
ニューギニアワイルドライム　253293
ニューサイラン　191317
ニューサンマー　93834
ニュージーマンドアサ　295600
ニュージーランド　328446
ニュージーランドイチビ属　77050
ニューシーランドマツ　10494
ニューメンラン　395865,404749
ニューヨークシオン　40923
ニョイスミレ　409656,410730
ニョホウチドリ　310897
ニョライ　304351
ニョロリ　272891
ニラ　15534,15843
ニラコ　294040
ニラネギ　15621
ニラハラン　254238
ニラバラン　254238
ニラハラン属　254226
ニラバラン属　254226
ニラミグサ　305334
ニラモ　418389
ニラモドキ　266959
ニラモドキ属　266957
ニリンオトギリ　201880,201892
ニリンサウ　23817
ニリンソウ　23817
ニリンソウ属　23696
ニレ　401489,401490,401581
ニレザクラ　19248
ニレッバズイショウ　385262
ニレッバスヰショウ　385262
ニレモミ　389
ニレ科　401415
ニレ属　401425
ニワアジサイ　200101
ニワウメ　83238
ニワウルシ　12559
ニワウルシ属　12558
ニワギキョウ　69942
ニワクサ　217361
ニワザクラ　83208,316485
ニワシロユリ　229789
ニワゼキショウ　365695,365743,365747
ニワゼキショウ属　365678
ニワタズ　345586
ニワタバコ　405667
ニワツゲ　64285
ニワトコ　345660,345679
ニワトコウライ　345592
ニワトコ属　345558
ニワナズナ　235090
ニワナズナ属　18313,235065
ニワナナカマド　369272
ニワハタザオ　30220
ニワハナビ　230655
ニワフジ　205876
ニワフジウツギ　61955
ニワホコリ　147823
ニワヤナギ　308816
ニンギョウニシキ　16687
ニンジン　123141,123164,280741,280793
ニンジンボク　411373
ニンジンボク属　411187
ニンジン属　123135,280712
ニンスーカ　59438
ニンドウ　235878
ニンドウバノヤドリギ　355319,385221
ニンドウモチノキ　203992
ニンドウモドキ　103862
ニンドオ　235878
ニンニク　15698,15726
ニンニンバ　1142
ニンフェア属　267627
ニンポウキンカン　167493
ヌイオスゲ　76665
ヌイヲスゲ　76665
ヌカイトナデシコ　183221
ヌカエ　290952
ヌカカゼクサ　147995
ヌカガラ　231334
ヌカキビ　281390
ヌガゴサウ　192851
ヌカスゲ　75391
ヌカススキ　12795
ヌカススキ属　12776
ヌカビ　85345
ヌカボ　12186
ヌカボガヤ　321236
ヌカボシサウ　238685,238690
ヌカボシソウ　238685
ヌカボシタデ　309054
ヌカボタデ　309853
ヌカボミチヤナギ　308791
ヌカボラン　12449
ヌカボラン属　12442
ヌカボ属　11968
ヌスビトノアシ　171918,200736
ヌスヒトハギ　200742
ヌスビトハギ属　126242
ヌソラーレア属　319116
ヌナワ　59148
ヌニシユリ　230038
ヌノマオ　300835
ヌノマオ属　300833
ヌノマヲ　300835
ヌノマヲ属　300833
ヌマアゼスゲ　74135,75357
ヌマイチゴツナギ　305804
ヌマカゼクサ　147496
ヌマガヤ　256632
ヌマガヤツリ　118957
ヌマガヤ属　256597
ヌマギシギシ　339925
ヌマキヌヤナギ　343119,343994,344034
ヌマクロネスゲ　75348
ヌマクロボスゲ　75348
ヌマシャジン　7728
ヌマスギ　385262
ヌマスギモドキ　252540
ヌマスギ属　385253
ヌマスゲ　76070
ヌマスノキ　403780
ヌマゼリ　365858
ヌマダイオウ　339925
ヌマダイコン　8012
ヌマダイコン属　8004
ヌマダイワン　339925
ヌマツルギク　4868,4869
ヌマツルギクモドキ　4863
ヌマドジョウツナギ　177661
ヌマトラノオ　239645
ヌマトラノヲ　239645
ヌマハコベ　258248
ヌマハッカ　250291
ヌマハリイ　143168,143200,143201,143271
ヌマハリヰ　143201,143271
ヌマヒノキ　85375
ヌマヒバ　85375
ヌマミズキ　267867
ヌマミズキ属　267849
ヌマヤナギ　343747
ヌマラフゲ　100257
ヌメゴマ　231942,232001
ヌメリグサ　341960
ヌメリグサ属　341937
ヌルデ　332509,332662
ヌルデアワブキ　249424
ヌンシュウツメクサ　31787
ネアポリタヌム　115949
ネイハ　167503
ネイハキンカン　167493
ネオアボッチア属　263559
ネオウエルデルマンニア属　264809
ネオキレニア属　263732
ネオダウソニア属　263800
ネオビンガミア属　263650
ネオベンタミア属　263641
ネオポルテリア属　264332
ネオマリカ属　264158
ネオライモンディア属　264386
ネオラウケア属　209024
ネオレゲリア属　264406
ネオロイディア属　264120
ネガモーレア属　264205
ネカラシナ　59461
ネギ　15289
ネギ属　15018
ネクタリン　20955

ネグンドカエデ 3218
ネコアサガオ 25316
ネコシデ 53388
ネコジャラシ 361935
ネコノシタ 413549
ネコノシタ属 413490
ネコノチチ 328550
ネコノチチ属 328544
ネコノヒゲソウ 95862,275639
ネコノヒゲソウ属 275624
ネコノメサウ属 90317
ネコノメソウ 90370
ネコノメソウ属 90317
ネコハギ 226924
ネゴヤタカスケ 259094
ネコヤナギ 343444
ネコヤマヒゴタイ 348536,348538
ネザサ 304010
ネジアヤメ 208543,208665,208667
ネジイトラン 416610
ネジキ 239392
ネジキ属 239369
ネジバナ 372247,372250
ネジバナ属 372176
ネジリカワツルモ 340472,340476
ネジレオニザミア 241465
ネジレモ 404556
ネズ 213896
ネズコ 390660
ネズサシ 213896
ネズミガヤ 259671,259672
ネズミガヤ属 259630
ネズミサシ 213896
ネズミサシ属 213606
ネズミシバ 398071
ネズミノオ 372659,372666
ネズミノオ属 372574
ネズミノシッポ 412465
ネズミノハナドオシ 122040
ネズミノハナヒゲ 333562
ネズミノヲ 372659,372666
ネズミノヲ属 372574
ネズミホソムギ 235301
ネズミムギ 235315
ネズミモチ 229485
ネズミユリ 73158
ネズミヨモギ 35282
ネズモドキ 226451
ネズモドキ属 226450
ネズモミ 80087
ネソバノサンジサウ 94140
ネヂアヤメ 208543,208667
ネヂイトラン 416610
ネヂキ 239391,239392
ネヂキ属 239369
ネヂバナ 372247
ネヂバナ属 372176
ネヂバリン 208667
ネヂレギ 242976
ネヂヰ 212890
ネツカオウギ 43024
ネッカキクアザミ 348538
ネッカキセカタ 295123
ネッカキセワタ 295123
ネッカグルミ 212595
ネッカタンポポ 384749
ネッカヌカボガヤ 321305
ネッカハナビゼキショウ 213553
ネッカヤナギ 343668
ネッタイモクレン属 383236
ネッタイラン 399634
ネッタイラン属 399633
ネテリカハツルモ 340472
ネナシカズラ 115050
ネナシカズラ科 115155
ネナシカズラ属 114947
ネナシカツラ 115050
ネナシカヅラ属 114947
ネヌナワ 59148
ネバリアズマヤマアザミ 92199
ネバリイヅハハコ 103516
ネバリオグルマ 181174
ネバリカワラサイコ 313103
ネバリクフュア 114614
ネバリコゴメグサ 160148
ネバリサウ 299759
ネバリサギサウ 183503
ネバリジナ 401542
ネバリタデ 309946
ネバリタニソバ 291872
ネバリノギク 380930
ネバリノギラン 14487
ネバリノボロギク 360321
ネバリノミノツヅリ 32222
ネバリハコベ 138482
ネバリハナハタザオ 136163
ネバリハナハタザホ 136163
ネバリハナヤナギ 114621
ネバリハンノキ 16447
ネバリビジョザクラ 405896
ネバリホロギク 293178
ネバリマガリバナ 61132
ネバリマンテマ 363641
ネバリミソハギ 114599
ネバリリンダウ 173379
ネバリリンドウ 173379
ネバリルリマガリバナ 61132
ネパールハンノキ 16421
ネビキグサ 240481
ネビキミヤコグサ 237713
ネフェラフィルルム属 265115
ネフェリウム属 265124
ネフェリューム属 265124
ネブカ 13578,15289
ネブタ 13578
ネフティティス属 265211
ネブノキ 13578
ネブラスカスゲ 75488
ネブリ 13578
ネブリノキ 13578
ネ-ブル 93765
ネ-ブルオレンジ 93797
ネペンテス属 264829
ネムカヅラ 1142
ネムチャ 78429
ネムノキ 13578
ネムノキ科 255141
ネムノキ属 13474
ネムリグサ 255098
ネムリグサ属 254979
ネムリハギ 366698
ネムロカハホネ 267320
ネムロガヤ 65471
ネムロカワホネ 267320
ネムロゴウソ 73554
ネムロコウホネ 267320
ネムロスゲ 74676
ネムロトド 463
ネムロトドマツ 463
ネムロブシダマ 235730,235733
ネメ-シア属 263374
ネモトシャクナゲ 330242
ネモフィラ属 263501
ネリ 229,401489,401490
ネリイマキ 306506
ネリネ属 265254
ネリマダイコン 326616
ネルテ-ラ属 265353
ネレ 401490
ネワタノキ 417247
ネンガヤシ属 263549
ネンガ属 263549
ネンドウ 217361
ネンパキンカン 167493
ネンバレムラサキ 66913
ネンボキンカン 167493
ノアサガオ 207891
ノアザミ 92022,92066
ノアズキ 138978
ノアズキ属 138966
ノアヅキ 138978
ノアヅキ属 138966
ノイシナラ 324190
ノイゼンカツラ科 54357
ノイネ 275957
ノイバラ 336783
ノウカウキンシバイ 201894
ノウカウグミ 141991
ノウカウチドリ 273481
ノウカウハグマ 12694
ノウカウマンネングサ 356964

ノウゴウイチゴ 167622
ノウコウウスユキソウ 21676
ノウコウキンシバイ 201894
ノウコウコゴメグサ 160210
ノウコギ 143588
ノウスユキソウ 224893
ノウゼン 70507
ノウセンカズラ 70507
ノウゼンカズラ 70507
ノウゼンカズラ科 54357
ノウゼンカズラ属 70502
ノウゼンカツラ 70507
ノウゼンカヅラ 70507
ノウゼンカツラ属 70502
ノウゼンハレン 399601
ノウゼンハレン科 399596
ノウゼンハレン属 399597
ノウリ 390128
ノウルシ 158397
ノエンドウ 408571
ノオケラ 44208
ノカイダウ 243710
ノカイドウ 243710
ノカミツレ 26746
ノカラマツ 388667,388668,388669
ノガリヤス 127197,127200,127209,127217
ノガリヤス属 65254
ノカンザウ 177932
ノカンゾウ 191289
ノギツネ 273029
ノキトウサゴヤシ 217920
ノキビ 151693,282162
ノギラン 14510
ノグイトウ 80381
ノグサ 352351
ノグサ属 352348
ノグルミ 302684
ノグルミ属 302674
ノグワ 259195
ノケイトウ 80381
ノゲイヌムギ 60974
ノゲエノコロ 33737
ノゲシ 368771
ノゲシバムギ 144667
ノゲシ属 368625
ノゲスゲ 75627
ノゲタイヌビエ 140466,140477
ノゲナガハネガヤ 376710
ノゲナシドクムギ 235303
ノゲヌカスゲ 75393
ノゲヤシチョウセンガリヤス 94591
ノコギリサウ 3921,4028
ノコギリサウ属 3913
ノコギリジヒ 78877
ノコギリシャジン 7739
ノコギリズイナ 210374
ノコギリスイバ 340019
ノコギリスミレ 410201,410531
ノコギリズヰナ 210374
ノコギリソウ 3921,4028
ノコギリソウモドキ 4052
ノコギリソウ属 3913
ノコギリダイゲキ 158758
ノコギリバ 203961
ノコギリバキスミレ 410327
ノコギリバケンチャヤシ属 138532
ノコギリパルメット 360646
ノコギリパルメット属 360645
ノコギリヒゴタイ 348794
ノコギリヒゴタヒ 348778
ノコギリボクチ 361029
ノコバチクセツニンジン 280793
ノコバツルグミ 141999,142207
ノコバツルシキミ 365961
ノコバニシキサウ 159092
ノコンギク 40849,41404
ノササゲ 138942
ノササゲ属 138928
ノザワツバキ 69213
ノシ 368771
ノジアオイ 249633
ノジアオイ属 248822,249627
ノシギク 89556
ノジスミレ 410412,410770
ノジトラノオ 239558
ノシバ 418426
ノシャクヤク 280207
ノシュンギク 41201
ノジヨモギ 36048
ノシラン 272086
ノスゲ 76491
ノスズメノテッポウ 17548
ノ-スポ-ルギク 246665
ノゼウ 70507
ノゼリ 24336
ノソリホシクサ 151398
ノソリリオン属 266893
ノダイオウ 340109
ノダイワウ 340109
ノダイヲウ 340109
ノダケ 24336
ノダケ属 292758
ノダナガフジ 414556
ノタヌキモ 403108
ノダフジ 414554
ノダフヂ 414554
ノヂアフヒ 249633
ノヂアフヒ属 248822,249627
ノヂシバリ 210666
ノヂシャ 404439,404454
ノヂシャ属 404394
ノヂスミレ 410770
ノチドメ 200329,200390
ノヂトラノヲ 239558
ノヂヨモギ 36048
ノチョロギ 373125
ノッポロガンクビソウ 77158
ノティリア属 267213
ノデボ 332509
ノテンツキ 166223,166224,166226
ノトカクタス属 267027
ノトロスゲ 73558
ノニガナ 210616,210654
ノニレ 401602
ノニンジン 35308
ノネ 275958
ノハカタカラクサ 394021
ノハギ 226721
ノハタザホ 30292
ノハッカ 250294
ノハナショウブ 208543,208552
ノバラ 336783
ノハラアカザ 87204
ノハラアザミ 92269,92425
ノハラカゼクサ 147737
ノハラガラシ 364557
ノハラキンポウゲ 325989
ノハラクサフジ 408278
ノハラクサフヂ 408278
ノハラコゴメナデシコ 183185
ノハラジャク 28023
ノハラスズメノテッポウ 17498
ノハラダイオウ 339884
ノハラツメクサ 370522
ノハラテンツキ 166435
ノハラナスビ 367014
ノハラナデシコ 127600
ノハラニンジン 77777
ノハラヒキオコシ 209848
ノハラヒジキ 344585
ノハラフウロ 174832
ノハラフタッブコムギ 398886
ノハラマツヨイグサ 269522
ノハラムラサキ 260760
ノハラワスレナグサ 260747
ノバ-ルホッキア属 266647
ノパレア属 266637
ノビエ 140408
ノヒネチドリ 182225
ノビネチドリ 182254
ノヒマハリ 188953
ノヒマワリ 188953
ノヒメユリ 229775
ノビユ 18670
ノビル 15450
ノビレダイオウ 329363
ノ-フオ-クマツ 30848
ノブキ 7433
ノブキ属 7427
ノブダウ 20354
ノブダウ属 20280

ノブドウ　20348,20354,20400
ノブドウ属　20280
ノブドオ　20321,20354
ノブノキ　302684
ノーブルモミ　453
ノボケ　84556
ノホソ　324384
ノボタン　248732
ノボタンカズラ　247564
ノボタンカツラ　247564
ノボタンカヅラ属　247526
ノボタン科　248806
ノボタン属　248726
ノボリクワガタ　407299
ノボリバナ　95796
ノボリフジ　238463,238473
ノボリフヂ　238463,238473
ノボリリュウ　340954
ノボロギク　360328
ノマアザミ　91862
ノマダケカンアオイ　37696
ノマメ　78429,177777
ノミノアッネ　340089
ノミノツヅリ　32015,32212
ノミノツヅリ属　31727
ノミノハゴロモグサ　13966
ノミノフスマ　374731,375126
ノミノフネ　340089
ノムラサキ　221685,221711
ノムラサキ属　221627
ノモキリス属　266548
ノモモ　20908
ノヤシ　97076
ノヤシ属　97072
ノヤナギ　343994,344155
ノヤマシロギク　39928
ノヤマスゲ　74406,74407
ノヤマテンツキ　166329
ノヤマトンボ　302439
ノヤマトンボソウ　302439
ノユリ　229872,229888,230058
ノヨモギ　35237
ノラエ　290952
ノラナ　266474
ノラナ科　266476
ノラナ属　266470
ノラビエ　143522
ノラマメ　408571
ノリアサ　225
ノリウツギ　200038
ノリクラアザミ　92242
ノリクライチゴ　338895
ノリノキ　200038
ノルウエーカエデ　3425
ノルウコン　114859
ノルゲスゲ　75253
ノルホオクアラウカリア　30848
ノルマンビーヤシ　266678
ノルマンビー属　266677
ノレンガヤ　220298
ノレンガヤ属　220295
バーケリア属　48039
ハアザミ　2695
ハアザミ属　2657
バアソブ　98424
ハアベンケイソウ属　250883
ハイアオイ　243823
ハイイヌガヤ　82532
ハイイヌツゲ　203734
ハイイバラ　337015
ハイイロキダチベンケイ　356776
ハイイロヨモギ　36286
ハイオオバコ　302181
ハイオオヘビイチゴ　312540
ハイオトギリ　201954
バイカアマチャ　302796
バイカアマチャ属　302795
バイカイカリソウ　146980
バィカイチグ　24071
バイカイチゲ　24103
バイカウツギ　294530
バイカウツギ科　294404
バイカウツギ属　294407
バイカオウレン　103843
ハイカカラマツソウ　388417
バイカシモツケ　161749
バイカツツジ　331791
バイカツメクサ　32004
バイカトラノオ　239781
ハイギ　116193
ハイキジムシロ　312357
ハイキビ　282162
ハイキンゴジカ　362622
ハイキンポウゲ　326287
ハイクサネム　126176
バイクワアマチャ　302796
バイクワアマチャ属　302795
バイクワイカリサウ　146980
バイクワウツギ属　294407
バイクワワウレン　103843
バイケイサウ　405598
バイケイソウ　405598
バイケイラン　106883
バイケワツツジ　331791
ハイゲンゲ　43012
ハイコトジソウ　345054
ハイコヌカグサ　12323
ハイシコタンソウ　349120
ハイジシバリ　210673
ハイシノブ科　307186
ハイシノブ属　307190
ハイシバ　226599
ハイシバ属　226585
ハイシマカンギク　89551
ハイシロノセンダングサ　54060
ハイスガモ　297182
ハイスギ　213739
ハイスグリ　334157
バイスタイ　158232
ハイゾウソウ　321643
ハイタウコギ　54168
ハイタツナミソウ　355683
ハイタネツケバナ　72942
ハイタムラソウ　345280
バイタラ　57122
ハイタワガタ　407299
ハイチゴザサ　209105
ハイッガザクラ　297033
ハイツボスミレ　410736
ハイツメクサ　255447
ハイツリガネニンジン　7867
ハイドジョウツナギ　177679
ハイトバ　125958
ハイトベニ　3300
ハイナス　266475
パイナップル　21479
パイナップル科　60557
バイナツプル属　21472
ハイナンエゴノキ　16295
ハイナンサカキ　8235
ハイナンワイルドカット　93480
ハイニガナ　210537
ハイニシキソウ　159634
ハイヌメリ　341960
ハイヌメリグサ　341960
ハイネオアギヤシ　57126
ハイネオウギヤシ　57126
ハイネズ　213739
ハイネズミモチ　229512
ハイノキ　381323
ハイノキ科　381067
ハイノキ属　381090
ハイハマボッス　345731
ハイハマボッス属　345728
ハイヒカゲツツジ　331000
バイビジョザクラ　710
ハイビジョザクラ属　687
バイビジョザクラ属　687
ハイヒメハッカ　250382
ハイビャクシン　213864
ハイビユ　18701
バイブカズラ　34171
パイプカズラ　34246
バイブバナ　34171
パイプバナ　34246
ハイベニギリ　147386
ハイマキエハギ　126651
ハイマツ　300163
ハイマツゲボタン　65855
ハイミチヤナギ　308788
ハイムロ　213739

ハイメドハギ 226745
ハイモ 65218
バイモ 168586
ハイモ属 65214
バイモ属 168332,229717
ハイユキソウ 31767
ハイヨウボタン属 79730
ハイリンドウ 173387
ハイルリソウ 270714
パイン 21479
パインアップル 21479
ハウ 411430
バウエラ属 48975
バウエルハケヤシ 332436
ハウォルティア属 186251
バウクワウマメ 100153
バウクワウマメ属 100149
バウコャブマメ 177789
ハウシャウ 91295
ハウチハノキ 135215
ハウチハノキ属 135192
ハウチハマメ 238473
ハウチハマメ属 238419
ハウチャクサウ 134467
ハウチャクモドキ 134405
ハウチャケモドキ属 134400
ハウチワカエデ 3034
ハウチワタヌキマメ 112601
ハウチワノキ 135215
ハウチワマメ 238473
ハウチワマメ属 238419
バウフウ 346448
バウマンチュウ 94556
ハウレン 371713
ハウレンサウ 371713
ハウレンサウ属 371710
ハウレンソウ 371713
ハウレンソウ属 371710
ハウロクイチゴ 339270
ハエジゴク属 131397
ハエドクソウ 295984,295986
ハエドクソウ科 296006
ハエドクソウ属 295978
ハエトリグサ属 131397
ハエトリソウ属 131397
ハエトリナズナ 260577
ハエトリナデシコ 363214
ハエマンサス・ムルテイフロルス 350312
ハエマンサス属 184345
ハエモドルム科 184524
バオバブ 7022
バオバブノキ 7022
バオバブノキ属 7018
パカエセンネンボク 137342,137514
パガエノモウセンゴケ 138299
ハガクレスゲ 74919
ハガクレツリフネ 205025
ハガクレナガミラン 390502
ハカタガラクサ 394093
ハカタユリ 229758
バカバヤシ 269374
ハカマオニゲシ 282522
ハカマカズラ 49140
ハカマカズラ属 48990
ハカマカヅラ 49140
ハカマカヅラ属 48990
ハガラシ 59438
ハカリノメ 32988
ハガワリトボシガラ 164006
ハガワリメダケ 304101
ハギ 226708
ハギカズラ 169657
ハギカズラ属 169645
ハギカヅラ 169657
ハギカヅラ属 169645
ハギクサウ 158857
ハギクソウ 158857,159492
パキコルムス属 279504
パキスタキス・ルテア 279769
パキセレウス属 279482
ハキダメガヤ 131224
ハキダメギク 170146
ハキダメギク属 170137
パキフィツム属 279635
パキポジューム属 279666
パキラ属 279380
ハギ属 226677
ハクウンカク 245252
ハクウンキスゲ 191301
ハクウンニシキ 273826
ハクウンボク 379414
ハクウンマル 244047
ハクウンラン 407612
ハクウンラン属 218294
ハクオウギョク 264348
ハクオウチュウ 263697
ハクオウマル 244182
ハクカ 250370
ハクカシャクナゲ 330534
ハクカ属 250287
ハクキュウデン 317689
ハクキュウマル 327294
ハクギョクト 244081
ハクギンサンゴ 159242
ハクギンリュウ 358925
ハククワイ 342424
ハククワヰ 342424
ハククンギョク 233572
ハクゲンカク 157169
ハクコウ 220494
ハクコウザン 354311
ハクコウマル 395653
ハクコウリュウ 171738
ハクサイ 59600
ハクサンアザミ 92190
ハクサンイチゲ 23901
ハクサンイチゴツナギ 305572
ハクサンオオバコ 301997
ハクサンオホバコ 301997
ハクサンオミナエシ 285874,285875
ハクサンカニコウモリ 283783
ハクサンカメバヒキオコシ 209855
ハクサンガヤ 297016
ハクサンコザクラ 314284
ハクサンサイコ 63764
ハクサンシャクナゲ 330240
ハクサンシャジン 7869
ハクサンスゲ 74004,74235
ハクサンタイゲキ 159980
ハクサンチドリ 121359
ハクサンチドリ属 121358,273317
ハクサンハタザオ 30111,30272,30286
ハクサンハタザホ 30286
ハクサンハタザヲ 30272
ハクサンフウロ 175029
ハクサンボウフウ 292942
ハクサンボク 407897
ハクサンモチ 203600
ハクサンモミジ 3718
ハクサンモミヂ 3718
ハクサンヨモギ 36225
ハクサンラン 82044
ハクシギョク 389218
ハクシマジンマル 284675
ハクシュマル 140866
ハクジュマル 244081
ハクシュラク 358561
ハクショウ 299830
ハクショウリュウ 299260
ハクジロ 10887
ハクズイホウギョク 43504
ハクセイマル 140843
ハクセイリュウ 171767
ハクセン 94567,129618,129629
ハクセンギョク 246595
ハクセンチュウ 94567
ハクセンナズナ 241250
ハクセンナズナ属 241248
ハクセンナヅナ 241250
ハクセンナヅナ属 241248
ハクセンヤシ属 59095,154571
ハクセン属 129615
ハクゾウマル 327259
バクタイカイ 376184
バクダンウリ 115991
バクダンウリ属 115990
ハクチク 11364
バクチノキ 223150
ハクチョウ 107852,244104
ハクチョウゲ 360933
ハクチョウゲ属 360923

ハクチョウソウ 172201
ハクチョウボク 360933
ハクチョオゲ 360933
ハクテイジョウ 163455
ハクテウク 360933
ハクテウゲ 360933
ハクテウゲ属 360923
ハクテンシ 340860
ハクテンマル 273823
ハクテンリュウ 299240
ハクトウイワガリヤス 65270
ハクトウクワガタ 407388
ハクトウセン 272981
ハクトウヤナギ 343901
ハクトジ 139993
ハクヒョウマル 244164
ハクホウ 181452
ハクマ 175496
ハクマウヒマハリ 188922
ハグマノキ 261399
ハクムオウ 273819
ハクモウアカツメクサ 397097
ハクモウヒマワリ 188922
ハクモウリュウ 299244
ハクモクレン 416694
ハクヤウ 311208
ハクヤギク 77441
ハクヨウ 311208
ハクヨウマル 243999
ハクラクオウ 155303
ハクラン 70989,117045
ハクランポウ 43524
ハクリュウマル 244037
ハクリョウリュウ 273812
ハクリン 82044
ハクリン属 82028,82113
ハグルミ 212636
ハクレイオク 45245
ハクレイマル 234923
ハクレンゲ 416694
ハクレンボク 242135
ハクロゥギョク 182410
ハクロウセキ 233637
ハクロギョク 263738
ハグロサウ 291161
ハグロサウ属 129220
ハクロソウ 291161
ハグロソウ 291162
ハグロソウ属 129220
ハクロバイ 312606
ハケア属 184586
バケイスゲ 75287
ハゲイトウ 18670,18779,18836
ハゲイトウモドギ 99711
ハーゲオセレウス属 183366
ハゲシバリ 16433
ハーゲセンネンボク 104357
バケヌカボ 11970
ハゲマノキ 107300
ハケヤシ属 332434
バケリエリイポプラ 311148
ハコウ 309494
ハコウリュウ 182477
ハコウリョウ 182489
ハコグリ 78782
ハコダテキリンソウ 294117
ハコツツジ 397860
ハコネイトスゲ 74736
ハコネウツギ 413576
ハコネウツギ属 413570
ハコネガマズミ 407791
ハコネギク 40805,41495
ハコネクサアジサイ 73128
ハコネグミ 142079
ハコネコメツツジ 399932
ハコネコメツツジ属 399930
ハコネシノ 347370
ハコネシモツケソウ 166099
ハコネスミレ 410571
ハコネツリガネツツジ 250519
ハコネナンブスズ 347300
ハコネニシキウツギ 413573
ハコネハナヒリノキ 228173
ハコネヒヨドリ 158136
ハコネユリ 229730
ハコネラン 146307
ハゴノキ 61935
ハコベ 374968
ハコベホオズキ 344372
バコベホオズキ属 344369
ハコベマンネングサ 356685,357174
ハコベラ 374968
ハコベ属 374724
ハコボレ 95358
ハコヤナギ 311480
ハコヤナギ属 311131
ハゴロモアオナラ 324389
ハゴロモアクシバ 403869
ハゴロモイヌホオズキ 367689
ハゴロモガシワ 323820
ハゴロモカンラン 59524,59538
ハゴロモギク属 31176
ハゴロモギヶ 31294
ハゴロモグサ 14065
ハゴロモグサ属 13951
ハゴロモサウ 3921,4028
ハゴロモソウ 3921,4028
ハゴロモナラ 324100
ハゴロモニワトコ 345672
ハゴロモノキ 180642
ハゴロモミズナラ 324178
ハゴロモモ 64496
ハゴロモモ科 64503
ハゴロモモ属 64491
ハゴロモルコウソウ 323457
ハザクラキブシ 373565
ハサコ 324384
パサニア属 285203
ハサミガヤ 209288
バシガイモ 16512
ハシカグサ 262960
ハシカグサモドキ 334338
ハシカノキ 57695
ハジガヘリ 61442
ハジカミ 417301
バジカミ 418010
ハシカン 59851
ハシカンボク 59851
ハシカンボク属 59822
ハシカン属 59822
バシクルモン属 29465
ハシコ 114427
ハシドイ 382276
ハシドイ属 382114
ハシナガアワボスゲ 73551
ハシナガカンスゲ 75771
ハシナガサギソウ 302404
ハシナガヤマサギソウ 302404
バージニアマツ 300296
ハシノキ 66808
ハジノキ 393479
ハシバミ 106736,106741
ハシバミ属 106696
バーシャフェルトベニオウギヤシ 222636
バジュウラシトロン 93550
ハシュクリ－ 93451
バショウ 260208
バショウ科 260284
バショウ属 260198
ハシリドコロ 354691
ハシリドコロ属 354678
ハシリビャクシン 213934
バシル 268438
ハス 263272
ハズ 113039
ハスイチゴ 339024
ハスイモ 16501,99919
ハスティングシア属 186143
パースニップ 285741
ハスノハイチゴ 339024
ハスノハカズラ 375870,375884
ハスノハカズラ属 375827
ハスノハカヅラ 375861,375870
ハズノハカヅラ 313197
ハスノハカヅラ属 375827
ハズノハカヅラ属 313186
ハスノハカラマツ 388463
ハスノハギリ 192913
ハスノハギリ科 192921
ハスノハギリ属 192909
ハスノハグサ 139629

ハスノハヒルガオ 250812
ハスノハベゴニア 50317
ハスノミカズラ 65034
ハスノミカヅラ 64971
ハズミダマ 272090
ハス科 263281
ハス属 263267
ハズ属 112829
ハゼ 393479,393485
バセウ 260208
ハゼウルシ 393479,393485
バセウ科 260284
バセウ属 260198
ハゼカツラ 9419
ハゼグリ 78780
ハゼナ 160954
ハゼノキ 393479
ハゼバナ 372050
ハゼラン 383345
ハゼラン属 383293
パセリ 292694
ハゼリサウ科 200476
ハゼリソウ 293176
ハゼリソウ科 200476
ハゼリソウ属 293151
ハセルトニア属 186105
バセヲ 260208
バタアンカエデ 3026
ハダカアキノキリンソウ 368106
ハダカアザミ 91699
ハダカエンバク 198293
ハダカガヤ 393081
ハダカコンロンソウ 72861
ハダカシャクジョウバナ 258019
バターカップノキ 98116
ハダカノキ 223150
ハダカハコベ 374970
ハタカハンニャ 43529
ハダカホオズキ 399998
ハダカホオズキ属 399995
ハダカホホヅキ 399998
ハダカホホヅキ属 399995
ハダカミヤマスミレ 410550
ハダカムギ 198293,198376,198391
ハタガヤ 63216
ハタガヤ属 63204
バタカンマユミ 157355
バタカンヨモギ 36303
ハタクイモ 99910
バタグルミ 212599
ハタケクガタ 406973
ハタケテンツキ 166511
ハタケニラ 266965
ハタザオ 400644
ハタザオガラシ 365398
ハタザオギキョウ 70255
ハタザオシホガマ 287659
ハタザオナズナ 203235
ハタザオモドモ 365585
ハタザオ属 30160
ハタササゲ 409092,409096
ハタザホ 400644
ハタザホ属 30160
ハダッカヅラ 13225
ハタツマリ 96466
ハタツモリ 96466
ハタニンジン 123141
ハタベスゲ 75079
ハタヨモギ 36474
ハダラ 273916
ハダンキョウ 316761
ハチク 297373
バチグサ 72038
ハチジョウアイノコチドリ 302242
ハチジョウアキノキリンソウ 368494
ハチジョウアザミ 92002
ハチジョウイタドリ 162538
ハチジョウイヌツゲ 203715
ハチジョウイノコヅチ 4274
ハチジョウイボタ 229580
ハチジョウオトギリ 201907
ハチジョウカンスゲ 74724
ハチジョウギボウシ 198645
ハチジョウクサイチゴ 338885
ハチジョウグワ 259097
ハチジョウコゴメグサ 160168
ハチジョウシュスラン 179625
ハチジョウショウマ 41813
ハチジョウススキ 255838,255886
ハチジョウスズタケ 347383
ハチジョウソウ 24384
ハチジョウチドリ 302409
ハチジョウツレサギ 302458
ハチジョウテンナンショウ 33343
ハチジョウナ 368635,368675
ハチジョウナ属 368625
ハチジョウネッタイラン 399651
ハチジョウモクセイ 276336
ハチジョウユリ 229992
ハチジョオナ 368675
ハチス 195269,263272
ハチダイリュオウ 160064
ハチヂャウシュスラン 179625
ハチヂャウナ 368635
ハチヂャウナ属 368625
パチナダケ 47402
ハチノジタデ 309570
パチパチグサ 136214
バチバナトチノキ 9680
ハチマンタイアザミ 92003
パチョリ 306964
ハヅ 113039
ハッカ 250331,250370,250376,250443
ハッカクウイキョウ 204603
ハッカクウヰキヤウ 204603
ハッカグサ 280286
ハッカクレン 139623
ハヅカシバナ 9419
ハツカダイコン 326616
ハッカ属 250287
ハッキンギョク 209509
ハックリ 110502,116880
ハックワホウサイ 117042
ハックンマル 234937
ハッケンギョク 242893
ハッコオダゴヨオ 300163
バッコクラン 148754
ハッコツジョウ 385878
パッコーメア属 46176
バッコヤナギ 343151
ハッサク 93482
ハツザメ 175527
ハツシマカンアオイ 37632
ハツシマラン 269025
ハツシモ 180269
ハッショウイモ 367696
ハッショウマメ 259520,259559
ハッショウマメ属 259479
パッション・フルート 285637
ハツタカ 359712
ハットクマメ 222724
ハツバキ 138635
ハツヒノデラン 79555
ハツヒメマル 167698
バッベムヒメバショウ 66160
ハッポウアザミ 92010
ハッポウウスユキソウ 224873
ハッポウタカネセンブリ 380381
ハツミドリ 10806
ハツュキカエデ 3531
ハツユキザクラ 83302
ハツユキソウ 159313
バツユクサ 100961
ハツユリ 168586
ハヅユリ 154926
ハティオラ属 186151
バテマニヤ属 48829
ハテルマカズラ 399303
ハテルマギリ 181743
ハテルマギリ属 181728
ハテンコウ 105439
ハーデンベルギア属 185758
ハト 213739
ハドイワガネ 273908
ハトガシ 116100
ハートカズラ 84305
ハトクサ 209217
ハトノキ 123245
ハドノキ 273916
ハドノキ属 273899
ハトノチャヒキ 60748

ハドハノキ 273916
ハトムギ 99124,99134
ハトヤバラ 336673
ハトラガラシ 4866
バトリカニア属 405191
ハナアオイ 13934,223393
ハナアオイ属 223350
ハナアカシア 1165,334965
ハナアスカ 123821
ハナアナナス 392015,392026
ハナアネモネ 23724
ハナアフヒ 13934,223393
ハナアフヒ属 223350
ハナアヤメ 208806
ハナイ 64184
ハナイカダ 191173
ハナイカダ属 191139
ハナイカリ 184691
ハナイカリ属 184689
ハナイソギク 89407
ハナイチゲ 23767
ハナイトナデシコ 183191
ハナイバナ 57685,57689
ハナイバナ属 57675
ハナイボタ 229610
ハナイワガラミ 351795
ハナインチンロウゲ 85648
ハナイ科 64173
ハナイ属 64180
ハナウジ 123823
ハナウスユキソウ 224822
ハナウド 192278,192312,192375
ハナウド属 192227
ハナウリグサ 392905
ハナエンジュ 334965
ハナオキナ 263802
ハナオトギリ 202099
ハナカイドウ 243623
ハナカエデ 3492
ハナガガブタ 267801
ハナカゴ 45994
ハナガサ 123867,405852
ハナガサギク 339581
ハナガササウ 284340
ハナガサシクナゲ 215395
ハナガサソウ 284340,405820
ハナガサノキ 258923,258926
ハナガサノキ属 258871
ハナガサマル 413678
ハナカズラ 5675
ハナカタノ 123959
ハナカタバミ 277702
ハナカツミ 418095
ハナカヅラ 5674
ハナカモノハシ 209276
ハナカラマツ 388615
ハナカンザシ 190819
ハナカンザシ属 189093,190791
ハナカンゼ 123963
ハナカンナ 71173
ハナキササゲ 79260
ハナキツネノボタン 325614
ハナキユウ 138205
ハナキョウソウ 138149
ハナギリ 285981
ハナギリソウ 4082
ハナギリソウ属 4070
ハナキリン 159363,159366
ハナキンバウゲ 325614
ハナキンポウゲ 325614
ハナクサアジサイ 73121
ハナクサキビ 281438
ハナグシ 159785
ハナクチナシ 171253,171374
ハナグリ 78783
ハナグルマ 175172
ハナグワイ 342421
ハナグワヰ 342421
ハナケマンソウ 128297
ハナゴ 264090
ハナゴショ 123969
ハナゴショウ 122532
ハナコセ 123904
ハナゴセウ 122532
ハナコゾメ 138237
ハナゴマ 205580
ハナコミカンボク 296641
ハナサイカチ 64990
ハナサカズキ 140214
ハナササゲ 293985
ハナサフラン 111625
ハナサワイ 64184
ハナサワヰ 64184
ハナジオオ 327450
ハナシキギク 340595
ハナシソウ属 155170
ハナシテンツキ 166550
ハナシノブ 307197,307215
ハナシノブ科 307186
ハナシノブ属 307190
ハナシバ 204583
ハナシホガマ 287259
ハナシャウガ 418031
ハナシャウブ 208547
ハナシャジン 7768
ハナショウガ 418031
ハナショウジョウ 138233
ハナショウブ 208543,208547
ハナシンボウギ 177822,177845,177846
ハナシンボウギ属 177817
ハナスオウ 83769
ハナズオウ 83769
ハナズオウ属 83752
ハナスゲ 23670,74513
ハナスゲ属 23669
ハナスズシロ 193417
ハナズハウ 83769,83771
ハナズハウ属 83752
ハナズホウ 83769
ハナズルソウ 29855
ハナズルソウ属 29853
ハナゼノ 243266
ハナセンナ 360429
ハナゾノツクバネウツギ 135
ハナダイゴ 123920
ハナタイコン 275876
ハナダイコン 193417
ハナダイコン属 193387
ハナダグサ 100961
ハナタチシバハギ 126596
ハナタチバナ 31396
ハナタデ 308907,309345,309624
ハナタネツケバナ 72934
ハナタバコ 266048,266054
ハナダマノキ 401761
ハナチゴ 247670
ハナヂシャ 90894
ハナチダケサシ 41808
ハナチャウジ 122482,341020
ハナチャウジ属 341016
ハナチョウジ 122482,341020
ハナチョウジ属 341016
パナックス属 280712
ハナツクハネ 135
ハナツヅキ 356590,356841
ハナツメクサ 295324
ハナツリフネソウ 204797
ハナヅル 5675
ハナツルクサ 9475,29855
ハナツルグサ属 9417
ハナツルグミ 142117
ハナヅルサウ 29855
ハナヅルソウ 29855
ハナヅルソウ属 29853
ハナツルボラン 39463
ハナテンジクアオイ 288297
ハナトラノオ 297984
ハナトラノオ属 297973
ハナトリカブト 5108,5185
バナナ 260253
ハナナギナタコウジュ 144079
ハナナズナ 53047
ハナナズナ属 53045
ハナナヅナ属 365386
ハナニガナ 210527,210532,210594
ハナニシキ 17452
ハナニラ 60526
ハナニラ属 60431
ハナヌカススキ 12816
ハナヌスビトハギ 126329
ハナネコノメ 90324

ハナノキ　3492,204583
ハナノボロギク　360306
ハナハギ　70833
ハナハクカ　274237
ハナハクカ属　274197
ハナバゴニヤ　49711
ハナハコベ　226621
ハナハシドイ　382329,382383
ハナハタザオ　136163
ハナハタザオ属　136156
ハナハタザホ　136163
ハナハタザホ属　136156
ハナハッカ　274237
ハナハッカ属　274197
ハナハツカ属　242792
ハナハマサジ　230761
ハナハマセンブリ　81539
ハナヒイラギ　89135
ハナビガヤ　249060
ハナビザクロソウ　256711
ハナビシサウ　155173
ハナビシソウ　155173
ハナビスゲ　74211
ハナビゼキシャウ　212816
ハナビゼキショウ　212816
ハナビゼリ　24376
ハナビソウ　288767
ハナビニラ　266970
ハナヒメウツギ　126948
ハナヒョウタンボク　235929
ハナヒョオタンボク　235929
ハナヒリグサ　81687
ハナヒリノキ　228166
ハナヒルガオ　103048
ハナフウロ　174593
ハナブサソウ　185226,380772
ハナフジウツギ　61954
ハナフスマ　32085
ハナフブキ　21097
ハナヘウタンボク　235929
バナベゴニヤ　50378
ハナホウキギ　217373
ハナホウショウ　138220
ハナボウラン　282934
ハナホシクサ　381835
ハナホトトギス　396598
パナボラヤシ　381807
ハナマガリスゲ　75785
ハナマキ　67253,67275
ハナマキアザミ　92007
バナマサウ　77043
パナマサウ科　115989
バナマサウ属　77038
パナマソウ　77043
パナマソウ科　115989
バナマソウ属　77038
パナマソウ属　77038
バナマハットサウ　77043
バナマハットソウ　77043
バナマハリネヤシ　113293
ハナマメ　293985
ハナマルコスゲ　75785
ハナミズキ　105023
ハナミョウガ　17695
ハナミョウガ属　17639
ハナムグラ　170342
ハナムラサキ　83769
ハナメウガ　17695
ハナメウガ属　17639
ハナモツヤク　64148
ハナモツヤクノキ　64148
ハナヤエムグラ　362097
ハナヤクシソウ　110608
ハナヤサイ　59529
ハナヤツシロソウ　70054
ハナヤナギ　114611,329699
ハナヤブジラミ　79664
ハナヤマゼリ　276770
ハナヤマツルリンドウ　398268
ハナヤマツルリンドオ　398268
ハナヤマボウシ　51140
バナヤマモモソウ　172203
ハナヤヨイ　138183
ハナユ　93869
バナラン　97218
ハナランザン　123854
ハナルリソウ　270723
ハナワオウジ　209507
ハナワギク　89466
ハナワラビ　90932
ハナヰ　64184
ハナヰ科　64173
ハナヰ属　64180
ハニガキ　364768
ハニシキ　65218
バニバナインゲン　293985
バニバナギリソウ　4076
バニバナフジウツギ　62000
バニバナフッドレア　62000
バニバナヰンゲン　293985
ハニヤミツバ　113879
パニュバン　93684
バニラ　405010,405017
バニラ属　404971
ハネガヤ　4134,4150
ハネガヤ属　4113,376687
ハネキビ　16231
ハネシンジュガヤ　354254
ハネズ　316485
ハネスゲ　74431
ハネスズメノヒエ　285435
ハネセンナ　360409
ハネタヌキマメ　111951
ハネノミカヅラ　121810
ハネノミマメ　121810
ハネバキンチャクソウ　66284
ハネバクンショウギク　172335
ハネミイヌエンジュ　240124
ハネミエンジュ　369130
ハネミギク　405915
ハネミササゲ　319112
ハネミヒメスイバ　340075
ハノジエビネ　65931
パパイア科　76820
パパイヤ　76813
パパイヤ属　76808
ハハカ　280023
ハハキイヌツゲ　203670
ハハキオオバコ　301871
ハハキオホバコ　301871
ハハキカサスゲ　76618
ハハキガヤツリ　118765
ハハキギ　217361
ハハキギク　41317,41404
ハハキギ属　217324
ハハキグサ　217361
ハハクリ　168586
ハハコクサ　178297
ハハコグサ　178062
ハハコグサ属　178056
ハハコヨモギ　35550
ハハジカミ　161373
ハハジマテンツキ　166388
ハハジマトベラ　301358
ハハジマノボタン　248787
ハハジマハナガサノキ　258925
ハハジマホザキラン　243042
ハハソ　323942,324384
ババノシリ　142252
パパーパル属　282496
パパベル・コンシュータム　282530
パパヤ　76813
ハバヤマボクチ　382073
バビア－ナ属　46022
ハヒアフヒ　243823
ハヒイノコヅチ　4319
ハヒイバラ　337015
ハヒイロヤナギ　343221
ハヒカヅラ　92920
ハヒカナビキサウ　389848
ハヒカナビキソウ　389848
ハヒギ　116193
ハヒキビ　282162
ハヒキンポウゲ　326287
ハヒシバ　226599
ハヒスガモ　297182
ハヒスグリ　334157
ハヒスズメノヒエ　285448
ハヒスミレ　409898
ハヒタネツケバナ　72942
ハヒチゴザサ　209105

ハヒッガザクラ 297033
ハヒヌメリ 341960
ハヒネズ 213739
ハヒノキ科 381067
ハヒノキ属 381090
ハビバアハブキ 249445
ハビバアワブキ 249445
ハヒバラモンジン 354930
ハヒヒメハギ 308123
ハヒビャクシ 213864
ハヒビユ 18701
ハヒマキエハギ 126396,126651
ハヒマツ 300163
ハヒミズ 288749
ハヒミヅ 288749
ハヒメソナ 252242
ハビロ 379414
ハビロモドキ 264222
ハヒヰノコヅチ 4319
ハブ 195311
パフイオペデイルム・ビローサム 282914
パフイオペデイルム・ロスチャイルデイアナム 282889
パフイオペデイルム属 282768
パフィニア属 282764
ハブカズラ 147349
ハブカヅラ 147349
ハブカツラ属 147336
ハブサウ 360463
ハブソウ 360463
ハブテコブラ 79257,309494
ハブランサス属 184249
パペダ 93523
ハヘドクサウ 295984
ハヘドクサウ科 296006
ハヘドクサウ属 295978
ハヘトリナデシコ 363214
バーベーナ 405812,405852
バーベナ属 405801
ハベルレア属 184222
パポー 38338
ハボソ 142396
ハボソノキ 142396,142399
ハボタン 59520,59524,59525,59532
ハボロシ 95358
ハマアオスゲ 73925
ハマアカザ 44659,44668,53249
ハマアカザ属 44298
ハマアキノキリンソウ 368491
ハマアザミ 92186
ハマアジサイ 199956
ハマアズキ 408957
ハマアラセイトウ 243209
ハマアリッスム 235090
ハマアヲスゲ 73925
ハマイ 213145
ハマイチビ 195311
ハマイチョウ 88977
ハマイヌビハ 165830
ハマイヌビワ 165847
ハマイバラ 337015
ハマイブキボウフウ 228572
ハマウツボ 275010
ハマウツボ科 274921
ハマウツボ属 274922
ハマウド 24380
ハマエゾリンドウ 174009
ハマエノコロ 361942
ハマエンドウ 222735,222743
ハマオオウシノケグサ 164286
ハマオオバコ 302091
ハマオガルカヤ 117185
ハマオギ 295888
ハマオトコヨモギ 35831
ハマオホバコ 302091
ハマオミナエシ 285860
ハマオモト 111167,111169,111170
ハマオモト属 111152
ハマカキネガラシ 365565
ハマカキラン 147197
ハマカヅラ 411430
ハマガヤ 226005
ハマガヤ属 132720
ハマガラシ 225360
ハマカンギク 124790
ハマカンザシ 34536
ハマカンザシ属 34488
ハマキ 67275
ハマギク 89646,266319
ハマキケマン 105976
ハマキジカクシ 38950
ハマギシギシ 340116
ハマギリ 192913
ハマクグ 76163
ハマクコ 239108
ハマクサ 297185
ハマクサギ 313692
ハマクサギ属 313593
ハマクサフジ 408435
ハマグス 91420
ハマクマツヅラ 405865
ハマグミ 411430
ハマグルマ 413514,413549
ハマグルマ属 413490
ハマグルミ 192494
ハマグワ 259100,411430
ハマクワガタ 407164
ハマゴウ 411430
ハマコウゾリナ 298594
ハマゴウ属 411187
ハマゴケ 64990
ハマゴバウ 92186
ハマコンギク 40847
ハマサオトメカズラ 280104
ハマザクラ 31038
ハマザクロ 368913
ハマザクロ科 368926
ハマササゲ 64990
ハマサジ 230797
ハマサワヒヨドリ 158210
ハマシキミ 411430
ハマシシウド 98777
ハマシバ 397159
ハマシャギク 406566
ハマジャコウソウ 391349
ハマジンチョウ 260711
ハマジンチョウ科 260700
ハマジンチョウ属 260706
ハマジンチョオ 260711
ハマスイバ 340116
ハマスゲ 119503
ハマススキ 341912,341915
ハマスベリヒユ 361667
ハマゼリ 97711
ハマゼリ属 97697
ハマセンダン 161335,161336,161356
ハマセンナ 274343
ハマセンブリ 380236
ハマダイゲキ 158490
ハマダイコン 326629
ハマタイセイ 209229,209238
ハマタガツラ 394093
ハマタカトウダイ 159217
ハマタデ 309262
ハマタバコ 229109
ハマタマボウキ 39051
ハマヂサ 230797
ハマヂシャ 387221
ハマチャヒキ 60745,60853
ハマヂンチャウ 260711
ハマヂンチャウ科 260700,296006
ハマヂンチャウ属 260706
ハマツバキ 411430
ハマツメクサ 342259
ハマデラソウ 168691
ハマテンツキ 166313
ハマトカクタス属 185155
ハマトラノオ 317963
ハマナ 108627
ハマナシ 333970,336901,404051
ハマナス 336901
ハマナタマメ 71055
ハマナツメ 280555
ハマナツメモドキ 415558
ハマナツメ属 280539
ハマナデシコ 127739
ハマナミソウ 355778
ハマナレン 110615
ハマナ属 108591
ハマニガナ 88977
ハマニンジン 97711

ハマニンドウ　235639
ハマニンドオ　235639
ハマニンニク　228369
ハマネズ　213775,213776
ハマネナシカズラ　114994
ハマネナシカヅラ　115072
ハマハイ　411430
ハマハイネズ　213963
ハマハギ　411430
ハマハコベ　198010
ハマハタザオ　30456,30458
ハマハタザヲ　30458
ハマバハギ　126329
ハマハヒ　411430
ハマバンケイサウ　250895
ハマバンケイサウ属　250883
ハマヒエガエリ　310125
ハマヒエガヘリ　310125
ハマヒサカキ　160458
ハマビシ　395146
ハマヒジキ　344496
ハマビシ科　418580
ハマビシ属　395064
ハマヒナノウスツボ　355132
ハマビハ　233949
ハマビハ属　233827
ハマヒルガオ　68737
ハマヒルガホ　68737
ハマビワ　233949
ハマビワ属　233827
ハマフウロ　175031
ハマフシグロ　363202
ハマフダンソウ　53254
ハマベノギク　193925,193939
ハマベノギク属　193914
ハマベハルシャギク　104545
ハマベブドウ　97865
ハマベマンテマ　364146
ハマベヨメナ　40835
ハマベンケイ　260711
ハマベンケイソウ　250895
ハマベンケイソウ属　250883
ハマボ　194911
ハマホウ　411430
ハマボウ　194911,195311,411430
ハマボウフウ　176923
ハマボウフウ属　176921
ハマボオ　194911
ハマボッス　239727
ハママツ　342859
ハママツナ　379499,379553
ハママンゴスチン　171115
ハママンネングサ　356751
ハマミズナ科　12911
ハマミヅナ　361667
ハマミヅナ属　361654
ハマムカシヨモギ　406566
ハマムギ　75259,75930,144271
ハマムギクサ　198319
ハマムギ属　144144
ハマムラサキ　190652,393269
ハマムラサキノキ　393242
ハマムラサキ属　393240
ハマモクコケ　329086,329114
ハマモチ　254303
ハマモッコク　329114
ハマモメン　111170
ハマヤブマオ　56096
ハマユウ　111170
ハマユリ　229915
ハマユンジュ属　274337
ハマヨモギ　35517,36232
ハマリンチョウ　260711
ハマレンゲ　220167
ハマレンゲ属　220155
ハマワスレナグサ　260791
ハマヲガルカヤ　117185
ハマヲギ　295888
ハマヲクロイゲ　342170
ハミズシクラメン　115953
ハミバミモドキ　276846
パミールギシギシ　340171
ハムグラ　170342
バメ　324283
バメガシ　324283
パメロ　93579
ハヤザキアワダチソウ　368188
ハヤザキウツギ　413621
ハヤサキギボウシ　198623
ハヤザキヒョウタンボク　236040,236041
ハヤザキヘウタンボク　236040
ハヤザキモモ　20908
ハヤタカンコノキ　177114
ハヤタマンネングサ　357138
ハヤチネウスユキソウ　224852
ハヤチネコウモリ　283827
ハヤチネミズタマソウ　91506
ハヤトウリ　356352
ハヤトウリ属　356350
ハヤトミツバツツジ　330588
バラ　336901
バラアカシア　334965
バライチゴ　338608,339194
ハライヌノヒゲ　151421
バライロキヌタソウ　170447
バライロモクセンナ　78338
バライロルエリア　339803
バラオキリハヤシ　181917
バラカヤシ属　46743
パラグアイチャ　204135
パラグラス　58128
パラグワイチャ　204135
パラゴムノキ　194453
パラゴムノキ属　194451
バラサン　265137
パラダイスナットノキ　223776
パラディセア属　283295
バラナマツ　30831
パラナマツ　30831
ハラヌメリ　341977
パラブアイオニバス　408735
パラファレノプシス属　283598
ハラミツ　61101
パラミツ　36920
バラモミ　298442
バラモンギク　394333
バラモンジン　394327
バラモンジン属　394257
バーラーヤシ属　85420
パラルゴニューム属　288045
ハラン　39532,39557
バラン　39532,39557
ハランアナナス　301106
ハランウチハ　28103
ハランウチワ　28103
バランガオレンジ　93708
ハラン属　39516
バラ科　337050
バラ属　336283
ハリアオイ　195119
ハリアカシア　1060
ハリアサガオ　208264
ハリアサガホ　208264
ハリイ　143100,143102,143298,143300
ハリイヌナズナ　136901
ハリイモ　131577
ハリイ属　143019
ハリウコギ　143657
ハリウメヅル　182683
ハリエニシダ　401388
ハリエニシダ属　401373
ハリエンジュ　334976
ハリエンジュ属　334946
ハリカウガイゼキシャウ　213569
ハリカガノアザミ　91986
ハリガネオトギリ　202132
ハリガネカズラ　96387,172092
ハリガネカズラ属　87668
ハリカネカヅラ　172092
ハリガネカツラ　96387
ハリガネカヅラ属　87668
ハリガネスゲ　74023
ハリガネツゲ　74023
ハリギ　302721
バリーギョク　233511
ハリギリ　154734,215442
ハリギリ属　215420
ハリクジャクソン属　12771
ハリクジャグヤシ属　12771
ハリクチナシ　79595
ハリグハ　240842

ハリグハ属　114319
ハリグワ　240842
ハリグワ属　114319
ハリゲコウゾリナ　298584
ハリゲタビラコ　20843
ハリゲナタネ　59651
ハリゲヤキ　191629
ハリゲヤキ属　191626
ハリコウガイゼキショウ　213569
ハリコザクラ　23294
ハリコモチトゲココヤシ　126689
ハリサキタフクグルミ　212656
ハリザクロ　79595
ハリシア属　185912
ハリシバ　418432
ハリジヒ　79027
ハリジヒノキ　233121
ハリスグリ　333929
ハリスゲ　75610
ハリスネ－リア属　404531
パリス属　284289
ハリセンボン　139681,385771
ハリタデ　291707,308922
ハリチドメ　200364
ハリヅス　83912
ハリツメクサ　31795
ハリツルマサキ　182683
ハリドリオシベ　111774
ハリドリヲシベ　111774
ハリナズナ　379650
ハリナスビ　367617
ハリナデシコ　292668
ハリナデシコ属　292656
ハリネズミマル　107021
ハリネヤシ属　113290
ハリノキ　16386,16389
ハリノキテンナンショウ　33363
ハリノホ　184581
ハリノホ属　226585
ハリハナムギ　371724
バリバリノキ　6762
ハリヒジキ　344743
ハリヒメハギ　307907
バリビヤキナノキ　91075
ハリビユ　18822
ハリフウテウサウ　95796
ハリブキ　272712
ハリブキ属　272704
ハリフタバ　57296,370744
ハリフタバ属　370736
ハリマツリ　139062,139065
ハリマツリ属　139060
ハリママムシグサ　33414
ハリミウム属　184776
ハリミコバンモチ　366047
ハリミコバンモチ属　366034
ハリミマユミ　157932
ハリモクシュク　271347
ハリモクシュク属　271290
ハリモデンドロン属　184832
ハリモミ　298442
ハリモミ属　298189
ハリヤシ属　329169,329173
ハリヰ　143102,143298
ハリヰ属　143019
バリン　191251,208667
バリンコロング　93406
ハルウコン　114859
ハルオキナグサ　24109
ハルオミナヘシ　404325
ハルガヤ　27969
ハルガヤ属　27930
ハルカラマツ　388432
ハルグミ　142152
ハルコガネバナ　105146
ハルコザクラ　314408
ハルゴマ　159641
ハルゴロモ　299262
バルサ　268344
ハルザキヤッシロラン　171944
ハルザキヤマガラシ　47964
ハルサザンカ　69740
ハルサフラン　111625
バルサミベラポプラ　311237
バルサムフヤ－　282
バルサムポプラ　311237
バルサムモミ　282
ハルサメソウ　350011
バルサ属　268341
ハルジオン　150862
ハルシャギク　104598
ハルシャギク属　104429
ハルジョオン　150862
パルストリス　324262
ハルタデ　291840,309570
ハルタマ　183066
ハルトラノオ　54821,309870
ハルニレ　401489,401490
ハルノコンギク　39980
ハルノタムラソウ　345338
ハルノノゲシ　368771
バルバドスアロエ　17381
バルバドスサクラ　243526
バルバドスチエリ－属　243519
バルバドスチユリ－　243526
ハルヒメソウ　94373
バルボフィラム属　62530
バルボフィルム属　62530
パルマロサグラス　117204
ハルユキソウ　197799
ハルユキノシタ　349697
ハルユリ　168586,230023
ハルライヌビハ　164985
ハルリンダウ　173974
ハルリンドウ　173974
ハルリンドオ　173974
バルレリア　48140
ハルヲミナヘシ　404325
パルンビ－ナ属　280655
バレイショ　367696
ハレ－シア属　184726
バレリアナ属　404213
バ－レ－リア属　48080
バ－レリア属　48080
バレン　39532,39557,208667
バレンギク属　326957
バレンシア　93785
バレンシバ　37389
バレンジバ　37379
バレンセキコク　166964
バロ－タ属　404575
パロチンベルガモット　93407
バロック　417247
パロディア属　284655
バンウコン　215011
バンウコン属　214995
バンオウカン　93471
ハンカイアザミ　287239
バンカイウ　417093
ハンカイシオガマ　287239
ハンカイソウ　229066
バンカジュ科　76820
ハンカチツリ－　123245
ハンカチノキ　123245
ハンカチノキ科　123253
バンギョク　267060
バンクシア属　47626
バンクスマツ　299818
パンクラチュ－ム属　280877
ハンクワイアザミ　287239
ハンクワイサウ　229066
ハンクワイシホガマ　287239
ハンクワイソウ　229066
バンクワジュ　76813
バンクワジュ科　76820
バンクワジュ属　76808
ハンゲ　299724
ハンゲシャウ　348087
ハンゲシャウ科　348003,348077
ハンゲシャウ属　348082
ハンゲショウ　348087
ハンゲショウ属　348082
ハンゲ属　299717
バンコウ　86517
バンコクハンゲ　401156
ハンゴンサウ　358500,359654
ハンゴンソウ　358500,359654
ハンサ　53465
バンサイカク　374031
バンサンジコ　111589
バンジ－　410677

パンジー 409624
ハンシジュ 56802
バンジュツ 215035
ハンショウヅル 95037
バンショウデン 385872
バンジラウ 318742
バンジラウ属 318733
バンジリア属 405059
ハンシレン 159222
バンジロウ 318742
バンジロウ属 318733
バンジロオ 318742
バンジンガンクビソウ 77160,77181
ハンスヤナギ 344268
バンソケイ 61304,281225
バンダ・セルーレア 404629
バンダイカエデ 3718
バンダイショウマ 41834
バンタイスギ 113700
バンダイスギ 113717
バンダイソウ属 358015
ハンダッカヅラ 13225
パンダーヌス属 280968
バンダ属 404620
バンテイシ 360933
ハンテンボク 232609
ハンドイチゴ 338516
バントウ 20952
ハントウギョク 163462
バンドプシス属 404745
バンナンクワ 114295
ハンニチバナ 188590,188846
ハンニチバナ科 93046
ハンニチバナ属 188583
ハンニャ 43526
ハンネマンニア属 199407
ハンノキ 16386,16389
パンノキ 36913
ハンノキエゴノキ 16304
ハンノキバエゴノキ 16295
ハンノキ科 53651
ハンノキ属 16309
パンノキ属 36902
ハンノハエゴノキ 16295
バンバ 142252
パンパス・グラス 105456
バンビカク 199057
バンブーサ属 47174
バンホウテコブダネヤシ 265204
バンホウテヤシ 265204
ハンボク 232609
バンボクベツ 378836
バンマツリ 61304,61316
バンマツリ属 61293
バンヤ 56802
ハンヤエウサギギク 34782
パンヤノキ 56802,80120
バンヤンジュ 164684
パンヤ科 56761
ハンヨウトキワラン 282797
バンリュウガン 310823
バンレイギョク 198501
バンレイシ 25898
バンレイシ科 25909
バンレイシ属 25828
ヒアシンス 199583
ヒアシンス属 199565
ヒアシント 199583
ビアソング 93634
ヒアフギ 50669
ヒアブギアヤメ 208819
ヒアフギモドキ 208524
ヒアフギ属 50667
ピアランッス属 298132
ヒイタマナ 67860
ヒイラギ 122040,276313,276327
ヒイラギガシ 223131,316824
ヒイラギギク 305100
ヒイラギサイコ 154327
ヒイラギズイナ 210402
ヒイラギソウ 13114
ヒイラギソヨゴ 203590
ヒイラギトラノオ 243523
ヒイラギトラノオ属 243519
ヒイラギナンテン 242563
ヒイラギナンテン属 242458
ヒイラギハギ 89136
ヒイラギハギ属 89134
ヒイラギマメ 89135
ヒイラギメギ 242478
ヒイラギモクセイ 276290
ヒイラギモチ 203763
ヒイラギヤブカラシ 79888
ヒイラチモチ 203660
ヒイランボタンヅル 94707
ビイリア属 54362
ヒイロサンジコ 196456
ヒウガウリ 114295
ヒウガミヅキ 106669
ヒウジ 56145
ヒウン 249586
ヒウンカク 279488
ヒエ 140367,140421,140423,140503
ヒエイボダイジュ 391850
ヒエガエリ 310109
ヒエガエリ属 310102
ヒエガヘリ 310109
ヒエガヘリ属 310102
ヒエスゲ 75189,75192
ヒエダンゴ 231334
ヒエボスゲ 73842
ヒエンサウ属 124008
ヒエンソウ 102833
ヒエンソウ属 102827,124008
ヒエンマル 234932
ヒエ属 140351
ピオイヒバ 114763
ヒオウギ 50669,208527
ヒオウギアヤメ 208819
ヒオウギズイセン 111458
ヒオウギズイゼン 413326
ヒオウギズイセン属 413315
ヒオウギラン 267958
ヒオウギ属 50667
ビオウレン 413480
ヒオオギ 50669
ヒオギモドキ 208524
ヒオドシナデシコ 127657
ビオフィタム属 54532
ヒカカク 140288
ヒカゲイノコズチ 4275
ヒカゲオトギリ 201967
ヒカゲカニツリ 398566
ヒカゲキセワタ 295215
ヒカゲシラスゲ 75827
ヒカゲスゲ 75048,76431
ヒカゲスミレ 410784
ヒカゲツツジ 330998
ヒカゲツルニンジン 98395
ヒカゲナミキソウ 355518
ヒカゲネコノメソウ 90400
ヒカゲハリスゲ 75610
ヒカゲヒメジソ 259284,259299
ヒカゲママコナ 248210
ヒカゲミズ 284160
ヒカゲミズ属 284131
ヒカゲミヅ 284160
ヒカゲミツバ 299448,299481
ヒカゲミツバ属 299342
ヒカゲミヅ属 284131
ヒカゲユキノシタ 350011
ヒカゲヰノコヅチ 4275
ヒガサギク 415317
ヒカチュウ 161922
ヒガヤ 37429
ヒガンガナ 239266
ヒカンザクラ 83158
ヒガンザクラ 83334
ヒガンバイザサ属 239253
ヒガンバナ 239266
ヒガンバナ科 18861
ヒガンバナ属 239253
ヒガンマムシグサ 33237
ヒキオコシ 209713,209717
ヒキオコシダマシ 345315
ヒキザクラ 242169,242171
ヒキノカサ 326431
ヒキヨモギ 365296
ヒキヨモギ属 365295
ヒギリ 96147
ピクェーリア属 300840

ピクナンサアカシア　1516
ビグノニア・カプレオラータ　54303
ビグミ－モウセンゴケ　138340
ヒグル　121561
ヒグルマ　188908
ヒグルマダリア　121547,121561
ヒグルマテンジクボタン　121547,121561
ビクンギョク　233570
ヒゲアオスゲ　73913
ビケイイチゴ　338425
ビケイチュウ　185919
ヒゲオシベ　306974
ヒゲオシベ属　306956
ヒゲガヤ　118319
ヒゲクサ　352351,372803
ヒゲクリノイガ　80815
ヒゲケマルバスミレ　410132
ヒゲコスミレ　410110
ヒゲサギサウ　183983
ヒゲシバ　88396,372737,372803
ヒゲシバ属　88316
ヒゲスゲ　76737,76740
ヒゲスズメノヒエ　285522
ヒゲナガキンギンザウ　179623
ヒゲナガコメススキ　376690
ヒゲナガコメススギ　321016
ヒゲナガコメススギ属　321004
ヒゲナガシナガリヤス　94632
ヒゲナガスズメノチャヒキ　60696,60935
ヒゲナガトンボ　183629
ヒゲナシイタチササゲ　222708
ヒゲナデシコ　127607
ヒゲネムリハギ　366654
ヒゲネワチガイソウ　318511
ヒゲノガリヤス　65413
ヒゲハリスゲ　217121,217234
ヒゲメヒシバ　281532
ヒゲリンドウ　174205
ヒゲレンリソウ　222788
ヒゴ　213962
ヒゴアオキ　44937
ビコウ　138171
ヒゴウカン　66669
ヒコウギョク　329580
ヒゴウコギ　143613
ヒゴオミナエシ　359565
ヒゴカンアオイ　37762
ヒコキミヅ　142658
ヒゴクサ　74941
ヒコサンヒメシャラ　376473
ヒゴシオン　40805
ヒゴシヲン　40805
ヒゴスゲ　74941
ヒゴスミレ　409816,409819
ヒゴタイ　140786
ヒゴタイサイコ　154317
ヒゴタイサイコ属　154276
ヒゴタイ属　140652
ヒゴヅタ　165630
ヒゴヒャクゼン　117364
ヒゴミズキ　106647
ヒゴメダケ　304034
ヒゴロモコンロンカ　260414
ヒゴロモサウ　345405
ヒゴロモソウ　345405
ヒゴヲミナヘシ　359565
ヒサウキヤウ　182526
ヒサウチソウ　50717
ヒサカキ　160503
ヒサカキサザンカ　322596
ヒサカキモドキ　69064
ヒサカキ属　160405
ヒサキ　243371
ヒザクラ　83158
ヒサゴナ　59512
ヒサマル　234947
ヒシ　394436
ビシ　394463
ヒシガタクリガシ　79027
ヒシガタマユミ　157912
ヒシガタヤッコサウ　256182
ヒシガタヤッコソウ　256182
ヒジキアカザ　379553
ヒシグリ　78933,79035,79054
ヒシバウオトリギ　180939
ヒシバカキドオシ　262247
ヒシバカキドホシ　262247
ヒシバザミア　417005
ヒシバシナガワハギ　249237
ヒシバデイコ　154632
ヒジハリノキ　278323
ヒシミガシ　233383
ヒシモドキ　394560
ヒシモドキ属　394559
ビシャコ　160503
ヒシュウギョク　284693
ヒシュウザサ　347343
ビジュナデシコ　127607
ビショウマル　140848
ヒショクギョク　404948
ビジョサクラ　405852
ビジョザクラ　405852
ビショップマツ　300098
ビジョナデシコ　127607
ビシンギョク　182414
ビジンサウ　282685
ビシンジュズネノキ　122048
ビジンショウ　260217
ビジンセウ　260217
ビジンソウ　282685
ヒシ科　394554
ヒシ属　394412
ビスイチュウ　182516
ヒスイホコ　340659
ヒスイヤガラ　171940
ヒスイラン　404629
ピスタシオノキ　301005
ピスタチオ　301005
ビスマルクヤシ属　54738
ピズム属　301053
ヒス卄ラン　404629
ヒス卄ラン属　404620
ヒセイマル　234929
ヒゼンガマズミ　407656
ビゼンギョク　275410
ビゼンナリヒラ　364820
ビゼンナリヒラ属　364783
ヒゼンマユミ　157365
ヒゼンモチ　203600
ビゼンモチ　203600
ビソウギョク　284665
ヒソップ　203095
ピゾ－ニア属　300941
ヒダアザミ　92432
ビタ－・ア－モンド　20892
ヒダイリオサラン　148790
ヒダイリヲサラン　148790
ヒダカアザミ　92026
ヒダカイワザクラ　314465
ヒダカゲンゲ　278936
ヒダカサウ属　66697
ヒダカソウ属　66697
ヒダカタチツボスミレ　410156
ヒダカトウヒレン　348442
ヒダカトックリヤシ属　201356
ヒダカノリウツギ　200051
ヒダカミセバヤ　200783,356607
ヒダカミツバツツジ　330580
ヒダカミネヤナギ　343752
ヒダカミヤマノエンドウ　279121
ヒダカヤエガワ　53402
ヒタチクマガイソウ　120374
ヒダフウロ　175027
ヒダブチエビネ　66099
ヒダベリクワクラン　293494
ヒダボタン　90413
ヒダリマキガヤ　393075,393080
ビタ－ルツバキ・ピタルデイ－　69494
ピタルツバキ・ユンナン　69504
ピタンガ　156385
ヒチノキ　261648
ヒヂハリノキ　278323
ヒツガタマユミ　157912
ビッキ　143573
ヒッコリ－　77920,77934
ヒツジグサ　267767,267770
ヒツジグサ科　267787
ヒツジグサ属　267627
ビッタディニア属　412022
ヒツヂグサ　267767
ビッチュウアザミ　91799

ビッチュウヒカゲスゲ 73884
ビッチュウフウロ 175034
ヒッチュウフウロサウ 175034
ビッチュウミヤコザサ 347230
ビッチュウヤマハギ 226805,226977
ヒッチョウカ 300377
ビッチリ 105810
ヒッツキ 54158
ヒッツキアザミ 91895
ヒツナヴ 26393
ヒッパリガヤ 201563
ヒッペアストラム属 196439
ヒッベルティア属 194661
ヒッポフェ属 196743
ヒデ 213962
ビテックス属 411187
ヒデリコ 166380,166402
ビトウクグ 245376
ヒトウバン 158653
ビトウマル 247668
ヒトエシジミバナ 372053
ヒトエノコクチナシ 171253
ヒトエノシジミバナ 372062
ピトカイルニア属 301099
ヒトシベサンザシ 109857
ヒトスジグサ 11341
ピトスポルム 301413
ヒトタバメヒシバ 130721
ヒトツノコシカニツリ 405429
ヒトッバ 134010
ヒトヅハ 306457
ヒトツバアメリカチャボヤシ 327543
ヒトツバイチヤクソウ 322915
ヒトツバエゾスミレ 409946
ヒトッバエニシダ 173082
ヒトッバエニシダ属 172898
ヒトツバオトコエシ 285826
ヒトツバオモダカ 342429
ヒトツバカエデ 2939
ヒトツバキジムシロ 312567
ヒトツバキソチドリ 302414
ヒトツバクマデヤシ属 212401
ヒトツバコンロンソウ 72682
ヒトツバショウマ 41851
ヒトツバタカヨモギ 36241
ヒトツバタゴ 87729
ヒトツバタゴ属 87694
ヒトツバテンナンショウ 33416
ヒトツバトウカエデ 2812
ヒトツバハギ 167092
ヒトツバハギ属 356392
ヒトツバハンゴンサウ 358504
ヒトツバハンゴンソウ 358504
ヒトツバマメ 185764
ヒトツバマメ属 185758
ヒトツバヨモギ 35648,35945
ヒトツブコムギ 398924
ヒトツボクロ 392349
ヒトツボクロモドキ 392350
ヒトツボクロモドキ属 129768
ヒトツボクロ属 392346
ヒトデ 325981
ヒトデカズラ 294839
ヒトハグサ 285981
ヒトハグハ 285981
ヒトハリヘビノボラズ 51301,52100
ヒトフサニワゼキショウ 365785
ヒトヘノコクチナシ 171253
ヒトマキ 306457
ヒトモジ 15289
ヒトモトススキ 93913,93915
ヒトモトススキ属 93901
ヒトモトメヒシバ 146015
ヒトモトメヒシバ属 145999
ヒトヨシテンナンショウ 33500
ヒドラスチス属 200173
ヒトリカヤツリ 118737
ヒトリシズカ 88289
ヒトリシヅカ 88289
ヒトリフウロ 174906
ヒドロカリス属 200225
ヒドロクレイス属 200241
ヒドロスメ属 20037
ヒナアズキ 408969,408976
ヒナイチャク 275482
ヒナウキクサ 224378
ヒナウスユキソウ 224833
ヒナウチワカエデ 3676
ヒナウツギ 127085
ヒナウンラン 84596
ヒナエンゴサク 106380,106385
ヒナカキササゲ 79243
ヒナガヤツリ 118889
ヒナカラスノエンドウ 408452
ヒナガリヤス 65434
ヒナカンアオイ 37698
ヒナギキャウ 412750
ヒナギキャウ属 412568
ヒナギキョウ 412750
ヒナキキョウソウ 397755
ヒナギキョウ属 412568
ヒナギキョオ 412750
ヒナギク 50825
ヒナギクモドキ 50809
ヒナギク属 50808
ヒナギョりュウ 383595
ヒナゲシ 282685
ヒナゲンゲ 181693
ヒナコゴメグサ 160294
ヒナコザクラ 23182
ヒナザクラ 314696
ヒナザサ 98451
ヒナザサ属 98446
ヒナシャジン 7695
ヒナスゲ 74697
ヒナスミレ 410667
ヒナセントウソウ 85483
ヒナソゥ 198708
ヒナソウ 187526
ヒナタイノコズチ 4286
ヒナタイノコヅチ 4286
ヒナタスゲ 74842,74851
ヒナタヰノコヅチ 4286
ヒナチドリ 310866
ヒナツメクサ 397050
ヒナナズナ 137151
ヒナノウスツボ 355106,355211
ヒナノカンザシ 344351
ヒナノカンザシ属 344345
ヒナノキンチャク 308397
ヒナノシャクジョウ 63960
ヒナノシャクジョウ科 64000
ヒナノシャクジョウ属 63949
ヒナノシャクヂャウ 63960
ヒナノシャクヂャウ科 64000
ヒナノシャクヂャウ属 63949
ヒナノボタン 59864
ヒナノボンボリ 278483
ヒナハギリ 1000
ヒナバショウ 260217
ヒナバセウ 260217
ヒナバト 102692
ヒナヒゴタイ 348403
ヒナビジョザクラ 405896
ヒナヒルガオ 102926
ヒナヒルガホ 102926
ヒナボウフウ 24397
ヒナマツヨイグサ 269486
ヒナミクリ 370088
ヒナミチヤナギ 309192
ヒナムラサキ 301627
ヒナモメンズル 42723
ヒナユリ 68803
ヒナヨウラク 308495
ヒナヨシ 37477
ヒナラン 19512
ヒナラン属 19498
ヒナリンドウ 173242
ヒナリンドオ 173242
ヒナワチガイソウ 318499
ビナンカズラ 214971
ビナンカヅラ 214971
ピナンガ属 299649
ヒネム 66669
ヒネリガヤツリ属 393185
ヒノカムリ 358997
ヒノカンムリ 358997
ヒノキ 85310,390680
ヒノキアスナロ 390684
ヒノキバヤドリギ 217906
ヒノキバヤドリギ属 217901

ヒノキンロウバイ　312421
ヒノキ科　114645
ヒノキ属　85263
ヒノデマル　163456
ヒノデラン　79546
ヒノミサキギク　89409,124836
ヒバ　85310,390680
ビハ　151162
ヒハツ　300446
ヒハツモドキ　300408,300495
ビハバハヒノキ　381412
ビハバミミズバヒ　381412
ビハモドキ　130919
ビハモドキ属　130913
ビハ属　151130
ヒビガヤ　82523
ピビダカイハマ　336938
ヒヒラギ　276313
ヒヒラギガシ　324436
ヒヒラギギク　305100
ヒヒラギギク属　305072
ヒヒラギズイナ　210374
ヒヒラギズヰナ　210374
ヒヒラギモチ　203545,203763
ヒヒラギ科　270182
ヒヒランイチゴ　339468
ヒヒランチク　351852
ヒヒランチク属　351844
ヒフガアフヒ　188908
ヒフキダケ　47178,47254
ビフレナ-リア属　54243
ビフレナリア属　54243
ヒホウマル　327278
ヒポエステス属　202495
ヒポキルタ属　202450,263320
ヒポクラテア科　196616
ヒボケ　84552
ヒボタン　182462
ヒボ属　390567
ヒマツバキ属　350904
ヒマハリ　188908
ヒマハリ属　188902
ヒマラアビロウ　234183
ヒマラヤ・シーダー　80087
ヒマラヤアオキ　44911
ヒマラヤアセビ　298722
ヒマラヤウバユリ　73159
ヒマラヤキブシ　373540
ヒマラヤキンシバイ　201925,202204
ヒマラヤゴヨウ　300297
ヒマラヤザクラ　83165
ヒマラヤサクラソウ　314212
ヒマラヤサザンカ　69235
ヒマラヤシーダー　80087
ヒマラヤスキ　80087
ヒマラヤスギ　80087
ヒマラヤスギ属　80074
ヒマラヤソケイ　211863
ヒマラヤトウヒ　298436
ヒマラヤドクウツギ　104701
ヒマラヤトベラ　301266
ヒマラヤナシ　323251
ヒマラヤニンジン　280722
ヒマラヤハシドイ　382153
ヒマラヤハッカクレン　365009
ヒマラヤハナイカダ　191157
ヒマラヤハリモミ　298436
ヒマラヤヒザクラ　83167
ヒマラヤピラカンサ　322458
ヒマラヤマツ　300128,300186
ヒマラヤミセバヤ　200788
ヒマラャモミ　480
ヒマラヤヤマボウシ　124891
ヒマラヤユキノシタ　52541
ヒマラヤリントウ　107454
ヒマワリ　188908
ヒマワリヒヨドリ　89299
ヒマワリ属　188902
ピーマン　72075
ビミギョク　233684
ヒムロ　85352,85353
ヒメアオガヤツリ　119461
ヒメアオキ　44932
ヒメアオゲイトウ　18661
ヒメアオスゲ　75136
ヒメアオミズ　298980
ヒメアカザ　139681
ヒメアカショウマ　41867
ヒメアカバナ　146697
ヒメアカボシタツナミ　355724
ヒメアギスミレ　410730,410743
ヒメアサガホ　208023
ヒメアサザ　267802
ヒメアサマスゲ　76778
ヒメアシホスゲ　76437
ヒメアシボソ　253997,254046
ヒメアスナロ　390660,390681
ヒメアゼスゲ　74438
ヒメアセビ　298751
ヒメアナナス　113365
ヒメアブラススキ　71608
ヒメアブラススキ属　22476
ヒメアフリカギク　31178
ヒメアフリカギク属　86025
ヒメアマ　231880
ヒメアマダマシ　266198
ヒメアマドコロ　308562
ヒメアマナ　15486,169446,169452
ヒメアマナズナ　68850
ヒメアマモドキ　266198
ヒメアミガサソウ　1876
ヒメアメリカアゼナ　231476
ヒメアメリカチャボヤシ　327541
ヒメアヤメ　208793
ヒメアラセイトウ　243209
ヒメアラセイトウ属　243156
ヒメアリアケカズラ　14881
ヒメアリアゲキズラ　14881
ヒメアリサンサンセウヅル　288762
ヒメアリドオシ　122041
ヒメイ　213042
ヒメイカリソウ　147068
ヒメイグマ　117425
ヒメイケマ　117425
ヒメイズイ　308562
ヒメイソツツジ　223911
ヒメイタビ　165515,165630
ヒメイチゲ　23780
ヒメイチゲソウ　23780
ヒメイチゴツナギ　306003
ヒメイチビ　905
ヒメイトフスマ　31968
ヒメイヌガラシ　336167
ヒメイヌハナヒゲ　333560
ヒメイヌビエ　140367
ヒメイバラモ　262108
ヒメイハラン　19512
ヒメイボクサ　260110
ヒメイヨカズラ　117721,400927
ヒメイラクサ　403039
ヒメイワカガミ　351441
ヒメイヮギボウシ　198619
ヒメイワシャジン　7712
ヒメイワタデ　308717
ヒメイワベンケイ　329836
ヒメイワヨモギ　35503
ヒメイワラン　19512
ヒメインチンロウゲ　85660
ヒメウイキャウ　24213
ヒメウイキョウ　24213,77784,114503
ヒメウイキョウ属　114502
ヒメウオトリキ　180925
ヒメウキガヤ　177586
ヒメウキクサ　221007
ヒメウキヤガラ　353707
ヒメウコギ　143677
ヒメウシオスゲ　76437
ヒメウシノシッペイ　193989,272350,388909
ヒメウシノシッペイ属　272345
ヒメウシノシャバイ属　272345
ヒメウシホスゲ　76437
ヒメウズ　357919
ヒメウズサバノオ　226341
ヒメウズサバノオ属　226339
ヒメウスノキ　403917,404066
ヒメウスユキソウ　224932
ヒメウズラヒトハラン　121436
ヒメウズ属　357918
ヒメウチワカエデ　3608
ヒメウヅ　357919

ヒメウツギ 126946
ヒメウツヅ属 357918
ヒメウマノミツバ 345978
ヒメウミヒルモ 184977
ヒメウメバチソウ 284503
ヒメウラシマソウ 33383
ヒメウラジロエゴノキ 379348
ヒメウロココゴヘイヤシ 225495
ヒメウワバミソウ 142696
ヒメウヰゼリ 77784
ヒメウヲトリキ 180925
ヒメウンラン 230889,230927
ヒメエゴノキ 379382
ヒメエゾネギ 15720
ヒメエニシダ 121013
ヒメエンコウソウ 68186
ヒメエンゴサク 106078
ヒメオウバイ 211956
ヒメオオイワボタン 90436
ヒメオオボウシバナ 100982
ヒメオガラバナ 3718
ヒメオダマキ 30031
ヒメオトギリ 201942
ヒメオドリコソウ 220416
ヒメオニササガヤ 128568
ヒメオニソテツ 145229
ヒメオノエスゲ 76219
ヒメオノオレ 53544
ヒメオモダカ 14734
ヒメカイウ 66600
ヒメカイウ属 66593
ヒメカイドウ 243717
ヒメカウガイゼキシャウ 212929
ヒメガガイモ 117469
ヒメガガブタ 267818
ヒメカカラ 366257
ヒメカクラン 293523
ヒメカジイチゴ 338064
ヒメカズラ 294827
ヒメカナリ-クサヨシ 293743
ヒメカノコサウ 398024
ヒメカノコサウ属 398022
ヒメカノコソウ 398024
ヒメカノコソウ属 398022
ヒメカハヅスゲ 73959
ヒメガマ 401094,401105
ヒメガマズミ 408019
ヒメカモジグサ 144661
ヒメカモノハシ 209321
ヒメガヤツリ 119687
ヒメカラスウリ 390128
ヒメカラフトイチゴツナギ 305942
ヒメカラマツ 388412
ヒメカリマタガヤ 131028
ヒメガルカヤ 29430
ヒメガルカヤ属 29422
ヒメカワズスゲ 73959,73961
ヒメカワヅスゲ 73959
ヒメカンアオイ 37741
ヒメカンガレイ 352223
ヒメカンガレヰ 352223
ヒメガンクビソウ 77185
ヒメカンザシ 34522
ヒメカンザブラウノキ 381148
ヒメカンスゲ 74174
ヒメカンゾウ 191272,191296
ヒメカンバ 53434,53454
ヒメガンピ 414222
ヒメキカシグサ 337351
ヒメギキョウ 70005
ヒメキクイモ 190505
ヒメキクイモ属 190502
ヒメキクタビラコ 261071
ヒメキクバスミレ 409610
ヒメキサワタ 220422
ヒメキジムシロ 312550
ヒメキツ 93483
ヒメキノヘラン 232117
ヒメキバナスズシロ 154019
ヒメキビ 402553
ヒメキブシ 373567
ヒメキヨシソウ 349851
ヒメキランサウ 13166
ヒメキランソウ 13166
ヒメギリソウ 377804
ヒメキリンギク 228511
ヒメキリンソウ 294124,294127
ヒメキンギョソウ 230911,231084
ヒメキンセンカ 66395
ヒメキンバイ 399524
ヒメキンポウゲ 325717
ヒメキンミズヒキ 11566
ヒメキンミヅヒキ 11566
ヒメキンレンカ 399604
ヒメクグ 218480
ヒメクグ属 218457
ヒメクサトベラ 350331
ヒメクジャクソウ 383107
ヒメクジラグサ 126120
ヒメクズ 138978
ヒメクチバシグサ 231587
ヒメクハガタ 407252
ヒメクマヤナギ 52436
ヒメクモマグサ 349121
ヒメクライタボ 165630
ヒメクリソラン 89936,185256
ヒメクリノイガ 80830
ヒメグルミ 212624,212655
ヒメクロアブラガヤ 353618
ヒメクロウメモドキ 328750
ヒメクロマメノキ 404030
ヒメクロモジ 231379
ヒメクワガタ 407252
ヒメクワクラン 293523
ヒメクワラン 82088
ヒメクワンザウ 191272
ヒメグンバイナズナ 225295,225340
ヒメケイヌホオズキ 367499
ヒメケイラン 274058
ヒメケフシグロ 248269
ヒメコイワカガミ 351447
ヒメコウオウソウ 383107
ヒメコウガイゼキショウ 212929
ヒメコウジ 172135
ヒメコウゾ 61103
ヒメゴウソ 75769
ヒメコウゾリナ 196043
ヒメコウボウ 27942
ヒメコウホネ 267331
ヒメコウモリ 283868
ヒメコガネナズナ 154466
ヒメコガネヒルガオ 250779
ヒメゴクラクチョウカ 190038
ヒメコケイラン 274058
ヒメコゴメグサ 160212
ヒメコザクラ 314604
ヒメコスモス 58671
ヒメコスモス属 58669
ヒメゴチョウ 293600
ヒメコチョウラン 293600
ヒメコナスビ 239693
ヒメコヌカグサ 12407
ヒメコハギ 126368
ヒメコハコベ 374986,375074
ヒメコバンサウ 60383
ヒメコバンソウ 60383
ヒメコバンノキ 60053,60083
ヒメコブシ 242331
ヒメコブナ 36710
ヒメコマツ 300130,300137
ヒメコマツナギ 206181
ヒメゴマノハグサ 355190
ヒメコメススキ 12104
ヒメゴヨウイチゴ 339115
ヒメコラノキ 99158
ヒメコレウス 99711
ヒメサイカチ 176886
ヒメサギゴケ 246980
ヒメザクラ 314408,314904
ヒメサクラサウ 314613
ヒメサクラマンテマ 364023
ヒメザクロ 321771
ヒメササガヤ 254012
ヒメササキビ 361709
ヒメサザンカ 69327
ヒメザゼンソウ 381079
ヒメサボンソウ 346421,346433
ヒメサヤバナ 99573
ヒメサユリ 230023
ヒメサルダヒコ 239238
ヒメサルトリ 366257

ヒメザンセウ 417340
ヒメシイン 400467
ヒメシオン 400467
ヒメシカクイ 143421
ヒメジガバチソウ 232211
ヒメジギタリス 17476
ヒメシキブ 301294
ヒメシキミ 301294
ヒメジソ 259284,259292
ヒメシタン 107549
ヒメシナガワハギモドキ 42697
ヒメシノ 347352
ヒメシバ 418453
ヒメシモツケ 371961
ヒメシャウジャウバカマ 191068
ヒメシャガ 208595
ヒメシャクナゲ 22445,330998
ヒメシャクナゲ属 22418
ヒメシャシャンポ 404060
ヒメシャジン 7706
ヒメジャモントウ 261091
ヒメシャラ 376457
ヒメシャリントウ 107549
ヒメシャリンバイ 329077
ヒメジュズスゲ 74539
ヒメショウジョウヤシ 120780
ヒメジョオン 150464
ヒメジョオン属 150414
ヒメジョヲン 150464
ヒメジヨヲン属 150414
ヒメシラスゲ 75400
ヒメシラタマソウ 363363
ヒメシラネアザミ 348505
ヒメシラネニンジン 391908
ヒメシロアサザ 267807
ヒメシロネ 239222,239224
ヒメシロビユ 18655
ヒメシンジュガヤ 132789,354175
ヒメスイカズラ 235887
ヒメズイナ 210407
ヒメスイバ 339897,339899
ヒメスギ 113696
ヒメスゲ 75649
ヒメススダケ 37196
ヒメスズムシソウ 232256
ヒメスズメガヤ 147891
ヒメスズメノテッポウ 17521
ヒメスズメノヒエ 58088
ヒメスズメラン 270832
ヒメスミレ 410105
ヒメスミレサイシン 410769
ヒメズヰナ 210407
ヒメセンサウ 252242
ヒメセンナリホオズキ 297720
ヒメセンナリホホヅキ 297703
ヒメセンニチモドキ 4876
ヒメセンブリ 235435
ヒメセンブリ属 235431
ヒメタイゲキ 160007
ヒメタイサンボク 242341
ヒメタイヌビエ 140367,140430
ヒメタウガラシ 72113
ヒメタウシャウブ属 399043
ヒメタカサゴヤマシロギク 40878
ヒメタガソデソウ 256444
ヒメタケウマヤシ属 208372
ヒメタケヤシ属 139300
ヒメタケラン 29814
ヒメタケラン属 29807
ヒメタコノキ 281114
ヒメタチホロギク 293171
ヒメタツナミサウ 355683
ヒメタデ 291744
ヒメタニタデ 309355
ヒメタヌキマメ 111856
ヒメタヌキモ 403252
ヒメタヌキラン 75387
ヒメタネツケバナ 72920
ヒメタマイチゴ 338281
ヒメタマサンゴ 367528
ヒメタマババハキ 81356
ヒメタマボウキ 81356
ヒメタムラソウ 345334
ヒメチゴザサ 120631,209098
ヒメヂシバリ 210673
ヒメチチコグサ 178481
ヒメチドメ 200393
ヒメチャルメルソウ 256018
ヒメヂンバリ 210539
ヒメツガ 399895
ヒメツクバネアサガオ 292753
ヒメツゲ 64285
ヒメツヅラフジ 364987
ヒメツバキ 69594,350932
ヒメツバキ属 350904
ヒメツユクサ 260110
ヒメツリガネヤナギ 289318,289332
ヒメツルアズキ 408969
ヒメツルアダン 168251
ヒメツルアリドオシ 256005
ヒメツルウンラン 216262
ヒメツルキジムシロ 313115
ヒメツルコケモモ 278360
ヒメツルソバ 308946
ヒメツルニチニチソウ 409339
ヒメツルニンジン 98341
ヒメツルリンダウ 398289
ヒメテイカカズラ 393650
ヒメテウセンガリヤス 306832
ヒメテキリスゲ 75638
ヒメテ-ブルヤシモドキ 201435
ヒメテンツキ 166228
ヒメテンマ 171930,171937
ヒメトウショウブ 399116
ヒメトウショウブ属 111455,399043
ヒメトウロウサウ 215160
ヒメトウ属 121486
ヒメトキホコリ 142898
ヒメドクゼリ 90935
ヒメトケイソウ 285671
ヒメトケンラン属 383119
ヒメドコロ 131873
ヒメトネリコ 167931
ヒメトベラ 301294
ヒメトモエサウ 201743
ヒメトモエソウ 201743,201745
ヒメトラノオ 317943,317966
ヒメトラノヲ 317966
ヒメトリカブト 5421
ヒメトレニア 392884
ヒメナエ 256154
ヒメナズナ 153964
ヒメナットウダイ 160007
ヒメナヅナ 153964
ヒメナツメヤシ 295468
ヒメナツメヤシモドキ 295478
ヒメナデシコ 127700
ヒメナベワリ 111677
ヒメナミキ 355428
ヒメナルコスゲ 75769
ヒメナンテンハギ 408650
ヒメナンバンギセル 8760
ヒメニガクサ 388299
ヒメニシキジソ 99711,99712
ヒメニシキノボタン 53136,53138
ヒメニシキハギ 226851
ヒメニセダイオウヤシ 347076
ヒメニチニチソウ 79424
ヒメニラ 15486
ヒメニレ 401602
ヒメヌカキビ 281733
ヒメヌカススキ 12818
ヒメヌカボ 12025
ヒメヌマハリイ 143397
ヒメヌマハリヰ 143397
ヒメネコノメサウ 90362
ヒメネジアヤメ 208657
ヒメネズ 213702
ヒメネズミノオ 372696
ヒメノアサガオ 208023
ヒメノアサガホ 208023
ヒメノアズキ 333323
ヒメノアヅキ 333323
ヒメノウスツボ 355148
ヒメノウゼンカズラ 385508
ヒメノウゼンカズラ属 385505
ヒメノウゼンカゼラ 385508
ヒメノウゼンハレン 399604
ヒメノカリス 200913
ヒメノカリス属 200906
ヒメノガリヤス 127243

ヒメノギネ属 201093
ヒメノコ 167700
ヒメノコギリソウ 4056
ヒメノダケ 24313,24314
ヒメノハギ 98157,126465
ヒメノボタン 53138,276090
ヒメノボタンサウ 346950
ヒメノボタン属 53130,276072
ヒメノマヘガミ 377973
ヒメノヤガラ 85455
ヒメノヤガラ属 193581
ヒメバイクワウツギ 294509
ヒメハイシノブ 307222
ヒメハイチゴザサ 209107
ヒメハイモ 65229
ヒメハウチハマメ 238467
ヒメハウチワマメ 238467
ヒメハギ 308123,308359
ヒメハギモドキ 418331
ヒメハギ科 308474
ヒメハギ属 307891
ヒメバショウ 260217
ヒメバセウ 260217
ヒメハゼザンセウ 417290
ヒメハッカ 250381
ヒメハナシノブ 175708,307195,307222
ヒメバナデシコ 127746
ヒメハナビシソウ 155171
ヒメハナヒリノキ 228171
ヒメハナヤナギ 114613
ヒメハニシキ 65229
ヒメハハキガヤツリ 118798
ヒメハブカズラ 329004
ヒメハマアカザ 87070
ヒメハマナシ 337006
ヒメハマナデシコ 127746
ヒメハマヒサカキ 160460
ヒメハマボッス 176775
ヒメハマヨモギ 35972
ヒメバラ 336490
ヒメバライチゴ 339194
ヒメバラモミ 298354
ヒメバラン 288658
ヒメハリイ 143178
ヒメハリイヌナズナ 137154
ヒメハリヰ 143178
ヒメハルガヤ 27934
ヒメハンシヨウヅル 95068
ヒメハンノキ 16433
ヒメビエ 281370
ヒメヒオウギズイセン 111464
ヒメヒカゲスゲ 74849
ヒメヒガンバナ 265299
ヒメヒゲシバ 88339,88340
ヒメヒゴタイ 348403,348690
ヒメヒサカキ 160629
ヒメビシ 394459
ヒメビジョザクラ 405896
ヒメビタイ 165630
ヒメヒダボタン 90414
ヒメヒマハリ 188936,188941
ヒメヒマワリ 188936,188941
ヒメヒャクニチソウ 418057
ヒメヒャクリコウ 391352
ヒメヒラテンツキ 166190,166228
ヒメビル 15486
ヒメヒレアザミ 73467
ヒメビロードカズラ 294825
ヒメビロードスゲ 75740
ヒメフウチョウソウ 34406
ヒメフウテウボク 71775
ヒメフウロ 174877
ヒメフジ 414570
ヒメフシグロ 248269
ヒメブシダマ 235730
ヒメフタツブダネヤシ 130065
ヒメブタナ 202404
ヒメフタバムグラ 125872
ヒメフタバラン 232919
ヒメブッソウゲ 243929
ヒメフトエクマテヤシ 315387
ヒメフトモモ 382514
ヒメフュイチゴ 339188
ヒメフユイチゴ 338508
ヒメフヨウ 243929,243932
ヒメフヨウ属 243927
ヒメフロックス 295267
ヒメヘイシソウ 347155
ヒメベゴニア 50287
ヒメベニサンゴアナナス 8578
ヒメヘビイチゴ 312446,312447
ヒメホウキガヤツリ 119284
ヒメホウワウチク 47351
ヒメホクシャ 114601
ヒメホシビジン 356660
ヒメホタルイ 352212,353557
ヒメホタルヰ 353557
ヒメボタンバラ 336462
ヒメボッス属 345728
ヒメホテイラン 68538
ヒメマイヅルソウ 242672
ヒメマサキ 157343
ヒメマツカサススキ 353510
ヒメマツバギク 220697
ヒメマツバシバ 33914
ヒメマツハダ 298329
ヒメマツバボタン 311915
ヒメマメラン 62571
ヒメマユミ 157345
ヒメマルバウツギ 126836
ヒメマンネングサ 357327
ヒメミカンサウ 296803
ヒメミカンソウ 296803
ヒメミクリ 370100,370106
ヒメミコシガヤ 75042
ヒメミズ 298945
ヒメミズトンボ 183799
ヒメミソハギ 19602
ヒメミソハギ属 19550
ヒメミゾホオズキ 255246
ヒメミチャナギ 309602
ヒメミヅ 142766,298945
ヒメミツバツツジ 331358
ヒメミナリトケイソウ 285645
ヒメミミカキグサ 403257
ヒメミヤガラシ 326842
ヒメミヤマウズラ 179685
ヒメミヤマウヅラ 179685
ヒメミヤマコナスビ 239719
ヒメミヤマジュズスゲ 76477
ヒメミヤマスミレ 409755
ヒメムカシヨモギ 103446
ヒメムギクサ 198297
ヒメムサシノ 385876
ヒメムシトリスミレ 299751
ヒメムシャ 106970
ヒメムツオレガヤツリ 118867
ヒメムョウラン 264651
ヒメムラサキ 321636
ヒメムラダチヒルガオ 103194
ヒメムレスズメ 72322,72351
ヒメムロ 85352,85353
ヒメモウズイカ 405691
ヒメモエギスゲ 76593
ヒメモチ 203977
ヒメモッコク 386698
ヒメモリサウ 85107
ヒメモロコシ 369652,369653
ヒメモンステラ 258177
ヒメヤエナリ 293989
ヒメヤエムグラ 170526,170545
ヒメヤガミスゲ 73782
ヒメヤシャブシ 16433
ヒメヤツシロラン 130041
ヒメヤツシロラン属 130037
ヒメヤナギラン 85884
ヒメヤブラン 232630
ヒメヤマイ 166471
ヒメヤマツツジ 330980
ヒメヤマハナソウ 349529
ヒメヤマヰ 166471
ヒメユズリハ 122727
ヒメユヅリハ 122679
ヒメユリ 229811,229813,229817,229818,229872
ヒメヨツバムグラ 170289
ヒメヨモギ 35483,35788,35794
ヒメラッキョウ 15064
ヒメラッギョウ 15813
ヒメラヤソケイ 211863
ヒメリンゴ 243580

ヒメルイヨウボタン　182634
ヒメルリザクラ　19045
ヒメルリトラノオ　406979
ピメレア属　299303
ヒメレモン　93546
ヒメレンゲ　357189
ヒメワキノテ　95031
ヒメワタスゲ　396005
ヒメワニクチ　259535
ヒメワレモコウ　345894
ヒメワンピ　94191
ヒメヰズヰ　308562
ヒメヲガラバナ　3718
ヒモゲイトウ　18687
ヒモサボテン　29715
ヒモサホテン属　29712
ヒモサボテン属　83855
ヒモズイトウ　18687
ビモードスゲ　75396
ヒモハノナンヨウスギ　30835
ヒモヒジキ　104765
ビモンギョク　233524
ビヤウヤナギ　202021
ビャクギフ　55575
ヒャクジッコウ　219933
ビャクジュッ　44218
ビャクジュツ　44205
ビャクシン　213634
ビャクシン属　213606
ビャクダン　85165,213634,346210
ビャクダン科　346181
ビャクダン属　346208
ヒャクニチサウ　418043
ヒャクニチサウ属　418035
ヒャクニチソウ　418043
ヒャクニチソウ属　418035
ビャクブ　375343,375351
ビャクブ科　375359
ビャクブ属　375336
ヒャクマントウ　105438
ビヤクレン　20408
ビャケレンゲ　279249
ヒヤシンス　199583
ビャッコイ　209983
ビャッコウアザミ　92085
ヒヤハサギゴケ属　191325
ヒヤンチン　392841
ヒユ　18776,18836
ヒュウガアオイ　188908
ヒュウガアジサイ　200099
ヒュウガギボウシ　198615,198619
ヒュウガセンキュウ　24407
ヒュウガタイゲキ　160077
ヒュウガトウキ　24486
ヒュウガナッ　93834
ヒュウガナツミカン　93834
ヒュウガヒロハテンナンショウ　33413
ヒュウガホシクサ　151486
ヒュウガミズキ　106669
ヒュウガミヅキ　106669
ヒュウガミツバツツジ　330909
ヒュウガモウセンゴケ　138362
ヒューゲリモクマオウ　79166
ビューティオブグレンリトリート　93692
ヒユナ　18776,18822
ヒユモドキ　18847
ヒユ科　18646
ヒユ属　18652
ビヨウカヅラ　398684
ビヨウカヅラ属　398681
ヒョウカン　93323
ビヨウサウ　201743
ヒヨウゼンギョク　103758
ビヨウソウ　201743
ヒョゥソゥギョク　45233
ビヨウタコノキ　281166
ヒョウタン　219843,219844
ヒョウタンギシギシ　340214
ヒョウタンボク　235970
ヒョウトウ　263747
ヒョウノコ　167703
ヒョウノセンカタバミ　277661
ヒョウビ　82499
ヒヨウビ　82523
ビヨウマル　234949
ヒョウモンウッボ　264834
ヒョウモンヨウショウ　245027
ヒョウモンラン　404680
ビヨウヤナギ　202021
ビヨウリッフヂ　125958
ビヨウリツヤブガラシ　387769
ヒョウレイ　86521
ヒョオタンカズラ　103862
ヒヨクサウ　407200
ヒヨクソウ　407200
ヒヨクヒバ　85348
ヒヨコマメ　90801
ヒヨコマメ属　90797
ヒヨス　201389
ヒヨス属　201370
ヒョソノキ　134946
ヒヨドリジャウコ　367322
ヒヨドリジョウゴ　367322
ヒヨドリジョオゴ　367322
ヒヨドリバナ　158161,158230
ヒヨドリバナ属　158021
ヒヨヒヨグサ　17525
ヒヨンカン　93333
ヒョンチク　297239
ヒラアンペライ　240478
ビラウ　234170,234194
ヒラウチハ　273103
ヒラウチハサボテン　273103
ビラウドイチゴ　338281
ビラウ属　234162
ヒラエサンキライ　194120
ヒラエブシダマ　235733
ビラカ　407865
ヒラカウガイゼキシャウ　213213
ヒラガマ　401112
ビラカンサ　322465
ピラカンサ　322450
ピラカンサス　322450,322454
ヒラギ　276313
ヒラギガシ　316819
ヒラギガン　223131,316824
ヒラギギク　55768
ヒラギシスゲ　73826
ヒラギシスゲモドキ　73547
ヒラギソヨゴ　203590
ヒラギナンテン　242563
ヒラギナンテン属　242458
ヒラギモクセイ　276287
ヒラギモチ　203545
ヒラコウガイゼキショウ　213213
ヒラスギ　113696
ビラディア属　409230
ヒラテンツキ　166223,166226
ヒラドツツジ　331581
ヒラナス　366920,367375
ヒラノニンジン　292892
ヒラバヒメマオ　56112
ヒラボガヤツリ　118436
ヒラマメ　224473
ヒラミガシ　233181
ヒラミカンコノキ　177170,177174
ヒラミスゲ　73877
ピラミダリスポプラ　311226
ヒラミレモン　93448
ヒラモ　404557
ヒラヰ　212890
ビラン　223150
ヒーランクリガシ　79032
ビランジ　363625
ビランジュ　223150
ビランジ属　363141
ビランヂ属　363141
ピリナッツツリ　71013
ヒリュウ　310852,331370
ヒリュウガシ　323957
ビリンビ　45724
ヒルガオ　68678,68692,68705,68708,68724
ビルガオ科　102892
ヒルガオ属　68671
ヒルガホ　68692
ヒルガホ科　101858
ヒルガホ属　68671
ヒルギ　215510
ヒルギカズラ　121647
ヒルギダマシ　45746
ヒルギダマシ属　45740

ヒルギモドキ　45746,238354
ヒルギモドキ属　238345
ヒルギ科　329776
ヒルザキツキミソウ　269504
ヒルゼンスゲ　74869
ビルベリ－　403916
ビルマガフクワン　13595
ビルマゴウカン　13595
ビルマコプシア　217869
ビルマシタン　320306
ビルマネム　13595
ビルマネムノキ　13595
ヒルムシロ　312090,312228
ヒルムシロシバ　200677
ヒルムシロ科　312300
ヒルムシロ属　312028
ヒルルシロ科　312300
ヒルレモン　93706
ヒレアザミ　73337,73339
ヒレアザミ属　73278
ヒレイガグリギョグ　167691
ヒレイマル　234901
ビレウ　234170
ピレオギク　89768,124862
ヒレギク　219996
ヒレギク属　219990
ヒレザンショウ　417177
ヒレタゴボウ　238166
ヒレタゴボオ　238166
ヒレタヌキマメ　112486
ピレナカンタ属　322517
ピレネ－イワタバコ　325225
ピレネ－フウロ　174856
ヒレハナガサヒゴタイ　348129
ヒレハリギク　81198
ヒレハリサウ　381035
ヒレハリサウ属　381024
ヒレハリソウ　381035
ヒレハリソウ属　381024
ヒレヒメヒゴタイ　348690
ヒレブダウ　92907
ヒレマオウ　146125
ヒレミゾホオズキ　255187
ヒレミヤガミスゲ　73933
ビロウ　234170,234194
ヒロウザサ　304071
ビロウドイチゴ　338281
ビロウドキスミレ　410466
ビロウドクサギ　96398
ビロウドゴセウ　290305
ビロウドスギ　113700,113717
ビロウドゼキシャウ　5809
ビロウドソバグリ　78932
ビロウドタツナミ　355510
ビロウドチョロギ　373321
ビロウドテンツキ　166473
ビロウトバライチゴ　338305
ビロウドヒメクヅ　65156
ビロウドヒメクヅ属　44824
ビロウドフウロ　175018
ヒロウドヘウタンボク　236076
ビロウドムラサキ　66829
ビロウドモウズイクワ　405788
ビロウドモウズヰクワ　405788
ビロウドヤマシロギク　39966
ビロウドラン　179708
ビロウドワンガシ　324490
ビロウボタンヅル　95077
ビロウモドキ　234175
ビロウ属　234162
ビロオドエノキグサ　1790
ビロオドシオガマ　287594
ビロオドムラサキ　66829
ビロオドモオズイカ　405788
ピロカクタス属　322986
ヒロガホ　68713
ヒロシマナツザボン　93485
ピロステギア属　322948
ヒロセレウズ属　200707
ピロソセレウス属　299238
ビロ－ドアオイ　18173
ビロ－ドアオイ属　18155
ビロ－ドアオダモ　168013
ビロ－ドアカツメクサ　396926
ビロ－ドアカネ　338021
ビロ－ドアサギリ　344855
ビロ－ドイチゴ　338281
ビロ－ドイボタ　229560
ビロ－ドイワギリ　16184
ビロトイワギリ属　16174
ビロ－ドイワギリ属　16174
ビロ－ドウチワ　28108
ビロ－ドウツギ　126898,413621,413624,413626
ビロ－ドウツボ　264858
ビロ－ドウマノスズクサ　34275
ビロ－ドウリノキ　13379
ビロ－ドエゾアザミ　92090
ビロ－ドエゾシオガマ　287838
ビロ－ドエゾニワトコ　345673
ビロ－ドオオウチワ　28138
ビロ－ドオトギリ　202185
ビロ－ドカズラ　294792
ビロ－トガンギク　124806
ビロ－トギジムシロ　312325
ビロ－ドキビ　58199
ビロ－ドギリ　366715
ビロ－ドギリ属　366714
ビロ－ドクサイチゴ　338060
ビロ－ドクサギ　96398,96400
ビロ－ドクサフジ　408693
ビロ－ドゴショウ　290305
ビロ－ドサワシバ　77278
ビロ－ドサンシチ　183056
ビロ－ドシオガマ　287594
ビロ－トシバ　418453
ビロ－ドスゲ　75396
ビロ－ドタツナミ　355510
ビロ－ドテンツキ　166473
ビロ－ドトラノオ　317938
ビロ－ドノリウツギ　200053
ビロ－ドハギ　226965
ビロ－ドヒメアナナス　113374
ビロ－ドヒメクズ　65156
ビロ－ドフウロ　175000
ヒロ－ドヘウタンボク　236076
ビロ－ドベゴニア　49617,50329
ビロ－ドホオズキ　297678
ビロ－ドボタンヅル　95077
ビロ－ドミヤコザサ　347205
ビロ－ドムラサキ　66829
ビロ－ドメヒシバ　130666
ビロ－ドモウズイカ　405788
ビロ－ドヤイトバナ　280107
ビロ－ドヤッシロソウ　70058
ビロ－ドヤマシロギク　39966
ビロ－ドラン　179708
ビロ－ドレンゲツツジ　331260
ヒロハアカガシ　323602
ヒロハアカザ　87112
ヒロハアキグミ　142219
ヒロハアマナ　400156
ヒロハアメリカシオン　40817
ヒロハイチゲ　23709
ヒロハイチゴツナギ　306010
ヒロハイッポンスゲ　75910
ヒロハイヌノヒゲ　151452
ヒロハウキガヤ　177648
ヒロハウバメガシ　324436
ヒロハウマノスズクサ　34268
ヒロハウラジロガシ　324359
ヒロハウラジロヨモギ　35742
ヒロバウロコザミア　225623
ヒロハエゾニワトコ　345675
ヒロハオオズミ　243649
ヒロハオガタマノキ　252855
ヒロハオキナグサ　321672
ヒロハオゼヌマスゲ　76570
ヒロバオニソテツ　145235
ヒロハオホゼリ　90937
ヒロハオランダフウロ　153769
ヒロハオリヅルラン　88565
ヒロハカスリソウ　130097
ヒロハカツラ　83743
ヒロハカハラサイコ　312815
ヒロハガマズミ　407905
ヒロハカワラナデシコ　127855
ヒロハカワラヒジキ　104772
ヒロハキエボシソゥ　5318
ヒロハキクヨモギ　35789
ヒロハキゲンゲ　279195

ヒロバギシギシ 340151
ヒロハキンバイザサ 256582
ヒロハクコ 239021
ヒロハクサフジ 408435
ヒロハクサフヂ 408435
ヒロハクサレダマ 239898,239901
ヒロハクマタケラン 17693
ヒロバクララ 369010
ヒロハクロイゲ 342191
ヒロハケニオイグサ 187697
ヒロハコアヤメ 208917
ヒロハコウゾリナ 195548
ヒロバコチョウラン 293643
ヒロハコックバネウツギ 168
ヒロバゴテフ 293643
ヒロハコリヤナギ 343529,343937
ヒロハコンロンカ 260499
ヒロハコンロンソウ 72685
ヒロハサギゴケ 191501
ヒロハサギゴケ属 191486
ヒロハサギスゲ 152765
ヒロバザミア 417010
ヒロハサンザシ 109936
ヒロハシヒノキカヅラ 125978
ヒロハシャゼンオモダカ 140545
ヒロハシラネニンジン 391906
ヒロハスギナモ 196816
ヒロバスゲ 74888
ヒロハスズサイコ 117645
ヒロハスズメノトウガラシ 231486
ヒロハセネガ 308348
ヒロハセンブリ 380238
ヒロハダイオウ 329318
ヒロハタイゲキ 158895
ヒロハタカトウダイ 158895
ヒロハタチャナギ 343680
ヒロハタマミズキ 204001
ヒロハダモ 233832
ヒロハチトセラン 346155
ヒロハチョウセンガリヤス 94604
ヒロハッツアナナス 54452
ヒロハツノミザミア 83683
ヒロハツリシュスラン 179673
ヒロハツリバナ 157705
ヒロバツルグミ 142001
ヒロハツルマサキ 157499
ヒロハツルマメ 177722
ヒロハテイショウソウ 12625
ヒロハテンナンショウ 33446
ヒロハドウダンツツジ 145710,145711
ヒロハドクゼリ 90937
ヒロハドジョウツナギ 177673
ヒロハトラノオ 317921,317933
ヒロバトラノヲ 317921
ヒロバトンボソウ 302513
ヒロハナアマナ 400156
ヒロハナミサキサウ 397159
ヒロハナミサキソウ 397159
ヒロハニワトコ 345620
ヒロバニンジン 7639
ヒロハヌマガヤ 128015
ヒロハヌマゼリ 365877
ヒロハネム 13565
ヒロハネムノキ 13565
ヒロハノアオキ 44921
ヒロハノイチゲ 24128
ヒロハノイヌノヒゲ 151219,151452
ヒロハノウシノケグサ 163819,164219
ヒロハノエビネ 65887
ヒロハノエビモ 312220
ヒロハノエビラフヂ 408374
ヒロハノオオウシノケグサ 164260
ヒロハノオオタマツリスゲ 74529
ヒロハノカウガイゼキシャウ 213045
ヒロバノカウガイゼキシャウ 213045
ヒロハノカエデ 3127
ヒロハノカラン 65869
ヒロハノカワラサイコ 312815
ヒロハノキハダ 294238,294251
ヒロハノキミツ 142654
ヒロハノクロタマガヤツリ 168903
ヒロハノコウガイゼキショウ 212816,213045
ヒロハノコジュズスゲ 75701
ヒロハノコヌカグサ 25343,25346
ヒロハノコヌカグサ属 25331
ヒロハノコメススキ 126039,126055
ヒロハノサヤヌカグサ 224003
ヒロハノセンニンモ 312032
ヒロハノツリバナ 157705
ヒロハノドジョウツナギ 177618
ヒロハノドゼウツナギ 177618,177629
ヒロハノナガハグサ 305875
ヒロハノナンヤウスギ 30835
ヒロハノナンヨウスギ 30835
ヒロハノハネガヤ 376754,376755
ヒロハノハネガヤ属 275607
ヒロハノハネミノキ 121730
ヒロハノハマサジ 230655
ヒロハノヒトツバヨモギ 36321
ヒロハノヒヨドリジョウゴ 367330
ヒロハノヘビノボラズ 51301,52100,52339
ヒロハノホロシ 367282
ヒロハノマツバボタン 311877
ヒロハノマンテマ 363411,363673
ヒロハノミズタマソウ 91580
ヒロハノミミズバイ 381429
ヒロハノヤガラ 156572
ヒロバノヤナギカモメズル 117721
ヒロバノヤハラスゲ 74915
ヒロハノユキヤナギ 371881
ヒロハノレンサウ 222750
ヒロハノレンソウ 222750
ヒロハハイノキ 381326
ヒロハハコヤナギ 311292,311565
ヒロハハナカズラ 5558
ヒロハハナカンザシ 329801
ヒロハハナヒリノキ 228167
ヒロハハマヨモギ 36232
ヒロハバラモンジン 354813
ヒロハハリブキ 272708
ヒロハハンゴンサウ 358500
ヒロハハンノキ 16416
ヒロハヒエスゲ 76651
ヒロバヒエスゲ 76773
ヒロハヒゲリンドウ 174212
ヒロハヒトツバヨモギ 36321
ヒロハヒトツバラン 265418,265420
ヒロハヒマワリ 392434
ヒロハヒメイチゲ 23705
ヒロハヒメウツギ 126947
ヒロハヒメジョオン 150972
ヒロハヒメチゴザサ 120632
ヒロハヒメハマアカザ 87133
ヒロバヒメマオ 56257
ヒロハヒョウタンボク 236040
ヒロハヒルガオ 68713,68716,68721
ヒロハヒルガホ 68713
ヒロハフウリンホオズキ 297650
ヒロハブタノマンヂャウ 115965
ヒロハフタバムグラ 57333,370802
ヒロハフタマハコベ 374836
ヒロハヘウタンボク 236040
ヒロハヘビノボラズ 51301,52100
ヒロハホウキギク 41328
ヒロハマツナ 379551
ヒロハミシマサイコ 63594
ヒロハムカシヨモギ 150421
ヒロハムギグワイ 400156
ヒロハムシトリナデシロ 363222
ヒロバモレノヤシ 85432
ヒロハヤブニッケイ 91255
ヒロハヤブマメ 177792
ヒロハヤブレガサモドキ 381824
ヒロハヤマカモジグサ 58596
ヒロハヤマトウバナ 97035
ヒロハヤマトオバナ 97008
ヒロハヤマトラノオ 317932
ヒロハヤマナラシ 311292
ヒロハヤマユリ 229732
ヒロハヤマヨモギ 36321
ヒロバユキザサ 242691
ヒロハヨノミ 235705
ヒロハラベンダ－ 223298
ヒロハレンギョウ 167452
ヒロマモウコシャジン 7826
ビロロ 93641
ビワ 151162
ビワコエビラフジ 408675
ビワモドキ 130919
ビワモドキ科 130928

ビワモドキ属　130913
ビワ属　151130
ビンォウジ　31680
ビンカ　203670
ピンカドウ　415693
ピンガドウ　415693
ビンキツ　93695
ビング　132230
ピンクダチュラ　61231
ピンクミンク　315810
ビンゴヌカスゲ　75394
ビンゴムグラ　170571
ヒンジガヤツリ　232405,232423
ヒンジガヤツリ属　232380
ヒンジモ　224396
ビンゼイ　244390
ビンセトキシカル属　409394
ビンッケズル　214971
ピンドジョオウヤシ　31707
ピンピカヅラ　97947
ビンボウガズラ　79850
ビンボフガヅラ　79850
ピンポン　376144,376150
ピンポンノキ　376144
ピンポン属　376063
ビンラウジ属　31677
ビンロウ　31680
ビンロウジ　31680
ビンロウジュ　31680
ビンロウジュ属　31677
ビンヲウジ　31680
ファウカリア属　162898
ファウル・メドウ・グラス　305804
ファケイロア属　161921
ファゴピラム属　162301
ファビアナ属　161896
フアーベルガシ　323881
ファレノプシス属　293577
フィカス属　164608
フィグーミウス属　296108
ブイサングミ　141999
ブイサンハナメウガ　17741
ブイサンヤマビハ　151145
フィジーノヤシモドキ属　264798
フィジーノヤシ属　405253
フィジーフトエクマテヤシ　315386
フィットーニア属　166706
フィテウーマ属　298054
フィトーニア属　166706
フイリアオキ　44919
フイリアオギリ　166628
フイリアスナロ　390682
フイリアブチロン　913
フイリイチョウ　175827
フイリイテフ　175827
フイリイナモリソウ　318186
フイリイヌコリヤナギ　343531
フイリイヌツゲ　203694
フイリイモ　28140
フイリインドゴムノキ　164940
フイリウコギ　2502,143679
フイリウスバサイシン　38319
フイリウヅラ　179650
フイリウリカエデ　2913
フイリウリハダカエデ　3531
フイリオオシマカンスゲ　75637
フイリオオバイボタ　229571
フイリオオバコ　301877
フイリオクタマスミレ　409619
フイリオニイタヤ　3398
フイリカキドオシ　176828
フイリカクレミノ　125645
フイリカンスゲ　75423
フイリクズウコン　245015
フイリクチナシ　171312
フイリクマタケラン　17675
フイリクマタゲラン　17711
フイリケイワタバコ　101684
フイリケンサキ　171253,171378
フイリゲンジスミレ　410721
フイリコウリカエデ　2913
フイリコクタン　132244
フイリコスミレ　410111
フイリサクララン　198831
フイリザゼンソウ　381082
フイリサンゴアナナス　8553
フイリシキンツユクサ　393999
フイリシハイスミレ　410753
フイリシャクナゲ　330537
フイリシュロガヤツリ　118478
フイリショウブ　5794
フイリジンジソウ　349211
フイリセイヨウキヅタ　187227
フイリセイヨウヤマハッカ　249506
フイリソシンカ　49247
フイリソヨゴ　204144
フイリタガネソウ　76266
フイリタコノキ　281169
フイリタラヨウ　203964
フイリダンチク　37469
フイリチシマセンブリ　380378
フイリツルニチニチソウ　409336
フイリトウモロコシ　417420
フイリドクダミ　198746
フイリナガバノスミレサイシン　409749
フイリナガバヒメアナナス　113381
フイリナワシログミ　142158
フイリニュウサイラン　295604
フイリネコヤナギ　343449
フイリノウゼンハレン　399603
フイリノセイヨウダンチク　37469
フイリハクチョウゲ　360934
フイリハラン　39533
フイリヒカゲスゲ　75051
フイリヒナスミレ　410670
フイリヒバ　390682
フイリヒメザゼンソウ　381080
フイリヒメバショウ　66181
フイリヒメフタバラン　232921
フイリヒロハオリヅルラン　88567
フィリビンクワズヤシ　273124
フィリピンスンキー　93710
フィリビンナコラン　9297
フイリフギレシハイスミレ　409619
フイリフッキソウ　279758
フイリフモトスミレ　410577
フイリブロメリア　60555
フイリホウワウチク　47350
フイリマキノスミレ　410755
フイリマサキ　157615,157616
フイリマルバウツギ　127076
フイリマルバチトセラン　346162
フイリミズキ　105003
フイリミチノクアゼスゲ　73569
フイリミヤマスミレ　410551
フイリミヤマフタバラン　232947
フイリミョウガ　418003
フイリヤエヤマスミレ　410642
フイリヤダケ　318313
フイリユズリハ　122701
フイリリュウキュウチク　304055
フイリロクワイ　17136
フィロデンドロン属　294785
フィロボルス属　296868
フィンブリアーム　125135
フウ　232557
フウキギク　290818,290821
フウキンギョク　272540
フウジュヤドリギ　411048
フウセンアカメガシワ　216642
フウセンアサガオ　103362
フウセンアサガオ属　250749
フウセンカズラ　73207
フウセンカズラ属　73194
フウセンカヅラ　73207
フウセンカヅラ属　73194
フウセンダマノキ　179032
フウセンツメクサ　397116
フウセントウワタ　179032,179096
フウセントウワタ属　178974
フウチウボク属　71676
フウチソウ　184656
フウチョウガシワ　121933
フウチョウソウ　95759,182915
フウチョウソウ科　71671
フウチョウボク属　71676
フウチョウラン　88553
フウテウサウ科　71671
フウテウボク　71760
フウテフボク　71760
フウトウカズラ　300427

フウトウカズラ属　300323
フウトウカヅラ　300427
フウトウカヅラ属　300323
フウライジン　10926
フウラン　263867,356457
フウラン属　24687,263865
フウリンアサガオ　208305
フウリンイワブクロ　289353
フウリンウメモドキ　203840
フウリンギョク　272516
フウリンサウ　70164
フウリンソウ　70164
フウリンツツジ　145677
フウリンハナシャジン　7768
フウリンブッサウゲ　195216
フウリンブッソウゲ　195216
フウロアオイ　67128
フウロケマン　106227
フウロサウ　174755
フウロサウ科　174431
フウロサウ属　174440
フウロソウ　174966,175023
フウロソウ科　174431
フウロソウ属　174440
フウ属　232550
フェイジョア　163117
フェイジョア属　163116
ブエイチュウ　279493
フェスツーカ属　163778
フェドラナッサ属　293252
フエニックス　295453
フェニックス属　295451
フェネストラリア属　163326
フェラーリア属　163510
フエリシア属　163121
フエルニア属　198969
フエロカクタス属　163417
フェロスペルマ属　294273
フェロデンドロン属　294230
フェローニア属　163499
フォエティダバラ　336563
フォーカリア属　162898
フォサーギラ属　167537
フォースターホエア　198808
フォステレラ属　167529
フォーチュネイー　242534
フォーチュン　242534
フォッケア属　167122
フォーリーアザミ　348313
フオーリーイチゴ　339390
フオリエゴノキ　16295
フォーリーガヤ　351254
フオーリーガヤ　351251
フオーリーガヤ属　351250
フオーリーササガャ　254005
フォリドータ属　295510
フォールズレモン　93707
フォンタネーシア属　167230
フカギレイタヤメイゲツ　3605
フカギレオオモミジ　2780
フガクスズムシソウ　232169
フカノキ　350706
フカノキ属　350646
フカミグサ　280286
フキ　292374
フキアゲ　10942,10943
フキエーラ属　167556
フキザクラ　290821
フキズツメソウ　127916,127919
フキスミレ　409895
フキズメサウ　127916,127919
フキタンポポ　400675
フキタンポポ属　400671
フキヅメサウ　127916,127919
フキブキ　290821
フキモドキ　41872
フキヤミツバ　346008
フキユキノシタ　349510
フギレアズキナシ　369306
フギレオオバキスミレ　409781
フギレキスミレ　409771
フギレサキガケスミレ　409665
フギレジンジソウ　349209
フギレヒルガオ　68688
フキ属　292342
フクイカサスゲ　75762
フクインギョク　233643
フクエジマカンアオイ　37686
フクオウソウ　313781
フクオウソウ属　313778
フクオウニガナ　313783
フクオウモドキ　313895
フクギ　171197,171201
フクギ属　171046
フクシア　168750
フクシア属　168735
フクシマシャジン　7639
フクシマヤジン　7639
フクシャ　168767
フクシュウキンカン　167511
フクジュウダケ　357939
フクジュギョク　233471
フクジュサウ　8331
フクジュサウ属　8324
フクジュソウ　8331,8378
フクジュソウ属　8324
フクジンソウ　107271
ブクスス属　64235
フクダンバラ　366284
フクド　35517
フクトメキンバイ　312714
フクトメブシ　5214
フクベ　219848
フクベノキ　111104,134946
フクベノキ属　111100
フクベラ　23817
フクベライチゲ　23817
フクボク　254303
フクボンシ　338292
フクマンギ　77067,141611
フクメンギョク　209506
フクライギョク　233558
フクライミカン　93460
フクラシバ　204138,204217
フクラモチ　204138,229513
ブクリャウサイ　129068
ブクリャウサイ属　129062
ブクリュウサイ　129068
ブクリュウサイ属　129062
ブクリュウサウ　312502
ブクリョウサイ　129068
ブクリョウサイ属　129062
ブクリョオサイ　129080
フクリンアカリフア　2014
フクリンアセビ　298749
フクリンカスリソウ　130096
フクリンキズタ　187319
フクリンキヅタ　187319
フクリンキブシ　373560
フクリンクチナシ　171253,171307
フクリンサカキ　96588
フクリンサンヒチ　171253
フクリンジガバチソウ　232207
フクリンシラン　55577
フクリンジンチョウゲ　122534
フクリンダイゲキ　159313
フクリンタラノキ　30638
フクリンチトセラン　346165
フクリンヂンチャウ　122534
フクリンナワシログミ　142158
フクリンネズミモチ　229492
フクリンマンネングサ　356885
フクリンヤツデ　162884
フクレウバメガシ　324285
フクレミカン　93859
フクロ　158456
フクロウンラン　263459
フクロカズラ　134044
フクロガンピ　363774
フクロギ　158456,159457
フクロスゲ　75403
フクロダガヤ　398091,398093
フクロナデシコ　363882
フクロネカヅラ　134037
フクロネジアヤメ　208921
フクロノキ　158456
フクロヒヨス　297882
フクロヒヨス属　297875
フクロモチ　229513
フクロヤシ属　244495
フクロユキノシタ　82561

フクロユキノシタ科　82490
フクロユキノシタ属　82560
フクワウサウ属　313778
フクワウドキ　313895
フクワバモクゲンジ　217613
フゲシザサ　347216
ブ-ゲンヴヰレヤ　57868
ブゲンカズラ　57864
ブ-ゲンビレア　57868
ブ-ケンビレア属　57852
ブコウカスミザクラ　83141
ブコウマメザクラ　83229
ブコウミツバフウロ　175007
フサアカシア　1165
フサウケバカラスウリ　396213
フサガヤ　91239,259636
フサガヤツリ　118433
フサガヤ属　91235
フサカンスゲ　76553
フサコメガヤ　249085
フサザガヤ　253992
フサザキシャリントウ　107637
フサザキズイセン　262457,262468
フサザキツヅラフヂ　116029
フサザキムニンヤッシロラン　171912
フサザクラ　160347
フサザクラ科　160349
フサザクラ属　160338
フサジュンサイ　64496
フサジュンサイ属　64491
フサスグリ　334179
フサスゲ　75347
フサタヌキモ　403156
フサナキリスゲ　76501
フサナリサンシチ　183097
フサナリタコノキ　280997
フサナンヨウスギ　30831
フサハリイ　353290
フサフジウツギ　62019
フサボタン　407897
フサモ　261379
フサモ属　261337
フサラ　93502
プサンモフォラ属　317026
フジ　414554,414576
ブジ　5193
フジアカショウマ　41861
フジアザミ　92332
フジイチゴツナギ　305584
フジイバラ　336578
フジイロテンジクボタン　121559
フジウツギ　62093,62099
フジウツキ科　235251
フジウツギ科　62208
フジウツギ属　61953
フジウメモドキ　204245
フジオトギリ　201865
フジガエソウ　408649,408652,408654
フジカサ　200736
フジカスミザクラ　83140
フジカンザウ　200736
フジカンゾウ　200736
フジキ　94030
フジギク　383103
フジキクザクラ　83225
フジキ属　94014
フジグルミ　320372
フシグロ　363451
フシグロセンノウ　363472,363755
フシグロ属　248235
フシゲクズ　321447
フシゲタチフウロ　174685
フシゲチガヤ　205517
フシザキサウ　381809
フシザキサウ属　381808
フシザキソウ　381809
フシザキソウ属　381808
フジザクラ　83224
フジサンシキウツギ　413571
フシシバ　418095
フジスミレ　410664
フジセンニンサウ　94818
フジセンニンソウ　94818
フジタイゲキ　160076
フシダカ　4275
フシダカフウロ　175006
フジチドリ　264769
フシチョウ　16657,256268
フジツツジ　331985
フジツリガネツツジ　250526
フジツリバナ　401773
フジテンニンソウ　228005
フジナ　384746
フシナシオサラン　299625
フシナシササハギ　18273
フシナシヲサラン　299625
フジナデシコ　127739
フシネキンエノコロ　361753,361773
フシノキ　332509
フシノハアワブキ　249424,249445
フジバカマ　158118,158161
フジバカマ属　158021
フジハタザオ　30435
フジハタザヲ　30435
ブジバヒエンソウ　124360
フジボグサ　402118
フジボゲンゲ　278757
フジボタン　95219,220729
フジボツルハギ　126411
フジマキ　323611
フジマツ　221894
フジマメ　218721
フジマメ属　135399,218715
フジマンヨウ　331286
フジムスメ　234941
フジモドキ　122438
フジヤナギ　342931
ブシュカン　93603,93604
プ-シュキニア属　321892
プシロカウロン属　318820
ブジンマル　284679
フジ属　414541
ブス　5193
フスマウツギ　294530
プセウデランテムム属　317245
プセウドアナナス属　317174
プセウドエスポストア属　317685
プセウドササ属　318277
プセウドロビビア属　317869
ブゼンテンツキ　166435
ブゼンノギク　40577
ブゾロイバナ　147084
ブソロイバナ属　25466
フタイロコリンシア　99838
フタイロマランタ　245016
ブダウ　411979
ブダウホホヅキ　297711
ブダウ科　411165
ブダウ属　411521
フタエアケビ　13226
フタエキキョウ　302754
フタエギキョウ　302753
フタエツリガネツツジ　250517
フタオモテ　302721
ブタクサ　19145
ブタクサモドキ　19179
ブタクサ属　19140
ブタグサ属　19140
フタゴシバ　235970
フタコロビ　235970
フタシベネズミノオ　372650,372666
フタシベネズミノヲ　372650,372666
フタタマオウ　146154
フタツブダネヤシ属　130057
ブタナ　202432
フタナミサウ　354930
フタナミサウ属　354797
フタナミソウ　354930
フタナミソウ属　354797
ブタノマンジュウ　115965
ブタノマンヂュウ　115949
ブタノマンヂュウ属　115931
フタバアオイ　37590
フタバアオイ属　37556
フタバアッサムヤシ　413093
フタバアフヒ　37590
フタバガキ科　133551
フタバガキ属　133552
フタハキノヘラン　232200
フタバツレサギ　302260
フタバナサザンクワ　68939,69411

フタバハギ 408648
フタバムグラ 187523,187565
フタバムグラ属 187497
フタバラン 232905
フタバラン属 232890
フタヒゲオオカニツリ 34934
フタヒゲカニツリ 398449
フタブツレサギ 302260
フタホセンナ 360431
フタマタイチゲ 23796
フタマタコゴメナデシコ 183191
フタマタタンポポ 110841
フタマタタンポポ属 110710
フタマタハコベ 374836
フタマタメヒシバ 130592,130763
フタマタモウセンゴケ 138270
ブタモロコシ 155878
ブタモロコシ属 155876
フタリシズカ 88297
フタリシヅカ 88297
フダンギク 89481
フダンサウ 53249
フダンサンショウ 417161
フダンザンショウ 417340
フダンソウ 53249,53257
フヂ 414554,414576
フヂイゼリ 299480
フヂイラン 38194
フヂウツギ 62093,62099
フヂウツキ科 235251
フヂウツギ属 61953
フヂキ 94030
フヂキ属 94014
フチグモイチゴ 338886
フヂゲイトウ 18788
フチゲオオバキスミレ 409777
フチゲハグロサウ 291161
フチゲハグロソウ 291161
フヂサキャウ 234715
フヂサザゲ 409113
フヂツツジ 331985
フチトリセッコク 125135
フチナシガクアジサイ 199990
フヂナデシコ 127607,127739
フヂバカマ 158070,158161
フヂバシデ 145517
フヂバシデ属 145504
フヂバナマメ 126471
フチベニシャクナゲ 330536
フチベニベンケイ 109229
フチベニホウサイラン 117042
フヂボグサ 402118
フヂボグサ属 402106
フヂボタン 220729
フヂボハギ 402118
フヂマメ 218721
フヂマメ属 135399
フヂモドキ 122438
プチロズティリス属 320862
フヂヰゼリ 299480
フヂヰラン 38194
フヂ属 414541
フッカ-ウツボ 264840
フッカ-メギ 51733
フッキサウ 279757
フッキサウ属 279743
フッキソウ 279757
フッキソウ属 279743
フックサ 95358
フッケンヒバ 167201
フッケンュサン 216142
フッコクカイガンショウ 300146
ブッサウゲ 195149
ブッソウゲ 195149
フッチンシア属 199480
ブットウ 61167
ブッバヒエンソウ 124360
ブツメンチク 297301
フデ 174095
ブティア 64161
フ-ディアプシス属 198079
フ-ディア属 198035
ブティア属 64159
フデオトギリ 202060
フデガタッチトリモチ 46863
フデクサ 75003,75259
フデノキ 302684
フデボテンナンシャウ 33337
フデリンドウ 174095
フデリンドオ 174095
プテロカクタス属 320258
プテロスティリス属 320871
プテロディスクス属 320525
フド 29304
フトイ 352278,352289
フトイガヤツリ 118520
ブドウ 411979
ブドウカズラ 387853
ブドウホオズキ 297711
ブドウホオヅキ 297711
ブドウ科 411165
ブドウ属 411521
フトエウチワヤシ属 103731
フトエクマデヤシ属 315376
フトヌマハリイ 143198
フトネアヤメ 208897
フ-トノキ 382582
フトバクレソラン 310798
フトヒルムシロ 312123
フトボアゼスゲ 74839
フトボタニガワスゲ 73545
フトボナガボソウ 373507
フトボナギナタコウジュ 144063
フトボノヌカボタデ 309859
フトボヤブマオ 56247
フトミノヒサカキ 160535
フトムギ 198376
フトモモ 382582
フトモモ科 261686
フトモモ属 156118
フトヰ 352278
ブナ 162368,162386
フナガタウチワ 28133
フナシホタルブクロ 70235
フナシミヤマウズラ 179694
ブナゾロ 77276
フナトウサゴヤシ 217925
フナドコ 93461
ブナノキ 162368
ブナノキ属 162359
フナバシソウ 210485
フナバハギ 35282
フナバラサウ 117385
フナバラソウ 117385
ブナ科 162033
ブナ属 162359
プニオセレウス属 288983
フノリカズラ 214971
ブバルディア属 57948
フヒメアナナス 113387
フブキショウマ 91008
フブキチュウ 94566
フブキノマッ 21136
フブキバナ 387032
フブキバナ属 386992
ブフタルムム属 63524
フモトクスミレ 410576
フモトスミレ 410474
フヤウ 195040
ファーリ-イチゴ 339390
ファーリ-ミズ 299037
ファーリ-ミヅ 299037
プヤ属 321936
フユアオイ 243862
フュアフイ 243862
フュアフヒ 243862
フユイチゴ 338205
ブユウマル 140242,140327
フュキ 15291
フュグミ 142214
フユグミ 141981
フユサンゴ 367522
フユザンショウ 417161,417163,417340
フユザンセウ 417161
フユズタ 187318
フユヅタ 187318
フユナ 59575
フユヌカボ 12139
フュネギ 15291
フユノヤンバルイチゴ 338698
フユボダイジュ 391691

フヨウ　195040,195269
フヨウカタバミ　278039,278137
フヨウホウ　331843
フヨウラン　245860
フヨウ属　194680
プライスハナメウガ　17741
ブライダルベール　175489
ブラウト－ニア属　61087
ブラウナンッス属　61135
ブラウ－ネア属　61154
ブラオンカヅラ　194157
ブラオンカヅラ属　194155
ブラオンカモメヅル　400945
ブラキカリキウム属　58301
ブラキステルマ属　58813
ブラキセレウス属　58320
フラグミペティラム属　295875
フラグモペティラム属　295875
フラク－ルティア属　166761
フラゲラリア属　166795
フラサバソウ　407144
プラジ　8563
フラシダ　98653
ブラシノキ　67298
ブラシリカクタス属　59158
ブラシリセレウス属　59167
ブラジルアラウカリア　30831
ブラジルウチワ　272818
ブラジルクマデヤシ属　398824
ブラジルゴムノキ　194453
ブラジルジャカランダ　211241
ブラジルジャケツイバラ　64975
ブラジルゾウゲヤシ　44807
ブラジルチドメグサ　200354
ブラジルナット　53057
ブラジルナットノキ　53057
ブラジルナットノキ属　53055
ブラジルハシカグサモドキ　334320
ブラジルバンジロウ　318734
フラジルヒメヤシ　380561
ブラジルマツ　30831
ブラジルロウヤシ　103737
フラセリアナモクマオウ　79164
フラセリアモクマオウ　79164
フラセリ－モミ　377
プラタナス属　302574
プラタヌス属　302574
フラック・カ－ラント　334117
ブラックウオルナット　212631
ブラックウッドアカシア　1384
ブラックスプル－ス　298349
ブラックソルガム　369593
ブラックチェリ－　83311
ブラックバ－ンサバル　341406
ブラッサボラ属　59254
ブラッシア属　59265
ブラッシノキ　67298
ブラッシノキ属　67248
プラティクリニス属　302740
プラティティラ属　303022
フラテルナ　69082
フラネルサウ　363370
フラネルソウ　363370
フラビマル　209508
ブラヘアヤシ属　59095
ブラヘア属　59095
フラベリギク　166824
フランケ－ニア属　167765
フランコ－ア属　167733
フランスオウギ　187842
フランスカイガンショウ　300146
フランスギク　227533
フランスゼリ　54240
フランスユリ　229789
フランツォシニイ　10855
ブランデゲ－ハクセンヤシ　59097
ブリオポプシア属　61549
フリ－ジア　168183
フリシア属　168328
フリ－ジア属　168153
フリ－セア属　412337
フリソデヤナギ　343602
プリムラ・ポリアンサ　314820
プリムロ－ス　315101
ブリュウマル　103740
ブリルランタイシア属　60295
プリンセ－ピア属　315171
ブリンチュウ　279490
ブル－・サルビア　345023
フルカワナスビ　367604
ブルケルリア属　63903
プルット　93492
プルネラ属　316095
ブル－ノニア科　61384
ブルノニア科　61384
ブル－ノニア属　61381
ブルノニア属　61381
ブル－ハイビスカス　18254
ブル－バ－ル　85347
ブルビネラ属　62471
ブルビ－ネ属　62334
ブルボコジュ－ム属　62518
ブ－ルボンセダカトックリヤシ　201359
ブ－ルボンベニオウギヤシ　222626
プルマ・マンゴ－　57847
プルムバ－ゴ属　305167
ブル－メンバッキア属　55851
プルモナ－リア属　321633
ブルンズウィギア属　61390
プレイオスピロズ属　304336
プレイオネ・フォレステイ　304248
プレイオネ属　304204
フレイシネチヤ属　168238
プレウロタリス属　304904
フレキシマツ　299942
プレクトメルミンツス属　303780
ブレッティア属　55529
フレップ　404051
ブレディア属　59822
ブレティラ属　55559
プレニア属　313909
フレモンティアデンドロン属　168215
フレモンティア属　168215
フレモンテイポプラ　311324
ブレヤガマズミ　407715
フレラン　295958
フレラン属　295952
フレ－レア属　167689,168221
ブロウニンギア属　61165
ブロスフェルディア属　55656
プロチアキナロィデス　315776
プロチア属　315702
フロックス属　295239
ブロッコリ－　59545
プロテア属　315702
フロ－ミス属　295049
プロメネ－ア属　315509
ブロ－メリア属　55644
ブロメ－リア属　60544
フロリダザミア　417006
フロリダソテツ　417006
フロリダダイオウヤシ　337879
フロリダロウバイ　68313
ブロワリア－スベシオ－サ　61131
ブンカンカ　415096
ブンカンカ属　415094
フンキコミズ　298919
フンキコミヅ　298919
ブンゲスゲ　74801
プンゲンストウヒ　298401
ブンゲンスマツ　300168
ブンゴウツギ　127127
ブンゴザサ　362131
ブンダイユリ　154926
ブンタン　93468,93579
ブンドウ　409025
フントレア属　199443
ブンパイクライニヤ　216654
ベアウフォルティア属　49347
ヘイケモリアザミ　92164
ヘイシソウ属　347144
ヘイシャナハヒノキ　381235
ベイスギ　390645
ベイヒバ　85301
ベイマツ　318580
ベイモミ　317
ペイレスキア属　290699
ペイレスキオプシス属　290717
ヘイワ　358062
ペウケダ－ヌム属　292758
ヘウタン　219844

ヘウタンカヅラ 103862
ヘウタングサ 407287
ヘウタンボク 103862,235970
ペウ－ムス属 293091
ヘウモンラン 404680
ベカン 77902
ペカン 77902
ペカン属 77874
ヘキカンギョク 32696
ヘキガンギョク 182447
ヘキギョク 175506
ヘキコウカン 257403
ヘキシギョク 233677
ヘキバンギョク 182485
ヘキルリランポウカク 43512
ヘキレイ 32691
ヘキロウギョク 233666
ベキンソウ 349895
ペキンバイカウツギ 294509
ペキンバイクワウツギ 294509
ペキンハシドイ 382278
ペキンヤナギ 343668
ペキンヨモギ 36468
ペグアセンヤク 1120
ベクインケンチャヤシ 138534
ヘクシセア属 350494
ヘクソカズラ 280097
ヘクソカズラ 280078
ヘクソカズラ属 280063
ヘクソカヅラ 280097
ヘクソカヅラ属 280063
ペグノキ 1120
ベゴニア属 49587
ベゴニヤ 50232
ペスカトレア属 292138
ヘスペランサ属 193232
ヘスペランタ属 193232
ヘスペリソウ 365532
ヘソクリ 299724
ヘソノヤシ属 119988
ペヂトオカクタス属 287881
ベチベルソウ 407585
ヘチマ 238261,238268
ヘチマカボチャ 114293
ヘチマ属 238257
ペチュニア属 292741
ベチュニヤ 295267
ヘツカコナスビ 239761
ヘツカニガキ 364768
ヘツカラン 116815
ベッカリフトエクマテヤシ 315379
ヘツカリンドウ 380368
ヘッケリア属 187156
ベッセラ属 53213
ヘッタマ 82523
ヘッピヤナギ 344255
ヘッレアンツス属 193109
ヘディサルム属 187753
ペディランツス属 287846
ペトラエア属 292517
ペトロセリ－ヌム属 292685
ベナフブキ 94568
ベニ 77748
ベニアヂッツアナナス 54456
ベニアマナ 231907
ベニアマモ 117308
ベニアミメグサ 166709
ベニイタヤ 3140,3198
ベニイチヤクサウ 322790
ベニイチヤクソウ 322790
ベニイトスゲ 76126
ベニイヌゴマ 373179
ベニウチハ 28119
ベニウチハ属 28071
ベニウチワ 28119
ベニウチワ属 26950,28071
ベニオウギ 303561
ベニオウギヤシ属 222625
ベニオオウチギョク 233619
ベニオグラコウホネ 267314
ベニカエデ 3505
ベニガク 199994,200093
ベニガクウツギ 200079
ベニガクエゴノキ 379384
ベニガクスミレ 410491
ベニガクッツアナナス 54451
ベニガクヒルギ 61263
ベニカタバミ 277706
ベニカノコソウ 81768
ベニカノコソウ属 81743
ベニカハ 333133
ベニガフクワン 66669
ベニガフクワン属 66660
ベニカヤラン 171869
ベニカワ 333133
ベニギリソウ 147386,217739
ベニキリン 158404
ベニクジャク 266648
ベニクロバナキハギ 226896
ベニケシヨウ 356628
ベニゴウカン 66669
ベニゴウカン属 66660
ベニコウジ 93737
ベニコウホネ 267287
ベニコスモス 107160
ベニコチョウ 17490
ベニサラサドウダン 145684
ベニサンゴバナ 279768
ベニサンゴバナ属 279765
ベニサンザシ 110014
ベニシオガマ 287315
ベニシタン 107486
ベニシタン属 107319
ベニシベウメバチソウ 284593
ベニシュスラン 179580,179644
ベニスギ 113696
ベニスジヒメバショウ 66185
ベニスズメガヤ 148028
ベニスヂサンジュ 196458
ベニタイゲキ 158809
ベニタカ 358997
ベニダマ 233642
ベニチカラシバ 289015
ベニチョウジ 84425,114606
ベニヅガ 399895
ベニツカサ 139998
ベニツクバネ 185
ベニヅルザクラ 316939
ベニドウダン 145690
ベニニガナ 144903,385892
ベニニガナ属 144884
ベニノキ 54862
ベニノキ科 54863
ベニノキ属 54859
ベニハエギリ 147386
ベニバスモモ 316302
ベニバナ 77748,77749
ベニバナアマ 231907,231908
ベニバナアワモリショウマ 41818
ベニバナイチゴ 339445
ベニバナイチヤクサウ 322790
ベニバナイチヤクソウ 322790
ベニバナイワウメ 127918
ベニバナイワハゼ 172054
ベニハナインゲン 293985
ベニバナインドソケイ 305225
ベニバナウツボグサ 316150
ベニバナウマゴヤシ 396934
ベニバナエニシダ 85401
ベニバナオオハナウド 192283
ベニバナオキナグサ 23767
ベニバナキジムシロ 312393
ベニバナクサギ 96078
ベニバナケイタドリ 328357
ベニバナゲッカコウ 59715
ベニバナゲッカコウ属 59714
ベニバナコックバネウツギ 170
ベニバナコバノイチヤクソウ 322781
ベニバナコメツツジ 332019
ベニバナザクラ 83301
ベニバナサナギイチゴ 339136
ベニバナサルビア 344980
ベニバナサワギキョウ 234351
ベニバナシュクシャ 187428
ベニバナスイカズラ 235878,235885
ベニバナスヒカヅラ 235878
ベニバナセキコク 125173
ベニバナセンブリ 81495
ベニバナソケイ 211749
ベニバナダイコンソウ 175400
ベニバナチャ 69641

ベニバナチョウセンアサガオ　61240
ベニバナツノゲシ　176728
ベニバナツメクサ　396934
ベニバナトケイソウ　285632
ベニバナトチノキ　9680
ベニバナニシキウツギ　413582
ベニバナニラ　15550
ベニバナノクロイチゴ　338810
ベニバナノツクバネウツギ　185
ベニバナノヘビイチゴ　167636
ベニバナハコネウツギ　413578
ベニバナハマハタザオ　30462
ベニバナハマヒルガオ　68740
ベニバナハマボッス　239728
ベニバナハリモミ　298444
ベニバナハルユキノシタ　349698
ベニバナハンショウヅル　95371
ベニバナヒイラギソウ　13116
ベニバナヒョウタンボク　236085
ベニバナフクジンソウ　107244
ベニバナベコニア　49729
ベニバナベゴニヤ　49845
ベニバナベンケイ　215102
ベニバナボロギク　108736
ベニバナマメアサガオ　207922
ベニバナマルバノイチヤクソウ　322860
ベニバナミゾホオズキ　255197
ベニバナミツモトソウ　312393
ベニバナミネズオウ　235287
ベニバナミヤマカタバミ　277880
ベニバナヤマシャクヤク　280241
ベニバナヨウラクラン　267959
ベニバナワタ　179884
ベニバナ属　77684
ベニヒ　85268
ベニヒメリンドウ　161531
ベニヒモノキ　1894
ベニブキ　292376
ベニフクリンセンネンボク　137360
ベニフデ　299261
ベニベンケイ　215102
ベニホウワウチク　47351
ベニマツリ　336104
ベニマツリ属　336096
ベニマメヅタラン　62697
ベニマメノキ　274392
ベニマンサク　134010
ベニマンサク属　134009
ベニモンジ　224336
ベニヤッコガズラ　294807
ベニヤマザクラ　83301
ベニヤマボウシ　51141
ベニユリ　229811
ベニラタンヤシ　222626
ベニラン　55575
ベニロイアルハッカ　250432
ベニワビスケ　69753
ヘネケン　10854
ヘノリンク　330538
ヘビイチゴ　138793,138796
ヘビイチゴ属　138790
ヘビイモ　20132,348036
ヘビウリ　396151
ヘヒティア属　187129
ヘビノダイハチ　33496
ヘビノボラズ　52151,52322
ヘベアゴムノキ　194453
ベベガヤ　82499,82523
ヘベヤゴムノキ　194453
ペペロミア属　290286
ヘーベ属　186925
ヘホウセイ　267046
ペポカボチャ　114300
ヘボガヤ　82499,82523
ヘマトキシルム属　184513
ヘミノキ　408034,408224
ヘメロカリス属　191261
ヘラオオバコ　302034
ヘラオホバコ　302034
ヘラオモダカ　14725
ヘラオモダカ属　14719
ヘラキノヘラン　232333
ヘラザクロサウ　256727
ベラドンナ　44708
ベラドンナ属　44704
ヘラナレン　110622
ヘラノキ　391749
ヘラハタザオ　30337
ヘラバトウロウサウ　215174
ヘラバヒメジョオン　150980
ヘラヒメアナナス　113372
ペラペラヨメナ　150724
ヘラマツバギク　136394
ヘラモ　404555
ヘリアントセレアス属　140842
ヘリオセレウス属　190237
ヘリオトロピューム属　190540
ヘリオトローブ　190559
ヘリオトロブ　190702
ヘリオフィラ属　190261
ヘリオプシス属　190502
ペリカンバナ　34198
ヘリコディケロス属　189982
ヘリコニア属　189986
ペリステリア属　291127
ペリストローフェ属　291134
ペリスフォルディア属　52947
ヘリドリオシベ　111724
ヘリドリオシベ属　111693
ヘリトリザサ　347324
ヘリドリヲシベ属　111693
ベリューム属　50841
ペリラ属　290932
ベルガモット　93407,93408
ベルガモットハッカ　250341
ヘルキア属　188365
ベルゲランツス属　52548
ベルゲロカケタス属　52571
ベルゲロセレウス属　52571
ベルサムモミ　282
ペルシアグルミ　212636
ペルシアナ　336568
ヘルシエリア属　193134
ペルシャコムギ　398869
ペルシャハシドイ　382241
ペルシャバッコヤナギ　342967
ペルセア属　291491
ベルタイ－ミア属　405356
ヘルチア属　193202
ベルックゼオセレウス属　413883
ペルティフィル－ム属　288840
ヘルトカ－ルス　363214
ベルトロ－ニア属　53130
ベルナンブコ　65002
ペルナンブコ　65002
ペルニ－ヒイラキ　204162
ペルネッチア属　291363
ベルノニア属　406040
ベルバナサルビア　344885
ペル－バルサムノキ　261559
ペルピ－ヒイラキ　204162
ベルベシ－ナ属　405911
ベルモアホエア　198806
ベルリンドロ　311144
ヘルレリア科　193119
ヘルレリア属　193117
ペレキフォラ属　288615
ペレスキオプシス属　290717
ベレッタ　93694
ペレニアル・ライグラス　235334
ヘレニュ－ム属　188393
ヘレロア属　192427
ベロニカ　317927
ヘンエフボク　98190
ヘンエフボク属　98177
ヘンカミカン　93703
ベンガルコ－ヒ－　98866
ベンガルコ－ヒ－ノキ　98866
ベンカルボダイジュ　164684
ベンガルヤハズカズラ　390787
ベンケイイモ　131501
ベンケイサウ科　109505
ベンケイサウ属　356467
ベンケイソウ　200784
ベンケイソウ科　109505
ベンケイソウ属　356467
ベンケイチュウ　77077
ベンケイナズナ　225398
ペンステモン属　289314
ペンタグロッティス属　289500
ペンタス属　289784

ベンティンキンブシス属 51165
ベンティンクヤシ属 51163
ヘンデルギ 14873,14874
ヘンデルギ属 14871
ベンテンツゲ 64303,64312
ヘントウ 20890,20897
ペントステモン属 289314
ペンペングサ 72038
ヘンヨウガマズミ 407790
ヘンヨウボク 98190,98191
ヘンリ-イモ 20088
ベンリイモ 207623
ヘンリエゴノキ 379347
ヘンリ-グリ 78795
ヘンリ-コクラン 232188
ヘンリ-ヅタ 285110
ヘンリ-メヒシバ 130590
ヘンルウダ 341064
ヘンルウダ科 341086
ヘンルウダ属 341039
ヘンル-ダ 341064
ヘンル-ダ属 341039
ホイ 410994
ポインセチヤ 159675
ボ-ウィエア属 57977
ホウイジユリ 229730
ホウオウガシワ 323821
ホウオウギョク 43498
ホウオウシャジン 7844
ホウオウチク 47347
ホウオウボク 123811
ホウオウボクモドキ 65014
ホウオウボク属 123801,307055
ホウオウヤナギ 343560
ホウオウラン 180126
ホウカ 243253
ホウガシラ 182424
ホウカゾウ 30032
ホウガンノキ 108195
ホウガンノキ属 108193
ホウガンヒルギ 415716
ホウカンボク 61157
ホウガンボク 108195
ホウカンマル 2101
ホウキアザミ 91997
ホウキアゼガヤ 226016
ホウキカサスゲ 76618
ホウキガヤツリ 118765
ホウキギ 217361
ホウキギク 41317
ホウキギョク 259726
ホウキギ属 217324
ホウキザクラ 83127
ホウキテンモンドウ 38983
ホウキヌカキビ 282214
ホウキヤシオ 145727
ホウギョク 284680
ホウゴウシデ 77314
ホウゴウシホデモドキ 194127,318450
ボウコウマメ 100153
ボウコウマメ属 100149
ホウコクサ 178297
ボウコツルマメ 177789
ホウコトウオモト 346173
ホウサイ 117042
ホウサイラン 117042
ホウザン 327283
ホウザンカラクサ 99361
ホウザンカラクサ属 99358
ホウザンスゲ 74824
ホウザンツヅラフジ 97938
ホウザンツヅラフヂ 97933
ボウシギョク 209511
ボウシバナ 100961
ホウシュウギョク 284661
ボウシュウボク 232483
ホウシュギョク 284660
ホウシュン 358030
ホウシュンマル 140844
ホウジョ 107053
ホウショウ 91289
ホウショウチク 47350
ホウスイギョク 233498
ボウズオトギリ 201925
ボウズヒメジョオン 150464
ホウセキコウ 257405
ボウセキコク 125383
ホウセンカ 204799
ホウセンカ属 204756
ホウセンクワ 204799
ホウセンクワ科 47077
ホウセンクワ属 204756
ホウソ 324384
ホウゾウバナ 43053
ボウダラ 215442
ボウチトセラン 346070
ホウチャクソウ 134467
ホウツイギョク 32712
ホウテンカ 364213
ホウテンクワ 364213
ホウトウ 109264
ホウノキ 198698
ホウビチク 47345
ホウビリュウ 358271
ボウフウ 346448
ボウブナ 114294
ボウフラ 114300
ボウブラ 114292,114294
ボウホウギョク 242883
ボウムギ 235310,235354
ホウメイデン 244152
ホウメイマル 244015
ホウヨカモメヅル 409467
ホウヨク 304367
ホウライアオイ 37674
ホウライアオカズラ 182384
ホウライアオカヅラ属 182361
ホウライアオキ 327069
ホウライアオキ属 326995
ホウライアフヒ 37674
ホウライアマ 266048
ホウライアヲカヅラ属 182361
ホウライアヲキ 327069
ホウライアヲキ属 326995
ホウライイケマ 117469
ホウライウヅラ 179685
ホウライオホケフチク 305226
ホウライカガミ 285056
ホウライカガミ属 285055
ホウライカズラ 171456
ホウライカズラ属 171439
ホウライカヅラ 171456
ホウライカヅラ属 171439
ホウライカン 93860
ホウライシソクサ 230322
ホウライショウ 258168
ホウライショウ属 258162
ホウライシンクサ 230320
ホウライセウ 258168
ホウライセウ属 258162
ホウライチク 39014,47345
ホウライチャヒキ 190130
ホウライツヅラフジ 290841
ホウライツヅラフヂ属 290840
ホウライツユクサ 100930
ホウライボタンヅル 94953
ホウライムラサキ 66779
ホウライリンダウ 173455
ホウライリンドウ 173455
ホウラン 116773,116774,116815,304340
ボウラン 238322
ホウランギョク 182456
ホウランジュリ 229730
ボウラン属 238301
ホウリュウ 192452
ホウリンギョク 159617
ホウレイマル 140862
ホウレンソウ 371713
ホウレンソウ属 371710
ホウロク 177465
ホウロクイチゴ 339270
ホウワウサウ 383103
ホウワウスゲ 74824
ホウワウチク 47345
ホウワウチク属 47174
ホウワウツゲ 360933
ボウンカク 140116
ホエア属 198804
ボエ-ニア属 57972
ホ-エンベルギア属 197143
ホオガシワ 283421

ホオガシワノキ 198698
ホオコグサ属 178056
ホオズキ 297643,297645
ホオズキハギ 89207
ホオズキハギ属 89201
ホオズキ属 297640
ホオヅキ属 297640
ホオナラ 283421
ホオノキ 198698
ホオノキ属 241951
ホオベニエニシダ 121005
ホオライアオカズラ 182384
ホオライカズラ 171456
ホオライシソクサ 230322
ホオライセンブリ 81502
ホオライムラサキ 66779
ホガエリガヤ 61442
ホガエリガヤ属 61441
ホガクレシバ 113316
ホカケサウ 78002
ホカケソウ 78002
ホカゴノキ 233967
ホカノキ 223131,316824
ホガヘリガヤ 61442
ホガヘリガヤ属 61441
ホカリヒゴクサ 73555
ホクシオミナエシ 285834
ホクシカラマツ 221939
ホクシタンポポ 384509
ホクシハネガヤ 376723
ホクシヒゴタイ 348585
ホクシヒョウタンボク 235781
ホクシマメナシ 323110
ホクシャ 168750,168767
ホクシヤマナシ 323330
ホクシャ属 168735
ホクセンイチゴツナギ 305858
ホクチアザミ 348349
ホクチキクアザミ 348094
ホクチグサ 283134
ホクチグサ属 56044
ホクトイ 212905
ホクトウリクサ 392903
ホクトカク 395658
ホクトガヤツリ 119419
ホクトクマタケラン 17674
ホクリ 116880
ホクリクアオウキクサ 224362
ホクリクネコノメ 90358
ホクリクムヨウラン 223645
ホクリョウアヤメ 208913
ホクリョウスゲ 74998
ホクリョウワレモコウ 345894
ホクロ 116880
ホクロクトウヒレン 348578
ボケ 84556,84573
ボケットメロン 114150
ボケ属 84549
ホコ 195269
ホコガタアカザ 44621
ホコガタギク 413535
ホコガタフウロ 174980
ホゴノキ 233967
ホコバイヌビハ 165762
ホコバガラシ 365529
ホザキアカメガシハ 243420
ホザキアケビ 13219,13241
ホザキアセビ 298734
ホザキアャメ 46148
ホザキアヤメ属 46022
ホザキイガカウゾリナ 143472
ホザキイガコウゾリナ 317233
ホザキイカリソウ 147048
ホザキイチョウラン 243084
ホザキウンラン 231183
ホザキカエデ 3732
ホザキカクラン 390917
ホザキカヘデ 3732
ホザキキカシグサ 337391
ホザキキケマン 106350
ホザキキンゴジカ 362670
ホザキクシノハラン 62806
ホザキゲンゲ 278883
ホザキザクラ 376683
ホザキザクラ属 376682
ホザキサルノオ 196824
ホザキサルノオ属 196821
ホザキサルノヲ 196824
ホザキサルノヲ属 196821
ホザキシオガマ 287689
ホザキシホガマ 287689
ホザキシモツケ 372075
ホザキシャウガ 187432
ホザキツキヌキソウ 397847
ホザキツチトリモチ 46859
ホザキツツジ 331239
ホザキツリガネツツジ 250516
ホザキツルウメモドキ 80203
ホザキトケイソウ 285695
ホザキナナカマド 369279,369282
ホザキナナカマド属 369259
ホザキニワヤナギ 309683
ホザキノイカリサウ 147048,147050
ホザキノイカリソウ 147048,147050
ホザキノキノヘラン 232122
ホザキノコケザクラ 376683
ホザキノトケイサウ 285695
ホザキノトケイソウ 285695
ホザキノナナカマド 369282
ホザキノフサモ 261364
ボザキノフサモ 261364
ホザキノミミカキグサ 403132,403299
ホザキハマビハ 295361
ホザキヒエンソウ 124151
ホザキヒトツバラン 125535
ホザキヒメラン 130185
ホザキブシ 5335
ホザキムシャリンダウ 137613
ホザキムシャリンドウ 137613
ホザキモクセイソウ 327879
ホザキヤドリギ 237077
ホザキヤブサンザシ 334055
ホザキヤブスグリ 334055
ホザキユリ 229789
ホザキヲサマン 148656
ホザクナナカマド 369279
ボサツソウ 396956
ボサツバラ 336789
ホサノキ 324384
ホシアオイ 243887
ホシアサガオ 208255
ホシオトメ 109266
ホシクサ 151257,151268
ホシクサ科 151206
ホシクサ属 151208
ホシケイ 293494
ホシケイラン 293494,293495
ホシケサ 151243
ホシザキイナモリソウ 318185
ホシザキカタバミ 277751
ホシザキカンアオイ 37734,194371
ホシザキキキョウ 70008
ホシザキシャクジョウ 278484
ホシザキスイレン 267723
ホシザキハマサオトメカズラ 280105
ホシザキフロックス 295272
ホシザキマンネングサ 356473
ホシザキユキノシタ 349938
ホシザクラ 83342
ホシサンゴ 116286
ホシセンネンボク 137409
ホシソケイ 211963
ホシダネヤシモドキ属 194609
ホシダネヤシ属 43372
ホシトウガラシ 72072
ホシナシアケボノソウ 380128
ホシナシゴウソ 75316
ホシナシサハヒヨドリ 158202
ホシナシサワヒヨドリ 158202
ホシナシヘツカリンドウ 380369
ホシノゲイトウ 18147
ホシノヤシ 321165
ホシバツルノゲイトウ 18147
ホシババンレイシ 25866
ホシバムカシヨモギ 58275
ホシハラン 39536
ホシビジン 279639
ホシベゴニヤ 50051
ホシヤシ属 41647
ホスゲ 74297,74299
ボスメン 143573

ボスリオキラス属 57545
ホソ 324384
ホソアオゲイトウ 18734,18795
ホソアカバナ 146605
ホソアヲゲイトウ 18795
ホソイ 213448
ホソイトスギ 114753
ホソイトヒバ 114753
ホソイリカンガレイ 352228
ホソエカエデ 2846
ホソエガラシ 365503
ホソエクマデヤシモドキ属 97883
ホソエクマデヤシ属 390451
ホソエダハヒノキ 381310
ホソエノアザミ 92439
ホソガタスズメウリ 417488
ホソカワヂシャ 407349
ホソグミ 141930,141932
ホソコウガイゼキショウ 213111
ホソスゲ 74342
ホソススキ 255886
ホソセイヨウヌカボ 28971
ホソツッチトセラン 346101
ホソテンツキ 144466
ホソテンニンギク 169589
ホソナルコビエ 151675
ホソヌカキビ 128483
ホソノゲムギ 198311
ホソノヤマスゲ 74453
ホソバ 225009
ホゾバアイアシ 293216
ホソバアイズシモツケ 371877
ホソバアオキ 44924
ホソバアオハダ 204008,204010
ホソバアオヨモギ 35416
ホソバアカザ 86893,86927
ホソバアカバナ 146812
ホソバアサガオ 415299
ホソバアジアタンポポ 384453
ホソバアブラギク 124806
ホソバアブラツツジ 145730
ホソバアマナ 234244
ホソバアミガサノキ 1923
ホソバアメリカシャクナゲ 215387
ホソバアラゲヒョウタンボク 236125
ホソバアラリア 310199
ホソバアリノトウグサ 185014
ホソバアリノトオグサ 184997
ホソバアヲダモ 168100
ホソバアンゲロンソウ 24522
ホソバイスノキ 134947
ホソバイスビハ 164956
ホソバイスビワ 164956
ホソバイズミケンチャ 200182
ホソバイソツツジ 223909,223911
ホソバイソノキ 328671
ホソバイトヒカゲスゲ 73992
ホソハィヌカャ 82528
ホソバイヌゴマ 373142
ホソバイヌタデ 291748,309909
ホソバイヌツゲ 203690
ホソハイヌメリ 341962
ホソバイブキトラノオ 309325
ホソバイラクサ 402847
ホソバイワウメ 127916
ホソバイワギキョウ 70279
ホソバイワベンケイ 329889
ホソバウキミクリ 370030
ホソバウジロガシ 116187
ホソバウチワ 28130
ホソバウマノスズクサ 34229
ホソバウメモドキ 204087
ホソバウラジロ 394635
ホソバウラジロエノキ 394628
ホソバウリハダカエデ 3530
ホソバウルップソウ 220196,220203
ホソバウロコザミア 225624
ホソバウロコヒジキ 104850
ホソバウロコマリ 225224
ホソバウンラン 231183
ホソバエゴノキ 379380
ホソバエゾキリンソウ 356857
ホソバエゾセンキュウ 24357
ホソバエゾニワトコ 345671
ホソバエゾノコギリソウ 4009
ホソバエゾボウフウ 8818
ホソバエゾリンダウ 174004
ホソバエゾリンドウ 174004
ホソバエノキグサ 1794
ホソバエビネ 82088
ホソバェンゴサク 105599
ホソバエンゴサク 105607
ホソバオオアリドオシ 122049
ホソバオオバコ 301864
ホソバオガタマノキ 252850
ホソバオカヒジキ 344743
ホソバオキナグサ 321716
ホソバオグルマ 207165
ホソバオケラ 44208
ホソバオゼヌマスゲ 75490
ホソバオニザミア 241457
ホソバオノオレ 53612
ホソバオホバコ 301864
ホソバオホハマニラ 274799
ホソバオモダカ 342409,342421
ホソバオヤマリンドウ 173623
ホソバオンタデ 309056
ホソバガシ 116153
ホソバガシワ 323591
ホソバカハラサイコ 312453
ホソバガマ 401094
ホソバカマツカ 295674
ホソバカモメヅル 400944
ホソバカンギク 124793
ホソバガンクビソウ 77157
ホソバカンスゲ 76503
ホソバキエボシソウ 5052
ホソバキカシグサ 337330,337353
ホソバギシギシ 340269,340270
ホソバキジュラン 194157
ホソバキスゲ 191312
ホソバキツネノメマゴ 337257
ホソバギョクシンクワ 384950
ホソバギリア 175669
ホソバキンゴジカ 362478
ホソバキンゴジクワ 362478
ホソバキンポウゲ 325571
ホソバキンマウツツジ 331839
ホソバギンヨウジュ 227345
ホソバグス 91351
ホソバクスノキ 91351
ホソバクチナシ 171253
ホソバグミ 141932
ホソバクモキリサウ 232229
ホソバクララ 369010
ホソバクリガシ 79032
ホソバクロウメ 328665
ホソバケンチャ 321174
ホソバコウゾリナ 298596
ホソバコゴメグサ 160183
ホソバコザクラ 23219
ホソバコツクバネウツギ 169
ホソバコナギ 257587
ホソバコラヤナギ 343414
ホソバコリヤナギ 344194
ホソバコンギク 40846
ホソバコンロンソウ 125863
ホソバサンキライ 366324
ホソバサンザシ 109937
ホソバシオガマ 287647
ホソバシオデ 366557
ホソバシバナ 397165
ホソバシマケンザン 139261
ホソバシモツケ 371944
ホソバシモツケソウ 166073
ホソバシモバシラ 215811
ホソバシャクナゲ 331187
ホソバシャクヤク 280318
ホソバシャリンバイ 329073
ホソバシュロサウ 405606
ホソバシュロソウ 405606
ホソバシュロチクヤシ 321174
ホソバシュンラン 116887
ホソバシラカシ 116139,116143
ホソバシラカンバ 53574
ホソバシラネセンキュウ 24434
ホソバシラネニンジン 391907
ホソバシロスミレ 410370
ホソバシロダモ 264091
ホソバシンジュガヤ 354022
ホソバスゲ 75012

ホソバスズタケ 347380
ホソバスズムシサウ 232318
ホソバスヒカヅラ 235633
ホソバセンキウ 276752
ホソバセンキュウ 276752
ホソバセンダイハギ 389533
ホソバセンダングサ 54042
ホソバセンナ 78227
ホソバセンネンボク 137337
ホソバソバナ 7799
ホソバタイセイ 209229
ホソバタイム 391369
ホソバタカサゴソウ 210572
ホソバタガネソウ 76271
ホソバタコノキ 280974
ホソバタゴボウ 238207
ホソバタゴボオ 238207
ホソバタチシオデ 366488
ホソバタヌキマメ 112340
ホソバタネツケバナ 125863
ホソバタブ 240604
ホソバタマミクリ 370067
ホソバダンドク 71196
ホソバチクセツニンジン 280751
ホソバチシバリ 210623
ホソバチヂミザサ 272691
ホソバチャ 69644
ホソバチョウジソウ 20846
ホソバチョウセンギク 124862
ホソバツヅラフヂ 97933
ホソバツメクサ 32306,255598
ホソバツユクサ 100972
ホソバヅルウズ 5674
ホソバツルグミ 142008
ホソバツルツゲ 204230,204276
ホソバツルノゲイトウ 18147
ホソバツルヒキノカサ 326278
ホソバツルマメ 177751
ホソバツルメヒシバ 45821,45833
ホソバテッパウユリ 229845
ホソバテッポウユリ 229845
ホソバテンジクアオイ 288468
ホソバテンツキ 166251,166272
ホソバテンナンシャウ 33258,33332
ホソバテンナンショウ 33258,33496
ホソバトウキ 24480
ホソバトウチク 364819
ホソバトウヌマゼリ 365872
ホソバトキワアワダチソウ 368387
ホソバトキワサンザシ 322450
ホソハドクゼリ 90940
ホソバドジョウツナギ 393100
ホソバトリアシショウマ 41835
ホソバトリカブト 5546
ホソバナエビネ 65952,65966,66074
ホソバナギナタコウジュ 144093
ホソバナソモソモ 306062
ホソバナツグミ 142088
ホソバナデシコ 127634
ホソバナナメノキ 203626
ホソバナナルコビエ 151703
ホソバナミキ 355711
ホソバナルコスゲ 76687
ホソバニガナ 210522
ホソバニセジュズネノキ 122049
ホソバニンジン 35132
ホソバヌマタデ 291746,309284
ホソバヌマヤナギ 344031
ホソバノアブラススキ 372431
ホソバノアマナ 234244
ホソバノイブキシモツケ 371884
ホソバノウナギツカミ 309629
ホソバノエノキグサ 1794
ホソバノエノコログサ 361946
ホソバノオオベンケイ 200804
ホソバノカウガイゼキシャウ 213355
ホソバノギク 41282
ホソバノキソチドリ 302536,302541
ホソバノキヌタサウ 170246
ホソバノキミズ 142719
ホソバノキリンサウ 294114
ホソバノキリンソウ 294114
ホソバノクルマムグラ 170430
ホソバノクロカハヅスゲ 75941
ホソバノコウガイゼキショウ 213355
ホソバノサンジソウ 94140
ホソバノシバナ 397165
ホソバノスズメガャ 147609
ホソバノスズメガヤ 147509
ホソバノセイタカギク 227453
ホソバノセンダングサ 54042
ホソバノタイトウサンザシ 322471
ホソバノダケ 24341,276752
ホソバノチシマフウロ 174587
ホソバノチチコグサモドキ 178101
ホソバノチョウセンエンゴサク 106568
ホソバノツルリンドウ 320927
ホソバノツルリンドウ属 320924
ホソバノトリカブト 5442
ホソバノナガハグサ 305328,306033
ホソバノハナイカダ 191174
ホソバノハマアカザ 44430
ホソバノヒイラギナンテン 242534
ホソバノヒメウスユキソウ 224834
ホソバノホロシ 367262
ホソバノマンシウハナウド 192320
ホソバノミシマサイコ 63562
ホソバノミヅキンバイ 238188
ホソバノミツバヒヨドリ 158347
ホソバノヤハズユンドウ 408284
ホソバノヤマハウコ 21616
ホソバノヤマハハコ 21607,21616
ホソバノヤマハハコモドキ 21630
ホソバノヤマホウコ 21616
ホソバノヤマホオコ 21616
ホソバノヨツバムグラ 170696,170697
ホソバノヨロイグサ 24457
ホソバノロクオンソウ 117385,117387
ホソバノヲケラ 44208
ホソバハカタユリ 230019
ホソバハグマ 12633
ホソバハゴロモノキ 180655
ホソバハナウド 192284
ホソバハナガサシャクナゲ 215405
ホソバハナカズラ 5403
ホソバハブサウ 360483
ホソバハブソウ 360483
ホソバハマアカザ 44575
ホソバハマナデシコ 127634
ホソバハママツナ 379598
ホソバハランウチワ 28107
ホソバハルシャギク 104509
ホソバハンゲ 299725
ホソバハンシャウヅル 95000
ホソバヒイラギギク 305123
ホソバヒイラギナンテン 242534
ホソバヒイラギマメ 89136
ホソバヒカゲスゲ 73992,74849
ホソバヒゲナデシコ 127608
ホソバヒサカキ 160504
ホソバヒナゲンゲ 181687
ホソバヒヒラギナンテン 242534
ホソバヒメオトギリ 201903
ホソバヒメトラノオ 317924
ホソバヒメハギ 308081,308124,308359
ホソバヒメハマアカザ 87073
ホソバヒメミソハギ 19575
ホソバヒャクニチソウ 418052
ホソバヒラギツバキ 69194
ホソバヒラギナンテン 242534
ホソバヒルムシロ 312037
ホソバフウテウボク 71775
ホソバフウリンホオズキ 297652
ホソバフジボグサ 402144
ホソバフタナミソウ 354930
ホソバフヂボグサ 402144
ホソバブラッシノキ 67278
ホソバベニゴチョウ 17481
ホソバヘラオモダカ 14727
ホソパヘラオモダカ 14727
ホソバホウチャクソウ 134478
ホソバボウフウ 292843
ホソバホトトギス 396588
ホソバマツムシソウ 350129
ホソバママコナ 248208
ホソバママツナ 379598
ホソバマユミ 157699
ホソバマユミ 157884
ホソバマンテマ 363516
ボソパミクリ 370100
ホソバミシマサイコ 63813

ホソバミズ　299048
ホソバミズヒキモ　312203
ホソバミチヤナギ　308832
ホソバミヅ　299048
ホソバミヅヒキモ　312200
ホソバムカシヨモギ　150423,150732
ホソバムクイヌビワ　164630
ホソバムニンネズミモチ　229547
ホソバムラサキ　66897
ホソバムレスズメ　72322,72351
ホソバメギ　52049
ホソバモウコヤナギ　343440
ホソバモクセイ　276344
ホソバモクセイソウ　327866
ホソバモクレイシ　254304
ホソバモンパミミナグサ　176955
ホソバヤナギハナシノブ　99859
ホソバヤハズエンドウ　408580
ホソバヤブコウジ　31480
ホソバヤブレガサ　381814
ホソバヤマカウバシ　231324
ホソバヤマグワ　259099
ホソバヤマジソ　259282
ホソバヤマツツジ　330968
ホソバヤマブキソウ　200759,200760
ホソバヤロオド　55212,161618
ホソバヤロ－ド　55212,161618
ホソバユリ　230005,230009
ホソバヨシ　295914
ホソバヨツバハギ　408672
ホソバヨツバヒヨドリ　158136
ホソバヨナメ　215350
ホソバヨモギモウゴホソバヨモギ　35916
ホソバラセンサウ　399296
ホソバラセンソウ　399296
ホソバラン　417795
ホソバラン属　417698
ホソバリンドウ　173852,173858
ホソバルリザクラ　153518
ホソバルリソウ　117908
ホソバワウレン　103835
ホソバワタスゲ　152769
ホソバワダン　110615
ホソバワチガイソウ　318515
ホソバヲグルマ　207165
ホソフデラン　154872
ホソフデラン属　154871
ホソベンギリア　175671
ホソボクサヨシ　293714
ホソボザミア　417006
ホソボチカラシバ　289205,289208
ホソボノクロカハヅスゲ　74230
ホソボヒメトウ　121487
ホソミエビスグサ　360493
ホソミキンガヤツリ　118803
ホソミナズナ　72038
ホソミハギ　126424
ホソムギ　235334
ホソムギクサ　198337
ホソムギ属　235300
ホソヤマアハ　65333
ホソリンダウ　173967
ホソリンドウ　173967
ホソヰ　213362,213448
ボダイジュ　165553,391773
ホタウ　382582
ホタウガシ　240604
ホダチハグロサウ　340344
ホタルイ　352200,352202
ホタルイ属　353179
ホタルカズラ　233786
ホタルカヅラ　233786
ホタルサイコ　63728,63735,63736,63807
ホタルサウ　63807
ホタルソウ　63728,233786
ホタルナデシコ　127747
ホタルブクロ　70234
ホタルブクロ属　69870
ホタルヰ　352200
ホタルヰ属　353179
ボタン　280286
ボタンイチゲ　23767
ボタンウキクサ　301025
ボタンウキクサ属　301021
ポダンギス属　306307
ボタンギョク　182463
ボタンキンバイ　399523
ボタングサ　280207
ボタンクサギ　95978
ボタンゲシ　282522,282655
ボタンサウ　79732
ボタンザキキンバイサウ　399519
ボタンザキキンバイソウ　399519
ボタンツツジ　332144
ボタンヅル　94740
ボタンナ　59520
ボタンニンジン　292892
ボタンネコノメソウ　90391
ボタンノキ　302588,302592
ボタンバウブウ　292892
ボタンバラ　336813
ボタンフウロ　67128
ボタンボウフウ　292892
ボタンボウブウ　292892
ボタンマル　140850
ボタン属　280143
ホチイアツモリ　120464
ボチャウジ　319810
ボチャウジ属　319392
ポ－チュラカリア属　311955
ホ－チュラナスモモ　316444
ボチョウジ　319810
ボチョウジ属　319392
ホッカイトウキ　24290
ホッギョクマル　107076
ポッキリヤナギ　343396
ホッコクアザミ　92191
ボッコニア属　56023
ホッスガヤ　65464
ホッスモ　262040
ボッセラ属　53213
ホツツジ　397862
ホツツジ属　397858
ホップ　199384,199386
ホップノキ　320071
ホップノキ属　320065
ホテイアオイ　141808
ホテイアオイ属　141805
ホテイアツモリ　120396,120407
ホテイアツモリソウ　120396
ホテイアフヒ　141808
ホテイアフヒ属　141805
ホテイカズラ　294797
ホテイサウ　141808
ホテイソウ　120371,141808
ホテイチク　297203
ホテイナ　59595
ホテイラン　68538,68542,68544,184372
ホテイラン属　68531,184345
ホド　29304
ボトァ属　56031
ホドイモ　29304
ホドイモ属　29295
ホトウ　382582
ホトケギ　311389
ホトケノザ　220355
ホトケノツヅレ　220355
ポトス　147338
ポドステモン科　306688
ポトス属　313186
ホドヅラ　375343
ホトトギス　396598
ホトトギス属　396583
ポドヒルマ　306636
ポドフィラム属　306609
ポドフィルム　306636
ポドレピス属　306564
ホド属　29295
ホナガアセビ　298750
ホナガイヌビユ　18848
ホナガカワヂシャ　406968
ホナガクマヤナギ　52439
ホナガコメガヤ　249110
ホナガサウ属　373490
ホナガソウ　373511
ホナガソウ属　373490
ホナガタツナミソウ　355559
ホナガナツハゼ　403998
ポネチ　236085
ホネヌキ　204799
ポネロルキス属　310859

ホバナヘナヘナヤシ 68617
ホプマン 304340
ホフマンセジア属 197109
ホフマンニア属 197101
ポプラ 311400
ポ－ポ－ 38338
ポポ－ 38338
ホホガシハ 198698
ホホガシハノキ 198698
ホホズキ 297645
ホホヅキ 297643
ホホヅキ属 297640
ホホノキ 198698
ホホノキ属 241951
ホホバ科 364451
ボマ－レア属 56755
ボマロケファラ属 197761
ホマロメナ・ルベッセンス 197796
ホメ－リア属 197823
ホヤ 410960,410992,410994
ホヤ属 198821
ホライサウ 120371
ホラガヒサウ 205175
ボラスキア属 307169
ボラビセレウス属 94551
ボラヤシ属 381806
ポ－ランドコムギ 398939
ポリキクニス属 307859
ホリサウ 350861
ホリシャアカマメ 274392
ホリシャアカマメ属 274375
ホリシャイチゴ 338281
ポリスキアス属 310169
ポリズダ－キア属 310309
ホリドハクタス属 198491
ホリバイヨカズラ 117722
ホリバルリカコソウ 13131
ボリビヤキナ 91075
ボリビヤキナノキ 91075
ボリンクエンダイオウヤシ 337875
ホルコグロッサム属 197253
ポルテア属 311780
ホルトカズラ 154245
ホルトカズラ属 154233
ホルトカヅラ 154245
ホルトカヅラ属 154233
ホルトサウ 159222
ホルトソウ 159222
ホルトノキ 142303,142399,270099
ホルトノギ 142396
ホルトノキ科 142267
ホルトノキ属 142269
ボルネオソケイ 211846,211912
ボルネオランブタン 265137
ホルムスキオ－ルディア属 197361
ホルリミノキ 222172
ボルレア属 56688
ボレボ－ルティア属 59982
ホロ 15621
ホロギク 143896
ボロギク 263515
ホロシ 367322
ポロシリトウウチソウ 345818
ボロヂノニシキサウ 159854
ホロテンナンショウ 33304
ポロナイカンバ 53527
ホロナイブキ 292403
ボロナイブキ 292403
ボロ－ニア属 57257
ホロビンソウ 311852
ボロボロノキ 352438
ボロボロノキ科 269640
ボロボロノキ属 352430
ホロマンノコギリソウ 3923
ホロムイイチゴ 338243
ホロムイクグ 75592
ホロムイコウガイ 213530
ホロムイサウ 350861
ホロムイサウ科 350864
ホロムイサウ属 350859
ホロムイスゲ 75370
ホロムイソウ 350861
ホロムイソウ科 350864
ホロムイソウ属 350859
ホロムイツツジ 85414,239395
ホロムイリンドウ 174010
ホロムイリンドオ 174010
ボロンカズラ 285623
ボロンカヅラ 285623
ホワイトオ－ク 323625
ホワイトスブル－ス 298286
ホワイトフヤ－ 317
ホンアズサ 53610
ホンアマリリス 18865
ホンウリ 2910
ホンオニク 93054
ホンカイドウ 243702
ホンガウサウ 352869
ホンガウサウ科 399365
ホンガウサウ属 352863
ホンガウソウ科 399365
ホンガヤ 393077
ボンガリャドリギ 411048
ポンカン 93717,93728
ポンキ－ 93698
ポンキツ 93698
ホンキリシマ 331370
ポンキン 114288
ホンキンセンカ 66395
ホンキンセンカ属 66380
ホンクワンザウ 191284
ホンゴウサウ属 352863
ホンゴウソウ 352869
ホンゴウソウ科 399365
ホンゴウソウ属 352863
ホンゴウロ 229872
ホンゴシュユ 161373,161376
ホンコブナグサ 36650
ホンコンカンコノキ 177192
ホンコンシュスラン属 238133
ホンコンツバキ 69135
ホンジソ 93732
ホンシチタウイ 119160
ホンシチタウヰ 119160
ホンシャクナゲ 330953
ボンシュウギョク 284682
ホンジュラスマホガニ－ 380527
ホンシロズイセン 262433
ホンセッコク 125288
ホンソネ 77323
ホンタデ 309199
ホンダパラ 130919
ポンチエバ属 311020
ポンチカプマウ 336736
ポンチカマウ 336736
ホンツガ 399923
ホンッゲ 64345
ホンツゲ 64303
ホンツツジ 332104
ポンデロ－ザ 93712
ポンデロ－サパイン 300153
ポンデロ－サマツ 300153
ボンデンカ 402267
ボンテンカ属 402238
ボンデンクワ 402267
ボンテンクワ属 402238
ボントクタデ 291916,309116,309644
ホンドミヤマネズ 213728
ホンバガシ 116187
ホンバコバンモチ 142351
ホンバサルナシ 6697
ホンバシデ 77346
ホンバツルアヅキ 293999
ホンバルリダマノキ 122028
ホンヒ 85310
ホンブナ 162368
ボンベイ麻 194779
ボンボリトウヒレン 348596
ホンマキ 306457,352856
ホンミカン 93724
ホンモンジスゲ 75792
ホンヤブラン 232631
ホンリュゥ 192442
ホンワスレグサ 191284
マアザミ 92390
マアニ 32336
マアブラキリ 212127
マイエロフィツム属 252697
マイカイ 336730
マイクサ 98159
マイグサ 98159

マイクジャク 3038
マイコアジサイ 200090
マイサギソウ 302404
マイズルテンナンショウ 33349
マイタジャク 3035
マイヅルソウ 242672,242680
マイヅルソウ属 242668
マイヅルテンナンショウ 33349
マイデンオ-ストラリアワイルドライム 253291
マイハギ 98159
マイフェニア属 242717
マイフェニオプシス属 242720
マイボシ 244005
マイヤ-レモン 93631
マイラチャ 2984
マイワブキ 349843
マウコシホガマ 287209
マウセングサ 138348
マウセンゴケ 138348
マウセンゴケ科 138369
マウセンゴケ属 138261
マウソウチク 297266,297306
マウンティン-サワ-ソップ 25866
マウンテンライム 93868
マエダカン 93457
マオ 56145,56229
マオウ 146253
マオウヒバ属 67405
マオウマル 140882
マオウ科 146270
マオウ属 146122
マオラン 191317,295600
マオ属 56094
マカイロフィルム属 240516
マカエロケレウス属 240511
マガオニザミア 241464
マカカヤ属 240735
マガクチヤシ属 97072
マカゴ 264090
マカススキ 12789,12794
マガタマ 233624
マガタマヤナギ 343027
マカダミア 240209
マカダミア属 240207
マカバ 53520
マガミバナ 203184
マガミバナ属 203181
マカヤキ 70507
マカラスムギ 45566
マガリナデシコ 61126
マガリバナ 203184
マガリバナ属 203181
マガリミイヌガラシ 336191
マ-ガレット 32563
マ-ガレットヒメトウ 121514
マカロニコムギ 398900
マカンバ 53400,53421
マカンマキ 306469
マキ 113685,306457,306469,324384,352856
マキエハギ 227009
マキエマル 234924
マギク 81020
マキグサ 217361
マキシミリヤンヤシ属 246731
マキシラ-リア属 246704
マギヅルザンセウ 417329
マキノスミレ 410754
マキノハヒゴタイ 348757
マキノハヨモギ 36348
マキバアスパラガス 39009
マキバクサギ 96028
マキバクロカワズスゲ 75690
マキバヌマハコベ 258258
マキバブラッシノキ 67291
マキバブラッシノキ属 67248
マキヤマザサ 347242
マキ科 306345
マキ属 306395
マクシミリァンヤシ属 246731
マクズ 321441
マクナブイトスギ 114721
マグハ 259067
マクラデニア属 240901
マクロレリア属 241144
マグワ 259067
マクワウリ 114213
マ-ケサズヤシ属 288038
マケンマル 140867
マコウギ 52436
マコデス属 240860
マゴノテ 138348
マコモ 418080,418095
マコモ属 418079
マゴヤシ 247425
マサキ 157601,157617,157631,157647
マサキカヅラ 157473
マサキズル 393616
マサキヅル 393621
マサキノカズラ 393616
マサキボク 171145
マサゴヤシ 252636
マシカクイ 143374
マシカクヰ 143374
マシケオトギリ 202224
マシケゲンゲ 279162
マシュウヨモギ 36428
マジョラム 274224
マシロラン 98634
マジンマル 284674
マスアラマル 140894
マスクサ 74644,119208
マスクサスゲ 74644
マスクメロン 114189
マズス属 246964
マスティクス 300994
マスデバリア属 246101
マスノグサ 268986,268999
マスノグサ属 268957
マスノスゲ 74333
マスノスミレ 410412
マスリマヘガミ 378024
マスリマヘガミ属 377969
マセイカク 198080
マセレウス属 240511
マダイオウ 340114
マタイセイ 209229
マダガスカルシタキソウ 245803
マダケ 297215,297435
マダケ属 297188
マタザキセキコク 125158
マタジイ 233193
マタジヒ 233193
マタタビ 6682
マタタビ科 6728
マタタビ属 6513
マタデ 309199
マダノキ 391739
マタバオニザミア 241454
マタバピナンガ 299660
マダラジソ 290956
マダラツボスミレ 410738
マダラナガバノタチツボスミレ 410334
マダラハウチワマメ 238468
マタラピナンガ 299667
マチク 125482
マチク属 125461
マ-チュ-スジュロ 393813
マチン 378836
マチン科 235251
マツアラシ 116648
マツカサアナナス 2625
マツガサギク 228519
マツカサススキ 353624
マツカサバ-レリア 48241
マツカサヒジキ 104755
マツガスミ 244198
マツカゼサウ 56392
マツカゼサウダ 56392
マツカゼサウ属 56380
マツカゼスゲ 76123
マツカゼソウ 56382,56383,56386,56392
マツカゼソウ属 56380
マツカナ属 246590
マツグミ 385202
マツグミ属 385186
マツグミ属 236507
マツゲイブキヌカボ 254517
マツゲカヤラン 342022
マツゲヒメハギ 308125

マツゲボタン 65835
マツゲボタン属 65827
マツダエビネ 65915
マツダオサラン 148754
マツダテンナンシャウ 33406
マツダヒメラン 110648
マツダミズ 298968
マツダミヅ 298968
マツダモクレイシ 254293
マツダヲサラン 148754
マツテニッケイ 264015
マツナ 379531
マツナモドキ 344538
マツナ属 379486
マツノキハダ 331593
マツノハマンネングサ 356778
マツノハラン 197265
マツノハラン属 197253
マツバイ 143022,143026,143029
マツバウド 39120
マツバウミジグサ 184940
マツバウンラン 230931,267347
マツバギク 220674,256719
マツバギク属 220464,251025
マツバギリア 175692
マツバキンポウゲ 326303
マツバコエンドロ 116384
マツバサワギク 358408
マツバシバ 33775,33817
マツバシバ属 33730
マツバスゲ 73885
マツバゼリ 116384
マツバゼリ属 29312
マツハダ 298226
マツバトウダイ 158730,158857
マツバナデシコ 231961
マツバニンジン 7765,231961
マツバハルシャギク 188437
マツバハルシャギク属 188393
マツバボタン 311852
マツバユリ 229803
マツバヰ 143022,143029
マツブサ 351073
マツブサ 351093
マツブサ科 351121
マツブサ属 351011
マツフジ 351073
マツボヤ 385202
マツマエスゲ 75189
マツマエベニソメイ 83123
マツマヘスゲ 75189
マツムシサウ 350175
マツムシサウ科 133444
マツムシサウ属 350086
マツムシソウ 350175
マツムシソウ科 133444
マツムシソウ属 350086
マツムライヌノヒゲ 151384
マツムラカヅラ属 113566
マツムラサウ 392424
マツムラサウ属 392423
マツムラソウ 392424
マツムラソウ属 392423
マツモ 83545
マツモ-スツリ- 360576
マツモト 364064
マツモトセンノウ 364064
マツモ科 83539
マツモ属 83540
マツユキオキナグサ 24071
マツユキソウ 169719
マツユキソウ属 169705
マツヨイグサ 269475,269509
マツヨイグサ属 269394
マツヨイセンノウ 363673
マツヨヒグサ 269475
マツヨヒグサ属 269394
マツヨヒセンノウ 363673
マツラコゴメグサ 160186
マツラニッケイ 91351
マツラン 171869,171870
マツラン属 342015
マツリ 211990
マツリカ 211990
マツリクワ 211990
マツ科 299594
マツ属 299780
マテ 204135
マデイラカズラ 26265
マテガシ 233193
マ-テコダケ 47516
マテチャ 204135
マテバシイ 233193
マテバシイ属 233099,285203
マテバシヒ 233193
マテバシヒ属 233099
マテンリュウ 182455
マテンロウ 34422
マトア 310823
マドカズラ 258170
マドリ-ドチャヒキ 60820
マナギョリュウ 383553
マナックヤシ属 68639
マナラ 324190
マニオク 244507
マニサンアザミ 91800
マニホットゴムノキ 244509
マニホットノキ 244507
マニホット属 244501
マニラアサ 260275
マニライトバショウ 260275
マニライトバセウ 260275
マニラエレミ 71004
マニラコチョウラン 293578
マニラコテフラン 293582
マニラサンユウカ 154192,382822
マニラヤシ 405258
マネキグサ 237958
マネキグサ属 237957
マネキシンジュガヤ 354175
マネッチア属 244358
マノセカワゴケソウ 93972
マハイカク 198044
マバマスゲ 76051
マヒエフハギ 98159
マヒグサ 98159
マヒヅルサウ 242672
マヒヅルサウ属 242668
マヒヅルテンナンシャウ 33349
マヒハギ 98159
マヒハギモドキ 98156
マベシイ 324283
マホガニ- 380528
マホガニ-属 380522
マボケ 84553
マホソ 324384
ママコナ 248201
ママコナ属 248155
ママコノキ 191173
ママコノシリヌグイ 309772
ママコノシリヌグヒ 309772
ママッコ 191173
ママン 116829
マムシアルム 37015
マムシカヅ 61107
マムシグサ 33369,33496,403028
マムシサウ 33496
マムシラヅ 61107
マムバサンキライ 366583
マメ 177777
マメアイ 206669
マメアカナス 239158
マメアサガオ 207921
マメアヰ 206669
マメイヌツゲ 203672
マメイバラ 64990
マメガキ 132264
マメカミツレ 107755
マメギ 61935
マメキリン 106958
マメキンカン 167499,167516
マメグミ 142083
マメグンバイナズナ 225475
マメグンバイナズナ属 225279
マメグンバイナヅナ 225475
マメコケ 356917
マメザクラ 83224
マメスゲ 75926
マメズタカズラ 134037
マメダオシ 114972
マメダフシ 114994

マメツゲ　203670,203672
マメヅタカズラ　134037
マメヅタカヅラ　134037
マメヅタカヅラ属　134027
マメヅタラン　62696
マメヅタラン属　62530
マメトマト　239158
マメナシ　323116
マメバス　373556
マメヒサカキ　160464
マメフジ　373556
マメブシ　373556
マメフヂ　373556
マメヤナギ　344045,373556
マメラン　62696
マメ科　224085
マヤビス　362226
マヤブシキ　368913
マヤプシキ科　368926
マヤブシキ属　368910
マヤヤシ属　272756
マヤラン　116975
マヤンテムム属　242668
マユハキグサ　152759,152792
マユハキサウ　88289
マユハケオモト　184347
マユハケオモト属　184345
マユミ　157879
マユミ　157559
マユミ属　157268
マヨラナ　274224
マヨラ－ナ属　242792
マヨワセアザミ　91704
マライキンカン　167513
マライピナンガ　299668
マラタケリグサ　147048,147050
マラッカノキ　296554
マラバルノボタン　248762
マラヤヒメヤシ属　203501
マランタ属　245013
マリ　211990
マリアナボウラン　238324
マリアビロウ　234185
マリカ　211990
マリシャジン　7707
マリヤマザミ　364360
マリュウギョク　264338
マリヨモギ　35220
マリランデイカポプラ　311155
マルガイ　310823
マルガタダマ　233603
マルキンカン　167503
マルグハ　259085
マルグワ　259152
マルザキオグルマ　207051
マルザキズイセン　262457
マルジクホエア属　188190
マルスグリ　334250
マルスゲ　352278
マルスピウムシタン　320307
マルチニア属　245977
マルドコロ　131501
マルバアアワユキセンダングサ　54061
マルバアオダモ　168100
マルバアカザ　86892,86893
マルバアカバナシキミ　204484
マルバアキグミ　142238
マルバアサガオ　208120
マルバアサガオガラクサ　161426
マルバアサガホカラクサ　161426
マルバアズキナシ　369308
マルバアヂサイ　199910
マルバアヂサヰ　199910
マルバアメリカアサガオ　207839
マルバアリドオシ　122056
マルバアルム　37027
マルバイチゴ　338516,339014
マルバイヌタデ　291827
マルバイノコヅチ　4287
マルバイワキンバイ　312500
マルバインドゴムノキ　164933
マルハウコギ　143685
マルバウスゴ　403996
マルハウチワヤシ　228741
マルバウツギ　127072,127099
マルバウメモドキ　204008
マルバエゾニュウ　24345
マルバエンゴサク　105602
マルバエンシュウハグマ　12628
マルバオオバクサフジ　408556
マルバオケラ　44205,44218
マルバオゴケ　409477
マルバオニウツギ　127049
マルバオモダカ　66343
マルバオモダカ属　66337
マルバカイドウ　243672
マルバカエデ　2939
マルバカモメヅル　409449
マルバカワヂシャ　407029
マルバガンピ　414250
マルバキカシグサ　337391
マルバギシギシ　278577
マルバギシギシ属　278576
マルバキッコウハグマ　12610
マルバキブシ　373562
マルバキンゴジカ　362523
マルバキンゴジクワ　362523
マルバギンバイソウ　123657
マルバキンレイカ　285830
マルバクサイチゴ　338518
マルバグス　91405,91429
マルバクスノキ　91429
マルバクチナシ　171253,171310
マルバクマツヅラ　405889
マルバグミ　142075
マルバクロウメモドキ　328752
マルバケスミレ　409834,410641
マルバゴウシュウアオギリ　376102
マルバコウツギ　126855
マルバコゴメグサ　160184
マルバコナラ　324395
マルバゴマギ　408134
マルバコマツヨイグサ　269457
マルバコンロンクワ　260499
マルバコンロンソウ　73000
マルバサイコ　63807
マルバサカキ　96598
マルバサッキ　331839
マルバサツキ　330629,330675,331843
マルバサンキライ　366583
マルバサンシュユ　105147
マルバシクラメン　115949
マルバシマザクラ　187592
マルバシモツケ　371823
マルバシャジン　7798
マルバシャリントウ　107504
マルバシロモジ　231457
マルバジンジソウ　349208
マルバスイバ　340254
マルバストラム属　243877
マルバスミレ　410131
マルバセウベンノキ　400536
マルバセスジグサ　11360
マルバダイオウ　329386,329388
マルバタイミンタチバナ　261632
マルバダイワウ　329388
マルバタウコギ　53805
マルバタケニグサ　240767
マルバタケハギ　18289
マルバダケブキ　229005,229016
マルバタチスミレ　410246
マルバタバコ　266053
マルバタマノカンザシ　198639
マルバダモ　113436
マルバタラヨウ　203538
マルバチシャノキ　141629,141633
マルバチトセラン　346163
マルバチャメルソウ　256032
マルバチャルメルソウ　256032
マルバックシハギ　226680
マルバックバネウツギ　416853
マルバツユクサ　100940
マルバツルグミ　142181
マルバツルツゲ　204226
マルバツルノゲイトウ　18135
マルバツルマサキ　157517
マルバデイコ　154648
マルバテイショウソウ　12635
マルバテイショオソウ　12636
マルバトウキ　229384,229385
マルバトウキ属　229283

マルバトキハアケビ 374425
マルバトゲチシャ 219508
マルバドコロ 131501
マルバナツグミ 142092
マルバニクケイ 91318
マルバニッケイ 91318
マルバニワトコ 345685
マルバニンジン 7830
マルバヌスヒトハギ 200739
マルバネコノメ 90442
マルバネコノメソウ 90442
マルバネズミモチ 229513
マルバノイチヤクサウ 322872
マルバノイヂャクサウ 322852
マルバノイチヤクソウ 322859,322872
マルバノウマノスズクサ 34154
マルバノウメモドキ 204008
マルバノウンラン 116736
マルバノキ 134010
マルバノキ属 134009
マルバノコゴメオウギ 42145
マルバノサカキ 96598
マルバノサワトウガラシ 123729
マルバノシャリンバイ 329086
マルバノチョウリョウソウ 229016
マルバノニンジン 7830
マルバノバツユクサ 100940
マルバノハマシャジン 7868
マルバノヒゴタイサイコ 154337
マルバノフナバラソウ 409483
マルバノホロシ 367120,367365
マルバノマテガシ 233219
マルバノミミカキグサ 403351
マルバハギ 226746
マルバハクカ 245770
マルバハダカホオズキ 400001
マルバハタケムシロ 234615
マルバハタザオ 30286
マルバハタザホ 30286
マルバハッカ 250457
マルバハナタマ 264222
マルバハマカキラン 147198
マルバハンノキ 16469
マルバヒイラギモチ 203668
マルバヒゴタイサイコ 154337
マルバヒメツバキ 350946
マルバヒメノアズキ 333324
マルバビユ 208349
マルバビユ属 208344
マルバビラウ 234194
マルバヒレアザミ 91999
マルバビロウ 234190
マルバフクオウソウ 313782
マルバフジバカマ 11154,158354
マルバフタバムグラ 370829
マルバフユイチゴ 339014
マルバブラッシノキ 67286
マルバマツグミ 385204
マルバママコナ 248204
マルバマンサク 185118
マルバマンネングサ 356917
マルバミゾカクシ 234884
マルハミツバアケビ 13238
マルバミツモト 312478
マルバミミカキグサ 403351
マルバミヤコアザミ 348516
マルバミヤマシグレ 408198
マルバムシトリスミレ 299743
マルバモグセイ 276313
マルバヤナギ 311480,343185
マルバヤナギザクラ 161749
マルバヤブニッケイ 91280
マルバヤブマオ 56302
マルバヤマズソウ 218345
マルバユウカリ 155710
マルバヨノミ 235696
マルバルコウ 323459
マルバルコウソウ 323459
マルバルリミノキ 222095,222305
マルバヲケラ 44205
マルピギア属 243519
マルブシュカン 93603
マルブッシュカン 93550
マルホハリイ 143261
マルミイチゴ 339290
マルミウマゴヤシ 247397
マルミオオバコ 301884
マルミカンアオイ 37736
マルミカンコノキ 177170
マルミキンカン 167503
マルミクダモノトケイ 285637
マルミケンチャ属 181915
マルミシウカイダウ 49626
マルミスズメノヒエ 285507
マルミスブタ 55913
マルミノウルシ 158809
マルミノキンシバイ 201899
マルミノクチナシ 171313
マルミノゴエフ 300130
マルミノシバナ 397159
マルミノダケカンバ 53427
マルミノハシカグサ 262961
マルミノヘビノボラズ 52151
マルミノヤマゴバウ 298114
マルミノヤマゴボウ 298114
マルミパンノキ 36913
マルメ 116546
マルメル 116546
マルメロ 116546
マルヤマカンコノキ 60189
マルヤマシウカイダウ 49848
マルヤマシュウカイドウ 49848
マルヤマシュウカイドオ 49848
マルヤマタニタデ 91504
マル属 243994
マレ-クサギ 96044
マレ-ジンコウ 29977
マレノプンティア属 245152
マレフォ-ラ属 243252
マレ-フトモモ 382606
マレ-ヤマバショウ 260199
マレ-ルュウガン 265136
マロニエ 9701
マロ-ペ属 243472
マワウ 146183,146253
マワウ属 146122
マヲ 56145,56229
マヲ属 56094
マンキチイヌツゲ 203670
マンキチスギ 113710
マンキツ 93844
マンキンアフヒ 195257
マンキンヤマサギサウ 58389
マングロ-ブ 329765
マンゴ- 244397
マンゴウ 244397
マンゴウノキ 244397
マンゴウノキ属 244386
マンゴウ属 244386
マンゴクドジョウツナギ 177557
マンゴスタン 171136
マンゴスタンノキ 171136
マンゴズチウ 171136
マンゴスチン 171136
マンゴスチン属 171046
マンゴ-メロン 114200
マンゴ-属 244386
マンザイギク 67314
マンサク 185104
マンサクヒャクニチソウ 418058
マンサク科 185089
マンサク属 185092
マンシウアカバナ 146606
マンシウアカマツ 300243
マンシウアキギリ 345450
マンシウアケボノソウ 380269
マンシウアサギリサウ 35505
マンシウアサギリソウ 35505
マンシウアンズ 34443
マンシウイタヤ 3714
マンシウイチゲ 23713
マンシウイチゴツナギ 305733
マンシウイヌノヒゲ 151536
マンシウイモ 16512
マンシウイワタデ 308730
マンシウイワヨモギ 36177
マンシウウコギ 143665
マンシウウマゴヤシ 247452
マンシウウマノスズクサ 34268
マンシウウリハダ 3665
マンシウウリハダカエデ 3665

マンシゥエンゴサク　105682
マンシウオキナグサ　321676
マンシウオトコヨモギ　35674
マンシウカエデ　3129
マンシウカキツバタ　208706
マンシウカヘデ　3129
マンシウカモジグサ　335461
マンシウカモノハシガヤ　57568
マンシウカラマツ　221919
マンシウカワホネ　267320
マンシウガンクビソウ　77191
マンシウギシギシ　339979
マンシウキスゲ　191284
マンシウクサイ　213502
マンシウクサイチゴ　167641
マンシウクルマバナ　96972
マンシウグルミ　212621
マンシウクロウメモドキ　328851,328853
マンシウクロカワスゲ　75737
マンシウクロスグリ　334147
マンシゥクロヒナスゲ　76628
マンシウクロマツ　300243,300251
マンシウコリヤナギ　343569
マンシウシホガマ　287411
マンシウシモツケ　371897
マンシウシャクヤク　280149
マンシウシラカバ　53581
マンシウシラカンバ　53503,53572
マンシウシラネアザミ　348355
マンシウシロカネソウ　210281
マンシウシロタンポポ　384760
マンシゥスズシロ　154424
マンシウスズメノテッポウ　17512
マンシウズミ　243649
マンシウダイゲキ　158857
マンシウタカサゴソウ　210567
マンシウタチャナギ　344211
マンシウタツノヒゲ　128019
マンシウタヌキモ　403382
マンシゥタネツケバナ　72971
マンシウタムラソウ　361023
マンシウダラ　30634
マンシウツリガネニンジン　7739
マンシウトリカブト　5294
マンシウナナカマド　369484
マンシウナミキソウ　355739
マンシウニガクサ　388298
マンシウニワトコ　345624
マンシウヌカボタデ　309121
マンシウネコノメソウ　90400
マンシウノガリヤス　65277
マンシウノブドウ　20348
マンシウバイカウツギ　294536
マンシウバイクワウツギ　294536
マンシウハイスグリ　334144
マンシウハシドイ　382277
マンシウハネガヤ　4148
マンシウハンノキ　16401
マンシウヒカゲスゲ　75055
マンシウヒカゲミツバ　299366
マンシウヒキオコシ　209717
マンシウヒゴタイ　348788
マンシウヒゴタヒ　348788
マンシウヒトツバヨモギ　35648
マンシウヒナノウスツボ　355182
マンシウヒメビシ　394496
マンシウヒョウタンボク　235945
マンシウヒルガオ　68701
マンシウフスマ　374858
マンシウボゥフウ　228597
マンシウボダイジュ　391760
マンシウホタルイ　352205
マンシウボタンヅル　94778
マンシウホトトギス　396601
マンシウマツモ　83568
マンシウミシマサイコ　63594
マンシウミチヤナギ　308850
マンシウミヅナラ　324173
マンジウミツバ　345998
マンシウミツバトリカブト　5639
マンシウムカシヨモギ　150732
マンシウムグラ　170474
マンシゥムレスズメ　72279
マンシウヤチマタブシ　5469
マンシウリンドウ　173625
マンシウレンギョウ　167450
マンシエウクロマツ　300243
マンシュウアヤメ　208797
マンシュウアンズ　34443
マンシュウウシタキソウ　91537
マンシュウウマノスズクサ　34268
マンシュウカメバヒキオコシ　209665
マンシュウカラマッ　221919
マンシュウクルマバナ　96972
マンシュウグルミ　212621
マンシュウクロカワスゲ　75737
マンシュウスイラン　195655
マンシュウズミ　243555
マンシュウスミレ　409914
マンシュウタカサゴソウ　210567
マンシュウタツノヒゲ　128019
マンシュウタムラソウ　361023
マンシュウダラ　30634
マンシュウツリガネニンジン　7739
マンシュウヌカボタデ　309121
マンシュウノガリヤス　127197
マンシュウバイカウツギ　294536
マンシュウハシドイ　382277
マンシュウハンノキ　16348
マンシュウヒキオコシ　209717
マンシュウヒトツバヨモギ　35648
マンシュウボダイジュ　391760
マンシュウマメナシ　323110
マンシュウヤマサギソウ　302290
マンジュカ　76813
マンジュギク　383090
マンジュギク属　383087
マンジュギョク　182481
マンジュクワ　76813
マンジュシャゲ　239266
マンショウハシドイ　382277
マンスウコマツナギ　206002
マンセンイブキトラノオ　309373
マンセンウド　30618
マンセンオオキヌタソウ　337916
マンセンカラマツ　388423
マンセンキヌタソウ　337916
マンセンスハマソウ　192138
マンセンビシ　394526
マンゾウ　186545
マンソニアナスモモ　316583
マンダリン　93717
マンテマ　363483
マンテマモドキ　363409
マンテマ属　363141
マンドラゴラ　244346
マンナ　14631
マンナノキ　168041
マンネングサ　356841,356884
マンネングサ属　356467
マンネンスギ　113700,113717
マンネンネギ　15293
マンネンラウ　337180
マンネンラン　10787
マンネンラン属　169242
マンネンロウ　337180
マンネンロウ属　337174
マンボウ　359880
マンポウ　359880
マンメア属　243979
マンリャウ　31396,31408
マンリャウイヌツケ　203571
マンリャウカナメモチ　295775
マンリャウ属　31347
マンリョウ　31396,31408
マンリョウ属　31347
マンルサウ　337180
マンルソウ　337180
ミアケザサ　347252
ミアライ　93633
ミイロスミレ　410677
ミウライボタ　229652
ミウラスズメノオゴケ　409557
ミエミツヤヤシ　263843
ミオソティディウム属　260734
ミカイダウ　243657
ミカイドウ　243657
ミガエリスゲ　75711
ミカエリソウ　228011
ミカエリヒヨドリ　158038
ミカヅキイトモ　417060

ミカヅキグサモドキ　333478
ミカドガシワ　241781
ミカドギリ　285960
ミカドニシキ　16894,159204
ミカドユリ　229828
ミカハスブタ　55924
ミガヘリスゲ　75711
ミカワイヌノヒゲ　151377
ミカワオオイトスゲ　73548
ミカワザサ　347277
ミカワシオガマ　287592,287593
ミカワショウマ　41836
ミカワシンジュガヤ　354159
ミカワスブタ　55924
ミカワタヌキモ　403191
ミカワチャルメルソウ　256020
ミカワツツジ　330975
ミカワナルコ　75639
ミカン　93717
ミカン科　341086
ミカン属　93314
ミギハハコベ　67376
ミギハヒゴタイ　348201
ミギワガラシ　336246
ミギワトダシバ　37424
ミクニテンナンショウ　33459
ミクラザサ　347228
ミクラジマトウヒレン　348881
ミクラトンボソウ　302440
ミクラントセレウス属　253092
ミクリ　370050,370075,370102
ミクリガヤ　333628
ミクリガヤツリ　118784
ミクリスゲ　74309
ミクリゼキショウ　213093
ミクリ科　370018
ミクリ属　370021
ミクロケリア属　253315
ミクロシカス属　253399
ミクロシトラス　253285
ミクロシトラス属　253284
ミクロプテルム属　253816
ミクロプンティア属　253832
ミコシガヤ　75500
ミコシギク　227453
ミコシグサ　174755
ミコニア属　253003
ミコミスゲ　75500
ミサオノキ　12519,12521,325309,325393
ミサオノキ属　325278
ミサキサウ　397165
ミサキシバ　91318
ミサキシラタマ　319833
ミサキソウ　397165
ミザクラ　83284
ミササゲ　409096
ミサヤマコシアブラ　143617
ミサヤマチャヒキ　190155
ミサヤマチャヒキ属　190129
ミサワアザミ　91702
ミサヲノキ　12519
ミサヲノキ属　325278
ミサンザシ　109933
ミシマサイコ　63625,63813,63821
ミシマサイコ属　63553
ミシマママコナ　248195
ミジンコウキクサ　414653,414664
ミジンコウキクサ属　414649
ミズ　142694,299024,299026
ミズアオイ　223131,257572,316824
ミズアオイ科　311019
ミズアオイ属　257564
ミズイ　213138
ミズイモ　66600
ミズイモ属　66593
ミズエノコロ　200827,200840
ミズオオバコ　277369
ミズオオバコ属　277361
ミズオグルマ　385908
ミズオジギソウ　265233
ミズオジギソウ属　265228
ミズオトギリ　394811
ミズオトギリ属　394807
ミズガシワ　250502
ミズガヤツリ　212778
ミズカンナ　388385
ミズカンナ属　388380
ミズガンピ　288927
ミズキ　57695,380461
ミズキカシグサ　337378,337389
ミズギク　207082
ミズギボウシ　198624,198634
ミズキンバイ　238152,238230
ミズキ科　104880
ミズキ属　104917
ミスクサ　401112
ミズクサ　57695
ミズコナラ　323584
ミズザゼン　66600
ミズサンザシ　29660
ミスジナガハグサ　305588,305605
ミズシマアジサイ　199789
ミズスギナ　337349
ミズズラン　22326
ミスズラン属　22320
ミズタカモジ　144336
ミズタガラシ　72873
ミスタシジューウル属　261772
ミズタンポポ　384572
ミズタネツケバナ　72979
ミズタビラコ　397386
ミズタビラコ属　397380
ミスタマアナナス　264421
ミズタマソウ　91575,151257
ミズタマソウ属　91499
ミスヂツヅラフヂ　290841
ミズチドリ　302355
ミスヂミズ　298971
ミスヂミヅ　298971
ミズトウズル　65697
ミズトラノオ　139585,139602
ミズトラノオ属　139544
ミズトンボ　183797
ミズトンボ属　183389
ミズナ　59438,142694
ミズナラ　324190
ミズネコノオ　139585
ミズハコベ　67376,67382
ミズハシカ　380461
ミズバショウ　239518
ミズバショウ属　239516
ミズヒキ　26624,291755
ミズヒキモ　312203
ミズヒナゲシ　200244
ミズヒマワリ　182537
ミズビワソウ　120509
ミズビワソウ属　120501
ミズブキ　160637
ミズマツバ　337368,337382
ミスミイ　143042,143149
ミスミギク　143464
ミスミギク属　143453
ミスミグサ　143464
ミスミソウ　192139
ミスミヰ　143149
ミズメ　53465
ミズユキノシタ　238191
ミズユキノシタ属　238150
ミズ-リマツヨイグサ　269464
ミズレンブ　156134,382474
ミズ属　298848
ミセバヤ　200798
ミセバヤベンケイ　200777
ミゼレモン　285660
ミセンアオスゲ　75139
ミセンギク　207219
ミゾイチゴツナギ　305278
ミゾカウシュ　259292
ミゾカウジュ　259292
ミゾカクシ　234363
ミゾカクシ科　234890
ミゾカクシ属　234275
ミゾカミツレ　397969
ミソガワソウ　265064
ミソクサ　269613
ミゾコウジュ　345310
ミゾコオジュ　345310
ミゾサデクサ　309356
ミゾソバ　309877
ミゾダイオウ　340082
ミゾダイワウ　340082

ミソナオシ　269613
ミゾナシオタフクグルミ　212622
ミソナヲシ　269613
ミソハギ　240030,240068
ミゾハギ　240030
ミソハギ科　240020
ミソハギ属　240025
ミゾハコベ　142578,142579
ミゾハコベ科　142552
ミゾハコベ属　142554
ミソブタ　125644
ミゾホオズキ　255253
ミゾホオズキ属　255184
ミソボシラン　405028,412388
ミソボシラン属　412385
ミゾホホヅキ　255253
ミゾホホヅキ属　255184
ミソマメ　177750,177777
ミソメキリシマ　330984
ミソ科　218697
ミタケウツヅ　5275
ミタケスゲ　75350
ミタニムプ　93536
ミタマスギ　113711
ミダレフブキ　94555
ミダレユキ　10852
ミチシバ　147671,249060
ミチタネツケバナ　72802
ミチノクアキグミ　142230
ミチノクエンゴサク　106217
ミチノクククワガタ　317957
ミチノクコゴメグサ　160216
ミチノクコザクラ　314461
ミチノクサイシン　37619
ミチノクスミレ　410517
ミチノクナシ　323330
ミチノクネコノメソウ　90389
ミチノクハリスゲ　74029
ミチノクホタルイ　352242
ミチノクホンモンジスゲ　76366
ミチノクメタカラコオ　229034
ミチノクヤマタバコ　229034
ミチノクヨロイグサ　24456
ミチハコベ　125872
ミチバタガラシ　336193
ミチヤナギ　308816,308863
ミヂンコウキクサ　414653
ミヂンコウキクサ属　414649
ミヂンコザクラ　291482,291483
ミヂンコザクラ属　291481
ミヅ　299024,299026
ミヅアサガホ　277369
ミヅアフヒ　257572
ミヅアフヒ科　311019
ミヅアフヒ属　257564
ミヅイモ　66600
ミヅイモ属　66593
ミヅウルシ　356917,357189
ミヅエノコロ　200827
ミヅオオバコ　277369
ミヅオオバコ属　277361
ミヅオトギリ　394811
ミヅオトギリ属　394807
ミヅオホバコ　277369,277397
ミヅオホバコ属　277361
ミヅガシハ　250502
ミヅガシハ属　250493
ミツガシワ　250502
ミツガシワ科　250492
ミツガシワ属　250493
ミツカドシカクイ　143415
ミヅガヤ　354583
ミヅガヤツリ　212778
ミヅガヤ属　354579
ミヅガンビ　288927
ミヅガンビ属　288926
ミヅキ　57695
ミヅキカシグサ　337368
ミヅキ科　104880
ミヅキ属　104917
ミヅクサ　138482
ミヅザクラ　83224
ミヅザゼン　66600
ミヅサハタウカラシ　180308
ミツシベマウセンゴケ　138350,138354
ミツジロ　348087
ミヅタカラシ　262722
ミヅタガラシ　72873
ミヅタハコベ　142578
ミヅタマサウ　91524,151268,151487
ミヅタマサウ属　91499
ミヅチドリ　302355
ミツデ　125644
ミツデイハガサ　372117
ミツデカエデ　2896
ミツデコトジソウ　345258
ミツデシモツケ　372117
ミツデモミジ　2896
ミヅトウヅル　65697
ミヅトラノヲ　139602
ミヅトラノヲ属　139544
ミヅトンボ　183921
ミヅナ　59438,393269
ミツナガシハ　125644
ミツナガシワ　125644
ミヅナギ　257583
ミヅネ　53465
ミヅネコノヲ　139585
ミツバ　113872,113879
ミツバアケビ　13238
ミツバイチゴ　338985
ミツバイヌノフグリ　407421
ミツバイハガサ　372117
ミツバイワガサ　371844,372117
ミツバウコギ　143694
ミヅバウシ　298878
ミツバウツギ　374089
ミツバウツギ科　374114
ミツバウツギ属　374087
ミツバオウレン　103850
ミツバオオハンゴンソウ　339622
ミツバオランダフウロ　153788
ミツバカイダウ　243717
ミツバカイドウ　243717
ミツバカシラ　387769
ミツバカシラ属　387740
ミツバカズラ　387766
ミツバガヤ　226005
ミツバカンチョウジ　57957,57958
ミヅハギ　288927
ミツバグサ　299395
ミツバグサ属　299342
ミツバゴシュユ　249168
ミヅハコベ　67393,142574
ミヅハコベ科　67334
ミヅハコベ属　67336
ミツバコンロンソウ　72680
ミツバザサ　242706
ミツバサワヒヨドリ　158207
ミツバシモツケソウ　175766
ミツバシモツケソウ属　175760
ミツバショウマ　91024
ミツバセイヨウソウフウチョウソウ　95687
ミツバゼリ　113872,113879
ミツバゼリ属　113868
ミツバチクセッニンジン　280793
ミツバツチグリ　312565
ミツバツツジ　330577
ミツバヅル　411918
ミツバテンナンショウ　33535
ミツバドコロ　131630
ミツバナ　3243
ミヅハナビ　118982,119687
ミツバナンテンハギ　408651
ミツバノコマツナギ　206684
ミツバノバイカオウレン　103853
ミツバノブドウ　20402
ミツバハマゴウ　411464
ミツバハンショウヅル　95061
ミツバヒヨドリ　158347
ミツバヒヨドリバナ　158347
ミツバヒルギ属　355908
ミツバビンボウカズラ　387774
ミツバビンボオズル　387774
ミツバビンボフカヅラ　387774
ミツバフウチョウソウ　95687,95807
ミツバフウロ　174684,175006
ミツバフサアサガホ　250853
ミヅバフヂマメ　135646
ミツバベンケイサウ　200816
ミツバベンケイソウ　200816

ミツバマツ　300181
ミツハル　93640
ミツバワウレン　103850
ミツバ属　113868
ミヅビエ　140356
ミヅヒキ　26624
ミヅヒキモ　312203
ミヅブキ　160637
ミツフデ　91024
ミヅホコリ　277369
ミツマタ　141470,141472
ミツマタスゲ　76618
ミツマタフュイチゴ　339015
ミツマタヤナギ　141470
ミツマタ属　141466
ミツミネモミ　495
ミヅメ　53465
ミツモト　312478
ミツモトサウ　312478
ミツモトソウ　312478,312480
ミツモリミミナグサ　82705
ミツヤヤシ属　263836
ミヅユキノシタ　238191
ミヅヰ　213138
ミヅ属　298848
ミドウシノ　347360
ミドリアカザ　86977
ミドリイソツツジ　223894
ミドリイモネヤガラ　157111
ミドリイラモミ　298227
ミドリウツボ　264850
ミドリゲンジスミレ　410721
ミドリコアジサイ　199899
ミドリザクラ　83228
ミトリササゲ　409096
ミドリサンゴ　159975
ミドリサンゴジュ　358621
ミドリシハイスミレ　410751
ミドリシャクジョウ　63964
ミドリスズムシ　232125
ミドリセンニンイモ　415191
ミドリツリバナ　157797
ミドリツルニンジン　98345
ミドリテンナンショウ　33487
ミドリドクダミ　198753
ミドリネナシカズラ　115051
ミドリノスズ　359924
ミドリノダケ　276780
ミドリノタマ　359924
ミドリバアソブ　98425
ミドリバイケイソウ　405627
ミドリハコベ　375007
ミドリハッカ　250450
ミドリハナヤサイ　59545
ミドリヒメザゼンソウ　381081
ミドリヒメフタバラン　232923
ミドリフジスミレ　410665
ミドリミツバウツギ　374090
ミドリミヤマフタバラン　232948
ミドリムヨウラン　223671
ミドリヤクシマコケリンドウ　174099
ミドリヨウラク　308565
ミドリヨシノ　83266
ミトロフィルム属　256257
ミナカミザサ　347295
ミナヅキ　200055
ミナトアカザ　87096
ミナトカラスムギ　45389
ミナトクマツヅラ　405815
ミナトタムラソウ　345471
ミナトマツヨイグサ　269452
ミナトムギクサ　198344
ミナトムグラ　170694
ミナミコミカンソウ　296553
ミナミススキノキ　415174
ミナミハマアカザ　44661
ミナモトサウ　312478
ミナリカイドウ　243717
ミネアザミ　91729
ミネウスユキソウ　224882
ミネオトギリ　202227
ミネカエデ　3718
ミネガラシ　72910
ミネザサ　347253
ミネズオウ　235283
ミネズオウ属　235279
ミネズオオ　235283
ミネスホウ　385355
ミネバリ　53610
ミネハリイ　353272,396010
ミネハリヰ　353272
ミネヨツバハギ　408678
ミノゴメ　49540,177560
ミノゴメ属　49532
ミノスゲ　74339
ミノチャルメルソウ　256012
ミノボロ　217441,217494
ミノボロスゲ　75547
ミノボロモドキ　217518
ミノボロ属　217416
ミバショウ　260253
ミビヨロギ　35868
ミフクラギ　83698,83700
ミフクラギ属　83690
ミフジン　233651
ミブナ　59434
ミブヨモギ　35868
ミマヒサカキ　160606
ミマミゾソバ　309877
ミミイヌガラシ　336182
ミミカウモリ　283792
ミミカキグサ　403119
ミミガタウチワ　28132
ミミガタテンナンショウ　33389
ミミコウモリ　283794
ミミスゲ　75785
ミミズノマクラ　381217
ミミズバイ　381217
ミミスベリ　381217
ミミヅバヒ　381217
ミミナグサ　82764,82824
ミミナグサ属　82629
ミミバフサアサガオ　250854
ミムラサキ　65670,66808
ミムラス　255223
ミモサアカシア　1168
ミモフリハリモミ　298358
ミヤウチソウ　73018
ミヤオギク　261069
ミヤオギク属　261060
ミヤオサウ　139623
ミヤオソウ　139623
ミヤオソウ属　139606
ミヤオヌカボ　12145
ミヤガラシ　326833
ミヤギザサ　347347
ミヤキノハギ　226977
ミャギノハギ　226916,226919
ミヤギノハギ　226969,226977
ミヤケスゲ　76446
ミヤケセキコク　125253
ミヤケセッコク　125253
ミヤケマユミ　157539
ミヤコアオイ　37568
ミヤコアザミ　348515
ミヤコイバラ　336842
ミヤコオトギリ　201964
ミヤコグサ　237539,237554
ミヤコグサ属　237476
ミヤコココケリンドウ　173955
ミヤコザサ　347259
ミヤコジシバリ　210649
ミヤコジマオヒシバ　121285
ミヤコジマソウ　191501
ミャコジマツヅラフジ　116020
ミャコジマツヅラフジ属　92523
ミャコジマツヅラフヂ　116020
ミャコジマツヅラフヂ属　92523
ミヤコジマツルマメ　177739
ミヤコジマニシキサウ　158534
ミヤコジマニシキソウ　158534
ミヤコジマハマアカザ　44520
ミャコジマボタンヅル　94814
ミヤコジマヲヒシバ　121285
ミャコダラ　215442
ミヤコダラ　215445
ミヤコツツジ　331943
ミヤコドリ　244254
ミヤコニシキ　264121
ミヤコミズ　298955
ミヤコワスレ　41201

ミヤジマママコナ 248177
ミヤトジマギク 89408
ミヤビカンアオイ 37595
ミヤマアオイ 37620
ミヤマアオスゲ 76121,76219
ミヤマアカヅラ 309711
ミヤマアカバナ 146734
ミヤマアカマツ 300269
ミヤマアキカラマツ 388654
ミヤマアキノキリンサウ 368073
ミヤマアキノキリンソウ 368073,368486
ミヤマアキノノゲシ 320523
ミヤマアクチ 31578
ミヤマアケボノソウ 380303
ミヤマアシクダシ 339047
ミヤマアシボソスゲ 76195
ミヤマアズマギク 150446,150732,151010
ミヤマアヅマギク 40016
ミヤマアツモリサウ 120357
ミヤマアツモリソウ 120357
ミヤマアブラススキ 372408
ミヤマアブラススキ属 372398
ミヤマアワガエリ 294962
ミヤマアヲスゲ 76219
ミヤマイ 212898
ミヤマイチゴ 338608,338698
ミヤマイチゴツナギ 177574,305970
ミヤマイヌノハナヒゲ 333715
ミヤマイボタ 229650
ミヤマイラクサ 221552
ミヤマイワスゲ 74125
ミヤマイワニガナ 210678
ミヤマウイキョウ 340463
ミヤマウグイスカグラ 235815
ミヤマウコギ 143693
ミヤマウシノケグサ 163817,164141
ミヤマウスユキソウ 224833,224932
ミヤマウズラ 179694
ミヤマウツギ 127118
ミヤマウツボグサ 316127,316136
ミヤマウヅラ 179694
ミヤマウド 30675
ミヤマウメモドキ 204087
ミヤマウラジロイチゴ 338572,338574
ミヤマウラジロヒゴタイ 348864
ミヤマエゾクロウスゴ 403939
ミヤマエンバク 190145
ミヤマエンレイサウ 397622
ミヤマエンレイソウ 397622
ミヤマオウギ 278722
ミヤマオオウシノケグサ 163836
ミヤマオグルマ 385902
ミヤマオタカラコウ 229078
ミヤマオタマキ 30031
ミヤマォダマキ 30043
ミヤマオトギリ 201942
ミヤマオトコヨモギ 36070
ミヤマカイドウ 243717
ミヤマカウゾリナ 195672
ミヤマカウバウ 27963
ミヤマカタバミ 277878
ミヤマカヂイチゴ 339188
ミヤマカナメモチ 295635
ミヤマカニツリ 398486
ミヤマカノコサウ 404292
ミヤマガマズミ 408224
ミヤマガラガラ 79130
ミヤマカラサナズナ 198480
ミヤマガラシ 47949
ミヤマカラナデシコ 127670
ミヤマカラマツ 388503,388703
ミヤマカワラハンノキ 16339
ミヤマガンクビサウ 77190
ミヤマガンクビソウ 77190
ミヤマカンスゲ 75443
ミヤマカンバ 53400
ミヤマガンピ 414123
ミヤマキクヨモギ 36363
ミヤマキケマン 106227,106238
ミヤマギシギシ 278577
ミヤマキスミレ 409765
ミヤマキタアザミ 348327
ミヤマキヌタソウ 170511
ミヤマキハダ 294234
ミヤマキリシマ 331010
ミヤマキンバイ 312745
ミヤマキンポウゲ 325518
ミヤマクサアジサイ 73124
ミヤマクサイチゴ 167653
ミヤマクスノキ 91351
ミヤマクハガタ 407387
ミヤマクマザサ 347221
ミヤマクマヤナギ 52448
ミヤマクリガシ 79027
ミヤマグルマ 138460,138461
ミヤマクルマバナ 97015
ミヤマクロスゲ 74569
ミヤマクロソヨゴ 204292,204299
ミヤマクワガタ 317958,317959
ミヤマゲンゲ 278710
ミヤマコアザミ 92075
ミヤマコウジュ 144002
ミヤマコウゾリナ 195672
ミヤマコウゾリナ属 195437
ミヤマコウボウ 27963
ミヤマコウモリソウ 283810
ミヤマコオゾリナ 195672
ミヤマコガンピ 414143
ミヤマコクラン 232322
ミヤマコケリンダウ 173260
ミヤマコケリンドウ 173260
ミヤマコゴメグサ 160176
ミヤマコシアブラ 143617
ミヤマコナスビ 239876
ミヤマコブシ 279249
ミヤマコメススキ 126039
ミヤマザクラ 83255
ミヤマサクラソウ 314519
ミヤマザサ 347297
ミヤマササガャ 254036
ミヤマサザンクワ 68953
ミヤマザンセウ 417330
ミヤマシオガマ 287012
ミヤマシキミ 365929,365974
ミヤマシキミ属 365916
ミヤマシグレ 408198,408199
ミヤマシシアクチ 31578
ミヤマシシウド 24445
ミヤマシャジン 7678,7706
ミヤマジュズスゲ 74343
ミヤマシュスラン 179708
ミヤマシラスゲ 74171,75596
ミヤマシロイヌナヅナ 30342
ミヤマシロバイ 381404
ミヤマシロバヒ 381355
ミヤマスイバ 278577
ミヤマスゲ 76571
ミヤマスズメノヒエ 238664
ミヤマスミレ 410547
ミヤマセウベンノキ 400522
ミヤマセンキウ 101837
ミヤマセンキウ属 101820
ミヤマセンキュウ 101825,101837
ミヤマセンキュウ属 101820
ミヤマゼンゴ 98781
ミヤマセンコ属 98772
ミヤマセントウソウ 85482
ミヤマタイゲキ 159544
ミヤマダイコンソウ 175390,175392
ミヤマタウキ 229485
ミヤマタゴボウ 239544
ミヤマタゴボオ 239544
ミヤマタツナミソウ 355675
ミヤマタニソバ 291732,309711
ミヤマタニタデ 91508
ミヤマタニワタシ 408648
ミヤマタネツケバナ 72910
ミヤマタムラサウ 345076
ミヤマタムラソウ 345181
ミヤマタンポポ 384434,384714
ミヤマチドメ 200394
ミヤマチドメグサ 200312
ミヤマチドリ 302533
ミヤマチャヒキ 60667
ミヤマチョウジザクラ 83147
ミヤマチングルマ 138460,138461
ミヤマツクバネウツギ 416841
ミヤマツチトリモチ 46848
ミヤマツツジ 330058,331791
ミヤマツボスミレ 410740
ミヤマツメクサ 32056

ミヤマトウキ　24284
ミヤマドウダン　145682
ミヤマトウバナ　97031
ミヤマトウヒレン　348645
ミヤマトサミズキ　106644
ミヤマトサミヅキ　106644
ミヤマトショウ　213934
ミヤマドジョウツナギ　177562
ミヤマドゼウツナギ　177574
ミヤマトベラ　155897
ミヤマトベラ属　155888
ミヤマナガハグサ　305303
ミヤマナガハシスミレ　410504
ミヤマナズナ　18317
ミヤマナデシコ　127834
ミヤマナナカマド　369513
ミヤマナミキ　355760
ミヤマナラ　323797
ミヤマナルコユリ　308582
ミヤマナワシロイチゴ　339047
ミヤマニガイチゴ　338678,338945,339318
ミヤマニガウリ　351762
ミヤマニガウリ属　351759
ミヤマニクケイ　91405
ミヤマニワトコ　345629
ミヤマニワナズナ　18317
ミヤマニンジン　276747
ミヤマヌカボ　12104
ミヤマヌカボシソウ　238633
ミヤマネズ　213735
ミヤマネズミガヤ　259646,259649
ミヤマノウルシ　159980
ミヤマノガリヤス　65486,65487
ミヤマノギク　150781
ミヤマノダケ　24322
ミヤマノボタン　48552
ミヤマノボタン属　48551
ミヤマハギ　226746
ミヤマハコベ　374780,375093
ミヤマハシカンボク　55147
ミヤマハシカンボク属　55135
ミヤマハタザオ　30121,30342,30345
ミヤマハタザホ　30342
ミヤマハナシノブ　307207
ミヤマハハソ　249468
ミヤマハブカツラ　147342
ミヤマハマナス　336805
ミヤマバラ　336736,336851
ミヤマハルガヤ　27969
ミヤマハンショウズル　95304
ミヤマハンショウヅル　94714,95304
ミヤマハンノキ　16328
ミヤマヒキオコシ　209830
ミヤマヒキノカサ　326273
ミヤマヒゴタイ　348882
ミヤマヒゴタイサイコ　154280
ミヤマヒサカキ　160606
ミヤマヒナゲシ　282506,282683
ミヤマヒナホシクサ　151394
ミヤマヒメツゲ　204418
ミヤマビャクシン　213696
ミヤマブキショウマ　37089
ミヤマフジキ　94032
ミヤマフタバラン　232946
ミヤマフヂキ　94032
ミヤマフユイチゴ　338698
ミヤマヘビノボラズ　52282
ミヤマホソエノアザミ　91998
ミヤマホソコウガイゼキショウ　213182
ミヤマホタルイ　352193
ミヤマホタルカズラ　233434
ミヤマホツツジ　397860
ミヤマママタタビ　6639
ミヤマママコナ　248175,248181
ミヤマママンネングサ　356847
ミヤマミズ　298854,298857
ミヤマミズタマソウ　91576
ミヤマミヅ　298854
ミヤマミミナグサ　83013
ミヤマムギラン　62813
ミヤマムグラ　170534,170535
ミヤマムラサキ　153491
ミヤマムラサキ属　153420
ミヤマメギ　52282
ミヤマモジズリ属　264757
ミヤマモミジ　2793
ミヤマモミジイチゴ　339111
ミヤマモミヂ　2793
ミヤマヤシャブシ　16342
ミヤマヤチヤナギ　343409
ミヤマヤナギ　343989
ミヤマヤブタバコ　77190
ミヤマヤブニンジン　276455,276460
ミヤマヤマブキショウマ　37064
ミヤマヨメナ　41201
ミヤマヨメナ属　256301
ミヤマラ　15779
ミヤマラッキョウ　15760
ミヤマリョウブ　96466
ミヤマリンドウ　173662
ミヤマリンドオ　173662
ミヤマルリミノキ　222133
ミヤマレンゲ　279249
ミヤマワスレナサウ　260892
ミヤマワスレナゾウ　260892
ミヤマワタスゲ　353606
ミヤマワレモコウ　345894
ミヤマヲダマキ　30043
ミヤミクロウスゴ　403937
ミヤラアオダモ　167923
ミヤヲギク　261069
ミヤヲギク属　261060
ミヤヲサウ　139623
ミヤヌカボ　12145
ミュイリア属　259725
ミュウケツジュ属　137330
ミユキカラマツ　388614
ミユキマル　244105
ミューラ-ビロウ　234188
ミョウガ　418002
ミョウギ　215198
ミョウギシャジン　7713
ミョウジンヤナギ　342933
ミョウトウセンニンソウ　95366
ミョウトウタンポポ　384629
ミラ　15843
ミラ属　254466,254562
ミリオカルパ属　261299
ミリス属　261576
ミルシネ属　261598
ミルスベリヒユ　361667
ミルチロカクタス属　261705
ミルト-ニア属　254924
ミルナ　344606,344718
ミルバアオダモ　168100
ミルバアヲダモ　168100
ミルマツナ　379542
ミルラノキ属　101285
ミロキシロン属　261549
ミロクチュウ　94563
ミロケマル　264357
ミロバラン　386504
ミロバランスモモ　316294
ミロバランノキ　386504
ミロマナデシコ　127833
ミンゲツイチゴ　338079
ミンブトキ　311890
ムイリア属　259725
ムエフラン　223647
ムエフラン属　223634
ムカゴイチゴツナギ　305422
ムカゴイラクサ　221538
ムカゴイラクサ属　221529
ムカゴサイシン　265412
ムカゴサイシン属　265366
ムカゴサウ　192851
ムカゴサウ属　192809
ムカゴソウ　192851
ムカゴソウ属　192809
ムカゴツヅリ　306106
ムカゴトラノオ　309954
ムカゴトラノヲ　309954
ムカゴトンボ　183625,291211
ムカゴトンボ属　291188
ムカゴニンジン　365871
ムカゴニンジン属　365835
ムカゴネコノメソウ　90411
ムカゴユキノシタ　349162
ムカシヨモギ　150419,150422
ムカシヨモギ属　150414
ムカデラン　288605

ムカデラン属　94425
ムギガラガヤツリ　322376
ムギカラグサ　49540
ムギクサ　198311,198327
ムギグワイ　18569
ムギグワヰ　18569
ムギコバノミツバツツジ　331625
ムギスゲ　74431,75700
ムギセンノウ　11947
ムギセンノウ属　11938
ムギナデシコ　11947,394327
ムギナデシコ属　11938
ムギラン　60379,62799
ムギワラギク　415386
ムギワラギク属　189093
ムクイヌビハ　165762
ムクイヌビワ　165166
ムクエノキ　29005
ムクゲ　195269
ムクゲアカシア　1488
ムクゲキンチャクズル　94933
ムクゲチャヒキ　60680
ムクノキ　29005
ムクノキ属　29004
ムクミカズラ　61102
ムクミコウゾ　61102
ムクレンジ　346338
ムクロジ　346338
ムクロジ科　346316
ムクロジ属　346323
ムゲンジョウ　244141
ムゲンラク　317688
ムゴマツ　300086
ムサシアブミ　33476
ムサシタイゲキ　159813
ムサシノササクサ　236285
ムサシモ　262018
ムシカリ　407865
ムシクサ　407275
ムシグサマムシグサ　33496
ムシグサマムシサウ　33496
ムシヅル　394560
ムシトリスミレ　299759,299764
ムシトリスミレ属　299736
ムシトリナデシコ　363214
ムシトリマンテマ　363181,364013
ムジナ　384746
ムジナオオバコ　301952
ムジナクグ　73574
ムジナスゲ　75065
ムジナノカミソリ　239276
ムジナモ　14273
ムジナモ属　14271
ムシャウチワ　106961
ムシャカゴノキ　6800
ムシャガマズミ　407930
ムシャザクラ　316809
ムシャスグ　74998
ムシャダモ　6800
ムシャツルウメモドキ　80279
ムシャバラン　39563
ムシャリンダウ　137551
ムシャリンダウ属　137545
ムシャリンドウ　137551
ムシャリンドウ属　137545
ムシュギョク　284659
ムジンサウ　89481
ムジンソウ　89481
ムスカリ属　260293
ムスビギ　141470
ムスビジャウ　396170
ムセンスゲ　75184
ムツアカバナ　146849
ムツイヌノヒゲ　151381
ムツオレガヤツリ　393194,393205
ムツオレグサ　177559,177560
ムツゲカヤラン　171836
ムッチャガラ　204025,204071
ムッチャギ　203636
ムツノガリヤス　65425,127268
ムツバアカネ　338038
ムツヲレガヤツリ　393194
ムツヲレグサ　177560
ムティシア属　260530
ムナガタイハウチハ　362251
ムニンアオガンピ　414248
ムニンイヌツゲ　204023
ムニンカラスウリ　396240
ムニンキケマン　105977
ムニンキヌラン　417713
ムニンクロガヤ　169550
ムニンクロキ　381124
ムニンゴシュユ　249156
ムニンシャシャンボ　403734
ムニンシュスラン　179585
ムニンススキ　255828
ムニンタイトゴメ　356843
ムニンタツナミソウ　355578
ムニンツツジ　330233
ムニンテイカカズラ　393656
ムニンテイカカヅラ　393616
ムニンテンツキ　166387
ムニンナキリスゲ　74768
ムニンネズミモチ　229546
ムニンノキ　301769
ムニンノボタン　248786
ムニンハダカホオズキ　400000
ムニンハナガサノキ　258924
ムニンハマウド　24381
ムニンヒサカキ　160510
ムニンビャクダン　346211
ムニンフトモモ　252609
ムニンボウラン　238307
ムニンホオズキ　238951
ムニンムラサキ　66931
ムニンモチ　204042,204044
ムニンヤッシロラン　171911
ムニンヤツデ　162891
ムヒギョク　175504
ムヒョウソウ　48762
ムベ　374409
ムベカヅラ　249649
ムベ属　374381
ムミピリア属　191582
ムユウゲ　234964
ムユウジュ　346504
ムユウジュ属　346497
ムヨウイチヤクソウ　322839
ムヨフラン　223647
ムヨフラン属　223634
ムラクモアオイ　37663
ムラクモアザミ　92189
ムラサキ　233731,233760,299726
ムラサキアカザ　87144
ムラサキアヂサイ　200031
ムラサキアヂサヰ　200031
ムラサキアムールテンナンショウ　33248
ムラサキイガヤグルマギク　80981
ムラサキイセハナビ　378166
ムラサキイソマツ　230816
ムラサキイヌナズナ　292531
ムラサキイヌナズナ属　292530
ムラサキイノコヅチ　4269
ムラサキイリス　208583
ムラサキイワオウギ　187769
ムラサキイワベンケイ　329888
ムラサキウズララン　418578
ムラサキウツギ　126896
ムラサキウツボカズラ　264845
ムラサキウマゴヤシ　247456
ムラサキウンラン　230911
ムラサキエゾソモソモ　305944
ムラサキエノコロ　361937,361959
ムラサキエンシュウハグマ　12629
ムラサキオウギ　42258
ムラサキオオハンゲ　299729
ムラサキオグルマ　358788
ムラサキオダマキ　30031
ムラサキオニシバリ　122587
ムラサキオバナ　341830
ムラサキオモト　394076
ムラサキオモト属　332293
ムラサキオンツツジ　332106
ムラサキカイウ　417118
ムラサキカタバミ　277776,278147
ムラサキカモソツル　117660
ムラサキギボウシ　198652
ムラサキギンバイソウ　123658
ムラサキクララ　369012
ムラサキクリンソウ　314157
ムラサキクワクカウアザミ　11199,11206

ムラサキクンシラン 10245
ムラサキクンシラン属 10244
ムラサキケマン 106004
ムラサキケマン属 105557
ムラサキコウキクサ 224369
ムラサキゴジアオイ 93219
ムラサキゴヤバラ 336789
ムラサキコンロンソウ 72876
ムラサキサギゴケ 247008
ムラサキサフラン 111625
ムラサキサルスベリ 219909
ムラサキサルビア 345474
ムラサキシキブ 66808
ムラサキシキブ属 66727
ムラサキシタイシャウ 202604
ムラサキシロウマリンドウ 174235
ムラサキジンジソウ 349207
ムラサキススキ 255873
ムラサキスズメノオゴケ 409395
ムラサキステビア 376414
ムラサキセンダイハギ 47859
ムラサキセンダイハギ属 47854
ムラサキセンニンイモ 415202
ムラサキセンブリ 380184,380319
ムラサキソシンカ 49211
ムラサキソシンクワ 49211
ムラサキソモソモ 305705
ムラサキタカオススキ 148873
ムラサキタデ 309201
ムラサキダンチク 37470
ムラサキタンボ 224110
ムラサキタンポポ 224110
ムラサキチョウジ 226125
ムラサキチョウセンガリヤス 94605
ムラサキチョウマメモドキ 81876
ムラサキックバネソウ 284406
ムラサキツツジ 331623
ムラサキツメクサ 397019
ムラサキツユクサ 394053
ムラサキツュクサ属 393987
ムラサキツリガネツツジ 250518
ムラサキツリバナ 157848,157930
ムラサキツリフネ 205369
ムラサキツルマサキ 157509
ムラサキツルマメ 185764
ムラサキテンニンサウ 228002
ムラサキテンニンソウ 228002
ムラサキナ 59603
ムラサキナギナタガヤ 412474
ムラサキナズナ 44868,154507
ムラサキナズナ属 44867
ムラサキナツフジ 66643
ムラサキナツフヂ 66643
ムラサキニガナ 283695
ムラサキニワホコリ 147883
ムラサキヌスビトハギ 126649
ムラサキネコハギ 226925
ムラサキネズミノオ 372666
ムラサキノキビ 151693
ムラサキノゲシ 219806
ムラサキノゲシ属 259746
ムラサキハシドイ 382329
ムラサキハチジョウナ 259772
ムラサキハナイカリ 184692
ムラサキハナナ属 275865
ムラサキハナニラ 60447
ムラサキバノカンアフヒ 37674
ムラサキハマエノコロ 361958
ムラサキハリナス 367042
ムラサキハリナスビ 367042
ムラサキバレンギク 140081
ムラサキハンゲ 299726
ムラサキヒキオコシ 209826
ムラサキヒゲシバ 146002
ムラサキヒメアナナス 113366
ムラサキヒメアブラススキ 71610,71613
ムラサキヒヨドリジョウゴ 367325
ムラサキビル 15822
ムラサキフイリバトケイソウ 285709
ムラサキフジ 414554
ムラサキフタッブダネヤシ 130067
ムラサキフタバラン 232919
ムラサキフトモモ 382522
ムラサキフロックス 16039
ムラサキヘイシソウ 347156
ムラサキベンケイサウ 200812
ムラサキベンケイソウ 200793,200812,357203
ムラサキホソバユリ 229803
ムラサキホタルブクロ 70237
ムラサキボタンボウフウ 292893
ムラサキマキエハギ 227010
ムラサキマムシゲサ 33496
ムラサキマユミ 157663
ムラサキマル 244260
ムラサキミズトラノオ 139602
ムラサキミズヒイラギ 2684
ムラサキミソハギ 114616
ムラサキミゾホホヅキ 392905
ムラサキミツバ 113881
ムラサキミミカキグサ 403098,403377
ムラサキムカシヨモギ 115595
ムラサキムヨウラン 223663
ムラサキモウズイカ 405748
ムラサキモクワンジュ 49211
ムラサキモメンズル 42594
ムラサキモメンヅル 42594
ムラサキヤシオ 330058
ムラサキヤシオツツジ 330058
ムラサキヤシホ 330058
ムラサキヤシホツツジ 330058
ムラサキヤハズカズラ 390699
ムラサキヤマツツジ 330975
ムラサキランタナ 221266
ムラサキリウキウツツジ 330883
ムラサキリュウキュウ 330883
ムラサキルエリア 339828
ムラサキワウギ 42264
ムラサキワセオバナ 262552
ムラサキワセヲバナ 262552
ムラサキヲダマキ 30031
ムラサキ科 57084
ムラサキ属 233692
ムラサヒゲシバ 88320
ムラスズメ属 72168
ムラダチヒルガオ 102980
ムラダチヤブマオ 56338
ムラマツノガリヤス 65257
ムリクサボタン 94995
ムリッ 228369
ムルティカウレ 89628
ムレ 379542
ムレイウズ 5302
ムレイカラマツ 221939
ムレイキリンソウ 329866
ムレイシャジン 7643
ムレイショウ 300243
ムレイセンキュウ 229344
ムレイソウ 193426
ムレイタウヒ 298358
ムレイナミキソウ 355757
ムレイヒョウタンボク 235905
ムレゴチョウ属 351343
ムレスギ 113710
ムレスズメ 72342
ムレスズメラン 270841
ムレスズメ属 72168
ムレナデシコ 183191
ムレホコ 340781
ムロ 213896
ムロウテンナンショウ 33569
メアオスゲ 75135
メアカンキンバイ 312782
メアカンフスマ 32070
メアゼガヤツリ 322235
メアゼテンツキ 166393,166501,166555
メイオウマル 103761
メイグンギョク 233554
メイケッカエデ 3034
メイゲッカエデ 3034
メイゲツソウ 162536
メイケンシ 266642
メイシュウギョク 284700
メイソウギョク 102699
メイソウチョウ 256270
メイヂサウ 103446
メイテンオトギリ 201859
メイヌナズナ 137136
メイヌヤマハッカ 209853
メイホウギョク 182449
メイロギョク 148141

メイワキンカン　167493
メウガ　418002
メウガサウ　307324
メウガ属　417964
メウマノチャヒキ　60992
メウリノキ　2910,2913
メオトアナナス　8573
メオトバナ　231680
メカ　418002
メガヤ　82523
メガルカヤ　389355
メガルカヤ属　389307
メギ　52225
メキシコ・ジニア　418038
メキシコアサガオ　208253
メキシコアナナス　8576
メキシコキョウチクトウ　389978
メキシコサバル　341418
メキシコサルビア　345043
メキシコジキタリス　387346
メキシコジギタリス属　387343
メキシコチモラン　416581
メキシコノボタン　193749
メキシコノボタン属　193746
メキシコハクセンヤシ　59099
メキシコハナヤナギ　114605
メキシコヒナギク　150788
メキシコヒマワリ　392434
メキシコヒメギク　150792
メキシコヒメノボタン　193749
メキシコヒャクニチソウ　418038
メキシコホシダネヤシモドキ　194610
メキシコマツ　299851
メキシコマホガニー　380526
メキシコマンネングサ　356934
メギヤナギ　343101
メキャベツ　59539
メキンコソリチャ　79916
メギ科　51261
メギ属　51268
メグサハッカ　250432
メグスリノキ　3492
メグズリノキ　3243
メクヌギ　324532
メクマイチゴ　338302
メクラフジ　414570
メグロ　3243
メゲヤキ　417559
メゴシツ　291161
メコノプシス属　247099
メザクラ　83255
メザマシグサ　69634
メシツブノキ　239392
メシバ　130489
メシャクナゲ　330998
メショウ　70507
メジロザクラ　83145
メジロスギ　113686
メジロホオズキ　238942
メジロホホヅキ　238942
メストクレマ属　252361
メズラ　280023
メソスピニディア属　252277
メダウ　30618
メタカラカウ　229202
メタガラカウ　229202
メタカラコウ　229202
メタカラコウ属　228969
メタカラコオ　229202
メダケ　304098
メダケ属　37127,303994
メタセコイア　252540
メタセコイア属　252538
メダラ　30643,30655
メタラヨウ　407980
メタルベゴニア　50076
メディォカクタス属　247648
メディオロビビア属　247664
メディコスマ属　247516
メディニラ属　247526
メドキ　226742
メドノキ　408009
メドハギ　226742,226860
メドラー　252305
メトロキシロン属　252631
メトロシデーロス属　252607
メナモミ　363093
メナモミ属　363073
メノウ　233639
メノマンネングサ　356841
メバウキ　268438
メハジキ　224989,225009
メハジキ属　224969
メハナヤサイ　59545
メバハキ　268438
メバハキ属　268418
メハリノキ　16447
メヒイラギ　223131,316824
メヒシバ　130489
メヒジハ　130745
メヒシバ属　130404
メヒラギ　223131,316819,316824
メヒルギ　215510,215511
メヒルギ属　215508
メヒルムシロ　312190
メホウエンドウ　408571
メボウキ　268438
メホタルイ　353516
メボタンヅル　94778
メマツ　299890
メマツヨイグサ　269404
メマツヨヒグサ　269404
メムクノキ　80739
メメズリ　381217
メメニラ　15486
メヤブマオ　56260,56343
メヤブマヲ　56260,56343
メヤマハッカ　209705
メラスフェヶルラ属　248718
メラネシアパペダ　93559
メラノクシロンアカシア　1384
メラローサ　93630
メラン　116832
メリケンガヤツリ　322342
メリケンカルカヤ　23077
メリケンサラダ　90894
メリケントキンソウ　368534,368536
メリケンニクキビ　402546
メリケンハギ　126278
メリケンマツ　318580
メリケンムグラ　131361
メリコッカ属　249124
メリッササウ　249504
メリッサソウ　249504
メリルビロウ　234187
メリロート　249232
メルクシマツ　300068
メルテンシア属　250883
メルテンスツガ　399916
メルテンモチ　204365
メレンデラ属　250627
メロカクタス属　249578
メロン　114189
メワレモコウ　345894
メヲトバナ　231680
メンガシ　116193
メンタ属　250287
メンツェーリア属　250481
メンツツジ　331985
メンテンササガヤ　254043
メンマツ　299890
モーアウロコケンチャ　225557
モイハシャジン　7739
モイハナズナ　137224
モイハナヅナ　137224
モイワシャジン　7739
モイワナズナ　136996,137222,137224
モイワボウフウ　293034
モイワボダイジュ　391768
モイワラン　110501
モウカ　53610
モウカンオウ　375919
モウカンチュウ　395638
モウキンシュウ　183378
モウコアカバナ　146842
モウコアカマツ　300240
モウゴアカマツ　300240
モウコアサガオ　250831
モウコアンズ　34470
モウコエビラハギ　249212
モウコオウギ　104640,187881

モウコオキナグサ　321659
モウコオナモミ　415031
モウコオミナエシ　285855
モウコガシワ　324173
モウコガマ　401103,401120
モウコキセワタ　295150
モウコギョク　103755,209510
モウコグハ　259165
モウコグワ　259165
モウコゲンゲ　278829
モウコシナノキ　391781
モウコスグ　74460
モウコスズシロ　154452
モウコスミレ　410269
モウコゼリ　97727
モウコタンポポ　384681
モウコドジョウツナギ　321241
モウコナズナ　18405
モウコナデシコ　127676
モウコナラ　324173
モウコネジアヤメ　208666
モウコノムラサキ　221727
モウコハマアカザ　44400
モウコハマサジ　230631
モウコハマタバコ　229109
モウコボクチ　361104
モウコマツナ　379509
モウコマユミ　157699
モウコマンテマ　363963
モウコムカシヨモギ　58265
モウコモメンヅル　42704
モウコヤナギ　343706
モウコヨナメ　215356
モウコラッキョウ　15632
モウシュウギョク　182467
モウズイカ　405667
モウズイカ属　405657
モウズイクワ　405667
モウズイクワ属　405657
モウズヰクワ　405667
モウズヰクワ属　405657
モウセンゴケ　138348
モウセンゴケ科　138369
モウセンゴケ属　138261
モウゼンツツジ　330457
モウソウチク　297301,297306
モウソウチグ　297367
モウソウチク属　297188
モウダ　391739
モウリンカ　211990
モウリンクワ　211990
モエカラ　346390
モエギアルム　37005
モエギウンゼンカンアオイ　37752
モエギオクエゾサイシン　38312
モエギカンアオイ　37756
モエギスゲ　76591
モエギベゴニア　49942
モーカザクラ　53465
モガシ　142303,142396,142399
モクアオイ　223355
モクアフヒ　223355
モクイラクサ　125545
モクカウイバラ　336366
モクカウクワ　211821
モクカウバラ　336366
モクキリン　290701
モクキリン属　290699
モククロ　76813
モクゲ　195269
モクゲンジ　217626
モクゲンジ属　217610
モクコク　386696
モクコクモドキ　329068
モクコク属　386689
モクシュンギク　32563
モクシュンク　32563
モクズキンポウゲ　326278
モクセイ　276257,276291
モクセイサウ　327896
モクセイサウ科　327934
モクセイサウ属　327793
モクセイソウ　327896
モクセイソウ属　327793
モクセイ科　270182
モクセイ属　276242
モクセキコク　125279
モクセンナ　78308,360490
モクタチバナ　31599
モクビヤクコウ　111826
モクビヤクコウ属　111823
モクビャッコ　111826
モクビャッコウ　111826
モクフヨウ　195040
モクベッシ　256804
モクベンケイ　162346,260711
モクマオ　56125,56254
モクマオウ　15953,79157,79161,79182
モクマオウ科　79193
モクマオウ属　79155
モクマワウ科　79193
モクマワウ属　79155
モクマヲ　56125,56254
モクラン　206669
モクレイシ　254303
モクレイシ属　254283
モクレン　416682,416707
モクレンゲ　416707
モクレンジ　217620,217626
モクレン科　242370
モクレン属　241951
モクワダン　110622
モクワンジュ　48993,259512
モケ　84573
モコキノガリヤス　65429
モシオグサ　397159,418392
モジズリ　372247
モシホホグサ　418392
モスカ－タバラ　336408
モスフロックス　295324
モスボ－ル　392039
モダマ　145899
モタマズル　145899
モタマヅル　145899
モダマ属　145857
モチアデク　123448,123456
モチイネ　275966
モチガシハ　323814
モチカシワ　323814
モチグサ　36097,36474
モチグハ　259085
モチゲワ　259152
モチゴメ　275958,275966
モチシバ　231355
モチヅキザクラ　83128
モチツツジ　331173
モチノキ　203908
モチノキ科　29968
モチノキ属　203527
モチババヒノキ　381390
モチバヒサカキ　160553
モチユ　93499
モッカ　76813
モッコウ　44881
モツコウカ　211821
モッコウバナ　302796
モッコウバラ　336366
モッコク　386696
モッコク属　386689
モッシア属　259328
モッスロ－ス　336476
モヅヒキモ　312288
モナデニウム属　256966
モナルダ属　257157
モナルドバナ　257169
モナンセス属　257069
モニラリア属　257395
モニワザクラ　83129
モノガタリナズナ　137144
モノタグマ属　257906
モノドラカンアオイ　37687
モノメリア属　257708
モ－パングリ　78823
モミ　365
モミジ　3300
モミジアオイ　194804
モミジイチゴ　338945,338946
モミジウリノキ　13376
モミジガサ　283804,283806
モミジカラマツ属　394583
モミジキンバイ　312540

モミジコウモリ 283833
モミジシモツケ 166101
モミジタマブキ 283810
モミジチドメ 200261
モミジドコロ 131840
モミジバアブラギリ 212186
モミジバキセワタ 224976
モミジハグマ 12600
モミジハグマ属 12599
モミジバゴキヅル 6907
モミジバショウマ 41839
モミジバスズカケ 302582
モミジバスズカケノキ 302582
モミジバセンダイソウ 349887
モミジバダイオウ 329372
モミジバダイモンジサウ 349604
モミジバダイモンジソウ 349204
モミジバヒメオドリコソウ 220392
モミジバフウ 232565
モミジバルコウソウ 323457
モミジヒルガオ 207659
モミジヘイシソウ 347163
モミジラン 267958
モミジルコウ 323457,323470
モミソ 365
モミヂ 3300
モミヂアオイ 194804
モミヂアフヒ 194804
モミヂイチゴ 338946
モミヂカラスウリ 396226
モミヂドコロ 131840
モミヂハグマ 12600
モミヂハグマ属 12599
モミヂバシウカイダウ 49890
モミヂバスズカケノキ 302582
モミヂバフウ 232565
モミチヒルガホ 207988
モミノキ 365
モミラン 171897
モミ属 269
モメンヅル 42417,42969
モモ 20935
モモイロカイウ 417118
モモイロカンアオイ 194369
モモイロキランソウ 13094
モモイロサギゴケ 247011
モモイロシロツメクサ 397047
モモイロタテヤマギク 40311
モモイロダンギク 78014
モモイロタンポポ 110985
モモイロテンジクアオイ 288268
モモイロナンバンサイカテ 78316
モモイロノジシャ 404408
モモイロノハラクサフジ 408280
モモイロハマオモト 111230
モモイロハマゴウ 411433
モモイロハリエンジュ 335013
モモイロヒルザキツキミソウ 269506
モモイロフイリイモ 28119
モモイロフウロ 174729
モモイロヘウタンボク 236146
モモイロホソバシャクナゲ 331192
モモイロマットギク 26579
モモイロラショウモンカズラ 247743
モモガサマル 234939
モモケラマ 331762
モモタマナ 386500
モモタマナ属 386448
モモチドリ 8302
モモノハギキョウ 70214
モモノミイヌビハ 164992,165527
モモハユリ 229947
モモミヤシ属 46374,181821
モモヤマアザミ 92037
モモルディカ属 256780
モヨウガラクサ 288767
モヨウビユ 18095
モラウェッチア属 273809
モラガシ 116150
モラスクマデヤシ 390462
モラン 11300,206626
モラン属 11273
モリアカネ 338032
モリアキノノグシ 320513
モリアザミ 91913
モリアラウカリア 30832
モリイチゲ 24105
モリイチゴ 167635,339047
モリイヌタデ 291747
モリイバラ 336830
モリオカシダレ 83114
モリオカハンノキ 16312
モリカ 211990
モリカサスゲ 74392
モリガシャドリギ 411048
モリカノコソウ 404316
モリカラマツ 388594
モリカンディア属 258791
モリギク 124829
モリキンバイ 312750
モリクリクリ 93865
モリクワ 211990
モリサウ 85107,359516
モリサウ属 85106
モリシマアカシア 1168,1380,1401
モリシャクナゲ 331276,331565
モ－リシャスパペダ 93492
モ－リシャセダカトックリヤシ 201363
モリスゲ 75417
モリゼンゴ 24411
モリソウ属 85106
モリソンガンピ 363774
モリソンヒラギナンデン 242608
モリハンジヤウヅル 95148
モリビランデ 248269
モリフレラン 295960
モリムグラ 170503
モリムグラモドキ 170661
モリメドハギ 226860
モリヤナギ 343321
モリヨシスゲ 73562
モリヨモギ 36048,36354
モリンカ 211990
モリンガ 258945
モリンガ属 258935
モルッカソテツ 115812
モルッカネム 162473
モルッカヤシ属 332375
モルトキア属 256746
モルモ－デス属 258987
モルモリカ属 259000
モレ－ア属 258380
モレイギョク 175526
モレノキ 350990
モレノヤシ属 85420
モ－レルッツアナナス 54453
モロコシ 369600,369720
モロコシガヤ 369677,369681
モロコシキビ 369600
モロコシサウ 239847
モロコシソウ 239847
モロコシ属 22476,369590
モロッコアヤメ 208899
モロビ 414
モンイキシア 210843
モンカタバミ 278124
モンカラクサ 263504
モンキ－ジャック 36928
モンゴリコナラ 323592,324404
モンゴリナラ 324173
モンジュブキ 292404
モンセテラ 258168
モンセンネンボク 137387
モンタチバナ 365940
モンタナアブラギリ 406019
モンタナマツ 300086
モンチア属 258237
モンチコラマツ 300082
モンツキウマゴヤシ 247246
モンツキガヤ 57548,57553
モンツキガヤ属 57545
モンツキシバ 203961
モンツキヒナゲシ 282530
モンティコ－ラマツ 300082
モンテンジクアオイ 288594
モンテンヂクアフヒ 288594
モンテンボク 194895
モントレイイトスギ 114723
モントレ－サイプレス 114723
モントレ－サイプレス 114723
モンナシノタヌキモ 403109

モンパイノコヅチ　4259,4273
モンパガシ　116118,324490
モンパノキ　393242
モンパモウズイ　405788
モンパモウズヰ　405788
モンパヤンバルクルマバナ　227577
モンパヰノコヅチ　4259
モンビレア属　258336
モンヨウショウ　245026
ヤイトバナ　280078,280097
ヤイニペシ　391766
ヤイマナスビ　367334
ヤウサイ　207590
ヤウジュカノコサウ　404316
ヤウシュシモツケ　166132
ヤウシュテウセンアサガホ　123077
ヤウシュヤマゴバウ　298094
ヤウラクボタン　220729
ヤウラクヤドリギ　411016
ヤウラクラン属　267922
ヤエイチゲサクラ　315102
ヤエイワウチワ　362274
ヤエオオダイコンソウ　175382
ヤエオグルマ　207052
ヤエオニユリ　230061
ヤエカイダウ　243624
ヤエカイドウ　243624
ヤエカザクルマ　95218,95221
ヤエガヤ　184323
ヤエガワ　235970
ヤエガワカンバ　53389,53400
ヤエキジムシロ　313015
ヤエキダチイナモリ　272222
ヤエキツネノカミソリ　239273
ヤエキバナシャクナゲ　330185
ヤエキョウチクトウ　265335
ヤエクチナシ　171253,171308,171339,
171374
ヤエコデマリ　371871
ヤエザキカイフウロ　174903
ヤエザキカジイチゴ　339395
ヤエザキクサイチゴ　338517
ヤエザキゲンノショウコ　174968
ヤエザキダイコンソウ　175419
ヤエザキチャボヤブデマリ　408044
ヤエザキニガイチゴ　338824
ヤエザキフヨウ　195052
ヤエザキヤナギラン　330971
ヤエズイセン　262458
ヤエセイヨウナツユキソウ　166127
ヤエセンブリ　380237
ヤエチョウセンアサガオ　123065,123066
ヤエチングルマ　363064
ヤエツワブキ　162622
ヤエテリハノイバラ　337017
ヤエドクダミ　198747
ヤエトラノオ　239595
ヤエナリ　409025
ヤエノキンキマメザクラ　83232
ヤエノツクバネウツギ　179
ヤエノミゾカクシ　234366
ヤエノミツバツチグリ　312566
ヤエノモミジイチゴ　338947
ヤエノユキツバキ　69211
ヤエハマナシ　336730
ヤエヒマワリ　188909
ヤエホソバシャクナゲ　331190
ヤエムグラ　170193,170199
ヤエムグラ属　170175
ヤエヤマアオキ　258882
ヤエヤマアオキ属　258871
ヤエヤマアブラスゲ　333530
ヤエヤマイナモリソウ　272282
ヤエヤマウツギ　127125
ヤエヤマカモノハシ　209345
ヤエヤマカンアオイ　37760
ヤエヤマギョクシンカ　384951
ヤエヤマキランソウ　13187
ヤエヤマクロバイ　381423
ヤエヤマクワズイモ　16492
ヤエヤマコウゾリナ　55766
ヤエヤマコクタン　132139,132153
ヤエヤマコンテリギ　199856
ヤエヤマコンロンカ　260475
ヤエヤマシキミ　204598
ヤエヤマシタン　320301
ヤエヤマジュウニヒトエ　13187
ヤエヤマスケロクラン　223652
ヤエヤマスズコオジュ　380045
ヤエヤマスミレ　410641
ヤエヤマセイシカ　330617
ヤエヤマタヌキマメ　112432,112433
ヤエヤマツツジ　330971
ヤエヤマナツメ　100061
ヤエヤマネコノチチ　328551
ヤエヤマネムノキ　13653
ヤエヤマノイバラ　336405
ヤエヤマノボタン　59888,59889
ヤエヤマハギカズラ　169658
ヤエヤマハシカグサ　187510
ヤエヤマハマゴウ　411208
ヤエヤマハマナツメ　100061
ヤエヤマヒイラギ　276341
ヤエヤマヒサカキ　160628
ヤエヤマヒトツボクロ　265373
ヤエヤマヒルギ　329765
ヤエヤマヒルギ属　329744
ヤエヤマブキ　216088
ヤエヤマメジロホオズキ　238956
ヤエヤマヤシ　347421
ヤエヤマヤマボウシ　124925
ヤエヤマラセイタソウ　56356
ヤエライラック　382371
ヤエラッパズイセン　262442
ヤオゼンタデ　309203
ヤオヤボウフウ　176923
ヤカイソウ　67736
ヤガミスゲ　75252
ヤガラ　56637,353587
ヤギ　98136
ヤギサウ　350344
ヤキバザサ　347324
ヤキバラン　116829,117104
ヤギムギ　8672
ヤキモチバラ　366284
ヤギュウカク　374018
ヤクシアゼトウナ　110603
ヤクシソウ　110606,283388
ヤクシマアオイ　37761
ヤクシマアカシュスラン　193592
ヤクシマアキノキリンソウ　368237
ヤクシマアザミ　92498
ヤクシマアジサイ　199792,199927
ヤクシマアセビ　298757
ヤクシマウスユキソウ　21698
ヤクシマウメバチソウ　284602
ヤクシマオオバコ　301885
ヤクシマオナガカエデ　3208
ヤクシマガクウツギ　199952
ヤクシマカラスザンショウ　417375
ヤクシマカワゴロモ　200206
ヤクシマカンスゲ　75425
ヤクシマキイチゴ　338950
ヤクシマギボウシ　198622
ヤクシマグミ　142251
ヤクシマコウモリ　283882
ヤクシマコオトギリ　201972
ヤクシマコケリンドウ　174079
ヤクシマコケリンドオ　174079
ヤクシマコナスビ　239693
ヤクシマコムラサキ　66927
ヤクシマサルスベリ　219922,219972
ヤクシマシソバタツナミ　355552
ヤクシマシャクナゲ　331237
ヤクシマシュスラン　179630
ヤクシマショウマ　41810
ヤクシマスミレ　410107
ヤクシマセントウソウ　85485
ヤクシマダイモンジソウ　349355
ヤクシマダケ　318336
ヤクシマチドリ　302248
ヤクシマツチトリモチ　46885
ヤクシマツバキ　69218
ヤクシマトウバナ　97037
ヤクシマトンボ　302408
ヤクシマニガナ　210546
ヤクシマネジバナ　372254
ヤクシマネッタイラン　399650
ヤクシマノガリヤス　65423
ヤクシマノギク　41511
ヤクシマノコンギク　41511

ヤクシマノダケ 24508
ヤクシマハコベ 374868
ヤクシマハシカグサ 262964
ヤクシマハマスゲ 76777
ヤクシマバライチゴ 338610
ヤクシマヒカゲツツジ 330998
ヤクシマヒゴタイ 348572
ヤクシマヒメアリドオシラン 218296
ヤクシマヒヨドリ 158373
ヤクシマヒロハテンナンショウ 33401
ヤクシマフウロ 174905
ヤクシマママコナ 248184
ヤクシマミツバツツジ 332134
ヤクシマムグラ 170438
ヤクシマヤマツツジ 332133
ヤクシマヤマムグラ 170564
ヤクシマヨウラクツツジ 250533
ヤクシマラン 29794
ヤクシマラン科 29796
ヤクシマラン属 29787
ヤクシマリンダウ 174080
ヤクシマリンドウ 174080
ヤクシマリンドオ 174080
ヤクシン 213647
ヤクタネゴヨウ 299789,299799
ヤクタネゴヨオ 299789
ヤクチ 17736
ヤクナガイヌムギ 60659
ヤクミマオナガカエテ 2847
ヤクムヨウラン 223659
ヤクモサウ 224989,225009
ヤクモソウ 225009
ヤクヨウカモメヅル 409463
ヤクヨウサルビヤ 345271
ヤクヨウニンジン 280741
ヤグラオオバコ 301871
ヤグラオホバコ 301871
ヤグラサクラ 314084
ヤグラバナ 19248
ヤグリマクワクカウ 257169
ヤグリマクワクカウ属 257157
ヤグルマアザミ 81150
ヤグルマギク 81020
ヤグルマギク属 80892
ヤグルマサウ 335151
ヤグルマサウ属 335141
ヤグルマセンノウ 363312
ヤグルマソウ 81020,335151
ヤグルマソウ属 335141
ヤグルマハッカ 257169
ヤグルマヤツデ 162886
ヤクワウクワ 84417
ヤクワウボク 84417
ヤケイシセンブリ 380382
ヤケマオナガカエデ 2846
ヤコウカ 84417
ヤコウボク 84417
ヤコビニア属 211338
ヤコブセニア属 211350
ヤコブボロギク 359158
ヤサイショウマ 91038
ヤシ 98136,320372
ヤシオ 333914
ヤシオネ属 211639
ヤジナ 401542
ヤジノ 318307
ヤシバダケ 304018
ヤシホ 98136
ヤシマスゲ 73566
ヤシャ 16433
ヤシャダケ 357959
ヤシャビシャク 333914
ヤシャブシ 16341
ヤシュウイヌノヒゲ 151476
ヤシ科 280608
ヤシ属 98134
ヤスノキ 320372
ヤセイカンラン 59520
ヤセウツボ 275135
ヤダケ 318307
ヤダケガヤ 391487
ヤダケガヤ属 391477
ヤダケ属 318277
ヤチアザミ 92388
ヤチイ 213476
ヤチイチゲ 23705
ヤチイヌツゲ 203734
ヤチエゾ 298298
ヤチエンドウ 222805
ヤチオウギ 43204
ヤチカバ 382276
ヤチカハツスゲ 75604
ヤチガヤ 256632
ヤチカワズスゲ 75604
ヤチカンバ 53544
ヤチコタヌキモ 403268
ヤチザクラ 109780
ヤチサンゴ 342859
ヤチシオガマ 287718
ヤチシャジン 7728
ヤチスゲ 75165
ヤチタデ 309338
ヤチダモ 167970,168023,168025
ヤチツツジ 85414,239395
ヤチバウズ 73826
ヤチハンノキ 16386,16389
ヤチブキ 68196
ヤチマタイカリソウ 146995
ヤチマタエンゴサク 105917
ヤチマタカラマツ 388617
ヤチムラサキ 397441
ヤチャナキ 261162
ヤチヤナギ 261164,261237
ヤチャナキ属 261120
ヤチヨニシキ 186546
ヤチラン 185200
ヤチラン属 242993
ヤチリンドウ 173751,174156
ヤツガシラ 275376
ヤツガタケキスミレ 409862
ヤツガタケザクラ 83126
ヤツガタケタンポポ 384862
ヤツガタケトウヒ 298323
ヤツガタケトオヒ 298323
ヤツガタケムグラ 170708
ヤッコカズラ 294830
ヤッコサウ 256180
ヤッコサウ科 325065
ヤッコサウ属 256175
ヤッコソウ 256171,256180,256182
ヤッコソウ科 256186,325065
ヤッコソウ属 256175
ヤツシロ 93873
ヤツシロサウ 70054
ヤツシロソウ 70054,70058,70061
ヤツシロラン 130043,171913
ヤツシロラン属 130037
ヤツタカネアザミ 92499
ヤツデ 162879
ヤツデアオギリ 376110
ヤツデアサガオ 207988,208051
ヤツデアサガホ 207988
ヤツデアヲギリ 376110
ヤツデイチゴ 338399
ヤツデウチワ 28113
ヤツデベゴニア 49915
ヤツデユキノシタ 349604
ヤツデ属 162876
ヤッブサ 72074
ヤツブサ 72074
ヤツブサウメ 316581
ヤツブサグリ 78778
ヤドクキリン 160064
ヤドリオホバトベラ 301253
ヤドリギ 410960,410992,410994
ヤドリギ科 236503,410792,410934
ヤドリギ属 410959
ヤドリコケモモ 403821
ヤドリサンゴジュ 407679
ヤドリタラノキ 289629
ヤドリタラノキ属 289627
ヤドリフカノキ 350654
ヤナギ 343070
ヤナギアカガシ 323601
ヤナギアカバナ 146812
ヤナギアザミ 91864,92132,406266
ヤナギイチゴ 123322,123332
ヤナギイチゴ属 123316
ヤナギイノコズチ 4304
ヤナギイノコヅチ 4304
ヤナギイボタ 229610

ヤナギイワタデ 308777
ヤナギガシ 116187
ヤナギサウ 85875,146609,146614
ヤナギザクラ 161755
ヤナギザクラ属 161741
ヤナギスイラン 195655
ヤナギスブタ 55922
ヤナギソウ 85875,146614
ヤナギタウコギ 53816
ヤナギタデ 309199
ヤナギタムラソウ 406631
ヤナギタンポポ 196057
ヤナギチョウジ 289324
ヤナギチョウジソウ 20860
ヤナギトウワタ 38147
ヤナギトラノオ 239881
ヤナギトラノヲ 239881
ヤナギニガナ 210543
ヤナギヌカボ 309125,309395
ヤナギノギク 193946
ヤナギバアカシア 1422
ヤナギバアンゲロン 24525
ヤナギバウロコマリ 225224
ヤナギバキリンソウ 368289
ヤナギハクカ 203095
ヤナギハクカ属 203079
ヤナギバグミ 141932
ヤナギハグロ 200652
ヤナギバザサ 209059,347214
ヤナギバサザンカ 69586
ヤナギバシッボウジュ 358271
ヤナギバシャジン 7649
ヤナギバシャリントウ 107668
ヤナギバスノキ 403716
ヤナギバチョウジソウ 20860
ヤナギハッカ 203095
ヤナギハッカ属 203079
ヤナギバツバキ 69586
ヤナギバテンモンドウ 39009
ヤナギバドクゼリ 90932
ヤナギバトラノオ 407424
ヤナギハナガサ 405812
ヤナギハナシノブ 99855
ヤナギバヒサカキ 160507
ヤナギバヒマワリ 189052
ヤナギバヒメギク 150464
ヤナギバヒメジョオン 150894
ヤナギバモクセイ 276338
ヤナギバモクマオ 56125
ヤナギバヤブマオ 56125,56254
ヤナギバヤブマヲ 56125,56254
ヤナギバラモンジン 354801
ヤナギバルイラソウ 339684
ヤナギバレンリソウ 222844
ヤナギヒゴタイ 348141,348201
ヤナギヒマワリ 188991
ヤナギヒョウタンボク 236161
ヤナギフナバラソウ 117387
ヤナギボウキ 38974
ヤナギモ 312206
ヤナギユウカリ 155624
ヤナギユ-カリ 155624
ヤナギヨモギ 150422
ヤナギラン 85875,146609,146614
ヤナギヰノコヅチ 4304
ヤナギ科 342841
ヤナギ属 342923
ヤニチシャノキ 141691
ヤニレ 401490
ヤネタビラコ 111050
ヤネバンダイソウ 358062
ヤノネアオイ 25926
ヤノネアオイ属 25922
ヤノネアザミ 348510
ヤノネグサ 309442
ヤノネジシバリ 221802
ヤノネシロバナスミレ 409716
ヤノネツハブキ 229065
ヤノネツワブキ 229065
ヤノネボンテンカ 286634
ヤノネボンテンカ属 286580
ヤハズアザミ 73337
ヤハズアザミ属 73278
ヤハズアジサイ 200105
ヤハズアメリカチャボヤシ 327540
ヤハズイグマ 117412
ヤハズイケマ 117412
ヤハズエンドウ 408284,408571,408578
ヤハズカズラ 390701
ヤバズカズラ属 390692
ヤハズカツラ 49046
ヤバズカツラ 390701
ヤバズカツラ属 390692
ヤハズカワツルモ 340491
ヤハズギク 188410
ヤハズキダチベンケイ 357080
ヤハズキンゴジカ 362490
ヤハズキンバイ 269475
ヤハズダンゴ 188410
ヤハズトウヒレン 348752
ヤハズニシキギ 157285
ヤハズハハコ 21682
ヤハズハンノキ 16412
ヤハズヒゴタイ 348872
ヤハズマメ 111874
ヤハズマンネングサ 357232
ヤハズミゾホオズキ 255187
ヤハズ属 321161
ヤバネイモ 415201
ヤバネオオムギ 198288
ヤバネカズラ 390701
ヤバネカズラ属 390692
ヤバネシハイヒメバショウ 66165
ヤバネダンゴギク 188410
ヤバネヒイラギモチ 203660
ヤバネホオコ 21682
ヤハラグサ 42417
ヤハラゲガキ 132145
ヤハラスギ 113696
ヤハラスゲ 76573
ヤヒコザサ 347328
ヤブイチゲ 23705,23925
ヤブイチゴ 338516
ヤブイバラ 336829,336954
ヤブイボタ 229522
ヤブウツギ 413588
ヤブウド 192278
ヤブウルシ 393485
ヤブエビネ 65928
ヤブエンゴサク 105810
ヤブオニザミア 241449
ヤブカウジ 31477
ヤブカウジ科 261596
ヤブカウジ属 31347
ヤブガヤツリ 212783
ヤブガラシ 79850
ヤブカラシ属 79831
ヤブガラシ属 79831,92586
ヤブガラツ 79850
ヤブカンサウ 200736
ヤブカンゾウ 191284,191291
ヤブクロウメモドキ 328851
ヤブクワンザウ 191284,191291
ヤブケマン 106004
ヤブケマン属 105557
ヤブコウジ 31477
ヤブコウジ科 261596
ヤブコウジ属 31347
ヤブコブダネヤシ 270998
ヤブザクラ 83122,316466
ヤブサンゴ 407980
ヤブサンザシ 333970
ヤブジラミ 392992
ヤブジラミ属 392959
ヤブシロジクトウモドキ 303099
ヤブスゲ 75993,76051
ヤブスミレ 410030
ヤブタチバナ 31477
ヤブタバコ 77146
ヤブタバコ属 77145
ヤブタビラコ 221792
ヤブタビラコ属 221780
ヤブチョロギ 373125
ヤブツアズキ 409080
ヤブツバキ 69156,69196,69201
ヤブツルアヅキ 408841
ヤブデマリ 408027,408034
ヤブテリハノイバラ 336313
ヤブナシ 6515
ヤブニクケイ 91351
ヤブニッケイ 91351,91437,91438

ヤブニンジン 276455
ヤブニンジン属 276450
ヤブハギ 126501,126527,200740,226838
ヤブハッショウマメ 259545
ヤブヒョウタンボク 235921
ヤブビワ 164947
ヤブヘビイチゴ 138796
ヤブマオ 56181,56195,273903
ヤブマメ 20566,20570
ヤブマメ属 20558
ヤブマヲ 56181,56195
ヤブミョウガ 307324
ヤブミョウガラン 179620
ヤブミョウガ属 307312
ヤフムグラ 170517
ヤブムラサキ 66873
ヤブメウガ 307324
ヤブメウガ属 307312
ヤブヨモギ 36162
ヤブラネギ 15293
ヤブラン 232623,232631,232640
ヤブラン属 232617
ヤブリンゴ 243717
ヤブレガサ 381820
ヤブレガサモドキ 381823
ヤブレガサ属 283781,381813
ヤヘオヒョウモモ 316884
ヤヘカハ 235970
ヤヘガハカンバ 53400
ヤヘクチナシ 171253,171339,171374
ヤヘクワンザウ 191291
ヤヘザキクサギ 96009
ヤヘジキオホハンゴンサウ 339581
ヤヘジキオホハンゴンソウ 339581
ヤヘナリ 409025
ヤヘムグラ 170193
ヤヘムグラ属 170175
ヤヘヤマアオキ 258882
ヤヘヤマアヲキ 258882
ヤヘヤマアヲキ属 258871
ヤヘヤマカウゾリナ 55766
ヤヘヤマカテンサウ 262249
ヤヘヤマカラクサ 161418
ヤヘヤマキツネノボタン 326365
ヤヘヤマクマガユサウ 265373
ヤヘヤマコウリンクワ 55754
ヤヘヤマシキミ 204598
ヤヘヤマジフニヒトヘ 13063
ヤヘヤマシャクヂャウ 63977
ヤヘヤマススキ 168903
ヤヘヤマセイシクワ 330617
ヤヘヤマセンニンサウ 95350
ヤヘヤマタヌキマメ 112340
ヤヘヤマノイバラ 336405
ヤヘヤマハマナツメ 100061
ヤヘヤマハマナツメ属 100058
ヤヘヤマヒルギ 329765
ヤヘヤマヒルギ属 329744
ヤヘライラック 382371
ャボオノオレ 53444
ヤマアイ 47845,250610
ヤマアイモドキ 178851
ヤマアイ属 250597
ヤマアカザ 87037
ヤマアキチョウジ 209631
ヤマアキチョオジ 209631
ヤマアサ 195311,359654,375821
ヤマアサクラザンショウ 417302
ヤマアザミ 92407
ヤマアジサイ 200085,200089
ヤマアゼスゲ 74797
ヤマアヂサイ 200089
ヤマアヂサヰ 200089
ヤマアデク 382512
ヤマアハ 65330
ヤマアブラガヤ 353676
ヤマアラシガヤ 376917
ヤマアララギ 242169
ヤマアワ 65330
ヤマアワモドキ 65258
ヤマアヰ 47845,250610
ヤマアヰモドキ 178851
ヤマアヰ属 250597
ヤマイ 166518
ヤマイチゴ 338300
ヤマイナモリ 272227
ヤマイバラ 336914
ヤマイワカガミ 351446
ヤマウグイスカグラ 235812
ヤマウコギ 86855,143682,143685
ヤマウスユキソウ 224874
ヤマウタイス 270696
ヤマウツギ 200038
ヤマウツボ 222645
ヤマウツボ属 222641
ヤマウルシ 393488
ヤマエビ 411623
ヤマエンゴサク 105594,106076
ヤマオウレン 214971
ヤマオオイトスゲ 74141
ヤマオオウシノケグサ 164139
ヤマオダマキ 30005
ヤマオホバコ 76264
ヤマカイダウ 243710
ヤマカウバシ 231355
ヤマカエデ 2910
ヤマカガミ 20408
ヤマガキ 80260,132230,132320
ヤマカジ 61103
ヤマカシウ 366573
ヤマカシュウ 366573
ヤマカタバミ 277878
ヤマカヅラ 58619
ヤマカノコサウ 404269
ヤマカノコサウ属 404213
ヤマカノコソウ属 404213
ヤマカノツメソウ 299365
ヤマカバ 83145
ヤマカハゴジクワ 249633
ヤマカムロザサ 304116
ヤマカモジグサ 58619,58629
ヤマカモジグサ属 58572
ヤマガラシ 47949
ヤマガラシ属 47931
ヤマカリヤス 255918
ヤマカリン 376466
ヤマキ 381390
ヤマキケマン 106021,106214
ヤマキジカクシ 38977
ヤマギハナガサ 405812
ヤマギバナシノブ 99854
ヤマキビ 281984
ヤマキマヒルギ 329765
ヤマギリ 215442,320372,406016
ヤマキリンサウ 294114
ヤマグチテンナンショウ 33501
ヤマクハ 124925
ヤマグハ 124925,160347,259067
ヤマクボスゲ 74857
ヤマグミ 105146,142092
ヤマグルマ 399433
ヤマクルマバナ 96979
ヤマグルマ属 399432
ヤマグルミ 212621,302684
ヤマクロウメモドキ 328689
ヤマクワ 124923
ヤマグワ 160347,259097
ヤマクワガタ 407162
ヤマコウズ 96398
ヤマコウバシ 231355
ヤマコオゾリナ 55766
ヤマコガメ 117597
ヤマコショウ 231355
ヤマゴバウ 298093
ヤマゴバウ科 298124
ヤマゴバウ属 298088
ヤマゴボウ 91913,298093
ヤマゴボウ科 298124
ヤマゴボウ属 298088
ヤマコンニャク 20088,20098
ヤマゴンニャク 33496
ヤマサギゴケ 247012
ヤマサギサウ 302404
ヤマサギソウ 302404
ヤマザクラ 83234,83314
ヤマサトイモ 99910
ヤマザトタンポポ 384447
ヤマサンザシ 333970
ヤマザンショウ 369401
ヤマザンセウ 417330
ヤマシ 23670

ヤマジオウ 220389
ヤマジオウギク 103509
ヤマシオジ 168109
ヤマシグレ 408198
ヤマジスゲ 73906
ヤマジソ 259300
ヤマシデコイヌシデ 77411
ヤマジノキク 40652
ヤマジノギク 193939
ヤマジノタツナミソウ 355360
ヤマジノテンナンショウ 33496
ヤマジノホトトギス 396584
ヤマシバ 418426
ヤマシバカエデ 2868
ヤマシバリ 16433
ヤマシマハヒノキ 381310
ヤマシャクヂャウ 280207
ヤマシャクヤク 280207,280241
ヤマシュロ 32336
ヤマシロ 66809
ヤマシロギク 39928,41226
ヤマシロネコノメ 90438
ヤマシンチョウ 122482
ヤマズエンドウ 408288
ヤマスグリ 334061
ヤマスズメノヒエ 238647
ヤマスズメノヤリ 238647
ヤマズソウ 218347
ヤマズソウ属 218344
ヤマズヒメバショウ 66190
ヤマズル 411903
ヤマゼリ 276768
ヤマゼリ属 276743
ヤマセンキウ 24433
ヤマタイミンガサ 283883
ヤマタウガラン 416382
ヤマタチバナ 31477,280286
ヤマタツナミソウ 355674
ヤマタニタデ 91524,91553,91557
ヤマタヌキラン 73699
ヤマタネッケバナ 72971
ヤマタバコ 199911,228982,229175
ヤマタマガサ 82107
ヤマタマガサ属 82091
ヤマチサ 379374
ヤマヂスゲ 73906
ヤマチドメ 200350
ヤマヂノキク 193939
ヤマヂノギク 193918
ヤマチャ 376457
ヤマヂワウギク 103509
ヤマヂワウギク属 103402
ヤマツクシハギ 226678
ヤマツゲ 203670
ヤマツゲモドキ 382535
ヤマヅサ 231403
ヤマヅタヒ 356917
ヤマツツジ 330966
ヤマツツミ 407434
ヤマツバ 28341
ヤマツバキ 69156
ヤマツリフネソウ 204969
ヤマヅル 411903
ヤマデキ 400546
ヤマテキリスゲ 74546
ヤマデマリ 408034
ヤマテラシ 407897
ヤマテリハノイバラ 337015
ヤマトアオダモ 168018
ヤマトアヲダモ 168018
ヤマドウシン 200077,375821
ヤマドウダン 145727
ヤマトウツバキ 69552
ヤマトウバナ 97034
ヤマトキサウ 306879
ヤマトキソウ 306879
ヤマトキホコリ 142703,142704
ヤマトグサ 389186
ヤマトグサ科 389181
ヤマトグサ属 389183
ヤマトコロ 23670
ヤマトザクラ 83365
ヤマトテンナンショウ 33496
ヤマトニシキ 140005
ヤマトフウロ 174904
ヤマトボシガラ 164025
ヤマトホシクサ 151336
ヤマトマメ 408393
ヤマトミクリ 370061
ヤマトミズキ 380461
ヤマトヤナギ 342952
ヤマトラノオ 317949,317966
ヤマトラノヲ 317966
ヤマトリカブト 5193,5298
ヤマトリクサ 147050
ヤマトリサウ 147048
ヤマドリサウ 78002
ヤマトリモチ 46848
ヤマトレンギョウ 167444
ヤマトレンギョオ 167444
ヤマトレンゲウ 167444
ヤマトロロ 231355
ヤマナシ 243717,323268
ヤマナシクロカンバ 328660
ヤマナシテンナンショウ 33437
ヤマナズナ 18433
ヤマナスビ 292892
ヤマナラシ 311480,311545
ヤマナラシ属 311131
ヤマナンテン 296565
ヤマナンバンギセル 8760,8767
ヤマニガキ 155897
ヤマニガナ 320511,320520
ヤマニシキギ 157879
ヤマニワナズナ 18433
ヤマニンジン 28044,293033,293038
ヤマヌカボ 12053,12186,12240
ヤマネクンボ 239847
ヤマネコノメサウ 90325,90383
ヤマネコノメソウ 90325,90383
ヤマネコヤナギ 343151
ヤマノイモ 131645
ヤマノイモ科 131918
ヤマノイモ属 131451
ヤマノコギリサウ 3921,4028
ヤマノコギリソウ 3930,4014,4028
ヤマハウコ 21596
ヤマバウシ 124925
ヤマハギ 226698,226708,226721
ヤマハコベ 375117
ヤマバショウ 260228,260258
ヤマハゼ 393485
ヤマバセウ 260228,260258
ヤマハタザオ 30292
ヤマハタザホ 30292
ヤマハッカ 209702
ヤマハッカ属 209610,303125,324712
ヤマハヅモドキ 28381
ヤマハナソウ 349869
ヤマハハコ 21596
ヤマハハコ属 21506
ヤマハマナス 336522,336736
ヤマバラモンジン 354897
ヤマハリノキ 16469
ヤマハルユキソウ 32301
ヤマハンゲ 299727
ヤマハンシャウヅル 94866
ヤマハンショウヅル 94866
ヤマハンノキ 16359,16362,16466,16469
ヤマヒジキ 344496
ヤマビハ 151144,249449
ヤマヒハツ 28341
ヤマヒハツ属 28309
ヤマヒョウタンボク 235963
ヤマヒヨドリ 158358
ヤマヒヨドリバナ 158358
ヤマビル 15868
ヤマビワ 249449
ヤマビワソウ 333735
ヤマビワソウ属 333733
ヤマブキ 93860,216085
ヤマブキサウ 200754
ヤマブキサウ属 200753
ヤマブキショウマ 37060,37069,37085
ヤマブキショウマ属 37053
ヤマブキソウ 200754
ヤマブキソウ属 200753
ヤマブキツメクサ 396908
ヤマブキミカン 93871
ヤマブキ属 216084
ヤマフジ 414546,414577

ヤマブダウ 411623
ヤマブドウ 411623
ヤマブドオ 411624
ヤマブヒルガオ 102933
ヤマフブキ 94559
ヤマフヨウ 195305
ヤマヘビイチゴ 138796
ヤマボウシ 124923,124925
ヤマホウレンソウ 44468
ヤマホオコ 21596
ヤマホオコ属 21506
ヤマホオズキ 297625
ヤマボクチ 382069,382078
ヤマボクチゾク 382069
ヤマボクチゾク属 382067
ヤマボクチ属 382067
ヤマホソ 324190
ヤマホタルブクロ 70241
ヤマホトトギス 396604
ヤマホホヅキ 297625,367416
ヤマホロシ 367133,367262
ヤマママキ 306506
ヤマママメザクラ 83119
ヤマママルバノホロシ 367365
ヤマミカン 93501,114325,240813
ヤマミコシガヤ 75101
ヤマミズ 298945
ヤマミゾイチゴツナギ 305584
ヤマミゾソバ 291885
ヤマミツ 298945
ヤマミヤギノハギ 226912
ヤマムギ 144279
ヤマムグラ 170294
ヤマムラサキ 66808,66829,66856,66873
ヤマムロ 19248
ヤマモガシ 189918
ヤマモガシ科 316007
ヤマモガシ属 189910
ヤマモクレン 242169
ヤマモミジ 2781
ヤマモモ 261212
ヤマモモサウ 172201
ヤマモモサウ属 172185
ヤマモモソウ 172201
ヤマモモソウ属 172185
ヤマモモ科 261243
ヤマモモ属 261120
ヤマヤグルマギク 81219
ヤマヤグルマソウ 81214
ヤマヤナギ 343151,343911,344060,344112
ヤマヤマナスミヤマバラ 336736
ヤマユキノシタ 383890
ヤマユリ 229730
ヤマユリテングユリ 229872
ヤマユンジュ 94030
ヤマヨナメ 380924
ヤマヨメナ 215352
ヤマヨモギ 35948
ヤマラッキヤウ 15381,15822
ヤマラッキョウ 15381,15822
ヤマリウガン 189931
ヤマリンチャウ 122482
ヤマリンチョウ 122482
ヤマルミトラノヲ 317927
ヤマルリソウ 270696
ヤマルリソウ属 270677
ヤマルリトラノオ 317927,317939
ヤマワキオゴケ 409572
ヤマヰ 166518
ヤマンワヤ 61666
ヤマンワヤ属 61665
ヤミウシ 263744
ヤメラガシハ 141595
ヤモメカズラ属 292517
ヤヨイ 28470,86537
ヤライコウ 385752
ヤラッパ 208117
ヤラニラ 15822
ヤラブ 67860
ヤラボ 67860
ヤラメガシワ 141683
ヤラメスゲ 74221
ヤリクサ 17525
ヤリクサヨシ 293729
ヤリゲイトウ 80396
ヤリズイセン 210843
ヤリズイセン属 210689
ヤリスゲ 74969
ヤリテンツキ 166402,166411,166428
ヤリノホアカザ 257687
ヤリハリイ 143107
ヤリンボ 17525
ヤロオド 263199
ヤロード 263199
ヤワゲフウロ 174739
ヤワタソウ 288848
ヤワタソウ属 288847
ヤワラケガキ 132145
ヤワラスギ 113696
ヤワラスゲ 76573
ヤワラミヤマカンスゲ 75445
ヤン 133553
ヤンバルアカメガシハ 243395
ヤンバルアカメガシワ 248500
ヤンバルアハブキ 249445
ヤンバルアリドオシ 122078
ヤンハルウメモドキ 203578
ヤンバルエボシグサ 237506
ヤンバルオヒシバ 121285
ヤンバルカゴノキ 233945
ヤンバルカナメモチ 295729
ヤンバルカマツカ 295729
ヤンバルガンピ 414193
ヤンバルキヌラン 417808
ヤンバルコマ 190094
ヤンバルコマ属 190092
ヤンバルサイシン 37674
ヤンバルジュズネノキ 122078
ヤンバルスズカウジュ 355464
ヤンバルセンニンサウ 95113
ヤンバルセンニンソウ 95113
ヤンバルタムラサウ 345076
ヤンバルタムラソウ 345076
ヤンバルツルハクカ 227657
ヤンバルツルハクカ属 227537
ヤンバルツルハッカ 227572
ヤンバルツルマオ 313483
ヤンバルツルマバナ 227657
ヤンバルナスビ 367146
ヤンバルナセビ 367241
ヤンバルナミキサウ 355391
ヤンバルノキク 115595
ヤンバルハグロサウ 129243
ヤンバルハグロソウ 129243
ヤンバルハコベ 138482
ヤンバルハコベ属 138470
ヤンバルヒゴタイ 115595
ヤンバルヒゴタイ属 406040
ヤンバルビハ 151144
ヤンバルマユミ 157912
ヤンバルミゾハコベ 52592
ヤンバルミチャナギ 309602
ヤンバルミミズバイ 381412
ヤンバルミョウガ 19491,19493
ヤンバルヲヒシバ 121285
ユ 93515
ユアヌルロア属 212553
ユ-アンテ属 155432
ユウカオ 67736
ユウガオ 219827,219851
ユウガオ属 219821
ユウガギク 40609
ユウカリ 155589
ユウカリジュ 155589
ユウカリノキ 155589
ユウカリノキ属 155468
ユウキシダレ 342954
ユウギリ 192445
ユウギリソウ 393608
ユウキリソウ属 393604
ユウゲシャウ 269498
ユウゲショウ 255711,269498
ユウコウギョク 352758
ユウコクラン 232164
ユウシウマル 45887
ユウシュンラン 82048
ユウスゲ 191318
ユウゼンギク 40923
ユゥソゥギョク 45236
ユウソウマル 163472
ユウチョウカ 409624

ユウヅルエビネ 65863
ユウナミ 123909
ユウバリアズマギク 151012
ユウバリカニツリ 126061
ユウバリキンバイ 312751
ユウバリクモマグサ 349122
ユウバリコザクラ 315136
ユウパリコザクラ 315136
ユウバリソウ 220200
ユウバリツガザクラ 297022
ユウバリリンドウ 174110
ユウパリリンドオ 174110
ユウヒギョク 86612
ユウフブキチュウ 94562
ユウホウマル 45886
ユウレイソウモトギ 258044
ユウレイラン 130043
ユウレツリュウ 395661
ユエフバイ 20970
ユーカリ 155589
ユーカリジュ 155589
ユカリッバキ 68920
ユーカリノキ 155589
ユーカリノキ属 155468
ユカン 296554
ユキイヌノヒゲ 151295
ユキイロハマエンドウ 222737
ユキグニカンアオイ 37649
ユキグニハリスゲ 76221
ユキクラヌカボ 12127
ユキゲユリ 87759,87765
ユキザサ 242691
ユキザサ属 366142
ユキツバキ 69207,69214
ユキナシ 323246
ユキノシタ 290573,349936
ユキノシタ科 350063
ユキノシタ属 349032
ユキノハナ 169719
ユキハギ 226914
ユキハタザオ 30164
ユキバトウヒレン 348954
ユキハナガタ 23209
ユキバヒゴタイ 348206
ユキマンテマ 363167
ユキミギク 359980
ユキミソウ 345310
ユキモチサウ 33505
ユキモチソウ 33505
ユキヤシオ 331464
ユキヤツデ 128115
ユキヤナギ 372101
ユキヤブケマン 106006
ユキョセサウ 215811
ユキョセソウ 215811
ユキヨモギ 35915
ユキワリイチゲ 23868
ユキワリコサクラ 314387
ユキワリサウ属 192120
ユキワリシホガマ 287802
ユキワリソウ 192139,314370,314377
ユギワリソウ 314366
ユキワリソウ属 192120
ユークニブ 93872
ユクノキ 94032
ユゲネザサ 304072
ユゲヒョウカン 93874
ユコウ 93875
ユーコミス属 156011
ユサン 216120,216125
ユサン属 216114
ユズ 93515
ユスノキ 134946
ユズノハカズラ 313197
ユズノハカヅラ 313197
ユズノハカヅラ属 313186
ユスラウメ 83347
ユスラバヤナギ 343063
ユスラヤシ 31013
ユスラヤシドキ 31015
ユスラヤシ属 31012
ユズリハ 122700
ユズリハワダン 110604
ユズリハ科 122656
ユソウボク 181505
ユソウボク属 181496
ユゾゴゼンタチバナ 105204
ユチャ 69411
ユーチャリス属 155857
ユッカ属 416535
ユッタディンテリア属 214911
ユヅヒハアヂサイ 199910
ユヅヒハアヂサヰ 199910
ユヅリハ 122700
ユヅリハ属 122661
ユデナシ 323268
ユバズサ 134010
ユバリタンポポ 384867
ユヒキツ 93677
ユフガオ属 219821
ユフガホ属 219821
ユフスゲ 191312
ユフダケオトギリ 201968
ユーホルビア属 158385
ユミグサ 308613
ユメノシマガヤツリ 118646
ユメルレア属 212695
ユメーレア属 212695
ユモトマムシグサ 33431
ユモトマユミ 157882
ユーユーヤシ属 46374
ユヨウバイ 20970
ユリアザミ 228511,228529
ユリアザミ属 228433
ユリアリア 64982
ユーリアレ属 160635
ユーリクニア属 157163
ユリグルマ 177251
ユリグルマ属 177230
ユリザキムクゲ 194971,195179
ユリズイセン 18076
ユリスチグマ属 160928
ユーリスティグマ属 160928
ユリッバキ 69199
ユリノキ 232609
ユリノキ属 232602
ユリバツバキ 69199
ユリラン 62985
ユリワサビ 161149
ユリワサビモドキ 97983
ユリ科 229700
ユリ属 229717
ユールアメリカナポプラ 311146
ユレハギ 98159
ユレバハギ 98159
ユーロフィア属 156521
ユーロフィエラ属 157145
ユワンギ 381404
ユンジュ 240114
ユンナンポプラ 311584
ユンベールティア属 199252
ヨーアニア属 416378
ヨウエン 82327
ヨウキヒ 107026
ヨウギョク 160012
ヨウサイ 207590
ヨウシュアキノキリンソウ 368480
ヨウシュイヌナズナ 137131
ヨウシュイブキジャコウソウ 391365
ヨウシュイボタ 229662
ヨウシュイボタノキ 229662
ヨウシュエンレイソウ 397570
ヨウシュオグルマ 207046
ヨウジュカノコソウ 404316
ヨウシュカラマツ 221855
ヨウシュカンボク 407989
ヨウシュキダチルリソウ 190622
ヨウシュギワリソウ 314366
ヨウシュクサノオウ 86755
ヨウシュクモマクサ 349856
ヨウシュコニスビ 239755
ヨウシュシモツケ 166132
ヨウシュジンチョウゲ 122515
ヨウシュタカネアズマギク 150454
ヨウシュチョウセンアサガオ 123077
ヨウシュトチノキ 9701
ヨゥシュトリカブト 5442
ヨウシュヌマガヤ 256601
ヨウシュネズ 213702
ヨウシュノツルキンバイ 312360
ヨウシュハクセン 129618

ヨウシュハッカ　250294
ヨウシュハナシノブ　307197
ヨウシュヒナギキョウ　412697
ヨウシュフクジュソウ　8387
ヨウシュホオズキ　297643
ヨウシュヤマゴボウ　298094
ヨウセイカク　373806
ヨウセイマル　234921
ヨウタケ　47345
ヨウトウ　45725
ヨウムギョク　55657
ヨウラククサアジサイ　73123
ヨウラクサウ　49886
ヨウラクソウ　49886
ヨウラクタマアジサイ　199915
ヨウラクツツアナナス　54454
ヨウラクツツジ　145677,250532
ヨウラクツツジ属　250505
ヨウラクニワトコ　345652
ヨウラクボク　19463
ヨウラクボク属　19462
ヨウラクボタン　220729
ヨウラクユリ　168430
ヨウラクラン　267958
ヨウラクラン属　267922
ヨウレイ　82399,264811
ヨウレイマル　234911,234913
ヨウロウ　140000
ヨオラクツツジ　250532
ョキミカゲソウ属　102856
ヨクセイ　242352
ヨグソアヅサ　53631
ヨグソカンバ　53631
ヨグソミネバリ　53465
ヨグンミネバリ　53631
ヨケロッパグリ　78811
ヨコイドオシ　52225
ヨコイドホシ　52225
ヨコグラノキ　52478
ヨコヅナ　107076
ヨコハマダケ　304069
ヨコメガシ　116106,323952
ヨコヤマリンドウ　173478
ヨコヤマリンドオ　173478
ヨゴレ　116926
ヨゴレタンポポ　384848
ヨゴレネコノメ　90407
ヨザキスイレン　267698
ヨサチマキ　347273
ヨシ　295888,295919
ヨシガシ　233228
ヨシガヤ　265916,265923
ヨシスス　341837
ヨシススキ　341837
ヨシタケ　37467
ヨシノアザミ　92241
ヨシノキスゲ　191312
ヨシノザクラ　83365
ヨシノシズカ　88289
ヨシノシヅカ　380131
ヨシノニンジン　280793
ヨシノヤナギ　344300
ヨシノユリ　229730
ヨシヒメマル　234940
ヨシュアノキ　416563
ヨシ属　295881
ヨソゾメ　407785
ヨッアミダケンチャ　6827
ヨッシベヤナギ　344199
ヨッシベヤマネコノメ　90384
ヨツズミ　407785
ヨッドゾメ　407785
ヨツバシオガマ　287079,287300
ヨツバセンナ　360434
ヨツバハギ　408508
ヨツバハコベ　307755,318184,318501
ヨツバヒヨドリ　158136
ヨツバヒヨドリバナ　158305
ヨツバムグラ　170678
ヨツバリキンギョモ　83575
ヨツマタモウセンゴケ　138326
ヨドガワツツジ　332144
ヨドボケ　84573
ヨナクニイソノギク　40081
ヨナクニカモメヅル　409573
ヨナクニトキホコリ　142901
ヨナバ　298734
ヨノミトリグサ　179438
ョヒュウガナツミカン　93834
ヨブスマソウ　283820,283825
ヨメナ　41516
ヨメナノキ　210389
ヨメナ属　215334
ヨメノツジラ　229569
ヨメフリ　311480
ヨモギ　35634,35638,36097,36474
ヨモギギク　383874
ヨモギギク属　383690
ヨモギダコチク　347359
ヨモギナ　35770
ヨモギ属　35084
ヨルガオ　67736,207570
ヨルガオ属　67734
ヨルガホ　67736
ヨルガホ属　67734
ヨルザキアラセイトウ　246443
ヨルザキマンテマ　363827
ヨルソケイ　267430
ヨルノジョオウ　357748
ヨレスギ　113703
ヨレハナビ　179384,230787
ヨレハユリ　229789
ヨロイグサ　24325
ヨロイドオシ　52225
ヨーロッパアカマツ　300223
ヨーロッパイチイ　385302
ヨーロッパキイチゴ　338557
ヨーロッパグリ　78811
ヨーロッパクロマツ　300117
ヨーロッパクロヤマナラシ　311398
ヨーロッパスモモ　316382
ヨーロッパナラ　324335
ヨーロッパニレ　401468,401593
ヨーロッパハイマツ　299842
ヨーロッパハンノキ　16352
ヨーロツパブドウ　411979
ヨーロッパブナ　162400
ヨーロッパベンケイソウ　356928
ヨーロッパポプラ　311414
ヨーロッパマンネングサ　356468
ヨロ-ッパミセバヤ　356535
ヨーロッパモミ　272
ヨーロッパヤマナラシ　311537
ヨロヒドホシ　52246
ライオンニシキ　273822
ライカク　374040,374041
ライケツマル　103756
ライコウ　10918
ライシャイモ　131761
ライシャエビネ　65887
ライジン　10925
ライジンカク　307170
ライチ　233078
ライチイ　233078
ライデンボク　392841
ライトウギョク　263748
ライトヤシ　4951
ライトヤシ属　286038
ライネッキア属　327524
ライホウマル　140889
ライム　93329
ライムギ　356237
ライムギモドキ　198267,198354
ライムギ属　356232
ライラック　382329
ラウススゲ　76413
ラウトウ　201392
ラカンマキ　306457,306469
ラクウショウ　385262
ラクウショウ属　385253
ラクエフショウ　221894
ラクエフスギ　385262
ラ-クスバ-　124301
ラクダガヤ　4163
ラクタギリ　285959
ラクトウ　109097
ラグナリア　220228
ラグナリア属　220226
ラクヨウショウ　221894
ラグルス属　220252
ラゲサカキ　160606

ラケナリアペンジユラ 218920
ラケナリア属 218824
ラジアータマツ 300173
ラジアタマツ 300173
ラシャウモンカヅラ 247739
ラシャウモンカヅラ属 247720
ラシャカキグサ 133478
ラシャキビ 58201
ラシャナス 367139
ラシュウギョク 284662
ラシュワコザクラ 314713,314948
ラショウモンカズラ 247739
ラショウモンカズラ属 247720
ラショウモンソウ 137647
ラショオモンカズラ 247739
ラスブニア属 375516
ラセイタソウ 56100
ラセイタタマアジサイ 199919
ラセイマル 182430
ラセンイ 213037,213067
ラセンサウ 399261
ラセンサウ属 399190
ラセンソウ 399261
ラセンソウ属 399190
ラセントウ 399261
ラセントウ属 399190
ラソッギョク 264366
ラッカセイ 30498
ラッキヤウ 15185
ラッキョウ 15185
ラッキョウアナナス 391985
ラッキョウアヤメ 208946
ラッキョウヤダケ 318309
ラッキョウラン 232122,232364
ラッパギリア 175672
ラッパグサ 81694
ラッパグサ属 81692
ラッパズイセン 262441
ラッパバナ 366866
ラッパバナ属 366863
ラデイアタマツ 300173
ラテイベス 93529
ラドー 93692
ラパゲーリア属 221340
ラバンジュラ属 223248
ラビエア属 324918
ラピダリア属 221515
ラフ・ストークド・メドウ・グラズ 306098
ラフィア属 326638
ラフィサムヌス属 329164
ラフィドフォラ属 328990
ラフゲ 312478
ラブドタムヌス属 328445
ラフノキ 393479
ラフバイ 87525
ラフバイモドキ 134010
ラフバイ科 68300
ラフバイ属 87510
ラフレシア 325062
ラフレシア科 325065
ラフレシア属 325061
ラフレモン 93505
ラベーニア属 327115
ラベニア属 327117
ラベンダー 223251
ラボウシ 383407
ラミー 56229,56233
ラミューム属 220345
ラムキョ 15185
ラモンダ属 325223
ラルディザバラ属 221815
ラワンデル属 223248
ランガク 158386
ランカンサルナシ 6530
ランカンシデ 77381
ランカンンボサウ 302519
ランギク 78012
ラングロイスキャベツヤシ 161056
ランサ 221226
ランザット 221226
ランサ属 221224
ランシウム属 221224
ランジョウ 304353
ランシンボク 300980
ランジンモドキ 264222
ランダイイチゴ 338425
ランダイイハウメ 362251
ランダイカウバシ 347411
ランダイカウバシ属 347402
ランダイガシ 79064
ランダイキミヅ 142794
ランタイグス 91429
ランダイグス 91429
ランダイコウバシ属 347402
ランダイサンキライ 366460
ランダイシキミ 204533,204575
ランダイシャシャンポ 403976
ランダイシュウカイドウ 49974
ランダイスギ 114539,114548
ランダイセキコク 125044
ランタイニクケイ 91429
ランダイニクケイ 91429
ランダイヒカゲツツジ 331420,331421
ランダイフタギ 381154
ランダイミズ 142793
ランダイムラサキ 66904
ランタイヤマサギサウ 58389
ランタナ 221238
ランタナ属 221229
ランチュウソウ 122759
ランチュウソウ属 122758
ランデンマ 416382
ランバイ 46193
ランフィーゴヘイヤシ 228764
ランブータン 265129
ランプランサス属 220464
ランポウカク 43510
ランポウギョク 43509
ランヨウアオイ 37571
ランリュウ 182464
ラン科 273267
リアトリス 228529
リアトリス属 228433
リウイストナ 234194
リウオウハゼ 346408
リウキウア 47845
リウキウアイ 47845
リウキウアハブキ 249445
リウキウアヰ 47845
リウキウアヲキ 319810
リウキウイチゴ 338487
リウキウイナモリ 272241
リウキウウキクサ 301025
リウキウエビネ 66101
リウキウカウガイ 215510
リウキウガキ 132291
リウキウカゴノキ 233945
リウキウガネブ 411735
リウキウカンコノキ 177153
リウキウコクラン 232164
リウキウコザクラ 23323
リウキウコスミレ 409843
リウキウシキミ 204598
リウキウシュスラン 179650
リウキウジュロ 393809
リウキウシュンギク 89704
リウキウシロスミレ 409716
リウキウセキコク 299625
リウキウソヨゴ 203871
リウキウチク 304054
リウキウツッジ 331284
リウキウトベラ 301294,301360
リウキウトロロアフヒ 235
リウキウハゼ 393479
リウキウバライチゴ 338305
リウキウハンゲ 401156
リウキウハンゲ属 401148
リウキウヒマゴ 337253
リウキウベンケイ 215174
リウキウベンケイ属 215086
リウキウマツ 300044
リウキウマッナ 379499
リヴキウマメ 65146
リウキウマユミ 157698
リウキウミヤマトベラ 155890
リウキウムクゲ 195149
リウキウモクカウ 229175
リウキウユリ 229900
リウキウルリミノキ 222133
リウキウワウバイ 211821

リウキウヰ 119160,119162
リウキンクワ 68189
リウキンクワ属 68157
リウクリンソウ 314624
リウトウンボサウ 302415
リウノヒゲモ 378991
リウヒガヤ 114539
リカステ属 238732
リ－ガル・リリ－ 230019
リ－キ 15621
リギダマツ 300180,300181
リキュウバイ 161749
リキュウマル 244090
リ－ク 15621
リクチュウダケ 357945
リクチュウナナカマド 369299
リグナムバイタ 181505
リコリス 239266
リコリス・スプリンゲリ 239278
リサンアラカシ 116200
リショウリュウ 395644
リシリイ 213114
リシリオウギ 42385,43030
リシリカニツリ 398531,398547
リシリゲンゲ 278765
リシリスゲ 76201
リシリソウ 417925
リシリソウ属 417880
リシリトウウチソウ 345835
リシリヒナグシ 282563
リシリビャクシン 213712,213931
リシリリンドウ 173540
リシリヰ 213114
リズサンザシ 29325
リステラ属 232890
リスレア属 233791
リセキシロバヒ 381235
リソスペルマム属 233692
リソ－ポメロ 93748
リチドフィルム属 333861
リ－チラン 116815
リッソキ－ルス属 156521
リッチ－チャボウチワヤシ 262235
リット－ニア属 234130
リッピア属 232457
リトウザンヨモギ 35136
リトウテン 264372
リトウトンボ 302415
リトウトンボサウ 302404
リトウトンボソウ 302404
リトウハクチョウゲ 360933
リトウハクテウゲ 360933
リトウヤドリギ 385208,385230
リトカルブス属 233099
リト－プス属 233459
リナム属 231856
リネフィルム属 329578
リネルミ－ザ属 329594
リバッツジ 331421
リプサリス属 329683
リプサリドプシス属 329679
リベットコムギ 398986
リベリアコ－ヒ－ノキ 98941
リベリャコ－ヒ－ 98941
リ－ベンボルシ－スゲ 75095
リボンガヤ 34936
リボングラス 293710
リボンザサ 272614
リムナンセマム属 267800
リムナンテス科 230183
リムナンテス属 230205
リムノカリス属 230244
リムノキ属 121084
リメッタ 93534
リメッタ－オ－ディネ－ル 93534
リモガソバニラ 405028
リモニア 93546
リモニュ－ム・ペレジ－ 230712
リモニュ－ム属 230508
リャウブ 96466
リャウブ科 96561
リャウブ属 96454
リュウエン 82380
リュウオウカク 199053
リュウオウマル 185158
リュウガギョク 103744
リュウカク 72425
リュウガン 131061,163479
リュウキュウアイ 47845
リュウキュウアオキ 319810
リュウギュウアオギ属 319392
リュウキュウアケボノソウ 380368
リュウキュウアセビ 298758
リュウキュウアマモ 117309
リュウキュウアリドオシ 122031
リュウキュウアワブキ 249424,249463
リュウキュウイ 119160,119162
リュウキュウイチゴ 338487
リュウキュウイナモリ 272241
リュゥキュウイモ 207623
リュウキュウイワウチワ 362251
リュウキュウウマノスズクサ 34247
リュウキュウウロコマリ 225178
リュウキュウエビネ 65865,66001,66101
リュウキュウオウバイ 211821
リュウキュウガキ 132291
リュウキュウガシワ 117571
リュウキュウカラスウリ 396224
リュゥキュウギク 89481
リュウキュウクルマバナ 227697
リュウキュウクロウメモドキ 328767
リュウキュウコオガイ 215510
リュウキュウコクタン 132139
リュウキュウコケリンドウ 173844
リュウキュウコザクラ 23323
リュウキュウコスミレ 410420
リュウキュウコマツナギ 206761
リュウキュウコンテリギ 199942
リュウキュウサンザシ 109924
リュウキュウシャマシキミ 365965
リュウキュウジュズネノキ 122078
リュウキュウジュロ 393816
リュウキュウシュンギク 89704
リュウキュウシロスミレ 409725
リュウキュウスガモ 388360
リュウキュウスガモ属 388357
リュウキュウスギ 114539
リュウキュウスゲリウギウスゲ 73636
リュウキュウスズカケ 407471
リュウキュウセッコク 148712,299625
リュウキュウソヨゴ 203860
リュウキュウタイゲキ 85768
リュウキュウタチスゲ 75280
リュウキュウタラノキ 30749
リュウキュウチク 304054
リュウキュウチシャノキ 141687
リュウキュウツチモチ 46826
リュウキュウツツジ 331284
リュウキュウツバキ 69327
リュウキュウツルウメモドキ 80221
リュウキュウツルグミ 142056
リュウキュウツルコウジ 31574
リュウキュウツワブキ 162627
リュウキュウテイカカズラ 393656
リュウキュウテリハノイバラ 336700
リュウキュウトロロアオイ 235
リュウキュウナガエサカキ 8270
リュウキュウヌスビトハギ 200727
リュウキュウハイノキ 381104
リュウキュウハギ 227013
リュウキュウハグマ 12611
リュウキュウバショウ 260206,260241
リュウキュウハゼ 346408,393479
リュウキュウハナイカダ 191185
リュウキュウハリギリ 215445
リュウギュウハリギリ 215442
リュウキュウハンゲ 401152,401156
リュウキュウハンゲ属 401148
リュウキュウヒエスゲ 74149
リュウキュウヒメアブラススキ 71614
リュウキュウヒメハギ 308170
リュウキュウフジウツギ 62110
リュウキュウフシグロ 363203
リュウキュウフヂウツギ 62110
リュウキュウベンケイ 215174
リュウキュウベンケイ属 215086
リュウキュウホウライカズラ 171456
リュウキュウマツ 300044
リュウキュウマメ 65146
リュウキュウマメガキ 132216
リュウキュウマメ属 65143

リュウキュウマユミ　157698
リュウキュウミヤマシキミ　365962,365963
リュウキュウムクゲ　195149
リュウキュウモクセイ　276268,276362
リュウキュウモチ　204365
リュウキュウモミ　114539
リュウキュウヤツデ　162889
リュウキュウヤノネグサ　309051
リュゥキュウヤブカラシ属　92586
リュウキュウヤブラン　232640
リュウキュウヤマツツジ　331839
リュウキュウユリ　229900
リュウキュウヨツバムグラ　170407
リュウキュウルリミノキ　222133
リュウギョク　352765
リュウキンカ　68189,68201
リュウキンカ属　68157
リュウグラジョウ　108828
リュウケツジュ　137382
リュウケツジュ属　137330
リュウコ　163438
リュウココリネ属　227813
リュウジュマル　264356
リュウショウカク　199020
リュウジンザンショウ　417306
リュウジンバク　261706
リュウジンボク　261706
リュウゼッサイ　219307
リュウゼッサウ　10788
リュウゼツラン　10788
リュウゼツラン科　10775
リュウゼツラン属　10778
リュウセン　496
リュウセンクワ　96257
リュウソウキョク　103749
リュウゾウジヤナギ　342930
リュウノウジュ　138548
リュウノウジュ属　138547
リュウノタマ　367522
リュウノツメガヤ　121285
リュウノツメビエ　121285
リュウノヒゲ　272090
リュウノヒゲモ　378991
リュウノヤブマオ　56342
リュウヒ　114539
リュウホウギョク　163442
リュウマギョク　103754
リュウラン　182428
リュウリンギョク　103764
リュウリンジュ　339869
リュウリンジュ属　339868
リュエリア属　339658
リュキュウヤナギ　367178
リュキュガキ　132291
リュキュクロキ　381327
リュキュツワブキ　162627
リュキュヨモギ　35237
リューコステレ属　228134
リューコスペルマム属　228029
リュセイギョク　264343
リュノウギク　89603
リューヒテンベルギア属　227755
リョウウン　86585
リョウウンカク　94556
リョウケツジュ　137382
リョウケンガリヤス　94608
リョウトウエンゴサク　106571
リョウトウニワトコ　345653
リョウトウブシ　5363
リョゥトウマンシウ　372038
リョウトウミチヤナギ　309334
リョゥトゥヨモギ　36453
リョウノウアザミ　91996
リョウブ　96466
リョウブ科　96561
リョウブ属　96454
リョウボウ　96466
リョウメイ　82295
リョウリギク　124826
リョウリュリ　229730
リョオブ　96466
リョクガクザクラ　83228
リョクガクバイ　34455
リョクキマル　247671
リョクチク　11362,47518,125453,125455
リョクチク属　11329
リョクトウ　109293,409025
リョクヨウカンラン　59524
リョクラン　70989
リョジョ　158809
リングア属　360582
リンゴ　243675
リンコステイリス・プレイオネッサ　333731
リンコスティリス属　333725
リンゴバショウ　260252
リンコレリァ属　333101
リンゴ属　243543
リンショウバイ　83238
リンダウ　173847,173852
リンダウ科　174100
リンダウ属　173181
リンチャウ　122532
リンチョウ　122532
リンデローフィア属　231243
リントウ　281138
リンドウ　173847,173852
リンドウツリガネヤナギ　289350
リンドウ科　174100
リンドウ属　173181
リンドレイクサギ　96180
リンネーア属　231676
リンネサウ　231679,231680
リンネサウ属　231676
リンネソウ　231679
リンネソウ属　231676
リンボウ　159305
リンボウキク　350175
リンポウラン　416538
リンボク　223131,316824
ルイヨウショウマ　6408,6414
ルイヨウショウマ属　6405
ルイヨウボタン　79732
ルイヨウボタン属　79730
ルイラサウ属　339658
ルイラソウ属　339658
ルウダサウ　139678
ルウダソウ　86934,139678
ルエルリア属　339658
ルカウサウ属　323450
ルカウソウ属　323450
ルガモットハッカ　250341
ルクカラクサ属　263501
ルクマ属　238110
ルクレア属　238100
ルコウソウ　323465
ルコウソウ属　323450
ルシア属　340548
ルスガス　122124,340990
ルスン　381404
ルゾンガマズミ　407853
ルゾンピンポン　376092
ルゾンホシクサ　151369
ルゾンヤマノイモ　131691
ルチョウギョク　233650
ルドベキア　339574
ルドベッキア属　339512
ルナア　238371
ルナア属　238369
ルナリア属　238369
ルバーブ　329388
ルビーガヤ　333024
ルピナス　238463
ルピヌス属　238419
ルベシベザサ　347272
ルベラナズナ　72059
ルミー　93554
ルリアザミ　81787,158263,158264
ルリイチゲ　23868
ルリイチゲサウ　23701,321667
ルリイチゲソウ　23701
ルリイロツルナス　367596
ルリカコソウ　13143
ルリカラクサ　263505
ルリギク　377288
ルリギク属　377286
ルリキョウ　272951
ルリコウ　159916
ルリゴクラクチョウカ　377563
ルリコシ　95219
ルリサウ　233786

ルリザクラ　153502,153534
ルリジサ　57101
ルリジソ　19459
ルリシャクジョウ　63977
ルリシャクヂャウ　63977
ルリスイレン　267658
ルリソウ　233786,270699
ルリソウ属　270677
ルリタマアザミ　140776
ルリダマノキ　222172
ルリヂサ属　57095
ルリヂシャ属　57095
ルリチョウチョウ　234363
ルリテフサウ　247739
ルリトウワタ　278547
ルリトウワタ属　278546
ルリトラノオ　317968
ルリトラノオ属　317904
ルリニガナ　79276
ルリニカナ属　79271,79295
ルリニガナ属　79271
ルリニワゼキショウ　365747
ルリネキ　15220
ルリハクカ　19459
ルリハクカ属　19458
ルリハコベ　21339,21340
ルリハコベ属　21335
ルリハッカ　19459
ルリハッカ属　19458
ルリハナガサ　148087
ルリバナカヅラ　390822
ルリハナガヤ属　148053
ルリハンショウズル　95304
ルリヒエンサウ　102833
ルリヒエンソウ　102833,124141
ルリヒゴタイ　140712,140786
ルリヒゴタイ属　140652
ルリヒナギク　163131
ルリヒナギク属　163121
ルリブクロ　333086,333089
ルリブクロ属　333083
ルリホホヅキ属　395708
ルリマガリバナ　61119
ルリマガリバナ属　61118
ルリマツリ　305172
ルリマツリモドキ　83646
ルリマツリモドキ属　83641
ルリマツリ属　305167
ルリマル　163425
ルリミゾカクシ　234363,234447
ルリミノウシコロシ　381137
ルリミノキ　222172
ルリミノキ属　222081
ルリムスカリ　260301
ルリヤナギ　367178
ルルニティディウム属　246704
ルロウマル　155220
ル丰エフショウマ　6414
ル丰エフショウマ属　6405
ル丰エフボタン　79732
ル丰エフボタン属　79730
ルンフソテツ　115888
レイオン　192454
レイギョク　86620
レイケオカクタス属　327498
レイケステーリア属　228319
レイコウギョク　233500
レイザン　272530
レイシ　233078
レイシギョグ　366825
レイシュウギョク　284683
レイショウカク　385151
レイジン　108905
レイジンカズラ　5021
レイシ属　233077
レイセイマル　327264
レイゼンギョク　275410
レイソウオウ　253093
レイダマル　182439
レイチイ　233078
レイテンギョク　233683
レイハイカク　198053
レイフギョク　182453
レイホウマル　327314
レイポチス属　224558
レイポルツティア属　224244
レイムギョク　263746
レイヨウマル　107059
レイリョウカウ　397208
レイリョウカウ属　397186
レイリョウコウ　397208
レイリョウコウ属　397186
レウィシア属　228256
レウボフ　96466
レウリバショウ　260253
レウリバセウ　260253
レウリユリ　229730
レオセレウス属　224521
レオヌールスゾク属　224969
レオノチス属　224558
レオボルドヤシ属　225037
レキシア属　329419
レゲネラータポプラ　311157
レジノーザマツ　300180
レスケナウルティア属　226669
レースソウ　29326,29663,29676
レースソウ科　29696
レストレピア属　328219
レダマ　370202
レダマキリン　159975
レダマ属　370191
レックステイネリア属　327322
レックスベゴニア　50235
レッドウッド　360565
レッドスプルース　298423
レツドラワン　362212
レヅム属　223888
レドゲリアナ　91075
レナギバガシ　233181
レナンセラ属　327681
レバノン・ジーダー　80100
レバノンジーダー　80100
レパンテス属　225061
レビスチカム属　228244
レピスミウム属　225691
レプティア属　327250
レプトスペルマム属　226450
レプトセレウス属　225974
レプトーテス属　226539
レブンアツモリソウ　120400
レブンイワレンゲ　275370
レブンウスユキソウ　224827
レブンコザクラ　314376
レブンサイコ　63558,63864
レブンシオガマ　287082
レブンソウ　279000
レブンタカネツメクサ　31745
レブントウヒレン　348719
レブンナニワズ　122471
レブンハナシノブ　307203
レブンフウロ　174586
レマイレオセレウス属　224335
レーマンニア属　327429
レモガンサルナシ　6530
レモン　93539
レモンエゴマ　290940
レモングラス　117153
レモンユウカリ　106793
レモンユーカリ　106793
レモンリアル　93456
レーリア属　219668
レンギヤウ　167456
レンギョウ　167456
レンギョウウツギ　167456
レンギョウエビネ　66001
レンギョウ属　167427
レングエ　263272
レンゲ　43053
レンゲイワヤナギ　343749
レンゲウ　167456
レンゲウ属　167427
レンゲギボウシ　198608
レンゲサウ　43053
レンゲショウマ属　24166,24177
レンゲソウ　43053
レンゲツツジ　331257,331258
レンゲバナ　43053
レンゲボク　232609
レンゲボタン　331263
レンゲボタンゲツツジ　331263
レンゲマル　234927

レンザアザミ 91958
レンザン 337157
レンシャウタウ 45725
レンジョウカク 83908
レンズマメ 224473
レンズ属 224471
レンバイ 46193
レンブ 382660
レンプクサウ 8399
レンプクサウ科 8406
レンプクサウ属 8396
レンプクソウ 8399
レンプクソウ科 8406
レンプクソウ属 8396
レンリサウ 222800,222807
レンリサウ属 222671
レンリソウ 222800,222828
レンリソウ属 222671
ロアサ科 234254
ロアサ属 234251
ロイルツリフネソウ 204982
ロウウン 249591
ロウエン 82345
ロウエンダイ 382992
ロウオウ 299266
ロウギョク 404946
ロウゲツ 86551
ロウココヤシ属 307538
ロウシウギョク 284706
ロウジュラク 155303
ロウシン 198499
ロゥソゥギョク 45232
ロウソクノキ 284497
ロウソクノキ属 284490
ロウテッオミナエシ 285821
ロウトウ 201392
ロウトウギョク 263749
ロウノウバニラ 405028
ロウノキ 346408,393479
ロウバイ 87525
ロウバイスケロクラン 223636
ロウバイモドキ 134010
ロウバイ科 68300
ロウマカミツレ 85526
ロウマカミツレ属 26738
ロウレルザクラ 223116
ロウレルノキ 223203
ロオレルカズラ 390822
ロガニア属 235246
ロクオンサウ 117364
ロクオンソウ 117364,117385
ロクカクアヤメ 208606
ロクカクサボテン 83890
ロクカクヤナギ 343070
ロクガツミカン 93752
ロクサントセレウス属 94551
ロクジャウコケモモカマツカ 107575
ロクジョウマンネングサ 357147
ロクダウボク 87729
ログハ 259085
ロクベンシモツケ 166132
ロクベンシモツケソウ 166132
ロクベンヤブツバキ 69193
ロクモギ 379374
ロクロギ 379374
ログワ 259152
ロクワイ 16817,17383
ロクワイ属 16540
ローケア属 335070
ロサウ 259085
ロシャハギ 20005
ロスコーエア属 337065
ロストラータユーカリ 155517
ロスラリア属 337292
ロセォカクタス属 337153
ローセーフローラ 69572
ローゼリサウ 195196
ローゼリソウ 195196
ローゼル 195196
ローゼルソウ 195196
ローセンベルギア属 337137
ロソウ 259085,259152
ロゾク 369700
ローソンヒノキ 85271
ロータス属 237476
ロタントウ 65782
ロチンベルガモット 93712
ロッカクソウ 360493
ロッカクヤナギ 343070
ログウッド 184518
ロックハーティア属 235131
ロッコウヤナギ 342928
ロッジポールペイン 299869
ロッシャーヤシ 337060
ロッシャーヤシ属 337059
ロッフォフォラ属 236423
ロディゲスベニオウギヤシ 222634
ロドーキトン属 329986
ロドスパサ属 332253
ロードヒポキシス属 332184
ロドフィアラ属 332224
ロドリゲーチア属 335161
ロドレイア属 332204
ローナス属 235474
ロニングルミ 212636,212640
ロビビア属 234900
ロビロビ 166776
ロビロリーパイン 300264
ロフォステモン属 236461
ロフォセレウス属 236362
ロフォフォラ属 236423
ロブスタコーヒーノキ 98872,98886
ロプスタポプラ 311158
ロブスタユーカリ 155722
ロブラーリア属 235065
ロブロリーパイン 300264
ロベージ 228250
ロベリアソウ 234547
ロベリア属 234275
ロベリヤ 234547
ロベリヤソウ 234547
ローベルヒメタケヤシ 139372
ロボウガラシ 133313
ローマカミツレ 85526
ローマカミツレ属 26738
ローマカミルレ 85526
ロマトフィム属 235474
ロマトフィルム属 235503
ロームセカミツレ 85526
ロムネヤ属 335869
ロムレア属 335879
ロリデュラ属 336149
ロリネンシスポプラ 311144
ローレル 223203
ローレルカズラ 390822
ローレルカヅラ 390822
ロワズルーリア属 235279
ロングリーフパイン 300128
ロンボフィルム属 332353
ワイゼイノボリフジ 238469
ワイゼイノボリフヂ 238469
ワイヘルディア属 404942
ワインニセダイオウヤシ 161059
ワインベリー 339047
ワウゴテフ 65055
ワウゴン 355387
ワウゴンガシハ 323630
ワウゴンサウ 202400
ワウゴンサウ属 202387
ワウゴンチク 297464
ワウゴンハギ属 105269
ワウシュクバイ 211931
ワウセイ 308538
ワウニクジュ 67529
ワウバイ 211931
ワウバイ属 211715
ワウラン 82051
ワウレン 103835
ワウレン属 103824
ワガキ 366284
ワカキノサクラ 83235
ワカサトウヒレン 348930
ワカサハマギク 89766
ワカバキャベツヤシ 161057
ワクヅル 94740
ワクノテ 94740
ワクラハ 403738
ワケギ 15170,15291
ワケノカワヤナギ 342940
ワサビ 161154
ワサビダイコン 34590

ワサビノキ 258945
ワサビノキ科 258961
ワサビノキ属 258935
ワサビ属 161120,413273
ワジキギク 89404,124780
ワシベッミヤマコウボウ 196122
ワジュロ 393809
ワシントンヤシ 413298
ワシントンヤシモドキ 413307
ワシントンヤシ属 413289
ワシントンルピナス 238481
ワスルナグサ 260868
ワスレグサ 191263,191286,191291,239280
ワスレグサ属 191261
ワスレタヌキモ 403191
ワスレナグサ 260747,260868
ワスレナグサ属 260737
ワスレナサウ 260868
ワスレナサウ属 260737
ワスレナソウ属 260737
ワセアキグミ 142151
ワセイチゴ 338516,338985
ワ－セウィッチア属 413261
ワ－セウィッチェラ属 413254
ワセオバナ 341912,341915
ワセオホイタビ 165086
ワセスイバ 340178
ワセビエ 140360
ワセボウブラ 114294
ワセワタスゲ 152759
ワセヲバナ 341912
ワタ 179878,179900
ワタイヌゴマ 373166
ワタエビ 411735
ワタカッコウ 373166
ワタガヤ 156513
ワタクヌギ 324532
ワタクワクカウ 373166
ワタゲカマツカ 295808
ワタゲスミレ 410075
ワタゲソモソモ 305687
ワタゲツルハナグルマ 31169
ワタゲトウヒレン 348344
ワタゲハナグルマ 31162
ワタゲベゴニア 49943
ワタゲミヤコグサ 237689
ワタシホガマ 287334
ワタスゲ 152791,152792
ワタスゲ属 152734
ワダソウ 318501
ワタタビ 6682
ワタチョロギ 373166
ワダツミノキ 266800
ワタドロ 311389
ワタナ 103509
ワタナベソウ 288850
ワタヌキ 324532
ワタネカズラ 294844
ワタノキ 56802
ワタフキノキ 218121
ワタフキヒゴタイ 348243
ワタフジ 351073
ワタムキアザミ 92431
ワタメフタギ 381310
ワタヨモギ 35527
ワタリスゲ 75792
ワタリミヤコグサ 237617
ワダン 110623
ワダンノキ 125433
ワタ属 179865
ワチガイサウ 318496
ワチガイサウ属 318489
ワチガイソウ 318496
ワチガイソウ属 318489
ワチソウ 318501
ワックスフラワ－ 122786
ワトソニア 413326
ワトソ－ニア属 413315
ワトソニア属 413315
ワニグチサウ 308567
ワニグチソウ 308567
ワニグモチダマ 259512
ワニナシ 291494
ワニナシ属 291491
ワビロウ 234194
ワ－ブルクドコロ 131593
ワラス 319114
ワラベナカセ 319833
ワラベノカンザシ 239268
ワリチャマラヤヒメヤシ 203513
ワリンゴ 243551,243675
ワルタビラコ 20837
ワルナスビ 367014
ワルブルギモチ 204396
ワルレア属 413243
ワレナ 104218
ワレバヒメトウ 121495
ワレモカウ 345881
ワレモカウ属 345816
ワレモコウ 345881,345888,345921
ワレモコウ属 345816
ワ－レンベルギア属 412568
ワンジュ 49140
ワンドスゲ 73751
ワンビ 94068
ワンピ 94207
ワンビ属 94169
ワンリュウヤナギ 343670
ワンンサザンカ 69694
ヰ 213036
ヰグサ 213036
ヰズヰ 308613
ヰゼキシャウ 213507
ヰノコヅチ 4273,4275
ヰノコヅチモドキ 115738
ヰノコヅチモドキ属 115697
ヰノコヅチ属 4249
ヰンゲサウ 383103
ヰンゲンササゲ 294056
ヰンゲンナ 53249
ヰンゲンマメ 294056
ヰンゲンマメ属 293966
ヰ科 212752,212753
ヰ属 212797
ヲカウコギ 143685
ヲカサダケ 57437
ヲガサハライチビ 228,934
ヲガサハラマツ 15953
ヲガサハラモクレイシ 172885
ヲガサハラモクレイシ属 172881
ヲガサハラヤブニクケイ 91438
ヲガサラッツジ 330233
ヲカスズメノヒエ 238675
ヲガタマ 252849
ヲガタマノキ 252849
ヲガタマノキ属 252803
ヲカトラノヲ 239594
ヲカトラノヲ属 239543
ヲカヒジキ 344718
ヲカヒジキ属 344425
ヲカボ 275958
ヲカミル 344718
ヲガラバナ 3732
ヲガルカヤ 117181
ヲガルカヤ属 117136
ヲカヲグルマ 385903
ヲギ 255873
ヲギノツメ 200652
ヲギノツメ属 200589
ヲギヨシ 255873
ヲグルマ 207046,207151
ヲグルマ属 207025
ヲグルミ 212626
ヲケラ 44208
ヲケラ属 44192
ヲゴケ 143022
ヲサラン 148754
ヲタカラカウ 229179
ヲタガラカウ 229179
ヲダマキ 30031
ヲダマキ属 29994
ヲタルスゲ 75638
ヲダンクワ 402267
ヲトコゼリ 326417
ヲトコビシ 394513
ヲトコヒルギ 61263
ヲトコヘシ 285880
ヲトコベシ 285880
ヲトコメシ 285880
ヲトコヨウゾメ 408024
ヲトコヨモギ 35674

ヲドリコサウ属 220345
ヲナガカエデ 2872,3058
ヲナガクマシデ 77268
ヲナガシ 379414
ヲナガタケ 304098
ヲナガヤブニンジン 276455
ヲナモミ 415057
ヲナモミ属 414990
ヲノオレ 53610
ヲノオレカンバ 53610
ヲノヘスゲ 76515
ヲノヘテンツキ 166329
ヲノマンネングサ 356884
ヲノヲレ 53610
ヲノヲレカンバ 53610
ヲバナ 255886
ヲヒシバ 143530
ヲヒシバ属 143512
ヲヒメクグ 218571
ヲヒルギ 61263
ヲヒルギ属 61250
ヲヒルムシロ 312190
ヲヘビイチゴ 312690
ヲボンテンクワ 402250
ヲマキザクラ 83224
ヲマツ 300281
ヲミナヘシ 285859
ヲミナベシ 285859
ヲミナヘシ科 404392
ヲミナヘシ属 285814
ヲミナメシ 285859
ヲメキグサ 354691
ヲヤマヲグリマ 207219
ヲンツツジ 332104
ンアンモック 392836
ンコウカク 158926
ンポイカク 198040
ンリエゴノキ 379347
ンリ－コクラン 232253
ンンチョウマル 163453
ンンパレッティア属 101626

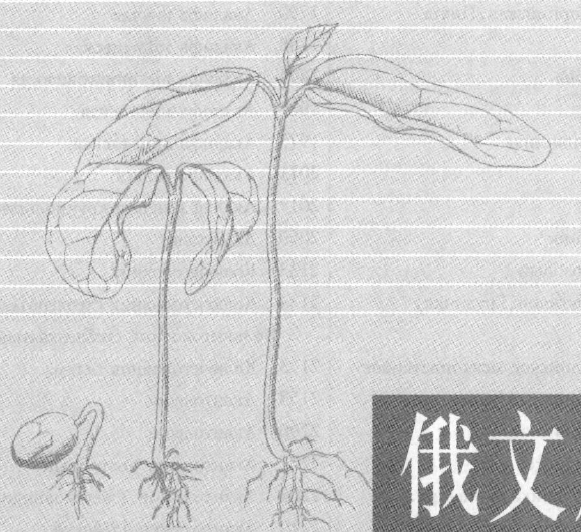

俄文序号—名称索引

Russian

A Dictionary of Seed Plant Names
种子植物名称

91 Абелия
113 Абелия китайская
117 Абелия корейская
119 Абелия щитковидная
167 Абелия пильчатая
177 Абелия японская
184 Абелия японская мелкоцветковая
200 Абелия мелколистная
212 Абелимош
219 Абелимош, Бамия, Бамья, Гибискус сабдарифа, Гибискус съедобный, Гомбо, Окра, Розелла
269 Пихта
272 Пихта белая, Пихта гребенчатая, Пихта европейская, Пихта обыкновенная, Пихта серебристая
280 Пихта миловидная
282 Бальзамник, Горечавка баварская, Пихта бальзамическая, Пихта канадская
299 Пихта ворнмюллвра
306 Пихта греческая
309 Пихта чаюйская
314 Пихта киликийская
317 Миристицевые, Мускатниковые, Пихта одноцветная
333 Пихта далиская
338 Пихта мотоская
362 Пихта факсона
365 Пихта крепкая, Пихта сильная
377 Пихта фразера
383 Пихта изящная, Пихта камчатская, Пихта тонкая
384 Пеларгоний попелечнополосатый
388 Клёновые, Пихта цельнолистная, Пихта черная
389 Пихта почкочешуйчатая, Пихта равночешуйчатая
396 Пихта каваками, Пихта тайваньская
398 Пихта корейская
403 Пихта аризонская
407 Пихта лоуа
409 Пихта великолепная, Пихта красивая
414 Пихта мариес
427 Пихта белокорая, Пихта почкочешуйная, Пихта почкочешуйчатая
429 Пихта кавказкая, Пихта нордманна, Пихта нордманнова
435 Пихта алжирская, Пихта нордмана, Пихта нубийская, Пихта нумидийская
439 Пихта западногималайская
442 Пихта испанская
453 Клён обыкновенный, Клён остролистный, Клён платановидный, Пихта благолодная
458 Пихта мексиканская, Пихта священная
460 Пихта сахалинская
463 Пихта сахалинская
464 Пихта вильсона
471 Пихта семенова
476 Пихта сибирская
480 Пихта гималайская, Пихта замечательная
496 Пихта вича, Пихта вура
507 Пихта калифорнийская, Пихта прелестная
510 Пихта вильсона
639 Абобра
641 Абобра тонколистная
672 Аброма
687 Аброния
738 Абрус, Чёточник
765 Абрус, Боб розовый
837 Абутилен, Абутилон, Грудника, Канатник
907 Абутилон индийское мелкоцветковое
913 Канатник гибридный
934 Абутилон индийское
989 Абутилон китайское
1000 Грудника, Грудника авиценны, Грудничник, Грудничник авиценны, Канан, Канатник, Канатник теофраста
1024 Акация
1049 Акация безжилковая, Мульга
1060 Акация вооруженная
1069 Акация бейли
1120 Акация дубильная, Акация катеху, Мимоза
1145 Акация тайваньская
1156 Акация ножевидная
1159 Акация синелистная
1165 Акация австралийская, Акация беловатая, Акация деальбада, Акация подбеленная, Акация серебристая, Мимоза
1168 Акация низбегающая
1176 Акация лицзянская
1177 Акация куэньминская
1219 Акация фарнеза, Акация фарнезе, Акация фарнеси
1268 Акация хайнаньская
1292 Дерево фиалковое, Пластинниковые, Пластинчатые, Шампиньоновые
1297 Акация устрашающая
1358 Акация длиннолистная
1384 Акация австралийская, Акация чернодревесная, Акация чернодревная, Дерево австралийское черное, Дерево чёрное, Дерево чёрное австралийское
1427 Акация аравийская, Акация нильская
1516 Акация густоцветная
1528 Акация камеденосная, Акация стойкая
1544 Акация красноватая
1572 Акация сенегальская, Рогористниковые
1686 Акация мутовчатая
1711 Акация юньнаньская
1739 Ацена
1769 Акалифа
1781 Акалифа тайваньская
1790 Акалифа южная
1888 Акалифа хайнаньская
1894 Акалифа шетинистоволосая
1900 Акалифа индийская
1975 Акалифа лицзянская
2011 Акалифа уилкса
2017 Акалифа уилкса крупнолистная
2060 Акантовые
2133 Колючеголовник
2134 Колючеголовник стеблеобъемлющий, Колючеголовник стеблеохватывающий
2135 Колючеголовник бетама
2158 Акантолепис
2206 Акантолепис
2207 Акантолепис восточный
2208 Акантолимон, Ежеголовниковые
2210 Акантолимон алайский
2211 Акантолимон алатавский
2213 Акантолимон альберта
2214 Акантолимон александра
2216 Акантолимон армянский
2217 Акантолимон аулиеатинский
2218 Акантолимон овсовый
2219 Акантолимон нежный
2220 Акантолимон бородина
2221 Акантолимон прицветниковый
2222 Акантолимон
2224 Акантолимон гвоздичный
2225 Акантолимон плодный
2226 Акантолимон
2227 Акантолимон диапенсиевидный
2230 Акантолимон екатерины
2231 Акантолимон иглистый
2232 Акантолимон красноватый
2233 Акантолимон
2234 Акантолимон фетикова
2235 Акантолимон фомина
2236 Акантолимон гауданский
2237 Акантолимон пленчатый
2238 Акантолимон гончарова
2239 Акантолимон дедина
2240 Акантолимон гиссарский
2241 Акантолимон гогенакера
2242 Акантолимон
2244 Акантолимон карелина
2245 Акантолимон карелина
2247 Акантолимон хорасанский
2248 Акантолимон кнорринг
2249 Акантолимон кокандский
2250 Акантолимон королькова

2252 Акантолимон рыхлый
2254 Акантолимон тонкохвостниковый
2257 Акантолимон плауновый
2259 Акантолимон маева
2260 Акантолимон маргариты
2262 Акантолимон микешина
2263 Акантолимон минжелкинский
2264 Акантолимон удивительный
2265 Акантолимон необычайный
2266 Акантолимон
2267 Акантолимон никитина
2268 Акантолимон нуратавский
2269 Акантолимон
2270 Акантолимон памирский
2271 Акантолимон мелкоцветковый
2274 Акантолимон распростертый
2275 Акантолимон пскемский
2276 Акантолимон крылоприцветниковый
2277 Акантолимон хорошенький
2278 Акантолимон пурпурный
2279 Акантолимон
2280 Акантолимон радде
2282 Акантолимон рупрехта
2283 Акантолимон сакена
2284 Акантолимон сахендский
2285 Акантолимон шемахинский
2286 Акантолимон
2287 Акантолимон
2288 Акантолимон растопыренный
2289 Акантолимон
2290 Акантолимон торчащий
2291 Акантолимон тарбагатайский
2292 Акантолимон татарский
2293 Акантолимон тонкоцветковый
2294 Акантолимон тяньшанский
2295 Акантолимон титова
2296 Акантолимон
2297 Акантолимон
2298 Акантолимон варивцевой
2299 Акантолимон бархатистый
2301 Акантолимон зеленый
2372 Акантопапакс тяньцюаньский
2395 Акантопапакс китайский
2412 Акантопапакс генри
2505 Акантопапакс узколистный
2521 Акантопапакс тибетский
2531 Акантопапакс чжэцзянский
2550 Акантофеникс
2555 Колючелистник
2565 Колючелистник колючий
2657 Акант, Акантус
2672 Акант мелкоцветковый
2684 Акант падуболистный
2695 Акант мягколистный
2711 Акант колючий
2765 Ацелидант антиклейный
2769 Клен, Клён
2785 Клен цзяньшуйский
2791 Клен аньхуйский
2798 Клен бородатожилковый, Клен бородатый
2811 Клен бюргера
2815 Клен тайваньский
2837 Граб сердцевидный, Клен полевой, Пакленок, Черноклениня
2844 Клен полевойпушистоплодный
2846 Клен волосистоножковый
2849 Клён весёлый, Клен кападокийский, Клён светрый
2866 Клен кападокийский мелколистный
2868 Клен граболистный
2884 Клен чанхуаский
2893 Клен завитой, Клён завитой
2896 Клен виноградолистный
2910 Клён боярышниколистный, Клен воярышниколистный
2920 Клен давиба
2923 Клен давиба крупнолистный
2941 Клен расходящийся
2943 Клён граболистный
2946 Клен эмэйский
2970 Клен лицзянский
2972 Клен франше
2978 Клен даньбаский
2980 Клен фупинский
2984 Гинала, Гиннала, Клен гиннала, Клён гиннала, Клен дальневосточный, Клен приречный, Клён приречный, Клен речной, Клён речной, Приречный клен
3003 Клён бумахный, Клен серый, Клён серый
3009 Клен гуйцзоуский
3010 Клен хайнаньский
3013 Клён гельдрейха
3023 Клен хуанпинский
3024 Клен ху
3025 Клен гиблидный
3029 Клен гирканский
3030 Клен гирканский, Клен грузинский, Клен иберийский
3034 Клен японский, Клён японский
3035 Клен японский аконитифолиум
3037 Клен японский мелколистный
3050 Клен японский крупнолистный
3055 Клен цзиндунский
3062 Клен цзянхиский
3063 Клен цюцзянский
3064 Клен комарова
3065 Клен гуншаньский
3067 Клен фэн
3070 Клен гуйлиньский
3075 Клен гладгий
3083 Клен ланьпинский
3089 Клен лаоюйский
3097 Клен лэйбоский
3101 Клен личуаньский
3102 Клен линьаньский
3103 Клен лин
3117 Клен чэнбуский
3119 Клен наньчуаньский
3127 Берёзовик, Берёзовик обыкновенная, Гриб подберёзовый, Гриб чёрный, Клен крупнолистный, Обабок, Подберёзовик, Подберёзовик обыкновенный, Подобабик, Подобабок
3129 Клен маньчжурский, Клён маньчжурский
3130 Клен ганьсуский
3131 Клен мабяньский
3137 Клен максимовича
3140 Клен майра
3145 Клен мяотайский
3149 Клен мийябе
3154 Клен мелколистный, Клен моно
3206 Клен монпельйский, Клён монпельйский, Клён трёхлопастный
3211 Клен мулиский
3216 Клен наюн
3217 Клен хунаньский
3218 Клен американский, Клён американский, Клен виргинский, Клен негундо, Клен перистолистный негундо, Клен перистый, Клен ясенелистный, Клён ясенелистный, Негундо, Неклён, Неклен вирдинский, Неклён вирдинский
3232 Клен ясенелистный вариегатум
3233 Клен ясенелистный фиолетовый
3236 Клён чёрный
3243 Ярофа, Ятропа
3255 Клен продолговатый
3266 Клен продолговатый эмэйский
3273 Клен корейский
3277 Клен оливера
3278 Клен оливера тайваньская
3286 Клен итальянский
3294 Клён критский
3297 Клен фунинский
3300 Клен веерный, Клён веерный, Клен дланевидный, Клён дланевидный, Клён пальмовидный
3301 Клен дланевидный атролинеаре
3302 Клен дланевидный 'Атропурпуреум'
3309 Клен дланевидныйрассеченный
3342 Клен дланевидный орнатум
3343 Клен дланевидный ретикулятум
3344 Клен дланевидный розео-маргинатум
3345 Клен дланевидный рубрум
3346 Клен дланевидный сангвинеум
3347 Клен дланевидный версиколор
3374 Клен пакс
3393 Клен пенсильванский, Клён

пенсильванский
3398 Клён красивый
3425 Клён обыкновенный, Клен остролистный, Клён остролистный, Клен платановидный, Клён платановидный
3462 Белый клен, Клён белый, Клен ложноплатановый, Клён ложноплатановый, Клен ложный платан, Клен явор, Клён явор, Сикамор, Сикомора, Явор
3479 Клен ложнозибольдов
3480 Клен ложнозибольдов мелкоплодный
3481 Клен опушеный
3484 Клен гуандунский
3494 Клен регеля
3498 Клен хэнаньский
3505 Клен красный, Клён красный
3529 Клён рыжеватожилковый, Клен рыжежилковый
3532 Клен сахаристый, Клён сахаристый, Клен серебристый, Клён серебристый
3549 Клен сахарный, Клён сахарный
3581 Клен яньюаньский
3587 Клен семенова
3593 Клен шансиский
3603 Клен сичоуский
3604 Клен зибольда, Клён зибольда
3615 Клен китайский
3622 Клен китайский мелкоплодный
3629 Клен колосистый
3645 Клен стевена
3648 Клен сычуаньский
3649 Клен тяньцюаньский
3656 Клен татарский, Неклен, Черноклен, Чина танжерская
3665 Клен зеленокорый, Клен покровный
3676 Клен мелколистный
3692 Клен чаюйский
3696 Клен гуансиский
3699 Высокогорный кавказский клен, Клён высокогорный, Клён кавказский, Клён кавказский высокогорный, Клен траутфеттера, Клён траутфеттера
3714 Клен усеченный
3718 Клен чоноски, Клен чоноского
3725 Клен тукурменский
3726 Клен туркестанский
3732 Дедюлэ, Клен березовый, Клен дедюлэ, Клен желтый, Клен укурундийский
3736 Клен бархатистый, Клён бархатистый, Клён бархатный, Клен велечественный, Клён величественный
3747 Клен цзиньюньский
3759 Клен яошаньский
3765 Кленовые, Клёновые
3913 Ахиллея, Дерганец, Деревей, Тысячелистник
3914 Тысячелистник заостренный
3921 Тысячелистник альпийский
3922 Тысячелистник камчатский
3934 Тысячелистник азиатский
3937 Тысячелистник биберштейна
3938 Тысячелистник дважпы-пильчатый
3939 Тысячелистник бореальный
3941 Тысячелистник хрящеватый
3943 Тысячелистник сжатый
3945 Тысячелистник клиновиднодольный
3946 Тысячелистник расставленный
3948 Тысячелистник таволговый, Тысячелистник таволголистый
3951 Тысячелистник голый
3953 Тысячелистник недотрога
3956 Тысячелистник керманский
3964 Тысячелистник широколопастный
3966 Тысячелистник ледебура
3967 Тысячелистник тонколистный
3973 Тысячелистник большой
3976 Тысячелистник мелкоцветковый
3977 Тысячелистник подовй
3978 Ахиллея тысячелистная, Миллефоль, Трава тысячелистника, Тысячелистник обыкновенный
3995 Пахлук, Тысячелистник мускатный, Тысячелистник мускусный
3998 Тысячелистник благородный
4000 Тысячелистник бледно-желтый
4004 Тысячелистник паннонский
4005 Ахиллеа-зоря, Птармика, Трава чихотная, Тысячелистник птармика, Тысячелистник чихотный, Чихотник обкновенный
4009 Тысячелистник крупноголовый
4014 Тысячелистник птармиковидный
4020 Тысячелистник сахокиа
4021 Тысячелистник иволистный
4024 Тысячелистник шишкина
4025 Тысячелистник шура
4026 Тысячелистник северный
4027 Тысячелистник щетинистый
4052 Тысячелистник торчаший
4055 Тысячелистник узколистный
4060 Тысячелистник червеобразный
4061 Тысячелистник вильгельмса
4070 Ахименес
4113 Ахнаерум, Чий
4119 Чий костеровидный
4123 Чий раскидистый
4134 Ковыль даленивосточный, Ковыль развесистый
4145 Чий длинностный
4160 Ковыль сибирский
4163 Чий, Чий блестящий
4249 Ахирантес, Самоцвет, Соломоцвет
4263 Ахирантес индийский
4273 Ахирантес двузучатый, Соломоцвет двузубый
4275 Ахирантес японский
4308 Ахирантес крупнолистный
4319 Ахирантес мелколистный
4661 Душевика, Душёвка альпийская, Щебрушка
4662 Душевик альпийский, Душевика альпийская, Душёвка обыкновенная, Чабер альпийский
4668 Душевика фомина
4669 Душевика пахучая
4866 Кресс бразильский, Кресс масляный, Шпилат, Шпилат трyванистый
4941 Акнистус
4986 Акомастилис
4995 Акомастилис росса
5014 Аконит, Борец
5019 Борец алатавский
5021 Борец бело-фиолетовый
5031 Аконит алтайский, Борец алтайский
5039 Аконит антора, Аконит противоядный, Борец волкобойник, Борец противоядный
5041 Борец анторовидный, Борец противоядновидный
5042 Борец мелколепестный
5043 Борец дуговидный
5048 Борец байкальский
5052 Аконит бороданый, Аконит бородатый, Борец бороданый
5065 Аконит двуцветковый
5066 Борец биробиджанский
5116 Аконит чаюйский
5127 Аконит цзиндунский
5144 Борец корейский
5151 Аконит толстолистный
5153 Борец ладьевидный
5154 Аконит чекановского
5160 Аконит десулави
5172 Аконит мотоский
5193 Аконит фишера
5195 Аконит извилистый
5198 Аконит плетевидный
5202 Аконит флерова
5204 Аконит тайваньский
5205 Аконит лицзянский
5208 Аконит франше
5283 Аконит хуйлиский
5284 Аконит хуйцзэский
5287 Аконит
5292 Аконит ялусский
5298 Аконит японский
5304 Аконит цзилунский
5314 Борец каракольский
5318 Борец гиринский

5326 Аконит комарова
5332 Аконит крылова
5333 Аконит гуншаньский
5335 Борец кусцнецова
5347 Аконит шерстистоустый，Волкобойник
5351 Борец мяньнинский
5359 Борец белоустый
5360 Борец хэбэйский
5403 Борец большеносый
5413 Аконит большой，Борец крупный
5422 Борец горный
5442 Аконит аптечный，Аконит жинокостнолистный，Аконит напелюс，Аконит реповидный，Аконит сборный，Аконит фиолетоный，Аконит ядовитый，Борец аптечный，Борец голубой，Борец синий
5451 Аконит носатый
5453 Борец лесной
5461 Аконит восточный
5466 Аконит теневой
5468 Аконит метельчатый
5484 Аконит мулиский
5486 Аконит чжундяньский
5498 Аконит мелкоцветковый
5502 Аконит лэйбоский
5506 Аконит опушенный
5525 Аконит радде
5539 Аконит фэн
5543 Борец круглолистный，Борец круглолистый
5546 Борец сахалинский
5548 Аконит сапожникова
5558 Борец щукина
5563 Аконит высокий，Аконит северный
5570 Аконит сиотинский
5572 Борец синьцзянский
5574 Борец горнокитайский
5581 Аконит смирнова
5585 Аконит джунгарский，Борец джунгарский
5593 Аконит ядунский
5600 Аконит штёрка
5611 Аконит сукачева
5618 Аконит таласский，Борец таласский
5619 Аконит илийский
5626 Аконит кандинский
5636 Аконит траншеля
5659 Аконит пестрый，Аконит пёстрый
5663 Аконит волосистый，Борец мохнатый
5674 Аконит вьющийся，Борец вьющийся
5676 Аконит волкобойный，Аконит волчий，Борец волкобойный，Борец лисий
5706 Аконит яньюаньский
5788 Аир，Лепеха
5793 Аир，Аир болотный，Аир многолетный，Аир обыкновенный，Аир тро стниковый，Ирный корень，Корень аирный，Корень ирный，Корень чёрный
5803 Аир травянистый，Аир элаковый
5809 Аир злаковый
6128 Акрокомия
6136 Акрокомия южноамериканская
6233 Акронихия
6237 Акронихия австралийская，Акронихия бауэра
6272 Горчак
6278 Горчак，Горчак ползучий，Горчак ядовитый，Калады-какря，Талхак
6405 Актеа，Воронец，Вороняжка
6408 Воронец заостренный
6414 Воронец азиатский
6422 Воронец красноплодный
6448 Воронец волосистый，Воронец колосистый，Воронец колосовидный
6513 Актинидия
6515 Актинидия аргут，Актинидия изогнутая，Актинидия крупная，Актинидия острая，Актинидия острозубиатая，Кишмиш крупный
6519 Актинидия джиральди
6523 Актинидия японская
6530 Актинидия мозолистая
6552 Актинидия чэнкоуская
6553 Актинидия киви，Актинидия китайская
6624 Актинидия гуйлиньская
6627 Актинидия хэнаньская
6628 Актинидия мэнцзыская
6632 Актинидия хубэйская
6638 Актинидия цзянкоуская
6639 Актинидия коломикта，Кишмиш обыкновенный，Коломикта，Крыжовник амурский，Максимовник，ползун
6669 Актинидия гуансиская
6682 Актинидия носатая，Актинидия полигама，Актинидия полигамная
6692 Актинидия красностовольная
6726 Актинидия чжэцзянская
6728 Актинидиевые
6830 Астинолема
6831 Астинолема синеголовниковая
6832 Астинолема крупнообертковая
6865 Фимбристилис китайский
6888 Актиностемма
6907 Актиностемма лопастная
7018 Адансония
7022 Адансония，Баобаб，Баобаб африканский，Дерево обезьяньe хлебное
7113 Аденандра
7178 Аденантера
7190 Аденантера павлинья，Дерево коралловое，Железняк，Пернамбук，Фернамбук
7427 Аденокаулон，Прилипало
7433 Аденокаулон сросшийся
7596 Аденофора，Бубенчик，Бубенчики
7649 Бубенчики гмелина，Короставник коронопусолистный
7657 Короставник голубинцевой
7661 Бубенчики гималайский
7672 Короставник якутский
7678 Бубенчики ламарка
7682 Бубенчик лилиелистный，Колокольчик льнюлистный
7695 Бубенчик максимовича
7697 Бубенчик мелкоцветковый
7702 Бубенчик тайваньский
7716 Бубенчики нинсяский
7739 Бубенчики широколистный
7806 Короставник скальный
7811 Бубенчик китайский
7816 Бубенчики узкоцветковый，Короставник курчавый
7846 Короставник крымский
7850 Бубенчик четырехлистный
7853 Бубенчики трахелиевидный
7855 Бубенчики трехконечный
8034 Аденостилес
8038 Аденостилес чесночная
8170 Адина
8199 Адина красноватая
8205 Адинандра
8298 Адлумия
8299 Адлумия азиатская
8324 Адонис，Горицвет，Желтоцвет，Златоцвет，Черногорка
8325 Адонис летний，Горицвет летний，Златоцвет летний
8331 Адонис амурский
8332 Адонис однолетний，Адонис осенний，Глазки павлиньи，Горец осенний，Горицвет осенний，Павлиньй глазки，Удо
8343 Адонис золотистый，Златоцвет золотистый
8362 Адонис огненный，Адонис пламенный，Горицвет огненый
8373 Златоцвет ийнепалск
8374 Златоцвет мелкоцветный
8381 Адонис сибирский，Златоцвет сибирский，Стародубка
8382 Адонис сычуаньский
8383 Адонис тяньшанский，Златоцвет тяньшанский
8385 Адонис туркестанский
8387 Адонис，Адонис весенний，Горицвет весенний，Горицвет черногорка，Желтоцвет весенний，Златоцвет весенний，Трава Горицвета，Черногорка

8389 Адонис пушистый, Златоцвет пушистый
8390 Адонис волжский
8396 Адокса, Мускусница
8399 Адокса, Адокса мускусная, Бесславник, Мускусница обыкновенная, Мускусница
8405 Адокса тибетская
8406 Адоксовые
8544 Коринкарпус, Эхмея
8645 Егицерас
8660 Бодлак, Коленница, Эгилёпс, Эгилопс
8665 Эгилёпс двуфостый, Эгилопс двхдюймовый
8671 Эгилёпс толстый, Эгилопс толстый
8672 Эгилопс цилиндрический
8683 Эгилопс жювенальский
8684 Эгилопс кочи
8695 Эгилёпс овальный, Эгилопс овальный
8727 Эгилопс оттопыренный
8728 Эгилёпс трехостый, Эгилопс трехостый
8732 Эгилёпс трехдюймовый, Эгилопс трехдюймовый
8810 Снедь-трава, Сныть
8811 Сныть альпийская, Сныть горная
8824 Сныть широколистная
8826 Снытка пухлая, Снытка съедобная, Сныть обыкновенная
8828 Сныть таджикстная
8848 Элления
8849 Элления ушастая
8850 Солянка сизая
8855 Прибрежница
8862 Ажрык, Прибрежница прибрежная, Прибрежница солончаковая, Шор-аджерик
8870 Прибрежница синьцзянская
8881 Прибрежница ползучая
8884 Прибрежница китайская
9417 Эшинантус
9489 Амбач, Эшиномена
9531 Сола
9675 Каштан конский, Конский каштан
9679 Каштан конский калифорнийский, Павия калифорнийская
9680 Конский каштан мясо-красный
9683 Каштан конский китайский, Конский каштан китайский
9684 Конский каштан чжэцзянский
9692 Каштан жёлтый американский, Каштан конский восьмитычнковый, Конский каштан восьмитычинковый, Павия жёлтая
9694 Каштан конский гладкий, Каштан конский голый, Конский каштан голый
9701 Каштан конский обыкновенный, Конский каштан белый, Конский каштан обыкновенный
9706 Конский каштан индийский
9715 Конский каштан крупнолистный
9719 Каштан конский мелкоцветный, Конский кашта мелкоцветковый
9720 Каштан конский красный, Каштан конский павия, Каштан конский розовоцветный, Конский каштан павия, Павия красная
9732 Каштан конский японский, Конский каштан японский
9736 Конский каштан ван
9765 Этеопаррус
9768 Этеопаррус кавказский
9769 Этеопаррус красивейщий
9770 Этеопаррус введенского
9785 Крылотычинник, Эвномия, Этионема
9787 Крылотычинник арабский
9788 Крылотычинник армянский
9789 Крылотычинник дерновинный
9790 Крылотычинник середцелистный
9791 Крылотычинник мясокрасный
9795 Крылотычинник складчатый
9796 Крылотычинник беззубый
9797 Крылотычинник длинный
9798 Крылотычинник крупноцветковый
9800 Крылотычинник разнолистный
9802 Крылотычинник левандовского
9803 Крылотычинник липского
9806 Крылотычинник перепончатый
9807 Крылотычинник красивый
9809 Крылотычинник салмасский, Крылотычинник стреловидный
9816 Крылотычинник осыпной
9817 Крылотычинник колючий
9818 Крылотычинник шовица
9820 Крылотычинник загирканский
9821 Крылотычинник трихжильный
9822 Крылотычинник воронова
9831 Кокорыш, Микания
9832 Зноиха, Кокорыш обыкновенный, Кокорыш собачья петрушка, Петрушка собачья
9856 Афлатуния
9858 Афлатуния вязолистая, Миндаль вязолистный
10105 Пагудия, Пахудия
10106 Афзелия африканская, Эвакс
10109 Афзелия бакери
10244 Агапант, Агапантус
10245 Агапантус африканский
10402 Многоколосник
10408 Многоколосник тайваньский
10414 Многоколосник морщинистый
10435 Агазиллис
10436 Агазиллис кавказская
10438 Агазиллис широколистная
10492 Агатис, Даммара, Каури
10494 Агатис новозеландский, Каури новозеландская
10497 Агатис ланцетный
10550 Гвоздичнник мадагаскарский
10557 Баросма
10775 Агавовые
10778 Агава, Алое американское
10787 Агава американская, Агава древовидная, Алоэ американское, Столетник
10854 Генекен, Икстль, Конопля юкатанская, Хенекен
10933 Агава магуэй, Агава сальма, Агава тёмно-зеленая
10940 Агава сизалевая, Агава сизаловая, Сизал, Сизаль
10955 Агава королев виктории
11194 Агератум, Долгоцветка, Целестина
11206 Агератум гаустона, Агератум мексиканский
11273 Аглайя, Аглая
11293 Аглайя тайваньская
11300 Аглайя душистая
11302 Аглайя душистая мелколистная
11329 Аглаонема
11332 Аглаонема переменчивая
11340 Аглаонема ребристая
11542 Агримониа, Приворот, Репейник, Репейничек, Репяшок
11546 Репейничек бархатистый
11549 Парило, Репейник обыкновенный, Репейничек аптечный, Репешок обыкновенный, Репяшок обыкновенный
11552 Репейничек азиатский
11557 Репейничек зернистый
11566 Репейник японский
11572 Репейник волосистый, Репейничек волосистый
11580 Репейничек японский
11587 Репейник непалский
11594 Агримониа душистая, Репейник душистый, Репейник пахучий, Репейничек пахучий
11609 Кумалчик, Кумарчик
11613 Верблюдка бокоцветковый, Кумарчик бокоцветковый
11614 Кумарчик широколистный
11616 Кумалчик малый, Кумарчик малый
11617 Кумарчик палецкого
11619 Кумарчик песчаный
11628 Житняк, Пырей, Пырей западный
11638 Пырей амгунский
11641 Пырей угловатый
11649 Пырей армянский
11650 Пырей оттянуточешуйный

11659 Пырей дернистый
11693 Пырей керченский
11696 Житняк, Житняк гребеневидный, Житняк гребенчатый, Пырей гребенчатый, Регнермей
11699 Житняк, Пырей гребневидный
11704 Пырей пушистоцетковый
11710 Житняк пустыный, Житняк узкоколосый, Житняк узколистный, Пырей пустынный
11732 Пырей крепкостебельный
11735 Житняк сибирьский, Пырей сибирский
11736 Пырей коленчатый
11737 Пырей сизый
11743 Пырей стройный
11749 Житняк, Житняк черепичатый, Пырей черепичатый
11754 Пырей якутов
11779 Пырей крылова
11783 Пырей плевеловидный
11824 Пырей понтийский
11830 Мортук пшеничный, Пырей простертый
11831 Пырей инееватый
11837 Пырей красивейший
11838 Пырей маленький
11841 Вострец, Острец, Пырей ветвистый, Пырей острец
11844 Пырей отогнутоостый
11861 Пырей рожевица
11871 Пырей щетинконосиный
11876 Пырей синьцзянский
11877 Пырей выемчатый
11882 Пырей сосновского
11886 Пырей ковылелистный
11888 Пырей щетинистый
11892 Пырей донской
11938 Агростемма, Горицвет, Куколь
11947 Куколь обыкновенный, Куколь посевной
11951 Куколь льняной
11968 Агростис, Полевица, Пятиостник
11972 Полевица белая, Полевица белая высокая, Полевица побегообразуюшая, Стелющаяся
11995 Полевица беловатая
11998 Полевица анадырская
12015 Полевица биберштейна
12025 Полевица собачья, Полевица собачья опущённая
12029 Полевица тайваньская
12034 Полевица волосовидная, Полевица тонкая
12053 Полевица булавовидная
12061 Полевица сычуаньская
12080 Полевица монгольская
12096 Полевица бороздчатая
12104 Полевица отклоненная
12118 Полевица гигантская
12139 Полевица зимующая
12151 Полевица якутская
12163 Полевица куэньминская
12168 Полевица лазистанская
12179 Полевица крупнометельчатая
12196 Полевица северная
12201 Агростис мелкоцветковый
12208 Агростис мулиский
12232 Полевица памирская
12235 Полевица паульсена
12256 Полевица плосколистная
12280 Полевица красная
12283 Полевица солончаковая
12292 Полевица лицзянская
12301 Полевица шаньдунская
12323 Полевица болотная, Полевица волосовидная, Полевица карская, Полевица обыкновенная, Полевица ползучая, Полевица приморская, Полевица сибирская, Полевица тонкая
12356 Полевица остистая
12357 Полевица гладконожковая
12365 Полевица далиская
12371 Полевица тебердинская
12386 Полевица закаспийская
12403 Полевица туркестанская
12410 Полевица мутовчатая
12413 Полевица тонколистная, Полевица триниуса
12558 Айлант
12559 Айлант высочайший, Айлант железистый, Айлант желёзковый, Айлант китайский ясень, Ясень китайский
12561 Айлант высочайший мелколистный
12564 Айлант тайваньский
12579 Айлант гуансиский
12588 Агава вильморена
12771 Айфанес
12776 Аира
12789 Аира волосовидная
12794 Аира гвоздичная
12839 Аира ранняя
12911 Аизоацневые, Аизовые
12917 Аизоон испанский
12988 Аяния
12990 Аяния лицзянская
12997 Аяния щитковая
12998 Аяния кустарничковая
13001 Аяния тонкая
13005 Аяния кокандская
13007 Аяния маньчжурская
13018 Аяния палласова
13021 Аяния мелкоцветковая
13032 Аяния шарнхорстая
13038 Аяния тибетская
13040 Аяния трехлопастная
13051 Дубница, Дубровка, Живучка
13069 Дубровка кандинская
13072 Живучка приземистая
13073 Дубровка елочковидная, Живучка елочковидная, Живучка ёлочковидная
13078 Дубровка хиосская, Живучка хиосская
13091 Дубровка лежачая
13104 Дубровка женевская, Живучка женевская
13130 Дубровка лаксманна, Живучка лаксманна
13143 Живучка многоцветковая
13146 Дубровка японская
13150 Дубровка продолговатая, Живучка продолговатая
13153 Дубровка восточная, Живучка восточная
13164 Живучка мелкоцветковая
13165 Живучка ложнохиосская
13166 Дубровка тайваньская
13167 Дубровка пирамидальная, Живучка пирамидальная
13174 Дубровка ползучая, Живучка ползучая, Трава живучая
13181 Живучка иволистная
13184 Живучка сикотанская
13190 Живучка туркестанская
13208 Акебия
13225 Акебия пятерная
13335 Эриантера уклоняющаяся
13345 Алангиум
13353 Алангиум китайский
13362 Алангиум мелкоцветковый
13374 Алангиум гуансиский
13376 Алангиум платанолистный
13393 Алангиум юньнаньский
13438 Альбертия
13443 Альбертия пленчатая
13474 Альбизия, Альбиция, Альбицция
13488 Альбиция противоглистная
13493 Альбиция хайнаньская
13505 Альбиция мэнцзыская
13515 Альбиция китайская
13526 Альбиция юньнаньская
13578 Акация шелковая, Акация шёлковая, Альбиция, Альбиция ленкоранская, Шелковая акация
13586 Альбиция калькора
13635 Альбиция душистая, Альбицция душистая, Мята лимонная аптечная
13653 Альбиция ланьюйская
13723 Альбовия трихраздельная
13911 шток-роза

13915 шток-роза угловатая
13916 шток-роза антонины
13917 шток-роза бальджуанская
13920 шток-роза фрейна
13921 шток-роза гроссгеима
13922 шток-роза гельдрейха
13923 шток-роза гирканская
13925 шток-роза каракалинская
13926 шток-роза карсская
13927 шток-роза копетдагская
13928 шток-роза кусаринская
13930 шток-роза ленкоранская
13931 шток-роза новопокровского
13934 Алетеа розеа, Алтей розовый, Алтей-штокроза, Вероника широколистная, Мальва махровая, Мальва садовая, Рожа, Штокроза, Шток-роза розовая
13940 Алтей морщинистный, Шток-роза моршинистая
13941 шток-роза сахсаханская
13942 шток-роза софии
13943 шток-роза сосховского
13945 шток-роза фиголистная
13946 шток-роза таласская
13947 шток-роза крымская
13948 шток-роза туркменская
13949 шток-роза туркевича
13950 шток-роза воронова
13951 Манжетка, Невзрачница
13952 Манжетка абхазская
13955 Манжетка остроугольная
13956 Манжетка подражающая
13957 Манжетка александра
13958 Манжетка приальпийская
13959 Манжетка альпийская
13960 Манжетка алтайская
13961 Манжетка неодинаковочерешковая
13962 Манжетка дуголопастная
13963 Манжетка тонкопильчатая
13966 Манжетка полевая, Невзрачница, Невзрачница полевая
13979 Манжетка бородатоцветковая
13980 Манжетка бородковатая
13982 Манжетка двухквадратная
13984 Манжетка короткозубая
13985 Манжетка коротколопастная
13986 Манжетка бунге
13987 Манжетка буша
13988 Манжетка изогнуточрешковая
13990 Манжетка кавказская
13991 Манжетка зелено-шелковая
13992 Манжетка черкесская
13996 Манжетка сжатая
13997 Манжетка шаровидноскученная
14000 Манжетка частозубая
14002 Манжетка волнистолистная
14003 Манжетка кривобокая
14004 Манжетка дагестанская
14005 Манжетка слабая
14006 Манжетка лысеющая
14007 Манжетка зубчиковатая
14009 Манжетка двуязычная
14010 Манжетка растопыривающаяся
14011 Манжетка разнорешковая
14012 Манжетка твертая
14013 Манжетка негостаточная
14014 Манжетка высокая
14016 Манжетка елизаветы
14023 Манжетка бескровная
14024 Манжетка сбрасывающая
14025 Манжетка тонкостебельная
14028 Манжетка желтеющая
14030 Манжетка ключевая
14031 Манжетка олиственная
14033 Манжетка грузинская
14036 Манжетка горбиковатая
14038 Манжетка голостебельная
14039 Манжетка клубочковая
14043 Манжетка крупеозубчатая
14044 Манжетка гроссгейма
14046 Манжетка гаральда
14048 Манжетка притупляющаяся
14050 Манжетка семиугольная
14051 Манжетка зияющая
14052 Манжетка жестковололистостебельная
14053 Манжетка самая жестковолосистая
14055 Манжетка волосистоцветоножковая
14056 Манжетка сплошь-волосистая
14057 Манжетка нивкостебельная
14058 Манжетка снизуиеленая
14059 Манжетка гирканская
14060 Манжетка безбородая
14062 Манжетка замечательная
14064 Манжетка яйлы
14065 Манжетка обыкновенный
14067 Манжетка юзепчука
14069 Манжетка козловского
14070 Манжетка крылова
14071 Манжетка светлая
14072 Манжетка светлозеленая
14073 Манжетка слабеющая
14074 Манжетка гололистная
14076 Манжетка линдберга
14077 Манжетка липшица
14079 Манжетка камнелюбивая
14080 Манжетка литвинова
14081 Манжетка лидин
14083 Манжетка сверкающая
14086 Манжетка мелкозубчатая
14087 Манжетка малая
14088 Манжетка мелковатоцветковая
14091 Манжетка мурбека
14093 Манжетка туманная
14094 Манжетка дубравная
14095 Манжетка ново-стевеновская
14097 Манжетка налегающая
14098 Манжетка тупая
14099 Манжетка туповидная
14100 Манжетка маловолосая
14101 Манжетка плосколистная
14102 Манжетка округленная
14103 Манжетка острочашелистковая
14104 Манжетка толстолистная
14106 Манжетка пастушья
14110 волосисто-скаладковая
14111 Манжетка тучная
14112 Манжетка складчатая
14113 Манжетка бородатая
14114 Манжетка прямозубая
14115 Манжетка близкая
14116 Манжетка лже-картвельская
14117 Манжетка лжемягкая
14118 Манжетка лысостебельная
14119 Манжетка гожилковая
14121 Манжетка багровеющая
14123 Манжетка густоцветковая
14124 Манжетка густоволосистая
14126 Манжетка радде
14128 Манжетка сетчатожилковая
14129 Манжетка твердеющая
14130 Манжетка жесткая
14133 Манжетка краснеющая
14134 Манжетка окровавленная
14135 Манжетка сарматская
14136 Манжетка шишкна
14137 Манжетка расщепленнолистная
14139 Манжетка полулунная
14140 Манжетка шелковистая
14141 Манжетка шелковая
14143 Манжетка сибирская
14144 Манжетка видная
14145 Манжетка звездчатая
14146 Манжетка звездочковая
14147 Манжетка стевена
14149 Манжетка городковатая
14151 Манжетка городковатовидная
14152 Манжетка почтипрямостоящеволосистая
14154 Манжетка почти-блестящая
14155 Манжетка почтищетинистая
14156 Манжетка приземистая
14157 Манжетка крымская
14158 Манжетка пепельно-шерковистая
14159 Манжетка тяньшанская
14161 Манжетка туруханская
14162 Манжетка мелкоцветковая
14163 Манжетка мешечковая
14164 Манжетка жилковатая
14166 Манжетка обыкновенная
14172 Манжетка ворoова

14177 Альхорнея
14271 Альдрованда
14273 Альдрованда пузырчатая
14359 Алектоион
14362 Алектоион высокий
14471 Алетрис
14491 Алетрис тайваньский
14538 Дерево масляное, Дерево тунговое, Тунг
14544 Гумбанг мягкий, Дерево лаковое, Дерево свечное, Лумбанг, Орех индейский, Орех чилийский, Тунг молуккский
14562 Александра
14593 Альфредия
14594 Альфредия колючечешуйная
14596 Альфредия пониклая, Альфредия поникшая
14599 Альфредия ледовая, Альфредия снежная
14623 Верблюжья колячка, Чагерак
14627 Велблюжья канадская сероватая
14629 Велблюжья канадская киргизская
14631 Велблюжья канадская персидская
14638 Велблюжья канадская обыкновенный, Велблюжья колючка обыкновенная, Велблюжья трава, Верблюжья колячка редколистная, Джантак, Жантак, Колючка верблюжья, Колючка верблюжья обыкновенная, Манна персидская, Трава верблюжья, Чагерак верблюжий
14719 Зуфи-оби, Частуха
14725 Частуха желобчатая
14734 Частуха валенберга, Частуха злаковая
14745 Частуха ланцетная, Частуха ланцетолистная
14749 Частуха лезеля, Частуха лёзеля
14751 Элизма плавающая
14754 Частуха восточная
14760 Частуха водяной подорожник, Частуха восточная, Частуха обыкновенная, Частуха подорожниковая
14789 Частуха валенберга
14790 Частуховые
14836 Алкана, Алканст, Алканна, Альканна
14837 Алкана селдселистная
14839 Алкана крупноплодная
14840 Алкана восточная
14841 Алканна красильная
14917 Вальдгеймия столички, Вальдгиймия столички
14953 Чесночник
14958 Чесночник коротколодный
14964 Чесночник аптечный, Чесночник лекарственный
15018 Лук, Лук зелёный, Лук-перо
15022 Лук родственный
15023 Лук афлатунский
15025 Лук акака
15027 Лук алайский
15030 Лук албанский, Лук восточнокавказский
15032 Лук беловатый
15036 Лук альбова
15039 Лук александры
15040 Лук алексея
15042 Лук алтайский
15043 Лук высочайший
15048 Лук виноградный, Лук жемчужный
15056 Лук смнительный
15060 Лук угловатый
15064 Лук неравнолучевой, Лук неравноногий
15073 Лук ароидный
15088 Лук однобратственный, Лук чернокрасный
15091 Лук черно-фиолетовый
15094 Лук ошера
15107 Лук барщевского
15112 Лук миленький
15113 Лук двузубый
15122 Лук боде
15128 Лук борщова
15131 Лук коротостебельный
15133 Лук коротозубый
15144 Лук синеголубой
15145 Лук дернистый
15146 Лук красивосетчатый
15149 Лук канадский
15156 Лук декандоля
15159 Лук осоковидный
15160 Лук килеватый
15161 Лук многолистный
15163 Лук каснийский
15165 Лук культурный, Лук многодетковый, Лук обыкновенный, Лук репчатый, Лук севок
15167 Лук картофельный
15170 Лук аскалона, Лук шалот, Лук-шалот, Сорокозубка, Шалот, Шарлот
15185 Лук бэкера, Лук китайский
15188 Лук христофа
15193 Лук решетчатый
15194 Лук голубой
15198 Лук густой, Лук сгущенный
15199 Лук неровный
15204 Лук лантышевый
15214 Лук кристаллоносный
15217 Лук чашеносный
15224 Лук дарвазский
15226 Лук обманываюший
15233 Лук привлекательный
15236 Лук дердернана
15241 Лук сетчатый
15242 Лук сетеносный
15244 Лук длинноцветоножковый
15245 Лук длинностолбиковый
15252 Лук серполистный
15253 Лук дробова
15260 Лук высокий
15262 Лук изящный
15266 Лук краснеющий
15267 Лук евгения
15271 Лук стареющий
15277 Лук ферганский
15278 Лук фетисова
15279 Лук волокинстый
15281 Лук нитезубный
15282 Лук нителистный
15285 Лук прочноодетый
15289 Батун, Катависса, Лук дудчатый, Лук зимний, Лук татарка, Лук трубчатый, Лук-батун, Лук-татарка, Татарка
15292 Лук японский
15295 Лук желтеющий
15296 Лук желтоватый
15300 Лук фомина
15308 Лук темнофиолетовый
15310 Лук молочноцветный
15319 Лук ледниковый
15324 Лук клубочный
15328 Лук стройный
15330 Лук крупный
15336 Лук гриффита, Лук неравнодольный
15340 Лук крапчатый
15341 Лук винтолистный
15352 Лук мулиский
15361 Лук влагалищнокорневищный, Лук плевокорневищный
15364 Лук илийский
15365 Лук неравный
15367 Лук незаметный
15368 Лук индерский
15371 Лук бесномощный
15375 Лук жакемонта, Лук красиенький
15380 Лук яйлинский
15384 Лук приятный
15389 Лук каратавский
15392 Лук карсский
15393 Лук кашский
15398 Лук коканский
15400 Лук комарова
15401 Лук копетдагский
15402 Лук королькова
15403 Лук куюкский
15408 Лук изорванный
15414 Лук мохнатолистный
15420 Лук ледебура
15421 Лук леманна
15423 Лук ленкоранский

15424 Лук кунта
15426 Лук белоцветный
15428 Лук белоголовый
15432 Лук линейный
15437 Лук длинноостроконечный
15438 Лук длиннолучевой
15442 Лук желтенький
15444 Лук маака
15445 Лук гигантский
15450 Лук крупнотычинковый
15461 Лук маргариты
15462 Лук марии
15464 Лук морской
15465 Лук маршалля
15467 Лук максимовича
15473 Лук мелколуковичный
15477 Лук мелкий
15480 Лук моли, Моли
15486 Лук одноцветковый
15493 Лук горный
15499 Лук мускатный
15513 Лук неринолистный
15523 Лук поникающий
15525 Лук косой
15534 Лук душистый
15540 Лук огородный
15541 Лук малоцветковый
15543 Лук эмэйский
15545 Лук змеелистный
15549 Лук высокогорный
15550 Лук горолюбивый
15551 Лук горцевидный
15552 Лук горцевидный
15553 Лук горноченочный
15555 Лук ошанина
15564 Лук палласа
15567 Лук метельчатый
15585 Лук странный
15590 Лук мелковатый
15597 Лук щебнистый
15602 Лук туполистый, Лук широкочехольный
15615 Лук многокорневой
15619 Лук понтийский
15620 Лук попова
15621 Лук порей, Лук поррей, Лук-поррей, Порей
15632 Лук стелющийся, Лук фишера
15634 Лук ложновиноградный
15637 Лук ложножелтый
15640 Лук ложно-зеравшанский
15642 Лук пскемский
15643 Лук хорошенький
15644 Лук малорослый
15651 Лук регелевский, Лук регеля
15652 Лук регеля
15662 Лук синьцзянский
15663 Лук мощный
15665 Лук розенбаха
15675 Лук круглый, Лук сорно-полевой
15680 Лук красиенький
15682 Лук красноватый
15686 Лук скальный
15689 Лук песчаный
15690 Лук комаровский, Лук мешочконосный
15698 Лук чеснок, Лук-чеснок, Чеснок, Чеснок посевный
15704 Лук скаловый, Лук шаровидный
15707 Лук шероховатостебельный
15708 Лук скородовида, Лук скородовидный
15709 Лук скорода, Лук-розенец, Лук-скорода, Лук-шнит, Резанец, Скорода, Трибулька, Шнит-лук
15717 Лук шероховатенький
15724 Лук шуберта
15726 Лук приречный, Лук причесночный, Лук-рокамболь, Рокамболь, Чеснок змеиный
15730 Лук ямцатый
15732 Лук семенова
15734 Лук стареющий, Лук стареющийся
15737 Лук зеравшанский
15738 Лук сергея
15740 Лук щечнолистный
15741 Лук северцова
15752 Лук круглоголовый
15760 Лук линейный
15764 Лук тычиночный
15767 Лук стеллера
15771 Лук венценосный
15774 Лук стебельчатый
15779 Лук короткозубчатый, Лук торчащий, Осока береговая
15791 Лук тончайший
15793 Лук суворова
15794 Лук сычуаньский
15796 Лук шовица
15797 Лук лентолепестный
15799 Лук таласский
15800 Лук талышский
15804 Лук татарский
15807 Лук пустынный
15810 Лук тонкостебельный
15813 Лук тонгий
15819 Лук вальковатолистный
15823 Лук тяньшанский
15828 Лук шероховатый
15829 Лук траутфеттера
15843 Лук душистый, Лук сибирский душистый
15850 Лук туркменский
15851 Лук туркестанский
15852 Лук мелкоцветный
15853 Лук мелкоголовый
15861 Лук медвежий, Лук-черемша, Черемша
15863 Лук вавилова
15866 Лук муговчатый
15867 Лук вешиякова
15868 Колба, Лук длинный, Лук победный, Лук сибирский колба, Черемша
15876 Лук виноградничный, Лук виноградный
15881 Лук зеленый
15885 Лук вальдштейна
15895 Лук винклера
15898 Лук мечелепестный
15953 Казуарина аргентинская
16295 Ольхолистник форчуна
16309 Ольха
16320 Ольха бородатая
16323 Ольха сердсевидная
16328 Ольха максимовича
16340 Ольха фердинанда кобурга
16345 Ольха тайваньская
16348 Ольха горная, Ольха кустарная, Ольха кустарниковая
16352 Ольха клейкая, Ольха липкая, Ольха черная, Ольха чёрная
16358 Ольха шерстистая
16362 Ольха сибирская
16368 Елоха, Ольха белая, Ольха серая
16386 Ольха японская
16398 Ольха камчатская
16399 Ольха кольская
16401 Ольха маньчжуьрская
16404 Ольха приморская
16421 Ольха непальская
16429 Ольха красная
16433 Ольха повислая
16437 Ольха красная
16440 Ольха морщинистая
16458 Ольха зибольда
16463 Ольха востчная, Ольха сердселистная
16466 Ольха красильная
16469 Ольха красильная голая
16472 Ольха горная, Ольха зеленая
16484 Алоказия
16498 Алоказия гуйцзоуская
16501 Алоказия индийская
16512 Алоказия крупнокорневая, Арум одоратум
16540 Алое, Алой, Алоэ, Бабушник
16592 Алое древовидное
16605 Алое остистое
16624 Алоэ барбадосское, Алоэ обыкновенное
16709 Алое ресничатое
16768 Алоэ раздвоенное

16817 Алоэ дикое, Алоэ колючее, Алоэ ферокс
16910 Алое индийское
17136 Алоэ пёстрое
17320 Алоэ сокотрина, Алоэ сокотринское
17381 Алоэ барбадосское, Алоэ настоящее, Алоэ обыкновенное
17383 Алое китайский
17475 Алонсоа
17497 Батлачёк, Батлачик, Батлачок, Батлячек, Лисехвостник, Лисохвост, Лисохвостник
17498 Батлачек равный, Лисохвост равный
17501 Лисохвост амурский
17506 Лисохвост альпийский
17512 Лисохвост блюшистый, Лисохвост вздутый, Лисохвост лисохвост вздутый, Лисохвост тростниковидный
17513 Лисохвост ошера
17515 Лисохвост коротколосый
17523 Лисохвост пушистоцветковый
17527 Батлачок коленчатый, Лисохвост коленчатый
17533 Лисохвост жерарда
17534 Лисохвост ледниковый
17535 Лисохвост сизый
17536 Лисохвост гимлайский
17537 Лисохвост японский
17538 Лисохвост рыхлоцветный
17541 Лисохвост длинноостный
17547 Лисохвост остроконечный
17548 Лисохвост мышехвостиковидный, Лисохвост мышехвостиковый, Лисохвост пашенный, Лисохвост полевой
17549 Батлачок непальский, Лисохвост непальский
17550 Лисохвост черный
17553 Батлачок луговый, Лисохвост джунгарский, Лисохвост луговой, Лисохвост луговый
17565 Лисохвост рожевица
17567 Лисохвост зеравшанский
17568 Лисохвост шелковистый
17571 Лисохвост штейнегера
17572 Лисохвост тонкой
17573 Лисохвост волокнистый
17574 Лисохвост тифлисский
17578 Лисохвост влагалищный
17613 Альфитония
17614 Альфитония высокая
17639 Алпиния, Альпиния
17653 Алпиния мелкоцветковая
17656 Алпиния китайская
17674 Алпиния тайваньская
17677 Альпиния галанга, Корень галанговый
17685 Алпиния хайнаньская
17715 Алпиния гуандунская
17721 Алпиния мэнхайская
17727 Алпиния напоская
17760 Алпиния янчуньская
18027 Альстония
18055 Альстония школьная
18060 Альстремерия
18089 Альтернантера, Очереднопыльник
18147 Очереднопыльник сидячий
18155 Алтей, Альтей, Шток-роза
18157 Алтей армянский
18160 Алтей конаплевидный, Алтей коноплевый
18166 Алтей жестковолосый, Алтей шершавоволосный, Алтей шершаволистный
18173 Алтей аптечный, Алтей аптечный, Алтей лекарственный , Дикая рожа, Корень алтейный, Рожа дикая, Трава алтейная
18174 Алтей бледный, Шток-роза бледная
18197 Расамала
18309 Бурачочек
18312 Бурачочек кочи
18313 Алиссум, Бурачек, Бурачок, Каменник, Конига, Плоскоплодник
18329 Бурачек андийский
18333 Бурачек армянский
18334 Бурачек артвинский
18341 Бурачек прицветничный
18342 Бурачек буша
18344 Бурачек чамечный, Бурачок чашечковый
18346 Бурачек чамечкоплодный
18349 Бурачек полевой
18369 Бурачек дагестанский
18370 Бурачек пушистоплодный, Бурачок пустынный, Бурачок пушистоплодный
18374 Бурачек пустынный
18381 Бурачек федченко
18385 Бурачек гмелина
18395 Бурачек шершавый
18403 Бурачек ленский, Бурачек седой, Бурачок ленский, Тунгусское назв
18407 Плоскоплодник льнолистный
18414 Бурачек длинностолбчатый
18419 Бурачек обвернутый
18432 Бурачек маленький
18433 Бурачок горный
18443 Бурачек мюллера
18444 Бурачек стенной
18451 Бурачек туполистный
18452 Бурачек мелкоцветковый
18457 Бурачек щитницевидный
18458 Бурачек персидский
18463 Бурачек носатый
18469 Аллисум сибирский, Бурачек двусемянный, Бурачок сибирский
18474 Бурачок колючий
18478 Бурачек прямой
18480 Бурачек шовица
18483 Бурачек извилистый, Бурачек искривленный, Бурачок извилистый
18484 Бурачек пушистый
18488 Бурачек зонтичный
18611 Амаракус
18646 Амарантовые, Щирицевые
18652 Аксамитник, Амарант, Бархатник, Щирица
18655 Амарант белый, Щирица белая, Щирица маскированная
18668 Амарант жминдовидный, Щирица жминдовидная
18670 Щирица жминда, Щирица синеватая
18683 Акнида, Конопля виргинская
18687 Аксамитник, Амарант пониклый, Амарант хвостатый, Бурачки, Вархатник, Хвост лисий, Щирица метельчатая, Щирица хвостатая
18701 Щирица согнутая
18734 Щирица гиблидный
18752 Щирица белосемянная
18763 Щирица белосемянная
18779 Амарант меланхолический
18788 Амарант метельчатый, Амарант хвостатый, Щирица американская, Щирица метельчатая, Щирица хвостатая
18810 Амарант колосистый, Подсвекольник, Щирица запрокинутая, Щирица колосистая
18822 Щирица колючая
18831 Щирица теллунга
18836 Щирица трехцветная
18861 Амариллисовые
18862 Амарилис, Гипеаструм, Гиппеаструм, Хипеаструм
18865 Амарилис белладонна
18946 Амбербоа
18948 Амбербоа обыкновенная
18951 Амбербоа бухарская
18954 Амбербоа сизая
18966 Василек дтшистый
18971 Амбербоа низкая
18980 Амбербоа сосновского
18983 Амбербоа туранская
19042 Круглоспинник
19045 Круглоспинник обратнояйцевидный, Незабудочник маака
19049 Амблиокарпум
19050 Амблиокарпум девясиловый
19140 Амброзия, Гертнериа
19143 Амброзия бескрылая
19145 Амброзия полыннолистная
19179 Амброзия горометельчатая

19191 Амброзия трёхнадрезная, Амброзия трехраздельная, Амброзия трёхраздельная
19240 Амеланхир, Ирга, Коринка
19248 Ирга азиатская
19251 Ирга канадская, Овсяица разнолистная
19271 Ирга цернокрайняя
19288 Ирга круглолистная, Ирга обыкновенная, Ирга овалолистная, Ирга овальная
19293 Ирга колосистая, Ирга метельчатая
19297 Ирга туркестанская
19458 Аметистеа
19459 Аметистеа голубая
19550 Аммания
19554 Аммания песчаная
19566 Аммания египетская
19602 Аммания многоцветковая
19616 Аммания пушистоцветковая
19639 Аммания мутовчатая
19641 Аммания зеленая
19659 Амми
19664 Амми большой
19672 Амми виснага, Амми зубная
19690 Аммобиум
19691 Аммобиум крылатый
19710 Песочник
19712 Песочник палестинский
19728 Акация печеная, Аммодендрон, Кандым, Печеная акация
19729 Аммодендрон серебристая, Печеная акация серебристая
19731 Аммодендрон конолли, Печеная акация конолли
19732 Аммодендрон эйхвальда, Печеная акация эйхвальда
19733 Аммодендрон карелина, Печеная акация карелина
19734 Аммодендрон лемана, Печеная акация лемана
19735 Аммодендрон длиннокистевая, Печеная акация длиннокистевая
19736 Аммодендрон персидская, Печеная акация персидская
19765 Аммофил, Аммофила, Песколюб
19766 Песколюб песчаный, Песколюбка песчаная
19836 Кардамон
19952 Амоора
19989 Аморфа
19994 Аморфа калифорнийская
19997 Аморфа седовадая
20005 Аморфа кустовидная, Аморфа кусторниковая, Амофа кустарниковая, Индиго бастардное, Индиго дикое, Индиго ложное, Крутик
20023 Аморфа карликовая
20037 Аморфофалл, Аморфофаллюс, Амрфофаллус
20051 Аморфофаллюс мэнхайский
20057 Амрфофаллус колокольчатый, Ямс слоновый
20125 Амрфофаллус колокольчатый, Ямс слоновый
20132 Аморфофаллюс ривьера, Амрфофаллус ривьера, Ямс слоновый ривьера
20136 Аморфофаллюс китайский
20145 Аморфофалл гигантский, Амрфофаллус гигантский
20153 Аморфофаллюс юньнаньский
20280 Ампелопсис, Виноградовик, Виноградовник
20348 Виноградовик коротоцветоножковый, Виноградовник коротоцветоножковый
20354 Виноградовник разнолистный
20356 Виноградовник коротоцветоножковый гулинский, Виноградовник разнолистный гулинский
20359 Виноградовник гуншаньский
20408 Виноградовник японский
20420 Виноградовник крупнолистный
20435 Виноград вьющийся, Виноград девичий пятилистный, Виноград дикий пятилистный, Партеноциссус пятилисточковый
20461 Виноград виноградолистный
20558 Фальиата
20570 Фальиата японская
20811 Амфорикарпус
20812 Амфорикарпус изящный
20827 Амзинцкия
20844 Амсония, Анабазис
20885 Миндаль, Персик
20889 Бодом, Миндаль бухарский, Таджиск
20890 Дерево миндальное, Миндаль, Миндаль настящий, Миндаль обыкновенный, Слива домашная, Слива европейская, Слива обыкновенная, Чернослив
20892 Миндаль горький
20897 Миндаль сладкий
20899 Миндаль ломкий
20908 Персик давида
20913 Миндаль фенцля
20921 Миндаль ганьсуский, Персик ганьсунский
20923 Миндаль ледебура, Миндаль ледебуровский
20926 Персик мира
20928 Абрикос монгольский
20930 Миндаль наирский
20931 Бобовник, Миндаль грузинский, Миндаль дикий, Миндаль изхий, Миндаль карликовый, Миндаль низкий, Миндаль степной, Миндальник, Орех заячий, Персик степной, Степной миндаль, Черёмуха карликовая
20933 Миндаль черешковый
20935 Дерево персиковое, Пави, Персик, Персик обыкновенный
20955 Нектарин, Персик нектарин, Персик-нектарин
20961 Миндаль петунникова
20964 Миндаль ложный-персик
20966 Миндаль метельчатый
20967 Миндаль колючейший
20970 Миндаль махровый, Миндаль трёхлопастный, Слива трёхлопастная
20976 Миндаль туркменский
20979 Миндаль урарту
20980 Миндаль вавилова
20983 Миндаль зангезурский
21003 Петрушка корневая
21005 Амирис бальзамический, Амирис ямайский, Дерево розовое
21023 Анабазис, Ежовник
21028 Ежовник безлистный
21032 Ежовник балхашский
21033 Ежовник канделябрный
21034 Ежовник аболна, Ежовник коротколистный, Ежовник смежный
21035 Биюргун, Ежовник меловой
21037 Ежовник высокий
21038 Ежовник шерстистоногий
21039 Ежовник ферганский
21042 Ежовник гипсолюбивый
21044 Ежовник щетинковолосый
21046 Ежовник сырдарьинский
21047 Ежовник коровина
21050 Ежовник крупнокрылый
21051 Ежовник мелкожелезистый
21053 Ежовник малоцветковый
21054 Ежовник пеллиота
21058 Анабазис солончаковый, Ежовник ветвистыи, Ежовник солончаковый
21064 Ежовник усеченный
21065 Ежовник тургайский
21066 Ежовник туркестанский
21162 Анакамптис
21176 Анакамптис пирамидальный
21181 Модестия
21190 Анакардиевые, Анакардовые, Сумаховые, Фисташковые
21191 Анакард, Анакардия, Дерево анакардовое
21195 Акажу, Анакард, Анакард акужа, Анакард западный, Анакардия акажу,

Анакардия западная,Кажу,Орех кешью
21230 Анациклус,Ромашка немецкая
21241 Анациклус реснитчатый
21263 Анациклус лекарственный,Немецкая ромашка
21329 Анагаллидиум
21335 Анагаллис,низманка,Очноцвет,Очный,Очный цвет,Просо воробьиное
21339 Анагаллис,Очноцвет пашенный,Очный цвет,Очный цвет багряный,Очный цвет голубой,Очный цвет пашеный,Очный цвет полевой,Полевой очный цвет,Просо воробьиное
21340 Голубой очный цвет,Очноцвет голубой,Очный цвет голубой
21394 низманка малая
21427 Очный цвет нежный
21433 Очный цвет мтовчатый
21445 Анагирис,Вонючка
21458 Анамирта
21472 Ананас
21475 Ананас посевной
21479 Ананас культурный,Ананас настоящий,Ананас посевной
21506 Анафалис
21540 Анафалис чжундяньский
21560 Анафалис дарвазский
21562 Анафалис обедненный
21588 Анафалис дэциньский
21592 Анафалис лицзянский
21596 Анафалис жемчужный
21616 Анафалис жемчужный,Бисерница
21637 Анафалис мулиский
21643 Анафалис непалский
21673 Анафалис кистеносная
21675 Анафалис белорозовый
21679 Анафалис утесистый
21682 Анафалис посьетский
21707 Анафалис сычуаньский
21710 Анафалис тонкостебельный
21711 Анафалис тибетский
21722 Анафалис бархатистый
21724 Анафалис прутьевидный
21733 Анафалис юньнаньский
21805 Роза иерихонская
21886 Анкафия
21888 Анкафия огненная,Анхуза
21889 Анхиета,Фиалка бразильская
21909 Анхоний
21910 Анхоний коротколплодный
21911 Анхоний суменицелистный
21912 Анхоний стеригмовидный
21914 Анхуза,Воловик,Филлокара,Язык волочий
21920 Кривоцвет пашенный,Кривоцвет полевой,Ликопсис полевой
21924 Воловик аушера,Филлокара оше
21928 Анхуза баррелиела,Анхуза баррелье,Воловик баррелиера
21940 Анхуза гмелина,Воловик гмелина
21949 Анхуза итальянская,Воловик итальянский
21953 Анхуза узколистная,Воловик уликостный
21957 Анхуза бледно-желтая,Воловик светложелтый
21959 Альканна,Анхуза лекарственная,Анхуза рослая,Воловик аптечный,Воловик лекарственный
21965 Кривоцвет восточный,Тамокука
21973 Анхуза ложно-бледножелтая
21974 Анхуза маленькая
21980 Анхуза длинностолбиковая,Воловик длинностолбиковый
21983 Анхуза фессалийская,Воловик фессалийский
22211 Андира
22218 Дерево капустное
22226 Андрахна
22234 Андрахна буша
22249 Андрахна федченко
22250 Андрахна нитевидная
22266 Андрахна крошечная
22267 Андрахна карликовая
22268 Андрахна круглолистная
22272 Андрахна узколистная
22273 Андрахна телефиевидная
22276 Андрахна тонгая
22418 Андромеда,Подбел
22445 Андромеда дубровниколистная,Андромеда-дубровник,Подбел дубровник,Подбел многолетний,Подбел многолистный
22476 Бородач,Осина
22569 Бородач китайский
22833 Бородач тибетский
23086 Бородач юньнаньский
23107 Проломник
23119 Проломник восточнокавказский
23121 Проломник альпийский
23122 Проломник арктический
23124 Проломник арминский
23127 Проломник бородчатый
23128 Проломник двузубчатый
23136 Проломник опадающий
23144 Проломник ресничатыймохнатый
23155 Проломник шерстистолистный
23163 Проломник удлиненный
23167 Проломник федчанко
23170 Проломник нитевидный
23182 Проломник гмелина
23199 Проломник ядунский
23202 Проломник седой
23205 Проломник промежуточный
23207 Проломник козо-полянского
23208 Проломник гуйцзоуский
23210 Проломник молочноцветковый
23213 Проломник горовчатый,Проломник лемана,Проломник лемановский
23215 Проломник бунге
23216 Проломник кандинский
23222 Проломник крупноцветковый
23224 Проломник тибетский
23227 Проломник большой,Проломник турчанинова
23234 Проломник мелколистный
23246 Проломник охотский
23247 Проломник ольги
23250 Проломник овчинникова
23252 Проломник эмэйский
23259 Проломник радде
23294 Проломник северный
23295 Проломник северный
23309 Проломник сычуаньский
23314 Проломник таврический
23321 Проломник трехцветковый
23323 Проломник зонтичный
23326 Проломник мохнатый
23347 Проломник чаюйский
23473 Андржеевския
23474 Андржеевския сердечниколистная
23696 Анемон,Анемона,Ветреница
23701 Ветреница алтайская
23705 Ветреница амурская
23709 Ветреница байкальская
23711 Ветреница ганьсуская
23722 Ветреница пермская
23724 Ветреница нежная
23734 Ветреница коротконожковая
23736 Ветреница бухарская
23739 Ветреница лысая
23750 Ветреница кавказская
23766 Ветреница голубая
23767 Анемона садовая
23773 Ветреница длинноволосая
23780 Ветреница слабая
23796 Ветреница вильчатая
23807 Ветреница весенниковая
23815 Ветреница пучковатая
23817 Ветреница гибгая
23823 Анемона блестящая
23827 Ветреница гладкая
23832 Ветреница горчакова
23850 Анемон хубэйский
23854 Анемон японский,Анемона японская,Ветреница японская
23866 Ветреница енисейская
23870 Ветреница кузнецова
23874 Ветреница рыхлая
23877 Ветреница прибрежная
23901 Ветреница нарциссовидная

23906 Ветреница вытянутая, Ветреница шренка
23917 Ветреница сахалинская
23919 Ветреница сибирская
23925 Ветреница дубравная
23939 Ветреница туподольчатая
23945 Ветреница туподольчатая чжэнькanская
23971 Ветреница охотская
23975 Ветреница малорассеченная
23980 Ветреница мелкоцветковая
23990 Ветреница тяньцюаньская
23998 Прострел луговой
24013 Ветреница радде
24018 Ветреница лютиковая, Ветреница лютичная
24019 Ветреница отогнутая
24020 Ветреница ричардсона
24058 Ветреница шаньдунская
24061 Ветреница лесная
24065 Ветреница видная
24071 Ветреница лесная
24077 Ветреница тэнчунская
24088 Ветреница тибетская
24100 Ветреница черняева
24103 Ветреница удская
24105 Ветреница теневая, Ветреница тенистая
24107 Ветреница уральская
24109 Ветреница весенняя, Прострел весенний
24111 Ветреница мохнатейшая
24132 Анемонелла
24133 Анемонелла василистниколистная
24208 Укроп, Укроп аптечный
24213 Укроп обертковый, Укроп огородный, Укроп пахучий, Укроп полевой, Уксусник
24236 Вострец
24238 Вострец алайский
24241 Вострец бальджуанский
24247 Вострец растопыренный
24248 Вострец пучковатый
24249 Вострец гибкий
24251 Вострец каратавский
24256 Вострец скальный
24267 Вострец угамский
24281 Ангелика, Дудник, Дягель, Дягиль
24291 Дудник амрский
24293 Дудник сахалинский, Дудник уклоняющийся, Дудник ялуйский
24318 Дудник опоясанный
24325 Дудник даурский
24326 Дудник тайваньский
24327 Дудник даурский
24336 Дудник низбегающий
24340 Дудник низбегающий
24342 Дудник чэнкоуский
24355 Дудник коленчатосогнутый, Дудник преломленный
24369 Дудник ичанский
24383 Дудник кандинский
24386 Дудник корейский
24391 Дудник лицзянский
24392 Дудник прибрежный
24406 Дудник крупнолистный
24410 Дудник горный
24424 Дудник эмэйский
24427 Дудник толстокрылый
24429 Маточник болотный, Остерикум болотный
24453 Дудник сахалинский
24465 Дудник сычуаньский
24475 Дудник китайский
24483 Ангелика лесная, Дудник лесной, Дягиль лесной
24489 Дудник тройчатый
24500 Дудник медвежий
24687 Ангрекум
25280 Анизанта
25316 Ипомея китайская
25379 Анизохилус
25438 Гималайская скополия
25447 Гималайская скополия буро-желтая
25450 Гималайская скополия лицзянская
25762 Кария китайская, Орешник китайский
25771 Аннеслея
25828 Аннона, Анона
25836 Анона черимойя, Анона черимола
25852 Анона обыкновенная
25868 Анона игольчатая, Анона колючая
25882 Пуходлевник, Хлопчатник древовидный
25898 Анона чешуйчатая, Канель, Яблоко сахарное
25909 Аноновые
25959 Анектохилюс
25978 Анектохилюс эмэйский
25980 Анектохилюс тайваньский
26015 Анектохилюс чжэцзянский
26172 Безшипник
26173 Безшипник сжатый
26264 Буссенгоя, Буссингоя
26299 Кошачья лапка, Лапка кошачья
26310 Кошачья лапка альпийская, Лапка кошачья альпийская
26362 Кошачья лапка карпатская, Лапка кошачья карпатская
26363 Кошачья лапка кавказская
26385 Бессмертник белый, Кошачья лапка двудомная, Лапка кошачья двдомная, Тиллея водная, Тиллея водяная
26393 Кошачья лапка двудомновидная
26413 Кошачья лапка фриса
26439 Кошачья лапка комарова
26474 Кошачья лапка одноголовчатая
26616 Кошачья лапка ворсоносная
26624 Горец нитевидный
26738 Антемис, Пупавка
26743 Пупавка высочайшая
26744 Пупавка анатолийская
26746 Пупавка полевая
26751 Пупавка австрийская
26752 Пупавка маршалла-биберштейна
26758 Пупавка известняковая
26759 Пупавка белейшая
26760 Пупавка карпатская
26768 Пупавка вонючая, Пупавка обыкновенная, Пупавка собачья, Ромашка вонючая, Ромашка собачья
26770 Пупавка меловая
26777 Пупавка слабая
26779 Пупавка пустынная
26780 Пупавка сомнительная
26781 Пупавка кустарниковая
26782 Пупавка звксинская
26787 Пупавка кустарничковая
26794 Пупавка гроссгейма
26797 Пупавка волосистая
26799 Пупавка грузинская
26811 Пупавка литовская
26815 Пупавка крупноязычковая
26818 Пупавка мархотская
26827 Пупавка темноокаймленная
26829 Пижма тысячелистная, Ромашник тысячелистный
26836 Пупавка однокорзиночная
26840 Пупавка зубчато-венечная
26871 Пупавка жестковадая
26874 Пупавка русская
26876 Пупавка сагурамская
26880 Пупавка шишкина
26885 Пупавка сосновского
26886 Пупавка бесплодная
26892 Пупавка светло-желтая
26893 Пупавка талышская
26900 Пупавка красильная, Ромашка жёлтая, Ромашка красильная
26906 Пупавка видемана
26907 Пупавка воронова
26909 Пупавка зефирова
26910 Пупавка зигийская
26950 Венечник
27190 Венечник лилейный, Венечник лилиаго, Венечник лилиецветковый
27320 Венечник ветвистый
27560 Антохламис
27561 Антохламис истодовая
27562 Антохламис туркменская
27664 Антолиза

27793 Антозахна
27798 Антозахна длинноостная
27930 Душистый колокок, Пахучеколосник
27934 Душистый колокок остистый, Колосок душистый остистый, Пахучеколосник остистый
27942 Зубровка гладкая, Зубровка голая
27947 Пахучеколосник тайваньский
27963 Зубровка альпийская
27967 Зубровка бунге, Зубровка душистая, Зубровка пахучая, Ликвидамбар, Лядник душистый, Лядник пахучий, Чаполоть
27969 Душистый колокок, Душистый колокок обыкновенный, Колос пахучий, Колосок душистый обыкновенный, Пахучеколосник
27971 Пахучеколосник японский
28013 Купырь
28023 Купырь обыкновенный
28024 Керверь, Керверь листовой, Керверь настоящий, Керверь обыкновенная, Керверь посевной, Керверь салатный, Купырь бутенелистный, Купырь садовый, Снедок
28027 Купырь приледниковый
28029 Купырь длинноносиковый
28030 Купырь дубравный
28035 Купырь лоснящийся
28037 Купырь рупрехта
28039 Купырь обыкновенный
28041 Купырь шмальгаузена
28042 Купырь сосновского
28044 Купырь лесной, Купырь похожий
28061 Купырь бархатистый
28066 Купырь юньнаньский
28071 Антуриум
28084 Антуриум андрэ
28093 Антуриум хрустальный
28118 Антуриум цепкий
28119 Антуриум шерцера
28137 Антуриум изменчивый
28138 Антуриум вейча
28143 Язвеник
28146 Язвеник родственный
28148 Язвеник приальпийский
28149 Язвеник песчаный
28154 Язвеник кавказский
28156 Язвеник яркокрасный
28157 Язвеник окрашенный
28171 Язвеник опушенный
28173 Язвеник лемана
28175 Язвеник приморский
28179 Язвеник горный
28189 Язвеник многолистный
28194 Язвеник крымский
28200 Клевер заячий, Трава зольная, Язвеник линнея, Язвеник обыкновенный
28248 Антемис африканский, Антемис ядовитый, Антияр, Анчар, Унас
28309 Антидесма
28316 Антидесма буниус, Дерево саламандровое
28576 Антиринум, Львиный зев
28617 Антиринум крупный, Антиринум майюс
28631 Антиринум горный, Зев львиный полевой, Львиный зев полевой
28808 Анура бледнозеленая
28963 Апейба, Тибурбу
28969 Метлица
28970 Метлица промежуточная
28971 Метлица прерывчатая
28972 Метлица обыкновенная, Метлица полевая
29004 Афананта
29005 Афананта шероховатая
29018 Невзрачница
29085 Неяснореберник, Совичия
29086 Неяснореберник волосолистый
29087 Неяснореберник тонкстебельный
29088 Неяснореберник шероховатоплодный
29121 Афеландра, Афеляндра
29123 Афеландра золотистая
29164 Одногнездка
29167 Одногнездка обвернутая
29169 Брайя альпийская, Брайя остроплодная
29172 Одногнездка мелкоплодный
29230 Зонтичные
29295 Апиос
29298 Апиос клубневый
29312 Сельдерей
29327 Сельдерей, Сельдерей душистый, Сельдерей обыкновенный, Сельдерей пахучий, Сельдерей черешковый
29328 Сельдерей листвентый, Сельдерей листовой, Сельдерей стеблевой, Сельдерей черешковый
29422 Мякинник
29430 Мякинник безостный, Мякинник тупоконечный
29462 Апоциновые, Кутровые
29465 Кендырь
29468 Кендырь плоромниковый
29470 кентырь армянский
29473 Апоцинум конопревидный, Кендырь коноплевидный, Кендырь конопляный, Хлопок дикий
29492 кентырь ланцетолистный
29506 Поацинум гендерсона, Поацинум пестрый
29509 кентырь русанова
29510 кентырь сарматский
29511 Кентырь шершавый
29522 Кентырь крымуский
29643 Апоногетон
29660 Апоногетон двуколосный
29663 Апоногетон решетчатый
29696 Апоногетоновые
29701 Апофиллум
29712 Апорокактус
29715 Апорокактус плетевидный, Цереус плетевидный
29727 Апороселла
29785 Апозерис
29786 Апозерис вонючий
29842 Аптандра
29956 Апулея
29968 Падубовые
29972 Дерево азиминовое, Каламбак
29994 Аквилегия, Акилей, Водосбор, Колокольчики водосборные, Орлики
29998 Водосбор амурский
30001 Водосбор темновинный, Водосбор темно-пурпуровый
30003 Водосбор бородина
30004 Водосбор короткошпорцевый
30007 Водосбор канадский
30031 Водосбор вееровидный
30034 Водосбор прекрасный
30038 Водосбор железистый
30041 Водосбор гибридный
30046 Водосбор каратавский
30047 Водосбор карелина
30048 Водосбор тайваньский
30049 Водосбор молочноцветный
30051 Водосбор тонкошпорцевый
30060 Водосбор олимпийский
30061 Водосбор остролепестный
30065 Водосбор ганьсуская
30068 Водосбор мелкоцветковый
30074 Водосбор сибирский
30075 Водосбор скиннера
30076 Водосбор тяньшанский
30077 Водосбор зеленоцветковый
30078 Водосбор темно-птрпуровый
30080 Водосбор виталия
30081 Акилей, Водосбор обыкновенный, Голубки, Колокольчики, Колокольчики садовые, Коломбина, Орлики обыкновенные, Садов
30105 Резушка
30111 Резуха максимовича
30144 Резуховидка шведская
30146 Резуховидка таля, Резушка таля
30155 Резушка одуванчиколистая
30157 Резушка ядунская
30160 Арабис, Резуха
30164 Арабис альпийский, Резуха

альпийская
30174 Арабис альпийский мелкоцветковый
30185 Резуха амурская
30186 Кардаминопсис, Резуха песчаная
30189 Резуха ушастая, Резуха ушковая
30199 Резуха бильярдье
30205 Резуха северная
30207 Резуха коротkoплодная
30211 Резуха бухарская
30214 Резуха канадская
30220 Резуха беловая, Резуха кавказская
30225 Резуха христиана
30256 Резуха желтушниковаю
30259 Резуха мучнистая
30267 Резуха желтая
30268 Резуха тайваньская
30270 Резуха кустарничковая
30277 Резуха жерара, Хвойник жерарда, Эфедра жерарда
30292 Резуха шершавая
30311 Резуха проежуточная
30313 Резуха фиолетовочашечная
30324 Резушка какандская
30325 Резуха каратегинская
30331 Резуха коканская
30336 Резуха рыхлая
30342 Резуха липовидная
30345 Резуха камчатская
30357 Резушка средняя
30360 Резушка монбретовская
30365 Резушка котовиковая
30367 Резушка нордманна
30380 Резушка толстокорневая
30387 Резуха висячая, Резуха повислая
30401 Резушка попова
30406 Резушка ядунская
30424 Резуха шершавая
30432 Резушка однобокая
30434 Резуха порярная
30454 Резушка выемчатая
30456 Резушка стеллера
30472 Резушка тибетская
30474 Резушка пушистоногая
30476 Резушка турчанинова
30477 Резуха башенная
30491 Ароидные, Аройниковые
30494 Арахис
30498 Арахис культурный, Арахис подземный, Земляной орех, Миндаль иерусалимский, Орех земляной, Орех подземный
30587 Аралия
30593 Аралия тайваньская
30604 Аралия китайская, Диморфант
30618 Аралия материковая
30619 Аралия сердцевидная, Спаржа японская, Ужовник
30634 Аралиа высокая, Аралиа маньчжурская, Аралия высокая, Шип-дерево
30666 Аралия мелколистная
30674 Аралия цзиндунская
30685 Аралия хубэйская
30691 Аралия ганьсуская
30753 Аралия шмидта
30760 Аралия колючая, Аралия шиповатая
30768 Аралия аньхуйская
30775 Аралия тэнчунская
30783 Аралия тибетская
30798 Аралия юньнаньская
30800 Аралиевые
30830 Араукария
30831 Араукария бразильская, Араукария узколистная
30832 Араукария чилийская
30835 Араукария бидвилла
30841 Араукария куннингама
30846 Араукария высокая
30848 Араукария высокая
30857 Араукариевые
30860 Аражия, Араушия
30862 Араушия шелковистая
30877 Дерево земляничное, Земляничник, Земляничное дерево
30880 Земляничное дерево, Земляничное дерево красное
30888 Дерево земляничное крупноплодное, Земляничник крупноплодный, Земляничное дерево крупноплодное
30906 Можжевелоягодник, Райграсс, Рузумовския
30912 Рузумовския жевельниковая
30929 Ангелика, Дягиль, Дядиль
30930 Дягиль короткостебельный
30931 Дягиль низбегающий, Дядиль низбегающий
30932 Ангелика, Ангелика аптечная, Арункус хангелика, Днгиль, Дудник лекаственный, Дягиль аптечный, Дягиль лекарственный, Дядиль аптечный
30937 Ангелика чимганский
31012 Архонтофеникс
31013 Архонтофеникс александра
31023 Арктагростис
31024 Арктагростис тростниковидная
31025 Арктагростис овсяницевидная
31026 Арктагростис мироколистная
31037 Арктерика
31038 Арктерика низкая
31044 Лопух, Лопушник
31051 Лопух большой, Лопух гладкосемянный, Лопух крупный, Лопух паутинистый, Лопух репейник, Лопушник большой, Лопушник съедобный, Репей, Репейник, Репейник исполинский, Собашник
31056 Лопух малый, Лопушник малый
31061 Лопух дубравный
31063 Лопух палладина
31067 Лопух скребница
31068 Лопух войлочный, Лопух паутинистый, Лопух паутинстый, Лопушник паутинистый
31079 Арктогерон
31080 Арктогерон злаковый
31085 Арктофила
31088 Арктофила рыжеватая
31102 Толокнянка
31142 Виноград медвежий, Медвежья ягода, Толокнянка аптечная, Толокнянка медвежья, Толокнянка обыкновенная, Ушко медвежье, Ягода медвежья
31176 Арктотис
31311 Арктоус
31314 Амприк альпийский, Арктоус альпийский, Арктоус японский, Толокнянка альпийская
31316 Арктоус японский
31319 Арктоус красноплодный
31321 Арктоус мелколистный
31322 Арктоус красный
31347 Ардизия
31362 Ардизия баотинская
31380 Ардизия китайская
31396 Ардизия городчатая
31455 Ардизия крупнолистная
31458 Ардизия мелкоцветковая
31477 Ардизия японская
31483 Ардизия цзиньюньская
31516 Ардизия малипоская
31519 Ардизия ланьюйская
31579 Ардизия хайнаньская
31598 Ардизия жуйлиская
31599 Ардизия зибольда
31644 Ардизия юньнаньская
31677 Пальма арековая
31680 Арека катеху, Бетель, Бетель пальма, Орех арековый, Орех нищих, Пальма арековая, Пальма бетель, Пальма бетельная, Пальма катеху, Пальма-бетель, Пинанг
31727 Песчанка
31758 Песчанка азиатская
31759 Песчанка дэциньская
31787 Песчанка волосовидная
31803 Песчанка головчатая
31842 Песчанка пузырниковидная
31848 Песчанка биверштейна
31849 Песчанка далиская
31852 Песчанка гвоздичковая
31874 Песчанка ферганская

31903 Песчанка красивая
31932 Песчанка злаковая
31933 Песчанка злаколистная, Песчанка траволистная
31936 Песчанка гриффита, Песчанка гриффица
31940 Песчанка качимовидная
31957 Песчанка приземистая
31961 Песчанка приметная
31967 Песчанка ситниковая
31971 Песчанка ганьсуская
31991 Песчанка корина
32014 Песчанка ледебра
32015 Песчанка тонкостебельная
32031 Песчанка длиннолистная
32039 Песчанка горицветная
32053 Песчанка крупноцветная
32058 Песчанка крупнолистная
32071 Песчанка мейера
32073 Песчанка луговостепная, Песчанка мелкозубчатая
32082 Песчанка монгольская
32089 Песчанка мулиская
32101 Песчанка чжэньканская
32114 Песчанка эмэйская
32124 Песчанка паульсена
32143 Песчанка полярная
32148 Песчанка потанина
32153 Песчанка ложнохолодная
32172 Песчанка цинхайская
32175 Песчанка ледовского
32178 Песчанка жесткая
32192 Песчанка круплолистная
32210 Альзина посевная
32212 Песчанка тимьянолистная
32240 Песчанка стевена
32266 Песчанка сырейшикова
32267 Песчанка сычуаньская
32268 Песчанка шовица
32288 Песчанка чукотская
32290 Песчанка туркестанская
32301 Мшанка шилолистная
32319 Песчанка юньнаньская
32328 Песчанка чжундяньская
32330 Аренга
32336 Аренга
32340 Аренга мелкоцветковая
32343 Аренга сахарная, Аренга сахароносная, Пальма сахарная
32371 Аретуза
32411 Аргемон, Аргемона, Аргемоне
32429 Аргемон мексиканский, Мак колючий
32548 Аргирантемум
32563 Аргирантемум куртаниковый, Ромашка, Хризантема куртаниковая
32600 Перламтровка
32769 Аргиролобий, Аргиролобиум
32792 Аргиролобий чашечный
32841 Аргиролобий лядвенцевидный
32923 Аргиролобий пажитниковый
32988 Мелкоплодник ольховолистный, Мелкоплодник ольхолистный, Рябина ольхолистная
33008 Рябина японская
33206 Ангалоний
33234 Аризема
33245 Аризема амурская
33310 Аризема далиская
33330 Аризема фаржа
33332 Аризема тайваньская
33335 Аризема франше
33353 Аризема хунаньский
33354 Аризема хуняский
33356 Аризема иланьский
33357 Аризема инцзянский
33369 Аризема японская
33375 Аризема цзиндунская
33388 Аризема лицзянская
33441 Аризема эмэйская
33458 Аризема пинбяньская
33512 Аризема яошаньская
33534 Аризема тэнчунская
33574 Аризема юньнаньская
33730 Аристида, Селин, Триостница
33737 Аристида вознесения, Селин, Триостница, Триостница вознесения
33760 Аристида паутинистая
33797 Триостница китайская
33898 Аристида карелина
33973 Аристида мелкоперистая
34097 Аристолохия, Кирказон
34126 Ольха ботта
34147 Кирказон сычуаньский
34149 Кирказон ломоносовидный, Кирказон обыкновенный, Кирказон понтийский
34154 Кирказон скрученный
34162 Аристолохия слабая, Кирказон слабый
34188 Аристолохия фуцзяньская
34198 Кирказон крупноцветтный
34200 Аристолохия тибетская
34201 Аристолохия гуйцзоуская
34211 Кирказон иберийский
34219 Аристолохия крупнолистная
34237 Аристолохия куэньминская
34238 Аристолохия гуансиская
34268 Кирказон маньчжурский
34300 Кирказон понтийский
34337 Аристолохия крупнолистная, Кирказон крупнолистный
34343 Ольха штейпа
34375 Аристолохия юньнаньская
34381 Аристолохия чжундяньская
34384 Кирказоновые
34394 Аристотелия
34425 Абрикос
34430 Абрикос альпийский
34431 Абрикос волосистоплодный, Абрикос фиолетовый, Абрикос черный, Абрикос чёрный, Абрикос шероховатоплодный, Абрикосо-слива, Слива канадская, Слива чёрная
34436 Абрикос тибетский, Абрикос шелковисто-бархатный
34443 Абрикос дикий, Абрикос маньчжурский, Абрикос монгольский, Маньчжурский абрикос, Прунус маньчжурский
34448 Абрикос китайский, Абрикос муме, Абрикос японский, Муме
34470 Абрикос сибирский, Прунус сибирский, Сибирский абрикос
34475 Абрикос, Абрикос дерево абрикосовое, Абрикос дикий, Абрикос культурный, Абрикос натоящий обыкновеный, Абрикос обыкновенный, Дерево абрикосовое, Урюк
34477 Абрикос ансу, Абрикос китайский, Ансу, Дерево абриковое
34488 Армерия, Гвоздичнник, Статице
34494 Армерия альпийская
34502 Армерия арктическая
34514 Армерия удлиненная
34536 Армерия приморская
34562 Армерия сибирская
34582 Ложечица, Хрен
34587 Жеруха хреновая, Хрен, Хрен деревенский, Хрен дикий, Хрен обыкновенный
34590 Жеруха хреновая, Хрен, Хрен деревенский, Хрен дикий, Хрен обыкновенный
34594 Хрен гулявниковыи, Хрен луговой
34600 Арнебия
34602 Арнебия бальджуанская
34605 Арнебия фиолетово-желтая
34609 Арнебия лежачая, Арнебия простертая
34615 Макротамия красящая
34622 Арнебия крупноцветковая
34624 Арнебия пятнистая, Арнебия тибетная
34628 Арнебия линейнолистная
34633 Арнебия обратнояйцевидная
34635 Макротамия синяковидная
34641 Арнебия томсона
34646 Арнебия закаспийская
34655 Арника, Баранник
34659 Арника альпийская

34712 Арника холодная
34724 Арника ильина
34725 Арника средня
34733 Арника лессинга
34752 Арника горная, Баранка, Баранник горный
34766 Арника сахалинская
34780 Арника уналашкинская
34815 Арнозерис
34816 Арнозерис малый
34845 Арония
34854 Арония арбутусолистная
34914 Аракача съедобная, Арракач, Арракача сьедобная
34920 Райграсс французский, Французский райграс
34930 Овес высокий, Овёс высокий, Райграсс итальяниский, Райграсс французский высокий, Французский райграс, Французский райграс высокий
34945 Райграсс французский кочи
34995 Артабоьрис
35084 Артемизия, Полынь
35088 Дерево божье, Полынь божье дерево, Полынь долинная, Полынь кустарниковая, Полынь лечебная
35090 Абсент, Абсинт, Вермут, Глистник, Нехворощ, Полынь горькая, Полынь настоящая, Полынь-абсент
35095 Полынь адамка
35105 Полынь беловосковая
35113 Полынь алтайская
35119 Полынь укрополистная
35132 Полынь однолетная, Полынь однолетняя
35144 Полынь аральская
35153 Полынь арктическая
35167 Полынь арги, Полынь аржий
35176 Полынь армянская
35178 Полынь ашурбаева
35187 Полынь золотистая
35188 Полынь австрийская
35193 Полынь бадхызская
35196 Полынь бальджуанская
35197 Полынь баргузинская
35211 Полынь островная, Полынь северная
35230 Полынь дернистая
35237 Нехворощ, Полынь полевая, Полынь равнинная
35273 Полынь канадская
35282 Полынь волосовидная
35313 Полынь каспийская
35314 Полынь кавказская, Полынь шерстистая
35323 Полынь ромашколистная, Полынь ромашниковая
35343 Полынь заменяющая, Полынь замещяющая
35361 Полынь заостренная
35362 Полынь дагестанская
35365 Полынь угнетенная
35369 Полынь обедненная
35370 Полынь пустынная
35373 Полынь пустынная
35405 Полынь эстрагоновидная
35411 Полынь непахучая, Полынь эстрагон, Трава драгун, Тургун, Эстрагон
35421 Полынь памирская
35422 Полынь цинхайская
35463 Полынь эмэйская
35464 Полынь волосистоцветная
35468 Полынь ганьсуская
35479 Полынь пучковатая
35501 Полынь душистая
35503 Полынь фрейна
35505 Полынь каменная, Полынь холодная
35520 Полынь вильчатая
35530 Полынь гладкая
35538 Полынь серая, Полынь сизая
35547 Полынь шаровницевая
35550 Полынь эдельвейсовая
35560 Полынь гмелина, Полынь сантолинолистая
35573 Полынь гуншаньская
35586 Полынь гипсовая
35595 Полынь солончаковая
35600 Полынь белотравная
35606 Полынь травянистая
35612 Полынь белая, Полынь бловойлочная
35627 Полынь седая
35634 Полынь индийская
35648 Полынь комарова, Полынь цельнолистная
35671 Полынь якутская
35674 Полынь японский
35701 Полынь японский хайнаньский
35719 Полынь цзилунский
35733 Полынь кейске
35740 Полынь·кнорринг
35742 Полынь коидзуми
35748 Полынь копетдагская
35752 Полынь кульбадская
35754 Полынь кушакевича
35759 Полынь надрезанная, Полынь раздельнолистная, Полынь рассеченная
35779 Полынь звёздчатая, Полынь стеллера
35785 Полынь куропаточья
35789 Полынь широколистная
35794 Полынь лавандолистная
35808 Полынь ледебура
35815 Полынь лессинга, Полынь лессинговская
35816 Полынь белолистная
35827 Полынь илийстая
35830 Полынь липского
35831 Полынь прбрежная
35833 Полынь лопастелистная
35856 Полынь худощавая
35857 Полынь крупноцветковая
35859 Полынь крупнокорзиночная
35860 Полынь крупнокорневищная
35868 Полынь морская, Полынь приморская
35888 Полынь маршалла, Полынь маршалловская
35895 Полынь максимовича
35897 Полынь промежуточная
35904 Полынь мексиканская
35916 Полынь монгольская
35944 Полынь однопестичная
35948 Полынь горная
35974 Полынь наманганская
35993 Полынь норвежская
36012 Полынь туподольчатая
36032 Полынь песчаная
36048 Полынь болотная
36054 Полынь лохматая
36059 Полынь мелкоцветковая
36072 Полынь персидская
36076 Полынь бурочешуйковая
36088 Полынь понтийская
36089 Полынь длинная
36091 Полынь лапчатколистная
36094 Полынь светлозеная
36105 Божедерево, Полынь высокая
36108 Полынь широкая
36122 Полынь точечная
36129 Полынь мощнокорневая
36134 Полынь пятидольчатая
36137 Полынь отставленнолопастая
36162 Полынь красноножковая
36166 Полынь каменная, Полынь скалистая, Полынь скальная, Чернобыльник каменистый
36173 Полынь рутолистая
36177 Полынь священная
36202 Полынь
36211 Полынь солянковая, Полынь солянковидная
36216 Полынь самоедов
36221 Полынь сапожникова
36225 Полынь щмидта
36232 Полынь веничная, Полынь метельчатая
36241 Полынь селенгинская
36259 Полынь сенявина
36262 Полынь шелковистая
36273 Полынь шелковистолистная
36276 Полынь поздная
36283 Полынь сычуаньская

36285 Полынь зибера
36286 Полынь сиверса, Полынь сиверсовская
36292 Полыньлесная
36304 Полынь джунгарская
36312 Полынь блестящая
36317 Полынь узкокорзиночная
36319 Полынь степная
36321 Полынь побегоносная
36348 Полынь шиловидная
36349 Полынь клейковатая
36350 Полынь мясистая
36354 Полынь лесная
36357 Полынь совича
36363 Полынь пижмолистная
36367 Полынь крымская, Полынь таврическая
36368 Полынь тонкорассеченная
36383 Полынь тилезиуса
36391 Полынь тонковойлочная
36394 Полынь турнефоровская
36396 Полынь траутфеттера
36426 Полынь черняевская
36429 Полынь туркменская
36440 Полынь уналашкская
36442 Полынь уссурийский
36443 Полынь ваханская
36444 Полынь крепкая
36474 Полынь обыкновенная, Полынь пышная, Полынь-чернобыль, Чернобыл, Чернобыль, Чернобыльник
36560 Полынь ядунский
36564 Полынь юньнаньский
36567 Полынь чаюйский
36570 Полынь чжундяньский
36612 Артраксон, Членистоостник
36625 Артраксон хайнаньский
36644 Артраксон гуйцзоуский
36647 Артраксон волосистый, Артраксон лангсдорфа, Артраксон отроняющаяся
36650 Артраксон лангсдорфа
36673 Артраксон ланцетовидный
36813 Саксаульчик
36822 Саксаульчик илийский
36902 Дерево хлебное
36913 Дерево хлебное
36919 Дерево гуншаньское
36920 Дерево хлебное, Дерево хлебное восточноиндийское, Дерево хлебное джек, Дерево хлебное индиское, Дерево хлебное разнолистное, Дерево хлебное цельнолистное, Джек, Джек-дерево, Джекфрут, Кемпедек, Хлебное дерево цельнолистное, Чембедек, Як-дерево
36924 Дерево хлебное восточноиндийское, Дерево хлебное джек, Дерево хлебное индиское, Дерево хлебное разнолистное, Дерево хлебное цельнолистное, Джек, Джек-дерево, Джекфрут, Кемпедек, Чембедек, Як-дерево
36928 Дерево хлебное лакооха, Хлебоплодник
36967 Аройник, Ароник, Аронник, Арум, Борода ааронова
36969 Аронник белокрылый
36991 Аронник удлиненный
37005 Аронник белокрылый, Аронник итальянский, Арум белокрылый, Арум итальянское
37013 Аройник королькова, Аронник королькова
37015 Аройник пятнистый, Аронник пятнистый, Арум пятнистый, Змеевик, Клещинец
37022 Аройник восточный
37053 Арункус, Волжанка, Таволжник
37056 Волжанка азиатская
37069 Волжанка камчатская
37079 Таволжник гуншаньский
37083 Волжанка малая
37085 Волжанка обыкновенная, Таволжник обыкновенный
37109 Арундинелла
37127 Арундинария, Камышовка
37162 Арундинария наньпинская
37229 Арундинария хайнаньская
37350 Арундинелла, Полевица обыкновенная
37352 Арундинелла уклоняющаяся
37379 Арундинелла уклоняющаяся
37398 Арундинелла лушаньская
37402 Арундинелла хубэйская
37417 Арундинелла мелкоцветковая
37431 Арундинелла тэнчунская
37442 Арундинелла юньнаньская
37450 Арундо
37467 Арундо донакс, Арундо тростниковый, Камыш, Камыш гигантский
37477 Арундо тайваньский
37556 Копытень
37574 Копытень канадский
37596 Копытень чэнкоуский
37597 Копытень китайский
37616 Копытень европейский, Ладан конский
37624 Копытень фуцзяньский
37647 Копытень грузинский, Копытень иберийский
37659 Копытень японский
37679 Копытень динхуский
37680 Копытень крупнолистный
37691 Копытень наньчуаньский
37695 Копытень фэнцзеский
37722 Копытень зибольда
37806 Асклепиадовые, Ластовневые, Латочниковые
37811 Асклепиас, Ваточник, Ластовень, Трава эскулапова
37888 Ваточник кюрасао, Ластовень курассавский
37967 Ваточник инкарнатный, Ваточник розово-малиновый
38135 Ваточник силийский, Латочник
38147 Ваточник клубневой, Ваточник клубненосный
38285 Асцирум, Асцирум эверобойниковидный
38322 Азимина, Дерево авраамово
38338 Азимина трёхлопастная
38904 Аспараг, Аспарагус, Спаржа
38908 Спаржа остролистная
38928 Спаржа джунгарская, Спаржа угловатонадломанная
38938 Спаржа спаржевидная
38950 Спаржа колотколистная
38951 Спаржа бреслера
38953 Спаржа бухарская
38977 Спаржа бугорчатая, Спаржа даурсая
38983 Аспарагус шпренгера, Спаржа шпренгера
39048 Спаржа ганьсуская
39049 Спаржа казахстанская
39061 Спаржа левиной
39064 Спаржа прибрежная
39082 Спаржа куэньминская
39095 Спаржа мищенко
39111 Спаржа пренебреженная
39120 Спаржа аптечная, Спаржа лекарственная, Спаржа многолистный, Спаржа обыкновенная, Спаржа овощная
39126 Спаржа маловетвистая
39141 Спаржа персидская
39154 Спаржа попова
39156 Аспарагус лицзянский
39157 Спаржа ложношероховатая
39163 Аспарагус цинхайский
39187 Спаржа шобериевидная
39195 Аспарагус перистый, Аспарагус пушистоперистый, Спаржа перистая, Спаржа плюмажная, Спаржа пушистая, Спаржа пушистоперистая
39196 Спаржа щетнновидная
39198 Аспарагус сычуаньский
39225 Аспарагус далиский
39229 Спаржа тонколистная
39234 Аспарагус тибетский
39242 Спаржа туркестанская
39251 Спаржа мутовчатая
39259 Аспарагус яньюаньский

39296 Бухингера
39297 Бухингера пазушная
39298 Асперуга, Острица
39299 Асперуга лежачая, Асперуга простертая, Острица лежачая, Острица простёртая
39300 Ясменник
39302 Ясменник абхазский
39303 Ясменник подоражающий
39304 Ясменник родотвенный
39305 Ясменник альбова
39306 Ясменник альпийский
39314 Ясменник полевой
39315 Ясменник утонченный
39317 Ясменник голубой
39320 Ясменник двузубый
39321 Ясменник биберштейна
39323 Ясменник кавказский
39324 Ясменник сжатый
39326 Ясменник меловой
39327 Ясменник гребенчатый
39328 Марзанка, Ясменник розоватый, Ясменник розовый
39337 Ясменник опушенноцветковый
39338 Ясменник слабый
39342 Ясменник данилевского, Ясменник уменьшенный
39343 Ясменник длиннолистный
39345 Ясменник шероховатый
39346 Ясменник ферганский
39347 Ясменник подмаренниковый, Ясменник сизый
39348 Ясменник скученый
39349 Ясменник изящный
39350 Ясменник гранитный
39351 Ясменник пахучий
39362 Ясменник жестковололистый
39365 Ясменник изломанный
39366 Ясменник неприятный
39367 Ясменник каратавский
39368 Ясменник кемулярии
39369 Ясменник крылова
39370 Ясменник гладкий
39371 Ясменник гладчайший
39372 Ясменник голоплодный
39373 Ясменник белопыльниковый
39374 Ясменник липского
39377 Ясменник маркотхский
39379 Ясменник моллюгообразный
39382 Ясменник супротивнолистный
39383 Ясменник восточный
39384 Ясменник памирский
39387 Ясменник цветкожковый
39393 Ясменник каменистый, Ясменник скальный
39395 Ясменник понтический
39396 Ясменник попова
39397 Ясменник переодетый
39398 Ясменник близкий
39399 Ясменник стелющийся
39402 Ясменник румелийский
39404 Ясменник полуодетый
39405 Ясменник щетинистый
39406 Ясменник щетинковый
39407 Ясменник стевена
39410 Ясменник низкий
39411 Ясменник крымский
39412 Ясменник сероплодный
39413 Ясменник красильный
39418 Ясменник туркменский
39419 Ясменник шерстеносный
39420 Ясменник воронова
39421 Ясменник ксерофитный
39435 Асфоделина
39436 Асфоделина древовидная
39438 Асфоделина жёлтая, Златок
39439 Асфоделина крымуская
39440 Асфоделина тонкоцветковая
39441 Асфоделина тонкая
39449 Асфодель, Асфоделюс, Асфодил
39454 Асфоделюс белый, Асфодил белый
39516 Аспидистра, Плектогине
39532 Аспидистра высокая, Плектогине
39534 Аспидистра тайваньская
39544 Аспидистра хайнаньская
39556 Аспидистра лодяньская
39557 Аспидистра грязнобурая
39560 Аспидистра мелкоцветковая
39567 Аспидистра эмэйская
39573 Аспидистра гуансиская
39575 Аспидистра сычуаньская
39583 Аспидистра инцзянская
39693 Аспидосперма
39700 Квебрахо белое
39874 Драцена астелия
39910 Астра, Грудница
39928 Астра агератовидная
39970 Астра агератовидная мелкоцветковая
40016 Астра альпийская
40021 Астра ложная
40022 Астра змеиногорская
40037 Астра ложноитальянская
40038 Астра дикая, Астра звёздчатая, Астра итальянская, Астра ромашковая, Астра степная
40425 Астра фори
40440 Мелколепестник обвислый, Мелколепестник разнощетинный
40474 Грудница тайваньская
40518 Астра глена
40598 Астра хунаньская
40607 Астра грузинская
40621 Астра индийская
40665 Астра коржинского
40754 Астра лицзянская
40803 Астра роскошнолистная
40805 Астра маака
40879 Астра мотоская
40880 Астра мупинская
40923 Астра виргинская, Астра новая бельгия, Астра новобельгийская
41025 Астра пенсауская
41133 Астра пиренейская
41150 Астра ричардсона
41190 Астра ивовая, Астра иволистная
41201 Гимнастер саватье
41224 Астра зее-буреинская
41244 Астра сычуаньская
41257 Астра сибирская
41285 Астра тибетская
41311 Астра почни цельная
41342 Астра татарская, Грудница татаркая
41351 Астра дэциньская
41376 Астра тяньцюаньская
41377 Астра толмачева
41520 Астра юньнаньская
41523 Астра чаюйская
41661 Астеролинум
41663 Астеролинум звездчатый
41747 Астеротамнус
41752 Астеротамнус кустарниковый
41755 Астеротамнус разнохохолковый
41757 Астеротамнус дубровнниколистный
41759 Астеротамнус шишкина
41786 Астильба, Астильбе
41795 Астильба китайская, Астильбе китайское
41822 Астильбе корейское
41828 Астильбе мелколистное
41852 Астильба тунберга
41876 Астомаеа, Астоматопсис
41906 Астрагал
41925 Астрагал пушковатый
41935 Астрагал аксуйский
41942 Астрагал алатавский
41948 Астрагал белостебелиный
41961 Астрагал лисохвостый
41962 Астрагал альпийский
41971 Астрагал горький
41974 Астрагал древообразный, Астрагал песчанодревесный
41975 Астрагал песочный
41976 Астрагал песколюбец
41989 Астрагал якорный
41998 Астрагал деревцовый, Астрагал дуговидный
42009 Астрагал аркалыкский
42014 Астрагал колючий, Астрагал колючковый
42021 Астрагал серповидный
42023 Астрагал шероховатый, Астрагал

шершавый
42047 Астрагал уральский, Астрагал южный
42048 Астрагал австрийский
42051 Астрагал южносибирский
42072 Астрагал бекетова
42086 Астрагал боетийский
42091 Астрагал бородина
42100 Астрагал короткодольный
42120 Астрагал бунге, Астрагал бунговский
42134 Астрагал лицзянский
42135 Астрагал хоботковый
42140 Астрагал беловойлочный
42164 Астрагал головчатый
42166 Астрагал роговой
42177 Астрагал китайский
42183 Астрагал цюцзянский
42186 Астрагал хомутова
42193 Астрагал гороховидный, Астрагал нутовый, Астрагал хлопунец, Хлопунец
42197 Астрагал клера
42203 Астрагал сродный
42207 Астрагал подложный, Астрагал смешанный
42211 Астрагал простертоветвистый
42218 Астрагал свернутый, Астрагал скрученный
42220 Астрагал рогатый, Астрагал рожковый
42221 Астрагал рогоплодный
42236 Астрагал крестовидный
42255 Астрагал вздуточашечковый
42258 Астрагал даурский, Остролодочник распростертый
42262 Астрагал датский, Астрагал луговой, Астрагал песчаный
42264 Астрагал пушистоцветковый, Астрагал шерстистоцветковый
42277 Астрагал древовидный
42280 Астрагал густоцветный
42294 Астрагал распластанный
42302 Астрагал длиннолистный, Астрагал долголистный
42303 Астрагал ядунский
42304 Астрагал джангартский
42322 Астрагал эллиптический
42352 Астрагал бесстебелиный, Астрагал бесстрелковый
42355 Астрагал серповидный, Астрагал серпоплодный
42366 Астрагал волокнистый, Астрагал тонкостебельный
42371 Астрагал согнутый
42382 Астрагал чжундяньский
42384 Астрагал холодный
42388 Астрагал кустарниковый
42391 Астрагал молочнобелый
42392 Астрагал галеговидный, Астрагал козлятниковиный
42409 Астрагал белесоватый, Астрагал сизый
42416 Астрагал ложносладколистный, Астрагал сладколистовиный
42417 Астрагал сладколистный, Астрагал солодколистный
42453 Астрагал камеденосный, Астрагал трагакантовый
42461 Астрагал крючковатый, Астрагал крючконосный
42485 Астрагал семиреченский
42494 Астрагал джимский
42503 Астрагал хуйнинский
42505 Астрагал мелколистный
42516 Астрагал илийский
42537 Астрагал японский
42544 Астрагал каркасинский
42556 Астрагал кроненбурга
42563 Астрагал курчумский
42566 Астрагал разорванный
42570 Астрагал заячий
42571 Астрагал заячий мелкоцветковый
42580 Астрагал синьцзянский, Астрагал шерстистый
42582 Астрагал лапландский
42585 Астрагал шелкоцветковый
42586 Астрагал мохнатолистный
42588 Астрагал мохнатый
42594 Астрагал приподнимающийся
42597 Астрагал лемана
42600 Астрагал лепсинский
42613 Астрагал беловетвистый
42617 Астрагал ганьсуский
42624 Астрагал каменный
42633 Астрагал длинноцветковый
42654 Астрагал крупнорогий
42655 Астрагал крупнодольный
42657 Астрагал длиннокрылый
42658 Астрагал длинноногий, Астрагал длинноцветоносный
42664 Астрагал длиннолодочный
42672 Астрагал маевского
42691 Астрагал средний
42694 Астрагал черноколосый
42696 Астрагал донниковый
42699 Астрагал перепончатый
42704 Астрагал монгольский
42713 Астрагал мелкоголовчатый
42723 Астрагал светлокрасный
42746 Астрагал монгутский
42757 Астрагал тяньцюаньский
42758 Астрагал мулиский
42784 Астрагал нитевидный
42801 Астрагал благородный
42821 Астрагал эспарцетный, Астрагал эспарцетовидный
42830 Астрагал округлолистковый
42834 Астрагал птицеклювый
42835 Астрагал горошковидный, Астрагал сочевичниковый
42840 Астрагал прямобобовый
42841 Астрагал прямодольный, Астрагал прямоплодный
42843 Астрагал остроязычковый, Астрагал остроплодный
42845 Астрагал палласа
42847 Астрагал бледнеющий, Астрагал бледный
42848 Астрагал кушакевича
42864 Астрагал цветоножечный
42878 Астрагал каменистый
42881 Астрагал волжский, Астрагал пузырчатоплодный, Астрагал пузырчатый
42883 Астрагал густоветвистый, Гуван, Гувен, Текен
42888 Астрагал плосколистный
42893 Астрагал татарский
42902 Астрагал понтийский
42923 Астрагал ложнокоротколодочный
42933 Астрагал ложкометельчатый
42948 Астрагал пушистоцветковый, Астрагал шерстистоцветковый
42967 Астрагал изогнутый, Астрагал отогнутый
42974 Астрагал сетчатый
42985 Астрагал розовый
42987 Астрагал камнеломный
42994 Астрагал мешечатый
42995 Астрагал сахалинский
43000 Астрагал солончаковый
43012 Астрагал острошерохватый
43024 Астрагал метельчатый
43039 Астрагал кунжутный
43040 Астрагал северцова
43044 Астрагал сычуаньский
43046 Астрагал сиверса
43053 Овения вишнелистная
43058 Астрагал скорнякова
43066 Астрагал соготинский
43074 Астрагал шароцветковый
43075 Астрагал шаровидный
43076 Астрагал шаровидновздутый
43085 Астрагал узкорогий
43096 Астрагал полосатый, Астрагал полосчатый
43100 Астрагал почти-дугообразный
43103 Астрагал шиловидный
43112 Астрагал бороздчатый, Астрагал желобчатый
43123 Астрагал обманчивый
43132 Астрагал кандинский

43139 Астрагал крымский
43151 Астрагал яичкоплодный, Астрагал яйцеплодный
43154 Астрагал тибетский
43175 Астрагал заилийский
43180 Астрагал якорцевидный, Астрагал якорцевый
43199 Астрагал мелкоплодный
43204 Астрагал болотный
43205 Астрагал зонтичный
43207 Астрагал однопарный
43209 Астрагал мешодчатый, Астрагал пузыристый
43223 Астрагал пузырчатый
43228 Астрагал ветвистый, Астрагал прутьевидный
43236 Астрагал лисий
43247 Астрагал владимира
43251 Астрагал
43275 Астрагал юньнаньский
43286 Астрагал зайсанский
43288 Астрагал чаюйский
43292 Астрагал цингера
43306 Астранция
43308 Астранция колхидская
43309 Астранция большая, Астранция крупная, Хвощ большой
43310 Астранция наибольшая
43311 Астранция малая
43312 Астранция понтийская
43313 Астранция урехнадрезная
43372 Астрокариум
43417 Морковица
43418 Морковица прибрежная
43419 Морковица восточная
43420 Морковица персидская
43472 Астрониум
43492 Астрофитум
43569 Азинеума
43570 Азинеума стебреобъемлющая
43571 Азинеума аньхуйская
43573 Азинеума острозубчатая
43574 Азинеума бальджуанская
43575 Азинеума колокольчиковидная
43576 Азинеума сероватая
43577 Азинеума китайская
43578 Азинеума цикориевидная
43580 Азинеума японская
43581 Азинеума ланцетная
43582 Азинеума гладкоцветковая
43583 Азинеума лобериевидная
43584 Азинеума красивая
43585 Азинеума ветвистая
43586 Азинеума жесткая
43588 Азинеума ивовая
43590 Азинеума тальшская
43593 Азинеума траутфеттера
43594 Азинеума кувшинчатая
43701 Аталантус
43702 Аталантус колючковидный
43718 Аталантия
43731 Аталантия однолистная
43765 Неполнопыльник
43768 Нопыльник карликовый
44114 Атрактилис
44192 Атрактилодес
44202 Атрактилодес корейский
44205 Атрактилодес яйцевидный
44208 Атрактилодес китайский
44247 Колючая греча, Курчавка
44249 Курчавка узколистная
44250 Курчавка бадхызская
44256 Курчавка сероватая
44257 Курчавка кавказская
44258 Курчавка скученная
44259 Курчавка обманчивая
44260 Курчавка кустарная
44265 Курчавка каратавская
44266 Курчавка яркозеленая
44270 Курчавка, Курчавка мушкетова
44271 Курчавка колючая
44272 Курчавка грушелистная
44273 Курчавка отогнутая
44275 Курчавка зеравшанская
44276 Курчавка шиповатая
44279 Курчавка вальковатолистная
44282 Курчавка турнефора
44284 Курчавка прутьевидная
44298 Лебеда, Лебедка
44324 Лебеда широкоплодная
44332 Лебеда красивоплодная
44333 Кокпек, Лебеда белая
44347 Лебеда центральноазиатская
44349 Лебеда крупноприцветничковая
44366 Лебеда толстолистная
44374 Лебеда диморфная
44400 Лебеда лебедадикая
44402 Лебеда веероплодная
44430 Лебеда гмелина
44468 Лебеда садовая, Лебеда саладная
44480 Лебеда кузеневой
44484 Лебеда гладкая
44510 Лебеда прибрежная
44520 Лебеда максимовича
44523 Лебеда разносемянная
44530 Лебеда монетоплодная
44537 Лебеда лоснящаяся, Лоснящаяся
44541 Лебеда голостебельная
44554 Лебеда продолговатолистная
44558 Лебеда украшенная
44564 Лебеда памирская
44575 Лебеда отклоненная
44576 Лебеда копьевидная, Лебеда копьелистная, Лебеда поникшая, Лебеда раскидистая
44601 Лебеда стебельчатая, Лебеда черешчатая
44619 Лебеда ранняя
44621 Лебеда копьевидная, Лебеда копьелистная, Лебеда обыкновенная
44624 Лебеда колючая
44629 Лебеда розовая
44640 Лебеда шугнанская
44646 Лебеда сибирская
44650 Лебеда гарообразная, Лебеда шарообразная
44668 Лебеда татарская
44674 Лебеда тунбергиелистная
44684 Лебеда туркменская
44688 Лебеда бородавчатая
44704 Белладонна, Красавка
44708 Белладонна, Красавка, Красавка белладонна, Красавка обыкновенная, Красавка одурь сонная, Одурь сонная
44709 Красавка кавказская
44711 Красавка комарова
44801 Атталея, Пальма-атталея
44867 Обриеция, Орех маньчжурский
44883 Аукумения
44887 Аукуба, Дерево золотое
44893 Аукуба китайская
44898 Аукуба китайская эмэйская
44911 Аукуба гималайская
44915 Аукуба японская, Дерево золотое
44919 Аукуба японская пестролистная
44949 Аукуба юньнаньская
45062 Болоздоплодник
45064 Болоздоплодник уклоняющийся
45065 Болоздоплодник дарвазский
45071 Болоздоплодник тяньшанский
45073 Болоздоплодник туркестанский
45165 Дазистома
45192 Бурачок скалистый, Бурачок скальный
45239 Кедр речной чилийский
45363 Овес, Овёс, Овсюг
45389 Овес бородатый, Овёс бородатый
45402 Овёс короткий
45417 Овес брунса
45424 Овёс византийский, Овёс византийский красный, Овёс посевной средиземноморский, Овёс средиземноморский
45431 Овес китайский
45433 Овес сомнительный
45437 Овес сходный
45439 Овес пустынный
45448 Гис, Овёс дикий, Овёс живой, Овес овсюг, Овес пустой, Овёс пустой, Овсец опушённый, Овсюг, Овсюг обыкновенный, Полетай, Хис

45481 Овес гиблидный
45495 Овес людовика
45503 Овес крупноцветковый
45513 Овес южный
45526 Овес волосистоузлый
45533 Овёс венгерский, Овёс восточный, Овёс египетский, Овёс косой, Овёс турецкий
45542 Овес волосисты
45546 Овес тяжелый
45548 Овес луговой, Овёс луговой, Овсец луговой
45558 Овес заячий, Овёс опушённый, Овес опушенный, Овёс пушистый
45566 Овес двухгривый, Овес обыкновенный, Овес посевной, Овёс посевной
45581 Овес северный, Поледай
45586 Галипея лекарственная, Нгустура, Овес бесплодный, Овёс бесплодный, Овёс декоративный, Овёс дикий красный, Овёс стерильный
45592 Овес волосистолистный
45602 Овёс калючий, Овёс тоший, Овес щетинистый, Овёс щетинистый
45619 Овес разноцветный
45621 Овес виеста
45724 Авероа-билимби, Билимби
45725 Авероа, Карамол, Карамола, Черёмуха поздняя иволистная
45740 Авиcения
45750 Куст соляной, Мангро поблескивающее, Медовик мангровый
45843 Аксирис
45844 Аксирис щирицевый
45851 Аксирис кавказский
45853 Аксирис гибридный
45856 Аксирис простертый
45859 Аксирис шароплодный
45906 Азадирахта
45908 Азадирахта индийская, Мелия индийская
45952 Азара
45955 Азара мелколистная
46176 Баккаурея
46210 Бакхарис
46227 Бакхарис галимолистный, Бакхарис лебедолистная
46348 Бакопа
46374 Бактрис, Пальма-бактрис
46376 Пальма персиковая
46888 Баланофоровые
46989 Белокудренник
46993 Белокудренник северный
46999 Белокудренник серый
47013 Белокудренник черный, Белокудренник чёрный, Кудренник чёрный, Кудри пестрые
47032 Белокудренник
47077 Бальзаминовые
47174 Бамбуза, Бамбук
47190 Бамбук обыкновенный
47222 Бамбуза хайнаньская
47225 Бамбуза бирманская
47345 Бамбук карликовый, Бамбук множественный
47430 Бамбук жесткий
47516 Бамбуза обыкновенная
47535 Бамбуза юньнаньская
47626 Банксия
47685 Бафия
47788 Бафия красивая
47854 Бантизия, Бантисия
47880 Бантизия красильная, Индиго дикое
47931 Сурепица, Сурепка
47934 Сурепка дуговидная
47944 Сурепка крупноцветковая
47948 Сурепка малая
47949 Сурепка прямая
47954 Сурепка подорожниковая
47959 Сурепка прижатая, Сурепка прямая
47962 Сурепка тайваньская
47963 Сурепица весенняя, Сурепка весенняя
47964 Сурепица обыкновенная, Сурепка обыкновеная
48472 Баросма
48509 Барингтония
48585 Бартония
48601 Бартсия, Бартшия, Зубчатка
48688 Безелла, Шпинат индийский, Шпинат малабарский
48689 Базелла-шпинат, Шпинат индийский, Шпинат малабарский
48698 Базеллевые
48753 Дерево масляное
48762 Эхинопсилон растопыреный
48767 Эхинопсилон волосистый
48769 Эхинопсилон иссополистная
48790 Эхинопсилон очитковидны
48880 Батис
48904 Водяной лютик, Шелковник
48907 Шелковник бунге
48910 Шелковник килеватый
48913 Шелковник дихотомический
48914 Водяной лютик расходящийся, Лютик расходящийся, Шелковник расходящийся
48915 Водяной лютик неуколеняющийся, Лютик нитчатый, Шелковник неуколеняющийся
48919 Водяной лютик, Лютик жестколистный, Лютик закрученный, Лютик фенхелевидный, Шелковник фенхелевидный
48920 Шелковник гилибера
48922 Водяной лютик кауфмана, Шелковник кауфмана
48923 Шелковник ланге
48924 Лютик бокочветковый
48926 Шелковник морской
48927 Водяной лютик монгольский
48928 Водяной лютик толстостебельный
48931 Водяной лютик риона, Шелковник риона
48933 Водяной лютик волосистый, Лютик волосолистный, Шелковник волосолистный
48944 Шелковник трехлистный
48990 Баугиния, Баухиния
48993 Баугиния остроконечная
49019 Баугиния лицзянская
49022 Баугиния фабера
49101 Баугиния хубэйская
49116 Баугиния хайнаньская
49140 Баугиния японская
49181 Баугиния бирманская
49241 Баугиния шерстистая
49268 Баугиния юньнаньская
49532 Бекманния
49540 Бекманния восточная, Бекманния обыкновенная
49587 Бегония
49603 Бегония белоточечная
49617 Бегония угловатая
49653 Бегония цзиньпинская
49662 Бегония бисмарка
49701 Бегония катаяна
49728 Бегония тэнчунская
49735 Бегония вьюнковая
49774 Бегония диадема
49778 Бегония наньчуаньская
49781 Бегония пальчатая
49808 Бегония эмэйская
49826 Бегония фэн
49827 Бегония ланьюйская
49844 Бегония лиственная
49877 Бегония сизолистная
49879 Бегония гоегская
49886 Бегония двансиана
49895 Бегония гуншаньская
49897 Бегония хайнаньская
49903 Бегония мотоская
49915 Бегония борщевиколистная
49925 Бегония волосистая
49926 Бегония волосистая колпачковая
49941 Бегония превосходная
49942 Бегония превосходная изумрудная
49943 Бегония седая, Бегония щитовидная
49944 Бегония мясокрасная
49966 Бегония гуйцзоуская

50056 Бегония малипоская
50059 Бегония рукавчатая
50071 Бегония крупнолистная
50076 Бегония металлическая
50079 Бегония мелкоцветковая
50087 Бегония юньнаньская
50092 Бегония лунчжоуская
50094 Бегония мулиская
50100 Бегония наньтоуская
50104 Бегония лотосолистная
50122 Бегония ольбия
50161 Бегония непалская
50181 Бегония лодяньская
50232 Бегония королевская, Бегония рекс, Царь-бегония
50243 Бегония клещевинолистная
50257 Бегония красноватая
50273 Бегония кровянокрасная
50278 Бегония лазящая
50298 Бегония вечкоцветушая, Бегония всегда цветущая, Бегония месячная
50317 Бегония полуклубневая
50352 Бегония тайваньская
50354 Бегония далиская
50374 Бегония пинбяньская
50378 Бегония клубневая
50379 Бегония ильмолистная
50382 Бегония волнистая
50400 Бегония ван
50405 Бегония вэньшаньская
50417 Бегония инцзянская
50433 Бегониевые
50465 Бейльшмидия
50667 Беламканда
50669 Беламканда китайская
50710 Беллардиа
50717 Беллардиа обыкновенная
50722 Беллардиахлоа
50724 Беллардиахлоа многокрасочная
50736 Белльвалия
50743 Белльвалия остролистная
50744 Белльвалия албанская
50745 Белльвалия араксинская
50746 Белльвалия темнофиолетовая
50747 Белльвалия оше
50759 Белльвалия фомина
50761 Белльвалия липского
50762 Белльвалия длинностолбчатая
50763 Белльвалия желтая
50767 Белльвалия горая
50769 Белльвалия густоцветковая
50773 Белльвалия сарматская
50774 Белльвалия савича
50776 Белльвалия самечательная
50778 Белльвалия туркестанская
50779 Белльвалия вильгельмса
50780 Белльвалия зигоморфная
50808 Белльс, Маргаритка
50818 Бакхарис гирканская
50825 Беллис многолетний, Бук энгрера, Маргаритка многолетняя, Маргаритка обыкновенная
50841 Беллиум
50994 Бенинказа
50998 Арбуз китайский, Тыква восковая
51261 Барбарисовые
51268 Барбарис
51279 Барбарис эмэйский
51280 Барбарис этненский
51284 Барбарис агрегатный, Барбарис скученный
51301 Барбарис амурский
51303 Барбарис бретшнейдера
51314 Барбарис аньхуйский
51327 Барбарис остистый
51333 Барбарис азиатский
51400 Барбарис бирманский
51401 Барбарис самшитолистный
51409 Барбарис быковский
51422 Барбарис канадский
51424 Барбарис сизобелый
51452 Барбарис китайский
51495 Барбарис дэциньский
51510 Барбарис боярышниковый
51518 Барбарис чэнкоуский
51520 Барбарис дарвина
51540 Барбарис густоцветковый
51563 Барбарис дильса
51586 Барбарис
51602 Барбарис наньчуаньский
51613 Барбарис фэн
51634 Барбарис франше
51642 Барбарис фуцзяньский
51643 Барбарис хубэйсий
51653 Барбарис хубэйсий эмэйский
51666 Барбарис цзилунский
51690 Барбарис гуйцзоуский
51711 Барбарис генри
51717 Барбарис разноножковый, Барбарис разноцветоножковый
51732 Барбарис хэнаньский
51733 Барбарис гукера
51757 Барбарис илийский
51766 Барбарис замечательный
51773 Барбарис сичанский
51782 Барбарис цельнокрайний
51802 Барбарис юлианы
51805 Барбарис ганьсуский
51807 Барбарис каркаралинский
51809 Барбарис кашгарский
51824 Барбарис корейский
51827 Барбарис куэньминский
51839 Барбарис лэйбоский
51856 Барбарис нежный
51860 Барбарис лицзянский
51891 Барбарис дерезовидный
51904 Барбарис малипоский
51916 Барбарис мяньнинский
51923 Барбарис мелколистный
51928 Барбарис мелкоцветковый
51943 Барбарис мулиский
51968 Барбарис монетный
51978 Барбарис продолговатый
51981 Барбарис восточный
52022 Барбарис перуский
52037 Барбарис пинбяньский
52049 Барбарис пуаре
52052 Барбарис многоцветковый
52066 Барбарис провансальский
52069 Барбарис сизый
52131 Барбарис саржента
52149 Барбарис сибирский
52150 Барбарис сычуаньский
52151 Барбарис зибольда
52169 Барбарис сулье, Барбарис узколистный
52182 Барбарис чжундяньский
52214 Барбарис далиский
52225 Барбарис тунберга, Барбарис японский
52227 Барбарис тунберга атропурпуреа
52246 Барбарис максимовича
52255 Барбарис тибетский
52258 Барбарис тишлера
52284 Барбарис туркменский
52306 Барбарис весенний
52308 Барбарис бородавчатый, Барбарис мелкобородавчатый
52313 Барбарис зеленоватый
52320 Барбарис чжэцзянский
52322 Барбарис обыкновенный
52355 Барбарис ван
52371 Барбарис вильсона
52391 Барбарис юньнаньский
52397 Барбарис чаюйский
52400 Берхемия
52466 Берхемия лазащая, Роза альберта
52503 Бадан, Бергения
52514 Бадан сердцелистный, Бадан толстолистный, Бадан толстолистый, Бадан узколистный, Чай могольский
52520 Бадан эмэйский
52530 Бадан тихоокеанский
52542 Бадан тяньцюаньский
52543 Бадан угамский
52587 Бергия
52592 Бергия амманиевая, Бергия амманиевидная
52597 Бергия водная
52843 Берлиния
53034 Икотник

53036 Икотник восходящий
53038 Икотник серозелёный, Икотник серый
53057 Орех американский, Орех бразильский
53154 Берула, Беруля
53157 Берула прямая, Беруля прямая, Поручейник узколистный
53161 Беруля восточная
53224 Бурак, Кызылша, Свекла, Свёкла
53231 Свекла кормовая, Свекла крупнокорневая
53242 Свекла многолетняя
53245 Свёкла кормовая
53248 Свекла трехстолбиковая
53249 Канткызылша, Кызылша, Салат мангольд, Свекла культурная, Свекла обыкновенная, Свёкла обыкновенная, Свекла сахарная
53254 Свёкла дикая, Свёкла приморская
53255 Свёкла сахарная, Свекловица сахарная
53257 Мангольд, Свекла лиственная, Свёкла листовая, Свекольник
53259 Свёкла столовая
53262 Свёкла кормовая
53269 Свёкла столовая красная
53293 Буквиза, Буквица, Буковица, Чистец
53295 Буквица абхазская
53298 Буквица олиственная
53301 Буквица крупноцветковый
53303 Буквица белоснежная
53304 Бетоника, Буквиза аптечная, Буквица аптечная, Буквица лекарственная
53307 Буквица восточная
53308 Буквица осетинская
53309 Береза, Берёза
53317 Береза аянская
53318 Береза алайская
53335 Береза белая китайская
53338 Береза аллеганская, Береза желтая, Берёза жёлтая
53352 Береза карликовая
53357 Береза байкальская
53370 Береза каяндера
53372 Береза мозолистная
53379 Береза китайская, Берёза китайская
53386 Береза стройная
53388 Береза японская
53389 Береза ребристая
53400 Береза даурская, Береза максимовича
53405 Береза делаве
53411 Береза ерманова, Береза желтая, Береза каменная, Берёза каменная, Береза эрмана, Берёза эрмана
53434 Береза тощая
53441 Берёза форрестая
53444 Береза овалинолистная
53461 Береза гмелина
53464 Береза крупнолистная
53466 Береза гуншаньская
53470 Береза низкая, Береза приземистая
53480 Береза иркутская
53483 Береза японская, Берёза японская
53494 Береза цзиньпинская
53497 Береза келлера
53499 Береза киргизская
53500 Береза коржинского
53501 Береза кузмичева
53505 Берёза вишенная, Береза вишневая, Берёза вишнёвая
53520 Берёза максимовича
53523 Береза медведева
53525 Береза мелколистная
53527 Береза миддендорфа, Берёза миддендорфа
53530 Береза карликовая, Берёза карликовая, Береза низкоросная, Ера пелярная, Ёра порярная, Ёрник, Ерник стелющийся, Стланец березовый
53536 Береза черная, Берёза чёрная
53544 Береза овальнолистная
53546 Береза памирская
53549 Береза бумажная, Берёза бумажная
53563 Берёза белая плакучая, Береза бородавчатая, Берёза бородавчатая, Берёза обыкновенная, Берёза плакучая, Берёза повислая, Береза поникляя
53572 Береза маньчжурская, Береза плосколистная
53586 Береза тополелистная
53587 Береза потанина, Берёза потанина
53589 Береза прохорова
53590 Береза пушистая, Берёза пушистая
53594 Берёза малорослая
53600 Береза радде
53603 Береза резниченко
53605 Береза круглолистная
53608 Береза сапожникова
53610 Береза шмидта, Берёза шмидта
53617 Береза субарктическая
53619 Береза сукачева
53625 Береза тяньшанская
53626 Береза извилистая
53628 Береза потанина эмэйская
53629 Береза туркестанская
53631 Береза ильмолистная
53651 Березовые, Берёзовые
53755 Череда
53797 Череда дваждыперистая, Череда двоякоперистая
53816 Череда поникшая
53903 Череда фэнцзеская
53913 Череда олиственная
53962 Череда камчатская
54001 Череда максимовича
54042 Череда мелкоцветковая
54081 Череда лучевая, Череда лучистая
54158 Трава золотушная, Череда трехраздельная, Череда трёхраздельная
54194 Биберштейния
54197 Биберштейния многораздельная
54198 Биберштейния душистая
54206 Бинерция
54208 Бинерция окружнокрылая
54237 Бифора, Двойчатка
54239 Бифора аучистая
54240 Бифора яйцеобразная
54292 Бигнония
54303 Бигнония усиковая
54357 Бигнониевые
54425 двукрайник бунге
54439 Билльбергия
54454 Билльбергия поникшая
54464 Билльбергия пирамидальнометельчатая
54532 Биофитум
54618 Бишофия
54620 Бишопия яванская, Бишофия, Бишофия трехлисточковая
54623 Бишофия многоплодная
54798 Горец змеиный, Горлец, Гречиха горлец, Змеевик, Корень змеиный, Остролист, Падуб востролистый, Шейки раковые, Шейки рачьи
54859 Викса
54862 Арнато, Бикса аннатовая, Бикса орлеанская, Дерево орлеанское, Лннато, Орлеан
54863 Биксовые
54891 Блейстония
54896 Блейстония пронзеннолистная
55202 Дербянка
55559 Блетилла
55563 Блетилла тайваньская
55574 Блетилла китайская
55592 Блигия
55596 Блигия африканская, Финган
55669 Блюмея
55693 Блюмея бальзамоносный
55738 Блюмея тайваньская
55889 Блисмус
55892 Блисмус сжатый
55895 Блисмус рыжый, Блисмус тощий, Камыш ржавый
56023 Бокконня, Маклейя, Смолосемянник
56094 Бемерия, Бомерия, Рами
56140 Бемерия зибольда, Бемерия тайваньская
56142 Бемерия фучжоуская
56179 Бемерия инцзянская

56181 Бемерия японская
56195 Бемерия крупнолистная
56206 Бемерия крупнолистная
56229 Конопля китайская, Кранива китайская, Рами, Рами белое, Рами белые
56336 Бемерия тибетская
56397 Боэргавия
56554 Бойсдувалия
56560 Буассьера
56563 Буассьера низкорослая
56572 Буассьера растопыреная
56633 Клбнекамыш
56700 Болтония
56724 Болтония лотюра
56755 Бомарея
56761 Баобабовые, Бомбаксовые
56770 Бомбакс
56802 Бомбакс малабарский
56827 Бомбицилена
56832 Бомбицилена прямостоящая
56941 Бонгардия
56944 Бонгардия золотистая
57084 Бурачниковые
57095 Бораго, Бурачик, Огуречная трава, Огуречник, Трава огуречная
57101 Бораго лекарственный, Бурачик лекарственный, Огуречник аптечный, Огуречник лекарственный, Трава огуречная лекарственная
57122 Борассус пальмира, Пальма борассус, Пальма лантар, Пальма пальмира, Пальма пальмировая, Пальма пальмирская, Пальма сахарная лантар-парьмира
57245 Бородиния
57246 Бородиния байкальская
57386 Борщовия
57387 Борщовия арало-каспийская
57429 Бошниакиа, Бошнякиа
57437 Бошниакиа русская, Бошнякиа русская
57516 Босвелия картера, Ладан
57539 Босвелия, Босвеллия, Босвеллия пильчатая
57545 Бородач
57551 Бородач кавказский
57567 Бородач кровоостанавливающий
57595 Бородач юньнаньский
57675 Кистесемянник
57678 Кистесемянник китайский
57685 Кистесемянник тонкий
57689 Асперуга тонкий
57693 Ботрокариум
57695 Ботрокариум спорный, Дерен спорный
57773 Ботриостеге
57852 Бугенвиллея, Бугенвиллия
57857 Бугенвиллия гладкая, Бугенвиллия голая
57903 Буррерия
57922 Бутелоа, Бутелоуа, Трава бизонья, Трава грама, Трава грамова, Трава пастбищная
57930 Бутелоа, Бутелоа изящная, Трава грама, Трава грамова, Трава пастбищная
58038 Брахантемим, Брахантемум
58041 Брахантемим кустарничковый, Брахантемум кустарничковый
58043 Брахантемим киргизский, Брахантемум киргизский
58048 Брахантем умтитова, Брахантемим титова
58052 Ветвянка
58088 Ветвянка гусеницевидная
58155 Просо ветвистое
58261 Брахиактис
58265 Брахиактис реснитчатая
58280 Брахиактис тибетский, Мелколепестник теневой
58299 Короткокистник
58300 Короткокистник воронеглазый
58335 Стеркулия плёнолистная
58348 Стеркулия утёсная
58447 Брахиэлитрум
58572 Коротконожка, Трахиния
58578 Трахиния двхколосковая
58588 Коротконожка тайваньская
58604 Коротконожка перестая
58616 Коротконожка скальная
58619 Коротконожка лесная, Овсяница лесная
58638 Коротконожка мохнатая
59095 Эритеа
59098 Пальма-брахея мексиканская
59099 Эритеа съедобная
59144 Бразения
59148 Бразения шребера
59265 Брассия
59278 Горчица, Капуста
59370 Капуста критская
59380 Капуста хреновидная
59438 Горчица сарептская, Горчица шпинатная
59493 Брюква, Бушма, Капуста брюкая, Рапс
59507 Брюква
59515 Горчинца черная, Горчица цёрная, Семена горчицы
59520 Браунколь, Капуста кормовая, Капуста кудрявая, Капуста листовая, Капуста овошная, Капуста огородная
59524 Капуста лиственная, Коллард, Тронхуда
59532 Грюнколь, Капуста белокачанная, Капуста кочанная, Капуста курчавая
59541 Голорепа, Капуста кольраби, Капуста кормовая мозговая, Капуста листовая культурная, Капуста репная, Кольраби, Репа венгерская
59575 Репа, Репа кормовая, Репа культурная, Репа овощная, Репа огородная, Репа столовая, Репа японская, Сурепица яровая, Турнепс
59595 Капуста китайская, Паккой, Пак-хой
59600 Капуста пенинская, Петсай, Пецай
59603 Капуста полевая, Репак, Сурепица
59651 Капуста турнефора
59714 Бравоа
59715 Бравоа
59727 Брайа, Брайя
59754 Брайя памирская
59757 Брайя багряная, Брайя багрянистая
59758 Брайа розовая, Брайя короткоплодная, Брайя меднокрасная, Брайя розовая, Брайя узколистная
59769 Брайя шарнгорста
59770 Брайа стручковая
59778 Брайа тибетская
60095 Колеостефус
60162 Бриделия
60372 Трясунка
60379 Стафилея, Трясунка большая
60382 Трясунка средняя
60383 Трясунка малая
60388 Трясунка колосовидная
60431 Бродиея, Броколи, Броколь, Капуста спаржевая, Капуста спаржевая цветная
60544 Бромелия
60557 Ананасные, Ананасовые, Бромелиевые
60612 Анизанта, Костер, Костёр, Стоколос, Хис-кахак
60626 Костер анатолийский
60628 Зерна ангренская, Костер ангренский
60634 Костер полевой, Костёр полевой
60643 Зерна бенекена, Костер бенекена
60646 Костер биберштейна
60653 Костер трясунковидный
60654 Костер канадский
60658 Костер каппадокийский
60680 Костёр ветвистый, Костер переменчивый, Костёр переменчивый, Костёр разветвлённый, Костёр снешанный
60692 Костер дантонии
60696 Костер двутычинковый
60707 Костер прямой, Костёр прямой, Костёр прямостоячий, Костёр стоячий, Стоколос ррямой
60724 Костер тайваньский

60731 Костер тонкий
60735 Костер тончайший, Невскиелла тончайшая
60760 Зерна аябезост, Костер безостый, Костёр безостый, Стоколос безостый
60776 Костер иркутский
60777 Костер японский, Костёр японский
60795 Костер копетдагский
60796 Костер короткого
60797 Костер крупноколосковый, Костер крупноколосый
60820 Костер мадридский, Костёр мадридский
60853 Костер мягкий, Костёр мягкий
60865 Костер непалский
60869 Костер укорашеннный
60871 Костер острозубый
60873 Зерна памирская, Костер памирский
60882 Зерна паульсена, Костер паульсена
60892 Костер попова
60912 Костер кистевидный, Костёр кистевидный
60918 Костер ветвистый, Костер кистистый
60931 Костер ричардсона
60935 Костер жестковатый
60942 Костер береговой
60943 Анизанта краснеющая, Костер краснеющий
60957 Костер жестковатый, Костер метельчатый
60963 Костер ржаной, Костёр ржаной
60968 Костер северцовя
60969 Костер сибирсий
60977 Костер растопыренный
60985 Анизанта бесплодная, Костер бесплодный, Костёр бесплодный, Стоколос бесплодный
60989 Анизанта кровельная, Костер кровельный, Костёр кровельный, Стоколос кровельный
60991 Анизанта шелковистая, Костер шелковистый
60999 Костер войлочковый
61001 Костер войлочный
61003 Зерна туркестанская, Костер туркестанский
61005 Костер тонкий
61006 Зерна мелкочешуйная, Костер мелкочешуйный
61008 Горовик униоловидный, Костер униоловидный, Костёр униоловидный, Роговик униоловидный
61016 Костер пестрый
61053 Бросимум
61057 Дерево коровье, Дерево молочное
61059 Пиратинера гвианская
61096 Бруссонетия, Бруссонеция
61107 Бруссонетия бумажная, Бруссонеция бумажная, Дерево бумажное, Шелковиза бумажная, Шелковица бумажная
61110 Бруссонетия наньчуаньская
61118 Броваллия
61221 Брукенталия
61223 Брукенталия остролистная
61231 Дурман древесный, Дурман древовидный
61242 Андира, Дурман душистый
61293 Брунфельзия, Францисцея
61297 Брунфельзия чашевидная
61306 Брунфельзия широколистная
61364 Брунера
61366 Брунера крупнолистная
61368 Брунера восточная
61369 Брунера сзбирская
61398 Брунсвигия жозефины, Дендробиум, Островския
61430 Мохоцветник
61432 Мохоцветник гмелина
61441 Брылкиния
61442 Брылкиния хвостатая
61464 Бриония, Переступень
61471 Бриония белая, Переступень белый
61497 Переступень двудомный
61518 Переступень однодомный
61538 Переступень заамударьинский
61561 Блиофиллюм
61572 Блиофиллюм перистый
61724 Бухлое, Бухлоэ
61953 Буддлейя, Буддлея
62019 Буддлейя изменчивая, Буддлея давида, Буддлея изменчивая
62059 Буддлея форрестая
62099 Буддлея японская
62110 Буддлея линдли
62178 Буддлея далиская
62244 Бюффония
62252 Бюффония крупноплодная
62261 Бюффония оливье
62264 Бюффония мелкоцветная
62518 Брандушка
62526 Брандушка разноцветная
62530 Бульбофиллум, Бульбофиллюм
62632 Дудник чэнкоуский
62739 Бульбофиллюм тайваньский
62745 Бульбофиллюм фунинский
62756 Бульбофиллюм гуншаньский
62767 Бульбофиллюм хайнаньский
62773 Бульбофиллюм хэнаньский
62813 Бульбофиллюм японский
62832 Бульбофиллюм гуандунский
62841 Бульбофиллюм лэдунский
62908 Бульбофиллюм мэнхайский
62963 Бульбофиллюм дэциньское
63197 Бульбофиллюм мэнцзыский
63204 Бульбостилис
63229 Бульбостилис волосовидный
63253 Бульбостилис волосовидный
63359 Бульбостилис тончайший
63380 Бульбостилис оронова
63405 Бульнезия
63440 Бунга
63442 Бунга трехраздельная
63443 Свербига, Тапсия, Трава злая
63446 Свербига ложкообразная, Свербига степная
63452 Свербига восточная
63467 Буниум, Шишечник, Шишник
63471 Буниум бадхыза
63472 Буниум буржо
63480 Буниум капю
63482 Буниум бутеневый
63485 Буниум цилиндрический
63488 Буниум изящный
63496 Буниум гинколюбивый
63497 Буниум гиссарский
63500 Буниум средный
63501 Буниум кугитанга
63502 Буниум длинноногий
63506 Буниум малолистный
63508 Буниум персидский
63509 Буниум шероховатый
63510 Буниум зеравшенский
63513 Буниум влагалищный
63524 Телекия
63553 Володушка
63554 Володушка абхазская
63557 Володушка родственная
63564 Володушка ясменниковая, Володушка ясменниковидная
63571 Володушка золотистая
63579 Володушка двстебельная
63582 Володушка буассье
63584 Володушка ветвистая
63594 Володушка китайская
63610 Володушка изменчивая, Володушка смешиваемая
63615 Володушка густоцветковая
63619 Володушка отклоненнолистная
63623 Володушка высокая
63625 Володушка серповидная, Кульбаба осенняя
63668 Володушка кустарниковая
63669 Володушка жерара
63675 Володушка сизая
63677 Володушка изащиная
63693 Володушка комарова
63694 Володушка козо-полянсксго
63695 Володушка крылова
63697 Володушка гуйцзоуская
63699 Володушка ланцетнолистная
63713 Володушка франше

63722 Володушка длиннолистная
63727 Володушка длиннообертковая
63728 Володушка длиннолучевая
63749 Володушка маршалла
63750 Володушка мартьянова
63761 Володушка многожильчатая
63769 Володушка нордмана
63776 Володушка бднолучевая
63786 Володушка многообразная
63787 Володушка многолистная
63792 Володушка карликовая
63796 Володушка лютиковидная
63803 Володушка лицзянская
63804 Володушка российская
63805 Володушка круглолистная, Ласковцы, Торичник
63807 Володушка сахалинская
63813 Володушка козелецелистая, Володушка козелецоволистная
63826 Володушка полусложная
63827 Володушка сибирская
63834 Володушка сосновского
63855 Володушка тончайшая
63859 Володушка тяньшанская
63864 Володушка трехлучевая
63869 Володушка витамана
63870 Володушка воронова
63873 Володушка юньнаньская
64063 Бурскария
64068 Бурзера, Бурсера
64088 Бурзеровые, Бурсеровые
64111 Бушия
64113 Бушия бокоцветная
64144 Бутеа
64159 бутия
64161 бутия головчатая
64173 Сусаковые
64180 Сусак
64181 Сусак ситниковидный
64184 Сусак зонтичный
64203 Дерево масляное
64223 Ши
64224 Самшитовые
64235 Буксус, Самшит
64240 Буксус балеарский
64251 Самшит колхидский
64253 Буксус хайнаньский
64263 Самшит гирканский
64264 Самшит ичанский
64284 Самшит крупнолистный
64285 Самшит мелколистный
64303 Самшит мелколистный японский, Самшит японский
64345 Буксус обыкновенный, Пальма кавказская, Самшит вечнозеленый, Самшит обыкновенный
64373 Самшит люцюский
64425 Бирсонима
64491 Кабомба
64496 Кабомба каролинская
64767 Каччиния
64769 Каччиния сомнительная
64772 Каччиния толстолистная
64776 Кахрис
64778 Кахрис альпийский
64783 Кахрис гердера
64788 Кахрис крупноплодная
64791 Кахрис опушенная
64822 Кактусовые
64932 Кадия
64965 Цезальпиния
64971 Бондук, Цезальпиния аптечная
64975 Дерево красное, Дерево красное бразилийское, Дерево фернамбуковое, Фернам буковые, Фернамбук
64982 Лапчатка ползучая
64983 Бондук, Цезальпиния аптечная
64990 Самшит японская, Цезальпиния заборная
64993 Цезальпиния японская
65002 Дерево красное фернамбфковое, Дерево пернамбуковое, Дерево фернамбуковое, Цезальпиния ежовая, Цезальпиния шиповатая
65014 Цезальпиния джиллиса
65033 Цезальпиния крупнолистная
65060 Дерево саппаовое, Цезальпиния ост-индская, Цезальпиния саппан
65084 Цезальпиниевые
65143 Кайанус, Каянус
65146 Айян, Горох голубиный, Кайанус, Кайанус индийский, Кайян, Кайянус, Катьянг, Каяну сголубиный горох, Каянус, Каянус индийский, Каянус каян, Каянус сголубиный горох
65164 Горчица морская, Морская горчица
65173 Морская горчица арктическая
65185 Морская горчица
65214 Каладиум
65218 Каладиум двуцветный
65254 Вейник
65266 Вейник алайский
65267 Вейник алексеенко
65269 Вейник алтайский
65270 Вейник узколистный
65274 Блестящеколосник желтоцветный, Вейник пахучеколосковый
65295 Вейник балансы
65301 Вейник бунге
65315 Вейник кавазский
65321 Вейник щучковидный
65324 Вейник высокий
65325 Вейник гималайский
65330 Вейник наземный, Вейник обыкновенный
65337 Вейник обыкновенный мелкоцветковый
65347 Вейник извилистый
65350 Вейник темный
65357 Вейник сизый
65358 Вейник крупноцветковый
65364 Вейник гиссарский
65366 Вейник хольма
65370 Вейник грузинский
65387 Вейник колымский
65388 Вейник короткого
65390 Вейник коржинского
65393 Вейник зайцехвостовидный
65397 Вейник ланцетный
65419 Вейник гигантский
65431 Вейник горный
65447 Вейник нинсяский
65448 Вейник тупоколосковый
65449 Вейник олимпийский
65452 Вейник своеобразный
65453 Вейник павлова
65454 Вейник персидский
65461 Вейник поплавской
65464 Вейник ложнотростниковый
65465 Вейник сомнительный
65470 Вейник краснеющий
65473 Вейник лангсдорфа
65476 Вейник ртпрехта
65477 Вейник сахалинский
65483 Вейник шугнанский
65488 Вейник сычуаньский
65502 Вейник тебердинский
65506 Вейник волосистый
65507 Вейник турчанинова
65509 Вейник лисохвостовидный, Вейник туркестанский
65511 Вейник уральский
65521 Вейник вильненский
65523 Вейник вилюйский
65528 Душевик, Клиноног, Пахучка
65531 Душёвка, Душёвка чапрецевидная, Пахучка острая, Пахучка остролистная, Пахучка полевая
65535 Душевик альпийский, Душёвка обыкновенная
65573 Душевик слабый
65581 Душевик круноцветковый
65603 Душевик аптечный, Душевик лекарственный, Пахучка аптечная
65637 Каламус
65674 Каламус шансиский
65687 Каламус бирманский
65688 Каламус мотоский
65697 Каламус тайваньский
65702 Каламус гуансиский
65703 Каламус хайнаньский

65713 Каламус яванский
65715 Каламус мэнлаский
65736 Каламус яошаньский
65747 Каламус инцзянский
65782 Пальма ползучая, Пальма ротанговая, Ротанг, Тростник индийский, Тростник испанский
65789 Каламус таиландский
65793 Каламус ланьюйский
65815 Каламус янчуньский
65816 Каламус юньнаньский
65818 Каламус пинбяньский
65827 Каландриния
65862 Каланта
65886 Каланта тайваньская
66101 Каланта чемерицелистная
66107 Каланта гуйцзоуская
66122 Каланта сычуаньская
66157 Калатея
66169 Калатея литце
66252 Кальцеолария, Кальцеолярия, Кошельки
66271 Кальцеолария гибридная, Кошельки гибридные
66273 Кошельки цельнолистные
66281 Кальцеолария мексиканская, Кошельки мексиканские
66337 Кальдезия
66343 Кальдезия березоролистная, Кальдезия почколистная
66380 Календула, Ноготки, Ногошок
66395 Календула полевая, Ноготки полевые
66422 Календула тонкая
66426 Календула каракалинская
66445 или ноготки, Календула лекарственная, Календула ноготки, Календула оффициналис, Ноготки аптечные, Ноготки лекарственные, Ноготки садовые
66453 Календула персидская
66504 Калепина
66506 Калепина неравномерная
66593 Белокрыльник, Калла
66600 Белокрыльник болотный, Калла болотная, Хлебница
66640 Миллетия эмэйская
66643 Миллетия сетчатая
66697 Каллиантемум, Красивоцвет
66698 Каллиантемум алатавский
66701 Каллиантемум узколистый, Красивоцвет узколистный
66710 Каллиантемум равноплодниковый
66717 Каллиантемум саянский
66727 Калликарпа
66733 Калликарпа амелликарпа
66743 Калликарпа боденье
66748 Калликарпа боденье наньчуаньская
66789 Калликарпа боденье джиральди
66790 Калликарпа цзиньюньская
66799 Каллика гуйцзоуская
66808 Калликарпа японская, Калликарпа японская цзилунская
66824 Калликарпа корейская
66826 Калликарпа мелкоплодная
66833 Калликарпа гуандунская
66835 Калликарпа индийская
66873 Калликарпа мягкая
66904 Калликарпа луаньдаская
66939 Калликарпа динхуский
66948 Калликарпа юньнаньская
66949 Каллицефалюс
66951 Каллицефалюс блестящий
66988 Джузгун, Жузгун, Кандум
66990 Жузгун колючекрылый
66991 Жузгун сродный
66994 Жузгун тонкокрыловидный
66995 Жузгун тонкокрылый
66996 Жузгун андросова
66998 Джузгун безлистый, Жузгун жесткокрылый, Кандум безлистный
67000 Жузгун аральский
67001 Жузгун древовидный
67002 Жузгун борщова
67003 Жузгун бубыра
67004 Жузгун известняковый
67008 Жузгун голова-медузы
67009 Жузгун хрящеватый
67010 Кандум китайский
67011 Жузгун змеиный
67013 Жузгун сердцевидный
67014 Жузгун кожистый
67017 Жузгун гребенчатый
67018 Жузгун густощетинковый
67021 Жузгун рассеченнокрылый
67023 Жузгун высокий
67024 Жузгун ежеплодный
67025 Жузгун евгения коровина
67026 Жузгун ферганский
67029 Жузгун стройный
67030 Жузгун сероватый
67032 Жузгун низкий
67033 Жузгун завернутый
67035 Жузгун ситниковый
67042 Жузгун блюдцевиднокрылый
67043 Жузгун белокорый
67044 Жузгун липского
67045 Жузгун литвинова
67046 Жузгун крупноплодный
67047 Жузгун перенончатый
67048 Жузгун мелкоплодный
67049 Жузгун мягкий
67050 Кандум
67052 Жузгун муравлянского
67053 Жузгун тупокрылый
67054 Жузгун прямощетинковый
67055 Жузгун палецкого
67057 Жузгун поникающий
67058 Жузгун прозрачноплодный
67059 Жузгун рерсидский
67060 Жузгун пеунникова
67061 Жузгун пузырчатокрылый
67062 Жузгун плоскощетинковый
67063 Жузгун складчатый
67064 Жузгун горецовидный
67069 Жузгун красивейший
67072 Жузгун квадратнокрылый
67075 Жузгун желтоплодный, Жузгун красноплодный, Жузгун курчавый
67076 Жузгун русанова
67077 Жузгун щетинистый
67079 Жузгун колючейший
67080 Жузгун растопыренный
67081 Жузгун тонкий
67082 Жузгун четырехкрылый
67084 Жузгун печальный
67085 Жузгун туркестанский
67248 Каллистемон
67291 Каллистемон жестколистный
67293 Каллистемон ивовый
67298 Каллистемон пышный
67312 Астра садовая, Каллистефус, Садовая астра
67314 Астра китайская, Астра садовая китайская, Каллистефус китайский, Садовая астра китайская
67334 Болотниковые, Красноволосковые
67336 Болотник, Водяная звездочка, Звёздочка водяная, Красноволоска, Красовласка
67345 Болотник ножкоплодный
67359 Красноволоска крючковатая
67361 Болотник осенний, Красноволоска осенная
67367 Красноволоска японская
67376 Болотник весений, Болотник весенний, Болотник изящный, Болотник сомнительный
67384 Красноволоска гуандунская
67389 Болотник изменчивый, Красноволоска изменчивая
67393 Болотник прудовой, Красавка прудовая, Красовласка прудовая, Шерстестебельник
67394 Болотник неясный
67401 Красноволоска весенная
67405 Каллитрис
67455 Вереск
67456 Вереск обыкновенный, Рыскун
67526 Калоцедрус
67530 Калоцедрус тайваньский
67554 Калохортус, Марипоза

67734 Луноцвет
67736 Луноцвет шиповатый
67771 Майкараган
67774 Майкараган китайский
67780 Майкараган крупноцветковый
67784 Майкараган шелковистыи
67786 Майкараган ховена
67787 Майкараган тяньшанский
67789 Майкараган волжский
67846 Дерево розовое, Калофиллум
67856 Калаба, Калофиллум обыкновенный
67860 Калофиллум круглоплодный
67882 Лимодорум
67894 Лимодорум крубневой
68086 Калотропис
68088 Калотропис гигантский
68157 Калужница, Кальта
68160 Калужница арктическая
68167 Калужница дернистая
68186 Калужница плавающая
68189 Калужница болотная, Кальта палюстрис
68196 Калужница дудчатая
68201 Калужница перенончатая
68213 Калужница многолепестная
68300 Каликантовые
68306 Каликант, Каликантус, Чашецветник
68308 Каликант китайский
68313 Каликант многоцветковый, Каликантус флоридский, Каликантус цветущий
68321 Каликантус плодовитый, Каликантус плодующий
68329 Каликант западный
68531 Калинсо луковичное, Калинсо луковичный, Калипсо
68538 Калинсо луковичная, Калинсо луковичное, Ясменник красильный
68542 Калипсо японская
68671 Вьюнок, Калистедия, Повой
68676 Калистедия даурская
68686 Вьюнок смолоносный, Вьюнок стрелолистный, Калистедия плющевидная, Повой плющевидный
68692 Калистедия японская, Повой японский
68701 Повой волосистый
68713 Вьюн белый, Калистедия заборная, Повой заборный
68724 Повой японский
68732 Повой лесной
68737 Повой круглолистный, Повой солданелевый
68746 Повой лесной
68789 Камассия, Квамассия
68794 Квамаш съедобный
68839 Рыжей, Рыжик
68840 Рыжик белоцветковый, Рыжик белоцветный
68844 Рыжик кавказский
68848 Рыжик рыхлый
68850 Рыжей мелкоплодный, Рыжик мелкоплодный
68857 Рыжик волосистый
68860 Мухоловка, Мухоловка рыжий, Рожь индийская, Рыжей яровой, Рыжик гладкий, Рыжик посевной, Рыжик яровой
68871 Рыжик дикий, Рыжик лесной
68873 Рыжик юньнаньский
68877 Зантедешия, Камелия, Рисовидка, Ричардия, Чай, Чайный куст
68906 Камелия аньлунская
68980 Камелия чжэцзянская
69049 Камелия маслоносная
69075 Камелия пинбяньская
69079 Камелия мэнцзыская
69137 Камелия ху
69142 Камелия хубэйская
69156 Камелия, Камелия обыкновенная, Камелия японская
69201 Камелия японская мелкоцветковая
69224 Камелия цзиньюньская
69244 Камелия гуансиская
69246 Камелия гуандунская
69248 Камелия гуйцзоуская
69286 Камелия либоская
69325 Камелия лунчжоуская
69355 Камелия мелкоцветковая
69389 Камелия наньчуаньская
69411 Камелия масличная, Камелия маслоносная
69423 Камелия эмэйская
69552 Камелия сетчатая
69594 Камелия гвоздичная, Камелия горная, Камелия евгенольная, Камелия масличная, Камелия сасанква, Камелия эвегенольная
69634 Дерево чайное, Куст чайный, Куст чайный китайский, Чай, Чай китайский, Чайный куст, Чайный куст китайский
69685 Камелия симаоская
69691 Камелия далиская
69706 Камелия наньтоуская
69760 Камелия вэньшаньская
69782 Камелия юньнаньская
69870 Кампанула, Колокольчик
69871 Колокольчик пихтовый
69873 Колокольчик
69881 Колокольчик альберта
69882 Колокольчик альбова
69883 Колокольчик
69885 Колокольчик чесночницелистный
69888 Колокольчик высокогорный
69889 Колокольчик альпийский
69891 Колокольчик алтайский
69896 Колокольчик анны
69897 Колокольчик необычный ардонский
69911 Колокольчик оше
69916 Колокольчик болодатый
69917 Колокольчик мергаритколистный
69919 Колокольчик березолистный
69921 Колокольчик болонский
69925 Колокольчик бротеруса
69928 Колокольчик шпорцеватый
69929 Колокольчик известняковый
69932 Колокольчик
69933 Колокольчик
69940 Колокольчик капю
69942 Кария северокаролинская, Колокольчик карпатский
69949 Колокольчиккавказский
69954 Колокольчик жестковолосый, Колокольчик олений
69956 Колокольчик шамиссо
69958 Колокольчик харкевича
69959 Колокольчик китайский
69961 Колокольчик хоцятовского
69963 Колокольчик ресничатый
69969 Колокольчик черкесский
69971 Колокольчик холмовой
69972 Колокольчик мелкоплодный
69977 Колокольчик сердцелистный
69980 Колокольчик курчавый
69982 Колокольчик дагестанский
69987 Колокольчик лицзянский
69991 Колокольчик
70001 Колокольчик
70003 Колокольчик
70004 Колокольчик доломитовый
70009 Колокольчик высокий
70010 Колокольчик
70011 Колокольчик изящный
70014 Колокольчик одонолетний
70019 Колокольчик евгеник
70021 Колокольчик мучнистый
70045 Колокольчик фомина
70046 Колокольчик фондервиза
70048 Колокольчик
70054 Колокольчик букетный, Колокольчик сборный, Колокольчик сжатый, Колокольчик скученный, Трава приточная
70074 Колокольчик гроссгейма
70075 Колокольчик хемшинский
70086 Колокольчик снизы-седой
70087 Колокольчик имеретинский
70088 Колокольчик сероватый
70089 Колокольчик
70092 Колокольчик ирины
70099 Колокольчик каракушский

70100 Колокольчик кемлярии
70103 Колокольчик колаковского
70104 Колокольчик комарова
70107 Колокольчик млечноцветный, Колокольчик молочноцветный
70116 Колокольчик лангсдорфа
70117 Колокольчик шершаваплодный
70119 Колокольчик широколистный
70135 Колокольчик
70143 Колокольчик лировидный
70155 Колокольчик макашвили
70161 Колокольчик масальского
70164 Кампанула медиум, Колокольчик корончатый, Колокольчик крупноцветный, Колокольчик средний
70168 Колокольчик мейера
70172 Колокольчик удивительный
70189 Колокольчик крупноколосый, Колокольчик многоцветковый
70198 Колокольчик продолговатолистный
70200 Колокольчик бледноохряный
70201 Колокольчик зубчаточашелистиковый
70202 Бубенчик эмэйский
70203 Колокольчик осетинский
70210 Колокольчик раскидистый
70214 Колокольчик персиколистный
70220 Колокольчик скальный
70226 Колокольчик полиморфный
70227 Колокольчик понтийский
70232 Колокольчик
70234 Колокольчик, Колокольчик точечный
70247 Колокольчик пирамидальный
70250 Колокольчик радде
70251 Колокольчик скребница
70254 Колокольчик рапунцель, Колокольчик репучатый, Колокольчик репчатовидный
70255 Колокольчик обыкновенный, Колокольчик рапунцелевидный, Колокольчик репчатовидный
70268 Колокольчик
70271 Колокольчик гизеке, Колокольчик кругролистный
70283 Колокольчик рупрехта
70288 Колокольчик камнеломка
70293 Колокольчик
70294 Колокольчик одноцветковый
70296 Колокольчик шишкина
70297 Колокольчик жестковолосый
70298 Колокольчик
70305 Колокольчик сибирский
70306 Колокольчик
70307 Колокольчик сомие
70308 Колокольчик сосновского
70314 Колокольчик стевена
70315 Колокольчик торчащий
70317 Колокольчик почти-головчатый
70320 Колокольчик тахтаджяна
70321 Колокольчик талиева
70322 Колокольчик крымский
70331 Колокольчик крапиволистный, Трава горляная
70338 Колокольчик
70341 Колокольчик трехзубый
70346 Колокольчик турчанинова
70351 Колокольчик одноцветковый
70353 Колокольчик вайды
70364 Колокольчик
70366 Колокольчик уральский
70367 Колокольчик воронова
70368 Кампанула юньнаньская
70372 Колокольчиковые
70460 Камфолосма, Трава камфарная
70461 Кампанула однолетная, Камфолосма однолетная
70463 Камфолосма марсельская, Караматау, Караматау монпелийская
70464 Кампанула лессинга, Камфолосма лессинга
70466 Кампанула джунгарская, Камфолосма джунгарская
70502 Кампсис
70507 Кампсис крупноцветковый
70512 Кампсис укореняющаяся, Кампсис укореняющийся, Текома укореняющаяся, Трубоцвет
70574 Камптотека
70575 Камптотека остроконечная
70704 Кривокрыльник мякокрасный
70712 Гомфия
70833 Леспедеза крупноплодная
70954 Кананга
70960 Иланг-иланг, Кананга, Кананга душистая
70988 Канариум
70996 Канариум обыкновенная
71015 Канариум черный
71036 Канавалия
71042 Ксерантемум однолетний, Сухоцвет однолетний
71050 Канавалия мечевидная, Канавалия мечелистная
71085 Канкриния
71086 Канкриния бочанцева
71088 Канкриния золотоглавая
71090 Канкриния безъязычковая
71091 Канкриния фергаская
71092 Канкриния голоскокова
71093 Канкриния каратавская
71096 Канкриния максимовича
71097 Канкриния остроконечная
71098 Канкриния невского
71099 Канкриния памироалайская
71101 Канкриния щетинконосная
71102 Канкриния неокаймленная
71103 Канкриния подобная
71104 Канкриния таджикская
71105 Канкриния тяньшанскя
71129 Канелла
71156 Канна
71166 Канна красная
71169 Аррорут квинслендский, Канна съедобная
71181 Канна индийская
71189 Канна ирикоцветковая
71202 Коноплевые
71209 Конопля
71218 Дербянка колосистая, Замашка, Конопля индийская, Конопля культурная, Конопля посевная, Посконь
71220 Конопля индийская
71228 Конопля сорная
71235 Канновые
71671 Каперсовые, Каперцовые
71676 Каперсник, Каперсы, Каперцы
71751 Каперцы тайваньский
71757 Каперцы хайнаньский
71762 Каперсы колючие, Каперцы колючие
71838 Каперсы розанова
71871 Каперсник колючий, Каперцы колючие
71933 Каперцы юньнаньский
71955 Жимолостные
72037 Пастушья сумка, Сумочник
72038 Пастушья сумка обыкновенная, Сумочник, Сумочник пастуший
72053 Пастушья сумка гирканская
72054 Пастушья сумка восточная
72068 Капсикум, Перец
72070 Перец ветвистый, Перец испанский, Перец красный, Перец кустарниковый, Перец многолетний, Перец овощной, Перец однолетний, Перец стручковый, Перец стручковыйоднолетний, Перец ягодный
72100 Перец ветвистый, Перец испанский, Перец красный, Перец кустарниковый, Перец многолетний, Перец овощной, Перец однолетний, Перец стручковый, Перец стручковыйоднолетний, Перец ягодный
72168 Карагана, Караганник, Чапышник, Чилига, Чилижник
72170 Карагана колючелистная
72175 Карагана алайская
72180 Акация желтая, Акация жёлтая, Желтая акация, Карагана высокая, Карагана древовидная, Карагана обыкновенная, Караганник древовидный, Челыжник, Черная

карагана, Чилига, Чилига древовидная, Чилига карагана
72187 Карагана оранжевая
72193 Карагана бонгарда
72199 Карагана бунге
72200 Карагана камилла шнейдера
72206 Карагана цинхайская
72215 Карагана пушистолистная
72227 Карагана франше
72228 Карагана цзилунская
72233 Дереза, Карагана кустарник, Карагана кустарниковая, Карагана степная, Чилига кустариниковая, Чилига степная, Чилижник кустарниковый
72237 Карагана кустарниковая
72241 Карагана крупноцветковая, Челига крупноцветковая
72244 Карагана всябелая
72249 Верблюжий хвост, Карагана гривастая
72251 Карагана гривастая сычуаньская
72257 Карагана киргизов
72267 Карагана красивая
72270 Карагана белокорая
72283 Карагана максимовича
72285 Карагана алтайская, Карагана мелколистная
72300 Карагана мягкая, Чилига пушистая
72311 Карагана многолистная
72316 Карагана прейна
72317 Карагана инееватая
72320 Карагана низкорослая
72322 Карагана карликовая, Чилига карликовая
72330 Карагана карликовая мелкоцветковая
72346 Карагана тибетская
72348 Карагана колючая
72351 Карагана узколистная
72364 Карагана трагакантовая
72370 Карагана турфанская
72371 Карагана туркестанская
72373 Карагана уссурийская
72634 Карапа
72640 Дерево крабовое, Карапа гвианская
72674 Сердечник
72679 Сердечник горький
72684 Сердечник аньхуйский
72698 Сердечник маргаритковый
72733 Сердечник густоцветковый
72734 Сердечник зубчатый
72749 Сердечник извилистый
72785 Сердечник греческий
72802 Сердечник шершавый
72809 Сердечник шершавый тайваньский
72812 Сердечник шершавый эмэйский
72829 Сердечник недотрога, Сердечник сычуаньский
72857 Сердечник лазистанский
72858 Сердечник белый
72873 Сердечник лировидный
72876 Сердечник крупнолистный
72897 Сердечник мелколистный
72900 Сердечник маленький
72901 Сердечник мулиский
72910 Сердечник японский
72920 Сердечник мелкоцветковый, Сердечник мелкоцветный
72925 Сердечник гребенчатый
72926 Сердечник стоповидный
72930 Сердечник каменный
72934 Сердечник луговой
72942 Сердечник ползучий
72946 Сердечник пурпурный
72963 Сердечник сахокиа
72971 Сердечник регеля
72985 Сердечник зейдлица
73003 Сердечник нежный
73018 Зубянка тонколистный, Сердечник тонколистный
73026 Сердечник болотнй
73027 Сердечник зонтичный
73033 Сердечник виктора
73044 Сердечник максимовича
73048 Сердечник юньнаньский
73052 Сердечник чжэцзянский
73085 Двугнёздка, Кардария, Сердечница
73090 Клоповник крупновидный, Кресс крупка, Сердечница крупновая
73092 Клоповник близкий, Сердечница ползучая
73098 Сердечница пушистая
73159 Лилия гигантская, Лилия исполинская
73194 Кардиосперм, Кардиоспермум
73278 Репейник, Чертогон, Чертополох
73280 Чертополох колючеголовый
73281 Чертополох акантовидный, Чертополох акантолистный, Чертополох колючий
73284 Чертополох прижатый
73288 Чертополох беловатый
73293 Чертополох аравийский
73310 Чертополох азербайджанский
73317 Чертополох беккеров
73329 Чертополох сереватый, Чертополох серый
73333 Чертополох холмовой
73337 Чертополох курчавый
73354 Чертополох сизый
73356 Чертополох хаястанский
73357 Чертополох крючечковый, Чертополох крючечный
73363 Чертополох поздный
73377 Чертополох кернера
73418 Чертополох многопарный
73425 Чертополох навашина
73426 Чертополох жилковатый
73427 Чертополох никитина
73430 Чертополох окрашенный, Чертополох поникающий, Чертополох поникший, Чертополох шишкина
73449 Чертополох татарниковый
73454 Чертополох замаскированный, Чертополох маскированный
73458 Чертополох серо-желтый
73459 Чертополох ложно-холмовый
73467 Чертополох мелкоголовчатый, Чертополох мелкоголовый
73486 Чертополох полуголый
73492 Чертополох джунгарский
73498 Чертополох узкоголовый
73508 Чертополох тёрмера
73514 Чертополох закаспийский
73516 Чертополох крючковатый
73543 Осока
73577 Осока аа
73587 Осока остролистная
73589 Осока острая, Осока стройная
73597 Осока ложно-острая, Осока островатая
73602 Осока неясноустая
73603 Осока гуандунская
73608 Осока двсмысленная
73620 Осока алайская
73622 Осока белая
73634 Осока алексеенковская
73652 Осока альпийская
73662 Осока алтайская
73667 Осока высокогорная
73668 Осока тупоносая
73673 Осока амгунская
73688 Осока нежилкоплодная
73690 Осока ангарская
73696 Осока суженная
73702 Осока неравножилковатая
73714 Осока открытая
73729 Осока сближенная, Осока своеобразная
73730 Осока сдвинутая
73732 Осока водяная
73740 Осока аркатская
73747 Осока песчаная
73748 Осока песколюбивая
73750 Осока аргунская
73753 Осока серебристочешуйная
73767 Осока алнелли, Осока алнелля, Осока арнелля
73770 Осока безносиковая
73774 Осока ичанская
73775 Осока шероховатая
73782 Осока прямоколосая

73788 Осока темная, Осока тёмная, Осока темноодетая
73792 Осока темнобурая
73814 Осока белозаостренная, Осока темнобурая, Осока чернобурая
73815 Осока белозаостренная
73826 Осока августиновича, Осока сойская
73873 Осока двуцветная
73875 Осока гиперборейская, Осока мечелистная
73877 Осока тайваньская
73888 Осока ресничатоплодная
73894 Осока богемская, Осока сытевидная
73898 Осока бонанцинская
73903 Осока бордзиловского
73906 Осока извилисторыльцевая, Осока курчаворыльцевая
73912 Осока парвуская
73940 Осока трясунковидная
73942 Осока бротертсов
73959 Осока буроватая
73969 Осока бухбрская
73971 Осока буека
73975 Осока бурятская
73987 Осока дернистая, Осока красная
73992 Осока красовлас
74004 Осока беловатая, Осока сероватая, Осока серозеленая
74023 Осока волосная
74029 Осока аоморийская
74032 Осока волосовидная, Осока зеленоколосая, Осока зеленоколосковая
74046 Осока головчатая
74049 Осока мелкоголовчатая
74051 Осока козерогая
74059 Осока гвоздичная
74068 Осока кавказская
74089 Осока шамисо
74098 Осока китайская
74101 Осока хинганская
74111 Осока зеленохолосая
74118 Осока длиннокорневищная, Осока струнокоренная
74121 Осока чозенская
74135 Осока сероватая
74138 Осока завитая
74148 Осока колхидская
74158 Осока сжатая
74182 Осока уплодненная
74187 Осока клопоносная
74221 Осока пильчатонлодная, Осока скрытоплодная
74230 Осока курайская
74244 Осока заостренная
74250 Осока чарвакская
74253 Осока даурская
74258 Осока дэвелла
74291 Осока обедненная
74295 Осока дэциньская
74296 Осока нисходящая
74304 Осока двутычинковая
74307 Осока двуцветная
74308 Осока двуцветноколосковая
74309 Осока корейская
74310 Осока лицзянская
74323 Осока калелина, Осока светлая, Осока солонцовая
74330 Осока двудомная
74333 Осока большеплодная
74337 Осока двуцветная, Осока пестрая
74339 Осока расходящаяся
74342 Осока двусеменная
74348 Осока расставленная, Осока расставлённая
74354 Осока двурядная, Осока средняя
74358 Осока разноцветная
74360 Осока раздельная, Осока скученная
74366 Осока прерванная
74372 Осока длинноплодная
74374 Осока длинноплодная
74392 Осока есолюбивая
74394 Осока амурская
74406 Осока твердоватая
74407 Осока жестковадая
74408 Осока ложнотвердоватая, Осока уральская
74423 Осока суженная
74435 Осока гуднозова, Осока омская, Репа, Репа культурная, Репа оголодная, Репа японская, Сурепица яровая, Турнепс
74438 Осока элевезиновидная
74442 Осока длинная, Осока удлиненая, Осока удлинённая
74444 Осока высокая
74453 Осока безжилковая, Осока тоненкая
74460 Осока мортуковидная
74461 Осока верещатниковая
74463 Осока радде
74476 Осока эвксинская
74481 Осока растянутая
74483 Осока серповидная
74491 Осока федийская
74495 Осока федчанко
74557 Осока желтая, Осока жёлтая
74568 Осока желтенькая
74584 Осока обильнолистная
74592 Осока уховертка
74606 Осока фуджиты
74614 Осока фунинская
74621 Осока меднобурая
74622 Осока буровлагалищная
74652 Осока ледниковая
74654 Осока галечная
74656 Осока сизая
74664 Осока ложно-сизая
74668 Осока глена
74672 Осока круграя
74676 Осока гмелина
74680 Осока гуншаньская
74681 Осока обыкновенная
74684 Осока городкова
74685 Осока сукачева
74696 Осока грэхэма, Осока узкочешуйная
74713 Осока гриффитса
74714 Осока гриолети
74718 Осока гросгейма
74722 Осока женосильная
74726 Осока кровяноколосая
74731 Осока хайнаньская
74733 Осока хаккоеская
74745 Осока галлеровская
74746 Осока галлера
74752 Осока хэнкока
74763 Осока выхолощеная, Осока выхолощенная, Осока гартмана, Осока цилиндрическая
74781 Осока болотолюбивая
74791 Осока хэпберна
74797 Осока разночешуйчатая
74806 Осока волосистая, Осока ворсистая, Осока коротковолосистая, Осока мохнатая
74818 Осока цельноротая, Осока цельноустая
74826 Осока ячменерядная
74830 Осока госта
74838 Осока юэта
74842 Осока низкая, Осока приземистая, Осока степная
74849 Осока низенькая
74856 Осока чащевная
74867 Осока ильина
74875 Осока индийская
74878 Осока вздутая
74884 Осока буреющая
74907 Осока затененная
74913 Осока заливная
74932 Осока якутская
74934 Осока ялусская
74937 Осока янковского
74941 Осока японсая
74967 Осока ситничек
74969 Осока кабанова
74972 Осока камчатская
74973 Осока ганьсуская
74974 Осока каро
74977 Осока карелина
74982 Осока каро
74986 Осока каттегатская
74992 Осока кенкольская

74998 Осока кирганикская
75000 Осока кирилова
75002 Осока кнорррнг
75003 Осока кобомуги
75010 Осока корагинская
75012 Осока коржинского
75013 Осока кожевникова
75015 Осока кочи
75026 Осока гуансиская
75038 Осока гладностбельная
75042 Осока гладчайшая
75048 Осока ланцетная
75062 Осока лапландская
75065 Осока волосистоплодная, Осока шерстистая
75078 Осока широколистная
75079 Осока широкочешуйная
75080 Осока рыхлая
75096 Осока ледебуровская
75101 Осока гладконосая
75115 Осока чешуеплодная
75116 Лапка заячья, Осока заячья
75131 Осока бледнозеленая
75165 Осока топяная
75176 Осока блестящая, Осока лоснящаяся
75178 Осока камнелюбивая
75180 Осока литвинова
75184 Осока свинцово-зеленая
75185 Осока плевельная
75189 Осока длипнноклювая
75227 Осока чэнкоуская
75230 Осока люмницера
75252 Осока маака
75253 Осока макензи
75259 Осока большеголовая
75265 Осока крупноостная
75268 Осока крупноколосковая
75275 Осока крупнорыльцевая
75276 Осока болошехвостая
75298 Осока маньчжурская
75303 Осока морская, Осока приморская
75304 Осока морская, Осока поморская
75314 Осока максимовича
75317 Осока суйфунская
75323 Осока медведева
75325 Осока мейнсгаузена
75326 Осока семноцветная, Осока черноцветковая
75328 Осока ложно-черноцветковая
75330 Осока черноплодная
75332 Осока черноголовая
75333 Осока черноколосковая
75334 Осока черноустая
75338 Осока перенончатая
75348 Осока мейера
75355 Осока микели
75357 Осока сероватая
75361 Осока мелкоостенник
75362 Осока коротконожчатая
75363 Осока ложнокоротконожчатая
75369 Осока мелковолосистая
75370 Осока миддендорфа
75378 Осока мелкая
75382 Осока слабошероховатая
75387 Осока мужененавистническая, Осока холодная
75396 Осока мийябе
75400 Осока мягковатая
75403 Осока мягчайшая
75409 Осока горная
75429 Осока мотоская
75431 Осока остриеносная
75438 Осока мулиская
75451 Осока остроконечная
75474 Осока наньчуаньская
75490 Осока невровская
75496 Осока жилковатая
75500 Осока жилкоплодная
75511 Осока чернеющая
75541 Осока новограбленова
75570 Осока притупленная
75572 Осока скрывающаяся
75575 Осока эдера
75588 Осока немногоцветковая
75595 Осока немногоколосковая
75601 Осока оливера
75602 Осока эмэйская
75604 Осока омианская
75608 Осока омская
75610 Осока головковидная, Осока оне
75618 Осока аркатская, Осока округлая
75623 Осока гололюбивая
75641 Осока медная
75649 Осока островерхая
75662 Осока толстоколосая
75664 Осока пустынная, Осока толстолбиковая
75672 Осока бледноватая
75675 Осока бледная, Осока возрастающая
75678 Осока памирская
75684 Осока просовидная, Осока просяная
75686 Осока метельчатая
75696 Осока поморская, Осока приморская
75697 Осока параллельная
75709 Осока малая
75711 Осока малоцветковая, Осока мелкоцветковая
75719 Осока необильная
75722 Осока лапчатая
75724 Осока большехвостая, Осока корневищная, Осока стоповидная, Осока сучанская
75734 Осока ножконосная
75737 Осока пейктусанская
75742 Осока висячая
75779 Осока листоколосая
75782 Илах, Осока вздутая, Осока песчаная, Осока ресчаная, Осока рянг, Рянг
75785 Осока волосистая, Осока ворсистая, Осока коротковолосистая, Осока кривоносая, Осока мохнатая
75788 Осока шариконосная
75827 Осока плоскостебельная
75834 Осока широконосая
75839 Осока свинцовая
75849 Осока многолистная
75855 Осока понтийская
75859 Клён завитой, Осока ранняя, Осока шребера
75878 Осока вытянутая
75880 Осока повислая
75894 Осока ложно-курайская
75895 Осока ложносыть
75899 Осока ложно-вонючая
75906 Осока ложнолансетная
75910 Осока ложноплевельная
75928 Осока красивенькая
75929 Осока блошиная
75930 Осока малорослая
75941 Осока густоколосая
75948 Осока цинхайская
75954 Осока пальчатая, Осока четырехцветковая
75957 Осока радде
75970 Осока раменского
75980 Осока редкоцветковая
75989 Осока редовского
75991 Осока регеля
75993 Осока раздвинутая
76005 Осока немногораздвинутая
76008 Осока ползучая
76010 Осока загнутая
76015 Осока возвратившаяся
76029 Осока вздутоносая
76031 Осока рабушинского
76042 Осока ложножесткая
76044 Осока береговая
76047 Осока ложнобереговая
76068 Осока носадая
76073 Осока кругловатая
76083 Осока далиская
76090 Осока морщинистая, Осока смирнова
76094 Осока скальная
76096 Осока русская
76098 Осока несчаная
76100 Осока зеленоватая, Осока шабинская
76126 Осока сикокийская
76132 Осока садоанская
76134 Осока саянская

76145 Осока сарептская
76153 Осока каменистая, Осока каменная
76162 Осока шершавошиноватая
76163 Осока шершаволистная
76165 Осока шероховатожилковатая
76178 Осока шмидта
76185 Осока камышевидная
76215 Осока ржаная
76219 Осока седакова
76222 Осока неполная
76230 Осока ключелюбивая, Осока эдера
76239 Осока щетинистая
76262 Осока сичоуская
76264 Осока ржавопятнистая
76273 Осока зигерта
76276 Осока лесная
76292 Осока слободова
76298 Осока сочавы
76299 Осока джунгарская
76301 Осока грязная
76306 Осока немногоплоданая
76328 Осока колосистая
76338 Осока прямая, Осока прямостоящая
76342 Осока звездчатая
76350 Осока узкоплодная
76354 Карабат, Осока узколистная, Осока уральская
76363 Осока ложноузколистная
76372 Осока глянцевитобурая
76375 Осока стесненная
76410 Осока сухощавая
76412 Осока стигийская
76413 Осока столбиконосная
76418 Осока малоприцветниковая
76435 Осока полувздттая
76437 Осока обертковидная
76446 Осока зонтиковидная
76456 Осока приземистая
76461 Осока сычуаньская
76465 Осока лесная
76469 Осока шовица
76482 Осока талдынская
76485 Осока хэбэйская
76490 Осока буксбаума
76513 Осока тонкоцветковая, Осока тонкоцветная
76515 Осока корейская
76545 Осока тунберга
76547 Осока придатконосная
76550 Осока тяньшанская
76552 Осока титова
76570 Осока траизискинская
76574 Осока траутфеттера
76585 Осока трехраздельная
76595 Осока печальная
76618 Осока туминганская
76622 Осока туркестанская
76626 Осока мочежинная
76628 Осока курчавая
76635 Осока теневая
76641 Осока унгурская
76647 Осока уральская
76650 Осока медвежья
76651 Осока уссурийская
76653 Осока мешечатая, Осока мешочковая
76654 Осока узонская
76656 Осока влагалищная
76665 Осока ван-хьюрка
76673 Осока весенняя
76677 Осока пузырчатая
76687 Осока пузыревая
76701 Осока зелененькая
76710 Осока спутанная
76716 Осока лисья
76751 Осока вэньшаньская
76760 Осока вилюйская
76773 Осока мечевидная
76789 Осока юньнаньская
76793 Осока чжэньканская
76808 Дерево дынное, Карика
76813 Дерево дынное, Дынное дерево, Карика, Карика папайа, Папайа, Папайя
76820 Дыниковые, Кариковые, Папаевые
76854 Карисса
76873 Карисса
76988 Колючелистник, Колючник
76990 Колючник бесстебельный
76994 Колючник биберштейна, Колючник узколистный
76998 Колючник стебельный
77000 Колючник артишоковидный
77008 Колючник шерсстистый
77011 Колючник татарниколистный
77022 Колючник обыкновенный
77038 Карллюдовика, Карлудовика
77067 Эреция мелколистный
77077 Цереус гигантский
77096 Тминоножка
77099 Тминоножка армянская
77100 Тминоножка уплощенноплодная
77145 Карпезиум
77146 Карпезиум польнный
77149 Карпезиум поникающий, Карпезиум поникший
77178 Карпезиум крупноголовый, Карпезиум особенный
77181 Карпезиум мелкоцветковый
77182 Карпезиум непалский
77187 Карпезиум сычуаньскй
77190 Карпезиум печальный
77252 Граб
77256 Бук белый, Граб европейский, Граб обыкновенный
77263 Граб американский, Граб каролинский
77271 Граб кавказский
77275 Граб сичоуский
77276 Граб приморский, Граб сердцевидный, Грибы сморчковые, Сморчковые
77283 Граб корейский
77306 Граб хубэйский
77312 Граб японский
77315 Граб гуйцзоуский
77323 Граб разреженноцветущий
77345 Граб мелколистный
77358 Граб эмэйский
77359 Граб восточный, Граб черный, Грабинник, Гравник
77397 Граб дэциньский
77401 Граб чанхуаский
77421 Граб гуншаньский
77468 Рогоплодник короткостолбчатый
77469 Рогоплодник длиннорогй
77472 Рогоплодник копьевидный
77474 Рогоплодник узкоплодный
77684 Сафлор
77716 Сафлор шерстистый
77734 Сафлор острошипый
77748 Сафлор красиьный, Сафлор культурный, Трава жетяница, Трава картамова
77766 Тмин
77768 Тмин альпийский
77774 Тмин темнокрасный
77775 Тмин хэбэйский
77777 Тмин бурятский
77784 Анис полевой, Тмин луговой, Тмин обыкновенный
77788 Тмин кавказский
77812 Тмин гроссгейма
77818 Тмин комарова
77836 Тмин скальный
77874 Гиккори, Кария
77881 Гиккори горикий, Кария водяная, Пекан горикий
77885 Кария северокаролинская
77889 Гиккори горький, Кария сердцевидная, Орех горький
77894 Гиккори гладкий, Гиккори свиной, Кария голая
77900 Кария хунаньская
77902 Кария пекан, Орех пекан, Пекан
77904 Кария гуйцзоуская
77905 Кария бахромчатая
77910 Кария мускатная
77912 Кария овальная
77920 Гиккори крупнопочечный, Гиккори яйцевидный, Кария косматая, Кария овальная, Кария яйцвевидная, Пекан яйцевидный твёрдокорый

77927 Кария мексиканская
77934 Гиккори белая, Гиккори белый, Гиккори косматый, Кария белая, Кария войлочная, Кария опушённая, Орех насмешливый
77946 Кариокар орехоносный, Орех савара
77980 Гвоздичные
77992 Кариоптерис
78012 Кариоптерис, Кариоптерис седой
78045 Кариота, Пальма, Пальма-Кариота
78047 Кариота мягкая
78056 Кариота жгучая, Пальма "рыбий хвост", Пальма винная, Пальма жгучая, Пальма-тодди, Тоддин-пальма
78172 Казимироа
78175 Казимироа съедобная, Казимироя
78204 Кассия, Сенна
78227 Кассия остролистная, Кассия узколистная, Лист александрийский, Сенна индийская, Сенна остролистная
78300 Кассия дудка, Кассия стручковидная, Кассия трубковидная, Кассия южноазиантская, Мёд дивий
78336 Кассия яванская
78436 Кассия овальнолистная, Кассия туполистная, Сенна итальянская
78540 Кассина
78622 Кассиопа, Кассиопея
78627 Кассиопея вересковидная
78629 Кассиопа фуцзяньская
78630 Кассиопа гипновидная
78631 Кассиопа плауновидная, Кассиопея плауновидная
78644 Кассиопея ледовского
78646 Галения стеллера, Кассиопея стеллера
78648 Кассиопа четырехрядная, Кассиопа четырёхрядная, Кассиопея четырехгранная, Кассиопея четырехгрянная
78766 Каштан
78777 Каштан городчатый, Каштан японский
78790 Каштан американский, Каштан высокий, Каштан зубчатый
78796 Каштан эмэйский
78802 Каштан мягчайший
78807 Каштан карликовый, Каштан низкорослый
78811 Дерево каштановое, Каштан благородный, Каштан европейский, Каштан настоящий, Каштан обыкновенная, Каштан посевной, Каштан съедобный
78848 Кастанопсис
78852 Кастанопсис наньнинский
78889 Кастанопсис китайский
78903 Кастанопсис длинно-остроконечный
78913 Кастанопсис люцюский
78934 Кастанопсис симаоский
78942 Кастанопсис тайваньский
78947 Кастанопсис хайнаньский
78952 Кастанопсис хубэйский
78959 Кастанопсис индийский
78969 Кастанопсис гуйцзоуский
78974 Кастанопсис лэдунский
78982 Кастанопсис лунчжоуский
79000 Кастанопсис пинбяньский
79004 Кастанопсис эмэйский
79067 Кастанопсис тэнчунский
79069 Кастанопсис сичоуский
79075 Ежевика лежачая, Ежевика плетевидная
79108 Кастилла, Кастиллоа
79110 Кастилла каучуконосная
79113 Касстиллея, Кастилейя, Кастиллейя
79117 Касстиллея арктическая
79125 Алоэ дикое, Алоэ колючее, Алоэ ферокс
79130 Касстиллея бледная
79155 Казуарина
79157 Казуарина каннингема
79161 Железное дерево, Казуарина, Казуарина хвощевидная, Казуарина хвощелистная
79185 Казуарина пробковая
79193 Казуариновые
79194 Поручейница
79198 Манник водный, Манник водяной, Поручейница водная, Поручейница водяная
79203 Поручейница капю
79214 колподим приземистый
79242 Катальпа
79243 Катальпа бигнониевая, Катальпа бигнониевидная, Катальпа сиренелистная
79257 Катальпа яйцевидная, Катальпа яйцевиднолистная, Катальпа японская
79260 Катальпа западная, Катальпа красивая, Катальпа прекрасная
79271 Катананхе
79387 Ката, Хат
79395 Кат, Ката, Ката съедобная, Катх съедобный
79418 Барвинок розовый
79526 Каттлея
79532 Каттлея цитрусовая
79595 Рандия
79614 Принцепник
79627 Принцепник бишофа
79664 Каукалис, Принцепник липучковый, Принцепник морковновидный, Принцепник морковный
79730 Каулофилум, Стеблелист
79732 Каулофилум, Каулофилум мощный, Стеблелист мощный
79902 Капуста цветная, Цеанотус
79905 Цеанотус американский
80003 Цекропия
80015 Цедрела
80025 Цедрела мексиканская
80030 Кедр остиндский, Цедрела душистая, Цедреладу шистая
80074 Дерево кедровое, Кедр, Орехи кедровые
80080 Кедр атласский
80083 Кедр атласский голубой
80085 Кедр короткохвойный
80087 Кедр гималайский
80100 Кедр ливанский
80114 Цейба
80120 Дерево капоковое, Дерево сырное, Дерево хлочатое, Дерево шелстяное, Капок, Сейба пятитычинковая, Сумаум, Цейба, Цейба пятитычинковая
80127 Бересклетовые, Краснопузырниковые
80129 Древогубец, Краснопузырник, Целяструс
80141 Краснопузырник угловатый
80189 Древогубец плетевидный, Краснопузырник плетеобразный
80191 Краснопузырник франше
80260 Древогубец круглолистный, Древогубец округлый, Краснопузырник круглолистный, Краснопузырник округлый
80285 Бересклет вьющийся, Краснопузырник древогубец
80295 Древогубец росторна
80296 Древогубец лёснера
80309 Древогубец лазящий
80327 Краснопузырник щетковидный
80378 Гребешок петуший, Петуший гребешки, Целозия
80381 Целозия серебристая
80395 Гребень петуший, Гребешок петуший, Петушьи гребешки, Целозия, Целозия гребенчатая
80521 Цельзия разнолистная
80540 Цельзия голостебьная
80542 Цельзия восточная
80555 Цельзия суволова
80569 Дерево камедное, Каркас
80575 Каркас шершавый
80577 Каркас турнефора, Каркас южный
80597 Каркас дэциньский
80600 Каркас иедзоский
80607 Каркас кавказский
80636 Каркас фэнцинский
80638 Каркас фунинский
80643 Каркас гладкий, Каркас голый

80645 Каркас гуншаньский
80663 Каркас люцюский
80664 Каркас корейский, Каркас южный
80668 Каркас куэньминский
80673 Каркас сглаженный
80698 Дерево железное, Каркас западный, Цельтис западный
80723 Каркас малорослый
80728 Каркас сетчатый
80737 Каркас шуннинский
80739 Каркас китайский
80764 Каркас тайваньский
80779 Каркас ван
80786 Каркас тибетский
80803 Ценхрус
80877 Пустореберник
80879 Пустореберник фишера
80892 Василек, Василёк, Центаурея
80894 Василек укороченный
80895 Василек абхазский
80897 Василек уклоняющийся
80902 Василек иглолистный
80904 Василек адамса
80907 Василек прижаточещунуй, Василек прижатый
80909 Василек подражающий
80912 Василек скученный
80913 Василек алайский
80915 Василек беловатый
80917 Василек альбова
80918 Василек александра
80922 Василек горький
80926 Василек тупочешуйчатый
80930 Василек андросова
80932 Василек шипиконосный
80933 Василек крупноридатковый
80935 Василек песчаный
80938 Василек армянский
80950 Василек темнокрасный
80957 Стизолофус бальзамический
80959 Василек беланже
80960 Василек красивый
80964 Василек барбеа
80965 Василек бессера
80978 Василек днепровский
80980 Василек короткоголовый
80981 Василёк голючеголовый, Василек колючеголовый
80983 Василек сборноголовый
80990 Василек козлиный
80991 Василек угольный
80993 Василек чертополоховидный
80995 Василек карпатский
80998 Василек каспийский
81002 Василек желтофиолевый
81005 Василек черкесский
81006 Василек предкавзский
81009 Василек колхидский
81010 Василек компера
81012 Василек рогатый, Живокость посевной
81014 Василек кожистый
81020 Василек, Василек голубой, Василек обыкновенный, Василек полевой, Василек посевной, Василек синий, Центаурея
81023 Василек дагестанский
81024 Василек подбеленный
81028 Василек незамеченный
81029 Василек наклоненный
81033 Василек низкий, Василек приплюснутый
81034 Василек раскидистый, Василек распростертый
81038 Василек дмитриевой
81052 Василек дубянского
81063 Василек ереванский
81069 Василек фишера
81076 Василек буроотороченный
81082 Василек грузинский
81083 Василек гербера
81086 Василек глена
81091 Василек гончарова
81098 Василек гроссгейма
81102 Василек гулисашвили
81104 Василек хаястанский
81107 Василек цельнолистный
81109 Василек пленчато-чешуйчатый
81110 Василек гирканский
81112 Василек грузинский, Василек иберийский
81113 Василек ильина
81145 Василек целнокрайнолистный
81150 Василёк горькуша, Василек луговой, Василёк луговой
81160 Василек карабахский
81161 Василек казахский
81162 Василек коктебельский
81163 Василек колковского
81164 Василек конки
81165 Василек копетдагский
81166 Василек кочи
81167 Василек кубанский
81168 Василек культиасова
81169 Василек шерстистоногий
81170 Василек широколопастный
81171 Василек лавренко
81172 Василек белопленчатый
81173 Василек белолистный
81174 Василек левзееподобный
81184 Василек крапчатый, Василек пятнистый, Василёк пятнистый
81188 Василек майорова
81193 Василек белая жемчужная
81194 Василек жемчужный
81196 Василек маршалла, Василек маршалля
81202 Василек месхетский
81204 Василек мейера
81209 Василек модеста
81210 Василек мягкий
81214 Василек горный, Василёк горный
81219 Белолист, Белолиственник, Василек дтшистый
81229 Василек натадзе
81237 Василек чёрный, Василёк чёрный
81244 Василек черноголовый
81245 Василек чернобахромчатый
81246 Василек новороссийский
81253 Василек ольтинский
81257 Василек восточный, Василёк восточный
81258 Василек осетинский
81259 Василек овечий
81260 Василек пачоского
81265 Василек паннонский
81269 Василек немного-лопастной
81270 Василек пехо
81272 Василек пелийский
81276 Василек феопаппусовидный
81277 Василек фригийский
81278 Василек черешочковый
81282 Василек боровой
81285 Василек многоножколистный
81290 Василек луговой
81292 Василек ложно-бледночешуйчатый
81293 Василек ложно-пятнисский
81294 Василек ложно-фригийский
81295 Василек ложно-скабиозовый
81323 Василек раздорского
81324 Василек отогнутый
81335 Василек корнецветковый
81336 Василек корнецветкововидный
81341 Василек красноватый
81342 Василек красноцветковый
81344 Василек рупрехта
81345 Василек русский, Василёк русский
81348 Василек иволистный
81352 Василек салоникский
81354 Василек шалфеелистный
81355 Василек сарандинаки
81356 Василек перистый, Василек скабиоза, Василёк скабиозный, Василек скабиозовидный, Василек скабиозовый, Василек шероховатый, Трава поясничная
81359 Василек шелковникова
81368 Василек сергея
81378 Василек сибирский
81380 Василек неразветвленный
81381 Василек синтеница

81382 Василек подколнечный, Василёк подсолнечный, Василек солнечный
81385 Василек софии
81386 Василек сосновского
81398 Василек растопыренный
81400 Василек узкопленчатый
81401 Василек жестколистный
81402 Василек бесплодный
81403 Василек стевеновский
81404 Василек стевена
81405 Василек мелкоцветковый
81407 Василек торчащий
81411 Василек сумской
81412 Василек совича
81415 Василек талиева
81416 Василек донской
81427 Василек тернопольский
81435 Василек закавказский
81436 Василек волосистоглавый, Василек волосистоголовый
81438 Василек трежжилковый
81439 Василек троицкого
81440 Василек тургайский
81441 Василек туркестанский
81445 Василек ванкова
81448 Василек горный
81450 Василек соседний
81453 Василек лозный
81460 Василек вильденова
81461 Василек воронова
81462 Василек желтоглавый
81463 Василек желтоглавовидный
81465 Василек зангезтрский
81466 Василек зувандский
81485 Золототысячник
81517 Золототысячник красивенькийй, Золототысячник красивый
81523 Золототысячник мейера
81536 Золототысячник колосистый, Золототысячник колосовидный
81539 Золототысячник тонкоцветковый
81544 Золототысячник зонтичный
81546 Золототысячник обыкновенный
81564 Центелла
81570 Центелла азиатская
81681 Стоножка, Центипеда
81687 Стоножка округлая, Центипеда малая
81743 Кентрантус, Центрантус
81751 Кентрантус валельяновидный
81758 Кентрантус длиноцветковый
81768 Кентрантус красный, Центрантус красный
81817 Центролобиум
81970 Ипекакуана, Корень рвотный, Свинушка бурая, Свинушка тонкая
82028 Пыльцеголовник, Цефаляптера
82036 Пыльцеголовник кавказский
82038 Пыльцеголовник крупноцветковый, Пыльцеголовник крупноцветный
82043 Пыльцеголовник дремликовидный
82058 Пыльцеголовник длиннорицветниковый
82060 Пыльцеголовник длиннолистный
82070 Пыльцеголовник красный
82075 Пыльцеголовник тайваньский
82091 Цефалантус, Цефаляптус
82098 Цефалант западный
82113 Головчатка, Пыльцеголовник, Цефалярия
82116 Головчатка албпийская
82117 Головчатка остистая
82118 Головчатка армянская
82126 Головчатка короткочешуйная
82129 Головчатка известняковая
82130 Головчатка кожистая
82131 Головчатка меловая
82133 Головчатка дмитрия
82139 Головчатка гигантская
82141 Головчатка гроссгейма
82145 Головчатка кочи
82151 Головчатка литвинова, Головчатка татарская
82160 Головчатка средная
82161 Головчатка мелкозубчатая
82165 Головчатка волосистая
82166 Головчатка высокая
82172 Головчатка шерстистая
82173 Головчатка сирийская, Махобели
82181 Головчатка чиначева
82182 Головчатка трансильванская
82183 Головчатка уральская
82186 Головчатка бархатистая
82204 Цефалоцереус
82407 Корнеглав
82408 Корнеглав туркменский
82409 Корнеглав яйценогий
82411 Цефалоринхус
82413 Цефалоринхус
82414 Цефалоринхус кирпичникова
82415 Цефалоринхус косинского
82420 Цефалоринхус многоветвистый
82422 Цефалоринхус джингарский
82423 Цефалоринхус почти-перестый
82424 Цефалоринхус тахтаджяна
82425 Цефалоринхус талышский
82426 Цефалоринхус клубневой
82493 Голпвчатотиссвые
82496 Головчатый тисс, Тис головчатый, Тисовник, Цефалотаксус
82499 Головчатый тисс костянковый, Тисовник головчато-костянковый, Тисовник костянковый, Цефалотаксус головчато-костянковый
82507 Головчатый тисс форчуна, Цефалотаксус фортуна
82521 Цефалотаксус хайнаньский
82523 Цефалотаксус японский
82539 Цефалотаксус гуншаньский
82550 Цефалотаксус тайваньский
82560 Камнеломка австоралийская
82629 Ясколка
82644 Ясколка алексеенко
82646 Ясколка альпийская
82664 Ясколка уклоняющаяся
82676 Ясколка серебристая
82679 Ясколка армянская
82680 Ясколка полевая
82732 Ясколка берингова, Ясколка беринговская
82738 Ясколка бялыницкого
82739 Ясколка биберштейна
82758 Ясколка дернистая
82772 Ясколка трехстолбиковая, Ясколка трёхстолбиковая, Ясколка ясколковидная
82776 Ясколка хлоролистная
82786 Ясколка даурская
82787 Ясколка зубчатая
82789 Ясколка вздутая, Ясколка вильчатая
82790 Ясколка вздутая
82808 Ясколка серповидная
82813 Ясколка фишера
82826 Ясколка костенецевидная
82834 Ясколка вильчатая
82845 Ясколка гладковатая
82849 Ясколка скученная, Ясколка скученноцветковая
82856 Ясколка липкая
82859 Ясколка городкова
82873 Ясколка хемшинская
82888 Ясколка кавказская
82902 Ясколка казбекская
82911 Ясколка воробейниколистная
82913 Ясколка длиннолистная
82918 Ясколка крупная
82921 Ясколка мейера
82922 Ясколка мелкосемянная
82924 Ясколка тайваньская
82926 Ясколка многоцветковая
82928 Ясколка лесная
82948 Ясколка гоная
82955 Ясколка малоцветковая
82968 Ясколка пронзеннолистная
82973 Ясколка полимофная
82974 Ясколка понтийская
82993 Ясколка пурпурная
82994 Ясколка мальнькая
82997 Ясколка регеля
83010 Ясколка сорная
83016 Ясколка шмальгаузена

83020 Ясколка пронзеннолистная, Ясколка пятитычинковая, Ясколка пятитычиночная
83030 Ясколка сосновского
83032 Ясколка стевена
83042 Ясколка сычуаньская
83043 Ясколка шовица
83049 Ясколка крымская
83059 Ясколка тяньшанская
83060 Ясколка шерстистая
83071 Ясколка волосистолистная
83110 Вишня
83143 Вишня алайская
83144 Вишня миндалецветная
83155 Вишня истанская, Вишня птичья, Вишня сладкая, Вишня черешня, Гини черешня, Черня, Черня птичья
83157 Вишня бессея, Вишня бесси
83174 Вишня холмовая
83177 Вишня конрадины
83201 Вишня краснoплодная
83202 Вишенник, Вишня дикая, Вишня кусгарная, Вишня кустарниковая, Вишня лесостепная, Вишня степная, Вишня шпанка кустарниковая, Слива куртарниковая
83207 Вишня железистолистная
83208 Вишня железистая, Прунус железистый
83215 Вишня хайнаньская
83216 Вишня мэнцзыская, Прунус мэнцзыский
83233 Вишня жакемона
83238 Вишня усурийская, Вишня японская
83253 Антипка, Вишня антипка, Вишня магалебская, Вишня чернильная, Кучина, Магалебка, Черемуха антипка, Черёмуха душистая, Черемуха магалепка, Черёмуха-антипка, Черёмуха-магалепка, Черня антипка
83255 Вишня максимовича, Прунус максимовича
83257 Вишня мелкоплодная
83271 Слива зибольда мелколистная
83274 Черемуха пенсильванская, Черёмуха пенсильванская
83285 Вишня ложнопростертая
83300 Вишня сахалинская
83301 Прунус сардента
83311 Вишня чёрная дикая, Черемуха оздняя, Черёмуха поздняя
83314 Вишня мелкопильцчатая, Вишня пильчатая, Вишня сакура, Сакура
83328 Вишня слабоопушенная восходящая
83340 Вишня сычуаньская, Прунус сычуаньский
83343 Вишня кандинская, Прунус кандинский
83345 Вишня тяньшанскя, Прунус тяньшанский
83347 Вишня войлочная, Войлочная вишня, Прунус войлочный
83353 Вишня туркменская
83360 Вишня бородавчатая
83361 Вишня обыкновенная, Вишня садовая, Вишня чёрная позднння
83365 Вишня иедоская
83368 Вишня юньнаньская
83418 Рогач, Устели-поле, Эбелек
83419 Рогач песчаный, Рогач туркестанский, Эбелек песчаный
83438 Рогоглавник
83439 Рогоглавник серповидный, Рогоглавник серпорогий
83443 Репяшок, Рогоглавник пряморогий
83466 Горовик
83523 Цератония
83527 Дерево рожковое, Дерево стручковое, Кароб, Каруба дикая, Рожечник, Рожки цареградские, Рожки царьградские, Рожковое дерево, Стручок цареградский, Хлеб дивий, Хлеб иванов, Цареградсие рожки, Цератония стручковая
83539 Рогористиковые, Рогористниковые
83540 Крапива водяная, Кушир, Рогористик, Рогористник
83545 Рогористик погруженный, Рогористник погруженный, Рогористник погружённый, Рогористник подводный, Рогористник темно-зеленый, Рогористник темно-зелёный
83564 Рогористник комарова
83573 Рогористник коссинского
83575 Рогористник японский
83577 Рогористик рисовый
83578 Рогористик полупогруженный, Рогористник полупогруженный
83583 Рогористик донской
83641 Цератостигма
83646 Цератостигма свинчатковая
83680 Цератозамия
83732 Баглянниковые
83734 Круглолистник
83736 Багрянник японский, Круглолистник японский
83743 Багрянник великолепный
83752 Багряник, Багрянник, Церцис
83757 Багрянник канадский, Церцис канадский
83769 Багрянник китайский, Багрянник японский, Церцис китайский
83788 Церцис гриффита
83809 Багрянник обыкновенный, Багрянник стручковатый, Багрянник стручковый, Дерево иудейское, Дерево иудино, Иудино дерево, Церцис европейский
83812 Церцис юньнаньский
83816 Церкокарпус
83855 Кактус колонновидный, Кактус-цереус, Цереус
83874 Кактус “царица ночи”, Кактус крупноцветный, Царица ночи, Цереус крупноцветный
83885 Цереус ямакару
83890 Цереус перуанский
83918 Восковник, Воскоцветник
83919 Восковник альпийский
83923 Восковник большой
83929 Воскоцветник малый
83975 Кактус свисающий, Кактус церопегия, Церопегия
84400 Цеструм
84404 Цеструм оранжевый
84419 Цеструм иволистный, Цеструм парка, Цеструм паркви
84549 Айва японская, Хеномелес, Цидония японская
84553 Хеномелес китайский
84556 Айва цветочная, Айва японская, Смородина японская, Хеномелес маулея, Хеномелес японский, Цидония японская
84573 Квит, Хеномелес китайский, Хеномелес легенария, Хеномелес прекрасный
84591 Хеномелес тибетский
84594 Хенорinum
84595 Хенориnum клокова
84615 Хенориnum колосовый
84619 Хенориnum клейкий
84723 Бутень
84726 Бутень ароматный
84729 Бутень боброва
84730 Бутень бородина
84731 Бутень клубненосный, Бутень луковичный, Керверь клубневой, Керверь корневой, Керверь ликовичный, Керверь репный, Репа кервельная
84736 Бутень кавказский
84738 Бутень смешиваемый
84739 Бутень длинноволосый
84744 Бутень низкий
84745 Бутень хорасанский
84746 Бутень кяпазский
84748 Бутень крупноплодный
84749 Бутень пятнистый
84750 Бутень мейера
84758 Бутень прескотта

84761 Бутень розовый
84762 Бутень красноваый
84768 Бутень одуряющий
84769 Бутень ложноодуряющий
84912 Хетолимон
84914 Хетолимон окаймленный
84915 Хетолимон щетинчатый
85054 Гривохвост, Хайтурус
85058 Гривохвост шандровый, Хайтурус шандровидный
85071 Медноцвет
85072 Медноцвет почколистый
85263 Кипарисник, Кипарисовик, Лжекипарис
85268 Кипарисовик формозский
85271 Кипарисовик лавсона, Кипарисовик лосона, Кипарисовик лоусона
85272 Кипарисовик лосона ‘Алюмии’
85278 Кипарисовик лосона эректа ‘Виридис
85282 Кипарисовик лосона ‘Флейера
85283 Кипарисовик лосона ‘Глука
85301 Кипарисовик нутканский
85310 Кипарисовик тупой, Кипарисовик туполистный
85316 Кипарисовик тупой ‘Филикоидес’
85330 Кипарисовик тупой ‘Тетрагона’
85345 Кипарисовик горохоплодный, Лжекипарис горохоносный
85375 Кипарисовик туеобразный
85401 Ракитник пупурный, Ракитник пурпуровый
85403 Бобовник альпийский, Дождь золотой альпийский, Ракитник альпийский, Ракитник головчатый
85412 Кассандра, Хамедафна, Хамедафне
85414 Хамедафна болотная, Хамедафне волотная, Хамедафне чашечная
85420 Хамедорея
85457 Хамегерон
85459 Хамегерон Бунге
85460 Хамегерон малоголовый
85510 Хамемелюм
85526 Пупавка благородная, Римская ромашка, Ромашка римская, Хамемелюм благолодный
85553 Хаменериум узкий
85554 Хаменериум кавказский
85555 Хаменериум колхидский
85559 Хаменериум лебедовый
85579 Дерен
85581 Дерен канадский
85646 Хамеродос
85647 Хамеродос алтайская
85651 Хамеродос прямой, Хамеродос прямостоящая
85652 Хамеродос крупноцветкнвая
85657 Хамеродос песчаная
85658 Вишня джунгарская
85660 Хамеродос трехнадоезная
85665 Хамеропс
85671 Пальма веерная европейская, Пальма веерная низкая, Пальма карликова, Пальма хамеропс, Пальмито, Хамеропс низкий, Хамеропс приземистый
85696 Низкозонтичник
85697 Низкозонтичник бесстебельный
85730 Хамесфакос, Шалфейчик
85731 Хамесфакос падуболистный, Шалфейчик падуболистый
85779 Молочай поникающий
85870 Кипрей, Кипрейник, Хаменерий, Хамениериум
85875 Иван-чай, Иван-чай узколистный, Кипрей, Кипрей узколистный, Конопля лесная, Копорский чай, Хаменерий узколистный, Хамениериум узколистный, Чай капорский
85884 Кипрей широколистный, Хамениериум широколистный
85928 Ятрышничк
85930 Ятрышничк альпийский
85944 Стробилантес японский
86012 Шардения
86013 Шардения крупноплодная
86014 Шардения восточная
86046 Хартолепис
86048 Хартолепис биберштейна
86049 Василек солончаковый, Хартолепис вайдолистный, Хартолепис средний
86051 Хартолепис крылатостебельный
86052 Бумагоплодник
86053 Бумагоплодник плоский
86409 Желтофиоль, Лакфиоль, Хейрантус
86427 Дерево восковое, Желтофиоль обыкновенный, Желтофиоль садовый, Лажфиоль, Лакфиоль садовый, Плющ восковой, Хейрантус садовый
86637 Хейролепис
86639 Хейролепис иранский
86733 Бородавочник, Честотел, Чистотел
86755 Бородавочник, Ластовица, Трава ластовичная, Честотел большой, Чистотел большой, Чистотел крупный
86781 Хелоне
86873 Маревые
86891 Лебеда, Марь
86892 Марь остроконечная
86901 Алабота, Марь белая, Марь обыкновенная, Шурок
86958 Марь берландье
86970 Марь доброго генриха
86977 Марь бриониелистная
86984 Жминда, Жминда головчатая, Марь головчатая, Марь переступнелистная
86990 Марь толстолистная
87016 Жминда, Марь многолистная
87019 Марь тайваньская
87023 Марь кустарничковая
87025 Марь гигантская
87029 Марь сизая
87045 Марь козлиная
87048 Марь гибридная
87057 Марь енисейская
87058 Марь лежачая, Марь стелюшаяся
87061 Марь клингреффа
87064 Марь коржинского
87096 Марь стенная
87112 Марь калинолистная
87130 Марь многосеменная, Марь многосемянная
87133 Марь луговая
87145 Квиноа, Квиноя, Киноа, Киноя, Лебеда кино, Лебеда перуанская, Лебеда рисовая, Марь киноа, Марь чилийская
87147 Марь красная
87158 Марь поздная
87162 Марь сосновского
87171 Марь торчащая
87182 Марь тибетская
87186 Марь городская
87191 Марь вахеля
87199 Марь зеленая
87200 Марь вонючая
87232 Чезнейя
87234 Чезнейя толстоногая
87237 Чезнейя изящная
87239 Чезнейя ганьсуская
87243 Чезнейя гиссарская
87244 Чезнейя чаюйская
87246 Чезнейя копетдагская
87247 Чезнейя линчевского
87264 Чезнейя якорцевая
87265 Чезнейя туркестанская
87266 Чезнейя юньнаньская
87271 Чезнейя ганьсуская, Чезнейя ферганская
87449 Губовник линейный
87480 Зимолюбка, Химафила
87485 Зимолюбка японская
87498 Зимолюбка зонтичная
87510 Химонант
87525 Химонант ранний, Химонант скороспелый
87545 Химонант чжэцзянский
87547 Химонобамбуза
87551 Химонобамбуза бирманская
87586 Химонобамбуза
87607 Бамбук четырехгранная, Химонобамбуза четырехгранная
87627 Химонобамбуза юньнаньская

87646 Юньнаньск тибетский
87661 Хиококка
87694 Дерево снежное, Хионант, Хионантус
87729 Хионант притупленный
87736 Дерево снежное, Хионантус виргинский
87757 Хионодокса
88022 Хирония
88242 Зелёнка, Хлора
88248 Зелёнка пронзённолистная
88263 Зеленоцветниковые, Хлорантовые
88264 Зеленоцвет, Хлорантус
88266 Хлорантус аньхуйский
88283 Хлорантус хубэйский
88289 Зеленоцвет японский, Хлорантус японский
88295 Хлорантус тайваньский
88297 Хлорантус пильчатый
88299 Хлорантус сычуаньский
88316 Хлорис
88351 Хлолис тайваньский
88421 Хлолис прутьевидная, Хлолис прутьевидный
88500 Хлорофора
88514 Дерево жёлтое, Хлорофора красильная
88527 Венечник живородящий, Хлорофитум
88546 Хлорофитум капский
88550 Хлорофитум китайский
88553 Хохлатый
88659 Хлороксилон
88663 Дерево апельсиновое, Дерево атласное, Дерево красное, Хлороксилон
88690 Хероспондиас
88691 Спондиас пазушный, Хероспондиас пазушный
88759 Хондрила, Хондрилла, Хондрилля
88760 Хондрилла колючечешуйная, Хондрилля щетинкопосная
88761 Хондрилла сомнительная, Хондрилля сомнительная
88763 Хондрилла шероховатая
88765 Хондрила боссэ
88766 Хондрилла коротноклювая, Хондрилля коротконосиковая
88767 Хондрилла седоватая, Хондрилля седоватая
88771 Хондрилля нителистная
88772 Хондрилла злаколистная, Хондрилля злаколистная
88773 Хондрилля каучуконосная
88776 Хондрилла обыкновенная, Хондрилла ситниковидная, Хондрилля ситниковидная
88779 Хондрилля косинского
88780 Хондрила кузнецова
88782 Хондрилла ширококорончатая
88783 Хондрилла широколистная
88784 Хондрилла гладкосемянная
88787 Хондрила тощая
88788 Хондрилла крупноплодная
88789 Хондрила самаркандская
88791 Хондрила украшенная
88793 Хондрилла малоцветковая
88794 Хондрилла темноголовая
88796 Хондрилла ломконосая, Хондрилла ломконосиковая
88799 Хондрилла рулье
88975 Хоризис
88977 Иксерис ползучий, Хоризис ползучий
88996 Хориспора
88997 Хориспора бунге, Хориспора бунговская
88998 Хориспора тонкая
89006 Хориспора грейга
89007 Хориспора грунинская, Хориспора иберийская
89008 Хориспора крупноногая
89013 Хориспора сибирская
89016 Хориспора джунгарская
89020 Многосемянниба нежная, Хориспора нежная
89159 Чозения
89160 Чозения земляничниколистная, Чозения крупночешуйная, Чозения толокнянколистная
89223 Христолея
89229 Христолея толстостенная
89231 Христолея вееровидная
89241 Христолея линейная
89246 Христолея майдантальская
89249 Христолея памирская
89253 Христолея парриевидная
89320 Трава лакмусовая, Хрозофора
89323 Хрозофора изящная
89324 Хрозофора косая
89328 Хрозофора песчаная
89330 Хрозофора красильная
89345 Хризалидокарпус
89360 Хризалидокарпус желтоватый
89402 Златоцвет, Нивяник, Пиретрум, Ромашка инсектисидная, Ромашник, Хризантема, Хризантема летние, Хризантемум
89421 Пиретрум щитковый, Пиретрум щитконосный, Поповник шитковидный, Поповник шитковый, Поповник шитконосный
89424 Дендрантема арктическая
89466 Хризантема килеватая
89481 Златоцвет венечный, Хризантема корончая, Хризантема увенчанная
89599 Пиретрум крупнолистный
89602 Калуфер, Кануфер, Колуфер, Конуфер, Пижма душистая, Пижма кануфер, Пижма пахучая, Пиретрум бальзамический, Хризантема бальзамная
89695 Нивяник круглолистный
89696 Хризантема роксбурга
89704 Златоцвет пашенный, Золотоцвет полевой, Нивяник посевной, Поповник посевной, Ромашка посевная, Хризантема межпосевной, Хризантема посевная
89723 Пиретрум сибирский, Поповник сибирский
89812 Слива кокосовая, Хризобаланус
89816 Икак, Икако, Слива золотая
89848 Златотравка
89850 Златотравка крупковидная
89952 Хризогонум
90007 Златолист, Хризофиллум, Хризофиллюм
90032 Златолист каймито, Каимито, Яблоко зведное
90107 Золотобородник, Поллиния
90108 Золотобородник игольчатый
90125 Золотобородник цикадовый
90133 Золотобородник восточный
90317 Селезеночник, Селезёночник
90325 Селезеночник обыкновенный, Селезёночник обыкновенный, Селезеночник очереднолистный, Селезёночник очереднолистный
90337 Селезеночник байкальский
90352 Селезеночник сомнительный
90361 Селезеночник тонгий
90362 Селезеночник плетеносный, Селезеночник усатый
90381 Селезеночник эмэйский
90382 Селезеночник гуандунский
90383 Селезеночник японский
90387 Селезеночник камчатский
90392 Селезеночник комарова
90396 Селезеночник тайваньский
90403 Селезеночник лушаньский
90416 Селезеночник непалский
90421 Селезеночник голостебельный
90423 Селезеночник супротивнолистный, Селезёночник супротивнолистный
90424 Селезеночник овальнолистный
90426 Селезеночник щитовидный
90429 Селезеночник волосистый
90434 Селезеночник ложный
90442 Селезеночник ветвистый
90447 Селезеночник седакова
90452 Селезеночник китайский
90460 Селезеночник четырехтычинковый
90461 Селезеночник тяньшанский
90465 Селезеночник шероховатосемянный

90471 Селезеночник райта
90629 Хускуеа
90709 Хымзыдия
90710 Хымзыдия агазиллевидная
90797 Нут
90800 Нут анатолийский
90801 Боб кофейный, Горох бараний, Горох турецкий, Материнка, Нагут, Нут, Нут бараний, Нут культурный, Нут обыкновенный, Нут полевой, Нут посевной, Нут рогообразный, Орех турецкий
90804 Нут балджуанский
90810 Нут чечевицеобразный
90812 Нут федченко
90813 Нут извилистый
90815 Нут жакемонта
90817 Нут копетдагский
90818 Нут коржинского
90819 Нут длинноколючий
90821 Нут мелколистный
90822 Нут крошечный
90823 Нут моголтавский
90827 Нут колючий
90828 Нут джунгарский
90835 Цицербита
90836 Цицербита альпийская
90837 Цицербита лазоревая
90839 Цицербита буржо
90845 Цицербита дельтовидная
90848 Цицербита ковалевткой
90851 Цицербита крупнолистная
90856 Цицербита далиская
90858 Цицербита пренантоидная
90859 Цицербита кистевидная
90862 Цицербита розовая
90864 Цицербита тибетская
90871 Цицербита тяньшанская
90872 Цицербита уральская
90873 Цицербита зеравшанская
90889 Цикорий, Цикорник
90894 Салат-эндивий, Цикорий корневой, Цикорий салатный, Цикорий эндивий, Эндивий
90900 Цикорий железистый
90901 Цикорий дикий, Цикорий корневой, Цикорий обыкновенный, Цикорий полевой, Цикорий садовий, Цикорий салатный
90910 Цикорий карликовый
90918 Вех, Цикута
90924 Блекота, Боликолов пятнистый, Омег пятнистый
90932 Бешеница водяная, Вех ядовитый, Омег болотный, Цикута пятнистая, Цикута ядовитая
90994 Клопогон, Цимицифуга
91008 Клопогон даурский, Цимицифуга даурская
91010 Клопогон европейский
91011 Клопогон вонючий, Ромашка клоповая, Ромашка клоповная
91023 Клопогон борщевиколистный
91024 Клопогон японский
91032 Клопогон наньчуаньский
91038 Клопогон простый, Цимицифуга простая
91042 Клопогон юньнаньский
91055 Дерево хинное, Хинное дерево, Цинхона
91071 Хинное гиблидное
91079 Хинное дерево мелкоцветное
91090 Дерево хинное, Хинное дерево красносоковое
91107 Пепериник, Цинерария
91235 Цинна
91239 Цинна широколистная
91252 Дерево коричное, Корица, Коричник, Лавр коричный, Лавр цейлонский, Циннамомум
91280 Циннамомум мелколистный
91287 Дерево кампешевое, Дерево камфарное, Корица, Коричник камфорный, Лавр камфорный, Циннамомум
91302 Дерево коричное китайское, Корифа китайская, Корица кассия, Корица китайская, Коричник кассия, Коричник китайский
91330 Коричник железконосный, Лавр ложнокамфарный, Ложнокамфорный лавр
91351 Коричник японский
91353 Циннамомум яванский
91357 Коричник ланьюйская
91366 Коричник лоурера
91407 Циннамомум пинбяньский
91409 Циннамомум чэнкоуский
91446 Дерево коричное, Дерево коричное цейлонское, Дерево кфичное, Корифа цейлонская, Корица цейлонская, Коричник цейлонский, Циннамомум цейланикум
91499 Двсемянник, Двулепестник, Колдун-трава, Трава колдуновая, Цирцея, Черноцвет
91508 Двулепестник альпийский, Двулепестник горный, Репей дикий, Цирцея альпейская
91524 Двулепестник четырехбороздчатый
91537 Двулепестник сердцелистный, Цирцея гиблидная, Цирцея сердцевидная
91553 Двулепестник средний
91557 Двулепестник обыкновенный, Двулепестник парижский, Цирцея обыкновенная, Цирцея парижская, Цирцея четырехбороздчатая
91575 Двулепестник мягкий, Цирцея мягкая
91697 Бодяк, Колютик, Хамепейце
91707 Бодяк абхазский
91713 Бодяк бесстебельный
91720 Бодяк крючковидный
91721 Бодяк скученный
91723 Бодяк крылатый, Бодяк щетинковидный
91724 Бодяк альберта
91738 Бодяк анатолийский
91744 Бодяк остроконечный
91746 Бодяк паутинистый
91749 Бодяк глинистый
91762 Бодяк армянский
91770 Бодяк желточешуйный, Бодяк полевой, Осот полевой, Осот розовый, Татарник полевой
91796 Бодяк биберштейна
91809 Бодяк борнмюллера
91815 Бодяк прицветниковый
91821 Бодяк коротхохохолковый
91827 Бодяк буша
91842 Бодяк серый
91852 Бодяк кавказский
91856 Бодяк головчатый
91864 Бодяк китайский
91870 Бодяк зереноверхий
91879 Бодяк ресчатый
91896 Бодяк собранный
91902 Бодяк дагестанский
91904 Бодяк беловатый
91939 Бодяк игольчатый
91943 Бодяк эльбрусский
91944 Бодяк колончаковый
91953 Бодяк гуншаньский
91955 Бодяк шершаво-липкий
91957 Бодяк красночешуйный
91958 Бодяк клейкий, Бодяк съедобный
91960 Бодяк черноморский
91962 Бодяк эмэйский
91978 Бодяк фомина
91984 Бодяк фрика
91988 Бодяк гагнидзе
91991 Бодяк гололистый
92012 Бодяк девясиловидный
92022 Бодяк разнолистный
92040 Бодяк хубэйский
92041 Бодяк болотистый
92044 Бодяк влаголюбивый
92045 Бодяк снизу-белый
92048 Бодяк имеретинский
92052 Бодяк седой, Бодяк серовойлочный

92066 Бодяк японский
92089 Бодяк камчатский
92098 Бодяк кемуларии
92099 Бодяк кецховели
92102 Бодяк комарова
92104 Бодяк кузнецова
92110 Бодяк ламировидный
92112 Анона сетчатая, Бодяк ланцетолистный
92120 Бодяк шерстистоцветковый
92121 Бодяк лопуховый
92131 Бодяк лицзянский
92160 Бодяк длинноцветковый
92167 Бодяк маака
92178 Бодяк крупнокистевой
92179 Бодяк крупноголовый
92211 Бодяк мулиский
92247 Бодяк плодолговатолистный
92248 Бодяк обвороченный, Бодяк окутанный
92268 Бодяк огородный
92278 Бодяк осетинский
92285 Бодяк болотный
92287 Бодяк паннонский
92293 Бодяк гребенчатый
92296 Бодяк поникший
92309 Бодяк польский
92321 Бодяк ложно-лопуховый
92348 Бодяк прямочешуйчатый
92349 Бодяк корнеголовчатый
92350 Бодяк поручейный, Бодяк речной
92356 Бодяк сайрамский
92367 Бодяк шантарский
92370 Бодяк шишкина
92375 Бодяк семенова
92380 Бодяк серпуховидный
92381 Бодяк зубчато-ресничатый, Бодяк мелкозубчатый
92384 Бодяк мягко-щетинистый, Бодяк щетинистый, Ляптхор, Наголоватьнь, Осот красный
92396 Бодяк сиверса
92398 Бодяк простой
92400 Бодяк выемчатый
92403 Бодяк кожистоглавый
92404 Бодяк сосновского
92412 Бодяк жесткощетинистый
92413 Бодяк слабо-вооруженный
92414 Бодяк почти-шерстистоцветковый
92420 Бодяк многоцветковый
92422 Бодяк совича
92433 Бодяк крымский
92448 Бодяк войлочный
92452 Бодяк волосистокаемчатый
92455 Бодяк туркестанский
92459 Бодяк украинский
92460 Бодяк топяной
92479 Бодяк власова
92485 Бодяк обыкновенный
92487 Бодяк вальдштейна
92492 Бодяк вейриха
92586 Виноград комнатный, Циссус
92729 Циссус
92777 Виноград комнатный, Циссус, Циссус двуцветный, Циссус разноцветный
92801 Циссус ланьюйский
93038 Циссус вэньшаньский
93046 Ладанниковые
93047 Цистанхе
93048 Цистанхе неясная
93058 Цистанхе рассеченная
93059 Цистанхе желтая
93068 Цистанхе монгольская
93069 Цистанхе нинсяская
93074 Цистанхе риджуэя
93075 Цистанхе солончаковая, Цистанхе солончаковое
93082 Цистанхе трубчатая
93120 Ладанник
93141 Ладанник кручавый
93152 Ладанник волосистый
93161 Лабданум
93170 Ладанник лавролистный
93194 Ладанник тополелистный
93202 Ладанник шалфеелистный
93216 Ладанник крымский
93232 Гитарник
93234 Гитарник леманна
93235 Гитарник весенний
93237 Лиродревесник, Цитарексилум
93286 Арбуз, Арбус
93297 Арбуз горький, Арбуз колоцинт, Арбуз пустотельт, Арбус дикий, Колоквинт, Колоцинт, Тыква горькая, Яблоко заморское
93304 Арбус, Арбус обыкновенный, Арбус столовый, Арбус съедобный, Кавун
93314 Агрум, Цитрон, Цитрус
93329 Лайм, Лайм настоящий, Лиметта, Лимон лиметта, Лимон сладкий, Ляйм, Цитрус лайм, Цитрус лиметта, Цитрус померанцелистный
93332 Апельсин горький, Апельсин кислый, Бигарадия, Бигарадия обыкновенный, Горький померанец, Дерево померанецевое, Кислый апельсин, Померанец, Померанец горький
93408 Бергамот, Цитрус бергамот
93414 Апельсин бигардия, Апельсин горький, Апельсин кислый, Бигардия, Померанец, Померанец горький, Цитрус бигардия
93446 Апельсин навель, Апельсин пупочный, Мандарин, Мандарин иволистный, Мандарин итальянский, Танжерин
93492 Дерево квассиевое, Лима, Папеда ежеиглистая
93495 Папеда китайская, Цитрус ичанский
93497 Цитрус индийский
93515 Цитрус юнос, Юзу, Юнос
93530 Цитрус кодзимикан
93534 Лайм, Лайм настоящий, Лиметта, Лимон лиметта, Лимон сладкий, Ляйм, Цитрус лиметта, Цитрус померанцелистный
93539 Дерево линонное, Лимон, Цитрон лимон, Цитрус лимон
93546 Дерево лимонное, Лимон
93579 Грейпфрут, Пампельмус, Помело, Поммело, Помпельмус, Пуммело, Цитрус помпельмус, Цитрус шеддок, Шеддок
93603 Лимон дикий, Цедрат, Цитрон, Цитрус мидийский
93604 Шитрон пальчатный
93636 Каламондин, Цитрус каламондин
93649 Мандарин благородный, Мандарин кинг
93690 Грейпфрут, Гроздевидный помпельмус, Помпельмус гроздевидный, Понтедериевые, Цитрус дивный
93717 Апельсин навель, Апельсин пупочный, Мандарин, Мандарин итальянский, Танжерин
93724 Цитрус кинокуни
93727 Мандарин кинг
93736 Мандарин, Мандарин уншиу, Мандарин японский, Уншиу сатсума, Цитрон уншиу, Цитрус уншиу, Японский мандарин
93748 Мандарин кислый
93765 Апельсин, Апельсин настоящий, Апельсин сладкий, Дерево анисовое, Дерево апельсиновое, Сладкий апельсин, Цитрус китайский
93771 Цитрус китайский ‘Гамлин
93774 Цитрус китайский ‘Яффа’
93785 Цитрус китайский ‘Валенсия’
93786 Цитрус китайский ‘Ващингтон-навель’
93837 Цитрус танжерин
93869 Цитрус вильсона
93901 Марискус, Меч-трава
93913 Меч-трава обыкновенная
93937 кладохета
93938 кладохета чистейшая
93939 кладохета каспийская
94014 Кладрастис

94022 Виргилия жёлтая, Кладрастис жёлтый
94026 Виргилия жёлтая, Кладрастис жёлтый
94132 Годециа, Кларкия, Эухаридиум
94138 Кларкия, Кларкия изящиная
94140 Кларкия красивая
94169 Цитроид
94232 Клаусия, Кляусия
94233 Кляусия солнцепечная
94236 Клаусия щетинистая
94237 Кляусия казахская
94238 Клаусия мягкая
94239 Клаусия ольги
94240 Клаусия сосочковаю
94292 Клайтония, Клейтония
94294 Клайтония остролистная
94299 Клайтония арктическая
94315 Клайтония эшольца
94322 Клайтония иоанна
94349 Клайтония пронзённолистный, Шпинат кубинский
94358 Клайтония отпрысковая
94362 Клайтония сибирная, Монция сибирская
94369 Клайтония клубневидная
94382 Клайтония васильева
94585 Змеевка
94591 Змеевка китайская
94604 Змеевка накаи
94605 Змеевка хэнса
94608 Змеевка китагавы
94622 Змеевка поздная
94631 Змеевка джунгарская, Змеевка торольда
94632 Змеевка растопыренная
94676 Жгунец, Клематис, Княжик, Лозинка, Ломонос
94704 Ломонос этузолистный
94714 Вьюнец, Клематис альпийский, Княжик альпийский, Ломонос альпийский
94745 Ломонос лицзянский
94748 Ломонос арманда
94763 Ломонос цзилунский
94778 Ломонос короткохвостый
94798 Ломонос бирманский
94811 Ломонос чжэцзянский
94814 Ломонос китайский
94816 Ломонос аньхуйский
94835 Ломонос усатый
94896 Ломонос фэн
94901 Ломонос жгучий
94910 Ломонос цветущий
94916 Ломонос тайваньский
94933 Ломонос бурый
94938 Ломонос бурый
94946 Ломонос сизый
94983 Ломонос гуйцзоуский
95000 Ломонос шестилепестный
95022 Ломонос усинский
95023 Ломонос хубэйский
95026 Ломонос илийский
95029 Клематис цельнолистный, Ломонос цельнолистный, Ломонос цельнолистый, Ломонос чельнолистный
95036 Клематис жакмана
95061 Ломонос корейский
95063 Ломонос гуйцзоуский
95067 Ломонос шерстистый
95084 Ломонос либоский
95100 Княжик крупнолепестковый
95111 Ломонос мэнлаский
95112 Ломонос мотоский
95127 Ломонос горный
95154 Ломонос непалский
95155 Ломонос напоский
95179 Ломонос восточный
95216 Ломонос раскидистый, Ломонос распростертый
95253 Ломонос ложножгучий
95261 Ломонос симаоский
95277 Ломонос прямой
95295 Ломонос пильчатолистный
95302 Клематис сибирский, Княжик сибирский
95304 Княжик охотский
95319 Ломонос асплениелистый, Ломонос джунгарский
95345 Ломонос тангутский
95358 Ломонос метельчатый
95366 Ломонос маньчжурский
95373 Княжик тяньшанский
95380 Ломонос динхуский
95423 Ломонос виорна
95426 Ломонос виргинский
95431 Лозинка белая, Лозинка обыкновенная, Ломонос виноградолистный, Ломонос вьющийся, Ломонос обыкновенный
95449 Ломонос фиолетовый
95454 Ломонос вэньшаньский
95463 Ломонос юньнаньский
95465 Ломонос юньнаньский цзиндунский
95606 Гинандропсис, Клеома, Клеоме, Паучник
95652 Клеоме колютеевбразная
95682 Клеоме гордягина
95723 Клеоме липского
95742 Клеоме ное, Клеоме ноевская
95746 Клеоме птиценогая, Клеоме птиценогое
95768 Клеоме пятижилковая
95771 Клеоме радде
95775 Клеоме клювообразная
95782 Клеоме хайнаньский
95806 Клеоме шерстистая
95810 Клеоме туркменская
95835 Клеоме юньнаньский
95934 Волкамерия, Клеродендрон, Сифонантус
95978 Клеродендрон бунге
96009 Клеродендрон китайский
96028 Клеродендрон криволистный
96029 Клеродендрон гуансиский
96078 Клеродендрон вонючий
96096 Клеродендрон таиландский
96114 Клеродендрон хайнаньский
96139 Клеродендрон индийское
96147 Клеродендрон японский
96157 Клеродендрон чжэцзянский
96167 Клеродендрон гуандунский
96392 Клеродендрон тибетский
96398 Клеродендрон трехнадрезный
96449 Волкамерия юньнаньская
96454 Клетра
96457 Клетра ольхолистная
96466 Клетра бородчатожилковая
96501 Клетра чэнкоуская
96536 Клетра наньчуаньская
96546 Клетра мелкоплодная
96561 Клетровые
96576 Клейера
96587 Клейера японская
96616 Клейера грушелистная
96627 Клейера мелколистная
96630 Клейера янчуньская
96633 Клиантус
96960 Душевга, Душевик, Клиноног, Пахучка
96964 Душевика тимьянная, Душёвка, Душёвка чапрецевидная, Пахучка острая, Пахучка остролистная, Пахучка полевая
96969 Душевик котовниковый
96970 Душевик китайский, Пахучка китайская
96981 Чабер пограничный
97002 Пахучка цернокрайняя
97031 Душевик сахалинский, Пахучка сахалинская
97041 Душевга эмэйская
97051 Пахучка теневая
97060 Пахучка обыкновенная
97086 Клинтония
97093 Клинтония удинская, Клинтония удская
97184 Клитория
97213 Кливия
97218 Кливия, Кливия оранжевая, Кливия суриковая, Кливия суриково-красная

97260 Клюзия
97266 Гуммигутовые, Зверобойные
97432 Щитница
97438 Щитница щетинистая
97442 Щитница дихотомическая
97444 Щитница изящная
97446 Щитница яруточная
97457 Щитница мелкоплодная
97491 Кнеорум, Оливник
97581 Бенедикт, Волчец, Кникус
97599 Бенедикт аптечный, Волчец благословленный, Волчец благословленный, Волчец кудрявый, Кардобенедикт, Кникус аптечный, Кникус благословенный, Кникус благословленный, Чертополох благословленный
97697 Жгун-корень, Жигунец
97703 Жгун-корень даурский
97705 Жгун-корень сомнительный, Жигунец сомнительный
97710 Жгун-корень гроссгейма
97711 Жгун-корень японский
97719 Жгун-корень монньe
97721 Жгун-корень многостебельный
97725 Жгун-корень восточный
97726 Жгун-корень малолучевой
97727 Жгун-корень солончаковый
97752 Кобея
97757 Кобея лазающая, Кобея лазящая, Кобея цепкая
97768 Кобрезия белларди
97770 Кобрезия нитевиднолистная
97771 Кобрезия низкая
97772 Кобрезия метельчатая
97773 Кобрезия ройля
97774 Кобрезия схонусовидная
97775 Кобрезия простая
97894 Коккулус, Коккулюс, Коломбо, Кукольван
97900 Коккулюс каролинский
97919 Коккулюс лавроволистный, Коккулюс лавролистный
97947 Коккулюс трехлопастный
97967 Лжеочток, Ложечница
97976 Ложечница английская
97977 Ложечница арктическая
97983 Ложечница тайваньская
97996 Ложечница гренландская
98002 Ложечница ху
98008 Ложечница ленская
98015 Ложечница мелкоплодная
98020 Ложечница аптечная, Ложечница арктическая, Ложечница лекарственная, Трава ложечная лекарственная, Трава лягушечная, Трава цынготная
98022 Ложечница продолговатолистная
98100 Кохлоспермум
98134 Кокос
98136 Кокос, Кокос орехоносный, Кокосовая пальма, Орех кокосовый, Пальма кокосовая
98159 Десмодиум движущийся, Трава терографная
98177 Кодиеум
98190 Кодиеум пестролистный, Кодиеум пестролистный расписная
98260 Кодоноцефалум
98263 Кодоноцефалум пекока
98273 Кодонопсис
98288 Кодонопсис бирманский
98290 Кодонопсис клематисовдный, Кодонопсис ломоносовидный
98343 Кодонопсис ланцетный, Кодонопсис ланцетолистный
98395 Кодонопсис мелковолосистый
98424 Кодонопсис уссурийский
98431 Кодонопсис тибетский
98560 Пололепестник
98586 Пололенестник зеленый, Пололепестник зелёный
98601 Целогина, Целогине
98634 Целогина гребенчатая
98653 Целогина поникшая
98664 Целогина гуншаньская
98694 Целогина малипоская
98743 Целогина красивая
98761 Целогина чжэньканская
98772 Пусторебрышник
98777 Пусторебрышник гмелина
98844 Дерево кофейное, Кофе
98857 Дерево кофейное, Дерево кофейное арабское, Дерево кофейное аравийское, Кофе, Кофе арабский, Кофе аравийское
98941 Дерево кофейное либерийское, Кофе либерийский, Кофе либерийское
99105 Коикс
99115 Коикс тайваньский
99124 Адлай, Коикс, Коикс слеза иова, Коикс слёзы Иовы, Слёзки богородицины, Слёзки Иова, Слёзник, Слёзы Иовы
99134 Коикс китайский
99157 Кола
99158 Кола блестящия, Кола заострённая, Кола курьтурная, Кола остролистная, Орех кола, Орешник-кола
99293 Безвременник, Зимовник, Колхикум, Осенник
99297 Безвременник осенний, Зимовник осенний, Крокус осенний, Осенчук, Песбой, Шафран логовой
99307 Безвременник биберштейна
99323 Безвременник регеля
99324 Безвременник веселый
99329 Безвременник желтый
99336 Безвременник подснежный
99343 Безвременник змесвиковый
99345 Безвременник великолепный
99348 Безвременник шовица
99350 Безвременник теневой
99355 Безвременник зангезурский
99411 Колеантус
99413 Колеантус тонкий, Колеантус тощий
99491 Колеостефус обыкновенный
99506 Колеус
99670 Картофель мадагаскарский
99711 Колеус блумего, Колеус пестрый, Колеус пёстрый
99712 Колеус ланьюйский
99804 Коллеция, Негниючник колёсивовидный
99815 Коллетия колючия
99833 Коллинзия, Коллинсия
99843 Коллинзония, Коллинсония
99850 Колломия
99858 Колломия крупноцветковая
99859 Колломия линейная
99899 Колоказия, Колокасия
99910 Колоказия древная, Колоказия съедобная, Таро
99918 Колоказия тайваньская
99922 Колоказия индийская
99987 Колподим
99988 колподим алтайский
99990 колподим араратский
99993 колподим золотистоцветковый
99994 колподим колхидский
99996 колподим волокнистый
100004 колподим бледночешуйный
100008 колподим украшенный
100009 колподим мелкоцветковый
100010 Колподим понтийский
100014 колподим валя
100015 колподим пестрый
100016 колподим разноцветный
100058 Колубрина
100132 Колюрия
100135 Колюрия гравилатовидная, Трава колюрин алтайская
100144 Колюрия эмэйская
100149 Пузырник
100152 Пузырник остролистный
100153 Дерево пузырное, Пузырник, Пузырник древовидный, Пузырник обыкновенная
100165 Пузырник армянский
100169 Пузырник бузе
100170 Пузырник седоватый
100171 Пузырник киликийский
100175 Пузырник тонгий

100177 Пузырник гибридный
100179 Пузырник ярмоленко
100180 Пузырник копетдагский
100185 Пузырник непалский
100187 Пузырник восточный, Пузырник кровавый
100188 Пузырник паульсена
100201 Пузыреплодник
100202 Пузыреплодник сетчатый
100254 Сабельник
100257 Лапчатка болотная, Пятилистник, Сабельник, Сабельник болотный
100259 Сабельник залесова, Сабельник залесовский
100271 Горечавка серповидная
100278 Горечавка распухлая
100284 Горечавка нежная
100296 Комбретовые
100309 Кольза, Комбретум
100892 Коммелина, Лазорник
100960 Коммелина мексиканская
100961 Коммелина обыкновенная, Коммелина синеглазка
101200 Коммелиновые
101285 Дерево мировое, Коммифора
101287 Коммифора абиссинская
101290 Дерево бальзамическое африканское, Коммифора африканская
101474 Комифора мира, Коммифора мирра
101488 Комифора бальзамная, Коммифора бальзамная
101633 Комперия
101635 Комперия крымская
101642 Астровые, Сложноцветные
101664 Комптония
101730 Кондалия
101820 Гирчевник, Гирчовник
101822 Гирчовник северный
101823 Гирчовник китайский
101827 Гирчовник камчатский
101829 Гирчовник широколистный
101830 Гирчовник длиннолистный
101831 Гирчовник тайваньский
101834 Гирчовник перистолистный
101839 Гирчовник влагалищный
101840 Гирчовник виктора
101845 Болиголов, Омег
101852 Болиголов крапчатый, Болиголов пятнистый, Омег пятнистый
102002 Конофолис
102808 Конрингия
102809 Конрингия австрийская
102811 Конрингия восточная
102812 Конрингия пронзеннолистая
102814 Конрингия персидская
102815 Конрингия плоскоплодная
102833 #######
102838 Живокость восточная
102842 Живокость мелкоморщинистая
102856 Ландыш
102867 Ландыш лесной, Ландыш майский, Ландыш обыкновенный, Трава майского ландыша
102883 Ландыш закавказский
102892 Вьюнковые
102903 Вьюнок, Калистедия, Повой
102926 Вьюнок аммана
102933 Березка, Берёзка, Вьюнок китайский, Вьюнок полевой, Печак, Чирмаук
102971 Вьюнок кальвера, Вьюнок карьверта
102980 Вьюнок кантабрийский
103025 Вьюнок растопыренный
103035 Вьюнок жестковетвистый
103053 Вьюнок кустарниковый
103065 Вьюнок горчакова
103072 Вьюнок пустыни
103080 Вьюнок волосистый
103084 Вьюнок шелковистный
103095 Вьюнок японский
103101 Вьюнок королькова
103102 Вьюнок краузе
103110 Вьюнок линейнолистный, Вьюнок узколистный, Вьюнок узколистый
103120 Вьюнок дерезовидный
103139 Вьюнок михельсова
103168 Вьюнок ольги
103188 Вьюнок персидский
103206 Вьюнок ашхабадский, Вьюнок ложнокантабрийский
103209 Вьюнок ложносмолоносный
103255 Вьюнок смолоносный, Вьюнок стрелолистный, Скандикс венерин гребень
103276 Вьюнок шерковистоголовый
103317 Вьюнок шерстистый
103319 Вьюнок седоватый
103329 Вьюнок крымский
103338 Вьюнок трагакантовый
103340 Вьюнок трёхцветный, Красавица дневная
103438 Мелколепестник буэносайресский, Мелколепестник курчавый
103446 Богатинка, Мелколепестник канадский
103651 конизантус
103653 конизантус злаколистный
103654 конизантус чешуйчатый
103691 Копайфера
103696 Копайфера копалоносная
103719 Зев львиный большой, Зев львиный садовый, Копайба, Копайфера лекарственная
103731 Коперниция
103737 Карнауба, Коперниция восконосная, Пальма бразильская, Пальма бразильская восковая, Пальма восковая бразильская, Пальма восконосная
103824 Коптис
103828 Коптис китайский
103835 Японск японский
103840 Японск эмэйский
103850 Коптис трехлистный
103976 Ладьян
104019 Ладьян трехлодвезный, Ладьян трёхлодвезный, Ладьян трехнадрезанный, Ладьян трехнадрезный
104056 Джут
104072 Джут длинноплодный, Джут короткоплодный, Джут круглоплодный, Джут огородный длинноплодный
104103 Джут длинноплодный, Джут короткоплодный
104147 Кордия
104334 Кордилина, Лилия-кордилина
104339 Драцена австралийская, Драцена южная, Кордилина австралийская, Кордилина южная
104350 Кордилина ветвистая
104366 Кордилина прямая
104367 Драцена южная, Кордилина верхушечная, Кордилина кустарниковая
104429 Кореопсис, Ленок, Лептосине
104448 Кореопсис друммонда
104468 Кореопсис сердечниколистный
104509 Кореопсис крупноцветконый, Кореопсис крупноцветный
104598 Кореопсис красильный, Ленок красильный
104687 Киндза, Киндзя, Кишнец, Кориандр
104690 Кишнец, Кишнец посевной, Кориандр, Кориандр огородный, Кориандр посевной
104691 Кориария
104696 Кориария японский
104700 Кориария миртолистная
104701 Кориария китайский, Кориария непалский
104710 Кориариевые
104750 Верблюдка
104751 Верблюдка зябкая
104752 Верблюдка алтайская
104754 Верблюдка арало-каспийская
104756 Верблюдка седоватая
104757 Верблюдка седоватая
104758 Верблюдка хинганская
104759 Верблюдка хинганская мелкоплодная
104762 Верблюдка густая
104763 Верблюдка толстолстная

104765 Верблюдка повислая
104770 Верблюдка бородавчатая
104771 Верблюдка вытянутая
104775 Верблюдка выгрызенная
104777 Верблюдка нителистная
104780 Верблюдка холодная
104783 Верблюдка семиречнская
104784 Верблюдка иларии
104786 Верблюдка иссополистная
104792 Верблюдка комарова
104793 Верблюдка коровина
104794 Верблюдка крылова
104796 Верблюдка рыхлоцветная
104797 Верблюдка лемана
104803 Верблюдка крупноплодная
104813 Верблюдка маршалла
104814 Верблюдка монгольская
104816 Верблюдка лоснящаяся
104819 Верблюдка восточная
104826 Верблюдка памирская
104828 Верблюдка бородавчатая
104839 Верблюдка редовского
104844 Верблюдка сибирская
104847 Верблюдка растопыренная
104849 Верблюдка стонтона
104856 Верблюдка тибетская
104860 Верблюдка курчавоплодный
104880 Деренные, Дерённые, Кизиловые
104909 Колнулака
104913 Колнулака коржинского
104917 Дерен, Дерён, Кизил, Свидина
104920 Дерен белый, Дерён белый, Кизил белый, Свида белая, Свидина белая
104930 Кизил сибирский, Кизил татарский, Свидина сибирская, Свидина татарская
104949 Дерен амом
104960 Дерен южный, Кизил южный, Свидина южная
104963 Дерен бейли
104994 Дерен китайский
105023 Дерево корнелиевое, Дерён съедобный, Дерён флолидский, Дерён цветточный, Дерен цветущий, Кизил флоридский, Циноксилон цветуший
105068 Свидина грузинская
105111 Дерен мощный, Дерён мужской, Дерен настоящий, Дерён отпрысковый, Кизил мужской, Кизил настоящий, Кизил обыкновенный, Кизил съедобный
105146 Дерен лекарственный, Кизил лекарсвенный
105169 Дерен кистевидный
105193 Дерён американский, Дерён обыкновенный, Дерен отпрысконосный, Кизил отпрысковый, Кизил побегоносный, Свидина отпрысковая, Свидина порослевая
105204 Дерен шведский, Дерён шведский, Кизил шведский
105207 Кизил сибирский, Кизил татарский
105229 Корнутия
105269 Вязель
105274 Вязель баланзы
105276 Вязель увенчанный
105277 Вязель критский
105279 Вязель эмеровый
105281 Вязель скутноцветный, Вязель эмеровый
105292 Вязель широколистный
105299 Вязель горный
105300 Вязель восточный
105301 Вязель мелкоцветкавый
105310 Вязель завитой
105324 Вязель изменчивый, Вязель обыкновенный, Вязель пестрый, Вязель пёстрый, Вязель разноцветный
105338 Воронья лапа, Лапа воронья, Лапка воронья
105340 Воронья лапа двойчатая
105350 Лапа воронья простёртая, Лапка воронья простёртая
105416 Корригиола береговая, Языки волобьиные
105450 Кортадерия
105456 Гинериум серебристый, Злак пампасовый, Кортадерия двдомная, Пампасова трава
105488 Кортуза
105489 Кортуза алтайская
105490 Кортуза лмурская
105491 Кортуза бротеруса
105496 Кортуза маттиоли, Кортуза маттиоля
105501 Кортуза пекинская, Кортуза хэбэйская
105505 Кортуза сибирская
105507 Кортуза туркестанская
105557 Хохлатка
105564 Хохлатка хубэйская
105581 Хохлатка эчкона
105587 Хохлатка алексеенко
105588 Хохлатка алпийская
105594 Хохлатка обманчивая, Хохлатка сомнительная
105615 Хохлатка узколистная
105617 Хохлатка узкоцветная
105618 Хохлатка узколистная
105624 Хохлатка арктическая
105663 Хохлатка крупноприцветниковая
105669 Хохлатка бухарская
105679 Хохлатка мулиская
105680 Хохлатка бунге
105682 Хохлатка буша
105688 Хохлатка сычуаньская
105699 Хохлатка бымянковидная, Хохлатка дымянкообразная
105719 Хохлатка кавказская
105721 Хохлатка полая
105738 Ганьсуск ганьсуский
105741 Хохлатка снеголюбивая
105752 Хохлатка коническикорневая
105803 Хохлатка дарвазская
105858 Хохлатка эмануеля
105860 Хохлатка эрдели
105872 Хохлатка федченковская
105881 Хохлатка мупинская
105888 Хохлатка мупинская мелкоцветковая
105914 Хохлатка франше
105917 Хохлатка дымянковолистная
105919 Хохлатка дымянковолистная
105929 Хохлатка гигантская
105936 Хохлатка сизоватая
105939 Хохлатка горчакова
105971 Хохлатка непалская
105999 Хохлатка недотрога
106004 Хохлатка надрезанная
106013 Хохлатка незаметная, Хохлатка тонкая
106017 Хохлатка промежуточная
106032 Хохлатка кашгарская
106040 Хохлатка цюцзянская
106066 Хохлатка ледебура
106117 Хохлатка жёлтая
106121 Хохлатка крупноцветная
106123 Хохлатка крупночашечная
106124 Хохлатка крупношпорцевая
106130 Хохлатка маршалла
106159 Хохлатка мелкоцветковая
106196 Хохлатка голостебельная
106197 Хохлатка охотская
106225 Хохлатка пачоского
106226 Хохлатка пионолистная
106227 Хохлатка бледная
106240 Хохлатка чжэцзянская
106242 Хохлатка бледноцветковая
106246 Хохлатка метельчатая
106255 Хохлатка малоцветковая
106268 Хохлатка персидская
106281 Хохлатка попова
106292 Хохлатка ложно-согнутая
106293 Хохлатка ложноалипийская
106346 Хохлатка цинхайская
106355 Хохлатка радде
106358 Хохлатка редовского
106380 Хохлатка ползучая
106414 Хохлатка шангина
106415 Хохлатка железнова
106427 Хохлатка семенова
106430 Хохлатка сизая
106431 Хохлатка северцева

106439 Хохлатка таиландская
106453 Хохлатка благородная, Хохлатка галлера, Хохлатка сибирская
106488 Хохлатка прямая
106531 Хохлатка тибетская
106557 Хохлатка чаюйская
106564 Хохлатка расставленная, Хохлатка удаленная
106613 Хохлатка юньнаньская
106620 Хохлатка чжундяньская
106623 Орешниковые
106628 Корилопсис
106656 Корилопсис тайваньский
106658 Корилопсис мелкоплодный
106668 Корилопсис эмэйский
106669 Корилопсис малоцветковый, Корилопсис слабоцветущий
106670 Корилопсис широколепестный
106675 Корилопсис китайский
106682 Корилопсис колосковый
106686 Корилопсис цюцзянский
106687 Корилопсис вича
106689 Корилопсис виллмотта, Корилопсис уилмотта
106695 Корилопсис юньнаньский
106696 Лещина, Орех лесной, Орешник
106698 Лещина американская, Лещина фундук
106703 Лесной орешник, Лещина обыкновенная, Лещина понтийская, Орех лесной американский, Орешник, Орешник лесной, Орешник обыкновенный, Орешник понтийский
106714 Лещина короткотрубчатая
106716 Лещина китайская, Орешник китайский
106719 Лещина колхидская
106720 Лещина древовидная, Орех мадвежий, Орех понтийская, Орешник медвежий
106725 Лещина калифорнийская
106734 Лещина тибетская
106735 Лещина тайваньская
106736 Лещина разнолистная, Орешник разнолистная
106748 Лещина гуйцзоуская
106751 Лещина маньчжурская, Орешник маньчжурсный
106757 Лещина высокая, Лещина крупная, Ломбардский орех, Орех ломбардский, Фундук, Фундук крупноплодный
106777 Лещина зибольда, Орешник зибольда
106784 Лещина ван
106786 Лещина юньнаньская
106793 Эвкалипт лимонноароматный, Эвкалипт лимонный
106800 Эвкалипт фиголистный
106805 Эвкалипт камеденосный
106809 Эвкалипт пятнистый
106924 Булавоносец
106925 Булавоносец суставчатый
106932 Булавоносец седой
106944 Каринокарпус
106984 Корифа, Корифа зонтичная
106999 Корифа зонтичная, Корифа теневаягвоздная, Пальма зонтичная, Пальма талипотовая, Пальма таллипотовая, Пальма тени
107158 Космея, Космос, Красотка
107161 Космея, Космея дваякоперистая, Космос, Космос дваякоперистый, Космос раздельнолистный, Красотка
107168 Космос разнолистный
107221 Костус
107271 Костус замечательный
107297 Дерево париковое, Желтинник, Скумпия
107300 Венецианский сумах, Дерево дымчатое, Дерево жёлтое, Дерево париковое, Дерево физетовое, Желтинник, Кожевенное дерево, Рай-дерево, Скумпия, Скумпия коггигрия, Скумпия кожевенная, Сумах венецианский, Сумах дубильный, Физетовое дерево
107308 Желтинник чэнкоуский
107315 Скумпия американская
107316 Скумпия сычуаньская
107319 Кизильник, Котонеастер
107322 Кизильник остроконечный
107336 Кизильник прижатый
107338 Кизильник разноцветный
107339 Кизильник сычуаньский
107340 Кизильник прелестный
107344 Кизильник антонины
107387 Кизильник чжэньканский
107416 Кизильник даммера
107424 Кизильник дильса
107435 Кизильник растопыренный
107440 Кизильник замечательный
107451 Кизильник франше
107454 Кизильник холодный
107472 Кизильник мэнцзыский
107474 Кизильник пушистолистый
107486 Кизильник горизонтальный
107504 Кизильник кистецветный, Кизильник цельнокрайний
107525 Кизильник чжундяньский
107526 Кизильник лицзянский
107531 Кизильник блестящий
107541 Кизильник крупноплодный
107543 Кизильник черноплодный
107549 Кизильник мелколистный, Кизильник тимьянолистный
107576 Кизильник мупинский
107581 Котонеастер мулиский
107583 Кизильник многоцветковый
107611 Кизильник малоцветковый
107615 Кизильник пленчатый
107624 Кизильник поярковой
107631 Кизильник ложномногоцветковый
107637 Кизильник кистецветный
107650 Кизильник роборовского
107668 Кизильник иволистный
107677 Кизильник скальный
107678 Кизильник
107688 Кизильник антониный
107689 Кизильник антониный мелкоплодный
107694 Кизильник приятный
107701 Кизильник крымский
107707 Кизильник тибетский
107710 Кизильник войлочный
107715 Кизильник одноцветковый
107726 Кизильник уорда
107729 Кизильник цабеля
107747 Котула
107766 Котула коронополистная, Котула коронопусолистная
107840 Котиледон
108217 Кузиния
108218 Кузиния угороченная
108219 Кузиния аболина
108220 Кузиния железконосная
108221 Кузиния родственная
108222 Кузиния головчатая
108223 Кузиния алайская
108224 Кузиния крылатая
108225 Кузиния алберта
108226 Кузиния белоцветковая
108227 Кузиния высокогорная
108228 Кузиния алпйская
108229 Кузиния обманывающая
108230 Кузиния прелестная
108231 Кузиния андросова
108232 Кузиния узкоголовая
108233 Кузиния неправильная
108234 Кузиния антонова
108235 Кузиния остроконечная
108236 Кузиния паутинистая
108237 Кузиния лопуховидная
108238 Кузиния армянская
108239 Кузиния астраханская, Кузиния волжская
108240 Кузиния золотистая
108241 Кузиния бадхизская
108242 Кузиния баталина
108243 Кузиния боброва
108244 Кузиния бонвало
108245 Кузиния бочанцева

108246 Кузиния короткокрылая
108247 Кузиния вунге
108248 Кузиния воловьеглазая
108249 Кузиния дернистая
108250 Кузиния лысая
108251 Кузиния согнутоколючковая
108252 Кузиния белесоватая
108253 Кузиния васильковидная
108254 Кузиния щетинкоголовая
108255 Кузиния зеленоватоцветковая
108256 Кузиния зеленоголовая
108257 Кузиния золотисто-цветковая
108258 Кузиния голубая
108259 Кузиния скученная
108260 Кузиния колонованная
108261 Кузиния шитковидная
108262 Кузиния артишоковидная
108263 Кузиния дарвазская
108264 Кузиния низбегающелистная
108265 Кузиния развилистая
108266 Кузиния двуцветковая
108267 Кузиния димо
108268 Кузиния рассеченная
108270 Кузиния рассеченнолистная
108271 Кузиния растопыренная
108272 Кузиния длинноветвистая
108273 Кузиния длинночешуйная
108274 Кузиния длиннолистная
108275 Кузиния подозрительная
108276 Кузиния отборная
108277 Кузиния торчащеколючковая
108279 Кузиния шерстисто-волосая
108280 Кузиния ереванская
108281 Кузиния синеголовниковидная
108282 Кузиния евгения
108284 Кузиния обманчивая
108285 Кузиния пучковатая
108286 Кузиния федченко
108287 Кузиния ферганская
108288 Кузиния ржавчинная
108289 Кузиния фетисова
108290 Кузиния франшэ
108291 Кузиния гладкощетинковая
108292 Кузиния желелистая
108293 Кузиния изящноголовая
108294 Кузиния меченосная
108295 Кузиния гнездиллы
108296 Кузиния гончарова
108297 Кузиния крупнолистная
108298 Кузиния григорьева
108299 Кузиния серая
108300 Кузиния гаматы
108301 Кузиния копьелистная
108302 Кузиния иларии
108303 Кузиния гогенакера
108304 Кузиния страшненькая
108305 Кузиния снизуседая
108306 Кузиния иглистая
108307 Кузиния ильина
108308 Кузиния целинолистная
108309 Кузиния каратавская
108310 Кузиния кнорринг
108311 Кузиния кокандская
108312 Кузиния комарава
108313 Кузиния королькова
108314 Кузиния коржинского
108315 Кузиния краузе
108316 Кузиния кюкенталя
108317 Кузиния ярко-зеленая
108318 Кузиния шерстистая
108319 Кузиния шерстистоголовая
108320 Кузиния репейниковая
108322 Кузиния трупкоопушенная
108323 Кузиния гладкоглавая, Кузиния гладкоголовая
108324 Кузиния тонкоколючковая
108325 Кузиния тонкосогнутая
108326 Кузиния тонкоголовая
108327 Кузиния тонковетвистая
108328 Кузиния почтитонковетвистая
108329 Кузиния бледноцветковая
108330 Кузиния линчевского
108331 Кузиния литвинова
108332 Кузиния ломакина
108333 Кузиния лировидная
108334 Кузиния крупноголовая
108335 Кузиния крупнохрылая
108336 Кузиния великолепная
108337 Кузиния самаркандская
108338 Кузиния маргариты
108339 Кузиния средная
108340 Кузиния мелкоплодная
108341 Кузиния мелкоголовчатая
108342 Кузиния минквиц
108343 Кузиния крошечная
108344 Кузиния моголтавская
108345 Кузиния мягкая
108346 Кузиния заплесневелая
108347 Кузиния многолопастная
108348 Кузиния незамеченная
108349 Кузиния невесского
108350 Кузиния нина
108351 Кузиния ольги
108352 Кузиния пупообразная
108353 Кузиния татарниковая
108354 Кузиния яйценогая
108355 Кузиния горная
108356 Кузиния горностхолюбивая
108357 Кузиния восточная
108358 Кузиния прямоколючковая
108359 Кузиния прямочешуйная
108360 Кузиния овчинникова
108361 Кузиния амударьинская
108362 Кузиния остронадрезная
108363 Кузиния лохматая
108364 Кузиния лохматовидная
108365 Кузиния маловетвистая
108366 Кузиния цветоносная
108367 Кузиния пятиколючковая
108368 Кузиния ложнопятиколючковая
108369 Кузиния широкочешуйная, Лопух широкочешуйный
108370 Кузиния широкооберточная
108371 Кузиния черешковая
108372 Кузиния многоглавая
108373 Кузиния ревосходная
108374 Кузиния главная
108375 Кузиния близкая
108376 Кузиния песколюбивая
108377 Кузиния ложнородственная
108378 Кузиния ложно-шерстистая
108379 Кузиния ложномягкая
108380 Кузиния перисточешуйная
108381 Кузиния красивенькая
108382 Кузиния красивая
108383 Кузиния колючая
108384 Кузиния пурпуровая
108385 Кузиния ничтожная
108386 Кузиния карликовая
108387 Кузиния радде
108388 Кузиния лучевая
108389 Кузиния ветвистая
108390 Кузиния желтовато-серая
108391 Кузиния преломленная
108392 Кузиния смолистая
108393 Кузиния красно-цветковая
108394 Кузиния жесткая
108395 Кузиния розовая
108396 Кузиния круглолистная
108397 Кузиния красно-бурая
108398 Кузиния шероховатая
108399 Кузиния шишкина
108401 Кузиния твердолистная
108402 Кузиния полунизбегающая
108403 Кузиния полуизорванная
108405 Кузиния сестринская
108406 Кузиния видная
108407 Кузиния спиридонова
108408 Кузиния блестящая
108409 Кузиния рассеянноголовая
108410 Кузиния шталя
108411 Кузиния звездчатая
108412 Кузиния узколистная
108413 Кузиния увенчанная
108414 Кузиния торчащая
108415 Кузиния шишкоголовая
108416 Кузиния почти-придаточная
108417 Кузиния почтибелесоватая
108418 Кузиния почти-тупоконечная
108419 Кузиния саксаульниковая
108420 Кузиния сырдарьинская

108421 Кузиния таласская
108422 Кузиния таммары
108423 Кузиния тоненькая
108424 Кузиния тонкорассеченная
108426 Кузиния тяньшанская
108427 Кузиния войлочненькая
108428 Кузиния шершаволистная
108429 Кузиния заилийская
108430 Кузиния заамударьинская
108432 Кузиния трехголовая
108433 Кузиния волосистая
108434 Кузиния трехцветковая
108435 Кузиния туркменская
108436 Кузиния туркестанская
108437 Кузиния курчаводольковая
108438 Кузиния пупочковая
108439 Кузиния теневая
108440 Кузиния вавилова
108441 Кузиния мутовчатая
108442 Кузиния викарная
108443 Кузиния введенского
108444 Кузиния вальдгейма
108448 Кузиния желтоватая
108449 Кузиния дтрнишникоголовая
108591 Капуста морская, Катран, Крамбе, Крамбекатран
108592 Крамбе абиссинская
108597 Катран армянский
108598 Катран шершавый
108599 Катран сердцелистный
108604 Катран беззубый
108607 Катран бугорчатый
108612 Катран гордягина
108613 Катран крупноцветковый
108618 Катран прутьевидный
108620 Катран коктебельский
108621 Катран котчиевский, Катран кочи
108626 Катран литвинова
108627 Капуста морская, Катран перистый, Катран приморский
108628 Катран восточный
108629 Катран кочи, Катран перистый
108630 Катран понтийский
108634 Катран шугнанский
108637 Катран стевена, Катран стевенова
108641 Катран татарский, Телимитра
108661 Черепоплодник
108662 Черепоплодник сероватый
108665 Черепоплодник ежистый
108666 Черепоплодник хлопьевидно-шерстистый, Черепоплодник щетинистый
108723 Гинура
108775 Крассула, Толокнянка
109505 Толстянка, Толстянковые
109509 Боярышник
109522 Боярышник алтайский
109524 Боярышник сомнительный
109535 Боярышник армянский
109536 Боярышник арнольда
109542 Боярышник темнокровавый
109545 Боярышник азароль, Боярышник испанский, Боярышник плодовый, Боярышник понтийский, Боярышник-азароль, Дуляна
109552 Боярышник беккера
109582 Боярышник канадский
109585 Боярышник кавказский
109597 Боярышник зеленомякотный, Боярышник зеленомясый
109606 Боярышник чжундяньский
109609 Боярышник американский, Боярышник шарлаховый, Боярышник ярко-красный
109612 Боярышник кокцинисвидный
109636 Боярышник петуший гребень, Боярышник петушья шпора, Боярышник шпорцевый
109650 Боярышник клиновидный
109662 Боярышник даурский
109671 Боярышник двукосточковый
109684 Боярышник дугласа
109685 Боярышник джунгарский
109693 Боярышник волосистоцветковый
109709 Боярышник жёлтыий
109750 Боярышник гельдрейха
109760 Боярышник хубэйский
109780 Боярышник хоккайдский
109781 Боярышник гансуйский
109785 Боярышник согнутостолбиковый
109788 Боярышник восточный
109790 Боярышник колючий, Боярышник обыкновенный
109796 Боярышник ламберта
109839 Боярышник максимовича
109845 Боярышник мейера
109846 Боярышник мелколистный
109850 Боярышник мягкий
109857 Боярышник однопестичный, Боярышник одностолбчатый, Гулявник
109868 Боярышник чёрный
109924 Боярышник пятипестичный
109933 Боярка бородавчатая, Боярышник перистонадрезанный, Боярышник перисторзадельный, Резаньглот
109964 Боярышник ложно-сомнительный
109965 Боярышник ложно-разнолистный
109966 Боярышник ложно-черноплодный
109994 Боярышник расставленнолопастный
110001 Боярышник робсона
110003 Боярышник круглолистный, Лиственница восточная американская, Орех чёрный, Орех чёрный восточный
110014 Боярышник кровавокрасный, Боярышник смбирский
110032 Боярышник шрадера
110039 Боярышник шаньдунский
110048 Боярышник джунгарский
110052 Боярышник клинолистный
110056 Боярышник стевенов
110076 Боярышник шовица
110084 Боярышник крымский
110087 Боярышник тяньшанский
110091 Боярышник закаспийский
110095 Боярышник туркменский
110096 Боярышник туркестанский
110097 Боярышник украинский
110110 Боярышник волжский
110116 Боярышник зангезурский
110292 Кравфурдия
110500 Кремастра
110608 Параиксерис лопастнолистный
110637 Малаксис ланьюйский
110644 Малаксис хайнаньский
110654 Малаксис сычуаньский
110710 Баркаузия, Зацинта, Скерда, Скерда болотная
110722 Скерда алайская
110726 Скерда аликера
110727 Скерда альпийская
110738 Скерда астраханская
110752 Скерда двулетняя, Скрипуха
110762 Скерда бунге
110763 Скерда буреинская
110767 Скерда волосовидная, Скерда двулетняя, Скерда зеленая, Скерда зелёная, Скерда колосовидная
110769 Скерда кавказская
110770 Скерда кавказородная
110775 Скерда золотистая, Скерда золотоцветковая
110777 Скерда реснитчатая
110783 Скерда рожковатая
110785 Скерда шафранно-желтая
110787 Скерда чуйская
110788 Скерда дарвазская
110806 Скерда извилистая
110808 Скерда вонючая
110820 Скерда голая
110827 Скерда гмелина
110841 Скерда хокайдская
110871 Скерда каракушская
110872 Скерда карелина
110874 Скерда хорасанская
110879 Скерда кочи
110880 Скерда млечная
110889 Скерда лировидная
110893 Скерда маршалла
110896 Скерда мелкоцветковая
110904 Скерда мягкая, Скерда

мягковолосистая
110906 Скерда многоцветбельная
110916 Скерда карликовая, Скерда низкая
110929 Скерда французская
110930 Скерда черноватая
110942 Скерда горная
110946 Скерда болотная, Скерда волосовидная
110947 Скерда венгерская
110954 Скерда многоволосистая
110956 Скерда обгрызанная, Скерда обгрызенная, Скерда тупокорневищная
110965 Скерда красивая
110975 Скерда разветвленная
110985 Скерда ксасная
111017 Скерда щетинистая
111019 Скерда сибирская
111050 Скерда кровельная
111059 Скерда тяньшанская
111066 Скерда туркменская
111091 Скерда вильденова
111092 Скерда вилеметиевидная
111096 Зацинта бородавчатая
111100 Дерево горлянковое, Дерево калабассовое, Кресченция
111104 Дерево голянковое обыкновенное, Дерево горлянковое
111111 Кресса
111113 Кресса критская
111152 Кринум
111158 Кринум американский, Лилия американская
111167 Кринум азиатский
111170 Японск японский
111180 Кринум длиннолистный
111343 Критмум, Серпник
111347 Критмум морской, Сосна скрученная
111480 Крокус, Шафран
111481 Шафран адама
111483 Шафран алатавский, Шафран алатауский
111487 Шафран артвинский
111492 Шафран отрана
111514 Шафран каспийский
111540 Шафран гейфела, Шафран гейфеля
111546 Шафран карский
111548 Шафран королькова
111556 Шафран михельсона
111558 Шафран золотистый
111573 Шафран палласа
111582 Шафран сетчатый
111584 Шафран рооп
111589 Крокус посевный, Шафран посевной
111590 Шафран шарояна
111603 Шафран красивый, Шафран прекрасный
111612 Шафран сузианский
111614 Шафран суворова
111615 Шафран крымский
111623 Крокус долинный, Шафран долиный
111624 Шафран полосатый
111625 Крокус весенний, Крокус голандский, Шафран весенний
111693 Кроссандра
111741 Кроссандра нильская
111823 Кроссостефиум
111826 Кроссостефиум китайский
111851 Кроталярия
112013 Кроталярия китайская
112207 Кроталярия хайнаньская
112248 Кроталярия промежуточная
112269 Звездовка, Кроталярия индийская, Кроталярия ситниковая, Кроталярия ситниковидная, Пенька индийская
112307 Кроталярия ланцетовидная
112491 Кроталярия остроконечная, Кроталярия полосадая
112563 Кроталярия цзиньпинская
112682 Кроталярия
112729 Кроталярия симаоская
112823 Кроталярия юньнаньская
112829 Дерево кротоновое, Кротон, Орешник проностый
112884 Каскариль
113033 Кротон таиландское
113039 Дерево кротоновое, Кротон, Кротон слабительный, Орешник проносный
113061 Кротон юньнаньский
113110 Крестовница
113135 Подмаренник херсонесский
113144 Крестоцветные
113151 Резушка мягкопушистая
113205 Крупина
113214 Крупина обыкновенная
113305 Скрытница
113306 Скрытница колючая
113310 Скрытница лизохностовидная
113312 Скрытница борщова
113316 Скрытница камышевидная
113318 Скрытница туркестанская
113364 Криптантус
113527 Криптокорина
113551 Скрытоподстолбник
113552 Скрытоподстолбник несколюбивый
113553 Скрытоподстолбник песчатый
113554 Скрытоподстолбник кахрисовидный
113555 Скрытоподстолбник двойчатый
113633 Криптомерия
113685 Криптомерия японская
113687 Криптомерия японская ‘Араукариоидес’
113693 Криптомерия японская ‘Дакридиоидес’
113696 Криптомерия японская ‘Элеганс’
113721 Криптомерия китайская
113815 Волосатик, Скрытосемянница
113817 Скрытосемянница серповидная
113868 Криптотения, Скрытница
113872 Скрытница канадская
113879 Криптотения японская, Скрытница японская
113944 Ктенанте
114057 Волдырник, Пузырник
114060 Волдырник ягодный, Волдырник ягодоплодный
114066 Волдырник японский
114108 Дыня, Кукумис, Огурец
114122 Ангурия, Коричник пикульный, Огурец антильский, Огурец вест-индский
114150 Дыня думаим, Дыня мелкоплодная, Сорная дыня
114189 Дыня, Дыня извилистая, Дыня мускатная, Дыня посевная, Дыня тарра, Огурец дыня
114199 Дыня десертная, Дыня канталупа, Дыня культурная, Канталупа
114209 Дыня змеевидная, Дыня извилистая, Дыня изогнутая, Дыня новислая, Дыня огуречная, Тарра
114214 Дыня обыкновенная, Дыня сетчатая
114221 Дыня декопативная красноплодная, Огурец африканский
114245 Огурец, Огурец грядовой, Огурец огородный, Огурец посевной
114246 Огурец английский тепличный
114247 Огурец серповидный
114250 Огурец иккимский
114251 Огурец чешуйчатый
114252 Огурец бугорочатый
114271 Кукурбита, Тыква
114277 Тыква черносеменная
114288 Выква большая столовая, Выква оплодная, Тыква гигантская, Тыква крупная, Тыква крупноплодная
114292 Выква китайская, Выква кускатная, Выква мускусная, Тыква мускатная
114300 Выква летняя, Выква овальная, Выква столовая, Выква тарелочная, Выква фестончатая, Кабачок, Патисон, Тыква обыкновенная
114301 Патиссон, Тыква кустовая, Тыква мозговая, Тыква фигурная
114304 Кабачок
114310 Тыква яйцевидная
114313 Тыквенные
114319 Кудрания
114427 Псоралея

114502 Кмин
114503 Горец щавелелистный, Кмин тминовый
114520 Кунила
114532 Куннингамия
114539 Куннингамия ланцетная, Куннингамия ланцетовидная
114540 Куннингамия ланцетовидная ‘Глаука’
114548 Куннингамия китайская, Куннингамия ланцетная, Куннингамия тайваньская
114574 Купания
114581 Блигия африканская, Финган
114595 Куфея, Парсония
114645 Кипарисовые
114649 Кипарис
114652 Кипарис аризонский, Кипарис аризонский бонита
114664 Кипарис португальский бентама
114668 Кипарис кашмирский
114669 Кипарис миньцзянский
114680 Кипарис гималайский, Кипарис дукло, Кипарис дюкло, Кипарис надутый
114690 Кипарис китайский плакучий, Кипарис плагучий
114695 Кипарис горибой говея
114698 Кипарис гауэна, Кипарис говена, Кипарис говея, Кипарис калифорнийский
114702 Кипарис гуадалупский
114709 Кипарис цзяньгэский
114714 Кипарис лузитанский, Кипарис португальский
114721 Кипарис макнаба
114723 Кипарис крупноплодный
114753 Кипарис вечнозеленый, Кипарис вечнозелёный, Кипарис настоящий, Кипарис пирамидальный
114760 Кипарис вечнозеленый, Кипарис горизонтальный
114763 Кипарис настоящий индийский
114765 Кипарис вечнозелёный пирамидальный, Кипарис пирамидальный
114775 Кипарис гималайский, Кипарис дукло, Кипарис надутый
114814 Куркулига, Куркулиго
114852 Куркума
114858 Аррорут восточноиндийский, Аррорут остиндский, Куркума узколистная
114868 Куркума гуансиская
114871 Имбирь жёлтый, Куркума длиная
114882 Куркума юньнаньская
114884 Корень цитварный, Куркума цитварная
114947 Зард-печак, Кускута, Плющ, Повилика, Трава войлочная
114953 Повилика белая
114961 Повилика бокальчатая, Повилика сближенная
114965 Повилика араратская
114972 Повилика желтая, Повилика перечная, Повилика южная
114973 Повилика вавилонская
114980 Повилика короткостолбиковая
114982 Повилика бухарская
114984 Повилика красностебельная
114986 Повилика полевая
114994 Повилика китайская
114997 Повилика ландышецветковая
115006 Повилика энгельмана
115007 Повилика креверная, Повилика льновая
115008 Повилика льняная, Повилика тимьяновая, Повилика чебрецовая
115031 Ластовица, Повилика гроновиуса, Повилика европейская, Трава ластовичная, Чистотел большой
115033 Повилика ферганская
115037 Повилика гигантская
115040 Повий, Повилика
115049 Повилика индийская
115050 Повилика окрашенная, Повилика японская
115056 Повилика каратавская
115061 Повилика кочи
115063 Повилика лемана
115065 Повилика гребневидная
115066 Повилика хмелевидная
115080 Повилика одностолбиковая
115096 Повилика памирская
115098 Повилика цветоножковая
115138 Повилика штапфа
115140 Повилика узкочашечная
115141 Повилика гронова
115143 Повилика тяньшанская
115146 Повилика душистая
115155 Повиликовые
115252 Кутандия
115259 Кутандия мемфийская
115278 Кювьера
115338 Цианантус
115793 Саговниковые, Цикадовые
115794 Деревосоговое, Саговник, Саговника шишка женская, Цикада, Цикас
115812 Саговник улитковидный
115830 Саговник гуйцзоуский
115833 Цикас хайнаньский
115884 Дерево соговое, Пальма пикадовая, Саговник, Саговник поникающий, Цикас, Цикас революта
115897 Цикас таиландский
115909 Цикас сычуаньский
115912 Цикас тайваньский
115931 Дряква, Фиалка альпийская, Цикламен
115932 Дряква абхазская, Дряква лджарская
115940 Дряква черкесская
115942 Дряква весенняя, Дряква косская
115948 Дряква изящная
115949 Дряква персидская, Цикламен альпийский, Цикламен европейский
115964 Дряква мелковетковая
115965 Дряквенник, Цикламен персидский
115970 Дряква понтийская
115981 Цикламен весенний, Цикламен весенняя
115989 Циклантовые
115990 Цикрантера
116100 Дуб сизый
116151 Дуб мотоский
116153 Дуб мирсинолистный
116216 Дуб тибетский
116240 Лапина палиурус
116516 Цикнохес
116532 Айва, дерево квитовое, Цидония
116546 Айва дикая, Айва низкая маулея, Айва обыкновенная, Айва продолговатая, Дерево айвовое
116547 Никандра пузыревидная, Никандра физалиевидная, Никандра физалисовидная
116716 Волноплодник
116717 Волноплодник гроссгейма
116718 Волноплодник разнолистный
116719 Волноплодник мохнатый
116732 Цимбалярия
116736 Цимбалярия постенная, Цимбалярия стенная
116745 Цимбария, Цимбохасма
116746 Цимбария днепровская
116747 Цимбария даурская
116769 Цимбидиум
116853 Цимбидиум эмэйский
116967 Цимбидиум лоу
117090 Цимбидиум тибетский
117100 Цимбидиум вэньшаньский
117117 Цимбохасма
117119 Цимбохасма днепровская
117121 цимболена
117122 цимболена длиннолистная
117153 Лемонграс, Сорго лимонное, Трава лимонная
117204 Мальмороза
117218 Нард, Сорго лимонное, Цитронелла
117334 Ластовенник, Ластовень, Латовник, Сиуноктонум, Цинанхум
117339 Ластовень заостренный, Цинанхум

остролистный
117345 Ластовень острый, Цинанхум острый
117346 Цинанхум сибирский
117357 Ластовень алебова
117364 Ластовень стеблеобъемлющий, Цинанхум стеблеобъемлющий
117385 Лавобень черноватый, Ластовень темный, Латовень черноватый
117424 Цинанхум чжэцзянский
117425 Цинанхум китайский
117469 Цинанхум тайваньский
117473 Цинанхум далиский
117482 Цинанхум эмэйский
117515 Цинанхум тибетский
117523 Ластовень неприятный
117537 Ластовень японский
117546 Цинанхум цзиндунский
117551 Цинанхум гуансиский
117555 Цинанхум ланьюйский
117564 Цинанхум лицзянский
117566 Цинанхум кандинский
117597 Ластовень максимовича, Цинанхум максимовича
117614 Цинанхум мулиский
117643 Пикностельма метельчатая
117660 Ластовень пурпуровый, Суноктонум пурпуровый, Циноктонум пурпуровый
117670 Ластовень русский
117675 Цинанхум тибетский
117679 Ластовень шмальгаузена
117713 Цинанхум сычуаньский
117716 Цинанхум таиландский
117721 Ластовень ленцовидный
117738 Цинанхум фунинский
117743 Ластовенник обыкновенный, Ластовень аптечный, Ластовень лекарственный, Ластовень обыкновенный, Цинанхум лекарственный
117745 Ластовень вьюшийся
117747 Цинанхум куэньминский
117748 Ластовень ван
117751 Ластовень вильфорда, Сейтера уилфорда
117765 Артишок
117770 Артишок испанский, Кардон
117787 Артишок, Артишок колючий, Артишок посевной
117852 Аджерик аджирык, Жиловник, Злак пальчатый, Зуб собачий, Лапка, Пальчатка, Свинорой, Свинорой пальчатый, Трава пальчатая
117859 Аджерик, Аджерик аджирык, Аджирык, Жиловник, Злак пальчатый, Зуб собачий, Лапка, Пальчатка, Свинорой, Свинорой пальчатый, Трава бермудская, Трава пальчатая, Трава собачья
117900 Бобы гиацинтовые, Мелколепестник канадский, Омфалодес, Пупочник, Циноглоссум, Чернокорень, Язык песий
117908 Циноглоссум амабиле, Чернокорень приятный
117948 Чернокорень пестрый
117956 Чернокорень растопыренный
117967 Циноглоссум ганьсуский
117968 Чернокорень немецкий
117978 Чернокорень шелковисто-войлочный
118004 Чернокорень горный
118008 Корень чёрный, Чернокорень аптечный, Чернокорень лекарственный
118010 Чернокорень распиканный
118017 Чернокорень тибетский
118018 Чернокорень зеравшанский
118022 Чернокорень тяньшанский
118033 Чернокорень зеленоцветковый
118124 Циноморневые
118127 Циноморий
118133 Циноморий джунгарский
118307 Гребенник, Гребневик, Гребник
118315 Гребневик гребенчатый, Гребневик обыкновенный
118319 Гребневик шиповадый
118325 Гребневик изящный
118393 Ципелла
118403 Осоковые
118428 Аирник, Сыть, Циперус
118473 Ситничек поздний
118476 Сыть разнолистная, Циперус очереднолистный
118491 Сыть амурская
118737 Ситовник щетиновидный
118744 Сыть разновидная, Сыть разнородная
118759 Циперу спинбяньский
118765 Аирник двухколосковый
118822 Миндаль земляной, Сыть золотистая, Сыть съедобная, Циперус съедобный, Чуфа
118844 Сыть хайнаньская
118928 Сыть чернобурая
118941 Сыть гладкая
118957 Сыть густоцветная, Сыть скученная
118982 Сыть гаспая
119041 Сыть ирия
119073 гладкий
119125 Сыть длинная
119127 Сыть темнокаштановая
119196 Дихостилис микели, Сыть микели
119262 Сыть ниппонская
119319 Сыть прямоколосая, Сыть усеченная
119347 Папирус
119503 Асселямалейкум, Донгузтопылак, Кага, Куга, Саламалейкум, Саламалик, Сыть клубненосная, Сыть кртглая, Сыть круглая, Сыть кругловатая
119578 Циперус шаньдунский
119593 Сыть джунгарская
119657 Циперус сычуаньский
119687 Сыть желтоватая
119973 Цифомандра
119974 Дерево томатное, Цифомандра, Цифомандра свекловидная
120287 Башмачек, Башмачок, Башмачок венерин, Винерин башмачек, Орхидные, Сапожки, Циприпедиум
120303 Винерин чжундяньский
120310 Башмачек настоящий, Башмачок венерин настоящий, Винерин башмачек настоящий, Циприпедиум башмачок
120336 Циприпедиум далиский
120353 Циприпедиум тайваньский
120355 Циприпедиум франше
120357 Башмачек пятнистый, Винерин башмачек пятнистый, Циприпедиум пятнистый
120371 Японск японский
120383 Циприпедиум лицзянский
120396 Башмачек крупноцветковый, Башмачек крупноцветный, Винерин, Циприпедиум крупноцветный
120411 Башмачек жемчужный
120414 Циприпедиум мелкоцветковый
120461 Циприпедиум тибетский
120464 Башмачек крупноцветный вздутый, Винерин башмачек вздутый
120473 Башмачек ятабе
120474 Циприпедиум юньнаньский
120532 Циртантус
120632 Просо верхоцветное
120779 Циртостакис, Циртостахис
120875 Подладанник
120882 Подладанник красный
120903 Жарновец, Острокильница, Ракитник, Цитисус
120909 Ракитник белый
120928 Ракитник австрийский
120942 Ракитник блоцкого
120943 Ракитник днепровский
120962 Ракитник волосистый
120970 Ракитник линдеманна
120972 Ракитник литвинова
120979 Ракитник многоцветковый
120983 Острокильница чернеюшая, Ракитник чернеющий
120986 Ракитник пачоского
120989 Ракитник подольский
120990 Ракитник многоволосистый,

Ракитник многоволосковый
120997 Ракитник регенсбургский
120998 Ракитник рошеля
121000 Ракитник русский
121001 Жарновец метельчатый, Ракитник метельчатый, Ракитник меттлистный, Цитисус прутьевидный
121007 Ракитник скробишевского
121012 Ракитник скробишегского
121030 Ракитник вульва, Ракитник вульфа
121031 Ракитник цингера
121045 Дудник черняева
121084 Дакридиум
121226 Ежа
121233 Ежа обыкновенная, Ежа сборная, Ежа скученная
121273 Ежа воронова
121281 Вороньянога
121359 Ятрышник остистый
121382 Ятрышник фукса
121389 Ятрышник кровавый
121422 Ятрышник теневой
121493 Каламус драконовый
121539 Георгина, Георгиния, Далия
121561 Георгина, Георгина изменчивая, Георгина перистая, Далия
121607 Дальбергия
121645 Дальбергия бирманская
121702 Дальбергия хайнаньская
121705 Дальбергия юньнаньская
121714 Дальбергия хубейская
121730 Дальбергия индийская, Дальбергия широколистная, Дерево розовое индийское
121750 Дальбергия сенегальская, Дальбергия чернодревесная
121760 Дальбергия мимозовидная
121770 Дерево палисандровое, Палисандр
121784 Дальбергия бирманская
121826 Дальбергия сиссоо, Дальбергия сиссу, Дерево палисандровое сиссоо
121859 Дальбергия юньнаньская
121860 Дальбергия куэньминская
121912 Далешампия
121993 Звездоплодник
121995 Звездоплодник частуховидный
122001 Звездоплодник перетянутый
122027 Дамнакант
122034 Дамнакант
122036 Дамнакант гуансиский
122037 Дамнакант хайнаньский
122040 Дамнакант индийский
122077 Дамнакант сычуаньский
122121 Вздутоплодник, Даная ветвистая, Пузыресеменник, Пузыресемянник
122124 Вздутоплодник, Даная ветвистая, Пузыресеменник, Пузыресемянник
122180 Дантония, Зиглингия
122200 Дантония чашечная
122210 Зиглингия лежачая, Зиглингия трехзубка
122227 Дантония форскаля
122242 Дантония промежуточная
122359 Волчеягодник, Волчник, Дафна, Лаврушка
122361 Дафна остролопастная
122363 Волчник альбова
122365 Волчник алтайский
122382 Волчник пазущноцветковый
122389 Волчник благая
122407 Волчеягодник кавказский, Волчник кавказский
122414 Волчник черкесский
122416 Волчеягодник-боровик, Волчник боровик, Волчник-боровик
122424 Дафна эмэйская
122432 Дафна далиская
122438 Дафна генква
122447 Волчник скученый
122472 Дафна цзиньюньская
122475 Волчея годникюлии, Волчни кюлии, Волчник юлия
122477 Волчник камчатский
122486 Дафна корейская
122492 Волчея годник лавролистный, Волчеягодник лавролистный, Волчник лавловый, Волчник лавролистный
122515 Волкобой, Волчеягодник волчьелыко, Волчеягодник обыкновенный, Волчник обыкновенный, Волчник смертельный, Волчьелыко обыкновенное, Корень волчий, Лыко волчье, Пережуй-лычко, Сирень лесная, Ягодки
122519 Дафна мийябе
122526 Волчник узколистный
122532 Дафна душистая
122534 Дафна душистая ‘маргината’, Дафна душистая отороченная
122585 Волчник понтийский, Дафна понтийская
122604 Волчея годниксофьи, Волчник софии
122614 Дафна ганьсуская
122622 Волчник закавказский
122634 Дафна сичоуская
122635 Дафна юньнаньская
122661 Дафнифиллум, Дафнифиллюм
122682 Дафнифиллум ланьюйский
122700 Дафнифиллум крупноножковый, Дафнифиллюм крупноножковый
122730 Дафнифиллюм юньнаньский
122758 Дарлингтония
122898 Дазилирион
122899 Дазилирион волосистоконечный
122912 Дазилирион вилера
122958 Гайнальдия
123027 Датиска
123028 Датиска коноплевая, Датиска коноплёвая, Пенька критская
123033 Датисковые
123036 Датура, Дурман
123061 Дурман безвредный, Дурман индийский
123065 Дурман индийский, Дурман метель
123077 Бангдевона, Дурман вонючий, Дурман обыкновенный, Дурман татуля, Зелье бешеное
123135 Морковь
123141 Баркан, Каротель, Морковь, Морковь двулетняя, Морковь дикая, Морковь культурная, Морковь культурная каротиновая, Морковь обыкновенная, Морковь-каротель
123164 Морковь огородная, Морковь посевная
123244 Давидия
123245 Давидия обернутая
123247 Давидия обернутая вильморена
123316 Дебрежазия
123322 Дебрежазия съедобная
123367 Декенея
123371 Декенея фаржа
123661 Дейнболлия
124008 Делет, Дельфиниум, Живокость, Разрыв-трава, Топорики, Шпорник
124012 Живокость редкоцветная
124023 Живокость белоокаймленная
124026 Живокость алтайская
124046 Живокость дуговидная
124053 Живокость бородатая
124056 Живокость баталина
124067 Живокость дваждытройчатая
124073 Живокость короткошпорцевая
124074 Живокость прицветничная
124076 Живокость брунона
124079 Живокость бухарская
124081 Живокость буша
124097 Живокость согнутоплодная
124110 Живокость кавказская
124127 Живокость губоцветковая, Живокость губоцветная
124140 Живокость спутанная
124141 Василёк рогатый, Васильки рогатые, Грабельки, Живокость полевая, Живокость посевная, Носики, Рогульки, Сокирки, Топорики, Червички
124147 Живокость ветвистая
124151 Живокость толстолистная
124153 Живокость курчавенькая
124154 Живокость холоднолюбивая

124156 Живокость клиновидная, Живокость русская
124158 Живокость синецветковая
124161 Живокость пушистоцветковая
124162 Живокость опушеноплодная
124181 Живокость сетчатоплодная
124183 Живокость растопыренная
124194 Живокость высокая, Царьзелье
124217 Живокость извилистая
124218 Живокость вонючая
124222 Живокость фрейна
124237 Дельфиниум крупноцветный, Живокость крупноцветковая, Живокость крупноцветная
124293 Живокость гогенакера
124294 Дельфиниум хэнаньский
124297 Живокость хуанчжунская
124301 Дельфиниум, Живокость гиблидная, Шпорник многолетний
124302 Живокость илийская
124304 Живокость незаметная
124305 Живокость неожиданная
124312 Живокость ганьсуская
124315 Живокость каратавская
124322 Живокость голоплодная
124324 Живокость кнорринг
124325 Живокость коржинского
124330 Живокость лакоста
124339 Живокость рыхловатая
124340 Живокость голоплодная
124344 Живокость тонкоплодная
124348 Живокость лицзянская
124352 Живокость липского
124355 Живокость длинноцветоножковая
124360 Живокость маака
124374 Живокость марии
124376 Живокость максимовича
124378 Живокость крупноцветковая
124386 Живокость чудесная
124394 Живокость мулиская
124424 Живокость охотская
124426 Живокость бледножелтая
124428 Живокость эмэйская
124432 Живокость горолюбивая
124439 Живокость осетинская
124454 Живокость палласа
124456 Живокость метьльчатая
124457 Живокость странная
124500 Живокость персидская
124502 Живокость полторацкого
124511 Живокость бливкая
124537 Живокость пунцовая
124548 Живокость пирамидальная
124549 Дельфиниум цинхайский
124550 Живокость дубняков
124555 Живокость отогнутоволосистая
124560 Живокость рупрехта
124563 Живокость саурская
124571 Живокость шмальгаузена
124572 Живокость полубородатая
124573 Живокость полубулавовидная
124589 Живокость джунгарская
124595 Живокость красивая
124605 Живокость стокса
124615 Живокость шовица
124616 Дельфиниум цзилунский
124618 Дельфиниум далиский
124630 Дельфиниум кандинский
124637 Живокость тройчатая
124643 Дельфиниум тяньшанский
124653 Живокость печальная
124659 Живокость туркменская
124669 Живокость уральская
124676 Живокость косматая
124708 Дельфиниум чжундяньский
124711 Дельфиниум юньнаньский
124767 Дендрантема
124778 Дендрантема краснеющая, Дендрантема сихоте-алинская
124785 Дендрантема шелковицелистная
124788 Дендрантема хультена
124790 Дендрантема индийская, Хризантема индийская
124804 Дендрантема курильская
124822 Дендрантема приморская
124824 Дендрантема максимовича
124825 Дендрантема монгольская, Хризантемум монгольский
124826 Дендрантема шелковицелистная
124832 Дендрантема нактонгенская, Хризантемум нактонгенский
124856 Дендрантема выемчатолистная
124862 Дендрантема вейриха, Дендрантема завадского, Хризантемум завадского
124891 Дерен головчатый, Кизил головчатый, Клубничноедерево, Циноксилон головчатый
124923 Деренкоуса японский
124925 Деренкоуса, Деренкоуса китайский
124957 Денгамия, Дендробиум
125048 Дендробиум золотоцветный
125058 Дендробиум золотистый
125173 Дендробиум хайнаньский
125179 Дендробиум хэнаньский
125233 Дендробиум лоддигеза
125264 Дендробиум мускустый
125279 Дендробиум благородный, Дендробиум тайваньский
125304 Дендробиум фаленопсис
125360 Дендробиум мэнхайский
125387 Дендробиум букетоцветный
125412 Дендробиум тэнчунский
125419 Дендробиум сичоуский
125461 Дендрокаламус
125526 Дендрохилюм
125531 Дендрохилюм пленчатый
125589 Дендропанакс
125596 Дендропанакс бирманский
125611 Дендропанакс хайнаньский
125623 Дендропанакс гуансиский
125627 Дендропанакс баотинский
125644 Дендропанакс трехнадрезный
125647 Дендропанакс юньнаньский
125763 Дендростеллера
125767 Дендростеллера песчаная
125769 Дендростеллера линейнолистная
125770 Дендростеллера крупноколосая
125771 Дендростеллера ольги
125772 Дендростеллера колоссобразная
125773 Дендростеллера туркменов
125802 Демьянка
125831 Зубянка
125834 Зубянка двояко-перистая
125836 Зубянка клубненосная, Зубянка луковичная
125854 Зубянка мелколистная
125856 Зубянка мелкоцветная
125858 Зубянка пятилисточковая
125861 Зубянка сибирская
125863 Зубянка тонколистная
125937 Деррис
125958 Деррис эллиптический
125970 Деррис далиский
125974 Деррис индийский
126012 Деррис юньнаньский
126025 Луговик, Щучка
126026 Луговик альпийский
126029 Луговик арктический
126034 Луговик беринговский
126035 Луговик северный
126039 Луговик ботнический, Луговик дернистый, Щучка дернистая
126047 Луговик рамирский
126071 Луговик двуцветный
126074 Луговик извилистый, Щучка извилистая
126086 Луговик тонконоговидный
126093 Луговик средний
126097 Луговик обский
126112 Дескурайния, Дескурения
126118 Дескурения коха
126127 Гулявник софьи, Гулявник струйчатый, Дескурения софии, Дескурения софия, Сарсальма
126131 Дескурения гулявниковая, Дескурения софиевидная
126151 Христолея вееровидная
126191 Десмацеря
126194 Десмацеря сжатая
126242 Десмодий, Десмодиум, Трилистник клещевой

126278 Десмодий канадский
126329 Десмодий колосковый, Десмодий липолистный
126338 Десмодиум яньюаньский
126453 Десмодиум маньчжурский
126465 Десмодиум мелколистный, Копеечник мелколистный
126675 Десмодиум юньнаньский
126832 Дейция
126839 Дейция ганьсуская
126841 Дейция амурская
126844 Дейция чжэньканская
126875 Дейция корейская
126891 Дейция городчатая
126933 Дейция гладкая
126940 Дейция скученноцветковая
126943 Дейция лицзянская
126946 Дейция изящная, Дейция красивая
126956 Дейция крупноцветковая
126971 Дейция тибетская
126976 Дейция сизоватая
126987 Дейция лэйбоская
126989 Дейция лемана
127005 Дейция великолепная
127012 Дейция максимовича
127018 Дейция мулиская
127022 Дейция наньчуаньская
127032 Дейция мелкоцветковая
127044 Дейция эмэйская
127072 Дейция шероховатая, Дейция шершавая
127088 Дейция шнейдера
127093 Дейция сычуаньская
127099 Дейция зибольда
127114 Дейция тайваньская
127124 Дейция вильсона
127126 Дейция юньнаньская
127128 Дейция чжундяньская
127197 Вейник тростниковидный
127217 Вейник торотковолосистый
127257 Вейник лапландский
127264 Вейник тонкостебельный
127276 Вейник незамечаемый
127314 Вейник тяньшанский
127375 Диалиум
127396 Диалиум гвинейский
127475 Мискантус непалский
127504 Дианелла
127505 Кордилина голубая
127573 Гвоздика
127577 Гвоздика акантолимоновидная
127578 Гвоздика иглолистная
127581 Гвоздика альпийская
127588 Гвоздика амурская
127591 Гвоздика андронаки
127592 Гвоздика анджеевского
127598 Гвоздика песчаная
127600 Гвоздика армериевидная
127607 Гвоздика бородатая, Гвоздика бородчатая, Гвоздика турецкая, Дерево мескитовое, Диантус барбатус
127612 Гвоздика двуцветная
127616 Гвоздика борбаша
127623 Гвоздика сизая
127629 Гвоздика красивоголовая
127630 Гвоздика равнинная
127631 Гвоздика седая
127632 Гвоздика головчатая
127633 Гвоздика угольная
127634 Гвоздика картузианская
127635 Гвоздика голландская, Гвоздика садовая
127654 Гвоздика китайская, Гвоздика фишера, Диантус китайский, Лоаза
127675 Гвоздика шаньдунская
127682 Гвоздика разноцветная, Гвоздика степная
127693 Гвоздика косматая
127697 Гвоздика бахромчатолепестная
127699 Гвоздика куринская
127700 Гвоздика травяная, Гвоздика травянка, Гвоздика трехгранная, Травянка
127705 Гвоздика пестрая
127707 Гвоздика высокая
127708 Гвоздика евгении
127712 Гвоздика многоцветковая
127713 Гвоздика душистая
127730 Гвоздика гроссгейма
127731 Гвоздика пятнистая
127733 Гвоздика гельцера
127735 Гвоздика низкая
127737 Гвоздика имеретинская
127739 Гвоздика японская
127743 Гвоздика каратавская
127744 Гвоздика киргизская
127748 Гвоздика кубанская
127749 Гвоздика кушакевича
127750 Гвоздика кузнецова
127752 Гвоздика ланцетная
127755 Гвоздика узколепестная
127763 Гвоздика длинноноготковая
127764 Гвоздика маршала биберштейна
127768 Гвоздика перепончатая
127778 Гвоздика многостебельная
127787 Гвоздика восточная
127789 Гвоздика бледноцветная
127792 Гвоздика скальная
127793 Гвоздика перистая, Гвоздика песчаная
127794 Гвоздика изменчивая
127795 Гвоздика луговая
127796 Гвоздика преображенского
127800 Гвоздика ложноармериевидная
127807 Гвоздика радде
127808 Гвоздика ветвистая
127809 Гвоздика ползучая
127811 Гвоздика жесткая
127812 Гвоздика роговича
127814 Гвоздика рупрехта
127820 Гвоздика семенова
127823 Гвоздика зеаавшанская
127839 Гвоздика джунгарская
127841 Гвоздика узкочашечная
127849 Гвоздика шилоносная
127852 Гвоздика пышная
127872 Гвоздика тайваньская
127880 Гвоздика сычуаньская
127884 Гвоздика талышская
127885 Гвоздика четырехчемуйная
127888 Гвоздика тяньшанская
127889 Гвоздика закавказкаская
127892 Гвоздика туркестанская
127893 Гвоздика уральская
127910 Диапензия, Диапенсия
127916 Диапензия лапландская, Диапенсия лапландская
127919 Диапензия обратнояйцевидная
127925 Диапенсия тибетская
127926 Диапензиевые, Диапенсиевые
128011 Диаррена
128015 Диаррена фори
128017 Диаррена японская
128019 Диаррена маньчжурская
128023 Двучленник, Дендростеллера, Диартрон
128026 Двучленник льнолистный, Диартрон льнолистный
128028 Двучленник пузырчатый
128029 Диасция
128032 Диасция
128288 Бикукулла, Диелитра, Дицентра, Диэритра, Сердечки
128291 Дицентра канадская
128310 Дицентра бродяжная
128589 Дихапеталиум, Дихапеталум
129010 Дихостилис
129018 Дихостилис крючковатый
129019 Дихостилис микели
129023 Дихостилис карликовый
129025 Дихотомант
129026 Дихотомант тристаниеплодный
129030 Дихроа, Дихроя
129033 Дихроа противолихорадочная, Дихроя противолихорадочная
129039 Дихроа мелкоцветковая
129051 Дихроя юньнаньская
129062 Дихроцефала
129068 Дихроцефала двуцветная
129489 Дикориния, Дикориния параензс
129568 Гляуциум бахромчатый

129615 Бадьян, Горюн-трава, Диктамнус, Купина неопалимая, Ясенец
129618 Бадьян-ясенец, Гриб чесночный, Диктамнус, Диктамнус белый, Диктамнус ясенелистный, Чесночник, Ясенец белый, Ясенец ясенелистный
129627 Ясенец узколистный
129628 Ясенец кавказский
129629 Ясенец мохнатоплодный
129631 Ясенец голостолбиковый
129633 Ясенец таджикский
129718 Корифа гвоздиная
129732 Диделотия африканская
130029 Двойчатка
130030 Двойчатка оше
130031 Двойчатка федченко
130083 Диффнбахия
130121 Диффнбахия пестрая
130187 Диерама
130242 Диервила, Диервилла
130344 Дигиталис, Наперстянка
130355 Наперстянка бахромчатая, Наперстянка ресничатая
130358 Наперстянка ржавая
130365 Наперстянка крупноцветковая, Наперстянка сомнительная
130369 Наперстянка шерстистая
130371 Наперстянка желтая, Наперстянка жёлтая
130374 Дигиталис мелкоцветковая
130375 Дигиталис жилковатая
130383 Дигиталис, Дигиталис пурпурный, Наперстянка, Наперстянка красная, Наперстянка лекарственная, Наперстянка пурпуровая
130389 Дигиталис шишкина
130404 Росичка
130481 Росичка китайская
130489 Росичка ресничатая
130598 Росичка горизонтальная
130612 Росичка кровеостанавливающая, Росичка линейная
130635 Росичка длинноцветковая
130745 Пальчатка, Просо крованое, Росичка крованая, Росичка кровяная, Росичка кровяно-красная
130825 Росичка фиолетовая
130913 Диления, Дилления
130956 Двукильник
130965 Двукильник солончаковый
131006 Димерия
131009 Димерия тайваньская
131014 Димерия гуансиская
131018 Димерия птиценогая
131137 Четочник железистый
131141 Донтостемон гребенчатый
131147 Диморфотека, Ноготки африканские
131397 Дионея, Диония
131418 Дионизия, Дионисия
131420 Дионисия гиссарская
131421 Дионисия обертковая
131422 Дионисия косинского
131425 Дионисия подушковидная
131451 Диоскорея, Тестудинария, Ямс
131458 Диоскорея крыратая, Ямс крыратый
131501 Диоскорея клубненосная, Ямс клубненосный
131515 Диоскорея кавкаская
131538 Диоскорея ланьюйская
131556 Диоскорея разноцветная
131577 Диоскорея съедобная, Ямс посевной, Ямс съедобный
131593 Ямс тайваньский
131597 Ямс фуцзяньский
131601 Картофель китайский, Ямс китайский
131645 Ямс японский
131710 Диоскорея мексиканская
131734 Диоскорея ниппонская
131772 Диоскорея батат, Диоскорея многокистевая, Диоскорея поликолосная, Ямс китайский
131846 Диоскорея мелкоцветковая
131880 Ямс куш-куш, Ямс трёхнадрезный
131909 Диоскорея инцзянская
131913 Диоскорея тибетская
131915 Диоскорея юньнаньская
131918 Диоскорейные
131938 Диосма
132030 Диоспирос, Персимон, Хурма
132061 Хурма далиская
132077 Хурма бирманская
132137 Дерево збеновое, Дерево черное
132152 Хурма фэн
132163 Хурма тэнчунская
132175 Хурма сизолистная
132186 Хурма хайнаньская
132219 Каки, Персймон, Слива финиковая, Хурма восточная, Хурма каки, Хурма субтропическая, Хурма японская
132238 Хурма цзиндунская
132241 Хурма ланьюйская
132264 Лотус, Слива турецкая, Слива финиковая, Финик дикий, Хурма кавказская, Хурма обыкновеная
132291 Хурма люцюская
132298 Хурма индийская
132359 Хурма баотинская
132371 Хурма ромболистная
132397 Хурма сичоуская
132466 Хурма виргинская
132478 Хурма юньнаньская
132677 Двулистник, Дифиллейя
132682 Дифиллейя японская
132685 Дифиллейя грея, Дифиллейя китайская
132720 Змеевка
132732 Змеевка болгарская
133226 Двурядка
133256 Двурядка меловая
133286 Двурядка стенная
133313 Двурядка тонколистная
133324 Двурядка прутяная
133394 Двурядка тонколистная
133444 Ворсянковые, Мориновые
133451 Ворсянка
133457 Ворсянка эмэйская
133461 Ворсянка голубая, Ворсянка лазолевая
133463 Ворсянка китайская
133466 Ворсянка далиская
133478 Ворсянка посевная, Ворсянка сукновалов, Ворсянка сукновальная, Часалка, Шишка ворсовальная
133486 Ворсянка гмелина
133490 Ворсянка японская
133492 Ворсянка кандинская
133494 Ворсянка раздельная, Ворсянка раздельнолистная, Ворсянка разрезная
133495 Ворсянка лицзянская
133497 Ворсянка лушаньская
133502 Ворсянка волосистая
133505 Ворсянка возделываемая, Ворсянка посевная, Кордовник
133508 Ворсянка лесная
133510 Ворсянка симаоская
133513 Ворсянка бесцветная, Ворсянка щетинистая
133517 Ворсянка лесная
133519 Ворсянка синьцзянская
133551 Диптерокарповые
133552 Двукрылоплодник, Двухкрылоплодник, Диптерокарпус
133587 Диптерокарпус шишковатый
133588 Диптерокарпус бальзамический
133626 Диптерикс
133629 Кумаруна
133640 Двоякоплодник
133641 Воякоплодник прижатый
133654 Дирка
133656 Дирка болотная
134027 Дисхидия
134032 Дисхидия китайская
134037 Дисхидия тайваньская
134181 Диоскостигма
134417 Диспорум
134441 Диспорум тайваньский
134453 Диспорум наньтоуский
134467 Диспорум сидячий
134486 Диспорум зеленоватый, Диспорум

смилациновый
134504 Диспорум зеленеющий, Диспорум зеленоватый
134918 Двупестфчник, Дистилиум
134922 Дистилиум китайский
134942 Дистилиум пинбяньский
134946 Двупестфчник кистевидный, Дистилиум кистевидный
135113 Дочиния
135114 Дочиния дераве
135120 Дочиния индийская
135132 Додартия, Додарция
135134 Додартия восточная, Додарция восточная, Машок, Теке-сакал
135153 Додекатеон, Дряква европейская, Дряквенник, Дуб
135161 Дряквенник холодный
135192 Додонея
135215 Додонея аптечная, Додонея клейкая
135253 Доллингерия
135264 Доллингерия шершавая
135399 Долихос
135412 Долихос лицзянский
135753 Домбея
135957 Домбея точечная
136156 Донтостемон
136163 Донтостемон зубчатый
136172 Донтостемон шершавый
136173 Донтостемон целенолистный
136179 Донтостемон мелкоцветковый
136184 Донтостемон многолетний
136209 Допатриум
136214 Допатриум ситниковый
136257 Дорема
136258 Дорема эчисона
136259 Дорема аммиачная, Дорема аммониакальная, Дорема камеденосная, Ошак
136261 Дорема голая
136262 Дорема смолоносная
136263 Дорема гирканская
136264 Дорема каратавская
136265 Дорема мелкоплодная
136266 Дорема наманганская
136267 Дорема инееватая
136268 Дорема песчаная
136333 Аароникум, Дороникум
136339 Дороникум алтайский
136344 Дороникум австрийский
136345 Дороникум баргузинский
136346 Дороникум кавказский
136348 Дороникум делеклюза
136355 Дороникум ганьсуский
136360 Дороникум длиннолистный
136361 Дороникум крупнолистный
136362 Дороникум продолговатолистный
136364 Дороникум восточный
136365 Дороникум ядовитый
136373 Дороникум шишкина
136376 Дороникум тибетский
136378 Дороникум тяньшанский
136379 Дороникум туркестанский
136417 Дорстения
136729 Дориантес
136740 Дорикниум
136746 Дорикниум средний, Дорикниум травянистый
136899 Крупка
136903 Крупка алайская
136905 Крупка альберта
136911 Крупка альпийская
136918 Крупка алтайская
136938 Крупка араратская
136940 Крупка арсеньева
136943 Крупка оше
136945 Крупка байкальская
136948 Крупка бародатая
136949 Крупка беринга
136951 Крупка северная
136956 Крупка бруниелистная
136963 Крупка сердечникоцветковая
136966 Крупка шамиссо
136969 Крупка серая
136973 Крупка вытянутолбиковая, Крупка остролистная
136976 Крупка дарвазская
136980 Крупка разнолистная
136984 Крупка войлочная
136988 Крупка елизаветы
136990 Крупка пушистая
136993 Крупка эшшольтца
136994 Крупка федченко
136995 Крупка фладницийская
136999 Крупка ледниковая
137003 Крупка круглоплодная
137019 Крупка волосистая, Крупка молочнобелая, Крупка мохнатая
137032 Крупка шетинистая
137033 Крупка гиссарская
137034 Крупка хюта
137038 Крупка имеретинская
137039 Крупка седая
137046 Крупка неубранная
137048 Крупка искривленная
137052 Крупка японская
137054 Крупка камчатская
137058 Крупка чьельмана
137059 Крупка коржинского
137061 Крупка курильская
137062 Крупка кузнецова
137063 Крупка молочнобелая
137069 Крупка ланцетная, Крупка ланцетоплодная
137083 Крупка волосистолистная
137088 Крупка лицзянская
137093 Крупка липского
137095 Крупка длинностручковая
137100 Крупка крупноплодная
137102 Крупка крупная
137107 Крупка высотная, Крупка черноногая
137108 Крупка черноногая
137110 Крупка мелкоплодная
137111 Крупка мелколепестная
137114 Крупка мягкоопушенная
137115 Крупка монгольская
137129 Крупка стенная
137133 Крупка дубравная, Крупка перелесковая
137146 Крупка снежная
137147 Крупка норвежская
137148 Крупка продолговатая, Крупка продолговатоплодная
137150 Крупка желто-белая
137152 Крупка ольги
137155 Крупка горная
137176 Крупка осетинская
137179 Крупка памирская
137180 Крупка мелкоцветковая
137181 Крупка мелкостручковая
137182 Крупка вздутоплодная
137185 Крупка волосистая
137189 Крупка поле
137191 Крупка многовласая
137194 Крупка первоцветовидная
137195 Крупка прозоровского
137196 Крупка ложноволосистя
137197 Крупка маленькая
137200 Крупка цинхайская
137224 Крупка сахалинская
137227 Крупка шершавая
137230 Крупка чжундяньская
137237 Крупка сибирская
137241 Крупка стручковая
137242 Крупка узкоплодная
137246 Крупка узколепестная
137248 Крупка столбиковая
137253 Крупка почти-стеблеобъемлющая
137255 Крупка почти-головчатая
137256 Крупка почти гладная
137257 Крупка однобокая
137259 Крупка таласская
137266 Крупка тибетская
137282 Крупка турчанинова
137286 Крупка уссурийская
137287 Крупка мощная
137298 Крупка юньнаньская
137321 Крупичка
137327 Крупичка весеная, Резушка весенняя
137330 Драцена

137337 Драцена узколистная
137367 Драцена деремская
137382 Дерево драконовое, Драконник, Драконово дерево, Драцена драконовая
137397 Драцена душистая
137409 Драцена годсефа
137425 Драцена грукера
137451 Драцена мэнлаская
137487 Драцена сандера
137545 Дракоцефалюм, Змеевик австрийский, Змееголовник
137551 Змеевик молдавский, Змеевик тимьянолистный, Змееголовник аргунский
137555 Змеевик аргунский, Змееголовник австрийский
137557 Змееголовник дважды-перистый
137562 Змееголовник кистевой
137565 Змееголовник бунге
137575 Змееголовник двуцветный
137577 Змееголовник разнообразнолистный
137581 Змееголовник вонючий
137583 Змееголовник красивый
137586 Змееголовник ломкий
137587 Змееголовник кусртарничковый
137591 Змееголовник крупноцветковый
137596 Змееголовник разнолистный
137599 Змееголовник безбородый
137601 Змееголовник цельнолистный
137606 Змееголовник комарова
137607 Змееголовник крылова
137611 Дракоцефалюм мелкоцветковое
137613 Змеевик руйша, Змееголовник молдавский, Маточник
137614 Змееголовник многостебельный
137615 Змееголовник многоцветный
137616 Змееголовник узловатый
137617 Змееголовник пониаюший, Змееголовник поникший
137619 Змееголовник продолговатолистный
137620 Змееголовник душицевидный
137622 Змееголовник дланевидный
137625 Змееголовник паульсена
137627 Змееголовник чужеземный
137630 Змееголовник перистый
137651 Змсевик молдавский, Змеевик тимьянолистный, Змееголовник руйша
137655 Змееголовник ямчатый
137665 Змееголовник колючий
137668 Змееголовник стеллера
137670 Змееголовник почтиголовчадый
137671 Змееголовник далиский
137677 Змееголовник тимьяноцветковый, Змееголовник тимьяноцветный, Змей-трава
137744 Арум, Дракункул
137747 Арум змейный, Левкой обыкновенный, Левкой садовый, Левкой седой, Маттиола садовая
137799 Хойя тайваньская
137828 Серпоплодник
137829 Серпоплодник северцова
138051 Дримис
138058 Дримис винтера, Корифа магелландская
138261 Росница, Росянка
138267 Росичка английская
138295 Росичка индийская
138299 Росичка промежуточная
138307 Росичка промежточная, Росичка шитовидная
138322 Росичка обратнояйцевидная
138328 Росичка щитовидная
138348 Ласковцы, Росичка круглолистая, Росичка круглолистная
138369 Росянковые
138375 Мухоловка
138408 Дриадоцвет
138411 Дриадоцвет четырехтычиночный, Сиббальдия четырехтычиночная
138445 Дриада, Дриада восьмилепестковая, Какавахом, Куропаточья трава, Трава куропаточья
138447 Дриада кавказская
138448 Дриада шамиссо
138449 Дриада мелкогородчатая
138451 Дриада бальшая
138460 Дриада аянская, Дриада восьмилепестная, Дриада чоноски
138461 Дриада, Дриада восьмилепестная, Дриада острозубчатая, Трава восъмилепестная, Трава кукушечная, Трава куропаточья восьмилепестная
138462 Дриада точечная
138466 Дриада клейкая
138470 Дримария
138547 Дриобаланопс
138548 Дерево камфарное, Дерево камфарное зондское, Дриобаланопс душистый
138739 Дубоизия
138744 Дубоизия
138790 Дюшенея
138796 Земляника индийская, Люшеная индийская
139037 Дюпонция
139038 Дюпонция фишера
139060 Дуранта
139089 Дуриан, Дурьян
139254 Дикия
139544 Дизофилла
139568 Дизофилла ятабе
139593 Дизофилла симаоская
139602 Дизофилла ятабе
139637 Дизоксилон
139678 Козыльник, Марь амброзиевидная, Марь благовонная, Чай иезуитский
139681 Марь остистая
139682 Кудрявец, Марь гроздевая, Марь душистая
139693 Марь пахучая
139763 Хурмовые, Эбеновые
139851 Бешеный огурец
139852 Бешеный огурец, Бешеный огурец выбрасывающий, Бешеный огурец обыкновенный, Деряба
139970 Эхеверия
140066 Эхинацея, Эхинация
140081 Рудбекия пурпурная, Эхинация пурпуровая
140099 Иглица
140105 Иглица головчатая
140111 Эхинокактус
140209 Эхиноцереус
140351 Ежовник, Куриное просо
140356 Ежовник хвостатый, Куриное просо хвостатое
140360 Ежовник крестьянский
140367 Ежовник куриное, Ежовник петушье, Ежовник петушье просо, Куриное просо обыкновенное, Пайза, Просо куриное, Просо мучнистое, Просо настоящее, Просо петушье, Просо посевное, Просо японское, Просяник, Просянка куриная
140382 Ежовник хвостатый
140403 Ежовник спиральный
140423 Ежовник хлебный, Пайза, Просо китайское, Просо японское, Пузыреплодник
140462 Ежовник рисовый, Куриное просо рисовидное, Просо рисовое
140466 Ежовник рисовый, Ежовник скученный, Крупноплодный, Курмак
140526 Эхиноцистис
140527 Эхиноцистис, Эхиноцистис шиповатый
140536 Эхинодорус
140640 колюченосник
140643 колюченосник сибторпа
140645 колюченосник волосолистный
140652 Мордовник, Эхинопс
140654 Мордовник белостебельный
140658 Мордовник бабатагский
140660 Мордовник банатский, Мордовник венгерский
140674 Мордовник короткокисточковый
140681 Мордовник хантавский
140684 Мордовник колхидский
140686 Мордовник сращенный
140687 Мордовник конрата

140692 Мордовник дагестанский
140696 Мордовник пушистоцветковый
140697 Мордовник рассеченный
140698 Мордовник дубянского
140704 Мордовник возвышеный, Мордовник русский
140705 Мордовник федченко
140706 Мордовник многолистный
140712 Мордовник гмелина
140722 Мордовник приземистый
140723 Мордовник ильина
140724 Мордовник цельнолистный
140725 Мордовник карабахский
140726 Мордовник каратавский
140727 Мордовник кнорринг
140732 Мордовник широколистный, Эхинопс широколистный
140734 Мордовник гладкорогий
140735 Мордовник белополосый
140737 Мордовник липского
140744 Мордовник клупнолистный
140745 Мордовник самаркандский
140749 Мордовник мейера
140751 Мордовник многостебельный
140753 Мордовник карликовый
140757 Мордовник косолопастный
140760 Мордовник матоволистный
140761 Мордовник восточный
140762 Мордовник осетинский
140765 Мордовник персидский
140766 Мордовник многоколючконосный
140767 Мордовник многоплосый
140768 Мордовник просмотренный
140772 Мордовник пушисто-чешуйчатый
140773 Мордовник колючий
140776 Крутай, Мордовник обыкновенный, Мордовник русский
140779 Мордовник крутай, Мордовник русский
140781 Мордовник зайсанский
140787 Мордовник севанский
140789 Мордовник круглоголовый, Мордовник шароголовый, Мордовник широколистный
140790 Мордовник колючконосный
140813 Мордовник почти-голый
140815 Мордовник совича
140816 Мордовник таласский
140820 Мордовник тяньшанский
140821 Мордовник турнефора
140822 Мордовник закаспийский
140823 Мордовник закаспийский
140824 Мордовник волосочешуйный
140825 Мордовник чимганский
140829 Мордовник зеренолистный
140830 Эхинопсилон
140842 Эхинопсис
141051 Эхитес
141087 Румянка, Синяк, Эхиум
141097 Синяк приятный
141202 Синяк итальянский
141258 Синяк подорожинковый
141289 Румянка, Синяк красный
141347 Румянка обкновенная, Синяк обыкновенный
141365 Зклипта, Румянка обкновенная, Синяк обкновенная
141366 Зклипта белая
141374 Зклипта белая, Зклипта простертая
141466 Эдгеворция, Эджевортия, Эджеворция
141470 Кислица, Магнолия звездчатая, Эджевортия бумажная
141508 Эдрайлнт
141510 Эдрайлнт оверина
141569 Зараза бодяная зубчатая
141592 Эреция
141618 Эреция лещинолистная
141629 Эреция диксона
141648 Эреция хайнаньская
141683 Эреция овальнолистная
141805 Эйхгорная, Эйхорния, Эйххорния
141808 Водяной гиацинт, Гиацит водный, Гиацит водяной, Эйхгорная толстоножковая, Эйхорния толсточерешковая
141928 Лоховые
141929 Джидда, Лох
141932 Джидда, Лох, Лох узколистный, Пшат
141936 Лох восточный
141941 Лох серебристный
141948 Лох мулиский
141959 Лох мэнхайский
141977 Лох сычуаньский
141991 Лох тайваньский
142021 Лох гуйцзоуский
142022 Лох ичанский
142037 Лох цзиндунский
142053 Лох ланьпинский
142081 Лох мелкоцветковый
142088 Лох многоцветковый
142118 Лох наньчуаньский
142140 Лох остроплодный
142152 Лох колючий, Лох колющий
142203 Лох далиский
142214 Лох зонтичный
142244 Джидда вэньшаньская
142248 Джидда сичоуская
142250 Джидда тибетская
142253 Джидда юньнаньская
142255 Пальма масличная, Пальма масляная
142257 Пальма гвинейская масличная, Пальма масличная, Пальма масличная гвинейская
142269 Элеокарпус
142271 Элеокарпус американский
142308 Элеокарпус зубчатый, Элеокарпус новозеландский
142422 Элеодендрон
142440 Элеодендрон канский
142552 Повойничковые
142554 Повойник, Повойничек
142555 Повойничек мокричный
142556 Повойничек сомнительный
142565 Повойничек болотниковый, Повойничек красновласковый
142568 Повойничек шеститычинковый
142569 Повойничек венгерский
142570 Повойничек согнутосемянный
142573 Повойничек прямосемянный
142574 Повойничек трехтычинковый, Повойничек трёхтычинковый
142578 Повойничек восточный
142579 Повойничек сомнительный
143019 Болотница, Ситняг
143022 Болотница игольчатая
143029 Болотница свенсона
143050 Болотница ложносеребристочешуйная
143051 Болотница серебристочешуйная
143053 Болотница чернопурпурная
143055 Болотница оттянутая
143109 Болотница тучная
143125 Болотница бескильная
143139 Болотница хвощевидная
143145 Болотница финская
143161 Болотница круглоколосковая
143168 Болотница промежуточная
143178 Болотница камчатская
143182 Болотница клинге
143183 Болотница комарова
143185 Болотница коржинского
143190 Болотница тонкостилоподная
143193 Болотница куэньминская
143198 Болотница сосочковая
143201 Болотница уссурийский
143205 Болотница жемчужная
143207 Болотница максимовича
143209 Болотница южная
143230 Болотница многощетинковая
143261 Болотница яйцевидная
143268 Болотница острочешуйная
143271 Болотница болотная, Ситняг болотный
143289 Болотница маленькая
143292 Болотница немногозубчатая
143298 Болотница прозрачная
143300 Болотница японская
143315 Болотница цинхайская

143319 Болотница малоцветковая, Болотница пятицветковая, Ситнят малоцветковый
143334 Болотница сахалинская
143343 Болотница северная
143384 Болотница закавказская
143395 Болотница туркменская
143397 Болотница одночешуйная
143414 Болотница вихуры, Болотница шляпконосная
143424 Болотница юньнаньская
143453 Элефантопус
143512 Елевзина, Елевзине, Элевзина
143522 Дагусса, Елевзина коракан, Коракан, Просо африканское, Раги, Такусса, Токусса
143530 Елевзине индийская
143575 Акантопапакс, Свободноягодник, Элеутерококк
143610 Акантопапакс генри
143625 Акантопапакс кандинский
143634 Акантопапакс сычуаньский
143657 Акантопапакс тернистый, Дикий перец, Перец дикий, Свободноягодник колючий
143665 Акантопапакс сидячецветковый, Акантопапакс спдячецвеный, Стосил
143667 Акантопапакс сидячецветковый мелкоплодный
143677 Акантопапакс зибольда
143682 Акантопапакс колючий
143685 Акантопапакс японский
143694 Акантопапакс трехлистный
143715 Свободносемянник
143716 Свободносемянник свободнолазский
143801 Элизанта
143807 Элизанта липкая
143914 бодяная Зараза, Зараза бодяная, Чима водяная, Элодея
143920 бодяная Зараза, Зараза бодяная канадская, Элодея канадская
143958 Элисгольция, Эльсгольция, Эльшольция
143974 Эльшольция реснитчатая
143993 Эльшольция гребенчатая
144002 Эльсгольция густоцветковая
144067 Эльшольция тайваньская
144090 Эльшольция поздняя
144144 Волоснец, Вострец, Клинэлимус, Колосняк, Элимус
144156 Пырей алайский
144159 Пырей алатавский
144168 Рэгнерия
144191 Колосняк бальджуанский
144216 Колосняк канадский
144228 Пырей собачий, Регнерия собачья, Рута собачья, Рэгнерия собачья
144271 Клинэлимус даурский, Колосняк даурский
144303 Клинэлимус высокий, Колосняк высокий
144312 Волоснец высокий, Волоснец гигантский, Кияк, Колосняк гигантский
144347 Антозахна жакемонта
144413 Пырей узловадый
144414 Клинэлимус поникший
144429 Пырей раннедерновинный
144459 Пырей скифский
144466 Волоснец сибирский, Клинэлимус сибирский, Колоняк сибирский
144474 Клинэлимус синьцзянский
144511 Колосняк чимганский
144517 Волоснец тургайский, Волоснец узкий, Вострец узкий
144522 Колосняк мохнатый
144624 Пырей, Элитригия
144629 Элитригия алайская
144635 Элитригия баталина
144643 Пырей русский, Пырей удлиненный, Пырей удлинённый
144648 Элитригия ферганская
144651 Пырей близкий
144652 Пырей промежуточный, Пырей средний, Элитригия средняя
144654 Пырей ситниковый
144661 Билаек, Житец, Пырей безостый, Пырей обыкновенный, Пырей ползучий, Ржанец, Элитригия ползучая
144665 Пырей длинноватый
144675 Пырей смида
144680 Пырей волосистый, Пырей волосоносный, Элитригия волосоносная
144716 Эмбелия
144847 Эмботриум
144884 Эмилия
144976 Эмилия яванская
145013 Эминиум
145015 Эминиум альберта
145016 Эминиум леманна
145052 Водяниковые, Ворониковые, Ёрникове, Шикшевые
145054 Багновка, Водяника, Водянка, Вороника, Шикша
145057 Шикша двуполая
145058 Шикша арктическая
145062 Шикша гермафродитная
145063 Шикша кардакова
145073 Багновка чёрная, Водяника черная, Вороника черная, Вороника чёрная, Ерник обыкновый, Медвежья ягода, Сыга, Шикша сибирская, Шикша чёрная, Шикща черная
145492 Энемион
145498 Энемион радде
145673 Энкиант
145693 Энкиант китайский
145694 Энкиант погнутый
145710 Энкиант почечный
145724 Энкиант сычуаньский
145756 Девятиостник
145766 Девятиостник северный, Хохлатник северный
145776 Девятиостник персидский
145857 Энтада
145899 Энтада ползучая
145948 Энтандрофрагма
145992 Энтеролобиум
146067 Эпакрис
146107 Эперуа
146122 Колча, Хвойник, Эфедра
146124 Хвойник эчисона
146134 Хвойник американский, Эфедра американская
146144 Хвойник калифорнийский
146148 Хвойник ресничатый
146154 Кузьмичева трава, Трава кузьчёва, Трава эфедра, Хвойник двухколосковый, Эфедра двуколосковая, Эфедра двухколосковая
146155 Хвойник хвощевидный, Хвойник хвощевый
146159 Хвойник федченко
146160 Хвойник ферганский
146183 Хвойник жерарда, Эфедра жерарда
146188 Хвойник сизый
146191 Хвойник разносемянный
146192 Хвойник крепкий, Хвойник персидский, Хвойник пустынный, Хвойник средний, Хвойник средный
146197 Хвойник средный тибетский, Хвойник тибетский
146203 Хвойник лицзянский
146209 Хвойник окаймленный
146212 Хвойник мелкосемянный
146215 Хвойник маленький
146219 Хвойник односемянный
146237 Хвойник высокий
146239 Хвойник кашгарский
146241 Хвойник подушечный
146243 Хвойник регелевский, Хвойник регеля
146255 Хвойник шишконосная
146270 Гнетовые, Хвойниковые, Эфедровые
146305 Седлоцвет
146308 Седлоцвет шмидта
146522 Эпигея
146524 Эпигея гаультериевидная
146565 Эпилазия
146567 Эпилазия верхушечношерстистая
146573 Эпилазия полушерстистая

146577 Эпилазия удивительная
146586 Кипрей, Кипрейник
146587 Кипрей железистостебельный
146595 Кипрей приросший, Кипрей сродный
146598 Кипрей альпийский
146599 Кипрей альпийский, Кипрейник альпийский
146601 Кипрей альсинолистный, Кипрейник мокричнолистный
146605 Кипрей амурский, Кипрей тонкий
146606 Кипрей головчаторыльцевый, Кипрей голоплодный, Кипрей цилиндрорыльцевый
146607 Кипрей головчато-рыльцевый, Кипрей чашучкоцветный
146609 Кипрей анагаллусолистный, Кипрейник очноцветолистный
146611 Кипрей анатолийский
146612 Кипрей угроватый
146623 Кипрей арктический, Кипрейник арктический
146629 Кипрей берингийский
146630 Кипрей двурядный
146635 Кипрей бонгарда
146669 Кипрей холмовый, Кипрейник холмовой
146671 Кипрей нечеткий
146680 Кипрей даурский, Кипрейник даурский
146696 Кипрей равноветвистый
146706 Кипрей холодостойкий
146709 Кипрей железистый
146724 Кипрей волосистый, Кипрей мохнатый, Кипрей шершавоволосистый, Кипрейник волосистый
146734 Кипрей горнеманна, Кипрей хорнеманна, Кипрейник горнемана
146738 Кипрей японский
146755 Кипрей белоцветковый
146760 Кипрей лансетолистный, Кипрейник лансетолистный
146763 Кипрей высокогорный
146775 Кипрей мелкоцветковый, Кипрей скромный
146782 Кипрей горный, Кипрейник горный
146795 Кипрей жилковатый, Кипрейник жилковатый
146803 Кипрей поникающий
146805 Кипрей тёмно-зелёный, Кипрей темный, Кипрейник темнозеленый, Кипрейник тёмно-зелёный
146812 Кипрей болотный, Кипрейник болотный
146834 Кипрей бедноцветковый, Кипрей бедноцветный, Кипрей мелкоцветковый, Кипрей мелкоцветный, Кипрейник мелкоцветковый
146841 Кипрей понтийский
146849 Кипрей красиохохолковый
146859 Кипрей розовый, Кипрейник розовый
146860 Кипрей алмматинский
146873 Кипрей скальный
146879 Кипрей гирляндообразный
146901 Кипрейник тайваньский
146905 Кипрей лами, Кипрейник лями, Кипрейник четырехгранный
146918 Кипрей теплолюбивый
146919 Кипрей тяньшанский
146928 Кипрей тундровый
146929 Кипрейник тундровый
146930 Кипрей уральский, Кипрейник уральский
146932 Кипрей бархатистый, Кипрей мохнатый
146953 Бесцветник, Горянка, Эпимедиум
146973 Эпимедиум колхидский
146992 Эпимедиум тяньцюаньский
146993 Эпимедиум франше
147008 Эпимедиум хунаньский
147013 Эпимедиум корейский
147021 Горянка японская
147023 Горянка крупночашелистиковая
147031 Эпимедиум мелколистный
147034 Эпимедиум перистый
147042 Эпимедиум опушенный
147067 Эпимедиум сычуаньский
147106 Дремлик
147121 Дремлик чемерицелистный
147149 Дремлик широкористный
147183 Дремлик мелколистный
147195 Дремлик болотный
147196 Дремлик сосочковый
147218 Дремлик ройля
147222 Дремлик ржавый
147238 Дремлик тунберга
147258 Дремлик юньнаньский
147283 Филлокактус, Эпифиллум, Эпифиллюм
147289 Эпифиллум гибридный
147291 Эпифиллум остролепестковый
147303 Надбородник, Эпиподиум
147307 Надбородник безлистный, Эпиподиум безлистный
147323 Дремлик безлистный, Надбородник безлистный
147468 Полевица, Полевичка, Тефф, Эрагростис
147475 Полевичка египетская
147593 Полевичка килианская, Полевичка крупноколосковая, Полевичка чилийская
147602 Полевичка тростниковая, Полевичка тростниковидная
147631 Полевичка дэцинская
147656 Полевичка удлиненная
147671 Полевичка ржавая
147673 Полевичка фуцзяньская
147703 Полевичка гуансиская
147708 Полевичка хайнаньская
147746 Полевичка японская
147754 Полевичка косинского
147797 Полевичка большой
147815 Мышей малый, Мышей мятликовидный, Полевичка малая
147844 Полевичка черная
147883 Мышей волосистый, Полевичка волосистая
147978 Полевичка старосельского
147986 Полевичка душистая
147994 Полевичка абиссинская, Полевичка тефф, Тефф абиссинский, Трава абиссинская
147995 Полевичка нежная, Полевичка приятная
148053 Эрантемум
148101 Весенник, Любник, Эрантис
148109 Весенник длинноножковый
148111 Весенник сибирский
148112 Весенник звездчатый
148148 Эрехтитес
148164 Эрехтитес валерианолистный
148265 Пустынноморковник
148266 Пустынноморковник лемана
148374 Пустынномятлик
148375 Пустынномятлик алтайски
148381 Пустынномятлик красивый
148385 Пустынномятлик персидский
148389 Пустынномятлик джунгарский
148398 Мортук
148400 Мортук бонапарта
148413 Мортук мохнадый
148415 Мортук, Мортук восточный
148420 Мортук пшеничный
148441 Эремоспартон
148444 Саярь-кафак, Эремоспартон безлистный
148446 Эремоспартон обвислый
148447 Эремоспартон джунгарский
148448 Эремоспартон туркестанвый
148466 Еремостахис, Пустынноколосник, Эремостахис
148468 Пустынноколосник родственный
148469 Пустынноколосник альберта
148470 Пустынноколосник аральский
148471 Пустынноколосник лопухолистный
148472 Пустынноколосник байсунский
148473 Пустынноколосник бальджуанский

148474 Пустынноколосник буасье
148475 Пустынноколосник седоватый
148476 Пустынноколосник головчатколистный
148477 Пустынноколосник сердцелистный
148478 Пустынноколосник пустынный
148479 Пустынноколосник бесприцветничковый
148480 Пустынноколосник шерстисточашечный
148481 Пустынноколосник фетисова
148482 Пустынноколосник блистающий
148483 Пустынноколосник железконосный
148484 Пустынноколосник голочашечковый
148485 Пустынноколосник гипсолюбивый
148486 Пустынноколосник армянский
148487 Пустынноколосник гиссарский
148488 Пустынноколосник гртзинский
148489 Пустынноколосник илийский
148490 Пустынноколосник кауфмана
148491 Пустынноколосник коровина
148492 Пустынноколосник губастый
148493 Пустынноколосник губастообразный
148494 Пустынноколосник надрезный
148495 Пустынноколосник лемана
148497 Пустынноколосник широкочашечный
148498 Пустынноколосник голый
148499 Пустынноколосник метельчатый
148500 Пустынноколосник гребенчатый
148501 Пустынноколосник зопниковый
148502 Пустынноколосник попова
148503 Пустынноколосник прекрасный
148504 Пустынноколосник регеля
148505 Пустынноколосник колесовидный
148506 Пустынноколосник шугнанский
148507 Пустынноколосник зеравшанский
148508 Пустынноколосник красивый
148510 Пустынноколосник почни-колосовый
148511 Пустынноколосник таджикский
148512 Пустынноколосник тяньшанский
148513 Пустынноколосник заамударьинский
148514 Пустынноколосник крувненосный, Эремостахис крубненосный
148515 Пустынноколосник одноцветковый
148517 Пустынноколосник зинаиды
148531 Череш, Шарыш, Ширяш, Эремурус
148532 Эремурус альберта
148533 Эремурус алтайский
148534 Эремурус неравнокрылый
148536 Эремурус бухарский
148538 Эремурус капю
148539 Ширяш китайский, Эремурус китайский
148540 Эремурус хохлатый
148542 Эремурус иларии
148544 Эремурус индерский
148546 Эремурус кауфмана
148547 Эремурус копетдагский
148548 Эремурус коровина
148549 Эремурус коржинского
148550 Эремурус млечноцветный
148551 Эремурус желтый
148552 Эремурус ольги
148553 Эремурус регеля
148554 Эремурус мощный
148555 Эремурус запрягаева
148556 Эремурус согдийский
148557 Эремурус величественный, Эремурус представительный
148558 Эремурус узколистный
148559 Эремурус крымский
148560 Эремурус туркестанский
148855 Шерстицвет, Шерстняк, Эриантус
148903 Эриантус краснеющий
148906 Камыс, Шерстицвет равеннский
148941 Ерика, Эрика
149017 Вереск древовидный, Вереск крупный средиземноморский, Ерика древовидная, Эрика древовидная
149147 Эрика мясо-красная, Эрика румяная
149205 Вереск серый, Ерика сизая, Эрика сизая
149398 Ерика средиземноморская
150027 Вереск метлистый, Эрика метловидная
150131 Ерика крестолистная, Эрика крестолистная
150286 Вересковые
150414 Блошник, Мелколепестник, Стенактис, Тонколистниковые, Эригерон
150419 Загадка, Мелколепестник едкий, Мелколепестник острый, Мелколепестник сахалинский
150422 Мелколепестник камчатский
150441 Мелколепестник алексеенио
150444 Мелколепестник иначе-крашенный
150454 Мелколепестник альпийский
150457 Мелколепестник алтайский
150461 Мелколепестник андриалевидный
150464 Мелколепестник однолетний, Стенактис однолетний
150484 Мелколепестник оранжевый
150488 Мелколепестник лазолевый
150489 Мелколепестник бадахшанский
150490 Мелколепестник байкальский
150497 Мелколепестник маргаритковидный
150501 Мелколепестник двукратноветвистый
150508 Мелколепестник северный
150510 Мелколепестник короткосемянный
150515 Мелколепестник тибетский
150521 Мелколепестник кабульский
150539 Мелколепестник кавказский
150597 Мелколепестник длинностолбиковый
150616 Мелколепестник удлиненный
150621 Мелколепестник пушисточашечный, Мелколепестник шерстисточашечный
150623 Мелколепестник пушистоголовый
150694 Мелколепестник гиссарский
150698 Мелколепестник низкий
150699 Мелколепестник гиблидный
150724 Мелколепестник мексиканский
150728 Мелколепестник хоросанский
150729 Мелколепестник цюцзянский
150732 Мелколепестник комарова
150735 Мелколепестник корагинский
150737 Мелколепестник крылова
150738 Мелколепестник гуншаньский
150740 Мелколепестник шерстистоголовый
150750 Мелколепестник гладко-горный
150756 Мелколепестник белолистнй
150763 Мелколепестник белолистный, Мелколепестник ланцетолистный
150827 Мелколепестник охары
150829 Мелколепестник олеги
150831 Мелколепестник горный
150834 Мелколепестник восточный
150841 Мелколепестник бледный
150851 Мелколепестник иноземный
150859 Мелколепестник черешковый
150877 Мелколепестник многолистный
150878 Мелколепестник подольский
150881 Мелколепестник многобразный
150883 Мелколепестник ророва
150888 Мелколепестник первоцветовидный
150892 Мелколепестник ложномелколепестник
150895 Мелколепестник ложноудлиненный
150897 Мелколепестник ложнозеравшанский
150951 Мелколепестник малбусский
150959 Мелколепестник зеравшанский
150961 Мелколепестник смолевколистный
150968 Мелколепестник синьцзянский
150980 Мелколепестник щетинистный
151008 Мелколепестник тунберга
151018 Мелколепестник тяньшанский
151036 Мелколепестник однокорзинчатый, Мелколепестник однокорзинчный, Мелколепестник одноцветковый
151054 Мелколепестник красивенький, Мелколепестник приятный
151058 Мелколепестник замещающий

151061 Мелколепестник фиолетовый
151130 Ериоботрия, Эриоботрия, Японская мушмула
151144 Эриоботрия тайваньская
151162 Башмала японская, Ериоботрия японская, Кизильцик японский, Локва, Мушмула субтропическая, Мушмула японская, Мушмула японская обыкновенная, Эриоботрия японская, Японская мушмула
151165 Эриоботрия малипоская
151206 Эриокаулиевые, Эриокаулоновые
151208 Шерстестебельник, Эриокаулон
151221 Шерстестебельник сычуаньский
151263 Шерстестебельник китайско-русский, Эриокаулон китайско-русский
151268 Шертестебельник зибольда
151288 Шерстестебельник десятицветковый, Эриокаулон десятицветковый
151322 Эриокаулон мэнцзыский
151400 Эриокаулон наньтоуский
151409 Эриокаулон непалский
151452 Шерстестебельник мощный, Эриокаулон мощный
151536 Шерстестебельник уссурийский, Эриокаулон уссурийский
151549 Эриокаулон яошаньский
151655 Шерстняк, Шерстяк
151699 Шерстяк перехваченный
151702 Шерстяк волосистый, Шерстяк моханатый
151792 Эриогонум
152734 Пушица
152744 Пушица короткопыльниковая
152746 Пушица красивощетниковая
152749 Пушица шамисо
152759 Пушица корейская, Пушица стройная
152763 Пушица низкая
152765 Пушица многокомоейовая, Пушица широколистная
152769 Пушица узколистная
152772 Пушица рыжеватая
152781 Пушица шейхцера
152791 Болотник влягалищный, Пушица влагалищная
152799 Эриофиллум
153358 пушистоспайник
153360 пушистоспайник длиннолистный
153420 Незабудочник, Эритрихиум
153426 Незабудочник подушковый
153437 Незабудочник кавказский
153439 Незабудочник чекановского
153446 Эритрихиум дэциньский
153448 Незабудочник сомнительный
153452 Незабудочник фетисова
153463 Незабудочник серный, Эритрихиум седой
153466 Незабудочник якутский
153469 Незабудочник енисейский
153470 Незабудочник камчатский
153471 Незабудочник кандинский
153473 Незабудочник широколистный, Незабудочник широколистый
153479 Незабудочник маньчжурский
153491 Незабудочник сахалинский, Эритрихиум ниппонский, Эритрихиум сахалинский
153497 Незабудочник памирский
153500 Незабудочник подушковидный
153502 Незабудочник гребенчатый
153509 Незабудочник ложно-шилоколистный
153518 Незабудочник скальный
153526 Незабудочник сахалинский
153531 Незабудочник шелковистый
153534 Незабудочник сихатэалинский
153535 Эритрихиум мелкоплодный
153542 Незабудочник почти-жакмонов
153546 Гакелия тимьянолистная
153549 Эритрихиум мелкоцветковое
153552 Незабудочник туркестанский
153555 Незабудочник мохнатый, Эритрихиум мохнатый
153711 Аистник, Грабельники, Журавельник
153723 Журавельник пупавковидный
153726 Журавельник армянский
153733 Журавельник бекетова
153754 Журавельник хиосский
153764 Аистник длинноклювый, Журавельник аистовый
153767 Аистник обыкновенный, Аистник цикутный, Аистник цикутовый, Аистник цикутолистный, Буськи, Грабельник цикутолистный, Журавельник веховый, Журавельник цикутовый, Журавельник цикутолистный
153800 Журавельник дымянковидный
153837 Журавельник литвинова
153839 Аистник цельнолистный, Журавельник мальвовидный
153883 Аистник геффта, Аистник остроклювый, Журавельник геффта, Журавельник остроносый
153907 Журавельник русский
153909 Аистник лоэбный
153914 Аистник стефана, Журавельник стефана
153916 Аистник стевена, Журавельник стевена
153917 Журавельник щетинистый
153920 Журавельник татарский
153923 Журавельник тибетский
153949 Веснянка
153956 Веснянка карликовая
153957 Веснянка ранняя
153964 Веснянка весенняя, Веснянка крокера, Крупка весенняя
153998 Индау, Эрука
154019 Горчица дикая, Индау посевной, Рокетсалат, Рокет-салат, Эрука посевная
154276 Синеголовник, Эрингиум
154280 Синеголовник альпийский
154291 Синеголовник биберштейна
154298 Синеголовник бунге
154300 Синеголовник голубой
154301 Синеголовник полевой, Синеголовник равнинный
154316 Синеголовник вонючий
154317 Синеголовник гигантский
154322 Синеголовник каратавский
154326 Синеголовник крупночашечковый
154327 Синеголовник морской, Синеголовник приморский
154330 Синеголовник удивительный
154331 Синеголовник карадагский
154332 Синеголовник ное
154337 Синеголовник плоский, Синеголовник плосколистный
154353 Синеголовник ванатура
154363 Желтушник, Хейриния, Эризимум
154367 Желтушник аксарский
154380 Желтушник амурский
154387 Желтушник сереброплодный
154394 Желтушник золотистый
154396 Желтушник бабатагский
154398 Желтушник сычуаньский
154409 Желтушник короткоплодный
154413 Желтушник короткостолбчатый
154414 Желтушник оранжевый
154417 Желтушник красивоплодный
154418 Желтушник сероватый, Желтушник серо-зелёный, Желтушник серый
154421 Желтушник каспийский
154422 Желтушник кавказский
154424 Желтушник левкойный
154433 Желтушник холмовый
154434 Желтушник скученный
154436 Желтушник толстоногий
154438 Желтушник меловой
154439 Желтушник шафранный
154440 Желтушник щитовидный
154441 Желтушник черняева
154451 Желтушник алтайский
154456 Желтушник франше
154458 Желтушник гауданский
154459 Желтушник холодный
154467 Желтушник маршалла, Желтушник прямой, Желтушник ястребинколистный

154471 Желтушник низкий
154474 Желтушник грузинский
154484 Желтушник узкостолбиковый
154485 Желтушник японский
154488 Желтушник крынкинский
154492 Желтушник узколистный
154493 Желтушник тонкостолбиковый
154494 Желтушник белоцветковый, Желтушник белоцветный
154495 Желтушник лиловый
154506 Желтушник мейера
154513 Желтушник венгерский, Желтушник душистый
154516 Желтушник палласа
154522 Желтушник персидский
154530 Желтушник выгрызенный, Желтушник выемчатый, Желтушник растопыренный
154544 Желтушник гулявниковый
154548 Желтушник прямостручковый
154551 Желтушник шершавый
154553 Желтушник лесной
154555 Желтушник шовица
154557 Желтушник заилийский
154560 Желтушник фиолетовый
154562 Желтушник желточный
154565 Эризимум юньнаньский
154617 Эритрина
154643 Дерево коралловое
154647 Эритрина петушиный гребень, Эритрина петушья
154734 Эритрина индийская
154746 Эритрина юньнаньская
154895 Зуб собачий, Кандык, Эритрониум
154906 Кандык кавказский
154909 Зуб собачий обыкновенный, Кандык собачий, Кандык южноевропейский, Кендик, Кендык
154926 Кандык японский
154946 Кандык сибирский
155048 Кокайновые
155051 Куст кокайновый, Эритроксилон
155067 Дерево кокаиноное, Кока, Кокайновый куст, Куст кокайновый, Кустарник кокаиновый, Эритроксилон кока, Эритроксилон яванское
155130 Эскаллония, Эскалония
155149 Эскаллония красная
155170 Полынёк, Полынок, Эсшольция, Эшолтциа, Эшольция, Эшшольция
155173 Мак калифорнийский, Полынок, Эшшольция, Эшшольция калифорнийская
155453 Евботриоидес
155468 Эвкалипт
155478 Эвкалипт миндальный
155505 Эвкалипт гроздевидный
155517 Эвкалипт камальдульский
155529 Эвкалипт пепельно-серый, Эвкалипт пепельный
155564 Эвкалипт великолепный
155579 Эвкалипт остроконечный
155585 Эвкалипт ясеневидный
155589 Эвкалипт бугорчатый, Эвкалипт голибой, Эвкалипт пепельный, Эвкалипт шариковый, Эвкалипт шариковый-голубой, Эвкалипт шаровидный
155594 Эвкалипт гвоздеголовый
155602 Эвкалипт ганна, Эвкалипт генна, Эвкалипт гэнна
155624 Эвкалипт белодревесный
155633 Эвкалипт макартура
155640 Эвкалипт мидена
155644 Эвкалипт западноавстралийский, Эвкалипт окаймленный
155649 Эвкалипт медомпахнущий
155652 Эвкалипт мелкоплодный
155668 Эвкалипт блестящий
155672 Эвкалипт косой
155688 Эвкалипт малоцветковый
155700 Эвкалипт многоцветковый
155716 Эвкалипт лучистый
155719 Эвкалипт царственный
155720 Эвкалипт смолоносный
155722 Эвкалипт исполинский, Эвкалипт мощный
155725 Эвкалипт красноватый
155727 Эвкалипт грубый
155730 Эвкалипт иволистный, Эвкалипт миндальный
155732 Эвкалипт ивовый
155743 Эвкалипт железнодревесный
155745 Эвкалипт смита
155756 Эвкалипт вальковаторогий
155772 Эвкалипт прутовидный, Эвкалипт прутьевидный
155857 Эвхарис
155876 Теозинт, Теозинта, Теозинте, Теосинт
155878 Теосинт мексиканский
155984 Крепкоплодник, Эвклидиум
155985 Крепкоплодник сирийский, Чанголак, Эвклидиум сирийский
156040 Эвкоммия
156041 Гуттаперчевое дерево, Дерево гуттаперчевое, Эвкоммия, Эвкоммия вязовидная, Эвкоммия вязолистная
156042 Эвкоммиевые
156118 Евгения
156148 Евгения бразильская
156385 Вишня суринамская, Вишня суринамская питанга, Питанга
156439 Эвлалия
156490 Эвлалия четырехжилковатая
156499 Эвлалия красивая
156510 Эвлалия юньнаньская
156521 Эулофия
156572 Эулофия тайваньская
157069 Эулофия туркестанская
157105 Эулофия юньнаньская
157195 Эвномия круглолистная
157268 Бересклет, Клещина
157285 Бересклет крылатый
157310 Бересклет священный
157315 Бересклет американский
157318 Бересклет далиский
157327 Бересклет тёмно-багряный, Бересклет тёмно-пурпуровый
157338 Бересклет наньчуаньский
157345 Бересклет бунге
157373 Бересклет цзиньюньский
157429 Бересклет европейский, Бересклет обыкновенный
157473 Бересклет форчуна
157499 Бересклет укореняющийся карьера, Бересклет форчуна карьера
157508 Бересклет укореняющийся серебристо-окаймленная
157547 Бересклет кртпноцветковый
157556 Бересклет хайнаньский
157559 Бересклет гамильтона
157561 Бересклет гамильтона мелкоплодный
157563 Бересклет гамильтона зияющий
157582 Бересклет хусюин
157587 Бересклет хубэйский
157601 Бересклет японский
157610 Бересклет японский 'Микрофиллюс'
157615 Бересклет японский золотисто-пестрая
157639 Бересклет яванский
157654 Бересклет коопмана
157658 Бересклет гуйцзоуский
157668 Вересклет широколистный
157683 Бересклет лицзянский
157695 Бересклет повислый
157698 Бересклет люцюский
157699 Бересклет маака
157705 Бересклет большекрылый
157711 Бересклет максимовича
157714 Бересклет мэнцзыский
157721 Бересклет мелкоплодный
157740 Бересклет карликовый, Бересклет низкий, Бересклет ниский
157755 Бересклет эмэйский
157761 Бересклет остролистный
157777 Бересклет малоцветковый
157791 Бересклет сичоуский
157797 Бересклет плоскочерешковый
157806 Бересклет баотинский

157848 Бересклет сахалинский, Бересклетокрас сахалинский
157876 Бересклет семенова
157879 Бересклет зибольда, Бересклет зибольдиев
157906 Бересклет сычуаньский
157913 Бересклет тэнчунский
157925 Бересклет тибетский
157946 Бересклет бархатистый
157947 Бересклет бородавчатый
157952 Бересклет бородавчатый, Бородавочник
157966 Бересклет вильсона
157974 Бересклет юньнаньский
158021 Посконник
158062 Посконник коноплевидный, Посконник коноплевый, Посконник конопляный, Седач, Седаш, Седаш коноплевидный
158070 Посконник китайский
158136 Посконник глена
158153 Посконник хиаляньский
158161 Японск японский
158182 Посконник цзилунский
158200 Посконник линдлея
158238 Посконник наньчуаньский
158243 Посконник эмэйский
158385 Молоко волчье, Молочай, Эйфорбия, Эуфорбия
158394 Молочай заостренный
158407 Молочай пашенный, Молочай плевой
158408 Молочай алайский
158409 Молочай алтавский
158414 Молочай алеппский
158417 Молочай альпийский
158420 Молочай алтайский
158433 Молочай миндалевидный
158443 Молочай адрахновидный
158447 Молочай ребристый, Молочай угловатый
158449 Молочай неравнолепестный
158471 Молочай остроконечный
158474 Молочай алмянский
158499 Молочай оше
158550 Молочай ресничнолистный
158553 Молочай буассье
158560 Молочай бородина
158561 Молочай борщова
158581 Молочай бухтарминский
158582 Молочай бузе
158585 Молочай бунге
158593 Молочай буша
158608 Молочай сереющий
158616 Молочай карпатский
158632 Молочай мелкосмоковник, Молочай приземистый
158642 Молочай дланечешуйный
158659 Молочай волосистый, Молочай мохнатый
158681 Молочай коническосеменный
158730 Молочай кипарисный, Молочай кипарисовидный, Молочай кипарисовый
158734 Молочай криволистный
158752 Молочай дельтоприцветничный
158756 Молочай густой
158762 Молочай мелкозубый
158805 Молочай сладкий
158844 Молочай шерстеносный
158857 Молочай двуцветный, Молочай лозный, Молочай маньчжурский, Молочай острый, Молочай полусердцевидный, Молочай скученный, Молочай слегка-сердцевидный, Молочай съедобный, Молочай сырдарьинский, Молочай эзуля
158865 Молочай евгении
158869 Молочай маленький, Молочай малый, Молочай ничтожный
158876 Молочай серповидный
158887 Молочай ферганский
158895 Молочай палласа
158906 Эйфорбия тайваньская
158909 Молочай форскаля
158915 Молочай франше
158948 Молочай оголенный
158952 Молочай хрящеватый
158958 Молочай шаровидный
158965 Молочай гмелина
158967 Молочай гольде
158977 Молочай греческий
158991 Молочай туркменский
159007 Молочай гроссгеймя
159018 Молочай хайнаньский
159027 Молочай солнцеглядный, Молочай солнцеглят, Солнцегляд
159046 Эйфорбия разнолистная
159092 Молочай приземистый, Молочай стелющийся
159095 Молочай приземистый
159101 Молочай комарова
159102 Молочай индийский
159112 Молочай грузинский
159129 Молочай индерский
159167 Молочай кандинский
159171 Молочай ганьсуский
159175 Молочай каро
159187 Молочай комалова
159191 Молочай копетдага
159199 Молочай кудряшова
159222 Молочай масличный, Молочай чеen
159223 Молочай широколистный
159226 Молочай ледебура
159236 Молочай тонкостебельный
159252 Молочай липского
159270 Молочай блестящий, Молочай глянцевитый
159272 Молочай рошевый
159285 Молочай длиннокорневой
159286 Молочай пятнистый
159305 Молочай сосочковый
159313 Эйфорбия маргината
159318 Молочай маршаллов
159342 Молочай дынеобразный
159357 Молочай мелкоплодный
159360 Молочай мелкосемянный
159363 Молочай яркий
159387 Молочай монгольский
159390 Молочай одноциатиевый
159391 Молочай одностолбиковый
159403 Молочай коротко остроконечный
159417 Молочай мирсинитский, Молочай миртолистный
159472 Молочай нормана
159503 Молочай восточный
159514 Молочай толстокорневой
159517 Молочай палласа
159522 Молочай болотный
159523 Молочай
159526 Молочай кожистый, Молочай прибрежный
159557 Молочай бутерлак, Молочай бутерлаковидный, Молочай бутерлаковый, Молочай садовый
159563 Молочай шерстистый
159567 Молочай камнелюбивый, Молочай скалолюбивый
159599 Молочай плосколистный, Молочай расширеннолистный
159603 Молочай
159615 Молочай многоцветный
159619 Молочай древнезеравшанский
159624 Молочай астраханский, Молочай ранний
159630 Молочай проханова
159640 Молочай ложно-полевой
159641 Молочай ложнокактусовый
159650 Молочай ложнополевой
159663 Молочай уральский, Молочай урарский
159670 Молочай пушистый
159675 Молочай красивейший, Молочай прекрасный, Поинсеттия, Поинсеция
159683 Молочай жесткий
159700 Молочай репка
159710 Молочай смолоносный
159726 Молочай двужелезковый
159733 Молочай русский
159736 Молочай розеточный

159751 Молочай скаловый
159759 Молочай иволистный
159770 Молочай донской, Молочай сарептский
159772 Молочай савари
159792 Молочай шугнанский
159794 Молочай твердобокальчатый
159799 Молочай исписанный
159808 Молочай сегиеров, Молочай сегюера
159811 Молочай полуволосистый, Молочай полумохнатый
159841 Молочай зибольда, Молочай зибольдиев
159851 Молочай джунгарский, Молочай светлоплодный
159853 Молочай сестричный
159863 Молочай колючезубый
159875 Молочай чешуйчатый, Молочай шершавый
159888 Молочай тибетский
159892 Молочай прямой, Молочай сжатый
159905 Молочай тонкий
159925 Молочай совича
159937 Молочай крымский
159967 Молочай тяньшанский
159974 Молочай тибетский
159975 Молочай тирукалли
159990 Молочай заамударьинский
160004 Молочай трехзубый
160019 Молочай турчанинова
160022 Молочай туркестанский
160028 Молочай волнистый
160062 Молочай лозный, Молочай лосный, Молочай острый, Молочай прутьевидный, Молочай съедобный, Молочай широковетвистый
160067 Молочай волжский
160091 Молочай виттмана
160093 Молочай воронова
160108 Молочай жигулевский
160110 Молочайные
160128 Очанка
160129 Очанка железистостебельная
160130 Очанка альбова
160131 Очанка алтайская
160132 Очанка тупозубая
160134 Очанка амурская
160136 Очанка бакурианская
160141 Очанка ботническая
160142 Очанка коротковолосистая
160144 Очанка кавказская
160145 Очанка короткая
160146 Очанка округлолистная
160147 Очанка дагестанская
160155 Очанка федченко
160156 Очанка финская
160161 Очанка холодная
160164 Очанка грузинская
160165 Очанка почти голая
160166 Очанка изящная
160167 Очанка гроссгейма
160169 Очанка волосистая
160188 Очанка ирины
160189 Очанка якутская
160193 Очанка юзепчука
160196 Очанка кемулярии
160197 Очанка кернера
160201 Очанка краснова
160203 Очанка широколистная
160204 Очанка чешуйчатая
160205 Очанка крупночашечковая
160206 Очанка крупнозубая
160214 Очанка максимовича
160219 Очанка мелкоцветковая
160222 Очанка маленькая
160224 Очанкамягкая
160225 Очанка лжемягкая
160226 Очанка горная
160227 Очанка многолистная
160230 Очанка мюрбека
160233 Очанка лекарственная, Очанка росткова, Очанка ростковиука, Очанка ростковская, Трава очная
160236 Очанка онежская
160237 Очанка осетинская
160238 Очанка коротноцветковая
160239 Очанка татарская
160240 Очанка сычуаньская
160241 Очанка максимовича
160243 Очанка цветоножковая
160245 Очанка раскрашенная
160246 Очанка летняя
160247 Очанка летнекоротная
160248 Очанка летнеростковская
160254 Очанка регеля
160256 Очанка рейтера
160258 Очанка ростковская
160259 Очанка саамская
160261 Очанка зальцбургская
160262 Очанка шугнанская
160265 Очанка севанская
160266 Очанка сибирская
160267 Очанка сосновского
160268 Очанка прямая
160270 Очанка шведская
160272 Очанка сванская
160273 Очанка сырейщикова
160277 Очанка татарская
160281 Очанка татры
160282 Очанка крымуская
160283 Очанка тонкая
160285 Очанка тоунсэнда
160286 Очанка тайваньская, Очанка черешковая
160288 Очанка траншеля
160289 Очанка ютрица
160290 Очанка уссурийская
160292 Очанка вилькома
160293 Очанка воронова
160338 Эвптелея
160347 Эвптелея многотычинковая
160405 Эврия
160442 Эврия китайская
160482 Эврия гуншаньская
160484 Эврия хайнаньская
160486 Эврия лицзянская
160492 Эврия мэнцзыская
160503 Эврия японская
160508 Эврия японская мелколистная
160523 Эврия цзиндунская
160525 Эврия гуйцзоуская
160542 Эврия малипоская
160626 Эврия вэньшаньская
160630 Эврия юньнаньская
160635 Эвриала, Эвриале
160637 Эвриала устрашающая, Эвриале устрашающая
160716 Эврикома
160946 Эвскафис
160949 Эвскафис фуцзяньский
160954 Эвскафис японская
161053 Пальма капустная
161120 Эвтрема, Эутрема
161125 Эутрема сердцелистная
161129 Эвтрема эдвардса, Эутрема эдвардса
161134 Эвтрема цельнолистная
161146 Эвтрема ложносердцелистная
161149 Эутрема японская
161158 Эутрема юньнаньская
161220 Эвакс
161222 Эвакс
161224 Эвакс анатолийский
161226 Эвакс песчаный
161238 Эвакс сжатый
161240 Эвакс жабниковый
161249 Эвакс мелконожковый
161251 Эвакс
161268 Эвакс
161294 Эверсманния
161295 Эверсманния гедизаровидная
161302 Эводия
161323 Эводия даньелла
161328 Эводия лицзянская
161333 Эводия ясенелистная
161336 Эводия сизая, Эводия японская
161340 Эводия генри
161376 Эводия лекарственная
161381 Эводия сычуаньская
161384 Эводия мотоская
161387 Эводия бархатистая

161528 Экзакум
161741 Экзохорда
161743 Экзохорда альберта
161757 Экзохорда тяньшанская
161896 Фабиана
161897 Фабиана
162033 Буковые
162301 Греча, Гречиха
162312 Гречиха культурная, Гречиха посевная, Гречиха съедобная
162334 Гречиха полукустарная
162335 Гречиха татарская, Кырлык
162359 Бук
162368 Бук зибольда, Бук зиебольда
162372 Бук энгрера
162375 Бук крупнолистный, Бук обыкновенный
162380 Бук тайваньский
162382 Бук японский
162395 Бук восточный
162400 Бук европейский, Бук лесной
162404 Бук лесной тёмно-пунцовый
162413 Бук лесной плакучий
162436 Бук крымский
162457 Резак
162460 Резак фалькариевидный
162465 Резак обыкновенный
162509 Гречишник бальджуанский
162519 Горец вьюнковый, Горец вьюшийся, Горец кустарниковый, Гречиха вьюнковая, Гречиха полевая, Повитель
162525 Горец зубчатокрылатый, Горец зубчатокрылый
162528 Горец кустарниковый
162530 Горец малоцветковый
162532 Горец остроконечный, Горец японский
162542 Горец многоцветковыи
162616 Лигулярия кемпфера
162635 Синарундинария
162718 Синарундинария блестящая
162813 Фарзетия, Фарсетия
162850 Фарзетия лопатчатая
162876 Фатсия, Фация
162879 Арария японская, Фатсия, Фатсия японская, Фация японская
163089 Змееголовник тычночный
163116 Фейхоа
163117 Трава ананасная, Фейхоа, Фейхоа селлоу
163121 Фелиция
163312 Фендлера
163499 Лимон персидский
163502 Ферония слоновая
163574 Смолоносница, Ферула
163576 Ферула эчисона
163577 Ферула акичкенская, Ферула переходная
163579 Ферула ангренская
163580 Ассафетида, Вонючка, Смолоносница вонючая, Ферула ассафетида, Ферула вонючая
163582 Ферула бадхызская
163584 Ферула жестковатая
163585 Ферула седовадая
163586 Смолоносница каспийская, Ферула каспийская
163587 Ферула кавказская
163588 Ферула роголистная
163589 Ферула ломоносолистная
163590 Ферула обыкновенная
163598 Ферула конусостебельная
163602 Ферула рассеченная
163604 Джауджамыр, Ферула джауджамыр
163605 Ферула хвощевая
163606 Ферула
163608 Ферула федченко
163609 Ферула ферганская
163610 Ферула метельчатая
163615 Ферула пахучая
163616 Ферула олиственная
163619 Ферула камеденосная
163620 Ферула гигантская
163621 Ферула гладная
163623 Ферула стройная
163624 Ферула григорьева
163626 Ферула гипсолюбивая
163630 Ферула илийская
163631 Ферула вздутая
163632 Ферула оберточковая
163633 Ферула иешке
163635 Ферула каракалинская
163636 Ферула бородлдайская
163637 Ферула каратавская
163638 Ферула каратегинская
163642 Ферула келифа
163643 Ферула келлера
163646 Ферула кокандская
163647 Ферула копетдагская
163649 Ферула коржинского
163650 Ферула козо-полянского
163651 Ферула крылова
163652 Ферула кугистанская
163654 Ферула каменная
163655 Ферула широколистная
163656 Ферула широкодольчатая
163658 Ферула лемана
163659 Ферула гладколистная
163662 Ферула язычковая
163663 Ферула линчевского
163665 Ферула липского
163666 Ферула литвинова
163676 Ферула мелкоплодная
163678 Ферула мелкодольчатая
163679 Ферула моголтавская
163680 Ферула мягкая
163681 Ферула мускусная
163684 Ферула невского
163685 Ферула голая
163688 Смолоносница восточная, Ферула восточная
163689 Ферула овечья, Ферула столбиковая
163690 Ферула
163691 Ферула толстолистная
163692 Ферула бледнозеленая
163694 Ферула перистонервная
163695 Ферула персидская
163700 Ферула многоканальцевая
163701 Ферула потанина
163702 Ферула прянголистная
163706 Ферула краснопесчаниковая
163707 Ферула самаркандская
163709 Ферула вонючая
163711 Ферула синьцзянская
163712 Ферула джунгарская
163717 Ферула тонкая
163727 Ферула сырейщикова
163728 Ферула совича
163729 Смолоносница татарская, Ферула татарская
163730 Ферула тонкорассеченная
163731 Ферула дурнопахнущая
163733 Ферула
163736 Ферула чимганская
163737 Ферула клубненосная
163740 Ферула угамская
163742 Ферула замещающая
163743 Ферула ксероморфная
163746 Ферульник
163750 Ферульник дагестанский
163751 Ферульник широкодольчатый
163764 Ферульник щетинколистный
163765 Ферульник лесной
163766 Ферульник крымский
163767 Ферульник туркменский
163778 Вульпия, Овсюг, Овсяница, Типчак, Фестука
163789 Овсяница алайская
163791 Овсяница алатавская, Овсяница тяньшанская
163802 Овсяница алтайская
163807 Овсяница тубоватая
163811 Овсяница аметистовая
163816 Овсяница песчаная
163817 Овсяница приземистая
163819 Овсяница луговая, Овсяница тростниковая, Овсяница тростниковидная
163827 Овсяница восточная
163834 Овсяница ужковатая
163836 Овсяница байкальская

163839 Овсяица беккера
163845 Овсяица боримюллера
163847 Овсяица коротколистная
163862 Овсяица чаюйская
163871 Овсяица поднебесная
163876 Овсяица меловая
163885 Овсяица даурская
163915 Овсяица жестковатая
163947 Овсяица пушистоцветковая
163950 Овсяница дальневосточная
163969 Овсяица тайваньская
163976 Овсяица ганешиная
163993 Овсяица высокая, Овсяица гигантская, Овсяица исполинская
164006 Ирга канадская, Овсяица разнолистная
164023 Овсяница якутская
164025 Фестука японская
164029 Овсяица ганьсуская
164030 Овсяица каратабский
164036 Овсяица кирилова
164039 Овсяица колымская
164040 Овсяица краузе
164041 Овсяица крыловская
164049 Овсяица ленская
164058 Овсяица длинноостная
164086 Овсяица мягчайшая
164089 Овсяица горная
164126 Овсяица бороздчатая, Овсяица овечья, Типчак, Трава овечья
164141 Овсяица рупрехта
164219 Овсяица выокая, Овсяица луговая
164224 Овсяица ложнобольшая
164226 Овсяица ложно-бороздчатая
164243 Овсяица красная
164249 Овсяица холодолюбивая
164280 Ирга канадская, Овсяица разнолистная
164295 Овсяица бороздчатая, Овсяица ганешиная, Овсяица желобчатая, Типчак
164329 Овсяица мелкочештйчатая
164357 Овсяица тяньшанская
164367 Овсяица печальная
164380 Овсяица влагалищная
164382 Овсяица ложноовечья
164386 Овсяица ложно-овечья
164393 Овсяица пестрая
164394 Овсяица прекрасная
164400 Фестука лицзянская
164401 Фестука юньнаньская
164445 Фибигия
164446 Фибигия щитовидная
164447 Фибигия мохнатоплодная
164449 Фибигия ширококрылая
164450 Фибигия полукустарниковая
164469 Чистяк
164473 Чистяк пучковатый
164474 Чистяк настоящий
164479 Чистяк весенний
164608 Инжир, Инжир смоковница, Смоква, Смоковница, Фикус
164635 Фикус наньтоуский
164684 Баниан, Баниян, Баньян, Смоковница бенгальская, Смоковница индийская, Фикус бенгальский, Фикус индийский
164751 Фикус капский
164762 Фикус лунчжоуский
164763 Дерево фиговое, Инжир, Инжир фига, Смоковница, Смоковница обыкновенная, Фига, Фиговое дерево, Фиговое смоковница, Фикус, Фикус инжир, Ягода винная
164925 Дерево каучуковое, Фикус, Фикус каучуконосный, Фикус эластика
164940 Фикус каучуконосный пестролистная
164947 Фикус прямостоячий
164964 Фикус лодяньский
164992 Фикус тайваньский
165079 Фикус гуансиский
165082 Фикус гуйцзоуский
165084 Фикус хайнаньский
165307 Фикус притупленный
165343 Фикус напоский
165442 Фикус ланьюйский
165515 Фикус карликовый, Фикус низкий
165520 Фикус карликовый малый
165553 Баниан, Баньян, Смоковница священая, Фикус благочестиный, Фикус священный
165628 Фикус японский
165711 Фикус булавовидный
165726 Инжир ослиный, Сикамор античный, Сикомор, Сикомора, Смоковница дикая, Фикус-сикомор
165871 Фикус сичоуский
165875 Фикус юньнаньский
165906 Жабник
165915 Жабник заячий, Жабник полевой
165922 Жабник борнмюллера
165940 Жабник шерстистоголовый
166005 Жабник горный
166025 Жабник лопатчатый
166038 Жабник германский, Жабник немецкий
166065 Нителистник
166066 Нителистник сибирский, Хризантемум триниевидный
166071 Лабазник
166073 Лабазник узкодольчатый, Лабазник узколопастный
166080 Лабазник обнаженный
166085 Лабазник голый
166088 Лабазник средний
166089 Лабазник камчатский
166092 Лабазник тайваньский
166097 Лабазник многопарный
166110 Лабазник дланевидный
166117 Лабазник пурпуровый
166123 Лабазник степной
166125 Лабазник вязолистный, Лабазник ильмовидный, Таболга вязолистная
166132 Земляные орешки, Лабазник шестилепестный, Орешки земляные, Таволга шестилепестная
166156 Фимбристилис
166164 Фимбристилис летний
166176 Фимбристилис однолетний
166248 Фимбристилис вильчатый
166268 Фимбристилис оходский
166313 Фимбристилис ржавый
166346 Фимбристилис хайнаньский
166348 Фимбристилис ичанский
166376 Фимбристилис гладкоплодный
166402 Фимбристилис пятигранный
166417 Фимбристилис наньнинский
166454 Фимбристилис пятигранный
166478 Фимбристилис зибера
166488 Фимбристилис симаоский
166499 Фимбристилис растопыренный
166518 Фимбристилис почти двухколосковый
166557 Фимбристилис бородавчатый
166565 Фимбристилис юньнаньский
166612 Фирмиана
166618 Фирмиана хайнаньская
166627 Стеркулия платанолистная, Фирмиана, Фирмиана платанолистная
166720 Фитцройя
166725 Фитцройя арчера, Фитцройя тасманийская
166726 Фитцройя патагонская
166761 Гребенщик пятитычинковый, Тамарикс патитычночный, Флакоуртия, Флакуртия
166773 Флакуртия индийская, Флакуртия рамонтша
166777 Флакоуртия кохинхинская
166786 Флакуртия рамонтша
166788 Флакуртия зондская
166793 Флакуртиевые, Флакуртовые
166814 Флаверия
166971 Флиндерсия
166974 Флиндерсия австралийская
167054 Флоуленсия
167067 Секуринега, Флюгея
167092 Бамбук полукустарниковый, Секуринега ветвецветная, Секуринега полукустарниковая
167094 Флюгея уссурийская

167141 Укроп душистый, Укроп конский, Фенхель
167146 Фенхель итальянский, Фенхель флорентинский
167156 Укроп водяной, Укроп сладкий, Фенхель аптечный, Фенхель душистый, Фенхель лекарственый, Фенхель обыкновенный
167230 Фонтанезия
167233 Фонтанезия форчуна
167427 Форзиция, Форсайтия, Форситня, Форсиция
167444 Форситня японская
167449 Форситня лицзянская
167452 Форсайтия яйцевидная, Форситня
167456 Форзиция пониклая, Форзиция свешивающаяся, Форсайтия висячия, Форситня повислая
167471 Форситня ярко-зеленая
167473 Форситня корейская
167487 Фортьюнария
167488 Фортьюнария китайская
167489 Кинкан, Кумкват
167499 Кинкан, Кинкан гонконгский дикий, Кинкан японский, Кумкват гонконгскийдикий, Кумкват дикий, Маруми
167503 Кемквот, Кинкан круглый, Кинкан японский, Кумкват круглый, Кумкват маруми, Кумкват японский, Маруми, Фортунелла японская
167506 Кемвот, Кемквот, Кинкан, Кинкан маргарита, Кинкан нагами, Кинкан овальный, Кумват, Кумкват, Кумкват овальный, Нагами, Нагами кинкан, Фортунелла жемчужная
167592 Земляника, Клубника
167597 Виктория ананасная, Земляника ананасная, Земляника крупноплодная, Земляника крупноцветковая, Земляника садовая, Клубника ананасная, Клубника ананасовая, Садовая земляника
167600 Земляника бухарская
167602 Земляника равнинная
167604 Земляника виргинская, Земляника чилийская
167614 Земляника мускуснаэ
167627 Земляника высокая, Земляника мускатная, Земляника мускусная, Клубника, Клубника высогая, Клубника настоящая
167640 Земляника тибетская
167641 Земляника восточная, Клубника восточная
167653 Земляника лесная
167663 Земляника виргинская, Клубника виргинская
167674 Земляника зеленая, Земляника зелёная, Земляника полуница, Полуница
167765 Сайгачья трава, Трава сайгачья, Франкения
167767 Франкения бухарская
167792 Сайгачья трава болосистая, Франкения жестковолосая
167802 Сайгачья трава гладкая
167813 Сайгачья трава мучнистая, Франкения порошистая
167827 Франкениевые
167893 Ясень
167897 Ясень американский, Ясень остороконечный
167908 Ясень американский орехолистный
167910 Ясень узколистный
167922 Боярышник однопестичный, Боярышник одностолбчатый, Кермек, Ясень неправильный
167934 Ясень калолинский
167940 Ясень китайский
167942 Ясень американский, Ясень остороконечный
167947 Ясень сумахолистный
167952 Ромашка ромашковидная, Ясень двойственный
167955 Ясень, Ясень высокий, Ясень европейский, Ясень обыкновенный
167958 Ясень золотистый
167970 Ясень японский
167975 Ясень обильноцветущий
167991 Ясень хубэйский
167999 Ясень японский
168008 Ясень зеленый, Ясень ланцетный
168018 Ясень длинно-остроконечный
168022 Ясень мягколистый
168023 Ясень маньчжурский
168030 Ясень наньчуаньский
168032 Ясень чёрный
168041 Ясень белый, Ясень маннитовый, Ясень манноносный, Ясень сахарный, Ясень цветочный
168043 Ясень остороплодный
168050 Ясень паллис
168053 Ясень кругролистный, Ясень мелколистный
168057 Ясень пенсильванский, Ясень пушистый
168070 Ясень поярковой
168071 Ясень туркестанский
168074 Ясень красный, Ясень опушенный, Ясень пенсильванский, Ясень пушистый
168079 Ясень голубой, Ясень четырёхгранный
168086 Ясень китайский клюволистный, Ясень носолистный
168093 Ясень кругролистный, Ясень мелколистный
168100 Ясень зибольда
168105 Ясень согдианский, Ясень согдийский
168114 Ясень сирийский
168120 Ясень войлочный
168124 Ясень мексиканская
168134 Ясень зеленый, Ясень зелёный, Ясень ланцетный, Ясень ланцетолистный
168136 Ясень бархатный
168153 Фреезия
168332 Корольковия, Ринопеталум, Рябчик, Фритиллярия
168344 Фритиллярия аньхуйская
168347 Ринопеталум ариаский
168348 Рябчик армянский
168360 Ринопеталум бухарский
168361 Рябчик камчатский
168364 Рябчик кавказский, Рябчик тюльпанолистный
168380 Рябчик цзилунский
168387 Рябчик дагана
168398 Рябчик эдуарда
168401 Рябчик ферганенский
168409 Ринопеталум горбатый
168416 Рябчик крупноцветный
168417 Рябчик гроссгейма
168430 Венец царский, Рябчик садовый, Рябчик турецкий, Рябчик царский
168438 Ринопеталум карелина
168442 Рябчик котчи
168443 Рябчик курдский
168450 Рябчик широколистный
168464 Рябчик желтый
168467 Рябчик камчатский, Рябчик максимовича
168469 Рябчик малый
168476 Рябчик шахматный
168489 Рябчик михайловского
168500 Рябчик ольги
168503 Рябчик восточный
168506 Рябчик бледноцветковый, Рябчик бледноцветный
168523 Рябчик ганьсуская
168547 Рябчик радде
168551 Рябчик регеля
168556 Рябчик русский
168566 Ринопеталум узкопыльниковый
168585 Рябчик горный, Рябчик тонкий
168612 Рябчик уссурийский
168614 Рябчик мумтвчатый, Рябчик мутовчатый, Фритиллярия мумтвчатая
168624 Рябчик валуева
168640 Рябчик тибетский
168717 Фролипия
168718 Фролипия почтиперистая

168735 Фуксия
168767 Фуксия
168780 Фуксия мелкоцветковая
168792 Фюрнрония
168793 Фюрнрония щетинолистная
168915 Фумана
168944 Фумана клейковатая
168959 Фумана лежащая
168964 Дымянка
169004 Дымянка бесчашечная
169030 Дымянка козья
169078 Дымянка кралика
169126 Дымянка аптечная, Дымянка лекарственная
169136 Дымянка мелкая, Дымянка мелкоцветковая
169155 Дымянка остроконечная
169168 Дымянка шлейхера
169183 Дымянка вайана
169194 Дымянковые
169196 Дымянока
169197 Дымянока туркестанская
169226 Фунтумия
169229 Фунтумия каучуконосная
169245 Конопля мавликийская, Конопля мавлициева
169313 Гертнерия
169375 Гусинй лук, Лук гусиный
169376 Гусинй лук афганский
169379 Гусинй лук альберта
169380 Гусинй лук алексеенко
169383 Гусинй лук алтайский
169384 Гусинй лук перавноцветковй
169385 Гусинй лук неравнолучевой
169388 Гусинй лук полевой
169389 Гусинй лук берга
169392 Гусинй лук клубненостый, Гусинй лук луковиценосный
169393 Гусинй лук калье
169394 Гусинй лук волосолистный
169395 Гусинй лук капю
169396 Гусинй лук коха
169397 Гусинй лук ханы
169398 Гусинй лук зеленовато-желтый
169399 Гусинй лук хомутовой
169400 Гусинй лук измененный
169401 Гусинй лук неясный
169409 Гусинй лук растопыренный
169410 Гусинй лук сомнительный
169416 Гусинй лук краснеющий
169417 Гусинй лук федченко
169420 Гусинй лук ложнокрасиноватый, Гусинй лук маленький, Гусинй лук малый, Гусинй лук нитевидный
169421 Гусинй лук дудчатый
169431 Гусинй лук выемчатый
169432 Гусинй лук персидский
169433 Гусинй лук жермены
169434 Гусинй лук леднйковый
169438 Гусинй лук злаколистный
169441 Гусинй лук гранателли
169443 Гусинй лук зелнистый
169444 Гусинй лук гельдрейха
169445 Гусинй лук елены
169446 Гусинй лук гиенский
169447 Гусинй лук гиссарский
169450 Гусинй лук илийский
169451 Гусинй лук памирский, Гусинй лук яшке
169452 Гусинй лук японский
169454 Гусинй лук копетдагский
169455 Гусинй лук коржинского
169459 Гусинй лук длинностебельный
169460 Гусинй лук желтый
169467 Гусинй лук мелкоцветковый
169468 Гусинй лук удивительный
169470 Гусинй лук влагалищный, Гусинй лук синьцзянский
169475 Гусинй лук ольги
169476 Гусинй лук яйцевидный
169477 Гусинй лук пачоского
169479 Гусинй лук небольшой
169480 Гусинй лук малоцветковый, Гусинй лук предвиденный
169483 Гусинй лук попова
169485 Гусинй лук луговой
169488 Гусинй лук ложносетчатый
169490 Гусинй лук низкий
169497 Гусинй лук сетчатый
169498 Гусинй лук жесткий
169499 Гусинй лук мешконосный
169501 Гусинй лук ненецкий
169503 Гусинй лук покрывалом, Гусинй лук покрывальцевый
169505 Гусинй лук стебельчатый
169507 Гусинй лук серножелтый
169508 Гусинй лук шовица
169509 Гусинй лук крымский
169510 Гусинй лук нежный
169511 Гусинй лук тонколистный
169512 Гусинй лук тончайший
169514 Гусинй лук косой, Гусинй лук поперецный
169519 Гусинй лук трехгранный
169520 Гусинй лук украинский
169568 Гайллардия
169575 Гайлярдия крупноцветная
169586 Гайллардия крупноцветная, Гайллардия остистая
169588 Гайллардия
169603 Гайллардия красивая
169705 Галантус, Подснежник
169706 Подснежник альпийский
169707 Подснежник кавказский
169717 Подснежник широколистный
169719 Подснежник белый, Подснежник снеговой, Подснежник снежный
169726 Подснежник складчатый
169731 Подснежник закавказский
169732 Подснежник воронова
169756 Солонечник
169763 Солонечник алтайский
169764 Солонечник узколистный
169767 Солонечник двухцветковый
169770 Солонечник окрашенно-хохолковый
169772 Солонечник кожистый
169773 Солонечник даурский
169775 Солонечник растопыренный
169777 Солонечник щитковидный
169780 Солонечник гаупта
169788 Солонечник истоговидный
169790 Солонечник точечный
169792 Солонечник редеря
169793 Солонечник русский
169795 Солонечник прутьевидный
169796 Солонечник синьцзянский
169803 Солонечник тяньшанский
169804 Солонечник трехжилковый
169806 Солонечник коротко-мохнатый
169817 Галакс
169887 Галеорхис круглогубый, Ятрышник круглогубый
169906 Галеорхис круглогубый, Ятрышник хэбэйская
169919 Галега, Козлятник
169935 Галега лекарственная, Козлятник аптечный, Козлятник лекарственный, Трава козья
169940 Галега восточная, Козлятник восточный
170015 Зеленчук
170024 Зеленчук гуандунский
170025 Зеленчук желный, Зеленчук жёлный
170026 Зеленчук сычуаньский
170056 Петушник, Пикульник
170058 Пикульник узколистный
170060 Пикульник двнадрезанный, Пикульник двнадрезный, Пикульник дврасщепленный
170066 Медунка, Пикульник ладанниковиднный, Пикульник ладанниковый, Пикульник ладанный
170073 Пикульник пушистый
170076 Забра, Зябра, Кадило медоволистное, Пикульник заметный, Пикульник красивый
170078 Жабрей, Медовик, Медовник, Пикульник жабрей, Пикульник колючий, Пикульник обыкновенный, Пикульник-жабрей
170137 Галинзога, Галинсога

170142 Галинсога мелкоцветковая, Галинсога мелкоцветная
170146 Галинзога щетинистая, Галинсога реснитчатая, Галинсога четырехлучевая
170164 Галипея
170167 Ангустура, Галипея лекарственная
170173 Икотник лопатчатый
170175 Подмаренник
170190 Подмаренник туполистный
170191 Подмаренник амурский
170192 Подмаренник разнолистный
170193 Лепчица, Липушник, Подмаренник лепчица, Подмаренник ложный, Подмаренник фальшивый, Подмаренник цепкий
170205 Подмаренник ложный
170216 Подмаренник членистый
170228 Подмаренник шершавы
170236 Подмаренник азербайджанский
170237 Подмаренник золотистый, Подмаренник золотой
170239 Подмаренник байкальский
170246 Подмаренник болеальный, Подмаренник северный
170281 Подмаренник коротколистный
170282 Подмаренник брауна
170288 Подмаренник пузырчатый
170301 Подмаренник изветняковый
170311 Подмаренник касрийский
170315 Подмаренник зеленовато-белый
170329 Подмаренник кожистй, Подмаренник сердцевидный
170330 Подмаренник венечный
170335 Подмаренник крестовидный
170337 Подмаренник черенанова
170340 Подмаренник даурский
170346 Подмаренник декена
170348 Подмаренник густоцветковый
170353 Подмаренник высокий
170367 Подмаренник вытянутый
170372 Подмаренник прямой
170378 Подмаренник дудчатый
170383 Подмаренник лицзянский
170391 Подмаренник коленчатый
170410 Подмаренник гроссгейма
170420 Подмаренник распростертый, Ясменник распростертый, Ясменник слабый, Ясменник стелющийся
170421 Подмаренник хубэйский
170423 Подмаренник иссополистный
170425 Подмаренник инсубрийский
170433 Подмаренник юзепчука
170434 Подмаренник камчатский
170440 Подмаренник каракульский
170442 Подмаренник кяпазский
170445 Подмаренник японский
170449 Подмаренник копетдагский
170450 Подмаренник крылова
170451 Подмаренник крымский
170452 Подмаренник кутцинга
170455 Подмаренник молочный
170459 Подмаренник гладколистный
170460 Подмаренник линчевского
170473 Подмаренник маймекенский
170477 Ясменник максимовича
170490 Подмаренник мягкий
170510 Подмаренник изменчивый
170523 Подмаренник бледножелтый
170524 Пахучка, Подмаренник дустый, Шарошница, Ясменник душистый, Ясменник пахучий
170529 Подмаренник болотный
170532 Подмаренник памиро-алайский
170533 Ясменник метельчатый
170534 Подмаренник необычный, Подмаренник удивительный
170536 Подмаренник парижский
170548 Подмаренник пьемонтский
170558 Подмаренник широкоподмаренниковый, Ясменник широкий
170566 Подмаренник поярковый
170568 Подмаренник кандинский
170579 Подмаренник маролослый
170588 Ясменник приручейный, Ясменник цепкий
170589 Подмаренник круглолистный
170597 Подмаренник мареновидный
170601 Подмаренник рупрехта
170603 Подмаренник русский
170609 Подмаренник чабрецелистный
170610 Подмаренник саурский
170611 Подмаренник герцинский, Подмаренник скальный
170614 Подмаренник шероховатый
170615 Подмаренник шишкина
170616 Подмаренник шультеса
170626 Подмаренник севанский
170631 Подмаренник лесной
170635 Подмаренник джунгарский
170650 Подмаренник иглолистный
170657 Подмаренник сырейщикова
170658 Подмаренник тайваньский
170662 Подмаренник тонкий, Подмаренник тончайщий
170667 Подмаренник тяньшанский
170690 Подмаренник закавказский
170692 Подмаренник волосоносный
170694 Пашмак, Подмаренник трехрогий, Подмаренник трёхрогий
170696 Подмаренник трёхдольный, Подмаренник трехнадрезанный, Подмаренник трехнадрезный, Подмаренник трёхнадрезный, Подмаренник трехраздельный, Подмаренник трёхраздельный
170705 Подмаренник трехцветкововидный
170708 Подмаренник трехцветковый, Подмаренник трёхцветковый
170716 Подмаренник туркестанский
170717 Подмаренник топяной
170724 Подмаренник уссурийский
170726 Подмаренник валантиевидный
170727 Подмаренник вайяна
170733 Подмаренник вартана
170734 Подмаренник васильченко
170737 Подмаренник весенний
170742 Подмаренник мутовчатый
170743 Подмаренник настоящий
170751 Подмаренник настоящий
170777 Подмаренник синьцзянский
170779 Подмаренник юньнаньский
170827 Гальтония
170859 Спайноцветник
170860 Спайноцветник ферганский
170861 Спайноцветник спайноплндный
170862 Спайноцветник келифский
170863 Спайноцветник волосистый
171046 Гарциния
171136 Манглис, Мангостан
171238 Гардения
171253 Гардения жасминовидная, Гардения обильноцветная, Гардения укореняющаяся
171301 Гардения хайнаньская
171333 Гардения жасминовидная крупноцветковая
171374 Гардения укореняющаяся
171475 Гарадиолюс, Рагадиолюс
171477 Гарадиолюс угловатый
171478 Гарадиолюс летучконосный
171545 Гаррия
171607 Гастерия
171767 Алоэ бородавчатое
171809 Пузадник, Пузатик
171818 Пузатик обыкновенный
171903 Гастрокотиле
171904 Гастрокотиле шершавый
171905 Гастродия
171918 Гастродия высокая
171939 Гастродия яванская
171943 Гастродия мэнхайская
171969 Гастролобиум
172032 Гаудиния
172034 Гаудиния ломкая
172047 Гаультерия
172059 Гаультерия тайваньская
172104 Гаультерия пинбяньская
172111 Гаультерия микели, Гаультерия микеля
172173 Гаультерия тибетская

172185 Гаура
172248 Гайлоссакия, Гейлюссакия, Гейлюссация
172276 Гацания
172778 Гельземий, Гельземиум
172779 Гельземиум изящный
172888 Генипа
172890 Генипап
172898 Дрок
172903 Дрок беловатый
172905 Дрок английский
172908 Дрок узколистный
172913 Дрок армянский
172914 Дрок артвинский
172945 Дрок плотный
172957 Дрок лежачий, Дрок прижатый
172972 Дрок плетевидный
172976 Дрок германский
172977 Дрок голый
172984 Дрок распростертый
173015 ложнодрок монпелийский
173028 Дрок отклоненный
173030 Дрок волосистый
173074 Дрок сванетский
173077 Дрок донской
173081 Дрок четырехгранный
173082 Дрок красильный
173089 Дрок закавказкаский
173181 Генциана, Горечавка
173184 Горечавка вырезанная
173198 Горечавка холодная
173225 Горечавка угроватая
173242 Горечавка водяная
173251 Горечавка арктическая
173264 Горечавка ластовневая, Горечавка ластовнная
173270 Горечавка черноватая
173280 Горечавка пазушная
173283 Горечавка балтийская
173287 Горечавка баварская
173293 Горечавка биберштейна
173297 Горечавка ресниценосная
173309 Горечавка бирманская
173318 Горечавка полевая
173325 Горечавка карпатская
173329 Горечавка кавказская
173336 Горечавка китайская
173341 Горечавка чжундяньская
173342 Горечавка ресничатая
173344 Горечавка клузия, Горечавка клюзия
173364 Горечавка крестовидная, Горечавка крестообразная, Горечавка перекрастолистная, Горечавка соколий перелет, Перелёт соколий
173373 Горечавка даурская
173379 Горечавка тайваньская
173383 Горечавка деши
173387 Горечавка лежачая, Горечавка приподнимающаяся
173402 Горечавка вильчатая
173408 Горечавка джимильская
173410 Горечавка долуханова
173412 Горечавка куэньминская
173430 Горечавка серповидная
173448 Горечавка фишера
173459 Горечавка франше
173461 Горечавка студеная
173470 Горечавка холодостойкая
173473 Горечавка германская, Горечавка немецкая
173478 Горечавка сизая
173486 Горечавка крупноцветковая
173491 Горечавка гроссгейма
173493 Горечавка цзилунская
173540 Горечавка джемся, Горечавка курильская
173547 Горечавка цзиндунская
173552 Горечавка кауфмана
173559 Горечавка колаковского
173560 Горечавка комарова
173561 Горечавка крылова
173569 Горечавка гуансиская
173571 Горечавка раздеренная
173580 Горечавка бело-черная
173598 Горечавка горькая, Горечавка язычковая, Стародубка
173600 Горечавка липского
173610 Горечавка желтая
173615 Горечавка крупнолистная
173617 Горечавка фетисова
173624 Горечавка макино
173627 Горечавка марковича
173655 Горечавка мотоская
173662 Горечавка ниппонская, Горечавка японская
173666 Горечавка снежная
173670 Горечавка поникающая
173677 Горечавка ольги
173679 Горечавка оливье
173680 Горечавка эмэйская
173695 Горечавка оверина
173698 Горечавка болотная
173700 Горечавка памирская
173701 Горечавка поннонская
173705 Горечавка особенная
173720 Горечавка краснолистная
173726 Горечавка простая
173728 Горечавка лазолевая, Горечавка легочая, Горечавка синне колокольчики, Горечавка синяя, Колокольчики синий
173730 Горечавка понтийская
173732 Горечавка скороспелая
173742 Горечавка похожная
173743 Горечавка водяная, Горечавка простертая, Горечавка раскидистая
173747 Горечавка карелина
173751 Горечавка ложноводяная
173776 Горечавка точечная
173781 Горечавка пурпурная
173785 Горечавка пиренейская
173786 Горечавка цюцзянская
173821 Горечавка прибрежная
173825 Горечавка романцова
173840 Горечавка мыльная
173847 Горечавка шероховатая
173886 Горечавка семираздельная
173893 Горечавка сибирская
173917 Горечавка растопыренная
173948 Горечавка сычуаньская
173956 Горечавка далиская
173960 Горечавка нежная, Горечавка тоненькая, Горечавка тонкая
173982 Горечавка тяньшанская
173988 Горечавка тибетская
174004 Горечавка трехцветковая
174007 Горечавка пазушноцветковая
174031 Горечавка зонтичная
174034 Горечавка мешочная
174049 Горечавка весенняя
174061 Горечавка введенского
174066 Горечавка валуева
174088 Горечавка юньнаньская
174095 Горечавка цоллингера
174100 Горечавковые
174103 Горечавка осторая, Горечавка острая
174119 Горечавка ушконосная
174120 Горечавка желтеющая, Горечавка золотистая, Горечавка лазолевая, Горечавка лазоревая, Горечавка лазурная
174122 Горечавка полевая
174128 Горечавка комарова
174147 Горечавка маленькая
174163 Горечавка туркестанцев
174164 Горечавка топяная
174205 Горечавка бородатая
174208 Горечавка бородатая китайская
174221 Горечавка желтая
174227 Горечавка бородатая
174431 Гераневые, Гераниевые, Журавельниковые
174440 Герань, Ерань, Журавельник
174453 Герань восточнокавказская
174454 Герань белоцветковая
174499 Герань двулистная
174503 Герань богемская
174530 Герань чарльза
174548 Герань скальная, Герань холмовая
174551 Герань голубиная
174559 Герань даурская
174569 Герань рассечённая

174570 Герань раскидистая, Герань растопыренная
174583 Герань волосистоцветковая, Егорьево копье пушистоцветковая
174589 Герань волосистотычинковая, Герань пушистотычинковая
174609 Герань ферганская
174619 Герань франше
174634 Герань стройная
174653 Герань миболда
174660 Герань грузинская
174682 Герань корейская
174683 Герань кочи
174684 Герань зибольда, Герань крамера
174706 Герань линейнолопастная
174711 Герань блестящая
174715 Герань крупнокорневищная
174729 Герань максимовича
174739 Герань мягкая, Герань нежная
174743 Герань горная
174755 Герань непалская
174796 Герань бледная
174799 Герань болотная
174811 Герань красно-бурая, Герань тёмная
174823 Герань плосколепестная
174829 Герань луговая
174832 Герань забайкальская, Герань луговая, Журавельник луговой
174836 Герань сходная
174846 Герань ложносибирская
174847 Герань мелкотычинковая
174852 Герань пурпурная
174854 Герань маленькая, Герань млкая
174856 Герань пиренейская
174865 Герань прямая
174872 Герань ренарда
174877 Герань роберта
174886 Герань круглолистая, Герань круглолистная
174889 Герань рупрехта
174891 Герань кроваво-расная, Герань кровянокрасная, Журавельник кровянокрасный
174906 Герань сибирская
174919 Герань отпрысконосная, Герань побегоносная
174924 Герань софии
174945 Герань почтибессбельная
174951 Герань лесная
174953 Герань голостебельная
174956 Герань крымуская
174966 Герань тунберга
174972 Герань поперечноклубневая
174986 Герань клубневая
175000 Герань власова
175006 Герань вильфорда, Герань уилфорда
175018 Герань власова
175023 Герань плоская
175107 Гербера, Лейбниция
175172 Гербера джемсона
175295 Геснерия, Горошек забослевый
175306 Геснериевые
175375 Геум, Гравилат
175378 Гравилат алеппский, Гравилат промежуточный, Гравилат прямой, Гравилат средний
175390 Лжегравилат калужницелистный
175400 Гравилат кровавокрасный
175410 Гравилат фори
175417 Гравилат японский
175425 Гравилат широколопастной
175426 Гравилат крупнолистный
175431 Гравилат сахалинский
175444 Гравилат прибрежный, Гравилат речной
175446 Гравилат прямой
175454 Гравилат городской, Гравилат обыкновенный, Любим-трава
175668 Гилия
175760 Гиления, Гилления
175812 Гинкго
175813 Гинкго двлопастной, Гинкго двлопастный, Гинкго двухлопастный, Гинкго китайское
175839 Гинкговые
175865 Жирардиния, Крапива
175868 Жирардиния чжэцзянская
175904 Гиргенсония
175907 Гиргенсония супротивноцветковая
175990 Гладиолус, Шпажник
176038 Шпажник темнофиолетовый
176112 Шпажник кавказский
176122 Шпажник большой, Шпажник обыкновенный
176222 Гладиолус, Гладиолус генский, Шпажник, Шпажник гентский
176246 Шпажник солелюбивый
176264 Шпажник черепитчатый, Шпажник черепичатый
176281 Шпажник итальянский, Шпажник посевный
176296 Шпажник кочи
176429 Шпажник болотный
176589 Шпажник тонкоцветный
176612 Шпажник туркменский
176724 Глауциум, Гляуциум, Мачёк, Мачок, Рогомак
176727 Гляуциум крупно-прицветниковый
176728 Глауциум рогатый, Гляуциум рогатый, Мак рогатый, Мачёк рогатый, Мачок рогатый
176734 Глауциум изяшный
176735 Глауциум бахромчатый, Глауциум жёлтый, Мачёк жёлтый
176738 Глауциум желтый, Гляуциум желтый, Мачок желтый
176739 Гляуциум крупноцветковый
176740 Гляуциум примерный
176743 Гляуциум гладкоплодный
176748 Глауциум остролопастный
176754 Глауциум чешуйчатый
176772 Глаукс, Млечник
176775 Глаукс морской, Глаукс приморский, Млечник морской
176812 Будра, Глехома
176821 Будра японская
176824 Будра плющевидная, Глехома хедерацеум, Котовник ползучий, Котяник
176837 Будра волосистая
176839 Будра длиннотрубчатая
176860 Гледичия
176862 Гледичия аргентинская
176864 Гледичия водяная
176867 Анакардиевые, Анакардовые, Гледичия каспийская, Сумаховые, Фисташковые
176870 Гледичия тайваньская
176881 Гледичия японская
176883 Гледичия японская делаве
176891 Гледичия крупноколючковая
176893 Гледичия мотоская
176901 Гледичия китайская
176903 Гледичия колючая, Гледичия обыкновенная, Гледичия сладкая, Гледичия трехколючковая, Гледичия трёхколючковая, Гледичия трёхшипная
176921 Гения, Гления
176923 Гения прибрежная, Гления прибрежная
176941 Глинус
176949 Глинус лядвенецевидный
177022 шаровница
177032 шаровница безлистноцветковая
177035 Шаровница сердцелистная
177052 шаровница волосоцветковая
177054 Шаровницевые
177096 Глохидион
177170 Глохидион форчуна
177230 Глориоза
177251 Лилия малабарская
177515 Глоксиния
177556 Манник
177560 Манник японский
177562 Манник ольховниковый
177574 Манник тростниковый
177577 Манник канадский
177580 Манник каспийский
177582 Манник китайский
177599 Манник наплывающий, Манник обыкновенный, Манник плавающий,

Манник ручьевой, Трава манная, Трава маточная, Трава утиная
177618 Манник тонкочешуйчатый, Манник уссурийский
177621 Манник тонокорневой
177626 Манник восточный, Манник литовский
177629 Глицерия водяная, Манник большой, Манник водный, Манник водяной, Манник наибольший, Поручейница водная, Поручейница водяная
177635 Манник дубравный
177648 Манник складчатый, Манник туркменский
177661 Манник заболачивающий, Манник колосковый
177672 Манник камчатский, Манник трехцветковый
177677 Манник уссурийский
177689 Глицина, Соя
177722 Соя изящная
177750 Боб сойный, Бобы соевые, Глицина посевная, Соя, Соя культурная, Соя щетинистая
177751 Глицина тайваньская
177777 Бобы соевые, Глицина соя, Соя культурная, Соя уссурийская, Соя щетинистая
177873 Лакрица, Лакричник, Раздельнолодочник, Солодка
177877 Солодка рыхлая, Солодка шершавая, Солодка шиповатая
177883 Солодка бухарская
177885 Солодка ежовая, Солодка щетинистая
177893 Лакричник голый, Солодка гладкая, Солодка голая
177913 Солодка гончарова
177916 Солодка коржинского
177918 Солодка кулябская
177925 Солодка македонская
177932 Солодка бледноцветковая, Солодка бледноцветная
177947 Солодка уральская
177950 Солодка юньнаньская
177951 Солодка зайсанская
178014 Глиптостробус, Кипарис азиатский болотный
178019 Глиптостробус повислый
178056 Гнафалиум, Сушеница
178062 Сушеница родственная
178090 Сушеница байкальская
178114 Сушеница кавказская
178237 Сушеница японская
178240 Сушеница казахстанская
178265 Сушеница желто-белая, Сушеница жёлто-белая, Сушеница желтоватобелая
178278 Сушеница маньчжурская
178303 Сушеница наньчуаньская
178311 Сушеница норвежская
178389 Сушеница русская
178409 Сушеница сибирская
178463 Сушеница лежачая, Сушеница приземистая, Фукус пильчатый
178465 Сушеница лесная
178475 Сушеница траншеля
178481 Сушеница болотная, Сушеница топяная
178522 Гнетовые, Хвойниковые, Эфедровые
178524 Гнетум
178535 Гнетум гнемон
178814 Гольдбахия
178821 Гольдбахия гладкая
178829 Гольдбахия пупырчатая
178831 Гольдбахия сетчатая
178835 Гольдбахия бородавчатая
178851 Стробилантес тайваньский
178974 Гомфокарпус
179032 Гомфокарпус кустарниковый
179221 Гомфрена
179236 Амарант шарообразный, Гомфрена головчатая, Гомфрена шаровидная
179368 Гониолимон
179369 Гониолимон бессера
179370 Гониолимон красивопучковый
179371 Гониолимон остроконечный
179372 Гониолимон джунгарский
179373 Гониолимон высокий
179374 Гониолимон превосходный, Гониолимон прямоветвистый
179375 Гониолимон злаколистный
179378 Гониолимон красноватый
179379 Гониолимон северцова
179380 Гониолимон красивый
179384 Гониолимон татарский
179515 Гончаровия
179516 Гончаровия попова
179571 Гудайера
179631 Гудайера максимовича
179653 Гудайера максимовича
179685 Гудайера ползучая
179694 Гудайера шелехтендаля
179716 Гудайера тяньцюаньская
179721 Гудайера ланьюйская
179722 Гудайера юньнаньская
179729 Гордония
179865 Хлопчатник
179876 Пуходлевник, Хлопчатник древовидный
179878 Хлопчатник индийский, Хлопчатник нанкинский
179884 Абасси, Сиайлэнд, Хлопчатник барбадоский, Хлопчатник египетский, Хлопчатник приморский, Хлопчатник южноамериканский
179900 Гуза, Обыкновенная гуза, Хлопчатник азиатский, Хлопчатник африканский, Хлопчатник афроазиатский, Хлопчатник среднеазиатский, Хлопчатник травянистый
179906 Упланд обыкновенный, Хлопчатник американский, Хлопчатник волосистый, Хлопчатник мохнатый, Хлопчатник нагорный, Хлопчатник средневолокнистый, Хлопчатник упланд, Хлопчатник шерстистый
179937 Хлопчатник зайцева
180046 Грелльсия, Кривоплодник
180050 Грелльсия камнеломколистная
180078 Злаки, Злаковые
180144 Граммосциадиум
180145 Граммосциадиум морковевидный
180292 Авран
180308 Авран японский
180321 Авран аптечный, Авран лекарственный, Зелье лихорадочное, Лихорадочник, Лихорадочник лекарственный, Трава лихорадочная
180561 Гревиллея
180570 Гревиллея банкса
180642 Гревиллея крепкая, Гревиллея сильная
180665 Гревия
180700 Гревия двлопастная
180702 Гревия двлопастная мелколистная
180703 Гревия мелкоцветковая
180835 Гревия гуандунская
180995 Гревия липолистная
181023 Гревия инцзянская
181024 Гревия юньнаньская
181092 Гринделия
181161 Гринделия высогая, Гринделия исполинская
181174 Гринделия растопыренная
181285 Гренландия
181286 Гренландия густая
181317 Гроссгей
181319 Гроссгей крупноголовая
181321 Гроссгей осетинская
181339 Крыжовниковые, Смородинные
181496 Бакаут, Гваякум, Гуаяк, Дерево бакаутовое, Дерево гваяковое
181505 Бакаут, Гваяк, Гваякум аптечный, Гваякум лекарственный, Дерево атласное индийское, Дерево бакаутовое, Дерево бакаутовое железное, Дерево гваяковое, Дерево железное
181544 Гварея
181578 Гварея, Гварея трёхтысячевидная

181626 Гюльденштедтия
181649 Гюльденштедтия ганьсуская
181661 Гюльденштедтия гуансиская
181673 Гюльденштедтия однолистная
181693 Гюльденштедтия мелкоцветковая, Гюльденштедтия ранняя
181728 Геттарда
181872 Гизоция, Нуг
181873 Гизоция абиссинская, Нуг абиссинская, Нуг масличтый, Подсолнечник абиссинский, Рамтиля
181929 Гунделия
181930 Гунделия турнефора
181946 Гунера
181958 Гунера чилийская
182149 Гуцмания
182213 Кокушник
182230 Кокушни, Кокушни комарниковый, Кокушник длиннопогий, Кокушник комалиный
182244 Кокушник эмэйский
182254 Кокушник камчатский, Неолидлейя камчатская
182261 Кокушник ароматнейший, Кокушник душистый
182355 Гимнастер юаньцюйский
182524 Бундук, Гимнокладус, Гимноклядус
182526 Гимнокладус китайский
182527 Бундук двудомный, Бундук канадский, Гимнокладус двудомный, Гимнокладус китайский, Дерево кентуккское кофейное, Шикот, Шипот
182529 Бундук гуансиский
182628 Голосеменные, Голосемянные
182632 Леонтица алтайская
182941 Гинериум, Трава памнаская, Трава памнасовая
182942 Гинериум серебристый, Злак пампасовый
182998 Гиностема
183000 Гиностема бирманская
183023 Гиностема пятилистная
183051 Гинура
183086 Гинура ланьюйская
183097 Гинура японская
183108 Гинура непалская
183157 Гипсолюбка, Гипсофила, Качим, Перекати-поле
183158 Качим остролистный
183167 Качим беловатые
183169 Качим мокоричниковидный
183170 Качим высокий
183171 Качим анатолийский
183172 Качим антонины
183173 Качим арециевидный
183175 Качим двуцветный
183176 Качим короткопепестный
183177 Качим бухарский
183179 Качим головчатый
183180 Качим головчатоцветковый
183181 Качим каппадокийский
183182 Качим густоцветный
183185 Качим даурский
183187 Качим пустыный
183188 Качим виличатый
183189 Качим раскидистый
183191 Гимсофила, Гипсофила изящная, Качим изящный
183194 Качим пушисточашечный
183195 Гипсолюбка пучковатая, Качим пучковатый
183197 Качим федченко
183198 Качим тонконожковый
183199 Качим разноцветный
183200 Качим железистый
183201 Качим сизый
183202 Качим шаровидный
183203 Качим скученный
183205 Качим крыжниковидный
183206 Качим разноножковый
183207 Качим волосистый
183209 Качим черепичатый
183210 Качим спутанный
183211 Качим крешенинникова
183213 Качим линейнолистный
183214 Качим липского
183215 Качим литвинова
183217 Качим мейера
183218 Качим мелколистный
183221 Гипсолюбка постенная, Качим постенный, Качим стенной
183223 Качим тихоокеанский
183225 Гипсолюбка метельчатая, Качим метельчатый, Кучерявка, Перекати-поле, Перекатун
183229 Качим патрэна
183231 Качим тихоокеанский
183232 Качим красивый
183234 Качим попова
183236 Качим вытянутый
183237 Качим плеображенского
183238 Качим ползучий
183240 Качим крепкий
183242 Качим самбука
183243 Качим козелецолистный
183244 Качим шелковистый
183245 Качим поздний, Качим постенный
183247 Качим смолевковидный
183249 Качим лопатчатолистный
183251 Качим степной
183252 Качим штейпа
183253 Качим стевена
183258 Качим шовица
183259 Качим узколистный
183260 Качим тяньшанский
183261 Качим триждывильчатый
183263 Качим хэбэйский
183264 Качим туркестанский
183265 Качим украинский
183266 Качим уральский
183268 Качим фиолетовый
183269 Качим прутьевидный
183270 Качим вязкий
183271 Качим иорский
183389 Поводник, Пололепестник
183565 Поводник гвоздичный
183797 Поводник линейнолистный
184114 Поводник сычуаньский
184136 Поводник тибетский
184234 Габлиция
184235 Габлиция тамусовидная
184291 Гакелия
184298 Гакелия повислоплодная
184326 Хакетия
184345 Гемантус, Хемантус
184396 Гемантус гиблидный
184518 Дерево каменное, Дерево кампешевое, Дерево сандаловое, Дерево сандальное, Сандал синий
184563 Гагения
184565 Гагения абиссинская, Куссо, Хагения абиссинская
184669 Соляноцветник
184670 Соляноцветник кульпский
184671 Соляноцветник редкоцветковый
184672 Соляноцветник розовый
184689 Галения
184691 Галения рогатая, Галения рожковая, Галения сибирская
184696 Галения эллиптическая
184702 Ползунок
184711 Ползунок русский
184713 Ползунок солончаковый
184726 Галезия, Халезия
184731 Галезия каролинская
184810 Галимокнемис
184811 Галимокнемис березина
184813 Галимокнемис голый
184815 Галимокнемис карелина
184816 Галимокнемис войлочноцветковый
184817 Галимокнемис широколистный
184818 Галимокнемис длиннолистный
184819 Галимокнемис крупнопыльниковый
184820 Галимокнемис мягковолосый
184824 Галимокнемис твердоплодный
184827 Галимокнемис смирнова
184829 Галимокнемис мохнатый
184832 Галимодендрон, Чемыш, Чингиль
184838 Чингиль серебристый
184919 Галохалис
184922 Галохалис щетинистоволосый

184923 Галохалис войлочноцветковый
184926 Галохалис туркменский
184930 Сарсазан
184933 Мурза-соранг, Сарсазан шишковатый, Сарсазан шищковатый, Темпексоранг
184938 Галодуле
184947 Галогетон
184950 Галогетон паутинистый
184951 Галогетон скученный, Галогетон тибетский
184952 Галогетон тибетский
184962 Соровник
184965 Соровник низкорослый
184993 Галлорагидовые, Галлораговые
184995 Сланоягодник, Халорагис
185034 Карабаркар, Солянколосник
185038 Карабаркар, Солянколосник, Солянколосник прикаспийский
185057 Галотис
185059 Галотис волосистый
185062 Саксаул
185064 Саксаул зайсанский, Саксаул черный
185072 Саксаул белый, Саксаул персидский, Саксаул песчаный
185089 Гамамелиговые, Гамамелидовые, Гамамелиевые
185092 Гамамелис, Орех волшеб, Хамамелис
185104 Гамамелис японский
185130 Гамамелис китайский, Гамамелис мягкий
185138 Гамамелис весенний
185141 Гамамелис виргинский
185164 Гамелия
185200 Мякотница болотная
185259 Хanделия
185260 Ханделия волосистолистая, Ханделия волосистолистная
185615 Простолистник, Цельнолистник
185619 Цельнолистник остролистный
185620 Цельнолистник близкий
185621 Цельнолистник альберта регеля
185625 Цельнолистник буржо
185629 Цельнолистник бухарский
185630 Цельнолистник бунге
185632 Цельнолистник ресничатый
185633 Цельнолистник предкавказский
185635 Цельнолистник даурский
185636 Цельнолистник джунгарский
185637 Цельнолистник сомнигельный
185639 Цельнолистник евгения коровина
185640 Цельнолистник ферганский
185643 Цельнолистник олиственный
185647 Цельнолистник коваленского
185648 Цельнолистник широколистный
185653 Цельнолистник однобратственный
185654 Цельнолистник многостебельный
185655 Цельнолистник туполистный
185657 Цельнолистник цветооножковый
185658 Цельнолистник исколотый
185662 Цельнолистник попова
185664 Цельнолистник ветвистый
185665 Цельнолистник мощный
185667 Цельнолистник шелковникова
185672 Простолистник крымский, Цельнолистник душистый, Цельнолистник крымский
185673 Цельнолистник тонгий
185674 Цельнолистник тонкорассеченный
185685 Цельнолистник разноцветный
185688 Цельнолистник мохнатый
185689 Цельнолистник введенского
185903 Галения
185904 Галения моховидная
186251 Гавортия
186868 Гайнальдия
186874 Гайнальдия мохнатаю
187176 Хедеома
187195 Плющ
187202 Плющ кавказский
187204 Плющ желтоплодный
187205 Плющ колхидский
187222 Блющ обыкновенный, Плющ обыкновенный
187292 Плющ гималайский
187306 Плющ непалская
187307 Плющ китайский
187311 Плющ пастухова
187318 Плющ японская
187325 Плющ японская тайваньская
187337 Плющ крымский
187361 Хединия
187373 Хединия тибетская
187417 Гедихиум, Хедихиум
187432 Хедихиум короновидный
187442 Хедихиум эмэйский
187455 Хедихиум гуансиский
187466 Хедихиум симаоский
187467 Хедихиум тибетский
187473 Хедихиум тэнчунский
187483 Хедихиум инцзянский
187484 Хедихиум юньнаньский
187709 гединоис
187718 гединоис критская
187734 гединоис персидская
187753 Гедизарум, Копеечник
187760 Копеечник, Копеечник алайский
187769 Копеечник альпийский, Копеечник сибирский
187775 Копеечник аманкутанский
187779 Копеечник арктический
187781 Копеечник серебристый
187783 Копеечник серебристолистный
187785 Копеечник армянский
187789 Копеечник азербайджанский
187790 Копеечник ушковидный
187792 Копеечник южносибирский
187793 Копеечник бабатагский
187794 Копеечник байкальский
187795 Копеечник бальджуанский
187801 Копеечник бордзиловского
187809 Копеечник бранта
187810 Копеечник бухарский
187815 Копеечник бледный, Копеечник блестящий
187823 Копеечник кавказский
187831 Копеечник китайский
187840 Копеечник родственный
187842 Копеечник венцевидный, Копеечник короноподобный, Сулла, Хедизарум, Эспарцет испанский
187843 Копеечник меловой
187848 Копеечник дагестанский
187849 Копеечник даурский
187851 Копеечник щетинистоплодный
187853 Копеечник зубчатый
187858 Копеечник дробова
187860 Копеечник изящный
187869 Копеечник федченко
187870 Копеечник ферганский
187871 Копеечник головчатый
187877 Копеечник извилистый
187879 Копеечник красивый
187881 Копеечник кустарниковый
187912 Копеечник гмелина
187918 Копеечник крупноцветковый, Копеечник крупноцветный
187939 Копеечник грузинский
187940 Копеечник илийский
187946 Копеечник затопляемый, Копеечник сросшийся
187948 Копеечник сырдарьинский
187952 Копеечник каратавский
187953 Копеечник киргизский
187955 Копеечник копетдагский
187956 Копеечник коржинского
187957 Копеечник краснова
187959 Копеечник крылова
187960 Копеечник кудряшева
187961 Копеечник кугитангский
187979 Копеечник лемана
187989 Копеечник липского
187995 Копеечник крупноцветковый
187997 Копеечник великолепный
188006 Копеечник минусинский
188012 Копеечник однолисточковый
188023 Копеечник забытый
188028 Копеечник копеечниковый, Копеечник темный, Копеечник тёмный

188034 Копеечник ольги
188071 Копеечник разомовского
188081 Копеечник сахалинский
188087 Копеечник шишкина
188088 Копеечник прутьевидный
188094 Копеечник семенова
188099 Копеечник щетинистый
188101 Копеечник северцова
188102 Копеечник горный
188104 Копеечник сибирский
188116 Копеечник джунгарский
188129 Копеечник блестящий
188136 Копеечник почтиголый
188140 Копеечник ташкентский
188141 Копеечник крымский
188142 Копеечник узколистный
188143 Копеечник чжундяньский
188155 Копеечник обрубленный
188160 Копеечник туркестанский
188161 Копеечник туркевича
188162 Копеечник украинский
188166 Копеечник уссурийский
188170 Копеечник пестрый
188177 Копеечник горошковидный, Копеечник съедобный
188186 Копеечник рослый
188187 Копеечник тибетский
188190 Хедисцепе
188250 Гегемона
188255 Гегемона мелкоцветная
188263 Геймия
188265 Геймия иволистная
188367 Гельдрейхия
188368 Гельдрейхия длиннолистная
188393 Гелениум
188410 Гелениум бигелова
188554 Гелиамфора
188583 Нежник, Солнцегляд, Солнцецвет
188586 Нежник альпийский, Солнцецвет альпийский, Солнцецвет приальпийский
188609 Солнцецвет буша
188613 Солнцецвет седой
188652 Солнцецвет дагестанский
188683 Солнцецвет крупноцветковый
188713 Солнцецвет щетинистоволосистый
188725 Солнцецвет мохнатоплодный
188726 Солнцецвет багульниколистный
188757 Нежник обыкновенный, Солнцецвет монетолистный, Солнцецвет обыкновенный
188773 Солнцецвет восточный
188828 Солнцецвет скалоломный
188829 Солнцецвет иволистный
188846 Солнцецвет джунгарский
188857 Солнцецвет войлочный
188902 Гелиантус, Подсолнечник
188908 Подсолнечник культурный масличный, Подсолнечник обыкновенный, Подсолнечник однолетний
188922 Подсолнечник серебристолистный
188926 Подсолнечник темно-красный
188936 Подсолнечник огурцеобразный, Подсолнечник слабый
188941 Подсолнечник огурцеобразный, Подсолнечник слабый
189073 Бульва, Груша земляная, Земляная груша, Подсолнечник клубневый, Подсолнечник клубненосный, Репа волошская, Тонконог сизый, Топинамбур
189093 Бессмертник, Гелихризум, Цмин
189147 Бессмертник песчаный, Гелиптерум жёлтый, Сушеница песчаная, Тмин песчаный, Цмин, Цмин жёлтый, Цмин песчаный
189162 Цмин армянский
189409 Цмин пахучий, Цмин сильнопахнущий
189481 Цмин копетдагский
189539 Цмин самаркандский
189583 Цмин мусы
189609 Цмин нуратавский
189638 Цмин палласа
189674 Цмин складчатый
189679 Цмин многочемйчатый
189681 Цмин многолистный
189859 Цмин тяньшанский
189880 Цмин волнистый
189982 Геликодисерос
189986 Геликония, Хеликония
189992 Хеликония бихаи
190092 Геликтерес
190129 Геликтотрихон, Овсец
190145 Овсец даурский
190149 Овсец федченко
190158 Овсец гиссарский
190159 Овсец азиатский
190175 Овсец монгольский
190187 Овсец опушеный
190199 Овсец тяньшанский
190502 Гелиопсис
190525 Гелиопсис шероховатый
190540 Гелиотроп
190542 Гелиотроп остроцветковый
190562 Гелиотроп аргузиевый, Гелиотроп синьцзянский
190577 Гелиотроп двухкольцевидный
190581 Гелиотроп бухарский
190586 Гелиотроп хороссанский
190601 Гелиотроп волосистоплодный
190614 Гелиотроп эллиптический
190622 Гелиотроп европейский
190625 Гелиотроп волосистоплодный
190628 Гелиотроп федчанко
190630 Гелиотроп тайваньский
190640 Гелиотроп крупны
190651 Гелиотроп индийский
190658 Гелиотроп коваленского
190666 Гелиотроп литвинова
190676 Гелиотроп мелкоцветковый
190683 Гелиотроп ольги
190692 Гелиотроп маленький
190702 Гелиотроп перувианский
190730 Гелиотроп зеравшанский
190746 Гелиотроп душистый
190753 Гелиотроп простертый
190755 Гелиотроп шовица
190757 Гелиотроп заамударьинский
190764 Гелиотроп туркменский
190791 Акроклиниум, Гелиптерум, Стеблеклонник
190925 Геллебор, Геллеборус, Зимовник, Морозник, Смелтоед
190928 Морозник абхазский
190932 Морозник кавказский
190936 Морозник вонючий
190938 Морозник крапчатый
190944 Морозник черный, Морозник чёрный
190949 Морозник восточный
190955 Морозник красноватый
190959 Морозник зелёный
191018 гельминция румянковидная
191046 Гелониас, Хелониас
191059 Гелониопсис
191065 Гелониопсис японский
191112 Болотнозонтичник
191117 Болотнозонтичник узлоцветковый
191225 Гемартрия
191232 Гемартрия пучковатая
191251 Гемартрия сирирская, Гемартрия японская
191261 Гемерокаллис, Красноднев, Красоднев, Лилейник, Лилия жёлтая, Рыжай, Сарана, Хемерокаллис
191266 Лилейник лимонножелтый
191268 Красоднев корейский
191272 Красоднев дюмортье
191278 Красоднев дюмортье, Красоднев съедобный
191284 Красноднев рыжий, Красоднев рыжий, Лилейник рыжеватый, Сарана
191302 Красоднев гиблидный
191304 Гемерокаллис, Красоднев желтый, Красоднев рыжий, Лилейник, Сарана
191309 Красоднев миддендорфа
191312 Красноднев малый, Красоднев малый
191318 Красоднев тунбера
191486 Хемиграфис

191501 Хемиграфис широковыемчатый
192027 Генрардия
192028 Генрардия голочешуйная
192029 Генрардия персидская
192120 Перелеска, Печёночник, Печеночница, Печёночница
192131 Печеночница фальконера
192134 Перелеска, Перелеска благолодная, Печёночница благолодная, Печеночница обыкновенная, Печёночница обыкновенная
192138 Печеночница азиатская
192148 Перелеска благолодная, Печёночница благолодная, Печёночница обыкновенная
192227 Борщевик, Гераклеум
192234 Борщевик альбова
192238 Борщевик жесткий
192239 Борщевик бородатый
192248 Борщевик известняковый
192252 Борщевик клауса
192253 Борщевик колхидский
192255 Борщевик рассеченнолистный, Борщевик рассеченный
192259 Борщевик сладкий
192262 Борщевик чэнкоуский
192263 Борщевик узколистный
192265 Борщевик чжундяньский
192266 Борщевик франше
192268 Борщевик крупноцветковый
192269 Борщевик гроссгейма
192271 Борщевик симаоский
192275 Борщевик ганьсуский
192276 Борщевик гуншаньский
192277 Борщевик выемчатолистный
192278 Борщевик шерстистый
192295 Борщевик лемана
192296 Борщевик лескова
192298 Борщевик лигустиколистный, Борщевик любистколистный
192299 Борщевик лицзянский
192302 Борщевик мантегацци
192306 Борщевик мантегацци
192312 Борщевик мелендорфа, Борщевик моллендорфа
192324 Борщевик непалский
192334 Борщевик ольги
192338 Борщевик осетинский
192339 Борщевик дланевидный
192342 Борщевик понтийский
192343 Борщевик пушийстый, Борщевик пушистый
192347 Борщевик розовый
192349 Борщевик шероховатый
192351 Борщевик шерковникова
192352 Борщевик сибирский, Борщевник сибирский
192355 Борщевик соммье
192356 Борщевик сосновского
192357 Борщевик кандинский
192358 Борщевик обыкновенный
192380 Борщевик тэнчунский
192382 Борщевик стевена
192390 Борщевик шероховато-окаймленный
192391 Борщевик закавказский
192394 Борщевик вымечатолистный, Борщевик мохнатый
192395 Борщевик ворошилова
192398 Борщевик вильгельмся, Борщевик пушистый
192403 Гераклеум юньнаньский
192809 Бровник
192862 Бровник одноклубневый
192889 Бровник юньнаньский
192926 Грыжник, Трава колосовая
192940 Грыжник кавказский
192965 Грыжник гладкий, Грыжник голый
192975 Грыжник волосистый
192981 Грыжник седоватый
193009 Грыжник многобрачный
193387 Вечерница, Гесперис, Фиалка ночная
193392 Вечерница армянская
193393 Вечерница двуострая
193403 Вечерница карсская
193417 Вечерница матроны, Вечерница ночная, Вечерница ночная фиалка
193423 Вечерница мейера
193428 Вечерница висячая
193429 Вечерница персидская
193435 Вечерница ложнобелоснежная
193444 Вечерница сибирская
193446 Вечерница лесная
193449 Вечерница стевена, Вечерница стевенова
193452 Вечерница мрачная, Вечерница печальная
193456 Вечерница фиолетовая
193457 Вечерница воронова
193657 Гетерация
193659 Гетерация шовица
193669 Гетерантериум
193671 Гетерантериум волосоносный
193676 Гетерантера
193736 Гетерокарий
193739 Гетерокарий еженосный
193741 Гетерокарий оголенный
193742 Гетерокарий крупноплодный
193743 Гетерокарий малошипиковый
193744 Гетерокарий грубый, Гетерокарий жесткий
193745 Гетерокарий шовица
193794 Гетеродерис
193796 Гетеродерис белоколовый
193797 Гетеродерис маленький
193897 Гетеропанакс
193914 Гетеропаппус
193916 Гетеропаппус альберта
193918 Гетеропаппус алтайский
193929 Гетеропаппус седеющий
193936 Гетеропаппус искривленный
193939 Астра щетинистоволосистая, Гетеропаппус сосняковый, Гетеропаппус щетинисто-волосистый, Гетеропаппус щетинистый
193970 Гетеропаппус средний
193973 Гетеропаппус мейендорфа
193983 Гетеропаппус татарский
194135 Гетероспата высокая
194411 Гейхера, Геухера, Хейхера
194451 Гевеа, Гевея, Сифония, Хевея
194453 Гевея бразильская, Дерево каучуковое, Хевея бразильская
194661 Гибертия
194664 Кандолея
194680 Гибиск, Гибискус, Хибиск, Хибискус
194684 Амбрет, Гибискус мускусный
194779 Гибискус кенаф, Гибискус коноплевый, Гибискус коноплёвый, Джут африканский, Джут яванский, Кенаф, Конопля декан, Пенька бомбейская
194850 Гибискус индийский
194881 Гибискус мотоский
194936 Гибискус индийский
195040 Гибискус переменчивый, Хибиск изменчивый
195089 Гибискус лушаньский
195111 Гибискус понтический
195149 Гибискус, Гибискус китайская роза, Гибискус розан китайский, Мальва сирийская, Роза китайская, Хибискус китайская роза
195196 Гибискус розелла, Гибискус сабдарифа, Кислица ямайская, Розелла, Розель, Салат-розель
195216 Хибискус расщепленнолепестный
195269 Азиат, Гибискус сирийский, Роза азиатская, Роза китайская, Роза сирийская, Роза шарона
195270 Гибискус сирийский, Хибиск сирийский
195305 Гибискус тайваньский
195311 Гибискус индийский, Гибискус липолистный, Дерево пробковое антильское
195326 Гибискус пройчатый, Гибискус северный, Гибискус тройчаный, Гибискус тройчиный
195359 Гибискус юньнаньский
195437 Ястребинка

195441 Ястребинка верхушечнозонтичная
195442 Ястребинка заостренноцветная
195443 Ястребинка остроуловатая
195444 Ястребинка острочешуйная
195445 Ястребинка железистоветочковая
195446 Ястребинка медно-зеленая
195447 Ястребинка акинфиева
195448 Ястребинка алатавская
195450 Ястребинка белесоватая
195452 Ястребинка белостопая
195453 Ястребинка белопепельная
195454 Ястребинка берокаемочная
195455 Ястребинка альпийская
195456 Ястребинка высокостебельная
195458 Ястребинка повсюду-пепельная
195465 Ястребинка остроконечная
195466 Ястребинка сиверковая
195467 Ястребинка изогнутозубчатая
195471 Ястребинка армянская
195472 Ястребинка артвинская
195474 Ястребинка полевая
195477 Ястребинка прежесткая
195478 Ястребинка оранжево-красная
195480 Ястребинка скороспелка
195481 Ястребинка авлорина
195482 Ястребинка бэница
195483 Ястребинка баланзы
195484 Ястребинка бородочковая
195485 Ястребинка основолистная
195486 Ястребинка белоосновная
195487 Ястребинка бессера
195488 Ястребинка биберштейна
195489 Ястребинка боброва
195492 Ястребинка бородина
195493 Ястребинка короткоглавая
195494 Ястребинка брандиса
195496 Ястребинка бриттатская
195497 Ястребинка бузе
195498 Ястребинка володушколистновидная
195499 Ястребинка володушколистная
195500 Ястребинка володушкообразная
195501 Ястребинка буша
195502 Ястребинка сизоцветковидная
195503 Ястребинка сизоветковая
195504 Ястребинка сизоватая
195505 Ястребинка сизая
195506 Ястребинка луговая
195507 Ястребинка прекраснообличная
195508 Ястребинка прекраснообразная
195509 Ястребинка прекрасная
195510 Ястребинка красивочешуйная
195518 Ястребинка седая
195520 Ястребинка сердцевидная
195521 Ястребинка сердцелистная
195523 Ястребинка каспари
195526 Ястребинка кавказская
195527 Ястребинка волосистосоцветноподобная
195528 Ястребинка волосистсоцветная
195530 Ястребинка зеленоцветная
195531 Ястребинка бледнозеленочешуйная
195533 Ястребинка зеленолюбивая
195535 Ястребинка зелененькая
195536 Ястребинка окрашенноешуйчатая
195538 Ястребинка кудрявая
195539 Ястребинка пепельнополосчатая
195540 Ястребинка холовообразная
195541 Ястребинка окрашенная
195544 Ястребинка хохлатая
195545 Ястребинка одноцветненькая
195546 Ястребинка конусообразная
195547 Ястребинка известная
195550 Ястребинка кудреватая
195551 Ястребинка курчавая
195553 Ястребинка кровянопятнистая
195554 Ястребинка кривоватая
195555 Ястребинка цимозная
195558 Ястребинка изогнутолистная
195559 Ястребинка дехи
195560 Ястребинка низбегающая
195561 Ястребинка зубчатая
195562 Ястребинка мелкозубчатая
195564 Ястребинка прозрачновидноглавая
195566 Ястребинка двуцветная
195568 Ястребинка двуцветненькая
195569 Ястребинка распущенная
195570 Ястребинка долотолистная
195571 Ястребинка дублицкого
195572 Ястребинка жесткошетинистая
195575 Ястребинка румянкоглавая
195576 Ястребинка румянковая
195577 Ястребинка эйхвальда
195578 Ястребинка красносемянковидная
195579 Ястребинка красноплодная
195581 Ястребинка отличная
195583 Ястребинка серпозубчатая
195584 Ястребинка обманчивая
195588 Ястребинка нитяная
195594 Ястребинка плетеобразная
195596 Ястребинка гибкостевельная
195597 Ястребинка звездчатохохлатая
195598 Ястребинка мелкохлопьевидная
195599 Ястребинка хлопьецветоносовая
195601 Ястребинка изобильноцветковая
195602 Ястребинка обильноцветковая
195603 Ястребинка многоцветковая, Ястребинка обильноцветущая
195604 Ястребинка цветстая
195605 Ястребинка листочконосная
195606 Ястребинка олиственнейшая
195607 Ястребинка фомина
195608 Ястребинка фрейна
195609 Ястребинка фрика
195611 Ястребинка хладолюбивая
195613 Ястребинка фрица
195614 Ястребинка олиственная
195615 Ястребинка сажистая
195616 Ястребинка рыжая
195617 Ястребинка отрубистая
195618 Ястребинка ганешина
195619 Ястребинка сродственная
195620 Ястребинка грузинская
195621 Ястребинка крупноватая
195622 Ястребинка крупностебельная
195623 Ястребинка гладкоязычковая
195625 Ястребинка голубовато-зеленая
195627 Ястребинка глена
195628 Ястребинка скученнообразная
195629 Ястребинка скученная
195630 Ястребинка сушеничная
195631 Ястребинка городкова
195637 Ястребинка крупнозубчатая
195641 Ястребинка сероватая
195645 Ястребинка гюнтера
195646 Ястребинка густава
195647 Ястребинка голая
195648 Ястребинка кровяноязычная
195649 Ястребинка гаральда
195651 Ястребинка разнозубчатая
195652 Ястребинка волосистейшая
195654 Ястребинка гогенакера
195657 Ястребинка однокровельная
195658 Ястребинка гоппе
195661 Ястребинка почти-сизая
195662 Ястребинка подъельниковая
195663 Ястребинка подбородчатая
195664 Ястребинка сероватоподобная
195665 Ястребинка седоватая
195666 Ястребинка необыкновенная
195668 Ястребинка посредническая
195669 Ястребинка тоше-железистая
195670 Ястребинка черноголовая
195672 Ястребинка японская
195673 Ястребинка юрская
195674 Ястребинка кабанова
195675 Ястребинка качурина
195676 Ястребинка каянская
195683 Ястребинка карягина
195684 Ястребинка карпинского
195685 Ястребинка кемуларии
195686 Ястребинка чильмана
195687 Ястребинка киргизская
195688 Ястребинка кнафа
195689 Ястребинка кенига
195690 Ястребинка кёрнике
195691 Ястребинка конжаковская
195692 Ястребинка коржинского
195693 Ястребинка косьвинская
195694 Ястребинка
195695 Ястребинка козловского

195698 Ястребинка крашана
195699 Ястребинка кречетовича
195700 Ястребинка крылова
195701 Ястребинка кубанская
195702 Ястребинка кубинская
195703 Ястребинка кулькова
195704 Ястребинка кумбельская
195705 Ястребинка купфера
195706 Ястребинка кузнецкая
195707 Ястребинка филандская
195708 Ястребинка обыкновенная
195709 Ястребинка лакшевица
195711 Ястребинка выровненная
195712 Ястребинка сглаженная
195713 Ястребинка блестащеволосоподобная
195714 Ястребинка блестащеволосистая
195716 Ястребинка ланцетнолистная
195717 Ястребинка копьезубая
195718 Ястребинка лапландская
195719 Ястребинка ляша
195720 Ястребинка густопушистая
195722 Ястребинка обильноопушенная
195723 Ястребинка скрытая
195724 Ястребинка однобочная
195725 Ястребинка бокоцветковая
195729 Ястребинка леберта
195731 Ястребинка ленкоранская
195732 Ястребинка миленькая
195733 Ястребинка чешуйчатообразная
195734 Ястребинка мелкожелезистая
195735 Ястребинка тонкостебельная
195736 Ястребинка тонковетвистая, Ястребинка узколистная
195737 Ястребинка мелкописаноподобная
195738 Ястребинка тонколослая
195739 Ястребинка тонколистная, Ястребинка узколистная
195740 Ястребинка леспинаса
195742 Ястребинка белокистеродная
195743 Ястребинка белокиствидная
195744 Ястребинка белокистевая
195745 Ястребинка голостебельная
195746 Ястребинка левье
195747 Ястребинка льнолистная
195748 Ястребинка липмаа
195749 Ястребинка липского
195750 Ястребинка гладкочешуйная
195751 Ястребинка литовская
195752 Ястребинка береговая
195753 Ястребинка литвинова
195755 Ястребинка побарчевского
195756 Ястребинка ломницкая
195758 Ястребинка длинностопая
195762 Ястребинка длиннолучистая
195763 Ястребинка длиннострелковая
195764 Ястребинка длиннощетинистая
195765 Ястребинка длинная
195766 Ястребинка желтожелезистая
195767 Ястребинка лидии
195768 Ястребинка лировидная
195769 Ястребинка крупнозонтичная
195770 Ястребинка крупноцветковая
195771 Ястребинка крупнощетинистая
195772 Ястребинка крупнозонтичная
195773 Ястребинка крупнопленчатая
195774 Ястребинка крупночешуевидная
195775 Ястребинка крупночешуйноподобная
195777 Ястребинка крупночешуйчатая
195778 Ястребинка крупноприкорнелистная
195779 Ястребинка крупнолучистая
195781 Ястребинка пятнисная
195783 Ястребинка крупно-скороспелковая
195784 Ястребинка мальме
195785 Ястребинка каемчатая
195787 Ястребинка мейнсглаузена
195788 Ястребинка черноватоглавая
195789 Ястребинка пленчатая
195790 Ястребинка менделя
195791 Ястребинка лоснящаяся
195792 Ястребинка мелкозвезчатая
195793 Ястребинка мелкоскрученная
195794 Ястребинка мелкошаровидная
195795 Ястребинка удивительная
195796 Ястребинка мягкощетинковая
195799 Ястребинка многоликая
195800 Ястребинка многожелезистая
195801 Ястребинка обильнощетинистая
195802 Ястребинка мурманская
195803 Ястребинка внешняя, Ястребинка постеная, Ястребинка стенная, Ястребинка стеночная
195805 Ястребинка нальчикская
195806 Ястребинка мелкоголовчатая
195807 Ястребинка нарымская
195808 Ястребинка ненюкова
195810 Ястребинка чернеющая
195811 Ястребинка чернощетинистая
195812 Ястребинка снежнокаемочная
195813 Ястребинка норрлиновидная
195814 Ястребинка норрлина
195815 Ястребинка норвежская
195817 Ястребинка темноприцветниковая
195818 Ястребинка неясная, Ястребинка темная
195819 Ястребинка олимрийская
195820 Ястребинка оманга
195821 Ястребинка лесная, Ястребинка онежская
195822 Ястребинка оносмовидная
195823 Ястребинка круглеющая
195825 Ястребинка украшенная
195826 Ястребинка прямостопая
195827 Ястребинка освальда
195828 Ястребинка овальноглавая
195829 Ястребинка овальноликая
195830 Ястребинка яйцеобразная
195831 Ястребинка плотная
195832 Ястребинка панша
195834 Ястребинка бледная
195837 Ястребинка панютина
195838 Ястребинка паннонскообразная
195840 Ястребинка лохматая
195841 Ястребинка уклончивая, Ястребинка утешающая
195842 Ястребинка парейса
195843 Ястребинка прозрлчная
195844 Ястребинка повислая
195845 Ястребинка прешероховатая
195846 Ястребинка пронзеннолистная
195847 Ястребинка большезубая
195849 Ястребинка черешчатая
195850 Ястребинка петергофская
195851 Ястребинка петунникова
195854 Ястребинка волосистостопая
195855 Ястребинка волосисточешуйчатая
195856 Ястребинка волосистоцветковая, Ястребинка волосистоцветная
195857 Ястребинка флорентийская
195858 Ястребинка волосистая
195862 Ястребинка сосновая
195863 Ястребинка белопленчатая
195865 Ястребинка складчатая
195866 Ястребинка многостебельная
195867 Ястребинка многолистная
195868 Ястребинка поле
195869 Ястребинка серозлачная
195870 Ястребинка разнолистная
195872 Ястребинка вытянутая
195873 Ястребинка плевысокая, Ястребинка стройная
195877 Ястребинка пропущенная
195879 Ястребинка лужаечная
195880 Ястребинка латуковидная
195881 Ястребинка высокорослая
195882 Ястребинка видная, Ястребинка высокая
195883 Ястребинка раскидистая
195884 Ястребинка вытянутая
195886 Ястребинка близкая
195887 Ястребинка инееносная
195888 Ястребинка прусская
195889 Ястребинка песколюбивая
195890 Ястребинка ушковатая
195891 Ястребинка ложноушковатая
195892 Ястребинка ложнолаздвоенная
195893 Ястребинка ложнокороткая
195894 Ястребинка ложносжатая
195895 Ястребинка ложно-юрская
195906 Ястребинка топяная

195907 Ястребинка головетвистая
195908 Ястребинка птшистая
195909 Ястребинка одутловатая
195910 Ястребинка пурпурноповязочная
195911 Ястребинка густосоцветная
195912 Ястребинка пятигорская
195913 Ястребинка радде
195914 Ястребинка ветвистая
195915 Ястребинка регелевская, Ястребинка регеля
195916 Ястребинка рэмана
195917 Ястребинка жесткая
195921 Ястребинка грубая, Ястребинка могучая, Ястребинка мощная
195922 Ястребинка ройовского
195924 Ястребинка рота
195928 Ястребинка рупрехта
195929 Ястребинка рузская
195931 Ястребинка савойская
195932 Ястребинка вересовая
195933 Ястребинка можжевеловая
195934 Ястребинка песчаная
195935 Ястребинка сахокия
195936 Ястребинка стреловидная, Ястребинка стрелолистная
195938 Ястребинка самурская
195939 Ястребинка хворостовидная
195940 Ястребинка камнеломковая
195941 Ястребинка негладкая
195950 Ястребинка скандинавская
195951 Ястребинка шелковникова
195952 Ястребинка шишкина
195953 Ястребинка шлякова
195955 Ястребинка шультеса
195957 Ястребинка зонтиконосная
195958 Ястребинка скальная
195963 Ястребинка полуранняя
195964 Ястребинка севирная
195965 Ястребинка шелковистостебельная
195966 Ястребинка пильчатолистная
195968 Ястребинка силена
195970 Ястребинка лесная
195971 Ястребинка ростостебельная
195975 Ястребинка сочавы
195980 Ястребинка рыжеющая
195981 Ястребинка сосновского
195983 Ястребинка лопатчатолистная
195984 Ястребинка оттопыренная
195985 Ястребинка ставропольская
195986 Ястребинка штейнберг
195988 Ястребинка звездчатая
195989 Ястребинка узкочешуйная
195991 Ястребинка побегоцветная, Ястребинка столоноцветковая
195992 Ястребинка скрученноволосая
195993 Ястребинка прямейшая
195994 Ястребинка сомнительная
195995 Ястребинка почти-северо-восточная
195996 Ястребинка почти-артвинская
195997 Ястребинка шероховатенькая
195998 Ястребинка ложноушастая
195999 Ястребинка голубоватая
196000 Ястребинка почти-толстолистная
196001 Ястребинка почти-прямая
196002 Ястребинка почти-многоцветковая
196003 Ястребинка почти-копьевидная
196004 Ястребинка жестковолосистейшая
196005 Ястребинка почти-имандоровая
196006 Ястребинка почти-густопушистая
196007 Ястребинка почти-бледная
196008 Ястребинка почти-пятнисная
196009 Ястребинка почти-окаймленная
196010 Ястребинка почти-черноватая
196011 Ястребинка черноватенькая
196012 Ястребинка почти-темноглавая
196013 Ястребинка почти-прозрачная
196014 Ястребинка почти-красноватая
196015 Ястребинка почти-простая
196016 Ястребинка почти-побежистая
196017 Ястребинка шведская
196018 Ястребинка сернистовидная
196019 Ястребинка сернистая
196021 Ястребинка серножелтая
196023 Ястребинка сырейщикова
196024 Ястребинка совича
196025 Ястребинка таненская
196028 Ястребинка тауша
196029 Ястребинка тебердинская
196030 Ястребинка пепельноцветная
196031 Ястребинка пепельносоцветная
196032 Ястребинка пеплолюбивая
196033 Ястребинка пепельностопая
196034 Ястребинка тепроухова
196035 Ястребинка терекская
196036 Ястребинка тилинга
196037 Ястребинка скрученногавая
196039 Ястребинка треугольная
196040 Ястребинка волосистоветвистая
196041 Ястребинка трехзубчатоглавая
196042 Ястребинка трехзубчатая
196043 Ястребинка траурная
196049 Ястребинка чхубианишвили
196051 Ястребинка трубчатая
196052 Ястребинка туркестанская
196053 Ястребинка черноватая
196054 Ястребинка угандийская
196055 Ястребинка
196057 Ястребинка зонтичная
196068 Ястребинка зонтоносная
196069 Ястребинка рыхлозонтичная
196070 Ястребинка теневая
196071 Ястребинка уральская
196072 Ястребинка усинская
196073 Ястребинка вайяна
196074 Ястребинка пестрочешуйная
196075 Ястребинка баскская
196076 Ястребинка ваги
196079 Ястребинка глянцевитая
196080 Ястребинка мелкобородавчатая
196081 Ястребинка шерстистая, Ястребинка ядовитая
196084 Ястребинка фиолетовостопая
196085 Ястребинка ветвистая
196086 Ястребинка ядовитая
196087 Ястребинка клейкая
196090 Ястребинка яичножелтая
196091 Ястребинка волжская
196092 Ястребинка обыкновенненькая
196093 Ястребинка обыкновенная
196095 Ястребинка виммера
196096 Ястребинка волчанская
196097 Ястребинка вологодская
196098 Ястребинка воронова
196100 Ястребинка желторыльцевая
196101 Ястребинка цинзерлинга
196102 Ястребинка цица, Ястребинка цицова
196116 Зубровка, Лядник
196124 Зубровка южная
196159 Зубровка малоцветковая
196372 Ремнелепестник
196374 Ремнелепестник прекрасный
196439 Амарилис, Гипеаструм, Гиппеаструм, Хипеаструм
196458 Гиппеаструм
196494 Конскокаштановые
196501 Гиппократия
196621 Гиппокрепис, Подковник
196627 Гиппокрепис двуцветковый
196629 Гиппокрепис ресничатый
196630 Гиппокрепис косматый
196675 Гиппокрепис одностручковый
196688 Ипполития
196692 Ипполития дарвазская
196698 Ипполития гердера
196703 Ипполития крупноглавая, Ипполития крупноголоавчатая
196706 Ипполития шугнанская
196712 Ипполития юньнаньская
196716 Гиппомана, Манцинелла
196723 Конский фенхель
196738 Конский фенхель длиннольчатый
196739 Конский фенхель мелкоплодный
196743 Облепиха
196757 Облепиха крушиновая, Облепиха крушиновидная, Облепиха обыкновенная
196780 Облепиха тибетская
196809 Воднососенковые, Хвостниковые
196810 Водяная сосенка, Сосенка водяная, Хвостник

196816 Хвостник четырехлистный
196818 Водяная сосенка обыкновенная, Сосенка водяная, Хвостник обыкновенная
196950 Гиршфельдия
196952 Гиршфельдия серая
197137 Гогенакерия
197139 Гогенакерия бесстебельная
197278 Бухарник
197301 Бухарник шерстистый, Вика кормовая, Трава модовая
197306 Бухарник мягкий, Просяник
197323 Бухарник щетинистый
197331 Сорго суданское, Суданска, Трава суданская
197468 Голосхенус
197469 Голосхенус римский
197486 Костенец
197490 Костенец окаймленный
197491 Костенец зонтичный, Костенец липкий
197619 Хомаляантус
197625 Хомаляантус тополевый
197633 Гомалиум
197943 Подбельник
197944 Подбельник альпийский
198003 Аммодения
198010 Аммодения бутерлаковидная
198126 Хопея
198176 Хопея душистая
198237 Горaниновия, Горяниновия, Сажереция
198238 Горaниновия улексовидная
198240 Горяниновия неправильная
198241 Горяниновия исключительная
198243 Горяниновия малая
198260 Критезион, Хордеум, Ячмень
198264 Ячмень богдана
198265 Ячмень северный
198267 Ячмень колончаковый, Ячмень короткоостистый, Ячмень короткоостый, Ячмень короткошиловидный, Ячмень луговой
198275 Халдумка, Ячмень дикий луковичный, Ячмень луковичный
198288 Ячмень двурядный, Ячмень двухрядный, Ячмень заячий, Ячмень крупный, Ячмень культурный двурядный, Ячмень пивоваренный, Ячмень яровой двурядный
198293 Овес голый, Овёс голый, Овёс мелкозёрный, Овес татарский, Овёс татарский, Ячмень голый
198297 Ячмень коленчатый, Ячмень ощетиненный
198307 Ячмень тощеостый
198311 Критезион гривастый, Ячмень гривастый, Ячмень гривоподобный
198316 Ячмень заячий
198327 Ячмень двурядный, Ячмень мышиный
198346 Ячмень рожевица
198354 Ячмень житяный, Ячмень ржаной
198360 Ячмень сибирский
198365 так-так, Ячмень дикий, Ячмень дикорастущий
198373 Ячмень туркестанский
198375 Ячмень фиолетовый
198376 #######
198483 Двсемянник скальный
198485 Многосемянник лежачий, Тонкостенник лежачий
198589 Госта, Функия, Хоста
198618 Госта японсая, Функия ланцетолистная
198624 Госта ланцетолистная, Функия ланцетолистная
198646 Хоста зибольда
198652 Функия яйцевидная
198676 Турча
198680 Перовския, Турча болотная
198698 Магнолия обратноовальная, Магнолия обратнояйцевидная, Магнолия туполистная, Магнолия японская
198703 Хоустония, Хустония
198743 Гуттуиния
198745 Гуттуиния сердцевидная
198766 Говения, Дерево конфетное, Ховения
198768 Говения цюцзянская
198769 Говения, Говения сладкая, Дерево конфетное, Конфетное дерево, Сладконожка, Сладконошник
198786 Говения тонковойлочная
198804 Ховея
198806 Ховея бельмора
198821 Плющ восковой, Хойя
198827 Дерево восковое, Плющ восковой, Хойя мясистая
198862 Хойя тибетская
198864 Хойя гуншаньская
198867 Хойя либоская
198875 Хойя мэнцзыская
199137 Гутония
199238 Дендрантема цельнолистная
199240 Гультемия, Хультемия
199243 Гультемия персидская, Хультемия иранская
199379 Ива гукера, Хмель
199382 Хмель японский
199384 Большеголовник альпийский, Хмелица, Хмель вьющийся, Хмель обыкновенный
199392 Хмель лазящий
199394 Хмель юньнаньский
199407 Хуннемания
199465 Хура трескающаяся
199480 Двсемянник
199545 Гиацинтик
199546 Гиацинтик беловатый
199547 Гиацинтик палласа
199555 Пролеска итальянская
199565 Гиацит
199577 Гиацит копетдагский
199580 Гиацит литвинова
199583 Гиацит восточный
199620 Гиалея
199624 Гиалея красивая
199626 Гиалея таджикская
199646 Гиалолена
199647 Гименолима володушковидная
199648 Гиалолена обедненная
199649 Гиалолена яксартская
199650 Гиалолена метельчатая
199740 Гигнокарпус, Хиднокарпус
199759 Гигнофитум
199787 Гидрангеа, Гидрангиа, Гортензия
199806 Гидрангея древовидная, Гортензия древовидная
199846 Гортензия бретшнейдера
199853 Гидрангеа китайская
199891 Гортензия разнонаправленная
199931 Гортензия гуансиская
199936 Гортензия гуандунская
199939 Гортензия личуаньская
199956 Гортензия крупнолистная
200038 Гидрангея метельчатая, Гидрангиа метельчатая, Гортензия метельчатая
200059 Гидрангиа черешчатая, Гортензия черешковая
200085 Гортензия пильчатая
200093 Гортензия пильчатая бело-розовая
200103 Гортензия шансиская
200121 Гидрангеа тайваньская
200148 Гортензия золотистожилковая
200157 Гидрангеа юньнаньская
200160 Гидранговые
200173 Гидрастис, Желтокорень, Печать золотая
200175 Гидрастис канадский, Желтокорень канадский
200184 Гидрилла
200190 Гидрилла мутовчатая
200225 Водокрас, Лягушатник, Лягушечник, Цикорий полевой
200228 Водокрас обыкновенный, Водокрас-лягушечник, Лягушечник
200230 Водокрас лягушечный, Водокрас обыкновенный
200235 Водокрасовые

200241 Гидроклеис
200253 Водолюб, Гидрокотиле, Гидрокотиль, Щитолистник
200267 Щитолистник бирманский
200274 Щитолистник китайский
200312 Гидрокотиле яванская
200350 Гидрокотиле веткоцветковая
200354 Щитолистник лютиковый
200388 Щитолистник обыкновенный
200476 Водолистниковые, Гидрофилляциевые
200482 Водолюб, Гидрофил, Гидрофиллум
200589 Гигрофила
200736 Десмодиум ольдама
200740 Десмодиум обманчивый
200745 Десмодиум сычуаньский
200753 Лесной мак, Лесной чистотел
200754 Лесной мак японский, Лесной чистотел японский
200775 Хилотелефиум
200788 Очиток эверса
200793 Очиток бледнеющий, Очиток бледноватый, Седум посконниковый
200798 Седум замечательный
200802 Седум зибольда
200812 Заячья капуста, Очиток пурпурный, Очиток пурпуровый, Седум пурпуровый, Скрипун
200816 Очиток мутовчатый, Седум мутовчатый
200820 Очиток живородящий
200906 Гименокаллис, Лилия нильская, Хименокаллис
200941 Хименокаллис прибрежный
201038 Гименократер
201039 Гименократер смолистый
201040 Гименократер изящный
201096 Гименолена
201098 Гименолена альпийская
201111 Гименолена бедренцелистная
201135 Многосемянник, Тонкостенник
201151 Многосемянник пушистый
201155 Гименолима
201157 Гименолима волосистая
201208 Двугнездка крупноплодная
201212 Двугнездка пушистая
201370 Белена
201372 Белена белая
201377 Белена богемсая, Белена чешская
201384 Белена копетдагская
201389 Белена черная, Белена чёрная, Мильг-девона
201401 Белена крошечная
201402 Белена сетчатая
201405 Белена туркменская
201436 Гипаканциум
201437 Гипаканциум мордовниковолистный
201597 Гипекоум, Житник
201605 Гипекоум прямой
201609 Гипекоум крупноцветковый
201612 Гипекоум китайский, Гипекоум тонкоплодная
201616 Гипекоум мелкоцветный
201618 Гипекоум вислоплодный, Житник повислый
201631 Гипекоум трехлопастной
201647 Генипа
201681 Зверобоецветные, Зверобойные
201705 Зверобой
201713 Зверобой крылатый, Зверобой осторый
201729 Зверобой аленийский, Зверобой приальпийский
201730 Зверобой аленийский
201732 Зверобой красильный
201738 Зверобой Ардасенова
201740 Зверобой армянский
201743 Зверобой большой, Зверобой геблера
201759 Зверобой ясменниковый
201761 Зверобой оттянутый
201785 Зверобой володушковидный
201786 Зверобой буша
201787 Зверобой чамчковидный, Зверобой чашечный, Зверобой чашечщый
201797 Зверобой кавказский
201812 Зверобой золотистокистевый, Зверобой иссополистный
201825 Зверобой курчавый
201840 Зверобой изящный
201848 Зверобой вытянутый, Зверобой удлиненный
201853 Зверобой прямой, Зверобой прямостоящий
201880 Зверобой тайваньский
201882 Зверобой прекрасный
201914 Зверобой солнцецветный
201920 Зверобой мэнцзыский
201922 Зверобой тибетский
201924 Зверобой жестковолосый, Зверобой опушенный, Зверобой шершавоволосый
201930 Зверобой распростертый, Зверобой стелющийся
201937 Зверобой непахнущий
201942 Зверобой японский
201954 Зверобой камчатский
201957 Зверобой
201960 Зверобой карсский
201969 Зверобой чаюйский
201973 Зверобой комарова
201974 Зверобой гуйцзоуский
202000 Зверобой лидийский
202003 Зверобой кандинский
202006 Зверобой пятнистый, Зверобой точечный, Зверобой четырехгранный
202011 Зверобой окаймленный
202021 Зверобой китайский
202026 Зверобой горный
202029 Зверобой недоразвитый
202049 Зверобой нордманна
202050 Зверобой округлолистный
202060 Зверобой олимпийский
202061 Зверобой душицелистный
202070 Зверобой растопыренный
202086 Зверобой исколотый, Зверобой обыкновенный, Зверобой плонзеннолистный, Зверобой пронзенный, Зверобой протырявленный
202098 Зверобой горецелистный
202100 Зверобой нонтийский, Зверобой черноморский
202142 Зверобой
202147 Зверобой шероховатый
202167 Зверобой торчащий
202179 Зверобой четырехгранный, Зверобой четырёхгранный
202180 Зверобой федора
202205 Зверобой красивый
202230 Зверобой иезский
202256 Думпальма, Пальма “дум”
202321 Гифене, Думпальма настоящая, Пальма-гифаена
202387 Гиохерис, Пазник, Прозанник
202400 Прозанник реснитчатый
202404 Гиохерис голый, Пазник гладкий, Пазник голый
202424 Прозанник крапчатый, Прозанник пятнистый
202432 Гиохерис короткокорневой, Гиохерис укореняющийся, Пазник уколенившийся, Пазник уколеняющийся
202444 Прозанник одноголовчатый
202649 Гипогомфия
202650 Гипогомфия высогая
202651 Гипогомфия туркестанская
202761 Подъельник
202767 Вертляница подъельник, Вертляница подъельниковая, Подъельник обкновенный
203079 Гиссон, Иссоп
203080 Тимьян алтайский
203081 Иссоп сомнительный
203082 Иссоп узколистный
203083 Иссоп меловой
203084 Иссоп остроконечный
203087 Иссоп ферганский
203092 Иссоп крупноцветный
203095 Иссоп аптечный, Иссоп лекарственный, Иссоп обыкновенный
203104 Иссоп зеравшанский

203105 Иссоп тяньшанский
203124 Гистрикс, Хистрикс, Чертополох курчавый, Шерохватка
203136 Гистрикс коморова
203142 Гистрикс сибириский
203175 Иберичка
203179 Иберичка трехжильная
203181 Иберийка, Иберис, Перечник, Разнолепестка, Разнолепестник, Разноорешник, Стенник
203184 Иберийка горькая, Иберис горький, Перечник горький, Разнолепестник горький, Разнолистник, Стенник горьгий
203208 Иберис корончатый
203234 Иберийка перистая
203236 Иберийка скальная
203239 Иберийка вечнозелёная, Стенник вечнозелённый
203248 Иберийка крымская
203249 Иберийка зонтичная, Иберис зонтикоцветный, Перечник зонтичный, Стенник зонтичный
203260 Иберис жёлтая
203303 Икациновые
203417 Идезия
203422 Идезия многоплодная, Ландышевое дерево
203424 Идезия фуцзяньская
203518 Иконниковия
203519 Иконниковия кауфмана, Иконниковия кауфмановская
203527 Илекс, Остролист, Падуб
203545 Остролист, Падуб востролистый, Падуб остролистный
203553 Падуб востролистый колючий
203584 Илекс ван
203621 Илекс чэнкоуский
203625 Илекс китайский, Падуб китайский, Падуб пурпурный
203643 Падуб колхидский
203645 Илекс гуансиский
203648 Падуб коралловый
203660 Илекс рогатый, Падуб рогатый
203670 Падуб городчатый
203789 Дрок колючий европейский, Урекс европейский
203798 Падуб фаржа
203807 Илекс фэнцинский
203821 Илекс тайваньский
203833 Илекс кандинский
203834 Илекс кандинский мелколистный
203835 Илекс фуцзяньский
203846 Илекс цзиндунский
203868 Илекс гуйцзоуский
203869 Илекс хайнаньский
203883 Падуб наньтоуский
203891 Падуб гуншаньский
203896 Падуб хусюин
203897 Падуб хунаньский
203904 Падуб гирканский
203908 Падуб цельнокрайний
203932 Падуб цзиньюньский
203950 Илекс куэньминский
203952 Илекс ланьюйский
203953 Илекс гуандунский
203961 Падуб широколистный
203971 Илекс лэйбоский
203980 Илекс баотинский
203991 Илекс лунчжоуский
204022 Илекс малипоский
204025 Падуб максимовича
204030 Илекс мотоский
204079 Илекс наньчуаньский
204080 Илекс наньнинский
204112 Илекс эмэйский
204114 Падуб тусклый
204135 Дерево чайное парагвайское, Илекс парагвайский, Йерба-матэ, Мате, Матэ, Падуб парагвайский, Парагвайский чай, Чай парагвайский, Чай-матэ, Чайный куст парагвайский
204154 Илекс шансиский
204184 Падуб опушенный
204226 Падуб морщинистый
204239 Падуб пильчатый
204275 Падуб узкоплодный
204276 Падуб узколистный
204292 Падуб сугероки
204311 Падуб сычуаньский
204341 Падуб трехцветковый
204376 Падуб мутовчатый
204391 Куст чайный апалашский, Падуб рвотный, Чай рвотный
204395 Падуб ван
204398 Падуб симаоский
204407 Илекс тибетский
204410 Илекс янчуньский
204413 Илекс юньнаньский
204418 Илекс мелколистный, Илекс юньнаньский мелколистный
204421 Илекс чжэцзянский
204434 Ильиниая
204435 Ильиниая ределя
204443 Хрущецветник
204469 Хрущецветник мутовчатый
204474 Бадьян, Иллициум
204484 Иллициум тайваньский
204490 Тайваньск бирманский
204491 Бадьян камбоджийский
204517 Бадьян флоридский
204524 Бирманск тибетский
204533 Анис звездчатый японский, Анис ядовитый
204558 Иллициум мелкоцветковое
204570 Бадьян мелкоцветковый
204571 Иллициум мелколистный
204575 Бадьян настоящий, Дерево анакардовое, Иллициум анисовый, Иллициум обрядный, Сикими
204583 Бадьян священный
204598 Иллициум луаньдаский
204601 Иллициум вэньшаньский
204603 Анис звездчатый, Анис китайский, Анис эвездчатый, Бадьян звездчатый, Бадьян китайский, Бадьян настоящий, Дерево анисовое, Звездчатый анис, Иллициум анисовый, Иллициум настоящий
204756 Бальзамин, Недотрога, Прыгун
204775 Прыгун аньхуйский
204788 Недотрога бирманская
204799 Бальзамин садовый, Недотрога балезаминовая, Недотрога бальзаминовидная
204825 Недотрога короткошпорцевая
204846 Недотрога чжэцзянская
204848 Недотрога китайская
204854 Недотрога чжундяньская
204881 Недотрога тибетская
204946 Недотрога фэнхуайская
204969 Недотрога вильчатая
204982 Недотрога железконосная
204987 Бальзамин гуншаньский
204997 Бальзамин гуйцзоуский
204998 Недотрога хайнаньская
205024 Недотрога хунаньская
205058 Недотрога комарова
205107 Недотрога маака
205129 Недотрога мотоская
205132 Недотрога мэнцзыская
205141 Недотрога мелкоцветковая
205154 Недотрога лунчжоуская
205155 Недотрога мулиская
205165 Недотрога напоская
205169 Недотрога невского
205175 Бальзамин недотрога, Недотрога "не тронь меня", Недотрога желтая, Недотрога обыкновенная, Нетронь меня, Не-тронь-меня, Прыгун
205197 Недотрога эмэйская
205212 Недотрога мелкоцветковая
205216 Недотрога мелколистная
205291 Недотрога железистая, Недотрога ройлевао, Недотрога ройля
205297 Недотрога жуйлиская
205327 Недотрога кандинская
205349 Недотрога сычуаньская
205352 Недотрога тайваньская
205355 Недотрога далиская
205369 Недотрога текстора

205381 Недотрога тяньцюаньская
205457 Недотрога инцзянская
205460 Недотрога юньнаньская
205470 Императа
205497 Императа, Императа цилиндрическая, Казаро
205535 Горичник настурциевый, Горичник царский корень, Корень царский, Царь-зелье
205546 Инкарвиллея
205552 Инкарвиллея сычуаньская
205574 Инкарвиллея олиги
205580 Инкарвиллеа китайская
205591 Инкарвиллеа тибетская
205611 Индиго, Индигоноска, Индигофера
205639 Индигофера тупоцветковая
205752 Индигофера хэбэйская
206127 Индигофера цзиндунская
206421 Индигофера ганьсуская
206445 Индигофера ложнокрасильная
206637 Индигоноска сычуаньская
206669 Индигоноска, Индигоноска красильная
206740 Индигофера хайнаньская
206918 Инга
207025 Девясил, Инуля
207026 Девясил бесстебельный
207038 Девясил шероховатый
207041 Девясил ушковатый
207046 Девясил британский
207075 Девясил каспийский
207087 Девясил блоший, Девясил растопыренный
207099 Девясил низбегающий
207105 Девясил мечелистный
207113 Девясил тибетский
207117 Девясил германский
207122 Девясил сизый
207131 Девясил кррупноцветковый
207132 Девясил большой
207134 Девясил громбчевского
207135 Девясил большой, Девясил высокий, Девясил обыкновенный
207140 Девясил волосистый, Девясил жестковолосый, Девясил шершавый
207145 Девясил хубэйский
207151 Девясил японский
207165 Девясил льнянколистный
207170 Девясил крупночешуйчатый
207173 Девясил великолепный
207175 Девясил марии
207181 Девясил монбре
207182 Девясил многостебельный
207189 Девясил глазковый, Девясил христово око
207193 Девясил восточный
207209 Девясил корнеглавый
207216 Девясил рубцова
207218 Девясил песчаный
207219 Девясил ивовый, Девясил иволистный
207222 Девясил китамуры
207224 Девясил солянковидные
207226 Девясил шмальгаузена
207228 Девясил зейдлитца
207240 Девясил песчаный
207241 Девясил коровяковый
207259 Девясил обыкновенный
207340 Иохрома
207546 Ипомея, Фарбитис
207623 Батат, Ипомея батат, Картофель батат, Картофель сладкий, Патат
207863 Ипомея щетинистая
207891 Ипмея индийская, Ипомея индийская
207901 Ипомея пурга, Ипомея слабительная, Ипомея ялана, Ялапа, Ялапа настоящая
207934 Ипомея леара
207988 Ипомея шаньдунская
208010 Ипомея мексиканская
208120 Вьюнок пурпурный, Вьюнок садовый, Ипомея пурпурная, Фарбитис пурпурный
208344 Илезина
208349 Илезина хербета
208366 Пальма-ириартея
208377 Ирисовые, Касатиковые
208433 Ирис, Касатик
208434 Касатик остродольный
208439 Касатик альберта
208447 Касатик безлистный, Касатик безлистый
208457 Касатик бальджуанский
208466 Касатик блудова
208473 Касатик бухарский
208477 Касатик камиллы
208481 Касатик кавказский
208492 Касатик тибетский
208494 Касатик голубой
208495 Касатик далиский
208516 Касатик дарвазский
208517 Касатик непалский
208524 Касатик вильчатый
208537 Касатик сепнолистный
208541 Касатик изящнейший
208543 Касатик кемпфера, Касатик мечевидный, Касатик остролепестковый, Касатик японский
208553 Касатик эвбанка
208555 Касатик серполистный
208562 Касатик желтейший, Касатик желтый, Касатик песчаный, Касатик ярко-желтый
208565 Касатик флорентийский
208566 Касатик вонючий
208572 Касатик фомина
208574 Касатик тайваньский
208576 Касатик фостера
208580 Касатик вильчатый
208583 Ирис германский, Ирис садовый, Касатик германский
208600 Касатик злаковидный, Касатик злаковый
208604 Касатик гроссгейма
208606 Касатик солелюбивый, Касатик солончаковый
208607 Касатик персидский, Касатик синьцзянский, Касатик согдийский
208627 Касатик гуга
208632 Касатик низкий, Касатик приземистый
208634 Касатик гирканский
208635 Касатик иберийский
208636 Касатик илийский
208638 Касатик черепичатый
208651 Касатик каратегинский
208657 Ирис японский
208659 Касатик колпаковского, Ксифиум колпаковского
208660 Касатик копетдагский
208662 Касатик королькова
208664 Касатик кушакевича
208672 Касатик вылощенный, Касатик гладкий
208685 Касатик лазистанский
208688 Касатик тонкокоренный
208689 Касатик линейнолистный
208690 Касатик тяньшанский
208696 Касатик людвига
208700 Касатик ложноаирный
208705 Касатик великолепный
208706 Касатик маньчжурский
208707 Касатик самаркандский
208708 Касатик мариковидный
208710 Касатик медведева
208723 Касатик мусульманский
208725 Касатик наньтоуский
208726 Касатик нарбута
208728 Касатик нарынский
208732 Касатик николая
208733 Касатик ненастоящий
208738 Касатик орхидейный, Касатик орхидный
208740 Касатик восточный
208744 Касатик бледный
208747 Касатик ганьсуский
208750 Касатик парадоксальный
208752 Касатик мелкий
208755 Ирис персидский
208764 Касатик попова

208765 Касатик потанина
208771 Ирис болотный, Касатик аировидный, Касатик болотный, Касатик водный, Касатик водяный, Касатик желтый, Касатик жёлтый, Касатик золотистый, Касатик ложноаирный, Касатик ложно-аировой, Касатик ложный
208777 Касатик ломнокавказский
208784 Касатик цинхайский
208788 Касатик сетчатый
208792 Касатик розенбаха
208797 Касатик русский
208806 Касатик кроваво-красный, Касатик родственный
208813 Касатик кожистый
208814 Касатик шелковниковая
208819 Касатик щетинистый, Касатик щетиновидный
208821 Касатик щетиновидный канадский
208829 Ирис сибирский, Касатик сибирский
208834 Касатик сычуаньский
208848 Касатик джунгарский
208853 Ирис мелкоцветковая
208862 Касатик побегоносный
208866 Касатик чжундяньский
208869 Касатик чешуйчатый
208870 Касатик сузианский
208872 Касатик таджикский
208874 Ирис пумила, Касатик крымский, Касатик маленикий, Касатик низкий, Касатик равнодольный, Касатик равнолопастный, Петушки степные, Петушок степной, Степные петушик
208882 Касатик тонколистный, Касатик узколистный
208897 Касатик тигровый
208911 Касатик тубердена
208917 Касатик одноцветковый
208919 Касатик пестрый, Касатик пёстрый, Петушки
208921 Касатик вздутый
208924 Касатик разноцветный
208928 Касатик замещающий
208930 Касатик фиолетовый
208936 Касатик ввведенского
208937 Касатик уорлийский
208940 Касатик уиллмотта
208942 Касатик винклера
208943 Касатик виноградова
209167 Вайда, Самерария
209171 Вайда обоюдоострая
209172 Вайда араратская
209174 Вайда арнольди
209175 Вайда бессера
209176 Вайда буассье
209177 Вайда короткоплодная
209179 Вайда бунге
209180 Вайда седая
209183 Вайда кавказская
209184 Вайда ребристая
209193 Вайда кустарниковая
209194 Вайда сизая
209195 Вайда гроссгейма
209196 Вайда пушисто-чашечная
209197 Вайда грузинская
209200 Вайда якутская
209201 Вайда японская
209204 Вайда гладкая
209205 Вайда пустоплодная, Вайда пушистоплодная
209208 Вайда широкоплодная
209209 Вайда беложилковая
209210 Вайда прибрежная
209213 Вайда крупнейшая
209215 Вайда маленькая
209217 Вайда продолговатая
209220 Вайда утконос
209223 Вайда тпнкоплоднаю
209224 Вайда сетчатая
209225 Вайда песчаная
209227 Вайда лучистая
209228 Вайда крымская
209229 Вайда красильная, Синильник
209236 Вайда пушистоплодная
209240 Вайда согнутоплодная
209241 Вайда выемчатая
209472 Исертия
209501 Искандера
209502 Искандера
209665 Плектрантус вырезной, Шпороцветник вырезной
209717 Плектрантус сизочашечный, Шпороцветник сизочашечный
209826 Плектрантус пильчатый, Шпороцветник пильчатый
210080 Камыш щединовидный
210251 Изопирум, Павноплодник, Равноплодник
210281 Изопирум маньчжурский
210291 Павноплодник василистниковый, Пукалка, Равноплодник василистниковый
210317 Изотома
210359 Итеа
210364 Итеа китайская
210389 Итеа японская
210390 Итеа цюцзянская
210391 Итеа гуансиская
210407 Итеа мелкоцветковая
210421 Итея виргинская
210422 Итеа янчуньская
210423 Итеа юньнаньская
210462 Лжедурмишник, Циклахена
210519 Иксеридиум
210525 Иксеридиум китайский, Иксерис китайский
210527 Иксеридиум зубчатолистный, Иксерис зубчатый
210540 Иксеридиум злаковидный
210541 Иксеридиум злаколистный
210549 Иксеридиум щетинистый
210552 Иксеридиум юньнаньский
210553 Иксерис
210623 Иксерис японский
210689 Иксия
210988 Иксиолирион
210992 Иксиолирион каратекинский
210999 Иксиолирион горный, Иксиолирион татарский
211029 Дерево железное, Иксора, Сидеродендрон
211067 Иксора китайская
211103 Иксора хайнаньская
211167 Иксора бирманская
211192 Иксора тибетская
211196 Иксора шансиская
211204 Иксора юньнаньская
211225 Дерево палисандровое, Якаранда
211226 Якаранда остролистная
211228 Дальбергия чёрная, Дерево бразильское палисандровое
211239 Якаранда минозолистная, Якаранда остролистная
211241 Якаранда остролистная
211273 Джексония
211550 Джемсия
211551 Джемсия американская, Лапчатка скальная
211639 Букашник
211660 Букашник хельдрейха
211670 Букашник многолетний
211672 Букашник горный
211693 Стефанотис
211695 Стефанотис люцюский, Стефанотис японский
211715 Жасмин
211749 Жасмин комнанин
211795 Жасмин двусемянный
211817 Жасмин инцзянский
211834 Жасмин желтый, Жасмин кустарниковый
211848 Жасмин крупноцветковый, Жасмин крупноцветный
211852 Жасмин гуансиский
211863 Жасмин низкий
211864 Жасмин отвороченный
211865 Жасмин низкий ганьсуский
211867 Жасмин опушенный
211893 Жасмин хайнаньский

211905 Жасмин месни
211931 Жасмин голоцветконый, Жасмин голоцветный
211938 Жасмин ароматный
211940 Жасмин антечный, Жасмин белый, Жасмин лекарственный, Жасмин настоящий
211952 Жасмин лекарственный тибетский
211987 Жасмин юньнаньский
211990 Жасмин арабский, Жасмин индийский, Жасмин самбак, Жасмин-самбак
212004 Жасмин китайский
212018 Жасмин ван
212071 Жасмин тибетский
212074 Жасмин юаньцзянский
212095 Колумба
212098 Ярофа, Ятропа
212127 Ятропа куркас, Ятрофа куркас
212296 Джефферсония
212433 Иорения
212434 Иорения малолисточковая
212541 Молодило отпрысковое
212553 Юануллоа
212556 Пальма слоновая, Юбея
212557 Пальма медовая, Юбея величавая, Юбея замечательная
212581 Ореховые
212582 Орех
212592 Орех калифорнийский
212595 Орех китайский
212599 Орех масляный, Орех серый
212602 Орех сердцевидный, Орех японский
212612 Орех хайндса
212614 Орех хэбэйский
212620 Орех большой
212621 Орех маньчжурский
212624 Орех сердцевидный, Орех японский
212626 Орех айлантолистный, Орех зибольда
212631 Лиственница восточная американская, Орех американский, Орех черный, Орех чёрный, Орех чёрный восточный
212636 Дерево ореховое, Орех волошский, Орех грецкий, Орех китайский, Орех обманчивый
212646 Орех скальный
212752 Ситниковые, Ситникоцветные
212753 Ситниковидные, Ситникоцветные
212763 Ситничек
212771 Ситовник илийстый
212774 Аирник венгерский
212778 Ситничек поздний
212797 Ситник
212808 Ситник острый
212819 Ситник алтайский
212822 Ситник альпийский
212832 Ситник неопределенный
212836 Ситник амурский
212844 Ситник арктический
212852 Ситник балтийский
212862 Ситник арийский
212866 Ситник блестящеплодный, Ситник членистый
212875 Ситник темноцветный
212876 Ситник темнобурый
212890 Ситник японский
212898 Ситник беринговский
212905 Ситник двухчешуйный, Ситник двухчешуйчатый
212915 Ситник короткооберточный, Ситник короткоприцветниковый
212929 Ситник жабий, Ситник лягушечий, Ситник лягушечный, Ситник мелкий
212955 Ситник луковичный
212970 Ситник канадский
212988 Ситник головчатый
212990 Ситник каштановый, Ситник трехглавый
213008 Ситник сплюснутый
213028 Ситник выдающийся
213036 Ситник сомнительный
213066 Ситник развесистый, Ситник раскидистый, Ситник расходящийся
213114 Ситник изогнутый, Ситник нитевидный
213127 Ситник герарда, Ситник жерара
213138 Ситник изящный, Ситник тончайший
213145 Ситник генке
213150 Ситник семиреченский
213153 Ситник гималайкий
213165 Ситник ероватый, Ситник искривлённый, Ситник короткоцветковый, Ситник склоненный, Ситник склонённый, Ситник склоняющийся
213176 Ситник затопляемый
213177 Ситник сырдарьинский
213180 Ситник юзепчука
213182 Ситник камчатский
213183 Ситник кандинский
213184 Ситник дэциньск
213185 Ситник капуский
213190 Ситник кочи
213191 Ситник крамера
213206 Ситник леерса
213213 Ситник лешено
213221 Ситник белооберточный
213226 Ситник береговой
213249 Ситник крупнопыльниковый
213263 Ситник морской, Ситник невского
213272 Ситник максимовича
213304 Ситник мюллера
213309 Ситник скученноцветковый
213322 Ситник ниппонский
213332 Ситник туполепестный
213341 Ситник тайваньский
213353 Ситник метельчатый
213355 Ситник сосочковый
213435 Ситник шишкина
213458 Ситник кучкоцветковый, Ситник кучкоцветный
213462 Ситник круглоплодный
213471 Ситник растопыренный
213476 Ситник стикийский
213492 Ситник шиловидный
213503 Ситник мелководный
213507 Ситник тонкий
213525 Ситник томсона
213527 Ситник тибетский
213534 Ситник трехглавый
213536 Ситник волосовидный
213537 Ситник трехраздельный
213545 Ситник трехчешуйный, Ситник трехчешуйчатый
213552 Ситник турчанинова
213554 Ситник туркестанский
213567 Ситник зеленоватый
213569 Ситник валиха
213579 Ситник юньнаньский
213606 Арча, Верес, Можжевельник
213623 Можжевельник барбадосский, Можжевельник бермудский
213625 Можжевельник карлифорнийский
213634 Арча китаиский, Арча красная, Можжевельник испанский, Можжевельник китайский
213696 Можжевельник саржента
213702 Можжевельник верес, Можжевельник китайский, Можжевельник обыкновенный
213739 Можжевельник приблежный, Можжевельник приморский, Можжевельник скученный
213752 Можжевельник даурский
213760 Можжевельник низкорослый
213763 Можжевельник косточковый
213767 Можжевельник высокий, Можжевельник древовадный
213773 Можжевельник повислый
213774 Можжевельник вонючий
213775 Можжевельник тайваньский, Можжевельник формозский
213780 Можжевельник куэньминский
213785 Можжевельник горизонтальный, Можжевельник распростёртый
213799 Можжевельник распростёртый опушённый
213809 Можжевельник равнолистный

213826 Можжевельник мексиканский, Можжевельник низкорослый
213833 Можжевельник длиннолистный
213834 Можжевельник американский, Можжевельник западный
213841 Арча красная, Дерево карандашное, Можжевельник испанский, Можжевельник красноватый, Можжевельник красный
213843 Можжевельник крупноплодный
213847 Можжевельник аллигаторовый, Можжевельник торстокорый
213849 Можжевельник красноплодный
213853 Можжевельник пинчота
213859 Можжевельник восточный, Можжевельник многоплодный
213863 Можжевельник стройный
213864 Можжевельник лежачий, Можжевельник распростертый
213873 Можжевельник, Можжевельник ложноказацкий, Можжевельник ложноказачий
213877 Балх-арча, Зардолу-арча, Можжевельник туркестанский, Тасва-арча, Урюк-арча, Хор-арча
213883 Можжевельник гималайский, Можжевельник отогнутый
213896 Можжевельник жесткий, Можжевельник твердолистный, Можжевельник твердый, Можжевельник твёрдый
213901 Можжевельник красноватый
213902 Можжевельник донской, Можжевельник казацкий, Можжевельник казачий
213921 Можжевельник шугнанский
213922 Можжевельник скальный
213932 Можжевельник полушаровидный, Саур-арча
213933 Кара-арча, махин-бура, Можжевельник зеравшанский
213934 Можжевельник карликовый, Можжевельник обыкновенный горный, Можжевельник сибирский, Потак-арча
213943 Можжевельник чешуйчатый
213963 Можжевельник люцюский
213969 Можжевельник ладанный
213984 Дерево карандашное, Карандашное дерево, Кедр красный, Красный кедр, Можжевельник американский, Можжевельник виргинский, Можжевельник казацкий
214028 Нагловадка, Наголоватка, Перплексия, Юринея
214029 Наголоватка аболина
214030 Наголовадка железистоплодная
214031 Наголовадка какрыладная, Наголоватка крылатая
214032 Наголоватка белостебельная
214033 Наголовадка холодная
214034 Наголоватка алтайская
214036 Наголовадка стеблеобъемлющая, Юринея многостебельная
214037 Наголоватка андросова
214038 Наголоватка анны
214039 Наголоватка безногая
214040 Наголовадка паутинистая, Юринея паутинистая
214043 Наголоватка армянская
214044 Наголоватка шершаволистная
214045 Наголоватка темнопурпуровая
214046 Наголоватка оше
214047 Наголоватка байсунская
214048 Наголоватка бальдшуанская
214049 Наголоватка маргаритковидная
214051 Наголоватка дважды-перистая
214052 Наголоватка привлекательная
214053 Наголоватка боброва
214054 Наголоватка бочанцева
214055 Наголоватка прицветничковая
214056 Наголоватка бухарская
214057 Наголоватка известняковая
214058 Наголоватка головоногая
214060 Наголовадка щетинкоплодная
214064 Наголоватка предкавказская
214066 Наголоватка вороньлапая
214067 Наголовадка меловая, Юринея меловая
214068 Наголоватка мелов
214070 Наголовадка васильковая, Наголовадка васильковидная, Юринея васильковая
214071 Наголоватка чиликиной
214072 Наголоватка дарвазская
214073 Наголоватка димитрия
214074 Наголовадка джунгарская
214078 Наголоватка изящная
214079 Наголоватка элегантная
214080 Наголовадка эверсмана, Юринея эверсмана
214081 Наголоватка превосходная
214082 Наголоватка федченко
214083 Наголоватка ферганская
214084 Наголоватка паноротниколистная
214087 Наголоватка олиственная
214090 Наголоватка голодкова
214091 Наголоватка изящная
214092 Наголоватка гранитовая
214093 Наголоватка гроссгейма
214094 Наголоватка пучковатая
214095 Наголоватка мелкоключковатая
214096 Наголоватка цминолистная
214100 Наголоватка ильина
214101 Наголоватка погруженно-жилковая
214102 Наголоватка капелькина
214103 Наголоватка карабугазская
214104 Наголоватка каратавская
214107 Наголоватка казахстанская
214108 Наголовадка киргизская, Юринея киргизская
214109 Наголоватка кнорринг
214110 Наголоватка кокандская
214112 Наголоватка крашенинникова
214113 Наголоватка культиасова
214114 Наголоватка кураминская
214115 Наголовадка шерстистоногая
214116 Наголоватка шерстистоногая
214117 Наголоватка рыхлая
214118 Наголоватка ледебура
214119 Наголоватка левье
214121 Наголовадка липского
214122 Наголоватка камнелюбивая
214123 Наголоватка лидии
214124 Наголоватка крупнокорзиночная
214126 Наголоватка маргалыкская
214127 Наголоватка марии
214128 Наголоватка крупная
214129 Наголоватка михельсона
214131 Юринея мягкая
214132 Наголоватка мягчайшая
214135 Наголоватка горная
214137 Наголоватка многоглавая
214138 Наголовадка многоцветковая, Эспарцет многоцветковая, Юринея сомнительная
214139 Наголоватка многолопастная
214140 Наголоватка снежная
214141 Наголоватка ольги
214142 Наголоватка восточная
214143 Наголоватка пачоского
214146 Наголоватка похожая
214149 Наголоватка сосняковая
214151 Наголоватка мятликовидная
214154 Наголоватка попова
214156 Наголоватка родственная
214157 Наголоватка песколюбивая
214158 Наголоватка ложно-васильковая
214159 Наголоватка крылостебельная
214160 Наголоватка хорошенькая
214161 Наголоватка низная
214162 Наголоватка корневищевидная
214163 Наголоватка мощная
214164 Наголоватка рупрехта
214165 Наголоватка иволистная
214168 Наголоватка шишкина
214169 Наголоватка змеестебельная
214170 Наголоватка серпуховидная
214171 Наголоватка шинтениша
214172 Наголоватка грязная
214173 Наголоватка сосновского
214175 Наголоватка примечательная

214176 Наголоватка спиридонова
214177 Наголоватка гтстая
214178 Наголоватка узколистная
214179 Наголовадка лавандолистная, Юринея узколистная
214180 Наголоватка полукустарниковая
214182 Наголовадка суйдунская
214183 Наголоватка таджикистанская
214184 Наголоватка талиеви
214185 Наголоватка донская
214186 Наголовадка тонколопастная, Юринея узколопастная
214188 Наголоватка рирамидально-метельчатоцветная
214189 Наголоватка тяньшанская
214191 Наголоватка загирканская
214192 Наголоватка зауральская
214193 Наголоватка траутфеттера
214194 Наголоватка трехвильчатая
214195 Наголоватка красивая
214197 Наголоватка винклера
214198 Наголоватка воронова
214199 Наголоватка бессмертниковая
214201 Юринелла
214203 Юринелла оттопыренная
214204 Юринелла бесстебельная
214304 Юстиция
214308 Атхатоба, Орешник индийский
214947 Кадзура
214964 Кадзура гуансиская
214969 Кадзура фэнцинская
214971 Кадзура японская
214975 Кадзура эмэйская
214995 Кемпферия
215086 Каланхое
215206 Каланхое мраморное
215263 Каланхое сомалиский
215272 Каланхое ланьюйское
215277 Каланхое войлочное
215322 Поташник
215324 Поташник каспийский
215327 Поташник олиственный
215330 Поташник шренка
215334 Калимерис
215339 Калимерис надрезанная, Калимерис надрезная
215343 Калимерис индийская
215350 Калимерис цельнолистная
215356 Калимерис монгольский
215386 Кальмия
215387 Кальмия узколистная
215393 Кальмия многолистная, Кальмия сизая
215395 Кальмия широколистная, Лавр американский, Лавр горный
215405 Кальмия многолистная, Кальмия сизая
215420 Диморфант, Калопанакс
215442 Диморфант, Калопанакс клещевинолистный, Калопанакс семилопастный, Орех белый, Шипдерево
215443 Калопанакс максимовича
215566 Карелиния
215567 Карелиния каспийская
215616 Кашгария
215617 Кашгария коротоцветковая
215618 Кашгария комарова
215639 Кауфманния
215640 Кауфманния короткопыльниковая
215641 Кауфманния семенова
215903 Пырей алатавский
215905 Пырей баталина
216074 Кернера
216084 Керия, Керрия
216085 Керия японская, Керрия японская
216092 Керрия японская плена
216114 Кетелеерия, Кетелерия
216120 Кетелеерия давида
216125 Кетелерия тайваньская
216134 Кетелерия эвелины
216142 Кетелеерия форчтна, Кетелерия форчуна
216147 Кетелерия хайнаньская
216205 Кайя иволензис, Кайя иворская
216214 Кайя сенегальская
216239 Киксия
216262 Киксия повойничек, Киксия повойничковая
216292 Киксия ложная
216313 Дерево колбасное, Кигелия
216473 Киренгешома
216486 Кириловия
216487 Кириловия пушиотоцветковая
216641 Клейнховия
216752 Короставник
216753 Короставник полевой
216764 Короставник ворсянколистный
216767 Короставник крупнообертковый
216771 Короставник горный
216772 Короставник восточный
216774 Короставник татарский
216923 Книфофия, Тритома
217117 Кобрезия
217130 Кобрезия волосолистная
217148 Кобрезия цзилунская
217195 Кобрезия низкая
217198 Кобрезия ганьсуская
217234 Кобрезия мышзхвостовидная
217236 Кобрезия непалская
217274 Кобрезия овальноколосая, Кобрезия схенусовидная
217280 Кобрезия сычуаньская
217299 Кобрезия тибетская
217307 Кобрезия мулиская
217310 Кобрезия ядунская
217324 Изень, Кипарис летний, Кохия
217331 Кохия песчаная, Кохия шерститоцветковая, Прутняк песчаный
217343 Кохия крылова
217344 Кохия шерститоцветковая
217347 Кохия темнокрылая
217353 Изень, Кохия простертая, Кохия стелющая, Кохия стелющаяся, Полынь красная, Прутняк, Прутняк распростертый, Солушец
217359 Кохия шленка
217361 Кипарис летний, Кохия, Кохия веничная, Кохия метельчатая, Неферощь веничная, Прутняк веничный
217372 Кохия сиверса
217379 Кохия иранская
217416 Келерия, Тонколучник, Тонконог
217419 Келерия альбова
217421 Тонконог алтайский
217424 Келерия азиатский
217427 Тонконог азиатский, Тонконог коленчатый, Тонконог темнофиолетовый
217431 Тонконог буша
217434 Тонконог кавказский
217441 Келерия гребенцатая, Келерия стерная, Келерия стройная, Тонконог гребенчатый, Тонконог обыкновенный, Тонконог степной, Тонконог стройный, Тонконог тонкий
217461 Тонконог дегена
217462 Тонконог делявиня
217469 Тонконог фомина
217473 Келерия сизая, Тонконог, Тонконог сизый
217474 Тонконог городкова
217483 Тонконог сомниельный
217485 Тонконог ледебура
217487 Тонконог литвинова
217493 Тонконог люерссена
217494 Келерия стерная мелкоцветковая, Тонконог токийский, Тонконог тонкий
217496 Тонконог одноцветковый
217499 Тонконог лоснящийся
217500 Лофохлоа тупоцветковая
217501 Лофохлоа тимофеевковидная
217518 Тонконог мятликовидный
217520 Тонконог поле
217522 Тонконог польский
217561 Тонконог жестнолистный
217562 Тонконог полуголый
217565 Тонконог сибирский
217568 Тонконог блестящий
217575 Тонконог тона
217600 Кельпиния

217604 Кельпиния линейная
217605 Кельпиния кртпноцветковая
217606 Кельпиния
217607 Кельпиния туранская
217610 Дерево мыльное, Кельрейтерия
217615 Кельрейтерия двухперистая, Кельрейтерия цельнолистная
217620 Кельрейтерия формозская
217626 Кельрейтерия метельчатая
217640 Кенигия
217648 Кенигия исландская
217785 Кольквиция
217787 Кольквиция прелестная
217826 Комаровия
217827 Комаровия разнореберная
217895 Коржинския
217896 Коржинския володушковая
217897 Коржинския ольги
217941 Козополянская
217942 Козополянская туркестанская
217944 Костелецкия
217955 Канап, Костелецкия пятиплодная
218116 Крашенинниковия, Терескен
218122 Терескен серый
218137 Крашенинниковия жесткая
218146 Красновия
218148 Красновия длиннодольчатая
218242 Крыловия
218245 Крыловия пустынно-степная
218246 Крыловия кермеколистная
218247 Крыловия новопокровского
218248 Крыловия попова
218274 Кудряшевия
218276 Кудряшевия разноволосый
218278 Кудряшевия якуба
218279 Кудряшевия коржинского
218280 Кудряшевия надины
218344 Куммеровия
218345 Куммеровия прилистниковая, Леспедеца корейская, Леспедеца прилистниковая
218347 Клевер японский, Куммеровия полосатая, Лесподеза полосатая, Леспедеца полосатая
218387 Ланция
218457 Киллинга
218480 Киллинга коротколистная
218556 Киллинга камчатская
218697 Губоцветные
218721 Боб гиацитовый, Боб индийский, Долихос лябляб, Долихос обыкновенный, Лобия, Лобия египетская
218754 Бобовник, Дождь золотой, Золотой дожль
218757 Бобовник альпийский, Дождь золотой альпийский, Ракитник альпийский
218761 Бобовник "золотый дождь", Бобовник анагировидный, Бобовник анагиролистный, Дождь золотой обыкновенный, Ракитник-золотой дождь
218948 Лахеналия, Лилия трёхцветная южноафриканская
219026 Лахнантес
219056 Шерстоплодник
219057 Шерстоплодник леманна
219059 Шерстистолистник
219060 Шерстистолистник хлопковидный
219206 Лактук, Латук, Молокан, Мульгедиум, Салат
219210 Латук алтайский
219221 Латук ушковатый
219249 Латук канадский
219270 Латук раздельнолистный, Латук шэ
219303 Латук ушковатый
219327 Латук грузинский
219336 Латук глауциелистный
219416 Латук удивительный
219444 Латук многолетний
219474 Латук розеточный
219480 Латук иволистный, Латук солончаковый
219485 Латук огородный, Латук посевной, Латук-салат, Малокан посевной, Салат культурный, Салат латук, Салат обыкновенный, Салат огородный, Салат посевной
219489 Латук-салат кочанный, Салат кочанный
219493 Латук римский, Латук-ромен, Латук-ромэн, Ромэн, Эндивий летний
219494 Пфлюк-салат, Салат латуксрывной, Салат листовой, Шнит-латук
219497 Латук дикий, Латук лесной, Малокан комнасный, Молокан комнасный, Салат дикий
219507 Латук дидий, Латук дикий, Латук комнастый, Латук лесной, Латук серриола, Молокан комнасный, Салат дикий
219513 Латук сибирский, Молокан сибирский
219533 Латук шиповатозубчатый
219547 Латук сжатый
219585 Латук клубненосный
219589 Латук волнистый
219600 Латук прутяной
219609 Латук дидий, Латук ядовитый, Молокан ядовитый, Салат вонючий
219614 Латук вильгельмся
219615 Латук винкрера
219660 Ладыгиния
219661 Ладыгиния бухарская
219806 Латук сибирский, Молокан сибирский
219821 Бутылочная тыква, Горлянка, Лагенарик, Лагенария, Тыква
219843 Горлянка, Горлянка обыкновенная, Калебаса, Лагенарик, Тыква бутылочная, Тыква бутылочная обыкновенная, Тыква горлянка, Тыква декоративная, Тыква посудная, Тыква фигурная, Тыква-горлянка, Ханка
219908 Лагерстремия, Лягерстремия, Сирень индийская
219927 Лагерстремия гуандунская
219931 Лагерстремия гуйлиньская
219933 Индийская сирень, Лагерстремия индийская, Сирень индийская
219952 Лагерстремия юньнаньская
219956 Лагерстремия фуцзяньская
219959 Лагерстремия мелкоцветковая
219965 Лагерстремия таиландская
219986 Дерево кружевное, Лагета
220050 Зайцегуб, Лягохилюс
220051 Зайцегуб остродольный
220053 Зайцегуб алтайский
220055 Зайцегуб андросова
220056 Зайцегуб валханский
220059 Зайцегуб бунге
220060 Зайцегуб кабульский
220063 Зайцегуб двиголчатый
220064 Лягохилюс голый, Яснотка голая
220068 Зайцегуб волосистый
220069 Зайцегуб падуболистный
220072 Зайцегуб опьяняющий
220073 Зайцегуб кашгарский
220074 Зайцегуб кнорринговский
220075 Зайцегуб кштутский
220077 Зайцегуб гладкоколючковый
220078 Зайцегуб длиннозубчатый
220079 Зайцегуб крупнозубый
220080 Зайцегуб невского
220084 Зайцегуб паульсена
220085 Зайцегуб плоскоколючковый
220086 Зайцегуб плоскочашечный
220087 Зайцегуб пушистый
220088 Зайцегуб красивый
220090 Зайцегуб колючий
220091 Зайцегуб зеравшанский
220092 Зайцегуб щетинистый
220093 Зайцегуб почтй-щетинистый
220094 Зайцегуб тяньшанский
220095 Зайцегуб узунахматский
220096 Зайцегуб синьцзянский
220103 мимозка
220105 мимозка выполненная
220117 Лагопсис
220121 Лагопсис мохнатоколосый
220124 Лагопсис желтый

220125 Лагопсис шандровый
220126 Лагопсис простертая, Шандра надрезанная
220131 Лагозериопсис
220132 Лагозериопсис попова
220135 Лагозерис
220137 Лагозерис аральский
220138 Лагозерис сизоватый
220139 Лагозерис крупноцветковый
220140 Лагозерис обрантояйцевидный
220141 Лагозерис
220142 Лагозерис пурпуровый
220143 Лагозерис мощный
220144 Лагозерис сахендский
220145 Лагозерис палестинский
220155 Лаготис, Ляготис
220161 Ляготис северный
220166 Ляготис лежачий
220167 Лаготис сизый, Ляготис сизый
220177 Лаготис иконникова
220180 Ляготис цельнолистный
220182 Лаготис королкова
220188 Лаготис малый
220190 Ляготис палласа
220196 Ляготис стеллдра
220199 Лаготис побегоносный
220201 Лаготис уральская
220226 Лагунария
220252 Зайцехвост, Лагурус
220281 Лаллеманция, Ляллеманция
220282 Ляллеманция бальджуанская
220283 Ляллеманция седоватая
220284 Ляллеманция грузинская
220285 Ляллеманция щитовидная
220286 Лаллеманция ройлевская, Ляллеманция ройля
220295 Ламаркия
220345 Зеленчук, Яснотка
220346 Глухая крапива, Яснотка белая
220355 Яснотка стеблеобъемлющая
220359 Зайцегуб бородатая
220367 Яснотка кавказская
220375 Яснотка кустарниковая
220387 Яснотка цзилунская
220388 Яснотка голая
220392 Яснотка гибридая
220401 Крапива глухая пятнистая, Яснотка крапчатая, Яснотка пятнистая
220407 Яснотка ганьсуская
220413 Яснотка пачоского
220416 Крапива глухая красная, Яснотка пурпуровая
220421 Яснотка войлочная
220729 Дикритра прекрасная, Дицентра великолепная, Дицентра зимная, Дицентра китайская
220775 Ламира
220779 Ламира колючеголовая
220780 Ламиропапус
220781 Ламиропапус шака-птарский
220814 Ландольфия
221224 Лансиум
221226 Лансиум домашний, Лансиум обыкновенный
221229 Крапивка цветная, Лантана
221238 Лантана сводчатая
221340 Лапагерия
221341 Лапагерия розовая
221353 Лаперузия
221529 Лапортея
221538 Лапортея клубненосная
221547 Лапортея канадская
221552 Жирардиния остроконечная
221561 Лапортея фуцзяньская
221576 Лапортея мотоская
221627 Липучка, Турица
221631 Липучка неравношипая, Липучка неравношипиковая
221634 Липучка балхашская
221635 Липучка бородчатая
221638 Липучка короткошипиковая
221639 Липучка бесшипиковая
221644 Липучка рогоносная
221645 Липучка родственная
221648 Липучка корончатая
221649 Липучка гребенчатая
221657 Липучка двуплодная
221659 Липучка дробова
221660 Липучка двоякоплодная
221672 Липучка ежеголовая
221675 Лепехиниелла ферганская
221678 Липучка ганьсуская
221680 Липучка оголенная
221685 Липучка разношипая
221689 Липучка промежуточная
221692 Липучка коржинского
221693 Липучка куликалонская
221697 Липучка липшица
221699 Липучка тошая
221700 Липучка тощая
221702 Липучка каемчатая, Липучка окаймленная
221705 Липучка мелкоплодная
221711 Дерябка, Липучка ежёвая, Липучка ежевидная, Липучка ежовая, Липучка незабудка, Липучка обыкновенная, Липучка репейчатая
221713 Липучка закрытоплодная
221715 Липучка пониклая, Липучка раскидистая
221724 Липучка попова
221735 Липучка скальная
221737 Липучка камнелюбивая
221741 Липучка полукрылатая
221742 Липучка полуголая
221744 Липучка синайская
221745 Липучка сидячецветковая
221748 Липучка синайская
221751 Липучка колючеплодная, Липучка шипоплодная
221758 Липучка прямая
221760 Липучка таджикская
221761 Липучка тонкая
221764 Липучка тяньшанская
221773 Липучка синьцзянская
221780 Бородавник
221787 Бородавник обыкновенный
221791 Бородавник крупный
221794 Бородавник средний
221802 Бородавник тайваньский
221815 Лардизабала
221818 Лардизабаловые
221829 Лиственница
221855 Лиственница европейская, Лиственница опадающая
221866 Лиственница гмелина, Лиственница даурская
221881 Лиственница гималайская, Лиственница гриффитка, Лиственница тибетская
221894 Лиственница тонкочешуйчатая, Лиственница японская
221904 Лиственница американская
221914 Лиственница сычуаньская
221917 Лиственница западная, Лиственница западная американская
221919 Лиственница ольгинская
221929 Лиственница польская
221934 Лиственница китайская
221944 Лиственница сибирская
221975 Куст креозотовый
221978 Ларрея
222020 Гладыш, Лазерпициум, Лазурник
222021 Гладыш близный
222022 Гладыш альпийский
222027 Гладыш щетинистоволосистый
222028 Гладыш широколистный
222031 Гладыш стевена
222038 Лазия
222454 Лазиопогон
222460 Лазиопогон моховидный
222625 Латания, Пальма веерная
222641 Крест петров, Петров крест
222645 Петров крест японский
222651 Крест петров чешуйчатый, Петров крест чешуйчатый, Чешуйник
222671 Латирус, Сочевичник, Чина
222676 Чина угловатая
222680 Чина однолетная
222681 Чина безлисточковая
222687 Чина азербайджанская

222688 Чина золотая
222693 Чина желтозеленая
222694 Чина красная
222704 Чина колхидская
222706 Чина голубая
222707 Чина давида
222711 Чина пальчатая
222713 Чина домина
222719 Чина фролова
222720 Сочевичник гмелина
222724 Чина танжерская
222726 Чина мохнатая
222727 Чина приземистная
222731 Чина незаметвая
222733 Чина согнутая
222735 Чина морская, Чина приморская, Чина японская
222743 Чина приморская
222747 Чина комарова, Чина крылатая
222748 Чина гладная, Чина крылова
222750 Чина широколистная
222756 Чина редкоцветная
222757 Чина ледебура
222758 Чина горная
222761 Чина литвинова
222775 Чина киноваревая
222782 Сочевичник чёрный, Чина черая, Чина чёрая
222784 Чина злаколлистная, Чина ниссолия
222788 Чина охряная
222789 Горошек душистый, Чина душистая, Чина душистый горошек
222799 Чина бледнеющая
222800 Чина болотная
222803 Чина волосистая
222815 Чина венгерская
222819 Чина гороховидная
222820 Чина луговая
222828 Чина пятижилковая
222830 Чина розовая
222831 Чина круглолистная
222832 Чина посевная
222836 Чина щетинолистная
222837 Чина шаровидная
222843 Чина кругловатая
222844 Чина лесная
222851 Сочевичник клубненосный, Чина клубневая, Чина клубненосная
222853 Чина синеватая
222860 Весенний горошик, Сочевичник весенний, Чина весенняя
222863 Чина воронова
222907 Рабдотэка
222947 Рабдотэка коровина
223006 Лавровые
223088 Лавровишня
223115 Лавровишня мэнхайская
223116 Дерево лавровишневое, Лавр дикий, Лавровишня аптечная, Лавровишня лекарственная, Лавровишня обыкновенная
223163 Лавр
223178 Лавр канарский
223203 Дерево лавровое, Лавр благородный
223248 Лаванда
223251 Колос благовонный, Лаванда аптечная, Лаванда колосовая, Лаванда настоящая, Лаванда обыкновенная
223298 Лаванда широколистная
223350 Лаватера, Хатьма
223355 Хатыма древовидная
223361 Хатьма кашмирская
223387 Хатьма точечная
223392 Рожа собачья, Хатыма тюрингенская
223393 Хатыма трёхмесячная, Хатьма трехмесячная
223451 Хенна, Хна
223454 Лавссония, Хенна, Хна
223476 Лайя, Лэйа
223497 Лэйа изящная
223764 Дерево горшечное, Лецитис
223825 Ледебуриелла
223827 Ледебуриелла многолестная
223888 Багульник
223897 Багульник подбел
223901 Багульник болотный
223905 Багульник подбел
223911 Багульник стелющийся
223973 Леерсия
223992 Леерсия посевная, Леерсия рисовидная
224002 Леерсия японская
224058 Зеркало девичье, Легузия, Легузия зеркало венеры, Легузия серповидная, Спекулярия
224067 Легузия либридная
224070 Легузия пятиугольная
224073 Легузия зеркало-венеры
224085 Бобовые
224106 Лейбниция
224110 Лейбниция бестычинковая, Лейбниция двликая
224115 Лейбниция кнорринг
224211 Паррия пушисточашечная
224212 Паррия бесстебельная
224275 Вигна, Лейтнерия
224359 Ряска, Чечевица водяная
224366 Ряска горбатая
224370 Ряска лэйбоская
224375 Ряска малаяРяска маленькая, Ряска меньшая
224396 Ряска прёхдольная, Ряска прехлопастная, Ряска тройчатая
224403 Рясковые
224471 Чечевица
224473 Чечевица культурная, Чечевица мелкозелная, Чечевица обкновенная, Чечевица пищевая, Чечевица съедобная
224483 Чечевица безусая, Чечевица линзообразная
224485 Чечевица чернеющая, Чечевица черноватая
224487 Чечевица восточная
224499 Пузырчатковые
224558 Леонотис
224625 Леонтица
224626 Леонтица альберта
224629 Леонтица дарвазская
224630 Леонтица еверсманна
224631 Леонтица сомнительная
224639 Леонтица малая
224640 Леонтица одесская
224642 Леонтица смирнова
224651 Кульбаба
224652 Кульбаба шероховатейшая
224655 Кульбаба осенняя
224662 Кульбаба кавказская
224666 Кульбаба шафранная
224668 Кульбаба дунайская
224698 Кульбаба щетинистая
224699 Кульбаба керетская
224700 Кульбаба кочи
224725 Кульбаба ложноодуванчиковаю
224726 Кульбаба ползучая
224736 Кульбаба шишкина
224767 Леонтоподиум, Сушеница, Эдельвейс
224772 Леонтоподиум альпийский, Лепка львиная, Сушеница альпийская, Эдельвейс альпийский
224806 Эдельвейс коротколучевой
224822 Эдельвейс скученный
224827 Эдельвейс разноцветный
224842 Леонтоподиум франше
224872 Эдельвейс японский
224890 Эдельвейс курильский
224893 Леонтоподиум сибирский, Эдельвейс обедненный, Эдельвейс эдельвейсовидный
224899 Эдельвейс эдельвейсовидный
224904 Леонтоподиум мелкоцветковое
224907 Эдельвейс мелколистный
224913 Эдельвейс низкорослый
224916 Эдельвейс бледно-желтый
224924 Леонтоподиум эмэйский
224925 Эдельвейс палибина
224969 Пустырник
224976 Пустырник обыкновенный, Пустырник сердечник, Трава сердечная
224981 Пустырник сизоватый, Пустырник

сизый
224983 Пустырник разнолистный
224989 Пустырник японский
224996 Пустырник крупноцветковый
225003 Пустырник монгольский
225004 Пустырник панцериевидный
225008 Пустырник пятилопастный
225009 Пустырник сибирский
225015 Пустырник татарский
225019 Пустырник туркестанский
225022 Пустырник мохнатый
225027 Леопольдия
225031 Леопольдия кавказсая
225035 Леопольдия длинная
225036 Леопольдия тонкоцветковая
225088 Лепехиниелла
225089 Лепехиниелла алайская
225092 Лепехиниелла коржинского
225095 Лепехиниелла маленькая
225096 Лепехиниелла пупковидкая
225097 Лепехиниелла
225098 Лепехиниелла зеравшанская
225099 Лепехиниелла заалайская
225279 Двугнездка, Двугнёздка, Кардария, Клоповник, Кресс, Чатыр
225293 Клоповник стебреобъемлющий
225295 Клоповник безлепестный, Клоповник безлистный
225303 Клоповник буассье
225305 Клоповник безлепестный
225308 Клоповник борщова
225315 Клоповник полевой, Кресс полевой
225326 Клоповник толстолистный
225333 Клоповник сердцевидный
225335 Клоповник воронцелистный, Клоповник воронцелистый
225336 Клоповник толстолистный
225340 Клоповник безлепестный, Клоповник безлистный, Клоповник густоцветковый
225344 Клоповник пустыни
225362 Клоповник пустынный
225364 Клоповник ферганский
225369 Клоповник злаколистный
225393 Клоповник каратавский
225397 Клоповник разрезной
225398 Клоповник широколистный, Кресс, Кресс широколистный, Солнечный хрен, Хрен солнечный, Хринок
225411 Клоповник лировидный
225414 Клоповник мейера
225425 Клоповник тупой
225428 Клоповник пронзеннолистный, Клоповник пронзённолистный, Клоповник пронзенный, Клоповник разнолистный, Ласковец, Перечник пронзеннолистный
225429 Клоповник персидский
225433 Клоповник перистый
225440 Клоповник низкий, Клоповник низкорослый
225443 Клоповник ползучий
225447 Веничник, Клоповник мусорный, Клоповник сорный, Клоровник, Кресс сорный
225450 Клоповник посевной, Кресс огородный, Кресс посевной, Кресс садовый, Кресс-салат, Перечник огородный, Перечник посевной
225455 Клоповник сибирский
225460 Клоповник джугарский
225472 Клоповник турчанинова
225473 Клоповник пузырчатый
225475 Клоповник виргинский
225519 лепидолофа
225520 лепидолофа федченко
225521 лепидолофа нителистная
225522 лепидолофа каратавская
225523 лепидолофа комарова
225524 лепидолофа моголтавская
225525 лепидолофа нуратавская
225678 Дерево мыльное сенегальское
225802 Лепталеум
225803 Лепталеум нителистный
226315 Андрахна, Лептопус
226320 Андрахна китайская
226330 Андрахна хайнаньская
226335 Андрахна мелколистная
226339 Лептопирум
226341 Лептопирум дымянковый
226367 Лепторабдос
226369 Лепторабдос мелкоцветковый
226450 Лептосперм, Лептоспермум
226569 Лептунис
226571 Лептунис волосовидный
226615 Пашенник
226621 Пашенник костенецовидный
226626 Пашенник звездочковидный
226677 Куммеровия, Лесспедеза, Леспедеца, Лесспедеция
226698 Лесспедеза двуцветная, Леспедеца двуцветковая, Леспедеца двуцветная
226708 Леспедеца японская
226739 Леспедеца китайская
226742 Копеечник шелковистый, Куммеровия клиновидная, Лесспедеза клиновидная, Леспедеца заострённая, Леспедеца клиновидная, Леспедеца шелковистая
226746 Леспедеца кривокистевая, Леспедеца плотная
226751 Леспедеца даурская
226840 Леспедеца хубэйская
226860 Леспедеца копеечниковая, Леспедеца ситниковая
226989 Леспедеца войлочная, Леспедеца мохнатая
227017 Везикария, Пузырник
227233 Дерево серебряное, Лейкодендрон серебристый
227420 Левкена
227452 Леукантемелла
227453 Леукантемелла линейная
227455 Леукантемелла поздняя
227457 Пиретрум альпийский, Хризантема алипийская
227467 Нивяник
227526 Нивяник поздний
227528 Нивяник сибирский
227532 Нивяник субальпийский
227533 Нивяник обыкновенный, Поповник, Поповник ниванка, Поповник обыкновенный, Хризантемум белоцветковый
227818 Леукокринум
227854 Амарилис альпийский, Белоцветник
227856 Белоцветник летний
227875 Белоцветник весенний
227916 Леукофиллум
227946 Беломятлик
227948 Беломятлик беловатый
227950 Беломятлик кавказский
227952 Беломятлик каратавский
227953 Беломятлик ольги
227956 Беломятлик жестколистный
228154 Лейкофоя
228171 Евботриоидес грея
228244 Любисток
228250 Зоря, Корень любистоковый, Любисток аптечный, Любисток лекарственный
228256 Льюзия
228319 Лейцестерия
228321 Лейцестерия красивая
228335 Лейцестерия тибетская
228337 Вострец, Мягкохвостник
228339 Вострец подражающий
228343 Вострец узкий
228346 Волоснец песчаный, Колосняк песчаный, Песчанк, Рожь полярная, Ячмень гигантский, Ячмень песчаный
228356 Вострец регеля, Колосняк китайский
228364 Вострец копетдагский
228369 Волоснец мягкий, Колосняк мягдий, Колосняк мягкий
228370 Вострец многостебельный, Колосняк многостебельный
228372 Вострец овальный
228373 Колосняк пабо
228374 Вострец пабо

228380 Вострец ветвистый
228384 Вострец пушистоколосый, Колосняк пушистоколосый
228392 Вострец тяньшанский
228433 Лиатрис
228529 Лиатрис колосковая
228556 Порезник
228565 Жабрица жабрицевидная
228568 Порезник бухторминский
228570 Порезник чамечный
228571 Порезник плотный
228578 Порезник длинностолбиковый
228581 Жабрица пушистоплодная, Порезник пушистоплодный
228583 Жабрица илийская, Порезник илийский
228584 Жабрица седоватая, Порезник седовадый
228585 Порезник промежуточный
228592 Порезник странный
228596 Порезник шренковский
228597 Порезник жаблицевидный
228598 Порезник щетинконосный
228599 Порезник сибирский
228605 Порезник закавказская
228625 Либертия
228635 Кедр ладанный, Кедр речной, Либоцедрус
228639 Калоцедрус калифорнийский, Калоцедрус сбежистый, Кедр калифорнийский, Кедр речной калифорнийский, Кедр речной сбежистый
228649 Кедр речной новозеландский, Кедр речной перистый
228663 Ликания
228729 Ликуала
228969 Бузульник, Лигулярия
228976 Бузульник альпийский, Бузульник высокогорный
228977 Бузульник алтайский
228981 Бузульник непалский
228986 Бузульник арктический
228990 Бузульник ядунский
228999 Бузульник калужницелистный
229000 Бузульник кавказский
229004 Лигулярия бирманская
229035 Бузульник фишера
229040 Бузульник франше
229043 Бузульник сизый
229046 Бузульник разнолистный
229049 Бузульник ходжсона
229058 Лигулярия хэбэйская
229063 Бузульник ялусский
229066 Лигулярия японская
229074 Бузульник кандинский
229076 Бузульник кратавский
229077 Бузульник кнорринг, Бузульник кнорринговский
229094 Бузульник лицзянский
229095 Бузульник ли
229101 Бузульник кнорринг
229104 Бузульник крупнолистный
229109 Бузульник монгольский
229116 Бузульник наньчуаньский
229117 Бузульник нарынский
229130 Лигулярия мелколистная
229131 Бузульник павлова
229144 Бузульник дллинproductногий
229160 Бузульник почковидный
229167 Бузульник сахалинский
229171 Бузульник шишкина
229175 Бузульник шмидта
229179 Бузульник сибирский
229195 Бузульник сихотинский
229197 Бузульник джунгарский
229201 Бузульник велъколепный
229211 Бузульник полустреловидный
229214 Бузульник таласский
229226 Бузульник томсона, Крестовник томсона
229227 Бузульник метельчатый, Бузульник пирамидально-метельчатый
229232 Бузульник волосистоголовый
229246 Бузульник
229252 Лигулярия синьцзянская
229259 Лигулярия юньнаньская
229283 Бородоплодник, Зоря, Лигустикум
229291 Лигустикум крылатый
229303 Лигустикум кавказский
229314 Лигустикум лицзянский
229318 Лигустикум разноцветный
229322 Лигустикум федченко
229329 Лигустикум франше
229333 Лигустикум цзилунский
229361 Лигустикум монгольский
229365 Толстореберник остроконечный
229378 Лигустикум низкий
229379 Лигустикум пурпуроволенестный
229384 Лигустикум северный, Лигустикум шотрандский
229385 Лигустикум хультена
229390 Лигустикум китайский
229398 Лигустикум мулиский
229415 Лигустикум тибетский
229416 Лигустикум яньюаньский
229417 Лигустикум юньнаньский
229428 Бирючина, Лигуструм
229454 Бирючина делаве
229463 Бирючина цзилунская
229472 Бирючина ибота
229485 Бирючина японская
229513 Бирючина японская ротундифолия
229522 Бирючина остолейшая
229523 Бирючина обыкновенная
229524 Лигуструм люцюский
229529 Бирючина блестящая, Лигуструм блестящий
229549 Бирючина мелкоплодная
229562 Бирючина амурская
229569 Бирючина овальнолистная
229593 Бирючина киу
229603 Бирючина чаойская
229617 Бирючина китайская
229625 Бирючина лодяньская
229634 Бирючина ичанская
229650 Бирючина чоноски
229651 Бирючина хоккайдинская
229662 Бирючина обыкновенная
229700 Лилейные
229717 Лилия
229723 Лилия корейская
229727 Лилия аньхуйская
229730 Коророва лилий, Лилия ауратум, Лилия золотистая, Лилия китайско-японская
229754 Лилия броуна
229766 Лилия бульбоносная, Лилия красная
229769 Лилия клубненосная красная, Лилия шафранная
229775 Лилия мозолистая
229776 Лилия мозолистая люцюская
229779 Лилия канадская
229789 Лилия белая, Лилия бледская, Лилия чисто-белая
229793 Лилия карниольская
229803 Лилия поникающая
229806 Лилия халцедонская
229822 Лилия
229828 Лилия даурсая, Лилия пенсиливанская
229829 Лилия сычуаньская
229834 Лилия двурядная
229837 Лилия эмэйская
229845 Лилия тайваньская
229857 Лилия хэнсона
229864 Лилия хуйдунская
229872 Лилия японская
229881 Лилия кессельринга
229886 Лилия ледебура
229887 Лилия максимовича
229888 Лилия максимовича
229890 Лилия ичанская
229898 Лилия лицзянская
229900 Лилия дллиноцветковая, Лилия дллиноцветная, Лилия лонгифлорум
229922 Кудри царские, Лилия кудреватая, Лилия мартагон, Лилия саранка, Лилия чалмовая, Мартагон, Сарана большая, Саранка, Царские кудри
229934 Лилия медеолевидная, Лилия

медеоловидная, Лилия овсяная, Сарана овсянка
229936 Лилия мотоская
229944 Лилия единобратственная, Лилия однобратственная
229996 Лилия понтийская
230006 Лилия красивенькая
230009 Лилия каприковая, Лилия узколистный
230019 Лилия королевская
230022 Лилия наньчуаньская
230046 Лилия шовица
230048 Лилия далиская
230055 Лилия терракотовая
230058 Лилия ланцетолистная, Лилия тигровая
230073 Лилия вэньшаньская
230205 Лимнантес
230219 Болодник
230220 Болодник стеллера
230221 Беломятлик верещагина
230244 Лимнохарис
230338 Лимнорхис
230342 Лимнорхис ландышелистный
230348 Лимнорхис расширенный
230369 Лимодорум
230378 Лимодорум недоразвитый
230506 Кермеквидка
230507 Кермеквидка оверина
230508 Гониолимон, Кермек, Лимониум
230524 Кермек золотой
230556 Кермек бунге
230565 Кермек мясистый
230572 Кермек каспкйский
230575 Кермек золотистый
230585 Кермек скученный
230586 Кермек коралловидный
230607 Кермек двуцветный
230610 Кермек серпоколосый
230620 Кермек файзиева
230623 Кермек ферганский
230625 Кермек фишера
230626 Кермек извилистый
230627 Кермек франше
230631 Кермек гмелина, Кермек гмелинов
230642 Кермек гельцера
230648 Кермек кашгарский
230655 Кермек плосколистный, Кермек широколистный
230657 Кермек узколопастный
230658 Кермек узколистный
230674 Кермек крупнокорневой
230678 Кермек мейера
230680 Кермек михельсона
230688 Кермек тысячецветковый
230700 Кермек ушколистный
230735 Кермек почковиднолистный
230737 Кермек резниченко
230748 Кермек прутяной
230753 Кермек семенова
230761 Кермек выемчатый, Лаванда морская
230782 Кермек кустарниковый, Кермек полукустарниковый
230787 Кермек татарский
230801 Кермек опушенный
230815 Кермек обыкновенная
230828 Лужайник, Лужница
230831 Лужайник водяной, Лужница водяная
230854 Лёновые, Льновые
230866 Лен дидий, Линария, Льнянка
230870 Льнянка остролопастная
230886 Льнянка белолобая
230887 Льнянка альпийская
230888 Льнянка алтайская
230897 Льнянка армянская
230901 Льнянка полевая
230910 Льнянка биберштейна
230911 Льнянка двураздельная
230920 Льнянка короткошпоровая
230925 Льнянка бунге, Льнянка заилийская
230927 Льнянка бурятская
230939 Льнянка меловая
230940 Льнянка верхнедонская
230945 Льнянка слабая
230951 Льнянка длинноплодная
230952 Льнянка шпороваядлинно
230953 Льнянка сладкая
230960 Льнянка элимаидская
230961 Льнянка черноморская
230976 Льнянка дроколистная
230978 Льнянка далматская
230987 Льнянка крупноцветковая
230989 Льнянка коричневая
231012 Льнянка неполноцветковая
231014 Льнянка итальянская
231015 Льнянка японская
231017 Льнянка кокандская
231018 Льнянка копетдагская
231019 Льнянка кулябская
231020 Льнянка курдская
231026 Льнянка ленкоранская
231027 Льнянка тонкошпоровая
231030 Льнянка линейнолистная
231033 Льнянка македонская
231036 Льнянка крупнолистная
231037 Льнянка длиннохвостая
231042 Льнянка марьянниковая
231043 Льнянка мейера
231044 Льнянка мелкоцветная
231061 Льнянка душистая
231062 Льнянка душистая
231074 Льнянка пелиссе
231080 Льнянка понтийская
231081 Льнянка попова
231090 Льнянка ветвистая
231093 Льнянка отогнутоножковая
231118 Льнянка русская
231120 Льнянка песчаная
231130 Льнянка сидячецветковая
231132 Льнянка простая
231139 Льнянка полосчатая
231165 Льнянка туркменская
231183 Бутея, Золотушник, Льнянка обыкновенная
231196 Льнянка зангезурская
231243 Линделофия
231244 Линделофия узкалистная
231246 Линделофия капю
231248 Линделофия длинностолбиковая
231249 Линделофия ольги
231252 Линделофия столбиковая
231253 Линделофия крылоплодная
231297 Дерево бензойное, Линдера
231303 Линдера узколистная
231322 Линдера динхуская
231324 Линдера обыкновенная
231331 Линдера гуншаньская
231366 Линдера гуансиская
231370 Линдера хэнаньская
231377 Линдера гуандунская
231395 Линдера мэнхайская
231396 Линдера тибетская
231403 Линдера туполопастная
231419 Линдера эмэйская
231433 Линдера шелковистая
231435 Линдера сычуаньская
231453 Линдера тяньцюаньская
231461 Линдера зонтичная
231473 Ванделлия, Линдерния
231509 Ванделлия раскидистая
231559 Линдерния стаканчатая, Льнянка лежачая
231600 Линдерния яошаньская
231676 Линнея
231679 Линнея северная
231818 Грудница, Линозирис
231828 Грудница фомина
231838 Грудница татарская
231842 Грудница мохнатая, Линозирис мохнатый
231844 Астра льнолистная, Грудница обыкновенная, Ленок, Ленок обыкновенный, Солонечник ленок
231856 Лен, Лён, Линум
231864 Лен алтайский
231865 Лен амурский
231872 Лен австрийский, Лён австрийский
231880 Лен узколистный, Лён узколистный
231884 Дерево дынное обыкновенное, Лён

слабетельный, Лен слабительный
231895 Лен щиткоцветковый, Лен щиточковатый
231909 Лен разночашелистиковый
231911 Лен волосистый, Лен естковолосый
231936 Лен жилковатый
231937 Лен узловатый
231941 Лен бледноватый
231942 Лен многолетний, Лён многолетний, Лен сибирский, Лён сибирский
231958 Лен чешуйчатый
231961 Лен стеллеровидный
231985 Лен крымский
231992 Лен тонколистный, Лён тонколистный
231996 Лен французский, Лён французский
232000 Лен украинский
232001 Лён белосемянный, Лен долгунец, Лён долгунец, Лён кудряш, Лен культурный, Лён культурный, Лен обыкновенный, Лён обыкновенный, Лён посевной
232002 Лён кудряш
232003 Лён-прыгунец, Лён-скакунец
232072 Липарис, Лосняк
232096 Липарис баотинский
232117 Липарис мелкоцветковая
232183 Липарис хайнаньский
232197 Липарис японская, Липарис японский
232215 Липарис кумокири
232217 Липарис гуандунский
232229 Липарис лёзеля, Лосняк лёзеля
232318 Липарис сахалинский
232322 Липарис макино
232457 Липия, Липпия
232483 Вербена лимонная, Липпия лимонная, Липпия трёхлистная, Ястребинка плетевая
232550 Ликвидамбар
232557 Ликвидамбар формозский
232562 Дерево стираксовое, Ликвидамбар восточный
232565 Амбровое дерево, Дерево амарантовое, Ликвидамбар смолоностый, Ликвидамбар стираксовый
232594 Магнолия коко
232595 Магнолия делаве
232598 Магнолия симаоская
232602 Дерево тюльпанное, Лиран, Лириодендрон, Тюльпанное дерево
232603 Дерево тюльпанное китайское, Лириодендрон китайский
232609 Дерево тюльпанное, Дерево тюльпанное виргинское, Лиран, Лириодендрон тюльпанный, Тюльпанное дерево
232627 Лириодендрон ганьсуский
232664 Лизея
232665 Лизея армянский
232667 Лизея разноплодная
232890 Листера, Офрис, Тайник
232901 Тайник коротковзубчатый
232905 Тайник сердцвидный
232919 Тайник японский
232950 Тайник овальнолистный, Тайник овальный
233077 Личжи, Нефелиум
233078 Личжи китайский, Нефелиум китайский
233099 Камнеплодник, Пазания
233692 Воробейник
233698 Воробейник узколистный
233700 Воробейник полевой
233731 Воробейник краснокорневишный, Воробейник краснокорневой
233760 Воробейник аптечный, Воробейник лекарственный
233772 Воробейник пурпурно-синий
233782 Воробейник тонкоцветковый
233784 Воробейник чимганский
234122 Литтледалия
234124 Литтледалия алайская
234130 Литтония
234146 Литторела, Прибрежник, Прибрежница
234150 Подорожник морской, Подорожник приморский, Прибрежник одноцветковый, Прибрежница одноцветковая
234154 Литвиновия
234158 Литвиновия тонкая
234162 Ливистона, Ливистония
234166 Ливистона австралийская, Ливистона южная
234170 Ливистона китайская, Ливистония китайская
234177 Ливистона фэнкайская
234181 Ливистона кохнихинская
234190 Ливистона круглолистная
234194 Ливистона кругловатая
234214 Ллоидия, Ллойдия
234229 Ллойдия красно-зеленая
234234 Ллойдия поздная, Ллойдия поздняя
234241 Ллойдия тибетская
234244 Ллойдия трехцветковая
234251 Лоаза
234269 Лобалия лёгочная, Лобалия лёгочнообразная
234275 Лобелия
234363 Лобелия китайская
234377 Лобелия гуйцзоуская
234405 Лобелия гуансиская
234407 Лобелия сычуаньская
234437 Лобелия дортмана, Лобелия дортманна
234520 Лобелия хайнаньская
234541 Лобелия гиблидная
234547 Зелье рвотное, Лобелия вздутая, Лобелия надутая, Лобелия одутлая
234766 Лобелия сидячелистная
234813 Лобелия сифилитическая
234816 Лобелия далиская
234827 Лобелия тибетская
234890 Лобелиевые
235065 Клоповник, Конига
235090 Алиссум душистый, Алиссум европейский, Алиссум морской, Бурачок приморский, Каменник норской, Конига морская, Конига приморская
235159 Орех малдивский, Пальма сейшельская
235251 Логаниевые, Чилибуховые
235262 Жабник галльский
235267 Жабник маленький, Жабник малый, Жабник мелкий
235283 Азалия порярная, Азалия стелющаяся, Лаузелеурия лежачая, Лаузелеурия стелющаяся, Лузалеурия стелющаяся
235297 Плевелок
235299 Плевелок восточный
235300 Плевел, Райграсс английский
235303 Плевел полевой
235310 Плевел южный
235315 Плевел итальянский, Плевел много, Плевел многоцветковый, Плевел многоцветный, Райграсс итальниский, Райграсс многолетний, Райграсс многоукосный
235334 Плевел многолетний, Райграсс английский, Райграсс высокий, Райграсс многолетний, Райграсс многолетний английский, Райграсс многолетний настбищный, Райграсс пастбищный
235351 Плевел персидский, Плевел пранский
235353 Плевел льняной, Плевел расставленный
235354 Плевел жесткий, Плевел южный
235373 Плевел, Плевел клиновидный, Плевел одуряющий, Плевел опьяняющий
235431 Ломатогониум, Плеврогина
235435 Ломатогониум каринтийский
235439 Ломатогониум ядунский
235450 Ломатогониум лицзянский
235455 Ломатогониум мелкоцветковая
235461 Ломатогониум колесовидный, Ломатогониум кольцевидный

235471 Ломатогониум чжундяньский
235513 Лонхокарпус
235614 Люндезия
235618 Люндезия пушистоцветная
235630 Жимолость, Каприфоль, Лоницера
235633 Жимолость генри
235649 Жимолость альберта
235651 Жимолость альпийская, Таволга альпийская
235663 Жимолость альтмана
235690 Жимолость бушей
235691 Жимолость, Жимолость голубая, Жимолость синяя
235699 Жимолость алтайская
235702 Жимолость съедобная
235710 Жимолость канадская
235713 Жимолость вьюизаяся, Жимолость душистая, Жимолость каприфол, Жимолость каприфоль, Жимолость козья, Жимолость лоницера, Каприфоль, Каприфоль обыкновенная, Лоницера каприфолиум
235715 Жимолость кавказская
235721 Жимолость шамиссо
235730 Жимолость золотистая, Жимолость золотистоцветковая
235740 Жимолость серая
235771 Жимолость этрусская
235780 Жимолость фэнкайская
235781 Жимолость фердинанда
235794 Жимолость цветущая
235796 Жимолость душистейшая
235798 Жимолость стандиша
235811 Жимолость глена
235841 Жимолость кандинская
235843 Жимолость раэнолистная
235845 Жимолость раэноволосая
235852 Жимолость щетинистая
235859 Жимолость низкая
235864 Жимолость грузинская
235865 Жимолость илийская
235874 Жимолость обернутая
235878 Жимолость японская
235890 Жимолость цзилунская
235894 Жимолость камчатская
235895 Жимолость ганьсуская
235903 Жимолость королькова
235909 Жимолость шерстистая
235929 Жимолость маака
235945 Жимолость максимовича
235949 Жимолость ланьпинская
235952 Жимолость мелколистная
235965 Жимолость лушаньская
235970 Жимолость морроу
235981 Жимолость жилковатая
235982 Жимолость черная, Жимолость чёрная
236000 Жимолость ольги
236012 Жимолость палласа
236013 Жимолость памирская
236015 Жимолость странная
236022 Жимолость немецкая
236035 Жимолость шапочная
236040 Жимолость раннецветущая, Жимолость ранняя, Жимолость раноцветущая
236070 Жимолость тибетская
236076 Жимолость рупрехта
236077 Жимолость рупрехта оголяющаяся
236085 Жимолость ахалинская
236088 Жимолость семенова
236089 Жимолость вечнозелёная
236106 Жимолость эмэйская
236115 Жимолость узкоцветковая
236130 Жимолость почтищетинистая, Жимолость щетинистоволосистая
236138 Жимолость сычуаньская
236146 Жимолость татарская
236158 Жимолость мелкоцветковая, Жимолость татарская мелкоцветковая
236174 Жимолость тяньшанская
236175 Жимолость толмачева
236198 Жимолость турчанинова
236214 Жимолость уэбба
236226 Жимолость лесная, Жимолость обыкновенная, Жимолость пушистая, Ягода волчья
236262 Лофант, Лофантус
236267 Лофант китайский, Лофантус китайский
236269 Лофант изящный
236271 Лофант крылова
236272 Лофант липского
236274 Лофант шренка
236277 Лофант приснежный
236278 Лофант тибетский
236279 Лофант чимганский
236334 Лофира
236336 Лофира африканская, Лофира крупнолистная, Лофира крылатая, Лофира пирамидальная
236425 Кактус лофофора, Лофофора вильямся
236462 Трипутник свёрнутая, Трипутник скрученная, Тристания сжатая
236503 Ремнецветниковые
236507 Ремнецветник
236704 Ремнецветник евронейский
236755 Ремнецветник гуйцзоуский
237190 Лоропеталум, Лоропеталюм
237191 Лоропеталюм китайский
237219 Лядвенестальник
237312 Лядвенестальник дроковидный
237476 Богатень, Лядвенец, Тетрагонолобус, Четырёхкрыльник
237485 Лядвенец узкий, Лядвенец узколистный
237506 Лядвенец ланьюйский
237526 Лядвенец кавказский
237539 Лядвенец рогатый
237554 Лядвенец
237609 Лядвенец густооблиственный
237615 Лядвенец гебелия
237617 Лядвенец тончий
237704 Лядвенец птиценогий
237709 Лядвенец болотный
237713 Лядвенец болотный, Лядвенец топяной
237725 Лядвенец просмотренный
237759 Лядвенец торчащий
237768 Лядвенец тонкий, Лядвенец тонколистнный
237771 Горох спаржевый, Лядвенец пурпурный, Лядвенец четырёхлопастный, Тетрагонолопус пурпурный, Четырёхкрыльник пурпуровый
238150 Людвигия
238192 Людвигия болотная
238211 Людвигия лежачая, Людвигия простертая
238230 Людвигия тайваньская
238242 Люхея
238257 Губка растительная, Губка-люффа, Люфа, Люффа
238258 Люффа гранистая, Люффа ребристая
238261 Губка растительная, Люфа цилиндрическая, Люффа мочалка, Люффа цилиндрическая, Тыква мочальная, Тыква цилиндрическая
238369 Лунник
238371 Лунник однолетний
238379 Лунник многолетный, Лунник оживающий
238419 Боб волчий, Лупин, Люпин
238421 Лупин белый, Люпин белый, Люпин древнеегипетский, Люпин тёплый
238422 Лупин узколистный, Люпин синий узколистный, Люпин узколистный
238430 Люпин древовидный
238448 Люпин мексиканский
238450 Люпин волосистый, Люпин жестковолосистый, Люпин жесткоопущённый, Люпин лохматый
238456 Люпин гибридный
238463 Лупин желтый, Люпин жёлтый
238468 Люпин изменчивый, Люпин пестроцветный
238469 Люпин низкий
238472 Лупин нуткинский

238473 Люпин многолестний
238481 Лупин многолистный, Люпин многолистный
238510 Элизма
238563 Ожика
238572 Ожика дуговидная, Ожика изогнутая
238573 Ожика камчадальская
238583 Ожика равнинная
238600 Ожика головчатая
238610 Ожика неясная, Ожика спутанная
238619 Ожика форстера
238629 Ожика японская
238631 Ожика тибетская
238637 Ожика чъельмана
238645 Ожика черноплодная
238647 Ожика многоцветковая
238650 Ожика холодная
238663 Ожика боровая, Ожика дубравная
238667 Ожика снеговая
238671 Ожика малоцветковая
238675 Ожика бледнеющая, Ожика бледноватая
238678 Ожика мелкоцветковая
238681 Ожика волосистая
238689 Ожика ложносудетская
238690 Ожика рыжеватая, Ожика рыжеющая
238691 Ожика крупноплодная
238694 Ожика сибирская
238695 Ожика сычуаньская
238697 Ожика колосистая
238702 Ожика слабоволосистая
238706 Ожика судетская
238709 Ожика лесная
238712 Ожика тайваньская
238717 Ожика валенберга
238765 Лихнис
238996 Дереза, Лиций
239011 Дереза берберов, Лиций обыкновенный
239021 Дереза китайская
239030 Дереза волосистотычинковая
239049 Дереза изогнутая
239055 Дереза лебедолистная, Дереза обыкновенная, Лиций обыкновенный
239068 Дереза копетдагская
239106 Дереза русская, Лиций русский
239128 Дереза туркменская
239154 Помидор, Томат
239157 Томат съедобный
239161 Томат гумбольдта
239163 Томат перуанский
239165 Томат смородинный
239167 Кривоцвет, Ликопсис
239184 Зюзник
239202 Зюзник европейский, Крапчатка
239205 Зюзник высокий
239211 Зюзник опушенный
239215 Зюзник блестящий, Зюзник дальневосточный
239224 Зюзник маака
239226 Зюзник маака тайваньская
239235 Зюзник мелкоцветковый
239369 Лиония
239391 Лиония овальнолистная
239392 Лиония овальнолистная эллиптическая
239516 Лизихитон
239518 Лизихитон камчатский
239532 Лизиелла малоцветковаю
239543 Вербейник, Наумбургия
239547 Вербейник гуансиский
239558 Вербейник густоколосый, Вербейник густоцветковый
239566 Вербейник белоснежный
239578 Вербейник чжэцзянский
239588 Вербейник чжундяньский
239594 Вербейник клетровидный, Вербейник ландышевый, Вербейник ландышный
239606 Вербейник японский
239623 Вербейник сомнительный
239626 Вербейник симаоский
239641 Вербейник фунинский
239645 Вербейник фортюна
239651 Вербейник фуцзяньский
239666 Вербейник франше
239668 Вербейник ичанский
239673 Вербейник ху
239687 Вербейник японский
239696 Вербейник яванский
239699 Вербейник цзиндунский
239702 Вербейник гуандунский
239712 Вербейник лицзянский
239731 Вербейник мотоский
239735 Вербейник мелкоплодный
239745 Вербейник наньчуаньский
239746 Вербейник наньпинский
239748 Вербейник хайнаньский
239750 Вербейник дубравный
239755 Вербейник луговой чай, Вербейник монетчатый, Луговой чай, Чай луговой
239762 Вербейник эмэйский
239775 Вербейник мелколистный
239781 Вербейник пятиленестный
239804 Вербейник пятнистный, Вербейник точечный
239822 Вербейник лушаньский
239846 Вербейник таиландский
239872 Наумбургия тайваньская
239874 Наумбургия далиская
239878 Наумбургия тэнчунская
239881 Кизляк кистецветковый, Кизляк кистецветный, Наумбургия кистецветная
239891 Вербейник мутовчатый
239898 Вербейник обыкновенный
239902 Вербейник даурский
240020 Дербенниковые
240025 Дербенка, Дербенник, Литрум, Плакун, Подбережник
240047 Дербенник иволистный, Дербенник иссополистный
240050 Дербенник промежуточный
240054 Дербенник комарова
240059 Дербенник карликовый, Дербенник низкий
240068 Дербенник securityзализняк, Дербенник иволистный, Дербенник лозный, Дербенник-плакн, Литрум обыкновенный, Плакун, Плакун иволистный, Плакун-трава, Подбережник
240086 Дербенник шелковникова
240087 Дербенник смолевковидный
240088 Дербенник федора
240089 Дербенник ленецевидный, Дербенник ленолистниковый
240090 Дербенник тимьянолистный
240094 Дербенник трехприцветниковый
240100 Дербенник иссополистный, Дербенник лозный, Дербенник плакун, Дербенник прутовидный, Дитрум лиловый, Плакун лозовый, Плакун пруныйтовид
240112 Акатик, Маакия
240114 Маакия амурская
240120 Маакия чжэцзянская
240128 Маакатик китайский, Маакия китайская
240130 Маакия тайваньская
240207 Макадамия
240210 Киндаль, Макадамия, Макадамия трехлистная, Орех австралийский
240550 Махил
240567 Махил
240568 Махил чжэцзянский
240570 Махил китайский
240587 Махил фуцзяньский
240592 Махил гуншаньский
240599 Махил ичанский
240604 Махил японский
240610 Махил гуандунский
240620 Махил личуаньский
240621 Махил либоский
240645 Махил мелкоплодный
240653 Махил наньчуаньский
240668 Махил мелкоцветковый
240698 Махил сичоуский
240699 Махил сычуаньский
240707 Махил тунберга

240725 Махил тибетский
240763 Боккония, Маклейя, Смолосемянник
240806 Лжеанельсин, Маклюра
240808 Кудрания цзиндунская
240828 Апельсин оседж, Апельсин-оседж, Дерево лжеапельсиновое, Лжеапельсин, Маклюра оранжевая, Маклюра яблоконосная
240842 Кудрания трехконечная
241248 Долгонот, Макроподиум
241249 Долгонот снеговой
241250 Долгонот крылатосемянный, Макроподиум круглосемянный, Макроподиум крылатосемянный
241386 Макротамия
241391 Макротамия густоцветковая
241398 Макротамия угамская
241448 Макрозамия
241514 Бассия широколистная, Мадука индийская, Опунция бассия
241523 Мадия
241543 Липучка масляная, Мадия посевная
241572 Меруя
241734 Меза
241755 Меза баотинская
241774 Меза хубэйская
241781 Меза японская
241795 Меза ланьюйская
241818 Меза тайваньская
241830 Меза морщинистая
241833 Меза иволистная
241951 Магнолия
241962 Дерево огуречное, Магнолия длиннозанстрённая, Магнолия заостренная, Магнолия остроконечная
242135 Лавр тюльпанный, Магнолия крупноцветковая, Магнолия крупноцветная
242152 Магнолия хэнаньская
242169 Магнолия кобус
242171 Магнолия кобус северная
242331 Магнолия звездчатая
242333 Магнолия зонтичная, Магнолия трехлепестковая, Магнолия трёхлепестковая
242341 Лавр берый, Лавр душистый, Магнолия виргинская, Магнолия сизая
242352 Магнолия уотсона
242370 Магнолиевые
242400 Оносма чжэнькanская
242458 Магония
242478 Магония падуболистная
242487 Магония била
242493 Магония сичанская
242502 Магония тибетская
242503 Магония чаюйская
242517 Магония куэньминская
242534 Магония форчуна
242550 Магония хайнаньская
242553 Магония мэнцзыская
242563 Магония японская
242597 Магония непалская
242608 Магония ломариелистная
242612 Магония цзиндунская
242660 Магония вагнера
242668 Майник
242672 Майник двулистный
242675 Майник канадский
242679 Смилацина даурская
242680 Майник широколистный
242687 Смилацина гуншаньская
242691 Смилацина волосистая, Смилацина японская
242693 Смилацина лицзянская
242694 Майоран наньчуаньский, Смилацина наньчуаньская
242698 Смилацина чжундяньская
242703 Смилацина сычуаньская
242706 Смилацина трехлистная
242804 Малабайла
242810 Малабайла пушистоцветковая
242813 Малабайла пахучая
242818 Малабайла
242825 Малабайла бороздчатая
242848 Яблоневые
242866 Малахра
242970 Мягкохвостник шерстистый
242993 Малаксис, Микростилис, Мякотница, Стагачка
243084 Малаксис однолистный
243156 Малькольмия
243164 Малькольмия африканская, Малькольмия одуванчколистая
243184 Малькольмия бухарская
243188 Малькольмия городчатая
243191 Малькольмия крупноцветная
243193 Малькольмия щетинистая
243195 Малькольмия карелина
243230 Малькольмия многостручковая, Малькольмия скорпионовидная
243243 Малькольмия вололистоплодная
243246 Малькольмия туркестанская
243315 Маллот, Маллотус
243320 Маллотус бесщитковый
243322 Маллотус гуансиский
243332 Маллотус хубэйский
243353 Маллотус наньпинский
243364 Маллотус хайнаньский
243371 Маллотус японский
243372 Маллот токолисный
243399 Маллотус непалский
243424 Маллот лушаньский
243472 Дыравка, Малопа
243475 Дыравка малаковидная
243494 Дыравка трехрассеченная
243519 Мальпигия, Мальпидия
243523 Мальпидия орехоносная
243526 Мальпигия гранатоволистная
243527 Мальпидия мексиканская
243530 Мальпигиевые
243543 Яблоня
243555 Крэб сибирский, Ранетка пурпурная, Яблоня дикая сибирская, Яблоня корнесобственная, Яблоня палласа, Яблоня палласова, Яблоня сибирская, Яблоня ягодная
243571 Яблоня гималайская
243583 Яблоня венечная, Яблоня душистая
243589 Яблоня войлочно-опушенная
243617 Яблоня многоцветковая, Яблоня обильноцветковая, Яблоня обильноцветущая
243629 Яблоня хэнаньская
243630 Яблоня хубейская
243642 Яблоня ганьсуская
243649 Яблоня маньчжуьрская
243657 Яблоня мелкоплодная
243658 Яблоня мулиская
243659 Яблоня красномясая, Яблоня недзведцкого, Яблоня тяньшаньская
243662 Яблоня восточная
243664 Яблоня ранняя
243667 Китайка, Яблоня китайская, Яблоня сливолистная
243675 Леснина, Яблоко-кислица, Яблоня дикая, Яблоня домашняя, Яблоня карликовая, Яблоня корнесобственная, Яблоня культурная, Яблоня лесная, Яблоня низкая, Яблоня низкорослая, Яблоня садовая
243682 Парадизка, Райка, Яблоко райское, Яблоня восточная, Яблоня парадизка, Яблоня райская
243687 Яблоня рока
243698 Дичка, Яблоня горная, Яблоня дикая, Яблоня сиверса
243702 Яблоня замечательная
243711 Крэб, Леснина, Яблоко-кислица, Яблоня дикая, Яблоня домашняя, Яблоня карликовая, Яблоня культурная, Яблоня лесная, Яблоня низкая, Яблоня низкорослая, Яблоня садовая
243717 Яблоня зибольда
243730 Яблоня туркменов
243732 Яблоня юньнаньская
243733 Яблоня юньнаньская вича
243738 Мальва, Просвирник, Просвирняк
243746 Просвирник штокрозовый, Просвирняк раздельнолистный
243747 Просвирник спорный

243750 Просвирник армянский
243758 Просвирник бухарский
243771 Мальва курчавая, Просвирник курчавый, Просвирняк курчавый
243781 Просвирник прямостоячий
243783 Просвирник вылезанный
243787 Просвирник гроссгейма
243793 Просвирник голоплодный
243797 Просвирняк мавританский
243801 Просвирник могилевский
243803 Мальва мускусная, Просвирняк мускусный
243805 Просвирник ницкий
243806 Алтей лекарственный, Просвирняк аптечный
243809 Просвирник памироалайский
243810 Просвирник мелкоцветковый
243818 Просвирняк плебейский
243821 Просвирник красивенький
243823 Мальва круглолистная, Мальва обыкновенная, Мальва приземистая, Мальва северная, Просвирник низкий, Просвирник пренебреженный, Просвирняк круглолистный, Просвирняк обыкновенный, Просвирняк приземистый
243833 Просвирняк китайский
243840 Гросвирняк дикий, Зензивер, Мальва лесная, Просвирник лесной
243862 Мальва красивая, Мальва мутовчатая, Просвирник мутовчатый, Просвирняк китайский, Просвирняк красивый, Просвирняк мутовчатый
243873 Мальвовые, Просвирниковые, Просвирняковые
243874 Мальвальтея
243875 Мальвальтея закавказкаская
243877 Мальваструм
243956 Мальвочка
243958 Мальвочка жерара
243979 Маммея
243982 Абрикос антильский, Абрикос тропический, Маммея американская
243994 Кактус маммилярия, Кактус сосочковый, Маммилярия
244012 Маммилярия бокасана
244203 Мамиллярия маленькая
244340 Зелье сонное, Мандрагора
244344 Мандрагора цинхайская
244349 Мандрагора туренская, Мандрагора туркменская, Селемелек
244358 Манеттия
244386 Дерево манговое, Манго
244393 Манго вонючее, Манго вонючий, Манго пахучее, Манго пахучий
244397 Дерево манговое, Манго, Манго индийское
244406 Манго душистое, Манго душистый
244409 Манго таиландское
244420 Манглиэция
244425 Манглиэция тибетская
244427 Манглиэция тонкоствольная
244435 Манглиэция хайнаньская
244440 Манглиэция таиландская
244451 Манглиэция гуандунская
244468 Манглиэция напоская
244471 Манглиэция сычуаньская
244501 Кассава, Маниок, Маниока, Маниот
244506 Маниок сладкий
244507 Аррорут бразильский, Кассава, Кассава горькая, Маниок горький, Маниок полезнейший, Маниок съедобный, Маниот съедобный, Тапиока
244509 Маниок глацийова, Маниока каучуконосная, Маниот глазиова
244512 Кассава сладкая, Маниок сладкий
244515 Аррорут бразильский, Кассава горькая, Маниок горький, Маниок полезнейший, Маниок съедобный, Маниот съедобный, Тапиока
244530 Балата
244605 Саподилла
244635 Маннагеттея
244637 Маннагеттея иркутская
245013 Аррорут, Маранта
245014 Аррорут вестидский, Маранта вестиндская, Маранта тростниковая, Маранта тростниковидный
245022 Маранта беложильчатая
245039 Марантовые
245157 Марезия
245164 Марезия низкорослая
245220 Филлантус индийский
245332 Марискус, Меч-трава
245723 Шандра
245726 Шандра разнозубая
245740 Шандра севанская
245748 Шандра пустырниковая, Шандра пустырниковидная
245754 Шандра карликовая
245756 Шандра малоцветковая
245757 Шандра чужеземная
245758 Шандра иранская
245759 Шандра перистая
245760 Шандра ранняя
245761 Шандра близкая
245764 Шандра пурпуровая
245769 Шандра туркевича
245770 Мята конская, Шандра белая, Шандра обыкновенная
245774 Шандра волонова
245777 Марсдения
245798 Кондуранго
245977 Мартиния, Роговник
245991 Мартиниевые
246307 Матрикария, Ромашка
246315 Ромашка сомнительная
246318 Ромашка золотистая
246336 Ромашка отмичная
246375 Ромашка безлепестная, Ромашка дисковидная, Ромашка душистая, Ромашка пахучая, Ромашка ромашковидная
246392 Ромашка ранняя
246396 Ромашка аптечная, Ромашка дикая, Ромашка лекарственная, Ромашка настоящая, Ромашка обнаженная, Ромашка ободранная
246399 Нивяник круглолистный
246411 Ромашка цвелева
246438 Левкой, Маттиола
246439 Левкой белостебельный
246443 Левкой двлогий
246446 Маттиола бухарская
246447 Левкой марелистый
246450 Левкой дагестанский
246455 Левкой мучнистый
246460 Левкой душистый, Левкой пахучий
246478 Левкой зимний, Левкой обыкновенный, Левкой однолетний, Левкой садовый, Левкой садой, Маттиола садовая
246487 Левкой цельнокрайний
246494 Левкой остророгий
246521 Левкой душистый
246544 Левкой грубый
246545 Левкой струговидный
246554 Левкой стоддарта
246555 Левкой великолепный
246556 Левкой татарский
246578 Паракариум гималайский
246651 Лофоспермум, Маурандия
246671 Пальма винная, Пальма маврикиева
246674 Бурути, Пальма бурути
246761 Майяка
246796 Майтенус
246817 Майтенус чилийский
246859 Майтенус падуболистный
246964 Мадзус, Мазус
246979 Мадзус фуцзяньский
247000 Мадзус гуйцзоуский
247024 Мадзус ван
247025 Мадзус японский, Мазус японский
247031 Мадзус лицзянский
247039 Мадзус чистецолистный
247043 Мадзус тибетский
247099 Мак голубой, Меконопсис
247113 Меконопсис кумберлендский, Меконопсис уэльский
247122 Меконопсис тибетский
247123 Меконопсис лицзянский

247139 Меконопсис цельнолистный
247157 Меконопсис непалский
247239 Люцерна, Рикшаи-одири
247242 Люцерна полевая
247246 Люцерна арабская, Люцерна аравийская, Люцерна пятнистая
247249 Люцерна древовидная
247257 Люцерна азиатская
247261 Люцерна решетчатая
247265 Люцерна голубая
247268 Люцерна увенчанная
247271 Люцерна меловая
247275 Люцерна дагестанская
247293 Люцерна
247298 Люцерна железистая, Люцерна кавказская
247304 Люцерна полузакрученная
247337 Люцерна раздельнолистная
247350 Люцерна шерстистая
247359 Люцерна прибрежная
247366 Буркунчик, Люцерна жмелевая, Люцерна жмелевдная, Люцерна хмелевая, Люцерна хмелевидная, Полуклевер, Трилистник жёлтый
247378 Люцерна приморская
247381 Люцерна маленькая
247407 Люцерна дисковидная, Люцерна круглая, Люцерна округлая
247418 Люцерна пушистая
247419 Люцерна пятицкличная
247422 Люцерна широкоплодная, Пажитник плоскоплодный
247424 Люцерна разноцветная
247425 Люцерна зубчатая, Люцерна маленькая, Люцерна мелкозубчатая, Люцерна шетинистая, Люцерна шетинистоволосистая
247437 Люцерна ранняя
247438 Люцерна лежачая
247443 Люцерна жестковатая
247451 Люцерна скалиная, Люцерна скалистая
247452 Люцерна русская, Пажитник русский
247456 Алфалфа, Альфальфа, Буркун красный, Люцерна обыкновенная, Люцерна посевная, Люцерна синяя
247457 Буркун, Джонушка, Люцерна желтая, Люцерна жёлтая, Люцерна серповидная, Юморка
247467 Люцерна каменистая
247468 Люцерна синьцзянская
247469 Люцерна щитковидная
247473 Люцерна солейроля
247486 Люцерна траутфеттера
247493 Люцерна якорцевидная
247507 Люцерна гибридная, Люцерна пестрая, Люцерна средняя, Люцерна тяньшанская
247510 Люцерна зеленоватая
247526 Мединилла
247720 Михения
247739 Михения крапиволистная
247767 Крупноплодник
247782 Крупноплодник гигантский
247783 Крупноплодник илийский
247786 Крупноплодник большеплодный, Крупноплодник крупноплодный, Крупноплодник мугоджарский
247788 Крупноплодник округлый
247791 Крупноплодник шугнанский
247850 Магадения
248072 Дерево каепутовое, Дерево кайюпутовое, Дерево каяпутовое, Мелалеука
248104 Дерево каяпутовое, Мелалеука бледнодревесная, Мелалеука цайная
248105 Мелалеука льнянколистная
248110 Мелалеука мелкоцветковая
248155 Марьянник
248156 Марьянник альбова
248161 Марьянник полевой
248164 Марьянник кавказский
248166 Марьянник зеленоколосый
248168 Марьянник гребенчатый
248169 Марьянник восокий
248170 Марьянник гербиха
248172 Марьянник разрезной
248193 Марьянник луговый
248195 Марьян розовый
248207 Марьянник скальный
248208 Марьянник щетинистый
248211 Марьянник сесной
248235 Дрема
248239 Дрема железистая
248241 Дрема родственная
248246 Дрема акинфиева
248248 Дрема белая
248252 Дрема узкоцветковая
248255 Дрема безлепестная
248269 Дрема пустынная
248284 Дрема астраханская
248291 Дрема баланзы
248294 Дрема буассы
248298 Дрема коротколепестнаю
248317 Дрема краснеющая
248318 Дрема федченко
248319 Дрема ферганская
248334 Дрема стройная
248382 Дрема ночная
248390 Дрема ольги
248393 Дрема овальнолистная
248413 Дрема сахалинская
248419 Дрема сочавы
248423 Дрема тусклая
248427 Дрема лесная
248429 Дрема таймырская
248435 Дрема тонкая
248440 Дрема траурная
248453 Дрема липкая
248806 Меластомовые
248893 Мелия, Ясенка
248895 Мелия ацедарах, Мелия гималайская, Мелия иранская, Мелия персндская
248903 Мелия японская
248929 Мелиациевые, Мелиевые
248936 Мелиант
248949 Перловник
248951 Перловник высокий
248960 Перловник азербайджанский
248962 Перловник короткоцветковый
248975 Перловник крымский
249008 Перловник желтоватый
249010 Перловник сизый
249012 Перловник гмелина
249043 Перловник малый
249054 Перловник поникающий
249065 Перловник кисточковидный
249066 Перловник жакемонта, Перловник неравночешуйный, Перловник персидский
249067 Перловник сероватый
249072 Перловник пестрый
249075 Перловник ганьсуский
249085 Перловник шероховотый
249090 Перловник однобокий
249104 Перловник трансильванский
249105 Перловник турчанинова
249106 Перловник одноцветковый
249110 Перловник прутьевидный
249124 Меликокка
249175 Мелицитус
249197 Гоу-ришка, Донник, Донник желтый
249203 Донник белый, Клевер бухарский
249205 Донник высокий, Донник рослый
249212 Донник зубчатый
249217 Донник волосистый
249218 Донник индийский, Донник мелкоцветковый
249228 Донник неаполитанский
249232 Бубрун желтый, Гоуришка, Донник аптечный, Донник ароманный, Донник аселтый, Донник душистый, Донник желтый, Донник жёлтый, Донник лекарственный
249240 Донник польский
249251 Донник бороздчатый
249260 Донник крымский
249263 Донник волжский

249303 Мелинис, Трава molасская, Трава паточная
249355 Мелиосма
249414 Мелиосма многоцветковая
249479 Мелиосма гуншаньская
249481 Мелиосма сичоуская
249482 Мелиосма юньнаньская
249491 Лимонная мята, Мелисса
249504 Лимонная мята аптечная, Мелисса аптечная, Мелисса лекарственная, Мята лимонная, Мята лимонная аптечная, Трава лимонная, Трава пчелиная, Цитрон-мелисса
249507 Мелисса мелколистная
249540 Кадило бабье
249542 Кадило, Кадило мелиссолистное
249578 Мелокактус
249760 Мелотрия
250217 Кукольвановые, Луносемянниковые, Мениспермовые
250218 Луносемянник, Мениспермум, Плющ дауский
250221 Луносемянник канадский, Мениспермум канадский, Плющ канадский
250225 Луносемянник китайский
250228 Луносемянник даурский, Плющ амурский
250287 Мента, Мята
250288 Мята алайская
250291 Мята водная, Мята водяная
250294 Мята полевая
250323 Мята австрийская, Мята азиатская
250326 Мята австрийская
250331 Мята канадская
250333 Мята сичанская
250339 Мята кавказская
250341 Мята бергамотная, Мята курчавая, Мята лимонная
250347 Мята курчавая
250349 Мята кудреватая, Мята кудрявая
250352 Мята дагестанская
250353 Мята даурская
250354 Мята дарвазская
250370 Мята полевая, Мята просто-чашечковая, Мята просточашечная
250380 Мята прерывистая
250381 Мята японская
250383 Мята копетдагская
250384 Мята лапландская
250385 Мята дикая, Мята длиннолистная
250400 Мята мелкоцветковая
250406 Мята памироалайская
250420 Мята ангрийская, Мята перечная, Мята холодка, Холодка
250432 Мята блошница, Мята болотная, Мята пурегиевая, Полей
250439 Мята круглолистная
250443 Мята сахалинская
250450 Мята зеленая, Мята зелёная, Мята колосовая
250463 Мята раскидистая
250481 Ментцелия
250492 Вахтовые
250493 Вахта, павун
250502 Вахта трехлистная, Вахта трёхлистная, Вахта трилистная, Третьина, Трефоль, Трилистник водяной, Трилистник голький, Трифоль
250505 Менцизия
250528 Менцизия пятитычинковая
250597 Пролеска, Пролесник
250600 Зелье курье, Пролеска однолетняя, Пролесник однолетний
250610 Пролесник гладкоплодный
250613 Пролеска яйцевиднолистная, Пролесник яйцевидный
250614 Капуста собачья, Пролеска многолетняя, Пролесник многолетный
250627 Мерендера
250631 Мерендера белейшая
250632 Мерендера кавказская
250633 Мерендера эйхлера
250636 Мерендера гиссарская
250637 Мерендера иоланты
250644 Мерендера радде
250645 Мерендера крупная
250647 Мерендера отпрысковая
250648 Мерендера трехстолбиковая
250730 Раздельнолодочник красноплодный
250733 Раздельнолодочник тройчатолистный
250734 Раздельнолодочник дурнишниковидный
250831 Ипомея сибирская
250883 Мертензия, Мертенсия
250888 Мертензия даурская
250890 Мертензия джагастайская
250892 Мертензия енисейская
250893 Мертензия камчатская
250894 Мертензия приморская
250898 Мертензия попова
250900 Мертензия палласа
250902 Мертензия волосистая
250905 Мертензия крылатоплодная
250909 Мертензия ручейковая
250910 Мертензия мелкопильчатая
250911 Мертензия сибирская
250912 Мертензия длинностолбиковая
250913 Мертензия тарбагатайская
251025 Мезембриантемум, Трава хрустальная
251265 Леденник, Мезембриантемум кристллический, Мезембриантемум хрустальный, Трава ледяная, Трава хрустальная
252291 Мушмула
252305 Мушмула германская
252370 Дерево железное, Мезуа железная
252508 Метаплексис
252514 Метаплексис китайский, Метаплексис staунтона, Метаплексис японская, Метаплексис японский
252538 Метасеквойя
252540 Метасеквойя глиртостробовидная, Метасеквойя глиртостробусовидная
252549 Метастахис
252550 Метастахис стреловидный
252631 Коелококкус, Целококкус
252634 Пальма саговая румфа
252636 Пальма саговая
252662 Корень медвежий, Меум, Ночецвет
252783 Мибора
252785 Мибора малая
252791 Мишоксия
252799 Мишоксия гладкая
252803 Микелия, Михелия
252845 Микелия малипоская
252849 Микелия сжатая
252857 Микелия сичоуская
252869 Микелия буроватая, Микелия ржавая, Микелия фиго
252878 Микелия фуцзяньская
252899 Микелия тибетская
252930 Микелия мелкоплодная
252974 Микелия эмэйская
253207 мелкоголовка
253208 мелкоголовка пластинчатая
253209 мелкоголовка кругловатая
253210 мелкоголовка туркменская
253631 Микромерия
253653 Микромерия изящная
253654 Микромерия эллиптическая
253661 Микромерия тайваньская
253666 Микромерия греческая
253698 Микромерия дубровниковая
253717 Микромерия тимьянолистная, Микромерия чабрецолистная
253734 Микромерия тибетская
253789 Мелкопузырник
253790 Мелкопузырник вытянутый
253841 микропус
253849 микропус лежащий
253980 Микростегиум
254020 Микростегиум японская
254034 Поллиния безбородая
254036 Поллиния голая
254046 Микростегиум лозный
254082 Микростигма
254085 Микростигма отогнутая
254094 Микростилтс, Мякотница, Стагачка

254387 Миддендорфия
254389 Миддендорфия днепровская
254414 Микания, Плющ комнатный
254426 Микания индийская
254443 Микания вьющаяся, Микания лазающая, Микания ползучая, Микания цепкая
254496 Бор, Просяник
254515 Бор развесистый, Бор раскидистый, Просяник развесистый
254542 Бор весенний
254562 Милла
254596 Миллетия
254741 Миллетия гуйцзоуская
254815 Миллетия цзиндунская
254924 Мильтония
254979 Мимоза
255098 Мимоза застенчивая, Мимоза стыдливая
255141 Мимозовые
255184 Губастик, Мимулус, Мимулюс
255197 Губастик темнокрасный, Мимулюс кардинальский, Мимулюс темнокрасный
255210 Губастик крапчатый, Мимулюс жёлтый, Мимулюс крапчатый
255218 Губастик желтый, Мимулюс желтый, Мимулюс крапчатый
255223 Губастик мускусный, Мимулюс мускусный
255235 Губастик слабоволосистый
255237 Губастик разинутый
255243 Губастик отполысковый
255244 Губастик сычуаньский
255246 Губастик тоненький
255256 Губастик тибетский
255260 Мимусопс
255435 Кверия, Минуартия, Минуарция
255437 Минуарция железистая
255438 Минуартия железистая
255439 Минуарция аицовидная
255440 Минуарция акинфиева
255442 Минуартия арктическая
255446 Минуарция биберштейна
255447 Минуартия двухцветковая, Минуарция двухцветковая
255449 Минуарция бротеруса
255450 Минуарция буша
255455 Минуарция кавказская
255461 Минуарция гвоздиколистная
255466 Минуарция изящная
255482 Минуартия скученная
255484 Минуарция стройная
255485 Минуарция зернышкоцветная
255489 Минуартия гельма
255490 Минуартия волосистая
255494 Минуарция гибридная
255497 Минуарция черепичатая
255498 Минуарция промежуточная
255499 Минуарция якутская
255503 Минуартия крашенинникова, Минуарция крашенинникова
255504 Минуарция крылова
255506 Минуарция лиственничная
255508 Минуарция полосчатая
255512 Минуарция литвинова
255514 Минуартия крупноплодная
255523 Минуарция мейера
255527 Минуарция мелкоцветковая
255549 Минуарция горная
255561 Минуарция маленькая
255564 Минуарция отогнутая
255565 Минуартия регеля
255566 Минуарция красночашечная
255569 Минуарция моховидная
255573 Минуартия красноватая
255578 Минуартия щетинковая
255582 Минуартия прямая
255584 Минуартия крымская
255595 Минуартия гибридная
255598 Минуартия весенняя
255606 Минуартия липкая
255608 Минуартия визнера
255609 Минуарция воронова
255672 Мирабилис
255694 Мирабилис вильчатая
255711 Красавица ночная, Мирабилис ялапа, Ночецветник, Ночная красавица, Ялапа
255727 Мирабилис длинноцветковая
255732 Мирабилис многоцветковая
255735 Мирабилис ночецветная
255750 Мирабилис липкая
255827 Мискантус, Трава серебряная
255873 Мискантус краснеющий, Мискантус сахароцветковый, Мискантус сахароцветный
255886 Мискантус китайский, Мискантус сахароцветный
256001 Мичелла
256005 Мичелла волнистая
256010 Мителла
256019 Мителла тайваньская
256024 Мителла японская
256032 Мителла голая
256107 Митрагина
256120 Митрагина крупнолистная, Митрагина прилистниковая
256142 Митрария
256301 Гимнастер
256310 Гимнастер сычуаньский
256424 Модестия
256426 Модестия дарвазская
256427 Модестия удивительная
256441 Мерингия
256443 Мерингия удлиненная
256444 Мерингия бокоцветковая, Мерингия бокоцветная
256456 Мерингия трехжилковая, Мерингия трехнервная
256459 Мерингия теневая
256515 Моголтавия
256516 Моголтавия северцова
256597 Молиния
256601 Молиния голубая, Молиния прибрежная, Синявк
256684 Моллюговые
256689 Моллуго, Моллюго, Мутовчатка
256695 Моллуго маленькая, Моллюго маленькое
256719 Мутовчатка пятилистная
256746 Мольткия
256750 Мольткия голубая
256758 Молюцелла, Трава молукская
256761 Лимонница гладкая
256780 Момордика
256786 Момордика, Момордика бальзамическая
256797 Момордика харантская, Момордика харанция, Огурчик пёсий виноградолистный, Тыква диковинная
257157 Монарда
257169 Монарда трубчатая
257371 Монезес, Одноцветка
257374 Груша, Дерево грушевое, Монезес одноцветковый, Одноцветка крупноцветковая, Одноцветка крупноцветная, Одноцветка одноцветковая
257564 Монохория
257572 Монохория корсакова
257583 Монохория подложниковая
257630 Монодора
257653 Монодора мускатная, Орех мускатный ямайский
257683 Однопокровник
257684 Однопокровник азиатский
257686 Однопокровник литвинова
258006 Вертляница, Подъельник
258041 Вертляница тайваньская
258044 Вертляница одноцветковая
258049 Вертляницевые
258162 Монстера
258168 Монстера любимая, Монстера привлекательная, Монстера пышная, Филодендрон пробитый
258237 Монция
258248 Монция блестящесемянная, Монция ручейная
258259 Монция маленькая
258378 Тутовые
258380 Морея

258831 Морина
258848 Морина кокансквя
258850 Морина лемана
258865 Молина мелкоцветковая
258867 Морина персидская
258870 Мориновые
258945 Дерево хренное, Моринга
259065 Дерево тутовое, Дерево шелковичное, Тут, Тута, Тутовник, Шелковица
259067 Дерево тутовое, Дерево шелковичное, Тут белый, Тута, Шелковица белая, Шовкун
259097 Шелковица атласная
259123 Тут гуншаньский
259137 Шелковица индийская
259154 Шелковица либоская
259164 Шелковица мелколистная
259165 Шелковица монгольская
259180 Тут черный, Тут чёрный, Шелковица черная, Шелковица чёрная
259186 Джамбоза, Помароза, Тут американский, Тут красный, Шедковица красная, Ямбоза
259193 Тут татарский
259278 Мосла, Ортодон
259281 Мосла мелкоцветковая
259282 Мосла китайская
259284 Мосла двпыльниковая
259292 Мосла крупнопильчатая, Ортодон крупнопильчатый
259300 Мосла японская
259479 Бархатные, Мукуна
259558 Бобы бархатные
259559 Бархатные бобы, Боб бархатный
259584 Мюленбекия
259630 Мюленбергия
259646 Мюленбергия короткоостая
259671 Мюленбергия гугела, Мюленбергия гюгеля
259672 Мюленбергия японская
259746 Лактук, Латук, Молокан, Мульгедиум, Салат
259763 Молокан цинхайский
259772 Латук татарский, Молокан синий, Молокан татарский, Чайшон
259910 Калабюра, Мунтингия, Мунтингия калабура
260102 Анейлема кейзака
260127 Мурезия
260129 Мурезия пахачая
260130 Мурезия желтая
260131 Мурезия эройланская
260132 Мурезия закаспийская
260133 Мурезия переходная
260158 Мурайя, Муррая, Муррея
260165 Мурайя экзотическая, Муррая экзотическая
260198 Банан, Дерево бархатное
260208 Банан базьо, Банан японский
260217 Сушеница лежачая
260224 Банан абиссинский, Банан декоративный, Банан эфиопский
260228 Банан базьо тайваньский
260236 Банан ланьюйский
260241 Банан люцюский
260250 Банан кавендиша, Банан карликовый, Банан китайский
260253 Банан браминов, Банан десертный, Банан кухонный, Банан мудрецов, Банан пизанг, Банан плодовой, Банан подоложник, Банан райский, Банан фруктовый, Банан яблочный, Пизанг, Смоковница райская
260258 Банан райский семеноносный
260275 Абака, Конопля манильская, Конопля маниписская, Манилла, Пенька манильская, Пенька маниписская
260284 Банановые
260293 Гадючий лук, Лук гадючий, Мускаримия, Мышиный Гиацинт
260301 Гадючий лук гроздевидный, Лук гадючий гроздевидный
260306 Гадючий лук синый
260307 Гадючий лук переменчивый
260308 Леопольдия хохлатая, Леопольдия хохолковая, Лук гадючий хохлатый
260311 Гадючий лук длинноцветковый
260313 Гадючий лук свободообразный
260318 Гадючий лук бело-зевный
260324 Гадючий лук пренебрегаемый
260326 Гадючий лук бледный
260327 Гадючий лук своеобразный
260329 Гадючий лук поникающий
260330 Гадючий лук многоцветковый
260333 Гадючий лук кистевидный, Кистецветвик обыкновенный, Лук гадючий кистевидный, Лук гадючий кистистый, Мускари
260337 Гадючий лук штейпа
260338 Гадючий лук шовица
260340 Гадючий лук туркевича
260341 Гадючий лук воронова
260345 Мускаримия
260347 Мускаримия мышиная
260359 Банан шертавоплодный
260569 Полевка, Полёвка
260577 Полевка пронзеннолистмная
260594 Муселис
260596 Латук стеной, Мицелис стенной
260700 Миопоровые
260706 Миопорум
260737 Миозотмс, Незабудка
260747 Незабудка альпийская
260757 Незабудка приятная
260760 Незабудка пашенная, Незабудка полевая, Незабудка средняя
260762 Незабудка азиатская
260766 Незабудка азорская
260772 Незабудка дернистая
260779 Незабудка жестковолосистая, Незабудка холмовая
260791 Незабудка разноцветная
260794 Незабудка крупноцветная
260803 Незабудка греческая, Незабудка идская
260807 Незабудка средняя
260809 Незабудка крылова
260815 Незабудка балтийская
260818 Незабудка лазистанская
260822 Незабудка воробейниколистная
260823 Незабудка литовская
260824 Незабудка береговая, Незабудка прибрежая
260828 Незабудка разноцветная
260833 Незабудка мелкоцветковая
260838 Незабудка горная, Незабудка лесная
260840 Незабудка дубравная
260842 Незабудка болотная
260848 Незабудка родственная
260849 Незабудка ложно-родственная
260850 Незабудка ложно-изменчивая
260860 Незабудка отогнутая
260868 Незабудка болотная
260877 Незабудка щетинистая
260878 Незабудка сицилийская
260880 Незабудка редкоцветковая
260890 Незабудка дужистая, Незабудка душистая
260892 Незабудка лесная, Незабудка сахалинская
260901 Незабудка туркменская
260902 Незабудка украинская
260909 Незабудка разноцветная
260920 Мягковолосник
260922 Мягковолосник водный, Мягковолосник водяной
260930 Мышехвостник
260940 Мышехвостник маленький
261060 Мириактис
261068 Мириактис гмелина
261120 Восковник, Восковница, Дерево восковое, Мирика
261132 Восковница калифорнийская
261137 Вереск восковой, Восковница восконосная, Мирт восконосный
261162 Болотный, Восковник обыкновенный, Восковница обыкновенная, Мирт болотный
261195 Восковница карликовая
261202 Восковница пенсильванская

261212 Восковник красный, Восковница красная, Мирт красный
261237 Восковница желтая
261243 Восковниковые
261246 Мирикария
261250 Мирикария лисохвостная, Мирикария лисоховостниковая
261254 Мирикария даурская
261259 Мирикария изящная
261262 Мирикария германская
261281 Мирикария длиннолистная
261293 Мирикария чешуйчатая
261337 Водоперица, Мириофиллюм, Уруть
261339 Уруть очевые, Уруть очередноцветковая
261342 Уруть бразильская
261364 Уруть колосистая, Уруть колосковая, Уруть колосовая
261377 Уруть уссурийская
261379 Водоперица мутовчая, Уруть мутовчатая
261402 Дерево мускатное, Мускат, Мускатник
261423 Орех мускатный фальшивый
261424 Дерево мускатное, Дерево мускатное душистое, Мускат, Мускатник душистый, Орех мускатный, Орех мушкатный
261461 Вирола суринамская
261464 Миристицевые, Мускатниковые
261467 Мирмехис
261469 Мирмехис китайская
261474 Мирмехис японский
261525 Мирокарпус
261527 Мирокарпус густолиственный
261549 Дерево бальзамвое, Мироксилон
261554 Дерево бальзамвое, Дерево бальзамическое толюйское, Дерево бальзамное толу
261561 Дерево бальзамвое, Дерево бальзамическое толюйское, Дерево бальзамное толу
261576 Миррис
261585 Донник, Керверь испанский, Керверь мускусный, Миррис душистый, Миррис пахучий
261592 Вздутостебельник
261596 Мирсиновые
261598 Мирзина, Мирсина
261601 Мирзина африканская, Мирсина африканская
261608 Мирзина африканская мелколистная
261619 Мирзина гуансиская
261626 Рапанея гуансиская
261632 Рапанея максимовича
261686 Миртовые
261726 Мирт, Мирта
261739 Мирт обыкновенный
262013 Наядовые
262015 Наяда, Резуха
262024 Наяда китайская
262028 Наяда гибкая
262035 Наяда ячеистая
262040 Наяда злаковая, Наяда злаковидная
262054 Наяда индийская
262067 Наяда морская
262089 Наяда малая
262110 Наяда тончайшая
262188 Нандина
262189 Нандина домашняя
262234 Нанноропс
262259 Нанофитон
262260 Нанофитон ежовый
262348 Нарцисс, Нарцисус
262391 Нарцисус кавказский
262404 Нарцисус гиблидный
262410 Жонкилия, Жонкилла, Жонкиль, Нарцисс жёлтый, Нарцисс-жонкиль
262415 Нарцисус большой
262438 Нарцисс белый, Нарцисус поэтический
262441 Лженарцисс, Лженарцисус, Нарцисс ложнонарциссовый, Нарцисс ложный, Нарцисс трубчатый, Нарцисус желтый, Нарцисус ложнонациссовый, Нарцисус ложный, Псевдонарцисс
262457 Нарцисс букетный, Нарцисс-тацетта, Нарцисус константинопольский
262509 Белоусник
262516 Белоусник краузе
262528 Белоусник тонкоцветковый
262532 Белоус
262537 Белоус, Белоус прямой, Белоус торчащий, Нард вытянутый, Щеть
262572 Костолом, Нартеций
262576 Нартеций кавказский
262645 Жеруха
262722 Брункресс, Жеруха аптечная, Жеруха водная, Жеруха лекарственная, Кресс водяной, Хрен водяной, Хрен лекарственный
262751 Жеруха тибетская
262776 Наталиелла
262777 Наталиелла алайская
263142 Нейллия
263161 Нейлия китайская
263172 Нейлия тибетская
263178 Нейллия метельчатая
263267 Лотос
263268 Лотос каспийский
263272 Лотос индийский, Лотос орехоносный
263374 Немезия
263501 Немофила
264006 Неолидлейя
264011 Неолицея
264024 Неолицея чжэцзянская
264052 Неолицея хайнаньская
264056 Неолицея баотинская
264061 Неолицея гуансиская
264062 Неолицея гуандунская
264070 Неолицея тибетская
264073 Неолицея мэнлаская
264084 Неолицея пинбяньская
264090 Неолицея шелковистая
264101 Неолицея сычуаньская
264104 Неолицея сычуаньская гуншаньская
264112 Неолицея ланьюйская
264241 Соя яванская
264248 Неопалассия
264249 Неопалассия гребенчатая, Полынь гребенчатая
264250 Полынь тибетская
264603 Торулярия, Четочник, Чёточник
264604 Четочник коротоплодный
264607 Четочник коротконогий
264609 Торулярия низкая, Четочник низкий, Чёточник низкий
264618 Четочник королькова, Чёточник королькова, Четочник серножелтый
264624 Четочник мягковолосый, Чёточник мягковолосый
264631 Четочник русский, Чёточник русский
264637 Четочник тибетский, Чёточник тибетский
264639 Четочник бугорчатый, Чёточник бугорчатый
264650 Гнездовка
264651 Гнездовка азиатская
264665 Гнездовка камчатская
264709 Гнездовка настоящая
264718 Гнездовка сосочковая, Гнездовка сосочконосная
264727 Тайник большой
264746 Тайник мелкоцветковая
264748 Тайник тяньшанский
264750 Гнездовка уссурийская
264757 Неоттианта
264765 Неоттианта клобучковая
264827 Непентесовые, Непентовые
264829 Непентес
264859 Котовик, Котовник, Непета
264862 Котовник алагезский
264867 Котовник прелестный
264869 Схизонепета однолетняя
264879 Котовник бадахшанский
264881 Котовник биберштейна
264890 Котовник прицветниковый
264891 Котовник короткολистный

264892 Котовник бухарский
264893 Котовник бузе
264894 Котовник буша
264897 Котовник кошачий, Котовник мятный, Кошачья мята, Мята кошачья
264903 Котовник родственный
264904 Котовник васильковый
264906 Котовник дагестанский
264909 Котовник густоцветковый
264917 Котовник ереванский
264921 Котовник федченко
264923 Котовник надрезанный
264926 Котовник прекрасный
264935 Котовник железистый
264936 Котовник гончароза
264939 Костер крупноцветковый
264940 Котовник гроссгейма
264941 Котовник гаястанский
264948 Котовник хэнаньская
264952 Котовник грузинский
264957 Котовник изфаганский
264958 Котовник японский
264962 Котовник кноринг
264963 Котовник кокамырский
264964 Котовник кокандский
264965 Котовник комарова
264966 Котовник копетдагский
264968 Котовник кубанский
264972 Котовник ладанный
264974 Котовник яснотколистный
264984 Котовник липского
264985 Котовник длинноприцветниковый
264986 Котовник длиннотрубчатый
264990 Котовник маньчжурский
264991 Котовник марии
264992 Котовник маусарифский
264994 Котовник мейера
264995 Котовник мелкоцветковый
264996 Котовник мелкоголовчатый
265001 Схизонепета многонадрезанная
265016 Костер голый, Котовник венгерский
265017 Котовник пахучий
265018 Котовник ольги
265019 Котовник бледный
265020 Котовник рамирский
265022 Котовник мелковатоцветковый
265027 Котовник ножкоколосый
265029 Котовник ложноклочковатый
265030 Котовник ложнококандский
265031 Котовник красивый
265032 Котовник колючий
265036 Котовник рейхенбаха
265041 Котовник сахаристый
265043 Котовник чаберовидный
265044 Котовник шишкива
265045 Котовник шугванский
265048 Котовник сибирский
265050 Котовник синтениза
265051 Котовник сосновского
265053 Котовник лопатчатый
265062 Котовник торчащий
265063 Котовник полукопьевидный
265068 Котовник серножелтый
265071 Котовник лежачий
265084 Котовник тибетский
265088 Котовник закавказский
265089 Котовник заилийский
265090 Котовник траутфеттера
265091 Котовник тройцкого
265094 Котовник украинский
265098 Котовник ваханский
265101 Котовник бархатистый
265129 Нефелиум пенейниковый, Рамбутан
265137 Герань пиренейская, Пулазан, Пулазан плодовый
265197 Фория гребневая
265198 Фория гребневая японская
265228 Нептуния
265254 Нерине
265310 Нириум, Олеандр
265327 Олеандр душистый, Олеандр индийский, Олеандр обыкновенный
265358 Нертера
265550 Неслия
265552 Неслия остроконечная
265553 Неслия метельчатая
265834 неуструевия каратавская
265856 Невиузия
265872 Невскиелла
265940 Никандра, Хмель огородный
265944 Адамова голова, Никандра, Никандра физалисовидная, Никандра физалисоподобная
266031 Растение табачное, Табак
266053 Махорка, Табак деревенский, Табак махорка, Табак тютюн, Табак-махорка, Тютюн
266060 Табак виргинский, Табак настоящий, Табак обыкновенный, Табак турецкий
266183 Недзвецкия
266185 Недзвецкия семиреченская
266194 Нирембергия
266215 Девица в зелени, Нигелла, Чернушка
266216 Чернушка полевная
266224 Чернушка бухарская
266225 Нигелла дамасская, Чернушка дамаскская, Чернушка дамасская
266229 Чернушка гарибелля, Чернушка мелкоцветковая
266230 Чернушка железистая
266235 Чернушка цельнолистая
266236 Чернушка восточная
266237 Чернушка остролепестная
266240 Чернушка персидская
266241 Тмин чёрный, Чернушка посевная
266242 Чернушка пашенная
266263 Никитиния
266264 Никитиния тонкостебельная
266355 Нитрария, Селитрянка
266358 Селитрянка комарова
266367 Заманиха, Селитрянка, Селитрянка шобера
266377 Селитрянка сибирская
266431 Ноэа
266433 Ноэа гладкостебельная
266434 Ноэа маленькая
266435 Ноэа остронечная
266470 Нолана
266482 Нолина
266593 Нонея, Ноннея
266594 Нонея альпийская
266597 Нонея каспийская
266598 Нонея темноплодная
266600 Нонея дагестанская
266601 Нонея низбегающая
266604 Нонея желтеющая
266606 Нонея промежуточная
266607 Нонея карсская
266608 Нонея желтая
266609 Нонея крупножковая
266618 Нонея бледная
266623 Нонея темнобурая
266624 Нонея розовая
266625 Нонея щетинистая
266626 Нонея туркменская
266627 Нонея вздудая
266628 Нонея разноцветная
266648 Эпифиллум аккерманна
266861 Нотофагус
266864 Бук антарктический, Нотофагус антарктический
267005 Нотобазис
267008 Нотобазис сурийский
267237 Новосиверрия
267238 Новосиверрия ледяная
267274 Кубышка
267276 Кубышка пришлая
267285 Кубышка средная
267286 Кубышка японская
267294 Кубышка желтая, Кубышка жёлтая
267320 Кубышка малая
267347 Льнянка канадская
267421 Никтагиновые, Ночецветные
267429 Никтантес
267627 Водяная роза, Кувшинка, Ненюфар, Нимфея
267639 Кувшинка гиблидная
267648 Кувшинка белая, Лилия водяная
267658 Кувшинка голубая

267664 Кувшинка чистобелая
267665 Кувшинка капская
267698 Кувшинка лотос, Лотос екипетский
267714 Кувшинка жёлтая, Кувшинка мексиканская
267730 Кувшинка душистая
267756 Кувшинка красная
267767 Кувшинка малая, Кувшинка четырехгранная, Кувшинка четырехугольная
267782 Кувшинка венцеля
267787 Кувшинковые, Нимфейные
267800 Болотноцветник, Болотоцветник, Лимнантемум, Нимфондес
267807 Нимфондес корейский
267819 Нимфондес индийский
267821 Болотноцветник корейский
267825 Болотноцветник кувшинковый, Болотноцветник щитолистный, Нимфондес щитовидный, Плавун нимфейный
267849 Нисса, Тупело
267862 Нисса шансиская
267863 Нисса жуйлиская
267864 Нисса китайский
267867 Нисса лесная
267876 Нисса вэньшаньская
267881 Ниссовые
268132 Охна
268261 Охна многоцветковая
268277 Охновые
268344 Бальза, Бальза заячья, Бальса, Дерево бальзовое, Охрома
268418 Базилик, Базилика
268438 Базилик душистый, Базилик камфорный, Базилик обыкновенный, Базилик оородный душки, Душки
268518 Базилик евгенольный, Базилик юбилейный
268688 Окотея
268768 Восьмитог
268769 Восьмитог леманновский
268957 Бартсия, Бартшия, Зубчатка
268966 Зубчатка железистая, Зубчатка клейкая
268970 Зубчатка прибрежная
268983 Зубчатка бартсиевидная, Зубчатка красная, Зубчатка осенняя
268985 Зубчатка солончаковая
268991 Зубчатка весенняя, Зубчатка ранняя
268999 Зубчатка осенняя, Зубчатка поздная
269274 Ойдибазис
269275 Ойдибазис остроконечный
269276 Ойдибазис бутеневый
269277 Ойдибазис каратавский
269278 Ойдибазис плоскоплодный
269292 Омежник
269293 Омежник абхазский
269299 Омежник водный, Омежник водяной, Раздулка, Укроп волошский
269307 Омежник шафранный, Омежник шафрановый, Омежник шафраноподобный
269316 Омежник федченко
269319 Омежник дучатый, Омежник трудубчатый
269326 Омежник лежачий, Омежник яванская
269337 Омежник мэнцзыский
269338 Омежник длиннолисточковый
269347 Омежник водный, Омежник водяной, Раздулка, Укроп волошский
269348 Омежник бедренецовидный
269358 Омежник морковниколистный
269360 Омежник китайский
269362 Омежник софии
269373 Энокарпус
269394 Онагра, Ослинник, Ослинник двулетний, Свеча ночная, Энотера
269403 Энотера мексиканская
269404 Ослинник двулетний, Ослинник двухлетний, Ослинник короткоиглый, Рапутник, Свеча ночная, Энотера двулетняя, Энотера двухлетняя
269475 Энотера душистая
269481 Энотера короткоиглая
270054 Ольдфильдия африканская
270058 Дерево оливковые, Маслина, Олива
270061 Маслина американская
270099 Дерево оливковое, Дерево оливковые, Маслина, Маслина европейская, Маслина культурная, Олеандр европейская, Олива, Оливка, Оливковое дерево
270100 Олеандр африканская
270110 Маслина дикая
270126 Олива гуансиская
270128 Олива хайнаньская
270182 Маслинные, Маслиновые, Масличные
270184 Олеария
270227 Ольгея
270228 Ольгея бальджуанская
270230 Ольгея шерстистоглавая
270233 Ольгея войлочноногая
270240 Ольгея длиннолистная
270241 Ольгея войлочногнездная
270242 Ольгея снежная
270243 Ольгея гребенчатая
270246 Ольгея колючконосная
270249 Ольгея введенского
270297 Олигохета
270300 Олигохета растопыренная
270301 Наголоватка комарова, Олигохета мелкая
270302 Олигохета войлочная
270461 Резушка карликовая
270462 Резушка карликовая, Резушка низкая
270563 Хомалянтус тополевый
270677 Омфалодес, Пупочник
270684 Пупочник каппадокийский
270703 Пупочник кузнецова
270705 Омфалодес линейнолистный, Пупочник льнолистный
270707 Пупочник лойки
270715 Пупочник скальный
270716 Пупочник ползучий
270723 Омфалодес настоящий, Пупочник весений
270757 Омфалотрикс
270758 Омфалотрикс длинножковый, Омфалотрикс длинноножковый
270773 Онагровые
270787 Онцидиум
271166 Эспарцет
271170 Эспарцет большой, Эспарцет высочайший
271171 Эспарцет приятный
271172 Эспарцет песчаный, Эспарцет полевой
271177 Эспарцет серебристный
271179 Эспарцет армянский
271181 Эспарцет биберштейна
271182 Эспарцет боброва
271183 Эспарцет бузе
271184 Эспарцет бунге
271185 Эспарцет педушья
271187 Эспарцет хорасанский
271190 Эспарцет рогообразный
271199 Эспарцет куринский
271200 Эспарцет дагестанский
271201 Эспарцет дарвазский
271204 Эспарцет дильса
271206 Эспарцет ехидна
271208 Эспарцет ферганский
271209 Эспарцет гончарова
271210 Эспарцет грациозный, Эспарцет изящный
271211 Эспарцет высокий
271212 Эспарцет гроссгейма
271213 Эспарцет североармянский
271214 Эспарцет крючковатый
271215 Эспарцет разнолистный
271216 Эспарцет гогенакера
271221 Эспарцет грузинский
271222 Эспарцет безостый, Эспарцет невооруженный
271224 Эспарцет комарова
271227 Эспарцет липского
271228 Эспарцет длинноиглистый

271229 Эспарцет большой
271230 Эспарцет майорова
271231 Эспарцет мишо
271232 Эспарцет мелкоцветный
271233 Эспарцет киноварнорасный
271234 Эспарцет горный
271236 Эспарцет немеца
271237 Эспарцет новопокровского
271239 Эспарцет острозубый
271240 Эспарцет остолодочниковый, Эспарцет сочевичниковый
271241 Эспарцет палласа
271254 Эспарцет скальный
271255 Эспарцет красивый
271256 Эспарцет лучстый
271259 Эспарцет рупрехта
271265 Эспарцет шугнанский
271266 Эспарцет зеравшанский
271267 Эспарцет сибирский
271268 Эспарцет синтениса
271269 Эспарцет сосновского
271271 Эспарцет короткостебельный
271272 Эспарцет лежачий
271273 Эспарцет донской
271274 Эспарцет шерстистый
271275 Эспарцет турнефора
271276 Эспарцет закаспийский
271277 Эспарцет закавказский
271278 Эспарцет влагальщный
271279 Эспарцет васильченко
271280 Клевер турецкий, Эспарцет виколистный, Эспарцет горошколистный, Эспарцет кормовой, Эспарцет обыкновенный, Эспарцет посевной
271290 Волчуг, Стальник, Трава воловья
271320 Стальник древник
271333 Стальник обыкновенный, Стальник олевой, Стальник пашенный, Стальник полевой
271337 Стальник австрийский
271347 Стальник колючий, Стальник колячий
271428 Стальник промежуточный
271436 Стальник глядкосемянный, Стальник голосемянный
271464 Стальник желтый, Стальник жёлтый, Стальник натрикс
271540 Стальник малеників
271563 Стальник ползучий
271663 Онопордон, Онопордум, Татаркик
271666 Онопордон колючий, Онопордум колючий, Онопордум обыкновенный, Татаркик колючий, Татаркик обыкновенный
271677 Онопордум армянский
271679 Онопордум снежно-белый
271680 Онопордум серый
271690 Онопордум фрика
271691 Онопордум разноколючковый
271693 Онопордон тонкочешуйный
271700 Онопордум пряхина
271701 Онопордум зеравшанский
271705 Онопордум крымский
271727 Громовик, Оносма
271731 Оносма расширенная
271733 Оносма песчаная
271734 Оносма армянская
271735 Оносма черно-синяя
271736 Оносма сине-голубая
271737 Оносма бальджуанская
271739 Оносма барщевского
271742 Оносма коротковолосистая
271743 Оносма чашечная
271745 Оносма кавказская
271759 Оносма мелкоцветковое
271766 Оносма ферганская
271772 Оносма гмелина
271773 Оносма изящная
271785 Оносма раздражающая
271788 Оносма белоплодная
271789 Оносма лицзянская
271793 Оносма ливанова
271795 Оносма длиннозубчатая
271800 Оносма крупнокорневая
271801 Оносма самаркандская
271805 Оносма мелкоплодная
271816 Оносма многоцветная
271817 Оносма многолистная
271820 Оносма жесткая
271821 Оносма скальная
271823 Оносма шелковистая
271824 Оносма щетинистая
271825 Оносма зауральская
271827 Оносма простая, Оносма простейшая
271830 Оносма тычиночная
271835 Оносма крымская
271837 Оносма красильная
271838 Оносма бугорчатоплодная
271841 Оносма куэньминская
271843 Оносма визиани
271846 Оносма тибетская
271849 Оносма чаюйская
272049 Ландышник, Офиопогон
272077 Офиопогон фунинская
272086 Офиопогон ябурая
272090 Офиопогон японский
272116 Офиопогон мотоский
272121 Офиопогон пинбяньский
272395 Офрис
272398 Офрис пчелоносная, Пчелка
272422 Офрис кавказская
272466 Офрис мухоносная
272481 Офрис оводоносная
272493 Офрис крымская
272497 Офрис трансгирканская
272611 Оплисменус, Остянка
272614 Оплисменус коротковолосистый
272619 Остянка бурманна
272645 Оплисменус фуцзяньський
272684 Остянка курчяволистная
272693 Оплисменус японский
272704 Заманиха, Эхинопанакс
272706 заманиха, Заманиха высокая, Эхинопанакс высокий
272730 Опопанакс
272731 Опопанакс армянский
272775 Кактус опунция, Опунция
272818 Опунция бразильская
272891 Опунция индийская, Опунция индийская фига, Смоква индийская, Тупа, Фига индийская
272987 Опунция обыкновенная, Опунция одоколючковая
273103 Опунция обыкновенная
273206 Стапелия пестрая
273247 Бабассу
273267 Орхидные, Осенник, Ятрышниковые
273317 Галеорхис, Комперия, Орхидея-пафиния, Ятрышник
273323 Ятрышник туполопастный
273333 Ятрышник балтийский
273359 Ятрышник зеленоватожелтый
273380 Ятрышник клопоносный
273420 Ятрышник черноморский
273429 Ятрышник федченко
273433 Ятрышник желтоватый
273467 Ятрышник иберийский
273469 Ятрышник мясокрасный
273501 Ятрышник солончаковый, Ятрышник широколистный
273524 Ятрышник патнистый, Ятрышник пятнистый
273529 Ятрышник
273531 Ятрышник мужской
273537 Ятрышник наибольший
273541 Певник болотный, Ятрышник шлемовидный, Ятрышник шлемоносный
273545 Ятрышник длемлик
273564 Ятрышник бледный
273565 Ятрышник болотный
273592 Ятрышник каскрашенный
273596 Ятрышник провансальский
273602 Ятрышник точечный
273604 Ятрышник пурпурный
273613 Ятрышник римский
273623 Ятрышник бузинный
273627 Ятрышник шелковникова

273633 Ятрышник обезьяний
273681 Ятрышник траунштейнера
273686 Ятрышник трехзубчатый
273688 Ятрышник трехлистный
273696 Ятрышник обожженный, Ятрышник обожжённый
273700 Ятрышник зеленобурый
273971 Стелигма войлочная, Стелигмостемум войлочный
274033 Ореорхис
274047 Ореорхис индийский
274058 Ореорхис раскидистый
274197 Душица, Майоран
274222 Душица копетдагская
274224 Душица, Майоран летний, Майоран садовый, Трава колбасная
274233 Амаракус круглолистный
274236 Душица мелкоцветковая
274237 Душица обыкновенная, Майоран дикий, Майоран зимний
274286 Орлайя
274287 Орлайя плоскоплодная
274288 Орлайя крупноцветковая
274370 Ожерельник
274372 Ожерельник туркменский
274373 Венечнозонтичник
274374 Венечнозонтичник красивый
274495 Ороксилум, Птицемлечник
274520 Птицемлечник дугообразный
274526 Птицемлечник балансы
274537 Птицемлечник буше
274538 Птицемлечник коротколосый
274614 Птицемлечник фишера, Птицемлечник фишеровский
274617 Лечник жетоватый, Птицемлечник желптицем, Птицемлечник жетоватый
274636 Птицемлечник тонкоцветковый
274651 Птицемлечник гирканский
274687 Птицемлечник крупный
274708 Птицемлечник поникший
274735 Птицемлечник плосколистный
274748 Птицемлечник пиренейский
274755 Птицемлечник преломленный
274769 Птицемлечник шолковникова
274770 Птицемлечник шишкина
274772 Птицемлечник шмальгаузена
274782 Птицемлечник синтениса
274799 Птицемлечник тонколистный
274813 Птицемлечник закавказский
274815 Птицемлечник бахромчатый, Птицемлечник ресничатый
274823 Белые брандушки, Брандушки белые, Птицемлечник зонтичный
274851 Птицемлечник воронова
274879 Сераделла
274880 Сераделла сжатая
274888 Сераделла маленькая, Сераделла очень маленькая
274895 Птиценожка, Сераделла посевная
274921 Заразиховые
274922 Заразиха
274927 Гули-харбуза, Заразиха бахчевая, Заразиха египетская, Шунгуя
274934 Заразиха белая, Заразиха бледноцветковая
274941 Заразиха эльзасская
274951 Заразиха красивая, Заразиха прелестная
274952 Заразиха амурская
274956 Заразиха андросова
274957 Заразиха песчаная
274958 Заразиха горчаковая
274971 Заразиха короткостебельная
274972 Заразиха капустная
274973 Заразиха короткозубчиковая
274977 Заразиха бунге
274985 Заразиха гвоздичная, Заразиха обыкновенная
274987 Заразиха нониклшая, Заразиха широкочешуйная
274991 Заразиха волчок, Заразиха подсолнечная, Заразиха поникшая
275000 Заразиха килинийская
275002 Заразиха кларке
275006 Заразиха лазурная
275010 Заразиха синеватая
275018 Заразиха коржинского
275021 Заразиха покрашеная
275027 Заразиха снаянная
275029 Заразиха городчатая, Заразиха зубчатая
275058 Заразиха желтая
275068 Заразиха сростночашелистиковая
275073 Заразиха голостебельная
275074 Заразиха сизоватая
275075 Заразиха тонкая
275079 Заразиха григольева
275081 Заразиха гроссгейма
275084 Заразиха ганса
275086 Заразиха плющевая
275090 Заразиха волосистоцветковая
275091 Заразиха хохенакера
275094 Заразиха перепончаточашечковая
275095 Заразиха иберийская
275100 Заразиха каратавская
275102 Заразиха келлера
275105 Заразиха кочи
275107 Заразиха крылова
275108 Заразиха курдская
275113 Заразиха голубая
275119 Заразиха линчевского
275126 Заразиха желтая, Заразиха люцерновая
275128 Заразиха большая
275135 Заразиха малая
275142 Заразиха монгольская
275148 Заразиха низкая
275155 Заразиха восточная, Заразиха остролопастная
275156 Заразиха оверина
275159 Заразиха бледноцветковая
275172 Заразиха красивенэкая
275173 Заразиха пурпуровая
275177 Заразиха толстоколосая, Заразиха толстоолосая
275182 Заразиха радде
275184 Заразиха бетвистая
275186 Заразиха мутеля
275195 Заразиха сетчатая
275196 Заразиха розовая
275201 Заразиха шелковникова
275208 Заразиха пильчаточашечковая
275211 Заразиха синтениса
275212 Заразиха соленантусовая
275213 Заразиха солмса
275214 Заразиха грязно-желтая
275222 Заразиха серножелдая
275225 Заразиха дубровниковая
275227 Заразиха закавказская
275242 Заразиха уральская
275251 Заразиха разноцветная
275257 Заразиха желтоватая
275258 Заразиха обыкновенная
275295 Ортосифон
275347 Горноколосник, Омфалодес, Оростахис, Пупочник
275356 Горноколосник хрящеватый, Очиток краснеющий
275363 Горноколосник бахромчатый
275367 Горноколосник шаньдунский
275375 Пупочник японский
275379 Горноколосник мягколистный, Очиток мягколистный
275394 Горноколосник колючий, Оростахис колючий, Очиток колючий
275396 Горноколосник пирамидальный, Горноколосник щитковый
275401 Оронтиум
275463 Оранта
275466 Оранта оше
275467 Оранта желтая
275480 Рамишия
275482 Рамишия тупая
275486 Грушанка однобокая, Раминия одновокая, Раминия одновочная, Рамишия однобокая
275624 Ортосифон
275761 Ортосифон хайнаньский
275781 Орхидейные, Чай почечный
275850 Прямохвостник
275852 Прямохвостник разноплодный

275853 Прямохвостник кокандский
275865 Спрыгиния
275869 Чесночник крупнолистный
275909 Рис
275958 Рис посевной, Шалы
275991 Оризопсис, Рисовидка
276001 Рисовидка китайская
276003 Рисовидка голубоватая
276008 Рисовидка ферганская
276020 Рисовидка бухарниковая
276028 Рисовидка копетдагская
276032 Рисовидка широколистная
276038 Рисовидка молиниевидная
276062 Рисовидка кокандская
276068 Рисовидка зеленоватая
276242 Османтус
276250 Маслина американская
276268 Османтус люцюский
276291 Османтус душистый
276292 Османтус оранжевый
276312 Османтус мэнцзыский
276313 Османтус разнолистный
276323 Османтус хубэйский
276374 Османтус наньчуаньский
276381 Османтус эмэйский
276450 Осмориза
276455 Осмориза остистая
276533 Костянкоплодник
276544 Костянкоплодник шверина
276743 Маточник
276752 Дудник максимовича
276768 Дудник зибольда
276780 Дудник зеленоцветковый
276795 Островския
276796 Островския величественная
276801 Хмелеграб
276803 Хмелеграб обыкновенный
276808 Хмелеграб япнский
276818 Хмелеграб американский, Хмелеграб виргинский
277326 Ототегия
277329 Ототегия бухарская
277333 Ототегия федченко
277345 Ототегия ольги
277361 Оттелия
277369 Оттелия частуховидная
277397 Оттелия японская
277636 Овения
277640 Овения вишнелистная
277643 Кисличные
277644 Кислица, Оксалис
277648 Заячья кислица, Кислица заячья, Кислица кисленькая, Кислица кисловатая, Кислица обыкновенная, Трилистник луговой
277650 Кислица заячья тайваньская
277747 Кислица поротая, Кислица рогатая, Кислица рожковая
277779 Кислица клубневая
277985 Кислица обратнотреугольная
278008 Кислица козья
278099 Кислица прямая, Кислица торчащая, Миллефоль, Тысячелистник обыкновенный
278147 Кислица фиолетовая
278351 Клюква
278360 Вакциниум мелкоплодный, Клюква мелкоплодная
278366 Вакциниум кислоплодный, Клюква болотная, Клюква обыкновенная, Клюква четырехлепестная
278388 Оксидендрон
278389 Андромеда древовидная, Оксидендрум древовидный
278466 Оксиграфис
278467 Оксиграфис шамиссо
278473 Оксиграфис ледниковый
278481 Оксиграфис обыкновенный
278576 Кисличник
278577 Кисличник двустолбиковый, Кисличник двухстолбчатый
278679 Остродка, Остролодочник
278680 Остролодочник колючий
278686 Остролодочник адамса
278687 Остролодочник восходящий
278689 Остролодочник аянский
278690 Остролодочник алайский
278691 Остролодочник албанский
278693 Остролодочник белоцветковый
278694 Остролодочник беломохнатый
278696 Остролодочник нижнеальпийский
278698 Остролодочник высокогорный
278700 Остролодочник альпийский
278701 Остролодочник алтайский
278702 Остролодочник сходный
278703 Остролодочник песколюбивый
278705 Остролодочник пузырчатоплодный
278706 Остролодочник анадырский
278715 Остролодочник близкий
278716 Остролодочник арктический
278719 Остролодочник серебристый
278723 Остролодочник жесткий
278724 Остролодочник ассийский
278725 Остролодочник астрагаловидный
278726 Остролодочник золотистожелтый
278728 Остролодочник южносахалинский
278729 Остролодочник птичий
278733 Остролодочник байсунский
278734 Остролодочник бальджуанский
278738 Остролодочник красивый
278739 Остролодочник белла
278745 Остролодочник двулопастной
278746 Остролодочник боброва
278748 Остролодочник богушский
278750 Остролодочник короткоплодный
278752 Остролодочник прицветничковый
278753 Остролодочник прицветничковый
278754 Остролодочник короткостебельный
278757 Остролодочник голубой
278760 Остролодочник дерновинный
278762 Остролодочник мелкодерновинный
278763 Остролодочник колокольчатый
278764 Остролодочник полевой
278767 Остролодочник седой
278768 Остролодочник беловатый
278769 Остролодочник капю
278772 Остролодочник карпатский
278773 Остролодочник кавказский
278774 Остролодочник хвостатый
278775 Остролодочник стеблевой
278777 Остролодочник хантенгринский
278779 Остролодочник чезнеевидный
278780 Остролодочник многолисточковый
278784 Остролодочник немногоцветковый, Остролодочник приснеговой
278785 Остролодочник снежнолистный
278786 Остролодочник хоргосский
278797 Остролодочник голубиный
278798 Остролодочник пузырниковидный
278799 Остролодочник смешиваемый
278800 Остролодочник толстоватый
278801 Остролодочник меловой
278803 Остролодочник остроконечный
278804 Остролодочник синий
278807 Остролодочник шерстеногий
278809 Остролодочник наклоненный, Остролодочник отклоненный
278814 Остролодочник двуцветный
278818 Остролодочник дологостайского
278822 Остролодочник шиповатый
278823 Остролодочник прямой
278824 Остролодочник волосистоплодный
278827 Остролодочник ферганский
278828 Остролодочник фетисова
278829 Остролодочник нитевидный
278831 Остролодочник многоцветковый, Остролодочник яркоцветный
278832 Остролодочник фомина
278835 Остролодочник хлодостойкий
278837 Остролодочник кустарничковый
278839 Остролодочник геблера
278842 Остролодочник голый
278851 Остролодочник железистый
278852 Остролодочник галечниковый
278854 Остролодочник летниковый, Остролодочник шароцветный
278855 Остролодочник гмелина
278857 Остролодочник горбунова
278862 Остролодочник крупноцветковый
278868 Остролодочник голопестичный
278869 Остролодочник хайларский

278877 Остролодочник разноножковый
278878 Остролодочник разноволосый
278880 Остролодочник ипполита
278881 Остролодочник жестковолосый
278882 Остролодочник жестковолосистый
278894 Остролодочник стелящийся
278899 Остролодочник иглистый
278906 Остролодочник цельнолепестный
278907 Остролодочник средний
278910 Остролодочник японский
278915 Остролодочник украшающий
278916 Остролодочник камчатский
278919 Остролодочник ганьсуский
278920 Остролодочник каратавский
278921 Остролодочник карягинб
278922 Остролодочник казбека
278923 Остролодочник хатангский
278924 Остролодочник кавасимский
278925 Остролодочник кетменский
278926 Остролодочник коморова
278930 Остролодочник копетдагский
278934 Остролодочник крылова
278935 Остролодочник кубанский
278938 Остролодочник кузнецова
278939 Остролодочник ладыгина
278941 Остролодочник шерстистый
278948 Остролодочник лапландский
278952 Остролодочник мохнатоплодный
278954 Остролодочник шишковидный
278956 Остролодочник лазистанский
278957 Остролодочник лемана
278958 Остролодочник тонколистный
278960 Остролодочник тонкопузырчатый
278961 Остролодочник бледноцветковый
278963 Остролодочник светлоголубой
278965 Остролодочник беловолосистый
278968 Остролодочник линчевского
278970 Остролодочник липского
278971 Остролодочник прибрежный
278972 Остролодочник литвинова
278974 Остролодочник длиннoприцветничный
278978 Остролодочник длинноногий
278979 Остролодочник длинносый
278980 Остролодочник лупиновый
278983 Остролодочник длиннокистевой
278984 Остролодочник крупноплодный
278985 Остролодочник крупнозубый
278987 Остролодочник майдантальский
278989 Остролодочник мягколистный
278990 Остролодочник маньчжурский
278997 Остролодочник приморский
278998 Остролодочник мартьянова
278999 Остролодочник майделиев
279002 Остролодочник крупнохоботковый
279003 Остролодочник мейнсхаузена
279004 Остролодочник чернобелый
279007 Остролодочник черноволосый
279009 Остролодочник меркенский
279010 Остролодочник мертенса
279011 Остролодочник мейера
279012 Остролодочник михельсона
279013 Остролодочник мелкоплодный
279014 Остролодочник мелколистный
279016 Остролодочник мелкосферический
279017 Остролодочник миддендорфа
279028 Остролодочник мягкоигольчатый
279029 Остролодочник бесчисленннолистный, Остролодочник тысячелистный
279033 Остролодочник недзвецкого
279034 Остролодочник чернеющий
279035 Остролодочник нинсяский
279037 Остролодочник блестящий
279039 Остролодочник белоснежный
279040 Остролодочник нагой
279041 Остролодочник поникший
279042 Остролодочник охотский
279055 Остролодочник бледножелтый
279058 Остролодочник оверина
279059 Остролодочник оверина
279060 Остролодочник остролистный
279067 Остролодочник палласа
279071 Остролодочник малоцветковый
279072 Остролодочник шубный
279073 Остролодочник повислоцветковый
279074 Остролодочник вздутоплодный
279075 Остролодочник ельниковый
279076 Остролодочник волосистый
279077 Остролодочник густоволосистый
279078 Остролодочник плосконоготковый
279079 Остролодочник широкопарусный
279080 Остролодочник бобоножковый
279083 Остролодочник многолистный
279084 Остролодочник понсэна
279087 Остролодочник распротертый
279088 Остролодочник ближайший
279097 Остролодочник ложнохолодостойкий
279102 Остролодочник пушистый
279104 Остролодочник подушковидный
279106 Остролодочник карликовый
279107 Остролодочник малорослый
279118 Остролодочник признанный
279121 Остролодочник притупленный
279122 Остролодочник завернутый
279126 Остролодочник вздутоносый
279129 Остролодочник мощный
279130 Остролодочник розовый
279131 Остролодочник розоватый
279132 Остролодочник лиловорозовый
279133 Остролодочник красноглинный
279134 Остролодочник красностержневой
279135 Остролодочник рюбзаамена
279136 Остролодочник рукутамский
279138 Остролодочник русский
279140 Остролодочник сахалинский
279141 Остролодочник солончаковый
279144 Остролодочник саркандский
279145 Остролодочник саурский
279146 Остролодочник савелланский
279147 Остролодочник щершавый
279149 Остролодочник шишкина
279150 Остролодочник шмидта
279151 Остролодочник шренка
279154 Остролодочник семенова
279155 Остролодочник зеравшанский
279156 Остролодочник шелковистый
279158 Остролодочник щетиноволосый
279159 Остролодочник северцова
279164 Остролодочник сычуаньский
279165 Остролодочник синьцзянский
279169 Остролодочник джунгарский
279170 Остролодочник арктический, Остролодочник грязноватый, Остролодочник грясный
279171 Остролодочник колосовый
279172 Остролодочник колючконосный
279175 Остролодочник чешуйчатый
279178 Остролодочник узколистный
279182 Остролодочник стукова
279183 Остролодочник приятский
279184 Остролодочник головчатый
279189 Остролодочник туповатый
279191 Остролодочник постимутовчатый
279192 Остролодочник серножелтый
279193 Остролодочник сусамырский
279195 Остролодочник лесно
279196 Остролодочник таласский
279201 Остролодочник тонкоклювый
279203 Остролодочник тонкий
279204 Остролодочник терекский
279208 Остролодочник подушечный, Остролодочник тяньшанский
279211 Остролодочник войлочный
279212 Остролодочник трагакантовый
279213 Остролодочник заалайский
279214 Остролодочник траутфеттера
279215 Остролодочник пушистозубчатый
279217 Остролодочник пушистопузырчатый
279218 Остролодочник пушистосферический
279220 Остролодочник трехлисточковый
279221 Остролодочник чимганский
279226 Остролодочник угамский
279228 Остролодочник уральский
279249 Магнолия зибольда
279250 Магнолия китайская
279476 Пахицентрия
279622 Толстостенка

279625 Толстостенкакрупнолистная
279644 Пахиплеурум, Толстореберник
279645 Пахиплеурум альпийский, Толстореберник альпийский
279651 Пахиплеурум
279653 Пахиплеурум тибетский
279703 Толстокрыл
279704 Толстокрыл коротконогий
279707 Толстокрыл густоцветковый
279714 Толстокрыл многостебельный
279743 Пахизандра
279757 Молочай японский, Пахизандра верхушечная
279899 Крестовник резедолистый
279998 Черемуха
280003 Глотуха, Прунус кистевой, Черемуха, Черёмуха, Черемуха кистевая, Черёмуха кистевая, Черемуха кистевидная, Черемуха обыкновенная, Черёмуха обыкновенная, Черемуха птичья
280004 Черемуха азиатская
280023 Черемуха грея
280028 Прунус маака, Черемуха маака, Черемуха японская
280036 Черемуха непалская
280044 Черёмуха, Черемуха кистевая, Черёмуха кистевая, Черемуха обыкновенная, Черёмуха обыкновенная
280053 Черемуха сьори
280061 Слива ватсона
280063 Пэдеря
280078 Куртарник хинный китайский, Лихорадочник китайский
280122 Педерота
280143 Марьин корень, Пеон, Пион
280146 Пион абхазский
280147 Пион белоцветковый
280152 Пеон алтайский
280154 Марьин корень, Пеон марьин-корень, Пион марьин-корень
280164 Пион биберштейна
280169 Пион кавказская
280204 Пион степной
280206 Пион степной
280207 Пион японский
280213 Пеон белоцветный, Пеон китаиский, Пеон мелкоцветковый, Пеон травянистый, Пион молочноцветковый
280227 Пион крупнолистный
280231 Ежовик коралловидный, Пеон коралловый
280236 Пион млокосевича
280241 Пион горный, Пион обратно-овальный, Пион обратнояйцевидный
280253 Пеон лекарственный, Пион аптечный, Пион лекарственный
280286 Дерево пеонное, Пеон древовидный, Пеон полустарниковый, Пион древовидный, Пион полукустарниковый
280318 Пион тонколистный
280319 Пион войлочноплодный
280320 Пион трехраздельный
280332 Пион весенний
280334 Пион виттмана
280437 Палаквиум
280441 Дерево гуттаперчевое, Паляквиум
280490 Налимбия
280491 Налимбия оживающая
280539 Держи-дерево, Палиурус, Пальма
280544 Держи-дерево, Христовы тернии
280548 Держи-дерево хемсли
280555 Держи-дерево многоответвистое
280559 Дерёновые, Держи-дерево, Держи-дерево обыкновенное, Держи-дерево христовы тернии, Драч, Палиурус держи-дерево, Тернии христовы, Христовы тернин
280578 Палленис
280585 Палленис колючий
280608 Пальмовые, Пальмы
280712 Женьшень, Панакс
280741 Желтокорень китайский, Женьшень, Жень-шень, Женьшень обыкновенный, Корень-человек, Панакс женьшень, Панакс жень-шень, Панакс корень-человик
280793 Жень-шень, Панакс ползучий
280799 Желтокорень китайский, Женьшень американский, Женьшень пятилистный, Стосил
280814 Панакс пинбяньский
280817 Бересклет карликовый, Бересклет низкий, Панакс трёхлистниковый
280877 Панкраций, Панкрациум
280900 Панкраций морской, Панкрациум приморский
280963 Пандановые, Панданусовые
280968 Дерево винтовое, Пандан, Панданус
281122 Панданус сандера
281166 Вакуа, Панданус полезный
281169 Панданус вейча
281173 Пандерия
281174 Пандерия волосистая
281175 Пандерия туркестанская
281245 Пангиум
281246 Пангиум съедобный
281274 Просо, Тары
281390 Просо двбороздчатое
281438 Просо власовидное, Просо волосовидное
281887 Просо гвинейское, Просо гигантское, Просо крупное, Трава гвинейская
281916 Арзан, Просо культурное, Просо метельчатое, Просо настоящее, Просо обыкновенное, Просо посевное, Тары
282366 Просо прутовидное, Просо прутьевидное
282477 Измодинь, Панцерия
282480 Панцерия сероватая
282481 Панцерия ганьсуская
282482 Панцерия шерстистая
282485 Панцерия серебристобелая
282496 Мак
282499 Мак аянский
282504 Мак белорозовый
282506 Мак альпийский
282508 Мак ложный
282511 Мак узкодольчатый
282514 Мак песчаный, Мак песчатый
282515 Мак аргемона, Мак колючий
282520 Мак беланже
282521 Мак дважтыперистый
282522 Мак прицветниковый
282524 Мак тяньшанский
282525 Мак сероватый
282526 Мак кавказский
282527 Мак чистотелолистный
282530 Мак спутанный
282534 Мак полярный
282536 Мак оранжевый
282546 Мак сомнительный
282574 Мак гиблидный
282579 Мак промежуточный
282580 Мак окутанный
282581 Мак югорский
282584 Мак разорванный
282585 Мак голый
282588 Мак лапландский
282592 Мак мохнатый
282593 Мак ледебура
282594 Мак голоплодный
282596 Мак лизы
282597 Мак литвинова
282603 Мак крупнокоробочный
282610 Мак мелкоплодный
282614 Мак одноцветковый
282616 Мак снежный
282617 Мак голостебельный, Мак оранжевный, Мак сибирский
282622 Мак амурский
282646 Мак горный
282647 Мак восточный
282658 Мак нмалолистный
282659 Мак глазчатый, Мак павлиний
282660 Мак павлиний
282662 Мак персидский
282666 Мак ложносероватый
282672 Мак ложный штубендорфа

282673 Мак подушковый
282674 Мак пиренейский
282676 Мак корневой, Мак полярный
282685 Мак полевой, Мак самосейка
282701 Мак роона
282705 Мак оранжево-красный
282713 Мак щетинконосный
282717 Мак обыкновенный, Мак опийный, Мак снотворный
282734 Мак щетинистый
282736 Мак штубендорфа
282739 Мак талышинский
282740 Мак тоненький
282743 Мак тяньшанский
282745 Мак толмачева
282751 Маковые
282768 Пафиопедилюм
282796 Пафиопедилюм бирманский
282844 Пафиопедилюм гиблидный
282845 Пафиопедилюм замечательный
282855 Пафиопедилюм яванский
282858 Пафиопедилюм малипоский
282914 Пафиопедилюм мохонадый
282926 Мотыльковые
282974 Хохлатник персидский
282979 Хохлатник
283171 Паракариум
283174 Паракариум бунге
283179 Паракариум изящный
283180 Паракариум гималайский
283182 Паракариум серый
283183 Паракариум промежуточный
283185 Паракариум каратавский
283186 Паракариум рыхлоцветковый
283190 Паракариум прямой
283191 Паракариум туркменский
283267 Парадиноглосс
283269 Парадиноглосс шершавейший
283270 Парадиноглосс зубчиковоплодный
283271 Парадиноглосс имеретинский
283285 Деррис хайнаньский
283295 Парадизея
283327 Лжегравилат
283385 Параиксерис
283386 Юнгия чистотелолистная
283388 Параиксерис зубчатолистный
283399 Параиксерис скальный
283400 Параиксерис поздний, Юнгия поздняя
283504 Парамикроринхус
283506 Парамикроринхус полегающий
283654 Парафорис
283656 Парафорис изогнутый, Чешуехвостник сонгутый
283720 Лжеводосбор
283721 Лжеводосбор ветреницевидный, Лжеводосбор крупноцветный
283723 Лжеводосбор дернистый
283726 Лжеводосбор мелколистый
283781 Недоспелка
283792 Недоспелка ушастая
283815 Какалия ганьсуская
283816 Какалия копьевидная, Какалия копьелистная
283835 Недоспелка комарова
283860 Недоспелка пропущенная
283921 Скафоспермум
284093 Парентучеллия
284094 Парентучеллия желтоцветковая
284096 Парентучеллия широколистная
284098 Парентучеллия клейкая
284131 Постенница
284133 Постенница мокричнолистная
284134 Постенница херсонская
284145 Постенница прямая
284152 Постенница иудейская
284160 Постенница меклоцветковая, Постенница слабая
284167 Мокрединник, Повой подстеный, Постенница апчечная, Постенница лекарственная, Стенница
284289 Вороний глаз, Воронятник, Одноягодник
284308 Воронятник сичоуский
284310 Воронятник далиский
284339 Вороний глаз неполный
284348 Вороний глаз маньчжуьрский
284368 Воронятник гуандунский
284387 Вороний глаз обыкновнный, Вороний глаз четырехлистный, Глаз вороний обыкновенный, Глаз вороний четырелистный, Глаз вороний четырёхлистный
284406 Вороний глаз мутовчатый, Вороний глаз шестилистный
284444 Паркия
284473 Паркия двхшаровидная, Паркия суданская
284475 Паркия африканская
284500 Белозор
284505 Белозор наньчуаньский
284510 Белозор двуристный, Белозор двуристый
284516 Белозор чэнкоуский
284517 Белозор китайский
284518 Белозор китайский сычуаньский
284528 Белозор дэциньский
284532 Белозор эмэйский
284544 Белозор ганьсуский
284549 Белозор кандинский
284550 Белозор боцебу
284556 Белозор лаксманна
284559 Белозор лицзянский
284591 Белозор болотный
284616 Белозор симаоский
284623 Белозор тибетский
284634 Белозор синаньский
284636 Белозор яньюаньский
284637 Белозор илянский
284639 Белозор юй
284640 Белозор юйлуншаньский
284641 Белозор юньнаньский
284647 Парохетус
284735 Приноготковник, Приноготовник
284808 Приноготовник головчатый
284866 Приноготовник курдский
284957 Железняк, Парротия
284961 Боккаут, Дерево железное, Парротия персидская
284969 Паррия
284971 Паррия беловатая
284973 Паррия шершавая
284974 Паррия бекетова
284990 Паррия кустарничковая
284991 Паррия голенкина
284993 Паррия каратавская
285001 Паррия мелкоплодная
285004 Паррия голостебельная, Паррия крупноплодная
285006 Паррия перистая
285008 Паррия перистая
285014 Паррия подушковая
285018 Паррия струговидная
285020 Паррия шугнанская
285021 Паррия стручковая
285022 Паррия узкоплодная
285029 Паррия узколистная
285030 Паррия стручковатая
285031 Паррия отпрысковая
285032 Паррия туркестанская
285055 Куфея, Парсония
285072 Гвайюла, Гваюла, Партениум
285076 Гвайюла, Гвайюла серебристая, Гваюла, Гваюла серебристая, Гуайюла, Гуаюла, Мариола, Партениум серебристый
285097 Виноград девичий, Виноград дикий, Девичий виноград, Дикий виноград, Партеноциссус
285102 Партеноциссус китайский
285110 Партеноциссус генри
285127 Девичий виноград садовый
285136 Виноград вьющийся, Виноград девичий пятилистный, Виноград дикий пятилистный, Девичий виноград пятилисточковый, Партеноциссус пятилисточковый
285157 Виноград девичий триостренный, Виноград дикий триостренный, Девичий виноград трехострконечный, Девичий виноград триостренный, Дикий

виноград триостренный
285390 Паспалум, Паспалюм
285417 Паспалюм двойчатосложный
285422 Гречка пальчатая, Паспалум пальчатый, Сухумка
285423 Гречка расширенная, Злак большой водяной, Паспалум расширенный, Паспалюм расширенный
285448 Паспалюм тайваньский
285460 Паспалюм длиннолистный
285469 Паспалум помеченый
285471 Паспалюм округлый
285507 Паспалюм ямчатый
285530 Паспалум тунберга, Паспалюм тунберга
285533 Паспалум
285607 Гренадилла, Звезда кавалерская, Звездочка, Пассифлора, Страстоцвет
285623 Звездочка голубая, Пассифлора олубая, Страстоцвет голубой, Страстоцвет светло-синий
285637 Гранадилла, Гренадилья, Пассифлора плодовая, Пассифлора съедобная, Страстоцвет съедобный
285639 Страстоцвет пахучий
285645 Страстоцвет изящный
285659 Страстоцвет гуандунский
285660 Пассифлора лавролистная, Страстоцвет лавлолистный
285694 Гранадилла гигантская, Пассифлора гигантская, Пассифлора крупная, Страстоцвет крупный, Страстоцвет настоящий, Страстоцвет четырехгранный, Страстоцвет четырёхугольный
285703 Страстоцвет пробковый
285725 Страстоцветные
285731 Пастернак
285732 Пастернак армянский
285734 Пастернак оранжевый
285738 Пастернак тусклый
285739 Пастернак бедренецолистный
285740 Пастернак луговой
285741 Пастернак посевной
285744 Пастернак дикий, Пастернак лесной
285745 Пастернак теневой
285748 Пастернаковник
285749 Пастернаковник ледниковый
285814 Валерьяна каменная, Патриния, Патрэния
285830 Патриния горбатая
285839 Патриния средняя
285854 Патриния каменистая, Патриния скальная, Патрэния скальная
285859 Патриния скабиозолистная, Патрэния скабиозолистная
285870 Патриния сибирская, Патрэния сибирская
285880 Патриния мохнатая
285928 Павлония, Пауллиния
285930 Гуарана, Пауллиния гуарана
285951 Павловния, Пауловния
285959 Павловния фаржа
285960 Павловния форчуна
285978 Павловния тайваньская
285981 Павловния войлочная, Павловния пушистая, Пауловния войлочная
285985 Павловния форчуна циньлинская
286073 Паветта
286754 Пайена
286918 Кунжутовые, Педалиевые, Сезамовые
286972 Быноица, Вшивица, Мытник, Трава вшивая
286973 Мытник полынелистный, Мытник полыннолистный
286974 Мытник тысячелистниковый
286975 Мытник адамса
286976 Мытник крючковатый
286984 Мытник алатавский
286985 Мытник альберта
286991 Мытник алтайский
286997 Мытник прелестноцветковый
287015 Мытник тонкопильчатый
287021 Мытник чернопурпуровый
287055 Мытник коротколосый
287067 Мытник голодчатый
287068 Мытник кавказский
287076 Мытник шамизо
287084 Мытник краекучниколистный
287090 Мытник китайский
287099 Мытник хохлатый
287100 Мытник густоцветковый, Мытник плотный
287102 Мытник сжатый
287120 Мытник толстоносый
287139 Мытник дагестанский
287142 Мытник мохнатоцветковый
287143 Мытник мохнатоколосый
287162 Мытник дэциньский
287175 Мытник длиннокорневой
287178 Мытник сомнительный
287182 Мытник высокий
287191 Мытник мохнатооденый
287194 Мытник очанковидный
287196 Мытник высокий, Мытник лослый
287201 Мытник фэн
287207 Мытник разрезной
287209 Мытник желтый
287222 Мытник франше
287243 Мытник гуншаньский
287259 Мытник крупноцветковый
287260 Мытник бальшой
287270 Мытник цзилунский
287279 Мытник волосистый
287282 Мытник хэнаньский
287290 Мытник мясокрасный
287298 Мытник прерванный
287300 Мытник японский
287301 Мытник кандинский
287302 Мытник ганьсуский
287307 Мытник каратавский
287310 Мытник кауфманна
287318 Мытник каролькова
287320 Мытник крылова
287321 Мытник кузнецова
287326 Мытник лабрадорский
287334 Мытник шерстистый
287336 Мытник лангсдорфа
287337 Мытник ланьпинский
287338 Мытник лапландский
287343 Мытник мохнатый, Мытник шелоховатоколосый
287356 Мытник лицзянский
287368 Мытник длинноцветковый
287382 Мытник людвига
287411 Мытник маньчжурский
287413 Мытник марин
287414 Мытник максимовича
287427 Мытник мелкоцветковая
287436 Мытник мупинский
287447 Мытник тысячелистный
287451 Мытник наньчуаньский
287455 Мытник носатый
287461 Мытник нордманна
287468 Мытник эдера
287481 Мытник ольги
287492 Мытник палласа
287494 Мытник болотный, Трава вшивая
287495 Мытник каро
287496 Мытник панютина
287507 Мытник цветоножковый
287508 Мытник пеннелля
287517 Мытник вздуточашечный
287520 Мытник ширококлювый
287532 Мытник понтийский
287533 Мытник попова
287541 Мытник хоботковый
287567 Мытник пушистоцветковый
287568 Мытник красивый
287569 Мытник густоцветковый
287584 Мытник запрокинутый, Мытник перевернутый
287617 Мытник погремковый
287645 Мытник краснеющий
287657 Мытник каролев-скипетр, Мытник королевский скипетр, Скипетр царский
287666 Мытник шугнанский
287668 Мытник семенова
287673 Мытник сибирский
287674 Мытник сибторпа

287689 Мытник колосистый, Мытник колосовый
287705 Мытник полосатый
287716 Мытник почти-носатый
287718 Мытник судетский
287721 Мытник лесной
287723 Мытник сычуаньский
287740 Мытник таласисский
287741 Мытник далиский
287760 Мытник тяньшанский
287761 Мытник тибетский
287779 Мытник печальный
287788 Мытник топяной
287790 Мытник уральский
287794 Мытник красивый, Мытник миловидный
287795 Мытник веры
287802 Мытник мутовчатый, Мытник прелестный
287814 Мытник мохнатый
287821 Мытник власова
287822 Мытник вальдгейма
287828 Мытник вильгельмся
287829 Мытник вильденова
287837 Мытник иезский
287842 Мытник чаюйский
287843 Мытник зеравшанский
287844 Мытник чжундяньский
287846 Педилантус
287975 Гармала, Могильник
287978 Адраспан, Гармала, Гармала обыкновенная, Могильник, Хутчик, Юзерлык
287983 Гармала чернушкообразная
288045 Пеларгониа, Пеларгоний, Пеларгониум, Пеларгония
288133 Пеларгониум головчатый
288216 Пеларгониум эндлихера
288264 Пеларгониум крупноцветный
288268 Герань душистая, Герань розовая, Пеларгоний розовый
288300 Пеларгониум гиблидная
288396 Пеларгония душистая
288430 Пеларгониа плющевидная, Пеларгониа ползучая
288594 Герань комнатная, Герань садовая, Пеларгоний попелечнополосатый, Пеларгония зональвая
288705 Пеллиония
288754 Пеллиония хайнаньская
288760 Пеллиония красивая
288800 Пельтандра
288818 Шит
288820 Шит оше
288824 Шит турменский
288825 Шит воронова
288827 Щитник
288829 Щитник гроссгейма
288840 Пельтифиллум
288863 Пельтогине
289011 Пеннизетум, Пеннисетум
289055 Пеннисетум ланьпинский
289096 Пеннисетум повислый
289116 Бажра, Дохн, Пеннисетум сизый, Просо африканское, Просо негритянское
289135 Пеннисетум японский
289141 Пеннисетум тибетский
289185 Пеннисетум восточный
289218 Пеннисетум красный, Трава слоновая
289258 Просо сычуаньский
289314 Пенстемон, Пентастемон
289341 Пентастемон кустарниковый
289407 Пентаце
289439 Пентаклетра крупнолистная
289476 Дерево масляное африканское, Дерево сальное, Пеннисетум масличная
289563 Пентанема
289566 Пентанема растопыренная
289567 Пентанема железистая
289568 Пентанема индийская
289571 Пентанема постенная
289572 Пентанема близкая
289574 Пентанема скальная
289708 Курильский чай дриадоцветовидный
289713 Курильский чай, Курильский чай кустарниковый, Лапчатка курильский, Лапчатка кустарниковая, Чай курильский, Чай курильский кустарниковый
289718 Курильский чай мелколистный, Лапчатка мелколистная
289720 Курильский чай листочашечный
289784 Пентас
290133 Петорум, Пятичленник
290134 Петорум китайский, Пятичленник китайский, Пятичленник низкий
290222 Кустарник овечий, Пентция
290286 Пеперомия
290357 Пеперомия мэнцзыская
290365 Пеперомия японская
290380 Пеперомия пятнистая
290382 Пеперомия магнолиелистная
290395 Пеперомия туполистная
290476 Бутерлак
290478 Бутерлак очереднолистный
290485 Бутерлак гирканский
290489 Бутерлак портулаковый
290572 Перакарпа
290573 Перакарпа двулепестниковидная
290701 Кактус пейреския, Крыжовник барбадосский, Переския колючая, Переския шиповатая
290821 Пеперигник багровый, Цинерария гибридная
290921 Стробилантес диера
290932 Пеллила, Судза
290940 Пеллила кустарниковая, Перилла базириковая, Перилла многолетняя
291031 Обвойник, Периплока
291063 Обвойник греческий
291082 Обвойник заборный, Периплока заборная
291134 Перисторофа
291363 Перо водяное
291427 Перовския, Персея
291428 Перовския полыная
291429 Перовския узколистная
291431 Перовския кудряшева
291432 Перовския линчевского
291433 Перовския памирская
291435 Перовския прутьевидная
291441 Перплексия мелкоголовчатая
291491 Авокадо, Персея
291494 Авокадо американское, Авокадо привлекательнейшее, Авокадо-груша, Груша аллигаторова, Персея приятнейшая
291495 Персея мексиканская
291505 Лавр красный, Персея, Персея индийская
291543 Персея индийская, Персея красная
291663 Башкирская капуста, Горец альпийский, Горец горный, Кислец, Таран
291818 Горец шероховатый
291932 Горец стрелолистный
292279 Клевер американский стенной, Петалостемум
292342 Белокопытник, Нардосмия, Подбел
292351 Нардосмия фомина
292352 Нардосмия тайваньская
292354 Нардосмия угловатая
292367 Нардосмия ледяная
292368 Нардосмия гмелина
292372 Белокопытник гибридный, Подбел гибридный
292374 Белокопытник белый, Нардосмия японская, Подбел белый, Подбел японский
292379 Нардосмия гладгая
292386 Белокопытник гибридный, Белокопытник лекарственный, Подбел гибридный, Подбел лекарственный, Подбел лечебный, Подбел обыкновенный
292394 Нардосмия скальная
292401 Белокопытник ложный, Белокопытник ненастоящий, Подбел

войлочный
292403 Белокопытник дланевидный, Нардосмия пальмолистная
292517 Петрея
292527 Петрокаллис
292645 Петрофитум
292656 Петорорагия, Туника
292658 Туника прямая
292665 Кольраушия побегоносная
292668 Петорорагия камнеломка, Туника камнеломка
292685 Петрушка
292694 Петрушка зеленная, Петрушка кудрявая, Петрушка обыкновенная, Петрушка огородная, Петрушка садовая
292701 Петрушка корневая
292714 Петросимония
292715 Петросимония супротивиолистная
292716 Петросимония коротколистная
292717 Петросимония толстолистная
292720 Петросимония сизая
292721 Петросимония сизоватая
292722 Петросимония жестковолосая
292723 Петросимония литвинова
292724 Петросимония однотычинковая
292726 Петросимония сибирская
292727 Петросимония растопыренная
292728 Петросимония трехтычинковая
292741 Петуния, Петунья
292745 Петунья гибридная, Петунья мелкоцветная
292748 Петунья вздутая
292751 Петунья ночецветковая
292754 Петунья фиолетовая
292758 Горечник, Горичник
292765 Горечник ады
292766 Горичник эльзасский
292774 Горичник песчный
292779 Горичник австрийский
292780 Горичник байкальский
292787 Горечник днепровский
292794 Горечник известняковый
292810 Горечник кавкажский
292813 Горичник олений
292814 Горечник оленелистный
292843 Горечник, Горичник изящный
292850 Горичник резаковый
292857 Горичник формозский
292873 Горичник гуансиский
292888 Горечник гиссарский
292892 Горичник японский
292908 Горичник широколистный
292916 Горечник любименко
292918 Горечник пышный
292929 Горечник крупнолистный
292937 Горичник морисона
292957 Горичник аптечный, Горичник лекарственный
292961 Горичник горный, Горная петрушка, Петрушка горная, Поречник, Трава прорезная
292965 Горечник палимбиевидный
292966 Горичник болотный, Петрушка болотная
292968 Горечник малоолиственный
292969 Горечник малолучевой
292975 Горечник подольский
292977 Горечник многоцветковый
292987 Горечник опушенный
292992 Горечник ленарда
293002 Горичник русский
293003 Горечник солончаковый
293007 Горичник шотта
293030 Горичник крымский
293032 Горечник
293033 Горичник, Горичник терпентинный
293052 Горечник туркменский
293059 Горечник влагалищный
293063 Горичник олений
293081 Горечник зедельмейеровский
293151 Фацелия, Эвтока
293176 Фацелия пижмолистная
293181 Пучкоцвет, Фаселантус
293186 Пучкоцвет трубкоцветковый, Фаселантус трубкоцветковый
293408 Фагналон
293411 Фагналон андросова
293423 Фагналон дарвазский
293553 Лысосемянник
293554 Лысосемянник лысый
293555 Лысосемянник девясиловидный
293577 Фаленопсис
293607 Фаленопсис хайнаньский
293611 Фаленопсис гиблидный
293633 Фаленопсис шиллера
293705 Двукисточник, Канареечник
293709 Двукисточник тростниковидный, Двукисточник тростниковидный житовник, Двукисточник тростниковый, Камыш стенной, Канареечник простниковидный, Канареечник тростниковый, Трава канареечная, Трава шелковая, Трава шёлковая
293722 Канареечник коротколосый
293723 Канареечник клубневидный, Канареечник клубненосный, Канареечник луковичный, Канареечник узловатый
293729 Каналейник, Канареечник канареечный, Канареечник канарский, Канареечник настоящий, Канареечник птичий, Семя канареечное, Трава канарская, Ятрориза драневидная
293743 Канареечник малый
293750 Канареечник своеобразный
293764 Канареечник клубневой
293765 Канареечник усеченный
293861 Ипмея
293966 Фасоль, Фасоль безволокновая, Фасоль ломкая
293969 Тепари, Фасоль остролистная
293985 Боб арабский, Боб турецкий, Бобы многоцветковые, Бобы огненные, Бобы турецкие, Фасоль арабская огненно-краская, Фасоль кроваво-красная, Фасоль многоцветковая, Фасоль огненая, Фасоль огнено-расная
294006 Бобы лимские, Лима, Фасоль лима, Фасоль лимская
294010 Боб лимский, Бобы каролинские, Бобы лимские мелкоплодные, Бобы сива, Лима, Лима мелкая, Лима мелкоплодная, Фасоль лимская, Фасоль полулунная, Фасоль сива
294056 Быбы турецкие, Фасоль волокнистая, Фасоль зерновая, Фасоль настоящая, Фасоль обыкновенная, Фасоль почечная, Фасоль почечновидная, Фасоль турецкая
294057 Фасоль обыкновенная кустовая
294114 Очиток живучий
294122 Очиток гиблидный
294123 Очиток камчатский
294124 Очиток миддендорфа
294126 Очиток сельского
294128 Очиток стевена
294230 Бархат, Дерево пробковое, Филодендрон
294231 Амурское пробковое дерево, Бархат амурский, Дерево абурское бархатное, Дерево амбровое, Дерево амурское пробковое, Дерево бархатное, Дерево пробковое амурское, Пробковое дерево
294233 Бархат амурский японский, Филодендрон японский
294238 Бархат китайский
294251 Бархат сахалинский
294407 Чубушник
294413 Чубушник короткокистевой
294419 Чубушник лицзянский
294422 Жасмин садовый, Чубушник кавказский
294425 Чобушник сердцелистный
294426 Жасмин глунтовой, Жасмин дикий, Жасмин ночной, Жасмин садовый, Чубушник венечный
294445 Чобушник культера
294451 Чобушник делаве
294459 Чобушник фальконера
294462 Чубушник гордона

294463 Чубушник крупноцветковый
294465 Чубушник генри
294468 Чубушник жестковолосый
294471 Чубушник седоватый, Чубушник седой
294478 Чубушник непахучий
294479 Жасмин непахучий, Чобушник непахучий
294484 Чубушник ганьсуский
294486 Чубушник куэньминский
294488 Чобушник широколистный
294493 Чубушник льюиса
294500 Чубушник мексиканский
294501 Чубушник мелколистный
294507 Жасмин обыкновенный, Чубушник душистый
294509 Чубушник пекинский
294517 Чобушник широколистный
294530 Жасмин японский
294536 Чубушник шренка
294542 Чубушник гулинский
294549 Чубушник полуседой
294556 Чубушник тонколистный
294560 Жасмин розовый
294564 Чобушник боротавчатый
294565 Чубушник девичий
294590 Чобушник цейера
294591 Чубушник чжэцзянский
294761 Липа каменная, Филирея
294762 Липа каменная узколистная, Филирея узколистная
294772 Липа каменная широколистная, Филирея широколистная
294785 Филодендрон
294792 Филодендрон андрэ
294795 Филодендрон шероховатый
294804 Филодендрон изящный
294805 Филодендрон красноватый
294830 Филодендрон рассеченный
294832 Филодендрон перистораздельный
294839 Филодендрон селло
294844 Филодендрон чешуйчатый
294850 Филодендрон бородавчатый
294907 Фиппсия
294909 Фиппсия холодолюбивая
294910 Фиппсия стройная
294960 Аржанец, Тимофеевка
294962 Тимофеевка альпийская
294966 Тимофеевка песчаная, Экзохорда
294972 Тимофеевка буассье
294976 Тимофеевка колючая
294979 Тимофеевка греческая
294985 Тимофеевка микели
294986 Тимофеевка горная
294988 Тимофеевка метельчатая
294989 Тимофеевка бёмора, Тимофеевка степная
294992 Аржанец, Аржанец луговоя, Палошник, Палошник луговой, Тимофеевка луговая, Трава тимофеева
295005 Тимофеевка тонкая
295035 Виздутоплодник, Флойодикарский
295038 Виздутоплодник сибирский, Флойодикарский сибирский
295041 Виздутоплодник мохнатый, Флойодикарский мохнатый
295045 Чистец мелкоцветковый
295048 Чистец мелкоцветковый
295049 Зопник
295051 Зопник нивяный
295052 Зопник алайский
295054 Зопник альпийский
295056 Зопник ангренский
295063 Зопник буквицелистный
295071 Зопник короткоприцветниковый
295077 Зопник бухарский
295080 Зопник решетчатый
295081 Зопник сероватый
295084 Зопник кавказский
295095 Зопник округлозубчатый
295098 Зопник дробова
295099 Зопник ферганский
295105 Зопник далиский
295111 Зопник кустарниковидный, Зопник кустарниковый, Шалфей иерусалимский
295116 Зопник гибридный
295117 Зопник снизу-белый
295124 Зопник ганьсуский
295126 Зопник кнорринговская
295127 Зопник копетдагский
295130 Зопник ленкоранский
295132 Зопник лицзянский
295133 Зопник линейнолистный
295139 Зопник майкопский
295142 Зопник максимовича
295150 Зопник монгольский
295153 Зопник мулиский
295162 Зопник ольги
295163 Зопник горный
295167 Зопник островского
295172 Зопник луговый
295173 Зопник ложно-колючий
295174 Зопник опушенный
295175 Зопник колючий
295181 Зопник регеля
295187 Зопник иволистный
295198 Зопник колючезубый
295203 Зопник кандинский
295205 Зопник крымский
295206 Зопник тонгий
295207 Зопник коровяковидный
295208 Зопник тибетский
295210 Зопник войлочный
295211 Зопник чимганский
295212 Зопник клубненосный
295223 Зопник хвостозубый
295225 Зопник вавилова
295228 Зопник зинаиды
295239 Пламенник, Флокс
295267 Флокс друммонда, Флокс друмонда, Флокс однолетний
295288 Флокс метельчатый
295318 Флокс сибирский
295451 Пальма финиковая, Феникс, Финик, Финикс
295459 Пальма финиковая канарская, Феникс канарский, Финик канарский, Финикс канарский
295461 Пальма финиковая, Пальма финиковая культурная, Феникс пальценосный, Финик пальчатый, Финик финиколистный, Финик финиконосный
295479 Финик изогнутый
295484 Финик робелена
295487 Финик лесной, Финикс лесной
295510 Фолидота
295518 Фолидота китайская
295523 Фолидота черепитчатая
295543 Фолидота вэньшаньская
295555 Чешуехвостник
295561 Чешуехвостник венгерский
295594 Лён новозеландский
295600 Конопля новозеландская, Лён новозеландский прядильный, Пенька новозеландская, Формиум
295609 Фотиния
295621 Фотиния аньлунская
295637 Фотиния бентама
295675 Фотиния давидсона
295686 Фотиния фуцзяньская
295688 Фотиния франше
295691 Фотиния голая
295715 Фотиния гуансиская
295767 Фотиния жуйлиская
295773 Фотиния пильчатая
295808 Фотиния мохнатая
295813 Фотиния мохнатая китайская
295820 Фотиния чжэцзянская
295881 Тростник
295888 Гала, Камыш обыкновенный, Курак, Тростник болотный, Тростник обыкновенный, Тростник южный, Урак
295909 Тростник гигантский, Тростник изиды
295911 Тростник поздный
295978 Фрима
295984 Фрима тонкокистевая, Фрима узкоколосая
296006 Фримовые
296011 Фрина

296012 Фрина хюта
296018 Фриниум
296033 Фриниум хайнаньский
296052 Фриниум головчатый
296061 Фтейроспермум
296068 Фтейроспермум китайский, Фтейроспермум японский
296070 Фтейроспермум мулиский
296086 Фуопсис
296091 Фуопсис столбиковый
296121 Липпия узлоццветковая
296465 Филлантус
296466 Филантус кислый, Филлантус кислый
296517 Филлантус чжэцзянский
296554 Миробалан, Филлантус лекарственный, Филлантус эмблика
296565 Филлантус извилистый
296573 Филлантус франше
296599 Филлантус хайнаньский
296803 Филлантус уссурийский
297015 Филлодоце
297016 Филлодоце алеутская, Филлодоце алеутское
297019 Филлодоце голубая, Филлодоце голубое, Филлодоце тиссолистое
297030 Филлодоце японский
297181 Филлоспадикс
297182 Филлоспадикс иватинский
297185 Филлоспадикс скаулера
297188 Листоколосник, Филлостахис
297194 Бамбук узкий, Листоколосник узкий
297203 Бамбук золотистый, Бамбук-листоколосник, Листоколосник золотистый, Листоколосник золотой
297215 Листоколосник бамбузовидный
297266 Листоколосник съедобный
297281 Листоколосник гибгий, Листоколосник извилистый
297282 Листоколосник тайваньский
297285 Листоколосник сизый
297289 Листоколосник гуйцзоуский
297306 Листоколосник пушистый
297359 Листоколосник гнездовый
297367 Бамбук чёрный, Листоколосник черный
297464 Листоколосник сернистый
297485 Листоколосник сизовато-зеленый
297486 Листоколосник зеленый
297624 Физалиаструм
297626 Физалиаструм иглистный
297628 Физалиаструм японский
297638 Кривоплодник
297639 Кривоплодник грелльсиелистный
297640 Вишня жидовская, Вишня пёсья, Вишня пузырная, Вишня пузырчатая, Можжуха, Мозжуха, Песья вишня, Физалис
297643 Вишня алкенинги, Вишня пёсья обыкновенная, Вишня пёсья франшетта, Вишня полевая, Вишня франшетти, Мохунка, Мошнуха, Песья вишня обыкновенная, Физалис обыкновенный, Физалис франшетта
297645 Вишня алкенинги, Вишня пёсья обыкновенная, Вишня пёсья франшетта, Вишня полевая, Вишня франшетти, Мохунка, Мошнуха, Физалис голоногий, Физалис обыкновенный, Физалис пропущенный, Физалис франше, Физалис франшетта
297686 Физалис клейкоплодный, Физалис мексиканский
297711 Вишня ананасная, Вишня перуанская, Вишня перувианская, Вишня пёсья сьедобная, Перуанская вишня, Томат, Физалис перуанский, Физалис съедовный, Физалис ягодный
297720 Земляничный томат, Физалис пушистый, Физалис рубчиковидный
297833 Пузыреплодник, Физокарпус
297837 Пузыреплодник амурский, Физокарпус амурский
297842 Пузыреплодник калинолистный, Физокарпус калинолистный
297851 Пузыреплодник смородинолистный
297854 Вздутостебельник
297856 Вздутостебельник узловатый
297875 Пузырница, Физохлайна
297876 Пузырница илийская
297881 Пузырница восточная
297882 Пузырница физалисовая, Физохлайна физалисовидная
297883 Пузырница тибетская
297885 Пузырница семенова
297934 Шарогнездка
297935 Шарогнездка сушеницевая
297973 Физостегия
298004 Физостигма
298009 Боб карабарский, Зеленчук, Физостигма, Физостигма ядовитая, Яснотка
298050 Фителефас
298051 Пальма костяная, Пальма слоновая, Фителефас крупноплодный
298054 Кольник, Фитеума
298064 Кольник черный, Кольник чёрный
298065 Кольник округлый, Кольник шаровидный
298069 Кольник колосистый, Кольник колосовидтый
298072 Кольник четырехлопастный
298082 Фитокренум
298088 Лаконос, Фитолакка
298094 Лаконос американский, Фитолакка американская, Фитолякка американская, Фитолякка десятитычинковая
298103 Лаконос двудомный
298114 Фитолакка японская
298124 Лаконосные, Фитоллякковые, Фитолякковые
298189 Ель
298191 Ель высокая, Ель европейская, Ель западная, Ель обыкновенный, Ель сизая, Ель стройная
298225 Ель алькокка, Ель двуцветная
298226 Ель алькокка, Ель двуцветная
298232 Ель шероховатая
298248 Ель балфура, Ель бальфура
298258 Ель саржента
298260 Ель саржента
298265 Ель вревера
298273 Ель энгельманна
298282 Ель финская
298286 Ель американская, Ель белая, Ель канадская, Ель серебристая, Ель сизая, Сосна канадская
298298 Ель глена
298307 Ель аянская, Ель иезская, Ель хоккайдская
298319 Ель камчатская
298323 Ель корейская
298331 Ель лицзянская
298349 Ель чёрная
298354 Ель максимовича
298357 Ель мексиканская
298368 Ель горная
298382 Ель обратнояйцевидная, Ель сибирская
298387 Ель сербская, Оморика
298391 Ель восточная, Ель кавказкая
298401 Ель голубая, Ель колючая
298406 Ель синяя
298417 Ель пурпурная
298423 Ель красная
298426 Ель тяньшаньская, Ель шренга, Ель шренка
298429 Ель шренга тяньшанская
298435 Ель ситкинская
298436 Ель гималайская, Ель индийская
298437 Ель восточно-гималайская, Ель шеповатая
298442 Ель изящная, Ель японская
298449 Ель вилькона
298469 Пикномон
298471 Пикномон колючий
298494 Пикрамния
298506 Пикразма, Пикрасма
298509 Пикрасма китайская
298510 Квассия ямайская, Пикрена высокая
298513 Пикрасма яванская

298516 Пикрасма квассиевидная
298550 Гельминция, Горечник, Горлюха, Горчак, Горчак желтый
298563 Горечник седоватая
298584 Горлюха румяниковидная, Горлюха румянковидная
298589 Горечник ястребинковый, Горлюха ястребинковая, Горлюха ястребинковидная, Горчак желтый ястребинковый, Горчак ястребинковидный
298602 Горлюха камчатская
298618 Горлюха японская
298625 Горлюха корейская
298627 Горлюха лицзянская
298633 Горлюха бедноцветковая, Горлюха малоцветковая
298636 Горечник твертая
298640 Горлюха похожая, Горлюха похожий
298643 Горлюха жестковолосая, Горлюха щетинистая
298703 Арктерика, Пиерис, Пиэрис
298721 Пиерис обилиноцветущий
298734 Пиерис японский, Пиэрис тайваньский, Пиэрис японский
298751 Пиэрис японский мелкоцветковый
298758 Пиэрис люцюский
298848 Пилеа, Пилея
298945 Ахудемия японская, Пилея японская
298970 Пилея мотоская
298972 Пилеа мэнхайская
298973 Пилеа гуанхиская
298974 Пилеа мелколистная
299006 Пилеа кртглолистная, Пилея бутерлаковидная
299024 Пилеа монгольская, Пилея зеленейшая, Пилея монгольская
299026 Пилея хамао
299167 Пилокарпус
299172 Пилокарпус перистолистный
299201 волосореберник
299203 волосореберник козо-полянского
299220 Ястребинка дернинная
299277 Волосопыльник
299278 Волосопыльник нителистный
299303 Пимелея
299325 Перец гвосдичный, Перец душистый, Перец ямайский, Пимента
299342 Альбовия, Анис, Бедренец, Реутера
299348 Бедренец близкий
299351 Анис, Анис обыновенный, Бедренец анис, Бедренец-анис, Ганус
299352 Бедренец купыревидный
299354 Бедренец армянский
299355 Бедренец ароматный
299360 Реутера золотистая
299365 Сныть короткоплодная
299382 Бедренец чжундяньск
299385 Бедренец смешиваемый
299390 Бедренец дагестанский
299394 Бедренец разрезной
299411 Бедренец чэнкоуский
299414 Бедренец тэнчунский
299423 Бедренец гроссгейма
299439 Бедренец иды
299446 Бедренец дэциньский
299448 Бедренец корейский
299450 Бедренец коржинского
299459 Бедренец цзиндунский
299462 Бедренец камнелюбивый
299463 Бедренец литвинова
299466 Бедренец большой
299467 Бедренец большой
299479 Бедренец черный
299484 Бедренец голостебельный
299490 Бедренец чужеземный
299493 Бедренец торичниколистный
299502 Бедренец опушенный
299506 Бедренец ветвистый
299511 Бедренец розовоцветный
299520 Бедренец лицзянский
299523 Бедренец камнеломка, Бедренец камнеломковый
299547 Бедренец мулиский
299551 Бедренец крымский
299555 Бедренец телунга
299563 Бедренец известколюбивый, Бедренец меловой
299564 Бедренец разрезаннолистный, Бедренец тимьянолистный
299573 Бедренец туркестанский
299583 Бедренец симаоский
299585 Бедренец тибетский
299586 Бедренец юньнаньский
299594 Сосновые
299649 Пинанга
299736 Жирянка
299737 Жирянка альпейская, Жирянка альпийская
299756 Жирянка пестрая
299757 Жирянка мохнатая
299759 Жирянка обыкновенная
299780 Сосна
299786 Сосна белоствольная
299797 Сосна аризонская, Сосна жёлтая аризонская
299799 Сосна арманда
299809 Сосна веймутова мексиканская, Сосна мексиканская веймутова
299818 Сосна банкса, Сосна банксова, Сосна растопыренная
299827 Сосна брута, Сосна калабрийская
299830 Сосна бунге, Сосна кружевнокорая
299835 Сосна канарская
299836 Сосна карибская, Сосна эллиота
299842 Кедр европейский, Сосна кедровая, Сосна кедровая европейская, Сосна шведская
299869 Сосна скрученная
299874 Сосна муррея, Сосна скрученная, Сосна скрученная широкохвойная, Сосна широкохвойная
299876 Сосна культера, Сосна куртера
299890 Сосна густоцветковая, Сосна густоцветная, Сосна японская
299925 Сосна ежовая, Сосна жёлтая
299926 Сосна съедобная
299927 Сосна эльдарская
299928 Сосна караибская, Сосна карибская, Сосна эллиота
299931 Сосна энгельмана
299942 Кедровая калифорнийская, Сосна гибкая, Сосна кедровая калифорнийская
299945 Сосна погребальная, Сосна похоронная
299953 Сосна джерарда, Сосна жерарда, Сосна жерардова
299964 Сосна алеппская, Сосна иерусалимская, Сосна калабрийская
299966 Сосна крючковатая
299968 Сосна гельдрейха
300006 Кедр корейский, Кедр маньчжурский, Кедровая сосна, Корейский кедр, Косна карейская, Маньчжурский кедр, Сосна кедровая корейская, Сосна кедровая маньчжурская, Сосна корейская
300009 Сосна гуандунская
300011 Сосна ламберта, Сосна сахарная, Сосна сахарная гигантская
300017 Сосна чёрная калабрийская
300054 Сосна китайская, Сосна массона, Сосна массонова, Сосна погребальная
300081 Сосна мексиканская, Сосна монтезумы
300082 Сосна горная веймутова
300086 Жереп, Сосна горная, Сосна кривая, Сосна муго
300098 Сосна мелкоигольчатая, Сосна мягкоигольчатая
300114 Сосна чёрная калабрийская
300116 Сосна чёрная калабрийская
300117 Сосна австрийская, Сосна столбовая, Сосна чёрная, Сосна чёрная австрийская
300126 Сосна яйцевидноплодная
300127 Сосна крымская, Сосна палласа, Сосна палласова
300128 Сосна болотная, Сосна

длиннохвойная, Сосна южная
300130 Сосна мелкоцветковая
300140 Сосна поникшая
300144 Сосна балканская, Сосна веймутова лумелийская, Сосна румелийская
300146 Кедр итальянский, Сосна итальянская, Сосна приморская
300151 Кедр итальянский, Сосна итальянская, Сосна пинния
300153 Дерево калабасовое, Желтая орегонская, Сосна желтая, Сосна жёлтая, Сосна жёлтая западная, Сосна тяжелтая
300163 Кедр капликовый, Кедр стланниковый, Кедровый стланик, Сосна низкорослая, Стланец кедровый
300173 Сосна замечательная, Сосна лучистая
300180 Сосна красная, Сосна норвежская, Сосна смолистая
300181 Сосна жесткая, Сосна жёсткая, Сосна смолистая
300186 Сосна длиннохвойная, Сосна роксбурга
300189 Белая калифорнийская, Сосна белая калифорнийская, Сосна сабина
300197 Кедр альпийсий, Кедр сибирский, Сосна кадровая, Сосна кедровая сибирская, Сосна сибирская
300209 Сосна станкевича
300211 Сосна белая, Сосна веймутова
300223 Сосна европейская, Сосна лесная, Сосна обыкновенная
300243 Сосна китайская, Сосна погребальная, Сосна ярусная
300264 Сосна, Сосна ладанная, Сосна скипидарная
300269 Сосна тайваньская
300281 Сосна тунберга, Сосна японская черная
300288 Сосна торрея
300296 Сосна виргинская
300297 Гималайская высокая, Сосна веймутова гималайская, Сосна гималайская веймутова, Сосна гриффита
300298 Сосна ван
300305 Сосна юньнаньская
300323 Дерево перечное, Перец, Перец настоящий
300333 Матико, Перец узколистный
300335 Перец ланьюйский
300354 Бетель, Бетель перечный
300372 Перец мэнхайский
300373 Перец китайский
300377 Кубеба
300385 Перец эмэйский
300407 Перец хайнаньский
300446 Перец длинный
300450 Перец тибетский
300458 Кава, Кава-перец, Перец кава
300462 Дерево непалское
300464 Дерево перечное, Перец черный, Перец чёрный
300473 Перец украшенный
300481 Перец пинбяньский
300488 Перец мексиканский
300524 Перец лесной
300526 Перец симаоский
300528 Перец тайваньский
300552 Перец инцзянский
300554 Перец юньнаньский
300557 Перечные
300607 Пиптадения
300609 Пиптадения африканская
300621 Пиптадения обыкновенная
300660 Пиптант, Пиптантус
300674 Пиптантус мэнцзыский
300677 Пиптантус бобовниколистный
300688 Пиптант карликовый
300691 Пиптантус непалский
300700 Рисовидка
300707 Рисовидка алпийская
300712 Рисовидка федченко
300713 Рисовидка ферганская
300717 Рисовидка илларии
300722 Рисовидка тупая
300724 Рисовидка боковая
300727 Рисовидка широколистная
300728 Рисовидка боковая
300737 Рисовидка памироалайская
300740 Рисовидка цирокоцветная
300741 Рисовидка краснеющая
300746 Рисовидка согдийская
300747 Рисовидка джунгарская
300755 Рисовидка дрожащая
300756 Рисовидка замещаящая
300793 Пиптоптера
300794 Пиптоптера туркестанская
300974 Фисташка
300977 Фисташка атлантическая
300980 Фисташка китайская
300994 Дерево мастиковое, Дерево мастичное, Кустарник мастиковый, Фисташка мастиковая, Фисташка мастичная, Фисташка-лентискус
300998 Кевовое дерево, Скипидарник, Скипидарное дерево, Фисташка дикая, Фисташка туполистная, Фисташник
301002 Дерево скипидарное, Дерево терпентинное, Дерево терпентиновое
301005 Бузгунча, Дерево фисташковое, Фисташка настоящая
301021 Пистия, Филодендрон писция
301053 Горох
301055 Горох полевой, Пелюшка
301060 Горох оше
301061 Горох обыкновенный, Горох посевной
301063 Горох красивый
301070 Горох обыкновенный, Горох посевной, Горох садовый, Горошек зелёный
301074 Горох высокий
301076 Горох рриземистый
301084 Горох закавказский
301203 Пнттоспоровые
301207 Питтоспорум, Смолосемянник
301253 Смолосемянник дафнифиллюмовидный
301266 Смолосемянник обильноцведущий
301279 Смолосемянник оголеный
301291 Смолосемянник разнолистный
301308 Питтоспорум куэньминский
301309 Питтоспорум гуансиский
301310 Питтоспорум гуйцзоуский
301340 Питтоспорум ланьюйский
301344 Питтоспорум непалский
301352 Питтоспорум эмэйский
301356 Питтоспорум мелкоплодный
301413 Питтоспорум тобира, Смолосемянник тобира
301433 Питтоспорум волнистый, Смолосемянник волнистый
301610 Плагиантус
301619 Пладиобазис
301620 Пладиобазис васильковый
301623 Разноорешек, Разноцвет
301626 Разноорешек восточный
301692 Джефферсония
301694 Джефферсония сомнительная
301801 Дзелква, Дзелкова, Зельква, Планера
301804 Планера водная
301821 Подорожниковые
301832 Подорожник
301853 Подорожник альпийский
301858 Подорожник паутинистый
301864 Колосеница, Подорожник блошный, Подорожник песчаный, Семя блошное
301871 Подорожник азиатский
301894 Подорожник камчатский
301897 Подорожник седоватый
301909 Подорожник корнута
301910 Подорожник перистный
301945 Подорожник собачья голова
301952 Подорожник прижатый, Подорожник приземистый
301980 Подорожник горечавковый
302009 Подорожник индийский
302014 Подорожник японский
302024 Подорожник комарова

302025 Подорожник войлочный
302026 Подорожник пушистоголовый
302034 Подорожник ланцетный, Подорожник ланцетовидный, Подорожник ланцетолистный
302054 Подорожник лёфлинга
302062 Подорожник крупноплодный
302068 Подорожник большой
302091 Мальваструм, Подорожник морской, Подорожник приморский
302101 Подорожник большой
302103 Подорожник средний
302107 Подорожник маленький, Подорожник малый
302112 Подорожник горный, Подорожник чернеющий
302114 Подорожник замечательный
302119 Подорожник яйцевидный
302132 Подорожник многосемянный
302169 Подорожник шварценберга
302183 Подорожник степной
302196 Подорожник мелкоцветковый
302199 Подорожник тибетский
302229 Ивоцветник, Платановые
302241 Лимнорхис, Любка, Перулярия
302260 Любка дальневосточная, Любка двулистная, Любка ложнодвулистная, Фиалка ночная
302280 Любка зеленоцветковая, Любка зеленоцветная
302290 Любка бычий рог
302326 Любка желтая
302331 Любка фрейна
302355 Лимнорхис цельногубый, Любка цельногубая
302377 Любка японская
302380 Любка гуансиская
302391 Любка лицзянская
302470 Любка офрисовидная
302502 Любка сахалинская
302513 Перулярия буреющая, Тулотис буреющий
302531 Любка тайваньская
302536 Любка комарниковая
302549 Перулярия уссурийская, Тулотис уссурийский
302574 Платан
302578 Платан клиновидный
302580 Платан пальчатый
302582 Платан испанский, Платан кленолистный, Платан клёнолистный
302586 Платан линдена
302587 Платан мексиканский
302588 Платан американский, Платан западный, Сикамор, Чинар американский, Чинар восточный, Чинар западный, Явор восточный
302592 Платан востока, Платан восточный, Платан обыкновенный, Чинар, Чинар восточный, Чинар кавказский, Явор восточный
302616 Платония
302622 Платония
302674 Платикария
302684 Платикария шишконосная
302718 Биота, Платикладус
302721 Биота восточная, Дерево жизненное, Платикладус восточный, Туя, Туя восточная
302734 Платикладус восточный 'Зиболидин'
302747 Валенбергая, Платикодон, Ширококолокольчик
302753 Валенбергая крупноцветная, Колокольчик китайский крупноцветный, Колокольчик крупноцветный, Колокольчик японский крупноцветный, Платикодон крупноцветконый, Платикодон крупноцветный, Ширококолокольчик крупноцветковый
302860 Платимисциум, Роза порицветковая
302961 Платистемон
302997 платитения бухарская
302999 платитения обедненная
303000 платитения разнозубая
303002 платитения комарова
303010 платитения памирская
303011 платитения бедренецонодобная
303013 Платитения рубцова
303125 Шпороцветник
303324 Крапивна
303994 Плейобластус
304040 Многобородник низкий
304051 Плейобластус гуансиский
304098 Арундинария симони
304695 Плевропогон, Плеуропогон
304752 Реброплодник
304765 Реброплодник австлийский
304766 Реброплодник уральский
304772 Реброплодник камчатский
304793 Реброплодник лицзянский
304795 Реброплодник франше
304817 Гименолена малорослая, Реброплодник тяньшанский
304831 Реброплодник кандинский
304846 Болоздоплодник псостой
305072 Плюхея
305162 Зубница
305163 Зубница мелкоцветковая
305165 Плюмбаговые, Свинчатковые
305167 Зубница, Корень свинцовый, Плюмбаго, Свинцовка, Свинчатка, Свинчатник
305172 Свинчатка капская
305183 Свинчатка европейская
305185 Свинчатка индийская
305202 Плюмбаго цейлонское
305206 Плюмрия
305274 Арктофиля, Мятлик
305275 Мятлик укороченный
305285 Мятлик остростебельный
305291 Мятлик альберта
305300 Мятлик высокогорный
305303 Мятлик альпийский
305316 Мятлик алтайский
305328 Мятлик узколистный
305331 Мятлик узкочешуйный
305334 Мятлик однолетний
305357 Мятлик араратский
305362 Мятлик арктический
305364 Мятлик толмачева
305368 Мятлик аргунский
305372 Мятлик членистый
305380 Мятлик даурский, Мятлик кистевидный, Мятлик оттянутый
305383 Мятлик даурский
305389 Мятлик бактрийский
305395 Мятлик бакунский
305409 Мятлик кистевидный
305411 Мятлик прицветниковый
305417 Мятлик бухарский
305418 Мятлик каратегинский
305419 Мятлик луковичный
305422 Мятлик луковичный живолодящий
305428 Мятлик бирманский
305431 Мятлик красивый
305433 Мятлик кавказский
305436 Мятлик ше, Мятлик шэ
305451 Мятлик сжадый, Мятлик сплюснутый
305474 Мятлик густой
305475 Мятлик густейший
305490 Овсяица джимильская
305498 Мятлик джилгинский
305506 Мятлик выдающийся
305521 Мятлик федченко
305539 Мятлик заостренный, Мятлик ломкий
305543 Мятлик ганешина
305544 Мятлик ганешина
305548 Мятлик гладкоцветковый
305551 Ель бальфура, Мятлик бальфура, Мятлик сизый
305559 Мятлик сизостебельный
305585 Мятлик гиссарский
305589 Мятлик гиблидный
305593 Мятлик грузинский
305599 Мятлик слабый
305601 Мятлик заметный
305604 Мятлик иркутский

305605 Мятлик заливной
305623 Мятлик комарова
305636 Мятлик скальный
305637 Мятлик шерстистоцветковый
305643 Мятлик лауданский
305646 Мятлик горный, Мятлик рыхлый
305664 Мятлик липского, Мятлик ложноразбросанный, Мятлик сжатый
305665 Мятлик джунгарский
305673 Мятлик альберта, Мятлик литвинова, Мятлик литвиновский, Мятлик окаймленный
305675 Мятлик длиннолистный
305680 Мятлик обманчиный
305686 Мятлик крупночашечный
305705 Мятлик мягкоцветковый
305708 Мятлик окаймленный
305711 Мятлик марии
305718 Мятлик мазендеранский
305721 Мятлик майделя
305728 Мятлик мейера
305733 Мятлик монгольский
305738 Мятлик нальчикский
305741 Мятлик ломнодубравный
305743 Мятлик боровой, Мятлик дубравный, Мятлик лесной
305764 Мятлик непалский
305771 Мятлик невского
305776 Мятлик японский
305804 Мятлик болотный, Мятлик поздний, Мятлик трхцветковыйё, Мятлик шафранный
305806 Мятлик памирский
305820 Мятлик снежный
305830 Мятлик скальный
305838 Мятлик пинежский
305840 Мятлик кисточный, Мятлик плоскоцветковый
305847 Мятлик подольский
305855 Мятлик луговой, Мятлик торфяной
305890 Мятлик инееватый
305895 Мятлик ложноуколоченный
305901 Мятлик ложноболотный
305906 Мятлик ложнодрожащий
305916 Мятлик шероховатый
305922 Мятлик расставленный
305926 Мятлик ревердатто
305928 Мятлик ромбический
305940 Мятлик песчаный
305945 Мятлик саянский
305966 Мятлик щетинистый
305973 Мятлик сибирский
305982 Мятлик лесной
305984 Мятлик синайский
305995 Мятлик смирнова
305998 Мятлик сочавы
306003 Мятлик веретеновидный, Мятлик охходский
306027 Мятлик бесплодный
306033 Мятлик колючий
306038 Мятлик широкометельчатый
306045 Мятлик раскинутый
306050 Мятлик сычуаньский
306054 Мятлик таймырский
306059 Мятлик корейский
306060 Мятлик танфильева
306071 Мятлик тяньшанский
306074 Мятлик тибетский
306082 Мятлик забайкальский
306087 Мятлик дрожащий
306094 Мятлик траурный
306097 Мятлик подобный
306098 Мятлик обыкновенный
306109 Мятлик торфяной
306113 Мятлик урянхайский
306119 Мятлик уссурийский
306131 Мятлик разноцветный, Мятлик расползающийся
306133 Мятлик веретеновидный, Мятлик охотский
306136 Мятлик расползающийся
306139 Мятлик степной
306146 Мятлик зелененький
306153 Мятлик вороновб
306159 Мятлик запрягаева
306163 Мятлик чжундяньский
306168 Поацинум
306345 Ногоплодниковые
306395 Ногоплодник, Подокарпус
306399 Ногоплодник альпийский
306419 Подокарпус ланьюйский
306424 Ногоплодник высокий, Подокарп
306431 Подокарпус тайваньский
306457 Подокарпус китайский, Подокарпус крупнолистный
306469 Подокарпус крупнолистный маки
306493 Ногоплодник нагея, Подокарпус наги
306510 Подокарпус чилийский
306522 Подокарпус тотара
306609 Ноголист, Ноголистник, Подофилл, Подофиллум, Подофиллюм
306636 Ноголистник пальчатый, Подофилл щитовидный
306847 Бородатка, Погония
306870 Бородатка японская, Погония японская
306964 Погостемон
306994 Калина канадская, Пачули
307076 Пуанзеция
307186 Синюховые
307190 Полемониум, Синюха
307195 Синюха северная
307197 Синюха голубая, Синюха лазолевая, Синюха лазурная, Синюха обыкновенная
307209 Синюха остролепестная, Синюха остроцветкоая
307214 Синюха кавказская
307215 Синюха китайская, Синюха льноцветковая, Синюха рыхлоцветковя
307222 Синюха низкая
307229 Синюха крупная
307233 Синюха тихоокеанская
307234 Синюха мелкоцветковая
307236 Синюха ложно-красивая
307237 Синюха красивая
307243 Синюха сибирская
307246 Синюха опушенная
307265 Полиантес, Тубероза
307269 Полиантес тубероза, Тубероза
307290 Полиотирсис
307291 Полиотирсис китайский
307355 Поллиния
307509 Полиальтия
307724 Многоплодник
307748 Многоплодник индийский
307755 Многоплодник четырехлистный
307830 Хруплявник
307832 Хруплявник полевой
307839 Хруплявник геуффеля
307840 Хруплявник крупный
307842 Хруплявник многолетний
307844 Хруплявник бородавчатый
307891 Истод
307903 Истод альпийский
307905 Истод австрийский, Истод горький, Истод горьковатый
307909 Истод прелестнейший
307910 Истод анаторийский
307911 Истод андрахновидный
307919 Истод тайваньский
307927 Истод мелкоцветковый
307981 Истод кавказский
307995 Истод колхидский
307998 Истод хохлатый
308010 Истод меловой
308081 Истод китайский
308095 Истод хайнаньский
308104 Истод гогенаккера
308116 Истод гибридный
308123 Истод японский
308135 Истод кемуларии
308140 Истод симаоский
308160 Истод белоцветный
308163 Истод лицзянский
308181 Истод большой
308183 Истод мариаме
308195 Истод молдавский
308248 Истод мотыльковый
308271 Истод подольский

308284 Истод инееватый
308347 Истод сенега
308359 Истод сибирский
308386 Истод штокса
308388 Истод сванский
308394 Истод лежачий, Истод простертый
308397 Истод татаринова
308403 Истод тонколистный
308412 Истод закавказский
308430 Истод урартийский
308454 Истод обыкновенный
308468 Истод вольфганга
308474 Истодовые
308481 Гречиха, Гречишные
308493 Купена, Печать соломонова, Соломонова печать
308495 Купена заостреннолистная, Купена остолистная
308503 Купена аньхуйская
308529 Купена многоцветковая, Купена многоцветная
308532 Купена десулави
308544 Купена франше
308551 Купена гладная
308558 Купена широколистная
308562 Купена низкая, Купена приземистая
308565 Купена вздутая
308567 Купена обвертковая, Купена оберткoвая
308584 Купена широколистная
308588 Купена лэйбоская
308598 Купена максимовича
308607 Купена многоцветковая
308612 Купена туполистная
308613 Купена аптечная, Купена душистая, Купена лекарственная, Купена максимовича, Купена обыкновенная, Купена японская
308616 Купена аптечная
308626 Купена яйцевиднолистная
308631 Купена кавказская, Купена тонкоцветковая
308632 Купена кандинская
308639 Купена розавая
308640 Купена северцова
308641 Купена сибирская
308647 Купена узколистная
308659 Купена мутовчатая
308666 Купена чжэцзянская
308690 Горец, Гречих, Гречишка, Гречишник, Полигонум
308692 Горец укороченный
308695 Горец игольчатый
308696 Горец кисленький
308697 Горец кислый
308717 Горец аянский
308726 Горец лисохвостовидный
308729 Горец альпийский
308737 Горец амманиевидный
308739 Горец земноводный, Гречиха земноводная, Трава щучья, Утевник
308777 Горец узколистный
308780 Горец араратский
308782 Горец песчаный
308788 Горец близкий, Горец спорыш, Горец топотун
308791 Горец серебристый
308803 Горец ашерсона
308816 Буркун, Горец птичий, Горец разнолистный, Гречиха придорожная, Гречиха птичья, Гусатница, Спорыш, Травка-муравка
308832 Горец незамечаемый
308850 Горец бурораструбовый, Горец рыже-желтоватый, Горец темнорасторубовый
308885 Горец двухостный
308893 Горец змеиный, Горлец, Гречиха горлец, Гречиха-горлец, Змеевик, Змеиный горлец, Корень змеиный, Раковые шейки, Рачьи шейки, Шейка, Шейки раковые, Шейки рачьи
308922 Горец бунге
308924 Горец борнмюллера
308925 Горец самшитолистный
308946 Горец горовчатый
308950 Горец мякокрасный
308952 Горец каспийский
308965 Полигонум китайский
309001 Горец скальный
309015 Горец бухарский, Горец дубильный, Таран бухарский, Таран дубильный
309016 Горец спорышевидный
309017 Горец меловой
309038 Горец опадающий
309054 Горец раздельноцветковый, Горец расставленноцветковый
309056 Горец растопыренный
309078 Горец эллиптический
309088 Горец хвощевидный
309098 Горец выступающий
309119 Горец многоцветковый
309121 Горец многоистый
309157 Горец злаколистный
309186 Горец гиссарский, Таран гиссарский
309187 Горец хэнаньский
309192 Горец распростертый
309199 Горец водянойперец, Горец перечный, Горчак, Гречиха перечная, Лягушечья трава, Перец водяной, Трава лягушечья, Трава манная
309247 Горец имеретинский
309252 Горец вогнутоветвистый
309253 Горец промежуточный
309258 Горец запутанный
309261 Горец янаты
309262 Полигонум японский
309271 Горец сидниковидный
309281 Горец китайбела
309284 Гречиха корейская
309298 Горец льняной, Горец шероховатый, Горец щавелелистный, Гречиха развесистая, Гречиха узловатая, Гречиха щавелелистная
309329 Горец ленкоранский
309335 Горец лицзянский
309338 Горец иловатый
309343 Горец прибрежный
309345 Горец длиннощетинковый, Горец длиннощетинный
309353 Горец пышный, Таран пышный
309354 Горец ожинколистный
309356 Горец маака
309374 Горец маньчжурский
309376 Горец морской
309395 Горец малый
309421 Горец моллиеобразный
309442 Полигонум мелкоцветковое
309446 Горец черничный
309459 Горец крылатый, Горец непальский, Гречиха крылатая
309466 Горец красивый
309468 Горец узловатый, Гречиха развесистая, Гречиха узловатая
309478 Горец охотский
309479 Горец лаксманна, Горец раструбистый
309494 Горец восточный, Персикария восточная
309500 Горец амударьинский
309503 Горец остроплодный, Горец рея
309505 Горец тихоокеанский
309515 Горец памироалайский
309526 Горец приноготковидный
309531 Горец отклоненный, Горец тонкий
309564 Горец пронзеннолистный, Горец сквоснолистный
309570 Блошная трава, Горец обыкновенный, Горец почечуйный, Гречиха птичья, Почечуй, Почечуйная трава, Почечуйник, Райдваки, Спорыш, Трава блошная, Трава гемолойная, Трава почечуйная
309607 Горец подземковидный
309624 Горец дерновинный, Горец посумбу
309650 Горец подушкообразный
309690 Горец регеля
309697 Горец роберта
309711 Горец стрговидный
309723 Горец сахалинский, Гречих сахалинская, Гречиха сахалинская,

Полигонум сахалинский
309749 Горец самарский
309766 Горец синьцзянский
309767 Горец шугнанский
309772 Горец тернистый, Горец терновый
309777 Горец мелковистый
309789 Горец щетинистый
309792 Горец сибирский
309796 Горец болотный, Горец зибольда
309808 Горец джунгарский
309831 Горец жесткощетинистый
309837 Горец полушастый
309839 Полигонум далиский
309877 Горец тунберга
309889 Горец тимьянолистный
309891 Горец тибетский
309892 Горец тифлисский
309893 Горец красильный, Гречиха красильная
309909 Горец треугольноплодный, Горец щетинистоплодный
309911 Горец трехкрылоплодный
309926 Горец уссурийский
309935 Горец жилковатый
309946 Горец клейкий, Горец кленносный
309951 Горец клейкий
309954 Горец живородящий, Гречиха живородящая, Корень сердечный, Сердечный корень
309966 Горец вейриха
309978 Горец зейский
310007 Многокрыльник
310008 Многокрыльник панютина
310021 Полимния
310102 Многобородник, Полипогон
310109 Многобородник низменый
310112 Полевица гиссарская, Пятиостник гиссарский
310121 Многобородник приморский
310125 Многобородник монпелиенский, Многобородник монпелийский, Полипогон монпелийский
310129 Многобородник полумутовчатый
310677 Политаксис лемана
310678 Политаксис винклера
310755 Помадерис
310848 Понцирус
310850 Апельсин трехлисточковый, Лимон трехлисточковый, Лимон трёхлисточковый, Понцирус трехлистный, Понцирус трехлисточковый, Понцирус трёхлисточковый, Понцирус трифолиата, Трехлисточковый апельсин, Трифолиата
310873 Чусва однобокая
310923 Ятрышник эмэйский
310933 Ятрышник сычуаньский
310939 Ятрышник тайваньская
310990 Понцирус
311019 Понтедериевые, Понтедерия
311037 поповиокодония
311039 поповиокодония узкоплодная
311040 поповиокодония уэмуры
311131 Тополь, Туранга
311144 Тополь берлинский
311146 Тополь дельтовидный, Тополь канадский
311182 Тополь ганьсуский
311204 Тополь узбекистанский
311206 Тополь таджикистанский
311208 Белолистка, Тополь белый, Тополь серебристый, Тополь-белолистка
311226 Тополь болле, Тополь самаркандский, Тополь туркестанский, Тополь туркестанский пирамидальный
311229 Тополь амурсий
311231 Тополь каролинский, Тополь ребристый
311237 Тополь бальзамический
311259 Тополь крупнолистный, Тополь седой
311261 Тополь седеющий, Тополь сереющий, Тополь серый
311265 Тополь китайский
311278 Тополь цзилунская
311281 Тополь давида, Тополь китайская
311292 Дерево чёрное, Дерево эбеновое, Тополь дельтовидный, Тополь канадский, Тополь треугольнолистный, Тополь трехгранный
311308 Тополь восточно-персидская, Тополь евфратская, Тополь литвинова, Тополь разнолистная
311324 Тополь фремонта
311340 Тополь дэциньский
311349 Тополь хэбэйский
311355 Тополь илийская
311362 Тополь кандинская
311368 Тополь корейский
311369 Тополь краузе
311375 Тополь лавролистный
311389 Тополь максимовича
311398 Осокорь, Осокорь Тополь, Тополь осокорь, Тополь черный, Тополь чёрный, Черный Тополь
311400 Итальяский Тополь, Пирамидальный Тополь, Тополь итальяский, Тополь пирамидальный, Украинский Тополь
311416 Тополь снежнобелый
311420 Тополь памирский
311423 Тополь волосистый
311433 Тополь сизолистная
311465 Тополь саржента
311480 Тополь зиболида
311509 Бравоа, Тополь, Тополь душистый, Топольник
311524 Тополь густолиственный, Тополь таласский
311537 Осина, Осина американская, Тополь дрожащий, Тополь обыкновенный, Тополь осина
311547 Осиновик, Тополь осинообразный
311554 Тополь волосистоплодный
311557 Тополь печальный
311562 Тополь уссурийский
311579 Тополь ядунская
311736 Порфира
311817 Портулак
311852 Портулак декоративный, Портулак крупноцветконый, Портулак крупноцветный
311857 Портулак хайнаньский
311890 Портулак овошной, Портулак овышной, Портулак огородный
311954 Портулаковые
311956 Портулакария
312028 Рдест
312036 Рдест остролистный
312037 Рдест альпийкий
312075 Рдест взморниколистный, Рдест зостеролистный
312079 Рдест курчавый
312083 Рдест гребенчатый
312086 Гренадилья густая, Рдест густолистный, Рдест плотный
312090 Рдест отльчный
312123 Рдест фриса
312126 Рдест злаковый
312136 Рдест геннинга
312138 Рдест злаковидный, Рдест разнолистный
312141 Рдест хубэйский
312155 Рдест прерванный
312166 Рдест длиннолистный
312168 Рдест блестящий
312170 Рдест маака
312171 Рдест малайский
312190 Рдест плавающий
312194 Рдест узловатый
312199 Рдест туполистный
312206 Рдест заостренный, Рдест остролистный
312208 Рдест памирский
312220 Рдест пронзеннолистный, Рдест пронзённолистный, Рдест стеблеобъемлющий
312228 Рдест гречихолистный
312232 Рдест длиннейший, Рдест длинный
312236 Рдест маленький, Рдест малый,

Рдест палермский
312259 Рдест красноватый
312274 Рдест выемчатый
312282 Рдест волосовидный
312285 Рдест влагалищный
312290 Рдест вольфганга
312292 Рдест сичанский
312300 Рдестовые
312322 Курильский чай, Лапчатка, Потентилла
312325 Лапчатка бесстебельная
312327 Лапчатка железолистная
312333 Лапчатка аджарская
312337 Лапчатка белая
312350 Лапчатка анадырская
312351 Лапчатка подражающая, Лапчатка траншеря
312357 Лапчатка лежачая
312358 Лапчатка узколистная
312360 Лапка гусиная, Лапчатка гусиная, Трава гусиная
312378 Лапчатка сближенная
312387 Лапчатка песчаная, Лапчатка серая
312388 Лапчатка песчанистая
312389 Лапчатка серебристая
312390 Лапчатка серебристовидная
312399 Лапчатка сильно-шероховатая
312401 Лапчатка астраханская
312402 Лапчатка астрагалолистная
312406 Гомфрена, Лапчатка золотистая
312409 Лапчатка двухцветковая
312412 Лапчатка вильчатая, Лапчатка двувильчатая, Лапчатка двунадрезанная
312417 Лапчатка муркрофта
312418 Лапчатка вильчатая
312427 Лапчатка борнмюллера
312428 Лапчатка коротколепестая
312430 Лапчатка коротностебельная
312431 Лапчатка бунге
312433 Лапчатка каллье
312434 Лапчатка камиллы
312441 Лапчатка кавказскаю
312446 Лапчатка стозернышковая, Лапчатка стоплодная
312450 Лапчатка китайская
312457 Лапчатка азиатская, Лапчатка золотистая, Лапчатка золотистоцветковая
312467 Лапчатка сжатая
312472 Лапчатка крандца, Лапчатка кранца
312473 Лапчатка толстая
312478 Лапчатка криптомерневая, Лапчатка криптотениевая
312496 Лапчатка прижатая
312497 Лапчатка пустынностепная
312502 Лапчатка двуцветная, Лапчатка разноцветная
312505 Лапчатка чудесная
312510 Лапчатка эгеде
312516 Лапчатка высокая
312517 Лапчатка изящная
312518 Лапчатка арктическая, Лапчатка выемчатая
312520 Дубровка, Завязник, Калган, Калган дикий, Лапчатка прямая, Лапчатка прямостоячая, Лапчатка прямостоящая, Лапчатка-узик, Узик, Укроп
312530 Лапчатка эверсманна
312531 Лапчатка неодетая
312535 Лапчатка федченковская
312538 Лапчатка веерная
312540 Лапчатка плетевидная
312544 Лапчатка облиственная
312550 Лапчатка земляниковидная, Лапчатка земляничная
312561 Лапчатка земляниковидная
312565 Лапчатка фрейна
312606 Курильский чай маньчжуьрский
312630 Лапчатка холодная
312633 Лапчатка гравилатовидная
312644 Лапчатка сизоватая
312645 Лапчатка гольдбаха
312656 Лапчатка гейденрейха
312658 Лапчатка семилисточковая, Лапчатка тюрингенская, Лапчатка тюрингская
312659 Лапчатка амурская
312667 Лапчатка сплошь белая, Лапчатка сплошь-белая
312668 Лапчатка цельнолепестная
312669 Лапчатка распростертая
312674 Лапчатка черепчатая
312675 Лапчатка неблестящая
312676 Лапчатка седоватая
312683 Лапчатка средняя
312687 Лапчатка якутская
312688 Лапчатка яйлы
312690 Лапчатка клейна
312695 Лапчатка комалова
312696 Лапчатка крылова
312697 Лапчатка кулябская
312698 Лапчатка кузнецова
312700 Лапчатка веселая
312703 Лапчатка лапландская
312721 Лапчатка белолистная
312726 Лапчатка вейссенбургская
312727 Лапчатка веловолосистая
312731 Лапчатка липского
312732 Лапчатка ломакина
312734 Лапчатка клейкая
312736 Лапчатка длинноцветоножковая
312739 Лапчатка крупноцветковая
312742 Лапчатка мягкволосая
312745 Лапчатка матзумуры
312757 Лапчатка мейера
312758 Лапчатка крупноцветковая
312762 Лапчатка мелкоцветковая
312764 Лапчатка мелколистная
312782 Лапчатка мийябе
312783 Лапчатка мягчайшая
312797 Лапчатка многонадрезная, Лапчатка многоразрезная
312811 Лапчатка жилковатая
312812 Лапчатка весенняя, Лапчатка табернемонтана
312818 Лапчатка снежная
312827 Лапчатка нордманна
312828 Лапчатка норвежская
312836 Лапчатка нурская
312837 Лапчатка темная
312840 Лапчатка округлая
312841 Лапчатка восточная
312851 Лапчатка памирская
312852 Лапчатка памироалайская
312861 Лапчатка стоповидная
312886 Лапчатка волосистая
312888 Лапчатка бедренцеволистная
312889 Лапчатка пиндская
312896 Лапчатка красноцветковая
312905 Лапчатка красивенькая
312911 Лапчатка радде
312912 Лапчатка прямая
312922 Лапчатка регелевская
312925 Лапчатка ползучая, Цезальпиния дубильная, Цезальпиния душистая, Цезальпиния кожевенная, Цезальпиния красильная, Цезальпиния центрально-американская
312934 Лапчатка жестковатая
312944 Лапчатка морщинистая, Лапчатка уссурийская
312946 Кларкия, Лапчатка пачкающая, Лапчатка скалистая, Лапчатка скальная, Эухаридиум
312948 Лапчатка рупрехта
312955 Лапчатка кровохлебковая
312963 Лапчатка шренковская
312964 Лапчатка шура
312965 Лапчатка зейдлица
312967 Лапчатка полуобнаженная
312968 Лапчатка полунадлезная
312969 Лапчатка шелковая
312975 Лапчатка многораздельная
313003 Лапчатка соммье
313005 Лапчатка джунгарская
313007 Лапчатка грясная
313009 Лапчатка клинолистная
313013 Лапчатка шпренгеля
313024 Лапчатка далиская
313027 Лапчатка прилистниковая
313028 Лапчатка побегоносная

313029 Лапчатка репейничковидная, Лапчатка щетинистая
313035 Лапчатка почти-дланевидная
313036 Лапчатка серножлтая
313039 Лапчатка низкая, Лапчатка распростертая, Лапчатка стелющаяся
313050 Лапчатка сванетская
313051 Лапчатка шовица
313054 Лапчатка лабазниковидная, Лапчатка лябинколистная, Лапчатка пижмолистная
313062 Лапчатка кандинская
313064 Лапчатка крымская
313067 Лапчатка пепельно-белая
313074 Лапчатка пирамидкоцветковая
313075 Лапчатка тяньшанская
313077 Лапчатка толля
313081 Лапчатка закаспийская
313090 Лапчатка теневая
313091 Лапчатка одноцветковая
313093 Лапчатка валя
313098 Лапчатка мутовчатая
313103 Лапчатка мохнатая
313104 Лапчатка белеющая
313113 Лапчатка вибеля
313114 Лапчатка тибетская
313140 Черноголовник мохнатоплодный
313516 Прангос
313517 Прангос бесстельный
313518 Прангос бухарский
313520 Прангос узкоплодный
313522 Прангос федченко
313524 Прангос широкоольчатый
313526 Прангос липского
313527 Прангос складчатокрылый
313528 Прангос кормовой
313529 Прангос чимганский
313530 Прангос курчавокрылый
313778 Косогорник, Пренантес
313779 Косогорник пихтовый
313787 Пренантес белый
313791 Косогорник узколистный
313802 Косогорник недоснелколистный
313850 Косогорник максимовича
313864 Косогорник понтийский
313866 Косогорник красный, Косогорник пурпуровый, Латук красный
313895 Косогорник татаринова, Пренантес татаринова
314083 Первоцвет, Примула
314086 Первоцвет абхазский
314106 Первоцвет холодный
314120 Первоцвет прелестный
314125 Первоцвет арктический
314143 Аурикула, Первоцвет-аврикула
314144 Первоцвет ушковатый
314147 Первоцвет бальджуанский
314156 Первоцвет байерна
314177 Первоцвет северный
314178 Примула мулиская
314221 Первоцвет карпатский
314242 Примула тэнчунская
314246 Примула чжундяньская
314259 Примула чжэньканская
314269 Первоцвет сердселистный
314271 Первоцвет кортусовидный
314280 Первоцвет клинолистный
314333 Первоцвет высокий, Примула высокая
314344 Первоцвет аньхуйский
314353 Первоцвет евгении
314356 Первоцвет отменный
314359 Первоцвет чэнкоуский
314365 Первоцвет мучнистолистный
314366 Первоцвет мучнистый, Примула мичнистая
314376 Первоцвет матсумуры
314388 Первоцвет федченко
314396 Первоцвет дудчатый
314404 Первоцвет извилистый
314449 Первоцвет галлера
314460 Первоцвет разноцветный
314490 Первоцвет ильинского
314499 Первоцвет цзиндунский
314501 Первоцвет промежуточныйторчащий
314508 Первоцвет японский, Примула японская
314514 Первоцвет иезский
314527 Первоцвет юлии
314542 Первоцвет кнорринг
314546 Первоцвет комалова
314548 Первоцвет кузнецова
314549 Первоцвет гуандунский
314550 Первоцвет гуйцзоуский
314595 Первоцвет длинногий
314599 Первоцвет длинностирелочный
314602 Первоцвет желтенький
314613 Примула мальвоподобная
314624 Примула максимовича
314637 Примула аньхуйская
314638 Первоцвет мейера
314646 Первоцвет маленький
314647 Первоцвет минквиц
314655 Примула мийябе
314670 Первоцвет муркрофта
314697 Первоцвет снежный
314700 Первоцвет туркестанская
314711 Первоцвет норвежский
314713 Первоцвет тяньшанский
314717 Первоцвет комнатный, Примула обконика
314722 Первоцвет фуцзяньский
314738 Примула мэнцзыская
314752 Мыльнянка коровья, Тысячеголов, Тысячеголовник
314756 Первоцвет ольги
314765 Первоцвет осетинский
314770 Первоцвет палласа
314772 Первоцвет памирский
314804 Первоцвет перистый
314862 Первоцвет порошковатый
314874 Примула цинхайская
314892 Первоцвет почколистный
314923 Первоцвет рупрехта
314925 Первоцвет сахалинский
314948 Первоцвет норвежский, Первоцвет поникающий, Первоцвет сибирский
314951 Первоцвет сибторпа
314952 Первоцвет отклоненный, Примула зибольда
314975 Первоцвет китайский, Примула китайская
314985 Примула фэнсянская
314991 Примула ядунская
315015 Первоцвет торчащий
315028 Примула сычуаньская
315030 Примула далиская
315049 Примула тибетская
315056 Первоцвет турнефора
315082 Баранчик, Ключики, Первоцвет аптечный, Первоцвет весенний, Первоцвет жёлтый, Первоцвет истиный, Первоцвет лекарственный, Примула аптечная, Примула настоящая
315083 Первоцвет крупночашечный
315101 Первоцвет бесстебельный, Первоцвет обыкновенный, Примула бесстебельная, Примула обыкновенная
315111 Примула ван
315113 Первоцвет варшеневского
315128 Первоцвет снизу-желтый
315138 Аврикуляриевые, Первоцветные
315154 Капуста кергуеленская
315171 Принсепия, Принцепия
315174 Принцепия тайваньская
315176 Принсепия китайская, Принцепия китайская
315246 Прионотрихон
315376 Притчардия
315430 Пробосцидеа, Пробосцидея
315434 Пробосцидеа луизианская
315436 Мартиния душистая
315547 Мескито, Мимозка, Прозопис
315551 Альгароба, Дерево москитовое, Мимозка серёжкоцветная, Прозопис серёжкоцветный
315560 Альгароба, Дерево москитовое, Мимозка серёжкоцветная, Прозопис серёжкоцветный
315702 Протея, Псевдобактерия

315937 Протея медоносная
316007 Протейные
316095 Селагинелловые, Селягинеллевые, Черноголовка
316104 Черноголовка крупноцветковая, Черноголовка крупноцветная
316112 Черноголовка средняя
316116 Черноголовка белоцветковая, Черноголовка разрезная
316127 Лойник, Черноголовка обыкновенная
316166 Прунус, Слива, Терновник
316206 Слива американская
316219 Слива узколистная, Слива чикаса
316220 Ватсония, Слива американская, Слива ватсона, Слива песчаная
316294 Алыча, Вишнеслива, Мирабель, Миробалан, Слива алыча обыкновенная, Слива алыча растопыренная, Слива вишнелистная, Слива дикая, Слива растопыренная
316318 Вишня мараскиновая, Вишня-гриот
316320 Вишня кислая
316333 Венгерка итальяская, Ренклод, Слива ленклод
316356 Слива курдистанская
316382 Венгерка, Слива венгерка, Слива домашная, Слива европейская, Слива обыкновенная, Угорка, Чернослив
316391 Венгерка итальяская, Ренклод, Слива ленклод
316416 Слива ферганская
316434 Слива изящная
316436 Прунус ланьюйский
316470 Слива малкая черная, Слива настоящая, Слива ненасточщая, Слива тернослива, Слива чёрная мелкая, Тернослив
316488 Прунус чжэцзянский
316518 Лавровишня лузитанская, Лавровишня португальская
316543 Слива морская
316553 Прунус мексиканский
316583 Слива дикая гусиная
316710 Вишня седая
316735 Вишня карликовая, Вишня песчаная
316761 Корейская слива, Прунус ивовый, Слива иволистная, Слива ивообразная, Слива китайская, Слива японская
316809 Слива зибольда
316813 Абрикоссовая слива, Слива абрикосовая, Слива абрикосовидная, Слива симона
316819 Слива колючая, Слива терн, Терн, Тёрн, Терн дикий, Терн терновник, Терновник
316833 Слива полусердцевидная
316884 Миндаль трёхлопастный, Слива трёхлопастная
316900 Слива уссурийская
316918 Вишня виргинская, Черемуха виргинская, Черёмуха виргинская
317006 Тескголюб
317009 Тесколюб седоватый
317012 Тесколюб щетинолистный
317073 Ломкоколосник
317077 Ломкоколосник дагестанский
317078 Ломкоколосник ломкий
317081 Колосняк ситниковый
317083 Ломкоколосник кроненбунга, Ячмень кроненбунга
317084 Ломкоколосник шерстистый
317086 Ломкоколосник скальный
317245 Псевдоерантемум
317290 Лжепустынноколосник
317291 Лжепустынноколосник северцова
317344 Гулявник ядовитый, Резушка стрелолистная
317414 Ложнобецкея
317415 Ложнобецкея кавказская
317485 лжерыжик
317487 лжерыжик шовица
317558 Ложноклаусия
317559 Ложноклаусия тончайшая
317560 Ложноклаусия щетинистоволосистая
317561 Ложноклаусия мягайшая
317563 Ложноклаусия сосочковая
317564 Ложноклаусия чимганская
317565 Клаусия туркестанская, Ложноклаусия туркестанская
317602 Айва китайская, Хеномелес китайский
317635 Лжетайник
317636 Лжетайник хориса
317790 Псевдохандения
317791 Псевдохандения зонтичная
317821 Лжелиственница, Лиственница ложная, Ложнолиственница
317822 Лжелиственница китайская, Лжелиственница приятная, Лжелиственница кемпфера
317839 псевдолинозирис
317842 псевдолинозирис гримна
317843 псевдолинозирис мелкоголовчатая
317844 псевдолинозирис синтениса
317907 Вероника алатавская
317909 Вероника даурская
317912 Вероника бедовойлочная, Вероника седая
317924 Вероника льнянколистная
317927 Вероника длиннолистная
317940 Вероника перистая, Вероника прекрасная
317947 Вероника корейская
317964 Андреев крест, Вероника колосистая, Крест андреев
317966 Вероника ложная, Вероника ненастоящая
317968 Вероника почни-сидячая
317987 неуструевия
318209 Левкорхис, Леукорхис
318210 Леукорхис беловатый
318277 Псевдосаза
318283 Арундинария приятная
318284 Псевдосаза фуцзяньская
318299 Псевдосаза хайнаньская
318303 Арундинария хайндса
318305 Лжесаса лушаньская
318307 Бамбук японский, Лжесаса японская, Псевдосаза японская, Саза японская
318388 Ложноочиток
318390 Очиток альберта
318391 Ложноочиток бухарский
318392 Ложноочиток колокольчиковый
318393 Ложноочиток плотный
318394 Ложноочиток федченко
318395 Ложноочиток ферганский
318396 Ложноочиток каратавский
318397 Ложноочиток ливена, Пупочная трава
318398 Ложноочиток длиннозубчатый
318399 Ложноочиток многостебельный
318432 лжеочток карсский
318489 Звездчаточка
318493 Звездчаточка давида, Крашенинниковия давида
318496 Звездчаточка разноцветковая
318501 Звездчаточка разнолистная
318506 Звездчаточка японская, Крашенинниковия японская
318507 Звездчаточка разноцветковая, Крашенинниковия максмовича
318513 Звездчаточка скальная, Крашенинниковия скальная
318515 Звездчаточка лесная, Крашенинниковия лесная
318517 Звездчаточка тибетская
318562 Дугласин, Дугласия, Лжетсуга, Псевдотсуга
318574 Дугласия сизая, Лжетсуга сизая
318575 Лжетсуга японская, Псевдотсуга японская
318579 Лжетсуга крупноплодная, Лжетсуга крупношишечная
318580 Дугласин, Дугласия тисолистная, Дугласия тиссолистная, Лжетсуга мензиса, Лжетсуга тисолистная, Лжетсуга тиссолистная, Пихта дугласова,

Псевдотсуга
318602 Псевдотсуга сичанская
318610 лжепузырник
318612 лжепузырник пальчатый
318733 Гуава, Гуайава, Гуайява, Гуаява, Гуйава, Гуйява, Псидиум
318737 Прибрежный
318740 Псидиум лавролистный
318742 Гойяба, Псидиум гваява
318747 Псидиум лавролистный
319062 Мелкохвостник
319064 Мелкохвостник остистый
319114 Бобы индийские
319116 Псолалея
319139 Псолалея смолоносная
319392 Психотрия
319477 Психотрия ланьюйская
319532 Психотрия тибетская
319569 Психотрия хайнаньская
319671 Психотрия люцюская
319810 Психотрия красная
319929 Мелколепестник карадагский
319930 Мелколепестник понсэ
319996 Подоложникоцветник
319997 Подоложникоцветник двсторонний
320000 Подоложникоцветник тонкоколосый
320001 Подоложникоцветник колосовидный
320002 Подоложникоцветник суворова
320032 Птероксилон
320065 Вязовик, Кожанка, Птелея
320068 Птелея желтоватая
320071 Вязовик трёхлистный, Кожанка, Кожанка трёхлистный, Птелея трехлистная, Птелея трёхлистная, Птелея трехлисточковая
320155 Крылоцветник
320156 Крылоцветник вильчатый
320258 Каркас крылатный, Птерокактус, Птероцелтис
320269 Падук, Птерокарпус
320288 Птерокарпус дальбергиевидный, Птерокарпус палисандровидный
320291 Дерево сандаловое антильское, Птерокарпус драконовый
320301 Дерево индийское розовое, Птерокарпус индийский, Сандал индийский
320306 Птерокарпус бирманский, Птерокарпус крупноплодный
320327 Дерево сандаловое, Дерево сандальное, Птерокарпус сандаловый, Птерокарпус санталинус, Птерокарпус санталовый, Сандал красный
320330 Птерокарпус кабонский, Птерокарпус сойакси, Рапс, Сабаль
320346 Крылоолешина, Крылоолешник, Лапина, Птерокария
320356 Лапина кавказская, Лапина крылатоплодная, Лапина ясенелистная, Орех кавказский, Птерокария рябинолистная
320358 Лапина хубэйская
320371 Лапина ледера
320372 Лапина сумахолистная
320377 Лапина узкокрылая
320411 Каркас крылатный
320412 Крылатый каркас татаринова
320421 Перистоголовник, Птероцефалюс
320423 Птероцефалюс афганский
320442 Птероцефалюс перистый
320511 Латук высокий
320515 Латук индийский
320520 Латук радде
320523 Латук треугольный
320827 Птероспермум
320828 Птероспермум кленолистный
320839 Птероспермум цзиндунский
320878 Птеростиракс
320880 Птеростиракс щщитковый
320885 Птеростиракс мелкоцветковый
320886 Птеростиракс щетинистый
320927 Кравфурдия вьющаяся
321004 Птилагростис
321008 Птилагростис красивый
321016 Птилагростис монгольский
321087 Птилотрихум
321089 Птилотрихум седоватый, Птилотрихум удлиненный
321099 Птилотрихум тибетский
321161 Птихосперма
321220 Бескильница
321226 Бескильница алтайская
321228 Бескильница суженная
321230 Бескильница неравноветвистая
321233 Бескильница луковичная
321236 Бескильница корейская
321237 Бескильница хорезмская
321240 Бескильница свернутолистная
321246 Бескильница раскидистая
321248 Бескильница расставленная, Бескильница сизая
321270 Бескильница удлиненночешуйная
321286 Бескильница коленчатая
321287 Бескильница крупная
321289 Бескильница сизая
321290 Бескильница гроссгейма
321291 Бескильница цзилунская
321295 Бескильница гаккелевская, Бескильница гаккеля
321296 Бескильница гаупта, Бескильница гауптовская
321298 Бескильница приземистая
321302 Бескильница илийская
321307 Бескильница енисейская
321309 Бескильница камчатская
321324 Бескильница курильская, Бескильница обедненая
321332 Бескильница крупнольниковая
321334 Бескильница большенот
321336 Бескильница морская, Пуцинелла приземистый
321343 Бескильница маловетвистая, Бескильница немноговетвистая
321357 Бескильница памирская
321358 Бескильница немноговетвистая
321361 Бескильница ползучая
321369 Бескильница пестроцветная
321377 Бескильница рожевица, Бескильница рожевицовская
321380 Бескильница шишкина
321382 Бескильница жесткая
321383 Бескильница севангская
321393 Бескильница колосовидная, Мятлик горбунова
321394 Бескильница шведская
321396 Бескильница аляскинская, Бескильница гладгая, Бескильница тонгая
321397 Бескильница тонкоцветковая
321401 Бескильница тончайшая
321404 Бескильница тяньшанская
321405 Бескильница ваханская
321419 Кудзу, Пуерария, Пуэрария
321441 Кудзу, Кудзу лопастная, Лиана верандная, Пуерария волосистая, Пуерария лопастная, Пузрария опушенная, Пуэрария волосистая, Пуэрария лопастная, Пуэрария опушенная, Пуэрария опушённая
321457 Пузрария эмэйская
321464 Пузрария мелкоцветковая
321509 Блошница, Пуликария
321554 Блошница дизентерийная, Зелье солодкое, Зелье солодьковое
321561 Блошница сушеницевидная
321595 Блошница обыкновенная, Блошница простертая, Блошница стелющаяся
321598 Блошница шалфеелистная
321607 Блошница болотная
321619 Блошница обыкновенная
321633 Лёгочница, Медуница, Трава легочная, Трава лёгочная
321636 Медуница узколистная
321639 Медуница мягкая, Медуница мягчайшая
321640 Медуница горная, Медуница красная
321641 Медуница неясная, Медуница

тёмная
321643 Медуница аптечная, Медуница лекарственная, Трава медовая
321646 Медуница красная
321651 Медуница филярского
321653 Прострел, Сон-трава
321655 Прострел аянский
321656 Прострел албанский
321658 Ветреница альпийская
321659 Прострел сомнительный
321661 Прострел армянский
321662 Прострел золотистый
321663 Прострел бунге
321665 Прострел колокольчатый
321667 Прострел поникающий
321672 Прострел китайский
321676 Прострел даурский
321679 Прострел грузинский
321681 Прострел крупный
321682 Прострел галлера
321688 Прострел костычева
321692 Прострел горный, Сон-трава горная
321696 Прострел чернеющий
321703 Прострел луговой раскрытый, Прострел поникающий, Прострел пониркший, Прострел раскрытый, Прострел распростёртый, Прострел сон-трава, Сон-трава
321707 Прострел желтелющий
321708 Прострел многонадрезной, Прострел многонадрецный, Прострел многораздельный
321713 Прострел сахалинский
321714 Прострел сукачева
321715 Прострел тонколопастный
321716 Прострел турчанинова
321720 Прострел фиолетовый
321721 Прострел грубoтерстистый, Прострел луговой раскрытый, Прострел обыкновенный, Прострел поникающий, Прострел раскрытый, Прострел распростёртый, Прострел сон-трава, Пульсатилла, Сон-трава
321763 Гранат, Гранатник
321764 Анар, Гранат, Гранат обыкновенный, Гранатник, Дерево гранатовое
321777 Гранатовые
321892 Пушкиния
321893 Пушкиния гиацинтовидная
321896 Пушкиния пролесковидная
321952 Мята горная, Пикнантемум
322165 Пикностельма
322179 Ситовник
322196 Ситовник чжэцзянский
322198 Ситовник колхидский
322223 Ситовник желтоватый
322233 Ситовник шаровидный
322271 Ситовник коржинского
322282 Ситовник лицзянский
322304 Ситовник нильгирийский
322323 Ситовник луговой
322336 Ситовник ремана
322342 Сыть краснопятнистая
322374 Ситовник дрожащий
322448 Пираканта
322450 Пираканта узколистная
322454 Пираканта колючая, Пираканта красная, Пираканта ярко-красная
322458 Пираканта мелкогородчатая
322459 Пираканта мелкогородчатая ганьсуская
322465 Пираканта городчато-зубчатая
322640 Златоцвет, Золотоцвет, Пиретрум, Поповник, Ромашник
322641 Пиретрум абротанолистный, Пиретрум абротанолистыый
322644 Пиретрум алатавский
322646 Пиретрум альпийский
322650 Пиретрум аррасанский
322654 Пиретрум оше
322655 Пиретрум бальзамоносновидный
322656 Пиретрум бальзамический
322658 Пиретрум красный, Пиретрум мюсокрасный, Пиретрум розовый, Пиретрум ярко-красивый, Ромашка кавказская, Ромашка персидская розовая, Ромашка розовая
322660 Далматская ромашка, Пиретрум красный, Пиретрум мюсокрасный, Пиретрум розовый, Пиретрум цинерариелистный, Пиретрум ярко-красивый, Поповник цинерариелистный, Ромашка далматская, Ромашка кавказская, Ромашка пепеленолистная, Ромашка персидская розовая, Ромашка розовая
322664 Пиретрум клюзия
322665 Пиретрум красный, Пиретрум мюсокрасный, Пиретрум розовый, Пиретрум ярко-красивый, Ромашка кавказская, Ромашка персидская розовая, Ромашка розовая
322669 Пиретрум щитковидный, Пиретрум щитковый
322675 Пиретрум дагестанский
322679 Пиретрум джилгинский
322683 Пиретрум кустарничковый
322685 Пиретрум гали
322688 Пиретрум железконосный
322689 Пиретрум гроссгейма
322690 Пиретрум хельдрейха
322691 Пиретрум гиссарский
322693 Пиретрум карелина
322697 Пиретрум келлера
322698 Пиретрум кочи
322699 Пиретрум крылова, Пиретрум крыловский
322700 Пиретрум кубинский
322701 Пиретрум шерстистый
322704 Пиретрум тонколистный, Пиретрум эдельвейсовиднй
322709 Кануфер, Колуфер, Пиретрум большой
322713 Пиретрум мариона
322714 Пиретрум микешина
322717 Пиретрум незамеченный
322723 Пиретрум гваюлолистный, Пиретрум девичьелистный
322724 Златоцвет девичий, Пиретрум девичий, Поповник девичий, Ромашка девичья, Трава медвежья
322729 Пиретрум горичнниколистшый
322730 Пиретрум красивый
322731 Пиретрум точечный
322732 Пиретрум пиретровидный
322741 Пиретрум семенова
322742 Пиретрум шелковистый
322745 Пиретрум джунгарский
322746 Пиретрум рябинолистный
322750 Пиретрум тяньшанский
322751 Пиретрум заилийский
322755 Пиретрум волосолопастный
322775 Грушанка
322787 Грушанка копытнелистная
322790 Грушанка красная
322802 Грушанка тибетская
322804 Грушанка зеленоватая, Грушанка зеленоцветковая
322810 Грушанка алтайская
322814 Грушанка даурская
322826 Грушанка далиская
322830 Грушанка крупноцветковая
322842 Грушанка японская
322852 Грушанка средняя
322853 Грушанка малая
322870 Грушанка почковидная, Грушанка почколистная
322872 Грушанка круглолистная
322919 Грушанка сычуаньская
322932 Грушанка синьцзянская
322935 Грушанковые
323076 Груша
323086 Груша миндалевидная
323095 Груша среднеазиатская
323107 Груша баланзы
323110 Груша берёзолистная
323112 Груша буассье
323116 Груша кальри
323129 Груша кавказская
323133 Груша дикая, Груша лесная, Груша

обыкновенная, Дерево грушевое
323165 Груша лохолистная
323185 Груша гроссгейма
323194 Груша хэбэйская
323207 Груша тайваньская
323214 Груша коржинского
323246 Груша снежная
323251 Груша пашия
323259 Груша темноплодная
323268 Груша китайския, Груша китайския песчаная, Груша песочная
323274 Груша радде
323276 Груша разнолистая, Груша регеля
323281 Груша иволистная
323283 Груша иволистная
323294 Груша пильчатая
323306 Груша синьцзянская
323307 Груша сосновского
323311 Груша сирийская
323329 Груша туркменская
323330 Груша уссурийская
323352 Груша зангезурская
323450 Квамоклит
323542 Квассия
323561 Дерево квебраховое, Квебрахо, Квебрачо, Кебрачо
323563 Квебрахо лоленца
323580 Дуб
323599 Дуб острый
323611 Дуб острейший
323621 Дуб траволистный
323625 Дуб белый
323630 Дуб чужой
323635 Дуб чужой остропильчатый
323649 Дуб ольхолистный
323707 Дуб двухцветный, Дуб двуцветный
323732 Дуб мирбека
323737 Дуб каштанолистный
323745 Валонея, Дуб австлийский, Дуб бургундский, Дуб турецкий
323762 Дуб золоточешуйчатый, Дуб калифорнийский вечнозелёный
323769 Дуб кермесовый, Дуб кермесоносный, Дуб хермесовый
323778 Дуб далиский
323780 Дуб кошенильный, Дуб пунцоволистный, Дуб шарлаховый, Дуб американский шарлаховый
323814 Дуб зубчатый, Дуб опушенный
323840 Дуб лицзянский
323876 Дуб эруколистный
323894 Дуб серповидный
323904 Дуб фэнчэнский
323917 Дуб густой, Дуб фрайнетто
323920 Дуб франше
323927 Дуб гамбелла
323996 Дуб гартвиса
324010 Дуб испанский, Дуб пробковый ложный
324012 Дуб хэбэйский
324020 Дуб золотистый
324025 Дуб грузинский
324027 Дуб вечнозелёный, Дуб каменный, Дуб падуболиский
324039 Дуб карликовый
324041 Дуб гонтовый, Дуб черепитчатый, Дуб черепичный
324043 Дуб имеретинский
324050 Дуб галловый
324079 Дуб гуншаньский
324086 Дуб козловского
324095 Дуб лавролистный, Тополь лавлолистный
324104 Дуб ливанский
324119 Дуб тибетский
324121 Дуб длиннолистный
324127 Дуб длинноножковый
324131 Дуб лэдунский
324134 Дуб лузитанский
324137 Дуб лиловидный, Дуб лилообразный
324140 Дуб мекедонский
324141 Дуб крупножелудевый, Дуб крупноплодный
324146 Дуб греческий, Дуб крупночешуйчатый
324155 Дуб мерилендский
324156 Дуб малипоский
324166 Дуб дубильный, Дуб каштановый, Дуб коровий, Дуб мишо
324173 Дуб монгольский
324190 Дуб курчатый
324201 Дуб горный
324225 Дуб черный, Дуб чёрный
324232 Дуб наттолла
324235 Дуб мексиканский
324247 Дуб пробковый западный
324262 Дуб болотный
324275 Дуб ножкоцветный
324278 Дуб зимний, Дуб сидячецветный, Дуб скалиный, Дуб-железняк, Дуб-зимняк, Зимняк
324282 Дуб ивовидный, Дуб иволистный
324283 Дуб мирзинолистный, Дуб филлиреевидный
324301 Дуб понтийский
324314 Дуб опушеный, Дуб пушистый
324319 Дуб пиренейский
324335 Дуб иволистный, Дуб летный, Дуб летный черешчатый, Дуб обыкновенный, Дуб черешчатый, Дуб черешчатый летний
324344 Дуб красный, Дуб северный, Дуб серповидный, Дуб серполистный
324383 Дуб мулиский
324409 Дуб фунинский
324443 Дуб звездчатый, Дуб малый
324453 Дуб пробковый
324467 Дуб тайваньский
324498 Дуб мекедонский
324532 Дуб изменчивый
324539 Дуб американский красильный, Дуб бархатистый, Дуб бархатный, Дуб красильный
324544 Дуб виргинский, Дуб каролинский
324558 Дуб воронова
324623 Квилайя, Квиллайя, Килайя
324624 Дерево мыльное, Квилайя мыльная, Квиллайя мыльная, Кора мыльная
324669 Квисквалис
325039 Радиола
325041 Радиола льновидная
325061 Раффлезия
325065 Рафлезиевые, Раффлезиевые
325185 Раяния
325223 Рамонда
325268 Рамондия
325494 Лютиковые
325498 Лютик, Ранупкул, Ранупкулюс
325504 Лютик абхазский
325515 Куросле п, Лютик едкий
325516 Лютик аконитолистный
325518 Куросле п, Лютик едкий
325535 Лютик острозубчатый
325536 Лютик острлопастный
325551 Лютик афганский
325553 Лютик агера
325555 Лютик алайский
325558 Лютик алеберта
325560 Лютик аллеманна
325562 Лютик альпийский
325563 Лютик алтайский
325571 Лютик амурский
325578 Лютик ветреницелистный
325580 Лютик водяной, Шелковник, Шелковник водяной
325608 Лютик паутинистый
325612 Лютик полевой
325614 Лютик азиатский
325627 Лютик золотолепестный
325628 Лютик золотистый
325636 Лютик бальджуанский
325656 Лютик бореальный, Лютик северный
325658 Лютик коротколопастный
325664 Лютик калабрийский
325665 Лютик бузе
325666 Лютик клубневой, Лютик клубненосный, Лютик луковичный, Носный

325686 Лютик буша
325714 Лютик кашубский
325715 Лютик кавказский
325718 Лютик китайский
325720 Лютик хиосский
325730 Лютик константинопольский
325740 Лютик толстолистный
325773 Лютик дихотомический
325774 Лютик кандинский
325777 Лютик лицзянский
325782 Лютик расширенный
325784 Лютик раздельный
325785 Лютик красивый, Лютик светлый
325788 Лютик обманчивый
325796 Лютик стройный
325800 Лютик эшшольца
325820 Лютик-чистяк, Чистяк весенний, Чистяк лютичный
325864 Лютик жгучий, Прыщенец, Прыщинец
325882 Лютик франше
325883 Лютик братский
325886 Лютик ледниковый
325898 Лютик ледниковый, Оксиграфис обыкновенный
325901 Лютик гмелина
325915 Лютик крупнолистный, Лютик крупнолистый
325920 Лютик нежелезистый
325926 Лютик елены
325960 Лютик гиперборейский
325964 Лютик гиперборейский
325971 Лютик иллирийский
325980 Лютик якутский
325981 Лютик японский
325989 Лютик близкий
326002 Лютик комарова
326003 Лютик копестдагский
326004 Лютик кочи
326005 Лютик краснова
326008 Лютик куэньминский
326009 Лютик куэньминский лэйбоский
326017 Лютик шерстевидный, Лютик шерстистовидный
326018 Лютик шерстевидтый, Лютик шерстистый
326020 Лютик лапландский
326021 Лютик волосистоплодный
326027 Лютик тонконосиковый
326033 Лютик линейнолопастной
326034 Курослеп болотный, Лютик длиннолистный, Лютик языковый, Лютик языколистный
326038 Лютик лойка
326039 Лютик окаймленный
326043 Лютик длиннолопастной
326067 Лютик крупноплодный
326069 Лютик мейнсхаузена
326078 Лютик мейера
326084 Лютик мелколистный
326093 Лютик крылова, Лютик однолистный, Лютик однолистый
326096 Лютик горный
326108 Лютик мягкоигольчатый, Лютик шиповатый
326113 Лютик борцовый
326114 Лютик плавающий
326116 Лютик неаполитанский
326118 Лютик дубравный, Лютик лесной
326128 Лютик снеговой
326133 Лютик тучный
326148 Лютик ужовниколистный
326154 Лютик горный
326156 Лютик восточный
326164 Лютик осетинский
326166 Лютик остроплодный
326172 Лютик палласа
326193 Лютик памира
326197 Лютик малозубчатый
326199 Лютик обедненный
326201 Лютик лапчатораздельный, Лютик сходный
326204 Лютик сходный
326206 Лютик стоповидный
326229 Лютик перисторассеченный
326237 Лютик плоскоплодный, Лютик плоскосемянный
326242 Лютик многоцветковый
326244 Лютик многолистный, Лютик многолистый
326245 Лютик многокореный, Лютик многокорневой
326248 Лютик попова
326253 Лютик ложноклубненосный
326254 Лютик мягколистный
326255 Лютик болотный
326261 Лютик изящный
326263 Лютик длинностебельный
326273 Лютик крошечный, Лютик пигмей
326277 Лютик радде
326278 Лютик укореняющийся
326281 Лютик изогнутый
326286 Лютик регелевский, Лютик регеля
326287 Лютик ползучий
326302 Лютик ползающий
326303 Лютик ползучий, Лютик простертый, Лютик распростертый
326310 Лютик жесткий
326317 Лютик красночашечный
326318 Лютик рыжечашечный
326324 Лютик ненецкий
326326 Лютик сардинский
326339 Лютик сарторнев
326340 Лютик ядовитый
326348 Лютик шафта
326361 Лютик северцова
326365 Лютик зибольда
326375 Лютик смирнова
326376 Лютик соммье
326377 Лютик джунгарский
326393 Лютик узколепестный
326397 Лютик стевена
326399 Лютик щетинковый
326405 Лютик окаймленный
326408 Лютик слабый
326411 Лютик серножелтый
326420 Ранупкулюс тайваньский
326422 Лютик бротеруса
326429 Лютик тонколопастной
326440 Лютик бугорчатый
326441 Лютик заилийский
326442 Лютик траутфеттера
326477 Лютик трехрассеченный
326481 Лютик туркестанский
326487 Лютик уссурийский
326494 Лютик волосистый
326496 Лютик вальтера
326497 Лютик ван
326504 Лютик чжундяньский
326513 Раулия
326523 Рапанея
326578 Редька
326601 Редька приморская
326609 Редька дикая, Редька полевная
326614 Редька носатая
326616 Редис, Редька европейская, Редька культурная, Редька летняя, Редька масличная, Редька обыкновенная, Редька огородная, Редька посевная
326629 Редька японская
326638 Рафия, Раффия
326639 Рафия мадгаскарская
326649 Пальма африканская, Пальма виная
326812 Редечник, Репник
326832 Репник многолетний
326833 Репник морщинистый
326995 Раувольфия
327001 Раувольфия кафрская
327058 Раувольфия змеевидная
327064 Раувольфия тайваньская
327067 Раувольфия седовадая
327069 Раувольфия мутовчатая
327070 Раувольфия хайнаньская
327093 Равенала
327095 Дерево путешественников, Равенала мадагаскарская
327194 Реомюрия, Хололахна
327199 Реомюрияская бадхызская
327201 Реомюрияская ладанниковая
327204 Реомюрияская кустарниковая
327212 Реомюрия кашгарская

327216 Реомюрияская кузнецова
327221 Реомюрияская амударьснская
327224 Реомюрияская персидская
327225 Реомюрияская отогнутая
327226 Хололахна джуграрская
327231 Реомюрия туркестанская
327234 Реомюрияская закирова
327344 Редька Редовския
327345 Редька Редовския двояко-перистая
327429 Ремания
327462 Рейхардия
327467 Рейхардия вильчатая
327524 Рейнекия
327525 Рейнекия мясистая
327548 Рейнвардтия
327670 Ремузация
327672 Ремузация тайваньская
327793 Резеда
327795 Резеда белая
327816 Резеда оше
327822 Резеда коротконогая
327823 Резеда вухарская
327855 Резеда шароплодная
327857 Резеда непахучая
327866 Резеда желтая, Резеда жёлтая
327879 Желтянка, Резеда желтеннькая, Резеда желтоватая, Резеда красильная, Трава желтая, Трава красильная, Трескучник, Церва, Черва
327887 Резеда мелкоплодная
327896 Резеда душистая, Резеда пахучая, Резеда цветмалиновый
327934 Резедовые
327958 Рестелла альберта
327963 Рестио
328328 Сахалинская гречиха
328345 Горец остроконечный, Горец тонкозаостренный, Горец японский
328507 Рагадиолюс
328512 Рагадиолюс съедобный
328513 Рагадиолюс опушенный
328543 Крушинные, Крушиновые
328592 Жестер, Жестёр, Жостер, Крушина, Крушинник
328595 Жестер вечнозеленый, Жостер вечнозеленый, Крушина вечнозелёная
328603 Крушина ольхолистная
328604 Крушина альпийская
328618 Жестер бальджуанский
328630 Крушина калифорнийская
328642 Жестер каролинский, Жестёр слабительный, Жостер каролинский, Жостер слабительный, Крушина слабительная, Крушина слабитьльная
328656 Жестер кожистолистный
328660 Жостер ребристый
328680 Жестер даурский, Жостер даурский, Крушина даурская
328688 Жестер прижатый
328689 Жестер диамантский, Жостер диамантский, Крушина диамантовая
328690 Жестер длиннолистный
328694 Жестер деревцо, Жостер красное деревцо
328700 Жостер тайваньский
328701 Крушина ломкая, Крушина ольховидная, Крушинник ломкий, Крушинник ольховидный
328719 Крушина крупнолистная
328720 Жостер хайнаньский
328731 Жосте хубэйский
328735 Жестер имеретинский
328737 Крушина красильная
328740 Буддлейя японская, Буддлея японская, Жосте японский, Крушина юпонская
328747 Жосте японский мелкоцветковый
328752 Жестер корейский
328753 Жестер гуансиский
328765 Крушина ливанская
328767 Жестер люцюский
328786 Жестер мелкоплодный
328788 Жестер крошечный
328812 Жестер палласа
328816 Жестер мелколистный
328838 Крушина американская, Крушина пурша
328842 Жостер росторна
328844 Крушина чжэцзянская
328846 Крушина скальная
328849 Крушина скалистая, Крушина скальная
328859 Жестер синтениса
328861 Жестер джунгарский
328862 Жестер лопатчатолистный
328874 Жестер красильный
328882 Жестер уссурийский, Крушина уссурийская
328883 Жостер полезный, Крушина китайская, Крушина полезная
328900 Жостер вильсона
328903 Жостер тибетский
328917 Рамфикарпа
329019 Рафидофитон
329020 Рафидофитон регеля
329056 Иглочешуйник
329068 Иглочешуйник индийский
329096 Иглочешуйник люцюский
329114 Иглочешуйник зонтичный
329169 Рапидофиллум, Рапидофиллюм
329173 Рапис
329176 Рапис высокий
329184 Рапис низкий, Рапис приземистый
329197 Большеголовник, Корень маралий, Рапонтикум
329205 Рапонтикум аулиеатинский
329217 Рапонтикум цельнолистный
329218 Рапонтикум каратавский
329222 Рапонтикум лировидный
329224 Рапонтикум наманганский
329225 Рапонтикум карликовый
329226 Рапонтикум блестящий
329227 Рапонтикум красивый
329231 Рапонтикум серпуховидный
329268 Реедия
329271 Реедия бразильская
329274 Реедия обыкновенная
329306 Ревень, Реум
329312 Ревень алтайский
329318 Ревень компактный, Ревень скученный
329319 Ревень сердцевидный
329320 Реум корейский
329323 Ревень дарвазский
329327 Ревень эмоди
329329 Ревень федченго
329341 Ревень гиссарский
329346 Ревень коржинского
329350 Ревень лицзянский
329351 Ревень лопастный
329352 Ревень блестящий
329353 Ревень крупноплодный
329356 Ревень максимовича
329361 Ревень низкий
329366 Ревень аптечный, Ревень китайский, Ревень лекарственный, Ревень московский, Ревень обыкновенный, Ревень русский
329368 Ревень восточный
329372 Ревень дланевидный, Ревень пальмовидный, Ревень пальчатый
329377 Ревень складчатый
329385 Ревень сетчатый
329386 Ревень волнистый
329388 Рапонтика, Ревень овощной, Ревень огородный, Ревень туполистный, Ревень черноморский
329389 Ревень бесстебельный
329391 Ревень смородинный
329392 Ревень скальный
329401 Ревень тангутский
329406 Ревень татарский
329407 Ревень тибетский
329408 Ревень туркестанский
329415 Ревень виттрока
329454 Погремок
329504 Погремок
329505 Погремок летний
329507 Погремок альпийский
329510 Погремок узколистный
329519 Погремок колхидский

329520 Погремок меловой
329521 Погремок петушиный гребешок
329525 Погремок ферганский
329527 Погремок весенний
329542 Погремок средиземноморский
329543 Погремок малый
329545 Погремок горный
329547 Погремок чернеющий
329549 Погремок олгз
329551 Погремок отклоненный
329552 Погремок гребенчатый
329553 Погремок понтийский
329556 Погремок румельский
329559 Погремок сельский
329560 Погремок сахалинский
329562 Погремок шишкина
329565 Погремок джингарский
329569 Погремок шиловидный
329683 Кактус пинсалис, Рипсалис
329738 Ризоцефалус
329739 Ризоцефалус туркестанский
329744 Древокорень, Ризофора
329745 Ризофора остроконечная
329761 Дерево мангровое, Древокорень мангле, Шандал-дерево
329826 Родиола
329829 Родиола холодная
329836 Родиола комарова
329839 Родиола арктическая
329842 Родиола дэциньская
329845 Родиола северная
329876 Родиола ганьнаньская
329877 Родиола холодная
329881 Родиола разнозубчатая
329894 Родиола ганьсуская
329896 Родиола кашгарская
329897 Родиола кириллова, Родиола линейнолистная
329901 Родиола куэньминская
329902 Родиола лицзянская
329905 Родиола литвинова
329917 Родиола памироалайская
329923 Родиола перистонадрезная
329930 Очиток четырехчленный, Родиола четырехчленная, Родиола ярко-красная
329932 Родиола прямостебельная
329935 Золотой корень, Корень розовый, Родиола розовая, Родиола темнокрасная, Родянка
329939 Родиола розовая мелколистная
329944 Родиола сахалинская
329950 Родиола семенова
329960 Родиола стефана
329961 Родиола стефана
329967 Родиола тибетская
329972 Родиола зеленоватая
330017 Азалея, Азалия, Рододендрон
330023 Рододендрон адамса
330031 Рододендрон мулиский
330058 Рододендрон альбрехта
330097 Рододендрон аньхуйское
330110 Рододендрон древовидный
330111 Рододендрон древовидный
330113 Рододендрон древесный розовый
330182 Кашкара, Кашкарник золотистый, Рододендрон золотистый
330240 Рододендрон короткоплодный
330267 Рододендрон бирманский
330276 Рододендрон календуловидный
330321 Рододендрон канадское
330331 Рододендрон катобский
330334 Рододендрон кавказский
330467 Рододендрон непалский
330495 Рододендрон даурский
330524 Рододендрон декоративный
330546 Рододендрон делаве
330591 Рододендрон тэнчунское
330592 Рододендрон разноцветный
330597 Рододендрон мотоское
330673 Рододендрон кандинский
330695 Рододендрон ржаволистный, Рододендрон ржавый
330722 Рододендрон тайваньский
330785 Рододендрон гуншаньский
330794 Рододендрон гриффита
330797 Рододендрон гуаннаньский
330798 Рододендрон гуйхайский
330799 Рододендрон гуйчжунский
330800 Рододендрон гуйцзоуский
330815 Рододендрон хайнаньский
330851 Рододендрон хэнаньский
330875 Рододендрон жестковолосистый
330876 Рододендрон тибетский
330886 Рододендрон ху
330891 Рододендрон ийхунаньск
330907 Рододендрон подъельниковый
330917 Азалия индийская, Рододендрон индийский
330939 Рододендрон увлажненный
330957 Рододендрон яванское
330961 Рододендрон цзиньпинский
330966 Рододендрон кемпфера
330989 Рододендрон камчатский
331018 Рододендрон кочи
331026 Рододендрон гуансиский
331031 Рододендрон гуандунский, Рододендрон фэнкайский
331057 Рододендрон лапландский
331073 Рододендрон ледебура
331079 Рододендрон лэйбоский
331161 Азалия жёлтая, Азалия понтийкая, Понтийсая азалия, Рододендрон желтый, Рододендрон жёлтый, Рододендрон понтийский
331194 Рододендрон малипоский
331207 Рододендрон большой, Рододендрон древовидный, Рододендрон крупнейший
331221 Рододендрон мэнцзыский
331238 Рододендрон мяньнинский
331240 Рододендрон мелкоцветковое
331257 Азалея моллис, Рододендрон мягкий, Рододендрон мяткий
331258 Рододендрон японский
331283 Рододендрон мупинский
331284 Рододендрон остоконечный
331289 Рододендрон остроконечный
331305 Рододендрон тимиянолистный
331307 Рододендрон дэциньский
331314 Рододендрон наньпинский
331341 Рододендрон японский
331370 Рододендрон тупой
331449 Рододендрон мелколистный
331487 Рододендрон чаюйский
331492 Рододендрон пинбяньский
331520 Азалея желтая, Азалея понтийская, Рододендрон понтийский, Рододендрон флабум
331592 Рододендрон цинхайский
331623 Рододендрон сетчатый
331759 Рододендрон люцюский
331768 Рододендрон шлиппенбаха
331805 Рододендрон тимиянолистный белоцветковый
331821 Рододендрон жуйлиский
331823 Рододендрон сихотинский
331832 Рододендрон фуцзяньский
331839 Рододендрон симса
331858 Рододендрон смирнова
331910 Рододендрон таиландский
331911 Рододендрон сычуаньский
331921 Рододендрон далиский
331971 Рододендрон томсона
331980 Рододендрон динхуский
332018 Рододендрон чоноски
332038 Рододендрон унгерна
332101 Рододендрон вэньшаньский
332104 Рододендрон корейский
332127 Рододендрон сичанский
332142 Рододендрон яошаньский
332144 Рододендрон иедоский
332155 Рододендрон юньнаньский
332162 Рододендрон чжундяньский
332218 Родомирт
332282 Родотип, Родотипус, Розовик
332284 Родотип, Родотипус красивый, Розовик керриевидный, Розовик японский, Розовнк цепкий
332293 Рео
332434 Ропалостилис
332452 Желтинник, Кожевник, Сумах,

Шмальница
332477 Сумах душистый
332509 Сумах китайский, Сумах яванский
332529 Сумах копалловый, Сумах копаловый
332533 Дерево кожевенное, Сумах дубильный, Сумах кожевеный, Сумах коиевенный, Сумах красильный, Сумах обыкновенный, Сумах сицилийский
332608 Сумах гладкий, Сумах голый
332662 Сумах китайский, Сумах яванский
332766 Сумах пятилистниковый, Фитцройя, Фитцроя
332916 Дерево уксусное, Сумах душистый, Сумах коротковолосый, Сумах оленерогий, Сумах оленьерогий, Сумах пушистый, Сумах уксусный, Уксусное дерево, Улекс
333131 Ринхозия
333182 Ринхозия китайская
333286 Ринхозия куэньминская
333477 Очеретник
333480 Очеретник белый
333510 Очеретник кавказский
333518 Очеретник китайский
333560 Очеретник фабера
333568 Очеретник фори
333576 Очеретник бурый
333628 Очеретник японский
333893 Крыжовник, Смородина
333894 Крыжовник игольчатый, Крыжовник колючий
333904 Смородина альпийская, Смородина глухая, Смородина горная
333913 Смородина высочайшая, Черная кислица
333916 Крыжовник миссурийский, Смородина американская
333918 Смородина кислица, Смородина темнопурпуровая
333919 Смородина белая, Смородина декоративная, Смородина золотая, Смородина золотистая
333924 Смородина биберштейна
333929 Буреинский крыжовник, Крыжовник буреинский, Крыжовник дальневосточный, Смородина буреинская
333940 Смородина колорадская
333955 Смородина двуиглая, Смородина скальная
333957 Алданский виноград, Смородина дикуша
333970 Смородина пучковатая, Смородина японская
333978 Смородина тайваньская
333981 Смородина душистая
333982 Смородина франше
334006 Смородина пахучая
334008 Крыжовник гуншаньский
334017 Смородина разноволосая
334030 Крыжовник шиповниковидный
334033 Кислица, Смородина щетинистая
334034 Смородина ощестиненная, Смородина щетинистая
334048 Смородина янчевского
334054 Смородина колымская
334055 Смородина комарова
334061 Смородина широколистная
334071 Смородина душистая
334078 Смородина блестящая, Смородина светлая
334083 Смородина мальволистная
334084 Смородина маньчжуьрская
334087 Смородина максимовича
334089 Смородина максимовича
334094 Смородина мейера
334104 Смородина мупинская
334109 Смородина мулиская
334117 Смородина черная, Смородина чёрная
334127 Смородина душистая
334132 Смородина восточная
334137 Смородина канадская
334144 Смородина пальчевского
334145 Смородина бледноцветная
334147 Смородина малоцветковая
334149 Колокольчик карпатский, Смородина каменная, Смородина камневая, Смородина карпатская
334157 Смородина лежачая
334159 Смородина железистая
334163 Смородина пушистая
334166 Смородина красивая
334176 Смородина наньчуаньская
334179 Смородина колосковая, Смородина красная, Смородина обыкновенная, Смородина пушистая, Смородина скандийская, Смородина скандинавская
334188 Смородина сахалинская
334189 Смородина кровянокрасная, Смородина кровяно-красная
334201 Смородина обыкновенная
334203 Смородина двуиглая, Смородина каменная, Смородина скальная, Смородина таранушка
334206 Смородина сычуаньская
334214 Смородина колосистая, Смородина колосковая, Смородина красная, Смородина обыкновенная, Смородина пушистая, Смородина скандинавская
334223 Смородина колосковая, Смородина красная, Смородина обыкновенная, Смородина пушистая, Смородина скандинавская
334236 Смородина тяньцюаньская
334240 Смородина высочайшая, Смородина печалица, Смородина печальная, Черная кислица
334247 Смородина кубарчатая
334249 Смородина тссурийская, Смородина уссурийский
334250 Крыжовник европейский, Крыжовник культурный, Крыжовник обыкновенный, Крыжовник отклоненный, Смородина отклоненный
334260 Смородина волосистая
334263 Смородина кандинская
334270 Смородина тибетская
334310 Зантедешия, Рисовидка, Ричардия
334411 Дерево касторовое африканское, Рицинодендрон африканский
334432 Клещевина, Рицин, Рицинус
334435 Боб касторовый, Клещевина занзибарская, Клещевина обыкновенная, Рициа, Рицинник клещевина, Рициннус обыкновенный, Рицинус
334540 Риндера
334541 Риндера южно-ежистая
334542 Риндера бальджуанская
334543 Риндера кругомзубчиковая
334544 Риндера ежистая
334545 Риндера ферганская
334548 Риндера сплошь-опушенная
334549 Риндера коржинского
334551 Риндера
334552 Риндера продолговатолистная
334553 Риндера светрожелтая
334554 Риндера ошская
334555 Риндера четырехостая, Риндера четырехщитковая
334556 Риндера тяньшанская
334557 Риндера уркестанская
334558 Риндера зонтичная
334892 Ривина
334897 Ривина низкая
334946 Белая аказия, Робиния
334965 Акация щетинистая, Робиния щетинистоволосая, Робиния щетинистоволосистая, Робиния щетиноволосая
334974 Робиния новомексикская, Робиния нью-мексикская
334976 Акация американская, Акация белая, Белая аказия ложная, Белая акация, Лжеакация, Псевдоакация, Робиния лжеакация, Робиния ложноакация, Робиния ложно-акация, Рябина лжеакация, Рябина ложноакация
335013 Робиния клейкая

335103 Рохелия
335107 Рохелия бунге, Рохелия согнутая
335108 Рохелия колокольчатая
335109 Рохелия сердцевидночашечная
335110 Рохелия двусемянная
335113 Рохелия карсская
335114 Рохелия голоплодная
335115 Рохелия крупночашечная
335118 Рохелия длинноножковая
335119 Рохелия персидская
335183 Рэгнерия
335184 Рэгнерия аболина
335200 Колосняк алтайский, Рэгнерия алтайская
335204 Рэгнерия амурская
335230 Рута северная, Рэгнерия северная
335233 Рэгнерия короткожковидная
335246 Рэгнерия буша
335254 Рэгнерия килеватая
335255 Рэгнерия кавказская
335258 Рэгнерия ресничатая
335278 Рэгнерия смешанная
335284 Рэгнерия искривленная
335287 Рэгнерия чимганская
335295 Рэгнерия дробова
335307 Рута волокнистая, Рэгнерия волокнистая
335312 Житняк тайваньский, Рэгнерия тайваньская
335338 Рэгнерия гималайская
335359 Рэгнерия прерывчатая
335364 Рэгнерия жакемонта
335366 Рэгнерия якутская
335370 Рэгнерия японская
335389 Пырей комарова, Рэгнерия комарова
335392 Рэгнерия пушистолистная
335403 Рэгнерия тонкохвостная
335404 Рэгнерия длинноостая
335407 Рэгнерия длинноостая
335415 Рэгнерия крупнощетинистая
335416 Рэгнерия длиниохвостая
335440 Колосняк переменчивый, Рута узкочешуйная, Рэгнерия узкочешуйная
335453 Рэгнерия ошская
335454 Рута палермская, Рэгнерия палелмская
335461 Рэгнерия повислая
335472 Рэгнерия пушистая
335487 Рэгнерия саянская
335490 Рута скандинавская, Рэгнерия скандинавская
335491 Пырей шренковский, Рэгнерия шренка
335493 Рэгнерия шугнанская
335494 Рэгнерия жестколистная
335501 Рэгнерия шаньдунская
335510 Рэгнерия синьцзянская
335526 Колосняк тяньшанский, Рэгнерия тяньшанская
335529 Рэгнерия тибетская
335543 Рэгнерия гиблидная
335547 Рэгнерия турчаннова, Элимус гмелин
335557 Рута туруханская, Рэгнерия туруханская
335558 Рэгнерия угамская
335560 Рута уральская, Рэгнерия уральская
335563 Рэгнерия зеленочешуйная
335624 Ремерия, Рёмерия
335638 Ремерия гибридная, Рёмерия гибридная
335650 Глауциум отогнутая, Ремерия отогнутая
335758 Родея
335760 Родея японская
336096 Ронделетия
336166 Жерушник
336172 Жерушник земноводный, Хрен водяной
336173 Жерушник обоюдоострый
336182 Жерушник австрийский
336185 Жерушник короткоплодный
336187 Жерушник кантонский
336192 Жерушник догадовой
336200 Жерушник шаровидный
336204 Жерушник щетинистый
336211 Жерушник индийский
336237 Жерушник мелкосемянный
336250 Жерушник болотный, Жерушник исландский
336277 Жерушник лесной
336283 Роза, Шиповник
336291 Роза гаррисона
336320 Роза ацикулярис, Роза иглистая, Шиповник иглистый, Шиповник приятный
336343 Роза белая, Хлопчатник, Шиповник белый, Шиповник холмовый
336345 Роза альберта, Шиповник альберта
336347 Роза алексеенко
336348 Роза альпийская, Роза повислая, Шиповник альпийский
336352 Роза высокогорная
336353 Роза тупоушковая
336356 Шиповник андржьевского
336364 Роза пашенная, Роза полевая
336366 Роза банкса, Роза банксиевая, Роза бэнкс, Роза бэнкса, Роза ванксиева, Шиповник бэнкса
336380 Роза беггера, Шиповник беггера, Шиповник беггеровский
336382 Роза красивая
336385 Хультемия барбарисолистная
336402 Роза буасье
336403 Роза бурбонская
336405 Роза макартова
336408 Роза мускусная
336409 Роза бунге
336411 Роза афцелиуса, Шиповник афцелиуса
336419 Роза канина, Роза собачья, Шиповник дикий, Шиповник кустарниковый, Шиповник обыкновенный, Шиповник собачий, Шиповник щитконосный
336462 Роза каролинская
336471 Роза хвостатая
336474 Роза столистная, Роза столрепестная, Роза центифолия, Центифолия
336476 Роза моховая, Роза моховидная
336477 Роза бургундская
336481 Роза помпонная
336484 Роза чэнкоуская
336485 Роза бенгальская, Роза индийская, Роза китайская, Роза чайная, Роза чайная китайская, Шиповник китайский
336490 Роза карликовая, Роза миниатюрная
336495 Ветреница дубравная, Роза месячная
336498 Роза коричная, Роза майская, Шиповник коричневый, Шиповник коричный, Шиповник северный
336504 Роза кожистолистная, Шиповник кожистолистный
336506 Роза щитковидная, Роза щитконосная, Шиповник щитконосный
336508 Шиповник острованый
336509 Роза мелкоплодная
336514 Роза дамасская, Роза казанлыкская, Роза полиантная, Роза полиантовая, Роза ремонтантная
336519 Роза давида
336522 Роза даурская, Шиповник даурский
336528 Роза дэциньская
336536 Роза эйчисон
336541 Роза плоскошипая
336546 Роза элимаитская
336554 Роза федченко, Шиповник федченковский
336558 Роза тонкочерешковая
336563 Роза вонючая, Роза желтая, Роза жёлтая, Роза пахучая, Шиповник вонючий
336581 Роза галлика, Роза французская, Шиповник галльский, Шиповник французский
336607 Роза джиральди
336610 Шиповник гололистый
336611 Роза сизая, Шиповник

краснолистный
336613 Роза клейкая
336618 Роза полушаровидная
336620 Роза елены
336625 Роза генри, Роза жантиля
336630 Роза щетинисто-волосистая
336632 Шиповник железистощетинистый
336633 Шиповник гиссарский
336636 Роза хьюго
336640 Роза гиблидная
336642 Роза грузинская
336643 Шиповник илийский
336649 Шиповник неколючий
336650 Роза гвосдичная, Роза полевая, Роза эллиптическолистная, Шиповник гвоздичный, Шиповник полевой, Шиповник эллинтический
336653 Роза якутская
336656 Роза камчатская
336658 Роза клука, Шиповник клюка
336659 Роза кокандская, Шиповник кокандский
336660 Шиповник корейский
336663 Роза коржинского, Шиповник коржинского
336664 Роза кугитангская
336665 Роза куэньминская
336667 Роза гуандунская
336671 Роза гуйцзоуская
336672 Роза рвущая
336675 Роза полированная
336685 Роза распростертая, Роза рыхлая, Шиповник рыхлый
336689 Роза белоцветковая
336690 Роза лицзянская
336692 Роза удлиненно-остроконечная
336734 Шиповник самаркандский
336735 Роза юндзилла, Шиповник юндзилла
336736 Роза маррэ
336738 Роза максимовича, Шиповник максимовича
336741 Роза мелкоцветковая, Шиповник мелкоцветковый
336762 Роза мягкая, Шиповник мягкий
336777 Роза мойза
336781 Роза многоприцветниковая
336783 Роза многоцветковая, Роза многоцветная, Роза многошипная, Роза мультифлора, Роза ползучая, Шиповник многоцветковый
336795 Роза наньнинская
336802 Роза многошипая, Шиповник многошиловатый
336804 Шиповник карликовый
336805 Роза ацикулярис мелкоцветковая
336810 Роза нутканская
336813 Роза душистая, Роза чайная, Роза чайная настоящая
336825 Роза оминская
336828 Роза оминская крылатошипная
336835 Роза остроиглая, Роза острошипная, Шиповник острошиповый
336836 Роза острозубая
336849 Гультемия барбарисолистая, Читыр, Шиповник персидсий, Шиповник розаперсидсий
336851 Роза колючейшая, Роза остроколючая, Роза шотлская, Шиповник бедренцоволистный, Шиповник бедренцовый, Шиповник колючейший, Шиповник колючий
336857 Роза широкошипая, Роза широкошиповая, Шиповник широкошиповый
336868 Роза чжундяньская
336879 Шиповник низкий
336885 Роза роксбурга
336892 Роза пжавая, Роза эглантерия, Шиповник эглянтерия
336901 Роза морщинистая, Роза ругоза, Роза страшная, Шиповник мерщинистый, Шиповник морщинистый, Шиповник страшный
336902 Роза морщинистая белая
336922 Шиповник шренковский
336925 Шиповник вечнозелёный
336930 Роза шелковистая
336944 Роза щетинистоножковая
336956 Роза илийская
336965 Роза сулье
336966 Роза сулье мелколистный
336980 Роза сванетская
336981 Роза звегинцова
336985 Роза цюцзянская
336990 Роза тибетская
336991 Роза войлочная, Шиповник опушенный
336993 Роза войлочная, Шиповник войлочный
337001 Роза кубарчатая
337002 Роза туркестанская
337003 Шиповник крючковатый
337005 Роза уссурийская
337006 Роза яблочная, Шиповник мохнатый, Шиповник яблочный
337007 Роза виргинская
337012 Роза уэбба
337015 Роза вихуры
337028 Роза воронова
337029 Роза желтоватая
337050 Розанные, Розовые, Розоцветные
337174 Розмарин
337180 Розмарин аптечный, Розмарин лекарственный
337253 Юстиция лежачая
337269 Лофохлоа
337292 Розеточница
337297 Розеточница альпийская
337298 Розеточница елимаитская
337299 Розеточница голая
337300 Розеточница гиссарская
337301 Розеточница кокандская
337302 Розеточница липского
337303 Розеточница желтая
337304 Розеточница метельчатая
337305 Розеточница персидская
337306 Розеточница волосистая
337307 Розеточница плосколистая, Розеточница плосколистная
337308 Розеточница корнецветковая
337310 Розеточница шишкина
337314 Розеточница вечнозеленая
337315 Розеточница живучковая
337316 Розеточница колосовидная
337317 Розеточница таджикская
337318 Розеточница туркестанская
337320 Ротала
337330 Ротала густоцветковая
337351 Ротала индийская
337368 Ротала мексиканская
337810 Роксбургия
337873 Пальма капустная
337880 Пальма королевская, Пальма олеодокса
337893 Малиноклен
337906 Марена
337916 Марена китайская
337923 Марена читралская
337925 Марена сердцевиднолистная, Марена сердцелистная
337948 Марена меловая
337949 Марена пустынная
337952 Марена длиннолистная
337957 Марена цветущая
337968 Марена грузинская
337970 Марена иезская
337971 Марена комарова
337972 Марена крашенинникова
337978 Марена рыхроцветковая
338022 Марена рехингера
338023 Марена регела
338024 Марена резниченко
338027 Марена шугнанская
338033 Марена татарская
338037 Марена тибетская
338038 Крап, Крапп, Марена красильная, Марсин
338045 Мареновые
338059 Ежевика, Костяника, Малина,

Морошка, Рубус
338074 Ежевика уклоняющаяся
338142 Княженика, Костяника арктическая, Малина арктическая, Мамура, Поляника, Рубус арктический
338209 Ежевика, Ежевика сизая, Ежевика синеватая, Куманника
338225 Рубус канадский
338232 Ежевика белесоватая
338236 Ежевика кавказская
338237 Ежевика кавказородная
338243 Морошка, Морошка приземистая, Рубус приземистый
338300 Малина боярышниколистная, Рубус боярышниколистный
338311 Ежевика куринская
338333 Ежевика длинноплодная
338384 Рубус эмэйский
338399 Ежевика лежачая, Ежевика плетевидная
338425 Рубус тайваньский
338435 Рубус ланьюйский
338447 Ежевика кустарниковая, Куманника, Росяника
338449 Рубус фуцзяньский
338459 Ежевика грузинская
338477 Рубус гуншаньский
338516 Ежевика щетинистая
338527 Ежевика щетинистая
338528 Ежевика щетиноволосная
338538 Рубус хуанпинский
338543 Рубус хмелелистный
338544 Рубус хунаньский
338552 Ежевика гирканская
338553 Ежевика иберийская
338554 Рубус ичанский
338557 Малина, Малина европейская, Малина красная, Малина красная европейская, Малина лесная, Малина обыкновенная, Малина садовая, Рубус идский
338566 Малина красная сибирская щетинистая, Малина обыкновенная черноворсистая
338679 Малина комарова, Рубус комарова
338683 Рубус гулинский
338687 Рубус гуансиский
338713 Ежевика шерстистая
338735 Ежевика мелковатая
338736 Ежевика мелкотычинковая
338742 Ежевика либоский
338757 Ежевика ллойда
338761 Ягода логанова
338787 Рубус малипоский
338806 Рубус мэнлаский
338808 Ежевика чёрная, Ежевика американская, Малина американская, Малина ежевикообразная, Малина западная, Малина североамериканская, Малина чёрная, Рубус наземный
338818 Рубус мотоский
338840 Ежевика мишенко
338878 Ежевика несская, Куманника
338886 Рубус мелкоцветковый
338917 Малина душистая, Малина пахучая, Малиноклен пахучий
338931 Ежевика осетинская
338985 Малина мелкоцветковая, Рубус мелколистный
339039 Ежевика персидская
339041 Ежевика сильно-согнутая
339047 Малина пурпурноволосистая
339049 Ежевика ельниковая
339072 Ежевика широколистная
339090 Ежевика понтийская
339101 Ежевика гигантская
339118 Рубус кандинский
339150 Ежевика радде
339162 Рубус отогнутый
339163 Рубус ху
339227 Малина сахалинская
339236 Ежевика кровавая
339240 Костяника, Костяника каменистая, Рубус скальный
339255 Ежевика змеевидная
339270 Рубус зибольда
339285 Малина великолепная, Малина замечательная, Малина превосходная
339298 Ежевика звездчатая
339370 Рубус тибетский
339383 Ежевика войлочная
339419 Ежевика вязолистная, Ежевика ильмолистная
339477 Ежевика воронова
339488 Ежевика сичоуская
339512 Рудбекия
339556 Рудбекия волосистая, Рудбекия шершавая
339572 Рудбекия гибрида
339574 Золотой шар, Рудбекия раздельнолистная, Рудбекия рассеченная, Рудбекия рассечённая, Рудбекия рассеянная, Шар золотой
339581 Рудбекия бахромчатая, Рудбекия раздельнолистная, Рудбекия разрезнолистная, Рудбекия рассеченная
339658 Руеллия, Руэллия
339731 Руэллия приятная
339880 Трава заячья, Щавель, Щульха
339887 Щавель кислый, Щавель настоящий, Щавель обыкновенный, Щавель огородный, Щавель стловый
339897 Щавелек, Щавелёк, Щавель воробьиный, Щавель кислелький, Щавель малый
339904 Щавель щавельковидный
339907 Щавель щитковидный
339914 Щавель альпийский, Щавель горный
339918 Щавель амурский
339923 Щавель узколистный
339925 Щавель водный, Щавель водяной
339937 Щавель арктический
339940 Щавель аройниколистный
339943 Щавель армянский
339944 Щавель ашхабадский
339947 Щавель золотистый
339987 Конятник, Шавель густой, Щавель конский
339989 Щавель клубковатый, Щавель сросшийся
339994 Щавель курчавый
340049 Щавель фори
340055 Щавель фишера
340058 Щавель ячеистый
340064 Щавель гмелина
340068 Щавель злаколистный
340071 Щавель галачи
340082 Щавель водनощавелевый, Щавель водяной, Щавель прибрежный
340084 Канегра, Щавель североамериканский
340089 Щавель японский
340093 Щавель камчадальский
340098 Щавель комарова
340103 Щавель лапландский
340109 Щавель длиннолистный, Щавель домашний
340116 Щавель морской, Щавель приморский, Щавель русский
340123 Щавель маршалля
340135 Щавель мелкоплодный
340141 Щавель непалский
340151 Щавель туполистный
340171 Щавель памирский
340178 Шпинат английский, Щавель английский, Щавель шпинатный
340194 Щавель паульсена
340209 Щавель ложносолончаковый, Щавель финский
340214 Лён красивый, Щавель красивый
340221 Щавель рехингера
340225 Щавель сетчатый
340249 Щавель крованой, Щавель кровано-краный
340253 Щавель шишкина
340254 Щавель щитковидный, Щавель щитковый
340256 Щавель сибирский
340269 Щавель узколистный
340270 Щавель уссурийский

340277 Щавель сирийский
340278 Щавель тонколистный
340279 Щавель пирамидальноцветковый, Щавель пирамидальный
340285 Щавель тяньшанский
340294 Щавель крубневый
340298 Щавель украинский
340322 Румия
340323 Румия критмолистная
340470 Руппия
340472 Руппия спиральная
340476 Руппия морская
340502 Руппиевые
340541 Рускусовые
340989 Иглица
340990 Иглица, Иглица колючая, Иглица понтийская, Иглица шиповатая, Иглица шиповая, Рускус колючий, Тёрн мышиный
340994 Иглица подъязычная, Иглица подъязычный
340996 Иглица подлистный
340997 Иглица гирканска
340998 Иглица понтийская, Иглица понтийский
341016 Русселия
341020 Русселия ситниковидная
341026 Руссовия
341028 Руссовия согдийская
341039 Рута
341059 Рута обыкновенная
341064 Выйнрута, Рута душистая, Рута обыкновенная, Рута пахучая, Рута содовая
341086 Рутовые
341401 Сабаль
341406 Сабаль блакбурна
341421 Сабаль адансона, Сабаль малый
341422 Капустная пальма, Пальма капустная американская, Пальметта-сабаль, Сабаль пальметто, Сабаль пальмовидный
341463 Сабатия
341765 Можжевельник полушаровидный
341829 Калам, Сахарный тростник
341837 Сахарный тростник тростниковидный
341858 Эриантус тайваньский
341887 Благородный сахарный тростник, Сахарный тростник лекарственный, Сахарный тростник, Тростник сахарный, Тростник сахарный брагородный, Тростник сахарный ктльтурный
341912 Калам, Сахарный тростник дикий
341960 Просо индийское
342155 Сажелеция
342181 Сажелеция мелкоцветковая
342182 Сажелеция эмэйская
342191 Сажелеция луаньдаская
342205 Сажелеция тайваньская
342221 Мшанка
342226 Мшанка безлепестковая, Мшанка безлепестная
342251 Мшанка японская
342259 Мшанка большая, Мшанка крупная, Мшанка прибрежная
342268 Мшанка промежуточная, Мшанка снежная
342271 Мшанка узловатая
342277 Мшанка острочашелистниковая
342279 Мшанка лежачая
342290 Мшанка линнеева, Мшанка моховидная, Мшанка мшанковидная
342308 Стрелолист
342338 Стрелолист злаковидный
342372 Стрелолист личуаньский
342385 Стрелолист плавающий
342400 Стрелолист китайский, Стрелолист обыкновенный, Стрелолист стрелолистный, Стрелолист стрелолистый
342417 Стрелолист шиловидный
342419 Стрелолист тэнчунский
342421 Стрелолист трехлистный, Стрелолист трилистный, Стрелолист трилистый
342424 Стрелолист китайский, Стрелолист обыкновенный, Стрелолист стрелолистный
342482 Сантпория
342493 Сантпория, Фиалка африканская
342841 Ивовые
342847 Солерос
342859 Солерос европейский, Солерос травянистый, Соранг
342887 Солерос поярковой
342923 Ива
342955 Ива скрытная
342961 Верба красная, Верболоз, Ива краснолистная, Ива остролистная, Ива шелюга, Ива шелюга красная, Красная верба, Краснотал, Шелюга, Шелюга красная
342967 Ива египетская
342969 Ива алатавская
342973 Белолоз, Ветла, Ива белая, Ива верба, Ива ветла, Ива прожащая, Ива серебристая
342982 Ветла блестяще-жёлтая, Ива жёлтая
342995 Ива альберта
343003 Ива альпийская
343013 Ива миндалевидная, Ива миндальная
343015 Ива анадерская
343032 Ива аномальная
343037 Ива безногая
343041 Ива деревовидная, Ива деревцовидная
343042 Ива деревцевидная
343045 Ива арктическая
343052 Ива серебристоберая
343063 Ива ушастая
343064 Ива южная, Кара-тал
343070 Ива бабилонская, Ива вавилонская, Ива китайская, Ива полакучая
343093 Ива барклая
343095 Ива бебба
343101 Ива барбарисолистная
343111 Ива блека, Ива линейнолистная
343116 Ива северная
343119 Ива коротконожковая
343128 Ива короткосережчатая
343133 Ива сизоватая
343138 Ива гуншаньская
343151 Бредина, Ива бредина, Ива вильгельмса, Ива козья, Ива хультена, Ива-бредина, Кубатал, Ракита
343167 Ива голубая, Ива недзвецкого
343168 Ива сердцевиднолистная
343174 Ива каспийская
343181 Ива кавказская
343190 Ива шамиссо
343200 Ива чжэцзянская
343216 Ива зеленосережчатая
343221 Ива пепельная, Ива серая
343231 Ива одновременная
343256 Ива клиновидная
343266 Ива дагестанская
343268 Ива даурская
343273 Ива далиская
343280 Верба, Ива волчниковая, Синетал, Шелюга, Шелюга желдая, Шелюга жёлдая, Шелюжник
343283 Ива длиннолистная, Ива пушистопобеговая, Ива шерстистопобегая, Ива шерстистопобеговая
343284 Ива девиса
343314 Ива растопыренная
343336 Ива сизоватая
343341 Ива изящная
343342 Ива длинностлбиковая
343355 Ива красноплодная
343358 Ива остроконечная
343360 Ива красная
343368 Ива ганьсуская
343373 Ива федченко
343378 Ива ферганская
343396 Ива ломкая, Ива хрупкая
343403 Ива дымчатая

343405 Ива буроватая
343409 Ива буреющая
343416 Ива голая
343417 Ива железистая
343425 Ива серая, Ива сизая
343444 Ива тонкостолбиковая
343458 Ива крупнолистная
343463 Ива цзилунская
343467 Ива копьевидная
343476 Ива травянистая
343495 Ива серебристая
343496 Ива гукера
343504 Ива хумаская
343509 Ива хубэйская
343523 Ива илийская
343529 Ива цельная
343543 Ива иссыская
343544 Ива якутская
343546 Ива японская
343553 Ива цзиндунская
343559 Ива кандинская
343560 Ива кангинская
343568 Ива кириллова
343569 Ива коха
343570 Ива кольская
343571 Ива колымская
343572 Ива комарова
343574 Ива корейская
343578 Ива шаньдунская
343579 Ива корианаги
343580 Ива коржинского
343581 Ива гуйцзоуская
343582 Ива крылова
343589 Ива кузнецова
343592 Ива мохнатая, Ива шерстистая
343595 Ива лапландская, Ива лопарская, Куропатник, Куропаточник
343600 Ива чемуесережчатая
343601 Ива тонковетвистая
343612 Ива липского
343613 Ива литвинова
343614 Ива сизовато-серая, Ива синевато-серая
343622 Ива длиннолистная
343628 Ива длинноножковая
343648 Ива тощая
343652 Ива крупноножковая
343653 Ива крупносережчатая
343667 Ива жемчугоносная
343668 Ива матсуды
343680 Ива максмовича
343681 Ива мотоская
343682 Ива медведева
343689 Ива михельсона
343691 Ива мелкосележчатая
343702 Ива мийябе
343706 Ива монгольская
343723 Ива мупинская
343734 Ива мулиская
343743 Ива миртолистная
343745 Ива черниковидная, Ива черничная
343752 Ива курильская
343767 Ива чёрная
343776 Ива чернеющая, Лоза волчья
343780 Ива монетная, Ива монетовидная
343784 Ива монетная мелкоцветковая, Ива монетовидная
343785 Ива ныйвинская
343787 Ива продолговатолистная
343802 Ива оленина
343803 Ива ольги
343804 Ива эмэйская
343809 Ива округлая
343820 Ива яйцевиднолистная
343822 Ива остроплодная
343825 Ива палибина
343826 Ива палласа
343832 Ива параллельноживковая
343839 Ива ядунская
343858 Верболоз, Ива ложнопятитычинковая, Ива пятитычинковая, Челнотал, Чернолоз, Чернотал
343867 Ива персидская
343884 Ива жилколистная
343886 Ива двуцветная, Ива филиколистная
343890 Ива пьеро
343899 Ива черешчатолистная
343900 Ива полярная
343913 Ива пржевальского
343915 Ива ложнобелая
343931 Ива головатая
343936 Ива красивая
343937 Желтолоз, Желтолозник, Ива пурпурная, Ива пурпуровая, Тальник
343962 Ива остроплодная
343968 Ива грушанколистная
343972 Ива аяцинхайск
343975 Ива радде, Ива скрытая
343982 Ива прямосережковая
343994 Ива ползучая, Ива стлющаяся, Ниселоз, Тальник ползучий
344000 Ива ползучая
344003 Ива сетчатая
344004 Ива туполистная
344005 Ива крушинолистная
344009 Ива ричардсона, Подорожник большой
344022 Ива росистая
344031 Ива розмаринолистная, Ива сибирская, Ива шугнанская
344035 Ива розмаринолистная ганьсуская
344037 Ива русская
344039 Ива наньчуаньская
344045 Ива круглолистная
344049 Ива рыжеватая
344053 Ива сахалинская, Ива темнолистная
344060 Ива саянская
344066 Ива скальная
344080 Ива зеемана
344081 Ива полупрутовидная
344096 Ива пильчатолистная
344112 Ива зибольда
344115 Ива силезская
344122 Ива ситхинская
344124 Ива сюзева
344127 Ива джунгарская
344132 Ива великолепная
344135 Ива колючепильчатая
344138 Ива блестящая
344148 Ива старка
344151 Ива длиннoприлистниковая, Ива прилистниковая
344152 Ива шишковидная
344172 Ива хэбэйская
344184 Ива тарайкинская
344193 Ива тэнчунская
344194 Ива тонколистная
344195 Ива тонкосережчатая
344203 Ива тяньшанская
344206 Ива тонтомуссирская
344208 Ива узловатая
344211 Заплатник, Ива миндалистнаяная, Ива миндальная, Ива трехтычинковая, Ива трёхтычинковая, Лоза, Лозина миндальная
344213 Ива ниппонская
344229 Ива тундровая
344230 Ива туранская
344231 Ива турчанинова
344233 Ива удская
344242 Ива настоящая прутьевидная
344243 Ива нарядная
344244 Белотал, Берболоз, Вязинник, Ива болотная, Ива гладколистная, Ива коноплянaя, Ива корзиночная, Ива прутовидная, Ива прутьевидная, Лоза, Лоза болотная
344245 Ива ложнолинейная
344251 Ива зеленоватая
344255 Ива волчья
344273 Ива вильгельмса, Кубатал
344286 Ива бебба, Ива сухолюбивая
344294 Ива тибетская
344307 Ива чаюйская
344381 Сальпиглоссис, Трубоязычник, Языкотруб
344425 Баялыч, Курай, Солянка
344434 Солянка пограничная
344441 Солянка андросова

344443 Солянка открытоплодная
344448 Солянка аральская
344452 Солянка боялыч, Солянка деревцевидная
344453 Солянка боялычевидная
344461 Солянка ошера
344466 Солянка баранова
344474 Солянка супоротивнолистная
344476 Солянка бухарская
344482 Солянка серая
344486 Солянка килеватая
344493 Солянка цинхайская
344494 Солянка хивинская
344496 Солянка холмовая
344499 Солянка мясистая
344506 Солянка дагестанская
344510 Солянка древовидная
344514 Солянка пустынная
344524 Солянка джунгарская
344526 Солянка вересковидная
344538 Солянка многолистная
344539 Солянка щипчиковая
344546 Солянка почечконосная
344564 Солянка семиреченская
344566 Солянка короткощетинковая
344574 Солянка илийская
344578 Солянка сероватая
344581 Солянка запутанная
344583 Солянка тяньшанская
344585 Солянка калийная
344606 Солянка комарова
344607 Солянка коржинского
344608 Солянка мерстистолистная
344609 Солянка шерстистая
344611 Солянка лиственничная, Солянка лиственничнолистная
344612 Солянка лиственничная
344618 Солянка длиниостолбиковая
344620 Солянка сухощавая
344623 Солянка самаркандская
344630 Солянка миквиц
344635 Солянка однокрылая
344636 Солянка горная
344639 Солянка туполистная
344642 Солянка непалская
344644 Солянка натронная
344645 Солянка узловатая
344647 Солянка ольги
344652 Солянка корявая
344654 Солянка палецкого
344658 Солянка паульсена
344662 Солянка прозрачная
344674 Солянка ранняя
344682 Солянка рихтера, Черкез
344686 Солянка розовая
344687 Солянка рожевица
344696 Солянка малолистная
344716 Солянка синьцзянская
344718 Солянка содоносная
344726 Солянка звездчатая
344728 Солянка толстоватая
344733 Солянка тахтаджана
344734 Солянка тамамшан
344735 Солянка тамарисковидная
344743 Солянка заразительная, Солянка русская, Солянка трагус, Урай
344747 Солянка закаспийская
344748 Солянка заамударьинская
344754 Солянка туркменская
344755 Солянка тургайская
344775 Солянка введенского
344798 Сальвадора
344821 Сальвия, Шалфей
344828 Шалфей железистоколосый
344833 Шалфей равнозубый
344836 Шалфей эфиопский
344843 Шалфей александра
344846 Шалфей амазийский
344849 Шалфей андрея
344863 Шалфей армянский
344883 Шалфей австрийский
344890 Шалфей бальджуанский
344899 Шалфей беккера
344913 Шалфей коротноцветковый
344922 Шалфей скобиозолистный
344935 Шалфей согнутозубый
344938 Шалфей седоватый
344949 Шалфей логолистный
344957 Сальвия китайская
344961 Шалфей зеленовато-белый
344984 Шалфей равный
344996 Шалфей дагестанский
344999 Шалфей пустынный
345013 Шалфей дробова
345014 Шалфей кустарниковый
345032 Шалфей фомина
345037 Шалфей форскаля
345042 Шалфей опадающий
345055 Шалфей голостебельный
345057 Шалфей железистый, Шалфей клейкий
345061 Шалфей гончарова
345064 Гортензия крупнолистная, Шалфей крупноцветковый, Шалфей крупноцветный
345065 Шалфей крупноцветковый
345071 Шалфей гроссгейма
345073 Шалфей гаястанский
345074 Шалфей мулиский
345086 Шалфей хэнаньский
345091 Шалфей хубэйский
345099 Шалфей замечательный
345102 Шалфей промежуточный
345108 Шалфей японский
345139 Шалфей карабахский
345146 Шалфей комарова
345147 Шалфей копетдагский
345149 Шалфей кузнецова
345173 Шалфей лиловосиний
345174 Шалфей окаймленный
345175 Шалфей линчевского
345176 Шалфей липского
345190 Шалфей длиннотрубчатый
345197 Шалфей маргариты
345205 Сальвия максимовича
345240 Шалфей наньчуаньский
345246 Шалфей дубравный, Шалфей лесной
345252 Шалфей японский
345254 Шалфей тайваньский, Шалфей японский тайваньский
345266 Шалфей поникающий, Шалфей поникший
345271 Шалфей аптечный, Шалфей лекарственный
345274 Шалфей эмэйский
345282 Шалфей толстоколосый
345298 Шалфей поникающий
345310 Шалфей обкновенный, Шалфей простой
345321 Шалфей луговой, Шалфей луговый
345322 Шалфей кандинский
345346 Шалфей раскрытый
345351 Шалфей розолистный
345363 Шалфей скабиозолистный
345378 Шалфей шмальгаузена
345379 Шалфей мускатный
345382 Шалфей полушерстистый
345384 Шалфей зеравшанский
345389 Шалфей сибторпа
345403 Шалфей колючий
345405 Сальвия блестящая, Шалфей блестящий, Шалфей великолепный, Шалфей голый
345412 Шалфей степной
345414 Шалфей туповатый
345422 Шалфей сиринский
345435 Шалфей траутфеттера
345438 Шалфей треугольночашечный
345445 Шалфей туркменский
345457 Шалфей коровяковый
345458 Шалфей вербеновидный, Шалфей вербеновый, Шалфей вербенолистный
345471 Шалфей мутовчатый
345473 Шалфей прутьевидный
345474 Шалфей горминовый, Шалфей зеленый, Шалфей хохлатый
345480 Шалфей тибетский
345483 Шалфей шертогубый
345525 Дерево дождевое, Саман, Энтеролобиум саман

345558 Бизина, Бузина, Самбук
345565 Бузина древовидная
345580 Бузина американская, Самбук канадский
345586 Самбук китайский
345592 Бизина корейская
345594 Бизина травянистая, Бузина карликовая, Бузинник
345620 Бизина широколисточковая
345624 Бизина маньчжурская
345627 Бизина мексиканская
345631 Бизина черная, Бузина чёрная
345633 Бузина золотистая
345655 Бузина перувианская
345660 Бизина кистистая, Бизина красная, Бизина пушистая, Бузина кистистая, Бузина красная, Бузина обыкновенная
345669 Бизина камчатская
345679 Бизина зибольда
345688 Бизина красная разлезнолистная
345690 Бизина сахалинская
345692 Бизина сибирская
345708 Бизина широколисточковая, Бузина вильямся, Бузина корейкая
345711 Самерария
345713 Самерария армянская
345715 Самерария пузырчатая
345716 Самерария желобчатая
345717 Самерария пустыная
345718 Самерария вайдолистная
345719 Самерария литвинова
345720 Самерария губчатоплодная
345721 Самерария туркменская
345728 Самолюс, Северница
345731 Самолюс мелкоцветковый
345733 Самолюс валеранда, Самолюс обыкновенный, Северница валеранда, Северница обыкновенная
345769 Санхезия
345770 Санхезия благолодная
345782 Лилия ландышавая натальская, Сандерсония
345785 Сандорик
345786 Сандорик индийский
345788 Сандорик индийский
345803 Сангвинария, Сангуинария
345805 Сангвинария канадская, Сонгвинария канадская
345816 Кровохлебка, Кровохлёбка, Черноголовник
345821 Кровохлебка альпийскя, Кровохлёбка альпийскя
345858 Кровохлебка великолепная
345860 Кровохлёбка малая, Пимпинель, Черноголовник кровохлёбковый
345881 Коточки, Кровохлёбка аптечная, Кровохлебка лекарственная, Кровохлебка лечебная, Кровохлёбка малая, Кровохлебка многобрачная, Кровохлёбка обыкновенная, Пимпинель, Черноголовник кровохбрачный, Черноголовник кровохлёбковый, Черноголовник многобрачный
345890 Кровохлебка железистая, Кровохлебка лекарственная
345906 Кровохлебка прибрежная
345911 Кровохлебка ситхинская, Кровохлёбка ситхинская
345917 Кровохлебка мелкоцветковая, Кровохлебка тонколистная, Кровохлёбка тонколистная
345921 Кровохлебка мелкоцветковая
345924 Кровохлебка мелкоцветковая
345940 Лечуха, Подлесник
345948 Подлесник китайский
345952 Подлесник высокий
345957 Подлесник европейский
345988 Подлесник мэнцзыский
345998 Подлесник красноцветковый
346047 Санзеверия, Сансевиерия, Сансевьера, Сансевьерия
346070 Сансевиерия цйлиндрическая
346094 Сансевиерия гвинейская, Сансевиерия кистецветковая
346155 Сансевиерия гвинейская, Сансевиерия кистецветковая
346173 Сансевиерия цейлонская, Сансевьера цейлонская
346181 Санталовые
346208 Дерево сандаловое, Дерево санталовое, Сандал
346210 Дерево сандаловое, Сандал, Сандал белый, Чефрас
346243 Глистогон, Сантолина
346250 Сантолина волосистовойлочная, Сантолина кипарисовидная, Сантолина кипарисовниковидная
346300 Санвиталия
346304 Санвиталия лежачая
346316 Каштановые, Сапиндовые
346323 Дерево мыльное, Сапиндус
346333 Сапиндус делаве
346338 Мельное дерево, Сапиндус макоросса, Сапиндус мукороси
346351 Дерево мыльное антильское, Дерево мыльное обыкновенное, Сапиндус мыльный, Ягода мельная
346361 Сапиум
346376 Сапиум гуйлиньский
346379 Сапиум разноцветный
346390 Сапиум японский
346408 Дерево восковое, Дерево восковое китайское, Дерево китайское восковое, Дерево китайское сальное, Дерево сальное, Дерево свечное, Сальное дерево, Сапиум салоносный
346418 Мыльница, Мыльнянка, Сапонария, Трава мыльная
346422 Мыльнянка ясколковидная
346425 Мыльнянка клейкая
346426 Мыльнянка гриффита
346434 Корень мыльный, Мыльнянка аптечная, Мыльнянка лекарственная, Сапонария аптечная, Сапонария лекарственная, Сапонария оффицциналис
346436 Мыльнянка восточная
346437 Мыльнянка маленькая
346438 Мыльнянка стелющаяся
346442 Мыльнянка северцова
346446 Мыльнянка липкая
346447 Сапожниковия
346448 Ледебуриелла жабрицевидная, Ледебуриелла растопыренная, Сапожниковия растопыренная
346460 Гуттаперчевые, Рододендрон
346497 Сарака
346525 Саркандра
346527 Саркандра голая
346532 Саркандра хайнаньская
346535 Саркандра пинбяньская
346723 Саркококка
346731 Саркококка гукера
346733 Саркококка гукера приземистая
346743 Саркококка иглицелистая
346747 Саркококка ивовая
347066 Парнолистник кашгарский
347117 Жарновец
347144 Саррацения
347156 Саррацения пурпурная
347166 Саррацениевые
347191 Саза, Саса
347218 Саза гуандунская
347219 Саза гуансиская
347220 Саза хайнаньская
347226 Саза хубэйская
347324 Саза вейтча, Саса вича
347402 Сассафрас
347407 Сассафрас лекарственный
347446 Сатурея, Чабер, Чабёр
347469 Чабер алтайский
347560 Чабер садовый, Чабёр садовый
347565 Чабер промежуточный
347579 Чабер рыхлоцветковый
347588 Чабер крупноцветковый
347597 Чабер горный, Чабёр горный
347599 Чабер тупоконечный
347610 Чабер толстолистный
347630 Чабер колосоносный
347631 Чабер мелкозубчатый
347632 Чабер кеымский

348003 Сауруровые
348022 Сауроматум
348036 Сауроматум капельный
348077 Сауруровые
348082 Саурурус
348087 Саурурус китайский
348089 Горькуша, Соссюрея, Сосюрея
348104 Горькуша, Соссюрея заостренная
348111 Соссюрея аянская
348113 Соссюрея крылатая
348118 Соссюрея альберта
348121 Горькуша альпийская, Соссюрея альпийская
348129 Горькуша горькая, Горькуша скученная, Соссюрея горькая, Соссюрея горькуша
348135 Горькуша горькая мелкоцветковая
348137 Соссюрея сомнительная
348141 Соссюрея амурская
348142 Соссюрея дуйская
348146 Соссюрея узколистная
348156 Соссюрея асбукина
348163 Соссюрея байкальская
348167 Соссюрея прелестная
348181 Соссюрея дернистая
348183 Соссюрея седая
348185 Соссюрея седоватая
348186 Соссюрея чертополохо-корзинчатая
348195 Соссюрея кандинская
348199 Соссюрея китайская
348208 Соссюрея хондрилловидная
348223 Соссюрея скученная
348226 Соссюрея спорная
348232 Соссюрея увенчанная
348243 Соссюрея даурская
348249 Соссюрея далиская
348256 Соссюрея тайваньская
348268 Соссюрея разноцветная
348273 Соссюрея дорогостайского
348274 Соссюрея чжундяньская
348275 Соссюрея сомнительная
348283 Соссюрея высокая
348284 Соссюрея изящная
348289 Соссюрея вытянутая
348308 Соссюрея фаминцина
348313 Соссюрея фори
348320 Соссюрея чэнкоуская
348322 Соссюрея густолистная
348328 Соссюрея фролова
348334 Соссюрея ледниковая
348342 Соссюрея суменицевидная
348355 Соссюрея крупнолистная
348381 Соссюрея серебристобелая
348392 Соссюрея обернутая
348401 Соссюрея испайская
348402 Соссюрея ядринцева
348403 Соссюрея японская
348418 Соссюрея ганьсуская
348420 Соссюрея кара-арча
348424 Соссюрея кашгарская
348434 Соссюрея китамуры
348441 Соссюрея крылова
348452 Соссюрея курильская
348454 Соссюрея кушакевича
348456 Соссюрея разрезная
348474 Соссюрея ларионова
348475 Соссюрея широколистная
348479 Соссюрея ленская
348488 Соссюрея серебристолистная
348492 Соссюрея лицзянская
348499 Соссюрея каемчато-чешуйчатая
348510 Соссюрея маньчжурская, Соссюрея треугольная
348513 Соссюрея мазарская
348515 Соссюрея максимовича
348533 Соссюрея микешина
348538 Соссюрея монгольская
348544 Соссюрея мулиская
348553 Соссюрея ново-хорошенькая
348554 Соссюрея ново-пильчатая, Соссюрея пильчатовидная
348555 Соссюрея непалская
348563 Соссюрея нины
348590 Соссюрея голая
348593 Соссюрея нупурипская
348604 Соссюрея зубчатая, Соссюрея зубчаточешуйчатая, Соссюрея зубчато-чешуйчатая
348614 Соссюрея яйцевидная
348616 Соссюрея острозубчатая
348624 Соссюрея особенная
348626 Соссюрея малоцветковая, Соссюрея мелкоцветковая
348654 Соссюрея полякова
348667 Соссюрея порчи
348671 Соссюрея прайса
348672 Горькуша бльзкая, Соссюрея сронная
348674 Соссюрея простертая
348677 Соссюрея ложно-узколистная
348683 Соссюрея ложно-оттопыренная
348684 Соссюрея ложно-сочавы
348690 Соссюрея хорошенькая
348695 Соссюрея хорошенькая
348697 Соссюрея подушечная
348706 Соссюрея отвороченная
348711 Соссюрея ридера
348727 Соссюрея твердая
348728 Соссюрея мощная
348742 Соссюрея скердолистная, Соссюрея струговидная
348749 Соссюрея сахалинская
348755 Соссюрея саянская
348756 Соссюрея залемана
348757 Соссюрея иволистная
348763 Горькуша солончаковая, Соссюрея солончаковая
348767 Соссюрея скальная
348770 Соссюрея шангиновская
348773 Соссюрея серпуховидная
348778 Горькуша пильчатая
348782 Соссюрея ширетокская
348788 Соссюрея выемчатая
348799 Соссюрея сочавы
348800 Соссюрея грязноцветковая
348808 Соссюрея советская
348810 Соссюрея блестящая
348811 Соссюрея оттопыренная
348821 Соссюрея штубендорфа
348823 Соссюрея почти-треугольная, Соссюрея треугольновидная
348830 Соссюрея сукачева
348831 Соссюрея бороздчатая
348832 Соссюрея сумневича
348835 Соссюрея сычуаньская
348838 Соссюрея таджиков
348861 Соссюрея тибетская
348863 Соссюрея тилезиева
348864 Соссюрея войлочная
348865 Соссюрея треугольная
348889 Горькуша тургайская, Соссюрея тургайская
348892 Соссюрея теневая
348900 Соссюрея уральская
348902 Соссюрея уссурийская
348925 Соссюрея липкая
348929 Соссюрея введенского
349032 Камнеломка, Разумовския, Саксифрага
349045 Камнеломка абхазская
349048 Камнеломка железистая
349049 Камнеломка восходящая
349052 Камнеломка жестколистная, Камнеломка жёстколистная
349054 Камнеломка вечнозелёная
349055 Камнеломка альберта
349058 Камнеломка анадырская
349060 Камнеломка проломниковая
349079 Камнеломка астильбовидная
349113 Камнеломка прицветниковая
349117 Камнеломка гребенчаторесничатая, Камнеломка колючая, Камнеломка реснитчатая
349118 Камнеломка мерлерневидная
349137 Камнеломка луковичная
349145 Камнеломка дернистая, Камнеломка дирнистая
349148 Камнеломка чамечная
349153 Камнеломка килеватая
349158 Камнеломка хрящеватая
349159 Камнеломка кавказская

349162 Камнеломка поникающая
349192 Камнеломка колхидская
349193 Камнеломка колончатая
349203 Камнеломка кожистолистная
349204 Камнеломка кортузолистная
349229 Камнеломка кимвальная
349234 Камнеломка даурская
349245 Камнеломка дэциньская
349246 Камнеломка десулави
349254 Камнеломка динника
349300 Камнеломка ешольтца
349303 Камнеломка рыхлая
349305 Камнеломка корневая
349315 Камнеломка крепная
349319 Камнеломка пеовионая, Камнеломка усатая
349339 Камнеломка листочковая
349340 Камнеломка листочковая
349344 Камнеломка форчуна
349411 Камнеломка гуншаньская
349418 Камнеломка крупнолепестная
349419 Камнеломка зернистая
349444 Камнеломка ястребинколистная
349452 Камнеломка болотная, Камнеломка козлиная, Очи царские, Очки царские, Царские очи
349505 Камнеломка крымская, Камнеломка орошенная
349511 Камнеломка цзиндунская
349514 Камнеломка можжевелолистная
349520 Камнеломка комарова
349523 Камнеломка коржинского
349524 Камнеломка кочи
349526 Камнеломка кузнецовская
349527 Камнеломка гуансиская
349532 Камнеломка молоцчнобелая
349533 Камнеломка кладная
349579 Камнеломка крупночашечковая
349607 Камнеломка маньчжуьрская
349617 Камнеломка мотоская
349622 Камнеломка черно-белая
349627 Камнеломка мэнцзыский
349631 Камнеломка мерка
349642 Камнеломка мелкоплодный
349655 Камнеломка мягкая
349671 Камнеломка мускусная
349680 Саксифрага моховидная
349691 Камнеломка нельсона, Камнеломка точечная
349700 Камнеломка снежная, Камнеломка тонкая
349701 Камнеломка голостебельная
349707 Камнеломка продолговатолистная
349719 Камнеломка супротивнолистная
349755 Камнеломка белозоровидная
349799 Камнеломка понтийская
349812 Камнеломка ложноголая
349832 Камнеломка пурпуовая
349835 Саксифрага цинхайская
349839 Камнеломка охотская
349840 Камнеломка редовского
349851 Камнеломка ручейная
349859 Камнеломка круглолистная
349869 Камнеломка сахалинская
349882 Камнеломка твердоногая
349888 Камнеломка чабрецелистная
349891 Камнеломка щетинистая, Камнеломка щетинчатая
349895 Камнеломка сибирская
349905 Камнеломка сиверса
349910 Камнеломка смолевкоцветная
349932 Кактус звездчатый, Камнеломка звездчатая
349934 Камнеломка узколистая, Камнеломка усатая
349936 Камнеломка корнеопрысковая, Камнеломка отпрысковая, Камнеломка плетеносная, Камнеломка побегоносная, Камнеломка ползучая
349956 Камнеломка сычуаньская
349988 Камнеломка моховидная, Камнеломка теректинская
349990 Камнеломка тибетская
349992 Камнеломка тилинга
349999 Камнеломка трехпалая, Камнеломка трёхпалая, Камнеломка трёхпальчатая
350011 Камнеломка тенистая
350027 Камнеломка мутовчатая
350050 Саксифрага чаойская
350055 Саксифрага чжэцзянская
350063 Камнеломковые
350086 Вдовушка, Вдовушки, Скабиоза, Скабиоса
350090 Скабиоза альпийская
350092 Скабиоса приятная
350096 Скабиоса серебристая
350099 Скабиоза темнопурпуровая, Скабиоза южная, Скабиоса темнопурпуровая
350111 Скабиоза южнолтайская
350113 Скабиоса дважды-перистая
350119 Скабиоза кавказская, Скабиоса кавказская
350124 Скабиоса колхидская
350125 Скабиоза голубиная, Скабиоса голубиная
350129 Скабиоза венечная
350132 Скабиоса шерстистолистная
350155 Скабиоза грузинская
350159 Скабиоза разветвленная, Скабиоса разветвленная
350168 Скабиоса гибридная, Скабиоса гирканская
350169 Скабиоса имеретинская
350173 Скабиоза исетская, Скабиоса исетская
350175 Скабиоза японская
350191 Скабиоса блестящатая
350203 Скабиоза мелкоцветковая, Скабиоса мелкоцветковая
350207 Скабиоза бледножелтая, Скабиоза бледно-жёлтая, Скабиоза желтоватая, Скабиоза светложелтая, Скабиоза тёмно-пурпурная, Скабиоса бледножелтая, Скабиоса елтоватая
350208 Скабиоса ольги
350209 Скабиоза оливье
350214 Скабиоса оверина
350222 Скабиоса персидская
350230 Скабиоза колесовидная, Скабиоса вращательная, Скабиоса колесовидная
350237 Скабиоса джунгарская
350238 Скабиоса сосновского
350279 Скабиоза украинская, Скабиоса украинская
350281 Скабиоса улугбека
350284 Скабиоса веленовского
350288 Скабиоса синьцзянская
350312 Гелиотроп мультифлорус, Гемантус многоцветковый
350313 Хемантус екатерины
350360 Скалигерия
350361 Скалигерия, Скалигерия алайская
350362 Скалигерия луковидная
350363 Скалигерия бухарская
350365 Скалигерия ферганская
350366 Скалигерия сизоватая
350367 Скалигерия шероховатая
350368 Скалигерия кнорринг
350369 Скалигерия копетдагская
350370 Скалигерия коржинского
350371 Скалигерия коровина
350372 Скалигерия липского
350374 Скалигерия плосколистная
350375 Скалигерия многоплодная
350376 Скалигерия самаркандская
350377 Скалигерия щетинковая
350378 Скалигерия закаспийская
350379 Скалигерия чимганская
350380 Скалигерия угамская
350402 Скандикс
350407 Скандикс оше
350421 Скандикс серповидный
350423 Скандикс грузинский
350425 Гребень венерин, Скандикс гребенчатый, Скерда
350428 Скандикс персидский
350431 Скандикс звездчатый
350486 Скафоспермум
350490 Скафоспермум азиатский
350504 Скариола

350506 Скариола альберторегерия
350507 Скариола восточная
350509 Скариола прутовидная
350859 Шейхцерия
350861 Шейхцерия болотная
350864 Шейхцериевые
350904 Шима
350949 Схима валлиха, Шима валлиша
350973 Дерево квебраховое, Квебрахо, Квебрачо, Кебрачо
350974 Квебрахо красное
350980 Схинус
350990 Схинус моле, Схинус мягкий
350995 Схинус теребинтолистный, Схинус терпентинолистный, Шинус терпентиннолистный
351011 Лимоник, Шизандра
351021 Лимонник китайский
351069 Лимоник мелкоцветковая
351103 Лимоник контснотычночный
351121 Лимонниковые, Схизеевые, Схицейные
351124 Шишкиния
351125 Шишкиния белощетинковая
351162 Схизмус, Схисмус
351163 Схизмус арабский
351167 Схизмус малый
351172 Схизмус чашечный
351245 Шиверекия
351247 Шиверекия подольская
351250 Схизахна
351251 Схизахна мозолистая
351253 Схизахна комарова
351261 Схизахне
351343 Схизантус, Шизантус
351439 Рдест фриса, Схизокодон
351742 Схизонепета
351759 Схизопепон
351762 Схизопепон переступенелистный
351767 Схизопепон сычуаньский
351774 Схизопепон тибетский
351780 Схизофрагма, Шизофрагма
351797 Схизофрагма гортензиевидная китайскя
351883 Схизостилис
351993 Шлюмбергерия
352004 Шмальгаузения
352006 Шмальгаузения гнездистая
352120 Сабалла
352124 Сабалла аптечная, Сабалла лекарственная, Семя вшивое
352200 Камыш ситниковидный
352205 Камыш комарова
352206 Камыш озерной, Камыш озёрный, Схеноплектус озёрный
352223 Камыш остроконечный
352238 Камыш обедненный
352249 Камыш американский
352276 Камыш прямой
352278 Камыш табернемонтана, Схеноплектус табернемонтана
352284 Камыш трехгранный, Куга трехгранная
352289 Камыш озерный, Куга, Куга озерная, Схеноплектус озерный
352348 Очерёт, Схенус
352385 Схенус ржавый
352408 Схенус чернеющий, Схенус черноватый
352409 Схенус чернеющий, Схенус черноватый
352485 Шотия
352558 Шрадерия блюдчатая
352559 Шрадерия бухарская
352560 Шрадерия змееговниковая
352561 Шрадерия королькова
352629 Шренкия
352631 Шренкия голике
352632 Шренкия замечательнаю
352633 Шренкия обертковая
352634 Шренкия культиасова
352635 Шренкия ворсинчатая
352636 Шренкия колючая
352637 Шренкия влагалищная
352643 Щуровския
352644 Щуровския тонкорассеченная
352645 Щуровския пятирогая
352688 Шульция
352689 Шульция белоцветковая
352690 Шульция косматая
352702 Шуманния
352703 Шуманния карелина
352722 Шумерия
352723 Шумерия широколистная
352724 Шумерия литвинова
352854 Сциадопитис
352856 Зонтичная сосна, Пихта японская зонтичная, Сциадопитис мутовчатый
352880 Пролеска, Ряст, Сцилла
352883 Подснежник армянская, Пролеска прелестная
352884 Подснежник азербайджанская
352885 Пролеска осенняя
352886 Пролеска двулистная, Сцилла двулистная
352889 Подснежник бухарская
352890 Подснежник кавказская
352895 Подснежник дизанская
352912 Подснежник гогенаккера
352920 Пролеска индийская
352960 Лук морской, Пролеска морская
352970 Подснежник мищенко
352972 Подснежник одноцветковая
352981 Подснежник снеговая
353001 Пролеска перуская
353036 Подснежник пушкиниевидная
353039 Подснежник раевского
353051 Подснежник розена
353057 Пролеска пролесковидная, Пролеска японская
353065 Подснежник сибирский, Пролеска сибирская, Сцилла сибирская
353114 Подснежник виноградова
353121 Сциндапсус
353127 Сциндапсус хайнаньский
353170 Голосхенус
353174 Голосхенус обыкновенный
353179 Голосхенус, Дихостилис, Изолепис, Камыш, Куга, Схеноплектус
353245 Камыш авечнский
353265 Камыш бухарский
353272 Камыш дернистый, Пухонос дернистый
353381 Камыш, Камыш эренберга
353448 Дихостилис хайнаньский
353557 Камыш линейчатый
353569 Камыш прибрежный, Камыш приморский
353574 Камыш хмелевый
353576 Камыш вихуры, Камыш лушаньский
353587 Камыш морской, Камыш приморский, Клубнекамыш морской, Нюнька
353606 Пушица японская
353608 Камыш черносемянный
353660 Камыш ниппонский
353676 Камыш восточный
353696 Камыш малоцветный
353707 Камыш плоскостебельный
353750 Камыш укореняющийся
353828 Клубнекамыш схожий
353845 Камыш лежачий, Камыш простерный, Камыш раскидистый
353856 Камыш лесной
353899 Камыш трехгранный, Схеноплектус трёхгранный
353986 Дивала, Дивало
353987 Дивало однолетнее
353994 Дивало многолетнее
353996 Дивала многоплодная
353997 Дивала крючковатая
353998 Склерия
354400 Жестковолосница, Жестколосница
354403 Жестковолосница твёрдая, Жестколосница твердая, Жестколосница твёрдая, Жестколосница трердая
354404 Жестколосница твердая
354477 Жесткомятлик

354485 Жесткомятлик жестковатый
354493 Жесткомятлик воронова
354579 Подостемон, Тростянка
354583 Тростянка овсяницевая
354648 Сколимус
354651 Блошак, Златокорень, Златокорень исканский, Кардуль, Сколимус испанский
354655 Сколимус пятнистый
354678 Скополия
354686 Корень пьяный, Скополия карниолийская
354691 Скополия японская
354755 Личинник, Скорпионница
354760 Личинник маленький
354761 Личинник полувойлочный
354797 Козелец, Скорцонера
354798 Козелец колючеветвистый
354800 Козелец альберта регеля
354801 Козелец белостебельный
354810 Козелец армянский
354813 Козелец австрийский, Козелец рупрехта
354823 Козелец бальджуанский
354824 Козелец двуцветный
354825 Козелец биберштейна
354826 Козелец прицветничный
354828 Козелец бунге
354834 Козелец хантавский
354835 Козелец завитой
354837 Козелец толстолистный
354838 Козелец курчавый
354840 Козелец сладкий корень
354841 Козелец искривленный
354845 Козелец гвоздиковидный
354854 Козелец мечелистный
354855 Козелец мохнатоплодный, Козелец пухосемянный
354858 Козелец ферганский
354859 Козелец франше
354862 Козелец изащный
354863 Козелец григорашвили
354864 Козелец гроссгейма
354870 Козелец испанский, Козелец сладкий корень, Корень сладгийцарский, Корень чёрный, Скорцонер, Скорцонер испанский
354871 Козелец гиссарский
354873 Козелец низкий, Козелец приземистый
354875 Козелец иды
354877 Козелец илийский
354878 Козелец неприметный
354880 Козелец кецховели
354881 Козелец разресной, Козелец рассечённый
354886 Козелец шерстистый
354887 Козелец широколистный
354888 Козелец тонколистный
354889 Козелец липского
354890 Козелец литвинова
354902 Козелец мейера
354903 Козелец мягкий
354904 Козелец монгольский
354910 Козелец яйцевидный
354911 Козелец амударьинский
354913 Козелец мелкоцветковый
354914 Козелец петрова
354917 Козелец луговой
354918 Козелец ложнорастопыренный
354920 Козелец ложношерстистый
354922 Козелец опушенный
354923 Козелец красивый
354925 Козелец пурпуровый
354926 Козелец крошечный, Козелец маленький
354929 Козелец радде
354930 Козелец лучетый
354936 Козелец многоветвистый
354940 Козелец
354941 Козелец розовый
354945 Козелец сафиева
354946 Козелец шишкина
354949 Козелец зейдлица
354950 Козелец шелковисто-шерстистый
354954 Козелец джунгарский
354955 Козелец прямой
354956 Козелец почти-бесстебельный
354957 Козелец пробкобый
354958 Козелец таджиков
354959 Козелец испанский, Козелец крымский
354961 Козелец тау-сагыз, Тау-сагыз
354962 Козелец тяньшанский
354965 Козелец козлобородниковый
354966 Козелец заиийский
354968 Козелец клубненосный
354970 Козелец туркменский
354971 Козелец туркестанский
354972 Козелец туркевича
355039 Буквица водяная, Норичник
355045 Норичник алтайский
355046 Норичник амгунский
355047 Норичник стеблеобъьмлющий
355057 Норичник армянский
355058 Норичник азербайджанский, Эспарцет североиранский
355061 Норичник водюной
355067 Норичник бюргера
355073 Норичник седоватый
355075 Норичник собачий
355084 Норичник харадзе
355088 Норичник зеленоцветковый
355089 Норичник золотистоцветковый
355091 Норичник меловой
355095 Норичник черняковской
355096 Норичник обманчивый
355102 Норичник рассеченный
355103 Норичник растопыренный
355114 Норичник тонкий
355115 Норичник чапандагский
355117 Норичник федченко
355122 Норичник франше
355124 Норичник холодный
355127 Норичник гольде
355128 Норичник гончарова
355129 Норичник грея
355133 Норичник гроссгейма
355134 Норичник кровянокоцветковый
355137 Норичник камнеломкоцветковый
355143 Норичник гирканский
355144 Норичник ильвенсибский
355145 Норичник вырезной
355146 Норичник цельнолистный
355147 Норичник кабадианский
355148 Норичник какудский, Норичник максимовича
355151 Норичник ганьсуский
355152 Норичник кирилова
355154 Норичник кочи
355166 Норичник бокоцветковый
355170 Норичник беловетвистый
355173 Норичник литвинова
355176 Норичник лунниколистный
355178 Норичник крупнокистевой
355181 Норичник маньчжурский
355182 Норичник маньчжурский
355183 Норичник мабяньский
355189 Норичник малый
355192 Норичник мягкий
355194 Норичник многостебельный
355199 Норичник жилковатый
355200 Норичник никитина
355202 Норичник узловатый, Норичник шишковатый
355204 Норичникольдгема
355206 Норичник ольги
355207 Норичник олимпийский
355208 Норичник восточный
355209 Норичник памироалайский
355216 Норичник чужеземный, Норичник чужестранный
355219 Норичник инееватый
355226 Норичник клювообразный
355227 Норичник скалистый, Норичник скальный
355228 Норичник рупрехта
355229 Норичник рутолистный
355234 Норичник зангтоденский
355236 Норичник скополи
355245 Норичник мелкоцветковый

355247 Норичник шпренгера
355249 Норичник полосатый
355253 Норичник таджикский
355256 Норичник лёнецевый
355260 Норичник туркменский
355261 Норичник крылатый
355264 Норичник пестрый
355265 Норичник весенний
355266 Норичник мутовчатый
355269 Норичник желтоязычковый
355274 Норичник заравшанский
355275 Норичник зувандский
355276 Норичниковые
355340 Шлемник
355344 Шлемник железисто-чешуйный
355346 Шлемник приподнимающийся
355351 Шлемник альберта
355353 Шлемник беловатый
355354 Шлемник алехеенкой
355355 Шлемник альпийский
355357 Шлемник алтайский
355359 Шлемник высокий, Шлемник высочайший
355366 Шлемник спрошь-зеленый
355367 Шлемник андрахновидный
355368 Шлемник андросова
355369 Шлемник ангренский
355375 Шлемник аньхуйский
355377 Шлемник аниты
355380 Шлемник араратский
355381 Шлемник араксинский
355382 Шлемник артвинский
355387 Шлемник байкальский
355389 Шлемник бальджуанский
355391 Шлемник чужеземный
355395 Шлемник бухарский
355401 Шлемник екатерины
355406 Шлемник чжэцзянский
355407 Шлемник марелистный
355410 Шлемник хитрово
355411 Шлемник чжундяньский
355414 Шлемник извилистый
355416 Шлемник хохлатый
355418 Шлемник сердцелиственный
355419 Шлемник меловой
355420 Шлемник гребнчатый
355422 Шлемник дагестанский
355423 Шлемник дарридагский
355424 Шлемник дарвазский
355428 Шлемник повислый
355432 Шлемник сомнительный
355438 Шлемник нитестебельный
355439 Шлемник веерный
355446 Шлемник мулиский
355448 Шлемник франше
355449 Шлемник колпаковидный, Шлемник колпаконосный, Шлемник красева, Шлемник обыкновенный
355455 Шлемник оголенный
355460 Шлемник мелкозернистый
355463 Шлемник гроссгеймовский
355465 Шлемник крапчатый
355466 Шлемник окровавлевно-зеленый
355467 Шлемник хайнаньский
355469 Шлемник копьелистный
355473 Шлемник елены
355474 Шлемник разноцветный
355475 Шлемник разноволосый
355477 Шлемник коротко-волосистый
355478 Шлемник гиссарский
355479 Шлемник спрошь-шелковистый
355480 Шлемник хэнаньский
355483 Шлемник хунаньский
355486 Шлемник снизу-белый
355489 Шлемник незапятнанный
355494 Шлемник индийская
355520 Шлемник средний
355521 Шлемник неправильный
355523 Шлемник искандера
355531 Шлемник яванский
355536 Шлемник юзепчука
355538 Шлемник каратавский
355539 Шлемник карягина
355540 Шлемник каркаралинский
355548 Шлемник кнорринг
355553 Шлемник курсанова
355561 Шлемник шерстисто-черешковый
355563 Шлемник боковоцветковый
355567 Шлемник тонкотрубковый
355568 Шлемник мелкочешуйный
355570 Шлемник лицзянский
355573 Шлемник линчевского
355575 Шлемник липского
355576 Шлемник литвинова
355579 Шлемник лодяньский
355580 Шлемник хмелевой
355582 Шлемник желто-синий
355586 Шлемник лэдунский
355601 Шлемник среднечешуйный
355604 Шлемник коротко-опуменный
355607 Шлемник мелкопузырный
355609 Шлемник малый
355614 Шлемник монгольский
355615 Шлемник четкокорневищный
355617 Шлемник лодочковый
355618 Шлемник котовниковидный
355621 Шлемник невского
355625 Шлемник новороссийский
355630 Шлемник глазховый
355632 Шлемник малозубый
355634 Шлемник эмэйский
355636 Шлемник кружсчковый
355637 Шлемник горолюбивый
355638 Шлемник восточный, Шлемник высокий, Шлемник крупноцветковый
355651 Шлемник остролистный
355652 Шлемник острочешуйный
355655 Шлемник бледный
355656 Шлемник памирский
355674 Шлемник переходный, Шлемник тихоокеанский
355675 Шлемник курилы, Шлемник уссурийский
355677 Шлемник олиственио колосый
355678 Шлемник пузырчаточашечный
355679 Шлемник раскрашенный
355680 Шлемник пинбяньский
355682 Шлемник широко-чешуйный
355688 Шлемник пестроцветковый
355690 Шлемник многозубый
355691 Шлемник многолистный
355692 Шлемник многоволосый
355693 Шлемник понтийский
355694 Шлемник попова
355699 Шлемник пржевальского
355705 Шлемник густовевистый
355709 Шлемник радде
355710 Шлемник ветвистейший
355711 Шлемник иконникова, Шлемник регелевский, Шлемник регеля
355712 Шлемник иконникова
355713 Шлемник ганьсуский
355716 Шлемник ромбовидный
355721 Шлемник краснопятнистый
355729 Шлемник шугнанский
355733 Шлемник скордиелистный
355758 Шлемник севанский
355760 Шлемник сикокский
355763 Шлемник жуйлиский
355764 Шлемник сичоуский
355765 Шлемник джунгарский, Шлемник заилийский, Шлемник крылова, Шлемник сиверса
355772 Шлемник сосновского
355774 Шлемник растопыренный
355776 Шлемник стевена
355777 Шлемник полосатенький
355778 Шлемник почти-дернистый, Шлемник такэ, Шлемник щетинковая
355782 Шлемник почти-дернистый
355783 Шлемник почтисердцевидный
355786 Шлемник приземистый, Шлемник простертый
355789 Шлемник тайваньский
355790 Шлемник таласский
355798 Шлемник татьяны
355799 Шлемник крымский
355810 Шлемник тяньцюаньский
355811 Шлемник титова
355812 Шлемник тогуз-тораукский
355813 Шлемник турнефора

355818 Шлемник чимганский
355820 Шлемник цзиньюньский
355824 Шлемник туминганский
355825 Шлемник тургайский
355826 Шлемник тувинский
355836 Шлемник весенний
355840 Шлемник мохнатейший
355847 Шлемник вэйшаньский
355848 Шлемник вэньшаньский
355850 Шлемник воронова
355854 Шлемник желтотрубковый
355855 Шлемник крапиволистный
355856 Шлемник иезский
356232 Рожь
356235 Рожь анатолийская
356237 Рожь культурная, Рожь обыкновенная, Рожь посевная
356244 Рожь куприянова
356248 Рожь сорно-полевая
356253 Рожь дикая
356255 Рожь вавилова
356352 Мексаканский огурец, Мирлитон, Огулец мексаканский, Сехиум, Чайот, Чайота, Шушу
356377 Меченосница, Секуригера, Тордириум
356387 Секуригера мечевидная
356392 Секуринега
356467 Очиток, Родиола, Седум
356468 Очиток едкий
356476 Очиток этнинкий
356503 Очиток белый
356542 Седум аньхуйский
356544 Очиток однолетний
356556 Седум черноватый
356577 Седум чэнкоуский
356589 Седум бухарский
356606 Седум кавказский
356619 Седум цзиндунский
356637 Седум щитковый
356655 Седум васильковый
356724 Седум еверса
356751 Седум тайваньский
356758 Седум франше
356774 Седум тонгий
356802 Очиток испанский
356828 Седум обертковый
356841 Очиток японский
356874 Седум ленкоранский
356895 Седум листонин
356896 Седум прибрежный
356904 Седум кандинский
356907 Седум лидийский
356928 Заячья капуста, Капуста заячья большая, Очиток большой, Скрипун
356958 Седум наньчуаньск
356959 Седум мелкий
356973 Седум туполистный
356998 Очиток бледный
357009 Седум мелкотычинковый
357013 Очиток пятилепестный
357036 Очиток моховидный, Очиток мохообразный
357039 Седум тополелистный
357107 Очиток краснеющий
357112 Очиток красный
357114 Очиток отогнутый
357147 Очиток шестирядный
357169 Очиток люжный
357181 Седум побегоносный
357191 Очиток шиловидный
357203 Капуста заячья, Очиток "заячья капуста", Очиток заячья капуста, Очиток телефиум, Скрипун, Трива живая
357219 Седум тоненький
357227 Очиток четырехмерный
357253 Очиток аньлунский
357255 Очиток цинхайский
357271 Очиток уссурийский
357282 Очиток волосистный
357294 Очиток дэцинький
357384 Зейдлиция
357386 Зейдлиция цветистая
357387 Зейдлиция розмариновая
357780 Гирча, Селин
357789 Гирча тминолистная
357802 Гирча культиасова
357814 Гирча попова
357828 Гирча тяньшанская
357905 Семеновия
357906 Малабайла пушистоцветковая
357912 Борщевик заилийский, Семеновия зайлийская
357918 Полуводосбор
357932 Семиарундинария
357939 Бамбук великолепный, Семиарундинария великолепная
358015 Живучка, Молодило, Семпервивум
358021 Живучка паутинная
358028 Молодило кавказское
358037 Молодило гололистное
358039 Молодило марообразное
358054 Молодило малорослое
358056 Молодило русское
358062 Живучка кровельная, Молодило кровельное
358159 Крестовник, Сенецио
358240 Крестовник амбровый
358253 Крестовник амурский
358280 Крестовник водяной, Крестовник эрратический
358282 Крестовник северный
358287 Крестовник арктический, Крестовник крученный
358292 Крестовник аргунский
358323 Крестовник темно-пурпуровый
358336 Крестовник золотистый
358369 Крестовник сурепколистный
358393 Крестовник бессера
358395 Крестовник акантолистный, Пепериник приморский, Цинерария ковровая, Цинерария летняя, Цинерария морская, Цинерария приморская
358424 Крестовник днепровский
358461 Крестовник буша
358484 Крестовник кальверта
358500 Крестовник коноплеволистный, Крестовник коноплелистный
358504 Крестовник литвинова
358510 Крестовник головчатый
358518 Крестовник карпатский
358523 Крестовник кавказский
358550 Сенецио чжундяньский
358717 Крестовник пустынный
358769 Крестовник сомнительный
358828 Желтозелье, Крестовник эруколистный
358848 Крестовник эмэйский
358866 Крестовник ферганский
358902 Крестовник приречный, Крестовник речной
358912 Крестовник франше
358995 Крестовник крупнозубчатый, Крестовник крупнолистный, Крестовник песчаный
359099 Крестовник ильина
359138 Крестовник целенолистный
359158 Желтуха, Крестовник желтуха, Крестовник луговой, Крестовник якова, Крестовник-желтуха
359165 Крестовник якутскик
359205 Крестовник карелиниевидный
359206 Крестовник карягина
359225 Крестовник киргизский
359236 Крестовник коленати
359240 Крестовник коржинского
359242 Крестовник крашенинникова
359244 Крестовник кубинский
359253 Крестовник аргунский
359270 Крестовник бородавниковый
359303 Крестовник ленский
359357 Крестовник лицзянский
359387 Крестовник лорента
359419 Крестовник крупнолистный
359450 Крестовник массагетова
359527 Крестовник мулиский
359565 Крестовник дубравный, Крестовник фукса
359619 Крестовник ольги
359634 Крестовник отонны

359654 Крестовник пальчатый
359660 Крестовник болотный
359667 Крестовник бандуролистный
359682 Крестовник малолистный
359691 Крестовник малолистный
359693 Крестовник малодольчатый
359696 Крестовник паульсена
359768 Крестовник плосколистный
359769 Крестовник широколистный
359777 Крестовник поярковой
359781 Крестовник многокорзиночный
359789 Крестовник пурпуровый
359802 Крестовник близкий
359811 Крестовник ложная арника, Крестовник ложноарниковый
359813 Крестовник лжеоранжевый
359844 Крестовник огненно-язычковый
359859 Крестовник разветвленный
359892 Крестовник ромболистный
359903 Крестовник приручейный
359938 Крестовник скалистый
359967 Крестовник сапожникова
359968 Крестовник сарацинский
359980 Крестовник цепкий
359991 Крестовник стрельчатый
360007 Крестовник швецова
360048 Крестовник сихотэа
360074 Крестовник сосновского
360129 Крестовник субальпийский
360137 Крестовник малозубчатый
360146 Крестовник клочковатый
360147 Крестовник холодный
360155 Крестовник
360160 Крестовник сивецолистный
360162 Крестовник сукачева
360166 Крестовник сумневича
360172 Крестовник лесной
360205 Крестовник тонколистный
360214 Крестовник тяньшанский
360220 Крестовник тибетский
360265 Крестовник тундровый
360281 Крестовник теневой
360306 Крестовник весенний
360321 Крестовник клейкий
360328 Крестовник обыкновенный
360453 Кассия мэрилендская, Кассия приморская
360561 Секвойя
360565 Дерево калифорнийское мамонтовое, Дерево красное, Калифорнийское мамонтовое дерево, Секвойя, Секвойя вечнозеленая, Секвойя вечнозелёная
360575 Секвойядендрон
360576 Веллингтония, Дерево мамонтовое, Мамонтово дерево, Секвойя гигантская, Секвойядендрон гигантский
360582 Серапиас
360642 Серапиас сошниковый
360653 Сергия
360654 Сергия регеря
360655 Сергия северцова
360812 Полынь приятная
360825 Дармина, Полынь цитварная
360827 Полынь плотная
360830 Полынь федченко
360831 Полынь ферганская
360835 Полынь бедноцветковая, Полынь малоцветковая, Полынь тонковатая
360837 Полынь белоземельная
360843 Полынь федченко
360845 Полынь ситниковая, Полынь ситниковидная
360846 Полынь ситниковидная
360847 Полынь каратавская, Полынь тонкорассеченная
360848 Полынь кашгарская
360850 Полынь коровина
360851 Полынь лемана, Полынь лемановская
360857 Полынь азотистая, Полынь селитряная
360866 Полынь розовоцветковая
360870 Полынь сантолинная
360872 Полынь шренка, Полынь шренковская
360873 Полынь прутьевидная
360874 Полынь полусухая
360878 Полынь лессинговидная
360879 Полынь белоземельная, Полынь белоземная, Полынь сероземная
360882 Полынь тибетская
360883 Полынь заилийская
360923 Серисса
360933 Серисса дурнопахнущая
360936 Серьяния
360982 Серпуха, Шумерия
360985 Серпуха алатавская
360987 Серпуха холодная
360992 Серпуха угловатая
360994 Серпуха голоногая
361002 Серпуха причветниковая
361004 Серпуха блестяшая, Серпуха чертополоховая
361010 Серпуха кавказская
361011 Серпуха лучевая
361017 Серпуха бумажистая
361018 Серпуха китайская
361021 Серпуха кожистая
361023 Крестовник увенчанная, Серпуха венечная, Серпуха венценосная
361039 Серпуха рассеченная
361043 Серпуха джунгарская
361045 Серпуха эриколистная
361057 Серпуха гмелина
361058 Серпуха копьелистная
361059 Серпуха разнолистная
361065 Серпуха красильная, Серпуха неколючая
361067 Серпуха равнолистная
361069 Серпуха киргизская
361070 Серпуха комаров
361074 Серпуха ланцетолистная
361076 Серпуха лиролистая, Серпуха модеста, Серпуха тяньшанская
361079 Серпуха окаймленная
361110 Серпуха простертая
361117 Серпуха пятилистная
361126 Серпуха серпуховидная
361134 Серпуха полукустарниковая
361135 Серпуха донская
361139 Серпуха красильная
361141 Серпуха
361144 Серпуха сухоцветная
361274 Астрокарпус
361294 Кунжут, Сезам
361317 Кунжут, Кунжут восточный, Кунжут индийский, Кунжут культурный, Сезам
361350 Сесбания
361357 Сесбания египетская
361456 Жабрица
361457 Жабрица аболина
361460 Жабрица подражаюшая
361462 Жабрица алексеенко
361464 Жабрица андронак
361465 Жабрица однолетняя
361466 Жабрица песчаная, Жабрица полевая
361467 Жабрица шероховатая
361472 Жабрица равнинная
361475 Жабрица увенчанная
361477 Жабрица клинолистная
361482 Жабрица виличатая
361483 Жабрица индийская
361485 Жабрица изящная
361487 Жабрица пушистоголовчатая
361488 Жабрица пучковая
361492 Жабрица многолистная
361493 Жабрица гигантская
361495 Жабрица гладковатая, Жабрица тонколистная
361496 Жабрица крупноканальцевая
361500 Жабрица камеденоская, Жабрица солоносная
361501 Жабрица конская
361511 Жабрица каратавская
361512 Жабрица коровина
361519 Жабрица ледебура
361520 Жабрица леманновская
361521 Жабрица леманна
361522 Жабрица тонковетвистая

361524 Жабрица лессинга
361526 Порезник горный
361537 Жабрица крупнолистная
361545 Жабрица тибетская
361547 Жабрица палласа
361548 Жабрица малолучевая
361550 Жабрица каменстая, Жабрица щебнистая
361551 Жабрица горичниколистная
361552 Жабрица горичниковидная
361553 Жабрица йомутская
361554 Жабрица понтийская
361559 Жабрица жесткая
361561 Жабрица скальная
361569 Жабрица сидячецветковая
361575 Жабрица джунгарская
361578 Жабрица растопыренная
361582 Жабрица прямостоячая, Жабрица торчашая
361585 Жабрица тонкорассеченная
361587 Жабрица растопыренная
361589 Жабрица чуилийская
361590 Жабрица коническая
361595 Жабрица валентины
361597 Жабрица варьирующая, Жабрица именчивая
361605 Жабрицевидка
361606 Жабрицевидка тяньшанская
361611 Сеслерия
361614 Сеслерия осенняя
361615 Сеслерия голубая
361621 Сеслерия гейфлера
361622 Сеслерия тимофеевковидная
361680 Мышей, Трехщетинник, Щетинник
361743 Щетинник фабера
361777 Щетинник гуйцзоуский
361789 Щетинник промежуточный
361794 Гоми, Могар, Просо головчатое, Просо итальянское, Просо татарское, Щетинник итальянский
361877 Зардкунак, Иткунак, Мышей сизый, Щетинник сизый
361930 Щетинник мутовчатый, Щетинник ципкий
361935 Брица, Гоми, Кунакисагак, Мысык-кайрык, Мышей зеленый, Мышей зелёный, Мышей сизый, Чумиза, Щетинник зеленый, Щетинник зелёный
362020 сейтера
362082 Шерардия
362083 Шерардия полевая
362089 Шефердия
362090 Ягода буйволова
362095 Жерардия, Шерардия
362097 Жерардия полевая, Шерардия полевая
362121 Сибатеа
362123 Сибатеа китайская
362131 Бамбук кумасаса, Сибатеа кумасаса
362140 Сибатеа наньпинская
362186 Дерево даммаровое, Сореа, Сорея, Шорея
362220 Сал, Сорея кистевая, Шорея мощная
362238 Дерево даммаровое
362242 Шортия
362339 Сиббальдиецвет, Сиббальдия
362341 Сиббальдиецвет прижатый
362354 Сиббальдия ольги
362355 Сиббальдия эмэйская
362356 Сиббальдия мелкоцветковая
362361 Сиббальдия крупнолистая, Сиббальдия крупнолистная, Сиббальдия лежачая, Сиббальдия распростертая
362372 Сиббальдия полуголая
362392 Сибирка
362396 Сибирка алтайская, Шомнольник
362398 Сибирка тяньшанская
362460 Огурец волосистый, Сициос
362461 Огурец волосистый угловатый, Огурец угроватый, Сициос угловатый
362473 Грудника, Просвирнячок, Сида
362514 Сида китайская
362617 Сида
362662 Грудника колючая
362702 Сидальцея
362728 Железница, Сидеритис
362746 Железница балансы
362765 Железница зеленочешуйная
362767 Железница хохлатая
362768 Железница скученная
362786 Железница черноморская
362807 Железница черепичатая
362835 Железница маршаллова
362838 Железница горная
362880 Железница таврическая
362902 Дерево железное, Сидероксирон
363002 Зибера
363006 Зибера колючая
363057 Сиверсия
363063 Сиверсия пятилепестная
363067 Сиверсия малая
363073 Зигесбекия, Сигезбекия
363080 Зигесбекия гладковатая
363090 Зигесбекия восточная, Сигезбекия восточная
363093 Зигесбекия пушистая
363123 Морковник
363124 Морковник бессера
363126 Морковник попова
363127 Морковник рубцова
363128 Морковник луговой
363141 Горицвет, Куколица, Лихнис, Силена, Силене, Смолевка, Смолёвка, Хлопушка
363144 Смолевка бесстебельная, Смолёвка бесстебельная
363153 Смолевка железистолепестная
363165 Смолевка александры
363167 Смолёвка альпийская
363168 Смолевка альпийская
363169 Смолевка алтайская
363180 Смолевка неравнолопастная
363193 Смолевка безлепестная
363206 Смолевка араратская
363209 Смолевка песчаная
363213 Смолевка армянская
363214 Смолевка армериевидная, Смолевка армерцевидная, Смолёвка армерцевидная
363217 Смолевка артвинская
363227 Смолевка ошера
363232 Смолевка бархашская
363233 Смолевка бальджуанская
363252 Смолевка боброва
363255 Смолевка борнмюллера
363259 Смолевка мелкоцветковая
363261 Смолевка коротколепестная
363263 Смолевка бротеруса
363264 Смолевка бухарская
363268 Смолевка володушковидная
363281 Смолевка дернистая
363295 Смолевка головчатая
363296 Смолевка мелкоголовчатая
363307 Смолевка кавказская
363308 Смолевка головчатая
363311 Смолевка щетинозубая
363312 Барская спесь, Зорька, Зорька "барская смесь", Зорька татарское мыло, Лихнис татарское мыло, Лихнис халцедонский, Мыло татарское
363313 Смолевка хамарская
363315 Смолевка зеленовацветковая, Смолёвка зелёновацветковая
363316 Смолевка хрололистная
363317 Смолевка зеленоватолепестная
363321 Смолевка чжундяньский
363326 Смолевка булавовидная
363331 Смолевка голубоваторозовая
363337 Лихнис родственный
363360 Смолевка замещающая
363361 Смолевка густоцветковая, Смолевка скученноцветковая
363363 Смолевка коническая, Смолёвка коническая
363366 Смолевка конусоцветковая
363368 Зорча, Смолевка конусовидная
363370 Горицвет кожистый, Горицвет короналиевый, Зорька-аксамитки, Лихнис кожистый
363381 Смолевка меловая

363383 Смолевка курчавая
363385 Смолевка сцера
363395 Смолевка куринская
363396 Смолевка чере
363397 Смолевка дагестанская
363399 Смолевка бесстебельная
363404 Смолевка густоцветковая
363406 Смолевка прижатая
363408 Смолевка гвоздинковидная
363409 Смолевка вильчатая, Смолёвка вильчатая
363442 Смолевка пустынная
363445 Смолевка черноморская
363451 Дрема крепкая, Смолевка крепкая, Смолевка твердая
363461 Горицвет кукушкин цвет, Горицвет-кукушкин цвет, Зорька-дрёма, Кокушник, Коронария кукущкин-цвет, Кукушник, Слёзки кукушкины, Цвет кукушкин
363463 Смолевка облиственная, Смолевка олиственная
363465 Смолевка олиственная мелкоцветковая
363472 Лихнис сверкающий
363477 Смолевка английская
363486 Смолевка газимайликская
363487 Смолевка гаврилова
363489 Смолевка, Смолевка геблера
363502 Смолевка сизоватый
363516 Смолевка злаколистная
363525 Смолевка гроссгейма
363528 Смолевка гунтская
363529 Смолевка цзилунская
363533 Смолевка гельмана
363534 Смолевка семиреченская
363551 Смолевка цельнолепестная
363557 Смолевка низкая
363559 Смолевка хубэйская
363561 Смолевка грузинская
363566 Смолевка изогнутолистная
363568 Смолевка индийская
363571 Куколица широколистная, Скабиоза жёлтая, Смолёвка-хлопушка, Щеркунец
363582 Смолевка испирская
363583 Дрёма липкая, Смолевка итальянская, Смолёвка итальянская
363606 Смолевка енисейская
363609 Смолевка енисейская мелкоцветковая
363620 Смолевка карачукурская
363640 Смолевка комарова
363641 Смолевка корейская
363642 Смолевка коржинского
363643 Смолевка кубанская
363644 Смолевка кудряшева
363646 Смолевка кунгесская
363647 Смолевка кушакевича
363648 Смолевка разрезная
363667 Смолевка опушенноцветковая
363671 Смолевка широколистная
363673 Дрема белая, Дрёма белая, Дрёма беловатая, Дрёма луговая, Лихнис белый, Смолевка белая
363680 Смолевка лазская
363682 Смолевка тонкостебельная
363684 Смолевка лицзянская
363686 Смолевка линейнолистная
363691 Лихнис сибирский, Смолевка сибирская, Смолёвка сибирская
363692 Смолевка камнелюбивая
363694 Смолевка литвинова
363700 Смолевка длиннозубая
363701 Смолевка длинноцветковая
363712 Смолевка горицветовидная
363718 Смолевка длинностолбиковая
363723 Смолевка марковича
363730 Смолевка маршара
363735 Смолевка промежуточная, Смолевка средная
363752 Смолевка мейера
363753 Смолевка михельсона
363778 Смолевка мулиская
363780 Смолевка многорассеченная
363781 Смолевка многоцветковая
363795 Смолевка мотоская
363796 Смолевка карликовая
363800 Дрема непалская
363804 Смолевка невского
363813 Дрема нинсяская
363815 Дрема ночецветная, Дрема ночная, Дрёма ночная, Элизанта ночная
363827 Смолёвка поникающая, Смолевка поникшая, Смолёвка поникшая
363833 Смолевка обратнояйцевидная
363834 Смолевка тупозубчатая
363840 Смолевка думистая
363843 Смолевка ольги
363850 Смолевка горная
363852 Силене восточное
363861 Смолевка толстожилковая
363865 Смолевка памирская
363879 Смолевка малолистная
363882 Смолевка повислая, Смолёвка повислая
363889 Смолевка вздуточашечная
363902 Смолевка многолистная
363903 Смолевка порярная
363912 Смолевка попова
363921 Смолевка приликоана
363926 Смолевка лежачая, Смолевка стелющаяся
363927 Смолевка бризкая
363928 Смолевка тонкоопушенная
363933 Смолевка ложнотонкая
363936 Смолевка ложноушковидная
363948 Смолевка маленькая
363950 Дрема четырехлопастная
363952 Смолевка радде
363958 Смолевка ползучая
364002 Смолевка скальная
364003 Смолёвка рупрехта
364007 Смолевка саманкардская
364010 Смолевка зеравшанская
364012 Дрема скальная
364020 Смолевка шероховатолистная
364023 Смолевка шафта
364024 Смолевка шугнанская
364047 Смолевка семенова
364071 Смолевка трубчатоцветная
364072 Дрема байкальская, Смолевка джунгарская
364079 Смолевка узколистная
364083 Дрема ароматная
364089 Смолевка кустарничковая
364091 Смолевка прилегающая
364094 Смолевка сычуаньская
364096 Смолевка тахтская
364098 Смолевка талышская
364100 Смолевка татарская, Смолёвка татарская
364108 Смолевка татьяны
364110 Смолевка стройная
364125 Смолевка тирке
364128 Смолевка тяньшанская
364131 Смолевка пушистая
364143 Смолевка туркменская
364144 Смолевка вздутая
364146 Смолевка одноцветковая
364158 Куколица широколистная, Скабиоза жёлтая, Смолёвка-хлопушка, Щеркунец
364178 Смолевка зеленоватая, Смолевка зеленовацветковая
364193 Куколица широколистная, Скабиоза жёлтая, Смолевка лопушка, Смолевка обыкновенная, Смолевка тилоколистная, Смолевка хлопушка, Смолевка широколистная, Смолёвка-хлопушка, Щеркунец
364213 Лихнис уилфорда
364216 Смолевка волжская
364226 Смолевка чаюйская
364255 Силифиум, Силифия
364353 Остро-пестро, Расторопша
364360 Остроло-пёстро, Остро-пестро, Расторопша остропёстрая, Расторопша пятнистая
364382 Кассия горькая, Квассия горькая
364384 Симаруба сизая
364387 Симарубовые
364544 Горчица

364545 Горчица английская, Горчица белая, Горчица белая английская
364557 Горчица дикая, Горчица полевая, Горький полевая
364631 Синарундинария
364737 Глоксиния
364754 Глоксиния
364951 Синоджекия
364957 Синоджекия редера
364959 Синоджекия дереванистоплодная
365002 Принсепия китайская, Принцепия китайская
365009 Подофилл гималайский, Подофилл эмода
365038 Крестовник наньчуаньский
365217 Сифонантус
365270 Сифонодон, Фителефас
365271 Сифонодон южный
365295 Сифоностегия
365296 Сифоностегия китайская
365357 Петрушечник, Сизон
365362 Петрушечник ароматный
365386 Гулявник
365398 Гулявник высокий, Гулявник высочайший
365412 Гулявник двулонастной
365416 Гулявник капустовидный
365433 Гулявник густой
365442 Гулявник дагестанский
365457 Гулявник высокий
365469 Гулявник фуцзяньский
365482 Гулявник вислоплодный
365490 Гулявник волосистый
365503 Гулявник ирио
365513 Гулявник исфаринский
365515 Гулявник японский
365528 Гулявник липского
365529 Гулявник лёзелиев, Гулявник лезеля
365532 Гулявник желтый
365564 Гулявник аптечный, Гулявник лекарственный, Сухоребрик
365568 Гулявник восточный
365585 Гулявник изменчивый
365606 Гулявник струговидный
365621 Гулявник сжатый
365624 Гулявник слабоколючий
365625 Гулявник тоненький
365643 Гулявник туркменский
365649 Гулявник волжский
365678 Сизюринхий, Сисюринхий
365747 Сизюринхий узколистный, Сисюринхий узколистный
365835 Поручейник, Почейник
365850 Поручейник ланцетолистный
365851 Поручейник широколистный, Почейник широколистный
365854 Поручейник средний, Почейник средний
365858 Поручейник японский
365871 Корень сахарный, Поручейник сахарный, Поручейник сахарный корень, Поручейник тонкий
365872 Поручейник привлекательный, Поручейник приятный, Ручейник приятный
365916 Скимия
365924 Скимия тайваньская
365929 Скимия японская
365965 Скимия люцюская
365970 Скимия перноплодный
365974 Скимия японская
365975 Скимия ползучая
366034 Слонея
366115 Смеловския
366116 Смеловския белая
366119 Смеловския папоротниколистая
366122 Смеловския чашечная
366127 Смеловския неожиданная
366141 Сассапариллевые, Смилаксовые
366142 Смилацина
366230 Павой, Сарсапарель, Сассапариль, Смилакс
366238 Смилакс алтайский
366284 Сассапарель китайская
366326 Смилакс эмэйский
366329 Сассапариль высокий, Экаль высокая
366334 Смилакс фунинский
366338 Сассапариль голая
366344 Смилакс тибетский
366447 Сассапариль мэра
366448 Смилакс малипоский
366481 Смилакс наньтоуский
366486 Смилакс японский
366537 Смилакс луаньдаский
366548 Сассапариль береговая, Сассапариль максимовича
366573 Смилакс зибольда, Смилакс ольдгема
366737 Смирниовитка
366738 Смирниовитка армянская
366739 Смирния
366741 Смирния сердсевиднолистная
366744 Смирния пронзеннолистная
366853 Пасленовые, Паслёновые
366902 Паслен, Паслён, Солянум
366934 Паслен безволосый
366953 Паслен среднеазиатский
367083 Паслен незамеченный
367120 Глистник, Паслён горько-сладгий, Паслен сладкогорький, Паслён сладко-горький
367205 Паслён арбузолистный
367241 Солянум индийское
367262 Солянум японский
367270 Паслен иудейский
367281 Паслен кизерицкого
367306 Паслен прибрежный
367311 Паслен желтый
367367 Паслен крупноплодный
367370 Бадиджан, Бадлажна, Баклажан, Баклажан обыкновенный, Дельфиниум гибридный, Демьянка, Паслён баклажан
367400 Груша дынная, Пепино
367416 Ангури-сагак, Паслен черный, Паслён чёрный, Поздника
367432 Паслен низкий
367468 Паслен ольги
367493 Паслен персидский
367518 Паслен хайнаньский
367522 Паслён лжеперечный, Паслён ложноперечный, Паслён перцеподобный
367529 Паслен ложножелтый
367531 Паслен ложноперсидский
367567 Паслен клювовидный, Паслён клювовидный, Паслён клювообразный
367604 Паслен семилопастный
367687 Паслен закавказский
367696 Картофель, Картофель культурный, Паслен клубненосный, Паслён клубненосный
367726 Паслён волосистый, Паслён жёлтый, Паслен крыладый, Паслён мохнатый
367748 Паслен воронова
367757 Паслен зеленецкого
367762 Сольданелла, Сольданелля
367763 Сольданелла альпийская
367766 Сольданелла горная
367771 Хельсиния
367772 Хельсиния
367804 Трубкоцвет, Трубкоцветник
367809 Трубкоцвет биберштейна, Трубкоцветник биберштейна
367810 Трубкоцвет короткотычиночный
367811 Трубкоцвет завитковый
367813 Трубкоцвет грубоволосистый
367815 Трубкоцвет каратегинский
367816 Трубкоцвет кокандский
367822 Трубкоцвет подорожниколистный
367825 Трубкоцвет тычиночный
367828 Трубкоцвет туркестанский
367939 Золотарник, Золотая розга, Солидаго, Солотянь
367948 Золотарник приальпийский
367965 Золотарник тонкозубчатый
368013 Золотарник канадский
368049 Золотарник сжатый
368056 Золотарник медный
368073 Золотарник низбегающий
368105 Золотарник геблера

368106 Золотарник поздний
368134 Золотарник злаколистный
368179 Солидаго гиблида
368183 Золотарник яйлинский
368188 Золотарник ситниковый
368196 Золотарник кухистанский
368208 Золотарник каучуконостый
368263 Золотарник незамеченный
368290 Золотарник пахучий
368297 Золотарник тихоокеанский
368377 Золотарник скальный
368442 Золотарник крымский
368453 Золотарник торфяный
368480 Золотарник "золотая розка", Золотарник золотая розга, Золотарник обыкновенный, Золотая розга, Золотель, Лоза золотая, Розга золотая
368502 Золотарник даурский
368625 Осот
368635 Осот жёлтый, Осот полевой, Хор
368649 Осот жёский, Осот жесткий, Осот колючий, Осот острый, Осот шероховатый, Сют-от
368771 Молочник, Молочник хробуст, Осот желтый, Осот овошной, Осот овыщной, Осот огородный, Хробуст
368781 Осот болотный
368813 Осот сосновского
368838 Осот закаспийский
368944 Софийка
368945 Софийка двулетная, Софийка однолетная
368947 Софийка желтейшая
368950 Софийка монгольская
368951 Софийка гулявниковая
368955 Гевелия, Софора
368962 Гёбелия лисохвостовидная, Горчак белоцвтеный, Ишек-мия, Софора лисохвостая, Софора лисохвостная, Софора обыкновенная, Софора осохвостная
368986 Софора китайская
368990 Софора мексиканская
368994 Софора виколистная
369004 Софора бирманская
369010 Софора желтеюшая, Софора жултеющая
369018 Софора франше
369037 Софора японская
369038 Софора японская
369044 Софора гиблидная
369051 Софора ичанская
369054 Софора корейская
369056 Аммртамнус лемана
369066 Софора мелкоплодная
369075 Софора гриффита
369095 Ишек-мия эгорчак белоцведный, Софора толстоплодная
369110 Софора продана
369122 Аммртамнус джунгарский
369130 Софора четырехкрылая
369254 Кучкоцветник
369256 Кучкоцветник мейера, Ферула
369259 Рябинник, Рябинолистник, Сорбария
369264 Рябинник амурская
369265 Рябинник древесный
369274 Рябинник ольги
369275 Рябинник палласа
369276 Рябинник сумахолистный
369277 Сорбария сибирская
369279 Рябинник рябинолистный, Рябинолистник обыкновенный, Спирея рябинолистная
369288 Рябинник линдлея, Сорбария тибетская
369290 Мелкоплодник, Рябина
369302 Рябина альбова
369313 Рябина американская
369317 Рябина анадырская
369322 Боярка белая, Рябина-ария
369334 Рябина армянская
369339 Бурка, Рябина ликёрная, Рябина обыкновенная
369349 Рябина балдаччи
369350 Рябина буассье
369351 Рябина буша
369357 Рябина кавказская
369361 Рябина колхидская
369362 Рябина смешанная
369367 Рябина смешанная
369379 Рябина двуцветная, Рябина китайская, Рябина пёстрая
369381 Рябина домашная, Рябина садовая
369391 Рябина непалский
369397 Рябина гладковатая
369403 Рябина греческая
369418 Рябина хунаньская
369420 Рябина хубейская
369429 Рябина гиблидная
369432 Рябина промежуточная, Рябина скандинавская
369436 Рябина камчатская
369439 Рябина цюцзянская
369444 Рябина кузнецова
369445 Рябина ланьпинская
369459 Арония черноплодная, Рябина черноплодная
369462 Рябина мелкоцветковая
369464 Рябина мелколистная
369466 Рябина мугеотта
369481 Рябина персидская
369484 Рябина амурская
369498 Рябина тайваньская
369512 Рябина бузинолистная
369514 Рябина бузинолистная
369518 Рябина шемахинская
369520 Рябина сычуаньская
369521 Рябина сибирская
369526 Рябина буроватая
369528 Рябина полувойлочная
369529 Рябина шаньдунская
369532 Рябина крымская
369536 Рябина тяньшанская
369541 Берек, Берека, Богорожник, Глоговина, Дерево красное, Рябина берека, Рябина глоговина, Рябина лечебная, Рябина-глоговина
369548 Рябина турецкая
369556 Рябина бархатистая
369566 Рябина воронова
369572 Рябина чаюйская
369590 Сорго
369600 Сорго двцветное
369601 Джугара, Дура белая, Дурра белая, Сорго поникшее
369636 Сорго веничное, Сорго метельчатое, Сорго развесистое, Сорго техническое
369640 Сорго зерновое, Сорго комовое, Сорго скученное
369652 Гумай, Джонсонова трава, Калами, Сорго алеппское, Сорго алепское, Сорго дикое, Сорго многолетнее, Сорго хлебное, Трава джонзонова, Чимай, Щалаппа
369671 Сорго китайское, Сорго японское
369700 Просо китайское, Сорго сахарное
369707 Сорго суданское, Суданска, Трава суданская
369720 Дурра, Дурра бурое, Дурро белое, Дурро бурое, Дурро египетское, Просо индийское, Сорго африканское зерновое, Сорго зерновое, Сорго метельчатое, Сорго обыкновенное, Сорго обыкновенный, Хлеб каффрский
369934 Соймида
369975 Спараксис
370018 Ежевиковые, Ежеголовковые, Ежеголовниковые
370021 Ежеголовка, Ежеголовник
370023 Ежеголовник родственный
370030 Ежеголовник узколистный
370050 Ежеголовка ветвистая, Ежеголовник ветвистый
370051 Ежеголовник незамеченный
370064 Ежеголовник глена
370066 Ежеголовник скученный
370068 Ежеголовник злаковый, Ежеголовник фриса
370071 Ежеголовник северный

370072 Ежеголовник японский
370075 Ежеголовник длиннолистный
370080 Ежеголовка мелкоплодная, Ежеголовник мелкоплодный
370088 Ежеголовка маленькая, Ежеголовник маленький, Ежеголовник малый
370090 Ежеголовник многогранный
370096 Ежеголовка простая, Ежеголовник выступающий, Ежеголовник простой
370100 Ежеголовник тонколистный, Ежеголовник узколистный
370102 Ежеголовка побегоносная, Ежеголовник побегоносный
370130 Спармания
370136 Пизанг, Спармания африканская
370156 Спартина
370191 Дрок испанский, Метельник
370202 Бобровик, Бобровник, Дрок безлистный, Дрок испанский, Дрок испанский обыкновенный, Испанский дрок, Метельник прутовый, Метельник прутьевидный, Прутняк ситниковидный, Прутовник, Шильник ситниковидный
370358 Дерево габонское тюльпанное
370521 Торица, Шпергель
370522 Торица крупная, Торица обыкновенная, Торица пашенная, Торица полевая, Шпергель
370546 Торица льняная
370568 Торица пятитычинковая
370595 Торица посевная
370604 Торица весенняя
370609 Торичник
370624 Торичник двутычинковый
370661 Торичник морской, Торичник солончаковый
370675 Торичник окаймленный, Торичник окаймлённый
370696 Торичник красивый, Торичник полевой
371160 Свайнсония, Сферофиза
371161 Сферофиза солонцовая
371237 Обманчивоплодник, Сфаллерокарпус
371240 Обманчивоплодник тонкий, Сфаллерокарпус тонкий
371400 Смирния Сфенокарпус
371403 Смирния Сфенокарпус синеголовниковый
371499 Булавоножка
371502 Булавоножка растопыренная
371601 Триния
371612 Спигелия
371627 Спирантес, Шпилат
371710 Шпинат
371713 Шпинат овошной, Шпинат огородный
371716 Шпинат четырехтычинковый
371717 Шпинат туркестанский
371785 Спирея, Таволга
371795 Таволга белая, Таволга шилоколистная
371798 Таволга альпийская
371807 Таволга водосборолистная
371815 Таволга острая
371818 Таволга бальджуанская
371823 Спирея березолистная, Таволга березолистная, Таволга берёзолистная
371828 Таволга бовера
371834 Таволга биллиарда, Таволга бильярда
371877 Спирея вязолистная, Спирея городчаталистная, Спирея городчатая, Спирея дубровколистая, Спирея дубровколистная, Таволга дубровколистная, Таволга уссурийская
371884 Таволга китайская
371890 Спирея городчатая, Таволга городчатая
371895 Таволга даурская
371902 Таволга дугласа
371903 Таволга изящная
371907 Таволга ферганская
371910 Спирея извилистая, Таволга извилистая
371912 Спирея тайваньская
371937 Таволга низкая
371938 Спирея зверобоелистая, Спирея зверобоелистная, Спирея зверобойнолистная, Таволга зверобоелистная
371944 Спирея мозолистая, Спирея японская, Таволга японская
371981 Таволга японская звездчатая
371988 Таволга гуансиская
371989 Спирея гуйцзоуская
371994 Таволга зверобоелистая
371997 Спирея лицзянская
372006 Спирея средняя, Товалга средняя
372012 Спирея мийябе
372018 Спирея монгольская
372022 Таволга мулиская
372027 Таволга нинсяская
372029 Таволга ниппонская
372047 Таволга волосистая
372050 Спирея сливолистная
372067 Спирея пушистая, Таволга пушистая
372072 Спирея наньчуаньская, Таволга ростгорна
372075 Спирея белая, Спирея иволистая, Спирея иволистная, Таволга белая, Таволга иволистная
372078 Таволга саржента
372081 Таволга шоха
372082 Таволга шренка
372083 Таволга шелковистая
372101 Спирея тунберга, Таволга тунберга
372104 Таволга тяньшанская
372106 Спирея волосистая, Таволга войлочная
372114 Таволга опушенноплодная
372117 Таволга трехлопастная
372134 Таволга виргинская
372135 Таволга вильсона
372136 Таволга тибетская
372141 Таволга цабеля
372151 Таволгоцвет
372153 Таволгоцвет шренковский
372176 Скрученник, Спирантес
372247 Скрученник китайский, Скрученник приятный, Спирантес китайский
372259 Скрученник осенний
372295 Многокоренник, Спиродела
372300 Многокоренник обыкновенный, Ряска многокорешковая, Спиродела многокоренная
372304 Спиродела сычуаньская
372333 Серпоносик
372335 Серпоносик песчаный
372354 Спиростегия
372355 Спиростегия бухарская
372398 Сербородник, Сподиопогон
372412 Серобородник тайваньский
372428 Серобородник сибирский, Сподиопогон сибирский
372459 Спондиас
372467 Спондиас цитеры
372477 Момбин жёлтый, Слива испанская тропическая, Спондиас жёлтый
372574 Споролобус, Спороболюс
372650 Спороболус индийский
372666 Спороболус индийский
372737 Спороболус японский
372912 Спрекелия
372934 Спрыгиния винкрева
372966 Срединския
372969 Срединския большая
373068 Стахиопсис
373069 Стахиопсис ясноткоцветковый
373070 Стахиопсис шандровый
373071 Стахиопсис продолговатолистный
373085 Буквица, Буковица, Медвежья лапа, Стахис, Чистец
373107 Чистец альпийский
373109 Чистец узколистный
373110 Чистец однолетний
373125 Чистец полевой
373139 Чистец байкальский, Чистец ридера
373151 Чистец балянзы

373157 Чистец буасы
373166 Живика пушистая, Чистец шерстистый
373173 Чистец китайский
373185 Чистец критский
373189 Чистец черняева
373212 Чистец фомина
373219 Чистец кустарничковый
373224 Чистец германский
373234 Чистец гросгейма
373238 Чистец разнозубый
373247 Чистец гиссарский
373256 Чистец грузинский
373258 Чистец вздутый
373260 Чистец промежуточный
373261 Чистец иранский
373262 Стахис японский, Чистец японский
373273 Чистец комарова
373276 Чистец франше
373288 Чистец лавандолистный
373300 Чистец крупнолистный
373303 Чистец приморский
373334 Чистец зубчатолистный
373346 Чистец болотный
373371 Чистец павла
373373 Чистец персидский
373381 Чистец ложнохлопьвидный
373386 Чистец пушистый
373389 Чистец прямой
373435 Чистец щетинистый
373439 Артишок китайский, Артишок японский, Стахис клубненосный, Хороги, Чистец клубневой, Чистец клубненосный, Чистец съедобный
373443 Чистец сосновского
373445 Чистец нарядный
373446 Чистец замечательный
373455 Чистец лесной
373456 Чистец далиский
373457 Чистец талышский
373463 Чистец терекский
373466 Чистец тибетский
373469 Чистец трапезунтский
373471 Чистец трехжилковый
373472 Чистец чаткальский
373475 Чистец туркменский
373476 Чистец туркестанский
373480 Чистец волжский
373481 Чистец ворована
373516 Стахиурус
373524 Стахиурус китайский
373556 Стахиурус ранний
373576 Стахиурус сычуаньский
373608 Бартсия альпийская
373673 Стангерия
373719 Кактус ластовневый, Стапелия
373827 Стапелия крупноцветковая
374087 Клекачка, Клокичка
374089 Клокичка бумальда
374098 Клекачка колхидская, Клокичка колхидская
374100 Клокичка гималайская
374107 Клекачка перистая, Клокичка перистая, Стафилея перистая
374114 Клекачковые, Клокичковые
374381 Стаунтония, Стонтония
374392 Стаунтония китайская
374400 Стонтония тайваньская
374403 Стаунтония хайнаньская
374409 Стаунтония шестилистная, Стонтония шестилисточковая
374440 Стонтония яошаньская
374724 Альзина, Алэина, Звездочка, Звездчатка, Мокрица алзина, Мокричник, Стеллария
374730 Звездочка алексеенко
374731 Звездочка топяная, Звездчатка альзиновидная
374739 Звездочка мокричная
374740 Звездочка тупочашелистная
374742 Звездочка курослепная
374743 Звездочка аньхуйская
374751 Звездочка арктиеская
374777 Звездочка коротколепестная
374780 Звездочка бунге, Звездчатка бунге
374783 Звездочка чашечкоцветковая, Звездчатка чашечкоцветковая
374794 Звездочка шерлериевидная
374802 Звездчатка китайская
374803 Звездочка ресничатая
374807 Звездочка толстолистная, Звездчатка толстолистная
374821 Звездочка даурская
374836 Звездочка развилистая, Звездчатка вильчатая
374846 Звездочка моховидная
374848 Звездочка раскидистая
374858 Звездочка двуцветная, Звездчатка двуцветная
374872 Звездочка бесприцветничковая, Звездчатка бесприцветничковая
374873 Звездочка эдвардса
374878 Звездочка ешшольца
374880 Звездочка фенцля
374882 Звездочка ялусская
374884 Звездочка фишера
374888 Звездочка прилучеймная
374897 Звездочка злачная, Звездчатка злаковая, Звездчатка злаковидная, Звездчатка злачная, Звездчатка травяная, Пьяная трава, Трава пьяная
374907 Звездчатка цзилунская
374913 Звездочка пушисточашечная, Звездчатка пушисточашечная
374914 Звездчатка хубэйская
374916 Звездочка жестколистная, Звездчатка ланцетовидная, Звездчатка ланцетолитная
374919 Звездочка приземистая, Звездчатка приземистая
374923 Звездочка черепичатая
374926 Звездочка орошаемая
374929 Звездочка каратавская
374932 Звездочка кочи
374940 Звездочка лаксманна
374944 Звездочка раскидистая, Звездчатка раскидистая
374951 Звездочка длинноцветоножковая, Звездочка эдвардса, Звездчатка длинноножковая, Звездчатка стебельчатая, Звездчатка эдвардса
374965 Звездочка мартьянова
374968 Звёздочка, Звездочка средная, Звездчатка средняя, Звездчатка-мокрица, Мокрица
374986 Звездчатка средная мелкоцветковая
375007 Звездочка незамеченная, Звездчатка незамеченная
375008 Звездочка дубравная, Звездчатка дубровная, Звездчатка лесная
375017 Звездчатка непалская
375028 Звездчатка эмэйская
375033 Звездочка бледная, Звездчатка бледная
375036 Звездочка болотная, Звездчатка болотная, Звездчатка сизая
375049 Звездочка персидская
375051 Звездочка каменная
375071 Звездочка лучистая, Звездчатка лучистая
375078 Звездочка иглицелистная
375089 Звездочка шугнанская
375094 Звездчатка сибирская
375100 Звездочка джунгарская
375101 Звездочка кандинская
375114 Звездочка тибетская
375116 Звездочка туркестанская
375128 Звездочка зонтичная, Звездчатка зонтичная
375144 Звездочка винклера
375179 Стеллера
375187 Стеллера, Стеллера карликовая
375198 Стеллера тайваньская
375207 Стеллеропсис
375209 Стеллеропсис алтайский
375210 Стеллеропсис антонины
375211 Стеллеропсис кавказский
375212 Стеллеропсис иранский
375213 Стеллеропсис иссыкульский
375214 Стеллеропсис магакяна
375215 Стеллеропсис тяньшанский

375216 Стеллеропсис туркменский
375268 Рапонтикум сафлоровидны
375274 Большеголовник дноцветковый, Рапонтикум одноцветковый
375451 Стенанциум
375459 Стенанциум сахалинский
375510 Стенокарпус
375511 Стенокарпус ивовый
375551 Стеноцелиум, Узколожбиник
375552 Узколожбиник атамантовидный
375555 Узколожбиник волосистоплодный
375725 Тонкотрубочник
375726 Тонкотрубочник скальный
375760 Стенотения
375761 Стенотения крапнолодная
375764 Стеноафрум
375765 Стеноафрум американский
375805 Пухополевица
375809 Пухополевица летучконоснаю
375811 Стефанандра
375812 Стефанандра китайская
375817 Стефанандра надрезаннолистная, Стефанандра надрезная
375821 Стефанандра танаки
375827 Стефания
375834 Стефания цзиндунская
375854 Стефания тибетская
375859 Стефания хайнаньская
375870 Стефания японская
375880 Стефания гуансиская
375889 Стефания ланьюйская
375890 Стефания мелкоцветковая
375896 Стефания округлая
375897 Стефания тайваньская
375903 Стефания сычуаньская
375916 Венцовник
375917 Венцовник ольги
376022 Стефанотис
376052 Степторамфус
376053 Степторамфус катранолистный
376054 Степторамфус толстостебельный
376055 Степторамфус черепанова
376056 Степторамфус линчевского
376057 Степторамфус персидский
376058 Степторамфус скальный
376059 Степторамфус клубненосный
376063 Стеркулия
376110 Стеркулия вонючая
376180 Стеркулия утёсная
376218 Стеркулиевые
376314 Стелигма, Стелигмостемум
376315 Стелигма иглоплодная
376322 Стелигма бугроватая
376350 Штернбергия
376353 Штернбергия зимовникоцветная
376354 Штернбергия фишера
376356 Штернбергия желтая, Штернбергия жёлтая
376380 Стеуднера
376382 Стеуднера колоказиелистная
376387 Стевения
376388 Стевения бурачковидная
376389 Стевения левкоевидная, Стевения левкойная
376390 Стевениелла
376391 Стевениелла
376405 Стевия
376410 Стевия яйцевидная
376428 Стюартия
376457 Стюартия однобратственная, Стюартия-монодерьфа
376466 Стюартия псевдокамелия
376536 Вьюнок колокольчатый
376660 Блестящеколосник
376661 Блестящеколосник желтоцветный
376662 Блестящеколосник зайцехвостовидный
376687 Ковыль, Чий перистый
376691 Ковыль уклоняющийся
376695 Ковыль тургайский
376701 Ковыль араксинский
376704 Ковыль алмянский
376709 Ковыль бадахшанский
376710 Ковыль байкальский
376718 Ковыль красивый
376720 Ковыль коротковетковый
376723 Ковыль бунге
376727 Ковыль седоватый
376728 Ковыль закручивающийся
376732 Ковыль волосатик, Ковыль тырса
376737 Ковыль кавказский
376753 Ковыль родственный
376757 Ковыль толстостебельный
376758 Ковыль дагестанский
376759 Ковыль опушенолистный
376762 Ковыль густоцветковый
376779 Ковыль гигантский
376785 Ковыль галечный
376793 Ковыль тонкий
376795 Ковыль большой
376798 Годенак керовский
376799 Ковыль шелковистый
376804 Ковыль ильина
376807 Ковыль ягнопский
376810 Ковыль каратавский
376814 Ковыль киргизский
376819 Ковыль коржинского
376821 Ковыль крашенинникова
376836 Ковыль лессинга
376840 Ковыль языкообразный
376841 Ковыль липского
376842 Ковыль камнелюбивый
376844 Ковыль длинноперистый
376845 Ковыль длинноязычковый
376846 Ковыль великолепный
376859 Ковыль восточный
376862 Ковыль овчинникова
376864 Ковыль памисский
376867 Ковыль своеобразный
376877 Ковыль обыкновенный, Ковыль перистый
376884 Ковыль понтийский
376888 Ковыль лже-волосатик
376890 Ковыль красивейший, Ковыль красивый
376893 Птилагростис пурпуровый
376899 Ковыль регеля
376902 Ковыль рихтеровский
376908 Ковыль сарептский
376910 Ковыль крылова
376920 Ковыль спиридонова
376924 Птилагростис сидячецветковый
376930 Ковыль шовица
376931 Альфа, Ковыль тончайщий, Спарто, Трава альфа, Эспарто
376934 Ковыль шовица
376936 Ковыль тяньшанский
376949 Ковыль волосовидный
376953 Ковыль туркменский
376956 Ковыль туркестанский
376958 Ковыль украинский
376962 Ковыль залесского, Ковыль красноватый
377011 Аристида перистая, Селин перистый, Триостница перистая
377137 Стизолофус
377140 Стизолофус коронопусолистный
377286 Стоксия
377433 Странвезия
377444 Странвезия давида
377452 Странвезия давида волнистая
377462 Странвезия нусия, Странвезия сизоватая
377482 Телорез
377485 Телорез алоэвидный, Телорез обыкновенный
377550 Стрелиция
377576 Стрелиция королевская
377652 Стрептокарпус
377888 Завиток
377889 Завиток пустыный
377902 Стрептопус
377930 Стрептопус стеблеобъемлющий
377933 Стрептопус мелкоцветковый
377942 Стрептопус стрептопусовидный
377969 Стрига
377973 Стрига азиатская, Стрига желтая
378079 Стробилантес
378287 Строгановия
378289 Строгановия афганская
378290 Строгановия коротkoухая

378292 Строгановия сердцелистая
378294 Строгановия тонкая
378295 Строгановия промежуточная
378296 Строгановия литвинова
378297 Строгановия метельчатая
378298 Строгановия персидская
378299 Строгановия коренстая
378300 Строгановия стрелолистая
378301 Строгановия субальпийская
378302 Строгановия трауфеттера
378319 Стромбозия
378363 Строфант
378386 Строфант расходящийся
378497 Строфостилес
378633 Орех рвотный, Чилибуха
378761 Бобы игнатьевские, Игнаций, Игнация
378834 Корень рвотный, Орех рвотный, Орешник рвотный, Стрихнос рвотный орех, Целибуха, Чилибуха
378836 Кучерява, Орех рвотный, Чилибуха
378959 Стрифнодендрон
378960 Стрифнодендрон-барбатимао
378976 Штубендорфия
378977 Штубендорфия бескрылая
378978 Штубендорфия липского
378979 Штубендорфия восточная
378980 Штубендорфия двойчатая
378983 Рдест туполистный
378986 Рдест нитевидный
378991 Рдест гребенчатый
379076 Стилидиум
379296 Стираксовые
379300 Дерево стираксовое, Стиракс
379312 Дерево бархатновое, Стиракс бензоин, Стиракс бензоиновый, Стирасбинзоин
379325 Стиракс китайский
379331 Стиракс густоцветковый
379347 Стиракс тайваньский
379357 Стиракс ланьюйский
379361 Стиракс хайнаньский
379370 Стиракс ху
379374 Стиракс японский
379411 Стиракс гуансиский
379414 Стиракс обасия
379419 Дерево стираксовое лекарственное, Стиракс аптечный, Стиракс лекарственный
379427 Стиракс жуйлиский
379474 Стиракс вильсона
379478 Стиракс чжэцзянский
379486 Свада, Сведа, Содник, Соранг, Суеда, Чорак, Шведка
379489 Сведа заостренная
379492 Сведа высокая
379495 Сведа дуголистная
379501 Сведа ягодоносная
379508 Сведа запутанная
379509 Сведа рожконосная
379510 Сведа мелкоплодная
379513 Сведа серполистная
379514 Сведа кустарничковая
379519 Сведа эльтонская
379531 Сведа сизая, Шведка сизая
379533 Сведа разнолистная
379542 Сведа японская
379543 Сведа косинского
379547 Сведа линейнолистная
379548 Сведа липского
379553 Свада морская, Свада приморская, Сведа приморская, Содник морской
379566 Сведа мелколистная
379567 Сведа мелкосемянная
379577 Сведа олуфсена
379580 Сведа странная
379582 Сведа вздутоплодная
379584 Сведа стелющаяся
379590 Сведа карликовая, Сведа крылоцветная
379594 Сведа туркестанская
379598 Сведа солончаковая, Сведа уссурийская
379613 Сведа заамударьинская
379649 Шильник
379650 Шильник водный, Шильник водяной
379655 Сивец
379657 Сивец южный, Синяк изогнутый, Скамоний
379660 Сивец луговой, Чертогон луговой
379662 Сукцизелла
379663 Сукцизелла когнутая
379673 Сухтеления
379674 Сухтеления чашечная
380065 Свайнсона
380105 Офелия, Свертия, Сверция, Трипутник
380123 Сверция оше
380165 Сверция сросшаяся
380179 Сверция лицзянская
380181 Анагаллидиум вильчатый
380184 Офелия китайская, Сверция бледная
380197 Сверция эмэйская
380202 Сверция выгрызенная
380212 Сверция франше
380217 Сверция тонкоцветвая
380234 Сверция японская
380245 Сверция гуйцзоуская
380250 Сверция молочнобелая
380252 Сверция мэнцзыская
380271 Сверция окаймленная
380291 Свертия тупая, Сверция тупая
380302 Свертия многолетная
380357 Сверция кандинская
380376 Сверция четырехлепестковая
380385 Офелия четырехлепестная
380387 Сверция тибетская
380400 Сверция чемерицевидная
380411 Офелия вильфорда, Сверция вильфорда
380419 Сверция чаюйская
380425 Свида
380444 Дерен корейский
380448 Свидина дарвазская
380451 Дерен хемсли
380461 Дерен крупнолистный, Кизил крупнолиствый, Свидина коротконогая
380484 Дерен мелкоцветковый
380488 Дерен белолистный
380499 Глот, Дерен кравано-красный, Дерен красная, Дерён красный, Дерён кровавокрасный, Кизил кровяной, Кизил свидина, Свидина, Свидина кровавокрасная, Спиж
380514 Дерен уолтера
380522 Акажу, Дерево красное, Свиетения, Свиетения махагона
380528 Акажу, Анлкардия западная, Дерево алойное, Дерево красное, Дерево махагониевое, Махагони
380586 Сикопсис
380590 Сикопсис тайваньский
380602 Сикопсис китайский
380681 Симпегма
380682 Симпегма регеря
380696 Симфония
380697 Симфония габонская
380734 Снежник, Снежноягодник, Ягода снежная
380736 Жимолость виргинская, Снежноягодник белый, Снежноягодник ветвистый, Снежноягодник кистевидный, Снежноягодник кистевой, Снежноягодник кистистый, Ягода снежная китайская
380746 Снежноягодник мягкий
380748 Снежноягодник западный
380749 Снежноягодник красноплодный, Снежноягодник круглолистный, Снежноягодник обыкновенный, Снежноягодник округлый
380753 Жимолость виргинская, Снежноягодник белый, Снежноягодник ветвистый, Снежноягодник кистевидный, Снежноягодник кистистый, Ягода снежная китайская
380769 Симфиандра
380771 Симфиандра армянская
380775 Симфиандра повислая

380777 Симфиандра закавказская
380779 Симфиандра зангезурскаялазская
380791 Симфилокарпус
380792 Симфилокарпус тощий
380809 Сростноплодник
380810 Сростноплодник пахучий
380907 Астра гладкая
380930 Астра американская
380945 Астра мелкоголовчатая
381024 Окопник
381026 Комфрей, Окопник жесткий, Окопник жёсткий, Окопник острый, Окопник шероховатый, Окопник шершавый
381027 Окопник богемский
381029 Окопник кавказский
381030 Окопник сердцевиднолистный, Окопник сердцевидный
381031 Окопник крупноцвтковый
381035 Окопник аптечный, Окопник лекарственный
381037 Окопник восточный
381038 Окопник бродячий
381039 Окопник крымский
381040 Окопник клубневой, Окопник клубненостный
381067 Симплоковые
381070 Симплокарпус
381073 Симплокарпус вонючий
381090 Симплокос
381143 Симплокос японский
381149 Симплокос ланьюйский
381206 Симплокос тайваньский
381210 Симплокос фуцзяньский
381233 Симплокос хайнаньский
381252 Симплокос японский
381288 Симплокос люцюский
381341 Симплокос китайский, Симплокос метельчатый
381390 Симплокос сычуаньский
381463 Симплокос яошаньский
381475 Синадениум
381813 Синейлезис
381814 Синейлезис асонитолистный
381965 Дерево розовое
382067 Синурус, Сростнохвостник
382069 Синурус дельтовидный, Синурус треугольный, Сростнохвостник дельтовидный
382096 Сырейщиковия
382097 Сырейщиковия шипиковатая
382098 Сырейщиковия тонкая
382101 Сирения
382103 Сирения узколистная
382104 Сирения крупноплодная
382105 Сирения сидячецветковая
382107 Сирения стручочковая
382108 Сирения талиева
382114 Куст сиреневый, Сирень, Трескун
382132 Сирень китайская
382153 Сирень гималайская, Сирень эмоди
382178 Сирень японская
382180 Сирень венгерская, Сирень угорская
382190 Сирень комарова, Сирень поникшая
382220 Сирень широколистная
382232 Сирень хубэйская
382241 Сирень персидская
382257 Сирень мохнатая, Сирень опушенная, Сирень опушённая
382266 Сирень опушенная мелколистная
382268 Сирень мохнатая
382276 Сирень сетчатая
382277 Сирень амурская, Трескун амурский
382278 Сирень пекинская
382282 Сирень крупная
382312 Сирень мохнатая, Сирень опушённая
382329 Сирень обыкновенная
382383 Сирень вольфа
382388 Сирень юньнаньская
382465 Евгения
382477 Дерево гвоздичное, Евгения
382481 Евгения миртолистная
382522 Джамболан, Евгения ямболана, Сизигиум яванский, Слива яванская, Слива явская, Ямболан
382535 Евгения тайваньская
382545 Евгения гуншаньская
382549 Евгения гуансиская
382569 Евгения хайнаньская
382570 Евгения мелкоцветковое
382582 Аблоко розовое, Дерево гвоздичное, Джамбоза, Евгения, Помароза, Яблоко розовое, Ямбоза
382591 Евгения гуандунская
382606 Евгения малаккская, Яблоня малайская
382664 Евгения ланьюйская
382670 Евгения сычуаньская
382671 Евгения симаоская
382672 Евгения тайваньская
382682 Евгения вэньшаньская
382684 Евгения тибетская
382685 Евгения юньнаньская
382696 Совичия красивоплодная
382704 Табебуйя
382852 Дерево молочное индийское
382908 Такка
382923 Аррорут индийский, Нианца, Цванга
382934 Такковые
383034 Лентоостник
383035 Лентоостник шероховатый
383040 Лентоостник длинноволосый
383087 Бархатцы, Тагетес
383090 Бархатцы высокие, Бархатцы крупноцветковые, Бархатцы крупноцветные, Бархатцы прямостоячие, Тагетес прямостоячий, Шапочки
383100 Бархатцы мелкие
383103 Бархатцы мелкоцветковые, Бархатцы низкие, Бархатцы отклоненные, Тагетес распростертые, Шапочки
383107 Тагетес мексиканская
383188 Тайвания
383189 Тайвания криртомериевидная
383233 Талассия
383235 Горечник заилийский, Талассия заилийская
383236 Талаума
383397 Бисерниковые, Гребенщиковые, Тамарксовые
383404 Тамаринд, Финик индейский
383407 Тамаринд индийский, Финик индийский
383412 Бисерник, Гребенчук, Гребенщик, Дерево бисерное, Жидовник, Тамарикс
383414 Гребенщик африканский, Смилаксовые
383431 Гребенщик андросова, Тамарикс андросова
383444 Гребенщик можжевеловый
383466 Бисерник бунге
383469 Гребенщик китайский, Тамарикс китайский
383477 Тамарикс одесский
383479 Гребенщик вытянутый, Гребенщик удлиненный, Тамарикс удлиненный
383484 Гребенщик эверсманна, Тамарикс эверсманна
383486 Бисерник яркий
383491 Гребенщик французский, Тамарикс французский
383514 Тамарикс ганьсуский
383516 Гребенщик изящный, Тамарикс изящный
383517 Гребенщик опушённый, Гребенщик щетинистоволосистый, Гребенщик щетинистоволосый, Тамарикс щетинистоволосистый, Тамарикс щетинистоволосисый
383519 Гребенщик гогенакера, Тамарикс гогенакера
383523 Бисерник каракалинский
383526 Гребенщик карелина
383528 Бисерник комарова
383530 Гребенщик королькова
383531 Бисерник араратсий, Бисерник кочи

383535 Гребенщик раскидистый, Гребенщик рыхлый, Тамарикс раскидистый
383542 Гребенщик тонкоколосый
383547 Бисерник литвинова
383555 Бисерник тонколепестный
383557 Гребенщик мейера, Тамарикс мейера
383559 Тамарикс монгольский
383564 Бисерник восимитычинковый
383581 Гребенщик четырехтычинковый, Гребенщик четырёхтычинковый, Тамарикс мелкоцветковая, Тамарикс четырехтычинковый, Тамарикс четырёхтычинковый
383585 Бисерник тимелеевый
383591 Гребенщик пятитычинковый, Тамарикс палласса, Тамарикс патитычночный
383595 Гребенщик многоветвистый, Гребенщик одесский, Тамарикс многоветвистый, Тамарикс одесский
383674 Тамус
383675 Тамус обыкновенный
383690 Пижма
383691 Пижма абротаноистная
383692 Пижма тысячелистниковая, Ромашник тысячелистниковый
383700 Пижма акинфиева
383712 Калуфер, Кануфер, Колуфер, Конуфер, Пижма душистая, Пижма кануфер, Пижма пахучая, Пиретрум бальзамический, Хризантема бальзамная
383714 Пижма алтайская, Пижма турланская
383716 Пижма дважты-перистая
383727 Пижма закавказская
383740 Пижма толстоножковая
383765 Пижма разнолистная
383780 Пижма карелина
383784 Пижма киттари
383793 Пижма длинноножковая
383806 Пижма многолистная
383810 Пижма малоголовчатая
383813 Пижма пачосского
383832 Пижма ложнотысячелистниковая
383842 Пижма сантолиновидная
383843 Пижма скальная
383845 Пижма жестколистная
383846 Пижма утесная
383857 Пижма пижмовидная
383867 Пижма туркменская
383869 Пижма улутавская
383871 Пижма одноцветковая
383872 Пижма уральская
383874 Дикая рябина, Дикая рябинка, Пижма дикая рябина, Пижма обыкновенная, Поповник обыкновенный, Пуговичник, Рябина дикая, Рябишник, Хризантемум обыкновенный
383876 Пижма северная
383989 Ямкосимяник
383990 Ямкосимяник алтайский
383996 Ямкосимяник широколепестный
384420 Одуванчик
384425 Одуванчик алайский
384427 Одуванчик алатавский
384429 Одуванчик беловатый
384437 Одуванчик алпийский
384438 Одуванчик алтайский
384443 Одуванчик андросова
384446 Одуванчик тяньцюаньский
384451 Одуванчик армянский
384452 Одуванчик ашхабадский
384460 Одуванчик черноватый
384462 Одуванчик почерневший
384466 Одуванчик балтийский
384467 Одуванчик бессарабский
384472 Одуванчик двурогий
384473 Одуванчик китайский, Одуванчик сизоцветковый
384474 Одуванчик бочанцева
384475 Одуванчик меркорогий
384479 Одуванчик капустнолистный
384480 Одуванчик коротксрогий
384482 Одуванчик коротконосиковый
384485 Одуванчик известняковый
384498 Одуванчик хирийский
384499 Одуванчик предкавказский
384500 Одуванчик холмовой
384501 Одуванчик скученный
384502 Одуванчик сложенный
384503 Одуванчик опутанный
384505 Одуванчик корейский
384507 Одуванчик скердоподобный
384508 Одуванчик шафранный
384513 Одуванчик чуйский
384515 Одуванчик лицзянская
384516 Одуванчик беловатоцветковый, Одуванчик беловатый
384518 Одуванчик обесцвеченный
384521 Одуванчик пустынный
384523 Одуванчик светрый
384524 Одуванчик байсальский, Одуванчик рассеченнолистнный, Одуванчик рассеченный
384526 Одуванчик непохожий
384529 Одуванчик безрогий
384533 Одуванчик удлиненный
384534 Одуванчик мохнатопазушный
384536 Одуванчик шерстистоногий
384540 Одуванчик красноплодный, Одуванчик красносемянный
384547 Одуванчик бесканальцевый
384551 Одуванчик федченко
384553 Одуванчик ключевой
384554 Одуванчик тайваньский
384559 Одуванчик почти-голый
384560 Одуванчик гладгий
384562 Одуванчик сизо-зеленый
384564 Одуванчик голоскокова
384565 Одуванчик крупнолисточковый
384567 Одуванчик гроссгейма
384569 Крым-сагыз, Одуванчик осенний
384576 Одуванчик хьелта
384581 Одуванчик хультена
384587 Одуванчик индийский
384590 Одуванчик промежутоный
384591 Одуванчик японский
384595 Одуванчик юзенпчука
384596 Одуванчик камчатский
384599 Одуванчик каратавский
384601 Одуванчик кетойский
384602 Одуванчик киргизский
384603 Одуванчик ключевской
384604 Одуванчик клокова
384605 Одуванчик койнмы
384606 Кок-сагыз, Одуванчик кок-сагыз, Одуванчик короткорожковый
384607 Одуванчик корагинский
384608 Одуванчик корагинец
384620 Одуванчик лапландский
384621 Одуванчик кирпичный
384624 Одуванчик широкочешуйный
384627 Одуванчик тонкорокий
384628 Одуванчик белоцветковый
384633 Одуванчик лиловый
384634 Одуванчик гнездилло
384636 Одуванчик липского
384637 Одуванчик гладкоплодный
384640 Одуванчик длиннорогий
384642 Одуванчик длинноножковый
384643 Одуванчик длинноносиковый, Одуванчик длиннопирамидковый
384647 Одуванчик бадахшанский, Одуванчик грязно-желтый, Одуванчик иконникова
384649 Одуванчик ировидный
384650 Одуванчик тоший
384653 Одуванчик крупночешуйный
384654 Одуванчик крупный
384655 Одуванчик большой
384656 Одуванчик малеза
384660 Одуванчик самаркандский
384669 Одуванчик мексиканский
384674 Одуванчик мелкосемянный
384676 Одуванчик мелколоастный
384677 Одуванчик ядунский
384678 Одуванчик миуаке
384681 Одуванчик монгольский

384683 Одуванчик, Одуванчик однопокровный, Одуванчик цельнолистный
384684 Одуванчик горный
384685 Одуванчик многокорзиночный
384687 Одуванчик остробугорчатый
384688 Одуванчик мурманский
384689 Одуванчик наиринский
384690 Одуванчик начикский
384692 Одуванчик новосахалинский
384693 Одуванчик невского
384694 Одуванчик чернеющий
384695 Одуванчик никитина
384696 Одуванчик снежный
384697 Одуванчик норвежский
384698 Одуванчик новаземельский
384699 Одуванчик нуратавский
384702 Одуванчик косой, Одуванчик неравнобокий
384714 Одуванчик аптестый, Одуванчик бугорчатый, Одуванчик лекарственный, Одуванчик обыкновенный, Одуванчик рогарый, Одуванчик рогоносный
384719 Одуванчик ови
384720 Одуванчик немногоцветковый
384721 Одуванчик ошский
384729 Одуванчик памирский
384732 Одуванчик мелкоцветковый
384736 Одуванчик богатомлечный
384737 Одуванчик маленький
384739 Одуванчик боровой
384740 Одуванчик жирный
384747 Одуванчик плоскорогий
384748 Одуванчик плоскочешуйный
384749 Одуванчик плосколисточковый
384752 Одуванчик порярковой
384753 Одуванчик пурпурный
384755 Одуванчик ранний
384756 Одуванчик луговой
384757 Одуванчик принтца
384763 Одуванчик ложноальпийский
384764 Одуванчик ложнопочерневший
384766 Одуванчик
384767 Одуванчик ложноголый
384769 Одуванчик ложномелколопастный
384770 Одуванчик ложночерноватый
384772 Одуванчик ложнорозовый
384778 Одуванчик загнутый
384782 Одуванчик красноватый
384784 Одуванчик рыжеватый
384785 Одуванчик сахалинский
384788 Одуванчик стреролистный
384789 Одуванчик стрелолистный
384792 Одуванчик шерковникова
384794 Одуванчик штганский
384797 Одуванчик зеравшанский
384799 Одуванчик поздний
384802 Одуванчик шикотанский
384803 Одуванчик симуширский
384804 Одуванчик шимшинский
384805 Одуванчик сибирский
384812 Одуванчик джунгарский
384814 Одуванчик станюковича
384817 Одуванчик узкочешуйный
384818 Одуванчик узколопастный
384819 Одуванчик стевена
384820 Одуванчик шишкоголовый
384823 Одуванчик приледниковый
384824 Одуванчик болотный
384825 Одуванчик сумневича
384827 Одуванчик сирийский
384828 Одуванчик тадшикский
384829 Одуванчик татеваки
384830 Одуванчик крымский
384831 Одуванчик тонкорассеченный
384832 Одуванчик тяньшанский
384833 Одуванчик тибетский
384836 Одуванчик туюксуйский
384837 Одуванчик тундровый
384839 Одуванчик тургайский
384840 Одуванчик турийский
384841 Одуванчик цвелева
384846 Одуванчик уссурийский
384849 Одуванчик варзобский
384850 Одуванчик васильченко
384854 Одуванчик воронова
384855 Одуванчик вулканный
384859 Одуванчик сухолюбивый
384861 Одуванчик ямамото
384864 Одуванчик йетрофуйский
385129 Таущерия
385132 Таущерия длинная, Таущерия опушенноплодная
385175 Тисовые, Тиссовые
385250 Таксодиевые, Тиссовые
385253 Болотный кипарис, Волотный кипалис, Таксодий
385262 Болотный кипарис обыкновенный, Волотный кипалис обыкновенный, Дерево жизненное, Дерево жизни, Кипарис болотный, Кипарис виргинский, Кипарис двураздельный, Кипарис лысый, Негниючка, Туя
385281 Болотный кипарис мексиканский, Болотный кипарис остроконечный, Волотный кипалис мексиканский, Кипарис болотный мексиканский
385301 Дерево тиковое, Негной-дерево, Таксус, Тис, Тисс
385302 Таксус баккатa, Таксус обыкновенный, Таксус ягодный, Тис европейский, Тис ягодный, Тисс европейский, Тисс ягодный
385342 Тис коротколистный, Тисс коротколистный
385344 Тис канадский, Тисс канадский
385355 Дерево розовое, Тис японский, Тисс восточный, Тисс дальневосточный, Тисс остроконечный, Тисс японский
385384 Тисс средний
385404 Таксус тибетский
385405 Тис китайский, Тисс китайский
385407 Тисс майра
385461 Текома
385505 Текомария
385508 Текомария капская
385529 Дерево тиковое, Тектона, Тик
385531 Дерево джатовое, Дерево тиковое, Тик
385549 Тисдайлия
385555 Тисдайлия голостебельная
385614 Телекия
385616 Телекия красивая
385629 Телефиум, Хлябник
385638 Телефиум восточный
385650 Тельфайрия
385655 Тельфайрия восточноафриканская, Тельфайрия стоповидная
385669 ложнодрок
385892 Крестовник пламенный
385902 Крестовник кавакамии
385903 Крестовник равинный
385908 Крестовник арктический
385915 Крестовник азиатский
385917 Крестовник головчатый
385920 Крестовник малозубчатый
385924 Крестовник турчанинова
385926 Тефрозия
386257 Тефрозия пурпурная
386448 Терминалия
386500 Миндаль тропический, Миробалан, Терминалия, Терминалия катаппа
386504 Дерево миробалановое
386652 Терминалия пышная, Терминалия хвалебная
386656 Терминалия шерстистая
386689 Тернстремия
386696 Тернстремия голопыльниковая, Тернстремия японская
386702 Тернстремия хайнаньская
386712 Тернстремия гуандунская
386720 Тернстремия мелколистная
386730 Тернстремия сычуаньская
386731 Тернстремия симаоская
386853 Тетрацера
386942 Дерево сандараковое, Тетраклинис членистый, Туя алжирская
386946 Четверозубец, Четырехзубчик
386948 Четырехзубчик бухарский
386952 Четверозубец памирский
386953 Четверозубец четырехрогий

386955 Четверозубец загнутый
387046 Тетрадиклис
387049 Тетрадиклис тонднький
387221 Тетрагония, Шпинат новозеландский
387239 Тетрагонолобус
387256 Тетрагонолобус струковый
387428 Тетрапапакс
387432 Бумажное дерево, Тетрапапакс бумажный, Чертово дерево
387435 Тетрапапакс тибетский
387512 Хлолис мохнатая
387989 Дубравник, Дубровник, Паклун
388000 Дубровник аньлунский
388015 Дубровник кистевидный, Дубровник метельчатый
388035 Дубровник седой
388042 Дубравник обыкновенный, Дубровник дубравный, Дубровник обыкновенный, Дубровник пурпуровый, Растигор
388076 Дубровник высокий
388078 Дубровник фишера
388104 Дубровник гирканский
388112 Дубровник яйлы
388114 Дубровник японский
388126 Дубравник крымский, Дубровник крымский
388147 Дубровник горький, Дубровник кошачий, Майоран кошачий
388154 Дубровник горный
388155 Дубровник многоузловый
388165 Дубравник нухинский, Дубровник нухинский
388172 Дубровник восточный
388177 Дубровник паннонский
388178 Дубровник мелкоцветковый
388184 Дубравник седой, Дубровник беловойлочный, Дубровник белый
388242 Дубровник предгорный
388246 Дубровник красивый
388270 Дубровник скордиевидный, Дубровник чесноч новидный
388272 Дубровник чесночный, Скордия, Чеснок дикий, Чеснок заячий
388275 Дубровник чесноковый, Дубровник шалфейный
388288 Дубровник испирский
388289 Дубровник тайваньский
388291 Дубровник тэйлора
388298 Дубровник уссурийский
388299 Дубровник верониковидный
388357 Талассия
388397 Василисник, Василистник
388404 Василисник альпийский, Василистник альпийский
388416 Василисник амурский, Василистник амурский
388421 Василисник узколистый
388423 Василисник водосборолистный, Василистник водосборолистный, Матренка
388432 Василисник байкальский, Василистник байкальский
388460 Василисник скрученный, Василистник скрученный
388503 Василисник нитчатый, Василистник тычинковый
388506 Василисник желтый, Василисник жёлтый, Василистник желтый
388510 Василисник вонючий, Василистник вонючий
388516 Василистник фриза
388533 Василисник хэнаньский
388539 Василисник изопироидный, Василистник изопироидный
388541 Василисник яванский
388556 Василистник блестящий, Василистник светлый
388569 Василисник мелкоплодный
388572 Василисник малый, Василисник обыкновенный, Василистник малый, Василистник холмовой
388584 Василистник кемский
388607 Василисник эмэйский
388614 Василисник ложнолепестковый, Василистник лепестковый, Василистник ложнолепестковый
388633 Василистник редкоцветковый
388654 Василисник сахалинский, Василистник сахалинский
388667 Василисник простой, Василистник простой
388673 Василисник редкоцветный, Василистник редкоцветный
388680 Василисник растопыренный, Василистник растопыренный
388683 Василисник прямой
388685 Василисник султанабадский
388701 Василисник триждысройчатый
388703 Василисник клубненосный, Василистник клубненосный
388710 Василистник тайваньский
388719 Василистник лицзянский
388722 Василистник иезский
388868 Тапсия, Трава злая
389046 Чайные
389153 Телесперма
389181 Телигоновые, Цинокрамбовые
389183 Капуста собачья, Телигонум
389184 Телигонум цинокрамба
389185 Телигонум тайваньский
389186 Телигонум японский
389197 Теллунгиэлла
389198 Теллунгиэлла солелюбка
389199 Резушка маленькая
389200 Теллунгиэлла солонцовая
389239 Свидина
389251 Свидина мейера
389258 Телимитра
389307 Темеда
389399 Дерево какаовое, Дерево шоколадное, Какао, Теоброма
389404 Дерево какаовое, Дерево шоколадное, Дерево-какао, Какао, Какао настоящее, Шоколадник, Шоколадное дерево
389505 Мышатник, Термопсис
389507 Термопсис альпийский
389510 Термопсис очередноцветковый
389518 Термопсис китайский
389520 Масти, Термопсис длинноплодный
389525 Термопсис цзилунский
389533 Мышатник ланцетовидный, Термопсис ланцетный, Термопсис ланцетолистный
389540 Термопсис бобовый, Термопсис лупиновый
389543 Термопсис монгольский
389557 Термопсис туркестанский
389572 Рододендрон камчатский
389577 Рододендрон редовского
389592 Ленец, Ленолистник
389597 Ленец алатавский
389600 Ленец альпийский
389623 Ленец коротколистный
389638 Ленец китайский
389672 Ленец дэциньский
389677 Ленец бесприцветничковый
389691 Ленец ферганский
389711 Ленец гончарова
389753 Ленец кочи
389763 Ленец льнолистный
389769 Ленец длиннолистный
389780 Ленец приморский
389792 Ленец минквиц
389795 Ленец многотебельный
389825 Ленец простертый
389838 Ленец сильно ветвистый
389840 Ленец ветвистый
389844 Ленец преломленный, Ленец сибирский
389848 Ленец ползучий
389856 Ленец каменистый
389857 Ленец щебнистый
389893 Ленец шовица
389967 Теведия
389978 Теведия олеандролистная
390107 Тладианта
390128 Тладианта дубия, Тладианта сомнительная

390202 Рогоплодник, Ярутка
390208 Ярутка альпийская
390210 Ярутка тибетская
390211 Ярутка однолетняя
390212 Ярутка армянская
390213 Вередник, Денежка широкопластинчатая, Денежник, Ярутка полевая
390223 Ярутка безухая, Ярутка ложечная
390231 Ярутка ферганская
390232 Ярутка сычуаньская
390233 Ярутка фрейна
390236 Ярутка хюта
390237 Ярутка японская
390239 Ярутка кочи
390241 Ярутка крупноцветковая
390242 Ярутка горная
390247 Ярутка округлая
390249 Ярутка пронзенная, Ярутка пронзённая
390255 Ярутка плоскоплодная
390256 Ярутка ранняя
390258 Ярутка карликовая
390259 Ярутка розовая
390260 Ярутка носатая
390264 Ярутка шовица
390268 Ярутка зонтичная
390424 Туиния
390460 Тринакс
390567 Биота, Дерево жизненное, Дерево жизни, Негниючка, Туя
390579 Туя корейская
390587 Дерево жизненное, Дерево жизни, Негниючка, Туя, Туя западная
390589 Туя западная ‘Ауреа’
390596 Туя западная ‘дугласин пирамидалис
390598 Туя западная ‘Эрикоидес
390602 Туя западная ‘Глобоза’
390617 Рекурва Нана, Туя западная
390645 Туя гигантская, Туя складчатая
390660 Туя стандиша, Туя стэндыша, Туя японсая
390661 Туевик сычуаньский
390673 Туевик
390680 Туевик поникающий, Туевик японский
390682 Туевик поникающий ‘вариегата’
390692 Тунбергия
390701 Тунбергия крылатая
390767 Тунбергия душистая
390793 Тунбергия хайнаньская
390916 Туния
390943 Туспейнанта
390944 Туспейнанта персидский
390967 Тилакоспермум
390968 Тилакоспермум плотнодернистый
390994 Тимелея, Тимелия
391015 Тимелея волобьиная, Тимелея обыкновенная, Тимелея однолетная
391030 Волчниковые, Тимелеевые, Ягодковые
391061 Тимиан, Тимиян, Тимьян, Травга богородичная, Чабрец, Чебрец
391064 Тимьян алатавский
391076 Тимьян попеременный
391077 Тимьян огетый
391078 Тимьян амурский
391079 Тимьян арараский
391080 Чебрец арктический
391083 Тимьян армянский
391084 Тимьян альсеньева
391085 Тимьян ашурбаева
391098 Тимьян башкирский, Чебрец башкирский
391101 Тимьян двжилковый, Чебрец двухнервный
391102 Тимьян смолистный
391106 Тимьян днепровский, Чебрец днепровский
391111 Тимьян бухарский
391112 Тимьян болгарский
391113 Тимьян буша
391115 Чебрец извесковый
391116 Чебрец каллье
391121 Тимьян кавказский
391151 Тимьян окаймленный
391155 Тимьян холмовый
391158 Тимьян кожелистный
391159 Тимьян частолистный
391160 Тимьян зазубренный
391161 Тимьян известковый, Чебрец меловой
391162 Чебрец сталинградский
391163 Тимьян клиновид
391164 Тимьян короткий
391165 Тимьян короткий, Чебрец черняева
391166 Тимьян дагестанский
391167 Тимьян даурский
391170 Тимьян уменишенный
391171 Тимьян двуформенный, Чебрец диморфный
391172 Тимьян разъединенный
391173 Тимьян разнолистный
391176 Тимьян дубянского, Чебрец дубянского
391177 Тимьян дзевановского, Чебрец дзевановского
391178 Тимьян елизаветы
391179 Тимьян эльтонский, Чебрец эльтонский
391180 Тимьян безжилковый
391181 Тимьян еравинский
391182 Тимьян пустынник
391183 Тимьян хлопчатый
391184 Тимьян байкальский
391185 Тимьян евпаторийский, Чебрец евпаторийский
391186 Тимьян федченко
391187 Тимьян гибкий
391188 Тимьян фомина
391193 Тимьян ледяной
391195 Тимьян гранитный, Чебрец гранитный
391196 Тимьян гроссгейма
391197 Тимьян губерлинский, Чебрец губерлинский
391199 Тимьян гаджиева
391200 Чебрец казахский
391203 Тимьян косматый, Чебрец волосистый
391204 Тимьян волосистотебельный
391218 Тимьян ильина
391220 Тимьян неравный
391221 Тимьян неверный
391223 Тимьян иртышский
391224 Тимьян яйлы, Чебрец яйлияский
391225 Тимьян японский
391227 Тимьян енисейский
391229 Тимьян кальмиусский, Чебрец кальмиусский
391230 Тимьян карамарьянский
391231 Тимьян тяньшанский
391232 Тимьян карягина
391233 Тимьян казахстанский
391234 Тимьян киргизский, Чебрец киргизский
391237 Тимьян клокова
391238 Тимьян комарова
391239 Тимьян кочиев
391240 Тимьян ладжанурский
391246 Тимьян шерстистый, Чебрец шерстистый
391247 Тимьян широколистный, Чебрец широколистный
391254 Тимьян липского
391255 Тимьян побережный, Чебрец прибрежный
391256 Тимьян лёви
391259 Тимьян лёви, Чебрец чешский
391263 Тимьян майкопский
391264 Тимьян маньчужский
391265 Тимьян мархотский
391274 Тимьян малшалла, Тимьян малшаллов, Чебрец маршалля
391278 Тимьян мигрийский
391279 Тимьян минусинский
391282 Тимьян молдавский, Чебрец молдавский
391284 Тимьян монгольский
391289 Тимьян мтгоджарский, Тимьян

мугоджарский, Чебрец мтгоджарский
391300 Тимьян нарымский
391301 Тимьян нерчинский
391302 Тимьян жилковатый
391306 Тимьян монетный
391307 Тимьян охотский
391310 Тимьян острозубый
391311 Тимьян палласов, Чебрец палласа
391321 Тимьян паннонский, Чабрец венгерский, Чебрец паннонский
391322 Тимьян пастуший
391323 Тимьян малолистный
391324 Тимьян каменистый
391325 Тимьян листоногий
391326 Тимьян подольский, Чебрец подольский
391334 Тимьян близкий
391338 Тимьян ложногранитный, Чебрец ложногранитный
391339 Тимьян ложногранитный, Тимьян ложноприземистый, Чебрец карликовый
391342 Тимьян ложномонетный
391344 Тимьян ложннблощийский
391345 Тимьян красивенький
391346 Тимьян блошиный, Чебрец блошницевидный
391347 Тимьян пржевальского, Тимьян пятижилковый, Тимьян пятиребрый
391351 Тимьян азиатский
391357 Тимьян редкоцветковый
391360 Тимьян розовый
391363 Тимьян шишкина
391364 Тимьян полуголый
391365 Тимьян обыкновенный, Тимьян полевой, Тимьян ползучий, Трава боголодская, Чабрец обыкновенная, Щебрец
391378 Тимьян сибирийский
391379 Тимьян соколова
391380 Тимьян сосновского
391381 Тимьян степной, Чебрец степной
391382 Тимьян субальпийский
391383 Тимьян субарктический, Чебрец субарктический
391384 Тимьян талиева, Чебрец талиева
391385 Тимьян крымский, Чебрец крымский
391387 Тимьян тбилисский
391388 Тимьян закаспийский
391389 Тимьян закавказский
391390 Тимьян траутфеттера
391391 Тимьян турчанинова
391392 Тимьян украинский
391393 Тимьян тссурийский
391397 Тимьян, Тимьян душистый, Тимьян обыкновенный, Травга богородичная, Травга богородская, Чабрец, Чабрец песчаный, Чебрец, Чебрец садовый
391399 Тимьян зеленецкого, Чебрец зеленецкого
391400 Чебрец жигулевский
391401 Тимьян зиаратинский
391477 Тизанолена
391517 Тяньшаночйа
391518 Тяньшаночйа зонтиконосная
391520 Тиарелла
391548 Гюльденштедтия юньнаньская
391622 Тигридия
391627 Тигридия павония
391646 Дерево липоное, Липа
391649 Липа американская, Липа чёрная
391661 Липа амурская
391666 Липа таке
391671 Липа венгерская, Липа войлочная, Липа серебристая
391682 Липа кавказская, Липа пушистостолбчатая
391685 Липа китайская
391691 Липа мелколистная, Липа сердцевидная
391699 Липа сердцелистная
391701 Липа кавказская, Липа опушенностолбиковая, Липа пушистостолбчатая
391705 Липа крымская, Липа ярко-зелёная
391706 Липа голландская, Липа европейская, Липа обыкновенная
391717 Липа франше
391730 Липа хубэйская
391739 Липа японская
391751 Липа гуйцзоуская
391759 Липа лицзянская
391760 Липа маньчжурская
391766 Липа максимовича
391781 Липа монгольская
391783 Липа многоцветковая
391784 Липа наньчуаньская
391797 Липа эмэйская
391816 Липа длинночешковая
391817 Липа крупнолистная, Липа плосколистная, Липа широколистная
391829 Липа бегониелистная, Липа красная
391832 Липа сибирская
391836 Липа белая, Липа венгерская, Липа войлочная, Липа пушистая, Липа серебристая
391860 Липоные
391903 Тиллея
391904 Жигунец аянсий
391918 Тиллея
391920 Тиллея крылатая
391923 Тиллея водная, Тиллея водяная
391939 Тиллея монгольская
391959 Тиллея вайяна
391966 Тилландсия
392052 Лишайник тилланзия
392091 Тимурия
392095 Тимурия сапожникова
392556 Тоддалия
392559 Тоддалия азиатская
392598 Тофилдия, Тофильдия
392601 Тофильдия цветная, Тофильдия чашечкоцветковая
392602 Тофильдия поникающая
392603 Тофилдия багряная, Тофилдия шарлаховая, Тофильдия понкшая
392622 Тофильдия японская
392629 Тофильдия болотная
392720 Томантеа
392722 Томантеа ошэ
392723 Томантеа даралагезская
392724 Томантеа темнохолковая
392725 Томантеа замечательная
392829 Цедрела сингапурская
392837 Цедрела мелкоплодная
392841 Цедрела китайская
392845 Цедрела китайская хубэйская
392863 Тордириум
392873 Тордириум комарова
392875 Тордириум крупный
392880 Торения
392905 Торения фурнье
392923 Торения наньтоуская
392927 Торения мелкоцветковая
392959 Купырник, Торилис, Чермеш
392968 Торилис полевой
392985 Купырник разнолистный, Торилис разнолистный
392989 Купырник надоеливый
392992 Кузырник купыревый, Пупырник японский, Торилис японский
392993 Торилис тонколитный
392995 Купырник пренебреженный
392996 Купырник узловатый, Торилис узловатый
393002 Торилис лучистый
393006 Торилис тонкий
393007 Купырник украинский, Торилис украинский
393009 Торилис желтоволосый
393047 Торрейя, Торрея
393057 Торрея калифорнийская
393061 Торрея большая, Торрея крупная, Торрея крупная дирла
393077 Торрея орехоносная, Торрея орехоплодная
393089 Торрея тисолистная, Торрея тиссолистная
393100 Манник плавающий
393136 Четочник скрученный, Чёточник скрученный

393137 Четочник прижатый
393175 Четочник сумбарский
393185 Членистник
393188 Членистник кавказский
393240 Турнефорция
393261 Турнефорция яйцевидная
393267 Турнефорция тайваньская
393269 Аргузия сибирская, Турнефорция сибирская
393276 Турнефорция согдийская
393441 Токсикодендрон
393449 Сумах делаве
393458 Токсикодендрон чжэньканский
393466 Сумах восточный, Сумах сомнительный, Токсикодендрон восточный
393470 Сумах канадский, Сумах укореняющийся, Сумах ядовитый, Токсикодендрон укореняющийся
393479 Дерево восковое, Дерево лаковое японское, Сумах замещающий, Сумах сочный, Сумах японский
393485 Сумах лесной
393488 Сумах волосистоплодный, Токсикодендрон волосистоплодный
393491 Дерево злаковое, Дерево лаковое, Дерево лаковое японское, Лаковое дерево, Сумах восковой, Сумах лаковый, Сумах лаконосный, Токсикодендрон лаконосный
393568 тоция
393570 тоция карпатская
393595 Трахелянт
393596 Трахелянт гиссарский
393597 Трахелянт королькова
393604 Трахелиум
393613 Трахелоспермум
393656 Трахелоспермум люцюск
393657 Трахелоспермум жасминовидный
393695 кентырь
393803 Пальма пеньковая, Трахикарпус
393809 Пальма веерная, Пальма веерная высокая, Трахикарпус высокий, Трахикарпус форчуна
393813 Трахикарпус мартиуса
393815 Трахикарпус такил
393816 Трахикарпус вагнера
393818 Бородавчатник, Трахидиум
393832 Бородавчатник вильчатый
393842 Бородавчатник копетдагский
393854 Трахидиум лицзянский
393864 Трахидиум тяньшанский
393872 Трахилобий
393882 Дидискус
393885 Трахиния
393937 Айован
393940 Ажгон, Айован, Айован душистый, Айован столовый
393952 Айован лицзянский
393959 Трахистемон
393961 Трахистемон восточный
393987 Зебрина, Традесканция
394021 Традесканция бассейновая, Традесканция белоцветковая
394076 Рео пестрое
394088 Традесканция виргинская
394093 Зебрина свизающая
394257 Козлобородник
394259 Козлобородник колючепродный
394260 Козлобородник алтайский
394261 Козлобородник абрикосово-желтый
394262 Козлобородник бадахшанский
394263 Козлобородник белорусский
394264 Козлобородник днепровский
394265 Козлобородник коротконосиковый, Козлобородник коротконосый
394266 Козлобородник буфтальмовидный
394268 Козлобородник головчатый
394269 Козлобородник харадзе
394270 Козлобородник колхидский
394271 Козлобородник холмовый
394272 Козлобородник окрашенный
394273 Козлобородник свернутый
394274 Козлобородник меловой
394276 Козлобородник дагестанский
394278 Козлобородник опушенноносый, Козлобородник шершавоносиковый
394279 Козлобородник донецкий, Козлобородник донской
394280 Козлобородник дубянского
394281 Козлобородник большой, Козлобородник сомнительный
394283 Козлобородник возвышеный, Козлобородник высокий
394284 Козлобородник, Козлобородник вытянутый
394285 Козлобородник нителистный
394286 Козлобородник шерстистый
394287 Козлобородник гауданский
394291 Козлобородник горского
394292 Козлобородник изящный
394293 Козлобородник злаколистный
394295 Козлобородник разносемянный
394297 Козлобородник иды
394298 Козлобородник карелина
394299 Козлобородник карягина
394300 Козлобородник казахстанский
394301 Козлобородник
394302 Козлобородник кемярии
394303 Козлобородник кецховели
394304 Козлобородник копетдагский
394305 Козлобородник крашенинникова
394306 Козлобородник культиасова
394308 Козлобородник широколистный
394309 Козлобородник литпвский
394311 Козлобородник крупнобороtый
394312 Козлобородник большой
394314 Козлобородник макашвили
394315 Козлобородник окаймленный
394316 Козлобородник гранистоплодный, Козлобородник окаймленнолистный
394317 Козлобородник скороспелый
394318 Козлобородник месхетский
394320 Одуванчик горный
394321 Козлобородник восточный
394323 Козлобородник особенный
394325 Козлобородник подорожниковый
394326 Козлобородник, Козлобородник подольский
394327 Козлобородник пореелистный, Козлобородник порейнолистный, Козлобородник порейолистный, Козлобородник поррейнолистный, Овсяный корень, Сальсифи
394332 Козлобородник
394333 Козлобородник луговой
394334 Козлобородник ложнокрупный
394335 Козлобородник крошечный
394336 Козлобородник сетчатыйК
394337 Козлобородник красный
394339 Козлобородник русский
394340 Козлобородник песчаный
394342 Козлобородник прутьевидный
394343 Козлобородник пашенный
394344 Козлобородник поздний
394345 Козлобородник сибирский
394346 Одуванчик джунгарский
394347 Козлобородник сосновского
394348 Козлобородник степной
394349 Козлобородник субальпийский
394350 Козлобородник донской
394351 Козлобородник пушистенький
394352 Козлобородник шероховатоплодный
394353 Козлобородник клубненосный
394354 Козлобородник туркестанский
394355 Козлобородник украинский
394358 Козлобородник волжский
394359 Козлобородник введенского
394372 Козелец, Козлец, Трагус
394390 Трагус кистевидный
394412 Водяной орех, Рогульник
394417 Водяной орех амурский
394423 Водяной орех астраханский
394426 Водяной орех двурогий
394430 Водяной орех потанина
394433 Водяной орех тайваньский
394447 Водяной орех каринтийский
394453 Водяной орех куполовидный
394454 Водяной орех крестообразный
394457 Водяной орех европейский

394458 Водяной орех гирканский
394459 Водяной орех выресной
394463 Водяной орех японский
394478 Водяной орех коржинского
394480 Водяной орех литвинова
394484 Водяной орех длиннорогий
394487 Водяной орех малеева
394489 Водяной орех маньчжурский
394496 Водяной орех максимовича
394500 Болотный орех, Водяной орех плавающий, Водяной орех северный, Каштан водяной, Орех болотный, Орех водяной, Орех водяной плавающий, Рогульник, Рогульник водяной орех, Рогульник плавающий, Рогурьки, Чилим
394523 Водяной орех гребенчатый
394526 Водяной орех комарова
394539 Водяной орех русский
394541 Водяной орех саянский
394542 Водяной орех сибирский
394547 Водяной орех спрыгина
394551 Водяной орех траншеля
394552 Водяной орех бугорчатый
394554 Рогульниковые
394559 Трапелла
394560 Трапелла китайская
394579 Траунштейнера
394580 Траунштейнера маровидная
394581 Траунштейнера сферическая
394583 Траутфеттерия
394588 Траутфеттерия японская
394601 Трекулия
394656 Трема восточная
394807 Триаденум, Трижелезник
394811 Триаденум японский, Трижелезник японский
394943 Мискантус сахароцветный
395064 Якорцы
395108 Якорцы крупнокрылые
395146 Маргелон, Якорцы земляные, Якорцы стелющиеся
395466 Трихилия
395668 Волосатик незаметный
395708 Триходесма
395759 Триходесма седая
395964 Власочешуйник
395968 Власочешуйник власоглавый
396003 Пухонос
396005 Пухонос альпийсйий
396010 Пухонос дернистый
396020 Пухонос приземистый
396150 Трихозант, Трихозантес, Трихозантус
396151 Огурец змеевидный, Огурец змеиный, Трихозант, Трихозантес
396583 Трициртис
396780 Седмичник, Троечница
396786 Седмичник еверопейский
396789 Седмичник арктический
396806 Клевер, Трилистник, Хмелёк
396817 Берсим, Клевер александрийский, Клевер берсим, Клевер египетский
396818 Клевер альпийский
396819 Клевер альпийский
396821 Клевер непостояный, Клевер сходный
396822 Клевер угловатый
396823 Клевер узколистный
396828 Дербенник, Клевер пашенный, Клевер полевой, Котики, Лапка заячьи
396845 Клевер бордзиловского
396854 Клевер полевой, Клевер равнинный
396858 Клевер седоватый
396860 Клевер кавказский
396881 Клевер паскидистый, Клевер паспростертый
396882 Клевер маленький, Клевер сомнительный
396885 Клевер иглистый
396893 Клевер превосходный
396895 Клевер венгерский, Клевер расширенный
396896 Клевер нитевидный, Клевер сомнительный, Клевер тонкостебельный
396903 Клевер ключевой
396905 Клевер земляниковидный, Клевер земляничный, Пустоягодник
396917 Клевер клубковидный, Клевер скрученный
396920 Люцерна гордеева, Пажитник гордеева
396926 Клевер мохнатый, Клевер шершавый
396930 Альзик, Клевер гиблидный, Клевер изящный, Клевер краснобелый, Клевер красно-белый, Клевер нарядный, Клевер розовый, Клевер шведский
396934 Клевер инкарнатный, Клевер маленовый, Клевер молинери, Клевер мясокрасный, Клевер пунцовый
396951 Клевер репейниковый
396953 Клевер белоцветковый, Клевер бледноцветковый
396956 Клевер лупиновидный, Клевер лупиновый, Клевер люпиновидный, Клевер люпиновый
396961 Клевер приморский
396967 Кашка лесная, Клевер средний
396981 Белоголовка, Клевер белоголовка, Клевер горный, Клевер темнокаштановый
396984 Клевер пренебреженный
396989 Клевер бледножелтый, Клевер бледно-жёлтый, Клевер светлоделтый, Клевер светложелтый, Клевер светло-жёлтый
396996 Клевер тихоокеанский
396998 Клевер бледный
396999 Клевер венгерский, Клевер жёлтый, Клевер паннонский
397000 Клевер палермский
397002 Клевер бедноцветковый, Клевер мегкоцветковый
397006 Клевер батлачковый
397013 Клевер вспухлый
397016 Клевер многолистный
397019 Дядлина белая, Дятлина красная, Кашка красная, Клевер красный, Клевер луговой, Конюшина, Трилистник, Трилистник луговой
397040 Клевер радде
397043 Дядлина белая, Кашка, Клевер белый, Клевер белый ползучий, Клевер ползучий, Трилистник белый
397050 Клевер опрокинутый, Клевер перевернутый, Клевер перевёрнутый, Клевер персидский, Клевер шабдар, Шабдар, Шафдар, Шафтал
397059 Клевер красноватый, Клевер красный
397065 Клевер культурный, Клевер скасный
397066 Клевер скасный Кудряш
397067 Клевер скасный позднеспелый
397068 Клевер жестковолосый, Клевер шероховатый, Клевер шершавый
397073 Клевер себастиана
397079 Клевер серавшанский
397083 Клевер горный, Клевер каштановый, Клевер темнокаштановый
397084 Клевер красивый
397086 Клевер пенистый
397087 Клевер открытозевый, Клевер растопыренный
397089 Клевер звездчатый
397091 Клевер плодоножковый
397095 Клевер полосатый
397096 Клевер пашенный, Клевер полевой, Клевер посевной, Клевер шуршащий, Хмелёк полевой, Хмелёк шуршащий
397097 Клевер полосатый
397106 Клевер подземный, Клевер подземный
397120 Клевер волосистоголовый
397122 Клевер вздутый
397130 Клевер пузырчатый
397142 Триостренник
397159 Триостренник морский, Триостренник морской, Триостренник приморский
397165 Триостренник болотный
397186 Пажитник, Пажитник греческий,

Сено греческое, Тригонелла
397187 Пажитник восходящий
397192 Пажитник дугообразный
397196 Пажитник лазулевый
397198 Пажитник бадахшанский
397200 Пажитник бессера
397202 Пажитник двухцветковый
397203 Пажитник короткоплодный
397206 Пажитник тонкий
397208 Гуньба, Донник голубой, Донник синий, Пажитник голубоватый, Пажитник голубой
397212 Пажитник меловой
397227 Пажитник фишера
397229 Клевер греческий, Пажитник греческое сено, Пажитник сенной, Сено греческое, Шамбала
397233 Пажитник мечевидный, Пажитник меченосный
397236 Пажитник крупноцветковая
397246 Пажитник рыхлоцветковая
397247 Пажитник липского
397254 Пажитник одноцветковый, Пажитник парноцветковый
397255 Пажитник ное
397261 Пажитник пряморогий
397263 Пажитник памирский
397271 Пажитник попова
397272 Пажитник простертый
397274 Пажитник лучевой
397278 Пажитник колосистый, Пажитник колосовый
397280 Пажитник четковидный
397281 Пажитник полосатый, Пажитник тонкий
397288 Пажитник бугорчатый
397289 Пажитник туркменский
397292 Пажитник вера
397338 Трехграннoплодник
397339 Трехграннoплодник окутанный
397380 Тригонотис
397393 Тригонотис чэнкоуский
397394 Тригонотис чусюнский
397410 Тригонотис тайваньский
397412 Тригонотис фунинский
397422 Тригонотис корейский
397423 Тригонотис наньчуаньский
397425 Тригонотис сичоуский
397439 Тригонотис мулиский
397441 Тригонотис незабудковый
397447 Тригонотис эмэйский
397451 Тригонотис булавовидный, Тригонотис длинноножковый, Тригонотис ножковый
397459 Тригонотис укореняющийся
397462 Тригонотис корейский
397473 Тригонотис тибетский
397533 Триллиевые
397540 Триллиум
397544 Триллиум камчатский
397568 Триллиум тибетский
397574 Триллиум камчатский
397604 Триллиум ромболистный
397616 Триллиум смолла
397620 Триллиум тайваньский
397622 Триллиум чоноский
397729 Триния
397732 Триния щетинистоволосая
397733 Триния китайбеля
397734 Триния гладкоплодная
397735 Триния многостебельная
397736 Триния шершавая
397737 Триния многоветвистая
397738 Триния станкова
397740 Триния украинская
397762 Трёхзубка, Триодия
397828 Трехкосточник, Трёхкосточник, Триостеум
397849 Трехкосточник выемчатый, Триостеум выемчатый
397860 Ботриостеге прицветниковая
397948 Трехреберник, Триплевроспермум
397949 Трехреберник сомнительный
397952 Трехреберник коротколучевой
397953 Трехреберник кавказский
397954 Трехреберник колхидский
397955 Трехреберник обманчивый
397956 Трехреберник дисковидный
397958 Трехреберник вытянутый
397960 Трехреберник гроссгейма
397964 Ромашка непахучая, Трехреберник непахучий, Триплевроспермум непахучий
397968 Трехреберник карягина
397969 Трехреберник илийстый, Триплевроспермум илийстый
397970 Ромашка приморская, Трехреберник приморский
397973 Трехреберник темноголовый
397975 Трехреберник мелкоцветковый
397981 Трехреберник скальный
397982 Трехреберник севанский
397983 Трехреберник совича
397984 Трехреберник чихачева
397986 Трехреберник четырехугольносемянный, Триплевроспермум четырехугольносемянный
397987 Трехреберник закавказский
398064 Трехбородник
398071 Трехбородник китайский
398126 Астра солончаковая, Триполиум
398134 Астра плавненая, Астра солончаковая, Астра солончаковая обыкновенная, Триполиум обыкновенный
398138 Трипсакум
398139 Трава гама
398273 Кравфурдия японская
398403 Трищетинник каванилля
398437 Трищетинник
398442 Трищетинник алтайский
398454 Трищетинник кандинский
398457 Трищетинник двряднолистный
398482 Трищетинник хубэйский
398488 Трищетинник линейный
398518 Овес желтеющий, Овёс желтеющий, Овёс желтоватый, Овес золотистый, Овёс золотистый, Трищетинник желтоватый, Трищетинник жёлтоватый, Трищетинник жёлтый, Трищетинник луговой
398524 Трищетинник жесткий
398528 Трищетинник сибирский
398531 Трищетинник колосистый, Трищетинник субальпийский
398565 Трищетинник тибетский
398666 Тристания
398824 Тритринакс
398826 Тритринакс бразильский
398834 Гандум, Пушеница, Пушиница, Пшеница
398836 Пшеница абисинская
398838 Пушеница эгилопсовая
398839 Пушеница летняя, Пушеница мягкая, Пшеница ветвистая, Пшеница летная, Пшеница мягкая, Пшеница обыкновенная
398857 Пшеница армянская, Пшеница полба армянская
398869 Пушеница карталинская, Пушеница прсидская, Пушиница кавказская
398877 Ежовка, Пушеница карликовая, Пушеница круглозёрная, Пушеница мягкая карликовая, Пушеница мягкая яровая, Пушеница плотная, Пушеница-ежовка, Пушиница ежовка
398886 Двузернянка дикая, Зандури, Пшеница двузернянковидная
398890 Двузернянка, Двузернянка культурная, Полба волжская, Полба двузёрная, Полба ненастоящая, Полба обыкновенная, Полуполба, Пшеница волжская, Пшеница двузернянка, Эммер
398900 Арнаутка, Гарновка, Кубанка, Пушеница выносливая, Пушеница стекловидная, Пушеница сьльная, Пушеница твердая, Пшеница твёрдая
398920 Пушеница маха
398924 Пушеница однозернянка, Пшеница

культурная однозернянка, Пшеница однозернянка культурная
398939 Пушеница польская, Пшеница полоникум, Рожь гималайская
398970 Полба, Полба настоящая, Польба, Польба настоящая, Пушеница полба, Пшеница спельта, Пшеница-спельта, Спелта, Спельта
398972 Пушеница круглозерная
398980 Пушеница таудар
398981 Пушеница тимофеева, Пушеница тимофеевская
398986 Пушеница английская, Пушеница варварийская, Пушеница тучная
399006 Пшеница волжская, Пшеница полба волжская
399043 Монтбреция, Тритония
399432 Троходендрон
399433 Троходендрон аралиевидный
399484 Авдотка, Купальница, Троллиус
399491 Купальница алтайская
399494 Купальница азиатская, Огоньки
399499 Купальница бумажночашечная
399500 Купальница китайская
399503 Купальница джунгарская
399504 Авдотка, Купава европейская, Купальница, Купальница европейская, Троллиус, Троллиус европейский
399510 Купальница японская
399515 Купальница ледебура
399518 Гегемона лиловая
399519 Купальница крупнолднестковая
399520 Купальница мелкоцветковая
399522 Купальница полуоткрытая
399540 Купальница ридера
399541 Купальница сибирская
399547 Купальница уральская
399596 Капуциновые
399597 Жеруха, Жерушник, Капуцин, Настурция, Тропелюм
399601 Капуцин большой, Красоля, Кресс испанский, Настурция, Настурция большая, Настурция высокая иноземная, Настурция садовая
399604 Капуцин малый, Настурция клубненосная, Настурция малая, Настурция мелкая, Настурция низкорослая, Настурция садовая
399606 Настурция канарская, Настурция карликовая
399611 Аньу, Настурция малая
399860 Гемлок, Тсуга, Цуга
399867 Гемлок восточный, Гемлок канадский, Тсуга канадская
399879 Тсуга китайская, Цуга китайская
399892 Тсуга разнолистная
399895 Тсуга разнолистная, Цуга разнолистная
399901 Тсуга тайваньская
399903 Тсуга лицзянская
399905 Тсуга западная, Цуга западная
399916 Цуга калифорнийская
399923 Тсуга зибольда, Тсуга японская, Цуга зибольда
399925 Тсуга чжэцзянская
400050 Лилия африканская
400110 Тюльпан
400116 Тюльпан алтайский
400117 Тюльпан аньхуйский
400118 Тюльпан неравнолистный
400130 Тюльпан беса
400131 Тюльпан биберштейна, Тюльпан биберштейновский
400132 Тюльпан бузе, Тюльпан бузовский, Тюльпан двуцветковый
400133 Тюльпан многоцветный
400134 Тюльпан ложнодвухцветковый
400135 Тюльпан боршова
400136 Тюльпан короткотычиночный
400138 Тюльпан каллье
400139 Тюльпан килеватый
400140 Тюльпан кавказский
400148 Тюльпан волосистотычиночный
400149 Тюльпан ложноволосистотычиночный
400151 Тюльпан сомнительный
400154 Тюльпан эйхлера
400157 Тюльпан ферганский
400158 Тюльпан флоренского
400159 Тюльпан фостера
400162 Тюльпан геснера, Тюльпан садовый, Тюльпан степной
400171 Тюльпан грейга
400173 Тюльпан разнолепестный
400174 Тюльпан разнолистный, Тюльпан разнолистый
400175 гиссарский
400176 Тюльпан хуга
400178 Тюльпан илийский
400179 Тюльпан великий
400180 Тюльпан юлии
400181 Тюльпан кауфмана, Тюльпан кауфмановский
400182 Тюльпан коктебельский
400183 Тюльпан колпаковского
400184 Тюльпан королькова
400185 Тюльпан коржинского
400186 Тюльпанкраузе
400187 Тюльпан кушкинский
400190 Тюльпан лемана, Тюльпан леманновский
400191 Тюльпан льнолистный
400193 Тюльпан максимовича
400194 Тюльпан микели
400195 Тюльпан моголтавский
400198 Тюльпан блестящий
400202 Тюльпан островского
400203 Тюльпан поникающий
400204 Тюльпан широкотычиночный
400207 Тюльпан превосходящий
400218 Тюльпан регели
400219 Тюльпан розовый
400221 Тюльпан шмидта
400222 Тюльпан шренка
400224 Тюльпан синьцзянский
400227 Тюльпан душистый
400228 Тюльпан превосходный
400229 Тюльпан лесной
400240 Тюльпан поздний
400241 Тюльпан четырехлистный
400242 Тюльпан тяньшанский
400244 Тюльпан тубергена
400245 Тюльпан туркменский
400246 Тюльпан туркестанский
400247 Тюльпан эйхлера
400248 Тюльпан одноцветковый
400250 Тюльпан фиолетовый
400253 Тюльпан вильсона
400254 Тюльпан зинаиды
400280 Тулотис
400343 Туника толстоpебристая
400466 Турчаниновия
400467 Турчаниновия верхушечная
400471 Тургеневия, Тургения
400473 Тургеневия широколистная, Тургения широколистная
400486 Турнера
400554 Туррея
400622 Пентастемон, Турраэнтус
400638 Башшеница, Вяжечка
400644 Башенница голая, Башшеница гладкая, Башшеница голая, Вяжечка гладкая, Вяжечка голая, Резуха гладкая
400671 Мать-и-мачеха
400675 Белокопытник, Камчужная трава, Мать-и-мачеха обыкновенная, Подонежник золотой, Трива камчужная
401085 Палочник, Рогоз, Ротоза, Чакан
401094 Мочёнка, Рогоз суженный, Рогоз узколистный, Рогоз узколистый, Ротоза узколистная
401105 Рогоз суженный, Рогоз южный
401107 Рогоз слоновый
401112 Рогоз широколистный, Рогоз широколистый, Тырса
401120 Рогоз лаксманна, Рогоз ляксманна
401128 Рогоз малый
401129 Рогоз восточный, Рогоза восточная, Ротоза, Ротоза восточная
401131 Рогоз бледный
401139 Рогоз верещагина

401140 Рогозоные
401210 Упака
401339 Юхтриница армянская
401340 Юхтриница коканская
401373 Золотохворост, Колючий дрок, Улекс, Улекс европейский, Утёсник
401388 Английский дрок, Дрок колючий европейский, Колючий дрок, Улекс европейский, Уллюко, Утесник европейский
401396 Утёсник карликовый, Утёсник малый
401411 Уллюко
401412 Картофель испанский гладкий, Уллюко клубневый, Уллюко клубненосный
401415 Вязовые, Ильмовые
401425 Берест, Вяз, Ильм
401430 Вяз крылатый, Вяз крыловаточерешковый, Ильм крылатый
401431 Вяз, Вяз американский, Вяз белый, Ильм американский, Ильм белый
401448 Вяз андросова
401468 Берест листоватый, Вяз граболистный, Вяз листоватый, Вяз полевой, Ильм листоватый
401475 Берест пробковый, Вяз пробковый, Ильм пробковый, Карагач
401478 Вяз каркасовый
401480 Вяз куэньминский
401485 Вяз граболистный роговидный
401490 Вяз сродный, Вяз японский
401499 Берест густой, Вяз густой
401501 Вяз раскидистый
401503 Вяз эллиптический
401508 Берест, Берест листоватый, Вяз граболистный, Вяз листоватый, Вяз полевой, Ильм листоватый, Карагач листоватый
401509 Вяз ржавый, Ильм сибирский
401512 Вяз голый, Вяз горный, Вяз шероховатый, Вяз шершавый, Ильм, Ильм горный, Ильм шершавый, Ильм шотландский
401542 Вяз лопастной, Вяз лопастный, Вяз разрезной
401549 Вяз гладкий, Вяз настоящий, Вяз обыкновенный, Вяз раскидистый, Ильм гладтый
401556 Вяз крупноплодный
401567 Вяз мексиканский
401569 Ильм мелкоплодный
401581 Астрагал китайский, Вяз мелколистный
401591 Вяз перисто-ветвистый
401593 Берест листоватый, Вяз граболистный, Вяз листоватый, Вяз полевой, Вяз равнинный, Илим, Илим полевой, Илом полевой, Ильм листоватый, Карагач полевой
401602 Берест приземистый, Вяз мелкоистный, Вяз низкий, Вяз приземистый, Ильм мелкоистный, Ильм приземистый, Ильм туркестанский, Ильмовник
401618 Вяз ветвистый, Вяз кистевидный, Вяз томаса
401633 Берест пробковый, Берест туркестанский, Вяз мелколистный, Вяз пробковый, Ильм пробковый, Карагач
401672 Лавр калифорнийский, Умбеллюлярия калифорнийская
401678 Трава пупочная, Умбиликус
401702 Седум супротивнолистный, Умбиликус супротивнолистный
401737 Ункария клюволистная
401754 Ункария тайваньская
401757 Гамбир
401817 Унгерния
401818 Унгерния ферганская
401820 Унгерния меньшая
401822 Унгерния
401824 Унгерния таджикская
401825 Унгерния трехсферная
402238 Урена
402245 Урена
402281 Урера
402314 Ургинея
402660 Уроспермум, Хвостосемянник
402663 Хвостосемянник горчаковидный
402743 Урзиния, Урсиния
402840 Крапива
402847 Крапива узколистная
402854 Крапива мелкоцветковая
402869 Крапива коноплевая
402886 Крапива большая, Крапива двудомная, Крапива обыкновенная
402889 Крапива двудомная ганьсуская
402893 Крапива синьцзянская
402943 Крапива киевская
402946 Крапива светро-зеленая
402947 Крапива синеющая
402992 Крапива шариконосная
402996 Крапива плосколистная
403001 Крапива опушенная
403016 Крапива сондена
403026 Крапива тайваньская
403028 Крапива тунбелга
403031 Крапива тибетская
403039 Крапива жгучая
403048 Крапива чаюйская
403050 Крапивные
403095 Пузырчатка
403110 Пузырчатка недосмотренная
403113 Пузырчатка хайнаньская
403130 Пузырчатка бремa
403191 Пузырчатка забытая
403220 Пузырчатка средная
403222 Пузырчатка японская
403246 Пузырчатка большая
403252 Пузырчатка малая
403382 Пузырчатка обыкновенная
403665 Увуляляия
403686 Коровница, Тысячеголов
403687 Геля, Коровница испанская, Мыльнянка коровья, Тысячеголов, Тысячеголов обыкновенный, Тысячеголов посевной, Тысячеголов пурамидальный, Тысячеголовник
403706 Брусничные
403710 Брусника, Вакциниум, Черника, Ягодник
403728 Кавказская черника, Черника кавказская, Черника кавказский
403738 Черника прицветниковая
403758 Голубика канадская, Черника канадская
403770 Черника сычуаньская
403780 Блюберри, Голубика американская, Голубика зонтичнокистевая, Черника щитоковая, Ягодник кистевой
403844 Черника гуандунская
403845 Черника хайнаньская
403854 Черника волосистый
403868 Вакциниум японский
403878 Вакциниум корейский
403894 Клюква крупноплодная, Черника крупноплодная
403912 Вакциниум мупинский
403916 Черника обыкновенная, Черница, Чирника, Ягодник черника
403929 Вакциниум эмэйский
403932 Вакциниум овальнолистный, Черник овальнолистный
403956 Вакциниум мелколистная
403967 Вакциниум выдающийся, Вакциниум красника
403976 Вакциниум луаньдаский
403998 Вакциниум зибольда
404001 Вакциниум гуансиский
404003 Вакциниум смолла
404027 Вакциниум болотный, Вахта, Голубика, Голубика обыкновенная, Гонобобель, Гоноболь, Павун, Пьяница
404051 Брусника, Брусника обыкновенная, Вакциниум обыкновенный
404063 Вакциниум яошаньский
404066 Вакциниум ятабе
404158 Валодеа
404160 Валодеа темнобагряная
404164 Валодеа парамуширская

404172 Вайянция
404181 Вайянция постенная
404213 Валериана, Валерьяна, Маун, Мяун
404217 Валерьяна аянская
404219 Валерьяна чесночниколистная
404220 Валерьяна приальпийская
404225 Валериана амурская
404236 Валерьяна головчатая
404242 Валерьяна снеголюбивая
404244 Валерьяна колхидская
404247 Валерьяна дагестанская
404251 Валерьяна болотная, Валерьяна двудомная
404252 Валерьяна сомнительная
404258 Валериана возвышенная
404259 Валериана высокая
404267 Валериана федченко
404284 Валерьяна японская
404291 Валерьяна еленевского
404306 Валериана мартьянова
404309 Валерьяна мелкоцветковая
404310 Валериана горная
404311 Валерьяна мулиская
404314 Валериана блестящая, Валерьяна лоснящаяся
404316 Болдырьян, Валериана аптечная, Валериана корейская, Валериана лекарственная, Валериана очереднолистная, Валерьяна аптечная, Валерьяна лекарственная, Маун аптечный
404330 Валерьяна камнелюбиая
404352 Валерьяна цельнолистная
404354 Валериана гулявниколистная
404370 Валерьяна линолистная
404373 Валерьяна трехлисточковая
404377 Валериана клубненосная, Валерьяна клубненосная
404379 Валериана турчанинова
404392 Валериановые, Валерьяновые, Мауновые
404394 Валерианелла, Валерианица, Валерианница, Валерьяница, Валерьянница
404396 Валерианелла тупая
404397 Валерианелла беззубая
404400 Валерианелла коротковенцовая
404402 Валерьянница килеватая
404403 Валерианелла кавказская
404408 Валерианелла увенчанная
404411 Валерианелла ладьеплодная
404413 Валерианелла зубчатая
404415 Валерианелла двузубая
404419 Валерианелла дюфлена
404421 Валерианелла ежовая
404422 Валерианелла бугорчатая, Валерьянница бугорчатая
404430 Валерианелла кочи
404431 Валерианелла кулябская
404433 Валерианелла пушистоплодная
404434 Валерианелла голоплодная
404437 Валерианелла липского
404439 Валерианелла колосковая, Валерианелла овощная, Валерианница овощная, Валерьянница овощная
404452 Валерианелла усеченная
404454 Валерианелла колосковая, Валерианелла малая, Валерьянница малая, Валерьянница огородная, Машсалат, Салат полевой
404457 Валерианелла остроносая
404462 Валерианелла косовенцовая
404463 Валерианелла плоскоплодная
404465 Валерианелла черноморская
404467 Валерианелла маленькая
404472 Валерианелла щелистая
404473 Валерианелла твердоплодная
404476 Валерианелла совича
404478 Валерианелла бугорчатая
404479 Валерианелла вздутая
404480 Валерианелла туркестанская
404481 Валерианелла крючковатая
404482 Валерианелла пузырчатая
404483 Валерианелла введенского
404531 Валлиснерия
404537 Валлиснерия аньхуйская
404555 Валлиснерия гигантская
404562 Валлиснерия спиральная
404563 Валлиснерия спиральная
404581 Валлота птрпурная
404620 Ванда
404632 Ванда одноцветная
404642 Ванда гуандунская
404647 Ванда гиблидная
404659 Ванда мелкоцветковая
404680 Ванда трехцветная
404971 Ваниль
404990 Ваниль душистая, Ваниль плосколистная
405010 Ваниль мексиканская
405017 Ваниль душистая, Ваниль плосколистная
405093 вартемия
405099 вартемия персидская
405153 Ватерия
405154 Ватерия индийская
405269 Велеция
405570 Ацелидант, Чемерица
405571 Чемерица белая
405583 Чемерица чашецветная
405585 Чемерица даурская
405593 Чемерица тайваньская
405597 Чемерица гуншаньская
405598 Чемерица крупноцветковая
405601 Чемерица японская
405604 Чемерица лобеля
405606 Чемерица маака
405614 Чемерица мэнцзыская
405615 Чемерица мелкоцветковое
405616 Чемерица арктическая
405617 Чемерица наньчуаньская
405618 Чемерица черная, Чемерица чёрная
405626 Чемерица эмэйская
405627 Чемерица остродольная
405633 Чемерица гулинская
405638 Чемерица далиская
405643 Чемерица американская, Чемерица белая американская, Чемерица зеленая, Чемерица зелёная
405657 Вербаскум, Коровяк, Скипетр царский, Ставрофрагма, Цельзия
405658 Коровяк приальпийский
405661 Коровяк артвинский
405667 Коровяк молевый, Коровяк тараканий, Коровяк темный, Коровяк тёмный, Трава молевая
405681 Коровяк восточный
405683 Коровяк левкоелистный
405685 Вербаскум китайский
405686 Коровяк холмовой
405694 Дивина, Коровяк высокий, Коровяк скипетровидный, Коровяк тапсовидый
405699 Коровяк ереванский
405703 Коровяк желтоватый
405704 Коровяк изогнутый
405706 Коровяк красивый
405708 Коровяк грузинский
405709 Коровяк скученый
405710 Коровяк белошерстистый, Коровяк черноморский
405711 Коровяк хлопковый
405713 Коровяк гаястанский
405724 Коровяк раскидистый
405727 Коровяк длинностный
405729 Коровяк метельчатый, Коровяк мучнистый
405731 Коровяк крупноплодный
405736 Коровяк крупнозопниковый
405737 Коровяк черный, Коровяк чёрный
405742 Коровяк горолюбивый
405744 Коровяк овальнолистный
405745 Коровяк метельчатый
405747 Коровяк зопниковидный, Коровяк лекарственный, Коровяк мохнатый, Коровяк шерстистый
405748 Коровяк фиолетовый
405749 Коровяк перистораздельный
405755 Коровяк пуналийский
405756 Коровяк пираидальный
405763 Коровяк краснобурый
405764 Коровяк мешечатый

405767 Коровяк сидячецветковый
405774 Коровяк выемчатый, Коровяк извилистый
405777 Коровяк джунгарский
405779 Коровяк великолепный, Коровяк красивый
405780 Коровяк видный
405781 Коровяк чистецовидный
405783 Коровяк шовица
405788 Коровяк медвежье, Коровяк медвежье ухо, Коровяк обыкновенный
405792 Коровяк закавказский
405793 Коровяк туркменский
405794 Коровяк туркестанский
405795 Коровяк изменчивый
405799 Коровяк вильгельмся
405801 Вербена
405817 Вербена прицветниковая, Вербена прицветничковая
405820 Вербена канадская
405844 Вербена копьевидная
405852 Вербена гибридная, Вербена плетистая
405872 Вербена аптечная, Вербена лекарственная
405893 Вербена лежачая, Вербена лежащая
405907 Вербеновые
405911 Вербезина, Вербесина
406016 Тунг сердцевидный, Тунг японский
406018 Дерево масляное китайское, Дерево тунговое, Масляное дерево, Тунг, Тунг китайский, Тунг форда, Тунг-ю
406040 Вернония, Трава железная
406966 Вероника
406971 Вероника ягодколистная
406973 Вероника пашенная
406978 Вероника восточнокавказская
406979 Вероника альпийская
406983 Вероника американская
406986 Вероника приятная
406988 Вероника анагалисовидная
406992 Вероника ключевая
407003 Вероника грязевая, Вероника ложноключевая
407008 Вероника безлистная
407010 Вероника песчаная
407011 Вероника борнмюллера, Вероника каратавская, Вероника остропильчатая
407012 Вероника армянская
407014 Вероника полевая
407018 Вероника австрийская
407019 Вероника крылова, Вероника широколиственная, Вероника широколистная
407022 Вероника бахофена
407024 Вероника баранецкого
407025 Вероника барелье
407028 Вероника баумгартена
407029 Вероника поручейная, Вероника поточная, Вероника ручейная, Вероника-поточник, Ибунка, Поточник
407037 Вероника ложнопоточная
407038 Вероника маргаритколистная
407039 Вероника маргаритковая
407040 Вероника двлопастная
407041 Вероника боброва
407045 Вероника бухарская
407048 Вероника болотниковидна
407049 Вероника кривоногая
407058 Вероника волосоногая
407063 Вероника сердцевидноплодная
407068 Вероника кавказская
407070 Вероника сливолистная
407071 Вероника рогоплодная
407072 Вероника дубравка, Вероника дубравная, Вероника дубровка, Дубравка, Дубровка
407074 Вероника хантавская
407075 Вероника харадзе
407076 Вероника чаюйская
407078 Вероника хэбэйская
407079 Вероника ресничатая
407081 Вероника ресничатая чжундяньская
407092 Вероника петушийгребень
407096 Вероника цимбаляриевая
407099 Вероника дегестанская
407106 Вероника густоцветковая
407107 Вероника зубчатая
407108 Вероника обнаженная
407109 Вероника двойчатая
407114 Вероника дилления
407120 Вероника чэнкоуская
407121 Вероника федченко
407123 Вероника нителистная
407124 Вероника нитевидная
407127 Вероника далиская
407129 Вероника кустящаяся
407130 Вероника кустарничковая
407135 Вероника голостная
407141 Вероника горбунова
407142 Вероника крупноцветковая
407144 Вероника плющелистная
407155 Вероника белая
407158 Вероника водная
407159 Вероника имеретинская
407161 Вероника промежуточная
407164 Вероника яванская
407167 Вероника кемулярии
407169 Вероника хоросанская
407184 Вероника комарова
407188 Вероника копетдагская
407192 Вероника курдская
407198 Вероника широколистная
407216 Вероника лидтке
407218 Вероника вербейниковая
407220 Вероника ложно-крупнотычнковая
407224 Вероника наибольшая
407227 Вероника мишо
407228 Вероника мелкоплодная
407229 Вероника мальнькая
407231 Вероника мелкая
407238 Вероника горная
407239 Вероника высокогорная
407240 Вероника монтиевидная
407245 Вероника многораздельная
407250 Вероника невского
407251 Вероника чернеющий
407252 Вероника японская
407256 Вероника аптечная, Вероника лекарственная, Купена аптечная, Купена лекарственная, Купена обыкновенная
407258 Вероника ольгинская
407260 Вероника ольтинская
407262 Вероника тусклая
407263 Вероника орхидная
407264 Вероника восточная
407267 Вероника остроплодная
407273 Вероника цветоножковая
407275 Вероника иноземная
407287 Вероника персидская, Вероника турнефора
407289 Вероника каменистая
407292 Вероника изящная
407295 Вероника порфирия
407297 Вероника ранняя
407298 Вероника близкая
407299 Вероника лежачая, Вероника простёртая
407306 Вероника крошечная
407308 Вероника многоветвистая
407329 Вероника краснолистная
407332 Вероника рупрехта
407333 Вероника сахалинская
407334 Вероника саянская
407337 Вероника шмидта
407349 Вероника шитковая
407357 Вероника северная
407358 Вероника тимьянная, Вероника тимьянолистная
407362 Вероника распростертая
407369 Вероника тимьянная
407371 Вероника сибирская
407387 Вероника стеллера
407393 Вероника столбиконосная
407400 Вероника сычуаньская
407403 Вероника тайваньская
407404 Вероника крымская
407405 Вероника тебердинская
407406 Вероника телефиелистная
407409 Вероника тончайшая
407415 Вероника тяньшанская

407416 Вероника тибетская
407420 Вероника трехраздельная
407421 Вероника трёхлистная
407424 Вероника трубкоцветковая
407426 Вероника туркменов
407430 Вероника береговая
407432 Вероника весенняя
407434 Вероника сибирская
407449 вероника
407549 Пузырник
407550 Пузырник прецеский
407551 Пузырник голоплодный
407579 Веттиверия
407645 Калина
407662 Калина кленолистная
407677 Горошек амурская
407696 Калина березолистная
407713 Калина буддлелистная
407715 Калина буреинская, Калина бурятская
407720 Калина мотоская
407727 Калина карлса, Калина корейская
407767 Калина скумпиелистная
407769 Калина цилиндрическая
407774 Калина давида
407778 Калина зазубренная
407785 Калина раскидистая, Калина расширенная
407795 Калина лушаньская
407802 Калина ичанская
407865 Калина вильчатая
407867 Калина люцюская
407877 Калина хайнаньская
407886 Калина хубейская
407897 Калина японская
407900 Калина ганьсуская
407905 Калина корейская
407907 Гордо обыкновенная, Гордовина, Калина гордовина, Калина цельнолистная
407917 Калина канадская
407939 Калина крупноголовчатая
407954 Калина монгольская
407977 Калина душистая
407980 Калина авабуки
407988 Калина эмэйская
407989 Калина красная, Калина обыкновенная, Крушина обыкновенная, Шар снежный
408009 Калина сарджента
408015 Калина восточная
408019 Калина мелколистная
408034 Калина войлочная, Калина складчатая войлочная
408061 Калина грушелистная, Калина сливолистная
408074 Калина крапчатая
408089 Калина морщинистолистная, Калина сморщеннолистная
408098 Калина сарджента
408120 Калина чжэцзянская
408130 Калина жуйлиская
408131 Калина зибольда, Калина зибольдова
408141 Калина яошаньская
408151 Калина висячая, Калина повислая
408156 Калина тайваньская
408165 Калина тибетская
408167 Калина вечнозелёная, Калина лавролистная
408224 Калина райта
408251 Вика, Вика мундж, Горошек, Мунч, Мшунки-гоуак
408252 Горошек укороченный
408255 Горошек горный
408262 Вика красивая, Горошек приятный
408274 Горошек круглоплодный
408278 Горошек амурский
408284 Вика узколистная, Горошек узколистный
408288 Горошек сорный
408295 Горошек древний
408302 Горошек байкальский
408303 Горошек чаюйский
408304 Горошек баланзы
408305 Вика бенгальская
408313 Горошек биберштейна
408314 Горошек двлетний
408322 Горошек вифинский
408325 Горошек обарта
408326 Горошек буасъе
408328 Горошек шпорцевый
408337 Горошек кассубский, Горошек кашубский
408339 Вика китайская
408343 Горошек ресничатый
408348 Горошек сердцевиднолистный, Горошек сердцевидный
408350 Вика синьцзянская, Горошек ребристый
408352 Горох мышиный, Горошек мышиный, Горошек оранжевый, Мышиный горошек
408361 Горошек японский
408370 Горошек далматский
408371 Вика пушистомлодная, Горошек шерстистоплодный
408381 Горошек зарослевый, Горошек кустарниковый
408383 Горошек безусиковый
408387 Горошек изящный
408392 Горошек горький, Горошек чёткообразный, Чечевица французская
408393 Боб конский, Боб огородный, Боб полевой, Бобы, Бобы кормовые, Бобы русские, Конские бобы
408394 Боб конский, Бобы конские, Бобы кормовые
408395 Бобы мелкоплодные
408401 Горошек федченко
408406 Горошек парноцветковый
408417 Горошек тонкий
408418 Горошек изящный
408419 Горошек крупноцветковый
408420 Горошек гроссгейма
408421 Горошек армянский
408423 Горошек волосистый, Горошек шершавоволосый
408427 Горошек плотноволосистый
408429 Горошек гиблидный, Горошек помесный
408431 Горошек гирканский
408432 Горошек грузинский
408433 Горошек надрезанный, Горошек надрезный
408434 Горошек иранский
408435 Горошек японский
408447 Горошек китайбеля
408448 Горошек коканский
408449 Горошек гулинский
408452 Горошек чиннобразный, Горошек чиновидный
408466 Горошек лиловый
408469 Горошек желтый, Горошек жёлтый
408478 Вика крупносемянная
408479 Горошек крупнолистный
408483 Горошек крупносемянный
408484 Горошек крупнолодочковый
408489 Горошек мейера
408490 Горошек мишо
408493 Горошек одноцветковый
408502 Горошек многостебельный
408505 Вика нарбонская, Вика римская, Вика французская, Горошек нарбонский
408517 Горошек ольвийский
408527 Вика бенгерская, Вика панноская, Горошек венгерский, Горошек паннонский
408535 Горошек малопарный
408537 Горошек иноземный, Горошек чужеземный
408540 Вика пестроцветная, Горошек вестрый, Горошек пестроцветный, Горошек раскрашенный
408541 Горошек волосистый, Горошек опушенный
408543 Вика гороховидная, Горошек гороховидный
408551 Горошек лжесочевичниковый, Горошек лжесочевниковый
408559 Горошек пушистый

408563 Горошек разветленный
408571 Вика кормовая, Вика обыкновенная, Вика плоско-семянная, Вика посевная, Вика сорнополевая, Вика яровая, Горох огородный, Горох полевой, Горошек зеленый, Горошек кормовой, Горошек полевой, Горошек посевной, Полюшка
408605 Горошек щебнистый
408609 Горошек семенова
408610 Горошек полуголый
408611 Горошек заборный, Горошек подзаборный
408612 Горошек зубчатолистный
408614 Вика лесная, Горошек лесной
408616 Горошек сосновского
408618 Горошек полосатый
408619 Горошек полуволосистый
408620 Горошек лесной
408624 Горошек тонколистный
408638 Горошек четырехсемянный, Горошек четырёхсемянный
408641 Горошек тибетский
408642 Горошек трехостроконечный
408646 Горошек обрубленный
408648 Горошек однопарный
408670 Горошек изменчивый
408671 Горошек пестрый
408672 Горошек жилковый
408686 Горошек зилковатый
408693 Вика мохнатая, Вика озимая, Вика озймая, Вика песчаная, Вика чёрная, Вика чёрная мохнатая, Горошек мохнатый, Горошек песчаный, Горошек шерстистоплодный
408733 Виктория
408825 Вигна
408839 Адзуки, Резеда, Фасоль адзуки, Фасоль лучистая, Фасоль угловатая
408982 Мунго, Урд, Фасоль азиатская, Фасоль маш, Фасоль много
409025 Маш, Фасоль золотистая, Фасоль кустовая однолистная
409036 Вигна люцюская
409069 Бобы земляные, Воандзея подземная, Горох земляной, Орех земляной бамбарский
409086 Бобы снаржевые, Вигна катьянг, Вигна китайская, Горох кормовой, Горох коровий, Дзян-доу, Долихос китайский, Фасоль спаржевая
409092 Вигна катьянг, Вигна коноплевидная
409096 Бобы снаржевые, Фасоль спаржевая
409251 Вилларезия
409325 Барвинок, Могильница
409329 Барвинок прямой
409335 Барвинок большой, Барвинок малый, Барвинок травянистый
409339 Барвинок малый, Гроб-трава
409345 Барвинок опушенный
409394 Ластовень
409431 Ластовень дарвазский
409608 Фиалка
409630 Фиалка заостренная, Фиалка остроконечная, Фиалка приостренная
409643 Фиалка остролистная
409655 Фиалка алайская
409661 Фиалка белая
409665 Фиалка корейская
409672 Фиалка александрова
409681 Фиалка альпийская
409682 Фиалка алтайская
409686 Фиалка изменчивая, Фиалка сомнительная
409687 Фиалка амурская
409698 Фиалка песчаная, Фиалка скальная
409703 Фиалка полевая
409706 Фиалка темнофиолетовая
409709 Фиалка авачинская
409711 Фиалка яньюаньская
409716 Фиалка непалская
409729 Фиалка двухцветковая, Фиалка двуцветковая
409762 Фиалка коротшпорцевая
409804 Фиалка собачья
409811 Фиалка мясистенькая
409816 Фиалка бутеневидная, Фиалка бутенелистная
409834 Фиалка холмовая
409858 Фиалка рогатая
409862 Фиалка толстая
409868 Фиалка толстошпорцевая
409878 Фиалка пальчатая
409883 Фиалка склоненная
409914 Фиалка рассеченная
409925 Фиалка надрезанная
409934 Фиалка длинношпорцевая
409942 Фиалка японская
409947 Фиалка высокая
409950 Фиалка елизаветы
409954 Фиалка эмэйская
409958 Фиалка кочкарная, Фиалка сверху голая
409979 Фиалка федченко
409993 Фиалка тайваньская
410010 Фиалка гмелина
410030 Фиалка орлиноклювая
410066 Фиалка коротковолосистая, Фиалка опушенная, Фиалка опушённая
410073 Фиалка волосистоногая, Фиалка волосисточерешковая
410080 Фиалка гиссарская
410090 Фиалка гультена
410091 Фиалка хунаньская
410106 Фиалка равнолепестная
410108 Фиалка японская
410117 Фиалка жордана
410126 Фиалка каракалинская
410140 Фиалка китайбелева, Фиалка китайбели, Фиалка китайбля
410149 Фиалка тибетская
410152 Фиалка купфера
410158 Фиалка кузнецова
410169 Фиалка лангсдорфа
410183 Фиалка прибрежная
410187 Фиалка жёлтая
410192 Фиалка крупношпорцевая
410201 Фиалка маньчжурская
410202 Фиалка манчьжурская
410232 Фиалка утренняя
410233 Фиалка морица
410246 Фиалка коротко-чашелистиковая, Фиалка удивительная
410266 Фиалка скромненькая
410269 Фиалка монгольская
410272 Фиалка одноцветная
410273 Фиалка горная
410279 Фиалка мюльдорфа
410318 Фиалка скрытая
410320 Фиалка душистая
410325 Фиалка крымский, Фиалка скальная
410326 Фиалка восточная
410329 Фиалка прямошнорая
410336 Фиалка острошпорцевая
410347 Фиалка болотная
410358 Фиалка крохотная
410360 Фиалка патрэна
410401 Фиалка прудовая
410403 Фиалка голоплодная, Фиалка лысоплодная
410452 Фиалка зубчатоцветковая
410474 Фиалка низкая, Фиалка низкорослая
410478 Фиалка радде
410483 Фиалка лесная
410493 Фиалка ривина, Фиалка ривиниуса, Фиалка ривинуса
410498 Фиалка росса
410513 Фиалка скальная
410516 Фиалка рупиуса
410517 Фиалка сахалинская
410547 Фиалка селькирка, Фиалка теневая
410579 Фиалка зиге, Фиалка сиэхеано
410584 Фиалка лесная
410598 Фиалка лушаньская
410605 Фиалка приятная
410622 Фиалка сычуаньская
410639 Фиалка донская
410640 Фиалка тарбагатайская
410647 Фиалка волосисточашечная
410662 Фиалка тяньшанская

410675 Фиалка ресничато-чашелистиковая
410677 Анютины глазки, Братки, Глазки анютины, Ивам-да-марья, Трёхцветка, Фиалка трехцветная, Фиалка трёхцветная
410700 Фиалка турчанинова
410702 Фиалка топяная
410704 Фиалка одноцветковая
410710 Фиалка пестрая
410721 Фиалка пестрая, Фиалка пестрололистная
410730 Фиалка застенчивая, Фиалка скромная
410770 Фиалка алисовой, Фиалка токийская
410787 Фиалка хэбэйская
410792 Фиалковые
410869 Виргилия
410950 Лихнис альпийский, Смолка альпийская
410953 Смолка клейкая
410957 Смолка клейкая, Смолка обыкновенная
410959 Омела
410960 Дубянка, Омела белая, Омела белая дубянка, Омела евронейская, Трава пресноцветущая
410972 Омела тайваньский
410992 Омела окрашенная
411140 Висмия
411165 Виноградные, Виноградовые
411187 Авраамово дерево, Витекс, Прутняк
411189 Авраамово дерево, Витекс священный, Деревей, Прутняк обыкновенный, Целомудренник
411310 Витекс гуансиский
411362 Прутняк китайский
411373 Прутняк китайский коноплеволистный
411421 Витекс лунчжоуский
411521 Виноград, Лоза виноградная
411540 Виноград амурский
411575 Виноград берландье, Виноград горный, Виноград зимний
411623 Виноград куанье
411631 Виноград лисий
411646 Виноград давида
411668 Виноград фэнцинский
411669 Виноград тунберга
411682 Виноград лунчжоуский
411742 Виноград лушаньский
411764 Виноград американский, Виноград изабелла, Виноград изабеллы, Виноград лабруска, Виноград ляблуска, Виноград североамериканский, Изабелла, Катауба
411796 Виноград мэнхайский
411797 Виноград мэнцзыский
411834 Виноград лапчатый
411868 Циссус четырехгранный
411885 Циссус ромболистный
411887 Виноград американский, Виноград лисий, Виноград прибрежный, Рипария
411890 Виноград романе
411893 Виноград круглолиственный, Виноград круглолистный, Мюскадина
411900 Виноград скальный
411979 Виноград, Виноград винный, Виноград виноносный, Виноград винородный, Виноград евроазиатский, Виноград европейский, Виноград культурный, Виноград настоящий, Виноград обыкновенный, Виноград старосветский, Лоза виноградная
411983 Виноград лесной
412010 Виноград чжэцзянский
412394 Вульпия, Зерна, Многолетний костер
412404 Вульпия альпийская
412407 Вульпия белкохвостая
412409 Вульпия ресничатая
412465 Вульпия мышехвостник, Овсяница мышехвостная
412479 Вульпия персидская
412495 Вульпия волосистая
412519 Вахендорфия
412568 Валенбергия
412750 Валенбергия изящная
413007 Вальдгеймия, Вальдгиймия
413009 Вальдгеймия голая, Вальдгиймия голая
413018 Вальдгеймия войлочная, Вальдгиймия войлочная
413020 Вальдгеймия заалайская
413021 Вальдгеймия трехлопастная, Вальдгиймия трехлопастная
413030 Вальдштейния
413034 Вальдштейния тройчатая
413059 Валления
413085 Валлихия
413090 Валлихия китайская
413091 Валлихия густоцветная
413094 Валлихия юньнаньская
413099 Валлихия таиландская
413289 Вашингтония
413298 Вашингтония нитевидная, Вашингтония нитеносная
413307 Вашингтония крепкая, Вашингтония мощная
413570 Вейгела, Вейгелия
413576 Вейгела корейская
413588 Вейгела обильноцветущая
413591 Вейгела цветущая
413611 Вейгела японская
413616 Вейгела максимовича
413617 Вейгела миддендорфа
413621 Вейгела ранняя
413624 Вейгела кроваво-красная
413628 Вейгела приятная
413741 Вельвичия
413765 Вендландия
414058 Виддрингтония
414065 Виддрингтония шварца
414066 Виддрингтония уайта
414072 Видеманния
414074 Видеманния многонадрезная
414075 Видеманния восточная
414092 Вигандия
414118 Викстремия, Рестелла
414133 Викстремия аньхуйская
414172 Викстремия чэнкоуская
414186 Викстремия хайнаньская
414191 Викстремия хуйдунская
414193 Викстремия индийская
414207 Викстремия лицзянская
414265 Викстремия тайваньская
414266 Викстремия дэциньская
414271 Викстремия японская
414312 Меркия пузырчатая
414385 Виллемеция
414389 Виллемеция крубненосная
414474 Винклера
414475 Винклера натриниевидная
414478 Винклера
414541 Вистария, Глициния
414554 Вистария обильноцветшая, Глициния обыкновенная, Глициния японская
414556 Глициния японская
414565 Вистария кустарниковая
414572 Глициния японская
414576 Вистария китайская, Глициния китайскя
414649 Вольфия
414653 Вольфия безкорневая
414772 Вороновия
414774 вороновия красивая
414829 Вульфения
414990 Дурнишник, Ксантиум
414999 Дурнишник калифорнийский
415003 Дурнишник канадский
415031 Ксантиум монгольский
415033 Дурнишник западный
415044 Дурнишник береговой
415046 Дурнишник сибирский, Ксантиум сибирский
415053 Дурнишник игольчатый, Дурнишник колючий, Ксантиум колючий
415057 Дурнишник американский, Дурнишник ежевидный, Дурнишник зобовидный, Дурнишник иглистый, Дурнишник обыкновенный, Дурнишник

шиповатый
415094 Ксантоцерас
415115 Ксантогалум
415117 Ксантогалум пурпуровый
415118 Ксантогалум сахокии
415119 Ксантогалум татьяны
415164 Ксанториза, Ксанторриза
415172 Акароидин, Дерево травяное, Желтосмолка, Ксанторрея
415190 Ксантозома, Ксантосома
415202 Ксантосома фиолетовая
415315 Бессмертник, Сухоцвет, Сухоцветник
415317 Ксерантемум однолетний, Сухоцвет однолетний
415386 Бессмертник, Бессмертник соломенный, Гелихризум болешой, Иммортель, Цмин прицветниковый
415558 Ксимения американская
415672 Ксилантемум
415673 Ксилантемум фишеры
415675 Ксилантемум памирский
415676 Ксилантемум скальный
415677 Ксилантемум тяньшанский
415714 Ксилосарпус
415863 Ксилосма
415869 Ксилосма японская
415990 Ксирис
416390 Юнгия
416394 Юнгия алтайская
416416 Юнгия разнолистная
416437 Юнгия тайваньская, Юнгия японская
416447 Юнгия кандинская
416475 Юнгия зеравшанская
416485 Юнгия узкая
416489 Юнгия тонкостебельная
416490 Юнгия тонколистная
416535 Юкка, Юкка мечевидная
416538 Юкка алоелистная, Юкка алойная, Юкка алоэлистная
416581 Пробосцидея, Юкка слоновая
416584 Игла адамова, Юкка виргинская, Юкка волокнистая, Юкка нитевидная, Юкка нитчатая
416607 Юкка величественная, Юкка прекрасная, Юкка сизая, Юкка славная
416610 Юкка пониклолистная
416682 Магнолия суланжа
416688 Магнолия кемпбелла, Магнолия кэмбелла
416694 Магнолия голая, Магнолия обнаженная, Магнолия обнажённая, Магнолия юлана
416707 Магнолия лилейная, Магнолия лилиецветковая, Магнолия лилиецветная, Магнолия пурпуровая
417002 Цамия
417045 Заникелия, Занникеллия
417051 Занникеллия большая
417053 Заникелия болотная, Занникеллия болотная, Занникеллия стебельчатая
417073 Занникелиевые
417092 Зантедесхия, Рисовидка, Ричардия
417093 Арум кукушечный, Зантедесхия абиссинская, Калла, Рисовидка африканская
417104 Зантедесхия эллиота
417132 Зантоксилум, Зантоксилюм, Ксантоксилум
417134 Зантоксилюм колюченожковый
417139 Зантоксилюм айлантовидный
417157 Дерево зубное, Зантоксилюм американский, Перец американский, Ясень колючий американский
417161 Зантоксилюм плоскошипный
417180 Перец бунге, Перец китайский
417216 Зантоксилюм рассеянный
417228 Цантоксилум мартиникский
417230 Дерево шёлковое
417231 Зантоксилум американский, Перец американский
417247 Зантоксилюм ланьюйский
417249 Зантоксилюм гуансиский
417254 Зантоксилюм лэйбоский
417258 Зантоксилюм либоский
417271 Зантоксилюм мелкоцветковое
417275 Зантоксилюм мотоский
417301 Зантоксилюм перечный, Перец японский
417330 Зантоксилюм схинусолистный
417340 Зантоксилюм подражающий, Зантоксилюм щетинистый, Перец бунге, Перец китайский
417374 Зантоксилюм сичоуский
417405 Цаущнерия
417412 Кукуруза, Кукуруза обыкновенная, Маис
417414 Кукуруза крахмалистая
417416 Кукуруза зубовидная
417417 Джугара, Кукуруза, Кукуруза обыкновенная, Маис, Макка-джугара
417423 Кукуруза зубовидная
417433 Кукуруза сахарная, Кукуруза сладкая столовая
417434 Кукуруза плёнчатая, Кукуруза чешуйчатая
417534 Дзелква, Дзелкова, Дзельква, Зельква, Планера
417540 Дзелква граболистная
417547 Дзелква тайваньская
417552 Дзелква шнейдера
417558 Дзелква пиличатая, Дзелква японская, Дзельква пиличатая
417563 Дзелква китайская
417606 Зефирантес
417640 Зеравшания
417641 Зеравшания регеля
417698 Зевксина
417795 Зевксина шлемовидная
417880 Зигаденус
417925 Зигаденус сибисский
417964 Имбирь, Цингибер
418001 Имбирь мэнхайский
418010 Имбирь, Имбирь лекарственный
418013 Имбирь эмэйский
418021 Имбирь симаоский
418029 Имбирь инцзянский
418030 Имбирь юньнаньский
418034 имбирные
418035 Циния, Цинния
418038 Цинния хагеа
418043 Циния изящная, Цинния изящная, Цинния обыкновенная красивая, Цинния стройная
418058 Цинния многоцветковая, Цинния перуская
418079 Цицания
418080 Рис болотный, Рис водяной, Рис дидий, Рис индийский, Рис канадский, Рис озёрный, Трава рисовая, Тускарора, Цицания водяная
418095 Цицания широколистная
418096 Рис индийский, Тускарора, Чалтык
418119 Зизифора
418120 Зизифора биберштейна
418121 Зизифора бранта
418122 Зизифора короткочашечная
418123 Зизифора бунге, Зизифора бунговская
418124 Зизифора головчатая
418125 Зизифора мелкоголовчатая
418126 Зизифора пахучковидная
418127 Зизифора мелкозубчатая
418128 Зизифора галины
418130 Зизифора прерваннар
418131 Зизифора памироалайская, Зизифора памиро-алайская
418132 Зизифора персидская
418135 Зизифора пушкина
418136 Зизифора радде
418137 Зизифора жесткая
418138 Зизифора тимьянниковая
418139 Зизифора крымская
418140 Зизифора тонкая
418141 Зизифора войлочная, Зизифора тяньшанская
418142 Зизифора туркменская
418143 Зизифора воронова
418144 Зизифус, Унаби, Юйюба, Ююба
418169 Зизифус, Зизифус обыкновенный,

Слива китайская, Унаби, Унаби обыкновенная, Унаби юйюба, Унаби ююба, Финик китайский, Юйюба, Ююба, Ююба унаби, Ягода грудная европейская, Ягода грудная французская
418175 Унаби колючий
418249 Зегея
418250 Зегея бальджуанская
418251 Зегея пурпуровая
418362 Зосимия
418364 Зосимия полыннолистная
418372 Зосимия тордилиевидная
418377 Взморник, Зостера
418381 Взморник азиатский
418389 Зостера японская
418392 Взморник, Взморник малый, Взморник морской, Взморник тикоокеанский, Зостера большая, Зостера морская, Камка, Морская трава, Трава морская
418396 Взморник низкий, Взморник японский
418405 Взморниковые, Зостеровые
418424 Цойзия, Цойсия
418426 Цойзия японская, Цойсия японская
418529 Зигокактус
418532 Зигокактус притупленный
418580 Парнолистниковые, Парнолистные
418585 Парнолистник
418599 Парнолистник лебедовый
418604 Парнолистник балхашский
418607 Парнолистник короткокрылый
418608 Парнолистник втхарский
418623 Парнолистник остроконечный
418626 Парнолистник дарвазский
418638 Парнолистник эйхвальда
418639 Парнолистник ширококрылый
418641 Парнолистник бобовидный, Парнолистник обокновенный
418646 Парнолистник фабаговидный
418648 Парнолистник ферганский
418653 Парнолистник вильчатый
418668 Парнолистник гопчарова
418677 Парнолистник илиййский
418680 Парнолистник дильса, Парнолистник сырдарьинский
418681 Парнолистник ганьсуский
418682 Парнолистник каратавский
418683 Парнолистник кашгарский
418685 Парнолистник кегенский
418686 Парнолистник копальский
418691 Парнолистник лемана
418701 Парнолистник крупноногий
418702 Парнолистник крупвокрыладый, Парнолистник крупнокрылый
418712 Парнолистник крупноплодный
418714 Парнолистник мелкоплодный
418720 Парнолистник суриковокрасный
418722 Парнолистник крупноногий, Парнолистник неравнобокий
418726 Парнолистник амудаьинский
418727 Парнолистник остоплодный
418730 Парнолистник портулаковидный
418732 Парнолистник потанина
418740 Парнолистник крылатоплодный
418753 Парнолистник розова
418754 Парнолистник широколистный
418760 Парнолистник синьцзянский
418768 Парнолистник узкокрылый
418769 Парнолистник почти трехпарный
418774 Парнолистник туркменский

俄文名称—序号索引

Russian

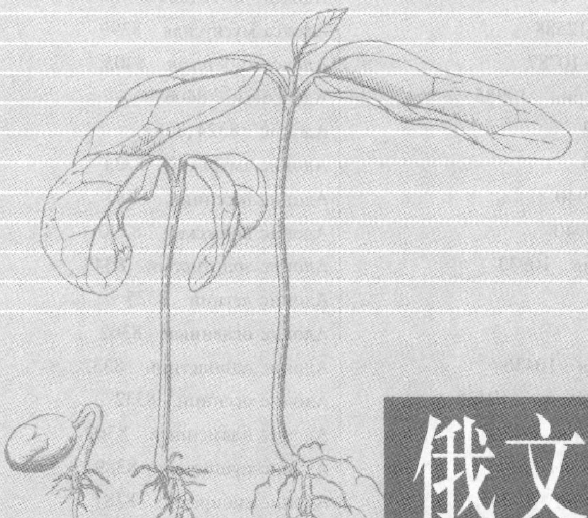

A Dictionary of Seed Plant Names
种子植物名称

Абака 260275
Абасси 179884
Абелимош 212,219
Абелия 91
Абелия китайская 113
Абелия корейская 117
Абелия мелколистная 200
Абелия пильчатая 167
Абелия щитковидная 119
Абелия японская 177
Абелия японская мелкоцветковая 184
Аблоко розовое 382582
Абобра 639
Абобра тонколистная 641
Абрикос 34425,34475
Абрикос альпийский 34430
Абрикос ансу 34477
Абрикос антильский 243982
Абрикос волосистоплодный 34431
Абрикос дерево абрикосовое 34475
Абрикос дикий 34443,34475
Абрикос китайский 34448,34477
Абрикос культурный 34475
Абрикос маньчжурский 34443
Абрикос монгольский 20928,34443
Абрикос муме 34448
Абрикос натоящий обыкновеный 34475
Абрикос обыкновенный 34475
Абрикос сибирский 34470
Абрикос тибетский 34436
Абрикос тропический 243982
Абрикос фиолетовый 34431
Абрикос черный 34431
Абрикос чёрный 34431
Абрикос шелковисто-бархатный 34436
Абрикос шероховатоплодный 34431
Абрикос японский 34448
Абрикосо-слива 34431
Абрикоссовая слива 316813
Аброма 672
Аброния 687
Абрус 738,765
Абсент 35090
Абсинт 35090
Абутилен 837
Абутилон 837
Абутилон индийское 934
Абутилон индийское мелкоцветковое 907
Абутилон китайское 989
Авдотка 399484,399504
Авероа 45725
Авероа-билимби 45724
Ависения 45740
Авокадо 291491
Авокадо американское 291494
Авокадо привлекательнейшее 291494
Авокадо-груша 291494
Авраамово дерево 411187,411189
Авран 180292
Авран аптечный 180321
Авран лекарственный 180321
Авран японский 180308
Аврикуляриевые 315138
Агава 10778
Агава американская 10787
Агава вильморена 12588
Агава древовидная 10787
Агава королев виктории 10955
Агава магуэй 10933
Агава сальма 10933
Агава сизалевая 10940
Агава сизаловая 10940
Агава тёмно-зеленая 10933
Агавовые 10775
Агазиллис 10435
Агазиллис кавказская 10436
Агазиллис широколистная 10438
Агапант 10244
Агапантус 10244
Агапантус африканский 10245
Агатис 10492
Агатис ланцетный 10497
Агатис новозеландский 10494
Агератум 11194
Агератум гаустона 11206
Агератум мексиканский 11206
Аглайя 11273
Аглайя душистая 11300
Аглайя душистая мелколистная 11302
Аглайя тайваньская 11293
Аглаонема 11329
Аглаонема переменчивая 11332
Аглаонема ребристая 11340
Аглая 11273
Агримониа 11542
Агримониа душистая 11594
Агростемма 11938
Агростис 11968
Агростис мелкоцветковый 12201
Агростис мулиский 12208
Агрум 93314
Адамова голова 265944
Адансония 7018,7022
Аденандра 7113
Аденантера 7178
Аденантера павлинья 7190
Аденокаулон 7427
Аденокаулон сросшийся 7433
Аденостилес 8034
Аденостилес чесночная 8038
Аденофора 7596
Аджерик 117859
Аджерик аджирык 117852,117859
Аджирык 117859
Адзуки 408839
Адина 8170
Адина красноватая 8199
Адинандра 8205
Адлай 99124
Адлумия 8298
Адлумия азиатская 8299
Адокса 8396,8399
Адокса мускусная 8399
Адокса тибетская 8405
Адоксовые 8406
Адонис 8324,8387
Адонис амурский 8331
Адонис весенний 8387
Адонис волжский 8390
Адонис золотистый 8343
Адонис летний 8325
Адонис огненный 8362
Адонис однолетний 8332
Адонис осенний 8332
Адонис пламенный 8362
Адонис пушистый 8389
Адонис сибирский 8381
Адонис сычуаньский 8382
Адонис туркестанский 8385
Адонис тяньшанский 8383
Адраспан 287978
Ажгон 393940
Ажрык 8862
Азадирахта 45906
Азадирахта индийская 45908
Азалея 330017
Азалея желтая 331520
Азалея моллис 331257
Азалея понтийская 331520
Азалия 330017
Азалия жёлтая 331161
Азалия индийская 330917
Азалия понтийкая 331161
Азалия порярная 235283
Азалия стелющаяся 235283
Азара 45952
Азара мелколистная 45955
Азиат 195269
Азимина 38322
Азимина трёхлопастная 38338
Азинеума 43569
Азинеума аньхуйская 43571
Азинеума бальджуанская 43574
Азинеума ветвистая 43585
Азинеума гладкоцветковая 43582
Азинеума жесткая 43586
Азинеума ивовая 43588
Азинеума китайская 43577
Азинеума колокольчиковидная 43575
Азинеума красивая 43584
Азинеума кувшинчатая 43594

Азинеума ланцетная 43581
Азинеума лобериевидная 43583
Азинеума острозубчатая 43573
Азинеума сероватая 43576
Азинеума стебреобъемлющая 43570
Азинеума тальшская 43590
Азинеума траутфеттера 43593
Азинеума цикориевидная 43578
Азинеума японская 43580
Аизоацневые 12911
Аизовые 12911
Аизоон испанский 12917
Аир 5788,5793
Аир болотный 5793
Аир злаковый 5809
Аир многолетный 5793
Аир обыкновенный 5793
Аир травянистый 5803
Аир тро стниковый 5793
Аир элаковый 5803
Аира 12776
Аира волосовидная 12789
Аира гвоздичная 12794
Аира ранняя 12839
Аирник 118428
Аирник венгерский 212774
Аирник двухколосковый 118765
Аистник 153711
Аистник геффта 153883
Аистник длинноклювый 153764
Аистник лоэбный 153909
Аистник обыкновенный 153767
Аистник остроклювый 153883
Аистник стевена 153916
Аистник стефана 153914
Аистник цельнолистный 153839
Аистник цикутный 153767
Аистник цикутовый 153767
Аистник цикутолистный 153767
Айва 116532
Айва дикая 116546
Айва китайская 317602
Айва низкая маулея 116546
Айва обыкновенная 116546
Айва продолговатая 116546
Айва цветочная 84556
Айва японская 84549,84556
Айлант 12558
Айлант высочайший 12559
Айлант высочайший мелколистный 12561
Айлант гуансиский 12579
Айлант железистый 12559
Айлант желёзковый 12559
Айлант китайский ясень 12559
Айлант тайваньский 12564
Айован 393937,393940
Айован душистый 393940
Айован лицзянский 393952
Айован столовый 393940
Айфанес 12771
Айян 65146
Акажу 21195,380522,380528
Акалифа 1769
Акалифа индийская 1900
Акалифа лицзянская 1975
Акалифа тайваньская 1781
Акалифа уилкса 2011
Акалифа уилкса крупнолистная 2017
Акалифа хайнаньская 1888
Акалифа шетинистоволосая 1894
Акалифа южная 1790
Акант 2657
Акант колючий 2711
Акант мелкоцветковый 2672
Акант мягколистный 2695
Акант падуболистный 2684
Акантовые 2060
Акантолепис 2158,2206
Акантолепис восточный 2207
Акантолимон 2208,2222,2226,2233,2242,2266,2269,2279,2286,2287,2289,2296,2297
Акантолимон алайский 2210
Акантолимон алатавский 2211
Акантолимон александра 2214
Акантолимон альберта 2213
Акантолимон армянский 2216
Акантолимон аулиеатинский 2217
Акантолимон бархатистый 2299
Акантолимон бородина 2220
Акантолимон варивцевой 2298
Акантолимон гауданский 2236
Акантолимон гвоздичный 2224
Акантолимон гиссарский 2240
Акантолимон гогенакера 2241
Акантолимон гончарова 2238
Акантолимон дедина 2239
Акантолимон диапенсиевидный 2227
Акантолимон екатерины 2230
Акантолимон зеленый 2301
Акантолимон иглистый 2231
Акантолимон карелина 2244,2245
Акантолимон кнорринг 2248
Акантолимон кокандский 2249
Акантолимон королькова 2250
Акантолимон красноватый 2232
Акантолимон крылоприцветниковый 2276
Акантолимон маева 2259
Акантолимон маргариты 2260
Акантолимон мелкоцветковый 2271
Акантолимон микешина 2262
Акантолимон минжелкинский 2263
Акантолимон нежный 2219
Акантолимон необычайный 2265
Акантолимон никитина 2267
Акантолимон нуратавский 2268
Акантолимон овсовый 2218
Акантолимон памирский 2270
Акантолимон плауновый 2257
Акантолимон пленчатый 2237
Акантолимон плодный 2225
Акантолимон прицветниковый 2221
Акантолимон пскемский 2275
Акантолимон пурпурный 2278
Акантолимон радде 2280
Акантолимон распростертый 2274
Акантолимон растопыренный 2288
Акантолимон рупрехта 2282
Акантолимон рыхлый 2252
Акантолимон сакена 2283
Акантолимон сахендский 2284
Акантолимон тарбагатайский 2291
Акантолимон татарский 2292
Акантолимон титова 2295
Акантолимон тонкохвостниковый 2254
Акантолимон тонкоцветковый 2293
Акантолимон торчащий 2290
Акантолимон тяньшанский 2294
Акантолимон удивительный 2264
Акантолимон фетикова 2234
Акантолимон фомина 2235
Акантолимон хорасанский 2247
Акантолимон хорошенький 2277
Акантолимон шемахинский 2285
Акантопапакс 143575
Акантопапакс генри 2412,143610
Акантопапакс зибольда 143677
Акантопапакс кандинский 143625
Акантопапакс китайский 2395
Акантопапакс колючий 143682
Акантопапакс сидячецветковый 143665
Акантопапакс сидячецветковый мелкоплодный 143667
Акантопапакс сычуаньский 143634
Акантопапакс спдячецвеный 143665
Акантопапакс тернистый 143657
Акантопапакс тибетский 2521
Акантопапакс трехлистный 143694
Акантопапакс тяньцюаньский 2372
Акантопапакс узколистный 2505
Акантопапакс чжэцзянский 2531
Акантопапакс японский 143685
Акантофеникс 2550
Акантус 2657
Акароидин 415172
Акатик 240112
Акация 1024
Акация австралийская 1165,1384
Акация американская 334976
Акация аравийская 1427
Акация безжилковая 1049

Акация бейли 1069
Акация белая 334976
Акация беловатая 1165
Акация вооруженная 1060
Акация густоцветная 1516
Акация деальбада 1165
Акация длиннолистная 1358
Акация дубильная 1120
Акация желтая 72180
Акация жёлтая 72180
Акация камеденосная 1528
Акация катеху 1120
Акация красноватая 1544
Акация куэньминская 1177
Акация лицзянская 1176
Акация мутовчатая 1686
Акация низбегающая 1168
Акация нильская 1427
Акация ножевидная 1156
Акация печеная 19728
Акация подбеленная 1165
Акация сенегальская 1572
Акация серебристая 1165
Акация синелистная 1159
Акация стойкая 1528
Акация тайваньская 1145
Акация устрашающая 1297
Акация фарнеза 1219
Акация фарнезе 1219
Акация фарнеси 1219
Акация хайнаньская 1268
Акация чернодревесная 1384
Акация чернодревная 1384
Акация шелковая 13578
Акация шёлковая 13578
Акация щетинистая 334965
Акация юньнаньская 1711
Аквилегия 29994
Акебия 13208
Акебия пятерная 13225
Акилей 29994,30081
Акнида 18683
Акнистус 4941
Акомастилис 4986
Акомастилис росса 4995
Аконит 5014,5287
Аконит алтайский 5031
Аконит антора 5039
Аконит аптечный 5442
Аконит большой 5413
Аконит бороданый 5052
Аконит бородатый 5052
Аконит волкобойный 5676
Аконит волосистый 5663
Аконит волчий 5676
Аконит восточный 5461
Аконит высокий 5563
Аконит вьющийся 5674
Аконит гуншаньский 5333
Аконит двуцветковый 5065
Аконит десулави 5160
Аконит джунгарский 5585
Аконит жинокостнолистный 5442
Аконит извилистый 5195
Аконит илийский 5619
Аконит кандинский 5626
Аконит комарова 5326
Аконит крылова 5332
Аконит лицзянский 5205
Аконит лэйбоский 5502
Аконит мелкоцветковый 5498
Аконит метельчатый 5468
Аконит мотоский 5172
Аконит мулиский 5484
Аконит напелюс 5442
Аконит носатый 5451
Аконит опушенный 5506
Аконит пестрый 5659
Аконит пёстрый 5659
Аконит плетевидный 5198
Аконит противоядный 5039
Аконит радде 5525
Аконит реповидный 5442
Аконит сапожникова 5548
Аконит сборный 5442
Аконит северный 5563
Аконит сиотинский 5570
Аконит смирнова 5581
Аконит сукачева 5611
Аконит тайваньский 5204
Аконит таласский 5618
Аконит теневой 5466
Аконит толстолистный 5151
Аконит траншеля 5636
Аконит фиолетоный 5442
Аконит фишера 5193
Аконит флерова 5202
Аконит франше 5208
Аконит фэн 5539
Аконит хуйлиский 5283
Аконит хуйцзэский 5284
Аконит цзилунский 5304
Аконит цзиндунский 5127
Аконит чаюйский 5116
Аконит чекановского 5154
Аконит чжундяньский 5486
Аконит шерстистоустый 5347
Аконит штёрка 5600
Аконит ядовитый 5442
Аконит ядунский 5593
Аконит ялусский 5292
Аконит яньюаньский 5706
Аконит японский 5298
Акроклиниум 190791
Акрокомия 6128
Акрокомия южноамериканская 6136
Акронихия 6233
Акронихия австралийская 6237
Акронихия бауэра 6237
Аксамитник 18652,18687
Аксирис 45843
Аксирис гибридный 45853
Аксирис кавказский 45851
Аксирис простертый 45856
Аксирис шароплодный 45859
Аксирис щирицевый 45844
Актея 6405
Актинидиевые 6728
Актинидия 6513
Актинидия аргут 6515
Актинидия гуансиская 6669
Актинидия гуйлиньская 6624
Актинидия джиральди 6519
Актинидия изогнутая 6515
Актинидия киви 6553
Актинидия китайская 6553
Актинидия коломикта 6639
Актинидия красноствольная 6692
Актинидия крупная 6515
Актинидия мозолистая 6530
Актинидия мэнцзыская 6628
Актинидия носатая 6682
Актинидия острая 6515
Актинидия острозубиатая 6515
Актинидия полигама 6682
Актинидия полигамная 6682
Актинидия хубэйская 6632
Актинидия хэнаньская 6627
Актинидия цзянкоуская 6638
Актинидия чжэцзянская 6726
Актинидия чэнкоуская 6552
Актинидия японская 6523
Актиностемма 6888
Актиностемма лопастная 6907
Алабота 86901
Алангиум 13345
Алангиум гуансиский 13374
Алангиум китайский 13353
Алангиум мелкоцветковый 13362
Алангиум платанолистный 13376
Алангиум юньнаньский 13393
Алданский виноград 333957
Александра 14562
Алектоион 14359
Алектоион высокий 14362
Алетея розеа 13934
Алетрис 14471
Алетрис тайваньский 14491
Алиссум 18313
Алиссум душистый 235090
Алиссум европейский 235090

Алиссум морской 235090
Алкана 14836
Алкана восточная 14840
Алкана крупноплодная 14839
Алкана селдселистная 14837
Алканет 14836
Алканна 14836
Алканна красильная 14841
Аллисум сибирский 18469
Алое 16540
Алое американское 10778
Алое древовидное 16592
Алое индийское 16910
Алое китайский 17383
Алое остистое 16605
Алое ресничатое 16709
Алой 16540
Алоказия 16484
Алоказия гуйцзоуская 16498
Алоказия индийская 16501
Алоказия крупнокорневая 16512
Алонсоа 17475
Алоэ 16540
Алоэ американское 10787
Алоэ барбадосское 16624,17381
Алоэ бородавчатое 171767
Алоэ дикое 16817,79125
Алоэ колючее 16817,79125
Алоэ настоящее 17381
Алоэ обыкновенное 16624,17381
Алоэ пёстрое 17136
Алоэ раздвоенное 16768
Алоэ сокотрина 17320
Алоэ сокотринское 17320
Алоэ ферокс 16817,79125
Алпиния 17639
Алпиния гуандунская 17715
Алпиния китайская 17656
Алпиния мелкоцветковая 17653
Алпиния мэнхайская 17721
Алпиния напоская 17727
Алпиния тайваньская 17674
Алпиния хайнаньская 17685
Алпиния янчуньская 17760
Алтей 18155
Алтей аптечнный 18173
Алтей аптечный 18173
Алтей армянский 18157
Алтей бледный 18174
Алтей жестковолосый 18166
Алтей конаплевидный 18160
Алтей коноплевый 18160
Алтей лекарственный 18173
Алтей лекарственный 243806
Алтей морщинистный 13940
Алтей розовый 13934
Алтей шершавоволосный 18166
Алтей шершаволистный 18166
Алтей – штокроза 13934
Алфалфа 247456
Алыча 316294
Альбертия 13438
Альбертия пленчатая 13443
Альбизия 13474
Альбиция 13474,13578
Альбиция душистая 13635
Альбиция калькора 13586
Альбиция китайская 13515
Альбиция ланьюйская 13653
Альбиция ленкоранская 13578
Альбиция мэнцзыская 13505
Альбиция противоглистная 13488
Альбиция хайнаньская 13493
Альбиция юньнаньская 13526
Альбицция 13474
Альбицция душистая 13635
Альбовия 299342
Альбовия трихраздельная 13723
Альгароба 315551,315560
Альдрованда 14271
Альдрованда пузырчатая 14273
Альзик 396930
Альзина 374724
Альзина посевная 32210
Альканна 14836,21959
Альпиния 17639
Альпиния галанга 17677
Альстония 18027
Альстония школьная 18055
Альстремерия 18060
Альтей 18155
Альтернантера 18089
Альфа 376931
Альфальфа 247456
Альфитония 17613
Альфитония высокая 17614
Альфредия 14593
Альфредия колючечешуйная 14594
Альфредия ледовая 14599
Альфредия пониклая 14596
Альфредия поникшая 14596
Альфредия снежная 14599
Альхорнея 14177
Алэина 374724
Амаракус 18611
Амаракус круглолистный 274233
Амарант 18652
Амарант белый 18655
Амарант жминдовидный 18668
Амарант колосистый 18810
Амарант меланхолический 18779
Амарант метельчатый 18788
Амарант поникльй 18687
Амарант хвостатый 18687,18788
Амарант шарообразный 179236
Амарантовые 18646
Амарилис 18862,196439
Амарилис альпийский 227854
Амарилис белладонна 18865
Амариллисовые 18861
Амбач 9489
Амбербоа 18946
Амбербоа бухарская 18951
Амбербоа низкая 18971
Амбербоа обыкновенная 18948
Амбербоа сизая 18954
Амбербоа сосновского 18980
Амбербоа туранская 18983
Амблиокарпум 19049
Амблиокарпум девясиловый 19050
Амбрет 194684
Амбровое дерево 232565
Амброзия 19140
Амброзия бескрылая 19143
Амброзия горометельчатая 19179
Амброзия полыннолистная 19145
Амброзия трёхнадрезная 19191
Амброзия трехраздельная 19191
Амброзия трёхраздельная 19191
Амеланхир 19240
Аметистеа 19458
Аметистеа голубая 19459
Амзинцкия 20827
Амирис бальзамический 21005
Амирис ямайский 21005
Аммания 19550
Аммания египетская 19566
Аммания зеленая 19641
Аммания многоцветковая 19602
Аммания мутовчатая 19639
Аммания песчаная 19554
Аммания пушистоцветковая 19616
Амми 19659
Амми большой 19664
Амми виснага 19672
Амми зубная 19672
Аммобиум 19690
Аммобиум крылатый 19691
Аммодендрон 19728
Аммодендрон длиннокистевая 19735
Аммодендрон карелина 19733
Аммодендрон конолли 19731
Аммодендрон лемана 19734
Аммодендрон персидская 19736
Аммодендрон серебристая 19729
Аммодендрон эйхвалъда 19732
Аммодения 198003
Аммодения бутерлаковидная 198010
Аммофил 19765
Аммофила 19765
Аммртамнус джунгарский 369122

Аммртамнус лемана 369056
Амоора 19952
Аморфа 19989
Аморфа калифорнийская 19994
Аморфа карликовая 20023
Аморфа кустовидная 20005
Аморфа кусторниковая 20005
Аморфа седовадая 19997
Аморфофалл 20037
Аморфофалл гигантский 20145
Аморфофаллюс 20037
Аморфофаллюс китайский 20136
Аморфофаллюс мэнхайский 20051
Аморфофаллюс риввера 20132
Аморфофаллюс юньнаньский 20153
Амофа кустарниковая 20005
Ампелопсис 20280
Амприк альпийский 31314
Амрфофаллус 20037
Амрфофаллус гигантский 20145
Амрфофаллус колокольчатый 20057,20125
Амрфофаллус риввера 20132
Амсония 20844
Амурское пробковое дерево 294231
Амфорикарпус 20811
Амфорикарпус изящный 20812
Анабазис 20844,21023
Анабазис солончаковый 21058
Анагаллидиум 21329
Анагаллидиум вильчатый 380181
Анагаллис 21335,21339
Анагирис 21445
Анакамптис 21162
Анакамптис пирамидальный 21176
Анакард 21191,21195
Анакард акужа 21195
Анакард западный 21195
Анакардиевые 21190,176867
Анакардия 21191
Анакардия акажу 21195
Анакардия западная 21195
Анакардовые 21190,176867
Анамирта 21458
Ананас 21472
Ананас культурный 21479
Ананас настоящий 21479
Ананас посевной 21475,21479
Ананасные 60557
Ананасовые 60557
Анар 321764
Анафалис 21506
Анафалис бархатистый 21722
Анафалис белорозовый 21675
Анафалис дарвазский 21560
Анафалис дэциньский 21588
Анафалис жемчужный 21596,21616
Анафалис кистеносная 21673
Анафалис лицзянский 21592
Анафалис мулиский 21637
Анафалис непалский 21643
Анафалис обедненный 21562
Анафалис посьетский 21682
Анафалис прутьевидный 21724
Анафалис сычуаньский 21707
Анафалис тибетский 21711
Анафалис тонкостебельный 21710
Анафалис утесистый 21679
Анафалис чжундяньский 21540
Анафалис юньнаньский 21733
Анациклус 21230
Анациклус лекарственный 21263
Анациклус реснитчатый 21241
Ангалоний 33206
Ангелика 24281,30929,30932
Ангелика аптечная 30932
Ангелика лесная 24483
Ангелика чимганский 30937
Английский дрок 401388
Ангрекум 24687
Ангури-сагак 367416
Ангурия 114122
Ангустура 170167
Андира 22211,61242
Андрахна 22226,226315
Андрахна буша 22234
Андрахна карликовая 22267
Андрахна китайская 226320
Андрахна крошечная 22266
Андрахна круглолистная 22268
Андрахна мелколистная 226335
Андрахна нитевидная 22250
Андрахна телефиевидная 22273
Андрахна тонгая 22276
Андрахна узколистная 22272
Андрахна федченко 22249
Андрахна хайнаньская 226330
Андреев крест 317964
Андржеевския 23473
Андржеевския сердечниколистная 23474
Андромеда 22418
Андромеда древовидная 278389
Андромеда дубровниколистная 22445
Андромеда-дубровник 22445
Анейлема кейзака 260102
Анектохилюс 25959
Анектохилюс тайваньский 25980
Анектохилюс чжэцзянский 26015
Анектохилюс эмэйский 25978
Анемон 23696
Анемон хубэйский 23850
Анемон японский 23854
Анемона 23696
Анемона блестящая 23823
Анемона садовая 23767
Анемона японская 23854
Анемонелла 24132
Анемонелла василистниколистная 24133
Анизанта 25280,60612
Анизанта бесплодная 60985
Анизанта краснеющая 60943
Анизанта кровельная 60989
Анизанта шелковистая 60991
Анизохилус 25379
Анис 299342,299351
Анис звездчатый 204603
Анис звездчатый японский 204533
Анис китайский 204603
Анис обыновенный 299351
Анис полевой 77784
Анис эвездчатый 204603
Анис ядовитый 204533
Анкафия 21886
Анкафия огненная 21888
Анлкардия западная 380528
Аннеслея 25771
Аннона 25828
Анона 25828
Анона игольчатая 25868
Анона колючая 25868
Анона обыкновенная 25852
Анона сетчатая 92112
Анона черимойя 25836
Анона черимола 25836
Анона чешуйчатая 25898
Аноновые 25909
Ансу 34477
Антемис 26738
Антемис африканский 28248
Антемис ядовитый 28248
Антидесма 28309
Антидесма буниус 28316
Антипка 83253
Антиринум 28576
Антиринум горный 28631
Антиринум крупный 28617
Антиринум майюс 28617
Антияр 28248
Антозахна 27793
Антозахна длинноостная 27798
Антозахна жакемонта 144347
Антолиза 27664
Антохламис 27560
Антохламис истодовая 27561
Антохламис туркменская 27562
Антуриум 28071
Антуриум андрэ 28084
Антуриум вейча 28138
Антуриум изменчивый 28137
Антуриум хрустальный 28093
Антуриум цепкий 28118
Антуриум шерцера 28119

Анура бледнозеленая 28808
Анхиета 21889
Анхоний 21909
Анхоний короткоплодный 21910
Анхоний стеригмовидный 21912
Анхоний суменицелистный 21911
Анхуза 21888,21914
Анхуза баррелиела 21928
Анхуза баррелье 21928
Анхуза бледно - желтая 21957
Анхуза гмелина 21940
Анхуза длинностолбиковая 21980
Анхуза итальянская 21949
Анхуза лекарственная 21959
Анхуза ложно - бледножелтая 21973
Анхуза маленькая 21974
Анхуза рослая 21959
Анхуза узколистная 21953
Анхуза фессалийская 21983
Анчар 28248
Аньу 399611
Анютины глазки 410677
Апейба 28963
Апельсин 93765
Апельсин бигардия 93414
Апельсин горький 93332,93414
Апельсин кислый 93332,93414
Апельсин навель 93446,93717
Апельсин настоящий 93765
Апельсин оседж 240828
Апельсин пупочный 93446,93717
Апельсин сладкий 93765
Апельсин трехлисточковый 310850
Апельсин - оседж 240828
Апиос 29295
Апиос клубневый 29298
Апозерис 29785
Апозерис вонючий 29786
Апоногетон 29643
Апоногетон двуколосный 29660
Апоногетон решетчатый 29663
Апоногетоновые 29696
Апорокактус 29712
Апорокактус плетевидный 29715
Апороселла 29727
Апофиллум 29701
Апоциновые 29462
Апоцинум конопревидный 29473
Аптандра 29842
Апулея 29956
Арабис 30160
Арабис альпийский 30164
Арабис альпийский мелкоцветковый 30174
Аракача съедобная 34914
Аралиа высокая 30634
Аралиа ганьсуская 30691
Аралиа маньчжурская 30634
Аралиа мелколистная 30666
Аралиа сердцевидная 30619
Аралиа хубэйская 30685
Аралиа цзиндунская 30674
Аралиевые 30800
Аралия 30587
Аралия аньхуйская 30768
Аралия высокая 30634
Аралия китайская 30604
Аралия колючая 30760
Аралия материковая 30618
Аралия тайваньская 30593
Аралия тибетская 30783
Аралия тэнчунская 30775
Аралия шиповатая 30760
Аралия шмидта 30753
Аралия юньнаньская 30798
Арария японская 162879
Араужия 30860
Араукариевые 30857
Араукария 30830
Араукария бидвилла 30835
Араукария бразильская 30831
Араукария высокая 30846,30848
Араукария куннингама 30841
Араукария узколистная 30831
Араукария чилийская 30832
Араушия 30860
Араушия шелковистая 30862
Арахис 30494
Арахис культурный 30498
Арахис подземный 30498
Арбуз 93286
Арбуз горький 93297
Арбуз китайский 50998
Арбуз колоцинт 93297
Арбуз пустотельт 93297
Арбус 93286,93304
Арбус дикий 93297
Арбус обыкновенный 93304
Арбус столовый 93304
Арбус съедобный 93304
Аргемон 32411
Аргемон мексиканский 32429
Аргемона 32411
Аргемоне 32411
Аргирантемум 32548
Аргирантемум куртаниковый 32563
Аргиролобий 32769
Аргиролобий лядвенцевидный 32841
Аргиролобий пажитниковый 32923
Аргиролобий чашечный 32792
Аргиролобиум 32769
Аргузия сибирская 393269
Ардизия 31347
Ардизия баотинская 31362
Ардизия городчатая 31396
Ардизия жуйлиская 31598
Ардизия зибольда 31599
Ардизия китайская 31380
Ардизия крупнолистная 31455
Ардизия ланьюйская 31519
Ардизия малипоская 31516
Ардизия мелкоцветковая 31458
Ардизия хайнаньская 31579
Ардизия цзиньюньская 31483
Ардизия юньнаньская 31644
Ардизия японская 31477
Арека катеху 31680
Аренга 32330,32336
Аренга мелкоцветковая 32340
Аренга сахарная 32343
Аренга сахароносная 32343
Аретуза 32371
Аржанец 294960,294992
Аржанец луговой 294992
Арзан 281916
Аризема 33234
Аризема амурская 33245
Аризема далиская 33310
Аризема иланьский 33356
Аризема инцзянский 33357
Аризема лицзянская 33388
Аризема пинбяньская 33458
Аризема тайваньская 33332
Аризема тэнчунская 33534
Аризема фаржа 33330
Аризема франше 33335
Аризема хунаньский 33353
Аризема хуняский 33354
Аризема цзиндунская 33375
Аризема эмэйская 33441
Аризема юньнаньская 33574
Аризема яошаньская 33512
Аризема японская 33369
Аристида 33730
Аристида вознесения 33737
Аристида карелина 33898
Аристида мелкоперистая 33973
Аристида паутинистая 33760
Аристида перистая 377011
Аристолохия 34097
Аристолохия гуансиская 34238
Аристолохия гуйцзоуская 34201
Аристолохия крупнолистная 34219,34337
Аристолохия куэньминская 34237
Аристолохия слабая 34162
Аристолохия тибетская 34200
Аристолохия фуцзяньская 34188
Аристолохия чжундяньская 34381
Аристолохия юньнаньская 34375
Аристотелия 34394
Арктагростис 31023
Арктагростис мироколистная 31026

Арктагростис овсяницевидная 31025
Арктагростис тростниковидная 31024
Арктерика 31037,298703
Арктерика низкая 31038
Арктогерон 31079
Арктогерон злаковый 31080
Арктотис 31176
Арктоус 31311
Арктоус альпийский 31314
Арктоус красноплодный 31319
Арктоус красный 31322
Арктоус мелколистный 31321
Арктоус японский 31314,31316
Арктофила 31085
Арктофила рыжеватая 31088
Арктофиля 305274
Армерия 34488
Армерия альпийская 34494
Армерия арктическая 34502
Армерия приморская 34536
Армерия сибирская 34562
Армерия удлиненная 34514
Арнато 54862
Арнаутка 398900
Арнебия 34600
Арнебия бальджуанская 34602
Арнебия закаспийская 34646
Арнебия крупноцветковая 34622
Арнебия лежачая 34609
Арнебия линейнолистная 34628
Арнебия обратнояйцевидная 34633
Арнебия простертая 34609
Арнебия пятнистая 34624
Арнебия тибетная 34624
Арнебия томсона 34641
Арнебия фиолетово－желтая 34605
Арника 34655
Арника альпийская 34659
Арника горная 34752
Арника ильина 34724
Арника лессинга 34733
Арника сахалинская 34766
Арника средня 34725
Арника уналашкинская 34780
Арника холодная 34712
Арнозерис 34815
Арнозерис малый 34816
Ароидные 30491
Аройник 36967
Аройник восточный 37022
Аройник королькова 37013
Аройник пятнистый 37015
Аройниковые 30491
Ароник 36967
Ароникум 136333
Арония 34845
Арония арбутусолистная 34854
Арония черноплодная 369459
Аронник 36967
Аронник белокрылый 36969,37005
Аронник итальянский 37005
Аронник королькова 37013
Аронник пятнистый 37015
Аронник удлиненный 36991
Арракач 34914
Арракача сьедобная 34914
Аррорут 245013
Аррорут бразильский 244507,244515
Аррорут восточноиндийский 114858
Аррорут вестидский 245014
Аррорут индийский 382923
Аррорут квинслендский 71169
Аррорут остиндский 114858
Артабоьрис 34995
Артемизия 35084
Артишок 117765,117787
Артишок испанский 117770
Артишок китайский 373439
Артишок колючий 117787
Артишок посевной 117787
Артишок японский 373439
Артраксон 36612
Артраксон волосистый 36647
Артраксон гуйцзоуский 36644
Артраксон лангсдорфа 36647,36650
Артраксон ланцетовидный 36673
Артраксон отроняющаяся 36647
Артраксон хайнаньский 36625
Арум 36967,137744
Арум белокрылый 37005
Арум змейный 137747
Арум итальянское 37005
Арум кукушечный 417093
Арум одоратум 16512
Арум пятнистый 37015
Арундинария 37127
Арундинария наньпинская 37162
Арундинария приятная 318283
Арундинария симони 304098
Арундинария хайнаньская 37229
Арундинария хайндса 318303
Арундинелла 37109,37350
Арундинелла лушаньская 37398
Арундинелла мелкоцветковая 37417
Арундинелла тэнчунская 37431
Арундинелла уклоняющаяся 37352,37379
Арундинелла хубэйская 37402
Арундинелла юньнаньская 37442
Арундо 37450
Арундо донакс 37467
Арундо тайваньский 37477
Арундо тростниковый 37467
Арункус 37053
Арункус хангелика 30932
Архонтофеникс 31012
Архонтофеникс александра 31013
Арча 213606
Арча китаиский 213634
Арча красная 213634,213841
Асклепиадовые 37806
Асклепиас 37811
Аспараг 38904
Аспарагус 38904
Аспарагус далиский 39225
Аспарагус лицзянский 39156
Аспарагус перистый 39195
Аспарагус пушистоперистый 39195
Аспарагус сычуаньский 39198
Аспарагус тибетский 39234
Аспарагус цинхайский 39163
Аспарагус шпренгера 38983
Аспарагус яньюаньский 39259
Асперуга 39298
Асперуга лежачая 39299
Асперуга простертая 39299
Асперуга тонкий 57689
Аспидистра 39516
Аспидистра высокая 39532
Аспидистра грязнобурая 39557
Аспидистра гуансиская 39573
Аспидистра инцзянская 39583
Аспидистра лодяньская 39556
Аспидистра мелкоцветковая 39560
Аспидистра сычуаньская 39575
Аспидистра тайваньская 39534
Аспидистра хайнаньская 39544
Аспидистра эмэйская 39567
Аспидосперма 39693
Ассафетида 163580
Асселямалейкум 119503
Астеролинум 41661
Астеролинум звездчатый 41663
Астеротамнус 41747
Астеротамнус дубровнниколистный 41757
Астеротамнус кустарниковый 41752
Астеротамнус разнохохолковый 41755
Астеротамнус шишкина 41759
Астильба 41786
Астильба китайская 41795
Астильба тунберга 41852
Астильбе 41786
Астильбе китайское 41795
Астильбе корейское 41822
Астильбе мелколистное 41828
Астинолема 6830
Астинолема крупнообертковая 6832
Астинолема синеголовниковая 6831
Астомаеа 41876
Астоматопсис 41876
Астра 39910
Астра агератовидная 39928

Астра агератовидная мелкоцветковая 39970
Астра альпийская 40016
Астра американская 380930
Астра виргинская 40923
Астра гладкая 380907
Астра глена 40518
Астра грузинская 40607
Астра дикая 40038
Астра дэциньская 41351
Астра звёздчатая 40038
Астра зее – буреинская 41224
Астра змеиногорская 40022
Астра ивовая 41190
Астра иволистная 41190
Астра индийская 40621
Астра итальянская 40038
Астра китайская 67314
Астра коржинского 40665
Астра лицзянская 40754
Астра ложная 40021
Астра ложноитальянская 40037
Астра льнолистная 231844
Астра маака 40805
Астра мелкоголовчатая 380945
Астра мотоская 40879
Астра мупинская 40880
Астра новая бельгия 40923
Астра новобельгийская 40923
Астра пенсауская 41025
Астра пиренейская 41133
Астра плавненая 398134
Астра почни цельная 41311
Астра ричардсона 41150
Астра ромашковая 40038
Астра роскошнолистная 40803
Астра садовая 67312
Астра садовая китайская 67314
Астра сибирская 41257
Астра сычуаньская 41244
Астра солончаковая 398126,398134
Астра солончаковая обыкновенная 398134
Астра степная 40038
Астра татарская 41342
Астра тибетская 41285
Астра толмачева 41377
Астра тяньцюаньская 41376
Астра фори 40425
Астра хунаньская 40598
Астра чаюйская 41523
Астра щетинистоволосистая 193939
Астра юньнаньская 41520
Астрагал 41906,43251
Астрагал австрийский 42048
Астрагал аксуйский 41935
Астрагал алатавский 41942
Астрагал альпийский 41962
Астрагал аркалыкский 42009
Астрагал бекетова 42072
Астрагал белесоватый 42409
Астрагал беловетвистый 42613
Астрагал беловойлочный 42140
Астрагал белостебелиный 41948
Астрагал бесстебелиный 42352
Астрагал бесстрелковый 42352
Астрагал благородный 42801
Астрагал бледнеющий 42847
Астрагал бледный 42847
Астрагал боетийский 42086
Астрагал болотный 43204
Астрагал бородина 42091
Астрагал бороздчатый 43112
Астрагал бунге 42120
Астрагал бунговский 42120
Астрагал ветвистый 43228
Астрагал вздуточашечковый 42255
Астрагал владимира 43247
Астрагал волжский 42881
Астрагал волокнистый 42366
Астрагал галеговидный 42392
Астрагал ганьсуский 42617
Астрагал головчатый 42164
Астрагал гороховидный 42193
Астрагал горошковидный 42835
Астрагал горький 41971
Астрагал густоветвистый 42883
Астрагал густоцветный 42280
Астрагал датский 42262
Астрагал даурский 42258
Астрагал деревцовый 41998
Астрагал джангартский 42304
Астрагал джимский 42494
Астрагал длиннокрылый 42657
Астрагал длиннолистный 42302
Астрагал длиннолодочный 42664
Астрагал длинноногий 42658
Астрагал длинноцветковый 42633
Астрагал длинноцветоносный 42658
Астрагал долголистный 42302
Астрагал донниковый 42696
Астрагал древовидный 42277
Астрагал древообразный 41974
Астрагал дуговидный 41998
Астрагал желобчатый 43112
Астрагал заилийский 43175
Астрагал зайсанский 43286
Астрагал заячий 42570
Астрагал заячий мелкоцветковый 42571
Астрагал зонтичный 43205
Астрагал изогнутый 42967
Астрагал илийский 42516
Астрагал камеденосный 42453
Астрагал каменистый 42878
Астрагал каменный 42624
Астрагал камнеломный 42987
Астрагал кандинский 43132
Астрагал каркасинский 42544
Астрагал китайский 42177,401581
Астрагал клера 42197
Астрагал козлятниковиный 42392
Астрагал колючий 42014
Астрагал колючковый 42014
Астрагал короткодольный 42100
Астрагал крестовидный 42236
Астрагал кроненбурга 42556
Астрагал крупнодольный 42655
Астрагал крупнорогий 42654
Астрагал крымский 43139
Астрагал крючковатый 42461
Астрагал крючконосный 42461
Астрагал кунжутный 43039
Астрагал курчумский 42563
Астрагал кустарниковый 42388
Астрагал кушакевича 42848
Астрагал лапландский 42582
Астрагал лемана 42597
Астрагал лепсинский 42600
Астрагал лисий 43236
Астрагал лисохвостый 41961
Астрагал лицзянский 42134
Астрагал ложкометельчатый 42933
Астрагал ложнокоротколодочный 42923
Астрагал ложносладколистный 42416
Астрагал луговой 42262
Астрагал маевского 42672
Астрагал мелкоголовчатый 42713
Астрагал мелколистный 42505
Астрагал мелкоплодный 43199
Астрагал метельчатый 43024
Астрагал мешечатый 42994
Астрагал мешодчатый 43209
Астрагал молочнобелый 42391
Астрагал монгольский 42704
Астрагал монгутский 42746
Астрагал мохнатолистный 42586
Астрагал мохнатый 42588
Астрагал мулиский 42758
Астрагал нитевидный 42784
Астрагал нутовый 42193
Астрагал обманчивый 43123
Астрагал однопарный 43207
Астрагал округлолистковый 42830
Астрагал остороязычковый 42843
Астрагал остроплодный 42843
Астрагал острошерохватый 43012
Астрагал отогнутый 42967
Астрагал палласа 42845
Астрагал перепончатый 42699
Астрагал песколюбец 41976
Астрагал песочный 41975
Астрагал песчанодревесный 41974
Астрагал песчаный 42262

Астрагал плосколистный 42888
Астрагал подложный 42207
Астрагал полосатый 43096
Астрагал полосчатый 43096
Астрагал понтийский 42902
Астрагал почти – дугообразный 43100
Астрагал приподнимающийся 42594
Астрагал простертоветвистый 42211
Астрагал прутьевидный 43228
Астрагал прямобобовый 42840
Астрагал прямодольный 42841
Астрагал прямоплодный 42841
Астрагал птицеклювый 42834
Астрагал пузыристый 43209
Астрагал пузырчатоплодный 42881
Астрагал пузырчатый 42881,43223
Астрагал пушистоцветковый 42264,42948
Астрагал пушковатый 41925
Астрагал разорванный 42566
Астрагал распластанный 42294
Астрагал рогатый 42220
Астрагал роговой 42166
Астрагал рогоплодный 42221
Астрагал рожковый 42220
Астрагал розовый 42985
Астрагал сахалинский 42995
Астрагал свернутый 42218
Астрагал светлокрасный 42723
Астрагал северцова 43040
Астрагал семиреченский 42485
Астрагал серповидный 42021,42355
Астрагал серпоплодный 42355
Астрагал сетчатый 42974
Астрагал сиверса 43046
Астрагал сизый 42409
Астрагал синьцзянский 42580
Астрагал сычуаньский 43044
Астрагал скорнякова 43058
Астрагал скрученный 42218
Астрагал сладколистный 42417
Астрагал сладколистовиный 42416
Астрагал смешанный 42207
Астрагал согнутый 42371
Астрагал соготинский 43066
Астрагал солодколистный 42417
Астрагал солончаковый 43000
Астрагал сочевичниковый 42835
Астрагал средний 42691
Астрагал сродный 42203
Астрагал татарский 42893
Астрагал тибетский 43154
Астрагал тонкостебельный 42366
Астрагал трагакантовый 42453
Астрагал тяньцюаньский 42757
Астрагал узкорогий 43085
Астрагал уральский 42047
Астрагал хлопунец 42193
Астрагал хоботковый 42135
Астрагал холодный 42384
Астрагал хомутова 42186
Астрагал хуйнинский 42503
Астрагал цветоножечный 42864
Астрагал цингера 43292
Астрагал цюцзянский 42183
Астрагал чаюйский 43288
Астрагал черноколосый 42694
Астрагал чжундяньский 42382
Астрагал шаровидновздутый 43076
Астрагал шаровидный 43075
Астрагал шароцветковый 43074
Астрагал шелкоцветковый 42585
Астрагал шероховатый 42023
Астрагал шерстистоцветковый 42264,42948
Астрагал шерстистый 42580
Астрагал шершавый 42023
Астрагал шиловидный 43103
Астрагал эллиптический 42322
Астрагал эспарцетный 42821
Астрагал эспарцетовидный 42821
Астрагал южносибирский 42051
Астрагал южный 42047
Астрагал юньнаньский 43275
Астрагал ядунский 42303
Астрагал яичкоплодный 43151
Астрагал яйцеплодный 43151
Астрагал якорный 41989
Астрагал якорцевидный 43180
Астрагал якорцевый 43180
Астрагал японский 42537
Астранция 43306
Астранция большая 43309
Астранция колхидская 43308
Астранция крупная 43309
Астранция малая 43311
Астранция наибольшая 43310
Астранция понтийская 43312
Астранция урехнадрезная 43313
Астровые 101642
Астрокариум 43372
Астрокарпус 361274
Астрониум 43472
Астрофитум 43492
Асфоделина 39435
Асфоделина древовидная 39436
Асфоделина жёлтая 39438
Асфоделина крымуская 39439
Асфоделина тонкая 39441
Асфоделина тонкоцветковая 39440
Асфодель 39449
Асфоделюс 39449
Асфоделюс белый 39454
Асфодил 39449
Асфодил белый 39454
Асцирум 38285
Асцирум эверобойниковидный 38285
Аталантия 43718
Аталантия однолистная 43731
Аталантус 43701
Аталантус колючковидный 43702
Атрактилис 44114
Атрактилодес 44192
Атрактилодес китайский 44208
Атрактилодес корейский 44202
Атрактилодес яйцевидный 44205
Атталея 44801
Атхатоба 214308
Аукуба 44887
Аукуба гималайская 44911
Аукуба китайская 44893
Аукуба китайская эмэйская 44898
Аукуба юньнаньская 44949
Аукуба японская 44915
Аукуба японская пестролистная 44919
Аукумения 44883
Аурикула 314143
Афананта 29004
Афананта шероховатая 29005
Афеландра 29121
Афеландра золотистая 29123
Афеляндра 29121
Афзелия африканская 10106
Афзелия бакери 10109
Афлатуния 9856
Афлатуния вязолистая 9858
Ахиллеа – зоря 4005
Ахиллея 3913
Ахиллея тысячелистная 3978
Ахименес 4070
Ахирантес 4249
Ахирантес двузучатый 4273
Ахирантес индийский 4263
Ахирантес крупнолистный 4308
Ахирантес мелколистный 4319
Ахирантес японский 4275
Ахнаерум 4113
Ахудемия японская 298945
Ацелидант 405570
Ацелидант антиклейный 2765
Ацена 1739
Аяния 12988
Аяния кокандская 13005
Аяния кустарничковая 12998
Аяния лицзянская 12990
Аяния маньчжурская 13007
Аяния мелкоцветковая 13021
Аяния палласова 13018
Аяния тибетская 13038
Аяния тонкая 13001
Аяния трехлопастная 13040
Аяния шарнхорстая 13032
Аяния щитковая 12997

Бабассу 273247
Бабушник 16540
Баглянниковые 83732
Багновка 145054
Багновка чёрная 145073
Багряник 83752
Багрянник 83752
Багрянник великолепный 83743
Багрянник канадский 83757
Багрянник китайский 83769
Багрянник обыкновенный 83809
Багрянник стручковатый 83809
Багрянник стручковый 83809
Багрянник японский 83736,83769
Багульник 223888
Багульник болотный 223901
Багульник подбел 223897,223905
Багульник стелющийся 223911
Бадан 52503
Бадан сердцелистный 52514
Бадан тихоокеанский 52530
Бадан толстолистный 52514
Бадан толстолистый 52514
Бадан тяньцюаньский 52542
Бадан угамский 52543
Бадан узколистный 52514
Бадан эмэйский 52520
Бадиджан 367370
Бадлажна 367370
Бадьян 129615,204474
Бадьян звездчатый 204603
Бадьян камбоджийский 204491
Бадьян китайский 204603
Бадьян мелкоцветковый 204570
Бадьян настоящий 204575,204603
Бадьян священный 204583
Бадьян флоридский 204517
Бадьян – ясенец 129618
Бажра 289116
Базелла – шпинат 48689
Базеллевые 48698
Базилик 268418
Базилик душистый 268438
Базилик евгенольный 268518
Базилик камфорный 268438
Базилик обыкновенный 268438
Базилик оородный душки 268438
Базилик юбилейный 268518
Базилика 268418
Бакаут 181496,181505
Баккаурея 46176
Баклажан 367370
Баклажан обыкновенный 367370
Бакопа 46348
Бактрис 46374
Бакхарис 46210
Бакхарис галимолистный 46227
Бакхарис гирканская 50818
Бакхарис лебедолистная 46227
Баланофоровые 46888
Балата 244530
Балх – арча 213877
Бальза 268344
Бальза заячья 268344
Бальзамин 204756
Бальзамин гуйцзоуский 204997
Бальзамин гуншаньский 204987
Бальзамин недотрога 205175
Бальзамин садовый 204799
Бальзаминовые 47077
Бальзамник 282
Бальса 268344
Бамбуза 47174
Бамбуза бирманская 47225
Бамбуза обыкновенная 47516
Бамбуза хайнаньская 47222
Бамбуза юньнаньская 47535
Бамбук 47174
Бамбук великолепный 357939
Бамбук жесткий 47430
Бамбук золотистый 297203
Бамбук карликовый 47345
Бамбук кумасаса 362131
Бамбук множественный 47345
Бамбук обыкновенный 47190
Бамбук полукустарниковый 167092
Бамбук узкий 297194
Бамбук чёрный 297367
Бамбук четырехгранная 87607
Бамбук японский 318307
Бамбук – листоколосник 297203
Бамия 219
Бамья 219
Банан 260198
Банан абиссинский 260224
Банан базьо 260208
Банан базьо тайваньский 260228
Банан браминов 260253
Банан декоративный 260224
Банан десертный 260253
Банан кавендиша 260250
Банан карликовый 260250
Банан китайский 260250
Банан кухонный 260253
Банан ланьюйский 260236
Банан люцюский 260241
Банан мудрецов 260253
Банан пизанг 260253
Банан плодовой 260253
Банан подоложник 260253
Банан райский 260253
Банан райский семеноносный 260258
Банан фруктовый 260253
Банан шертавоплодный 260359
Банан эфиопский 260224
Банан яблочный 260253
Банан японский 260208
Банановые 260284
Бангдевона 123077
Баниан 164684,165553
Банияn 164684
Банксия 47626
Бантизия 47854
Бантизия красильная 47880
Бантисия 47854
Баньян 164684,165553
Баобаб 7022
Баобаб африканский 7022
Баобабовые 56761
Баранка 34752
Баранник 34655
Баранник горный 34752
Баранчик 315082
Барбарис 51268,51586
Барбарис агрегатный 51284
Барбарис азиатский 51333
Барбарис амурский 51301
Барбарис аньхуйский 51314
Барбарис бирманский 51400
Барбарис бородавчатый 52308
Барбарис боярышниковый 51510
Барбарис бретшнейдера 51303
Барбарис быковский 51409
Барбарис ван 52355
Барбарис весенний 52306
Барбарис вильсона 52371
Барбарис восточный 51981
Барбарис ганьсуский 51805
Барбарис генри 51711
Барбарис гуйцзоуский 51690
Барбарис гукера 51733
Барбарис густоцветковый 51540
Барбарис далиский 52214
Барбарис дарвина 51520
Барбарис дерезовидный 51891
Барбарис дильса 51563
Барбарис дэцинький 51495
Барбарис замечательный 51766
Барбарис зеленоватый 52313
Барбарис зибольда 52151
Барбарис илийский 51757
Барбарис канадский 51422
Барбарис каркаралинский 51807
Барбарис кашгарский 51809
Барбарис китайский 51452
Барбарис корейский 51824
Барбарис куэньминский 51827
Барбарис лицзянский 51860
Барбарис лэйбоский 51839
Барбарис максимовича 52246
Барбарис малипоский 51904

Барбарис мелкобородавчатый 52308
Барбарис мелколистный 51923
Барбарис мелкоцветковый 51928
Барбарис многоцветковый 52052
Барбарис монетный 51968
Барбарис мулиский 51943
Барбарис мяньнинский 51916
Барбарис наньчуаньский 51602
Барбарис нежный 51856
Барбарис обыкновенный 52322
Барбарис остистый 51327
Барбарис перуский 52022
Барбарис пинбяньский 52037
Барбарис провансальский 52066
Барбарис продолговатый 51978
Барбарис пуаре 52049
Барбарис разноножковый 51717
Барбарис разноцветоножковый 51717
Барбарис самшитолистный 51401
Барбарис саржента 52131
Барбарис сибирский 52149
Барбарис сизобелый 51424
Барбарис сизый 52069
Барбарис сичанский 51773
Барбарис сычуаньский 52150
Барбарис скученный 51284
Барбарис сулье 52169
Барбарис тибетский 52255
Барбарис тишлера 52258
Барбарис тунберга 52225
Барбарис тунберга атропурпуреа 52227
Барбарис туркменский 52284
Барбарис узколистный 52169
Барбарис франше 51634
Барбарис фуцзяньский 51642
Барбарис фэн 51613
Барбарис хубэйсий 51643
Барбарис хубэйсий эмэйский 51653
Барбарис хэнаньский 51732
Барбарис цельнокрайний 51782
Барбарис цзилунский 51666
Барбарис чаюйский 52397
Барбарис чжундяньский 52182
Барбарис чжэцзянский 52320
Барбарис чэнкоуский 51518
Барбарис эмэйский 51279
Барбарис этненский 51280
Барбарис юлианы 51802
Барбарис юньнаньский 52391
Барбарис японский 52225
Барбарисовые 51261
Барвинок 409325
Барвинок большой 409335
Барвинок малый 409335,409339
Барвинок опушенный 409345
Барвинок прямой 409329
Барвинок розовый 79418
Барвинок травянистый 409335
Барингтония 48509
Баркан 123141
Баркаузия 110710
Баросма 10557,48472
Барская спесь 363312
Бартония 48585
Бартсия 48601,268957
Бартсия альпийская 373608
Бартшия 48601,268957
Бархат 294230
Бархат амурский 294231
Бархат амурский японский 294233
Бархат китайский 294238
Бархат сахалинский 294251
Бархатник 18652
Бархатные 259479
Бархатные бобы 259559
Бархатцы 383087
Бархатцы высокие 383090
Бархатцы крупноцветковые 383090
Бархатцы крупноцветные 383090
Бархатцы мелкие 383100
Бархатцы мелкоцветковые 383103
Бархатцы низкие 383103
Бархатцы отклоненные 383103
Бархатцы прямостоячие 383090
Бассия широколистная 241514
Батат 207623
Батис 48880
Батлачёк 17497
Батлачек равный 17498
Батлачик 17497
Батлачок 17497
Батлачок коленчатый 17527
Батлачок луговый 17553
Батлачок непальский 17549
Батлячек 17497
Батун 15289
Баугиния 48990
Баугиния бирманская 49181
Баугиния лицзянская 49019
Баугиния остроконечная 48993
Баугиния фабера 49022
Баугиния хайнаньская 49116
Баугиния хубэйская 49101
Баугиния шерстистая 49241
Баугиния юньнаньская 49268
Баугиния японская 49140
Баухиния 48990
Бафия 47685
Бафия красивая 47788
Башенница голая 400644
Башкирская капуста 291663
Башмала японская 151162
Башмачек 120287
Башмачек жемчужный 120411
Башмачек крупноцветковый 120396
Башмачек крупноцветный 120396
Башмачек крупноцветный вздутый 120464
Башмачек настоящий 120310
Башмачек пятнистый 120357
Башмачек ятабе 120473
Башмачок 120287
Башмачок венерин 120287
Башмачок венерин настоящий 120310
Башшеница 400638
Башшеница гладкая 400644
Башшеница голая 400644
Баялыч 344425
Бегониевые 50433
Бегония 49587
Бегония белоточечная 49603
Бегония бисмарка 49662
Бегония борщевиколистная 49915
Бегония ван 50400
Бегония вечкоцветушая 50298
Бегония волнистая 50382
Бегония волосистая 49925
Бегония волосистая колпачковая 49926
Бегония всегда цветущая 50298
Бегония вьюнковая 49735
Бегония вэньшаньская 50405
Бегония гоегская 49879
Бегония гуйцзоуская 49966
Бегония гуншаньская 49895
Бегония далиская 50354
Бегония двансиана 49886
Бегония диадема 49774
Бегония ильмолистная 50379
Бегония инцзянская 50417
Бегония катаяна 49701
Бегония клещевинолистная 50243
Бегония клубневая 50378
Бегония королевская 50232
Бегония красноватая 50257
Бегония кровянокрасная 50273
Бегония крупнолистная 50071
Бегония лазящая 50278
Бегония ланьюйская 49827
Бегония лиственная 49844
Бегония лодяньская 50181
Бегония лотосолистная 50104
Бегония лунчжоуская 50092
Бегония малипоская 50056
Бегония мелкоцветковая 50079
Бегония месячная 50298
Бегония металлическая 50076
Бегония мотоская 49903
Бегония мулиская 50094
Бегония мясокрасная 49944
Бегония наньтоуская 50100
Бегония наньчуаньская 49778
Бегония непалская 50161

Бегония ольбия 50122
Бегония пальчатая 49781
Бегония пинбяньская 50374
Бегония полуклубневая 50317
Бегония превосходная 49941
Бегония превосходная изумрудная 49942
Бегония рекс 50232
Бегония рукавчатая 50059
Бегония седая 49943
Бегония сизолистная 49877
Бегония тайваньская 50352
Бегония тэнчунская 49728
Бегония угловатая 49617
Бегония фэн 49826
Бегония хайнаньская 49897
Бегония цзиньпинская 49653
Бегония щитовидная 49943
Бегония эмэйская 49808
Бегония юньнаньская 50087
Бедренец 299342
Бедренец анис 299351
Бедренец армянский 299354
Бедренец ароматный 299355
Бедренец близкий 299348
Бедренец большой 299466,299467
Бедренец ветвистый 299506
Бедренец голостебельный 299484
Бедренец гроссгейма 299423
Бедренец дагестанский 299390
Бедренец дэциньский 299446
Бедренец иды 299439
Бедренец известколюбивый 299563
Бедренец камнеломка 299523
Бедренец камнеломковый 299523
Бедренец камнелюбивый 299462
Бедренец корейский 299448
Бедренец коржинского 299450
Бедренец крымский 299551
Бедренец купыревидный 299352
Бедренец литвинова 299463
Бедренец лицзянский 299520
Бедренец меловой 299563
Бедренец мулиский 299547
Бедренец опушенный 299502
Бедренец разрезаннолистный 299564
Бедренец разрезной 299394
Бедренец розовоцветный 299511
Бедренец симаоский 299583
Бедренец смешиваемый 299385
Бедренец телунга 299555
Бедренец тибетский 299585
Бедренец тимьянолистный 299564
Бедренец торичниколистный 299493
Бедренец туркестанский 299573
Бедренец тэнчунский 299414
Бедренец цзиндунский 299459
Бедренец черный 299479
Бедренец чжундяньск 299382
Бедренец чужеземных 299490
Бедренец чэнкоуский 299411
Бедренец юньнаньский 299586
Бедренец – анис 299351
Безвременник 99293
Безвременник биберштейна 99307
Безвременник великолепный 99345
Безвременник веселый 99324
Безвременник желтый 99329
Безвременник зангезурский 99355
Безвременник змесвиковый 99343
Безвременник осенний 99297
Безвременник подснежный 99336
Безвременник регеля 99323
Безвременник теневой 99350
Безвременник шовица 99348
Безелла 48688
Безшипник 26172
Безшипник сжатый 26173
Бейльшмидия 50465
Бекманния 49532
Бекманния восточная 49540
Бекманния обыкновенная 49540
Беламканда 50667
Беламканда китайская 50669
Белая аказия 334946
Белая аказия ложная 334976
Белая акация 334976
Белая калифорнийская 300189
Белена 201370
Белена белая 201372
Белена богемская 201377
Белена копетдагская 201384
Белена крошечная 201401
Белена сетчатая 201402
Белена туркменская 201405
Белена черная 201389
Белена чёрная 201389
Белена чешская 201377
Белладонна 44704,44708
Беллардиа 50710
Беллардия обыкновенная 50717
Беллардиахлоа 50722
Беллардиахлоа многокрасочная 50724
Беллис многолетний 50825
Беллиум 50841
Белльвалия 50736
Белльвалия албанская 50744
Белльвалия араксинская 50745
Белльвалия вильгельмса 50779
Белльвалия горая 50767
Белльвалия густоцветковая 50769
Белльвалия длинностолбчатая 50762
Белльвалия желтая 50763
Белльвалия зигоморфная 50780
Белльвалия липского 50761
Белльвалия остролистная 50743
Белльвалия оше 50747
Белльвалия савича 50774
Белльвалия самечательная 50776
Белльвалия сарматская 50773
Белльвалия темнофиолетовая 50746
Белльвалия туркестанская 50778
Белльвалия фомина 50759
Белльс 50808
Белоголовка 396981
Белозор 284500
Белозор болотный 284591
Белозор боцебу 284550
Белозор ганьсуский 284544
Белозор двуристный 284510
Белозор двуристый 284510
Белозор дэциньский 284528
Белозор илянский 284637
Белозор кандинский 284549
Белозор китайский 284517
Белозор китайский сычуаньский 284518
Белозор лаксманна 284556
Белозор лицзянский 284559
Белозор наньчуаньский 284505
Белозор симаоский 284616
Белозор синаньский 284634
Белозор тибетский 284623
Белозор чэнкоуский 284516
Белозор эмэйский 284532
Белозор юй 284639
Белозор юйлуншаньский 284640
Белозор юньнаньский 284641
Белозор яньюаньский 284636
Белокопытник 292342,400675
Белокопытник белый 292374
Белокопытник гибридный 292372,292386
Белокопытник дланевидный 292403
Белокопытник лекарственный 292386
Белокопытник ложный 292401
Белокопытник ненастоящий 292401
Белокрыльник 66593
Белокрыльник болотный 66600
Белокудренник 46989,47032
Белокудренник северный 46993
Белокудренник серый 46999
Белокудренник черный 47013
Белокудренник чёрный 47013
Белолист 81219
Белолиственник 81219
Белолистка 311208
Белолоз 342973
Беломятлик 227946
Беломятлик беловатый 227948
Беломятлик верещагина 230221
Беломятлик жестколистный 227956
Беломятлик кавказский 227950
Беломятлик каратавский 227952

Беломятлик ольги 227953
Белотал 344244
Белоус 262532,262537
Белоус прямой 262537
Белоус торчащий 262537
Белоусник 262509
Белоусник краузе 262516
Белоусник тонкоцветковый 262528
Белоцветник 227854
Белоцветник весенний 227875
Белоцветник летний 227856
Белые брандущки 274823
Белый клен 3462
Бемерия 56094
Бемерия зибольда 56140
Бемерия инцзянская 56179
Бемерия крупнолистная 56195,56206
Бемерия тайваньская 56140
Бемерия тибетская 56336
Бемерия фучжоуская 56142
Бемерия японская 56181
Бенедикт 97581
Бенедикт аптечный 97599
Бенинказа 50994
Берболоз 344244
Бергамот 93408
Бергения 52503
Бергия 52587
Бергия амманиевая 52592
Бергия амманиевидная 52592
Бергия водная 52597
Береза 53309
Берёза 53309
Береза алайская 53318
Береза аллеганская 53338
Береза аянская 53317
Береза байкальская 53357
Береза белая китайская 53335
Берёза белая плакучая 53563
Береза бородавчатая 53563
Берёза бородавчатая 53563
Береза бумажная 53549
Берёза бумажная 53549
Берёза вишенная 53505
Береза вишневая 53505
Берёза вишнёвая 53505
Береза гмелина 53461
Береза гуншаньская 53466
Береза даурская 53400
Береза делаве 53405
Береза ерманова 53411
Береза желтая 53338,53411
Берёза жёлтая 53338
Береза извилистая 53626
Береза ильмолистная 53631
Береза иркутская 53480
Береза каменная 53411
Берёза каменная 53411
Береза карликовая 53352,53530
Берёза карликовая 53530
Береза каяндера 53370
Береза келлера 53497
Береза киргизская 53499
Береза китайская 53379
Берёза китайская 53379
Береза коржинского 53500
Береза круглолистная 53605
Береза крупнолистная 53464
Береза кузмичева 53501
Береза максимовича 53400
Берёза максимовича 53520
Берёза малорослая 53594
Береза маньчжурская 53572
Береза медведева 53523
Береза мелколистная 53525
Береза миддендорфа 53527
Берёза миддендорфа 53527
Береза мозолистная 53372
Береза низкая 53470
Береза низкоросная 53530
Берёза обыкновенная 53563
Береза овалинолистная 53444
Береза овальнолистная 53544
Береза памирская 53546
Берёза плакучая 53563
Береза плосколистная 53572
Берёза повислая 53563
Береза пониклая 53563
Береза потанина 53587
Берёза потанина 53587
Береза потанина эмэйская 53628
Береза приземистая 53470
Береза прохорова 53589
Береза пушистая 53590
Берёза пушистая 53590
Береза радде 53600
Береза ребристая 53389
Береза резниченко 53603
Береза сапожникова 53608
Береза стройная 53386
Береза субарктическая 53617
Береза сукачева 53619
Береза тополелистная 53586
Береза тощая 53434
Береза туркестанская 53629
Береза тяньшанская 53625
Берёза форрестая 53441
Береза цзиньпинская 53494
Береза черная 53536
Берёза чёрная 53536
Береза шмидта 53610
Берёза шмидта 53610
Береза эрмана 53411
Берёза эрмана 53411
Береза японская 53388,53483
Берёза японская 53483
Березка 102933
Берёзка 102933
Берёзовик 3127
Берёзовик обыкновенная 3127
Березовые 53651
Берёзовые 53651
Берек 369541
Берека 369541
Бересклет 157268
Бересклет американский 157315
Бересклет баотинский 157806
Бересклет бархатистый 157946
Бересклет большекрылый 157705
Бересклет бородавчатый 157947,157952
Бересклет бунге 157345
Бересклет вильсона 157966
Бересклет вьющийся 80285
Бересклет гамильтона 157559
Бересклет гамильтона зияющий 157563
Бересклет гамильтона мелкоплодный 157561
Бересклет гуйцзоуский 157658
Бересклет далиский 157318
Бересклет европейский 157429
Бересклет зибольда 157879
Бересклет зибольдиев 157879
Бересклет карликовый 157740,280817
Бересклет коопмана 157654
Бересклет кртпноцветковый 157547
Бересклет крылатый 157285
Бересклет люцюский 157698
Бересклет маака 157699
Бересклет максимовича 157711
Бересклет малоцветковый 157777
Бересклет мелкоплодный 157721
Бересклет мэнцзыский 157714
Бересклет наньчуаньский 157338
Бересклет низкий 157740,280817
Бересклет ниский 157740
Бересклет обыкновенный 157429
Бересклет остролистный 157761
Бересклет плоскочерешковый 157797
Бересклет повислый 157695
Бересклет сахалинский 157848
Бересклет священный 157310
Бересклет семенова 157876
Бересклет сичоуский 157791
Бересклет сычуаньский 157906
Бересклет тёмно – багряный 157327
Бересклет тёмно – пурпуровый 157327
Бересклет тибетский 157925
Бересклет тэнчунский 157913
Бересклет укореняющийся каррьера 157499
Бересклет укореняющийся серебристо – окаймленная 157508

Бересклет форчуна 157473
Бересклет форчуна карръера 157499
Бересклет хайнаньский 157556
Бересклет хубэйский 157587
Бересклет хусюин 157582
Бересклет цзиньюньский 157373
Бересклет эмэйский 157755
Бересклет юньнаньский 157974
Бересклет яванский 157639
Бересклет японский 157601
Бересклет японский ‘Микрофиллюс’ 157610
Бересклет японский золотисто – пестрая 157615
Бересклетовые 80127
Бересклетокрас сахалинский 157848
Берест 401425,401508
Берест густой 401499
Берест листоватый 401468,401508,401593
Берест приземистый 401602
Берест пробковый 401475,401633
Берест туркестанский 401633
Берлиния 52843
Берсим 396817
Берула 53154
Берула прямая 53157
Беруля 53154
Беруля восточная 53161
Беруля прямая 53157
Берхемия 52400
Берхемия лазащая 52466
Бескильница 321220
Бескильница алтайская 321226
Бескильница аляскинская 321396
Бескильница большенот 321334
Бескильница ваханская 321405
Бескильница гаккелевская 321295
Бескильница гаккеля 321295
Бескильница гаупта 321296
Бескильница гауптовская 321296
Бескильница гладгая 321396
Бескильница гроссгейма 321290
Бескильница енисейская 321307
Бескильница жесткая 321382
Бескильница илийская 321302
Бескильница камчатская 321309
Бескильница коленчатая 321286
Бескильница колосовидная 321393
Бескильница корейская 321236
Бескильница крупная 321287
Бескильница крупнольниковая 321332
Бескильница курильская 321324
Бескильница луковичная 321233
Бескильница маловетвистая 321343
Бескильница морская 321336
Бескильница немноговетвистая 321343, 321358
Бескильница неравноветвистая 321230
Бескильница обедненая 321324
Бескильница памирская 321357
Бескильница пестроцветная 321369
Бескильница ползучая 321361
Бескильница приземистая 321298
Бескильница раскидистая 321246
Бескильница расставленная 321248
Бескильница рожевица 321377
Бескильница рожевицовская 321377
Бескильница свернутолистная 321240
Бескильница севангская 321383
Бескильница сизая 321248,321289
Бескильница суженная 321228
Бескильница тонгая 321396
Бескильница тонкоцветковая 321397
Бескильница тончайшая 321401
Бескильница тяньшанская 321404
Бескильница удлиненночешуйная 321270
Бескильница хорезмская 321237
Бескильница цзилунская 321291
Бескильница шведская 321394
Бескильница шишкина 321380
Бесславник 8399
Бессмертник 189093,415315,415386
Бессмертник белый 26385
Бессмертник песчаный 189147
Бессмертник соломенный 415386
Бесцветник 146953
Бетель 31680,300354
Бетель пальма 31680
Бетель перечный 300354
Бетоника 53304
Бешеница водяная 90932
Бешеный огурец 139851,139852
Бешеный огурец выбрасывающий 139852
Бешеный огурец обыкновенный 139852
Биберштейния 54194
Биберштейния душистая 54198
Биберштейния многораздельная 54197
Бигарадия 93332
Бигарадия обыкновенный 93332
Бигардия 93414
Бигнониевые 54357
Бигнония 54292
Бигнония усиковая 54303
Бизина 345558
Бизина зибольда 345679
Бизина камчатская 345669
Бизина кистистая 345660
Бизина корейская 345592
Бизина красная 345660
Бизина красная разлезнолистная 345688
Бизина маньчжурская 345624
Бизина мексиканская 345627
Бизина пушистая 345660
Бизина сахалинская 345690
Бизина сибирская 345692
Бизина травянистая 345594
Бизина черная 345631
Бизина широколисточковая 345620,345708
Бикса аннатовая 54862
Бикса орлеанская 54862
Биксовые 54863
Бикукулла 128288
Билаек 144661
Билимби 45724
Билльбергия 54439
Билльбергия пирамидальнометельчатая 54464
Билльбергия поникшая 54454
Бинерция 54206
Бинерция окружнокрылая 54208
Биота 302718,390567
Биота восточная 302721
Биофитум 54532
Бирманск тибетский 204524
Бирсонима 64425
Бирючина 229428
Бирючина амурская 229562
Бирючина блестящая 229529
Бирючина делаве 229454
Бирючина ибота 229472
Бирючина ичанская 229634
Бирючина китайская 229617
Бирючина киу 229593
Бирючина лодяньская 229625
Бирючина мелкоплодная 229549
Бирючина обыкновенная 229523,229662
Бирючина овальнолистная 229569
Бирючина остолейшая 229522
Бирючина хоккайдинская 229651
Бирючина цзилунская 229463
Бирючина чаюйская 229603
Бирючина чоноски 229650
Бирючина японская 229485
Бирючина японская ротундифолия 229513
Бисерник 383412
Бисерник араратсий 383531
Бисерник бунге 383466
Бисерник восимитычинковый 383564
Бисерник каракалинский 383523
Бисерник комарова 383528
Бисерник кочи 383531
Бисерник литвинова 383547
Бисерник тимелеевый 383585
Бисерник тонколепестный 383555
Бисерник яркий 383486
Бисерниковые 383397
Бисерница 21616
Бифора 54237
Бифора аучистая 54239
Бифора яйцеобразная 54240
Бишопия яванская 54620

Бишофия 54618,54620
Бишофия многоплодная 54623
Бишофия трехлисточковая 54620
Биюргун 21035
Благородный сахарный тростник 341887
Блейстония 54891
Блейстония пронзеннолистная 54896
Блекота 90924
Блестящеколосник 376660
Блестящеколосник желтоцветный 65274, 376661
Блестящеколосник зайцехвостовидный 376662
Блетилла 55559
Блетилла китайская 55574
Блетилла тайваньская 55563
Блигия 55592
Блигия африканская 55596,114581
Блиофиллюм 61561
Блиофиллюм перистый 61572
Блисмус 55889
Блисмус рыжый 55895
Блисмус сжатый 55892
Блисмус тощий 55895
Блошак 354651
Блошная трава 309570
Блошник 150414
Блошница 321509
Блошница болотная 321607
Блошница дизентерийная 321554
Блошница обыкновенная 321595,321619
Блошница простертая 321595
Блошница стелющаяся 321595
Блошница сушеницевидная 321561
Блошница шалфеелистная 321598
Блюберри 403780
Блюмея 55669
Блюмея бальзамоносный 55693
Блюмея тайваньская 55738
Блющ обыкновенный 187222
Боб арабский 293985
Боб бархатный 259559
Боб волчий 238419
Боб гиацитовый 218721
Боб индийский 218721
Боб карабарский 298009
Боб касторовый 334435
Боб конский 408393,408394
Боб кофейный 90801
Боб лимский 294010
Боб огородный 408393
Боб полевой 408393
Боб розовый 765
Боб сойный 177750
Боб турецкий 293985
Бобовник 20931,218754
Бобовник "золотый дождь" 218761
Бобовник альпийский 85403,218757
Бобовник анагировидный 218761
Бобовник анагиролистный 218761
Бобовые 224085
Бобровик 370202
Бобровник 370202
Бобы 408393
Бобы бархатные 259558
Бобы гиацинтовые 117900
Бобы земляные 409069
Бобы игнатьевские 378761
Бобы индийские 319114
Бобы каролинские 294010
Бобы конские 408394
Бобы кормовые 408393,408394
Бобы лимские 294006
Бобы лимские мелкоплодные 294010
Бобы мелкоплодные 408395
Бобы многоцветковые 293985
Бобы огненные 293985
Бобы русские 408393
Бобы сива 294010
Бобы снаржевые 409086,409096
Бобы соевые 177750,177777
Бобы турецкие 293985
Богатень 237476
Богатинка 103446
Богорожник 369541
Бодлак 8660
Бодом 20889
Бодяк 91697
Бодяк абхазский 91707
Бодяк альберта 91724
Бодяк анатолийский 91738
Бодяк армянский 91762
Бодяк беловатый 91904
Бодяк бесстебельный 91713
Бодяк биберштейна 91796
Бодяк болотистый 92041
Бодяк болотный 92285
Бодяк борнмюллера 91809
Бодяк буша 91827
Бодяк вальдштейна 92487
Бодяк вейриха 92492
Бодяк влаголюбивый 92044
Бодяк власова 92479
Бодяк войлочный 92448
Бодяк волосистокаемчатый 92452
Бодяк выемчатый 92400
Бодяк гагнидзе 91988
Бодяк глинистый 91749
Бодяк головчатый 91856
Бодяк護лолистный 91991
Бодяк гребенчатый 92293
Бодяк гуншаньский 91953
Бодяк дагестанский 91902
Бодяк девясиловидный 92012
Бодяк длинноцветковый 92160
Бодяк желточешуйный 91770
Бодяк жесткощетинистый 92412
Бодяк зереноверхий 91870
Бодяк зубчато-ресничатый 92381
Бодяк игольчатый 91939
Бодяк имеретинский 92048
Бодяк кавказский 91852
Бодяк камчатский 92089
Бодяк кемуларии 92098
Бодяк кецховели 92099
Бодяк китайский 91864
Бодяк клейкий 91958
Бодяк кожистоглавый 92403
Бодяк колончаковый 91944
Бодяк комарова 92102
Бодяк корнеголовчатый 92349
Бодяк короткохохолковый 91821
Бодяк красночешуйный 91957
Бодяк крупноголовый 92179
Бодяк крупнокистевой 92178
Бодяк крылатый 91723
Бодяк крымский 92433
Бодяк крючковидный 91720
Бодяк кузнецова 92104
Бодяк ламировидный 92110
Бодяк ланцетолистный 92112
Бодяк лицзянский 92131
Бодяк ложно-лопуховый 92321
Бодяк лопуховый 92121
Бодяк маака 92167
Бодяк мелкозубчатый 92381
Бодяк многоцветковый 92420
Бодяк мулиский 92211
Бодяк мягко-щетинистый 92384
Бодяк обвороченный 92248
Бодяк обыкновенный 92485
Бодяк огородный 92268
Бодяк окутанный 92248
Бодяк осетинский 92278
Бодяк остроконечный 91744
Бодяк паннонский 92287
Бодяк паутинистый 91746
Бодяк плодолговатолистный 92247
Бодяк полевой 91770
Бодяк польский 92309
Бодяк поникший 92296
Бодяк поручейный 92350
Бодяк почти-шерстистоцветковый 92414
Бодяк прицветниковый 91815
Бодяк простой 92398
Бодяк прямочешуйчатый 92348
Бодяк разнолистный 92022
Бодяк ресчатый 91879
Бодяк речной 92350
Бодяк сайрамский 92356
Бодяк седой 92052

Бодяк семенова 92375
Бодяк серовойлочный 92052
Бодяк серпуховидный 92380
Бодяк серый 91842
Бодяк сиверса 92396
Бодяк скученный 91721
Бодяк слабо – вооруженный 92413
Бодяк снизу – белый 92045
Бодяк собранный 91896
Бодяк совича 92422
Бодяк сосновского 92404
Бодяк съедобный 91958
Бодяк топяной 92460
Бодяк туркестанский 92455
Бодяк украинский 92459
Бодяк фомина 91978
Бодяк фрика 91984
Бодяк хубэйский 92040
Бодяк черноморский 91960
Бодяк шантарский 92367
Бодяк шерстистоцветковый 92120
Бодяк шершаво – липкий 91955
Бодяк шишкина 92370
Бодяк щетинистый 92384
Бодяк щетинковидный 91723
Бодяк эльбрусский 91943
Бодяк эмэйский 91962
Бодяк японский 92066
бодяная Зараза 143914,143920
Божедерево 36105
Бойсдувалия 56554
Боккаут 284961
Боккония 56023,240763
Болдырьян 404316
Болиголов 101845
Болиголов крапчатый 101852
Болиголов пятнистый 101852
Боликолов пятнистый 90924
Болодник 230219
Болодник стеллера 230220
Болоздоплодник 45062
Болоздоплодник дарвазский 45065
Болоздоплодник псостой 304846
Болоздоплодник туркестанский 45073
Болоздоплодник тяньшанский 45071
Болоздоплодник уклоняющийся 45064
Болотник 67336
Болотник весений 67376
Болотник весенний 67376
Болотник влягалищный 152791
Болотник изменчивый 67389
Болотник изящный 67376
Болотник неясный 67394
Болотник ножкоплодный 67345
Болотник осенний 67361
Болотник прудовой 67393
Болотник сомнительный 67376
Болотниковые 67334
Болотница 143019
Болотница бескильная 143125
Болотница болотная 143271
Болотница вихуры 143414
Болотница жемчужная 143205
Болотница закавказская 143384
Болотница игольчатая 143022
Болотница камчатская 143178
Болотница клинге 143182
Болотница комарова 143183
Болотница коржинского 143185
Болотница круглоколосковая 143161
Болотница куэньминская 143193
Болотница ложносеребристочешуйная 143050
Болотница максимовича 143207
Болотница маленькая 143289
Болотница малоцветковая 143319
Болотница многощетинковая 143230
Болотница немногозубчатая 143292
Болотница одночешуйная 143397
Болотница острочешуйная 143268
Болотница оттянутая 143055
Болотница прозрачная 143298
Болотница промежуточная 143168
Болотница пятицветковая 143319
Болотница сахалинская 143334
Болотница свенсона 143029
Болотница северная 143343
Болотница серебристочешуйная 143051
Болотница сосочковая 143198
Болотница тонкостилоподная 143190
Болотница туркменская 143395
Болотница тучная 143109
Болотница уссурийская 143201
Болотница финская 143145
Болотница хвощевидная 143139
Болотница цинхайская 143315
Болотница чернопурпурная 143053
Болотница шляпконосная 143414
Болотница южная 143209
Болотница юньнаньская 143424
Болотница яйцевидная 143261
Болотница японская 143300
Болотнозонтичник 191112
Болотнозонтичник узлоцветковый 191117
Болотноцветник 267800
Болотноцветник корейский 267821
Болотноцветник кувшинковый 267825
Болотноцветник щитолистный 267825
Болотный 261162
Болотный кипарис 385253
Болотный кипарис мексиканский 385281
Болотный кипарис обыкновенный 385262
Болотный кипарис остроконечный 385281
Болотный орех 394500
Болотоцветник 267800
Болтония 56700
Болтония лотюра 56724
Большеголовник 329197
Большеголовник альпийский 199384
Большеголовник дноцветковый 375274
Бомарея 56755
Бомбакс 56770
Бомбакс малабарский 56802
Бомбаксовые 56761
Бомбицилена 56827
Бомбицилена прямостоящая 56832
Бомерия 56094
Бонгардия 56941
Бонгардия золотистая 56944
Бондук 64971,64983
Бор 254496
Бор весенний 254542
Бор развесистый 254515
Бор раскидистый 254515
Бораго 57095
Бораго лекарственный 57101
Борассус пальмира 57122
Борец 5014
Борец алатавский 5019
Борец алтайский 5031
Борец анторовидный 5041
Борец аптечный 5442
Борец байкальский 5048
Борец белоустый 5359
Борец бело – фиолетовый 5021
Борец биробиджанский 5066
Борец большеносый 5403
Борец бороданый 5052
Борец волкобойник 5039
Борец волкобойный 5676
Борец вьющийся 5674
Борец гиринский 5318
Борец голубой 5442
Борец горнокитайский 5574
Борец горный 5422
Борец джунгарский 5585
Борец дуговидный 5043
Борец каракольский 5314
Борец корейский 5144
Борец круглолистный 5543
Борец круглолистый 5543
Борец крупный 5413
Борец куснецова 5335
Борец ладьевидный 5153
Борец лесной 5453
Борец лисий 5676
Борец мелколепестный 5042
Борец мохнатый 5663
Борец мяньнинский 5351
Борец противоядновидный 5041
Борец противоядный 5039

Борец сахалинский 5546
Борец синий 5442
Борец синьцзянский 5572
Борец таласский 5618
Борец хэбэйский 5360
Борец щукина 5558
Борода ааронова 36967
Бородавник 221780
Бородавник крупный 221791
Бородавник обыкновенный 221787
Бородавник средний 221794
Бородавник тайваньский 221802
Бородавочник 86733,86755,157952
Бородавчатник 393818
Бородавчатник вильчатый 393832
Бородавчатник копетдагский 393842
Бородатка 306847
Бородатка японская 306870
Бородач 22476,57545
Бородач кавказский 57551
Бородач китайский 22569
Бородач кровоостанавливающий 57567
Бородач тибетский 22833
Бородач юньнаньский 23086,57595
Бородиния 57245
Бородиния байкальская 57246
Бородоплодник 229283
Борщевик 192227
Борщевик альбова 192234
Борщевик бородатый 192239
Борщевик вильгельмся 192398
Борщевик ворошилова 192395
Борщевик выемчатолистный 192277
Борщевик вымечатолистный 192394
Борщевик ганьсуский 192275
Борщевик гроссгейма 192269
Борщевик гуншаньский 192276
Борщевик дланевидный 192339
Борщевик жесткий 192238
Борщевик заилийский 357912
Борщевик известняковый 192248
Борщевик кандинский 192357
Борщевик клауса 192252
Борщевик колхидский 192253
Борщевик крупноцветковый 192268
Борщевик лемана 192295
Борщевик лескова 192296
Борщевик лигустиколистный 192298
Борщевик лицзянский 192299
Борщевик любистколистный 192298
Борщевик мантегацци 192302,192306
Борщевик мелендорфа 192312
Борщевик моллендорфа 192312
Борщевик мохнатый 192394
Борщевик непалский 192324
Борщевик обыкновенный 192358
Борщевик ольги 192334
Борщевик осетинский 192338
Борщевик понтийский 192342
Борщевик пушийстый 192343
Борщевик пушистый 192343,192398
Борщевик рассеченнолистный 192255
Борщевик рассеченный 192255
Борщевик розовый 192347
Борщевик сибирский 192352
Борщевик симаоский 192271
Борщевик сладкий 192259
Борщевик соммье 192355
Борщевик стевена 192382
Борщевик тэнчунский 192380
Борщевик узколистный 192263
Борщевик франше 192266
Борщевик чжундяньский 192265
Борщевик чэнкоуский 192262
Борщевик шерковникова 192351
Борщевик шероховато – окаймленный 192390
Борщевик шероховатый 192349
Борщевик шерстистый 192278
Борщевик закавказский 192391
Борщевик сосновского 192356
Борщовия 57386
Борщовия арало – каспийская 57387
Босвелия 57539
Босвелия картера 57516
Босвеллия 57539
Босвеллия пильчатая 57539
Ботриостеге 57773
Ботриостеге прицветниковая 397860
Ботрокариум 57693
Ботрокариум спорный 57695
Бошниакиа 57429
Бошниакиа русская 57437
Бошнякиа 57429
Бошнякиа русская 57437
Боэргавия 56397
Боярка белая 369322
Боярка бородавчатая 109933
Боярышник 109509
Боярышник азароль 109545
Боярышник алтайский 109522
Боярышник американский 109609
Боярышник армянский 109535
Боярышник арнольда 109536
Боярышник беккера 109552
Боярышник волжский 110110
Боярышник волосистоцветковый 109693
Боярышник восточный 109788
Боярышник гансуйский 109781
Боярышник гельдрейха 109750
Боярышник даурский 109662
Боярышник двукосточковый 109671
Боярышник джунгарский 109685,110048
Боярышник дугласа 109684
Боярышник жёлтый 109709
Боярышник закаспийский 110091
Боярышник зангезурский 110116
Боярышник зеленомякотный 109597
Боярышник зеленомясый 109597
Боярышник испанский 109545
Боярышник кавказский 109585
Боярышник канадский 109582
Боярышник клиновидный 109650
Боярышник клинолистный 110052
Боярышник кокцинисвидный 109612
Боярышник колючий 109790
Боярышник кровавокрасный 110014
Боярышник круглолистный 110003
Боярышник крымский 110084
Боярышник ламберта 109796
Боярышник ложно – разнолистный 109965
Боярышник ложно – сомнительный 109964
Боярышник ложно – черноплодный 109966
Боярышник максимовича 109839
Боярышник мейера 109845
Боярышник мелколистный 109846
Боярышник мягкий 109850
Боярышник обыкновенный 109790
Боярышник однопестичный 109857,167922
Боярышник одностолбчатый 109857,167922
Боярышник перистонадрезанный 109933
Боярышник перисторзадельный 109933
Боярышник петуший гребень 109636
Боярышник петушья шпора 109636
Боярышник плодовый 109545
Боярышник понтийский 109545
Боярышник пятипестичный 109924
Боярышник расставленнолонностный 109994
Боярышник робсона 110001
Боярышник смбирский 110014
Боярышник согнутостолбиковый 109785
Боярышник сомнительный 109524
Боярышник стевенов 110056
Боярышник темнокровавый 109542
Боярышник туркестанский 110096
Боярышник туркменский 110095
Боярышник тяньшанский 110087
Боярышник украинский 110097
Боярышник хоккайдский 109780
Боярышник хубэйский 109760
Боярышник чёрный 109868
Боярышник чжундяньский 109606
Боярышник шаньдунский 110039
Боярышник шарлаховый 109609
Боярышник шовица 110076
Боярышник шпорцевый 109636
Боярышник шрадера 110032
Боярышник ярко – красный 109609
Боярышник – азароль 109545

Бравоа 59714,59715,311509
Бразения 59144
Бразения шребера 59148
Брайа 59727
Брайа розовая 59758
Брайа стручковая 59770
Брайа тибетская 59778
Брайя 59727
Брайя альпийская 29169
Брайя багряная 59757
Брайя багрянистая 59757
Брайя короткоплодная 59758
Брайя меднокрасная 59758
Брайя остроплодная 29169
Брайя памирская 59754
Брайя розовая 59758
Брайя узколистная 59758
Брайя шарнгорста 59769
Брандушка 62518
Брандушка разноцветная 62526
Брандушки белые 274823
Брассия 59265
Братки 410677
Браунколь 59520
Брахантем умтитова 58048
Брахантемим 58038
Брахантемим киргизский 58043
Брахантемим кустарничковый 58041
Брахантемим титова 58048
Брахантемум 58038
Брахантемум киргизский 58043
Брахантемум кустарничковый 58041
Брахиактис 58261
Брахиактис реснитчатая 58265
Брахиактис тибетский 58280
Брахиэлитрум 58447
Бредина 343151
Бриделия 60162
Бриония 61464
Бриония белая 61471
Брица 361935
Броваллия 61118
Бровник 192809
Бровник одноклубневый 192862
Бровник юньнаньский 192889
Бродиея 60431
Броколи 60431
Броколь 60431
Бромелиевые 60557
Бромелия 60544
Бросимум 61053
Брукенталия 61221
Брукенталия остролистная 61223
Брунера 61364
Брунера восточная 61368
Брунера крупнолистная 61366
Брунера сзбирская 61369
Брункресс 262722
Брунсвигия жозефины 61398
Брунфельзия 61293
Брунфельзия чашевидная 61297
Брунфельзия широколистная 61306
Брусника 403710,404051
Брусника обыкновенная 404051
Брусничные 403706
Бруссонетия 61096
Бруссонетия бумажная 61107
Бруссонетия наньчуаньская 61110
Бруссонеция 61096
Бруссонеция бумажная 61107
Брылкиния 61441
Брылкиния хвостатая 61442
Брюква 59493,59507
Буассьера 56560
Буассьера низкорослая 56563
Буассьера растопыреная 56572
Бубенчик 7596
Бубенчик китайский 7811
Бубенчик лилиелистный 7682
Бубенчик максимовича 7695
Бубенчик мелкоцветковый 7697
Бубенчик тайваньский 7702
Бубенчик четырехлистный 7850
Бубенчик эмэйский 70202
Бубенчики 7596
Бубенчики гималайский 7661
Бубенчики гмелина 7649
Бубенчики ламарка 7678
Бубенчики нинсяский 7716
Бубенчики трахелиевидный 7853
Бубенчики трехконечный 7855
Бубенчики узкоцветковый 7816
Бубенчики широколистный 7739
Бубрун желтый 249232
Бугенвиллеея 57852
Бугенвиллия 57852
Бугенвиллия гладкая 57857
Бугенвиллия голая 57857
Буддлейя 61953
Буддлейя изменчивая 62019
Буддлейя японская 328740
Буддлея 61953
Буддлея давида 62019
Буддлея далиская 62178
Буддлея изменчивая 62019
Буддлея линдли 62110
Буддлея форрестая 62059
Буддлея японская 62099,328740
Будра 176812
Будра волосистая 176837
Будра длиннотрубчатая 176839
Будра плющевидная 176824
Будра японская 176821
Бузгунча 301005
Бузина 345558
Бузина американская 345580
Бузина вильямся 345708
Бузина древовидная 345565
Бузина золотистая 345633
Бузина карликовая 345594
Бузина кистистая 345660
Бузина корейкая 345708
Бузина красная 345660
Бузина обыкновенная 345660
Бузина перувианская 345655
Бузина чёрная 345631
Бузинник 345594
Бузульник 228969,229246
Бузульник алтайский 228977
Бузульник альпийский 228976
Бузульник арктический 228986
Бузульник велколепный 229201
Бузульник волосистоголовый 229232
Бузульник высокогорный 228976
Бузульник джунгарский 229197
Бузульник длиноногий 229144
Бузульник кавказский 229000
Бузульник калужницелистный 228999
Бузульник кандинский 229074
Бузульник кнорринг 229077,229101
Бузульник кнорринговский 229077
Бузульник кратавский 229076
Бузульник крупнолистный 229104
Бузульник ли 229095
Бузульник лицзянский 229094
Бузульник метельчатый 229227
Бузульник монгольский 229109
Бузульник наньчуаньский 229116
Бузульник нарынский 229117
Бузульник непалский 228981
Бузульник павлова 229131
Бузульник пирамидально – метельчатый 229227
Бузульник полустреловидный 229211
Бузульник почковидный 229160
Бузульник разнолистный 229046
Бузульник сахалинский 229167
Бузульник сибирский 229179
Бузульник сизый 229043
Бузульник сихотинский 229195
Бузульник таласский 229214
Бузульник томсона 229226
Бузульник фишера 229035
Бузульник франше 229040
Бузульник ходжсона 229049
Бузульник шишкина 229171
Бузульник шмидта 229175
Бузульник ядунский 228990
Бузульник ялусский 229063
Бук 162359
Бук антарктический 266864

Бук белый 77256
Бук восточный 162395
Бук европейский 162400
Бук зибольда 162368
Бук зиебольда 162368
Бук крупнолистный 162375
Бук крымский 162436
Бук лесной 162400
Бук лесной плакучий 162413
Бук лесной тёмно－пунцовый 162404
Бук обыкновенный 162375
Бук тайваньский 162380
Бук энгрера 50825,162372
Бук японский 162382
Букашник 211639
Букашник горный 211672
Букашник многолетний 211670
Букашник хельдрейха 211660
Буквиза 53293
Буквиза аптечная 53304
Буквица 53293,373085
Буквица абхазская 53295
Буквица аптечная 53304
Буквица белоснежная 53303
Буквица водяная 355039
Буквица восточная 53307
Буквица крупноцветковый 53301
Буквица лекарственная 53304
Буквица олиственная 53298
Буквица осетинская 53308
Буковица 53293,373085
Буковые 162033
Буксус 64235
Буксус балеарский 64240
Буксус обыкновенный 64345
Буксус хайнаньский 64253
Булавоножка 371499
Булавоножка растопыренная 371502
Булавоносец 106924
Булавоносец седой 106932
Булавоносец суставчатый 106925
Бульбостилис 63204
Бульбостилис волосовидный 63229,63253
Бульбостилис оронова 63380
Бульбостилис тончайший 63359
Бульбофиллум 62530
Бульбофиллюм 62530
Бульбофиллюм гуандунский 62832
Бульбофиллюм гуншаньский 62756
Бульбофиллюм дэциньское 62963
Бульбофиллюм лэдунский 62841
Бульбофиллюм мэнхайский 62908
Бульбофиллюм мэнцзыский 63197
Бульбофиллюм тайваньский 62739
Бульбофиллюм фунинский 62745
Бульбофиллюм хайнаньский 62767
Бульбофиллюм хэнаньский 62773
Бульбофиллюм японский 62813
Бульва 189073
Бульнезия 63405
Бумагоплодник 86052
Бумагоплодник плоский 86053
Бумажное дерево 387432
Бунга 63440
Бунга трехраздельная 63442
Бундук 182524
Бундук гуансиский 182529
Бундук двудомный 182527
Бундук канадский 182527
Буниум 63467
Буниум бадхыза 63471
Буниум буржо 63472
Буниум бутеневый 63482
Буниум влагалищный 63513
Буниум гинколюбивый 63496
Буниум гиссарский 63497
Буниум длинноногий 63502
Буниум зеравшенский 63510
Буниум изящный 63488
Буниум капю 63480
Буниум кугитанга 63501
Буниум малолистный 63506
Буниум персидский 63508
Буниум средный 63500
Буниум цилиндрический 63485
Буниум шероховатый 63509
Бурак 53224
Бурачек 18313
Бурачек андийский 18329
Бурачек армянский 18333
Бурачек артвинский 18334
Бурачек буша 18342
Бурачек гмелина 18385
Бурачек дагестанский 18369
Бурачек двусемянный 18469
Бурачек длинностолбчатый 18414
Бурачек зонтичный 18488
Бурачек извилистый 18483
Бурачек искривленный 18483
Бурачек ленский 18403
Бурачек маленький 18432
Бурачек мелкоцветковый 18452
Бурачек мюллера 18443
Бурачек носатый 18463
Бурачек обвернутый 18419
Бурачек персидский 18458
Бурачек полевой 18349
Бурачек прицветничный 18341
Бурачек прямой 18478
Бурачек пустынный 18374
Бурачек пушистоплодный 18370
Бурачек пушистый 18484
Бурачек седой 18403
Бурачек стенной 18444
Бурачек туполистный 18451
Бурачек федченко 18381
Бурачек чамечкоплодный 18346
Бурачек чамечный 18344
Бурачек шершавый 18395
Бурачек шовица 18480
Бурачек щитницевидный 18457
Бурачик 57095
Бурачик лекарственный 57101
Бурачки 18687
Бурачниковые 57084
Бурачок 18313
Бурачок горный 18433
Бурачок извилистый 18483
Бурачок колючий 18474
Бурачок ленский 18403
Бурачок приморский 235090
Бурачок пустынный 18370
Бурачок пушистоплодный 18370
Бурачок сибирский 18469
Бурачок скалистый 45192
Бурачок скальный 45192
Бурачок чашечковый 18344
Бурачочек 18309
Бурачочек кочи 18312
Буреинский крыжовник 333929
Бурзера 64068
Бурзеровые 64088
Бурка 369339
Буркун 247457,308816
Буркун красный 247456
Буркунчик 247366
Буррерия 57903
Бурсера 64068
Бурсеровые 64088
Бурскария 64063
Бурути 246674
Буссенгоя 26264
Буссингоя 26264
Буськи 153767
Бутеа 64144
Бутелоа 57922,57930
Бутелоа изящная 57930
Бутелоуа 57922
Бутень 84723
Бутень ароматный 84726
Бутень бородина 84730
Бутень воброва 84729
Бутень длинноволосый 84739
Бутень кавказский 84736
Бутень клубненосный 84731
Бутень красноватый 84762
Бутень крупноплодный 84748
Бутень кяпазский 84746
Бутень ложноодуряющий 84769
Бутень луковичный 84731
Бутень мейера 84750

Бутень низкий 84744
Бутень одуряющий 84768
Бутень прескотта 84758
Бутень пятнистый 84749
Бутень розовый 84761
Бутень смешиваемый 84738
Бутень хорасанский 84745
Бутерлак 290476
Бутерлак гирканский 290485
Бутерлак очередонлистный 290478
Бутерлак портулаковый 290489
Бутея 231183
бутия 64159
бутия головчатая 64161
Бутылочная тыква 219821
Бухарник 197278
Бухарник мягкий 197306
Бухарник шерстистый 197301
Бухарник щетинистый 197323
Бухингера 39296
Бухингера пазушная 39297
Бухлое 61724
Бухлоэ 61724
Бушия 64111
Бушия бокоцветная 64113
Бушма 59493
Быбы турецкие 294056
Быноица 286972
Бюффония 62244
Бюффония крупноплодная 62252
Бюффония мелкоцветная 62264
Бюффония оливье 62261
Вайда 209167
Вайда араратская 209172
Вайда арнольди 209174
Вайда беложилковая 209209
Вайда бессера 209175
Вайда буассье 209176
Вайда бунге 209179
Вайда выемчатая 209241
Вайда гладкая 209204
Вайда гроссгейма 209195
Вайда грузинская 209197
Вайда кавказская 209183
Вайда короткоплодная 209177
Вайда красильная 209229
Вайда крупнейшая 209213
Вайда крымская 209228
Вайда кустарниковая 209193
Вайда лучистая 209227
Вайда маленькая 209215
Вайда обоюдоострая 209171
Вайда песчаная 209225
Вайда прибрежная 209210
Вайда продолговатая 209217
Вайда пустоплодная 209205
Вайда пушистоплодная 209205,209236
Вайда пушисто-чашечная 209196
Вайда ребристая 209184
Вайда седая 209180
Вайда сетчатая 209224
Вайда сизая 209194
Вайда согнутоплодная 209240
Вайда тпнкоплоднаю 209223
Вайда утконос 209220
Вайда широкоплодная 209208
Вайда якутская 209200
Вайда японская 209201
Вайянция 404172
Вайянция постенная 404181
Вакуа 281166
Вакциниум 403710
Вакциниум болотный 404027
Вакциниум выдающийся 403967
Вакциниум гуансиский 404001
Вакциниум зибольда 403998
Вакциниум кислоплодный 278366
Вакциниум корейский 403878
Вакциниум красника 403967
Вакциниум луаньдаский 403976
Вакциниум мелколистная 403956
Вакциниум мелкоплодный 278360
Вакциниум мупинский 403912
Вакциниум обыкновенный 404051
Вакциниум овальнолистный 403932
Вакциниум смолла 404003
Вакциниум эмэйский 403929
Вакциниум яошаньский 404063
Вакциниум японский 403868
Вакциниум ятабе 404066
Валенбергая 302747
Валенбергая крупноцветная 302753
Валенбергия 412568
Валенбергия изящная 412750
Валериана 404213
Валериана амурская 404225
Валериана аптечная 404316
Валериана блестящая 404314
Валериана возвышенная 404258
Валериана высокая 404259
Валериана горная 404310
Валериана гулявниколистная 404354
Валериана клубненосная 404377
Валериана корейская 404316
Валериана лекарственная 404316
Валериана мартьянова 404306
Валериана очередонлистная 404316
Валериана турчанинова 404379
Валериана федченко 404267
Валерианелла 404394
Валерианелла беззубая 404397
Валерианелла бугорчатая 404422,404478
Валерианелла введенского 404483
Валерианелла вздутая 404479
Валерианелла голоплодная 404434
Валерианелла двузубая 404415
Валерианелла дюфлена 404419
Валерианелла ежовая 404421
Валерианелла зубчатая 404413
Валерианелла кавказская 404403
Валерианелла колосковая 404439,404454
Валерианелла коротковенцовая 404400
Валерианелла косовенцовая 404462
Валерианелла кочи 404430
Валерианелла крючковатая 404481
Валерианелла кулябская 404431
Валерианелла ладьеплодная 404411
Валерианелла липского 404437
Валерианелла малая 404454
Валерианелла маленькая 404467
Валерианелла овощная 404439
Валерианелла остроносая 404457
Валерианелла плоскоплодная 404463
Валерианелла пузырчатая 404482
Валерианелла пушистоплодная 404433
Валерианелла совича 404476
Валерианелла твердоплодная 404473
Валерианелла тупая 404396
Валерианелла туркестанская 404480
Валерианелла увенчанная 404408
Валерианелла усеченная 404452
Валерианелла черноморская 404465
Валерианелла щелистая 404472
Валерианица 404394
Валериинница 404394
Валериинница овощная 404439
Валериановые 404392
Валерьяна 404213
Валерьяна аптечная 404316
Валерьяна аянская 404217
Валерьяна болотная 404251
Валерьяна головчатая 404236
Валерьяна дагестанская 404247
Валерьяна двудомная 404251
Валерьяна еленевского 404291
Валерьяна каменная 285814
Валерьяна камнелюбиая 404330
Валерьяна клубненосная 404377
Валерьяна колхидская 404244
Валерьяна лекарственная 404316
Валерьяна линолистная 404370
Валерьяна лоснящаяся 404314
Валерьяна мелкоцветковая 404309
Валерьяна мулиская 404311
Валерьяна приальпийская 404220
Валерьяна снеголюбивая 404242
Валерьяна сомнительная 404252
Валерьяна трехлисточковая 404373
Валерьяна цельнолистная 404352
Валерьяна чесночниколистная 404219
Валерьяна японская 404284

Валерьяница 404394
Валерьянница 404394
Валерьянница бугорчатая 404422
Валерьянница килеватая 404402
Валерьянница малая 404454
Валерьянница овощная 404439
Валерьянница огородная 404454
Валерьяновые 404392
Валления 413059
Валлисинерия 404531
Валлиснерия аньхуйская 404537
Валлиснерия гигантская 404555
Валлиснерия спиральная 404562,404563
Валлихия 413085
Валлихия густоцветная 413091
Валлихия китайская 413090
Валлихия таиландская 413099
Валлихия юньнаньская 413094
Валлота птрпурная 404581
Валодеа 404158
Валодеа парамуширская 404164
Валодеа темнобагряная 404160
Валонея 323745
Вальдгеймия 413007
Вальдгеймия войлочная 413018
Вальдгеймия голая 413009
Вальдгеймия заалайская 413020
Вальдгеймия столички 14917
Вальдгеймия трехлопастная 413021
Вальдгиймия 413007
Вальдгиймия войлочная 413018
Вальдгиймия голая 413009
Вальдгиймия столички 14917
Вальдгиймия трехлопастная 413021
Вальдштейния 413030
Вальдштейния тройчатая 413034
Ванда 404620
Ванда гиблидная 404647
Ванда гуандунская 404642
Ванда мелкоцветковая 404659
Ванда одноцветная 404632
Ванда трехцветная 404680
Ванделлия 231473
Ванделлия раскидистая 231509
Ваниль 404971
Ваниль душистая 404990,405017
Ваниль мексиканская 405010
Ваниль плосколистная 404990,405017
вартемия 405093
вартемия персидская 405099
Вархатник 18687
Василек 80892,81020
Василёк 80892
Василек абхазский 80895
Василек адамса 80904
Василек алайский 80913
Василек александра 80918
Василек альбова 80917
Василек андросова 80930
Василек армянский 80938
Василек барбеа 80964
Василек беланже 80959
Василек белая жемчужная 81193
Василек беловатый 80915
Василек белолистный 81173
Василек белопленчатый 81172
Василек бесплодный 81402
Василек бессера 80965
Василек боровой 81282
Василек бурооторoченный 81076
Василек ванкова 81445
Василек вильденова 81460
Василек волосистоглавый 81436
Василек волосистоголовый 81436
Василек воронова 81461
Василек восточный 81257
Василёк восточный 81257
Василек гербера 81083
Василек гирканский 81110
Василек глена 81086
Василек голубой 81020
Василёк голючеголовый 80981
Василек гончарова 81091
Василек горный 81214,81448
Василёк горный 81214
Василек горький 80922
Василёк горькуша 81150
Василек гроссгейма 81098
Василек грузинский 81082,81112
Василек гулисашвили 81102
Василек дагестанский 81023
Василек дминтриевой 81038
Василек днепровский 80978
Василек донской 81416
Василек дтшистый 18966,81219
Василек дубянского 81052
Василек ереванский 81063
Василек желтоглавовидный 81463
Василек желтоглавый 81462
Василек желтофиолевый 81002
Василек жемчужный 81194
Василек жестколистный 81401
Василек закавказский 81435
Василек зангезтрский 81465
Василек зувандский 81466
Василек иберийский 81112
Василек иволистный 81348
Василек иглолистный 80902
Василек ильина 81113
Василек казахский 81161
Василек карабахский 81160
Василек карпатский 80995
Василек каспийский 80998
Василек кожистый 81014
Василек козлиный 80990
Василек коктебельский 81162
Василек колковского 81163
Василек колхидский 81009
Василек колючеголовый 80981
Василек компера 81010
Василек конки 81164
Василек копетдагский 81165
Василек корнецветкововидный 81336
Василек корнецветковый 81335
Василек короткоголовый 80980
Василек кочи 81166
Василек крапчатый 81184
Василек красивый 80960
Василек красноватый 81341
Василек красноцветковый 81342
Василек крупноридатковый 80933
Василек кубанский 81167
Василек культиасова 81168
Василек лавренко 81171
Василек левзееподобный 81174
Василек ложно-бледночешуйчатый 81292
Василек ложно-пятнисский 81293
Василек ложно-скабиозовый 81295
Василек ложно-фригийский 81294
Василек лозный 81453
Василек луговой 81150,81290
Василёк луговой 81150
Василек майорова 81188
Василек маршалла 81196
Василек маршалля 81196
Василек мейера 81204
Василек мелкоцветковый 81405
Василек месхетский 81202
Василек многоножколистный 81285
Василек модеста 81209
Василек мягкий 81210
Василек наклоненный 81029
Василек натадзе 81229
Василек незамеченный 81028
Василек немного-лопастной 81269
Василек неразветвленный 81380
Василек низкий 81033
Василек новороссийский 81246
Василек обыкновенный 81020
Василек овечий 81259
Василек ольтинский 81253
Василек осетинский 81258
Василек отогнутый 81324
Василек паннонский 81265
Василек пачоского 81260
Василек пелийский 81272
Василек перистый 81356
Василек песчаный 80935
Василек пехо 81270
Василек пленчато-чешуйчатый 81109
Василек подбеленный 81024

Василек подколнечный 81382
Василек подражающий 80909
Василёк подсолнечный 81382
Василек полевой 81020
Василек посевной 81020
Василек предказский 81006
Василек прижаточещунуй 80907
Василек прижатый 80907
Василек приплюснутый 81033
Василек пятнистый 81184
Василёк пятнистый 81184
Василек раздорского 81323
Василек раскидистый 81034
Василек распростертый 81034
Василек растопыренный 81398
Василек рогатый 81012,102833
Василёк рогатый 124141
Василек рупрехта 81344
Василек русский 81345
Василёк русский 81345
Василек салоникский 81352
Василек сарандинаки 81355
Василек сборноголовый 80983
Василек сергея 81368
Василек сибирский 81378
Василек синий 81020
Василек синтеница 81381
Василек скабиоза 81356
Василёк скабиозный 81356
Василек скабиозовидный 81356
Василек скабиозовый 81356
Василек скученный 80912
Василек совича 81412
Василек солнечный 81382
Василек солончаковый 86049
Василек соседний 81450
Василек сосновского 81386
Василек софии 81385
Василек стевена 81404
Василек стевеновский 81403
Василек сумской 81411
Василек талиева 81415
Василек темнокрасный 80950
Василек тернопольский 81427
Василек торчащий 81407
Василек трехжилковый 81438
Василек троицкого 81439
Василек тупочешуйчатый 80926
Василек тургайский 81440
Василек туркестанский 81441
Василек угольный 80991
Василек узкопленчатый 81400
Василек уклоняющийся 80897
Василек укороченный 80894
Василек феопаппусовидный 81276
Василек фишера 81069
Василек фригийский 81277
Василек хаястанский 81104
Василек целнокрайнолистный 81145
Василек цельнолистный 81107
Василек черешочковый 81278
Василек черкесский 81005
Василек чернобахромчатый 81245
Василек черноголовый 81244
Василек чёрный 81237
Василёк чёрный 81237
Василек чертополоховидный 80993
Василек шалфеелистный 81354
Василек шелковникова 81359
Василек шероховатый 81356
Василек шерстистоногий 81169
Василек шипиконосный 80932
Василек широколопастный 81170
Василисник 388397
Василисник альпийский 388404
Василисник амурский 388416
Василисник байкальский 388432
Василисник водосборолистный 388423
Василисник вонючий 388510
Василисник желтый 388506
Василисник жёлтый 388506
Василисник изопироидный 388539
Василисник клубненосный 388703
Василисник ложнолепестковый 388614
Василисник малый 388572
Василисник мелкоплодный 388569
Василисник нитчатый 388503
Василисник обыкновенный 388572
Василисник простой 388667
Василисник прямой 388683
Василисник растопыренный 388680
Василисник редкоцветный 388673
Василисник сахалинский 388654
Василисник скрученный 388460
Василисник султанабадский 388685
Василисник триждысройчатый 388701
Василисник узколистый 388421
Василисник хэнаньский 388533
Василисник эмэйский 388607
Василисник яванский 388541
Василистник 388397
Василистник альпийский 388404
Василистник амурский 388416
Василистник байкальский 388432
Василистник блестящий 388556
Василистник водосборолистный 388423
Василистник вонючий 388510
Василистник желтый 388506
Василистник иезский 388722
Василистник изопироидный 388539
Василистник кемский 388584
Василистник клубненосный 388703
Василистник лепестковый 388614
Василистник лицзянский 388719
Василистник ложнолепестковый 388614
Василистник малый 388572
Василистник простой 388667
Василистник растопыренный 388680
Василистник редкоцветковый 388633
Василистник редкоцветный 388673
Василистник сахалинский 388654
Василистник светлый 388556
Василистник скрученный 388460
Василистник тайваньский 388710
Василистник тычинковый 388503
Василистник фриза 388516
Василистник холмовой 388572
Васильки 102833
Васильки рогатые 124141
Ватерия 405153
Ватерия индийская 405154
Ваточник 37811
Ваточник инкарнатный 37967
Ваточник клубневой 38147
Ваточник клубненосный 38147
Ваточник кюрасао 37888
Ваточник розово－малиновый 37967
Ваточник силийский 38135
Ватсония 316220
Вахендорфия 412519
Вахта 250493,404027
Вахта трехлистная 250502
Вахта трёхлистная 250502
Вахта трилистная 250502
Вахтовые 250492
Вашингтония 413289
Вашингтония крепкая 413307
Вашингтония мощная 413307
Вашингтония нитевидная 413298
Вашингтония нитеносная 413298
Вдовушка 350086
Вдовушки 350086
Везикария 227017
Вейгела 413570
Вейгела корейская 413576
Вейгела кроваво－красная 413624
Вейгела максимовича 413616
Вейгела миддендорфа 413617
Вейгела обильноцветущая 413588
Вейгела приятная 413628
Вейгела ранняя 413621
Вейгела цветущая 413591
Вейгела японская 413611
Вейгелия 413570
Вейник 65254
Вейник алайский 65266
Вейник алексеенко 65267
Вейник алтайский 65269
Вейник балансы 65295
Вейник бунге 65301
Вейник вильненский 65521

Вейник вилюйский 65523
Вейник волосистый 65506
Вейник высокий 65324
Вейник гигантский 65419
Вейник гималайский 65325
Вейник гиссарский 65364
Вейник горный 65431
Вейник грузинский 65370
Вейник зайцехвостовидный 65393
Вейник извилистый 65347
Вейник кавазский 65315
Вейник колымский 65387
Вейник коржинского 65390
Вейник короткого 65388
Вейник краснеющий 65470
Вейник крупноцветковый 65358
Вейник лангсдорфа 65473
Вейник ланцетный 65397
Вейник лапландский 127257
Вейник лисохвостовидный 65509
Вейник ложнотростниковый 65464
Вейник наземный 65330
Вейник незамечаемый 127276
Вейник нинсяский 65447
Вейник обыкновенный 65330
Вейник обыкновенный мелкоцветковый 65337
Вейник олимпийский 65449
Вейник павлова 65453
Вейник пахучеколосковый 65274
Вейник персидский 65454
Вейник поплавской 65461
Вейник ртпрехта 65476
Вейник сахалинский 65477
Вейник своеобразный 65452
Вейник сизый 65357
Вейник сычуаньский 65488
Вейник сомнительный 65465
Вейник тебердинский 65502
Вейник темный 65350
Вейник тонкостебельный 127264
Вейник торотковолосистый 127217
Вейник тростниковидный 127197
Вейник тупоколосковый 65448
Вейник туркестанский 65509
Вейник турчанинова 65507
Вейник тяньшанский 127314
Вейник узколистный 65270
Вейник уральский 65511
Вейник хольма 65366
Вейник шугнанский 65483
Вейник щучковидный 65321
Велблюжья канадская киргизская 14629
Велблюжья канадская обыкновенный 14638
Велблюжья канадская персидская 14631
Велблюжья канадская сероватая 14627
Велблюжья колючка обыкновенная 14638
Велблюжья трава 14638
Велеция 405269
Веллингтония 360576
Вельвичия 413741
Венгерка 316382
Венгерка итальяская 316333,316391
Вендландия 413765
Венец царский 168430
Венецианский сумах 107300
Венечник 26950
Венечник ветвистый 27320
Венечник живородящий 88527
Венечник лилейный 27190
Венечник лилиаго 27190
Венечник лилиецветковый 27190
Венечнозонтичник 274373
Венечнозонтичник красивый 274374
Веничник 225447
Венцовник 375916
Венцовник ольги 375917
Верба 343280
Верба красная 342961
Вербаскум 405657
Вербаскум китайский 405685
Вербезина 405911
Вербейник 239543
Вербейник белоснежный 239566
Вербейник гуандунский 239702
Вербейник гуансиский 239547
Вербейник густоколосый 239558
Вербейник густоцветковый 239558
Вербейник даурский 239902
Вербейник дубравный 239750
Вербейник ичанский 239668
Вербейник клетровидный 239594
Вербейник ландышевый 239594
Вербейник ландышный 239594
Вербейник лицзянский 239712
Вербейник луговой чай 239755
Вербейник лушаньский 239822
Вербейник мелколистный 239775
Вербейник мелкоплодный 239735
Вербейник монетчатый 239755
Вербейник мотоский 239731
Вербейник мутовчатый 239891
Вербейник наньпинский 239746
Вербейник наньчуаньский 239745
Вербейник обыкновенный 239898
Вербейник пятиленестный 239781
Вербейник пятнистный 239804
Вербейник симаоский 239626
Вербейник сомнительный 239623
Вербейник таиландский 239846
Вербейник точечный 239804
Вербейник фортюна 239645
Вербейник франше 239666
Вербейник фунинский 239641
Вербейник фуцзяньский 239651
Вербейник хайнаньский 239748
Вербейник ху 239673
Вербейник цзиндунский 239699
Вербейник чжундяньский 239588
Вербейник чжэцзянский 239578
Вербейник эмэйский 239762
Вербейник яванский 239696
Вербейник японский 239606,239687
Вербена 405801
Вербена аптечная 405872
Вербена гибридная 405852
Вербена канадская 405820
Вербена копьевидная 405844
Вербена лежачая 405893
Вербена лежащая 405893
Вербена лекарственная 405872
Вербена лимонная 232483
Вербена плетистая 405852
Вербена прицветниковая 405817
Вербена прицветничковая 405817
Вербеновые 405907
Вербесина 405911
Верблюдка 104750
Верблюдка алтайская 104752
Верблюдка арало－каспийская 104754
Верблюдка бокоцветковый 11613
Верблюдка бородавчатая 104770,104828
Верблюдка восточная 104819
Верблюдка выгрызенная 104775
Верблюдка вытянутая 104771
Верблюдка густая 104762
Верблюдка зябкая 104751
Верблюдка иларии 104784
Верблюдка иссополистная 104786
Верблюдка комарова 104792
Верблюдка коровина 104793
Верблюдка крупноплодная 104803
Верблюдка крылова 104794
Верблюдка курчавоплодный 104860
Верблюдка лемана 104797
Верблюдка лоснящаяся 104816
Верблюдка маршалла 104813
Верблюдка монгольская 104814
Верблюдка нителистная 104777
Верблюдка памирская 104826
Верблюдка повислая 104765
Верблюдка растопыренная 104847
Верблюдка редовского 104839
Верблюдка рыхлоцветная 104796
Верблюдка седоватая 104756,104757
Верблюдка семиреченская 104783
Верблюдка сибирская 104844
Верблюдка стонтона 104849
Верблюдка тибетская 104856
Верблюдка толстолстная 104763
Верблюдка хинганская 104758

Верблюдка хинганская мелкоплодная 104759
Верблюдка холодная 104780
Верблюжий хвост 72249
Верблюжья колючка 14623
Верблюжья колючка редколистная 14638
Верболоз 342961,343858
Вередник 390213
Верес 213606
Вереск 67455
Вереск восковой 261137
Вереск древовидный 149017
Вереск крупный средиземноморский 149017
Вереск метлистый 150027
Вереск обыкновенный 67456
Вереск серый 149205
Вересклет лицзянский 157683
Вересклет широколистный 157668
Вересковые 150286
Вермут 35090
Вернония 406040
Вероника 406966
вероника 407449
Вероника австрийская 407018
Вероника алатавская 317907
Вероника альпийская 406979
Вероника американская 406983
Вероника анагалисовидная 406988
Вероника аптечная 407256
Вероника армянская 407012
Вероника баранецкого 407024
Вероника барелье 407025
Вероника баумгартена 407028
Вероника бахофена 407022
Вероника бедовойлочная 317912
Вероника безлистная 407008
Вероника белая 407155
Вероника береговая 407430
Вероника близкая 407298
Вероника боброва 407041
Вероника болотниковидна 407048
Вероника борнмюллера 407011
Вероника бухарская 407045
Вероника вербейниковая 407218
Вероника весенняя 407432
Вероника водная 407158
Вероника волосоногая 407058
Вероника восточная 407264
Вероника восточнокавказская 406978
Вероника высокогорная 407239
Вероника гололистная 407135
Вероника горбунова 407141
Вероника горная 407238
Вероника грязевая 407003
Вероника густоцветковая 407106
Вероника далиская 407127
Вероника даурская 317909
Вероника двлопастная 407040
Вероника двойчатая 407109
Вероника дегестанская 407099
Вероника диллениа 407114
Вероника длиннолистная 317927
Вероника дубравка 407072
Вероника дубравная 407072
Вероника дубровка 407072
Вероника зубчатая 407107
Вероника изящная 407292
Вероника имеретинская 407159
Вероника иноземная 407275
Вероника кавказская 407068
Вероника каменистая 407289
Вероника каратавская 407011
Вероника кемулярии 407167
Вероника ключевая 406992
Вероника колосистая 317964
Вероника комарова 407184
Вероника копетдагская 407188
Вероника корейская 317947
Вероника краснолистная 407329
Вероника кривоногая 407049
Вероника крошечная 407306
Вероника крупноцветковая 407142
Вероника крылова 407019
Вероника крымская 407404
Вероника курдская 407192
Вероника кустарничковая 407130
Вероника кустящаяся 407129
Вероника лежачая 407299
Вероника лекарственная 407256
Вероника лидтке 407216
Вероника ложная 317966
Вероника ложноключевая 407003
Вероника ложно-крупнотычнковая 407220
Вероника ложнопоточная 407037
Вероника льнянколистная 317924
Вероника мальнькая 407229
Вероника маргаритковая 407039
Вероника маргаритколистная 407038
Вероника мелкая 407231
Вероника мелкоплодная 407228
Вероника мишо 407227
Вероника многоветвистая 407308
Вероника многораздельная 407245
Вероника монтиевидная 407240
Вероника наибольшая 407224
Вероника невского 407250
Вероника ненастоящая 317966
Вероника нитевидная 407124
Вероника нителистная 407123
Вероника обнаженная 407108
Вероника ольгинская 407258
Вероника ольтинская 407260
Вероника орхидная 407263
Вероника остропильчатая 407011
Вероника остроплодная 407267
Вероника пашенная 406973
Вероника перистая 317940
Вероника персидская 407287
Вероника песчаная 407010
Вероника петушийгребень 407092
Вероника плющелистная 407144
Вероника полевая 407014
Вероника поручейная 407029
Вероника порфирия 407295
Вероника поточная 407029
Вероника почни-сидячая 317968
Вероника прекрасная 317940
Вероника приятная 406986
Вероника промежуточная 407161
Вероника простёртая 407299
Вероника ранняя 407297
Вероника распростертая 407362
Вероника ресничатая 407079
Вероника ресничатая чжундяньская 407081
Вероника рогоплодная 407071
Вероника рупрехта 407332
Вероника ручейная 407029
Вероника сахалинская 407333
Вероника саянская 407334
Вероника северная 407357
Вероника седая 317912
Вероника сердцевидноплодная 407063
Вероника сибирская 407371,407434
Вероника сычуаньская 407400
Вероника сливолистная 407070
Вероника стеллера 407387
Вероника столбиконосная 407393
Вероника тайваньская 407403
Вероника тебердинская 407405
Вероника телефиелистная 407406
Вероника тибетская 407416
Вероника тимьянная 407358,407369
Вероника тимьянолистная 407358
Вероника тончайшая 407409
Вероника трёхлистная 407421
Вероника трехраздельная 407420
Вероника трубкоцветковая 407424
Вероника туркменов 407426
Вероника турнефора 407287
Вероника тусклая 407262
Вероника тяньшанская 407415
Вероника федченко 407121
Вероника хантавская 407074
Вероника харадзе 407075
Вероника хоросанская 407169
Вероника хэбэйская 407078
Вероника цветоножковая 407273
Вероника цимбаляриевая 407096
Вероника чаюйская 407076

Вероника чернеющий 407251
Вероника чэнкоуская 407120
Вероника широколиственная 407019
Вероника широколистная 13934,407019,407198
Вероника шитковая 407349
Вероника шмидта 407337
Вероника яванская 407164
Вероника ягодколистная 406971
Вероника японская 407252
Вероника – поточник 407029
Вертляница 258006
Вертляница одноцветковая 258044
Вертляница подъельник 202767
Вертляница подъельниковая 202767
Вертляница тайваньская 258041
Вертляницевые 258049
Весенний горошик 222860
Весенник 148101
Весенник длинноножковый 148109
Весенник звездчатый 148112
Весенник сибирский 148111
Веснянка 153949
Веснянка весенняя 153964
Веснянка карликовая 153956
Веснянка крокера 153964
Веснянка ранняя 153957
Ветвянка 58052
Ветвянка гусеницевидная 58088
Ветла 342973
Ветла блестяще – жёлтая 342982
Ветреница 23696
Ветреница алтайская 23701
Ветреница альпийская 321658
Ветреница амурская 23705
Ветреница байкальская 23709
Ветреница бухарская 23736
Ветреница весенниковая 23807
Ветреница весенняя 24109
Ветреница видная 24065
Ветреница вильчатая 23796
Ветреница вытянутая 23906
Ветреница ганьсуская 23711
Ветреница гибгая 23817
Ветреница гладкая 23827
Ветреница голубая 23766
Ветреница горчакова 23832
Ветреница длинноволосая 23773
Ветреница дубравная 23925,336495
Ветреница енисейская 23866
Ветреница кавказская 23750
Ветреница коротконожковая 23734
Ветреница кузнецова 23870
Ветреница лесная 24061,24071
Ветреница лысая 23739
Ветреница лютиковая 24018
Ветреница лютичная 24018
Ветреница малорассеченная 23975
Ветреница мелкоцветковая 23980
Ветреница мохнатейшая 24111
Ветреница нарциссовидная 23901
Ветреница нежная 23724
Ветреница отогнутая 24019
Ветреница охотская 23971
Ветреница пермская 23722
Ветреница прибрежная 23877
Ветреница пучковатая 23815
Ветреница радде 24013
Ветреница ричардсона 24020
Ветреница рыхлая 23874
Ветреница сахалинская 23917
Ветреница сибирская 23919
Ветреница слабая 23780
Ветреница теневая 24105
Ветреница тенистая 24105
Ветреница тибетская 24088
Ветреница туподольчатая 23939
Ветреница туподольчатая чжэньканская 23945
Ветреница тэнчунская 24077
Ветреница тяньцюаньская 23990
Ветреница удская 24103
Ветреница уральская 24107
Ветреница черняева 24100
Ветреница шаньдунская 24058
Ветреница шренка 23906
Ветреница японская 23854
Веттиверия 407579
Вех 90918
Вех ядовитый 90932
Вечерница 193387
Вечерница армянская 193392
Вечерница висячая 193428
Вечерница воронова 193457
Вечерница двуострая 193393
Вечерница карсская 193403
Вечерница лесная 193446
Вечерница ложнобелоснежная 193435
Вечерница матроны 193417
Вечерница мейера 193423
Вечерница мрачная 193452
Вечерница ночная 193417
Вечерница ночная фиалка 193417
Вечерница персидская 193429
Вечерница печальная 193452
Вечерница сибирская 193444
Вечерница стевена 193449
Вечерница стевенова 193449
Вечерница фиолетовая 193456
Вздутоплодник 122121,122124
Вздутостебельник 261592,297854
Вздутостебельник узловатый 297856
Взморник 418377,418392
Взморник азиатский 418381
Взморник малый 418392
Взморник морской 418392
Взморник низкий 418396
Взморник тикоокеанский 418392
Взморник японский 418396
Взморниковые 418405
Вигандия 414092
Вигна 224275,408825
Вигна катьянг 409086,409092
Вигна китайская 409086
Вигна коноплевидная 409092
Вигна люцюская 409036
Виддрингтония 414058
Виддрингтония уайта 414066
Виддрингтония шварца 414065
Видеманния 414072
Видеманния восточная 414075
Видеманния многонадрезная 414074
Виздутоплодник 295035
Виздутоплодник мохнатый 295041
Виздутоплодник сибирский 295038
Вика 408251
Вика бенгальская 408305
Вика бенгерская 408527
Вика гороховидная 408543
Вика китайская 408339
Вика кормовая 197301,408571
Вика красивая 408262
Вика крупносемянная 408478
Вика лесная 408614
Вика мохнатая 408693
Вика мундж 408251
Вика нарбонская 408505
Вика обыкновенная 408571
Вика озимая 408693
Вика озймая 408693
Вика панноская 408527
Вика пестроцветная 408540
Вика песчаная 408693
Вика плоско – семянная 408571
Вика посевная 408571
Вика пушистомлодная 408371
Вика римская 408505
Вика синьцзянская 408350
Вика сорнополевая 408571
Вика узколистная 408284
Вика французская 408505
Вика чёрная 408693
Вика чёрная мохнатая 408693
Вика яровая 408571
Викса 54859
Викстремия 414118
Викстремия аньхуйская 414133
Викстремия дэцинььская 414266
Викстремия индийская 414193
Викстремия лицзянская 414207
Викстремия тайваньская 414265

Викстремия хайнаньская 414186
Викстремия хуйдунская 414191
Викстремия чэнкоуская 414172
Викстремия японская 414271
Виктория 408733
Виктория ананасная 167597
Вилларезия 409251
Виллемеция 414385
Виллемеция крубненосная 414389
Винерин 120396
Винерин башмачек 120287
Винерин башмачек вздутый 120464
Винерин башмачек настоящий 120310
Винерин башмачек пятнистый 120357
Винерин чжундяньский 120303
Винклера 414474,414478
Винклера натриниевидная 414475
Виноград 411521,411979
Виноград американский 411764,411887
Виноград амурский 411540
Виноград берландье 411575
Виноград винный 411979
Виноград виноградолистный 20461
Виноград виноносный 411979
Виноград винородный 411979
Виноград вьющийся 20435,285136
Виноград горный 411575
Виноград давида 411646
Виноград девичий 285097
Виноград девичий пятилистный 20435, 285136
Виноград девичий триостренный 285157
Виноград дикий 285097
Виноград дикий пятилистный 20435, 285136
Виноград дикий триостренный 285157
Виноград евроазиатский 411979
Виноград европейский 411979
Виноград зимний 411575
Виноград изабелла 411764
Виноград изабеллы 411764
Виноград комнатный 92586,92777
Виноград круглолиственный 411893
Виноград круглолистный 411893
Виноград куанье 411623
Виноград культурный 411979
Виноград лабруска 411764
Виноград лапчатый 411834
Виноград лесной 411983
Виноград лисий 411631,411887
Виноград лунчжоуский 411682
Виноград лушаньский 411742
Виноград ляблуска 411764
Виноград медвежий 31142
Виноград мэнхайский 411796
Виноград мэнцзыский 411797
Виноград настоящий 411979
Виноград обыкновенный 411979
Виноград прибрежный 411887
Виноград романе 411890
Виноград североамериканский 411764
Виноград скальный 411900
Виноград старосветский 411979
Виноград тунберга 411669
Виноград фэнцинский 411668
Виноград чжэцзянский 412010
Виноградные 411165
Виноградовик 20280
Виноградовик короткоцветоножковый 20348
Виноградовник 20280
Виноградовник гуншаньский 20359
Виноградовник короткоцветоножковый 20348
Виноградовник короткоцветоножковый гулинский 20356
Виноградовник крупнолистный 20420
Виноградовник разнолистный 20354
Виноградовник разнолистный гулинский 20356
Виноградовник японский 20408
Виноградовые 411165
Виргилия 410869
Виргилия жёлтая 94022,94026
Вирола суринамская 261461
Висмия 411140
Вистария 414541
Вистария китайская 414576
Вистария кустарниковая 414565
Вистария обильноцветшая 414554
Витекс 411187
Витекс гуансиский 411310
Витекс лунчжоуский 411421
Витекс священный 411189
Вишенник 83202
Вишнеслива 316294
Вишня 83110
Вишня алайская 83143
Вишня алкенинги 297643,297645
Вишня ананасная 297711
Вишня антипка 83253
Вишня бессея 83157
Вишня бесси 83157
Вишня бородавчатая 83360
Вишня виргинская 316918
Вишня войлочная 83347
Вишня джунгарская 85658
Вишня дикая 83202
Вишня жакемона 83233
Вишня железистая 83208
Вишня железистолистная 83207
Вишня жидовская 297640
Вишня иедоская 83365
Вишня истанская 83155
Вишня кандинская 83343
Вишня карликовая 316735
Вишня кислая 316320
Вишня конрадины 83177
Вишня красноплодная 83201
Вишня кусгарная 83202
Вишня кустарниковая 83202
Вишня лесостепная 83202
Вишня ложнопростертая 83285
Вишня магалебская 83253
Вишня максимовича 83255
Вишня мараскиновая 316318
Вишня мелкопильцчатая 83314
Вишня мелкоплодная 83257
Вишня миндалецветная 83144
Вишня мэнцзыская 83216
Вишня обыкновенная 83361
Вишня перуанская 297711
Вишня перувианская 297711
Вишня песчаная 316735
Вишня пёсья 297640
Вишня пёсья обыкновенная 297643,297645
Вишня пёсья съедобная 297711
Вишня пёсья франшетта 297643,297645
Вишня пильчатая 83314
Вишня полевая 297643,297645
Вишня птичья 83155
Вишня пузырная 297640
Вишня пузырчатая 297640
Вишня садовая 83361
Вишня сакура 83314
Вишня сахалинская 83300
Вишня седая 316710
Вишня сычуаньская 83340
Вишня слабоопушенная восходящая 83328
Вишня сладкая 83155
Вишня степная 83202
Вишня суринамская 156385
Вишня суринамская питанга 156385
Вишня туркменская 83353
Вишня тяньшанскя 83345
Вишня усурийская 83238
Вишня франшетти 297643,297645
Вишня хайнаньская 83215
Вишня холмовая 83174
Вишня черешня 83155
Вишня чёрная дикая 83311
Вишня чёрная позднян 83361
Вишня чернильная 83253
Вишня шпанка кустарниковая 83202
Вишня юньнаньская 83368
Вишня японская 83238
Вишня – гриот 316318
Власочешуйник 395964
Власочешуйник власоглавый 395968
Воандзея подземная 409069
Воднососенковые 196809

Водокрас 200225
Водокрас лягушечный 200230
Водокрас обыкновенный 200228,200230
Водокрас – лягушечник 200228
Водокрасовые 200235
Водолистниковые 200476
Водолюб 200253,200482
Водоперица 261337
Водоперица мутовчая 261379
Водосбор 29994
Водосбор амурский 29998
Водосбор бородина 30003
Водосбор вееровидный 30031
Водосбор виталия 30080
Водосбор ганьсуская 30065
Водосбор гибридный 30041
Водосбор железистый 30038
Водосбор зеленоцветковый 30077
Водосбор канадский 30007
Водосбор каратавский 30046
Водосбор карелина 30047
Водосбор короткошпорцевый 30004
Водосбор мелкоцветковый 30068
Водосбор молочноцветный 30049
Водосбор обыкновенный 30081
Водосбор олимпийский 30060
Водосбор остролепестный 30061
Водосбор прекрасный 30034
Водосбор сибирский 30074
Водосбор скиннера 30075
Водосбор тайваньский 30048
Водосбор темновинный 30001
Водосбор темно – птрпуровый 30078
Водосбор темно – пурпуровый 30001
Водосбор тонкошпорцевый 30051
Водосбор тяньшанский 30076
Водяная звездочка 67336
Водяная роза 267627
Водяная сосенка 196810
Водяная сосенка обыкновенная 196818
Водяника 145054
Водяника черная 145073
Водяниковые 145052
Водянка 145054
Водяной гиацинт 141808
Водяной лютик 48904,48919
Водяной лютик волосистый 48933
Водяной лютик кауфмана 48922
Водяной лютик монгольский 48927
Водяной лютик неуколеняющийся 48915
Водяной лютик расходящийся 48914
Водяной лютик риона 48931
Водяной лютик толстостебельный 48928
Водяной орех 394412
Водяной орех амурский 394417
Водяной орех астраханский 394423
Водяной орех бугорчатый 394552
Водяной орех вcomponents 394459
Водяной орех гирканский 394458
Водяной орех гребенчатый 394523
Водяной орех двурогий 394426
Водяной орех длиннорогий 394484
Водяной орех европейский 394457
Водяной орех каринтийский 394447
Водяной орех комарова 394526
Водяной орех коржинского 394478
Водяной орех крестообразный 394454
Водяной орех куполовидный 394453
Водяной орех литвинова 394480
Водяной орех максимовича 394496
Водяной орех малеева 394487
Водяной орех маньчжурский 394489
Водяной орех плавающий 394500
Водяной орех потанина 394430
Водяной орех русский 394539
Водяной орех саянский 394541
Водяной орех северный 394500
Водяной орех сибирский 394542
Водяной орех спрыгина 394547
Водяной орех тайваньский 394433
Водяной орех траншеля 394551
Водяной орех японский 394463
Войлочная вишня 83347
Волдырник 114057
Волдырник ягодный 114060
Волдырник ягодоплодный 114060
Волдырник японский 114066
Волжанка 37053
Волжанка азиатская 37056
Волжанка камчатская 37069
Волжанка малая 37083
Волжанка обыкновенная 37085
Волкамерия 95934
Волкамерия юньнаньская 96449
Волкобой 122515
Волкобойник 5347
Волноплодник 116716
Волноплодник гроссгейма 116717
Волноплодник мохнатый 116719
Волноплодник разнолистный 116718
Воловик 21914
Воловик аптечный 21959
Воловик аушера 21924
Воловик баррелиера 21928
Воловик гмелина 21940
Воловик длинностолбиковый 21980
Воловик итальянский 21949
Воловик лекарственный 21959
Воловик светложелтый 21957
Воловик уликостный 21953
Воловик фессалийский 21983
Володушка 63553
Володушка абхазская 63554
Володушка бднолучевая 63776
Володушка буассье 63582
Володушка ветвистая 63584
Володушка витамана 63869
Володушка воронова 63870
Володушка высокая 63623
Володушка гуйцзоуская 63697
Володушка густоцветковая 63615
Володушка двстебельная 63579
Володушка длиннолистная 63722
Володушка длиннолучевая 63728
Володушка длиннообертковая 63727
Володушка жерара 63669
Володушка золотистая 63571
Володушка изащиная 63677
Володушка изменчивая 63610
Володушка карликовая 63792
Володушка китайская 63594
Володушка козелецелистая 63813
Володушка козелецоволистная 63813
Володушка козо – полянскго 63694
Володушка комарова 63693
Володушка круглолистная 63805
Володушка крылова 63695
Володушка кустарниковая 63668
Володушка ланцетнолистная 63699
Володушка лицзянская 63803
Володушка лютиковидная 63796
Володушка мартьянова 63750
Володушка маршалла 63749
Володушка многожильчатая 63761
Володушка многолистная 63787
Володушка многообразная 63786
Володушка нордмана 63769
Володушка отклоненнолистная 63619
Володушка полуслозкная 63826
Володушка родственная 63557
Володушка российская 63804
Володушка сахалинская 63807
Володушка серповидная 63625
Володушка сибирская 63827
Володушка сизая 63675
Володушка смешиваемая 63610
Володушка сосновского 63834
Володушка тончайшая 63855
Володушка трехлучевая 63864
Володушка тяньшанская 63859
Володушка франше 63713
Володушка юньнаньская 63873
Володушка ясменниковая 63564
Володушка ясменниковидная 63564
Волосатик 113815
Волосатик незаметный 395668
волосисто – скаладковая 14110
Волоснец 144144
Волоснец высокий 144312
Волоснец гигантский 144312
Волоснец мягкий 228369

Волоснец песчаный 228346
Волоснец сибирский 144466
Волоснец тургайский 144517
Волоснец узкий 144517
Волосопыльник 299277
Волосопыльник нителистный 299278
волосореберник 299201
волосореберник козо - полянского 299203
Волотный кипалис 385253
Волотный кипалис мексиканский 385281
Волотный кипалис обыкновенный 385262
Волчец 97581
Волчец благословенный 97599
Волчец благословленный 97599
Волчец кудрявый 97599
Волчея годник лавролистный 122492
Волчея годниксофьи 122604
Волчея годникюлии 122475
Волчеягодник 122359
Волчеягодник волчьелыко 122515
Волчеягодник кавказский 122407
Волчеягодник лавролистный 122492
Волчеягодник обыкновенный 122515
Волчеягодник - боровик 122416
Волчни кюлии 122475
Волчник 122359
Волчник алтайский 122365
Волчник альбова 122363
Волчник благая 122389
Волчник боровик 122416
Волчник закавказский 122622
Волчник кавказский 122407
Волчник камчатский 122477
Волчник лавловый 122492
Волчник лавролистный 122492
Волчник обыкновенный 122515
Волчник пазущноцветковый 122382
Волчник понтийский 122585
Волчник скученый 122447
Волчник смертельный 122515
Волчник софии 122604
Волчник узколистный 122526
Волчник черкесский 122414
Волчник юлия 122475
Волчник - боровик 122416
Волчниковые 391030
Волчуг 271290
Волчьелыко обыкновенное 122515
Вольфия 414649
Вольфия безкорневая 414653
Вонючка 21445,163580
Воробейник 233692
Воробейник аптечный 233760
Воробейник краснокорневишный 233731
Воробейник краснокорневой 233731
Воробейник лекарственный 233760
Воробейник полевой 233700
Воробейник пурпурно - синий 233772
Воробейник тонкоцветковый 233782
Воробейник узколистный 233698
Воробейник чимганский 233784
Воронец 6405
Воронец азиатский 6414
Воронец волосистый 6448
Воронец заостренный 6408
Воронец колосистый 6448
Воронец колосовидный 6448
Воронец красноплодный 6422
Вороний глаз 284289
Вороний глаз маньчжуьрский 284348
Вороний глаз мутовчатый 284406
Вороний глаз неполный 284339
Вороний глаз обыкновнный 284387
Вороний глаз четырехлистный 284387
Вороний глаз шестилистный 284406
Вороника 145054
Вороника черная 145073
Вороника чёрная 145073
Ворониковые 145052
Вороновия 414772
вороновия красивая 414774
Воронья лапа 105338
Воронья лапа двойчатая 105340
Вороньянога 121281
Вороняжка 6405
Воронятник 284289
Воронятник гуандунский 284368
Воронятник далиский 284310
Воронятник сичоуский 284308
Ворсянка 133451
Ворсянка бесцветная 133513
Ворсянка возделываемая 133505
Ворсянка волосистая 133502
Ворсянка гмелина 133486
Ворсянка голубая 133461
Ворсянка далиская 133466
Ворсянка кандинская 133492
Ворсянка китайская 133463
Ворсянка лазолевая 133461
Ворсянка лесная 133508,133517
Ворсянка лицзянская 133495
Ворсянка лушаньская 133497
Ворсянка посевная 133478,133505
Ворсянка раздельная 133494
Ворсянка раздельнолистная 133494
Ворсянка разрезная 133494
Ворсянка симаоская 133510
Ворсянка синьцзянская 133519
Ворсянка сукновалов 133478
Ворсянка сукновальная 133478
Ворсянка щетинистая 133513
Ворсянка эмэйская 133457
Ворсянка японская 133490
Ворсянковые 133444
Восковник 83918,261120
Восковник альпийский 83919
Восковник большой 83923
Восковник красный 261212
Восковник обыкновенный 261162
Восковниковые 261243
Восковница 261120
Восковница восконосная 261137
Восковница желтая 261237
Восковница калифорнийская 261132
Восковница карликовая 261195
Восковница красная 261212
Восковница обыкновенная 261162
Восковница пенсильванская 261202
Воскоцветник 83918
Воскоцветник малый 83929
Вострец 11841,24236,144144,228337
Вострец алайский 24238
Вострец бальджуанский 24241
Вострец ветвистый 228380
Вострец гибкий 24249
Вострец каратавский 24251
Вострец копетдагский 228364
Вострец многостебельный 228370
Вострец овальный 228372
Вострец пабо 228374
Вострец подражающий 228339
Вострец пучковатый 24248
Вострец пушистоколосый 228384
Вострец растопыренный 24247
Вострец регеля 228356
Вострец скальный 24256
Вострец тяньшанский 228392
Вострец угамский 24267
Вострец узкий 144517,228343
Восьмитог 268768
Восьмитог леманновский 268769
Воякоплодник прижатый 133641
Вульпия 163778,412394
Вульпия альпийская 412404
Вульпия белкохвостая 412407
Вульпия волосистая 412495
Вульпия мышехвостник 412465
Вульпия персидская 412479
Вульпия ресничатая 412409
Вульфения 414829
Вшивица 286972
Выйнрута 341064
Выква большая столовая 114288
Выква китайская 114292
Выква кускатная 114292
Выква летняя 114300
Выква мускусная 114292
Выква овальная 114300
Выква оплодная 114288
Выква столовая 114300
Выква тарелочная 114300

Выква фестончатая 114300
Высокогорный кавказский клен 3699
Вьюн белый 68713
Вьюнец 94714
Вьюнковые 102892
Вьюнок 68671,102903
Вьюнок аммана 102926
Вьюнок ашхабадский 103206
Вьюнок волосистый 103080
Вьюнок горчакова 103065
Вьюнок дерезовидный 103120
Вьюнок жестковетвистый 103035
Вьюнок кальвера 102971
Вьюнок кантабрийский 102980
Вьюнок карьверта 102971
Вьюнок китайский 102933
Вьюнок колокольчатый 376536
Вьюнок королькова 103101
Вьюнок краузе 103102
Вьюнок крымский 103329
Вьюнок кустарниковый 103053
Вьюнок линейнолистный 103110
Вьюнок ложнокантабрийский 103206
Вьюнок ложносмолоносный 103209
Вьюнок михельсова 103139
Вьюнок ольги 103168
Вьюнок персидский 103188
Вьюнок полевой 102933
Вьюнок пурпурный 208120
Вьюнок пустыни 103072
Вьюнок растопыренный 103025
Вьюнок садовый 208120
Вьюнок седоватый 103319
Вьюнок смолоносный 68686,103255
Вьюнок стрелолистный 68686,103255
Вьюнок трагакантовый 103338
Вьюнок трёхцветный 103340
Вьюнок узколистный 103110
Вьюнок узколистый 103110
Вьюнок шелковистный 103084
Вьюнок шерковистоголовый 103276
Вьюнок шерстистый 103317
Вьюнок японский 103095
Вяжечка 400638
Вяжечка гладкая 400644
Вяжечка голая 400644
Вяз 401425,401431
Вяз американский 401431
Вяз андросова 401448
Вяз белый 401431
Вяз ветвистый 401618
Вяз гладкий 401549
Вяз голый 401512
Вяз горный 401512
Вяз граболистный 401468,401508,401593
Вяз граболистный роговидный 401485
Вяз густой 401499
Вяз каркасовый 401478
Вяз кистевидный 401618
Вяз крупноплодный 401556
Вяз крылатый 401430
Вяз крыловаточерешковый 401430
Вяз куэньминский 401480
Вяз листоватый 401468,401508,401593
Вяз лопастной 401542
Вяз лопастный 401542
Вяз мексиканский 401567
Вяз мелкоистный 401602
Вяз мелколистный 401581,401633
Вяз настоящий 401549
Вяз низкий 401602
Вяз обыкновенный 401549
Вяз перисто - ветвистый 401591
Вяз полевой 401468,401508,401593
Вяз приземистый 401602
Вяз пробковый 401475,401633
Вяз равнинный 401593
Вяз разрезной 401542
Вяз раскидистый 401501,401549
Вяз ржавый 401509
Вяз сродный 401490
Вяз томаса 401618
Вяз шероховатый 401512
Вяз шершавый 401512
Вяз эллиптический 401503
Вяз японский 401490
Вязель 105269
Вязель баланзы 105274
Вязель восточный 105300
Вязель горный 105299
Вязель завитой 105310
Вязель изменчивый 105324
Вязель критский 105277
Вязель мелкоцветкавый 105301
Вязель обыкновенный 105324
Вязель пестрый 105324
Вязель пёстрый 105324
Вязель разноцветный 105324
Вязель скутноцветный 105281
Вязель увенчанный 105276
Вязель широколистный 105292
Вязель эмеровый 105279,105281
Вязинник 344244
Вязовик 320065
Вязовик трёхлистный 320071
Вязовые 401415
Габлиция 184234
Габлиция тамусовидная 184235
Гавортия 186251
Гагения 184563
Гагения абиссинская 184565
Гадючий лук 260293
Гадючий лук бело - зевный 260318
Гадючий лук бледный 260326
Гадючий лук воронова 260341
Гадючий лук гроздевидный 260301
Гадючий лук длинноцветковый 260311
Гадючий лук кистевидный 260333
Гадючий лук многоцветковый 260330
Гадючий лук переменчивый 260307
Гадючий лук поникающий 260329
Гадючий лук пренебрегаемый 260324
Гадючий лук свободообразный 260313
Гадючий лук своеобразный 260327
Гадючий лук синый 260306
Гадючий лук туркевича 260340
Гадючий лук шовица 260338
Гадючий лук штейпа 260337
Гайллардия 169568,169588
Гайллардия красивая 169603
Гайллардия крупноцветная 169586
Гайллардия остистая 169586
Гайлоссакия 172248
Гайлярдия крупноцветная 169575
Гайнальдия 122958,186868
Гайнальдия мохнатаю 186874
Гакелия 184291
Гакелия повислоплодная 184298
Гакелия тимьянолистная 153546
Гала 295888
Галакс 169817
Галантус 169705
Галега 169919
Галега восточная 169940
Галега лекарственная 169935
Галезия 184726
Галезия каролинская 184731
Галения 184689,185903
Галения моховидная 185904
Галения рогатая 184691
Галения рожковая 184691
Галения сибирская 184691
Галения стеллера 78646
Галения эллиптическая 184696
Галеорхис 273317
Галеорхис круглогубый 169887,169906
Галимодендрон 184832
Галимокнемис 184810
Галимокнемис березина 184811
Галимокнемис войлочноцветковый 184816
Галимокнемис голый 184813
Галимокнемис длиннолистный 184818
Галимокнемис карелина 184815
Галимокнемис крупнопылыниковый 184819
Галимокнемис мохнатый 184829
Галимокнемис мягковолосый 184820
Галимокнемис смирнова 184827
Галимокнемис твердоплодный 184824
Галимокнемис широколистный 184817
Галинзога 170137
Галинзога щетинистая 170146

Галинсога 170137
Галинсога мелкоцветковая 170142
Галинсога мелкоцветная 170142
Галинсога реснитчатая 170146
Галинсога четырехлучевая 170146
Галипея 170164
Галипея лекарственная 45586,170167
Галлорагидовые 184993
Галлораговые 184993
Галогетон 184947
Галогетон паутинистый 184950
Галогетон скученный 184951
Галогетон тибетский 184951,184952
Галодуле 184938
Галотис 185057
Галотис волосистый 185059
Галохалис 184919
Галохалис войлочноцветковый 184923
Галохалис туркменский 184926
Галохалис щетинистоволосый 184922
Гальтония 170827
Гамамелиговые 185089
Гамамелидовые 185089
Гамамелиевые 185089
Гамамелис 185092
Гамамелис весенний 185138
Гамамелис виргинский 185141
Гамамелис китайский 185130
Гамамелис мягкий 185130
Гамамелис японский 185104
Гамбир 401757
Гамелия 185164
Гандум 398834
Ганус 299351
Ганьсуск ганьсуский 105738
Гарадиолюс 171475
Гарадиолюс летучконосный 171478
Гарадиолюс угловатый 171477
Гардения 171238
Гардения жасминовидная 171253
Гардения жасминовидная крупноцветковая 171333
Гардения обильноцветная 171253
Гардения укореняющаяся 171253,171374
Гардения хайнаньская 171301
Гармала 287975,287978
Гармала обыкновенная 287978
Гармала чернушкообразная 287983
Гарновка 398900
Гаррия 171545
Гарциния 171046
Гастерия 171607
Гастродия 171905
Гастродия высокая 171918
Гастродия мэнхайская 171943
Гастродия яванская 171939
Гастрокотиле 171903
Гастрокотиле шершавый 171904
Гастролобиум 171969
Гаудиния 172032
Гаудиния ломкая 172034
Гаультерия 172047
Гаультерия микели 172111
Гаультерия микеля 172111
Гаультерия пинбяньская 172104
Гаультерия тайваньская 172059
Гаультерия тибетская 172173
Гаура 172185
Гацания 172276
Гвайюла 285072,285076
Гвайюла серебристая 285076
Гварея 181544,181578
Гварея трёхтысячевидная 181578
Гваюла 285072,285076
Гваюла серебристая 285076
Гваяк 181505
Гваякум 181496
Гваякум аптечный 181505
Гваякум лекарственный 181505
Гвоздика 127573
Гвоздика акантолимоновидная 127577
Гвоздика альпийская 127581
Гвоздика амурская 127588
Гвоздика андржеевского 127592
Гвоздика андронаки 127591
Гвоздика армериевидная 127600
Гвоздика бахромчатолепестная 127697
Гвоздика бледноцветная 127789
Гвоздика борбаша 127616
Гвоздика бородатая 127607
Гвоздика бородчатая 127607
Гвоздика ветвистая 127808
Гвоздика восточная 127787
Гвоздика высокая 127707
Гвоздика гельцера 127733
Гвоздика голландская 127635
Гвоздика головчатая 127632
Гвоздика гроссгейма 127730
Гвоздика двуцветная 127612
Гвоздика джунгарская 127839
Гвоздика длинноноготковая 127763
Гвоздика душистая 127713
Гвоздика евгении 127708
Гвоздика жесткая 127811
Гвоздика закавказкаская 127889
Гвоздика зеаавшанская 127823
Гвоздика иглолистная 127578
Гвоздика изменчивая 127794
Гвоздика имеретинская 127737
Гвоздика каратавская 127743
Гвоздика картузианская 127634
Гвоздика киргизская 127744
Гвоздика китайская 127654
Гвоздика косматая 127693
Гвоздика красивоголовая 127629
Гвоздика кубанская 127748
Гвоздика кузнецова 127750
Гвоздика куринская 127699
Гвоздика кушакевича 127749
Гвоздика ланцетная 127752
Гвоздика ложноармериевидная 127800
Гвоздика луговая 127795
Гвоздика маршала биберштейна 127764
Гвоздика многостебельная 127778
Гвоздика многоцветковая 127712
Гвоздика низкая 127735
Гвоздика перенончатая 127768
Гвоздика перистая 127793
Гвоздика пестрая 127705
Гвоздика песчаная 127598,127793
Гвоздика ползучая 127809
Гвоздика преображенского 127796
Гвоздика пышная 127852
Гвоздика пятнистая 127731
Гвоздика равнинная 127630
Гвоздика радде 127807
Гвоздика разноцветная 127682
Гвоздика роговича 127812
Гвоздика рупрехта 127814
Гвоздика садовая 127635
Гвоздика седая 127631
Гвоздика семенова 127820
Гвоздика сизая 127623
Гвоздика сычуаньская 127880
Гвоздика скальная 127792
Гвоздика степная 127682
Гвоздика тайваньская 127872
Гвоздика талышская 127884
Гвоздика травяная 127700
Гвоздика травянка 127700
Гвоздика трехгранная 127700
Гвоздика турецкая 127607
Гвоздика туркестанская 127892
Гвоздика тяньшанская 127888
Гвоздика угольная 127633
Гвоздика узколепестная 127755
Гвоздика узкочашечная 127841
Гвоздика уральская 127893
Гвоздика фишера 127654
Гвоздика четырехчемуйная 127885
Гвоздика шаньдунская 127675
Гвоздика шилоносная 127849
Гвоздика японская 127739
Гвоздичнник 34488
Гвоздичнник мадагаскарский 10550
Гвоздичные 77980
Гёбелия лисохвостовидная 368962
Гевеа 194451
Гевелия 368955
Гевея 194451
Гевея бразильская 194453

Гегемона 188250
Гегемона лиловая 399518
Гегемона мелкоцветная 188255
Гедизарум 187753
гединоис 187709
гединоис критская 187718
гединоис персидская 187734
Гедихиум 187417
Гейлюссакия 172248
Гейлюссация 172248
Геймия 188263
Геймия иволистная 188265
Гейхера 194411
Гелениум 188393
Гелениум бигелова 188410
Гелиамфора 188554
Гелиантус 188902
Геликодисерос 189982
Геликония 189986
Геликтерес 190092
Геликтотрихон 190129
Гелиопсис 190502
Гелиопсис шероховатый 190525
Гелиотроп 190540
Гелиотроп аргузиевый 190562
Гелиотроп бухарский 190581
Гелиотроп волосистоплодный 190601, 190625
Гелиотроп двухкольцевидный 190577
Гелиотроп душистый 190746
Гелиотроп европейский 190622
Гелиотроп заамударьинский 190757
Гелиотроп зеравшанский 190730
Гелиотроп индийский 190651
Гелиотроп коваленского 190658
Гелиотроп крупны 190640
Гелиотроп литвинова 190666
Гелиотроп маленький 190692
Гелиотроп мелкоцветковый 190676
Гелиотроп мультифлорус 350312
Гелиотроп ольги 190683
Гелиотроп остроцветковый 190542
Гелиотроп перувианский 190702
Гелиотроп простертый 190753
Гелиотроп синьцзянский 190562
Гелиотроп тайваньский 190630
Гелиотроп туркменский 190764
Гелиотроп федчанко 190628
Гелиотроп хороссанский 190586
Гелиотроп шовица 190755
Гелиотроп эллиптический 190614
Гелиптерум 190791
Гелиптерум жёлтый 189147
Гелихризум 189093
Гелихризум болешой 415386
Геллебор 190925
Геллеборус 190925
Гелониас 191046
Гелониопсис 191059
Гелониопсис японский 191065
Гельдрейхия 188367
Гельдрейхия длиннолистная 188368
Гельземий 172778
Гельземиум 172778
Гельземиум изящный 172779
Гельминция 298550
гельминция румянковидная 191018
Геля 403687
Гемантус 184345
Гемантус гиблидный 184396
Гемантус многоцветковый 350312
Гемартрия 191225
Гемартрия пучковатая 191232
Гемартрия сирирская 191251
Гемартрия японская 191251
Гемерокаллис 191261, 191304
Гемлок 399860
Гемлок восточный 399867
Гемлок канадский 399867
Генекен 10854
Генипа 172888, 201647
Генипап 172890
Гения 176921
Гения прибрежная 176923
Генрардия 192027
Генрардия голочешуйная 192028
Генрардия персидская 192029
Генциана 173181
Георгина 121539, 121561
Георгина изменчивая 121561
Георгина перистая 121561
Георгиния 121539
Гераклеум 192227
Гераклеум юньнаньский 192403
Гераневые 174431
Гераниевые 174431
Герань 174440
Герань белоцветковая 174454
Герань бледная 174796
Герань блестящая 174711
Герань богемская 174503
Герань болотная 174799
Герань вильфорда 175006
Герань власова 175000, 175018
Герань волосистотычинковая 174589
Герань волосистоцветковая 174583
Герань восточнокавказская 174453
Герань голостебельная 174953
Герань голубиная 174551
Герань горная 174743
Герань грузинская 174660
Герань даурская 174559
Герань двулистная 174499
Герань душистая 288268
Герань забайкальская 174832
Герань зибольда 174684
Герань клубневая 174986
Герань комнатная 288594
Герань корейская 174682
Герань кочи 174683
Герань крамера 174684
Герань красно-бурая 174811
Герань кроваво-расная 174891
Герань кровянокрасная 174891
Герань круглолистная 174886
Герань крупнокорневищная 174715
Герань крымуская 174956
Герань лесная 174951
Герань линейнолопастная 174706
Герань ложносибирская 174846
Герань луговая 174829, 174832
Герань максимовича 174729
Герань маленькая 174854
Герань мелкотычинковая 174847
Герань миболда 174653
Герань млкая 174854
Герань мягкая 174739
Герань нежная 174739
Герань непалская 174755
Герань отпрысконосная 174919
Герань пиренейская 174856, 265137
Герань плоская 175023
Герань плосколепестная 174823
Герань побегоносная 174919
Герань поперечноклубневая 174972
Герань почтибессбельная 174945
Герань прямая 174865
Герань пурпурная 174852
Герань пушистотычинковая 174589
Герань раскидистая 174570
Герань рассечённая 174569
Герань растопыренная 174570
Герань ренарда 174872
Герань роберта 174877
Герань розовая 288268
Герань рупрехта 174889
Герань садовая 288594
Герань сибирская 174906
Герань скальная 174548
Герань софии 174924
Герань стройная 174634
Герань сходная 174836
Герань тёмная 174811
Герань тунберга 174966
Герань уилфорда 175006
Герань ферганская 174609
Герань франше 174619
Герань холмовая 174548
Герань чарльза 174530
Гербера 175107
Гербера джемсона 175172

Гертнериа 19140,169313
Геснериевые 175306
Геснерия 175295
Гесперис 193387
Гетерантера 193676
Гетерантериум 193669
Гетерантериум волосоносный 193671
Гетерация 193657
Гетерация шовица 193659
Гетеродерис 193794
Гетеродерис белоколовый 193796
Гетеродерис маленький 193797
Гетерокарий 193736
Гетерокарий грубый 193744
Гетерокарий еженосный 193739
Гетерокарий жесткий 193744
Гетерокарий крупноплодный 193742
Гетерокарий малошипиковый 193743
Гетерокарий оголенный 193741
Гетерокарий шовица 193745
Гетеропанакс 193897
Гетеропаппус 193914
Гетеропаппус алтайский 193918
Гетеропаппус альберта 193916
Гетеропаппус искривленный 193936
Гетеропаппус мейендорфа 193973
Гетеропаппус седеющий 193929
Гетеропаппус сосняковый 193939
Гетеропаппус средний 193970
Гетеропаппус татарский 193983
Гетеропаппус щетинисто – волосистый 193939
Гетеропаппус щетинистый 193939
Гетероспата высокая 194135
Геттарда 181728
Геум 175375
Геухера 194411
Гиалея 199620
Гиалея красивая 199624
Гиалея таджикская 199626
Гиалолена 199646
Гиалолена метельчатая 199650
Гиалолена обедненная 199648
Гиалолена яксартская 199649
Гиацинтик 199545
Гиацинтик беловатый 199546
Гиацинтик палласа 199547
Гиацит 199565
Гиацит водный 141808
Гиацит водяной 141808
Гиацит восточный 199583
Гиацит копетдагский 199577
Гиацит литвинова 199580
Гибертия 194661
Гибиск 194680
Гибискус 194680,195149
Гибискус индийский 194850,194936,195311
Гибискус кенаф 194779
Гибискус китайская роза 195149
Гибискус коноплевый 194779
Гибискус коноплёвый 194779
Гибискус липолистный 195311
Гибискус лушаньский 195089
Гибискус мотоский 194881
Гибискус мускусный 194684
Гибискус переменчивый 195040
Гибискус понтический 195111
Гибискус пройчатый 195326
Гибискус розан китайский 195149
Гибискус розелла 195196
Гибискус сабдарифа 219,195196
Гибискус северный 195326
Гибискус сирийский 195269,195270
Гибискус съедобный 219
Гибискус тайваньский 195305
Гибискус тройчаный 195326
Гибискус тройчиный 195326
Гибискус юньнаньский 195359
Гигнокарпус 199740
Гигнофитум 199759
Гигрофила 200589
Гидрангеа 199787
Гидрангеа китайская 199853
Гидрангеа тайваньская 200121
Гидрангеа юньнаньская 200157
Гидрангея древовидная 199806
Гидрангея метельчатая 200038
Гидрангиа 199787
Гидрангиа метельчатая 200038
Гидрангиа черешчатая 200059
Гидранговые 200160
Гидрастис 200173
Гидрастис канадский 200175
Гидрилла 200184
Гидрилла мутовчатая 200190
Гидроклеис 200241
Гидрокотиле 200253
Гидрокотиле веткоцветковая 200350
Гидрокотиле яванская 200312
Гидрокотиль 200253
Гидрофил 200482
Гидрофиллум 200482
Гидрофилляциевые 200476
Гизоция 181872
Гизоция абиссинская 181873
Гиккори 77874
Гиккори белая 77934
Гиккори белый 77934
Гиккори гладкий 77894
Гиккори горикий 77881
Гиккори горький 77889
Гиккори косматый 77934
Гиккори крупнопочечный 77920
Гиккори свиной 77894
Гиккори яйцевидный 77920
Гиления 175760
Гилия 175668
Гилления 175760
Гималайская высокая 300297
Гималайская скополия 25438
Гималайская скополия буро – желтая 25447
Гималайская скополия лицзянская 25450
Гименокаллис 200906
Гименократер 201038
Гименократер изящный 201040
Гименократер смолистый 201039
Гименолена 201096
Гименолена альпийская 201098
Гименолена бедренцелистная 201111
Гименолена малорослая 304817
Гименолима 201155
Гименолима володушковидная 199647
Гименолима волосистая 201157
Гимнастер 256301
Гимнастер саватье 41201
Гимнастер сычуаньский 256310
Гимнастер юаньцюйсий 182355
Гимнокладус 182524
Гимнокладус двудомный 182527
Гимнокладус китайский 182526,182527
Гимнокльдус 182524
Гимсофила 183191
Гинала 2984
Гинандропсис 95606
Гинериум 182941
Гинериум серебристый 105456,182942
Гини черешня 83155
Гинкго 175812
Гинкго двлопастной 175813
Гинкго двлопастный 175813
Гинкго двухлопастный 175813
Гинкго китайское 175813
Гинкговые 175839
Гиннала 2984
Гиностема 182998
Гиностема бирманская 183000
Гиностема пятилистная 183023
Гинура 108723,183051
Гинура ланьюйская 183086
Гинура непалская 183108
Гинура японская 183097
Гиохерис 202387
Гиохерис голый 202404
Гиохерис короткокорневой 202432
Гиохерис укореняющийся 202432
Гипаканциум 201436
Гипаканциум мордовниковолистный 201437
Гипеаструм 18862,196439
Гипекоум 201597
Гипекоум вислоплодный 201618

Гипекоум китайский 201612
Гипекоум крупноцветковый 201609
Гипекоум мелкоцветный 201616
Гипекоум прямой 201605
Гипекоум тонкоплодная 201612
Гипекоум трехлопастной 201631
Гипогомфия 202649
Гипогомфия высогая 202650
Гипогомфия туркестанская 202651
Гиппеаструм 18862,196439,196458
Гиппократия 196501
Гиппокрепис 196621
Гиппокрепис двуцветковый 196627
Гиппокрепис косматый 196630
Гиппокрепис однострючковый 196675
Гиппокрепис ресничатый 196629
Гиппомана 196716
Гипсолюбка 183157
Гипсолюбка метельчатая 183225
Гипсолюбка постенная 183221
Гипсолюбка пучковатая 183195
Гипсофила 183157
Гипсофила изящная 183191
Гиргенсония 175904
Гиргенсония супротивноцветковая 175907
Гирча 357780
Гирча культиасова 357802
Гирча попова 357814
Гирча тминолистная 357789
Гирча тяньшанская 357828
Гирчевник 101820
Гирчовник 101820
Гирчовник виктора 101840
Гирчовник влагалищный 101839
Гирчовник длиннолистный 101830
Гирчовник камчатский 101827
Гирчовник китайский 101823
Гирчовник перистолистный 101834
Гирчовник северный 101822
Гирчовник тайваньский 101831
Гирчовник широколистный 101829
Гиршфельдия 196950
Гиршфельдия серая 196952
Гис 45448
гиссарский 400175
Гиссон 203079
Гистрикс 203124
Гистрикс коморова 203136
Гистрикс сибириский 203142
Гитарник 93232
Гитарник весенний 93235
Гитарник леманна 93234
Гифене 202321
Гладиолус 175990,176222
Гладиолус генский 176222
гладкий 119073
Гладыш 222020
Гладыш альпийский 222022
Гладыш близный 222021
Гладыш стевена 222031
Гладыш широколистный 222028
Гладыш щетинистоволосистый 222027
Глаз вороний обыкновенный 284387
Глаз вороний четырелистный 284387
Глаз вороний четырёхлистный 284387
Глазки анютины 410677
Глазки павлиньи 8332
Глаукс 176772
Глаукс морской 176775
Глаукс приморский 176775
Глауциум 176724
Глауциум бахромчатый 176735
Глауциум желтый 176738
Глауциум жёлтый 176735
Глауциум изяшный 176734
Глауциум остролопастный 176748
Глауциум отогнутая 335650
Глауциум рогатый 176728
Глауциум чешуйчатый 176754
Гледичия 176860
Гледичия аргентинская 176862
Гледичия водяная 176864
Гледичия каспийская 176867
Гледичия китайская 176901
Гледичия колючая 176903
Гледичия крупноколючковая 176891
Гледичия мотоская 176893
Гледичия обыкновенная 176903
Гледичия сладкая 176903
Гледичия тайваньская 176870
Гледичия трехколючковая 176903
Гледичия трёхколючковая 176903
Гледичия трёхшипная 176903
Гледичия японская 176881
Гледичия японская делаве 176883
Гления 176921
Гления прибрежная 176923
Глехома 176812
Глехома хедерацеум 176824
Глинус 176941
Глинус лядвенецевидный 176949
Глиптостробус 178014
Глиптостробус повислый 178019
Глистник 35090,367120
Глистогон 346243
Глицерия водяная 177629
Глицина 177689
Глицина посевная 177750
Глицина соя 177777
Глицина тайваньская 177751
Глициния 414541
Глициния китайская 414576
Глициния обыкновенная 414554
Глициния японская 414554,414556,414572
Глоговина 369541
Глоксиния 177515,364737,364754
Глориоза 177230
Глот 380499
Глотуха 280003
Глохидион 177096
Глохидион форчуна 177170
Глухая крапива 220346
Гляуциум 176724
Гляуциум бахромчатый 129568
Гляуциум гладкоплодный 176743
Гляуциум желтый 176738
Гляуциум крупно－прицветниковый 176727
Гляуциум крупноцветковый 176739
Гляуциум примерный 176740
Гляуциум рогатый 176728
Гнафалиум 178056
Гнездовка 264650
Гнездовка азиатская 264651
Гнездовка камчатская 264665
Гнездовка настоящая 264709
Гнездовка сосочковая 264718
Гнездовка сосочконосная 264718
Гнездовка уссурийская 264750
Гнетовые 146270,178522
Гнетум 178524
Гнетум гнемон 178535
Говения 198766,198769
Говения сладкая 198769
Говения тонковойлочная 198786
Говения цюцзянская 198768
Гогенакерия 197137
Гогенакерия бесстебельная 197139
Годенак керовский 376798
Годеция 94132
Гойяба 318742
Головчатка 82113
Головчатка албпийская 82116
Головчатка армянская 82118
Головчатка бархатистая 82186
Головчатка волосистая 82165
Головчатка высокая 82166
Головчатка гигантская 82139
Головчатка гроссгейма 82141
Головчатка дмитрия 82133
Головчатка известняковая 82129
Головчатка кожистая 82130
Головчатка короткочешуйная 82126
Головчатка кочи 82145
Головчатка литвинова 82151
Головчатка мелкозубчатая 82161
Головчатка меловая 82131
Головчатка остистая 82117
Головчатка сирийская 82173
Головчатка средная 82160
Головчатка татарская 82151

Головчатка трансильванская 82182
Головчатка уральская 82183
Головчатка чиначева 82181
Головчатка шерстистая 82172
Головчатый тисс 82496
Головчатый тисс костянковый 82499
Головчатый тисс форчуна 82507
Голорепа 59541
Голосеменные 182628
Голосемянные 182628
Голосхенус 197468,353170,353179
Голосхенус обыкновенный 353174
Голосхенус римский 197469
Голпвчатотиссвые 82493
Голубика 404027
Голубика американская 403780
Голубика зонтичнокистевая 403780
Голубика канадская 403758
Голубика обыкновенная 404027
Голубки 30081
Голубой очный цвет 21340
Гольдбахия 178814
Гольдбахия бородавчатая 178835
Гольдбахия гладкая 178821
Гольдбахия пупырчатая 178829
Гольдбахия сетчатая 178831
Гомалиум 197633
Гомбо 219
Гоми 361794,361935
Гомфия 70712
Гомфокарпус 178974
Гомфокарпус кустарниковый 179032
Гомфрена 179221,312406
Гомфрена головчатая 179236
Гомфрена шаровидная 179236
Гониолимон 179368,230508
Гониолимон бессера 179369
Гониолимон высокий 179373
Гониолимон джунгарский 179372
Гониолимон злаколистный 179375
Гониолимон красивопучковый 179370
Гониолимон красивый 179380
Гониолимон красноватый 179378
Гониолимон остроконечный 179371
Гониолимон превосходный 179374
Гониолимон прямоветвистый 179374
Гониолимон северцова 179379
Гониолимон татарский 179384
Гонобобель 404027
Гоноболь 404027
Гончаровия 179515
Гончаровия попова 179516
Гораниновия 198237
Гораниновия улексовидная 198238
Гордо обыкновенная 407907
Гордовина 407907
Гордония 179729
Горец 308690
Горец альпийский 291663,308729
Горец амманиевидный 308737
Горец амударьинский 309500
Горец араратский 308780
Горец ашерсона 308803
Горец аянский 308717
Горец близкий 308788
Горец болотный 309796
Горец борнмюллера 308924
Горец бунге 308922
Горец бурораструбовый 308850
Горец бухарский 309015
Горец вейриха 309966
Горец вогнутоветвистый 309252
Горец водянойперец 309199
Горец восточный 309494
Горец выступающий 309098
Горец вьюнковый 162519
Горец вьюшийся 162519
Горец гиссарский 309186
Горец горный 291663
Горец горовчатый 308946
Горец двухостный 308885
Горец дерновинный 309624
Горец джунгарский 309808
Горец длинноцетинковый 309345
Горец длинноцетинный 309345
Горец дубильный 309015
Горец жесткоцетинистый 309831
Горец живородящий 309954
Горец жилковатый 309935
Горец запутанный 309258
Горец зейский 309978
Горец земноводный 308739
Горец зибольда 309796
Горец злаколистный 309157
Горец змеиный 54798,308893
Горец зубчатокрылатый 162525
Горец зубчатокрылый 162525
Горец игольчатый 308695
Горец иловатый 309338
Горец имеретинский 309247
Горец каспийский 308952
Горец кисленький 308696
Горец кислый 308697
Горец китайбела 309281
Горец клейкий 309946,309951
Горец кленносный 309946
Горец красивый 309466
Горец красильный 309893
Горец крылатый 309459
Горец кустарниковый 162519,162528
Горец лаксманна 309479
Горец ленкоранский 309329
Горец лисохвостовидный 308726
Горец лицзянский 309335
Горец льняной 309298
Горец маака 309356
Горец малоцветковый 162530
Горец малый 309395
Горец маньчжурский 309374
Горец мелковистый 309777
Горец меловой 309017
Горец многоистый 309121
Горец многоцветковыи 162542
Горец многоцветковый 309119
Горец моллиеобразный 309421
Горец морской 309376
Горец мякокрасный 308950
Горец незамечаемый 308832
Горец непальский 309459
Горец нитевидный 26624
Горец обыкновенный 309570
Горец ожинколистный 309354
Горец опадающий 309038
Горец осенний 8332
Горец остроконечный 162532,328345
Горец остроплодный 309503
Горец отклоненный 309531
Горец охотский 309478
Горец памироалайский 309515
Горец перечный 309199
Горец песчаный 308782
Горец подземковидный 309607
Горец подушкообразный 309650
Горец полушастый 309837
Горец посумбу 309624
Горец почечуйный 309570
Горец прибрежный 309343
Горец приноготковидный 309526
Горец промежуточный 309253
Горец пронзеннолистный 309564
Горец птичий 308816
Горец пышный 309353
Горец раздельноцветковый 309054
Горец разнолистный 308816
Горец распростертый 309192
Горец расставленноцветковый 309054
Горец растопыренный 309056
Горец раструбистый 309479
Горец регеля 309690
Горец рея 309503
Горец роберта 309697
Горец рыже-желтоватый 308850
Горец самарский 309749
Горец самшитолистный 308925
Горец сахалинский 309723
Горец серебристый 308791
Горец сибирский 309792
Горец сидниковидный 309271
Горец синьцзянский 309766
Горец скальный 309001
Горец сквоснолистный 309564

Горец спорыш 308788
Горец спорышевидный 309016
Горец стрговидный 309711
Горец стрелолистный 291932
Горец темнорасторубовый 308850
Горец тернистый 309772
Горец терновый 309772
Горец тибетский 309891
Горец тимьянолистный 309889
Горец тифлисский 309892
Горец тихоокеанский 309505
Горец тонкий 309531
Горец тонкозаостренный 328345
Горец топотун 308788
Горец треугольноплодный 309909
Горец трехкрылоплодный 309911
Горец тунберга 309877
Горец узколистный 308777
Горец узловатый 309468
Горец укороченный 308692
Горец уссурийский 309926
Горец хвощевидный 309088
Горец хэнаньский 309187
Горец черничный 309446
Горец шероховатый 291818,309298
Горец шугнанский 309767
Горец щавелелистный 114503,309298
Горец щетинистоплодный 309909
Горец щетинистый 309789
Горец эллиптический 309078
Горец янаты 309261
Горец японский 162532,328345
Горечавка 173181
Горечавка арктическая 173251
Горечавка баварская 282,173287
Горечавка балтийская 173283
Горечавка бело－черная 173580
Горечавка биберштейна 173293
Горечавка бирманская 173309
Горечавка болотная 173698
Горечавка бородатая 174205,174227
Горечавка бородатая китайская 174208
Горечавка валуева 174066
Горечавка введенского 174061
Горечавка весенняя 174049
Горечавка вильчатая 173402
Горечавка водяная 173242,173743
Горечавка вырезанная 173184
Горечавка германская 173473
Горечавка горькая 173598
Горечавка гроссгейма 173491
Горечавка гуансиская 173569
Горечавка далиская 173956
Горечавка даурская 173373
Горечавка деши 173383
Горечавка джемся 173540
Горечавка джимильская 173408
Горечавка долуханова 173410
Горечавка желтая 173610,174221
Горечавка желтеющая 174120
Горечавка золотистая 174120
Горечавка зонтичная 174031
Горечавка кавказская 173329
Горечавка карелина 173747
Горечавка карпатская 173325
Горечавка кауфмана 173552
Горечавка китайская 173336
Горечавка клузия 173344
Горечавка клюзия 173344
Горечавка колаковского 173559
Горечавка комарова 173560,174128
Горечавка краснолистная 173720
Горечавка крестовидная 173364
Горечавка крестообразная 173364
Горечавка крупнолистная 173615
Горечавка крупноцветковая 173486
Горечавка крылова 173561
Горечавка курильская 173540
Горечавка куэньминская 173412
Горечавка лазолевая 173728,174120
Горечавка лазоревая 174120
Горечавка лазурная 174120
Горечавка ластовневая 173264
Горечавка ластовнная 173264
Горечавка легочая 173728
Горечавка лежачая 173387
Горечавка липского 173600
Горечавка ложноводяная 173751
Горечавка макино 173624
Горечавка маленькая 174147
Горечавка марковича 173627
Горечавка мешочная 174034
Горечавка мотоская 173655
Горечавка мыльная 173840
Горечавка нежная 100284,173960
Горечавка немецкая 173473
Горечавка ниппонская 173662
Горечавка оверина 173695
Горечавка оливье 173679
Горечавка ольги 173677
Горечавка особенная 173705
Горечавка осторая 174103
Горечавка острая 174103
Горечавка пазушная 173280
Горечавка пазушноцветковая 174007
Горечавка памирская 173700
Горечавка перекрастолистная 173364
Горечавка пиренейская 173785
Горечавка полевая 173318,174122
Горечавка поникающая 173670
Горечавка понноская 173701
Горечавка понтийская 173730
Горечавка похожная 173742
Горечавка прибрежная 173821
Горечавка приподнимающаяся 173387
Горечавка простая 173726
Горечавка простертая 173743
Горечавка пурпурная 173781
Горечавка раздеренная 173571
Горечавка раскидистая 173743
Горечавка распухлая 100278
Горечавка растопыренная 173917
Горечавка ресниценосная 173297
Горечавка ресничатая 173342
Горечавка романцова 173825
Горечавка семираздельная 173886
Горечавка серповидная 100271,173430
Горечавка сибирская 173893
Горечавка сизая 173478
Горечавка синие колокольчики 173728
Горечавка синяя 173728
Горечавка сычуаньская 173948
Горечавка скороспелая 173732
Горечавка снежная 173666
Горечавка соколий перелет 173364
Горечавка студеная 173461
Горечавка тайваньская 173379
Горечавка тибетская 173988
Горечавка тоненькая 173960
Горечавка тонкая 173960
Горечавка топяная 174164
Горечавка точечная 173776
Горечавка трехцветковая 174004
Горечавка туркестанцев 174163
Горечавка тяньшанская 173982
Горечавка угроватая 173225
Горечавка ушконосная 174119
Горечавка фетисова 173617
Горечавка фишера 173448
Горечавка франше 173459
Горечавка холодная 173198
Горечавка холодостойкая 173470
Горечавка цзилунская 173493
Горечавка цзиндунская 173547
Горечавка цоллингера 174095
Горечавка цюцзянская 173786
Горечавка черноватая 173270
Горечавка чжундяньская 173341
Горечавка шероховатая 173847
Горечавка эмэйская 173680
Горечавка юньнаньская 174088
Горечавка язычковая 173598
Горечавка японская 173662
Горечавковые 174100
Горечник 292758,292843,293032,298550
Горечник ады 292765
Горечник влагалищный 293059
Горечник гиссарский 292888
Горечник днепровский 292787
Горечник заилийский 383235
Горечник зедельмейеровский 293081

Горечник известняковый 292794
Горечник кавкажский 292810
Горечник крупнолистный 292929
Горечник ленарда 292992
Горечник любименко 292916
Горечник малолучевой 292969
Горечник малоолиственный 292968
Горечник многоцветковый 292977
Горечник оленелистный 292814
Горечник опушенный 292987
Горечник палимбиевидный 292965
Горечник подольский 292975
Горечник пышный 292918
Горечник седоватая 298563
Горечник солончаковый 293003
Горечник твертая 298636
Горечник туркменский 293052
Горечник ястребинковый 298589
Горицвет 8324,11938,363141
Горицвет весенний 8387
Горицвет кожистый 363370
Горицвет короналиевый 363370
Горицвет кукушкин цвет 363461
Горицвет летний 8325
Горицвет огненый 8362
Горицвет осенний 8332
Горицвет черногорка 8387
Горицвет – кукушкин цвет 363461
Горичник 292758,293033
Горичник австрийский 292779
Горичник аптечный 292957
Горичник байкальский 292780
Горичник болотный 292966
Горичник горный 292961
Горичник гуансиский 292873
Горичник изящный 292843
Горичник крымский 293030
Горичник лекарственный 292957
Горичник морисона 292937
Горичник настурциевый 205535
Горичник олений 292813,293063
Горичник песчный 292774
Горичник резаковый 292850
Горичник русский 293002
Горичник терпентинный 293033
Горичник формозский 292857
Горичник царский корень 205535
Горичник широколистный 292908
Горичник шотта 293007
Горичник эльзасский 292766
Горичник японский 292892
Горлец 54798,308893
Горлюха 298550
Горлюха бедноцветковая 298633
Горлюха жестковолосая 298643
Горлюха камчатская 298602
Горлюха корейская 298625
Горлюха лицзянская 298627
Горлюха малоцветковая 298633
Горлюха похожая 298640
Горлюха похожий 298640
Горлюха румяниковидная 298584
Горлюха румянковидная 298584
Горлюха щетинистая 298643
Горлюха японская 298618
Горлюха ястребинковая 298589
Горлюха ястребинковидная 298589
Горлянка 219821,219843
Горлянка обыкновенная 219843
Горная петрушка 292961
Горноколосник 275347
Горноколосник бахромчатый 275363
Горноколосник колючий 275394
Горноколосник мягколистный 275379
Горноколосник пирамидальный 275396
Горноколосник хрящеватый 275356
Горноколосник шаньдунский 275367
Горноколосник щитковый 275396
Горовик 83466
Горовик униоловидный 61008
Горох 301053
Горох бараний 90801
Горох высокий 301074
Горох голубиный 65146
Горох закавказский 301084
Горох земляной 409069
Горох кормовой 409086
Горох коровий 409086
Горох красивый 301063
Горох мышиный 408352
Горох обыкновенный 301061,301070
Горох огородный 408571
Горох оше 301060
Горох полевой 301055,408571
Горох посевной 301061,301070
Горох рриземистый 301076
Горох садовый 301070
Горох спаржевый 237771
Горох турецкий 90801
Горошек 408251
Горошек амурская 407677
Горошек амурский 408278
Горошек армянский 408421
Горошек байкальский 408302
Горошек баланзы 408304
Горошек безусиковый 408383
Горошек биберштейна 408313
Горошек буасье 408326
Горошек венгерский 408527
Горошек вестрый 408540
Горошек вифинский 408322
Горошек волосистый 408423,408541
Горошек гиблидный 408429
Горошек гирканский 408431
Горошек горный 408255
Горошек гороховидный 408543
Горошек горький 408392
Горошек гроссгейма 408420
Горошек грузинский 408432
Горошек гулинский 408449
Горошек далматский 408370
Горошек двлетний 408314
Горошек древний 408295
Горошек душистый 222789
Горошек желтый 408469
Горошек жёлтый 408469
Горошек жилковый 408672
Горошек заборный 408611
Горошек забослевый 175295
Горошек зарослевый 408381
Горошек зеленый 408571
Горошек зелёный 301070
Горошек зилковатый 408686
Горошек зубчатолистный 408612
Горошек изменчивый 408670
Горошек изящный 408387,408418
Горошек иноземный 408537
Горошек иранский 408434
Горошек кассубский 408337
Горошек кашубский 408337
Горошек китайбеля 408447
Горошек коканский 408448
Горошек кормовой 408571
Горошек круглоплодный 408274
Горошек крупнолистный 408479
Горошек крупнолодочковый 408484
Горошек крупносемянный 408483
Горошек крупноцветковый 408419
Горошек кустарниковый 408381
Горошек лесной 408614,408620
Горошек лжесочевичниковый 408551
Горошек лжесочевниковый 408551
Горошек лиловый 408466
Горошек малопарный 408535
Горошек мейера 408489
Горошек мишо 408490
Горошек многостебельный 408502
Горошек мохнатый 408693
Горошек мышиный 408352
Горошек надрезанный 408433
Горошек надрезный 408433
Горошек нарбонский 408505
Горошек обарта 408325
Горошек обрубленный 408646
Горошек однопарный 408648
Горошек одноцветковый 408493
Горошек ольвийский 408517
Горошек опушенный 408541
Горошек оранжевый 408352
Горошек паннонский 408527
Горошек парноцветковый 408406

Горошек пестроцветный 408540
Горошек пестрый 408671
Горошек песчаный 408693
Горошек плотноволосистый 408427
Горошек подзаборный 408611
Горошек полевой 408571
Горошек полосатый 408618
Горошек полуволосистый 408619
Горошек полуголый 408610
Горошек помесный 408429
Горошек посевной 408571
Горошек приятный 408262
Горошек пушистый 408559
Горошек разветленный 408563
Горошек раскрашенный 408540
Горошек ребристый 408350
Горошек ресничатый 408343
Горошек семенова 408609
Горошек сердцевиднолистный 408348
Горошек сердцевидный 408348
Горошек сорный 408288
Горошек сосновского 408616
Горошек тибетский 408641
Горошек тонкий 408417
Горошек тонколистный 408624
Горошек трехостроконечный 408642
Горошек узколистный 408284
Горошек укороченный 408252
Горошек федченко 408401
Горошек чаюйский 408303
Горошек чёткообразный 408392
Горошек четырехсемянный 408638
Горошек четырёхсемянный 408638
Горошек чиннобразный 408452
Горошек чиновидный 408452
Горошек чужеземный 408537
Горошек шерстистоплодный 408371, 408693
Горошек шершавоволосый 408423
Горошек шпорцевый 408328
Горошек щебнистый 408605
Горошек японский 408361,408435
Гортензия 199787
Гортензия бретшнейдера 199846
Гортензия гуандунская 199936
Гортензия гуансиская 199931
Гортензия древовидная 199806
Гортензия золотистожилковая 200148
Гортензия крупнолистная 199956,345064
Гортензия личуаньская 199939
Гортензия метельчатая 200038
Гортензия пильчатая 200085
Гортензия пильчатая бело - розовая 200093
Гортензия разнонаправленная 199891
Гортензия черешковая 200059
Гортензия шансиская 200103
Горчак 6272,6278,298550,309199
Горчак белоцвтеный 368962
Горчак желтый 298550
Горчак желтый ястребинковый 298589
Горчак ползучий 6278
Горчак ядовитый 6278
Горчак ястребинковидный 298589
Горчинца черная 59515
Горчица 59278,364544
Горчица английская 364545
Горчица белая 364545
Горчица белая английская 364545
Горчица дикая 154019,364557
Горчица морская 65164
Горчица полевая 364557
Горчица сарептская 59438
Горчица цёрная 59515
Горчица шпинатная 59438
Горький полевая 364557
Горький померанец 93332
Горькуша 348089,348104
Горькуша альпийская 348121
Горькуша бльзкая 348672
Горькуша горькая 348129
Горькуша горькая мелкоцветковая 348135
Горькуша пильчатая 348778
Горькуша скученная 348129
Горькуша солончаковая 348763
Горькуша тургайская 348889
Горюн - трава 129615
Горяниновия 198237
Горяниновия исключительная 198241
Горяниновия малая 198243
Горяниновия неправильная 198240
Горянка 146953
Горянка крупночашелистиковая 147023
Горянка японская 147021
Госта 198589
Госта ланцетолистная 198624
Госта японсая 198618
Гоуришка 249232
Гоу - ришка 249197
Граб 77252
Граб американский 77263
Граб восточный 77359
Граб гуйцзоуский 77315
Граб гуншаньский 77421
Граб дэциньский 77397
Граб европейский 77256
Граб кавказский 77271
Граб каролинский 77263
Граб корейский 77283
Граб мелколистный 77345
Граб обыкновенный 77256
Граб приморский 77276
Граб разреженноцветущий 77323
Граб сердцевидный 2837,77276
Граб сичоуский 77275
Граб хубэйский 77306
Граб чанхуаский 77401
Граб черный 77359
Граб эмэйский 77358
Граб японский 77312
Грабельки 102833,124141
Грабельник цикутолистный 153767
Грабельники 153711
Грабинник 77359
Гравилат 175375
Гравилат алеппский 175378
Гравилат городской 175454
Гравилат кровавокрасный 175400
Гравилат крупнолистный 175426
Гравилат обыкновенный 175454
Гравилат прибрежный 175444
Гравилат промежуточный 175378
Гравилат прямой 175378,175446
Гравилат речной 175444
Гравилат сахалинский 175431
Гравилат средний 175378
Гравилат фори 175410
Гравилат широколопастной 175425
Гравилат японский 175417
Гравник 77359
Граммосциадиум 180144
Граммосциадиум морковевидный 180145
Гранадилла 285637
Гранадилла гигантская 285694
Гранат 321763,321764
Гранат обыкновенный 321764
Гранатник 321763,321764
Гранатовые 321777
Гребенник 118307
Гребенчук 383412
Гребенщик 383412
Гребенщик андросова 383431
Гребенщик африканский 383414
Гребенщик вытянутый 383479
Гребенщик гогенакера 383519
Гребенщик изящный 383516
Гребенщик карелина 383526
Гребенщик китайский 383469
Гребенщик королькова 383530
Гребенщик мейера 383557
Гребенщик многоветвистый 383595
Гребенщик можжевеловый 383444
Гребенщик одесский 383595
Гребенщик опушённый 383517
Гребенщик пятитычинковый 166761, 383591
Гребенщик раскидистый 383535
Гребенщик рыхлый 383535
Гребенщик тонкоколосый 383542
Гребенщик удлиненный 383479
Гребенщик французский 383491
Гребенщик четырехтычинковый 383581

Гребенщик четырёхтычинковый 383581
Гребенщик щетинистоволосистый 383517
Гребенщик щетинистоволосый 383517
Гребенщик эверсманна 383484
Гребенщиковые 383397
Гребень венерин 350425
Гребень петуший 80395
Гребешок петуший 80378,80395
Гребневик 118307
Гребневик гребенчатый 118315
Гребневик изящный 118325
Гребневик обыкновенный 118315
Гребневик шиповадый 118319
Гребник 118307
Гревиллея 180561
Гревиллея банкса 180570
Гревиллея крепкая 180642
Гревиллея сильная 180642
Гревия 180665
Гревия гуандунская 180835
Гревия двлопастная 180700
Гревия двлопастная мелколистная 180702
Гревия инцзянская 181023
Гревия липолистная 180995
Гревия мелкоцветковая 180703
Гревия юньнаньская 181024
Грейпфрут 93579,93690
Грелльсия 180046
Грелльсия камнеломколистная 180050
Гренадилла 285607
Гренадилья 285637
Гренадилья густая 312086
Гренландия 181285
Гренландия густая 181286
Греча 162301
Гречих 308690
Гречих сахалинская 309723
Гречиха 162301,308481
Гречиха вьюнковая 162519
Гречиха горлец 54798,308893
Гречиха живородящая 309954
Гречиха земноводная 308739
Гречиха корейская 309284
Гречиха красильная 309893
Гречиха крылатая 309459
Гречиха культурная 162312
Гречиха перечная 309199
Гречиха полевая 162519
Гречиха полукустарная 162334
Гречиха посевная 162312
Гречиха придорожная 308816
Гречиха птичья 308816,309570
Гречиха развесистая 309298,309468
Гречиха сахалинская 309723
Гречиха съедобная 162312
Гречиха татарская 162335
Гречиха узловатая 309298,309468
Гречиха щавелелистная 309298
Гречиха – горлец 308893
Гречишка 308690
Гречишник 308690
Гречишник бальджуанский 162509
Гречишные 308481
Гречка пальчатая 285422
Гречка расширенная 285423
Гриб подберёзовый 3127
Гриб чёрный 3127
Гриб чесночный 129618
Грибы сморчковые 77276
Гривохвост 85054
Гривохвост шандровый 85058
Гринделия 181092
Гринделия высогая 181161
Гринделия исполинская 181161
Гринделия растопыренная 181174
Гроб – трава 409339
Гроздевидный помпельмус 93690
Громовик 271727
Просвирняк дикий 243840
Гроссгей 181317
Гроссгей крупноголовая 181319
Гроссгей осетинская 181321
Грудника 837,1000,362473
Грудника авиценны 1000
Грудника колючая 362662
Грудница 39910,231818
Грудница мохнатая 231842
Грудница обыкновенная 231844
Грудница тайваньская 40474
Грудница татарка 41342
Грудница татарская 231838
Грудница фомина 231828
Грудничник 1000
Грудничник авиценны 1000
Груша 257374,323076
Груша аллигаторова 291494
Груша баланзы 323107
Груша берёзолистная 323110
Груша буассье 323112
Груша гроссгейма 323185
Груша дикая 323133
Груша дынная 367400
Груша зангезурская 323352
Груша земляная 189073
Груша иволистная 323281,323283
Груша кавказская 323129
Груша кальри 323116
Груша китайскя 323268
Груша китайская песчаная 323268
Груша коржинского 323214
Груша лесная 323133
Груша лохолистная 323165
Груша миндалевидная 323086
Груша обыкновенная 323133
Груша пашия 323251
Груша песочная 323268
Груша пильчатая 323294
Груша радде 323274
Груша разнолистая 323276
Груша регеля 323276
Груша синьцзянская 323306
Груша сирийская 323311
Груша снежная 323246
Груша сосновского 323307
Груша среднеазиатская 323095
Груша тайваньская 323207
Груша темноплодная 323259
Груша туркменская 323329
Груша уссурийская 323330
Груша хэбэйская 323194
Грушанка 322775
Грушанка алтайская 322810
Грушанка далиская 322826
Грушанка даурская 322814
Грушанка зеленоватая 322804
Грушанка зеленоцветковая 322804
Грушанка копытнелистная 322787
Грушанка красная 322790
Грушанка круглолистная 322872
Грушанка крупноцветковая 322830
Грушанка малая 322853
Грушанка однобокая 275486
Грушанка почковидная 322870
Грушанка почколистная 322870
Грушанка синьцзянская 322932
Грушанка сычуаньская 322919
Грушанка средняя 322852
Грушанка тибетская 322802
Грушанка японская 322842
Грушанковые 322935
Грыжник 192926
Грыжник волосистый 192975
Грыжник гладкий 192965
Грыжник голый 192965
Грыжник кавказский 192940
Грыжник многобрачный 193009
Грыжник седоватый 192981
Грюнколь 59532
Гуава 318733
Гуайава 318733
Гуайюла 285076
Гуайява 318733
Гуарана 285930
Гуаюла 285076
Гуаява 318733
Гуаяк 181496
Губастик 255184
Губастик желтый 255218
Губастик крапчатый 255210
Губастик мускусный 255223
Губастик отполысковый 255243

Губастик разинутый　255237
Губастик сычуаньский　255244
Губастик слабоволосистый　255235
Губастик темнокрасный　255197
Губастик тибетский　255256
Губастик тоненький　255246
Губка растительная　238257,238261
Губка – люффа　238257
Губовник линейный　87449
Губоцветные　218697
Гуван　42883
Гувен　42883
Гугония　199137
Гудайера　179571
Гудайера ланьюйская　179721
Гудайера максимовича　179631,179653
Гудайера ползучая　179685
Гудайера тяньцюаньская　179716
Гудайера шелехтендаля　179694
Гудайера юньнаньская　179722
Гуза　179900
Гуйава　318733
Гуйява　318733
Гули – харбуза　274927
Гультемия　199240
Гультемия барбарисолистая　336849
Гультемия персидская　199243
Гулявник　109857,365386
Гулявник аптечный　365564
Гулявник вислоплодный　365482
Гулявник волжский　365649
Гулявник волосистый　365490
Гулявник восточный　365568
Гулявник высокий　365398,365457
Гулявник высочайший　365398
Гулявник густой　365433
Гулявник дагестанский　365442
Гулявник двулонастной　365412
Гулявник желтый　365532
Гулявник изменчивый　365585
Гулявник ирио　365503
Гулявник исфаринский　365513
Гулявник капустовидный　365416
Гулявник лёзелиев　365529
Гулявник лезеля　365529
Гулявник лекарственный　365564
Гулявник липского　365528
Гулявник сжатый　365621
Гулявник слабоколючий　365624
Гулявник софьи　126127
Гулявник струговидный　365606
Гулявник струйчатый　126127
Гулявник тоненький　365625
Гулявник туркменский　365643
Гулявник фуцзяньский　365469
Гулявник ядовитый　317344
Гулявник японский　365515
Гумай　369652
Гумбанг мягкий　14544
Гуммигутовые　97266
Гунделия　181929
Гунделия турнефора　181930
Гунера　181946
Гунера чилийская　181958
Гуньба　397208
Гусатница　308816
Гусиный лук　169375
Гусиный лук алексеенко　169380
Гусиный лук алтайский　169383
Гусиный лук альберта　169379
Гусиный лук афганский　169376
Гусиный лук берга　169389
Гусиный лук влагалищный　169470
Гусиный лук волосолистный　169394
Гусиный лук выемчатый　169431
Гусиный лук гельдрейха　169444
Гусиный лук гиенский　169446
Гусиный лук гиссарский　169447
Гусиный лук гранателли　169441
Гусиный лук длинностебельный　169459
Гусиный лук дудчатый　169421
Гусиный лук елены　169445
Гусиный лук желтый　169460
Гусиный лук жермены　169433
Гусиный лук жесткий　169498
Гусиный лук зеленовато – желтый　169398
Гусиный лук зелнистый　169443
Гусиный лук злаколистный　169438
Гусиный лук измененный　169400
Гусиный лук илийский　169450
Гусиный лук калье　169393
Гусиный лук капю　169395
Гусиный лук клубненостый　169392
Гусиный лук копетдагский　169454
Гусиный лук коржинского　169455
Гусиный лук косой　169514
Гусиный лук коха　169396
Гусиный лук краснеющий　169416
Гусиный лук крымский　169509
Гусиный лук ледниковый　169434
Гусиный лук ложнокрасиноватый　169420
Гусиный лук ложносетчатый　169488
Гусиный лук луговой　169485
Гусиный лук луковиценосный　169392
Гусиный лук маленький　169420
Гусиный лук малоцветковый　169480
Гусиный лук малый　169420
Гусиный лук мелкоцветковый　169467
Гусиный лук мешконосный　169499
Гусиный лук небольшой　169479
Гусиный лук нежный　169510
Гусиный лук ненецкий　169501
Гусиный лук неравнолучевой　169385
Гусиный лук неясный　169401
Гусиный лук низкий　169490
Гусиный лук нитевидный　169420
Гусиный лук ольги　169475
Гусиный лук памирский　169451
Гусиный лук пачоского　169477
Гусиный лук перавноцветковй　169384
Гусиный лук персидский　169432
Гусиный лук покрывалом　169503
Гусиный лук покрывальцевый　169503
Гусиный лук полевой　169388
Гусиный лук поперецный　169514
Гусиный лук попова　169483
Гусиный лук предвиденный　169480
Гусиный лук растопыренный　169409
Гусиный лук серножелтый　169507
Гусиный лук сетчатый　169497
Гусиный лук синьцзянский　169470
Гусиный лук сомнительный　169410
Гусиный лук стебельчатый　169505
Гусиный лук тонколистный　169511
Гусиный лук тончайший　169512
Гусиный лук трехгранный　169519
Гусиный лук удивительный　169468
Гусиный лук украинский　169520
Гусиный лук федченко　169417
Гусиный лук ханы　169397
Гусиный лук хомутовой　169399
Гусиный лук шовица　169508
Гусиный лук яйцевидный　169476
Гусиный лук японский　169452
Гусиный лук яшке　169451
Гуттаперчевое дерево　156041
Гуттаперчевые　346460
Гуттуиния　198743
Гуттуиния сердцевидная　198745
Гуцмания　182149
Гюльденштедтия　181626
Гюльденштедтия ганьсуская　181649
Гюльденштедтия гуансиская　181661
Гюльденштедтия мелкоцветковая　181693
Гюльденштедтия однолистная　181673
Гюльденштедтия ранняя　181693
Гюльденштедтия юньнаньская　391548
Давидия　123244
Давидия обернутая　123245
Давидия обернутая вильморена　123247
Дагусса　143522
Дазилирион　122898
Дазилирион вилера　122912
Дазилирион волосистоконечный　122899
Дазистома　45165
Дакридиум　121084
Далешампия　121912
Далия　121539,121561
Далматская ромашка　322660
Дальбергия　121607
Дальбергия бирманская　121645,121784

Дальбергия индийская 121730
Дальбергия куэньминская 121860
Дальбергия мимозовидная 121760
Дальбергия сенегальская 121750
Дальбергия сиссоо 121826
Дальбергия сиссу 121826
Дальбергия хайнаньская 121702
Дальбергия хубейская 121714
Дальбергия чёрная 211228
Дальбергия чернодревесная 121750
Дальбергия широколистная 121730
Дальбергия юньнаньская 121705,121859
Даммара 10492
Дамнакант 122027,122034
Дамнакант гуансиский 122036
Дамнакант индийский 122040
Дамнакант сычуаньский 122077
Дамнакант хайнаньский 122037
Даная ветвистая 122121,122124
Дантония 122180
Дантония промежуточная 122242
Дантония форскаля 122227
Дантония чашечная 122200
Дарлингтония 122758
Дармина 360825
Датиска 123027
Датиска коноплевая 123028
Датиска коноплёвая 123028
Датисковые 123033
Датура 123036
Дафна 122359
Дафна ганьсуская 122614
Дафна генква 122438
Дафна далиская 122432
Дафна душистая 122532
Дафна душистая ‘маргината’ 122534
Дафна душистая отороченная 122534
Дафна корейская 122486
Дафна мийябе 122519
Дафна остролопастная 122361
Дафна понтийская 122585
Дафна сичоуская 122634
Дафна цзиньюньская 122472
Дафна эмэйская 122424
Дафна юньнаньская 122635
Дафнифиллум 122661
Дафнифиллум крупноножковый 122700
Дафнифиллум ланьюйский 122682
Дафнифиллюм 122661
Дафнифиллюм крупноножковый 122700
Дафнифиллюм юньнаньский 122730
Двойчатка 54237,130029
Двойчатка оше 130030
Двойчатка федченко 130031
Двоякоплодник 133640
Двсемянник 91499,199480
Двсемянник скальный 198483
Двугнездка 225279
Двугнёздка 73085,225279
Двугнездка крупноплодная 201208
Двугнездка пушистая 201212
Двузернянка 398890
Двузернянка дикая 398886
Двузернянка культурная 398890
Двукильник 130956
Двукильник солончаковый 130965
Двукисточник 293705
Двукисточник тростниковидный 293709
Двукисточник тростниковидный житовник 293709
Двукисточник тростниковый 293709
двукрайник бунге 54425
Двукрылоплодник 133552
Двулепестник 91499
Двулепестник альпийский 91508
Двулепестник горный 91508
Двулепестник мягкий 91575
Двулепестник обыкновенный 91557
Двулепестник парижский 91557
Двулепестник сердцелистный 91537
Двулепестник средний 91553
Двулепестник четырехбороздчатый 91524
Двулистник 132677
Двупестфчник 134918
Двупестфчник кистевидный 134946
Двурядка 133226
Двурядка меловая 133256
Двурядка прутяная 133324
Двурядка стенная 133286
Двурядка тонколистная 133313,133394
Двухкрылоплодник 133552
Двучленник 128023
Двучленник льнолистный 128026
Двучленник пузырчатый 128028
Дебрежазия 123316
Дебрежазия съедобная 123322
Девица в зелени 266215
Девичий виноград 285097
Девичий виноград пятилисточковый 285136
Девичий виноград садовый 285127
Девичий виноград трехостроконечный 285157
Девичий виноград триостренный 285157
Девясил 207025
Девясил бесстебельный 207026
Девясил блоший 207087
Девясил большой 207132,207135
Девясил британский 207046
Девясил великолепный 207173
Девясил волосистый 207140
Девясил восточный 207193
Девясил высокий 207135
Девясил германский 207117
Девясил глазковый 207189
Девясил громбчевского 207134
Девясил жестковолосый 207140
Девясил зейдлитца 207228
Девясил ивовый 207219
Девясил иволистный 207219
Девясил каспийский 207075
Девясил китамуры 207222
Девясил корнеглавый 207209
Девясил коровяковый 207241
Девясил крупноцветковый 207131
Девясил крупночешуйчатый 207170
Девясил льнянколистный 207165
Девясил марии 207175
Девясил мечелистный 207105
Девясил многостебельный 207182
Девясил монбре 207181
Девясил низбегающий 207099
Девясил обыкновенный 207135,207259
Девясил песчаный 207218,207240
Девясил растопыренный 207087
Девясил рубцова 207216
Девясил сизый 207122
Девясил солянковидные 207224
Девясил тибетский 207113
Девясил ушковатый 207041
Девясил христово око 207189
Девясил хубэйский 207145
Девясил шероховатый 207038
Девясил шершавый 207140
Девясил шмальгаузена 207226
Девясил японский 207151
Девятиостник 145756
Девятиостник персидский 145776
Девятиостник северный 145766
Дедюлэ 3732
Дейнболлия 123661
Дейция 126832
Дейция амурская 126841
Дейция великолепная 127005
Дейция вильсона 127124
Дейция ганьсуская 126839
Дейция гладкая 126933
Дейция городчатая 126891
Дейция зибольда 127099
Дейция изящная 126946
Дейция корейская 126875
Дейция красивая 126946
Дейция крупноцветковая 126956
Дейция лемана 126989
Дейция лицзянская 126943
Дейция лэйбоская 126987
Дейция максимовича 127012
Дейция мелкоцветковая 127032
Дейция мулиская 127018
Дейция наньчуаньская 127022
Дейция сизоватая 126976

Дейция сычуаньская 127093
Дейция скученноцветковая 126940
Дейция тайваньская 127114
Дейция тибетская 126971
Дейция чжундяньская 127128
Дейция чжэньканская 126844
Дейция шероховатая 127072
Дейция шершавая 127072
Дейция шнейдера 127088
Дейция эмэйская 127044
Дейция юньнаньская 127126
Декенея 123367
Декенея фаржа 123371
Делет 124008
Дельфиниум 102833,124008,124301
Дельфиниум аякса 102833
Дельфиниум гибридный 367370
Дельфиниум далиский 124618
Дельфиниум кандинский 124630
Дельфиниум крупноцветный 124237
Дельфиниум полевой 102833
Дельфиниум тяньшанский 124643
Дельфиниум хэнаньский 124294
Дельфиниум цзилунский 124616
Дельфиниум цинхайский 124549
Дельфиниум чжундяньский 124708
Дельфиниум юньнаньский 124711
Демьянка 125802,367370
Денгамия 124957
Дендрантема 124767
Дендрантема арктическая 89424
Дендрантема вейриха 124862
Дендрантема выемчатолистная 124856
Дендрантема завадского 124862
Дендрантема индийская 124790
Дендрантема краснеющая 124778
Дендрантема курильская 124804
Дендрантема максимовича 124824
Дендрантема монгольская 124825
Дендрантема нактонгенская 124832
Дендрантема приморская 124822
Дендрантема сихоте-алинская 124778
Дендрантема хультена 124788
Дендрантема цельнолистная 199238
Дендрантема шелковицелистная 124785, 124826
Дендробиум 61398,124957
Дендробиум благородный 125279
Дендробиум букетоцветный 125387
Дендробиум золотистый 125058
Дендробиум золотоцветный 125048
Дендробиум лоддигеза 125233
Дендробиум мускустый 125264
Дендробиум мэнхайский 125360
Дендробиум сичоуский 125419
Дендробиум тайваньский 125279
Дендробиум тэнчунский 125412
Дендробиум фаленопсис 125304
Дендробиум хайнаньский 125173
Дендробиум хэнаньский 125179
Дендрокаламус 125461
Дендропанакс 125589
Дендропанакс баотинский 125627
Дендропанакс бирманский 125596
Дендропанакс гуансиский 125623
Дендропанакс трехнадрезный 125644
Дендропанакс хайнаньский 125611
Дендропанакс юньнаньский 125647
Дендростеллера 125763,128023
Дендростеллера колоссобразная 125772
Дендростеллера крупноколосая 125770
Дендростеллера линейнолистная 125769
Дендростеллера ольги 125771
Дендростеллера песчаная 125767
Дендростеллера туркменов 125773
Дендрохилюм 125526
Дендрохилюм пленчатый 125531
Денежка широкопластинчатая 390213
Денежник 390213
Дербенка 240025
Дербенник 240025,396828
Дербенник züализняк 240068
Дербенник иволистный 240047,240068
Дербенник иссополистный 240047,240100
Дербенник карликовый 240059
Дербенник комарова 240054
Дербенник ленецевидный 240089
Дербенник ленолистниковый 240089
Дербенник лозный 240068,240100
Дербенник низкий 240059
Дербенник плакун 240100
Дербенник промежуточный 240050
Дербенник прутовидный 240100
Дербенник смолевковидный 240087
Дербенник тимьянолистный 240090
Дербенник трехприцветниковый 240094
Дербенник федора 240088
Дербенник шелковникова 240086
Дербенниковые 240020
Дербенник-плакн 240068
Дербянка 55202
Дербянка колосистая 71218
Дерганец 3913
Деревей 3913,411189
Дерево абриковое 34477
Дерево абрикосовое 34475
Дерево абурское бархатное 294231
Дерево авраамово 38322
Дерево австралийское черное 1384
Дерево азиминовое 29972
Дерево айвовое 116546
Дерево алойное 380528
Дерево амарантовое 232565
Дерево амбровое 294231
Дерево амурское пробковое 294231
Дерево анакардовое 21191,204575
Дерево анисовое 93765,204603
Дерево апельсиновое 88663,93765
Дерево атласное 88663
Дерево атласное индийское 181505
Дерево бакаутовое 181496,181505
Дерево бакаутовое железное 181505
Дерево бальзамвое 261549,261554,261561
Дерево бальзамическое африканское 101290
Дерево бальзамическое толуйское 261554, 261561
Дерево бальзамное толу 261554,261561
Дерево бальзовое 268344
Дерево бархатновое 379312
Дерево бархатное 260198,294231
Дерево бензойное 231297
Дерево бисерное 383412
Дерево божье 35088
Дерево бразильское палисандровое 211228
Дерево бумажное 61107
Дерево винтовое 280968
Дерево восковое 86427,198827,261120, 346408,393479
Дерево восковое китайское 346408
Дерево габонское тюльпанное 370358
Дерево гваяковое 181496,181505
Дерево гвоздичное 382477,382582
Дерево голянковое обыкновенное 111104
Дерево горлянковое 111100,111104
Дерево горшечное 223764
Дерево гранатовое 321764
Дерево грушевое 257374,323133
Дерево гуншаньское 36919
Дерево гуттаперчевое 156041,280441
Дерево даммаровое 362186,362238
Дерево джатовое 385531
Дерево дождевое 345525
Дерево драконовое 137382
Дерево дымчатое 107300
Дерево дынное 76808,76813
Дерево дынное обыкновенное 231884
Дерево железное 80698,181505,211029, 252370,284961,362902
Дерево жёлтое 88514,107300
Дерево жизненное 302721,385262,390567, 390587
Дерево жизни 385262,390567,390587
Дерево збеновое 132137
Дерево земляничное 30877
Дерево земляничное крупноплодное 30888
Дерево злаковое 393491
Дерево золотое 44887,44915
Дерево зубное 417157
Дерево индийское розовое 320301
Дерево иудейское 83809
Дерево иудино 83809

Дерево каепутовое 248072
Дерево кайюпутовое 248072
Дерево какаовое 389399,389404
Дерево калабасовое 300153
Дерево калабассовое 111100
Дерево калифорнийское мамонтовое 360565
Дерево камедное 80569
Дерево каменное 184518
Дерево кампешевое 91287,184518
Дерево камфарное 91287,138548
Дерево камфарное зондское 138548
Дерево капоковое 80120
Дерево капустное 22218
Дерево карандашное 213841,213984
Дерево касторовое африканское 334411
Дерево каучуковое 164925,194453
Дерево каштановое 78811
Дерево каяпутовое 248072,248104
Дерево квассиевое 93492
Дерево квебраховое 323561,350973
дерево квитовое 116532
Дерево кедровое 80074
Дерево кентуккское кофейное 182527
Дерево китайское восковое 346408
Дерево китайское сальное 346408
Дерево кожевенное 332533
Дерево кокаиноное 155067
Дерево колбасное 216313
Дерево конфетное 198766,198769
Дерево коралловое 7190,154643
Дерево коричное 91252,91446
Дерево коричное китайское 91302
Дерево коричное цейлонское 91446
Дерево корнелиевое 105023
Дерево коровье 61057
Дерево кофейное 98844,98857
Дерево кофейное арабское 98857
Дерево кофейное аравийское 98857
Дерево кофейное либерийское 98941
Дерево крабовое 72640
Дерево красное 64975,88663,360565,369541,380522,380528
Дерево красное бразилийское 64975
Дерево красное фернамбфковое 65002
Дерево кротоновое 112829,113039
Дерево кружевное 219986
Дерево кфичное 91446
Дерево лавровишневое 223116
Дерево лавровое 223203
Дерево лаковое 14544,393491
Дерево лаковое японское 393479,393491
Дерево лжеапельсиновое 240828
Дерево лимонное 93546
Дерево линонное 93539
Дерево липоное 391646
Дерево мамонтовое 360576
Дерево манговое 244386,244397
Дерево мангровое 329761
Дерево масляное 14538,48753,64203
Дерево масляное африканское 289476
Дерево масляное китайское 406018
Дерево мастиковое 300994
Дерево мастичное 300994
Дерево махагониевое 380528
Дерево мескитовое 127607
Дерево миндальное 20890
Дерево миробалановое 386504
Дерево мировое 101285
Дерево молочное 61057
Дерево молочное индийское 382852
Дерево москитовое 315551,315560
Дерево мускатное 261402,261424
Дерево мускатное душистое 261424
Дерево мыльное 217610,324624,346323
Дерево мыльное антильское 346351
Дерево мыльное обыкновенное 346351
Дерево мыльное сенегальское 225678
Дерево непалское 300462
Дерево обезьянье хлебное 7022
Дерево огуречное 241962
Дерево оливковое 270099
Дерево оливковые 270058,270099
Дерево ореховое 212636
Дерево орлеанское 54862
Дерево палисандровое 121770,211225
Дерево палисандровое сиссоо 121826
Дерево париковое 107297,107300
Дерево пеонное 280286
Дерево перечное 300323,300464
Дерево пернамбуковое 65002
Дерево персиковое 20935
Дерево померанецевое 93332
Дерево пробковое 294230
Дерево пробковое амурское 294231
Дерево пробковое антильское 195311
Дерево пузырное 100153
Дерево путешественников 327095
Дерево рожковое 83527
Дерево розовое 21005,67846,381965,385355
Дерево розовое индийское 121730
Дерево саламандоровое 28316
Дерево сальное 289476,346408
Дерево сандаловое 184518,320327,346208,346210
Дерево сандаловое антильское 320291
Дерево сандальное 184518,320327
Дерево сандараковое 386942
Дерево санталовое 346208
Дерево саппаоовое 65060
Дерево свечное 14544,346408
Дерево серебряное 227233
Дерево скипидарное 301002
Дерево снежное 87694,87736
Дерево соговое 115884
Дерево стираксовое 232562,379300
Дерево стираксовое лекарственное 379419
Дерево стручковое 83527
Дерево сырное 80120
Дерево терпентинное 301002
Дерево терпентиновое 301002
Дерево тиковое 385301,385529,385531
Дерево томатное 119974
Дерево травяное 415172
Дерево тунговое 14538,406018
Дерево тутовое 259065,259067
Дерево тюльпанное 232602,232609
Дерево тюльпанное виргинское 232609
Дерево тюльпанное китайское 232603
Дерево уксусное 332916
Дерево фернамбуковое 64975,65002
Дерево фиалковое 1292
Дерево фиговое 164763
Дерево физетовое 107300
Дерево фисташковое 301005
Дерево хинное 91055,91090
Дерево хлебное 36902,36913,36920
Дерево хлебное восточноиндийское 36920,36924
Дерево хлебное джек 36920,36924
Дерево хлебное индиское 36920,36924
Дерево хлебное лакооха 36928
Дерево хлебное разнолистное 36920,36924
Дерево хлебное цельнолистное 36920,36924
Дерево хлочатое 80120
Дерево хренное 258945
Дерево чайное 69634
Дерево чайное парагвайское 204135
Дерево черное 132137
Дерево чёрное 1384,311292
Дерево чёрное австралийское 1384
Дерево шелковичное 259065,259067
Дерево шёлковое 417230
Дерево шелстяное 80120
Дерево шоколадное 389399,389404
Дерево эбеновое 311292
Дерево – какао 389404
Деревосоговое 115794
Дереза 72233,238996
Дереза берберов 239011
Дереза волосистотычинковая 239030
Дереза изогнутая 239049
Дереза китайская 239021
Дереза копетдагская 239068
Дереза лебедолистная 239055
Дереза обыкновенная 239055
Дереза русская 239106
Дереза туркменская 239128
Дерен 85579,104917
Дерён 104917

Дерён американский 105193
Дерен амом 104949
Дерен бейли 104963
Дерен белолистный 380488
Дерен белый 104920
Дерён белый 104920
Дерен головчатый 124891
Дерен канадский 85581
Дерен кистевидный 105169
Дерен китайский 104994
Дерен корейский 380444
Дерен кравано – красный 380499
Дерен красная 380499
Дерён красный 380499
Дерён кроваво – красный 380499
Дерен крупнолистный 380461
Дерен лекарственный 105146
Дерен мелкоцветковый 380484
Дерен мощный 105111
Дерён мужской 105111
Дерен настоящий 105111
Дерён обыкновенный 105193
Дерён отпрысковый 105111
Дерен отпрысконосный 105193
Дерен спорный 57695
Дерён съедобный 105023
Дерен уолтера 380514
Дерён флолидский 105023
Дерен хемсли 380451
Дерён цветточный 105023
Дерен цветущий 105023
Дерен шведский 105204
Дерён шведский 105204
Дерен южный 104960
Деренкоуса 124925
Деренкоуса китайский 124925
Деренкоуса японский 124923
Деренные 104880
Дерённые 104880
Дерёновые 280559
Держи – дерево 280539,280544,280559
Держи – дерево многоветвистое 280555
Держи – дерево обыкновенное 280559
Держи – дерево хемсли 280548
Держи – дерево христовы терний 280559
Деррис 125937
Деррис далиский 125970
Деррис индийский 125974
Деррис хайнаньский 283285
Деррис эллиптический 125958
Деррис юньнаньский 126012
Деряба 139852
Дерябка 221711
Дескурайния 126112
Дескурения 126112
Дескурения гулявниковая 126131
Дескурения коха 126118
Дескурения софиевидная 126131
Дескурения софии 126127
Дескурения софия 126127
Десмацеря 126191
Десмацеря сжатая 126194
Десмодий 126242
Десмодий канадский 126278
Десмодий колосковый 126329
Десмодий липолистный 126329
Десмодиум 126242
Десмодиум движущийся 98159
Десмодиум маньчжурский 126453
Десмодиум мелколистный 126465
Десмодиум обманчивый 200740
Десмодиум ольдама 200736
Десмодиум сычуаньский 200745
Десмодиум юньнаньский 126675
Десмодиум яньюаньский 126338
Джамбоза 259186,382582
Джамболан 382522
Джантак 14638
Джауджамыр 163604
Джек 36920,36924
Джек – дерево 36920,36924
Джексония 211273
Джекфрут 36920,36924
Джемсия 211550
Джемсия американская 211551
Джефферсония 212296,301692
Джефферсония сомнительная 301694
Джидда 141929,141932
Джидда вэньшаньская 142244
Джидда сичоуская 142248
Джидда тибетская 142250
Джидда юньнаньская 142253
Джонсонова трава 369652
Джонушка 247457
Джугара 369601,417417
Джузгун 66988
Джузгун безлистый 66998
Джут 104056
Джут африканский 194779
Джут длиннoплодный 104072,104103
Джут короткоплодный 104072,104103
Джут круглоплодный 104072
Джут огородный длинноплодный 104072
Джут яванский 194779
Дзелква 301801,417534
Дзелква граболистная 417540
Дзелква китайская 417563
Дзелква пиличатая 417558
Дзелква тайваньская 417547
Дзелква шнейдера 417552
Дзелква японская 417558
Дзелкова 301801,417534
Дзельква 417534
Дзельква пиличатая 417558
Дзян – доу 409086
Диалиум 127375
Диалиум гвинейский 127396
Дианелла 127504
Диантус барбатус 127607
Диантус китайский 127654
Диапензиевые 127926
Диапензия 127910
Диапензия лапландская 127916
Диапензия обратнояйцевидная 127919
Диапенсиевые 127926
Диапенсия 127910
Диапенсия лапландская 127916
Диапенсия тибетская 127925
Диаррена 128011
Диаррена маньчжурская 128019
Диаррена фори 128015
Диаррена японская 128017
Диартрон 128023
Диартрон льнолистный 128026
Диасция 128029,128032
Дивала 353986
Дивала крючковатая 353997
Дивала многоплодная 353996
Дивало 353986
Дивало многолетнее 353994
Дивало однолетнее 353987
Дивина 405694
Дигиталис 130344,130383
Дигиталис жилковатая 130375
Дигиталис мелкоцветковая 130374
Дигиталис пурпурный 130383
Дигиталис шишкина 130389
Диделотия африканская 129732
Дидискус 393882
Диелитра 128288
Диерама 130187
Диервила 130242
Диервилла 130242
Дизоксилон 139637
Дизофилла 139544
Дизофилла симаоская 139593
Дизофилла ятабе 139568,139602
Дикая рожа 18173
Дикая рябина 383874
Дикая рябинка 383874
Дикий виноград 285097
Дикий виноград триостренный 285157
Дикий перец 143657
Дикия 139254
Дикориния 129489
Дикориния параензс 129489
Дикритра прекрасная 220729
Диктамнус 129615,129618
Диктамнус белый 129618
Диктамнус ясенелистный 129618
Диления 130913

Дилления 130913
Димерия 131006
Димерия гуансиская 131014
Димерия птиценогая 131018
Димерия тайваньская 131009
Диморфант 30604,215420,215442
Диморфотека 131147
Дионея 131397
Дионизия 131418
Дионисия 131418
Дионисия гиссарская 131420
Дионисия косинского 131422
Дионисия оберткова 131421
Дионисия подушковидная 131425
Диония 131397
Диоскорейные 131918
Диоскорея 131451
Диоскорея батат 131772
Диоскорея инцзянская 131909
Диоскорея кавкаская 131515
Диоскорея клубненосная 131501
Диоскорея крыратая 131458
Диоскорея ланьюйская 131538
Диоскорея мексиканская 131710
Диоскорея мелкоцветковая 131846
Диоскорея многокистевая 131772
Диоскорея ниппонская 131734
Диоскорея поликолосная 131772
Диоскорея разноцветная 131556
Диоскорея съедобная 131577
Диоскорея тибетская 131913
Диоскорея юньнаньская 131915
Диоскостигма 134181
Диосма 131938
Диоспирос 132030
Диптерикс 133626
Диптерокарповые 133551
Диптерокарпус 133552
Диптерокарпус бальзамический 133588
Диптерокарпус шишковатый 133587
Дирка 133654
Дирка болотная 133656
Диспорум 134417
Диспорум зеленеющий 134504
Диспорум зеленоватый 134486,134504
Диспорум наньтоуский 134453
Диспорум сидячий 134467
Диспорум смилациновый 134486
Диспорум тайваньский 134441
Дистилиум 134918
Дистилиум кистевидный 134946
Дистилиум китайский 134922
Дистилиум пинбяньский 134942
Дисхидия 134027
Дисхидия китайская 134032
Дисхидия тайваньская 134037
Дитрум лиловый 240100
Дифиллейя 132677
Дифиллейя грея 132685
Дифиллейя китайская 132685
Дифиллейя японская 132682
Диффнбахия 130083
Диффнбахия пестрая 130121
Дихапеталиум 128589
Дихапеталум 128589
Дихостилис 129010,353179
Дихостилис карликовый 129023
Дихостилис крючковатый 129018
Дихостилис микели 119196,129019
Дихостилис хайнаньский 353448
Дихотомант 129025
Дихотомант тристаниеплодный 129026
Дихроа 129030
Дихроа мелкоцветковая 129039
Дихроа противолихорадочная 129033
Дихроцефала 129062
Дихроцефала двуцветная 129068
Дихроя 129030
Дихроя противолихорадочная 129033
Дихроя юньнаньская 129051
Дицентра 128288
Дицентра бродяжная 128310
Дицентра великолепная 220729
Дицентра зимная 220729
Дицентра канадская 128291
Дицентра китайская 220729
Дичка 243698
Диэритра 128288
Днгиль 30932
Додартия 135132
Додартия восточная 135134
Додарция 135132
Додарция восточная 135134
Додекатеон 135153
Додонея 135192
Додонея аптечная 135215
Додонея клейкая 135215
Дождь золотой 218754
Дождь золотой альпийский 85403,218757
Дождь золотой обыкновенный 218761
Долгонот 241248
Долгонот крылатосемянный 241250
Долгонот снеговой 241249
Долгоцветка 11194
Долихос 135399
Долихос китайский 409086
Долихос лицзянский 135412
Долихос лябляб 218721
Долихос обыкновенный 218721
Доллингерия 135253
Доллингерия шершавая 135264
Домбея 135753
Домбея точечная 135957
Донгузтопылак 119503
Донник 249197,261585
Донник аптечный 249232
Донник ароманный 249232
Донник аселтый 249232
Донник белый 249203
Донник бороздчатый 249251
Донник волжский 249263
Донник волосистый 249217
Донник высокий 249205
Донник голубой 397208
Донник душистый 249232
Донник желтый 249197,249232
Донник жёлтый 249232
Донник зубчатый 249212
Донник индийский 249218
Донник крымский 249260
Донник лекарственный 249232
Донник мелкоцветковый 249218
Донник неаполитанский 249228
Донник польский 249240
Донник рослый 249205
Донник синий 397208
Донтостемон 136156
Донтостемон гребенчатый 131141
Донтостемон зубчатый 136163
Донтостемон мелкоцветковый 136179
Донтостемон многолетний 136184
Донтостемон целенолистный 136173
Донтостемон шершавый 136172
Допатриум 136209
Допатриум ситниковый 136214
Дорема 136257
Дорема аммиачная 136259
Дорема аммониакальная 136259
Дорема гирканская 136263
Дорема голая 136261
Дорема инееватая 136267
Дорема камеденосная 136259
Дорема каратавская 136264
Дорема мелкоплодная 136265
Дорема наманганская 136266
Дорема песчаная 136268
Дорема смолоносная 136262
Дорема эчисона 136258
Дориантес 136729
Дорикниум 136740
Дорикниум средний 136746
Дорикниум травянистый 136746
Дороникум 136333
Дороникум австрийский 136344
Дороникум алтайский 136339
Дороникум баргузинский 136345
Дороникум восточный 136364
Дороникум ганьсуский 136355
Дороникум делеклюза 136348
Дороникум длиннолистный 136360
Дороникум кавказский 136346

Дороникум крупнолистный 136361
Дороникум продолговатолистный 136362
Дороникум тибетский 136376
Дороникум туркестанский 136379
Дороникум тяньшанский 136378
Дороникум шишкина 136373
Дороникум ядовитый 136365
Дорстения 136417
Дохн 289116
Дочиния 135113
Дочиния дераве 135114
Дочиния индийская 135120
Драконник 137382
Драконово дерево 137382
Дракоцефалюм 137545
Дракоцефалюм мелкоцветковое 137611
Дракункул 137744
Драцена 137330
Драцена австралийская 104339
Драцена астелия 39874
Драцена годсефа 137409
Драцена грукера 137425
Драцена деремская 137367
Драцена драковован 137382
Драцена душистая 137397
Драцена мэнласкан 137451
Драцена сандера 137487
Драцена узколистная 137337
Драцена южная 104339,104367
Драч 280559
Древогубец 80129
Древогубец круглолистный 80260
Древогубец лазящий 80309
Древогубец лёснера 80296
Древогубец округлый 80260
Древогубец плетевидный 80189
Древогубец росторна 80295
Древокорень 329744
Древокорень мангле 329761
Дрема 248235
Дрема акинфиева 248246
Дрема ароматная 364083
Дрема астраханская 248284
Дрема байкальская 364072
Дрема баланзы 248291
Дрема безлепестная 248255
Дрема белая 248248,363673
Дрёма белая 363673
Дрёма беловатая 363673
Дрема буассы 248294
Дрема железистая 248239
Дрема коротколепестнаю 248298
Дрема краснеющая 248317
Дрема крепкая 363451
Дрема лесная 248427
Дрема липкая 248453
Дрёма липкая 363583
Дрёма луговая 363673
Дрема непалская 363800
Дрема нинсяская 363813
Дрема ночецветная 363815
Дрема ночная 248382,363815
Дрёма ночная 363815
Дрема овальнолистная 248393
Дрема ольги 248390
Дрема пустынная 248269
Дрема родственная 248241
Дрема сахалинская 248413
Дрема скальная 364012
Дрема сочавы 248419
Дрема стройная 248334
Дрема таймырская 248429
Дрема тонкая 248435
Дрема траурная 248440
Дрема тусклая 248423
Дрема узкоцветковая 248252
Дрема федченко 248318
Дрема ферганская 248319
Дрема четырехлопастная 363950
Дремлик 147106
Дремлик безлистный 147323
Дремлик болотный 147195
Дремлик мелколистный 147183
Дремлик ржавый 147222
Дремлик ройля 147218
Дремлик сосочковый 147196
Дремлик тунберга 147238
Дремлик чемерицелистный 147121
Дремлик широкористный 147149
Дремлик юньнаньский 147258
Дриада 138445,138461
Дриада аянская 138460
Дриада бальшая 138451
Дриада восьмилепестковая 138445
Дриада восьмилепестная 138460,138461
Дриада кавказская 138447
Дриада клейкая 138466
Дриада мелкогородчатая 138449
Дриада острозубчатая 138461
Дриада точечная 138462
Дриада чоноски 138460
Дриада шамиссо 138448
Дриадоцвет 138408
Дриадоцвет четырехтычиночный 138411
Дримария 138470
Дримис 138051
Дримис винтера 138058
Дриобаланопс 138547
Дриобаланопс душистый 138548
Дрок 172898
Дрок английский 172905
Дрок армянский 172913
Дрок артвинский 172914
Дрок безлистный 370202
Дрок беловатый 172903
Дрок волосистый 173030
Дрок германский 172976
Дрок голый 172977
Дрок донской 173077
Дрок закавказкаский 173089
Дрок испанский 370191,370202
Дрок испанский обыкновенный 370202
Дрок колючий европейский 203789,401388
Дрок красильный 173082
Дрок лежачий 172957
Дрок отклоненный 173028
Дрок плетевидный 172972
Дрок плотный 172945
Дрок прижатый 172957
Дрок распростертый 172984
Дрок сванетский 173074
Дрок узколистный 172908
Дрок четырехгранный 173081
Дрякба 115931
Дряква абхазская 115932
Дряква весенняя 115942
Дряква европейская 135153
Дряква изящная 115948
Дряква косская 115942
Дряква лджарская 115932
Дряква мелковетковая 115964
Дряква персидская 115949
Дряква понтийская 115970
Дряква черкесская 115940
Дрякvenник 115965,135153
Дрякbенник холодный 135161
Дуб 135153,323580
Дуб австлийский 323745
Дуб американский красильный 324539
Дуб американский шарлаховый 323780
Дуб бархатистый 324539
Дуб бархатный 324539
Дуб белый 323625
Дуб болотный 324262
Дуб бургундский 323745
Дуб вечнозелёный 324027
Дуб виргинский 324544
Дуб воронова 324558
Дуб галловый 324050
Дуб гамбелла 323927
Дуб гартвиса 323996
Дуб гонтовый 324041
Дуб горный 324201
Дуб греческий 324146
Дуб грузинский 324025
Дуб гуншаньский 324079
Дуб густой 323917
Дуб далиский 323778
Дуб двухцветный 323707
Дуб двуцветный 323707
Дуб длиннолистный 324121

Дуб длинноножковый 324127
Дуб дубильный 324166
Дуб звездчатый 324443
Дуб зимний 324278
Дуб золотистый 324020
Дуб золоточешуйчатый 323762
Дуб зубчатый 323814
Дуб ивовидный 324282
Дуб иволистный 324282,324335
Дуб изменчивый 324532
Дуб имеретинский 324043
Дуб испанский 324010
Дуб калифорнийский вечнозелёный 323762
Дуб каменный 324027
Дуб карликовый 324039
Дуб каролинский 324544
Дуб каштановый 324166
Дуб каштанолистный 323737
Дуб кермесовый 323769
Дуб кермесоносный 323769
Дуб козловского 324086
Дуб коровий 324166
Дуб кошенильный 323780
Дуб красильный 324539
Дуб красный 324344
Дуб крупножелудевый 324141
Дуб крупноплодный 324141
Дуб крупночешуйчатый 324146
Дуб курчатый 324190
Дуб лавролистный 324095
Дуб летный 324335
Дуб летный черешчатый 324335
Дуб ливанский 324104
Дуб лиловидный 324137
Дуб лилообразный 324137
Дуб лицзянский 323840
Дуб лузитанский 324134
Дуб лэдунский 324131
Дуб малипоский 324156
Дуб малый 324443
Дуб мекедонский 324140,324498
Дуб мексиканский 324235
Дуб мерилендский 324155
Дуб мирбека 323732
Дуб мирзинолистный 324283
Дуб мирсинолистный 116153
Дуб мишо 324166
Дуб монгольский 324173
Дуб мотоский 116151
Дуб мулиский 324383
Дуб наттолла 324232
Дуб ножкоцветный 324275
Дуб обыкновенный 324335
Дуб ольхолистный 323649
Дуб опушенный 323814
Дуб опушеный 324314
Дуб острейший 323611
Дуб острый 323599
Дуб падуболиский 324027
Дуб пиренейский 324319
Дуб понтийский 324301
Дуб пробковый 324453
Дуб пробковый западный 324247
Дуб пробковый ложный 324010
Дуб пунцоволистный 323780
Дуб пушистый 324314
Дуб северный 324344
Дуб серповидный 323894,324344
Дуб серполистный 324344
Дуб сидячецветный 324278
Дуб сизый 116100
Дуб скалиный 324278
Дуб тайваньский 324467
Дуб тибетский 116216,324119
Дуб траволистный 323621
Дуб турецкий 323745
Дуб филлиреевидный 324283
Дуб фрайнетто 323917
Дуб франше 323920
Дуб фунинский 324409
Дуб фэнчэнский 323904
Дуб хермесовый 323769
Дуб хэбэйский 324012
Дуб черепитчатый 324041
Дуб черепичный 324041
Дуб черешчатый 324335
Дуб черешчатый летний 324335
Дуб черный 324225
Дуб чёрный 324225
Дуб чужой 323630
Дуб чужой острипильчатый 323635
Дуб шарлаховый 323780
Дуб эруколистный 323876
Дуб－железняк 324278
Дуб－зимняк 324278
Дубница 13051
Дубоизия 138739,138744
Дубравка 407072
Дубравник 387989
Дубравник крымский 388126
Дубравник нухинский 388165
Дубравник обыкновенный 388042
Дубравник седой 388184
Дубровка 13051,312520,407072
Дубровка восточная 13153
Дубровка елочковидная 13073
Дубровка женевская 13104
Дубровка кандинская 13069
Дубровка лаксманна 13130
Дубровка лежачая 13091
Дубровка пирамидальная 13167
Дубровка ползучая 13174
Дубровка продолговатая 13150
Дубровка тайваньская 13166
Дубровка хиосская 13078
Дубровка японская 13146
Дубровник 387989
Дубровник аньлунский 388000
Дубровник беловойлочный 388184
Дубровник белый 388184
Дубровник верониковидный 388299
Дубровник восточный 388172
Дубровник высокий 388076
Дубровник гирканский 388104
Дубровник горный 388154
Дубровник горький 388147
Дубровник дубравный 388042
Дубровник испирский 388288
Дубровник кистевидный 388015
Дубровник кошачий 388147
Дубровник красивый 388246
Дубровник крымский 388126
Дубровник мелкоцветковый 388178
Дубровник метельчатый 388015
Дубровник многоузловый 388155
Дубровник нухинский 388165
Дубровник обыкновенный 388042
Дубровник паннонский 388177
Дубровник предгорный 388242
Дубровник пурпуровый 388042
Дубровник седой 388035
Дубровник скордиевидный 388270
Дубровник тайваньский 388289
Дубровник тэйлора 388291
Дубровник уссурийский 388298
Дубровник фишера 388078
Дубровник чесноковый 388275
Дубровник чесноч новидный 388270
Дубровник чесночный 388272
Дубровник шалфейный 388275
Дубровник яйлы 388112
Дубровник японский 388114
Дубянка 410960
Дугласин 318562,318580
Дугласия 318562
Дугласия сизая 318574
Дугласия тисолистная 318580
Дугласия тиссолистная 318580
Дудник 24281
Дудник амрский 24291
Дудник горный 24410
Дудник даурский 24325,24327
Дудник зеленоцветковый 276780
Дудник зибольда 276768
Дудник ичанский 24369
Дудник кандинский 24383
Дудник китайский 24475
Дудник коленчатосогнутый 24355
Дудник корейский 24386
Дудник крупнолистный 24406
Дудник лекаственный 30932

Дудник лесной 24483
Дудник лицзянский 24391
Дудник максимовича 276752
Дудник медвежий 24500
Дудник низбегающий 24336,24340
Дудник опоясанный 24318
Дудник преломленный 24355
Дудник прибрежный 24392
Дудник сахалинский 24293,24453
Дудник сычуаньский 24465
Дудник тайваньский 24326
Дудник толстокрылый 24427
Дудник тройчатый 24489
Дудник уклоняющийся 24293
Дудник черняева 121045
Дудник чэнкоуский 24342,62632
Дудник эмэйский 24424
Дудник ялуйский 24293
Дуляна 109545
Думпальма 202256
Думпальма настоящая 202321
Дура белая 369601
Дуранта 139060
Дуриан 139089
Дурман 123036
Дурман безвредный 123061
Дурман вонючий 123077
Дурман древесный 61231
Дурман древовидный 61231
Дурман душистый 61242
Дурман индийский 123061,123065
Дурман метель 123065
Дурман обыкновенный 123077
Дурман татуля 123077
Дурнишник 414990
Дурнишник американский 415057
Дурнишник береговой 415044
Дурнишник ежевидный 415057
Дурнишник западный 415033
Дурнишник зобовидный 415057
Дурнишник иглистый 415057
Дурнишник игольчатый 415053
Дурнишник калифорнийский 414999
Дурнишник канадский 415003
Дурнишник колючий 415053
Дурнишник обыкновенный 415057
Дурнишник сибирский 415046
Дурнишник шиповатый 415057
Дурра 369720
Дурра белая 369601
Дурра бурое 369720
Дурро белое 369720
Дурро бурое 369720
Дурро египетское 369720
Дурьян 139089
Душевга 96960
Душевга эмэйская 97041
Душевик 65528,96960
Душевик альпийский 4662,65535
Душевик аптечный 65603
Душевик китайский 96970
Душевик котовниковый 96969
Душевик круноцветковый 65581
Душевик лекарственный 65603
Душевик сахалинский 97031
Душевик слабый 65573
Душевика 4661
Душевика альпийская 4662
Душевика пахучая 4669
Душевика тимьянная 96964
Душевика фомина 4668
Душёвка 65531,96964
Душёвка альпийская 4661
Душёвка обыкновенная 4662,65535
Душёвка чапрецевидная 65531,96964
Душистый колокок 27930,27969
Душистый колокок обыкновенный 27969
Душистый колокок остистый 27934
Душица 274197,274224
Душица копетдагская 274222
Душица мелкоцветковая 274236
Душица обыкновенная 274237
Душки 268438
Дымянка 168964
Дымянка аптечная 169126
Дымянка бесчашечная 169004
Дымянка вайана 169183
Дымянка козья 169030
Дымянка кралика 169078
Дымянка лекарственная 169126
Дымянка мелкая 169136
Дымянка мелкоцветковая 169136
Дымянка остроконечная 169155
Дымянка шлейхера 169168
Дымянковые 169194
Дымянока 169196
Дымянока туркестанская 169197
Дыниковые 76820
Дынное дерево 76813
Дыня 114108,114189
Дыня декопативная краснoплодная 114221
Дыня десертная 114199
Дыня думаим 114150
Дыня змеевидная 114209
Дыня извилистая 114189,114209
Дыня изогнутая 114209
Дыня канталупа 114199
Дыня культурная 114199
Дыня мелкоплодная 114150
Дыня мускатная 114189
Дыня новислая 114209
Дыня обыкновенная 114214
Дыня огуречная 114209
Дыня посевная 114189
Дыня сетчатая 114214
Дыня тарра 114189
Дыравка 243472
Дыравка малаковидная 243475
Дыравка трехрассеченная 243494
Дюпонция 139037
Дюпонция фишера 139038
Дюшенея 138790
Дягель 24281
Дягиль 24281,30929
Дягиль аптечный 30932
Дягиль короткостебельный 30930
Дягиль лекарственный 30932
Дягиль лесной 24483
Дягиль низбегающий 30931
Дядиль 30929
Дядиль аптечный 30932
Дядиль низбегающий 30931
Дядлина белая 397019,397043
Дятлина красная 397019
Евботриоидес 155453
Евботриоидес грея 228171
Евгения 156118,382465,382477,382582
Евгения бразильская 156148
Евгения вэньшаньская 382682
Евгения гуандунская 382591
Евгения гуансиская 382549
Евгения гуншаньская 382545
Евгения ланьюйская 382664
Евгения малаккская 382606
Евгения мелкоцветковое 382570
Евгения миртолистная 382481
Евгения симаоская 382671
Евгения сычуаньская 382670
Евгения тайваньская 382535,382672
Евгения тибетская 382684
Евгения хайнаньская 382569
Евгения юньнаньская 382685
Евгения ямболана 382522
Егицерас 8645
Егорьево копье пушистоцветковая 174583
Ежа 121226
Ежа воронова 121273
Ежа обыкновенная 121233
Ежа сборная 121233
Ежа скученная 121233
Ежевика 338059,338209
Ежевика американская 338808
Ежевика белесоватая 338232
Ежевика войлочная 339383
Ежевика воронова 339477
Ежевика вязолистная 339419
Ежевика гигантская 339101
Ежевика гирканская 338552
Ежевика грузинская 338459
Ежевика длинноплодная 338333
Ежевика ельниковая 339049

Ежевика звездчатая 339298
Ежевика змеевидная 339255
Ежевика иберийская 338553
Ежевика ильмолистная 339419
Ежевика кавказородная 338237
Ежевика кавказская 338236
Ежевика кровавая 339236
Ежевика куринская 338311
Ежевика кустарниковая 338447
Ежевика лежачая 79075,338399
Ежевика либоский 338742
Ежевика ллойда 338757
Ежевика мелковатая 338735
Ежевика мелкотычинковая 338736
Ежевика мишенко 338840
Ежевика несская 338878
Ежевика осетинская 338931
Ежевика персидская 339039
Ежевика плетевидная 79075,338399
Ежевика понтийская 339090
Ежевика радде 339150
Ежевика сизая 338209
Ежевика сильно－согнутая 339041
Ежевика синеватая 338209
Ежевика сичоуская 339488
Ежевика уклоняющаяся 338074
Ежевика чёрная 338808
Ежевика шерстистая 338713
Ежевика широколистная 339072
Ежевика щетинистая 338516,338527
Ежевика щетиноволосная 338528
Ежевиковые 370018
Ежеголовка 370021
Ежеголовка ветвистая 370050
Ежеголовка маленькая 370088
Ежеголовка мелкоплодная 370080
Ежеголовка побегоносная 370102
Ежеголовка простая 370096
Ежеголовковые 370018
Ежеголовник 370021
Ежеголовник ветвистый 370050
Ежеголовник выступающий 370096
Ежеголовник глена 370064
Ежеголовник длиннолистный 370075
Ежеголовник злаковый 370068
Ежеголовник маленький 370088
Ежеголовник малый 370088
Ежеголовник мелкоплодный 370080
Ежеголовник многогранный 370090
Ежеголовник незамеченный 370051
Ежеголовник побегоносный 370102
Ежеголовник простой 370096
Ежеголовник родственный 370023
Ежеголовник северный 370071
Ежеголовник скученный 370066
Ежеголовник тонколистный 370100
Ежеголовник узколистный 370030,370100
Ежеголовник фриса 370068
Ежеголовник японский 370072
Ежеголовниковые 2208,370018
Ежовик коралловидный 280231
Ежовка 398877
Ежовник 21023,140351
Ежовник аболна 21034
Ежовник балхашский 21032
Ежовник безлистный 21028
Ежовник ветвистыи 21058
Ежовник высокий 21037
Ежовник гипсолюбивый 21042
Ежовник канделябрный 21033
Ежовник коровина 21047
Ежовник коротколистный 21034
Ежовник крестьянский 140360
Ежовник крупнокрылый 21050
Ежовник куриное 140367
Ежовник малоцветковый 21053
Ежовник мелкожелезистый 21051
Ежовник меловой 21035
Ежовник пеллиота 21054
Ежовник петушье 140367
Ежовник петушье просо 140367
Ежовник рисовый 140462,140466
Ежовник скученный 140466
Ежовник смежный 21034
Ежовник солончаковый 21058
Ежовник спиральный 140403
Ежовник сырдарьинский 21046
Ежовник тургайский 21065
Ежовник туркестанский 21066
Ежовник усеченный 21064
Ежовник ферганский 21039
Ежовник хвостатый 140356,140382
Ежовник хлебный 140423
Ежовник шерстистоногий 21038
Ежовник щетинковолосый 21044
Елевзина 143512
Елевзина коракан 143522
Елевзине 143512
Елевзине индийская 143530
Елоха 16368
Ель 298189
Ель алькокка 298225,298226
Ель американская 298286
Ель аянская 298307
Ель балфура 298248
Ель бальфура 298248,305551
Ель белая 298286
Ель вилькона 298449
Ель восточная 298391
Ель восточно－гималайская 298437
Ель вревера 298265
Ель высокая 298191
Ель гималайская 298436
Ель глена 298298
Ель голубая 298401
Ель горная 298368
Ель двуцветная 298225,298226
Ель европейская 298191
Ель западная 298191
Ель иезская 298307
Ель изящная 298442
Ель индийская 298436
Ель кавказкая 298391
Ель камчатская 298319
Ель канадская 298286
Ель колючая 298401
Ель корейская 298323
Ель красная 298423
Ель лицзянская 298331
Ель максимовича 298354
Ель мексиканская 298357
Ель обратнояйцевидная 298382
Ель обыкновенный 298191
Ель пурпурная 298417
Ель саржента 298258,298260
Ель сербская 298387
Ель серебристая 298286
Ель сибирская 298382
Ель сизая 298191,298286
Ель синяя 298406
Ель ситкинская 298435
Ель стройная 298191
Ель тяньшаньская 298426
Ель финская 298282
Ель хоккайдская 298307
Ель чёрная 298349
Ель шеповатая 298437
Ель шероховатая 298232
Ель шренга 298426
Ель шренга тяньшанская 298429
Ель шренка 298426
Ель энгельманна 298273
Ель японская 298442
Ера пелярная 53530
Ёра порярная 53530
Ерань 174440
Еремостахис 148466
Ерика 148941
Ерика древовидная 149017
Ерика крестолистная 150131
Ерика сизая 149205
Ерика средиземноморская 149398
Ериоботрия 151130
Ериоботрия японская 151162
Ёрник 53530
Ерник обыкновый 145073
Ерник стелющийся 53530
Ёрникове 145052
Жабник 165906
Жабник борнмюллера 165922
Жабник галльский 235262

Жабник германский 166038
Жабник горный 166005
Жабник заячий 165915
Жабник лопатчатый 166025
Жабник маленький 235267
Жабник малый 235267
Жабник мелкий 235267
Жабник немецкий 166038
Жабник полевой 165915
Жабник шерстистоголовый 165940
Жабрей 170078
Жабрица 361456
Жабрица аболина 361457
Жабрица алексеенко 361462
Жабрица андронак 361464
Жабрица валентины 361595
Жабрица варьирующая 361597
Жабрица виличатая 361482
Жабрица гигантская 361493
Жабрица гладковатая 361495
Жабрица горичниковидная 361552
Жабрица горичниколистная 361551
Жабрица джунгарская 361575
Жабрица жабрицевидная 228565
Жабрица жесткая 361559
Жабрица изящная 361485
Жабрица илийская 228583
Жабрица именчивая 361597
Жабрица индийская 361483
Жабрица йомутская 361553
Жабрица камеденоская 361500
Жабрица каменстая 361550
Жабрица каратавская 361511
Жабрица клинолистная 361477
Жабрица коническая 361590
Жабрица конская 361501
Жабрица коровина 361512
Жабрица крупноканальцевая 361496
Жабрица крупнолистная 361537
Жабрица ледебура 361519
Жабрица леманна 361521
Жабрица леманновская 361520
Жабрица лессинга 361524
Жабрица малолучевая 361548
Жабрица многолистная 361492
Жабрица однолетняя 361465
Жабрица палласа 361547
Жабрица песчаная 361466
Жабрица подражаюшая 361460
Жабрица полевая 361466
Жабрица понтийская 361554
Жабрица прямостоячая 361582
Жабрица пучковая 361488
Жабрица пушистоголовчатая 361487
Жабрица пушистоплодная 228581
Жабрица равнинная 361472
Жабрица растопыренная 361578,361587
Жабрица седоватая 228584
Жабрица сидячецветковая 361569
Жабрица скальная 361561
Жабрица солоносная 361500
Жабрица тибетская 361545
Жабрица тонковетвистая 361522
Жабрица тонколистная 361495
Жабрица тонкорассеченная 361585
Жабрица торчашая 361582
Жабрица увенчанная 361475
Жабрица чуилийская 361589
Жабрица шероховатая 361467
Жабрица щебнистая 361550
Жабрицевидка 361605
Жабрицевидка тяньшанская 361606
Жантак 14638
Жарновец 120903,347117
Жарновец метельчатый 121001
Жасмин 211715
Жасмин антечный 211940
Жасмин арабский 211990
Жасмин ароматный 211938
Жасмин белый 211940
Жасмин ван 212018
Жасмин глунтовой 294426
Жасмин голоцветконый 211931
Жасмин голоцветный 211931
Жасмин гуансиский 211852
Жасмин двусемянный 211795
Жасмин дикий 294426
Жасмин желтый 211834
Жасмин индийский 211990
Жасмин инцзянский 211817
Жасмин китайский 212004
Жасмин комнанин 211749
Жасмин крупноцветковый 211848
Жасмин крупноцветный 211848
Жасмин кустарниковый 211834
Жасмин лекарственный 211940
Жасмин лекарственный тибетский 211952
Жасмин месни 211905
Жасмин настоящий 211940
Жасмин непахучий 294479
Жасмин низкий 211863
Жасмин низкий ганьсуский 211865
Жасмин ночной 294426
Жасмин обыкновенный 294507
Жасмин опушенный 211867
Жасмин отвороченный 211864
Жасмин розовый 294560
Жасмин садовый 294422,294426
Жасмин самбак 211990
Жасмин тибетский 212071
Жасмин хайнаньский 211893
Жасмин юаньцзянский 212074
Жасмин юньнаньский 211987
Жасмин японский 294530
Жасмин－самбак 211990
Жгунец 94676
Жгун－корень 97697
Жгун－корень восточный 97725
Жгун－корень гроссгейма 97710
Жгун－корень даурский 97703
Жгун－корень малолучевой 97726
Жгун－корень многостебельный 97721
Жгун－корень моннье 97719
Жгун－корень солончаковый 97727
Жгун－корень сомнительный 97705
Жгун－корень японский 97711
Железница 362728
Железница балансы 362746
Железница горная 362838
Железница зеленочешуйная 362765
Железница маршаллова 362835
Железница скученная 362768
Железница таврическая 362880
Железница хохлатая 362767
Железница черепичатая 362807
Железница черноморская 362786
Железное дерево 79161
Железняк 7190,284957
Желтая акация 72180
Желтая орегонская 300153
Желтинник 107297,107300,332452
Желтинник чэнкоуский 107308
Желтозелье 358828
Желтокорень 200173
Желтокорень канадский 200175
Желтокорень китайский 280741,280799
Желтолоз 343937
Желтолозник 343937
Желтосмолка 415172
Желтофиоль 86409
Желтофиоль обыкновенный 86427
Желтофиоль садовый 86427
Желтоцвет 8324
Желтоцвет весенний 8387
Желтуха 359158
Желтушник 154363
Желтушник аксарский 154367
Желтушник алтайский 154451
Желтушник амурский 154380
Желтушник бабатагский 154396
Желтушник белоцветковый 154494
Желтушник белоцветный 154494
Желтушник венгерский 154513
Желтушник выгрызенный 154530
Желтушник выемчатый 154530
Желтушник гауданский 154458
Желтушник грузинский 154474
Желтушник гулявниковый 154544
Желтушник душистый 154513
Желтушник желточный 154562
Желтушник заилийский 154557

Желтушник золотистый 154394
Желтушник кавказский 154422
Желтушник каспийский 154421
Желтушник коротколодный 154409
Желтушник короткостолбчатый 154413
Желтушник красивоплодный 154417
Желтушник крынкинский 154488
Желтушник левкойный 154424
Желтушник лесной 154553
Желтушник лиловый 154495
Желтушник маршалла 154467
Желтушник мейера 154506
Желтушник меловой 154438
Желтушник низкий 154471
Желтушник оранжевый 154414
Желтушник палласа 154516
Желтушник персидский 154522
Желтушник прямой 154467
Желтушник прямостручковый 154548
Желтушник растопыренный 154530
Желтушник сереброплодный 154387
Желтушник сероватый 154418
Желтушник серо－зелёный 154418
Желтушник серый 154418
Желтушник сычуаньский 154398
Желтушник скученный 154434
Желтушник толстоногий 154436
Желтушник тонкостолбиковый 154493
Желтушник узколистный 154492
Желтушник узкостолбиковый 154484
Желтушник фиолетовый 154560
Желтушник франше 154456
Желтушник холмовый 154433
Желтушник холодный 154459
Желтушник черняева 154441
Желтушник шафранный 154439
Желтушник шершавый 154551
Желтушник шовица 154555
Желтушник щитовидный 154440
Желтушник японский 154485
Желтушник ястребинколистный 154467
Желтянка 327879
Женьшень 280712,280741
Жень－шень 280741,280793
Женьшень американский 280799
Женьшень обыкновенный 280741
Женьшень пятилистный 280799
Жерардия 362095
Жерардия полевая 362097
Жереп 300086
Жеруха 262645,399597
Жеруха аптечная 262722
Жеруха водная 262722
Жеруха лекарственная 262722
Жеруха тибетская 262751
Жеруха хреновая 34587,34590
Жерушник 336166,399597
Жерушник австрийский 336182
Жерушник болотный 336250
Жерушник догадовой 336192
Жерушник земноводный 336172
Жерушник индийский 336211
Жерушник исландский 336250
Жерушник кантонский 336187
Жерушник короткоплодный 336185
Жерушник лесной 336277
Жерушник мелкосемянный 336237
Жерушник обоюдоострый 336173
Жерушник шаровидный 336200
Жерушник щетинистый 336204
Жестер 328592
Жестёр 328592
Жестер бальджуанский 328618
Жестер вечнозеленый 328595
Жестер гуансиский 328753
Жестер даурский 328680
Жестер деревцо 328694
Жестер джунгарский 328861
Жестер диамантский 328689
Жестер длиннолистный 328690
Жестер имеретинский 328735
Жестер каролинский 328642
Жестер кожистолистный 328656
Жестер корейский 328752
Жестер красильный 328874
Жестер крошечный 328788
Жестер лопатчатолистный 328862
Жестер люцюский 328767
Жестер мелколистный 328816
Жестер мелкоплодный 328786
Жестер палласа 328812
Жестер прижатый 328688
Жестер синтениса 328859
Жестёр слабительный 328642
Жестер тибетский 328903
Жестер уссурийский 328882
Жесткоковолосница 354400
Жесткоковолосница твёрдая 354403
Жесткоколосница 354400
Жесткоколосница твердая 354403,354404
Жесткоколосница твёрдая 354403
Жесткоколосница трердая 354403
Жесткомятлик 354477
Жесткомятлик воронова 354493
Жесткомятлик жестковатый 354485
Живика пушистая 373166
Живокость 102833,124008
Живокость алтайская 124026
Живокость аякса 102833
Живокость аяксова 102833
Живокость баталина 124056
Живокость белоокаймленная 124023
Живокость бледножелтая 124426
Живокость бливкая 124511
Живокость бородатая 124053
Живокость брунона 124076
Живокость бухарская 124079
Живокость буша 124081
Живокость ветвистая 124147
Живокость вонючая 124218
Живокость восточная 102838
Живокость высокая 124194
Живокость ганьсуская 124312
Живокость гиблидная 124301
Живокость гогенакера 124293
Живокость голоплодная 124322,124340
Живокость горолюбивая 124432
Живокость губоцветковая 124127
Живокость губоцветная 124127
Живокость дваждытройчатая 124067
Живокость джунгарская 124589
Живокость длинноцветоножковая 124355
Живокость дубняков 124550
Живокость дуговидная 124046
Живокость извилистая 124217
Живокость илийская 124302
Живокость кавказская 124110
Живокость каратавская 124315
Живокость клиновидная 124156
Живокость кнорринг 124324
Живокость коржинского 124325
Живокость короткошпорцевая 124073
Живокость косматая 124676
Живокость красивая 124595
Живокость крупноцветковая 124237, 124378
Живокость крупноцветная 124237
Живокость курчавенькая 124153
Живокость лакоста 124330
Живокость липского 124352
Живокость лицзянская 124348
Живокость маака 124360
Живокость максимовича 124376
Живокость марии 124374
Живокость мелкоморщинистая 102842
Живокость метльчатая 124456
Живокость мулиская 124394
Живокость незаметная 124304
Живокость неожиданная 124305
Живокость опушеноплодная 124162
Живокость осетинская 124439
Живокость отогнутоволосистая 124555
Живокость охотская 124424
Живокость палласа 124454
Живокость персидская 124500
Живокость печальная 124653
Живокость пирамидальная 124548
Живокость плотная 102833
Живокость полевая 102833,124141
Живокость полторацкого 124502
Живокость полубородатая 124572

Живокость полубулавовидная 124573
Живокость посевная 102833,124141
Живокость посевной 81012
Живокость прицветничная 124074
Живокость пунцовая 124537
Живокость пушистоцветковая 124161
Живокость растопыренная 124183
Живокость редкоцветная 124012
Живокость рогатые 102833
Живокость рупрехта 124560
Живокость русская 124156
Живокость рыхловатая 124339
Живокость саурская 124563
Живокость сетчатоплодная 124181
Живокость синецветковая 124158
Живокость согнутоплодная 124097
Живокость спутанная 124140
Живокость стокса 124605
Живокость странная 124457
Живокость толстолистная 124151
Живокость тонкоплодная 124344
Живокость тройчатая 124637
Живокость туркменская 124659
Живокость уральская 124669
Живокость фрейна 124222
Живокость холоднолюбивая 124154
Живокость хуанчжунская 124297
Живокость чудесная 124386
Живокость шмальгаузена 124571
Живокость шовица 124615
Живокость эмэйская 124428
Живучка 13051,358015
Живучка восточная 13153
Живучка елочковидная 13073
Живучка ёлочковидная 13073
Живучка женевская 13104
Живучка иволистная 13181
Живучка кровельная 358062
Живучка лаксманна 13130
Живучка ложнохиосская 13165
Живучка мелкоцветковая 13164
Живучка многоцветковая 13143
Живучка паутинная 358021
Живучка пирамидальная 13167
Живучка ползучая 13174
Живучка приземистая 13072
Живучка продолговатая 13150
Живучка сикотанская 13184
Живучка туркестанская 13190
Живучка хиосская 13078
Жигунец 97697
Жигунец аянсий 391904
Жигунец сомнительный 97705
Жидовник 383412
Жиловник 117852,117859
Жимолостные 71955
Жимолость 235630,235691
Жимолость алтайская 235699
Жимолость альберта 235649
Жимолость альпийская 235651
Жимолость альтмана 235663
Жимолость ахалинская 236085
Жимолость бушей 235690
Жимолость вечнозелёная 236089
Жимолость виргинская 380736,380753
Жимолость вьющаяся 235713
Жимолость ганьсуская 235895
Жимолость генри 235633
Жимолость глена 235811
Жимолость голубая 235691
Жимолость грузинская 235864
Жимолость душистая 235713
Жимолость душистейшая 235796
Жимолость жилковатая 235981
Жимолость золотистая 235730
Жимолость золотистоцветковая 235730
Жимолость илийская 235865
Жимолость кавказская 235715
Жимолость камчатская 235894
Жимолость канадская 235710
Жимолость кандинская 235841
Жимолость каприфол 235713
Жимолость каприфоль 235713
Жимолость козья 235713
Жимолость королькова 235903
Жимолость ланьпинская 235949
Жимолость лесная 236226
Жимолость лоницера 235713
Жимолость лушаньская 235965
Жимолость маака 235929
Жимолость максимовича 235945
Жимолость мелколистная 235952
Жимолость мелкоцветковая 236158
Жимолость морроу 235970
Жимолость немецкая 236022
Жимолость низкая 235859
Жимолость обернутая 235874
Жимолость обыкновенная 236226
Жимолость ольги 236000
Жимолость палласа 236012
Жимолость памирская 236013
Жимолость почтищетинистая 236130
Жимолость пушистая 236226
Жимолость раннецветущая 236040
Жимолость ранняя 236040
Жимолость раноцветущая 236040
Жимолость разноволосая 235845
Жимолость разнолистная 235843
Жимолость рупрехта 236076
Жимолость рупрехта оголяющаяся 236077
Жимолость семенова 236088
Жимолость серая 235740
Жимолость синяя 235691
Жимолость сычуаньская 236138
Жимолость стандиша 235798
Жимолость странная 236015
Жимолость съедобная 235702
Жимолость татарская 236146
Жимолость татарская мелкоцветковая 236158
Жимолость тибетская 236070
Жимолость толмачева 236175
Жимолость турчанинова 236198
Жимолость тяньшанская 236174
Жимолость узкоцветковая 236115
Жимолость уэбба 236214
Жимолость фердинанда 235781
Жимолость фэнкайская 235780
Жимолость цветущая 235794
Жимолость цзилунская 235890
Жимолость черная 235982
Жимолость чёрная 235982
Жимолость шамиссо 235721
Жимолость шапочная 236035
Жимолость шерстистая 235909
Жимолость щетинистая 235852
Жимолость щетинистоволосистая 236130
Жимолость эмэйская 236106
Жимолость этрусская 235771
Жимолость японская 235878
Жирардиния 175865
Жирардиния остроконечная 221552
Жирардиния чжэцзянская 175868
Жирянка 299736
Жирянка альпейская 299737
Жирянка альпийская 299737
Жирянка мохнатая 299757
Жирянка обыкновенная 299759
Жирянка пестрая 299756
Житец 144661
Житник 201597
Житник повислый 201618
Житняк 11628,11696,11699,11749
Житняк гребеневидный 11696
Житняк гребенчатый 11696
Житняк пустынный 11710
Житняк сибирський 11735
Житняк тайваньский 335312
Житняк узкоколосый 11710
Житняк узколистный 11710
Житняк черепичатый 11749
Жминда 86984,87016
Жминда головчатая 86984
Жонкилия 262410
Жонкилла 262410
Жонкиль 262410
Жосте хубэйский 328731
Жосте японский 328740
Жосте японский мелкоцветковый 328747
Жостер 328592
Жостер вечнозеленый 328595

Жостер вильсона 328900
Жостер даурский 328680
Жостер диамантский 328689
Жостер каролинский 328642
Жостер красное деревцо 328694
Жостер полезный 328883
Жостер ребристый 328660
Жостер росторна 328842
Жостер слабительный 328642
Жостер тайваньский 328700
Жостер хайнаньский 328720
Жузгун 66988
Жузгун андросова 66996
Жузгун аральский 67000
Жузгун белокорый 67043
Жузгун блюдцевиднокрылый 67042
Жузгун борщова 67002
Жузгун бубыра 67003
Жузгун высокий 67023
Жузгун голова – медузы 67008
Жузгун горецовидный 67064
Жузгун гребенчатый 67017
Жузгун густощетинковый 67018
Жузгун древовидный 67001
Жузгун евгения коровина 67025
Жузгун ежеплодный 67024
Жузгун желтоплодный 67075
Жузгун жесткокрылый 66998
Жузгун завернутый 67033
Жузгун змеиный 67011
Жузгун известняковый 67004
Жузгун квадратнокрылый 67072
Жузгун кожистый 67014
Жузгун колючейший 67079
Жузгун колючекрылый 66990
Жузгун красивейший 67069
Жузгун красноплодный 67075
Жузгун крупноплодный 67046
Жузгун курчавый 67075
Жузгун липского 67044
Жузгун литвинова 67045
Жузгун мелкоплодный 67048
Жузгун муравлянского 67052
Жузгун мягкий 67049
Жузгун низкий 67032
Жузгун палецкого 67055
Жузгун перепончатый 67047
Жузгун пеунникова 67060
Жузгун печальный 67084
Жузгун плоскощетинковый 67062
Жузгун поникающий 67057
Жузгун прозрачноплодный 67058
Жузгун прямощетинковый 67054
Жузгун пузырчатокрылый 67061
Жузгун рассеченнокрылый 67021
Жузгун растопыренный 67080
Жузгун ререидский 67059
Жузгун русанова 67076
Жузгун сердцевидный 67013
Жузгун сероватый 67030
Жузгун ситниковый 67035
Жузгун складчатый 67063
Жузгун сродный 66991
Жузгун стройный 67029
Жузгун тонкий 67081
Жузгун тонкокрыловидный 66994
Жузгун тонкокрылый 66995
Жузгун тупокрылый 67053
Жузгун туркестанский 67085
Жузгун ферганский 67026
Жузгун хрящеватый 67009
Жузгун четырехкрылый 67082
Жузгун щетинистый 67077
Журавельник 153711,174440
Журавельник аистовый 153764
Журавельник армянский 153726
Журавельник бекетова 153733
Журавельник веховый 153767
Журавельник геффта 153883
Журавельник дымянковидный 153800
Журавельник кровяно – красный 174891
Журавельник литвинова 153837
Журавельник луговой 174832
Журавельник мальвовидный 153839
Журавельник остроносый 153883
Журавельник пупавковидный 153723
Журавельник русский 153907
Журавельник стевена 153916
Журавельник стефана 153914
Журавельник татарский 153920
Журавельник тибетский 153923
Журавельник хиосский 153754
Журавельник цикутовый 153767
Журавельник цикутолистный 153767
Журавельник щетинистый 153917
Журавельниковые 174431
Забра 170076
Завиток 377888
Завиток пустынный 377889
Завязник 312520
Загадка 150419
Зайцегуб 220050
Зайцегуб алтайский 220053
Зайцегуб андросова 220055
Зайцегуб бородатая 220359
Зайцегуб бунге 220059
Зайцегуб валханский 220056
Зайцегуб волосистый 220068
Зайцегуб гладкоколючковый 220077
Зайцегуб двигольчатый 220063
Зайцегуб длиннозубчатый 220078
Зайцегуб зеравшанский 220091
Зайцегуб кабульский 220060
Зайцегуб кашгарский 220073
Зайцегуб кнорринговский 220074
Зайцегуб колючий 220090
Зайцегуб красивый 220088
Зайцегуб крупнозубый 220079
Зайцегуб кштутский 220075
Зайцегуб невского 220080
Зайцегуб опьяняющий 220072
Зайцегуб остродольный 220051
Зайцегуб падуболистный 220069
Зайцегуб паульсена 220084
Зайцегуб плоскоколючковый 220085
Зайцегуб плоскочашечный 220086
Зайцегуб почтй – щетинистый 220093
Зайцегуб пушистый 220087
Зайцегуб синьцзянский 220096
Зайцегуб тяньшанский 220094
Зайцегуб узунахматский 220095
Зайцегуб щетинистый 220092
Зайцехвост 220252
Заманиха 266367,272704
заманиха 272706
Заманиха высокая 272706
Замашка 71218
Зандури 398886
Заникелия 417045
Заникелия болотная 417053
Занникелиевые 417073
Занникеллия 417045
Занникеллия болотная 417053
Занникеллия большая 417051
Занникеллия стебельчатая 417053
Зантедесхия 417092
Зантедесхия абиссинская 417093
Зантедесхия эллиота 417104
Зантедешия 68877,334310
Зантоксилум 417132
Зантоксилум американский 417231
Зантоксилюм 417132
Зантоксилюм айлантовидный 417139
Зантоксилюм американский 417157
Зантоксилюм гуансиский 417249
Зантоксилюм колюченожковый 417134
Зантоксилюм ланьюйский 417247
Зантоксилюм либоский 417258
Зантоксилюм лэйбоский 417254
Зантоксилюм мелкоцветковое 417271
Зантоксилюм мотоский 417275
Зантоксилюм перечный 417301
Зантоксилюм плоскошипный 417161
Зантоксилюм подражающий 417340
Зантоксилюм рассеянный 417216
Зантоксилюм сичоуский 417374
Зантоксилюм схинусолистный 417330
Зантоксилюм щетинистый 417340
Заплатник 344211
Зараза бодяная 143914
Зараза бодяная зубчатая 141569

Зараза бодяная канадская 143920
Заразиха 274922
Заразиха амурская 274952
Заразиха андросова 274956
Заразиха бахчевая 274927
Заразиха белая 274934
Заразиха бетвистая 275184
Заразиха бледноцветковая 274934,275159
Заразиха большая 275128
Заразиха бунге 274977
Заразиха волосистоцветковая 275090
Заразиха волчок 274991
Заразиха восточная 275155
Заразиха ганса 275084
Заразиха гвоздичная 274985
Заразиха голостебельная 275073
Заразиха голубая 275113
Заразиха городчатая 275029
Заразиха горчаковая 274958
Заразиха григольева 275079
Заразиха гроссгейма 275081
Заразиха грязно-желтая 275214
Заразиха дубровниковая 275225
Заразиха египетская 274927
Заразиха желтая 275058,275126
Заразиха желтоватая 275257
Заразиха закавказская 275227
Заразиха зубчатая 275029
Заразиха иберийская 275095
Заразиха капустная 274972
Заразиха каратавская 275100
Заразиха келлера 275102
Заразиха килинийская 275000
Заразиха кларке 275002
Заразиха коржинского 275018
Заразиха короткозубчиковая 274973
Заразиха короткостебельная 274971
Заразиха кочи 275105
Заразиха красивая 274951
Заразиха красивенэкая 275172
Заразиха крылова 275107
Заразиха курдская 275108
Заразиха лазурная 275006
Заразиха линчевского 275119
Заразиха люцерновая 275126
Заразиха малая 275135
Заразиха монгольская 275142
Заразиха мутеля 275186
Заразиха низкая 275148
Заразиха ноникшая 274987
Заразиха обыкновенная 274985,275258
Заразиха оверина 275156
Заразиха остролопастная 275155
Заразиха перепончаточашечковая 275094
Заразиха песчаная 274957
Заразиха пильчаточашечковая 275208
Заразиха плющевая 275086
Заразиха подсолнечная 274991
Заразиха покрашеная 275021
Заразиха поникшая 274991
Заразиха прелестная 274951
Заразиха пурпуровая 275173
Заразиха радде 275182
Заразиха разноцветная 275251
Заразиха розовая 275196
Заразиха серножелдая 275222
Заразиха сетчатая 275195
Заразиха сизоватая 275074
Заразиха синеватая 275010
Заразиха синтениса 275211
Заразиха снаянная 275027
Заразиха соленантусовая 275212
Заразиха солмса 275213
Заразиха сростночашелистиковая 275068
Заразиха толстоколосая 275177
Заразиха толстоолосая 275177
Заразиха тонкая 275075
Заразиха уральская 275242
Заразиха хохенакера 275091
Заразиха шелковникова 275201
Заразиха широкочешуйная 274987
Заразиха эльзасская 274941
Заразиховые 274921
Зардкунак 361877
Зардолу-арча 213877
Зард-печак 114947
Зацинта 110710
Зацинта бородавчатая 111096
Заячья капуста 200812,356928
Заячья кислица 277648
Звезда кавалерская 285607
Звездовка 112269
Звездоплодник 121993
Звездоплодник перетянутый 122001
Звездоплодник частуховидный 121995
Звездочка 285607,374724
Звёздочка 374968
Звездочка алексеенко 374730
Звездочка аньхуйская 374743
Звездочка арктиеская 374751
Звездочка бесприцветничковая 374872
Звездочка бледная 375033
Звездочка болотная 375036
Звездочка бунге 374780
Звездочка винклера 375144
Звёздочка водяная 67336
Звездочка голубая 285623
Звездочка даурская 374821
Звездочка двуцветная 374858
Звездочка джунгарская 375100
Звездочка длинноцветоножковая 374951
Звездочка дубравная 375008
Звездочка ешшольца 374878
Звездочка жестколистная 374916
Звездочка злачная 374897
Звездочка зонтичная 375128
Звездочка иглицелистная 375078
Звездочка каменная 375051
Звездочка кандинская 375101
Звездочка каратавская 374929
Звездочка коротколепестная 374777
Звездочка кочи 374932
Звездочка курослепная 374742
Звездочка лаксманна 374940
Звездочка лучистая 375071
Звездочка мартьянова 374965
Звездочка мокричная 374739
Звездочка моховидная 374846
Звездочка незамеченная 375007
Звездочка орошаемая 374926
Звездочка персидская 375049
Звездочка приземистая 374919
Звездочка прилучеймная 374888
Звездочка пушисточашечная 374913
Звездочка развилистая 374836
Звездочка раскидистая 374848,374944
Звездочка ресничатая 374803
Звездочка средная 374968
Звездочка тибетская 375114
Звездочка толстолистная 374807
Звездочка топяная 374731
Звездочка тупочашелистная 374740
Звездочка туркестанская 375116
Звездочка фенцля 374880
Звездочка фишера 374884
Звездочка чашечкоцветковая 374783
Звездочка черепичатая 374923
Звездочка шерлериевидная 374794
Звездочка шугнанская 375089
Звездочка эдвардса 374873,374951
Звездочка ялусская 374882
Звездчатка 374724
Звездчатка альзиновидная 374731
Звездчатка бесприцветничковая 374872
Звездчатка бледная 375033
Звездчатка болотная 375036
Звездчатка бунге 374780
Звездчатка вильчатая 374836
Звездчатка двуцветная 374858
Звездчатка длинноножковая 374951
Звездчатка дубровная 375008
Звездчатка злаковая 374897
Звездчатка злаковидная 374897
Звездчатка злачная 374897
Звездчатка зонтичная 375128
Звездчатка китайская 374802
Звездчатка ланцетовидная 374916
Звездчатка ланцетолитная 374916
Звездчатка лесная 375008
Звездчатка лучистая 375071
Звездчатка незамеченная 375007

Звездчатка непалская 375017
Звездчатка приземистая 374919
Звездчатка пушисточашечная 374913
Звездчатка раскидистая 374944
Звездчатка сибирская 375094
Звездчатка сизая 375036
Звездчатка средная мелкоцветковая 374986
Звездчатка средняя 374968
Звездчатка стебельчатая 374951
Звездчатка толстолистная 374807
Звездчатка травяная 374897
Звездчатка хубэйская 374914
Звездчатка цзилунская 374907
Звездчатка чашечкоцветковая 374783
Звездчатка эдварда 374951
Звездчатка эмэйская 375028
Звездчатка – мокрица 374968
Звездчаточка 318489
Звездчаточка давида 318493
Звездчаточка лесная 318515
Звездчаточка разнолистная 318501
Звездчаточка разноцветковая 318496, 318507
Звездчаточка скальная 318513
Звездчаточка тибетская 318517
Звездчаточка японская 318506
Звездчатый анис 204603
Зверобоецветные 201681
Зверобой 201705, 201957, 202142
Зверобой аленийский 201729, 201730
Зверобой Ардасенова 201738
Зверобой армянский 201740
Зверобой большой 201743
Зверобой буша 201786
Зверобой володушковидный 201785
Зверобой вытянутый 201848
Зверобой геблера 201743
Зверобой горецелистный 202098
Зверобой горный 202026
Зверобой гуйцзоуский 201974
Зверобой душицелистный 202061
Зверобой жестковолосый 201924
Зверобой золотистокистевый 201812
Зверобой иезский 202230
Зверобой изящный 201840
Зверобой исколотый 202086
Зверобой иссополистный 201812
Зверобой кавказский 201797
Зверобой камчатский 201954
Зверобой кандинский 202003
Зверобой карсский 201960
Зверобой китайский 202021
Зверобой комарова 201973
Зверобой красивый 202205
Зверобой красильный 201732
Зверобой крылатый 201713
Зверобой курчавый 201825
Зверобой лидийский 202000
Зверобой мэнцзыский 201920
Зверобой недоразвитый 202029
Зверобой непахнущий 201937
Зверобой нонтийский 202100
Зверобой нордманна 202049
Зверобой обыкновенный 202086
Зверобой окаймленный 202011
Зверобой округлолистный 202050
Зверобой олимпийский 202060
Зверобой опушенный 201924
Зверобой осторый 201713
Зверобой оттянутый 201761
Зверобой плонзенолистный 202086
Зверобой прекрасный 201882
Зверобой приальпийский 201729
Зверобой пронзенный 202086
Зверобой протырявленный 202086
Зверобой прямой 201853
Зверобой прямостоящий 201853
Зверобой пятнистый 202006
Зверобой распростертый 201930
Зверобой растопыренный 202070
Зверобой солнцецветный 201914
Зверобой стелющийся 201930
Зверобой тайваньский 201880
Зверобой тибетский 201922
Зверобой торчащий 202167
Зверобой точечный 202006
Зверобой удлиненный 201848
Зверобой федора 202180
Зверобой чамчковидный 201787
Зверобой чашечный 201787
Зверобой чашечцый 201787
Зверобой чаюйский 201969
Зверобой черноморский 202100
Зверобой четырехгранный 202006, 202179
Зверобой четырёхгранный 202179
Зверобой шероховатый 202147
Зверобой шершавоволосый 201924
Зверобой японский 201942
Зверобой ясменниковый 201759
Зверобойные 97266, 201681
Зебрина 393987
Зебрина свизающая 394093
Зев львиный большой 103719
Зев львиный полевой 28631
Зев львиный садовый 103719
Зевксина 417698
Зевксина шлемовидная 417795
Зегея 418249
Зегея бальджуанская 418250
Зегея пурпуровая 418251
Зейдлиция 357384
Зейдлиция розмариновая 357387
Зейдлиция цветистая 357386
Зелёнка 88242
Зелёнка пронзённолистная 88248
Зеленоцвет 88264
Зеленоцвет японский 88289
Зеленоцветниковые 88263
Зеленчук 170015, 220345, 298009
Зеленчук гуандунский 170024
Зеленчук желный 170025
Зеленчук жёлный 170025
Зеленчук сычуаньский 170026
Зелье бешеное 123077
Зелье курье 250600
Зелье лихорадочное 180321
Зелье рвотное 234547
Зелье солодкое 321554
Зелье солодьковое 321554
Зелье сонное 244340
Зельква 301801, 417534
Земляная груша 189073
Земляника 167592
Земляника ананасная 167597
Земляника бухарская 167600
Земляника виргинская 167604, 167663
Земляника восточная 167641
Земляника высокая 167627
Земляника зеленая 167674
Земляника зелёная 167674
Земляника индийская 138796
Земляника крупноплодная 167597
Земляника крупноцветковая 167597
Земляника лесная 167653
Земляника мускатная 167627
Земляника мускуснаэ 167614
Земляника мускусная 167627
Земляника полуница 167674
Земляника равнинная 167602
Земляника садовая 167597
Земляника тибетская 167640
Земляника чилийская 167604
Земляничник 30877
Земляничник крупноплодный 30888
Земляничное дерево 30877, 30880
Земляничное дерево красное 30880
Земляничное дерево крупноплодное 30888
Земляничный томат 297720
Земляной орех 30498
Земляные орешки 166132
Зензивер 243840
Зеравшания 417640
Зеравшания регеля 417641
Зеркало девичье 224058
Зерна 412394
Зерна ангренская 60628
Зерна аябезост 60760
Зерна бенекена 60643
Зерна мелкочешуйная 61006
Зерна памирская 60873
Зерна паульсена 60882

Зерна туркестанская 61003
Зефирантес 417606
Зибера 363002
Зибера колючая 363006
Зигаденус 417880
Зигаденус сибисский 417925
Зигесбекия 363073
Зигесбекия восточная 363090
Зигесбекия гладковатая 363080
Зигесбекия пушистая 363093
Зиглингия 122180
Зиглингия лежачая 122210
Зиглингия трехзубка 122210
Зигокактус 418529
Зигокактус притупленный 418532
Зизифора 418119
Зизифора биберштейна 418120
Зизифора бранта 418121
Зизифора бунге 418123
Зизифора бунговская 418123
Зизифора войлочная 418141
Зизифора воронова 418143
Зизифора галины 418128
Зизифора головчатая 418124
Зизифора жесткая 418137
Зизифора короткочашечная 418122
Зизифора крымская 418139
Зизифора мелкоголовчатая 418125
Зизифора мелкозубчатая 418127
Зизифора памироалайская 418131
Зизифора памиро－алайская 418131
Зизифора пахучковидная 418126
Зизифора персидская 418132
Зизифора прерваннар 418130
Зизифора пушкина 418135
Зизифора радде 418136
Зизифора тимьянниковая 418138
Зизифора тонкая 418140
Зизифора туркменская 418142
Зизифора тяньшанская 418141
Зизифус 418144,418169
Зизифус обыкновенный 418169
Зимняк 324278
Зимовник 99293,190925
Зимовник осенний 99297
Зимолюбка 87480
Зимолюбка зонтичная 87498
Зимолюбка японская 87485
Зклипта 141365
Зклипта белая 141366,141374
Зклипта простертая 141374
Злак большой водяной 285423
Злак пальчатый 117852,117859
Злак пампасовый 105456,182942
Злаки 180078
Злаковые 180078
Златок 39438
Златокорень 354651
Златокорень исканский 354651
Златолист 90007
Златолист каймито 90032
Златотравка 89848
Златотравка крупковидная 89850
Златоцвет 8324,89402,322640
Златоцвет венечный 89481
Златоцвет весенний 8387
Златоцвет девичий 322724
Златоцвет золотистый 8343
Златоцвет ийнепалск 8373
Златоцвет летний 8325
Златоцвет мелкоцветный 8374
Златоцвет пашенный 89704
Златоцвет пушистый 8389
Златоцвет сибирский 8381
Златоцвет тяньшанский 8383
Змеевик 37015,54798,308893
Змеевик австрийский 137545
Змеевик аргунский 137555
Змеевик молдавский 137551,137651
Змеевик руйша 137613
Змеевик тимьянолистный 137551,137651
Змеевка 94585,132720
Змеевка болгарская 132732
Змеевка джунгарская 94631
Змеевка китагавы 94608
Змеевка китайская 94591
Змеевка накаи 94604
Змеевка поздная 94622
Змеевка растопыренная 94632
Змеевка торольда 94631
Змеевка хэнса 94605
Змееголовник 137545
Змееголовник австрийский 137555
Змееголовник аргунский 137551
Змееголовник безбородый 137599
Змееголовник бунге 137565
Змееголовник вонючий 137581
Змееголовник далиский 137671
Змееголовник дважды－перистый 137557
Змееголовник двуцветный 137575
Змееголовник дланевидный 137622
Змееголовник душицевидный 137620
Змееголовник кистевой 137562
Змееголовник колючий 137665
Змееголовник комарова 137606
Змееголовник красивый 137583
Змееголовник крупноцветковый 137591
Змееголовник крылова 137607
Змееголовник кустарничковый 137587
Змееголовник ломкий 137586
Змееголовник многостебельный 137614
Змееголовник многоцветный 137615
Змееголовник молдавский 137613
Змееголовник паульсена 137625
Змееголовник перистый 137630
Змееголовник пониаюший 137617
Змееголовник поникший 137617
Змееголовник почтиголовчадый 137670
Змееголовник продолговатолистный 137619
Змееголовник разнолистный 137596
Змееголовник разнообразнолистный 137577
Змееголовник руйша 137651
Змееголовник стеллера 137668
Змееголовник тимьяноцветковый 137677
Змееголовник тимьяноцветный 137677
Змееголовник тычночный 163089
Змееголовник узловатый 137616
Змееголовник цельнолистный 137601
Змееголовник чужеземный 137627
Змееголовник ямчатый 137655
Змеиный горлец 308893
Змей－трава 137677
Зноиха 9832
Золотарник 367939
Золотарник “золотая розка” 368480
Золотарник геблера 368105
Золотарник даурский 368502
Золотарник злаколистный 368134
Золотарник золотая розга 368480
Золотарник канадский 368013
Золотарник каучуконостый 368208
Золотарник крымский 368442
Золотарник кухистанский 368196
Золотарник медный 368056
Золотарник незамеченный 368263
Золотарник низбегающий 368073
Золотарник обыкновенный 368480
Золотарник пахучий 368290
Золотарник поздний 368106
Золотарник приальпийский 367948
Золотарник сжатый 368049
Золотарник ситниковый 368188
Золотарник скальный 368377
Золотарник тихоокеанский 368297
Золотарник тонкозубчатый 367965
Золотарник торфяный 368453
Золотарник яйлинский 368183
Золотая розга 367939,368480
Золотель 368480
Золотобородник 90107
Золотобородник восточный 90133
Золотобородник игольчатый 90108
Золотобородник цикадовый 90125
Золотой дожль 218754
Золотой корень 329935
Золотой шар 339574
Золототысячник 81485
Золототысячник зонтичный 81544
Золототысячник колосистый 81536
Золототысячник колосовидный 81536
Золототысячник красивенькийй 81517

Золототысячник красивый 81517
Золототысячник мейера 81523
Золототысячник обыкновенный 81546
Золототысячник тонкоцветковый 81539
Золотохворост 401373
Золотоцвет 322640
Золотоцвет полевой 89704
Золотушник 231183
Зонтичная сосна 352856
Зонтичные 29230
Зопник 295049
Зопник алайский 295052
Зопник альпийский 295054
Зопник ангренский 295056
Зопник буквицелистный 295063
Зопник бухарский 295077
Зопник вавилова 295225
Зопник войлочный 295210
Зопник ганьсуский 295124
Зопник гибридный 295116
Зопник горный 295163
Зопник далиский 295105
Зопник дробова 295098
Зопник зинаиды 295228
Зопник иволистный 295187
Зопник кавказский 295084
Зопник кандинский 295203
Зопник клубненосный 295212
Зопник кнорринговская 295126
Зопник колючезубый 295198
Зопник колючий 295175
Зопник копетдагский 295127
Зопник коровяковидный 295207
Зопник короткоприцветниковый 295071
Зопник крымский 295205
Зопник кустарниковидный 295111
Зопник кустарниковый 295111
Зопник ленкоранский 295130
Зопник линейнолистный 295133
Зопник лицзянский 295132
Зопник ложно－колючий 295173
Зопник луговый 295172
Зопник майкопский 295139
Зопник максимовича 295142
Зопник монгольский 295150
Зопник мулиский 295153
Зопник нивяный 295051
Зопник округлозубчатый 295095
Зопник ольги 295162
Зопник опушенный 295174
Зопник островского 295167
Зопник регеля 295181
Зопник решетчатый 295080
Зопник сероватый 295081
Зопник снизу－белый 295117
Зопник тибетский 295208
Зопник тонгий 295206
Зопник ферганский 295099
Зопник хвостозубый 295223
Зопник чимганский 295211
Зорча 363368
Зорька 363312
Зорька “барская смесь” 363312
Зорька татарское мыло 363312
Зорька－аксамитки 363370
Зорька－дрёма 363461
Зоря 228250,229283
Зосимия 418362
Зосимия полыннолистная 418364
Зосимия тордилиевидная 418372
Зостера 418377
Зостера большая 418392
Зостера морская 418392
Зостера японская 418389
Зостеровые 418405
Зуб собачий 117852,117859,154895
Зуб собачий обыкновенный 154909
Зубница 305162,305167
Зубница мелкоцветковая 305163
Зубровка 196116
Зубровка альпийская 27963
Зубровка бунге 27967
Зубровка гладкая 27942
Зубровка голая 27942
Зубровка душистая 27967
Зубровка малоцветковая 196159
Зубровка пахучая 27967
Зубровка южная 196124
Зубчатка 48601,268957
Зубчатка бартсиевидная 268983
Зубчатка весенняя 268991
Зубчатка железистая 268966
Зубчатка клейкая 268966
Зубчатка красная 268983
Зубчатка осенняя 268983,268999
Зубчатка поздная 268999
Зубчатка прибрежная 268970
Зубчатка ранняя 268991
Зубчатка солончаковая 268985
Зубянка 125831
Зубянка двояко－перистая 125834
Зубянка клубненосная 125836
Зубянка луковичная 125836
Зубянка мелколистная 125854
Зубянка мелкоцветная 125856
Зубянка пятилисточковая 125858
Зубянка сибирская 125861
Зубянка тонколистная 125863
Зубянка тонколистный 73018
Зуфи－оби 14719
Зюзник 239184
Зюзник блестящий 239215
Зюзник высокий 239205
Зюзник дальневосточный 239215
Зюзник европейский 239202
Зюзник маака 239224
Зюзник маака тайваньская 239226
Зюзник мелкоцветковый 239235
Зюзник опушенный 239211
Зябра 170076
Иберичка 203175
Иберичка трехжильная 203179
Иберийка 203181
Иберийка вечнозелёная 203239
Иберийка горькая 203184
Иберийка зонтичная 203249
Иберийка крымская 203248
Иберийка перистая 203234
Иберийка скальная 203236
Иберис 203181
Иберис горький 203184
Иберис жёлтая 203260
Иберис зонтикоцветный 203249
Иберис корончатый 203208
Ибунка 407029
Ива 342923
Ива алатавская 342969
Ива альберта 342995
Ива альпийская 343003
Ива анадерская 343015
Ива аномальная 343032
Ива арктическая 343045
Ива аяцинхайск 343972
Ива бабилонская 343070
Ива барбарисолистная 343101
Ива барклая 343093
Ива бебба 343095,344286
Ива безногая 343037
Ива белая 342973
Ива блека 343111
Ива блестящая 344138
Ива болотная 344244
Ива бредина 343151
Ива буреющая 343409
Ива буроватая 343405
Ива вавилонская 343070
Ива великолепная 344132
Ива верба 342973
Ива ветла 342973
Ива вильгельмса 343151,344273
Ива волчниковая 343280
Ива волчья 344255
Ива ганьсуская 343368
Ива гладколистная 344244
Ива голая 343416
Ива головатая 343931
Ива голубая 343167
Ива грушанколистная 343968
Ива гуйцзоуская 343581
Ива гукера 199379,343496
Ива гуншаньская 343138

Ива дагестанская 343266
Ива далиская 343273
Ива даурская 343268
Ива двуцветная 343886
Ива девиса 343284
Ива деревовидная 343041
Ива деревцевидная 343042
Ива деревцовидная 343041
Ива джунгарская 344127
Ива длиннолистная 343283,343622
Ива длинноножковая 343628
Ива длиннoприлистниковая 344151
Ива длинностлбиковая 343342
Ива дымчатая 343403
Ива египетская 342967
Ива железистая 343417
Ива жёлтая 342982
Ива жемчугоносная 343667
Ива жилколистная 343884
Ива зеемана 344080
Ива зеленоватая 344251
Ива зеленосережчатая 343216
Ива зибольда 344112
Ива изящная 343341
Ива илийская 343523
Ива иссыкская 343543
Ива кавказская 343181
Ива кангинская 343560
Ива кандинская 343559
Ива каспийская 343174
Ива кириллова 343568
Ива китайская 343070
Ива клиновидная 343256
Ива козья 343151
Ива колымская 343571
Ива кольская 343570
Ива колючепильчатая 344135
Ива комарова 343572
Ива конопляная 344244
Ива копьевидная 343467
Ива корейская 343574
Ива коржинского 343580
Ива корзиночная 344244
Ива корианаги 343579
Ива коротконожковая 343119
Ива короткосережчатая 343128
Ива коха 343569
Ива красивая 343936
Ива красная 343360
Ива краснолистная 342961
Ива красноплодная 343355
Ива круглолистная 344045
Ива крупнолистная 343458
Ива крупноножковая 343652
Ива крупносережчатая 343653
Ива крушинолистная 344005
Ива крылова 343582
Ива кузнецова 343589
Ива курильская 343752
Ива лапландская 343595
Ива линейнолистная 343111
Ива липского 343612
Ива литвинова 343613
Ива ложнобелая 343915
Ива ложнолинейная 344245
Ива ложнопятитычинковая 343858
Ива ломкая 343396
Ива лопарская 343595
Ива максмовича 343680
Ива матсуды 343668
Ива медведева 343682
Ива мелкосележчатая 343691
Ива мийябе 343702
Ива миндалевидная 343013
Ива миндалистнаяная 344211
Ива миндальная 343013,344211
Ива миртолистная 343743
Ива михельсона 343689
Ива монгольская 343706
Ива монетная 343780
Ива монетная мелкоцветковая 343784
Ива монетовидная 343780,343784
Ива мотоская 343681
Ива мохнатая 343592
Ива мулиская 343734
Ива мупинская 343723
Ива наньчуаньская 344039
Ива нарядная 344243
Ива настоящая прутьевидная 344242
Ива недзвецкого 343167
Ива ниппонская 344213
Ива ныйвинская 343785
Ива одновременная 343231
Ива округлая 343809
Ива оленина 343802
Ива ольги 343803
Ива остроконечная 343358
Ива остролистная 342961
Ива остроплодная 343822,343962
Ива палибина 343825
Ива палласа 343826
Ива параллельноживковая 343832
Ива пепельная 343221
Ива персидская 343867
Ива пильчатолистная 344096
Ива полакучая 343070
Ива ползучая 343994,344000
Ива полупрутовидная 344081
Ива полярная 343900
Ива пржевальского 343913
Ива прилистниковая 344151
Ива продолговатолистная 343787
Ива прожацая 342973
Ива прутовидная 344244
Ива прутьевидная 344244
Ива прямосережковая 343982
Ива пурпурная 343937
Ива пурпуровая 343937
Ива пушистопобеговая 343283
Ива пьеро 343890
Ива пятитычинковая 343858
Ива радде 343975
Ива растопыренная 343314
Ива ричардсона 344009
Ива розмаринолистная 344031
Ива розмаринолистная ганьсуская 344035
Ива росистая 344022
Ива русская 344037
Ива рыжеватая 344049
Ива сахалинская 344053
Ива саянская 344060
Ива северная 343116
Ива серая 343221,343425
Ива сердцевиднолистная 343168
Ива серебристая 342973,343495
Ива серебристоберая 343052
Ива сетчатая 344003
Ива сибирская 344031
Ива сизая 343425
Ива сизоватая 343133,343336
Ива сизовато－серая 343614
Ива силезская 344115
Ива синевато－серая 343614
Ива ситхинская 344122
Ива скальная 344066
Ива скрытая 343975
Ива скрытная 342955
Ива старка 344148
Ива стлющаяся 343994
Ива сухолюбивая 344286
Ива сюзева 344124
Ива тарайкинская 344184
Ива темнолистная 344053
Ива тибетская 344294
Ива тонковетвистая 343601
Ива тонколистная 344194
Ива тонкосережчатая 344195
Ива тонкостолбиковая 343444
Ива тонтомуссирская 344206
Ива тощая 343648
Ива травянистая 343476
Ива трехтычинковая 344211
Ива трёхтычинковая 344211
Ива тундровая 344229
Ива туполистная 344004
Ива туранская 344230
Ива турчанинова 344231
Ива тэнчунская 344193
Ива тяньшанская 344203
Ива удская 344233
Ива узловатая 344208

Ива ушастая 343063
Ива федченко 343373
Ива ферганская 343378
Ива филиколистная 343886
Ива хрупкая 343396
Ива хубэйская 343509
Ива хультена 343151
Ива хумаская 343504
Ива хэбэйская 344172
Ива цельная 343529
Ива цзилунская 343463
Ива цзиндунская 343553
Ива чаюйская 344307
Ива чемуесережчатая 343600
Ива черешчатолистная 343899
Ива чёрная 343767
Ива чернеющая 343776
Ива черниковидная 343745
Ива черничная 343745
Ива чжэцзянская 343200
Ива шамиссо 343190
Ива шаньдунская 343578
Ива шелюга 342961
Ива шелюга красная 342961
Ива шерстистая 343592
Ива шерстистопобегая 343283
Ива шерстистопобеговая 343283
Ива шишковидная 344152
Ива шугнанская 344031
Ива эмэйская 343804
Ива южная 343064
Ива ядунская 343839
Ива яйцевиднолистная 343820
Ива якутская 343544
Ива японская 343546
Ива – бредина 343151
Ивам – да – марья 410677
Иван – чай 85875
Иван – чай узколистный 85875
Ивовые 342841
Ивоцветник 302229
Игла адамова 416584
Иглица 140099,340989,340990
Иглица гирканска 340997
Иглица головчатая 140105
Иглица колючая 340990
Иглица подлистный 340996
Иглица подъязычная 340994
Иглица подъязычный 340994
Иглица понтийская 340990,340998
Иглица понтийский 340998
Иглица шиповатая 340990
Иглица шиповая 340990
Иглочешуйник 329056
Иглочешуйник зонтичный 329114
Иглочешуйник индийский 329068
Иглочешуйник люцюский 329096
Игнаций 378761
Игнация 378761
Идезия 203417
Идезия многоплодная 203422
Идезия фуцзяньская 203424
Изабелла 411764
Изень 217324,217353
Измодинь 282477
Изолепис 353179
Изопирум 210251
Изопирум маньчжурский 210281
Изотома 210317
Икак 89816
Икако 89816
Икациновые 203303
Иконниковия 203518
Иконниковия кауфмана 203519
Иконниковия кауфмановская 203519
Икотник 53034
Икотник восходящий 53036
Икотник лопатчатый 170173
Икотник серозелёный 53038
Икотник серый 53038
Иксеридиум 210519
Иксеридиум злаковидный 210540
Иксеридиум злаколистный 210541
Иксеридиум зубчатолистный 210527
Иксеридиум китайский 210525
Иксеридиум щетинистый 210549
Иксеридиум юньнаньский 210552
Иксерис 210553
Иксерис зубчатый 210527
Иксерис китайский 210525
Иксерис ползучий 88977
Иксерис японский 210623
Иксиолирион 210988
Иксиолирион горный 210999
Иксиолирион каратекинский 210992
Иксиолирион татарский 210999
Иксия 210689
Иксора 211029
Иксора бирманская 211167
Иксора китайская 211067
Иксора тибетская 211192
Иксора хайнаньская 211103
Иксора шансиская 211196
Иксора юньнаньская 211204
Икстль 10854
Иланг – иланг 70960
Илах 75782
Илезина 208344
Илезина хербета 208349
Илекс 203527
Илекс баотинский 203980
Илекс ван 203584
Илекс гуандунский 203953
Илекс гуансиский 203645
Илекс гуйцзоуский 203868
Илекс кандинский 203833
Илекс кандинский мелколистный 203834
Илекс китайский 203625
Илекс куэньминский 203950
Илекс ланьюйский 203952
Илекс лунчжоуский 203991
Илекс лэйбоский 203971
Илекс малипоский 204022
Илекс мелколистный 204418
Илекс мотоский 204030
Илекс наньнинский 204080
Илекс наньчуаньский 204079
Илекс парагвайский 204135
Илекс рогатый 203660
Илекс тайваньский 203821
Илекс тибетский 204407
Илекс фуцзяньский 203835
Илекс фэнцинский 203807
Илекс хайнаньский 203869
Илекс цзиндунский 203846
Илекс чжэцзянский 204421
Илекс чэнкоуский 203621
Илекс шансиский 204154
Илекс эмэйский 204112
Илекс юньнаньский 204413
Илекс юньнаньский мелколистный 204418
Илекс янчуньский 204410
или ноготки 66445
Илим 401593
Илим полевой 401593
Иллициум 204474
Иллициум анисовый 204575,204603
Иллициум вэньшаньский 204601
Иллициум луаньдаский 204598
Иллициум мелколистный 204571
Иллициум мелкоцветковое 204558
Иллициум настоящий 204603
Иллициум обрядный 204575
Иллициум тайваньский 204484
Илом полевой 401593
Ильиниая 204434
Ильиниая ределя 204435
Ильм 401425,401512
Ильм американский 401431
Ильм белый 401431
Ильм гладтый 401549
Ильм горный 401512
Ильм крылатый 401430
Ильм листоватый 401468,401508,401593
Ильм мелкоистный 401602
Ильм мелкоплодный 401569
Ильм приземистый 401602
Ильм пробковый 401475,401633
Ильм сибирский 401509
Ильм туркестанский 401602
Ильм шершавый 401512

Ильм шотландский 401512
Ильмовник 401602
Ильмовые 401415
имбирные 418034
Имбирь 417964,418010
Имбирь жёлтый 114871
Имбирь инцзянский 418029
Имбирь лекарственный 418010
Имбирь мэнхайский 418001
Имбирь симаоский 418021
Имбирь эмэйский 418013
Имбирь юньнаньский 418030
Иммортель 415386
Императа 205470,205497
Императа цилиндрическая 205497
Инга 206918
Индау 153998
Индау посевной 154019
Индиго 205611
Индиго бастардное 20005
Индиго дикое 20005,47880
Индиго ложное 20005
Индигоноска 205611,206669
Индигоноска красильная 206669
Индигоноска сычуаньская 206637
Индигофера 205611
Индигофера ганьсуская 206421
Индигофера ложнокрасильная 206445
Индигофера тупоцветковая 205639
Индигофера хайнаньская 206740
Индигофера хэбэйская 205752
Индигофера цзиндунская 206127
Индийская сирень 219933
Инжир 164608,164763
Инжир ослиный 165726
Инжир смоковница 164608
Инжир фига 164763
Инкарвиллеа китайская 205580
Инкарвиллеа тибетская 205591
Инкарвиллея 205546
Инкарвиллея олиги 205574
Инкарвиллея сычуаньская 205552
Инуля 207025
Иорения 212433
Иорения малолисточковая 212434
Иохрома 207340
Ипекакуана 81970
Ипмея 293861
Ипмея индийская 207891
Ипомея 207546
Ипомея батат 207623
Ипомея индийская 207891
Ипомея китайская 25316
Ипомея леара 207934
Ипомея мексиканская 208010
Ипомея пурга 207901
Ипомея пурпурная 208120
Ипомея сибирская 250831
Ипомея слабительная 207901
Ипомея шаньдунская 207988
Ипомея щетинистая 207863
Ипомея ялана 207901
Ипполития 196688
Ипполития гердера 196698
Ипполития дарвазская 196692
Ипполития крупноглавая 196703
Ипполития крупноголоавчатая 196703
Ипполития шугнанская 196706
Ипполития юньнаньская 196712
Ирга 19240
Ирга азиатская 19248
Ирга канадская 19251,164006,164280
Ирга колосистая 19293
Ирга круглолистная 19288
Ирга метельчатая 19293
Ирга обыкновенная 19288
Ирга овалолистная 19288
Ирга овальная 19288
Ирга туркестанская 19297
Ирга цернокрайняя 19271
Ирис 208433
Ирис болотный 208771
Ирис германский 208583
Ирис мелкоцветковая 208853
Ирис персидский 208755
Ирис пумила 208874
Ирис садовый 208583
Ирис сибирский 208829
Ирис японский 208657
Ирисовые 208377
Ирный корень 5793
Исертия 209472
Искандера 209501,209502
Испанский дрок 370202
Иссоп 203079
Иссоп аптечный 203095
Иссоп зеравшанский 203104
Иссоп крупноцветный 203092
Иссоп лекарственный 203095
Иссоп меловой 203083
Иссоп обыкновенный 203095
Иссоп остроконечный 203084
Иссоп сомнительный 203081
Иссоп тяньшанский 203105
Иссоп узколистный 203082
Иссоп ферганский 203087
Истод 307891
Истод австрийский 307905
Истод альпийский 307903
Истод анаторийский 307910
Истод андрахновидный 307911
Истод белоцветный 308160
Истод большой 308181
Истод вольфганга 308468
Истод гибридный 308116
Истод гогенаккера 308104
Истод горький 307905
Истод горьковатый 307905
Истод закавказский 308412
Истод инееватый 308284
Истод кавказский 307981
Истод кемуларии 308135
Истод китайский 308081
Истод колхидский 307995
Истод лежачий 308394
Истод лицзянский 308163
Истод мариаме 308183
Истод мелкоцветковый 307927
Истод меловой 308010
Истод молдавский 308195
Истод мотыльковый 308248
Истод обыкновенный 308454
Истод подольский 308271
Истод прелестнейший 307909
Истод простертый 308394
Истод сванский 308388
Истод сенега 308347
Истод сибирский 308359
Истод симаоский 308140
Истод тайваньский 307919
Истод татаринова 308397
Истод тонколистный 308403
Истод урартийский 308430
Истод хайнаньский 308095
Истод хохлатый 307998
Истод штокса 308386
Истод японский 308123
Истодовые 308474
Итальяский Тополь 311400
Итеа 210359
Итеа гуансиская 210391
Итеа китайская 210364
Итеа мелкоцветковая 210407
Итеа цюцзянская 210390
Итеа юньнаньская 210423
Итеа янчуньская 210422
Итеа японская 210389
Итея виргинская 210421
Иткунак 361877
Иудино дерево 83809
Ишек – мия 368962
Ишек – мия эгорчак белоцведный 369095
Йерба – матэ 204135
Кабачок 114300,114304
Кабомба 64491
Кабомба каролинская 64496
Кава 300458
Кава – перец 300458
Кавказская черника 403728
Кавун 93304
Кага 119503

Кадзура 214947
Кадзура гуансиская 214964
Кадзура фэнцинская 214969
Кадзура эмэйская 214975
Кадзура японская 214971
Кадило 249542
Кадило бабье 249540
Кадило медоволистное 170076
Кадило мелиссолистное 249542
Кадия 64932
Кажу 21195
Казаро 205497
Казимироа 78172
Казимироа съедобная 78175
Казимироя 78175
Казуарина 79155,79161
Казуарина аргентинская 15953
Казуарина каннингема 79157
Казуарина пробковая 79185
Казуарина хвощевидная 79161
Казуарина хвощелистная 79161
Казуариновые 79193
Каимито 90032
Кайанус 65143,65146
Кайанус индийский 65146
Кайя иволензис 216205
Кайя иворская 216205
Кайя сенегальская 216214
Кайян 65146
Кайянус 65146
Какавахом 138445
Какалия ганьсуская 283815
Какалия копьевидная 283816
Какалия копьелистная 283816
Какао 389399,389404
Какао настоящее 389404
Каки 132219
Кактус "царица ночи" 83874
Кактус звездчатый 349932
Кактус колонновидный 83855
Кактус крупноцветный 83874
Кактус ластовневый 373719
Кактус лофофора 236425
Кактус маммилярия 243994
Кактус опунция 272775
Кактус пейреския 290701
Кактус пинсалис 329683
Кактус свисающий 83975
Кактус сосочковый 243994
Кактус церопегия 83975
Кактусовые 64822
Кактус – цереус 83855
Калаба 67856
Калабюра 259910
Каладиум 65214
Каладиум двуцветный 65218
Калады – какря 6278
Калам 341829,341912
Каламбак 29972
Калами 369652
Каламондин 93636
Каламус 65637
Каламус бирманский 65687
Каламус гуансиский 65702
Каламус драконовый 121493
Каламус инцзянский 65747
Каламус ланьюйский 65793
Каламус мотоский 65688
Каламус мэнлаский 65715
Каламус пинбяньский 65818
Каламус таиландский 65789
Каламус тайваньский 65697
Каламус хайнаньский 65703
Каламус шансиский 65674
Каламус юньнаньский 65816
Каламус яванский 65713
Каламус янчуньский 65815
Каламус яошаньский 65736
Каландриния 65827
Каланта 65862
Каланта гуйцзоуская 66107
Каланта сычуаньская 66122
Каланта тайваньская 65886
Каланта чемерицелистная 66101
Каланхое 215086
Каланхое войлочное 215277
Каланхое ланьюйское 215272
Каланхое мраморное 215206
Каланхое сомалиский 215263
Калатея 66157
Калатея литце 66169
Калган 312520
Калган дикий 312520
Калебаса 219843
Календула 66380
Календула каракалинская 66426
Календула лекарственная 66445
Календула ноготки 66445
Календула оффициналис 66445
Календула персидская 66453
Календула полевая 66395
Календула тонкая 66422
Калепина 66504
Калепина неравномерная 66506
Каликант китайский 68308
Каликант 68306
Каликант западный 68329
Каликант многоцветковый 68313
Каликантовые 68300
Каликантус 68306
Каликантус плодовитый 68321
Каликантус плодующий 68321
Каликантус флоридский 68313
Каликантус цветущий 68313
Калимерис 215334
Калимерис индийская 215343
Калимерис монгольский 215356
Калимерис надрезанная 215339
Калимерис надрезная 215339
Калимерис цельнолистная 215350
Калина 407645
Калина авабуки 407980
Калина березолистная 407696
Калина буддлелистная 407713
Калина буреинская 407715
Калина бурятская 407715
Калина вечнозелёная 408167
Калина вильчатая 407865
Калина висячая 408151
Калина войлочная 408034
Калина восточная 408015
Калина ганьсуская 407900
Калина гордовина 407907
Калина грушелистная 408061
Калина давида 407774
Калина душистая 407977
Калина жуйлиская 408130
Калина зазубренная 407778
Калина зибольда 408131
Калина зибольдова 408131
Калина ичанская 407802
Калина канадская 306994,407917
Калина карлса 407727
Калина кленолистная 407662
Калина корейская 407727,407905
Калина крапчатая 408074
Калина красная 407989
Калина крупноголовчатая 407939
Калина лавролистная 408167
Калина лушаньская 407795
Калина люцюская 407867
Калина мелколистная 408019
Калина монгольская 407954
Калина морщинистолистная 408089
Калина мотоская 407720
Калина обыкновенная 407989
Калина повислая 408151
Калина райта 408224
Калина раскидистая 407785
Калина расширенная 407785
Калина сарджента 408009,408098
Калина складчатая войлочная 408034
Калина скумпиелистная 407767
Калина сливолистная 408061
Калина сморщеннолистная 408089
Калина тайваньская 408156
Калина тибетская 408165
Калина хайнаньская 407877
Калина хубейская 407886
Калина цельнолистная 407907
Калина цилиндрическая 407769

Калина чжэцзянская 408120
Калина эмэйская 407988
Калина яошаньская 408141
Калина японская 407897
Калинсо луковичная 68538
Калинсо луковичное 68531,68538
Калинсо луковичный 68531
Калипсо 68531
Калипсо японская 68542
Калистедия 68671,102903
Калистедия даурская 68676
Калистедия заборная 68713
Калистедия плющевидная 68686
Калистедия японская 68692
Калифорнийское мамонтовое дерево 360565
Калла 66593,417093
Калла болотная 66600
Каллиантемум 66697
Каллиантемум алатавский 66698
Каллиантемум равноплодниковый 66710
Каллиантемум саянский 66717
Каллиантемум узколистый 66701
Каллика гуйцзоуская 66799
Калликарпа 66727
Калликарпа амелликарпа 66733
Калликарпа боденье 66743
Калликарпа боденье джиральди 66789
Калликарпа боденье наньчуаньская 66748
Калликарпа гуандунская 66833
Калликарпа динхуский 66939
Калликарпа индийская 66835
Калликарпа корейская 66824
Калликарпа луаньдаская 66904
Калликарпа мелкоплодная 66826
Калликарпа мягкая 66873
Калликарпа цзиньюньская 66790
Калликарпа юньнаньская 66948
Калликарпа японская 66808
Калликарпа японская цзилунская 66808
Каллистемон 67248
Каллистемон жестколистный 67291
Каллистемон ивовый 67293
Каллистемон пышный 67298
Каллистефус 67312
Каллистефус китайский 67314
Каллитрис 67405
Каллицефалюс 66949
Каллицефалюс блестящий 66951
Калопанакс 215420
Калопанакс клещевинолистный 215442
Калопанакс максимовича 215443
Калопанакс семилопастный 215442
Калотропис 68086
Калотропис гигантский 68088
Калофиллум 67846
Калофиллум круглоплодный 67860
Калофиллум обыкновенный 67856
Калохортус 67554
Калоцедрус 67526
Калоцедрус калифорнийский 228639
Калоцедрус сбежистый 228639
Калоцедрус тайваньский 67530
Калужница 68157
Калужница арктическая 68160
Калужница болотная 68189
Калужница дернистая 68167
Калужница дудчатая 68196
Калужница многолепестная 68213
Калужница перепончатая 68201
Калужница плавающая 68186
Калуфер 89602,383712
Кальдезия 66337
Кальдезия березоролистная 66343
Кальдезия почколистная 66343
Кальмия 215386
Кальмия многолистная 215393,215405
Кальмия сизая 215393,215405
Кальмия узколистная 215387
Кальмия широколистная 215395
Кальта 68157
Кальта палюстрис 68189
Кальцеолария 66252
Кальцеолария гибридная 66271
Кальцеолария мексиканская 66281
Кальцеолярия 66252
Камассия 68789
Камелия 68877,69156
Камелия аньлунская 68906
Камелия вэньшаньская 69760
Камелия гвоздичная 69594
Камелия горная 69594
Камелия гуандунская 69246
Камелия гуансиская 69244
Камелия гуйцзоуская 69248
Камелия далиская 69691
Камелия евгенольная 69594
Камелия либоская 69286
Камелия лунчжоуская 69325
Камелия масличная 69411,69594
Камелия маслоносная 69049,69411
Камелия мелкоцветковая 69355
Камелия мэнцзыская 69079
Камелия наньтоуская 69706
Камелия наньчуаньская 69389
Камелия обыкновенная 69156
Камелия пинбяньская 69075
Камелия сасанква 69594
Камелия сетчатая 69552
Камелия симаоская 69685
Камелия ху 69137
Камелия хубэйская 69142
Камелия цзиньюньская 69224
Камелия чжэцзянская 68980
Камелия эвегенольная 69594
Камелия эмэйская 69423
Камелия юньнаньская 69782
Камелия японская 69156
Камелия японская мелкоцветковая 69201
Каменник 18313
Каменник норской 235090
Камка 418392
Камнеломка 349032
Камнеломка абхазская 349045
Камнеломка австоралийская 82560
Камнеломка альберта 349055
Камнеломка анадырская 349058
Камнеломка астильбовидная 349079
Камнеломка белозоровидная 349755
Камнеломка болотная 349452
Камнеломка вечнозелёная 349054
Камнеломка восходящая 349049
Камнеломка голостебельная 349701
Камнеломка гребенчаторесничатая 349117
Камнеломка гуансиская 349527
Камнеломка гуншаньская 349411
Камнеломка даурская 349234
Камнеломка дернистая 349145
Камнеломка десулави 349246
Камнеломка динника 349254
Камнеломка дирнистая 349145
Камнеломка дэциньская 349245
Камнеломка ешольтца 349300
Камнеломка железистая 349048
Камнеломка жестколистная 349052
Камнеломка жёстколистная 349052
Камнеломка звездчатая 349932
Камнеломка зернистая 349419
Камнеломка кавказская 349159
Камнеломка килеватая 349153
Камнеломка кимвальная 349229
Камнеломка кладная 349533
Камнеломка кожистолистная 349203
Камнеломка козлиная 349452
Камнеломка колончатая 349193
Камнеломка колхидская 349192
Камнеломка колючая 349117
Камнеломка комарова 349520
Камнеломка коржинского 349523
Камнеломка корневая 349305
Камнеломка корнеопрысковая 349936
Камнеломка кортузолистная 349204
Камнеломка кочи 349524
Камнеломка крепная 349315
Камнеломка круглолистная 349859
Камнеломка крупнолепестная 349418
Камнеломка крупночашечковая 349579
Камнеломка крымская 349505
Камнеломка кузнецовская 349526
Камнеломка листочковая 349339,349340
Камнеломка ложноголая 349812

Камнеломка луковичная 349137
Камнеломка маньчжуьрская 349607
Камнеломка мелкоплодный 349642
Камнеломка мерка 349631
Камнеломка мерлерневидная 349118
Камнеломка можжевелолистная 349514
Камнеломка молоцчнобелая 349532
Камнеломка мотоская 349617
Камнеломка моховидная 349988
Камнеломка мускусная 349671
Камнеломка мутовчатая 350027
Камнеломка мэнцзыский 349627
Камнеломка мягкая 349655
Камнеломка нельсона 349691
Камнеломка орошенная 349505
Камнеломка отпрысковая 349936
Камнеломка охотская 349839
Камнеломка пеовионая 349319
Камнеломка плетеносная 349936
Камнеломка побегоносная 349936
Камнеломка ползучая 349936
Камнеломка поникающая 349162
Камнеломка понтийская 349799
Камнеломка прицветниковая 349113
Камнеломка продолговатолистная 349707
Камнеломка проломниковая 349060
Камнеломка пурпуовая 349832
Камнеломка редовского 349840
Камнеломка реснитчатая 349117
Камнеломка ручейная 349851
Камнеломка рыхлая 349303
Камнеломка сахалинская 349869
Камнеломка сибирская 349895
Камнеломка сиверса 349905
Камнеломка сычуаньская 349956
Камнеломка смолевкоцветная 349910
Камнеломка снежная 349700
Камнеломка супротивнолистная 349719
Камнеломка твердоногая 349882
Камнеломка тенистая 350011
Камнеломка теректинская 349988
Камнеломка тибетская 349990
Камнеломка тилинга 349992
Камнеломка тонкая 349700
Камнеломка точечная 349691
Камнеломка трехпалая 349999
Камнеломка трёхпалая 349999
Камнеломка трёхпальчатая 349999
Камнеломка узколистая 349934
Камнеломка усатая 349319,349934
Камнеломка форчуна 349344
Камнеломка хрящеватая 349158
Камнеломка цзиндунская 349511
Камнеломка чабрецелистная 349888
Камнеломка чамечная 349148
Камнеломка черно–белая 349622
Камнеломка щетинистая 349891
Камнеломка щетинчатая 349891
Камнеломка ястребинколистная 349444
Камнеломковые 350063
Камнеплодник 233099
Кампанула 69870
Кампанула джунгарская 70466
Кампанула лессинга 70464
Кампанула медиум 70164
Кампанула однолетная 70461
Кампанула юньнаньская 70368
Кампсис 70502
Кампсис крупноцветковый 70507
Кампсис укореняющаяся 70512
Кампсис укореняющийся 70512
Камптотека 70574
Камптотека остроконечная 70575
Камфолосма 70460
Камфолосма джунгарская 70466
Камфолосма лессинга 70464
Камфолосма марсельская 70463
Камфолосма однолетная 70461
Камчужная трава 400675
Камыс 148906
Камыш 37467,353179,353381
Камыш авечнский 353245
Камыш американский 352249
Камыш бухарский 353265
Камыш вихуры 353576
Камыш восточный 353676
Камыш гигантский 37467
Камыш дернистый 353272
Камыш комарова 352205
Камыш лежачий 353845
Камыш лесной 353856
Камыш линейчатый 353557
Камыш лушаньский 353576
Камыш малоцветный 353696
Камыш морской 353587
Камыш ниппонский 353660
Камыш обедненный 352238
Камыш обыкновенный 295888
Камыш озерной 352206
Камыш озерный 352289
Камыш озёрный 352206
Камыш остроконечный 352223
Камыш плоскостебельный 353707
Камыш прибрежный 353569
Камыш приморский 353569,353587
Камыш простерный 353845
Камыш прямой 352276
Камыш раскидистый 353845
Камыш ржавый 55895
Камыш ситниковидный 352200
Камыш стенной 293709
Камыш табернемонтана 352278
Камыш трехгранный 352284,353899
Камыш укореняющийся 353750
Камыш хмелевый 353574
Камыш черносемянный 353608
Камыш щединовидный 210080
Камыш эренберга 353381
Камышовка 37127
Канавалия 71036
Канавалия мечевидная 71050
Канавалия мечелистная 71050
Каналейник 293729
Канан 1000
Кананга 70954,70960
Кананга душистая 70960
Канап 217955
Канареечник 293705
Канареечник канареечный 293729
Канареечник канарский 293729
Канареечник клубневидный 293723
Канареечник клубневой 293764
Канареечник клубненосный 293723
Канареечник коротколосый 293722
Канареечник луковичный 293723
Канареечник малый 293743
Канареечник настоящий 293729
Канареечник простниковидный 293709
Канареечник птичий 293729
Канареечник своеобразный 293750
Канареечник тростниковый 293709
Канареечник узловатый 293723
Канареечник усеченный 293765
Канариум 70988
Канариум обыкновенная 70996
Канариум черный 71015
Канатник 837,1000
Канатник гибридный 913
Канатник теофраста 1000
Кандолея 194664
Кандум 66988,67050
Кандум безлистный 66998
Кандум китайский 67010
Кандык 154895
Кандык кавказский 154906
Кандык сибирский 154946
Кандык собачий 154909
Кандык южноевропейский 154909
Кандык японский 154926
Кандым 19728
Канегра 340084
Канелла 71129
Канель 25898
Канкриния 71085
Канкриния безъязычковая 71090
Канкриния бочанцева 71086
Канкриния голоскокова 71092
Канкриния золотоглавая 71088
Канкриния каратавская 71093
Канкриния максимовича 71096
Канкриния невского 71098

Канкриния неокаймленная 71102
Канкриния остроконечная 71097
Канкриния памироалайская 71099
Канкриния подобная 71103
Канкриния таджикская 71104
Канкриния тяньшанскя 71105
Канкриния фергаская 71091
Канкриния щетинконосная 71101
Канна 71156
Канна индийская 71181
Канна ирикоцветковая 71189
Канна красная 71166
Канна съедобная 71169
Канновые 71235
Канталупа 114199
Канткызылша 53249
Кануфер 89602,322709,383712
Каперсник 71676
Каперсник колючий 71871
Каперсовые 71671
Каперсы 71676
Каперсы колючие 71762
Каперсы розанова 71838
Каперцовые 71671
Каперцы 71676
Каперцы колючие 71762,71871
Каперцы тайваньский 71751
Каперцы хайнаньский 71757
Каперцы юньнаньский 71933
Капок 80120
Каприфоль 235630,235713
Каприфоль обыкновенная 235713
Капсикум 72068
Капуста 59278
Капуста белокачанная 59532
Капуста брюкая 59493
Капуста заячья 357203
Капуста заячья большая 356928
Капуста кергуеленская 315154
Капуста китайская 59595
Капуста кольраби 59541
Капуста кормовая 59520
Капуста кормовая мозговая 59541
Капуста кочанная 59532
Капуста критская 59370
Капуста кудрявая 59520
Капуста курчавая 59532
Капуста лиственная 59524
Капуста листовая 59520
Капуста листовая культурная 59541
Капуста морская 108591,108627
Капуста овошная 59520
Капуста огородная 59520
Капуста пенинская 59600
Капуста полевая 59603
Капуста репная 59541
Капуста собачья 250614,389183
Капуста спаржевая 60431
Капуста спаржевая цветная 60431
Капуста турнефора 59651
Капуста хреновидная 59380
Капуста цветная 79902
Капустная пальма 341422
Капуцин 399597
Капуцин большой 399601
Капуцин малый 399604
Капуцин овые 399596
Кара－арча 213933
Карабаркар 185034,185038
Карабат 76354
Карагана 72168
Карагана алайская 72175
Карагана алтайская 72285
Карагана белокорая 72270
Карагана бонгарда 72193
Карагана бунге 72199
Карагана всябелая 72244
Карагана высокая 72180
Карагана гривастая 72249
Карагана гривастая сычуаньская 72251
Карагана древовидная 72180
Карагана инееватая 72317
Карагана камилла шнейдера 72200
Карагана карликовая 72322
Карагана карликовая мелкоцветковая 72330
Карагана киргизов 72257
Карагана колючая 72348
Карагана колючелистная 72170
Карагана красивая 72267
Карагана крупноцветковая 72241
Карагана кустарник 72233
Карагана кустарниковая 72233,72237
Карагана максимовича 72283
Карагана мелколистная 72285
Карагана многолистная 72311
Карагана мягкая 72300
Карагана низкорослая 72320
Карагана обыкновенная 72180
Карагана оранжевая 72187
Карагана прейна 72316
Карагана пушистолистная 72215
Карагана степная 72233
Карагана тибетская 72346
Карагана трагакантовая 72364
Карагана туркестанская 72371
Карагана турфанская 72370
Карагана узколистная 72351
Карагана уссурийская 72373
Карагана франше 72227
Карагана цзилунская 72228
Карагана цинхайская 72206
Караганник 72168
Караганник древовидный 72180
Карагач 401475,401633
Карагач листоватый 401508
Карагач полевой 401593
Караматау 70463
Караматау монпелийская 70463
Карамол 45725
Карамола 45725
Карандашное дерево 213984
Карапа 72634
Карапа гвианская 72640
Кара－тал 343064
Кардаминопсис 30186
Кардамон 19836
Кардария 73085,225279
Кардиосперм 73194
Кардиоспермум 73194
Кардобенедикт 97599
Кардон 117770
Кардуль 354651
Карелиния 215566
Карелиния каспийская 215567
Карика 76808,76813
Карика папайа 76813
Кариковые 76820
Каринокарпус 106944
Кариокар орехоносный 77946
Кариоптерис 77992,78012
Кариоптерис седой 78012
Кариота 78045
Кариота жгучая 78056
Кариота мягкая 78047
Карисса 76854,76873
Кария 77874
Кария бахромчатая 77905
Кария белая 77934
Кария водяная 77881
Кария войлочная 77934
Кария голая 77894
Кария гуйцзоуская 77904
Кария китайская 25762
Кария косматая 77920
Кария мексиканская 77927
Кария мускатная 77910
Кария овальная 77912,77920
Кария опушённая 77934
Кария пекан 77902
Кария северокаролинская 69942,77885
Кария сердцевидная 77889
Кария хунаньская 77900
Кария яйцевидная 77920
Каркас 80569
Каркас ван 80779
Каркас гладкий 80643
Каркас голый 80643
Каркас гуншаньский 80645
Каркас дэциньский 80597
Каркас западный 80698
Каркас иедзоский 80600

Каркас кавказский 80607
Каркас китайский 80739
Каркас корейский 80664
Каркас крылатный 320258,320411
Каркас куэньминский 80668
Каркас люцюский 80663
Каркас малорослый 80723
Каркас сглаженный 80673
Каркас сетчатый 80728
Каркас тайваньский 80764
Каркас тибетский 80786
Каркас турнефора 80577
Каркас фунинский 80638
Каркас фэнцинский 80636
Каркас шершавый 80575
Каркас шуннинский 80737
Каркас южный 80577,80664
Карллюдовика 77038
Карлудовика 77038
Карнауба 103737
Кароб 83527
Каротель 123141
Карпезиум 77145
Карпезиум крупноголовый 77178
Карпезиум мелкоцветковый 77181
Карпезиум непалский 77182
Карпезиум особенный 77178
Карпезиум печальный 77190
Карпезиум польнный 77146
Карпезиум поникающий 77149
Карпезиум поникший 77149
Карпезиум сычуаньскй 77187
Картофель 367696
Картофель батат 207623
Картофель испанский гладкий 401412
Картофель китайский 131601
Картофель культурный 367696
Картофель мадагаскарский 99670
Картофель сладкий 207623
Каруба дикая 83527
Касатик 208433
Касатик аировидный 208771
Касатик альберта 208439
Касатик бальджуанский 208457
Касатик безлистный 208447
Касатик безлистый 208447
Касатик бледный 208744
Касатик блудова 208466
Касатик болотный 208771
Касатик бухарский 208473
Касатик ввведенского 208936
Касатик великолепный 208705
Касатик вздутый 208921
Касатик вильчатый 208524,208580
Касатик винклера 208942
Касатик виноградова 208943
Касатик водный 208771
Касатик водяный 208771
Касатик вонючий 208566
Касатик восточный 208740
Касатик вылощенный 208672
Касатик ганьсуский 208747
Касатик германский 208583
Касатик гирканский 208634
Касатик гладкий 208672
Касатик голубой 208494
Касатик гроссгейма 208604
Касатик гуга 208627
Касатик далиский 208495
Касатик дарвазский 208516
Касатик джунгарский 208848
Касатик желтейший 208562
Касатик желтый 208562,208771
Касатик жёлтый 208771
Касатик замещающий 208928
Касатик злаковидный 208600
Касатик злаковый 208600
Касатик золотистый 208771
Касатик иберийский 208635
Касатик изящнейший 208541
Касатик илийский 208636
Касатик кавказский 208481
Касатик камиллы 208477
Касатик каратегинский 208651
Касатик кемпфера 208543
Касатик кожистый 208813
Касатик колпаковского 208659
Касатик копетдагский 208660
Касатик королькова 208662
Касатик кроваво-красный 208806
Касатик крымский 208874
Касатик кушакевича 208664
Касатик лазистанский 208685
Касатик линейнолистный 208689
Касатик ложноаирный 208700,208771
Касатик ложно-аировой 208771
Касатик ложный 208771
Касатик ломнокавказский 208777
Касатик людвига 208696
Касатик маленикий 208874
Касатик маньчжурский 208706
Касатик мариковидный 208708
Касатик медведева 208710
Касатик мелкий 208752
Касатик мечевидный 208543
Касатик мусульманский 208723
Касатик наньтоуский 208725
Касатик нарбута 208726
Касатик нарынский 208728
Касатик ненастоящий 208733
Касатик непалский 208517
Касатик низкий 208632,208874
Касатик николая 208732
Касатик одноцветковый 208917
Касатик орхидейный 208738
Касатик орхидный 208738
Касатик остродольный 208434
Касатик остролепестковый 208543
Касатик парадоксальный 208750
Касатик персидский 208607
Касатик пестрый 208919
Касатик пёстрый 208919
Касатик песчаный 208562
Касатик побегоносный 208862
Касатик попова 208764
Касатик потанина 208765
Касатик приземистый 208632
Касатик равнодольный 208874
Касатик равнолопастный 208874
Касатик разноцветный 208924
Касатик родственный 208806
Касатик розенбаха 208792
Касатик русский 208797
Касатик самаркандский 208707
Касатик сепнолистный 208537
Касатик серполистный 208555
Касатик сетчатый 208788
Касатик сибирский 208829
Касатик синьцзянский 208607
Касатик сычуаньский 208834
Касатик согдийский 208607
Касатик солелюбивый 208606
Касатик солончаковый 208606
Касатик сузианский 208870
Касатик таджикский 208872
Касатик тайваньский 208574
Касатик тибетский 208492
Касатик тигровый 208897
Касатик тонкокоренный 208688
Касатик тонколистный 208882
Касатик тубердена 208911
Касатик тяньшанский 208690
Касатик узколистный 208882
Касатик уиллмотта 208940
Касатик уорлийский 208937
Касатик фиолетовый 208930
Касатик флорентийский 208565
Касатик фомина 208572
Касатик фостера 208576
Касатик цинхайский 208784
Касатик черепичатый 208638
Касатик чешуйчатый 208869
Касатик чжундяньский 208866
Касатик шелковниковая 208814
Касатик щетинистый 208819
Касатик щетиновидный 208819
Касатик щетиновидный канадский 208821
Касатик эвбанка 208553
Касатик японский 208543
Касатик ярко-желтый 208562
Касатиковые 208377

Каскариль 112884
Кассава 244501,244507
Кассава горькая 244507,244515
Кассава сладкая 244512
Кассандра 85412
Кассина 78540
Кассиопа 78622
Кассиопа гипновидная 78630
Кассиопа плауновидная 78631
Кассиопа фуцзяньская 78629
Кассиопа четырехрядная 78648
Кассиопа четырёхрядная 78648
Кассиопея 78622
Кассиопея вересковидная 78627
Кассиопея ледовского 78644
Кассиопея плауновидная 78631
Кассиопея стеллера 78646
Кассиопея четырехгранная 78648
Кассиопея четырехгрянная 78648
Кассия 78204
Кассия горькая 364382
Кассия дудка 78300
Кассия мэрилендская 360453
Кассия овальнолистная 78436
Кассия остролистная 78227
Кассия приморская 360453
Кассия стручковидная 78300
Кассия трубковидная 78300
Кассия туполистная 78436
Кассия узколистная 78227
Кассия южноазиантская 78300
Кассия яванская 78336
Касстиллея 79113
Касстиллея арктическая 79117
Касстиллея бледная 79130
Кастанопсис 78848
Кастанопсис гуйцзоуский 78969
Кастанопсис длинно – остроконечный 78903
Кастанопсис индийский 78959
Кастанопсис китайский 78889
Кастанопсис лунчжоуский 78982
Кастанопсис лэдунский 78974
Кастанопсис люцюский 78913
Кастанопсис наньнинский 78852
Кастанопсис пинбяньский 79000
Кастанопсис симаоский 78934
Кастанопсис сичоуский 79069
Кастанопсис тайваньский 78942
Кастанопсис тэнчунский 79067
Кастанопсис хайнаньский 78947
Кастанопсис хубэйский 78952
Кастанопсис эмэйский 79004
Кастилейя 79113
Кастилла 79108
Кастилла каучуконосная 79110
Кастиллейя 79113
Кастиллоа 79108
Кат 79395
Ката 79387,79395
Ката съедобная 79395
Катависса 15289
Катальпа 79242
Катальпа бигнониевая 79243
Катальпа бигнониевидная 79243
Катальпа западная 79260
Катальпа красивая 79260
Катальпа прекрасная 79260
Катальпа сиренелистная 79243
Катальпа яйцевидная 79257
Катальпа яйцевиднолистная 79257
Катальпа японская 79257
Катананхе 79271
Катауба 411764
Катран 108591
Катран армянский 108597
Катран беззубый 108604
Катран бугорчатый 108607
Катран восточный 108628
Катран гордягина 108612
Катран коктебельский 108620
Катран котчиевский 108621
Катран кочи 108621,108629
Катран крупноцветковый 108613
Катран литвинова 108626
Катран перистый 108627,108629
Катран понтийский 108630
Катран приморский 108627
Катран прутьевидный 108618
Катран сердцелистный 108599
Катран стевена 108637
Катран стевенова 108637
Катран татарский 108641
Катран шершавый 108598
Катран шугнанский 108634
Каттлея 79526
Каттлея цитрусовая 79532
Катх съедобный 79395
Катьянг 65146
Каукалис 79664
Каулофилум 79730,79732
Каулофилум мощный 79732
Каури 10492
Каури новозеландская 10494
Кауфманния 215639
Кауфманния короткопыльниковая 215640
Кауфманния семенова 215641
Кахрис 64776
Кахрис альпийский 64778
Кахрис гердера 64783
Кахрис крупноплодная 64788
Кахрис опушенная 64791
Качим 183157
Качим анатолийский 183171
Качим антонины 183172
Качим арециевидный 183173
Качим беловатые 183167
Качим бухарский 183177
Качим виличатый 183188
Качим волосистый 183207
Качим высокий 183170
Качим вытянутый 183236
Качим вязкий 183270
Качим головчатоцветковый 183180
Качим головчатый 183179
Качим густоцветный 183182
Качим даурский 183185
Качим двуцветный 183175
Качим железистый 183200
Качим изящный 183191
Качим иорский 183271
Качим каппадокийский 183181
Качим козелецолистный 183243
Качим коротколепестный 183176
Качим красивый 183232
Качим крепкий 183240
Качим крешенинникова 183211
Качим крыжниковидный 183205
Качим линейнолистный 183213
Качим липского 183214
Качим литвинова 183215
Качим лопатчатолистный 183249
Качим мейера 183217
Качим мелколистный 183218
Качим метельчатый 183225
Качим мокоричниковидный 183169
Качим остролистный 183158
Качим патрэна 183229
Качим плеображенского 183237
Качим поздний 183245
Качим ползучий 183238
Качим попова 183234
Качим постенный 183221,183245
Качим прутьевидный 183269
Качим пустыный 183187
Качим пучковатый 183195
Качим пушисточашечный 183194
Качим разноножковый 183206
Качим разноцветный 183199
Качим раскидистый 183189
Качим самбука 183242
Качим сизый 183201
Качим скученный 183203
Качим смолевковидный 183247
Качим спутанный 183210
Качим стевена 183253
Качим стенной 183221
Качим степной 183251
Качим тихоокеанский 183223,183231
Качим тонконожковый 183198
Качим триждывильчатый 183261

Качим туркестанский 183264
Качим тяньшанский 183260
Качим узколистный 183259
Качим украинский 183265
Качим уральский 183266
Качим федченко 183197
Качим фиолетовый 183268
Качим хэбэйский 183263
Качим черепичатый 183209
Качим шаровидный 183202
Качим шелковистый 183244
Качим шовица 183258
Качим штейпа 183252
Каччиния 64767
Каччиния сомнительная 64769
Каччиния толстолистная 64772
Кашгария 215616
Кашгария комарова 215618
Кашгария коротксцветковая 215617
Кашка 397043
Кашка красная 397019
Кашка лесная 396967
Кашкара 330182
Кашкарник золотистый 330182
Каштан 78766
Каштан американский 78790
Каштан благородный 78811
Каштан водяной 394500
Каштан высокий 78790
Каштан городчатый 78777
Каштан европейский 78811
Каштан жёлтый американский 9692
Каштан зубчатый 78790
Каштан карликовый 78807
Каштан конский 9675
Каштан конский восьмитычнковый 9692
Каштан конский гладкий 9694
Каштан конский голый 9694
Каштан конский калифорнийский 9679
Каштан конский китайский 9683
Каштан конский красный 9720
Каштан конский мелкоцветный 9719
Каштан конский обыкновенный 9701
Каштан конский павия 9720
Каштан конский розовоцветный 9720
Каштан конский японский 9732
Каштан мягчайший 78802
Каштан настоящий 78811
Каштан низкорослый 78807
Каштан обыкновенная 78811
Каштан посевной 78811
Каштан съедобный 78811
Каштан эмэйский 78796
Каштан японский 78777
Каштановые 346316
Каяну сголубиный горох 65146
Каянус 65143,65146
Каянус индийский 65146
Каянус каян 65146
Каянус сголубиный горох 65146
Квамассия 68789
Квамаш съедобный 68794
Квамоклит 323450
Квассия 323542
Квассия горькая 364382
Квассия ямайская 298510
Квебрахо 323561,350973
Квебрахо белое 39700
Квебрахо красное 350974
Квебрахо лоленца 323563
Квебрачо 323561,350973
Кверия 255435
Квилайя 324623
Квилайя мыльная 324624
Квиллайя 324623
Квиллайя мыльная 324624
Квиноа 87145
Квиноя 87145
Квисквалис 324669
Квит 84573
Кебрачо 323561,350973
Кевовое дерево 300998
Кедр 80074
Кедр альпийсий 300197
Кедр атласский 80080
Кедр атласский голубой 80083
Кедр гималайский 80087
Кедр европейский 299842
Кедр итальянский 300146,300151
Кедр калифорнийский 228639
Кедр капликовый 300163
Кедр корейский 300006
Кедр коротхохвойный 80085
Кедр красный 213984
Кедр ладанный 228635
Кедр ливанский 80100
Кедр маньчжурский 300006
Кедр остиндский 80030
Кедр речной 228635
Кедр речной калифорнийский 228639
Кедр речной новозеландский 228649
Кедр речной перистый 228649
Кедр речной сбежистый 228639
Кедр речной чилийский 45239
Кедр сибирский 300197
Кедр стланниковый 300163
Кедровая калифорнийская 299942
Кедровая сосна 300006
Кедровый стланик 300163
Келерия 217416
Келерия азиатский 217424
Келерия альбова 217419
Келерия гребенцатая 217441
Келерия сизая 217473
Келерия стерная 217441
Келерия стерная мелкоцветковая 217494
Келерия стройная 217441
Кельпиния 217600,217606
Кельпиния кртпноцветковая 217605
Кельпиния линейная 217604
Кельпиния туранская 217607
Кельрейтерия 217610
Кельрейтерия двухперистая 217615
Кельрейтерия метельчатая 217626
Кельрейтерия формозская 217620
Кельрейтерия цельнолистная 217615
Кембот 167506
Кемквот 167503,167506
Кемпедек 36920,36924
Кемпферия 214995
Кенаф 194779
Кендик 154909
Кендык 154909
Кендырь 29465
Кендырь коноплевидный 29473
Кендырь конопляный 29473
Кендырь плоромниковый 29468
Кенигия 217640
Кенигия исландская 217648
Кентрантус 81743
Кентрантус валельяновидный 81751
Кентрантус длиноцветковый 81758
Кентрантус красный 81768
кентырь 393695
кентырь армянский 29470
Кентырь крымуский 29522
кентырь ланцетолистный 29492
кентырь русанова 29509
кентырь сарматский 29510
Кентырь шершавый 29511
Керверь 28024
Керверь испанский 261585
Керверь клубневой 84731
Керверь корневой 84731
Керверь ликовичный 84731
Керверь листовой 28024
Керверь мускусный 261585
Керверь настоящий 28024
Керверь обыкновенная 28024
Керверь посевной 28024
Керверь репный 84731
Керверь салатный 28024
Керия 216084
Керия японская 216085
Кермек 167922,230508
Кермек бунге 230556
Кермек выемчатый 230761
Кермек гельцера 230642
Кермек гмелина 230631
Кермек гмелинов 230631
Кермек двуцветный 230607

Кермек золотой 230524
Кермек золотистый 230575
Кермек извилистый 230626
Кермек каспкйский 230572
Кермек кашгарский 230648
Кермек коралловидный 230586
Кермек крупнокорневой 230674
Кермек кустарниковый 230782
Кермек мейера 230678
Кермек михельсона 230680
Кермек мясистый 230565
Кермек обыкновенная 230815
Кермек опушенный 230801
Кермек плосколистный 230655
Кермек полукустарниковый 230782
Кермек почковиднолистный 230735
Кермек прутяной 230748
Кермек резниченко 230737
Кермек семенова 230753
Кермек серпоколосый 230610
Кермек скученный 230585
Кермек татарский 230787
Кермек тысячецветковый 230688
Кермек узколистный 230658
Кермек узколопастный 230657
Кермек ушколистный 230700
Кермек файзиева 230620
Кермек ферганский 230623
Кермек фишера 230625
Кермек франше 230627
Кермек широколистный 230655
Кермеквидка 230506
Кермеквидка оверина 230507
Кернера 216074
Керрия 216084
Керрия японская 216085
Керрия японская плена 216092
Кетелеерия 216114
Кетелеерия давида 216120
Кетелеерия форчтна 216142
Кетелерия 216114
Кетелерия тайваньская 216125
Кетелерия форчуна 216142
Кетелерия хайнаньская 216147
Кетелерия эвелины 216134
Кигелия 216313
Кизил 104917
Кизил белый 104920
Кизил головчатый 124891
Кизил кровяной 380499
Кизил крупнолиствый 380461
Кизил лекарсвенный 105146
Кизил мужской 105111
Кизил настоящий 105111
Кизил обыкновенный 105111
Кизил отпрысковый 105193
Кизил побегоносный 105193
Кизил свидина 380499
Кизил сибирский 104930,105207
Кизил съедобный 105111
Кизил татарский 104930,105207
Кизил флоридский 105023
Кизил шведский 105204
Кизил южный 104960
Кизиловые 104880
Кизильник 107319,107678
Кизильник антонины 107344
Кизильник антониный 107688
Кизильник антониный мелкоплодный 107689
Кизильник блестящий 107531
Кизильник войлочный 107710
Кизильник горизонтальный 107486
Кизильник даммера 107416
Кизильник дильса 107424
Кизильник замечательный 107440
Кизильник иволистный 107668
Кизильник кистецветный 107504,107637
Кизильник крупноплодный 107541
Кизильник крымский 107701
Кизильник лицзянский 107526
Кизильник ложномногоцветковый 107631
Кизильник малоцветковый 107611
Кизильник мелколистный 107549
Кизильник многоцветковый 107583
Кизильник мупинский 107576
Кизильник мэнцзыский 107472
Кизильник одноцветковый 107715
Кизильник остроконечный 107322
Кизильник пленчатый 107615
Кизильник поярковой 107624
Кизильник прелестный 107340
Кизильник прижатый 107336
Кизильник приятный 107694
Кизильник пушистолистый 107474
Кизильник разноцветный 107338
Кизильник растопыренный 107435
Кизильник роборовского 107650
Кизильник сычуаньский 107339
Кизильник скальный 107677
Кизильник тибетский 107707
Кизильник тимьянолистный 107549
Кизильник уорда 107726
Кизильник франше 107451
Кизильник холодный 107454
Кизильник цабеля 107729
Кизильник цельнокрайний 107504
Кизильник черноплодный 107543
Кизильник чжундяньский 107525
Кизильник чжэньканский 107387
Кизильцик японский 151162
Кизляк кистецветковый 239881
Кизляк кистецветный 239881
Киксия 216239
Киксия ложная 216292
Киксия повойничек 216262
Киксия повойничковая 216262
Килайя 324623
Киллинга 218457
Киллинга камчатская 218556
Киллинга коротколистная 218480
Киндаль 240210
Киндза 104687
Киндзя 104687
Кинкан 167489,167499,167506
Кинкан гонконгский дикий 167499
Кинкан круглый 167503
Кинкан маргарита 167506
Кинкан нагами 167506
Кинкан овальный 167506
Кинкан японский 167499,167503
Киноа 87145
Киноя 87145
Кипарис 114649
Кипарис азиатский болотный 178014
Кипарис аризонский 114652
Кипарис аризонский бонита 114652
Кипарис болотный 385262
Кипарис болотный мексиканский 385281
Кипарис вечнозеленый 114753,114760
Кипарис вечнозелёный 114753
Кипарис вечнозелёный пирамидальный 114765
Кипарис виргинский 385262
Кипарис гауэна 114698
Кипарис гималайский 114680,114775
Кипарис говена 114698
Кипарис говея 114698
Кипарис горибой говея 114695
Кипарис горизонтальный 114760
Кипарис гуадалупский 114702
Кипарис двураздельный 385262
Кипарис дукло 114680,114775
Кипарис дюкло 114680
Кипарис калифорнийский 114698
Кипарис кашмирский 114668
Кипарис китайский плакучий 114690
Кипарис крупноплодный 114723
Кипарис летний 217324,217361
Кипарис лузитанский 114714
Кипарис лысый 385262
Кипарис макнаба 114721
Кипарис миньцзянский 114669
Кипарис надутый 114680,114775
Кипарис настоящий 114753
Кипарис настоящий индийский 114763
Кипарис пирамидальный 114753,114765
Кипарис плагучий 114690
Кипарис португальский 114714
Кипарис португальский бентама 114664
Кипарис цзяньгэский 114709

Кипарисник 85263
Кипарисовик 85263
Кипарисовик горохоплодный 85345
Кипарисовик лавсона 85271
Кипарисовик лосона 85271
Кипарисовик лосона 'Алюмии' 85272
Кипарисовик лосона 'Глука 85283
Кипарисовик лосона 'Флейера 85282
Кипарисовик лосона эректа 'Виридис 85278
Кипарисовик лоусона 85271
Кипарисовик нутканский 85301
Кипарисовик туеобразный 85375
Кипарисовик тупой 85310
Кипарисовик тупой 'Тетрагона' 85330
Кипарисовик тупой 'Филикоидес' 85316
Кипарисовик туполистный 85310
Кипарисовик формозский 85268
Кипарисовые 114645
Кипрей 85870,85875,146586
Кипрей алмматинский 146860
Кипрей алыпийский 146598
Кипрей альпийский 146599
Кипрей альсинолистный 146601
Кипрей амурский 146605
Кипрей анагаллусолистный 146609
Кипрей анатолийский 146611
Кипрей арктический 146623
Кипрей бархатистый 146932
Кипрей бедноцветковый 146834
Кипрей бедноцветный 146834
Кипрей белоцветковый 146755
Кипрей берингийский 146629
Кипрей болотный 146812
Кипрей бонгарда 146635
Кипрей волосистый 146724
Кипрей высокогорный 146763
Кипрей гирляндообразный 146879
Кипрей головчаторыльцевый 146606
Кипрей головчато-рыльцевый 146607
Кипрей голоплодный 146606
Кипрей горнеманна 146734
Кипрей горный 146782
Кипрей даурский 146680
Кипрей двурядный 146630
Кипрей железистый 146709
Кипрей железистостебельный 146587
Кипрей жилковатый 146795
Кипрей красиохохолковый 146849
Кипрей лами 146905
Кипрей лансетолистный 146760
Кипрей мелкоцветковый 146775,146834
Кипрей мелкоцветный 146834
Кипрей мохнатый 146724,146932
Кипрей нечеткий 146671
Кипрей поникающий 146803
Кипрей понтийский 146841
Кипрей приросший 146595
Кипрей равноветвистый 146696
Кипрей розовый 146859
Кипрей скальный 146873
Кипрей скромный 146775
Кипрей сродный 146595
Кипрей тёмно-зелёный 146805
Кипрей темный 146805
Кипрей теплолюбивый 146918
Кипрей тонкий 146605
Кипрей тундровый 146928
Кипрей тяньшанский 146919
Кипрей угроватый 146612
Кипрей узколистный 85875
Кипрей уральский 146930
Кипрей холмовый 146669
Кипрей холодостойкий 146706
Кипрей хорнеманна 146734
Кипрей цилиндрорыльцевый 146606
Кипрей чашучкоцветный 146607
Кипрей шершавоволосистый 146724
Кипрей широколистный 85884
Кипрей японский 146738
Кипрейник 85870,146586
Кипрейник альпийский 146599
Кипрейник арктический 146623
Кипрейник болотный 146812
Кипрейник волосистый 146724
Кипрейник горнемана 146734
Кипрейник горный 146782
Кипрейник даурский 146680
Кипрейник жилковатый 146795
Кипрейник лансетолистный 146760
Кипрейник лями 146905
Кипрейник мелкоцветковый 146834
Кипрейник мокричнолистный 146601
Кипрейник очноцветолистный 146609
Кипрейник розовый 146859
Кипрейник тайваньский 146901
Кипрейник темнозеленый 146805
Кипрейник тёмно-зелёный 146805
Кипрейник тундровый 146929
Кипрейник уральский 146930
Кипрейник холмовой 146669
Кипрейник четырехгранный 146905
Киренгешома 216473
Кириловия 216486
Кириловия пушиотоцветковая 216487
Кирказон 34097
Кирказон иберийский 34211
Кирказон крупнолистный 34337
Кирказон крупноцветтный 34198
Кирказон ломоносовидный 34149
Кирказон маньчжурский 34268
Кирказон обыкновенный 34149
Кирказон понтийский 34149,34300
Кирказон сычуаньский 34147
Кирказон скрученный 34154
Кирказон слабый 34162
Кирказоновые 34384
Кислец 291663
Кислица 141470,277644,334033
Кислица заячья 277648
Кислица заячья тайваньская 277650
Кислица кисленькая 277648
Кислица кисловатая 277648
Кислица клубневая 277779
Кислица козья 278008
Кислица обратнотреугольная 277985
Кислица обыкновенная 277648
Кислица поротая 277747
Кислица прямая 278099
Кислица рогатая 277747
Кислица рожковая 277747
Кислица торчащая 278099
Кислица фиолетовая 278147
Кислица ямайская 195196
Кисличник 278576
Кисличник двустолбиковый 278577
Кисличник двухстолбчатый 278577
Кисличные 277643
Кислый апельсин 93332
Кистесемянник 57675
Кистесемянник китайский 57678
Кистесемянник тонкий 57685
Кистецветвик обыкновенный 260333
Китайка 243667
Кишмиш крупный 6515
Кишмиш обыкновенный 6639
Кишнец 104687,104690
Кишнец посевной 104690
Кияк 144312
кладохета 93937
кладохета каспийская 93939
кладохета чистейшая 93938
Кладрастис 94014
Кладрастис жёлтый 94022,94026
Клайтония 94292
Клайтония арктическая 94299
Клайтония васильева 94382
Клайтония иоанна 94322
Клайтония клубневидная 94369
Клайтония остролистная 94294
Клайтония отпрысковая 94358
Клайтония пронзённолистный 94349
Клайтония сибирная 94362
Клайтония эшольца 94315
Кларкия 94132,94138,312946
Кларкия изящиная 94138
Кларкия красивая 94140
Клаусия 94232
Клаусия мягкая 94238
Клаусия ольги 94239
Клаусия сосочковаю 94240

Клаусия туркестанская 317565
Клаусия щетинистая 94236
Клбнекамыш 56633
Клевер 396806
Клевер александрийский 396817
Клевер альпийский 396818,396819
Клевер американский стенной 292279
Клевер батлачковый 397006
Клевер бедноцветковый 397002
Клевер белоголовка 396981
Клевер белоцветковый 396953
Клевер белый 397043
Клевер белый ползучий 397043
Клевер берсим 396817
Клевер бледножелтый 396989
Клевер бледно - жёлтый 396989
Клевер бледноцветковый 396953
Клевер бледный 396998
Клевер бордзиловского 396845
Клевер бухарский 249203
Клевер венгерский 396895,396999
Клевер вздутый 397122
Клевер волосистоголовый 397120
Клевер вспухлый 397013
Клевер гиблидный 396930
Клевер горный 396981,397083
Клевер греческий 397229
Клевер египетский 396817
Клевер жёлтый 396999
Клевер жестковолосый 397068
Клевер заячий 28200
Клевер звездчатый 397089
Клевер земляниковидный 396905
Клевер земляничный 396905
Клевер иглистый 396885
Клевер изящный 396930
Клевер инкарнатный 396934
Клевер кавказский 396860
Клевер каштановый 397083
Клевер клубковидный 396917
Клевер ключевой 396903
Клевер красивый 397084
Клевер краснобелый 396930
Клевер красно - белый 396930
Клевер красноватый 397059
Клевер красный 397019,397059
Клевер культурный 397065
Клевер луговой 397019
Клевер лупиновидный 396956
Клевер лупиновый 396956
Клевер люпиновидный 396956
Клевер люпиновый 396956
Клевер маленовый 396934
Клевер маленький 396882
Клевер мегкоцветковый 397002
Клевер многолистный 397016
Клевер молинери 396934
Клевер мохнатый 396926
Клевер мясокрасный 396934
Клевер нарядный 396930
Клевер непостояный 396821
Клевер нитевидный 396896
Клевер опрокинутый 397050
Клевер открытозевый 397087
Клевер палермский 397000
Клевер паннонский 396999
Клевер паскидистый 396881
Клевер паспростертый 396881
Клевер пашенный 396828,397096
Клевер пенистый 397086
Клевер перевернутый 397050
Клевер перевёрнутый 397050
Клевер персидский 397050
Клевер плодоножковый 397091
Клевер подземнный 397106
Клевер подземный 397106
Клевер полевой 396828,396854,397096
Клевер ползучий 397043
Клевер полосатый 397095,397097
Клевер посевной 397096
Клевер превосходный 396893
Клевер пренебреженный 396984
Клевер приморский 396961
Клевер пузырчатый 397130
Клевер пунцовый 396934
Клевер равнинный 396854
Клевер радде 397040
Клевер растопыренный 397087
Клевер расширенный 396895
Клевер репейниковый 396951
Клевер розовый 396930
Клевер светлоделтый 396989
Клевер светложелтый 396989
Клевер светло - жёлтый 396989
Клевер себастиана 397073
Клевер седоватый 396858
Клевер серавшанский 397079
Клевер скасный 397065
Клевер скасный Кудряш 397066
Клевер скасный позднеспелый 397067
Клевер скрученный 396917
Клевер сомнительный 396882,396896
Клевер средний 396967
Клевер сходный 396821
Клевер темнокаштановый 396981,397083
Клевер тихоокеанский 396996
Клевер тонкостебельный 396896
Клевер турецкий 271280
Клевер угловатый 396822
Клевер узколистный 396823
Клевер шабдар 397050
Клевер шведский 396930
Клевер шероховатый 397068
Клевер шершавый 396926,397068
Клевер шуршащий 397096
Клевер японский 218347
Клейера 96576
Клейера грушелистная 96616
Клейера мелколистная 96627
Клейера янчуньская 96630
Клейера японская 96587
Клейнховия 216641
Клейтония 94292
Клекачка 374087
Клекачка колхидская 374098
Клекачка перистая 374107
Клекачковые 374114
Клематис 94676
Клематис альпийский 94714
Клематис жакмана 95036
Клематис сибирский 95302
Клематис цельнолистный 95029
Клен 2769
Клён 2769
Клен американский 3218
Клён американский 3218
Клен аньхуйский 2791
Клен бархатистый 3736
Клён бархатистый 3736
Клён бархатный 3736
Клён белый 3462
Клен березовый 3732
Клен бородатожилковый 2798
Клен бородатый 2798
Клён боярышниколистный 2910
Клён бумахный 3003
Клен бюргера 2811
Клен веерный 3300
Клён веерный 3300
Клен велечественный 3736
Клён величественный 3736
Клён весёлый 2849
Клен виноградолистный 2896
Клен виргинский 3218
Клен волосистоножковый 2846
Клен воярышниколистный 2910
Клён высокогорный 3699
Клен ганьсуский 3130
Клён гельдрейха 3013
Клен гиблидный 3025
Клен гиннала 2984
Клён гиннала 2984
Клен гирканский 3029,3030
Клен гладгий 3075
Клен граболистный 2868
Клён граболистный 2943
Клен грузинский 3030
Клен гуандунский 3484
Клен гуансиский 3696
Клен гуйлиньский 3070
Клен гуйцзоуский 3009

Клен гуншаньский 3065
Клен давиба 2920
Клен давиба крупнолистный 2923
Клен дальневосточный 2984
Клен даньбаский 2978
Клен дедюлэ 3732
Клен дланевидный 3300
Клён дланевидный 3300
Клен дланевидный ‘Атропурпуреум’ 3302
Клен дланевидный атролинеаре 3301
Клен дланевидный версиколор 3347
Клен дланевидный орнатум 3342
Клен дланевидный ретикулятум 3343
Клен дланевидный розео – маргинатум 3344
Клен дланевидный рубрум 3345
Клен дланевидный сангвинеум 3346
Клен дланевидныйрассеченный 3309
Клен желтый 3732
Клен завитой 2893
Клён завитой 2893,75859
Клен зеленокорый 3665
Клен зибольда 3604
Клён зибольда 3604
Клен иберийский 3030
Клен итальянский 3286
Клён кавказский 3699
Клён кавказский высокогорный 3699
Клен кападокийский 2849
Клен кападокийский мелколистный 2866
Клен китайский 3615
Клен китайский мелкоплодный 3622
Клен колосистый 3629
Клен комарова 3064
Клен корейский 3273
Клён красивый 3398
Клен красный 3505
Клён красный 3505
Клён критский 3294
Клен крупнолистный 3127
Клен ланьпинский 3083
Клен лаоюйский 3089
Клен лин 3103
Клен линьаньский 3102
Клен лицзянский 2970
Клен личуаньский 3101
Клен ложнозибольдов 3479
Клен ложнозибольдов мелкоплодный 3480
Клен ложноплатановый 3462
Клён ложноплатановый 3462
Клен ложный платан 3462
Клен лэйбоский 3097
Клен мабяньский 3131
Клен майра 3140
Клен максимовича 3137
Клен маньчжурский 3129
Клён маньчжурский 3129
Клен мелколистный 3154,3676
Клен мийябе 3149
Клен моно 3154
Клен монпельйский 3206
Клён монпельйский 3206
Клен мулиский 3211
Клен мяотайский 3145
Клен наньчуаньский 3119
Клен наюн 3216
Клен негундо 3218
Клён обыкновенный 453,3425
Клен оливера 3277
Клен оливера тайваньская 3278
Клен опушеный 3481
Клен остролистный 3425
Клён остролистный 453,3425
Клен пакс 3374
Клён пальмовидный 3300
Клен пенсильванский 3393
Клён пенсильванский 3393
Клен перистолистный негундо 3218
Клен перистый 3218
Клен платановидный 3425
Клён платановидный 453,3425
Клен покровный 3665
Клен полевой 2837
Клен полевойпушистоплодный 2844
Клен приречный 2984
Клён приречный 2984
Клен продолговатый 3255
Клен продолговатый эмэйский 3266
Клен расходящийся 2941
Клен регеля 3494
Клен речной 2984
Клён речной 2984
Клён рыжеватожилковый 3529
Клен рыжежилковый 3529
Клен сахаристый 3532
Клён сахаристый 3532
Клен сахарный 3549
Клён сахарный 3549
Клён светрый 2849
Клен семенова 3587
Клен серебристый 3532
Клён серебристый 3532
Клен серый 3003
Клён серый 3003
Клен сичоуский 3603
Клен сычуаньский 3648
Клен стевена 3645
Клен тайваньский 2815
Клен татарский 3656
Клен траутфеттера 3699
Клён траутфеттера 3699
Клён трёхлопастный 3206
Клен тукурменский 3725
Клен туркестанский 3726
Клен тяньцюаньский 3649
Клен укурундийский 3732
Клен усеченный 3714
Клен франше 2972
Клен фунинский 3297
Клен фупинский 2980
Клен фэн 3067
Клен хайнаньский 3010
Клен ху 3024
Клен хуанпинский 3023
Клен хунаньский 3217
Клен хэнаньский 3498
Клен цзиндунский 3055
Клен цзиньюньский 3747
Клен цзянхиский 3062
Клен цзяньшуйский 2785
Клен цюцзянский 3063
Клен чанхуаский 2884
Клен чаюйский 3692
Клён чёрный 3236
Клен чоноски 3718
Клен чоноского 3718
Клен чэнбуский 3117
Клен шансиский 3593
Клен эмэйский 2946
Клен явор 3462
Клён явор 3462
Клен яньюаньский 3581
Клен яошаньский 3759
Клен японский 3034
Клён японский 3034
Клен японский аконитифолиум 3035
Клен японский крупнолистный 3050
Клен японский мелколистный 3037
Клен ясенелистный 3218
Клён ясенелистный 3218
Клен ясенелистный вариегатум 3232
Клен ясенелистный фиолетовый 3233
Кленовые 3765
Клёновые 388,3765
Клеома 95606
Клеоме 95606
Клеоме гордягина 95682
Клеоме клювообразная 95775
Клеоме колютееобразная 95652
Клеоме липского 95723
Клеоме ное 95742
Клеоме ноевская 95742
Клеоме птиценогая 95746
Клеоме птиценогое 95746
Клеоме пятижилковая 95768
Клеоме радде 95771
Клеоме туременская 95810
Клеоме хайнаньский 95782
Клеоме шерстистая 95806
Клеоме юньнаньский 95835
Клеродендрон 95934

Клеродендрон бунге 95978
Клеродендрон вонючий 96078
Клеродендрон гуандунский 96167
Клеродендрон гуансиский 96029
Клеродендрон индийское 96139
Клеродендрон китайский 96009
Клеродендрон криволистный 96028
Клеродендрон таиландский 96096
Клеродендрон тибетский 96392
Клеродендрон трехнадрезный 96398
Клеродендрон хайнаньский 96114
Клеродендрон чжэцзянский 96157
Клеродендрон японский 96147
Клетра 96454
Клетра бородчатожилковая 96466
Клетра мелкоплодная 96546
Клетра наньчуаньская 96536
Клетра ольхолистная 96457
Клетра чэнкоуская 96501
Клетровые 96561
Клещевина 334432
Клещевина занзибарская 334435
Клещевина обыкновенная 334435
Клещина 157268
Клещинец 37015
Клиантус 96633
Кливия 97213,97218
Кливия оранжевая 97218
Кливия суриковая 97218
Кливия сурикова-красная 97218
Клиноног 65528,96960
Клинтония 97086
Клинтония удинская 97093
Клинтония удская 97093
Клинэлимус 144144
Клинэлимус высокий 144303
Клинэлимус даурский 144271
Клинэлимус поникший 144414
Клинэлимус сибирский 144466
Клинэлимус синьцзянский 144474
Клитория 97184
Клокичка 374087
Клокичка бумальда 374089
Клокичка гималайская 374100
Клокичка колхидская 374098
Клокичка перистая 374107
Клокичковые 374114
Клоповник 225279,235065
Клоповник безлепестный 225295,225305,225340
Клоповник безлистный 225295,225340
Клоповник близкий 73092
Клоповник борщова 225308
Клоповник буассье 225303
Клоповник виргинский 225475
Клоповник воронцелистный 225335
Клоповник воронцелистый 225335
Клоповник густоцветковый 225340
Клоповник джугарский 225460
Клоповник злаколистный 225369
Клоповник каратавский 225393
Клоповник крупновидный 73090
Клоповник лировидный 225411
Клоповник мейера 225414
Клоповник мусорный 225447
Клоповник низкий 225440
Клоповник низкорослый 225440
Клоповник перистый 225433
Клоповник персидский 225429
Клоповник полевой 225315
Клоповник ползучий 225443
Клоповник посевной 225450
Клоповник пронзеннолистный 225428
Клоповник пронзённолистный 225428
Клоповник пронзенный 225428
Клоповник пузырчатый 225473
Клоповник пустыни 225344
Клоповник пустынный 225362
Клоповник разнолистный 225428
Клоповник разрезной 225397
Клоповник сердцевидный 225333
Клоповник сибирский 225455
Клоповник сорный 225447
Клоповник стебреобъемлющий 225293
Клоповник толстолистный 225326,225336
Клоповник тупой 225425
Клоповник турчанинова 225472
Клоповник ферганский 225364
Клоповник широколистный 225398
Клопогон 90994
Клопогон борщевиколистный 91023
Клопогон вонючий 91011
Клопогон даурский 91008
Клопогон европейский 91010
Клопогон наньчуаньский 91032
Клопогон простый 91038
Клопогон юньнаньский 91042
Клопогон японский 91024
Клоровник 225447
Клубнекамыш морской 353587
Клубнекамыш схожий 353828
Клубника 167592,167627
Клубника ананасная 167597
Клубника ананасовая 167597
Клубника виргинская 167663
Клубника восточная 167641
Клубника высогая 167627
Клубника настоящая 167627
Клубничноедерево 124891
Клюзия 97260
Клюква 278351
Клюква болотная 278366
Клюква крупноплодная 403894
Клюква мелкоплодная 278360
Клюква обыкновенная 278366
Клюква четырехлепестная 278366
Ключики 315082
Кляусия 94232
Кляусия казахская 94237
Кляусия солнцепечная 94233
Кмин 114502
Кмин тминовый 114503
Кнеорум 97491
Кникус 97581
Кникус аптечный 97599
Кникус благословенный 97599
Кникус благословленный 97599
Книфофия 216923
Княженика 338142
Княжик 94676
Княжик альпийский 94714
Княжик крупнолепестковый 95100
Княжик охотский 95304
Княжик сибирский 95302
Княжик тяньшанский 95373
Кобея 97752
Кобея лазающая 97757
Кобея лазящая 97757
Кобея цепкая 97757
Кобрезия 217117
Кобрезия белларди 97768
Кобрезия волосолистная 217130
Кобрезия ганьсуская 217198
Кобрезия метельчатая 97772
Кобрезия мулиская 217307
Кобрезия мышзхвостовидная 217234
Кобрезия непалская 217236
Кобрезия низкая 97771,217195
Кобрезия нитевиднолистная 97770
Кобрезия овальноколосая 217274
Кобрезия простая 97775
Кобрезия ройля 97773
Кобрезия сычуаньская 217280
Кобрезия схенусовидная 217274
Кобрезия схонусовидная 97774
Кобрезия тибетская 217299
Кобрезия цзилунская 217148
Кобрезия ядунская 217310
Ковыль 376687
Ковыль алмянский 376704
Ковыль араксинский 376701
Ковыль бадахшанский 376709
Ковыль байкальский 376710
Ковыль большой 376795
Ковыль бунге 376723
Ковыль великолепный 376846
Ковыль волосатик 376732
Ковыль волосовидный 376949
Ковыль восточный 376859
Ковыль галечный 376785
Ковыль гигантский 376779

Ковыль густоцветковый 376762
Ковыль дагестанский 376758
Ковыль даленивосточный 4134
Ковыль длинноперистый 376844
Ковыль длинноязычковый 376845
Ковыль закручивающийся 376728
Ковыль залесского 376962
Ковыль ильина 376804
Ковыль кавказский 376737
Ковыль камнелюбивый 376842
Ковыль каратавский 376810
Ковыль киргизский 376814
Ковыль коржинского 376819
Ковыль коротκοцветковый 376720
Ковыль красивейший 376890
Ковыль красивый 376718,376890
Ковыль красноватый 376962
Ковыль крашенинникова 376821
Ковыль крылова 376910
Ковыль лессинга 376836
Ковыль лже – волосатик 376888
Ковыль липского 376841
Ковыль обыкновенный 376877
Ковыль овчинникова 376862
Ковыль опушенолистный 376759
Ковыль памисский 376864
Ковыль перистый 376877
Ковыль понтийский 376884
Ковыль развесистый 4134
Ковыль регеля 376899
Ковыль рихтеровский 376902
Ковыль родственный 376753
Ковыль сарептский 376908
Ковыль своеобразный 376867
Ковыль седоватый 376727
Ковыль сибирский 4160
Ковыль спиридонова 376920
Ковыль толстостебельный 376757
Ковыль тонкий 376793
Ковыль тончайщий 376931
Ковыль тургайский 376695
Ковыль туркестанский 376956
Ковыль туркменский 376953
Ковыль тырса 376732
Ковыль тяньшанский 376936
Ковыль уклоняющийся 376691
Ковыль украинский 376958
Ковыль шелковистый 376799
Ковыль шовица 376930,376934
Ковыль ягнопский 376807
Ковыль языкообразный 376840
Кодиеум 98177
Кодиеум пестролистный 98190
Кодиеум пестролистный расписная 98190
Кодонопсис 98273
Кодонопсис бирманский 98288
Кодонопсис клематисовдный 98290
Кодонопсис ланцетный 98343
Кодонопсис ланцетолистный 98343
Кодонопсис ломоносовидный 98290
Кодонопсис мелковолосистый 98395
Кодонопсис тибетский 98431
Кодонопсис уссурийский 98424
Кодоноцефалум 98260
Кодоноцефалум пекока 98263
Коелококкус 252631
Кожанка 320065,320071
Кожанка трёхлистный 320071
Кожевенное дерево 107300
Кожевник 332452
Козелец 354797,354940,394372
Козелец австрийский 354813
Козелец альберта регеля 354800
Козелец амударьинский 354911
Козелец армянский 354810
Козелец бальджуанский 354823
Козелец белостебельный 354801
Козелец биберштейна 354825
Козелец бунге 354828
Козелец гвоздиковидный 354845
Козелец гиссарский 354871
Козелец григорашвили 354863
Козелец гроссгейма 354864
Козелец двуцветный 354824
Козелец джунгарский 354954
Козелец завитой 354835
Козелец заиийский 354966
Козелец зейдлица 354949
Козелец иды 354875
Козелец изащный 354862
Козелец илийский 354877
Козелец искривленный 354841
Козелец испанский 354870,354959
Козелец кецховели 354880
Козелец клубненосный 354968
Козелец козлобородниковый 354965
Козелец колючеветвистый 354798
Козелец красивый 354923
Козелец крошечный 354926
Козелец крымский 354959
Козелец курчавый 354838
Козелец липского 354889
Козелец литвинова 354890
Козелец ложнорастопыренный 354918
Козелец ложношерстистый 354920
Козелец луговой 354917
Козелец лучетый 354930
Козелец маленький 354926
Козелец мейера 354902
Козелец мелкоцветковый 354913
Козелец мечелистный 354854
Козелец многоветвистый 354936
Козелец монгольский 354904
Козелец мохнатоплодный 354855
Козелец мягкий 354903
Козелец неприметный 354878
Козелец низкий 354873
Козелец опушенный 354922
Козелец петрова 354914
Козелец почти – бесстебельный 354956
Козелец приземистый 354873
Козелец прицветничный 354826
Козелец пробкобый 354957
Козелец прямой 354955
Козелец пурпуровый 354925
Козелец пухосемянный 354855
Козелец радде 354929
Козелец разресной 354881
Козелец рассечённый 354881
Козелец розовый 354941
Козелец рупрехта 354813
Козелец сафиева 354945
Козелец сладкий корень 354840,354870
Козелец таджиков 354958
Козелец тау – сагыз 354961
Козелец толстолистный 354837
Козелец тонколистный 354888
Козелец туркевича 354972
Козелец туркестанский 354971
Козелец туркменский 354970
Козелец тяньшанский 354962
Козелец ферганский 354858
Козелец франше 354859
Козелец хантавский 354834
Козелец шелковисто – шерстистый 354950
Козелец шерстистный 354886
Козелец широколистный 354887
Козелец шишкина 354946
Козелец яйцевидный 354910
Козлец 394372
Козлобородник 394257,394284,394301, 394326,394332
Козлобородник абрикосово – желтый 394261
Козлобородник алтайский 394260
Козлобородник бадахшанский 394262
Козлобородник белорусский 394263
Козлобородник большой 394281
Козлобородник большой 394312
Козлобородник буфтальмовидный 394266
Козлобородник введенского 394359
Козлобородник возвышеный 394283
Козлобородник волжский 394358
Козлобородник восточный 394321
Козлобородник высокий 394283
Козлобородник вытянутый 394284
Козлобородник гауданский 394287
Козлобородник головчатый 394268
Козлобородник горского 394291
Козлобородник гранистоплодный 394316
Козлобородник дагестанский 394276

Козлобородник днепровский 394264
Козлобородник донецкий 394279
Козлобородник донской 394279,394350
Козлобородник дубянского 394280
Козлобородник злаколистный 394293
Козлобородник иды 394297
Козлобородник изящный 394292
Козлобородник казахстанский 394300
Козлобородник карелина 394298
Козлобородник карягина 394299
Козлобородник кемрярии 394302
Козлобородник кецховели 394303
Козлобородник клубненосный 394353
Козлобородник колхидский 394270
Козлобородник колючепродный 394259
Козлобородник копетдагский 394304
Козлобородник коротконосиковый 394265
Козлобородник коротконосый 394265
Козлобородник красный 394337
Козлобородник крашенинникова 394305
Козлобородник крошечный 394335
Козлобородник крупноборотый 394311
Козлобородник культиасова 394306
Козлобородник литпвский 394309
Козлобородник ложнокрупный 394334
Козлобородник луговой 394333
Козлобородник макашвили 394314
Козлобородник меловой 394274
Козлобородник месхетский 394318
Козлобородник нителистный 394285
Козлобородник окаймленнолистный 394316
Козлобородник окаймленный 394315
Козлобородник окрашенный 394272
Козлобородник опушенноносый 394278
Козлобородник особенный 394323
Козлобородник пашенный 394343
Козлобородник песчаный 394340
Козлобородник подольский 394326
Козлобородник подорожниковый 394325
Козлобородник поздний 394344
Козлобородник пореелистный 394327
Козлобородник порейнолистный 394327
Козлобородник порейолистный 394327
Козлобородник поррейнолистный 394327
Козлобородник прутьевидный 394342
Козлобородник пушистенький 394351
Козлобородник разносемянный 394295
Козлобородник русский 394339
Козлобородник свернутый 394273
Козлобородник сетчатыйК 394336
Козлобородник сибирский 394345
Козлобородник скороспелый 394317
Козлобородник сомнительный 394281
Козлобородник сосновского 394347
Козлобородник степной 394348
Козлобородник субальпийский 394349
Козлобородник туркестанский 394354
Козлобородник украинский 394355
Козлобородник харадзе 394269
Козлобородник холмовый 394271
Козлобородник шероховатоплодный 394352
Козлобородник шерстистый 394286
Козлобородник шершавоносиковый 394278
Козлобородник широколистный 394308
Козлятник 169919
Козлятник аптечный 169935
Козлятник восточный 169940
Козлятник лекарственный 169935
Козополянская 217941
Козополянская туркестанская 217942
Козыльник 139678
Коикс 99105,99124
Коикс китайский 99134
Коикс слеза иова 99124
Коикс слёзы Иовы 99124
Коикс тайваньский 99115
Кока 155067
Кокайновые 155048
Кокайновый куст 155067
Коккулус 97894
Коккулюс 97894
Коккулюс каролинский 97900
Коккулюс лавроволистный 97919
Коккулюс лавролистный 97919
Коккулюс трехлопастный 97947
Кокорыш 9831
Кокорыш обыкновенный 9832
Кокорыш собачья петрушка 9832
Кокос 98134,98136
Кокос орехоносный 98136
Кокосовая пальма 98136
Кокпек 44333
Кок – сагыз 384606
Кокушни 182230
Кокушни комарниковый 182230
Кокушник 182213,363461
Кокушник ароматнейший 182261
Кокушник длиннопогий 182230
Кокушник душистый 182261
Кокушник камчатский 182254
Кокушник комалиный 182230
Кокушник эмэйский 182244
Кола 99157
Кола блестящия 99158
Кола заострённая 99158
Кола курьтурная 99158
Кола остролистная 99158
Колба 15868
Колдун – трава 91499
Колеантус 99411
Колеантус тонкий 99413
Колеантус тощий 99413
Коленница 8660
Колеостефус 60095
Колеостефус обыкновенный 99491
Колеус 99506
Колеус блумего 99711
Колеус ланьюйский 99712
Колеус пестрый 99711
Колеус пёстрый 99711
Коллард 59524
Коллетия колючия 99815
Коллеция 99804
Коллинзия 99833
Коллинзония 99843
Коллинсия 99833
Коллинсония 99843
Колломия 99850
Колломия крупноцветковая 99858
Колломия линейная 99859
Колнулака 104909
Колнулака коржинского 104913
Колоказия 99899
Колоказия древная 99910
Колоказия индийская 99922
Колоказия съедобная 99910
Колоказия тайваньская 99918
Колокасия 99899
Колоквинт 93297
Колокольчик 69870,69873,69883,69932,69933,69991,70001,70003,70010,70048,70089,70135,70232,70234,70268,70293,70298,70306,70338,70364
Колокольчик алтайский 69891
Колокольчик альберта 69881
Колокольчик альбова 69882
Колокольчик альпийский 69889
Колокольчик анны 69896
Колокольчик березолистный 69919
Колокольчик бледноохряный 70200
Колокольчик болодатый 69916
Колокольчик болонский 69921
Колокольчик бротеруса 69925
Колокольчик букетный 70054
Колокольчик вайды 70353
Колокольчик воронова 70367
Колокольчик высокий 70009
Колокольчик высокогорный 69888
Колокольчик гизеке 70271
Колокольчик гроссгейма 70074
Колокольчик дагестанский 69982
Колокольчик доломитовый 70004
Колокольчик евгеник 70019
Колокольчик жестковолосый 69954,70297
Колокольчик зубчаточашелистиковый 70201
Колокольчик известняковый 69929
Колокольчик изящный 70011
Колокольчик имеретинский 70087
Колокольчик ирины 70092

Колокольчик камнеломка 70288
Колокольчик капю 69940
Колокольчик каракушский 70099
Колокольчик карпатский 69942,334149
Колокольчик кемлярии 70100
Колокольчик китайский 69959
Колокольчик китайский крупноцветный 302753
Колокольчик колаковского 70103
Колокольчик комарова 70104
Колокольчик корончатый 70164
Колокольчик крапиволистный 70331
Колокольчик кругролистный 70271
Колокольчик крупноколосый 70189
Колокольчик крупноцветный 70164,302753
Колокольчик крымский 70322
Колокольчик курчавый 69980
Колокольчик лангсдорфа 70116
Колокольчик лировидный 70143
Колокольчик лицзянский 69987
Колокольчик льнюлистный 7682
Колокольчик макашвили 70155
Колокольчик масальского 70161
Колокольчик мейера 70168
Колокольчик мелкоплодный 69972
Колокольчик мергаритколистный 69917
Колокольчик млечноцветный 70107
Колокольчик многоцветковый 70189
Колокольчик молочноцветный 70107
Колокольчик мучнистый 70021
Колокольчик необычный ардонский 69897
Колокольчик обыкновенный 70255
Колокольчик одноцветковый 70294,70351
Колокольчик одонолетний 70014
Колокольчик олений 69954
Колокольчик осетинский 70203
Колокольчик оше 69911
Колокольчик персиколистный 70214
Колокольчик пирамидальный 70247
Колокольчик пихтовый 69871
Колокольчик полиморфный 70226
Колокольчик понтийский 70227
Колокольчик почти－головчатый 70317
Колокольчик продолговатолистный 70198
Колокольчик радде 70250
Колокольчик рапунцелевидный 70255
Колокольчик рапунцель 70254
Колокольчик раскидистый 70210
Колокольчик репучатый 70254
Колокольчик репчатовидный 70254,70255
Колокольчик ресничатый 69963
Колокольчик рупрехта 70283
Колокольчик сборный 70054
Колокольчик сердцелистный 69977
Колокольчик car сероватый 70088
Колокольчик сжатый 70054
Колокольчик сибирский 70305
Колокольчик скальный 70220
Колокольчик скребница 70251
Колокольчик скученный 70054
Колокольчик снизы－седой 70086
Колокольчик сомие 70307
Колокольчик сосновского 70308
Колокольчик средний 70164
Колокольчик стевена 70314
Колокольчик талиева 70321
Колокольчик тахтаджяна 70320
Колокольчик торчащий 70315
Колокольчик точечный 70234
Колокольчик трехзубый 70341
Колокольчик турчанинова 70346
Колокольчик удивительный 70172
Колокольчик уральский 70366
Колокольчик фомина 70045
Колокольчик фондервиза 70046
Колокольчик харкевича 69958
Колокольчик хемшинский 70075
Колокольчик холмовой 69971
Колокольчик хоцятовского 69961
Колокольчик черкесский 69969
Колокольчик чесночницелистный 69885
Колокольчик шамиссо 69956
Колокольчик шершаваплодный 70117
Колокольчик широколистный 70119
Колокольчик шишкина 70296
Колокольчик шпорцеватый 69928
Колокольчик японский крупноцветный 302753
Колокольчики 30081
Колокольчики водосборные 29994
Колокольчики садовые 30081
Колокольчики синий 173728
Колокольчиккавказский 69949
Колокольчиковые 70372
Коломбина 30081
Коломбо 97894
Коломикта 6639
Колоняк сибирский 144466
Колос благовонный 223251
Колос пахучий 27969
Колосеница 301864
Колосняк 144144
Колосняк алтайский 335200
Колосняк бальджуанский 144191
Колосняк высокий 144303
Колосняк гигантский 144312
Колосняк даурский 144271
Колосняк канадский 144216
Колосняк китайский 228356
Колосняк многостебельный 228370
Колосняк мохнатый 144522
Колосняк мягдий 228369
Колосняк мягкий 228369
Колосняк пабо 228373
Колосняк переменчивый 335440
Колосняк песчаный 228346
Колосняк пушистоколосый 228384
Колосняк ситниковый 317081
Колосняк тяньшанский 335526
Колосняк чимганский 144511
Колосок душистый обыкновенный 27969
Колосок душистый остистый 27934
Колоцинт 93297
Колподим 99987
колподим алтайский 99988
колподим араратский 99990
колподим бледночешуйный 100004
колподим валя 100014
колподим волокнистый 99996
колподим золотистоцветковый 99993
колподим колхидский 99994
колподим мелкоцветковый 100009
колподим пестрый 100015
Колподим понтийский 100010
колподим приземистый 79214
колподим разноцветный 100016
колподим украшенный 100008
Колубрина 100058
Колумба 212095
Колуфер 89602,322709,383712
Колхикум 99293
Колча 146122
Кольза 100309
Кольквиция 217785
Кольквиция прелестная 217787
Кольник 298054
Кольник колосистый 298069
Кольник колосовидтый 298069
Кольник округлый 298065
Кольник черный 298064
Кольник чёрный 298064
Кольник четырехлопастный 298072
Кольник шаровидный 298065
Кольраби 59541
Кольраушия побегоносная 292665
Колюрия 100132
Колюрия гравилатовидная 100135
Колюрия эмэйская 100144
Колютик 91697
Колючая греча 44247
Колючеголовник 2133
Колючеголовник бетама 2135
Колючеголовник стеблеобъемлющий 2134
Колючеголовник стеблеохватывающий 2134
Колючелистник 2555,76988
Колючелистник колючий 2565
колюченосник 140640
колюченосник волосолистный 140645
колюченосник сибторпа 140643
Колючий дрок 401373,401388
Колючка верблюжья 14638

Колючка верблюжья обыкновенная 14638
Колючник 76988
Колючник артишоковидный 77000
Колючник бесстебельный 76990
Колючник биберштейна 76994
Колючник обыкновенный 77022
Колючник стебельный 76998
Колючник татарниколистный 77011
Колючник узколистный 76994
Колючник шерсстистый 77008
Комаровия 217826
Комаровия разнореберная 217827
Комаровыносики 102833
Комбретовые 100296
Комбретум 100309
Комифора бальзамная 101488
Комифора мира 101474
Коммелина 100892
Коммелина мексиканская 100960
Коммелина обыкновенная 100961
Коммелина синеглазка 100961
Коммелиновые 101200
Коммифора 101285
Коммифора абиссинская 101287
Коммифора африканская 101290
Коммифора бальзамная 101488
Коммифора мирра 101474
Комперия 101633,273317
Комперия крымская 101635
Комптония 101664
Комфрей 381026
Кондалия 101730
Кондуранго 245798
Конига 18313,235065
Конига морская 235090
Конига приморская 235090
конизантус 103651
конизантус злаколистный 103653
конизантус чешуйчатый 103654
Коноплевые 71202
Конопля 71209
Конопля виргинская 18683
Конопля декан 194779
Конопля индийская 71218,71220
Конопля китайская 56229
Конопля культурная 71218
Конопля лесная 85875
Конопля мавликийская 169245
Конопля мавлициева 169245
Конопля манильская 260275
Конопля манипиская 260275
Конопля новозеландская 295600
Конопля посевная 71218
Конопля сорная 71228
Конопля юкатанская 10854
Конофолис 102002
Конрингия 102808
Конрингия австрийская 102809
Конрингия восточная 102811
Конрингия персидская 102814
Конрингия плоскоплодная 102815
Конрингия пронзеннолистая 102812
Конские бобы 408393
Конский кашта мелкоцветковый 9719
Конский каштан 9675
Конский каштан белый 9701
Конский каштан ван 9736
Конский каштан восьмитычинковый 9692
Конский каштан голый 9694
Конский каштан индийский 9706
Конский каштан китайский 9683
Конский каштан крупнолистный 9715
Конский каштан мясо – красный 9680
Конский каштан обыкновенный 9701
Конский каштан павия 9720
Конский каштан чжэцзянский 9684
Конский каштан японский 9732
Конский фенхель 196723
Конский фенхель длиннольчатый 196738
Конский фенхель мелкоплодный 196739
Конскокаштановые 196494
Конуфер 89602,383712
Конфетное дерево 198769
Конюшина 397019
Конятник 339987
Копайба 103719
Копайфера 103691
Копайфера копалоносная 103696
Копайфера лекарственная 103719
Копеечник 187753,187760
Копеечник азербайджанский 187789
Копеечник алайский 187760
Копеечник альпийский 187769
Копеечник аманкутанский 187775
Копеечник арктический 187779
Копеечник армянский 187785
Копеечник бабатагский 187793
Копеечник байкальский 187794
Копеечник бальджуанский 187795
Копеечник бледный 187815
Копеечник блестящий 187815,188129
Копеечник бордзиловского 187801
Копеечник бранта 187809
Копеечник бухарский 187810
Копеечник великолепный 187997
Копеечник венцевидный 187842
Копеечник гмелина 187912
Копеечник головчатый 187871
Копеечник горный 188102
Копеечник горошковидный 188177
Копеечник грузинский 187939
Копеечник дагестанский 187848
Копеечник даурский 187849
Копеечник джунгарский 188116
Копеечник дробова 187858
Копеечник забытый 188023
Копеечник затопляемый 187946
Копеечник зубчатый 187853
Копеечник извилистый 187877
Копеечник изящный 187860
Копеечник илийский 187940
Копеечник кавказский 187823
Копеечник каратавский 187952
Копеечник киргизский 187953
Копеечник китайский 187831
Копеечник копеечниковый 188028
Копеечник копетдагский 187955
Копеечник коржинского 187956
Копеечник короноподобный 187842
Копеечник красивый 187879
Копеечник краснова 187957
Копеечник крупноцветковый 187918, 187995
Копеечник крупноцветный 187918
Копеечник крылова 187959
Копеечник крымский 188141
Копеечник кугитангский 187961
Копеечник кудряшева 187960
Копеечник кустарниковый 187881
Копеечник лемана 187979
Копеечник липского 187989
Копеечник мелколистный 126465
Копеечник меловой 187843
Копеечник минусинский 188006
Копеечник обрубленный 188155
Копеечник однолисточковый 188012
Копеечник ольги 188034
Копеечник пестрый 188170
Копеечник почтиголый 188136
Копеечник прутьевидный 188088
Копеечник разомовского 188071
Копеечник родственный 187840
Копеечник рослый 188186
Копеечник сахалинский 188081
Копеечник северцова 188101
Копеечник семенова 188094
Копеечник серебристолистный 187783
Копеечник серебристый 187781
Копеечник сибирский 187769,188104
Копеечник сросшийся 187946
Копеечник съедобный 188177
Копеечник сырдарьинский 187948
Копеечник ташкентский 188140
Копеечник темный 188028
Копеечник тёмный 188028
Копеечник тибетский 188187
Копеечник туркевича 188161
Копеечник туркестанский 188160
Копеечник узколистный 188142
Копеечник украинский 188162
Копеечник уссурийский 188166

Копеечник ушковидный 187790
Копеечник федченко 187869
Копеечник ферганский 187870
Копеечник чжундяньский 188143
Копеечник шелковистый 226742
Копеечник шишкина 188087
Копеечник щетинистоплодный 187851
Копеечник щетинистый 188099
Копеечник южносибирский 187792
Копернициа 103731
Коперниция восконосная 103737
Копорский чай 85875
Коптис 103824
Коптис китайский 103828
Коптис трехлистный 103850
Копытень 37556
Копытень грузинский 37647
Копытень динхуский 37679
Копытень европейский 37616
Копытень зибольда 37722
Копытень иберийский 37647
Копытень канадский 37574
Копытень китайский 37597
Копытень крупнолистный 37680
Копытень наньчуаньский 37691
Копытень фуцзяньский 37624
Копытень фэнцзеский 37695
Копытень чэнкоуский 37596
Копытень японский 37659
Кора мыльная 324624
Коракан 143522
Кордилина 104334
Кордилина австралийская 104339
Кордилина верхушечная 104367
Кордилина ветвистая 104350
Кордилина голубая 127505
Кордилина кустарниковая 104367
Кордилина прямая 104366
Кордилина южная 104339
Кордия 104147
Кордовник 133505
Корейская слива 316761
Корейский кедр 300006
Корень аирный 5793
Корень алтейный 18173
Корень волчий 122515
Корень галанговый 17677
Корень змеиный 54798,308893
Корень ирный 5793
Корень любистоковый 228250
Корень маралий 329197
Корень медвежий 252662
Корень мыльный 346434
Корень пьяный 354686
Корень рвотный 81970,378834
Корень розовый 329935
Корень сахарный 365871
Корень свинцовый 305167
Корень сердечный 309954
Корень сладгийцарский 354870
Корень царский 205535
Корень цитварный 114884
Корень чёрный 5793,118008,354870
Корень – человек 280741
Кореопсис 104429
Кореопсис друммонда 104448
Кореопсис красильный 104598
Кореопсис крупноцветконый 104509
Кореопсис крупноцветный 104509
Кореопсис сердечниколистный 104468
Коржинския 217895
Коржинския володушковая 217896
Коржинския ольги 217897
Кориандр 104687,104690
Кориандр огородный 104690
Кориандр посевной 104690
Кориариевые 104710
Кориария 104691
Кориария китайский 104701
Кориария миртолистная 104700
Кориария непалский 104701
Кориария японский 104696
Корилопсис 106628
Корилопсис вилльмотта 106689
Корилопсис вича 106687
Корилопсис китайский 106675
Корилопсис колосковый 106682
Корилопсис малоцветковый 106669
Корилопсис мелкоплодный 106658
Корилопсис слабоцветущий 106669
Корилопсис тайваньский 106656
Корилопсис уилмотта 106689
Корилопсис цюцзянский 106686
Корилопсис широколепестный 106670
Корилопсис эмэйский 106668
Корилопсис юньнаньский 106695
Коринка 19240
Коринкарпус 8544
Корифа 106984
Корифа гвоздиная 129718
Корифа зонтичная 106984,106999
Корифа китайская 91302
Корифа магелландская 138058
Корифа теневаятвоздная 106999
Корифа цейлонская 91446
Корица 91252,91287
Корица кассия 91302
Корица китайская 91302
Корица цейлонская 91446
Коричник 91252
Коричник железконосный 91330
Коричник камфорный 91287
Коричник кассия 91302
Коричник китайский 91302
Коричник ланьюйская 91357
Коричник лоурера 91366
Коричник пикульный 114122
Коричник цейлонский 91446
Коричник японский 91351
Корнеглав 82407
Корнеглав туркменский 82408
Корнеглав яйценогий 82409
Корнутия 105229
Коровница 403686
Коровница испанская 403687
Коровяк 405657
Коровяк артвинский 405661
Коровяк белошерстистый 405710
Коровяк великолепный 405779
Коровяк видный 405780
Коровяк вильгельмся 405799
Коровяк восточный 405681
Коровяк выемчатый 405774
Коровяк высокий 405694
Коровяк гаястанский 405713
Коровяк горолюбивый 405742
Коровяк грузинский 405708
Коровяк джунгарский 405777
Коровяк длинностный 405727
Коровяк ереванский 405699
Коровяк желтоватый 405703
Коровяк закавказский 405792
Коровяк зопниковидный 405747
Коровяк извилистый 405774
Коровяк изменчивый 405795
Коровяк изогнутый 405704
Коровяк красивый 405706,405779
Коровяк краснобурый 405763
Коровяк крупнозопниковый 405736
Коровяк крупноплодный 405731
Коровяк левкоелистный 405683
Коровяк лекарственный 405747
Коровяк медвежье 405788
Коровяк медвежье ухо 405788
Коровяк метельчатый 405729,405745
Коровяк мешечатый 405764
Коровяк молевый 405667
Коровяк мохнатый 405747
Коровяк мучнистый 405729
Коровяк обыкновенный 405788
Коровяк овальнолистный 405744
Коровяк перистораздельный 405749
Коровяк пираидальный 405756
Коровяк приальпийский 405658
Коровяк пуналийский 405755
Коровяк раскидистый 405724
Коровяк сидячецветковый 405767
Коровяк скипетровидный 405694
Коровяк скученый 405709
Коровяк тапсовидый 405694
Коровяк тараканий 405667

Коровяк темный 405667
Коровяк тёмный 405667
Коровяк туркестанский 405794
Коровяк туркменский 405793
Коровяк фиолетовый 405748
Коровяк хлопковый 405711
Коровяк холмовой 405686
Коровяк черноморский 405710
Коровяк черный 405737
Коровяк чёрный 405737
Коровяк чистецовидный 405781
Коровяк шерстистый 405747
Коровяк шовица 405783
Корольковия 168332
Коронария кукущкин－цвет 363461
Корорева лилий 229730
Короставник 216752
Короставник ворсянколистный 216764
Короставник восточный 216772
Короставник голубинцевой 7657
Короставник горный 216771
Короставник коронопусолистный 7649
Короставник крупнообертковый 216767
Короставник крымский 7846
Короставник курчавый 7816
Короставник полевой 216753
Короставник скальный 7806
Короставник татарский 216774
Короставник якутский 7672
Короткокистник 58299
Короткокистник воронеглазый 58300
Коротконожка 58572
Коротконожка лесная 58619
Коротконожка мохнатая 58638
Коротконожка перестая 58604
Коротконожка скальная 58616
Коротконожка тайваньская 58588
Корригиола береговая 105416
Кортадерия 105450
Кортадерия двдомная 105456
Кортуза 105488
Кортуза алтайская 105489
Кортуза бротеруса 105491
Кортуза лмурская 105490
Кортуза маттиоли 105496
Кортуза маттиоля 105496
Кортуза пекинская 105501
Кортуза сибирская 105505
Кортуза туркестанская 105507
Кортуза хэбэйская 105501
Космея 107158,107161
Космея дваякоперистая 107161
Космос 107158,107161
Космос дваякоперистый 107161
Космос раздельнолистный 107161
Космос разнолистный 107168
Косна карейская 300006
Косогорник 313778
Косогорник красный 313866
Косогорник максимовича 313850
Косогорник недоснелколистный 313802
Косогорник пихтовый 313779
Косогорник понтийский 313864
Косогорник пурпуровый 313866
Косогорник татаринова 313895
Косогорник узколистный 313791
Костелецкия 217944
Костелецкия пятиплодная 217955
Костенец 197486
Костенец зонтичный 197491
Костенец липкий 197491
Костенец окаймленный 197490
Костер 60612
Костёр 60612
Костер анатолийский 60626
Костер ангренский 60628
Костер безостый 60760
Костёр безостый 60760
Костер бенекена 60643
Костер береговой 60942
Костер бесплодный 60985
Костёр бесплодный 60985
Костер биберштейна 60646
Костер ветвистый 60918
Костёр ветвистый 60680
Костер войлочковый 60999
Костер войлочный 61001
Костер голый 265016
Костер дантонии 60692
Костер двутычинковый 60696
Костер жестковатый 60935,60957
Костер иркутский 60776
Костер канадский 60654
Костер каппадокийский 60658
Костер кистевидный 60912
Костёр кистевидный 60912
Костер кистистый 60918
Костер копетдагский 60795
Костер короткого 60796
Костер краснеющий 60943
Костер кровельный 60989
Костёр кровельный 60989
Костер крупноколосковый 60797
Костер крупноколосый 60797
Костер крупноцветковый 264939
Костер мадридский 60820
Костёр мадридский 60820
Костер мелкочешуйный 61006
Костер метельчатый 60957
Костер мягкий 60853
Костёр мягкий 60853
Костер непалский 60865
Костер острозубый 60871
Костер памирский 60873
Костер паульсена 60882
Костер переменчивый 60680
Костёр переменчивый 60680
Костер пестрый 61016
Костер полевой 60634
Костёр полевой 60634
Костер попова 60892
Костер прямой 60707
Костёр прямой 60707
Костёр прямостоячий 60707
Костёр разветвлённый 60680
Костер растопыренный 60977
Костер ржаной 60963
Костёр ржаной 60963
Костер ричардсона 60931
Костер северцовя 60968
Костер сибирсий 60969
Костёр снешанный 60680
Костёр стоячий 60707
Костер тайваньский 60724
Костер тонкий 60731,61005
Костер тончайший 60735
Костер трясунковидный 60653
Костер туркестанский 61003
Костер укорашеннный 60869
Костер униоловидный 61008
Костёр униоловидный 61008
Костер шелковистый 60991
Костер японский 60777
Костёр японский 60777
Костолом 262572
Костус 107221
Костус замечательный 107271
Костяника 338059,339240
Костяника арктическая 338142
Костяника каменистая 339240
Костянкоплодник 276533
Костянкоплодник шверина 276544
Котики 396828
Котиледон 107840
Котовик 264859
Котовник 264859
Котовник алагезский 264862
Котовник бадахшанский 264879
Котовник бархатистый 265101
Котовник биберштейна 264881
Котовник бледный 265019
Котовник бузе 264893
Котовник бухарский 264892
Котовник буша 264894
Котовник васильковый 264904
Котовник ваханский 265098
Котовник венгерский 265016
Котовник гаястанский 264941
Котовник гончароза 264936
Котовник гроссгейма 264940
Котовник грузинский 264952

Котовник густоцветковый 264909
Котовник дагестанский 264906
Котовник длинноприцветниковый 264985
Котовник длиннотрубчатый 264986
Котовник ереванский 264917
Котовник железистый 264935
Котовник заилийский 265089
Котовник закавказский 265088
Котовник изфаганский 264957
Котовник кноринг 264962
Котовник кокамырский 264963
Котовник кокандский 264964
Котовник колючий 265032
Котовник комарова 264965
Котовник копетдагский 264966
Котовник коротколистный 264891
Котовник кошачий 264897
Котовник красивый 265031
Котовник кубанский 264968
Котовник ладанный 264972
Котовник лежачий 265071
Котовник липского 264984
Котовник ложноклочковатый 265029
Котовник ложнококандский 265030
Котовник лопатчатый 265053
Котовник маньчжурский 264990
Котовник марии 264991
Котовник маусарифский 264992
Котовник мейера 264994
Котовник мелковатоцветковый 265022
Котовник мелкоголовчатый 264996
Котовник мелкоцветковый 264995
Котовник мятный 264897
Котовник надрезанный 264923
Котовник ножкоколосый 265027
Котовник ольги 265018
Котовник пахучий 265017
Котовник ползучий 176824
Котовник полукопьевидный 265063
Котовник прекрасный 264926
Котовник прелестный 264867
Котовник прицветниковый 264890
Котовник рамирский 265020
Котовник рейхенбаха 265036
Котовник родственный 264903
Котовник сахаристый 265041
Котовник серножелтый 265068
Котовник сибирский 265048
Котовник синтениза 265050
Котовник сосновского 265051
Котовник тибетский 265084
Котовник торчащий 265062
Котовник траутфеттера 265090
Котовник тройцкого 265091
Котовник украинский 265094
Котовник федченко 264921
Котовник хэнаньская 264948
Котовник чаберовидный 265043
Котовник шишкива 265044
Котовник шугванский 265045
Котовник японский 264958
Котовник яснотколистный 264974
Котонеастер 107319
Котонеастер мулиский 107581
Коточки 345881
Котула 107747
Котула коронополистная 107766
Котула коронопусолистная 107766
Котяник 176824
Кофе 98844,98857
Кофе арабский 98857
Кофе аравийское 98857
Кофе либерийский 98941
Кофе либерийское 98941
Кохия 217324,217361
Кохия веничная 217361
Кохия иранская 217379
Кохия крылова 217343
Кохия метельчатая 217361
Кохия песчаная 217331
Кохия простертая 217353
Кохия сиверса 217372
Кохия стелющая 217353
Кохия стелющаяся 217353
Кохия темнокрылая 217347
Кохия шерстистоцветковая 217331,217344
Кохия шленка 217359
Кохлоспермум 98100
Кошачья лапка 26299
Кошачья лапка альпийская 26310
Кошачья лапка ворсоносная 26616
Кошачья лапка двудомная 26385
Кошачья лапка двудомновидная 26393
Кошачья лапка кавказская 26363
Кошачья лапка карпатская 26362
Кошачья лапка комарова 26439
Кошачья лапка одноголовчатая 26474
Кошачья лапка фриса 26413
Кошачья мята 264897
Кошельки 66252
Кошельки гибридные 66271
Кошельки мексиканские 66281
Кошельки цельнолистные 66273
Кравфурдия 110292
Кравфурдия вьющаяся 320927
Кравфурдия японская 398273
Крамбе 108591
Крамбе абиссинская 108592
Крамбекатран 108591
Кранива китайская 56229
Крап 338038
Крапива 175865,402840
Крапива большая 402886
Крапива водяная 83540
Крапива глухая красная 220416
Крапива глухая пятнистая 220401
Крапива двудомная 402886
Крапива двудомная ганьсуская 402889
Крапива жгучая 403039
Крапива киевская 402943
Крапива коноплевая 402869
Крапива мелкоцветковая 402854
Крапива обыкновенная 402886
Крапива опушенная 403001
Крапива плосколистная 402996
Крапива светро-зеленая 402946
Крапива синеющая 402947
Крапива синьцзянская 402893
Крапива сондена 403016
Крапива тайваньская 403026
Крапива тибетская 403031
Крапива тунбелга 403028
Крапива узколистная 402847
Крапива чаюйская 403048
Крапива шариконосная 402992
Крапивка цветная 221229
Крапивна 303324
Крапивные 403050
Крапп 338038
Крапчатка 239202
Красавица дневная 103340
Красавица ночная 255711
Красавка 44704,44708
Красавка белладонна 44708
Красавка кавказская 44709
Красавка комарова 44711
Красавка обыкновенная 44708
Красавка одурь сонная 44708
Красавка прудовая 67393
Красивоцвет 66697
Красивоцвет узколистный 66701
Красная верба 342961
Красновия 218146
Красновия длиннодольчатая 218148
Красноволоска 67336
Красноволоска весенная 67401
Красноволоска гуандунская 67384
Красноволоска изменчивая 67389
Красноволоска крючковатая 67359
Красноволоска осенная 67361
Красноволоска японская 67367
Красноволосковые 67334
Красноднев 191261
Красноднев малый 191312
Красноднев рыжий 191284
Краснопузырник 80129
Краснопузырник древогубец 80285
Краснопузырник круглолистный 80260
Краснопузырник округлый 80260
Краснопузырник плетеобразный 80189
Краснопузырник угловатый 80141

Краснопузырник франше 80191
Краснопузырник щетковидный 80327
Краснопузырниковые 80127
Краснотал 342961
Красный кедр 213984
Красовласка 67336
Красовласка прудовая 67393
Красоднев 191261
Красоднев гиблидный 191302
Красоднев дюмортье 191272,191278
Красоднев желтый 191304
Красоднев корейский 191268
Красоднев малый 191312
Красоднев миддендорфа 191309
Красоднев рыжий 191284,191304
Красоднев съедобный 191278
Красоднев тунбера 191318
Красоля 399601
Красотка 107158,107161
Крассула 108775
Крашенинниковия 218116
Крашенинниковия давида 318493
Крашенинниковия жесткая 218137
Крашенинниковия лесная 318515
Крашенинниковия максмовича 318507
Крашенинниковия скальная 318513
Крашенинниковия японская 318506
Кремастра 110500
Крепкоплодник 155984
Крепкоплодник сирийский 155985
Кресс 225279,225398
Кресс бразильский 4866
Кресс водяной 262722
Кресс испанский 399601
Кресс крупка 73090
Кресс масляный 4866
Кресс огородный 225450
Кресс полевой 225315
Кресс посевной 225450
Кресс садовый 225450
Кресс сорный 225447
Кресс широколистный 225398
Кресса 111111
Кресса критская 111113
Кресс－салат 225450
Крест андреев 317964
Крест петров 222641
Крест петров чешуйчатый 222651
Крестовник 358159,360155
Крестовник азиатский 385915
Крестовник акантолистный 358395
Крестовник амбровый 358240
Крестовник амурский 358253
Крестовник аргунский 358292,359253
Крестовник арктический 358287,385908
Крестовник бандуролистный 359667
Крестовник бессера 358393
Крестовник близкий 359802
Крестовник болотный 359660
Крестовник бородавниковый 359270
Крестовник буша 358461
Крестовник весенний 360306
Крестовник водяной 358280
Крестовник головчатый 358510,385917
Крестовник днепровский 358424
Крестовник дубравный 359565
Крестовник желтуха 359158
Крестовник золотистый 358336
Крестовник ильина 359099
Крестовник каваками 385902
Крестовник кавказский 358523
Крестовник кальверта 358484
Крестовник карелиниевидный 359205
Крестовник карпатский 358518
Крестовник карягина 359206
Крестовник киргизский 359225
Крестовник клейкий 360321
Крестовник клочковатый 360146
Крестовник коленати 359236
Крестовник коноплеволистный 358500
Крестовник коноплелистный 358500
Крестовник коржинского 359240
Крестовник крашенинникова 359242
Крестовник крупнозубчатый 358995
Крестовник крупнолистный 358995,359419
Крестовник крученный 358287
Крестовник кубинский 359244
Крестовник ленский 359303
Крестовник лесной 360172
Крестовник лжеоранжевый 359813
Крестовник литвинова 358504
Крестовник лицзянский 359357
Крестовник ложная арника 359811
Крестовник ложноарниковый 359811
Крестовник лорента 359387
Крестовник луговой 359158
Крестовник малодольчатый 359693
Крестовник малозубчатый 360137,385920
Крестовник малолистный 359682,359691
Крестовник массагетова 359450
Крестовник многокорзиночный 359781
Крестовник мулиский 359527
Крестовник наньчуаньский 365038
Крестовник обыкновенный 360328
Крестовник огненно－язычковый 359844
Крестовник ольги 359619
Крестовник отонны 359634
Крестовник пальчатый 359654
Крестовник паульсена 359696
Крестовник песчаный 358995
Крестовник пламенный 385892
Крестовник плосколистный 359768
Крестовник поярковой 359777
Крестовник приречный 358902
Крестовник приручейный 359903
Крестовник пурпуровый 359789
Крестовник пустынный 358717
Крестовник равнинный 385903
Крестовник разветвленный 359859
Крестовник резедолистый 279899
Крестовник речной 358902
Крестовник ромболистный 359892
Крестовник сапожникова 359967
Крестовник сарацинский 359968
Крестовник северный 358282
Крестовник сивецолистный 360160
Крестовник сихотэа 360048
Крестовник скалистый 359938
Крестовник сомнительный 358769
Крестовник сосновского 360074
Крестовник стрельчатый 359991
Крестовник субальпийский 360129
Крестовник сукачева 360162
Крестовник сумневича 360166
Крестовник сурепколистный 358369
Крестовник темно－пурпуровый 358323
Крестовник теневой 360281
Крестовник тибетский 360220
Крестовник томсона 229226
Крестовник тонколистный 360205
Крестовник тундровый 360265
Крестовник турчанинова 385924
Крестовник тяньшанский 360214
Крестовник увенчанная 361023
Крестовник ферганский 358866
Крестовник франше 358912
Крестовник фукса 359565
Крестовник холодный 360147
Крестовник целенолистный 359138
Крестовник цепкий 359980
Крестовник швецова 360007
Крестовник широколистный 359769
Крестовник эмэйский 358848
Крестовник эрратический 358280
Крестовник эруколистный 358828
Крестовник якова 359158
Крестовник якутский 359165
Крестовник－желтуха 359158
Крестовница 113110
Крестоцветные 113144
Кресченция 111100
Кривокрыльник мякокрасный 70704
Кривоплодник 180046,297638
Кривоплодник грелльсиелистный 297639
Кривоцвет 239167
Кривоцвет восточный 21965
Кривоцвет пашенный 21920
Кривоцвет полевой 21920
Кринум 111152
Кринум азиатский 111167
Кринум американский 111158

Кринум длиннолистный 111180
Криптантус 113364
Криптокорина 113527
Криптомерия 113633
Криптомерия китайская 113721
Криптомерия японская 113685
Криптомерия японская ‘Араукариоидес’ 113687
Криптомерия японская ‘Дакридиоидес’ 113693
Криптомерия японская ‘Элеганс’ 113696
Криптотения 113868
Криптотения японская 113879
Критезион 198260
Критезион гривастый 198311
Критмум 111343
Критмум морской 111347
Кровохлебка 345816
Кровохлёбка 345816
Кровохлебка альпийскя 345821
Кровохлёбка альпийскя 345821
Кровохлебка аптечная 345881
Кровохлебка великолепная 345858
Кровохлебка железистая 345890
Кровохлебка лекарственная 345881,345890
Кровохлебка лечебная 345881
Кровохлёбка малая 345860,345881
Кровохлебка мелкоцветковая 345917,345921,345924
Кровохлебка многобрачная 345881
Кровохлёбка обыкновенная 345881
Кровохлебка прибрежная 345906
Кровохлебка ситхинская 345911
Кровохлёбка ситхинская 345911
Кровохлебка тонколистная 345917
Кровохлёбка тонколистная 345917
Крокус 111480
Крокус весенний 111625
Крокус голандский 111625
Крокус долинный 111623
Крокус осенний 99297
Крокус посевный 111589
Кроссандра 111693
Кроссандра нильская 111741
Кроссостефиум 111823
Кроссостефиум китайский 111826
Кроталярия 111851,112682
Кроталярия индийская 112269
Кроталярия китайская 112013
Кроталярия ланцетовидная 112307
Кроталярия остроконечная 112491
Кроталярия полосадая 112491
Кроталярия промежуточная 112248
Кроталярия симаоская 112729
Кроталярия ситниковая 112269
Кроталярия ситниковидная 112269
Кроталярия хайнаньская 112207
Кроталярия цзиньпинская 112563
Кроталярия юньнаньская 112823
Кротон 112829,113039
Кротон слабительный 113039
Кротон таиландское 113033
Кротон юньнаньский 113061
Круглолистник 83734
Круглолистник японский 83736
Круглоспинник 19042
Круглоспинник обратнояйцевидный 19045
Крупина 113205
Крупина обыкновенная 113214
Крупичка 137321
Крупичка весеная 137327
Крупка 136899
Крупка алайская 136903
Крупка алтайская 136918
Крупка альберта 136905
Крупка альпийская 136911
Крупка араратская 136938
Крупка арсеньева 136940
Крупка байкальская 136945
Крупка бародатая 136948
Крупка беринга 136949
Крупка брунаелистная 136956
Крупка весенняя 153964
Крупка вздутоплодная 137182
Крупка войлочная 136984
Крупка волосистая 137019,137185
Крупка волосистолистная 137083
Крупка высотная 137107
Крупка вытянутолбиковая 136973
Крупка гиссарская 137033
Крупка горная 137155
Крупка дарвазская 136976
Крупка длинностручковая 137095
Крупка дубравная 137133
Крупка елизаветы 136988
Крупка желто-белая 137150
Крупка имеретинская 137038
Крупка искривленная 137048
Крупка камчатская 137054
Крупка коржинского 137059
Крупка круглоплодная 137003
Крупка крупная 137102
Крупка крупноплодная 137100
Крупка кузнецова 137062
Крупка курильская 137061
Крупка ланцетная 137069
Крупка ланцетоплодная 137069
Крупка ледниковая 136999
Крупка липского 137093
Крупка лицзянская 137088
Крупка ложноволосистя 137196
Крупка маленькая 137197
Крупка мелколепестная 137111
Крупка мелкоплодная 137110
Крупка мелкостручковая 137181
Крупка мелкоцветковая 137180
Крупка многовласая 137191
Крупка молочнобелая 137019,137063
Крупка монгольская 137115
Крупка мохнатая 137019
Крупка мощная 137287
Крупка мягкоопушенная 137114
Крупка неубранная 137046
Крупка норвежская 137147
Крупка однобокая 137257
Крупка ольги 137152
Крупка осетинская 137176
Крупка остролистная 136973
Крупка оше 136943
Крупка памирская 137179
Крупка первоцветовидная 137194
Крупка перелесковая 137133
Крупка поле 137189
Крупка почти гладная 137256
Крупка почти-головчатая 137255
Крупка почти-стеблеобъемлющая 137253
Крупка продолговатая 137148
Крупка продолговатоплодная 137148
Крупка прозоровского 137195
Крупка пушистая 136990
Крупка разнолистная 136980
Крупка сахалинская 137224
Крупка северная 136951
Крупка седая 137039
Крупка серая 136969
Крупка сердечникоцветковая 136963
Крупка сибирская 137237
Крупка снежная 137146
Крупка стенная 137129
Крупка столбиковая 137248
Крупка стручковая 137241
Крупка таласская 137259
Крупка тибетская 137266
Крупка турчанинова 137282
Крупка узколепестная 137246
Крупка узкоплодная 137242
Крупка уссурийская 137286
Крупка федченко 136994
Крупка фладницийская 136995
Крупка хюта 137034
Крупка цинхайская 137200
Крупка черноногая 137107,137108
Крупка чжундяньская 137230
Крупка чьельмана 137058
Крупка шамиссо 136966
Крупка шершавая 137227
Крупка шетинистая 137032
Крупка эшшольтца 136993
Крупка юньнаньская 137298
Крупка японская 137052
Крупноплодник 247767

Крупноплодник большеплодный 247786
Крупноплодник гигантский 247782
Крупноплодник илийский 247783
Крупноплодник крупноплодный 247786
Крупноплодник муголжарский 247786
Крупноплодник округлый 247788
Крупноплодник шугнанский 247791
Крупноплодный 140466
Крутай 140776
Крутик 20005
Крушина 328592
Крушина альпийская 328604
Крушина американская 328838
Крушина вечнозелёная 328595
Крушина даурская 328680
Крушина диамантовая 328689
Крушина калифорнийская 328630
Крушина китайская 328883
Крушина красильная 328737
Крушина крупнолистная 328719
Крушина ливанская 328765
Крушина ломкая 328701
Крушина обыкновенная 407989
Крушина ольховидная 328701
Крушина ольхолистная 328603
Крушина полезная 328883
Крушина пурша 328838
Крушина скалистая 328849
Крушина скальгная 328846
Крушина скальная 328849
Крушина слабительная 328642
Крушина слабитьльная 328642
Крушина уссурийская 328882
Крушина чжэцзянская 328844
Крушина юпонская 328740
Крушинник 328592
Крушинник ломкий 328701
Крушинник ольховидный 328701
Крушинные 328543
Крушиновые 328543
Крыжовник 333893
Крыжовник амурский 6639
Крыжовник барбадосский 290701
Крыжовник буреинский 333929
Крыжовник гуншаньский 334008
Крыжовник дальневосточный 333929
Крыжовник европейский 334250
Крыжовник игольчатый 333894
Крыжовник колючий 333894
Крыжовник культурный 334250
Крыжовник миссурийский 333916
Крыжовник обыкновенный 334250
Крыжовник отклоненный 334250
Крыжовник шиповниковидный 334030
Крыжовниковые 181339
Крылатый каркас татаринова 320412
Крыловия 218242
Крыловия кермеколистная 218246
Крыловия новопокровского 218247
Крыловия попова 218248
Крыловия пустынно－степная 218245
Крылоолешина 320346
Крылоолешник 320346
Крылотычинник 9785
Крылотычинник арабский 9787
Крылотычинник армянский 9788
Крылотычинник беззубый 9796
Крылотычинник воронова 9822
Крылотычинник дерновинный 9789
Крылотычинник длинный 9797
Крылотычинник загирканский 9820
Крылотычинник колючий 9817
Крылотычинник красивый 9807
Крылотычинник крупноцветковый 9798
Крылотычинник левандовского 9802
Крылотычинник липского 9803
Крылотычинник мясокрасный 9791
Крылотычинник осыпной 9816
Крылотычинник перепончатый 9806
Крылотычинник разнолистный 9800
Крылотычинник салмасский 9809
Крылотычинник середцелистный 9790
Крылотычинник складчатый 9795
Крылотычинник стреловидный 9809
Крылотычинник трихжильный 9821
Крылотычинник шовица 9818
Крылоцветник 320155
Крылоцветник вильчатый 320156
Крым－сагыз 384569
Крэб 243711
Крэб сибирский 243555
Ксантиум 414990
Ксантиум колючий 415053
Ксантиум монгольский 415031
Ксантиум сибирский 415046
Ксантогалум 415115
Ксантогалум пурпуровый 415117
Ксантогалум сахокии 415118
Ксантогалум татьяны 415119
Ксантозома 415190
Ксантоксилум 417132
Ксанториза 415164
Ксанторрея 415172
Ксанторриза 415164
Ксантосома 415190
Ксантосома фиолетовая 415202
Ксантоцерас 415094
Ксерантемум однолетний 71042,415317
Ксилантемум 415672
Ксилантемум памирский 415675
Ксилантемум скальный 415676
Ксилантемум тяньшанский 415677
Ксилантемум фишеры 415673
Ксилосарпус 415714
Ксилосма 415863
Ксилосма японская 415869
Ксимения американская 415558
Ксирис 415990
Ксифиум колпаковского 208659
Ктенанте 113944
Кубанка 398900
Кубатал 343151,344273
Кубеба 300377
Кубышка 267274
Кубышка желтая 267294
Кубышка жёлтая 267294
Кубышка малая 267320
Кубышка пришлая 267276
Кубышка средная 267285
Кубышка японская 267286
Кувшинка 267627
Кувшинка белая 267648
Кувшинка венцеля 267782
Кувшинка гиблидная 267639
Кувшинка голубая 267658
Кувшинка душистая 267730
Кувшинка жёлтая 267714
Кувшинка капская 267665
Кувшинка красная 267756
Кувшинка лотос 267698
Кувшинка малая 267767
Кувшинка мексиканская 267714
Кувшинка четырехгранная 267767
Кувшинка четырехугольная 267767
Кувшинка чистобелая 267664
Кувшинковые 267787
Куга 119503,352289,353179
Куга озерная 352289
Куга трехгранная 352284
Кудзу 321419,321441
Кудзу лопастная 321441
Кудрания 114319
Кудрания трехконечная 240842
Кудрания цзиндунская 240808
Кудренник чёрный 47013
Кудри пестрые 47013
Кудри царские 229922
Кудрявец 139682
Кудряшевия 218274
Кудряшевия коржинского 218279
Кудряшевия надины 218280
Кудряшевия разноволосый 218276
Кудряшевия якуба 218278
Кузиния 108217
Кузиния аболина 108219
Кузиния алайская 108223
Кузиния алберта 108225
Кузиния алпйская 108228
Кузиния амударьинская 108361
Кузиния андросова 108231
Кузиния антонова 108234

Кузиния армянская 108238
Кузиния артищоковидная 108262
Кузиния астраханская 108239
Кузиния бадхизская 108241
Кузиния баталина 108242
Кузиния белесоватая 108252
Кузиния белоцветковая 108226
Кузиния бледноцветковая 108329
Кузиния блестящая 108408
Кузиния близкая 108375
Кузиния боброва 108243
Кузиния бонвало 108244
Кузиния бочанцева 108245
Кузиния вавилова 108440
Кузиния вальдгейма 108444
Кузиния васильковидная 108253
Кузиния введенского 108443
Кузиния великолепная 108336
Кузиния ветвистая 108389
Кузиния видная 108406
Кузиния викарная 108442
Кузиния войлочненькая 108427
Кузиния волжская 108239
Кузиния воловьеглазая 108248
Кузиния волосистая 108433
Кузиния восточная 108357
Кузиния вунге 108247
Кузиния высокогорная 108227
Кузиния гаматы 108300
Кузиния главная 108374
Кузиния гладкоглавая 108323
Кузиния гладкоголовая 108323
Кузиния гладкощетинковая 108291
Кузиния гнездиллы 108295
Кузиния гогенакера 108303
Кузиния головчатая 108222
Кузиния голубая 108258
Кузиния гончарова 108296
Кузиния горная 108355
Кузиния горностхолюбивая 108356
Кузиния григорьева 108298
Кузиния дарвазская 108263
Кузиния двуцветковая 108266
Кузиния дернистая 108249
Кузиния димо 108267
Кузиния длинноветвистая 108272
Кузиния длиннолистная 108274
Кузиния длинночешуйная 108273
Кузиния дтрнишникоголовая 108449
Кузиния евгения 108282
Кузиния ереванская 108280
Кузиния желелистая 108292
Кузиния железконосная 108220
Кузиния желтоватая 108448
Кузиния желтовато – серая 108390
Кузиния жесткая 108394
Кузиния заамударьинская 108430
Кузиния заилийская 108429
Кузиния заплезневелая 108346
Кузиния звездчатая 108411
Кузиния зеленоватоцветковая 108255
Кузиния зеленоголовая 108256
Кузиния золотистая 108240
Кузиния золотисто – цветковая 108257
Кузиния иглистая 108306
Кузиния изящноголовая 108293
Кузиния иларии 108302
Кузиния ильина 108307
Кузиния каратавская 108309
Кузиния карликовая 108386
Кузиния кнорринг 108310
Кузиния кокандская 108311
Кузиния колонованная 108260
Кузиния колючая 108383
Кузиния комарава 108312
Кузиния копьелистная 108301
Кузиния коржинского 108314
Кузиния королькова 108313
Кузиния короткокрылая 108246
Кузиния красивая 108382
Кузиния красивенькая 108381
Кузиния красно – бурая 108397
Кузиния красно – цветковая 108393
Кузиния краузе 108315
Кузиния крошечная 108343
Кузиния круглолистная 108396
Кузиния крупноголовая 108334
Кузиния крупнолистная 108297
Кузиния крупнохрылая 108335
Кузиния крылатая 108224
Кузиния курчаводольковая 108437
Кузиния кюкентaля 108316
Кузиния линчевского 108330
Кузиния лировидная 108333
Кузиния литвинова 108331
Кузиния ложномягкая 108379
Кузиния ложнопятиколючковая 108368
Кузиния ложнородственная 108377
Кузиния ложно – шерстистая 108378
Кузиния ломакина 108332
Кузиния лопуховидная 108237
Кузиния лохматая 108363
Кузиния лохматовидная 108364
Кузиния лучевая 108388
Кузиния лысая 108250
Кузиния маловетвистая 108365
Кузиния маргариты 108338
Кузиния мелкоголовчатая 108341
Кузиния мелкоплодная 108340
Кузиния меченосная 108294
Кузиния минквиц 108342
Кузиния многоглавая 108372
Кузиния многолопастная 108347
Кузиния моголтавская 108344
Кузиния мутовчатая 108441
Кузиния мягкая 108345
Кузиния невесского 108349
Кузиния незамеченная 108348
Кузиния неправильная 108233
Кузиния низбегающелистная 108264
Кузиния нина 108350
Кузиния ничтожная 108385
Кузиния обманчивая 108284
Кузиния обманывающая 108229
Кузиния овчинникова 108360
Кузиния ольги 108351
Кузиния остроконечная 108235
Кузиния остронадрезная 108362
Кузиния отборная 108276
Кузиния паутинистая 108236
Кузиния перисточешуйная 108380
Кузиния песколюбивая 108376
Кузиния подозрительная 108275
Кузиния полуизорванная 108403
Кузиния полунизбегающая 108402
Кузиния почтибелесоватая 108417
Кузиния почти – придаточная 108416
Кузиния почтитонковетвистая 108328
Кузиния почти – тупоконечная 108418
Кузиния прелестная 108230
Кузиния преломленная 108391
Кузиния прямоколючковая 108358
Кузиния прямочешуйная 108359
Кузиния пупообразная 108352
Кузиния пупочковая 108438
Кузиния пурпуровая 108384
Кузиния пучковатая 108285
Кузиния пятиколючковая 108367
Кузиния радде 108387
Кузиния развилистая 108265
Кузиния рассеченная 108268
Кузиния рассеченнолистная 108270
Кузиния рассеянноголовая 108409
Кузиния растопыренная 108271
Кузиния репейниковая 108320
Кузиния ржавчинная 108288
Кузиния родственная 108221
Кузиния розовая 108395
Кузиния рревосходная 108373
Кузиния саксаульниковая 108419
Кузиния самаркандская 108337
Кузиния серая 108299
Кузиния сестринская 108405
Кузиния синеголовниковидная 108281
Кузиния скученная 108259
Кузиния смолистая 108392
Кузиния снизуседая 108305
Кузиния согнутоколючковая 108251
Кузиния спиридонова 108407
Кузиния средная 108339
Кузиния страшненькая 108304

Кузиния сырдарьинская 108420
Кузиния таласская 108421
Кузиния таммары 108422
Кузиния татарниковая 108353
Кузиния твердолистная 108401
Кузиния теневая 108439
Кузиния тоненькая 108423
Кузиния тонковетвистая 108327
Кузиния тонкоголовая 108326
Кузиния тонкоколючковая 108324
Кузиния тонкорассеченная 108424
Кузиния тонкосогнутая 108325
Кузиния торчащая 108414
Кузиния торчащеколючковая 108277
Кузиния трехголовая 108432
Кузиния трехцветковая 108434
Кузиния трупкоопушенная 108322
Кузиния туркестанская 108436
Кузиния туркменская 108435
Кузиния тяньшанская 108426
Кузиния увенчанная 108413
Кузиния угороченная 108218
Кузиния узкоголовая 108232
Кузиния узколистная 108412
Кузиния федченко 108286
Кузиния ферганская 108287
Кузиния фетисова 108289
Кузиния франшэ 108290
Кузиния цветоносная 108366
Кузиния целинолистная 108308
Кузиния черешковая 108371
Кузиния шероховатая 108398
Кузиния шерстистая 108318
Кузиния шерстисто - волосая 108279
Кузиния шерстистоголовая 108319
Кузиния шершаволистная 108428
Кузиния широкооберточная 108370
Кузиния широкочешуйная 108369
Кузиния шитковидная 108261
Кузиния шишкина 108399
Кузиния шишкоголовая 108415
Кузиния шталя 108410
Кузиния щетинкоголовая 108254
Кузиния яйценогая 108354
Кузиния ярко - зеленая 108317
Кузырник купыревый 392992
Кузьмичева трава 146154
Куколица 363141
Куколица широколистная 363571,364158,364193
Куколь 11938
Куколь льняной 11951
Куколь обыкновенный 11947
Куколь посевной 11947
Кукольван 97894
Кукольвановые 250217
Кукумис 114108
Кукурбита 114271
Кукуруза 417412,417417
Кукуруза зубовидная 417416,417423
Кукуруза крахмалистая 417414
Кукуруза обыкновенная 417412,417417
Кукуруза плёнчатая 417434
Кукуруза сахарная 417433
Кукуруза сладкая столовая 417433
Кукуруза чешуйчатая 417434
Кукушник 363461
Кульбаба 224651
Кульбаба дунайская 224668
Кульбаба кавказская 224662
Кульбаба керетская 224699
Кульбаба кочи 224700
Кульбаба ложноодуванчиковаю 224725
Кульбаба осенняя 63625,224655
Кульбаба ползучая 224726
Кульбаба шафранная 224666
Кульбаба шероховатейшая 224652
Кульбаба шишкина 224736
Кульбаба щетинистая 224698
Кумалчик 11609
Кумалчик малый 11616
Куманника 338209,338447,338878
Кумаруна 133629
Кумарчик 11609
Кумарчик бокоцветковый 11613
Кумарчик малый 11616
Кумарчик палецкого 11617
Кумарчик песчаный 11619
Кумарчик широколистный 11614
Кумват 167506
Кумкват 167489,167506
Кумкват гонконгскийдикий 167499
Кумкват дикий 167499
Кумкват круглый 167503
Кумкват маруми 167503
Кумкват овальный 167506
Кумкват японский 167503
Куммеровия 218344,226677
Куммеровия клиновидная 226742
Куммеровия полосатая 218347
Куммеровия прилистниковая 218345
Кунакисагак 361935
Кунжут 361294,361317
Кунжут восточный 361317
Кунжут индийский 361317
Кунжут культурный 361317
Кунжутовые 286918
Кунила 114520
Куннингамия 114532
Куннингамия китайская 114548
Куннингамия ланцетная 114539,114548
Куннингамия ланцетовидная 114539
Куннингамия ланцетовидная 'Глаука' 114540
Куннингамия тайваньская 114548
Купава европейская 399504
Купальница 399484,399504
Купальница азиатская 399494
Купальница алтайская 399491
Купальница бумажночашечная 399499
Купальница джунгарская 399503
Купальница европейская 399504
Купальница китайская 399500
Купальница крупнолднестковая 399519
Купальница ледебура 399515
Купальница мелкоцветковая 399520
Купальница полуоткрытая 399522
Купальница ридера 399540
Купальница сибирская 399541
Купальница уральская 399547
Купальница японская 399510
Купания 114574
Купена 308493
Купена аньхуйская 308503
Купена аптечная 308613,308616,407256
Купена вздутая 308565
Купена гладная 308551
Купена десулави 308532
Купена душистая 308613
Купена заостреннолистная 308495
Купена кавказская 308631
Купена кандинская 308632
Купена лекарственная 308613,407256
Купена лэйбоская 308588
Купена максимовича 308598,308613
Купена многоцветковая 308529,308607
Купена многоцветная 308529
Купена мутовчатая 308659
Купена низкая 308562
Купена оберткoвая 308567
Купена обертковая 308567
Купена обыкновенная 308613,407256
Купена остолистная 308495
Купена приземистая 308562
Купена розавая 308639
Купена северцова 308640
Купена сибирская 308641
Купена тонкоцветковая 308631
Купена туполистная 308612
Купена узколистная 308647
Купена франше 308544
Купена чжэцзянская 308666
Купена широколистная 308558,308584
Купена яйцевиднолистная 308626
Купена японская 308613
Купина неопалимая 129615
Купырник 392959
Купырник надоеливый 392989
Купырник пренебреженный 392995
Купырник разнолистный 392985
Купырник узловатый 392996

Купырник украинский 393007
Купырь 28013
Купырь бархатистый 28061
Купырь бутенелистный 28024
Купырь длинноносиковый 28029
Купырь дубравный 28030
Купырь лесной 28044
Купырь лоснящийся 28035
Купырь обыкновенный 28023,28039
Купырь похожий 28044
Купырь приледниковый 28027
Купырь рупрехта 28037
Купырь садовый 28024
Купырь сосновского 28042
Купырь шмальгаузена 28041
Купырь юньнаньский 28066
Курай 344425
Курак 295888
Курильский чай 289713,312322
Курильский чай дриадоцветовидный 289708
Курильский чай кустарниковый 289713
Курильский чай листочашечный 289720
Курильский чай маньчжурьрский 312606
Курильский чай мелколистный 289718
Куриное просо 140351
Куриное просо обыкновенное 140367
Куриное просо рисовидное 140462
Куриное просо хвостатое 140356
Куркулига 114814
Куркулиго 114814
Куркума 114852
Куркума гуансиская 114868
Куркума длиная 114871
Куркума узколистная 114858
Куркума цитварная 114884
Куркума юньнаньская 114882
Курмак 140466
Куропатник 343595
Куропаточник 343595
Куропаточья трава 138445
Курослеп 325515,325518
Курослеп болотный 326034
Куртарник хинный китайский 280078
Курчавка 44247,44270
Курчавка бадхызская 44250
Курчавка валькоьатолистная 44279
Курчавка грушелистная 44272
Курчавка зеравшанская 44275
Курчавка кавказская 44257
Курчавка каратавская 44265
Курчавка колючая 44271
Курчавка кустарная 44260
Курчавка мушкетова 44270
Курчавка обманчивая 44259
Курчавка отогнутая 44273
Курчавка прутьевидная 44284
Курчавка сероватая 44256
Курчавка скученная 44258
Курчавка турнефора 44282
Курчавка узколистная 44249
Курчавка шиповатая 44276
Курчавка яркозеленая 44266
Кускута 114947
Куссо 184565
Куст кокайновый 155051,155067
Куст креозотовый 221975
Куст сиреневый 382114
Куст соляной 45750
Куст чайный 69634
Куст чайный апалашский 204391
Куст чайный китайский 69634
Кустарник кокаиновый 155067
Кустарник мастиковый 300994
Кустарник овечий 290222
Кутандия 115252
Кутандия мемфийская 115259
Кутровые 29462
Куфея 114595,285055
Кучерява 378836
Кучерявка 183225
Кучина 83253
Кучкоцветник 369254
Кучкоцветник мейера 369256
Кушир 83540
Кызылша 53224,53249
Кырлык 162335
Кювьера 115278
Лабазник 166071
Лабазник вязолистный 166125
Лабазник голый 166085
Лабазник дланевидный 166110
Лабазник ильмовидный 166125
Лабазник камчатский 166089
Лабазник многопарный 166097
Лабазник обнаженный 166080
Лабазник пурпуровый 166117
Лабазник средний 166088
Лабазник степной 166123
Лабазник тайваньский 166092
Лабазник узкодольчатый 166073
Лабазник узколопастный 166073
Лабазник шестилепестный 166132
Лабданум 93161
Лаванда 223248
Лаванда аптечная 223251
Лаванда колосовая 223251
Лаванда морская 230761
Лаванда настоящая 223251
Лаванда обыкновенная 223251
Лаванда широколистная 223298
Лаватера 223350
Лавобень черноватый 117385
Лавр 223163
Лавр американский 215395
Лавр берый 242341
Лавр благородный 223203
Лавр горный 215395
Лавр дикий 223116
Лавр душистый 242341
Лавр калифорнийский 401672
Лавр камфорный 91287
Лавр канарский 223178
Лавр коричный 91252
Лавр красный 291505
Лавр ложнокамфарный 91330
Лавр тюльпанный 242135
Лавр цейлонский 91252
Лавровишня 223088
Лавровишня аптечная 223116
Лавровишня лекарственная 223116
Лавровишня лузитанская 316518
Лавровишня мэнхайская 223115
Лавровишня обыкновенная 223116
Лавровишня португальская 316518
Лавровые 223006
Лаврушка 122359
Лавссония 223454
Лагенарик 219821,219843
Лагенария 219821
Лагерстремия 219908
Лагерстремия гуандунская 219927
Лагерстремия гуйлиньская 219931
Лагерстремия индийская 219933
Лагерстремия мелкоцветковая 219959
Лагерстремия таиландская 219965
Лагерстремия фуцзяньская 219956
Лагерстремия юньнаньская 219952
Лагета 219986
Лагозериопсис 220131
Лагозериопсис попова 220132
Лагозерис 220135,220141
Лагозерис аральский 220137
Лагозерис крупноцветковый 220139
Лагозерис мощный 220143
Лагозерис обрантояйцевидный 220140
Лагозерис палестинский 220145
Лагозерис пурпуровый 220142
Лагозерис сахендский 220144
Лагозерис сизоватый 220138
Лагопсис 220117
Лагопсис желтый 220124
Лагопсис мохнатоколосый 220121
Лагопсис простертая 220126
Лагопсис шандровый 220125
Лаготис 220155
Лаготис иконникова 220177
Лаготис королкова 220182
Лаготис малый 220188
Лаготис побегоносный 220199
Лаготис сизый 220167

Лаготис уральская 220201
Лагунария 220226
Лагурус 220252
Ладан 57516
Ладан конский 37616
Ладанник 93120
Ладанник волосистый 93152
Ладанник кручавый 93141
Ладанник крымский 93216
Ладанник лавролистный 93170
Ладанник тополелистный 93194
Ладанник шалфеелистный 93202
Ладанниковые 93046
Ладыгиния 219660
Ладыгиния бухарская 219661
Ладьян 103976
Ладьян трехлодвезный 104019
Ладьян трёхлодвезный 104019
Ладьян трехнадрезанный 104019
Ладьян трехнадрезный 104019
Лажфиоль 86427
Лазерпициум 222020
Лазиопогон 222454
Лазиопогон моховидный 222460
Лазия 222038
Лазорник 100892
Лазурник 222020
Лайм 93329,93534
Лайм настоящий 93329,93534
Лайя 223476
Лаковое дерево 393491
Лаконос 298088
Лаконос американский 298094
Лаконос двудомный 298103
Лаконосные 298124
Лакрица 177873
Лакричник 177873
Лакричник голый 177893
Лактук 219206,259746
Лакфиоль 86409
Лакфиоль садовый 86427
Лаллеманция 220281
Лаллеманция ройлевская 220286
Ламаркия 220295
Ламира 220775
Ламира колючеголовая 220779
Ламиропапус 220780
Ламиропапус шака – птарский 220781
Ландольфия 220814
Ландыш 102856
Ландыш закавказский 102883
Ландыш лесной 102867
Ландыш майский 102867
Ландыш обыкновенный 102867
Ландышевое дерево 203422
Ландышник 272049
Лансиум 221224
Лансиум домашний 221226
Лансиум обыкновенный 221226
Лантана 221229
Лантана сводчатая 221238
Ланция 218387
Лапа воронья 105338
Лапа воронья простёртая 105350
Лапагерия 221340
Лапагерия розовая 221341
Лаперузия 221353
Лапина 320346
Лапина кавказская 320356
Лапина крылатоплодная 320356
Лапина ледера 320371
Лапина палиурус 116240
Лапина сумахолистная 320372
Лапина узкокрылая 320377
Лапина хубэйская 320358
Лапина ясенелистная 320356
Лапка 117852,117859
Лапка воронья 105338
Лапка воронья простёртая 105350
Лапка гусиная 312360
Лапка заячьи 396828
Лапка заячья 75116
Лапка кошачья 26299
Лапка кошачья альпийская 26310
Лапка кошачья двдомная 26385
Лапка кошачья карпатская 26362
Лапортея 221529
Лапортея канадская 221547
Лапортея клубненосная 221538
Лапортея мотоская 221576
Лапортея фуцзяньская 221561
Лапчатка 312322
Лапчатка аджарская 312333
Лапчатка азиатская 312457
Лапчатка амурская 312659
Лапчатка анадырская 312350
Лапчатка арктическая 312518
Лапчатка астрагалолистная 312402
Лапчатка астраханская 312401
Лапчатка бедренцеволистная 312888
Лапчатка белая 312337
Лапчатка белеющая 313104
Лапчатка белолистная 312721
Лапчатка бесстебельная 312325
Лапчатка болотная 100257
Лапчатка борнмюллера 312427
Лапчатка бунге 312431
Лапчатка валя 313093
Лапчатка веерная 312538
Лапчатка вейссенбургская 312726
Лапчатка веловолосистая 312727
Лапчатка веселая 312700
Лапчатка весенняя 312812
Лапчатка вибеля 313113
Лапчатка вильчатая 312412,312418
Лапчатка волосистая 312886
Лапчатка восточная 312841
Лапчатка выемчатая 312518
Лапчатка высокая 312516
Лапчатка гейденрейха 312656
Лапчатка гольдбаха 312645
Лапчатка гравилатовидная 312633
Лапчатка грясная 313007
Лапчатка гусиная 312360
Лапчатка далиская 313024
Лапчатка двувильчатая 312412
Лапчатка двунадрезанная 312412
Лапчатка двухцветковая 312409
Лапчатка двуцветная 312502
Лапчатка джунгарская 313005
Лапчатка длинноцветоножковая 312736
Лапчатка железтолистная 312327
Лапчатка жестковатая 312934
Лапчатка жилковатая 312811
Лапчатка закаспийская 313081
Лапчатка зейдлица 312965
Лапчатка земляниковидная 312550,312561
Лапчатка земляничная 312550
Лапчатка золотистая 312406,312457
Лапчатка золотистоцветковая 312457
Лапчатка изящная 312517
Лапчатка кавказскаю 312441
Лапчатка калльe 312433
Лапчатка камиллы 312434
Лапчатка кандинская 313062
Лапчатка китайская 312450
Лапчатка клейкая 312734
Лапчатка клейна 312690
Лапчатка клинолистная 313009
Лапчатка комалова 312695
Лапчатка коротколепестая 312428
Лапчатка коротностебельная 312430
Лапчатка крандца 312472
Лапчатка кранца 312472
Лапчатка красивенькая 312905
Лапчатка красноцветковая 312896
Лапчатка криптомерневая 312478
Лапчатка криптотениевая 312478
Лапчатка кровохлебковая 312955
Лапчатка крупноцветковая 312739,312758
Лапчатка крылова 312696
Лапчатка крымская 313064
Лапчатка кузнецова 312698
Лапчатка кулябская 312697
Лапчатка курильский 289713
Лапчатка кустарниковая 289713
Лапчатка лабазниковидная 313054
Лапчатка лапландская 312703
Лапчатка лежачая 312357
Лапчатка липского 312731
Лапчатка ломакина 312732

Лапчатка лябинколистная 313054
Лапчатка матзумуры 312745
Лапчатка мейера 312757
Лапчатка мелколистная 289718,312764
Лапчатка мелкоцветковая 312762
Лапчатка мийябе 312782
Лапчатка многонадрезная 312797
Лапчатка многораздельная 312975
Лапчатка многоразрезная 312797
Лапчатка морщинистая 312944
Лапчатка мохнатая 313103
Лапчатка муркрофта 312417
Лапчатка мутовчатая 313098
Лапчатка мягковолосая 312742
Лапчатка мягчайшая 312783
Лапчатка неблестящая 312675
Лапчатка неодетая 312531
Лапчатка низкая 313039
Лапчатка норвежская 312828
Лапчатка нордманна 312827
Лапчатка нурская 312836
Лапчатка облиственная 312544
Лапчатка одноцветковая 313091
Лапчатка округлая 312840
Лапчатка памироалайская 312852
Лапчатка памирская 312851
Лапчатка пачкающая 312946
Лапчатка пепельно - белая 313067
Лапчатка песчаная 312387
Лапчатка песчанистая 312388
Лапчатка пижмолистная 313054
Лапчатка пиндская 312889
Лапчатка пирамидкоцветковая 313074
Лапчатка плетевидная 312540
Лапчатка побегоносная 313028
Лапчатка подражающая 312351
Лапчатка ползучая 64982,312925
Лапчатка полунадлезная 312968
Лапчатка полуобнаженная 312967
Лапчатка почти - дланевидная 313035
Лапчатка прижатая 312496
Лапчатка прилистниковая 313027
Лапчатка прямая 312520,312912
Лапчатка прямостоячая 312520
Лапчатка прямостоящая 312520
Лапчатка пустынностепная 312497
Лапчатка радде 312911
Лапчатка разноцветная 312502
Лапчатка распростертая 312669,313039
Лапчатка регелевская 312922
Лапчатка репейничковидная 313029
Лапчатка рупрехта 312948
Лапчатка сближенная 312378
Лапчатка сванетская 313050
Лапчатка седоватая 312676
Лапчатка семилисточковая 312658
Лапчатка серая 312387
Лапчатка серебристая 312389
Лапчатка серебристовидная 312390
Лапчатка серножлтая 313036
Лапчатка сжатая 312467
Лапчатка сизоватая 312644
Лапчатка сильно - шероховатая 312399
Лапчатка скалистая 312946
Лапчатка скальная 211551,312946
Лапчатка снежная 312818
Лапчатка соммье 313003
Лапчатка сплошь белая 312667
Лапчатка сплошь - белая 312667
Лапчатка средняя 312683
Лапчатка стелющаяся 313039
Лапчатка стозернышковая 312446
Лапчатка стоплодная 312446
Лапчатка стоповидная 312861
Лапчатка табернемонтана 312812
Лапчатка темная 312837
Лапчатка теневая 313090
Лапчатка тибетская 313114
Лапчатка толля 313077
Лапчатка толстая 312473
Лапчатка траншеря 312351
Лапчатка тюрингенская 312658
Лапчатка тюрингская 312658
Лапчатка тяньшанская 313075
Лапчатка узколистная 312358
Лапчатка уссурийская 312944
Лапчатка федченковская 312535
Лапчатка фрейна 312565
Лапчатка холодная 312630
Лапчатка цельнолепестная 312668
Лапчатка черепчатая 312674
Лапчатка чудесная 312505
Лапчатка шелковая 312969
Лапчатка шовица 313051
Лапчатка шпренгеля 313013
Лапчатка шренковская 312963
Лапчатка шура 312964
Лапчатка щетинистая 313029
Лапчатка эверсманна 312530
Лапчатка эгеде 312510
Лапчатка яйлы 312688
Лапчатка якутская 312687
Лапчатка - узик 312520
Лардизабала 221815
Лардизабаловые 221818
Ларрея 221978
Ласковец 225428
Ласковцы 63805,138348
Ластовенник 117334
Ластовенник обыкновенный 117743
Ластовень 37811,117334,409394
Ластовень алебова 117357
Ластовень аптечный 117743
Ластовень ван 117748
Ластовень вильфорда 117751
Ластовень вьюшийся 117745
Ластовень дарвазский 409431
Ластовень заостренный 117339
Ластовень куррассавский 37888
Ластовень лекарственный 117743
Ластовень ленцовидный 117721
Ластовень максимовича 117597
Ластовень неприятный 117523
Ластовень обыкновенный 117743
Ластовень острый 117345
Ластовень пурпуровый 117660
Ластовень русский 117670
Ластовень стеблеобъемлющий 117364
Ластовень темный 117385
Ластовень шмальгаузена 117679
Ластовень японский 117537
Ластовица 86755,115031
Ластовневые 37806
Латания 222625
Латирус 222671
Латовень черноватый 117385
Латовник 117334
Латочник 38135
Латочниковые 37806
Латук 219206,259746
Латук алтайский 219210
Латук вильгельмся 219614
Латук винкрера 219615
Латук волнистый 219589
Латук высокий 320511
Латук глауциелистный 219336
Латук грузинский 219327
Латук дидий 219507,219609
Латук дикий 219497,219507
Латук иволистный 219480
Латук индийский 320515
Латук канадский 219249
Латук клубненосный 219585
Латук комнастый 219507
Латук красный 313866
Латук лесной 219497,219507
Латук многолетний 219444
Латук огородный 219485
Латук посевной 219485
Латук прутяной 219600
Латук радде 320520
Латук раздельнолистный 219270
Латук римский 219493
Латук розеточный 219474
Латук серриола 219507
Латук сжатый 219547
Латук сибирский 219513,219806
Латук солончаковый 219480
Латук стеной 260596
Латук татарский 259772
Латук треугольный 320523

Латук удивительный 219416
Латук ушковатый 219221,219303
Латук шиповатозубчатый 219533
Латук шэ 219270
Латук ядовитый 219609
Латук – ромен 219493
Латук – ромэн 219493
Латук – салат 219485
Латук – салат кочанный 219489
Лаузелеурия лежачая 235283
Лаузелеурия стелющаяся 235283
Лахеналия 218948
Лахнантес 219026
Лебеда 44298,86891
Лебеда белая 44333
Лебеда бородавчатая 44688
Лебеда веероплодная 44402
Лебеда гарообразная 44650
Лебеда гладкая 44484
Лебеда гмелина 44430
Лебеда голостебельная 44541
Лебеда диморфная 44374
Лебеда кино 87145
Лебеда колючая 44624
Лебеда копьевидная 44576,44621
Лебеда копьелистная 44576,44621
Лебеда красивоплодная 44332
Лебеда крупноприцветничковая 44349
Лебеда кузеневой 44480
Лебеда лебедадикая 44400
Лебеда лоснящаяся 44537
Лебеда максимовича 44520
Лебеда монетоплодная 44530
Лебеда обыкновенная 44621
Лебеда отклоненная 44575
Лебеда памирская 44564
Лебеда перуанская 87145
Лебеда поникшая 44576
Лебеда прибрежная 44510
Лебеда продолговатолистная 44554
Лебеда разносемянная 44523
Лебеда ранняя 44619
Лебеда раскидистая 44576
Лебеда рисовая 87145
Лебеда розовая 44629
Лебеда садовая 44468
Лебеда саладная 44468
Лебеда сибирская 44646
Лебеда стебельчатая 44601
Лебеда татарская 44668
Лебеда толстолистная 44366
Лебеда тунбергиелистная 44674
Лебеда туркменская 44684
Лебеда украшенная 44558
Лебеда центральноазиатская 44347
Лебеда черешчатая 44601
Лебеда шарообразная 44650
Лебеда широкоплодная 44324
Лебеда шугнанская 44640
Лебедка 44298
Левкена 227420
Левкой 246438
Левкой белостебельный 246439
Левкой великолепный 246555
Левкой грубый 246544
Левкой дагестанский 246450
Левкой двлогий 246443
Левкой душистый 246460,246521
Левкой зимний 246478
Левкой марелистый 246447
Левкой мучнистый 246455
Левкой обыкновенный 137747,246478
Левкой однолетний 246478
Левкой остророгий 246494
Левкой пахучий 246460
Левкой садовый 137747,246478
Левкой садой 246478
Левкой седой 137747
Левкой стоддарта 246554
Левкой струговидный 246545
Левкой татарский 246556
Левкой цельнокрайний 246487
Левкорхис 318209
Лёгочница 321633
Легузия 224058
Легузия зеркало венеры 224058
Легузия зеркало – венеры 224073
Легузия либридная 224067
Легузия пятиугольная 224070
Легузия серповидная 224058
Ледебуриелла 223825
Ледебуриелла жабрицевидная 346448
Ледебуриелла многолестная 223827
Ледебуриелла растопыренная 346448
Ледянник 251265
Леерсия 223973
Леерсия посевная 223992
Леерсия рисовидная 223992
Леерсия японская 224002
Лейбниция 175107,224106
Лейбниция бестычинковая 224110
Лейбниция двликая 224110
Лейбниция кнорринг 224115
Лейкодендрон серебристый 227233
Лейкофоя 228154
Лейтнерия 224275
Лейцестерия 228319
Лейцестерия красивая 228321
Лейцестерия тибетская 228335
Лемонграс 117153
Лен 231856
Лён 231856
Лен австрийский 231872
Лён австрийский 231872
Лен алтайский 231864
Лен амурский 231865
Лён белосемянный 232001
Лен бледноватый 231941
Лен волосистый 231911
Лен дидий 230866
Лен долгунец 232001
Лён долгунец 232001
Лен естковолосый 231911
Лен жилковатый 231936
Лён красивый 340214
Лен крымский 231985
Лён кудряш 232001,232002
Лен культурный 232001
Лён культурный 232001
Лен многолетний 231942
Лён многолетний 231942
Лён новозеландский 295594
Лён новозеландский прядильный 295600
Лен обыкновенный 232001
Лён обыкновенный 232001
Лён посевной 232001
Лен разночашелистиковый 231909
Лен сибирский 231942
Лён сибирский 231942
Лён слабетельный 231884
Лен слабительный 231884
Лен стеллеровидный 231961
Лен тонколистный 231992
Лён тонколистный 231992
Лен узколистный 231880
Лён узколистный 231880
Лен узловатый 231937
Лен украинский 232000
Лен французский 231996
Лён французский 231996
Лен чешуйчатый 231958
Лен щиткоцветковый 231895
Лен щиточковатый 231895
Ленец 389592
Ленец алатавский 389597
Ленец альпийский 389600
Ленец бесприцветничковый 389677
Ленец ветвистый 389840
Ленец гончарова 389711
Ленец длиннолистный 389769
Ленец дэцинький 389672
Ленец каменистый 389856
Ленец китайский 389638
Ленец коротколистный 389623
Ленец кочи 389753
Ленец льнолистный 389763
Ленец минквиц 389792
Ленец многотебельный 389795
Ленец ползучий 389848
Ленец преломленный 389844
Ленец приморский 389780

Ленец простертый 389825
Ленец сибирский 389844
Ленец сильно ветвистый 389838
Ленец ферганский 389691
Ленец шовица 389893
Ленец щебнистый 389857
Лёновые 230854
Ленок 104429,231844
Ленок красильный 104598
Ленок обыкновенный 231844
Ленолистник 389592
Лён – прыгунец 232003
Лён – скакунец 232003
Лентоостник 383034
Лентоостник длинноволосый 383040
Лентоостник шероховатый 383035
Леонотис 224558
Леонтица 224625
Леонтица алтайская 182632
Леонтица альберта 224626
Леонтица дарвазская 224629
Леонтица еверсманна 224630
Леонтица малая 224639
Леонтица одесская 224640
Леонтица смирнова 224642
Леонтица сомнительная 224631
Леонтоподиум 224767
Леонтоподиум альпийский 224772
Леонтоподиум мелкоцветковое 224904
Леонтоподиум сибирский 224893
Леонтоподиум франше 224842
Леонтоподиум эмэйский 224924
Леопольдия 225027
Леопольдия длинная 225035
Леопольдия кавказсая 225031
Леопольдия тонкоцветковая 225036
Леопольдия хохлатая 260308
Леопольдия хохолковая 260308
Лепеха 5788
Лепехиниелла 225088,225097
Лепехиниелла алайская 225089
Лепехиниелла заалайская 225099
Лепехиниелла зеравшанская 225098
Лепехиниелла коржинского 225092
Лепехиниелла маленькая 225095
Лепехиниелла пупковидкая 225096
Лепехиниелла ферганская 221675
лепидолофа 225519
лепидолофа каратавская 225522
лепидолофа комарова 225523
лепидолофа моголтавская 225524
лепидолофа нителистная 225521
лепидолофа нуратавская 225525
лепидолофа федченко 225520
Лепка львиная 224772
Лепталеум 225802
Лепталеум нителистный 225803
Лептопирум 226339
Лептопирум дымянковый 226341
Лептопус 226315
Лепторабдос 226367
Лепторабдос мелкоцветковый 226369
Лептосине 104429
Лептосперм 226450
Лептоспермум 226450
Лептунис 226569
Лептунис волосовидный 226571
Лепчица 170193
Леснина 243675,243711
Лесной мак 200753
Лесной мак японский 200754
Лесной орешник 106703
Лесной чистотел 200753
Лесной чистотел японский 200754
Леспедеза 226677
Леспедеза двуцветная 226698
Леспедеза клиновидная 226742
Леспедеза крупноплодная 70833
Леспедеза полосатая 218347
Леспедеца 226677
Леспедеца войлочная 226989
Леспедеца даурская 226751
Леспедеца двуцветковая 226698
Леспедеца двуцветная 226698
Леспедеца заострённая 226742
Леспедеца китайская 226739
Леспедеца клиновидная 226742
Леспедеца копеечниковая 226860
Леспедеца корейская 218345
Леспедеца кривокистевая 226746
Леспедеца мохнатая 226989
Леспедеца плотная 226746
Леспедеца полосатая 218347
Леспедеца прилистниковая 218345
Леспедеца ситниковая 226860
Леспедеца хубэйская 226840
Леспедеца шелковистая 226742
Леспедеца японская 226708
Леспедеция 226677
Леукантемелла 227452
Леукантемелла линейная 227453
Леукантемелла поздняя 227455
Леукокринум 227818
Леукорхис 318209
Леукорхис беловатый 318210
Леукофиллум 227916
Лецитис 223764
Лечник жетоватый 274617
Лечуха 345940
Лещина 106696
Лещина американская 106698
Лещина ван 106784
Лещина высокая 106757
Лещина гуйцзоуская 106748
Лещина древовидная 106720
Лещина зибольда 106777
Лещина калифорнийская 106725
Лещина китайская 106716
Лещина колхидская 106719
Лещина короткотрубчатая 106714
Лещина крупная 106757
Лещина маньчжурская 106751
Лещина обыкновенная 106703
Лещина понтийская 106703
Лещина разнолистная 106736
Лещина тайваньская 106735
Лещина тибетская 106734
Лещина фундук 106698
Лещина юньнаньская 106786
Лжеакация 334976
Лжеанельсин 240806
Лжеапельсин 240828
Лжеводосбор 283720
Лжеводосбор ветреницевидный 283721
Лжеводосбор дернистый 283723
Лжеводосбор крупноцветный 283721
Лжеводосбор мелколистый 283726
Лжегравилат 283327
Лжегравилат калужницелистный 175390
Лжедурмишник 210462
Лжекипарис 85263
Лжекипарис гороохоносный 85345
Лжелиственница 317821
Лжелиственница кемпфера 317822
Лжелиственница китайская 317822
Лжелиственница приятная 317822
Лженарцисс 262441
Лженарцисус 262441
Лжеочток 97967
лжеочток карсский 318432
лжепузырник 318610
лжепузырник пальчатый 318612
Лжепустынноколосник 317290
Лжепустынноколосник северцова 317291
лжерыжик 317485
лжерыжик шовица 317487
Лжесаса лушаньская 318305
Лжесаса японская 318307
Лжетайник 317635
Лжетайник хориса 317636
Лжетсуга 318562
Лжетсуга крупноплодная 318579
Лжетсуга крупношишечная 318579
Лжетсуга мензиса 318580
Лжетсуга сизая 318574
Лжетсуга тисолистная 318580
Лжетсуга тиссолистная 318580
Лжетсуга японская 318575
Лиана верандная 321441
Лиатрис 228433
Лиатрис колосковая 228529

Либертия 228625
Либоцедрус 228635
Ливистона 234162
Ливистона австралийская 234166
Ливистона китайская 234170
Ливистона кохнихинская 234181
Ливистона кругловатая 234194
Ливистона круглолистная 234190
Ливистона фэнкайская 234177
Ливистона южная 234166
Ливистония 234162
Ливистония китайская 234170
Лигулярия 228969
Лигулярия бирманская 229004
Лигулярия кемпфера 162616
Лигулярия мелколистная 229130
Лигулярия синьцзянская 229252
Лигулярия хэбэйская 229058
Лигулярия юньнаньская 229259
Лигулярия японская 229066
Лигустикум 229283
Лигустикум кавказский 229303
Лигустикум китайский 229390
Лигустикум крылатый 229291
Лигустикум лицзянский 229314
Лигустикум монгольский 229361
Лигустикум мулиский 229398
Лигустикум низкий 229378
Лигустикум пурпуроволенестный 229379
Лигустикум разноцветный 229318
Лигустикум северный 229384
Лигустикум тибетский 229415
Лигустикум федченко 229322
Лигустикум франше 229329
Лигустикум хультена 229385
Лигустикум цзилунский 229333
Лигустикум шотрандский 229384
Лигустикум юньнаньский 229417
Лигустикум яньюаньский 229416
Лигуструм 229428
Лигуструм блестящий 229529
Лигуструм люцюский 229524
Лизея 232664
Лизея армянский 232665
Лизея разноплодная 232667
Лизиелла малоцветковаю 239532
Лизихитон 239516
Лизихитон камчатский 239518
Ликания 228663
Ликвидамбар 27967,232550
Ликвидамбар восточный 232562
Ликвидамбар смолоностый 232565
Ликвидамбар стираксовый 232565
Ликвидамбар формозский 232557
Ликопсис 239167
Ликопсис полевой 21920
Ликуала 228729
Лилейник 191261,191304
Лилейник лимонножелтый 191266
Лилейник рыжеватый 191284
Лилейные 229700
Лилия 229717,229822
Лилия американская 111158
Лилия аньхуйская 229727
Лилия ауратум 229730
Лилия африканская 400050
Лилия белая 229789
Лилия бледская 229789
Лилия броуна 229754
Лилия бульбоносная 229766
Лилия водяная 267648
Лилия вэньшаньская 230073
Лилия гигантская 73159
Лилия далиская 230048
Лилия даурсая 229828
Лилия двурядная 229834
Лилия дллиноцветковая 229900
Лилия дллиноцветная 229900
Лилия единобратственная 229944
Лилия жёлтая 191261
Лилия золотистая 229730
Лилия исполинская 73159
Лилия ичанская 229890
Лилия канадская 229779
Лилия каприковая 230009
Лилия карниольская 229793
Лилия кессельринга 229881
Лилия китайско－японская 229730
Лилия клубненосная красная 229769
Лилия корейская 229723
Лилия королевская 230019
Лилия красивенькая 230006
Лилия красная 229766
Лилия кудреватая 229922
Лилия ландышавая натальская 345782
Лилия ланцетолистная 230058
Лилия ледебура 229886
Лилия лицзянская 229898
Лилия лонгифлорум 229900
Лилия максимовича 229887,229888
Лилия малабарская 177251
Лилия мартагон 229922
Лилия медеолевидная 229934
Лилия медеоловидная 229934
Лилия мозолистая 229775
Лилия мозолистая люцюская 229776
Лилия мотоская 229936
Лилия наньчуаньская 230022
Лилия нильская 200906
Лилия овсяная 229934
Лилия однобратственная 229944
Лилия пенсиливанская 229828
Лилия поникающая 229803
Лилия понтийская 229996
Лилия саранка 229922
Лилия сычуаньская 229829
Лилия тайваньская 229845
Лилия терракотовая 230055
Лилия тигровая 230058
Лилия трёхцветная южноафриканская 218948
Лилия узколистный 230009
Лилия халцедонская 229806
Лилия хуйдунская 229864
Лилия хэнсона 229857
Лилия чалмовая 229922
Лилия чисто－белая 229789
Лилия шафранная 229769
Лилия шовица 230046
Лилия эмэйская 229837
Лилия японская 229872
Лилия－кордилина 104334
Лима 93492,294006,294010
Лима мелкая 294010
Лима мелкоплодная 294010
Лиметта 93329,93534
Лимнантемум 267800
Лимнантес 230205
Лимнорхис 230338,302241
Лимнорхис ландышелистный 230342
Лимнорхис расширенный 230348
Лимнорхис цельногубый 302355
Лимнохарис 230244
Лимодорум 67882,230369
Лимодорум крубневой 67894
Лимодорум недоразвитый 230378
Лимон 93539,93546
Лимон дикий 93603
Лимон лиметта 93329,93534
Лимон персидский 163499
Лимон сладкий 93329,93534
Лимон трехлисточковый 310850
Лимон трёхлисточковый 310850
Лимоник 351011
Лимоник контснотычночный 351103
Лимоник мелкоцветковая 351069
Лимониум 230508
Лимонная мята 249491
Лимонная мята аптечная 249504
Лимонник китайский 351021
Лимонниковые 351121
Лимонница гладкая 256761
Линария 230866
Линделофия 231243
Линделофия длинностолбиковая 231248
Линделофия капю 231246
Линделофия крылоплодная 231253
Линделофия ольги 231249
Линделофия столбиковая 231252
Линделофия узкалистная 231244
Линдера 231297

Линдера гуандунская 231377
Линдера гуансиская 231366
Линдера гуншаньская 231331
Линдера динхуская 231322
Линдера зонтичная 231461
Линдера мэнхайская 231395
Линдера обыкновенная 231324
Линдера сычуаньская 231435
Линдера тибетская 231396
Линдера туполопастная 231403
Линдера тяньцюаньская 231453
Линдера узколистная 231303
Линдера хэнаньская 231370
Линдера шелковистая 231433
Линдера эмэйская 231419
Линдерния 231473
Линдерния стаканчатая 231559
Линдерния яошаньская 231600
Линнея 231676
Линнея северная 231679
Линозирис 231818
Линозирис мохнатый 231842
Линум 231856
Лиония 239369
Лиония овальнолистная 239391
Лиония овальнолистная эллиптическая 239392
Липа 391646
Липа американская 391649
Липа амурская 391661
Липа бегониелистная 391829
Липа белая 391836
Липа венгерская 391671,391836
Липа войлочная 391671,391836
Липа голландская 391706
Липа гуйцзоуская 391751
Липа длинночешковая 391816
Липа европейская 391706
Липа кавказская 391682,391701
Липа каменная 294761
Липа каменная узколистная 294762
Липа каменная широколистная 294772
Липа китайская 391685
Липа красная 391829
Липа крупнолистная 391817
Липа крымская 391705
Липа лицзянская 391759
Липа максимовича 391766
Липа маньчжурская 391760
Липа мелколистная 391691
Липа многоцветковая 391783
Липа монгольская 391781
Липа наньчуаньская 391784
Липа обыкновенная 391706
Липа опушенностолбиковая 391701
Липа плосколистная 391817
Липа пушистая 391836
Липа пушистостолбчатая 391682,391701
Липа сердцевидная 391691
Липа сердцелистная 391699
Липа серебристая 391671,391836
Липа сибирская 391832
Липа таке 391666
Липа франше 391717
Липа хубэйская 391730
Липа чёрная 391649
Липа широколистная 391817
Липа эмэйская 391797
Липа японская 391739
Липа ярко-зелёная 391705
Липарис 232072
Липарис баотинский 232096
Липарис гуандунский 232217
Липарис кумокири 232215
Липарис лёзеля 232229
Липарис макино 232322
Липарис мелкоцветковая 232117
Липарис сахалинский 232318
Липарис хайнаньский 232183
Липарис японская 232197
Липарис японский 232197
Липия 232457
Липоные 391860
Липпия 232457
Липпия лимонная 232483
Липпия трёхлистная 232483
Липпия узлоццветковая 296121
Липучка 221627
Липучка балхашская 221634
Липучка бесшипиковая 221639
Липучка бородчатая 221635
Липучка ганьсуская 221678
Липучка гребенчатая 221649
Липучка двоякоплодная 221660
Липучка двуплодная 221657
Липучка дробова 221659
Липучка ежёвая 221711
Липучка ежевидная 221711
Липучка ежеголовая 221672
Липучка ежовая 221711
Липучка закрытоплодная 221713
Липучка каемчатая 221702
Липучка камнелюбивая 221737
Липучка колючеплодная 221751
Липучка коржинского 221692
Липучка корончатая 221648
Липучка коротпошипиковая 221638
Липучка куликалонская 221693
Липучка липшица 221697
Липучка масляная 241543
Липучка мелкоплодная 221705
Липучка незабудка 221711
Липучка неравношипая 221631
Липучка неравношипиковая 221631
Липучка обыкновенная 221711
Липучка оголенная 221680
Липучка окаймленная 221702
Липучка полуголая 221742
Липучка полукрылатая 221741
Липучка пониклая 221715
Липучка попова 221724
Липучка промежуточная 221689
Липучка прямая 221758
Липучка разношипая 221685
Липучка раскидистая 221715
Липучка репейчатая 221711
Липучка рогоносная 221644
Липучка родственная 221645
Липучка сидячецветковая 221745
Липучка синайская 221744,221748
Липучка синьцзянская 221773
Липучка скальная 221735
Липучка таджикская 221760
Липучка тонкая 221761
Липучка тошая 221699
Липучка тощая 221700
Липучка тяньшанская 221764
Липучка шипоплодная 221751
Липушник 170193
Лиран 232602,232609
Лириодендрон 232602
Лириодендрон ганьсуский 232627
Лириодендрон китайский 232603
Лириодендрон тюльпанный 232609
Лиродревесник 93237
Лисехвостник 17497
Лисохвост 17497
Лисохвост альпийский 17506
Лисохвост амурский 17501
Лисохвост блюшистый 17512
Лисохвост вздутый 17512
Лисохвост влагалищный 17578
Лисохвост волокнистый 17573
Лисохвост гималайский 17536
Лисохвост джунгарский 17553
Лисохвост длинноостный 17541
Лисохвост жерарда 17533
Лисохвост зеравшанский 17567
Лисохвост коленчатый 17527
Лисохвост короткоколосый 17515
Лисохвост ледниковый 17534
Лисохвост лисохвост вздутый 17512
Лисохвост луговой 17553
Лисохвост луговый 17553
Лисохвост мышехвостиковидный 17548
Лисохвост мышехвостиковый 17548
Лисохвост непальский 17549
Лисохвост остроконечный 17547
Лисохвост ошера 17513
Лисохвост пашенный 17548
Лисохвост полевой 17548

Лисохвост пушистоцветковый 17523
Лисохвост равный 17498
Лисохвост рожевица 17565
Лисохвост рыхлоцветный 17538
Лисохвост сизый 17535
Лисохвост тифлисский 17574
Лисохвост тонкой 17572
Лисохвост тростниковидный 17512
Лисохвост черный 17550
Лисохвост шелковистый 17568
Лисохвост штейнегера 17571
Лисохвост японский 17537
Лисохвостник 17497
Лист александрийский 78227
Лиственница 221829
Лиственница американская 221904
Лиственница восточная американская 110003,212631
Лиственница гималайская 221881
Лиственница гмелина 221866
Лиственница гриффитка 221881
Лиственница даурская 221866
Лиственница европейская 221855
Лиственница западная 221917
Лиственница западная американская 221917
Лиственница китайская 221934
Лиственница ложная 317821
Лиственница ольгинская 221919
Лиственница опадающая 221855
Лиственница польская 221929
Лиственница сибирская 221944
Лиственница сычуаньская 221914
Лиственница тибетская 221881
Лиственница тонкочешуйчатая 221894
Лиственница японская 221894
Листера 232890
Листоколосник 297188
Листоколосник бамбузовидный 297215
Листоколосник гибгий 297281
Листоколосник гнездовый 297359
Листоколосник гуйцзоуский 297289
Листоколосник зеленый 297486
Листоколосник золотистый 297203
Листоколосник золотой 297203
Листоколосник извилистый 297281
Листоколосник пушистый 297306
Листоколосник сернистый 297464
Листоколосник сизовато－зеленый 297485
Листоколосник сизый 297285
Листоколосник съедобный 297266
Листоколосник тайваньский 297282
Листоколосник узкий 297194
Листоколосник черный 297367
Литвиновия 234154
Литвиновия тонкая 234158
Литрум 240025
Литрум обыкновенный 240068
Литтледалия 234122
Литтледалия алайская 234124
Литтония 234130
Литторела 234146
Лихнис 238765,363141
Лихнис альпийский 410950
Лихнис белый 363673
Лихнис кожистый 363370
Лихнис родственный 363337
Лихнис сверкающий 363472
Лихнис сибирский 363691
Лихнис татарское мыло 363312
Лихнис уилфорда 364213
Лихнис халцедонский 363312
Лихорадочник 180321
Лихорадочник китайский 280078
Лихорадочник лекарственный 180321
Лиций 238996
Лиций обыкновенный 239011,239055
Лиций русский 239106
Личжи 233077
Личжи китайский 233078
Личинник 354755
Личинник маленький 354760
Личинник полувойлочный 354761
Лишайник тилланзия 392052
Ллоидия 234214
Ллойдия 234214
Ллойдия красно－зеленая 234229
Ллойдия поздная 234234
Ллойдия поздняя 234234
Ллойдия тибетская 234241
Ллойдия трехцветковая 234244
Лннато 54862
Лоаза 127654,234251
Лобалия лёгочная 234269
Лобалия лёгочнообразная 234269
Лобелиевые 234890
Лобелия 234275
Лобелия вздутая 234547
Лобелия гиблидная 234541
Лобелия гуансиская 234405
Лобелия гуйцзоуская 234377
Лобелия далиская 234816
Лобелия дортмана 234437
Лобелия дортманна 234437
Лобелия китайская 234363
Лобелия надутая 234547
Лобелия одутлая 234547
Лобелия сидячелистная 234766
Лобелия сифилитическая 234813
Лобелия сычуаньская 234407
Лобелия тибетская 234827
Лобелия хайнаньская 234520
Лобия 218721
Лобия египетская 218721
Логаниевые 235251
Ложечица 34582
Ложечница 97967
Ложечница английская 97976
Ложечница аптечная 98020
Ложечница арктическая 97977,98020
Ложечница гренландская 97996
Ложечница лекарственная 98020
Ложечница ленская 98008
Ложечница мелкоплодная 98015
Ложечница продолговатолистная 98022
Ложечница тайваньская 97983
Ложечница ху 98002
Ложнобецкея 317414
Ложнобецкея кавказская 317415
ложнодрок 385669
ложнодрок монпелийский 173015
Ложнокамфорный лавр 91330
Ложноклясусия 317558
Ложноклясусия мягайшая 317561
Ложноклясусия сосочковая 317563
Ложноклясусия тончайшая 317559
Ложноклясусия туркестанская 317565
Ложноклясусия чимганская 317564
Ложноклясусия щетинистоволосистая 317560
Ложнолиственница 317821
Ложноочиток 318388
Ложноочиток бухарский 318391
Ложноочиток длиннозубчатый 318398
Ложноочиток каратавский 318396
Ложноочиток колокольчиковый 318392
Ложноочиток ливена 318397
Ложноочиток многостебельный 318399
Ложноочиток плотный 318393
Ложноочиток федченко 318394
Ложноочиток ферганский 318395
Лоза 344211,344244
Лоза болотная 344244
Лоза виноградная 411521,411979
Лоза волчья 343776
Лоза золотая 368480
Лозина миндальная 344211
Лозинка 94676
Лозинка белая 95431
Лозинка обыкновенная 95431
Лойник 316127
Локва 151162
Ломатогониум 235431
Ломатогониум каринтийский 235435
Ломатогониум колесовидный 235461
Ломатогониум кольцевидный 235461
Ломатогониум лицзянский 235450
Ломатогониум мелкоцветковая 235455
Ломатогониум чжундяньский 235471
Ломатогониум ядунский 235439
Ломбардский орех 106757

Ломколосник 317073
Ломколосник дагестанский 317077
Ломколосник кроненбунга 317083
Ломколосник ломкий 317078
Ломколосник скальный 317086
Ломколосник шерстистый 317084
Ломонос 94676
Ломонос альпийский 94714
Ломонос аньхуйский 94816
Ломонос арманда 94748
Ломонос асплениелистый 95319
Ломонос бирманский 94798
Ломонос бурый 94933,94938
Ломонос виноградолистный 95431
Ломонос виорна 95423
Ломонос виргинский 95426
Ломонос восточный 95179
Ломонос вьющийся 95431
Ломонос вэньшаньский 95454
Ломонос горный 95127
Ломонос гуйцзоуский 94983,95063
Ломонос джунгарский 95319
Ломонос динхуский 95380
Ломонос жгучий 94901
Ломонос илийский 95026
Ломонос китайский 94814
Ломонос корейский 95061
Ломонос короткохвостый 94778
Ломонос либоский 95084
Ломонос лицзянский 94745
Ломонос ложножгучий 95253
Ломонос маньчжурский 95366
Ломонос метельчатый 95358
Ломонос мотоский 95112
Ломонос мэнлаский 95111
Ломонос напоский 95155
Ломонос непалский 95154
Ломонос обыкновенный 95431
Ломонос пильчатолистный 95295
Ломонос прямой 95277
Ломонос раскидистый 95216
Ломонос распростертый 95216
Ломонос сизый 94946
Ломонос симаоский 95261
Ломонос тайваньский 94916
Ломонос тангутский 95345
Ломонос усатый 94835
Ломонос усинский 95022
Ломонос фиолетовый 95449
Ломонос фэн 94896
Ломонос хубэйский 95023
Ломонос цветущий 94910
Ломонос цельнолистный 95029
Ломонос цельнолистый 95029
Ломонос цзилунский 94763
Ломонос чельнолистный 95029
Ломонос чжэцзянский 94811
Ломонос шерстистый 95067
Ломонос шестилепестный 95000
Ломонос этузолистный 94704
Ломонос юньнаньский 95463
Ломонос юньнаньский цзиндунский 95465
Лоницера 235630
Лоницера каприфолиум 235713
Лонхокарпус 235513
Лопух 31044
Лопух большой 31051
Лопух войлочный 31068
Лопух гладкосемянный 31051
Лопух дубравный 31061
Лопух крупный 31051
Лопух малый 31056
Лопух палладина 31063
Лопух паутинистый 31051,31068
Лопух паутинстый 31068
Лопух репейник 31051
Лопух скребница 31067
Лопух широкочешуйный 108369
Лопушник 31044
Лопушник большой 31051
Лопушник малый 31056
Лопушник паутинистый 31068
Лопушник съедобный 31051
Лоропеталум 237190
Лоропеталюм 237190
Лоропеталюм китайский 237191
Лосняк 232072
Лосняк лёзеля 232229
Лоснящаяся 44537
Лотос 263267
Лотос екипетский 267698
Лотос индийский 263272
Лотос каспийский 263268
Лотос орехоносный 263272
Лотус 132264
Лофант 236262
Лофант изящный 236269
Лофант китайский 236267
Лофант крылова 236271
Лофант липского 236272
Лофант приснежный 236277
Лофант тибетский 236278
Лофант чимганский 236279
Лофант шренка 236274
Лофантус 236262
Лофантус китайский 236267
Лофира 236334
Лофира африканская 236336
Лофира крупнолистная 236336
Лофира крылатая 236336
Лофира пирамидальная 236336
Лофоспермум 246651
Лофофора вильямся 236425
Лофохлоа 337269
Лофохлоа тимофеевковидная 217501
Лофохлоа тупоцветковая 217500
Лох 141929,141932
Лох восточный 141936
Лох гуйцзоуский 142021
Лох далиский 142203
Лох зонтичный 142214
Лох иванский 142022
Лох колючий 142152
Лох колющий 142152
Лох ланьпинский 142053
Лох мелкоцветковый 142081
Лох многоцветковый 142088
Лох мулиский 141948
Лох мэнхайский 141959
Лох наньчуаньский 142118
Лох остроплодный 142140
Лох серебристный 141941
Лох сычуаньский 141977
Лох тайваньский 141991
Лох узколистный 141932
Лох цзиндунский 142037
Лоховые 141928
Лутовик 126025
Луговик альпийский 126026
Луговик арктический 126029
Луговик беринговский 126034
Луговик ботнический 126039
Луговик двуцветный 126071
Луговик дернистый 126039
Луговик извилистый 126074
Луговик обский 126097
Луговик рамирский 126047
Луговик северный 126035
Луговик средний 126093
Луговик тонконоговидный 126086
Луговой чай 239755
Лужайник 230828
Лужайник водяной 230831
Лужница 230828
Лужница водяная 230831
Лузалеурия стелющаяся 235283
Лук 15018
Лук акака 15025
Лук алайский 15027
Лук албанский 15030
Лук александры 15039
Лук алексея 15040
Лук алтайский 15042
Лук альбова 15036
Лук ароидный 15073
Лук аскалона 15170
Лук афлатунский 15023
Лук барщевского 15107
Лук беловатый 15032
Лук белоголовый 15428
Лук белоцветный 15426

Лук бесномощный 15371
Лук боде 15122
Лук борщова 15128
Лук бэкера 15185
Лук вавилова 15863
Лук вальдштейна 15885
Лук вальковатолистный 15819
Лук венценосный 15771
Лук вешиякова 15867
Лук винклера 15895
Лук виноградничный 15876
Лук виноградный 15048,15876
Лук винтолистный 15341
Лук влагалищнокорневищный 15361
Лук волокинстый 15279
Лук восточнокавказский 15030
Лук высокий 15260
Лук высокогорный 15549
Лук высочайший 15043
Лук гадючий 260293
Лук гадючий гроздевидный 260301
Лук гадючий кистевидный 260333
Лук гадючий кистистый 260333
Лук гадючий хохлатый 260308
Лук гигантский 15445
Лук голубой 15194
Лук горноченочный 15553
Лук горный 15493
Лук горолюбивый 15550
Лук горцевидный 15551,15552
Лук гриффита 15336
Лук гусиный 169375
Лук густой 15198
Лук дарвазский 15224
Лук двузубый 15113
Лук декандоля 15156
Лук дердернана 15236
Лук дернистый 15145
Лук длиннолучевой 15438
Лук длинноостроконечный 15437
Лук длинностолбиковый 15245
Лук длинноцветоножковый 15244
Лук длинный 15868
Лук дробова 15253
Лук дудчатый 15289
Лук душистый 15534,15843
Лук евгения 15267
Лук жакемонта 15375
Лук желтенький 15442
Лук желтеющий 15295
Лук желтоватый 15296
Лук жемчужный 15048
Лук зеленый 15881
Лук зелёный 15018
Лук зеравшанский 15737
Лук зимний 15289
Лук змеелистный 15545
Лук изорванный 15408
Лук изящный 15262
Лук илийский 15364
Лук индерский 15368
Лук канадский 15149
Лук каратавский 15389
Лук карсский 15392
Лук картофельный 15167
Лук каснийский 15163
Лук кашский 15393
Лук килеватый 15160
Лук китайский 15185
Лук клубочный 15324
Лук коканский 15398
Лук комарова 15400
Лук комаровский 15690
Лук копетдагский 15401
Лук королькова 15402
Лук короткозубчатый 15779
Лук короткозубый 15133
Лук короткостебельный 15131
Лук косой 15525
Лук крапчатый 15340
Лук красивосетчатый 15146
Лук красиенький 15375,15680
Лук краснеющий 15266
Лук красноватый 15682
Лук кристаллоносный 15214
Лук круглоголовый 15752
Лук круглый 15675
Лук крупнотычинковый 15450
Лук крупный 15330
Лук культурный 15165
Лук кунта 15424
Лук куюкский 15403
Лук лантышевый 15204
Лук ледебура 15420
Лук ледниковый 15319
Лук леманна 15421
Лук ленкоранский 15423
Лук лентолепестный 15797
Лук линейный 15432,15760
Лук ложновиноградный 15634
Лук ложножелтый 15637
Лук ложно－зеравшанский 15640
Лук маака 15444
Лук максимовича 15467
Лук малорослый 15644
Лук малоцветковый 15541
Лук маргариты 15461
Лук марии 15462
Лук маршалля 15465
Лук медвежий 15861
Лук мелкий 15477
Лук мелковатый 15590
Лук мелкоголовый 15853
Лук мелколуковичный 15473
Лук мелкоцветный 15852
Лук метельчатый 15567
Лук мечелепестный 15898
Лук мешочконосный 15690
Лук миленький 15112
Лук многодетковый 15165
Лук многокорневой 15615
Лук многолистный 15161
Лук моли 15480
Лук молочноцветный 15310
Лук морской 15464,352960
Лук мохнатолистный 15414
Лук мощный 15663
Лук муговчатый 15866
Лук мулиский 15352
Лук мускатный 15499
Лук незаметный 15367
Лук неравнодольный 15336
Лук неравнолучевой 15064
Лук неравноногий 15064
Лук неравный 15365
Лук нериволистный 15513
Лук неровный 15199
Лук нитезубный 15281
Лук нителистный 15282
Лук обманываюший 15226
Лук обыкновенный 15165
Лук огородный 15540
Лук однобратственный 15088
Лук одноцветковый 15486
Лук осоковидный 15159
Лук ошанина 15555
Лук ошера 15094
Лук палласа 15564
Лук песчаный 15689
Лук плевокорневищный 15361
Лук победный 15868
Лук поникающий 15523
Лук попова 15620
Лук порей 15621
Лук поррей 15621
Лук понтийский 15619
Лук привлекательный 15233
Лук приречный 15726
Лук причесночный 15726
Лук приятный 15384
Лук прочноодетый 15285
Лук пскемский 15642
Лук пустыный 15807
Лук регелевский 15651
Лук регеля 15651,15652
Лук репчатый 15165
Лук решетчатый 15193
Лук родственный 15022
Лук розенбаха 15665
Лук сгущенный 15198
Лук северцова 15741

Лук севок 15165
Лук семенова 15732
Лук сергея 15738
Лук серполистный 15252
Лук сетеносный 15242
Лук сетчатый 15241
Лук сибирский душистый 15843
Лук сибирский колба 15868
Лук синеголубой 15144
Лук синьцзянский 15662
Лук сычуаньский 15794
Лук скаловый 15704
Лук скальный 15686
Лук скорода 15709
Лук скородовида 15708
Лук скородовидный 15708
Лук смнительный 15056
Лук сорно－полевой 15675
Лук стареющий 15271,15734
Лук стареющийся 15734
Лук стебельчатый 15774
Лук стеллера 15767
Лук стелющийся 15632
Лук странный 15585
Лук стройный 15328
Лук суворова 15793
Лук таласский 15799
Лук талышский 15800
Лук татарка 15289
Лук татарский 15804
Лук темнофиолетовый 15308
Лук тонгий 15813
Лук тонкостебельный 15810
Лук тончайший 15791
Лук торчащий 15779
Лук траутфеттера 15829
Лук трубчатый 15289
Лук туполистый 15602
Лук туркестанский 15851
Лук туркменский 15850
Лук тычиночный 15764
Лук тяньшанский 15823
Лук угловатый 15060
Лук ферганский 15277
Лук фетисова 15278
Лук фишера 15632
Лук фомина 15300
Лук хорошенький 15643
Лук христофа 15188
Лук чашеносный 15217
Лук чернокрасный 15088
Лук черно－фиолетовый 15091
Лук чеснок 15698
Лук шалот 15170
Лук шаровидный 15704
Лук шероховатенький 15717
Лук шероховатостебельный 15707
Лук шероховатый 15828
Лук широкочехольный 15602
Лук шовица 15796
Лук шуберта 15724
Лук щебнистый 15597
Лук щечнолистный 15740
Лук эмэйский 15543
Лук яйлинский 15380
Лук ямцатый 15730
Лук японский 15292
Лук－батун 15289
Лук－перо 15018
Лук－поррей 15621
Лук－розенец 15709
Лук－рокамболь 15726
Лук－скорода 15709
Лук－татарка 15289
Лук－черемша 15861
Лук－чеснок 15698
Лук－шалот 15170
Лук－шнит 15709
Лумбанг 14544
Лунник 238369
Лунник многолетный 238379
Лунник однолетний 238371
Лунник оживающий 238379
Луносемянник 250218
Луносемянник даурский 250228
Луносемянник канадский 250221
Луносемянник китайский 250225
Луносемянниковые 250217
Луноцвет 67734
Луноцвет шиповатый 67736
Лупин 238419
Лупин белый 238421
Лупин желтый 238463
Лупин многолистный 238481
Лупин нуткинский 238472
Лупин узколистный 238422
Лыко волчье 122515
Лысосемянник 293553
Лысосемянник девясиловидный 293555
Лысосемянник лысый 293554
Львиный зев 28576
Львиный зев полевой 28631
Льновые 230854
Льнянка 230866
Льнянка алтайская 230888
Льнянка альпийская 230887
Льнянка армянская 230897
Льнянка белолобая 230886
Льнянка биберштейна 230910
Льнянка бунге 230925
Льнянка бурятская 230927
Льнянка верхнедонская 230940
Льнянка ветвистая 231090
Льнянка далматская 230978
Льнянка двураздельная 230911
Льнянка длинноплодная 230951
Льнянка длиннохвостая 231037
Льнянка дроколистная 230976
Льнянка душистая 231061,231062
Льнянка заилийская 230925
Льнянка зангезурская 231196
Льнянка итальянская 231014
Льнянка канадская 267347
Льнянка кокандская 231017
Льнянка копетдагская 231018
Льнянка коричневая 230989
Льнянка коротшпоровая 230920
Льнянка крупнолистная 231036
Льнянка крупноцветковая 230987
Льнянка кулябская 231019
Льнянка курдская 231020
Льнянка лежачая 231559
Льнянка ленкоранская 231026
Льнянка линейнолистная 231030
Льнянка македонская 231033
Льнянка марьянниковая 231042
Льнянка мейера 231043
Льнянка мелкоцветная 231044
Льнянка меловая 230939
Льнянка неполноцветковая 231012
Льнянка обыкновенная 231183
Льнянка остролопастная 230870
Льнянка отогнутоножковая 231093
Льнянка пелиссе 231074
Льнянка песчаная 231120
Льнянка полевая 230901
Льнянка полосчатая 231139
Льнянка понтийская 231080
Льнянка попова 231081
Льнянка простая 231132
Льнянка русская 231118
Льнянка сидячецветковая 231130
Льнянка слабая 230945
Льнянка сладкая 230953
Льнянка тонкошпоровая 231027
Льнянка туркменская 231165
Льнянка черноморская 230961
Льнянка шпороваядлинно 230952
Льнянка элимаидская 230960
Льнянка японская 231015
Льюзия 228256
Лэйа 223476
Лэйа изящная 223497
Любим－трава 175454
Любисток 228244
Любисток аптечный 228250
Любисток лекарственный 228250
Любка 302241
Любка бычий рог 302290
Любка гуансиская 302380
Любка дальневосточная 302260

Любка двулистная 302260
Любка желтая 302326
Любка зеленоцветковая 302280
Любка зеленоцветная 302280
Любка комарниковая 302536
Любка лицзянская 302391
Любка ложнодвулистная 302260
Любка офрисовидная 302470
Любка сахалинская 302502
Любка тайваньская 302531
Любка фрейна 302331
Любка цельногубая 302355
Любка японская 302377
Любник 148101
Людвигия 238150
Людвигия болотная 238192
Людвигия лежачая 238211
Людвигия простертая 238211
Людвигия тайваньская 238230
Люндезия 235614
Люндезия пушистоцветная 235618
Люпин 238419
Люпин белый 238421
Люпин волосистый 238450
Люпин гибридный 238456
Люпин древнеегипетский 238421
Люпин древовидный 238430
Люпин жёлтый 238463
Люпин жестковолосистый 238450
Люпин жесткоопушённый 238450
Люпин изменчивый 238468
Люпин лохматый 238450
Люпин мексиканский 238448
Люпин многолестний 238473
Люпин многолистный 238481
Люпин низкий 238469
Люпин пестроцветный 238468
Люпин синий узколистный 238422
Люпин тёплый 238421
Люпин узколистный 238422
Лютик 325498
Лютик абхазский 325504
Лютик агера 325553
Лютик азиатский 325614
Лютик аконитолистный 325516
Лютик алайский 325555
Лютик алеберта 325558
Лютик аллеманна 325560
Лютик алтайский 325563
Лютик альпийский 325562
Лютик амурский 325571
Лютик афганский 325551
Лютик бальджуанский 325636
Лютик близкий 325989
Лютик бокочветковый 48924
Лютик болотный 326255
Лютик бореальный 325656
Лютик борцовый 326113
Лютик братский 325883
Лютик бротеруса 326422
Лютик бугорчатый 326440
Лютик бузе 325665
Лютик буша 325686
Лютик вальтера 326496
Лютик ван 326497
Лютик ветреницелистный 325578
Лютик водяной 325580
Лютик волосистоплодный 326021
Лютик волосистый 326494
Лютик волосолистный 48933
Лютик восточный 326156
Лютик гиперборейский 325960,325964
Лютик гмелина 325901
Лютик горный 326096,326154
Лютик джунгарский 326377
Лютик дихотомический 325773
Лютик длиннолистный 326034
Лютик длиннолопастной 326043
Лютик длинностебельный 326263
Лютик дубравный 326118
Лютик едкий 325515,325518
Лютик елены 325926
Лютик жгучий 325864
Лютик жесткий 326310
Лютик жестколистный 48919
Лютик заилийский 326441
Лютик закрученный 48919
Лютик зибольда 326365
Лютик золотистый 325628
Лютик золотолепестный 325627
Лютик изогнутый 326281
Лютик изящный 326261
Лютик иллирийский 325971
Лютик кавказский 325715
Лютик калабрийский 325664
Лютик кандинский 325774
Лютик кашубский 325714
Лютик китайский 325718
Лютик клубневой 325666
Лютик клубненосный 325666
Лютик комарова 326002
Лютик константинопольский 325730
Лютик копестдагский 326003
Лютик коротколопастный 325658
Лютик кочи 326004
Лютик красивый 325785
Лютик краснова 326005
Лютик красночашечный 326317
Лютик крошечный 326273
Лютик крупнолистный 325915
Лютик крупнолистый 325915
Лютик крупноплодный 326067
Лютик крылова 326093
Лютик куэньминский 326008
Лютик куэньминский лэйбоский 326009
Лютик лапландский 326020
Лютик лапчатораздельный 326201
Лютик ледниковый 325886,325898
Лютик лесной 326118
Лютик линейнолопастной 326033
Лютик лицзянский 325777
Лютик ложноклубненосный 326253
Лютик лойка 326038
Лютик луковичный 325666
Лютик малозубчатый 326197
Лютик мейера 326078
Лютик мейнсхаузена 326069
Лютик мелколистный 326084
Лютик многокореный 326245
Лютик многокорневой 326245
Лютик многолистный 326244
Лютик многолистый 326244
Лютик многоцветковый 326242
Лютик мягкоигольчатый 326108
Лютик мягколистный 326254
Лютик неаполитанский 326116
Лютик нежелезистый 325920
Лютик ненецкий 326324
Лютик нитчатый 48915
Лютик обедненный 326199
Лютик обманчивый 325788
Лютик однолистный 326093
Лютик однолистый 326093
Лютик окаймленный 326039,326405
Лютик осетинский 326164
Лютик острозубчатый 325535
Лютик остролопастный 325536
Лютик остроплодный 326166
Лютик палласа 326172
Лютик памира 326193
Лютик паутинистый 325608
Лютик периссторассеченный 326229
Лютик пигмей 326273
Лютик плавающий 326114
Лютик плоскоплодный 326237
Лютик плоскосемянный 326237
Лютик полевой 325612
Лютик ползающий 326302
Лютик ползучий 326287,326303
Лютик попова 326248
Лютик простертый 326303
Лютик радде 326277
Лютик раздельный 325784
Лютик распростертый 326303
Лютик расходящийся 48914
Лютик расширенный 325782
Лютик регелевский 326286
Лютик регеля 326286
Лютик рыжечашечный 326318
Лютик сардинский 326326
Лютик сарторнев 326339

Лютик светльй 325785
Лютик северный 325656
Лютик северцова 326361
Лютик серножелтый 326411
Лютик слабый 326408
Лютик смирнова 326375
Лютик снеговой 326128
Лютик соммье 326376
Лютик стевена 326397
Лютик стоповидный 326206
Лютик стройный 325796
Лютик сходный 326201,326204
Лютик толстолистный 325740
Лютик тонколопастной 326429
Лютик тонконосиковый 326027
Лютик траутфеттера 326442
Лютик трехрассеченный 326477
Лютик туркестанский 326481
Лютик тучный 326133
Лютик ужовниколистный 326148
Лютик узколепестный 326393
Лютик укореняющийся 326278
Лютик уссурийский 326487
Лютик фенхелевидный 48919
Лютик франше 325882
Лютик хиосский 325720
Лютик чжундяньский 326504
Лютик шафта 326348
Лютик шерстевидный 326017
Лютик шерстевидтый 326018
Лютик шерстистовидный 326017
Лютик шерстистый 326018
Лютик шиповатый 326108
Лютик щетинковый 326399
Лютик эшшольца 325800
Лютик ядовитый 326340
Лютик языковый 326034
Лютик языколистный 326034
Лютик якутский 325980
Лютик японский 325981
Лютиковые 325494
Лютик－чистяк 325820
Люфа 238257
Люфа цилиндрическая 238261
Люффа 238257
Люффа гранистая 238258
Люффа мочалка 238261
Люффа ребристая 238258
Люффа цилиндрическая 238261
Люхея 238242
Люцерна 247239,247293
Люцерна азиатская 247257
Люцерна арабская 247246
Люцерна аравийская 247246
Люцерна гибридная 247507
Люцерна голубая 247265
Люцерна гордеева 396920
Люцерна дагестанская 247275
Люцерна дисковидная 247407
Люцерна древовидная 247249
Люцерна железистая 247298
Люцерна желтая 247457
Люцерна жёлтая 247457
Люцерна жестковатая 247443
Люцерна жмелевая 247366
Люцерна жмелевдная 247366
Люцерна зеленоватая 247510
Люцерна зубчатая 247425
Люцерна кавказская 247298
Люцерна каменистая 247467
Люцерна круглая 247407
Люцерна лежачая 247438
Люцерна маленькая 247381,247425
Люцерна мелкозубчатая 247425
Люцерна меловая 247271
Люцерна обыкновенная 247456
Люцерна округлая 247407
Люцерна пестрая 247507
Люцерна полевая 247242
Люцерна полузакрученная 247304
Люцерна посевная 247456
Люцерна прибрежная 247359
Люцерна приморская 247378
Люцерна пушистая 247418
Люцерна пятицкличная 247419
Люцерна пятнистая 247246
Люцерна раздельнолистная 247337
Люцерна разноцветная 247424
Люцерна ранняя 247437
Люцерна решетчатая 247261
Люцерна русская 247452
Люцерна серповидная 247457
Люцерна синьцзянская 247468
Люцерна синяя 247456
Люцерна скалиная 247451
Люцерна скалистая 247451
Люцерна солейроля 247473
Люцерна средняя 247507
Люцерна траутфеттера 247486
Люцерна тяньшанская 247507
Люцерна увенчанная 247268
Люцерна хмелевая 247366
Люцерна хмелевидная 247366
Люцерна шерстистая 247350
Люцерна шетинистая 247425
Люцерна шетинистоволосистая 247425
Люцерна широкоплодная 247422
Люцерна щитковидная 247469
Люцерна якорцевидная 247493
Люшеная индийская 138796
Лягерстремия 219908
Ляготис 220155
Ляготис лежачий 220166
Ляготис палласа 220190
Ляготис северный 220161
Ляготис сизый 220167
Ляготис стеллдра 220196
Ляготис цельнолистный 220180
Лягохилюс 220050
Лягохилюс голый 220064
Лягушатник 200225
Лягушечник 200225,200228
Лягушечья трава 309199
Лядвенестальник 237219
Лядвенестальник дроковидный 237312
Лядвенец 237476,237554
Лядвенец болотный 237709,237713
Лядвенец гебелия 237615
Лядвенец густооблиственный 237609
Лядвенец кавказский 237526
Лядвенец ланьюйский 237506
Лядвенец просмотренный 237725
Лядвенец птиценогий 237704
Лядвенец пурпурный 237771
Лядвенец рогатый 237539
Лядвенец тонкий 237768
Лядвенец тонколистнный 237768
Лядвенец тончий 237617
Лядвенец топяной 237713
Лядвенец торчащий 237759
Лядвенец узкий 237485
Лядвенец узколистный 237485
Лядвенец четырёхлопастный 237771
Лядник 196116
Лядник душистый 27967
Лядник пахучий 27967
Ляйм 93329,93534
Ляллеманция 220281
Ляллеманция бальджуанская 220282
Ляллеманция грузинская 220284
Ляллеманция ройля 220286
Ляллеманция седоватая 220283
Ляллеманция щитовидная 220285
Ляптхор 92384
Маакатик китайский 240128
Маакия 240112
Маакия амурская 240114
Маакия китайская 240128
Маакия тайваньская 240130
Маакия чжэцзянская 240120
Магадения 247850
Магалебка 83253
Магнолиевые 242370
Магнолия 241951
Магнолия виргинская 242341
Магнолия голая 416694
Магнолия делаве 232595
Магнолия длиннозаострённая 241962
Магнолия заостренная 241962
Магнолия звездчатая 141470,242331
Магнолия зибольда 279249

Магнолия зонтичная 242333
Магнолия кемпбелла 416688
Магнолия китайская 279250
Магнолия кобус 242169
Магнолия кобус северная 242171
Магнолия коко 232594
Магнолия крупноцветковая 242135
Магнолия крупноцветная 242135
Магнолия кэмбелла 416688
Магнолия лилейная 416707
Магнолия лилиецветковая 416707
Магнолия лилиецветная 416707
Магнолия обнаженная 416694
Магнолия обнажённая 416694
Магнолия обратноовальная 198698
Магнолия обратнояйцевидная 198698
Магнолия остроконечная 241962
Магнолия пурпуровая 416707
Магнолия сизая 242341
Магнолия симаоская 232598
Магнолия суланжа 416682
Магнолия трехлепестковая 242333
Магнолия трёхлепестковая 242333
Магнолия туполистная 198698
Магнолия уотсона 242352
Магнолия хэнаньская 242152
Магнолия юлана 416694
Магнолия японская 198698
Магония 242458
Магония била 242487
Магония вагнера 242660
Магония куэньминская 242517
Магония ломариелистная 242608
Магония мэнцзыская 242553
Магония непалская 242597
Магония падуболистная 242478
Магония сичанская 242493
Магония тибетская 242502
Магония форчуна 242534
Магония хайнаньская 242550
Магония цзиндунская 242612
Магония чаойская 242503
Магония японская 242563
Мадзус 246964
Мадзус ван 247024
Мадзус гуйцзоуский 247000
Мадзус лицзянский 247031
Мадзус тибетский 247043
Мадзус фуцзяньский 246979
Мадзус чистецолистный 247039
Мадзус японский 247025
Мадия 241523
Мадия посевная 241543
Мадука индийская 241514
Мазус 246964
Мазус японский 247025
Маис 417412,417417
Майкараган 67771
Майкараган волжский 67789
Майкараган китайский 67774
Майкараган крупноцветковый 67780
Майкараган тяньшанский 67787
Майкараган ховена 67786
Майкараган шелковистыи 67784
Майник 242668
Майник двулистный 242672
Майник канадский 242675
Майник широколистный 242680
Майоран 274197
Майоран дикий 274237
Майоран зимний 274237
Майоран кошачий 388147
Майоран летний 274224
Майоран наньчуаньский 242694
Майоран садовый 274224
Майтенус 246796
Майтенус падуболистный 246859
Майтенус чилийский 246817
Майяка 246761
Мак 282496
Мак альпийский 282506
Мак амурский 282622
Мак аргемона 282515
Мак аянский 282499
Мак беланже 282520
Мак белорозовый 282504
Мак восточный 282647
Мак гиблидный 282574
Мак глазчатый 282659
Мак голоплодный 282594
Мак голостебельный 282617
Мак голубой 247099
Мак голый 282585
Мак горный 282646
Мак дважтыперистый 282521
Мак кавказский 282526
Мак калифорнийский 155173
Мак колючий 32429,282515
Мак корневой 282676
Мак крупнокоробочный 282603
Мак лапландский 282588
Мак ледебура 282593
Мак лизы 282596
Мак литвинова 282597
Мак ложносероватый 282666
Мак ложный 282508
Мак ложный штубендорфа 282672
Мак мелкоплодный 282610
Мак мохнатый 282592
Мак нмалолистный 282658
Мак обыкновенный 282717
Мак одноцветковый 282614
Мак окутанный 282580
Мак опийный 282717
Мак оранжевный 282617
Мак оранжево-красный 282705
Мак оранжевый 282536
Мак павлиний 282659,282660
Мак персидский 282662
Мак песчаный 282514
Мак песчатый 282514
Мак пиренейский 282674
Мак подушковый 282673
Мак полевой 282685
Мак полярный 282534,282676
Мак прицветниковый 282522
Мак промежуточный 282579
Мак разорванный 282584
Мак рогатый 176728
Мак роона 282701
Мак самосейка 282685
Мак сероватый 282525
Мак сибирский 282617
Мак снежный 282616
Мак снотворный 282717
Мак сомнительный 282546
Мак спутанный 282530
Мак талышинский 282739
Мак толмачева 282745
Мак тоненький 282740
Мак тяньшанский 282524,282743
Мак узкодольчатый 282511
Мак чистотелолистный 282527
Мак штубендорфа 282736
Мак щетинистый 282734
Мак щетинконосный 282713
Мак югорский 282581
Макадамия 240207,240210
Макадамия трехлистная 240210
Макка-джугара 417417
Маклейя 56023,240763
Маклюра 240806
Маклюра оранжевая 240828
Маклюра яблоконосная 240828
Маковые 282751
Макрозамия 241448
Макроподиум 241248
Макроподиум круглосемянный 241250
Макроподиум крылатосемянный 241250
Макротамия 241386
Макротамия густоцветковая 241391
Макротамия красящая 34615
Макротамия синяковидная 34635
Макротамия угамская 241398
Максимовник 6639
Малабайла 242804,242818
Малабайла бороздчатая 242825
Малабайла пахучая 242813
Малабайла пушистоцветковая 242810,357906
Малаксис 242993

Малаксис ланьюйский 110637
Малаксис однолистный 243084
Малаксис сычуаньский 110654
Малаксис хайнаньский 110644
Малахра 242866
Малина 338059,338557
Малина американская 338808
Малина арктическая 338142
Малина боярышниколистная 338300
Малина великолепная 339285
Малина душистая 338917
Малина европейская 338557
Малина ежевикообразная 338808
Малина замечательная 339285
Малина западная 338808
Малина комарова 338679
Малина красная 338557
Малина красная европейская 338557
Малина красная сибирская щетинистая 338566
Малина лесная 338557
Малина мелкоцветковая 338985
Малина обыкновенная 338557
Малина обыкновенная черноворсистая 338566
Малина пахучая 338917
Малина превосходная 339285
Малина пурпурноволосистая 339047
Малина садовая 338557
Малина сахалинская 339227
Малина североамериканская 338808
Малина чёрная 338808
Малиноклен 337893
Малиноклен пахучий 338917
Маллот 243315
Маллот лушаньский 243424
Маллот токолисный 243372
Маллотус 243315
Маллотус бесщитковый 243320
Маллотус гуансиский 243322
Маллотус наньпинский 243353
Маллотус непалскик 243399
Маллотус хайнаньский 243364
Маллотус хубэйский 243332
Маллотус японский 243371
Малокан комнасный 219497
Малокан посевной 219485
Малопа 243472
Мальва 243738
Мальва красивая 243862
Мальва круглолистная 243823
Мальва курчавая 243771
Мальва лесная 243840
Мальва махровая 13934
Мальва мускусная 243803
Мальва мутовчатая 243862
Мальва обыкновенная 243823
Мальва приземистая 243823
Мальва садовая 13934
Мальва северная 243823
Мальва сирийская 195149
Мальвальтея 243874
Мальвальтея закавказкаская 243875
Мальваструм 243877,302091
Мальвовые 243873
Мальвочка 243956
Мальвочка жерара 243958
Малькольмия 243156
Малькольмия африканская 243164
Малькольмия бухарская 243184
Малькольмия вололистоплодная 243243
Малькольмия городчатая 243188
Малькольмия карелина 243195
Малькольмия крупноцветная 243191
Малькольмия многостручковая 243230
Малькольмия одуванчколистая 243164
Малькольмия скорпионовидная 243230
Малькольмия туркестанская 243246
Малькольмия щетинистая 243193
Мальмороза 117204
Мальпигиевые 243530
Мальпигия 243519
Мальпигия гранатоволистная 243526
Мальпидия 243519
Мальпидия мексиканская 243527
Мальпидия орехоносная 243523
Мамиллярия маленькая 244203
Маммея 243979
Маммея американская 243982
Маммилярия 243994
Маммилярия бокасана 244012
Мамонтово дерево 360576
Мамура 338142
Манглис 171136
Манглиэция 244420
Манглиэция гуандунская 244451
Манглиэция напоская 244468
Манглиэция сычуаньская 244471
Манглиэция таиландская 244440
Манглиэция тибетская 244425
Манглиэция тонкоствольная 244427
Манглиэция хайнаньская 244435
Манго 244386,244397
Манго вонючее 244393
Манго вонючий 244393
Манго душистое 244406
Манго душистый 244406
Манго индийское 244397
Манго пахучее 244393
Манго пахучий 244393
Манго таиландское 244409
Мангольд 53257
Мангостан 171136
Мангро поблескивающее 45750
Мандарин 93446,93717,93736
Мандарин благородный 93649
Мандарин иволистный 93446
Мандарин итальянский 93446,93717
Мандарин кинг 93649,93727
Мандарин кислый 93748
Мандарин уншиу 93736
Мандарин японский 93736
Мандрагора 244340
Мандрагора туренская 244349
Мандрагора туркменская 244349
Мандрагора цинхайская 244344
Манеттия 244358
Манжетка 13951
Манжетка абхазская 13952
Манжетка александра 13957
Манжетка алтайская 13960
Манжетка альпийская 13959
Манжетка багровеющая 14121
Манжетка безбородая 14060
Манжетка бескровная 14023
Манжетка близкая 14115
Манжетка бородатая 14113
Манжетка бородатоцветковая 13979
Манжетка бородковатая 13980
Манжетка бунге 13986
Манжетка буша 13987
Манжетка видная 14144
Манжетка волнистолистная 14002
Манжетка волосистоцветоножковая 14055
Манжетка вороова 14172
Манжетка высокая 14014
Манжетка гаральда 14046
Манжетка гирканская 14059
Манжетка гожилковая 14119
Манжетка гололистная 14074
Манжетка голостебельная 14038
Манжетка горбиковатая 14036
Манжетка городковатая 14149
Манжетка городковатовидная 14151
Манжетка гроссгейма 14044
Манжетка грузинская 14033
Манжетка густоволосистая 14124
Манжетка густоцветковая 14123
Манжетка дагестанская 14004
Манжетка двухквадратная 13982
Манжетка двуязычная 14009
Манжетка дубравная 14094
Манжетка дуголопастная 13962
Манжетка елизаветы 14016
Манжетка желтеющая 14028
Манжетка жесткая 14130
Манжетка жестковолоистостебельная 14052
Манжетка жилковатая 14164
Манжетка замечательная 14062
Манжетка звездочковая 14146

Манжетка звездчатая 14145
Манжетка зелено – шелковая 13991
Манжетка зияющая 14051
Манжетка зубчиковатая 14007
Манжетка изогнуточрешковая 13988
Манжетка кавказская 13990
Манжетка камнелюбивая 14079
Манжетка клубочковая 14039
Манжетка ключевая 14030
Манжетка козловского 14069
Манжетка короткозубая 13984
Манжетка коротколопастная 13985
Манжетка краснеющая 14133
Манжетка кривобокая 14003
Манжетка крупеозубчатая 14043
Манжетка крылова 14070
Манжетка крымская 14157
Манжетка лже – картвельская 14116
Манжетка лжемягкая 14117
Манжетка лидин 14081
Манжетка линдберга 14076
Манжетка липшица 14077
Манжетка литвинова 14080
Манжетка лысеющая 14006
Манжетка лысостебельная 14118
Манжетка малая 14087
Манжетка маловолосая 14100
Манжетка мелковатоцветковая 14088
Манжетка мелкозубчатая 14086
Манжетка мелкоцветковая 14162
Манжетка мешечковая 14163
Манжетка мурбека 14091
Манжетка налегающая 14097
Манжетка негостаточная 14013
Манжетка неодинаковочерешковая 13961
Манжетка нивкостебельная 14057
Манжетка ново – стевеновская 14095
Манжетка обыкновенная 14166
Манжетка обыкновенный 14065
Манжетка окровавленная 14134
Манжетка округленная 14102
Манжетка олиственная 14031
Манжетка остроугольная 13955
Манжетка острочашелистковая 14103
Манжетка пастушья 14106
Манжетка пепельно – шерковистая 14158
Манжетка плосколистная 14101
Манжетка подражающая 13956
Манжетка полевая 13966
Манжетка полулунная 14139
Манжетка почти – блестящая 14154
Манжетка почтипрямостоящеволосистая 14152
Манжетка почтищетинистая 14155
Манжетка приальпийская 13958
Манжетка приземистая 14156
Манжетка притупляющаяся 14048
Манжетка прямозубая 14114
Манжетка радде 14126
Манжетка разнорешковая 14011
Манжетка растопыривающаяся 14010
Манжетка расщепленнолистная 14137
Манжетка самая жестковолосистая 14053
Манжетка сарматская 14135
Манжетка сбрасывающая 14024
Манжетка сверкающая 14083
Манжетка светлая 14071
Манжетка светлозеленая 14072
Манжетка семиугольная 14050
Манжетка сетчатожилковая 14128
Манжетка сжатая 13996
Манжетка сибирская 14143
Манжетка складчатая 14112
Манжетка слабая 14005
Манжетка слабеющая 14073
Манжетка снизуиеленая 14058
Манжетка сплошь – волосистая 14056
Манжетка стевена 14147
Манжетка твердеющая 14129
Манжетка твертая 14012
Манжетка толстолистная 14104
Манжетка тонкопильчатая 13963
Манжетка тонкостебельная 14025
Манжетка туманная 14093
Манжетка тупая 14098
Манжетка туповидная 14099
Манжетка туруханская 14161
Манжетка тучная 14111
Манжетка тяньшанская 14159
Манжетка частозубая 14000
Манжетка черкесская 13992
Манжетка шаровидноскученная 13997
Манжетка шелковая 14141
Манжетка шелковистая 14140
Манжетка шишкна 14136
Манжетка юзепчука 14067
Манжетка яйлы 14064
Манилла 260275
Маниок 244501
Маниок глацийова 244509
Маниок горький 244507,244515
Маниок полезнейший 244507,244515
Маниок сладкий 244506,244512
Маниок съедобный 244507,244515
Маниока 244501
Маниока каучуконосная 244509
Маниот 244501
Маниот глазиова 244509
Маниот съедобный 244507,244515
Манна персидская 14638
Маннагеттея 244635
Маннагеттея иркутская 244637
Манник 177556
Манник большой 177629
Манник водный 79198,177629
Манник водяной 79198,177629
Манник восточный 177626
Манник дубравный 177635
Манник заболачивающий 177661
Манник камчатский 177672
Манник канадский 177577
Манник каспийский 177580
Манник китайский 177582
Манник колосковый 177661
Манник литовский 177626
Манник найбольший 177629
Манник наплывающий 177599
Манник обыкновенный 177599
Манник ольховниковый 177562
Манник плавающий 177599,393100
Манник ручьевой 177599
Манник складчатый 177648
Манник тонкочешуйчатый 177618
Манник тонокорневой 177621
Манник трехцветковый 177672
Манник тростниковый 177574
Манник туркменский 177648
Манник уссурийский 177618,177677
Манник японский 177560
Манцинелла 196716
Маньчжурский абрикос 34443
Маньчжурский кедр 300006
Маранта 245013
Маранта беложильчатая 245022
Маранта вестиндская 245014
Маранта тростниковая 245014
Маранта тростниковидный 245014
Марантовые 245039
Маргаритка 50808
Маргаритка многолетняя 50825
Маргаритка обыкновенная 50825
Маргелон 395146
Маревые 86873
Марезия 245157
Марезия низкорослая 245164
Марена 337906
Марена грузинская 337968
Марена длиннолистная 337952
Марена иезская 337970
Марена китайская 337916
Марена комарова 337971
Марена красильная 338038
Марена крашенинникова 337972
Марена меловая 337948
Марена пустынная 337949
Марена регела 338023
Марена резниченко 338024
Марена рехингера 338022
Марена рыхроцветковая 337978
Марена сердцевиднолистная 337925
Марена сердцелистная 337925

Марена татарская 338033
Марена тибетская 338037
Марена цветущая 337957
Марена читралская 337923
Марена шугнанская 338027
Мареновые 338045
Марзанка 39328
Мариола 285076
Марипоза 67554
Марискус 93901,245332
Марсдения 245777
Марсин 338038
Мартагон 229922
Мартиниевые 245991
Мартиния 245977
Мартиния душистая 315436
Маруми 167499,167503
Марь 86891
Марь амброзиевидная 139678
Марь белая 86901
Марь берландье 86958
Марь благовонная 139678
Марь бриониелистная 86977
Марь вахеля 87191
Марь вонючая 87200
Марь гибридная 87048
Марь гигантская 87025
Марь головчатая 86984
Марь городская 87186
Марь гроздевая 139682
Марь доброго генриха 86970
Марь душистая 139682
Марь енисейская 87057
Марь зеленая 87199
Марь калинолистная 87112
Марь киноа 87145
Марь клингреффа 87061
Марь козлиная 87045
Марь коржинского 87064
Марь красная 87147
Марь кустарничковая 87023
Марь лежачая 87058
Марь луговая 87133
Марь многолистная 87016
Марь многосеменная 87130
Марь многосемянная 87130
Марь обыкновенная 86901
Марь остистая 139681
Марь остроконечная 86892
Марь пахучая 139693
Марь переступнелистная 86984
Марь поздная 87158
Марь сизая 87029
Марь сосновского 87162
Марь стелющаяся 87058
Марь стенная 87096
Марь тайваньская 87019
Марь тибетская 87182
Марь толстолистная 86990
Марь торчащая 87171
Марь чилийская 87145
Марьин корень 280143,280154
Марьян розовый 248195
Марьянник 248155
Марьянник альбова 248156
Марьянник восокий 248169
Марьянник гербиха 248170
Марьянник гребенчатый 248168
Марьянник зеленоколосый 248166
Марьянник кавказский 248164
Марьянник луговый 248193
Марьянник полевой 248161
Марьянник разрезной 248172
Марьянник сесной 248211
Марьянник скальный 248207
Марьянник щетинистый 248208
Маслина 270058,270099
Маслина американская 270061,276250
Маслина дикая 270110
Маслина европейская 270099
Маслина культурная 270099
Маслинные 270182
Маслиновые 270182
Масличные 270182
Масляное дерево 406018
Масти 389520
Мате 204135
Материнка 90801
Матико 300333
Маточник 137613,276743
Маточник болотный 24429
Матренка 388423
Матрикария 246307
Маттиола 246438
Маттиола бухарская 246446
Маттиола садовая 137747,246478
Мать – и – мачеха 400671
Мать – и – мачеха обыкновенная 400675
Матэ 204135
Маун 404213
Маун аптечный 404316
Мауновые 404392
Маурандия 246651
Махагони 380528
Махил 240550,240567
Махил гуандунский 240610
Махил гуншаньский 240592
Махил иčанский 240599
Махил китайский 240570
Махил либоский 240621
Махил личуаньский 240620
Махил мелкоплодный 240645
Махил мелкоцветковый 240668
Махил наньчуаньский 240653
Махил сичоуский 240698
Махил сычуаньский 240699
Махил тибетский 240725
Махил тунберга 240707
Махил фуцзяньский 240587
Махил чжэцзянский 240568
Махил японский 240604
махин – бура 213933
Махобели 82173
Махорка 266053
Мачёк 176724
Мачёк жёлтый 176735
Мачёк рогатый 176728
Мачок 176724
Мачок желтый 176738
Мачок рогатый 176728
Маш 409025
Машок 135134
Маш – салат 404454
Мёд дивий 78300
Медвежья лапа 373085
Медвежья ягода 31142,145073
Мединилла 247526
Медноцвет 85071
Медноцвет почколистый 85072
Медовик 170078
Медовик мангровый 45750
Медовник 170078
Медуница 321633
Медуница аптечная 321643
Медуница горная 321640
Медуница красная 321640,321646
Медуница лекарственная 321643
Медуница мягкая 321639
Медуница мягчайшая 321639
Медуница неясная 321641
Медуница тёмная 321641
Медуница узколистная 321636
Медуница филярского 321651
Медунка 170066
Меза 241734
Меза баотинская 241755
Меза иволистная 241833
Меза ланьюйская 241795
Меза морщинистая 241830
Меза тайваньская 241818
Меза хубэйская 241774
Меза японская 241781
Мезембриантемум 251025
Мезембриантемум кристллический 251265
Мезембриантемум хрустальный 251265
Мезуа железная 252370
Меконопсис 247099
Меконопсис кумберлендский 247113
Меконопсис лицзянский 247123
Меконопсис непалский 247157
Меконопсис тибетский 247122

Меконопсис уэльский 247113
Меконопсис цельнолистный 247139
Мексаканский огурец 356352
Мелалеука 248072
Мелалеука бледнодревесная 248104
Мелалеука льнянколистная 248105
Мелалеука мелкоцветковая 248110
Мелалеука цайная 248104
Меластомовые 248806
Мелиант 248936
Мелиациевые 248929
Мелиевые 248929
Меликокка 249124
Мелинис 249303
Мелиосма 249355
Мелиосма гуншаньская 249479
Мелиосма многоцветковая 249414
Мелиосма сичоуская 249481
Мелиосма юньнаньская 249482
Мелисса 249491
Мелисса аптечная 249504
Мелисса лекарственная 249504
Мелисса мелколистная 249507
Мелицитус 249175
Мелия 248893
Мелия ацедарах 248895
Мелия гималайская 248895
Мелия индийская 45908
Мелия иранская 248895
Мелия перснդская 248895
Мелия японская 248903
мелкоголовка 253207
мелкоголовка кругловатая 253209
мелкоголовка пластинчатая 253208
мелкоголовка туркменская 253210
Мелколепестник 150414
Мелколепестник алексеенио 150441
Мелколепестник алтайский 150457
Мелколепестник альпийский 150454
Мелколепестник андриалевидный 150461
Мелколепестник бадахшанский 150489
Мелколепестник байкальский 150490
Мелколепестник белолистнй 150756
Мелколепестник белолистный 150763
Мелколепестник бледный 150841
Мелколепестник буэносайресский 103438
Мелколепестник восточный 150834
Мелколепестник гиблидный 150699
Мелколепестник гиссарский 150694
Мелколепестник гладко – горный 150750
Мелколепестник горный 150831
Мелколепестник гуншаньский 150738
Мелколепестник двукратноветвистый 150501
Мелколепестник длинностолбиковый 150597
Мелколепестник едкий 150419
Мелколепестник замещающий 151058
Мелколепестник зеравшанский 150959
Мелколепестник иначе – крашенный 150444
Мелколепестник иноземный 150851
Мелколепестник кабульский 150521
Мелколепестник кавказский 150539
Мелколепестник камчатский 150422
Мелколепестник канадский 103446,117900
Мелколепестник карадагский 319929
Мелколепестник комарова 150732
Мелколепестник корагинский 150735
Мелколепестник короткосемянный 150510
Мелколепестник красивенький 151054
Мелколепестник крылова 150737
Мелколепестник курчавый 103438
Мелколепестник лазолевый 150488
Мелколепестник ланцетолистный 150763
Мелколепестник ложнозеравшанский 150897
Мелколепестник ложномелколепестник 150892
Мелколепестник ложноудлиненный 150895
Мелколепестник малбусский 150951
Мелколепестник маргаритковидный 150497
Мелколепестник мексиканский 150724
Мелколепестник многобразный 150881
Мелколепестник многолистный 150877
Мелколепестник низкий 150698
Мелколепестник обвислый 40440
Мелколепестник однокорзинчатый 151036
Мелколепестник однокорзинчный 151036
Мелколепестник однолетний 150464
Мелколепестник одноцветковый 151036
Мелколепестник олеги 150829
Мелколепестник оранжевый 150484
Мелколепестник острый 150419
Мелколепестник охары 150827
Мелколепестник первоцветовидный 150888
Мелколепестник подольский 150878
Мелколепестник понсэ 319930
Мелколепестник приятный 151054
Мелколепестник пушистоголовый 150623
Мелколепестник пушисточашечный 150621
Мелколепестник разноцетинный 40440
Мелколепестник ророва 150883
Мелколепестник сахалинский 150419
Мелколепестник северный 150508
Мелколепестник синьцзянский 150968
Мелколепестник смолевколистный 150961
Мелколепестник теневой 58280
Мелколепестник тибетский 150515
Мелколепестник тунберга 151008
Мелколепестник тяньшанский 151018
Мелколепестник удлиненный 150616
Мелколепестник фиолетовый 151061
Мелколепестник хоросанский 150728
Мелколепестник цюцзянский 150729
Мелколепестник черешковый 150859
Мелколепестник шерстистоголовый 150740
Мелколепестник шерстисточашечный 150621
Мелколепестник щетинистный 150980
Мелкоплодник 369290
Мелкоплодник ольховолистный 32988
Мелкоплодник ольхолистный 32988
Мелкопузырник 253789
Мелкопузырник вытянутый 253790
Мелкохвостник 319062
Мелкохвостник остистый 319064
Мелокактус 249578
Мелотрия 249760
Мельное дерево 346338
Мениспермовые 250217
Мениспермум 250218
Мениспермум канадский 250221
Мента 250287
Ментцелия 250481
Менцизия 250505
Менцизия пятитычинковая 250528
Мерендера 250627
Мерендера белейшая 250631
Мерендера гиссарская 250636
Мерендера иоланты 250637
Мерендера кавказская 250632
Мерендера крупная 250645
Мерендера отпрысковая 250647
Мерендера радде 250644
Мерендера трехстолбиковая 250648
Мерендера эйхлера 250633
Мерингия 256441
Мерингия бокоцветковая 256444
Мерингия бокоцветная 256444
Мерингия теневая 256459
Мерингия трехжилковая 256456
Мерингия трехнервная 256456
Мерингия удлиненная 256443
Меркия пузырчатая 414312
Мертензия 250883
Мертензия волосистая 250902
Мертензия даурская 250888
Мертензия джагастайская 250890
Мертензия длинностолбиковая 250912
Мертензия енисейская 250892
Мертензия камчатская 250893
Мертензия крылатоплодная 250905
Мертензия мелкопильчатая 250910
Мертензия палласа 250900
Мертензия попова 250898
Мертензия приморская 250894
Мертензия ручейковая 250909
Мертензия сибирская 250911
Мертензия тарбагатайская 250913
Мертенсия 250883

Меруя 241572
Мескито 315547
Метаплексис 252508
Метаплексис китайский 252514
Метаплексис стаунтона 252514
Метаплексис японская 252514
Метаплексис японский 252514
Метасеквойя 252538
Метасеквойя глиртостробовидная 252540
Метасеквойя глиртостробусовидная 252540
Метастахис 252549
Метастахис стреловидный 252550
Метельник 370191
Метельник прутовый 370202
Метельник прутьевидный 370202
Метлица 28969
Метлица обыкновенная 28972
Метлица полевая 28972
Метлица прерывчатая 28971
Метлица промежуточная 28970
Меум 252662
Меченосница 356377
Меч – трава 93901,245332
Меч – трава обыкновенная 93913
Мибора 252783
Мибора малая 252785
Миддендорфия 254387
Миддендорфия днепровская 254389
Микания 9831,254414
Микания вьющаяся 254443
Микания индийская 254426
Микания лазающая 254443
Микания ползучая 254443
Микания цепкая 254443
Микелия 252803
Микелия буроватая 252869
Микелия малипоская 252845
Микелия мелкоплодная 252930
Микелия ржавая 252869
Микелия сжатая 252849
Микелия сичоуская 252857
Микелия тибетская 252899
Микелия фиго 252869
Микелия фуцзяньская 252878
Микелия эмэйская 252974
Микромерия 253631
Микромерия греческая 253666
Микромерия дубровниковая 253698
Микромерия изящная 253653
Микромерия тайваньская 253661
Микромерия тибетская 253734
Микромерия тимьянолистная 253717
Микромерия чабрецолистная 253717
Микромерия эллиптическая 253654
микропус 253841
микропус лежащий 253849
Микростегиум 253980
Микростегиум лозный 254046
Микростегиум японская 254020
Микростигма 254082
Микростигма отогнутая 254085
Микростилис 242993
Микростилтс 254094
Милла 254562
Миллетия 254596
Миллетия гуйцзоуская 254741
Миллетия сетчатая 66643
Миллетия цзиндунская 254815
Миллетия эмэйская 66640
Миллефоль 3978,278099
Мильг – девона 201389
Мильтония 254924
Мимоза 1120,1165,254979
Мимоза застенчивая 255098
Мимоза стыдливая 255098
мимозка 220103
Мимозка 315547
мимозка выполненная 220105
Мимозка серёжкоцветная 315551,315560
Мимозовые 255141
Мимулус 255184
Мимулюс 255184
Мимулюс желтый 255218
Мимулюс жёлтый 255210
Мимулюс кардинальский 255197
Мимулюс крапчатый 255210,255218
Мимулюс мускусный 255223
Мимулюс темнокрасный 255197
Мимусопс 255260
Миндаль 20885,20890
Миндаль бухарский 20889
Миндаль вавилова 20980
Миндаль вязолистный 9858
Миндаль ганьсуский 20921
Миндаль горький 20892
Миндаль грузинский 20931
Миндаль дикий 20931
Миндаль зангезурский 20983
Миндаль земляной 118822
Миндаль иерусалимский 30498
Миндаль изхий 20931
Миндаль карликовый 20931
Миндаль колючейший 20967
Миндаль ледебура 20923
Миндаль ледебуровский 20923
Миндаль ложный – персик 20964
Миндаль ломкий 20899
Миндаль махровый 20970
Миндаль метельчатый 20966
Миндаль наирский 20930
Миндаль настящий 20890
Миндаль низкий 20931
Миндаль обыкновенный 20890
Миндаль петунникова 20961
Миндаль сладкий 20897
Миндаль степной 20931
Миндаль трёхлопастный 20970,316884
Миндаль тропический 386500
Миндаль туркменский 20976
Миндаль урарту 20979
Миндаль фенцля 20913
Миндаль черешковый 20933
Миндальник 20931
Минуартия 255435
Минуартия арктическая 255442
Минуартия весенняя 255598
Минуартия визнера 255608
Минуартия волосистая 255490
Минуартия гельма 255489
Минуартия гибридная 255595
Минуартия двухцветковая 255447
Минуартия железистая 255438
Минуартия красноватая 255573
Минуартия крашенинникова 255503
Минуартия крупноплодная 255514
Минуартия крымская 255584
Минуартия липкая 255606
Минуартия прямая 255582
Минуартия регеля 255565
Минуартия скученная 255482
Минуартия щетинковая 255578
Минуарция 255435
Минуарция аицовидная 255439
Минуарция акинфиева 255440
Минуарция биберштейна 255446
Минуарция бротеруса 255449
Минуарция буша 255450
Минуарция воронова 255609
Минуарция гвоздиколистная 255461
Минуарция гибридная 255494
Минуарция горная 255549
Минуарция двухцветковая 255447
Минуарция железистая 255437
Минуарция зернышкоцветная 255485
Минуарция изящная 255466
Минуарция кавказская 255455
Минуарция красночашечная 255566
Минуарция крашенинникова 255503
Минуарция крылова 255504
Минуарция лиственничная 255506
Минуарция литвинова 255512
Минуарция маленькая 255561
Минуарция мейера 255523
Минуарция мелкоцветковая 255527
Минуарция моховидная 255569
Минуарция отогнутая 255564
Минуарция полосчатая 255508
Минуарция промежуточная 255498
Минуарция стройная 255484
Минуарция черепичатая 255497
Минуарция якутская 255499

Миозотмс 260737
Миопоровые 260700
Миопорум 260706
Мирабель 316294
Мирабилис 255672
Мирабилис вильчатая 255694
Мирабилис длинноцветковая 255727
Мирабилис липкая 255750
Мирабилис многоцветковая 255732
Мирабилис ночецветная 255735
Мирабилис ялапа 255711
Мирзина 261598
Мирзина африканская 261601
Мирзина африканская мелколистная 261608
Мирзина гуансиская 261619
Мириактис 261060
Мириактис гмелина 261068
Мирика 261120
Мирикария 261246
Мирикария германская 261262
Мирикария даурская 261254
Мирикария длиннолистная 261281
Мирикария изящная 261259
Мирикария лисохвостная 261250
Мирикария лисохвостниковая 261250
Мирикария чешуйчатая 261293
Мириофиллюм 261337
Миристицевые 317,261464
Мирлитон 356352
Мирмехис 261467
Мирмехис китайская 261469
Мирмехис японский 261474
Миробалан 296554,316294,386500
Мирокарпус 261525
Мирокарпус густолиственный 261527
Мироксилон 261549
Миррис 261576
Миррис душистый 261585
Миррис пахучий 261585
Мирсина 261598
Мирсина африканская 261601
Мирсиновые 261596
Мирт 261726
Мирт болотный 261162
Мирт восконосный 261137
Мирт красный 261212
Мирт обыкновенный 261739
Мирта 261726
Миртовые 261686
Мискантус 255827
Мискантус китайский 255886
Мискантус краснеющий 255873
Мискантус непалский 127475
Мискантус сахароцветковый 255873
Мискантус сахароцветный 255873,255886, 394943
Мителла 256010
Мителла голая 256032
Мителла тайваньская 256019
Мителла японская 256024
Митрагина 256107
Митрагина крупнолистная 256120
Митрагина прилистниковая 256120
Митрария 256142
Михелия 252803
Михения 247720
Михения крапиволистная 247739
Мицелис стенной 260596
Мичелла 256001
Мичелла волнистая 256005
Мишоксия 252791
Мишоксия гладкая 252799
Млечник 176772
Млечник морской 176775
Многобородник 310102
Многобородник монпелиенский 310125
Многобородник монпелийский 310125
Многобородник низкий 304040
Многобородник низменый 310109
Многобородник полумутовчатый 310129
Многобородник приморский 310121
Многоколосник 10402
Многоколосник морщинистый 10414
Многоколосник тайваньский 10408
Многокоренник 372295
Многокоренник обыкновенный 372300
Многокрыльник 310007
Многокрыльник панютина 310008
Многолетний костер 412394
Многоплодник 307724
Многоплодник индийский 307748
Многоплодник четырехлистный 307755
Многосемянниба нежная 89020
Многосемянник 201135
Многосемянник лежачий 198485
Многосемянник пушистый 201151
Могар 361794
Могильник 287975,287978
Могильница 409325
Моголтавия 256515
Моголтавия северцова 256516
Модестия 21181,256424
Модестия дарвазская 256426
Модестия удивительная 256427
Можжевелоягодник 30906
Можжевельник 213606,213873
Можжевельник аллигаторовый 213847
Можжевельник американский 213834, 213984
Можжевельник барбадосский 213623
Можжевельник бермудский 213623
Можжевельник верес 213702
Можжевельник виргинский 213984
Можжевельник вонючий 213774
Можжевельник восточный 213859
Можжевельник высокий 213767
Можжевельник гималайский 213883
Можжевельник горизонтальный 213785
Можжевельник даурский 213752
Можжевельник длиннолистный 213833
Можжевельник донской 213902
Можжевельник древовадный 213767
Можжевельник жесткий 213896
Можжевельник западный 213834
Можжевельник зеравшанский 213933
Можжевельник испанский 213634,213841
Можжевельник казацкий 213902,213984
Можжевельник казачий 213902
Можжевельник карликовый 213934
Можжевельник карлифорнийский 213625
Можжевельник китайский 213634,213702
Можжевельник косточковый 213763
Можжевельник красноватый 213841, 213901
Можжевельник красноплодный 213849
Можжевельник красный 213841
Можжевельник крупноплодный 213843
Можжевельник куэньминский 213780
Можжевельник ладанный 213969
Можжевельник лежачий 213864
Можжевельник ложноказацкий 213873
Можжевельник ложноказачий 213873
Можжевельник люцюский 213963
Можжевельник мексиканский 213826
Можжевельник многоплодный 213859
Можжевельник низкорослый 213760, 213826
Можжевельник обыкновенный 213702
Можжевельник обыкновенный горный 213934
Можжевельник отогнутый 213883
Можжевельник пинчота 213853
Можжевельник повислый 213773
Можжевельник полушаровидный 213932, 341765
Можжевельник приблежный 213739
Можжевельник приморский 213739
Можжевельник равнолистный 213809
Можжевельник распростертый 213864
Можжевельник распростёртый 213785
Можжевельник распростёртый опушённый 213799
Можжевельник саржента 213696
Можжевельник сибирский 213934
Можжевельник скальный 213922
Можжевельник скученный 213739
Можжевельник стройный 213863
Можжевельник тайваньский 213775
Можжевельник твердолистный 213896
Можжевельник твердый 213896

Можжевельник твёрдый 213896
Можжевельник торстокорый 213847
Можжевельник туркестанский 213877
Можжевельник формозский 213775
Можжевельник чешуйчатый 213943
Можжевельник шугнанский 213921
Можжуха 297640
Мозжуха 297640
Мокрединник 284167
Мокрица 374968
Мокрица алзина 374724
Мокричник 374724
Моли 15480
Молина мелкоцветковая 258865
Молиния 256597
Молиния голубая 256601
Молиния прибрежная 256601
Моллуго 256689
Моллуго маленькая 256695
Моллюго 256689
Моллюго маленькое 256695
Моллюговые 256684
Молодило 358015
Молодило гололистное 358037
Молодило кавказское 358028
Молодило кровельное 358062
Молодило малорослое 358054
Молодило марообразное 358039
Молодило отпрысковое 212541
Молодило русское 358056
Молокан 219206,259746
Молокан комнасный 219497,219507
Молокан сибирский 219513,219806
Молокан синий 259772
Молокан татарский 259772
Молокан цинхайский 259763
Молокан ядовитый 219609
Молоко волчье 158385
Молочай 158385,159523,159603
Молочай адрахновидный 158443
Молочай алайский 158408
Молочай алеппский 158414
Молочай алмянский 158474
Молочай алтавский 158409
Молочай алтайский 158420
Молочай альпийский 158417
Молочай астраханский 159624
Молочай блестящий 159270
Молочай болотный 159522
Молочай бородина 158560
Молочай борцова 158561
Молочай буассье 158553
Молочай бузе 158582
Молочай бунге 158585
Молочай бутерлак 159557
Молочай бутерлаковидный 159557
Молочай бутерлаковый 159557
Молочай бухтарминский 158581
Молочай буша 158593
Молочай виттмана 160091
Молочай волжский 160067
Молочай волнистый 160028
Молочай волосистый 158659
Молочай воронова 160093
Молочай восточный 159503
Молочай ганьсуский 159171
Молочай глянцевитый 159270
Молочай гмелина 158965
Молочай гольде 158967
Молочай греческий 158977
Молочай гроссгейма 159007
Молочай грузинский 159112
Молочай густой 158756
Молочай двужелезковый 159726
Молочай двуцветный 158857
Молочай дельтоприцветничный 158752
Молочай джунгарский 159851
Молочай дланечешуйный 158642
Молочай длиннокорневой 159285
Молочай донской 159770
Молочай древнезеравшанский 159619
Молочай дынеобразный 159342
Молочай евгении 158865
Молочай жесткий 159683
Молочай жигулевский 160108
Молочай заамударьинский 159990
Молочай заостренный 158394
Молочай зибольда 159841
Молочай зибольдиев 159841
Молочай иволистный 159759
Молочай индерский 159129
Молочай индийский 159102
Молочай исписанный 159799
Молочай камнелюбивый 159567
Молочай кандинский 159167
Молочай каро 159175
Молочай карпатский 158616
Молочай кипарисный 158730
Молочай кипарисовидный 158730
Молочай кипарисовый 158730
Молочай кожистый 159526
Молочай колючезубый 159863
Молочай комалова 159187
Молочай комарова 159101
Молочай коническосеменный 158681
Молочай копетдага 159191
Молочай коротко остроконечный 159403
Молочай красивейший 159675
Молочай криволистный 158734
Молочай крымский 159937
Молочай кудряшова 159199
Молочай ледебура 159226
Молочай липского 159252
Молочай ложнокактусовый 159641
Молочай ложнополевой 159650
Молочай ложно－полевой 159640
Молочай лозный 158857,160062
Молочай лосный 160062
Молочай маленький 158869
Молочай малый 158869
Молочай маньчжурский 158857
Молочай маршаллов 159318
Молочай масличный 159222
Молочай мелкозубый 158762
Молочай мелкоплодный 159357
Молочай мелкосемянный 159360
Молочай мелкосмоковник 158632
Молочай миндалевидный 158433
Молочай мирсинитский 159417
Молочай миртолистный 159417
Молочай многоцветный 159615
Молочай монгольский 159387
Молочай мохнатый 158659
Молочай неравнолепестный 158449
Молочай ничтожный 158869
Молочай нормана 159472
Молочай оголенный 158948
Молочай одностолбиковый 159391
Молочай одноциатиевый 159390
Молочай остроконечный 158471
Молочай острый 158857,160062
Молочай оше 158499
Молочай палласа 158895,159517
Молочай пашенный 158407
Молочай плевой 158407
Молочай плосколистный 159599
Молочай полуволосистый 159811
Молочай полумохнатый 159811
Молочай полусердцевидный 158857
Молочай поникающий 85779
Молочай прекрасный 159675
Молочай прибрежный 159526
Молочай приземистый 158632,159092,159095
Молочай проханова 159630
Молочай прутьевидный 160062
Молочай прямой 159892
Молочай пушистый 159670
Молочай пятнистый 159286
Молочай ранний 159624
Молочай расширеннолистный 159599
Молочай ребристый 158447
Молочай репка 159700
Молочай ресничнолистный 158550
Молочай розеточный 159736
Молочай рошевый 159272
Молочай русский 159733
Молочай савари 159772
Молочай садовый 159557
Молочай сарептский 159770
Молочай светлоплодный 159851

Молочай сегиеров 159808
Молочай сегюера 159808
Молочай сереющий 158608
Молочай серповидный 158876
Молочай сестричный 159853
Молочай сжатый 159892
Молочай скаловый 159751
Молочай скалолюбивый 159567
Молочай скученный 158857
Молочай сладкий 158805
Молочай слегка－сердцевидный 158857
Молочай смолоносный 159710
Молочай совича 159925
Молочай солнцеглядный 159027
Молочай солнцеглят 159027
Молочай сосочковый 159305
Молочай стелющийся 159092
Молочай съедобный 158857,160062
Молочай сырдарьинский 158857
Молочай твердобокальчатый 159794
Молочай тибетский 159888,159974
Молочай тирукалли 159975
Молочай толстокорневой 159514
Молочай тонкий 159905
Молочай тонкостебельный 159236
Молочай трехзубый 160004
Молочай туркестанский 160022
Молочай туркменский 158991
Молочай турчанинова 160019
Молочай тяньшанский 159967
Молочай угловатый 158447
Молочай уральский 159663
Молочай урарский 159663
Молочай ферганский 158887
Молочай форскаля 158909
Молочай франше 158915
Молочай хайнаньский 159018
Молочай хрящеватый 158952
Молочай ченовидный 159222
Молочай чешуйчатый 159875
Молочай чиный 159222
Молочай шаровидный 158958
Молочай шерстеносный 158844
Молочай шерстистый 159563
Молочай шершавый 159875
Молочай широковетвистый 160062
Молочай широколистный 159223
Молочай шугнанский 159792
Молочай эзуля 158857
Молочай японский 279757
Молочай яркий 159363
Молочайные 160110
Молочник 368771
Молочник хробуст 368771
Мольткия 256746
Мольткия голубая 256750
Молюцелла 256758
Момбин жёлтый 372477
Момордика 256780,256786
Момордика бальзамическая 256786
Момордика харантская 256797
Момордика харанция 256797
Монарда 257157
Монарда трубчатая 257169
Монезес 257371
Монезес одноцветковый 257374
Монодора 257630
Монодора мускатная 257653
Монохория 257564
Монохория корсакова 257572
Монохория подложниковая 257583
Монстера 258162
Монстера любимая 258168
Монстера привлекательная 258168
Монстера пышная 258168
Монтбреция 399043
Монция 258237
Монция блестящесемянная 258248
Монция маленькая 258259
Монция ручейная 258248
Монция сибирская 94362
Мордовник 140652
Мордовник бабатагский 140658
Мордовник банатский 140660
Мордовник белополосый 140735
Мордовник белостебельный 140654
Мордовник венгерский 140660
Мордовник возвышеный 140704
Мордовник волосочешуйный 140824
Мордовник восточный 140761
Мордовник гладкорогий 140734
Мордовник гмелина 140712
Мордовник дагестанский 140692
Мордовник дубянского 140698
Мордовник зайсанский 140781
Мордовник закаспийский 140822,140823
Мордовник зеренолистный 140829
Мордовник ильина 140723
Мордовник карабахский 140725
Мордовник каратавский 140726
Мордовник карликовый 140753
Мордовник клупнолистный 140744
Мордовник кнорринг 140727
Мордовник колхидский 140684
Мордовник колючий 140773
Мордовник колючконосный 140790
Мордовник конрата 140687
Мордовник короткокисточковый 140674
Мордовник косолопастный 140757
Мордовник круглоголовый 140789
Мордовник крутай 140779
Мордовник липского 140737
Мордовник матоволистный 140760
Мордовник мейера 140749
Мордовник многоколючконосный 140766
Мордовник многолистный 140706
Мордовник многоплосый 140767
Мордовник многостебельный 140751
Мордовник обыкновенный 140776
Мордовник осетинский 140762
Мордовник персидский 140765
Мордовник почти－голый 140813
Мордовник приземистый 140722
Мордовник просмотренный 140768
Мордовник пушистоцветковый 140696
Мордовник пушисто－чешуйчатый 140772
Мордовник рассеченный 140697
Мордовник русский 140704,140776,140779
Мордовник самаркандский 140745
Мордовник севанский 140787
Мордовник совича 140815
Мордовник сращеннный 140686
Мордовник таласский 140816
Мордовник турнефора 140821
Мордовник тяньшанский 140820
Мордовник федченко 140705
Мордовник хантавский 140681
Мордовник цельнолистный 140724
Мордовник чимганский 140825
Мордовник шароголовый 140789
Мордовник широколистный 140732,140789
Морея 258380
Морина 258831
Морина кокансквя 258848
Морина лемана 258850
Морина персидская 258867
Моринга 258945
Мориновые 133444,258870
Морковица 43417
Морковица восточная 43419
Морковица персидская 43420
Морковица прибрежная 43418
Морковник 363123
Морковник бессера 363124
Морковник луговой 363128
Морковник попова 363126
Морковник рубцова 363127
Морковь 123135,123141
Морковь двулетняя 123141
Морковь дикая 123141
Морковь культурная 123141
Морковь культурная каротиновая 123141
Морковь обыкновенная 123141
Морковь огородная 123164
Морковь посевная 123164
Морковь－каротель 123141
Морозник 190925
Морозник абхазский 190928
Морозник вонючий 190936
Морозник восточный 190949

Морозник зелёный 190959
Морозник кавказский 190932
Морозник крапчатый 190938
Морозник красноватый 190955
Морозник черный 190944
Морозник чёрный 190944
Морошка 338059,338243
Морошка приземистая 338243
Морская горчица 65164,65185
Морская горчица арктическая 65173
Морская трава 418392
Мортук 148398,148415
Мортук бонапарта 148400
Мортук восточный 148415
Мортук мохнадый 148413
Мортук пшеничный 11830,148420
Мосла 259278
Мосла двпыльниковая 259284
Мосла китайская 259282
Мосла крупнопильчатая 259292
Мосла мелкоцветковая 259281
Мосла японская 259300
Мотыльковые 282926
Мохоцветник 61430
Мохоцветник гмелина 61432
Мохунка 297643,297645
Мочёнка 401094
Мошнуха 297643,297645
Мукуна 259479
Мульга 1049
Мульгедиум 219206,259746
Муме 34448
Мунго 408982
Мунтингия 259910
Мунтингия калабура 259910
Мунч 408251
Мурайя 260158
Мурайя экзотическая 260165
Мурезия 260127
Мурезия желтая 260130
Мурезия закаспийская 260132
Мурезия пахачая 260129
Мурезия переходная 260133
Мурезия эройланская 260131
Мурза – соранг 184933
Муррая 260158
Муррая экзотическая 260165
Муррея 260158
Муселис 260594
Мускари 260333
Мускаримия 260293,260345
Мускаримия мышиная 260347
Мускат 261402,261424
Мускатник 261402
Мускатник душистый 261424
Мускатниковые 317,261464
Мускусница 8396,8399
Мускусница обыкновенная 8399
Мутовчатка 256689
Мутовчатка пятилистная 256719
Мухоловка 68860,138375
Мухоловка рыжий 68860
Мушмула 252291
Мушмула германская 252305
Мушмула субтропическая 151162
Мушмула японская 151162
Мушмула японская обыкновенная 151162
Мшанка 342221
Мшанка безлепестковая 342226
Мшанка безлепестная 342226
Мшанка большая 342259
Мшанка крупная 342259
Мшанка лежачая 342279
Мшанка линнеева 342290
Мшанка моховидная 342290
Мшанка мшанковидная 342290
Мшанка острочашелистниковая 342277
Мшанка прибрежная 342259
Мшанка промежуточная 342268
Мшанка снежная 342268
Мшанка узловатая 342271
Мшанка шилолистная 32301
Мшанка японская 342251
Мшунки – гоуак 408251
Мыло татарское 363312
Мыльница 346418
Мыльнянка 346418
Мыльнянка аптечная 346434
Мыльнянка восточная 346436
Мыльнянка гриффита 346426
Мыльнянка клейкая 346425
Мыльнянка коровья 314752,403687
Мыльнянка лекарственная 346434
Мыльнянка липкая 346446
Мыльнянка маленькая 346437
Мыльнянка северцова 346442
Мыльнянка стелющаяся 346438
Мыльнянка ясколковидная 346422
Мысык – кайрык 361935
Мытник 286972
Мытник адамса 286975
Мытник алатавский 286984
Мытник алтайский 286991
Мытник альберта 286985
Мытник бальшой 287260
Мытник болотный 287494
Мытник вальдгейма 287822
Мытник веры 287795
Мытник вздуточашечный 287517
Мытник вильгельмся 287828
Мытник вильденова 287829
Мытник власова 287821
Мытник волосистый 287279
Мытник высокий 287182,287196
Мытник ганьсуский 287302
Мытник голодчатый 287067
Мытник гуншаньский 287243
Мытник густоцветковый 287100,287569
Мытник дагестанский 287139
Мытник далиский 287741
Мытник длиннокорневой 287175
Мытник длинноцветковый 287368
Мытник дэциньский 287162
Мытник желтый 287209
Мытник запрокинутый 287584
Мытник зеравшанский 287843
Мытник иезский 287837
Мытник кавказский 287068
Мытник кандинский 287301
Мытник каратавский 287307
Мытник каро 287495
Мытник каролев – скипетр 287657
Мытник каролькова 287318
Мытник кауфманна 287310
Мытник китайский 287090
Мытник колосистый 287689
Мытник колосовый 287689
Мытник королевский скипетр 287657
Мытник коротколосый 287055
Мытник краекучниколистный 287084
Мытник красивый 287568,287794
Мытник краснеющий 287645
Мытник крупноцветковый 287259
Мытник крылова 287320
Мытник крючковатый 286976
Мытник кузнецова 287321
Мытник лабрадорский 287326
Мытник лангсдорфа 287336
Мытник ланьпинский 287337
Мытник лапландский 287338
Мытник лесной 287721
Мытник лицзянский 287356
Мытник лослый 287196
Мытник людвига 287382
Мытник максимовича 287414
Мытник маньчжурский 287411
Мытник марин 287413
Мытник мелкоцветковая 287427
Мытник миловидный 287794
Мытник мохнатоколосый 287143
Мытник мохнатоoдeный 287191
Мытник мохнатоцветковый 287142
Мытник мохнатый 287343,287814
Мытник мупинский 287436
Мытник мутовчатый 287802
Мытник мясокрасный 287290
Мытник наньчуаньский 287451
Мытник нордманна 287461
Мытник носатый 287455
Мытник ольги 287481
Мытник очанковидный 287194

Мытник палласа 287492
Мытник панютина 287496
Мытник пеннелля 287508
Мытник перевернутый 287584
Мытник печальный 287779
Мытник плотный 287100
Мытник погремковый 287617
Мытник полосатый 287705
Мытник полынелистный 286973
Мытник полыннолистный 286973
Мытник понтийский 287532
Мытник попова 287533
Мытник почти – носатый 287716
Мытник прелестноцветковый 286997
Мытник прелестный 287802
Мытник прерванный 287298
Мытник пушистоцветковый 287567
Мытник разрезной 287207
Мытник семенова 287668
Мытник сжатый 287102
Мытник сибирский 287673
Мытник сиборпа 287674
Мытник сычуаньский 287723
Мытник сомнительный 287178
Мытник судетский 287718
Мытник таласисский 287740
Мытник тибетский 287761
Мытник толстоносый 287120
Мытник тонкопильчатый 287015
Мытник топяной 287788
Мытник тысячелистниковый 286974
Мытник тысячелистный 287447
Мытник тяньшанский 287760
Мытник уральский 287790
Мытник франше 287222
Мытник фэн 287201
Мытник хоботковый 287541
Мытник хохлатый 287099
Мытник хэнаньский 287282
Мытник цветоножковый 287507
Мытник цзилунский 287270
Мытник чаюйский 287842
Мытник чернопурпуровый 287021
Мытник чжундяньский 287844
Мытник шамизо 287076
Мытник шелоховатоколосый 287343
Мытник шерстистый 287334
Мытник ширококлювый 287520
Мытник шугнанский 287666
Мытник эдера 287468
Мытник японский 287300
Мышатник 389505
Мышатник ланцетовидный 389533
Мышей 361680
Мышей волосистый 147883
Мышей зеленый 361935
Мышей зелёный 361935
Мышей малый 147815
Мышей мятликовидный 147815
Мышей сизый 361877,361935
Мышехвостник 260930
Мышехвостник маленький 260940
Мышиный Гиацинт 260293
Мышиный горошек 408352
Мюленбекия 259584
Мюленбергия 259630
Мюленбергия гугела 259671
Мюленбергия гюгеля 259671
Мюленбергия коротоостая 259646
Мюленбергия японская 259672
Мюскадина 411893
Мягковолосник 260920
Мягковолосник водный 260922
Мягковолосник водяной 260922
Мягкохвостник 228337
Мягкохвостник шерстистый 242970
Мякинник 29422
Мякинник безостный 29430
Мякинник тупоконечный 29430
Мякотница 242993,254094
Мякотница болотная 185200
Мята 250287
Мята австрийская 250323,250326
Мята азиатская 250323
Мята алайская 250288
Мята ангрийская 250420
Мята бергамотная 250341
Мята блошница 250432
Мята болотная 250432
Мята водная 250291
Мята водяная 250291
Мята горная 321952
Мята дагестанская 250352
Мята дарвазская 250354
Мята даурская 250353
Мята дикая 250385
Мята длиннолистная 250385
Мята зеленая 250450
Мята зелёная 250450
Мята кавказская 250339
Мята канадская 250331
Мята колосовая 250450
Мята конская 245770
Мята копетдагская 250383
Мята кошачья 264897
Мята круглолистная 250439
Мята кудревадая 250349
Мята кудрявая 250349
Мята курчавая 250341,250347
Мята лапландская 250384
Мята лимонная 249504,250341
Мята лимонная аптечная 13635,249504
Мята мелкоцветковая 250400
Мята памироалайская 250406
Мята перечная 250420
Мята полевая 250294,250370
Мята прерывистая 250380
Мята просто – чашечковая 250370
Мята просточашечная 250370
Мята пурегиевая 250432
Мята раскидистая 250463
Мята сахалинская 250443
Мята сичанская 250333
Мята холодка 250420
Мята японская 250381
Мятлик 305274
Мятлик алтайский 305316
Мятлик альберта 305291,305673
Мятлик альпийский 305303
Мятлик араратский 305357
Мятлик аргунский 305368
Мятлик арктический 305362
Мятлик бактрийский 305389
Мятлик бакунский 305395
Мятлик бальфура 305551
Мятлик бесплодный 306027
Мятлик бирманский 305428
Мятлик болотный 305804
Мятлик боровой 305743
Мятлик бухарский 305417
Мятлик веретеновидный 306003,306133
Мятлик вороновб 306153
Мятлик выдающийся 305506
Мятлик высокогорный 305300
Мятлик ганешина 305543,305544
Мятлик гиблидный 305589
Мятлик гиссарский 305585
Мятлик гладкоцветковый 305548
Мятлик горбунова 321393
Мятлик горный 305646
Мятлик грузинский 305593
Мятлик густейший 305475
Мятлик густой 305474
Мятлик даурский 305380,305383
Мятлик джилгинский 305498
Мятлик джунгарский 305665
Мятлик длиннолистный 305675
Мятлик дрожащий 306087
Мятлик дубравный 305743
Мятлик забайкальский 306082
Мятлик заливной 305605
Мятлик заметный 305601
Мятлик заостренный 305539
Мятлик запрягаева 306159
Мятлик зелененький 306146
Мятлик инееватый 305890
Мятлик иркутский 305604
Мятлик кавказский 305433
Мятлик каратегинский 305418
Мятлик кистевидный 305380,305409
Мятлик кисточный 305840

Мятлик колючий 306033
Мятлик комарова 305623
Мятлик корейский 306059
Мятлик красивый 305431
Мятлик крупночашечный 305686
Мятлик лауданский 305643
Мятлик лесной 305743,305982
Мятлик липского 305664
Мятлик литвинова 305673
Мятлик литвиновский 305673
Мятлик ложноболотный 305901
Мятлик ложнодрожащий 305906
Мятлик ложноразбросанный 305664
Мятлик ложноуколоченный 305895
Мятлик ломкий 305539
Мятлик ломнодубравный 305741
Мятлик луговой 305855
Мятлик луковичный 305419
Мятлик луковичный живолодящий 305422
Мятлик мазендеранский 305718
Мятлик майделя 305721
Мятлик марии 305711
Мятлик мейера 305728
Мятлик монгольский 305733
Мятлик мягкоцветковый 305705
Мятлик нальчикский 305738
Мятлик невского 305771
Мятлик непалский 305764
Мятлик обманчиный 305680
Мятлик обыкновенный 306098
Мятлик однолетний 305334
Мятлик окаймленный 305673,305708
Мятлик остростебельный 305285
Мятлик оттянутый 305380
Мятлик оходский 306003
Мятлик охотский 306133
Мятлик памирский 305806
Мятлик песчаный 305940
Мятлик пинежский 305838
Мятлик плоскоцветковый 305840
Мятлик подобный 306097
Мятлик подольский 305847
Мятлик поздний 305804
Мятлик прицветниковый 305411
Мятлик разноцветный 306131
Мятлик раскинутый 306045
Мятлик расползающийся 306131,306136
Мятлик расставленный 305922
Мятлик ревердатто 305926
Мятлик ромбический 305928
Мятлик рыхлый 305646
Мятлик саянский 305945
Мятлик сжадый 305451
Мятлик сжатый 305664
Мятлик сибирский 305973
Мятлик сизостебельный 305559
Мятлик сизый 305551
Мятлик синайский 305984
Мятлик сычуаньский 306050
Мятлик скальный 305636,305830
Мятлик слабый 305599
Мятлик смирнова 305995
Мятлик снежный 305820
Мятлик сочавы 305998
Мятлик сплюснутый 305451
Мятлик степной 306139
Мятлик таймырский 306054
Мятлик танфильева 306060
Мятлик тибетский 306074
Мятлик толмачева 305364
Мятлик торфяной 305855,306109
Мятлик траурный 306094
Мятлик трхцветковыйё 305804
Мятлик тяньшанский 306071
Мятлик узколистный 305328
Мятлик узкочешуйный 305331
Мятлик укороченный 305275
Мятлик урянхайский 306113
Мятлик уссурийский 306119
Мятлик федченко 305521
Мятлик чжундяньский 306163
Мятлик членистый 305372
Мятлик шафранный 305804
Мятлик ше 305436
Мятлик шероховатый 305916
Мятлик шерстистоцветковый 305637
Мятлик широкометельчатый 306038
Мятлик шэ 305436
Мятлик щетинистый 305966
Мятлик японский 305776
Мяун 404213
Нагами 167506
Нагами кинкан 167506
Нагловадка 214028
Наголовадка васильковая 214070
Наголовадка васильковидная 214070
Наголовадка джунгарская 214074
Наголовадка железистоплодная 214030
Наголовадка какрыладная 214031
Наголовадка киргизская 214108
Наголовадка лавандолистная 214179
Наголовадка липского 214121
Наголовадка меловая 214067
Наголовадка многоцветковая 214138
Наголовадка паутинистая 214040
Наголовадка стеблеобъемлющая 214036
Наголовадка суйдунская 214182
Наголовадка тонколопастная 214186
Наголовадка холодная 214033
Наголовадка шерстистоногая 214115
Наголовадка щетинкоплодная 214060
Наголовадка эверсмана 214080
Наголоватка 214028
Наголоватка аболина 214029
Наголоватка алтайская 214034
Наголоватка андросова 214037
Наголоватка анны 214038
Наголоватка армянская 214043
Наголоватка байсунская 214047
Наголоватка бальдшуанская 214048
Наголоватка безногая 214039
Наголоватка белостебельная 214032
Наголоватка бессмертниковая 214199
Наголоватка боброва 214053
Наголоватка бочанцева 214054
Наголоватка бухарская 214056
Наголоватка винклера 214197
Наголоватка воронова 214198
Наголоватка вороньлапая 214066
Наголоватка восточная 214142
Наголоватка головоногая 214058
Наголоватка голодкова 214090
Наголоватка горная 214135
Наголоватка гранитовая 214092
Наголоватка гроссгейма 214093
Наголоватка грязная 214172
Наголоватка гтстая 214177
Наголоватка дарвазская 214072
Наголоватка дважды – перистая 214051
Наголоватка димитрия 214073
Наголоватка донская 214185
Наголоватка загирканская 214191
Наголоватка зауральская 214192
Наголоватка змеестебельная 214169
Наголоватка иволистная 214165
Наголоватка известняковая 214057
Наголоватка изящная 214078,214091
Наголоватка ильина 214100
Наголоватка казахстанская 214107
Наголоватка камнелюбивая 214122
Наголоватка капелькина 214102
Наголоватка карабугазская 214103
Наголоватка каратавская 214104
Наголоватка кнорринг 214109
Наголоватка кокандская 214110
Наголоватка комарова 270301
Наголоватка корневищевидная 214162
Наголоватка красивая 214195
Наголоватка крашенинникова 214112
Наголоватка крупная 214128
Наголоватка крупнокорзиночная 214124
Наголоватка крылатая 214031
Наголоватка крылостебельная 214159
Наголоватка культиасова 214113
Наголоватка кураминская 214114
Наголоватка левье 214119
Наголоватка ледебура 214118
Наголоватка лидии 214123
Наголоватка ложно – васильковая 214158
Наголоватка маргалыкская 214126
Наголоватка маргаритковидная 214049

Наголоватка марии 214127
Наголоватка мелкоключковатая 214095
Наголоватка мелов 214068
Наголоватка михельсона 214129
Наголоватка многоглавая 214137
Наголоватка многолопастная 214139
Наголоватка мощная 214163
Наголоватка мягчайшая 214132
Наголоватка мятликовидная 214151
Наголоватка низная 214161
Наголоватка олиственная 214087
Наголоватка ольги 214141
Наголоватка оше 214046
Наголоватка паноротниколистная 214084
Наголоватка пачоского 214143
Наголоватка песколюбивая 214157
Наголоватка погруженно－жилковая 214101
Наголоватка полукустарниковая 214180
Наголоватка попова 214154
Наголоватка похожая 214146
Наголоватка превосходная 214081
Наголоватка предкавказская 214064
Наголоватка привлекательная 214052
Наголоватка примечательная 214175
Наголоватка прицветничковая 214055
Наголоватка пучковатая 214094
Наголоватка рирамидально－метельчатоцветная 214188
Наголоватка родственная 214156
Наголоватка рупрехта 214164
Наголоватка рыхлая 214117
Наголоватка серпуховидная 214170
Наголоватка снежная 214140
Наголоватка сосновского 214173
Наголоватка сосняковая 214149
Наголоватка спиридонова 214176
Наголоватка таджикистанская 214183
Наголоватка талиеви 214184
Наголоватка темнопурпуровая 214045
Наголоватка траутфеттера 214193
Наголоватка трехвильчатая 214194
Наголоватка тяньшанская 214189
Наголоватка узколистная 214178
Наголоватка федченко 214082
Наголоватка ферганская 214083
Наголоватка хорошенькая 214160
Наголоватка цминолистная 214096
Наголоватка чиликиной 214071
Наголоватка шерстистоногая 214116
Наголоватка шершаволистная 214044
Наголоватка шинтенища 214171
Наголоватка шишкина 214168
Наголоватка элегантная 214079
Наголоватьнь 92384
Нагут 90801
Надбородник 147303
Надбородник безлистный 147307,147323
Налимбия 280490
Налимбия оживающая 280491
Нандина 262188
Нандина домашняя 262189
Наннoропс 262234
Нанофитон 262259
Нанофитон ежовый 262260
Наперстянка 130344,130383
Наперстянка бахромчатая 130355
Наперстянка желтая 130371
Наперстянка жёлтая 130371
Наперстянка красная 130383
Наперстянка крупноцветковая 130365
Наперстянка лекарственная 130383
Наперстянка пурпуровая 130383
Наперстянка ресничатая 130355
Наперстянка ржавая 130358
Наперстянка сомнительная 130365
Наперстянка шерстистая 130369
Нард 117218
Нард вытянутый 262537
Нардосмия 292342
Нардосмия гладгая 292379
Нардосмия гмелина 292368
Нардосмия ледяная 292367
Нардосмия пальмолистная 292403
Нардосмия скальная 292394
Нардосмия тайваньская 292352
Нардосмия угловатая 292354
Нардосмия фомина 292351
Нардосмия японская 292374
Нартеций 262572
Нартеций кавказский 262576
Нарцисс 262348
Нарцисс белый 262438
Нарцисс букетный 262457
Нарцисс жёлтый 262410
Нарцисс ложнонарциссовый 262441
Нарцисс ложный 262441
Нарцисс трубчатый 262441
Нарцисс－жонкиль 262410
Нарцисс－тацетта 262457
Нарцисус 262348
Нарцисус большой 262415
Нарцисус гиблидный 262404
Нарцисус желтый 262441
Нарцисус кавказский 262391
Нарцисус константинопольский 262457
Нарцисус ложнонацисcовый 262441
Нарцисус ложный 262441
Нарцисус поэтический 262438
Настурция 399597,399601
Настурция большая 399601
Настурция высокая иноземная 399601
Настурция канарская 399606
Настурция карликовая 399606
Настурция клубненосная 399604
Настурция малая 399604,399611
Настурция мелкая 399604
Настурция низкорослая 399604
Настурция садовая 399601,399604
Наталиелла 262776
Наталиелла алайская 262777
Наумбургия 239543
Наумбургия далиская 239874
Наумбургия кистецветная 239881
Наумбургия тайваньская 239872
Наумбургия тэнчунская 239878
Наяда 262015
Наяда гибкая 262028
Наяда злаковая 262040
Наяда злаковидная 262040
Наяда индийская 262054
Наяда китайская 262024
Наяда малая 262089
Наяда морская 262067
Наяда тончайшая 262110
Наяда ячеистая 262035
Наядовые 262013
Нгустура 45586
Невзрачница 13951,13966,29018
Невзрачница полевая 13966
Невиузия 265856
Невскиелла 265872
Невскиелла тончайшая 60735
Негниючка 385262,390567,390587
Негниючник колёсивовидный 99804
Негной－дерево 385301
Негундо 3218
Недзвецкия 266183
Недзвецкия семиреченская 266185
Недоспелка 283781
Недоспелка комарова 283835
Недоспелка пропущенная 283860
Недоспелка ушастая 283792
Недотрога 204756
Недотрога "не тронь меня" 205175
Недотрога балезаминовая 204799
Недотрога бальзаминовидная 204799
Недотрога бирманская 204788
Недотрога вильчатая 204969
Недотрога далиская 205355
Недотрога железистая 205291
Недотрога железконосная 204982
Недотрога желтая 205175
Недотрога жуйлиская 205297
Недотрога инцзянская 205457
Недотрога кандинская 205327
Недотрога китайская 204848
Недотрога комарова 205058
Недотрога коротконшпорцевая 204825
Недотрога лунчжоуская 205154
Недотрога маака 205107

Недотрога мелколистная 205216
Недотрога мелкоцветковая 205141,205212
Недотрога мотоская 205129
Недотрога мулиская 205155
Недотрога мэнцзыская 205132
Недотрога напоская 205165
Недотрога невского 205169
Недотрога обыкновенная 205175
Недотрога ройлевао 205291
Недотрога ройля 205291
Недотрога сычуаньская 205349
Недотрога тайваньская 205352
Недотрога текстора 205369
Недотрога тибетская 204881
Недотрога тяньцюаньская 205381
Недотрога фэнхуайская 204946
Недотрога хайнаньская 204998
Недотрога хунаньская 205024
Недотрога чжундяньская 204854
Недотрога чжэцзянская 204846
Недотрога эмэйская 205197
Недотрога юньнаньская 205460
Нежник 188583
Нежник альпийский 188586
Нежник обыкновенный 188757
Незабудка 260737
Незабудка азиатская 260762
Незабудка азорская 260766
Незабудка альпийская 260747
Незабудка балтийская 260815
Незабудка береговая 260824
Незабудка болотная 260842,260868
Незабудка воробейниколистная 260822
Незабудка горная 260838
Незабудка греческая 260803
Незабудка дернистая 260772
Незабудка дубравная 260840
Незабудка дужистая 260890
Незабудка душистая 260890
Незабудка жестковолосистая 260779
Незабудка идская 260803
Незабудка крупноцветная 260794
Незабудка крылова 260809
Незабудка лазистанская 260818
Незабудка лесная 260838,260892
Незабудка литовская 260823
Незабудка ложно-изменчивая 260850
Незабудка ложно-родственная 260849
Незабудка мелкоцветковая 260833
Незабудка отогнутая 260860
Незабудка пашенная 260760
Незабудка полевая 260760
Незабудка прибрежая 260824
Незабудка приятная 260757
Незабудка разноцветная 260791,260828,260909
Незабудка редкоцветковая 260880
Незабудка родственная 260848
Незабудка сахалинская 260892
Незабудка сицилийская 260878
Незабудка средняя 260760,260807
Незабудка туркменская 260901
Незабудка украинская 260902
Незабудка холмовая 260779
Незабудка щетинистая 260877
Незабудочник 153420
Незабудочник гребенчатый 153502
Незабудочник енисейский 153469
Незабудочник кавказский 153437
Незабудочник камчатский 153470
Незабудочник кандинский 153471
Незабудочник ложно-шилоколистный 153509
Незабудочник маака 19045
Незабудочник маньчжурский 153479
Незабудочник мохнатый 153555
Незабудочник памирский 153497
Незабудочник подушковидный 153500
Незабудочник подушковый 153426
Незабудочник почти-жакмонов 153542
Незабудочник сахалинский 153491,153526
Незабудочник серный 153463
Незабудочник сихатэалинский 153534
Незабудочник скальный 153518
Незабудочник сомнительный 153448
Незабудочник туркестанский 153552
Незабудочник фетисова 153452
Незабудочник чекановского 153439
Незабудочник шелковистый 153531
Незабудочник широколистный 153473
Незабудочник широколистый 153473
Незабудочник якутский 153466
Нейлия китайская 263161
Нейлия тибетская 263172
Нейллия 263142
Нейллия метельчатая 263178
Неклен 3656
Неклён 3218
Неклен вирдинский 3218
Неклён вирдинский 3218
Нектарин 20955
Немезия 263374
Немецкая ромашка 21263
Немофила 263501
Ненюфар 267627
Неолидлейя 264006
Неолидлейя камчатская 182254
Неолицея 264011
Неолицея баотинская 264056
Неолицея гуандунская 264062
Неолицея гуансиская 264061
Неолицея ланьюйская 264112
Неолицея мэнлаская 264073
Неолицея пинбяньская 264084
Неолицея сычуаньская 264101
Неолицея сычуаньская гуншаньская 264104
Неолицея тибетская 264070
Неолицея хайнаньская 264052
Неолицея чжэцзянская 264024
Неолицея шелковистая 264090
Неопалассия 264248
Неопалассия гребенчатая 264249
Неоттианта 264757
Неоттианта клобучковая 264765
Непентес 264829
Непентесовые 264827
Непентовые 264827
Непета 264859
Неполнопыльник 43765
Нептуния 265228
Нерине 265254
Нертера 265358
Неслия 265550
Неслия метельчатая 265553
Неслия остроконечная 265552
Нетронь меня 205175
Не-тронь-меня 205175
неуструевия 317987
неуструевия каратавская 265834
Нефелиум 233077
Нефелиум китайский 233078
Нефелиум пенейниковый 265129
Нефорощь веничная 217361
Нехворощ 35090,35237
Неяснореберник 29085
Неяснореберник волосолистый 29086
Неяснореберник тонкстебельный 29087
Неяснореберник шероховатоплодный 29088
Нианца 382923
Нивяник 89402,227467
Нивяник круглолистный 89695,246399
Нивяник обыкновенный 227533
Нивяник поздний 227526
Нивяник посевной 89704
Нивяник сибирский 227528
Нивяник субальпийский 227532
Нигелла 266215
Нигелла дамасская 266225
Низкозонтичник 85696
Низкозонтичник бесстебельный 85697
низманка 21335
низманка малая 21394
Никандра 265940,265944
Никандра пузыревидная 116547
Никандра физалиевидная 116547
Никандра физалисовидная 116547,265944
Никандра физалисоподобная 265944
Никитиния 266263
Никитиния тонкостебельная 266264
Никтагиновые 267421
Никтантес 267429

Нимфейные 267787
Нимфея 267627
Нимфондес 267800
Нимфондес индийский 267819
Нимфондес корейский 267807
Нимфондес щитовидный 267825
Ниремберфия 266194
Нириум 265310
Ниселоз 343994
Нисса 267849
Нисса вэньшанськая 267876
Нисса жуйлиская 267863
Нисса китайский 267864
Нисса лесная 267867
Нисса шансиская 267862
Ниссовые 267881
Нителистник 166065
Нителистник сибирский 166066
Нитрария 266355
Новосиверрия 267237
Новосиверрия ледяная 267238
Ноголист 306609
Ноголистник 306609
Ноголистник пальчатый 306636
Ногоплодник 306395
Ногоплодник альпийский 306399
Ногоплодник высокий 306424
Ногоплодник нагея 306493
Ногоплодниковые 306345
Ноготки 66380
Ноготки аптечные 66445
Ноготки африканские 131147
Ноготки лекарственные 66445
Ноготки полевые 66395
Ноготки садовые 66445
Ногошок 66380
Нолана 266470
Нолина 266482
Нонея 266593
Нонея альпийская 266594
Нонея бледная 266618
Нонея вздудая 266627
Нонея дагестанская 266600
Нонея желтая 266608
Нонея желтеющая 266604
Нонея карсская 266607
Нонея каспийская 266597
Нонея крупножковая 266609
Нонея низбегающая 266601
Нонея промежуточная 266606
Нонея разноцветная 266628
Нонея розовая 266624
Нонея темнобурая 266623
Нонея темноплодная 266598
Нонея туркменская 266626
Нонея щетинистая 266625
Ноннея 266593
Нопыльник карликовый 43768
Норичник 355039
Норичник азербайджанский 355058
Норичник алтайский 355045
Норичник амгунский 355046
Норичник армянский 355057
Норичник беловетвистый 355170
Норичник бокоцветковый 355166
Норичник бюргера 355067
Норичник весенний 355265
Норичник водюной 355061
Норичник восточный 355208
Норичник вырезной 355145
Норичник ганьсуский 355151
Норичник гирканский 355143
Норичник гольде 355127
Норичник гончарова 355128
Норичник грея 355129
Норичник гроссгейма 355133
Норичник желтоязычковый 355269
Норичник жилковатый 355199
Норичник занãтоденский 355234
Норичник заравшанский 355274
Норичник зеленоцветковый 355088
Норичник золотистоцветковый 355089
Норичник зувандский 355275
Норичник ильвенсибский 355144
Норичник инееватый 355219
Норичник кабадианский 355147
Норичник какудский 355148
Норичник камнеломкоцветковый 355137
Норичник кирилова 355152
Норичник клювообразный 355226
Норичник кочи 355154
Норичник кровянокоцветковый 355134
Норичник крупнокистевой 355178
Норичник крылатый 355261
Норичник лёнецевый 355256
Норичник литвинова 355173
Норичник лунниколистный 355176
Норичник мабяньский 355183
Норичник максимовича 355148
Норичник малый 355189
Норичник маньчжурский 355181,355182
Норичник мелкоцветковый 355245
Норичник меловой 355091
Норичник многостебельный 355194
Норичник мутовчатый 355266
Норичник мягкий 355192
Норичник никитина 355200
Норичник обманчивый 355096
Норичник олимпийский 355207
Норичник ольги 355206
Норичник памироалайский 355209
Норичник пестрый 355264
Норичник полосатый 355249
Норичник рассеченный 355102
Норичник растопыренный 355103
Норичник рупрехта 355228
Норичник рутолистный 355229
Норичник седоватый 355073
Норичник скалистый 355227
Норичник скальный 355227
Норичник скополи 355236
Норичник собачий 355075
Норичник стеблеобъьмлющий 355047
Норичник таджикский 355253
Норичник тонкий 355114
Норичник туркменский 355260
Норичник узловатый 355202
Норичник федченко 355117
Норичник франше 355122
Норичник харадзе 355084
Норичник холодный 355124
Норичник цельнолистный 355146
Норичник чапандагский 355115
Норичник черняковской 355095
Норичник чужеземный 355216
Норичник чужестранный 355216
Норичник шишковатый 355202
Норичник шпренгера 355247
Норичниковые 355276
Норичникольдгема 355204
Носики 124141
Носный 325666
Нотобазис 267005
Нотобазис сурийский 267008
Нотофагус 266861
Нотофагус антарктический 266864
Ночецвет 252662
Ночецветник 255711
Ночецветные 267421
Ночная красавица 255711
Ноэа 266431
Ноэа гладкостебельная 266433
Ноэа маленькая 266434
Ноэа остронечная 266435
Нут 181872
Нут абиссинская 181873
Нут масличтый 181873
Нут 90797,90801
Нут анатолийский 90800
Нут балджуанский 90804
Нут бараний 90801
Нут джунгарский 90828
Нут длинноколючий 90819
Нут жакемонта 90815
Нут извилистый 90813
Нут колючий 90827
Нут копетдагский 90817
Нут коржинского 90818
Нут крошечный 90822
Нут культурный 90801
Нут мелколистный 90821

Нут моголтавский 90823
Нут обыкновенный 90801
Нут полевой 90801
Нут посевной 90801
Нут рогообразный 90801
Нут федченко 90812
Нут чечевицеобразный 90810
Нюнька 353587
Обабок 3127
Обвойник 291031
Обвойник греческий 291063
Обвойник заборный 291082
Облепиха 196743
Облепиха крушиновая 196757
Облепиха крушиновидная 196757
Облепиха обыкновенная 196757
Облепиха тибетская 196780
Обманчивоплодник 371237
Обманчивоплодник тонкий 371240
Обриеция 44867
Обыкновенная гуза 179900
Овения 277636
Овения вишнелистная 43053,277640
Овес 45363
Овёс 45363
Овес бесплодный 45586
Овёс бесплодный 45586
Овес бородатый 45389
Овёс бородатый 45389
Овес брунса 45417
Овёс венгерский 45533
Овес виеста 45621
Овёс византийский 45424
Овёс византийский красный 45424
Овес волосистолистный 45592
Овес волосистоузлый 45526
Овес волосисты 45542
Овёс восточный 45533
Овес высокий 34930
Овёс высокий 34930
Овес гиблидный 45481
Овес голый 198293
Овёс голый 198293
Овес двухгривый 45566
Овёс декоративный 45586
Овёс дикий 45448
Овёс дикий красный 45586
Овёс египетский 45533
Овес желтеющий 398518
Овёс желтеющий 398518
Овёс желтоватый 398518
Овёс живой 45448
Овес заячий 45558
Овес золотистый 398518
Овёс золотистый 398518
Овёс калючий 45602
Овес китайский 45431
Овёс короткий 45402
Овёс косой 45533
Овес крупноцветковый 45503
Овес луговой 45548
Овёс луговой 45548
Овес людовика 45495
Овёс мелкозёрный 198293
Овес обыкновенный 45566
Овес овсюг 45448
Овёс опушённый 45558
Овес опушенный 45558
Овес посевной 45566
Овёс посевной 45566
Овёс посевной средиземноморский 45424
Овес пустой 45448
Овёс пустой 45448
Овес пустынный 45439
Овёс пушистый 45558
Овес разноцветный 45619
Овес северный 45581
Овес сомнительный 45433
Овёс средиземноморский 45424
Овёс стерильный 45586
Овес сходный 45437
Овес татарский 198293
Овёс татарский 198293
Овёс тоший 45602
Овёс турецкий 45533
Овес тяжелый 45546
Овес щетинистый 45602
Овёс щетинистый 45602
Овес южный 45513
Овсец 190129
Овсец азиатский 190159
Овсец гиссарский 190158
Овсец даурский 190145
Овсец луговой 45548
Овсец монгольский 190175
Овсец опушённый 45448
Овсец опушеный 190187
Овсец тяньшанский 190199
Овсец федченко 190149
Овсюг 45363,45448,163778
Овсюг обыкновенный 45448
Овсяица алайская 163789
Овсяица алатавская 163791
Овсяица аметистовая 163811
Овсяица байкальская 163836
Овсяица беккера 163839
Овсяица боримюллера 163845
Овсяица бороздчатая 164126,164295
Овсяица влагалищная 164380
Овсяица восточная 163827
Овсяица выокая 164219
Овсяица высокая 163993
Овсяица ганешиная 163976,164295
Овсяица ганьсуская 164029
Овсяица гигантская 163993
Овсяица горная 164089
Овсяица даурская 163885
Овсяица джимильская 305490
Овсяица длинноостная 164058
Овсяица желобчатая 164295
Овсяица жестковатая 163915
Овсяица исполинская 163993
Овсяица каратабский 164030
Овсяица кирилова 164036
Овсяица колымская 164039
Овсяица коротколистная 163847
Овсяица красная 164243
Овсяица краузе 164040
Овсяица крыловская 164041
Овсяица ленская 164049
Овсяица лесная 58619
Овсяица ложнобольшая 164224
Овсяица ложно – бороздчатая 164226
Овсяица ложноовечья 164382
Овсяица ложно – овечья 164386
Овсяица луговая 163819,164219
Овсяица мелкочешйчатая 164329
Овсяица меловая 163876
Овсяица мышехвостная 412465
Овсяица мягчайшая 164086
Овсяица овечья 164126
Овсяица пестрая 164393
Овсяица песчаная 163816
Овсяица печальная 164367
Овсяица поднебесная 163871
Овсяица прекрасная 164394
Овсяица приземистая 163817
Овсяица пушистоцветковая 163947
Овсяица разнолистная 19251,164006,164280
Овсяица рупрехта 164141
Овсяица тростниковая 163819
Овсяица тростниковидная 163819
Овсяица тяньшанская 163791,164357
Овсяица ужковатая 163834
Овсяица холодолюбивая 164249
Овсяица чаюйская 163862
Овсяница 163778
Овсяница алтайская 163802
Овсяница дальневосточная 163950
Овсяница тайваньская 163969
Овсяница тубоватая 163807
Овсяница якутская 164023
Овсяный корень 394327
Огоньки 399494
Огурец 114108,114245
Огурец английский тепличный 114246
Огурец антильский 114122
Огурец африканский 114221
Огурец бугорочатый 114252
Огурец вест – индский 114122

Огурец волосистый 362460
Огурец волосистый угловатый 362461
Огурец грядовой 114245
Огурец дыня 114189
Огурец змеевидный 396151
Огурец змеиный 396151
Огурец иккимский 114250
Огурец мексаканский 356352
Огурец огородный 114245
Огурец посевной 114245
Огурец серповидный 114247
Огурец угроватый 362461
Огурец чешуйчатый 114251
Огуречная трава 57095
Огуречник 57095
Огуречник аптечный 57101
Огуречник лекарственный 57101
Огурчик пёсий виноградолистный 256797
Одногнездка 29164
Одногнездка мелкоплодный 29172
Одногнездка обвернутая 29167
Однопокровник 257683
Однопокровник азиатский 257684
Однопокровник литвинова 257686
Одноцветка 257371
Одноцветка крупноцветковая 257374
Одноцветка крупноцветная 257374
Одноцветка одноцветковая 257374
Одноягодник 284289
Одуванчик 384420,384683,384766
Одуванчик алайский 384425
Одуванчик алатавский 384427
Одуванчик алпийский 384437
Одуванчик алтайский 384438
Одуванчик андросова 384443
Одуванчик аптестый 384714
Одуванчик армянский 384451
Одуванчик ашхабадский 384452
Одуванчик бадахшанский 384647
Одуванчик байсальский 384524
Одуванчик балтийский 384466
Одуванчик безрогий 384529
Одуванчик беловатоцветковый 384516
Одуванчик беловатый 384429,384516
Одуванчик белоцветковый 384628
Одуванчик бесканальцевый 384547
Одуванчик бессарабский 384467
Одуванчик богатомлечный 384736
Одуванчик болотный 384824
Одуванчик большой 384655
Одуванчик боровой 384739
Одуванчик бочанцева 384474
Одуванчик бугорчатый 384714
Одуванчик варзобский 384849
Одуванчик васильченко 384850
Одуванчик воронова 384854
Одуванчик вулканный 384855
Одуванчик гладгий 384560
Одуванчик гладкоплодный 384637
Одуванчик гнездилло 384634
Одуванчик голоскокова 384564
Одуванчик горный 384684,394320
Одуванчик гроссгейма 384567
Одуванчик грязно－желтый 384647
Одуванчик двурогий 384472
Одуванчик джунгарский 384812,394346
Одуванчик длинноножковый 384642
Одуванчик длинноносиковый 384643
Одуванчик длиннопирамидковый 384643
Одуванчик длиннорогий 384640
Одуванчик жирный 384740
Одуванчик загнутый 384778
Одуванчик зеравшанский 384797
Одуванчик известняковый 384485
Одуванчик иконникова 384647
Одуванчик индийский 384587
Одуванчик ировидный 384649
Одуванчик йетрофуйский 384864
Одуванчик камчатский 384596
Одуванчик капустнолистный 384479
Одуванчик каратавский 384599
Одуванчик кетойский 384601
Одуванчик киргизский 384602
Одуванчик кирпичный 384621
Одуванчик китайский 384473
Одуванчик клокова 384604
Одуванчик ключевой 384553
Одуванчик ключевской 384603
Одуванчик койнмы 384605
Одуванчик кок－сагыз 384606
Одуванчик корагинец 384608
Одуванчик корагинский 384607
Одуванчик корейский 384505
Одуванчик коротконосиковый 384482
Одуванчик короткорогий 384480
Одуванчик короткорожковый 384606
Одуванчик косой 384702
Одуванчик красноватый 384782
Одуванчик красноплодный 384540
Одуванчик красносемянный 384540
Одуванчик крупнолисточковый 384565
Одуванчик крупночешуйный 384653
Одуванчик крупный 384654
Одуванчик крымский 384830
Одуванчик лапландский 384620
Одуванчик лекарственный 384714
Одуванчик лиловый 384633
Одуванчик липского 384636
Одуванчик лицзянская 384515
Одуванчик ложноальпийский 384763
Одуванчик ложноголый 384767
Одуванчик ложномелколопастный 384769
Одуванчик ложнопочерневший 384764
Одуванчик ложнорозовый 384772
Одуванчик ложночерноватый 384770
Одуванчик луговой 384756
Одуванчик малеза 384656
Одуванчик маленький 384737
Одуванчик мексиканский 384669
Одуванчик мелколоастный 384676
Одуванчик мелкосемянный 384674
Одуванчик мелкоцветковый 384732
Одуванчик меркорогий 384475
Одуванчик миуаке 384678
Одуванчик многокорзиночный 384685
Одуванчик монгольский 384681
Одуванчик мохнатопазушный 384534
Одуванчик мурманский 384688
Одуванчик наиринский 384689
Одуванчик начикский 384690
Одуванчик невского 384693
Одуванчик немногоцветковый 384720
Одуванчик непохожий 384526
Одуванчик неравнобокий 384702
Одуванчик никитина 384695
Одуванчик новаземельский 384698
Одуванчик новосахалинский 384692
Одуванчик норвежский 384697
Одуванчик нуратавский 384699
Одуванчик обесцвеченный 384518
Одуванчик обыкновенный 384714
Одуванчик ови 384719
Одуванчик однопокровный 384683
Одуванчик опутанный 384503
Одуванчик осенний 384569
Одуванчик остробугорчатый 384687
Одуванчик ошский 384721
Одуванчик памирский 384729
Одуванчик плосколисточковый 384749
Одуванчик плоскорогий 384747
Одуванчик плоскочешуйный 384748
Одуванчик поздний 384799
Одуванчик порярковой 384752
Одуванчик почерневший 384462
Одуванчик почти－голый 384559
Одуванчик предкавказский 384499
Одуванчик приледниковый 384823
Одуванчик принтца 384757
Одуванчик промежутоный 384590
Одуванчик пурпурный 384753
Одуванчик пустынный 384521
Одуванчик ранний 384755
Одуванчик рассеченнолистный 384524
Одуванчик рассеченный 384524
Одуванчик рогарый 384714
Одуванчик рогоносный 384714
Одуванчик рыжеватый 384784
Одуванчик самаркандский 384660
Одуванчик сахалинский 384785
Одуванчик светрый 384523
Одуванчик сибирский 384805

Одуванчик сизо－зеленый 384562
Одуванчик сизоцветковый 384473
Одуванчик симуширский 384803
Одуванчик сирийский 384827
Одуванчик скердоподобный 384507
Одуванчик скученный 384501
Одуванчик сложенный 384502
Одуванчик снежный 384696
Одуванчик станюковича 384814
Одуванчик стевена 384819
Одуванчик стрелолистный 384789
Одуванчик стреролистный 384788
Одуванчик сумневича 384825
Одуванчик сухолюбивый 384859
Одуванчик тадшикский 384828
Одуванчик тайваньский 384554
Одуванчик татеваки 384829
Одуванчик тибетский 384833
Одуванчик тонкорассеченный 384831
Одуванчик тонкорокий 384627
Одуванчик тоший 384650
Одуванчик тундровый 384837
Одуванчик тургайский 384839
Одуванчик турийский 384840
Одуванчик туюксуйский 384836
Одуванчик тяньцюаньский 384446
Одуванчик тяньшанский 384832
Одуванчик удлиненный 384533
Одуванчик узколопастный 384818
Одуванчик узкочешуйный 384817
Одуванчик уссурийский 384846
Одуванчик федченко 384551
Одуванчик хирийский 384498
Одуванчик холмовой 384500
Одуванчик хультена 384581
Одуванчик хьелта 384576
Одуванчик цвелева 384841
Одуванчик цельнолистный 384683
Одуванчик чернеющий 384694
Одуванчик черноватый 384460
Одуванчик чуйский 384513
Одуванчик шафранный 384508
Одуванчик шерковникова 384792
Одуванчик шерстистоногий 384536
Одуванчик шикотанский 384802
Одуванчик шимшинский 384804
Одуванчик широкочешуйный 384624
Одуванчик шишкоголовый 384820
Одуванчик штгнанский 384794
Одуванчик юзенпчука 384595
Одуванчик ядунский 384677
Одуванчик ямамото 384861
Одуванчик японский 384591
Одурь сонная 44708
Ожерельник 274370
Ожерельник туркменский 274372
Ожика 238563
Ожика бледнеющая 238675
Ожика бледноватая 238675
Ожика боровая 238663
Ожика валенберга 238717
Ожика волосистая 238681
Ожика головчатая 238600
Ожика дубравная 238663
Ожика дуговидная 238572
Ожика изогнутая 238572
Ожика камчадальская 238573
Ожика колосистая 238697
Ожика крупноплодная 238691
Ожика лесная 238709
Ожика ложносудетская 238689
Ожика малоцветковая 238671
Ожика мелкоцветковая 238678
Ожика многоцветковая 238647
Ожика неясная 238610
Ожика равнинная 238583
Ожика рыжеватая 238690
Ожика рыжеющая 238690
Ожика сибирская 238694
Ожика сычуаньская 238695
Ожика слабоволосистая 238702
Ожика снеговая 238667
Ожика спутанная 238610
Ожика судетская 238706
Ожика тайваньская 238712
Ожика тибетская 238631
Ожика форстера 238619
Ожика холодная 238650
Ожика черноплодная 238645
Ожика чъельмана 238637
Ожика японская 238629
Ойдибазис 269274
Ойдибазис бутеневый 269276
Ойдибазис каратавский 269277
Ойдибазис остроконечный 269275
Ойдибазис плоскоплодный 269278
Окопник 381024
Окопник аптечный 381035
Окопник богемский 381027
Окопник бродячий 381038
Окопник восточный 381037
Окопник жесткий 381026
Окопник жёсткий 381026
Окопник кавказский 381029
Окопник клубневой 381040
Окопник клубненостный 381040
Окопник крупноцвтковый 381031
Окопник крымский 381039
Окопник лекарственный 381035
Окопник острый 381026
Окопник сердцевиднолистный 381030
Окопник сердцевидный 381030
Окопник шероховатый 381026
Окопник шершавый 381026
Окотеа 268688
Окра 219
Оксалис 277644
Оксиграфис 278466
Оксиграфис ледниковый 278473
Оксиграфис обыкновенный 278481,325898
Оксиграфис шамиссо 278467
Оксидендрон 278388
Оксидендрум древовидный 278389
Олеандр 265310
Олеандр африканская 270100
Олеандр душистый 265327
Олеандр европейская 270099
Олеандр индийский 265327
Олеандр обыкновенный 265327
Олеария 270184
Олива 270058,270099
Олива гуансиская 270126
Олива хайнаньская 270128
Оливка 270099
Оливковое дерево 270099
Оливник 97491
Олигохета 270297
Олигохета войлочная 270302
Олигохета мелкая 270301
Олигохета растопыренная 270300
Ольгея 270227
Ольгея бальджуанская 270228
Ольгея введенского 270249
Ольгея войлочногнездная 270241
Ольгея войлочноногая 270233
Ольгея гребенчатая 270243
Ольгея длиннолистная 270240
Ольгея колючконосная 270246
Ольгея снежная 270242
Ольгея шерстистоглавая 270230
Ольдфильдия африканская 270054
Ольха 16309
Ольха белая 16368
Ольха бородатая 16320
Ольха ботта 34126
Ольха востная 16463
Ольха горная 16348,16472
Ольха зеленая 16472
Ольха зибольда 16458
Ольха камчатская 16398
Ольха клейкая 16352
Ольха кольская 16399
Ольха красильная 16466
Ольха красильная голая 16469
Ольха красная 16429,16437
Ольха кустарная 16348
Ольха кустарниковая 16348
Ольха липкая 16352
Ольха максимовича 16328
Ольха маньчжуърская 16401
Ольха морщинистая 16440

Ольха непальская 16421
Ольха повислая 16433
Ольха приморская 16404
Ольха серая 16368
Ольха сердсевидная 16323
Ольха сердселистная 16463
Ольха сибирская 16362
Ольха тайваньская 16345
Ольха фердинанда кобурга 16340
Ольха черная 16352
Ольха чёрная 16352
Ольха шерстистая 16358
Ольха штейпа 34343
Ольха японская 16386
Ольхолистник форчуна 16295
Омег 101845
Омег болотный 90932
Омег пятнистый 90924,101852
Омежник 269292
Омежник абхазский 269293
Омежник бедренецовидный 269348
Омежник водный 269299,269347
Омежник водяной 269299,269347
Омежник длиннолисточковый 269338
Омежник дучатый 269319
Омежник китайский 269360
Омежник лежачий 269326
Омежник морковниколистный 269358
Омежник мэнцзыский 269337
Омежник софии 269362
Омежник трудубчатый 269319
Омежник федченко 269316
Омежник шафранный 269307
Омежник шафрановый 269307
Омежник шафраноподобный 269307
Омежник яванская 269326
Омела 410959
Омела белая 410960
Омела белая дубянка 410960
Омела евронейская 410960
Омела окрашенная 410992
Омела тайваньский 410972
Оморика 298387
Омфалодес 117900,270677,275347
Омфалодес линейнолистный 270705
Омфалодес настоящий 270723
Омфалотрикс 270757
Омфалотрикс длинножковый 270758
Омфалотрикс длинноножковый 270758
Онагра 269394
Онагровые 270773
Онопордон 271663
Онопордон колючий 271666
Онопордон тонкочешуйный 271693
Онопордум 271663
Онопордум армянский 271677
Онопордум зеравшанский 271701
Онопордум колючий 271666
Онопордум крымский 271705
Онопордум обыкновенный 271666
Онопордум пряхина 271700
Онопордум разноколючковый 271691
Онопордум серый 271680
Онопордум снежно－белый 271679
Онопордум фрика 271690
Оносма 271727
Оносма армянская 271734
Оносма бальджуанская 271737
Оносма барщевского 271739
Оносма белоплодная 271788
Оносма бугорчатоплодная 271838
Оносма визиани 271843
Оносма гмелина 271772
Оносма длиннозубчатая 271795
Оносма жесткая 271820
Оносма зауральская 271825
Оносма изящная 271773
Оносма кавказская 271745
Оносма коротковолосистая 271742
Оносма красильная 271837
Оносма крупнокорневая 271800
Оносма крымская 271835
Оносма куэньминская 271841
Оносма ливанова 271793
Оносма лицзянская 271789
Оносма мелкоплодная 271805
Оносма мелкоцветковое 271759
Оносма многолистная 271817
Оносма многоцветная 271816
Оносма песчаная 271733
Оносма простая 271827
Оносма простейшая 271827
Оносма раздражающая 271785
Оносма расширенная 271731
Оносма самаркандская 271801
Оносма сине－голубая 271736
Оносма скальная 271821
Оносма тибетская 271846
Оносма тычиночная 271830
Оносма ферганская 271766
Оносма чашечная 271743
Оносма чаойская 271849
Оносма черно－синяя 271735
Оносма чжэньканская 242400
Оносма шелковистая 271823
Оносма щетинистая 271824
Онцидиум 270787
Оплисменус 272611
Оплисменус коротковолосистый 272614
Оплисменус фуцзяньский 272645
Оплисменус японский 272693
Опопанакс 272730
Опопанакс армянский 272731
Опунция 272775
Опунция бассия 241514
Опунция бразильская 272818
Опунция индийская 272891
Опунция индийская фига 272891
Опунция обыкновенная 272987,273103
Опунция одноколючковая 272987
Оранта 275463
Оранта желтая 275467
Оранта оше 275466
Ореорхис 274033
Ореорхис индийский 274047
Ореорхис раскидистый 274058
Орех 212582
Орех австралийский 240210
Орех айлантолистный 212626
Орех американский 53057,212631
Орех арековый 31680
Орех белый 215442
Орех болотный 394500
Орех большой 212620
Орех бразильский 53057
Орех водяной 394500
Орех водяной плавающий 394500
Орех волошский 212636
Орех волшеб 185092
Орех горький 77889
Орех грецкий 212636
Орех заячий 20931
Орех земляной 30498
Орех земляной бамбарский 409069
Орех зибольда 212626
Орех индейский 14544
Орех кавказский 320356
Орех калифорнийский 212592
Орех кешью 21195
Орех китайский 212595,212636
Орех кокосовый 98136
Орех кола 99158
Орех лесной 106696
Орех лесной американский 106703
Орех ломбардский 106757
Орех мадвежий 106720
Орех малдивский 235159
Орех маньчжурский 44867,212621
Орех масляный 212599
Орех мускатный 261424
Орех мускатный фальшивый 261423
Орех мускатный ямайский 257653
Орех мушкатный 261424
Орех насмешливый 77934
Орех нищих 31680
Орех обманчивый 212636
Орех пекан 77902
Орех подземный 30498
Орех понтийская 106720
Орех рвотный 378633,378834,378836
Орех савара 77946

Орех сердцевидный 212602,212624
Орех серый 212599
Орех скальный 212646
Орех турецкий 90801
Орех хайндса 212612
Орех хэбэйский 212614
Орех черный 212631
Орех чёрный 110003,212631
Орех чёрный восточный 110003,212631
Орех чилийский 14544
Орех японский 212602,212624
Орехи кедровые 80074
Ореховые 212581
Орешки земляные 166132
Орешник 106696,106703
Орешник зибольда 106777
Орешник индийский 214308
Орешник китайский 25762,106716
Орешник лесной 106703
Орешник маньчжурсный 106751
Орешник медвежий 106720
Орешник обыкновенный 106703
Орешник понтийский 106703
Орешник проносный 113039
Орешник проностый 112829
Орешник разнолистная 106736
Орешник рвотный 378834
Орешник－кола 99158
Орешниковые 106623
Оризопсис 275991
Орлайя 274286
Орлайя крупноцветковая 274288
Орлайя плоскоплодная 274287
Орлеан 54862
Орлики 29994
Орлики обыкновенные 30081
Ороксилум 274495
Оронтиум 275401
Оростахис 275347
Оростахис колючий 275394
Ортодон 259278
Ортодон крупнопильчатый 259292
Ортосифон 275295,275624
Ортосифон хайнаньский 275761
Орхидейные 275781
Орхидея－пафиния 273317
Орхидные 120287,273267
Осенник 99293,273267
Осенчук 99297
Осина 22476,311537
Осина американская 311537
Осиновик 311547
Ослинник 269394
Ослинник двулетний 269394,269404
Ослинник двухлетний 269404
Ослинник короткоиглый 269404
Османтус 276242
Османтус душистый 276291
Османтус люцюский 276268
Османтус мэнцзыский 276312
Османтус наньчуаньский 276374
Османтус оранжевый 276292
Османтус разнолистный 276313
Османтус хубэйский 276323
Османтус эмэйский 276381
Осмориза 276450
Осмориза остистая 276455
Осока 73543
Осока аа 73577
Осока августиновича 73826
Осока алайская 73620
Осока алексеенковская 73634
Осока алнелли 73767
Осока алнелля 73767
Осока алтайская 73662
Осока альпийская 73652
Осока амгунская 73673
Осока амурская 74394
Осока ангарская 73690
Осока аоморийская 74029
Осока аргунская 73750
Осока аркатская 73740,75618
Осока арнелля 73767
Осока безжилковая 74453
Осока безносиковая 73770
Осока белая 73622
Осока беловатая 74004
Осока белозаостренная 73814,73815
Осока береговая 15779,76044
Осока бледная 75675
Осока бледноватая 75672
Осока бледнозеленая 75131
Осока блестящая 75176
Осока блошиная 75929
Осока богемская 73894
Осока болотолюбивая 74781
Осока болошехвостая 75276
Осока большеголовая 75259
Осока большеплодная 74333
Осока большехвостая 75724
Осока бонанцинская 73898
Осока бордзиловского 73903
Осока бротертсов 73942
Осока буека 73971
Осока буксбаума 76490
Осока буреющая 74884
Осока буроватая 73959
Осока буровлагалищная 74622
Осока бурятская 73975
Осока бухбрская 73969
Осока ван－хьюрка 76665
Осока верещатниковая 74461
Осока весенняя 76673
Осока вздутая 74878,75782
Осока вздутоносая 76029
Осока вилюйская 76760
Осока висячая 75742
Осока влагалищная 76656
Осока водяная 73732
Осока возвратившаяся 76015
Осока возрастающая 75675
Осока волосистая 74806,75785
Осока волосистоплодная 75065
Осока волосная 74023
Осока волосовидная 74032
Осока ворсистая 74806,75785
Осока высокая 74444
Осока высокогорная 73667
Осока вытянутая 75878
Осока выхолощеная 74763
Осока выхолощенная 74763
Осока вэньшаньская 76751
Осока галечная 74654
Осока галлера 74746
Осока галлеровская 74745
Осока ганьсуская 74973
Осока гартмана 74763
Осока гвоздичная 74059
Осока гиперборейская 73875
Осока гладконосая 75101
Осока гладкостбельная 75038
Осока гладчайшая 75042
Осока глена 74668
Осока глянцевитобурая 76372
Осока гмелина 74676
Осока головковидная 75610
Осока головчатая 74046
Осока гололюбивая 75623
Осока горная 75409
Осока городкова 74684
Осока госта 74830
Осока гриолети 74714
Осока гриффитса 74713
Осока гросгейма 74718
Осока грэхэма 74696
Осока грязная 76301
Осока гуандунская 73603
Осока гуансиская 75026
Осока гуднозова 74435
Осока гуншаньская 74680
Осока густоколосая 75941
Осока далиская 76083
Осока даурская 74253
Осока двсмысленная 73608
Осока двудомная 74330
Осока двурядная 74354
Осока двусеменная 74342
Осока двутычинковая 74304
Осока двуцветная 73873,74307,74337
Осока двуцветноколосковая 74308
Осока дернистая 73987

Осока джунгарская 76299
Осока длинная 74442
Осока длиннокорневищная 74118
Осока длинноплодная 74372,74374
Осока длипнноклювая 75189
Осока дэвелла 74258
Осока дэциньская 74295
Осока есолюбивая 74392
Осока желтая 74557
Осока жёлтая 74557
Осока желтенькая 74568
Осока женосильная 74722
Осока жестковадая 74407
Осока жилковатая 75496
Осока жилкоплодная 75500
Осока завитая 74138
Осока загнутая 76010
Осока заливная 74913
Осока заостренная 74244
Осока затененная 74907
Осока заячья 75116
Осока звездчатая 76342
Осока зелененькая 76701
Осока зеленоватая 76100
Осока зеленоколосая 74032
Осока зеленоколосковая 74032
Осока зеленохолосая 74111
Осока зигерта 76273
Осока зонтиковидная 76446
Осока извилисторыльцевая 73906
Осока ильина 74867
Осока индийская 74875
Осока ичанская 73774
Осока кабанова 74969
Осока кавказская 74068
Осока калелина 74323
Осока каменистая 76153
Осока каменная 76153
Осока камнелюбивая 75178
Осока камчатская 74972
Осока камышевидная 76185
Осока карелина 74977
Осока каро 74974,74982
Осока каттегатская 74986
Осока кенкольская 74992
Осока кирганикская 74998
Осока кирилова 75000
Осока китайская 74098
Осока клопоносная 74187
Осока ключелюбивая 76230
Осока кнорррнг 75002
Осока кобомуги 75003
Осока кожевникова 75013
Осока козерогая 74051
Осока колосистая 76328
Осока колхидская 74148
Осока корагинская 75010
Осока корейская 74309,76515
Осока коржинского 75012
Осока корневищная 75724
Осока коротковолосистая 74806,75785
Осока коротконожчатая 75362
Осока кочи 75015
Осока красивенькая 75928
Осока красная 73987
Осока красовлас 73992
Осока кривоносая 75785
Осока кровяноколосая 74726
Осока кругловатая 76073
Осока круграя 74672
Осока крупноколосковая 75268
Осока крупноостная 75265
Осока крупнорыльцевая 75275
Осока курайская 74230
Осока курчавая 76628
Осока курчаворыльцевая 73906
Осока ланцетная 75048
Осока лапландская 75062
Осока лапчатая 75722
Осока ледебуровская 75096
Осока ледниковая 74652
Осока лесная 76276,76465
Осока листоколосая 75779
Осока лисья 76716
Осока литвинова 75180
Осока лицзянская 74310
Осока ложнобереговая 76047
Осока ложно – вонючая 75899
Осока ложножесткая 76042
Осока ложнокоротконожчатая 75363
Осока ложно – курайская 75894
Осока ложнолансетная 75906
Осока ложно – острая 73597
Осока ложноплевельная 75910
Осока ложно – сизая 74664
Осока ложносыть 75895
Осока ложнотвердоватая 74408
Осока ложноузколистная 76363
Осока ложно – черноцветковая 75328
Осока лоснящаяся 75176
Осока люмницера 75230
Осока маака 75252
Осока макензи 75253
Осока максимовича 75314
Осока малая 75709
Осока малоприцветниковая 76418
Осока малорослая 75930
Осока малоцветковая 75711
Осока маньчжурская 75298
Осока медведева 75323
Осока медвежья 76650
Осока медная 75641
Осока меднобурая 74621
Осока мейера 75348
Осока мейнсгаузена 75325
Осока мелкая 75378
Осока мелковолосистая 75369
Осока мелкоголовчатая 74049
Осока мелкоостенник 75361
Осока мелкоцветковая 75711
Осока метельчатая 75686
Осока мечевидная 76773
Осока мечелистная 73875
Осока мешечатая 76653
Осока мешочковая 76653
Осока миддендорфа 75370
Осока мийябе 75396
Осока микели 75355
Осока многолистная 75849
Осока морская 75303,75304
Осока мортуковидная 74460
Осока морщинистая 76090
Осока мотоская 75429
Осока мохнатая 74806,75785
Осока мочежинная 76626
Осока мужененавистническая 75387
Осока мулиская 75438
Осока мягковатая 75400
Осока мягчайшая 75403
Осока наньчуаньская 75474
Осока невровская 75490
Осока нежилкоплодная 73688
Осока немногоколосковая 75595
Осока немногоплоданая 76306
Осока немногораздвинутая 76005
Осока немногоцветковая 75588
Осока необильная 75719
Осока неполная 76222
Осока неравножилковатая 73702
Осока несчаная 76098
Осока неясноустая 73602
Осока низенькая 74849
Осока низкая 74842
Осока нисходящая 74296
Осока новограбленова 75541
Осока ножконосная 75734
Осока носадая 76068
Осока обедненная 74291
Осока обертковидная 76437
Осока обильнолистная 74584
Осока обыкновенная 74681
Осока округлая 75618
Осока оливера 75601
Осока омианская 75604
Осока омская 74435,75608
Осока оне 75610
Осока острая 73589
Осока остриеносная 75431
Осока островатая 73597
Осока островерхая 75649
Осока остроконечная 75451

Осока остролистная 73587
Осока открытая 73714
Осока пальчатая 75954
Осока памирская 75678
Осока параллельная 75697
Осока парвуская 73912
Осока пейктусанская 75737
Осока перенончатая 75338
Осока песколюбивая 73748
Осока пестрая 74337
Осока песчаная 73747,75782
Осока печальная 76595
Осока пильчатонлодная 74221
Осока плевельная 75185
Осока плоскостебельная 75827
Осока повислая 75880
Осока ползучая 76008
Осока полувздттая 76435
Осока поморская 75304,75696
Осока понтийская 75855
Осока прерванная 74366
Осока придатконосная 76547
Осока приземистая 74842,76456
Осока приморская 75303,75696
Осока притупленная 75570
Осока просовидная 75684
Осока просяная 75684
Осока прямая 76338
Осока прямоколосая 73782
Осока прямостоящая 76338
Осока пузыревая 76687
Осока пузырчатая 76677
Осока пустынная 75664
Осока рабушинского 76031
Осока радде 74463,75957
Осока раздвинутая 75993
Осока раздельная 74360
Осока разноцветная 74358
Осока разночешуйчатая 74797
Осока раменского 75970
Осока ранняя 75859
Осока расставленная 74348
Осока расставлённая 74348
Осока растянутая 74481
Осока расходящаяся 74339
Осока регеля 75991
Осока редкоцветковая 75980
Осока редовского 75989
Осока ресничатоплодная 73888
Осока ресчаная 75782
Осока ржавопятнистая 76264
Осока ржаная 76215
Осока русская 76096
Осока рыхлая 75080
Осока рянг 75782
Осока садоанская 76132
Осока сарептская 76145
Осока саянская 76134
Осока сближенная 73729
Осока светлая 74323
Осока свинцовая 75839
Осока свинцово－зеленая 75184
Осока своеобразная 73729
Осока сдвинутая 73730
Осока седакова 76219
Осока семноцветная 75326
Осока серебристочешуйная 73753
Осока сероватая 74004,74135,75357
Осока серозеленая 74004
Осока серповидная 74483
Осока сжатая 74158
Осока сизая 74656
Осока сикокийская 76126
Осока ситничек 74967
Осока сичоуская 76262
Осока сычуаньская 76461
Осока скальная 76094
Осока скрывающаяся 75572
Осока скрытоплодная 74221
Осока скученная 74360
Осока слабошероховатая 75382
Осока слободова 76292
Осока смирнова 76090
Осока сойская 73826
Осока солонцовая 74323
Осока сочавы 76298
Осока спутанная 76710
Осока средняя 74354
Осока степная 74842
Осока стесненная 76375
Осока стигийская 76412
Осока столбиконосная 76413
Осока стоповидная 75724
Осока стройная 73589
Осока струнокоренная 74118
Осока суженная 73696,74423
Осока суйфунская 75317
Осока сукачева 74685
Осока сухощавая 76410
Осока сучанская 75724
Осока сытевидная 73894
Осока тайваньская 73877
Осока талдынская 76482
Осока твердоватая 74406
Осока темная 73788
Осока тёмная 73788
Осока темнобурая 73792,73814
Осока темноодетая 73788
Осока теневая 76635
Осока титова 76552
Осока толстоколосая 75662
Осока толстолбиковая 75664
Осока тоненкая 74453
Осока тонкоцветковая 76513
Осока тонкоцветная 76513
Осока топяная 75165
Осока траизискинская 76570
Осока траутфеттера 76574
Осока трехраздельная 76585
Осока трясунковидная 73940
Осока туминганская 76618
Осока тунберга 76545
Осока тупоносая 73668
Осока туркестанская 76622
Осока тяньшанская 76550
Осока удлиненая 74442
Осока удлинённая 74442
Осока узколистная 76354
Осока узкоплодная 76350
Осока узкочешуйная 74696
Осока узонская 76654
Осока унгурская 76641
Осока уплодненная 74182
Осока уральская 74408,76354,76647
Осока уссурийская 76651
Осока уховертка 74592
Осока федийская 74491
Осока федчанко 74495
Осока фуджиты 74606
Осока фунинская 74614
Осока хайнаньская 74731
Осока хаккоеская 74733
Осока хинганская 74101
Осока холодная 75387
Осока хэбэйская 76485
Осока хэнкока 74752
Осока хэпберна 74791
Осока цельноротая 74818
Осока цельноустая 74818
Осока цилиндрическая 74763
Осока цинхайская 75948
Осока чарвакская 74250
Осока чащевная 74856
Осока чернеющая 75511
Осока чернобурая 73814
Осока черноголовая 75332
Осока черноколосковая 75333
Осока черноплодная 75330
Осока черноустая 75334
Осока черноцветковая 75326
Осока четырехцветковая 75954
Осока чешуеплодная 75115
Осока чжэнькканская 76793
Осока чозенская 74121
Осока чэнкоуская 75227
Осока шабинская 76100
Осока шамисо 74089
Осока шариконосная 75788
Осока шероховатая 73775
Осока шероховатожилковатая 76165
Осока шерстистая 75065

Осока шершаволистная 76163
Осока шершавошиноватая 76162
Осока широколистная 75078
Осока широконосая 75834
Осока широкочешуйная 75079
Осока шмидта 76178
Осока шовица 76469
Осока шребера 75859
Осока щетинистая 76239
Осока эвксинская 74476
Осока эдера 75575,76230
Осока элевезиновидная 74438
Осока эмэйская 75602
Осока юньнаньская 76789
Осока юэта 74838
Осока якутская 74932
Осока ялусская 74934
Осока янковского 74937
Осока японсая 74941
Осока ячменерядная 74826
Осоковые 118403
Осокорь 311398
Осокорь Тополь 311398
Осот 368625
Осот болотный 368781
Осот желтый 368771
Осот жёлтый 368635
Осот жёский 368649
Осот жесткий 368649
Осот закаспийский 368838
Осот колючий 368649
Осот красный 92384
Осот овошной 368771
Осот овыщной 368771
Осот огородный 368771
Осот острый 368649
Осот полевой 91770,368635
Осот розовый 91770
Осот сосновского 368813
Осот шероховатый 368649
Остерикум болотный 24429
Острец 11841
Острица 39298
Острица лежачая 39299
Острица простёртая 39299
Островския 61398,276795
Островския величественная 276796
Острокильница 120903
Острокильница чернеюшая 120983
Остролист 54798,203527,203545
Остролодка 278679
Остролодочник 278679
Остролодочник адамса 278686
Остролодочник алайский 278690
Остролодочник албанский 278691
Остролодочник алтайский 278701
Остролодочник альпийский 278700
Остролодочник анадырский 278706
Остролодочник арктический 278716, 279170
Остролодочник ассийский 278724
Остролодочник астрагаловидный 278725
Остролодочник аянский 278689
Остролодочник байсунский 278733
Остролодочник бальджуанский 278734
Остролодочник белла 278739
Остролодочник беловатый 278768
Остролодочник беловолосистый 278965
Остролодочник беломохнатый 278694
Остролодочник белоснежный 279039
Остролодочник белоцветковый 278693
Остролодочник бесчисленнолистнный 279029
Остролодочник бледножелтый 279055
Остролодочник бледноцветковый 278961
Остролодочник блестящий 279037
Остролодочник ближайший 279088
Остролодочник близкий 278715
Остролодочник бобоножковый 279080
Остролодочник боброва 278746
Остролодочник богушский 278748
Остролодочник вздутоносый 279126
Остролодочник вздутоплодный 279074
Остролодочник войлочный 279211
Остролодочник волосистоплодный 278824
Остролодочник волосистый 279076
Остролодочник восходящий 278687
Остролодочник высокогорный 278698
Остролодочник галечниковый 278852
Остролодочник ганьсуский 278919
Остролодочник геблера 278839
Остролодочник гмелина 278855
Остролодочник головчатый 279184
Остролодочник голопестичный 278868
Остролодочник голубиный 278797
Остролодочник голубой 278757
Остролодочник голый 278842
Остролодочник горбунова 278857
Остролодочник грязноватый 279170
Остролодочник грясный 279170
Остролодочник густоволосистый 279077
Остролодочник двулопастной 278745
Остролодочник двуцветный 278814
Остролодочник дерновинный 278760
Остролодочник джунгарский 279169
Остролодочник длиннокистевой 278983
Остролодочник длинноногий 278978
Остролодочник длинноприцветничный 278974
Остролодочник длинносый 278979
Остролодочник дологостайского 278818
Остролодочник ельниковый 279075
Остролодочник железистый 278851
Остролодочник жесткий 278723
Остролодочник жестковолосистый 278882
Остролодочник жестковолосый 278881
Остролодочник заалайский 279213
Остролодочник завернутый 279122
Остролодочник зеравшанский 279155
Остролодочник золотистожелтый 278726
Остролодочник иглистый 278899
Остролодочник ипполита 278880
Остролодочник кавасимский 278924
Остролодочник кавказский 278773
Остролодочник казбека 278922
Остролодочник камчатский 278916
Остролодочник капю 278769
Остролодочник каратавский 278920
Остролодочник карликовый 279106
Остролодочник карпатский 278772
Остролодочник карягинб 278921
Остролодочник кетменский 278925
Остролодочник колокольчатый 278763
Остролодочник колосовый 279171
Остролодочник колючий 278680
Остролодочник колючконосный 279172
Остролодочник коморова 278926
Остролодочник копетдагский 278930
Остролодочник коротколлодный 278750
Остролодочник короткостебельный 278754
Остролодочник красивый 278738
Остролодочник красноглинный 279133
Остролодочник красностержневой 279134
Остролодочник крупнозубый 278985
Остролодочник крупноплодный 278984
Остролодочник крупнохоботковый 279002
Остролодочник крупноцветковый 278862
Остролодочник крылова 278934
Остролодочник кубанский 278935
Остролодочник кузнецова 278938
Остролодочник кустарничковый 278837
Остролодочник ладыгина 278939
Остролодочник лазистанский 278956
Остролодочник лапландский 278948
Остролодочник лемана 278957
Остролодочник лесно 279195
Остролодочник летниковый 278854
Остролодочник лиловорозовый 279132
Остролодочник линчевского 278968
Остролодочник липского 278970
Остролодочник литвинова 278972
Остролодочник ложнохолодостойкий 279097
Остролодочник лупиновый 278980
Остролодочник майдантальский 278987
Остролодочник майделиев 278999
Остролодочник малорослый 279107
Остролодочник малоцветковый 279071
Остролодочник маньчжурский 278990
Остролодочник мартьянова 278998
Остролодочник мейера 279011

Остролодочник мейнсхаузена 279003
Остролодочник мелкодерновинный 278762
Остролодочник мелколистный 279014
Остролодочник мелкоплодный 279013
Остролодочник мелкосферический 279016
Остролодочник меловой 278801
Остролодочник меркенский 279009
Остролодочник мертенса 279010
Остролодочник миддендорфа 279017
Остролодочник михельсона 279012
Остролодочник многолистный 279083
Остролодочник многолисточковый 278780
Остролодочник многоцветковый 278831
Остролодочник мохнатоплодный 278952
Остролодочник мощный 279129
Остролодочник мягкоигольчатый 279028
Остролодочник мягколистный 278989
Остролодочник нагой 279040
Остролодочник наклоненный 278809
Остролодочник недзвецкого 279033
Остролодочник немногоцветковый 278784
Остролодочник нижнеальпийский 278696
Остролодочник нинсяский 279035
Остролодочник нитевидный 278829
Остролодочник оверина 279058,279059
Остролодочник остроконечный 278803
Остролодочник остролистный 279060
Остролодочник отклоненный 278809
Остролодочник охотский 279042
Остролодочник палласа 279067
Остролодочник песколюбивый 278703
Остролодочник плосконоготковый 279078
Остролодочник повислоцветковый 279073
Остролодочник подушечный 279208
Остролодочник подушковидный 279104
Остролодочник полевой 278764
Остролодочник поникший 279041
Остролодочник понсэна 279084
Остролодочник постимутовчатый 279191
Остролодочник прибрежный 278971
Остролодочник признанный 279118
Остролодочник приморский 278997
Остролодочник приснеговой 278784
Остролодочник притупленный 279121
Остролодочник прицветничковый 278752, 278753
Остролодочник приятский 279183
Остролодочник прямой 278823
Остролодочник птичий 278729
Остролодочник пузырниковидный 278798
Остролодочник пузырчатоплодный 278705
Остролодочник пушистозубчатый 279215
Остролодочник пушистопузырчатый 279217
Остролодочник пушистосферический 279218
Остролодочник пушистый 279102
Остролодочник разноволосый 278878
Остролодочник разноножковый 278877
Остролодочник распростертый 42258
Остролодочник распротертый 279087
Остролодочник розоватый 279131
Остролодочник розовый 279130
Остролодочник рукутамский 279136
Остролодочник русский 279138
Остролодочник рюбзаамена 279135
Остролодочник савелланский 279146
Остролодочник саркандский 279144
Остролодочник саурский 279145
Остролодочник сахалинский 279140
Остролодочник светлоголубой 278963
Остролодочник северцова 279159
Остролодочник седой 278767
Остролодочник семенова 279154
Остролодочник серебристый 278719
Остролодочник серножелтый 279192
Остролодочник синий 278804
Остролодочник синьцзянский 279165
Остролодочник сычуаньский 279164
Остролодочник смешиваемый 278799
Остролодочник снежнолистный 278785
Остролодочник солончаковый 279141
Остролодочник средний 278907
Остролодочник стеблевой 278775
Остролодочник стелящийся 278894
Остролодочник стукова 279182
Остролодочник сусамырский 279193
Остролодочник сходный 278702
Остролодочник таласский 279196
Остролодочник терекский 279204
Остролодочник толстоватый 278800
Остролодочник тонкий 279203
Остролодочник тонкоклювый 279201
Остролодочник тонколистный 278958
Остролодочник тонкопузырчатый 278960
Остролодочник трагакантовый 279212
Остролодочник траутфеттера 279214
Остролодочник трехлисточковый 279220
Остролодочник туповатый 279189
Остролодочник тысячелистный 279029
Остролодочник тяньшанский 279208
Остролодочник угамский 279226
Остролодочник узколистный 279178
Остролодочник украшающий 278915
Остролодочник уральский 279228
Остролодочник ферганский 278827
Остролодочник фетисова 278828
Остролодочник фомина 278832
Остролодочник хайларский 278869
Остролодочник хантенгринский 278777
Остролодочник хатангский 278923
Остролодочник хвостатый 278774
Остролодочник хлодостойкий 278835
Остролодочник хоргосский 278786
Остролодочник цельнолепестный 278906
Остролодочник чезнеевидный 278779
Остролодочник чернеющий 279034
Остролодочник чернобелый 279004
Остролодочник черноволосый 279007
Остролодочник чешуйчатый 279175
Остролодочник чимганский 279221
Остролодочник шароцветный 278854
Остролодочник шелковистый 279156
Остролодочник шерстеногий 278807
Остролодочник шерстистый 278941
Остролодочник шиповатый 278822
Остролодочник широкопарусный 279079
Остролодочник шишкина 279149
Остролодочник шишковидный 278954
Остролодочник шмидта 279150
Остролодочник шренка 279151
Остролодочник шубный 279072
Остролодочник щершавый 279147
Остролодочник щетиноволосый 279158
Остролодочник южносахалинский 278728
Остролодочник японский 278910
Остролодочник яркоцветный 278831
Остроло – пёстро 364360
Остро – пестро 364353,364360
Остянка 272611
Остянка бурманна 272619
Остянка курчяволистная 272684
Ототегия 277326
Ототегия бухарская 277329
Ототегия ольги 277345
Ототегия федченко 277333
Оттелия 277361
Оттелия частуховидная 277369
Оттелия японская 277397
Офелия 380105
Офелия вильфорда 380411
Офелия китайская 380184
Офелия четырехлепестная 380385
Офиопогон 272049
Офиопогон мотоский 272116
Офиопогон пинбяньский 272121
Офиопогон фунинская 272077
Офиопогон ябурая 272086
Офиопогон японский 272090
Офрис 232890,272395
Офрис кавказская 272422
Офрис крымская 272493
Офрис мухоносная 272466
Офрис оводоносная 272481
Офрис пчелоносная 272398
Офрис трансгирканская 272497
Охна 268132
Охна многоцветковая 268261
Охновые 268277
Охрома 268344
Очанка 160128

Очанка алтайская 160131
Очанка альбова 160130
Очанка амурская 160134
Очанка бакурианская 160136
Очанка ботническая 160141
Очанка вилькома 160292
Очанка волосистая 160169
Очанка воронова 160293
Очанка горная 160226
Очанка гроссгейма 160167
Очанка грузинская 160164
Очанка дагестанская 160147
Очанка железистостебельная 160129
Очанка зальцбургская 160261
Очанка изящная 160166
Очанка ирины 160188
Очанка кавказская 160144
Очанка кемулярии 160196
Очанка кернера 160197
Очанка короткая 160145
Очанка коротковолосистая 160142
Очанка короткоцветковая 160238
Очанка краснова 160201
Очанка крупнозубая 160206
Очанка крупночашечковая 160205
Очанка крымуская 160282
Очанка лекарственная 160233
Очанка летнекоротная 160247
Очанка летнеростковская 160248
Очанка летняя 160246
Очанка лжемягкая 160225
Очанка максимовича 160214,160241
Очанка маленькая 160222
Очанка мелкоцветковая 160219
Очанка многолистная 160227
Очанка мюрбека 160230
Очанка округлолистная 160146
Очанка онежская 160236
Очанка осетинская 160237
Очанка почти голая 160165
Очанка прямая 160268
Очанка раскрашенная 160245
Очанка регеля 160254
Очанка рейтера 160256
Очанка росткова 160233
Очанка ростковиука 160233
Очанка ростковская 160233,160258
Очанка саамская 160259
Очанка сванская 160272
Очанка севанская 160265
Очанка сибирская 160266
Очанка сычуаньская 160240
Очанка сосновского 160267
Очанка сырейщикова 160273
Очанка тайваньская 160286
Очанка татарская 160239,160277
Очанка татры 160281
Очанка тонкая 160283
Очанка тоунсэнда 160285
Очанка траншеля 160288
Очанка тупозубая 160132
Очанка уссурийская 160290
Очанка федченко 160155
Очанка финская 160156
Очанка холодная 160161
Очанка цветоножковая 160243
Очанка черешковая 160286
Очанка чешуйчатая 160204
Очанка шведская 160270
Очанка широколистная 160203
Очанка шугнанская 160262
Очанка юзепчука 160193
Очанка юхтрица 160289
Очанка якутская 160189
Очанкамягкая 160224
Очереднопыльник 18089
Очереднопыльник сидячий 18147
Очерёт 352348
Очеретник 333477
Очеретник белый 333480
Очеретник бурый 333576
Очеретник кавказский 333510
Очеретник китайский 333518
Очеретник фабера 333560
Очеретник фори 333568
Очеретник японский 333628
Очи царские 349452
Очиток 356467
Очиток “заячья капуста” 357203
Очиток альберта 318390
Очиток аньлунский 357253
Очиток белый 356503
Очиток бледнеющий 200793
Очиток бледноватый 200793
Очиток бледный 356998
Очиток большой 356928
Очиток волосистный 357282
Очиток гиблидный 294122
Очиток дэциньский 357294
Очиток едкий 356468
Очиток живородящий 200820
Очиток живучий 294114
Очиток заячья капуста 357203
Очиток испанский 356802
Очиток камчатский 294123
Очиток колючий 275394
Очиток краснеющий 275356,357107
Очиток красный 357112
Очиток люжный 357169
Очиток миддендорфа 294124
Очиток моховидный 357036
Очиток мохообразный 357036
Очиток мутовчатый 200816
Очиток мягколистный 275379
Очиток однолетний 356544
Очиток отогнутый 357114
Очиток пурпурный 200812
Очиток пурпуровый 200812
Очиток пятилепестный 357013
Очиток сельского 294126
Очиток стевена 294128
Очиток телефиум 357203
Очиток уссурийский 357271
Очиток цинхайский 357255
Очиток четырехмерный 357227
Очиток четырехчленный 329930
Очиток шестирядный 357147
Очиток шиловидный 357191
Очиток эверса 200788
Очиток этнинкий 356476
Очиток японский 356841
Очки царские 349452
Очноцвет 21335
Очноцвет голубой 21340
Очноцвет пашенный 21339
Очный 21335
Очный цвет 21335,21339
Очный цвет багряный 21339
Очный цвет голубой 21339,21340
Очный цвет мтовчатый 21433
Очный цвет нежный 21427
Очный цвет пашеный 21339
Очный цвет полевой 21339
Ошак 136259
Паветта 286073
Пави 20935
Павия жёлтая 9692
Павия калифорнийская 9679
Павия красная 9720
Павлиньй глазки 8332
Павловния 285951
Павловния войлочная 285981
Павловния пушистая 285981
Павловния тайваньская 285978
Павловния фаржа 285959
Павловния форчуна 285960
Павловния форчуна циньлинская 285985
Павлония 285928
Павноплодник 210251
Павноплодник василистниковый 210291
Павой 366230
павун 250493
Павун 404027
Пагудия 10105
Падуб 203527
Падуб ван 204395
Падуб востролистый 54798,203545
Падуб востролистый колючий 203553
Падуб гирканский 203904
Падуб городчатый 203670
Падуб гуншаньский 203891

Падуб китайский 203625
Падуб колхидский 203643
Падуб коралловый 203648
Падуб максимовича 204025
Падуб морщинистый 204226
Падуб мутовчатый 204376
Падуб наньтоуский 203883
Падуб опушенный 204184
Падуб остролистный 203545
Падуб парагвайский 204135
Падуб пильчатый 204239
Падуб пурпурный 203625
Падуб рвотный 204391
Падуб рогатый 203660
Падуб симаоский 204398
Падуб сычуаньский 204311
Падуб сугероки 204292
Падуб трехцветковый 204341
Падуб тусклый 204114
Падуб узколистный 204276
Падуб узкоплодный 204275
Падуб фаржа 203798
Падуб хунаньский 203897
Падуб хусюин 203896
Падуб цельнокрайний 203908
Падуб цзиньюньский 203932
Падуб широколистный 203961
Падубовые 29968
Падук 320269
Пажитник 397186
Пажитник бадахшанский 397198
Пажитник бессера 397200
Пажитник бугорчатый 397288
Пажитник вера 397292
Пажитник восходящий 397187
Пажитник голубоватый 397208
Пажитник голубой 397208
Пажитник гордеева 396920
Пажитник греческий 397186
Пажитник греческое сено 397229
Пажитник двухцветковый 397202
Пажитник дугообразный 397192
Пажитник колосистый 397278
Пажитник колосовый 397278
Пажитник короткоплодный 397203
Пажитник крупноцветковая 397236
Пажитник лазулевый 397196
Пажитник липского 397247
Пажитник лучевой 397274
Пажитник меловой 397212
Пажитник мечевидный 397233
Пажитник меченосный 397233
Пажитник ное 397255
Пажитник одноцветковый 397254
Пажитник памирский 397263
Пажитник парноцветковый 397254
Пажитник плоскоплодный 247422
Пажитник полосатый 397281
Пажитник попова 397271
Пажитник простертый 397272
Пажитник пряморогий 397261
Пажитник русский 247452
Пажитник рыхлоцветковая 397246
Пажитник сенной 397229
Пажитник тонкий 397206,397281
Пажитник туркменский 397289
Пажитник фишера 397227
Пажитник четковидный 397280
Пазания 233099
Пазник 202387
Пазник гладкий 202404
Пазник голый 202404
Пазник уколенившийся 202432
Пазник уколеняющийся 202432
Пайена 286754
Пайза 140367,140423
Паккой 59595
Пакленок 2837
Паклун 387989
Пак–хой 59595
Палаквиум 280437
Палисандр 121770
Палиурус 280539
Палиурус держи–дерево 280559
Палленис 280578
Палленис колючий 280585
Палочник 401085
Палошник 294992
Палошник луговой 294992
Пальма 78045,280539
Пальма “дум” 202256
Пальма “рыбий хвост” 78056
Пальма арековая 31677,31680
Пальма африканская 326649
Пальма бетель 31680
Пальма бетельная 31680
Пальма борассус 57122
Пальма бразильская 103737
Пальма бразильская восковая 103737
Пальма бурути 246674
Пальма веерная 222625,393809
Пальма веерная высокая 393809
Пальма веерная европейская 85671
Пальма веерная низкая 85671
Пальма виная 326649
Пальма винная 78056,246671
Пальма восковая бразильская 103737
Пальма восконосная 103737
Пальма гвинейская масличная 142257
Пальма жгучая 78056
Пальма зонтичная 106999
Пальма кавказская 64345
Пальма капустная 161053,337873
Пальма капустная американская 341422
Пальма карликова 85671
Пальма катеху 31680
Пальма кокосовая 98136
Пальма королевская 337880
Пальма костяная 298051
Пальма лантар 57122
Пальма маврикиева 246671
Пальма масличная 142255,142257
Пальма масличная гвинейская 142257
Пальма масляная 142255
Пальма медовая 212557
Пальма олеодокса 337880
Пальма пальмира 57122
Пальма пальмировая 57122
Пальма пальмирская 57122
Пальма пеньковая 393803
Пальма персиковая 46376
Пальма пикадовая 115884
Пальма ползучая 65782
Пальма ротанговая 65782
Пальма саговая 252636
Пальма саговая румфа 252634
Пальма сахарная 32343
Пальма сахарная лантар–парьмира 57122
Пальма сейшельская 235159
Пальма слоновая 212556,298051
Пальма талипотовая 106999
Пальма таллипотовая 106999
Пальма тени 106999
Пальма финиковая 295451,295461
Пальма финиковая канарская 295459
Пальма финиковая культурная 295461
Пальма хамеропс 85671
Пальма–атталея 44801
Пальма–бактрис 46374
Пальма–бетль 31680
Пальма–брахея мексиканская 59098
Пальма–гифаена 202321
Пальма–ириартея 208366
Пальма–Кариота 78045
Пальма–тодди 78056
Пальметта–сабаль 341422
Пальмито 85671
Пальмовые 280608
Пальмы 280608
Пальчатка 117852,117859,130745
Паляквиум 280441
Пампасова трава 105456
Пампельмус 93579
Панакс 280712
Панакс женьшень 280741
Панакс жень–шень 280741
Панакс корень–человик 280741
Панакс пинбяньский 280814
Панакс ползучий 280793
Панакс трёхлистниковый 280817
Пангиум 281245

Пангиум съедобный 281246
Пандан 280968
Пандановые 280963
Пандануc 280968
Пандануc вейча 281169
Пандануc полезный 281166
Пандануc сандера 281122
Панданусовые 280963
Пандерия 281173
Пандерия волосистая 281174
Пандерия туркестанская 281175
Панкраций 280877
Панкраций морской 280900
Панкрациум 280877
Панкрациум приморский 280900
Панцерия 282477
Панцерия ганьсуская 282481
Панцерия серебристобелая 282485
Панцерия сероватая 282480
Панцерия шерстистая 282482
Папаевые 76820
Папайа 76813
Папайя 76813
Папеда ежеиглистая 93492
Папеда китайская 93495
Папирус 119347
Парагвайский чай 204135
Парадизея 283295
Парадизка 243682
Парадиноглосс 283267
Парадиноглосс зубчиковоплодный 283270
Парадиноглосс имеретинский 283271
Парадиноглосс шершавейший 283269
Параиксерис 283385
Параиксерис зубчатолистный 283388
Параиксерис лопастнолистный 110608
Параиксерис поздний 283400
Параиксерис скальный 283399
Паракариум 283171
Паракариум бунге 283174
Паракариум гималайский 246578,283180
Паракариум изящный 283179
Паракариум каратавский 283185
Паракариум промежуточный 283183
Паракариум прямой 283190
Паракариум рыхлоцветковый 283186
Паракариум серый 283182
Паракариум туркменский 283191
Парамикроринхус 283504
Парамикроринхус полегающий 283506
Парафорис 283654
Парафорис изогнутый 283656
Парентучеллия 284093
Парентучеллия желтоцветковая 284094
Парентучеллия клейкая 284098
Парентучеллия широколистная 284096
Парило 11549
Паркия 284444
Паркия африканская 284475
Паркия двхшаровидная 284473
Паркия суданская 284473
Парнолистник 418585
Парнолистник амудаьинский 418726
Парнолистник балхашский 418604
Парнолистник бобовидный 418641
Парнолистник вильчатый 418653
Парнолистник втхарский 418608
Парнолистник ганьсуский 418681
Парнолистник гончарова 418668
Парнолистник дарвазский 418626
Парнолистник дильса 418680
Парнолистник илийский 418677
Парнолистник каратавский 418682
Парнолистник кашгарский 347066,418683
Парнолистник кегенский 418685
Парнолистник копальский 418686
Парнолистник короткокрылый 418607
Парнолистник крупвокрыладый 418702
Парнолистник крупнокрылый 418702
Парнолистник крупноногий 418701,418722
Парнолистник крупноплодный 418712
Парнолистник крылатоплодный 418740
Парнолистник лебедовый 418599
Парнолистник лемана 418691
Парнолистник мелкоплодный 418714
Парнолистник неравнобокий 418722
Парнолистник обокновенный 418641
Парнолистник остоплодный 418727
Парнолистник остроконечный 418623
Парнолистник портулаковидный 418730
Парнолистник потанина 418732
Парнолистник почти трехпарный 418769
Парнолистник розова 418753
Парнолистник синьцзянский 418760
Парнолистник суриковокрасный 418720
Парнолистник сырдарьинский 418680
Парнолистник туркменский 418774
Парнолистник узкокрылый 418768
Парнолистник фабаговидный 418646
Парнолистник ферганский 418648
Парнолистник ширококрылый 418639
Парнолистник широколистный 418754
Парнолистник эйхвальда 418638
Парнолистниковые 418580
Парнолистные 418580
Парохетус 284647
Паррия 284969
Паррия бекетова 284974
Паррия беловатая 284971
Паррия бесстебельная 224212
Паррия голенкина 284991
Паррия голостебельная 285004
Паррия каратавская 284993
Паррия крупноплодная 285004
Паррия кустарничковая 284990
Паррия мелкоплодная 285001
Паррия отпрысковая 285031
Паррия перистая 285006,285008
Паррия подушковая 285014
Паррия пушисточашечная 224211
Паррия струговидная 285018
Паррия стручковатая 285030
Паррия стручковая 285021
Паррия туркестанская 285032
Паррия узколистная 285029
Паррия узкоплодная 285022
Паррия шершавая 284973
Паррия шугнанская 285020
Парротия 284957
Парротия персидская 284961
Парсония 114595,285055
Партениум 285072
Партениум серебристый 285076
Партеноциссус 285097
Партеноциссус генри 285110
Партеноциссус китайский 285102
Партеноциссус пятилисточковый 20435,
285136
Паслен 366902
Паслён 366902
Паслён арбузолистный 367205
Паслён баклажан 367370
Паслен безволосый 366934
Паслён волосистый 367726
Паслен воронова 367748
Паслён горько－сладгий 367120
Паслен желтый 367311
Паслён жёлтый 367726
Паслен закавказский 367687
Паслен зеленецкого 367757
Паслен иудейский 367270
Паслен кизерицкого 367281
Паслен клубненосный 367696
Паслён клубненосный 367696
Паслен клювовидный 367567
Паслён клювовидный 367567
Паслён клювообразный 367567
Паслен крупноплодный 367367
Паслен крыладый 367726
Паслён лжеперечный 367522
Паслен ложножелтый 367529
Паслён ложноперечный 367522
Паслен ложноперсидский 367531
Паслён мохнатый 367726
Паслен незамеченный 367083
Паслен низкий 367432
Паслен ольги 367468
Паслен персидский 367493
Паслён перцеподобный 367522
Паслен прибрежный 367306
Паслен семилопастный 367604

Паслен сладкогорький 367120
Паслён сладко－горький 367120
Паслен среднеазиатский 366953
Паслен хайнаньский 367518
Паслен черный 367416
Паслён чёрный 367416
Пасленовые 366853
Паслёновые 366853
Паспалум 285390，285533
Паспалум пальчатый 285422
Паспалум помеченый 285469
Паспалум расширенный 285423
Паспалум тунберга 285530
Паспалюм 285390
Паспалюм двойчатосложный 285417
Паспалюм длиннолистный 285460
Паспалюм округлый 285471
Паспалюм расширенный 285423
Паспалюм тайваньский 285448
Паспалюм тунберга 285530
Паспалюм ямчатый 285507
Пассифлора 285607
Пассифлора гигантская 285694
Пассифлора крупная 285694
Пассифлора лавролистная 285660
Пассифлора олубая 285623
Пассифлора плодовая 285637
Пассифлора съедобная 285637
Пастернак 285731
Пастернак армянский 285732
Пастернак бедренецолистный 285739
Пастернак дикий 285744
Пастернак лесной 285744
Пастернак луговой 285740
Пастернак оранжевый 285734
Пастернак посевной 285741
Пастернак теневой 285745
Пастернак тусклый 285738
Пастернаковник 285748
Пастернаковник ледниковый 285749
Пастушья сумка 72037
Пастушья сумка восточная 72054
Пастушья сумка гирканская 72053
Пастушья сумка обыкновенная 72038
Патат 207623
Патисон 114300
Патиссон 114301
Патриния 285814
Патриния горбатая 285830
Патриния каменистая 285854
Патриния мохнатая 285880
Патриния сибирская 285870
Патриния скабиозолистная 285859
Патриния скальная 285854
Патриния средняя 285839
Патрэния 285814
Патрэния сибирская 285870
Патрэния скабиозолистная 285859
Патрэния скальная 285854
Пауллиния 285928
Пауллиния гуарана 285930
Пауловния 285951
Пауловния войлочная 285981
Паучник 95606
Пафиопедилюм 282768
Пафиопедилюм бирманский 282796
Пафиопедилюм гиблидный 282844
Пафиопедилюм замечательный 282845
Пафиопедилюм малипоский 282858
Пафиопедилюм мохонадый 282914
Пафиопедилюм яванский 282855
Пахизандра 279743
Пахизандра верхушечная 279757
Пахиплеурум 279644，279651
Пахиплеурум альпийский 279645
Пахиплеурум тибетский 279653
Пахицентрия 279476
Пахлук 3995
Пахудия 10105
Пахучеколосник 27930，27969
Пахучеколосник остистый 27934
Пахучеколосник тайваньский 27947
Пахучеколосник японский 27971
Пахучка 65528，96960，170524
Пахучка аптечная 65603
Пахучка китайская 96970
Пахучка обыкновенная 97060
Пахучка острая 65531，96964
Пахучка остролистная 65531，96964
Пахучка полевая 65531，96964
Пахучка сахалинская 97031
Пахучка теневая 97051
Пахучка цернокрайняя 97002
Пачули 306994
Пашенник 226615
Пашенник звездочковидный 226626
Пашенник костенецовидный 226621
Пашмак 170694
Певник болотный 273541
Педалиевые 286918
Педерота 280122
Педилантус 287846
Пекан 77902
Пекан горикий 77881
Пекан яйцевидный твёрдокорый 77920
Пеларгониа 288045
Пеларгониа плющевидная 288430
Пеларгониа ползучая 288430
Пеларгоний 288045
Пеларгоний попелечнополосатый 384，288594
Пеларгоний розовый 288268
Пеларгониум 288045
Пеларгониум гиблидная 288300
Пеларгониум головчатый 288133
Пеларгониум крупноцветный 288264
Пеларгониум эндлихера 288216
Пеларгония 288045
Пеларгония душистая 288396
Пеларгония зональвая 288594
Пеллила 290932
Пеллила кустарниковая 290940
Пеллиония 288705
Пеллиония красивая 288760
Пеллиония хайнаньская 288754
Пельтандра 288800
Пельтифиллум 288840
Пельтогине 288863
Пелюшка 301055
Пеннизетум 289011
Пеннисетум 289011
Пеннисетум восточный 289185
Пеннисетум красный 289218
Пеннисетум ланьпинский 289055
Пеннисетум масличная 289476
Пеннисетум повислый 289096
Пеннисетум сизый 289116
Пеннисетум тибетский 289141
Пеннисетум японский 289135
Пенстемон 289314
Пентаклетра крупнолистная 289439
Пентанема 289563
Пентанема близкая 289572
Пентанема железистая 289567
Пентанема индийская 289568
Пентанема постенная 289571
Пентанема растопыренная 289566
Пентанема скальная 289574
Пентас 289784
Пентастемон 289314，400622
Пентастемон кустарниковый 289341
Пентаце 289407
Пентция 290222
Пенька бомбейская 194779
Пенька индийская 112269
Пенька критская 123028
Пенька манильская 260275
Пенька манипиская 260275
Пенька новозеландская 295600
Пеон 280143
Пеон алтайский 280152
Пеон белоцветный 280213
Пеон древовидный 280286
Пеон китаиский 280213
Пеон коралловый 280231
Пеон лекарственный 280253
Пеон марьин－корень 280154
Пеон мелкоцветковый 280213
Пеон полустарниковый 280286
Пеон травянистый 280213
Пеперинник 91107

Пепериник багровый 290821
Пепериник приморский 358395
Пеперомия 290286
Пеперомия магнолиелистная 290382
Пеперомия мэнцзыская 290357
Пеперомия пятнистая 290380
Пеперомия туполистная 290395
Пеперомия японская 290365
Пепино 367400
Перакарпа 290572
Перакарпа двулепестниковидная 290573
Первоцвет 314083
Первоцвет абхазский 314086
Первоцвет аньхуйский 314344
Первоцвет аптечный 315082
Первоцвет арктический 314125
Первоцвет байерна 314156
Первоцвет бальджуанский 314147
Первоцвет бесстебельный 315101
Первоцвет варшеневского 315113
Первоцвет весенний 315082
Первоцвет высокий 314333
Первоцвет галлера 314449
Первоцвет гуандунский 314549
Первоцвет гуйцзоуский 314550
Первоцвет длинногий 314595
Первоцвет длиннострелочный 314599
Первоцвет дудчатый 314396
Первоцвет евгении 314353
Первоцвет желтенький 314602
Первоцвет жёлтый 315082
Первоцвет иезский 314514
Первоцвет извилистый 314404
Первоцвет ильинского 314490
Первоцвет истиный 315082
Первоцвет карпатский 314221
Первоцвет китайский 314975
Первоцвет клинолистный 314280
Первоцвет кнорринг 314542
Первоцвет комалова 314546
Первоцвет комнатный 314717
Первоцвет кортусовидный 314271
Первоцвет крупночашечный 315083
Первоцвет кузнецова 314548
Первоцвет лекарственный 315082
Первоцвет маленький 314646
Первоцвет матсумуры 314376
Первоцвет мейера 314638
Первоцвет минквиц 314647
Первоцвет муркрофта 314670
Первоцвет мучнистолистный 314365
Первоцвет мучнистый 314366
Первоцвет норвежский 314711,314948
Первоцвет обыкновенный 315101
Первоцвет ольги 314756
Первоцвет осетинский 314765
Первоцвет отклоненный 314952
Первоцвет отменный 314356
Первоцвет палласа 314770
Первоцвет памирский 314772
Первоцвет перистый 314804
Первоцвет поникающий 314948
Первоцвет порошковатый 314862
Первоцвет почколистный 314892
Первоцвет прелестный 314120
Первоцвет промежуточныйторчащий 314501
Первоцвет разноцветный 314460
Первоцвет рупрехта 314923
Первоцвет сахалинский 314925
Первоцвет северный 314177
Первоцвет сердселистный 314269
Первоцвет сибирский 314948
Первоцвет сиборпа 314951
Первоцвет снежный 314697
Первоцвет снизу-желтый 315128
Первоцвет торчащий 315015
Первоцвет туркестанская 314700
Первоцвет турнефора 315056
Первоцвет тяньшанский 314713
Первоцвет ушковатый 314144
Первоцвет федченко 314388
Первоцвет фуцзяньский 314722
Первоцвет холодный 314106
Первоцвет цзиндунский 314499
Первоцвет чэнкоуский 314359
Первоцвет юлии 314527
Первоцвет японский 314508
Первоцвет-аврикула 314143
Первоцветные 315138
Пережуй-лычко 122515
Перекати-поле 183157,183225
Перекатун 183225
Перелеска 192120,192134
Перелеска благолодная 192134,192148
Перелёт соколий 173364
Пересксия колючая 290701
Пересксия шиповатая 290701
Переступень 61464
Переступень белый 61471
Переступень двудомный 61497
Переступень заамударьинский 61538
Переступень однодомный 61518
Перец 72068,300323
Перец американский 417157,417231
Перец бунге 417180,417340
Перец ветвистый 72070,72100
Перец водяной 309199
Перец гвосдичный 299325
Перец дикий 143657
Перец длинный 300446
Перец душистый 299325
Перец инцзянский 300552
Перец испанский 72070,72100
Перец кава 300458
Перец китайский 300373,417180,417340
Перец красный 72070,72100
Перец кустарниковый 72070,72100
Перец ланьюйский 300335
Перец лесной 300524
Перец мексиканский 300488
Перец многолетний 72070,72100
Перец мэнхайский 300372
Перец настоящий 300323
Перец овощной 72070,72100
Перец однолетний 72070,72100
Перец пинбяньский 300481
Перец симаоский 300526
Перец стручковый 72070,72100
Перец стручковыйоднолетний 72070,72100
Перец тайваньский 300528
Перец тибетский 300450
Перец узколистный 300333
Перец украшенный 300473
Перец хайнаньский 300407
Перец черный 300464
Перец чёрный 300464
Перец эмэйский 300385
Перец юньнаньский 300554
Перец ягодный 72070,72100
Перец ямайский 299325
Перец японский 417301
Перечник 203181
Перечник горький 203184
Перечник зонтичный 203249
Перечник огородный 225450
Перечник посевной 225450
Перечник пронзеннолистный 225428
Перечные 300557
Перилла базириковая 290940
Перилла многолетняя 290940
Периплока 291031
Периплока заборная 291082
Перистоголовник 320421
Перисторофа 291134
Перламтровка 32600
Перловник 248949
Перловник азербайджанский 248960
Перловник высокий 248951
Перловник ганьсуский 249075
Перловник гмелина 249012
Перловник жакемонта 249066
Перловник желтоватый 249008
Перловник кисточковидный 249065
Перловник короткоцветковый 248962
Перловник крымский 248975
Перловник малый 249043
Перловник неравночешуйный 249066
Перловник однобокий 249090
Перловник одноцветковый 249106
Перловник персидский 249066

Перловник пестрый 249072
Перловник поникающий 249054
Перловник прутьевидный 249110
Перловник сероватый 249067
Перловник сизый 249010
Перловник трансильванский 249104
Перловник турчанинова 249105
Перловник шероховотый 249085
Пернамбук 7190
Перо водяное 291363
Перовския 198680,291427
Перовския кудряшева 291431
Перовския линчевского 291432
Перовския памирская 291433
Перовския полыная 291428
Перовския прутьевидная 291435
Перовския узколистная 291429
Перплексия 214028
Перплексия мелкоголовчатая 291441
Персея 291427,291491,291505
Персея индийская 291505,291543
Персея красная 291543
Персея мексиканская 291495
Персея приятнейшая 291494
Персик 20885,20935
Персик ганьсунский 20921
Персик давида 20908
Персик мира 20926
Персик нектарин 20955
Персик обыкновенный 20935
Персик степной 20931
Персикария восточная 309494
Персик – нектарин 20955
Персимон 132030
Персймон 132219
Перуанская вишня 297711
Перулярия 302241
Перулярия буреющая 302513
Перулярия уссурийская 302549
Песбой 99297
Песколюб 19765
Песколюб песчаный 19766
Песколюбка песчаная 19766
Песочник 19710
Песочник палестинский 19712
Песчанк 228346
Песчанка 31727
Песчанка азиатская 31758
Песчанка биверштейна 31848
Песчанка волосовидная 31787
Песчанка ганьсуская 31971
Песчанка гвоздичковая 31852
Песчанка головчатая 31803
Песчанка горицветная 32039
Песчанка гриффита 31936
Песчанка гриффица 31936
Песчанка далиская 31849
Песчанка длиннолистная 32031
Песчанка дэциньская 31759
Песчанка жесткая 32178
Песчанка злаковая 31932
Песчанка злаколистная 31933
Песчанка качимовидная 31940
Песчанка корина 31991
Песчанка красивая 31903
Песчанка круплолистная 32192
Песчанка крупнолистная 32058
Песчанка крупноцветная 32053
Песчанка ледебра 32014
Песчанка ледовского 32175
Песчанка ложнохолодная 32153
Песчанка луговостепная 32073
Песчанка мейера 32071
Песчанка мелкозубчатая 32073
Песчанка монгольская 32082
Песчанка мулиская 32089
Песчанка паульсена 32124
Песчанка полярная 32143
Песчанка потанина 32148
Песчанка приземистая 31957
Песчанка приметная 31961
Песчанка пузырниковидная 31842
Песчанка ситниковая 31967
Песчанка сычуаньская 32267
Песчанка стевена 32240
Песчанка сырейщикова 32266
Песчанка тимьянолистная 32212
Песчанка тонкостебельная 32015
Песчанка траволистная 31933
Песчанка туркестанская 32290
Песчанка ферганская 31874
Песчанка цинхайская 32172
Песчанка чжундяньская 32328
Песчанка чжэньканская 32101
Песчанка чукотская 32288
Песчанка шовица 32268
Песчанка эмэйская 32114
Песчанка юньнаньская 32319
Песья вишня 297640
Песья вишня обыкновенная 297643
Петалостемум 292279
Петорорагия 292656
Петорорагия камнеломка 292668
Петорум 290133
Петорум китайский 290134
Петрея 292517
Петров крест 222641
Петров крест чешуйчатый 222651
Петров крест японский 222645
Петрокаллис 292527
Петросимония 292714
Петросимония жестковолосая 292722
Петросимония коротколистная 292716
Петросимония литвинова 292723
Петросимония однотычинковая 292724
Петросимония растопыренная 292727
Петросимония сибирская 292726
Петросимония сизая 292720
Петросимония сизоватая 292721
Петросимония супротивиолистная 292715
Петросимония толстолистная 292717
Петросимония трехтычинковая 292728
Петрофитум 292645
Петрушечник 365357
Петрушечник ароматный 365362
Петрушка 292685
Петрушка болотная 292966
Петрушка горная 292961
Петрушка зеленная 292694
Петрушка корневая 21003,292701
Петрушка кудрявая 292694
Петрушка обыкновенная 292694
Петрушка огородная 292694
Петрушка садовая 292694
Петрушка собачья 9832
Петсай 59600
Петуния 292741
Петунья 292741
Петунья вздутая 292748
Петунья гибридная 292745
Петунья мелкоцветная 292745
Петунья ночецветковая 292751
Петунья фиолетовая 292754
Петуший гребешки 80378
Петушки 208919
Петушки степные 208874
Петушник 170056
Петушок степной 208874
Петушьи гребешки 80395
Пецай 59600
Печак 102933
Печать золотая 200173
Печать соломонова 308493
Печеная акация 19728
Печеная акация длиннокистевая 19735
Печеная акация карелина 19733
Печеная акация конолли 19731
Печеная акация лемана 19734
Печеная акация персидская 19736
Печеная акация серебристая 19729
Печеная акация эйхвальда 19732
Печёночник 192120
Печеночница 192120
Печёночница 192120
Печеночница азиатская 192138
Печёночница благолодная 192134,192148
Печеночница обыкновенная 192134
Печёночница обыкновенная 192134,192148
Печеночница фальконера 192131
Пиерис 298703
Пиерис обилиноцветущий 298721

Пиерис японский 298734
Пижма 383690
Пижма абротаноистная 383691
Пижма акинфиева 383700
Пижма алтайская 383714
Пижма дважты – перистая 383716
Пижма дикая рябина 383874
Пижма длинноножковая 383793
Пижма душистая 89602,383712
Пижма жестколистная 383845
Пижма закавказская 383727
Пижма канупер 89602,383712
Пижма карелина 383780
Пижма киттари 383784
Пижма ложнотысячелистниковая 383832
Пижма малоголовчатая 383810
Пижма многолистная 383806
Пижма обыкновенная 383874
Пижма одноцветковая 383871
Пижма пахучая 89602,383712
Пижма пачосского 383813
Пижма пижмовидная 383857
Пижма разнолистная 383765
Пижма сантолиновидная 383842
Пижма северная 383876
Пижма скальная 383843
Пижма толстоножковая 383740
Пижма туркменская 383867
Пижма турланская 383714
Пижма тысячелистная 26829
Пижма тысячелистниковая 383692
Пижма улутавская 383869
Пижма уральская 383872
Пижма утесная 383846
Пизанг 260253,370136
Пикнантемум 321952
Пикномон 298469
Пикномон колючий 298471
Пикностельма 322165
Пикностельма метельчатая 117643
Пикразма 298506
Пикрамния 298494
Пикрасма 298506
Пикрасма квассиевидная 298516
Пикрасма китайская 298509
Пикрасма яванская 298513
Пикрена высокая 298510
Пикульник 170056
Пикульник двнадрезанный 170060
Пикульник двнадрезный 170060
Пикульник двурасщепленный 170060
Пикульник жабрей 170078
Пикульник заметный 170076
Пикульник колючий 170078
Пикульник красивый 170076
Пикульник ладанниковидный 170066
Пикульник ладанниковый 170066
Пикульник ладанный 170066
Пикульник обыкновенный 170078
Пикульник пушистый 170073
Пикульник узколистный 170058
Пикульник – жабрей 170078
Пилеа 298848
Пилеа гуанхиская 298973
Пилеа кртглолистная 299006
Пилеа мелколистная 298974
Пилеа монгольская 299024
Пилеа мэнхайская 298972
Пилея 298848
Пилея бутерлаковидная 299006
Пилея зеленейшая 299024
Пилея монгольская 299024
Пилея мотоская 298970
Пилея хамао 299026
Пилея японская 298945
Пилокарпус 299167
Пилокарпус перистолистный 299172
Пимелея 299303
Пимента 299325
Пимпинель 345860,345881
Пинанг 31680
Пинанга 299649
Пион 280143
Пион абхазский 280146
Пион аптечный 280253
Пион белоцветковый 280147
Пион биберштейна 280164
Пион весенний 280332
Пион виттмана 280334
Пион войлочноплодный 280319
Пион горный 280241
Пион древовидный 280286
Пион кавказская 280169
Пион крупнолистный 280227
Пион лекарственный 280253
Пион марьин – корень 280154
Пион млокосевича 280236
Пион молочноцветковый 280213
Пион обратно – овальный 280241
Пион обратнояйцевидный 280241
Пион полукустарниковый 280286
Пион степной 280204,280206
Пион тонколистный 280318
Пион трехраздельный 280320
Пион японский 280207
Пиптадения 300607
Пиптадения африканская 300609
Пиптадения обыкновенная 300621
Пиптант 300660
Пиптант карликовый 300688
Пиптантус 300660
Пиптантус бобовниколистный 300677
Пиптантус мэнцзыский 300674
Пиптантус непалский 300691
Пиптоптера 300793
Пиптоптера туркестанская 300794
Пираканта 322448
Пираканта городчато – зубчатая 322465
Пираканта колючая 322454
Пираканта красная 322454
Пираканта мелкогородчатая 322458
Пираканта мелкогородчатая ганьсуская 322459
Пираканта узколистная 322450
Пираканта ярко – красная 322454
Пирамидальный Тополь 311400
Пиратинера гвианская 61059
Пиретрум 89402,322640
Пиретрум абротанолистный 322641
Пиретрум абротанолистый 322641
Пиретрум алатавский 322644
Пиретрум альпийский 227457,322646
Пиретрум аррасанский 322650
Пиретрум бальзамический 89602,322656,383712
Пиретрум бальзамоносновидный 322655
Пиретрум большой 322709
Пиретрум волосолопастный 322755
Пиретрум гали 322685
Пиретрум гваюлолистный 322723
Пиретрум гиссарский 322691
Пиретрум горичнниколистшый 322729
Пиретрум гроссгейма 322689
Пиретрум дагестанский 322675
Пиретрум девичий 322724
Пиретрум девичьелистный 322723
Пиретрум джилгинский 322679
Пиретрум джунгарский 322745
Пиретрум железконосный 322688
Пиретрум заилийский 322751
Пиретрум карелина 322693
Пиретрум келлера 322697
Пиретрум клюзия 322664
Пиретрум кочи 322698
Пиретрум красивый 322730
Пиретрум красный 322658,322660,322665
Пиретрум крупнолистный 89599
Пиретрум крылова 322699
Пиретрум крыловский 322699
Пиретрум кубинский 322700
Пиретрум кустарничковый 322683
Пиретрум мариона 322713
Пиретрум микешина 322714
Пиретрум мюсокрасный 322658,322660,322665
Пиретрум незамеченный 322717
Пиретрум оше 322654
Пиретрум пиретровидный 322732
Пиретрум розовый 322658,322660,322665
Пиретрум рябинолистный 322746
Пиретрум семенова 322741

Пиретрум сибирский 89723
Пиретрум тонколистный 322704
Пиретрум точечный 322731
Пиретрум тяньшанский 322750
Пиретрум хельдрейха 322690
Пиретрум цинерариелистный 322660
Пиретрум шелковистый 322742
Пиретрум шерстистый 322701
Пиретрум щитковидный 322669
Пиретрум щитковый 89421,322669
Пиретрум щитконосный 89421
Пиретрум эдельвейсовиднй 322704
Пиретрум ярко－красивый 322658,322660,322665
Пистия 301021
Питанга 156385
Питтоспорум 301207
Питтоспорум волнистый 301433
Питтоспорум гуансиский 301309
Питтоспорум гуйцзоуский 301310
Питтоспорум куэньминский 301308
Питтоспорум ланьюйский 301340
Питтоспорум мелкоплодный 301356
Питтоспорум непалский 301344
Питтоспорум тобира 301413
Питтоспорум эмэйский 301352
Пихта 269
Пихта алжирская 435
Пихта аризонская 403
Пихта бальзамическая 282
Пихта белая 272
Пихта белокорая 427
Пихта благолодная 453
Пихта великолепная 409
Пихта вильсона 464,510
Пихта вича 496
Пихта ворнмюллвра 299
Пихта вура 496
Пихта гималайская 480
Пихта гребенчатая 272
Пихта греческая 306
Пихта далиская 333
Пихта дугласова 318580
Пихта европейская 272
Пихта замечательная 480
Пихта западногималайская 439
Пихта изящная 383
Пихта испанская 442
Пихта каваками 396
Пихта кавказкая 429
Пихта калифорнийская 507
Пихта камчатская 383
Пихта канадская 282
Пихта киликийская 314
Пихта корейская 398
Пихта красивая 409
Пихта крепкая 365
Пихта лоуа 407
Пихта мариес 414
Пихта мексиканская 458
Пихта миловидная 280
Пихта мотоская 338
Пихта нордмана 435
Пихта нордманна 429
Пихта нордманнова 429
Пихта нубийская 435
Пихта нумидийская 435
Пихта обыкновенная 272
Пихта одноцветная 317
Пихта почкочешуйная 427
Пихта почкочешуйчатая 389,427
Пихта прелестная 507
Пихта равночешуйчатая 389
Пихта сахалинская 460,463
Пихта священная 458
Пихта семенова 471
Пихта серебристая 272
Пихта сибирская 476
Пихта сильная 365
Пихта тайваньская 396
Пихта тонкая 383
Пихта факсона 362
Пихта фразера 377
Пихта цельнолистная 388
Пихта чаюйская 309
Пихта черная 388
Пихта японская зонтичная 352856
Пиэрис 298703
Пиэрис люцюский 298758
Пиэрис тайваньский 298734
Пиэрис японский 298734
Пиэрис японский мелкоцветковый 298751
Плавун нимфейный 267825
Плагиантус 301610
Пладиобазис 301619
Пладиобазис васильковый 301620
Плакун 240025,240068
Плакун иволистный 240068
Плакун лозовый 240100
Плакун пруныйтовид 240100
Плакун－трава 240068
Пламенник 295239
Планера 301801,417534
Планера водная 301804
Пластинниковые 1292
Пластинчатые 1292
Платан 302574
Платан американский 302588
Платан востока 302592
Платан восточный 302592
Платан западный 302588
Платан испанский 302582
Платан кленолистный 302582
Платан клёнолистный 302582
Платан клиновидный 302578
Платан линдена 302586
Платан мексиканский 302587
Платан обыкновенный 302592
Платан пальчатый 302580
Платановые 302229
Платикария 302674
Платикария шишконосная 302684
Платикладус 302718
Платикладус восточный 302721
Платикладус восточный 'Зиболидин' 302734
Платикодон 302747
Платикодон крупноцветконый 302753
Платикодон крупноцветный 302753
Платимисциум 302860
Платистемон 302961
платитения бедренецонодобная 303011
платитения бухарская 302997
платитения комарова 303002
платитения обедненная 302999
платитения памирская 303010
платитения разнозубая 303000
Платитения рубцова 303013
Платония 302616,302622
Плевел 235300,235373
Плевел жесткий 235354
Плевел итальянский 235315
Плевел клиновидный 235373
Плевел льняной 235353
Плевел много 235315
Плевел многолетний 235334
Плевел многоцветковый 235315
Плевел многоцветный 235315
Плевел одуряющий 235373
Плевел опьяняющий 235373
Плевел персидский 235351
Плевел полевой 235303
Плевел пранский 235351
Плевел расставленный 235353
Плевел южный 235310,235354
Плевелок 235297
Плевелок восточный 235299
Плеврогина 235431
Плевропогон 304695
Плейобластус 303994
Плейобластус гуансиский 304051
Плектогине 39516,39532
Плектрантус вырезной 209665
Плектрантус пильчатый 209826
Плектрантус сизочашечный 209717
Плеуропогон 304695
Плоскоплодник 18313
Плоскоплодник льнолистный 18407
Плюмбаго 305167
Плюмбаго цейлонское 305202
Плюмбаговые 305165

Плюмрия 305206
Плюхея 305072
Плюш дауский 250218
Плюш комнатный 254414
Плющ 114947,187195
Плющ амурский 250228
Плющ восковой 86427,198821,198827
Плющ гималайский 187292
Плющ желтоплодный 187204
Плющ кавказский 187202
Плющ канадский 250221
Плющ китайский 187307
Плющ колхидский 187205
Плющ крымский 187337
Плющ непалская 187306
Плющ обыкновенный 187222
Плющ пастухова 187311
Плющ японская 187318
Плющ японская тайваньская 187325
Пнттоспоровые 301203
Поацинум 306168
Поацинум гендерсона 29506
Поацинум пестрый 29506
Повий 115040
Повилика 114947,115040
Повилика араратская 114965
Повилика белая 114953
Повилика бокальчатая 114961
Повилика бухарская 114982
Повилика вавилонская 114973
Повилика гигантская 115037
Повилика гребневидная 115065
Повилика гронова 115141
Повилика гроновиуса 115031
Повилика душистая 115146
Повилика европейская 115031
Повилика желтая 114972
Повилика индийская 115049
Повилика каратавская 115056
Повилика китайская 114994
Повилика короткостолбиковая 114980
Повилика кочи 115061
Повилика красностебельная 114984
Повилика креверная 115007
Повилика ландышецветковая 114997
Повилика лемана 115063
Повилика льновая 115007
Повилика льняная 115008
Повилика одностолбиковая 115080
Повилика окрашенная 115050
Повилика памирская 115096
Повилика перечная 114972
Повилика полевая 114986
Повилика сблиясанная 114961
Повилика тимьяновая 115008
Повилика тяньшанская 115143
Повилика узкочашечная 115140
Повилика ферганская 115033
Повилика хмелевидная 115066
Повилика цветоножковая 115098
Повилика чебрецовая 115008
Повилика штапфа 115138
Повилика энгельмана 115006
Повилика южная 114972
Повилика японская 115050
Повиликовые 115155
Повитель 162519
Поводник 183389
Поводник гвоздичный 183565
Поводник линейнолистный 183797
Поводник сычуаньский 184114
Поводник тибетский 184136
Повой 68671,102903
Повой волосистый 68701
Повой заборный 68713
Повой круглолистный 68737
Повой лесной 68732,68746
Повой плющевидный 68686
Повой подстеный 284167
Повой солданелевый 68737
Повой японский 68692,68724
Повойник 142554
Повойничек 142554
Повойничек болотниковый 142565
Повойничек венгерский 142569
Повойничек восточный 142578
Повойничек красновласковый 142565
Повойничек мокричный 142555
Повойничек прямосемянный 142573
Повойничек согнутосемянный 142570
Повойничек сомнительный 142556,142579
Повойничек трехтычинковый 142574
Повойничек трёхтычинковый 142574
Повойничек шеститычинковый 142568
Повойничковые 142552
Погония 306847
Погония японская 306870
Погостемон 306964
Погремок 329454,329504
Погремок альпийский 329507
Погремок весенний 329527
Погремок горный 329545
Погремок гребенчатый 329552
Погремок джингарский 329565
Погремок колхидский 329519
Погремок летний 329505
Погремок малый 329543
Погремок меловой 329520
Погремок олгз 329549
Погремок отклоненный 329551
Погремок петушиный гребешок 329521
Погремок понтийский 329553
Погремок румельский 329556
Погремок сахалинский 329560
Погремок сельский 329559
Погремок средиземноморский 329542
Погремок узколистный 329510
Погремок ферганский 329525
Погремок чернеющий 329547
Погремок шиловидный 329569
Погремок шишкина 329562
Подбел 22418,292342
Подбел белый 292374
Подбел войлочный 292401
Подбел гибридный 292372,292386
Подбел дубровник 22445
Подбел лекарственный 292386
Подбел лечебный 292386
Подбел многолетний 22445
Подбел многолистный 22445
Подбел обыкновенный 292386
Подбел японский 292374
Подбельник 197943
Подбельник альпийский 197944
Подбережник 240025,240068
Подберёзовик 3127
Подберёзовик обыкновенный 3127
Подковник 196621
Подладанник 120875
Подладанник красный 120882
Подлесник 345940
Подлесник высокий 345952
Подлесник европейский 345957
Подлесник китайский 345948
Подлесник красноцветковый 345998
Подлесник мэнцзыский 345988
Подмаренник 170175
Подмаренник азербайджанский 170236
Подмаренник амурский 170191
Подмаренник байкальский 170239
Подмаренник бледножелтый 170523
Подмаренник болеальный 170246
Подмаренник болотный 170529
Подмаренник брауна 170282
Подмаренник вайяна 170727
Подмаренник валантиевидный 170726
Подмаренник вартана 170733
Подмаренник васильченко 170734
Подмаренник венечный 170330
Подмаренник весенний 170737
Подмаренник волосоносный 170692
Подмаренник высокий 170353
Подмаренник вытянутый 170367
Подмаренник герцинский 170611
Подмаренник гладколистный 170459
Подмаренник гроссгейма 170410
Подмаренник густоцветковый 170348
Подмаренник даурский 170340
Подмаренник декена 170346
Подмаренник джунгарский 170635
Подмаренник дудчатый 170378

Подмаренник дустый 170524
Подмаренник закавказский 170690
Подмаренник зеленовато – белый 170315
Подмаренник золотистый 170237
Подмаренник золотой 170237
Подмаренник иглолистный 170650
Подмаренник изветняковый 170301
Подмаренник изменчивый 170510
Подмаренник инсубрийский 170425
Подмаренник иссополистный 170423
Подмаренник камчатский 170434
Подмаренник кандинский 170568
Подмаренник каракульский 170440
Подмаренник касрийский 170311
Подмаренник кожистй 170329
Подмаренник коленчатый 170391
Подмаренник копетдагский 170449
Подмаренник коротколистный 170281
Подмаренник крестовидный 170335
Подмаренник круглолистный 170589
Подмаренник крылова 170450
Подмаренник крымский 170451
Подмаренник кутцинга 170452
Подмаренник кяпазский 170442
Подмаренник лепчица 170193
Подмаренник лесной 170631
Подмаренник линчевского 170460
Подмаренник лицзянский 170383
Подмаренник ложный 170193,170205
Подмаренник маймекенский 170473
Подмаренник мареновидный 170597
Подмаренник маролослый 170579
Подмаренник молочный 170455
Подмаренник мутовчатый 170742
Подмаренник мягкий 170490
Подмаренник настоящий 170743,170751
Подмаренник необычный 170534
Подмаренник памиро – алайский 170532
Подмаренник парижский 170536
Подмаренник поярковый 170566
Подмаренник прямой 170372
Подмаренник пузырчатый 170288
Подмаренник пьемонтский 170548
Подмаренник разнолистный 170192
Подмаренник распростертый 170420
Подмаренник рупрехта 170601
Подмаренник русский 170603
Подмаренник саурский 170610
Подмаренник севанский 170626
Подмаренник северный 170246
Подмаренник сердцевидный 170329
Подмаренник синьцзянский 170777
Подмаренник скальный 170611
Подмаренник сырейщикова 170657
Подмаренник тайваньский 170658
Подмаренник тонкий 170662
Подмаренник тончайщий 170662
Подмаренник топяной 170717
Подмаренник трёхдольный 170696
Подмаренник трехнадрезанный 170696
Подмаренник трехнадрезный 170696
Подмаренник трёхнадрезный 170696
Подмаренник трехраздельный 170696
Подмаренник трёхраздельный 170696
Подмаренник трехрогий 170694
Подмаренник трёхрогий 170694
Подмаренник трехцветкововидный 170705
Подмаренник трехцветковый 170708
Подмаренник трёхцветковый 170708
Подмаренник туполистный 170190
Подмаренник туркестанский 170716
Подмаренник тяньшанский 170667
Подмаренник удивительный 170534
Подмаренник уссурийский 170724
Подмаренник фальшивый 170193
Подмаренник херсонесский 113135
Подмаренник хубэйский 170421
Подмаренник цепкий 170193
Подмаренник чабрецелистный 170609
Подмаренник черенанова 170337
Подмаренник членистый 170216
Подмаренник шероховатый 170614
Подмаренник шершавы 170228
Подмаренник широкоподмаренниковый 170558
Подмаренник шишкина 170615
Подмаренник шультеса 170616
Подмаренник юзепчука 170433
Подмаренник юньнаньский 170779
Подмаренник японский 170445
Подобабик 3127
Подобабок 3127
Подокарп 306424
Подокарпус 306395
Подокарпус китайский 306457
Подокарпус крупнолистный 306457
Подокарпус крупнолистный маки 306469
Подокарпус ланьюйский 306419
Подокарпус наги 306493
Подокарпус тайваньский 306431
Подокарпус тотара 306522
Подокарпус чилийский 306510
Подоложникоцветник 319996
Подоложникоцветник двсторонний 319997
Подоложникоцветник колосовидный 320001
Подоложникоцветник суворова 320002
Подоложникоцветник тонкоколосый 320000
Подонежник золотой 400675
Подорожник 301832
Подорожник азиатский 301871
Подорожник альпийский 301853
Подорожник блошный 301864
Подорожник большой 302068,302101,344009
Подорожник войлочный 302025
Подорожник горечавковый 301980
Подорожник горный 302112
Подорожник замечательный 302114
Подорожник индийский 302009
Подорожник камчатский 301894
Подорожник комарова 302024
Подорожник корнута 301909
Подорожник крупноплодный 302062
Подорожник ланцетный 302034
Подорожник ланцетовидный 302034
Подорожник ланцетолистный 302034
Подорожник лёфлинга 302054
Подорожник маленький 302107
Подорожник малый 302107
Подорожник мелкоцветковый 302196
Подорожник многосемянный 302132
Подорожник морской 234150,302091
Подорожник паутинистый 301858
Подорожник перистный 301910
Подорожник песчаный 301864
Подорожник прижатый 301952
Подорожник приземистый 301952
Подорожник приморский 234150,302091
Подорожник пушистоголовый 302026
Подорожник седоватый 301897
Подорожник собачья голова 301945
Подорожник средний 302103
Подорожник степной 302183
Подорожник тибетский 302199
Подорожник чернеющий 302112
Подорожник шварценберга 302169
Подорожник яйцевидный 302119
Подорожник японский 302014
Подорожниковые 301821
Подостемон 354579
Подофилл 306609
Подофилл гималайский 365009
Подофилл щитовидный 306636
Подофилл эмода 365009
Подофиллум 306609
Подофиллюм 306609
Подсвекольник 18810
Подснежник 169705
Подснежник азербайджанская 352884
Подснежник альпийский 169706
Подснежник армянская 352883
Подснежник белый 169719
Подснежник бухарская 352889
Подснежник виноградова 353114
Подснежник воронова 169732
Подснежник гогенаккера 352912
Подснежник дизанская 352895
Подснежник закавказский 169731
Подснежник кавказская 352890

Подснежник кавказский 169707
Подснежник мищенко 352970
Подснежник одноцветковая 352972
Подснежник пушкиниевидная 353036
Подснежник раевского 353039
Подснежник розена 353051
Подснежник сибирский 353065
Подснежник складчатый 169726
Подснежник снеговая 352981
Подснежник снеговой 169719
Подснежник снежный 169719
Подснежник широколистный 169717
Подсолнечник 188902
Подсолнечник абиссинский 181873
Подсолнечник клубневый 189073
Подсолнечник клубненосный 189073
Подсолнечник культурный масличный 188908
Подсолнечник обыкновенный 188908
Подсолнечник огурцеобразный 188936, 188941
Подсолнечник однолетний 188908
Подсолнечник серебристолистный 188922
Подсолнечник слабый 188936,188941
Подсолнечник темно-красный 188926
Подъельник 202761,258006
Подъельник обкновенный 202767
Поздника 367416
Поинсеттия 159675
Поинсеция 159675
Полба 398970
Полба волжская 398890
Полба двузёрная 398890
Полба настоящая 398970
Полба ненастоящая 398890
Полба обыкновенная 398890
Полевица 11968,147468
Полевица анадырская 11998
Полевица белая 11972
Полевица белая высокая 11972
Полевица беловатая 11995
Полевица биберштейна 12015
Полевица болотная 12323
Полевица бороздчатая 12096
Полевица булавовидная 12053
Полевица волосовидная 12034,12323
Полевица гигантская 12118
Полевица гиссарская 310112
Полевица гладконожковая 12357
Полевица далиская 12365
Полевица закаспийская 12386
Полевица зимующая 12139
Полевица карская 12323
Полевица красная 12280
Полевица крупнометельчатая 12179
Полевица куэньминская 12163
Полевица лазистанская 12168
Полевица лицзянская 12292
Полевица монгольская 12080
Полевица мутовчатая 12410
Полевица обыкновенная 12323,37350
Полевица остистая 12356
Полевица отклоненная 12104
Полевица памирская 12232
Полевица паульсена 12235
Полевица плосколистная 12256
Полевица побегообразуюшая 11972
Полевица ползучая 12323
Полевица приморская 12323
Полевица северная 12196
Полевица сибирская 12323
Полевица сычуаньская 12061
Полевица собачья 12025
Полевица собачья опущённая 12025
Полевица солончаковая 12283
Полевица тайваньская 12029
Полевица тебердинская 12371
Полевица тонкая 12034,12323
Полевица тонколистная 12413
Полевица триниуса 12413
Полевица туркестанская 12403
Полевица шаньдунская 12301
Полевица якутская 12151
Полевичка 147468
Полевичка абиссинская 147994
Полевичка большой 147797
Полевичка волосистая 147883
Полевичка гуансиская 147703
Полевичка душистая 147986
Полевичка дэциньская 147631
Полевичка египетская 147475
Полевичка килианская 147593
Полевичка косинского 147754
Полевичка крупноколосковая 147593
Полевичка малая 147815
Полевичка нежная 147995
Полевичка приятная 147995
Полевичка ржавая 147671
Полевичка старосельского 147978
Полевичка тефф 147994
Полевичка тростниковая 147602
Полевичка тростниковидная 147602
Полевичка удлиненная 147656
Полевичка фуцзяньская 147673
Полевичка хайнаньская 147708
Полевичка черная 147844
Полевичка чилийская 147593
Полевичка японская 147746
Полевка 260569
Полёвка 260569
Полевка пронзеннолистмная 260577
Полевой очный цвет 21339
Поледай 45581
Полей 250432
Полемониум 307190
Полетай 45448
ползун 6639
Ползунок 184702
Ползунок русский 184711
Ползунок солончаковый 184713
Полиальтия 307509
Полиантес 307265
Полиантес тубероза 307269
Полигонум 308690
Полигонум далиский 309839
Полигонум китайский 308965
Полигонум мелкоцветковое 309442
Полигонум сахалинский 309723
Полигонум японский 309262
Полимния 310021
Полиотирсис 307290
Полиотирсис китайский 307291
Полипогон 310102
Полипогон монпелийский 310125
Политаксис винклера 310678
Политаксис лемана 310677
Поллиния 90107,307355
Поллиния безбородая 254034
Поллиния голая 254036
Пололенестник зеленый 98586
Пололепестник 98560,183389
Пололепестник зелёный 98586
Полуводосбор 357918
Полуклевер 247366
Полуница 167674
Полуполба 398890
Полынёк 155170
Полынок 155170,155173
Полынь 35084,36202
Полынь австрийская 35188
Полынь адамка 35095
Полынь азотистая 360857
Полынь алтайская 35113
Полынь аральская 35144
Полынь арги 35167
Полынь аржий 35167
Полынь арктическая 35153
Полынь армянская 35176
Полынь ашурбаева 35178
Полынь бадхызская 35193
Полынь бальджуанская 35196
Полынь баргузинская 35197
Полынь бедноцветковая 360835
Полынь белая 35612
Полынь беловосковая 35105
Полынь белоземельная 360837,360879
Полынь белоземная 360879
Полынь белолистная 35816
Полынь белотравная 35600
Полынь блестящая 36312
Полынь бловойлочная 35612

Полынь божье дерево 35088
Полынь болотная 36048
Полынь бурочешуйковая 36076
Полынь ваханская 36443
Полынь веничная 36232
Полынь вильчатая 35520
Полынь волосистоцветная 35464
Полынь волосовидная 35282
Полынь высокая 36105
Полынь ганьсуская 35468
Полынь гипсовая 35586
Полынь гладкая 35530
Полынь гмелина 35560
Полынь горная 35948
Полынь горькая 35090
Полынь гребенчатая 264249
Полынь гуншаньская 35573
Полынь дагестанская 35362
Полынь дернистая 35230
Полынь джунгарская 36304
Полынь длинная 36089
Полынь долинная 35088
Полынь душистая 35501
Полынь заилийская 360883
Полынь заменяющая 35343
Полынь замещающая 35343
Полынь заостренная 35361
Полынь звёздчатая 35779
Полынь зибера 36285
Полынь золотистая 35187
Полынь илийстая 35827
Полынь индийская 35634
Полынь кавказская 35314
Полынь каменная 35505,36166
Полынь канадская 35273
Полынь каратавская 360847
Полынь каспийская 35313
Полынь кашгарская 360848
Полынь кейске 35733
Полынь клейковатая 36349
Полынь кнорринг 35740
Полынь коидзуми 35742
Полынь комарова 35648
Полынь копетдагская 35748
Полынь коровина 360850
Полынь красная 217353
Полынь красноножковая 36162
Полынь крепкая 36444
Полынь крупнокорзиночная 35859
Полынь крупнокорневищная 35860
Полынь крупноцветковая 35857
Полынь крымская 36367
Полынь кульбадская 35752
Полынь куропаточья 35785
Полынь кустарниковая 35088
Полынь кушакевича 35754
Полынь лавандолистная 35794
Полынь лапчатколистная 36091
Полынь ледебура 35808
Полынь лемана 360851
Полынь лемановская 360851
Полынь лесная 36354
Полынь лессинга 35815
Полынь лессинговидная 360878
Полынь лессинговская 35815
Полынь лечебная 35088
Полынь липского 35830
Полынь лопастелистная 35833
Полынь лохматая 36054
Полынь максимовича 35895
Полынь малоцветковая 360835
Полынь маршалла 35888
Полынь маршалловская 35888
Полынь мексиканская 35904
Полынь мелкоцветковая 36059
Полынь метельчатая 36232
Полынь монгольская 35916
Полынь морская 35868
Полынь мощнокорневая 36129
Полынь мясистая 36350
Полынь надрезанная 35759
Полынь намангакская 35974
Полынь настоящая 35090
Полынь непахучая 35411
Полынь норвежская 35993
Полынь обедненная 35369
Полынь обыкновенная 36474
Полынь однолетная 35132
Полынь однолетняя 35132
Полынь однопестичная 35944
Полынь островная 35211
Полынь отставленнолопастая 36137
Полынь памирская 35421
Полынь персидская 36072
Полынь песчаная 36032
Полынь пижмолистная 36363
Полынь плотная 360827
Полынь побегоносная 36321
Полынь поздная 36276
Полынь полевая 35237
Полынь полусухая 360874
Полынь понтийская 36088
Полынь прбрежная 35831
Полынь приморская 35868
Полынь приятная 360812
Полынь промежуточная 35897
Полынь прутьевидная 360873
Полынь пустынная 35370,35373
Полынь пучковатая 35479
Полынь пышная 36474
Полынь пятидольчатая 36134
Полынь равнинная 35237
Полынь раздельнолистная 35759
Полынь рассеченная 35759
Полынь розовоцветковая 360866
Полынь ромашколистная 35323
Полынь ромашниковая 35323
Полынь рутолистая 36173
Полынь самоедов 36216
Полынь сантолинная 360870
Полынь сантолинолистая 35560
Полынь сапожникова 36221
Полынь светлозеная 36094
Полынь священная 36177
Полынь северная 35211
Полынь седая 35627
Полынь селенгинская 36241
Полынь селитряная 360857
Полынь сенявина 36259
Полынь серая 35538
Полынь сероземная 360879
Полынь сиверса 36286
Полынь сиверсовская 36286
Полынь сизая 35538
Полынь ситниковая 360845
Полынь ситниковидная 360845,360846
Полынь сычуаньская 36283
Полынь скалистая 36166
Полынь скальная 36166
Полынь совича 36357
Полынь солончаковая 35595
Полынь солянковая 36211
Полынь солянковидная 36211
Полынь стеллера 35779
Полынь степная 36319
Полынь таврическая 36367
Полынь тибетская 264250,360882
Полынь тилезиуса 36383
Полынь тонковатая 360835
Полынь тонковойлочная 36391
Полынь тонкорассеченная 36368,360847
Полынь точечная 36122
Полынь травянистая 35606
Полынь траутфеттера 36396
Полынь туподольчатая 36012
Полынь туркменская 36429
Полынь турнефоровская 36394
Полынь угнетенная 35365
Полынь узкокорзиночная 36317
Полынь укрополистная 35119
Полынь уналашкская 36440
Полынь уссурийский 36442
Полынь федченко 360830,360843
Полынь ферганская 360831
Полынь фрейна 35503
Полынь холодная 35505
Полынь худощавая 35856
Полынь цельнолистная 35648
Полынь цзилунский 35719
Полынь цинхайская 35422
Полынь цитварная 360825

Полынь чаюйский 36567
Полынь черняевская 36426
Полынь чжундяньский 36570
Полынь шаровницевая 35547
Полынь шелковистая 36262
Полынь шелковистолистная 36273
Полынь шерстистая 35314
Полынь шиловидная 36348
Полынь широкая 36108
Полынь широколистная 35789
Полынь шренка 360872
Полынь шренковская 360872
Полынь щмидта 36225
Полынь эдельвейсовая 35550
Полынь эмэйская 35463
Полынь эстрагон 35411
Полынь эстрагоновидная 35405
Полынь юньнаньский 36564
Полынь ядунский 36560
Полынь якутская 35671
Полынь японский 35674
Полынь японский хайнаньский 35701
Полынь – абсент 35090
Полыньлесная 36292
Полынь – чернобыль 36474
Польба 398970
Польба настоящая 398970
Полюшка 408571
Поляника 338142
Помадерис 310755
Помароза 259186,382582
Помело 93579
Померанец 93332,93414
Померанец горький 93332,93414
Помидор 239154
Поммело 93579
Помпельмус 93579
Помпельмус гроздевидный 93690
Понтедериевые 93690,311019
Понтедерия 311019
Понтийсая азалия 331161
Понцирус 310848,310990
Понцирус трехлистный 310850
Понцирус трехлисточковый 310850
Понцирус трёхлисточковый 310850
Понцирус трифолиата 310850
поповиокодония 311037
поповиокодония узкоплодная 311039
поповиокодония уэмуры 311040
Поповник 227533,322640
Поповник девичий 322724
Поповникниванка 227533
Поповник обыкновенный 227533,383874
Поповник посевной 89704
Поповник сибирский 89723
Поповник цинерариелистный 322660
Поповник шитковидный 89421
Поповник шитковый 89421
Поповник шитконосный 89421
Порезник 228556
Порезник бухторминский 228568
Порезник горный 361526
Порезник длинностолбиковый 228578
Порезник жаблицевидный 228597
Порезник закавказская 228605
Порезник илийский 228583
Порезник плотный 228571
Порезник промежуточный 228585
Порезник пушистоплодный 228581
Порезник седовадый 228584
Порезник сибирский 228599
Порезник странный 228592
Порезник чамечный 228570
Порезник шренковский 228596
Порезник щетинконосный 228598
Порей 15621
Поречник 292961
Портулак 311817
Портулак декоративный 311852
Портулак крупноцветконый 311852
Портулак крупноцветный 311852
Портулак овошной 311890
Портулак овышной 311890
Портулак огородный 311890
Портулак хайнаньский 311857
Портулакария 311956
Портулаковые 311954
Поручейник 365835
Поручейник ланцетолистный 365850
Поручейник привлекательный 365872
Поручейник приятный 365872
Поручейник сахарный 365871
Поручейник сахарный корень 365871
Поручейник средний 365854
Поручейник тонкий 365871
Поручейник узколистный 53157
Поручейник широколистный 365851
Поручейник японский 365858
Поручейница 79194
Поручейница водная 79198,177629
Поручейница водяная 79198,177629
Поручейница капю 79203
Порфира 311736
Посконник 158021
Посконник глена 158136
Посконник китайский 158070
Посконник коноплевидный 158062
Посконник коноплевый 158062
Посконник конопляный 158062
Посконник линдлея 158200
Посконник наньчуаньский 158238
Посконник хиаляньский 158153
Посконник цзилунский 158182
Посконник эмэйский 158243
Посконь 71218
Постенница 284131
Постенница апчечная 284167
Постенница иудейская 284152
Постенница лекарственная 284167
Постенница меклоцветковая 284160
Постенница мокричнолистная 284133
Постенница прямая 284145
Постенница слабая 284160
Постенница херсонская 284134
Потак – арча 213934
Поташник 215322
Поташник каспийский 215324
Поташник олиственный 215327
Поташник шренка 215330
Потентилла 312322
Поточник 407029
Почейник 365835
Почейник средний 365854
Почейник широколистный 365851
Почечуй 309570
Почечуйная трава 309570
Почечуйник 309570
Прангос 313516
Прангос бесстельный 313517
Прангос бухарский 313518
Прангос кормовой 313528
Прангос курчавокрылый 313530
Прангос липского 313526
Прангос складчатокрылый 313527
Прангос узкоплодный 313520
Прангос федченко 313522
Прангос чимганский 313529
Прангос широкоольчатый 313524
Пренантес 313778
Пренантес белый 313787
Пренантес татаринова 313895
Прибрежник 234146
Прибрежник одноцветковый 234150
Прибрежница 8855,234146
Прибрежница китайская 8884
Прибрежница одноцветковая 234150
Прибрежница ползучая 8881
Прибрежница прибрежная 8862
Прибрежница синьцзянская 8870
Прибрежница солончаковая 8862
Прибрежный 318737
Приворот 11542
Прилипало 7427
Примула 314083
Примула аньхуйская 314637
Примула аптечная 315082
Примула бесстебельная 315101
Примула ван 315111
Примула высокая 314333
Примула далиская 315030
Примула зибольда 314952

Примула китайская 314975
Примула максимовича 314624
Примула мальвоподобная 314613
Примула мийябе 314655
Примула мичнистая 314366
Примула мулиская 314178
Примула мэнцзыская 314738
Примула настоящая 315082
Примула обконика 314717
Примула обыкновенная 315101
Примула сычуаньская 315028
Примула тибетская 315049
Примула тэнчунская 314242
Примула фэнсянская 314985
Примула цинхайская 314874
Примула чжундяньская 314246
Примула чжэнькканская 314259
Примула ядунская 314991
Примула японская 314508
Приноготковник 284735
Приноготовник 284735
Приноготовник головчатый 284808
Приноготовник курдский 284866
Принсепия 315171
Принсепия китайская 315176,365002
Принцепия 315171
Принцепия китайская 315176,365002
Принцепия тайваньская 315174
Принцепник 79614
Принцепник бишофа 79627
Принцепник липучковый 79664
Принцепник морковновидный 79664
Принцепник морковный 79664
Прионотрихон 315246
Приречный клен 2984
Притчардия 315376
Пробковое дерево 294231
Пробосцидеа 315430
Пробосцидеа луизианская 315434
Пробосцидея 315430,416581
Прозанник 202387
Прозанник крапчатый 202424
Прозанник одноголовчатый 202444
Прозанник пятнистый 202424
Прозанник реснитчатый 202400
Прозопис 315547
Прозопис серёжкоцветный 315551,315560
Пролеска 250597,352880
Пролеска двулистная 352886
Пролеска индийская 352920
Пролеска итальянская 199555
Пролеска многолетняя 250614
Пролеска морская 352960
Пролеска однолетняя 250600
Пролеска осенняя 352885
Пролеска перуская 353001
Пролеска прелестная 352883
Пролеска пролесковидная 353057
Пролеска сибирская 353065
Пролеска яйцевиднолистная 250613
Пролеска японская 353057
Пролесник 250597
Пролесник гладкоплодный 250610
Пролесник многолетный 250614
Пролесник однолетний 250600
Пролесник яйцевидный 250613
Проломник 23107
Проломник альпийский 23121
Проломник арктический 23122
Проломник арминский 23124
Проломник большой 23227
Проломник бородчатый 23127
Проломник бунге 23215
Проломник восточнокавказский 23119
Проломник гмелина 23182
Проломник горовчатый 23213
Проломник гуйцзоуский 23208
Проломник двузубчатый 23128
Проломник зонтичный 23323
Проломник кандинский 23216
Проломник козо－полянского 23207
Проломник крупноцветковый 23222
Проломник лемана 23213
Проломник лемановский 23213
Проломник мелколистный 23234
Проломник молочноцветковый 23210
Проломник мохнатый 23326
Проломник нитевидный 23170
Проломник овчинникова 23250
Проломник ольги 23247
Проломник опадающий 23136
Проломник охотский 23246
Проломник промежуточный 23205
Проломник радде 23259
Проломник ресничатыймохнатый 23144
Проломник северный 23294,23295
Проломник седой 23202
Проломник сычуаньский 23309
Проломник таврический 23314
Проломник тибетский 23224
Проломник трехцветковый 23321
Проломник турчанинова 23227
Проломник удлиненный 23163
Проломник федчанко 23167
Проломник чаюйский 23347
Проломник шерстистолистный 23155
Проломник эмэйский 23252
Проломник ядунский 23199
Просвирник 243738
Просвирник армянский 243750
Просвирник бухарский 243758
Просвирник вылезанный 243783
Просвирник голоплодный 243793
Просвирник гроссгейма 243787
Просвирник красивенький 243821
Просвирник курчавый 243771
Просвирник лесной 243840
Просвирник мелкоцветковый 243810
Просвирник могилевский 243801
Просвирник мутовчатый 243862
Просвирник низкий 243823
Просвирник ницций 243805
Просвирник памироалайский 243809
Просвирник пренебреженный 243823
Просвирник прямостоячий 243781
Просвирник спорный 243747
Просвирник штокрозовый 243746
Просвирниковые 243873
Просвирняк 243738
Просвирняк аптечный 243806
Просвирняк китайский 243833,243862
Просвирняк красивый 243862
Просвирняк круглолистный 243823
Просвирняк курчавый 243771
Просвирняк мавританский 243797
Просвирняк мускусный 243803
Просвирняк мутовчатый 243862
Просвирняк обыкновенный 243823
Просвирняк плебейский 243818
Просвирняк приземистый 243823
Просвирняк раздельнолистный 243746
Просвирняковые 243873
Просвирнячок 362473
Просо 281274
Просо африканское 143522,289116
Просо верхоцветное 120632
Просо ветвистое 58155
Просо власовидное 281438
Просо волосовидное 281438
Просо воробьиное 21335,21339
Просо гвинейское 281887
Просо гигантское 281887
Просо головчатое 361794
Просо двбороздчатое 281390
Просо индийское 341960,369720
Просо итальянское 361794
Просо китайское 140423,369700
Просо крованое 130745
Просо крупное 281887
Просо культурное 281916
Просо куриное 140367
Просо метельчатое 281916
Просо мучнистое 140367
Просо настоящее 140367,281916
Просо негритянское 289116
Просо обыкновенное 281916
Просо петушье 140367
Просо посевное 140367,281916
Просо прутовидное 282366
Просо прутьевидное 282366
Просо рисовое 140462

Просо сычуаньский　289258
Просо татарское　361794
Просо японское　140367,140423
Простолистник　185615
Простолистник крымский　185672
Прострел　321653
Прострел албанский　321656
Прострел армянский　321661
Прострел аянский　321655
Прострел бунге　321663
Прострел весенний　24109
Прострел галлера　321682
Прострел горный　321692
Прострел груботерстистый　321721
Прострел грузинский　321679
Прострел даурский　321676
Прострел желтелющий　321707
Прострел золотистый　321662
Прострел китайский　321672
Прострел колокольчатый　321665
Прострел костычева　321688
Прострел крупный　321681
Прострел луговой　23998
Прострел луговой раскрытый　321703,321721
Прострел многонадрезной　321708
Прострел многонадрецный　321708
Прострел многораздельный　321708
Прострел обыкновенный　321721
Прострел поникающий　321667,321703,321721
Прострел пониркший　321703
Прострел раскрытый　321703,321721
Прострел распростёртый　321703,321721
Прострел сахалинский　321713
Прострел сомнительный　321659
Прострел сон－трава　321703,321721
Прострел сукачева　321714
Прострел тонколопастный　321715
Прострел турчанинова　321716
Прострел фиолетовый　321720
Прострел чернеющий　321696
Просяник　140367,197306,254496
Просяник развесистый　254515
Просянка куриная　140367
Протейные　316007
Протея　315702
Протея медоносная　315937
Прунус　316166
Прунус войлочный　83347
Прунус железистый　83208
Прунус ивовый　316761
Прунус кандинский　83343
Прунус кистевой　280003
Прунус ланьюйский　316436
Прунус маака　280028
Прунус максимовича　83255
Прунус маньчжурский　34443
Прунус мексиканский　316553
Прунус мэнцзыский　83216
Прунус сардента　83301
Прунус сибирский　34470
Прунус сычуаньский　83340
Прунус тяньшанский　83345
Прунус чжэцзянский　316488
Прутняк　217353,411187
Прутняк веничный　217361
Прутняк китайский　411362
Прутняк китайский коноплеволистный　411373
Прутняк обыкновенный　411189
Прутняк песчаный　217331
Прутняк распростертый　217353
Прутняк ситниковидный　370202
Прутовник　370202
Прыгун　204756,205175
Прыгун аньхуйский　204775
Прыщенец　325864
Прыщинец　325864
Прямохвостник　275850
Прямохвостник кокандский　275853
Прямохвостник разноплодный　275852
Псевдоакация　334976
Псевдобактерия　315702
Псевдоерантемум　317245
псевдолинозирис　317839
псевдолинозирис гримна　317842
псевдолинозирис мелкоголовчатая　317843
псевдолинозирис синтениса　317844
Псевдонарцисс　262441
Псевдосаза　318277
Псевдосаза фуцзяньская　318284
Псевдосаза хайнаньская　318299
Псевдосаза японская　318307
Псевдотсуга　318562,318580
Псевдотсуга сичанская　318602
Псевдотсуга японская　318575
Псевдохандалия　317790
Псевдохандалия зонтичная　317791
Псидиум　318733
Псидиум гваява　318742
Псидиум лавролистный　318740,318747
Психотрия　319392
Психотрия красная　319810
Психотрия ланьюйская　319477
Психотрия люцюская　319671
Психотрия тибетская　319532
Психотрия хайнаньская　319569
Псолалея　319116
Псолалея смолоносная　319139
Псоралея　114427
Птармика　4005
Птелея　320065
Птелея желтоватая　320068
Птелея трехлистная　320071
Птелея трёхлистная　320071
Птелея трехлисточковая　320071
Птерокактус　320258
Птерокария　320346
Птерокария рябинолистная　320356
Птерокарпус　320269
Птерокарпус бирманский　320306
Птерокарпус дальбергиевидный　320288
Птерокарпус драконовый　320291
Птерокарпус индийский　320301
Птерокарпус кабонский　320330
Птерокарпус крупноплодный　320306
Птерокарпус палисандровидный　320288
Птерокарпус сандаловый　320327
Птерокарпус санталинус　320327
Птерокарпус санталовый　320327
Птерокарпус сойакси　320330
Птероксилон　320032
Птероспермум　320827
Птероспермум кленолистный　320828
Птероспермум цзиндунский　320839
Птеростиракс　320878
Птеростиракс мелкоцветковый　320885
Птеростиракс щетинистый　320886
Птеростиракс щщитковый　320880
Птероцелтис　320258
Птероцефалюс　320421
Птероцефалюс афганский　320423
Птероцефалюс перистый　320442
Птилагростис　321004
Птилагростис красивый　321008
Птилагростис монгольский　321016
Птилагростис пурпуровый　376893
Птилагростис сидячецветковый　376924
Птилотрихум　321087
Птилотрихум седоватый　321089
Птилотрихум тибетский　321099
Птилотрихум удлиненный　321089
Птихосперма　321161
Птицемлечник　274495
Птицемлечник балансы　274526
Птицемлечник бахромчатый　274815
Птицемлечник буше　274537
Птицемлечник воронова　274851
Птицемлечник гирканский　274651
Птицемлечник дугообразный　274520
Птицемлечник желптицем　274617
Птицемлечник жетоватый　274617
Птицемлечник закавказский　274813
Птицемлечник зонтичный　274823
Птицемлечник коротколосый　274538
Птицемлечник крупный　274687
Птицемлечник пиренейский　274748
Птицемлечник плосколистный　274735
Птицемлечник поникший　274708
Птицемлечник преломленный　274755

Птицемлечник ресничатый 274815
Птицемлечник синтениса 274782
Птицемлечник тонколистный 274799
Птицемлечник тонкоцветковый 274636
Птицемлечник фишера 274614
Птицемлечник фишеровский 274614
Птицемлечник шишкина 274770
Птицемлечник шмальгаузена 274772
Птицемлечник шолковникова 274769
Птиценожка 274895
Пуанзеция 307076
Пуговичник 383874
Пуерария 321419
Пуерария волосистая 321441
Пуерария лопастная 321441
Пузадник 171809
Пузатик 171809
Пузатик обыкновенный 171818
Пузрария мелкоцветковая 321464
Пузрария опушенная 321441
Пузрария эмэйская 321457
Пузыреплодник 100201,140423,297833
Пузыреплодник амурский 297837
Пузыреплодник калинолистный 297842
Пузыреплодник сетчатый 100202
Пузыреплодник смородинолистный 297851
Пузыресеменник 122121,122124
Пузыресемянник 122121,122124
Пузырник 100149,100153,114057,227017,407549
Пузырник армянский 100165
Пузырник бузе 100169
Пузырник восточный 100187
Пузырник гибридный 100177
Пузырник голоплодный 407551
Пузырник древовидный 100153
Пузырник киликийский 100171
Пузырник копетдагский 100180
Пузырник кровавый 100187
Пузырник непалский 100185
Пузырник обыкновенная 100153
Пузырник остролистный 100152
Пузырник паульсена 100188
Пузырник прецеский 407550
Пузырник седоватый 100170
Пузырник тонгий 100175
Пузырник ярмоленко 100179
Пузырница 297875
Пузырница восточная 297881
Пузырница илийская 297876
Пузырница семенова 297885
Пузырница тибетская 297883
Пузырница физалисовая 297882
Пузырчатка 403095
Пузырчатка большая 403246
Пузырчатка брема 403130
Пузырчатка забытая 403191
Пузырчатка малая 403252
Пузырчатка недосмотренная 403110
Пузырчатка обыкновенная 403382
Пузырчатка средная 403220
Пузырчатка хайнаньская 403113
Пузырчатка японская 403222
Пузырчатковые 224499
Пукалка 210291
Пулазан 265137
Пулазан плодовый 265137
Пуликария 321509
Пульсатилла 321721
Пуммело 93579
Пупавка 26738
Пупавка австрийская 26751
Пупавка анатолийская 26744
Пупавка белейшая 26759
Пупавка бесплодная 26886
Пупавка благородная 85526
Пупавка видемана 26906
Пупавка волосистая 26797
Пупавка вонючая 26768
Пупавка воронова 26907
Пупавка высочайшая 26743
Пупавка гроссгейма 26794
Пупавка грузинская 26799
Пупавка жестковадая 26871
Пупавка звксинская 26782
Пупавка зефирова 26909
Пупавка зигийская 26910
Пупавка зубчато-венечная 26840
Пупавка известняковая 26758
Пупавка карпатская 26760
Пупавка красильная 26900
Пупавка крупноязычковая 26815
Пупавка кустарниковая 26781
Пупавка кустарничковая 26787
Пупавка литовская 26811
Пупавка мархотская 26818
Пупавка маршалла-биберштейна 26752
Пупавка меловая 26770
Пупавка обыкновенная 26768
Пупавка однокорзиночная 26836
Пупавка полевая 26746
Пупавка пустынная 26779
Пупавка русская 26874
Пупавка сагурамская 26876
Пупавка светло-желтая 26892
Пупавка слабая 26777
Пупавка собачья 26768
Пупавка сомнительная 26780
Пупавка сосновского 26885
Пупавка талышская 26893
Пупавка темноокаймленная 26827
Пупавка шишкина 26880
Пупочная трава 318397
Пупочник 117900,270677,275347
Пупочник весений 270723
Пупочник каппадокийский 270684
Пупочник кузнецова 270703
Пупочник лойки 270707
Пупочник льнолистный 270705
Пупочник ползучий 270716
Пупочник скальный 270715
Пупочник японский 275375
Пупырник японский 392992
Пустореберник 80877
Пустореберник фишера 80879
Пусторебрышник 98772
Пусторебрышник гмелина 98777
Пустоягодник 396905
Пустынноколосник 148466
Пустынноколосник альберта 148469
Пустынноколосник аральский 148470
Пустынноколосник армянский 148486
Пустынноколосник байсунский 148472
Пустынноколосник бальджуанский 148473
Пустынноколосник бесприцветничковый 148479
Пустынноколосник блистающий 148482
Пустынноколосник буасье 148474
Пустынноколосник гипсолюбивый 148485
Пустынноколосник гиссарский 148487
Пустынноколосник головчатколистный 148476
Пустынноколосник голочашечковый 148484
Пустынноколосник голый 148498
Пустынноколосник гребенчатый 148500
Пустынноколосник гртзинский 148488
Пустынноколосник губастообразный 148493
Пустынноколосник губастый 148492
Пустынноколосник железконосный 148483
Пустынноколосник заамударьинский 148513
Пустынноколосник зеравшанский 148507
Пустынноколосник зинаиды 148517
Пустынноколосник зопниковый 148501
Пустынноколосник илийский 148489
Пустынноколосник кауфмана 148490
Пустынноколосник колесовидный 148505
Пустынноколосник коровина 148491
Пустынноколосник красивый 148508
Пустынноколосник крувненосный 148514
Пустынноколосник лемана 148495
Пустынноколосник лопухолистный 148471
Пустынноколосник метельчатый 148499
Пустынноколосник надрезный 148494
Пустынноколосник одноцветковый 148515
Пустынноколосник попова 148502
Пустынноколосник почни-колосовый 148510
Пустынноколосник прекрасный 148503

Пустынноколосник пустынный 148478
Пустынноколосник регеля 148504
Пустынноколосник родственный 148468
Пустынноколосник седоватый 148475
Пустынноколосник сердцелистный 148477
Пустынноколосник таджикский 148511
Пустынноколосник тяньшанский 148512
Пустынноколосник фетисова 148481
Пустынноколосник шерстисточашечный 148480
Пустынноколосник широкочашечный 148497
Пустынноколосник шугнанский 148506
Пустынноморковник 148265
Пустынноморковник лемана 148266
Пустынномятлик 148374
Пустынномятлик алтайски 148375
Пустынномятлик джунгарский 148389
Пустынномятлик красивый 148381
Пустынномятлик персидский 148385
Пустырник 224969
Пустырник крупноцветковый 224996
Пустырник монгольский 225003
Пустырник мохнатый 225022
Пустырник обыкновенный 224976
Пустырник панцериевидный 225004
Пустырник пятилопастный 225008
Пустырник разнолистный 224983
Пустырник сердечник 224976
Пустырник сибирский 225009
Пустырник сизоватый 224981
Пустырник сизый 224981
Пустырник татарский 225015
Пустырник туркестанский 225019
Пустырник японский 224989
Походлевник 25882,179876
Пухонос 396003
Пухонос альпийсийи 396005
Пухонос дернистый 353272,396010
Пухонос приземистый 396020
Пухополевица 375805
Пухополевица летучконоснаю 375809
Пуцинелла приземистый 321336
Пучкоцвет 293181
Пучкоцвет трубкоцветковый 293186
Пушеница 398834
Пушеница английская 398986
Пушеница варварийская 398986
Пушеница выносливая 398900
Пушеница карликовая 398877
Пушеница карталинская 398869
Пушеница круглозерная 398972
Пушеница круглозёрная 398877
Пушеница летняя 398839
Пушеница маха 398920
Пушеница мягкая 398839
Пушеница мягкая карликовая 398877
Пушеница мягкая яровая 398877
Пушеница однозернянка 398924
Пушеница плотная 398877
Пушеница полба 398970
Пушеница польская 398939
Пушеница прсидская 398869
Пушеница стекловидная 398900
Пушеница сьльная 398900
Пушеница таудар 398980
Пушеница твердая 398900
Пушеница тимофеева 398981
Пушеница тимофеевская 398981
Пушеница тучная 398986
Пушеница эгилопсовая 398838
Пушеница－ежовка 398877
Пушиница 398834
Пушиница ежовка 398877
Пушиница кавказская 398869
пушистоспайник 153358
пушистоспайник длиннолистный 153360
Пушица 152734
Пушица влагалищная 152791
Пушица корейская 152759
Пушица короткопыльниковая 152744
Пушица красивощетниковая 152746
Пушица многокомоейовая 152765
Пушица низкая 152763
Пушица рыжеватая 152772
Пушица стройная 152759
Пушица узколистная 152769
Пушица шамисо 152749
Пушица шейхцера 152781
Пушица широколистная 152765
Пушица японская 353606
Пушкиния 321892
Пушкиния гиацинтовидная 321893
Пушкиния пролесковидная 321896
Пуэрария 321419
Пуэрария волосистая 321441
Пуэрария лопастная 321441
Пуэрария опушенная 321441
Пуэрария опушённая 321441
Пфлюк－салат 219494
Пчелка 272398
Пшат 141932
Пшеница 398834
Пшеница абисинская 398836
Пшеница армянская 398857
Пшеница ветвистая 398839
Пшеница волжская 398890,399006
Пшеница двузернянка 398890
Пшеница двузернянковидная 398886
Пшеница культурная однозернянка 398924
Пшеница летная 398839
Пшеница мягкая 398839
Пшеница обыкновенная 398839
Пшеница однозернянка культурная 398924
Пшеница полба армянская 398857
Пшеница полба волжская 399006
Пшеница полоникум 398939
Пшеница спельта 398970
Пшеница твёрдая 398900
Пшеница－спельта 398970
Пыльцеголовник 82028,82113
Пыльцеголовник длиннолистный 82060
Пыльцеголовник длинноприцветниковый 82058
Пыльцеголовник дремликовидный 82043
Пыльцеголовник кавказский 82036
Пыльцеголовник красный 82070
Пыльцеголовник крупноцветковый 82038
Пыльцеголовник крупноцветный 82038
Пыльцеголовник тайваньский 82075
Пырей 11628,144624
Пырей алайский 144156
Пырей алатавский 144159,215903
Пырей амгунский 11638
Пырей армянский 11649
Пырей баталина 215905
Пырей безостый 144661
Пырей близкий 144651
Пырей ветвистый 11841
Пырей волосистый 144680
Пырей волосоносный 144680
Пырей выемчатый 11877
Пырей гребенчатый 11696
Пырей гребневидный 11699
Пырей дернистый 11659
Пырей длинноватый 144665
Пырей донской 11892
Пырей западный 11628
Пырей инееватый 11831
Пырей керченский 11693
Пырей ковылелистный 11886
Пырей коленчатый 11736
Пырей комарова 335389
Пырей красивейший 11837
Пырей крепкостебельный 11732
Пырей крылова 11779
Пырей маленький 11838
Пырей обыкновенный 144661
Пырей острец 11841
Пырей отогнутоостый 11844
Пырей оттянуточешуйный 11650
Пырей плевеловидный 11783
Пырей ползучий 144661
Пырей понтийский 11824
Пырей промежуточный 144652
Пырей простертый 11830
Пырей пустынный 11710
Пырей пушистоцетковый 11704
Пырей раннедерновинный 144429
Пырей рожевица 11861
Пырей русский 144643

Пырей сибирский 11735
Пырей сизый 11737
Пырей синьцзянский 11876
Пырей ситниковый 144654
Пырей скифский 144459
Пырей смида 144675
Пырей собачий 144228
Пырей сосновского 11882
Пырей средний 144652
Пырей стройный 11743
Пырей угловатый 11641
Пырей удлиненный 144643
Пырей удлинённый 144643
Пырей узловадый 144413
Пырей черепичатый 11749
Пырей шренковский 335491
Пырей щетинистый 11888
Пырей щетинконосиный 11871
Пырей якутов 11754
Пьяная трава 374897
Пьяница 404027
Пэдеря 280063
Пятилистник 100257
Пятиостник 11968
Пятиостник гиссарский 310112
Пятичленник 290133
Пятичленник китайский 290134
Пятичленник низкий 290134
Рабдотэка 222907
Рабдотэка коровина 222947
Равенала 327093
Равенала мадагаскарская 327095
Равноплодник 210251
Равноплодник василистниковый 210291
Рагадиолюс 171475,328507
Рагадиолюс опушенный 328513
Рагадиолюс съедобный 328512
Раги 143522
Радиола 325039
Радиола льновидная 325041
Раздельнолодочник 177873
Раздельнолодочник дурнишниковидный 250734
Раздельнолодочник красноплодный 250730
Раздельнолодочник тройчатолистный 250733
Раздулка 269299,269347
Разнолепестка 203181
Разнолепестник 203181
Разнолепестник горький 203184
Разнолистник 203184
Разноорешек 301623
Разноорешек восточный 301626
Разноорешник 203181
Разноцвет 301623
Разрыв – трава 124008
Разумовския 349032
Райграсс 30906
Райграсс английский 235300,235334
Райграсс высокий 235334
Райграсс итальниский 235315
Райграсс итальяниский 34930
Райграсс многолетний 235315,235334
Райграсс многолетний английский 235334
Райграсс многолетний настбищный 235334
Райграсс многоукосный 235315
Райграсс пастбищный 235334
Райграсс французский 34920
Райграсс французский высокий 34930
Райграсс французский кочи 34945
Райдваки 309570
Рай – дерево 107300
Райка 243682
Ракита 343151
Ракитник 120903
Ракитник австрийский 120928
Ракитник альпийский 85403,218757
Ракитник белый 120909
Ракитник блоцкого 120942
Ракитник волосистый 120962
Ракитник вульва 121030
Ракитник вульфа 121030
Ракитник головчатый 85403
Ракитник днепровский 120943
Ракитник линдеманна 120970
Ракитник литвинова 120972
Ракитник метельчатый 121001
Ракитник метглистный 121001
Ракитник многоволосистый 120990
Ракитник многоволосковый 120990
Ракитник многоцветковый 120979
Ракитник пачоского 120986
Ракитник подольский 120989
Ракитник пупурный 85401
Ракитник пурпуровый 85401
Ракитник регенсбургский 120997
Ракитник рошеля 120998
Ракитник русский 121000
Ракитник скробишевского 121007
Ракитник скробишегского 121012
Ракитник цингера 121031
Ракитник чернеющий 120983
Ракитник – золотой дождь 218761
Раковые шейки 308893
Рамбутан 265129
Рами 56094,56229
Рами белое 56229
Рами белые 56229
Раминия одновокая 275486
Раминия одновочная 275486
Рамишия 275480
Рамишия однобокая 275486
Рамишия тупая 275482
Рамонда 325223
Рамондия 325268
Рамтиля 181873
Рамфикарпа 328917
Рандия 79595
Ранетка пурпурная 243555
Ранупкул 325498
Ранупкулюс 325498
Ранупкулюс тайваньский 326420
Рапанея 326523
Рапанея гуансиская 261626
Рапанея максимовича 261632
Рапидофиллум 329169
Рапидофиллюм 329169
Рапис 329173
Рапис высокий 329176
Рапис низкий 329184
Рапис приземистый 329184
Рапонтика 329388
Рапонтикум 329197
Рапонтикум аулиеатинский 329205
Рапонтикум блестящий 329226
Рапонтикум каратавский 329218
Рапонтикум карликовый 329225
Рапонтикум красивый 329227
Рапонтикум лировидный 329222
Рапонтикум наманганский 329224
Рапонтикум одноцветковый 375274
Рапонтикум сафлоровидны 375268
Рапонтикум серпуховидный 329231
Рапонтикум цельнолистный 329217
Рапс 59493,320330
Рапутник 269404
Расамала 18197
Растение табачное 266031
Растигор 388042
Расторопша 364353
Расторопша остропёстрая 364360
Расторопша пятнистая 364360
Раувольфия 326995
Раувольфия змеевидная 327058
Раувольфия кафрская 327001
Раувольфия мутовчатая 327069
Раувольфия седовадая 327067
Раувольфия тайваньская 327064
Раувольфия хайнаньская 327070
Раулия 326513
Рафидофитон 329019
Рафидофитон регеля 329020
Рафия 326638
Рафия мадгаскарская 326639
Рафлезиевые 325065
Раффия 326638
Раффлезиевые 325065
Раффлезия 325061
Рачьи шейки 308893
Раяния 325185
Рдест 312028

Рдест альпийкий　312037
Рдест блестящий　312168
Рдест взморниколистный　312075
Рдест влагалищный　312285
Рдест волосовидный　312282
Рдест вольфганга　312290
Рдест выемчатый　312274
Рдест геннинга　312136
Рдест гребенчатый　312083,378991
Рдест гречихолистный　312228
Рдест густолистный　312086
Рдест длиннейший　312232
Рдест длиннолистный　312166
Рдест длинный　312232
Рдест заостренный　312206
Рдест злаковидный　312138
Рдест злаковый　312126
Рдест зостеролистный　312075
Рдест красноватый　312259
Рдест курчавый　312079
Рдест маака　312170
Рдест малайский　312171
Рдест маленький　312236
Рдест малый　312236
Рдест нитевидный　378986
Рдест остролистный　312036,312206
Рдест отльчный　312090
Рдест палермский　312236
Рдест памирский　312208
Рдест плавающий　312190
Рдест плотный　312086
Рдест прерванный　312155
Рдест пронзеннолистный　312220
Рдест пронзённолистный　312220
Рдест разнолистный　312138
Рдест сичанский　312292
Рдест стеблеобъемлющий　312220
Рдест туполистный　312199,378983
Рдест узловатый　312194
Рдест фриса　312123,351439
Рдест хубэйский　312141
Рдестовые　312300
Ребраплодник　304752
Ребраплодник австлийский　304765
Ребраплодник камчатский　304772
Ребраплодник кандинский　304831
Ребраплодник лицзянский　304793
Ребраплодник тяньшанский　304817
Ребраплодник уральский　304766
Ребраплодник франше　304795
Ревень　329306
Ревень алтайский　329312
Ревень аптечный　329366
Ревень бесстебельный　329389
Ревень блестящий　329352
Ревень виттрока　329415
Ревень волнистый　329386
Ревень восточный　329368
Ревень гиссарский　329341
Ревень дарвазский　329323
Ревень дланевидный　329372
Ревень китайский　329366
Ревень компактный　329318
Ревень коржинского　329346
Ревень крупноплодный　329353
Ревень лекарственный　329366
Ревень лицзянский　329350
Ревень лопастный　329351
Ревень максимовича　329356
Ревень московский　329366
Ревень низкий　329361
Ревень обыкновенный　329366
Ревень овощной　329388
Ревень огородный　329388
Ревень пальмовидный　329372
Ревень пальчатый　329372
Ревень русский　329366
Ревень сердцевидный　329319
Ревень сетчатый　329385
Ревень скальный　329392
Ревень складчатый　329377
Ревень скученный　329318
Ревень смородинный　329391
Ревень тангутский　329401
Ревень татарский　329406
Ревень тибетский　329407
Ревень туполистный　329388
Ревень туркестанский　329408
Ревень федченго　329329
Ревень черноморский　329388
Ревень эмоди　329327
Регнерия собачья　144228
Регнермей　11696
Редечник　326812
Редис　326616
Редька　326578
Редька дикая　326609
Редька европейская　326616
Редька культурная　326616
Редька летняя　326616
Редька масличная　326616
Редька носатая　326614
Редька обыкновенная　326616
Редька огородная　326616
Редька полевная　326609
Редька посевная　326616
Редька приморская　326601
Редька Редовския　327344
Редька Редовския двояко – перистая　327345
Редька японская　326629
Редия　329268
Редия бразильская　329271
Редия обыкновенная　329274
Резак　162457
Резак обыкновенный　162465
Резак фалькариевидный　162460
Резанец　15709
Резаньглот　109933
Резеда　327793,408839
Резеда белая　327795
Резеда вухарская　327823
Резеда душистая　327896
Резеда желтая　327866
Резеда жёлтая　327866
Резеда желтенњкая　327879
Резеда желтоватая　327879
Резеда коротконогая　327822
Резеда красильная　327879
Резеда мелкоплодная　327887
Резеда непахучая　327857
Резеда оше　327816
Резеда пахучая　327896
Резеда цветмалиновый　327896
Резеда шароплодная　327855
Резедовые　327934
Резуха　30160,262015
Резуха альпийская　30164
Резуха амурская　30185
Резуха башенная　30477
Резуха беловая　30220
Резуха бильярдье　30199
Резуха бухарская　30211
Резуха висячая　30387
Резуха гладкая　400644
Резуха желтая　30267
Резуха желтушниковаю　30256
Резуха жерара　30277
Резуха кавказская　30220
Резуха камчатская　30345
Резуха канадская　30214
Резуха каратегинская　30325
Резуха коканская　30331
Резуха коротколодная　30207
Резуха кустарничковая　30270
Резуха липовидная　30342
Резуха максимовича　30111
Резуха мучнистая　30259
Резуха песчаная　30186
Резуха повислая　30387
Резуха порярная　30434
Резуха проежуточная　30311
Резуха рыхлая　30336
Резуха северная　30205
Резуха тайваньская　30268
Резуха ушастая　30189
Резуха ушковая　30189
Резуха фиолетовочашечная　30313
Резуха христиана　30225
Резуха шершавая　30292,30424
Резуховидка таля　30146
Резуховидка шведская　30144

Резушка 30105
Резушка весенняя 137327
Резушка выемчатая 30454
Резушка какандская 30324
Резушка карликовая 270461,270462
Резушка котовиковая 30365
Резушка маленькая 389199
Резушка монбретовская 30360
Резушка мягкопушистая 113151
Резушка низкая 270462
Резушка нордманна 30367
Резушка однобокая 30432
Резушка одуванчиколистая 30155
Резушка попова 30401
Резушка пушистоногая 30474
Резушка средняя 30357
Резушка стеллера 30456
Резушка стрелолистная 317344
Резушка таля 30146
Резушка тибетская 30472
Резушка толстокорневая 30380
Резушка турчанинова 30476
Резушка ядунская 30157,30406
Рейнвардтия 327548
Рейнекия 327524
Рейнекия мясистая 327525
Рейхардия 327462
Рейхардия вильчатая 327467
Рекурва Нана 390617
Ремания 327429
Ремерия 335624
Рёмерия 335624
Ремерия гибридная 335638
Рёмерия гибридная 335638
Ремерия отогнутая 335650
Ремнелепестник 196372
Ремнелепестник прекрасный 196374
Ремнецветник 236507
Ремнецветник гуйцзоуский 236755
Ремнецветник евронейский 236704
Ремнецветниковые 236503
Ремузация 327670
Ремузация тайваньская 327672
Ренклод 316333,316391
Рео 332293
Рео пестрое 394076
Реомюрия 327194
Реомюрия кашгарская 327212
Реомюрия туркестанская 327231
Реомюрияская амударьснская 327221
Реомюрияская бадхызская 327199
Реомюрияская закирова 327234
Реомюрияская кузнецова 327216
Реомюрияская кустарниковая 327204
Реомюрияская ладанниковая 327201
Реомюрияская отогнутая 327225
Реомюрияская персидская 327224
Репа 59575,74435
Репа венгерская 59541
Репа волошская 189073
Репа кервельная 84731
Репа кормовая 59575
Репа культурная 59575,74435
Репа овощная 59575
Репа оголодная 74435
Репа огородная 59575
Репа столовая 59575
Репа японская 59575,74435
Репак 59603
Репей 31051
Репей дикий 91508
Репейник 11542,31051,73278
Репейник волосистый 11572
Репейник душистый 11594
Репейник исполинский 31051
Репейник непалский 11587
Репейник обыкновенный 11549
Репейник пахучий 11594
Репейник японский 11566
Репейничек 11542
Репейничек азиатский 11552
Репейничек аптечный 11549
Репейничек бархатистый 11546
Репейничек волосистый 11572
Репейничек зернистый 11557
Репейничек пахучий 11594
Репейничек японский 11580
Репешок обыкновенный 11549
Репник 326812
Репник многолетний 326832
Репник морщинистый 326833
Репяшок 11542,83443
Репяшок обыкновенный 11549
Рестелла 414118
Рестелла альберта 327958
Рестио 327963
Реум 329306
Реум корейский 329320
Реутера 299342
Реутера золотистая 299360
Ржанец 144661
Ривина 334892
Ривина низкая 334897
Ризофора 329744
Ризофора остроконечная 329745
Ризоцефалус 329738
Ризоцефалус туркестанский 329739
Рикшаи – одири 247239
Римская ромашка 85526
Риндера 334540,334551
Риндера бальджуанская 334542
Риндера ежистая 334544
Риндера зонтичная 334558
Риндера коржинского 334549
Риндера кругомзубчиковая 334543
Риндера ошская 334554
Риндера продолговатолистная 334552
Риндера светрожелтая 334553
Риндера сплошь – опушенная 334548
Риндера тяньшанская 334556
Риндера уркестанская 334557
Риндера ферганская 334545
Риндера четырехостая 334555
Риндера четырехщитковая 334555
Риндера южно – ежистая 334541
Ринопеталум 168332
Ринопеталум ариаский 168347
Ринопеталум бухарский 168360
Ринопеталум горбатый 168409
Ринопеталум карелина 168438
Ринопеталум узкопыльниковый 168566
Ринхозия 333131
Ринхозия китайская 333182
Ринхозия куэньминская 333286
Рипария 411887
Рипсалис 329683
Рис 275909
Рис болотный 418080
Рис водяной 418080
Рис дидий 418080
Рис индийский 418080,418096
Рис канадский 418080
Рис озёрный 418080
Рис посевной 275958
Рисовидка 68877,275991,300700,334310,417092
Рисовидка алпийская 300707
Рисовидка африканская 417093
Рисовидка боковая 300724,300728
Рисовидка бухарниковая 276020
Рисовидка голубоватая 276003
Рисовидка джунгарская 300747
Рисовидка дрожащая 300755
Рисовидка замещающая 300756
Рисовидка зеленоватая 276068
Рисовидка иллларии 300717
Рисовидка китайская 276001
Рисовидка кокандская 276062
Рисовидка копетдагская 276028
Рисовидка краснеющая 300741
Рисовидка молиниевидная 276038
Рисовидка памироалайская 300737
Рисовидка согдийская 300746
Рисовидка тупая 300722
Рисовидка федченко 300712
Рисовидка ферганская 276008,300713
Рисовидка цирокоцветная 300740
Рисовидка широколистная 276032,300727
Рициа 334435
Рицин 334432
Рицинник клещевина 334435

Рицинус обыкновенный 334435
Рицинодендрон африканский 334411
Рицинус 334432,334435
Ричардия 68877,334310,417092
Робиния 334946
Робиния клейкая 335013
Робиния лжеакация 334976
Робиния ложноакация 334976
Робиния ложно－акация 334976
Робиния новомексикская 334974
Робиния нью－мексикская 334974
Робиния щетинистоволосая 334965
Робиния щетинистоволоситая 334965
Робиния щетиноволосая 334965
Рогач 83418
Рогач песчаный 83419
Рогач туркестанский 83419
Роговик униоловидный 61008
Роговник 245977
Рогоглавник 83438
Рогоглавник пряморогий 83443
Рогоглавник серповидный 83439
Рогоглавник серпорогий 83439
Рогоз 401085
Рогоз бледный 401131
Рогоз верещагина 401139
Рогоз восточный 401129
Рогоз лаксманна 401120
Рогоз ляксманна 401120
Рогоз малый 401128
Рогоз слоновый 401107
Рогоз суженный 401094,401105
Рогоз узколистный 401094
Рогоз узколистый 401094
Рогоз широколистный 401112
Рогоз широколистый 401112
Рогоз южный 401105
Рогоза восточная 401129
Рогозоные 401140
Рогомак 176724
Рогоплодник 390202
Рогоплодник длиннорогий 77469
Рогоплодник копьевидный 77472
Рогоплодник короткостолбчатый 77468
Рогоплодник узкоплодный 77474
Рогористик 83540
Рогористик донской 83583
Рогористик погруженный 83545
Рогористик полупогруженный 83578
Рогористик рисовый 83577
Рогористиковые 83539
Рогористник 83540
Рогористник комарова 83564
Рогористник коссинского 83573
Рогористник погруженный 83545
Рогористник погружённый 83545
Рогористник подводный 83545
Рогористник полупогруженный 83578
Рогористник темно－зеленый 83545
Рогористник темно－зелёный 83545
Рогористник японский 83575
Рогористниковые 1572,83539
Рогульки 124141
Рогульник 394412,394500
Рогульник водяной орех 394500
Рогульник плавающий 394500
Рогульниковые 394554
Рогурики 102833
Рогурьки 394500
Родея 335758
Родея японская 335760
Родиола 329826,356467
Родиола арктическая 329839
Родиола ганьнаньская 329876
Родиола ганьсуская 329894
Родиола дэциньская 329842
Родиола зеленоватая 329972
Родиола кашгарская 329896
Родиола кириллова 329897
Родиола комарова 329836
Родиола куэньминская 329901
Родиола линейнолистная 329897
Родиола литвинова 329905
Родиола лицзянская 329902
Родиола памироалайская 329917
Родиола перистонадрезная 329923
Родиола прямостебельная 329932
Родиола разнозубчатая 329881
Родиола розовая 329935
Родиола розовая мелколистная 329939
Родиола сахалинская 329944
Родиола северная 329845
Родиола семенова 329950
Родиола стефана 329960,329961
Родиола темнокрасная 329935
Родиола тибетская 329967
Родиола холодная 329829,329877
Родиола четырехчленная 329930
Родиола ярко－красная 329930
Рододендрон 330017,346460
Рододендрон адамса 330023
Рододендрон альбрехта 330058
Рододендрон аньхуйское 330097
Рододендрон бирманский 330267
Рододендрон большой 331207
Рододендрон вэньшаньский 332101
Рододендрон гриффита 330794
Рододендрон гуандунский 331031
Рододендрон гуаннаньский 330797
Рододендрон гуансиский 331026
Рододендрон гуйхайский 330798
Рододендрон гуйцзоуский 330800
Рододендрон гуйчжунский 330799
Рододендрон гуншаньский 330785
Рододендрон далиский 331921
Рододендрон даурский 330495
Рододендрон декоративный 330524
Рододендрон делаве 330546
Рододендрон динхуский 331980
Рододендрон древесный розовый 330113
Рододендрон древовидный 330110,330111,331207
Рододендрон дэциньский 331307
Рододендрон желтый 331161
Рододендрон жёлтый 331161
Рододендрон жестковолосистый 330875
Рододендрон жуйлиский 331821
Рододендрон золотистый 330182
Рододендрон иедоский 332144
Рододендрон ийхунаньск 330891
Рододендрон индийский 330917
Рододендрон кавказский 330334
Рододендрон календуловидный 330276
Рододендрон камчатский 330989,389572
Рододендрон канадское 330321
Рододендрон кандинский 330673
Рододендрон катобский 330331
Рододендрон кемпфера 330966
Рододендрон корейский 332104
Рододендрон короткоплодный 330240
Рододендрон кочи 331018
Рододендрон крупнейший 331207
Рододендрон лапландский 331057
Рододендрон ледебура 331073
Рододендрон лэйбоский 331079
Рододендрон люцюский 331759
Рододендрон малипоский 331194
Рододендрон мелколистный 331449
Рододендрон мелкоцветковое 331240
Рододендрон мотоское 330597
Рододендрон мулиский 330031
Рододендрон мупинский 331283
Рододендрон мэнцзыский 331221
Рододендрон мягкий 331257
Рододендрон мяньнинский 331238
Рододендрон мяткий 331257
Рододендрон наньпинский 331314
Рододендрон непалский 330467
Рододендрон остоконечный 331284
Рододендрон остроконечный 331289
Рододендрон пинбяньский 331492
Рододендрон подъельниковый 330907
Рододендрон понтийский 331161,331520
Рододендрон разноцветный 330592
Рододендрон редовского 389577
Рододендрон ржаволистный 330695
Рододендрон ржавый 330695
Рододендрон сетчатый 331623
Рододендрон симса 331839
Рододендрон сихотинский 331823
Рододендрон сичанский 332127

Рододендрон сычуаньский 331911
Рододендрон смирнова 331858
Рододендрон таиландский 331910
Рододендрон тайваньский 330722
Рододендрон тибетский 330876
Рододендрон тимиянолистный 331305
Рододендрон тимиянолистный белоцветковый 331805
Рододендрон томсона 331971
Рододендрон тупой 331370
Рододендрон тэнчунское 330591
Рододендрон увлажненный 330939
Рододендрон унгерна 332038
Рододендрон флабум 331520
Рододендрон фуцзяньский 331832
Рододендрон фэнкайский 331031
Рододендрон хайнаньский 330815
Рододендрон ху 330886
Рододендрон хэнаньский 330851
Рододендрон цзиньпинский 330961
Рододендрон цинхайский 331592
Рододендрон чаюйский 331487
Рододендрон чжундяньский 332162
Рододендрон чоноски 332018
Рододендрон шлиппенбаха 331768
Рододендрон юньнаньский 332155
Рододендрон яванское 330957
Рододендрон яошаньский 332142
Рододендрон японский 331258,331341
Родомирт 332218
Родотип 332282,332284
Родотипус 332282
Родотипус красивый 332284
Родянка 329935
Рожа 13934
Рожа дикая 18173
Рожа собачья 223392
Рожечник 83527
Рожки цареградские 83527
Рожки царьградские 83527
Рожковое дерево 83527
Рожь 356232
Рожь анатолийская 356235
Рожь вавилова 356255
Рожь гималайская 398939
Рожь дикая 356253
Рожь индийская 68860
Рожь культурная 356237
Рожь куприянова 356244
Рожь обыкновенная 356237
Рожь полярная 228346
Рожь посевная 356237
Рожь сорно－полевая 356248
Роза 336283
Роза азиатская 195269
Роза алексеенко 336347
Роза альберта 52466,336345
Роза альпийская 336348
Роза афцелиуса 336411
Роза ацикулярис 336320
Роза ацикулярис мелкоцветковая 336805
Роза банкса 336366
Роза банксиевая 336366
Роза беггера 336380
Роза белая 336343
Роза белоцветковая 336689
Роза бенгальская 336485
Роза буасье 336402
Роза бунге 336409
Роза бурбонская 336403
Роза бургундская 336477
Роза бэнкс 336366
Роза бэнкса 336366
Роза ванксиева 336366
Роза виргинская 337007
Роза вихуры 337015
Роза войлочная 336991,336993
Роза вонючая 336563
Роза воронова 337028
Роза высокогорная 336352
Роза галлика 336581
Роза гаррисона 336291
Роза гвосдичная 336650
Роза генри 336625
Роза гиблидная 336640
Роза грузинская 336642
Роза гуандунская 336667
Роза гуйцзоуская 336671
Роза давида 336519
Роза дамасская 336514
Роза даурская 336522
Роза джиральди 336607
Роза душистая 336813
Роза дэциньская 336528
Роза елены 336620
Роза жантиля 336625
Роза желтая 336563
Роза жёлтая 336563
Роза желтоватая 337029
Роза звегинцова 336981
Роза иглистая 336320
Роза иерихонская 21805
Роза илийская 336956
Роза индийская 336485
Роза казанлыкская 336514
Роза камчатская 336656
Роза канина 336419
Роза карликовая 336490
Роза каролинская 336462
Роза китайская 195149,195269,336485
Роза клейкая 336613
Роза клука 336658
Роза кожистолистная 336504
Роза кокандская 336659
Роза колючейшая 336851
Роза коржинского 336663
Роза коричная 336498
Роза красивая 336382
Роза кубарчатая 337001
Роза кугитангская 336664
Роза куэньминская 336665
Роза лицзянская 336690
Роза майская 336498
Роза макартова 336405
Роза максимовича 336738
Роза маррэ 336736
Роза мелкоплодная 336509
Роза мелкоцветковая 336741
Роза месячная 336495
Роза миниатюрная 336490
Роза многоприцветниковая 336781
Роза многоцветковая 336783
Роза многоцветная 336783
Роза многошипая 336802
Роза многошипная 336783
Роза мойза 336777
Роза морщинистая 336901
Роза морщинистая белая 336902
Роза моховая 336476
Роза моховидная 336476
Роза мультифлора 336783
Роза мускусная 336408
Роза мягкая 336762
Роза наньнинская 336795
Роза нутканская 336810
Роза оминская 336825
Роза оминская крылатошипная 336828
Роза острозубая 336836
Роза остроиглая 336835
Роза остроколючая 336851
Роза острошипная 336835
Роза пахучая 336563
Роза пашенная 336364
Роза пжавая 336892
Роза плоскошипая 336541
Роза повислая 336348
Роза полевая 336364,336650
Роза ползучая 336783
Роза полиантная 336514
Роза полиантовая 336514
Роза полированная 336675
Роза полушаровидная 336618
Роза помпонная 336481
Роза порицветковая 302860
Роза распростертая 336685
Роза рвущая 336672
Роза ремонтантная 336514
Роза роксбурга 336885
Роза ругоза 336901
Роза рыхлая 336685
Роза сванетская 336980

Роза сизая 336611
Роза сирийская 195269
Роза собачья 336419
Роза столистная 336474
Роза столрепестная 336474
Роза страшная 336901
Роза сулье 336965
Роза сулье мелколистный 336966
Роза тибетская 336990
Роза тонкочерешковая 336558
Роза тупоушковая 336353
Роза туркестанская 337002
Роза удлиненно－остроконечная 336692
Роза уссурийская 337005
Роза уэбба 337012
Роза федченко 336554
Роза французская 336581
Роза хвостатая 336471
Роза хьюго 336636
Роза центифолия 336474
Роза цюцзянская 336985
Роза чайная 336485,336813
Роза чайная китайская 336485
Роза чайная настоящая 336813
Роза чжундяньская 336868
Роза чэнкоуская 336484
Роза шарона 195269
Роза шелковистая 336930
Роза широкошипая 336857
Роза широкошиповая 336857
Роза шотлская 336851
Роза щетинисто－волосистая 336630
Роза щетинистоножковая 336944
Роза щитковидная 336506
Роза щитконосная 336506
Роза эглантерия 336892
Роза эйчисон 336536
Роза элимаитская 336546
Роза эллиптическолистная 336650
Роза юндзилла 336735
Роза яблочная 337006
Роза якутская 336653
Розанные 337050
Розга золотая 368480
Розелла 219,195196
Розель 195196
Розеточница 337292
Розеточница альпийская 337297
Розеточница вечнозеленая 337314
Розеточница волосистая 337306
Розеточница гиссарская 337300
Розеточница голая 337299
Розеточница елимаитская 337298
Розеточница желтая 337303
Розеточница живучковая 337315
Розеточница кокандская 337301
Розеточница колосовидная 337316
Розеточница корнецветковая 337308
Розеточница липского 337302
Розеточница метельчатая 337304
Розеточница персидская 337305
Розеточница плосколистая 337307
Розеточница плосколистная 337307
Розеточница таджикская 337317
Розеточница туркестанская 337318
Розеточница шишкина 337310
Розмарин 337174
Розмарин аптечный 337180
Розмарин лекарственный 337180
Розовик 332282
Розовик керриевидный 332284
Розовик японский 332284
Розовнк цепкий 332284
Розовые 337050
Розоцветные 337050
Рокамболь 15726
Рокетсалат 154019
Рокет－салат 154019
Роксбургия 337810
Ромашка 32563,246307
Ромашка аптечная 246396
Ромашка безлепестная 246375
Ромашка вонючая 26768
Ромашка далматская 322660
Ромашка девичья 322724
Ромашка дикая 246396
Ромашка дисковидная 246375
Ромашка душистая 246375
Ромашка жёлтая 26900
Ромашка золотистая 246318
Ромашка инсектисидная 89402
Ромашка кавказская 322658,322660,322665
Ромашка клоповая 91011
Ромашка клоповная 91011
Ромашка красильная 26900
Ромашка лекарственная 246396
Ромашка настоящая 246396
Ромашка немецкая 21230
Ромашка непахучая 397964
Ромашка обнаженная 246396
Ромашка ободранная 246396
Ромашка отмичная 246336
Ромашка пахучая 246375
Ромашка пепеленолистная 322660
Ромашка персидская розовая 322658,
322660,322665
Ромашка посевная 89704
Ромашка приморская 397970
Ромашка ранняя 246392
Ромашка римская 85526
Ромашка розовая 322658,322660,322665
Ромашка ромашковидная 167952,246375
Ромашка собачья 26768
Ромашка сомнительная 246315
Ромашка цвелева 246411
Ромашник 89402,322640
Ромашник тысячелистниковый 383692
Ромашник тысячелистный 26829
Ромэн 219493
Ронделетия 336096
Ропалостилис 332434
Росичка 130404
Росичка английская 138267
Росичка горизонтальная 130598
Росичка длинноцветковая 130635
Росичка индийская 138295
Росичка китайская 130481
Росичка крованая 130745
Росичка кровеостанавливающая 130612
Росичка кровяная 130745
Росичка кровяно－красная 130745
Росичка круглолистая 138348
Росичка круглолистная 138348
Росичка линейная 130612
Росичка обратнояйцевидная 138322
Росичка промежточная 138307
Росичка промежуточная 138299
Росичка ресничатая 130489
Росичка фиолетовая 130825
Росичка шитовидная 138307
Росичка щитовидная 138328
Росница 138261
Росяника 338447
Росянка 138261
Росянковые 138369
Ротала 337320
Ротала густоцветковая 337330
Ротала индийская 337351
Ротала мексиканская 337368
Ротанг 65782
Ротоза 401085,401129
Ротоза восточная 401129
Ротоза узколистная 401094
Рохелия 335103
Рохелия бунге 335107
Рохелия голоплодная 335114
Рохелия двусемянная 335110
Рохелия длинноножковая 335118
Рохелия карсская 335113
Рохелия колокольчатая 335108
Рохелия крупночашечная 335115
Рохелия персидская 335119
Рохелия сердцевидночашечная 335109
Рохелия согнутая 335107
Рубус 338059
Рубус арктический 338142
Рубус боярышниколистный 338300
Рубус гуансиский 338687
Рубус гулинский 338683
Рубус гуншаньский 338477
Рубус зибольда 339270

Рубус идский 338557
Рубус ичанский 338554
Рубус канадский 338225
Рубус кандинский 339118
Рубус комарова 338679
Рубус ланьюйский 338435
Рубус малипоский 338787
Рубус мелколистный 338985
Рубус мелкоцветковый 338886
Рубус мотоский 338818
Рубус мэнлаский 338806
Рубус наземный 338808
Рубус отогнутый 339162
Рубус приземистый 338243
Рубус скальный 339240
Рубус тайваньский 338425
Рубус тибетский 339370
Рубус фуцзяньский 338449
Рубус хмелелистный 338543
Рубус ху 339163
Рубус хуанпинский 338538
Рубус хуналньский 338544
Рубус эмэйский 338384
Рудбекия 339512
Рудбекия бахромчатая 339581
Рудбекия волосистая 339556
Рудбекия гибрида 339572
Рудбекия пурпурная 140081
Рудбекия раздельнолистная 339574,339581
Рудбекия разрезнолистная 339581
Рудбекия рассеченная 339574,339581
Рудбекия рассечённая 339574
Рудбекия рассеянная 339574
Рудбекия шершавая 339556
Руеллия 339658
Рузумовския 30906
Рузумовския жевельниковая 30912
Румия 340322
Румия критмолистная 340323
Румянка 141087,141289
Румянка обкновенная 141347,141365
Руппиевые 340502
Руппия 340470
Руппия морская 340476
Руппия спиральная 340472
Рускус колючий 340990
Рускусовые 340541
Русселия 341016
Русселия ситниковидная 341020
Руссовия 341026
Руссовия согдийская 341028
Рута 341039
Рута волокнистая 335307
Рута душистая 341064
Рута обыкновенная 341059,341064
Рута палермская 335454
Рута пахучая 341064
Рута северная 335230
Рута скандинавская 335490
Рута собачья 144228
Рута содовая 341064
Рута туруханская 335557
Рута узкочешуйная 335440
Рута уральская 335560
Рутовые 341086
Ручейник приятный 365872
Руэллия 339658
Руэллия приятная 339731
Рыжай 191261
Рыжей 68839
Рыжей мелкоплодный 68850
Рыжей яровой 68860
Рыжик 68839
Рыжик белоцветковый 68840
Рыжик белоцветный 68840
Рыжик волосистый 68857
Рыжик гладкий 68860
Рыжик дикий 68871
Рыжик кавказский 68844
Рыжик лесной 68871
Рыжик мелкоплодный 68850
Рыжик посевной 68860
Рыжик рыхлый 68848
Рыжик юньнаньский 68873
Рыжик яровой 68860
Рыскун 67456
Рэгнерия 144168,335183
Рэгнерия аболина 335184
Рэгнерия алтайская 335200
Рэгнерия амурская 335204
Рэгнерия буша 335246
Рэгнерия волокнистая 335307
Рэгнерия гиблидная 335543
Рэгнерия гималайская 335338
Рэгнерия длиниохвостая 335416
Рэгнерия длинноостая 335404,335407
Рэгнерия дробова 335295
Рэгнерия жакемонта 335364
Рэгнерия жестколистная 335494
Рэгнерия зеленочешуйная 335563
Рэгнерия искривленная 335284
Рэгнерия кавказская 335255
Рэгнерия килеватая 335254
Рэгнерия комарова 335389
Рэгнерия короткожковидная 335233
Рэгнерия крупнощетинистая 335415
Рэгнерия ошская 335453
Рэгнерия палелмская 335454
Рэгнерия повислая 335461
Рэгнерия прерывчатая 335359
Рэгнерия пушистая 335472
Рэгнерия пушистолистная 335392
Рэгнерия ресничатая 335258
Рэгнерия саянская 335487
Рэгнерия северная 335230
Рэгнерия синьцзянская 335510
Рэгнерия скандинавская 335490
Рэгнерия смешанная 335278
Рэгнерия собачья 144228
Рэгнерия тайваньская 335312
Рэгнерия тибетская 335529
Рэгнерия тонкохвостная 335403
Рэгнерия туруханская 335557
Рэгнерия турчаннова 335547
Рэгнерия тяньшанская 335526
Рэгнерия угамская 335558
Рэгнерия узкочешуйная 335440
Рэгнерия уральская 335560
Рэгнерия чимганская 335287
Рэгнерия шаньдунская 335501
Рэгнерия шренка 335491
Рэгнерия шугнанская 335493
Рэгнерия якутская 335366
Рэгнерия японская 335370
Рябина 369290
Рябина альбова 369302
Рябина американская 369313
Рябина амурская 369484
Рябина анадырская 369317
Рябина армянская 369334
Рябина балдаччи 369349
Рябина бархатистая 369556
Рябина берека 369541
Рябина буассье 369350
Рябина бузинолистная 369512,369514
Рябина буроватая 369526
Рябина буша 369351
Рябина воронова 369566
Рябина гиблидная 369429
Рябина гладковатая 369397
Рябина глоговина 369541
Рябина греческая 369403
Рябина двуцветная 369379
Рябина дикая 383874
Рябина домашная 369381
Рябина кавказская 369357
Рябина камчатская 369436
Рябина китайская 369379
Рябина колхидская 369361
Рябина крымская 369532
Рябина кузнецова 369444
Рябина ланьпинская 369445
Рябина лечебная 369541
Рябина лжеакация 334976
Рябина ликёрная 369339
Рябина ложноакация 334976
Рябина мелколистная 369464
Рябина мелкоцветковая 369462
Рябина мугеотта 369466
Рябина непалский 369391
Рябина обыкновенная 369339

Рябина ольхолистная 32988
Рябина персидская 369481
Рябина пёстрая 369379
Рябина полувойлочная 369528
Рябина промежуточная 369432
Рябина садовая 369381
Рябина сибирская 369521
Рябина сычуаньская 369520
Рябина скандинавская 369432
Рябина смешанная 369362,369367
Рябина тайваньская 369498
Рябина турецкая 369548
Рябина тяньшанская 369536
Рябина хубейская 369420
Рябина хунаньская 369418
Рябина цюцзянская 369439
Рябина чаюйская 369572
Рябина черноплодная 369459
Рябина шаньдунская 369529
Рябина шемахинская 369518
Рябина японская 33008
Рябина – ария 369322
Рябина – глоговина 369541
Рябинник 369259
Рябинник амурская 369264
Рябинник древесный 369265
Рябинник линдлея 369288
Рябинник ольги 369274
Рябинник палласа 369275
Рябинник рябинолистный 369279
Рябинник сумахолистный 369276
Рябинолистник 369259
Рябинолистник обыкновенный 369279
Рябишник 383874
Рябчик 168332
Рябчик армянский 168348
Рябчик бледноцветковый 168506
Рябчик бледноцветный 168506
Рябчик валуева 168624
Рябчик восточный 168503
Рябчик ганьсуская 168523
Рябчик горный 168585
Рябчик гроссгейма 168417
Рябчик дагана 168387
Рябчик желтый 168464
Рябчик кавказский 168364
Рябчик камчатский 168361,168467
Рябчик котчи 168442
Рябчик крупноцветный 168416
Рябчик курдский 168443
Рябчик максимовича 168467
Рябчик малый 168469
Рябчик михайловского 168489
Рябчик мумтвчатый 168614
Рябчик мутовчатый 168614
Рябчик ольги 168500
Рябчик радде 168547
Рябчик регеля 168551
Рябчик русский 168556
Рябчик садовый 168430
Рябчик тибетский 168640
Рябчик тонкий 168585
Рябчик турецкий 168430
Рябчик тюльпанолистный 168364
Рябчик уссурийский 168612
Рябчик ферганенский 168401
Рябчик царский 168430
Рябчик цзилунский 168380
Рябчик шахматный 168476
Рябчик широколистный 168450
Рябчик эдуарда 168398
Рянг 75782
Ряска 224359
Ряска горбатая 224366
Ряска лэйбоская 224370
Ряска малаяРяска маленькая 224375
Ряска меньшая 224375
Ряска многокорешковая 372300
Ряска прёходльная 224396
Ряска прехлопастная 224396
Ряска тройчатая 224396
Рясковые 224403
Ряст 352880
Сабалла 352120
Сабалла аптечная 352124
Сабалла лекарственная 352124
Сабаль 320330,341401
Сабаль адансона 341421
Сабаль блакбурна 341406
Сабаль малый 341421
Сабаль пальметто 341422
Сабаль пальмовидный 341422
Сабатия 341463
Сабельник 100254,100257
Сабельник болотный 100257
Сабельник залесова 100259
Сабельник залесовский 100259
Саговник 115794,115884
Саговник гуйцзоуский 115830
Саговник поникающий 115884
Саговник улитковидный 115812
Саговника шишка женская 115794
Саговниковые 115793
Садов 30081
Садовая астра 67312
Садовая астра китайская 67314
Садовая земляника 167597
Сажелеция 342155
Сажелеция луаньдаская 342191
Сажелеция мелкоцветковая 342181
Сажелеция тайваньская 342205
Сажелеция эмэйская 342182
Сажереция 198237
Саза 347191
Саза вейтча 347324
Саза гуандунская 347218
Саза гуансиская 347219
Саза хайнаньская 347220
Саза хубэйская 347226
Саза японская 318307
Сайгачья трава 167765
Сайгачья трава болосистая 167792
Сайгачья трава гладкая 167802
Сайгачья трава мучнистая 167813
Саксаул 185062
Саксаул белый 185072
Саксаул зайсанский 185064
Саксаул персидский 185072
Саксаул песчаный 185072
Саксаул черный 185064
Саксаульчик 36813
Саксаульчик илийский 36822
Саксифрага 349032
Саксифрага моховидная 349680
Саксифрага цинхайская 349835
Саксифрага чаюйская 350050
Саксифрага чжэцзянская 350055
Сакура 83314
Сал 362220
Саламалейкум 119503
Саламалик 119503
Салат 219206,259746
Салат вонючий 219609
Салат дикий 219497,219507
Салат кочанный 219489
Салат культурный 219485
Салат латук 219485
Салат латукрывной 219494
Салат листовой 219494
Салат мангольд 53249
Салат обыкновенный 219485
Салат огородный 219485
Салат полевой 404454
Салат посевной 219485
Салат – розель 195196
Салат – эндивий 90894
Сальвадора 344798
Сальвия 344821
Сальвия блестящая 345405
Сальвия китайская 344957
Сальвия максимовича 345205
Сальное дерево 346408
Сальпиглоссис 344381
Сальсифи 394327
Саман 345525
Самбук 345558
Самбук канадский 345580
Самбук китайский 345586
Самерария 209167,345711
Самерария армянская 345713
Самерария вайдолистная 345718

Самерария губчатоплодная 345720
Самерария желобчатая 345716
Самерария литвинова 345719
Самерария пузырчатая 345715
Самерария пустыная 345717
Самерария туркменская 345721
Самолюс 345728
Самолюс валеранда 345733
Самолюс мелкоцветковый 345731
Самолюс обыкновенный 345733
Самоцвет 4249
Самшит 64235
Самшит вечнозеленый 64345
Самшит гирканский 64263
Самшит ичанский 64264
Самшит колхидский 64251
Самшит крупнолистный 濮学擐瞍 64284
Самшит люцюский 64373
Самшит мелколистный 64285
Самшит мелколистный японский 64303
Самшит обыкновенный 64345
Самшит японская 64990
Самшит японский 64303
Самшитовые 64224
Санвиталия 346300
Санвиталия лежачая 346304
Сангвинария 345803
Сангвинария канадская 345805
Сангуинария 345803
Сандал 346208,346210
Сандал белый 346210
Сандал индийский 320301
Сандал красный 320327
Сандал синий 184518
Сандерсония 345782
Сандорик 345785
Сандорик индийский 345786,345788
Санзеверия 346047
Сансевиерия 346047
Сансевиерия гвинейская 346094,346155
Сансевиерия кистецветковая 346094, 346155
Сансевиерия цейлонская 346173
Сансевиерия цйлиндрическая 346070
Сансевьера 346047
Сансевьера цейлонская 346173
Сансевьерия 346047
Санталовые 346181
Сантолина 346243
Сантолина волосистовойлочная 346250
Сантолина кипарисовидная 346250
Сантолина кипарисовниковидная 346250
Сантпория 342482,342493
Санхезия 345769
Санхезия благолодная 345770
Сапиндовые 346316
Сапиндус 346323
Сапиндус делаве 346333
Сапиндус макоросса 346338
Сапиндус мукороси 346338
Сапиндус мыльный 346351
Сапиум 346361
Сапиум гуйлиньский 346376
Сапиум разноцветный 346379
Сапиум салоносный 346408
Сапиум японский 346390
Сاподилла 244605
Сапожки 120287
Сапожниковия 346447
Сапожниковия растопыренная 346448
Сапонария 346418
Сапонария аптечная 346434
Сапонария лекарственная 346434
Сапонария оффициналис 346434
Сарака 346497
Сарана 191261,191284,191304
Сарана большая 229922
Сарана овсянка 229934
Саранка 229922
Саркандра 346525
Саркандра голая 346527
Саркандра пинбяньская 346535
Саркандра хайнаньская 346532
Саркококка 346723
Саркококка гукера 346731
Саркококка гукера приземистая 346733
Саркококка ивовая 346747
Саркококка иглицелистая 346743
Саррацениевые 347166
Саррацения 347144
Саррацения пурпурная 347156
Сарсазан 184930
Сарсазан шишковатый 184933
Сарсазан шищковатый 184933
Сарсальма 126127
Сарсапарель 366230
Саса 347191
Саса вича 347324
Сассапарель китайская 366284
Сассапариллевые 366141
Сассапариль 366230
Сассапариль береговая 366548
Сассапариль высокий 366329
Сассапариль голая 366338
Сассапариль максимовича 366548
Сассапариль мэра 366447
Сассафрас 347402
Сассафрас лекарственный 347407
Сатурея 347446
Саур－арча 213932
Сауроматум 348022
Сауроматум капельный 348036
Сауруровые 348003,348077
Саурурус 348082
Саурурус китайский 348087
Сафлор 77684
Сафлор красиьный 77748
Сафлор культурный 77748
Сафлор острошипый 77734
Сафлор шерстистый 77716
Сахалинская гречиха 328328
Сахарный тростник лекарственный 341887
Сахарный тростник 341829,341887
Сахарный тростник дикий 341912
Сахарный тростник тростниковидный 341837
Саярь－кафак 148444
Свада 379486
Свада морская 379553
Свада приморская 379553
Свайнсона 380065
Свайнсония 371160
Сведа 379486
Сведа вздутоплодная 379582
Сведа высокая 379492
Сведа дуголистная 379495
Сведа заамударьинская 379613
Сведа заостренная 379489
Сведа запутанная 379508
Сведа карликовая 379590
Сведа косинского 379543
Сведа крылоцветная 379590
Сведа кустарничковая 379514
Сведа линейнолистная 379547
Сведа липского 379548
Сведа мелколистная 379566
Сведа мелкоплодная 379510
Сведа мелкосемянная 379567
Сведа олуфсена 379577
Сведа приморская 379553
Сведа разнолистная 379533
Сведа рожконосная 379509
Сведа серполистная 379513
Сведа сизая 379531
Сведа солончаковая 379598
Сведа стелющаяся 379584
Сведа странная 379580
Сведа туркестанская 379594
Сведа уссурийская 379598
Сведа эльтонская 379519
Сведа ягодоносная 379501
Сведа японская 379542
Свекла 53224
Свёкла 53224
Свёкла дикая 53254
Свекла кормовая 53231
Свёкла кормовая 53245,53262
Свекла крупнокорневая 53231
Свекла культурная 53249
Свекла лиственная 53257
Свёкла листовая 53257

Свекла многолетняя 53242
Свекла обыкновенная 53249
Свёкла обыкновенная 53249
Свёкла приморская 53254
Свекла сахарная 53249
Свёкла сахарная 53255
Свёкла столовая 53259
Свёкла столовая красная 53269
Свекла трехстолбиковая 53248
Свекловица сахарная 53255
Свекольник 53257
Свербига 63443
Свербига восточная 63452
Свербига ложкообразная 63446
Свербига степная 63446
Свертия 380105
Свертия многолетная 380302
Свертия тупая 380291
Сверция 380105
Сверция бледная 380184
Сверция вильфорда 380411
Сверция выгрызенная 380202
Сверция гуйцзоуская 380245
Сверция кандинская 380357
Сверция лицзянская 380179
Сверция молочнобелая 380250
Сверция мэнцзыская 380252
Сверция окаймленная 380271
Сверция оше 380123
Сверция сросшаяся 380165
Сверция тибетская 380387
Сверция тонкоцветвая 380217
Сверция тупая 380291
Сверция франше 380212
Сверция чаюйская 380419
Сверция чемерицевидная 380400
Сверция четырехлепестковая 380376
Сверция эмэйская 380197
Сверция японская 380234
Свеча ночная 269394,269404
Свида 380425
Свида белая 104920
Свидина 104917,380499,389239
Свидина белая 104920
Свидина грузинская 105068
Свидина дарвазская 380448
Свидина коротконогая 380461
Свидина кроваво－красная 380499
Свидина мейера 389251
Свидина отпрысковая 105193
Свидина порослевая 105193
Свидина сибирская 104930
Свидина татарская 104930
Свидина южная 104960
Свиетения 380522
Свиетения махагона 380522
Свинорой 117852,117859
Свинорой пальчатый 117852,117859
Свинушка бурая 81970
Свинушка тонкая 81970
Свинцовка 305167
Свинчатка 305167
Свинчатка европейская 305183
Свинчатка индийская 305185
Свинчатка капская 305172
Свинчатковые 305165
Свинчатник 305167
Свободносемянник 143715
Свободносемянник свободнолазский 143716
Свободноягодник 143575
Свободноягодник колючий 143657
Северница 345728
Северница валеранда 345733
Северница обыкновенная 345733
Седач 158062
Седаш 158062
Седаш коноплевидный 158062
Седлоцвет 146305
Седлоцвет шмидта 146308
Седмичник 396780
Седмичник арктический 396789
Седмичник европейский 396786
Седум 356467
Седум аньхуйский 356542
Седум бухарский 356589
Седум васильковый 356655
Седум еверса 356724
Седум замечательный 200798
Седум зибольда 200802
Седум кавказский 356606
Седум кандинский 356904
Седум ленкоранский 356874
Седум лидийский 356907
Седум листонин 356895
Седум мелкий 356959
Седум мелкотычинковый 357009
Седум мутовчатый 200816
Седум наньчуаньск 356958
Седум обертковый 356828
Седум побегоносный 357181
Седум посконниковый 200793
Седум прибрежный 356896
Седум пурпуровый 200812
Седум супротивнолистный 401702
Седум тайваньский 356751
Седум тонгий 356774
Седум тоненький 357219
Седум тополелистный 357039
Седум туполистный 356973
Седум франше 356758
Седум цзиндунский 356619
Седум черноватый 356556
Седум чэнкоуский 356577
Седум щитковый 356637
Сезам 361294,361317
Сезамовые 286918
Сейба пятитычинковая 80120
сейтера 362020
Сейтера уилфорда 117751
Секвойя 360561,360565
Секвойя вечнозеленая 360565
Секвойя вечнозелёная 360565
Секвойя гигантская 360576
Секвойядендрон 360575
Секвойядендрон гигантский 360576
Секуригера 356377
Секуригера мечевидная 356387
Секуринега 167067,356392
Секуринега ветвецветная 167092
Секуринега полукустарниковая 167092
Селагинелловые 316095
Селезеночник 90317
Селезёночник 90317
Селезеночник байкальский 90337
Селезеночник ветвистый 90442
Селезеночник волосистый 90429
Селезеночник голостебельный 90421
Селезеночник гуандунский 90382
Селезеночник камчатский 90387
Селезеночник китайский 90452
Селезеночник комарова 90392
Селезеночник ложный 90434
Селезеночник лушаньский 90403
Селезеночник непалский 90416
Селезеночник обыкновенный 90325
Селезёночник обыкновенный 90325
Селезеночник овальнолистный 90424
Селезеночник очереднолистный 90325
Селезёночник очереднолистный 90325
Селезеночник плетеносный 90362
Селезеночник райта 90471
Селезеночник седакова 90447
Селезеночник сомнительный 90352
Селезеночник супротивнолистный 90423
Селезёночник супротивнолистный 90423
Селезеночник тайваньский 90396
Селезеночник тонгий 90361
Селезеночник тяньшанский 90461
Селезеночник усатый 90362
Селезеночник четырехтычинковый 90460
Селезеночник шероховатосемянный 90465
Селезеночник щитовидный 90426
Селезеночник эмэйский 90381
Селезеночник японский 90383
Селемелек 244349
Селин 33730,33737,357780
Селин перистый 377011
Селитрянка 266355,266367
Селитрянка комарова 266358
Селитрянка сибирская 266377

Селитрянка шобера 266367
Сельдерей 29312,29327
Сельдерей душистый 29327
Сельдерей листвентый 29328
Сельдерей листовой 29328
Сельдерей обыкновенный 29327
Сельдерей пахучий 29327
Сельдерей стеблевой 29328
Сельдерей черешковый 29327,29328
Селягинеллевые 316095
Семена горчицы 59515
Семеновия 357905
Семеновия зайлийская 357912
Семиарундинария 357932
Семиарундинария великолепная 357939
Семпервивум 358015
Семя блошное 301864
Семя вшивое 352124
Семя канареечное 293729
Сенецио 358159
Сенецио чжундяньский 358550
Сенна 78204
Сенна индийская 78227
Сенна итальянская 78436
Сенна остролистная 78227
Сено греческое 397186,397229
Сераделла 274879
Сераделла маленькая 274888
Сераделла очень маленькая 274888
Сераделла посевная 274895
Сераделла сжатая 274880
Серапиас 360582
Серапиас сошниковый 360642
Сергия 360653
Сергия регеря 360654
Сергия северцова 360655
Сердечки 128288
Сердечник 72674
Сердечник аньхуйский 72684
Сердечник белый 72858
Сердечник болотнй 73026
Сердечник виктора 73033
Сердечник горький 72679
Сердечник гребенчатый 72925
Сердечник греческий 72785
Сердечник густоцветковый 72733
Сердечник зейдлица 72985
Сердечник зонтичный 73027
Сердечник зубчатый 72734
Сердечник извилистый 72749
Сердечник каменный 72930
Сердечник крупнолистный 72876
Сердечник лазистанский 72857
Сердечник лировидный 72873
Сердечник луговой 72934
Сердечник максимовича 73044
Сердечник маленький 72900
Сердечник маргаритковый 72698
Сердечник мелколистный 72897
Сердечник мелкоцветковый 72920
Сердечник мелкоцветный 72920
Сердечник мулиский 72901
Сердечник недотрога 72829
Сердечник нежный 73003
Сердечник ползучий 72942
Сердечник пурпурный 72946
Сердечник регеля 72971
Сердечник сахокия 72963
Сердечник сычуаньский 72829
Сердечник стоповидный 72926
Сердечник тонколистный 73018
Сердечник чжэцзянский 73052
Сердечник шершавый 72802
Сердечник шершавый тайваньский 72809
Сердечник шершавый эмэйский 72812
Сердечник юньнаньский 73048
Сердечник японский 72910
Сердечница 73085
Сердечница крупновая 73090
Сердечница ползучая 73092
Сердечница пушистая 73098
Сердечный корень 309954
Серисса 360923
Серисса дурнопахнущая 360933
Серобородник 372398
Серобородник сибирский 372428
Серобородник тайваньский 372412
Серпник 111343
Серпоносик 372333
Серпоносик песчаный 372335
Серпоплодник 137828
Серпоплодник северцова 137829
Серпуха 360982,361141
Серпуха алатавская 360985
Серпуха блестяшая 361004
Серпуха бумажистая 361017
Серпуха венечная 361023
Серпуха венценосная 361023
Серпуха гмелина 361057
Серпуха голоногая 360994
Серпуха джунгарская 361043
Серпуха донская 361135
Серпуха кавказская 361010
Серпуха киргизская 361069
Серпуха китайская 361018
Серпуха кожистая 361021
Серпуха комаров 361070
Серпуха копьелистная 361058
Серпуха красильная 361065,361139
Серпуха ланцетолистная 361074
Серпуха лиролистая 361076
Серпуха лучевая 361011
Серпуха модеста 361076
Серпуха неколючая 361065
Серпуха окаймленная 361079
Серпуха полукустарниковая 361134
Серпуха причветниковая 361002
Серпуха простертая 361110
Серпуха пятилистная 361117
Серпуха равнолистная 361067
Серпуха разнолистная 361059
Серпуха рассеченная 361039
Серпуха серпуховидная 361126
Серпуха сухоцветная 361144
Серпуха тяньшанская 361076
Серпуха угловатая 360992
Серпуха холодная 360987
Серпуха чертополоховая 361004
Серпуха эриколистная 361045
Серьяния 360936
Сесбания 361350
Сесбания египетская 361357
Сеслерия 361611
Сеслерия гейфлера 361621
Сеслерия голубая 361615
Сеслерия осенняя 361614
Сеслерия тимофеевковидная 361622
Сехиум 356352
Сиайлэнд 179884
Сибатеа 362121
Сибатеа китайская 362123
Сибатеа кумасаса 362131
Сибатеа наньпинская 362140
Сиббальдиецвет 362339
Сиббальдиецвет прижатый 362341
Сиббальдия 362339
Сиббальдия крупнолистая 362361
Сиббальдия крупнолистная 362361
Сиббальдия лежачая 362361
Сиббальдия мелкоцветковая 362356
Сиббальдия ольги 362354
Сиббальдия полуголая 362372
Сиббальдия распростертая 362361
Сиббальдия четырехтычиночная 138411
Сиббальдия эмэйская 362355
Сибирка 362392
Сибирка алтайская 362396
Сибирка тяньшанская 362398
Сибирский абрикос 34470
Сиверсия 363057
Сиверсия малая 363067
Сиверсия пятилепестная 363063
Сивец 379655
Сивец луговой 379660
Сивец южный 379657
Сигезбекия 363073
Сигезбекия восточная 363090
Сида 362473,362617
Сида китайская 362514
Сидальцея 362702
Сидеритис 362728

Сидеродендрон 211029
Сидероксирон 362902
Сизал 10940
Сизаль 10940
Сизигиум яванский 382522
Сизон 365357
Сизюринхий 365678
Сизюринхий узколистный 365747
Сикамор 3462,302588
Сикамор античный 165726
Сикими 204575
Сикомор 165726
Сикомора 3462,165726
Сикопсис 380586
Сикопсис китайский 380602
Сикопсис тайваньский 380590
Силена 363141
Силене 363141
Силене восточное 363852
Силифиум 364255
Силифия 364255
Симаруба сизая 364384
Симарубовые 364387
Симпегма 380681
Симпегма регеря 380682
Симплокарпус 381070
Симплокарпус вонючий 381073
Симплоковые 381067
Симплокос 381090
Симплокос китайский 381341
Симплокос ланьюйский 381149
Симплокос люцюский 381288
Симплокос метельчатый 381341
Симплокос сычуаньский 381390
Симплокос тайваньский 381206
Симплокос фуцзяньский 381210
Симплокос хайнаньский 381233
Симплокос яошаньский 381463
Симплокос японский 381143,381252
Симфиандра 380769
Симфиандра армянская 380771
Симфиандра закавказская 380777
Симфиандра зангезурскаялазская 380779
Симфиандра повислая 380775
Симфилокарпус 380791
Симфилокарпус тощий 380792
Симфония 380696
Симфония габонская 380697
Синадениум 381475
Синарундинария 162635,364631
Синарундинария блестящая 162718
Синеголовник 154276
Синеголовник альпийский 154280
Синеголовник биберштейна 154291
Синеголовник бунге 154298
Синеголовник ванатура 154353
Синеголовник вонючий 154316
Синеголовник гигантский 154317
Синеголовник голубой 154300
Синеголовник карадагский 154331
Синеголовник каратавский 154322
Синеголовник крупночашечковый 154326
Синеголовник морской 154327
Синеголовник ное 154332
Синеголовник плоский 154337
Синеголовник плосколистный 154337
Синеголовник полевой 154301
Синеголовник приморский 154327
Синеголовник равнинный 154301
Синеголовник удивительный 154330
Синейлезис 381813
Синейлезис асонитолистный 381814
Синетал 343280
Синильник 209229
Синоджекия 364951
Синоджекия деревянистоплодная 364959
Синоджекия редера 364957
Синурус 382067
Синурус дельтовидный 382069
Синурус треугольный 382069
Синюха 307190
Синюха голубая 307197
Синюха кавказская 307214
Синюха китайская 307215
Синюха красивая 307237
Синюха крупная 307229
Синюха лазолевая 307197
Синюха лазурная 307197
Синюха ложно-красивая 307236
Синюха льноцветковая 307215
Синюха мелкоцветковая 307234
Синюха низкая 307222
Синюха обыкновенная 307197
Синюха опушенная 307246
Синюха остролепестная 307209
Синюха остроцветкоая 307209
Синюха рыхлоцветковя 307215
Синюха северная 307195
Синюха сибирская 307243
Синюха тихоокеанская 307233
Синюховые 307186
Синявк 256601
Синяк 141087
Синяк изогнутый 379657
Синяк итальянский 141202
Синяк красный 141289
Синяк обкновенная 141365
Синяк обыкновенный 141347
Синяк подорожинковый 141258
Синяк приятный 141097
Сирения 382101
Сирения крупноплодная 382104
Сирения сидячецветковая 382105
Сирения стручочковая 382107
Сирения талиева 382108
Сирения узколистная 382103
Сирень 382114
Сирень амурская 382277
Сирень венгерская 382180
Сирень вольфа 382383
Сирень гималайская 382153
Сирень индийская 219908,219933
Сирень китайская 382132
Сирень комарова 382190
Сирень крупная 382282
Сирень лесная 122515
Сирень мохнатая 382257,382268,382312
Сирень обыкновенная 382329
Сирень опушенная 382257
Сирень опушённая 382257,382312
Сирень опушенная мелколистная 382266
Сирень пекинская 382278
Сирень персидская 382241
Сирень поникшая 382190
Сирень сетчатая 382276
Сирень угорская 382180
Сирень хубэйская 382232
Сирень широколистная 382220
Сирень эмоди 382153
Сирень юньнаньская 382388
Сирень японская 382178
Сисюринхий 365678
Сисюринхий узколистный 365747
Ситник 212797
Ситник алтайский 212819
Ситник альпийский 212822
Ситник амурский 212836
Ситник арийский 212862
Ситник арктический 212844
Ситник балтийский 212852
Ситник белооберточный 213221
Ситник береговой 213226
Ситник беринговский 212898
Ситник блестящеплодный 212866
Ситник валиха 213569
Ситник волосовидный 213536
Ситник выдающийся 213028
Ситник генке 213145
Ситник герарда 213127
Ситник гималайкий 213153
Ситник головчатый 212988
Ситник двухчешуйный 212905
Ситник двухчешуйчатый 212905
Ситник дэциньск 213184
Ситник ероватый 213165
Ситник жабий 212929
Ситник жерара 213127
Ситник затопляемый 213176
Ситник зеленоватый 213567
Ситник изогнутый 213114
Ситник изящный 213138

Ситник искривлённый 213165
Ситник камчатский 213182
Ситник канадский 212970
Ситник кандинский 213183
Ситник канпуский 213185
Ситник каштановый 212990
Ситник короткооберточный 212915
Ситник короткоприцветниковый 212915
Ситник коротцветковый 213165
Ситник кочи 213190
Ситник крамера 213191
Ситник круглоплодный 213462
Ситник крупнопыльниковый 213249
Ситник кучкоцветковый 213458
Ситник кучкоцветный 213458
Ситник леерса 213206
Ситник лешено 213213
Ситник луковичный 212955
Ситник лягушечий 212929
Ситник лягушечный 212929
Ситник максимовича 213272
Ситник мелкий 212929
Ситник мелководный 213503
Ситник метельчатый 213353
Ситник морской 213263
Ситник мюллера 213304
Ситник невского 213263
Ситник неопределенный 212832
Ситник ниппонский 213322
Ситник нитевидный 213114
Ситник острый 212808
Ситник развесистый 213066
Ситник раскидистый 213066
Ситник растопыренный 213471
Ситник расходящийся 213066
Ситник семиреченский 213150
Ситник склоненный 213165
Ситник склонённый 213165
Ситник склоняющийся 213165
Ситник скученноцветковый 213309
Ситник сомнительный 213036
Ситник сосочковый 213355
Ситник сплюснутый 213008
Ситник стикийский 213476
Ситник сырдарьинский 213177
Ситник тайваньский 213341
Ситник темнобурый 212876
Ситник темноцветный 212875
Ситник тибетский 213527
Ситник томсона 213525
Ситник тонкий 213507
Ситник тончайший 213138
Ситник трехглавый 212990,213534
Ситник трехраздельный 213537
Ситник трехчешуйный 213545
Ситник трехчешуйчатый 213545
Ситник туполепестный 213332
Ситник туркестанский 213554
Ситник турчанинова 213552
Ситник членистый 212866
Ситник шиловидный 213492
Ситник шишкина 213435
Ситник юзепчука 213180
Ситник юньнаньский 213579
Ситник японский 212890
Ситниковидные 212753
Ситниковые 212752
Ситникоцветные 212752,212753
Ситничек 212763
Ситничек поздний 118473,212778
Ситняг 143019
Ситняг болотный 143271
Ситнят малоцветковый 143319
Ситовник 322179
Ситовник дрожащий 322374
Ситовник желтоватый 322223
Ситовник илийстый 212771
Ситовник колхидский 322198
Ситовник коржинского 322271
Ситовник лицзянский 322282
Ситовник луговой 322323
Ситовник нильгирийский 322304
Ситовник ремана 322336
Ситовник чжэцзянский 322196
Ситовник шаровидный 322233
Ситовник щетиновидный 118737
Сифонантус 95934,365217
Сифония 194451
Сифонодон 365270
Сифонодон южный 365271
Сифоностегия 365295
Сифоностегия китайская 365296
Сициос 362460
Сициос угловатый 362461
Скабиоза 350086
Скабиоза альпийская 350090
Скабиоза бледножелтая 350207
Скабиоза бледно-жёлтая 350207
Скабиоза венечная 350129
Скабиоза голубиная 350125
Скабиоза жёлтая 363571,364158,364193
Скабиоза желтоватая 350207
Скабиоза исетская 350173
Скабиоза кавказская 350119
Скабиоза колесовидная 350230
Скабиоза мелкоцветковая 350203
Скабиоза оливье 350209
Скабиоза разветленная 350159
Скабиоза светложелтая 350207
Скабиоза тёмно-пурпурная 350207
Скабиоза темнопурпуровая 350099
Скабиоза украинская 350279
Скабиоза южная 350099
Скабиоза южнолтайская 350111
Скабиоза японская 350175
Скабиоса 350086
Скабиоса бледножелтая 350207
Скабиоса блестящатая 350191
Скабиоса веленовского 350284
Скабиоса вращательная 350230
Скабиоса гибридная 350168
Скабиоса гирканская 350168
Скабиоса голубиная 350125
Скабиоса грузинская 350155
Скабиоса дважды-перистая 350113
Скабиоса джунгарская 350237
Скабиоса елтоватая 350207
Скабиоса имеретинская 350169
Скабиоса исетская 350173
Скабиоса кавказская 350119
Скабиоса колесовидная 350230
Скабиоса колхидская 350124
Скабиоса мелкоцветковая 350203
Скабиоса оверина 350214
Скабиоса ольги 350208
Скабиоса персидская 350222
Скабиоса приятная 350092
Скабиоса разветленная 350159
Скабиоса серебристая 350096
Скабиоса синьцзянская 350288
Скабиоса сосновского 350238
Скабиоса темнопурпуровая 350099
Скабиоса украинская 350279
Скабиоса улугбека 350281
Скабиоса шерстистолистная 350132
Скалигерия 350360,350361
Скалигерия алайская 350361
Скалигерия бухарская 350363
Скалигерия закаспийская 350378
Скалигерия кнорринг 350368
Скалигерия копетдагская 350369
Скалигерия коржинского 350370
Скалигерия коровина 350371
Скалигерия липского 350372
Скалигерия луковидная 350362
Скалигерия многоплодная 350375
Скалигерия плосколистная 350374
Скалигерия самаркандская 350376
Скалигерия сизоватая 350366
Скалигерия угамская 350380
Скалигерия ферганская 350365
Скалигерия чимганская 350379
Скалигерия шероховатая 350367
Скалигерия щетинковая 350377
Скамоний 379657
Скандикс 350402
Скандикс венерин гребень 103255
Скандикс гребенчатый 350425
Скандикс грузинский 350423
Скандикс звездчатый 350431
Скандикс оше 350407

Скандикс персидский 350428
Скандикс серповидный 350421
Скариола 350504
Скариола альберторегерия 350506
Скариола восточная 350507
Скариола прутовидная 350509
Скафоспермум 283921,350486
Скафоспермум азиатский 350490
Скерда 110710,350425
Скерда алайская 110722
Скерда аликера 110726
Скерда альпийская 110727
Скерда астраханская 110738
Скерда болотная 110710,110946
Скерда бунге 110762
Скерда буреинская 110763
Скерда венгерская 110947
Скерда вилеметиевидная 111092
Скерда вильденова 111091
Скерда волосовидная 110767,110946
Скерда вонючая 110808
Скерда гмелина 110827
Скерда голая 110820
Скерда горная 110942
Скерда дарвазская 110788
Скерда двулетняя 110752,110767
Скерда зеленая 110767
Скерда зелёная 110767
Скерда золотистая 110775
Скерда золотоцветковая 110775
Скерда извилистая 110806
Скерда кавказородная 110770
Скерда кавказская 110769
Скерда каракушская 110871
Скерда карелина 110872
Скерда карликовая 110916
Скерда колосовидная 110767
Скерда кочи 110879
Скерда красивая 110965
Скерда кровельная 111050
Скерда ксасная 110985
Скерда лировидная 110889
Скерда маршалла 110893
Скерда мелкоцветковая 110896
Скерда млечная 110880
Скерда многоволосистая 110954
Скерда многоцветбельная 110906
Скерда мягкая 110904
Скерда мягковолосистая 110904
Скерда низкая 110916
Скерда обгрызанная 110956
Скерда обгрызенная 110956
Скерда разветвленная 110975
Скерда реснитчатая 110777
Скерда рожковатая 110783
Скерда сибирская 111019
Скерда тупокорневищная 110956
Скерда туркменская 111066
Скерда тяньшанская 111059
Скерда французская 110929
Скерда хокайдская 110841
Скерда хорасанская 110874
Скерда черноватая 110930
Скерда чуйская 110787
Скерда шафранно-желтая 110785
Скерда щетинистая 111017
Скимия 365916
Скимия люцюская 365965
Скимия перноплодный 365970
Скимия ползучая 365975
Скимия тайваньская 365924
Скимия японская 365929,365974
Скипетр царский 287657,405657
Скипидарник 300998
Скипидарное дерево 300998
Склерия 353998
Сколимус 354648
Сколимус испанский 354651
Сколимус пятнистый 354655
Скополия 354678
Скополия карниолийская 354686
Скополия японская 354691
Скордия 388272
Скорода 15709
Скорпионница 354755
Скорцонер 354870
Скорцонер испанский 354870
Скорцонера 354797
Скрипун 200812,356928,357203
Скрипуха 110752
Скрученник 372176
Скрученник китайский 372247
Скрученник осенний 372259
Скрученник приятный 372247
Скрытница 113305,113868
Скрытница борщова 113312
Скрытница камышевидная 113316
Скрытница канадская 113872
Скрытница колючая 113306
Скрытница лизохностовидная 113310
Скрытница туркестанская 113318
Скрытница японская 113879
Скрытоподстолбник 113551
Скрытоподстолбник двойчатый 113555
Скрытоподстолбник кахрисовидный 113554
Скрытоподстолбник несколючбивый 113552
Скрытоподстолбник песчатый 113553
Скрытосемянница 113815
Скрытосемянница серповидная 113817
Скумпия 107297,107300
Скумпия американская 107315
Скумпия коггигрия 107300
Скумпия кожевенная 107300
Скумпия сычуаньская 107316
Сладкий апельсин 93765
Сладконожка 198769
Сладконошник 198769
Сланоягодник 184995
Слёзки богородицины 99124
Слёзки Иова 99124
Слёзки кукушкины 363461
Слёзник 99124
Слёзы Иовы 99124
Слива 316166
Слива абрикосовая 316813
Слива абрикосовидная 316813
Слива алыча обыкновенная 316294
Слива алыча растопыренная 316294
Слива американская 316206,316220
Слива ватсона 280061,316220
Слива венгерка 316382
Слива вишнелистная 316294
Слива дикая 316294
Слива дикая гусиная 316583
Слива домашная 20890,316382
Слива европейская 20890,316382
Слива зибольда 316809
Слива зибольда мелколистная 83271
Слива золотая 89816
Слива иволистная 316761
Слива ивообразная 316761
Слива изящная 316434
Слива испанская тропическая 372477
Слива канадская 34431
Слива китайская 316761,418169
Слива кокосовая 89812
Слива колючая 316819
Слива курдистанская 316356
Слива куртарниковая 83202
Слива ленклод 316333,316391
Слива малкая черная 316470
Слива морская 316543
Слива настоящая 316470
Слива ненасточщая 316470
Слива обыкновенная 20890,316382
Слива песчаная 316220
Слива полусердцевидная 316833
Слива растопыренная 316294
Слива симона 316813
Слива терн 316819
Слива тернослива 316470
Слива трёхлопастная 20970,316884
Слива турецкая 132264
Слива узколистная 316219
Слива уссурийская 316900
Слива ферганская 316416
Слива финиковая 132219,132264
Слива чёрная 34431
Слива чёрная мелкая 316470
Слива чикаса 316219
Слива яванская 382522

Слива явская 382522
Слива японская 316761
Сложноцветные 101642
Слонея 366034
Смеловския 366115
Смеловския белая 366116
Смеловския неожиданная 366127
Смеловския папоротниколистая 366119
Смеловския чашечная 366122
Смелтоед 190925
Смилакс 366230
Смилакс алтайский 366238
Смилакс зибольда 366573
Смилакс луаньдаский 366537
Смилакс малипоский 366448
Смилакс наньтоуский 366481
Смилакс ольдгема 366573
Смилакс тибетский 366344
Смилакс фунинский 366334
Смилакс эмэйский 366326
Смилакс японский 366486
Смилаксовые 366141,383414
Смилацина 366142
Смилацина волосистая 242691
Смилацина гуншаньская 242687
Смилацина даурская 242679
Смилацина лицзянская 242693
Смилацина наньчуаньская 242694
Смилацина сычуаньская 242703
Смилацина трехлистная 242706
Смилацина чжундяньская 242698
Смилацина японская 242691
Смирниовитка 366737
Смирниовитка армянская 366738
Смирния 366739
Смирния пронзеннолистная 366744
Смирния сердсевиднолистная 366741
Смирния Сфенокарпус 371400
Смирния Сфенокарпус синеголовниковый 371403
Смоква 164608
Смоква индийская 272891
Смоковница 164608,164763
Смоковница бенгальская 164684
Смоковница дикая 165726
Смоковница индийская 164684
Смоковница обыкновенная 164763
Смоковница райская 260253
Смоковница священая 165553
Смолевка 363141,363489
Смолёвка 363141
Смолевка александры 363165
Смолевка алтайская 363169
Смолевка альпийская 363168
Смолёвка альпийская 363167
Смолевка английская 363477
Смолевка араратская 363206
Смолевка армериевидная 363214
Смолевка армерцевидная 363214
Смолёвка армерцевидная 363214
Смолевка армянская 363213
Смолевка артвинская 363217
Смолевка бальджуанская 363233
Смолевка бархашская 363232
Смолевка безлепестная 363193
Смолевка белая 363673
Смолевка бесстебельная 363144,363399
Смолёвка бесстебельная 363144
Смолевка боброва 363252
Смолевка борнмюллера 363255
Смолевка бризкая 363927
Смолевка бротеруса 363263
Смолевка булавовидная 363326
Смолевка бухарская 363264
Смолевка вздутая 364144
Смолевка вздуточашечная 363889
Смолевка вильчатая 363409
Смолёвка вильчатая 363409
Смолевка волжская 364216
Смолевка володушковидная 363268
Смолевка гаврилова 363487
Смолевка газимайликская 363486
Смолевка гвозднковидная 363408
Смолевка геблера 363489
Смолевка гельмана 363533
Смолевка головчатая 363295,363308
Смолевка голубоватороэовая 363331
Смолевка горицветовидная 363712
Смолевка горная 363850
Смолевка гроссгейма 363525
Смолевка грузинская 363561
Смолевка гунтская 363528
Смолевка густоцветковая 363361,363404
Смолевка дагестанская 363397
Смолевка дернистая 363281
Смолевка джунгарская 364072
Смолевка длиннозубая 363700
Смолевка длинностолбиковая 363718
Смолевка длинноцветковая 363701
Смолевка думистая 363840
Смолевка енисейская 363606
Смолевка енисейская мелкоцветковая 363609
Смолевка железистолепестная 363153
Смолевка замещающая 363360
Смолевка зеленоватая 364178
Смолевка зеленоватолепестная 363317
Смолевка зеленовацветковая 363315, 364178
Смолёвка зелёновацветковая 363315
Смолевка зеравшанская 364010
Смолевка злаколистная 363516
Смолевка изогнутолистная 363566
Смолевка индийская 363568
Смолевка испирская 363582
Смолевка итальянская 363583
Смолёвка итальянская 363583
Смолевка кавказская 363307
Смолевка камнелюбивая 363692
Смолевка карачукурская 363620
Смолевка карликовая 363796
Смолевка комарова 363640
Смолевка коническая 363363
Смолёвка коническая 363363
Смолевка конусовидная 363368
Смолевка конусоцветковая 363366
Смолевка корейская 363641
Смолевка коржинского 363642
Смолевка коротколепестная 363261
Смолевка крепкая 363451
Смолевка кубанская 363643
Смолевка кудряшева 363644
Смолевка кунгесская 363646
Смолевка куринская 363395
Смолевка курчавая 363383
Смолевка кустарничковая 364089
Смолевка кушакевича 363647
Смолевка лазская 363680
Смолевка лежачая 363926
Смолевка линейнолистная 363686
Смолевка литвинова 363694
Смолевка лицзянская 363684
Смолевка ложнотонкая 363933
Смолевка ложноушковидная 363936
Смолевка лопушка 364193
Смолевка маленькая 363948
Смолевка малолистная 363879
Смолевка марковича 363723
Смолевка маршара 363730
Смолевка мейера 363752
Смолевка мелкоголовчатая 363296
Смолевка мелкоцветковая 363259
Смолевка меловая 363381
Смолевка михельсона 363753
Смолевка многолистная 363902
Смолевка многорассеченная 363780
Смолевка многоцветковая 363781
Смолевка мотоская 363795
Смолевка мулиская 363778
Смолевка невского 363804
Смолевка неравнолопастная 363180
Смолевка низкая 363557
Смолевка облиственная 363463
Смолевка обратнояйцевидная 363833
Смолевка обыкновенная 364193
Смолевка одноцветковая 364146
Смолевка олиственная 363463
Смолевка олиственная мелкоцветковая 363465
Смолевка ольги 363843
Смолевка опушенноцветковая 363667

Смолевка ошера 363227
Смолевка памирская 363865
Смолевка песчаная 363209
Смолевка повислая 363882
Смолёвка повислая 363882
Смолевка ползучая 363958
Смолёвка поникающая 363827
Смолевка поникшая 363827
Смолёвка поникшая 363827
Смолевка попова 363912
Смолевка порярная 363903
Смолевка прижатая 363406
Смолевка прилегающая 364091
Смолевка приликоана 363921
Смолевка промежуточная 363735
Смолевка пустынная 363442
Смолевка пушистая 364131
Смолевка радде 363952
Смолевка разрезная 363648
Смолёвка рупрехта 364003
Смолевка саманкардская 364007
Смолевка семенова 364047
Смолевка семиреченская 363534
Смолевка сибирская 363691
Смолёвка сибирская 363691
Смолевка сизоватый 363502
Смолевка сычуаньская 364094
Смолевка скальная 364002
Смолевка скученноцветковая 363361
Смолевка средная 363735
Смолевка стелющаяся 363926
Смолевка стройная 364110
Смолевка сцера 363385
Смолевка талышская 364098
Смолевка татарская 364100
Смолёвка татарская 364100
Смолевка татьяны 364108
Смолевка тахтская 364096
Смолевка твердая 363451
Смолевка тилоколистная 364193
Смолевка тирке 364125
Смолевка толстожилковая 363861
Смолевка тонкоопушенная 363928
Смолевка тонкостебельная 363682
Смолевка трубчатоцветная 364071
Смолевка тупозубчатая 363834
Смолевка туркменская 364143
Смолевка тяньшанская 364128
Смолевка узколистная 364079
Смолевка хамарская 363313
Смолевка хлопушка 364193
Смолевка хрололистная 363316
Смолевка хубэйская 363559
Смолевка цельнолепестная 363551
Смолевка цзилунская 363529
Смолевка чаюйская 364226
Смолевка чере 363396
Смолевка черноморская 363445
Смолевка чжундяньский 363321
Смолевка шафта 364023
Смолевка шероховатолистная 364020
Смолевка широколистная 363671,364193
Смолевка шугнанская 364024
Смолевка щетинозубая 363311
Смолёвка – хлопушка 363571,364158,364193
Смолка альпийская 410950
Смолка клейкая 410953,410957
Смолка обыкновенная 410957
Смолоносница 163574
Смолоносница вонючая 163580
Смолоносница восточная 163688
Смолоносница каспийская 163586
Смолоносница татарская 163729
Смолосемянник 56023,240763,301207
Смолосемянник волнистый 301433
Смолосемянник дафнифиллюмовидный 301253
Смолосемянник обильноцведущий 301266
Смолосемянник оголеный 301279
Смолосемянник разнолистный 301291
Смолосемянник тобира 301413
Смородина 333893
Смородина альпийская 333904
Смородина американская 333916
Смородина белая 333919
Смородина биберштейна 333924
Смородина бледноцветная 334145
Смородина блестящая 334078
Смородина буреинская 333929
Смородина волосистая 334260
Смородина восточная 334132
Смородина высочайшая 333913,334240
Смородина глухая 333904
Смородина горная 333904
Смородина двуиглая 333955,334203
Смородина декоративная 333919
Смородина дикуша 333957
Смородина душистая 333981,334071,334127
Смородина железистая 334159
Смородина золотая 333919
Смородина золотистая 333919
Смородина каменная 334149,334203
Смородина камневая 334149
Смородина канадская 334137
Смородина кандинская 334263
Смородина карпатская 334149
Смородина кислица 333918
Смородина колорадская 333940
Смородина колосистая 334214
Смородина колосковая 334179,334214,334223
Смородина колымская 334054
Смородина комарова 334055
Смородина красивая 334166
Смородина красная 334179,334214,334223
Смородина кровянокрасная 334189
Смородина кровяно – красная 334189
Смородина кубарчатая 334247
Смородина лежачая 334157
Смородина максимовича 334087,334089
Смородина малоцветковая 334147
Смородина мальволистная 334083
Смородина маньчжуърская 334084
Смородина мейера 334094
Смородина мулиская 334109
Смородина мупинская 334104
Смородина наньчуаньская 334176
Смородина обыкновенная 334179,334201,334214,334223
Смородина отклоненный 334250
Смородина ощестиненная 334034
Смородина пальчевского 334144
Смородина пахучая 334006
Смородина печалица 334240
Смородина печальная 334240
Смородина пучковатая 333970
Смородина пушистая 334163,334179,334214,334223
Смородина разноволосая 334017
Смородина сахалинская 334188
Смородина светлая 334078
Смородина сычуаньская 334206
Смородина скальная 333955,334203
Смородина скандийская 334179
Смородина скандинавская 334179,334214,334223
Смородина тайваньская 333978
Смородина таранушка 334203
Смородина темнопурпуровая 333918
Смородина тибетская 334270
Смородина тссурийская 334249
Смородина тяньцюаньская 334236
Смородина уссурийский 334249
Смородина франше 333982
Смородина черная 334117
Смородина чёрная 334117
Смородина широколистная 334061
Смородина щетинистая 334033,334034
Смородина янчевского 334048
Смородина японская 84556,333970
Смородинные 181339
Сморчковые 77276
Снедок 28024
Снедь – трава 8810
Снежник 380734
Снежноягодник 380734
Снежноягодник белый 380736,380753
Снежноягодник ветвистый 380736,380753
Снежноягодник западный 380748

Снежноягодник кистевидный 380736, 380753
Снежноягодник кистевой 380736
Снежноягодник кистистый 380736,380753
Снежноягодник красноплодный 380749
Снежноягодник круглолистный 380749
Снежноягодник мягкий 380746
Снежноягодник обыкновенный 380749
Снежноягодник округлый 380749
Снытка пухлая 8826
Снытка съедобная 8826
Сныть 8810
Сныть альпийская 8811
Сныть горная 8811
Сныть коротколодная 299365
Сныть обыкновенная 8826
Сныть таджикстная 8828
Сныть широколистная 8824
Собашник 31051
Совичия 29085
Совичия красивоплодная 382696
Содник 379486
Содник морской 379553
Соймида 369934
Сокирки 102833,124141
Сокирки аяксовы 102833
Сокирки полевые 102833
Сола 9531
Солерос 342847
Солерос европейский 342859
Солерос поярковой 342887
Солерос травянистый 342859
Солидаго 367939
Солидаго гиблида 368179
Солнечный хрен 225398
Солнцегляд 159027,188583
Солнцецвет 188583
Солнцецвет альпийский 188586
Солнцецвет багульниколистный 188726
Солнцецвет буша 188609
Солнцецвет войлочный 188857
Солнцецвет восточный 188773
Солнцецвет дагестанский 188652
Солнцецвет джунгарский 188846
Солнцецвет иволистный 188829
Солнцецвет крупноцветковый 188683
Солнцецвет монетолистный 188757
Солнцецвет мохнатоплодный 188725
Солнцецвет обыкновенный 188757
Солнцецвет приальпийский 188586
Солнцецвет седой 188613
Солнцецвет скалоломный 188828
Солнцецвет щетинистоволосистый 188713
Солодка 177873
Солодка бледноцветковая 177932
Солодка бледноцветная 177932
Солодка бухарская 177883
Солодка гладкая 177893
Солодка голая 177893
Солодка гончарова 177913
Солодка ежовая 177885
Солодка зайсанская 177951
Солодка коржинского 177916
Солодка кулябская 177918
Солодка македонская 177925
Солодка рыхлая 177877
Солодка уральская 177947
Солодка шершавая 177877
Солодка шиповатая 177877
Солодка щетинистая 177885
Солодка юньнаньская 177950
Соломонова печать 308493
Соломоцвет 4249
Соломоцвет двузубый 4273
Солонечник 169756
Солонечник алтайский 169763
Солонечник гаупта 169780
Солонечник даурский 169773
Солонечник двухцветковый 169767
Солонечник истоговидный 169788
Солонечник кожистый 169772
Солонечник коротко－мохнатый 169806
Солонечник ленок 231844
Солонечник окрашенно－хохолковый 169770
Солонечник прутьевидный 169795
Солонечник растопыренный 169775
Солонечник редеря 169792
Солонечник русский 169793
Солонечник синьцзянский 169796
Солонечник точечный 169790
Солонечник трехжилковый 169804
Солонечник тяньшанский 169803
Солонечник узколистный 169764
Солонечник щитковидный 169777
Солотьнь 367939
Солушец 217353
Сольданелла 367762
Сольданелла альпийская 367763
Сольданелла горная 367766
Сольданелля 367762
Солянка 344425
Солянка андросова 344441
Солянка аральская 344448
Солянка баранова 344466
Солянка боялыч 344452
Солянка боялычевидная 344453
Солянка бухарская 344476
Солянка введенского 344775
Солянка вересковидная 344526
Солянка горная 344636
Солянка дагестанская 344506
Солянка деревцевидная 344452
Солянка джунгарская 344524
Солянка длиностолбиковая 344618
Солянка древовидная 344510
Солянка заамударьинская 344748
Солянка закаспийская 344747
Солянка запутанная 344581
Солянка заразительная 344743
Солянка звездчатая 344726
Солянка илийская 344574
Солянка калийная 344585
Солянка килеватая 344486
Солянка комарова 344606
Солянка коржинского 344607
Солянка короткощетинковая 344566
Солянка корявая 344652
Солянка лиственничная 344611,344612
Солянка лиственничнолистная 344611
Солянка малолистная 344696
Солянка мерстистолистная 344608
Солянка миквиц 344630
Солянка многолистная 344538
Солянка мясистая 344499
Солянка натронная 344644
Солянка непалская 344642
Солянка однокрылая 344635
Солянка ольги 344647
Солянка открытоплодная 344443
Солянка ошера 344461
Солянка палецкого 344654
Солянка паульсена 344658
Солянка пограничная 344434
Солянка почечконосная 344546
Солянка прозрачная 344662
Солянка пустынная 344514
Солянка ранняя 344674
Солянка рихтера 344682
Солянка рожевица 344687
Солянка розовая 344686
Солянка русская 344743
Солянка самаркандская 344623
Солянка семиреченская 344564
Солянка серая 344482
Солянка сероватая 344578
Солянка сизая 8850
Солянка синьцзянская 344716
Солянка содоносная 344718
Солянка супоротивнолистная 344474
Солянка сухощавая 344620
Солянка тамамшан 344734
Солянка тамарисковидная 344735
Солянка тахтаджана 344733
Солянка толстоватая 344728
Солянка трагус 344743
Солянка туполистная 344639
Солянка тургайская 344755
Солянка туркменская 344754
Солянка тяньшанская 344583
Солянка узловатая 344645

Солянка хивинская 344494
Солянка холмовая 344496
Солянка цинхайская 344493
Солянка шерстистая 344609
Солянка щипчиковая 344539
Соляноколосник 185034,185038
Соляноколосник прикаспийский 185038
Соляноцветник 184669
Соляноцветник кульпский 184670
Соляноцветник редкоцветковый 184671
Соляноцветник розовый 184672
Солянум 366902
Солянум индийское 367241
Солянум японский 367262
Сонгвинария канадская 345805
Сон – трава 321653,321703,321721
Сон – трава горная 321692
Соранг 342859,379486
Сорбария 369259
Сорбария сибирская 369277
Сорбария тибетская 369288
Сорго 369590
Сорго алеппское 369652
Сорго алепское 369652
Сорго африканское зерновое 369720
Сорго веничное 369636
Сорго двцветное 369600
Сорго дикое 369652
Сорго зерновое 369640,369720
Сорго китайское 369671
Сорго комовое 369640
Сорго лимонное 117153,117218
Сорго метельчатое 369636,369720
Сорго многолетнее 369652
Сорго обыкновенное 369720
Сорго обыкновенный 369720
Сорго поникшее 369601
Сорго развесистое 369636
Сорго сахарное 369700
Сорго скученное 369640
Сорго суданское 197331,369707
Сорго техническое 369636
Сорго хлебное 369652
Сорго японское 369671
Сореа 362186
Сорея 362186
Сорея кистевая 362220
Сорная дыня 114150
Соровник 184962
Соровник низкорослый 184965
Сорокозубка 15170
Сосенка водяная 196810,196818
Сосна 299780,300264
Сосна австрийская 300117
Сосна алеппская 299964
Сосна аризонская 299797
Сосна арманда 299799
Сосна балканская 300144
Сосна банкса 299818
Сосна банксова 299818
Сосна белая 300211
Сосна белая калифорнийская 300189
Сосна белоствольная 299786
Сосна болотная 300128
Сосна брута 299827
Сосна бунге 299830
Сосна ван 300298
Сосна веймутова 300211
Сосна веймутова гималайская 300297
Сосна веймутова лумелийская 300144
Сосна веймутова мексиканская 299809
Сосна виргинская 300296
Сосна гельдрейха 299968
Сосна гибкая 299942
Сосна гималайская веймутова 300297
Сосна горная 300086
Сосна горная веймутова 300082
Сосна гриффита 300297
Сосна гуандунская 300009
Сосна густоцветковая 299890
Сосна густоцветная 299890
Сосна джерарда 299953
Сосна длиннохвойная 300128,300186
Сосна европейская 300223
Сосна ежовая 299925
Сосна желтая 300153
Сосна жёлтая 299925,300153
Сосна жёлтая аризонская 299797
Сосна жёлтая западная 300153
Сосна жерарда 299953
Сосна жерардова 299953
Сосна жесткая 300181
Сосна жёсткая 300181
Сосна замечательная 300173
Сосна иерусалимская 299964
Сосна итальянская 300146,300151
Сосна кадровая 300197
Сосна калабрийская 299827,299964
Сосна канадская 298286
Сосна канарская 299835
Сосна караибская 299928
Сосна карибская 299836,299928
Сосна кедровая 299842
Сосна кедровая европейская 299842
Сосна кедровая калифорнийская 299942
Сосна кедровая корейская 300006
Сосна кедровая маньчжурская 300006
Сосна кедровая сибирская 300197
Сосна китайская 300054,300243
Сосна корейская 300006
Сосна красная 300180
Сосна кривая 300086
Сосна кружевнокорая 299830
Сосна крымская 300127
Сосна крючковатая 299966
Сосна культера 299876
Сосна куртера 299876
Сосна ладанная 300264
Сосна ламберта 300011
Сосна лесная 300223
Сосна лучистая 300173
Сосна массона 300054
Сосна массонова 300054
Сосна мексиканская 300081
Сосна мексиканская веймутова 299809
Сосна мелкоигольчатая 300098
Сосна мелкоцветковая 300130
Сосна монтезумы 300081
Сосна муго 300086
Сосна муррея 299874
Сосна мягкоигольчатая 300098
Сосна низкорослая 300163
Сосна норвежская 300180
Сосна обыкновенная 300223
Сосна палласа 300127
Сосна палласова 300127
Сосна пинния 300151
Сосна погребальная 299945,300054,300243
Сосна поникшая 300140
Сосна похоронная 299945
Сосна приморская 300146
Сосна растопыренная 299818
Сосна роксбурга 300186
Сосна румелийская 300144
Сосна сабина 300189
Сосна сахарная 300011
Сосна сахарная гигантская 300011
Сосна сибирская 300197
Сосна скипидарная 300264
Сосна скрученная 111347,299869,299874
Сосна скрученная широкохвойная 299874
Сосна смолистая 300180,300181
Сосна станкевича 300209
Сосна столбовая 300117
Сосна съедобная 299926
Сосна тайваньская 300269
Сосна торрея 300288
Сосна тунберга 300281
Сосна тяжелтая 300153
Сосна чёрная 300117
Сосна чёрная австрийская 300117
Сосна чёрная калабрийская 300017,
300114,300116
Сосна шведская 299842
Сосна широкохвойная 299874
Сосна эллиота 299836,299928
Сосна эльдарская 299927
Сосна энгельмана 299931
Сосна южная 300128
Сосна юньнаньская 300305
Сосна яйцевидноплодная 300126

Сосна японская 299890
Сосна японская черная 300281
Сосна ярусная 300243
Сосновые 299594
Соссюрея 348089
Соссюрея альберта 348118
Соссюрея альпийская 348121
Соссюрея амурская 348141
Соссюрея асбукина 348156
Соссюрея аянская 348111
Соссюрея байкальская 348163
Соссюрея блестящая 348810
Соссюрея бороздчатая 348831
Соссюрея введенского 348929
Соссюрея войлочная 348864
Соссюрея выемчатая 348788
Соссюрея высокая 348283
Соссюрея вытянутая 348289
Соссюрея ганьсуская 348418
Соссюрея голая 348590
Соссюрея горькая 348129
Соссюрея горькуша 348129
Соссюрея грязноцветковая 348800
Соссюрея густолистная 348322
Соссюрея далиская 348249
Соссюрея даурская 348243
Соссюрея дернистая 348181
Соссюрея дорогостайского 348273
Соссюрея дуйская 348142
Соссюрея залемана 348756
Соссюрея заостренная 348104
Соссюрея зубчатая 348604
Соссюрея зубчаточешуйчатая 348604
Соссюрея зубчато－чешуйчатая 348604
Соссюрея иволистная 348757
Соссюрея изящная 348284
Соссюрея испайская 348401
Соссюрея каемчато－чешуйчатая 348499
Соссюрея кандинская 348195
Соссюрея кара－арча 348420
Соссюрея кашгарская 348424
Соссюрея китайская 348199
Соссюрея китамуры 348434
Соссюрея крупнолистная 348355
Соссюрея крылатая 348113
Соссюрея крылова 348441
Соссюрея курильская 348452
Соссюрея кушакевича 348454
Соссюрея ларионова 348474
Соссюрея ледниковая 348334
Соссюрея ленская 348479
Соссюрея липкая 348925
Соссюрея лицзянская 348492
Соссюрея ложно－оттопыренная 348683
Соссюрея ложно－сочавы 348684
Соссюрея ложно－узколистная 348677
Соссюрея мазарская 348513
Соссюрея максимовича 348515
Соссюрея малоцветковая 348626
Соссюрея маньчжурская 348510
Соссюрея мелкоцветковая 348626
Соссюрея микешина 348533
Соссюрея монгольская 348538
Соссюрея мощная 348728
Соссюрея мулиская 348544
Соссюрея непалская 348555
Соссюрея нины 348563
Соссюрея ново－пильчатая 348554
Соссюрея ново－хорошенькая 348553
Соссюрея нупурипская 348593
Соссюрея обернутая 348392
Соссюрея особенная 348624
Соссюрея острозубчатая 348616
Соссюрея отвороченная 348706
Соссюрея оттопыренная 348811
Соссюрея пильчатовидная 348554
Соссюрея подушечная 348697
Соссюрея полякова 348654
Соссюрея порчи 348667
Соссюрея почти－треугольная 348823
Соссюрея прайса 348671
Соссюрея прелестная 348167
Соссюрея простертая 348674
Соссюрея разноцветная 348268
Соссюрея разрезная 348456
Соссюрея ридера 348711
Соссюрея сахалинская 348749
Соссюрея саянская 348755
Соссюрея седая 348183
Соссюрея седоватая 348185
Соссюрея серебристобелая 348381
Соссюрея серебристолистная 348488
Соссюрея серпуховидная 348773
Соссюрея сычуаньская 348835
Соссюрея скальная 348767
Соссюрея скердолистная 348742
Соссюрея скученная 348223
Соссюрея советская 348808
Соссюрея солончаковая 348763
Соссюрея сомнительная 348137,348275
Соссюрея сочавы 348799
Соссюрея спорная 348226
Соссюрея срорная 348672
Соссюрея струговидная 348742
Соссюрея сукачева 348830
Соссюрея суменицевидная 348342
Соссюрея сумневича 348832
Соссюрея таджиков 348838
Соссюрея тайваньская 348256
Соссюрея твердая 348727
Соссюрея теневая 348892
Соссюрея тибетская 348861
Соссюрея тилезиева 348863
Соссюрея треугольная 348510,348865
Соссюрея треугольновидная 348823
Соссюрея тургайская 348889
Соссюрея увенчанная 348232
Соссюрея узколистная 348146
Соссюрея уральская 348900
Соссюрея уссурийская 348902
Соссюрея фаминцина 348308
Соссюрея фори 348313
Соссюрея фролова 348328
Соссюрея хондрилловидная 348208
Соссюрея хорошенькая 348690,348695
Соссюрея чертополохо－корзинчатая 348186
Соссюрея чжундяньская 348274
Соссюрея чэнкоуская 348320
Соссюрея шангиновская 348770
Соссюрея ширетокская 348782
Соссюрея широколистая 348475
Соссюрея штубендорфа 348821
Соссюрея ядринцева 348402
Соссюрея яйцевидная 348614
Соссюрея японская 348403
Сосюрея 348089
Софийка 368944
Софийка гулявниковая 368951
Софийка двулетная 368945
Софийка желтейшая 368947
Софийка монгольская 368950
Софийка однолетная 368945
Софора 368955
Софора бирманская 369004
Софора виколистная 368994
Софора гиблидная 369044
Софора гриффита 369075
Софора желтеюшая 369010
Софора жултеющая 369010
Софора ичанская 369051
Софора китайская 368986
Софора корейская 369054
Софора лисохвостая 368962
Софора лисохвостная 368962
Софора мексиканская 368990
Софора мелкоплодная 369066
Софора обыкновенная 368962
Софора осохвостная 368962
Софора продана 369110
Софора толстоплодная 369095
Софора франше 369018
Софора четырехкрылая 369130
Софора японская 369037,369038
Сочевичник 222671
Сочевичник весенний 222860
Сочевичник гмелина 222720
Сочевичник клубненосный 222851
Сочевичник чёрный 222782
Соя 177689,177750
Соя изящная 177722

Соя культурная 177750,177777
Соя уссурийская 177777
Соя щетинистая 177750,177777
Соя яванская 264241
Спайноцветник 170859
Спайноцветник волосистый 170863
Спайноцветник келифский 170862
Спайноцветник спайноплндный 170861
Спайноцветник ферганский 170860
Спараксис 369975
Спаржа 38904
Спаржа аптечная 39120
Спаржа бреслера 38951
Спаржа бугорчатая 38977
Спаржа бухарская 38953
Спаржа ганьсуская 39048
Спаржа даурсая 38977
Спаржа джунгарская 38928
Спаржа казахстанская 39049
Спаржа колотколистная 38950
Спаржа куэньминская 39082
Спаржа левиной 39061
Спаржа лекарственная 39120
Спаржа ложношероховатая 39157
Спаржа маловетвистая 39126
Спаржа мищенко 39095
Спаржа многолистный 39120
Спаржа мутовчатая 39251
Спаржа обыкновенная 39120
Спаржа овощная 39120
Спаржа остролистная 38908
Спаржа перистая 39195
Спаржа персидская 39141
Спаржа плюмажная 39195
Спаржа попова 39154
Спаржа пренебреженная 39111
Спаржа прибрежная 39064
Спаржа пушистая 39195
Спаржа пушистоперистая 39195
Спаржа спаржевидная 38938
Спаржа тонколистная 39229
Спаржа туркестанская 39242
Спаржа угловатонадломанная 38928
Спаржа шобериевидная 39187
Спаржа шпренгера 38983
Спаржа щетнновидная 39196
Спаржа японская 30619
Спармания 370130
Спармания африканская 370136
Спартина 370156
Спарто 376931
Спекулярия 224058
Спелта 398970
Спельта 398970
Спигелия 371612
Спиж 380499
Спирантес 371627,372176
Спирантес китайский 372247
Спирея 371785
Спирея белая 372075
Спирея березолистная 371823
Спирея волосистая 372106
Спирея вязолистная 371877
Спирея городчаталистная 371877
Спирея городчатая 371877,371890
Спирея гуйцзоуская 371989
Спирея дубровколистая 371877
Спирея дубровколистная 371877
Спирея зверобоелистая 371938
Спирея зверобоелистная 371938
Спирея зверобой - нолистная 371938
Спирея иволистая 372075
Спирея иволистная 372075
Спирея извилистая 371910
Спирея лицзянская 371997
Спирея мийябе 372012
Спирея мозолистая 371944
Спирея монгольская 372018
Спирея наньчуаньская 372072
Спирея пушистая 372067
Спирея рябинолистная 369279
Спирея сливолистная 372050
Спирея средняя 372006
Спирея тайваньская 371912
Спирея тунберга 372101
Спирея японская 371944
Спиродела 372295
Спиродела многокоренная 372300
Спиродела сычуаньская 372304
Спиростегия 372354
Спиростегия бухарская 372355
Сподиопогон 372398
Сподиопогон сибирский 372428
Спондиас 372459
Спондиас жёлтый 372477
Спондиас пазушный 88691
Спондиас цитеры 372467
Спороболус 372574
Спороболус индийский 372650,372666
Спороболус японский 372737
Спороболюс 372574
Спорыш 308816,309570
Спрекелия 372912
Спрыгиния 275865
Спрыгиния винкрева 372934
Срединския 372966
Срединския большая 372969
Сростноплодник 380809
Сростноплодник пахучий 380810
Сростнохвостник 382067
Сростнохвостник дельтовидный 382069
Ставрофрагма 405657
Стагачка 242993,254094
Стальник 271290
Стальник австрийский 271337
Стальник глядкосемянный 271436
Стальник голосемянный 271436
Стальник древник 271320
Стальник желтый 271464
Стальник жёлтый 271464
Стальник колючий 271347
Стальник колячий 271347
Стальник маленикий 271540
Стальник натрикс 271464
Стальник обыкновенный 271333
Стальник олевой 271333
Стальник пашенный 271333
Стальник полевой 271333
Стальник ползучий 271563
Стальник промежуточный 271428
Стангерия 373673
Стапелия 373719
Стапелия крупноцветковая 373827
Стапелия пестрая 273206
Стародубка 8381,173598
Статице 34488
Стаунтония 374381
Стаунтония китайская 374392
Стаунтония хайнаньская 374403
Стаунтония шестилистная 374409
Стафилея 60379
Стафилея перистая 374107
Стахиопсис 373068
Стахиопсис продолговатолистный 373071
Стахиопсис шандровый 373070
Стахиопсис ясноткоцветковый 373069
Стахис 373085
Стахис клубненосный 373439
Стахис японский 373262
Стахиурус 373516
Стахиурус китайский 373524
Стахиурус ранний 373556
Стахиурус сычуаньский 373576
Стеблеклонник 190791
Стеблелист 79730
Стеблелист мощный 79732
Стевениелла 376390,376391
Стевения 376387
Стевения бурачковидная 376388
Стевения левкоевидная 376389
Стевения левкойная 376389
Стевия 376405
Стевия яйцевидная 376410
Стелигма 376314
Стелигма бугроватая 376322
Стелигма войлочная 273971
Стелигма иглоплодная 376315
Стелигмостемум 376314
Стелигмостемум войлочный 273971
Стелларня 374724
Стеллера 375179,375187

Стеллера карликовая 375187
Стеллера тайваньская 375198
Стеллеропсис 375207
Стеллеропсис алтайский 375209
Стеллеропсис антонины 375210
Стеллеропсис иранский 375212
Стеллеропсис иссыкульский 375213
Стеллеропсис кавказский 375211
Стеллеропсис магакяна 375214
Стеллеропсис туркменский 375216
Стеллеропсис тяньшанский 375215
Стелющаяся 11972
Стенактис 150414
Стенактис однолетний 150464
Стенанциум 375451
Стенанциум сахалинский 375459
Стенник 203181
Стенник вечнозелёный 203239
Стенник горьгий 203184
Стенник зонтичный 203249
Стенница 284167
Стеноафрум 375764
Стеноафрум американский 375765
Стенокарпус 375510
Стенокарпус ивовый 375511
Стенотения 375760
Стенотения крапнолодная 375761
Стеноцелиум 375551
Степной миндаль 20931
Степные петушик 208874
Степторамфус 376052
Степторамфус катранолистный 376053
Степторамфус клубненосный 376059
Степторамфус линчевского 376056
Степторамфус персидский 376057
Степторамфус скальный 376058
Степторамфус толстостебельный 376054
Степторамфус черепанова 376055
Стеркулиевые 376218
Стеркулия 376063
Стеркулия вонючая 376110
Стеркулия платанолистная 166627
Стеркулия плёнолистная 58335
Стеркулия утёсная 58348,376180
Стеуднера 376380
Стеуднера колоказиелистная 376382
Стефанандра 375811
Стефанандра китайская 375812
Стефанандра надрезаннолистная 375817
Стефанандра надрезная 375817
Стефанандра танаки 375821
Стефания 375827
Стефания гуансиская 375880
Стефания ланьюйская 375889
Стефания мелкоцветковая 375890
Стефания округлая 375896
Стефания сычуаньская 375903
Стефания тайваньская 375897
Стефания тибетская 375854
Стефания хайнаньская 375859
Стефания цзиндунская 375834
Стефания японская 375870
Стефанотис 211693,376022
Стефанотис люцюский 211695
Стефанотис японский 211695
Стизолофус 377137
Стизолофус бальзамический 80957
Стизолофус коронопусолистный 377140
Стилидиум 379076
Стиракс 379300
Стиракс аптечный 379419
Стиракс бензоин 379312
Стиракс бензоиновый 379312
Стиракс вильсона 379474
Стиракс гуансиский 379411
Стиракс густоцветковый 379331
Стиракс жуйлиский 379427
Стиракс китайский 379325
Стиракс ланьюйский 379357
Стиракс лекарственный 379419
Стиракс обасия 379414
Стиракс тайваньский 379347
Стиракс хайнаньский 379361
Стиракс ху 379370
Стиракс чжэцзянский 379478
Стиракс японский 379374
Стираксовые 379296
Стирас – бинзоин 379312
Стланец березовый 53530
Стланец кедровый 300163
Стоколос 60612
Стоколос безостый 60760
Стоколос бесплодный 60985
Стоколос кровельный 60989
Стоколос рррямой 60707
Стоксия 377286
Столетник 10787
Стоножка 81681
Стоножка округлая 81687
Стонтония 374381
Стонтония тайваньская 374400
Стонтония шестилисточковая 374409
Стонтония яошаньская 374440
Стосил 143665,280799
Странвезия 377433
Странвезия давида 377444
Странвезия давида волнистая 377452
Странвезия нусия 377462
Странвезия сизоватая 377462
Страстоцвет 285607
Страстоцвет голубой 285623
Страстоцвет гуандунский 285659
Страстоцвет изящный 285645
Страстоцвет крупный 285694
Страстоцвет лавлолистный 285660
Страстоцвет настоящий 285694
Страстоцвет пахучий 285639
Страстоцвет пробковый 285703
Страстоцвет светло – синий 285623
Страстоцвет съедобный 285637
Страстоцвет четырехгранный 285694
Страстоцвет четырёхугольный 285694
Страстоцветные 285725
Стрелиция 377550
Стрелиция королевская 377576
Стрелолист 342308
Стрелолист злаковидный 342338
Стрелолист китайский 342400,342424
Стрелолист личуаньский 342372
Стрелолист обыкновенный 342400,342424
Стрелолист плавающий 342385
Стрелолист стрелолистный 342400,342424
Стрелолист стрелолистый 342400
Стрелолист трехлистный 342421
Стрелолист трилистный 342421
Стрелолист трилистый 342421
Стрелолист тэнчунский 342419
Стрелолист шиловидный 342417
Стрептокарпус 377652
Стрептопус 377902
Стрептопус мелкоцветковый 377933
Стрептопус стеблеобъемлющий 377930
Стрептопус стрептопусовидный 377942
Стрига 377969
Стрига азиатская 377973
Стрига желтая 377973
Стрифнодендрон 378959
Стрифнодендрон – барбатимао 378960
Стрихнос рвотный орех 378834
Стробилантес 378079
Стробилантес диера 290921
Стробилантес тайваньский 178851
Стробилантес японский 85944
Строгановия 378287
Строгановия афганская 378289
Строгановия коренстая 378299
Строгановия короткоухая 378290
Строгановия литвинова 378296
Строгановия метельчатая 378297
Строгановия персидская 378298
Строгановия промежуточная 378295
Строгановия сердцелистая 378292
Строгановия стрелолистая 378300
Строгановия субальпийская 378301
Строгановия тонкая 378294
Строгановия трауфеттера 378302
Стромбозия 378319
Строфант 378363
Строфант расходящийся 378386
Строфостилес 378497
Стручок цареградский 83527

Стюартия 376428
Стюартия однобратственная 376457
Стюартия псевдокамелия 376466
Стюартия – монодерьфа 376457
Суданска 197331,369707
Судза 290932
Суеда 379486
Сукцизелла 379662
Сукцизелла когнутая 379663
Сулла 187842
Сумаум 80120
Сумах 332452
Сумах венецианский 107300
Сумах волосистоплодный 393488
Сумах восковой 393491
Сумах восточный 393466
Сумах гладкий 332608
Сумах голый 332608
Сумах делаве 393449
Сумах дубильный 107300,332533
Сумах душистый 332477,332916
Сумах замещающий 393479
Сумах канадский 393470
Сумах китайский 332509,332662
Сумах кожевеный 332533
Сумах коиевенный 332533
Сумах копалловый 332529
Сумах копаловый 332529
Сумах коротковолосый 332916
Сумах красильный 332533
Сумах лаковый 393491
Сумах лаконосный 393491
Сумах лесной 393485
Сумах обыкновенный 332533
Сумах оленерогий 332916
Сумах оленьерогий 332916
Сумах пушистый 332916
Сумах пятилистниковый 332766
Сумах сицилийский 332533
Сумах сомнительный 393466
Сумах сочный 393479
Сумах укореняющийся 393470
Сумах уксусный 332916
Сумах яванский 332509,332662
Сумах ядовитый 393470
Сумах японский 393479
Сумаховые 21190,176867
Сумочник 72037,72038
Сумочник пастуший 72038
Суноктонум 117334
Суноктонум пурпуровый 117660
Сурепица 47931,59603
Сурепица весенняя 47963
Сурепица обыкновенная 47964
Сурепица яровая 59575,74435
Сурепка 47931
Сурепка весенняя 47963
Сурепка дуговидная 47934
Сурепка крупноцветковая 47944
Сурепка малая 47948
Сурепка обыкновеная 47964
Сурепка подорожниковая 47954
Сурепка прижатая 47959
Сурепка прямая 47949,47959
Сурепка тайваньская 47962
Сусак 64180
Сусак зонтичный 64184
Сусак ситниковидный 64181
Сусаковые 64173
Сухоребрик 365564
Сухоцвет 415315
Сухоцвет однолетний 71042,415317
Сухоцветник 415315
Сухтеления 379673
Сухтеления чашечная 379674
Сухумка 285422
Сушеница 178056,224767
Сушеница альпийская 224772
Сушеница байкальская 178090
Сушеница болотная 178481
Сушеница желто – белая 178265
Сушеница жёлто – белая 178265
Сушеница желтоватобелая 178265
Сушеница кавказская 178114
Сушеница казахстанская 178240
Сушеница лежачая 178463,260217
Сушеница лесная 178465
Сушеница маньчжурская 178278
Сушеница наньчуаньская 178303
Сушеница норвежская 178311
Сушеница песчаная 189147
Сушеница приземистая 178463
Сушеница родственная 178062
Сушеница русская 178389
Сушеница сибирская 178409
Сушеница топяная 178481
Сушеница траншеля 178475
Сушеница японская 178237
Сфаллерокарпус 371237
Сфаллерокарпус тонкий 371240
Сферофиза 371160
Сферофиза солонцовая 371161
Схеноплектус 353179
Схеноплектус озерный 352289
Схеноплектус озёрный 352206
Схеноплектус табернемонтана 352278
Схеноплектус трёхгранный 353899
Схенус 352348
Схенус ржавый 352385
Схенус чернеющий 352408,352409
Схенус черноватый 352408,352409
Схизантус 351343
Схизахна 351250
Схизахна комарова 351253
Схизахна мозолистая 351251
Схизахне 351261
Схизеевые 351121
Схизмус 351162
Схизмус арабский 351163
Схизмус малый 351167
Схизмус чашечный 351172
Схизокодон 351439
Схизонепета 351742
Схизонепета многонадрезанная 265001
Схизонепета однолетняя 264869
Схизопепон 351759
Схизопепон переступенелистный 351762
Схизопепон сычуаньский 351767
Схизопепон тибетский 351774
Схизостилис 351883
Схизофрагма 351780
Схизофрагма гортензиевидная китайскя 351797
Схима валлиха 350949
Схинус 350980
Схинус моле 350990
Схинус мягкий 350990
Схинус теребинтолистный 350995
Схинус терпентинолистный 350995
Схисмус 351162
Схицейные 351121
Сциадопитис 352854
Сциадопитис мутовчатый 352856
Сцилла 352880
Сцилла двулистная 352886
Сцилла сибирская 353065
Сциндапсус 353121
Сциндапсус хайнаньский 353127
Сыга 145073
Сырейщиковия 382096
Сырейщиковия тонкая 382098
Сырейщиковия шипиковатая 382097
Сыть 118428
Сыть амурская 118491
Сыть гаспая 118982
Сыть гладкая 118941
Сыть густоцветная 118957
Сыть джунгарская 119593
Сыть длинная 119125
Сыть желтоватая 119687
Сыть золотистая 118822
Сыть ирия 119041
Сыть клубненосная 119503
Сыть краснопятнистая 322342
Сыть кртглая 119503
Сыть круглая 119503
Сыть кругловатая 119503
Сыть микели 119196
Сыть ниппонская 119262
Сыть прямоколосая 119319
Сыть разновидная 118744

Сыть разнолистная 118476
Сыть разнородная 118744
Сыть скученная 118957
Сыть съедобная 118822
Сыть темнокаштановая 119127
Сыть усеченная 119319
Сыть хайнаньская 118844
Сыть чернобурая 118928
Сют – от 368649
Табак 266031
Табак виргинский 266060
Табак деревенский 266053
Табак махорка 266053
Табак настоящий 266060
Табак обыкновенный 266060
Табак турецкий 266060
Табак тютюн 266053
Табак – махорка 266053
Табебуйя 382704
Таболга вязолистная 166125
Таволга 371785
Таволга альпийская 235651,371798
Таволга бальджуанская 371818
Таволга белая 371795,372075
Таволга березолистная 371823
Таволга берёзолистная 371823
Таволга биллиарда 371834
Таволга бильярда 371834
Таволга бовера 371828
Таволга вильсона 372135
Таволга виргинская 372134
Таволга водосборолистная 371807
Таволга войлочная 372106
Таволга волосистая 372047
Таволга городчатая 371890
Таволга гуансиская 371988
Таволга даурская 371895
Таволга дубровколистная 371877
Таволга дугласа 371902
Таволга зверобоелистая 371994
Таволга зверобоелистная 371938
Таволга иволистная 372075
Таволга извилистая 371910
Таволга изящная 371903
Таволга китайская 371884
Таволга мулиская 372022
Таволга низкая 371937
Таволга нинсяская 372027
Таволга ниппонская 372029
Таволга опушенноплодная 372114
Таволга острая 371815
Таволга пушистая 372067
Таволга ростгорна 372072
Таволга саржента 372078
Таволга тибетская 372136
Таволга трехлопастная 372117
Таволга тунберга 372101
Таволга тяньшанская 372104
Таволга уссурийская 371877
Таволга ферганская 371907
Таволга цабеля 372141
Таволга шелковистая 372083
Таволга шестилепестная 166132
Таволга шилоколистная 371795
Таволга шоха 372081
Таволга шренка 372082
Таволга японская 371944
Таволга японская звездчатая 371981
Таволгоцвет 372151
Таволгоцвет шренковский 372153
Таволжник 37053
Таволжник гуншаньский 37079
Таволжник обыкновенный 37085
Тагетес 383087
Тагетес мексиканская 383107
Тагетес прямостоячий 383090
Тагетес распростертые 383103
Таджиск 20889
Тайвания 383188
Тайвания криртомериевидная 383189
Тайваньск бирманский 204490
Тайник 232890
Тайник большой 264727
Тайник короткозубчатый 232901
Тайник мелкоцветковая 264746
Тайник овальнолистный 232950
Тайник овальный 232950
Тайник сердцвидный 232905
Тайник тяньшанский 264748
Тайник японский 232919
Такка 382908
Такковые 382934
Таксодиевые 385250
Таксодий 385253
Таксус 385301
Таксус бакката 385302
Таксус обыкновенный 385302
Таксус тибетский 385404
Таксус ягодный 385302
так – так 198365
Такусса 143522
Талассия 383233,388357
Талассия заилийская 383235
Талаума 383236
Талхак 6278
Тальник 343937
Тальник ползучий 343994
Тамарикс 383412
Тамарикс андросова 383431
Тамарикс ганьсуский 383514
Тамарикс гогенакера 383519
Тамарикс изящный 383516
Тамарикс китайский 383469
Тамарикс мейера 383557
Тамарикс мелкоцветковая 383581
Тамарикс многоветвистый 383595
Тамарикс монгольский 383559
Тамарикс одесский 383477,383595
Тамарикс палласса 383591
Тамарикс патитычночный 166761,383591
Тамарикс раскидистый 383535
Тамарикс удлиненный 383479
Тамарикс французский 383491
Тамарикс четырехтычинковый 383581
Тамарикс четырёхтычинковый 383581
Тамарикс щетинистоволосистый 383517
Тамарикс щетинистоволосисый 383517
Тамарикс эверсманна 383484
Тамаринд 383404
Тамаринд индийский 383407
Тамарксовые 383397
Тамокука 21965
Тамус 383674
Тамус обыкновенный 383675
Танжерин 93446,93717
Тапиока 244507,244515
Тапсия 63443,388868
Таран 291663
Таран бухарский 309015
Таран гиссарский 309186
Таран дубильный 309015
Таран пышный 309353
Таро 99910
Тарра 114209
Тары 281274,281916
Тасва – арча 213877
Татарка 15289
Татаркик 271663
Татаркик колючий 271666
Татаркик обыкновенный 271666
Татарник полевой 91770
Тау – сагыз 354961
Таущерия 385129
Таущерия длинная 385132
Таущерия опушенноплодная 385132
Теведия 389967
Теведия олеандролистная 389978
Текен 42883
Теке – сакал 135134
Текома 385461
Текома укореняющаяся 70512
Текомария 385505
Текомария капская 385508
Тектона 385529
Телекия 63524,385614
Телекия красивая 385616
Телесперма 389153
Телефиум 385629
Телефиум восточный 385638
Телигоновые 389181
Телигонум 389183

Телигонум тайваньский 389185
Телигонум цинокрамба 389184
Телигонум японский 389186
Телимитра 108641,389258
Теллунгиэлла 389197
Теллунгиэлла солелюбка 389198
Теллунгиэлла солонцовая 389200
Телорез 377482
Телорез алоэвидный 377485
Телорез обыкновенный 377485
Тельфайрия 385650
Тельфайрия восточноафриканская 385655
Тельфайрия стоповидная 385655
Темеда 389307
Темпексоранг 184933
Теоброма 389399
Теозинт 155876
Теозинта 155876
Теозинте 155876
Теосинт 155876
Теосинт мексиканский 155878
Тепари 293969
Терескен 218116
Терескен серый 218122
Терминалия 386448,386500
Терминалия катаппа 386500
Терминалия пышная 386652
Терминалия хвалебная 386652
Терминалия шерстистая 386656
Термопсис 389505
Термопсис альпийский 389507
Термопсис бобовый 389540
Термопсис длинноплодный 389520
Термопсис китайский 389518
Термопсис ланцетный 389533
Термопсис ланцетолистный 389533
Термопсис лупиновый 389540
Термопсис монгольский 389543
Термопсис очередноцветковый 389510
Термопсис туркестанский 389557
Термопсис цзилунский 389525
Терн 316819
Тёрн 316819
Терн дикий 316819
Тёрн мышиный 340990
Терн терновник 316819
Тернии христовы 280559
Терновник 316166,316819
Тернослив 316470
Тернстремия 386689
Тернстремия голопыльниковая 386696
Тернстремия гуандунская 386712
Тернстремия мелколистная 386720
Тернстремия симаоская 386731
Тернстремия сычуаньская 386730
Тернстремия хайнаньская 386702
Тернстремия японская 386696
Тесколюб 317006
Тесколюб седоватый 317009
Тесколюб щетинолистный 317012
Тестудинария 131451
Тетрагония 387221
Тетрагонолобус 237476,387239
Тетрагонолобус струковый 387256
Тетрагонолопус пурпурный 237771
Тетрадиклис 387046
Тетрадиклис тондньский 387049
Тетраклинис членистый 386942
Тетрапапакс 387428
Тетрапапакс бумажный 387432
Тетрапапакс тибетский 387435
Тетрацера 386853
Тефрозия 385926
Тефрозия пурпурная 386257
Тефф 147468
Тефф абиссинский 147994
Тиарелла 391520
Тибурбу 28963
Тигридия 391622
Тигридия павония 391627
Тизанолена 391477
Тик 385529,385531
Тилакоспермум 390967
Тилакоспермум плотнодернистый 390968
Тилландсия 391966
Тиллея 391903,391918
Тиллея вайяна 391959
Тиллея водная 26385,391923
Тиллея водяная 26385,391923
Тиллея крылатая 391920
Тиллея монгольская 391939
Тимелеевые 391030
Тимелея 390994
Тимелея волобьиная 391015
Тимелея обыкновенная 391015
Тимелея однолетная 391015
Тимелия 390994
Тимиан 391061
Тимиян 391061
Тимофеевка 294960
Тимофеевка альпийская 294962
Тимофеевка бёмора 294989
Тимофеевка буассье 294972
Тимофеевка горная 294986
Тимофеевка греческая 294979
Тимофеевка колючая 294976
Тимофеевка луговая 294992
Тимофеевка метельчатая 294988
Тимофеевка микели 294985
Тимофеевка песчаная 294966
Тимофеевка степная 294989
Тимофеевка тонкая 295005
Тимурия 392091
Тимурия сапожникова 392095
Тимьян 391061,391397
Тимьян азиатский 391351
Тимьян алатавский 391064
Тимьян алтайский 203080
Тимьян альсеньева 391084
Тимьян амурский 391078
Тимьян арараский 391079
Тимьян армянский 391083
Тимьян ашурбаева 391085
Тимьян байкальский 391184
Тимьян башкирский 391098
Тимьян безжилковый 391180
Тимьян близкий 391334
Тимьян блошиный 391346
Тимьян болгарский 391112
Тимьян бухарский 391111
Тимьян буша 391113
Тимьян волосистотебельный 391204
Тимьян гаджиева 391199
Тимьян гибкий 391187
Тимьян гранитный 391195
Тимьян гроссгейма 391196
Тимьян губерлинский 391197
Тимьян дагестанский 391166
Тимьян даурский 391167
Тимьян двжилковый 391101
Тимьян двуформенный 391171
Тимьян дзевановского 391177
Тимьян днепровский 391106
Тимьян дубянского 391176
Тимьян душистый 391397
Тимьян евпаторийский 391185
Тимьян елизаветы 391178
Тимьян енисейский 391227
Тимьян еравинский 391181
Тимьян жилковатый 391302
Тимьян зазубренный 391160
Тимьян закавказский 391389
Тимьян закаспийский 391388
Тимьян зеленецкого 391399
Тимьян зиаратинский 391401
Тимьян известковый 391161
Тимьян ильина 391218
Тимьян иртышский 391223
Тимьян кавказский 391121
Тимьян казахстанский 391233
Тимьян кальмиусский 391229
Тимьян каменистый 391324
Тимьян карамарьянский 391230
Тимьян карягина 391232
Тимьян киргизский 391234
Тимьян клиновид 391163
Тимьян клокова 391237
Тимьян кожелистный 391158
Тимьян комарова 391238
Тимьян короткий 391164,391165
Тимьян косматый 391203

Тимьян кочиев 391239
Тимьян красивенький 391345
Тимьян крымский 391385
Тимьян ладжанурский 391240
Тимьян лёви 391256,391259
Тимьян ледяной 391193
Тимьян липского 391254
Тимьян листоногий 391325
Тимьян ложннблощийский 391344
Тимьян ложногранитный 391339
Тимьян ложнограранитный 391338
Тимьян ложномонетный 391342
Тимьян ложноприземистый 391339
Тимьян майкопский 391263
Тимьян малолистный 391323
Тимьян малшалла 391274
Тимьян малшаллов 391274
Тимьян маньчужский 391264
Тимьян мархотский 391265
Тимьян мигрийский 391278
Тимьян минусинский 391279
Тимьян молдавский 391282
Тимьян монгольский 391284
Тимьян монетный 391306
Тимьян мтгоджарский 391289
Тимьян мугоджарский 391289
Тимьян нарымский 391300
Тимьян неверный 391221
Тимьян неравный 391220
Тимьян нерчинский 391301
Тимьян обыкновенный 391365,391397
Тимьян огетый 391077
Тимьян окаймленный 391151
Тимьян острозубый 391310
Тимьян охотский 391307
Тимьян палласов 391311
Тимьян паннонский 391321
Тимьян пастуший 391322
Тимьян побережный 391255
Тимьян подольский 391326
Тимьян полевой 391365
Тимьян ползучий 391365
Тимьян полуголый 391364
Тимьян попеременный 391076
Тимьян пржевальского 391347
Тимьян пустынник 391182
Тимьян пятижилковый 391347
Тимьян пятиребрый 391347
Тимьян разнолистный 391173
Тимьян разъединенный 391172
Тимьян редкоцветковый 391357
Тимьян розовый 391360
Тимьян сибирийский 391378
Тимьян смолистный 391102
Тимьян соколова 391379
Тимьян сосновского 391380
Тимьян степной 391381
Тимьян субальпийский 391382
Тимьян субарктический 391383
Тимьян талиева 391384
Тимьян тбилисский 391387
Тимьян траутфеттера 391390
Тимьян тссурийский 391393
Тимьян турчанинова 391391
Тимьян тяньшанский 391231
Тимьян украинский 391392
Тимьян уменишенный 391170
Тимьян федченко 391186
Тимьян фомина 391188
Тимьян хлопчатый 391183
Тимьян холмовый 391155
Тимьян частолистный 391159
Тимьян шерстистый 391246
Тимьян широколистный 391247
Тимьян шишкина 391363
Тимьян эльтонский 391179
Тимьян яйлы 391224
Тимьян японский 391225
Типчак 163778,164126,164295
Тис 385301
Тис головчатый 82496
Тис европейский 385302
Тис канадский 385344
Тис китайский 385405
Тис коротколистный 385342
Тис ягодный 385302
Тис японский 385355
Тисдайлия 385549
Тисдайлия голостебельная 385555
Тисовник 82496
Тисовник головчато – костянковый 82499
Тисовник костянковый 82499
Тисовые 385175
Тисс 385301
Тисс восточный 385355
Тисс дальневосточный 385355
Тисс европейский 385302
Тисс канадский 385344
Тисс китайский 385405
Тисс коротколистный 385342
Тисс майра 385407
Тисс остроконечный 385355
Тисс средний 385384
Тисс ягодный 385302
Тисс японский 385355
Тиссовые 385175,385250
Тладианта 390107
Тладианта дубия 390128
Тладианта сомнительная 390128
Тмин 77766
Тмин альпийский 77768
Тмин бурятский 77777
Тмин гроссгейма 77812
Тмин кавказский 77788
Тмин комарова 77818
Тмин луговой 77784
Тмин обыкновенный 77784
Тмин песчаный 189147
Тмин скальный 77836
Тмин темнокрасный 77774
Тмин хэбэйский 77775
Тмин чёрный 266241
Тминоножка 77096
Тминоножка армянская 77099
Тминоножка уплощенноплодная 77100
Товалга средняя 372006
Тоддалия 392556
Тоддалия азиатская 392559
Тоддин – пальма 78056
Токсикодендрон 393441
Токсикодендрон волосистоплодный 393488
Токсикодендрон восточный 393466
Токсикодендрон лаконосный 393491
Токсикодендрон укореняющийся 393470
Токсикодендрон чжэньканский 393458
Токусса 143522
Толокнянка 31102,108775
Толокнянка альпийская 31314
Толокнянка аптечная 31142
Толокнянка медвежья 31142
Толокнянка обыкновенная 31142
Толстокрыл 279703
Толстокрыл густоцветковый 279707
Толстокрыл коротконогий 279704
Толстокрыл многостебельный 279714
Толстореберник 279644
Толстореберник альпийский 279645
Толстореберник остроконечный 229365
Толстостенка 279622
Толстостенкакрупнолистная 279625
Толстянка 109505
Толстянковые 109505
Томантеа 392720
Томантеа даралагезская 392723
Томантеа замечательная 392725
Томантеа ошэ 392722
Томантеа темнохолковая 392724
Томат 239154,297711
Томат гумбольдта 239161
Томат перуанский 239163
Томат смородинный 239165
Томат съедобный 239157
Тонколистниковые 150414
Тонколучник 217416
Тонконог 217416,217473
Тонконог азиатский 217427
Тонконог алтайский 217421
Тонконог блестящий 217568
Тонконог буша 217431
Тонконог городкова 217474
Тонконог гребенчатый 217441

Тонконог дегена 217461
Тонконог делявиня 217462
Тонконог жестнолистный 217561
Тонконог кавказский 217434
Тонконог коленчатый 217427
Тонконог ледебура 217485
Тонконог литвинова 217487
Тонконог лоснящийся 217499
Тонконог люерссена 217493
Тонконог мятликовидный 217518
Тонконог обыкновенный 217441
Тонконог одноцветковый 217496
Тонконог поле 217520
Тонконог полуголый 217562
Тонконог польский 217522
Тонконог сибирский 217565
Тонконог сизый 189073,217473
Тонконог сомниельный 217483
Тонконог степной 217441
Тонконог стройный 217441
Тонконог темнофиолетовый 217427
Тонконог токийский 217494
Тонконог тона 217575
Тонконог тонкий 217441,217494
Тонконог фомина 217469
Тонкостенник 201135
Тонкостенник лежачий 198485
Тонкотрубочник 375725
Тонкотрубочник скальный 375726
Топинамбур 189073
Тополь 311131,311509
Тополь амурсий 311229
Тополь бальзамический 311237
Тополь белый 311208
Тополь берлинский 311144
Тополь болле 311226
Тополь волосистоплодный 311554
Тополь волосистый 311423
Тополь восточно－персидская 311308
Тополь ганьсуский 311182
Тополь густолиственный 311524
Тополь давида 311281
Тополь дельтовидный 311146,311292
Тополь дрожащий 311537
Тополь душистый 311509
Тополь дэциньский 311340
Тополь евфратская 311308
Тополь зиболида 311480
Тополь илийская 311355
Тополь итальяский 311400
Тополь канадский 311146,311292
Тополь кандинская 311362
Тополь каролинский 311231
Тополь китайская 311281
Тополь китайский 311265
Тополь корейский 311368
Тополь краузе 311369
Тополь крупнолистный 311259
Тополь лавролистный 311375,324095
Тополь литвинова 311308
Тополь максимовича 311389
Тополь обыкновенный 311537
Тополь осина 311537
Тополь осинообразный 311547
Тополь осокорь 311398
Тополь памирский 311420
Тополь печальный 311557
Тополь пирамидальный 311400
Тополь разнолистная 311308
Тополь ребристый 311231
Тополь самаркандский 311226
Тополь саржента 311465
Тополь седеющий 311261
Тополь седой 311259
Тополь серебристый 311208
Тополь сереющий 311261
Тополь серый 311261
Тополь сизолистная 311433
Тополь снежнобелый 311416
Тополь таджикистанский 311206
Тополь таласский 311524
Тополь треугольнолистный 311292
Тополь трехгранный 311292
Тополь туркестанский 311226
Тополь туркестанский пирамидальный 311226
Тополь узбекистанский 311204
Тополь уссурийский 311562
Тополь фремонта 311324
Тополь хэбэйский 311349
Тополь цзилунская 311278
Тополь черный 311398
Тополь чёрный 311398
Тополь ядунская 311579
Тополь－белолистка 311208
Топольник 311509
Топорики 102833,124008,124141
Тордириум 356377,392863
Тордириум комарова 392873
Тордириум крупный 392875
Торения 392880
Торения мелкоцветковая 392927
Торения наньтоуская 392923
Торения фурнье 392905
Торилис 392959
Торилис желтоволосый 393009
Торилис лучистый 393002
Торилис полевой 392968
Торилис разнолистный 392985
Торилис тонкий 393006
Торилис тонколитный 392993
Торилис узловатый 392996
Торилис украинский 393007
Торилис японский 392992
Торица 370521
Торица весенняя 370604
Торица крупная 370522
Торица льняная 370546
Торица обыкновенная 370522
Торица пашенная 370522
Торица полевая 370522
Торица посевная 370595
Торица пятитычинковая 370568
Торичник 63805,370609
Торичник двутычинковый 370624
Торичник красивый 370696
Торичник морской 370661
Торичник окаймленный 370675
Торичник окаймлённый 370675
Торичник полевой 370696
Торичник солончаковый 370661
Торрейя 393047
Торрея 393047
Торрея большая 393061
Торрея калифорнийская 393057
Торрея крупная 393061
Торрея крупная дирла 393061
Торрея орехоносная 393077
Торрея орехоплодная 393077
Торрея тисолистная 393089
Торрея тиссолистная 393089
Торулярия 264603
Торулярия низкая 264609
Тофилдия 392598
Тофилдия багряная 392603
Тофилдия шарлаховая 392603
Тофильдия 392598
Тофильдия болотная 392629
Тофильдия поникающая 392602
Тофильдия понкшая 392603
Тофильдия цветная 392601
Тофильдия чашечкоцветковая 392601
Тофильдия японская 392622
тоция 393568
тоция карпатская 393570
Трава абиссинская 147994
Трава алтейная 18173
Трава альфа 376931
Трава ананасная 163117
Трава бермудская 117859
Трава бизонья 57922
Трава блошная 309570
Трава боголодская 391365
Трава верблюжья 14638
Трава войлочная 114947
Трава воловья 271290
Трава восьмилепестная 138461
Трава вшивая 286972,287494
Трава гама 398139
Трава гвинейская 281887
Трава гемолойная 309570

Трава Горицвета 8387
Трава горляная 70331
Трава грама 57922,57930
Трава грамова 57922,57930
Трава гусиная 312360
Трава джонзонова 369652
Трава драгун 35411
Трава железная 406040
Трава желтая 327879
Трава жетяница 77748
Трава живучая 13174
Трава заячья 339880
Трава злая 63443,388868
Трава золотушная 54158
Трава зольная 28200
Трава камфарная 70460
Трава канареечная 293709
Трава канарская 293729
Трава картамова 77748
Трава козья 169935
Трава колбасная 274224
Трава колдуновая 91499
Трава колосовая 192926
Трава колюрин алтайская 100135
Трава красильная 327879
Трава кузьчёва 146154
Трава кукушечная 138461
Трава куропаточья 138445
Трава куропаточья восьмилепестная 138461
Трава лакмусовая 89320
Трава ластовичная 86755,115031
Трава легочная 321633
Трава лёгочная 321633
Трава ледяная 251265
Трава лимонная 117153,249504
Трава лихорадочная 180321
Трава ложечная лекарственная 98020
Трава лягушечная 98020
Трава лягушечья 309199
Трава майского ландыша 102867
Трава манная 177599,309199
Трава маточная 177599
Трава медвежья 322724
Трава медовая 321643
Трава модовая 197301
Трава молассkая 249303
Трава молевая 405667
Трава молукская 256758
Трава морская 418392
Трава мыльная 346418
Трава овечья 164126
Трава огуречная 57095
Трава огуречная лекарственная 57101
Трава очная 160233
Трава пальчатая 117852,117859
Трава памнаская 182941
Трава памнасовая 182941
Трава пастбищная 57922,57930
Трава паточная 249303
Трава почечуйная 309570
Трава поясничная 81356
Трава пресноцветущая 410960
Трава приточная 70054
Трава прорезная 292961
Трава пупочная 401678
Трава пчелиная 249504
Трава пьяная 374897
Трава рисовая 418080
Трава сайгачья 167765
Трава сердечная 224976
Трава серебряная 255827
Трава слоновая 289218
Трава собачья 117859
Трава суданская 197331,369707
Трава терерафная 98159
Трава тимофеева 294992
Трава тысячелистника 3978
Трава утиная 177599
Трава хрустальная 251025,251265
Трава цынготная 98020
Трава чихотная 4005
Трава шелковая 293709
Трава шёлковая 293709
Трава щучья 308739
Трава эскулапова 37811
Трава эфедра 146154
Травга богородичная 391061,391397
Травга богородская 391397
Травка – муравка 308816
Травянка 127700
Трагус 394372
Трагус кистевидный 394390
Традесканция 393987
Традесканция бассейновая 394021
Традесканция белоцветковая 394021
Традесканция виргинская 394088
Трапелла 394559
Трапелла китайская 394560
Траунштейнера 394579
Траунштейнера маровидная 394580
Траунштейнера сферическая 394581
Траутфеттерия 394583
Траутфеттерия японская 394588
Трахелиум 393604
Трахелоспермум 393613
Трахелоспермум жасминовидный 393657
Трахелоспермум люцюск 393656
Трахелянт 393595
Трахелянт гиссарский 393596
Трахелянт королькова 393597
Трахидиум 393818
Трахидиум лицзянский 393854
Трахидиум тяньшанский 393864
Трахикарпус 393803
Трахикарпус вагнера 393816
Трахикарпус высокий 393809
Трахикарпус мартиуса 393813
Трахикарпус такил 393815
Трахикарпус форчуна 393809
Трахилобий 393872
Трахиния 58572,393885
Трахиния двхколосковая 58578
Трахистемон 393959
Трахистемон восточный 393961
Трекулия 394601
Трема восточная 394656
Трескун 382114
Трескун амурский 382277
Трескучник 327879
Третьина 250502
Трефоль 250502
Трехбородник 398064
Трехбородник китайский 398071
Трехгранноплодник 397338
Трехгранноплодник окутанный 397339
Трёхзубка 397762
Трехкосточник 397828
Трёхкосточник 397828
Трехкосточник выемчатый 397849
Трехлисточковый апельсин 310850
Трехреберник 397948
Трехреберник вытянутый 397958
Трехреберник гроссгейма 397960
Трехреберник дисковидный 397956
Трехреберник закавказский 397987
Трехреберник илийстый 397969
Трехреберник кавказский 397953
Трехреберник карягина 397968
Трехреберник колхидский 397954
Трехреберник коротколучевой 397952
Трехреберник мелкоцветковый 397975
Трехреберник непахучий 397964
Трехреберник обманчивый 397955
Трехреберник приморский 397970
Трехреберник севанский 397982
Трехреберник скальный 397981
Трехреберник совича 397983
Трехреберник сомнительный 397949
Трехреберник темноголовый 397973
Трехреберник четырехугольносемянный 397986
Трехреберник чихачева 397984
Трёхцветка 410677
Трехщетинник 361680
Триаденум 394807
Триаденум японский 394811
Трибулька 15709
Трива живая 357203
Трива камчужная 400675
Тригонелла 397186

Тригонотис 397380
Тригонотис булавовидный 397451
Тригонотис длинноножковый 397451
Тригонотис корейский 397422,397462
Тригонотис мулиский 397439
Тригонотис наньчуаньский 397423
Тригонотис незабудковый 397441
Тригонотис ножковый 397451
Тригонотис сичоуский 397425
Тригонотис тайваньский 397410
Тригонотис тибетский 397473
Тригонотис укореняющийся 397459
Тригонотис фунинский 397412
Тригонотис чусюнский 397394
Тригонотис чэнкоуский 397393
Тригонотис эмэйский 397447
Трижелезник 394807
Трижелезник японский 394811
Трилистник 396806,397019
Трилистник белый 397043
Трилистник водяной 250502
Трилистник голький 250502
Трилистник жёлтый 247366
Трилистник клещевой 126242
Трилистник луговой 277648,397019
Триллиевые 397533
Триллиум 397540
Триллиум камчатский 397544,397574
Триллиум ромболистный 397604
Триллиум смолла 397616
Триллиум тайваньский 397620
Триллиум тибетский 397568
Триллиум чоноский 397622
Тринакс 390460
Триния 371601,397729
Триния гладкоплодная 397734
Триния китайбеля 397733
Триния многоветвистая 397737
Триния многостебельная 397735
Триния станкова 397738
Триния украинская 397740
Триния шершавая 397736
Триния щетинистоволосая 397732
Триодия 397762
Триостеум 397828
Триостеум выемчатый 397849
Триостница 33730,33737
Триостница вознесения 33737
Триостница китайская 33797
Триостница перистая 377011
Триостренник 397142
Триостренник болотный 397165
Триостренник морский 397159
Триостренник морской 397159
Триостренник приморский 397159
Триплевроспермум 397948
Триплевроспермум илийстый 397969
Триплевроспермум непахучий 397964
Триплевроспермум четырехугольносемянный 397986
Триполиум 398126
Триполиум обыкновенный 398134
Трипсакум 398138
Трипутник 380105
Трипутник свёрнутая 236462
Трипутник скрученная 236462
Тристания 398666
Тристания сжатая 236462
Тритома 216923
Тритония 399043
Тритринакс 398824
Тритринакс бразильский 398826
Трифолиата 310850
Трифоль 250502
Трихилия 395466
Триходесма 395708
Триходесма седая 395759
Трихозант 396150,396151
Трихозантес 396150,396151
Трихозантус 396150
Трициртис 396583
Трищетинник 398437
Трищетинник алтайский 398442
Трищетинник двряднолистный 398457
Трищетинник желтоватый 398518
Трищетинник жёлтоватый 398518
Трищетинник жёлтый 398518
Трищетинник жесткий 398524
Трищетинник каванилля 398403
Трищетинник кандинский 398454
Трищетинник колосистый 398531
Трищетинник линейный 398488
Трищетинник луговой 398518
Трищетинник сибирский 398528
Трищетинник субальпийский 398531
Трищетинник тибетский 398565
Трищетинник хубэйский 398482
Троечница 396780
Троллиус 399484,399504
Троллиус европейский 399504
Тронхуда 59524
Тропелюм 399597
Тростник 295881
Тростник болотный 295888
Тростник гигантский 295909
Тростник изиды 295909
Тростник индийский 65782
Тростник испанский 65782
Тростник обыкновенный 295888
Тростник поздный 295911
Тростник сахарный 341887
Тростник сахарный брагородный 341887
Тростник сахарный ктльтурный 341887
Тростник южный 295888
Тростянка 354579
Тростянка овсяницевая 354583
Троходендрон 399432
Троходендрон аралиевидный 399433
Трубкоцвет 367804
Трубкоцвет биберштейна 367809
Трубкоцвет грубоволосистый 367813
Трубкоцвет завитковый 367811
Трубкоцвет каратегинский 367815
Трубкоцвет кокандский 367816
Трубкоцвет короткотычиночный 367810
Трубкоцвет подорожниколистный 367822
Трубкоцвет туркестанский 367828
Трубкоцвет тычиночный 367825
Трубкоцветник 367804
Трубкоцветник биберштейна 367809
Трубоцвет 70512
Трубоязычник 344381
Трясунка 60372
Трясунка большая 60379
Трясунка колосовидная 60388
Трясунка малая 60383
Трясунка средняя 60382
Тсуга 399860
Тсуга западная 399905
Тсуга зибольда 399923
Тсуга канадская 399867
Тсуга китайская 399879
Тсуга лицзянская 399903
Тсуга разнолистная 399892,399895
Тсуга тайваньская 399901
Тсуга чжэцзянская 399925
Тсуга японская 399923
Тубероза 307265,307269
Туевик 390673
Туевик поникающий 390680
Туевик поникающий ‘вариегата’ 390682
Туевик сычуаньский 390661
Туевик японский 390680
Туиния 390424
Тулотис 400280
Тулотис буреющий 302513
Тулотис уссурийский 302549
Тунбергия 390692
Тунбергия душистая 390767
Тунбергия крылатая 390701
Тунбергия хайнаньская 390793
Тунг 14538,406018
Тунг китайский 406018
Тунг молуккский 14544
Тунг сердцевидный 406016
Тунг форда 406018
Тунг японский 406016
Тунгусское назв 18403
Тунг – ю 406018
Туника 292656
Туника камнеломка 292668

Туника прямая 292658
Туника толсторебристая 400343
Туния 390916
Тупа 272891
Тупело 267849
Туранга 311131
Тургеневия 400471
Тургеневия широколистная 400473
Тургения 400471
Тургения широколистная 400473
Тургун 35411
Турица 221627
Турнепс 59575,74435
Турнера 400486
Турнефорция 393240
Турнефорция сибирская 393269
Турнефорция согдийская 393276
Турнефорция тайваньская 393267
Турнефорция яйцевидная 393261
Турраэнтус 400622
Туррея 400554
Турча 198676
Турча болотная 198680
Турчаниновия 400466
Турчаниновия верхушечная 400467
Тускарора 418080,418096
Туспейнанта 390943
Туспейнанта персидский 390944
Тут 259065
Тут американский 259186
Тут белый 259067
Тут гуншаньский 259123
Тут красный 259186
Тут татарский 259193
Тут черный 259180
Тут чёрный 259180
Тута 259065,259067
Тутовник 259065
Тутовые 258378
Туя 302721,385262,390567,390587
Туя алжирская 386942
Туя восточная 302721
Туя гигантская 390645
Туя западная 390587,390617
Туя западная ‘Ауреа’ 390589
Туя западная ‘Глобоза’ 390602
Туя западная ‘дугласин пирамидалис 390596
Туя западная ‘Эрикоидес 390598
Туя корейская 390579
Туя складчатая 390645
Туя стандиша 390660
Туя стэндыша 390660
Туя японсая 390660
Тыква 114271,219821
Тыква бутылочная 219843
Тыква бутылочная обыкновенная 219843
Тыква восковая 50998
Тыква гигантская 114288
Тыква горлянка 219843
Тыква горькая 93297
Тыква декоративная 219843
Тыква диковинная 256797
Тыква крупная 114288
Тыква крупноплодная 114288
Тыква кустовая 114301
Тыква мозговая 114301
Тыква мочальная 238261
Тыква мускатная 114292
Тыква обыкновенная 114300
Тыква посудная 219843
Тыква фигурная 114301,219843
Тыква цилиндрическая 238261
Тыква черносеменная 114277
Тыква яйцевидная 114310
Тыква – горлянка 219843
Тыквенные 114313
Тырса 401112
Тысячеголов 314752,403686,403687
Тысячеголов обыкновенный 403687
Тысячеголов посевной 403687
Тысячеголов пурамидальный 403687
Тысячеголовник 314752,403687
Тысячелистник 3913
Тысячелистник азиатский 3934
Тысячелистник альпийский 3921
Тысячелистник биберштейна 3937
Тысячелистник благородный 3998
Тысячелистник бледно – желтый 4000
Тысячелистник большой 3973
Тысячелистник бореальный 3939
Тысячелистник вильгельмса 4061
Тысячелистник голый 3951
Тысячелистник дважпы – пильчатый 3938
Тысячелистник заостренный 3914
Тысячелистник иволистный 4021
Тысячелистник камчатский 3922
Тысячелистник керманский 3956
Тысячелистник клиновиднодольный 3945
Тысячелистник крупноголовый 4009
Тысячелистник ледебура 3966
Тысячелистник мелкоцветковый 3976
Тысячелистник мускатный 3995
Тысячелистник мускусный 3995
Тысячелистник недотрога 3953
Тысячелистник обыкновенный 3978, 278099
Тысячелистник паннонский 4004
Тысячелистник подовй 3977
Тысячелистник птармика 4005
Тысячелистник птармиковидный 4014
Тысячелистник расставленный 3946
Тысячелистник сахокиа 4020
Тысячелистник северный 4026
Тысячелистник сжатый 3943
Тысячелистник таволговый 3948
Тысячелистник таволголистый 3948
Тысячелистник тонколистный 3967
Тысячелистник торчаший 4052
Тысячелистник узколистный 4055
Тысячелистник хрящеватый 3941
Тысячелистник червеобразный 4060
Тысячелистник чихотный 4005
Тысячелистник широколопастный 3964
Тысячелистник шишкина 4024
Тысячелистник шура 4025
Тысячелистник щетинистый 4027
Тюльпан 400110
Тюльпан алтайский 400116
Тюльпан аньхуйский 400117
Тюльпан беса 400130
Тюльпан биберштейна 400131
Тюльпан биберштейновский 400131
Тюльпан блестящий 400198
Тюльпан боршова 400135
Тюльпан бузе 400132
Тюльпан бузовский 400132
Тюльпан великий 400179
Тюльпан вильсона 400253
Тюльпан волосистотычиночный 400148
Тюльпан геснера 400162
Тюльпан грейга 400171
Тюльпан двуцветковый 400132
Тюльпан душистый 400227
Тюльпан зинаиды 400254
Тюльпан илийский 400178
Тюльпан кавказский 400140
Тюльпан каллье 400138
Тюльпан кауфмана 400181
Тюльпан кауфмановский 400181
Тюльпан килеватый 400139
Тюльпан коктебельский 400182
Тюльпан колпаковского 400183
Тюльпан коржинского 400185
Тюльпан королькова 400184
Тюльпан короткотычиночный 400136
Тюльпан кушкинский 400187
Тюльпан лемана 400190
Тюльпан леманновский 400190
Тюльпан лесной 400229
Тюльпан ложноволосистотычиночный 400149
Тюльпан ложнодвухцветковый 400134
Тюльпан льнолистный 400191
Тюльпан максимовича 400193
Тюльпан микели 400194
Тюльпан многоцветный 400133
Тюльпан моголтавский 400195
Тюльпан неравнолистный 400118
Тюльпан одноцветковый 400248
Тюльпан островского 400202

Тюльпан поздний 400240
Тюльпан поникающий 400203
Тюльпан превосходный 400228
Тюльпан превосходящий 400207
Тюльпан разнолепестный 400173
Тюльпан разнолистный 400174
Тюльпан разнолистый 400174
Тюльпан регели 400218
Тюльпан розовый 400219
Тюльпан садовый 400162
Тюльпан синьцзянский 400224
Тюльпан сомнительный 400151
Тюльпан степной 400162
Тюльпан тубергена 400244
Тюльпан туркестанский 400246
Тюльпан туркменский 400245
Тюльпан тяньшанский 400242
Тюльпан ферганский 400157
Тюльпан фиолетовый 400250
Тюльпан флоренского 400158
Тюльпан фостера 400159
Тюльпан хуга 400176
Тюльпан четырехлистный 400241
Тюльпан широкотычиночный 400204
Тюльпан шмидта 400221
Тюльпан шренка 400222
Тюльпан эйхлера 400154,400247
Тюльпан юлии 400180
Тюльпанкраузе 400186
Тюльпанное дерево 232602,232609
Тютюн 266053
Тяньшаночйа 391517
Тяньшаночйа зонтиконосная 391518
Уапака 401210
Увулялия 403665
Угорка 316382
Удо 8332
Ужовник 30619
Узик 312520
Узколожбиник 375551
Узколожбиник атамантовидный 375552
Узколожбиник волосистоплодный 375555
Украинский Тополь 311400
Укроп 24208,312520
Укроп аптечный 24208
Укроп водяной 167156
Укроп волошский 269299,269347
Укроп душистый 167141
Укроп конский 167141
Укроп обертковый 24213
Укроп огородный 24213
Укроп пахучий 24213
Укроп полевой 24213
Укроп сладкий 167156
Уксусник 24213
Уксусное дерево 332916
Улекс 332916,401373
Улекс европейский 401373,401388
Уллюко 401388,401411
Уллюко клубневый 401412
Уллюко клубненосный 401412
Умбеллюлярия калифорнийская 401672
Умбиликус 401678
Умбиликус супротивнолистный 401702
Унаби 418144,418169
Унаби колючий 418175
Унаби обыкновенная 418169
Унаби юйюба 418169
Унабиююба 418169
Унас 28248
Унгерния 401817,401822
Унгерния меньшая 401820
Унгерния таджикская 401824
Унгерния трехсферная 401825
Унгерния ферганская 401818
Species Ункария клюволистная 401737
Ункария тайваньская 401754
Уншиу сатсума 93736
Упланд обыкновенный 179906
Урай 344743
Урак 295888
Ургинея 402314
Урд 408982
Урекс европейский 203789
Урена 402238,402245
Урера 402281
Урзиния 402743
Уроспермум 402660
Урсиния 402743
Уруть 261337
Уруть бразильская 261342
Уруть колосистая 261364
Уруть колосковая 261364
Уруть колосовая 261364
Уруть мутовчатая 261379
Уруть очевые 261339
Уруть очередноцветковая 261339
Уруть уссурийская 261377
Урюк 34475
Урюк - арча 213877
Устели - поле 83418
Утевник 308739
Утёсник 401373
Утесник европейский 401388
Утёсник карликовый 401396
Утёсник малый 401396
Ушко медвежье 31142
Фабиана 161896,161897
Фагналон 293408
Фагналон андросова 293411
Фагналон дарвазский 293423
Фаленопсис 293577
Фаленопсис гиблидный 293611
Фаленопсис хайнаньский 293607
Фаленопсис шиллера 293633
Фальиата 20558
Фальиата японская 20570
Фарбитис 207546
Фарбитис пурпурный 208120
Фарзетия 162813
Фарзетия лопатчатая 162850
Фарсетия 162813
Фаселантус 293181
Фаселантус трубкоцветковый 293186
Фасоль 293966
Фасоль адзуки 408839
Фасоль азиатская 408982
Фасоль арабская огненно - краская 293985
Фасоль безволокновая 293966
Фасоль волокнистая 294056
Фасоль зерновая 294056
Фасоль золотистая 409025
Фасоль кроваво - красная 293985
Фасоль кустовая однолистная 409025
Фасоль лима 294006
Фасоль лимская 294006,294010
Фасоль ломкая 293966
Фасоль лучистая 408839
Фасоль маш 408982
Фасоль много 408982
Фасоль многоцветковая 293985
Фасоль настоящая 294056
Фасоль обыкновенная 294056
Фасоль обыкновенная кустовая 294057
Фасоль огненая 293985
Фасоль огнено - расная 293985
Фасоль остролистная 293969
Фасоль полулунная 294010
Фасоль почечная 294056
Фасоль почечновидная 294056
Фасоль сива 294010
Фасоль спаржевая 409086,409096
Фасоль турецкая 294056
Фасоль угловатая 408839
Фатсия 162876,162879
Фатсия японская 162879
Фацелия 293151
Фацелия пижмолистная 293176
Фация 162876
Фация японская 162879
Фейхоа 163116,163117
Фейхоа селлоу 163117
Фелиция 163121
Фендлера 163312
Феникс 295451
Феникс канарский 295459
Феникс пальценосный 295461
Фенхель 167141
Фенхель аптечный 167156
Фенхель душистый 167156
Фенхель итальянский 167146

Фенхель лекарственый 167156
Фенхель обыкновенный 167156
Фенхель флорентинский 167146
Фернам буковые 64975
Фернамбук 7190,64975
Ферония слоновая 163502
Ферула 163574,163606,163690,163733,369256
Ферула акичкенская 163577
Ферула ангренская 163579
Ферула ассафетида 163580
Ферула бадхызская 163582
Ферула бледнозеленая 163692
Ферула бородлайская 163636
Ферула вздутая 163631
Ферула вонючая 163580,163709
Ферула восточная 163688
Ферула гигантская 163620
Ферула гипсолюбивая 163626
Ферула гладколистная 163659
Ферула гладная 163621
Ферула голая 163685
Ферула григорьева 163624
Ферула джауджамыр 163604
Ферула джунгарская 163712
Ферула дурнопахнущая 163731
Ферула жестковатая 163584
Ферула замещающая 163742
Ферула иешке 163633
Ферула илийская 163630
Ферула кавказская 163587
Ферула камеденосная 163619
Ферула каменная 163654
Ферула каракалинская 163635
Ферула каратавская 163637
Ферула каратегинская 163638
Ферула каспийская 163586
Ферула келифа 163642
Ферула келлера 163643
Ферула клубненосная 163737
Ферула козо－полянского 163650
Ферула кокандская 163646
Ферула конусостебельная 163598
Ферула копетдагская 163647
Ферула коржинского 163649
Ферула краснопесчаниковая 163706
Ферула крылова 163651
Ферула ксероморфная 163743
Ферула кугистанская 163652
Ферула лемана 163658
Ферула линчевского 163663
Ферула липского 163665
Ферула литвинова 163666
Ферула ломоносолистная 163589
Ферула мелкодольчатая 163678
Ферула мелкоплодная 163676
Ферула метельчатая 163610
Ферула многоканальцевая 163700
Ферула моголтавская 163679
Ферула мускусная 163681
Ферула мягкая 163680
Ферула невского 163684
Ферула оберточковая 163632
Ферула обыкновенная 163590
Ферула овечья 163689
Ферула олиствинная 163616
Ферула пахучая 163615
Ферула переходная 163577
Ферула перистонервная 163694
Ферула персидская 163695
Ферула потанина 163701
Ферула пранголистная 163702
Ферула рассеченная 163602
Ферула роголистная 163588
Ферула самаркандская 163707
Ферула седовадая 163585
Ферула синьцзянская 163711
Ферула совича 163728
Ферула столбиковая 163689
Ферула стройная 163623
Ферула сырейщикова 163727
Ферула татарская 163729
Ферула толстолистная 163691
Ферула тонкая 163717
Ферула тонкорассеченная 163730
Ферула угамская 163740
Ферула федченко 163608
Ферула ферганская 163609
Ферула хвощевая 163605
Ферула чимганская 163736
Ферула широкодольчатая 163656
Ферула широколистная 163655
Ферула эчисона 163576
Ферула язычковая 163662
Ферульник 163746
Ферульник дагестанский 163750
Ферульник крымский 163766
Ферульник лесной 163765
Ферульник туркменский 163767
Ферульник широкодольчатый 163751
Ферульник щетинколистный 163764
Фестука 163778
Фестука лицзянская 164400
Фестука юньнаньская 164401
Фестука японская 164025
Фиалка 409608
Фиалка авачинская 409709
Фиалка алайская 409655
Фиалка александрова 409672
Фиалка алисовой 410770
Фиалка алтайская 409682
Фиалка альпийская 115931,409681
Фиалка амурская 409687
Фиалка африканская 342493
Фиалка белая 409661
Фиалка болотная 410347
Фиалка бразильская 21889
Фиалка бутеневидная 409816
Фиалка бутенелистная 409816
Фиалка волосистонотая 410073
Фиалка волосисточашечная 410647
Фиалка волосисточерешковая 410073
Фиалка восточная 410326
Фиалка высокая 409947
Фиалка гиссарская 410080
Фиалка гмелина 410010
Фиалка голоплодная 410403
Фиалка горная 410273
Фиалка гультена 410090
Фиалка двухцветковая 409729
Фиалка двуцветковая 409729
Фиалка длинношпорцевая 409934
Фиалка донская 410639
Фиалка душистая 410320
Фиалка елизаветы 409950
Фиалка жёлтая 410187
Фиалка жордана 410117
Фиалка заостренная 409630
Фиалка застенчивая 410730
Фиалка зиге 410579
Фиалка зубчатоцветковая 410452
Фиалка изменчивая 409686
Фиалка каракалинская 410126
Фиалка китайбелева 410140
Фиалка китайбели 410140
Фиалка китайбля 410140
Фиалка корейская 409665
Фиалка коротковолосистая 410066
Фиалка коротко－чашелистиковая 410246
Фиалка короткошпорцевая 409762
Фиалка кочкарная 409958
Фиалка крохотная 410358
Фиалка крупношпорцевая 410192
Фиалка крымский 410325
Фиалка кузнецова 410158
Фиалка купфера 410152
Фиалка лангсдорфа 410169
Фиалка лесная 410483,410584
Фиалка лушаньская 410598
Фиалка лысоплодная 410403
Фиалка манчьжурская 410202
Фиалка маньчжурская 410201
Фиалка монгольская 410269
Фиалка морица 410233
Фиалка мюльдорфа 410279
Фиалка мясистенькая 409811
Фиалка надрезанная 409925
Фиалка непалская 409716
Фиалка низкая 410474
Фиалка низкорослая 410474
Фиалка ночная 193387,302260

Фиалка одноцветковая 410704
Фиалка одноцветная 410272
Фиалка опушенная 410066
Фиалка опушённая 410066
Фиалка орлиноклювая 410030
Фиалка остроконечная 409630
Фиалка остролистная 409643
Фиалка острошпорцевая 410336
Фиалка пальчатая 409878
Фиалка патрэна 410360
Фиалка пестрая 410710,410721
Фиалка пестрололистная 410721
Фиалка песчаная 409698
Фиалка полевая 409703
Фиалка прибрежная 410183
Фиалка приостренная 409630
Фиалка приятная 410605
Фиалка прудовая 410401
Фиалка прямошпорая 410329
Фиалка равнолепестная 410106
Фиалка радде 410478
Фиалка рассеченная 409914
Фиалка ресничато – чашелистиковая 410675
Фиалка ривина 410493
Фиалка ривиниуса 410493
Фиалка ривинуса 410493
Фиалка рогатая 409858
Фиалка росса 410498
Фиалка рупиуса 410516
Фиалка сахалинская 410517
Фиалка сверху голая 409958
Фиалка селькирка 410547
Фиалка сычуаньская 410622
Фиалка сиэхеано 410579
Фиалка скальная 409698,410325,410513
Фиалка склоненная 409883
Фиалка скромная 410730
Фиалка скромненькая 410266
Фиалка скрытая 410318
Фиалка собачья 409804
Фиалка сомнительная 409686
Фиалка тайваньская 409993
Фиалка тарбагатайская 410640
Фиалка темнофиолетовая 409706
Фиалка теневая 410547
Фиалка тибетская 410149
Фиалка токийская 410770
Фиалка толстая 409862
Фиалка толстошпорцевая 409868
Фиалка топяная 410702
Фиалка трехцветная 410677
Фиалка трёхцветная 410677
Фиалка турчанинова 410700
Фиалка тяньшанская 410662
Фиалка удивительная 410246
Фиалка утренняя 410232
Фиалка федченко 409979
Фиалка холмовая 409834
Фиалка хунаньская 410091
Фиалка хэбэйская 410787
Фиалка эмэйская 409954
Фиалка яньюаньская 409711
Фиалка японская 409942,410108
Фиалковые 410792
Фибигия 164445
Фибигия мохнатоплодная 164447
Фибигия полукустарниковая 164450
Фибигия ширококрылая 164449
Фибигия щитовидная 164446
Фига 164763
Фига индийская 272891
Фиговое дерево 164763
Фиговое смоковница 164763
Физалиаструм 297624
Физалиаструм иглистный 297626
Физалиаструм японский 297628
Физалис 297640
Физалис голоногий 297645
Физалис клейкоплодный 297686
Физалис мексиканский 297686
Физалис обыкновенный 297643,297645
Физалис перуанский 297711
Физалис пропущенный 297645
Физалис пушистый 297720
Физалис рубчиковидный 297720
Физалис съедовный 297711
Физалис франше 297645
Физалис франшетта 297643,297645
Физалис ягодный 297711
Физетовое дерево 107300
Физокарпус 297833
Физокарпус амурский 297837
Физокарпус калинолистный 297842
Физостегия 297973
Физостигма 298004,298009
Физостигма ядовитая 298009
Физохлайна 297875
Физохлайна физалисовидная 297882
Фикус 164608,164763,164925
Фикус бенгальский 164684
Фикус благочестиный 165553
Фикус булавовидный 165711
Фикус гуансиский 165079
Фикус гуйцзоуский 165082
Фикус индийский 164684
Фикус инжир 164763
Фикус капский 164751
Фикус карликовый 165515
Фикус карликовый малый 165520
Фикус каучуконосный 164925
Фикус каучуконосный пестролистная 164940
Фикус ланьюйский 165442
Фикус лодяньский 164964
Фикус лунчжоуский 164762
Фикус наньтоуский 164635
Фикус напоский 165343
Фикус низкий 165515
Фикус притупленный 165307
Фикус прямостоячий 164947
Фикус священный 165553
Фикус сичоуский 165871
Фикус тайваньский 164992
Фикус хайнаньский 165084
Фикус эластика 164925
Фикус юньнаньский 165875
Фикус японский 165628
Фикус – сикомор 165726
Филантус кислый 296466
Филирея 294761
Филирея узколистная 294762
Филирея широколистная 294772
Филлантус 296465
Филлантус извилистый 296565
Филлантус индийский 245220
Филлантус кислый 296466
Филлантус лекарственный 296554
Филлантус уссурийский 296803
Филлантус франше 296573
Филлантус хайнаньский 296599
Филлантус чжэцзянский 296517
Филлантус эмблика 296554
Филлодоце 297015
Филлодоце алеутская 297016
Филлодоце алеутское 297016
Филлодоце голубая 297019
Филлодоце голубое 297019
Филлодоце тиссолистое 297019
Филлодоце японский 297030
Филлокактус 147283
Филлокара 21914
Филлокара оше 21924
Филлоспадикс 297181
Филлоспадикс иватинский 297182
Филлоспадикс скаулера 297185
Филлостахис 297188
Филодендрон 294230,294785
Филодендрон андрэ 294792
Филодендрон бородавчатый 294850
Филодендрон изящный 294804
Филодендрон красноватый 294805
Филодендрон перистораздельный 294832
Филодендрон писция 301021
Филодендрон пробитый 258168
Филодендрон рассеченный 294830
Филодендрон селло 294839
Филодендрон чешуйчатый 294844
Филодендрон шероховатый 294795
Филодендрон японский 294233
Фимбристилис 166156
Фимбристилис бородавчатый 166557

Фимбристилис вильчатый 166248
Фимбристилис гладкоплодный 166376
Фимбристилис зибера 166478
Фимбристилис ичанский 166348
Фимбристилис китайский 6865
Фимбристилис летний 166164
Фимбристилис наньнинский 166417
Фимбристилис однолетний 166176
Фимбристилис оходский 166268
Фимбристилис почти двухколосковый 166518
Фимбристилис пятигранный 166402, 166454
Фимбристилис растопыренный 166499
Фимбристилис ржавый 166313
Фимбристилис симаоский 166488
Фимбристилис хайнаньский 166346
Фимбристилис юньнаньский 166565
Финган 55596,114581
Финик 295451
Финик дикий 132264
Финик изогнутый 295479
Финик индейский 383404
Финик индийский 383407
Финик канарский 295459
Финик китайский 418169
Финик лесной 295487
Финик пальчатый 295461
Финик робелена 295484
Финик финиколистный 295461
Финик финиконосный 295461
Финикс 295451
Финикс канарский 295459
Финикс лесной 295487
Фиппсия 294907
Фиппсия стройная 294910
Фиппсия холодолюбивая 294909
Фирмиана 166612,166627
Фирмиана платанолистная 166627
Фирмиана хайнаньская 166618
Фисташка 300974
Фисташка атлантическая 300977
Фисташка дикая 300998
Фисташка китайская 300980
Фисташка мастиковая 300994
Фисташка мастичная 300994
Фисташка настоящая 301005
Фисташка туполистная 300998
Фисташка - лентискус 300994
Фисташковые 21190,176867
Фисташник 300998
Фителефас 298050,365270
Фителефас крупноплодный 298051
Фитеума 298054
Фитокренум 298082
Фитолакка 298088
Фитолакка американская 298094
Фитолакка японская 298114
Фитоллякковые 298124
Фитолякка американская 298094
Фитолякка десятитычинковая 298094
Фитолякковые 298124
Фитцройя 166720,332766
Фитцройя арчера 166725
Фитцройя патагонская 166726
Фитцройя тасманийская 166725
Фитцроя 332766
Флаверия 166814
Флакоуртия 166761
Флакоуртия кохинхинская 166777
Флакуртиевые 166793
Флакуртия 166761
Флакуртия зондская 166788
Флакуртия индийская 166773
Флакуртия рамонтша 166773,166786
Флакуртовые 166793
Флиндерсия 166971
Флиндерсия австралийская 166974
Флойодикарский 295035
Флойодикарский мохнатый 295041
Флойодикарский сибирский 295038
Флокс 295239
Флокс друммонда 295267
Флокс друмонда 295267
Флокс метельчатый 295288
Флокс однолетний 295267
Флокс сибирский 295318
Флоуленсия 167054
Флюгея 167067
Флюгея уссурийская 167094
Фолидота 295510
Фолидота вэньшаньская 295543
Фолидота китайская 295518
Фолидота черепитчатая 295523
Фонтанезия 167230
Фонтанезия форчуна 167233
Форзиция 167427
Форзиция поникля 167456
Форзиция свешивающаяся 167456
Фория гребневая 265197
Фория гребневая японская 265198
Формиум 295600
Форсайтия 167427
Форсайтия висячия 167456
Форсайтия яйцевидная 167452
Форситня 167427,167452
Форситня корейская 167473
Форситня лицзянская 167449
Форситня повислая 167456
Форситня японская 167444
Форситня ярко - зеленая 167471
Форсиция 167427
Фортунелла жемчужная 167506
Фортунелла японская 167503
Фортьюнария 167487
Фортьюнария китайская 167488
Фотиния 295609
Фотиния аньлунская 295621
Фотиния бентама 295637
Фотиния голая 295691
Фотиния гуансиская 295715
Фотиния давидсона 295675
Фотиния жуйлиская 295767
Фотиния мохнатая 295808
Фотиния мохнатая китайская 295813
Фотиния пильчатая 295773
Фотиния франше 295688
Фотиния фуцзяньская 295686
Фотиния чжэцзянская 295820
Франкениевые 167827
Франкения 167765
Франкения бухарская 167767
Франкения жестковолосая 167792
Франкения порошистая 167813
Францисцея 61293
Французский райграс 34920,34930
Французский райграс высокий 34930
Фреезия 168153
Фрима 295978
Фрима тонкокистевая 295984
Фрима узкоколосая 295984
Фримовые 296006
Фрина 296011
Фрина хюта 296012
Фриниум 296018
Фриниум головчатый 296052
Фриниум хайнаньский 296033
Фритиллярия 168332
Фритиллярия аньхуйская 168344
Фритиллярия мумтвчатая 168614
Фролипия 168717
Фролипия почтиперистая 168718
Фтейроспермум 296061
Фтейроспермум китайский 296068
Фтейроспермум мулиский 296070
Фтейроспермум японский 296068
Фуксия 168735,168767
Фуксия мелкоцветковая 168780
Фукус пильчатый 178463
Фумана 168915
Фумана клейковатая 168944
Фумана лежащая 168959
Фундук 106757
Фундук крупноплодный 106757
Функия 198589
Функия ланцетолистная 198618,198624
Функия яйцевидная 198652
Фунтумия 169226
Фунтумия каучуконосная 169229
Фуопсис 296086
Фуопсис столбиковый 296091

Фюрнрония 168792
Фюрнрония щетинолистная 168793
Хагения абиссинская 184565
Хайтурус 85054
Хайтурус шандровидный 85058
Хакетия 184326
Халдумка 198275
Халезия 184726
Халорагис 184995
Хамамелис 185092
Хамегерон 85457
Хамегерон Бунге 85459
Хамегерон малоголовый 85460
Хамедафна 85412
Хамедафна болотная 85414
Хамедафне 85412
Хамедафне волотная 85414
Хамедафне чашечная 85414
Хамедорея 85420
Хамемелюм 85510
Хамемелюм благолодный 85526
Хаменерий 85870
Хаменерий узколистный 85875
Хаменериум 85870
Хаменериум кавказский 85554
Хаменериум колхидский 85555
Хаменериум лебедовый 85559
Хаменериум узкий 85553
Хаменериум узколистный 85875
Хаменериум широколистный 85884
Хамепейце 91697
Хамеродос 85646
Хамеродос алтайская 85647
Хамеродос крупноцветкная 85652
Хамеродос песчаная 85657
Хамеродос прямой 85651
Хамеродос прямостоящая 85651
Хамеродос трехнадоезная 85660
Хамеропс 85665
Хамеропс низкий 85671
Хамеропс приземистый 85671
Хамесфакос 85730
Хамесфакос падуболистный 85731
Ханделия 185259
Ханделия волосистолистая 185260
Ханделия волосистолистная 185260
Ханка 219843
Хартолепис 86046
Хартолепис биберштейна 86048
Хартолепис вайдолистный 86049
Хартолепис крылатостебельный 86051
Хартолепис средний 86049
Хат 79387
Хатыма древовидная 223355
Хатыма трёхмесячная 223393
Хатыма тюрингенская 223392
Хатьма 223350
Хатьма кашмирская 223361
Хатьма точечная 223387
Хатьма трехмесячная 223393
Хвойник 146122
Хвойник американский 146134
Хвойник высокий 146237
Хвойник двухколосковый 146154
Хвойник жерарда 30277,146183
Хвойник калифорнийский 146144
Хвойник кашгарский 146239
Хвойник крепкий 146192
Хвойник лицзянский 146203
Хвойник маленький 146215
Хвойник мелкосемянный 146212
Хвойник односемянный 146219
Хвойник окаймленный 146209
Хвойник персидский 146192
Хвойник подушечный 146241
Хвойник пустынный 146192
Хвойник разносемянный 146191
Хвойник регелевский 146243
Хвойник регеля 146243
Хвойник ресничатый 146148
Хвойник сизый 146188
Хвойник средний 146192
Хвойник средный 146192
Хвойник средный тибетский 146197
Хвойник тибетский 146197
Хвойник федченко 146159
Хвойник ферганский 146160
Хвойник хвощевидный 146155
Хвойник хвощевый 146155
Хвойник шишконосная 146255
Хвойник эчисона 146124
Хвойниковые 146270,178522
Хвост лисий 18687
Хвостник 196810
Хвостник обыкновенная 196818
Хвостник четырехлистный 196816
Хвостниковые 196809
Хвостосемянник 402660
Хвостосемянник горчаковидный 402663
Хвощ большой 43309
Хевея 194451
Хевея бразильская 194453
Хедеома 187176
Хедизарум 187842
Хединия 187361
Хединия тибетская 187373
Хедисцепе 188190
Хедихиум 187417
Хедихиум гуансиский 187455
Хедихиум инцзянский 187483
Хедихиум короновидный 187432
Хедихиум симаоский 187466
Хедихиум тибетский 187467
Хедихиум тэнчунский 187473
Хедихиум эмэйский 187442
Хедихиум юньнаньский 187484
Хейрантус 86409
Хейрантус садовый 86427
Хейриния 154363
Хейролепис 86637
Хейролепис иранский 86639
Хейхера 194411
Хеликония 189986
Хеликония бихаи 189992
Хелоне 86781
Хелониас 191046
Хельсиния 367771,367772
Хемантус 184345
Хемантус екатерины 350313
Хемерокаллис 191261
Хемиграфис 191486
Хемиграфис широковыемчатый 191501
Хенекен 10854
Хенна 223451,223454
Хеномелес 84549
Хеномелес китайский 84553,84573,317602
Хеномелес легенария 84573
Хеномелес маулея 84556
Хеномелес прекрасный 84573
Хеномелес тибетский 84591
Хеномелес японский 84556
Хеноринум 84594
Хеноринум клейкий 84619
Хеноринум клокова 84595
Хеноринум колосовый 84615
Хероспондиас 88690
Хероспондиас пазушный 88691
Хетолимон 84912
Хетолимон окаймленный 84914
Хетолимон щетинчатый 84915
Хибиск 194680
Хибиск изменчивый 195040
Хибиск сирийский 195270
Хибискус 194680
Хибискус китайская роза 195149
Хибискус расщепленнолепестный 195216
Хиднокарпус 199740
Хилотелефиум 200775
Химафила 87480
Хименокаллис 200906
Хименокаллис прибрежный 200941
Химонант 87510
Химонант ранний 87525
Химонант скороспелый 87525
Химонант чжэцзянский 87545
Химонобамбуза 87547,87586
Химонобамбуза бирманская 87551
Химонобамбуза четырехгранная 87607
Химонобамбуза юньнаньская 87627
Хинное гиблидное 91071
Хинное дерево 91055

Хинное дерево красносоковое 91090
Хинное дерево мелкоцветное 91079
Хиококка 87661
Хионант 87694
Хионант притупленный 87729
Хионантус 87694
Хионантус виргинский 87736
Хионодокса 87757
Хипеаструм 18862,196439
Хирония 88022
Хис 45448
Хис－кахак 60612
Хистрикс 203124
Хлеб дивий 83527
Хлеб иванов 83527
Хлеб каффрский 369720
Хлебница 66600
Хлебное дерево цельнолистное 36920
Хлебоплодник 36928
Хлолис мохнатая 387512
Хлолис прутьевидная 88421
Хлолис прутьевидный 88421
Хлолис тайваньский 88351
Хлопок дикий 29473
Хлопунец 42193
Хлопушка 363141
Хлопчатник 179865,336343
Хлопчатник азиатский 179900
Хлопчатник американский 179906
Хлопчатник африканский 179900
Хлопчатник афроазиатский 179900
Хлопчатник барбадоский 179884
Хлопчатник волосистый 179906
Хлопчатник древовидный 25882,179876
Хлопчатник египетский 179884
Хлопчатник зайцева 179937
Хлопчатник индийский 179878
Хлопчатник мохнатый 179906
Хлопчатник нагорный 179906
Хлопчатник нанкинский 179878
Хлопчатник приморский 179884
Хлопчатник среднеазиатский 179900
Хлопчатник средневолокнистый 179906
Хлопчатник травянистый 179900
Хлопчатник упланд 179906
Хлопчатник шерстистый 179906
Хлопчатник южноамериканский 179884
Хлора 88242
Хлорантовые 88263
Хлорантус 88264
Хлорантус аньхуйский 88266
Хлорантус пильчатый 88297
Хлорантус сычуаньский 88299
Хлорантус тайваньский 88295
Хлорантус хубэйский 88283
Хлорантус японский 88289
Хлорис 88316
Хлороксилон 88659,88663
Хлорофитум 88527
Хлорофитум капский 88546
Хлорофитум китайский 88550
Хлорофора 88500
Хлорофора красильная 88514
Хлябник 385629
Хмелеграб 276801
Хмелеграб американский 276818
Хмелеграб виргинский 276818
Хмелеграб обыкновенный 276803
Хмелеграб япнский 276808
Хмелёк 396806
Хмелёк полевой 397096
Хмелёк шуршащий 397096
Хмелица 199384
Хмель 199379
Хмель вьющийся 199384
Хмель лазящий 199392
Хмель обыкновенный 199384
Хмель огородный 265940
Хмель юньнаньский 199394
Хмель японский 199382
Хна 223451,223454
Ховения 198766
Ховея 198804
Ховея бельмора 198806
Хойя 198821
Хойя гуншаньская 198864
Хойя либоская 198867
Хойя мэнцзыская 198875
Хойя мясистая 198827
Хойя тайваньская 137799
Хойя тибетская 198862
Холодка 250420
Хололахна 327194
Хололахна джугрnative 327226
Хомалянтус 197619
Хомалянтус тополевый 197625,270563
Хондрила 88759
Хондрила боссэ 88765
Хондрила кузнецова 88780
Хондрила самаркандская 88789
Хондрила тощая 88787
Хондрила украшенная 88791
Хондрилла 88759
Хондрилла гладкосемянная 88784
Хондрилла злаколистная 88772
Хондрилла колючечешуйная 88760
Хондрилла коротоклювая 88766
Хондрилла крупноплодная 88788
Хондрилла ломконосая 88796
Хондрилла ломконосиковая 88796
Хондрилла малоцветковая 88793
Хондрилла обыкновенная 88776
Хондрилла рулье 88799
Хондрилла седоватая 88767
Хондрилла ситниковидная 88776
Хондрилла сомнительная 88761
Хондрилла темноголовая 88794
Хондрилла шероховатая 88763
Хондрилла ширококорончатая 88782
Хондрилла широколистная 88783
Хондрилля 88759
Хондрилля злаколистная 88772
Хондрилля каучуконосная 88773
Хондрилля коротконосиковая 88766
Хондрилля косинского 88779
Хондрилля нителистная 88771
Хондрилля седоватая 88767
Хондрилля ситниковидная 88776
Хондрилля сомнительная 88761
Хондрилля щетинконосная 88760
Хопея 198126
Хопея душистая 198176
Хор 368635
Хор－арча 213877
Хордеум 198260
Хоризис 88975
Хоризис ползучий 88977
Хориспора 88996
Хориспора бунге 88997
Хориспора бунговская 88997
Хориспора грейга 89006
Хориспора грунинская 89007
Хориспора джунгарская 89016
Хориспора иберийская 89007
Хориспора крупноногая 89008
Хориспора нежная 89020
Хориспора сибирская 89013
Хориспора тонкая 88998
Хороги 373439
Хоста 198589
Хоста зибольда 198646
Хоустония 198703
Хохлатка 105557
Хохлатка алексеенко 105587
Хохлатка алпийская 105588
Хохлатка арктическая 105624
Хохлатка благородная 106453
Хохлатка бледная 106227
Хохлатка бледноцветковая 106242
Хохлатка бунге 105680
Хохлатка бухарская 105669
Хохлатка буша 105682
Хохлатка бымянковидная 105699
Хохлатка галлера 106453
Хохлатка гигантская 105929
Хохлатка голостебельная 106196
Хохлатка горчакова 105939
Хохлатка дарвазская 105803
Хохлатка дымянколистная 105917,105919
Хохлатка дымянкообразная 105699

Хохлатка железнова 106415
Хохлатка жёлтая 106117
Хохлатка кавказская 105719
Хохлатка кашгарская 106032
Хохлатка коническикорневая 105752
Хохлатка крупноприцветниковая 105663
Хохлатка крупноцветная 106121
Хохлатка крупночашечная 106123
Хохлатка крупношпорцевая 106124
Хохлатка ледебура 106066
Хохлатка ложноалипийская 106293
Хохлатка ложно－согнутая 106292
Хохлатка малоцветковая 106255
Хохлатка маршалла 106130
Хохлатка мелкоцветковая 106159
Хохлатка метельчатая 106246
Хохлатка мулиская 105679
Хохлатка мупинская 105881
Хохлатка мупинская мелкоцветковая 105888
Хохлатка надрезанная 106004
Хохлатка недотрога 105999
Хохлатка незаметная 106013
Хохлатка непалская 105971
Хохлатка обманчивая 105594
Хохлатка охотская 106197
Хохлатка пачоского 106225
Хохлатка персидская 106268
Хохлатка пионолистная 106226
Хохлатка полая 105721
Хохлатка ползучая 106380
Хохлатка попова 106281
Хохлатка промежуточная 106017
Хохлатка прямая 106488
Хохлатка радде 106355
Хохлатка расставленная 106564
Хохлатка редовского 106358
Хохлатка северцева 106431
Хохлатка семенова 106427
Хохлатка сибирская 106453
Хохлатка сизая 106430
Хохлатка сизоватая 105936
Хохлатка сычуаньская 105688
Хохлатка снеголюбивая 105741
Хохлатка сомнительная 105594
Хохлатка таиландская 106439
Хохлатка тибетская 106531
Хохлатка тонкая 106013
Хохлатка удаленная 106564
Хохлатка узколистная 105615,105618
Хохлатка узкоцветная 105617
Хохлатка федченковская 105872
Хохлатка франше 105914
Хохлатка хубэйская 105564
Хохлатка цинхайская 106346
Хохлатка цюцзянская 106040
Хохлатка чаюйская 106557
Хохлатка чжундяньская 106620
Хохлатка чжэцзянская 106240
Хохлатка шангина 106414
Хохлатка эмануеля 105858
Хохлатка эрдели 105860
Хохлатка эчкона 105581
Хохлатка юньнаньская 106613
Хохлатник 282979
Хохлатник персидский 282974
Хохлатник северный 145766
Хохлатый 88553
Хрен 34582,34587,34590
Хрен водяной 262722,336172
Хрен гулявниковыи 34594
Хрен деревенский 34587,34590
Хрен дикий 34587,34590
Хрен лекарственный 262722
Хрен луговой 34594
Хрен обыкновенный 34587,34590
Хрен солнечный 225398
Хризалидокарпус 89345
Хризалидокарпус желтоватый 89360
Хризантема 89402
Хризантема алипийская 227457
Хризантема бальзамная 89602,383712
Хризантема индийская 124790
Хризантема килеватая 89466
Хризантема корончая 89481
Хризантема куртаниковая 32563
Хризантема летние 89402
Хризантема межпосевной 89704
Хризантема посевная 89704
Хризантема роксбурга 89696
Хризантема увенчанная 89481
Хризантемум 89402
Хризантемум белоцветковый 227533
Хризантемум завадского 124862
Хризантемум монгольский 124825
Хризантемум нактонгенский 124832
Хризантемум обыкновенный 383874
Хризантемум триниевидный 166066
Хризобаланус 89812
Хризогонум 89952
Хризофиллум 90007
Хризофиллюм 90007
Хринок 225398
Христовы тернии 280544
Христовы тернин 280559
Христолея 89223
Христолея веероведная 89231,126151
Христолея линейная 89241
Христолея майдантальская 89246
Христолея памирская 89249
Христолея парриевидная 89253
Христолея толстостенная 89229
Хробуст 368771
Хрозофора 89320
Хрозофора изящная 89323
Хрозофора косая 89324
Хрозофора красильная 89330
Хрозофора песчаная 89328
Хруплявник 307830
Хруплявник бородавчатый 307844
Хруплявник геуффеля 307839
Хруплявник крупный 307840
Хруплявник многолетний 307842
Хруплявник полевой 307832
Хрущецветник 204443
Хрущецветник мутовчатый 204469
Хультемия 199240
Хультемия барбарисолистная 336385
Хультемия иранская 199243
Хуннемания 199407
Хура трескающаяся 199465
Хурма 132030
Хурма баотинская 132359
Хурма бирманская 132077
Хурма виргинская 132466
Хурма восточная 132219
Хурма далиская 132061
Хурма индийская 132298
Хурма кавказская 132264
Хурма каки 132219
Хурма ланьюйская 132241
Хурма люцюская 132291
Хурма обыкновеная 132264
Хурма ромболистная 132371
Хурма сизолистная 132175
Хурма сичоуская 132397
Хурма субтропическая 132219
Хурма тэнчунская 132163
Хурма фэн 132152
Хурма хайнаньская 132186
Хурма цзиндунская 132238
Хурма юньнаньская 132478
Хурма японская 132219
Хурмовые 139763
Хускуеа 90629
Хустония 198703
Хутчик 287978
Хымзыдия 90709
Хымзыдия агазиллевидная 90710
Цамия 417002
Цантоксилум мартиникский 417228
Цареградсие рожки 83527
Царица ночи 83874
Царские кудри 229922
Царские очи 349452
Царь－бегония 50232
Царьзелье 124194
Царь－зелье 205535
Цаущнерия 417405
Цванга 382923
Цвет кукушкин 363461

Цеанотус 79902
Цеанотус американский 79905
Цедрат 93603
Цедрела 80015
Цедрела душистая 80030
Цедрела китайская 392841
Цедрела китайская хубэйская 392845
Цедрела мексиканская 80025
Цедрела мелкоплодная 392837
Цедрела сингапурская 392829
Цедреладу шистая 80030
Цезальпиниевые 65084
Цезальпиния 64965
Цезальпиния аптечная 64971,64983
Цезальпиния джиллися 65014
Цезальпиния дубильная 312925
Цезальпиния душистая 312925
Цезальпиния ежовая 65002
Цезальпиния заборная 64990
Цезальпиния кожевенная 312925
Цезальпиния красильная 312925
Цезальпиния крупнолистная 65033
Цезальпиния ост – индская 65060
Цезальпиния саппан 65060
Цезальпиния центрально – американская 312925
Цезальпиния шиповатая 65002
Цезальпиния японская 64993
Цейба 80114,80120
Цейба пятитычинковая 80120
Цекропия 80003
Целестина 11194
Целибуха 378834
Целогина 98601
Целогина гребенчатая 98634
Целогина гуншаньская 98664
Целогина красивая 98743
Целогина малипоская 98694
Целогина поникшая 98653
Целогина чжэньканская 98761
Целогине 98601
Целозия 80378,80395
Целозия гребенчатая 80395
Целозия серебристая 80381
Целококкус 252631
Целомудренник 411189
Цельзия 405657
Цельзия восточная 80542
Цельзия голостебьная 80540
Цельзия разнолистная 80521
Цельзия суволова 80555
Цельнолистник 185615
Цельнолистник альберта регеля 185621
Цельнолистник близкий 185620
Цельнолистник бунге 185630
Цельнолистник буржо 185625
Цельнолистник бухарский 185629
Цельнолистник введенского 185689
Цельнолистник ветвистый 185664
Цельнолистник даурский 185635
Цельнолистник джунгарский 185636
Цельнолистник душистый 185672
Цельнолистник евгения коровина 185639
Цельнолистник исколотый 185658
Цельнолистник коваленского 185647
Цельнолистник крымский 185672
Цельнолистник многостебельный 185654
Цельнолистник мохнатый 185688
Цельнолистник мощный 185665
Цельнолистник однобратственный 185653
Цельнолистник олиственный 185643
Цельнолистник остролистный 185619
Цельнолистник попова 185662
Цельнолистник предкавказский 185633
Цельнолистник разноцветный 185685
Цельнолистник ресничатый 185632
Цельнолистник сомнительный 185637
Цельнолистник тонгий 185673
Цельнолистник тонкорассеченный 185674
Цельнолистник туполистный 185655
Цельнолистник ферганский 185640
Цельнолистник цветоножковый 185657
Цельнолистник шелковникова 185667
Цельнолистник широколистный 185648
Цельтис западный 80698
Целяструс 80129
Центауреа 80892,81020
Центелла 81564
Центелла азиатская 81570
Центипеда 81681
Центипеда малая 81687
Центифолия 336474
Центрантус 81743
Центрантус красный 81768
Центролобиум 81817
Ценхрус 80803
Цератозамия 83680
Цератония 83523
Цератония стручковая 83527
Цератостигма 83641
Цератостигма свинчатковая 83646
Церва 327879
Цереус 83855
Цереус гигантский 77077
Цереус крупноцветный 83874
Цереус перуанский 83890
Цереус плетевидный 29715
Цереус ямакару 83885
Церкокарпус 83816
Церопегия 83975
Церцис 83752
Церцис гриффита 83788
Церцис европейский 83809
Церцис канадский 83757
Церцис китайский 83769
Церцис юньнаньский 83812
Цеструм 84400
Цеструм иволистный 84419
Цеструм оранжевый 84404
Цеструм парка 84419
Цеструм паркви 84419
Цефалант западный 82098
Цефалантус 82091
Цефалоринхус 82411,82413
Цефалоринхус джингарский 82422
Цефалоринхус кирпичникова 82414
Цефалоринхус клубневой 82426
Цефалоринхус косинского 82415
Цефалоринхус многоветвистый 82420
Цефалоринхус почти – перестый 82423
Цефалоринхус талышский 82425
Цефалоринхус тахтаджяна 82424
Цефалотаксус 82496
Цефалотаксус головчато – костянковый 82499
Цефалотаксус гуншаньский 82539
Цефалотаксус тайваньский 82550
Цефалотаксус фортуна 82507
Цефалотаксус хайнаньский 82521
Цефалотаксус японский 82523
Цефалоцереус 82204
Цефалянтера 82028
Цефалянтус 82091
Цефалярия 82113
Цианантус 115338
Цидония 116532
Цидония японская 84549,84556
Цикада 115794
Цикадовые 115793
Цикас 115794,115884
Цикас революта 115884
Цикас сычуаньский 115909
Цикас таиландский 115897
Цикас тайваньский 115912
Цикас хайнаньский 115833
Цикламен 115931
Цикламен альпийский 115949
Цикламен весенний 115981
Цикламен весенняя 115981
Цикламен европейский 115949
Цикламен персидский 115965
Циклантовые 115989
Циклахена 210462
Цикнохес 116516
Цикорий 90889
Цикорий дикий 90901
Цикорий железистый 90900
Цикорий карликовый 90910
Цикорий корневой 90894,90901
Цикорий обыкновенный 90901
Цикорий полевой 90901,200225

Цикорий садовий 90901
Цикорий салатный 90894,90901
Цикорий эндивий 90894
Цикорник 90889
Цикрантера 115990
Цикута 90918
Цикута пятнистая 90932
Цикута ядовитая 90932
Цимбалярия 116732
Цимбалярия постенная 116736
Цимбалярия стенная 116736
Цимбария 116745
Цимбария даурская 116747
Цимбария днепровская 116746
Цимбидиум 116769
Цимбидиум вэньшаньский 117100
Цимбидиум лоу 116967
Цимбидиум тибетский 117090
Цимбидиум эмэйский 116853
цимболена 117121
цимболена длиннолистная 117122
Цимбохасма 116745,117117
Цимбохасма днепровская 117119
Цимицифуга 90994
Цимицифуга даурская 91008
Цимицифуга простая 91038
Цинанхум 117334
Цинанхум гуансиский 117551
Цинанхум далиский 117473
Цинанхум кандинский 117566
Цинанхум китайский 117425
Цинанхум куэньминский 117747
Цинанхум ланьюйский 117555
Цинанхум лекарственный 117743
Цинанхум лицзянский 117564
Цинанхум максимовича 117597
Цинанхум мулиский 117614
Цинанхум остролистный 117339
Цинанхум острый 117345
Цинанхум сибирский 117346
Цинанхум сычуаньский 117713
Цинанхум стеблеобъемлющий 117364
Цинанхум таиландский 117716
Цинанхум тайваньский 117469
Цинанхум тибетский 117515,117675
Цинанхум фунинский 117738
Цинанхум цзиндунский 117546
Цинанхум чжэцзянский 117424
Цинанхум эмэйский 117482
Цингибер 417964
Цинерария 91107
Цинерария гибридная 290821
Цинерария ковровая 358395
Цинерария летняя 358395
Цинерария морская 358395
Цинерария приморская 358395
Циния 418035
Циния изящная 418043
Цинна 91235
Цинна широколистная 91239
Циннамомум 91252,91287
Циннамомум мелколистный 91280
Циннамомум пинбяньский 91407
Циннамомум цейланикум 91446
Циннамомум чэнкоуский 91409
Циннамомум яванский 91353
Цинния 418035
Цинния изящная 418043
Цинния многоцветковая 418058
Цинния обыкновенная красивая 418043
Цинния перуская 418058
Цинния стройная 418043
Цинния хагеа 418038
Циноглоссум 117900
Циноглоссум амабиле 117908
Циноглоссум ганьсуский 117967
Цинокрамбовые 389181
Циноксилон головчатый 124891
Циноксилон цветуший 105023
Циноктонум пурпуровый 117660
Циноморииевые 118124
Циноморий 118127
Циноморий джунгарский 118133
Цинхона 91055
Ципелла 118393
Циперу спинбяньский 118759
Циперус 118428
Циперус очереднолистный 118476
Циперус сычуаньский 119657
Циперус съедобный 118822
Циперус шаньдунский 119578
Циприпедиум 120287
Циприпедиум башмачок 120310
Циприпедиум далиский 120336
Циприпедиум крупноцветный 120396
Циприпедиум лицзянский 120383
Циприпедиум мелкоцветковый 120414
Циприпедиум пятнистый 120357
Циприпедиум тайваньский 120353
Циприпедиум тибетский 120461
Циприпедиум франше 120355
Циприпедиум юньнаньский 120474
Циртантус 120532
Циртостакис 120779
Циртостахис 120779
Цирцея 91499
Цирцея альпейская 91508
Цирцея гиблидная 91537
Цирцея мягкая 91575
Цирцея обыкновенная 91557
Цирцея парижская 91557
Цирцея сердцевидная 91537
Цирцея четырехбороздчатая 91557
Циссус 92586,92729,92777
Циссус вэньшаньский 93038
Циссус двуцветный 92777
Циссус ланьюйский 92801
Циссус разноцветный 92777
Циссус ромболистный 411885
Циссус четырехгранный 411868
Цистанхе 93047
Цистанхе желтая 93059
Цистанхе монгольская 93068
Цистанхе неясная 93048
Цистанхе нинсяская 93069
Цистанхе рассеченная 93058
Цистанхе риджуэя 93074
Цистанхе солончаковая 93075
Цистанхе солончаковое 93075
Цистанхе трубчатая 93082
Цитарексилум 93237
Цитисус 120903
Цитисус прутьевидный 121001
Цитроид 94169
Цитрон 93314,93603
Цитрон лимон 93539
Цитрон уншиу 93736
Цитронелла 117218
Цитрон – мелисса 249504
Цитрус 93314
Цитрус бергамот 93408
Цитрус бигардия 93414
Цитрус вильсона 93869
Цитрус дивный 93690
Цитрус индийский 93497
Цитрус ичанский 93495
Цитрус каламондин 93636
Цитрус кинокуни 93724
Цитрус китайский 93765
Цитрус китайский ‘Валенсия’ 93785
Цитрус китайский ‘Вашингтон – навель’ 93786
Цитрус китайский ‘Гамлин 93771
Цитрус китайский ‘Яффа’ 93774
Цитрус кодзимикан 93530
Цитрус лайм 93329
Цитрус лиметта 93329,93534
Цитрус лимон 93539
Цитрус мидийский 93603
Цитрус померанцелистный 93329,93534
Цитрус помпельмус 93579
Цитрус танжерин 93837
Цитрус уншиу 93736
Цитрус шеддок 93579
Цитрус юнос 93515
Цифомандра 119973,119974
Цифомандра свекловидная 119974
Цицания 418079
Цицания водяная 418080
Цицания широколистная 418095
Цицербита 90835

Цицербита альпийская 90836
Цицербита буржо 90839
Цицербита далиская 90856
Цицербита дельтовидная 90845
Цицербита зеравшанская 90873
Цицербита кистевидная 90859
Цицербита ковалевткой 90848
Цицербита крупнолистная 90851
Цицербита лазоревая 90837
Цицербита пренантоидная 90858
Цицербита розовая 90862
Цицербита тибетская 90864
Цицербита тяньшанская 90871
Цицербита уральская 90872
Цмин 189093,189147
Цмин армянский 189162
Цмин волнистый 189880
Цмин жёлтый 189147
Цмин копетдагский 189481
Цмин многолистный 189681
Цмин многочемичатый 189679
Цмин мусы 189583
Цмин нуратавский 189609
Цмин палласа 189638
Цмин пахучий 189409
Цмин песчаный 189147
Цмин прицветниковый 415386
Цмин самаркандский 189539
Цмин сильнопахнущий 189409
Цмин складчатый 189674
Цмин тяньшанский 189859
Цойзия 418424
Цойзия японская 418426
Цойсия 418424
Цойсия японская 418426
Цуга 399860
Цуга западная 399905
Цуга зибольда 399923
Цуга калифорнийская 399916
Цуга китайская 399879
Цуга разнолистная 399895
Чабер 347446
Чабёр 347446
Чабер алтайский 347469
Чабер альпийский 4662
Чабер горный 347597
Чабёр горный 347597
Чабер кеымский 347632
Чабер колосоносный 347630
Чабер крупноцветковый 347588
Чабер мелкозубчатый 347631
Чабер пограничный 96981
Чабер промежуточный 347565
Чабер рыхлоцветковый 347579
Чабер садовый 347560
Чабёр садовый 347560
Чабер толстолистный 347610
Чабер тупоконечный 347599
Чабрец 391061,391397
Чабрец венгерский 391321
Чабрец обыкновенная 391365
Чабрец песчаный 391397
Чагерак 14623
Чагерак верблюжий 14638
Чай 68877,69634
Чай иезуитский 139678
Чай капорский 85875
Чай китайский 69634
Чай курильский 289713
Чай курильский кустарниковый 289713
Чай луговой 239755
Чай многольский 52514
Чай парагвайский 204135
Чай почечный 275781
Чай рвотный 204391
Чай – матэ 204135
Чайные 389046
Чайный куст 68877,69634
Чайный куст китайский 69634
Чайный куст парагвайский 204135
Чайот 356352
Чайота 356352
Чайшон 259772
Чакан 401085
Чалтык 418096
Чанголак 155985
Чаполоть 27967
Чапышник 72168
Часалка 133478
Частуха 14719
Частуха валенберга 14734,14789
Частуха водяной подорожник 14760
Частуха восточная 14754,14760
Частуха желобчатая 14725
Частуха злаковая 14734
Частуха ланцетная 14745
Частуха ланцетолистная 14745
Частуха лезеля 14749
Частуха лёзеля 14749
Частуха обыкновенная 14760
Частуха подорожниковая 14760
Частуховые 14790
Чатыр 225279
Чашецветник 68306
Чебрец 391061,391397
Чебрец арктический 391080
Чебрец башкирский 391098
Чебрец блошницевидный 391346
Чебрец волосистый 391203
Чебрец гранитный 391195
Чебрец губерлинский 391197
Чебрец двухнервный 391101
Чебрец дзевановского 391177
Чебрец диморфный 391171
Чебрец днепровский 391106
Чебрец дубянского 391176
Чебрец евпаторийский 391185
Чебрец жигулевский 391400
Чебрец зеленецкого 391399
Чебрец извесковый 391115
Чебрец казахский 391200
Чебрец каллье 391116
Чебрец кальмиусский 391229
Чебрец карликовый 391339
Чебрец киргизский 391234
Чебрец крымский 391385
Чебрец ложногранитный 391338
Чебрец маршалля 391274
Чебрец меловой 391161
Чебрец молдавский 391282
Чебрец мтгоджарский 391289
Чебрец палласа 391311
Чебрец паннонский 391321
Чебрец подольский 391326
Чебрец прибрежный 391255
Чебрец садовый 391397
Чебрец сталинградский 391162
Чебрец степной 391381
Чебрец субарктический 391383
Чебрец талиева 391384
Чебрец черняева 391165
Чебрец чешский 391259
Чебрец шерститый 391246
Чебрец широколистный 391247
Чебрец эльтонский 391179
Чебрец яйлияский 391224
Чезнейя 87232
Чезнейя ганьсуская 87239,87271
Чезнейя гиссарская 87243
Чезнейя изящная 87237
Чезнейя копетдагская 87246
Чезнейя линчевского 87247
Чезнейя толстоногая 87234
Чезнейя туркестанская 87265
Чезнейя ферганская 87271
Чезнейя чаюйская 87244
Чезнейя юньнаньская 87266
Чезнейя якорцевая 87264
Челига крупноцветковая 72241
Челнотал 343858
Челыжник 72180
Чембедек 36920,36924
Чемерица 405570
Чемерица американская 405643
Чемерица арктическая 405616
Чемерица белая 405571
Чемерица белая американская 405643
Чемерица гулинская 405633
Чемерица гуншаньская 405597
Чемерица далиская 405638
Чемерица даурская 405585

Чемерица зеленая 405643
Чемерица зелёная 405643
Чемерица крупноцветковая 405598
Чемерица лобеля 405604
Чемерица маака 405606
Чемерица мелкоцветковое 405615
Чемерица мэнцзыская 405614
Чемерица наньчуаньская 405617
Чемерица остродольная 405627
Чемерица тайваньская 405593
Чемерица чашецветная 405583
Чемерица черная 405618
Чемерица чёрная 405618
Чемерица эмэйская 405626
Чемерица японская 405601
Чемыш 184832
Черва 327879
Червички 124141
Череда 53755
Череда дваждыперистая 53797
Череда двоякоперистая 53797
Череда камчатская 53962
Череда лучевая 54081
Череда лучистая 54081
Череда максимовича 54001
Череда мелкоцветковая 54042
Череда олиственная 53913
Череда поникшая 53816
Череда трехраздельная 54158
Череда трёхраздельная 54158
Череда фэнцзеская 53903
Черемуха 279998,280003
Черёмуха 280003,280044
Черемуха азиатская 280004
Черемуха антипка 83253
Черемуха виргинская 316918
Черёмуха виргинская 316918
Черемуха грея 280023
Черёмуха душистая 83253
Черёмуха карликовая 20931
Черемуха кистевая 280003,280044
Черёмуха кистевая 280003,280044
Черемуха кистевидная 280003
Черемуха маака 280028
Черемуха магалепка 83253
Черемуха непалская 280036
Черемуха обыкновенная 280003,280044
Черёмуха обыкновенная 280003,280044
Черемуха оздняя 83311
Черемуха пенсильванская 83274
Черёмуха пенсильванская 83274
Черёмуха поздняя 83311
Черёмуха поздняя иволистная 45725
Черемуха птичья 280003
Черемуха сьори 280053
Черемуха японская 280028
Черёмуха – антипка 83253
Черёмуха – магалепка 83253
Черемша 15861,15868
Черепоплодник 108661
Черепоплодник ежистый 108665
Черепоплодник сероватый 108662
Черепоплодник хлопьевидно – шерстистый 108666
Черепоплодник щетинистый 108666
Череш 148531
Черкез 344682
Чермеш 392959
Черная карагана 72180
Черная кислица 333913,334240
Черник овальнолистный 403932
Черника 403710
Черника волосистый 403854
Черника гуандунская 403844
Черника кавказская 403728
Черника кавказский 403728
Черника канадская 403758
Черника крупноплодная 403894
Черника обыкновенная 403916
Черника прицветниковая 403738
Черника сычуаньская 403770
Черника хайнаньская 403845
Черника щитковая 403780
Черница 403916
Чернобыл 36474
Чернобыль 36474
Чернобыльник 36474
Чернобыльник каменистый 36166
Черноголовка 316095
Черноголовка белоцветковая 316116
Черноголовка крупноцветковая 316104
Черноголовка крупноцветная 316104
Черноголовка обыкновенная 316127
Черноголовка разрезная 316116
Черноголовка средняя 316112
Черноголовник 345816
Черноголовник кровохбрачный 345881
Черноголовник кровохлёбковый 345860, 345881
Черноголовник многобрачный 345881
Черноголовник мохнатоплодный 313140
Черногорка 8324,8387
Черноклен 3656
Черноклёнина 2837
Чернокорень 117900
Чернокорень аптечный 118008
Чернокорень горный 118004
Чернокорень зеленоцветковый 118033
Чернокорень зеравшанский 118018
Чернокорень лекарственный 118008
Чернокорень немецкий 117968
Чернокорень пестрый 117948
Чернокорень приятный 117908
Чернокорень распиканный 118010
Чернокорень растопыренный 117956
Чернокорень тибетский 118017
Чернокорень тяньшанский 118022
Чернокорень шелковисто – войлочный 117978
Чернолоз 343858
Чернослив 20890,316382
Чернотал 343858
Черноцвет 91499
Чернушка 266215
Чернушка бухарская 266224
Чернушка восточная 266236
Чернушка гарибелля 266229
Чернушка дамаскская 266225
Чернушка дамасская 266225
Чернушка железистая 266230
Чернушка мелкоцветковая 266229
Чернушка остролепестная 266237
Чернушка пашенная 266242
Чернушка персидская 266240
Чернушка полевная 266216
Чернушка посевная 266241
Чернушка цельнолистая 266235
Черный Тополь 311398
Черня 83155
Черня антипка 83253
Черня птичья 83155
Чертово дерево 387432
Чертогон 73278
Чертогон луговой 379660
Чертополох 73278
Чертополох азербайджанский 73310
Чертополох акантовидный 73281
Чертополох акантолистный 73281
Чертополох аравийский 73293
Чертополох беккеров 73317
Чертополох беловатый 73288
Чертополох благословленный 97599
Чертополох джунгарский 73492
Чертополох жилковатый 73426
Чертополох закаспийский 73514
Чертополох замаскированный 73454
Чертополох кернера 73377
Чертополох колючеголовый 73280
Чертополох колючий 73281
Чертополох крючечковый 73357
Чертополох крючечный 73357
Чертополох крючковатый 73516
Чертополох курчавый 73337,203124
Чертополох ложно – холмовый 73459
Чертополох маскированный 73454
Чертополох мелкоголовчатый 73467
Чертополох мелкоголовый 73467
Чертополох многопарный 73418
Чертополох навашина 73425
Чертополох никитина 73427
Чертополох окрашенный 73430

Чертополох поздный 73363
Чертополох полуголый 73486
Чертополох поникающий 73430
Чертополох поникший 73430
Чертополох прижатый 73284
Чертополох сереватый 73329
Чертополох серо - желтый 73458
Чертополох серый 73329
Чертополох сизый 73354
Чертополох татарниковый 73449
Чертополох тёрмера 73508
Чертополох узкоголовый 73498
Чертополох хаястанский 73356
Чертополох холмовой 73333
Чертополох шишкина 73430
Чеснок 15698
Чеснок дикий 388272
Чеснок заячий 388272
Чеснок змеиный 15726
Чеснок посевный 15698
Чесночник 14953,129618
Чесночник аптечный 14964
Чесночник коротколплодный 14958
Чесночник крупнолистный 275869
Чесночник лекарственный 14964
Честотел 86733
Честотел большой 86755
Четверозубец 386946
Четверозубец загнутый 386955
Четверозубец памирский 386952
Четверозубец четырехрогий 386953
Четочник 264603
Чёточник 738,264603
Четочник бугорчатый 264639
Чёточник бугорчатый 264639
Четочник железистый 131137
Четочник королькова 264618
Чёточник королькова 264618
Четочник коротконогий 264607
Четочник короткоплодный 264604
Четочник мягковолосый 264624
Чёточник мягковолосый 264624
Четочник низкий 264609
Чёточник низкий 264609
Четочник прижатый 393137
Четочник русский 264631
Чёточник русский 264631
Четочник серножелтый 264618
Четочник скрученный 393136
Чёточник скрученный 393136
Четочник сумбарский 393175
Четочник тибетский 264637
Чёточник тибетский 264637
Четырехзубчик 386946
Четырехзубчик бухарский 386948
Четырёхкрыльник 237476
Четырёхкрыльник пурпуровый 237771
Чефрас 346210
Чечевица 224471
Чечевица безусая 224483
Чечевица водяная 224359
Чечевица восточная 224487
Чечевица культурная 224473
Чечевица линзообразная 224483
Чечевица мелкозелная 224473
Чечевица обкновенная 224473
Чечевица пищевая 224473
Чечевица съедобная 224473
Чечевица французская 408392
Чечевица чернеющая 224485
Чечевица черноватая 224485
Чешуехвостник 295555
Чешуехвостник венгерский 295561
Чешуехвостник сонгутый 283656
Чешуйник 222651
Чий 4113,4163
Чий блестящий 4163
Чий длинностный 4145
Чий костеровидный 4119
Чий перистый 376687
Чий раскидистый 4123
Чилибуха 378633,378834,378836
Чилибуховые 235251
Чилига 72168,72180
Чилига древовидная 72180
Чилига карагана 72180
Чилига карликовая 72322
Чилига кустариниковая 72233
Чилига пушистая 72300
Чилига степная 72233
Чилижник 72168
Чилижник кустарниковый 72233
Чилим 394500
Чима водяная 143914
Чимай 369652
Чина 222671
Чина азербайджанская 222687
Чина безлисточковая 222681
Чина бледнеющая 222799
Чина болотная 222800
Чина венгерская 222815
Чина весенняя 222860
Чина волосистая 222803
Чина воронова 222863
Чина гладная 222748
Чина голубая 222706
Чина горная 222758
Чина гороховидная 222819
Чина давида 222707
Чина домина 222713
Чина душистая 222789
Чина душистый горошек 222789
Чина желтозеленая 222693
Чина злаколлистная 222784
Чина золотая 222688
Чина киноваревая 222775
Чина клубневая 222851
Чина клубненосная 222851
Чина колхидская 222704
Чина комарова 222747
Чина красная 222694
Чина кругловатая 222843
Чина круглолистная 222831
Чина крылатая 222747
Чина крылова 222748
Чина ледебура 222757
Чина лесная 222844
Чина литвинова 222761
Чина луговая 222820
Чина морская 222735
Чина мохнатая 222726
Чина незаметвая 222731
Чина ниссолия 222784
Чина однолетная 222680
Чина охряная 222788
Чина пальчатая 222711
Чина посевная 222832
Чина приземистная 222727
Чина приморская 222735,222743
Чина пятижилковая 222828
Чина редкоцветная 222756
Чина розовая 222830
Чина синеватая 222853
Чина согнутая 222733
Чина танжерская 3656,222724
Чина угловатая 222676
Чина фролова 222719
Чина черая 222782
Чина чёрая 222782
Чина шаровидная 222837
Чина широколистная 222750
Чина щетинолистная 222836
Чина японская 222735
Чинар 302592
Чинар американский 302588
Чинар восточный 302588,302592
Чинар западный 302588
Чинар кавказский 302592
Чингиль 184832
Чингиль серебристый 184838
Чирмаук 102933
Чирника 403916
Чистец 53293,373085
Чистец альпийский 373107
Чистец байкальский 373139
Чистец балянзы 373151
Чистец болотный 373346
Чистец буасы 373157
Чистец вздутый 373258
Чистец волжский 373480
Чистец ворована 373481

Чистец германский 373224
Чистец гиссарский 373247
Чистец гросгейма 373234
Чистец грузинский 373256
Чистец далиский 373456
Чистец замечательный 373446
Чистец зубчатолистный 373334
Чистец иранский 373261
Чистец китайский 373173
Чистец клубневой 373439
Чистец клубненосный 373439
Чистец комарова 373273
Чистец критский 373185
Чистец крупнолистный 373300
Чистец кустарничковый 373219
Чистец лавандолистный 373288
Чистец лесной 373455
Чистец ложнохлопьвидный 373381
Чистец мелкоцветковый 295045,295048
Чистец нарядный 373445
Чистец однолетний 373110
Чистец павла 373371
Чистец персидский 373373
Чистец полевой 373125
Чистец приморский 373303
Чистец промежуточный 373260
Чистец прямой 373389
Чистец пушистый 373386
Чистец разнозубый 373238
Чистец ридера 373139
Чистец сосновского 373443
Чистец съедобный 373439
Чистец талышский 373457
Чистец терекский 373463
Чистец тибетский 373466
Чистец трапезунтский 373469
Чистец трехжилковый 373471
Чистец туркестанский 373476
Чистец туркменский 373475
Чистец узколистный 373109
Чистец фомина 373212
Чистец франше 373276
Чистец чаткальский 373472
Чистец черняева 373189
Чистец шерстистый 373166
Чистец щетинистый 373435
Чистец японский 373262
Чистотел 86733
Чистотел большой 86755,115031
Чистотел крупный 86755
Чистяк 164469
Чистяк весенний 164479,325820
Чистяк лютичный 325820
Чистяк настоящий 164474
Чистяк пучковатый 164473
Читыр 336849
Чихотник обкновенный 4005
Членистник 393185
Членистник кавказский 393188
Членистоостник 36612
Чобушник боротавчатый 294564
Чобушник культера 294445
Чобушник непахучий 294479
Чобушник сердцелистный 294425
Чобушник фальконера 294459
Чобушник цейера 294590
Чобушник широколистный 294488,294517
Чозения 89159
Чозения земляничниколистная 89160
Чозения крупночешуйная 89160
Чозения толокнянколистная 89160
Чорак 379486
Чубушник 294407
Чубушник венечный 294426
Чубушник ганьсуский 294484
Чубушник генри 294465
Чубушник гордона 294462
Чубушник гулинский 294542
Чубушник девичий 294565
Чубушник делаве 294451
Чубушник душистый 294507
Чубушник жестковолосый 294468
Чубушник кавказский 294422
Чубушник короткокистевой 294413
Чубушник крупноцветковый 294463
Чубушник куэньминский 294486
Чубушник лицзянский 294419
Чубушник льюиса 294493
Чубушник мексиканский 294500
Чубушник мелколистный 294501
Чубушник непахучий 294478
Чубушник пекинский 294509
Чубушник полуседой 294549
Чубушник седоватый 294471
Чубушник седой 294471
Чубушник тонколистный 294556
Чубушник чжэцзянский 294591
Чубушник шренка 294536
Чумиза 361935
Чусва однобокая 310873
Чуфа 118822
Шабдар 397050
Шавель густой 339987
Шавель конский 339987
Шалот 15170
Шалфей 344821
Шалфей австрийский 344883
Шалфей александра 344843
Шалфей амазийский 344846
Шалфей андрея 344849
Шалфей аптечный 345271
Шалфей армянский 344863
Шалфей бальджуанский 344890
Шалфей беккера 344899
Шалфей блестящий 345405
Шалфей великолепный 345405
Шалфей вербеновидный 345458
Шалфей вербеновый 345458
Шалфей вербенолистный 345458
Шалфей гаястанский 345073
Шалфей голостебельный 345055
Шалфей голый 345405
Шалфей гончарова 345061
Шалфей горминовый 345474
Шалфей гроссгейма 345071
Шалфей дагестанский 344996
Шалфей длиннотрубчатый 345190
Шалфей дробова 345013
Шалфей дубравный 345246
Шалфей железистоколосый 344828
Шалфей железистый 345057
Шалфей замечательный 345099
Шалфей зеленовато-белый 344961
Шалфей зеленый 345474
Шалфей зеравшанский 345384
Шалфей иерусалимский 295111
Шалфей кандинский 345322
Шалфей карабахский 345139
Шалфей клейкий 345057
Шалфей колючий 345403
Шалфей комарова 345146
Шалфей копетдагский 345147
Шалфей коровяковый 345457
Шалфей коротцветковый 344913
Шалфей крупноцветковый 345064,345065
Шалфей крупноцветный 345064
Шалфей кузнецова 345149
Шалфей кустарниковый 345014
Шалфей лекарственный 345271
Шалфей лесной 345246
Шалфей лиловосиний 345173
Шалфей линчевского 345175
Шалфей липского 345176
Шалфей логолистный 344949
Шалфей луговой 345321
Шалфей луговый 345321
Шалфей маргариты 345197
Шалфей мулиский 345074
Шалфей мускатный 345379
Шалфей мутовчатый 345471
Шалфей наньчуаньский 345240
Шалфей обкновенный 345310
Шалфей окаймленный 345174
Шалфей опадающий 345042
Шалфей полушерстистый 345382
Шалфей поникающий 345266,345298
Шалфей поникший 345266
Шалфей промежуточный 345102
Шалфей простой 345310
Шалфей прутьевидный 345473
Шалфей пустынный 344999

Шалфей равнозубый 344833
Шалфей равный 344984
Шалфей раскрытый 345346
Шалфей розолистный 345351
Шалфей седоватый 344938
Шалфей сиборпа 345389
Шалфей сиринский 345422
Шалфей скабиозолистный 345363
Шалфей скобиозолистный 344922
Шалфей согнутозубый 344935
Шалфей степной 345412
Шалфей тайваньский 345254
Шалфей тибетский 345480
Шалфей толстоколосый 345282
Шалфей траутфеттера 345435
Шалфей треугольночашечный 345438
Шалфей туповатый 345414
Шалфей туркменский 345445
Шалфей фомина 345032
Шалфей форскаля 345037
Шалфей хохлатый 345474
Шалфей хубэйский 345091
Шалфей хэнаньский 345086
Шалфей шертогубый 345483
Шалфей шмальгаузена 345378
Шалфей эмэйский 345274
Шалфей эфиопский 344836
Шалфей японский 345108,345252
Шалфей японский тайваньский 345254
Шалфейчик 85730
Шалфейчик падуболистый 85731
Шалы 275958
Шамбала 397229
Шампиньоновые 1292
Шандал - дерево 329761
Шандра 245723
Шандра белая 245770
Шандра близкая 245761
Шандра волонова 245774
Шандра иранская 245758
Шандра карликовая 245754
Шандра малоцветковая 245756
Шандра надрезанная 220126
Шандра обыкновенная 245770
Шандра перистая 245759
Шандра пурпуровая 245764
Шандра пустырниковая 245748
Шандра пустырниковидная 245748
Шандра разнозубая 245726
Шандра ранняя 245760
Шандра севанская 245740
Шандра туркевича 245769
Шандра чужеземная 245757
Шапочки 383090,383103
Шар золотой 339574
Шар снежный 407989
Шардения 86012
Шардения восточная 86014
Шардения крупноплодная 86013
Шарлот 15170
шаровница 177022
шаровница безлистноцветковая 177032
шаровница волосоцветковая 177052
Шаровница сердцелистная 177035
Шаровницевые 177054
Шарогнездка 297934
Шарогнездка сушеницевая 297935
Шарошница 170524
Шарыш 148531
Шафдар 397050
Шафран 111480
Шафран адама 111481
Шафран алатавский 111483
Шафран алатауский 111483
Шафран артвинский 111487
Шафран вессенний 111625
Шафран гейфела 111540
Шафран гейфеля 111540
Шафран долиный 111623
Шафран золотистый 111558
Шафран карский 111546
Шафран каспийский 111514
Шафран королькова 111548
Шафран красивый 111603
Шафран крымский 111615
Шафран логовой 99297
Шафран михельсона 111556
Шафран отрана 111492
Шафран палласа 111573
Шафран полосатый 111624
Шафран посевной 111589
Шафран прекрасный 111603
Шафран рооп 111584
Шафран сетчатый 111582
Шафран суворова 111614
Шафран сузианский 111612
Шафран шарояна 111590
Шафтал 397050
Шведка 379486
Шведка сизая 379531
Шеддок 93579
Шедковица красная 259186
Шейка 308893
Шейки раковые 54798,308893
Шейки рачьи 54798,308893
Шейхцериевые 350864
Шейхцерия 350859
Шейхцерия болотная 350861
Шелковая акация 13578
Шелковиза бумажная 61107
Шелковица 259065
Шелковица атласная 259097
Шелковица белая 259067
Шелковица бумажная 61107
Шелковица индийская 259137
Шелковица либоская 259154
Шелковица мелколистная 259164
Шелковица монгольская 259165
Шелковица черная 259180
Шелковица чёрная 259180
Шелковник 48904,325580
Шелковник бунге 48907
Шелковник водяной 325580
Шелковник волосолистный 48933
Шелковник гилибера 48920
Шелковник дихотомический 48913
Шелковник кауфмана 48922
Шелковник килеватый 48910
Шелковник ланге 48923
Шелковник морской 48926
Шелковник неуколеняющийся 48915
Шелковник расходящийся 48914
Шелковник риона 48931
Шелковник трехлистный 48944
Шелковник фенхелевидный 48919
Шелюга 342961,343280
Шелюга желдая 343280
Шелюга жёлдая 343280
Шелюга красная 342961
Шелюжник 343280
Шерардия 362082,362095
Шерардия полевая 362083,362097
Шерохватка 203124
Шерстестебельник 67393,151208
Шерстестебельник десятицветковый 151288
Шерстестебельник китайско - русский 151263
Шерстестебельник мощный 151452
Шерстестебельник сычуаньский 151221
Шерстестебельник уссурийский 151536
Шерстистолистник 219059
Шерстистолистник хлопковидный 219060
Шерстицвет 148855
Шерстицвет равеннский 148906
Шерстняк 148855,151655
Шерстоплодник 219056
Шерстоплодник леманна 219057
Шерстяк 151655
Шерстяк волосистый 151702
Шерстяк моханатый 151702
Шерстяк перехваченный 151699
Шертестебельник зибольда 151268
Шефердия 362089
Ши 64223
Шиверекия 351245
Шиверекия подольская 351247
Шизандра 351011
Шизантус 351343
Шизофрагма 351780
Шикот 182527
Шикша 145054

Шикша арктическая 145058
Шикша гермафродитная 145062
Шикша двуполая 145057
Шикша кардакова 145063
Шикша сибирская 145073
Шикша чёрная 145073
Шикшевые 145052
Шикща черная 145073
Шильник 379649
Шильник водный 379650
Шильник водяной 379650
Шильник ситниковидный 370202
Шима 350904
Шима валлиша 350949
Шинус терпентиннолистный 350995
Шипдерево 215442
Шип－дерево 30634
Шиповник 336283
Шиповник альберта 336345
Шиповник альпийский 336348
Шиповник андржьевского 336356
Шиповник афцелиуса 336411
Шиповник беггера 336380
Шиповник беггеровский 336380
Шиповник бедренцоволистный 336851
Шиповник бедренцовый 336851
Шиповник белый 336343
Шиповник бэнкса 336366
Шиповник вечнозелёный 336925
Шиповник войлочный 336993
Шиповник вонючий 336563
Шиповник галльский 336581
Шиповник гвоздичный 336650
Шиповник гиссарский 336633
Шиповник гололистый 336610
Шиповник даурский 336522
Шиповник дикий 336419
Шиповник железистощетинистый 336632
Шиповник иглистый 336320
Шиповник илийский 336643
Шиповник карликовый 336804
Шиповник китайский 336485
Шиповник клюка 336658
Шиповник кожистолистный 336504
Шиповник кокандский 336659
Шиповник колючейший 336851
Шиповник колючий 336851
Шиповник корейский 336660
Шиповник коржинского 336663
Шиповник коричневый 336498
Шиповник коричный 336498
Шиповник краснолистный 336611
Шиповник крючковатый 337003
Шиповник кустарниковый 336419
Шиповник максимовича 336738
Шиповник мелкоцветковый 336741
Шиповник мерщинистый 336901
Шиповник многоцветковый 336783
Шиповник многошиловатый 336802
Шиповник морщинистый 336901
Шиповник мохнатый 337006
Шиповник мягкий 336762
Шиповник неколючий 336649
Шиповник низкий 336879
Шиповник обыкновенный 336419
Шиповник опушенный 336991
Шиповник острованый 336508
Шиповник острошиповый 336835
Шиповник персидсий 336849
Шиповник полевой 336650
Шиповник приятный 336320
Шиповник розаперсидсий 336849
Шиповник рыхлый 336685
Шиповник самаркандский 336734
Шиповник северный 336498
Шиповник собачий 336419
Шиповник страшный 336901
Шиповник федченковский 336554
Шиповник французский 336581
Шиповник холмовый 336343
Шиповник широкошиповый 336857
Шиповник шренковский 336922
Шиповник щитконосный 336419,336506
Шиповник эглянтерия 336892
Шиповник эллинтический 336650
Шиповник юндзилла 336735
Шиповник яблочный 337006
Шипот 182527
Ширококолокольчик 302747
Ширококолокольчик крупноцветковый 302753
Ширяш 148531
Ширяш китайский 148539
Шит 288818
Шит воронова 288825
Шит оше 288820
Шит турменский 288824
Шитрон пальчатный 93604
Шишечник 63467
Шишка ворсовальная 133478
Шишкиния 351124
Шишкиния белощетинковая 351125
Шишник 63467
Шлемник 355340
Шлемник алехеенкой 355354
Шлемник алтайский 355357
Шлемник альберта 355351
Шлемник альпийский 355355
Шлемник ангренский 355369
Шлемник андрахновидный 355367
Шлемник андросова 355368
Шлемник аниты 355377
Шлемник аньхуйский 355375
Шлемник араксинский 355381
Шлемник араратский 355380
Шлемник артвинский 355382
Шлемник байкальский 355387
Шлемник бальджуанский 355389
Шлемник беловатый 355353
Шлемник бледный 355655
Шлемник боковоцветковый 355563
Шлемник бухарский 355395
Шлемник веерный 355439
Шлемник весенний 355836
Шлемник ветвистейший 355710
Шлемник воронова 355850
Шлемник восточный 355638
Шлемник высокий 355359,355638
Шлемник высочайший 355359
Шлемник вэйшаньский 355847
Шлемник вэньшаньский 355848
Шлемник ганьсуский 355713
Шлемник гиссарский 355478
Шлемник глазховый 355630
Шлемник горолюбивый 355637
Шлемник гребнчатый 355420
Шлемник гроссгеймовский 355463
Шлемник густовевистый 355705
Шлемник дагестанский 355422
Шлемник дарвазский 355424
Шлемник дарридагский 355423
Шлемник джунгарский 355765
Шлемник екатерины 355401
Шлемник елены 355473
Шлемник железисто－чешуйный 355344
Шлемник желто－синий 355582
Шлемник желтотрубковый 355854
Шлемник жуйлиский 355763
Шлемник заилийский 355765
Шлемник иезский 355856
Шлемник извилистый 355414
Шлемник иконникова 355711,355712
Шлемник индийская 355494
Шлемник искандера 355523
Шлемник каратавский 355538
Шлемник каркаралинский 355540
Шлемник карягина 355539
Шлемник кнорринг 355548
Шлемник колпаковидный 355449
Шлемник колпаконосный 355449
Шлемник копьелистный 355469
Шлемник коротко－волосистый 355477
Шлемник коротко－опуменный 355604
Шлемник котовниковидный 355618
Шлемник крапиволистный 355855
Шлемник крапчатый 355465
Шлемник красева 355449
Шлемник краснопятнистый 355721
Шлемник кружсчковый 355636
Шлемник крупноцветковый 355638
Шлемник крылова 355765

Шлемник крымский 355799
Шлемник курилы 355675
Шлемник курсанова 355553
Шлемник линчевского 355573
Шлемник липского 355575
Шлемник литвинова 355576
Шлемник лицзянский 355570
Шлемник лодочковый 355617
Шлемник лодяньский 355579
Шлемник лэдунский 355586
Шлемник малозубый 355632
Шлемник малый 355609
Шлемник марелистный 355407
Шлемник мелкозернистый 355460
Шлемник мелкопузырный 355607
Шлемник мелкочешуйный 355568
Шлемник меловой 355419
Шлемник многоволосый 355692
Шлемник многозубый 355690
Шлемник многолистный 355691
Шлемник монгольский 355614
Шлемник мохнатейший 355840
Шлемник мулиский 355446
Шлемник невского 355621
Шлемник незапятнанный 355489
Шлемник неправильный 355521
Шлемник нитестебельный 355438
Шлемник новороссийский 355625
Шлемник обыкновенный 355449
Шлемник оголенный 355455
Шлемник окровавлевно－зеленый 355466
Шлемник олиствениоколосый 355677
Шлемник остролистный 355651
Шлемник острочешуйный 355652
Шлемник памирский 355656
Шлемник переходный 355674
Шлемник пестроцветковый 355688
Шлемник пинбяньский 355680
Шлемник повислый 355428
Шлемник полосатенький 355777
Шлемник понтийский 355693
Шлемник попова 355694
Шлемник почти－дернистый 355778, 355782
Шлемник почтисердцевидный 355783
Шлемник пржевальского 355699
Шлемник приземистый 355786
Шлемник приподнимающийся 355346
Шлемник простертый 355786
Шлемник пузырчаточашечный 355678
Шлемник радде 355709
Шлемник разноволосый 355475
Шлемник разноцветный 355474
Шлемник раскрашенный 355679
Шлемник растопыренный 355774
Шлемник регелевский 355711
Шлемник регеля 355711
Шлемник ромбовидный 355716
Шлемник севанский 355758
Шлемник сердцелиственный 355418
Шлемник сиверса 355765
Шлемник сикокский 355760
Шлемник сичоуский 355764
Шлемник скордиелистный 355733
Шлемник снизу－белый 355486
Шлемник сомнительный 355432
Шлемник сосновского 355772
Шлемник спрошь－зеленый 355366
Шлемник спрошь－шелковистый 355479
Шлемник среднечешуйный 355601
Шлемник средний 355520
Шлемник стевена 355776
Шлемник тайваньский 355789
Шлемник такэ 355778
Шлемник таласский 355790
Шлемник татьяны 355798
Шлемник титова 355811
Шлемник тихоокеанский 355674
Шлемник тогуз－тораусқий 355812
Шлемник тонкотрубковый 355567
Шлемник тувинский 355826
Шлемник туминганский 355824
Шлемник тургайский 355825
Шлемник турнефора 355813
Шлемник тяньцюаньский 355810
Шлемник уссурийский 355675
Шлемник франше 355448
Шлемник хайнаньский 355467
Шлемник хитрово 355410
Шлемник хмелевой 355580
Шлемник хохлатый 355416
Шлемник хунаньский 355483
Шлемник хэнаньский 355480
Шлемник цзиньюньский 355820
Шлемник четкокорневищный 355615
Шлемник чжундяньский 355411
Шлемник чжэцзянский 355406
Шлемник чимганский 355818
Шлемник чужеземный 355391
Шлемник шерстисто－черешковый 355561
Шлемник широко－чешуйный 355682
Шлемник шугнанский 355729
Шлемник щетинковая 355778
Шлемник эмэйский 355634
Шлемник юзепчука 355536
Шлемник яванский 355531
Шлюмбергерия 351993
Шмальгаузения 352004
Шмальгаузения гнездистая 352006
Шмальница 332452
Шнит－латук 219494
Шнит－лук 15709
Шовкун 259067
Шоколадник 389404
Шоколадное дерево 389404
Шомнольник 362396
Шор－аджерик 8862
Шорея 362186
Шорея мощная 362220
Шортия 362242
Шотия 352485
Шпажник 175990,176222
Шпажник болотный 176429
Шпажник большой 176122
Шпажник гентский 176222
Шпажник итальянский 176281
Шпажник кавказский 176112
Шпажник кочи 176296
Шпажник обыкновенный 176122
Шпажник посевный 176281
Шпажник солелюбивый 176246
Шпажник темнофиолетовый 176038
Шпажник тонкоцветный 176589
Шпажник туркменский 176612
Шпажник черепитчатый 176264
Шпажник черепичатый 176264
Шпергель 370521,370522
Шпилат 4866,371627
Шпилат трянистый 4866
Шпинат 371710
Шпинат английский 340178
Шпинат индийский 48688,48689
Шпинат кубинский 94349
Шпинат малабарский 48688,48689
Шпинат новозеландский 387221
Шпинат овошной 371713
Шпинат огородный 371713
Шпинат туркестанский 371717
Шпинат четырехтычинковый 371716
Шпорник 102833,124008
Шпорник аякса 102833
Шпорник многолетний 124301
Шпорник посевной 102833
Шпороцветник 303125
Шпороцветник вырезной 209665
Шпороцветник пильчатый 209826
Шпороцветник сизочашечный 209717
Шрадерия блюдчатая 352558
Шрадерия бухарская 352559
Шрадерия змееговниковая 352560
Шрадерия королькова 352561
Шренкия 352629
Шренкия влагалищная 352637
Шренкия ворсинчатая 352635
Шренкия голике 352631
Шренкия замечательнаю 352632
Шренкия колючая 352636
Шренкия культиасова 352634
Шренкия обертковая 352633
Штернбергия 376350
Штернбергия желтая 376356

Штернбергия жёлтая 376356
Штернбергия зимовникоцветная 376353
Штернбергия фишера 376354
Штокроза 13934
шток – роза 13911
Шток – роза 18155
шток – роза антонины 13916
шток – роза бальджуанская 13917
Шток – роза бледная 18174
шток – роза воронова 13950
шток – роза гельдрейха 13922
шток – роза гирканская 13923
шток – роза гроссгеима 13921
шток – роза каракалинская 13925
шток – роза карсская 13926
шток – роза копетдагская 13927
шток – роза крымская 13947
шток – роза кусаринская 13928
шток – роза ленкоранская 13930
Шток – роза морщинистая 13940
шток – роза новопокровского 13931
Шток – роза розовая 13934
шток – роза сахсаханская 13941
шток – роза сосховского 13943
шток – роза софии 13942
шток – роза таласская 13946
шток – роза туркевича 13949
шток – роза туркменская 13948
шток – роза угловатая 13915
шток – роза фиголистная 13945
шток – роза фрейна 13920
Штубендорфия 378976
Штубендорфия бескрылая 378977
Штубендорфия восточная 378979
Штубендорфия двойчатая 378980
Штубендорфия липского 378978
Шульция 352688
Шульция белоцветковая 352689
Шульция косматая 352690
Шуманния 352702
Шуманния карелина 352703
Шумерия 352722,360982
Шумерия литвинова 352724
Шумерия широколистная 352723
Шунгуя 274927
Шурок 86901
Шучка 126025
Шучка извилистая 126074
Шушу 356352
Щавелек 339897
Щавелёк 339897
Щавель 339880
Щавель альпийский 339914
Щавель амурский 339918
Щавель английский 340178
Щавель арктический 339937
Щавель армянский 339943
Щавель аройниколистный 339940
Щавель ашхабадский 339944
Щавель воднощавелевый 340082
Щавель водный 339925
Щавель водяной 339925,340082
Щавель воровьиный 339897
Щавель галачи 340071
Щавель гмелина 340064
Щавель горный 339914
Щавель длиннолистный 340109
Щавель домашний 340109
Щавель злаколистный 340068
Щавель золотистый 339947
Щавель камчадальский 340093
Щавель кислелький 339897
Щавель кислый 339887
Щавель клубковатый 339989
Щавель комарова 340098
Щавель красивый 340214
Щавель крованой 340249
Щавель кровано – краный 340249
Щавель крубневый 340294
Щавель курчавый 339994
Щавель лапландский 340103
Щавель ложносолончаковый 340209
Щавель малый 339897
Щавель маршалля 340123
Щавель мелкоплодный 340135
Щавель морской 340116
Щавель настоящий 339887
Щавель непальский 340141
Щавель обыкновенный 339887
Щавель огородный 339887
Щавель памирский 340171
Щавель паульсена 340194
Щавель пирамидальноцветковый 340279
Щавель пирамидальный 340279
Щавель прибрежный 340082
Щавель приморский 340116
Щавель рехингера 340221
Щавель русский 340116
Щавель североамериканский 340084
Щавель сетчатый 340225
Щавель сибирский 340256
Щавель сирийский 340277
Щавель сросшийся 339989
Щавель стловый 339887
Щавель тонколистный 340278
Щавель туполистный 340151
Щавель тяньшанский 340285
Щавель узколистный 339923,340269
Щавель украинский 340298
Щавель уссурийский 340270
Щавель финский 340209
Щавель фишера 340055
Щавель фори 340049
Щавель шишкина 340253
Щавель шпинатный 340178
Щавель щавельковидный 339904
Щавель щитковидный 339907,340254
Щавель щитковый 340254
Щавель японский 340089
Щавель ячеистый 340058
Щалаппа 369652
Щебрец 391365
Щебрушка 4661
Щеркунец 363571,364158,364193
Щетинник 361680
Щетинник гуйцзоуский 361777
Щетинник зеленый 361935
Щетинник зелёный 361935
Щетинник итальянский 361794
Щетинник мутовчатый 361930
Щетинник промежуточный 361789
Щетинник сизый 361877
Щетинник фабера 361743
Щетинник ципкий 361930
Щеть 262537
Щирица 18652
Щирица американская 18788
Щирица белая 18655
Щирица белосемянная 18752,18763
Щирица гиблидный 18734
Щирица жминда 18670
Щирица жминдовидная 18668
Щирица запрокинутая 18810
Щирица колосистая 18810
Щирица колючая 18822
Щирица маскированная 18655
Щирица метельчатая 18687,18788
Щирица синеватая 18670
Щирица согнутая 18701
Щирица теллунга 18831
Щирица трехцветная 18836
Щирица хвостатая 18687,18788
Щирицевые 18646
Щитник 288827
Щитник гроссгейма 288829
Щитница 97432
Щитница дихотомическая 97442
Щитница изящная 97444
Щитница мелкоплодная 97457
Щитница щетинистая 97438
Щитница яруточная 97446
Щитолистник 200253
Щитолистник бирманский 200267
Щитолистник китайский 200274
Щитолистник лютиковый 200354
Щитолистник обыкновенный 200388
Щульха 339880
Щуровския 352643
Щуровския пятирогая 352645
Щуровския тонкорассеченная 352644
Щучка дернистая 126039

Эбелек 83418
Эбелек песчаный 83419
Эбеновые 139763
Эвакс 10106,161220,161222,161251,161268
Эвакс анатолийский 161224
Эвакс жабниковый 161240
Эвакс мелконожковый 161249
Эвакс песчаный 161226
Эвакс сжатый 161238
Эверсманния 161294
Эверсманния гедизаровидная 161295
Эвкалипт 155468
Эвкалипт белодревесный 155624
Эвкалипт блестящий 155668
Эвкалипт бугорчатый 155589
Эвкалипт вальковатторогий 155756
Эвкалипт великолепный 155564
Эвкалипт ганна 155602
Эвкалипт гвоздеголовый 155594
Эвкалипт генна 155602
Эвкалипт голибой 155589
Эвкалипт гроздевидный 155505
Эвкалипт грубый 155727
Эвкалипт гэнна 155602
Эвкалипт железнодревесный 155743
Эвкалипт западноавстралийский 155644
Эвкалипт ивовый 155732
Эвкалипт иволистный 155730
Эвкалипт исполинский 155722
Эвкалипт камальдульский 155517
Эвкалипт камеденосный 106805
Эвкалипт косой 155672
Эвкалипт красноватый 155725
Эвкалипт лимонноароматный 106793
Эвкалипт лимонный 106793
Эвкалипт лучистый 155716
Эвкалипт макартура 155633
Эвкалипт малоцветковый 155688
Эвкалипт медомпахнущий 155649
Эвкалипт мелкоплодный 155652
Эвкалипт миндальный 155478,155730
Эвкалипт мйдена 155640
Эвкалипт многоцветковый 155700
Эвкалипт мощный 155722
Эвкалипт окаймленный 155644
Эвкалипт остроконечный 155579
Эвкалипт пепельно – серый 155529
Эвкалипт пепельный 155529,155589
Эвкалипт прутовидный 155772
Эвкалипт прутьевидный 155772
Эвкалипт пятнистый 106809
Эвкалипт смита 155745
Эвкалипт смолоносный 155720
Эвкалипт фиголистный 106800
Эвкалипт царственный 155719
Эвкалипт шариковый 155589
Эвкалипт шариковый – голубой 155589
Эвкалипт шаровидный 155589
Эвкалипт ясеневидный 155585
Эвклидиум 155984
Эвклидиум сирийский 155985
Эвкоммиевые 156042
Эвкоммия 156040,156041
Эвкоммия вязовидная 156041
Эвкоммия вязолистная 156041
Эвлалия 156439
Эвлалия красивая 156499
Эвлалия четырехжилковатая 156490
Эвлалия юньнаньская 156510
Эвномия 9785
Эвномия круглолистная 157195
Эводия 161302
Эводия бархатистая 161387
Эводия генри 161340
Эводия даньелла 161323
Эводия лекарственная 161376
Эводия лицзянская 161328
Эводия мотоская 161384
Эводия сизая 161336
Эводия сычуаньская 161381
Эводия японская 161336
Эводия ясенелистная 161333
Эвптелея 160338
Эвптелея многотычинковая 160347
Эвриала 160635
Эвриала устрашающая 160637
Эвриале 160635
Эвриале устрашающая 160637
Эврикома 160716
Эврия 160405
Эврия вэньшаньская 160626
Эврия гуйцзоуская 160525
Эврия гуншаньская 160482
Эврия китайская 160442
Эврия лицзянская 160486
Эврия малипоская 160542
Эврия мэнцзыская 160492
Эврия хайнаньская 160484
Эврия цзиндунская 160523
Эврия юньнаньская 160630
Эврия японская 160503
Эврия японская мелколистная 160508
Эвскафис 160946
Эвскафис фуцзяньский 160949
Эвскафис японская 160954
Эвтока 293151
Эвтрема 161120
Эвтрема ложносердцелистная 161146
Эвтрема цельнолистная 161134
Эвтрема эдвардса 161129
Эвхарис 155857
Эгилёпс 8660
Эгилёпс двуфостый 8665
Эгилёпс овальный 8695
Эгилёпс толстый 8671
Эгилёпс трехдюймовый 8732
Эгилёпс трехостый 8728
Эгилопс 8660
Эгилопс двхдюймовый 8665
Эгилопс жювенальский 8683
Эгилопс кочи 8684
Эгилопс овальный 8695
Эгилопс оттопыренный 8727
Эгилопс толстый 8671
Эгилопс трехдюймовый 8732
Эгилопс трехостый 8728
Эгилопс цилиндрический 8672
Эдгеворция 141466
Эдельвейс 224767
Эдельвейс альпийский 224772
Эдельвейс бледно – желтый 224916
Эдельвейс коротколучевой 224806
Эдельвейс курильский 224890
Эдельвейс мелколистный 224907
Эдельвейс низкорослый 224913
Эдельвейс обедненный 224893
Эдельвейс палибина 224925
Эдельвейс разноцветный 224827
Эдельвейс скученный 224822
Эдельвейс эдельвейсовидный 224893,224899
Эдельвейс японский 224872
Эджевортия 141466
Эджевортия бумажная 141470
Эджеворция 141466
Эдрайлнт 141508
Эдрайлнт оверина 141510
Эйфорбия 158385
Эйфорбия маргината 159313
Эйфорбия разнолистная 159046
Эйфорбия тайваньская 158906
Эйхгорная 141805
Эйхгорная толстоножковая 141808
Эйхорния 141805
Эйхорния толсточерешковая 141808
Эйххорния 141805
Экаль высокая 366329
Экзакум 161528
Экзохорда 161741,294966
Экзохорда альберта 161743
Экзохорда тяньшанская 161757
Элевзина 143512
Элеодендрон 142422
Элеодендрон канский 142440
Элеокарпус 142269
Элеокарпус американский 142271
Элеокарпус зубчатый 142308
Элеокарпус новозеландский 142308
Элеутерококк 143575
Элефантопус 143453

Элизанта 143801
Элизанта липкая 143807
Элизанта ночная 363815
Элизма 238510
Элизма плавающая 14751
Элимус 144144
Элимус гмелин 335547
Элисгольция 143958
Элитригия 144624
Элитригия алайская 144629
Элитригия баталина 144635
Элитригия волосоносная 144680
Элитригия ползучая 144661
Элитригия средняя 144652
Элитригия ферганская 144648
Эллениа 8848
Эллениа ушастая 8849
Элодея 143914
Элодея канадская 143920
Эльсгольция 143958
Эльсгольция густоцветковая 144002
Эльшольция 143958
Эльшольция гребенчатая 143993
Эльшольция поздняя 144090
Эльшольция реснитчатая 143974
Эльшольция тайваньская 144067
Эмбелия 144716
Эмботриум 144847
Эмилия 144884
Эмилия яванская 144976
Эминиум 145013
Эминиум альберта 145015
Эминиум леманна 145016
Эммер 398890
Эндивий 90894
Эндивий летний 219493
Энемион 145492
Энемион радде 145498
Энкиант 145673
Энкиант китайский 145693
Энкиант погнутый 145694
Энкиант почечный 145710
Энкиант сычуаньский 145724
Энокарпус 269373
Энотера 269394
Энотера двулетняя 269404
Энотера двухлетняя 269404
Энотера душистая 269475
Энотера короткоиглая 269481
Энотера мексиканская 269403
Энтада 145857
Энтада ползучая 145899
Энтандрофрагма 145948
Энтеролобиум 145992
Энтеролобиум саман 345525
Эпакрис 146067
Эперуа 146107
Эпигея 146522
Эпигея гаультериевидная 146524
Эпилазия 146565
Эпилазия верхушечношерстистая 146567
Эпилазия полушерстистая 146573
Эпилазия удивительная 146577
Эпимедиум 146953
Эпимедиум колхидский 146973
Эпимедиум корейский 147013
Эпимедиум мелколистный 147031
Эпимедиум опушенный 147042
Эпимедиум перистый 147034
Эпимедиум сычуаньский 147067
Эпимедиум тяньцюаньский 146992
Эпимедиум франше 146993
Эпимедиум хунаньский 147008
Эпиподиум 147303
Эпиподиум безлистный 147307
Эпифиллум 147283
Эпифиллум аккерманна 266648
Эпифиллум гибридный 147289
Эпифиллум остролепестковый 147291
Эпифиллюм 147283
Эрагростис 147468
Эрантемум 148053
Эрантис 148101
Эремоспартон 148441
Эремоспартон безлистный 148444
Эремоспартон джунгарский 148447
Эремоспартон обвислый 148446
Эремоспартон туркестанвый 148448
Эремостахис 148466
Эремостахис крубненосный 148514
Эремурус 148531
Эремурус алтайский 148533
Эремурус альберта 148532
Эремурус бухарский 148536
Эремурус величественный 148557
Эремурус желтый 148551
Эремурус запрягаева 148555
Эремурус иларии 148542
Эремурус индерский 148544
Эремурус капю 148538
Эремурус кауфмана 148546
Эремурус китайский 148539
Эремурус копетдагский 148547
Эремурус коржинского 148549
Эремурус коровина 148548
Эремурус крымский 148559
Эремурус млечноцветный 148550
Эремурус мощный 148554
Эремурус неравнокрылый 148534
Эремурус ольги 148552
Эремурус представительный 148557
Эремурус регеля 148553
Эремурус согдийский 148556
Эремурус туркестанский 148560
Эремурус узколистный 148558
Эремурус хохлатый 148540
Эрехтитес 148148
Эрехтитес валерианолистный 148164
Эреция 141592
Эреция диксона 141629
Эреция лещинолистная 141618
Эреция мелколистный 77067
Эреция овальнолистная 141683
Эреция хайнаньская 141648
Эриантера уклоняющаяся 13335
Эриантус 148855
Эриантус краснеющий 148903
Эриантус тайваньский 341858
Эригерон 150414
Эризимум 154363
Эризимум юньнаньский 154565
Эрика 148941
Эрика древовидная 149017
Эрика крестолистная 150131
Эрика метловидная 150027
Эрика мясо – красная 149147
Эрика румяная 149147
Эрика сизая 149205
Эрингиум 154276
Эриоботрия 151130
Эриоботрия малипоская 151165
Эриоботрия тайваньская 151144
Эриоботрия японская 151162
Эриогонум 151792
Эриокаулиевые 151206
Эриокаулон 151208
Эриокаулон десятицветковый 151288
Эриокаулон китайско – русский 151263
Эриокаулон мощный 151452
Эриокаулон мэнцзыский 151322
Эриокаулон наньтоуский 151400
Эриокаулон непалский 151409
Эриокаулон уссурийский 151536
Эриокаулон яошаньский 151549
Эриокаулоновые 151206
Эриофиллум 152799
Эритеа 59095
Эритеа съедобная 59099
Эритрина 154617
Эритрина индийская 154734
Эритрина петушиный гребень 154647
Эритрина петушья 154647
Эритрина юньнаньская 154746
Эритрихиум 153420
Эритрихиум дэциньский 153446
Эритрихиум мелкоплодный 153535
Эритрихиум мелкоцветковое 153549
Эритрихиум мохнатый 153555
Эритрихиум ниппонский 153491
Эритрихиум сахалинский 153491
Эритрихиум седой 153463

Эритроксилон 155051
Эритроксилон кока 155067
Эритроксилон яванское 155067
Эритрониум 154895
Эрука 153998
Эрука посевная 154019
Эскаллония 155130
Эскаллония красная 155149
Эскалония 155130
Эспарто 376931
Эспарцет 271166
Эспарцет армянский 271179
Эспарцет безостый 271222
Эспарцет биберштейна 271181
Эспарцет боброва 271182
Эспарцет большой 271170,271229
Эспарцет бузе 271183
Эспарцет бунге 271184
Эспарцет васильченко 271279
Эспарцет виколистный 271280
Эспарцет влагальщный 271278
Эспарцет высокий 271211
Эспарцет высочайший 271170
Эспарцет гогенакера 271216
Эспарцет гончарова 271209
Эспарцет горный 271234
Эспарцет горошколистный 271280
Эспарцет грациозный 271210
Эспарцет гроссгейма 271212
Эспарцет грузинский 271221
Эспарцет дагестанский 271200
Эспарцет дарвазский 271201
Эспарцет дильса 271204
Эспарцет длинноиглистый 271228
Эспарцет донской 271273
Эспарцет ехидна 271206
Эспарцет закавказский 271277
Эспарцет закаспийский 271276
Эспарцет зеравшанский 271266
Эспарцет изящный 271210
Эспарцет испанский 187842
Эспарцет киноварнорасный 271233
Эспарцет комарова 271224
Эспарцет кормовой 271280
Эспарцет короткостебельный 271271
Эспарцет красивый 271255
Эспарцет крючковатый 271214
Эспарцет куринский 271199
Эспарцет лежачий 271272
Эспарцет липского 271227
Эспарцет лучстый 271256
Эспарцет майорова 271230
Эспарцет мелкоцветный 271232
Эспарцет мишо 271231
Эспарцет многоцветковая 214138
Эспарцет невооруженный 271222
Эспарцет немеца 271236
Эспарцет новопокровского 271237
Эспарцет обыкновенный 271280
Эспарцет остолодочниковый 271240
Эспарцет острозубый 271239
Эспарцет палласа 271241
Эспарцет педушья 271185
Эспарцет песчаный 271172
Эспарцет полевой 271172
Эспарцет посевной 271280
Эспарцет приятный 271171
Эспарцет разнолистный 271215
Эспарцет рогообразный 271190
Эспарцет рупрехта 271259
Эспарцет североармянский 271213
Эспарцет североиранский 355058
Эспарцет серебристный 271177
Эспарцет сибирский 271267
Эспарцет синтениса 271268
Эспарцет скальный 271254
Эспарцет сосновского 271269
Эспарцет сочевичниковый 271240
Эспарцет турнефора 271275
Эспарцет ферганский 271208
Эспарцет хорасанский 271187
Эспарцет шерстистый 271274
Эспарцет шугнанский 271265
Эстрагон 35411
Эсшольция 155170
Этеопаррус 9765
Этеопаррус введенского 9770
Этеопаррус кавказский 9768
Этеопаррус красивейщий 9769
Этионема 9785
Эулофия 156521
Эулофия тайваньская 156572
Эулофия туркестанская 157069
Эулофия юньнаньская 157105
Эутрема 161120
Эутрема сердцелистная 161125
Эутрема эдвардса 161129
Эутрема юньнаньская 161158
Эутрема японская 161149
Эуфорбия 158385
Эухаридиум 94132,312946
Эфедра 146122
Эфедра американская 146134
Эфедра двуколосковая 146154
Эфедра двухколосковая 146154
Эфедра жерарда 30277,146183
Эфедровые 146270,178522
Эхеверия 139970
Эхинацея 140066
Эхинация 140066
Эхинация пурпуровая 140081
Эхинодорус 140536
Эхинокактус 140111
Эхинопанакс 272704
Эхинопанакс высокий 272706
Эхинопс 140652
Эхинопс широколистный 140732
Эхинопсилон 140830
Эхинопсилон волосистый 48767
Эхинопсилон иссополистная 48769
Эхинопсилон очитковидны 48790
Эхинопсилон растопыреный 48762
Эхинопсис 140842
Эхиноцереус 140209
Эхиноцистис 140526,140527
Эхиноцистис шиповатый 140527
Эхитес 141051
Эхиум 141087
Эхмея 8544
Эшинантус 9417
Эшиномена 9489
Эшолтциа 155170
Эшольция 155170
Эшшольция 155170,155173
Эшшольция калифорнийская 155173
Юануллоа 212553
Юбея 212556
Юбея величавая 212557
Юбея замечательная 212557
Юзерлык 287978
Юзу 93515
Юйюба 418144,418169
Юкка 416535
Юкка алоелистная 416538
Юкка алойная 416538
Юкка алоэлистная 416538
Юкка величественная 416607
Юкка виргинская 416584
Юкка волокнистая 416584
Юкка мечевидная 416535
Юкка нитевидная 416584
Юкка нитчатая 416584
Юкка пониклолистная 416610
Юкка прекрасная 416607
Юкка сизая 416607
Юкка славная 416607
Юкка слоновая 416581
Юморка 247457
Юнгия 416390
Юнгия алтайская 416394
Юнгия зеравшанская 416475
Юнгия кандинская 416447
Юнгия поздняя 283400
Юнгия разнолистная 416416
Юнгия тайваньская 416437
Юнгия тонколистная 416490
Юнгия тонкостебельная 416489
Юнгия узкая 416485
Юнгия чистотелолистная 283386
Юнгия японская 416437
Юнос 93515

Юньнаньск тибетский 87646
Юринелла 214201
Юринелла бесстебельная 214204
Юринелла оттопыренная 214203
Юринея 214028
Юринея васильковая 214070
Юринея киргизская 214108
Юринея меловая 214067
Юринея многостебельная 214036
Юринея мягкая 214131
Юринея паутинистая 214040
Юринея сомнительная 214138
Юринея узколистная 214179
Юринея узколопастная 214186
Юринея эверсмана 214080
Юстиция 214304
Юстиция лежачая 337253
Юхтриница армянская 401339
Юхтриница коканская 401340
Ююба 418144,418169
Ююба унаби 418169
Яблоко заморское 93297
Яблоко зведное 90032
Яблоко райское 243682
Яблоко розовое 382582
Яблоко сахарное 25898
Яблоко – кислица 243675,243711
Яблоневые 242848
Яблоня 243543
Яблоня венечная 243583
Яблоня войлочно – опушенная 243589
Яблоня восточная 243662,243682
Яблоня ганьсуская 243642
Яблоня гималайская 243571
Яблоня горная 243698
Яблоня дикая 243675,243698,243711
Яблоня дикая сибирская 243555
Яблоня домашняя 243675,243711
Яблоня душистая 243583
Яблоня замечательная 243702
Яблоня зибольда 243717
Яблоня карликовая 243675,243711
Яблоня китайская 243667
Яблоня корнесобственная 243555,243675
Яблоня красномясая 243659
Яблоня культурная 243675,243711
Яблоня лесная 243675,243711
Яблоня малайская 382606
Яблоня маньчжуьрская 243649
Яблоня мелкоплодная 243657
Яблоня многоцветковая 243617
Яблоня мулиская 243658
Яблоня недзведцкого 243659
Яблоня низкая 243675,243711
Яблоня низкорослая 243675,243711
Яблоня обильноцветковая 243617
Яблоня обильноцветущая 243617
Яблоня палласа 243555
Яблоня палласова 243555
Яблоня парадизка 243682
Яблоня райская 243682
Яблоня ранняя 243664
Яблоня рока 243687
Яблоня садовая 243675,243711
Яблоня сибирская 243555
Яблоня сиверса 243698
Яблоня сливолистная 243667
Яблоня туркменнов 243730
Яблоня тяньшаньская 243659
Яблоня хубейская 243630
Яблоня хэнаньская 243629
Яблоня юньнаньская 243732
Яблоня юньнаньская вича 243733
Яблоня ягодная 243555
Явор 3462
Явор восточный 302588,302592
Ягода буйволова 362090
Ягода винная 164763
Ягода волчья 236226
Ягода грудная европейская 418169
Ягода грудная французская 418169
Ягода логанова 338761
Ягода медвежья 31142
Ягода мельная 346351
Ягода снежная 380734
Ягода снежная китайская 380736,380753
Ягодки 122515
Ягодковые 391030
Ягодник 403710
Ягодник кистевой 403780
Ягодник черника 403916
Язвеник 28143
Язвеник горный 28179
Язвеник кавказский 28154
Язвеник крымский 28194
Язвеник лемана 28173
Язвеник линнея 28200
Язвеник многолистный 28189
Язвеник обыкновенный 28200
Язвеник окрашенный 28157
Язвеник опушенный 28171
Язвеник песчаный 28149
Язвеник приальпийский 28148
Язвеник приморский 28175
Язвеник родственный 28146
Язвеник яркокрасный 28156
Язык волочий 21914
Язык песий 117900
Языки волобьиные 105416
Языкотруб 344381
Якаранда 211225
Якаранда минозолистная 211239
Якаранда остролистная 211226,211239,211241
Як – дерево 36920,36924
Якорцы 395064
Якорцы земляные 395146
Якорцы крупнокрылые 395108
Якорцы стелющиеся 395146
Ялапа 207901,255711
Ялапа настоящая 207901
Ямбоза 259186,382582
Ямболан 382522
Ямкосимяник 383989
Ямкосимяник алтайский 383990
Ямкосимяник широколепестный 383996
Ямс 131451
Ямс китайский 131601,131772
Ямс клубненосный 131501
Ямс крыратый 131458
Ямс куш – куш 131880
Ямс посевной 131577
Ямс слоновый 20057,20125
Ямс слоновый ривьера 20132
Ямс съедобный 131577
Ямс тайваньский 131593
Ямс трёхнадрезный 131880
Ямс фуцзяньский 131597
Ямс японский 131645
Японск эмэйский 103840
Японск японский 103835,111170,120371,158161
Японская мушмула 151130,151162
Японский мандарин 93736
Ярофа 3243,212098
Ярутка 390202
Ярутка альпийская 390208
Ярутка армянская 390212
Ярутка безухая 390223
Ярутка горная 390242
Ярутка зонтичная 390268
Ярутка карликовая 390258
Ярутка кочи 390239
Ярутка крупноцветковая 390241
Ярутка ложечная 390223
Ярутка носатая 390260
Ярутка однолетняя 390211
Ярутка округлая 390247
Ярутка плоскоплодная 390255
Ярутка полевая 390213
Ярутка пронзенная 390249
Ярутка пронзённая 390249
Ярутка ранняя 390256
Ярутка розовая 390259
Ярутка сычуаньская 390232
Ярутка тибетская 390210
Ярутка ферганская 390231
Ярутка фрейна 390233
Ярутка хюта 390236
Ярутка шовица 390264
Ярутка японская 390237

Ясенец 129615
Ясенец белый 129618
Ясенец голостолбиковый 129631
Ясенец кавказский 129628
Ясенец мохнатоплодный 129629
Ясенец таджикский 129633
Ясенец узколистный 129627
Ясенец ясенелистный 129618
Ясенка 248893
Ясень 167893,167955
Ясень американский 167897,167942
Ясень американский ореxолистный 167908
Ясень бархатный 168136
Ясень белый 168041
Ясень войлочный 168120
Ясень высокий 167955
Ясень голубой 168079
Ясень двойственный 167952
Ясень длинно－остроконечный 168018
Ясень европейский 167955
Ясень зеленый 168008,168134
Ясень зелёный 168134
Ясень зибольда 168100
Ясень золотистый 167958
Ясень калолинский 167934
Ясень китайский 12559,167940
Ясень китайский клюволистный 168086
Ясень колючий американский 417157
Ясень красный 168074
Ясень круглолистный 168053,168093
Ясень ланцетный 168008,168134
Ясень ланцетолистный 168134
Ясень маннитовый 168041
Ясень манноносный 168041
Ясень маньчжурский 168023
Ясень мексиканская 168124
Ясень мелколистный 168053,168093
Ясень мягколистый 168022
Ясень наньчуаньский 168030
Ясень неправильный 167922
Ясень носолистный 168086
Ясень обильноцветущий 167975
Ясень обыкновенный 167955
Ясень опушенный 168074
Ясень остороконечный 167897,167942
Ясень остороплодный 168043
Ясень паллис 168050
Ясень пенсильванский 168057,168074
Ясень поярковой 168070
Ясень пушистый 168057,168074
Ясень сахарный 168041
Ясень сирийский 168114
Ясень согдианский 168105
Ясень согдийский 168105
Ясень сумахолистный 167947
Ясень туркестанский 168071
Ясень узколистный 167910
Ясень хубэйский 167991
Ясень цветочный 168041
Ясень чёрный 168032
Ясень четырёхгранный 168079
Ясень японский 167970,167999
Ясколка 82629
Ясколка алексеенко 82644
Ясколка альпийская 82646
Ясколка армянская 82679
Ясколка берингова 82732
Ясколка беринговская 82732
Ясколка биберштейна 82739
Ясколка бялыницкого 82738
Ясколка вздутая 82789,82790
Ясколка вильчатая 82789,82834
Ясколка воробейниколистная 82911
Ясколка волосистолистная 83071
Ясколка гладковатая 82845
Ясколка гоная 82948
Ясколка городкова 82859
Ясколка даурская 82786
Ясколка дернистая 82758
Ясколка длиннолистная 82913
Ясколка зубчатая 82787
Ясколка кавказская 82888
Ясколка казбекская 82902
Ясколка костенецевидная 82826
Ясколка крупная 82918
Ясколка крымская 83049
Ясколка лесная 82928
Ясколка липкая 82856
Ясколка малоцветковая 82955
Ясколка мальнькая 82994
Ясколка мейера 82921
Ясколка мелкосемянная 82922
Ясколка многоцветковая 82926
Ясколка полевая 82680
Ясколка полимофная 82973
Ясколка понтийская 82974
Ясколка пронзеннолистная 82968,83020
Ясколка пурпурная 82993
Ясколка пятитычинковая 83020
Ясколка пятитычиночная 83020
Ясколка регеля 82997
Ясколка серебристая 82676
Ясколка серповидная 82808
Ясколка сычуаньская 83042
Ясколка скученная 82849
Ясколка скученноцветковая 82849
Ясколка сорная 83010
Ясколка сосновского 83030
Ясколка стевена 83032
Ясколка тайваньская 82924
Ясколка трехстолбиковая 82772
Ясколка трёхстолбиковая 82772
Ясколка тяньшанская 83059
Ясколка уклоняющаяся 82664
Ясколка фишера 82813
Ясколка хемшинская 82873
Ясколка хлоролистная 82776
Ясколка шерстистая 83060
Ясколка шмальгаузена 83016
Ясколка шовица 83043
Ясколка ясколковидная 82772
Ясменник 39300
Ясменник абхазский 39302
Ясменник альбова 39305
Ясменник альпийский 39306
Ясменник белопыльниковый 39373
Ясменник биберштейна 39321
Ясменник близкий 39398
Ясменник воронова 39420
Ясменник восточный 39383
Ясменник гладкий 39370
Ясменник гладчайший 39371
Ясменник голоплодный 39372
Ясменник голубой 39317
Ясменник гранитный 39350
Ясменник гребенчатый 39327
Ясменник данилевского 39342
Ясменник двузубый 39320
Ясменник длиннолистный 39343
Ясменник душистый 170524
Ясменник жестковололистый 39362
Ясменник изломанный 39365
Ясменник изящный 39349
Ясменник кавказский 39323
Ясменник каменистый 39393
Ясменник каратавский 39367
Ясменник кемулярии 39368
Ясменник красильный 39413,68538
Ясменник крылова 39369
Ясменник крымский 39411
Ясменник ксерофитный 39421
Ясменник липского 39374
Ясменник максимовича 170477
Ясменник маркотхский 39377
Ясменник меловой 39326
Ясменник метельчатый 170533
Ясменник моллюгообразный 39379
Ясменник неприятный 39366
Ясменник низкий 39410
Ясменник опушенноцветковый 39337
Ясменник памирский 39384
Ясменник пахучий 39351,170524
Ясменник переодетый 39397
Ясменник подмаренниковый 39347
Ясменник подоражающий 39303
Ясменник полевой 39314
Ясменник полуодетый 39404
Ясменник понтический 39395
Ясменник попова 39396
Ясменник приручейный 170588
Ясменник распростертый 170420

Ясменник родотвенный 39304
Ясменник розоватый 39328
Ясменник розовый 39328
Ясменник румелийский 39402
Ясменник сероплодный 39412
Ясменник сжатый 39324
Ясменник сизый 39347
Ясменник скальный 39393
Ясменник скученый 39348
Ясменник слабый 39338,170420
Ясменник стевена 39407
Ясменник стелющийся 39399,170420
Ясменник супротивнолистный 39382
Ясменник туркменский 39418
Ясменник уменьшенный 39342
Ясменник утонченный 39315
Ясменник ферганский 39346
Ясменник цветкожковый 39387
Ясменник цепкий 170588
Ясменник шероховатый 39345
Ясменник шерстеносный 39419
Ясменник широкий 170558
Ясменник щетинистый 39405
Ясменник щетинковый 39406
Яснотка 220345,298009
Яснотка белая 220346
Яснотка войлочная 220421
Яснотка ганьсуская 220407
Яснотка гибридая 220392
Яснотка голая 220064,220388
Яснотка кавказская 220367
Яснотка крапчатая 220401
Яснотка кустарниковая 220375
Яснотка пачоского 220413
Яснотка пурпуровая 220416
Яснотка пятнистая 220401
Яснотка стеблеобъемлющая 220355
Яснотка цзилунская 220387
Ястребинка 195437,195694,196055
Ястребинка авлорина 195481
Ястребинка акинфиева 195447
Ястребинка алатавская 195448
Ястребинка альпийская 195455
Ястребинка армянская 195471
Ястребинка артвинская 195472
Ястребинка баланзы 195483
Ястребинка баскская 196075
Ястребинка белесоватая 195450
Ястребинка белокистевая 195744
Ястребинка белокистевидная 195743
Ястребинка белокистеродная 195742
Ястребинка белоосновная 195486
Ястребинка белопепельная 195453
Ястребинка белопленчатая 195863
Ястребинка белостопая 195452
Ястребинка береговая 195752
Ястребинка берокаемочная 195454
Ястребинка бессера 195487
Ястребинка биберштейна 195488
Ястребинка бледная 195834
Ястребинка бледнозеленочешуйная 195531
Ястребинка блестащеволосистая 195714
Ястребинка блестащеволосоподобная 195713
Ястребинка близкая 195886
Ястребинка боброва 195489
Ястребинка бокоцветковая 195725
Ястребинка большезубая 195847
Ястребинка бородина 195492
Ястребинка бородочковая 195484
Ястребинка брандиса 195494
Ястребинка бриттатская 195496
Ястребинка бузе 195497
Ястребинка буша 195501
Ястребинка бэница 195482
Ястребинка ваги 196076
Ястребинка вайяна 196073
Ястребинка вересовая 195932
Ястребинка верхушечнозонтичная 195441
Ястребинка ветвистая 195914,196085
Ястребинка видная 195882
Ястребинка виммера 196095
Ястребинка внешняя 195803
Ястребинка волжская 196091
Ястребинка вологодская 196097
Ястребинка володушколистная 195499
Ястребинка володушколистновидная 195498
Ястребинка володушкообразная 195500
Ястребинка волосистая 195858
Ястребинка волосистейшая 195652
Ястребинка волосистоветвистая 196040
Ястребинка волосистосоцветноподобная 195527
Ястребинка волосистостопая 195854
Ястребинка волосистоцветковая 195856
Ястребинка волосистоцветная 195856
Ястребинка волосисточешуйчатая 195855
Ястребинка волосистсоцветная 195528
Ястребинка волчанская 196096
Ястребинка воронова 196098
Ястребинка выровненная 195711
Ястребинка высокая 195882
Ястребинка высокорослая 195881
Ястребинка высокостебельная 195456
Ястребинка вытянутая 195872,195884
Ястребинка ганешина 195618
Ястребинка гаральда 195649
Ястребинка гибкостевельная 195596
Ястребинка гладкочешуйная 195750
Ястребинка гладкоязычковая 195623
Ястребинка глена 195627
Ястребинка глянцевитая 196079
Ястребинка гогенакера 195654
Ястребинка голая 195647
Ястребинка головетвистая 195907
Ястребинка голостебельная 195745
Ястребинка голубоватая 195999
Ястребинка голубовато – зеленая 195625
Ястребинка гоппе 195658
Ястребинка городкова 195631
Ястребинка грубая 195921
Ястребинка грузинская 195620
Ястребинка густава 195646
Ястребинка густопушистая 195720
Ястребинка густосоцветная 195911
Ястребинка гюнтера 195645
Ястребинка двуцветная 195566
Ястребинка двуцветненькая 195568
Ястребинка дернинная 299220
Ястребинка дехи 195559
Ястребинка длинная 195765
Ястребинка длиннолучистая 195762
Ястребинка длинностопая 195758
Ястребинка длиннострелковая 195763
Ястребинка длиннощетинистая 195764
Ястребинка долотолистная 195570
Ястребинка дублицкого 195571
Ястребинка железистоветочковая 195445
Ястребинка желтожелезистая 195766
Ястребинка желторыльцевая 196100
Ястребинка жесткая 195917
Ястребинка жестковолосистейшая 196004
Ястребинка жесткошетинистая 195572
Ястребинка заостренносоцветная 195442
Ястребинка звездчатая 195988
Ястребинка звездчатохохлатая 195597
Ястребинка зелененькая 195535
Ястребинка зеленолюбивая 195533
Ястребинка зеленоцветная 195530
Ястребинка зонтиконосная 195957
Ястребинка зонтичная 196057
Ястребинка зонтоносная 196068
Ястребинка зубчатая 195561
Ястребинка известная 195547
Ястребинка изобильноцветковая 195601
Ястребинка изогнутозубчатая 195467
Ястребинка изогнутолистная 195558
Ястребинка инеенсная 195887
Ястребинка кабанова 195674
Ястребинка кавказская 195526
Ястребинка каемчатая 195785
Ястребинка камнеломковая 195940
Ястребинка карпинского 195684
Ястребинка карягина 195683
Ястребинка каспари 195523
Ястребинка качурина 195675
Ястребинка каянская 195676
Ястребинка кемуларии 195685
Ястребинка кенига 195689
Ястребинка кёрнике 195690

Ястребинка киргизская 195687
Ястребинка клейкая 196087
Ястребинка кнафа 195688
Ястребинка козловского 195695
Ястребинка конжаковская 195691
Ястребинка конусообразная 195546
Ястребинка копьезубая 195717
Ястребинка коржинского 195692
Ястребинка короткоглавая 195493
Ястребинка косьвинская 195693
Ястребинка красивочешуйная 195510
Ястребинка красноплодная 195579
Ястребинка красносемянковидная 195578
Ястребинка крашана 195698
Ястребинка кречетовича 195699
Ястребинка кривоватая 195554
Ястребинка кровянопятнистая 195553
Ястребинка кровяноязычная 195648
Ястребинка круглеющая 195823
Ястребинка крупноватая 195621
Ястребинка крупнозонтичная 195769, 195772
Ястребинка крупнозубчатая 195637
Ястребинка крупнолучистая 195779
Ястребинка крупнопленчатая 195773
Ястребинка крупноприкорнелистная 195778
Ястребинка крупно－скороспелковая 195783
Ястребинка крупностебельная 195622
Ястребинка крупноцветковая 195770
Ястребинка крупночешуевидная 195774
Ястребинка крупночешуйноподобная 195775
Ястребинка крупночешуйчатая 195777
Ястребинка крупнощетинистая 195771
Ястребинка крылова 195700
Ястребинка кубанская 195701
Ястребинка кубинская 195702
Ястребинка кудреватая 195550
Ястребинка кудрявая 195538
Ястребинка кузнецкая 195706
Ястребинка кулькова 195703
Ястребинка кумбельская 195704
Ястребинка купфера 195705
Ястребинка курчавая 195551
Ястребинка лакшевица 195709
Ястребинка ланцетнолистная 195716
Ястребинка лапландская 195718
Ястребинка латуковидная 195880
Ястребинка леберта 195729
Ястребинка левье 195746
Ястребинка ленкоранская 195731
Ястребинка лесная 195821, 195970
Ястребинка леспинаса 195740
Ястребинка лидии 195767
Ястребинка липмаа 195748
Ястребинка липского 195749
Ястребинка лировидная 195768
Ястребинка листочконосная 195605
Ястребинка литвинова 195753
Ястребинка литовская 195751
Ястребинка ложнокороткая 195893
Ястребинка ложнолаздвоенная 195892
Ястребинка ложносжатая 195894
Ястребинка ложноушастая 195998
Ястребинка ложноушковатая 195891
Ястребинка ложно－юрская 195895
Ястребинка ломницкая 195756
Ястребинка лопатчатолистная 195983
Ястребинка лоснящаяся 195791
Ястребинка лохматая 195840
Ястребинка луговая 195506
Ястребинка лужаечная 195879
Ястребинка льнолистная 195747
Ястребинка ляша 195719
Ястребинка мальме 195784
Ястребинка медно－зеленая 195446
Ястребинка мейнсглаузена 195787
Ястребинка мелкобородавчатая 196080
Ястребинка мелкоголовчатая 195806
Ястребинка мелкожелезистая 195734
Ястребинка мелкозвезчатая 195792
Ястребинка мелкозубчатая 195562
Ястребинка мелкописаноподобная 195737
Ястребинка мелкоскрученная 195793
Ястребинка мелкохлопьевидная 195598
Ястребинка мелкошаровидная 195794
Ястребинка менделя 195790
Ястребинка миленькая 195732
Ястребинка многожелезистая 195800
Ястребинка многоликая 195799
Ястребинка многолистная 195867
Ястребинка многостебельная 195866
Ястребинка многоцветковая 195603
Ястребинка могучая 195921
Ястребинка можжевеловая 195933
Ястребинка мощная 195921
Ястребинка мурманская 195802
Ястребинка мягкощетинковая 195796
Ястребинка нальчикская 195805
Ястребинка нарымская 195807
Ястребинка негладкая 195941
Ястребинка ненюкова 195808
Ястребинка необыкновенная 195666
Ястребинка неясная 195818
Ястребинка низбегающая 195560
Ястребинка нитяная 195588
Ястребинка норвежская 195815
Ястребинка норрлина 195814
Ястребинка норрлиновидная 195813
Ястребинка обильноопушенная 195722
Ястребинка обильноцветковая 195602
Ястребинка обильноцветущая 195603
Ястребинка обильнощетинистая 195801
Ястребинка обманчивая 195584
Ястребинка обыкновенная 195708, 196093
Ястребинка обыкновенненькая 196092
Ястребинка овальноглавая 195828
Ястребинка овальноликая 195829
Ястребинка однобочная 195724
Ястребинка однокровельная 195657
Ястребинка одноцветненькая 195545
Ястребинка одутловатая 195909
Ястребинка окрашенная 195541
Ястребинка окрашенноешуйчатая 195536
Ястребинка олимрийская 195819
Ястребинка олиственная 195614
Ястребинка олиственнейшая 195606
Ястребинка оманга 195820
Ястребинка онежская 195821
Ястребинка оносмовидная 195822
Ястребинка оранжево－красная 195478
Ястребинка освальда 195827
Ястребинка основолистная 195485
Ястребинка остроконечная 195465
Ястребинка остроуловатая 195443
Ястребинка острочешуйная 195444
Ястребинка отличная 195581
Ястребинка отрубистая 195617
Ястребинка оттопыренная 195984
Ястребинка паннонскообразная 195838
Ястребинка панша 195832
Ястребинка панютина 195837
Ястребинка парейса 195842
Ястребинка пепельнополосчатая 195539
Ястребинка пепельносоцветная 196031
Ястребинка пепельностопая 196033
Ястребинка пепельноцветная 196030
Ястребинка пеплолюбивая 196032
Ястребинка песколюбивая 195889
Ястребинка пестрочешуйная 196074
Ястребинка песчаная 195934
Ястребинка петергофская 195850
Ястребинка петунникова 195851
Ястребинка пильчатолистная 195966
Ястребинка плевысокая 195873
Ястребинка пленчатая 195789
Ястребинка плетевая 232483
Ястребинка плетеобразная 195594
Ястребинка плотная 195831
Ястребинка побарчевского 195755
Ястребинка побегоцветная 195991
Ястребинка повислая 195844
Ястребинка повсюду－пепельная 195458
Ястребинка подбородчатая 195663
Ястребинка подъельниковая 195662
Ястребинка поле 195868
Ястребинка полевая 195474
Ястребинка полуранняя 195963
Ястребинка посредническая 195668
Ястребинка постеная 195803

Ястребинка почти – артвинская 195996
Ястребинка почти – бледная 196007
Ястребинка почти – густопушистая 196006
Ястребинка почти – имандоровая 196005
Ястребинка почти – копьевидная 196003
Ястребинка почти – красноватая 196014
Ястребинка почти – многоцветковая 196002
Ястребинка почти – окаймленная 196009
Ястребинка почти – побежистая 196016
Ястребинка почти – прозрачная 196013
Ястребинка почти – простая 196015
Ястребинка почти – прямая 196001
Ястребинка почти – пятнисная 196008
Ястребинка почти – северо – восточная 195995
Ястребинка почти – сизая 195661
Ястребинка почти – темноглавая 196012
Ястребинка почти – толстолистная 196000
Ястребинка почти – черноватая 196010
Ястребинка прежесткая 195477
Ястребинка прекрасная 195509
Ястребинка прекраснообличная 195507
Ястребинка прекраснообразная 195508
Ястребинка прешероховатая 195845
Ястребинка прозрачновидноглавая 195564
Ястребинка прозрлчная 195843
Ястребинка пронзеннолистная 195846
Ястребинка пропущенная 195877
Ястребинка прусская 195888
Ястребинка прямейшая 195993
Ястребинка прямостопая 195826
Ястребинка птшистая 195908
Ястребинка пурпурноповязочная 195910
Ястребинка пятигорская 195912
Ястребинка пятнисная 195781
Ястребинка радде 195913
Ястребинка разнозубчатая 195651
Ястребинка разнолистная 195870
Ястребинка раскидистая 195883
Ястребинка распущенная 195569
Ястребинка регелевская 195915
Ястребинка регеля 195915
Ястребинка ройовского 195922
Ястребинка ростостебельная 195971
Ястребинка рота 195924
Ястребинка рузская 195929
Ястребинка румянковая 195576
Ястребинка румянкоглавая 195575
Ястребинка рупрехта 195928
Ястребинка рыжая 195616
Ястребинка рыжеющая 195980
Ястребинка рыхлозонтичная 196069
Ястребинка рэмана 195916
Ястребинка савойская 195931
Ястребинка сажистая 195615
Ястребинка самурская 195938
Ястребинка сахокия 195935
Ястребинка сглаженная 195712
Ястребинка севирная 195964
Ястребинка седая 195518
Ястребинка седоватая 195665
Ястребинка сердцевидная 195520
Ястребинка сердцелистная 195521
Ястребинка сернистая 196019
Ястребинка сернистовидная 196018
Ястребинка серножелтая 196021
Ястребинка сероватая 195641
Ястребинка сероватоподобная 195664
Ястребинка серозлачная 195869
Ястребинка серпозубчатая 195583
Ястребинка сиверковая 195466
Ястребинка сизая 195505
Ястребинка сизоватая 195504
Ястребинка сизоветковая 195503
Ястребинка сизоцветковидная 195502
Ястребинка силена 195968
Ястребинка скальная 195958
Ястребинка скандинавская 195950
Ястребинка складчатая 195865
Ястребинка скороспелка 195480
Ястребинка скрученноволосая 195992
Ястребинка скрученногавая 196037
Ястребинка скрытая 195723
Ястребинка скученная 195629
Ястребинка скученнообразная 195628
Ястребинка снежнокаемочная 195812
Ястребинка совича 196024
Ястребинка сомнительная 195994
Ястребинка сосновая 195862
Ястребинка сосновского 195981
Ястребинка сочавы 195975
Ястребинка сродственная 195619
Ястребинка ставропольская 195985
Ястребинка стенная 195803
Ястребинка стеночная 195803
Ястребинка столоноцветковая 195991
Ястребинка стреловидная 195936
Ястребинка стрелолистная 195936
Ястребинка стройная 195873
Ястребинка сушеничная 195630
Ястребинка сырейщикова 196023
Ястребинка таненская 196025
Ястребинка тауша 196028
Ястребинка тебердинская 196029
Ястребинка темная 195818
Ястребинка темноприцветниковая 195817
Ястребинка теневая 196070
Ястребинка тепроухова 196034
Ястребинка терекская 196035
Ястребинка тилинга 196036
Ястребинка тонковетвистая 195736
Ястребинка тонколистная 195739
Ястребинка тонколосая 195738
Ястребинка тонкостебельная 195735
Ястребинка топяная 195906
Ястребинка тоше – железистая 195669
Ястребинка траурная 196043
Ястребинка треугольная 196039
Ястребинка трехзубчатая 196042
Ястребинка трехзубчатоглавая 196041
Ястребинка трубчатая 196051
Ястребинка туркестанская 196052
Ястребинка угандийская 196054
Ястребинка удивительная 195795
Ястребинка узколистная 195736,195739
Ястребинка узкочешуйная 195989
Ястребинка уклончивая 195841
Ястребинка украшенная 195825
Ястребинка уральская 196071
Ястребинка усинская 196072
Ястребинка утешающая 195841
Ястребинка ушковатая 195890
Ястребинка филандская 195707
Ястребинка фиолетовостопая 196084
Ястребинка флорентийская 195857
Ястребинка фомина 195607
Ястребинка фрейна 195608
Ястребинка фрика 195609
Ястребинка фрица 195613
Ястребинка хворостовидная 195939
Ястребинка хладолюбивая 195611
Ястребинка хлопьецветоносовая 195599
Ястребинка холовообразная 195540
Ястребинка хохлатая 195544
Ястребинка цветстая 195604
Ястребинка цимозная 195555
Ястребинка цинзерлинга 196101
Ястребинка цица 196102
Ястребинка цицова 196102
Ястребинка черещчатая 195849
Ястребинка чернеющая 195810
Ястребинка черноватая 196053
Ястребинка черноватенькая 196011
Ястребинка черноватоглавая 195788
Ястребинка черноголовая 195670
Ястребинка черноцетинистая 195811
Ястребинка чешуйчатообразная 195733
Ястребинка чильмана 195686
Ястребинка чхубианишвили 196049
Ястребинка шведская 196017
Ястребинка шелковистостебельная 195965
Ястребинка шелковникова 195951
Ястребинка шероховатенькая 195997
Ястребинка шерстистая 196081
Ястребинка шишкина 195952
Ястребинка шлякова 195953
Ястребинка штейнберг 195986
Ястребинка шультеса 195955
Ястребинка эйхвальда 195577
Ястребинка юрская 195673
Ястребинка ядовитая 196081,196086

Ястребинка яичножелтая　196090
Ястребинка яйцеобразная　195830
Ястребинка японская　195672
Ятропа　3243,212098
Ятропа куркас　212127
Ятрориза драневидная　293729
Ятрофа куркас　212127
Ятрышник　273317,273529
Ятрышник балтийский　273333
Ятрышник бледный　273564
Ятрышник болотный　273565
Ятрышник бузинный　273623
Ятрышник длемлик　273545
Ятрышник желтоватый　273433
Ятрышник зеленобурый　273700
Ятрышник зеленоватожелтый　273359
Ятрышник иберийский　273467
Ятрышник каскрашенный　273592
Ятрышник клопоносный　273380
Ятрышник кровавый　121389
Ятрышник круглогубый　169887
Ятрышник мужской　273531
Ятрышник мясокрасный　273469
Ятрышник наибольший　273537
Ятрышник обезьяний　273633
Ятрышник обожженный　273696
Ятрышник обожжённый　273696
Ятрышник остистый　121359
Ятрышник патнистый　273524
Ятрышник провансальский　273596
Ятрышник пурпурный　273604
Ятрышник пятнистый　273524
Ятрышник римский　273613
Ятрышник сычуаньский　310933
Ятрышник солончаковый　273501
Ятрышник тайваньская　310939
Ятрышник теневой　121422
Ятрышник точечный　273602
Ятрышник траунштейнера　273681
Ятрышник трехзубчатый　273686
Ятрышник трехлистный　273688
Ятрышник туполопастный　273323
Ятрышник федченко　273429
Ятрышник фукса　121382
Ятрышник хэбэйская　169906
Ятрышник черноморский　273420
Ятрышник шелковникова　273627
Ятрышник широколистный　273501
Ятрышник шлемовидный　273541
Ятрышник шлемоносный　273541
Ятрышник эмэйский　310923
Ятрышниковые　273267
Ятрышничк　85928
Ятрышничк альпийский　85930
Ячмень　198260
Ячмень богдана　198264
Ячмень гигантский　228346
Ячмень голый　198293
Ячмень гривастый　198311
Ячмень гривоподобный　198311
Ячмень двурядный　198288,198327
Ячмень двухрядный　198288
Ячмень дикий　198365
Ячмень дикий луковичный　198275
Ячмень дикорастущий　198365
Ячмень житяный　198354
Ячмень заячий　198288,198316
Ячмень коленчатый　198297
Ячмень колончаковый　198267
Ячмень короткоостистый　198267
Ячмень короткоостый　198267
Ячмень короткошиловидный　198267
Ячмень кроненбунга　317083
Ячмень крупный　198288
Ячмень культурный двурядный　198288
Ячмень культурный многорядный　198376
Ячмень культурный шестирядный　198376
Ячмень луговой　198267
Ячмень луковичный　198275
Ячмень многолядный　198376
Ячмень мышиный　198327
Ячмень обыкновенный　198376
Ячмень обыкновенный многорядный　198376
Ячмень озимый　198376
Ячмень ощетиненный　198297
Ячмень песчаный　228346
Ячмень пивоваренный　198288
Ячмень посевной　198376
Ячмень ржаной　198354
Ячмень рожевица　198346
Ячмень северный　198265
Ячмень сибирский　198360
Ячмень тощеостый　198307
Ячмень туркестанский　198373
Ячмень фиолетовый　198375
Ячмень четытёхрядный　198376
Ячмень четытёхстолонний　198376
Ячмень шестигранный　198376
Ячмень шестирядный　198376
Ячмень яровой двурядный　198288
Ячмень яровой четырехрядный　198376